e of the Elements

10	11	12	13	14	15	16	17	18
								2 He helium 4.0026
			5 **B** boron 10.81 [10.806, 10.821]	6 **C** carbon 12.011 [12.009, 12.012]	7 **N** nitrogen 14.007 [14.006, 14.008]	8 **O** oxygen 15.999 [15.999, 16.000]	9 **F** fluorine 18.998	10 **Ne** neon 20.180
			13 **Al** aluminium 26.982	14 **Si** silicon 28.085 [28.084, 28.086]	15 **P** phosphorus 30.974	16 **S** sulfur 32.06 [32.059, 32.076]	17 **Cl** chlorine 35.45 [35.446, 35.457]	18 **Ar** argon 39.948
28 **Ni** nickel 58.693	29 **Cu** copper 63.546(3)	30 **Zn** zinc 65.38(2)	31 **Ga** gallium 69.723	32 **Ge** germanium 72.630(8)	33 **As** arsenic 74.922	34 **Se** selenium 78.971(8)	35 **Br** bromine 79.904 [79.901, 79.907]	36 **Kr** krypton 83.798(2)
46 **Pd** alladium 106.42	47 **Ag** silver 107.87	48 **Cd** cadmium 112.41	49 **In** indium 114.82	50 **Sn** tin 118.71	51 **Sb** antimony 121.76	52 **Te** tellurium 127.60(3)	53 **I** iodine 126.90	54 **Xe** xenon 131.29
78 **Pt** platinum 195.08	79 **Au** gold 196.97	80 **Hg** mercury 200.59	81 **Tl** thallium 204.38 [204.38, 204.39]	82 **Pb** lead 207.2	83 **Bi** bismuth 208.98	84 **Po** polonium	85 **At** astatine	86 **Rn** radon
110 **Ds** rmstadtium	111 **Rg** roentgenium	112 **Cn** copernicium	113 **Nh** nihonium	114 **Fl** flerovium	115 **Mc** moscovium	116 **Lv** livermorium	117 **Ts** tennessine	118 **Og** oganesson

63 **Eu** europium 151.96	64 **Gd** gadolinium 157.25(3)	65 **Tb** terbium 158.93	66 **Dy** dysprosium 162.50	67 **Ho** holmium 164.93	68 **Er** erbium 167.26	69 **Tm** thulium 168.93	70 **Yb** ytterbium 173.05	71 **Lu** lutetium 174.97
95 **Am** mericium	96 **Cm** curium	97 **Bk** berkelium	98 **Cf** californium	99 **Es** einsteinium	100 **Fm** fermium	101 **Md** mendelevium	102 **No** nobelium	103 **Lr** lawrencium

Perry's Chemical Engineers' Handbook

ABOUT THE EDITORS

Dr. Don W. Green is Emeritus Distinguished Professor of Chemical and Petroleum Engineering at the University of Kansas (KU). He holds a B.S. in petroleum engineering from the University of Tulsa, and M.S. and Ph.D. degrees in chemical engineering from the University of Oklahoma. He is the coeditor of the sixth edition of *Perry's Chemical Engineers' Handbook*, and editor of the seventh and eighth editions. He has authored/coauthored 70 refereed publications, over 100 technical meeting presentations, and is coauthor of the first and second editions of the SPE textbook *Enhanced Oil Recovery*. Dr. Green has won numerous teaching awards at KU, including the *Honors for Outstanding Progressive Educator* (HOPE) Award and the Chancellor's Club Career Teaching Award, the highest teaching recognitions awarded at the University. He has also been featured as an outstanding educator in ASEE's Chemical Engineering Education Journal. He received the KU School of Engineering *Distinguished Engineering Service Award* (DESA), and has been designated an *Honorary Member* of both SPE and AIME and a Fellow of the AIChE.

Dr. Marylee Z. Southard is Associate Professor of Chemical and Petroleum Engineering at the University of Kansas. She holds B.S., M.S., and Ph.D. degrees in chemical engineering from the University of Kansas. Dr. Southard's research deals with small molecule drug formulations; but her industrial background is in production and process development of inorganic chemical intermediates. Dr. Southard's work in inorganic chemicals production has included process engineering, design, and product development. She has consulted for industrial and pharmaceutical chemical production and research companies. She teaches process design and project economics, and has won several university-wide teaching awards, including the *Honors for Outstanding Progressive Educator* (HOPE) Award and the Kemper Teaching Fellowship. She has authored 1 patent, 15 refereed publications, and numerous technical presentations. Her research interests are in biological and pharmaceutical mass transport. She is a senior member of AIChE and ASEE.

PERRY'S CHEMICAL ENGINEERS' HANDBOOK

NINTH EDITION

McGraw Hill Education

New York
Chicago
San Francisco
Athens
London
Madrid
Mexico City
Milan
New Delhi
Singapore
Sydney
Toronto

Editor-in-Chief
Don W. Green
Emeritus Distinguished Professor of
Chemical and Petroleum Engineering,
University of Kansas

Associate Editor
Marylee Z. Southard
Associate Professor of Chemical & Petroleum
Engineering, University of Kansas

Library of Congress Control Number: 2017963193

McGraw-Hill Education books are available at special quantity discounts to use as premiums and sales promotions, or for use in corporate training programs. To contact a representative please visit the Contact Us page at www.mhprofessional.com.

Perry's Chemical Engineers' Handbook, Ninth Edition

3 4 5 6 7 8 9 LWI 23 22

ISBN 978-0-07-183408-7
MHID 0-07-183408-7

The pages within this book were printed on acid-free paper.

Sponsoring Editor
Robert Argentieri

Editorial Supervisor
Donna M. Martone

Production Supervisor
Pamela A. Pelton

Acquisitions Coordinator
Elizabeth Houde

Project Manager
Tania Andrabi,
Cenveo® Publisher Services

Copy Editors
Patti Scott
Kevin Campbell

Proofreader
Cenveo Publisher Services

Indexer
Cenveo Publisher Services

Art Director, Cover
Jeff Weeks

Composition
Cenveo Publisher Services

Contents

For the detailed contents of any section, consult the title page of that section.
See also the alphabetical index in the back of the handbook.

Contributors

D. Shabbir Ahmed, Ph.D. *Chemical Engineer, Chemical Sciences and Engineering Division, Argonne National Laboratory* (Section Editor, Sec. 24, Energy Resources, Conversion, and Utilization)

Brooke Albin, M.S.E. *Chemical Engineer, MATRIC (Mid-Atlantic Technology, Research and Innovation Center), Charleston, WV; Member, American Institute of Chemical Engineers, American Filtration Society (Crystallization from the Melt)* (Sec. 18, Liquid-Solid Operations and Equipment)

John Alderman, M.S., P.E., C.S.P. *Managing Partner, Hazard and Risk Analysis, LLC (Electrical Area Classification, Fire Protection Systems)* (Sec. 23, Process Safety)

Paul Amyotte, Ph.D., P.Eng. *Professor of Chemical Engineering and C.D. Howe Chair in Process Safety, Dalhousie University; Fellow, Chemical Institute of Canada; Fellow, Canadian Academy of Engineering (Dust Explosions)* (Sec. 23, Process Safety)

Frank A. Baczek, B.S. *Sr. Research Advisor, FLSmidth USA, Inc. (Gravity Sedimentation Operations)* (Sec. 18, Liquid-Solid Operations and Equipment)

Wayne E. Beimesch, Ph.D. *Technical Associate Director (Retired), Corporate Engineering, The Procter & Gamble Company (Drying Equipment, Operation and Troubleshooting)* (Sec. 12, Psychrometry, Evaporative Cooling, and Solids Drying)

Ray Bennett, Ph.D., P.E., CEFEI *Senior Principal Engineer, Baker Engineering and Risk Consultants, Inc.; Member, American Petroleum Institute 752, 753, and 756 (Estimation of Damage Effects)* (Sec. 23, Process Safety)

B. Wayne Bequette, Ph.D. *Professor of Chemical and Biological Engineering, Rensselaer Polytechnic Institute (Unit Operations Control, Advanced Control Systems)* (Sec. 8, Process Control)

Patrick M. Bernhagen, P.E., B.S. *Director of Sales—Fired Heater, Amec Foster Wheeler North America Corp.; API Subcommittee on Heat Transfer Equipment API 530, 536, 560, and 561 (Compact and Nontubular Heat Exchangers)* (Sec. 11, Heat-Transfer Equipment)

Michael J. Betenbaugh, Ph.D. *Professor of Chemical and Biomolecular Engineering, Johns Hopkins University; Member, American Institute of Chemical Engineers (Emerging Biopharmaceutical and Bioprocessing Technologies and Trends)* (Sec. 20, Bioreactions and Bioprocessing)

Lorenz T. Biegler, Ph.D. *Bayer Professor of Chemical Engineering, Carnegie Mellon University; Member, National Academy of Engineering* (Sec. 3, Mathematics)

Meherwan P. Boyce, Ph.D., P.E. *(Deceased) Chairman and Principal Consultant, The Boyce Consultancy Group, LLC; Fellow, American Society of Mechanical Engineers (U.S.); Fellow, National Academy Forensic Engineers (U.S.); Fellow, Institution of Mechanical Engineers (U.K.); Fellow, Institution of Diesel and Gas Turbine Engineers (U.K.); Registered Professional Engineer (Texas), Chartered Engineer (U.K.); Sigma Xi, Tau Beta Pi, Phi Kappa Phi.* (Section Coeditor, Sec. 10, Transport and Storage of Fluids)

Jeffrey Breit, Ph.D. *Principal Scientist, Capsugel; Member, American Association of Pharmaceutical Scientists (Product Attribute Control)* (Sec. 20, Bioreactions and Bioprocessing)

Laurence G. Britton, Ph.D. *Process Safety Consultant; Fellow, American Institute of Chemical Engineers; Fellow, Energy Institute; Member, Institute of Physics (U.K.) (Flame Arresters)* (Sec. 23, Process Safety)

Nathan Calzadilla, M.S.E. *Research Program Assistant, Johns Hopkins Medicine, Chemical and Biomolecular Engineering, Johns Hopkins University; Member, American Institute of Chemical Engineers (Emerging Biopharmaceutical and Bioprocessing Technologies and Trends)* (Sec. 20, Bioreactions and Bioprocessing)

John W. Carson, Ph.D. *President, Jenike & Johanson, Inc., Founding member and past chair of ASTM Subcommittee D18.24, "Characterization and Handling of Powders and Bulk Solids" (Bulk Solids Flow and Hopper Design)* (Sec. 21, Solids Processing and Particle Technology)

Giorgio Carta, Ph.D. *Lawrence R. Quarles Professor, Department of Chemical Engineering, University of Virginia; Member, American Institute of Chemical Engineers, American Chemical Society* (Section Coeditor, Sec. 16, Adsorption and Ion Exchange)

Jeffrey Chalmers, Ph.D. *Professor of Chemical and Biomolecular Engineering, The Ohio State University; Member, American Institute of Chemical Engineers; American Chemical Society; Fellow, American Institute for Medical and Biological Engineering* (Section Coeditor, Sec. 20, Bioreactions and Bioprocessing)

J. Wayne Chastain, B.S., P.E., CCPSC *Engineering Associate, Eastman Chemical Company; Member, American Institute of Chemical Engineers (Layer of Protection Analysis)* (Sec. 23, Process Safety)

Wu Chen, Ph.D. *Principal Research Scientist, The Dow Chemical Company; Fellow, American Filtration and Separations Society (Expression)* (Sec. 18, Liquid-Solid Operations and Equipment)

Martin P. Clouthier, M.Sc., P.Eng. *Director, Jensen Hughes Consulting Canada Ltd. (Dust Explosions)* (Sec. 23, Process Safety)

James R. Couper, D.Sc. *Professor Emeritus, The Ralph E. Martin Department of Chemical Engineering, University of Arkansas—Fayetteville* (Section Editor, Sec. 9, Process Economics)

Daniel A. Crowl, Ph.D., CCPSC *AIChE/CCPS Staff Consultant; Adjunct Professor, University of Utah; Professor Emeritus of Chemical Engineering, Michigan Technological University; Fellow, American Institute of Chemical Engineers; Fellow, AIChE Center for Chemical Process Safety* (Section Coeditor, Sec. 23, Process Safety)

Rita D'Aquino, M.E. *Consultant, Member, American Institute of Chemical Engineers (Pollution Prevention)* (Sec. 22, Waste Management)

Michael Davies, Ph.D. *President and CEO, Braunschweiger Flammenfilter GmbH (PROTEGO), Member, American Institute of Chemical Engineers; Member, National Fire Protection Association (Flame Arresters)* (Sec. 23, Process Safety)

Sheldon W. Dean, Jr., ScD, P.E. *President, Dean Corrosion Technology, Inc.; Fellow, Air Products and Chemicals, Inc., Retired; Fellow, ASTM; Fellow, NACE; Fellow, AIChE; Fellow, Materials Technology Institute (Corrosion Fundamentals, Corrosion Prevention)* (Sec. 25, Materials of Construction)

Dennis W. Dees, Ph.D. *Senior Electrochemical Engineer, Chemical Sciences and Engineering Division, Argonne National Laboratory (Electrochemical Energy Storage)* (Sec. 24, Energy Resources, Conversion, and Utilization)

Vinay P. Deodeshmukh, Ph.D. *Sr. Applications Development Manager—High Temperature and Corrosion Resistant Alloys, Haynes International Inc. (Corrosion Fundamentals, High-Temperature Corrosion, Nickel Alloys)* (Sec. 25, Materials of Construction)

Shrikant Dhodapkar, Ph.D. *Fellow, The Dow Chemical Company; Fellow, American Institute of Chemical Engineers (Gas-Solids Separations)* (Sec. 17, Gas–Solid Operations and Equipment); *(Feeding, Metering, and Dosing)* (Sec. 21, Solids Processing and Particle Technology)

David S. Dickey, Ph.D. *Consultant, MixTech, Inc.; Fellow, American Institute of Chemical Engineers; Member, North American Mixing Forum (NAMF); Member, American Chemical Society; Member, American Society of Mechanical Engineers; Member, Institute of Food Technology (Mixing and Processing of Liquids and Solids & Mixing of Viscous Fluids, Pastes, and Doughs)* (Sec. 18, Liquid-Solid Operations and Equipment)

Michael F. Doherty, Ph.D. *Professor of Chemical Engineering, University of California—Santa Barbara* (Section Editor, Sec. 13, Distillation)

Arthur M. Dowell, III, P.E., B.S. *President, A M Dowell III PLLC; Fellow, American Institute of Chemical Engineers; Senior Member, Instrumentation, Systems and Automation Society (Risk Analysis)* (Sec. 23, Process Safety)

Brandon Downey, B.A.Sc. *Principal Engineer, R&D, Lonza; Member, American Institute of Chemical Engineers (Product Attribute Control)* (Sec. 20, Bioreactions and Bioprocessing)

Karin Nordström Dyvelkov, Ph.D. *GEA Process Engineering A/S Denmark (Drying Equipment, Fluidized Bed Dryers, Spray Dryers)* (Sec. 12, Psychrometry, Evaporative Cooling, and Solids Drying)

Thomas F. Edgar, Ph.D. *Professor of Chemical Engineering, University of Texas—Austin* (Section Editor, Sec. 8, Process Control)

Victor H. Edwards, Ph.D., P.E. *Principal, VHE Technical Analysis; Fellow and Life Member, American Institute of Chemical Engineers; Member, American Association for the Advancement of Science, American Chemical Society, National Society of Professional Engineers; Life Member, New York Academy of Sciences; Registered Professional Engineer (Texas), Phi Lambda Upsilon, Sigma Tau* (Section Coeditor, Sec. 10, Transport and Storage of Fluids)

J. Richard Elliott, Ph.D. *Professor, Department of Chemical and Biomolecular Engineering, University of Akron; Member, American Institute of Chemical Engineers; Member, American Chemical Society; Member, American Society of Engineering Educators* (Section Coeditor, Sec. 4, Thermodynamics)

Dirk T. Van Essendelft, Ph.D. *Chemical Engineer, National Energy Technology Laboratory, U.S. Department of Energy (Coal)* (Sec. 24, Energy Resources, Conversion, and Utilization)

James R. Fair, Ph.D., P.E. *(Deceased) Professor of Chemical Engineering, University of Texas; Fellow, American Institute of Chemical Engineers; Member, American Chemical Society, American Society for Engineering Education, National Society of Professional Engineers (Section Editor of the 7th edition and major contributor to the 5th, 6th, and 7th editions)* (Sec. 14, Equipment for Distillation, Gas Absorption, Phase Dispersion, and Phase Separation)

Yi Fan, Ph.D. *Associate Research Scientist, The Dow Chemical Company (Solids Mixing)* (Sec. 21, Solids Processing and Particle Technology)

Paul S. Farber, P.E., M.S. *Principal, P. Farber & Associates, LLC, Willowbrook, Illinois; Member, American Institute of Chemical Engineers, Air & Waste Management Association* (Section Coeditor, Sec. 22, Waste Management)

Hans K. Fauske, D.Sc. *Emeritus President and Regent Advisor, Fauske and Associates, LLC; Fellow, American Institute of Chemical Engineers; Fellow, American Nuclear Society; Member, National Academy of Engineering (Pressure Relief Systems)* (Sec. 23, Process Safety)

Zbigniew T. Fidkowski, Ph.D. *Process Engineer, Evonik Industries (Distillation Systems, Batch Distillation)* (Sec. 13, Distillation)

Bruce A. Finlayson, Ph.D. *Rehnberg Professor Emeritus, Department of Chemical Engineering, University of Washington; Member, National Academy of Engineering* (Section Editor, Sec. 3, Mathematics)

Emory A. Ford, Ph.D. *Associate Director, Materials Technology Institute, Chief Scientist and Director of Research, Lyondell/Bassel Retired, Fellow Materials Technology Institute* (Section Coeditor, Sec. 25, Materials of Construction)

Gregory Frank, Ph.D. *Principal Engineer, Amgen Inc.; Fellow, American Institute of Chemical Engineers; Member, Society of Biological Engineering; North American Mixing Forum; Pharmaceutical Discovery, Development, and Manufacturing Forum* (Section Coeditor, Sec. 20, Bioreactions and Bioprocessing)

Timothy C. Frank, Ph.D. *Fellow, The Dow Chemical Company; Fellow, American Institute of Chemical Engineers* (Section Coeditor, Sec. 4, Thermodynamics; Sec. 15, Liquid-Liquid Extraction and Other Liquid-Liquid Operations and Equipment)

Walter L. Frank, B.S., P.E., CCPSC *President, Frank Risk Solutions, Inc.; AIChE/CCPS Staff Consultant; Fellow, American Institute of Chemical Engineers; Fellow, AIChE Center for Chemical Process Safety (Hazards of Vacuum, Hazards of Inerts)* (Sec. 23, Process Safety)

Ben J. Freireich, Ph.D. *Technical Director, Particulate Solid Research, Inc. (Solids Mixing, Size Enlargement)* (Sec. 21, Solids Processing and Particle Technology)

James D. Fritz, Ph.D. *Consultant, NACE International certified Material Selection Design Specialist; Member of the Metallic Materials and Materials Joining Subcommittees of the ASME Bioprocessing Equipment Standard, the Ferrous Specifications Subcommittee of the ASME Boiler & Pressure Vessel Code, and ASM International (Stainless Steels)* (Sec. 25, Materials of Construction)

Kevin L. Ganschow, B.S., P.E. *Senior Staff Materials Engineer, Chevron Corporation; Registered Professional Mechanical Engineer (California) (Ferritic Steels)* (Sec. 25, Materials of Construction)

Wayne J. Genck, Ph.D. *President, Genck International; consultant on crystallization and precipitation; Member, American Chemical Society, American Institute of Chemical Engineers, Association for Crystallization Technology, International Society of Pharmaceutical Engineers (ISPE)* (Section Editor, Sec. 18, Liquid-Solid Operations and Equipment)

Craig G. Gilbert, B.Sc. *Global Product Manager-Paste, FLSmidth USA, Inc.; Member, Society for Mining, Metallurgy, and Exploration; Mining and Metallurgical Society of America; Registered Professional Engineer (Gravity Sedimentation Operations)* (Sec. 18, Liquid-Solid Operations and Equipment)

Roy A. Grichuk, P.E. *Piping Director, Fluor, BSME, P.E.; Member, American Society of Mechanical Engineers, B31 Main Committee, B31MTC Committee, and B31.3 Committee; Registered Professional Engineer (Texas) (Piping)* (Sec. 10, Transport and Storage of Fluids)

Juergen Hahn, Ph.D. *Professor of Biomedical Engineering, Rensselaer Polytechnic Institute (Advanced Control Systems, Bioprocess Control)* (Sec. 8, Process Control)

Roger G. Harrison, Ph.D. *Professor of Chemical, Biological, and Materials Engineering and Professor of Biomedical Engineering, University of Oklahoma; Member, American Institute of Chemical Engineers, American Chemical Society, American Society for Engineering Education, Oklahoma Higher Education Hall of Fame; Fellow, American Institute for Medical and Biological Engineering (Downstream Processing: Primary Recovery and Purification)* (Sec. 20, Bioreactions and Bioprocessing)

John P. Hecht, Ph.D. *Technical Section Head, Drying and Particle Processing, The Procter & Gamble Company; Member, American Institute of Chemical Engineers* (Section Editor, Sec. 12, Psychrometry, Evaporative Cooling, and Solids Drying)

Matthew K. Heermann, P.E., B.S. *Consultant—Fossil Power Environmental Technologies, Sargent & Lundy LLC, Chicago, Illinois (Introduction to Waste Management and Regulatory Overview)* (Sec. 22, Waste Management)

Dennis C. Hendershot, M.S. *Process Safety Consultant; Fellow, American Institute of Chemical Engineers (Inherently Safer Design and Related Concepts, Hazard Analysis, Key Procedures)* (Sec. 23, Process Safety)

Taryn Herrera, B.S. *Process Engineer, Manager Separations Laboratory, FLSmidth USA, Inc. (Gravity Sedimentation Operations)* (Sec. 18, Liquid-Solid Operations and Equipment)

Darryl W. Hertz, B.S. *Senior Manager, Value Improvement Group, KBR, Houston, Texas (Front-End Loading, Value-Improving Practices)* (Sec. 9, Process Economics)

Bruce S. Holden, M.S. *Principal Research Scientist, The Dow Chemical Company; Fellow, American Institute of Chemical Engineers* (Sec. 15, Liquid-Liquid Extraction and Other Liquid-Liquid Operations and Equipment)

Predrag S. Hrnjak, Ph.D. *Will Stoecker Res. Professor of Mechanical Science and Engineering, University of Illinois at Urbana-Champaign; Principal Investigator—U of I Air Conditioning and Refrigeration Center; Assistant Professor, University of Belgrade; International Institute of Chemical Engineers; American Society of Heat, Refrigerating, and Air Conditioning Engineers (Refrigeration)* (Sec. 11, Heat-Transfer Equipment)

Lindell R. Hurst, Jr., M.S., P.E. *Senior Materials and Corrosion Engineer, Shell Global Solutions (US) Inc. Retired, Registered Professional Metallurgical Engineer (Alabama, Ohio, North Dakota)* (Section Coeditor, Sec. 25, Materials of Construction)

Karl V. Jacob, B.S. *Fellow, The Dow Chemical Company; Lecturer, University of Michigan; Fellow, American Institute of Chemical Engineers* (Section Editor, Sec. 21, Solids Processing and Particle Technology)

Pradeep Jain, M.S. *Senior Fellow, The Dow Chemical Company (Feeding, Metering, and Dosing)* (Sec. 21, Solids Processing and Particle Technology)

David Johnson, P.E., M.Ch.E. *Retired (Thermal Design of Heat Exchangers, Condensers, Reboilers)* (Sec. 11, Heat-Transfer Equipment)

Robert W. Johnson, M.S.Ch.E. *President, Unwin Company; Fellow, American Institute of Chemical Engineers* (Section Coeditor, Sec. 23, Process Safety)

Hugh D. Kaiser, P.E., B.S., M.B.A. *Principal Engineer, WSP USA; Fellow, American Institute of Chemical Engineers; Registered Professional Engineer (Indiana, Nebraska, Oklahoma, and Texas) (Storage and Process Vessels)* (Sec. 10, Transport and Storage of Fluids)

Ian C. Kemp, M.A. (Cantab) *Scientific Leader, GlaxoSmithKline; Fellow, Institution of Chemical Engineers; Associate Member, Institution of Mechanical Engineers (Psychrometry, Solids-Drying Fundamentals, Freeze Dryers)* (Sec. 12, Psychrometry, Evaporative Cooling, and Solids Drying); *(Pinch Analysis)* (Sec. 24, Energy Resources, Conversion, and Utilization)

Pradip R. Khaladkar, M.S., P.E. *Principal Consultant, Materials Engineering Group, Dupont Company (Retired), Registered Professional Engineer (Delaware), Fellow, Materials Technology Institute, St. Louis (Nonmetallic Materials)* (Sec. 25, Materials of Construction)

Henry Z. Kister, M.E., C.Eng., C.Sc. *Senior Fellow and Director of Fractionation Technology, Fluor Corporation; Member, National Academy of Engineering (NAE); Fellow, American Institute of Chemical Engineers; Fellow, Institution of Chemical Engineers (U.K.); Member, Institute of Energy* (Section Editor, Sec. 14, Equipment for Distillation, Gas Absorption, Phase Dispersion, and Phase Separation)

Kent S. Knaebel, Ph.D. *President, Adsorption Research, Inc.; Member, American Institute of Chemical Engineers, International Adsorption Society; Professional Engineer (Ohio)* (Mass Transfer Coeditor, Sec. 5, Heat and Mass Transfer)

Ted M. Knowlton, Ph.D. *Technical Consultant and Fellow, Particulate Solid Research, Inc.; Member, American Institute of Chemical Engineers* (Section Editor, Sec. 17, Gas–Solid Operations and Equipment)

James F. Koch, M.S. *Senior Process Engineering Specialist, The Dow Chemical Company (Size Reduction, Screening)* (Sec. 21, Solids Processing and Particle Technology)

Tim Langrish, D. Phil. *School of Chemical and Biomolecular Engineering, The University of Sydney, Australia (Solids-Drying Fundamentals, Cascading Rotary Dryers)* (Sec. 12, Psychrometry, Evaporative Cooling, and Solids Drying)

Tim J. Laros, M.S. *Owner, Filtration Technologies, LLC, Park City, UT; Member, Society for Mining, Metallurgy, and Exploration (Filtration)* (Sec. 18, Liquid-Solid Operations and Equipment)

Tiberiu M. Leib, Ph.D. *Principal Consultant, The Chemours Company (retired); Fellow, American Institute of Chemical Engineers* (Section Coeditor, Sec. 7, Reaction Kinetics; Sec. 19, Reactors)

M. Douglas LeVan, Ph.D. *J. Lawrence Wilson Professor of Engineering Emeritus, Department of Chemical and Biomolecular Engineering, Vanderbilt University; Member, American Institute of Chemical Engineers, American Chemical Society, International Adsorption Society* (Section Coeditor, Sec. 16, Adsorption and Ion Exchange)

Wenping Li, Ph.D. *R&D Director, Agrilectric Research Company; Member, American Filtration and Separations Society, American Institute of Chemical Engineers (Expression)* (Sec. 18, Liquid-Solid Operations and Equipment)

Eugene L. Liening, M.S., P.E. *Manufacturing & Engineering Technology Fellow, The Dow Chemical Company Retired; Fellow, Materials Technology Institute; Registered Professional Metallurgical Engineer (Michigan) (Corrosion Testing)* (Sec. 25, Materials of Construction)

Dirk Link, Ph.D. *Chemist, National Energy Technology Laboratory, U.S. Department of Energy (Nonpetroleum Liquid Fuels)* (Sec. 24, Energy Resources, Conversion, and Utilization)

Carl T. Lira, Ph.D. *Associate Professor, Department of Chemical and Materials Engineering, Michigan State University; Member, American Institute of Chemical Engineers; Member, American Chemical Society; Member, American Society of Engineering Educators* (Section Coeditor, Sec. 4, Thermodynamics)

Peter J. Loftus, D. Phil. *Chief Scientist, Primaira LLC, Member, American Society of Mechanical Engineers (Heat Generation)* (Sec. 24, Energy Resources, Conversion, and Utilization)

Michael F. Malone, Ph.D. *Professor of Chemical Engineering and Vice-Chancellor for Research and Engagement, University of Massachusetts—Amherst (Batch Distillation, Enhanced Distillation)* (Sec. 13, Distillation)

Paul E. Manning, Ph.D. *Director CRA Marketing and Business Development, Haynes International (Nickel Alloys)* (Sec. 25, Materials of Construction)

Chad V. Mashuga, Ph.D., P.E. *Assistant Professor of Chemical Engineering, Texas A&M University (Flammability, Combustion and Flammability Hazards, Explosions, Vapor Cloud Explosions, Boiling-Liquid Expanding-Vapor Explosions)* (Sec. 23, Process Safety)

Paul M. Mathias, Ph.D. *Senior Fellow and Technical Director, Fluor Corporation; Fellow, American Institute of Chemical Engineers* (Section Coeditor, Sec. 4, Thermodynamics); *(Design of Gas Absorption Systems)* (Sec. 14, Equipment for Distillation, Gas Absorption, Phase Dispersion, and Phase Separation)

Paul McCurdie, B.S. *Product Manager-Vacuum Filtration, FLSmidth USA, Inc. (Filtration)* (Sec. 18, Liquid-Solid Operations and Equipment)

James K. McGillicuddy, B.S. *Product Specialist, Centrifuges, Andritz Separation Inc.; Member, American Institute of Chemical Engineers (Centrifuges)* (Sec. 18, Liquid-Solid Operations and Equipment)

John D. McKenna, Ph.D. *Principal, ETS, Inc.; Member, American Institute of Chemical Engineers, Air and Waste Management Association (Air Pollution Management of Stationary Sources)* (Sec. 22, Waste Management)

Terence P. McNulty, Ph.D. *President, T. P. McNulty and Associates, Inc.; consultants in mineral processing and extractive metallurgy; Member, National Academy of Engineering; Member, American Institute of Mining, Metallurgical, and Petroleum Engineers; Member, Society for Mining, Metallurgy, and Exploration; Member, The Metallurgical Society; Member Mining and Metallurgical Society of America (Leaching)* (Sec. 18, Liquid-Solid Operations and Equipment)

Greg Mehos, Ph.D., P.E. *Senior Project Engineer, Jenike & Johanson, Inc. (Bulk Solids Flow and Hopper Design)* (Sec. 21, Solids Processing and Particle Technology)

Georges A. Melhem, Ph.D. *President and CEO, IoMosaic; Fellow, American Institute of Chemical Engineers (Emergency Relief Device Effluent Collection and Handling)* (Sec. 23, Process Safety)

Valerie S. Monical, B.S. *Fellow, Ascend Performance Materials, Inc. (Phase Separation)* (Sec. 14, Equipment for Distillation, Gas Absorption, Phase Dispersion, and Phase Separation)

Ronnie Montgomery *Technical Manager, Process Control Systems, IHI Engineering and Construction International Corporation; Member, Process Industries Practices, Process Controls Function Team; Member, International Society of Automation (Flow Measurement)* (Sec. 10, Transport and Storage of Fluids)

David A. Moore, B.Sc., M.B.A., P.E., C.S.P. *President, AcuTech Consulting Group; Member, ASSE, ASIS, NFPA (Security)* (Sec. 23, Process Safety)

Charles G. Moyers, Ph.D. *Senior Chemical Engineering Consultant, MATRIC (Mid-Atlantic Technology, Research and Innovation Center), Charleston, WV; Fellow, American Institute of Chemical Engineers (Crystallization from the Melt)* (Sec. 18, Liquid-Solid Operations and Equipment)

William E. Murphy, Ph.D., P.E. *Professor of Mechanical Engineering, University of Kentucky; American Society of Heating, Refrigerating, and Air-Conditioning Engineers; American Society of Mechanical Engineers; International Institute of Refrigeration (Air Conditioning)* (Sec. 11, Heat-Transfer Equipment)

Edward R. Naylor, B.S., M.S. *Senior Materials Engineering Associate, AkzoNobel; Certified API 510, 570, 653 and Fixed Equipment Source Inspector* (Section Coeditor, Sec. 25, Materials of Construction)

James J. Noble, Ph.D., P.E., Ch.E. [U.K.] *Research Affiliate, Department of Chemical Engineering, Massachusetts Institute of Technology; Fellow, American Institute of Chemical Engineers; Member, New York Academy of Sciences* (Heat Transfer Coeditor, Sec. 5, Heat and Mass Transfer)

W. Roy Penney, Ph.D., P.E. *Professor Emeritus, Department of Chemical Engineering, University of Arkansas; Fellow, American Institute of Chemical Engineers (Gas-in-Liquid Dispersions)* (Sec. 14, Equipment for Distillation, Gas Absorption, Phase Dispersion, and Phase Separation)

Clint Pepper, Ph.D. *Director, Lonza; Member, American Institute of Chemical Engineers (Product Attribute Control)* (Sec. 20, Bioreactions and Bioprocessing)

Carmo J. Pereira, Ph.D., M.B.A. *DuPont Fellow, E. I. du Pont de Nemours and Company; Fellow, American Institute of Chemical Engineers* (Section Coeditor, Sec. 7, Reaction Kinetics; Sec. 19, Reactors)

Demetri P. Petrides, Ph.D. *President, Intelligen, Inc.; Member, American Institute of Chemical Engineers, American Chemical Society (Downstream Processing: Primary Recovery and Purification)* (Sec. 20, Bioreactions and Bioprocessing)

Thomas H. Pratt, Ph.D., P.E., C.S.P. *Retired; Emeritus Member, NFPA 77 (Static Electricity)* (Sec. 23, Process Safety)

Richard W. Prugh, M.S., P.E., C.S.P. *Principal Process Safety Consultant, Chilworth Technology, Inc., a Dekra Company; Fellow, American Institute of Chemical Engineers; Member, National Fire Protection Association (Toxicity)* (Sec. 23, Process Safety)

Massood Ramezan, Ph.D., P.E. *Sr. Technical Advisor, KeyLogic Systems, Inc. (Coal Conversion)* (Sec. 24, Energy Resources, Conversion, and Utilization)

George A. Richards, Ph.D. *Mechanical Engineer, National Energy Technology Laboratory, U.S. Department of Energy (Natural Gas, Liquefied Petroleum Gas, Other Gaseous Fuels)* (Sec. 24, Energy Resources, Conversion, and Utilization)

John R. Richards, Ph.D. *Research Fellow, E. I. du Pont de Nemours and Company* (*retired*)*; Fellow, American Institute of Chemical Engineers* (*Polymerization Reactions*) (Sec. 7, Reaction Kinetics)

James A. Ritter, Ph.D. *L. M. Weisiger Professor of Engineering and Carolina Distinguished Professor, Department of Chemical Engineering, University of South Carolina; Member, American Institute of Chemical Engineers, American Chemical Society, International Adsorption Society (Sorption Equilibrium, Process Cycles, Equipment)* (Sec. 16, Adsorption and Ion Exchange)

Richard L. Rowley, Ph.D. *Professor Emeritus of Chemical Engineering, Brigham Young University* (Section Coeditor, Sec. 2, Physical and Chemical Data)

Scott R. Rudge, Ph.D. *Chief Operating Officer and Chairman, RMC Pharmaceutical Solutions, Inc.; Adjunct Professor, Chemical and Biological Engineering, University of Colorado; Vice President, Margaux Biologics, Scientific Advisory Board, Sundhin Biopharma (Downstream Processing: Primary Recovery and Purification); Member, American Chemical Society, International Society of Pharmaceutical Engineers, American Association for the Advancement of Science, Parenteral Drug Association (Downstream Processing: Primary Recovery and Purification)* (Sec. 20, Bioreactions and Bioprocessing)

Adel F. Sarofim, Sc.D. *Deceased; Presidential Professor of Chemical Engineering, Combustion, and Reactors, University of Utah; Member, American Institute of Chemical Engineers, American Chemical Society, Combustion Institute (Radiation)* (Sec. 5, Heat and Mass Transfer)

David K. Schmalzer, Ph.D., P.E. *Argonne National Laboratory (Retired), Member, American Chemical Society, American Institute of Chemical Engineers (Resources and Reserves, Liquid Petroleum Fuels)* (Sec. 24, Energy Resources, Conversion, and Utilization)

Fred Schoenbrunn, B.S. *Director-Sedimentation Products, Member, Society of Metallurgical and Exploration Engineers of the American Institute of Minting, Metallurgical and Petroleum Engineers; Registered Professional Engineer (Gravity Sedimentation Operations)* (Sec. 18, Liquid-Solid Operations and Equipment)

A. Frank Seibert, Ph.D., P.E. *Technical Manager, Separations Research Program, The University of Texas at Austin; Fellow, American Institute of Chemical Engineers* (Sec. 15, Liquid-Liquid Extraction and Other Liquid-Liquid Operations and Equipment)

Yongkoo Seol, Ph.D. *Geologist, National Energy Technology Laboratory, U.S. Department of Energy (Natural Gas)* (Sec. 24, Energy Resources, Conversion, and Utilization)

Lawrence J. Shadle, Ph.D. *Mechanical Engineer, National Energy Technology Laboratory, U.S. Department of Energy (Coke)* (Sec. 24, Energy Resources, Conversion, and Utilization)

Robert R. Sharp, P.E., Ph.D. *Environmental Consultant; Professor of Environmental Engineering, Manhattan College; Member, American Water Works Association; Water Environment Federation Section Director (Wastewater Management)* (Sec. 22, Waste Management)

Dushyant Shekhawat, Ph.D., P.E. *Chemical Engineer, National Energy Technology Laboratory, U.S. Department of Energy (Natural Gas, Fuel and Energy Costs)* (Sec. 24, Energy Resources, Conversion, and Utilization)

Richard L. Shilling, P.E., B.E.M.E. *Senior Engineering Consultant, Heat Transfer Research, Inc.; American Society of Mechanical Engineers* (Section Editor, Sec. 11, Heat-Transfer Equipment)

Nicholas S. Siefert, Ph.D., P.E. *Mechanical Engineer, National Energy Technology Laboratory, U.S. Department of Energy (Other Solid Fuels)* (Sec. 24, Energy Resources, Conversion, and Utilization)

Geoffrey D. Silcox, Ph.D. *Professor of Chemical Engineering, University of Utah; Member, American Institute of Chemical Engineers, American Chemical Society* (Heat Transfer Section Coeditor, Sec. 5, Heat and Mass Transfer)

Cecil L. Smith, Ph.D. *Principal, Cecil L. Smith Inc. (Batch Process Control, Telemetering and Transmission, Digital Technology for Process Control, Process Control and Plant Safety)* (Sec. 8, Process Control)

(Francis) Lee Smith, Ph.D. *Principal, Wilcrest Consulting Associates, LLC, Katy, Texas; Partner and General Manager, Albutran USA, LLC, Katy, Texas (Front-End Loading, Value-Improving Practices)* (Sec. 9, Process Economics); *(Evaporative Cooling)* (Sec. 12, Psychrometry, Evaporative Cooling, and Solids Drying); *(Energy Recovery)* (Sec. 24, Energy Resources, Conversion, and Utilization)

Joseph D. Smith, Ph.D. *Professor of Chemical and Biochemical Engineering, Missouri University of Science and Technology (Thermal Energy Conversion and Utilization)* (Sec. 24, Energy Resources, Conversion, and Utilization)

Daniel J. Soeder, M.S. *Director, Energy Resources Initiative, South Dakota School of Mines & Technology (Gaseous Fuels)* (Sec. 24, Energy Resources, Conversion, and Utilization)

Marylee Z. Southard, Ph.D. *Associate Professor of Chemical and Petroleum Engineering, University of Kansas; Senior Member, American Institute of Chemical Engineers; Member, American Society for Engineering Education* (Section Editor, Sec. 1, Unit Conversion Factors and Symbols); (Section Editor, Sec. 2, Physical and Chemical Data)

Thomas O. Spicer III, Ph.D., P.E. *Professor; Maurice E. Barker Chair in Chemical Engineering, Chemical Hazards Research Center Director, Ralph E. Martin Department of Chemical Engineering, University of Arkansas; Fellow, American Institute of Chemical Engineers (Atmospheric Dispersion)* (Sec. 23, Process Safety)

Jason A. Stamper, M. Eng. *Technology Leader, Drying and Particle Processing, The Procter & Gamble Company; Member, Institute for Liquid Atomization and Spray Systems (Drying Equipment, Fluidized Bed Dryers, Spray Dryers)* (Sec. 12, Psychrometry, Evaporative Cooling, and Solids Drying)

Daniel E. Steinmeyer, P.E., M.S. *Distinguished Science Fellow, Monsanto Company (retired); Fellow, American Institute of Chemical Engineers; Member, American Chemical Society (Phase Dispersion, Liquid in Gas Systems)* (Sec. 14, Equipment for Distillation, Gas Absorption, Phase Dispersion, and Phase Separation)

Gary J. Stiegel, P.E., M.S. *Technology Manager (Retired), National Energy Technology Laboratory, U.S. Department of Energy (Coal Conversion)* (Sec. 24, Energy Resources, Conversion, and Utilization)

Angela Summers, Ph.D., P.E. *President, SIS-TECH; Adjunct Professor, Department of Environmental Management, University of Houston–Clear Lake; Fellow, International Society of Automation; Fellow, American Institute of Chemical Engineers; Fellow, AIChE Center for Chemical Process Safety (Safety Instrumented Systems)* (Sec. 23, Process Safety)

Richard C. Sutherlin, B.S., P.E. *Richard Sutherlin, PE, Consulting, LLC; Registered Professional Metallurgical Engineer (Oregon) (Reactive Metals)* (Sec. 25, Materials of Construction)

Ross Taylor, Ph.D. *Distinguished Professor of Chemical Engineering, Clarkson University (Simulation of Distillation Processes)* (Sec. 13, Distillation)

Louis Theodore, Eng.Sc.D. *Consultant, Theodore Tutorials, Professor of Chemical Engineering, Manhattan College; Member, Air and Waste Management Association* (Section Coeditor, Sec. 22, Waste Management)

Susan A. Thorneloe, M.S. *U.S. EPA/Office of Research & Development, National Risk Management Research Laboratory; Member, Air and Waste Management Association, International Waste Working Group* (Sec. 22, Waste Management)

James N. Tilton, Ph.D., P.E. *DuPont Fellow, Chemical and Bioprocess Engineering, E. I. du Pont de Nemours & Co.; Member, American Institute of Chemical Engineers; Registered Professional Engineer (Delaware)* (Section Editor, Sec. 6, Fluid and Particle Dynamics)

Paul W. Todd, Ph.D. *Chief Scientist Emeritus, Techshot, Inc.; Member, American Institute of Chemical Engineers (Downstream Processing: Primary Recovery and Purification)* (Sec. 20, Bioreactions and Bioprocessing)

Krista S. Walton, Ph.D. *Professor and Robert "Bud" Moeller Faculty Fellow, School of Chemical & Biomolecular Engineering, Georgia Institute of Technology; Member, American Institute of Chemical Engineers, American Chemical Society, International Adsorption Society (Adsorbents)* (Sec. 16, Adsorption and Ion Exchange)

Phillip C. Wankat, Ph.D. *Clifton L. Lovell Distinguished Professor of Chemical Engineering Emeritus, Purdue University; Member, American Institute of Chemical Engineers* (Mass Transfer Coeditor, Sec. 5, Heat and Mass Transfer)

Kenneth N. Weiss, P.E., BCEE, B.Ch.E, M.B.A. *Managing Partner, ERM; Member, Air and Waste Management Association (Introduction to Waste Management and Regulatory Overview)* (Sec. 22, Waste Management)

W. Vincent Wilding, Ph.D. *Professor of Chemical Engineering, Brigham Young University; Fellow, American Institute of Chemical Engineers* (Section Coeditor, Sec. 2, Physical and Chemical Data)

Ronald J. Willey, Ph.D., P.E. *Professor, Department of Chemical Engineering, Northeastern University; Fellow, American Institute of Chemical Engineers (Case Histories)* (Sec. 23, Process Safety)

Todd W. Wisdom, M.S. *Director-Separations Technology, FLSmidth USA, Inc.; Member, American Institute of Chemical Engineers (Filtration)* (Sec. 18, Liquid-Solid Operations and Equipment)

John L. Woodward, Ph.D. *Senior Principal Consultant, Baker Engineering and Risk Consultants, Inc.; Fellow, American Institute of Chemical Engineers (Discharge Rates from Punctured Lines and Vessels)* (Sec. 23, Process Safety)

Preface to the Ninth Edition

"This handbook is intended to supply both the practicing engineer and the student with an authoritative reference work that covers comprehensively the field of chemical engineering as well as important related fields."

—*John H. Perry, 1934*

Chemical engineering is generally accepted to have had its origin in the United Kingdom (U.K.) during the latter part of the nineteenth century, largely in response to the industrial revolution and growth in the demand for industrial chemicals. To answer this demand, chemical companies began to mass-produce their products, which meant moving from batch processing to continuous operation. New processes and equipment, in turn, called for new methods. Initially, continuous reactions and processing were implemented largely by plant operators, mechanical engineers, and industrial chemists. Chemical engineering evolved from this advancement of the chemical industry, creating engineers who were trained in chemistry as well as the fundamentals of engineering, physics, and thermodynamics.

As an academic discipline, the earliest reported chemical engineering lectures were given in the United Kingdom. George Davis is generally recognized as the first chemical engineer, lecturing at the Manchester Technical School (later the University of Manchester) in 1887. The first American chemical engineering courses were taught at MIT in 1888. Davis also proposed an appropriate professional society that evolved with the industrial and academic profession, ultimately called the Society of Chemical Industry (1881). His initial proposal was for a society of chemical engineers but the name was changed because so few chemical engineers existed at that time. From there, the American Institute of Chemical Engineers, AIChE (1908), and the U.K.-origin Institution of Chemical Engineers, IChemE (1922), were created.

As the discipline advanced, important approaches to describing and designing chemical and physical processes developed. George Davis is credited with an early description of what came to be termed "unit operations," although he did not use that specific term. Arthur D. Little coined the phrase in 1908 in a report to the president of MIT and developed the concept and applications with William H. Walker. Walker later defined "unit operations" in his 1923 seminal textbook published by McGraw-Hill, *Principles of Chemical Engineering*, coauthored with Warren K. Lewis and William H. McAdams. Other concepts developed over time, including chemical reactor engineering, transport phenomena, and use of computers to enhance mathematical simulation, have increased our ability to understand and design chemical/physical industrial processes. Chemical engineering concepts and methods have been applied in increasingly diverse fields, including environmental engineering, pharmaceutical processing, microelectronics, and biological/biosimilar engineering.

The first known handbook of chemical engineering was in two volumes, written by George Davis, and published in the United Kingdom in 1901. A second edition followed in 1904. The emphasis was on materials and their properties; laboratory equipment and techniques; steam production and distribution; power and its applications; moving solids, liquids, and gases; and solids handling. In the preface, Davis acknowledged the advances in industrial chemistry made in Germany, especially in commercial organic chemistry. He also noted the "severe competition" coming from America "in the ammonia-soda industry." The first US handbook was edited by Donald M. Liddell and published by McGraw-Hill in 1922. It was a two-volume book with thirty-one contributing writers. It dealt with many of the same topics as in the Davis handbook, but also had significantly more emphasis on operations such as leaching, crystallization, evaporation, and drying.

Perry's Chemical Engineers' Handbook originated from a decision by McGraw-Hill in 1930 (during the Great Depression) to develop a new handbook of chemical engineering. Receiving support for the project from DuPont Company, they selected John H. Perry to be the editor. Perry had earned a Ph.D. from MIT in 1922 in Physical Chemistry and Chemical Engineering. He subsequently worked for the US Bureau of Mines, next as a chemist for a DuPont subsidiary in Cleveland, OH, then moved to Wilmington, DE, to work for DuPont as a chemist in the company's experimental station, and back to

Cleveland, still with DuPont. Family lore says that Perry was a very hard worker, dedicated to chemical engineering, and willing to basically live two lives: one as a full-time engineer for DuPont and the other as editor of the handbook. On weekends he would hitchhike to New York, go to the Chemist's Club with a packet of galley proofs and a carton of cigarettes, and work all weekend, sometimes for 24 hours at a time. His work on the book extended through 1933, leading to publication of the first edition in January 1934. There were 63 contributors, 14 from the DuPont Company and 21 from different universities, all experts in their respective technical areas. The first sentence in the preface was applicable then as well as for this ninth edition: "This handbook is intended to supply both the practicing engineer and the student with an authoritative reference work that covers comprehensively the field of chemical engineering as well as important related fields."

Several chemical engineers, serving as editor or coeditor, have guided the preparation of the different editions over the years. John H. Perry was editor of the first (1934), second (1941), and third (1950) editions before his untimely death in 1953. The position of editor passed to his only child, Robert H. Perry (Bob), a notable chemical engineer in his own right. Bob had a Ph.D. in chemical engineering from the University of Delaware and was working in industry at the time of his father's death. In 1958, he took a position as professor and later chair of the Department of Chemical Engineering at the University of Oklahoma. He was the editor of the fourth (1963) edition, coedited with Cecil H. Chilton and assisted by Sidney D. Kirkpatrick, and the fifth (1973) edition, coedited with Chilton.

For the sixth edition, Bob asked Don W. Green, his first Ph.D. student and now a professor of Chemical and Petroleum Engineering at the University of Kansas, to assist him. Tragically, Bob Perry's work on the handbook ceased when he was killed in an accident south of London in November 1978. Green assumed responsibility as editor and completed the sixth edition (1984), assisted by a colleague at KU, James O. Maloney. The first five editions were titled *The Chemical Engineers' Handbook*. Beginning with the sixth edition, the book was renamed *Perry's Chemical Engineers' Handbook* in honor of the father and son. Green was also editor of the seventh (1997) and eighth (2008) editions, with Maloney assisting on the seventh edition. Robert H. Perry was listed as the "late editor" for the seventh and eighth editions; honoring his ideas that carried over to these recent editions. To create the ninth edition, Green brought on Marylee Z. Southard, a colleague with industrial, consulting, and academic experience in chemical engineering.

The organization of this ninth edition replicates the logic of the eighth edition, although content changes are extensive. The first group of sections includes comprehensive tables with unit conversions and fundamental constants, physical and chemical data, methods to predict properties, and basics of mathematics most useful to engineers. The second group, comprising the fourth through the ninth sections, covers fundamentals of chemical engineering. The third and largest group of sections deals with processes, including heat transfer operations, distillation, gas–liquid processes, chemical reactors, and liquid–liquid processes. The last group of sections covers auxiliary information, including waste management, safety and handling of hazardous materials, energy sources, and materials of construction.

In 2012, McGraw-Hill launched Access Engineering (ACE), an electronic engineering reference tool for professionals, academics, and students. This edition of *Perry's Chemical Engineers' Handbook* is a part of ACE, as was the eighth edition. Beyond the complete text of the handbook, ACE provides:

- Interactive graphs
- Video tutorials for example problems given in the handbook
- Excel spreadsheets to solve guided and user-defined problems in different areas, such as heat transfer or fluid flow
- Curriculum maps for use in complementing engineering course content

All 25 sections have been updated to cover the latest advances in technology related to chemical engineering. Notable updates and completely new materials include:

- Sec. 2 includes new and updated chemical property data produced by the Design Institute for Physical Properties (DIPPR) of AIChE
- Sec. 4 on thermodynamics fundamentals has been redesigned to be more practical, and less theoretical than in earlier editions, to suit the practicing engineer and student pursuing applications
- A new Sec. 20, "Bioreactions and Bioprocessing," has been added in response to the significant, large-scale growth of commercial processes for nonfood products since the end of the twentieth century
- Sec. 21 on solids handling operations and equipment has been rewritten by industrial experts in their field

A group of 147 professionals, serving as section editors and contributors, has worked on this ninth edition. Their names, affiliations, and writing responsibilities are listed herein as part of the front material and on the title page of their respective sections. These authors are known experts in their field, with many having received professional awards and named as Fellows of their professional societies.

Since the publication of the eighth edition, we have lost two major contributors to *Perry's Chemical Engineers' Handbook*. Dr. Adel F. Sarofim died in December 2011. He was a section coeditor/contributor in the radiation subsection from the fifth edition (1973) through this current ninth edition. Dr. Sarofim, a Professor Emeritus at MIT, was a recognized pioneer in the development of combustion science and radiation heat transfer. He received numerous U.S. and international prizes for his work.

Dr. Meherwan P. Boyce died in December 2017. He was the editor for the "Transport and Storage of Fluids" section in the seventh edition and co-section editor for the eighth and current editions. Dr. Boyce was founder of Boyce Engineering International. He was also known for his role as the first director of the Turbomachinery Laboratory and founding member of the Turbomachinery Symposium.

On this 85th anniversary of *Perry's Chemical Engineers' Handbook*, we celebrate the memory of its creators, Dr. John H. Perry and Dr. Robert H. Perry. Often referred to as "the Bible of Chemical Engineering," this handbook is the gold standard as a source of valuable information to innumerable chemical engineers.

We dedicate this ninth edition to chemical engineers who carry on the profession, creating solutions, products, and processes needed in the challenging world ahead. We hope this edition will provide information and focus for you—to work for the quality and improvement of human life and the earth we inhabit.

DON W. GREEN
Editor-in-Chief

MARYLEE Z. SOUTHARD
Associate Editor

Section 1

Unit Conversion Factors and Symbols

Marylee Z. Southard, Ph.D. *Associate Professor of Chemical and Petroleum Engineering, University of Kansas; Senior Member, American Institute of Chemical Engineers; Member, American Society for Engineering Education*

UNITS AND SYMBOLS

CONVERSION FACTORS

UNITS AND SYMBOLS

TABLE 1-1 Standard SI Quantities and Units

Quantity or "dimension"	SI unit	SI unit symbol ("abbreviation")
Base quantity or "dimension"		
length	meter	m
mass	kilogram	kg
time	second	s
electric current	ampere	A
thermodynamic temperature	kelvin	K
amount of substance	mole*	mol
luminous intensity	candela	cd
Supplementary quantity or "dimension"		
plane angle	radian	rad
solid angle	steradian	sr

*When the mole is used, the elementary entities must be specified; they may be atoms, molecules, ions, electrons, other particles, or specified groups of such particles.

TABLE 1-2a Common Derived Units of SI

Quantity	Unit	Symbol
acceleration	meter per second squared	m/s^2
angular acceleration	radian per second squared	rad/s^2
angular velocity	radian per second	rad/s
area	square meter	m^2
concentration (mass)	kilogram per cubic meter	kg/m^3
concentration (molar)	mole per cubic meter	mol/m^3
current density	ampere per square meter	A/m^2
density, mass	kilogram per cubic meter	kg/m^3
electric charge density	coulomb per cubic meter	C/m^3
electric field strength	volt per meter	V/m
electric flux density	coulomb per square meter	C/m^2
energy density	joule per cubic meter	J/m^3
entropy	joule per kelvin	J/K
heat capacity	joule per kelvin	J/K
heat flux density, irradiance	watt per square meter	W/m^2
luminance	candela per square meter	cd/m^2
magnetic field strength	ampere per meter	A/m
molar energy	joule per mole	J/mol
molar entropy	joule per mole-kelvin	J/(mol·K)
molar heat capacity	joule per mole-kelvin	J/(mol·K)
moment of force	newton-meter	N·m
permeability	henry per meter	H/m
permittivity	farad per meter	F/m
radiance	watt per square meter-steradian	$W/(m^2 \cdot sr)$
radiant intensity	watt per steradian	W/sr
specific energy	joule per kilogram	J/kg
specific entropy	joule per kilogram-kelvin	J/(kg·K)
specific heat capacity	joule per kilogram-kelvin	J/(kg·K)
specific volume	cubic meter per kilogram	m^3/kg
surface tension	newton per meter	N/m
thermal conductivity	watt per meter-kelvin	W/(m·K)
velocity	meter per second	m/s
viscosity, dynamic	pascal-second	Pa·s
viscosity, kinematic	square meter per second	m^2/s
volume	cubic meter	m^3
wave number	reciprocal meter	1/m

TABLE 1-2b Derived Units of SI That Have Special Names

Quantity	Unit	Symbol	Formula
absorbed dose	gray	Gy	J/kg
activity (of radionuclides)	becquerel	Bq	1/s
capacitance	farad	F	C/V
conductance	siemens	S	A/V
electric potential, potential difference, electromotive force	volt	V	W/A
electric resistance	ohm	Ω	V/A
energy, work, quantity of heat	joule	J	N·m
force	newton	N	$(kg \cdot m)/s^2$
frequency (of a periodic phenomenon)	hertz	Hz	1/s
illuminance	lux	lx	lm/m^2
inductance	henry	H	Wb/A
luminous flux	lumen	lm	Cd·sr
magnetic flux	weber	Wb	V·s
magnetic flux density	tesla	T	Wb/m^2
power, radiant flux	watt	W	J/s
pressure, stress	pascal	Pa	N/m^2
quantity of electricity, electric charge	coulomb	C	A·s

TABLE 1-3 SI Prefixes

Multiplication factor	Prefix	Symbol
$1\ 000\ 000\ 000\ 000\ 000\ 000 = 10^{18}$	exa	E
$1\ 000\ 000\ 000\ 000\ 000 = 10^{15}$	peta	P
$1\ 000\ 000\ 000\ 000 = 10^{12}$	tera	T
$1\ 000\ 000\ 000 = 10^{9}$	giga	G
$1\ 000\ 000 = 10^{6}$	mega	M
$1\ 000 = 10^{3}$	kilo	k
$100 = 10^{2}$	hecto*	h
$10 = 10^{1}$	deka*	da
$0.1 = 10^{-1}$	deci*	d
$0.01 = 10^{-2}$	centi	c
$0.001 = 10^{-3}$	milli	m
$0.000\ 001 = 10^{-6}$	micro	μ
$0.000\ 000\ 001 = 10^{-9}$	nano	n
$0.000\ 000\ 000\ 001 = 10^{-12}$	pico	p
$0.000\ 000\ 000\ 000\ 001 = 10^{-15}$	femto	f
$0.000\ 000\ 000\ 000\ 000\ 001 = 10^{-18}$	atto	a

*Generally to be avoided.

TABLE 1-4 Greek Alphabet

alpha = Α, α	nu = Ν, ν
beta = Β, β	xi = Ξ, ξ
gamma = Γ, γ	omicron = Ο, ο
delta = Δ, δ	pi = Π, π
epsilon = Ε, ε	rho = Ρ, ρ
zeta = Ζ, ζ	sigma = Σ, σ
eta = Η, η	tau = Τ, τ
theta = Θ, θ	upsilon = Υ, υ
iota = Ι, ι	phi = Φ, ϕ
kappa = Κ, κ	chi = Χ, χ
lambda = Λ, λ	psi = Ψ, ψ
mu = Μ, μ	omega = Ω, ω

TABLE 1-5 United States Traditional System of Weights and Measures

Linear Measure

12 inches (in) or (″) = 1 foot (ft) or (′)
3 feet = 1 yard (yd)

16.5 feet, 5.5 yards = 1 rod (rd)

5280 feet, 320 rods = 1 mile (mi)

1 mil = 0.001 in

Nautical:

6080.2 feet = 1 nautical mile
6 feet = 1 fathom
120 fathoms = 1 cable length
1 knot (kn) = 1 nautical mile per hour
60 nautical miles = 1° of latitude

Square Measure

144 square inches (sq in) or (in^2) = 1 sq ft (ft^2)
9 sq ft (ft^2) = 1 sq yd (yd^2)
30.25 sq yd = 1 sq rod, pole, or perch
160 sq rods = {10 sq chains, 43.560 sq ft} = 1 acre
640 acres = 1 sq mi = 1 section
1 circular in (area of circle of 1-in diameter) = 0.7854 sq in
1 sq in = 1.2732 circular in
1 circular mil = area of circle of 0.001-in diameter
1,000,000 circular mils = 1 circular in

Circular Measure

60 seconds (″) = 1 minute or (′)
60 minutes (′) = 1 degree (1°)
90 degrees (90°) = 1 quadrant
360 degrees (360°) = 1 circumference
57.29578 degrees = {1 radian (rad), 57°17′44.81″}

Volume Measure

Solid:

1728 cubic in (cu in) (in^3) = 1 cubic foot (cu ft) (ft^3)
27 cu ft = 1 cubic yard (cu yd) (yd^3)

Dry Measure:

2 pints = 1 quart
8 quarts = 1 peck
4 pecks = 1 bushel
1 U.S. Winchester bushel = 2150.42 cubic inches (in^3)

Liquid:

4 gills = 1 pint (pt)
2 pints = 1 quart (qt)
4 quarts = 1 gallon (gal)
7.4805 gallons = 1 cubic foot (ft^3)

Apothecaries' Liquid:

60 minims (min. or ♏) = 1 fluid dram or drachm
8 drams (ʒ) = 1 fluid ounce
16 ounces (oz. ℥) = 1 pint

Avoirdupois Weight

16 drams = 437.5 grains (gr) = 1 ounce (oz)
16 ounces = 7000 grains = 1 pound (lb)
100 pounds = 1 hundredweight (cwt)
2000 pounds = 1 short ton; 2240 pounds = 1 long ton

Troy Weight

24 grains (gr) = 1 pennyweight (dwt)
20 pennyweights = 1 ounce (oz)
12 ounces = 1 pound (lb)

Apothecaries' Weight

20 grains (gr) = 1 scruple (℈)
3 scruples = 1 dram (ʒ)
8 drams = 1 ounce (℥)
12 ounces = 1 pound (lb)

CONVERSION FACTORS

TABLE 1-6 Common Units and Conversion Factors*

Mass (M)		
	1 pound mass	= 453.5924 grams
		= 0.45359 kilogram
		= 7000 grains
	1 slug	= 32.174 pounds mass
	1 ton (short)	= 2000 pounds mass
	1 ton (long)	= 2240 pounds mass
	1 ton (metric)	= 1000 kilograms
		= 2204.62 pounds mass
	1 pound-mole	= 453.59 gram-moles
Length (L)		
	1 foot	= 30.480 centimeters
		= 0.3048 meter
	1 inch	= 2.54 centimeters
		= 0.0254 meter
	1 mile (U.S.)	= 1.60935 kilometers
	1 yard	= 0.9144 meter
Area (L^2)		
	1 square foot	= 929.0304 square centimeters
		= 0.09290304 square meter
	1 square inch	= 6.4516 square centimeters
	1 square yard	= 0.836127 square meter
Volume (L^3)		
	1 cubic foot	= 28,316.85 cubic centimeters
		= 0.02831685 cubic meter
		= 28.31685 liters
		= 7.481 gallons (U.S.)
	1 gallon	= 3.7853 liters
		= 231 cubic inches
Time (θ)		
	1 hour (h)	= 60 minutes (min)
		= 3600 seconds (s)
Temperature (T)		
	1 centigrade or Celsius degree	= 1.8 Fahrenheit degrees
	Temperature, Kelvin	= T°C + 273.15
	Temperature, Rankine	= T°F + 459.7
	Temperature, Fahrenheit	= 9/5 T°C + 32
	Temperature, Celsius or centigrade	= 5/9 (T°F − 32)
	Temperature, Rankine	= 1.8T K
Force (F)		
	1 pound force	= 444,822.2 dynes
		= 4.448222 newtons (N)
		= 32.174 poundals
Pressure (F/L^2)		
	Normal atmospheric pressure	
	1 atm	= 760 millimeters of mercury at 0°C (density 13.5951 g/cm^3)
		= 29.921 inches of mercury at 32°F
		= 14.696 pounds force/square inch
		= 33.899 feet of water at 39.1°F
		= 1.01325 × 10^6 dynes/square centimeter
		= 1.01325 × 10^5 newtons/square meter
Density (M/L^3)		
	1 pound mass/cubic foot	= 0.01601846 gram/cubic centimeter
		= 16.01846 kilograms/cubic meter
Energy (H or FL)		
	1 British thermal unit	= 251.98 calories
		= 1054.4 joules
		= 777.97 foot-pounds force
		= 10.409 liter-atmospheres
		= 0.2930 watthour
Diffusivity (L^2/θ)		
	1 square foot/hour	= 0.258 cm^2/s
		= 2.58 × 10^{-5} m^2/s
Viscosity ($M/L\theta$)		
	1 pound mass/foot-hour	= 0.00413 g/cm s
		= 0.000413 kg/m s
	1 centipoise (cP)	= 0.01 poise (P)
		= 0.01 g/cm s
		= 0.001 kg/m s
		= 0.000672 lbm/ft s
		= 0.0000209 lb$_f$-s/ft^2
Thermal conductivity [$H/\theta\ L^2(T/L)$]		
	1 Btu/h ft^2 (°F/ft)	= 0.00413 cal/s cm^2 (°C/cm)
		= 1.728 J/s m^2 (°C/m)
Heat transfer coefficient		
	1 Btu/h ft^2 °F	= 5.678 J/s m^2 °C
Heat capacity (H/MT)		
	1 Btu/lbm °F	= 1 cal/g °C
		= 4184 J/kg °C
Gas constant		
	1.987 Btu/lbm mol °R	= 1.987 cal/mol K
		= 82.057 atm cm^3/mol K
		= 0.7302 atm ft^3/lbmol °F
		= 10.73 (lb$_f$/in^2)(ft^3)/lb mol °R
		= 1545 (lb$_f$/ft^2)(ft^3)/lb mol °R
		= 8.314 (N/m^2)(m^3)/mol K
Gravitational acceleration		
		g = 9.8066 m/s^2
		= 32.174 ft/s^2

NOTE: U.S. Customary units, or British units, on left and SI units on right.
*Adapted from Faust et al., *Principles of Unit Operations,* John Wiley & Sons, 1980.

TABLE 1-7 Alphabetical Listing of Common Unit Conversions

To Convert from	To	Multiply by
acres	square feet	43,560
acres	square meters	4074
acres	square miles	0.001563
acre-feet	cubic meters	1233
ampere-hours (absolute)	Coulombs (absolute)	3600
angstrom units	inches	3.937×10^{-9}
angstrom units	meters	1×10^{-10}
angstrom units	microns or micrometers	1×10^{-4}
atmospheres	millimeters of mercury at 32°F	760
atmospheres	dynes per square centimeter	1.0133×10^{6}
atmospheres	newtons per square meter	101,325
atmospheres	feet of water at 39.1°F	33.90
atmospheres	grams per square centimeter	1033.3
atmospheres	inches of mercury at 32°F	29.921
atmospheres	pounds per square foot	2116.3
atmospheres	pounds per square inch	14.696
bags (cement)	pounds (cement)	94
barrels (cement)	pounds (cement)	376
barrels (oil)	cubic meters	0.15899
barrels (oil)	gallons	42
barrels (U.S. liquid)	cubic meters	0.11924
barrels (U.S. liquid)	gallons	31.5
barrels per day	gallons per minute	0.02917
bars	atmospheres	0.9869
bars	newtons per square meter	1×10^{5}
bars	pounds per square inch	14.504
board feet	cubic feet	$\frac{1}{12}$
boiler horsepower	Btu per hour	33,480
boiler horsepower	kilowatts	9.803
Btu	calories (gram)	252
Btu	celsius heat units (chu or pcu)	0.55556
Btu	foot-pounds	777.9
Btu	horsepower-hours	3.929×10^{-4}
Btu	joules	1055.1
Btu	liter-atmospheres	10.41
Btu	pounds carbon to CO_2	6.88×10^{-5}
Btu	pounds water evaporated from and at 212°F	0.001036
Btu	cubic foot–atmospheres	0.3676
Btu	kilowatt-hours	2.930×10^{-4}
Btu per cubic foot	joules per cubic meter	37,260
Btu per hour	watts	0.29307
Btu per minute	horsepower	0.02357
Btu per pound	joules per kilogram	2326
Btu per pound per degree Fahrenheit	calories per gram per degree celsius	1
Btu per pound per degree Fahrenheit	joules per kilogram per degree kelvin	4186.8
Btu per second	watts	1054.4
Btu per square foot per hour	joules per square meter per second	3.1546
Btu per square foot per minute	kilowatts per square foot	0.1758
Btu per square foot per second for a temperature gradient of 1°F per inch	calories, gram (15°C), per square centimeter per second for a temperature gradient of 1°C per centimeter	1.2405
Btu (60°F) per degree Fahrenheit	calories per degree Celsius	453.6
Bushels (U.S. dry)	cubic feet	1.2444
Bushels (U.S. dry)	cubic meters	0.03524
calories, gram	Btu	3.968×10^{-3}
calories, gram	foot-pounds	3.087
calories, gram	joules	4.1868
calories, gram	liter-atmospheres	4.130×10^{-2}
calories, gram	horsepower-hours	1.5591×10^{-6}
calories, gram, per gram per degree C	joules per kilogram per kelvin	4186.8
calories, kilogram	kilowatt-hours	0.0011626
calories, kilogram per second	kilowatts	4.185
candle power (spherical)	lumens	12.556
carats (metric)	grams	0.2
centigrade heat units	Btu	1.8
centimeters	Angstrom units	1×10^{8}
centimeters	feet	0.03281
centimeters	inches	0.3937
centimeters	meters	0.01
centimeters	microns or micrometers	10,000
centimeters of mercury at 0°C	atmospheres	0.013158
centimeters of mercury at 0°C	feet of water at 39.1°F	0.4460
centimeters of mercury at 0°C	newtons per square meter	1333.2
centimeters of mercury at 0°C	pounds per square foot	27.845
centimeters of mercury at 0°C	pounds per square inch	0.19337
centimeters per second	feet per minute	1.9685
centimeters of water at 4°C	newtons per square meter	98.064
centistokes	square meters per second	1×10^{-6}
circular mils	square centimeters	5.067×10^{-6}
circular mils	square inches	7.854×10^{-7}
circular mils	square mils	0.7854
cords	cubic feet	128
cubic centimeters	cubic feet	3.532×10^{-5}
cubic centimeters	gallons	2.6417×10^{-4}
cubic centimeters	ounces (U.S. fluid)	0.03381
cubic centimeters	quarts (U.S. fluid)	0.0010567
cubic feet	Bushels (U.S.)	0.8036
cubic feet	cubic centimeters	28,317
cubic feet	cubic meters	0.028317
cubic feet	cubic yards	0.03704
cubic feet	gallons	7.481
cubic feet	liters	28.316
cubic foot–atmospheres	foot-pounds	2116.3
cubic foot–atmospheres	liter-atmospheres	28.316
cubic feet of water (60°F)	pounds	62.37
cubic feet per minute	cubic centimeters per second	472.0
cubic feet per minute	gallons per second	0.1247
cubic feet per second	gallons per minute	448.8
cubic feet per second	million gallons per day	0.64632
cubic inches	cubic meters	1.6387×10^{-5}
cubic yards	cubic meters	0.76456
curies	disintegrations per minute	2.2×10^{12}
curies	coulombs per minute	1.1×10^{12}
degrees	radians	0.017453
drams (apothecaries' or troy)	grams	3.888
drams (avoirdupois)	grams	1.7719
dynes	newtons	1×10^{-5}
ergs	joules	1×10^{-7}
Faradays	Coulombs (abs.)	96,500
fathoms	feet	6
feet	meters	0.3048
feet per minute	centimeters per second	0.5080
feet per minute	miles per hour	0.011364
feet per (second)^2	meters per (second)^2	0.3048
feet of water at 39.2°F	newtons per square meter	2989
foot-poundals	Btu	3.995×10^{-5}
foot-poundals	joules	0.04214
foot-poundals	liter-atmospheres	4.159×10^{-4}
foot-pounds	Btu	0.0012856
foot-pounds	calories, gram	0.3239
foot-pounds	foot-poundals	32.174
foot-pounds	horsepower-hours	5.051×10^{-7}
foot-pounds	kilowatt-hours	3.766×10^{-7}
foot-pounds	liter-atmospheres	0.013381
foot-pounds force	joules	1.3558
foot-pounds per second	horsepower	0.0018182
foot-pounds per second	kilowatts	0.0013558
furlongs	miles	0.125
gallons (U.S. liquid)	barrels (U.S. liquid)	0.03175

(Continued)

TABLE 1-7 Alphabetical Listing of Common Unit Conversions *(Continued)*

To Convert from	To	Multiply by
gallons	cubic meters	0.003785
gallons	cubic feet	0.13368
gallons	gallons (imperial)	0.8327
gallons	liters	3.785
gallons	ounces (U.S. fluid)	128
gallons per minute	cubic feet per hour	8.021
gallons per minute	cubic feet per second	0.002228
grains	grams	0.06480
grains	pounds	$\frac{1}{7000}$
grains per cubic foot	grams per cubic meter	2.2884
grains per gallon	parts per million	17.118
grams	drams (avoirdupois)	0.5644
grams	drams (troy)	0.2572
grams	grains	15.432
grams	kilograms	0.001
grams	pounds (avoirdupois)	0.0022046
grams	pounds (troy)	0.002679
grams per cubic centimeter	pounds per cubic foot	62.43
grams per cubic centimeter	pounds per gallon	8.345
grams per liter	grains per gallon	58.42
grams per liter	pounds per cubic foot	0.0624
grams per square centimeter	pounds per square foot	2.0482
grams per square centimeter	pounds per square inch	0.014223
hectares	acres	2.471
hectares	square meters	10,000
horsepower (British)	btu per minute	42.42
horsepower (British)	btu per hour	2545
horsepower (British)	foot-pounds per minute	33,000
horsepower (British)	foot-pounds per second	550
horsepower (British)	watts	745.7
horsepower (British)	horsepower (metric)	1.0139
horsepower (British)	pounds carbon to CO_2 per hour	0.175
horsepower (British)	pounds water evaporated per hour at 212°F	2.64
horsepower (metric)	foot-pounds per second	542.47
horsepower (metric)	kilogram-meters per second	75.0
hours (mean solar)	seconds	3600
inches	meters	0.0254
inches of mercury at 60°F	newtons per square meter	3376.9
inches of water at 60°F	newtons per square meter	248.84
joules (absolute)	Btu (mean)	9.480×10^{-4}
joules (absolute)	calories, gram (mean)	0.2389
joules (absolute)	cubic foot–atmospheres	0.3485
joules (absolute)	foot-pounds	0.7376
joules (absolute)	kilowatt-hours	2.7778×10^{-7}
joules (absolute)	liter-atmospheres	0.009869
kilocalories	joules	4186.8
kilograms	pounds (avoirdupois)	2.2046
kilograms force	newtons	9.807
kilograms per square centimeter	pounds per square inch	14.223
kilometers	miles	0.6214
kilowatt-hours	Btu	3414
kilowatt-hours	foot-pounds	2.6552×10^{6}
kilowatts	horsepower	1.3410
knots (international)	meters per second	0.5144
knots (nautical miles per hour)	miles per hour	1.1516
lamberts	candles per square inch	2.054
liter-atmospheres	cubic foot–atmospheres	0.03532
liter-atmospheres	foot-pounds	74.74
liters	cubic feet	0.03532
liters	cubic meters	0.001
liters	gallons	0.26418
lumens	watts	0.001496
micromicrons	microns or micrometers	1×10^{-6}
microns	angstrom units	1×10^{4}
microns	meters	1×10^{-6}
miles (nautical)	feet	6080
miles (nautical)	miles (U.S. statute)	1.1516
miles	feet	5280
miles	meters	1609.3
miles per hour	feet per second	1.4667
miles per hour	meters per second	0.4470
milliliters	cubic centimeters	1
millimeters	meters	0.001
millimeters of mercury at 0°C	newtons per square meter	133.32
millimicrons	microns	0.001
mils	inches	0.001
mils	meters	2.54×10^{-5}
minims (U.S.)	cubic centimeters	0.06161
minutes (angle)	radians	2.909×10^{-4}
minutes (mean solar)	seconds	60
newtons	kilograms	0.10197
ounces (avoirdupois)	kilograms	0.02835
ounces (avoirdupois)	ounces (troy)	0.9115
ounces (U.S. fluid)	cubic meters	2.957×10^{-5}
ounces (troy)	ounces (apothecaries')	1.000
pints (U.S. liquid)	cubic meters	4.732×10^{-4}
poundals	newtons	0.13826
pounds (avoirdupois)	grains	7000
pounds (avoirdupois)	kilograms	0.45359
pounds (avoirdupois)	pounds (troy)	1.2153
pounds per cubic foot	grams per cubic centimeter	0.016018
pounds per cubic foot	kilograms per cubic meter	16.018
pounds per square foot	atmospheres	4.725×10^{-4}
pounds per square foot	kilograms per square meter	4.882
pounds per square inch	atmospheres	0.06805
pounds per square inch	kilograms per square centimeter	0.07031
pounds per square inch	newtons per square meter	6894.8
pounds force	newtons	4.4482
pounds force per square foot	newtons per square meter	47.88
pounds water evaporated from and at 212°F	horsepower-hours	0.379
pound-celsius units (pcu)	Btu	1.8
quarts (U.S. liquid)	cubic meters	9.464×10^{-4}
radians	degrees	57.30
revolutions per minute	radians per second	0.10472
seconds (angle)	radians	4.848×10^{-6}
slugs	*g* pounds	1
slugs	kilograms	14.594
slugs	pounds	32.17
square centimeters	square feet	0.0010764
square feet	square meters	0.0929
square feet per hour	square meters per second	2.581×10^{-5}
square inches	square centimeters	6.452
square inches	square meters	6.452×10^{-4}
square yards	square meters	0.8361
stokes	square meters per second	1×10^{-4}
tons (long)	kilograms	1016
tons (long)	pounds	2240
tons (metric)	kilograms	1000
tons (metric)	pounds	2204.6
tons (metric)	tons (short)	1.1023
tons (short)	kilograms	907.18
tons (short)	pounds	2000
tons (refrigeration)	Btu per hour	12,000
tons (British shipping)	cubic feet	42.00
tons (U.S. shipping)	cubic feet	40.00
torr (mm mercury, 0°C)	newtons per square meter	133.32
watts	Btu per hour	3.413
watts	joules per second	1
watts	kilogram-meters per second	0.10197
watthours	joules	3600
yards	meters	0.9144

TABLE 1-8 Conversion Factors: Commonly Used and Traditional Units to SI Units

The following unit symbols are used in the table:

Unit symbol	Name	Unit symbol	Name	Unit symbol	Name
A	ampere	Gy	gray	Oe	oersted
a	annum (year)	H	henry	Ω	ohm
Bq	becquerel	h	hour	Pa	pascal
C	coulomb	ha	hectare	rad	radian
cd	candela	Hz	hertz	r	revolution
Ci	curie	J	joule	S	siemens
d	day	K	kelvin	s	second
°C	degree Celsius	L, ℓ, l	liter	″	second
°	degree	lm	lumen	sr	steradian
dyn	dyne	lx	lux	St	stokes
F	farad	m	meter	T	tesla
fc	footcandle	min	minute	t	tonne
G	gauss	′	minute	V	volt
g	gram	N	newton	W	watt
gr	grain	naut mi	U.S. nautical mile	Wb	weber

NOTE: Copyright SPE-AIME, *The SI Metric System of Units and SPE's Tentative Metric Standard*, Society of Petroleum Engineers, Dallas, 1977.

Quantity	Customary or commonly used unit	SI unit	Alternate SI unit	Conversion factor; multiply customary unit by factor to obtain SI unit	
		Space, time			
Length	naut mi	km		1.852*	E + 00
	mi	km		1.609 344*	E + 00
	chain	m		2.011 68*	E + 01
	link	m		2.011 68*	E − 01
	fathom	m		1.828 8*	E + 00
	yd	m		9.144*	E − 01
	ft	m		3.048*	E − 01
	in	cm		3.048*	E + 01
	in	mm		2.54*	E + 01
	mil	cm		2.54	E + 00
		μm		2.54*	E + 01
Length/length	ft/mi	m/km		1.893 939	E − 01
Length/volume	ft/U.S. gal	m/m^3		8.051 964	E + 01
	ft/ft^3	m/m^3		1.076 391	E + 01
	ft/bbl	m/m^3		1.917 134	E + 00
Area	mi^2	km^2		2.589 988	E + 00
	section	ha		2.589 988	E + 02
	acre	ha		4.046 856	E − 01
	ha	m^2		1.000 000*	E + 04
	yd^2	m^2		8.361 274	E − 01
	ft^2	m^2		9.290 304*	E − 02
	in^2	mm^2		6.451 6*	E + 02
		cm^2		6.451 6*	E + 00
Area/volume	ft^2/in^3	m^2/cm^3		5.699 291	E − 03
	ft^2/ft^3	m^2/m^3		3.280 840	E + 00
Volume	m^3	km^3		4.168 182	E + 00
	acre · ft	m^3		1.233 482	E + 03
		ha · m		1.233 482	E − 01
	yd^3	m^3		7.645 549	E + 01
	bbl (42 U.S. gal)	m^3		1.589 873	E − 01
	ft^3	m^3		2.831 685	E − 02
		dm^3	L	2.831 685	E + 01
	U.K. gal	m^3		4.546 092	E − 03
		dm^3	L	4.546 092	E + 00
	U.S. gal	m^3		3.785 412	E − 03
		dm^3	L	3.785 412	E + 00
	U.K. qt	dm^3	L	1.136 523	E + 00
	U.S. qt	dm^3	L	9.463 529	E − 01
	U.S. pt	dm^3	L	4.731 765	E − 01
	U.K. fl oz	cm^3		2.841 307	E + 01
	U.S. fl oz	cm^3		2.957 353	E + 01
	in^3	cm^3		1.638 706	E + 01
Volume/length (linear displacement)	bbl/in	m^{3}m		6.259 342	E + 00
	bbl/ft	m^3/m		5.216 119	E − 01
	ft^3/ft	m^3/m		9.290 304*	E − 02
	U.S. gal/ft	m^3/m		1.241 933	E − 02
		L/m		1.241 933	E + 01
Plane angle	rad	rad		1	
	deg (°)	rad		1.745 329	E − 02
	min (′)	rad		2.908 882	E − 04
	sec (″)	rad		4.848 137	E − 06
Solid angle	sr	sr		1	

*An asterisk indicates that the conversion factor is exact.

(*Continued*)

TABLE 1-8 Conversion Factors: Commonly Used and Traditional Units to SI Units *(Continued)*

Quantity	Customary or commonly used unit	SI unit	Alternate SI unit	Conversion factor; multiply customary unit by factor to obtain SI unit	
Time	year	a		1	
	week	d		7.0*	E + 00
	h	s		3.6*	E + 03
		min		6.0*	E + 01
	min	s		6.0*	E + 01
		h		1.666 667	E − 02
Mass, amount of substance					
Mass	U.K. ton	Mg	t	1.016 047	E + 00
	U.S. ton	Mg	t	9.071 847	E − 01
	U.K. cwt	kg		5.080 234	E + 01
	U.S. cwt	kg		4.535 924	E + 01
	lbm	kg		4.535 924	E − 01
	oz (troy)	g		3.110 348	E + 01
	oz (av)	g		2.834 952	E + 01
	gr	mg		6.479 891	E + 01
Amount of substance	lbmmol	kmol		4.535 924	E − 01
	std m^3 (0°C, 1 atm)	kmol		4.461 58	E − 02
	std ft^3 (60°F, 1 atm)	kmol		1.195 30	E − 03
Enthalpy, calorific value, heat, entropy, heat capacity					
Caloric value, enthalpy (mass basis)	Btu/lbm	MJ/kg		2.326 000	E − 03
		kJ/kg	J/g	2.326 000	E + 00
		kWh/kg		6.461 112	E − 04
	cal/g	kJ/kg	J/g	4.184*	E + 00
	cal/lbm	J/kg		9.224 141	E + 00
Caloric value, enthalpy (mole basis)	kcal/(g · mol)	kJ/kmol		4.184*	E + 03
	Btu/(lb · mol)	kJ/kmol		2.326 000	E + 00
Caloric value (volume basis—solids and liquids)	Btu/U.S. gal	MJ/m^3	kJ/dm^3	2.787 163	E − 01
		kJ/m^3		2.787 163	E + 02
		kWh/m^3		7.742 119	E − 02
	Btu/U.K. gal	MJ/m^3	kJ/dm^3	2.320 800	E − 01
		kJ/m^3		2.320 800	E + 02
	Btu/ft^3	kWh/m^3		6.446 667	E − 02
		MJ/m^3	kJ/dm^3	3.725 895	E − 02
		kJ/m^3		3.725 895	E + 01
		kWh/m^3		1.034 971	E − 02
	cal/mL	MJ/m^3		4.184*	E + 00
	(ft · lbf)/U.S. gal	kJ/m^3		3.581 692	E − 01
Caloric value (volume basis—gases)	cal/mL	kJ/m^3	J/dm^3	4.184*	E + 03
	kcal/m^3	kJ/m^3	J/dm^3	4.184*	E + 00
	Btu/ft^3	kJ/m^3	J/dm^3	3.725 895	E + 01
		kWh/m^3		1.034 971	E − 02
Specific entropy	Btu/(lbm · °R)	kJ/(kg · K)	J/(g · K)	4.186 8*	E + 00
	cal/(g · K)	kJ/(kg · K)	J/(g · K)	4.184*	E + 00
	kcal/(kg · °C)	kJ/(kg · K)	J/(g · K)	4.184*	E + 00
Specific heat capacity (mass basis)	kWh/(kg · °C)	kJ/(kg · K)	J/(g · K)	3.6*	E + 03
	Btu/(lbm · °F)	kJ/(kg · K)	J/(g · K)	4.186 8*	E + 00
	kcal/(kg · °C)	kJ/(kg · K)	J/(g · K)	4.184*	E + 00
Specific heat capacity (mole basis)	Btu/(lb · mol · °F)	kJ/(kmol · K)		4.186 8*	E + 00
	cal/(g · mol · °C)	kJ/(kmol · K)		4.184*	E + 00
Temperature, pressure, vacuum					
Temperature (absolute)	°R	K		5/9	
	K	K		1	
Temperature (traditional)	°F	°C		5/9(°F + 32)	
Temperature (difference)	°F	K, °C		5/9	
Pressure	atm (760 mmHg at 0°C or 14,696 psi)	MPa		1.013 250*	E − 01
		kPa		1.013 250*	E + 02
		bar		1.013 250*	E + 00
	bar	MPa		1.0*	E + 01
		kPa		1.0*	E + 02
	mmHg (0°C) = torr	MPa		6.894 757	E − 03
		kPa		6.894 757	E + 00
		bar		6.894 757	E − 02
	μmHg (0°C)	kPa		3.376 85	E + 00
	μ bar	kPa		2.488 4	E − 01
	mmHg = torr (0°C)	kPa		1.333 224	E − 01
	cmH_2O (4°C)	kPa		9.806 38	E − 02
	lb_f/ft^2 (psf)	kPa		4.788 026	E − 02
	mHg (0°C)	Pa		1.333 224	E − 01
	bar	Pa		1.0*	E + 05
	dyn/cm^2	Pa		1.0*	E − 01

*An asterisk indicates that the conversion factor is exact.

TABLE 1-8 Conversion Factors: Commonly Used and Traditional Units to SI Units *(Continued)*

Quantity	Customary or commonly used unit	SI unit	Alternate SI unit	Conversion factor; multiply customary unit by factor to obtain SI unit	
Vacuum, draft	inHg (60°F)	kPa		3.376 85	E + 00
	inH_2O (39.2°F)	kPa		2.490 82	E − 01
	inH_2O (60°F)	kPa		2.488 4	E − 01
	mmHg (0°C) = torr	kPa		1.333 224	E − 01
	cmH_2O (4°C)	kPa		9.806 38	E −02
Liquid head	ft	m		3.048*	E − 01
	in	mm		2.54*	E + 01
		cm		2.54*	E + 00
Pressure drop/length	psi/ft	kPa/m		2.262 059	E + 01
	Density, specific volume, concentration, dosage				
Density	lbm/ft^3	kg/m^3		1.601 846	E + 01
		g/m^3		1.601 846	E + 04
	lbm/U.S. gal	kg/m^3		1.198 264	E + 02
		g/cm^3		1.198 264	E − 01
	lbm/U.K. gal	kg/m^3		9.977 633	E + 01
	lbm/ft^3	kg/m^3		1.601 846	E + 01
		g/cm^3		1.601 846	E − 02
	g/cm^3	kg/m^3		1.0*	E + 03
	lbm/ft^3	kg/m^3		1.601 846	E + 01
Specific volume	ft^3/lbm	m^3/kg		6.242 796	E − 02
		m^3/g		6.242 796	E − 05
	ft^3/lbm	dm^3/kg		6.242 796	E + 01
	U.K. gal/lbm	dm^3/kg	cm^3/g	1.002 242	E + 01
	U.S. gal/lbm	dm^3/kg	cm^3/g	8.345 404	E + 00
Specific volume (mole basis)	L/(gmol)	$m^3/kmol$		1	
	$ft^3/(lbmol)$	$m^3/kmol$		6.242 796	E − 02
Specific volume	bbl/U.S. ton	m^3/t		1.752 535	E − 01
	bbl/U.K. ton	m^3/t		1.564 763	E − 01
Yield	bbl/U.S. ton	dm^3/t	L/t	1.752 535	E + 02
	bbl/U.K. ton	dm^3/t	L/t	1.564 763	E + 02
	U.S. gal/U.S. ton	dm^3/t	L/t	4.172 702	E + 00
	U.S. gal/U.K. ton	dm^3/t	L/t	3.725 627	E + 00
Concentration (mass/mass)	wt %	kg/kg		1.0*	E − 02
		g/kg		1.0*	E + 01
	wt ppm	mg/kg		1	
Concentration (mass/volume)	lbm/bbl	kg/m^3	g/dm^3	2.853 010	E + 00
	g/U.S. gal	kg/m^3		2.641 720	E − 01
	g/U.K. gal	kg/m^3	g/L	2.199 692	E − 01
	lbm/1000 U.S. gal	g/m^3	mg/dm^3	1.198 264	E + 02
	lbm/1000 U.K. gal	g/m^3	mg/dm^3	9.977 633	E + 01
	gr/U.S. gal	g/m^3	mg/dm^3	1.711 806	E + 01
	gr/ft^3	mg/m^3		2.288 351	E + 03
	lbm/1000 bbl	g/m^3	mg/dm^3	2.853 010	E + 00
	mg/U.S. gal	g/m^3	mg/dm^3	2.641 720	E − 01
	$gr/100\ ft^3$	mg/m^3		2.288 351	E + 01
Concentration (volume/volume)	ft^3/ft^3	m^3/m^3		1	
	bbl/(acreft)	m^3/m^3		1.288 931	E − 04
	vol %	m^3/m^3		1.0*	E − 02
	$U.K.\ gal/ft^3$	dm^3/m^3	L/m^3	1.605 437	E + 02
	$U.S.\ gal/ft^3$	dm^3/m^3	L/m^3	1.336 806	E + 02
	mL/U.S. gal	dm^3/m^3	L/m^3	2.641 720	E − 01
	mL/U.K. gal	dm^3/m^3	L/m^3	2.199 692	E − 01
	vol ppm	cm^3/m^3		1	
		dm^3/m^3	L/m^3	1.0*	E − 03
	U.K. gal/1000 bbl	cm^3/m^3		2.859 403	E + 01
	U.S. gal/1000 bbl	cm^3/m^3		2.380 952	E + 01
	U.K. pt/1000 bbl	cm^3/m^3		3.574 253	E + 00
Concentration (mole/volume)	(lbmol)/U.S. gal	$kmol/m^3$		1.198 264	E + 02
	(lbmol)/U.K. gal	$kmol/m^3$		9.977 644	E + 01
	$(lbmol)/ft^3$	$kmol/m^3$		1.601 846	E + 01
	std ft^3 (60°F, 1 atm)/bbl	$kmol/m^3$		7.518 21	E − 03
Concentration (volume/mole)	U.S. gal/1000 std ft^3 (60°F/60°F)	$dm^3/kmol$	L/kmol	3.166 91	E + 00
	bbl/million std ft^3 (60°F/60°F)	$dm^3/kmol$	L/kmol	1.330 10	E − 01

*An asterisk indicates that the conversion factor is exact.

(Continued)

TABLE 1-8 Conversion Factors: Commonly Used and Traditional Units to SI Units *(Continued)*

Quantity	Customary or commonly used unit	SI unit	Alternate SI unit	Conversion factor; multiply customary unit by factor to obtain SI unit	
		Facility throughput, capacity			
Throughput (mass basis)	U.K. ton/yr	t/a		1.016 047	E + 00
	U.S. ton/yr	t/a		9.071 847	E − 01
	U.K. ton/day	t/d		1.016 047	E + 00
		t/h		4.233 529	E − 02
	U.S. ton/day	t/d		9.071 847	E − 01
		t/h		3.779 936	E − 02
	U.K. ton/h	t/h		1.016 047	E + 00
	U.S. ton/h	t/h		9.071 847	E − 01
	lbm/h	kg/h		4.535 924	E − 01
Throughput (volume basis)	bbl/day	t/a		5.803 036	E + 01
		m^3/d		1.589 873	E − 01
	ft^3/day	m^3/h		1.179 869	E − 03
	bbl/h	m^3/h		1.589 873	E − 01
	ft^3/h	m^3/h		2.831 685	E − 02
	U.K. gal/h	m^3/h		4.546 092	E − 03
		L/s		1.262 803	E − 03
	U.S. gal/h	m^3/h		3.785 412	E − 03
		L/s		1.051 503	E − 03
	U.K. gal/min	m^3/h		2.727 655	E − 01
		L/s		7.576 819	E − 02
	U.S. gal/min	m^3/h		2.271 247	E − 01
		L/s		6.309 020	E − 02
Throughput (mole basis)	(lbmmol)/h	kmol/h		4.535 924	E − 01
		kmol/s		1.259 979	E − 04
		Flow rate			
Flow rate (mass basis)	U.K. ton/min	kg/s		1.693 412	E + 01
	U.S. ton/min	kg/s		1.511 974	E + 01
	U.K. ton/h	kg/s		2.822 353	E − 01
	U.S. ton/h	kg/s		2.519 958	E − 01
	U.K. ton/day	kg/s		1.175 980	E − 02
	U.S. ton/day	kg/s		1.049 982	E − 02
	million lbm/yr	kg/s		5.249 912	E + 00
	U.K. ton/yr	kg/s		3.221 864	E − 05
	U.S. ton/yr	kg/s		2.876 664	E − 05
	lbm/s	kg/s		4.535 924	E − 01
	lbm/min	kg/s		7.559 873	E − 03
	lbm/h	kg/s		1.259 979	E − 04
Flow rate (volume basis)	bbl/day	m^3/d		1.589 873	E − 01
		L/s		1.840 131	E − 03
	ft^3/day	m^3/d		2.831 685	E − 02
		L/s		3.277 413	E − 04
	bbl/h	m^3/s		4.416 314	E − 05
		L/s		4.416 314	E − 02
	ft^3/h	m^3/s		7.865 791	E − 06
		L/s		7.865 791	E − 03
	U.K. gal/h	dm^3/s	L/s	1.262 803	E − 03
	U.S. gal/h	dm^3/s	L/s	1.051 503	E − 03
	U.K. gal/min	dm^3/s	L/s	7.576 820	E − 02
	U.S. gal/min	dm^3/s	L/s	6.309 020	E − 02
	ft^3/min	dm^3/s	L/s	4.719 474	E − 01
	ft^3/s	dm^3/s	L/s	2.831 685	E + 01
Flow rate (mole basis)	(lbmol)/s	kmol/s		4.535 924	E − 01
	(lbmol)/h	kmol/s		1.259 979	E − 04
	million scf/D	kmol/s		1.383 45	E − 02
Flow rate/length (mass basis)	lbm/(sft)	kg/(sm)		1.488 164	E + 00
	lbm/(hft)	kg/(sm)		4.133 789	E − 04
Flow rate/length (volume basis)	U.K. gal/(min · ft)	m^2/s	m^3/(s · m)	2.485 833	E − 04
	U.S. gal/(min · ft)	m^2/s	m^3/(s · m)	2.069 888	E − 04
	U.K. gal/(h · in)	m^2/s	m^3/(s · m)	4.971 667	E − 05
	U.S. gal/(h · in)	m^2/s	m^3/(s · m)	4.139 776	E − 05
	U.K. gal/(h · ft)	m^2/s	m^3/(s · m)	4.143 055	E − 06
	U.S. gal/(h · ft)	m^2/s	m^3/(s · m)	3.449 814	E − 06
Flow rate/area (mass basis)	lbm/(s · ft^2)	kg/(s · m^2)		4.882 428	E + 00
	lbm/(h · ft^2)	kg/(s · m^2)		1.356 230	E − 03
Flow rate/area (volume basis)	ft^3/(s · ft^2)	m/s	m^3/(s · m^2)	3.048*	E − 01
	ft^3/(min · ft^2)	m/s	m^3/(s · m^2)	5.08*	E − 03
	U.K. gal/(h · in^2)	m/s	m^3/(s · m^2)	1.957 349	E − 03
	U.S. gal/(h · in^2)	m/s	m^3/(s · m^2)	1.629 833	E − 03
	U.K. gal/(min · ft^2)	m/s	m^3/(s · m^2)	8.155 621	E − 04
	U.S. gal/(min · ft^2)	m/s	m^3/(s · m^2)	6.790 972	E − 04
	U.K. gal/(h · ft^2)	m/s	m^3/(s · m^2)	1.359 270	E − 05
	U.S. gal/(h · ft^2)	m/s	m^3/(s · m^2)	1.131 829	E − 05

*An asterisk indicates that the conversion factor is exact.

TABLE 1-8 Conversion Factors: Commonly Used and Traditional Units to SI Units *(Continued)*

Quantity	Customary or commonly used unit	SI unit	Alternate SI unit	Conversion factor; multiply customary unit by factor to obtain SI unit	
		Energy, work, power			
Energy, work	therm	MJ		1.055 056	E + 02
		kJ		1.055 056	E + 05
		kWh		2.930 711	E + 01
	U.S. tonf · mi	MJ		1.431 744	E + 01
	hp · h	MJ		2.684 520	E + 00
		kJ		2.684 520	E + 03
		kWh		7.456 999	E − 01
	ch · h or CV · h	MJ		2.647 780	E + 00
		kJ		2.647 780	E + 03
		kWh		7.354 999	E − 01
	kWh	MJ		3.6*	E + 00
		kJ		3.6*	E + 03
	Chu	kJ		1.899 101	E + 00
		kWh		5.275 280	E − 04
	Btu	kJ		1.055 056	E + 00
		kWh		2.930 711	E − 04
	kcal	kJ		4.184*	E + 00
	cal	kJ		4.184*	E − 03
	ft · lbf	kJ		1.355 818	E − 03
	lbf · ft	kJ		1.355 818	E − 03
	J	kJ		1.0*	E − 03
	$(\text{lbf}\cdot\text{ft}^2)/\text{s}^2$	kJ		4.214 011	E − 05
	erg	J		1.0*	E − 07
Impact energy	kgf · m	J		9.806 650*	E + 00
	lbf · ft	J		1.355 818	E + 00
Surface energy	erg/cm^2	mJ/m^2		1.0*	E + 00
Specific-impact energy	$(\text{kgf}\cdot\text{m})/\text{cm}^2$	J/cm^2		9.806 650*	E − 02
	$(\text{lbf}\cdot\text{ft})/\text{in}^2$	J/cm^2		2.101 522	E − 03
Power	million Btu/h	MW		2.930 711	E − 01
	tons of refrigeration	kW		3.516 853	E + 00
	Btu/s	kW		1.055 056	E + 00
	kW	kW		1	
	hydraulic horsepower (hhp)	kW		7.460 43	E − 01
	hp (electric)	kW		7.46*	E − 01
	hp [(550 ft · lbf)/s]	kW		7.456 999	E − 01
	ch or CV	kW		7.354 999	E − 01
	Btu/min	kW		1.758 427	E − 02
	(ft · lbf)/s	kW		1.355 818	E − 03
	kcal/h	W		1.162 222	E + 00
	Btu/h	W		2.930 711	E − 01
	(ft · lbf)/min	W		2.259 697	E − 02
Power/area	$\text{Btu/(s}\cdot\text{ft}^2)$	kW/m^2		1.135 653	E + 01
	$\text{cal/(h}\cdot\text{cm}^2)$	kW/m^2		1.162 222	E − 02
	$\text{Btu/(h}\cdot\text{ft}^2)$	kW/m^2		3.154 591	E − 03
Heat-release rate, mixing power	hp/ft^3	kW/m^3		2.633 414	E + 01
	$\text{cal/(h}\cdot\text{cm}^3)$	kW/m^3		1.162 222	E + 00
	$\text{Btu/(s}\cdot\text{ft}^3)$	kW/m^3		3.725 895	E + 01
	$\text{Btu/(h}\cdot\text{ft}^3)$	kW/m^3		1.034 971	E − 02
Cooling duty (machinery)	Btu/(bhp · h)	W/kW		3.930 148	E − 01
Specific fuel consumption (mass basis)	lbm/(hp · h)	mg/J	kg/MJ	1.689 659	E − 01
		kg/kWh		6.082 774	E − 01
Specific fuel consumption (volume basis)	$\text{m}^3\text{/kWh}$	$\text{dm}^3\text{/MJ}$	$\text{mm}^3\text{/J}$	2.777 778	E + 02
	U.S. gal/(hp · h)	$\text{dm}^3\text{/MJ}$	$\text{mm}^3\text{/J}$	1.410 089	E + 00
	U.K. pt/(hp · h)	$\text{dm}^3\text{/MJ}$	$\text{mm}^3\text{/J}$	2.116 806	E − 01
Fuel consumption	U.K. gal/mi	$\text{dm}^3\text{/100 km}$	L/100 km	2.824 807	E + 02
	U.S. gal/mi	$\text{dm}^3\text{/100 km}$	L/100 km	2.352 146	E + 02
	mi/U.S. gal	km/dm^3	km/L	4.251 437	E − 01
	mi/U.K. gal	km/dm^3	km/L	3.540 064	E − 01
Velocity (linear), speed	knot	km/h		1.852*	E + 00
	mi/h	km/h		1.609 344*	E + 00
	ft/s	m/s		3.048*	E − 01
		cm/s		3.048*	E + 01
	ft/min	m/s		5.08*	E − 03
	ft/h	mm/s		8.466 667	E − 02
	ft/day	mm/s		3.527 778	E − 03
		m/d		3.048*	E − 01
	in/s	mm/s		2.54*	E + 01
	in/min	mm/s		4.233 333	E − 01

*An asterisk indicates that the conversion factor is exact.

(Continued)

TABLE 1-8 Conversion Factors: Commonly Used and Traditional Units to SI Units *(Continued)*

Quantity	Customary or commonly used unit	SI unit	Alternate SI unit	Conversion factor; multiply customary unit by factor to obtain SI unit	
Corrosion rate	in/yr (ipy)	mm/a		2.54*	E + 01
	mil/yr	mm/a		2.54*	E − 02
Rotational frequency	r/min	r/s		1.666 667	E + 02
		rad/s		1.047 198	E − 01
Acceleration (linear)	ft/s^2	m/s^2		3.048*	E − 01
		cm/s^2		3.048*	E + 01
Acceleration (rotational)	rpm/s	rad/s^2		1.047 198	E − 01
Momentum	(lbm · ft)/s	(kg · m)/s		1.382 550	E − 01
Force	U.K. tonf	kN		9.964 016	E + 00
	U.S. tonf	kN		8.896 443	E + 00
	kgf	N		9.806 650*	E + 00
	lbf	N		4.448 222	E + 00
	dyn	mN		1.0	E − 02
Bending moment, torque	U.S. tonf · ft	kN · m		2.711 636	E + 00
	kgf · m	N · m		9.806 650*	E + 00
	lbf · ft	N · m		1.355 818	E + 00
	lbf · in	N · m		1.129 848	E − 01
Bending moment/length	(lbf · ft)/in	(N · m)/m		5.337 866	E + 01
	(lbf · in)/in	(N · m)/m		4.448 222	E + 00
Moment of inertia	lbm · ft^2	kg · m^2		4.214 011	E − 02
Stress	U.S. $tonf/in^2$	MPa	N/mm^2	1.378 951	E + 01
	kgf/mm^2	MPa	N/mm^2	9.806 650*	E + 00
	U.S. $tonf/ft^2$	MPa	N/mm^2	9.576 052	E − 02
	lbf/in^2 (psi)	MPa	N/mm^2	6.894 757	E − 03
	lbf/ft^2 (psf)	kPa		4.788 026	E − 02
	dyn/cm^2	Pa		1.0*	E − 01
Mass/length	lbm/ft	kg/m		1.488 164	E + 00
Mass/area structural loading,	U.S. ton/ft^2	Mg/m^2		9.764 855	E + 00
bearing capacity (mass basis)	lbm/ft^2	kg/m^2		4.882 428	E + 00
	Miscellaneous transport properties				
Diffusivity	ft^2/s	m^2/s		9.290 304*	E − 02
	m^2/s	mm^2/s		1.0*	E + 06
	ft^2/h	m^2/s		2.580 64*	E − 05
Thermal resistance	(°C · m^2 · h)/kcal	(K · m^2)/kW		8.604 208	E + 02
	(°F · ft^2 · h)/Btu	(K · m^2)/kW		1.761 102	E + 02
Heat flux	Btu/(h · ft^2)	kW/m^2		3.154 591	E − 03
Thermal conductivity	(cal · cm)/(s · cm^2 · °C)	W/(m · K)		4.184*	E + 02
	(Btu · ft)/(h · ft^2 · °F)	W/(m · K)		1.730 735	E + 00
		(kJ · m)/(h · m^2 · K)		6.230 646	E + 00
	(kcal · m)/(h · m^2 · °C)	W/(m · K)		1.162 222	E + 00
	(Btu · in)/(h · ft^2 · °F)	W/(m · K)		1.442 279	E − 01
	(cal · cm)/(h · cm^2 · °C)	W/(m · K)		1.162 222	E − 01
Heat-transfer coefficient	cal/(s · cm^2 · °C)	kW/(m^2 · K)		4.184*	E + 01
	Btu/(s · ft^2 · °F)	kW/(m^2 · K)		2.044 175	E + 01
	cal/(h · cm^2 · °C)	kW/(m^2 · K)		1.162 222	E − 02
	Btu/(h · ft^2 · °F)	kW/(m^2 · K)		5.678 263	E − 03
		kJ/(h · m^2 · K)		2.044 175	E + 01
	Btu/(h · ft^2 · °R)	kW/(m^2 · K)		5.678 263	E − 03
	kcal/(h · m^2 · °C)	kW/(m^2 · K)		1.162 222	E − 03
Volumetric heat-transfer	Btu/(s · ft^3 · °F)	kW/(m^3 · K)		6.706 611	E + 01
coefficient	Btu/(h · ft^3 · °F)	kW/(m^3 · K)		1.862 947	E − 02
Surface tension	dyn/cm	mN/m		1	
Viscosity (dynamic)	(lbf · s)/in^2	Pa · s	(N · s)/m^2	6.894 757	E + 03
	(lbf · s)/ft^2	Pa · s	(N · s)/m^2	4.788 026	E + 01
	(kgf · s)/m^2	Pa · s	(N · s)/m^2	9.806 650*	E + 00
	lbm/(ft · s)	Pa · s	(N · s)/m^2	1.488 164	E + 00
	(dyn · s)/cm^2	Pa · s	(N · s)/m^2	1.0*	E − 01
	cP	Pa · s	(N · s)/m^2	1.0*	E − 03
	lbm/(ft · h)	Pa · s	(N · s)/m^2	4.133 789	E − 04

*An asterisk indicates that the conversion factor is exact.

TABLE 1-8 Conversion Factors: Commonly Used and Traditional Units to SI Units *(Continued)*

Quantity	Customary or commonly used unit	SI unit	Alternate SI unit	Conversion factor; multiply customary unit by factor to obtain SI unit	
Viscosity (kinematic)	ft^2/s	m^2/s		9.290 304*	E − 02
	in^2/s	mm^2/s		6.451 6*	E + 02
	m^2/h	mm^2/s		2.777 778	E + 02
	ft^2/h	m^2/s		2.580 64*	E − 05
	cSt	mm^2/s		1	
Permeability	darcy	μm^2		9.869 233	E − 01
	millidarcy	μm^2		9.869 233	E − 04
Thermal flux	$Btu/(h \cdot ft^2)$	W/m^2		3.152	E + 00
	$Btu/(s \cdot ft^2)$	W/m^2		1.135	E + 04
	$cal/(s \cdot cm^2)$	W/m^2		4.184	E + 04
Mass-transfer coefficient	$(lbmol)/[h \cdot ft^2(lbmol/ft^3)]$	m/s		8.467	E − 05
	$(gmol)/[s \cdot m^2(gmol/L)]$	m/s		1.0	E + 01
	Electricity, magnetism				
Admittance	S	S		1	
Capacitance	μF	μF		1	
Charge density	C/mm^3	C/mm^3		1	
Conductance	S	S		1	
	℧ (mho)	S		1	
Conductivity	S/m	S/m		1	
	℧/m	S/m		1	
	m℧/m	mS/m		1	
Current density	A/mm^2	A/mm^2		1	
Displacement	C/cm^2	C/cm^2		1	
Electric charge	C	C		1	
Electric current	A	A		1	
Electric-dipole moment	C · m	C · m		1	
Electric-field strength	V/m	V/m		1	
Electric flux	C	C		1	
Electric polarization	C/cm^2	C/cm^2		1	
Electric potential	V	V		1	
	mV	mV		1	
Electromagnetic moment	$A \cdot m^2$	$A \cdot m^2$		1	
Electromotive force	V	V		1	
Flux of displacement	C	C		1	
Frequency	cycles/s	Hz		1	
Impedance	Ω	Ω		1	
Linear-current density	A/mm	A/mm		1	
Magnetic-dipole moment	Wb · m	Wb · m		1	
Magnetic-field strength	A/mm	A/mm		1	
	Oe	A/m		7.957 747	E + 01
	gamma	A/m		7.957 747	E + 04
Magnetic flux	mWb	mWb		1	
Magnetic-flux density	mT	mT		1	
	G	T		1.0*	E − 04
	gamma	nT		1	
Magnetic induction	mT	mT		1	
Magnetic moment	$A \cdot m^2$	$A \cdot m^2$		1	
Magnetic polarization	mT	mT		1	
Magnetic potential difference	A	A		1	
Magnetic-vector potential	Wb/mm	Wb/mm		1	
Magnetization	A/mm	A/mm		1	
Modulus of admittance	S	S		1	

*An asterisk indicates that the conversion factor is exact.

(Continued)

TABLE 1-8 Conversion Factors: Commonly Used and Traditional Units to SI Units *(Continued)*

Quantity	Customary or commonly used unit	SI unit	Alternate SI unit	Conversion factor; multiply customary unit by factor to obtain SI unit
Modulus of impedance	Ω	Ω		1
Mutual inductance	H	H		1
Permeability	μH/m	μH/m		1
Permeance	H	H		1
Permittivity	μF/m	μF/m		1
Potential difference	V	V		1
Quantity of electricity	C	C		1
Reactance	Ω	Ω		1
Reluctance	H^{-1}	H^{-1}		1
Resistance	Ω	Ω		1
Resistivity	Ω · cm Ω · m	Ω · cm Ω · m		1 1
Self-inductance	mH	mH		1
Surface density of change	mC/m^2	mC/m^2		1
Susceptance	S	S		1
Volume density of charge	C/mm^3	C/mm^3		1
Acoustics, light, radiation				
Absorbed dose	rad	Gy		1.0* E − 02
Acoustical energy	J	J		1
Acoustical intensity	W/cm^2	W/m^2		1.0* E + 04
Acoustical power	W	W		1
Sound pressure	N/m^2	N/m^2		1.0*
Illuminance	fc	lx		1.076 391 E + 01
Illumination	fc	lx		1.076 391 E + 01
Irradiance	W/m^2	W/m^2		1
Light exposure	fc · s	lx · s		1.076 391 E + 01
Luminance	cd/m^2	cd/m^2		1
Luminous efficacy	lm/W	lm/W		1
Luminous exitance	lm/m^2	lm/m^2		1
Luminous flux	lm	lm		1
Luminous intensity	cd	cd		1
Radiance	W/m^2 · sr	W/m^2 · sr		1
Radiant energy	J	J		1
Radiant flux	W	W		1
Radiant intensity	W/sr	W/sr		1
Radiant power	W	W		1
Wavelength	Å	nm		1.0* E − 01
Capture unit	10^{-3} cm^{-1}	m^{-1}		1.0* E + 01
			10^{-3} cm^{-1}	1
	m^{-1}	m^{-1}		1
Radioactivity	Ci	Bq		3.7* E + 10

*An asterisk indicates that the conversion factor is exact.

TABLE 1-9 Other Conversion Factors to SI Units

The first two digits of each numerical entry represent a power of 10. For example, the entry "–02 2.54" expresses the fact that 1 in = 2.54 × 10^{-2} m.

To Convert from	To	Multiply by
abampere	ampere	+01 1.00
abcoulomb	coulomb	+01 1.00
abfarad	farad	+09 1.00
abhenry	henry	–09 1.00
abmho	mho	+09 1.00
abohm	ohm	–09 1.00
abvolt	volt	–08 1.00
acre	meter^2	+03 4.046 856
ampere (international of 1948)	ampere	–01 9.998 35
angstrom	meter	–10 1.00
are	meter^2	+02 1.00
astronomical unit	meter	+11 1.495 978
atmosphere	newton/meter^2	+05 1.013 25
bar	newton/meter^2	+05 1.00
barn	meter^2	–28 1.00
barrel (petroleum 42 gal)	meter^3	–01 1.589 873
barye	newton/meter^2	–01 1.00
British thermal unit (ISO/TC 12)	joule	+03 1.055 06
British thermal unit (International Steam Table)	joule	+03 1.055 04
British thermal unit (mean)	joule	+03 1.055 87
British thermal unit (thermochemical)	joule	+03 1.054 350
British thermal unit (39°F)	joule	+03 1.059 67
British thermal unit (60°F)	joule	+03 1.054 68
bushel (U.S.)	meter^3	–02 3.523 907
cable	meter	+02 2.194 56
caliber	meter	–04 2.54
calorie (International Steam Table)	joule	+00 4.1868
calorie (mean)	joule	+00 4.190 02
calorie (thermochemical)	joule	+00 4.184
calorie (15°C)	joule	+00 4.185 80
calorie (20°C)	joule	+00 4.181 90
calorie (kilogram, International Steam Table)	joule	+03 4.186 8
calorie (kilogram, mean)	joule	+03 4.190 02
calorie (kilogram, thermochemical)	joule	+03 4.184
carat (metric)	kilogram	–04 2.00
Celsius (temperature)	kelvin	$t_K = t_C + 273.15$
centimeter of mercury (0°C)	newton/meter^2	+03 1.333 22
centimeter of water (4°C)	newton/meter^2	+01 9.806 38
chain (engineer's)	meter	+01 3.048
chain (surveyor's or Gunter's)	meter	+01 2.011 68
circular mil	meter^2	–10 5.067 074
cord	meter^3	+00 3.624 556
coulomb (international of 1948)	coulomb	–01 9.998 35
cubit	meter	–01 4.572
cup	meter^3	–04 2.365 882
curie	disintegration/second	+10 3.70
day (mean solar)	second (mean solar)	+04 8.64
day (sidereal)	second (mean solar)	+04 8.616 409
degree (angle)	radian	–02 1.745 329
denier (international)	kilogram/meter	–07 1.111 111
dram (avoirdupois)	kilogram	–03 1.771 845
dram (troy or apothecary)	kilogram	–03 3.887 934
dram (U.S. fluid)	meter^3	–06 3.696 691
dyne	newton	–05 1.00
electron volt	joule	–19 1.602 10
erg	joule	–07 1.00
Fahrenheit (temperature)	kelvin	$t_K = (5/9)(t_F + 459.67)$
Fahrenheit (temperature)	Celsius	$t_C = (5/9)(t_F - 32)$
farad (international of 1948)	farad	–01 9.995 05
faraday (based on carbon 12)	coulomb	+04 9.648 70
faraday (chemical)	coulomb	+04 9.649 57
faraday (physical)	coulomb	+04 9.652 19
fathom	meter	+00 1.828 8
fermi (femtometer)	meter	–15 1.00
fluid ounce (U.S.)	meter^3	–05 2.957 352
foot	meter	–01 3.048
foot (U.S. survey)	meter	–01 3.048 006
foot of water (39.2°F)	newton/meter^2	+03 2.988 98
footcandle	lumen/meter^2	+01 1.076 391
footlambert	candela/meter^2	+00 3.426 259
furlong	meter	+02 2.011 68
galileo	meter/second^2	–02 1.00
gallon (U.K. liquid)	meter^3	–03 4.546 087
gallon (U.S. dry)	meter^3	–03 4.404 883
gallon (U.S. liquid)	meter^3	–03 3.785 411
gamma	tesla	–09 1.00
gauss	tesla	–04 1.00
gilbert	ampere turn	–01 7.957 747
gill (U.K.)	meter^3	–04 1.420 652
gill (U.S.)	meter^3	–04 1.182 941
grad	degree (angular)	–01 9.00
grad	radian	–02 1.570 796
grain	kilogram	–05 6.479 891
gram	kilogram	–03 1.00
hand	meter	–01 1.016
hectare	meter^2	+04 1.00
henry (international of 1948)	henry	+00 1.000 495
hogshead (U.S.)	meter^3	–01 2.384 809
horsepower (550 ft lbf/s)	watt	+02 7.456 998
horsepower (boiler)	watt	+03 9.809 50
horsepower (electric)	watt	+02 7.46
horsepower (metric)	watt	+02 7.354 99
horsepower (U.K.)	watt	+02 7.457
horsepower (water)	watt	+02 7.460 43
hour (mean solar)	second (mean solar)	+03 3.60
hour (sidereal)	second (mean solar)	+03 3.590 170
hundredweight (long)	kilogram	+01 5.080 234
hundredweight (short)	kilogram	+01 4.535 923
inch	meter	–02 2.54
inch of mercury (32°F)	newton/meter^2	+03 3.386 389
inch of mercury (60°F)	newton/meter^2	+03 3.376 85
inch of water (39.2°F)	newton/meter^2	+02 2.490 82
inch of water (60°F)	newton/meter^2	+02 2.4884
joule (international of 1948)	joule	+00 1.000 165
kayser	1/meter	+02 1.00
kilocalorie (International Steam Table)	joule	+03 4.186 74
kilocalorie (mean)	joule	+03 4.190 02
kilocalorie (thermochemical)	joule	+03 4.184
kilogram mass	kilogram	+00 1.00
kilogram-force (kgf)	newton	+00 9.806 65
kilopound-force	newton	+00 9.806 65
kip	newton	+03 4.448 221
knot (international)	meter/second	–01 5.144 444
lambert	candela/meter^2	+04 $1/\pi$
lambert	candela/meter^2	+03 3.183 098
langley	joule/meter^2	+04 4.184
lbf (pound-force, avoirdupois)	newton	+00 4.448 221
lbm (pound-mass, avoirdupois)	kilogram	–01 4.535 923
league (British nautical)	meter	+03 5.559 552
league (international nautical)	meter	+03 5.556
league (statute)	meter	+03 4.828 032
light-year	meter	+15 9.460 55
link (engineer's)	meter	–01 3.048
link (surveyor's or Gunter's)	meter	–01 2.011 68
liter	meter^3	–03 1.00
lux	lumen/meter^2	+00 1.00
maxwell	weber	–08 1.00
meter	wavelengths Kr 86	+06 1.650 763
micrometer	meter	–06 1.00
mil	meter	–05 2.54
mile (U.S. statute)	meter	+03 1.609 344
mile (U.K. nautical)	meter	+03 1.853 184
mile (international nautical)	meter	+03 1.852
mile (U.S. nautical)	meter	+03 1.852
millibar	newton/meter^2	+02 1.00
millimeter of mercury (0°C)	newton/meter^2	+02 1.333 224

(Continued)

TABLE 1-9 Other Conversion Factors to SI Units *(Continued)*

The first two digits of each numerical entry represent a power of 10. For example, the entry "–02 2.54" expresses the fact that 1 in = 2.54×10^{-2} m.

To Convert from	To	Multiply by
minute (angle)	radian	–04 2.908 882
minute (mean solar)	second (mean solar)	+01 6.00
minute (sidereal)	second (mean solar)	+01 5.983 617
month (mean calendar)	second (mean solar)	+06 2.628
nautical mile (international)	meter	+03 1.852
nautical mile (U.S.)	meter	+03 1.852
nautical mile (U.K.)	meter	+03 1.853 184
oersted	ampere/meter	+01 7.957 747
ohm (international of 1948)	ohm	+00 1.000 495
ounce-force (avoirdupois)	newton	–01 2.780 138
ounce-mass (avoirdupois)	kilogram	–02 2.834 952
ounce-mass (troy or apothecary)	kilogram	–02 3.110 347
ounce (U.S. fluid)	meter^3	–05 2.957 352
pace	meter	–01 7.62
parsec	meter	+16 3.083 74
pascal	newton/meter^2	+00 1.00
peck (U.S.)	meter^3	–03 8.809 767
pennyweight	kilogram	–03 1.555 173
perch	meter	+00 5.0292
phot	lumen/meter^2	+04 1.00
pica (printer's)	meter	–03 4.217 517
pint (U.S. dry)	meter^3	–04 5.506 104
pint (U.S. liquid)	meter^3	–04 4.731 764
point (printer's)	meter	–04 3.514 598
poise	(newton-second)/meter^2	–01 1.00
pole	meter	+00 5.0292
pound-force (lbf avoirdupois)	newton	+00 4.448 221
pound-mass (lbm avoirdupois)	kilogram	–01 4.535 923
pound-mass (troy or apothecary)	kilogram	–01 3.732 417
poundal	newton	–01 1.382 549
quart (U.S. dry)	meter^3	–03 1.101 220
quart (U.S. liquid)	meter^3	–04 9.463 529
rad (radiation dose absorbed)	joule/kilogram	–02 1.00
Rankine (temperature)	kelvin	$t_K = (5/9)t_R$
rayleigh (rate of photon emission)	1/second-meter^2	+10 1.00
rhe	meter^2/(newton-second)	+01 1.00
rod	meter	+00 5.0292
roentgen	coulomb/kilogram	–04 2.579 76
rutherford	disintegration/second	+06 1.00
second (angle)	radian	–06 4.848 136

To Convert from	To	Multiply by
second (ephemeris)	second	+00 1.000 000
second (mean solar)	second (ephemeris)	Consult *American Ephemeris and Nautical Almanac*
second (sidereal)	second (mean solar)	–01 9.972 695
section	meter^2	+06 2.589 988
scruple (apothecary)	kilogram	–03 1.295 978
shake	second	–08 1.00
skein	meter	+02 1.097 28
slug	kilogram	+01 1.459 390
span	meter	–01 2.286
statampere	ampere	–10 3.335 640
statcoulomb	coulomb	–10 3.335 640
statfarad	farad	–12 1.112 650
stathenry	henry	+11 8.987 554
statmho	mho	–12 1.112 650
statohm	ohm	+11 8.987 554
statute mile (U.S.)	meter	+03 1.609 344
statvolt	volt	+02 2.997 925
stere	meter^3	+00 1.00
stilb	candela/meter^2	+04 1.00
stoke	meter^2/second	–04 1.00
tablespoon	meter^3	–05 1.478 676
teaspoon	meter^3	–06 4.928 921
ton (assay)	kilogram	–02 2.916 666
ton (long)	kilogram	+03 1.016 046
ton (metric)	kilogram	+03 1.00
ton (nuclear equivalent of TNT)	joule	+09 4.20
ton (register)	meter^3	+00 2.831 684
ton (short, 2000 lb)	kilogram	+02 9.071 847
tonne	kilogram	+03 1.00
torr (0°C)	newton/meter^2	+02 1.333 22
township	meter^2	+07 9.323 957
unit pole	weber	–07 1.256 637
volt (international of 1948)	volt	+00 1.000 330
watt (international of 1948)	watt	+00 1.000 165
yard	meter	–01 9.144
year (calendar)	second (mean solar)	+07 3.1536
year (sidereal)	second (mean solar)	+07 3.155 815
year (tropical)	second (mean solar)	+07 3.155 692
year 1900, tropical, Jan., day 0, hour 12	second (ephemeris)	+07 3.155 692
year 1900, tropical, Jan., day 0, hour 12	second	+07 3.155 692

TABLE 1-10 Temperature Conversion Formulas

°F = (°C × 5/9) + 32
°C = (°F − 32) × 5/9
°R = °F + 459.67
K = °C + 273.15
K = °R × 5/9

Temperature difference ΔT:
°F = °C × 9/5

TABLE 1-11 Density Conversion Formulas

$$°\text{Bé} = 145 - \frac{145}{\text{sp gr}} \text{(heavier than } H_2O) \qquad °\text{Bé} = \frac{140}{\text{sp gr}} - 130 \text{ (lighter than } H_2O)$$

$$°\text{Tw} = \frac{\text{sp gr } 60°/60°\text{F} - 1}{0.005}$$

$$°\text{API} = \frac{141.5}{\text{sp gr}} - 131.5$$

$$\left.\frac{\text{lb}}{\text{gal}}\right|_{T,P} = \text{sp gr}\big|_{T,P} \times 8.345406$$

$$\left.\frac{\text{lb}}{\text{ft}^3}\right|_{T,P} = \text{sp gr}\big|_{T,P} \times 62.42797$$

TABLE 1-12 Kinematic Viscosity Conversion Formulas

Viscosity scale	Range of t, s	Kinematic viscosity, stokes*
Saybolt Universal	$32 < t < 100$	$0.00226t - 1.95/t$
	$t > 100$	$0.00220t - 1.35/t$
Saybolt Furol	$25 < t < 40$	$0.0224t - 1.84/t$
	$t > 40$	$0.0216t - 0.60/t$
Redwood No. 1	$34 < t < 100$	$0.00260t - 1.79/t$
	$t > 100$	$0.00247t - 0.50/t$
Redwood Admiralty		$0.027t - 20/t$
Engler		$0.00147t - 3.74/t$

*1 stoke (St) = 1 $cm^2/s = 10^{-4}\ m^2/s$

TABLE 1-13 Values of the Ideal Gas Constant

Temp. scale	Pressure units	Volume units	Weight units	Energy units*	R Energy / (Weight · Temp)
Kelvin			g mol	calories	1.9872
			g mol	joules (abs)	8.3144
			g mol	joules (int)	8.3130
	atm	cm^3	g mol	atm · cm^3	82.057
	atm	liters	g mol	atm · liters	0.08205
	mmHg	liters	g mol	mmHg · liters	62.361
	bar	liters	g mol	bar · liters	0.08314
	kg/cm^2	liters	g mol	$kg/(cm^2)$(liters)	0.08478
	atm	ft^3	lb mol	atm · ft^3	1.314
	mmHg	ft^3	lb mol	mmHg · ft^3	998.9
			lb mol	chu or pcu	1.9872
Rankine			lb mol	Btu	1.9872
			lb mol	hph	0.0007805
			lb mol	kWh	0.0005819
	atm	ft^3	lb mol	atm · ft^3	0.7302
	in Hg	ft^3	lb mol	in Hg · ft^3	21.85
	mmHg	ft^3	lb mol	mmHg · ft^3	555.0
	lb/in^2 abs	ft^3	lb mol	(lb)(ft^3)/in^2	10.73
	lb/ft^2 abs	ft^3	lb mol	ft · lbf	1545.0

*Energy units are the product of pressure units and volume units.

TABLE 1-14 Fundamental Physical Constants

1 sec = 1.00273791 sidereal seconds	sec = mean solar second
g_0 = 9.80665 m/s^2	Definition: g_0 = standard gravity
1 liter = 0.001 cu m	
1 atm = 101,325 newtons/sq m	Definition: atm = standard atmosphere
1 mmHg (pressure) = ($\frac{1}{760}$) atm	mmHg (pressure) = standard millimeter mercury
= 133.3224 newtons/sq m	
1 int ohm = 1.000495 ± 0.000015 abs ohm	int = international; abs = absolute
1 int amp = 0.999835 ± 0.000025 abs amp	amp = ampere
1 int coul = 0.999835 ± 0.000025 abs coul	coul = coulomb
1 int volt = 1.000330 ± 0.000029 abs volt	
1 int watt = 1.000165 ± 0.000052 abs watt	
1 int joule = 1.000165 ± 0.000052 abs joule	
$T_{0°C}$ = 273.150 ± 0.010 K	Absolute temperature of the ice point, 0°C
$(PV)_{0°C}^{P=0} = (RT)_{0°C}$ = 2271.16 ± 0.04 abs joule/mole	
= 22,414.6 ± 0.4 cu cm atm/mole	PV = product for ideal gas at 0°C
= 22.4146 ± 0.0004 liter atm/mole	
R = 8.31439 ± 0.00034 abs joule/deg mole	R = gas constant per mole
= 1.98719 ± 0.00013 cal/deg mole	
= 82.0567 ± 0.0034 cu cm atm/deg mole	
= 0.0820567 ± 0.0000034 liter atm/deg mole	
ln 10 = 2.302585	ln = natural logarithm (base e)
R ln 10 = 19.14460 ± 0.00078 abs joule/deg mole	
= 4.57567 ± 0.00030 cal/deg mole	
N = (6.02283 ± 0.0022) × 10^{23}/mole	N = Avogadro number
h = (6.6242 ± 0.0044) × 10^{-34} joule s	h = Planck constant
c = (2.99776 ± 0.00008) × 10^8 m/s	c = velocity of light
$(h^2/8\pi^2k)$ = (4.0258 ± 0.0037) × 10^{-39} g sq cm deg	Constant in rotational partition function of gases
$(h/8\pi^2c)$ = (2.7986 ± 0.0018) × 10^{-39} g cm	Constant relating wave number and moment of inertia
$Z = Nhc$ = 11.9600 ± 0.0036 abs joule cm/mole	Z = constant relating wave number and energy per mole
= 2.85851 ± 0.0009 cal cm/mole	
$Z/R = hc/k = c_2$ = 1.43847 ± 0.00045 cm deg	c_2 = second radiation constant
f = 96,501.2 ± 10.0 int coul/g-equiv or int joule/int volt g-equiv	$\mathcal{F}$ = Faraday constant
= 96,485.3 ± 10.0 abs coul/g-equiv or abs joule/abs volt g-equiv	
= 23,068.1 ± 2.4 cal/int volt g-equiv	
= 23,060.5 ± 2.4 cal/abs volt g-equiv	e = electronic charge
e = (1.60199 ± 0.00060) × 10^{-19} abs coul	emu = electromagnetic unit of charge
= (1.60199 ± 0.00060) × 10^{-20} abs emu	esu = electrostatic unit of charge
= (4.80239 ± 0.00180) × 10^{-10} abs esu	
1 int electron-volt/molecule = 96,501.2 ± 10 int joule/mole	
= 23,068.1 ± 2.4 cal/mole	
1 abs electron-volt/molecule = 96,485.3 ± 10. abs joule/mole	
= 23,060.5 ± 2.4 cal/mole	
1 int electron-volt = (1.60252 ± 0.00060) × 10^{-12} erg	
1 abs electron-volt = (1.60199 ± 0.00060) × 10^{-12} erg	
hc = (1.23916 ± 0.00032) × 10^{-4} int electron-volt cm	Constant relating wave number and energy per molecule
= (1.23957 ± 0.00032) × 10^{-4} abs electron-volt cm	
k = (8.61442 ± 0.00100) × 10^{-5} int electron-volt/deg	k = Boltzmann constant
= (8.61727 ± 0.00100) × 10^{-5} abs electron-volt/deg	
= R/N = (1.38048 ± 0.00050) × 10^{-23} joule/deg	
1 IT cal = ($\frac{1}{860}$) = 0.00116279 int watt-h	Definition of IT cal: IT = International steam tables
= 4.18605 int joule	
= 4.18674 abs joule	
= 1.000654 cal	cal = thermochemical calorie
1 cal = 4.1840 abs joule	Definition: cal = thermochemical calorie
= 4.1833 int joule	
= 41.2929 ± 0.0020 cu cm atm	
= 0.0412929 ± 0.0000020 liter atm	
1 IT cal/g = 1.8 Btu/lb	Definition of Btu: Btu = IT British thermal unit
1 Btu = 251.996 IT cal	
= 0.293018 int watt-h	
= 1054.866 int joule	
= 1055.040 abs joule	
= 252.161 cal	cal = thermochemical calorie
1 horsepower = 550 ft-lb$_f$(wt)/s	Definition of horsepower (mechanical): lb (wt) = weight of 1 lb
= 745.578 int watt	at standard gravity
= 745.70 abs watt	Definition of inch: in = U.S. inch
1 in = (1/0.3937) = 2.54 cm	ft = U.S. foot (1 ft = 12 in)
1 ft = 0.304800610 m	
1 lb = 453.5924277 g	Definition: lb = avoirdupois pound
1 gal = 231 cu in	Definition: gal = U.S. gallon
= 0.133680555 cu ft	
= 3.785412 × 10^{-3} cu m	
= 3.785412 liter	

Section 2

Physical and Chemical Data

Marylee Z. Southard, Ph.D. *Associate Professor of Chemical and Petroleum Engineering, University of Kansas; Senior Member, American Institute of Chemical Engineers; Member, American Society for Engineering Education (Section Coeditor, Physical and Chemical Data)*

Richard L. Rowley, Ph.D. *Department of Chemical Engineering, Emeritus, Brigham Young University (Section Coeditor, Prediction and Correlation of Physical Properties)*

W. Vincent Wilding, Ph.D. *Professor of Chemical Engineering, Brigham Young University; Fellow, American Institute of Chemical Engineers (Section Coeditor, Prediction and Correlation of Physical Properties)*

DENSITIES OF MISCELLANEOUS MATERIALS

LATENT HEATS

SPECIFIC HEATS

PROPERTIES OF FORMATION AND COMBUSTION REACTIONS

HEATS OF SOLUTION

THERMAL EXPANSION AND COMPRESSIBILITY

THERMODYNAMIC PROPERTIES

TRANSPORT PROPERTIES

PREDICTION AND CORRELATION OF PHYSICAL PROPERTIES

GENERAL REFERENCES

Considerations of reader interest, space availability, the system or systems of units employed, copyright issues, etc., have all influenced the revision of material in previous editions for the present edition. Reference is made at numerous places to various specialized works and, when appropriate, to more general works. A listing of general works may be useful to readers in need of further information.

ASHRAE Handbook—Fundamentals, SI edition, ASHRAE, Atlanta, 2005; Benedek, P., and F. Olti, *Computer-Aided Chemical Thermodynamics of Gases and Liquids,* Wiley, New York, 1985; Brule, M. R., L. L. Lee, and K. E. Starling, *Chem. Eng.*, **86,** 25, Nov. 19, 1979, pp. 155–164; Cox, J. D., and G. Pilcher, *Thermochemistry of Organic and Organometallic Compounds,* Academic Press, New York, 1970; Cox, J. D., D. D. Wagman, and V. A. Medvedev, *CODATA Key Values for Thermodynamics,* Hemisphere Publishing Corp., New York, 1989; Daubert, T. E., R. P. Danner, H. M. Sibel, and C. C. Stebbins, *Physical and Thermodynamic Properties of Pure Chemicals: Data Compilation,* Taylor & Francis, Washington, 1997; Domalski, E. S., and E. D. Hearing, Heat capacities and entropies of organic compounds in the condensed phase, vol. 3, *J. Phys. Chem. Ref. Data* 25(1):1–525, Jan-Feb 1996; Dykyj, J., and M. Repas, Saturated vapor pressures of organic compounds, *Veda,* Bratislava, 1979 (Slovak); Dykyj, J., M. Repas, and J. Svoboda, Saturated vapor pressures of organic compounds, *Veda,* Bratislava, 1984 (Slovak); Glushko, V. P., ed., *Thermal Constants of Compounds,* Issues I–X, Moscow, 1965–1982 (Russian only); Gmehling, J., *Azeotropic Data,* 2 vols., VCH Weinheim, Germany, 1994; Gmehling, J., and U. Onken, *Vapor-Liquid Equilibrium Data Collection,* Dechema Chemistry Data Series, Frankfurt, 1977–1978; International Data Series, *Selected Data on Mixtures,* Series A: Thermodynamics Research Center, National Institute of Standards and Technology, Boulder, Colo.; Kaye, S. M., *Encyclopedia of Explosives and Related Items,* U.S. Army R&D command, Dover, N.J., 1980; King, M. B., *Phase Equilibrium in Mixtures,* Pergamon, Oxford, 1969; Landolt-Boernstein, Numerical Data and Functional Relationships in Science and Technology (New Series), http://www.springeronline.com/sgw/cda/frontpage/0,11855,4-10113-2-95859-0,00.html; Lide, D. R., *CRC Handbook of Chemistry and Physics,* 86th ed., CRC Press, Boca Raton, Fla., 2005; Lyman, W. J., W. F. Reehl, and D. H. Rosenblatt, *Handbook of Chemical Property Estimation Methods,* McGraw-Hill, New York, 1990; Majer, V., and V. Svoboda, *Enthalpies of Vaporization of Organic Compounds: A Critical Review and Data Compilation,* Blackwell Science, 1985; Majer V., V. Svoboda, and J. Pick, *Heats of Vaporization of Fluids,* Elsevier, Amsterdam, 1989 (general discussion); Marsh, K. N., *Recommended Reference Materials for the Realization of Physicochemical Properties,* Blackwell Science, 1987; *NIST-IUPAC Solubility Data Series,* Pergamon Press, http://www.iupac.org/publications/ci/1999/march/solubility.html; Ohse, R. W., and H. von Tippelskirch, *High Temp.—High Press.*, **9:**367–385, 1977; Ohse, R. W., *Handbook of Thermodynamic and Transport Properties of Alkali Metals,* Blackwell Science Pubs., Oxford, England, 1985; Pedley, J. B., R. D. Naylor, and S. P. Kirby, *Thermochemical Data of Organic Compounds,* Chapman and Hall, New York, 1986; *Physical Property Data for the Design Engineer,* Hemisphere, New York, 1989; Poling, B. E., J. M. Prausnitz, and J. P. O'Connell, *The Properties of Gases and Liquids,* 5th ed., McGraw-Hill, New York, 2001; Rothman, D., et al., Max Planck Inst. f. Stromungsforschung, Ber 6, 1978; Smith, B. D., and R. Srivastava, *Thermodynamic Data for Pure Compounds,* Part A: *Hydrocarbons and Ketones,* Elsevier, Amsterdam, 1986, Physical sciences data 25, http://www.elsevier.com/wps/find/bookseriesdescription.librarians/BS_PSD/description; Sterbacek, Z., B. Biskup, and P. Tausk, *Calculation of Properties Using Corresponding States Methods,* Elsevier, Amsterdam, 1979; Stull, D. R., E. F. Westrum, and G. C. Sink, *The Chemical Thermodynamics of Organic Compounds,* Wiley, New York, 1969; *TRC Thermodynamic Tables—Hydrocarbons,* Thermodynamics Research Center, National Institute of Standards and Technology, Boulder, Colo.; *TRC Thermodynamic Tables—Non-Hydrocarbons,* Thermodynamics Research Center, National Institute of Standards and Technology, Boulder, Colo.; Young, D. A., "Phase Diagrams of the Elements," UCRL Rep. 51902, 1975 republished in expanded form by the University of California Press, 1991; Zabransky, M., V. Ruzicka, Jr., V. Majer, and E. S. Domalski, Heat Capacity of Liquids: Critical Review and Recommended Values, *J. Phys. Chem. Ref. Data,* Monograph No. 6, 1996.

CRITICAL DATA SOURCES

Ambrose, D., "Vapor-Liquid Critical Properties," N. P. L. Teddington, Middlesex, Rep. 107, 1980; Kudchaker, A. P., G. H. Alani, and B. J. Zwolinski, *Chem. Revs.* **68:** 659–735, 1968; Matthews, J. F., *Chem. Revs.* **72:** 71–100, 1972; Simmrock, K., R. Janowsky, and A. Ohnsorge, *Critical Data of Pure Substances,* Parts 1 and 2, Dechema Chemistry Data Series, 1986. Other recent references for critical data can be found in Lide, D. R., *CRC Handbook of Chemistry and Physics,* 86th ed., CRC Press, Boca Raton, Fla., 2005.

PUBLICATIONS ON THERMOCHEMISTRY

Pedley, J. B., *Thermochemical Data and Structures of Organic Compounds,* 1, Thermodynamic Research Center, Texas A&M Univ., 1994 (976 pp., 3000 cpds.); Frenkel, M., et al., *Thermodynamics of Organic Compounds in the Gas State,* 2 vols., Thermodynamic Research Center, Texas A&M Univ., 1994 (1825 pp., 2000 cpds.); Barin, I., *Thermochemical Data of Pure Substances,* 2nd ed., 2 vols., VCH Weinheim, Germany, 1993 (1834 pp., 2400 substances); Gurvich, L. V., et al., *Thermodynamic Properties of Individual Substances,* 4th ed., 3 vols., Hemisphere, New York, 1989, 1990, and 1993 (2520 pp.); Lide, D. R., and G. W. A. Milne, *Handbook of Data on Organic Compounds,* 3rd ed., 7 vols., Chemical Rubber, Miami, 1993 (7000 pp.); Daubert, T. E., et al., *Physical and Thermodynamic Properties of Pure Chemicals: Data Compilation,* extant 1995, Taylor & Francis, Bristol, Pa., 1995; Database 11, NIST, Gaithersburg, Md. U.S. Bureau of Mines publications include Bulletins 584, 1960 (232 pp.); 592, 1961 (149 pp.); 595, 1961 (68 pp.); 654, 1970 (26 pp.); Chase, M. W., et al., *JANAF Thermochemical Tables,* 3d ed., *J. Phys. Chem. Ref. Data* 14 suppl. 1, 1986 (1896 pp.); *Journal of Physical and Chemical Reference Data* is available online at http://listserv.nd.edu/cgi-bin/wa?×A2=ind0501&L=pamnet&F=&S=&P=8490 and at http://www.nist.gov/srd/reprints.htm

PHYSICAL PROPERTIES OF PURE SUBSTANCES

TABLE 2-1 Physical Properties of the Elements and Inorganic Compounds*

Abbreviations Used in the Table

a., acid
A., specific gravity with reference to air = 1
abs., absolute
ac., acetic acid
act., acetone
al., 95 percent ethyl alcohol
alk, alkali (i.e., aq. NaOH or KOH)
am., amyl (C_5H_{11})
amor., amorphous
anh., anhydrous
aq., aqueous or water
aq. reg., aqua regia
atm., atmosphere or 760 mm. of mercury pressure
bk., black
brn., brown
bz., benzene
c., cold
cb., cubic
cc, cubic centimeter
chl., chloroform
col., colorless or white
conc., concentrated
cr., crystals or crystalline
d., decomposes
D., specific gravity with reference to hydrogen = 1
d. 50, decomposes at 50°C; 50 d., melts at 50°C with decomposition
delq., deliquescent
dil., dilute
dk., dark
eff., effloresces or efflorescent
et., ethyl ether
expl., explodes
gel., gelatinous
gly., glycerol (glycerin)
gn., green
h., hot
hex., hexagonal
hyg., hygroscopic
i., insoluble
ign., ignites
lq., liquid
lt., light
m. al., methyl alcohol
mn., monoclinic
nd., needles
NH_3, liquid ammonia
NH_4OH, ammonium hydroxide solution
oct., octahedral
or., orange
pd., powder
pl., plates
pr., prisms or prismatic
pyr., pyridine
rhb., rhombic (orthorhombic)
s., soluble
satd., saturated
sl., slightly
soln., solution
subl., sublimes
sulf., sulfides
tart. a., tartaric acid
tet., tetragonal
tr., transition
tri., triclinic
trig., trigonal
v., very
vac., in vacuo
vl., violet
volt., volatile or volatilizes
wh., white
yel., yellow
∞, soluble in all proportions
<, less than
>, greater than
42±, about or near 42
$-3H_2O$, 100, loses 3 moles of water per formula weight at 100°C

Formula weights are based upon the International Atomic Weights in "Atomic Weights of the Elements 2001," *Pure Appl. Chem.,* **75,** 1107, 2003, and are computed to the nearest hundredth.

Refractive index, where given for a uniaxial crystal, is for the ordinary (ω) ray; where given for a biaxial crystal, the index given is for the median (β) value. Unless otherwise specified, the index is given for the sodium D-line (λ = 589.3 μm).

Specific gravity values are given at room temperatures (15 to 20°C) unless otherwise indicated by the small figures which follow the value: thus, $5.6^{18°}_{4}$ indicates a specific gravity of 5.6 for the substance at 18°C referred to water at 4°C. In this table the values for the specific gravity of gases are given with reference to air (A) = 1, or hydrogen (D) = 1.

Melting point is recorded in a certain case as 82 d. and in some other case as d. 82, the distinction being made in this manner to indicate that the former is a melting point with decomposition at 82°C, while in the latter decomposition only occurs at 82°C. Where a value such as $-2H_2O$, 82 is given, it indicates loss of 2 moles of water per formula weight of the compound at a temperature of 82°C.

Boiling point is given at atmospheric pressure (760 mm of mercury) unless otherwise indicated; thus, $82^{15\,mm}$ indicates the boiling point is 82°C when the pressure is 15 mm.

Solubility is given in parts by weight (of the formula shown at the extreme left) per 100 parts by weight of the solvent; the small superscript indicates the temperature. In the case of gases the solubility is often expressed in some manner as $5^{10°}$ cc which indicates that at 10°C, 5 cc of the gas are soluble in 100 g of the solvent. The symbols of the common mineral acids: H_2SO_4, HNO_3, HCl, etc., represent dilute aqueous solutions of these acids. See also special tables on Solubility.

REFERENCES: The information given in this table has been collected mainly from the following sources: Mellor, *A Comprehensive Treatise on Inorganic and Theoretical Chemistry,* Longmans, New York, 1922. Abegg, *Handbuch der anorganischen Chemie,* S. Hirzel, Leipzig, 1905. Gmelin-Kraut, *Handbuch der anorganischen Chemie,* 7th ed., Carl Winter, Heidelberg; 8th ed., Verlag Chemie, Berlin, 1924. Friend, *Textbook of Inorganic Chemistry,* Griffin, London, 1914. Winchell, *Microscopic Character of Artificial Inorganic Solid Substances or Artificial Minerals,* Wiley, New York, 1931. *International Critical Tables,* McGraw-Hill, New York, 1926. *Tables annuelles internationales de constants et donnes numeriques,* McGraw-Hill, New York. *Annual Tables of Physical Constants and Numerical Data,* National Research Council, Princeton, N.J., 1943. Comey and Hahn, *A Dictionary of Chemical Solubilities,* Macmillan, New York, 1921. Seidell, *Solubilities of Inorganic and Metal Organic Compounds,* Van Nostrand, New York, 1940.

Name	Formula	Formula weight	Color, crystalline form, and refractive index	Specific gravity	Melting point, °C	Boiling point, °C	Solubility in 100 parts: Cold water	Hot water	Other reagents
Aluminum	Al	26.98	silv., cb.	$2.70^{20°}$	660	2056	i.	i.	s. HCl, H_2SO_4, alk.
acetate, normal	$Al(C_2H_3O_2)_3$	204.11	wh. pd.		d. 200		s.	d.	
acetate, basic	$Al(OH)(C_2H_3O_2)_2$	162.08	wh., amor.		d.		i.		s.a.; i. NH_4 salts
bromide	$AlBr_3$	266.69	trig.	$3.01^{25°}_{4}$	97.5	268	s.		s.al., act., CS_2
bromide	$AlBr_3·6H_2O$	374.78	col., delq. cr.		d. 100		s.	s.	s. al., CS_2
carbide	Al_4C_3	143.96	yel., hex., 2.70	2.95	d. >2200		d. to CH_4		s. a.; i. act.
chloride	$AlCl_3$	133.34	wh., delq., hex.	$2.44^{25°}_{4}$	$194^{5.2\,atm.}$	$182.7^{752\,mm}$; subl. 178	$69.87^{15°}$	s. d.	s. et., chl., CCl_4; i. bz.
chloride	$AlCl_3·6H_2O$	241.43	col., delq., trig., 1.560				400	v. s.	50 al.; s. et.
fluoride (fluellite)	$AlF_3·H_2O$	101.99	col., rhb., 1.490	2.17	d.		sl. s.		
fluoride	$Al_2F_6·7H_2O$	294.06	wh., cr. pd.		$-4H_2O$, 120	$-6H_2O$, 250	i.	sl. s.	
hydroxide	$Al(OH)_3$	78.00	wh., mn.	2.42	$-2H_2O$, 300		$0.000104^{18°}$	i.	s. a., alk.; i. a.
nitrate	$Al(NO_3)_3·9H_2O$	375.13	rhb., delq.		73	d. 134	v. s.	v. s. d.	s. al., CS_2
nitride	Al_2N_2	81.98	yel., hex.	$3.05^{25°}_{4}$	$2150^{4\,atm.}$	d. >1400	d. slowly		s. alk. d.
oxide	Al_2O_3	101.96	col., hex., 1.67–8	3.99	1999 to 2032		i.	i.	v. sl. s. a., alk.
oxide (corundum)	Al_2O_3	101.96	wh., trig., 1.768	4.00	1999 to 2032	2210	i.	i.	v. sl. s. a., alk.
phosphate	$AlPO_4$	121.95	col., hex.	2.59			i.	i.	s. a., alk.; i. ac.

*By N. A. Lange, Ph.D., Handbook Publishers, Inc., Sandusky, Ohio. Abridged from table of Physical Constants of Inorganic Compounds in *Lange's Handbook of Chemistry.*

(*Continued*)

TABLE 2-1 Physical Properties of the Elements and Inorganic Compounds *(Continued)*

Name	Formula	Formula weight	Color, crystalline form, and refractive index	Specific gravity	Melting point, °C	Boiling point, °C	Solubility in 100 parts: Cold water	Hot water	Other reagents
Aluminum (*Cont.*)									
potassium silicate (muscovite)	$3Al_2O_3 \cdot K_2O \cdot 6SiO_2 \cdot 2H_2O$	796.61	mn., 1.590	2.9	d.		i.		
potassium silicate (orthoclase)	$Al_2O_3 \cdot K_2O \cdot 6SiO_2$	556.66	col., mn., 1.524	2.56	1450 (1150)		i.		
Aluminum potassium tartrate	$AlK(C_4H_4O_6)_2$	362.22	col.				s.	s.	
sodium fluoride (cryolite)	$AlF_3 \cdot 3NaF$	209.94	wh., mn., 1.3389	2.90	1000		sl. s.		i. HCl
sodium silicate	$Al_2O_3 \cdot Na_2O \cdot 6SiO_2$	524.44	col., tri., 1.529	2.61	1100		i.	i.	d. a.
sulfate	$Al_2(SO_4)_3$	342.15	wh. cr.	2.71	d. 770		$31.3^{0°}$	$89^{100°}$	
Alum, ammonium (tschermigite)	$Al_2(SO_4)_3 \cdot (NH_4)_2SO_4 \cdot 24H_2O$	906.66	col., oct., 1.4594	$1.64_{4}^{20°}$	93.5	$-20H_2O$, 120; $-24H_2O$, 200	$3.9^{0°}$	∞ 100°	i. al.
ammonium chrome	$Cr_2(SO_4)_3 \cdot (NH_4)_2SO_4 \cdot 24H_2O$	956.69	gn. or vl., oct., 1.4842	1.72	100 d.		$21.2^{25°}$		s. al.
ammonium iron	$Fe_2(SO_4)_3 \cdot (NH_4)_2SO_4 \cdot 24H_2O$	964.38	vl., oct., 1.485	1.71	40		$124^{25°}$		i. al.
potassium (kalinite)	$Al_2(SO_4)_3 \cdot K_2SO_4 \cdot 24H_2O$	948.78	col., mn., 1.4564	$1.76_{4}^{26°}$	92	$-18H_2O$, 64.5	$5.7^{0°}$	$\infty^{93°}$	
potassium chrome	$Cr_2(SO_4)_3 \cdot K_2SO_4 \cdot 24H_2O$	998.81	red or gn., cb., 1.4814	1.83	89		20	50	i. al.
sodium	$Al_2(SO_4)_3 \cdot Na_2SO_4 \cdot 24H_2O$	916.56	col., oct., 1.4388	$1.675_{4}^{20°}$	61		$106.4^{0°}$	$121.7^{45°}$	i. al.
Ammonia†	NH_3	17.03	col. gas, 1.325 (lq.)	$0.817^{-79°}$ 0.5971 (A)	−77.7	−33.4	$89.9^{0°}$	$7.4^{96°}$	$14.8^{20°}$ al.; s. et.
Ammonium acetate	$NH_4C_2H_3O_2$	77.08	wh., hyg. cr.	1.073	114	d.	$148^{4°}$		s. al.; sl. s. act.
auricyanide	$NH_4CN \cdot Au(CN)_3 \cdot H_2O$	337.09	pl.		d. 200		s.	v. s.	i. al.
bicarbonate	NH_4HCO_3	79.06	mn. or rhb., 1.5358	1.573	d. 35–60		$11.9^{0°}$	$27^{30°}$	i. al.
bromide	NH_4Br	97.94	col., cb., 1.7108	$2.327_{4}^{15°}$	subl. 542		$68^{10°}$	$145.6^{100°}$	s. al., et., act.
carbonate	$(NH_4)_2CO_3 \cdot H_2O$	114.10	col. pl.		d. 58		$100^{15°}$		i. al., CS_2, NH_3
carbonate, carbamate	$NH_4HCO_3 \cdot NH_2CO_2NH_4$‡	157.13	wh. cr.		subl.		$25^{15°}$	$67^{65°}$	
carbonate, sesqui-	$(NH_4)_2CO_3 \cdot 2NH_4HCO_3 \cdot H_2O$	272.21	wh.		d.		$20^{15°}$	$50^{49°}$	
chloride (salammoniac)	NH_4Cl	53.49	wh., cb., 1.639, 1.6426	$1.53^{17°}$	d. 350	subl. 520	$29.4^{0°}$	$77.3^{100°}$	s. NH_3; sl. s. al., m. al.
chloroplatinate	$(NH_4)_2PtCl_6$	443.87	yel., cb.	3.065	d.		$0.7^{15°}$	$1.25^{100°}$	0.005 al.
chloroplatinite	$(NH_4)_2PtCl_4$	372.97	tet.		d.		s.	v. s.	
chlorostannate	$(NH_4)_2SnCl_6$	367.50	pink., cb.	2.4			$33.3^{15°}$		
chromate	$(NH_4)_2CrO_4$	152.07	yel., mn.	$1.917^{12°}$	d. 180		$40.5^{30°}$	d.	sl. s. act., NH_3; i. al.
cyanide	NH_4CN	44.06	col., cb.	$0.79^{100°}$ (A)	36		s.	v. s.	s. al.
dichromate	$(NH_4)_2Cr_2O_7$	252.06	or., mn.	2.15	d. 185		$47.2^{30°}$	v. s.	s. al.; i. act.
ferrocyanide	$(NH_4)_4Fe(CN)_6 \cdot 6H_2O$	392.19	mn.		d.		s.		i. al.
fluoride	NH_4F	37.04	wh., hex.				v. s.	d.	s. al.; i. NH_3
fluoride, acid	$NH_4F \cdot HF$	57.04	wh., rhb., 1.390	$2.21_{12}^{12°}$			v. s.		
formate	HCO_2NH_4	63.06	col., mn., delq.	1.266	114–116	d. 180; subl. in vac.	$102^{0°}$	$531^{80°}$	s. al.
hydrosulfide	NH_4HS	51.11	col., rhb.		d.	subl. 120	v. s.		s. al.
hydroxide	NH_4OH	35.05	in soln. only				s.		
molybdate	$(NH_4)_2MoO_4$	196.01	mn.	2.27	d.		d.	d.	i. al., NH_3
molybdate, hepta-	$(NH_4)_6Mo_7O_{24} \cdot 4H_2O$‡	1235.86	col., mn.				$44^{25°}$		i. al.
nitrate (α), stable −16° to 32°	NH_4NO_3	80.04	col., tet., 1.611	$1.66_{4}^{25°}$	169.6	d. 210	$118.3^{0°}$	$241.8^{30°}$	
nitrate (β), stable 32° to 84°	NH_4NO_3	80.04	col., rhb. or mn.	$1.725_{4}^{25°}$		d. 210	$365.8^{35°}$	$580^{80°}$	$3.8^{20°}$ al., $17.1^{20°}$ m. al.; v. s. NH_3
nitrite	NH_4NO_2	64.04	wh. nd.	1.69	expl.		s.	d.	s. al.
osmochloride	$(NH_4)_2OsCl_6$	439.02	cb.	$2.93_{4}^{20°}$					
oxalate	$(NH_4)_2C_2O_4 \cdot H_2O$	142.11	col., rhb.	1.501			$2.5^{0°}$	$11.8^{50°}$	sl. s. al.; i. NH_3
oxalate, acid	$NH_4HC_2O_4 \cdot H_2O$	125.08	col., trimetric	1.556	d.		s.		
perchlorate	NH_4ClO_4	117.49	col., rhb., 1.4833	1.95	d.		$10.9^{0°}$	$46.9^{100°}$	$2^{20°}$ al.; s. act.; i. et.
persulfate	$(NH_4)_2S_2O_8$	228.20	wh., mn., 1.5016	1.98	d. 120		$58.2^{0°}$	d.	
phosphate, monobasic	$NH_4H_2PO_4$	115.03	col., tet., 1.5246	$1.803_{4}^{19°}$			$22.7^{0°}$	$173.2^{100°}$	i. ac.
phosphate, dibasic	$(NH_4)_2HPO_4$	132.06	col., mn., 1.53	1.619			$131^{15°}$		i. act.
phosphate, meta-	$(NH_4)_4P_4O_{12}$	388.04	col., mn.	2.21			s.		

Ammonium phosphomolybdate	$(NH_4)_3PO_4 \cdot 12MoO_3 \cdot 3H_2O$ (?)	1930.39	yel.		d.		$0.03^{15°}$	i.	s. alk.; i. al., HNO_3
silicofluoride	$(NH_4)_2SiF_6$	178.15	cb., 1.3696	2.01		subl.	$18.5^{17.5°}$	55.5	s. al.; i. act.
sulfamate	$NH_4 \cdot SO_3NH_2$	114.12	col. pl.		132	d. 160	$134^{0°}$	$357^{50°}$	
sulfate (mascagnite)	$(NH_4)_2SO_4$	132.14	col., rhb., 1.5230	$1.769\frac{20°}{4}$	235 d.		$70.6^{0°}$	$103.3^{100°}$	i. al., act., CS_2
sulfate, acid	NH_4HSO_4	115.11	col., rhb., 1.480	1.78	146.9	490	100		v. sl. s. al.; i. act.
sulfide	$(NH_4)_2S$	68.14	yel.-wh.		d.		v. s.		$120^{25°}$ NH_3
sulfide, penta-	$(NH_4)_2S_5$	196.40	or.-red pr.				s.		
sulfite	$(NH_4)_2SO_3 \cdot H_2O$	134.16	col., mn.	1.41	d.		$100^{12°}$		i. al., act.
sulfite, acid	NH_4HSO_3	99.11	rhb.	$2.03\frac{12°}{4}$	d.		s.		
tartrate	$(NH_4)_2C_4H_4O_6$	184.15	col., mn.	1.60	d.		$45^{0°}$	$87^{60°}$	sl. s. al.
thiocyanate	NH_4CNS	76.12	col., mn., 1.685±	1.305	149.6	d. 170	$120^{0°}$	$170^{20°}$	s. al., act., NH_3, SO_2
vanadate, meta-	NH_4VO_3	116.98	col. cr.	2.326	d.		$0.44^{18°}$	$3.05^{70°}$	i. al., NH_4Cl
Antimony	Sb	121.76	tin wh., trig.	$6.684^{25°}$	630.5	1380	i.	i.	s. aq. reg., h. conc. H_2SO_4
chloride, tri- (butter of antimony)*	$SbCl_3$	228.12	col., rhb., delq.	$3.14\frac{20°}{4}$	73.4	220.2	$601.6^{0°}$	$\infty^{72°}$	s. al., HCl, HBr, $H_2C_4H_4O_6$
oxide, tri- (valentinite)	Sb_2O_3	291.52	rhb., 2.35	5.67	656	1570	v. sl. s.	sl. s.	s. HCl, KOH, $H_2C_4H_4O_6$
oxide, tri- (senarmontite)	Sb_2O_3	291.52	cb., 2.087	5.2	652				
sulfide, tri- (stibnite)	Sb_2S_3	339.72	bk., rhb., 4.046	4.64	550		$0.00017^{18°}$	d.	s. HCl; alk., NH_4HS, K_2S; i. ac.
sulfide, penta-	Sb_2S_5	403.85	golden	$4.120^{0°}$	−2S, 135		i.	i.	s. HCl, alk., NH_4HS
telluride, tri-	Sb_2Te_3	626.32	gray		629				
Antimonyl potassium tartrate (tartar emetic)	$(SbO)KC_4H_4O_6 \cdot \frac{1}{2}H_2O$	333.94	wh., rhb.	2.60	−½H_2O, 100		$5.26^{8.7°}$	$35.7^{100°}$	s. gly.; i. al.
sulfate, normal	$(SbO)_2SO_4$	371.58	wh. pd.	4.89			d.	d.	
sulfate, basic	$(SbO)_2SO_4 \cdot Sb_2(OH)_4$	683.20	wh. pd.				i.	d.	$5.15^{15°}$ gly.
Argon	Ar	39.95	col. gas	$1.65^{-288°}$; $1.402^{-185.7°}$; 1.38 (A)	−189.2	−185.7	$5.6^{0°}$ cc	$2.23^{50°}$ cc	$24^{25°}$ cc al.
Arsenic (crystalline) (α)	As_4	299.69	met., hex.	$5.727^{14°}$	$814^{36atm.}$	subl. 615	i.	i.	s. HNO_3
Arsenic (black) (β)	As_4	299.69	bk., amor.	$4.7^{20°}$			i.	i.	s. HNO_3, aq. reg., aq. Cl_2, h. alk.
Arsenic (yellow) (γ)	As_4	299.69	yel., cb.	$2.0^{20°}$	d. 358				
acid, ortho-	$H_3AsO_4 \cdot \frac{1}{2}H_2O$	150.95	col., hyg.	2.0–2.5	35.5	−H_2O, 160	16.7	50	s. alk.
acid, meta-	$HAsO_3$	123.93	wh., hyg.		d.		d. to form	H_3AsO_4	
acid, pyro-	$H_4As_2O_7$	265.87	col.		d. 206		d. to form	H_3AsO_4	
pentoxide	As_2O_5	229.84	wh., amor.	4.086		d.	$59.5^{0°}$	$76.7^{100°}$	s. alk., al.
sulfide, di- (realgar)	As_2S_2	213.97	red, mn., 2.68	(α)$3.506^{19°}$; (β)$3.254^{19°}$	(α)tr. 267; (β)307	565	i.	d.	s. K_2S, $NaHCO_3$
sulfide, penta-	As_2S_5	310.17	yel.			d. 500	$0.000136^{0°}$	i.	s. HNO_3, alk.
Arsenious chloride (butter of arsenic)	$AsCl_3$	181.28	oily lq.	lq. 2.163	−18	130	d.	d.	s. HCl, HBr, PCl_3
hydride (arsine)	AsH_3	77.95	col. gas	2.695 (A)	−113.5	−55; d. 230	20 cc	sl. s.	sl. s. alk.
oxide (arsenolite)	As_2O_3	197.84	col., cb., fibrous, 1.755	$3.865\frac{25°}{4}$	subl.		sl. s.	sl. s.	i. al., et.
oxide (claudetite)	As_2O_3	197.84	col., mn., 1.92	3.85	subl.		sl. s.	sl. s.	i. al., et.
oxide	As_2O_3	197.84	amor. or vitreous	3.738	315		$1.21^{0°}$	$2.93^{40°}$	s. HCl, alk., Na_2CO_3; i. al., et.
Auric chloride	$AuCl_3 \cdot 2H_2O$	339.36	or. cr.		d.		v. s.	v. s.	s. HCl, al., et.; sl. s. NH_3
cyanide	$Au(CN)_3 \cdot 6H_2O$	383.11			d. 50		v. s.	v. s.	s. al.
Aurous chloride	AuCl	232.42	yel. cr.	7.4	$AuCl_3$, 170	d. 290	d.	d.	s. HCl, HBr; d. al.
cyanide	AuCN	222.98	yel. cr.		d.		i.	i.	s. KCN; i. al., et.
Cf. also under *Gold*									
Barium	Ba	137.33	silv. met.	3.5	850	1140	d.	d.	s. a.; d. al.
acetate	$Ba(C_2H_3O_2)_2$	255.42	col.	2.468			$58.8^{0°}$	$75.0^{100°}$	
acetate	$Ba(C_2H_3O_2)_2 \cdot H_2O$	273.43	wh., tri. pr., 1.517	2.19	−H_2O, 41		$75^{30°}$(anh.)	$79^{40°}$(anh.)	i. al.
bromide	$BaBr_2$	297.14	col.	$4.781\frac{24°}{4}$	847	d.	$98^{0°}$	$149^{100°}$	v. s. m. al.; v. sl. s. act.

*Usually the solution.
†See special tables.
‡Usual commercial form.

(*Continued*)

TABLE 2-1 Physical Properties of the Elements and Inorganic Compounds *(Continued)*

Name	Formula	Formula weight	Color, crystalline form, and refractive index	Specific gravity	Melting point, °C	Boiling point, °C	Solubility in 100 parts: Cold water	Hot water	Other reagents
Barium (*Cont.*)									
bromide	$BaBr_2 \cdot 2H_2O$	333.17	col., mn., 1.7266	3.69	$-2H_2O$, 100	d.	v. s.	v. s.	s. al.
carbonate (witherite)	$BaCO_3$	197.34	wh., rhb., 1.676	4.29	tr. 811 to α	d. 1450	0.0022^{18°	0.0065^{100°	s. a.; i. al.
carbonate (α)	$BaCO_3$	197.34	wh., hex.		tr. 982 to β				
carbonate (β)	$BaCO_3$	197.34	wh.		$1740^{90 atm.}$		0.0022^{18°	0.0065^{100°	s. a.; i. al.
Barium chlorate	$Ba(ClO_3)_2$	304.23	col.		414		20.35^{0°	84.8^{80°	
chlorate	$Ba(ClO_3)_2 \cdot H_2O$*	322.24	col., mn., 1.577	3.179	d. 120		s.	s.	sl. s. al., act.
chloride	$BaCl_2$	208.23	col., mn., 1.7361	$3.856^{\frac{24^\circ}{4}}$	tr. 925	1560	31^{0°	59^{100°	sl. s. HCl, HNO_3; i. al.
chloride	$BaCl_2$	208.23	col., cb.		962	1560			
chloride	$BaCl_2 \cdot 2H_2O$†	244.26	col., mn., 1.646	$3.097^{\frac{24^\circ}{4}}$	$-2H_2O$, 100		39.3^{0°	76.8^{100°	sl. s. HCl, HNO_3; i. al.
hydroxide	$Ba(OH)_2$	171.34	col., mn.	4.495			1.67^{0°	101.4^{80°	
hydroxide	$Ba(OH)_2 \cdot 8H_2O$	315.46	col., mn., 1.5017	2.188^{16°	77.9	$-8H_2O$, 550	5.6^{15°		v. sl. s. al.; i. et.
nitrate (nitrobarite)	$Ba(NO_3)_2$	261.34	col., cb., 1.572	3.244^{28°	592	d.	5.0^{0°	34.2^{100°	sl. s. a.; i. al.
oxalate	BaC_2O_4	225.35	wh. cr.	2.658			0.0016^{8°	0.0024^{24°	s. a., NH_4Cl; i. al.
oxide	BaO	153.33	col., cb., 1.98	5.72	1923	2000±	1.5^{0°	90.8^{80°	s. HCl, HNO_3, abs. al.; i. NH_3, act.
peroxide	BaO_2*	169.33	gray or wh. pd.	4.958	–O, 800		v. sl. s.	d.	s. dil. a.; i. act.
peroxide	$BaO_2 \cdot 8H_2O$	313.45	pearly sc.		$-8H_2O$, 100		0.168	d.	s. dil. a.; i. al., et., act.
phosphate, monobasic	$BaH_4(PO_4)_2$	331.30	tri.	2.9^{4°			d.	d.	s. a.
phosphate, dibasic	$BaHPO_4$	233.31	wh., rhb. nd., 1.635	4.165^{15°			0.015		s. a., NH_4 salts
phosphate, tribasic	$Ba_3(PO_4)_2$	601.92	wh., cb.	4.1^{16°			i.		s. a.
phosphate, pyro-	$Ba_2P_2O_7$	448.60	wh., rhb.	3.9^{20°			0.01		s. a., NH_4 salts
silicofluoride	$BaSiF_6$	279.40	pr.	4.279^{15°			0.026^{17°	0.09^{100°	sl. s. HCl, NH_4Cl; i. al.
sulfate (barite, barytes)	$BaSO_4$	233.39	col., rhb., 1.636	4.499^{15°	1580 d.	tr. to mn. 1149	0.000115^{0°	0.000285^{30°	s. conc. H_2SO_4; 0.006, 3% HCl
sulfide, mono-	BaS	169.39	col., cb., 2.155	4.25^{15°			d.	d.	d. HCl; i. al.
sulfide, tri-	BaS_3	233.52	yel.-gn.		d. 400		s.	s.	
sulfide, tetra-	$BaS_4 \cdot 2H_2O$	301.62	red, rhb.	2.988^{20°	d. 200		41^{15°	v. s.	i. al., CS_2
Beryllium (glucinum)	Be(Gl)	9.01	gray, met., hex.	1.816	1284	2767	i.	sl. s. d.	s. dil. a., alk.
Bismuth	Bi	208.98	silv. wh. or reddish, hex.	9.80^{20°	271	1450	i.	i.	s. aq. reg., conc. H_2SO_4, HNO_3
carbonate, sub-	$Bi_2O_3 \cdot CO_2 \cdot H_2O$	527.98	wh. pd.	6.86	d.		i.	i.	s. a.
chloride, di-	$BiCl_2$	279.89	bk. nd.	4.86	163	d. 300	d.		
chloride, tri-	$BiCl_3$*	315.34	wh. cr.	4.75	230	447	d.		s. al.
nitrate	$Bi(NO_3)_3 \cdot 5H_2O$	485.07	col., tri.	2.82	d. 30	$-5H_2O$, 80	d.		42^{19° act.; s. a.; i. al.
nitrate, sub-	$BiONO_3 \cdot H_2O$	305.00	hex. pl.	4.928^{15°	d. 260		i.	i.	s. a.
oxide, tri-	Bi_2O_3	465.96	yel., rhb.	8.9	820	1900±	i.	i.	s. a.
oxide, tri-	Bi_2O_3	465.96	yel., tet.	8.55	860		i.	i.	s. a.
oxide, tri-	Bi_2O_3	465.96	yel., cb.	8.20	tr. 704		i.	i.	s. a.
oxychloride	$BiOCl$	260.43	wh., amor.	7.72^{15°			sl. s.	sl. s.	s. a.; i. act., NH_3, $H_2C_4H_4O_6$
Boric acid	H_3BO_3	61.83	wh., tri.	1.435^{15°	185 d.		2.66^{0°	40.2^{100°	22.2^{20° gly., 0.24^{25° et.; s. al.
Boron	B	10.81	gray or bk., amor. or mn.	2.32	2300	2550	i.	i.	s. HNO_3; i. al.
carbide	B_4C	55.25	bk. cr.	2.54	2450	>3500	i.	i.	i. a.
oxide	B_2O_3	69.62	col. glass, 1.459	1.85	577	>1500	1.1^{0°	15.7^{100°	s. a., al., gly.
oxide (sassolite)	$B_2O_3 \cdot 3H_2O$	123.67	tri., 1.456	1.49	d.		sl. s.	s.	
Bromic acid	$HBrO_3$	128.91	col.; in soln. only		d. 100		v. s.	d.	
Bromine	Br_2	159.81	rhb., or red lq.	3.119^{20°; 5.87 (A)	–7.2	58.78	4.22^{0°	3.13^{30°	s. al., et., alk., CS_2
hydrate	$Br_2 \cdot 10H_2O$	339.96	red, oct.		d. 6.8		s.		
Cadmium	Cd	112.41	silv. met., hex.	8.65^{20°	320.9	767	i.	i.	s. a., NH_4NO_3
acetate	$Cd(C_2H_3O_2)_2$	230.50	col.	2.341	256	d.	v. s.		s. m. al.
acetate	$Cd(C_2H_3O_2)_2 \cdot 2H_2O$*	266.53	col., mn.	2.01	$-H_2O$, 130		v. s.		s. al.
carbonate	$CdCO_3$	172.42	wh., trig.	4.258^{4°	d. <500		i.	i.	s. a., KCN, NH_4 salts; i. NH_3
chloride	$CdCl_2$	183.32	wh., cb.	$4.047^{\frac{25^\circ}{4}}$	568	960	90^{0°	147^{100°	1.52^{15° al.; i. et., act.

chloride	$CdCl_2·2½H_2O$	228.36	col., mn., 1.6513	3.327	tr. 34		$168^{20°}$	$180^{100°}$	$2.05^{15°}$ m. al.
cyanide	$Cd(CN)_2$	164.45			d. >200		$0.0247^{18°}$		s. a.; NH_4OH, KCN
hydroxide	$Cd(OH)_2$	146.43	wh., trig.	$4.79^{15°}_{4}$	d. 300		$0.00026^{25°}$		s. a., NH_4 salts; i. alk.
nitrate	$Cd(NO_3)_2$	236.42	col.		350		$109.7^{0°}$	$326^{59.5°}$	v. s. a.
nitrate	$Cd(NO_3)_2·4H_2O$*	308.48	col. nd.	$2.455^{17°}_{4}$	59.4	132	$215^{0°}$		s. al., NH_3; i. HNO_3
oxide	CdO	128.41	brn., cb.	8.15			i.	i.	s. a., NH_4 salts; i. alk.
oxide	CdO	128.41	brn., amor, 2.49	6.95	d. 900–1000		i.	i.	s. a., NH_4 salts; i. alk.
oxide, sub-	Cd_2O	240.82	gn., amor.	$8.192^{18°}_{4}$	d.				d. a., alk.
Cadmium sulfate	$CdSO_4$	208.47	rhb.	$4.691^{24°}_{4}$	1000		$76.5^{0°}$	$60.8^{100°}$	i.act., NH_3
sulfate	$CdSO_4·H_2O$	226.49	mn.	$3.786^{20°}$	tr. 108		s.	s.	
sulfate	$3CdSO_4·8H_2O$*	769.54	col., mn., 1.565	3.09	tr. 41.5		$114.2^{0°}$	$127.6^{60°}$	i. al.
sulfate	$CdSO_4·4H_2O$	280.53	col.	3.05			s.	s.	i. al.
sulfate	$CdSO_4·7H_2O$	334.58	mn.	$2.48^{20°}_{4}$	tr. 4		$350^{-5°}$		i. al.
sulfide (greenockite)	CdS	144.48	yel.-or., hex., 2.506	4.58	$1750^{100atm.}$	subl. in N_2, 980	0.000001	Colloidal	s. a.; v. s. NH_4OH
Calcium	Ca	40.08	silv. met., cb.	$1.55^{20°}$	810	1200 ± 30	d.	d.	s, a.; sl. s. al.
acetate	$Ca(C_2H_3O_2)_2·H_2O$	176.18	wh. nd.		d.		$52^{0°}$	$45.5^{80°}$	sl. s. al.
aluminate	$Ca(AlO_2)_2$	158.04	col., rhb. or mn.	$3.67^{20°}$	1600		d.		s. HCl
aluminum silicate (anorthite)	$CaO·Al_2O_3·2SiO_2$	278.21	tri., 1.5832	2.765	1551				
arsenate	$Ca_3(AsO_4)_2$	398.07	wh. pd.				$0.013^{25°}$	i.	s. dil. a.
bromide	$CaBr_2$	199.89	delq. nd.	$3.353^{25°}_{4}$	760	1810	$125^{0°}$	$312^{105°}$	s. al., act.; sl. s. NH_3
carbonate (aragonite)	$CaCO_3$	100.09	col., rhb., 1.6809	2.93	d. 825		$0.0012^{20°}$†	$0.002^{100°}$	s. a., NH_4Cl
carbonate (calcite)	$CaCO_3$	100.09	col., hex., 1.550	$2.711^{25°}_{4}$	$1339^{103atm.}$		$0.0014^{25°}$	$0.002^{100°}$	s. a., NH_4Cl
chloride (hydrophilite)	$CaCl_2$*	110.98	wh., delq., cb, 1.52	$2.152^{15°}_{4}$	772	>1600	$59.5^{0°}$	$347^{260°}$	s. al.
chloride	$CaCl_2·H_2O$	129.00	col., delq.				s.	s.	s. al.
chloride	$CaCl_2·6H_2O$	219.08	col., trig., 1.417	$1.68^{17°}$	29.92	$-6H_2O$, 200	v. s.	v. s.	s. al.
citrate	$Ca_3(C_6H_5O_7)_2·4H_2O$	570.49	col. nd.		$-2H_2O$, 130	$-4H_2O$, 185	$0.085^{18°}$	$0.096^{26°}$	$0.0065^{18°}$ al.
cyanamide	$CaCN_2$	80.10	col., rhombohedral				s. d.	d.	
ferrocyanide	$Ca_2Fe(CN)_6·12H_2O$	508.29	yel., tri., 1.5818	1.7			s.	$150^{90°}$	i. al.
fluoride (fluorite)	CaF_2	78.07	wh., cb., 1.4339	$3.180^{20°}$	1330		$0.0016^{18°}$	$0.0017^{26°}$	sl. s. a.
formate	$Ca(HCO_2)_2$	130.11	col., rhb.	2.015	d.		$16.1^{0°}$	18.4^{100}	i. al., et.
hydride	CaH_2	42.09	wh. cr. or pd.	1.7	d. 675		d.		d. a.; i. bz.
hydroxide	$Ca(OH)_2$	74.09	col., hex., 1.574	2.2	$-H_2O$, 580		$0.185^{0°}$	$0.077^{100°}$	s. NH_4Cl
hypochlorite	$Ca(ClO)_2·4H_2O$	215.04	wh., feathery cr.		d.		delq.; d.	d.	d. a.
hypophosphate	$Ca_2P_2O_6·2H_2O$	274.13	granular		$-2H_2O$, 200		i.		s. HCl, $H_4P_2O_6$
lactate	$Ca(C_3H_5O_3)_2·5H_2O$	308.29	col., eff.		$-3H_2O$, 100		10.5	∞	∞h. al.; i. et.
magnesium carbonate (dolomite)	$CaO·MgO·2CO_2$	184.40	trig., 1.68174	2.872	d. 730–760		$0.032^{18°}$		
magnesium silicate (diopside)	$CaO·MgO·2SiO_2$	216.55	wh., mn.	3.3	1391		i.	i.	
nitrate (nitrocalcite)	$Ca(NO_3)_2$	164.09	col., cb.	2.36	561		$102^{0°}$	$376^{151°}$	$14^{15°}$ al.; s. amyl al., NH_3
nitrate	$Ca(NO_3)_2·4H_2O$*	236.15	col., mn., 1.498	1.82	42.7		$266^{0°}$	v. s.	
nitride	Ca_3N_2	148.25	brn. cr.	$2.63^{17°}$	900		d.	d.	s. dil. a.; i. abs. al.
nitrite	$Ca(NO_2)_2·H_2O$	150.10	delq., hex.	$2.23^{34°}$			$77^{0°}$	$417^{90°}$	s. 90% al.
oxalate	CaC_2O_4	128.10	col., cb.	$2.2^{4°}$	d.		$0.00067^{13°}$	$0.0014^{95°}$	s. a.; i. ac.
oxalate	$CaC_2O_4·H_2O$	146.11	col.	2.2	$-H_2O$, 200		i.	i.	s. a.; i. ac
oxide	CaO	56.08	col., cb., 1.837	3.32	2570	2850	Forms $Ca(OH)_2$		s. a.; i. al.
peroxide	$CaO_2·8H_2O$	216.20	pearly, tet.		$-8H_2O$, 100	expl. 275	sl. s.	d.	s. a. d.; i. al., et.
phosphate, monobasic	$CaH_4(PO_4)_2·H_2O$	252.07	wh., tri.	$2.220^{16°}_{4}$	$-H_2O$, 100	d. 200		d.	
phosphate, dibasic	$CaHPO_4·2H_2O$	172.09	wh., mn. pl.	$2.306^{16°}_{4}$	d.		$0.02^{24.5°}$	$0.075^{100°}$	
phosphate, tribasic	$Ca_3(PO_4)_2$	310.18	wh., amor.	3.14	1670		0.0025	d.	s. a.; i. al., ac.
phosphate, meta-	$Ca(PO_3)_2$	198.02	wh., tet., 1.588	2.82	975		i.	i.	i. a.
phosphate, pyro-	$Ca_2P_2O_7$	254.10	col., biaxial, 1.60	3.09	1230		i.		s. a.
phosphate, pyro- (brushite)	$Ca_2P_2O_7·5H_2O$	344.18	wh., mn.	2.25			sl. s.		s. a.; i. NH_4Cl
phosphide	Ca_3P_2	182.18	red cr.	$2.51^{15°}$	>1600		d.		s. dil. a.; i. al., et.
silicate (α) (pseudowollastonite)	$CaSiO_3$	116.16	col., pseudo hex., 1.6150 or mn.	2.905	1540		$0.0095^{17°}$		s. HCl
silicate (β) (wollastonite)	$CaSiO_3$	116.16	col., mn., 1.610	2.915	tr. 1190 to α				
sulfate (anhydrite)	$CaSO_4$	136.14	col., rhb., 1.576, or mn., 1.50	2.96	1450(mn.)	tr. 1193 to rhb.	$0.298^{20°}$	$0.1619^{100°}$	s. a., $Na_2S_2O_3$, NH_4 salts

*Usual commercial form.
†The solubility of $CaCO_3$ in H_2O is greatly increased by increasing the amount of CO_2 in the H_2O.

(Continued)

TABLE 2-1 Physical Properties of the Elements and Inorganic Compounds *(Continued)*

Name	Formula	Formula weight	Color, crystalline form, and refractive index	Specific gravity	Melting point, °C	Boiling point, °C	Solubility in 100 parts: Cold water	Hot water	Other reagents
Calcium (*Cont.*)									
sulfate (gypsum)	$CaSO_4·2H_2O$	172.17	col., mn., 1.5226	2.32	$-1½H_2O$, 128	$-2H_2O$, 163	$0.223^{0°}$	$0.257^{50°}$	s. a., gly., $Na_2S_2O_3$, NH_4 salts
sulfhydrate	$Ca(SH)_2·6H_2O$	214.32	col. pr.		d. 15		v. s.	v. s.	s. al.
sulfide (oldhamite)	CaS	72.14	col., cb.	$2.8^{15°}$			d.	d.	s. a.
sulfite	$CaSO_3·2H_2O$	156.17	wh., cr., 1.595		$-2H_2O$, 100	d. 650	$0.0043^{18°}$	$0.0027^{90°}$	s. H_2SO_3
tartrate	$CaC_4H_4O_6·4H_2O$	260.21	col., rhb.		d.		$0.037^{0°}$	$0.22^{85°}$	sl. s. al.
thiocyanate	$Ca(CNS)_2·3H_2O$	210.29	wh., delq. cr.				s.	v. s.	v. s. al.
thiosulfate	$CaS_2O_3·6H_2O$	260.30	col., tri., 1.56	$1.873^{16°}$	d.		$71.2^{9°}$	d.	i. al.
tungstate (scheelite)	$CaWO_4$	287.92	wh., tet., 1.9200	6.06			0.2		s. NH_4Cl; i. a.
Carbon, *cf.* table of organic compounds									
Carbon, amorphous	C	12.01	bk., amor.	1.8–2.1	>3500	4200	i.	i.	i. a., alk.
Carbon, diamond	C	12.01	col., cb., 2.4195	$3.51^{20°}$	>3500	4200	i.	i.	i. a., alk.
Carbon, graphite	C	12.01	bk., hex.	$2.26^{20°}$	>3500	4200	i.	i.	i. a., alk.
dioxide	CO_2	44.01	col. gas	lq. $1.101^{-87°}$; 1.53 (A); solid $1.56^{-79°}$	$-56.6^{5.2atm.}$	subl. −78.5	$179.7^{0°}$ cc	$90.1^{20°}$ cc	s. a., alk.
disulfide	CS_2	76.14	col. lq.	lq. $1.261\frac{22°}{20}$; 2.63 (A)	−108.6	46.3	$0.2^{0°}$	$0.014^{50°}$	s. al.; et.
monoxide	CO	28.01	col., poisonous, odorless gas	lq. $0.814\frac{-195°}{4}$; 0.968 (A)	−207	−192	$0.0044^{0°}$; $3.5^{0°}$ cc	$0.0018^{50°}$; $2.32^{20°}$ cc	s. al., Cu_2Cl_2
oxychloride (phosgene)	$COCl_2$	98.92	poisonous gas	$1.392\frac{19°}{4}$	−104	8.2^{756mm}	v. s. sl. d.	d.	s. ac., CCl_4, bs.; d.a.
oxysulfide	COS	60.08	gas	lq. $1.24^{-87°}$; 2.10 (A)	−138.2	-50.2^{760mm}	$133^{0°}$ cc	$40.3^{30°}$ cc	v. s. alk., al.
suboxide	C_3O_2	68.03	gas	lq. $1.114^{0°}$	−107	7^{761mm}	d.		s. et.
thionyl chloride	$CSCl_2$	114.98	yel.-red lq.	$1.509^{15°}$		73.5			
Ceric hydroxide	$2CeO_2·3H_2O$	398.28	yel., gelatinous						s. a.; sl. s. alk. carb.; i. alk
hydroxynitrate	$Ce(OH)(NO_3)_3·3H_2O$	397.18	red, mn.				d.		
oxide	CeO_2	172.11	wh. or pa. yel., cb.	7.3	1950		i.	i.	s. H_2SO_4, HCl
sulfate	$Ce(SO_4)_2·4H_2O$	404.30	yel., rhb.	3.91			s. d.		s. dil. H_2SO_4
Cerium	Ce	140.12	steel gray, cb. or hex.	$6.9^{20°}$ cb.; 6.7 hex.	645	1400	i.	Slowly oxidized	s. dil. a.; i. al.
Cerous sulfate	$Ce_2(SO_4)_3$	568.42	wh., mn. or rhb.	3.91			$18.98^{0°}$	$0.4^{100°}$	
sulfate	$Ce_2(SO_4)_3·8H_2O$	712.54	tri.	$2.886^{17°}$	$-8H_2O$, 630		$25^{0°}$	$7.6^{40°}$	
Cesium	Cs	132.91	silv. met., hex.	$1.90^{20°}$	28.5	670	d.		s. a., al., NH_3
Chloric acid	$HClO_3·7H_2O$	210.57	lq.	$1.282^{14.2°}$	<−20	d. 40	v. s.		
Chlorine	Cl_2	70.91	rhb., or gn.-yel. gas	lq. $1.56^{-33.6°}$; $2.49^{0°}$ (A)	−101.6	−34.6	$1.46^{0°}$; $310^{10°}$ cc	$0.57^{30°}$; $177^{30°}$ cc	s. alk.
hydrate	$Cl_2·8H_2O$	215.03	rhb.	1.23	d. 9.6		s.		s. alk.
Chloroplatinic acid	$H_2PtCl_6·6H_2O$	517.90	red-brn., delq.	2.431	60		v. s.	v. s.	s. al., et.
Chlorostannic acid	$H_2SnCl_6·6H_2O$	441.54	delq.	$1.971^{28°}$	19.2		s.		
Chlorosulfonic acid	$HO·SO_2·Cl$	116.52	col. lq.	$1.787^{25°}$	−80	151.5^{765mm}	d.		d. al.; i. CS_2
Chromic acetate	$Cr_2(C_2H_3O_2)_6·2H_2O$	494.29	gn.				s.		$4.76^{15°}$ m. al.
chloride	$CrCl_3$	158.36	pink, trig.	$2.757^{15°}$		1200–1500 d.	i.[§]	sl. s.	i. a., act., CS_2
chloride	$CrCl_3·6H_2O$*	266.45	vl. or gn., hex. pl.	$1.835\frac{25°}{4}$	subl. 83		v. s. d.		s. al.; i. et.
fluoride	CrF_3	108.99	gn., rhb.	3.8	>1000	d.	i.		sl. s. a.; i. al., NH_3
hydroxide	$Cr(OH)_3$	103.02	gn. or blue, gelatinous				i.		s. a., alk.; sl. s. NH_3
hydroxide	$Cr(OH)_3·2H_2O$	139.05	gn.		$-2H_2O$, 100		i.	i.	s. a., alk.
nitrate	$Cr(NO_3)_3·9H_2O$*	400.15	purple pr.		36.5	d. 100	s.	s.	s. a., alk., al., act.
nitrate	$Cr(NO_3)_3·7½H_2O$	373.13	purple, mn.		100	d.	s.	s.	
oxide	Cr_2O_3	151.99	dark gn., hex.	5.21	1900		i.	i.	sl. s. a.
sulfate	$Cr_2(SO_4)_3$	392.18	rose pd.	3.012			i.[†]		i. a.
sulfate	$Cr_2(SO_4)_3·5H_2O$	482.26	gn.				s.		s. al., H_2SO_4
sulfate	$Cr_2(SO_4)_3·15H_2O$	662.41	vl.	$1.867^{17°}$	100	$-10H_2O$, 100	s.	d. 67°	sl. s. al.
sulfate	$Cr_2(SO_4)_3·18H_2O$	716.46	vl., cb., 1.564	$1.7^{22°}$		$-12H_2O$, 100	$120^{20°}$	d.	s. al.
sulfide	Cr_2S_3	200.19	brn.-bk. pd.	$3.77^{19°}$	−S, 1350		i.	d.	s. h. HNO_3

Chromium	Cr	52.00	gray, met., cb.	7.1	1615	2200	i.	i.	s. HCl, dil. H_2SO_4; i. HNO_3
trioxide (chromic acid)	CrO_3	99.99	red, rhb.	2.70	197 d.		$164.9^{0°}$	$206.7^{100°}$	s. H_2SO_4, al., et.
Chromous chloride	$CrCl_2$	122.90	wh., delq.	2.75			v. s.	v. s.	sl. s. al.; i. et.
hydroxide	$Cr(OH)_2$	86.01	yel.-brn.		d.		d.		s. conc. a.
oxide	CrO	68.00	bk. pd.				i.	i.	i. dil. HNO_3
sulfate	$CrSO_4{\cdot}7H_2O$	274.17	blue				$12.35^{0°}$		sl. s. al.
sulfide (daubrelite)	CrS	84.06	bk. pd.	3.97	1550		i.		v. s. a.
Chromyl chloride	CrO_2Cl_2	154.90	dark red lq.	1.92	−96.5	117.6	d.		s. et.
Cobalt	Co	58.93	silv. met., cb.	$8.9^{20°}$	1480	2900	i.	i.	s. a.
carbonyl	$Co(CO)_4$	170.97	or. cr.	$1.73^{18°}$	51	d. 52	i.	d.	s. al., et., CS_2
sulfide, di-	CoS_2	123.06	bk., cb.	4.269			i.		s. HNO_3, aq. reg.
Cobaltic chloride	$CoCl_3$	165.29	red cr.	2.94	subl.		s.	s.	
chloride, dichro	$Co(NH_3)_3Cl_3{\cdot}H_2O$	234.40					s.		s. a.; al.
chloride, luteo	$Co(NH_3)_6Cl_3$	267.48	or., mn.	$1.7016^{20°}$			$4.26^{0°}$	$12.74^{46.5°}$	i. al., NH_4OH
chloride, praseo	$Co(NH_3)_4Cl_3{\cdot}H_2O$	251.43	gn., rhb.	1.847			v. s.		s. a.; i. al.
Cobaltic chloride, purpureo	$Co(NH_3)_5Cl_3$	250.44	rhb.	$1.819\frac{25°}{25}$			$0.232^{0°}$	$1.031^{46.5°}$	i. al.
chloride, roseo	$Co(NH_3)_5Cl_3{\cdot}H_2O$	268.46	brick red		d. 100		$16.12^{0°}$	$24.87^{16°}$	sl. s. HCl
hydroxide	$Co(OH)_3$	109.96	bk.		$-1\frac{1}{2}H_2O$, 100		i.	i.	s. a.; i. al.
oxide	Co_2O_3	165.86	bk.	5.18	d. 900		i.	i.	s. a.
sulfate	$Co_2(SO_4)_3$	406.05	blue cr.				d.		s. H_2SO_4
sulfide	Co_2S_3	214.06	bk. cr.	4.8			i.		d. a.
Cobalto-cobaltic oxide	Co_3O_4	240.80	bk., cb.	6.07			i.	i.	s. H_2SO_4; i. HCl, HNO_3
Cobaltous acetate	$Co(C_2H_3O_2)_2{\cdot}4H_2O$	249.08	red-vl., mn., 1.542	$1.7053^{18.7°}$	$-4H_2O$, 140		s.	s.	s. a., al.
chloride	$CoCl_2$	129.84	blue cr.	3.356	subl.	1049	$45^{7°}$	$105^{96°}$	31 al.; 8.6 act.
chloride	$CoCl_2{\cdot}6H_2O$*	237.93	red, mn.	$1.924\frac{25°}{25}$	86	$-6H_2O$, 110	$116.5^{0°}$	$177^{80°}$	v. s. et., act.
nitrate	$Co(NO_3)_2{\cdot}6H_2O$	291.03	red, mn., 1.4	$1.883\frac{25°}{25}$	<100	d.	$84.03^{0°}$(anh.)	$334.9^{90°}$ (anh.)	$100^{12.5°}$ al.; s. act.; sl. s. NH_3
oxide	CoO	74.93	brn., cb.	5.68	d. 1800		i.	i.	s. a., NH_4OH; i. al.
sulfate	$CoSO_4$	155.00	red pd.	$3.710^{25°}$	d. 880		$25.6^{0°}$	$83^{100°}$	$1.04^{18°}$ m. al.; i. NH_8
sulfate	$CoSO_4{\cdot}H_2O$	173.01	red pd., mn.(?), 1.639	3.13	d.		s.	s.	
sulfate (biebeorite)	$CoSO_4{\cdot}7H_2O$*	281.10	red, mn., 1.483	$1.948\frac{25°}{25}$	96.8	$-7H_2O$, 420	$33^{80°}$	s.	$2.5^{8°}$ al.
sulfide (syeporite)	CoS	91.00	brn. nd.	$5.45^{18°}$	>1100		$0.00038^{18°}$		s. a., aq. reg.
Copper	Cu	63.55	yel.-red met., cb.	$8.92^{20°}$	1083	2300	i.	i.	s. HNO_3, h. H_2SO_4
Cupric	$Cu(C_2H_3O_2)_2$	181.63		$1.930\frac{20°}{4}$			s.		
acetate	$Cu(C_2H_3O_2)_2{\cdot}H_2O$	199.65	dark gn., mn.	1.882	115	240 d.	7.2	20	7 al.; s. et.; gly.
aceto-arsenite (Paris green)	$(CuOAs_2O_3)_3{\cdot}Cu(C_2H_3O_2)_2$*	1013.79	gn.				i.		s. a., NH_4OH
ammonium chloride	$CuCl_2{\cdot}2NH_4Cl{\cdot}2H_2O$	277.47	blue, tet., 1.670, 1.744	1.98	d. 110		$33.8^{0°}$	$99.3^{80°}$	s. a.
ammonium sulfate	$CuSO_4{\cdot}4NH_3{\cdot}H_2O$	245.75	blue, rhb.	1.81	d. 150		$18.05^{21.5°}$	d.	i. al.
carbonate, basic (azurite)	$2CuCO_3{\cdot}Cu(OH)_2$	344.67	blue, mn., 1.758	3.88	d. 220		i.	d.	s. NH_4OH, h. aq. $NaHCO_3$
carbonate, basic (malachite)	$CuCO_3{\cdot}Cu(OH)_2$	221.12	dark gn., mn., 1.875	3.9	d.		i.	d.	s. KCN; 0.03 aq. CO
chloride (eriochalcite)	$CuCl_2$	134.45	brn.-yel. pd.	3.054	498	Forms Cu_2Cl_2 993	$70.7^{0°}$	$107.9^{100°}$	$53^{15°}$ al.; $68^{15°}$ m. al.
chloride	$CuCl_2{\cdot}2H_2O$	170.48	gn., rhb., 1.684	$2.39^{22.4°}$	$-2H_2O$, 110	d.	$110.4^{0°}$	$192.4^{100°}$	s. al.; et., NH_4Cl
chromate, basic	$CuCrO_4{\cdot}2CuO{\cdot}2H_2O$	374.66	yel.-brn.		$-2H_2O$, 260		i.		s. HNO_3, NH_4OH
cyanide	$Cu(CN)_2$	115.58	yel.-gn.		d.		i.		s. KCN, C_5H_5N
dichromate	$CuCr_2O_7{\cdot}2H_2O$	315.56	bk., tri.	$2.286^{18°}$	$-2H_2O$, 100		sl. s.	d.	s. a.; NH_4OH
ferricyanide	$Cu_3[Fe(CN)_6]_2$	614.54	yel.-gn.				i.		s. NH_4OH; i. HCl
ferrocyanide	$Cu_2Fe(CN)_6{\cdot}7H_2O$	465.15	red-brn.				i.	i.	s. NH_4OH; i. a., NH_8
formate	$Cu(HCO_2)_2$	153.58	blue, mn.	1.831			12.5	d.	0.25 al.
hydroxide	$Cu(OH)_2$	97.56	blue, gelatinous	3.368	$-H_2O$		i.	d.	s. a., NH_4OH, KCN, al.
lactate	$Cu(C_3H_5O_3)_2{\cdot}2H_2O$	277.72	dark blue, mn.				16.7	$45^{100°}$	sl. s. al.
nitrate	$Cu(NO_3)_2{\cdot}3H_2O$*	241.60	blue, delq.	$2.047^{3.9°}$	114.5	$-HNO_3$, 170	$381^{40°}$	$666^{80°}$	$100^{12.5°}$ al.
nitrate	$Cu(NO_3)_2{\cdot}6H_2O$	295.65	blue, rhb.	2.074	$-3H_2O$, 26.4		$243.7^{0°}$	∞	s. al.

*Usual commercial form.
†Also a soluble modification.

(Continued)

TABLE 2-1 Physical Properties of the Elements and Inorganic Compounds *(Continued)*

Name	Formula	Formula weight	Color, crystalline form, and refractive index	Specific gravity	Melting point, °C	Boiling point, °C	Solubility in 100 parts: Cold water	Hot water	Other reagents
Cupric (*Cont.*)									
oxide (paramelaconite)	CuO	79.55	bk., cb.	6.40	d. 1026		i.	i.	s. a.; KCN, NH_4Cl
oxide (tenorite)	CuO	79.55	bk., tri., 2.63	6.45	d. 1026		i.	i.	s. a., KCN, NH_4Cl
oxychloride	$CuCl_2{\cdot}2CuO{\cdot}4H_2O$	365.60	blue-gn.		$-3H_2O$, 140		i.		s. a.
phosphide	Cu_3P_2	252.59	bk.	6.35	d.		i.		s. HNO_3; i. HCl
sulfate (hydrocyanite)	$CuSO_4$	159.61	gn.-wh., rhb., 1.733	$3.606^{15°}$	d. >600	Forms CuO, 650	$14.3^{0°}$	$75.4^{100°}$	i. al.
sulfate (blue vitriol or chalcanthite)	$CuSO_4{\cdot}5H_2O$*	249.69	blue, tri., 1.5368	$2.286\frac{15.6°}{4}$	$-4H_2O$, 110	$-5H_2O$, 250	$24.3^{0°}$	$205^{100°}$	$1.1^{8°}$ al.
sulfide (covellite)	CuS	95.61	blue, hex. or mn., 1.45	4.6	tr. 103	d. 220	$0.000033^{18°}$		s. HNO_3, KCN
tartate	$CuC_4H_4O_6{\cdot}3H_2O$	265.66	1 gn. pd.		d.		$0.02^{15°}$	$0.14^{85°}$	s. a., KOH
Cuprous ammonium iodide	$CuI{\cdot}NH_4I{\cdot}H_2O$	353.41	rhb. pl.				d.		s. NH_4I
carbonate	Cu_2CO_3	187.10	yel.	4.4	d.		i.	i.	s. a., NH_4OH
chloride (nantokite)	Cu_2Cl_2	198.00	wh., cb., 1.973	3.53	422	1366	$1.52^{25°}$		s. HCl, NH_4OH, al.
cyanide	$Cu_2(CN)_2$	179.13	wh., mn.	2.9	474.5	d.	i.	i.	s. KCN, HCl, NH_4OH; sl. s. NH_3
ferricyanide	$Cu_3Fe(CN)_6$	402.59	brn.-red				i.		s. NH_4OH; i. HCl
ferrocyanide	$Cu_4Fe(CN)_6$	466.13	brn.-red				i.		s. NH_4OH; i. NH_4Cl
fluoride	Cu_2F_2	165.09	red cr.		908	subl. 1100	i.		s. HF, HCl, HNO_3; i. al.
hydroxide	$CuOH$	80.55	yel.	3.4	$-½H_2O$, 360		i.	i.	s. a., NH_4OH
oxide (cuprite)	Cu_2O	143.09	red, cb., 2.705	6.0	1235	−O, 1800	i.	i.	s. HCl, NH_4Cl, NH_4OH
Cuprous phosphide	Cu_6P_2	443.22	gray-bk.	6.4 to 6.8			i.		s. HNO_3; i. HCl
sulfide (chalcocite)	Cu_2S	159.16	bk., rhb.	5.6	1100		$0.0005^{18°}$		s. HNO_3, NH_4OH; i. act.
sulfide	Cu_2S	159.16	bk., cb.	5.80	1130		$0.0005^{18°}$		s. HNO_3, NH_4OH; i. act.
Cyanogen	C_2N_2	52.03	poisonous gas	lq. $0.866^{-17.2°}$; 1.806 (A)	−34.4	−20.5	$450^{20°}$ cc		$2300^{20°}$ cc al.; $500^{18°}$ cc et.
Cyanogen compounds, *cf.* table of organic compounds									
Ferric acetate, basic	$Fe(OH)(C_2H_3O_2)_2$	190.94	brn., amor.				i.		s. a.; al.
ammonium sulfate, *cf.* Alum									
chloride (molysite)	$FeCl_3$	162.20	bk.-brn., hex. delq.	$2.804^{11°}$	282	315	$74.4^{0°}$	$535.8^{100°}$	v. s. al.; et. +HCl
chloride	$FeCl_3{\cdot}6H_2O$*	270.30	red-yel., delq.		37	280	$246^{0°}$	∞	s. al., act., gly.
ferrocyanide (Prussian blue)	$Fe_4[Fe(CN)_6]_3$	859.23	dark blue		d.		i.	d.	s. HCl, conc. H_2SO_4; i. al., et.
hydroxide	$Fe(OH)_3$	106.87	red-brn.	3.4 to 3.9	$-1½H_2O$, 500		i.	i.	s. a.; i. al., et.
lactate	$Fe(C_3H_5O_3)_3$	323.06	brn., amor., delq.				v. s.	v. s.	i. et.
nitrate	$Fe(NO_3)_3{\cdot}6H_2O$	349.95	rhb., delq.	$1.684^{20°}$	35	d.	$150^{0°}$	∞	s. al., act.
oxide (hematite)	Fe_2O_3	159.69	red or bk., trig., 3.042	5.12	1560 d.		i.		s. HCl
sulfate	$Fe_2(SO_4)_3$	399.88	rhb., 1.814	$3.097^{18°}$	d. 480		sl. s.	d.	i. H_2SO_4, NH_3
sulfate (coquimbite)	$Fe_2(SO_4)_3{\cdot}9H_2O$	562.02	yel., trig.	2.1			440	d.	s. abs. al.
Ferroso-ferric chloride	$FeCl_2{\cdot}2FeCl_3{\cdot}18H_2O$	775.43	yel., delq.		d. 50		s.	s.	
ferricyanide (Prussian green)	$Fe_4Fe_3[Fe(CN)_6]_6$	1662.61	gn.		d. 180		i.		s. d. h. HCl
oxide (magnetite; magnetic iron oxide)	Fe_3O_4	231.53	bk., cb., 2.42	5.2	1538 d.		i.	i.	i. al.
oxide, hydrated	$Fe_3O_4{\cdot}4H_2O$	303.59	bk.		d.		i.	i.	s. a.
Ferrous ammonium sulfate	$FeSO_4{\cdot}(NH_4)_2SO_4{\cdot}6H_2O$	392.14	blue-gn., mn., 1.4915	1.864	d.		$18^{0°}$	$100^{75°}$	i. al.
chloride (lawrencite)	$FeCl_2$	126.75	gn.-yel., hex., 1.567	2.7		delq.	$64.4^{10°}$	$105.7^{100°}$	100 al.; s. act.; i. et.
chloroplatinate	$FePtCl_6{\cdot}6H_2O$	571.73	yel., hex.	2.714			v. s.	v. s.	
ferricyanide (Turnbull's blue)	$Fe_3[Fe(CN)_6]_2$	591.43	dark blue		d.		i.		i. dil. a., al.
ferrocyanide	$Fe_2Fe(CN)_6$	323.64	blue-wh., amor.				i.		
formate	$Fe(HCO_2)_2{\cdot}2H_2O$	181.91			d.		sl. s.		
hydroxide	$Fe(OH)_2$	89.86	lt. gn.	3.4			0.00067		s. a., NH_4Cl
nitrate	$Fe(NO_3)_2{\cdot}6H_2O$	287.95	cr.		60.5		$200^{0°}$	$300^{25°}$	
oxide	FeO	71.84	bk.	5.7	1420		i.	i.	s. a.; i. alk.

phosphate (vivianite)	$Fe_3(PO_4)_2 \cdot 8H_2O$	501.60	blue, mn., 1.592, 1.603	2.58			i.	i.	s. a.; i. ac.
silicate	$FeSiO_3$	131.93	mn.	3.5	1550				
sulfate (siderotilate)	$FeSO_4 \cdot 5H_2O$	241.98	gn., tri., 1.536	2.2		$-5H_2O$, 300	s.	s.	i. al.
sulfate (copperas)	$FeSO_4 \cdot 7H_2O$*	278.01	blue-gn., mn.	$1.899^{14.8°}$	64	$-7H_2O$, 300	$32.8^{0°}$	$149^{50°}$	i. al.
sulfide	FeS	87.91	bk., hex.	4.84	1193	d.	$0.000616^{18°}$		s. a.; i. NH_3
cf. also under iron									
Fluoboric acid	HBF_4	87.81	col. lq.			130 d.	∞	∞	s. al.
Fluorine	F_2	38.00	gn.-yel. gas	lq. $1.51^{-187°}$; $1.31^{15°}$ (A)	−223	−187	d.		
Fluosilicic acid	H_2SiF_6	144.09					s.	s.	
Gadolinium	Gd	157.25							
Gallium bromide	$GaBr_3$	309.44	delq. cr.				s.	s.	
Glucinum *cf.* Beryllium									
Gold	Au	196.97	yel. met., cb.	$19.3^{20°}$	1063	2600	i.	i.	s. aq. reg., KCN; i. a.
Gold, colloidal	Au	196.97	blue to vl.				s.		s. aq. reg., KCN; i. a.
Gold salts *cf.* under Auric and Aurous									
Hafnium	Hf	178.49	hex.	12.1	>1700	>3200(?)			
Helium	He	4.00	col. gas	0.1368 (A)	<−272.2	−268.9	$0.97^{0°}$ cc	$1.08^{50°}$ cc	Absorbed by Pt
Hydrazine	N_2H_4	32.05	col. lq.	$1.011\frac{15°}{4}$	1.4	113.5	∞	∞	s. al.
formate	$N_2H_4 \cdot 2HCO_2H$	124.10	cb.		128		s.		
hydrate	$N_2H_4 \cdot H_2O$	50.06	col.	$1.03^{21°}$	−40	$118.5^{739.5mm}$	∞	∞	∞ al.; i. et.
hydrochloride	$N_2H_4 \cdot HCl$	68.51	yel. lq.				v. s.	v. s.	sl. s. al.
hydrochloride, di-	$N_2H_4 \cdot 2HCl$	104.97	wh., cb.	1.42	198		s.	v. s.	s. al.
nitrate	$N_2H_4 \cdot HNO_3$	95.06	cr.		70.7	subl. 140	$174.9^{10°}$	v. s.	
nitrate, di-	$N_2H_4 \cdot 2HNO_3$	158.07	nd.		104	d.	v. s.		
sulfate	$N_2H_4 \cdot ½H_2SO_4$	81.08	delq. pl.		85		v. s.		i. al.
sulfate	$N_2H_4 \cdot H_2SO_4$	130.12	rhb.	1.378	254		$3.055^{22°}$	$27.65^{60°}$	v. sl. s. abs. al.
Hydrazoic acid (azoimide)	HN_3	43.03	col. lq.		−80	37	∞	∞	∞ al.
Hydriodic acid	HI	127.91	col. gas	$4.4^{0°}$ (A)	−50.8	−35.5	$42{,}500^{10°}$ cc	v. s.	s. al.
Hydriodic acid	$HI \cdot H_2O$	145.93	col. lq.	$1.7^{15°}$		127^{774mm}	∞		∞ al.
Hydriodic acid	$HI \cdot 2H_2O$	163.94	col. lq.		−43		∞		s. al.
Hydriodic acid	$HI \cdot 3H_2O$	181.96	col. lq.		−48		∞		s. al.
Hydriodic acid	$HI \cdot 4H_2O$	199.97	col. lq.		−36.5		∞		s. al.
Hydrobromic acid	HBr	80.91	col. gas; 1.325 (lq.)	$2.71^{0°}$ (A)	−86	−67	$221^{0°}$	$130^{100°}$	s. al.
Hydrobromic acid	$HBr \cdot H_2O$	98.93	col. lq.	1.78					Stable at −15.5° and 1 atm., and at −11.3° and 2.5 atm.
Hydrobromic acid	HBr (47.8% in H_2O)	80.91	col. lq.	1.486		126	∞		s. al.
Hydrobromic acid	$HBr \cdot 2H_2O$	118.96	wh. cr.	$2.11^{-15°}$	−11		s.	s.	
Hydrochloric acid	HCl†	36.46	col. gas; 1.256 (lq.)	$1.268^{0°}$ (A)	−111	−85	$82.3^{0°}$	$56.1^{60°}$	s. al., et.
Hydrochloric acid	HCl (45.2% in H_2O)	36.46	col. lq.	1.48	−15.35		∞		s. al.
Hydrochloric acid	$HCl \cdot 2H_2O$	72.49	col. lq.	$1.46\frac{-18.3°}{4}$	0	d.	∞		s. al.
Hydrochloric acid	$HCl \cdot 3H_2O$	90.51	col. lq.		−24.4	d.	∞		s. al.
Hydrocyanic acid (prussic acid)	HCN	27.03	poisonous gas or col. lq., 1.254	$0.697^{18°}$	−14	26	∞		∞ al., et.
Hydrofluoric acid	HF	20.01	gas or col. lq.	$0.988^{13.6°}$	−83	19.4	∞ 0° to 19.4°	v. s.	
Hydrofluoric acid	HF (35.35% in H_2O)	20.01	col. lq.	1.15	−35	120	v. s.		
Hydrogen	H_2	2.02	col. gas or cb.	lq. $0.0709^{-252.7°}$ 0.06948 (A)	−259.1	−252.7	$2.1^{0°}$ cc	$0.85^{80°}$ cc	sl. s. Fe, Pd, Pt
peroxide	H_2O_2‡	34.01	col. lq., 1.333	$1.438\frac{20°}{4}$	−0.89	151.4^{760mm}	∞		s. a., et.; i. petr. et
selenide	H_2Se	80.98	col. gas	$2.12^{-42°}$	−64	−42	$377^{4°}$ cc	$270^{22.5°}$ cc	s. CS_2, $COCl_2$
sulfide	H_2S	34.08	col. gas	1.1895 (A)	−82.9	−59.6	$437^{0°}$ cc	$186^{40°}$ cc	$9.54^{15°}$ cc al.; s. CS_2
Hydroxylamine	NH_2OH	33.03	rhb., delq.	$1.35^{18°}$	34	56.5^{22mm}	s.	d.	s. a., al.
hydrochloride	$NH_2OH \cdot HCl$	69.49	col., mn.	$1.67^{17°}$	151	d.	$83.3^{17°}$	v. s.	s. al.; i. et.
nitrate	$NH_2OH \cdot HNO_3$	96.04	col. cr.		48	d. <100	v. s.	d.	v. s. abs. al.
sulfate	$NH_2OH \cdot ½H_2SO_4$	82.07	col., mn.		170 d.		$32.9^{0°}$	$68.5^{90°}$	v. sl. s. al.; i. et., abs. al.

*Usual commercial form.
†Usual commercial form about 31 percent.
‡Usual commercial forms 3 or 30 percent.

(Continued)

TABLE 2-1 Physical Properties of the Elements and Inorganic Compounds *(Continued)*

Name	Formula	Formula weight	Color, crystalline form, and refractive index	Specific gravity	Melting point, °C	Boiling point, °C	Solubility in 100 parts		
							Cold water	Hot water	Other reagents
Hypobromous acid	HBrO	96.91	yel.			40^{50mm}	s.	d.	
Indium	In	114.82	soft, tet. met.	7.3^{20°	155	1450	i.	i.	s. a.
Iodic acid	HIO_3	175.91	col., rhb.	4.629^{0°	110 d.		286^{0°	576^{101°	v. s. 87% al.; i. abs. al. et., chl.
Iodine	I_2	253.81	blue-bk., rhb.	4.93^{20°	113.5	184.35	0.0162^{0°	0.09566^{60°	s. al., KI, et.
oxide, penta-	I_2O_5	333.81	wh., trimetric	$4.799\frac{25^\circ}{4}$	d. 300		187.4^{12°		i. abs. al., et., chl.
Iodoplatinic acid	$H_2PtI_6 \cdot 9H_2O$	1120.66	brn., delq. mn.				s. d.		
Iridium	Ir	192.22	wh. met., cb.	22.4^{20°	2350	>4800	i.	i.	sl. s. aq. reg., aq. Cl_2
Iron, cast†	Fe	55.85	gray	7.03	1275		i.	i.	s. a.; i. alk.
pure	Fe	55.85	silv. met., cb.	7.86^{20°	1535	3000	i.	i.	s. a.; i. alk.
steel	Fe	55.85	silv. gray	7.6 to 7.8	1375		i.	i.	s. a.; i. alk.
white pig	Fe	55.85	gray	7.6 to 7.8	1075		i.	i.	s. a.; i. alk.
wrought	Fe	55.85	gray	7.86	1505		i.	i.	s. a.; i. alk.
carbide (cementite)	Fe_3C	179.55	pseudo hex.	7.4	1837		i.	i.	s. a.
carbonyl	$Fe(CO)_5$	195.90	pa. yel. lq.	1.457^{21°	−21	102.5^{760mm}	i.		s. al., H_2SO_4, alk.
nitride	Fe_2N	125.70	gray	6.35	d. >560		d.		s. HCl, H_2SO_4
silicide	FeSi	83.93	yel.-gray, oct.	$6.1\frac{20^\circ}{4}$			i.	i.	i. aq. reg.
sulfide, di- (marcasite)	FeS_2	119.98	yel., rhb.	4.87	tr. 450	d.	0.00049		i. dil. a.
sulfide, di- (pyrite)	FeS_2	119.98	yel., cb.	5.0	1171	d.	0.0005		i. dil. a.
sulfide (pyrrhotite)	Fe_7S_8	647.44	hex.	$4.6\frac{20^\circ}{4}$	d. >700		i.		
Cf. also under ferric and ferrous									
Krypton	Kr	83.80	col. gas	2.818 (A)	−169	−151.8	11.05^{0° cc	3.57^{60° cc	sl. s. al., bz.
Lanthanum	La	138.91	lead gray	6.15^{20°	826	1800	d.		s. a.
Lead	Pb	207.20	silv. met., cb.	$11.337\frac{20^\circ}{20}$	327.5	1620	i.	i.	s. HNO_3; i. c. HCl, H_2SO_4
acetate	$Pb(C_2H_3O_2)_2$	325.29	wh. cr.	$3.251\frac{20^\circ}{4}$	280		19.7^{0°	221^{50°	s. gly.; v. sl. s. al.
acetate (sugar of lead)	$Pb(C_2H_3O_2)_2 \cdot 3H_2O$†	379.33	wh., mn.	2.55	$-3H_2O$, 75		45.64^{15°	200^{100°	s. gly.; sl. s. al.
acetate	$Pb(C_2H_3O_2)_2 \cdot 10H_2O$	505.44	wh., rhb.	1.689	22		s.	s.	
acetate, basic	$Pb_2(C_2H_3O_2)_3OH$	608.54	wh.				v. s.		sl. s. al.
acetate, basic	$Pb(C_2H_3O_2)_2 \cdot Pb(OH)_2 \cdot H_2O$	584.52	wh. nd.				v. s.		s. al.
acetate, basic	$Pb(C_2H_3O_2)_2 \cdot 2Pb(OH)_2$	807.72	wh. nd.				5.55	18.2	s. al.
arsenate, monobasic	$PbH_4(AsO_4)_2$	489.07	tri., 1.82	4.46^{15°	d. 140		d.		s. HNO_3
arsenate, dibasic (schultenite)	$PbHAsO_4$	347.13	wh., mn., 1.9097	5.94	d. >200	$-H_2O$, 280	i.	sl. s.	s. HNO_3, NaOH
arsenate, meta-	$Pb(AsO_3)_2$	453.04	hex.	6.42^{15°			d.		s. HNO_3
arsenate, pyro-	$Pb_2As_2O_7$	676.24	rhb., 2.03	$6.85\frac{15^\circ}{15}$	802		i.	d.	s. HCl, HNO_3; i. sc.
azide	PbN_6	291.24	col. nd.		expl. 350		i.	0.05^{100°	v. s. ac.; i. NH_4OH
bromide	$PbBr_2$	367.01	col., rhb.	6.66	373	918	0.4554^{0°	4.75^{100°	s. a., KBr.; sl. s. NH_3; i. al.
carbonate (cerussite)	$PbCO_3$	267.21	wh., rhb., 2.0763	6.6	d. 315		0.00011^{20°	d.	s. a., alk.; i. NH_3, al.
carbonate, basic (hydrocerussite; white lead)	$2PbCO_3 \cdot Pb(OH)_2$†	775.63	wh., hex.	6.14	d. 400		i.	i.	s. ac.; sl. s. aq. CO_2
chloride (cotunnite)	$PbCl_2$	278.11	wh., rhb., 2.2172	5.80	501	954^{760mm}	0.673^{0°	3.34^{100°	sl. s. dil. HCl, NH_3, i. al.
chromate (crocoite)	$PbCrO_4$	323.19	yel., mn., 2.42	6.12	844	d.	0.000007^{20°	i.	s. a., alk.; i. NH_3, ac.
chromate, basic	$PbCrO_4 \cdot PbO$	546.39	or.-yel. nd.				i.	i.	s. a., alk.
formate	$Pb(HCO_2)_2$	297.23	wh., rhb.	4.56	d. 190		1.6^{16°	18^{100° d.	i. al.
hydroxide	$3PbO \cdot H_2O$	687.61	cb.	7.592	$-H_2O$, 130		0.014		s. a., alk.
nitrate	$Pb(NO_3)_2$	331.21	col., cb. or mn., 1.7815	4.53	d. 470		38.8^{0°	138.8^{100°	8.8^{22° al.
oxide, sub-	Pb_2O	430.40	bk., amor.	8.34	d. red heat		i.	i.	s. a., alk.
oxide, mono- (litharge)	PbO	223.20	yel., tet.	9.53	888		0.0068^{18°		s. alk., PbAc, NH_4Cl, $CaCl_2$
oxide, mono (massicotite)	PbO	223.20	yel., rhb., 2.61	8.0					

oxide, mono-	PbO	223.20	amor.	9.2 to 9.5			i.	i.	s. alk., PbAc, NH_4Cl, $CaCl_2$
oxide, red (minium)	Pb_3O_4	685.60	red, amor.	9.1	d. 500		i.	i.	s. ac., h. HCl
oxide, sesqui-	Pb_2O_3	462.40	red-yel., amor.		d. 360		i.	i.	s. a., alk.
oxide, di- (plattnerite)	PbO_2	239.20	brn., tet., 2.229	9.375	d. 290		i.	i.	s. ac., h. alk.; i. al.
silicate	$PbSiO_3$	283.28	col., mn., 1.961	6.49	766		i.		s. a.
sulfate (anglesite)	$PbSO_4$	303.26	wh., mn. or rhb., 1.8823	6.2	1170		$0.0028^{0°}$	$0.0056^{40°}$	s. conc. a., NH_4 salts; i. al.
sulfate, acid	$Pb(HSO_4)_2{\cdot}H_2O$	419.36	cr.		d.		$0.0001^{18°}$		sl. s. H_2SO_4
sulfate, basic (lanarkite)	$PbSO_4{\cdot}PbO$	526.46	col., mn.	6.92	977		$0.0044^{18°}$		sl. s. H_2SO_4
sulfide (galena)	PbS	239.27	lead gray, cb., 3.912	7.5	1120		$0.00009^{18°}$	i.	s. a.; i. alk.
thiocyanate	$Pb(CNS)_2$	323.36	col., mn.	3.82	d. 190		$0.05^{20°}$	s.	s. KCNS, HNO_3
Lithium	Li	6.94	silv. met. cb.	$0.53^{20°}$	186	1336 ± 5	d.	d.	s. a., NH_3
benzoate	$LiC_7H_5O_2$	128.05	wh. leaflets				$33^{25°}$	$40^{100°}$	$7.7^{25°}$, $10^{78°}$ al.
bromide	LiBr	86.85	wh., delq., cb., 1.784	$3.464\frac{25°}{4}$	547	1265	$143^{0°}$ ($2H_2O$)	$266^{100°}$ ($1H_2O$)	s. al., act.
bromide	$LiBr{\cdot}2H_2O$	122.88	wh. pr.		44		$246^{20°}$		s. al.
carbonate	Li_2CO_3	73.89	col., mn., 1.567	$2.11^{0°}$	618	d.	$1.54^{0°}$	$0.72^{100°}$	s. dil. a.; i. al., act., NH_3
chloride	LiCl	42.39	wh., delq., cb., 1.662	$2.068\frac{25°}{4}$	614	1360	$67^{0°}$	$127.5^{100°}$	$2.48^{15°}$ al.; s. et.
citrate	$Li_3C_6H_5O_7{\cdot}4H_2O$	281.98	wh. cr.		d.		$61.2^{15°}$	$66.7^{100°}$	sl. s. al., et.
fluoride	LiF	25.94	wh., cb., 1.3915	$2.295^{21.5°}$	870	1670	$0.27^{18°}$	$0.135^{35°}$	s. HF; i. act.
formate	$LiHCO_2{\cdot}H_2O$	69.97	col., rhb.	1.46	$-H_2O$, 94		$49.2^{0°}$	$346.6^{104°}$	sl. s. al., et.
hydride	LiH	7.95	wh., cb.	0.820	680		d.		i. et.
hydroxide	LiOH	23.95	wh. cr.	2.54	445	925±	$12.7^{0°}$	$17.5^{100°}$	sl. s. al.
hydroxide	$LiOH{\cdot}H_2O$	41.96	col., mn.	1.83		d.	$22.3^{10°}$	$26.8^{80°}$	sl. s. al.
nitrate	$LiNO_3$	68.95	col., trig., 1.735	2.38	261		$53.4^{0°}$	$194^{70°}$	s. al., NH_3
nitrate	$LiNO_3{\cdot}3H_2O$	122.99	col.		29.88		v. s.	∞	
oxide	Li_2O	29.88	col., 1.644	$2.013\frac{25°}{4}$		subl. <1000	forms LiOH		
phosphate, monobasic	LiH_2PO_4	103.93	col.	2.461	>100				
phosphate, tribasic	Li_3PO_4	115.79	wh., rhb.	$2.537^{17.5°}$	837		$0.034^{18°}$	v. sl. s.	s. a., NH_4Cl; i. act.
phosphate, tribasic	$Li_3PO_4{\cdot}12H_2O$	331.98	wh., trig.	1.645	100		v. sl. s.	v. sl. s.	
salicylate	$LiC_7H_5O_3$	144.05	col.		d.		$128^{26°}$		v. s. al.
sulfate	Li_2SO_4	109.94	col., mn., 1.465	2.22	860		$35.34^{0°}$	$29.9^{100°}$	i. act., 80% al.
sulfate	$Li_2SO_4{\cdot}H_2O$†	127.96	col., mn., 1.477	2.06	$-H_2O$, 130		$43.6^{0°}$	$35^{100°}$	i. 80% al.
sulfate, acid	$LiHSO_4$	104.01	pr.	$2.123^{13°}$	170.5		d.		
Lutecium	Lu	174.97							
Magnesium	Mg	24.31	silv. met., hex.	$1.74^{20°}$	651	1110	i.	sl. s. d.	s. a., NH_4 salts
acetate	$Mg(C_2H_3O_2)_2$	142.39	wh.	1.42	323		v. s.	v. s.	$5.25^{15°}$ m. al.
acetate	$Mg(C_2H_3O_2)_2{\cdot}4H_2O$†	214.45	wh., mn. pr., 1.491	1.454	80		v. s.	v. s.	v. s. al.
aluminate (spinel)	$MgO{\cdot}Al_2O_3$	142.26	col. cb., 1.718–23	3.6	2135		i.		v. sl. s. dil. HCl; i. dil. HNO_3
ammonium chloride	$MgCl_2{\cdot}NH_4Cl{\cdot}6H_2O$	256.79	wh., rhb., delq.	1.456	$-4H_2O$, 195		16.7	s.	
ammonium phosphate (struvite)	$MgNH_4PO_4{\cdot}6H_2O$	245.41	col., rhb., 1.496	1.715	d. 100		$0.0231^{0°}$	$0.0195^{80°}$	s. a.; i. al.
ammonium sulfate (boussingaultite)	$MgSO_4{\cdot}(NH_4)_2SO_4{\cdot}6H_2O$	360.60	col., mn.	1.72	>120		$16.86^{0°}$	$130^{100°}$	
benzoate	$Mg(C_7H_5O_2)_2{\cdot}3H_2O$	320.58	wh. pd.		$-3H_2O$, 110		$4.5^{25°}$ (anh.)	s.	s. act.
carbonate (magnesite)	$MgCO_3$	84.31	wh., trig. 1.700	3.037	d. 350		0.0106		s. a., aq. CO_2; i. act., NH_3
carbonate (nesquehonite)	$MgCO_3{\cdot}3H_2O$	138.36	col., rhb., 1.501	1.852	$-H_2O$, 100		$0.1518^{19°}$	d.	s. a., aq. CO_2
carbonate, basic (hydromagnesite)	$3MgCO_3{\cdot}Mg(OH)_2{\cdot}3H_2O$	365.31	wh., rhb., 1.530	2.16	d.		0.04	0.011	s. a., NH_4 salts; i. al.
Magnesium chloride (chloromagnesite)	$MgCl_2$	95.21	col., hex., 1.675	$2.325^{25°}$	712	1412	$52.8^{0°}$	$73^{100°}$	50 al.
chloride (bischofite)	$MgCl_2{\cdot}6H_2O$†	203.30	wh., delq., mn., 1.507	1.56	118 d.	d.	$281^{0°}$	$918^{100°}$	50 al.
hydroxide (brucite)	$Mg(OH)_2$	58.32	wh., trig., 1.5617	2.4	d.		$0.0009^{18°}$		s. NH_4 salts, dil. a.
nitride	Mg_3N_2	100.93	gn.-yel., amor.		d.		i.	d.	s. a.; i. al.
oxide (magnesia; periclase)	MgO	40.30	col., cb., 1.7364	3.65	2800	3600	0.00062		s. a., NH_4 salts; i. al.
perchlorate	$Mg(ClO_4)_2$†	223.21	wh., delq.	$2.60^{25°}$	d.		$99.6^{25°}$	v. s.	24^{25} al., $51.8^{25°}$ m. al.; 0.29 et.

*See also a table of alloys.
†Usual commercial form.

(Continued)

TABLE 2-1 Physical Properties of the Elements and Inorganic Compounds *(Continued)*

Name	Formula	Formula weight	Color, crystalline form, and refractive index	Specific gravity	Melting point, °C	Boiling point, °C	Solubility in 100 parts: Cold water	Hot water	Other reagents
Magnesium (*Cont.*)									
peroxide	MgO_2	56.30	wh. pd.		expl. 275		i.	i.	s. a.
phosphate, pyro-	$Mg_2P_2O_7$	222.55	col., mn., 1.604	$2.598^{22°}$	1383		i.	i.	s. a.; i. alk.
phosphate, pyro-	$Mg_2P_2O_7 \cdot 3H_2O$	276.60	wh., amor.	2.56	$-3H_2O$, 100		i.	sl. s.	s. a.; i. al.
potassium chloride (carnallite)	$MgCl_2 \cdot KCl \cdot 6H_2O$	277.85	delq., rhb., 1.475	$1.60\frac{19.4°}{4}$	265		$64.5^{19°}$ d.	d.	d. al.
potassium sulfate (picromerite)	$MgSO_4 \cdot K_2SO_4 \cdot 6H_2O$	402.72	mn., 1.4629	2.15	d. 72		$19.26^{0°}$	$81.7^{75°}$	
silicofluoride	$MgSiF_6 \cdot 6H_2O$	274.47	col., trig., 1.3439	$1.788\frac{17.5°}{4}$	d.		$64.8^{17.5°}$	s.	d. HF
sodium chloride	$MgCl_2 \cdot NaCl \cdot H_2O$	171.67	col.				s.	s.	
sulfate	$MgSO_4$	120.37	col.	2.66	1185		$26.9^{0°}$	$68.3^{100°}$	s. al.
sulfate (epsom salt; epsomite)	$MgSO_4 \cdot 7H_2O$*	246.47	col., rhb., 1.4554	1.68	70 d.		$72.4^{0°}$	$178^{40°}$	s. al.
Manganese	Mn	54.94	gray-pink met.	$7.2^{20°}$	1260	1900	d.		s. dil. a.
acetate	$Mn(C_2H_3O_2)_2$	173.03		$1.74\frac{20°}{4}$			s.	s.	
acetate	$Mn(C_2H_3O_2)_2 \cdot 4H_2O$*	245.09	pa. pink, mn.	1.589			s.	$64.5^{50°}$	s. al., m. al.
carbonate (rhodocrosite)	$MnCO_3$	114.95	rose, trig., 1.817	3.125	d.		$0.0065^{25°}$		s. aq. CO_2, dil. a.; l. NH_3, al.
chloride (scacchite)	$MnCl_2$	125.84	rose, delq., cb.	$2.977\frac{25°}{4}$	650	1190	$63.4^{0°}$	$123.8^{100°}$	s. al.; i. et., NH_3
chloride	$MnCl_2 \cdot 4H_2O$*	197.91	rose red, delq., mn. 1.575	2.01	58.0	$-H_2O$, 106; $-4H_2O$, 200	$151^{8°}$	∞	s. al.; i. et.
chloride, per-	$MnCl_4$	196.75	gn.				s.	s.	s. al., et.
hydroxide (ous) (pyrochroite)	$Mn(OH)_2$	88.95	wh., trig.	$3.258^{18°}$	d.		$0.002^{20°}$	i.	s. a., NH_4 salts; i. alk.
hydroxide (ic) (manganite)	$Mn_2O_3 \cdot H_2O$	175.89	brn., rhb., 2.24	3.258	d.		i.	i.	s. h. H_2SO_4
nitrate	$Mn(NO_3)_2 \cdot 6H_2O$	287.04	rose red, mn.	$1.82^{21°}$	25.8	129.5	$426^{0°}$	∞	v. s. al.
oxide (ous) (manganosite)	MnO	70.94	gray-gn., cb., 2.16	5.18	1650		i.	i.	s. a., NH_4Cl
oxide (ic)	Mn_2O_3	157.87	brn.-bk., cb.	4.81	–0, 1080		i.	i.	s. a.; i. act.
oxide, di- (pyrolusite; polianite)	MnO_2*	86.94	bk., rhb.	5.026	–0, >230		i.	i.	s. HCl; i. HNO_3, act.
sulfate (ous)	$MnSO_4$	151.00	red-wh.	3.235	700	d. 850	$53^{0°}$	$73^{50°}$	s. al.; i. et.
sulfate (ous) (szmikite)	$MnSO_4 \cdot H_2O$	169.02	pa. pink, mn., 1.595	2.87	Stable 57 to 117		$98.47^{48°}$	$79.77^{100°}$	
sulfate (ous)	$MnSO_4 \cdot 2H_2O$	187.03		$2.526^{15°}$	Stable 40 to 57		$85.27^{35°}$	$106.8^{55°}$	
sulfate (ous)	$MnSO_4 \cdot 3H_2O$	205.05		$2.356^{15°}$	Stable 30 to 40		$74.22^{5°}$	$99.31^{57°}$	
sulfate (ous)	$MnSO_4 \cdot 4H_2O$*	223.06	pink, rhb. or mn., 1.518	2.107	Stable 18 to 30	$-4H_2O$, 450	$136^{16°}$	$169^{50°}$	i. al.
sulfate (ous)	$MnSO_4 \cdot 5H_2O$	241.08	pink, tri., 1.508	$2.103^{15°}$	Stable 8 to 18		$142^{5°}$	$200^{35°}$	
sulfate (ous)	$MnSO_4 \cdot 6H_2O$	259.09			Stable –5 to +8		$204^{0°}$	$247^{9°}$	
sulfate (ous)	$MnSO_4 \cdot 7H_2O$	277.11	pink, mn. or rhb.	2.092	Stable –10 to –5; 19 d.	$-7H_2O$, 280	$176^{0°}$	$251^{14°}$	
sulfate (ic)	$Mn_2(SO_4)_3$	398.06	gn., delq. cr.	3.24	d. 160		v. s.	d.	s. HCl, dil. H_2SO_4; l.
Mercuric acetate	$Hg(C_2H_3O_2)_2$	318.68	wh. pl.	3.270	d.		$25^{10°}$	$100^{100°}$	s. al. sl. d.
bromide	$HgBr_2$	360.40	wh., rhb.	6.053	237	322	$0.5^{20°}$	$25^{100°}$	$25.2^{0°}$ al.; v. sl. s. et.
carbonate, basic	$HgCO_3 \cdot 2HgO$	693.78	brn.-red				i.		s. aq. CO_2, NH_4Cl
chloride (corrosive sublimate)	$HgCl_2$	271.50	wh., rhb., 1.859	5.44	277	304	$3.6^{0°}$	$61.3^{100°}$	$33^{25°}$ 99% al.; 33 et.
fulminate	$Hg(CNO)_2$	284.62	cb.	4.42	expl.		sl. s.		s. NH_4OH, al.
hydroxide	$Hg(OH)_2$	234.60			$-H_2O$, 175		i.	i.	s. a.
oxide (montroydite)	HgO	216.59	yel. or red, rhb., 2.5	11.14	d. 100		$0.0052^{25°}$	$0.041^{100°}$	s. a.; i. al.
oxychloride (kleinite)	$HgCl_2 \cdot 3HgO$	921.26	yel., hex.	7.93	d. 260		i.	d.	s. HCl
silicofluoride, basic	$HgSiF_6 \cdot HgO \cdot 3H_2O$	613.30	yel. nd.				d.		s. a.
sulfate	$HgSO_4$	296.65	wh., rhb.	6.47	d.		d.		s. a.; i. al., act., NH_8
sulfate, basic (turpeth)	$HgSO_4 \cdot 2HgO$	729.83	yel., tet.	6.44			0.005	$0.167^{100°}$	s. a.; i. al.
Mercurous acetate	$HgC_2H_3O_2$	259.63	wh. sc.		d.		$0.75^{13°}$	d.	s. H_2SO_4, HNO_3; i. al.
bromide	HgBr	280.49	wh., tet.	7.307	subl. 345		7×10^{-9}	i.	s. a.; i. al., act.
carbonate	Hg_2CO_3	461.19	yel. pd.		d. 130		i.	d.	s. NH_4Cl

chloride (calomel)	HgCl	236.04	wh., tet., 1.9733	7.150	302	383.7	$0.0014^{0°}$	$0.0007^{43°}$	s. aq. reg., $Hg(NO_3)_2$; sl. s. HNO_3, HCl; i. al., etc.
iodide	HgI	327.49	yel., tet.	7.70	290 d.	subl. 140; 310d.	2×10^{-8}	v. sl. s.	s. KI; i. al.
nitrate	$HgNO_3 \cdot H_2O$	280.61	wh. mn.	$4.785^{3.9°}$	70	expl.	v. s.	d.	s. HNO_3; i. al., et.
Mercurous oxide	Hg_2O	417.18	bk.	9.8	d. 100		i.	0.0007	s. h. ac.; i. alk., dil. HCl, NH_3
sulfate	Hg_2SO_4	497.24	wh., mn.	7.56	d.		$0.055^{16.5°}$	$0.092^{100°}$	s. H_2SO_4, HNO_3
Mercury[†]	Hg	200.59	silv. lq. or hex.(?)	$13.546^{20°}$	−38.87	356.9	i.	i.	s. HNO_3; i. HCl
Molybdenum	Mo	95.94	gray, cb.	10.2	2620 ± 10	3700	i.	i.	s. h. conc. H_2SO_4; i. HCl, HF, NH_3, dil. H_2SO_4, Hg
chloride, di-	$MoCl_2$	166.85	yel., amor.	$3.714^{25°}_{4}$	d.		i.	i.	s. HCl, H_2SO_4, NH_4OH, al., et.
chloride, tri-	$MoCl_3$	202.30	dark red pd.	$3.578^{25°}_{4}$	d.		i.	d.	s. HNO_3, H_2SO_4; v. sl. s. al., et.
chloride, tetra-	$MoCl_4$	237.75	brn., delq.		volt.	d.	s.	d.	s. HNO_3, H_2SO_4; sl. s. al., et.
chloride, penta-	$MoCl_5$	273.21	bk. cr.	$2.928^{25°}_{4}$	194	268	s.	d.	s. HNO_3, H_2SO_4; i. abs. al., et.
oxide, tri- (molybdite)	MoO_3	143.94	col., rhb.	$4.50^{19.5°}$	795	subl.	$0.107^{18°}$	$2.106^{79°}$	s. a., NH_4OH
sulfide, di- (molybdenite)	MoS_2	160.07	bk., hex., 4.7	$4.801^{14°}$	1185		i.	i.	s. H_2SO_4, aq. reg.
sulfide, tri-	MoS_3	192.14	red-brn.		d.		sl. s.	s.	s. alk. sulfides
sulfide, tetra-	MoS_4	224.20	brn. pd.		d.		i.	i.	s. alk. sulfides; i. NH_3
Molybdic acid	H_2MoO_4	161.95	yel-wh., hex.		d. 115		v. sl. s.	sl. s.	s. NH_4OH, H_2SO_4; i. NH
Molybdic acid	$H_2MoO_4 \cdot H_2O$	161.95	yel., mn.	$3.124^{15°}$	$-H_2O$, 70	$-2H_2O$, 200	$0.133^{18°}$	$2.13^{70°}$	s. a., NH_4OH, NH_4, salts
Neodymium	Nd	144.24	yellowish	$6.9^{20°}$	840		d.		
Neon	Ne	20.18	col. gas	lq. $1.204^{-245.9°}$ 0.674 (A)	−248.67	−245.9	$2.6^{0°}$ cc	$1.1^{45°}$ cc	s. lq. O_2, al., act., bz.
Neptunium	Np^{239}	239.05			Produced by Neutron bombardment of U^{238}				
Nickel	Ni	58.69	silv. met., cb.	8.90^{20}	1452	2900	i.	i.	s. dil. HNO_3; sl. s. H_2SO_4, HCl; i. NH_3
acetate	$Ni(C_2H_3O_2)_2$	176.78	gn. pr.	1.798	d.		16.6		i. al.
ammonium chloride	$NiCl_2 \cdot NH_4Cl \cdot 6H_2O$	291.18	gn., delq., mn.	1.645			$150^{25°}$	v. s.	
ammonium sulfate	$NiSO_4 \cdot (NH_4)_2SO_4 \cdot 6H_2O$	394.99	blue-gn., mn., 1.5007	1.923			$2.5^{3.5°}$	$39.2^{88°}$	v. sl. s. $(NH_4)_2SO_4$
bromate	$Ni(BrO_3)_2 \cdot 6H_2O$	422.59	gn., cb.	2.575	d.		28		s. NH_4OH
bromide	$NiBr_2$	218.50	yel., delq.	$4.64^{28°}_{4}$	d.		$112.8^{0°}$	$156^{100°}$	s. al., et., NH_4OH
bromide	$NiBr_2 \cdot 3H_2O$	272.55	gn., delq.		$-3H_2O$, 200		$199^{0°}$	$316^{100°}$	s. al., et., NH_4OH
bromide, ammonia	$NiBr_2 \cdot 6NH_3$	320.68	vl. pd.	1.837			v. s.	d.	i. c. NH_4OH
bromoplatinate	$NiPtBr_6 \cdot 6H_2O$	841.29	trig.	3.715					
carbonate	$NiCO_3$	118.70	lt. gn., rhb.		d.		$0.0093^{25°}$	i.	s. a.
carbonate, basic	$2NiCO_3 \cdot 3Ni(OH)_2 \cdot 4H_2O$	587.59	lt. gn.		d.		i.	d.	s. a., NH_4 salts
carbonyl	$Ni(CO)_4$	170.73	lq.	$1.31^{17°}$	−25	43^{751mm}	$0.018^{9.8°}$	i.	s. aq. reg., HNO_3, al., et.
chloride	$NiCl_2$	129.60	yel., delq.	3.544	subl.	973	$53.8^{0°}$	$87.6^{100°}$	s. NH_4OH, al.; i. NH_3
chloride	$NiCl_2 \cdot 6H_2O$*	237.69	gn., delq., mn., 1.57±				180	v. s.	v. s. al.
chloride, ammonia	$NiCl_2 \cdot 6NH_3$	231.78					s.	d.	s. NH_4OH; i. al.
cyanide	$Ni(CN)_2 \cdot 4H_2O$	182.79	gn. pl.		$-4H_2O$, 200	d.	i.	i.	s. KCN; i. dil. KCl
dimethylglyoxime	$NiC_8H_{14}O_4N_4$	288.91	scarlet red cr.		subl. 250		i.	i.	s. abs. al., a.; i. ac., NH_4OH
formate	$Ni(HCO_2)_2 \cdot 2H_2O$	184.76	gn. cr.	2.154	d.		s.		
hydroxide (ic)	$Ni(OH)_3$	109.72	bk.		d.		i.	i.	s. a., NH_4OH, NH_4Cl
hydroxide (ous)	$Ni(OH)_2 \cdot \frac{1}{4}H_2O$	97.21	lt. gn.	4.36	d.		v. sl. s.	v. sl. s.	s. a., NH_4OH; i. alk.
nitrate	$Ni(NO_3)_2 \cdot 6H_2O$	290.79	gn., mn.	2.05	56.7	136.7	$243.0^{0°}$	$\infty^{56.7°}$	s. NH_4OH; i. abs. al.
nitrate, ammonia	$Ni(NO_3)_2 \cdot 4NH_3 \cdot 2H_2O$	286.86					v. s.		i. al.
oxide, mono- (bunsenite)	NiO	74.69	gn.-bk., cb., 2.37	7.45	Forms Ni_2O_3 at 400		i.	i.	s. a., NH_4OH
potassium cyanide	$Ni(CN)_2 \cdot 2KCN \cdot H_2O$	258.97	red yel., mn.	$1.875^{11°}$	$-H_2O$, 100		s.		d. a.
sulfate	$NiSO_4$	154.76	yel., cb.	3.68	$-SO_3$, 840		$27.2^{0°}$	$76.7^{100°}$	i. al., et., act.

*Usual commercial form.
†See also Tables 2-28 and 2-280.

(*Continued*)

TABLE 2-1 Physical Properties of the Elements and Inorganic Compounds *(Continued)*

Name	Formula	Formula weight	Color, crystalline form, and refractive index	Specific gravity	Melting point, °C	Boiling point, °C	Solubility in 100 parts: Cold water	Hot water	Other reagents
Nickel (*Cont.*)									
sulfate	$NiSO_4{\cdot}6H_2O$*	262.85	gn. mn. or blue, tet., 1.5109	2.07	tr. 53.3	$-6H_2O$, 280	$131^{50°}$	$280^{100°}$	v. s. NH_4OH, al.
sulfate (morenosite)	$NiSO_4{\cdot}7H_2O$	280.86	gn., rhb., 1.4893	1.948	98–100	$-6H_2O$, 103	$63.5^{0°}$	$117.8^{30°}$	s. al.
Nitric acid	HNO_3	63.01	col. lq.	1.502	−42	86	∞	∞	expl. with al.
Nitric acid	$HNO_3{\cdot}H_2O$	81.03	col. lq.		−38		∞	∞	d. al.
Nitric acid	$HNO_3{\cdot}3H_2O$	117.06	col. lq.		−18.5		$263^{-20°}$	∞	d. al.
Nitro acid sulfite	NO_2HSO_3	127.08	col., rhb.		73 d.		d.		s. H_2SO_4
Nitrogen	N_2	28.01	col. gas or cb. cr.	$1.026^{-252.5°}$ $0.808^{-195.8°}$ $12.5^{0°}$ (D)	−209.86	−195.8	$2.35^{0°}$ cc	$1.55^{20°}$ cc	sl. s. al.
Nitrogen oxide, mono- (ous)	N_2O	44.01	col. gas	lq. $1.226^{-89°}$ 1.530 (A)	−102.3	−90.7	$130.52^{0°}$ cc	$60.82^{24°}$ cc	s. H_2SO_4, al.
oxide, di- (ic)	NO or $(NO)_2$	30.01 60.01	col. gas	lq. $1.269^{-150.2°}$ 1.0367 (A)	−161	−151	$7.34^{0°}$ cc	$0.0^{100°}$	26.6 cc al.; 3.5 cc H_2SO_4; s. aq. $FeSO_4$
oxide, tri-	N_2O_3	76.01	red-brn. gas or blue lq. or solid	$1.447^{2°}$	−102	3.5	s.		s. a., et.
oxide, tetra- (per- or di-)	NO_2 or $(NO_2)_2$	46.01 92.01	yel. lq., col. solid, red-brn. gas	$1.448^{20°}$	−9.3	21.3	d.		s. HNO_3, H_2SO_4, chl., CS_2
oxide, penta-	N_2O_5	108.01	wh., rhb.	$1.63^{18°}$	30	47	s.	Forms HNO_3	
oxybromide	NOBr	109.91	brn. lq.	>1.0	−55.5	−2	d.		
oxychloride	NOCl	65.46	red-yel. lq. or gas	$1.417^{-12°}$ 2.31 (A)	−64.5	−5.5	d.		s. fuming H_2SO_4
Nitroxyl chloride	NO_2Cl	81.46	yel.-brn. gas	lq. $1.32^{14°}$	<−30	5	d		
Osmium	Os	190.23	blue, hex.	$22.48^{20°}$	2700	>5300	i.	i.	sl. s. aq. reg., HNO_3; i. NH_3
chloride, di-	$OsCl_2$	261.14	gn., delq.				s. d.		s. NaCl, al., et.
chloride, tri-	$OsCl_3$	296.59	brn., cb.		d. 560–600		sl. s.		s. a., alk., al.; sl. s. et.
chloride, tetra-	$OsCl_4$	332.04	red-yel. nd.				s. d.		s. HCl, al.
Oxygen	O_2	32.00	col. gas or hex. solid	$1.14^{-188°}$ $1.426^{-252.5°}$ 1.1053 (A)	−218.4	−183	$4.89^{0°}$ cc	$2.6^{30°}$ cc $1.7^{100°}$ cc	sl. s. al., s. fused Ag
Ozone	O_3	48.00	col. gas	$1.71^{-183°}$ $3.03^{-80°}$ 1.658 (A)	−251	−112	$0.494^{0°}$ cc	$0^{60°}$ cc	s. oil turp., oil cinn.
Palladium	Pd	106.42	silv. met., cb.	$12.0^{20°}$ $11^{1550°}$	1555	2200	i.	i.	s. aq. reg., h. H_2SO_4; i. NH_3
bromide (ous)	$PdBr_2$	266.23	brn.				i.	i.	s. HBr
chloride	$PdCl_2$	177.33	brn., cb.		500 d.		s.	s.	s. HCl, act., al.
chloride	$PdCl_2{\cdot}2H_2O$	213.36	brn. pr.				s.	s.	s. HCl, act., al.
cyanide	$Pd(CN)_2$	158.45	yel.		d.		i.	i.	s. HCN, KCN, NH_4OH; i. dil. a.
hydride	Pd_2H	213.85	met.	11.06	d.				
Palladous dichlorodiammine	$Pd(NH_3)_2Cl_2$	211.39	red or yel., tet.	2.5			s.		s. a., NH_4OH
Perchloric acid	$HClO_4$	100.46	unstable, col. lq	$1.768\frac{22°}{4}$	−112	16^{18mm}	s.		
Perchloric acid	$HClO_4{\cdot}H_2O$	118.47	fairly stable nd.	1.88	50	d.	s.		
Perchloric acid	$HClO_4{\cdot}2H_2O$* 73.6% anh.	136.49	stable lq., col.	$1.71\frac{25°}{4}$	−17.8	200	v. s.		s. al.
Periodic acid	HIO_4	191.91	wh. cr.		d. 138	subl. 110	s.		
Periodic acid	$HIO_4{\cdot}2H_2O$	227.94	delq., mn.		d. 110		v. s.	v. s.	sl. s. al., et.
Permanganic acid	$HMnO_4$	119.94	exists only in solution				v. s.	d.	d. al.
Permolybdic acid	$HMoO_4{\cdot}2H_2O$	196.98	wh. cr.				v. s.	v. s.	
Persulfuric acid	$H_2S_2O_8$	194.14	hyg. cr.		<60		v. s.	d.	
Phosphamic acid	$PONH_2{\cdot}(OH)_2$	97.01	cb.		d.		v. s.	v. s. d.	i. al.
Phosphatomolybdic acid	$H_7P(Mo_2O_7)_6{\cdot}28H_2O$	2365.71	yel. cb.		78	$-25H_2O$, 140	s.		s. HNO_8
Phosphine	PH_3	34.00	col. gas	lq. $0.746^{-90°}$ 1.146 (A)	−132.5	−85	$26^{17°}$ cc	$i.^{100°}$	s. Cu_2Cl_2, al., et.
Phosphonium chloride	PH_4Cl	70.46	wh., cb.		$28^{46atm.}$	subl.	d.		

Phosphoric acid, hypo-	$H_4P_2O_6$	161.98	cr.		55	d. 70	s.	$450^{62°}$	
Phosphoric acid, meta-	HPO_3	79.98	vitreous, delq.	2.2–2.5	subl.		s.	Forms H_3PO_4	i. lq. CO_2
Phosphoric acid, ortho-	H_3PO_4†	98.00	col., rhb.	$1.834^{18.2°}$	42.35	$-½H_2O$, 213	$2340^{26°}$	v. s.	s. al.
Phosphoric acid, pyro-	$H_4P_2O_7$	177.98	wh. nd.		61		$800^{28°}$	Forms H_3PO_4	v. s. al., et.
Phosphorous acid, hypo-	H_3PO_2	66.00	syrupy	$1.493^{18.8°}$	26.5	d.	∞	∞	
Phosphorous acid, ortho-	H_3PO_3	82.00	col.	$1.651^{21.2°}$	74	d. 200	$307.3^{0°}$	$730^{40°}$	
Phosphorous acid, pyro-	$H_4P_2O_5$	145.98	nd.		38	d. 130	d.		
Phosphorus, black	P_4	123.90	rhombohedral	2.69		ign. in air, 400	i.	i.	i. CS_2
Phosphorus, red	P_4	123.90	red, cb.	$2.20^{20°}$	$590^{43atm.}$	ign. in air, 725	i.	i.	s. alk.; i. CS_2, NH_3, et.
Phosphorus, yellow	P_4	123.90	yel., hex., 2.1168	$1.82^{20°}$; lq. $1.745^{44.5°}$	44.1; ign. 34	280	0.0003	sl. s.	0.4 al.; $1000^{10°}$ CS_2; $1.5^{0°}$, $10^{81°}$ bs.; s. NH_3
chloride, tri-	PCl_3	137.33	col., fuming lq.	$1.574\frac{20.8°}{4}$	−111.8	75.95^{760mm}	d.		s. et., chl., CS_2
chloride, penta-	PCl_5	208.24	delq., tet.	solid 1.6; $3.60^{295°}$ (A)	148 under pressure	subl. 160	d.		s. CS_2, C_6H_5COCl
oxide, penta-	P_2O_5	141.94	wh., delq., amor.	2.387	subl. 250		Forms H_3PO_4	v. s.	s. H_2SO_4; i. NH_3, act.
oxychloride	$POCl_3$	153.33	col., fuming lq.	1.675	2	107.2^{760mm}	d.		d. al.
Phosphotungstic acid	$H_3PO_4{\cdot}12WO_3{\cdot}xH_2O$	2880.05	yel.-gn. cr.				s.		s. al., et.
Platinum	Pt	195.08	silv. met., cb.	$21.45^{20°}$ lq. $19^{1755°}$	1755	4300	i.	i.	s. aq. reg., fused alk.
chloride (ic)	$PtCl_4$	336.89	brn.		d. 370		$140^{25°}$	v. s.	s. al., act.; sl. s. NH_2; i. et.
chloride (ous)	$PtCl_2$	265.98	brn.	$5.87^{11°}$	d. 581		i.	i.	s. HCl, NH_4OH; sl. s. NH_3; i. al., et.
chloride (ic)	$PtCl_4{\cdot}8H_2O$	481.01	red, mn.	2.43	$-4H_2O$, 100		v. s.	v. s.	s. al., et.
cyanide (ous)	$Pt(CN)_2$	247.11	yel.-brn.				i.	i.	i. alk.
Plutonium	Pu	238.05		Produced by deuteron bombardment on U^{238}					
Plutonium	Pu	239.05		Produced by neutron bombardment on U^{238}					
Potassium	K	39.10	silv. met., cb.	$0.86^{20°}$ lq. $0.83^{42°}$	62.3	760	d.	Forms KOH	s. a., al., Hg
acetate	$KC_2H_3O_2$	98.14	wh. pd.	1.8	292		$217^{0°}$	$396^{90°}$	33 al.; i. et.
acetate, acid	$KH(C_2H_3O_2)_2$	158.19	delq. nd. or pl.		148	d. 200	d.		s. ac.
aluminate	$K_2(AlO_2)_2{\cdot}3H_2O$	250.20	cr.				s.	d.	s. alk.; i al.
amide	KNH_2	55.12	yel.-grn.		338	subl. 400	d.		d. al.; $3.6^{25°}$ NH_3
arsenate (monobasic)	KH_2AsO_4	180.03	col., tet., 1.5674	2.867	288		$18.87^{6°}$	v. s.	i. al.
auricyanide	$KAu(CN)_4{\cdot}1{\cdot}5H_2O$	367.16	pl.		d. 200		s.	v. s.	s. al.
aurocyanide	$KAu(CN)_2$	288.10	rhb.				14.3	$200^{100°}$	sl. s. al.; i. et.
bicarbonate	$KHCO_3$	100.12	mn., 1.482	2.17	d. 100–200		$22.4^{0°}$	$60^{60°}$	i. satd. K_2CO_3, al.
bisulfate	$KHSO_4$	136.17	rhb., or mn., 1.480	2.35	210	d.	$36.3^{0°}$	$121.6^{100°}$	d. al.
bromate	$KBrO_3$	167.00	trig.	$3.27^{17.5°}$	370 d.		$3.11^{0°}$	$49.75^{100°}$	sl. s. al.; i. act.
bromide	KBr	119.00	col., cb., 1.5594	$2.75^{25°}$	730	1380	$53.5^{0°}$	$104^{100°}$	sl. s. al., et.
carbonate	K_2CO_3	138.21	wh., delq. pd., 1.531	2.29	891	d.	$105.5^{0°}$	$156^{100°}$	i. al.
carbonate	$K_2CO_3{\cdot}2H_2O$	174.24	rhb.	2.043			$183^{0°}$	$331^{100°}$	
carbonate	$2K_2CO_3{\cdot}3H_3O$	330.46	mn.	2.13			$129.4^{0°}$	$268^{100°}$	
chlorate	$KClO_3$	122.55	col., mn., 1.5167	2.32	368	d. 400	$3.3^{0°}$	$57^{100°}$	0.83 al.; s. alk.
chloride (sylvite)	KCl	74.55	col., cb., 1.4904	1.988	790	1500	$27.6^{0°}$	$56.7^{100°}$	s. al., alk.
chloroplatinate	K_2PtCl_6	485.99	yel., cb., 1.825±	3.499	d. 250		$0.74^{0°}$	$5.2^{100°}$	i. al., et.
chromate (tarapacaite)	K_2CrO_4	194.19	yel., rhb., 1.7261	$2.732^{18°}$	975		$58.0^{0°}$	$75.6^{100°}$	i. al.
cyanate	KCNO	81.12	wh., tet.	2.048			s.	d.	v. sl. s. al.
cyanide	KCN	65.12	wh., cb., delq., 1.410	$1.52^{16°}$	634.5		s.	$122.2^{108.8°}$	s. gly.; $0.9^{19.5°}$ al.; 1.3 h. al.
dichromate	$K_2Cr_2O_7$	294.18	red, tri.	2.69	398	d.	$4.9^{0°}$	$80^{100°}$	i. al.
ferricyanide	$K_3Fe(CN)_6$	329.24	red, mn. pr., 1.5689	1.84	d.		$33^{4.4°}$	$77.5^{100°}$	s. act.; sl. s. al.; i. NH_3
ferrocyanide	$K_4Fe(CN)_6{\cdot}3H_2O$	422.39	yel., mn., 1.5772	$1.853^{17°}$	$-3HO_2$, 70		$27.8^{12.2°}$	$90.6^{96.8°}$	s. act.; i. NH_3, al., et.
formate	$KHCO_2$	84.12	col., rhb.	1.91	167.5	d.	$331^{18°}$	$657^{90°}$	sl. s. al.; i. et.
hydride	KH	40.11	cb., 1.453	0.80	d.		d.		i. et., bz., CS_2
hydrosulfide	KHS	72.17	wh., delq., rhb.	2.0	455		s.	s. d.	s. al.
hydroxide	KOH	56.11	wh., delq., rhb.	2.044	380	1320	$97^{0°}$	$178^{100°}$	v. s. al., et.; i. NH_3
iodate	KIO_3	214.00	col., mn.	3.89	560		$4.73^{0°}$	$32.2^{100°}$	s. KI; i. al., NH_3
iodide	KI	166.00	wh., cb., 1.6670	3.13	723	1330	$127.5^{0°}$	$208^{100°}$	$4^{20°}$ al.; s. NH_3; sl. s. et.

*One commercial form 70 to 72 per cent.
†Common commercial form 85 per cent H_3PO_4 in aqueous solution.

(Continued)

TABLE 2-1 Physical Properties of the Elements and Inorganic Compounds *(Continued)*

Name	Formula	Formula weight	Color, crystalline form, and refractive index	Specific gravity	Melting point, °C	Boiling point, °C	Solubility in 100 parts: Cold water	Hot water	Other reagents
Potassium (*Cont.*)									
iodide, tri-	KI_3	419.81	dark blue, delq., mn.	3.498	45	d. 225	v. s.		s. KI, al.
iodoplatinate	K_2PtI_6	1034.70	cb.	5.18			s.		
manganate	K_2MnO_4	197.13	gn., rhb.		d. 190		d.		s. KOH
metabisulfite	$K_2S_2O_5$	222.32	mn., pl.		d. 150		$25^{0°}$	$120^{94°}$	sl. s. al.; i. et.
nitrate (saltpeter)	KNO_3	101.10	col., rhb., 1.5038	$2.11^{10.6°}$	tr. 129; 333	d. 400	$13.3^{0°}$	$246^{100°}$	$0.1^{30°}$ al.; i. et.
nitrite	KNO_2	85.10	pr.	1.915	297	d. 350	$281^{0°}$	$413^{100°}$	v. s. NH_3; sl. s. al.
oxalate	$K_2C_2O_4 \cdot H_2O$	184.23	wh., mn.	2.13	d.		$28.7^{0°}$	$83.2^{100°}$	
oxalate, acid	KHC_2O_4*	128.13	mn., 1.545	2.0	d.		$14.3^{50°}$	$48.1^{100°}$	
oxalate, acid	$KHC_2O_4 \cdot \frac{1}{2}H_2O$	137.13	trimetric		d.		$2.2^{0°}$	$51.5^{100°}$	
oxide	K_2O	94.20	wh., cb.	$2.32_{4}^{20°}$			Forms KOH	v. s.	s. al., et.
perchlorate	$KClO_4$	138.55	col., rhb., 1.4737	$2.524_{4}^{11°}$	d. 400		$0.75^{0°}$	$21.8^{100°}$	$0.105^{20°}$ m. al.; i. et.
permanganate	$KMnO_4$	158.03	purple, rhb.	2.703	d. <240		$2.83^{0°}$	$32.35^{75°}$	s. H_2SO_4; d. al.
persulfate	$K_2S_2O_3$	190.32	wh., tri., 1.4669		d. <100		$1.77^{0°}$	$10^{40°}$	i. al.
phosphate, monobasic	KH_2PO_4	136.09	col., delq., tet., 1.5095	2.338	256		$14.8^{0°}$	$83.5^{90°}$	i. al.
phosphate, dibasic	K_2HPO_4	174.18	wh., delq.		d.		$33^{25°}$	v. s.	sl. s. al.
phosphate, tribasic	K_3PO_4	212.27	wh., rhb.	$2.564^{17°}$	1340		$193.1^{25°}$	v. s.	i. al.
phosphate, meta-	KPO_3	118.07	wh. pd.	$2.258^{14.5°}$	tr. 450; 798	1320	s.	s.	
phosphate, meta-	$K_4P_4O_{12} \cdot 2H_2O$	508.31	amor.	$2.264^{14.5°}$	$-2H_2O$, 100	d.	s.	83	s. a.
phosphate, pyro-	$K_4P_2O_7 \cdot 3H_2O$	384.38	delq.	2.33	$-2H_2O$, 180	$-3H_2O$, 300	s.	v. s.	i. al.
phthalate, acid	$KHC_8H_4O_4$	204.22	wh. cr.	1.63	d.		$10.2^{25°}$	36	
platinocyanide	$K_2Pt(CN)_4 \cdot 3H_2O$	431.39	yel., rhb., 1.62±	$2.45^{16°}$			sl. s.	v. s.	s. al., et.
silicate	K_2SiO_3	154.28	hyg. 1.521±		976		s.	s.	i. al.
silicate, tetra-	$K_2Si_4O_9 \cdot H_2O$	352.55	rhb., 1.530	2.417	d. 400		s.	s.	i. al.
sulfate (arcanite)	K_2SO_4	174.26	col., rhb., 1.4947	2.662	tr. 588		$7.35^{0°}$	$24.1^{100°}$	i. al., act., CS_2
sulfate, pyro-	$K_2S_2O_7$	254.32	col.	2.277	300		s.	d.	
sulfide, mono-	$K_2S \cdot 5H_2O$	200.34	rhb., delq.		60	$-3H_2O$, 150	s.		s. al., gly.; i. et.
sulfite	$K_2SO_3 \cdot 2H_2O$	194.29	wh., rhb.		d.		100	>100	sl. s. al.; i. NH_3
sulfite, acid	$KHSO_3$	120.17	wh., mn.		d. 190		$45.5^{15°}$	$91.5^{75°}$	i. abs. al.
tartrate	$K_2C_4H_4O_6 \cdot \frac{1}{2}H_2O$	235.28	col., mn., 1.526	1.98		d.	$12.5^{17.5°}$	$278^{100°}$	sl. s. al.
tartrate, acid	$KHC_4H_4O_6$*	188.18	col., rhb.	1.956			$0.37^{0°}$	$6.1^{100°}$	s. a., alk.; i. al., ac.
thiocyanate	KCNS	97.18	col., delq., mn., 1.660±	1.886	172.3	d. 500	$177^{0°}$	$217^{20°}$	$20.8^{22°}$ act.; s. al.
thiosulfate	$K_2S_2O_3$	190.32	col., cb.		d. 400		$96.1^{0°}$	$311.2^{90°}$	
thiosulfate	$3K_2S_2O_3 \cdot H_2O$	588.99	delq., mn.	2.23	$-H_2O$, 180	d.			i. al.
Praseodymium	Pr	140.91	yel.	$6.5^{20°}$	940		d.		
Radium	Ra	226.03	wh., met.	5?	960	1140	d. +H_2		d. a.
bromide	$RaBr_2$	385.83	wh., mn.	5.79	728	subl. 900	$70^{20°}$	s.	s. al.
Radon (Niton)	Rn	222.02	gas	lq. 5.5; 111 (D)	−71	−62	$51^{0°}$ cc	$8.5^{60°}$ cc	
Rhenium	Re	186.21	hex.		3440				i. HF, HCl; s. H_2SO_4; HNO_3
Rhodium	Rh	102.91	gray-wh., cb.	12.5	1955	>2500	i.	i.	sl. s. aq. reg., a.
chloride	$RhCl_3$	209.26	red		d. 450	subl. 800±	i.	i.	v. sl. s. alk.; i. aq. reg., a.
chloride	$RhCl_3 \cdot 4H_2O$	281.33	dark red				v. s.		s. HCl, al.; i. et.
Rubidium	Rb	85.47	silv. wh.	lq. $1.475^{88.5}$; $1.53^{20°}$	38.5	700	d.		s. a., al.
Ruthenium	Ru	101.07	bk., porous	8.6	>1950		i.	i.	sl. s. aq. reg., a.
Ruthenium	Ru	101.07	gray, hex.	$12.2^{20°}$	2450	>2700	i.	i.	
Samarium	Sm (also Sa)	150.36		7.7	>1300				
Scandium	Sc	44.96		2.5?	1200	2400			
Selenic acid	H_2SeO_4	144.97	hex. pr.	$2.950_{4}^{15°}$	58	260	$1300^{30°}$	$\infty^{60°}$	s. H_2SO_4; d. al.; i. NH_3
Selenic acid	$H_2SeO_4 \cdot H_2O$	162.99	nd.	$2.627_{4}^{15°}$	26	205	v. s.		
Selenium	Se_8	631.68	red pd., amor., 2.92	$4.26^{25°}$	50	688	i.	i.	s. CS_2, H_2SO_4, CH_2I_2
Selenium	Se_8	631.68	gray, trig., 3.00; red, hex.	4.80; 4.50	220	688	i.	i.	s. CS_2, H_2SO_4

Selenium	Se_8	631.68	steel gray	$4.8^{25°}$	217	688	i.	i.	i. CS_2; s. H_2SO_4
Selenous acid	H_2SeO_3	128.97	hex.	$3.004\frac{15°}{4}$	d.		$90^{0°}$	$400^{90°}$	v. s. al.; i. NH_3
Silicic acid, meta-	H_2SiO_3	78.10	amor., 1.41	2.1–2.3			i.	i.	s. alk.; i. NH_4Cl
Silicic acid, ortho-	H_4SiO_4	96.11	amor.	$1.576^{17°}$			sl. s.	sl. s.	s. alk.; i. NH_4Cl
Silicon, crystalline	Si	28.09	gray, cb., 3.736	$2.4^{20°}$	1420	2600	i.	i.	s. HNO_3 + HF, Ag; sl. s. Pb, Zn; i. HF
Silicon, graphitic	Si	28.09	cr.	2.0–2.5		2600	i.	i.	s. HNO_3 + HF, fused alk.; i. HF.
Silicon, amorphous	Si	28.09	brn., amor.	2		2600	i.	i.	s. HF, KOH
carbide	SiC	40.10	blue-bk., trig., 2.654	3.17	>2700	subl. 2200	i.	i.	s. fused alk.; i. a.
chloride, tri-	Si_2Cl_6	268.89	lf. or lq.	$1.58^{0°}$	−1	144^{760mm}	d.		d. alk.
chloride, tetra-	$SiCl_4$	169.90	col., fuming lq., 1.412	1.50	−70	57.6	d.		d. conc. H_2SO_4, al.
fluoride	SiF_4	104.08	gas	3.57 (A)	−95.7	-65^{1810mm}	v. s. d.		s. HNO_3, al., et.
hydride (silane)	SiH_4	32.12	col. gas	lq. $0.68^{-185°}$	−185	-112^{760mm}	i.		i. al., et.; d. KOH
oxide, di- (opal)	$SiO_2 \cdot xH_2O$		iridescent, amor.	2.2	1600–1750	subl. 1750	i.	i.	s. HF, h. alk., fused $CaCl_2$
oxide, di- (cristobalite)	SiO_2	60.08	col., cb. or tet., 1.487	2.32	1710	2230	i.	i.	s. HF; i. alk.
oxide, di- (lechatelierite)	SiO_2	60.08		2.20		2230	i.	i.	s. HF; i. alk.
oxide, di- (quartz)	SiO_2	60.08	hex., 1.5442	$2.650^{20°}$	tr. <1425	2230	i.	i.	s. HF; i. alk.
oxide, di- (tridymite)	SiO_2	60.08	trig., rhb., 1.469	2.26	tr. 1670	2230	i.		s. HF; i. alk.
Silver	Ag	107.87	silv. met., cb.	$10.5^{20°}$	960.5	1950	i.	i.	s. HNO_3, h. H_2SO_4; i. alk.
bromide (bromyrite)	AgBr	187.77	pa. yel., cb., 2.252	$6.473\frac{25°}{4}$	434	d. 700	$0.00002^{20°}$	$0.00037^{100°}$	$0.51^{18°}$ NH_4OH; s. KCN, $Na_2S_2O_3$
carbonate	Ag_2CO_3	275.75	yel. pd.	6.077	218 d.		$0.003^{20°}$	$0.05^{100°}$	s. NH_4OH, $Na_2S_2O_3$; i. al.
chloride (cerargyrite)	AgCl	143.32	wh., cb., 2.071	5.56	455	1550	$0.000089^{10°}$	$0.00217^{100°}$	s. NH_4OH, KCN; sl. s. HCl
cyanide	AgCN	133.89	wh., 1.685±	3.95	$-(CN)_2$, 320		$0.000022^{20°}$		s. NH_4OH, KCN, HNO_3
nitrate (lunar caustic)	$AgNO_3$	169.87	col., rhb., 1.744	$4.352\frac{19°}{4}$	212	444 d.	$122^{0°}$	$952^{100°}$	s. gly.; v. sl. s. al.
Sodium	Na	22.99	silv. met, cb.	$0.97^{20°}$	97.5	880	d., forms NaOH		i. bz.; d. al.
acetate	$NaC_2H_3O_2$	82.03	wh., mn., 1.464	1.528	324		$46.5^{20°}$	$170^{100°}$	$2.1^{18°}$ al.
acetate	$NaC_2H_3O_2 \cdot 3H_2O$	136.08	wh., mn.	1.45	58	$-3H_2O$, 120	v. s.	v. s.	$7.8^{25°}$ abs. al.
aluminate	$NaAlO_2$	81.97	amor.		1650		s.	v. s.	i. al.
amide	$NaNH_2$	39.01	olive gn.		210	400	d.		d. al.
ammonium phosphate	$NaNH_4HPO_4 \cdot 4H_2O$	209.07	col., mn.	1.574	79 d.		16.7	100	i. al.
antimonate, meta-	$2NaSbO_3 \cdot 7H_2O$	511.60	cb.				$0.031^{12.8°}$		sl. s. al., NH_4 salts; i. ac.
arsenate	$Na_3AsO_4 \cdot 12H_2O$	424.07	hex., 1.4589	1.759	86.3		$26.7^{17°}$		1.67 al., $50^{15°}$ gly.
arsenate, acid (monobasic)	$NaH_2AsO_4 \cdot H_2O$	181.94	rhb., 1.5535	2.535	d. 100		s.		
arsenate, acid (dibasic)	$Na_2HAsO_4 \cdot 7H_2O$*	312.01	col., mn., 1.4658	1.871	125	$-7H_2O$, 100	$61^{15°}$	v. s.	sl. s. al.
arsenate, acid (dibasic)	$Na_2HAsO_4 \cdot 12H_2O$	402.09	mn., 1.4496	1.72	28	$-12H_2O$, 100	$5.59^{0.1°}$	$140.7^{30°}$	sl. s. al.
arsenite, acid	Na_2HAsO_3	169.91	col.	1.87			v. s.		
benzoate	$NaC_7H_5O_2$	144.10	col. cr.				$62.5^{25°}$	$76.9^{100°}$	$2.3^{25°}$, $8.3^{78°}$ al.
bicarbonate	$NaHCO_3$	84.01	wh., mn., 1.500	2.20	$-CO_2$, 270		$6.9^{0°}$	$16.4^{60°}$	i. al.
bifluoride	$NaHF_2$	61.99	col. cr.		d.		$3.7^{20°}$	s.	
bisulfate	$NaHSO_4$	120.06	col., tri.	2.742	>315	d., $-H_2O$	$50^{0°}$	$100^{100°}$	d. al.; i. NH_3
bisulfite	$NaHSO_3$	104.06	col., mn., 1.526	1.48	d.		sl. s.	s.	i. al., act.
borate, tetra-	$Na_2B_4O_7$	201.22		2.367	741		$1.3^{0°}$	$8.79^{40°}$	i. al.
borate, tetra	$Na_2B_4O_7 \cdot 5H_2O$	291.30	col., rhb., 1.461	1.815			$22^{62°}$ (anh.)	$52.3^{100°}$ (anh.)	
borate, tetra- (borax)	$Na_2B_4O_7 \cdot 10H_2O$*	381.37	wh., mn., 1.4694	1.73	75	$-10H_2O$, 200	$1.3^{0.5}$ (anh.)	$20.3^{80°}$ (anh.)	s. gly.; i. abs. al.
bromate	$NaBrO_3$	150.89	col., cb.	$3.339^{17.5°}$	381		$27.5^{0°}$	$90.9^{100°}$	i. al.
bromide	NaBr	102.89	col., cb., 1.6412	$3.205^{17.5°}$	755	1390	$90^{20°}$	$121^{100°}$	sl. s. al.
bromide	$NaBr \cdot 2H_2O$	138.92	col., mn.	2.176	50.7		$79.5^{0°}$ (anh.)	$118.3^{80°}$ (anh.)	sl. s. al.
carbonate (soda ash)	Na_2CO_3	105.99	wh. pd., 1.535	2.533	851	d.	$7.1^{0°}$	$48.5^{104°}$	i. al., et.
carbonate	$Na_2CO_3 \cdot H_2O$	124.00	wh., rhb., 1.506–1.509	1.55	$-H_2O$, 100		s.	s.	s. gly.; i. al., et.
carbonate	$Na_2CO_3 \cdot 7H_2O$	232.10	rhb. or trig.	1.51	d. 35.1		s.	s.	
carbonate (sal soda)	$Na_2CO_3 \cdot 10H_2O$	286.14	wh., mn., 1.425	1.46			$21.5^{0°}$	$238^{30°}$	i. al.

*Usual commercial form.

(*Continued*)

TABLE 2-1 Physical Properties of the Elements and Inorganic Compounds *(Continued)*

Name	Formula	Formula weight	Color, crystalline form, and refractive index	Specific gravity	Melting point, °C	Boiling point, °C	Solubility in 100 parts: Cold water	Hot water	Other reagents
Sodium (*Cont.*)									
carbonate, sesqui- (trona)	$Na_3H(CO_3)_2{\cdot}2H_2O$	226.03	wh., mn., 1.5073	2.112	d.		$13^{0°}$	$42^{100°}$	
chlorate	$NaClO_3$	106.44	wh., cb., or trig., 1.5151	$2.490^{15°}$	248	d.	$79^{0°}$	$230^{100°}$	s. al.
chloride	$NaCl$	58.44	col., cb., 1.5443	2.163	800.4	1413	$35.7^{0°}$	$39.8^{100°}$	sl. s. al.; i. conc. HCl
chromate	Na_2CrO_4	161.97	yel., rhb.	2.723	392		$32^{0°}$	$126^{100°}$	
chromate	$Na_2CrO_4{\cdot}10H_2O$	342.13	yel., delq., mn.	1.483	19.9		v. s.	∞	sl. s. al.
citrate	$2Na_3C_6H_5O_7{\cdot}11H_2O$	714.31	wh., rhb.	$1.857^{\frac{23.5°}{4}}$	$-11H_2O$, 150	d.	$91^{25°}$	$250^{100°}$	i. al.
cyanide	$NaCN$	49.01	wh., cb., 1.452		563.7	1496	$48^{10°}$	$82^{35°}$	s. NH_3; sl. s. al.
dichromate	$Na_2Cr_2O_7{\cdot}2H_2O$	298.00	red, mn., 1.6994	$2.52^{18°}$	$-2H_2O$, 84.6; 356 (anh.)	d. 400	$238^{0°}$	$508^{80°}$	
ferricyanide	$Na_3Fe(CN)_6{\cdot}H_2O$	298.93	red, delq.				$18.9^{0°}$	$67^{100°}$	i. al.
ferrocyanide	$Na_4Fe(CN)_6{\cdot}10H_2O$	484.06	yel., mn.	1.458			$17.9^{20°}$ (anh.)	$63^{98.5°}$ (anh.)	i. al.
fluoride (villiaumite)	NaF	41.99	tet., 1.3258	2.79	992		$4^{0°}$	$5^{100°}$	v. sl. s. al.
formate	$NaHCO_2$	68.01	wh., mn.	1.919	253		$44^{0°}$	$160^{100°}$	sl. s. al.; i. et.
hydride	NaH	24.00	silv. nd., 1.470	0.92	d. 800		d.		i. bz., CS_2, CCl_4, NH_3; s. molten metal
hydrosulfide	$NaSH{\cdot}2H_2O$	92.09	col., delq., nd.		d.		s.	s.	s. al.; d. a.
hydrosulfide	$NaSH{\cdot}3H_2O$	110.11	rhb.		22	d.	s.	s.	s. al.; d. a.
hydrosulfite	$Na_2S_2O_4{\cdot}2H_2O$	210.14	col. cr.		d.		$22^{20°}$	d.	i. al.
hydroxide	$NaOH$	40.00	wh., delq.	2.130	318.4	1390	$42^{0°}$	$347^{100°}$	v. s. al., et., gly.; i. act.
hydroxide	$NaOH{\cdot}3\frac{1}{2}H_2O$	103.05	mn.		15.5		s.	v. s.	
hypochlorite	$NaOCl$	74.44	pa. yel., in soln. only		d.		$26^{0°}$	$158^{56°}$	
iodide	NaI*	149.89	col., cb., 1.7745	$3.667^{0°}$	651	1300	$158.7^{0°}$	$302^{100°}$	v. s. al., act.
iodide	$NaI{\cdot}2H_2O$	185.92	col., mn.	2.448			v. s.	v. s.	v. s. NH_3
lactate	$NaC_3H_5O_3$	112.06	col., amor.		d.		v. s.	v. s.	s. al.; i. et.
nitrate (soda niter)	$NaNO_3$	84.99	col., trig., 1.5874	2.257	308	d. 380	$73^{0°}$	$180^{100°}$	s. NH_3; sl. s. gly., al.
nitrite	$NaNO_2$	69.00	pa. yel., rhb.	$2.168^{0°}$	271	d. 320	$72.1^{0°}$	$163.2^{100°}$	$0.3^{20°}$ et.; 0.3 abs. al.; $4.4^{20°}$ m. al.; v. s. NH_3
oxide	Na_2O	61.98	wh., delq.	2.27	subl.		Forms NaOH		d. al.
perborate	$NaBO_3{\cdot}H_2O$	99.81	wh. pd.		d. 40		sl. s.	d.	s. gly., alk.
perchlorate	$NaClO_4$	122.44	rhb., 1.4617		482 d.		$170^{0°}$	$320^{100°}$	s. al.; 51 m. al.; 52 act.; i. et.
perchlorate	$NaClO_4{\cdot}H_2O$	140.46	hex.	2.02	d. 130		$209^{15°}$	$284^{50°}$	s. al.
peroxide	Na_2O_2*	77.98	yel.-wh. pd.	2.805	d.		s. d.	d.	s. dil. a.
peroxide	$Na_2O_2{\cdot}8H_2O$	222.10	wh., hex.		d. 30		s. d.	d.	
phosphate, monobasic	$NaH_2PO_4{\cdot}H_2O$*	137.99	col., rhb., 1.4852	2.040	$-H_2O$, 100	d. 200	$71^{0°}$	$390^{83°}$	i. al.
phosphate, monobasic	$NaH_2PO_4{\cdot}2H_2O$	156.01	col., rhb., 1.4629	1.91	60		$91.1^{0°}$	$308^{40°}$	
phosphate, dibasic	$Na_2HPO_4{\cdot}7H_2O$	268.07	col., mn., 1.4424	1.679	d.		$185^{40°}$	$2000^{100°}$	
phosphate, dibasic	$Na_2HPO_4{\cdot}12H_2O$	358.14	col., mn., 1.4361	1.52	34.6	$-12H_2O$, 180	$4.3^{0°}$	$76.7^{30°}$	i. al.
phosphate, tribasic	Na_3PO_4	163.94	wh.	$2.537^{17.5°}$	1340		$4.5^{0°}$	$77^{100°}$	
phosphate, tribasic	$Na_3PO_4{\cdot}12H_2O$*	380.12	wh., trig., 1.4458	1.62	73.4	$-11H_2O$, 100	$28.3^{15°}$	∞	i. CS_2
phosphate, meta-	$Na_4P_4O_{12}$	407.85	col.	2.476	616 d.		s.	s.	s. a., alk.
phosphate, pyro-	$Na_4P_2O_7$*	265.90	wh.	2.45	988		$2.26^{0°}$	$45^{96°}$	d. a.
phosphate, pyro-	$Na_4P_2O_7{\cdot}10H_2O$	446.06	mn., 1.4525	1.82	d.		$5.4^{0°}$	$93^{100°}$	i. al., NH_3
phosphate (pyrodisodium)	$Na_2H_2P_2O_7$	221.94	col., mn., 1.510	1.862	d. 220		$4.5^{0°}$	$21^{40°}$	
phosphate (pyrodisodium)	$Na_2H_2P_2O_7{\cdot}6H_2O$	330.03	col., mn., 1.4645	1.848			$6.9^{0°}$	$36^{40°}$	
potassium tartrate	$NaKC_4H_4O_6{\cdot}4H_2O$	282.22	rhb., 1.493	1.790	70 to 80	$-4H_2O$, 215	$26^{0°}$	$66^{26°}$	sl. s. al.
silicate, meta-	Na_2SiO_3	122.06	col., rhb., 1.520		1088		s.	s. d.	i. Na or K salts, al.
Sodium silicate, meta-	$Na_2SiO_3{\cdot}9H_2O$	284.20	rhb.		47	$-6H_2O$, 100	v. s.	v. s.	$29^{18°}$, aN NaOH
silicate, ortho-	Na_4SiO_4	184.04	col., hex., 1.530		1018		s.	s.	
silicofluoride	Na_2SiF_6	188.06	wh., hex., 1.312	2.679	d.		$0.44^{0°}$	$2.45^{100°}$	i. al.
stannate	$Na_2SnO_3{\cdot}3H_2O$	266.73	hex. tablets		d. 140		$50^{0°}$	$67^{50°}$	i. al., act.
sulfate (thenardite)	Na_2SO_4	142.04	col., rhb., 1.477	2.698	tr. 100 to mn.		$5^{0°}$	$42^{100°}$	i. al.
sulfate	Na_2SO_4	142.04	col., mn.		tr. 500 to hex.		$48.8^{40°}$	$42.5^{100°}$	d. HI; s. H_2SO_4

sulfate	Na_2SO_4	142.04	col., hex.		884		$19.4^{20°}$	$45.3^{60°}$	
sulfate	$Na_2SO_4·7H_2O$	268.15	tet.				$44.9^{0°}$	$202.6^{26°}$	
sulfate (Glauber's salt)	$Na_2SO_4·10H_2O$	322.19	col., mn., 1.396	1.464	32.4	$-10H_2O$, 100	$36^{15°}$	$412^{34°}$	i. al.
sulfide, mono-	Na_2S	78.04	pink or wh., amor.	1.856			$15.4^{10°}$	$57.3^{90°}$	sl. s. al.; i. et.
sulfide, tetra-	Na_2S_4	174.24	yel., cb.		275		s.	s.	s. al.
sulfide, penta-	Na_2S_5	206.30	yel.		251.8		s.	s.	s. al.
sulfite	Na_2SO_3	126.04	hex. pr., 1.565	$2.633_{4}^{15°}$	d.		$13.9^{0°}$	$28.3^{84°}$	i. al., NH
sulfite	$Na_2SO_3·7H_2O$	252.15	mn.	1.561	$-7H_2O$, 150	d.	$34.7^{2°}$	$67.8^{18°}$	i. al.
tartrate	$Na_2C_4H_4O_6·2H_2O$	230.08	rhb.	1.818			$29^{6°}$	$66^{43°}$	i. al.
thiocyanate	NaCNS	81.07	delq., rhb., 1.625±		287		$110^{10°}$	$225^{100°}$	v. s. al.
thiosulfate	$Na_2S_2O_3$	158.11	mn.	1.667			$50^{0°}$	$231^{80°}$	
thiosulfate (hypo)	$Na_2S_2O_3·5H_2O$*	248.18	mn. pr., 1.5079	1.685	d. 48.0		$74.7^{0°}$	$301.8^{60°}$	s. NH_3; v. sl. s. al.
tungstate	Na_2WO_4	293.82	wh., rhb.	4.179	692		$57.58^{0°}$	$97^{100°}$	
tungstate	$Na_2WO_4·2H_2O$*	329.85	wh., rhb.	3.245	$-2H_2O$, 100		$88^{0°}$	$123.5^{100°}$	sl. s. NH_3; i. a., al.
tungstate, para-	$Na_6W_7O_{24}·16H_2O$	2097.05	wh., tri.	$3.987^{14°}$	$-16H_2O$, 300		8	d.	
uranate	Na_2UO_4	348.01	yel.				i.	i.	s. alk. carb., dil. a.
vanadate	$Na_3VO_4·16H_2O$	472.15	col. nd.		866 (anh.)		v. s.	d.	i. al.
vanadate, pyro-	$Na_4V_2O_7$	305.84	hex.		654		s.		i. al.
Stannic chloride	$SnCl_4$	260.52	col., fuming lq.	2.226	−30.2	114.1	s.	d.	s. abs. al., act., NH_3; s. ∞ CS_2
oxide (cassiterite)	SnO_2	150.71	wh., tet., 1.9968	7.0	1127		i.	i.	s. conc. H_2SO_4; i. alk.; NH_4OH, NH_3
sulfate	$Sn(SO_4)_2·2H_2O$	346.87	col., delq., hex.				v. s.	d.	s. dil. H_2SO_4, HCl; d. abs. al.
Stannous bromide	$SnBr_2$	278.52	yel., rhb.	$5.12^{17°}$	215.5	620	s.	d.	s. C_6H_5N
chloride	$SnCl_2$	189.62	wh., rhb.		246.8	623	$83.9^{0°}$	$269.8^{15°}$	s. alk., abs. al., et.
chloride (tin salt)	$SnCl_2·2H_2O$*	225.65	wh., tri.	$2.71^{15.5°}$	37.7	d.	$118.7^{0°}$	∞	s. tart. a., alk., al.
sulfate	$SnSO_4$	214.77	wh. cr.		$-SO_2$, 360		$19^{19°}$	$18^{100°}$	s. H_2SO_4
Strontium	Sr	87.62	silv. met.	2.6	800	1150	d.	Forms $Sr(OH)_2$	s. al., a.
acetate	$Sr(C_2H_3O_2)_2$	205.71	wh. cr.	2.099		d.	$36.9^{0°}$	$36.4^{97°}$	$0.26^{15°}$ m. al.
carbonate (strontianite)	$SrCO_3$	147.63	wh., rhb., 1.664	3.70	$1497^{60atm.}$	$-CO_2$, 1350	$0.0011^{18°}$	$0.065^{100°}$	s. a., NH_4 salts, aq. CO_2
chloride	$SrCl_2$	158.53	wh., cb., 1.6499	3.052	873		$43.5^{0°}$	$100.8^{100°}$	v. sl. s. act., abs. al.; i. NH_3
chloride	$SrCl_2·6H_2O$*	266.62	wh., rhb., 1.5364	$1.933^{17°}$	$-4H_2O$, 61	$-6H_2O$, 100	$104^{0°}$	$198^{40°}$	
hydroxide	$Sr(OH)_2$	121.63	wh., delq.	3.625	375		$0.41^{0°}$	$21.83^{100°}$	s. NH_4Cl
hydroxide	$Sr(OH)_2·8H_2O$*	265.76	col., tet., 1.499	1.90	$-7H_2O$ in dry air		$0.90^{0°}$	$47.7^{100°}$	s. NH_4Cl; i. act.
nitrate	$Sr(NO_3)_2$*	211.63	col., cb., 1.5878	2.986	570		$40^{0°}$	$100^{89°}$	s. NH_3; 0.012 abs. al.
nitrate	$Sr(NO_3)_2·4H_2O$	283.69	wh., mn.	2.2			$62.2^{0°}$	$124^{20°}$	i. HNO_3
oxide (strontia)	SrO	103.62	col., cb., 1.870	4.7	2430		Forms $Sr(OH)_2$		sl. s. al.; i. et.
peroxide	SrO_2	119.62	wh. pd.		d.		$0.008^{20°}$	d.	s. al., NH_4Cl; i. act.
peroxide	$SrO_2·8H_2O$	263.74	wh. cr.		$-8H_2O$, 100	d.	$0.018^{20°}$	d.	s. al.; i. NH_4OH
sulfate (celestite)	$SrSO_4$	183.68	col., rhb., 1.6237	3.96	1580 d.		$0.0113^{0°}$	$0.0114^{32°}$	sl. s. a.; i. dil. H_2SO_4, al.
sulfate, acid	$Sr(HSO_4)_2$	281.76	col., granular		d.		d.		$14^{70°}$ H_2SO_4
Sulfamic acid	NH_2SO_3H	97.09	wh., rhb.	$2.03_{4}^{12°}$	205 d.		$20^{0°}$	$40^{70°}$	sl. s. al., act.; i. et.
Sulfur, amorphous	S	32.07	pa. yel. pd., 2.0–2.9	2.046	120	444.6	i.	i.	sl. s. CS_2
Sulfur, monoclinic	S_8	256.52	pa. yel., mn.	1.96	119.0	444.6	i.	i.	s. CS_2, al.
Sulfur, rhombic	S_8	256.52	pa. yel., rhb.	2.07	112.8	444.6	i.	i.	$24^{0°}$, $181^{55°}$ CS_2
Sulfur bromide, mono-	S_2Br_2	223.94	red, fuming lq.	2.635	−46	$54^{0.18mm}$	d.		
chloride, mono-	S_2Cl_2	135.04	red-yel. lq.	1.687	−80	138	d.		s. CS_2, et., bz.
chloride, di-	SCl_2	102.97	dark red fuming lq.	$1.621_{15}^{15°}$	−78	59	d.		d. al.
chloride, tetra-	SCl_4	173.88	yel.-brn. lq.		−30	d. > −20	d.		
oxide, di-	SO_2	64.06	col. gas	lq., $1.434^{0°}$; 2.264 (A)	−75.5	−10.0	$22.8^{0°}$	$4.5^{50°}$	s. H_2SO_4; al., ac.
oxide, tri-(α)	SO_3	80.06	col. pr.	lq., 1.923; 2.75 (A)	16.83	44.6	d.		s. H_2SO_4
oxide, tri-(β)	$(SO_3)_2$	160.13	col., silky, nd.	$1.97^{20°}$	50		Forms H_2SO_4		s. H_2SO_4
Sulfuric acid	H_2SO_4*	98.08	col., viscous lq.	$1.834_{4}^{18°}$	10.49	d. 340	∞	∞	d. al.
Sulfuric acid	$H_2SO_4·H_2O$	116.09	pr. or lq.	$1.842_{4}^{15°}$	8.62	290	∞	∞	d. al.

*Usual commercial form.

(*Continued*)

TABLE 2-1 Physical Properties of the Elements and Inorganic Compounds *(Continued)*

Name	Formula	Formula weight	Color, crystalline form, and refractive index	Specific gravity	Melting point, °C	Boiling point, °C	Solubility in 100 parts		
							Cold water	Hot water	Other reagents
Sulfuric acid	$H_2SO_4 \cdot 2H_2O$	134.11	col. lq.	$1.650\frac{0°}{4}$	−38.9	167	∞	∞	d. al.
Sulfuric acid, pyro-	$H_2S_2O_7$	178.14	cr.	$1.9^{20°}$	35	d.	d.		d. al.
Sulfuric oxychloride	SO_2Cl_2	134.97	col. lq.	$1.667\frac{20°}{4}$	−54.1	69.1^{760mm}	d.		s. ac.; d. al.
Sulfurous oxybromide	$SOBr_2$	207.87	or.-yel. lq.	$2.68^{18°}$	−50	68^{40mm}	d.		s. bz., CS_2, CCl_4; d. act.
oxychloride	$SOCl_2$	118.97	yel. fuming lq.	1.631	−104.5	75.6	d.		s. bz., chl.
Tantalum	Ta	180.95	bk.-gray, cb.	16.6	2850	>4100	i.	i.	s. fused alk., HF; i. HCl, HNO_3, H_2SO_4
Tellurium	Te	127.60	met., hex.	(α) 6.24; (β) 6.00	452	1390	i.	i.	s. H_2SO_4, HNO_3, KCN, KOH, aq. reg.; i. CS_2
Terbium	Tb	158.93							
Thallium	Tl	204.38	blue-wh., tet.	11.85	303.5	1650	i.	i.	s. HNO_3, H_2SO_4; i. NH_3
acetate	$TlC_2H_3O_2$	263.43	silky nd.	3.68	110		v. s.		v. s. al.
chloride, mono-	TlCl	239.84	wh., cb.	7.00	430	806	$0.21^{0°}$	$1.8^{100°}$	sl. s. HCl; i. al., NH_4OH
chloride, sesqui-	Tl_2Cl_3	515.13	yel., hex.	5.9	400–500	d.	$0.26^{15°}$	$1.9^{100°}$	
chloride, tri-	$TlCl_3$	310.74	hex. pl.		25	d.	v. s.		s. al., et.
chloride, tri-	$TlCl_3 \cdot 4H_2O$	382.80	nd.		37	$-4H_2O$, 100	$86.2^{17°}$	d.	s. al., et.
sulfate (ic)	$Tl_2(SO_4)_3 \cdot 7H_2O$	823.06	lf.		$-6H_2O$, 200	d.	d.	d.	s. dil. H_2SO_4
sulfate (ous)	Tl_2SO_4	504.83	col., rhb., 1.8671	6.77	632	d.	$2.70^{0°}$	$18.45^{100°}$	
sulfate, acid	$TlHSO_4$	301.45	trimorphous		115 d.				v. sl. s. dil. H_2SO_4
Thio, *cf.* sulfo or sulfur									
Thorium	Th	232.04	cb.	11.2	1845	>3000	i.	i.	s. HCl, H_2SO; sl. s. HNO_3; i. HF, alk.
oxide, di- (thorianite)	ThO_2	264.04	wh., cb.	9.69	>2800	4400	i.		s. h. H_2SO_4; i. alk.
sulfate	$Th(SO_4)_2$	424.16		$4.225^{17°}$			$0.74^{0°}$	$5.22^{50°}$	
sulfate	$Th(SO_4)_2 \cdot 9H_2O$	586.30	mn. pr.	2.77	$-9H_2O$, 400		sl. s.	sl. s.	
Thulium	Tm	168.93					i.	i.	
Tin	Sn	118.71	silv. met., tet.	7.31	231.85	2260	i.	i.	s. HCl, H_2SO_4, dil. HNO_3 h. aq KOH
Tin	Sn	118.71	gray, cb.	5.750	Stable −163 to +18	2260	i.	i.	s. a., h. alk. solns.
Tin salts, *cf.* stannic and stannous									
Titanic acid	H_2TiO_3	97.88	wh. pd.				i.	i.	s. alk.; v. sl. s. dil. a.; i. al.
Titanium	Ti	47.87	dark gray, cb.	$4.50^{17.5°}$	1800	>3000	i.	d.	s. a.
chloride, di-	$TiCl_2$	118.77	bk., delq.		Unstable in air		d.		i. CS_2, et., chl.
chloride, tri-	$TiCl_3$	154.23	vl., delq.		d. 440		s.	s.	
chloride, tetra-	$TiCl_4$*	189.68	col. lq.	lq., 1.726	−30	136.4	s.	d.	s. dil. HCl
oxide, di- (anatase)	TiO_2	79.87	brn. or bk., tet., 2.534–2.564	3.84			i.	i.	sl. s. alk.
oxide, di- (brookite)	TiO_2	79.87	brn. or bk., rhb., 2.586	4.17			i.	i.	
oxide, di- (rutile)	TiO_2	79.87	col. if pure, tet., 2.615	4.26	1640 d.	<3000	i.	i.	s. H_2SO_4, alk.
Tungsten	W	183.84	gray-bk., cb.	19.3	3370	5900	i.	i.	s. h. conc. KOH; sl. s. NH_3, HNO_3, aq. reg.
carbide	WC	195.85	gray pd., cb.	$15.7^{18°}$	2777	6000	i.	i.	s. F_2; i. a.
carbide	W_2C	379.69	iron gray	$16.06^{18°}$	2877	6000	i.	i.	s. h. HNO_3; sl. s. HCl, H_2SO_4
oxide, tri-	WO_3	231.84	yel., rhb.	7.16	>2130		i.	i.	s. alk.; i. a.
Tungstic acid (tungstite)	H_2WO_4	249.85	yel., rhb. 2.24	5.5	$-\frac{1}{2}H_2O$, 100; 1473		i.	sl. s.	s. HF, alk., NH_3
Uranic acid	H_2UO_4	304.04	yel. pd.	$5.926^{15°}$	$-H_2O$, 250 to 300		i.	i.	s. a., alk. carb.; i. alk.
Uranium	U	238.03	wh. cr.	$18.485\frac{13°}{4}$	1133	3500	i.	i.	s. a.; i. alk.
carbide	U_2C_3	512.09	cr.	11.28	2400		d.		d. a.
oxide, di- (uraninite)	UO_2	270.03	bk., rhb.	10.9	2176		i.	i.	s. HNO_3, conc. H_2SO_4

oxide (pitchblende)	U_3O_8	842.08	olive gn.	7.31	d.		i.	i.	s. HNO_3, H_2SO_4
sulfate (ous)	$U(SO_4)_2 \cdot 4H_2O$	502.22	gn., rhb.		$-4H_2O$, 300		$23^{11°}$	$9^{63°}$	s. dil. a.
Uranyl acetate	$UO_2(C_2H_3O_2)_2 \cdot 2H_2O$	424.15	yel., rhb.	$2.89^{15°}$	$-2H_2O$, 110		$9.2^{17°}$	d.	s. al., act.
carbonate (rutherfordine)	UO_2CO_3	330.04	tet.	5.6					
nitrate	$UO_2(NO_3)_2 \cdot 6H_2O$	502.13	yel., rhb., 1.4967	2.807	60.2	118	$170.3^{0°}$	$\infty^{60°}$	v. s. ac., al., et.; i. dil., alk.
sulfate	$UO_2SO_4 \cdot 3H_2O$	420.14	yel. cr.	$3.28^{16.5°}$	d. 100		$18.9^{13.2°}$	$230^{25°}$	4 al.; s. a.
Vanadic acid, meta-	HVO_3	99.95	yel. scales				i.		s. a., alk.; i. NH_3
Vanadic acid, pyro-	$H_4V_2O_7$	217.91	pa. yel., amor.				i.		s. a., alk., NH_4OH
Vanadium	V	50.94	lt. gray, cb.	5.96	1710	3000	i.	i.	s. HNO_3, H_2SO_4; i. aq., alk.
chloride, di-	VCl_2	121.85	gn., hex., delq.	$3.23^{18°}$			s.	d.	s. al., et.
chloride, tri-	VCl_3	157.30	pink, tabular, delq.	$3.00^{18°}$	d.		s.	d.	s. abs. al., et.
chloride, tetra-	VCl_4	192.75	red lq.	$1.816^{30°}$	−109	148.5^{755mm}	s. d.		s. abs. al., et., chl., ac.
oxide, di-	V_2O_2	133.88	lt. gray cr.	3.64	ign.		i.	i.	s. a.
oxide, tri-	V_2O_3	149.88	bk. cr.	$4.87^{18°}_{4}$	1970		sl. s.	s.	s. HNO_3, HF, alk.
oxide, tetra-	V_2O_4	165.88	blue cr.	4.399	1967		i.	i.	s. a., alk.
oxide, penta-	V_2O_5	181.88	red-yel., rhb.	$3.357^{18°}_{4}$	800	d. 1750	$0.8^{20°}$		s. a., alk.; i. abs. al.
oxychloride, mono-	VOCl	86.39	brn. pd.	2.824			i.		v. s. HNO_3
Vanadyl chloride	$(VO)_2Cl$	169.33	yel. cr.	3.64	d. in air		i.		s. HNO_3
chloride, di-	$VOCl_2$	137.85	gn., delq.	$2.88^{13°}$			d.		s. abs. al., dil. HNO_3
chloride, tri-	$VOCl_3$	173.30	yel. lq.	1.829	<−15	127.19	s. d.		s. al., et., ∞Br_2
Water†	H_2O	18.02	col. lq., $1.33300^{20°}$; hex. solid, 1.309	$1.00^{4°}$ (lq.); $0.915^{0°}$ (ice)	0	100			∞ al.; sl. s. et.
Water, heavy	D_2O	20.029	col. lq., $1.32844^{20°}$	$1.107^{20°}$	3.82	101.42	∞	∞	∞ al.; sl. s. et.
Xenon	Xe	131.29	col. gas	lq., $3.06^{-109.1}$ $2.7^{-140°}$ 4.53 (A)	−140	−109.1	$24.2^{0°}$ cc	$7.3^{50°}$ cc	
Ytterbium	Yb	173.04							
Yttrium	Y	88.91	dark gray, hex.	5.51	1490	2500	sl. d.	d.	v. s. dil. a., h. KOH
Zinc	Zn	65.41	silv. met., hex.	7.140	419.4	907	i.	i.	s. a., ac., alk.
acetate	$Zn(C_2H_3O_2)_2$	183.50	mn.	1.840	242	subl. in vac.	$30^{25°}$	$44.6^{100°}$	$2.8^{25°}$, $166^{79°}$ al.
acetate	$Zn(C_2H_3O_2)_2 \cdot 2H_2O$*	219.53	wh., mn., 1.494	1.735	237	$-2H_2O$, 100	$40^{25°}$	$66.6^{100°}$	v. s. al.
bromide	$ZnBr_2$	225.22	rhb.	$4.219^{4°}$	394	650	$390^{0°}$	$670^{100°}$	v. s. NH_4OH, al., et.
carbonate	$ZnCO_3$	125.42	wh., trig., 1.818	4.42	$-CO_2$, 300		$0.001^{15°}$		s. a., alk., NH_4 salts; i. act., NH_3
chloride	$ZnCl_2$	136.32	wh., delq., 1.687, uniaxial	$2.91^{25°}_{4}$	283	732	$432^{25°}$	$615^{100°}$	$100^{12.5°}$ al.; v. s. et.; i. NH_3
cyanide	$Zn(CN)_2$	117.44	col., rhb.		d. 80		$0.0005^{18°}$	sl. s.	s. KCN, NH_3, alk.; i. al.
hydroxide	$Zn(OH)_2$	99.42	col., rhb.	3.053	d. 125		$0.00052^{18°}$		s. a., alk., NH_4OH
iodide	ZnI_2	319.22	cb.	$4.666^{14.2°}_{4}$	446	624	$430^{0°}$	$510^{100°}$	s. a., al., NH_3, aq. $(NH_4)_2CO_3$
nitrate	$Zn(NO_3)_2 \cdot 6H_2O$	297.51	col., tet.	$2.065^{14°}_{4}$	36.4	$-6H_2O$, 105	324.5	$\infty^{36.4°}$	v. s. al.
oxide (zincite)	ZnO	81.41	wh., hex., 2.004	5.606	>1800		$0.00042^{18°}$		s. a., alk., NH_4Cl; i. NH_3
oxide	ZnO	81.41	wh., amor.	5.47	>1800		$0.00042^{18°}$		
peroxide	ZnO_2	97.41	yel.	1.571	expl. 212		0.0022		i. NH_4OH; d. a.
phosphide	Zn_3P_2	258.17	steel gray, cb.	$4.55^{13°}_{4}$	>420	1100	i.		s. dil. a.
silicate	$ZnSiO_3$	141.49	hex. or rhb.; glass, 1.650	3.52	1437		i.		
sulfate (zincosite)	$ZnSO_4$	161.47	wh., rhb., 1.669	$3.74^{15°}_{4}$	d. 740		$42^{0°}$	$61^{100°}$	sl. s. al.; s. gly.
sulfate	$ZnSO_4 \cdot H_2O$	179.49	col.	$3.28^{15°}_{4}$	d. 238		s.	$89.5^{100°}$	
sulfate	$ZnSO_4 \cdot 6H_2O$	269.56	mn.	$2.072^{15°}_{4}$	$-5H_2O$, 70		s.	s.	sl. s. al.; i. act.; NH_3
sulfate (goslarite)	$ZnSO_4 \cdot 7H_2O$*	287.58	rhb., 1.4801	$1.966^{16.5°}$	tr. 39	$-7H_2O$, 280	$115.2^{0°}$	$653.6^{100°}$	sl. s. al.; i. act.; NH_3
sulfide (α) (wurzite)	ZnS	97.47	wh., hex., 2.356	4.087	$1850^{150atm.}$	subl. 1185	$0.00069^{18°}$	i.	v. s. a.; i. ac.
sulfide (β) (sphalerite)	ZnS	97.47	wh., cb.; glass (?) 2.18–2.25	$4.102^{25°}_{4}$	tr. 1020		i.	i.	s. a.
sulfide (blende)	ZnS	97.47	wh., granular	4.04			i.	i.	v. s. a.; i. ac.
sulfite	$ZnSO_3 \cdot 2½H_2O$	190.51	mn.		$-2½H_2O$, 100	d. 200	0.16	d.	s. H_2SO_3, NH_4OH; i. al.
Zirconium	Zr	91.22	cb., pd. ign. easily	6.4	1700	>2900	i.	i.	s. HF, aq. reg.; sl. s. a.
oxide, di- (baddeleyite)	ZrO_2	123.22	yel. or brn., mn., 2.19	5.49	2700		i.	i.	s. H_2SO_4, HF
oxide, di- (free from Hf)	ZrO_2	123.22	wh., mn.	5.73		4300	i.	i.	s. H_2SO_4, HF

*Usual commercial form.

†*Cf.* special tables on water and steam, Tables 2-3, 2-4, and 2-5.

NOTE: °F = 9/5°C + 32.

TABLE 2-2 Physical Properties of Organic Compounds*

Abbreviations Used in the Table

(A), density referred to air	cr., crystalline	i-, iso-, containing the group $(CH_3)_2CH$-	nd., needles	s-, sec-, secondary	v. s., very soluble
al., ethyl alcohol	d., decomposes	i., insoluble	o-, ortho	silv., silvery	v. sl. s., very slightly soluble
amor., amorphous	d-, dextrorotatory	ign., ignites	or., orange	sl., slightly	wh., white
aq., aqua, water	dl-, dextro-laevorotatory	l-, laevorotatory	p-, para	subl., sublimes	yel., yellow
brn., brown	et., ethyl ether	lf., leaflets	pd., powder	sym., symmetrical	(+), right rotation
bz., benzene	expl., explodes	lq., liquid	pet., petroleum ether	t-, tertiary	>, greater than
c., cubic	gn., green	m-, meta	pl., plates	tet., tetragonal	<, less than
cc., cubic centimeter	h., hot	mn., monoclinic	pr., prisms	tri., triclinic	∞, infinitely
chl., chloroform	hex., hexagonal	n-, normal	rhb., rhombic	uns., unsymmetrical	
col., colorless			s., soluble	v., very	

This table of the physical properties includes the organic compounds of most general interest. For the properties of other organic compounds, reference must be made to larger tables in *Lange's Handbook of Chemistry* (Handbook Publishers), *Handbook of Chemistry and Physics* (Chemical Rubber Publishing Co.), Van Nostrand's *Chemical Annual, International Critical Tables* (McGraw-Hill), and similar works.

The **molecular weights** are based on the atomic weight values in "Atomic weights of the Elements 2001," *PURE Appl. Chem.*, **75,** 1107, 2003. The **densities** are given for the temperature indicated and are usually referred to water at 4°C, *e.g.,* $1.028^{95/4}$ a density of 1.028 at 95°C referred to water at 4°C, the 4 being omitted when it is not clear whether the reference is to water at 4°C or at the temperature indicated by the upper figure. The melting and boiling points given have been selected from available data as probably the most accurate. The **solubility** is given in grams of the substance in 100 of the solvent. In the case of gases, the solubility is often expressed in some manner as "5^{10} cc." which indicates that, at 10°C, 5 cc. of the gas are soluble in 100 of the solvent.

Name	Synonym	Formula	Formula weight	Form and color	Specific gravity	Melting point, °C	Boiling point, °C	Solubility in 100 parts		
								Water	Alcohol	Ether
Abietic acid	sylvic acid, abietinic acid	$C_{20}H_{30}O_2$	302.45	lf.		182		i.	v. s.	v. s.
Acenaphthene	naphthylene ethylene	$C_{10}H_6(CH_2)_2$	154.21	rhb./al.	$1.069^{95/95}$	95	278–9	i.	s. h.	s. chl.
Acetal	acetaldehyde diethylacetal	$CH_3CH(OC_2H_5)_2$	118.17	lq.	$0.821^{22/4}$		102.2	6^{25}	∞	∞
Acet-aldehyde	ethanal	CH_3CHO	44.05	col. lq.	$0.783^{18/4}$	−123.5	20.2	∞	∞	∞
-aldehyde, par-	paraldehyde	$(C_2H_4O)_3$	132.16	col. cr.	$0.994^{20/4}$	10.5–12	124.4^{752}	12^{13}	∞	∞
-aldehyde ammonia		$CH_3CHOHNH_2$	61.08	col. cr.		97	100–10 d.	v. s.	v. s.	sl. s.
-amide	ethanamide	CH_3CONH_2	59.07	col. cr.	1.159	81(69.4)	222	s.	s.	v. sl. s.
-anilide	antifebrin	$C_6H_5NHCOCH_3$	135.16	rhb./al.	1.21^{4}	113–4	305	0.5^{6}	21^{20}	7^{25}
-phenetidide (*o*-)	*o*-ethoxyacetanilide	$CH_3CONHC_6H_4OC_2H_5$	179.22	lf./al.		79	>250	i.	s.	
(*m*-)	acetyl-*m*-phenetidine	$CH_3CONHC_6H_4OC_2H_5$	179.22	lf./al.		96–7		sl. s.	s.	
-toluidide (*o*-)	N-tolylacetamide	$CH_3C_6H_4NHCOCH_3$	149.19	rhb.	1.168^{15}	110	296	0.86^{19}	s.	s.
(*p*-)	N-tolylacetamide	$CH_3C_6H_4NHCOCH_3$	149.19	rhb. or mn.	1.212^{15}	153	306–7	0.09^{22}	10^{25}	s.
Acetic acid	ethanoic acid, vinegar acid	CH_3CO_2H	60.05	col. lq.	$1.049^{20/4}$	16.7	118.1	∞	∞	∞
anhydride	acetyl oxide, acetic oxide	$(CH_3CO)_2O$	102.09	col. lq.	$1.082^{20/4}$	−73	139.6	12 c.	∞	∞
nitrile	methyl cyanide	CH_3CN	41.05	col. lq.	$0.783^{20/4}$	−41	81.6–2.0	∞	∞	∞
Acetone	propanone, dimethyl ketone	CH_3COCH_3	58.08	col. lq.	$0.792^{20/4}$	−94.6	56.5	∞	∞	∞
Acetonyl urea	dimethyl hydantoin	<NHCONHCOC>$(CH_3)_2$	128.13	tri./al.		175	subl.	s.	s.	s.
Acetophenone benzoyl hydride	methyl-phenyl ketone	$CH_3COC_6H_5$	120.15	lf.	$1.033^{15/15}$	20.5	202.3^{749}	i.	s.	s.
Acetyl-chloride	ethanoyl chloride	CH_3COCl	78.50	col. lq.	$1.105^{20/4}$	−112.0	51–2	d.	d.	∞
-phenylenediamine (-*p*)	amino-acetanilide (*p*)	$C_2H_3ONHC_6H_4NH_2$	150.18	nd./aq.		162		s. h.	v. s.	v. s.
Acetylene	ethyne; ethine	$HC{:}CH$	26.04	col. gas	(A) 0.906	-81.5^{891}	-84^{760}	100 $cc.^{18}$	600 $cc.^{18}$	
dichloride (*cis*)	1,2-dichloroethene	$CHCl{:}CHCl$	96.94	col. lq.	$1.291^{15/4}$	−80.5	60.3	0.35^{20}	∞	∞
(*trans*)	dioform	$CHCl{:}CHCl$	96.94	col. lq.	$1.265^{15/4}$	−50	48.4	0.63^{20}	∞	∞
Aconitic acid	equisetic acid; citridic acid	$C_3H_3(CO_2H)_3$	174.11	cr./aq.		192 d.		33^{15}	sl. s.	v. sl. s.
Acridine		$C_6H_4<(CH)(N)>C_6H_4$	179.22	rhb./aq. al.		110–1	346	sl. s. h.	s.	s.
Acrolein ethylene aldehyde	acrylic aldehyde, propenal	$CH_2{:}CH{\cdot}CHO$	56.06	col. lq.	$0.841^{20/4}$	−87.7	52.5	40	s.	s.
Acrylic acid	propenoic acid	$CH_2{:}CH{\cdot}CO_2H$	72.06	col. lq.	$1.062^{16/4}$	12–13	141–2	∞	∞	
nitrile	vinyl cyanide	$CH_2{:}CH{\cdot}CN$	53.06	col. lq.	0.811^{20}	−82	78–9	s.		
Adipic acid	hexandioc acid, adipinic acid	$(CH_2CH_2CO_2H)_2$	146.14	mn. pr.	$1.360^{25/4}$	151–3	265^{10}	1.4^{15}	v. s.	0.6^{15}
amide		$(CH_2CH_2CONH_2)_2$	144.17	cr. pd.		226–7		0.4^{12}		
nitrile	tetramethylene	$(CH_2CH_2CN)_2$	108.14	col. oil	$0.951^{19/19}$	1	295	v. sl. s.	s.	v. sl. s.
Adrenaline (1-) (3,4,1)	1-suprarenine	$C_6H_3(OH)_2(CHOHCH_2NHCH_3)$	183.20	col. pd.		d. 207–11		0.03^{20}	v. sl. s.	i.
Alanine (α) (*dl*-)		$CH_3CH(NH_2)CO_2H$	89.09	nd./aq.		295 d.	subl. >200	22^{17}	v. sl. s.	i.
Aldol acetaldol	2-hydroxybutyraldehyde	$CH_3CH(OH)CH_2COH$	88.11	col. lq.	$1.103^{20/4}$		83^{20}	∞	∞	s.
Alizarin	Anthraquinoic acid	$C_6H_4(CO)_2C_6H_2(OH)_2$	240.21	red rhb.		289–90	430	0.03^{100}	v. s.	v. s.
Allyl alcohol	propen-1-ol-3,propenyl alcohol	$CH_2{:}CH{\cdot}CH_2OH$	58.08	col. lq.	$0.854^{20/4}$	−129	96.6	∞	∞	∞
bromide	3-bromo-propene-1	$CH_2{:}CH{\cdot}CH_2Br$	120.98	lq.	$1.398^{20/4}$	−119.4	$70-1^{753}$	i.	∞	∞
chloride	3-chloro-propene-1	$CH_2{:}CH{\cdot}CH_2Cl$	76.52	col. lq.	$0.938^{20/4}$	−136.4	44.6	<0.1	∞	∞
thiocyanate (*i*)	mustard oil	$CH_2{:}CH{\cdot}CH_2NCS$	99.15	col. oil	$1.013^{20/4}$	−80	152	0.2	∞	∞
thiourea	thiosinamide	$CH_2{:}CH{\cdot}CH_2NHCSNH_2$	116.18	col. pr.	$1.219^{20/20}$	77–8		3^{0}	s.	v. sl. s.
Aluminum ethoxide		$Al(OCH_2CH_3)_3$	162.16	pd.	$1.142^{20/0}$	150–60	$200-5^{10}$	d.	i.	v. sl. s.
Amino-anthraquinone (α)		$C_6H_4(CO)_2C_6H_3NH_2$	223.23	red nd.		256	subl.	i.	s.	s.
(β)		$C_6H_4(CO)_2C_6H_3NH_2$	223.23	red nd.		302	subl.	i.	s.	i.
-azobenzene		$C_6H_5{\cdot}N{:}N{\cdot}C_6H_4NH_2$	197.24	yel. mn.		126–7	225^{120}	sl. s. h.	s. h.	s.
-benzoic acid (*m*-)		$H_2N{\cdot}C_6H_4CO_2H$	137.14	nd./aq.	$1.511^{4°}$	173–4		v. sl. s.	2^{10}	1.8^{6}
(*p*-)	aminodracylic acid	$H_2N{\cdot}C_6H_4CO_2H$	137.14	mn. pr.		187–8		0.3^{13}	11^{10}	8.2^{6}

Amino-diphenylamine (*p*-)		$H_2N{\cdot}C_6H_4NH{\cdot}C_6H_5$	184.24	nd./aq. al.		67	354	sl. s.	s.	s.
-G-acid (2-)(6-,8-), Na_2 salt		$C_{10}H_5(NH_2)(SO_3Na)_2$	347.28					v. sl. s.		
-mono-potassium salt		$C_{10}H_5(NH_2)S_2O_6HK$	341.40					12.8^{20}		
-sodium salt		$C_{10}H_5(NH_2)S_2O_6HNa$	325.29					2.7^{18}		
-J-acid (2-)(5-,7-)		$C_{10}H_5(NH_2)(SO_3H)_2$	303.31					10.0^{20}		
-mono-potassium salt		$C_{10}H_5(NH_2)S_2O_6HK$	341.40					3.4^{18}		
-naphthol sulfonic (1-,2-,4-)(α-)		$C_{10}H_5OHNH_2SO_3H½H_2O$	248.26					v. s.		
(1-,8-,4-)		$NH_2(OH)C_{10}H_5SO_3H$	239.25					v. sl. s.		
-phenol (*o*-)	2-aminophenol	$H_2N{\cdot}C_6H_4{\cdot}OH$	109.13	col. nd.		173	subl.	1.7^0	4.3^0	v. s.
(*m*-)	3-aminophenol	$H_2N{\cdot}C_6H_4{\cdot}OH$	109.13	pr.		122–3		2.6^0	s.	sl. s.
(*p*-)	*p*-hydroxyaniline	$H_2N{\cdot}C_6H_4{\cdot}OH$	109.13	lf.		184–6 d.	subl.	1.1^0	4^0	i. bz.
-toluene sulfonic acid (1-,2-,3-)		$C_6H_3(CH_3)(NH_2)SO_3H$	187.22	nd.				0.97^{11}		
(1-,4-,2-)		$C_6H_3(CH_3)(NH_2)SO_3H{\cdot}H_2O$	205.23	mn.		d.		0.5^{20}	i.	
(1-,4-,3-)		$C_6H_3(CH_3)(NH_2)SO_3H{\cdot}½H_2O$	196.22	nd.				0.47		
(1-,2-,5-)		$C_6H_3(CH_3)(NH_2)SO_3H{\cdot}H_2O$	205.23	tri./aq.		$-H_2O$, 120		3^{11}	i.	
Amyl acetate (*n*-)		$CH_3CO_2CH_2(CH_2)_3CH_3$	130.18	col. lq.	$0.879^{20/20}$	−70.8	148.4^{737}	v. sl. s.	∞	∞
(*i*-)	common amyl acetate	$CH_3CO_2CH_2CH_2CH(CH_3)_2$	130.18	col. lq.	$0.876^{15/4}$		142^{757}	0.3^{15}	∞	∞
		$CH_3CO_2CH_2CH(CH_3)C_2H_5$	130.18	col. lq.	0.880^{13}		141–2	v. sl. s.	∞	∞
(*s*-)	α-Me-Bu-acetate	$CH_3CO_2CH(CH_3)CH_2C_2H_5$	130.18	col. lq.	0.922^0		133.5	sl. s.	∞	∞
(*s*-)	di Et-carbinol acetate	$CH_3CO_2CH(C_2H_5)_2$	130.18	col. lq.	$0.871^{20/4}$		133	sl. s.	∞	∞
(*t*-)		$CH_3CO_2C(CH_3)_2C_2H_5$	130.18	col. lq.	0.874^{19}		124.5^{749}	v. sl. s.	∞	∞
alcohol (*n*-) fusel oil,	pentanol-1	$CH_3(CH_2)_3CH_2OH$	88.15	col. lq.	$0.817^{20/20}$	−78.5	137.9	2.7^{22}	∞	∞
(*s*-,*n*-) methyl-propyl carbinol,	pentanol-2	$C_2H_5CH_2CH(OH)CH_3$	88.15	col. lq.	$0.810^{20/20}$		119.5	4^{20}	∞	∞
(prim.-,*i*-) isobutyl carbinol,	2-methyl-butanol-4	$(CH_3)_2CHCH_2CH_2OH$	88.15	col. lq.	$0.813^{15/4}$	−117.2	132.0	2^{14}	∞	∞
		$(C_2H_5)_2CHOH$	88.15	col. lq.	$0.815^{25/4}$		115.6	5.5^{30}	∞	∞
(*s*-,*i*-)	2-methyl-butanol-3	$(CH_3)_2CHCH(OH)CH_3$	88.15	col. lq.	0.819^{19}		113–4	2.8^{30}	∞	∞
(*t*-)	2-methyl-butanol-2	$(CH_3)_2C(OH)C_2H_5$	88.15	col. lq.	$0.809^{20/4}$	−11.9	102	sl. s.	s.	s.
		$(CH_3)_3CCH_2OH$	88.15	cr.		52–3	113–4	sl. s.	∞	∞
(*d*-)	active amyl alcohol	$C_2H_5CH(CH_3)CH_2OH$	88.15	col. lq.	$0.816^{20/4}$		128	3.6^{30}	∞	∞
-amine (*n*-)		$CH_3(CH_2)_4NH_2$	87.16	col. lq.	0.766^{19}	−55	103–4	s.	s.	s.
(*s*-,*n*-)		$(C_3H_7)(CH_3)CHNH_2$	87.16	col. lq.	$0.749^{20/4}$		91–2	∞	∞	∞
(*i*-)		$(CH_3)_2CH(CH_2)_2NH_2$	87.16	col. lq.	$0.751^{18/4}$		95	∞	∞	∞
(*t*-)		$(C_2H_5)(CH_3)_2CNH_2$	87.16	col. lq.	$0.731^{25/4}$	−105	77–8	∞	∞	∞
	$1\text{-}NH_2\text{-}2\text{-Me-butane}$	$C_2H_5CH(CH_3)CH_2NH_2$	87.16	col. lq.	0.755^{18}		95–6	∞	∞	∞
	3-amino pentane	$(C_2H_5)_2CHNH_2$	87.16	col. lq.	$0.749^{20/4}$		90–1	∞	∞	∞
	$3\text{-}NH_2\text{-}2\text{-Me-butane}$	$(CH_3)_2CHCH(CH_3)NH_2$	87.16	col. lq.	0.757^{18}		83–4	∞	∞	∞
aniline (*i*-)		$C_6H_5NHC_5H_{11}$	163.26	lq.	$0.928^{15/4}$		254.5	i.		
benzoate (*i*-)		$C_6H_5CO_2C_5H_{11}$	192.25	col. lq.	$0.992^{14/14}$		261^{746}	i.	∞	∞
bromide (*n*-)	1-bromopentane	$CH_3(CH_2)_3CH_2Br$	151.04	col. lq.	$1.218^{20/4}$	−95	129.7	i.	s.	
(*i*-)	4-Br-2-Me-butane	$(CH_3)_2CH(CH_2)_2Br$	151.04	col. lq.	$1.220^{17/15}$		120^{745}	0.02^{16}	s.	s.
(*t*-)	2-Br-2-Me-butane	$(CH_3)_2C(Br)C_2H_5$	151.04	lq.	$1.216^{19/0}$		108^{765}	i.	s.	s.
n-butyrate (*n*-)		$C_2H_5CH_2CO_2(CH_2)_4CH_3$	158.24	col. lq.	$0.871^{15/4}$	−73.2	186.4	0.05^{50}	∞	∞
(*i*-)		$C_2H_5CH_2CO_2{\cdot}C_5H_{11}$	158.24	col. lq.	$0.866^{19/15}$		178.6	i.	∞	∞
(*t*-)		$C_3H_7CO_2C(CH_3)_2C_2H_5$	158.24	col. lq.	$0.865^{15/0}$		164	sl. s.	∞	∞
i-butyrate (*i*-)		$(CH_3)_2CHCO_2C_5H_{11}$	158.24	lq.	$0.876^{0/4}$		168.8	i.	s.	s.
chloride (*n*-)	1-chloropentane	$CH_3(CH_2)_3CH_2Cl$	106.59	col. lq.	$0.878^{20/4}$	−99	108.4	i.	s.	s.
(*s*-)	2-chloropentane	$C_2H_5CH_2CHClCH_3$	106.59	lq.	$0.870^{20/4}$		96.7	i.	s.	s.
(*s*-)	3-chloropentane	$(C_2H_5)_2CHCl$	106.59	col. lq.	0.895^{21}		97.3	i.	∞	∞
(*i*-)	4-Cl-2-Me-butane	$(CH_3)_2CH(CH_2)_2Cl$	106.59	col. lq.	$0.893^{20/4}$		99.7^{758}	i.	s.	∞
(*s*-,*i*-)	3-Cl-2-Me-butane	$(CH_3)CHCHClCH_3$	106.59	lq.	0.883^0		91^{753}	i.	s.	s.
(*t*-)	2-Cl-2-Me-butane	$(CH_3)_2CClC_2H_5$	106.59	lq.	$0.871^{20/4}$	−72.9	85.7	i.	s.	s.
	1-Cl-2-Me-butane	$(CH_3)(C_2H_5)CHCH_2Cl$	106.59	lq.	$0.881^{17.5}$		98–9	i.	s.	s.
i-cyanide (*i*-)	iso-caproic iso-nitrile	$(CH_3)_2CH(CH_2)_2NC$	97.16	lq.			137–9	i.	s.	s.
formate (*n*-)		$HCO_2CH_2(CH_2)_3CH_3$	116.16	lq.	0.902^0	−73.5	132	v. sl. s.	∞	∞
(*i*-)		$HCO_2CH_2CH_2CH(CH_3)_2$	116.16	lq.	$0.882^{20/4}$	−93.5	123.5	0.3^{22}	∞	∞
iodide (*n*-)	1-iodopentane	$CH_3(CH_2)_3CH_2I$	198.05	lq.	$1.510^{20/4}$	−86	157.0	i.	s.	∞
(*i*-)	4-I-2-Me-butane	$(CH_3)_2CHCH_2CH_2I$	198.05	lq.	$1.515^{18/4}$		147^{765}	i.	∞	∞
(*s*-,*n*-)	2-iodopentane	$C_2H_5CH_2CHICH_3$	198.05	lq.	$1.507^{17/4}$		144–5	i.	∞	∞
(*t*-)	2-I-2-Me-butane	$(CH_3)_2CIC_2H_5$	198.05	lq.	$1.471^{19/15}$		127^{765}	i.	∞	∞
		$C_2H_5CH(CH_3)CH_2I$	198.05	lq.	$1.524^{20/4}$		148	i.	∞	∞
mercaptan (*n*-)	pentanthiol-1	$CH_3(CH_2)_3CH_2SH$	104.21	lq.	0.857^{20}		126^{767}	i.	∞	∞
(*n*-)	pentanthiol-3	$(C_2H_5)_2CHSH$	104.21	col. lq.			105	i.	∞	∞
(*i*-)	2-Me-butanthiol-4	$(CH_3)_2CH(CH_2)_2SH$	104.21	lq.	$0.835^{20/4}$		120	i.	∞	∞
phenol (*t*-)(*p*-)	pentaphen	$C_5H_{11}{\cdot}C_6H_4OH$	164.24	cr.		93	265–7	sl. s.	s.	s.
propionate (*n*-)		$C_2H_5CO_2(CH_2)_4CH_3$	144.21	lq.	$0.876^{15/4}$	−73.1	168.7	i.	∞	∞
(*i*-)		$C_2H_5CO_2(CH_2)_2CH(CH_3)_2$	144.21	col. lq.	$0.870^{20/4}$		160.2	0.1^{25}	∞	∞
(act.)		$C_2H_5CO_2C_5H_{11}$	144.21	col. lq.	$0.866^{20/4}$		58^{16}	v. sl. s.	∞	∞
salicylate (*n*-)		$HOC_6H_4CO_2C_5H_{11}$	208.25	lq.	1.065^{15}		265	i.	∞	∞
Amyl *i*-valerate (*i*)		$C_4H_9CO_2C_5H_{11}$	172.26	col. lq.	$0.858^{20/15}$		194	v. sl. s.	∞	∞
(*t*-)		$C_4H_9CO_2C_5H_{11}$	172.26	col. lq.	$0.861^{14/0}$		173–4	sl. s.	s.	s.

*By N. A. Lange, Ph.D., Handbook Publishers, Inc., Sandusky, Ohio. Abridged from table of Physical Constants of Organic Compounds in *Lange's Handbook of Chemistry.*

(*Continued*)

TABLE 2-2 Physical Properties of Organic Compounds *(Continued)*

Name	Synonym	Formula	Formula weight	Form and color	Specific gravity	Melting point, °C	Boiling point, °C	Solubility in 100 parts: Water	Alcohol	Ether
Amylene (*n*-)(α-)	pentene-1	$C_2H_5CH_2CH{:}CH_2$	70.13	lq.	0.644^{20}		30–1	i.	∞	∞
(*i*-)	2-methyl-butene-3	$(CH_3)_2CHCH{:}CH_2$	70.13	col. lq.	0.632^{15}	−135	20.5^{771}	i.	∞	∞
(α-)	2-methyl-butene-1	$(C_2H_5)(CH_3)C{:}CH_2$	70.13	col. lq.	$0.667^{0/0}$		$31–2^{758}$	i.	∞	∞
(-*n*)(β-)	pentene-2	$C_2H_5CH{:}CHCH_3$	70.13	col. lq.	$0.650^{20/4}$	−139	36.4	v. sl. s.	∞	∞
(*i*-)(β-)	2-methyl-butene-2	$(CH_3)_2C{:}CHCH_3$	70.13	col. lq.	$0.663^{19/4}$	−124	37–8	i.	s.	∞
Anethole (*p*-)	*p*-propenyl anisole	$CH_3CH{:}CH{\cdot}C_6H_4OCH_3$	148.20	lf./al.	$0.991^{20/20}$	22.5	235.3	v. sl. s.	s.	∞
Anhydroformald-aniline	methylene aniline	$(CH_2NC_6H_5)_3$	315.41	pr./al.		143	185	i.	sl. s.	s.
Aniline	amino benzene, phenyl amine, cyanol	$C_6H_5NH_2$	93.13	col. oil	$1.022^{20/4}$	−6.2	184.4	3.6^{18}	∞	∞
hydrochloride	aniline salt, aniline chloride	$C_6H_5NH_2{\cdot}HCl$	129.59	cr.	1.222^{4}	198	245	18^{15}	s.	i.
nitrate		$C_6H_5NH_2{\cdot}HNO_3$	156.14	rhb.	1.356^{4}	d. 190		s.	s.	sl. s.
sulfate		$(C_6H_5NH_2)_2{\cdot}H_2SO_4$	284.33	lf./al.	1.377^{4}	d.		5^{14}	sl. s.	i.
Anisal-acetone (*p*-)	MeO-benzalacetone	$CH_3OC_6H_4CH{:}CHCOCH_3$	176.21	lf./et.		73–4		i.	v. s.	v. s.
Anisic acid (*p*-)		$CH_3OC_6H_4CO_2H$	152.15	mn./aq.	1.385^{4}	184.2	275–80	0.03^{19}	v. s.	v. s.
aldehyde (*p*-)		$CH_3OC_6H_4CHO$	136.15	col. oil	$1.123^{20/4}$	2.5	247–8	v. sl. s.	∞	∞
Anisidine (*o*-)	2-amino-anisole	$CH_3OC_6H_4NH_2$	123.15	col. lq.	$1.098^{15/15}$	5.2	225	v. sl. s.	∞	∞
(*m*-)	MeO-aniline(*m*)	$CH_3OC_6H_4NH_2$	123.15	oil	$1.096^{20/4}$	<−12	251	v. sl. s.	s.	s.
(*p*-)	4-amino anisole	$CH_3OC_6H_4NH_2$	123.15	pl./aq.	$1.089^{55/55}$	57.2	243	s. h.	s.	s.
Anisole	methyl phenyl ether	$CH_3OC_6H_5$	108.14	col. lq.	$0.990^{22/4}$	−37.3	154–5	i.	s.	s.
Anthracene	paranaphthalene, anthracin green oil	$C_6H_4{:}(CH)_2{:}C_6H_4$	178.23	col. mn.	$1.25^{27/4}$	217–8	340–2	i.	1.5^{20}	
Anthramine (α)	α-amino-anthracene	$C_6H_4{:}(CH)_2{:}C_6H_3NH_2$	193.24	yel./al.		130±		i.	s.	
(β)	β-amino-anthracene	$C_6H_4{:}(CH)_2{:}C_6H_3NH_2$	193.24	yel./al.		238	subl.	i.	sl. s.	sl. s.
Anthranil		$C_6H_4{:}(NH)CO$	119.12	col. oil	$1.187^{15/4}$	<−18	d. >215	sl. s. h.	s.	s.
Anthranilic acid (*o*-)		$H_2NC_6H_4CO_2H$	137.14	col. rhb.		144–5	subl.	0.35^{14}	11^{10}	16^{7}
Anthrapurpurin (1-,2-,7-)		$C_{14}H_5O_2(OH)_3$	256.21	or. nd./al.		369	462	sl. s. h.	v. s. h.	sl. s.
Anthraquinone	diphenyleneketone, dihydrodiketoanthracene	$C_6H_4{:}(CO)_2{:}C_6H_4$	208.21	yel. rhb.	$1.438^{20/4}$	286	379–81	i.	0.05^{18}	v. sl. s.
disulfonate Na_2 (1-,5-)	ρ-anthraquinone disulfonate	$C_{14}H_6O_2(SO_3Na)_2{\cdot}5H_2O$	502.38	yel. lf.				v. s.	i.	i.
(1-,8-)	*x*-anthraquinone disulfonate	$C_{14}H_6O_2(SO_3Na)_2{\cdot}4H_2O$	484.36	yel. pr.				sl. s.		
(2-,6-)		$C_{14}H_6O_2(SO_3Na)_2{\cdot}7H_2O$	538.41	col. cr.				3.9^{20}		
(2-,7-)		$C_{14}H_6O_2(SO_3Na)_2{\cdot}4H_2O$	484.36	cr.				30.5^{20}	v. sl. s.	i.
sulfonate Na (1-)		$C_{14}H_7O_2SO_3Na$	310.26	yel. lf.				0.53^{20}	i.	i.
(2-)		$C_{14}H_7O_2SO_3Na$	310.26	silv. lf.				0.84^{25}	i.	i.
Anthrarufin (1-,5-)		$C_{14}H_6O_2(OH)_2$	240.21	yel. lf.		280	subl.	i.	sl. s.	s.
Antipyrene	1-ph-2,3-diMepyrazolone-5	$C_{11}H_{12}ON_2$	188.23	mn./aq.	$1.088^{113/4}$	113(109)	319^{174}	100^{25}	100	sl. s.
Apiole	1-allyl-2, 5-diMeO-3,4 methylenedioxybenzene	$C_{12}H_{14}O_4$	222.24	col. nd.	$1.02^{20/4}$	30	294	i	s.	s.
Arabinose (α)(*d*- or *l*-)		$CH_2OH(CHOH)_3CHO$	150.13	rhb. pr.	$1.585^{20/4}$	159.5		46^{0}	$0.5^{9°}$	i.
(*dl*-)		$CH_2OH(CHOH)_3CHO$	150.13			164.5		16.9^{10}		
Arachidic acid	eicosanoic acid	$CH_3(CH_2)_{18}CO_2H$	312.53	col. lf.		77	328	i.	s. h.	v. s.
Arsanilic acid (*p*-)		$H_2N{\cdot}C_6H_4{\cdot}AsO_3H_2$	217.05	nd./aq.		232		v. s. h.	v. s. h.	i.
Asparagine (*l*-)		$HO_2C{\cdot}C_2H_3(NH_2){\cdot}CONH_2$	132.12	rhb.	$1.543^{15/4}$	227–35	d. 235	3.1^{28}	i. c.	
Aspirin (*o*-)		$CH_3CO_2{\cdot}C_6H_4{\cdot}CO_2H$	180.16	nd./aq.		135–6		1^{37}	s.	5^{20}
Atropic acid	α-phenyl acrylic acid	$C_6H_5C({:}CH_2){\cdot}CO_2H$	148.16	nd./aq.		106–7	267 d.	0.1 c.	s.	s.
Auramine	4,4′-dimethylaminobenzophenomide	$[(CH_3)_2NC_6H_4]_2C{:}NH$	267.37	col./al.		136		i.	7^{20}	2.3^{20}
Aurine, coralline (4-,4′-)		$(HOC_6H_4)_2C{:}C_6H_4{:}O$	290.31	red		310 d.		i.	s.	s.
Azo-anisole (2-,2′-)	diMeO-azobenzene	$(CH_3O{\cdot}C_6H_4N{:})_2$	242.27	or. pr.		153		i.	s.	s.
benzene	diphenyldiimide	$C_6H_5N{:}N{\cdot}C_6H_5$	182.22	or. mn.	$1.203^{20/4}$	68	297	i.	4.2^{20}	
Azoxybenzene		$(C_6H_5)_2N_2O$	198.22	yel. rhb.	$1.248^{20/20}$	36	d.	i.	11.4^{15}	
Barbituric acid	malonyl urea	$CO{:}(NHCO)_2{:}CH_2{\cdot}2H_2O$	164.12	col./aq.		d. 245		s. h.	sl. s.	s.
Benzal acetone	Me-cinnamyl ketone	$C_6H_5CH{:}CHCOCH_3$	146.19	pl.	$1.035^{20/20}$	41–2	260–2	i.	s.	s.
Benzaldehyde	artificial almond oil	C_6H_5CHO	106.12	col. lq.	$1.046^{20/4}$	−26	179	0.3	∞	∞
Benzamide		$C_6H_5CONH_2$	121.14	col. pr.	1.341	130	290	1.35^{25}	17^{25}	sl. s.
Benzanilide		$C_6H_5CONHC_6H_5$	197.23	lf./al.	1.31^{4}	163	$117–9^{10}$	i.	4^{30}	sl. s.
Benzene	benzol, phenyl hydride, cyclohexatriene	C_6H_6	78.11	col. lq.	$0.879^{20/4}$	5.5	80.1	0.07^{22}	s.	∞
sulfinic acid		$C_6H_5SO_2H$	142.18	pr./aq.		83–4	d. > 100	v. s. h.	v. s.	v. s.
sulfonic acid		$C_6H_5SO_3H$	158.18	col. nd.		65–6	d.	v. s.	v. s.	i.
sulfonic amide	benzene sulfonamide	$C_6H_5SO_2NH_2$	157.19	mn./aq.		156		0.43^{16}	v. s.	v. s.
sulfonic chloride	benzene sulfonyl chloride	$C_6H_5SO_2Cl$	176.62	cr.	$1.384^{15/15}$	14.5	251.5	i.	v. s.	s.
Benzidine (4-,4′-)		$NH_2{\cdot}C_6H_4{\cdot}C_6H_4{\cdot}NH_2$	184.24	cr./aq.		128–9	400^{740}	1 h.	1 h.	2
disulfonic acid (2-,2′-)		$({\cdot}C_6H_3(NH_2)SO_3H)_2{\cdot}3H_2O$	398.41	pr./aq.		d. >175		0.09^{25}	i.	i.
(3-,3′-)		$({\cdot}C_6H_3(NH_2)SO_3H)_2$	344.36					v. sl. s.		
Benzil	dibenzoyl	$C_6H_5CO{\cdot}COC_6H_5$	210.23	pr.	1.23^{15}	95	348 d.	i.	v. s.	v. s.
Benzoic acid		$C_6H_5CO_2H$	122.12	mn. pr.	$1.266^{15/4}$	121.7	249.2	0.2^{17}	46^{15}	66^{15}
anhydride		$(C_6H_5CO)_2O$	226.23	rhb./et.	$1.199^{15/4}$	42	360	i.	s.	s.
nitrile	phenyl cyanide	C_6H_5CN	103.12	col. lq.	$1.001^{25/6}$	−12.9	190.7	1^{100}	∞	∞

Benzoin (*dl*-)		$C_6H_5CO{\cdot}CHOHC_6H_5$	212.24	mn.		133–7	344^{768}	v. sl. s.	s. h.	sl. s.
Benzophenone	diphenyl ketone	$C_6H_5COC_6H_5$	182.22	col. rhb.	1.083^{54}	48.5	305.4	i.	6.5^{15}	15^{13}
Benzotrichloride	phenyl chloroform	$C_6H_5CCl_3$	195.47	col. lq.	1.380^{14}	−4.75	220.7	i.	s.	s.
Benzoyl-benzoic acid (*o*-)		$C_6H_5COC_6H_4CO_2H{\cdot}H_2O$	244.24	tri./aq.		93(128)		sl. s.		
-chloride		C_6H_5COCl	140.57	col. lq.	$1.212^{20/4}$	−0.5	197.2	d.	d. h.	∞
-peroxide		$(C_6H_5CO)_2O_2$	242.23	rhb./et.		108 d.	expl.	i.	s. h.	s.
Benzyl acetate		$CH_3CO_2CH_2C_6H_5$	150.17	col. lq.	1.057^{17}	−51.5	213.5	i.	∞	∞
alcohol	phenyl carbinol	$C_6H_5CH_2OH$	108.14	col. lq.	$1.043^{20/4}$	−15.3	204.7	4^{17}	∞	∞
amine	ω-amino toluene	$C_6H_5CH_2NH_2$	107.15	lq.	$0.982^{20/4}$		184.5	∞	∞	∞
aniline	phenyl-benzylamine	$C_6H_5CH_2NHC_6H_5$	183.25	mn. pr.	$1.065^{25/25}$	37–8	306^{750}	i.		s.
benzoate		$C_6H_5CO_2CH_2C_6H_5$	212.24	nd.	$1.12^{20/4}$	21	323–4	i.	∞	∞
butyrate		$C_2H_5CH_2CO_2CH_2C_6H_5$	178.23	col. lq.	$1.016^{16/18}$	238–40	i.	v. s.	v. s.	
chloride	ω-chlorotoluene	$C_6H_5CH_2Cl$	126.58	col. lq.	$1.100^{20/20}$	−39	179.4	i.	∞	∞
ether	dibenzyl ether	$(C_6H_5CH_2)_2O$	198.26	lq.	1.036^{16}		295–8	i.	s. h.	s.
formate		$HCO_2CH_2C_6H_5$	136.15	col. lq.	1.081^{23}	3.6	$202–3^{747}$	i.	s.	∞
propionate		$C_2H_5CO_2CH_2C_6H_5$	164.20	lq.	$1.036^{16/17}$		220–2	i.		
Berberonic acid (2-,4-,5-)		$C_5H_2N(CO_2H)_3{\cdot}2H_2O$	247.16	tri.		243		v. sl. s.	sl. s. h.	i.
Biuret	allophanamide	$NH(CONH_2)_2$	103.08	nd./al.		192–3 d.		1.3^{0}	s.	
Borneol (*dl*-)		$C_{10}H_{17}OH$	154.25	col. cr.	$1.011^{20/4}$	210.5	subl.	v. sl. s.		
(*d*- or *l*-)		$C_{10}H_{17}OH$	154.25	col. cr.	$1.011^{20/4}$	208–9	212–3	v. sl. s.	v. s.	v. s.
(iso-)		$C_{10}H_{17}OH$	154.25	col. cr.		212		i.		
Bornyl acetate (*d*-)		$CH_3CO_2C_{10}H_{17}$	196.29	rhb./pet.	0.991^{15}	29	226–7	i.	s.	s.
Bromo-aniline (*p*-)		$BrC_6H_4NH_2$	172.02	rhb.	1.8^{20}	63–4		i. c.	v. s.	v. s.
-benzene	phenyl bromide	C_6H_5Br	157.01	col. lq.	$1.495^{20/4}$	−30.6	156.2	i.	s.	∞
-camphor (3-)(*d*-)	α-bromocamphor	$BrC_{10}H_{15}O$	231.13	cr.	$1.449^{20/4}$	77–8	274	i.	20^{26}	v. s.
-diphenyl (*p*-)		$BrC_6H_4{\cdot}C_6H_5$	233.10	cr./al.		90–1	310	i.	s.	34^{25}
-naphthalene (α-)	α-naphthyl bromide	$C_{10}H_7Br$	207.07	col. oil	$1.482^{20/4}$	5–6	281.1	i.	s.	∞
(β-)	β-naphthyl bromide	$C_{10}H_7Br$	207.07	lf./al.	1.605^{0}	59	281–2	i.	6^{20}	v. s.
-phenol (*o*-)		BrC_6H_4OH	173.01	col. lq.	1.553^{80}	5.6	194–5	s.	s.	∞
(*m*-)		BrC_6H_4OH	173.01	cr.		32–3	236–7		s.	s.
(*p*-)		BrC_6H_4OH	173.01	tet. cr.	1.588^{80}	63.5	238	1.4^{15}	v. s.	v. s.
-styrene (ω)(1)		$C_6H_5CH{:}CHBr$	183.05	lq.	$1.422^{20/4}$	7	221	i.	∞	∞
(2)		$C_6H_5CH{:}CHBr$	183.05	lq.	$1.427^{20/4}$	−7.5	108^{26}	i.	∞	∞
-toluene (*o*-)	*o*-tolyl bromide	$CH_3{\cdot}C_6H_4Br$	171.03	col. lq.	$1.422^{20/4}$	−28	181.8	i.	s.	$∞^{25}$
(*m*-)		$CH_3{\cdot}C_6H_4Br$	171.03	col. lq.	$1.410^{20/4}$	−39.8	183.7	i.	s.	s.
(*p*-)		$CH_3{\cdot}C_6H_4Br$	171.03	cr./al.	$1.390^{20/4}$	28.5	184–5	i.	s.	$∞^{25}$
Bromoform	tribromo-methane	$CHBr_3$	252.73	col. lq.	$2.890^{20/4}$	8–9	150.5	0.1 c.	∞	∞
Butadiene (1-,2-)	methyl-allene	$CH_3CH{:}C{:}CH_2$	54.09	lq.			18–9	i.	∞	∞
(1-,3-)	erythrene	$CH_2{:}CHCH{:}CH_2$	54.09	col. gas	$0.621^{20/4}$	−108.9	−4.41	i.	∞	∞
Butadienyl acetylene		$CH_2{:}(CH)_2{:}CH{\cdot}C{\vdots}CH$	78.11	col. lq.	$0.773^{20/4}$		83–6	i.		
Butane	diethyl	$CH_3CH_2CH_2CH_3$	58.12	col. gas	0.60^{0}	−135	−0.6	i.	s.	s.
(*i*-)	trimethyl-methane	$(CH_3)_2CHCH_3$	58.12	col. gas	0.60^{0}	−145	−10	i.	s.	s.
Butyl acetate (*n*-)		$CH_3CO_2(CH_2)_2C_2H_5$	116.16	col. lq.	0.882^{20}	−76.3	125^{740}	0.7	∞	∞
(*s*-)		$CH_3CO_2CH(CH_3)C_2H_5$	116.16	col. lq.	$0.865^{25/4}$		112^{744}	i.	∞	∞
(*i*-)		$CH_3CO_2CH_2CH(CH_3)_2$	116.16	col. lq.	$0.871^{20/4}$	−98.9	118	0.6^{25}	∞	∞
(*tert*-)		$CH_3CO_2C(CH_3)_3$	116.16	col. lq.	$0.866^{20/4}$		$95–6^{760}$	i.	∞	∞
alcohol (*n*-)	butanol-1	$C_2H_5CH_2CH_2OH$	74.12	col. lq.	$0.810^{20/4}$	−79.9	117	9^{15}	∞	∞
(*s*-)	butanol-2	$C_2H_5CH(OH)CH_3$	74.12	col. lq.	$0.808^{20/4}$	−114.7	99.5	12.5^{20}	∞	∞
(*i*-)	2-methyl-propanol-1	$(CH_3)_2CHCH_2OH$	74.12	col. lq.	$0.805^{17.5}$	−108	107–8	10^{15}	∞	∞
(*tert*-)	2-methyl-propanol-2	$(CH_3)_3COH$	74.12	lq.	0.779^{26}	25.5	82.9	∞	∞	∞
amine (*n*-)		$C_2H_5CH_2CH_2NH_2$	73.14	col. lq.	$0.739^{25/4}$	−50	77.8	∞	∞	∞
(*s*-)		$C_2H_5CH(NH_2)CH_3$	73.14	col. lq.	$0.724^{20/4}$	−104	66^{772}	∞	∞	∞
(*i*-)		$(CH_3)_2CHCH_2NH_2$	73.14	col. lq.	$0.732^{20/20}$	−85	68–9	∞	∞	∞
(*t*-)		$(CH_3)_3CNH_2$	73.14	col. lq.	$0.698^{18/4}$	−67.5	45.2		∞	
p-aminophenol (*N*)(*n*)		$C_4H_9NH{\cdot}C_6H_4{\cdot}OH$	165.23			71		i.		
(*N*)(*i*-)		$C_4H_9NH{\cdot}C_6H_4{\cdot}OH$	165.23			79		i.		
aniline (*n*-)		$C_4H_9NHC_6H_5$	149.23	lq.			235^{720}	i.	v. s.	v. s.
(*i*-)		$C_4H_9NHC_6H_5$	149.23	oil	$0.940^{20/4}$		231–2	0.01^{15}	v. s.	v. s.
arsonic acid (*n*-)		$C_4H_9AsO(OH)_2$	182.05	col. lf.		158–9		s.	s.	i.
benzoate (*n*-)		$C_6H_5CO_2C_4H_9$	178.23	col. oil	$1.005^{25/25}$	−22	249–50	i.	s.	s.
(*i*-)		$C_6H_5CO_2C_4H_9$	178.23	col. oil	$0.997^{25/25}$		241.5	i.	∞	∞
bromide (*n*-)	1-bromo-butane	$C_2H_5CH_2CH_2Br$	137.02	lq.	$1.277^{20/4}$	−112.4	101.6	0.06^{16}	∞	∞
(*s*-)	2-bromo-butane	$C_2H_5CH(Br)CH_3$	137.02	lq.	$1.251^{25/4}$	−112	91.3	i.		
(*i*-)	1-Br-2-Me-propane	$(CH_3)_2CHCH_2Br$	137.02	lq.	$1.258^{25/4}$	−118.5	91.5	0.06^{18}	∞	∞
(*t*-)	2-Br-2-Me-propane	$(CH_3)_3CBr$	137.02	lq.	$1.211^{20/4}$	−16.2	73.3	i.	∞	∞
butyrate (*n*-)(*n*-)		$C_2H_5CH_2CO_2CH_2CH_2C_2H_5$	144.21	col. lq.	$0.872^{20/20}$		165.7^{736}	i.	∞	∞
(*n*-)(*i*-)		$C_2H_5CH_2CO_2CH_2CH(CH_3)_2$	144.21	col. lq.	$0.863^{18/4}$		156.9	i.	∞	∞
(*i*-)(*i*-)		$(CH_3)_2CHCO_2CH_2CH(CH_3)_2$	144.21	col. lq.	$0.875^{0/4}$	−80.7	148–9	i.	∞	∞
caproate		$CH_3(CH_2)_4CO_2C_4H_9$	172.26	col. lq.	$0.882^{0/0}$		204.3	i.		
carbamate (*i*-)		$NH_2CO_2CH_2CH(CH_3)_2$	117.15	col. lf.	$0.956^{76/4}$	65	206–7	i.	s.	s.
cellosolve (*n*-)	2-BuO-ethanol-1	$C_4H_9OCH_2CH_2OH$	118.17	col. lq.	$0.903^{20/4}$		171.2	∞	∞	∞

(*Continued*)

TABLE 2-2 Physical Properties of Organic Compounds *(Continued)*

Name	Synonym	Formula	Formula weight	Form and color	Specific gravity	Melting point, °C	Boiling point, °C	Solubility in 100 parts		
								Water	Alcohol	Ether
chloride (*n*-)	1-chloro-butane	$C_2H_5CH_2CH_2Cl$	92.57	col. lq.	0.887^{20}	−123.1	77.9^{763}	0.07^{18}	∞	∞
(*s*-)	2-chloro-butane	$C_2H_5\cdot CHCl\cdot CH_3$	92.57	col. lq.	$0.871^{20/4}$	−131	67.8^{767}	i.	∞	∞
(*i*-)	1-Cl_2-2-Me-propane	$(CH_3)_2CHCH_2Cl$	92.57	col. lq.	0.884^{15}	−131.2	68.9	i.	∞	∞
(*t*-)	2-Cl_2-2-Me-propane	$(CH_3)_3CCl$	92.57	col. lq.	0.847^{15}	−26.5	51–2	i.	∞	∞
dimethylbenzene (*t*-)(1-,3-,5-)		$(CH_3)_3C\cdot C_6H_3{:}(CH_3)_2$	162.27	col. lq.			$200–2^{147}$	i.		
formate (*n*-)		$HCO_2CH_2CH_2C_2H_5$	102.13	lq.	0.911^{0}		106.9	v. sl. s.	∞	∞
(*s*-)		$HCO_2CH(CH_3)C_2H_5$	102.13	lq.	$0.882^{20/4}$		97	sl. s.	∞	∞
(*i*-)		$HCO_2CH_2CH(CH_3)_2$	102.13	lq.	$0.885^{20/4}$	−95.3	98.2	1.1^{22}	∞	∞
furoate (*n*-)		$OC_4H_3CO_2C_4H_9$	168.19	col. lq.	$1.056^{20/4}$		$118–20^{25}$	i.	∞	∞
iodide (*n*-)	1-iodo-butane	$C_2H_5CH_2CH_2I$	184.02	lq.	$1.617^{20/4}$	−103.5	129.9	i.	∞	∞
(*s*-)	2-iodo-butane	$C_2H_5CHICH_3$	184.02	lq.	1.595^{20}	−104	118–9	i.	∞	∞
(*i*-)	1-iodo-2-Me-propane	$(CH_3)_2CHCH_2I$	184.02	lq.	$1.606^{20/4}$	−90.7	120	i.	∞	∞
(*t*-)	2-iodo-2-Me-propane	$(CH_3)_3CI$	184.02	lq.	$1.370^{19/15}$	−34	99	i.	∞	∞
lactate (*n*-)		$CH_3CH(OH)CO_2C_4H_9$	146.18	col. lq.	0.968		$75–6^{6}$	sl. s.	∞	∞
mercaptan (*n*-)	butanthiol-1	$C_2H_5CH_2CH_2SH$	90.19	col. lq.	$0.837^{25/4}$	−116	97–8	sl. s.	v. s.	v. s.
(*i*-)	2-Me-propanthiol-1	$(CH_3)_2CHCH_2SH$	90.19	lq.	$0.836^{20/4}$	<−79	88	v. sl. s.	s.	s.
(*t*-)		$(CH_3)_3CSH$	90.19	lq.			65–7			
methacrylate (*n*-)		$CH_2{:}C(CH_3)CO_2C_4H_9$	142.20	lq.	$0.889^{15.6}$		155	i.		
(*i*-)		$CH_2{:}C(CH_3)CO_2C_4H_9$	142.20	lq.	$0.889^{15.6}$		155	i.		
phenol (*p*-)(*t*-)		$(CH_3)_3C\cdot C_6H_4\cdot OH$	150.22	nd./aq.	$0.908^{112/4}$	99	236–8	sl. s.	s.	s.
propionate (*n*-)		$C_2H_5CO_2C_4H_9$	130.18	col. lq.	0.883^{15}	−89.55	146	i.	∞	∞
(*s*-)		$C_2H_5CO_2C_4H_9$	130.18	col. lq.	$0.866^{20/4}$		132.5	i.	∞	∞
(*i*-)		$C_2H_5CO_2C_4H_9$	130.18	col. lq.	$0.888^{0/4}$	−71.4	136.8	i.	∞	∞
stearate (*n*-)		$CH_3(CH_2)_{16}CO_2C_4H_9$	340.58	col. lq.	$0.855^{25/25}$	27.5	$220–5^{25}$	0.3^{25}	s.	s.
(*i*-)		$CH_3(CH_2)_{16}CO_2C_4H_9$	340.58	wax		25		i.		
iso-thiocyanate (*n*-)	butyl mustard oil	$C_2H_5CH_2CH_2\cdot N{:}CS$	115.20	lq.	0.956^{11}		165^{724}	i.	s.	s.
(*i*-)	iso-Bu mustard oil	$(CH_3)_2CHCH_2\cdot N{:}CS$	115.20	lq.	$0.964^{14/4}$		162	i.	s.	s.
(*s*-)(*d*-)		$C_4H_9\cdot N{:}CS$	115.20	lq.	$0.943^{20/4}$		159–63	i.	s.	s.
(*t*-)		$(CH_3)_3C\cdot N{:}CS$	115.20	lq.	0.919^{10}	10.5	140^{770}	i.	s.	s.
valerate (*n*-)(*n*-)		$CH_3(CH_2)_3CO_2(CH_2)_3CH_3$	158.24	lq.	$0.870^{15/4}$	−93	186	v. sl. s.	∞	∞
(*i*-)(*n*-)		$(CH_3)_2CHCH_2CO_2(CH_2)_3CH_3$	158.24	lq.	$0.862^{25/4}$		168.8	i.	∞	∞
(*i*-)(*s*-)		$(CH_3)_2CHCH_2CO_2C_4H_9$	158.24	col. lq.	$0.848^{20/4}$		$163–4^{752}$	i.	∞	∞
(*i*-)(*i*-)		$C_4H_9CO_2C_4H_9$	158.24	col. lq.	$0.874^{0/4}$		168.7	i.	∞	∞
Butylene (α-)	butene-1	$C_2H_5CH{:}CH_2$	56.11	col. gas	0.6^{9}	−130	-5^{758}	i.	v. s.	v. s.
(β-)	butene-2	$CH_3CH{:}CHCH_3$	56.11	col. gas		−127	3^{746}			
Butyraldehyde (*n*-)		$CH_3CH_2CH_2CHO$	72.11	col. lq.	$0.817^{20/4}$	−99	75.7	4	∞	∞
(*i*-)	2-Me-propanol	$(CH_3)_2CHCHO$	72.11	col. lq.	$0.794^{20/4}$	−65.9	64^{757}	11^{20}	∞	∞
Butyric acid (*n*-)	butanoic acid	$C_2H_5CH_2CO_2H$	88.11	col. lq.	$0.964^{20/4}$	−4.7	163.5^{757}	∞	∞	∞
(*i*-)	2-Me-propanoic acid	$(CH_3)_2CHCO_2H$	88.11	col. lq.	$0.949^{20/4}$	−47	154.5	20^{20}	∞	∞
amide (*n*-)	*n*-butyramide	$C_2H_5CH_2CONH_2$	87.12	rhb.	1.032	115–6	216	16.3^{15}	s.	sl. s.
(*i*-)	iso-butyramide	$(CH_3)_2CHCONH_2$	87.12	mn. pl.	1.013	129–30	216–20	v. s.	s.	sl. s.
anhydride (*n*-)		$(C_2H_5CH_2CO)_2O$	158.19	col. lq.	$0.968^{20/20}$	−75	199.5	d.	d	∞
(*i*-)		$[(CH_3)_2CHCO]_2O$	158.19	col. lq.	$0.950^{25/4}$	−53.5	181.5^{734}	d.	d	∞
anilide (*n*-)	*n*-butyranilide	$C_3H_7CONHC_6H_5$	163.22	mn. pr.	1.134	92	189^{15}	i.	s.	s.
Caffeic acid (3-,4-)		$(HO)_2C_6H_3C_2H_2CO_2H$	180.16	yel./aq.		195–213	d.	s. h.	s.	sl. s.
Caffeine		$C_8H_{10}O_2N_4\cdot H_2O$	212.21	nd./al.	1.23^{19}	237	subl.	2	2	0.3
Camphene (*dl*-)		$C_{10}H_{16}$	136.23	cr.	0.822^{78}	50	160	i.	s.	s.
(*d*- or *l*-)		$C_{10}H_{16}$	136.23	cr.	$0.845^{50/4}$	42.7	159.6	i.	s.	s.
Camphor (*d*-)		$C_{10}H_{16}O$	152.23	trig.	$0.999^{9/9}$	178–9	209.1^{759}	0.1	120^{12}	v. s.
Camphoric acid (*d*-)		$C_8H_{14}(CO_2H)_2$	200.23	mn.	1.186	187		0.6^{12}	s.	
Cantharidine		$C_{10}H_{12}O_4$	196.20	cr.		212		0.003		
Capric acid	decanoic acid	$CH_3(CH_2)_8CO_2H$	172.26	col. nd.	0.889^{87}	31.5	268–70	0.003	s.	s.
Caproic acid (*n*-)	hexanoic acid	$CH_3(CH_2)_4CO_2H$	116.16	oily lq.	$0.922^{20/4}$	−1.5	202^{761}	1.1^{20}	s.	s.
(*i*-)	2-Me-pentanoic-5 acid	$(CH_3)_2CH(CH_2)_2\cdot CO_2H$	116.16	col. oil	$0.925^{20/4}$	−35	207.7	v. sl. s.	s.	s.
Caprylic acid (*n*-)	octanoic acid	$CH_3(CH_2)_6CO_2H$	144.21	col. lf.	$0.910^{20/4}$	16	237.5	0.07^{15}	s.	s.
Carbazole	diphenylenelimine, dibenzopyrrole	$(C_6H_4)_2NH$	167.21	lf.		244.8	354.8	i.	0.92^{14}	sl. s.
Carbitol	diethylene glycol mono-Et ether	$C_2H_5O(CH_2)_2O(CH_2)_2OH$	134.17	col. lq.	$0.990^{20/20}$		201.9	∞	v. s.	s.
Carbon disulfide		CS_2	76.14	col. lq.	$1.263^{20/4}$	−108.6	46.3	0.2^{0}	∞	∞
monoxide		CO	28.01	col. gas	$0.81^{-195/4}$	−207	−192	3.5^{0} cc.	s.	
suboxide		$OC{:}C{:}CO$	68.03	gas	1.114^{0}	−107	7^{761}	d.		s.
tetrabromide	tetrabromomethane	CBr_4	331.63	col. mn.	3.42	90.1(48)	189.5	0.02^{30}	s.	s.
tetrachloride	tetrachloromethane	CCl_4	153.82	col. lq.	$1.595^{20/4}$	−22.6	76.8	0.08^{20}	∞	∞
tetrafluoride	tetrafluoromethane	CF_4	88.00	gas			−128	sl. s.		
Carbonyl sulfide		COS	60.08	col. gas	1.24^{-87}	−138.2	-50.2^{760}	80^{14} cc.	s.	s.
Carminic acid		$C_{22}H_{20}O_{13}$	492.39	red pd.		d. 136		s.	s.	v. sl. s.
Carvacrol (1-,2-,4-)		$CH_3C_6H_3(OH)CH(CH_3)_2$	150.22	col. lq.	$0.977^{20/4}$	0.5	238	v. sl. s.	∞	∞

Carvacrylamine (2-,1-,4-)	$H_2NC_6H_3(CH_3)C_3H_7$	149.23	oil	0.994^{20}	−16	241	v. sl. s.	s.	s.
Carvone (*d*-)	$C_{10}H_{14}O$	150.22	col. lq.	$0.961^{20/4}$		230^{766}	i.	∞	∞
Cellosolve	$C_2H_5O(CH_2)_2OH$	90.12	col. lq.	$0.931^{20/4}$	−70	135.1	∞	∞	∞
acetate	$CH_3CO_2CH_2CH_2OC_2H_5$	132.16	col. lq.	$0.975^{20/4}$		156.3	22	∞	∞
Cellulose	$(C_6H_{10}O_5)x$	162.14	amor.	1.3–1.4		i.	i.	i.	
Cetyl acetate	$CH_3CO_2(CH_2)_{15}CH_3$	284.48	nd.	0.858^{20}	22–3	200^{15}	i.	v. sl. s. c.	
alcohol	$CH_3(CH_2)_{14}CH_2OH$	242.44	lf.	$0.818^{50/4}$	49–50	189.5^{15}	i.	s.	s.
Chloral	$CCl_3{\cdot}CHO$	147.39	col. lq.	$1.505^{25/4}$	−57	97.6^{768}	v. s.	∞	∞
hydrate	$CCl_3{\cdot}CH(OH)_2$	165.40	mn. pr.	$1.619^{50/4}$	51.7	d. 98	474^{17}	v. s.	s.
Chloranil	$OC{:}(CCl{\cdot}CCl)_2{:}CO$	245.88	yel./bz.		290	subl.	i.	i. c.	i. c.
Chloretone	$Cl_3C{\cdot}C(OH)(CH_3)_2$	177.46	col. cr.		97	167	0.8 c.	111	s.
Chloro-acetanilide (*p*-)	$CH_3CONHC_6H_4Cl$	169.61	rhb.	1.385^{22}	175–6		sl. s.	s.	v. s.
-acetic acid	$ClCH_2CO_2H$	94.50	col. cr.	$1.58^{20/20}$	61.2	189.5	v. s.	s.	s.
-acetone	CH_3COCH_2Cl	92.52	col. lq.	1.162^{16}	−44.5	121	∞	∞	∞
-acetophenone (ω-)	$C_6H_5COCH_2Cl$	154.59	rhb.	1.324^{15}	58–9	245–7	0.11	v. s.	v. s.
-acetyl chloride	$ClCH_2COCl$	112.94	col. lq.	$1.498^{20/20}$		105	d.	d.	
-aniline (*o*-)	$ClC_6H_4NH_2$	127.57	lq.	$1.213^{20/4}$	0	210.5	i.		s.
(*m*-)	$ClC_6H_4NH_2$	127.57	lq.	$1.216^{20/4}$	−10.4	230^{767}	i.		s.
(*p*-)	$ClC_6H_4NH_2$	127.57	rhb.	1.427^{19}	70–1	230–1	s. h.	s.	s.
-anthraquinone (1-)	$C_6H_4(CO)_2C_6H_3Cl$	242.66	yel. nd.		162	subl.	i.	sl. s. h.	
(2-)	$C_6H_4(CO)_2C_6H_3Cl$	242.66	nd./al.		208–9		i.		
-benzaldehyde (*o*-)	ClC_6H_4CHO	140.57	nd.	1.29^{8}	11	208^{748}	v. sl. s.	v. s.	v. s.
(*m*-)	ClC_6H_4CHO	140.57	pr.	1.250^{15}	17–8	213–4	v. sl. s.	v. s.	v. s.
(*p*-)	ClC_6H_4CHO	140.57	pr.	1.196^{61}	47.8	213^{748}	s. h.	v. s.	v. s.
-benzene	C_6H_5Cl	112.56	col. lq.	$1.107^{20/4}$	−45.2	132.1	0.049^{20}	∞	∞
-benzoic acid (*o*-)	$ClC_6H_4CO_2H$	156.57	mn./aq.	$1.544^{25/4}$	141–2		0.208^{25}	s.	s.
(*m*-)	$ClC_6H_4CO_2H$	156.57	pr.	$1.496^{25/4}$	158		0.041^{25}	s.	s.
(*p*-)	$ClC_6H_4CO_2H$	156.57	tri.	1.541^{24}	242–3	subl.	0.008^{25}	s.	s.
-buta-1,3-diene (2-)	$CH_2{:}CCl{\cdot}CH{:}CH_2$	88.54	col. lq.	$0.958^{20/20}$		59.4	v. sl. s.	∞	∞
(1-)	$CH_2{:}CH{\cdot}CH{:}CHCl$	88.54	col. lq.	$0.965^{20/20}$		69	v. sl. s.	∞	∞
-buta-1,2-diene (4-)	$CH_2{:}C{:}CH{\cdot}CH_2Cl$	88.54	col. lq.	$0.991^{20/20}$		88	d.		
-dimethylhydantoin	$—C(CH_3)_2N(Cl)CON(Cl)CO—$	197.02		$1.5^{20/20}$	130		0.21^{25}		
-dinitrobenzene (α)(1-,2-)(4-)	$ClC_6H_3(NO_2)_2$	202.55	cr./et.		39(36)	315 d.	i.	v. s. h.	v. s.
(α)(1-,3-)(4-)	$ClC_6H_3(NO_2)_2$	202.55	rhb./et.	1.697^{22}	53(43)	315 d.	i.	s. h.	s.
-diphenyl (*o*-)	$C_6H_5{\cdot}C_6H_4Cl$	188.65	cr.		34	267–8	i.		
(*m*-)	$C_6H_5{\cdot}C_6H_4Cl$	188.65	cr.		89	284–5	i.		
(*p*-)	$C_6H_5{\cdot}C_6H_4Cl$	188.65	lf.		77.5	282	i.		
-hydroquinone	$ClC_6H_3(OH)_2$	144.56	mn.		106	263 sl. d.	v. s.	v. s.	v. s.
-naphthalene (α-)	$C_{10}H_7Cl$	162.62	col. lq.	$1.194^{20/4}$	−20	259.3	i.	s.	∞
(β-)	$C_{10}H_7Cl$	162.62	lf./al.	1.266^{16}	56–7	264^{751}	i.	v. s.	v. s.
-nitrobenzene (*o*-)	$ClC_6H_4NO_2$	157.55	mn. nd.	$1.305^{80/4}$	32.5	245.5^{753}	i.	s. h.	s.
(*m*-)	$ClC_6H_4NO_2$	157.55	yel./al.	$1.343^{50/4}$	44.4(24)	235.6	i.	v. s. h.	v. s.
(*p*-)	$ClC_6H_4NO_2$	157.55	mn. pr.	1.298^{91}	83–4	242^{761}	i.	v. s. h.	v. s.
-nitrotoluene (2-,4-)	$CH_3C_6H_3(NO_2)(Cl)$	171.58	cr.	1.256^{80}	38.2	240^{718}	i.		
(2-,6-)	$CH_3C_6H_3(NO_2)(Cl)$	171.58	cr.		37.5	238	i.		
-phenol (*o*-)	ClC_6H_4OH	128.56	col. lq.	$1.241^{18/15}$	7(0)	175–6	2.85^{20}	s.	s.
(*m*-)	ClC_6H_4OH	128.56	nd.	1.268^{25}	32–3	214	2.60^{20}	s.	s.
(*p*-)	ClC_6H_4OH	128.56	nd.	$1.306^{20/4}$	41–3	217	2.71^{20}	v. s.	v. s.
-propionic acid (α)(dl-)	$CH_3{\cdot}CHCl{\cdot}CO_2H$	108.52	col. lq.	1.306^{9}	<−20	186	∞	∞	∞
-toluene (*o*-)	$CH_3{\cdot}C_6H_4Cl$	126.58	col. lq.	$1.082^{20/4}$	−34	159.5	i.	s.	∞
(*m*-)	$CH_3{\cdot}C_6H_4Cl$	126.58	col. lq.	$1.072^{20/4}$	−47.8	161.6	i.	s.	∞
(*p*-)	$CH_3{\cdot}C_6H_4Cl$	126.58	col. lq.	$1.070^{20/4}$	7.5	162.2	i.	s.	∞
Chloroform	$CHCl_3$	119.38	col. lq.	1.489^{20}	−63.5	61.2	0.82^{20}	∞	∞
Chlorophyll (α-)	$C_{55}H_{72}O_5N_4Mg$	893.49			d.	i.	s.		
Chloropicrin	Cl_3CNO_2	164.38	lq.	$1.651^{23/4}$	−64	112.3^{766}	0.17^{18}	s.	s.
Cholesterol	$C_{27}H_{45}OH{\cdot}H_2O$	404.67	rhb./al.	1.067	149–51	subl.	0.26^{20}	1.1^{17}	18
Chrysene	$C_{18}H_{12}$	228.29	col. rhb.		253–4	448	i.	0.1^{16}	v. sl. s.
Chrysoidine (2-,4-)	$C_6H_5{\cdot}N{:}N{\cdot}C_6H_3(NH_2)_2$	212.25	yel. cr.		117.5		sl. s. h.	s.	s.
Chrysophanic acid	$C_{14}H_5(OH)_2(CH_3)O_2$	254.24	yel./al.		195	subl.	i. c.	s. h.	sl. s.
Cinchomeronic acid (3-,4-)	$C_5H_3N(CO_2H)_2$	167.12	cr./HCl		258–9 d.	subl. d.	v. sl. s.	sl. s.	i.
Cineole, eucalyptole	$C_{10}H_{18}O$	154.25	col. oil	0.927^{20}	1.5	176–7	1.9^{15}	∞	∞
Cinnamic acid (*cis*-)	$C_6H_5CH{:}CHCO_2H$	148.16	mn. pr.	1.284^{4}	68	125^{19}			
(*trans*-)	$C_6H_5CH{:}CHCO_2H$	148.16	mn. pr.	1.245	133	300	0.04^{18}	24^{20}	v. s.
aldehyde	$C_6H_5CH{:}CHCHO$	132.16	lq.	$1.110^{20/20}$	−7.5	252 sl. d.	v. sl. s.	s.	∞
Cinnamyl alcohol	$C_6H_5CH{:}CHCH_2OH$	134.18	nd.	$1.040^{35/35}$	33	257.5	sl. s.	v. s.	v. s.
cinnamate	$C_8H_7CO_2C_9H_9$	264.32	nd. or pr.	$1.085^{16.5}$	44		i.	4 c.	33
Citraconic acid (*cis*-)	$CH_3C(CO_2H){:}CHCO_2H$	130.10	nd.	1.617	92–3		360^{25}	s.	s.
Citral (α)	$C_9H_{15}CHO$	152.23	col. oil	$0.890^{17/4}$		229	i.	∞	∞
Citric acid	$C_3H_4(OH)(CO_2H)_3$	192.12	cr.	$1.542^{20/4}$	153	d.	207.7^{25}	76^{15}	2^{15}
Citronellal (*d*-)	$C_9H_{17}{\cdot}CHO$	154.25	col. oil	$0.855^{17.5}$		204–8	v. sl. s.	∞	∞
Citronellol (*d*-)	$C_{10}H_{20}O$	156.27	col. oil	$0.848^{20/4}$		224–5	v. sl. s.	∞	∞
Coniine (*d*-)(2-)	$C_3H_7{\cdot}C_5H_{10}N$	127.23	col. lq.	0.847^{17}	−2	166–7	1.1	v. s.	v. s.

(*Continued*)

TABLE 2-2 Physical Properties of Organic Compounds *(Continued)*

Name	Formula	Formula weight	Form and color	Specific gravity	Melting point, °C	Boiling point, °C	Solubility in 100 parts: Water	Alcohol	Ether
Coumaric acid (*o*-)	$HOC_6H_4CH:CHCO_2H$	164.16	nd./aq.		207–8	subl.	sl. s. c.	s.	v. sl. s.
(*p*-)	$HOC_6H_4CH:CHCO_2H$	164.16	cr./aq.		206–7 d.		s. h.	v. s. h.	v. s.
Coumarin	$C_9H_6O_2$	146.14	rhb./et.	$0.935^{20/4}$	70	290–1	0.3 c.	v. s.	s.
Coumarone	C_8H_6O	118.13	oil	$1.078^{15/15}$	<–18	173–4	i.		s.
Creatine	$C_4H_9N_3O_2{\cdot}H_2O$	149.15	mn./aq.		295		1.4^{18}	0.01^{17}	i.
Creatinine	$C_4H_7N_3O$	113.12	mn.		260 d.		8.7^{16}	1^{16}	
Creosol (3-,1-,4-)	$CH_3O{\cdot}C_6H_3(CH_3)OH$	138.16	pr.	$1.092^{20/20}$	5.5	$221–2^{765}$	v. sl. s.	∞	∞
Cresidine (1-,2-,4-)	$CH_3(NH_2)C_6H_3{\cdot}OCH_3$	137.18	nd./pet.		93–4	235	v. sl. s.	s.	s.
Cresol (*o*-)	$CH_3C_6H_4OH$	108.14	cr.	$1.048^{20/4}$	30.8	190.8	2.5	∞^{30}	∞^{30}
(*m*-)	$CH_3C_6H_4OH$	108.14	lq.	$1.034^{20/4}$	10.9	202.8	0.5	∞	∞
(*p*-)	$CH_3C_6H_4OH$	108.14	pr.	$1.035^{20/4}$	35–6	202	1.8	∞^{36}	∞^{36}
Cresyl benzoate (*o*-)	$C_6H_5CO_2C_6H_4CH_3$	212.24	lq.			308	i.		
(*m*-)	$C_6H_5CO_2C_6H_4CH_3$	212.24	cr.		55	314	i.		
(*p*-)	$C_6H_5CO_2C_6H_4CH_3$	212.24	cr.		71.5	316	i.		
Crotonic acid (α-)	$CH_3CH:CHCO_2H$	86.09	col. mn.	$0.964^{79.7}$	72	189	8.3^{15}		
acid (β-)(*cis*-)	$CH_3CH:CHCO_2H$	86.09	nd.	$1.031^{15/4}$	15.5	170–1 d.	∞^{25}	s.	
aldehyde (α)	$CH_3CH:CHCHO$	70.09	col. lq.	$0.853^{20/20}$	–69	102.2	18	∞	∞
Cumene	$C_6H_5CH(CH_3)_2$	120.19	col. lq.	$0.862^{20/4}$	–96.9	152.5	i.	∞	∞
Cumic acid (*p*-)	$(CH_3)_2CH{\cdot}C_6H_4CO_2H$	164.20	tri.	1.162^{4}	116–7	subl.	0.02^{25}	s.	s.
Cumidine (*p*-)	$(CH_3)_2CH{\cdot}C_6H_4NH_2$	135.21	lq.	0.953	<–20	225^{761}	i.		
Cyanamide	$H_2N{\cdot}CN$	42.04	col. nd.	$1.073^{48/4}$	44–5	140^{19}	v. s.	v. s.	v. s.
Cyanic acid	HOCN or HNCO	43.02	gas	1.140^{0}	–80	$–64^{0}$	sl. s.		s.
Cyanoacetic acid	$CH_2(CN)CO_2H$	85.06	col. lq.		65–6	$108^{0.2}$	s.	s.	s.
Cyanogen	$(CN)_2$	52.03	col. gas	0.866^{17}	–34.4	–21	450^{20} cc.	2300^{20} cc.	500^{20} cc.
bromide	BrCN	105.92	nd.	$2.015^{20/4}$	52	61.3^{750}	s.	s.	s.
chloride	ClCN	61.47	gas	1.222^{0}	–6.5	12.5–13	2500^{20} cc.	v. s.	5000^{20} cc.
Cyanuric acid	$C_3H_3O_3N_3{\cdot}2H_2O$	165.10	mn./aq.	$1.768^{0/4}$	>360	d.	0.27^{17}	0.1^{22}	
Cyclo-butane	$CH_2<(CH_2)_2>CH_2$	56.11	col. gas	$0.703^{0/4}$	–50	$11–12^{726}$	i.	v. s.	
-heptane	$CH_2<(CH_2CH_2CH_2)_2>$	98.19	oil	$0.810^{20/4}$	–12	118–20	i.		
-hexane	$CH_2<(CH_2CH_2)_2>CH_2$	84.16	col. lq.	$0.779^{20/4}$	6.5	80–1	i.	∞	∞
-hexanol	$CH_2<(CH_2CH_2)_2>CHOH$	100.16	col. nd.	$0.962^{20/4}$	23.9	160–1	3.6^{20}	s.	s.
-hexanone	$CH_2<(CH_2CH_2)_2>CO$	98.14	col. oil	$0.947^{19/4}$	–45	155–6	s.	s.	s.
-hexene	$({\cdot}CH_2{\cdot}CH_2CH:)_2$	82.14	lq.	$0.810^{20/4}$	–103.7	83.3	v. sl. s.	v. s.	v. s.
-hexyl acetate	$CH_3CO_2C_6H_{11}$	142.20	oil	$0.985^{0/4}$		174^{750}	i.	∞	∞
amine	$CH_2<(CH_2CH_2)_2>CHNH_2$	99.17	col. lq.	$0.865^{20/0}$		134	i.	s.	
bromide	$CH_2<(CH_2CH_2)_2>CHBr$	163.06	col. lq.	$1.324^{20/20}$		165^{714}	i.	s.	s.
chloride	$CH_2<(CH_2CH_2)_2>CHCl$	118.60	col. lq.	$0.977^{18/4}$	–43.9	142	i.		∞
-pentadiene (1-,3-)	$CH_2<(CH:CH)_2>$	66.10	col. lq.	$0.805^{19/4}$	–85	41–2	i.	∞	∞
-pentane	$CH_2<(CH_2CH_2)_2>$	70.13	col. oil	$0.745^{20/4}$	–93.3	49–50	i.		
-pentanone	$<(CH_2CH_2)_2>CO$	84.12	col. oil	0.948^{20}	–58.2	129–30	v. sl. s.		
-propane	$<CH_2CH_2CH_2>$	42.08	col. gas	0.720^{-79}	–126.6	$–34^{749}$	i.	s.	s.
Cymene (*o*-)	$CH_3{\cdot}C_6H_4CH(CH_3)_2$	134.22	col. lq.	$0.875^{20/4}$		177	i.	s.	s.
(*m*-)	$CH_3{\cdot}C_6H_4CH(CH_3)_2$	134.22	col. lq.	0.862^{20}	<–25	175–6	i.	s.	s.
(*p*-)	$CH_3{\cdot}C_6H_4CH(CH_3)_2$	134.22	col. lq.	$0.857^{20/4}$	–73.5	176–7	i.	s.	s.
Cystine (*l*-)	$[{\cdot}SCH_2CH(NH_2)CO_2H]_2$	240.30	pl.		d. 258–61		0.01^{19}	i.	
Dambose	$C_6H_6(OH)_6$	180.16	mn./aq.	1.752	253	319^{15}	2^{12}	i.	i.
Decahydronaphthalene (*cis*-)	$C_{10}H_{18}$	138.25	lq.	$0.895^{18/4}$	–51	193.3	i.	s.	s.
(*trans*-)	$C_{10}H_{18}$	138.25	lq.	$0.872^{20/4}$	–32	185.3	i.	s.	s.
Decane (*n*-)	$CH_3(CH_2)_8CH_3$	142.28	col. lq.	0.730^{2}	–29.7	174.0	i.	∞	∞
Decyl alcohol	$CH_3(CH_2)_8CH_2OH$	158.28	col. oil	$0.830^{20/4}$	7	232.9	i.	s.	
Dextrin	$(C_6H_{10}O_5)x$	162.14	amor.	1.038			s.	i.	i.
Diacetone alcohol	$(CH_3)_2C(OH){\cdot}CH_2COCH_3$	116.16	lq.	0.931^{25}	–47	167.9	∞	∞	∞
Diamino-benzophenone (4-,4′-)	$H_2NC_6H_4COC_6H_4NH_2$	212.25	yel. nd.		237–9		sl. s. h.	s.	s.
-diphenylamine (4-,4′-)	$H_2NC_6H_4NHC_6H_4NH_2$	199.25	lf./aq.		158	d.	sl. s.	s.	s.
-diphenylmethane (4-,4′-)	$H_2NC_6H_4CH_2C_6H_4NH_2$	198.26	nd./aq.		93–4	$249–53^{15}$	sl. s. c.	s.	s.
-diphenylurea (4-,4′-)	$(H_2NC_6H_4NH)_2CO$	242.28	cr.		subl. 310		v. sl. s.		
Diamyl-amine (*i*-)	$[(CH_3)_2CHCH_2CH_2]_2NH$	157.30	col. lq.	$0.767^{21/4}$	–44	188–90	sl. s.	s.	∞
ether (*n*-)	$(C_2H_5CH_2CH_2CH_2)_2O$	158.28	col. lq.	$0.774^{20/4}$	–69	190	i.	∞	∞
(*i*-)	$[(CH_3)_2CH(CH_2)_2]_2O$	158.28	col. lq.	$0.777^{20/4}$		173.4	i.	∞	∞
Diamyl ketone (*i*-)	$[(CH_3)_2CHCH_2CH_2]_2CO$	170.29	yel. oil	$0.821^{25/4}$	14.6	228	i.	s.	s.
phthalate (*n*-)	$C_6H_4(CO_2C_5H_{11})_2$	306.40	col. lq.			$204–6^{11}$			
(*i*-)	$C_6H_4(CO_2C_5H_{11})_2$	306.40	col. lq.	1.03		225^{40}	i.	s.	s.
tartrate (*i*-)	$(HOCH{\cdot}CO_2C_5H_{11})_2$	290.35	lq.	$1.063^{15/4}$		195^{16}	i.		
Dianisidine (*o*-)(4-,3-)$_2$	$[NH_2(OCH_3)C_6H_3{\cdot}]_2$	244.29	col. lf.		131.5		i.	s.	s.
Diazo-aminobenzene	$C_6H_5N:N{\cdot}NHC_6H_5$	197.24	yel. lf.		96–8	expl.	i.	s. h.	v. s.
-aminotoluene (2-,2′-)	$C_7H_7N:N{\cdot}NHC_7H_7$	225.29	or. cr.		51		0.05		
-methane	$CH_2:N_2$	42.04	gas		–145	–23	d.		s.

Dibenzothiazyl-disulfide (2-,2′-)	$(C_6H_4NSC)_2S_2$	332.49	cr.	1.50	180	d.	i.		
Dibensoyl methane	$(C_6H_5CO)_2CH_2$	224.25	rhb./al.		78	$219–21^{18}$	i.	4.4^{20}	s.
Dibensyl-amine	$(C_6H_5CH_2)_2NH$	197.28	col. oil	$1.028^{25/25}$	–26	$268–71^{250}$	i.	s.	s.
-aniline	$C_6H_5N(CH_2C_6H_5)_2$	273.37	pr./al.		70–1	>300	i.	v. s. h.	s.
ketone	$(C_6H_5CH_2)_2CO$	210.27	cr.		34–5	330.6	i.		
phthalate (*o*-)	$C_6H_4(CO_2CH_2C_6H_5)_2$	346.38	pr./al.		42–3	274^{12}	v. sl. s.	s.	s.
succinate	$(\cdot CH_2CO_2CH_2C_6H_5)_2$	298.33	lf./al.		45–6	238^{14}	i.	s.	s.
Dibromo-benzene (*o*-)	$C_6H_4Br_2$	235.90	col. lq.	$1.956^{20/4}$	1.8	221–2	i.	s.	s.
(*m*-)	$C_6H_4Br_2$	235.90	col. lq.	$1.952^{20/4}$	–6.9	219^{755}	i.	s.	s.
(*p*-)	$C_6H_4Br_2$	235.90	pl./al.	2.261^{18}	87–8	218.6^{758}	i.	1.6	71^{25}
-diphenyl (4-,4′-)	$BrC_6H_4\cdot C_6H_4Br$	312.00	mn. pr.	1.897	164–5	355–60	i.	v. sl. s. h.	
Dibutyl-adipate (*n*-)	$(\cdot CH_2CH_2CO_2C_4H_9)_2$	258.35	col. lq.	$0.965^{20/4}$	–38	183^{14}	i.	∞	∞
(*i*-)	$(\cdot CH_2CH_2CO_2C_4H_9)_2$	258.35	col. lq.	0.950^{25}	–20	278–80	i.		
-amine (*n*-)	$(C_2H_5CH_2CH_2)_2NH$	129.24	col. lq.	$0.768^{20/20}$		159^{761}	∞	∞	∞
(*i*-)	$[(CH_3)_2CHCH_2]_2NH$	129.24	col. lq.	$0.741^{25/4}$	–70	139–40	v. sl. s.	s.	s.
-*p*-aminophenol (*s*-)	$(C_4H_9)_2N\cdot C_6H_4OH$	221.34	lq.			170^{10}	i.		
-aniline (*n*-)	$C_6H_5N(C_4H_9)_2$	205.34	lq.			262.8	i.	∞	∞
carbonate (*n*-)	$CO(OC_4H_9)_2$	174.24	col. lq.	$0.924^{20/4}$		207^{740}	i.	s.	
(*i*-)	$CO(OC_4H_9)_2$	174.24	col. lq.	0.919^{15}		190	i.		
(*s*-)	$CO(OC_4H_9)_2$	174.24	col. lq.			178–80			
ether (*n*-)	$(C_2H_5CH_2CH_2)_2O$	130.23	lq.	$0.769^{20/20}$	–98	142.4	<0.05	∞	∞
(*i*-)	$[(CH_3)_2CHCH_2]_2O$	130.23	lq.	0.762^{15}		122.5	i.	∞	∞
(*s*-)	$[C_2H_5(CH_3)CH]_2O$	130.23	lq.	0.756^{21}		121	i.	∞	∞
ketone (*n*-)	$(C_2H_5CH_2CH_2)_2CO$	142.24	lq.	$0.827^{18/4}$	–5.9	187.7	i.	s.	v. s.
(*i*-)	$[(CH_3)_2CHCH_2]_2CO$	142.24	oil	$0.805^{21/4}$		168.1	<0.06	∞	∞
malate (*l*-)(*n*-)	$C_2H_4O(CO_2C_4H_9)_2$	246.30	lq.	$1.038^{20/4}$		$170–1^{18}$	v. sl. s.		
oxalate (*n*-)	$(\cdot CO_2C_4H_9)_2$	202.25	col. lq.	$0.986^{20/4}$	–29.6	245.5	i.	s.	s.
phthalate (*n*-)	$C_6H_4(CO_2C_4H_9)_2$	278.34	col. lq.	1.045^{21}		340	0.04^{25}	∞	∞
tartrate (*d*-)(*n*-)	$(CHOHCO_2C_4H_9)_2$	262.30	pr.	1.098^{15}	22–2.5	$200–3^{18}$	i.		
(*d*-)(*i*-)	$(CHOHCO_2C_4H_9)_2$	262.30	cr.	$1.031^{75/4}$	73–4	323–5	v. sl. s.		
Dichloro-acetic acid	$Cl_2CH\cdot CO_2H$	128.94	lq.	$1.560^{25/25}$	9.7(–4)	194.4	∞	∞	∞
-acetone (αα-)	$Cl_2CHCOCH_3$	126.97	lq.	1.234^{15}		120	v. sl. s.	s.	s.
-aniline (2-,5-)	$Cl_2C_6H_3NH_2$	162.02	nd.		50	251	v. sl. s.	s.	s.
-anthraquinone (1-,3-)	$C_6H_4{:}(CO)_2{:}C_6H_2Cl_2$	277.10	yel. nd.		208–9		i.	i.	
(1-,4-)	$C_6H_4{:}(CO)_2{:}C_6H_2Cl_2$	277.10	yel. nd.		187.5		i.	v. sl. s.	v. sl. s.
(1-,5-)	$C_6H_3Cl{:}(CO)_2{:}C_6H_3Cl$	277.10	yel. nd.		251		i.	sl. s.	
(1-,6-)	$C_6H_3Cl{:}(CO)_2{:}C_6H_3Cl$	277.10	yel. nd.		203–4		i.		
(1-,8-)	$C_6H_3Cl{:}(CO)_2{:}C_6H_3Cl$	277.10	yel. nd.		202–3		i.	sl. s.	
(2-,3-)	$C_6H_4{:}(CO)_2{:}C_6H_2Cl_2$	277.10	yel. nd.		268–70		i.	sl. s.	
(2-,6-)	$C_6H_3Cl{:}(CO)_2{:}C_6H_3Cl$	277.10	yel. nd.		282		i.		
(2-,7-)	$C_6H_3Cl{:}(CO)_2{:}C_6H_3Cl$	277.10	yel. nd.		210–11		i.		
-benzene (*o*-)	$C_6H_4Cl_2$	147.00	col. lq.	$1.305^{20/4}$	–17.6	179	i.	∞	∞
(*m*-)	$C_6H_4Cl_2$	147.00	col. lq.	$1.288^{20/4}$	–24.8	172^{766}	i.	s.	s.
(*p*-)	$C_6H_4Cl_2$	147.00	col. mn.	1.458^{21}	53	174^{764}	i.	v. s.	v. s.
-butane (*n*-)(1-,4-)	$ClCH_2(CH_2)_2CH_2Cl$	127.01	lq.		–38.7	161–3			
-diphenyl (4-,4′-)	$ClC_6H_4\cdot C_6H_4Cl$	223.10	pr.	$1.442^{0/4}$	148	315–9	i.	v. sl. s.	4^{25}
-ethane (1-,2-)	$ClCH_2\cdot CH_2Cl$	98.96	col. lq.	$1.256^{20/20}$	–35.3	83.7	0.9^{0}	∞	∞
-naphthalene (β-)(1-,4-)	$C_{10}H_6Cl_2$	197.06	nd./al.	$1.300^{76/4}$	67–8	$286–7^{740}$	i.	v. sl. s.	
(γ-)(1-,5-)	$C_{10}H_6Cl_2$	197.06	lf./al.		107	subl.	i.	s.	s.
-nitrobenzene (2-,5-)	$Cl_2C_6H_3NO_2$	192.00	tri./al.	1.669^{22}	54.6	266	i.	v. s. h.	
-pentane (1-,5-)	$ClCH_2(CH_2)_3CH_2Cl$	141.04	col. lq.	$1.094^{25/4}$		180–1	i.	s.	s.
-phenol (2-,4-)	$Cl_2C_6H_3OH$	163.00	nd.	$1.383^{60/25}$	45	209–10	0.45^{20}	v. s.	v. s.
Dichloramine T (*p*-)	$CH_3C_6H_4SO_2NCl_2$	240.11	cr.		83		sl. s.		
Dicyandiamide	$H_2N\cdot C({:}NH)\cdot NH\cdot CN$	84.08	mn. pl.	1.40^{14}	207–8	d.	2.3^{18}	1.3^{18}	0.01^{18}
Diethanolamine	$HN(CH_2CH_2OH)_2$	105.14	pr.	$1.097^{20/4}$	28	270^{748}	∞	∞	v. sl. s.
Diethyl adipate	$(\cdot CH_2CH_2CO_2C_2H_5)_2$	202.25	col. lq.	$1.009^{20/4}$	–21	$239–41^{761}$	0.43^{80}	s.	s.
-amine	$(C_2H_5)_2NH$	73.14	col. lq.	$0.712^{15/15}$	–38.9	55.5^{759}	v. s.	∞	∞
-aminophenol (*m*-)	$(C_2H_5)_2N\cdot C_6H_4\cdot OH$	165.23	rhb.		78	276–80	s.		
-aniline	$(C_2H_5)_2NC_6H_5$	149.23	oil	$0.934^{20/4}$	–34.4	216	1.4^{12}	s.	s.
sulfonic acid (*m*-)	$(C_2H_5)_2NC_6H_4SO_3H$	229.30	cr.		270 d.		s.		
carbonate	$OC(OC_2H_5)_2$	118.13	col. lq.	$0.975^{20/4}$	–43	126^{759}	i.	∞	∞
diethyl malonate	$(C_2H_5)_2C(CO_2C_2H_5)_2$	216.27	col. lq.	$0.985^{20/4}$		230	i.	∞	∞
Diethyl dimethyl malonate	$(CH_3)_2C(CO_2C_2H_5)_2$	188.22	col. lq.	$0.994^{25/25}$		196.7	i.	∞	∞
glutarate	$CH_2(CH_2CO_2C_2H_5)_2$	188.22	syrup	1.025^{21}	–24	237	0.88^{20}	v. s.	s.
ketone	$(C_2H_5)_2CO$	86.13	col. lq.	$0.816^{19/4}$	–42	101.7	4.7^{20}	∞	∞
malonate	$CH_2(CO_2C_2H_5)_2$	160.17	col. lq.	$1.055^{20/4}$	–49.8	198.9	2.08^{20}	∞	∞
-malonic acid	$(C_2H_5)_2C(CO_2H)_2$	160.17	pr./aq.		125	d. 170–80	65^{16}	v. s.	v. s.
-naphthylamine (α-)	$C_{10}H_7N(C_2H_5)_2$	199.29	col. oil	1.005		285–90	i.	∞	∞
(β-)	$C_{10}H_7N(C_2H_5)_2$	199.29	col. oil	1.026		318	i.	∞	∞
oxalate	$(\cdot CO_2C_2H_5)_2$	146.14	col. lq.	$1.079^{20/4}$	–40.6	186	v. sl. s.	∞	∞
phthalate (*o*-)	$C_6H_4(CO_2C_2H_5)_2$	222.24	col. lq.	$1.121^{25/25}$		298–9	i.	∞	∞
sulfate	$O_2S(OC_2H_5)_2$	154.18	col. lq.	$1.172^{25/4}$	–25	210	i.	s.	∞
sulfide	$(C_2H_5)_2S$	90.19	col. lq.	$0.837^{20/4}$	–99.5	$92–3^{754}$	0.31^{20}	∞	∞

(*Continued*)

TABLE 2-2 Physical Properties of Organic Compounds *(Continued)*

Name	Formula	Formula weight	Form and color	Specific gravity	Melting point, °C	Boiling point, °C	Solubility in 100 parts		
							Water	Alcohol	Ether
tartrate (*d*-)	$(CHOH\cdot CO_2C_2H_5)_2$	206.19	lq.	$1.204^{20/4}$	17	280	sl. s.	∞	∞
-toluidine (*o*-)	$CH_3\cdot C_6H_4\cdot N(C_2H_5)_2$	163.26	lq.			$208–9^{755}$	i.	s.	s.
(*m*-)	$CH_3\cdot C_6H_4\cdot N(C_2H_5)_2$	163.26	lq.			231–2	i.	s.	s.
(*p*-)	$CH_3\cdot C_6H_4\cdot N(C_2H_5)_2$	163.26	lq.	$0.924^{15.5}$		228–9	i.		
Diethyleneglycol dinitrate	$O(CH_2CH_2ONO_2)_2$	196.12	lq.	$1.377^{25/4}$	−11.3		i.		
Difluorodichloromethane	F_2CCl_2	120.91	gas	1.486^{-30}	−155	−29.2	5.7 cc.^{26}	s.	s.
Diglycerol	$[(HO)_2C_3H_5]_2O$	166.17	lq.			$220–30^{10}$	s. h.		i.
Dihydroxy-dinaphthyl (α-)	$(HO\cdot C_{10}H_6\cdot)_2$	286.32	pl./al.		300		i.	s.	v. s.
(-2,-2′,-1,-1′)	$(HO\cdot C_{10}H_6\cdot)_2$	286.32	nd./al.		218	subl.	i.	s.	v. s.
-diphenyl (4-,4′-)	$(HO\cdot C_6H_4\cdot)_2$	186.21	rhb./al.	1.25	270–2	subl.	sl. s.	v. s.	v. s.
-ethyl formal (β-)	$CH_2(OCH_2CH_2OH)_2$	136.15	lq.	1.154^{25}	−5.3	264	∞		
-naphthalene (1-,5-)	$C_{10}H_6(OH)_2$	160.17	pr./aq.		258–60	d.	sl. s.	s.	v. s.
(1-,8-)	$C_{10}H_6(OH)_2$	160.17	nd.		140		sl. s. h.		v. s.
Dimethoxy-benzene (*p*-)	$(CH_3O)_2C_6H_4$	138.16	lf.	$1.053^{55/55}$	56	212.6	v. sl. s.	v. s.	v. s.
-diphenylamine (4-,4′-)	$HN(C_6H_4OCH_3)_2$	229.27	cr.		103		i.		
-ethyl adipate	$(CH_2)_4(CO_2C_2H_4OCH_3)_2$	262.30	lq.	$1.075^{15.6}$		$145–50^{2}$	5		
Dimethyl adipate	$[(CH_2)_2CO_2CH_3]_2$	174.19	col. lq.	$1.063^{20/4}$	10–1	115^{18}	i.		
-amine	$(CH_3)_2NH$	45.08	col. lq.	$0.680^{0/4}$	−96	7.4	v. s.	s.	s.
-aminoasobenzene (*p*-)	$C_6H_5N{:}N\cdot C_6H_4N(CH_3)_2$	225.29	yel./al.		116–7	d.	i.	s.	s.
-aminoethanol	$(CH_3)_2NCH_2CH_2OH$	89.14	col. lq.	$0.887^{20/4}$		135^{756}	∞		
-aminophenol (*m*-)	$(CH_3)_2NC_6H_4OH$	137.18	nd.		85	265–8	sl. s. h.	s.	s.
-aniline	$(CH_3)_2NC_6H_5$	121.18	yel. lq.	$0.956^{20/4}$	2.5	193	i.	s.	s.
sulfonic acid (*m*-)	$(CH_3)_2NC_6H_4SO_3H$	201.24	cr.		d. 266		s.		
(*p*-)	$(CH_3)_2NC_6H_4SO_3H\cdot H_2O$	219.26	pr.		257		s. h.	v. sl. s.	v. sl. s.
carbonate	$OC(OCH_3)_2$	90.08	col. lq.	$1.070^{20/4}$	0.5	89–90	i.	∞	∞
ether	CH_3OCH_3	46.07	gas		−138.5	−23.7	3700 cc.^{18}	s.	s.
-formamide	$HCON(CH_3)_2$	73.09	lq.	0.945^{25}	−58.3	152.8	∞		
fumarate	$({:}CHCO_2CH_3)_2$	144.13	col. tri.		102	192	i.	sl. s.	sl. s.
glutarate	$(CH_2)_3(CO_2CH_3)_2$	160.17	lq.	$1.089^{15.6}$	−37	130^{50}			
glyoxime	$(CH_3\cdot C{:}NOH)_2$	116.12	col. cr.		240–6		0.06^{20}	v. s.	v. s.
-naphthalene (1-,4-)	$C_{10}H_6(CH_3)_2$	156.22	lq.	$1.016^{20/4}$	<−18	264–6	i.		
(2-,3-)	$C_{10}H_6(CH_3)_2$	156.22	lf./al.		104	265^{767}	i.	sl. s.	
-naphthylamine (α-)	$C_{10}H_7N(CH_3)_2$	171.24	col. oil	1.042^{20}		274.5^{711}	i.	s.	s.
(β-)	$C_{10}H_7N(CH_3)_2$	171.24	col. cr.	$1.039^{70/70}$	46	304–5	i.	s.	s.
oxalate	$(\cdot CO_2CH_3)_2$	118.09	col. mn.	1.148^{54}	54	163.3	6	s.	s.
phthalate (*o*-)	$C_6H_4(CO_2CH_3)_2$	194.18	col. lq.	$1.189^{25/25}$		280^{734}	0.43		
sulfate	$(CH_3O)_2SO_2$	126.13	col. oil	$1.352^{0/4}$	−26.8	188.3	v. sl. s.	∞	∞
sulfide	$(CH_3)_2S$	62.13	oil	$0.846^{21/4}$	−83.2	37.3	i.	s.	s.
tartrate (*d*-)	$(CHOH\cdot CO_2CH_3)_2$	178.14	cr.	$1.328^{20/4}$	61.5	280	s.	200^{15}	
-vinyl-ethenyl carbinol	$(CH_3)_2COH\cdot C⋮C\cdot CH{:}CH_2$	110.15	lq.	$0.887^{20/4}$		150	6^{20}		
Dinaphthyl (αα-)	$C_{10}H_7\cdot C_{10}H_7$	254.33	lf./al.		160	$240–4^{12}$	i.	s. h.	s.
-methane (αα′-)	$(C_{10}H_7)_2CH_2$	268.35	pr./al.		109	>360	i.	0.8 c.	v. s.
(β,β′-)	$(C_{10}H_7)_2CH_2$	268.35	nd./al.		92		i.	s.	
Dinitro-anisole (1-)(2-,4-)	$CH_3OC_6H_3(NO_2)_2$	198.13	col. mn.	1.341^{20}	94–5		sl. s. h.	1.5^{20}	
-benzene (*o*-)	$C_6H_4(NO_2)_2$	168.11	col. mn.	1.59^{18}	117–8	319^{774}	0.01 c.	1.9^{21}	
(*m*-)	$C_6H_4(NO_2)_2$	168.11	col. rhb.	$1.575^{20/4}$	89.8	300–2	0.3^{99}	3^{20}	
(*p*-)	$C_6H_4(NO_2)_2$	168.11	col. mn.	1.625^{18}	173–4	299^{777}	0.18^{100}	0.18^{21}	
sulfonic acid (2-,4-)(1-)	$(NO_2)_2C_6H_3SO_3H\cdot 3H_2O$	302.22	pr.		106–8		s.	s.	v. sl. s.
-benzoic acid (2-,4-)	$(NO_2)_2C_6H_3CO_2H$	212.12	cr./aq.		179–80		1.85^{25}	s.	
(3-,5-)	$(NO_2)_2C_6H_3CO_2H$	212.12	mn. pr.		204–5	subl.	s h.	v. s.	sl. s.
-benzophenone (4-,4′-)	$(NO_2C_6H_4)_2CO$	272.21	col. nd.		189		i.		
-diphenyl (4-,4′-)	$(NO_2C_6H_4)_2$	244.20	nd./al.	1.445	233		i.	1.5^{20}	
(2-,4′-)	$(NO_2C_6H_4)_2$	244.20	mn.	1.474	93.5		i.	v. s. h.	
-naphthalene (1-,5-)	$C_{10}H_6(NO_2)_2$	218.17	nd.		216	subl.	i.		
(1-,8-)	$C_{10}H_6(NO_2)_2$	218.17	rhb.		170–2	d.	i.	0.2 c.	
Dinitro-phenol (2-,3-)	$(NO_2)_2C_6H_3OH$	184.11	yel. mn.	1.681^{20}	144–5		sl. s.	v. s. h.	v. s.
(2-,4-)	$(NO_2)_2C_6H_3OH$	184.11	yel. rhb.	1.683^{24}	114–5	subl.	0.5 c.	4^{20}	v. s. h.
(2-,6-)	$(NO_2)_2C_6H_3OH$	184.11	yel. rhb.		63–4		s. h.	s. h.	s.
-salicylic acid (3-,5-)	$(NO_2)_2C_6H_2(OH)CO_2H\cdot H_2O$	246.13	pl./aq.		173 d.		s. c.	v. s.	v. s.
-stilbene (4-,4′-)	$(NO_2C_6H_4CH{:})_2$	270.24	yel. lf.		210–6		i.	v. sl. s.	v. sl. s.
-toluene (2-,4-)	$(NO_2)_2C_6H_3CH_3$	182.13	nd.	1.321^{71}	70	300	0.03^{22}	1.2^{15}	9^{16}
(3-,4-)	$(NO_2)_2C_6H_3CH_3$	182.13	nd.	1.259^{111}	60–1		i.		
(3-,5-)	$(NO_2)_2C_6H_3CH_3$	182.13	mn. pr.	1.277^{111}	92–3	subl.	sl. s.	s. h.	s.
Dioxane	$O<(CH_2\cdot CH_2)_2>O$	88.11	col. lq.	$1.033^{20/4}$	9.5–10.5	101.1	∞	s.	s.
Dipentene	$C_{10}H_{16}$	136.23	col. lq.	0.865^{18}		178	i.		

Diphenyl	$C_6H_5{\cdot}C_6H_5$	154.21	col. mn.	$0.992^{73/4}$	69–70	254.9	i.	10^{20}	6.6^{20}
-amine	$C_6H_5NHC_6H_5$	169.22	col. mn.	$1.160^{20/20}$	52.9	302	0.03^{25}	$56^{19.5}$	s.
carbonate	$CO(OC_6H_5)_2$	214.22	nd./al.	1.272^{14}	80	302–6	i.	v. s.	s.
-chloroarsine	$(C_6H_5)_2AsCl$	264.58	rhb.	1.583^{40}	43–4	d. 327	0.2 d.	20	s.
-ethane	$(C_6H_5CH_2{\cdot})_2$	182.26	col. pr.	$0.978^{50/50}$	52–3	284	i.	s.	v. s.
ether	$C_6H_5OC_6H_5$	170.21	col. rhb.	1.073^{20}	27	259	v. sl. s.	s.	∞
guanidine	$(C_6H_5NH)_2C{:}NH$	211.26	mn./al.		147–8	d. > 170	v. sl. s.	9^{20}	sl. s.
-methane	$(C_6H_5)_2CH_2$	168.23	col. pr.	$1.001^{26/4}$	26–7	265	i.	v. s.	v. s.
phenylenediamine (*p*-)	$(C_6H_5NH)_2C_6H_4$	260.33	cr.		152		i.		
succinate	$({\cdot}CH_2CO_2C_6H_5)_2$	270.28	lf./al.		122–3	330	i.		s.
sulfide	$(C_6H_5)_2S$	186.27	col. lq.	$1.119^{15/15}$	<–40	296–7	i.	s. h.	∞
sulfone	$(C_6H_5)_2SO_2$	218.27	nd./aq.	$1.248^{25/4}$	128–9	379	sl. s. h.	s. h.	
urea (uns.)	$(C_6H_5)_2NCONH_2$	212.25	rhb.	1.276	189		v. sl. s.	s.	s.
Diphenylene oxide	$<(C_6H_4)_2O$	168.19	lf./al.		86–7	287–8	i.	s. h.	v. s.
Dipropyl adipate (*n*-)	$({\cdot}CH_2CH_2CO_2C_3H_7)_2$	230.30	col. lq.	$0.979^{20/4}$	–20.3	$143–5^{10}$	i.	s.	s.
-amine (*n*-)	$(C_2H_5CH_2)_2NH$	101.19	col. lq.	$0.739^{20/4}$	–39.6	110–1	s.	∞	∞
(*i*-)	$[(CH_3)_2CH]_2NH$	101.19	col. lq.	0.722^{22}	–61	83.5^{743}	s.	s.	
aniline (*n*-)	$C_6H_5N(C_3H_7)_2$	177.29	yel. oil	0.910^{20}		245.4	i.	s.	s.
carbonate (*n*-)	$CO(OCH_2C_2H_5)_2$	146.18	col. lq.	0.968^{22}		168.2	v. sl. s.		
ether (*n*-)	$(C_2H_5CH_2)_2O$	102.17	col. lq.	$0.744^{21/0}$	–122	91	sl. s.	∞	∞
(*i*-)	$[(CH_3)_2CH]_2O$	102.17	col. lq.	$0.725^{21/0}$	–60	69	0.2	∞	∞
ketone (*n*-)	$(C_2H_5CH_2)_2CO$	114.19	col. lq.	$0.822^{20/4}$	–32.6	144.2	0.43	∞	∞
(*i*-)	$[(CH_3)_2CH]_2CO$	114.19	col. lq.	$0.806^{20/4}$		123.7	v. sl. s.	∞	∞
oxalate (*n*-)	$(CO_2CH_2C_2H_5)_2$	174.19	col. lq.	$1.038^{0/0}$	–51.7	213.5	d. h.		
(*i*-)	$[CO_2CH(CH_3)_2]_2$	174.19	col. lq.			190			
Disalicylal ethylenediamine	$[HOC_6H_4CH{:}NCH_2{\cdot}]_2$	268.31	cr.	1.34	125–6		0.03^{28}		
Ditolyl guanidine (*o*-)	$(C_7H_7NH)_2C{:}NH$	239.32	cr.	$1.10^{20/4}$	178–9		v. sl. s.	s. h.	s.
Divinyl acetylene	$(H_2C{:}CH{\cdot}C{:})_2$	78.11	lq.	$0.776^{20/4}$		85	i.		
Docosane (*n*-)	$CH_3(CH_2)_{20}CH_3$	310.60	cr.	$0.778^{44/4}$	44.5	224.5^{15}	i.	4 h.	v. s.
Dodecane (*n*-)	$CH_3(CH_2)_{10}CH_3$	170.33	lq.	$0.751^{20/4}$	–9.6	214.5	i.	v. s.	v. s.
Dulcitol	$CH_2OH(CHOH)_4CH_2OH$	182.17	mn.	1.466^{15}	189	$290–5^{3}$	3.2^{15}	v. sl. s.	i.
Durene (1-,2-,4-,5-)	$(CH_3)_4C_6H_2$	134.22	mn.	$0.838^{81/4}$	79–80	193–5	i.	s.	s.
Elaidic acid	$C_8H_{17}CH{:}CH(CH_2)_7CO_2H$	282.46	lf./al.	$0.851^{79/4}$	51–2	288^{100}	i.	v. s.	v. s.
Eosine	$C_{20}H_8O_5Br_4$	647.89	col. cr.				i.	s.	
Ephedrine (*l*-)	$C_6H_5CHOHCH(CH_3)NHCH_3$	165.23	cr./et.		40	255	5	500	s.
Epichlorhydrin (α-)	$C_2H_3O{\cdot}CH_2Cl$	92.52	lq.	$1.183^{25/25}$	–25.6	117^{756}	<5	∞	∞
Epidichlorohydrin (α-)	$CH_2{:}CCl{\cdot}CH_2Cl$	110.97	col. lq.	1.204^{25}		94	i.	∞	∞
Erythritol (*dl*-)	$CH_2OH(CHOH)_2CH_2OH$	122.12	tet. pr.	$1.451^{20/4}$	126	329–31	60	sl. s. c.	i.
tetranitrate	$C_4H_6(ONO_2)_4$	302.11	lf./al.		61	expl.	i. c.	s.	s.
Ethane	CH_3CH_3	30.07	col. gas	0.546^{-88}	–172	–88.6	4.7 cc.^{20}	150 cc.	
Ethanol-amine	$HOCH_2CH_2NH_2$	61.08	col. oil	1.022^{20}	10.5	171^{757}	∞	∞	1
formamide	$HCONHCH_2CH_2OH$	89.09	lq.	1.169^{25}	<–40	d.	∞		
Ether	$(CH_3CH_2)_2O$	74.12	col. lq.	$0.708^{25/4}$	–116.3	34.6	7.5^{20}	∞	
Ethyl abietate	$C_{19}H_{29}CO_2C_2H_5$	330.50	lq.	$1.020^{20/20}$		200^{4}	i.		
acetate	$CH_3CO_2C_2H_5$	88.11	col. lq.	$0.901^{20/4}$	–82.4	77.1	8.5^{15}	∞	∞
acetoacetate	$CH_3COCH_2CO_2C_2H_5$	130.14	col. lq.	$1.025^{20/4}$	–45	180^{755}	13^{17}	∞	∞
alcohol	CH_3CH_2OH	46.07	col. lq.	$0.789^{20/4}$	–112	78.4	∞		∞
-amine	$C_2H_5NH_2$	45.08	col. lq.	$0.689^{15/15}$	–80.6	16.6	∞	∞	∞
hydrochloride	$C_2H_5NH_2{\cdot}HCl$	81.54	mn.	1.216	108–9		240^{17}	v. s.	i.
aniline	$C_6H_5NHC_2H_5$	121.18	lq.	$0.963^{20/4}$	–63.5	204	i.	∞	∞
sulfonic acid (*m*-)	$C_2H_5NHC_6H_4SO_3H$	201.24	nd./aq.		d. 294		2.15^{15}		
anisate (*p*-)	$CH_3OC_6H_4CO_2C_2H_5$	180.20	lq.	$1.103^{25/25}$	7–8	269–70	i.	s.	s.
anthranilate (*o*-)	$NH_2C_6H_4CO_2C_2H_5$	165.19	cr.	$1.117^{20/4}$	13	266–8	v. sl. s.	s.	s.
benzene	$C_6H_5{\cdot}C_2H_5$	106.17	col. lq.	$0.867^{20/4}$	–94.4	136.2	0.01^{15}	∞	∞
benzoate	$C_6H_5CO_2C_2H_5$	150.17	col. lq.	$1.052^{15/15}$	–34.6	211–2	0.08^{20}	∞	∞
-benzyl-aniline	$C_6H_5N(C_2H_5)CH_2C_6H_5$	211.30	yel. oil	$1.034^{18.5}$		285^{10}	i.	18	∞
bromide	C_2H_5Br	108.97	col. lq.	$1.431^{20/4}$	–117.8	38.4	1.06^{0}	∞	∞
butyrate (*n*-)	$C_2H_5CH_2CO_2C_2H_5$	116.16	col. lq.	$0.879^{20/4}$	–93.3	120–1	0.68^{25}	∞	∞
(*i*-)	$(CH_3)_2CHCO_2C_2H_5$	116.16	col. lq.	$0.871^{20/4}$	–88.2	110–1	sl. s.	∞	∞
caprate (*n*-)	$CH_3(CH_2)_8CO_2C_2H_5$	200.32	lq.	0.859^{28}	–20	244.6^{758}	0.002^{20}	∞	∞
Ethyl caproate (*n*-)	$CH_3(CH_2)_4CO_2C_2H_5$	144.21	col. lq.	$0.873^{20/20}$	–67.5	$165–6^{736}$	i.	∞	∞
caprylate (*n*-)	$CH_3(CH_2)_6CO_2C_2H_5$	172.26	col. lq.	0.878^{17}	–45	$207–8^{753}$	i.	∞	∞
chloride	CH_3CH_2Cl	64.51	col. lq.	$0.917^{6/6}$	–139	13	0.45^{0}	∞	∞
chloroacetate	$ClCH_2CO_2C_2H_5$	122.55	col. lq.	$1.159^{20/4}$	–26	144	i.	∞	∞
chlorocarbonate	$ClCO_2CH_2CH_3$	108.52	col. lq.	$1.138^{20/4}$	–80.6	94–5	d.	∞	∞
cinnamate (*trans*-)	$C_6H_5CH{:}CHCO_2C_2H_5$	176.21	col. lq.	$1.049^{20/4}$	12	271	i.	∞	∞
cyanoacetate	$CH_2(CN)CO_2C_2H_5$	113.11	col. lq.	$1.062^{20/4}$	–22.5	208^{753}	2^{25}	∞	∞
formate	$HCO_2CH_2CH_3$	74.08	col. lq.	$0.923^{20/4}$	–79	54^{760}	11^{18}	∞	∞
furoate (α)	$OC_4H_3CO_2C_2H_5$	140.14	lf.	$1.117^{21/4}$	34	195^{766}	i.	∞	∞
heptoate	$CH_3(CH_2)_5CO_2C_2H_5$	158.24	col. lq.	$0.872^{20/20}$	–66.1	187–8	0.029^{20}	∞	∞
hypochlorite	$ClOCH_2CH_3$	80.51	yel. lq.	$1.013^{-6/4}$	expl.	36^{752}			∞
iodide	CH_3CH_2I	155.97	col. lq.	$1.933^{20/4}$	–105	72.4	0.4^{20}	∞	∞
lactate	$CH_3CH(OH)CO_2C_2H_5$	118.13	oil	$1.030^{25/4}$		155	∞	∞	∞

(*Continued*)

TABLE 2-2 Physical Properties of Organic Compounds *(Continued)*

Name	Formula	Formula weight	Form and color	Specific gravity	Melting point, °C	Boiling point, °C	Solubility in 100 parts: Water	Alcohol	Ether
laurate	$CH_3(CH_2)_{10}CO_2C_2H_5$	228.37	oil	$0.868^{13/4}$	−10.7	269	i.	s.	∞
mercaptan	CH_3CH_2SH	62.13	lq.	$0.839^{20/4}$	−121	36–7	1.5	s.	s.
methacrylate	$CH_2{:}C(CH_3)CO_2C_2H_5$	114.14	col. lq.	$0.913^{15.6}$		118	i.	s.	s.
naphthylamine (α-)	$C_{10}H_7NHC_2H_5$	171.24	oil	$1.060^{20/4}$		303^{723}	i.	s.	s.
naphthyl ether (α-)	$C_{10}H_7OC_2H_5$	172.22	cr.	$1.061^{20/20}$	5.5	276.4	i.	s.	s.
nitrate	$C_2H_5ONO_2$	91.07	col. lq.	$1.100^{25/4}$	−102	87–8	1.3^{55}	∞	∞
nitrite	C_2H_5ONO	75.07	lq.	$0.900^{15.5}$		17	v. sl. s.	∞	∞
oleate	$C_{17}H_{33}CO_2C_2H_5$	310.51	oil	0.867^{25}	<−15	$216–8^{15}$	i.	∞	∞
palmitate	$CH_3(CH_2)_{14}CO_2C_2H_5$	284.48	col. nd.	$0.858^{25/4}$	24–5	191^{10}	i.	s.	s.
pelargonate	$CH_3(CH_2)_7CO_2C_2H_5$	186.29	col. lq.	$0.866^{17.5}$	−44.5	$227–8^{757}$	i.	∞	∞
propionate	$CH_3CH_2CO_2C_2H_5$	102.13	col. lq.	$0.891^{20/4}$	−72.6	99.1	2.4^{20}	∞	∞
salicylate (*o*-)	$HOC_6H_4CO_2C_2H_5$	166.17	col. lq.	$1.136^{15/4}$	1.3	233–4	i.	∞	∞
stearate	$CH_3(CH_2)_{16}CO_2C_2H_5$	312.53	col. cr.	$0.848^{36.3}$	33.4(31)	201^{10}	i.	s.	s.
toluate (*o*-)	$CH_3{\cdot}C_6H_4CO_2C_2H_5$	164.20	lq.	$1.032^{25/25}$	<−10	227	i.	∞	∞
(*m*-)	$CH_3{\cdot}C_6H_4CO_2C_2H_5$	164.20	lq.	$1.030^{20/20}$		231^{750}	i.	∞	∞
toluene sulfonate (*p*-)	$CH_3{\cdot}C_6H_4SO_3C_2H_5$	200.25	pr./al.	$1.166^{48/4}$	33–4	221.3	i.	s.	s.
toluidine (*o*-)	$CH_3{\cdot}C_6H_4NHC_2H_5$	135.21	lq.	$0.948^{25/4}$	<−15	215–6	i.		
(*p*-)	$CH_3{\cdot}C_6H_4NHC_2H_5$	135.21	lq.	$0.942^{25/4}$		217	i.		
urea	$C_2H_5NH{\cdot}CO{\cdot}NH_2$	88.11	nd.	1.213^{18}	92		v. s.	80	i.
valerate (*n*-)	$CH_3(CH_2)_3CO_2C_2H_5$	130.18	col. lq.	0.877^{20}	−91.2	145.5	0.24^{25}	∞	∞
(*i*-)	$(CH_3)_2CH(CH_2)CO_2C_2H_5$	130.18	col. lq.	$0.867^{20/4}$	−99.3	135	0.17^{20}	∞	∞
Ethylal	$CH_2(OC_2H_5)_2$	104.15	lq.	$0.824^{25/4}$	−66.5	89	9^{18}	∞	∞
Ethylene	$H_2C{:}CH_2$	28.05	col. gas	$0.57^{-102/4}$	−169	−103.9	26 cc.^{0}	360 cc.	s.
bromide	$BrCH_2{\cdot}CH_2Br$	187.86	col. lq.	$2.180^{20/4}$	10	131.5	0.43^{80}	∞	∞
bromohydrin	$BrCH_2{\cdot}CH_2OH$	124.96	col. lq.	$1.772^{20/4}$		150.3	sl. s.	s.	
chlorobromide	$ClCH_2{\cdot}CH_2Br$	143.41	lq.	1.689^{19}	−16.6	106.7	0.69^{80}		
chlorohydrin	$ClCH_2{\cdot}CH_2OH$	80.51	col. lq.	$1.213^{20/4}$	−69	128.8	∞	∞	∞
diamine	$H_2NCH_2{\cdot}CH_2NH_2$	60.10	col. lq.	$0.900^{20/20}$	8.5	117.2	∞	∞	0.3
oxide	$<(CH_2)_2>O$	44.05	lq.	$0.887^{7/4}$	−111.3	13.5^{747}	∞	∞	v. s.
Ethylidene diacetate	$CH_3CH(O_2CCH_3)_2$	146.14	col. lq.	1.061^{12}	18.85	168^{740}	sl. s.	∞	
Eugenol (1-,4-,3-)	$C_3H_5{\cdot}C_6H_3(OH)OCH_3$	164.20	oil	$1.070^{15/15}$	10.3	253.5	v. sl. s.	∞	∞
(*i*-)(1-,3-,4-)	$C_3H_5{\cdot}C_6H_3(OCH_3)OH$	164.20	oil	$1.091^{15/15}$	−10	267.5	v. sl. s.	∞	∞
Fenchyl alcohol (*dl*-)	$C_{10}H_{17}OH$	154.25	col. cr.	0.935^{40}	35	201	sl. s.		
(*d*-)(α-)	$C_{10}H_{17}OH$	154.25	col. pr.	$0.964^{20/4}$	45–7	201–2	sl. s.	s.	s.
(*i*-)(*l*-)	$C_{10}H_{17}OH$	154.25	col. cr.	0.961	61–2	201–2	i.		
Ferric dimethyl-dithiocarbamate	$Fe[SSCN(CH_3)_2]_3$	416.49	cr.		d. 100–30	ign. >150	v. sl. s.		
Fluorene	$(C_6H_4)_2>CH_2$	166.22	cr./al.	$1.203^{0/4}$	115–6	293–5	i.	s. h.	s.
Fluorescein	$C_{20}H_{12}O_5$	332.31	yel. red		d. > 290		v. sl. s. h.	s. h.	
Fluoro-dichloromethane	$FCHCl_2$	102.92	gas	1.426^{0}	−127	14.5	i.	s.	s.
-trichloromethane	Cl_3CF	137.37	col. lq.	$1.494^{17.2}$		24.9	i.	∞	∞
Formaldehyde	HCHO	30.03	gas	0.815^{-20}	−92	−21	v. s.	v. s.	v. s.
(*m*-)	$(CH_2O)_3$	90.08	wh.	1.17^{65}	64	114.5^{759}	21^{25}	s.	s.
(*p*-)	$(CH_2O)_x{\cdot}xH_2O$	(30.03)	amor.		150–60	subl.	$20–30^{18}$	i.	i.
Formamide	$HCONH_2$	45.04	lq.	$1.139^{20/4}$	2	193	∞	∞	v. sl. s.
Formanilide	$HCONHC_6H_5$	121.14	mn.	$1.147^{15/15}$	47	216^{120}	sl. s.	v. s.	s.
Formic acid	HCO_2H	46.03	col. lq.	$1.220^{20/4}$	8.6	100.8	∞	∞	
Fructose	$CH_2OH(CHOH)_3COCH_2OH$	180.16	nd./aq.	$1.669^{17.5}$	95–105		v. s.	8^{18}	
Fuchsin	$C_{20}H_{19}N_3HCl$	337.85	red	1.22	d. >200		0.3	s.	i.
Fulminic acid	C:NOH	43.02							
Fumaric acid (*trans*-)	$HO_2CCH{:}CHCO_2H$	116.07	col. pr.	$1.635^{20/4}$	286–7	290	0.7^{17}	5.8^{30}	0.7^{25}
Furfural	$C_4H_3O{\cdot}CHO$	96.08	lq.	$1.159^{20/4}$	−38.7	161.7^{760}	9.1^{13}	∞	∞
Furfuran	C_4H_4O	68.07	col. lq.	$0.937^{20/4}$		$31–2^{756}$	i.	s.	s.
Furfuryl acetate	$CH_3CO_2CH_2C_4H_3O$	140.14	col. oil	$1.118^{20/4}$		175–7	i.	s.	s.
alcohol	$C_4H_3O{\cdot}CH_2OH$	98.10	oil	$1.129^{25/4}$		169.5^{752}	∞	s.	s.
butyrate	$C_3H_7CO_2CH_2{\cdot}C_4H_3O$	168.19	col. lq.	$1.053^{20/4}$		212–3	v. sl. s.	s.	∞
propionate	$C_2H_5CO_2CH_2{\cdot}C_4H_3O$	154.16	col. lq.	$1.109^{20/4}$		195–6	v. sl. s.	s.	∞
Furoic acid	$C_4H_3O{\cdot}CO_2H$	112.08	mn. pr.		133–4	230–2	3.6^{15}	s.	s.
G-acid, K salt (2-)(6-,8-)	$HOC_{10}H_5(SO_3K)_2$	380.48	cr.				8^{25}		
Na salt (2-)(6-,8-)	$HOC_{10}H_5(SO_3Na)_2$	348.26	cr.				34^{20}		
Galactose (*d*-)(α-)	$C_5H_{11}O_5{\cdot}CHO$	180.16	pr.		165.5		10.3^{0}	0.6^{40}	
Gallic acid (3-,4-,5-)	$(HO)_3C_6H_2CO_2H{\cdot}H_2O$	188.13	mn./aq.	$1.694^{4/4}$	d. 220		1^{13}	28^{15}	2.5^{15}
Gamma acid (2-,8-,6-)	$C_{10}H_5(NH_2)(OH)SO_3H$	239.25	cr.						
Geraniol	$C_9H_{15}CH_2OH$	154.25	col. lq.	0.883^{15}	<−15	230	i.	∞	∞
Glucose (*d*-)(α-)	$C_5H_{11}O_5{\cdot}CHO$	180.16	rhb.	1.544^{25}	146		$82^{17.5}$	sl. s.	i.
(*d*-)(β-)	$C_6H_{12}O_6{\cdot}H_2O$	198.17	cr.	$1.562^{18/4}$	150		154^{15}		
Glucuronic acid	$CHO(CHOH)_4CO_2H$	194.14	cr.		154	d.	v. s.		
Glutam(in)ic acid (*dl*-)	$[{\cdot}CHNH_2(CH_2)_2{\cdot}](CO_2H)_2$	147.13	cr./aq.	1.460	199 d.		1.5^{20}	v. sl. s.	v. sl. s.

Glutaric acid	$CH_2(CH_2CO_2H)_2$	132.11	col. cr.	1.429^{15}	97.5	200^{20}	63.9^{20}	v. s.	v. s.
Glycerol	$CH_2OH \cdot CHOH \cdot CH_2OH$	92.09	col. lq.	$1.260^{50/4}$	17.9	290	∞	∞	i.
acetate (mono-)	$C_5H_{10}O_4$	134.13	col. oil	$1.20^{20/4}$		158^{165}	v. s.	v. s.	sl. s.
(di-)	$(CH_3CO_2)_2C_3H_5OH$	176.17	col. lq.	$1.178^{15/15}$	40	$175–6^{40}$	s.	s.	sl. s.
nitrate (mono-) (α-)	$CH_2OH \cdot CHOH \cdot CH_2NO_3$	137.09	col. pr.	1.40^{15}	58–9	155–60	70^{15}	v. s.	v. sl. s.
(β-)	$CH_2OH \cdot CHNO_3 \cdot CH_2OH$	137.09	lf.	1.40^{15}	54	155–60		v. s.	sl. s.
dinitrate (1-,3-)	$CHOH(CH_2ONO_2)_2$	182.09	oil	1.47^{15}	<−30	$146–8^{15}$		v. s.	v. s.
Glyceryl triacetate	$(CH_3CO_2)_3C_3H_5$	218.20	col. lq.	$1.161^{17/4}$	−78	258–9	7.17^{15}	∞	∞
tribenzoate	$(C_6H_5CO_2)_3C_3H_5$	404.41	nd.	1.228^{12}	75–6	d.	i.	s. h.	s.
tributyrate	$(C_2H_5CH_2CO_2)_3C_3H_5$	302.36	col. lq.	$1.032^{20/4}$	<−75	305–9	i.	s.	s.
tricaprate	$[CH_3(CH_2)_8CO_2]_3C_3H_5$	554.84	col. cr.	$0.921^{40/4}$	31(25)		i.	s. h.	v. s.
tricaproate	$[CH_3(CH_2)_4CO_2]_3C_3H_5$	386.52	col. lq.	$0.987^{20/4}$	−25		i.	s.	s.
tricaprylate	$[CH_3(CH_2)_6CO_2]_3C_3H_5$	470.68	col. lq.	$0.954^{20/4}$	8.3(−21)		i.	s.	s.
trilaurate	$[CH_3(CH_2)_{10}CO_2]_3C_3H_5$	639.00	col. nd.	$0.894^{60/4}$	45–6		i.	sl. s. c.	v. s.
trimyristate	$[CH_3(CH_2)_{12}CO_2]_3C_3H_5$	723.16	lf.	$0.885^{60/6}$	56.5		i.	s.	
trinitrate	$CH_2NO_3 \cdot CHNO_3 \cdot CH_2NO_3$	227.09	yel. oil	1.601^{15}	13.3(2)	160^{15}	0.18^{20}	50^{20}	∞
trinitrite	$CH_2NO_2 \cdot CHNO_2 \cdot CH_2NO_2$	179.09	yel. lq.	$1.291^{10/16}$		150 sl. d.	d.	d.	s.
trioleate	$(C_{17}H_{33}CO_2)_3C_3H_5$	885.43	col. oil	0.915^{15}	−4	240^{18}	i.	sl. s.	v. s.
tripalmitate	$[CH_3(CH_2)_{14}CO_2]_3C_3H_5$	807.32	col. nd.	$0.866^{80/4}$	65.1	$310–20^{0.1}$	i.	0.004^{21}	v. s.
tristearate	$[CH_3(CH_2)_{16}CO_2]_3C_3H_5$	891.48	col. pr.	$0.862^{80/4}$	70.8(55)		i.	s. h.	s. h.
Glycide	$C_2H_3O \cdot CH_2OH$	74.08	col. lq.	$1.114^{16/16}$		166 sl. d.	∞	∞	∞
Glycine, Glycocoll	$NH_2CH_2 \cdot CO_2H$	75.07	mn.	1.161	232–6 d.		23 c.	0.1 c.	i.
Glycol	$CH_2OH \cdot CH_2OH$	62.07	col. lq.	$1.113^{19/4}$	−15.6	197.4	∞	∞	1.0
diacetate	$(CH_3CO_2CH_2 \cdot)_2$	146.14	col. lq.	$1.109^{14/4}$	−31	190.5	14.3^{22}	∞	∞
dibenzoate	$(C_6H_5CO_2CH_2 \cdot)_2$	270.28	rhb./et.		73–4	>360	i.		s.
dibutyrate	$(C_3H_7CO_2CH_2 \cdot)_2$	202.25	col. lq.	1.024^{0}		240	i.	v. s.	v. s.
dicaprylate	$(C_7H_{15}CO_2CH_2 \cdot)_2$	314.46	lq.		22		i.		
diformate	$(HCO_2CH_2 \cdot)_2$	118.09	lq.			174	v. sl. s.		
dilaurate	$(C_{11}H_{23}CO_2CH_2 \cdot)_2$	426.67	amor.		52–4	188^{20}	i.	v. s.	v. s.
dinitrate	$(O_2NO \cdot CH_2 \cdot)_2$	152.06	yel. lq.	$1.482^{21/2}$	−20	expl. 114	0.92^{25}	s.	∞
dinitrite	$(ONO \cdot CH_2 \cdot)_2$	120.06	lq.	1.216^{0}	<−15	96–8	i.	s. d.	s.
dipalmitate	$(C_{15}H_{31}CO_2CH_2 \cdot)_2$	538.89	nd.		71–2	$260^{0.1}$	i.	s.	s.
dipropionate	$(C_2H_5CO_2CH_2 \cdot)_2$	174.19	lq.	1.045^{25}		211–2	sl. s.	∞	∞
ether	$(HO \cdot CH_2CH_2)_2O$	106.12	lq.	$1.118^{20/20}$	−10.5	244.8	∞	∞	i.
formal	$< O \cdot CH_2CH_2OCH_2 >$	74.08	lq.	$1.060^{20/4}$		75–6	∞		
formate (mono-)	$HCO_2CH_2CH_2OH$	90.08	lq.	$1.199^{15/4}$		180	∞		
Glycolic acid	$HOCH_2CO_2H$	76.05	nd./aq.		79(63)	d.	∞	90^{25}	v. s.
Guaiacol (*o*-)	$CH_3O \cdot C_6H_4OH$	124.14	pr.	$1.140^{15/15}$	28.3	205	1.7^{15}	v. s.	v. s.
Guanidine	$NH:C(NH_2)_2$	59.07	col. cr.		50		v. s.	s.	
H-acid, Na salt (1-,8-,3-,6-)	$C_{10}H_8O_7NS_2Na \cdot 1½H_2O$	368.32	cr.				0.17^{20}		
Heptacosane (*n*-)	$CH_3(CH_2)_{25}CH_3$	380.73	col. cr.	$0.780^{60/4}$	59.5	270^{15}	i.		
Heptane (*n*-)	$CH_3(CH_2)_5CH_3$	100.20	col. lq.	$0.684^{20/4}$	−90.6	98.4^{760}	0.005^{15}	sl. s.	∞
(*i*-)	$(CH_3)_2CH(CH_2)_3CH_3$	100.20	col. lq.	$0.679^{20/4}$	−118.2	90.0	i.	s.	∞
	$C_3H_7 \cdot CH(CH_3) \cdot C_2H_5$	100.20	col. lq.	$0.687^{20/4}$	−119.4	91.8	i.	s.	∞
	$(CH_3)_3C \cdot CH_2 \cdot C_2H_5$	100.20	col. lq.	$0.674^{20/4}$	−125	79.1	i.	s.	∞
	$[(CH_3)_2CH]_2CH_2$	100.20	col. lq.	$0.675^{20/4}$	−119.4	80.8	i.	s.	∞
	$(CH_3)_2C(C_2H_5)_2$	100.20	col. lq.	$0.693^{20/4}$	−135.0	86.0	i.	s.	∞
	$(C_2H_5)_3CH$	100.20	col. lq.	$0.698^{20/4}$	−118.7	93.5	i.	s.	∞
	$(CH_3)_3C \cdot CH(CH_3)_2$	100.20	col. lq.	$0.690^{20/4}$	−25	80.8	i.	s.	∞
Heptoic acid	$CH_3(CH_2)_5CO_2H$	130.18	col. lq.	0.918^{20}	−10	221–2	0.25^{15}	s.	s.
aldehyde	$CH_3(CH_2)_5CHO$	114.19	col. lq.	$0.850^{20/\ell}$	−42	155	0.02^{20}	∞	∞
Heptyl acetate (*n*-)	$CH_3CO_2CH_2(CH_2)_5CH_3$	158.24	col. lq.	$0.874^{16/16}$		191.5^{759}	i.	s.	s.
alcohol (*n*-)	$CH_3(CH_2)_5CH_2OH$	116.20	col. lq.	$0.824^{20/4}$	−34	175^{756}	0.18^{25}	∞	∞
	$[(CH_3)_2CH]_2CHOH$	116.20	col. lq.	$0.829^{20/4}$		140	v. sl. s.	∞	∞
	$(C_2H_5 \cdot CH_2)_2CHOH$	116.20	lq.	$0.820^{20/4}$	−37	156	i.	s.	s.
mercaptan	$CH_3CH(SH) \cdot C_5H_{11}$	132.27	lq.	0.835^{20}		$174–5^{765}$	i.		
Hexachloro-benzene	C_6Cl_6	284.78	mn.	2.044^{24}	228–31	309^{742}	i.	v. sl. s. h.	s. h.
-ethane	$CCl_3 \cdot CCl_3$	236.74	rhb.	$2.091^{20/4}$	186–7	186^{777}	0.005^{22}	v. s.	v. s.
Hexacosane (*n*-)	$CH_3(CH_2)_{24}CH_3$	366.71	cr.	$0.779^{57/4}$	56.6	262^{15}	i.	v. sl. s.	
Hexadecane (*n*-)	$CH_3(CH_2)_{14}CH_3$	226.44	lf.	$0.774^{20/4}$	18.5	287.5	i.	∞	∞
Hexaethylbenzene	$C_6(C_2H_5)_6$	246.43	pr./al.	$0.831^{130/4}$	130	298.3	i.	0.75^{25}	8^{25}
Hexamethylbenzene	$C_6(CH_3)_6$	162.27	pl./al.		166	265	i.	0.2^{0}	v. s.
Hexamethylene-diamine	$NH_2(CH_2)_6NH_2$	116.20	lf.		42	204–5	v. s.	s.	
-diisocyanate	$OCN(CH_2)_6NCO$	168.19	lq.	1.04^{28}		$143–4^{20}$	d.	d.	
-glycol	$HO(CH_2)_6OH$	118.17	nd./aq.		42	250	s.	s.	sl. s. h.
tetramine	$(CH_2)_6N_4$	140.19	col. rhb.		subl.		81^{12}	3	v. sl. s.
Hexane (*n*-)	$CH_3(CH_2)_4CH_3$	86.18	col. lq.	$0.659^{20/4}$	−94	69	0.014^{15}	50^{33}	∞
(*i*-)	$(CH_3)_2CH(CH_2)_2CH_3$	86.18	lq.	$0.654^{20/4}$	−153.7	60.2	i.		s.
(neo-)	$(CH_3)_3C \cdot C_2H_5$	86.18	lq.	$0.649^{20/20}$	−98.2	49.7	i.		s.
	$(CH_3)_2CH \cdot CH(CH_3)_2$	86.18	lq.	$0.662^{20/4}$	−129.8	58.0^{760}	i.		s.
	$(C_2H_5)_2CHCH_3$	86.18	lq.	$0.664^{20/4}$	−118	63.2	i.		s.

(*Continued*)

TABLE 2-2 Physical Properties of Organic Compounds *(Continued)*

Name	Formula	Formula weight	Form and color	Specific gravity	Melting point, °C	Boiling point, °C	Solubility in 100 parts		
							Water	Alcohol	Ether
Hexyl acetate (*n*-)	$CH_3CO_2(CH_2)_5CH_3$	144.21	col. lq.	$0.890^{0/0}$		169.2	i.	v. s.	v. s.
alcohol (*n*-)	$CH_3(CH_2)_4CH_2OH$	102.17	col. lq.	$0.820^{20/20}$	−51.6	157.2	0.6^{20}	∞	∞
	$(CH_3)_2CH\cdot C(CH_3)_2OH$	102.17	lq.	$0.821^{20/0}$	−14	120–1	v. sl. s.	∞	∞
	$(CH_3)_2COH\cdot CH_2C_2H_5$	102.17	lq.	$0.809^{20/4}$	−107	123^{762}	v. sl. s.	∞	∞
formate (*n*-)	$HCO_2CH_2(CH_2)_4CH_3$	130.18	lq.	0.898^{0}		153.6		∞	∞
resorcinol (2-,4-)	$CH_3(CH_2)_5C_6H_3(OH)_2$	194.27	col. nd.		68–70	179^{7}	0.05	v. s.	s.
Hippuric acid	$C_6H_5CONHCH_2CO_2H$	179.17	rhb.	$1.371^{20/4}$	187–8	d.	0.4^{20}	s. h.	0.25^{18}
Histidine (*l*-)	$C_6H_9O_2N_3$	155.15	lf./aq.		d. 287		s.	v. sl. s.	i.
Homophthalic acid (*o*-)	$HO_2C\cdot C_6H_4\cdot CH_2CO_2H$	180.16	cr./aq.		175–80		s. h.	v. s.	sl. s.
Hydracrylic acid	$HOCH_2CH_2CO_2H$	90.08	syrup			d.			
Hydro-cyanic acid	HCN	27.03	lq.	0.697^{18}	−12	25–6	∞	∞	∞
-quinone (*p*-)	$C_6H_4(OH)_2$	110.11	cr.	1.332^{15}	170.3	285^{730}	6^{15}	v. s.	v. s.
Hydroxy-benzaldehyde (*p*-)	$HO\cdot C_6H_4\cdot CHO$	122.12	nd./aq.	1.129^{130}	116–7	subl.	1.38^{31}		
-benzanilide (*o*-)	$HO\cdot C_6H_4\cdot CONHC_6H_5$	213.23	pr./al.		135	d.	v. sl. s. h.	s.	s.
-quinoline (2-)(α-)	$C_9H_6N\cdot OH$	145.16	pr./al.		199–200	subl.	s. h.	v. s.	v. s.
(8-)(*o*-)	$C_9H_6N\cdot OH$	145.16	pr.		75–6	266.6^{752}	v. sl. s. c.	s.	sl. s.
Indigo	$[C_6H_4(CO)(NH)C:]_2$	262.26	cr.	1.35	390–2	subl.	i.	i.	i.
White	$C_{16}H_{12}O_2N_2$	264.28	gray				i.	s.	s.
Indole	C_8H_7N	117.15	lf./aq.		52	253–4	s. h.	s. h.	s.
Indoxyl	C_8H_6NOH	133.15	yel. pr.		85	110	s.	s.	s.
Iodo-benzene	C_6H_5I	204.01	col. lq.	$1.824^{25/4}$	−28.5	188.6	0.034^{20}	s.	∞
-phenol (*p*-)	IC_6H_4OH	220.01	nd./aq.	1.857^{112}	93–4	d.	sl. s.	v. s.	v. s.
Iodoform	HCI_3	393.73	yel. hex.	4.008^{17}	119	subl.	0.01^{25}	1.5^{17}	13.6^{25}
Ionone (α-)	$C_{10}H_{16}:CHCOCH_3$	192.30	col. oil	0.930^{20}		136.1^{17}	sl. s.	∞	∞
(β-)	$C_{10}H_{16}:CHCOCH_3$	192.30	col. oil	0.944^{20}		140^{18}	sl. s.	∞	∞
Irone (β-)	$C_{14}H_{22}O$	206.32	col. oil	0.939^{20}		144^{16}	v. sl. s.	v. s.	v. s.
Isatin	$C_6H_4<(CO)(N)>COH$	147.13	yel. red		200–1	subl.	s. h.	v. s. h.	sl. s.
Isoprene	$CH_2:CH\cdot C(CH_3):CH_2$	68.12	col. lq.	$0.681^{20/4}$	−120	34	i.	∞	∞
Ketene	$H_2C:CO$	42.04	col. gas		−151	−56	d.	d.	s.
Koch acid (1-)(3-,6-,8-)	$C_{10}H_4(NH_2)S_3O_9HNa_2$	427.34	cr.				7.2^{20}		
Lactic acid (*dl*-)	$CH_3CH(OH)CO_2H$	90.08	hyg.	$1.249^{15/4}$	16.8	122^{14}	∞	∞	∞
anhydride	$C_6H_{10}O_5$	162.14	yel. oil			d. 250	v. sl. s.	s.	s.
Lactide (*dl*-)	$C_6H_8O_4$	144.13	tri./al.	$0.862^{10/4}$	124.5	255^{757}	v. sl. s.	v. sl. s. c.	
Lactose	$C_{12}H_{22}O_{11}\cdot H_2O$	360.31	col. rhb.	1.525^{20}	202	d.	17^{10}	i.	i.
Lauric acid	$CH_3(CH_2)_{10}CO_2H$	200.32	col. nd.	$0.869^{50/4}$	48(44)	225^{100}	i.	s.	s.
Laurone	$[CH_3(CH_2)_{10}]_2CO$	338.61	pl.	$0.809^{69/4}$	69–70		i.	i. c.	
Lauryl alcohol	$CH_3(CH_2)_{10}CH_2OH$	186.33	lf.	$0.831^{24/4}$	24	255–9	i.	s.	s.
Lead tetraethyl	$Pb(CH_2CH_3)_4$	323.44	col. lq.	$1.659^{18/4}$	−136	152^{291}	i.	sl. s.	∞
tetramethyl	$Pb(CH_3)_4$	267.34	col. lq.	$1.995^{20/4}$	−27.5	110^{760}	i.	∞	∞
Lepidine (*py*-4)	$C_9H_6N\cdot CH_3$	143.19	lq.	1.086^{20}	9–10	261–3	sl. s.	∞	∞
Leucine (*l*-)	$(CH_3)_2CHCH_2CH(NH_2)CO_2H$	131.17	cr.	1.293^{18}	295	subl.	2.2^{18}		
Levulinic acid	$CH_3CO(CH_2)_2CO_2H$	116.12	lf.	$1.140^{20/20}$	33.5	245–6	v. s.	v. s.	v. s.
Limonene (*d*- or *l*-)	$C_{10}H_{16}$	136.23	lq.	$0.842^{20/4}$	−96.9	177	i.	∞	∞
Linalool (*d*- or *l*-)	$C_{10}H_{17}OH$	154.25	col. oil	0.868^{20}		198–200	v. sl. s.	s.	∞
Linalyl acetate	$CH_3CO_2C_{10}H_{17}$	196.29	col. lq.	0.895^{20}		220^{762} d.	v. sl. s.	∞	∞
Linoleic acid	$C_{17}H_{31}CO_2H$	280.45	yel. oil	$0.903^{18/4}$	−9.5	$229–30^{16}$	i.	∞	∞
Maleic acid	$HO_2C\cdot CH:CH\cdot CO_2H$	116.07	mn.	1.609	130.5	135 d.	79^{25}	70^{30}	8^{25}
anhydride	$<(\cdot CHCO)_2>O$	98.06	cr.	1.5	57–60	202	16.3^{80}		
Malic acid (*dl*-)	$HO_2CCH_2CH(OH)CO_2H$	134.09	col. cr.	$1.601^{20/4}$	128–9	150 d.	144^{26}	v. s.	v. s.
(*d*- or *l*-)	$HO_2CCH_2CH(OH)CO_2H$	134.09	col. cr.	$1.595^{20/4}$	99–100	140 d.	v. s.	v. s.	8.4^{15}
Malonic acid	$H_2C(CO_2H)_2$	104.06	col. tri.	1.631^{15}	130–5 d.		138^{16}	42^{25}	8^{15}
Maltose	$C_{12}H_{22}O_{11}\cdot H_2O$	360.31	col. nd.	1.540^{17}	d.		108^{25}	v. sl. s. c.	i.
Mandelic acid (*dl*-)	$C_6H_5CH(OH)CO_2H$	152.15	rhb./aq.	$1.300^{20/4}$	118.1	d.	16^{20}	s.	s.
Mannitol (*d*-)	$CH_2OH(CHOH)_4CH_2OH$	182.17	col. rhb.	$1.489^{20/4}$	166	$290–5^{3}$	13^{14}	0.01^{14}	i.
Mannose (*d*-)	$CH_2OH(CHOH)_4CHO$	180.16	rhb.	$1.539^{20/4}$	132		248^{17}	v. sl. s.	i.
Margaric acid	$CH_3(CH_2)_{15}CO_2H$	270.45	col. pl.	0.853^{60}	60–1	227^{100}	i.	32^{28}	v. s.
Mellitic acid	$C_6(CO_2H)_6$	342.17	nd./al.		286–8	d.	v. s.	v. s.	
Menthol (l-)(α-)	$C_{10}H_{19}OH$	156.27	col. cr.	$0.890^{15/15}$	42–3	212	0.04 c.	v. s.	v. s.
Mercapto-benzothiazole (2-)	$<C_6H_4N:C(SH)S>$	167.25	nd.	$1.42^{20/4}$	179	d.	i.	s.	sl. s.
-thiazoline (2-)	$<CH_2N:C(SH)SCH_2>$	119.21	cr.	1.50	106		1.6^{60}		
Mercuric cyanide	$Hg(CN)_2$	252.62	cr.	4.003^{22}	d. 320		12.5^{15}		
fulminate	$Hg(ONC)_2\cdot \frac{1}{2}H_2O$	293.63	cr./aq.	4.4	expl.		0.07^{12}	s.	
Mesityl oxide	$(CH_3)_2C:CHCOCH_3$	98.14	lq.	$0.858^{20/4}$	−59	130^{750}	3^{20}	∞	∞
Mesitylene (1-,3-,5-)	$C_6H_3(CH_3)_3$	120.19	col. lq.	$0.865^{20/4}$	−45(−52)	164.8	i.	s.	∞
Metanilic acid (*m*-)	$H_2NC_6H_4SO_3H$	173.19	col. nd.		d.		2^{15}	v. sl. s.	v. sl. s.
Methane	CH_4	16.04	gas	0.415^{-164}	−182.6	−161.4	0.4^{20} cc.	47^{20} cc.	104^{10} cc.

Methoxy-methoxyethanol	$CH_3(OCH_2)_2CH_2OH$	106.12	lq.	1.038^{25}	<−70	167.5	∞		
Methyl acetate	$CH_3CO_2CH_3$	74.08	col. lq.	$0.924^{20/4}$	−98.7	57.1	33^{22}	∞	∞
acrylic acid (α-)	$CH_2{:}C(CH_3)CO_2H$	86.09	pr.	$1.015^{20/4}$	15–16	161–3	s. h.	∞	∞
alcohol	CH_3OH	32.04	col. lq.	$0.792^{20/4}$	−97–8	64.7	∞	∞	∞
-amine	CH_3NH_2	31.06	col. gas	0.699^{-11}	−92.5	-6.7^{758}	v. s.	v. s.	
-amine hydrochloride	$CH_3NH_2{\cdot}HCl$	67.52	pl./al.	1.23	226–8	230^{15}	v. s.	23 h.	i.
aniline	$C_6H_5NHCH_3$	107.15	lq.	$0.989^{20/4}$	−57	195.5	0.01^{25}	s.	∞
anthracene (α-)	$C_6H_4{:}(CH)_2{:}C_6H_3CH_3$	192.26	lf./al.	$1.047^{99.4}$	86		i.		
(β-)	$C_6H_4{:}(CH)_2{:}C_6H_3CH_3$	192.26	col. lf.	$1.181^{0/4}$	207		i.	v. sl. s.	v. sl. s.
anthranilate (*o*-)	$NH_2C_6H_4CO_2CH_3$	151.16	col. lq.	$1.168^{19/4}$	24	135.5^{15}	sl. s.	s.	s.
anthraquinone (2-)	$C_6H_4{:}(CO)_2{:}C_6H_3CH_3$	222.24	col. nd.		176–7	subl.	i.	s.	s.
benzoate	$C_6H_5CO_2CH_3$	136.15	col. lq.	$1.087^{25/25}$	−12.5	198–9	0.02^{30}	∞	∞
benzylaniline	$C_6H_5N(CH_3)CH_2C_6H_5$	197.28	lq.		9.2	305–6	i.	s.	s.
bromide	CH_3Br	94.94	gas	$1.732^{0/0}$	−93	4.5^{758}	v. sl. s.	s.	s.
butyrate (*n*-)	$CH_3(CH_2)_2CO_2CH_3$	102.13	col. lq.	$0.898^{20/4}$	<−95	102.3	1.7	∞	∞
(*i*-)	$(CH_3)_2CHCO_2CH_3$	102.13	col. lq.	$0.891^{20/4}$	−84.7	92.6	v. sl. s.	∞	∞
caprate	$CH_3(CH_2)_8CO_2CH_3$	186.29	lq.		−18	223–4	i.	∞	∞
caproate (*n*-)	$CH_3(CH_2)_4CO_2CH_3$	130.18	col. lq.	$0.904^{0/0}$		149.5	i.	∞	∞
caprylate	$CH_3(CH_2)_6CO_2CH_3$	158.24	col. lq.	0.887^{18}	−40	192–4	i.	∞	∞
cellosolve	$CH_3OCH_2CH_2OH$	76.09	col. lq.	$0.965^{20/4}$		124–5	∞	∞	∞
chloride	CH_3Cl	50.49	gas	0.952^{0}	−97.7	−24	280^{16} cc.	v. s.	v. s.
chloroacetate	$ClCH_2CO_2CH_3$	108.52	col. lq.	$1.236^{20/4}$	−32.7	130^{740}	v. sl. s.	∞	∞
chloroformate	$ClCO_2CH_3$	94.50	col. lq.	1.236^{15}		71–2	d.	∞	∞
cinnamate	$C_6H_5CH{:}CHCO_2CH_3$	162.19	cr.	$1.042^{36/0}$	33.4	263	i.	v. s.	v. s.
cyclohexane	$CH_2 < (CH_2CH_2)_2 > CHCH_3$	98.19	col. lq.	$0.769^{20/4}$	−126.3	101	i.	s.	s.
ethyl carbonate	$CH_3O{\cdot}CO{\cdot}OC_2H_5$	104.10	lq.	1.002^{27}	−14.5	109.2	i.	∞	∞
ethyl ketone	$CH_3.CO{\cdot}C_2H_5$	72.11	col. lq.	$0.805^{20/4}$	−85.9	79.6	35^{10}	∞	∞
ethyl oxalate	$CH_3OCO{\cdot}CO_2C_2H_5$	132.11	lq.	$1.156^{0/0}$		173.7	i.	v. s.	v. s.
formate	HCO_2CH_3	60.05	lq.	$0.974^{20/4}$	−99.8	32	30^{20}	∞	
furoate	$C_4H_3O{\cdot}CO_2CH_3$	126.11	col. lq.	$1.179^{21/4}$		181.3	i.	∞	∞
glucamine	$CH_2OH(CHOH)_4CH_2NHCH_3$	195.21							
glycolate	$HOCH_2CO_2CH_3$	90.08	lq.	1.168^{18}		151.2			
heptoate	$CH_3(CH_2)_5CO_2CH_3$	144.21	lq.	$0.881^{15/4}$		172–3	i.		
hypochlorite	$ClOCH_3$	66.49	gas			12^{726}			
iodide	CH_3I	141.94	col. lq.	$2.279^{20/4}$	−64.4	42.4	1.8^{15}	∞	∞
lactate	$CH_3CH(OH)CO_2CH_3$	104.10	lq.	1.090^{19}		144.8	∞	s.	s.
laurate	$CH_3(CH_2)_{10}CO_2CH_3$	214.34	lq.		5	148^{18}	i.		
mercaptan	CH_3SH	48.11	gas	0.896^{0}	−121	5.8^{752}	s.	v. s.	v. s.
methacrylate	$CH_2{:}C(CH_3)CO_2CH_3$	100.12	lq.	$0.950^{15.6}$	−48	100.3	i.		
myristate	$CH_3(CH_2)_{12}CO_2CH_3$	242.40	cr./al.		18–9	295^{715}	i.		
naphthalene (α-)	$C_{10}H_7CH_3$	142.20	oil	$1.025^{14/4}$	−19	244.6	i.	v. s.	v. s.
(β-)	$C_{10}H_7CH_3$	142.20	mn.	$0.994^{40/4}$	35–6	241–2	i.	v. s.	v. s.
nitrate	CH_3ONO_2	77.04	lq.	1.203^{25}	expl.	65	sl. s.	s.	s.
nitrite	CH_3ONO	61.04	gas	0.991^{15}		−12		s.	s.
nonyl ketone (*n*-)	$CH_3(CH_2)_8COCH_3$	170.29	col. oil	$0.828^{20/20}$	13.5	228	i.	s.	s.
oleate	$C_{17}H_{33}CO_2CH_3$	296.49	oil	0.879^{18}		$190\text{–}1^{10}$	i.	∞	∞
orange	$(CH_3)_2NC_6H_4N_2C_6H_4SO_3Na$	327.33	red pd.				0.2 c.		
palmitate	$CH_3(CH_2)_{14}CO_2CH_3$	270.45	col. cr.		30–1	196^{15}	i.	s.	s.
phosphine	CH_3PH_2	48.02	gas			-14^{759}	i.	sl. s.	
propionate	$CH_3CH_2CO_2CH_3$	88.11	col. lq.	$0.915^{20/4}$	−87.5	79.7	0.5^{20}	∞	∞
propyl ketone (*n*-)	$CH_3COCH_2CH_2CH_3$	86.13	col. lq.	$0.812^{15/15}$	−77.8	102	v. sl. s.	∞	∞
salicylate (*o*-)	$HO{\cdot}C_6H_4CO_2CH_3$	152.15	col. lq.	$1.182^{25/25}$	−8.3	222.2	0.07^{30}	∞	∞
stearate	$CH_3(CH_2)_{16}CO_2CH_3$	298.50	col. cr.		38–9	215^{15}	i.	s.	s.
toluate (*o*-)	$CH_3{\cdot}C_6H_4CO_2CH_3$	150.17	col. lq.	1.073^{15}	<−50	213	i.	∞	∞
(*m*-)	$CH_3{\cdot}C_6H_4CO_2CH_3$	150.17	col. lq.	1.066^{15}		215	i.		
(*p*-)	$CH_3{\cdot}C_6H_4CO_2CH_3$	150.17	cr.		33–4	217	i.	v. s.	v. s.
Methyl toluidine (*o*-)	$CH_3{\cdot}C_6H_4NHCH_3$	121.18	lq.	0.973^{15}		206–7	i.	∞	∞
(*m*-)	$CH_3{\cdot}C_6H_4NHCH_3$	121.18	lq.			206–7	i.	∞	∞
(*p*-)	$CH_3{\cdot}C_6H_4NHCH_3$	121.18	lq.	$0.935^{55/4}$		211^{761}	i.	∞	∞
valerate (*n*-)	$CH_3(CH_2)_3CO_2CH_3$	116.16	lq.	$0.895^{15/4}$	−91	127.3	v. sl. s.	∞	∞
(*i*-)	$(CH_3)_2CHCH_2CO_2CH_3$	116.16	col. lq.	$0.881^{20/4}$		116.7^{764}	v. sl. s.	∞	∞
vinyl ketone	$CH_3COCH{:}CH_2$	70.09	lq.	$0.836^{20/4}$		81	>85		
Methylal	$HCH(OCH_3)_2$	76.09	col. lq.	$0.866^{15/4}$	−104.8	42–3	33	∞	∞
Methylene-bis-(phenyl-4-isocyanate)	$(OCN{\cdot}C_6H_4)_2CH_2$	250.25	lq.	1.222^{30}		$210\text{–}2^{13}$	d.	d.	
bromide	CH_2Br_2	173.83	col. lq.	$2.495^{20/4}$	−52.8	98.5^{756}	1.17^{0}	∞	∞
chloride	CH_2Cl_2	84.93	col. lq.	$1.336^{20/4}$	−96.7	40–1	2^{20}	∞	∞
dianiline	$(C_6H_5NH)_2CH_2$	198.26	cr.		65	208–9 d.	i.	s.	s.
iodide	CH_2I_2	267.84	col. lq.	$3.325^{20/4}$	5.7	180 d.	1.4^{20}	∞	∞
Michler's hydrol (*p*-,*p*′-)	$[(CH_3)_2NC_6H_4]_2CHOH$	270.37	gn.		96–7		i.	s. h.	s.
ketone	$[(CH_3)_2NC_6H_4]_2CO$	268.35	lf./al.		174	>360 d.	i.	sl. s.	v. sl. s.
Morphine	$C_{17}H_{19}O_3N{\cdot}H_2O$	303.35	pr./al.	1.317	254 d.		0.02^{20}	sl. s.	s.
Mucic acid	$({\cdot}CHOHCHOHCO_2H)_2$	210.14	pd.		206–14		0.33^{14}	i.	i.

(*Continued*)

TABLE 2-2 Physical Properties of Organic Compounds *(Continued)*

Name	Formula	Formula weight	Form and color	Specific gravity	Melting point, °C	Boiling point, °C	Solubility in 100 parts: Water	Alcohol	Ether
Mustard gas	$(ClCH_2{\cdot}CH_2)_2S$	159.08	oil	$1.275^{20/4}$	13–4	217	0.07^{25}	s.	s.
Myricyl alcohol	$C_{31}H_{63}OH(?)$	452.84	cr.	0.777^{95}	88		i.	v. sl. s.	v. s.
Myristic acid	$CH_3(CH_2)_{12}CO_2H$	228.37	col. lf.	$0.853^{70/4}$	57–8	250.5^{100}	i.	v. s.	v. s.
Myristyl alcohol	$CH_3(CH_2)_{12}CH_2OH$	214.39	cr.	$0.824^{38/4}$	38	167^{15}	<0.02	sl. s.	s.
Naphthalene	$C_{10}H_8$	128.17	pl./al.	$1.145^{20/4}$	80.2	217.9	0.003^{25}	9.5^{20}	v. s.
disulfonic acid (1-,5-)	$C_{10}H_6(SO_3H)_2$	288.30	lf.		d.		102^{20}	s.	i.
(1-,6-)	$C_{10}H_6(SO_3H)_2$	288.30	cr.		d. 125		164^{20}	s.	i.
sulfonic acid (α-)	$C_{10}H_7SO_3H{\cdot}2H_2O$	244.26	cr.		90		v. s.	v. s.	sl. s.
(β-)	$C_{10}H_7SO_3H{\cdot}H_2O$	226.25	cr.		125		77^{30}		
Naphthasultam (1-,8-)	$C_{10}H_7O_2NS$	205.23	nd.		177–8		s. h.	sl. s.	s.
disulfonate Na (1-,8-)	$C_{10}H_5O_8NS_3Na_2{\cdot}2H_2O$	445.35	cr.				v. s.		
(2-,4-)	$C_{10}H_4O_8NS_3Na_3{\cdot}8½H_2O$	584.43	lf.				v. s.	sl. s.	
Naphthoic acid (α-)	$C_{10}H_7CO_2H$	172.18	nd.		160–1	300	v. sl. s. h.	s. h.	s.
(β-)	$C_{10}H_7CO_2H$	172.18	mn.	$1.077^{100/4}$	184	>300	0.007^{25}	s.	s.
Naphthol (α-)	$C_{10}H_7OH$	144.17	mn.	1.224^{4}	96	278–80	sl. s. h.	v. s.	v. s.
(β-)	$C_{10}H_7OH$	144.17	mn.	1.217^{4}	122–3	285–6	0.074^{25}	v. s.	v. s.
sulfonic acid (α-)(1-,2-)	$HO{\cdot}C_{10}H_6SO_3H$	224.23	pl./aq.		>250		v. s. h.		i.
(β-)(2-,6-)	$HO{\cdot}C_{10}H_6SO_3H$	224.23	lf.		125		v. s.	v. s.	
Naphthyl acetate (α-)	$CH_3CO_2C_{10}H_7$	186.21	nd./al.		46–9		sl. s. h.	s.	s.
(β-)	$CH_3CO_2C_{10}H_7$	186.21	nd./al.		69–70		i.	s.	s.
amine (α-)	$C_{10}H_7NH_2$	143.19	rhb.	$1.123^{25/25}$	50	300.8	0.17 c.	v. s.	v. s.
(β-)	$C_{10}H_7NH_2$	143.19	lf./aq.	$1.061^{98/4}$	111–2	306.1	v. s. h.	s.	s.
amine hydrochloride (α-)	$C_{10}H_7NH_2{\cdot}HCl$	179.65	nd.			subl.	3.8^{20}	s.	s.
(β-)	$C_{10}H_7NH_2{\cdot}HCl$	179.65	lf.				v. s.	v. s.	
amine sulfonic acid (1-,4-)	$NH_2{\cdot}C_{10}H_6{\cdot}SO_3H$	223.25	nd.		d.		0.2^{100}	i.	i.
(1-,5-)	$NH_2{\cdot}C_{10}H_6{\cdot}SO_3H{\cdot}H_2O$	241.26	cr.				sl. s.		
(1-,7-)	$NH_2{\cdot}C_{10}H_6{\cdot}SO_3H{\cdot}H_2O$	241.26	cr.				0.46^{25}		
(1-,8-)	$NH_2{\cdot}C_{10}H_6{\cdot}SO_3H{\cdot}H_2O$	241.26	cr.				0.42^{100}		
(2-,5-)	$NH_2{\cdot}C_{10}H_6{\cdot}SO_3H$	223.25	cr.				0.08		
(2-,6-)	$NH_2{\cdot}C_{10}H_6{\cdot}SO_3H{\cdot}H_2O$	241.26	cr.				0.38^{100}		
(2-,7-)	$NH_2{\cdot}C_{10}H_6{\cdot}SO_3H{\cdot}H_2O$	241.26	cr.				0.28^{100}		
isocyanate (α-)	$C_{10}H_7N{:}CO$	169.18	col. lq.	1.18		269–70	d.	s.	s.
Nicotine	$C_{10}H_{14}N_2$	162.23	oil	$1.009^{20/4}$	<–80	246^{730}	s.	∞	∞
Nicotinic acid (3-)	$C_5H_4NCO_2H$	123.11	nd./al.		235.2	subl.	s. h.	s. h.	v. sl. s.
(*i*-)(4-)	$C_5H_4NCO_2H$	123.11	nd./aq.		317	d.	s. h.	sl. s. h.	v. sl. s.
Nitro-acetanilide (*p*-)	$CH_3CONHC_6H_4NO_2$	180.16	rhb.		215–6		s. h.	s.	s.
-acetophenone (*m*-)	$CH_3COC_6H_4NO_2$	165.15	nd.		80–1	202	i.	s.	
-aminoanisole (4-,1-,2-)	$NO_2{\cdot}C_6H_3(OCH_3)NH_2$	168.15	red nd.	1.207^{156}	118		i.	s.	
(5-,1-,2-)	$NO_2{\cdot}C_6H_3(OCH_3)NH_2$	168.15	yel. nd.	1.211^{156}	139–40				
(3-,1-,4-)	$NO_2{\cdot}C_6H_3(OCH_3)NH_2$	168.15	red		123		sl. s.	s.	s.
-aminophenol (4-,2-,1-)	$NO_2{\cdot}C_6H_3(NH_2)OH$	154.12	or. pr.		142–3		sl. s. c.	v. s.	v. s.
-aniline (*o*-)	$NO_2{\cdot}C_6H_4NH_2$	138.12	yel. rhb.	1.442^{15}	71.5	284.1	s. h.	v. s.	v. s.
(*m*-)	$NO_2{\cdot}C_6H_4NH_2$	138.12	yel. rhb.	1.43	114	306.4	0.11^{20}	7.1^{20}	7.9^{20}
(*p*-)	$NO_2{\cdot}C_6H_4NH_2$	138.12	yel. mn.	1.437^{14}	146–7	331.7	0.08^{19}	5.8^{20}	6.1^{20}
-anisole (*o*-)	$CH_3OC_6H_4NO_2$	153.14	col. cr.	$1.254^{20/4}$	9.4	272–3	0.17^{30}	∞	∞
(*p*-)	$CH_3OC_6H_4NO_2$	153.14	pr./al.	1.233^{20}	54	274	0.06^{30}	v. s.	v. s.
-anthraquinone (α-)	$C_6H_4{:}(CO)_2{:}C_6H_3NO_2$	253.21	nd.		230	270^{7}	i.	sl. s.	v. sl. s.
-anthraquinone sulfonic acid (1-,5-)	$NO_2{\cdot}C_{14}H_6O_2{\cdot}SO_3H$	333.27	yel. cr.				s.	i.	i.
-benzal chloride (*m*-)	$NO_2{\cdot}C_6H_4{\cdot}CHCl_2$	206.03	mn.		65		i.	v. s. h.	v. s.
-benzaldehyde (*m*-)	$NO_2{\cdot}C_6H_4CHO$	151.12	nd./aq.		58	164^{23}	1.95^{112}	v. s. h.	v. s.
Nitro-benzene	$C_6H_5NO_2$	123.11	yel. lq.	$1.205^{18/4}$	5.7	210.9	0.19^{20}	v. s.	∞
-benzidine (2-)	$NH_2C_6H_4C_6H_3(NH_2)NO_2$	229.23	red nd.		143		sl. s. h.		
-benzoic acid (*o*-)	$NO_2{\cdot}C_6H_4{\cdot}CO_2H$	167.12	tri./aq.	$1.575^{4/4}$	147.5		0.65^{20}	28^{11}	22^{11}
(*m*-)	$NO_2{\cdot}C_6H_4{\cdot}CO_2H$	167.12	mn.	$1.494^{4/4}$	140–1		0.24^{165}	31^{12}	25^{10}
(*p*-)	$NO_2{\cdot}C_6H_4{\cdot}CO_2H$	167.12	yel. mn.	$1.550^{22/4}$	240–2	subl.	0.02^{15}	0.9^{10}	2.2^{18}
-benzyl alcohol (*m*-)	$NO_2{\cdot}C_6H_4{\cdot}CH_2OH$	153.14	cr.		27	$175–80^{3}$			
-benzyl bromide (*p*-)	$NO_2{\cdot}C_6H_4CH_2Br$	216.03	nd./al.		99–100		i.	2^{19}	v. s.
-chlorotoluene (1-,2-,6-)	$CH_3{\cdot}C_6H_3(NO_2)Cl$	171.58	cr.		37.5	238	i.		
-cresol (1-,3-,4-)	$CH_3{\cdot}C_6H_3(NO_2)OH$	153.14	yel.	$1.240^{89/4}$	32	125^{22}	v. sl. s.	v. s.	v. s.
-cymene (1-,2-,4-)	$CH_3{\cdot}C_6H_3(NO_2)CH(CH_3)_2$	179.22	oil	$1.067^{20/4}$		152^{15}	i.		
-dimethylaniline (*o*-)	$NO_2{\cdot}C_6N_4N(CH_3)_2$	166.18	yel. oil	$1.179^{20/4}$		$151–3^{80}$	v. sl. s.	v. s.	v. s.
(*m*-)	$NO_2{\cdot}C_6H_4N(CH_3)_2$	166.18	red mn.	1.313^{17}	60–1	280–5	i.	s.	s.
(*p*-)	$NO_2{\cdot}C_6H_4N(CH_3)_2$	166.18	yel. nd.		163–4		i.	s. h.	
-diphenyl (*o*-)	$C_6H_5{\cdot}C_6H_4NO_2$	199.21	rhb.	1.44	37	320	i.	s.	v. s.
(*p*-)	$C_6H_5{\cdot}C_6H_4NO_2$	199.21	nd./al.		113–4	340	i.	sl. s. c.	v. s.
-diphenylamine (*o*-)	$C_6H_5{\cdot}NH{\cdot}C_6H_4NO_2$	214.22	or. cr.		75–6				
-guanidine	$H_2NC(NH)NHNO_2$	104.07	nd./aq.		246–7		9^{100}	sl. s.	v. sl. s.

-naphthalene (α-)	$C_{10}H_7NO_2$	173.17	yel./al.	1.223^{62}	59–60	304	i.	s.	s.
(β-)	$C_{10}H_7NO_2$	173.17	col./al.		79	165^{15}	i.	v. s.	v. s.
-phenol (*o*-)	$NO_2{\cdot}C_6H_4{\cdot}OH$	139.11	yel. mn.	1.295^{45}	44–5	214.5	1.08^{100}	v. s.	v. s.
(*m*-)	$NO_2{\cdot}C_6H_4{\cdot}OH$	139.11	col. mn.	1.485^{20}	96–7	194^{70}	1.35^{20}	v. s.	s.
(*p*-)	$NO_2{\cdot}C_6H_4{\cdot}OH$	139.11	yel. pr.	1.479^{20}	113–4	subl.	1.6^{25}	v. s.	v. s.
-phenol sulfonic acid (1-,4-,2-)	$HO{\cdot}C_6H_3(NO_2)SO_3H{\cdot}3H_2O$	273.22	nd.		d. 110		v. s.	v. s.	sl. s.
(1-,2-,4-)	$HO{\cdot}C_6H_3(NO_2)SO_3H{\cdot}3H_2O$	273.22	nd./aq.		51.5		v. s.	v. s.	
-phthalic acid (3-)	$NO_2{\cdot}C_6H_3(CO_2H)_2$	211.13	yel./aq.		222		2.05^{25}	v. s. h.	sl. s.
(4-)	$NO_2{\cdot}C_6H_3(CO_2H)_2$	211.13	yel. cr.		164–5		v. s.	v. s.	s.
-toluene (*o*-)	$CH_3{\cdot}C_6H_4NO_2$	137.14	yel. lq.	$1.163^{20/4}$	−4.1	222.3	0.07^{80}	∞	∞
(*m*-)	$CH_3{\cdot}C_6H_4NO_2$	137.14	lq.	$1.160^{18/4}$	15–16	230–1	0.05^{80}	∞	∞
(*p*-)	$CH_3{\cdot}C_6H_4NO_2$	137.14	rhb.	$1.139^{55/55}$	51.9	237.7	0.04^{80}	8.6^{15}	80.8^{15}
-toluene sulfonic acid (1-,4-,2-)	$CH_3{\cdot}C_6H_3(NO_2)SO_3H{\cdot}2H_2O$	253.23	pl./aq.		130		47.7^{28}	v. s.	v. s.
-toluidine (4-,1-,2-)	$NO_2{\cdot}C_6H_3(CH_3)NH_2$	152.15	yel. mn.	1.365^{15}	105–7		v. sl. s.	s.	s.
(3-,1-,4-)	$NO_2{\cdot}C_6H_3(CH_3)NH_2$	152.15	red mn.	1.312^{17}	116–7		sl. s. h.	s.	
Nitron	$C_{20}H_{16}N_4$	312.37	yel. lf.		189–90 d.		i.	s. h.	v. sl. s.
Nitroso-dimethylaniline (*p*-)	$ON{\cdot}C_6H_4N(CH_3)_2$	150.18	gn. tri.		86–7		i.	s.	s.
-naphthol (β-)(1-)	$ON{\cdot}C_{10}H_6OH$	173.17	brn. pr.		109.5		0.1^{20}	2.4^{18}	
Nonadecane (*n*-)	$CH_3(CH_2)_{17}CH_3$	268.52	cr.	$0.777^{32/4}$	32	330	i.	sl. s.	s.
Nonane (*n*-)	$CH_3(CH_2)_7CH_3$	128.26	col. lq.	$0.718^{20/4}$	−53.7	150.5^{759}	i.	sl. s.	s.
Octadecane (*n*-)	$CH_3(CH_2)_{16}CH_3$	254.49	cr.	$0.775^{28/4}$	28	317	i.	sl. s.	s.
Octane (*n*-)	$CH_3(CH_2)_6CH_3$	114.23	col. lq.	$0.703^{20/4}$	−56.5	125.7	0.002^{16}	sl. s.	s.
(iso-)	$(CH_3)_3CCH_2CH(CH_3)_2$	114.23	col. lq.	$0.692^{20/4}$	−107.4	99.3^{760}	i.	sl. s.	s.
Octyl acetate (*n*-)	$CH_3CO_2CH_2(CH_2)_6CH_3$	172.26	col. lq.	$0.885^{0/4}$	−38.5	210	i.	s.	s.
(*sec*-)	$CH_3CO_2CH(CH_3)C_6H_{13}$	172.26	col. lq.	$0.863^{14/4}$		195	i.	s.	s.
alcohol (*n*-)	$CH_3(CH_2)_6CH_2OH$	130.23	col. lq.	$0.827^{20/4}$	−16	194–5	0.054^{25}	∞	∞
(*sec*-)	$CH_3(CH_2)_5CH(OH)CH_3$	130.23	col. lq.	$0.822^{20/4}$	−38.6	179–80	0.096^{25}	∞	∞
Octylene (*n*-)	$CH_3(CH_2)_5CH{:}CH_2$	112.21	lq.	$0.721^{18/4}$		126	i.	∞	∞
Oleic acid	$C_8H_{17}CH{:}CH(CH_2)_7CO_2H$	282.46	col. nd.	$0.854^{78/4}$	14	$285–6^{100}$	i.	∞	∞
Orcinol (1-,3-,5-)	$(HO)_2C_6H_3{\cdot}CH_3$	124.14	pr./bz.	1.290^{4}	107–8	287–90	v. s.	v. s.	v. s.
Oxalic acid	$HO_2C{\cdot}CO_2H{\cdot}2H_2O$	126.07	col. mn.	$1.653^{19/4}$	101.5	subl.	s.	s.	1.3
Palmitic acid	$CH_3(CH_2)_{14}CO_2H$	256.42	col. pl.	$0.849^{70/4}$	63–4	271.5^{100}	i.	9^{20}	s.
Pelargonic acid	$CH_3(CH_2)_7CO_2H$	158.24	col. oil	$0.906^{20/4}$	12.5	253–4	v. sl. s.	s.	s.
Penta-chloroethane	$CHCl_2{\cdot}CCl_3$	202.29	col. lq.	$1.671^{25/4}$	−22	162	0.05^{20}	∞	∞
-decane (*n*-)	$CH_3(CH_2)_{13}CH_3$	212.41	col. lq.	$0.770^{20/4}$	10	270.5	i.	v. s.	v. s.
-erythritol	$C(CH_2OH)_4$	136.15	cr.		262	276^{30}	5.6^{15}	v. sl. s.	i.
Pentandiol	$HOCH_2(CH_2)_3CH_2OH$	104.15	lq.	$0.994^{20/4}$		239.4	∞		
Pentane (*n*-)	$CH_3(CH_2)_3CH_3$	72.15	col. lq.	$0.630^{18/4}$	−129.7	36.3	0.036^{16}	∞	∞
(*i*-)	$(CH_3)_2CHCH_2CH_3$	72.15	col. lq.	0.621^{19}	−160.0	27.95	i.	∞	∞
(neo-)	$(CH_3)_2C(CH_3)_2$	72.15	col. lq.	$0.613^{20/4}$	−20	9.5	i.	s.	s.
Phenacetin	$C_2H_5OC_6H_4NHCOCH_3$	179.22	col. mn.		134–5	d.	0.7^{20}	40 h.	1.6^{25}
Phenanthrene	$<(C_6H_4CH)_2>$	178.23	pl./al.	1.179^{25}	99–100	340	i.	10 h.	v. s.
Phenetidine (*o*-)	$C_2H_5O{\cdot}C_6H_4{\cdot}NH_2$	137.18	oil		<−21	228–9	i.	s.	s.
(*p*-)	$C_2H_5O{\cdot}C_6H_4{\cdot}NH_2$	137.18	lq.	1.061^{15}	3–4	254–5	i.	s.	s.
Phenetole	$C_2H_5O{\cdot}C_6H_5$	122.16	col. lq.	$0.967^{20/4}$	−30.2	172	i.	∞	∞
Phenol	C_6H_5OH	94.11	col. nd.	$1.071^{25/4}$	42–3	181.4	8.2^{15}	∞	∞
-phthalein	$C_{20}H_{14}O_4$	318.32	col. rhb.	$1.299^{25/4}$	261–2		0.2^{20}	10^{25}	5.9 c.
-sulfonic acid (*o*-)	$HO{\cdot}C_6H_4SO_3H{\cdot}\frac{3}{4}H_2O$	187.69	cr.		50 d.		v. s.	v. s.	
Phenyl acetaldehyde	$C_6H_5CH_2CHO$	120.15	lq.	1.025^{20}		193–4	v. sl. s.	∞	∞
acetic acid	$C_6H_5CH_2CO_2H$	136.15	lf.	$1.081^{80/4}$	76–7	265.5	1.66^{20}	v. s.	v. s.
-acetylene	$C_6H_5C{:}CH$	102.13	col. lq.	$0.930^{20/4}$	−43	142–3	i.	∞	∞
aniline (*o*-)	$C_6H_5{\cdot}C_6H_4{\cdot}NH_2$	169.22	cr.		45–6	299^{760}	v. sl. s.	s.	s.
(*p*-)	$C_6H_5{\cdot}C_6H_4{\cdot}NH_2$	169.22	lf.		50–2	302	s. h.	s.	s.
Phenyl-ethyl alcohol	$C_6H_5CH_2CH_2OH$	122.16	col. oil	$1.023^{18/4}$		$219–21^{750}$	1.6^{20}	s.	∞
-glycine	$C_6H_5NHCH_2CO_2H$	151.16	cr.		127		s.	s.	sl. s.
-hydrazine	$C_6H_5NH{\cdot}NH_2$	108.14	yel. oil	$1.097^{23/4}$	19.6	243.5	sl. s. h.	∞	∞
-hydrazine sulfonic acid (*p*-)	$H_2NNHC_6H_4SO_3H$	188.20	cr./al.		286		0.6^{12}	sl. s.	
isocyanate	$C_6H_5N{:}CO$	119.12	lq.	$1.096^{20/4}$		166^{769}	d.	d.	v. s.
-methylpyrazolone (3-)(*N*-)	$C_4H_5ON_2{\cdot}C_6H_5$	174.20	pr./aq.		128	191^{17}	1^{20}	v. s. h.	v. sl. s.
-mustard oil	$C_6H_5N{:}CS$	135.19	col. lq.	$1.138^{15/15}$	−21	219–20	i.	s.	s.
naphthalene (α-)	$C_{10}H_7{\cdot}C_6H_5$	204.27	waxy		45	336–7	i.	v. s.	v. s.
(β-)	$C_{10}H_7{\cdot}C_6H_5$	204.27	lf./al.		102.5	345–6	i.	sl. s.	sl. s.
naphthylamine (α-)	$C_{10}H_7NHC_6H_5$	219.28	pr./al.	1.17	62	335^{258}	0.08^{60}	s.	s.
(β-)	$C_{10}H_7NHC_6H_5$	219.28	rhb.	1.18	107–8	399.5	0.4^{60}	v. s. h.	v. s. h.
phenol (*o*-)	$C_6H_5{\cdot}C_6H_4OH$	170.21	nd.		56–7	275	i.	s.	s.
(*p*-)	$C_6H_5{\cdot}C_6H_4OH$	170.21	nd.		164–5	305–8	i.	s.	s.
propyl alcohol (γ-)	$C_6H_5(CH_2)_3OH$	136.19	oil	$1.008^{20/4}$	<−18	235–7	sl. s.	∞	∞
quinoline (2-)(α-)	$C_6H_5{\cdot}C_9H_6N$	205.25	nd.		86	363	sl. s.	s. h.	s.
(8-)(0-)	$C_6H_5{\cdot}C_9H_6N$	205.25	lq.			283^{187}	sl. s.	s.	s.
salicylate, salol	$HO{\cdot}C_6H_4CO_2C_6H_5$	214.22	rhb./al.	$1.250^{20/4}$	42–3	$172–3^{12}$	0.015^{25}	v. s.	s.
stearate	$CH_3(CH_2)_{16}CO_2C_6H_5$	360.57	cr.		52	267^{15}	i.		
urethane	$C_6H_5NHCO_2C_2H_5$	165.19	pl./al.	$1.106^{30/4}$	52–3	237–8	i. c.	s.	s.

(*Continued*)

TABLE 2-2 Physical Properties of Organic Compounds *(Continued)*

Name	Formula	Formula weight	Form and color	Specific gravity	Melting point, °C	Boiling point, °C	Solubility in 100 parts		
							Water	Alcohol	Ether
Phenylene-diamine (*o*-)	$C_6H_4(NH_2)_2$	108.14	lf./aq.		103–4	256–8	733^{81}	v. s.	v. s.
(*m*-)	$C_6H_4(NH_2)_2$	108.14	rhb.	$1.139^{15/15}$	62.8	284–7	35.1^{25}	v. s.	s.
(*p*-)	$C_6H_4(NH_2)_2$	108.14	mn.		140	267	669^{107}	s.	s.
Phloroglucinol (1-,3-,5-)	$C_6H_3(OH)_3{\cdot}2H_2O$	162.14	rhb.		117	subl.	1.13^{25}	v. s.	v. s.
Phorone	$[(CH_3)_2C{:}CH]_2CO$	138.21	yel. pr.	$0.885^{20/4}$	28	197.2^{743}	0.1^{50}	s.	s.
Phosgene	$OCCl_2$	98.92	gas	$1.392^{19/4}$	−104	8.2^{756}	v. sl. s.		
Phthalic acid (*o*-)	$C_6H_4(CO_2H)_2$	166.13	mn./aq.	$1.593^{20/4}$	208	d.	0.70^{25}	12^{18}	0.68^{15}
(*m*-)(iso-)	$C_6H_4(CO_2H)_2$	166.13	nd./aq.		330	subl.	0.2^{100}	s.	
anhydride (*o*-)	$C_6H_4<(CO)_2>O$	148.12	rhb.	1.527^{4}	130.8	284.5	v. sl. s.	s.	sl. s.
nitrile (*o*-)	$C_6H_4(CN)_2$	128.13	cr.		141		sl. s. c.		
Phthalide	$C_6H_4(CH_2)(CO)>O$	134.13	nd./aq.	$1.164^{99/4}$	73(65)	290	v. sl. s.	s.	
Phthalimide (*o*-)	$C_6H_4<(CO)_2>NH$	147.13	cr./et.		238	subl.	0.04^{25}	5	s. h.
Picoline (α-)	$C_5H_4N{\cdot}CH_3$	93.13	col. lq.	$0.950^{15/4}$	−70	128.8	v. s.	∞	∞
(β-)	$C_5H_4N{\cdot}CH_3$	93.13	col. lq.	$0.961^{15/4}$		143.5	∞	∞	∞
(γ-)	$C_5H_4N{\cdot}CH_3$	93.13	lq.	$0.957^{15/4}$		143.1	∞	∞	∞
Picramic acid (1-,2-,4-,6-)	$HO{\cdot}C_6H_2(NH_2)(NO_2)_2$	199.12	red nd.		169		0.14^{22}	s.	sl. s.
Picric acid (2-,4-,6-)	$HO{\cdot}C_6H_2(NO_2)_3$	229.10	yel. rhb.	$1.763^{20/4}$	121.8	expl.	1.23^{20}	6^{20}	1^{13}
Picryl chloride (2-,4-,6-)	$ClC_6H_2(NO_2)_3$	247.55	yel. mn.	1.797^{20}	83	d.	0.018^{15}	4.8^{17}	7^{17}
Pinacol	$[(CH_3)_2C{\cdot}OH]_2$	118.17	col. nd.	0.967^{15}	43(38)	$171–2^{789}$	sl. s. c.	v. s.	v. s.
Pinacoline	$CH_3COC(CH_3)_3$	100.16	col. lq.	0.800^{16}	−52.5	106.2	2.5^{15}	s.	s.
Pinene (α-)(*dl*-)	$C_{10}H_{16}$	136.23	col. lq.	$0.878^{20/4}$	−55	154–6	v. sl. s.	s.	∞
hydrochloride	$C_{10}H_{17}Cl$	172.69	lf.		131–2	207–8	i.	33	s.
Pinol (*dl*-)	$C_{10}H_{16}O$	152.23	lq.	$0.953^{20/20}$		183–4		s.	s.
Piperidine	$CH_2<(CH_2CH_2)_2>NH$	85.15	lq.	$0.860^{20/4}$	−9	106	∞	∞	
carboxylic acid (α-)(*dl*-)	$HO_2C{\cdot}CH<(CH_2CH_2)_2>NH$	129.16	cr.		264		s.		
Piperidinium pentamethylene dithiocarbamate	$(CH_2)_5CS_2H{\cdot}HN(CH_2)_5$	232.43	cr.	1.13	175		6^{28}		
Propane	$CH_3CH_2CH_3$	44.10	gas	$0.585^{-45/4}$	−187.1	−42.2	6.5^{18} cc.	s.	v. s.
Propionic acid	$CH_3CH_2CO_2H$	74.08	col. lq.	$0.992^{20/4}$	−22	141.1	∞	∞	∞
aldehyde	CH_3CH_2CHO	58.08	col. lq.	$0.807^{20/4}$	−81	49.5^{740}	20^{20}	∞	∞
anhydride	$(CH_3CH_2CO)_2O$	130.14	col. lq.	$1.012^{20/4}$	−45	168.8^{780}	d.	d.	
Propyl acetate (*n*-)	$CH_3CO_2CH_2CH_2CH_3$	102.13	col. lq.	$0.886^{20/4}$	−92.5	101.6	1.6^{16}	∞	∞
(*i*-)	$CH_3CO_2CH(CH_3)_2$	102.13	col. lq.	$0.874^{20/20}$	−73.4	88.4	3^{20}	∞	∞
alcohol (*n*-)	$CH_3CH_2CH_2OH$	60.10	col. lq.	$0.804^{20/4}$	−127	97.8	∞	∞	∞
(*i*-)	$(CH_3)_2CHOH$	60.10	col. lq.	$0.789^{20/4}$	−85.8	82.5	∞	∞	∞
amine (*n*-)	$CH_3CH_2CH_2NH_2$	59.11	col. lq.	$0.718^{20/20}$	−83	$49–50^{761}$	∞	∞	∞
(*i*-)	$(CH_3)_2CHNH_2$	59.11	col. lq.	$0.694^{15/4}$	−101	33–4	∞	∞	∞
aniline (*n*-)	$C_6H_5NHCH_2CH_2CH_3$	135.21	lq.	0.949^{18}		222	i.	v. s.	v. s.
benzoate (*n*-)	$C_6H_5CO_2CH_2CH_2CH_3$	164.20	col. lq.	$1.021^{25/25}$	−51.6	231	i.	s.	s.
(*i*-)	$C_6H_5CO_2CH(CH_3)_2$	164.20	col. lq.	$1.010^{25/25}$		218.5	i.	s.	s.
bromide (*n*-)	$CH_3CH_2CH_2Br$	122.99	col. lq.	$1.353^{20/4}$	−109.9	70.8	0.25^{20}	∞	∞
(*i*-)	$(CH_3)_2CHBr$	122.99	col. lq.	$1.310^{20/4}$	−89	60	0.32^{20}	∞	∞
n-butyrate (*n*-)	$C_2H_5CH_2CO_2CH_2C_2H_5$	130.18	col. lq.	0.879^{15}	−95.2	142.7	0.17^{17}	∞	∞
i-butyrate (*n*-)	$(CH_3)_2CHCO_2CH_2C_2H_5$	130.18	col. lq.	$0.884^{0/4}$		134–5	v. sl. s.		
n-butyrate (*i*-)	$C_2H_5CH_2CO_2CH(CH_3)_2$	130.18	col. lq.	0.865^{18}		128	v. sl. s.		
i-butyrate (*i*-)	$(CH_3)_2CHCO_2CH(CH_3)_2$	130.18	col. lq.	$0.869^{0/4}$		120.8	v. sl. s.		
chloride (*n*-)	$CH_3CH_2CH_2Cl$	78.54	col. lq.	$0.890^{20/4}$	−122.8	46.4	0.27^{20}	∞	∞
(*i*-)	$(CH_3)_2CHCl$	78.54	col. lq.	0.859^{20}	−117	36.5	0.31^{20}	∞	∞
Propyl formate (*n*-)	$HCO_2CH_2CH_2CH_3$	88.11	col. lq.	$0.901^{20/4}$	−92.9	81.3	12.2^{22}	∞	∞
(*i*-)	$HCO_2CH(CH_3)_2$	88.11	col. lq.	$0.873^{20/4}$		$68–71^{751}$	2.1^{22}	∞	∞
furoate (*n*-)	$C_4H_3O{\cdot}CO_2C_3H_7$	154.16	col. lq.	$1.075^{26/4}$		211	v. sl. s.	s.	∞
lactate (*n*-)	$CH_3CH(OH)CO_2CH_2C_2H_5$	132.16	col. lq.			$122–3^{150}$	s.	s.	s.
(*i*-)	$CH_3CH(OH)CO_2CH(CH_3)_2$	132.16	col. lq.			167.5	s.	s.	s.
mercaptan (*n*-)	$CH_3CH_2CH_2SH$	76.16	lq.	$0.836^{25/4}$	−112	67–8	v. sl. s.	s.	s.
(*i*-)	$(CH_3)_2CHSH$	76.16	lq.	$0.809^{25/4}$	−130.7	58–60	v. sl. s.	∞	∞
propionate (*n*-)	$C_2H_5CO_2CH_2C_2H_5$	116.16	col. lq.	$0.883^{20/4}$	−76	122–3	0.56^{25}	∞	∞
(*i*-)	$C_2H_5CO_2CH(CH_3)_2$	116.16	col. lq.	0.893^{0}		$109–11^{750}$	0.6^{25}	∞	∞
thiocyanate (*i*-)	$(CH_3)_2CH{\cdot}CNS$	101.17	lq.	0.963^{20}		$152–3^{754}$	i.	∞	∞
n-valerate (*n*-)	$CH_3(CH_2)_3CO_2CH_2C_2H_5$	144.21	lq.	0.874^{15}	−70.7	67.5	i.	∞	∞
i-valerate (*n*-)	$(CH_3)_2CHCH_2CO_2C_3H_7$	144.21	col. lq.	$0.863^{20/4}$		155.9	i.	∞	∞
i-valerate (*i*-)	$(CH_3)_2CHCH_2CO_2C_3H_7$	144.21	col. lq.	0.854^{17}		142^{756}			
Propylene	$CH_3CH{:}CH_2$	42.08	gas	$0.609^{-47/4}$	−185	-48^{749}	44.6 cc.	1200 cc.	
bromide	$CH_3CHBrCH_2Br$	201.89	col. lq.	$1.933^{20/4}$	−55.5	141.6	0.25^{20}	s.	v. s.
chlorohydrin	$CH_3CHClCH_2OH$	94.54	col. lq.	1.103^{20}		133–4	s.	s.	s.
chloride	$CH_3CHClCH_2Cl$	112.99	col. lq.	$1.159^{20/20}$	<−70	96.8	0.27^{20}	v. s.	v. s.
glycol	$CH_3CH(OH)CH_2OH$	76.09	col. oil	$1.040^{19.4}$		188–9	∞	∞	8
oxide	$CH_3(CHCH_2)O$	58.08	col. lq.	$0.831^{20/20}$		35	33^{20}	∞	∞
Protocatechuic acid (3-,4-)	$(HO)_2C_6H_3CO_2H{\cdot}H_2O$	172.14	nd./aq.	$1.542^{4/4}$	199 d.		1.82^{14}	v. s.	s.

Pulegol (iso-)(*d*-)	$C_{10}H_{17}OH$	154.25	col. lq.	$0.911^{20/4}$		$86–9^{10}$	v. sl. s.		
Pulegone	$C_{10}H_{16}O$	152.23	col. lq.	$0.932^{20/20}$		224^{754}	i.	∞	∞
Pyrazole	—NH·N:CH·CH:CH—	68.08	nd./et.		70	186–8	s.	s.	s.
Pyrazoline	$—NH·N:CH·CH_2CH_2—$	70.09	lq.			144	∞	∞	sl. s.
Pyrazolone	$—NH·CO·CH_2CH:N—$	84.08	nd.		165	subl. d.	s.	v. s.	v. sl. s.
Pyrene	$C_{16}H_{10}$	202.25	yel. pr.	$1.277^{0/4}$	149–50	>360	i.	3 h.	v. s.
Pyridazine	$N_2<(CHCH)_2>$	80.09	lq.	$1.107^{20/4}$	−8	208	∞	s.	s.
Pyridine	$CH<(CHCH)_2>N$	79.10	col. lq.	$0.982^{20/4}$	−42	115–6	∞	∞	s.
Pyrocatechol (*o*-)	$C_6H_4(OH)_2$	110.11	nd./aq.	1.344^{4}	104–5	240–5	45.1^{20}	v. s.	v. s.
Pyrogallol (1-,2-,3-)	$C_6H_3(OH)_3$	126.11	nd.	1.453^{4}	133–4	309	40^{13}	s.	s.
Pyrone	$CO<(CHCH)_2>O$	96.08	cr.	$1.190^{40.3}$	32.5	215–7	v. sl. s.	s.	v. s.
Pyrrole	$<(CH:CH)_2>NH$	67.09	lq.	$0.948^{20/4}$		131	i.	s.	s.
Pyrrolidine	$<(CH_2·CH_2)_2>NH$	71.12	lq.	$0.852^{22.5}$		87–8	∞	∞	∞
Pyrroline	$<(CH·CH_2)_2>NH$	69.11	lq.	$0.910^{20/4}$		90–1	v. s.	∞	∞
Pyruvic acid	CH_3COCO_2H	88.06	col. lq.	$1.267^{20/4}$	13.6	165	∞	∞	∞
Quercitrin	$C_{21}H_{20}O_{11}·2H_2O$	484.41	yel. nd.		182–5		0.04^{20}	s.	sl. s.
Quinaldine (py-2)	$CH_3·C_9H_6N$	143.19	lq.	$1.059^{20/4}$	−1	$244–5^{750}$	v. sl. s.		s.
Quinoline	C_9H_7N	129.16	lq.	1.095^{20}	−15	237.1^{747}	6	∞	∞
(iso-)	C_9H_7N	129.16	pl.	$1.099^{21/4}$	24.6	240.5^{763}	sl. s.		s.
-diol (1-,3-)	—C_6H_4CH:C(OH)N:C(OH)—	161.16	cr.		237		v. sl. s.		
Quinone (*p*-)	$CO<(CHCH)_2>CO$	108.09	yel. mn.	$1.318^{20/4}$	115.7	subl.	sl. s. h.	s.	s.
R-acid Ca salt (2-)(3-,6-)	$HOC_{10}H_5(SO_3)_2Ca$	342.36	cr.				30.6^{25}		
K salt	$HOC_{10}H_5(SO_3K)_2$	380.48	cr.				29.5^{25}		
Na salt	$HOC_{10}H_5(SO_3Na)_2$	348.26	cr.				25.2^{25}		
Raffinose	$C_{18}H_{32}O_{16}·5H_2O$	594.51	cr./aq.	1.465^{0}	119	d. 130	14.3^{20}	0.1^{20}	
Resorcinol (*m*-)	$C_6H_4(OH)_2$	110.11	col. rhb.	1.272^{15}	110.7	276.5	147^{12}	v. s.	v. s.
Retene	$C_{18}H_{18}$	234.34	lf./al.	1.13^{16}	98–9	390–4	i.	69 h.	v. s. h.
Rhamnose (β-)	$CH_3(CHOH)_4CHO·H_2O$	182.17	col. mn.	$1.471^{20/4}$	126		60.8^{21}		i.
Ricinoleic acid	$C_{17}H_{32}(OH)CO_2H$	298.46	lq.	0.954^{16}	4–5	$226–8^{10}$	i.	∞	∞
Rosaniline	$C_{20}H_{21}ON_3$	319.40	col. nd.		186 d.		v. sl. s.	sl. s.	i.
Rosolic acid	$C_{20}H_{16}O_3$	304.34	red lf.		308–10 d.		0.12^{25}	v. s. h.	sl. s.
Saccharin	$C_6H_4(CO)(SO_2)>NH$	183.18	mn.		225–8	subl.	0.4^{25}	3.1 c.	1.05 c.
Safrole (1-,3-,4-)	$CH_2:CHCH_2·C_6H_3:O_2CH_2$	162.19	col. mn.	$1.100^{20/4}$	11.2	233–4	i.	s.	∞
(iso-)(1-,3-,4-)	$CH_3·CH:CH·C_6H_3:O_2CH_2$	162.19	col. lq.	$1.122^{20/4}$	6–7	252–3	i.	∞	∞
Salicylic acid (*o*-)	$HO·C_6H_4·CO_2H$	138.12	mn.	$1.443^{20/4}$	159	211^{20}	0.2^{23}	49^{15}	51^{15}
aldehyde (*o*-)	$HO·C_6H_4·CHO$	122.12	col. oil	$1.153^{25/4}$	−7	196.5	1.7^{86}	∞	∞
Saligenin	$HO·C_6H_4·CH_2OH$	124.14	rhb./aq.	1.161^{25}	86–7	subl.	6.6^{15}	v. s.	v. s.
Schaeffer's salt, Ca	$(HOC_{10}H_6SO_3)_2Ca·5H_2O$	576.60	cr.				4.76^{20}		
K	$HOC_{10}H_6SO_3K$	262.32	cr.				3.46^{25}		
Na	$HOC_{10}H_6SO_3Na$	246.21	cr.				6.29^{25}		
Semicarbazide	$NH_2·CO·NH·NH_2$	75.07	pr./al.		96		v. s.	v. s.	i.
hydrochloride	$NH_2·CO·NH·NH_3Cl$	111.53	pr.		173 d.		v. s.	sl. s.	i.
Skatole (3-)	$CH_3·C_8H_6N$	131.17	lf.		95	$265–6^{755}$	0.05 c.	s.	s.
Sodium methylate	CH_3ONa	54.02	pd.		d. 300		d.		
Sorbitol	$[CH_2OH(CHOH)_2]_2$	182.17	cr.		110–2		v. s.	v. s. h.	
Sorbose (*d*- or *l*-)	$C_6H_{12}O_6$	180.16	rhb.	1.654^{15}	165		55^{17}	sl. s.	
Starch	$(C_6H_{10}O_5)x$	162.14	amor.	1.50^{21}	d.		i.	i.	i.
Stearic acid	$CH_3(CH_2)_{16}CO_2H$	284.48	mn.	$0.847^{69.3}$	70–1	291^{110}	0.03^{25}	2^{20}	6^{8}
amide	$CH_3(CH_2)_{16}CONH_2$	283.49	col. cr.		108–9	251^{12}	i.	s. h.	s. h:
Styrene	$C_6H_5CH:CH_2$	104.15	col. lq.	$0.903^{20/4}$	−31	145–6	v. sl. s.	∞	∞
Suberic acid	$HO_2C(CH_2)_6CO_2H$	174.19	nd./aq.	$1.266^{25/4}$	140–4	279^{100}	0.14^{16}	s.	0.8^{15}
Succinic acid	$HO_2C(CH_2)_2CO_2H$	118.09	col. mn.	$1.572^{25/4}$	189–90	235 d.	6.8^{20}	9.9^{15}	1.2^{15}
Sucrose	$C_{12}H_{22}O_{11}$	342.30	col. mn.	1.588^{15}	170–86 d.		179^{0}	0.9	i.
Sulfanilic acid (*p*-)	$H_2N·C_6H_4·SO_3H$	173.19	col. cr.		d. > 280		0.8^{10}	v. sl. s.	v. sl. s.
Sylvestrene (*d*-)	$C_{10}H_{16}$	136.23	lq.	$0.863^{20/4}$		176–7			
Tartaric acid (meso-)	$(CHOHCO_2H)_2$	150.09	cr.	1.737	159–60		120^{15}		
(racemic)	$(CHOHCO_2H)_2·H_2O$	168.10	tri.	$1.697^{20/4}$	205–6		20.6^{20}	2^{0}	0.09
(*d*- or *l*-)	$(CHOHCO_2H)_2$	150.09	mn.	$1.760^{20/4}$	168–70	d.	139^{20}	25^{15}	0.4^{15}
Tartronic acid	$CH(OH)(CO_2H)_2·½H_2O$	129.07	pr./aq.		d. 155–8	subl.	v. s.	v. s.	i.
Terephthalic acid (*p*-)	$C_6H_4(CO_2H)_2$	166.13	cr.	1.510	subl.		0.001 c.	sl. s. h.	i.
Terpin hydrate (*cis*-)	$C_{10}H_{20}O_2·H_2O$	190.28	rhb.		117	d.	0.4^{15}	10^{15}	1^{15}
Terpineol (α-)(*d*- or *l*-)	$C_{10}H_{18}O$	154.25	col. cr.	0.935^{15}	38–40	219–21	i.	v. s.	v. s.
(*dl*-)	$C_{10}H_{18}O$	154.25	col. cr.	$0.935^{20/20}$	35	$218–9^{752}$	i.	v. s.	v. s.
Terpinyl acetate (α-)(*dl*-)	$CH_3CO_2·C_{10}H_{17}$	196.29	lq.	$0.966^{20/4}$	< −50	220 d.	i.	20	
Tetrabromo-ethane (sym)	$Br_2CH·CHBr_2$	345.65	col. lq.	$2.964^{20/4}$	−1.0	151^{54}	i.	∞	∞
(uns)	$Br_3C·CH_2Br$	345.65	col. lq.	$2.875^{20/4}$	0	104^{13}		s.	
Tetrachloro-ethane (sym)	$Cl_2CH·CHCl_2$	167.85	col. lq.	$1.600^{20/4}$	−36	146.3	0.29^{20}	∞	∞
(uns)	$Cl_3C·CH_2Cl$	167.85	lq.	$1.588^{20/4}$		129–30	i.	∞	∞
-ethylene	$Cl_2C:CCl_2$	165.83	col. lq.	$1.624^{15/4}$	−19	120.8	0.02^{20}	∞	∞
Tetracosane (*n*-)	$CH_3(CH_2)_{22}CH_3$	338.65	cr.	$0.779^{51/4}$	51.1	324			s.
Tetradecane (*n*-)	$CH_3(CH_2)_{12}CH_3$	198.39	col. lq.	$0.765^{20/4}$	5.5	252.5	i.	v. s.	v. s.
Tetraethyl-thiuram disulfide	$[(C_2H_5)_2NCS]_2S_2$	296.54	cr.	1.17	70		i.		

(*Continued*)

TABLE 2-2 Physical Properties of Organic Compounds *(Continued)*

Name	Formula	Formula weight	Form and color	Specific gravity	Melting point, °C	Boiling point, °C	Solubility in 100 parts		
							Water	Alcohol	Ether
Tetrafluoro-ethylene	$F_2C{:}CF_2$	100.02	gas	1.58^{-78}	−142.5	−76.3	0.01^{30}		
Tetrahydro-furan	$—CH_2(CH_2)_2CH_2{\cdot}O—$	72.11	col. lq.	$0.888^{21/4}$	−65	65–6	s.	s.	s.
-furfuryl alcohol	$C_4H_7O{\cdot}CH_2OH$	102.13	col. lq.	$1.050^{20/4}$		$177–8^{743}$	∞	∞	∞
-pyran	$—CH_2(CH_2)_3CH_2{\cdot}O—$	86.13	lq.	$0.881^{20/4}$		88	s.		
Tetralin	$—C_6H_4CH_2(CH_2)_2CH_2—$	132.20	col. lq.	$0.973^{18/4}$	−31	206^{764}	i.	s.	s.
Tetramethyl-thiuram disulfide	$[(CH_3)_2NCS]_2S_2$	240.43	cr.	1.29	155–6		i.		
Tetryl (2-,4-,6-)	$(NO_2)_3C_6H_2{\cdot}N(CH_3)NO_2$	287.14	yel. mn.	1.57^{19}	130.5	expl.	i.	s. h.	s.
Theobromine	$C_7H_8O_2N_4$	180.16	rhb.		330		0.06^{15}	0.06 c.	0.03 h.
Thio-acetic acid	$CH_3{\cdot}CO{\cdot}SH$	76.12	yel. lq.	1.074^{10}	< −17	93	s.	∞	∞
-aniline (4-,4′-)	$(NH_2{\cdot}C_6H_4)_2S$	216.30	nd./aq.		108		sl. s. h.	s.	s.
-carbanilide	$(C_6H_5{\cdot}NH)_2CS$	228.31	rhb./al.	1.3^{24}	154	d.	i.	v. s.	v. s.
-naphthol (β-)	$C_{10}H_7{\cdot}SH$	160.24	cr./al.		81	286–8	v. sl. s.	v. s.	v. s.
-phenol	$C_6H_5{\cdot}SH$	110.18	col. lq.	$1.074^{23/4}$		168–9	v. sl. s.	v. s.	∞
-salicylic acid (*o*-)	$HS{\cdot}C_6H_4{\cdot}CO_2H$	154.19	yel. nd.		164	subl.	sl. s. h.	s.	
-urea	$NH_2{\cdot}CS{\cdot}NH_2$	76.12	rhb./al.	$1.405^{20/4}$	180–2	d.	9.2^{13}	s.	sl. s.
Thiophene	$<(CH{:}CH)_2>S$	84.14	col. lq.	$1.070^{15/4}$	−30	84	i.	s.	
Thymol (5-,2-,1-)	$(CH_3)(C_3H_7)C_6H_3OH$	150.22	cr.	$0.972^{25/25}$	51.5	232^{752}	0.09^{19}	v. s.	v. s.
Tolidine (0-)(3-,3′-,4-,4′-)	$[CH_3(NH_2)C_6H_3]_2$	212.29	lf.		128–9		v. sl. s.	s.	s.
Toluene	$C_6H_5{\cdot}CH_3$	92.14	col. lq.	$0.866^{20/4}$	−95	110.8	0.05^{16}	s.	∞
sulfonic acid (*o*-)	$CH_3{\cdot}C_6H_4SO_3H{\cdot}2H_2O$	208.23	cr.		d.	128.8^{0}	v. s.	s.	
(*p*-)	$CH_3{\cdot}C_6H_4SO_3H{\cdot}H_2O$	190.22	mn.		104–5	$146–7^{0}$	v. s.	s.	
sulfonic amide (*p*-)	$CH_3{\cdot}C_6H_4SO_2NH_2$	171.22	mn.		137		0.2^{9}	7.4^{5}	
sulfonic chloride (*p*-)	$CH_3{\cdot}C_6H_4{\cdot}SO_2Cl$	190.65	tri.		69	134.5^{10}	i.	s.	s.
Toluic acid (*o*-)	$CH_3{\cdot}C_6H_4{\cdot}CO_2H$	136.15	cr./aq.	$1.062^{115/4}$	104–5	259^{751}	2.17^{100}	v. s.	
(*m*-)	$CH_3{\cdot}C_6H_4{\cdot}CO_2H$	136.15	pr./aq.	$1.054^{112/4}$	110–1	263	1.6^{100}	v. s.	v. s.
(*p*-)	$CH_3{\cdot}C_6H_4{\cdot}CO_2H$	136.15	cr./aq.		179–80	274–5	1.3^{100}	v. s.	v. s.
Toluidine (*o*-)	$CH_3{\cdot}C_6H_4{\cdot}NH_2$	107.15	col. lq.	$0.999^{20/4}$	−16.3	199.7	1.5^{25}	∞	∞
(*m*-)	$CH_3{\cdot}C_6H_4{\cdot}NH_2$	107.15	col. lq.	$0.989^{20/4}$	−31.5	203.3	sl. s.	∞	∞
(*p*-)	$CH_3{\cdot}C_6H_4{\cdot}NH_2$	107.15	cr.	$1.046^{20/4}$	44–5	200.3	0.74^{21}	v. s.	v. s.
hydrochloride (*o*-)	$CH_3{\cdot}C_6H_4{\cdot}NH_3Cl$	143.61	mn. pr.		218–20	242	s.	sl. s.	
sulfonic acid (1-,2-,3-)	$CH_3(NH_2)C_6H_3SO_3H$	187.22	cr.				0.97^{11}		
Toluylenediamine (1-,2-,4-)	$CH_3{\cdot}C_6H_3(NH_2)_2$	122.17	rhb.		99	283–5	s. h.	s.	s.
Tolylene diisocyanate (1-,2-,4-)	$CH_3{\cdot}C_6H_3(NCO)_2$	174.16	lq.	1.23^{28}		134.5^{20}	d.	d.	
Trehalose	$C_{12}H_{22}O_{11}{\cdot}2H_2O$	378.33	rhb./al.		97		s. h.	sl. s. h.	i.
Triamylamine (*n*-)	$[CH_3(CH_2)_3CH_2]_3N$	227.43	lq.			240–5	i.		
(*i*-)	$[(CH_3)_2CH(CH_2)_2]_3N$	227.43	col. lq.	$0.786^{20/4}$		235	i.		
Tributyl-amine (*n*-)	$[CH_3(CH_2)_2CH_2]_3N$	185.35	col. lq.	$0.778^{20/20}$		216.5^{761}	i.	s.	∞
phosphite	$[CH_3(CH_2)_3O]_3P$	250.31	lq.	$0.925^{20/4}$		$122–3^{12}$	i.		
Trichloro-acetic acid	$Cl_3C{\cdot}CO_2H$	163.39	cr.	$1.617^{46/15}$	58	195.5^{754}	120^{25}	s.	s.
-benzene (*s*-)(1-,3-,5-)	$C_6H_3Cl_3$	181.45	nd.		63.5	208.5^{764}	i.	sl. s.	
-ethane (1-,1-,1-)	$Cl_3C{\cdot}CH_3$	133.40	lq.	$1.325^{26/4}$		74.1	i.	∞	∞
-ethylene	$Cl_2C{:}CHCl$	131.39	col. lq.	$1.466^{20/20}$	−73	87.2	0.1^{25}	∞	∞
-phenol	$Cl_3C_6H_2OH$	197.45	nd.	$1.490^{75/4}$	68–9	246	0.09^{25}	v. s.	v. s.
Tricosane (*n*-)	$CH_3(CH_2)_{21}CH_3$	324.63	lf.	$0.779^{48/4}$	47.7	234^{15}	i.		
Tricresyl phosphate (*o*-)	$OP(OC_6H_4CH_3)_3$	368.36	lq.				i.		
Tridecane (*n*-)	$CH_3(CH_2)_{11}CH_3$	184.36	col. lq.	$0.757^{20/4}$	−6.2	234	i.	v. s.	v. s.
Triethanol amine	$(HOCH_2CH_2)_3N$	149.19	col. lq.	$1.126^{20/20}$	20–1	$277–9^{150}$	∞	∞	sl. s.
Triethyl-amine	$(CH_3CH_2)_3N$	101.19	col. oil	$0.729^{20/20}$	−114.8	89.4	$∞ > 19^{0}$	∞	∞
-benzene (1-,3-,5-)	$(C_2H_5)_3C_6H_3$	162.27	lq.	$0.861^{20/4}$		215	i.	s.	s.
(1-,2-,4-)	$(C_2H_5)_3C_6H_3$	162.27	lq.	$0.882^{17/4}$		$217–8^{755}$	i.	s.	s.
borate	$B(OCH_2CH_3)_3$	145.99	lq.	$0.864^{20/20}$		120	d.		
citrate	$HOC_3H_4(CO_2C_2H_5)_3$	276.28	oil	$1.137^{20/4}$		294	i.	∞	∞
Triethylene glycol	$({\cdot}CH_2OCH_2CH_2OH)_2$	150.17	col. lq.	$1.125^{20/20}$	−5	290	∞	∞	v. sl. s.
Trifluoro-chloromethane	CF_3Cl	104.46	gas	1.726^{-130}	−182	−80			
-chloroethylene	$F_2C{:}CFCl$	116.47	gas		−157.5	−27.9	d.		
-trichloroethane	$Cl_2CF{\cdot}CClF_2$	187.38	lq.	$1.576^{20/4}$	−35	47.6	i.	∞	∞
Trimethoxybutane (1-,3-,3-)	$CH_2(OCH_3)CH_2C(OCH_3)_2CH_3$	148.20	lq.	0.932		$63–5^{25}$	d.		
Trimethylamine	$(CH_3)_3N$	59.11	gas	0.662^{-5}	−124	3.5	41^{19}	s.	s.
Trimethylene bromide	$BrCH_2CH_2CH_2Br$	201.89	lq.	$1.987^{15/4}$	−34.4	167.5	0.17^{30}	s.	s.
chloride	$ClCH_2CH_2CH_2Cl$	112.99	lq.	1.201^{15}		123–5	0.27^{25}	s.	s.
glycol	$HOCH_2CH_2CH_2OH$	76.09	oil	$1.060^{20/4}$		214	∞	∞	

Trinitro-benzene (1-,3-,5-)	$C_6H_3(NO_2)_3$	213.10	col. rhb.	$1.688^{20/4}$	121	d.	0.03^{15}	1.9^{18}	1.5^{18}
-benzoic acid (2-,4-,6-)	$(NO_2)_3C_6H_2CO_2H$	257.11	rhb./aq.		210–20 d.		2.05^{24}		
-tert-butylxylene	$(NO_2)_3C_6(CH_3)_2C_4H_9$	297.26	nd./al.		110		i.	sl. s.	s.
-naphthalene (α-)(1-,3-,5-)	$C_{10}H_5(NO_2)_3$	263.16	rhb.		122–3		i.	s.	
(β-)(1-,3-,8-)	$C_{10}H_5(NO_2)_3$	263.16	cr./al.		218–9		0.02^{100}	0.05^{23}	0.13^{15}
(γ-)(1-,4-,5-)	$C_{10}H_5(NO_2)_3$	263.16	yel. cr.		148–9		i.	0.11^{19}	0.4^{19}
-phenol (2-,3-,6-)	$(NO_2)_3C_6H_2OH$	229.10	nd.		117–8		s. h.	v. s.	v. s.
-toluene (β-)(2-,3-,4-)	$CH_3C_6H_2(NO_2)_3$	227.13	cr.	$1.620^{20/4}$	112	expl.	i.	sl. s. c.	s.
(γ-)(2-,4-,5-)	$CH_3C_6H_2(NO_2)_3$	227.13	yel. pl.	$1.620^{20/4}$	104	expl.	i.	s. h.	v. s.
(α-)(2-,4-,6-)	$CH_3C_6H_3(NO_2)_3$	227.13	cr./al.	1.654	80.8	expl.	0.01^{20}	1.5^{22}	5^{33}
Trional	$(C_2H_5SO_2C_2H_4)_2$	242.36	pl./al.	$1.199^{85/4}$	76	d.	0.3^{15}	5^{0}	6.6^{15}
Triphenyl-arsine	$(C_6H_5)_3As$	306.23	pl.	1.306	59–60	>360	i.	s.	v. s.
carbinol	$(C_6H_5)_3COH$	260.33	cr.	$1.188^{20/4}$	162.5	>360	i.	v. s.	v. s.
guanidine (α-)	$C_6H_5N{:}C(NHC_6H_5)_2$	287.36	rhb./al.	1.13	144–5	d.	i.	4^{0}	
methane	$(C_6H_5)_3CH$	244.33	cr.	$1.014^{99/4}$	93.4	359^{754}	i.	v. s. h.	v. s.
methyl	$(C_6H_5)_3C\ldots$	243.32	col. cr.		145–7	d.	i.	sl. s. h.	
phosphate	$OP(OC_6H_5)_3$	326.28	pr./al.	$1.206^{58/4}$	49–50	245^{11}	i.	155^{25}	v. s.
Tripropylamine (*n*-)	$(CH_3CH_2CH_2)_3N$	143.27	col. lq.	$0.757^{20/4}$	−93.5	156.5	v. sl. s.	∞	∞
Undecane (*n*-)	$CH_3(CH_2)_9CH_3$	156.31	col. lq.	$0.741^{20/4}$	−25.6	194.5	i.	∞	∞
Urea	$H_2N{\cdot}CO{\cdot}NH_2$	60.06	col. pr.	$1.335^{20/4}$	132.7	d.	100^{17}	20^{20}	sl. s.
nitrate	$CO(NH_2)_2{\cdot}HNO_3$	123.07	col. mn.		152 d.		v. s. h.	s.	
Uric acid	$C_5H_4O_3N_4$	168.11	cr.	1.893^{20}	d.		0.06 h.	i.	i.
Valeric acid (*n*-)	$C_2H_5CH_2CH_2CO_2H$	102.13	col. lq.	$0.939^{20/4}$	−34.5	187	3.3^{16}	∞	∞
(*i*-)	$(CH_3)_2CHCH_2CO_2H$	102.13	col. lq.	$0.931^{20/20}$	−37.6	176	4.2^{20}	∞	∞
aldehyde (*n*-)	$C_2H_5CH_2CH_2CHO$	86.13	lq.	0.819^{11}	−92	103.4	v. sl. s.	s.	s.
(*i*-)	$(CH_3)_2CHCH_2CHO$	86.13	col. lq.	0.803^{17}	−51	92.5	sl. s.	s.	s.
amide (*n*-)	$C_2H_5CH_2CH_2CONH_2$	101.15	mn. pl.	1.023	106		v. s.	v. s.	v. s.
(*i*-)	$(CH_3)_2CHCH_2CONH_2$	101.15	mn.	$0.965^{20/4}$	135–7	232	s.	s.	s.
Vanillic acid (3-,4-,1-)	$CH_3O(OH)C_6H_3CO_2H$	168.15	nd./aq.		207	subl.	0.12^{14}	v. s.	v. s.
alcohol (3-,4-,1-)	$CH_3O(OH)C_6H_3CH_2OH$	154.16	mn./aq.		115	d.	v. s. h.	v. s.	v. s.
hyl-thiuram disulfide	$[(C_2H_5)_2NCS]_2S_2$	296.54	cr.	1.17	70		i.		
Vanillin (3-,4-,1-)	$CH_3O(OH)C_6H_3CHO$	152.15	mn.	1.056	81–2	285	1^{14}	v. s.	v. s.
Veratrole (*o*-)	$C_6H_4(OCH_3)_2$	138.16	cr.	$1.091^{15/15}$	22.5	207.1	v. sl. s.	s.	s.
Vinyl acetate	$CH_3CO_2CH{:}CH_2$	86.09	col. lq.	$0.932^{20/4}$	<−60	72–3	2^{20}	∞	∞
(poly-)	$(CH_3CO_2CH{:}CH_2)x$	(86.09)		1.19^{20}	100–25		i.		
acetic acid	$CH_2{:}CH{\cdot}CH_2CO_2H$	86.09	col. lq.	$1.013^{15/15}$	−39	163	s.	∞	∞
acetylene	$CH_2{:}CH{\cdot}C{:}CH$	52.07	gas	$0.705^{1.5}$		5.5	$0.67^{0.6}$		
alcohol	$CH_2{:}CHOH$	44.05							
(poly-)	$(CH_2{:}CHOH)x$	(44.05)		1.3^{20}	d. >200		s.		
chloride	$CH_2{:}CHCl$	62.50	gas	$0.908^{25/25}$	−160	−12	sl. s.	s.	v. s.
propionate	$C_2H_5CO_2CH{:}CH_2$	100.12	lq.			93–5	v. sl. s.		
Xylene (*o*-)	$C_6H_4(CH_3)_2$	106.17	col. lq.	$0.881^{20/4}$	−25	144	i.	s.	∞
(*m*-)	$C_6H_4(CH_3)_2$	106.17	col. lq.	$0.867^{17/4}$	−47.4	139.3	i.	s.	∞
(*p*-)	$C_6H_4(CH_3)_2$	106.17	col. lq.	$0.861^{20/4}$	13.2	138.5	i.	s.	v. s.
sulfonic acid (1-,4-,2-)	$(CH_3)_2C_6H_3SO_3H{\cdot}2H_2O$	222.26	col. lf.		86	$149^{0.1}$	s.		
Xylidine (1:2)(3-)	$(CH_3)_2C_6H_3NH_2$	121.18	lq.	0.991^{15}	<−15	223	v. sl. s.	s.	s.
(1:2)(4-)	$(CH_3)_2C_6H_3NH_2$	121.18	pr.	$1.076^{17.5}$	49–50	224–6	v. sl. s.		
(1:3)(2-)	$(CH_3)_2C_6H_3NH_2$	121.18	lq.	0.980^{15}		216–7	v. sl. s.		
(1:3)(4-)	$(CH_3)_2C_6H_3NH_2$	121.18	lq.	$0.978^{20/4}$		213–4	v. sl. s.		
(1:3)(5-)	$(CH_3)_2C_6H_3NH_2$	121.18	oil	$0.972^{20/4}$		221–2	v. sl. s.		
(1:4)(2-)	$(CH_3)_2C_6H_3NH_2$	121.18	oil	$0.979^{21/4}$	15.5	215^{789}	v. sl. s.		
Xylose (*l*-)(+)	$CH_2OH(CHOH)_3CHO$	150.13	nd.	1.535^{0}	153–4		117^{20}	v. sl. s.	i.
Xylylene dichloride (*p*-)	$C_6H_4(CH_2Cl)_2$	175.06	mn.	1.417^{0}	100.5	240–5 d.	i.	s.	v. sl. s.
Zinc diethyl	$Zn(CH_2CH_3)_2$	123.53	col. lq.	1.182^{18}	−28	118	d.	d.	
dimethyl	$Zn(CH_3)_2$	95.48	col. lq.	1.386^{11}	−40	46	d.	d.	
dimethyl-dithiocarbamate	$Zn[S_2CN(CH_3)_2]_2$	305.84		$2.00^{40/4}$	248–50		i.		

NOTE: °F = $\frac{9}{5}$°C + 32.

VAPOR PRESSURES

TABLE 2-3 Vapor Pressure of Water Ice from 0 to −40°C

	Vapor pressure			Vapor pressure			Vapor pressure	
t, °C	mmHg	kPa	*t*, °C	mmHg	kPa	*t*, °C	mmHg	kPa
0	4.584	0.6112	−13.5	1.423	0.1897	−27.0	0.3881	0.05174
−0.5	4.399	0.5865	−14.0	1.359	0.1812	−27.5	0.3688	0.04918
−1.0	4.220	0.5627	−14.5	1.298	0.1731	−28.0	0.3505	0.04673
−1.5	4.049	0.5398	−15.0	1.240	0.1653	−28.5	0.3330	0.04439
−2.0	3.883	0.5177	−15.5	1.184	0.1578	−29.0	0.3162	0.04216
−2.5	3.724	0.4965	−16.0	1.130	0.1507	−29.5	0.3003	0.04004
−3.0	3.571	0.4761	−16.5	1.079	0.1438	−30.0	0.2851	0.03801
−3.5	3.423	0.4564	−17.0	1.029	0.1372	−30.5	0.2706	0.03608
−4.0	3.281	0.4375	−17.5	0.9822	0.1310	−31.0	0.2568	0.03424
−4.5	3.145	0.4193	−18.0	0.9370	0.1249	−31.5	0.2437	0.03249
−5.0	3.013	0.4018	−18.5	0.8937	0.1191	−32.0	0.2311	0.03082
−5.5	2.887	0.3849	−19.0	0.8522	0.1136	−32.5	0.2192	0.02923
−6.0	2.766	0.3687	−19.5	0.8125	0.1083	−33.0	0.2078	0.02771
−6.5	2.649	0.3532	−20.0	0.7745	0.1033	−33.5	0.1970	0.02627
−7.0	2.537	0.3382	−20.5	0.7381	0.09841	−34.0	0.1867	0.02490
−7.5	2.429	0.3238	−21.0	0.7034	0.09377	−34.5	0.1769	0.02359
−8.0	2.325	0.3100	−21.5	0.6701	0.08934	−35.0	0.1676	0.02235
−8.5	2.225	0.2967	−22.0	0.6383	0.08510	−35.5	0.1587	0.02116
−9.0	2.130	0.2839	−22.5	0.6078	0.08104	−36.0	0.1503	0.02004
−9.5	2.038	0.2717	−23.0	0.5787	0.07716	−36.5	0.1423	0.01897
−10.0	1.949	0.2599	−23.5	0.5509	0.07345	−37.0	0.1347	0.01796
−10.5	1.865	0.2486	−24.0	0.5243	0.06991	−37.5	0.1274	0.01699
−11.0	1.783	0.2377	−24.5	0.4989	0.06652	−38.0	0.1206	0.01607
−11.5	1.705	0.2273	−25.0	0.4747	0.06329	−38.5	0.1140	0.01520
−12.0	1.630	0.2173	−25.5	0.4515	0.06020	−39.0	0.1078	0.01437
−12.5	1.558	0.2077	−26.0	0.4294	0.05725	−39.5	0.1019	0.01359
−13.0	1.489	0.1985	−26.5	0.4083	0.05443	−40.0	0.0963	0.01284

SOURCE: Formulation of Wagner, Saul, and Pruss, *J. Phys. Chem. Ref. Data*, **23**, 515 (1994), implemented in Harvey, Peskin, and Klein, *NIST/ASME Steam Properties*, NIST Standard Reference Database 10, Version 2.2, National Institute of Standards and Technology, Gaithersburg, Md., 2000. This source provides data down to 190 K (−83.15°C). A formula extending to 110 K may be found in Murphy and Koop, *Q. J. R. Meteorol. Soc.*, **131**, 1539 (2005).

TABLE 2-4 Vapor Pressure of Supercooled Liquid Water from 0 to −40°C*

	Vapor pressure			Vapor pressure			Vapor pressure	
t, °C	mmHg	kPa	*t*, °C	mmHg	kPa	*t*, °C	mmHg	kPa
0	4.584	0.6112	−13.5	1.623	0.2163	−27.0	0.5051	0.06734
−0.5	4.421	0.5894	−14.0	1.558	0.2077	−27.5	0.4824	0.06431
−1.0	4.262	0.5682	−14.5	1.495	0.1993	−28.0	0.4606	0.06141
−1.5	4.108	0.5477	−15.0	1.435	0.1913	−28.5	0.4397	0.05862
−2.0	3.959	0.5279	−15.5	1.377	0.1836	−29.0	0.4197	0.05595
−2.5	3.816	0.5087	−16.0	1.321	0.1761	−29.5	0.4005	0.05339
−3.0	3.676	0.4901	−16.5	1.267	0.1689	−30.0	0.3820	0.05094
−3.5	3.542	0.4722	−17.0	1.215	0.1620	−30.5	0.3644	0.04858
−4.0	3.411	0.4548	−17.5	1.165	0.1553	−31.0	0.3475	0.04633
−4.5	3.285	0.4380	−18.0	1.117	0.1489	−31.5	0.3313	0.04417
−5.0	3.163	0.4218	−18.5	1.070	0.1427	−32.0	0.3158	0.04210
−5.5	3.046	0.4061	−19.0	1.026	0.1367	−32.5	0.3009	0.04012
−6.0	2.932	0.3909	−19.5	0.9827	0.1310	−33.0	0.2867	0.03822
−6.5	2.822	0.3762	−20.0	0.9414	0.1255	−33.5	0.2731	0.03640
−7.0	2.715	0.3620	−20.5	0.9016	0.1202	−34.0	0.2600	0.03467
−7.5	2.612	0.3483	−21.0	0.8633	0.1151	−34.5	0.2476	0.03300
−8.0	2.513	0.3351	−21.5	0.8265	0.1102	−35.0	0.2356	0.03141
−8.5	2.417	0.3223	−22.0	0.7911	0.1055	−35.5	0.2242	0.02989
−9.0	2.324	0.3099	−22.5	0.7571	0.1009	−36.0	0.2133	0.02844
−9.5	2.235	0.2980	−23.0	0.7244	0.0965	−36.5	0.2029	0.02705
−10.0	2.149	0.2865	−23.5	0.6930	0.0923	−37.0	0.1929	0.02572
−10.5	2.065	0.2753	−24.0	0.6628	0.08836	−37.5	0.1834	0.02445
−11.0	1.985	0.2646	−24.5	0.6337	0.08449	−38.0	0.1743	0.02324
−11.5	1.907	0.2542	−25.0	0.6059	0.08078	−38.5	0.1656	0.02208
−12.0	1.832	0.2442	−25.5	0.5791	0.07721	−39.0	0.1573	0.02098
−12.5	1.760	0.2346	−26.0	0.5534	0.07379	−39.5	0.1494	0.01992
−13.0	1.690	0.2253	−26.5	0.5288	0.07050	−40.0	0.1419	0.01891

*SOURCE: Murphy and Koop, *Q. J. R. Meteorol. Soc.*, **131**, 1552 (2005). The formula in the reference extends down to 123 K (−150.15°C), although in practice pure liquid water cannot be supercooled below about 235 K.

VAPOR PRESSURES OF PURE SUBSTANCES

Unit Conversions For this subsection, the following unit conversions are applicable: °F = 9/5°C + 32.

To convert millimeters of mercury to pounds-force per square inch, multiply by 0.01934. To convert cubic feet to cubic meters, multiply by 0.02832. To convert bars to pounds-force per square inch, multiply by 14.504. To convert bars to kilopascals, multiply by 1×10^2.

Additional References Additional vapor-pressure data may be found in major thermodynamic property databases, such as those produced by the AIChE's DIPPR program (aiche.org/dippr), NIST's Thermodynamics Research Center (trc.nist.gov), the Dortmund Databank (ddbst.de), and the Physical Property Data Service (ppds.co.uk). Additional sources include the *NIST Chemistry Webbook* (webbook.nist.gov/chemistry/); Boublik, T., V. Fried, and E. Hala, *The Vapor Pressures of Pure Substances,* 2d ed., Elsevier, Amsterdam, 1984; Bruce Poling, JohnPrausnitz, and John O'Connell, *The Properties of Gases and Liquids,* 5th ed., McGraw-Hill, New York, 2001; *Vapor Pressure of Chemicals* (subvolumes A, B, and C), vol. IV/20 in Landolt-Bornstein: *Numerical Data and Functional Relationships in Science and Technology—New Series,* Springer-Verlag, Berlin, 1999–2001. The most recent work on water may be found at The International Association for the Properties of Water and Steam website http://iapws.org.

TABLE 2-5 Vapor Pressure (MPa) of Liquid Water from 0 to 100°C

t, °C	P_{vp}, MPa	*t*, °C	P_{vp}, MPa	*t*, °C	P_{vp}, MPa
0.01	0.00061165	34	0.0053251	68	0.028599
1	0.00065709	35	0.0056290	69	0.029876
2	0.00070599	36	0.0059479	70	0.031201
3	0.00075808	37	0.0062823	71	0.032575
4	0.00081355	38	0.0066328	72	0.034000
5	0.00087258	39	0.0070002	73	0.035478
6	0.00093536	40	0.0073849	74	0.037009
7	0.0010021	41	0.0077878	75	0.038595
8	0.0010730	42	0.0082096	76	0.040239
9	0.0011483	43	0.0086508	77	0.041941
10	0.0012282	44	0.0091124	78	0.043703
11	0.0013130	45	0.0095950	79	0.045527
12	0.0014028	46	0.010099	80	0.047414
13	0.0014981	47	0.010627	81	0.049367
14	0.0015990	48	0.011177	82	0.051387
15	0.0017058	49	0.011752	83	0.053476
16	0.0018188	50	0.012352	84	0.055635
17	0.0019384	51	0.012978	85	0.057867
18	0.0020647	52	0.013631	86	0.060173
19	0.0021983	53	0.014312	87	0.062556
20	0.0023393	54	0.015022	88	0.065017
21	0.0024882	55	0.015762	89	0.067558
22	0.0026453	56	0.016533	90	0.070182
23	0.0028111	57	0.017336	91	0.072890
24	0.0029858	58	0.018171	92	0.075684
25	0.0031699	59	0.019041	93	0.078568
26	0.0033639	60	0.019946	94	0.081541
27	0.0035681	61	0.020888	95	0.084608
28	0.0037831	62	0.021867	96	0.087771
29	0.0040092	63	0.022885	97	0.091030
30	0.0042470	64	0.023943	98	0.094390
31	0.0044969	65	0.025042	99	0.097852
32	0.0047596	66	0.026183	100	0.10142
33	0.0050354	67	0.027368		

From E. W. Lemmon, M. O. McLinden, and D. G. Friend, "Thermophysical Properties of Fluid Systems" in *NIST Chemistry WebBook,* NIST Standard Reference Database Number 69, Eds. P. J. Linstrom and W. G. Mallard, June 2005, National Institute of Standards and Technology, Gaithersburg, Md. (http://webbook.nist.gov) and Wagner, W., and A., Pruss, "The IAPWS Formulation 1995 for the Thermodynamic Properties of Ordinary Water Substance for General and Scientific Use," *J. Phys. Chem. Ref. Data* **31**(2):387–535, 2002.

The website mentioned above allows users to generate their own tables of thermodynamic properties. The user can select the units as well as the temperatures and/or pressures for which properties are to be generated. The results can then be copied into spreadsheets or other files.

TABLE 2-6 Substances in Tables 2-8, 2-22, 2-32, 2-69, 2-72, 2-74, 2-75, 2-95, 2-106, 2-139, 2-140, 2-146, and 2-148 Sorted by Chemical Family

Name	Cmpd. no.	Formula
Paraffins		
Methane	193	CH_4
Ethane	125	C_2H_6
Propane	295	C_3H_8
Butane	31	C_4H_{10}
Pentane	279	C_5H_{12}
Hexane	171	C_6H_{14}
Heptane	160	C_7H_{16}
Octane	265	C_8H_{18}
Nonane	256	C_9H_{20}
Decane	74	$C_{10}H_{22}$
Undecane	336	$C_{11}H_{24}$
Dodecane	123	$C_{12}H_{26}$
Tridecane	327	$C_{13}H_{28}$
Tetradecane	319	$C_{14}H_{30}$
Pentadecane	277	$C_{15}H_{32}$
Hexadecane	169	$C_{16}H_{34}$
Heptadecane	158	$C_{17}H_{36}$
Octadecane	263	$C_{18}H_{38}$
Nonadecane	254	$C_{19}H_{40}$
Eicosane	124	$C_{20}H_{42}$
2-Methylpropane	236	C_4H_{10}
2-Methylbutane	202	C_5H_{12}
2,3-Dimethylbutane	107	C_6H_{14}
2-Methylpentane	234	C_6H_{14}
2,3-Dimethylpentane	114	C_7H_{16}
2,2,3,3-Tetramethylbutane	323	C_8H_{18}
2,2,4-Trimethylpentane	332	C_8H_{18}
2,3,3-Trimethylpentane	333	C_8H_{18}
Cyclopropane	71	C_3H_6
Cyclobutane	64	C_4H_8
Cyclopentane	69	C_5H_{10}
Cyclohexane	65	C_6H_{12}
Methylcyclopentane	217	C_6H_{12}
Ethylcyclopentane	134	C_7H_{14}
Methylcyclohexane	213	C_7H_{14}
1,1-Dimethylcyclohexane	108	C_8H_{16}
cis-1,2-Dimethylcyclohexane	109	C_8H_{16}
trans-1,2-Dimethylcyclohexane	110	C_8H_{16}
Ethylcyclohexane	133	C_8H_{16}
Olefins		
Ethylene	135	C_2H_4
Propylene	305	C_3H_6
1-Butene	36	C_4H_8
cis-2-Butene	37	C_4H_8
trans-2-Butene	38	C_4H_8
Cyclopentene	70	C_5H_8
1-Pentene	285	C_5H_{10}
Cyclohexene	68	C_6H_{10}
1-Hexene	177	C_6H_{12}
1-Heptene	166	C_7H_{14}
1-Octene	271	C_8H_{16}
1-Nonene	260	C_9H_{18}
1-Decene	77	$C_{10}H_{20}$
2-Methyl propene	238	C_4H_8
2-Methyl-1-butene	205	C_5H_{10}
2-Methyl-2-butene	206	C_5H_{10}
1-Methylcyclopentene	218	C_6H_{10}
3-Methylcyclopentene	219	C_6H_{10}
Propenylcyclohexene	298	C_9H_{14}
Propadiene	294	C_3H_4
1,2-Butadiene	29	C_4H_6
1,3-Butadiene	30	C_4H_6
3-Methyl-1,2-butadiene	201	C_5H_8
Acetylenes		
Acetylene	7	C_2H_2
1-Butyne	43	C_4H_6
1-Pentyne	288	C_5H_8
2-Pentyne	289	C_5H_8
3-Hexyne	178	C_6H_{10}
1-Hexyne	180	C_6H_{10}
2-Hexyne	181	C_6H_{10}
1-Heptyne	168	C_7H_{12}
1-Octyne	273	C_8H_{14}
1-Nonyne	262	C_9H_{16}
1-Decyne	79	$C_{10}H_{18}$
Methyl acetylene	197	C_3H_4
Vinyl acetylene	339	C_4H_4
Dimethyl acetylene	105	C_4H_6
2-Methyl -1-butene-3-yne	207	C_5H_6
3-Methyl-1-butyne	210	C_5H_8
Aromatics		
Benzene	16	C_6H_6
Toluene	325	C_7H_8
Styrene	312	C_8H_8
Ethylbenzene	129	C_8H_{10}
m-Xylene	343	C_8H_{10}
o-Xylene	344	C_8H_{10}
p-Xylene	345	C_8H_{10}
alpha-Methyl styrene	243	C_9H_{10}
Cumene	62	C_9H_{12}
Propylbenzene	304	C_9H_{12}
1,2,3-Trimethylbenzene	330	C_9H_{12}
1,2,4-Trimethylbenzene	331	C_9H_{12}
Naphthalene	246	$C_{10}H_8$
1,2,3,4-Tetrahydronaphthalene	321	$C_{10}H_{12}$
Butylbenzene	40	$C_{10}H_{14}$
Biphenyl	24	$C_{12}H_{10}$
Phenanthrene	290	$C_{14}H_{10}$
o-Terphenyl	318	$C_{18}H_{14}$
Aldehydes		
Formaldehyde	153	CH_2O
Acetaldehyde	1	C_2H_4O
Propionaldehyde	299	C_3H_6O
Butyraldehyde	44	C_4H_8O
Pentanal	278	$C_5H_{10}O$
Hexanal	170	$C_6H_{12}O$
Heptanal	159	$C_7H_{14}O$
Octanal	264	$C_8H_{16}O$
Nonanal	255	$C_9H_{18}O$
Decanal	73	$C_{10}H_{20}O$
Ketones		
Acrolein	8	C_3H_4O
Acetone	5	C_3H_6O
Methylethyl ketone	222	C_4H_8O
Methylisopropyl ketone	229	$C_5H_{10}O$
2-Pentanone	283	$C_5H_{10}O$
3-Pentanone	284	$C_5H_{10}O$
Quinone	310	$C_6H_4O_2$
Cyclohexanone	67	$C_6H_{10}O$
Ethylisopropyl ketone	144	$C_6H_{12}O$
2-Hexanone	175	$C_6H_{12}O$
3-Hexanone	176	$C_6H_{12}O$
Methylisobutyl ketone	226	$C_6H_{12}O$
Diisopropyl ketone	102	$C_7H_{14}O$
3-Heptanone	164	$C_7H_{14}O$
2-Heptanone	165	$C_7H_{14}O$
2-Octanone	269	$C_8H_{16}O$
3-Octanone	270	$C_8H_{16}O$
Benzophenone	20	$C_{13}H_{10}O$
Heterocyclics		
Furan	156	C_4H_4O
Thiophene	324	C_4H_4S
Tetrahydrofuran	320	C_4H_8O
Tetrahydrothiophene	322	C_4H_8S
Elements		
Argon	14	Ar
Bromine	25	Br_2
Chlorine	52	Cl_2
Deuterium	80	D_2
Fluorine	149	F_2

(*Continued*)

TABLE 2-6 Substances in Tables 2-8, 2-22, 2-32, 2-69, 2-72, 2-74, 2-75, 2-95, 2-106, 2-139, 2-140, 2-146, and 2-148 Sorted by Chemical Family *(Continued)*

Name	Cmpd. no.	Formula
Elements		
Hydrogen	183	H_2
Helium-4	157	He
Nitrogen	249	N_2
Neon	247	Ne
Oxygen	275	O_2
Alcohols		
Methanol	194	CH_4O
Ethanol	126	C_2H_6O
1-Propanol	296	C_3H_8O
2-Propanol	297	C_3H_8O
1-Butanol	34	$C_4H_{10}O$
2-Butanol	35	$C_4H_{10}O$
1-Pentanol	281	$C_5H_{12}O$
2-Pentanol	282	$C_5H_{12}O$
Cyclohexanol	66	$C_6H_{12}O$
1-Hexanol	173	$C_6H_{14}O$
2-Hexanol	174	$C_6H_{14}O$
1-Heptanol	162	$C_7H_{16}O$
2-Heptanol	163	$C_7H_{16}O$
1-Octanol	267	$C_8H_{18}O$
2-Octanol	268	$C_8H_{18}O$
1-Nonanol	258	$C_9H_{20}O$
2-Nonanol	259	$C_9H_{20}O$
1-Decanol	76	$C_{10}H_{22}O$
1-Undecanol	337	$C_{11}H_{24}O$
2-Methyl-2-propanol	237	$C_4H_{10}O$
3-Methyl-1-butanol	204	$C_5H_{12}O$
Benzyl alcohol	21	C_7H_8O
1-Methylcyclohexanol	214	$C_7H_{14}O$
cis-2-Methylcyclohexanol	215	$C_7H_{14}O$
trans-2-Methylcyclohexanol	216	$C_7H_{14}O$
Ethylene glycol	137	$C_2H_6O_2$
1,2-Propylene glycol	309	$C_3H_8O_2$
1,2-Butanediol	32	$C_4H_{10}O_2$
1,3-Butanediol	33	$C_4H_{10}O_2$
Phenols		
Phenol	291	C_6H_6O
m-Cresol	59	C_7H_8O
o-Cresol	60	C_7H_8O
p-Cresol	61	C_7H_8O
Ethers		
Dimethyl ether	112	C_2H_6O
Methyl vinyl ether	245	C_3H_6O
Methylethyl ether	221	C_3H_8O
1,4-Dioxane	120	$C_4H_8O_2$
Diethyl ether	95	$C_4H_{10}O$
Methylpropyl ether	240	$C_4H_{10}O$
Methylisopropyl ether	228	$C_4H_{10}O$
1,1-Dimethoxyethane	103	$C_4H_{10}O_2$
Methylbutyl ether	208	$C_5H_{12}O$
Methylisobutyl ether	225	$C_5H_{12}O$
Methyl *tert*-butyl ether	244	$C_5H_{12}O$
Ethylpropyl ether	147	$C_5H_{12}O$
Ethylisopropyl ether	143	$C_5H_{12}O$
1,2-Dimethoxypropane	104	$C_5H_{12}O_2$
Di-isopropyl ether	101	$C_6H_{14}O$
Methyl pentyl ether	235	$C_6H_{14}O$
Anisole	13	C_7H_8O
Dibutyl ether	84	$C_8H_{18}O$
Ethylhexyl ether	142	$C_8H_{18}O$
Benzyl ethyl ether	22	$C_9H_{12}O$
Diphenyl ether	121	$C_{12}H_{10}O$
Acids		
Formic acid	155	CH_2O_2
Oxalic acid	274	$C_2H_2O_4$
Acetic acid	3	$C_2H_4O_2$
Acrylic acid	9	$C_3H_4O_2$
Malonic acid	191	$C_3H_4O_4$
Propionic acid	300	$C_3H_6O_2$
Methacrylic acid	192	$C_4H_6O_2$
Acetic anhydride	4	$C_4H_6O_3$
Succinic acid	313	$C_4H_6O_4$
Butyric acid	45	$C_4H_8O_2$
Isobutyric acid	189	$C_4H_8O_2$
2-Methylbutanoic acid	203	$C_5H_{10}O_2$
Pentanoic acid	280	$C_5H_{10}O_2$
2-Ethyl butanoic acid	131	$C_6H_{12}O_2$
Hexanoic acid	172	$C_6H_{12}O_2$
Benzoic acid	18	$C_7H_6O_2$
Heptanoic acid	161	$C_7H_{14}O_2$
Phthalic anhydride	293	$C_8H_4O_3$
Terephthalic acid	317	$C_8H_6O_4$
2-Ethyl hexanoic acid	141	$C_8H_{16}O_2$
Octanoic acid	266	$C_8H_{16}O_2$
2-Methyloctanoic acid	233	$C_9H_{18}O_2$
Nonanoic acid	257	$C_9H_{18}O_2$
Decanoic acid	75	$C_{10}H_{20}O_2$
Esters		
Methyl formate	224	$C_2H_4O_2$
Ethyl formate	140	$C_3H_6O_2$
Methyl acetate	196	$C_3H_6O_2$
Methyl acrylate	198	$C_4H_6O_2$
Vinyl acetate	338	$C_4H_6O_2$
Ethyl acetate	127	$C_4H_8O_2$
Methyl propionate	239	$C_4H_8O_2$
Propyl formate	306	$C_4H_8O_2$
Methyl methacrylate	232	$C_5H_8O_2$
Ethyl propionate	146	$C_5H_{10}O_2$
Methyl butyrate	211	$C_5H_{10}O_2$
Propyl acetate	302	$C_5H_{10}O_2$
Butyl acetate	39	$C_6H_{12}O_2$
Ethyl butyrate	132	$C_6H_{12}O_2$
Methyl benzoate	200	$C_8H_8O_2$
Ethyl benzoate	130	$C_9H_{10}O_2$
Dimethyl phthalate	115	$C_{10}H_{10}O_4$
Dimethyl terephthalate	119	$C_{10}H_{10}O_4$
Amines		
Methyl amine	199	CH_5N
Ethyleneimine	138	C_2H_5N
Dimethyl amine	106	C_2H_7N
Ethyl amine	128	C_2H_7N
Ethylenediamine	136	$C_2H_8N_2$
Isopropyl amine	190	C_3H_9N
Propyl amine	303	C_3H_9N
Trimethyl amine	329	C_3H_9N
Diethyl amine	94	$C_4H_{11}N$
Diethanol amine	93	$C_4H_{11}NO_2$
Di-isopropyl amine	100	$C_6H_{15}N$
Dipropyl amine	122	$C_6H_{15}N$
Triethyl amine	328	$C_6H_{15}N$
Amides		
Formamide	154	CH_3NO
Acetamide	2	C_2H_5NO
N,N-Dimethyl formamide	113	C_3H_7NO
N-Methyl acetamide	195	C_3H_7NO
Benzamide	15	C_7H_7NO
Nitriles		
Acetonitrile	6	C_2H_3N
Cyanogen	63	C_2N_2
Acrylonitrile	10	C_3H_3N
Propionitrile	301	C_3H_5N
Butyronitrile	46	C_4H_7N
Benzonitrile	19	C_7H_5N
Nitro Compounds		
Nitromethane	251	CH_3NO_2
Nitroethane	248	$C_2H_5NO_2$

TABLE 2-6 Substances in Tables 2-8, 2-22, 2-32, 2-69, 2-72, 2-74, 2-75, 2-95, 2-106, 2-139, 2-140, 2-146, and 2-148 Sorted by Chemical Family *(Continued)*

Name	Cmpd. no.	Formula
Nitro Compounds		
1,3,5-Trinitrobenzene	334	$C_6H_3N_3O_6$
2,4,6-Trinitrotoluene	335	$C_7H_5N_3O_6$
Isocyanates		
Methyl isocyanate	227	C_2H_3NO
Phenyl isocyanate	292	C_7H_5NO
Mercaptans		
Methyl mercaptan	231	CH_4S
Ethyl mercaptan	145	C_2H_6S
Propyl mercaptan	308	C_3H_8S
2-Propyl mercaptan	307	C_3H_8S
Butyl mercaptan	41	$C_4H_{10}S$
sec-Butyl mercaptan	42	$C_4H_{10}S$
Pentyl mercaptan	287	$C_5H_{12}S$
2-Pentyl mercaptan	286	$C_5H_{12}S$
Benzenethiol	17	C_6H_6S
Cyclohexyl mercaptan	72	$C_6H_{12}S$
Hexyl mercaptan	179	$C_6H_{14}S$
Benzyl mercaptan	23	C_7H_8S
Heptyl mercaptan	167	$C_7H_{16}S$
Octyl mercaptan	272	$C_8H_{18}S$
Nonyl mercaptan	261	$C_9H_{20}S$
Decyl mercaptan	78	$C_{10}H_{22}S$
Sulfides		
Dimethyl sulfide	117	C_2H_6S
Dimethyl disulfide	111	$C_2H_6S_2$
Methylethyl sulfide	223	C_3H_8S
Diethyl sulfide	96	$C_4H_{10}S$
Methylisopropyl sulfide	230	$C_4H_{10}S$
Methylpropyl sulfide	241	$C_4H_{10}S$
Methylbutyl sulfide	209	$C_5H_{12}S$
Halogenated Hydrocarbons		
Carbon tetrachloride	50	CCl_4
Carbon tetrafluoride	51	CF_4
Chloroform	55	$CHCl_3$
Dibromomethane	83	CH_2Br_2
Dichloromethane	90	CH_2Cl_2
Difluoromethane	99	CH_2F_2
Bromomethane	28	CH_3Br
Chloromethane	56	CH_3Cl
Fluoromethane	152	CH_3F
Vinyl chloride	340	C_2H_3Cl
1,1,2-Trichloroethane	326	$C_2H_3Cl_3$
1,1-Dibromoethane	81	$C_2H_4Br_2$
1,2-Dibromoethane	82	$C_2H_4Br_2$
1,1-Dichloroethane	88	$C_2H_4Cl_2$

Name	Cmpd. no.	Formula
Halogenated Hydrocarbons		
1,2-Dichloroethane	89	$C_2H_4Cl_2$
1,1-Difluoroethane	97	$C_2H_4F_2$
1,2-Difluoroethane	98	$C_2H_4F_2$
Bromoethane	27	C_2H_5Br
Chloroethane	54	C_2H_5Cl
Fluoroethane	151	C_2H_5F
1,1-Dichloropropane	91	$C_3H_6Cl_2$
1,2-Dichloropropane	92	$C_3H_6Cl_2$
1-Chloropropane	57	C_3H_7Cl
2-Chloropropane	58	C_3H_7Cl
m-Dichlorobenzene	85	$C_6H_4Cl_2$
o-Dichlorobenzene	86	$C_6H_4Cl_2$
p-Dichlorobenzene	87	$C_6H_4Cl_2$
Bromobenzene	26	C_6H_5Br
Chlorobenzene	53	C_6H_5Cl
Fluorobenzene	150	C_6H_5F
Silanes		
Methylsilane	242	CH_6Si
Methylchlorosilane	212	CH_5ClSi
Methyldichlorosilane	220	CH_4Cl_2Si
Vinyl trichlorosilane	341	$C_2H_3Cl_3Si$
Ethyltrichlorosilane	148	$C_2H_5Cl_3Si$
Dimethylsilane	116	C_2H_8Si
Silicon tetrafluoride	311	F_4Si
Light Gases		
Hydrogen cyanide	186	CHN
Carbon monoxide	49	CO
Carbon dioxide	47	CO_2
Carbon disulfide	48	CS_2
Nitrogen trifluoride	250	F_3N
Sulfur hexafluoride	315	F_6S
Hydrogen bromide	184	HBr
Hydrogen chloride	185	HCl
Hydrogen fluoride	187	HF
Hydrogen sulfide	188	H_2S
Ammonia	12	H_3N
Hydrazine	182	H_4N_2
Nitric oxide	253	NO
Nitrous oxide	252	N_2O
Sulfur dioxide	314	O_2S
Ozone	276	O_3
Sulfur trioxide	316	O_3S
Others		
Air	11	Mixture
Ethylene oxide	139	C_2H_4O
Dimethyl sulfoxide	118	C_2H_6OS
Water	342	H_2O

TABLE 2-7 Formula Index of Substances in Tables 2-8, 2-22, 2-32, 2-69, 2-72, 2-74, 2-75, 2-95, 2-106, 2-139, 2-140, 2-146, and 2-148

Formula	No.	Name
	11	Air
Ar	14	Argon
Br_2	25	Bromine
CCl_4	50	Carbon tetrachloride
CF_4	51	Carbon tetrafluoride
$CHCl_3$	55	Chloroform
CHN	186	Hydrogen cyanide
CH_2Br_2	83	Dibromomethane
CH_2Cl_2	90	Dichloromethane
CH_2F_2	99	Difluoromethane
CH_2O	153	Formaldehyde
CH_2O_2	155	Formic acid
CH_3Br	28	Bromomethane
CH_3Cl	56	Chloromethane
CH_3F	152	Fluoromethane
CH_3NO	154	Formamide
CH_3NO_2	251	Nitromethane
CH_4	193	Methane
CH_4Cl_2Si	220	Methyldichlorosilane
CH_4O	194	Methanol
CH_4S	231	Methyl mercaptan
CH_5ClSi	212	Methylchlorosilane
CH_5N	199	Methyl amine
CH_6Si	242	Methylsilane
CO	49	Carbon monoxide
CO_2	47	Carbon dioxide
CS_2	48	Carbon disulfide
C_2H_2	7	Acetylene
$C_2H_2O_4$	274	Oxalic acid
C_2H_3Cl	340	Vinyl chloride
$C_2H_3Cl_3$	326	1,1,2-Trichloroethane
$C_2H_3Cl_3Si$	341	Vinyl trichlorosilane
C_2H_3N	6	Acetonitrile
C_2H_3NO	227	Methyl Isocyanate
C_2H_4	135	Ethylene
$C_2H_4Br_2$	81	1,1-Dibromoethane
$C_2H_4Br_2$	82	1,2-Dibromoethane
$C_2H_4Cl_2$	88	1,1-Dichloroethane
$C_2H_4Cl_2$	89	1,2-Dichloroethane
$C_2H_4F_2$	97	1,1-Difluoroethane
$C_2H_4F_2$	98	1,2-Difluoroethane
C_2H_4O	1	Acetaldehyde
C_2H_4O	139	Ethylene oxide
$C_2H_4O_2$	3	Acetic acid
$C_2H_4O_2$	224	Methyl formate
C_2H_5Br	27	Bromoethane
C_2H_5Cl	54	Chloroethane
$C_2H_5Cl_3Si$	148	Ethyltrichlorosilane
C_2H_5F	151	Fluoroethane
C_2H_5N	138	Ethyleneimine
C_2H_5NO	2	Acetamide
$C_2H_5NO_2$	248	Nitroethane
C_2H_6	125	Ethane
C_2H_6O	112	Dimethyl ether
C_2H_6O	126	Ethanol
$C_2H_6O_2$	137	Ethylene glycol
C_2H_6OS	118	Dimethyl sulfoxide
C_2H_6S	117	Dimethyl sulfide
C_2H_6S	145	Ethyl mercaptan
$C_2H_6S_2$	111	Dimethyl disulfide
C_2H_7N	106	Dimethyl amine
C_2H_7N	128	Ethyl amine
$C_2H_8N_2$	136	Ethylenediamine
C_2H_8Si	116	Dimethylsilane
C_2N_2	63	Cyanogen
C_3H_3N	10	Acrylonitrile
C_3H_4	197	Methyl acetylene
C_3H_4	294	Propadiene
C_3H_4O	8	Acrolein
$C_3H_4O_2$	9	Acrylic acid
$C_3H_4O_4$	191	Malonic acid
C_3H_5N	301	Propionitrile
C_3H_6	71	Cyclopropane
C_3H_6	305	Propylene
$C_3H_6Cl_2$	91	1,1-Dichloropropane
$C_3H_6Cl_2$	92	1,2-Dichloropropane
C_3H_6O	5	Acetone
C_3H_6O	245	Methyl vinyl ether
C_3H_6O	299	Propionaldehyde
$C_3H_6O_2$	140	Ethyl formate
$C_3H_6O_2$	196	Methyl acetate
$C_3H_6O_2$	300	Propionic acid
C_3H_7Cl	57	1-Chloropropane
C_3H_7Cl	58	2-Chloropropane
C_3H_7NO	113	*N,N*-Dimethyl formamide
C_3H_7NO	195	*N*-Methyl acetamide
C_3H_8	295	Propane
C_3H_8O	221	Methylethyl ether
C_3H_8O	296	1-Propanol
C_3H_8O	297	2-Propanol
$C_3H_8O_2$	309	1,2-Propylene glycol
C_3H_8S	223	Methylethyl sulfide
C_3H_8S	308	Propyl mercaptan
C_3H_8S	307	2-Propyl mercaptan
C_3H_9N	190	Isopropyl amine
C_3H_9N	303	Propyl amine
C_3H_9N	329	Trimethyl amine
C_4H_4	339	Vinyl acetylene
C_4H_4O	156	Furan
C_4H_4S	324	Thiophene
C_4H_6	29	1,2-Butadiene
C_4H_6	30	1,3-Butadiene
C_4H_6	43	1-Butyne
C_4H_6	105	Dimethyl acetylene
$C_4H_6O_2$	192	Methacrylic acid
$C_4H_6O_2$	198	Methyl acrylate
$C_4H_6O_2$	338	Vinyl acetate
$C_4H_6O_3$	4	Acetic anhydride
$C_4H_6O_4$	313	Succinic acid
C_4H_7N	46	Butyronitrile
C_4H_8	36	1-Butene
C_4H_8	37	*cis*-2-Butene
C_4H_8	38	*trans*-2-Butene
C_4H_8	64	Cyclobutane
C_4H_8	238	2-Methyl propene
C_4H_8O	44	Butyraldehyde
C_4H_8O	222	Methylethyl ketone
C_4H_8O	320	Tetrahydrofuran
$C_4H_8O_2$	45	Butyric acid
$C_4H_8O_2$	120	1,4-Dioxane
$C_4H_8O_2$	127	Ethyl acetate
$C_4H_8O_2$	189	Isobutyric acid
$C_4H_8O_2$	239	Methyl propionate
$C_4H_8O_2$	306	Propyl formate
C_4H_8S	322	Tetrahydrothiophene
C_4H_{10}	31	Butane
C_4H_{10}	236	2-Methylpropane
$C_4H_{10}O$	34	1-Butanol
$C_4H_{10}O$	35	2-Butanol
$C_4H_{10}O$	95	Diethyl ether
$C_4H_{10}O$	237	2-Methyl-2-propanol
$C_4H_{10}O$	240	Methylpropyl ether
$C_4H_{10}O$	228	Methylisopropyl ether
$C_4H_{10}O_2$	32	1,2-Butanediol
$C_4H_{10}O_2$	33	1,3-Butanediol
$C_4H_{10}O_2$	103	1,1-Dimethoxyethane
$C_4H_{10}S$	41	Butyl mercaptan
$C_4H_{10}S$	42	*sec*-Butyl mercaptan
$C_4H_{10}S$	96	Diethyl sulfide
$C_4H_{10}S$	230	Methylisopropyl sulfide
$C_4H_{10}S$	241	Methylpropyl sulfide
$C_4H_{11}N$	94	Diethyl amine
$C_4H_{11}NO_2$	93	Diethanol amine
C_5H_6	207	2-Methyl-1-butene-3-yne
C_5H_8	70	Cyclopentene
C_5H_8	201	3-Methyl-1,2-butadiene
C_5H_8	210	3-Methyl-1-butyne
C_5H_8	288	1-Pentyne
C_5H_8	289	2-Pentyne
$C_5H_8O_2$	232	Methyl methacrylate

TABLE 2-7 Formula Index of Substances in Tables 2-8, 2-22, 2-32, 2-69, 2-72, 2-74, 2-75, 2-95, 2-106, 2-139, 2-140, 2-146, and 2-148 *(Continued)*

Formula	No.	Name	Formula	No.	Name
C_5H_{10}	69	Cyclopentane	C_7H_8	325	Toluene
C_5H_{10}	205	2-Methyl-1-butene	C_7H_8O	13	Anisole
C_5H_{10}	206	2-Methyl-2-butene	C_7H_8O	21	Benzyl alcohol
C_5H_{10}	285	1-Pentene	C_7H_8O	59	*m*-Cresol
$C_5H_{10}O$	229	Methylisopropyl ketone	C_7H_8O	60	*o*-Cresol
$C_5H_{10}O$	278	Pentanal	C_7H_8O	61	*p*-Cresol
$C_5H_{10}O$	283	2-Pentanone	C_7H_8S	23	Benzyl mercaptan
$C_5H_{10}O$	284	3-Pentanone	C_7H_{12}	168	1-Heptyne
$C_5H_{10}O_2$	146	Ethyl propionate	C_7H_{14}	134	Ethylcyclopentane
$C_5H_{10}O_2$	203	2-Methylbutanoic acid	C_7H_{14}	166	1-Heptene
$C_5H_{10}O_2$	211	Methyl butyrate	C_7H_{14}	213	Methylcyclohexane
$C_5H_{10}O_2$	280	Pentanoic acid	$C_7H_{14}O$	102	Di-isopropyl ketone
$C_5H_{10}O_2$	302	Propyl acetate	$C_7H_{14}O$	159	Heptanal
C_5H_{12}	202	2-Methylbutane	$C_7H_{14}O$	164	3-Heptanone
C_5H_{12}	279	Pentane	$C_7H_{14}O$	165	2-Heptanone
$C_5H_{12}O$	143	Ethylisopropyl ether	$C_7H_{14}O$	214	1-Methylcyclohexanol
$C_5H_{12}O$	147	Ethylpropyl ether	$C_7H_{14}O$	215	*cis*-2-Methylcyclohexanol
$C_5H_{12}O$	204	3-Methyl-1-butanol	$C_7H_{14}O$	216	*trans*-2-Methylcyclohexanol
$C_5H_{12}O$	208	Methylbutyl ether	$C_7H_{14}O_2$	161	Heptanoic acid
$C_5H_{12}O$	225	Methylisobutyl ether	C_7H_{16}	114	2,3-Dimethylpentane
$C_5H_{12}O$	244	Methyl *tert*-butyl ether	C_7H_{16}	160	Heptane
$C_5H_{12}O$	281	1-Pentanol	$C_7H_{16}O$	162	1-Heptanol
$C_5H_{12}O$	282	2-Pentanol	$C_7H_{16}O$	163	2-Heptanol
$C_5H_{12}O_2$	104	1,2-Dimethoxypropane	$C_7H_{16}S$	167	Heptyl mercaptan
$C_5H_{12}S$	209	Methylbutyl sulfide	$C_8H_4O_3$	293	Phthalic anhydride
$C_5H_{12}S$	286	2-Pentyl mercaptan	$C_8H_6O_4$	317	Terephthalic acid
$C_5H_{12}S$	287	Pentyl mercaptan	C_8H_8	312	Styrene
$C_6H_3N_3O_6$	334	1,3,5-Trinitrobenzene	$C_8H_8O_2$	200	Methyl benzoate
$C_6H_4Cl_2$	85	*m*-Dichlorobenzene	C_8H_{10}	129	Ethylbenzene
$C_6H_4Cl_2$	86	*o*-Dichlorobenzene	C_8H_{10}	343	*m*-Xylene
$C_6H_4Cl_2$	87	*p*-Dichlorobenzene	C_8H_{10}	344	*o*-Xylene
$C_6H_4O_2$	310	Quinone	C_8H_{10}	345	*p*-Xylene
C_6H_5Br	26	Bromobenzene	C_8H_{14}	273	1-Octyne
C_6H_5Cl	53	Chlorobenzene	C_8H_{16}	108	1,1-Dimethylcyclohexane
C_6H_5F	150	Fluorobenzene	C_8H_{16}	109	*cis*-1,2-Dimethylcyclohexane
C_6H_6	16	Benzene	C_8H_{16}	110	*trans*-1,2-Dimethylcyclohexane
C_6H_6O	291	Phenol	C_8H_{16}	133	Ethylcyclohexane
C_6H_6S	17	Benzenethiol	C_8H_{16}	271	1-Octene
C_6H_{10}	218	1-Methylcyclopentene	$C_8H_{16}O$	264	Octanal
C_6H_{10}	68	Cyclohexene	$C_8H_{16}O$	269	2-Octanone
C_6H_{10}	178	3-Hexyne	$C_8H_{16}O$	270	3-Octanone
C_6H_{10}	180	1-Hexyne	$C_8H_{16}O_2$	141	2-Ethyl hexanoic acid
C_6H_{10}	181	2-Hexyne	$C_8H_{16}O_2$	266	Octanoic acid
C_6H_{10}	219	3-Methylcyclopentene	C_8H_{18}	265	Octane
$C_6H_{10}O$	67	Cyclohexanone	C_8H_{18}	323	2,2,3,3-Tetramethylbutane
C_6H_{12}	65	Cyclohexane	C_8H_{18}	332	2,2,4-Trimethylpentane
C_6H_{12}	177	1-Hexene	C_8H_{18}	333	2,3,3-Trimethylpentane
C_6H_{12}	217	Methylcyclopentane	$C_8H_{18}O$	84	Dibutyl ether
$C_6H_{12}O$	66	Cyclohexanol	$C_8H_{18}O$	142	Ethylhexyl ether
$C_6H_{12}O$	144	Ethylisopropyl ketone	$C_8H_{18}O$	267	1-Octanol
$C_6H_{12}O$	170	Hexanal	$C_8H_{18}O$	268	2-Octanol
$C_6H_{12}O$	175	2-Hexanone	$C_8H_{18}S$	272	Octyl mercaptan
$C_6H_{12}O$	176	3-Hexanone	C_9H_{10}	243	alpha-Methyl styrene
$C_6H_{12}O$	226	Methylisobutyl ketone	$C_9H_{10}O_2$	130	Ethyl benzoate
$C_6H_{12}O_2$	39	Butyl acetate	C_9H_{12}	62	Cumene
$C_6H_{12}O_2$	131	2-Ethyl butanoic acid	C_9H_{12}	304	Propylbenzene
$C_6H_{12}O_2$	132	Ethyl butyrate	C_9H_{12}	330	1,2,3-Trimethylbenzene
$C_6H_{12}O_2$	172	Hexanoic acid	C_9H_{12}	331	1,2,4-Trimethylbenzene
$C_6H_{12}S$	72	Cyclohexyl mercaptan	$C_9H_{12}O$	22	Benzyl ethyl ether
C_6H_{14}	107	2,3-Dimethylbutane	C_9H_{14}	298	Propenylcyclohexene
C_6H_{14}	171	Hexane	C_9H_{16}	262	1-Nonyne
C_6H_{14}	234	2-Methylpentane	C_9H_{18}	260	1-Nonene
$C_6H_{14}O$	101	Di-isopropyl ether	$C_9H_{18}O$	255	Nonanal
$C_6H_{14}O$	173	1-Hexanol	$C_9H_{18}O_2$	233	2-Methyloctanoic acid
$C_6H_{14}O$	174	2-Hexanol	$C_9H_{18}O_2$	257	Nonanoic acid
$C_6H_{14}O$	235	Methyl pentyl ether	C_9H_{20}	256	Nonane
$C_6H_{14}S$	179	Hexyl mercaptan	$C_9H_{20}O$	258	1-Nonanol
$C_6H_{15}N$	100	Di-isopropyl amine	$C_9H_{20}O$	259	2-Nonanol
$C_6H_{15}N$	122	Dipropyl amine	$C_9H_{20}S$	261	Nonyl mercaptan
$C_6H_{15}N$	328	Triethyl amine	$C_{10}H_8$	246	Naphthalene
C_7H_5N	19	Benzonitrile	$C_{10}H_{10}O_4$	115	Dimethyl phthalate
$C_7H_5N_3O_6$	335	2,4,6-Trinitrotoluene	$C_{10}H_{10}O_4$	119	Dimethyl terephthalate
C_7H_5NO	292	Phenyl isocyanate	$C_{10}H_{12}$	321	1,2,3,4-Tetrahydronaphthalene
$C_7H_6O_2$	18	Benzoic acid	$C_{10}H_{14}$	40	Butylbenzene
C_7H_7NO	15	Benzamide	$C_{10}H_{18}$	79	1-Decyne

(Continued)

TABLE 2-7 Formula Index of Substances in Tables 2-8, 2-22, 2-32, 2-69, 2-72, 2-74, 2-75, 2-95, 2-106, 2-139, 2-140, 2-146, and 2-148 *(Continued)*

Formula	No.	Name	Formula	No.	Name
$C_{10}H_{20}$	77	1-Decene	D_2	80	Deuterium
$C_{10}H_{20}O$	73	Decanal	F_2	149	Fluorine
$C_{10}H_{20}O_2$	75	Decanoic acid	F_3N	250	Nitrogen trifluoride
$C_{10}H_{22}$	74	Decane	F_4Si	311	Silicon tetrafluoride
$C_{10}H_{22}O$	76	1-Decanol	F_6S	315	Sulfur hexafluoride
$C_{10}H_{22}S$	78	Decyl mercaptan	HBr	184	Hydrogen bromide
$C_{11}H_{24}$	336	Undecane	HCl	185	Hydrogen chloride
$C_{11}H_{24}O$	337	1-Undecanol	HF	187	Hydrogen fluoride
$C_{12}H_{10}$	24	Biphenyl	H_2	183	Hydrogen
$C_{12}H_{10}O$	121	Diphenyl ether	H_2O	342	Water
$C_{12}H_{26}$	123	Dodecane	H_2S	188	Hydrogen sulfide
$C_{13}H_{10}O$	20	Benzophenone	H_3N	12	Ammonia
$C_{13}H_{28}$	327	Tridecane	H_4N_2	182	Hydrazine
$C_{14}H_{10}$	290	Phenanthrene	He	157	Helium-4
$C_{14}H_{30}$	319	Tetradecane	NO	253	Nitric oxide
$C_{15}H_{32}$	277	Pentadecane	N_2	249	Nitrogen
$C_{16}H_{34}$	169	Hexadecane	N_2O	252	Nitrous oxide
$C_{17}H_{36}$	158	Heptadecane	Ne	247	Neon
$C_{18}H_{14}$	318	*o*-Terphenyl	O_2	275	Oxygen
$C_{18}H_{38}$	263	Octadecane	O_2S	314	Sulfur dioxide
$C_{19}H_{40}$	254	Nonadecane	O_3	276	Ozone
$C_{20}H_{42}$	124	Eicosane	O_3S	316	Sulfur trioxide
Cl_2	52	Chlorine			

TABLE 2-8 Vapor Pressure of Inorganic and Organic Liquids, $\ln P = C_1 + C_2/T + C_3 \ln T + C_4 T^{C_5}$, P in Pa, T in K

Cmpd. no.*	Name	Formula	CAS	C_1	C_2	C_3	C_4	C_5	T_{min}, K	P at T_{min}	T_{max}, K	P at T_{max}
1	Acetaldehyde	C_2H_4O	75-07-0	52.9107	−4643.14	−4.50683	2.70E-17	6	149.78	5.15E-01	466	5.570E+06
2	Acetamide	C_2H_5NO	60-35-5	125.8 1	−12,376	−14.589	5.0824E-06	2	353.33	3.36E+02	761	6.569E+06
3	Acetic acid	$C_2H_4O_2$	64-19-7	53.27	−6304.5	−4.2985	8.89E-18	6	289.81	1.28E+03	591.95	5.739E+06
4	Acetic anhydride	$C_4H_6O_3$	108-24-7	67.1818	−7463.47	−6.24388	6.86E-18	6	200.15	4.10E-02	606	4.000E+06
5	Acetone	C_3H_6O	67-64-1	69.006	−5599.6	−7.0985	6.2237E-06	2	178.45	2.79E+00	508.2	4.709E+06
6	Acetonitrile	C_2H_3N	75-05-8	46.735	−5126.18	−3.54064	1.40E-17	6	229.32	1.71E+02	545.5	4.850E+06
7	Acetylene	C_2H_2	74-86-2	39.63	−2552.2	−2.78	2.39E-16	6	192.4	1.27E+05	308.3	6.106E+06
8	Acrolein	C_3H_4O	107-02-8	138.4	−7122.7	−19.638	0.026447	1	185.45	1.03E+01	506	5.020E+06
9	Acrylic acid	$C_3H_4O_2$	79-10-7	46.745	−6587.1	−3.2208	5.2253E-07	2	286.15	2.57E+02	615	5.661E+06
10	Acrylonitrile	C_3H_3N	107-13-1	57.3157	−5662.2	−5.06221	1.51E-17	6	189.63	2.47E+00	540	4.660E+06
11	Air	Mixture	132259-10-0	21.662	−692.39	−0.39208	0.0047574	1	59.15	5.64E+03	132.45	3.793E+06
12	Ammonia	H_3N	7664-41-7	90.483	−4669.7	−11.607	0.017194	1	195.41	6.11E+03	405.65	1.130E+07
13	Anisole	C_7H_8O	100-66-3	128.06	−9307.7	−16.693	0.014919	1	235.65	2.45E+00	645.6	4.273E+06
14	Argon	Ar	7440-37-1	42.127	−1093.1	−4.1425	0.000057254	2	83.78	6.87E+04	150.86	4.896E+06
15	Benzamide	C_7H_7NO	55-21-0	85.474	−11,932	−8.3348	1.29E-18	6	403	3.55E+02	824	5.047E+06
16	Benzene	C_6H_6	71-43-2	83.107	−6486.2	−9.2194	6.9844E-06	2	278.68	4.76E+03	562.05	4.875E+06
17	Benzenethiol	C_6H_6S	108-98-5	77.765	−8455.1	−7.7404	4.31E-18	6	258.27	7.68E+00	689	4.728E+06
18	Benzoic acid	$C_7H_6O_2$	65-85-0	88.513	−11,829	−8.6826	2.32E-19	6	395.45	7.96E+02	751	4.469E+06
19	Benzonitrile	C_7H_5N	100-47-0	55.0403	−7363.83	−4.50612	1.95E-18	6	260.28	5.40E+00	702.3	4.215E+06
20	Benzophenone	$C_{13}H_{10}O$	119-61-9	88.404	−11,769	−8.9014	1.93E-18	6	321.35	1.49E+00	830	3.357E+06
21	Benzyl alcohol	C_7H_8O	100-51-6	100.68	−11,059	−10.709	3.06E-18	6	257.85	1.88E-01	720.15	4.372E+06
22	Benzyl ethyl ether	$C_9H_{12}O$	539-30-0	68.541	−7886.2	−6.5804	2.4285E-06	2	275.65	2.31E+01	662	3.113E+06
23	Benzyl mercaptan	C_7H_8S	100-53-8	118.02	−10,527	−13.91	6.4794E-06	2	243.95	2.98E-01	718	4.074E+06
24	Biphenyl	$C_{12}H_{10}$	92-52-4	77.314	−9910.4	−7.5079	2.24E-18	6	342.2	9.42E+01	773	3.407E+06
25	Bromine	Br_2	7726-95-6	108.26	−6592	−14.16	0.016043	1	265.85	5.85E+03	584.15	1.028E+07
26	Bromobenzene	C_6H_5Br	108-86-1	63.749	−7130.2	−5.879	5.21E-18	6	242.43	7.84E+00	670.15	4.520E+06
27	Bromoethane	C_2H_5Br	74-96-4	57.3242	−4931.2	−5.2244	3.08E-17	6	154.25	3.80E-01	503.8	5.565E+06
28	Bromomethane	CH_3Br	74-83-9	44.7643	−3907.8	−3.4016	2.95E-17	6	179.44	2.07E+02	464	6.929E+06
29	1,2-Butadiene	C_4H_6	590-19-2	39.714	−3769.9	−2.6407	6.94E-18	6	136.95	4.47E-01	452	4.361E+06
30	1,3-Butadiene	C_4H_6	106-99-0	75.572	−4621.9	−8.5323	0.000012269	2	164.25	6.92E+01	425	4.303E+06
31	Butane	C_4H_{10}	106-97-8	66.343	−4363.2	−7.046	9.4509E-06	2	134.86	6.74E-01	425.12	3.770E+06
32	1,2-Butanediol	$C_4H_{10}O_2$	584-03-2	103.28	−11,548	−10.925	4.26E-18	6	220	2.93E-04	680	5.202E+06
33	1,3-Butanediol	$C_4H_{10}O_2$	107-88-0	123.22	−12,620	−13.986	0.000003926	2	196.15	3.74E-07	676	4.033E+06
34	1-Butanol	$C_4H_{10}O$	71-36-3	106.29483	−9866.35511	−11.6553	1.08E-17	6	183.85	2.91E-04	563.1	4.414E+06
35	2-Butanol	$C_4H_{10}O$	78-92-2	122.552	−10,236.2	−14.125	2.36E-17	6	158.45	1.24E-06	535.9	4.190E+06
36	1-Butene	C_4H_8	106-98-9	51.836	−4019.2	−4.5229	4.88E-17	6	87.8	6.94E-07	419.5	4.021E+06
37	*cis*-2-Butene	C_4H_8	590-18-1	72.541	−4691.2	−7.9776	0.000010368	2	134.26	2.72E-01	435.5	4.238E+06
38	*trans*-2-Butene	C_4H_8	624-64-6	71.704	−4563.1	−7.9053	0.000011319	2	167.62	7.45E+01	428.6	4.100E+06
39	Butyl acetate	$C_6H_{12}O_2$	123-86-4	122.82	−9253.2	−14.99	0.00001047	2	199.65	8.17E-02	575.4	3.087E+06
40	Butylbenzene	$C_{10}H_{14}$	104-51-8	101.22	−9255.4	−11.538	5.9208E-06	2	185.3	1.54E-04	660.5	2.882E+06
41	Butyl mercaptan	$C_4H_{10}S$	109-79-5	65.382	−6262.4	−6.2585	1.49E-17	6	157.46	2.35E-03	570.1	3.973E+06
42	*sec*-Butyl mercaptan	$C_4H_{10}S$	513-53-1	60.649	−5785.9	−5.6113	1.59E-17	6	133.02	3.40E-05	554	4.060E+06
43	1-Butyne	C_4H_6	107-00-6	77.004	−5054.5	−8.5665	0.000010161	2	147.43	1.18E+00	440	4.599E+06
44	Butyraldehyde	C_4H_8O	123-72-8	51.648	−5301.36	−4.2559	1.14E-17	6	176.8	6.97E-01	537.2	4.410E+06
45	Butyric acid	$C_4H_8O_2$	107-92-6	78.1171	−8924.37	−7.59929	7.39E-18	6	267.95	1.03E+01	615.7	4.060E+06
46	Butyronitrile	C_4H_7N	109-74-0	60.6576	−6404.32	−5.49286	1.13E-17	6	161.3	9.41E-04	585.4	3.880E+06
47	Carbon dioxide	CO_2	124-38-9	47.0169	−2839	−3.86388	2.81E-16	6	216.58	5.18E+05	304.21	7.384E+06
48	Carbon disulfide	CS_2	75-15-0	67.114	−4820.4	−7.5303	0.0091695	1	161.11	1.49E+00	552	8.041E+06
49	Carbon monoxide	CO	630-08-0	45.698	−1076.6	−4.8814	0.000075673	2	68.15	1.54E+04	132.92	3.494E+06
50	Carbon tetrachloride	CCl_4	56-23-5	78.441	−6128.1	−8.5766	6.8465E-06	2	250.33	1.12E+03	556.35	4.544E+06
51	Carbon tetrafluoride	CF_4	75-73-0	61.89	−2296.3	−7.086	0.000034687	2	89.56	1.08E+02	227.51	3.742E+06
52	Chlorine	Cl_2	7782-50-5	71.334	−3855	−8.5171	0.012378	1	172.12	1.37E+03	417.15	7.793E+06
53	Chlorobenzene	C_6H_5Cl	108-90-7	54.144	−6244.4	−4.5343	4.70E-18	6	227.95	8.45E+00	632.35	4.529E+06
54	Chloroethane	C_2H_5Cl	75-00-3	44.677	−4026	−3.371	2.27E-17	6	136.75	2.61E-01	460.35	5.267E+06
55	Chloroform	$CHCl_3$	67-66-3	146.43	−7792.3	−20.614	0.024578	1	207.15	5.25E+01	536.4	5.554E+06
56	Chloromethane	CH_3Cl	74-87-3	44.555	−3521.3	−3.4258	5.63E-17	6	175.45	8.84E+02	416.25	6.759E+06
57	1-Chloropropane	C_3H_7Cl	540-54-5	58.3592	−5111.33	−5.35261	2.47E-17	6	150.35	8.47E-02	503.15	4.425E+06
58	2-Chloropropane	C_3H_7Cl	75-29-6	46.854	−4445.5	−3.6533	1.33E-17	6	155.97	9.08E-01	489	4.510E+06
59	*m*-Cresol	C_7H_8O	108-39-4	95.403	−10,581	−10.004	4.30E-18	6	285.39	5.86E+00	705.85	4.522E+06
60	*o*-Cresol	C_7H_8O	95-48-7	210.88	−13,928	−29.483	0.025182	1	304.19	6.53E+01	697.55	5.058E+06

(*Continued*)

TABLE 2-8 Vapor Pressure of Inorganic and Organic Liquids, $\ln P = C_1 + C_2/T + C_3 \ln T + C_4 T^{C_5}$, P in Pa, T in K (*Continued*)

Cmpd. no.*	Name	Formula	CAS	C_1	C_2	C_3	C_4	C_5	T_{min}, K	P at T_{min}	T_{max}, K	P at T_{max}
61	*p*-Cresol	C_7H_8O	106-44-5	118.53	−11,957	−13.293	8.70E-18	6	307.93	3.45E+01	704.65	5.151E+06
62	Cumene	C_9H_{12}	98-82-8	102.81	−8674.6	−11.922	7.0048E-06	2	177.14	4.71E-04	631	3.226E+06
63	Cyanogen	C_2N_2	460-19-5	39.0596	−3473.98	−2.48683	2.86E-17	6	245.25	7.44E+04	400.15	5.924E+06
64	Cyclobutane	C_4H_8	287-23-0	85.899	−4884.4	−10.883	0.014934	1	182.48	1.80E+02	459.93	4.991E+06
65	Cyclohexane	C_6H_{12}	110-82-7	51.087	−5226.4	−4.2278	9.76E-18	6	279.69	5.36E+03	553.8	4.093E+06
66	Cyclohexanol	$C_6H_{12}O$	108-93-0	189.19	−14,337	−24.148	0.00001074	2	296.6	7.65E+01	650.1	4.265E+06
67	Cyclohexanone	$C_6H_{10}O$	108-94-1	85.424	−7944.4	−9.2862	4.9957E-06	2	242	6.80E+00	653	3.989E+06
68	Cyclohexene	C_6H_{10}	110-83-8	88.184	−6624.9	−10.059	8.2566E-06	2	169.67	1.04E-01	560.4	4.392E+06
69	Cyclopentane	C_5H_{10}	287-92-3	66.341	−5198.5	−6.8103	0.000006193	2	179.28	9.07E+00	511.7	4.513E+06
70	Cyclopentene	C_5H_8	142-29-0	67.952	−5187.5	−7.0785	6.8165E-06	2	138.13	1.28E-02	507	4.799E+06
71	Cyclopropane	C_3H_6	75-19-4	40.608	−3179.6	−2.8937	5.61E-17	6	145.59	7.80E+01	398	5.494E+06
72	Cyclohexyl mercaptan	$C_6H_{12}S$	1569-69-3	85.146	−7843.7	−9.2982	5.1788E-06	2	189.64	8.24E-03	664	3.970E+06
73	Decanal	$C_{10}H_{20}O$	112-31-2	93.5742	−10,403.8	−9.79483	4.57E-18	6	285	5.51E+00	674	2.600E+06
74	Decane	$C_{10}H_{22}$	124-18-5	112.73	−9749.6	−13.245	7.1266E-06	2	243.51	1.39E+00	617.7	2.091E+06
75	Decanoic acid	$C_{10}H_{20}O_2$	334-48-5	126.405	−14,864.6	−13.9067	2.51E-18	6	304.55	1.45E-01	722.1	2.280E+06
76	1-Decanol	$C_{10}H_{22}O$	112-30-1	156.23933	−15,212.33492	−18.42393	8.50E-18	6	280.05	1.50E-01	688	2.308E+06
77	1-Decene	$C_{10}H_{20}$	872-05-9	68.401	−7776.9	−6.4637	6.38E-18	6	206.89	2.59E-02	616.6	2.223E+06
78	Decyl mercaptan	$C_{10}H_{22}S$	143-10-2	91.91	−10,565	−9.5957	5.70E-18	6	247.56	2.59E-02	696	2.130E+06
79	1-Decyne	$C_{10}H_{18}$	764-93-2	142.94	−11,119	−17.818	0.00001102	2	229.15	1.60E-01	619.85	2.363E+06
80	Deuterium	D_2	7782-39-0	18.947	−154.47	−0.57226	0.038899	1	18.73	1.72E+04	38.35	1.663E+06
81	1,1-Dibromoethane	$C_2H_4Br_2$	557-91-5	62.711	−6503.5	−5.7669	1.0427E-06	2	210.15	2.64E+00	628	6.034E+06
82	1,2-Dibromoethane	$C_2H_4Br_2$	106-93-4	43.751	−5587.7	−3.0891	8.2664E-07	2	282.85	7.53E+02	650.15	5.375E+06
83	Dibromomethane	CH_2Br_2	74-95-3	86.295	−7010.3	−9.5972	6.7794E-06	2	220.6	2.13E+01	611	7.170E+06
84	Dibutyl ether	$C_8H_{18}O$	142-96-1	72.227	−7537.6	−7.0596	9.14E-18	6	175.3	7.14E-04	584.1	2.459E+06
85	*m*-Dichlorobenzene	$C_6H_4Cl_2$	541-73-1	53.187	−6827.5	−4.3233	2.31E-18	6	248.39	6.41E+00	683.95	4.070E+06
86	*o*-Dichlorobenzene	$C_6H_4Cl_2$	95-50-1	77.105	−8111.1	−7.8886	2.7267E-06	2	256.15	6.49E+00	705	4.074E+06
87	*p*-Dichlorobenzene	$C_6H_4Cl_2$	106-46-7	88.31	−8463.4	−9.6308	4.5833E-06	2	326.14	1.23E+03	684.75	4.070E+06
88	1,1-Dichloroethane	$C_2H_4Cl_2$	75-34-3	66.611	−5493.1	−6.7301	5.3579E-06	2	176.19	2.21E+00	523	5.106E+06
89	1,2-Dichloroethane	$C_2H_4Cl_2$	107-06-2	92.355	−6920.4	−10.651	9.1426E-06	2	237.49	2.37E+02	561.6	5.318E+06
90	Dichloromethane	CH_2Cl_2	75-09-2	101.6	−6541.6	−12.247	0.000012311	2	178.01	5.93E+00	510	6.093E+06
91	1,1-Dichloropropane	$C_3H_6Cl_2$	78-99-9	83.495	−6661.4	−9.2386	6.7652E-06	2	192.5	1.72E+00	560	4.239E+06
92	1,2-Dichloropropane	$C_3H_6Cl_2$	78-87-5	65.955	−6015.6	−6.5509	4.3172E-06	2	172.71	8.25E-02	572	4.232E+06
93	Diethanol amine	$C_4H_{11}NO_2$	111-42-2	106.38	−13,714	−11.06	3.26E-18	6	301.15	1.02E-01	736.6	4.260E+06
94	Diethyl amine	$C_4H_{11}N$	109-89-7	49.314	−4949	−3.9256	9.20E-18	6	223.35	3.74E+02	496.6	3.674E+06
95	Diethyl ether	$C_4H_{10}O$	60-29-7	136.9	−6954.3	−19.254	0.024508	1	156.85	3.95E-01	466.7	3.641E+06
96	Diethyl sulfide	$C_4H_{10}S$	352-93-2	46.705	−5177.4	−3.5985	1.7147E-06	2	169.2	9.93E-02	557.15	3.961E+06
97	1,1-Difluoroethane	$C_2H_4F_2$	75-37-6	73.491	−4385.9	−8.1851	0.000012978	2	154.56	6.45E+01	386.44	4.507E+06
98	1,2-Difluoroethane	$C_2H_4F_2$	624-72-6	84.625	−5217.4	−9.871	0.00001305	2	179.6	1.17E+02	445	4.372E+06
99	Difluoromethane	CH_2F_2	75-10-5	69.132	−3847.7	−7.5868	0.000015065	2	136.95	5.43E+01	351.26	5.761E+06
100	Di-sopropyl amine	$C_6H_{15}N$	108-18-9	462.84	−18,227	−73.734	0.092794	1	176.85	4.47E-03	523.1	3.199E+06
101	Di-sopropyl ether	$C_6H_{14}O$	108-20-3	41.631	−4668.7	−2.8551	0.00063693	1	187.65	6.86E+00	500.05	2.869E+06
102	Di-sopropyl ketone	$C_7H_{14}O$	565-80-0	50.868	−6036.5	−4.066	1.1326E-06	2	204.81	8.21E-01	576	3.017E+06
103	1,1-Dimethoxyethane	$C_4H_{10}O_2$	534-15-6	53.637	−5251.2	−4.5649	1.68E-17	6	159.95	9.45E-02	507.8	3.773E+06
104	1,2-Dimethoxypropane	$C_5H_{12}O_2$	7778-85-0	62.097	−6174.9	−5.715	1.23E-17	6	226.1	4.50E+01	543	3.447E+06
105	Dimethyl acetylene	C_4H_6	503-17-3	66.592	−4999.8	−6.8387	6.6793E-06	2	240.91	6.12E+03	473.2	4.870E+06
106	Dimethyl amine	C_2H_7N	124-40-3	71.738	−5302	−7.3324	6.42E-17	6	180.96	7.56E+01	437.2	5.258E+06
107	2,3-Dimethylbutane	C_6H_{14}	79-29-8	77.161	−5691.1	−8.501	8.0325E-06	2	145.19	1.52E-02	500	3.130E+06
108	1,1-Dimethylcyclohexane	C_8H_{16}	590-66-9	81.184	−6927	−8.8498	0.000005458	2	239.66	6.06E+01	591.15	2.939E+06
109	*cis*-1,2-Dimethylcyclohexane	C_8H_{16}	2207-01-4	78.952	−7075.4	−8.4344	4.5035E-06	2	223.16	6.41E+00	606.15	2.939E+06
110	*trans*-1,2-Dimethylcyclohexane	C_8H_{16}	6876-23-9	78.429	−6882.1	−8.4129	4.9831E-06	2	184.99	8.04E-02	596.15	2.938E+06
111	Dimethyl disulfide	$C_2H_6S_2$	624-92-0	81.045	−6941.3	−8.777	5.5501E-06	2	188.44	2.07E-01	615	5.363E+06
112	Dimethyl ether	C_2H_6O	115-10-6	44.704	−3525.6	−3.4444	5.46E-17	6	131.65	3.05E+00	400.1	5.274E+06
113	*N,N*-Dimethyl formamide	C_3H_7NO	68-12-2	82.762	−7955.5	−8.8038	4.2431E-06	2	212.72	1.95E-01	649.6	4.365E+06
114	2,3-Dimethylpentane	C_7H_{16}	565-59-3	78.335	−6348.7	−8.5105	6.4311E-06	2	160	1.26E-02	537.3	2.882E+06
115	Dimethyl phthalate	$C_{10}H_{10}O_4$	131-11-3	72.517	−10,415	−6.755	1.3269E-06	2	274.18	3.72E-02	766	2.779E+06
116	Dimethylsilane	C_2H_8Si	1111-74-6	63.08	−4062.3	−6.425	1.51E-16	6	122.93	4.15E-01	402	3.561E+06
117	Dimethyl sulfide	C_2H_6S	75-18-3	84.39	−5740.6	−9.6454	0.000010073	2	174.88	7.86E+00	503.04	5.533E+06
118	Dimethyl sulfoxide	C_2H_6OS	67-68-5	56.273	−7620.6	−4.6279	4.3819E-07	2	291.67	5.02E+01	729	5.648E+06
119	Dimethyl terephthalate	$C_{10}H_{10}O_4$	120-61-6	66.1795	−9870.41	−5.85599	1.47E-18	6	413.79	1.15E+03	777.4	2.759E+06
120	1,4-Dioxane	$C_4H_8O_2$	123-91-1	44.494	−5406.7	−3.1287	2.89E-18	6	284.95	2.53E+03	587	5.158E+06

121	Diphenyl ether	$C_{12}H_{10}O$	101-84-8	59.969	−8585.5	−5.1538	2.00E-18	6	300.03	7.09E+00	766.8	3.097E+06
122	Dipropyl amine	$C_6H_{15}N$	142-84-7	54	−6018.5	−4.4981	9.97E-18	6	210.15	3.69E+00	550	3.111E+06
123	Dodecane	$C_{12}H_{26}$	112-40-3	137.47	−11,976	−16.698	8.0906E-06	2	263.57	6.15E-01	658	1.822E+06
124	Eicosane	$C_{20}H_{42}$	112-95-8	203.66	−19,441	−25.525	8.8382E-06	2	309.58	9.26E-03	768	1.175E+06
125	Ethane	C_2H_6	74-84-0	51.857	−2598.7	−5.1283	0.000014913	2	90.35	1.13E+00	305.32	4.852E+06
126	Ethanol	C_2H_6O	64-17-5	73.304	−7122.3	−7.1424	2.8853E-06	2	159.05	4.96E-04	514	6.109E+06
127	Ethyl acetate	$C_4H_8O_2$	141-78-6	66.824	−6227.6	−6.41	1.79E-17	6	189.6	1.43E+00	523.3	3.850E+06
128	Ethyl amine	C_2H_7N	75-04-7	81.56	−5596.9	−9.0779	0.000008792	2	192.15	1.52E+02	456.15	5.594E+06
129	Ethylbenzene	C_8H_{10}	100-41-4	89.063	−7733.7	−9.917	0.000005986	2	178.2	3.91E-03	617.15	3.590E+06
130	Ethyl benzoate	$C_9H_{10}O_2$	93-89-0	52.923	−7531.7	−4.2347	1.1835E-06	2	238.45	1.69E-01	698	3.203E+06
131	2-Ethyl butanoic acid	$C_6H_{12}O_2$	88-09-5	90.464	−10,243	−9.2836	5.26E-18	6	258.15	4.63E-01	655	3.403E+06
132	Ethyl butyrate	$C_6H_{12}O_2$	105-54-4	57.661	−6346.5	−5.032	8.25E-18	6	175.15	1.04E-02	571	2.935E+06
133	Ethylcyclohexane	C_8H_{16}	1678-91-7	80.208	−7203.2	−8.6023	4.5901E-06	2	161.84	3.57E-04	609.15	3.041E+06
134	Ethylcyclopentane	C_7H_{14}	1640-89-7	88.671	−7012.7	−10.045	7.4578E-06	2	134.71	3.71E-06	569.5	3.412E+06
135	Ethylene	C_2H_4	74-85-1	53.963	−2443	−5.5643	0.000019079	2	104	1.26E+02	282.34	5.032E+06
136	Ethylenediamine	$C_2H_8N_2$	107-15-3	73.51	−7572.7	−7.1435	1.21E-17	6	284.29	6.78E+02	593	6.290E+06
137	Ethylene glycol	$C_2H_6O_2$	107-21-1	84.09	−10,411	−8.1976	1.65E-18	6	260.15	2.19E-01	720	8.257E+06
138	Ethyleneimine	C_2H_5N	151-56-4	66.51	−6019.2	−6.3332	1.04E-17	6	195.2	9.71E+00	537	6.850E+06
139	Ethylene oxide	C_2H_4O	75-21-8	91.944	−5293.4	−11.682	0.014902	1	160.65	7.79E+00	469.15	7.255E+06
140	Ethyl formate	$C_3H_6O_2$	109-94-4	73.833	−5817	−7.809	0.00000632	2	193.55	1.81E+01	508.4	4.708E+06
141	2-Ethyl hexanoic acid	$C_8H_{16}O_2$	149-57-5	122.364	−13,308.8	−13.5709	6.42E-18	6	155.15	1.44E-14	674.6	2.780E+06
142	Ethylhexyl ether	$C_8H_{18}O$	5756-43-4	77.523	−7978.8	−7.7757	1.01E-17	6	180	7.60E-04	583	2.460E+06
143	Ethylisopropyl ether	$C_5H_{12}O$	625-54-7	57.723	−5236.9	−5.2136	2.30E-17	6	140	4.31E-03	489	3.414E+06
144	Ethylisopropyl ketone	$C_6H_{12}O$	565-69-5	57.459	−6356.8	−4.9545	5.20E-18	6	204.15	9.70E-01	567	3.293E+06
145	Ethyl mercaptan	C_2H_6S	75-08-1	65.551	−5027.4	−6.6853	6.3208E-06	2	125.26	1.14E-03	499.15	5.492E+06
146	Ethyl propionate	$C_5H_{10}O_2$	105-37-3	105.64	−8007	−12.477	0.000009	2	199.25	7.80E-01	546	3.336E+06
147	Ethylpropyl ether	$C_5H_{12}O$	628-32-0	86.898	−6646.4	−9.5758	5.96E-17	6	145.65	1.61E-03	500.23	3.372E+06
148	Ethyltrichlorosilane	$C_2H_5Cl_3Si$	115-21-9	61.6271	−6095.88	−5.69714	1.06E-17	6	167.55	1.96E-02	559.95	3.321E+06
149	Fluorine	F_2	7782-41-4	42.393	−1103.3	−4.1203	0.000057815	2	53.48	2.53E+02	144.12	5.167E+06
150	Fluorobenzene	C_6H_5F	462-06-6	51.915	−5439	−4.2896	8.75E-18	6	230.94	1.51E+02	560.09	4.544E+06
151	Fluoroethane	C_2H_5F	353-36-6	38.593	−3123.34	−2.53014	5.30E-17	6	129.95	9.43E+00	375.31	4.980E+06
152	Fluoromethane	CH_3F	593-53-3	41.2744	−2676.65	−3.03914	2.45E-16	6	131.35	4.34E+02	317.42	5.875E+06
153	Formaldehyde	CH_2O	50-00-0	49.3632	−3847.87	−4.09834	4.64E-17	6	155.15	4.89E+01	420	6.590E+06
154	Formamide	CH_3NO	75-12-7	100.3	−10,763	−10.946	3.8503E-06	2	275.6	1.04E+00	771	7.751E+06
155	Formic acid	CH_2O_2	64-18-6	43.8066	−5131.03	−3.18777	2.37819E-06	2	281.45	2.41E+03	588	5.810E+06
156	Furan	C_4H_4O	110-00-9	74.738	−5417	−8.0636	0.00000747	2	187.55	5.00E+01	490.15	5.550E+06
157	Helium-4	He	7440-59-7	11.533	−8.99	0.6724	0.2743	1	1.76	1.46E+03	5.2	2.284E+05
158	Heptadecane	$C_{17}H_{36}$	629-78-7	156.95	−15,557	−18.966	6.4559E-06	2	295.13	4.65E-02	736	1.344E+06
159	Heptanal	$C_7H_{14}O$	111-71-7	55.3058	−6694.68	−4.64122	5.28E-18	6	229.8	2.56E+00	620	3.160E+06
160	Heptane	C_7H_{16}	142-82-5	87.829	−6996.4	−9.8802	7.2099E-06	2	182.57	1.83E-01	540.2	2.719E+06
161	Heptanoic acid	$C_7H_{14}O_2$	111-14-8	112.372	−12,660.1	−12.147	4.39E-18	6	265.83	4.66E-02	677.3	3.042E+06
162	1-Heptanol	$C_7H_{16}O$	111-70-6	147.41	−13,466	−17.353	1.13E-17	6	239.15	1.95E-02	632.3	3.013E+06
163	2-Heptanol	$C_7H_{16}O$	543-49-7	153.088	−12,618.7	−18.7479	7.45073E-06	2	220	6.55E-03	608.3	3.000E+06
164	3-Heptanone	$C_7H_{14}O$	106-35-4	78.463	−8077.2	−7.9062	8.05E-18	6	234.15	2.30E+00	606.6	2.919E+06
165	2-Heptanone	$C_7H_{14}O$	110-43-0	75.494	−7896.5	−7.5047	8.91E-18	6	238.15	3.54E+00	611.4	2.946E+06
166	1-Heptene	C_7H_{14}	592-76-7	65.922	−6189	−6.3629	2.01E-17	6	154.12	1.86E-03	537.4	2.921E+06
167	Heptyl mercaptan	$C_7H_{16}S$	1639-09-4	79.858	−8501.8	−8.1043	8.15E-18	6	229.92	3.05E-01	645	2.772E+06
168	1-Heptyne	C_7H_{12}	628-71-7	59.083	−6031.8	−5.3072	1.44E-17	6	192.22	8.15E-01	547	3.209E+06
169	Hexadecane	$C_{16}H_{34}$	544-76-3	156.06	−15,015	−18.941	6.8172E-06	2	291.31	9.23E-02	723	1.411E+06
170	Hexanal	$C_6H_{12}O$	66-25-1	58.7734	−6529.3	−5.17151	6.95E-18	6	214.93	1.86E+00	594	3.460E+06
171	Hexane	C_6H_{14}	110-54-3	104.65	−6995.5	−12.702	0.000012381	2	177.83	9.02E-01	507.6	3.045E+06
172	Hexanoic acid	$C_6H_{12}O_2$	142-62-1	98.3767	−11,394	−10.2239	3.29E-18	6	269.25	3.17E-01	660.2	3.309E+06
173	1-Hexanol	$C_6H_{14}O$	111-27-3	135.42149	−12,288.40621	−15.73191	1.27E-17	6	228.55	2.25E-02	611.3	3.446E+06
174	2-Hexanol	$C_6H_{14}O$	626-93-7	122.695	−10,870	−14.192	0.000003871	2	223	7.46E-02	585.3	3.323E+06
175	2-Hexanone	$C_6H_{12}O$	591-78-6	107.44	−8528.6	−12.679	8.4606E-06	2	217.35	1.45E+00	587.61	3.286E+06
176	3-Hexanone	$C_6H_{12}O$	589-38-8	73.155	−7242.9	−7.2569	1.27E-17	6	217.5	2.22E+00	582.82	3.322E+06
177	1-Hexene	C_6H_{12}	592-41-6	51.9766	−5104.66	−4.34844	1.17E-17	6	133.39	5.16E-04	504	3.210E+06
178	3-Hexyne	C_6H_{10}	928-49-4	47.091	−5104	−3.6371	0.00051621	1	170.05	2.20E-01	544	3.540E+06
179	Hexyl mercaptan	$C_6H_{14}S$	111-31-9	68.467	−7390.5	−6.5456	7.76E-18	6	192.62	1.31E-02	623	3.079E+06
180	1-Hexyne	C_6H_{10}	693-02-7	133.2	−7492.9	−18.405	0.022062	1	141.25	3.92E-04	516.2	3.635E+06
181	2-Hexyne	C_6H_{10}	764-35-2	123.71	−7639	−16.451	0.016495	1	183.65	5.40E-01	549	3.530E+06
182	Hydrazine	H_4N_2	302-01-2	76.858	−7245.2	−8.22	0.0061557	1	274.69	4.08E+02	653.15	1.473E+07
183	Hydrogen	H_2	1333-74-0	12.69	−94.896	1.1125	0.00032915	2	13.95	7.21E+03	33.19	1.315E+06

(Continued)

TABLE 2-8 Vapor Pressure of Inorganic and Organic Liquids, $\ln P = C_1 + C_2/T + C_3 \ln T + C_4 T^{C_5}$, P in Pa, T in K *(Continued)*

Cmpd. no.*	Name	Formula	CAS	C_1	C_2	C_3	C_4	C_5	T_{min}, K	P at T_{min}	T_{max}, K	P at T_{max}
184	Hydrogen bromide	BrH	10035-10-6	29.315	−2424.5	−1.1354	2.38E-18	6	185.15	2.95E+04	363.15	8.463E+06
185	Hydrogen chloride	ClH	7647-01-0	104.27	−3731.2	−15.047	0.03134	1	158.97	1.35E+04	324.65	8.356E+06
186	Hydrogen cyanide	CHN	74-90-8	36.75	−3927.1	−2.1245	3.89E-17	6	259.83	1.87E+04	456.65	5.353E+06
187	Hydrogen fluoride	FH	7664-39-3	59.544	−4143.8	−6.1764	0.000014161	2	189.79	3.37E+02	461.15	6.487E+06
188	Hydrogen sulfide	H_2S	7783-06-4	85.584	−3839.9	−11.199	0.018848	1	187.68	2.29E+04	373.53	8.999E+06
189	Isobutyric acid	$C_4H_8O_2$	79-31-2	110.38	−10,540	−12.262	1.43E-17	6	227.15	7.82E-02	605	3.683E+06
190	Isopropyl amine	C_3H_9N	75-31-0	136.66	−7201.5	−18.934	0.022255	1	177.95	7.73E+00	471.85	4.540E+06
191	Malonic acid	$C_3H_4O_4$	141-82-2	119.172	−15,688.8	−12.6757	1.55E-18	6	409.15	9.97E+01	834	6.097E+06
192	Methacrylic acid	$C_4H_6O_2$	79-41-4	109.53	−10,410	−12.289	0.000003199	2	288.15	5.86E+01	662	4.812E+06
193	Methane	CH_4	74-82-8	39.205	−1324.4	−3.4366	0.000031019	2	90.69	1.17E+04	190.56	4.590E+06
194	Methanol	CH_4O	67-56-1	82.718	−6904.5	−8.8622	7.4664E-06	2	175.47	1.11E-01	512.5	8.145E+06
195	*N*-Methyl acetamide	C_3H_7NO	79-16-3	79.128	−9523.9	−7.7355	3.16E-18	6	301.15	2.86E+01	718	4.997E+06
196	Methyl acetate	$C_3H_6O_2$	79-20-9	61.267	−5618.6	−5.6473	2.11E-17	6	175.15	1.02E+00	506.55	4.695E+06
197	Methyl acetylene	C_3H_4	74-99-7	50.242	−3811.9	−4.2526	6.53E-17	6	170.45	4.15E+02	402.4	5.619E+06
198	Methyl acrylate	$C_4H_6O_2$	96-33-3	107.69	−7027.2	−13.916	0.015185	1	196.32	4.07E+00	536	4.277E+06
199	Methyl amine	CH_5N	74-89-5	75.206	−5082.8	−8.0919	0.000008113	2	179.69	1.77E+02	430.05	7.414E+06
200	Methyl benzoate	$C_8H_8O_2$	93-58-3	84.828	−9334.7	−8.7063	6.17E-18	6	260.75	1.81E+00	693	3.589E+06
201	3-Methyl-1,2-butadiene	C_5H_8	598-25-4	66.575	−5213.4	−6.7693	4.8106E-06	2	159.53	7.28E-01	490	3.831E+06
202	2-Methylbutane	C_5H_{12}	78-78-4	71.308	−4976	−7.7169	8.7271E-06	2	113.25	1.21E-04	460.4	3.366E+06
203	2-Methylbutanoic acid	$C_5H_{10}O_2$	116-53-0	85.383	−9575.4	−8.6164	5.61E-18	6	193	6.94E-05	643	3.886E+06
204	3-Methyl-1-butanol	$C_5H_{12}O$	123-51-3	117.074	−10,743.2	−13.1654	1.17E-17	6	155.95	1.14E-08	577.2	3.933E+06
205	2-Methyl-1-butene	C_5H_{10}	563-46-2	93.131	−5525.4	−11.852	0.014205	1	135.58	2.05E-02	465	3.465E+06
206	2-Methyl-2-butene	C_5H_{10}	513-35-9	83.927	−5640.5	−9.6453	0.000011121	2	139.39	1.94E-02	470	3.394E+06
207	2-Methyl-1-butene-3-yne	C_5H_6	78-80-8	95.453	−5448.8	−12.384	0.015643	1	160.15	2.92E+00	492	4.469E+06
208	Methylbutyl ether	$C_5H_{12}O$	628-28-4	60.164	−5621.7	−5.53	1.86E-17	6	157.48	2.99E-02	512.74	3.377E+06
209	Methylbutyl sulfide	$C_5H_{12}S$	628-29-5	96.344	−7856.3	−11.058	0.000007308	2	175.3	4.61E-03	593	3.464E+06
210	3-Methyl-1-butyne	C_5H_8	598-23-2	69.459	−5250	−7.1125	7.93E-17	6	183.45	4.36E+01	463.2	4.199E+06
211	Methyl butyrate	$C_5H_{10}O_2$	623-42-7	71.87	−6885.7	−7.0944	1.49E-17	6	187.35	1.34E-01	554.5	3.480E+06
212	Methylchlorosilane	CH_5ClSi	993-00-0	95.984	−5401.7	−11.829	0.000018092	2	139.05	4.12E-01	442	4.170E+06
213	Methylcyclohexane	C_7H_{14}	108-87-2	92.684	−7080.8	−10.695	8.1366E-06	2	146.58	1.52E-04	572.1	3.486E+06
214	1-Methylcyclohexanol	$C_7H_{14}O$	590-67-0	134.63	−10,682	−16.511	8.4427E-06	2	299.15	2.57E+02	686	3.994E+06
215	*cis*-2-Methylcyclohexanol	$C_7H_{14}O$	7443-70-1	125.1	−10,288	−15.157	0.000010918	2	280.15	4.56E+01	614	3.807E+06
216	*trans*-2-Methylcyclohexanol	$C_7H_{14}O$	7443-52-9	54.179	−7477.2	−4.22	3.52E-18	6	269.15	1.62E+01	617	3.767E+06
217	Methylcyclopentane	C_6H_{12}	96-37-7	55.368	−5149.8	−5.0136	0.000003222	2	130.73	2.25E-04	532.7	3.759E+06
218	1-Methylcyclopentene	C_6H_{10}	693-89-0	52.732	−5286.9	−4.4509	1.09E-17	6	146.62	3.98E-03	542	4.130E+06
219	3-Methylcyclopentene	C_6H_{10}	1120-62-3	52.601	−5120.3	−4.4554	1.33E-17	6	168.54	5.37E-01	526	4.129E+06
220	Methyldichlorosilane	CH_4Cl_2Si	75-54-7	79.788	−5420	−9.0702	0.000011489	2	182.55	2.58E+01	483	3.964E+06
221	Methylethyl ether	C_3H_8O	540-67-0	78.586	−5176.3	−8.7501	9.1727E-06	2	160	7.85E+00	437.8	4.433E+06
222	Methylethyl ketone	C_4H_8O	78-93-3	72.698	−6143.6	−7.5779	5.6476E-06	2	186.48	1.39E+00	535.5	4.120E+06
223	Methylethyl sulfide	C_3H_8S	624-89-5	79.07	−6114.1	−8.631	6.5333E-06	2	167.23	2.25E-01	533	4.261E+06
224	Methyl formate	$C_2H_4O_2$	107-31-3	77.184	−5606.1	−8.392	7.8468E-06	2	174.15	6.88E+00	487.2	5.983E+06
225	Methylisobutyl ether	$C_5H_{12}O$	625-44-5	57.984	−5339.6	−5.2362	2.08E-17	6	188	8.70E+00	497	3.416E+06
226	Methylisobutyl ketone	$C_6H_{12}O$	108-10-1	80.503	−7421.8	−8.379	1.81E-17	6	189.15	6.99E-02	574.6	3.272E+06
227	Methyl Isocyanate	C_2H_3NO	624-83-9	57.612	−5197.9	−5.1269	2.17E-17	6	256.15	7.28E+03	488	5.480E+06
228	Methylisopropyl ether	$C_4H_{10}O$	598-53-8	53.867	−4701	−4.7052	2.88E-17	6	127.93	3.32E-03	464.48	3.764E+06
229	Methylisopropyl ketone	$C_5H_{10}O$	563-80-4	45.242	−5324.4	−3.2551	3.04E-18	6	180.15	2.95E-01	553.4	3.792E+06
230	Methylisopropyl sulfide	$C_4H_{10}S$	1551-21-9	52.82	−5437.7	−4.442	9.51E-18	6	171.64	1.80E-01	553.1	4.022E+06
231	Methyl mercaptan	CH_4S	74-93-1	54.15	−4337.7	−4.8127	4.50E-17	6	150.18	3.15E+00	469.95	7.231E+06
232	Methyl methacrylate	$C_5H_8O_2$	80-62-6	107.36	−8085.3	−12.72	8.3307E-06	2	224.95	1.91E+01	566	3.674E+06
233	2-Methyloctanoic acid	$C_9H_{18}O_2$	3004-93-1	105.7	−12,458	−11.234	4.46E-18	6	240	4.19E-04	694	2.545E+06
234	2-Methylpentane	C_6H_{14}	107-83-5	53.579	−5041.2	−4.6404	1.94E-17	6	119.55	2.07E-05	497.7	3.044E+06
235	Methyl pentyl ether	$C_6H_{14}O$	628-80-8	61.907	−6188.9	−5.706	1.18E-17	6	176	6.33E-02	546.49	3.041E+06
236	2-Methylpropane	C_4H_{10}	75-28-5	108.43	−5039.9	−15.012	0.022725	1	113.54	1.21E-02	407.8	3.630E+06
237	2-Methyl-2-propanol	$C_4H_{10}O$	75-65-0	172.27	−11,589	−22.113	0.000013703	2	298.97	5.88E+03	506.2	3.957E+06
238	2-Methyl propene	C_4H_8	115-11-7	78.01	−4634.1	−8.9575	0.000013413	2	132.81	6.45E-01	417.9	4.004E+06
239	Methyl propionate	$C_4H_8O_2$	554-12-1	70.717	−6439.7	−6.9845	2.01E-17	6	185.65	6.34E-01	530.6	4.028E+06
240	Methylpropyl ether	$C_4H_{10}O$	557-17-5	67.942	−5419.1	−6.8067	4.78E-17	6	133.97	2.90E-03	476.25	3.802E+06
241	Methylpropyl sulfide	$C_4H_{10}S$	3877-15-4	83.711	−6786.9	−9.2526	6.6666E-06	2	160.17	4.26E-03	565	3.972E+06
242	Methylsilane	CH_6Si	992-94-9	37.205	−2590.3	−2.5993	6.0508E-06	2	116.34	1.43E+01	352.5	4.702E+06

243	alpha-Methyl styrene	C_9H_{10}	98-83-9	56.485	−6954.2	−4.7889	2.78E-18	6	249.95	9.23E+00	654	3.341E+06
244	Methyl tert-butyl ether	$C_5H_{12}O$	1634-04-4	57.1299	−5200.7	−5.13976	1.65E-17	6	164.55	4.94E-01	497.1	3.286E+06
245	Methyl vinyl ether	C_3H_6O	107-25-5	51.085	−4271	−4.307	3.05E-17	6	151.15	3.37E+00	437	4.583E+06
246	Naphthalene	$C_{10}H_8$	91-20-3	62.964	−8137.5	−5.6317	2.27E-18	6	353.43	9.91E+02	748.4	4.069E+06
247	Neon	Ne	7440-01-9	29.755	−271.06	−2.6081	0.000527	2	24.56	4.38E+04	44.4	2.665E+06
248	Nitroethane	$C_2H_5NO_2$	79-24-3	75.632	−7202.3	−7.6464	1.83E-17	6	183.63	3.18E-02	593	5.159E+06
249	Nitrogen	N_2	7727-37-9	58.282	−1084.1	−8.3144	0.044127	1	63.15	1.25E+04	126.2	3.391E+06
250	Nitrogen trifluoride	F_3N	7783-54-2	68.149	−2257.9	−8.9118	0.023233	1	66.46	1.86E-01	234	4.500E+06
251	Nitromethane	CH_3NO_2	75-52-5	57.278	−6089	−4.9821	1.22E-17	6	244.6	1.47E+02	588.15	6.309E+06
252	Nitrous oxide	N_2O	10024-97-2	96.512	−4045	−12.277	0.00002886	2	182.3	8.69E+04	309.57	7.278E+06
253	Nitric oxide	NO	10102-43-9	72.974	−2650	−8.261	9.70E-15	6	109.5	2.20E+04	180.15	6.516E+06
254	Nonadecane	$C_{19}H_{40}$	629-92-5	182.54	−17,897	−22.498	7.4008E-06	2	305.04	1.59E-02	758	1.208E+06
255	Nonanal	$C_9H_{18}O$	124-19-6	80.3832	−9096.15	−8.03581	4.71E-18	6	267.3	4.25E+00	658.5	2.680E+06
256	Nonane	C_9H_{20}	111-84-2	109.35	−9030.4	−12.882	7.8544E-06	2	219.66	4.31E-01	594.6	2.305E+06
257	Nonanoic acid	$C_9H_{18}O_2$	112-05-0	123.374	−14,215.3	−13.5607	3.17E-18	6	285.55	4.58E-02	710.7	2.513E+06
258	1-Nonanol	$C_9H_{20}O$	143-08-8	162.854	−15,204.55331	−19.42436	1.07E-17	6	268.15	8.58E-02	670.9	2.528E+06
259	2-Nonanol	$C_9H_{20}O$	628-99-9	213.069	−16,246	−27.6195	1.31827E-05	2	238.15	3.85E-03	649.5	2.540E+06
260	1-Nonene	C_9H_{18}	124-11-8	63.313	−7040.4	−5.8055	7.58E-18	6	191.91	2.04E-02	593.1	2.427E+06
261	Nonyl mercaptan	$C_9H_{20}S$	1455-21-6	106.2	−10,982	−11.696	8.90E-18	6	253.05	1.47E-01	681	2.330E+06
262	1-Nonyne	C_9H_{16}	3452-09-3	114.77	−9430.8	−13.631	8.1918E-06	2	223.15	4.50E-01	598.05	2.619E+06
263	Octadecane	$C_{18}H_{38}$	593-45-3	157.68	−16,093	−18.954	5.9272E-06	2	301.31	3.39E-02	747	1.255E+06
264	Octanal	$C_8H_{16}O$	124-13-0	74.0298	−8302.12	−7.19776	5.31E-18	6	251.65	3.49E+00	638.9	2.960E+06
265	Octane	C_8H_{18}	111-65-9	96.084	−7900.2	−11.003	7.1802E-06	2	216.38	2.11E+00	568.7	2.467E+06
266	Octanoic acid	$C_8H_{16}O_2$	124-07-2	116.477	−13,300.4	−12.6746	3.98E-18	6	289.65	2.76E-01	694.26	2.779E+06
267	1-Octanol	$C_8H_{18}O$	111-87-5	144.11083	−13,667.15667	−16.82611	9.37E-18	6	257.65	9.60E-02	652.3	2.781E+06
268	2-Octanol	$C_8H_{18}O$	123-96-6	185.828	−14,520.2	−23.6236	1.08854E-05	2	241.55	3.79E-02	629.8	2.749E+06
269	2-Octanone	$C_8H_{16}O$	111-13-7	63.775	−7711.3	−5.7359	3.09E-18	6	252.85	4.68E+00	632.7	2.647E+06
270	3-Octanone	$C_8H_{16}O$	106-68-3	72.382	−8054.8	−7.0002	5.83E-18	6	255.55	7.84E+00	627.7	2.705E+06
271	1-Octene	C_8H_{16}	111-66-0	74.936	−7155.9	−7.5843	1.71E-17	6	171.45	2.98E-03	566.9	2.663E+06
272	Octyl mercaptan	$C_8H_{18}S$	111-88-6	78.368	−8855.4	−7.8202	5.66E-18	6	223.95	3.05E-02	667.3	2.523E+06
273	1-Octyne	C_8H_{14}	629-05-0	64.612	−6802.5	−6.0261	1.10E-17	6	193.55	1.04E-01	574	2.880E+06
274	Oxalic acid	$C_2H_2O_4$	144-62-7	107.476	−12,833.4	−11.3837	1.34E-18	6	462.65	1.97E+04	828	8.203E+06
275	Oxygen	O_2	7782-44-7	51.245	−1200.2	−6.4361	0.028405	1	54.36	1.48E+02	154.58	5.021E+06
276	Ozone	O_3	10028-15-6	40.067	−2204.8	−2.9351	7.75E-16	6	80.15	7.35E-01	261	5.566E+06
277	Pentadecane	$C_{15}H_{32}$	629-62-9	135.57	−13,478	−16.022	5.6136E-06	2	283.07	1.29E-01	708	1.474E+06
278	Pentanal	$C_5H_{10}O$	110-62-3	28.3041	−4657.56	−0.732149	−8.31E-18	6	191.59	1.16E+00	566.1	3.845E+06
279	Pentane	C_5H_{12}	109-66-0	78.741	−5420.3	−8.8253	9.6171E-06	2	143.42	6.86E-02	469.7	3.364E+06
280	Pentanoic acid	$C_5H_{10}O_2$	109-52-4	93.2079	−10,470.5	−9.61345	5.62E-18	6	239.15	3.97E-02	639.16	3.630E+06
281	1-Pentanol	$C_5H_{12}O$	71-41-0	114.74801	−10,643.3	−12.85754	1.25E-17	6	195.56	5.47E-04	588.1	3.897E+06
282	2-Pentanol	$C_5H_{12}O$	6032-29-7	116.828	−10,453	−13.1768	1.07E-17	6	200	5.24E-03	561	3.699E+06
283	2-Pentanone	$C_5H_{10}O$	107-87-9	84.635	−7078.4	−9.3	6.2702E-06	2	196.29	7.52E-01	561.08	3.706E+06
284	3-Pentanone	$C_5H_{10}O$	96-22-0	44.286	−5415.1	−3.0913	1.86E-18	6	234.18	7.34E+01	560.95	3.699E+06
285	1-Pentene	C_5H_{10}	109-67-1	46.994	−4289.5	−3.7345	2.54E-17	6	108.02	3.71E-05	464.8	3.562E+06
286	2-Pentyl mercaptan	$C_5H_{12}S$	2084-19-7	58.985	−6193.1	−5.2746	7.40E-18	6	160.75	1.77E-03	584.3	3.537E+06
287	Pentyl mercaptan	$C_5H_{12}S$	110-66-7	67.309	−6880.8	−6.4449	1.01E-17	6	197.45	2.01E-01	598	3.473E+06
288	1-Pentyne	C_5H_8	627-19-0	82.805	−5683.8	−9.4301	0.000010767	2	167.45	2.40E+00	481.2	4.170E+06
289	2-Pentyne	C_5H_8	627-21-4	137.29	−7447.1	−19.01	0.021415	1	163.83	2.05E-01	519	4.020E+06
290	Phenanthrene	$C_{14}H_{10}$	85-01-8	72.958	−10,943	−6.7902	1.09E-18	6	372.38	2.93E+01	869	2.902E+06
291	Phenol	C_6H_6O	108-95-2	95.444	−10,113	−10.09	6.76E-18	6	314.06	1.88E+02	694.25	6.058E+06
292	Phenyl isocyanate	C_7H_5NO	103-71-9	86.779	−8101.8	−9.5303	6.1367E-06	2	243.15	4.33E+00	653	4.063E+06
293	Phthalic anhydride	$C_8H_4O_3$	85-44-9	126.5	−12,551	−15.002	7.7521E-06	2	404.15	7.90E+02	791	4.734E+06
294	Propadiene	C_3H_4	463-49-0	57.069	−3682.7	−5.5662	6.5133E-06	2	136.87	1.82E+01	394	5.218E+06
295	Propane	C_3H_8	74-98-6	59.078	−3492.6	−6.0669	0.000010919	2	85.47	1.68E-04	369.83	4.213E+06
296	1-Propanol	C_3H_8O	71-23-8	84.66416	−8307.24422	−8.57673	7.51E-18	6	146.95	4.27E-07	536.8	5.169E+06
297	2-Propanol	C_3H_8O	67-63-0	110.717	−9040	−12.676	0.000005538	2	185.26	1.69E-02	508.3	4.771E+06
298	Propenylcyclohexene	C_9H_{14}	13511-13-2	64.268	−7298.9	−5.9109	4.85E-18	6	199	2.48E-02	636	3.130E+06
299	Propionaldehyde	C_3H_6O	123-38-6	50.8769	−4931	−4.16673	1.67E-17	6	165	7.54E-01	503.6	5.040E+06
300	Propionic acid	$C_3H_6O_2$	79-09-4	54.552	−7149.4	−4.2769	1.18E-18	6	252.45	1.31E+01	600.81	4.608E+06
301	Propionitrile	C_3H_5N	107-12-0	59.9958	−6006.16	−5.46004	1.70E-17	6	180.37	1.89E-01	561.3	4.260E+06
302	Propyl acetate	$C_5H_{10}O_2$	109-60-4	115.16	−8433.9	−13.934	0.000010346	2	178.15	1.71E-02	549.73	3.366E+06
303	Propyl amine	C_3H_9N	107-10-8	58.398	−5312.7	−5.2876	1.9913E-06	2	188.36	1.30E+01	496.95	4.738E+06
304	Propylbenzene	C_9H_{12}	103-65-1	91.379	−8276.8	−10.176	0.000005624	2	173.55	1.81E-04	638.35	3.202E+06
305	Propylene	C_3H_6	115-07-1	43.905	−3097.8	−3.4425	1.00E-16	6	87.89	1.17E-03	364.85	4.599E+06

(Continued)

TABLE 2-8 Vapor Pressure of Inorganic and Organic Liquids, $\ln P = C_1 + C_2/T + C_3 \ln T + C_4 T^{C_5}$, P in Pa, T in K *(Continued)*

Cmpd. no.*	Name	Formula	CAS	C_1	C_2	C_3	C_4	C_5	T_{min}, K	P at T_{min}	T_{max}, K	P at T_{max}
306	Propyl formate	$C_4H_8O_2$	110-74-7	104.08	−7535.9	−12.348	0.000009602	2	180.25	2.11E-01	538	4.031E+06
307	2-Propyl mercaptan	C_3H_8S	75-33-2	60.43	−5276.9	−5.6572	2.60E-17	6	142.61	9.73E-03	517	4.752E+06
308	Propyl mercaptan	C_3H_8S	107-03-9	62.165	−5624	−5.8595	2.06E-17	6	159.95	6.51E-02	536.6	4.627E+06
309	1,2-Propylene glycol	$C_3H_8O_2$	57-55-6	212.8	−15,420	−28.109	0.000021564	2	213.15	9.29E-05	626	6.041E+06
310	Quinone	$C_6H_4O_2$	106-51-4	48.651	−7289.5	−3.4453	1.01E-18	6	388.85	1.17E+04	683	5.925E+06
311	Silicon tetrafluoride	F_4Si	7783-61-1	272.85	−9548.9	−40.089	6.37E-15	6	186.35	2.21E+05	259	3.748E+06
312	Styrene	C_8H_8	100-42-5	105.93	−8685.9	−12.42	7.5583E-06	2	242.54	1.06E+01	636	3.823E+06
313	Succinic acid	$C_4H_6O_4$	110-15-6	165.977	−19,914.4	−18.9344	1.91E-18	6	460.85	7.78E+02	838	5.001E+06
314	Sulfur dioxide	O_2S	7446-09-5	47.365	−4084.5	−3.6469	1.80E-17	6	197.67	1.67E+03	430.75	7.860E+06
315	Sulfur hexafluoride	F_6S	2551-62-4	29.16	−2383.6	−1.1342			223.15	2.30E+05	318.69	3.771E+06
316	Sulfur trioxide	O_3S	7446-11-9	180.99	−12,060	−22.839	7.24E-17	6	289.95	2.09E+04	490.85	8.192E+06
317	Terephthalic acid	$C_8H_6O_4$	100-21-0	124.004	−17,894.4	−13.156	1.18E-18	6	700.15	2.42E+05	883.6	3.487E+06
318	*o*-Terphenyl	$C_{18}H_{14}$	84-15-1	110.52	−14,045	−11.861	2.21E-18	6	329.35	4.14E-01	857	2.974E+06
319	Tetradecane	$C_{14}H_{30}$	629-59-4	140.47	−13,231	−16.859	6.5877E-06	2	279.01	2.53E-01	693	1.569E+06
320	Tetrahydrofuran	C_4H_8O	109-99-9	54.898	−5305.4	−4.7627	1.43E-17	6	164.65	1.96E-01	540.15	5.203E+06
321	1,2,3,4-Tetrahydronaphthalene	$C_{10}H_{12}$	119-64-2	137.23	−10,620	−17.908	0.014506	1	237.38	1.33E-01	720	3.624E+06
322	Tetrahydrothiophene	C_4H_8S	110-01-0	75.881	−6910.6	−7.9499	4.4315E-06	2	176.99	1.54E-02	631.95	5.117E+06
323	2,2,3,3-Tetramethylbutane	C_8H_{18}	594-82-1	57.963	−5901.5	−5.2048	9.13E-18	6	373.96	8.69E+04	568	2.871E+06
324	Thiophene	C_4H_4S	110-02-1	93.193	−7001.5	−10.738	8.2308E-06	2	234.94	1.86E+02	579.35	5.702E+06
325	Toluene	C_7H_8	108-88-3	76.945	−6729.8	−8.179	5.3017E-06	2	178.18	4.75E-02	591.75	4.080E+06
326	1,1,2-Trichloroethane	$C_2H_3Cl_3$	79-00-5	54.153	−6041.8	−4.5383	4.98E-18	6	236.5	4.47E+01	602	4.447E+06
327	Tridecane	$C_{13}H_{28}$	629-50-5	137.45	−12,549	−16.543	7.1275E-06	2	267.76	2.51E-01	675	1.679E+06
328	Triethyl amine	$C_6H_{15}N$	121-44-8	56.55	−5681.9	−4.9815	1.24E-17	6	158.45	1.06E-02	535.15	3.037E+06
329	Trimethyl amine	C_3H_9N	75-50-3	134.68	−6055.8	−19.415	0.028619	1	156.08	9.92E+00	433.25	4.102E+06
330	1,2,3-Trimethylbenzene	C_9H_{12}	526-73-8	78.341	−8019.8	−8.1458	3.8971E-06	2	247.79	3.71E+00	664.5	3.447E+06
331	1,2,4-Trimethylbenzene	C_9H_{12}	95-63-6	85.301	−8215.9	−9.2166	4.7979E-06	2	229.33	6.93E-01	649.1	3.211E+06
332	2,2,4-Trimethylpentane	C_8H_{18}	540-84-1	84.912	−6722.2	−9.5157	7.2244E-06	2	165.78	1.71E-02	543.8	2.550E+06
333	2,3,3-Trimethylpentane	C_8H_{18}	560-21-4	83.105	−6903.7	−9.1858	6.4703E-06	2	172.22	1.68E-02	573.5	2.812E+06
334	1,3,5-Trinitrobenzene	$C_6H_3N_3O_6$	99-35-4	506.33	−37,483	−69.22	0.000027381	2	398.4	8.50E+00	846	3.410E+06
335	2,4,6-Trinitrotoluene	$C_7H_5N_3O_6$	118-96-7	302	−24,324	−40.13	0.000017403	2	354	9.36E-01	828	3.019E+06
336	Undecane	$C_{11}H_{24}$	1120-21-4	131	−11,143	−15.855	8.1871E-06	2	247.57	4.08E-01	639	1.949E+06
337	1-Undecanol	$C_{11}H_{24}O$	112-42-5	182.57122	−17,112.47062	−22.1251	1.13E-17	6	288.45	1.25E-01	703.9	2.119E+06
338	Vinyl acetate	$C_4H_6O_2$	108-05-4	57.406	−5702.8	−5.0307	1.10E-17	6	180.35	7.06E-01	519.13	3.930E+06
339	Vinyl acetylene	C_4H_4	689-97-4	55.682	−4439.3	−5.0136	1.97E-17	6	173.15	6.69E+01	454	4.887E+06
340	Vinyl chloride	C_2H_3Cl	75-01-4	91.432	−5141.7	−10.981	0.000014318	2	119.36	1.92E-02	432	5.749E+06
341	Vinyl trichlorosilane	$C_2H_3Cl_3Si$	75-94-5	54.571	−5561.5	−4.712	1.07E-17	6	178.35	3.54E-01	543.15	3.058E+06
342	Water	H_2O	7732-18-5	73.649	−7258.2	−7.3037	4.1653E-06	2	273.16	6.11E+02	647.1	2.193E+07
343	*m*-Xylene	C_8H_{10}	108-38-3	85.099	−7615.9	−9.3072	5.5643E-06	2	225.3	3.18E+00	617	3.528E+06
344	*o*-Xylene	C_8H_{10}	95-47-6	90.405	−7955.2	−10.086	5.9594E-06	2	247.98	2.18E+01	630.3	3.741E+06
345	*p*-Xylene	C_8H_{10}	106-42-3	88.72	−7741.2	−9.8693	0.000006077	2	286.41	5.76E+02	616.2	3.501E+06

Vapor pressure P_s is calculated by $P_s = \exp(C_1 + C_2/T + C_3 \ln(T) + C_4 T^{C_5})$ where P_s is in Pa and T is in K.
*All substances and their numbers are listed by chemical family in Table 2-6 and by formula in Table 2-7.
Values in this table were taken from the Design Institute for Physical Properties (DIPPR) of the American Institute of Chemical Engineers (AIChE), 801 Critically Evaluated Gold Standard™ Database, copyright 2016 AIChE, and reproduced with permission of AIChE and of the DIPPR Evaluated Process Design Data Project Steering Committee. Their source should be cited as "R. L. Rowley, W. V. Wilding, J. L. Oscarson, T. A. Knotts, N. F. Giles, *DIPPR® Data Compilation of Pure Chemical Properties*, Design Institute for Physical Properties, AIChE, New York, NY (2016)".

TABLE 2-9 Vapor Pressures of Inorganic Compounds, up to 1 atm*

Compound		Pressure, mmHg										Melting point, °C
		1	5	10	20	40	60	100	200	400	760	
Name	Formula	Temperature, °C										
Aluminum	Al	1284	1421	1487	1555	1635	1684	1749	1844	1947	2056	660
borohydride	$Al(BH_4)_3$		−52.2	−42.9	−32.5	−20.9	−13.4	−3.9	+11.2	28.1	45.9	−64.
bromide	$AlBr_3$	81.3	103.8	118.0	134.0	150.6	161.7	176.1	199.8	227.0	256.3	97.
chloride	Al_2Cl_6	100.0	116.4	123.8	131.8	139.9	145.4	152.0	161.8	171.6	180.2	192.4
fluoride	AlF_3	1238	1298	1324	1350	1378	1398	1422	1457	1496	1537	1040
iodide	AlI_3	178.0	207.7	225.8	244.2	265.0	277.8	294.5	322.0	354.0	385.5	
oxide	Al_2O_3	2148	2306	2385	2465	2549	2599	2665	2766	2874	2977	2050
Ammonia	NH_3	−109.1	−97.5	−91.9	−85.8	−79.2	−74.3	−68.4	−57.0	−45.4	−33.6	−77.7
heavy	ND_3						−74.0	−67.4	−57.0	−45.4	−33.4	−74.0
Ammonium bromide	NH_4Br	198.3	234.5	252.0	270.6	290.0	303.8	320.0	345.3	370.8	396.0	
carbamate	$N_2H_6CO_2$	−26.1	−10.4	−2.9	+5.3	14.0	19.6	26.7	37.2	48.0	58.3	
chloride	NH_4Cl	160.4	193.8	209.8	226.1	245.0	256.2	271.5	293.2	316.5	337.8	520
cyanide	NH_4CN	−50.6	−35.7	−28.6	−20.9	−12.6	−7.4	−0.5	+9.6	20.5	31.7	36
hydrogen sulfide	NH_4HS	−51.1	−36.0	−28.7	−20.8	−12.3	−7.0	0.0	+10.5	21.8	33.3	
iodide	NH_4I	210.9	247.0	263.5	282.8	302.8	316.0	331.8	355.8	381.0	404.9	
Antimony	Sb	886	984	1033	1084	1141	1176	1223	1288	1364	1440	630.5
tribromide	$SbBr_3$	93.9	126.0	142.7	158.3	177.4	188.1	203.5	225.7	250.2	275.0	96.6
trichloride	$SbCl_3$	49.2	71.4	85.2	100.6	117.8	128.3	143.3	165.9	192.2	219.0	73.4
pentachloride	$SbCl_5$	22.7	48.6	61.8	75.8	91.0	101.0	114.1				2.8
triiodide	SbI_3	163.6	203.8	223.5	244.8	267.8	282.5	303.5	333.8	368.5	401.0	167
trioxide	Sb_4O_6	574	626	666	729	812	873	957	1085	1242	1425	656
Argon	A	−218.2	−213.9	−210.9	−207.9	−204.9	−202.9	−200.5	−195.6	−190.6	−185.6	−189.2
Arsenic	As	372	416	437	459	483	498	518	548	579	610	814
Arsenic tribromide	$AsBr_3$	41.8	70.6	85.2	101.3	118.7	130.0	145.2	167.7	193.6	220.0	
trichloride	$AsCl_3$	−11.4	+11.7	+23.5	36.0	50.0	58.7	70.9	89.2	109.7	130.4	−18
trifluoride	AsF_3					−2.5	+4.2	13.2	26.7	41.4	56.3	−5.9
pentafluoride	AsF_5	−117.9	−108.0	−103.1	−98.0	−92.4	−88.5	−84.3	−75.5	−64.0	−52.8	−79.8
trioxide	As_2O_3	212.5	242.6	259.7	279.2	299.2	310.3	332.5	370.0	412.2	457.2	312.8
Arsine	AsH_3	−142.6	−130.8	−124.7	−117.7	−110.2	−104.8	−98.0	−87.2	−75.2	−62.1	−116.3
Barium	Ba		984	1049	1120	1195	1240	1301	1403	1518	1638	850
Beryllium borohydride	$Be(BH_4)_2$	+1.0	19.8	28.1	36.8	46.2	51.7	58.6	69.0	79.7	90.0	123
bromide	$BeBr_2$	289	325	342	361	379	390	405	427	451	474	490
chloride	$BeCl_2$	291	328	346	365	384	395	411	435	461	487	405
iodide	BeI_2	283	322	341	361	382	394	411	435	461	487	488
Bismuth	Bi	1021	1099	1136	1177	1217	1240	1271	1319	1370	1420	271
tribromide	$BiBr_3$		261	282	305	327	340	360	392	425	461	218
trichloride	$BiCl_3$		242	264	287	311	324	343	372	405	441	230
Diborane hydrobromide	B_2H_5Br	−93.3	−75.3	−66.3	−56.4	−45.4	−38.2	−29.0	−15.4	0.0	+16.3	−104.2
Borine carbonyl	BH_3CO	−139.2	−127.3	−121.1	−114.1	−106.6	−101.9	−95.3	−85.5	−74.8	−64.0	−137.0
triamine	$B_3N_3H_6$	−63.0	−45.0	−35.3	−25.0	−13.2	−5.8	+4.0	18.5	34.3	50.6	−58.2
Boron hydrides												
dihydrodecaborane	$B_{10}H_{14}$	60.0	80.8	90.2	100.0	117.4	127.8	142.3	163.8			99.6
dihydrodiborane	B_2H_6	−159.7	−149.5	−144.3	−138.5	−131.6	−127.2	−120.9	−111.2	−99.6	−86.5	−169
dihydropentaborane	B_5H_9		−40.4	−30.7	−20.0	−8.0	−0.4	+9.6	24.6	40.8	58.1	−47.0
tetrahydropentaborane	B_5H_{11}	−50.2	−29.9	−19.9	−9.2	+2.7	10.2	20.1	34.8	51.2	67.0	
tetrahydrotetraborane	B_4H_{10}	−90.9	−73.1	−64.3	−54.8	−44.3	−37.4	−28.1	−14.0	+0.8	16.1	−119.9
Boron tribromide	BBr_3	−41.4	−20.4	−10.1	+1.5	14.0	22.1	33.5	50.3	70.0	91.7	−45
trichloride	BCl_3	−91.5	−75.2	−66.9	−57.9	−47.8	−41.2	−32.4	−18.9	−3.6	+12.7	−107
trifluoride	BF_3	−154.6	−145.4	−141.3	−136.4	−131.0	−127.6	−123.0	−115.9	−108.3	−100.7	−126.8
Bromine	Br_2	−48.7	−32.8	−25.0	−16.8	−8.0	−0.6	+9.3	24.3	41.0	58.2	−7.3
pentafluoride	BrF_5	−69.3	−51.0	−41.9	−32.0	−21.0	−14.0	−4.5	+9.9	25.7	40.4	−61.4
Cadmium	Cd	394	455	484	516	553	578	611	658	711	765	320.9
chloride	$CdCl_2$		618	656	695	736	762	797	847	908	967	568
fluoride	CdF_2	1112	1231	1286	1344	1400	1436	1486	1561	1651	1751	520
iodide	CdI_2	416	481	512	546	584	608	640	688	742	796	385
oxide	CdO	1000	1100	1149	1200	1257	1295	1341	1409	1484	1559	
Calcium	Ca		926	983	1046	1111	1152	1207	1288	1388	1487	851
Carbon (graphite)	C	3586	3828	3946	4069	4196	4273	4373	4516	4660	4827	
dioxide	CO_2	−134.3	−124.4	−119.5	−114.4	−108.6	−104.8	−100.2	−93.0	−85.7	−78.2	−57.5
disulfide	CS_2	−73.8	−54.3	−44.7	−34.3	−22.5	−15.3	−5.1	+10.4	28.0	46.5	−110.8
monoxide	CO	−222.0	−217.2	−215.0	−212.8	−210.0	−208.1	−205.7	−201.3	−196.3	−191.3	−205.0
oxyselenide	COSe	−117.1	−102.3	−95.0	−86.3	−76.4	−70.2	−61.7	−49.8	−35.6	−21.9	
oxysulfide	COS	−132.4	−119.8	−113.3	−106.0	−98.3	−93.0	−85.9	−75.0	−62.7	−49.9	−138.8
selenosulfide	CSeS	−47.3	−26.5	−16.0	−4.4	+8.6	17.0	28.3	45.7	65.2	85.6	−75.2
subsulfide	C_3S_2	14.0	41.2	54.9	69.3	85.6	96.0	109.9	130.8			+0.4
tetrabromide	CBr_4					96.3	106.3	119.7	139.7	163.5	189.5	90.1
tetrachloride	CCl_4	−50.0	−30.0	−19.6	−8.2	+4.3	12.3	23.0	38.3	57.8	76.7	−22.6
tetrafluoride	CF_4	−184.6	−174.1	−169.3	−164.3	−158.8	−155.4	−150.7	−143.6	−135.5	−127.7	−183.7
Cesium	Cs	279	341	375	409	449	474	509	561	624	690	28.5
bromide	CsBr	748	838	887	938	993	1026	1072	1140	1221	1300	636
chloride	CsCl	744	837	884	934	989	1023	1069	1139	1217	1300	646
fluoride	CsF	712	798	844	893	947	980	1025	1092	1170	1251	683
iodide	CsI	738	828	873	923	976	1009	1055	1124	1200	1280	621

*Compiled from the extended tables published by D. R. Stull in *Ind. Eng. Chem.*, **39**, 517 (1947).

(*Continued*)

TABLE 2-9 Vapor Pressures of Inorganic Compounds, up to 1 atm *(Continued)*

Compound		Pressure, mmHg										Melting point, °C
		1	5	10	20	40	60	100	200	400	760	
Name	Formula	Temperature, °C										
Chlorine	Cl_2	−118.0	−106.7	−101.6	−93.3	−84.5	−79.0	−71.7	−60.2	−47.3	−33.8	−100.7
fluoride	ClF		−143.4	−139.0	−134.3	−128.8	−125.3	−120.8	−114.4	−107.0	−100.5	−145
trifluoride	ClF_3		−80.4	−71.8	−62.3	−51.3	−44.1	−34.7	−20.7	−4.9	+11.5	−83
monoxide	Cl_2O	−98.5	−81.6	−73.1	−64.3	−54.3	−48.0	−39.4	−26.5	−12.5	+2.2	−116
dioxide	ClO_2			−59.0	−51.2	−42.8	−37.2	−29.4	−17.8	−4.0	+11.1	−59
heptoxide	Cl_2O_7	−45.3	−23.8	−13.2	−2.1	+10.3	+18.2	29.1	44.6	62.2	78.8	−91
Chlorosulfonic acid	HSO_3Cl	32.0	53.5	64.0	75.3	87.6	95.2	105.3	120.0	136.1	151.0	−80
Chromium	Cr	1616	1768	1845	1928	2013	2067	2139	2243	2361	2482	1615
carbonyl	$Cr(CO)_6$	36.0	58.0	68.3	79.5	91.2	98.3	108.0	121.8	137.2	151.0	
oxychloride	CrO_2Cl_2	−18.4	+3.2	13.8	25.7	38.5	46.7	58.0	75.2	95.2	117.1	
Cobalt chloride	$CoCl_2$					770	801	843	904	974	1050	735
nitrosyl tricarbonyl	$Co(CO)_3NO$				−1.3	+11.0	18.5	29.0	44.4	62.0	80.0	−11
Columbium fluoride	CbF_5			86.3	103.0	121.5	133.2	148.5	172.2	198.0	225.0	75.5
Copper	Cu	1628	1795	1879	1970	2067	2127	2207	2325	2465	2595	1083
Cuprous bromide	Cu_2Br_2	572	666	718	777	844	887	951	1052	1189	1355	504
chloride	Cu_2Cl_2	546	645	702	766	838	886	960	1077	1249	1490	422
iodide	Cu_2I_2		610	656	716	786	836	907	1018	1158	1336	605
Cyanogen	C_2N_2	−95.8	−83.2	−76.8	−70.1	−62.7	−57.9	−51.8	−42.6	−33.0	−21.0	−34.4
bromide	$CNBr$	−35.7	−18.3	−10.0	−1.0	+8.6	14.7	22.6	33.8	46.0	61.5	58
chloride	$CNCl$	−76.7	−61.4	−53.8	−46.1	−37.5	−32.1	−24.9	−14.1	−2.3	+13.1	−6.5
fluoride	CNF	−134.4	−123.8	−118.5	−112.8	−106.4	−102.3	−97.0	−89.2	−80.5	−72.6	
Deuterium cyanide	DCN	−68.9	−54.0	−46.7	−38.8	−30.1	−24.7	−17.5	−5.4	+10.0	26.2	−12
Fluorine	F_2	−223.0	−216.9	−214.1	−211.0	−207.7	−205.6	−202.7	−198.3	−193.2	−187.9	−223
oxide	F_2O	−196.1	−186.6	−182.3	−177.8	−173.0	−170.0	−165.8	−159.0	−151.9	−144.6	−223.9
Germanium bromide	$GeBr_4$		43.3	56.8	71.8	88.1	98.8	113.2	135.4	161.6	189.0	26.1
chloride	$GeCl_4$	−45.0	−24.9	−15.0	−4.1	+8.0	16.2	27.5	44.4	63.8	84.0	−49.5
hydride	GeH_4	−163.0	−151.0	−145.3	−139.2	−131.6	−126.7	−120.3	−111.2	−100.2	−88.9	−165
Trichlorogermane	$GeHCl_3$	−41.3	−22.3	−13.0	−3.0	+8.8	16.2	26.5	41.6	58.3	75.0	−71.1
Tetramethylgermane	$Ge(CH_3)_4$	−73.2	−54.6	−45.2	−35.0	−23.4	−16.2	−6.3	+8.8	26.0	44.0	−88
Digermane	Ge_2H_6	−88.7	−69.8	−60.1	−49.9	−38.2	−30.7	−20.3	−4.7	+13.3	31.5	−109
Trigermane	Ge_3H_8	−36.9	−12.8	−0.9	+11.8	26.3	35.5	47.9	67.0	88.6	110.8	−105.6
Gold	Au	1869	2059	2154	2256	2363	2431	2521	2657	2807	2966	1063
Helium	He	−271.7	−271.5	−271.3	−271.1	−270.7	−270.6	−270.3	−269.8	−269.3	−268.6	
para-Hydrogen	H_2	−263.3	−261.9	−261.3	−260.4	−259.6	−258.9	−257.9	−256.3	−254.5	−252.5	−259.1
Hydrogen bromide	HBr	−138.8	−127.4	−121.8	−115.4	−108.3	−103.8	−97.7	−88.1	−78.0	−66.5	−87.0
chloride	HCl	−150.8	−140.7	−135.6	−130.0	−123.8	−119.6	−114.0	−105.2	−95.3	−84.8	−114.3
cyanide	HCN	−71.0	−55.3	−47.7	−39.7	−30.9	−25.1	−17.8	−5.3	+10.2	25.9	−13.2
fluoride	H_2F_2		−74.7	−65.8	−56.0	−45.0	−37.9	−28.2	−13.2	+2.5	19.7	−83.7
iodide	HI	−123.3	−109.6	−102.3	−94.5	−85.6	−79.8	−72.1	−60.3	−48.3	−35.1	−50.9
oxide (water)	H_2O	−17.3	+1.2	11.2	22.1	34.0	41.5	51.6	66.5	83.0	100.0	0.0
sulfide	H_2S	−134.3	−122.4	−116.3	−109.7	−102.3	−97.9	−91.6	−82.3	−71.8	−60.4	−85.5
disulfide	$HSSH$	−43.2	−24.4	−15.2	−5.1	+6.0	12.8	22.0	35.3	49.6	64.0	−89.7
selenide	H_2Se	−115.3	−103.4	−97.9	−91.8	−84.7	−80.2	−74.2	−65.2	−53.6	−41.1	−64
telluride	H_2Te	−96.4	−82.4	−75.4	−67.8	−59.1	−53.7	−45.7	−32.4	−17.2	−2.0	−49.0
Iodine	I_2	38.7	62.2	73.2	84.7	97.5	105.4	116.5	137.3	159.8	183.0	112.9
heptafluoride	IF_7	−87.0	−70.7	−63.0	−54.5	−45.3	−39.4	−31.9	−20.7	−8.3	+4.0	5.5
Iron	Fe	1787	1957	2039	2128	2224	2283	2360	2475	2605	2735	1535
pentacarbonyl	$Fe(CO)_5$		−6.5	+4.6	16.7	30.3	39.1	50.3	68.0	86.1	105.0	−21
Ferric chloride	Fe_2Cl_6	194.0	221.8	235.5	246.0	256.8	263.7	272.5	285.0	298.0	319.0	304
Ferrous chloride	$FeCl_2$			700	737	779	805	842	897	961	1026	
Krypton	Kr	−199.3	−191.3	−187.2	−182.9	−178.4	−175.7	−171.8	−165.9	−159.0	−152.0	−156.7
Lead	Pb	973	1099	1162	1234	1309	1358	1421	1519	1630	1744	327.5
bromide	$PbBr_2$	513	578	610	646	686	711	745	796	856	914	373
chloride	$PbCl_2$	547	615	648	684	725	750	784	833	893	954	501
fluoride	PbF_2		861	904	950	1003	1036	1080	1144	1219	1293	855
iodide	PbI_2	479	540	571	605	644	668	701	750	807	872	402
oxide	PbO	943	1039	1085	1134	1189	1222	1265	1330	1402	1472	890
sulfide	PbS	852	928	975	1005	1048	1074	1108	1160	1221	1281	1114
Lithium	Li	723	838	881	940	1003	1042	1097	1178	1273	1372	186
bromide	$LiBr$	748	840	888	939	994	1028	1076	1147	1226	1310	547
chloride	$LiCl$	783	880	932	987	1045	1081	1129	1203	1290	1382	614
fluoride	LiF	1047	1156	1211	1270	1333	1372	1425	1503	1591	1681	870
iodide	LiI	723	802	841	883	927	955	993	1049	1110	1171	446
Magnesium	Mg	621	702	743	789	838	868	909	967	1034	1107	651
chloride	$MgCl_2$	778	877	930	988	1050	1088	1142	1223	1316	1418	712
Manganese	Mn	1292	1434	1505	1583	1666	1720	1792	1900	2029	2151	1260
chloride	$MnCl_2$		736	778	825	879	913	960	1028	1108	1190	650
Mercury	Hg	126.2	164.8	184.0	204.6	228.8	242.0	261.7	290.7	323.0	357.0	−38.9
Mercuric bromide	$HgBr_2$	136.5	165.3	179.8	194.3	211.5	221.0	237.8	262.7	290.0	319.0	237
chloride	$HgCl_2$	136.2	166.0	180.2	195.8	212.5	222.2	237.0	256.5	275.5	304.0	277
iodide	HgI_2	157.5	189.2	204.5	220.0	238.2	249.0	261.8	291.0	324.2	354.0	259
Molybdenum	Mo	3102	3393	3535	3690	3859	3964	4109	4322	4553	4804	2622
hexafluoride	MoF_6	−65.5	−49.0	−40.8	−32.0	−22.1	−16.2	−8.0	+4.1	17.2	36.0	17
oxide	MoO_3	734	785	814	851	892	917	955	1014	1082	1151	795

TABLE 2-9 Vapor Pressures of Inorganic Compounds, up to 1 atm *(Continued)*

Compound		Pressure, mmHg										Melting point, °C
		1	5	10	20	40	60	100	200	400	760	
Name	Formula	Temperature, °C										
Neon	Ne	−257.3	−255.5	−254.6	−253.7	−252.6	−251.9	−251.0	−249.7	−248.1	−246.0	−248.7
Nickel	Ni	1810	1979	2057	2143	2234	2289	2364	2473	2603	2732	1452
carbonyl	$Ni(CO)_4$					−23.0	−15.9	−6.0	+8.8	25.8	42.5	−25
chloride	$NiCl_2$	671	731	759	789	821	840	866	904	945	987	1001
Nitrogen	N_2	−226.1	−221.3	−219.1	−216.8	−214.0	−212.3	−209.7	−205.6	−200.9	−195.8	−210.0
Nitric oxide	NO	−184.5	−180.6	−178.2	−175.3	−171.7	−168.9	−166.0	−162.3	−156.8	−151.7	−161
Nitrogen dioxide	NO_2	−55.6	−42.7	−36.7	−30.4	−23.9	−19.9	−14.7	−5.0	+8.0	21.0	−9.3
Nitrogen pentoxide	N_2O_5	−36.8	−23.0	−16.7	−10.0	−2.9	+1.8	7.4	15.6	24.4	32.4	30
Nitrous oxide	N_2O	−143.4	−133.4	−128.7	−124.0	−118.3	−114.9	−110.3	−103.6	−96.2	−85.5	−90.9
Nitrosyl chloride	NOCl					−60.2	−54.2	−46.3	−34.0	−20.3	−6.4	−64.5
fluoride	NOF	−132.0	−120.3	−114.3	−107.8	−100.3	−95.7	−88.8	−79.2	−68.2	−56.0	−134
Osmium tetroxide (yellow)	OsO_4	3.2	22.0	31.3	41.0	51.7	59.4	71.5	89.5	109.3	130.0	56
(white)	OsO_4	−5.6	+15.6	26.0	37.4	50.5	59.4	71.5	89.5	109.3	130.0	42
Oxygen	O_2	−219.1	−213.4	−210.6	−207.5	−204.1	−201.9	−198.8	−194.0	−188.8	−183.1	−218.7
Ozone	O_3	−180.4	−168.6	−163.2	−157.2	−150.7	−146.7	−141.0	−132.6	−122.5	−111.1	−251
Phosgene	$COCl_2$	−92.9	−77.0	−69.3	−60.3	−50.3	−44.0	−35.6	−22.3	−7.6	+8.3	−104
Phosphorus (yellow)	P	76.6	111.2	128.0	146.2	166.7	179.8	197.3	222.7	251.0	280.0	44.1
(violet)	P	237	271	287	306	323	334	349	370	391	417	590
tribromide	PBr_3	7.8	34.4	47.8	62.4	79.0	89.8	103.6	125.2	149.7	175.3	−40
trichloride	PCl_3	−51.6	−31.5	−21.3	−10.2	+2.3	10.2	21.0	37.6	56.9	74.2	−111.8
pentachloride	PCl_5	55.5	74.0	83.2	92.5	102.5	108.3	117.0	131.3	147.2	162.0	
Phosphine	PH_3					−129.4	−125.0	−118.8	−109.4	−98.3	−87.5	−132.5
Phosphonium bromide	PH_4Br	−43.7	−28.5	−21.2	−13.3	−5.0	+0.3	7.4	17.6	28.0	38.3	
chloride	PH_4Cl	−91.0	−79.6	−74.0	−68.0	−61.5	−57.3	−52.0	−44.0	−35.4	−27.0	−28.5
iodide	PH_4I	−25.2	−9.0	−1.1	+7.3	16.1	21.9	29.3	39.9	51.6	62.3	
Phosphorus trioxide	P_4O_6		39.7	53.0	67.8	84.0	94.2	108.3	129.0	150.3	173.1	22.5
pentoxide	P_4O_{10}	384	424	442	462	481	493	510	532	556	591	569
oxychloride	$POCl_3$			2.0	13.6	27.3	35.8	47.4	65.0	84.3	105.1	2
thiobromide	$PSBr_3$	50.0	72.4	83.6	95.5	108.0	116.0	126.3	141.8	157.8	175.0	38
thiochloride	$PSCl_3$	−18.3	+4.6	16.1	29.0	42.7	51.8	63.8	82.0	102.3	124.0	−36.2
Platinum	Pt	2730	3007	3146	3302	3469	3574	3714	3923	4169	4407	1755
Potassium	K	341	408	443	483	524	550	586	643	708	774	62.3
bromide	KBr	795	892	940	994	1050	1087	1137	1212	1297	1383	730
chloride	KCl	821	919	968	1020	1078	1115	1164	1239	1322	1407	790
fluoride	KF	885	988	1039	1096	1156	1193	1245	1323	1411	1502	880
hydroxide	KOH	719	814	863	918	976	1013	1064	1142	1233	1327	380
iodide	KI	745	840	887	938	995	1030	1080	1152	1238	1324	723
Radon	Rn	−144.2	−132.4	−126.3	−119.2	−111.3	−106.2	−99.0	−87.7	−75.0	−61.8	−71
Rhenium heptoxide	Re_2O_7	212.5	237.5	248.0	261.0	272.0	280.0	289.0	307.0	336.0	362.4	296
Rubidium	Rb	297	358	389	422	459	482	514	563	620	679	38.5
bromide	RbBr	781	876	923	975	1031	1066	1114	1186	1267	1352	682
chloride	RbCl	792	887	937	990	1047	1084	1133	1207	1294	1381	715
fluoride	RbF	921	982	1016	1052	1096	1123	1168	1239	1322	1408	760
iodide	RbI	748	839	884	935	991	1026	1072	1141	1223	1304	642
Selenium	Se	356	413	442	473	506	527	554	594	637	680	217
dioxide	SeO_2	157.0	187.7	202.5	217.5	234.1	244.6	258.0	277.0	297.7	317.0	340
hexafluoride	SeF_6	−118.6	−105.2	−98.9	−92.3	−84.7	−80.0	−73.9	−64.8	−55.2	−45.8	−34.7
oxychloride	$SeOCl_2$	34.8	59.8	71.9	84.2	98.0	106.5	118.0	134.6	151.7	168.0	8.5
tetrachloride	$SeCl_4$	74.0	96.3	107.4	118.1	130.1	137.8	147.5	161.0	176.4	191.5	
Silicon	Si	1724	1835	1888	1942	2000	2036	2083	2151	2220	2287	1420
dioxide	SiO_2			1732	1798	1867	1911	1969	2053	2141	2227	1710
tetrachloride	$SiCl_4$	−63.4	−44.1	−34.4	−24.0	−12.1	−4.8	+5.4	21.0	38.4	56.8	−68.8
tetrafluoride	SiF_4	−144.0	−134.8	−130.4	−125.9	−120.8	−117.5	−113.3	−170.2	−100.7	−94.8	−90
Trichlorofluorosilane	$SiFCl_3$	−92.6	−76.4	−68.3	−59.0	−48.8	−42.2	−33.2	−19.3	−4.0	+12.2	−120.8
Iodosilane	SiH_3I		−53.0	−47.7	−33.4	−21.8	−14.3	−4.4	+10.7	27.9	45.4	−57.0
Diiodosilane	SiH_2I_2		3.8	18.0	34.1	52.6	64.0	79.4	101.8	125.5	149.5	−1.0
Disiloxan	$(SiH_3)_2O$	−112.5	−95.8	−88.2	−79.8	−70.4	−64.2	−55.9	−43.5	−29.3	−15.4	−144.2
Trisilane	Si_3H_8	−68.9	−49.7	−40.0	−29.0	−16.9	−9.0	+1.6	17.8	35.5	53.1	−117.2
Trisilazane	$(SiH_3)_3N$	−68.7	−49.9	−40.4	−30.0	−18.5	−11.0	−1.1	+14.0	31.0	48.7	−105.7
Tetrasilane	Si_4H_{10}	−27.7	−6.2	+4.3	15.8	28.4	36.6	47.4	63.6	81.7	100.0	−93.6
Octachlorotrisilane	Si_3Cl_3	46.3	74.7	89.3	104.2	121.5	132.0	146.0	166.2	189.5	211.4	
Hexachlorodisiloxane	$(SiCl_3)_2O$	−5.0	17.8	29.4	41.5	55.2	63.8	75.4	92.5	113.6	135.6	−33.2
Hexachlorodisilane	Si_2Cl_6	+4.0	27.4	38.8	51.5	65.3	73.9	85.4	102.2	120.6	139.0	−1.2
Tribromosilane	$SiHBr_3$	−30.5	−8.0	+3.4	16.0	30.0	39.2	51.6	70.2	90.2	111.8	−73.5
Trichlorosilane	$SiHCl_3$	−80.7	−62.6	−53.4	−43.8	−32.9	−25.8	−16.4	−1.8	+14.5	31.8	−126.6
Trifluorosilane	$SiHF_3$	−152.0	−142.7	−138.2	−132.9	−127.3	−123.7	−118.7	−111.3	−102.8	−95.0	−131.4
Dibromosilane	SiH_2Br_2	−60.9	−40.0	−29.4	−18.0	−5.2	+3.2	14.1	31.6	50.7	70.5	−70.2
Difluorosilane	SiH_2F_2	−146.7	−136.0	−130.4	−124.3	−117.6	−113.3	−107.3	−98.3	−87.6	−77.8	
Monobromosilane	SiH_3Br		−85.7	−77.3	−68.3	−57.8	−51.1	−42.3	−28.6	−13.3	+2.4	−93.9
Monochlorosilane	SiH_3Cl	−117.8	−104.3	−97.7	−90.1	−81.8	−76.0	−68.5	−57.0	−44.5	−30.4	
Monofluorosilane	SiH_3F	−153.0	−145.5	−141.2	−136.3	−130.8	−127.2	−122.4	−115.2	−106.8	−98.0	
Tribromofluorosilane	$SiFBr_3$	−46.1	−25.4	−15.1	−3.7	+9.2	17.4	28.6	45.7	64.6	83.8	−82.5
Dichlorodifluorosilane	SiF_2Cl_2	−124.7	−110.5	−102.9	−94.5	−85.0	−78.6	−70.3	−58.0	−45.0	−31.8	−139.7
Trifluorobromosilane	SiF_3Br								−69.8	−55.9	−41.7	−70.5

(Continued)

TABLE 2-9 Vapor Pressures of Inorganic Compounds, up to 1 atm *(Continued)*

Compound		Pressure, mmHg										Melting point, °C
		1	5	10	20	40	60	100	200	400	760	
Name	Formula	Temperature, °C										
Trifluorochlorosilane	SiF_3Cl	−144.0	−133.0	−127.0	−120.5	−112.8	−108.2	−101.7	−91.7	−81.0	−70.0	−142
Hexafluorodisilane	Si_2F_6	−81.0	−68.8	−63.1	−57.0	−50.6	−46.7	−41.7	−34.2	−26.4	−18.9	−18.6
Dichlorofluorobromosilane	$SiFCl_2Br$	−86.5	−68.4	−59.0	−48.8	−37.0	−29.0	−19.5	−3.2	+15.4	35.4	−112.3
Dibromochlorofluorosilane	$SiFClBr_2$	−65.2	−45.5	−35.6	−24.5	−12.0	−4.7	+6.3	23.0	43.0	59.5	−99.3
Silane	SiH_4	−179.3	−168.6	−163.0	−156.9	−150.3	−146.3	−140.5	−131.6	−122.0	−111.5	−185
Disilane	Si_2H_6	−114.8	−99.3	−91.4	−82.7	−72.8	−66.4	−57.5	−44.6	−29.0	−14.3	−132.6
Silver	Ag	1357	1500	1575	1658	1743	1795	1865	1971	2090	2212	960.5
chloride	AgCl	912	1019	1074	1134	1200	1242	1297	1379	1467	1564	455
iodide	AgI	820	927	983	1045	1111	1152	1210	1297	1400	1506	552
Sodium	Na	439	511	549	589	633	662	701	758	823	892	97.5
bromide	NaBr	806	903	952	1005	1063	1099	1148	1220	1304	1392	755
chloride	NaCl	865	967	1017	1072	1131	1169	1220	1296	1379	1465	800
cyanide	NaCN	817	928	983	1046	1115	1156	1214	1302	1401	1497	564
fluoride	NaF	1077	1186	1240	1300	1363	1403	1455	1531	1617	1704	992
hydroxide	NaOH	739	843	897	953	1017	1057	1111	1192	1286	1378	318
iodide	NaI	767	857	903	952	1005	1039	1083	1150	1225	1304	651
Strontium	Sr		847	898	953	1018	1057	1111	1192	1285	1384	800
Strontium oxide	SrO	2068	2198	2262	2333	2410						2430
Sulfur	S	183.8	223.0	243.8	264.7	288.3	305.5	327.2	359.7	399.6	444.6	112.8
monochloride	S_2Cl_2	−7.4	+15.7	27.5	40.0	54.1	63.2	75.3	93.5	115.4	138.0	−80
hexafluoride	SF_6	−132.7	−120.6	−114.7	−108.4	−101.5	−96.8	−90.9	−82.3	−72.6	−63.5	−50.2
Sulfuryl chloride	SO_2Cl_2		−35.1	−24.8	−13.4	−1.0	+7.2	17.8	33.7	51.3	69.2	−54.1
Sulfur dioxide	SO_2	−95.5	−83.0	−76.8	−69.7	−60.5	−54.6	−46.9	−35.4	−23.0	−10.0	−73.2
trioxide (α)	SO_3	−39.0	−23.7	−16.5	−9.1	−1.0	+4.0	10.5	20.5	32.6	44.8	16.8
trioxide (β)	SO_3	−34.0	−19.2	−12.3	−4.9	+3.2	8.0	14.3	23.7	32.6	44.8	32.3
trioxide (γ)	SO_3	−15.3	−2.0	+4.3	11.1	17.9	21.4	28.0	35.8	44.0	51.6	62.1
Tellurium	Te	520	605	650	697	753	789	838	910	997	1087	452
chloride	$TeCl_4$			233	253	273	287	304	330	360	392	224
fluoride	TeF_6	−111.3	−98.8	−92.4	−86.0	−78.4	−73.8	−67.9	−57.3	−48.2	−38.6	−37.8
Thallium	Tl	825	931	983	1040	1103	1143	1196	1274	1364	1457	3035
Thallous bromide	TlBr		490	522	559	598	621	653	703	759	819	460
chloride	TlCl		487	517	550	589	612	645	694	748	807	430
iodide	TlI	440	502	531	567	607	631	663	712	763	823	440
Thionyl bromide	$SOBr_2$	−6.7	+18.4	31.0	44.1	58.8	68.3	80.6	99.0	119.2	139.5	−52.2
Thionyl chloride	$SOCl_2$	−52.9	−32.4	−21.9	−10.5	+2.2	10.4	21.4	37.9	56.5	75.4	−104.5
Tin	Sn	1492	1634	1703	1777	1855	1903	1968	2063	2169	2270	231.9
Stannic bromide	$SnBr_4$		58.3	72.7	88.1	105.5	116.2	131.0	152.8	177.7	204.7	31.0
Stannous chloride	$SnCl_2$	316	366	391	420	450	467	493	533	577	623	246.8
Stannic chloride	$SnCl_4$	−22.7	−1.0	+10.0	22.0	35.2	43.5	54.7	72.0	92.1	113.0	−30.2
iodide	SnI_4		156.0	175.8	196.2	218.8	234.2	254.2	283.5	315.5	348.0	144.5
hydride	SnH_4	−140.0	−125.8	−118.5	−111.2	−102.3	−96.6	−89.2	−78.0	−65.2	−52.3	−149.9
Tin tetramethyl	$Sn(CH_3)_4$	−51.3	−31.0	−20.6	−9.3	+3.5	11.7	22.8	39.8	58.5	78.0	
trimethyl-ethyl	$Sn(CH_3)_3{\cdot}C_2H_5$	−30.0	−7.6	+3.8	16.1	30.0	38.4	50.0	67.3	87.6	108.8	
trimethyl-propyl	$Sn(CH_3)_3{\cdot}C_3H_7$	−12.0	+10.7	21.8	34.0	48.5	57.5	69.8	88.0	109.6	131.7	
Titanium chloride	$TiCl_4$	−13.9	+9.4	21.3	34.2	48.4	58.0	71.0	90.5	112.7	136.0	−30
Tungsten	W	3990	4337	4507	4690	4886	5007	5168	5403	5666	5927	3370
Tungsten hexafluoride	WF_6	−71.4	−56.5	−49.2	−41.5	−33.0	−27.5	−20.3	−10.0	+1.2	17.3	−0.5
Uranium hexafluoride	UF_6	−38.8	−22.0	−13.8	−5.2	+4.4	10.4	18.2	30.0	42.7	55.7	69.2
Vanadyl trichloride	$VOCl_3$	−23.2	+0.2	12.2	26.6	40.0	49.8	62.5	82.0	103.5	127.2	
Xenon	Xe	−168.5	−158.2	−152.8	−147.1	−141.2	−137.7	−132.8	−125.4	−117.1	−108.0	−111.6
Zinc	Zn	487	558	593	632	673	700	736	788	844	907	419.4
chloride	$ZnCl_2$	428	481	508	536	566	584	610	648	689	732	365
fluoride	ZnF_2	970	1055	1086	1129	1175	1207	1254	1329	1417	1497	872
diethyl	$Zn(C_2H_5)_2$	−22.4	0.0	+11.7	24.2	38.0	47.2	59.1	77.0	97.3	118.0	−28
Zirconium bromide	$ZrBr_4$	207	237	250	266	281	289	301	318	337	357	450
chloride	$ZrCl_4$	190	217	230	243	259	268	279	295	312	331	437
iodide	ZrI_4	264	297	311	329	344	355	369	389	409	431	499

TABLE 2-10 Vapor Pressures of Organic Compounds, up to 1 atm*

Compound		Pressure, mmHg										Melting point, °C
		1	5	10	20	40	60	100	200	400	760	
Name	Formula	Temperature, °C										
Acenaphthalene	$C_{12}H_{10}$		114.8	131.2	148.7	168.2	181.2	197.5	222.1	250.0	277.5	95
Acetal	$C_6H_{14}O_2$	−23.0	−2.3	+8.0	19.6	31.9	39.8	50.1	66.3	84.0	102.2	
Acetaldehyde	C_2H_4O	−81.5	−65.1	−56.8	−47.8	−37.8	−31.4	−22.6	−10.0	+4.9	20.2	−123.5
Acetamide	C_2H_5NO	65.0	92.0	105.0	120.0	135.8	145.8	158.0	178.3	200.0	222.0	81
Acetanilide	C_8H_9NO	114.0	146.6	162.0	180.0	199.6	211.8	227.2	250.5	277.0	303.8	113.5
Acetic acid	$C_2H_4O_2$	−17.2	+6.3	17.5	29.9	43.0	51.7	63.0	80.0	99.0	118.1	16.7
anhydride	$C_4H_6O_3$	1.7	24.8	36.0	48.3	62.1	70.8	82.2	100.0	119.8	139.6	−73
Acetone	C_3H_6O	−59.4	−40.5	−31.1	−20.8	−9.4	−2.0	+7.7	22.7	39.5	56.5	−94.6
Acetonitrile	C_2H_3N	−47.0	−26.6	−16.3	−5.0	+7.7	15.9	27.0	43.7	62.5	81.8	−41
Acetophenone	C_8H_8O	37.1	64.0	78.0	92.4	109.4	119.8	133.6	154.2	178.0	202.4	20.5
Acetyl chloride	C_2H_3OCl	−50.0	−35.0	−27.6	−19.6	−10.4	−4.5	+3.2	16.1	32.0	50.8	−112.0
Acetylene	C_2H_2	−142.9	−133.0	−128.2	−122.8	−116.7	−112.8	−107.9	−100.3	−92.0	−84.0	−81.5
Acridine	$C_{13}H_9N$	129.4	165.8	184.0	203.5	224.2	238.7	256.0	284.0	314.3	346.0	110.5
Acrolein (2-propenal)	C_3H_4O	−64.5	−46.0	−36.7	−26.3	−15.0	−7.5	+2.5	17.5	34.5	52.5	−87.7
Acrylic acid	$C_3H_4O_2$	+3.5	27.3	39.0	52.0	66.2	75.0	86.1	103.3	122.0	141.0	14
Adipic acid	$C_6H_{10}O_4$	159.5	191.0	205.5	222.0	240.5	251.0	265.0	287.8	312.5	337.5	152
Allene (propadiene)	C_3H_4	−120.6	−108.0	−101.0	−93.4	−85.2	−78.8	−72.5	−61.3	−48.5	−35.0	−136
Allyl alcohol (propen-1-ol-3)	C_3H_6O	−20.0	+0.2	10.5	21.7	33.4	40.3	50.0	64.5	80.2	96.6	−129
chloride (3-chloropropene)	C_3H_5Cl	−70.0	−52.0	−42.9	−32.8	−21.2	−14.1	−4.5	10.4	27.5	44.6	−136.4
isopropyl ether	$C_6H_{12}O$	−43.7	−23.1	−12.9	−1.8	+10.9	18.7	29.0	44.3	61.7	79.5	
isothiocyanate	C_4H_5NS	−2.0	+25.3	38.3	52.1	67.4	76.2	89.5	108.0	129.8	150.7	−80
n-propyl ether	$C_6H_{12}O$	−39.0	−18.2	−7.9	+3.7	16.4	25.0	35.8	52.6	71.4	90.5	
4-Allylveratrole	$C_{11}H_{14}O_2$	85.0	113.9	127.0	142.8	158.3	169.6	183.7	204.0	226.2	248.0	
iso-Amyl acetate	$C_7H_{14}O_2$	0.0	+23.7	35.2	47.8	62.1	71.0	83.2	101.3	121.5	142.0	
n-Amyl alcohol	$C_5H_{12}O$	+13.6	34.7	44.9	55.8	68.0	75.5	85.8	102.0	119.8	137.8	
iso-Amyl alcohol	$C_5H_{12}O$	+10.0	30.9	40.8	51.7	63.4	71.0	80.7	95.8	113.7	130.6	−117.2
sec-Amyl alcohol (2-pentanol)	$C_5H_{12}O$	+1.5	22.1	32.2	42.6	54.1	61.5	70.7	85.7	102.3	119.7	
tert-Amyl alcohol	$C_5H_{12}O$	−12.9	+7.2	17.2	27.9	38.8	46.0	55.3	69.7	85.7	101.7	−11.9
sec-Amylbenzene	$C_{11}H_{16}$	29.0	55.8	69.2	83.8	100.0	110.4	124.1	145.2	168.0	193.0	
iso-Amyl benzoate	$C_{12}H_{16}O_2$	72.0	104.5	121.6	139.7	158.3	171.4	186.8	210.2	235.8	262.0	
bromide (1-bromo-3-methylbutane)	$C_5H_{11}Br$	−20.4	+2.1	13.6	26.1	39.8	48.7	60.4	78.7	99.4	120.4	
n-butyrate	$C_9H_{18}O_2$	21.2	47.1	59.9	74.0	90.0	99.8	113.1	133.2	155.3	178.6	
formate	$C_6H_{12}O_2$	−17.5	+5.4	17.1	30.0	44.0	53.3	65.4	83.2	102.7	123.3	
iodide (1-iodo-3-methylbutane)	$C_5H_{11}I$	−2.5	+21.9	34.1	47.6	62.3	71.9	84.4	103.8	125.8	148.2	
isobutyrate	$C_9H_{18}O_2$	14.8	40.1	52.8	66.6	81.8	91.7	104.4	124.2	146.0	168.8	
Amyl isopropionate	$C_8H_{16}O_2$	+8.5	33.7	46.3	60.0	75.5	85.2	97.6	117.3	138.4	160.2	
iso-Amyl isovalerate	$C_{10}H_{20}O_2$	27.0	54.4	68.6	83.8	100.6	110.3	125.1	146.1	169.5	194.0	
n-Amyl levulinate	$C_{10}H_{18}O_3$	81.3	110.0	124.0	139.7	155.8	165.2	180.5	203.1	227.4	253.2	
iso-Amyl levulinate	$C_{10}H_{18}O_3$	75.6	104.0	118.8	134.4	151.7	162.6	177.0	198.1	222.7	247.9	
nitrate	$C_5H_{11}NO_3$	+5.2	28.8	40.3	53.5	67.6	76.3	88.6	106.7	126.5	147.5	
4-*tert*-Amylphenol	$C_{11}H_{16}O$		109.8	125.5	142.3	160.3	172.6	189.0	213.0	239.5	266.0	93
Anethole	$C_{10}H_{12}O$	62.6	91.6	106.0	121.8	139.3	149.8	164.2	186.1	210.5	235.3	22.5
Angelonitrile	C_5H_7N	−8.0	+15.0	28.0	41.0	55.8	65.2	77.5	96.3	117.7	140.0	
Aniline	C_6H_7N	34.8	57.9	69.4	82.0	96.7	106.0	119.9	140.1	161.9	184.4	−6.2
2-Anilinoethanol	$C_8H_{11}NO$	104.0	134.3	149.6	165.7	183.7	194.0	209.5	230.6	254.5	279.6	
Anisaldehyde	$C_8H_8O_2$	73.2	102.6	117.8	133.5	150.5	161.7	176.7	199.0	223.0	248.0	2.5
o-Anisidine (2-methoxyaniline)	C_7H_9NO	61.0	88.0	101.7	116.1	132.0	142.1	155.2	175.3	197.3	218.5	5.2
Anthracene	$C_{14}H_{10}$	145.0	173.5	187.2	201.9	217.5	231.8	250.0	279.0	310.2	342.0	217.5
Anthraquinone	$C_{14}H_8O_2$	190.0	219.4	234.2	248.3	264.3	273.3	285.0	314.6	346.2	379.9	286
Azelaic acid	$C_9H_{16}O_4$	178.3	210.4	225.5	242.4	260.0	271.8	286.5	309.6	332.8	356.5	106.5
Azelaldehyde	$C_9H_{18}O$	33.3	58.4	71.6	85.0	100.2	110.0	123.0	142.1	163.4	185.0	
Azobenzene	$C_{12}H_{10}N_2$	103.5	135.7	151.5	168.3	187.9	199.8	216.0	240.0	266.1	293.0	68
Benzal chloride (α,α-Dichlorotoluene)	$C_7H_6Cl_2$	35.4	64.0	78.7	94.3	112.1	123.4	138.3	160.7	187.0	214.0	−16.1
Benzaldehyde	C_7H_6O	26.2	50.1	62.0	75.0	90.1	99.6	112.5	131.7	154.1	179.0	−26
Benzanthrone	$C_{17}H_{10}O$	225.0	274.5	297.2	322.5	350.0	368.8	390.0	426.5			174
Benzene	C_6H_6	−36.7	−19.6	−11.5	−2.6	+7.6	15.4	26.1	42.2	60.6	80.1	+5.5
Benzenesulfonylchloride	$C_6H_5ClO_2S$	65.9	96.5	112.0	129.0	147.7	158.2	174.5	198.0	224.0	251.5	14.5
Benzil	$C_{14}H_{10}O_2$	128.4	165.2	183.0	202.8	224.5	238.2	255.8	283.5	314.3	347.0	95
Benzoic acid	$C_7H_6O_2$	96.0	119.5	132.1	146.7	162.6	172.8	186.2	205.8	227.0	249.2	121.7
anhydride	$C_{14}H_{10}O_3$	143.8	180.0	198.0	218.0	239.8	252.7	270.4	299.1	328.8	360.0	42
Benzoin	$C_{14}H_{12}O_2$	135.6	170.2	188.1	207.0	227.6	241.7	258.0	284.4	313.5	343.0	132
Benzonitrile	C_7H_5N	28.2	55.3	69.2	83.4	99.6	109.8	123.5	144.1	166.7	190.6	−12.9
Benzophenone	$C_{13}H_{10}O$	108.2	141.7	157.6	175.8	195.7	208.2	224.4	249.8	276.8	305.4	48.5
Benzotrichloride (α,α,α-Trichlorotoluene)	$C_7H_5Cl_3$	45.8	73.7	87.6	102.7	119.8	130.0	144.3	165.6	189.2	213.5	−21.2
Benzotrifluoride (α,α,α-Trifluorotoluene)	$C_7H_5F_3$	−32.0	−10.3	−0.4	12.2	25.7	34.0	45.3	62.5	82.0	102.2	−29.3
Benzoyl bromide	C_7H_5BrO	47.0	75.4	89.8	105.4	122.6	133.4	147.7	169.2	193.7	218.5	0
chloride	C_7H_5ClO	32.1	59.1	73.0	87.6	103.8	114.7	128.0	149.5	172.8	197.2	−0.5
nitrile	C_8H_5NO	44.5	71.7	85.5	100.2	116.6	127.0	141.0	161.3	185.0	208.0	33.5
Benzyl acetate	$C_9H_{10}O_2$	45.0	73.4	87.6	102.3	119.6	129.8	144.0	165.5	189.0	213.5	−51.5
alcohol	C_7H_8O	58.0	80.8	92.6	105.8	119.8	129.3	141.7	160.0	183.0	204.7	−15.3

*Compiled from the extended tables published by D. R. Stull in *Ind. Eng. Chem.*, **39**, 517 (1947). For information on fuels see Hibbard, N.A.C.A. Research Mem. E56I21, 1956. For methane see Johnson (ed.), WADD-TR-60-56, 1960.

(*Continued*)

TABLE 2-10 Vapor Pressures of Organic Compounds, up to 1 atm *(Continued)*

Compound		Pressure, mmHg										Melting point, °C
		1	5	10	20	40	60	100	200	400	760	
Name	Formula	Temperature, °C										
Benzylamine	C_7H_9N	29.0	54.8	67.7	81.8	97.3	107.3	120.0	140.0	161.3	184.5	
Benzyl bromide (α-bromotoluene)	C_7H_7Br	32.2	59.6	73.4	88.3	104.8	115.6	129.8	150.8	175.2	198.5	−4
chloride (α-chlorotoluene)	C_7H_7Cl	22.0	47.8	60.8	75.0	90.7	100.5	114.2	134.0	155.8	179.4	−39
cinnamate	$C_{16}H_{14}O_2$	173.8	206.3	221.5	239.3	255.8	267.0	281.5	303.8	326.7	350.0	39
Benzyldichlorosilane	$C_7H_8Cl_2Si$	45.3	70.2	83.2	96.7	111.8	121.3	133.5	152.0	173.0	194.3	
Benzyl ethyl ether	$C_9H_{12}O$	26.0	52.0	65.0	79.6	95.4	105.5	118.9	139.6	161.5	185.0	
phenyl ether	$C_{13}H_{12}O$	95.4	127.7	144.0	160.7	180.1	192.6	209.2	233.2	259.8	287.0	
isothiocyanate	C_8H_7NS	79.5	107.8	121.8	137.0	153.0	163.8	177.7	198.0	220.4	243.0	
Biphenyl	$C_{12}H_{10}$	70.6	101.8	117.0	134.2	152.5	165.2	180.7	204.2	229.4	254.9	69.5
1-Biphenyloxy-2,3-epoxypropane	$C_{15}H_{14}O_2$	135.3	169.9	187.2	205.8	226.3	239.7	255.0	280.4	309.8	340.0	
d-Bornyl acetate	$C_{12}H_{20}O_2$	46.9	75.7	90.2	106.0	123.7	135.7	149.8	172.0	197.5	223.0	29
Bornyl *n*-butyrate	$C_{14}H_{24}O_2$	74.0	103.4	118.0	133.8	150.7	161.8	176.4	198.0	222.2	247.0	
formate	$C_{11}H_{18}O_2$	47.0	74.8	89.3	104.0	121.2	131.7	145.8	166.4	190.2	214.0	
isobutyrate	$C_{14}H_{24}O_2$	70.0	99.8	114.0	130.0	147.2	157.6	172.2	194.2	218.2	243.0	
propionate	$C_{13}H_{22}O_2$	64.6	93.7	108.0	123.7	140.4	151.2	165.7	187.5	211.2	235.0	
Brassidic acid	$C_{22}H_{42}O_2$	209.6	241.7	256.0	272.9	290.0	301.5	316.2	336.8	359.6	382.5	61.5
Bromoacetic acid	$C_2H_3BrO_2$	54.7	81.6	94.1	108.2	124.0	133.8	146.3	165.8	186.7	208.0	49.5
4-Bromoanisole	C_7H_7BrO	48.8	77.8	91.9	107.8	125.0	136.0	150.1	172.7	197.5	223.0	12.5
Bromobenzene	C_6H_5Br	+2.9	27.8	40.0	53.8	68.6	78.1	90.8	110.1	132.3	156.2	−30.7
4-Bromobiphenyl	$C_{12}H_9Br$	98.0	133.7	150.6	169.8	190.8	204.5	221.8	248.2	277.7	310.0	90.5
1-Bromo-2-butanol	C_4H_9BrO	23.7	45.4	55.8	67.2	79.5	87.0	97.6	112.1	128.3	145.0	
1-Bromo-2-butanone	C_4H_7BrO	+6.2	30.0	41.8	54.2	68.2	77.3	89.2	107.0	126.3	147.0	
cis-1-Bromo-1-butene	C_4H_7Br	−44.0	−23.2	−12.8	−1.4	+11.5	19.8	30.8	47.8	66.8	86.2	
trans-1-Bromo-1-butene	C_4H_7Br	−38.4	−17.0	−6.4	+5.4	18.4	27.2	38.1	55.7	75.0	94.7	−100.3
2-Bromo-1-butene	C_4H_7Br	−47.3	−27.0	−16.8	−5.3	+7.2	15.4	26.3	42.8	61.9	81.0	−133.4
cis-2-Bromo-2-butene	C_4H_7Br	−39.0	−17.9	−7.2	+4.6	17.7	26.2	37.5	54.5	74.0	93.9	−111.2
trans-2-Bromo-2-butene	C_4H_7Br	−45.0	−24.1	−13.8	−2.4	+10.5	18.7	29.9	46.5	66.0	85.5	−114.6
1,4-Bromochlorobenzene	C_6H_4BrCl	32.0	59.5	72.7	87.8	103.8	114.8	128.0	149.5	172.6	196.9	
1-Bromo-1-chloroethane	C_2H_4BrCl	−36.0	−18.0	−9.4	0.0	+10.4	17.0	28.0	44.7	63.4	82.7	16.6
1-Bromo-2-chloroethane	C_2H_4BrCl	−28.8	−7.0	+4.1	16.0	29.7	38.0	49.5	66.8	86.0	106.7	−16.6
2-Bromo-4,6-dichlorophenol	$C_6H_3BrCl_2O$	84.0	115.6	130.8	147.7	165.8	177.6	193.2	216.5	242.0	268.0	68
1-Bromo-4-ethyl benzene	C_8H_9Br	30.4	42.5	74.0	90.2	108.5	121.0	135.5	156.5	182.0	206.0	−45.0
(2-Bromoethyl)-benzene	C_8H_9Br	48.0	76.2	90.5	105.8	123.2	133.8	148.2	169.8	194.0	219.0	
2-Bromoethyl 2-chloroethyl ether	C_4H_8BrClO	36.5	63.2	76.3	90.8	106.6	116.4	129.8	150.0	172.3	195.8	
(2-Bromoethyl)-cyclohexane	$C_8H_{15}Br$	38.7	66.6	80.5	95.8	113.0	123.7	138.0	160.0	186.2	213.0	
1-Bromoethylene	C_2H_3Br	−95.4	−77.8	−68.8	−58.8	−48.1	−41.2	−31.9	−17.2	−1.1	+15.8	−138
Bromoform (tribromomethane)	$CHBr_3$		22.0	34.0	48.0	63.6	73.4	85.9	106.1	127.9	150.5	8.5
1-Bromonaphthalene	$C_{10}H_7Br$	84.2	117.5	133.6	150.2	170.2	183.5	198.8	224.2	252.0	281.1	5.5
2-Bromo-4-phenylphenol	$C_{12}H_9BrO$	100.0	135.4	152.3	171.8	193.8	207.0	224.5	251.0	280.2	311.0	95
3-Bromopyridine	C_5H_4BrN	16.8	42.0	55.2	69.1	84.1	94.1	107.8	127.7	150.0	173.4	
2-Bromotoluene	C_7H_7Br	24.4	49.7	62.3	76.0	91.0	100.0	112.0	133.6	157.3	181.8	−28
3-Bromotoluene	C_7H_7Br	14.8	50.8	64.0	78.1	93.9	104.1	117.8	138.0	160.0	183.7	39.8
4-Bromotoluene	C_7H_7Br	10.3	47.5	61.1	75.2	91.8	102.3	116.4	137.4	160.2	184.5	28.5
3-Bromo-2,4,6-trichlorophenol	$C_6H_2BrCl_3O$	112.4	146.2	163.2	181.8	200.5	213.0	229.3	253.0	278.0	305.8	
2-Bromo-1,4-xylene	C_8H_9Br	37.5	65.0	78.8	94.0	110.6	121.6	135.7	156.4	181.0	206.7	+9.5
1,2-Butadiene (methyl allene)	C_4H_6	−89.0	−72.7	−64.2	−54.9	−44.3	−37.5	−28.3	−14.2	+1.8	18.5	
1,3-Butadiene	C_4H_6	−102.8	−87.6	−79.7	−71.0	−61.3	−55.1	−46.8	−33.9	−19.3	−4.5	−108.9
n-Butane	C_4H_{10}	−101.5	−85.7	−77.8	−68.9	−59.1	−52.8	−44.2	−31.2	−16.3	−0.5	−135
iso-Butane (2-methylpropane)	C_4H_{10}	−109.2	−94.1	−86.4	−77.9	−68.4	−62.4	−54.1	−41.5	−27.1	−11.7	−145
1,3-Butanediol	$C_4H_{10}O_2$	22.2	67.5	85.3	100.0	117.4	127.5	141.2	161.0	183.8	206.5	77
1,2,3-Butanetriol	$C_4H_{10}O_3$	102.0	132.0	146.0	161.0	178.0	188.0	202.5	222.0	243.5	264.0	
1-Butene	C_4H_8	−104.8	−89.4	−81.6	−73.0	−63.4	−57.2	−48.9	−36.2	−21.7	−6.3	−130
cis-2-Butene	C_4H_8	−96.4	−81.1	−73.4	−64.6	−54.7	−48.4	−39.8	−26.8	−12.0	+3.7	−138.9
trans-2-Butene	C_4H_8	−99.4	−84.0	−76.3	−67.5	−57.6	−51.3	−42.7	−29.7	−14.8	+0.9	−105.4
3-Butenenitrile	C_4H_5N	−19.6	+2.9	14.1	26.6	40.0	48.8	60.2	78.0	98.0	119.0	
iso-Butyl acetate	$C_6H_{12}O_2$	−21.2	+1.4	12.8	25.5	39.2	48.0	59.7	77.6	97.5	118.0	−98.9
n-Butyl acrylate	$C_7H_{12}O_2$	−0.5	+23.5	35.5	48.6	63.4	72.6	85.1	104.0	125.2	147.4	−64.6
alcohol	$C_4H_{10}O$	−1.2	+20.0	30.2	41.5	53.4	60.3	70.1	84.3	100.8	117.5	−79.9
iso-Butyl alcohol	$C_4H_{10}O$	−9.0	+11.6	21.7	32.4	44.1	51.7	61.5	75.9	91.4	108.0	−108
sec-Butyl alcohol	$C_4H_{10}O$	−12.2	+7.2	16.9	27.3	38.1	45.2	54.1	67.9	83.9	99.5	−114.7
tert-Butyl alcohol	$C_4H_{10}O$	−20.4	−3.0	+5.5	14.3	24.5	31.0	39.8	52.7	68.0	82.9	25.3
iso-Butyl amine	$C_4H_{11}N$	−50.0	−31.0	−21.0	−10.3	+1.3	8.8	18.8	32.0	50.7	68.6	−85.0
n-Butylbenzene	$C_{10}H_{14}$	22.7	48.8	62.0	76.3	92.4	102.6	116.2	136.9	159.2	183.1	−88.0
iso-Butylbenzene	$C_{10}H_{14}$	14.1	40.5	53.7	67.8	83.3	93.3	107.0	127.2	149.6	172.8	−51.5
sec-Butylbenzene	$C_{10}H_{14}$	18.6	44.2	57.0	70.6	86.2	96.0	109.5	128.8	150.3	173.5	−75.5
tert-Butylbenzene	$C_{10}H_{14}$	13.0	39.0	51.7	65.6	80.8	90.6	103.8	123.7	145.8	168.5	−58
iso-Butyl benzoate	$C_{11}H_{14}O_2$	64.0	93.6	108.6	124.2	141.8	152.0	166.4	188.2	212.8	237.0	
n-Butyl bromide (1-bromobutane)	C_4H_9Br	−33.0	−11.2	−0.3	+11.6	24.8	33.4	44.7	62.0	81.7	101.6	−112.4
iso-Butyl *n*-butyrate	$C_8H_{16}O_2$	+4.6	30.0	42.2	56.1	71.7	81.3	94.0	113.9	135.7	156.9	
carbamate	$C_5H_{11}NO_2$		83.7	96.4	110.1	125.3	134.6	147.2	165.7	186.0	206.5	65
Butyl carbitol (diethylene glycol butyl ether)	$C_8H_{18}O_3$	70.0	95.7	107.8	120.5	135.5	146.0	159.8	181.2	205.0	231.2	
n-Butyl chloride (1-chlorobutane)	C_4H_9Cl	−49.0	−28.9	−18.6	−7.4	+5.0	13.0	24.0	40.0	58.8	77.8	−123.1
iso-Butyl chloride	C_4H_9Cl	−53.8	−34.3	−24.5	−13.8	−1.9	+5.9	16.0	32.0	50.0	68.9	−131.2

TABLE 2-10 Vapor Pressures of Organic Compounds, up to 1 atm *(Continued)*

Compound		Pressure, mmHg										Melting point, °C
		1	5	10	20	40	60	100	200	400	760	
Name	Formula	Temperature, °C										
sec-Butyl chloride (2-Chlorobutane)	C_4H_9Cl	−60.2	−39.8	−29.2	−17.7	−5.0	+3.4	14.2	31.5	50.0	68.0	−131.3
tert-Butyl chloride	C_4H_9Cl					−19.0	−11.4	−1.0	+14.6	32.6	51.0	−26.5
sec-Butyl chloroacetate	$C_6H_{11}ClO_2$	17.0	41.8	54.6	68.2	83.6	93.0	105.5	124.1	146.0	167.8	
2-*tert*-Butyl-4-cresol	$C_{11}H_{16}O$	70.0	98.0	112.0	127.2	143.9	153.7	167.0	187.8	210.0	232.6	
4-*tert*-Butyl-2-cresol	$C_{11}H_{16}O$	74.3	103.7	118.0	134.0	150.8	161.7	176.2	197.8	221.8	247.0	
iso-Butyl dichloroacetate	$C_6H_{10}Cl_2O_2$	28.6	54.3	67.5	81.4	96.7	106.6	119.8	139.2	160.0	183.0	
2,3-Butylene glycol (2,3-butanediol)	$C_4H_{10}O_2$	44.0	68.4	80.3	93.4	107.8	116.3	127.8	145.6	164.0	182.0	22.5
2-Butyl-2-ethylbutane-1,3-diol	$C_{10}H_{22}O_2$	94.1	122.6	136.8	151.2	167.8	178.0	191.9	212.0	233.5	255.0	
2-*tert*-Butyl-4-ethylphenol	$C_{12}H_{15}O$	76.3	106.2	121.0	137.0	154.0	165.4	179.0	200.3	223.8	247.8	
n-Butyl formate	$C_5H_{10}O_2$	−26.4	−4.7	+6.1	18.0	31.6	39.8	51.0	67.9	86.2	106.0	
iso-Butyl formate	$C_5H_{10}O_2$	−32.7	−11.4	−0.8	+11.0	24.1	32.4	43.4	60.0	79.0	98.2	−95.3
sec-Butyl formate	$C_5H_{10}O_2$	−34.4	−13.3	−3.1	+8.4	21.3	29.6	40.2	56.8	75.2	93.6	
sec-Butyl glycolate	$C_6H_{12}O_3$	28.3	53.6	66.0	79.8	94.2	104.0	116.4	135.5	155.6	177.5	
iso-Butyl iodide (1-iodo-2-methylpropane)	C_4H_9I	−17.0	+5.8	17.0	29.8	42.8	51.8	63.5	81.0	100.3	120.4	−90.7
isobutyrate	$C_8H_{16}O_2$	+4.1	28.0	39.9	52.4	67.2	75.9	88.0	106.3	126.3	147.5	−80.7
isovalerate	$C_9H_{18}O_2$	16.0	41.2	53.8	67.7	82.7	92.4	105.2	124.8	146.4	168.7	
levulinate	$C_9H_{16}O_3$	65.0	92.1	105.9	120.2	136.2	147.0	160.2	181.8	205.5	229.9	
naphthylketone (1-isovaleronaphthone)	$C_{15}H_{16}O$	136.0	167.9	184.0	201.6	219.7	231.5	246.7	269.7	294.0	320.0	
2-*sec*-Butylphenol	$C_{10}H_{14}O$	57.4	86.0	100.8	116.1	133.4	143.9	157.3	179.7	203.8	228.0	
2-*tert*-Butylphenol	$C_{10}H_{14}O$	56.6	84.2	98.1	113.0	129.2	140.0	153.5	173.8	196.3	219.5	
4-iso-Butylphenol	$C_{10}H_{14}O$	72.1	100.9	115.5	130.3	147.2	157.0	171.2	192.1	214.7	237.0	
4-*sec*-Butylphenol	$C_{10}H_{14}O$	71.4	100.5	114.8	130.3	147.8	157.9	172.4	194.3	217.6	242.1	
4-*tert*-Butylphenol	$C_{10}H_{14}O$	70.0	99.2	114.0	129.5	146.0	156.0	170.2	191.5	214.0	238.0	99
2-(4-*tert*-Butylphenoxy)ethyl acetate	$C_{14}H_{20}O_3$	118.0	150.0	165.8	183.3	201.5	212.8	228.0	250.3	277.6	304.4	
4-*tert*-Butylphenyl dichlorophosphate	$C_{10}H_{13}Cl_2O_2P$	96.0	129.6	146.0	164.0	184.3	197.2	214.3	240.0	268.2	299.0	
tert-Butyl phenyl ketone (pivalophenone)	$C_{11}H_{14}O$	57.8	85.7	99.0	114.3	130.4	140.8	154.0	175.0	197.7	220.0	
iso-Butyl propionate	$C_7H_{14}O_2$	−2.3	+20.9	32.3	44.8	58.5	67.6	79.5	97.0	116.4	136.8	−71
4-*tert*-Butyl-2,5-xylenol	$C_{12}H_{18}O$	88.2	119.8	135.0	151.0	169.8	180.3	195.0	217.5	241.3	265.3	
4-*tert*-Butyl-2,6-xylenol	$C_{12}H_{18}O$	74.0	103.9	119.0	135.0	152.2	163.6	176.0	196.0	217.8	239.8	
6-*tert*-Butyl-2,4-xylenol	$C_{12}H_{18}O$	70.3	100.2	115.0	131.0	148.5	158.2	172.0	192.3	214.2	236.5	
6-*tert*-Butyl-3,4-xylenol	$C_{12}H_{18}O$	83.9	113.6	127.0	143.0	159.7	170.0	184.0	204.5	226.7	249.5	
Butyric acid	$C_4H_8O_2$	25.5	49.8	61.5	74.0	88.0	96.5	108.0	125.5	144.5	163.5	−74
iso-Butyric acid	$C_4H_8O_2$	14.7	39.3	51.2	64.0	77.8	86.3	98.0	115.8	134.5	154.5	−47
Butyronitrile	C_4H_7N	−20.0	+2.1	13.4	25.7	38.4	47.3	59.0	76.7	96.8	117.5	
iso-Valerophenone	$C_{11}H_{14}O$	58.3	87.0	101.4	116.8	133.8	144.6	158.0	180.1	204.2	228.0	
Camphene	$C_{10}H_{16}$			47.2	60.4	75.7	85.0	97.9	117.5	138.7	160.5	50
Campholenic acid	$C_{10}H_{16}O_2$	97.6	125.7	139.8	153.9	170.0	180.0	193.7	212.7	234.0	256.0	
d-Camphor	$C_{10}H_{16}O$	41.5	68.6	82.3	97.5	114.0	124.0	138.0	157.9	182.0	209.2	178.5
Camphylamine	$C_{10}H_{19}N$	45.3	74.0	83.7	97.6	112.5	122.0	134.6	153.0	173.8	195.0	
Capraldehyde	$C_{10}H_{20}O$	51.9	78.8	92.0	106.3	122.2	132.0	145.3	164.8	186.3	208.5	
Capric acid	$C_{10}H_{20}O_2$	125.0	142.0	152.2	165.0	179.9	189.8	200.0	217.1	240.3	268.4	31.5
n-Caproic acid	$C_6H_{12}O_2$	71.4	89.5	99.5	111.8	125.0	133.3	144.0	160.8	181.0	202.0	−1.5
iso-Caproic acid	$C_6H_{12}O_2$	66.2	83.0	94.0	107.0	120.4	129.6	141.4	158.3	181.0	207.7	−35
iso-Caprolactone	$C_6H_{10}O_2$	38.3	66.4	80.3	95.7	112.3	123.2	137.2	157.8	182.1	207.0	
Capronitrile	$C_6H_{11}N$	9.2	34.6	47.5	61.7	76.9	86.8	99.8	119.7	141.0	163.7	
Capryl alcohol (2-octanol)	$C_8H_{18}O$	32.8	57.6	70.0	83.3	98.0	107.4	119.8	138.0	157.5	178.5	−38.6
Caprylaldehyde	$C_8H_{16}O$	73.4	92.0	101.2	110.2	120.0	126.0	133.9	145.4	156.5	168.5	
Caprylic acid (octanoic acid)	$C_8H_{16}O_2$	92.3	114.1	124.0	136.4	150.6	160.0	172.2	190.3	213.9	237.5	16
Caprylonitrile	$C_8H_{15}N$	43.0	67.6	80.4	94.6	110.6	121.2	134.8	155.2	179.5	204.5	
Carbazole	$C_{12}H_9N$						248.2	265.0	292.5	323.0	354.8	244.8
Carbon dioxide	CO_2	−134.3	−124.4	−119.5	−114.4	−108.6	−104.8	−100.2	−93.0	−85.7	−78.2	−57.5
disulfide	CS_2	−73.8	−54.3	−44.7	−34.3	−22.5	−15.3	−5.1	+10.4	28.0	46.5	−110.8
monoxide	CO	−222.0	−217.2	−215.0	−212.8	−210.0	−208.1	−205.7	−201.3	−196.3	−191.3	−205.0
oxyselenide (carbonyl selenide)	COSe	−117.1	−102.3	−95.0	−86.3	−76.4	−70.2	−61.7	−49.8	−35.6	−21.9	
oxysulfide (carbonyl sulfide)	COS	−132.4	−119.8	−113.3	−106.0	−98.3	−93.0	−85.9	−75.0	−62.7	−49.9	−138.8
tetrabromide	CBr_4					96.3	106.3	119.7	139.7	163.5	189.5	90.1
tetrachloride	CCl_4	−50.0	−30.0	−19.6	−8.2	+4.3	12.3	23.0	38.3	57.8	76.7	−22.6
tetrafluoride	CF_4	−184.6	−174.1	−169.3	−164.3	−158.8	−155.4	−150.7	−143.6	−135.5	−127.7	−183.7
Carvacrol	$C_{10}H_{14}O$	70.0	98.4	113.2	127.9	145.2	155.3	169.7	191.2	213.8	237.0	+0.5
Carvone	$C_{10}H_{14}O$	57.4	86.1	100.4	116.1	133.0	143.8	157.3	179.6	203.5	227.5	
Chavibetol	$C_{10}H_{12}O_2$	83.6	113.3	127.0	143.2	159.8	170.7	185.5	206.8	229.8	254.0	
Chloral (trichloroacetaldehyde)	C_2HCl_3O	−37.8	−16.0	−5.0	+7.2	20.2	29.1	40.2	57.8	77.5	97.7	−57
hydrate (trichloroacetaldehyde hydrate)	$C_2H_3Cl_3O_2$	−9.8	+10.0	19.5	29.2	39.7	46.2	55.0	68.0	82.1	96.2	51.7
Chloranil	$C_6Cl_4O_2$	70.7	89.3	97.8	106.4	116.1	122.0	129.5	140.3	151.3	162.6	290
Chloroacetic acid	$C_2H_3ClO_2$	43.0	68.3	81.0	94.2	109.2	118.3	130.7	149.0	169.0	189.5	61.2
anhydride	$C_4H_4Cl_2O_3$	67.2	94.1	108.0	122.4	138.2	148.0	159.8	177.8	197.0	217.0	46
2-Chloroaniline	C_6H_6ClN	46.3	72.3	84.8	99.2	115.6	125.7	139.5	160.0	183.7	208.8	0
3-Chloroaniline	C_6H_6ClN	63.5	89.8	102.0	116.7	133.6	144.1	158.0	179.5	203.5	228.5	−10.4
4-Chloroaniline	C_6H_6ClN	59.3	87.9	102.1	117.8	135.0	145.8	159.9	182.3	206.6	230.5	70.5
Chlorobenzene	C_6H_5Cl	−13.0	+10.6	22.2	35.3	49.7	58.3	70.7	89.4	110.0	132.2	−45.2
2-Chlorobenzotrichloride (2-α,α,α-tetrachlorotoluene)	$C_7H_4Cl_4$	69.0	101.8	117.9	135.8	155.0	167.8	185.0	208.0	233.0	262.1	28.7

(Continued)

TABLE 2-10 Vapor Pressures of Organic Compounds, up to 1 atm *(Continued)*

Compound		Pressure, mmHg										Melting point, °C
		1	5	10	20	40	60	100	200	400	760	
Name	Formula	Temperature, °C										
2-Chlorobenzotrifluoride (2-chloro-α,α,α-trifluorotoluene)	$C_7H_4ClF_3$	0.0	24.7	37.1	50.6	65.9	75.4	88.3	108.3	130.0	152.2	−6.0
2-Chlorobiphenyl	$C_{12}H_9Cl$	89.3	109.8	134.7	151.2	169.9	182.1	197.0	219.6	243.8	267.5	34
4-Chlorobiphenyl	$C_{12}H_9Cl$	96.4	129.8	146.0	164.0	183.8	196.0	212.5	237.8	264.5	292.9	75.5
α-Chlorocrotonic acid	$C_4H_5ClO_2$	70.0	95.6	108.0	121.2	135.6	144.4	155.9	173.8	193.2	212.0	
Chlorodifluoromethane	$CHClF_2$	−122.8	−110.2	−103.7	−96.5	−88.6	−83.4	−76.4	−65.8	−53.6	−40.8	−160
Chlorodimethylphenylsilane	$C_8H_{11}ClSi$	29.8	56.7	70.0	84.7	101.2	111.5	124.7	145.5	168.6	193.5	
1-Chloro-2-ethoxybenzene	C_8H_9ClO	45.8	72.8	86.5	101.5	117.8	127.8	141.8	162.0	185.5	208.0	
2-(2-Chloroethoxy) ethanol	$C_4H_9ClO_2$	53.0	78.3	90.7	104.1	118.4	127.5	139.5	157.2	176.5	196.0	
bis-2-Chloroethyl acetacetal	$C_6H_{12}Cl_2O_2$	56.2	83.7	97.6	112.2	127.8	138.0	150.7	169.8	190.5	212.6	
1-Chloro-2-ethylbenzene	C_8H_9Cl	17.2	43.0	56.1	70.3	86.2	96.4	110.0	130.2	152.2	177.6	−80.2
1-Chloro-3-ethylbenzene	C_8H_9Cl	18.6	45.2	58.1	73.0	89.2	99.6	113.6	133.8	156.7	181.1	−53.3
1-Chloro-4-ethylbenzene	C_8H_9Cl	19.2	46.4	60.0	75.5	91.8	102.0	116.0	137.0	159.8	184.3	−62.6
2-Chloroethyl chloroacetate	$C_4H_6Cl_2O_2$	46.0	72.1	86.0	100.0	116.0	126.2	140.0	159.8	182.2	205.0	
2-Chloroethyl 2-chloroisopropyl ether	$C_5H_{10}Cl_2O$	24.7	50.1	63.0	77.2	92.4	102.2	115.8	135.7	156.5	180.0	
2-Chloroethyl 2-chloropropyl ether	$C_5H_{10}Cl_2O$	29.8	56.5	70.0	84.8	101.5	111.8	125.6	146.3	169.8	194.1	
2-Chloroethyl α-methylbenzyl ether	$C_{10}H_{13}ClO$	62.3	91.4	106.0	121.8	139.6	150.0	164.8	186.3	210.8	235.0	
Chloroform (trichloromethane)	$CHCl_3$	−58.0	−39.1	−29.7	−19.0	−7.1	+0.5	10.4	25.9	42.7	61.3	−63.5
1-Chloronaphthalene	$C_{10}H_7Cl$	80.6	104.8	118.6	134.4	153.2	165.6	180.4	204.2	230.8	259.3	−20
4-Chlorophenethyl alcohol	C_8H_9ClO	84.0	114.3	129.0	145.0	162.0	173.5	188.1	210.0	234.5	259.3	
2-Chlorophenol	C_6H_5ClO	12.1	38.2	51.2	65.9	82.0	92.0	106.0	126.4	149.8	174.5	7
3-Chlorophenol	C_6H_5ClO	44.2	72.0	86.1	101.7	118.0	129.4	143.0	164.8	188.7	214.0	32.5
4-Chlorophenol	C_6H_5ClO	49.8	78.2	92.2	108.1	125.0	136.1	150.0	172.0	196.0	220.0	42
2-Chloro-3-phenylphenol	$C_{12}H_9ClO$	118.0	152.2	169.7	186.7	207.4	219.6	237.0	261.3	289.4	317.5	+6
2-Chloro-6-phenylphenol	$C_{12}H_9ClO$	119.8	153.7	170.7	189.8	208.2	220.0	237.1	261.6	289.5	317.0	
Chloropicrin (trichloronitromethane)	CCl_3NO_2	−25.5	−3.3	+7.8	20.0	33.8	42.3	53.8	71.8	91.8	111.9	−64
1-Chloropropene	C_3H_5Cl	−81.3	−63.4	−54.1	−44.0	−32.7	−25.1	−15.1	+1.3	18.0	37.0	−99.0
2-Chloropyridine	C_5H_4ClN	13.3	38.8	51.7	65.8	81.7	91.6	104.6	125.0	147.7	170.2	
3-Chlorostyrene	C_8H_7Cl	25.3	51.3	65.2	80.0	96.5	107.2	121.2	142.2	165.7	190.0	
4-Chlorostyrene	C_8H_7Cl	28.0	54.5	67.5	82.0	98.0	108.5	122.0	143.5	166.0	191.0	−15.0
1-Chlorotetradecane	$C_{14}H_{29}Cl$	98.5	131.8	148.2	166.2	187.0	199.8	215.5	240.3	267.5	296.0	+0.9
2-Chlorotoluene	C_7H_7Cl	+5.4	30.6	43.2	56.9	72.0	81.8	94.7	115.0	137.1	159.3	
3-Chlorotoluene	C_7H_7Cl	+4.8	30.3	43.2	57.4	73.0	83.2	96.3	116.6	139.7	162.3	
4-Chlorotoluene	C_7H_7Cl	+5.5	31.0	43.8	57.8	73.5	83.3	96.6	117.1	139.8	162.3	+7.3
Chlorotriethylsilane	$C_6H_{15}ClSi$	−4.9	+19.8	32.0	45.5	60.2	69.5	82.3	101.6	123.6	146.3	
1-Chloro-1,2,2-trifluoroethylene	C_2ClF_3	−116.0	−102.5	−95.9	−88.2	−79.7	−74.1	−66.7	−55.0	−41.7	−27.9	−157.5
Chlorotrifluoromethane	$CClF_3$	−149.5	−139.2	−134.1	−128.5	−121.9	−117.3	−111.7	−102.5	−92.7	−81.2	
Chlorotrimethylsilane	C_3H_9ClSi	−62.8	−43.6	−34.0	−23.2	−11.4	−4.0	+6.0	21.9	39.4	57.9	
trans-Cinnamic acid	$C_9H_8O_2$	127.5	157.8	173.0	189.5	207.1	217.8	232.4	253.3	276.7	300.0	133
Cinnamyl alcohol	$C_9H_{10}O$	72.6	102.5	117.8	133.7	151.0	162.0	177.8	199.8	224.6	250.0	33
Cinnamylaldehyde	C_9H_8O	76.1	105.8	120.0	135.7	152.2	163.7	177.7	199.3	222.4	246.0	−7.5
Citraconic anhydride	$C_5H_4O_3$	47.1	74.8	88.9	103.8	120.3	131.3	145.4	165.8	189.8	213.5	
cis-α-Citral	$C_{10}H_{16}O$	61.7	90.0	103.9	119.4	135.9	146.3	160.0	181.8	205.0	228.0	
d-Citronellal	$C_{10}H_{18}O$	44.0	71.4	84.8	99.8	116.1	126.2	140.1	160.0	183.8	206.5	
Citronellic acid	$C_{10}H_{18}O_2$	99.5	127.3	141.4	155.6	171.9	182.1	195.4	214.5	236.6	257.0	
Citronellol	$C_{10}H_{20}O$	66.4	93.6	107.0	121.5	137.2	147.2	159.8	179.8	201.0	221.5	
Citronellyl acetate	$C_{12}H_{22}O_2$	74.7	100.2	113.0	126.0	140.5	149.7	161.0	178.8	197.8	217.0	
Coumarin	$C_9H_6O_2$	106.0	137.8	153.4	170.0	189.0	200.5	216.5	240.0	264.7	291.0	70
o-Cresol (2-cresol; 2-methylphenol)	C_7H_8O	38.2	64.0	76.7	90.5	105.8	115.5	127.4	146.7	168.4	190.8	30.8
m-Cresol (3-cresol; 3-methylphenol)	C_7H_8O	52.0	76.0	87.8	101.4	116.0	125.8	138.0	157.3	179.0	202.8	10.9
p-Cresol (4-cresol; 4-methylphenol)	C_7H_8O	53.0	76.5	88.6	102.3	117.7	127.0	140.0	157.3	179.4	201.8	35.5
cis-Crotonic acid	$C_4H_6O_2$	33.5	57.4	69.0	82.0	96.0	104.5	116.3	133.9	152.2	171.9	15.5
trans-Crotonic acid	$C_4H_6O_2$			80.0	93.0	107.8	116.7	128.0	146.0	165.5	185.0	72
cis-Crotononitrile	C_4H_5N	−29.0	−7.1	+4.0	16.4	30.0	38.5	50.1	68.0	88.0	108.0	
trans-Crotononitrile	C_4H_5N	−19.5	+3.5	15.0	27.8	41.8	50.9	62.8	81.1	101.5	122.8	
Cumene	C_9H_{12}	+2.9	26.8	38.3	51.5	66.1	75.4	88.1	107.3	129.2	152.4	−96.0
4-Cumidene	$C_9H_{13}N$	60.0	88.2	102.2	117.8	134.2	145.0	158.0	180.0	203.2	227.0	
Cuminal	$C_{10}H_{12}O$	58.0	87.3	102.0	117.9	135.2	146.0	160.0	182.8	206.7	232.0	
Cuminyl alcohol	$C_{10}H_{14}O$	74.2	103.7	118.0	133.8	150.3	161.7	176.2	197.9	221.7	246.6	
2-Cyano-2-*n*-butyl acetate	$C_7H_{11}NO_2$	42.0	68.7	82.0	96.2	111.8	121.5	133.8	152.2	173.4	195.2	
Cyanogen	C_2N_2	−95.8	−83.2	−76.8	−70.1	−62.7	−57.9	−51.8	−42.6	−33.0	−21.0	−34.4
bromide	$CBrN$	−35.7	−18.3	−10.0	−1.0	+8.6	14.7	22.6	33.8	46.0	61.5	58
chloride	$CClN$	−76.7	−61.4	−53.8	−46.1	−37.5	−32.1	−24.9	−14.1	−2.3	+13.1	−6.5
iodide	CIN	25.2	47.2	57.7	68.6	80.3	88.0	97.6	111.5	126.1	141.1	
Cyclobutane	C_4H_8	−92.0	−76.0	−67.9	−58.7	−48.4	−41.8	−32.8	−18.9	−3.4	+12.9	−50
Cyclobutene	C_4H_6	−99.1	−83.4	−75.4	−66.6	−56.4	−50.0	−41.2	−27.8	−12.2	+2.4	
Cyclohexane	C_6H_{12}	−45.3	−25.4	−15.9	−5.0	+6.7	14.7	25.5	42.0	60.8	80.7	+6.6
Cyclohexaneethanol	$C_8H_{16}O$	50.4	77.2	90.0	104.0	119.8	129.8	142.7	161.7	183.5	205.4	
Cyclohexanol	$C_6H_{12}O$	21.0	44.0	56.0	68.8	83.0	91.8	103.7	121.7	141.4	161.0	23.9
Cyclohexanone	$C_6H_{10}O$	+1.4	26.4	38.7	52.5	67.8	77.5	90.4	110.3	132.5	155.6	−45.0
2-Cyclohexyl-4,6-dinitrophenol	$C_{12}H_{14}N_2O_5$	132.8	161.8	175.9	191.2	206.7	216.0	229.0	248.7	269.8	291.5	
Cyclopentane	C_5H_{10}	−68.0	−49.6	−40.4	−30.1	−18.6	−11.3	−1.3	+13.8	31.0	49.3	−93.7
Cyclopropane	C_3H_6	−116.8	−104.2	−97.5	−90.3	−82.3	−77.0	−70.0	−59.1	−46.9	−33.5	−126.6
Cymene	$C_{10}H_{14}$	17.3	43.9	57.0	71.1	87.0	97.2	110.8	131.4	153.5	177.2	−68.2

TABLE 2-10 Vapor Pressures of Organic Compounds, up to 1 atm *(Continued)*

Compound		Pressure, mmHg										Melting point, °C
		1	5	10	20	40	60	100	200	400	760	
Name	Formula	Temperature, °C										
cis-Decalin	$C_{10}H_{18}$	22.5	50.1	64.2	79.8	97.2	108.0	123.2	145.4	169.9	194.6	−43.3
trans-Decalin	$C_{10}H_{18}$	−0.8	+30.6	47.2	65.3	85.7	98.4	114.6	136.2	160.1	186.7	−30.7
Decane	$C_{10}H_{22}$	16.5	42.3	55.7	69.8	85.5	95.5	108.6	128.4	150.6	174.1	−29.7
Decan-2-one	$C_{10}H_{20}O$	44.2	71.9	85.8	100.7	117.1	127.8	142.0	163.2	186.7	211.0	+3.5
1-Decene	$C_{10}H_{20}$	14.7	40.3	53.7	67.8	83.3	93.5	106.5	126.7	149.2	172.0	
Decyl alcohol	$C_{10}H_{22}O$	69.5	97.3	111.3	125.8	142.1	152.0	165.8	186.2	208.8	231.0	+7
Decyltrimethylsilane	$C_{13}H_{30}Si$	67.4	96.4	111.0	126.5	144.0	154.3	169.5	191.0	215.5	240.0	
Dehydroacetic acid	$C_8H_8O_4$	91.7	122.0	137.3	153.0	171.0	181.5	197.5	219.5	244.5	269.0	
Desoxybenzoin	$C_{14}H_{12}O$	123.3	156.2	173.5	192.0	212.0	224.5	241.3	265.2	293.0	321.0	60
Diacetamide	$C_4H_7NO_2$	70.0	95.0	108.0	122.6	138.2	148.0	160.6	180.8	202.0	223.0	78.5
Diacetylene (1,3-butadiyne)	C_4H_2	−82.5	−68.0	−61.2	−53.8	−45.9	−41.0	−34.0	−20.9	−6.1	+9.7	−34.9
Diallyldichlorosilane	$C_6H_{10}Cl_2Si$	+9.5	34.8	47.4	61.3	76.4	86.3	99.7	119.4	142.0	165.3	
Diallyl sulfide	$C_6H_{10}S$	−9.5	+14.4	26.6	39.7	54.2	63.7	75.8	94.8	116.1	138.6	−83
Diisoamyl ether	$C_{10}H_{22}O$	18.6	44.3	57.0	70.7	86.3	96.0	109.6	129.0	150.3	173.4	
oxalate	$C_{12}H_{22}O_4$	85.4	116.0	131.4	147.7	165.7	177.0	192.2	215.0	240.0	265.0	
sulfide	$C_{10}H_{22}S$	43.0	73.0	87.6	102.7	120.0	130.6	145.3	166.4	191.0	216.0	
Dibenzylamine	$C_{14}H_{15}N$	118.3	149.8	165.6	182.2	200.2	212.2	227.3	249.8	274.3	300.0	−26
Dibenzyl ketone (1,3-diphenyl-2-propanone)	$C_{15}H_{14}O$	125.5	159.8	177.6	195.7	216.6	229.4	246.6	272.3	301.7	330.5	34.5
1,4-Dibromobenzene	$C_6H_4Br_2$	61.0	79.3	87.7	103.6	120.8	131.6	146.5	168.5	192.5	218.6	87.5
1,2-Dibromobutane	$C_4H_8Br_2$	7.5	33.2	46.1	60.0	76.0	86.0	99.8	120.2	143.5	166.3	−64.5
dl-2,3-Dibromobutane	$C_4H_8Br_2$	+5.0	30.0	41.6	56.4	72.0	82.0	95.3	115.7	138.0	160.5	
meso-2,3-Dibromobutane	$C_4H_8Br_2$	+1.5	26.6	39.3	53.2	68.0	78.0	91.7	111.8	134.2	157.3	−34.5
1,2-Dibromodecane	$C_{10}H_{20}Br_2$	95.7	123.6	137.3	151.0	167.4	177.5	190.2	209.6	229.8	250.4	
Di(2-bromoethyl) ether	$C_4H_8Br_2O$	47.7	75.3	88.5	103.6	119.8	130.0	144.0	165.0	188.0	212.5	
α,β-Dibromomaleic anhydride	$C_4H_2Br_2O_3$	50.0	78.0	92.0	106.7	123.5	133.8	147.7	168.0	192.0	215.0	
1,2-Dibromo-2-methylpropane	$C_4H_8Br_2$	−28.8	−3.0	+10.5	25.7	42.3	53.7	68.8	92.1	119.8	149.0	−70.3
1,3-Dibromo-2-methylpropane	$C_4H_8Br_2$	14.0	40.0	53.0	67.5	83.5	93.7	107.4	117.8	150.6	174.6	
1,2-Dibromopentane	$C_5H_{10}Br_2$	19.8	45.4	58.0	72.0	87.4	97.4	110.1	130.2	151.8	175.0	
1,2-Dibromopropane	$C_3H_6Br_2$	−7.0	+17.3	29.4	42.3	57.2	66.4	78.7	97.8	118.5	141.6	−55.5
1,3-Dibromopropane	$C_3H_6Br_2$	+9.7	35.4	48.0	62.1	77.8	87.8	101.3	121.7	144.1	167.5	−34.4
2,3-Dibromopropene	$C_3H_4Br_2$	−6.0	+17.9	30.0	43.2	57.8	67.0	79.5	98.0	119.5	141.2	
2,3-Dibromo-1-propanol	$C_3H_6Br_2O$	57.0	84.5	98.2	113.5	129.8	140.0	153.0	173.8	196.0	219.0	
Diisobutylamine	$C_8H_{19}N$	−5.1	+18.4	30.6	43.7	57.8	67.0	79.2	97.6	118.0	139.5	−70
2,6-Ditert-butyl-4-cresol	$C_{15}H_{24}O$	85.8	116.2	131.0	147.0	164.1	175.2	190.0	212.8	237.6	262.5	
4,6-Ditert-butyl-2-cresol	$C_{15}H_{24}O$	86.2	117.3	132.4	149.0	167.4	179.0	194.0	217.5	243.4	269.3	
4,6-Ditert-butyl-3-cresol	$C_{15}H_{24}O$	103.7	135.2	150.0	167.0	185.3	196.1	211.0	233.0	257.1	282.0	
2,6-Ditert-butyl-4-ethylphenol	$C_{16}H_{26}O$	89.1	121.4	137.0	154.0	172.1	183.9	198.0	220.0	244.0	268.6	
4,6-Ditert-butyl-3-ethylphenol	$C_{16}H_{26}O$	111.5	142.6	157.4	174.0	192.3	204.4	218.0	241.7	264.6	290.0	
Diisobutyl oxalate	$C_{10}H_{18}O_4$	63.2	91.2	105.3	120.3	137.5	147.8	161.8	183.5	205.8	229.5	
2,4-Ditert-butylphenol	$C_{14}H_{22}O$	84.5	115.4	130.0	146.0	164.3	175.8	190.0	212.5	237.0	260.8	
Dibutyl phthalate	$C_{16}H_{22}O_4$	148.2	182.1	198.2	216.2	235.8	247.8	263.7	287.0	313.5	340.0	
sulfide	$C_8H_{18}S$	+21.7	51.8	66.4	80.5	96.0	105.8	118.6	138.0	159.0	182.0	−79.7
Diisobutyl *d*-tartrate	$C_{12}H_{22}O_6$	117.8	151.8	169.0	188.0	208.5	221.6	239.5	264.7	294.0	324.0	73.5
Dicarvacryl-mono-(6-chloro-2-xenyl) phosphate	$C_{32}H_{34}ClO_4P$	204.2	234.5	249.3	264.5	280.5	290.7	304.9	323.8	342.0	361.0	
Dicarvacryl-2-tolyl phosphate	$C_{27}H_{33}O_4P$	180.2	209.3	221.8	237.0	251.5	260.3	272.5	290.0	309.8	330.0	
Dichloroacetic acid	$C_2H_2Cl_2O_2$	44.0	69.8	82.6	96.3	111.8	121.5	134.0	152.3	173.7	194.4	9.7
1,2-Dichlorobenzene	$C_6H_4Cl_2$	20.0	46.0	59.1	73.4	89.4	99.5	112.9	133.4	155.8	179.0	−17.6
1,3-Dichlorobenzene	$C_6H_4Cl_2$	12.1	39.0	52.0	66.2	82.0	92.2	105.0	125.9	149.0	173.0	−24.2
1,4-Dichlorobenzene	$C_6H_4Cl_2$			54.8	69.2	84.8	95.2	108.4	128.3	150.2	173.9	53.0
1,2-Dichlorobutane	$C_4H_8Cl_2$	−23.6	−0.3	+11.5	24.5	37.7	47.8	60.2	79.7	100.8	123.5	
2,3-Dichlorobutane	$C_4H_8Cl_2$	−25.2	−3.0	+8.5	21.2	35.0	43.9	56.0	74.0	94.2	116.0	−80.4
1,2-Dichloro-1,2-difluoroethylene	$C_2Cl_2F_2$	−82.0	−65.6	−57.3	−48.3	−38.2	−31.8	−23.0	−10.0	+5.0	20.9	−112
Dichlorodifluoromethane	CCl_2F_2	−118.5	−104.6	−97.8	−90.1	−81.6	−76.1	−68.6	−57.0	−43.9	−29.8	
Dichlorodiphenyl silane	$C_{12}H_{10}Cl_2Si$	109.6	142.4	158.0	176.0	195.5	207.5	223.8	248.0	275.5	304.0	
Dichlorodiisopropyl ether	$C_6H_{12}Cl_2O$	29.6	55.2	68.2	82.2	97.3	106.9	119.7	139.0	159.8	182.7	
Di(2-chloroethoxy) methane	$C_5H_{10}Cl_2O_2$	53.0	80.4	94.0	109.5	125.5	135.8	149.6	170.0	192.0	215.0	
Dichloroethoxymethylsilane	$C_3H_8Cl_2OSi$	−33.8	−12.1	−1.3	+11.3	24.4	32.6	44.1	61.0	80.3	100.6	
1,2-Dichloro-3-ethylbenzene	$C_8H_8Cl_2$	46.0	75.0	90.0	105.9	123.8	135.0	149.8	172.0	197.0	222.1	−40.8
1,2-Dichloro-4-ethylbenzene	$C_8H_8Cl_2$	47.0	77.2	92.3	109.6	127.5	139.0	153.3	176.0	201.7	226.6	−76.4
1,4-Dichloro-2-ethylbenzene	$C_8H_8Cl_2$	38.5	68.0	83.2	99.8	118.0	129.0	144.0	166.2	191.5	216.3	−61.2
cis-1,2-Dichloroethylene	$C_2H_2Cl_2$	−58.4	−39.2	−29.9	−19.4	−7.9	−0.5	+9.5	24.6	41.0	59.0	−80.5
trans-1,2-Dichloro ethylene	$C_2H_2Cl_2$	−65.4	−47.2	−38.0	−28.0	−17.0	−10.0	−0.2	+14.3	30.8	47.8	−50.0
Di(2-chloroethyl) ether	$C_4H_8Cl_2O$	23.5	49.3	62.0	76.0	91.5	101.5	114.5	134.0	155.4	178.5	
Dichlorofluoromethane	$CHCl_2F$	−91.3	−75.5	−67.5	−58.6	−48.8	−42.6	−33.9	−20.9	−6.2	+8.9	−135
1,5-Dichlorohexamethyltrisiloxane	$C_6H_{18}Cl_2O_2Si_3$	26.0	52.0	65.1	79.0	94.8	105.0	118.2	138.3	160.2	184.0	−53.0
Dichloromethylphenylsilane	$C_7H_8Cl_2Si$	35.7	63.5	77.4	92.4	109.5	120.0	134.2	155.5	180.2	205.5	
1,1-Dichloro-2-methylpropane	$C_4H_8Cl_2$	−31.0	−8.4	+2.6	14.6	28.2	37.0	48.2	65.8	85.4	106.0	
1,2-Dichloro-2-methylpropane	$C_4H_8Cl_2$	−25.8	−4.2	+6.7	18.7	32.0	40.2	51.7	68.9	87.8	108.0	
1,3-Dichloro-2-methylpropane	$C_4H_8Cl_2$	−3.0	+20.6	32.0	44.8	58.6	67.5	78.8	96.1	115.4	135.0	
2,4-Dichlorophenol	$C_6H_4Cl_2O$	53.0	80.0	92.8	107.7	123.4	133.5	146.0	165.2	187.5	210.0	45.0
2,6-Dichlorophenol	$C_6H_4Cl_2O$	59.5	87.6	101.0	115.5	131.6	141.8	154.6	175.5	197.7	220.0	

(Continued)

TABLE 2-10 Vapor Pressures of Organic Compounds, up to 1 atm *(Continued)*

Compound		Pressure, mmHg										Melting point, °C
		1	5	10	20	40	60	100	200	400	760	
Name	Formula	Temperature, °C										
α,α-Dichlorophenylacetonitrile	$C_8H_5Cl_2N$	56.0	84.0	98.1	113.8	130.0	141.0	154.5	176.2	199.5	223.5	
Dichlorophenylarsine	$C_6H_5AsCl_2$	61.8	100.0	116.0	133.1	151.0	163.2	178.9	202.8	228.8	256.5	
1,2-Dichloropropane	$C_3H_6Cl_2$	−38.5	−17.0	−6.1	+6.0	19.4	28.0	39.4	57.0	76.0	96.8	
2,3-Dichlorostyrene	$C_8H_6Cl_2$	61.0	90.1	104.6	120.5	137.8	149.0	163.5	185.7	210.0	235.0	
2,4-Dichlorostyrene	$C_8H_6Cl_2$	53.5	82.2	97.4	111.8	129.2	140.0	153.8	176.0	200.0	225.0	
2,5-Dichlorostyrene	$C_8H_6Cl_2$	55.5	83.9	98.2	114.0	131.0	142.0	155.8	178.0	202.5	227.0	
2,6-Dichlorostyrene	$C_8H_6Cl_2$	47.8	75.7	90.0	105.5	122.4	133.3	147.6	169.0	193.5	217.0	
3,4-Dichlorostyrene	$C_8H_6Cl_2$	57.2	86.0	100.4	116.2	133.7	144.6	158.2	181.5	205.7	230.0	
3,5-Dichlorostyrene	$C_8H_6Cl_2$	53.5	82.2	97.4	111.8	129.2	140.0	153.8	176.0	200.0	225.0	
1,2-Dichlorotetraethylbenzene	$C_{14}H_{20}Cl_2$	105.6	138.7	155.0	172.5	192.2	204.8	220.7	245.6	272.8	302.0	
1,4-Dichlorotetraethylbenzene	$C_{14}H_{20}Cl_2$	91.7	126.1	143.8	162.0	183.2	195.8	212.0	238.5	265.8	296.5	
1,2-Dichloro-1,1,2,2-tetrafluoroethane	$C_2Cl_2F_4$	−95.4	−80.0	−72.3	−63.5	−53.7	−47.5	−39.1	−26.3	−12.0	+3.5	−94
Dichloro-4-tolylsilane	$C_7H_8Cl_2Si$	46.2	71.7	84.2	97.8	113.2	122.6	135.5	153.5	175.2	196.3	
3,4-Dichloro-α,α,α-trifluorotoluene	$C_7H_3Cl_2F_3$	11.0	38.3	52.2	67.3	84.0	95.0	109.2	129.0	150.5	172.8	−12.1
Dicyclopentadiene	$C_{10}H_8$		34.1	47.6	62.0	77.9	88.0	101.7	121.8	144.2	166.6	32.9
Diethoxydimethylsilane	$C_6H_{16}O_2Si$	−19.1	+2.4	13.3	25.3	38.0	46.3	57.6	74.2	93.2	113.5	
Diethoxydiphenylsilane	$C_{16}H_{20}O_2Si$	111.5	142.8	157.6	174.3	193.2	205.0	220.0	243.8	259.7	296.0	
Diethyl adipate	$C_{10}H_{18}O_4$	74.0	106.6	123.0	138.3	154.6	165.8	179.0	198.2	219.1	240.0	−21
Diethylamine	$C_4H_{11}N$			−33.0	−22.6	−11.3	−4.0	+6.0	21.0	38.0	55.5	−38.9
N-Diethylaniline	$C_{10}H_{15}N$	49.7	78.0	91.9	107.2	123.6	133.8	147.3	168.2	192.4	215.5	−34.4
Diethyl arsanilate	$C_{10}H_{16}AsNO_3$	38.0	62.6	74.8	88.0	102.6	111.8	123.8	141.9	161.0	181.0	
1,2-Diethylbenzene	$C_{10}H_{14}$	22.3	48.7	62.0	76.4	92.5	102.6	116.2	136.7	159.0	183.5	−31.4
1,3-Diethylbenzene	$C_{10}H_{14}$	20.7	46.8	59.9	74.5	90.4	100.7	114.4	134.8	156.9	181.1	−83.9
1,4-Diethylbenzene	$C_{10}H_{14}$	20.7	47.1	60.3	74.7	91.1	101.3	115.3	136.1	159.0	183.8	−43.2
Diethyl carbonate	$C_5H_{10}O_3$	−10.1	+12.3	23.8	36.0	49.5	57.9	69.7	86.5	105.8	125.8	−43
cis-Diethyl citraconate	$C_9H_{14}O_4$	59.8	88.3	103.0	118.2	135.7	146.2	160.0	182.3	206.5	230.3	
Diethyl dioxosuccinate	$C_8H_{10}O_6$	70.0	98.0	112.0	126.8	143.8	153.7	167.7	188.0	210.8	233.5	
Diethylene glycol	$C_4H_{10}O_3$	91.8	120.0	133.8	148.0	164.3	174.0	187.5	207.0	226.5	244.8	
Diethyleneglycol-bis-chloroacetate	$C_8H_{12}Cl_2O_5$	148.3	180.0	195.8	212.0	229.0	239.5	252.0	271.5	291.8	313.0	
Diethylene glycol dimethyl ether												
Di(2-methoxyethyl) ether	$C_6H_{14}O_3$	13.0	37.6	50.0	63.0	77.5	86.8	99.5	118.0	138.5	159.8	
glycol ethyl ether	$C_6H_{14}O_3$	45.3	72.0	85.8	100.3	116.7	126.8	140.3	159.0	180.3	201.9	
Diethyl ether	$C_4H_{10}O$	−74.3	−56.9	−48.1	−38.5	27.7	−21.8	−11.5	+2.2	17.9	34.6	−116.3
ethylmalonate	$C_9H_{16}O_4$	50.8	77.8	91.6	106.0	122.4	132.4	146.0	166.0	188.7	211.5	
fumarate	$C_8H_{12}O_4$	53.2	81.2	95.3	110.2	126.7	137.7	151.1	172.2	195.8	218.5	+0.6
glutarate	$C_9H_{16}O_4$	65.6	94.7	109.7	125.4	142.8	153.2	167.8	189.5	212.8	237.0	
Diethylhexadecylamine	$C_{20}H_{43}N$	139.8	175.8	194.0	213.5	235.0	248.5	265.5	292.8	324.6	355.0	
Diethyl itaconate	$C_9H_{14}O_4$	51.3	80.2	95.2	111.0	128.2	139.9	154.3	177.5	203.1	227.9	
ketone (3-pentanone)	$C_5H_{10}O$	−12.7	+7.5	17.2	27.9	39.4	46.7	56.2	70.6	86.3	102.7	−42
malate	$C_8H_{14}O_5$	80.7	110.4	125.3	141.2	157.8	169.0	183.9	205.3	229.5	253.4	
maleate	$C_8H_{12}O_4$	57.3	85.6	100.0	115.3	131.8	142.4	156.0	177.8	201.7	225.0	
malonate	$C_7H_{12}O_4$	40.0	67.5	81.3	95.9	113.3	123.0	136.2	155.5	176.8	198.9	−49.8
mesaconate	$C_9H_{14}O_4$	62.8	91.0	105.3	120.3	137.3	147.9	161.6	183.2	205.8	229.0	
oxalate	$C_6H_{10}O_4$	47.4	71.8	83.8	96.8	110.6	119.7	130.8	147.9	166.2	185.7	−40.6
phthalate	$C_{12}H_{14}O_4$	108.8	140.7	156.0	173.6	192.1	204.1	219.5	243.0	267.5	294.0	
sebacate	$C_{14}H_{26}O_4$	125.3	156.2	172.1	189.8	207.5	218.4	234.4	255.8	280.3	305.5	1.3
2,5-Diethylstyrene	$C_{12}H_{16}$	49.7	78.4	92.6	108.5	125.8	136.8	151.0	173.2	198.0	223.0	
Diethyl succinate	$C_8H_{14}O_4$	54.6	83.0	96.6	111.7	127.8	138.2	151.1	171.7	193.8	216.5	−20.8
isosuccinate	$C_8H_{14}O_4$	39.8	66.7	80.0	94.7	111.0	121.4	134.8	155.1	177.7	201.3	
sulfate	$C_4H_{10}O_4S$	47.0	74.0	87.7	102.1	118.0	128.6	142.5	162.5	185.5	209.5	−25.0
sulfide	$C_4H_{10}S$	−39.6	−18.6	−8.0	+3.5	16.1	24.2	35.0	51.3	69.7	88.0	−99.5
sulfite	$C_4H_{10}O_3S$	10.0	34.2	46.4	59.7	74.2	83.8	96.3	115.8	137.0	159.0	
d-Diethyl tartrate	$C_8H_{14}O_6$	102.0	133.0	148.0	164.2	182.3	194.0	208.5	230.4	254.8	280.0	17
dl-Diethyl tartrate	$C_8H_{14}O_6$	100.0	131.7	147.2	163.8	181.7	193.2	208.0	230.0	254.3	280.0	
3,5-Diethyltoluene	$C_{11}H_{16}$	34.0	61.5	75.3	90.2	107.0	117.7	131.7	152.4	176.5	200.7	
Diethylzinc	$C_4H_{10}Zn$	−22.4	0.0	+11.7	24.2	38.0	47.2	59.1	77.0	97.3	118.0	−28
1-Dihydrocarvone	$C_{10}H_{16}O$	46.6	75.5	90.0	106.0	123.7	134.7	149.7	171.8	197.0	223.0	
Dihydrocitronellol	$C_{10}H_{22}O$	68.0	91.7	103.0	115.0	127.6	136.7	145.9	160.2	176.8	193.5	
1,4-Dihydroxyanthraquinone	$C_{14}H_8O_4$	196.7	239.8	259.8	282.0	307.4	323.3	344.5	377.8	413.0	450.0	194
Dimethylacetylene (2-butyne)	C_4H_6	−73.0	−57.9	−50.5	−42.5	−33.9	−27.8	−18.8	−5.0	+10.6	27.2	−32.5
Dimethylamine	C_2H_7N	−87.7	−72.2	−64.6	−56.0	−46.7	−40.7	−32.6	−20.4	−7.1	+7.4	−96
N,N-Dimethylaniline	$C_8H_{11}N$	29.5	56.3	70.0	84.8	101.6	111.9	125.8	146.5	169.2	193.1	+2.5
Dimethyl arsanilate	$C_8H_{12}AsNO_3$	15.0	39.6	51.8	65.0	79.7	88.6	101.0	119.8	140.3	160.5	
Di(α-methylbenzyl) ether	$C_{16}H_{18}O$	96.7	128.3	144.0	160.3	179.6	191.5	206.8	229.7	254.8	281.0	
2,2-Dimethylbutane	C_6H_{14}	−69.3	−50.7	−41.5	−31.1	−19.5	−12.1	−2.0	+13.4	31.0	49.7	−99.8
2,3-Dimethylbutane	C_6H_{14}	−63.6	−44.5	−34.9	−24.1	−12.4	−4.9	+5.4	21.1	39.0	58.0	−128.2
Dimethyl citraconate	$C_7H_{10}O_4$	50.8	78.2	91.8	106.5	122.6	132.7	145.8	165.8	188.0	210.5	
1,1-Dimethylcyclohexane	C_8H_{16}	−24.4	−1.4	+10.3	23.0	37.3	45.7	57.9	76.2	97.2	119.5	−34
cis-1,2-Dimethylcyclohexane	C_8H_{16}	−15.9	+7.3	18.4	31.1	45.3	54.4	66.8	85.6	107.0	129.7	−50.0
trans-1,2-Dimethylcyclohexane	C_8H_{16}	−21.1	+1.7	13.0	25.6	39.7	48.7	61.0	79.6	100.9	123.4	−88.0
trans-1,3-Dimethylcyclohexane	C_8H_{16}	−19.4	+3.4	14.9	27.4	41.4	50.4	62.5	81.0	102.1	124.4	−92.0
cis-1,3-Dimethylcyclohexane	C_8H_{16}	−22.7	0.0	+11.2	23.6	37.5	46.4	58.5	76.9	97.8	120.1	−76.2
cis-1,4-Dimethylcyclohexane	C_8H_{16}	−20.0	+3.2	14.5	27.1	41.1	50.1	62.3	80.8	101.9	124.3	−87.4
trans-1,4-Dimethylcyclohexane	C_8H_{16}	−24.3	−1.7	+10.1	22.6	36.5	45.4	57.6	76.0	97.0	119.3	−36.9

TABLE 2-10 Vapor Pressures of Organic Compounds, up to 1 atm *(Continued)*

Compound		Pressure, mmHg										Melting point, °C
		1	5	10	20	40	60	100	200	400	760	
Name	Formula	Temperature, °C										
Dimethyl ether	C_2H_6O	−115.7	−101.1	−93.3	−85.2	−76.2	−70.4	−62.7	−50.9	−37.8	−23.7	−138.5
2,2-Dimethylhexane	C_8H_{18}	−29.7	−7.9	+3.1	15.0	28.2	36.7	48.2	65.7	85.6	106.8	
2,3-Dimethylhexane	C_8H_{18}	−23.0	−1.1	+9.9	22.1	35.6	44.2	56.0	73.8	94.1	115.6	
2,4-Dimethylhexane	C_8H_{18}	−26.9	−5.3	+5.2	17.2	30.5	39.0	50.6	68.1	88.2	109.4	
2,5-Dimethylhexane	C_8H_{18}	−26.7	−5.5	+5.3	17.2	30.4	38.9	50.5	68.0	87.9	109.1	−90.7
3,3-Dimethylhexane	C_8H_{18}	−25.8	−4.4	+6.1	18.2	31.7	40.4	52.5	70.0	90.4	112.0	
3,4-Dimethylhexane	C_8H_{18}	−22.1	+0.2	11.3	23.5	37.1	45.8	57.7	75.6	96.0	117.7	
Dimethyl itaconate	$C_7H_{10}O_4$	69.3	94.0	106.6	119.7	133.7	142.6	153.7	171.0	189.8	208.0	38
1-Dimethyl malate	$C_6H_{10}O_5$	75.4	104.0	118.3	133.8	150.1	160.4	175.1	196.3	219.5	242.6	
Dimethyl maleate	$C_6H_8O_4$	45.7	73.0	86.4	101.3	117.2	127.1	140.4	160.0	182.2	205.0	
malonate	$C_5H_8O_4$	35.0	59.8	72.0	85.0	100.0	109.7	121.9	140.0	159.8	180.7	−62
trans-Dimethyl mesaconate	$C_7H_{10}O_4$	46.8	74.0	87.8	102.1	118.0	127.8	141.5	161.0	183.5	206.0	
2,7-Dimethyloctane	$C_{10}H_{22}$	+6.3	30.5	42.3	55.8	71.2	80.8	93.9	114.0	136.0	159.7	−52.8
Dimethyl oxalate	$C_4H_6O_4$	20.0	44.0	56.0	69.4	83.6	92.8	104.8	123.3	143.3	163.3	
2,2-Dimethylpentane	C_7H_{16}	−49.0	−28.7	−18.7	−7.5	+5.0	13.0	23.9	40.3	59.2	79.2	−123.7
2,3-Dimethylpentane	C_7H_{16}	−42.0	−20.8	−10.3	+1.1	13.9	22.1	33.3	50.1	69.4	89.8	−135
2,4-Dimethylpentane	C_7H_{16}	−48.0	−27.4	−17.1	−5.9	+6.5	14.5	25.4	41.8	60.6	80.5	−119.5
3,3-Dimethylpentane	C_7H_{16}	−45.9	−25.0	−14.4	−2.9	+9.9	18.1	29.3	46.2	65.5	86.1	−135.0
2,3-Dimethylphenol (2,3-xylenol)	$C_8H_{10}O$	56.0	83.8	97.6	112.0	129.2	139.5	152.2	173.0	196.0	218.0	75
2,4-Dimethylphenol (2,4-xylenol)	$C_8H_{10}O$	51.8	78.0	91.3	105.0	121.5	131.0	143.0	161.5	184.2	211.5	25.5
2,5-Dimethylphenol (2,5-xylenol)	$C_8H_{10}O$	51.8	78.0	91.3	105.0	121.5	131.0	143.0	161.5	184.2	211.5	74.5
3,4-Dimethylphenol (3,4-xylenol)	$C_8H_{10}O$	66.2	93.8	107.7	122.0	138.0	148.0	161.0	181.5	203.6	225.2	62.5
3,5-Dimethylphenol (3,5-xylenol)	$C_8H_{10}O$	62.0	89.2	102.4	117.0	133.3	143.5	156.0	176.2	197.8	219.5	68
Dimethylphenylsilane	$C_8H_{12}Si$	+5.3	30.3	42.6	56.2	71.4	81.3	94.2	114.2	136.4	159.3	
Dimethyl phthalate	$C_{10}H_{10}O_4$	100.3	131.8	147.6	164.0	182.8	194.0	210.0	232.7	257.8	283.7	
3,5-Dimethyl-1,2-pyrone	$C_7H_8O_2$	78.6	107.6	122.0	136.4	152.7	163.8	177.5	198.0	221.0	245.0	51.5
4,6-Dimethylresorcinol	$C_8H_{10}O_2$	49.0	76.8	90.7	105.8	122.5	133.2	147.3	167.8	192.0	215.0	
Dimethyl sebacate	$C_{12}H_{22}O_4$	104.0	139.8	156.2	175.8	196.0	208.0	222.6	245.0	269.6	293.5	38
2,4-Dimethylstyrene	$C_{10}H_{12}$	34.2	61.9	75.8	90.8	107.7	118.0	132.3	153.2	177.5	202.0	
2,5-Dimethylstyrene	$C_{10}H_{12}$	29.0	55.9	69.0	84.0	100.2	110.7	124.7	145.6	168.7	193.0	
α,α-Dimethylsuccinic anhydride	$C_6H_8O_3$	61.4	88.1	102.0	116.3	132.3	142.4	155.3	175.8	197.5	219.5	
Dimethyl sulfide	C_2H_6S	−75.6	−58.0	−49.2	−39.4	−28.4	−21.4	−12.0	+2.6	18.7	36.0	−83.2
d-Dimethyl tartrate	$C_6H_{10}O_6$	102.1	133.2	148.2	164.3	182.4	193.8	208.8	230.5	255.0	280.0	61.5
dl-Dimethyl tartrate	$C_6H_{10}O_6$	100.4	131.8	147.5	164.0	182.4	193.8	209.5	232.3	257.4	282.0	89
N,N-Dimethyl-2-toluidine	$C_9H_{13}N$	28.8	54.1	66.2	80.2	95.0	105.2	118.1	138.3	161.5	184.8	−61
N,N-Dimethyl-4-toluidine	$C_9H_{13}N$	50.1	74.3	86.7	100.0	116.3	126.4	140.3	161.6	185.4	209.5	
Di(nitrosomethyl) amine	$C_2H_5N_3O_2$	+3.2	27.8	40.0	53.7	68.2	77.7	90.3	110.0	131.3	153.0	
Diosphenol	$C_{10}H_{16}O_2$	66.7	95.4	109.0	124.0	141.2	151.3	165.6	186.2	209.5	232.0	
1,4-Dioxane	$C_4H_8O_2$	−35.8	−12.8	−1.2	+12.0	25.2	33.8	45.1	62.3	81.8	101.1	10
Dipentene	$C_{10}H_{16}$	14.0	40.4	53.8	68.2	84.3	94.6	108.3	128.2	150.5	174.6	
Diphenylamine	$C_{12}H_{11}N$	108.3	141.7	157.0	175.2	194.3	206.9	222.8	247.5	274.1	302.0	52.9
Diphenyl carbinol (benzhydrol)	$C_{13}H_{12}O$	110.0	145.0	162.0	180.9	200.0	212.0	227.5	250.0	275.6	301.0	68.5
chlorophosphate	$C_{12}H_{10}ClPO_3$	121.5	160.5	182.0	203.8	227.9	244.2	265.0	299.5	337.2	378.0	
disulfide	$C_{12}H_{10}S_2$	131.6	164.0	180.0	197.0	214.8	226.2	241.3	262.6	285.8	310.0	61
1,2-Diphenylethane (dibenzyl)	$C_{14}H_{14}$	86.8	119.8	136.0	153.7	173.7	186.0	202.8	227.8	255.0	284.0	51.5
Diphenyl ether	$C_{12}H_{10}O$	66.1	97.8	114.0	130.8	150.0	162.0	178.8	203.3	230.7	258.5	27
1,1-Diphenylethylene	$C_{14}H_{12}$	87.4	119.6	135.0	151.8	170.8	183.4	198.6	222.8	249.8	277.0	
trans-Diphenylethylene	$C_{14}H_{12}$	113.2	145.8	161.0	179.8	199.0	211.5	227.4	251.7	278.3	306.5	124
1,1-Diphenylhydrazine	$C_{12}H_{12}N_2$	126.0	159.3	176.1	194.0	213.5	225.9	242.5	267.2	294.0	322.2	44
Diphenylmethane	$C_{13}H_{12}$	76.0	107.4	122.8	139.8	157.8	170.2	186.3	210.7	237.5	264.5	26.5
Diphenyl sulfide	$C_{12}H_{10}S$	96.1	129.0	145.0	162.0	182.8	194.8	211.8	236.8	263.9	292.5	
Diphenyl-2-tolyl thiophosphate	$C_{18}H_{17}O_3PS$	159.7	179.8	201.6	215.5	230.6	240.4	252.5	270.3	290.0	310.0	
1,2-Dipropoxyethane	$C_8H_{18}O_2$	−38.8	−10.3	+5.0	22.3	42.3	55.8	74.2	103.8	140.0	180.0	
1,2-Diisopropylbenzene	$C_{12}H_{18}$	40.0	67.8	81.8	96.8	114.0	124.3	138.7	159.8	184.3	209.0	
1,3-Diisopropylbenzene	$C_{12}H_{18}$	34.7	62.3	76.0	91.2	107.9	118.2	132.3	153.7	177.6	202.0	−105
Dipropylene glycol	$C_6H_{14}O_3$	73.8	102.1	116.2	131.3	147.4	156.5	169.9	189.9	210.5	231.8	
Dipropyleneglycol monobutyl ether	$C_{10}H_{22}O_3$	64.7	92.0	106.0	120.4	136.3	146.3	159.8	180.0	203.8	227.0	
isopropyl ether	$C_9H_{20}O_3$	46.0	72.8	86.2	100.8	117.0	126.8	140.3	160.0	183.1	205.6	
Di-*n*-propyl ether	$C_6H_{14}O$	−43.3	−22.3	−11.8	0.0	+13.2	21.6	33.0	50.3	69.5	89.5	−122
Diisopropyl ether	$C_6H_{14}O$	−57.0	−37.4	−27.4	−16.7	−4.5	+3.4	13.7	30.0	48.2	67.5	−60
Di-*n*-propyl ketone (4-heptanone)	$C_7H_{14}O$	23.0	44.4	55.0	66.2	78.1	85.8	96.0	111.2	127.3	143.7	−32.6
Di-*n*-propyl oxalate	$C_8H_{14}O_4$	53.4	80.2	93.9	108.6	124.6	134.8	148.1	168.0	190.3	213.5	
Diisopropyl oxalate	$C_8H_{14}O_4$	43.2	69.0	81.9	95.6	110.5	120.0	132.6	151.2	171.8	193.5	
Di-*n*-propyl succinate	$C_{10}H_{18}O_4$	77.5	107.6	122.2	138.0	154.8	166.0	180.3	202.5	226.5	250.8	
Di-*n*-propyl *d*-tartrate	$C_{10}H_{18}O_6$	115.6	147.7	163.5	180.4	199.7	211.7	227.0	250.1	275.6	303.0	
Diisopropyl *d*-tartrate	$C_{10}H_{18}O_6$	103.7	133.7	148.2	164.0	181.8	192.6	207.3	228.2	251.8	275.0	
Divinyl acetylene (1,5-hexadiene-3-yne)	C_6H_6	−45.1	−24.4	−14.0	−2.8	+10.0	18.1	29.5	46.0	64.4	84.0	
1,3-Divinylbenzene	$C_{10}H_{10}$	32.7	60.0	73.8	88.7	105.5	116.0	130.0	151.4	175.2	199.5	−66.9
Docosane	$C_{22}H_{46}$	157.8	195.4	213.0	233.5	254.5	268.3	286.0	314.2	343.5	376.0	44.5
n-Dodecane	$C_{12}H_{26}$	47.8	75.8	90.0	104.6	121.7	132.1	146.2	167.2	191.0	216.2	−9.6
1-Dodecene	$C_{12}H_{24}$	47.2	74.0	87.8	102.4	118.6	128.5	142.3	162.2	185.5	208.0	−31.5
n-Dodecyl alcohol	$C_{12}H_{26}O$	91.0	120.2	134.7	150.0	167.2	177.8	192.0	213.0	235.7	259.0	24
Dodecylamine	$C_{12}H_{27}N$	82.8	111.8	127.8	141.6	157.4	168.0	182.1	203.0	225.0	248.0	
Dodecyltrimethylsilane	$C_{15}H_{34}Si$	91.2	122.1	137.7	153.8	172.1	184.2	199.5	222.0	248.0	273.0	
Elaidic acid	$C_{18}H_{34}O_2$	171.3	206.7	223.5	242.3	260.8	273.0	288.0	312.4	337.0	362.0	51.5

(Continued)

TABLE 2-10 Vapor Pressures of Organic Compounds, up to 1 atm *(Continued)*

Compound		Pressure, mmHg										Melting point, °C
		1	5	10	20	40	60	100	200	400	760	
Name	Formula	Temperature, °C										
Epichlorohydrin	C_3H_5ClO	−16.5	+5.6	16.6	29.0	42.0	50.6	62.0	79.3	98.0	117.9	−25.6
1,2-Epoxy-2-methylpropane	C_4H_8O	−69.0	−50.0	−40.3	−29.5	−17.3	−9.7	+1.2	17.5	36.0	55.5	
Erucic acid	$C_{22}H_{42}O_2$	206.7	239.7	254.5	270.6	289.1	300.2	314.4	336.5	358.8	381.5	33.5
Estragole (*p*-methoxy allyl benzene)	$C_{10}H_{12}O$	52.6	80.0	93.7	108.4	124.6	135.2	148.5	168.7	192.0	215.0	
Ethane	C_2H_6	−159.5	−148.5	−142.9	−136.7	−129.8	−125.4	−119.3	−110.2	−99.7	−88.6	−183.2
Ethoxydimethylphenylsilane	$C_{10}H_{16}OSi$	36.3	63.1	76.2	91.0	107.2	127.5	131.4	151.5	175.0	199.5	
Ethoxytrimethylsilane	$C_5H_{14}OSi$	−50.9	−31.0	−20.7	−9.8	+3.7	11.5	22.1	38.1	56.3	75.7	
Ethoxytriphenylsilane	$C_{20}H_{20}OSi$	167.0	198.2	213.5	230.0	247.0	258.3	273.5	295.0	319.5	344.0	
Ethyl acetate	$C_4H_8O_2$	−43.4	−23.5	−13.5	−3.0	+9.1	16.6	27.0	42.0	59.3	77.1	−82.4
acetoacetate	$C_6H_{10}O_3$	28.5	54.0	67.3	81.1	96.2	106.0	118.5	138.0	158.2	180.8	−45
Ethylacetylene (1-butyne)	C_4H_6	−92.5	−76.7	−68.7	−59.9	−50.0	−43.4	−34.9	−21.6	−6.9	+8.7	−130
Ethyl acrylate	$C_5H_8O_2$	−29.5	−8.7	+2.0	13.0	26.0	33.5	44.5	61.5	80.0	99.5	−71.2
α-Ethylacrylic acid	$C_5H_8O_2$	47.0	70.7	82.0	94.4	108.1	116.7	127.5	144.0	160.7	179.2	
α-Ethylacrylonitrile	C_5H_7N	−29.0	−6.4	+5.0	17.7	31.8	40.6	53.0	71.6	92.2	114.0	
Ethyl alcohol (ethanol)	C_2H_6O	−31.3	−12.0	−2.3	+8.0	19.0	26.0	34.9	48.4	63.5	78.4	−112
Ethylamine	C_2H_7N	−82.3	−66.4	−58.3	−48.6	−39.8	−33.4	−25.1	−12.3	+2.0	16.6	−80.6
4-Ethylaniline	$C_8H_{11}N$	52.0	80.0	93.8	109.0	125.7	136.0	149.8	170.6	194.2	217.4	−4
N-Ethylaniline	$C_8H_{11}N$	38.5	66.4	80.6	96.0	113.2	123.6	137.3	156.9	180.8	204.0	−63.5
2-Ethylanisole	$C_9H_{12}O$	29.7	55.9	69.0	83.1	98.8	109.0	122.3	142.1	164.2	187.1	
3-Ethylanisole	$C_9H_{12}O$	33.7	60.3	73.9	88.5	104.8	115.5	129.2	149.7	172.8	196.5	
4-Ethylanisole	$C_9H_{12}O$	33.5	60.2	73.9	88.5	104.7	115.4	128.4	149.2	172.3	196.5	
Ethylbenzene	C_8H_{10}	−9.8	+13.9	25.9	38.6	52.8	61.8	74.1	92.7	113.8	136.2	−94.9
Ethyl benzoate	$C_9H_{10}O_2$	44.0	72.0	86.0	101.4	118.2	129.0	143.2	164.8	188.4	213.4	−34.6
benzoylacetate	$C_{11}H_{12}O_3$	107.6	136.4	150.3	166.8	181.8	191.9	205.0	223.8	244.7	265.0	
bromide	C_2H_5Br	−74.3	−56.4	−47.5	−37.8	−26.7	−19.5	−10.0	+4.5	21.0	38.4	−117.8
α-bromoisobutyrate	$C_6H_{11}BrO_2$	10.6	35.8	48.0	61.8	77.0	86.7	99.8	119.7	141.2	163.6	
n-butyrate	$C_6H_{12}O_2$	−18.4	+4.0	15.3	27.8	41.5	50.1	62.0	79.8	100.0	121.0	−93.3
isobutyrate	$C_6H_{12}O_2$	−24.3	−2.4	+8.4	20.6	33.8	42.3	53.5	71.0	90.0	110.0	−88.2
Ethylcamphoronic anhydride	$C_{11}H_{16}O_5$	118.2	149.8	165.0	181.8	199.8	211.5	226.6	248.5	272.8	298.0	
Ethyl isocaproate	$C_8H_{16}O_2$	11.0	35.8	48.0	61.7	76.3	85.8	98.4	117.8	139.2	160.4	
carbamate	$C_3H_7NO_2$		65.8	77.8	91.0	105.6	114.8	126.2	144.2	164.0	184.0	49
carbanilate	$C_9H_{11}NO_2$	107.8	131.8	143.7	155.5	168.8	177.3	187.9	203.8	220.0	237.0	52.5
Ethylcetylamine	$C_{18}H_{39}N$	133.2	168.2	186.0	205.5	226.5	239.8	256.8	283.3	313.0	342.0	
Ethyl chloride	C_2H_5Cl	−89.8	−73.9	−65.8	−56.8	−47.0	−40.6	−32.0	−18.6	−3.9	+12.3	−139
chloroacetate	$C_4H_7ClO_2$	+1.0	25.4	37.5	50.4	65.2	74.0	86.0	103.8	123.8	144.2	−26
chloroglyoxylate	$C_4H_5ClO_3$	−5.1	+18.0	29.9	42.0	56.0	65.2	76.6	94.5	114.7	135.0	
α-chloropropionate	$C_5H_9ClO_2$	+6.6	30.2	41.9	54.3	68.2	77.3	89.3	107.2	126.2	146.5	
trans-cinnamate	$C_{11}H_{12}O_2$	87.6	108.5	134.0	150.3	169.2	181.2	196.0	219.3	245.0	271.0	12
3-Ethylcumene	$C_{11}H_{16}$	28.3	55.5	68.8	83.6	99.9	110.2	124.3	145.4	168.2	193.0	
4-Ethylcumene	$C_{11}H_{16}$	31.5	58.4	72.0	86.7	103.3	113.8	127.2	148.3	171.8	195.8	
Ethyl cyanoacetate	$C_5H_7NO_2$	67.8	93.5	106.0	119.8	133.8	142.1	152.8	169.8	187.8	206.0	
Ethylcyclohexane	C_8H_{16}	−14.5	+9.2	20.6	33.4	47.6	56.7	69.0	87.8	109.1	131.8	−111.3
Ethylcyclopentane	C_7H_{14}	−32.2	−10.8	−0.1	+11.7	25.0	33.4	45.0	62.4	82.3	103.4	−138.6
Ethyl dichloroacetate	$C_4H_6Cl_2O_2$	9.6	34.0	46.3	59.5	74.0	83.6	96.1	115.2	135.9	156.5	
N,N-diethyloxamate	$C_8H_{15}NO_3$	76.0	106.3	121.7	137.7	154.4	166.0	180.3	202.8	226.5	252.0	
N-Ethyldiphenylamine	$C_{14}H_{15}N$	98.3	130.2	146.0	162.8	182.0	193.7	209.8	233.0	258.8	286.0	
Ethylene	C_2H_4	−168.3	−158.3	−153.2	−147.6	−141.3	−137.3	−131.8	−123.4	−113.9	−103.7	−169
Ethylene-bis-(chloroacetate)	$C_6H_8Cl_2O_4$	112.0	142.4	158.0	173.5	191.0	201.8	215.0	237.3	259.5	283.5	
Ethylene chlorohydrin (2-chloroethanol)	C_2H_5ClO	−4.0	+19.0	30.3	42.5	56.0	64.1	75.0	91.8	110.0	128.8	−69
diamine (1,2-ethanediamine)	$C_2H_8N_2$	−11.0	+10.5	21.5	33.0	45.8	53.8	62.5	81.0	99.0	117.2	8.5
dibromide (1,2-dibromethane)	$C_2H_4Br_2$	−27.0	+4.7	18.6	32.7	48.0	57.9	70.4	89.8	110.1	131.5	10
dichloride (1,2-dichloroethane)	$C_2H_4Cl_2$	−44.5	−24.0	−13.6	−2.4	+10.0	18.1	29.4	45.7	64.0	82.4	−35.3
glycol (1,2-ethanediol)	$C_2H_6O_2$	53.0	79.7	92.1	105.8	120.0	129.5	141.8	158.5	178.5	197.3	−15.6
glycol diethyl ether (1,2-diethoxyethane)	$C_6H_{14}O_2$	−33.5	−10.2	+1.6	14.7	29.7	39.0	51.8	71.8	94.1	119.5	
glycol dimethyl ether (1,2-dimethoxyethane)	$C_4H_{10}O_2$	−48.0	−26.2	−15.3	−3.0	+10.7	19.7	31.8	50.0	70.8	93.0	
glycol monomethyl ether (2-methoxyethanol)	$C_3H_8O_2$	−13.5	+10.2	22.0	34.3	47.8	56.4	68.0	85.3	104.3	124.4	
oxide	C_2H_4O	−89.7	−73.8	−65.7	−56.6	−46.9	−40.7	−32.1	−19.5	−4.9	+10.7	−111.3
Ethyl α-ethylacetoacetate	$C_8H_{14}O_3$	40.5	67.3	80.2	94.6	110.3	120.6	133.8	153.2	175.6	198.0	
fluoride	C_2H_5F	−117.0	−103.8	−97.7	−90.0	−81.8	−76.4	−69.3	−58.0	−45.5	−32.0	
formate	$C_3H_6O_2$	−60.5	−42.2	−33.0	−22.7	−11.5	−4.3	−5.4	20.0	37.1	54.3	−79
2-furoate	$C_7H_8O_3$	37.6	63.8	77.1	91.5	107.5	117.5	130.4	150.1	172.5	195.0	34
glycolate	$C_4H_8O_3$	14.3	38.8	50.5	63.9	78.1	87.6	99.8	117.8	138.0	158.2	
3-Ethylhexane	C_8H_{18}	−20.0	+2.1	12.8	25.0	38.5	47.1	58.9	76.7	97.0	118.5	
2-Ethylhexyl acrylate	$C_{11}H_{20}O_2$	50.0	77.7	91.8	106.3	123.7	134.0	147.9	168.2	192.2	216.0	
Ethylidene chloride (1,1-dichloroethane)	$C_2H_4Cl_2$	−60.7	−41.9	−32.3	−21.9	−10.2	−2.9	+7.2	22.4	39.8	57.4	−96.7
fluoride (1,1-difluoroethane)	$C_2H_4F_2$	−112.5	−98.4	−91.7	−84.1	−75.8	−70.4	−63.2	−52.0	−39.5	−26.5	−117
Ethyl iodide	C_2H_5I	−54.4	−34.3	−24.3	−13.1	−0.9	+7.2	18.0	34.1	52.3	72.4	−105
Ethyl *l*-leucinate	$C_8H_{17}NO_2$	27.8	57.3	72.1	88.0	106.0	117.8	131.8	149.8	167.3	184.0	
Ethyl levulinate	$C_7H_{12}O_3$	47.3	74.0	87.3	101.8	117.7	127.6	141.3	160.2	183.0	206.2	
Ethyl mercaptan (ethanethiol)	C_2H_6S	−76.7	−59.1	−50.2	−40.7	−29.8	−22.4	−13.0	+1.5	17.7	35.0	−121
Ethyl methylcarbamate	$C_4H_9NO_2$	26.5	51.0	63.2	76.1	91.0	100.0	112.0	130.0	149.8	170.0	
Ethyl methyl ether	C_3H_8O	−91.0	−75.6	−67.8	−59.1	−49.4	−43.3	−34.8	−22.0	−7.8	+7.5	

TABLE 2-10 Vapor Pressures of Organic Compounds, up to 1 atm *(Continued)*

Compound		Pressure, mmHg										Melting point, °C
		1	5	10	20	40	60	100	200	400	760	
Name	Formula	Temperature, °C										
1-Ethylnaphthalene	$C_{12}H_{12}$	70.0	101.4	116.8	133.8	152.0	164.1	180.0	204.6	230.8	258.1	−27
Ethyl α-naphthyl ketone (1-propionaphthone)	$C_{13}H_{12}O$	124.0	155.5	171.0	188.1	206.9	218.2	233.5	255.5	280.2	306.0	
Ethyl 3-nitrobenzoate	$C_9H_9NO_4$	108.1	140.2	155.0	173.6	192.6	205.0	220.3	244.6	270.6	298.0	47
3-Ethylpentane	C_7H_{16}	−37.8	−17.0	−6.8	+4.7	17.5	25.7	36.9	53.8	73.0	93.5	−118.6
4-Ethylphenetole	$C_{10}H_{14}O$	48.5	75.7	89.5	103.8	119.8	129.8	143.5	163.2	185.7	208.0	
2-Ethylphenol	$C_8H_{10}O$	46.2	73.4	87.0	101.5	117.9	127.9	141.8	161.6	184.5	207.5	−45
3-Ethylphenol	$C_8H_{10}O$	60.0	86.8	100.2	114.5	130.0	139.8	152.0	171.8	193.3	214.0	−4
4-Ethylphenol	$C_8H_{10}O$	59.3	86.5	100.2	115.0	131.3	141.7	154.2	175.0	197.4	219.0	46.5
Ethyl phenyl ether (phenetole)	$C_8H_{10}O$	18.1	43.7	56.4	70.3	86.6	95.4	108.4	127.9	149.8	172.0	−30.2
Ethyl propionate	$C_5H_{10}O_2$	−28.0	−7.2	+3.4	14.3	27.2	35.1	45.2	61.7	79.8	99.1	−72.6
Ethyl propyl ether	$C_5H_{12}O$	−64.3	−45.0	−35.0	−24.0	−12.0	−4.0	+6.8	23.3	41.6	61.7	
Ethyl salicylate	$C_9H_{10}O_3$	61.2	90.0	104.2	119.3	136.7	147.6	161.5	183.7	207.0	231.5	1.3
3-Ethylstyrene	$C_{10}H_{12}$	28.3	55.0	68.3	82.8	99.2	109.6	123.2	144.0	167.2	191.5	
4-Ethylstyrene	$C_{10}H_{12}$	26.0	52.7	66.3	80.8	97.3	107.6	121.5	142.0	165.0	189.0	
Ethylisothiocyanate	C_3H_5NS	−13.2	+10.6	22.8	36.1	50.8	59.8	71.9	90.0	110.1	131.0	−5.9
2-Ethyltoluene	C_9H_{12}	9.4	34.8	47.6	61.2	76.4	86.0	99.0	119.0	141.4	165.1	
3-Ethyltoluene	C_9H_{12}	7.2	32.3	44.7	58.2	73.3	82.9	95.9	115.5	137.8	161.3	−95.5
4-Ethyltoluene	C_9H_{12}	7.6	32.7	44.9	58.5	73.6	83.2	96.3	116.1	136.4	162.0	
Ethyl trichloroacetate	$C_4H_5Cl_3O_2$	20.7	45.5	57.7	70.6	85.5	94.4	107.4	125.8	146.0	167.0	
Ethyltrimethylsilane	$C_5H_{14}Si$	−60.6	−41.4	−31.8	−21.0	−9.0	−1.2	+9.2	25.0	42.8	62.0	
Ethyltrimethyltin	$C_5H_{14}Sn$	−30.0	−7.6	+3.8	16.1	30.0	38.4	50.0	67.3	87.6	108.8	
Ethyl isovalerate	$C_7H_{14}O_2$	−6.1	+17.0	28.7	41.3	55.2	64.0	75.9	93.8	114.0	134.3	−99.3
2-Ethyl-1,4-xylene	$C_{10}H_{14}$	25.7	52.0	65.6	79.8	96.0	106.2	120.0	140.2	163.1	186.9	
4-Ethyl-1,3-xylene	$C_{10}H_{14}$	26.3	53.0	66.4	80.6	97.2	107.4	121.2	141.8	164.4	188.4	
5-Ethyl-1,3-xylene	$C_{10}H_{14}$	22.1	48.8	62.1	76.5	92.6	103.0	116.5	137.4	159.6	183.7	
Eugenol	$C_{10}H_{12}O_2$	78.4	108.1	123.0	138.7	155.8	167.3	182.2	204.7	228.3	253.5	
iso-Eugenol	$C_{10}H_{12}O_2$	86.3	117.0	132.4	149.0	167.0	178.2	194.0	217.2	242.3	267.5	−10
Eugenyl acetate	$C_{12}H_{14}O_3$	101.6	132.3	148.0	164.2	183.0	194.0	209.7	232.5	257.4	282.0	295
Fencholic acid	$C_{10}H_{16}O_2$	101.7	128.7	142.3	155.8	171.8	181.5	194.0	215.0	237.8	264.1	19
d-Fenchone	$C_{10}H_{16}O$	28.0	54.7	68.3	83.0	99.5	109.8	123.6	144.0	166.8	191.0	5
dl-Fenchyl alcohol	$C_{10}H_{18}O$	45.8	70.3	82.1	95.6	110.8	120.2	132.3	150.0	173.2	201.0	35
Fluorene	$C_{13}H_{10}$		129.3	146.0	164.2	185.2	197.8	214.7	240.3	268.6	295.0	113
Fluorobenzene	C_6H_5F	−43.4	−22.8	−12.4	−1.2	+11.5	19.6	30.4	47.2	65.7	84.7	−42.1
2-Fluorotoluene	C_7H_7F	−24.2	−2.2	+8.9	21.4	34.7	43.7	55.3	73.0	92.8	114.0	−80
3-Fluorotoluene	C_7H_7F	−22.4	−0.3	+11.0	23.4	37.0	45.8	57.5	75.4	95.4	116.0	−110.8
4-Fluorotoluene	C_7H_7F	−21.8	+0.3	11.8	24.0	37.8	46.5	58.1	76.0	96.1	117.0	
Formaldehyde	CH_2O			−88.0	−79.6	−70.6	−65.0	−57.3	−46.0	−33.0	−19.5	−92
Formamide	CH_3NO	70.5	96.3	109.5	122.5	137.5	147.0	157.5	175.5	193.5	210.5	
Formic acid	CH_2O_2	−20.0	−5.0	+2.1	10.3	24.0	32.4	43.8	61.4	80.3	100.6	8.2
trans-Fumaryl chloride	$C_4H_2Cl_2O_2$	+15.0	38.5	51.8	65.0	79.5	89.0	101.0	120.0	140.0	160.0	
Furfural (2-furaldehyde)	$C_5H_4O_2$	18.5	42.6	54.8	67.8	82.1	91.5	103.4	121.8	141.8	161.8	
Furfuryl alcohol	$C_5H_6O_2$	31.8	56.0	68.0	81.0	95.7	104.0	115.9	133.1	151.8	170.0	
Geraniol	$C_{10}H_{18}O$	69.2	96.8	110.0	125.6	141.8	151.5	165.3	185.6	207.8	230.0	
Geranyl acetate	$C_{12}H_{20}O_2$	73.5	102.7	117.9	133.0	150.0	160.3	175.2	196.3	219.8	243.3	
Geranyl *n*-butyrate	$C_{14}H_{24}O_2$	96.8	125.2	139.0	153.8	170.1	180.2	193.8	214.0	235.0	257.4	
Geranyl isobutyrate	$C_{14}H_{24}O_2$	90.9	119.6	133.0	147.9	164.0	174.0	187.7	207.6	228.5	251.0	
Geranyl formate	$C_{11}H_{18}O_2$	61.8	90.3	104.3	119.8	136.2	147.2	160.7	182.6	205.8	230.0	
Glutaric acid	$C_5H_8O_4$	155.5	183.8	196.0	210.5	226.3	235.5	247.0	265.0	283.5	303.0	97.5
Glutaric anhydride	$C_5H_6O_3$	100.8	133.3	149.5	166.0	185.5	196.2	212.5	236.5	261.0	287.0	
Glutaronitrile	$C_5H_6N_2$	91.3	123.7	140.0	156.5	176.4	189.5	205.5	230.0	257.3	286.2	
Glutaryl chloride	$C_5H_6Cl_2O_2$	56.1	84.0	97.8	112.3	128.3	139.1	151.8	172.4	195.3	217.0	
Glycerol	$C_3H_8O_3$	125.5	153.8	167.2	182.2	198.0	208.0	220.1	240.0	263.0	290.0	17.9
Glycerol dichlorohydrin (1,3-dichloro-2-propanol)	$C_3H_6Cl_2O$	28.0	52.2	64.7	78.0	93.0	102.0	114.8	133.3	153.5	174.3	
Glycol diacetate	$C_6H_{10}O_4$	38.3	64.1	77.1	90.8	106.1	115.8	128.0	147.8	168.3	190.5	−31
Glycolide (1,4-dioxane-2,6-dione)	$C_4H_4O_4$		103.0	116.6	132.0	148.6	158.2	173.2	194.0	217.0	240.0	97
Guaicol (2-methoxyphenol)	$C_7H_8O_2$	52.4	79.1	92.0	106.0	121.6	131.0	144.0	162.7	184.1	205.0	28.3
Heneicosane	$C_{21}H_{44}$	152.6	188.0	205.4	223.2	243.4	255.3	272.0	296.5	323.8	350.5	40.4
Heptacosane	$C_{27}H_{56}$	211.7	248.6	266.8	284.6	305.7	318.3	333.5	359.4	385.0	410.6	59.5
Heptadecane	$C_{17}H_{36}$	115.0	145.2	160.0	177.7	195.8	207.3	223.0	247.8	274.5	303.0	22.5
Heptaldehyde (enanthaldehyde)	$C_7H_{14}O$	12.0	32.7	43.0	54.0	66.3	74.0	84.0	102.0	125.5	155.0	−42
n-Heptane	C_7H_{16}	−34.0	−12.7	−2.1	+9.5	22.3	30.6	41.8	58.7	78.0	98.4	−90.6
Heptanoic acid (enanthic acid)	$C_7H_{14}O_2$	78.0	101.3	113.2	125.6	139.5	148.5	160.0	179.5	199.6	221.5	−10
1-Heptanol	$C_7H_{16}O$	42.4	64.3	74.7	85.8	99.8	108.0	119.5	136.6	155.6	175.8	34.6
Heptanoyl chloride (enanthyl chloride)	$C_7H_{13}ClO$	34.2	54.6	64.6	75.0	86.4	93.5	102.7	116.3	130.7	145.0	
2-Heptene	C_7H_{14}	−35.8	−14.1	−3.5	+8.3	21.5	30.0	41.3	58.6	78.1	98.5	
Heptylbenzene	$C_{13}H_{20}$	64.0	94.6	110.0	126.0	144.0	154.8	170.2	193.3	217.8	244.0	
Heptyl cyanide (enanthonitrile)	$C_7H_{13}N$	21.0	47.8	61.6	76.3	92.6	103.0	116.8	137.7	160.0	184.6	
Hexachlorobenzene	C_6Cl_6	114.4	149.3	166.4	185.7	206.0	219.0	235.5	258.5	283.5	309.4	230
Hexachloroethane	C_2Cl_6	32.7	49.8	73.5	87.6	102.3	112.0	124.2	143.1	163.8	185.6	186.6
Hexacosane	$C_{26}H_{54}$	204.0	240.0	257.4	275.8	295.2	307.8	323.2	348.4	374.6	399.8	56.6
Hexadecane	$C_{16}H_{34}$	105.3	135.2	149.8	164.7	181.3	193.2	208.5	231.7	258.3	287.5	18.5
1-Hexadecene	$C_{16}H_{32}$	101.6	131.7	146.2	162.0	178.8	190.8	205.3	226.8	250.0	274.0	4
n-Hexadecyl alcohol (cetyl alcohol)	$C_{16}H_{34}O$	122.7	158.3	177.8	197.8	219.8	234.3	251.7	280.2	312.7	344.0	49.3

(Continued)

TABLE 2-10 Vapor Pressures of Organic Compounds, up to 1 atm *(Continued)*

Compound		Pressure, mmHg										Melting point, °C
		1	5	10	20	40	60	100	200	400	760	
Name	Formula	Temperature, °C										
n-Hexadecylamine (cetylamine)	$C_{16}H_{35}N$	123.6	157.8	176.0	195.7	215.7	228.8	245.8	272.2	300.4	330.0	
Hexaethylbenzene	$C_{18}H_{30}$		134.3	150.3	168.0	187.7	199.7	216.0	241.7	268.5	298.3	130
n-Hexane	C_6H_{14}	−53.9	−34.5	−25.0	−14.1	−2.3	+5.4	15.8	31.6	49.6	68.7	−95.3
1-Hexanol	$C_6H_{14}O$	24.4	47.2	58.2	70.3	83.7	92.0	102.8	119.6	138.0	157.0	−51.6
2-Hexanol	$C_6H_{14}O$	14.6	34.8	45.0	55.9	67.9	76.0	87.3	103.7	121.8	139.9	
3-Hexanol	$C_6H_{14}O$	+2.5	25.7	36.7	49.0	62.2	70.7	81.8	98.3	117.0	135.5	
1-Hexene	C_6H_{12}	−57.5	−38.0	−28.1	−17.2	−5.0	+2.8	13.0	29.0	46.8	66.0	−98.5
n-Hexyl levulinate	$C_{11}H_{20}O_3$	90.0	120.0	134.7	150.2	167.8	179.0	193.6	215.7	241.0	266.8	
n-Hexyl phenyl ketone (enanthophenone)	$C_{13}H_{18}O$	100.0	130.3	145.5	161.0	178.9	189.8	204.2	225.0	248.3	271.3	
Hydrocinnamic acid	$C_9H_{10}O_2$	102.2	133.5	148.7	165.0	183.3	194.0	209.0	230.8	255.0	279.8	48.5
Hydrogen cyanide (hydrocyanic acid)	CHN	−71.0	−55.3	−47.7	−39.7	−30.9	−25.1	−17.8	−5.3	+10.2	25.9	−13.2
Hydroquinone	$C_6H_6O_2$	132.4	153.3	163.5	174.6	192.0	203.0	216.5	238.0	262.5	286.2	170.3
4-Hydroxybenzaldehyde	$C_7H_6O_2$	121.2	153.2	169.7	186.8	206.0	217.5	233.5	256.8	282.6	310.0	115.5
α-Hydroxyisobutyric acid	$C_4H_8O_3$	73.5	98.5	110.5	123.8	138.0	146.4	157.7	175.2	193.8	212.0	79
α-Hydroxybutyronitrile	C_5H_9NO	41.0	65.8	77.8	90.7	104.8	113.9	125.0	142.0	159.8	178.8	
4-Hydroxy-3-methyl-2-butanone	$C_5H_{10}O_2$	44.6	69.3	81.0	94.0	108.2	117.4	129.0	146.5	165.5	185.0	
4-Hydroxy-4-methyl-2-pentanone	$C_6H_{12}O_2$	22.0	46.7	58.8	72.0	86.7	96.0	108.2	126.8	147.5	167.9	−47
3-Hydroxypropionitrile	C_3H_5NO	58.7	87.8	102.0	117.9	134.1	144.7	157.7	178.0	200.0	221.0	
Indene	C_9H_8	16.4	44.3	58.5	73.9	90.7	100.8	114.7	135.6	157.8	181.6	−2
Iodobenzene	C_6H_5I	24.1	50.6	64.0	78.3	94.4	105.0	118.3	139.8	163.9	188.6	−28.5
Iodononane	$C_9H_{19}I$	70.0	96.2	109.0	123.0	138.1	147.7	159.8	179.0	199.3	219.5	
2-Iodotoluene	C_7H_7I	37.2	65.9	79.8	95.6	112.4	123.8	138.1	160.0	185.7	211.0	
α-Ionone	$C_{13}H_{20}O$	79.5	108.8	123.0	139.0	155.6	166.3	181.2	202.5	225.2	250.0	
Isoprene	C_5H_8	−79.8	−62.3	−53.3	−43.5	−32.6	−25.4	−16.0	−1.2	+15.4	32.6	−146.7
Lauraldehyde	$C_{12}H_{24}O$	77.7	108.4	123.7	140.2	157.8	168.7	184.5	207.8	231.8	257.0	44.5
Lauric acid	$C_{12}H_{24}O_2$	121.0	150.6	166.0	183.6	201.4	212.7	227.5	249.8	273.8	299.2	48
Levulinaldehyde	$C_5H_8O_2$	28.1	54.9	68.0	82.7	98.3	108.4	121.8	142.0	164.0	187.0	
Levulinic acid	$C_5H_8O_3$	102.0	128.1	141.8	154.1	169.5	178.0	190.2	208.3	227.4	245.8	33.5
d-Limonene	$C_{10}H_{16}$	14.0	40.4	53.8	68.2	84.3	94.6	108.3	128.5	151.4	175.0	−96.9
Linalyl acetate	$C_{12}H_{20}O_2$	55.4	82.5	96.0	111.4	127.7	138.1	151.8	173.3	196.2	220.0	
Maleic anhydride	$C_4H_2O_3$	44.0	63.4	78.7	95.0	111.8	122.0	135.8	155.9	179.5	202.0	58
Menthane	$C_{10}H_{20}$	+9.7	35.7	48.3	62.7	78.3	88.6	102.1	122.7	146.0	169.5	
1-Menthol	$C_{10}H_{20}O$	56.0	83.2	96.0	110.3	126.1	136.1	149.4	168.3	190.2	212.0	42.5
Menthyl acetate	$C_{12}H_{22}O_2$	57.4	85.8	100.0	115.4	132.1	143.2	156.7	178.8	202.8	227.0	
benzoate	$C_{17}H_{24}O_2$	123.2	154.2	170.0	186.3	204.3	215.8	230.4	253.2	277.1	301.0	54.5
formate	$C_{11}H_{20}O_2$	47.3	75.8	90.0	105.8	123.0	133.8	148.0	169.8	194.2	219.0	
Mesityl oxide	$C_6H_{10}O$	−8.7	+14.1	26.0	37.9	51.7	60.4	72.1	90.0	109.8	130.0	−59
Methacrylic acid	$C_4H_6O_2$	25.5	48.5	60.0	72.7	86.4	95.3	106.6	123.9	142.5	161.0	15
Methacrylonitrile	C_4H_5N	−44.5	−23.3	−12.5	−0.6	+12.8	21.5	32.8	50.0	70.3	90.3	
Methane	CH_4	−205.9	−199.0	−195.5	−191.8	−187.7	−185.1	−181.4	−175.5	−168.8	−161.5	−182.5
Methanethiol	CH_4S	−90.7	−75.3	−67.5	−58.8	−49.2	−43.1	−34.8	−22.1	−7.9	+6.8	−121
Methoxyacetic acid	$C_3H_6O_3$	52.5	79.3	92.0	106.5	122.0	131.8	144.5	163.5	184.2	204.0	
N-Methylacetanilide	$C_9H_{11}NO$		103.8	118.6	135.1	152.2	164.2	179.8	202.3	227.4	253.0	102
Methyl acetate	$C_3H_6O_2$	−57.2	−38.6	−29.3	−19.1	−7.9	−0.5	+9.4	24.0	40.0	57.8	−98.7
acetylene (propyne)	C_3H_4	−111.0	−97.5	−90.5	−82.9	−74.3	−68.8	−61.3	−49.8	−37.2	−23.3	−102.7
acrylate	$C_4H_6O_2$	−43.7	−23.6	−13.5	−2.7	+9.2	17.3	28.0	43.9	61.8	80.2	
alcohol (methanol)	CH_4O	−44.0	−25.3	−16.2	−6.0	+5.0	12.1	21.2	34.8	49.9	64.7	−97.8
Methylamine	CH_5N	−95.8	−81.3	−73.8	−65.9	−56.9	−51.3	−43.7	−32.4	−19.7	−6.3	−93.5
N-Methylaniline	C_7H_9N	36.0	62.8	76.2	90.5	106.0	115.8	129.8	149.3	172.0	195.5	−57
Methyl anthranilate	$C_8H_9NO_2$	77.6	109.0	124.2	141.5	159.7	172.0	187.8	212.4	238.5	266.5	24
benzoate	$C_8H_8O_2$	39.0	64.4	77.3	91.8	107.8	117.4	130.8	151.4	174.7	199.5	−12.5
2-Methylbenzothiazole	C_8H_7NS	70.0	97.5	111.2	125.5	141.2	150.4	163.9	183.2	204.5	225.5	15.4
α-Methylbenzyl alcohol	$C_8H_{10}O$	49.0	75.2	88.0	102.1	117.8	127.4	140.3	159.0	180.7	204.0	
Methyl bromide	CH_3Br	−96.3	−80.6	−72.8	−64.0	−54.2	−48.0	−39.4	−26.5	−11.9	+3.6	−93
2-Methyl-1-butene	C_5H_{10}	−89.1	−72.8	−64.3	−54.8	−44.1	−37.3	−28.0	−13.8	+2.5	20.2	−135
2-Methyl-2-butene	C_5H_{10}	−75.4	−57.0	−47.9	−37.9	−26.7	−19.4	−9.9	+4.9	21.6	38.5	−133
Methyl isobutyl carbinol (2-methyl-4-pentanol)	$C_6H_{14}O$	−0.3	+22.1	33.3	45.4	58.2	67.0	78.0	94.9	113.5	131.7	
n-butyl ketone (2-hexanone)	$C_6H_{12}O$	+7.7	28.8	38.8	50.0	62.0	69.8	79.8	94.3	111.0	127.5	−56.9
isobutyl ketone (4-methyl-2-pentanone)	$C_6H_{12}O$	−1.4	+19.7	30.0	40.8	52.8	60.4	70.4	85.6	102.0	119.0	−84.7
n-butyrate	$C_5H_{10}O_2$	−26.8	−5.5	+5.0	16.7	29.6	37.4	48.0	64.3	83.1	102.3	
isobutyrate	$C_5H_{10}O_2$	−34.1	−13.0	−2.9	+8.4	21.0	28.9	39.6	55.7	73.6	92.6	−84.7
caprate	$C_{11}H_{22}O_2$	63.7	93.5	108.0	123.0	139.0	148.6	161.5	181.6	202.9	224.0	−18
caproate	$C_7H_{14}O_2$	+5.0	30.0	42.0	55.4	70.0	79.7	91.4	109.8	129.8	150	
caprylate	$C_9H_{18}O_2$	34.2	61.7	74.9	89.0	105.3	115.3	128.0	148.1	170.0	193.0	−40
chloride	CH_3Cl		−99.5	−92.4	−84.8	−76.0	−70.4	−63.0	−51.2	−38.0	−24.0	−97.7
chloroacetate	$C_3H_5ClO_2$	−2.9	19.0	30.0	41.5	54.5	63.0	73.5	90.5	109.5	130.3	−31.9
cinnamate	$C_{10}H_{10}O_2$	77.4	108.1	123.0	140.0	157.9	170.0	185.8	209.6	235.0	263.0	33.4
α-Methylcinnamic acid	$C_{10}H_{10}O_2$	125.7	155.0	169.8	185.2	201.8	212.0	224.8	245.0	266.8	288.0	
Methylcyclohexane	C_7H_{14}	−35.9	−14.0	−3.2	+8.7	22.0	30.5	42.1	59.6	79.6	100.9	−126.4
Methylcyclopentane	C_6H_{12}	−53.7	−33.8	−23.7	−12.8	−0.6	+7.2	17.9	34.0	52.3	71.8	−142.4
Methylcyclopropane	C_4H_8	−96.0	−80.6	−72.8	−64.0	−54.2	−48.0	−39.3	−26.0	−11.3	+4.5	
Methyl *n*-decyl ketone (*n*-dodecan-2-one)	$C_{12}H_{24}O$	77.1	106.0	120.4	136.0	152.4	163.8	177.5	199.0	222.5	246.5	
dichloroacetate	$C_3H_4Cl_2O_2$	3.2	26.7	38.1	50.7	64.7	73.6	85.4	103.2	122.6	143.0	
N-Methyldiphenylamine	$C_{13}H_{13}N$	103.5	134.0	149.7	165.8	184.0	195.4	210.1	232.8	257.0	282.0	−7.6

TABLE 2-10 Vapor Pressures of Organic Compounds, up to 1 atm *(Continued)*

Compound		Pressure, mmHg										Melting point, °C
		1	5	10	20	40	60	100	200	400	760	
Name	Formula	Temperature, °C										
Methyl *n*-dodecyl ketone (2-tetradecanone)	$C_{14}H_{28}O$	99.3	130.0	145.5	161.3	179.8	191.4	206.0	228.2	253.3	278.0	
Methylene bromide (dibromomethane)	CH_2Br_2	−35.1	−13.2	−2.4	+9.7	23.3	31.6	42.3	58.5	79.0	98.6	−52.8
chloride (dichloromethane)	CH_2Cl_2	−70.0	−52.1	−43.3	−33.4	−22.3	−15.7	−6.3	+8.0	24.1	40.7	−96.7
Methyl ethyl ketone (2-butanone)	C_4H_8O	−48.3	−28.0	−17.7	−6.5	+6.0	14.0	25.0	41.6	60.0	79.6	−85.9
2-Methyl-3-ethylpentane	C_8H_{18}	−24.0	−1.8	+9.5	21.7	35.2	43.9	55.7	73.6	94.0	115.6	−114.5
3-Methyl-3-ethylpentane	C_8H_{18}	−23.9	−1.4	+9.9	22.3	36.2	45.0	57.1	75.3	96.2	118.3	−90
Methyl fluoride	CH_3F	−147.3	−137.0	−131.6	−125.9	−119.1	−115.0	−109.0	−99.9	−89.5	−78.2	
formate	$C_2H_4O_2$	−74.2	−57.0	−48.6	−39.2	−28.7	−21.9	−12.9	+0.8	16.0	32.0	−99.8
α-Methylglutaric anhydride	$C_6H_8O_3$	93.8	125.4	141.8	157.7	177.5	189.9	205.0	229.1	255.5	282.5	
Methyl glycolate	$C_3H_6O_3$	+9.6	33.7	45.3	58.1	72.3	81.8	93.7	111.8	131.7	151.5	
2-Methylheptadecane	$C_{18}H_{38}$	119.8	152.0	168.7	186.0	204.8	216.3	231.5	254.5	279.8	306.5	
2-Methylheptane	C_8H_{18}	−21.0	+1.3	12.3	24.4	37.9	46.6	58.3	76.0	96.2	117.6	−109.5
3-Methylheptane	C_8H_{18}	−19.8	+2.6	13.3	25.4	38.9	47.6	59.4	77.1	97.4	118.9	−120.8
4-Methylheptane	C_8H_{18}	−20.4	+1.5	12.4	24.5	38.0	46.6	58.3	76.1	96.3	117.7	−121.1
2-Methyl-2-heptene	C_8H_{16}	−16.1	+6.7	17.8	30.4	44.0	52.8	64.6	82.3	102.2	122.5	
6-Methyl-3-hepten-2-ol	$C_8H_{16}O$	41.6	65.0	76.7	89.3	102.7	111.5	122.6	139.5	156.6	175.5	
6-Methyl-5-hepten-2-ol	$C_8H_{16}O$	41.9	66.0	77.8	90.4	104.0	112.8	123.8	140.0	156.6	174.3	
2-Methylhexane	C_7H_{16}	−40.4	−19.5	−9.1	+2.3	14.9	23.0	34.1	50.8	69.8	90.0	−118.2
3-Methylhexane	C_7H_{16}	−39.0	−18.1	−7.8	+3.6	16.4	24.5	35.6	52.4	71.6	91.9	
Methyl iodide	CH_3I		−55.0	−45.8	−35.6	−24.2	−16.9	−7.0	+8.0	25.3	42.4	−64.4
laurate	$C_{13}H_{26}O_2$	87.8	117.9	133.2	149.0	166.0	176.8	190.8				5
levulinate	$C_6H_{10}O_3$	39.8	66.4	79.7	93.7	109.5	119.3	133.0	153.4	175.8	197.7	
methacrylate	$C_5H_8O_2$	−30.5	−10.0	+1.0	11.0	25.5	34.5	47.0	63.0	82.0	101.0	
myristate	$C_{15}H_{30}O_2$	115.0	145.7	160.8	177.8	195.8	207.5	222.6	245.3	269.8	295.8	18.5
α-naphthyl ketone (1-acetonaphthone)	$C_{12}H_{10}O$	115.6	146.3	161.5	178.4	196.8	208.6	223.8	246.7	270.5	295.5	
β-naphthyl ketone (2-acetonaphthone)	$C_{12}H_{10}O$	120.2	152.3	168.5	185.7	203.8	214.7	229.8	251.6	275.8	301.0	55.5
n-nonyl ketone (undecan-2-one)	$C_{11}H_{22}O$	68.2	95.5	108.9	123.1	139.0	148.6	161.0	181.2	202.3	224.0	15
palmitate	$C_{17}H_{34}O_2$	134.3	166.8	184.3	202.0							30
n-pentadecyl ketone (2-heptdecanone)	$C_{17}H_{34}O$	129.6	161.6	178.0	196.4	214.3	226.7	242.0	265.8	291.7	319.5	
2-Methylpentane	C_6H_{14}	−60.9	−41.7	−32.1	−21.4	−9.7	−1.9	+8.1	24.1	41.6	60.3	−154
3-Methylpentane	C_6H_{14}	−59.0	−39.8	−30.1	−19.4	−7.3	+0.1	10.5	26.5	44.2	63.3	−118
2-Methyl-1-pentanol	$C_6H_{14}O$	15.4	38.0	49.6	61.6	74.7	83.4	94.2	111.3	129.8	147.9	
2-Methyl-2-pentanol	$C_6H_{14}O$	−4.5	+16.8	27.6	38.8	51.3	58.8	69.2	85.0	102.6	121.2	−103
Methyl *n*-pentyl ketone (2-heptanone)	$C_7H_{14}O$	19.3	43.6	55.5	67.7	81.2	89.8	100.0	116.1	133.2	150.2	
phenyl ether (anisole)	C_7H_8O	+5.4	30.0	42.2	55.8	70.7	80.1	93.0	112.3	133.8	155.5	−37.3
2-Methylpropene	C_4H_8	−105.1	−96.5	−81.9	−73.4	−63.8	−57.7	−49.3	−36.7	−22.2	−6.9	−140.3
Methyl propionate	$C_4H_8O_2$	−42.0	−21.5	−11.8	−1.0	+11.0	18.7	29.0	44.2	61.8	79.8	−87.5
4-Methylpropiophenone	$C_{10}H_{12}O$	59.6	89.3	103.8	120.2	138.0	149.3	164.2	187.4	212.7	238.5	
2-Methylpropionyl bromide	C_4H_7BrO	13.5	38.4	50.6	64.1	79.4	88.8	101.6	120.5	141.7	163.0	
Methyl propyl ether	$C_4H_{10}O$	−72.2	−54.3	−45.4	−35.4	−24.3	−17.4	−8.1	+6.0	22.5	39.1	
n-propyl ketone (2-pentanone)	$C_5H_{10}O$	−12.0	+8.0	17.9	28.5	39.8	47.3	56.8	71.0	86.8	103.3	−77.8
isopropyl ketone (3-Methyl-2-butanone)	$C_5H_{10}O$	−19.9	−1.0	+8.3	18.3	29.6	36.2	45.5	59.0	73.8	88.9	−92
2-Methylquinoline	$C_{10}H_9N$	75.3	104.0	119.0	134.0	150.8	161.7	176.2	197.8	211.7	246.5	−1
Methyl salicylate	$C_8H_8O_3$	54.0	81.6	95.3	110.0	126.2	136.7	150.0	172.6	197.5	223.2	−8.3
α-Methyl styrene	C_9H_{10}	7.4	34.0	47.1	61.8	77.8	88.3	102.2	121.8	143.0	165.4	−23.2
4-Methyl styrene	C_9H_{10}	16.0	42.0	55.1	69.2	85.0	95.0	108.6	128.7	151.2	175.0	
Methyl *n*-tetradecyl ketone (2-hexadecanone)	$C_{16}H_{32}O$	109.8	151.5	167.3	184.6	203.7	215.0	230.5	254.4	279.8	307.0	
thiocyanate	C_2H_3NS	−14.0	+9.8	21.6	34.5	49.0	58.1	70.4	89.8	110.8	132.9	−51
isothiocyanate	C_2H_3NS	−34.7	−8.3	+5.4	20.4	38.2	47.5	59.3	77.5	97.8	119.0	35.5
undecyl ketone (2-tridecanone)	$C_{13}H_{26}O$	86.8	117.0	131.8	147.8	165.7	176.6	191.5	214.0	238.3	262.5	28.5
isovalerate	$C_6H_{12}O_2$	−19.2	+2.9	14.0	26.4	39.8	48.2	59.8	77.3	96.7	116.7	
Monovinylacetylene (butenyne)	C_4H_4	−93.2	−77.7	−70.0	−61.3	−51.7	−45.3	−37.1	−24.1	−10.1	+5.3	
Myrcene	$C_{10}H_{16}$	14.5	40.0	53.2	67.0	82.6	92.6	106.0	126.0	148.3	171.5	
Myristaldehyde	$C_{14}H_{28}O$	99.0	132.0	148.3	166.2	186.0	198.3	214.5	240.4	267.9	297.8	23.5
Myristic acid (tetradecanoic acid)	$C_{14}H_{28}O_2$	142.0	174.1	190.8	207.6	223.5	237.2	250.5	272.3	294.6	318.0	57.5
Naphthalene	$C_{10}H_8$	52.6	74.2	85.8	101.7	119.3	130.2	145.5	167.7	193.2	217.9	80.2
1-Naphthoic acid	$C_{11}H_8O_2$	156.0	184.0	196.8	211.2	225.0	234.5	245.8	263.5	281.4	300.0	160.5
2-Naphthoic acid	$C_{11}H_8O_2$	160.8	189.7	202.8	216.9	231.5	241.3	252.7	270.3	289.5	308.5	184
1-Naphthol	$C_{10}H_8O$	94.0	125.5	142.0	158.0	177.8	190.0	206.0	229.6	255.8	282.5	96
2-Naphthol	$C_{10}H_8O$		128.6	145.5	161.8	181.7	193.7	209.8	234.0	260.6	288.0	122.5
1-Naphthylamine	$C_{10}H_9N$	104.3	137.7	153.8	171.6	191.5	203.8	220.0	244.9	272.2	300.8	50
2-Naphthylamine	$C_{10}H_9N$	108.0	141.6	157.6	175.8	195.7	208.1	224.3	249.7	277.4	306.1	111.5
Nicotine	$C_{10}H_{14}N_2$	61.8	91.8	107.2	123.7	142.1	154.7	169.5	193.8	219.8	247.3	
2-Nitroaniline	$C_6H_6N_2O_2$	104.0	135.7	150.4	167.7	186.0	197.8	213.0	236.3	260.0	284.5	71.5
3-Nitroaniline	$C_6H_6N_2O_2$	119.3	151.5	167.8	185.5	204.2	216.5	232.1	255.3	280.2	305.7	114
4-Nitroaniline	$C_6H_6N_2O_2$	142.4	177.6	194.4	213.2	234.2	245.9	261.8	284.5	310.2	336.0	146.5
2-Nitrobenzaldehyde	$C_7H_5NO_3$	85.8	117.7	133.4	150.0	168.8	180.7	196.2	220.0	246.8	273.5	40.9
3-Nitrobenzaldehyde	$C_7H_5NO_3$	96.2	127.4	142.8	159.0	177.7	189.5	204.3	227.4	252.1	278.3	58
Nitrobenzene	$C_6H_5NO_2$	44.4	71.6	84.9	99.3	115.4	125.8	139.9	161.2	185.8	210.6	+5.7
Nitroethane	$C_2H_5NO_2$	−21.0	+1.5	12.5	24.8	38.0	46.5	57.8	74.8	94.0	114.0	−90
Nitroglycerin	$C_3H_5N_3O_9$	127	167	188	210	235	251					11
Nitromethane	CH_3NO_2	−29.0	−7.9	+2.8	14.1	27.5	35.5	46.6	63.5	82.0	101.2	−29
2-Nitrophenol	$C_6H_5NO_3$	49.3	76.8	90.4	105.8	122.1	132.6	146.4	167.6	191.0	214.5	45
2-Nitrophenyl acetate	$C_8H_7NO_4$	100.0	128.0	142.0	155.8	172.8	181.7	194.1	213.0	233.5	253.0	

(Continued)

TABLE 2-10 Vapor Pressures of Organic Compounds, up to 1 atm *(Continued)*

Compound		Pressure, mmHg										Melting point, °C
		1	5	10	20	40	60	100	200	400	760	
Name	Formula	Temperature, °C										
1-Nitropropane	$C_3H_7NO_2$	−9.6	+13.5	25.3	37.9	51.8	60.5	72.3	90.2	110.6	131.6	−108
2-Nitropropane	$C_3H_7NO_2$	−18.8	+4.1	15.8	28.2	41.8	50.3	62.0	80.0	99.8	120.3	−93
2-Nitrotoluene	$C_7H_7NO_2$	50.0	79.1	93.8	109.6	126.3	137.6	151.5	173.7	197.7	222.3	−4.1
3-Nitrotoluene	$C_7H_7NO_2$	50.2	81.0	96.0	112.8	130.7	142.5	156.9	180.3	206.8	231.9	15.5
4-Nitrotoluene	$C_7H_7NO_2$	53.7	85.0	100.5	117.7	136.0	147.9	163.0	186.7	212.5	238.3	51.9
4-Nitro-1,3-xylene (4-nitro-*m*-xylene)	$C_8H_9NO_2$	65.6	95.0	109.8	125.8	143.3	153.8	168.5	191.7	217.5	244.0	+2
Nonacosane	$C_{29}H_{60}$	234.2	269.8	286.4	303.6	323.2	334.8	350.0	373.2	397.2	421.8	63.8
Nonadecane	$C_{19}H_{40}$	133.2	166.3	183.5	200.8	220.0	232.8	248.0	271.8	299.8	330.0	32
n-Nonane	C_9H_{20}	+1.4	25.8	38.0	51.2	66.0	75.5	88.1	107.5	128.2	150.8	−53.7
1-Nonanol	$C_9H_{20}O$	59.5	86.1	99.7	113.8	129.0	139.0	151.3	170.5	192.1	213.5	−5
2-Nonanone	$C_9H_{18}O$	32.1	59.0	72.3	87.2	103.4	113.8	127.4	148.2	171.2	195.0	−19
Octacosane	$C_{28}H_{58}$	226.5	260.3	277.4	295.4	314.2	326.8	341.8	364.8	388.9	412.5	61.6
Octadecane	$C_{18}H_{38}$	119.6	152.1	169.6	187.5	207.4	219.7	236.0	260.6	288.0	317.0	28
n-Octane	C_8H_{18}	−14.0	+8.3	19.2	31.5	45.1	53.8	65.7	83.6	104.0	125.6	−56.8
n-Octanol (1-octanol)	$C_8H_{18}O$	54.0	76.5	88.3	101.0	115.2	123.8	135.2	152.0	173.8	195.2	−15.4
2-Octanone	$C_8H_{16}O$	23.6	48.4	60.9	74.3	89.8	99.0	111.7	130.4	151.0	172.9	−16
n-Octyl acrylate	$C_{11}H_{20}O_2$	58.5	87.7	102.0	117.8	135.6	145.6	159.1	180.2	204.0	227.0	
iodide (1-Iodooctane)	$C_8H_{17}I$	45.8	74.8	90.0	105.9	123.8	135.4	150.0	173.3	199.3	225.5	−45.9
Oleic acid	$C_{18}H_{34}O_2$	176.5	208.5	223.0	240.0	257.2	269.8	286.0	309.8	334.7	360.0	14
Palmitaldehyde	$C_{16}H_{32}O$	121.6	154.6	171.8	190.0	210.0	222.6	239.5	264.1	292.3	321.0	34
Palmitic acid	$C_{16}H_{32}O_2$	153.6	188.1	205.8	223.8	244.4	256.0	271.5	298.7	326.0	353.8	64.0
Palmitonitrile	$C_{16}H_{31}N$	134.3	168.3	185.8	204.2	223.8	236.6	251.5	277.1	304.5	332.0	31
Pelargonic acid	$C_9H_{18}O_2$	108.2	126.0	137.4	149.8	163.7	172.3	184.4	203.1	227.5	253.5	12.5
Pentachlorobenzene	C_6HCl_5	98.6	129.7	144.3	160.0	178.5	190.1	205.5	227.0	251.6	276.0	85.5
Pentachloroethane	C_2HCl_5	+1.0	27.2	39.8	53.9	69.9	80.0	93.5	114.0	137.2	160.5	−22
Pentachloroethylbenzene	$C_8H_5Cl_5$	96.2	130.0	148.0	166.0	186.2	199.0	216.0	241.8	269.3	299.0	
Pentachlorophenol	C_6HCl_5O				192.2	211.2	223.4	239.6	261.8	285.0	309.3	188.5
Pentacosane	$C_{25}H_{52}$	194.2	230.0	248.2	266.1	285.6	298.4	314.0	339.0	365.4	390.3	53.3
Pentadecane	$C_{15}H_{32}$	91.6	121.0	135.4	150.2	167.7	178.4	194.0	216.1	242.8	270.5	10
1,3-Pentadiene	C_5H_8	−71.8	−53.8	−45.0	−34.8	−23.4	−16.5	−6.7	+8.0	24.7	42.1	
1,4-Pentadiene	C_5H_8	−83.5	−66.2	−57.1	−47.7	−37.0	−30.0	−20.6	−6.7	+8.3	26.1	
Pentaethylbenzene	$C_{16}H_{26}$	86.0	120.0	135.8	152.4	171.9	184.2	200.0	224.1	250.2	277.0	
Pentaethylchlorobenzene	$C_{16}H_{25}Cl$	90.0	123.8	140.7	158.1	178.2	191.0	208.0	230.3	257.2	285.0	
n-Pentane	C_5H_{12}	−76.6	−62.5	−50.1	−40.2	−29.2	−22.2	−12.6	+1.9	18.5	36.1	−129.7
iso-Pentane (2-methylbutane)	C_5H_{12}	−82.9	−65.8	−57.0	−47.3	−36.5	−29.6	−20.2	−5.9	+10.5	27.8	−159.7
neo-Pentane (2,2-dimethylpropane)	C_5H_{12}	−102.0	−85.4	−76.7	−67.2	−56.1	−49.0	−39.1	−23.7	−7.1	+9.5	−16.6
2,3,4-Pentanetriol	$C_5H_{12}O_3$	155.0	189.3	204.5	220.5	239.6	249.8	263.5	284.5	307.0	327.2	
1-Pentene	C_5H_{10}	−80.4	−63.3	−54.5	−46.0	−34.1	−27.1	−17.7	−3.4	+12.8	30.1	
α-Phellandrene	$C_{10}H_{16}$	20.0	45.7	58.0	72.1	87.8	97.6	110.6	130.6	152.0	175.0	
Phenanthrene	$C_{14}H_{10}$	118.2	154.3	173.0	193.7	215.8	229.9	249.0	277.1	308.0	340.2	99.5
Phenethyl alcohol (phenyl cellosolve)	$C_8H_{10}O_2$	58.2	85.9	100.0	114.8	130.5	141.2	154.0	175.0	197.5	219.5	
2-Phenetidine	$C_8H_{11}NO$	67.0	94.7	108.6	123.7	139.9	149.8	163.5	184.0	207.0	228.0	
Phenol	C_6H_6O	40.1	62.5	73.8	86.0	100.1	108.4	121.4	139.0	160.0	181.9	40.6
2-Phenoxyethanol	$C_8H_{10}O_2$	78.0	106.6	121.2	136.0	152.2	163.2	176.5	197.6	221.0	245.3	11.6
2-Phenoxyethyl acetate	$C_{10}H_{12}O_3$	82.6	113.5	128.0	144.5	162.3	174.0	189.2	211.3	235.0	259.7	−6.7
Phenyl acetate	$C_8H_8O_2$	38.2	64.8	78.0	92.3	108.1	118.1	131.6	151.2	173.5	195.9	
Phenylacetic acid	$C_8H_8O_2$	97.0	127.0	141.3	156.0	173.6	184.5	198.2	219.5	243.0	265.5	76.5
Phenylacetonitrile	C_8H_7N	60.0	89.0	103.5	119.4	136.3	147.7	161.8	184.2	208.5	233.5	−23.8
Phenylacetyl chloride	C_8H_7ClO	48.0	75.3	89.0	103.6	119.8	129.8	143.5	163.8	186.0	210.0	
Phenyl benzoate	$C_{13}H_{10}O_2$	106.8	141.5	157.8	177.0	197.6	210.8	227.8	254.0	283.5	314.0	70.5
4-Phenyl-3-buten-2-one	$C_{10}H_{10}O$	81.7	112.2	127.4	143.8	161.3	172.6	187.8	211.0	235.4	261.0	41.5
Phenyl isocyanate	C_7H_5NO	10.6	36.0	48.5	62.5	77.7	87.7	100.6	120.8	142.7	165.6	
isocyanide	C_7H_5N	12.0	37.0	49.7	63.4	78.3	88.0	101.0	120.8	142.3	165.0	
Phenylcyclohexane	$C_{12}H_{16}$	67.5	96.5	111.3	126.4	144.0	154.2	169.3	191.3	214.6	240.0	+7.5
Phenyl dichlorophosphate	$C_6H_5Cl_2O_2P$	66.7	95.9	110.0	125.9	143.4	153.6	168.0	189.8	213.0	239.5	
m-Phenylene diamine (1,3-phenylenediamine)	$C_6H_8N_2$	99.8	131.2	147.0	163.8	182.5	194.0	209.9	233.0	259.0	285.5	62.8
Phenylglyoxal	$C_8H_6O_2$		75.0	87.8	100.7	115.5	124.2	136.2	153.8	173.5	193.5	73
Phenylhydrazine	$C_6H_8N_2$	71.8	101.6	115.8	131.5	148.2	158.7	173.5	195.4	218.2	243.5	19.5
N-Phenyliminodiethanol	$C_{10}H_{15}NO_2$	145.0	179.2	195.8	213.4	233.0	245.3	260.6	284.5	311.3	337.8	
1-Phenyl-1,3-pentanedione	$C_{11}H_{12}O_2$	98.0	128.5	144.0	159.9	178.0	189.8	204.5	226.7	251.2	276.5	
2-Phenylphenol	$C_{12}H_{10}O$	100.0	131.6	146.2	163.3	180.3	192.2	205.9	227.9	251.8	275.0	56.5
4-Phenylphenol	$C_{12}H_{10}O$			176.2	193.8	213.0	225.3	240.9	263.2	285.5	308.0	164.5
3-Phenyl-1-propanol	$C_9H_{12}O$	74.7	102.4	116.0	131.2	147.4	156.8	170.3	191.2	212.8	235.0	
Phenyl isothiocyanate	C_7H_5NS	47.2	75.6	89.8	115.5	122.5	133.3	147.7	169.6	194.0	218.5	−21.0
Phorone	$C_9H_{14}O$	42.0	68.3	81.5	95.6	111.3	121.4	134.0	153.5	175.3	197.2	28
iso-Phorone	$C_9H_{14}O$	38.0	66.7	81.2	96.8	114.5	125.6	140.6	163.3	188.7	215.2	
Phosgene (carbonyl chloride)	CCl_2O	−92.9	−77.0	−69.3	−60.3	−50.3	−44.0	−35.6	−22.3	−7.6	+8.3	−104
Phthalic anhydride	$C_8H_4O_3$	96.5	121.3	134.0	151.7	172.0	185.3	202.3	228.0	256.8	284.5	130.8
Phthalide	$C_8H_6O_2$	95.5	127.7	144.0	161.3	181.0	193.5	210.0	234.5	261.8	290.0	73
Phthaloyl chloride	$C_8H_4Cl_2O_2$	86.3	118.3	134.2	151.0	170.0	182.2	197.8	222.0	248.3	275.8	88.5
2-Picoline	C_6H_7N	−11.1	+12.6	24.4	37.4	51.2	59.9	71.4	89.0	108.4	128.8	−70
Pimelic acid	$C_7H_{12}O_4$	163.4	196.2	212.0	229.3	247.0	258.2	272.0	294.5	318.5	342.1	103
α-Pinene	$C_{10}H_{16}$	−1.0	+24.6	37.3	51.4	66.8	76.8	90.1	110.2	132.3	155.0	−55
β-Pinene	$C_{10}H_{16}$	+4.2	30.0	42.3	58.1	71.5	81.2	94.0	114.1	136.1	158.3	

TABLE 2-10 Vapor Pressures of Organic Compounds, up to 1 atm *(Continued)*

Compound		Pressure, mmHg										Melting point, °C
		1	5	10	20	40	60	100	200	400	760	
Name	Formula	Temperature, °C										
Piperidine	$C_5H_{11}N$		−7.0	+3.9	15.8	29.2	37.7	49.0	66.2	85.7	106.0	−9
Piperonal	$C_8H_6O_3$	87.0	117.4	132.0	148.0	165.7	177.0	191.7	214.3	238.5	263.0	37
Propane	C_3H_8	−128.9	−115.4	−108.5	−100.9	−92.4	−87.0	−79.6	−68.4	−55.6	−42.1	−187.1
Propenylbenzene	C_9H_{10}	17.5	43.8	57.0	71.5	87.7	97.8	111.7	132.0	154.7	179.0	−30.1
Propionamide	C_3H_7NO	65.0	91.0	105.0	119.0	134.8	144.3	156.0	174.2	194.0	213.0	79
Propionic acid	$C_3H_6O_2$	4.6	28.0	39.7	52.0	65.8	74.1	85.8	102.5	122.0	141.1	−22
anhydride	$C_6H_{10}O_3$	20.6	45.3	57.7	70.4	85.6	94.5	107.2	127.8	146.0	167.0	−45
Propionitrile	C_3H_5N	−35.0	−13.6	−3.0	+8.8	22.0	30.1	41.4	58.2	77.7	97.1	−91.9
Propiophenone	$C_9H_{10}O$	50.0	77.9	92.2	107.6	124.3	135.0	149.3	170.2	194.2	218.0	21
n-Propyl acetate	$C_5H_{10}O_2$	−26.7	−5.4	+5.0	16.0	28.8	37.0	47.8	64.0	82.0	101.8	−92.5
iso-Propyl acetate	$C_5H_{10}O_2$	−38.3	−17.4	−7.2	+4.2	17.0	25.1	35.7	51.7	69.8	89.0	
n-Propyl alcohol (1-propanol)	C_3H_8O	−15.0	+5.0	14.7	25.3	36.4	43.5	52.8	66.8	82.0	97.8	−127
iso-Propyl alcohol (2-propanol)	C_3H_8O	−26.1	−7.0	+2.4	12.7	23.8	30.5	39.5	53.0	67.8	82.5	−85.8
n-Propylamine	C_3H_9N	−64.4	−46.3	−37.2	−27.1	−16.0	−9.0	+0.5	15.0	31.5	48.5	−83
Propylbenzene	C_9H_{12}	6.3	31.3	43.4	56.8	71.6	81.1	94.0	113.5	135.7	159.2	−99.5
Propyl benzoate	$C_{10}H_{12}O_2$	54.6	83.8	98.0	114.3	131.8	143.3	157.4	180.1	205.2	231.0	−51.6
n-Propyl bromide (1-bromopropane)	C_3H_7Br	−53.0	−33.4	−23.3	−12.4	−0.3	+7.5	18.0	34.0	52.0	71.0	−109.9
iso-Propyl bromide (2-bromopropane)	C_3H_7Br	−61.8	−42.5	−32.8	−22.0	−10.1	−2.5	+8.0	23.8	41.5	60.0	−89.0
n-Propyl n-butyrate	$C_7H_{14}O_2$	−1.6	+22.1	34.0	47.0	61.5	70.3	82.6	101.0	121.7	142.7	−95.2
isobutyrate	$C_7H_{14}O_2$	−6.2	+16.8	28.3	40.6	54.3	63.0	73.9	91.8	112.0	133.9	
iso-Propyl isobutyrate	$C_7H_{14}O_2$	−16.3	+5.8	17.0	29.0	42.4	51.4	62.3	80.2	100.0	120.5	
Propyl carbamate	$C_4H_9NO_2$	52.4	77.6	90.0	103.2	117.7	126.5	138.3	155.8	175.8	195.0	
n-Propyl chloride (1-chloropropane)	C_3H_7Cl	−68.3	−50.0	−41.0	−31.0	−19.5	−12.1	−2.5	+12.2	29.4	46.4	−122.8
iso-Propyl chloride (2-chloropropane)	C_3H_7Cl	−78.8	−61.1	−52.0	−42.0	−31.0	−23.5	−13.7	+1.3	18.1	36.5	−117
iso-Propyl chloroacetate	$C_5H_9ClO_2$	+3.8	28.1	40.2	53.9	68.7	78.0	90.3	108.8	128.0	148.6	
Propyl chloroglyoxylate	$C_5H_7ClO_3$	9.7	32.3	43.5	55.6	68.8	77.2	88.0	104.7	123.0	150.0	
Propylene	C_3H_6	−131.9	−120.7	−112.1	−104.7	−96.5	−91.3	−84.1	−73.3	−60.9	−47.7	−185
Propylene glycol (1,2-Propanediol)	$C_3H_8O_2$	45.5	70.8	83.2	96.4	111.2	119.9	132.0	149.7	168.1	188.2	
Propylene oxide	C_3H_6O	−75.0	−57.8	−49.0	−39.3	−28.4	−21.3	−12.0	+2.1	17.8	34.5	−112.1
n-Propyl formate	$C_4H_8O_2$	−43.0	−22.7	−12.6	−1.7	+10.8	18.8	29.5	45.3	62.6	81.3	−92.9
iso-Propyl formate	$C_4H_8O_2$	−52.0	−32.7	−22.7	−12.1	−0.2	+7.5	17.8	33.6	50.5	68.3	
4,4′-iso-Propylidenebisphenol	$C_{15}H_{16}O_2$	193.0	224.2	240.8	255.5	273.0	282.9	297.0	317.5	339.0	360.5	
n-Propyl iodide (1-iodopropane)	C_3H_7I	−36.0	−13.5	−2.4	+10.0	23.6	32.1	43.8	61.8	81.8	102.5	−98.8
iso-Propyl iodide (2-iodopropane)	C_3H_7I	−43.3	−22.1	−11.7	0.0	+13.2	21.6	32.8	50.0	69.5	89.5	−90
n-Propyl levulinate	$C_8H_{14}O_3$	59.7	86.3	99.9	114.0	130.1	140.6	154.0	175.6	198.0	221.2	
iso-Propyl levulinate	$C_8H_{14}O_3$	48.0	74.5	88.0	102.4	118.1	127.8	141.8	161.6	185.2	208.2	
Propyl mercaptan (1-propanethiol)	C_3H_8S	−56.0	−36.3	−26.3	−15.4	−3.2	+4.6	15.3	31.5	49.2	67.4	−112
2-iso-Propylnaphthalene	$C_{13}H_{14}$	76.0	107.9	123.4	140.3	159.0	171.4	187.6	211.8	238.5	266.0	
iso-Propyl β-naphthyl ketone (2-isobutyronaphthone)	$C_{14}H_{14}O$	133.2	165.4	181.0	197.7	215.6	227.0	242.3	264.0	288.2	313.0	
2-iso-Propylphenol	$C_9H_{12}O$	56.6	83.8	97.0	111.7	127.5	137.7	150.3	170.1	192.6	214.5	15.5
3-iso-Propylphenol	$C_9H_{12}O$	62.0	90.3	104.1	119.8	136.2	146.6	160.2	182.0	205.0	228.0	26
4-iso-Propylphenol	$C_9H_{12}O$	67.0	94.7	108.0	123.4	139.8	149.7	163.3	184.0	206.1	228.2	61
Propyl propionate	$C_6H_{12}O_2$	−14.2	+8.0	19.4	31.6	45.0	53.8	65.2	82.7	102.0	122.4	−76
4-iso-Propylstyrene	$C_{11}H_{14}$	34.7	62.3	76.0	91.2	108.0	118.4	132.8	153.9	178.0	202.5	
Propyl isovalerate	$C_8H_{16}O_2$	+8.0	32.8	45.1	58.0	72.8	82.3	95.0	113.9	135.0	155.9	
Pulegone	$C_{10}H_{16}O$	58.3	82.5	94.0	106.8	121.7	130.2	143.1	162.5	189.8	221.0	
Pyridine	C_5H_5N	−18.9	+2.5	13.2	24.8	38.0	46.8	57.8	75.0	95.6	115.4	−42
Pyrocatechol	$C_6H_6O_2$		104.0	118.3	134.0	150.6	161.7	176.0	197.7	221.5	245.5	105
Pyrocaltechol diacetate (1,2-phenylene diacetate)	$C_{10}H_{10}O_4$	98.0	129.8	145.7	161.8	179.8	191.6	206.5	228.7	253.3	278.0	
Pyrogallol	$C_6H_6O_3$		151.7	167.7	185.3	204.2	216.3	232.0	255.3	281.5	309.0	133
Pyrotartaric anhydride	$C_5H_6O_3$	69.7	99.7	114.2	130.0	147.8	158.6	173.8	196.1	221.0	247.4	
Pyruvic acid	$C_3H_4O_3$	21.4	45.8	57.9	70.8	85.3	94.1	106.5	124.7	144.7	165.0	13.6
Quinoline	C_9H_7N	59.7	89.6	103.8	119.8	136.7	148.1	163.2	186.2	212.3	237.7	−15
iso-Quinoline	C_9H_7N	63.5	92.7	107.8	123.7	141.6	152.0	167.6	190.0	214.5	240.5	24.6
Resorcinol	$C_6H_6O_2$	108.4	138.0	152.1	168.0	185.3	195.8	209.8	230.8	253.4	276.5	110.7
Safrole	$C_{10}H_{10}O_2$	63.8	93.0	107.6	123.0	140.1	150.3	165.1	186.2	210.0	233.0	11.2
Salicylaldehyde	$C_7H_6O_2$	33.0	60.1	73.8	88.7	105.2	115.7	129.4	150.0	173.7	196.5	−7
Salicylic acid	$C_7H_6O_3$	113.7	136.0	146.2	156.8	172.2	182.0	193.4	210.0	230.5	256.0	159
Sebacic acid	$C_{10}H_{18}O_4$	183.0	215.7	232.0	250.0	268.2	279.8	294.5	313.2	332.8	352.3	134.5
Selenophene	C_4H_4Se	−39.0	−16.0	−4.0	+9.1	24.1	33.8	47.0	66.7	89.8	114.3	
Skatole	C_9H_9N	95.0	124.2	139.6	154.3	171.9	183.6	197.4	218.8	242.5	266.2	95
Stearaldehyde	$C_{18}H_{36}O$	140.0	174.6	192.1	210.6	230.8	244.2	260.0	285.0	313.8	342.5	63.5
Stearic acid	$C_{18}H_{36}O_2$	173.7	209.0	225.0	243.4	263.3	275.5	291.0	316.5	343.0	370.0	69.3
Stearyl alcohol (1-octadecanol)	$C_{18}H_{38}O$	150.3	185.6	202.0	220.0	240.4	252.7	269.4	293.5	320.3	349.5	58.5
Styrene	C_8H_8	−7.0	+18.0	30.8	44.6	59.8	69.5	82.0	101.3	122.5	145.2	−30.6
Styrene dibromide [(1,2-dibromoethyl) benzene]	$C_8H_8Br_2$	86.0	115.6	129.8	145.2	161.8	172.2	186.3	207.8	230.0	254.0	
Suberic acid	$C_8H_{14}O_4$	172.8	205.5	219.5	238.2	254.6	265.4	279.8	300.5	322.8	345.5	142
Succinic anhydride	$C_4H_4O_3$	92.0	115.0	128.2	145.3	163.0	174.0	189.0	212.0	237.0	261.0	119.6
Succinimide	$C_4H_5NO_2$	115.0	143.2	157.0	174.0	192.0	203.0	217.4	240.0	263.5	287.5	125.5
Succinyl chloride	$C_4H_4Cl_2O_2$	39.0	65.0	78.0	91.8	107.5	117.2	130.0	149.3	170.0	192.5	17
α-Terpineol	$C_{10}H_{18}O$	52.8	80.4	94.3	109.8	126.0	136.3	150.1	171.2	194.3	217.5	35
Terpenoline	$C_{10}H_{16}$	32.3	58.0	70.6	84.8	100.0	109.8	122.7	142.0	163.5	185.0	

(Continued)

TABLE 2-10 Vapor Pressures of Organic Compounds, up to 1 atm *(Continued)*

Compound		Pressure, mmHg										Melting point, °C
		1	5	10	20	40	60	100	200	400	760	
Name	Formula	Temperature, °C										
1,1,1,2-Tetrabromoethane	$C_2H_2Br_4$	58.0	83.3	95.7	108.5	123.2	132.0	144.0	161.5	181.0	200.0	
1,1,2,2-Tetrabromoethane	$C_2H_2Br_4$	65.0	95.5	110.0	126.0	144.0	155.1	170.0	192.5	217.5	243.5	
Tetraisobutylene	$C_{16}H_{32}$	63.8	93.7	108.5	124.5	142.2	152.6	167.5	190.0	214.6	240.0	
Tetracosane	$C_{24}H_{50}$	183.8	219.6	237.6	255.3	276.3	288.4	305.2	330.5	358.0	386.4	51.1
1,2,3,4-Tetrachlorobenzene	$C_6H_2Cl_4$	68.5	99.6	114.7	131.2	149.2	160.0	175.7	198.0	225.5	254.0	46.5
1,2,3,5-Tetrachlorobenzene	$C_6H_2Cl_4$	58.2	89.0	104.1	121.6	140.0	152.0	168.0	193.7	220.0	246.0	54.5
1,2,4,5-Tetrachlorobenzene	$C_6H_2Cl_4$					146.0	157.7	173.5	196.0	220.5	245.0	139
1,1,2,2-Tetrachloro-1,2-difluoroethane	$C_2Cl_4F_2$	−37.5	−16.0	−5.0	+6.7	19.8	28.1	38.6	55.0	73.1	92.0	26.5
1,1,1,2-Tetrachloroethane	$C_2H_2Cl_4$	−16.3	+7.4	19.3	32.1	46.7	56.0	68.0	87.2	108.2	130.5	−68.7
1,1,2,2-Tetrachloroethane	$C_2H_2Cl_4$	−3.8	+20.7	33.0	46.2	60.8	70.0	83.2	102.2	124.0	145.9	−36
1,2,3,5-Tetrachloro-4-ethylbenzene	$C_8H_6Cl_4$	77.0	110.0	126.0	143.7	162.1	175.0	191.6	215.3	243.0	270.0	
Tetrachloroethylene	C_2Cl_4	−20.6	+2.4	13.8	26.3	40.1	49.2	61.3	79.8	100.0	120.8	−19.0
2,3,4,6-Tetrachlorophenol	$C_6H_2Cl_4O$	100.0	130.3	145.3	161.0	179.1	190.0	205.2	227.2	250.4	275.0	69.5
3,4,5,6-Tetrachloro-1,2-xylene	$C_8H_6Cl_4$	94.4	125.0	140.3	156.0	174.2	185.8	200.5	223.0	248.3	273.5	
Tetradecane	$C_{14}H_{30}$	76.4	106.0	120.7	135.6	152.7	164.0	178.5	201.8	226.8	252.5	5.5
Tetradecylamine	$C_{14}H_{31}N$	102.6	135.8	152.0	170.0	189.0	200.2	215.7	239.8	264.6	291.2	
Tetradecyltrimethylsilane	$C_{17}H_{38}Si$	120.0	150.7	166.2	183.5	201.5	213.3	227.8	250.0	275.0	300.0	
Tetraethoxysilane	$C_8H_{20}O_4Si$	16.0	40.3	52.6	65.8	81.1	90.7	103.6	123.5	146.2	168.5	
1,2,3,4-Tetraethylbenzene	$C_{14}H_{22}$	65.7	96.2	111.6	127.7	145.8	156.7	172.4	196.0	221.4	248.0	11.6
Tetraethylene glycol	$C_8H_{18}O_5$	153.9	183.7	197.1	212.3	228.0	237.8	250.0	268.4	288.0	307.8	
Tetraethylene glycol chlorohydrin	$C_8H_{17}ClO_4$	110.1	141.8	156.1	172.6	190.0	200.5	214.7	236.5	258.2	281.5	
Tetraethyllead	$C_8H_{20}Pb$	38.4	63.6	74.8	88.0	102.4	111.7	123.8	142.0	161.8	183.0	−136
Tetraethylsilane	$C_8H_{20}Si$	−1.0	+23.9	36.3	50.0	65.3	74.8	88.0	108.0	130.2	153.0	
Tetralin	$C_{10}H_{12}$	38.0	65.3	79.0	93.8	110.4	121.3	135.3	157.2	181.8	207.2	−31.0
1,2,3,4-Tetramethylbenzene	$C_{10}H_{14}$	42.6	68.7	81.8	95.8	111.5	121.8	135.7	155.7	180.0	204.4	−6.2
1,2,3,5-Tetramethylbenzene	$C_{10}H_{14}$	40.6	65.8	77.8	91.0	105.8	115.4	128.3	149.9	173.7	197.9	−24.0
1,2,4,5-Tetramethylbenzene	$C_{10}H_{14}$	45.0	65.0	74.6	88.0	104.2	114.8	128.1	149.5	172.1	195.9	79.5
2,2,3,3-Tetramethylbutane	C_8H_{18}	−17.4	+3.2	13.5	24.6	36.8	44.5	54.8	70.2	87.4	106.3	−102.2
Tetramethylene dibromide (1,4-dibromobutane)	$C_4H_8Br_2$	32.0	58.8	72.4	87.6	104.0	115.1	128.7	149.8	173.8	197.5	−20
Tetramethyllead	$C_4H_{12}Pb$	−29.0	−6.8	+4.4	16.6	30.3	39.2	50.8	68.8	89.0	110.0	−27.5
Tetramethyltin	$C_4H_{12}Sn$	−51.3	−31.0	−20.6	−9.3	+3.5	11.7	22.8	39.8	58.5	78.0	
Tetrapropylene glycol monoisopropyl ether	$C_{15}H_{32}O_5$	116.6	147.8	163.0	179.8	197.7	209.0	223.3	245.0	268.3	292.7	
Thioacetic acid (mercaptoacetic acid)	$C_2H_4O_2S$	60.0	87.7	101.5	115.8	131.8	142.0	154.0				−16.5
Thiodiglycol (2,2′-thiodiethanol)	$C_4H_{10}O_2S$	42.0	96.0	128.0	165.0	210.0	240.5	285				
Thiophene	C_4H_4S	−40.7	−20.8	−10.9	0.0	+12.5	20.1	30.5	46.5	64.7	84.4	−38.3
Thiophenol (benzenethiol)	C_6H_6S	18.6	43.7	56.0	69.7	84.2	93.9	106.6	125.8	146.7	168.0	
α-Thujone	$C_{10}H_{16}O$	38.3	65.7	79.3	93.7	110.0	120.2	134.0	154.2	177.8	201.0	
Thymol	$C_{10}H_{14}O$	64.3	92.8	107.4	122.6	139.8	149.8	164.1	185.5	209.2	231.8	51.5
Tiglaldehyde	C_5H_8O	−25.0	−1.6	+10.0	23.2	37.0	45.8	57.7	75.4	95.5	116.4	
Tiglic acid	$C_5H_8O_2$	52.0	77.8	90.2	103.8	119.0	127.8	140.5	158.0	179.2	198.5	64.5
Tiglonitrile	C_5H_7N	−25.5	−2.4	+9.2	22.1	36.7	46.0	58.2	77.8	99.7	122.0	
Toluene	C_7H_8	−26.7	−4.4	+6.4	18.4	31.8	40.3	51.9	69.5	89.5	110.6	−95.0
Toluene-2,4-diamine	$C_7H_{10}N_2$	106.5	137.2	151.7	167.9	185.7	196.2	211.5	232.8	256.0	280.0	99
2-Toluic nitrile (2-tolunitrile)	C_8H_7N	36.7	64.0	77.9	93.0	110.0	120.8	135.0	156.0	180.0	205.2	−13
4-Toluic nitrile (4-tolunitrile)	C_8H_7N	42.5	71.3	85.8	101.7	109.5	130.0	145.2	167.3	193.0	217.6	29.5
2-Toluidine	C_7H_9N	44.0	69.3	81.4	95.1	110.0	119.8	133.0	153.0	176.2	199.7	−16.3
3-Toluidine	C_7H_9N	41.0	68.0	82.0	96.7	113.5	123.8	136.7	157.6	180.6	203.3	−31.5
4-Toluidine	C_7H_9N	42.0	68.2	81.8	95.8	111.5	121.5	133.7	154.0	176.9	200.4	44.5
2-Tolyl isocyanide	C_8H_7N	25.2	51.0	64.0	78.2	94.0	104.0	117.7	137.8	159.9	183.5	
4-Tolylhydrazine	$C_7H_{10}N_2$	82.4	110.0	123.8	138.6	154.1	165.0	178.0	198.0	219.5	242.0	65.5
Tribromoacetaldehyde	C_2HBr_3O	18.5	45.0	58.0	72.1	87.8	97.5	110.2	130.0	151.6	174.0	
1,1,2-Tribromobutane	$C_4H_7Br_3$	45.0	73.5	87.8	103.2	120.2	131.6	146.0	167.8	192.0	216.2	
1,2,2-Tribromobutane	$C_4H_7Br_3$	41.0	69.0	83.2	98.6	116.0	127.0	141.8	163.5	188.0	213.8	
2,2,3-Tribromobutane	$C_4H_7Br_3$	38.2	66.0	79.8	94.6	111.8	122.2	136.3	157.8	182.2	206.5	
1,1,2-Tribromoethane	$C_2H_3Br_3$	32.6	58.0	70.6	84.2	100.0	110.0	123.5	143.5	165.4	188.4	−26
1,2,3-Tribromopropane	$C_3H_5Br_3$	47.5	75.8	90.0	105.8	122.8	134.0	148.0	170.0	195.0	220.0	16.5
Triisobutylamine	$C_{12}H_{27}N$	32.3	57.4	69.8	83.0	97.8	107.3	119.7	138.0	157.8	179.0	−22
Triisobutylene	$C_{12}H_{24}$	18.0	44.0	56.5	70.0	86.7	96.7	110.0	130.2	153.0	179.0	
2,4,6-Tritertbutylphenol	$C_{18}H_{30}O$	95.2	126.1	142.0	158.0	177.4	188.0	203.0	226.2	250.6	276.3	
Trichloroacetic acid	$C_2HCl_3O_2$	51.0	76.0	88.2	101.8	116.3	125.9	137.8	155.4	175.2	195.6	57
Trichloroacetic anhydride	$C_4Cl_6O_3$	56.2	85.3	99.6	114.3	131.2	141.8	155.2	176.2	199.8	223.0	
Trichloroacetyl bromide	C_2BrCl_3O	−7.4	+16.7	29.3	42.1	57.2	66.7	79.5	98.4	120.2	143.0	
2,4,6-Trichloroaniline	$C_6H_4Cl_3N$	134.0	157.8	170.0	182.6	195.8	204.5	214.6	229.8	246.4	262.0	78
1,2,3-Trichlorobenzene	$C_6H_3Cl_3$	40.0	70.0	85.6	101.8	119.8	131.5	146.0	168.2	193.5	218.5	52.5
1,2,4-Trichlorobenzene	$C_6H_3Cl_3$	38.4	67.3	81.7	97.2	114.8	125.7	140.0	162.0	187.7	213.0	17
1,3,5-Trichlorobenzene	$C_6H_3Cl_3$		63.8	78.0	93.7	110.8	121.8	136.0	157.7	183.0	208.4	63.5
1,2,3-Trichlorobutane	$C_4H_7Cl_3$	+0.5	27.2	40.0	55.0	71.5	82.0	96.2	118.0	143.0	169.0	
1,1,1-Trichloroethane	$C_2H_3Cl_3$	−52.0	−32.0	−21.9	−10.8	+1.6	9.5	20.0	36.2	54.6	74.1	−30.6
1,1,2-Trichloroethane	$C_2H_3Cl_3$	−24.0	−2.0	+8.3	21.6	35.2	44.0	55.7	73.3	93.0	113.9	−36.7
Trichloroethylene	C_2HCl_3	−43.8	−22.8	−12.4	−1.0	+11.9	20.0	31.4	48.0	67.0	86.7	−73
Trichlorofluoromethane	CCl_3F	−84.3	−67.6	−59.0	−49.7	−39.0	−32.3	−23.0	−9.1	+6.8	23.7	
2,4,5-Trichlorophenol	$C_6H_3Cl_3O$	72.0	102.1	117.3	134.0	151.5	162.5	178.0	201.5	226.5	251.8	62
2,4,6-Trichlorophenol	$C_6H_3Cl_3O$	76.5	105.9	120.2	135.8	152.2	163.5	177.8	199.0	222.5	246.0	68.5

TABLE 2-10 Vapor Pressures of Organic Compounds, up to 1 atm *(Continued)*

Compound		Pressure, mmHg										Melting point, °C
		1	5	10	20	40	60	100	200	400	760	
Name	Formula	Temperature, °C										
Tri-2-chlorophenylthiophosphate	$C_{18}H_{12}Cl_3O_3PS$	188.2	217.2	231.2	246.7	261.7	271.5	283.8	302.8	322.0	341.3	
1,1,1-Trichloropropane	$C_3H_5Cl_3$	−28.8	−7.0	+4.2	16.2	29.9	38.3	50.0	67.7	87.5	108.2	−77.7
1,2,3-Trichloropropane	$C_3H_5Cl_3$	+9.0	33.7	46.0	59.3	74.0	83.6	96.1	115.6	137.0	158.0	−14.7
1,1,2-Trichloro-1,2,2-trifluoroethane	$C_2Cl_3F_3$	−68.0	−49.4	−40.3	−30.0	−18.5	−11.2	−1.7	+13.5	30.2	47.6	−35
Tricosane	$C_{23}H_{48}$	170.0	206.3	223.0	242.0	261.3	273.8	289.8	313.5	339.8	366.5	47.7
Tridecane	$C_{13}H_{28}$	59.4	98.3	104.0	120.2	137.7	148.2	162.5	185.0	209.4	234.0	−6.2
Tridecanoic acid	$C_{13}H_{26}O_2$	137.8	166.3	181.0	195.8	212.4	222.0	236.0	255.2	276.5	299.0	41
Triethoxymethylsilane	$C_7H_{18}O_3Si$	−1.5	+22.8	34.6	47.2	61.7	70.4	82.7	101.0	121.8	143.5	
Triethoxyphenylsilane	$C_{12}H_{20}O_3Si$	71.0	98.8	112.6	127.2	143.5	153.2	167.5	188.0	210.5	233.5	
1,2,4-Triethylbenzene	$C_{12}H_{18}$	46.0	74.2	88.5	104.0	121.7	132.2	146.8	168.3	193.7	218.0	
1,3,4-Triethylbenzene	$C_{12}H_{18}$	47.9	76.0	90.2	105.8	122.6	133.4	147.7	168.3	193.2	217.5	
Triethylborine	$C_6H_{15}B$			−148.0	−140.6	−131.4	−125.2	−116.0	−101.0	−81.0	−56.2	
Triethyl camphoronate	$C_{15}H_{26}O_6$		150.2	166.0	183.6	201.8	213.5	228.6	250.8	276.0	301.0	135
citrate	$C_{12}H_{20}O_7$	107.0	138.7	144.0	171.1	190.4	202.5	217.8	242.2	267.5	294.0	
Triethyleneglycol	$C_6H_{14}O_4$	114.0	144.0	158.1	174.0	191.3	201.5	214.6	235.2	256.6	278.3	
Triethylheptylsilane	$C_{13}H_{30}Si$	70.0	99.8	114.6	130.3	148.0	158.2	174.0	196.0	221.0	247.0	
Triethyloctylsilane	$C_{14}H_{32}Si$	73.7	104.8	120.6	137.7	155.7	168.0	184.3	208.0	235.0	262.0	
Triethyl orthoformate	$C_7H_{16}O_3$	+5.5	29.2	40.5	53.4	67.5	76.0	88.0	106.0	125.7	146.0	
phosphate	$C_6H_{15}O_4P$	39.6	67.8	82.1	97.8	115.7	126.3	141.6	163.7	187.0	211.0	
Triethylthallium	$C_6H_{15}Tl$	+9.3	37.6	51.7	67.7	85.4	95.7	112.1	136.0	163.5	192.1	−63.0
Trifluorophenylsilane	$C_6H_5F_3Si$	−31.0	−9.7	+0.8	12.3	25.4	33.2	44.2	60.1	78.7	98.3	
Trimethallyl phosphate	$C_{12}H_{21}PO_4$	93.7	131.0	149.8	169.8	192.0	207.0	225.7	255.0	288.5	324.0	
2,3,5-Trimethylacetophenone	$C_{11}H_{14}O$	79.0	108.0	122.3	137.5	154.2	165.7	179.7	201.3	224.3	247.5	
Trimethylamine	C_3H_9N	−97.1	−81.7	−73.8	−65.0	−55.2	−48.8	−40.3	−27.0	−12.5	+2.9	−117.1
2,4,5-Trimethylaniline	$C_9H_{13}N$	68.4	95.9	109.0	123.7	139.8	149.5	162.0	182.3	203.7	234.5	67
1,2,3-Trimethylbenzene	C_9H_{12}	16.8	42.9	55.9	69.9	85.4	95.3	108.8	129.0	152.0	176.1	−25.5
1,2,4-Trimethylbenzene	C_9H_{12}	13.6	38.3	50.7	64.5	79.8	89.5	102.8	122.7	145.4	169.2	−44.1
1,3,5-Trimethylbenzene	C_9H_{12}	9.6	34.7	47.4	61.0	76.1	85.8	98.9	118.6	141.0	164.7	−44.8
2,2,3-Trimethylbutane	C_7H_{16}			−18.8	−7.5	+5.2	13.3	24.4	41.2	60.4	80.9	−25.0
Trimethyl citrate	$C_9H_{14}O_7$	106.2	146.2	160.4	177.2	194.2	205.5	219.6	241.3	264.2	287.0	78.5
Trimethyleneglycol (1,3-propanediol)	$C_3H_8O_2$	59.4	87.2	100.6	115.5	131.0	141.1	153.4	172.8	193.8	214.2	
1,2,4-Trimethyl-5-ethylbenzene	$C_{11}H_{16}$	43.7	71.2	84.6	99.7	106.0	126.3	140.3	160.3	184.5	208.1	
1,3,5-Trimethyl-2-ethylbenzene	$C_{11}H_{16}$	38.8	67.0	80.5	96.0	113.2	123.8	137.9	158.4	183.5	208.0	
2,2,3-Trimethylpentane	C_8H_{18}	−29.0	−7.1	+3.9	16.0	29.5	38.1	49.9	67.8	88.2	109.8	−112.3
2,2,4-Trimethylpentane	C_8H_{18}	−36.5	−15.0	−4.3	+7.5	20.7	29.1	40.7	58.1	78.0	99.2	−107.3
2,3,3-Trimethylpentane	C_8H_{18}	−25.8	−3.9	+6.9	19.2	33.0	41.8	53.8	72.0	92.7	114.8	−101.5
2,3,4-Trimethylpentane	C_8H_{18}	−26.3	−4.1	+7.1	19.3	32.9	41.6	53.4	71.3	91.8	113.5	−109.2
2,2,4-Trimethyl-3-pentanone	$C_8H_{16}O$	14.7	36.0	46.4	57.6	69.8	77.3	87.6	102.2	118.4	135.0	
Trimethyl phosphate	$C_3H_9O_4P$	26.0	53.7	67.8	83.0	100.0	110.0	124.0	145.0	167.8	192.7	
2,4,5-Trimethylstyrene	$C_{11}H_{14}$	48.1	77.0	91.6	107.1	124.2	135.5	149.8	171.8	196.1	221.2	
2,4,6-Trimethylstyrene	$C_{11}H_{14}$	37.5	65.7	79.7	94.8	111.8	122.3	136.8	157.8	182.3	207.0	
Trimethylsuccinic anhydride	$C_7H_{10}O_3$	53.5	82.6	97.4	113.8	131.0	142.2	156.5	179.8	205.5	231.0	
Triphenylmethane	$C_{19}H_{16}$	169.7	188.4	197.0	206.8	215.5	221.2	228.4	239.7	249.8	259.2	93.4
Triphenylphosphate	$C_{18}H_{15}O_4P$	193.5	230.4	249.8	269.7	290.3	305.2	322.5	349.8	379.2	413.5	49.4
Tripropyleneglycol	$C_9H_{20}O_4$	96.0	125.7	140.5	155.8	173.7	184.6	199.0	220.2	244.3	267.2	
Tripropyleneglycol monobutyl ether	$C_{13}H_{28}O_4$	101.5	131.6	147.0	161.8	179.8	190.2	204.4	224.4	247.0	269.5	
Tripropyleneglycol monoisopropyl ether	$C_{12}H_{26}O_4$	82.4	112.4	127.3	143.7	161.4	173.2	187.8	209.7	232.8	256.6	
Tritolyl phosphate	$C_{21}H_{21}O_4P$	154.6	184.2	198.0	213.2	229.7	239.8	252.2	271.8	292.7	313.0	
Undecane	$C_{11}H_{24}$	32.7	59.7	73.9	85.6	104.4	115.2	128.1	149.3	171.9	195.8	−25.6
Undecanoic acid	$C_{11}H_{22}O_2$	101.4	133.1	149.0	166.0	185.6	197.2	212.5	237.8	262.8	290.0	29.5
10-Undecenoic acid	$C_{11}H_{20}O_2$	114.0	142.8	156.3	172.0	188.7	199.5	213.5	232.8	254.0	275.0	24.5
Undecan-2-ol	$C_{11}H_{24}O$	71.1	99.0	112.8	127.5	143.7	153.7	167.2	187.7	209.8	232.0	
n-Valeric acid	$C_5H_{10}O_2$	42.2	67.7	79.8	93.1	107.8	116.6	128.3	146.0	165.0	184.4	−34.5
iso-Valeric acid	$C_5H_{10}O_2$	34.5	59.6	71.3	84.0	98.0	107.3	118.9	136.2	155.2	175.1	−37.6
γ-Valerolactone	$C_5H_8O_2$	37.5	65.8	79.8	95.2	101.9	122.4	136.5	157.7	182.3	207.5	
Valeronitrile	C_5H_9N	−6.0	+18.1	30.0	43.3	57.8	66.9	78.6	97.7	118.7	140.8	
Vanillin	$C_8H_8O_3$	107.0	138.4	154.0	170.5	188.7	199.8	214.5	237.3	260.0	285.0	81.5
Vinyl acetate	$C_4H_6O_2$	−48.0	−28.0	−18.0	−7.0	+5.3	13.0	23.3	38.4	55.5	72.5	
2-Vinylanisole	$C_9H_{10}O$	41.9	68.0	81.0	94.7	110.0	119.8	132.3	151.0	172.1	194.0	
3-Vinylanisole	$C_9H_{10}O$	43.4	69.9	83.0	97.2	112.5	122.3	135.3	154.0	175.8	197.5	
4-Vinylanisole	$C_9H_{10}O$	45.2	72.0	85.7	100.0	116.0	126.1	139.7	159.0	182.0	204.5	
Vinyl chloride (1-chloroethylene)	C_2H_3Cl	−105.6	−90.8	−83.7	−75.7	−66.8	−61.1	−53.2	−41.3	−28.0	−13.8	−153.7
cyanide (acrylonitrile)	C_3H_3N	−51.0	−30.7	−20.3	−9.0	+3.8	11.8	22.8	38.7	58.3	78.5	−82
fluoride (1-fluoroethylene)	C_2H_3F	−149.3	−138.0	−132.2	−125.4	−118.0	−113.0	−106.2	−95.4	−84.0	−72.2	−160.5
Vinylidene chloride (1,1-dichloroethene)	$C_2H_2Cl_2$	−77.2	−60.0	−51.2	−41.7	−31.1	−24.0	−15.0	−1.0	+14.8	31.7	−122.5
4-Vinylphenetole	$C_{10}H_{12}O$	64.0	91.7	105.6	120.3	136.3	146.4	159.8	180.0	202.8	225.0	
2-Xenyl dichlorophosphate	$C_{12}H_9Cl_2PO$	138.2	171.1	187.0	205.0	223.8	236.0	251.5	275.3	301.5	328.5	
2,4-Xyaldehyde	$C_9H_{10}O$	59.0	85.9	99.0	114.0	129.7	139.8	152.2	172.3	194.1	215.5	75
2-Xylene (2-xylene)	C_8H_{10}	−3.8	+20.2	32.1	45.1	59.5	68.8	81.3	100.2	121.7	144.4	−25.2
3-Xylene (3-xylene)	C_8H_{10}	−6.9	+16.8	28.3	41.1	55.3	64.4	76.8	95.5	116.7	139.1	−47.9
4-Xylene (4-xylene)	C_8H_{10}	−8.1	+15.5	27.3	40.1	54.4	63.5	75.9	94.6	115.9	138.3	+13.3
2,4-Xylidine	$C_8H_{11}N$	52.6	79.8	93.0	107.6	123.8	133.7	146.8	166.4	188.3	211.5	
2,6-Xylidine	$C_8H_{11}N$	44.0	72.6	87.0	102.7	120.2	131.5	146.0	168.0	193.7	217.9	

VAPOR PRESSURES OF SOLUTIONS

TABLE 2-11 Partial Pressures of Water over Aqueous Solutions of HCl*

$\log_{10} p$mm $= A - B/T$, (T in K), which, however, agrees only approximately with the table. The table is more nearly correct.

			Partial pressure of H_2O, mmHg, °C																
% HCl	*A*	*B*	0°	5°	10°	15°	20°	25°	30°	35°	40°	45°	50°	60°	70°	80°	90°	100°	110°
6	8.99156	2282	4.18	6.04	8.45	11.7	15.9	21.8	29.1	39.4	50.6	66.2	86.0	139	220	333	492	715	
10	8.99864	2295	3.84	5.52	7.70	10.7	14.6	20.0	26.8	35.5	47.0	61.5	80.0	130	204	310	463	677	960
14	8.97075	2300	3.39	4.91	6.95	9.65	13.1	18.0	24.1	31.9	42.1	55.3	72.0	116	185	273	425	625	892
18	8.98014	2323	2.87	4.21	5.92	8.26	11.3	15.4	20.6	27.5	36.4	47.9	62.5	102	162	248	374	550	783
20	8.97877	2334	2.62	3.83	5.40	7.50	10.3	14.1	19.0	25.1	33.3	43.6	57.0	93.5	150	230	345	510	729
22	9.02708	2363	2.33	3.40	4.82	6.75	9.30	12.6	17.1	22.8	30.2	39.8	52.0	85.6	138	211	317	467	670
24	8.96022	2356	2.05	3.04	4.31	6.03	8.30	11.4	15.4	20.4	27.1	35.7	46.7	77.0	124	194	290	426	611
26	9.01511	2390	1.76	2.60	3.71	5.21	7.21	9.95	13.5	18.0	24.0	31.7	41.5	69.0	112	173	261	387	555
28	8.97611	2395	1.50	2.24	3.21	4.54	6.32	8.75	11.8	15.8	21.1	27.9	36.5	60.7	99.0	154	234	349	499
30	9.00117	2422	1.26	1.90	2.73	3.88	5.41	7.52	10.2	13.7	18.4	24.3	32.0	53.5	87.5	136	207	310	444
32	9.03317	2453	1.04	1.57	2.27	3.25	4.55	6.37	8.70	11.7	15.7	21.0	27.7	46.5	76.5	120	184	275	396
34	9.07143	2487	0.85	1.29	1.87	2.70	3.81	5.35	7.32	9.95	13.5	18.1	24.0	40.5	66.5	104	161	243	355
36	9.11815	2526	0.68	1.03	1.50	2.19	3.10	4.41	6.08	8.33	11.4	15.4	20.4	34.8	57.0	90.0	140	212	311
38	9.20783	2579	0.53	0.81	1.20	1.75	2.51	3.60	5.03	6.92	9.52	13.0	17.4	29.6	49.1	77.5	120	182	266
40	9.33923	2647	0.41	0.63	0.94	1.37	2.00	2.88	4.09	5.68	7.85	10.7	14.5	25.0	42.1	67.3	105	158	230
42	9.44953	2709	0.31	0.48	0.72	1.06	1.56	2.30	3.28	4.60	6.45	8.90	12.1	21.2	35.8	57.2	89.2	135	195

*Uncertainty, ca. 2 percent for solutions of 15 to 30 percent HCl between 0 and 100°; for solutions of > 30 percent HCl the accuracy is ca. 5 percent at the lower temperatures and ca. 15 percent at the higher temperatures. Below 15 percent HCl, the uncertainty is ca. 5 percent at the lower temperatures and higher strengths to ca. 15 to 20 percent at the lower strengths and perhaps 15 to 20 percent at the higher temperatures and lower strengths.

International Critical Tables, vol. 3, p. 301.

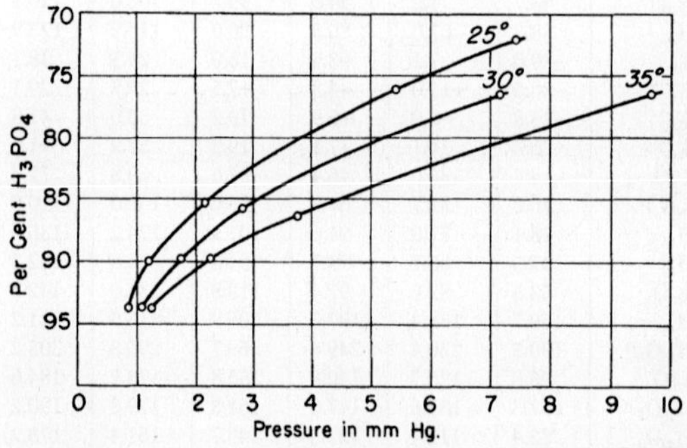

FIG. 2-1 Vapor pressures of H_3PO_4 aqueous: partial pressure of H_2O vapor. (*Courtesy of Victor Chemical Works, Stauffer Chemical Company; measurements by W. H. Woodstock.*)

TABLE 2-12 Water Partial Pressure, Bar, over Aqueous Sulfuric Acid Solutions*

	Weight percent, H_2SO_4									
°C	10.0	20.0	30.0	40.0	50.0	60.0	70.0	75.0	80.0	85.0
0	.582E–02	.534E–02	.448E–02	.326E–02	.193E–02	.836E–03	.207E–03	.747E–04	.197E–04	.343E–05
10	.117E–01	.107E–01	.909E–02	.670E–02	.405E–02	.180E–02	.467E–03	.175E–03	.490E–04	.952E–05
20	.223E–01	.205E–01	.174E–01	.130E–01	.802E–02	.367E–02	.995E–03	.388E–03	.115E–04	.245E–04
30	.404E–01	.373E–01	.319E–01	.241E–01	.151E–01	.710E–02	.201E–02	.811E–03	.253E–03	.589E–04
40	.703E–01	.649E–01	.558E–01	.427E–01	.272E–01	.131E–01	.387E–02	.162E–02	.531E–03	.133E–03
50	.117	.109	.939E–01	.725E–01	.470E–01	.232E–01	.715E–02	.309E–02	.106E–02	.286E–03
60	.189	.175	.152	.119	.782E–01	.395E–01	.127E–01	.565E–02	.204E–02	.584E–03
70	.296	.275	.239	.188	.126	.651E–01	.217E–01	.997E–02	.376E–02	.114E–02
80	.449	.417	.365	.290	.196	.104	.360E–01	.170E–01	.668E–02	.213E–02
90	.664	.617	.542	.434	.298	.161	.578E–01	.281E–01	.115E–01	.383E–02
100	.957	.891	.786	.634	.441	.244	.905E–01	.452E–01	.192E–01	.666E–02
110	1.349	1.258	1.113	.904	.638	.360	.138	.708E–01	.312E–01	.112E–01
120	1.863	1.740	1.544	1.264	.903	.519	.206	.108	.493E–01	.183E–01
130	2.524	2.361	2.101	1.732	1.253	.734	.301	.162	.760E–01	.291E–01
140	3.361	3.149	2.810	2.333	1.708	1.020	.481	.236	.115	.451E–01
150	4.404	4.132	3.697	3.090	2.289	1.392	.605	.339	.170	.682E–01
160	5.685	5.342	4.793	4.031	3.021	1.870	.837	.478	.246	.101
170	7.236	6.810	6.127	5.185	3.930	2.475	1.138	.662	.350	.147
180	9.093	8.571	7.731	6.584	5.045	3.233	1.525	.902	.489	.208
190	11.289	10.658	9.640	8.259	6.397	4.169	2.017	1.212	.673	.291
200	13.861	13.107	11.887	10.245	8.020	5.312	2.632	1.606	.913	.401
210	16.841	15.951	14.505	12.576	9.948	6.696	3.395	2.101	1.220	.542
220	20.264	19.225	17.529	15.287	12.217	8.354	4.331	2.714	1.609	.724
230	24.160	22.960	20.992	18.414	14.864	10.322	5.466	3.467	2.096	.952
240	28.561	27.188	24.927	21.992	17.929	12.641	6.831	4.381	2.699	1.237
250	33.494	31.939	29.364	26.056	21.452	15.351	8.458	5.480	3.435	1.587
260	38.984	37.240	34.334	30.642	25.472	18.496	10.382	6.788	4.326	2.012
270	45.055	43.116	39.865	35.784	30.030	22.121	12.640	8.333	5.395	2.525
280	51.726	49.590	45.984	41.514	35.168	26.274	15.269	10.142	6.663	3.136
290	59.015	56.681	52.715	47.865	40.926	31.003	18.311	12.242	8.155	3.857
300	66.934	64.407	60.081	54.868	47.346	36.360	21.808	14.665	9.897	4.701
310	75.495	72.781	68.100	62.553	54.470	42.395	25.804	17.438	11.912	5.680
320	84.705	81.816	76.792	70.947	62.337	49.164	30.343	20.591	14.227	6.806
330	94.567	91.518	86.172	80.077	70.988	56.721	35.473	24.153	16.867	8.093
340	105.083	101.894	96.252	89.969	80.463	65.123	41.240	28.154	19.855	9.551
350	116.251	112.946	107.043	100.646	90.802	74.426	47.692	32.622	23.217	11.193

	Weight percent, H_2SO_4									
°C	90.0	92.0	94.0	96.0	97.0	98.0	98.5	99.0	99.5	100.0
0	.518E–06	.242E–06	.107E–06	.401E–07	.218E–07	.980E–08	.569E–08	.268E–08	.775E–09	.196E–09
10	.159E–05	.762E–06	.344E–06	.130E–06	.713E–07	.323E–07	.188E–07	.888E–08	.258E–08	.655E–09
20	.448E–05	.220E–05	.101E–05	.390E–06	.215E–06	.978E–07	.572E–07	.271E–07	.789E–08	.201E–08
30	.117E–04	.587E–05	.275E–05	.108E–05	.598E–06	.275E–06	.161E–06	.766E–07	.224E–07	.575E–08
40	.285E–04	.146E–04	.696E–05	.278E–05	.155E–05	.720E–06	.424E–06	.202E–06	.595E–07	.153E–07
50	.652E–04	.341E–04	.166E–04	.672E–05	.379E–05	.177E–05	.105E–05	.503E–06	.149E–06	.384E–07
60	.141E–03	.754E–04	.372E–04	.154E–04	.875E–05	.413E–05	.245E–05	.118E–05	.350E–06	.910E–07
70	.290E–03	.158E–03	.795E–04	.334E–04	.192E–04	.912E–05	.544E–05	.263E–05	.784E–06	.205E–06
80	.569E–03	.316E–03	.162E–03	.691E–04	.400E–04	.192E–04	.115E–04	.559E–05	.168E–05	.439E–06
90	.107E–02	.606E–03	.315E–03	.137E–03	.801E–04	.388E–04	.234E–04	.114E–04	.343E–05	.903E–06
100	.194E–02	.112E–02	.590E–03	.261E–03	.154E–03	.752E–04	.455E–04	.223E–04	.674E–05	.178E–05
110	.338E–02	.198E–02	.107E–02	.479E–03	.285E–03	.141E–03	.855E–04	.420E–04	.128E–04	.339E–05
120	.571E–02	.341E–02	.186E–02	.851E–03	.511E–03	.254E–03	.155E–03	.766E–04	.233E–04	.623E–05
130	.938E–02	.569E–02	.315E–02	.146E–02	.886E–03	.445E–03	.278E–03	.135E–03	.414E–04	.111E–04
140	.150E–01	.923E–02	.519E–02	.245E–02	.149E–02	.757E–03	.467E–03	.232E–03	.711E–04	.191E–04
150	.233E–01	.146E–01	.832E–02	.399E–02	.245E–02	.125E–02	.776E–03	.387E–03	.119E–03	.321E–04
160	.354E–01	.225E–01	.130E–01	.633E–02	.393E–02	.202E–02	.126E–02	.629E–03	.194E–03	.526E–04
170	.526E–01	.340E–01	.199E–01	.983E–02	.614E–02	.319E–02	.199E–02	.999E–03	.309E–03	.840E–04
180	.766E–01	.502E–01	.298E–01	.149E–01	.941E–02	.492E–02	.309E–02	.155E–02	.482E–03	.131E–03
190	.110	.729E–01	.438E–01	.222E–01	.141E–01	.744E–02	.469E–02	.236E–02	.735E–03	.201E–03
200	.154	.104	.631E–01	.325E–01	.208E–01	.110E–01	.698E–02	.352E–02	.110E–02	.300E–03
210	.213	.146	.894E–01	.467E–01	.300E–01	.161E–01	.102E–01	.516E–02	.161E–02	.442E–03
220	.290	.201	.125	.660E–01	.427E–01	.230E–01	.147E–01	.743E–02	.232E–02	.638E–03
230	.389	.273	.171	.918E–01	.598E–01	.325E–01	.208E–01	.105E–01	.329E–02	.906E–03
240	.514	.366	.232	.126	.825E–01	.451E–01	.290E–01	.147E–01	.460E–02	.127E–02
250	.673	.485	.310	.170	.112	.618E–01	.398E–01	.202E–01	.633E–02	.174E–02
260	.870	.635	.409	.227	.151	.835E–01	.540E–01	.274E–01	.858E–02	.237E–02
270	1.112	.822	.534	.300	.200	.111	.723E–01	.366E–01	.115E–01	.317E–02
280	1.407	1.052	.689	.391	.263	.147	.957E–01	.485E–01	.152E–01	.420E–02
290	1.763	1.335	.880	.505	.341	.192	.125	.634E–01	.199E–01	.548E–02
300	2.190	1.676	1.112	.646	.437	.248	.162	.820E–01	.257E–01	.708E–02
310	2.696	2.088	1.394	.817	.556	.316	.208	.105	.328E–01	.905E–02
320	3.292	2.578	1.732	1.025	.701	.400	.264	.133	.415E–01	.114E–01
330	3.990	3.159	2.133	1.274	.875	.502	.331	.167	.520E–01	.143E–01
340	4.801	3.843	2.608	1.571	1.083	.624	.413	.208	.646E–01	.178E–01
350	5.738	4.641	3.164	1.922	1.331	.770	.511	.256	.795E–01	.218E–01

*Vermeulen, Dong, Robinson, Nguyen, and Gmitro, AIChE meeting, Anaheim, Calif., 1982; and private communication from Prof. Theodore Vermeulen, Chemical Engineering Dept., University of California, Berkeley.

TABLE 2-13 Partial Vapor Pressure of Sulfur Dioxide over Water, mmHg

$g\ SO_2/100\ g\ H_2O$	Temperature, °C 0	10	20	30	40	50	60	90	120
0.01	0.02	0.04	0.07	0.12	0.19	0.29	0.43	1.21	2.82
0.05	0.38	0.66	1.07	1.68	2.53	3.69	5.24	12.9	27.0
0.10	1.15	1.91	3.03	4.62	6.80	9.71	13.5	31.7	63.9
0.15	2.10	3.44	5.37	8.07	11.7	16.5	22.7	52.2	104
0.20	3.17	5.13	7.93	11.8	17.0	23.8	32.6	73.7	145
0.25	4.34	6.93	10.6	15.7	22.5	31.4	42.8	95.8	186
0.30	5.57	8.84	13.5	19.8	28.2	39.2	53.3	118	229
0.40	8.17	12.8	19.4	28.3	40.1	55.3	74.7	164	316
0.50	10.9	17.0	25.6	37.1	52.3	72.0	96.8	211	404
1.00	25.8	39.5	58.4	83.7	117	159	212	454	856
2.00	58.6	88.5	129	183	253	342	453	955	
3.00	93.2	139	202	285	393	530	700		
4.00	129	192	277	389	535	720			
5.00	165	245	353	496	679				
6.00	202	299	430	602	824				
8.00	275	407	585	818					
10.00	351	517	741						
15.00	542	796							
20.00	735								

Condensed from Rabe, A. E. and Harris, J. F., *J. Chem. Eng. Data*, **8** (3), 333–336, 1963.

TABLE 2-14 Partial Pressures of HNO_3 and H_2O over Aqueous Solutions of HNO_3*

mmHg
Percentages are weight % HNO_3 in solution.

	20%		25%		30%		35%		40%		45%		50%	
°C	HNO_3	H_2O	HNO_3	H_2O	HNO_3	H_2O	HNO_3	H_2O	HNO_3	H_2O	HNO_3	H_2O	HNO_3	H_2O
0		4.1		3.8		3.6		3.3		3.0		2.6		2.1
5		5.7		5.4		5.0		4.6		4.2		3.6		3.0
10		8.0		7.6		7.1		6.5		5.8		5.0	0.12	4.2
15		10.9		10.3		9.7		8.9		8.0	0.10	6.9	.18	5.8
20		15.2		14.2		13.2		12.0		10.8	.15	9.4	.27	7.9
25		20.6		19.2		17.8		16.2	0.12	14.6	.23	12.7	.39	10.7
30		27.6		25.7		23.8	0.09	21.7	.17	19.5	.33	16.9	.56	14.4
35		36.5		33.8		31.1	.13	28.3	.25	25.5	.48	22.3	.80	19.0
40		47.5		44	0.11	41	.20	37.7	.36	33.5	.68	29.3	1.13	25.0
45		62	0.09	57.5	.17	53	.28	48	.52	43	.96	38.0	1.57	32.5
50		80	.13	75	.25	69	.42	63	.75	56	1.35	49.5	2.18	42.5
55	0.09	100	.18	94	.35	87	.59	79	1.04	71	1.83	62.5	2.95	54
60	.13	128	.28	121	.51	113	.85	102	1.48	90	2.54	80	4.05	70
65	.19	162	.40	151	.71	140	1.18	127	2.05	114	3.47	100	5.46	88
70	.27	200	.54	187	1.00	174	1.63	159	2.80	143	4.65	126	7.25	110
75	.38	250	.77	234	1.38	217	2.26	198	3.80	178	6.20	158	9.6	138
80	.53	307	1.05	287	1.87	267	3.07	243	5.10	218	8.15	195	12.5	170
85	.74	378	1.44	352	2.53	325	4.15	297	6.83	268	10.7	240	16.3	211
90	1.01	458	1.95	426	3.38	393	5.50	359	9.0	325	13.7	292	20.9	258
95	1.37	555	2.62	517	4.53	478	7.32	436	11.7	394	17.8	355	26.8	315
100	1.87	675	3.50	628	6.05	580	9.7	530	15.5	480	23.0	430	34.2	383
105	2.50	800	4.65	745	7.90	690	12.7	631	20.0	573	29.2	520	43.0	463
110							16.5	755	25.7	688	37.0	625	54.5	560
115									32.5	810	46	740	67	665
120													84	785

**International Critical Tables*, vol. 3, pp. 304–305.

(Continued)

TABLE 2-14 Partial Pressures of HNO_3 and H_2O over Aqueous Solutions of HNO_3 *(Continued)*

mmHg
Percentages are weight % HNO_3 in solution.

	55%		60%		65%		70%		80%		90%		100%
°C	HNO_3	H_2O	HNO_3	H_2O	HNO_3	H_2O	HNO_3	H_2O	HNO_3	H_2O	HNO_3	H_2O	HNO_3
0		1.8	0.19	1.5	0.41	1.3	0.79	1.1	2		5.5		11
5	0.14	2.5	.28	2.1	.60	1.8	1.12	1.6	3		8		15
10	.21	3.5	.41	3.0	.86	2.6	1.58	2.2	4	1.2	11		22
15	.31	4.9	.59	4.1	1.21	3.5	2.18	3.0	6	1.7	15		30
20	.45	6.7	.84	5.6	1.68	4.9	3.00	4.1	8	2.4	20		42
25	.66	9.1	1.21	7.7	2.32	6.6	4.10	5.5	10.5	3.2	27	1	57
30	.93	12.2	1.66	10.3	3.17	8.8	5.50	7.4	14	4	36	1.3	77
35	1.30	16.1	2.28	13.6	4.26	11.6	7.30	9.8	18.5	5.5	47	1.8	102
40	1.82	21.3	3.10	18.1	5.70	15.5	9.65	12.8	24.5	7	62	2.4	133
45	2.50	28.0	4.20	23.7	7.55	20.0	12.6	16.7	32	9.5	80	3	170
50	3.41	36.3	5.68	31	10.0	26.0	16.5	21.8	41	12	103	4	215
55	4.54	46	7.45	39	12.8	33.0	21.0	27.3	52	15	127	5	262
60	6.15	60	9.9	51	16.8	43.0	27.1	35.3	67	20	157	6.5	320
65	8.18	76	13.0	64	21.7	54.5	34.5	44.5	85	25	192	8	385
70	10.7	95	16.8	81	27.5	68	43.3	56	106	31	232	10	460
75	13.9	120	21.8	102	35.0	86	54.5	70	130	38	282	13	540
80	18.0	148	27.5	126	43.5	106	67.5	86	158	48	338	16	625
85	23.0	182	34.8	156	54.5	131	83	107	192	60	405	20	720
90	29.4	223	43.7	192	67.5	160	103	130	230	73	480	24	820
95	37.3	272	55.0	233	83.5	195	125	158	278	89	570	29	
100	47	331	69.5	285	103	238	152	192	330	108	675	35	
105	58.5	400	84.5	345	124	288	183	231	392	129	790	42	
110	73	485	103	417	152	345	221	278	465	155			
115	90	575	126	495	181	410	262	330	545	185			
120	110	685	156	590	218	490	312	393	640	219			
125			187	700	260	580	372	469					

TABLE 2-15 Total Vapor Pressures of Aqueous Solutions of CH_3COOH*

Percentages of weight % acetic acid in the solution
mmHg

°C	25%	50%	75%
20	16.3	15.7	15.3
25	22.1	21.4	20.8
30	29.6	28.8	27.8
35	39.4	38.3	36.6
40	51.7	50.2	48.1
45	67.0	65.0	62.0
50	87.2	85.0	80.1
55	110	107	102
60	141	138	130
65	178	172	162
70	223	216	203
75	277	269	251
80	342	331	310
85	419	407	376
90	510	497	458
95	618	602	550
100	743	725	666

**International Critical Tables*, vol. 3, p. 306.

TABLE 2-16 Partial Pressure of H_2O over Aqueous Solutions of NH_3 (psia)

	Liquid mole percent NH_3 (liquid weight percent NH_3)																			
	0	5	10	15	20	25	30	35	40	45	50	55	60	65	70	75	80	85	90	95
t, °F	(0)	(4.74)	(9.5)	(14.29)	(19.1)	(23.94)	(28.81)	(33.71)	(38.64)	(43.59)	(48.57)	(53.58)	(58.62)	(63.69)	(68.79)	(73.91)	(79.07)	(84.26)	(89.47)	(94.72)
32	0.089	0.083	0.077	0.071	0.063	0.055	0.047	0.039	0.031	0.025	0.019	0.014	0.011	0.008	0.006	0.004	0.003	0.002	0.002	0.001
40	0.122	0.115	0.106	0.097	0.087	0.077	0.065	0.054	0.044	0.035	0.027	0.021	0.016	0.012	0.009	0.007	0.005	0.004	0.002	0.001
50	0.178	0.168	0.156	0.143	0.129	0.113	0.097	0.081	0.066	0.053	0.041	0.032	0.025	0.019	0.014	0.011	0.008	0.006	0.004	0.002
60	0.256	0.242	0.225	0.207	0.186	0.164	0.142	0.119	0.098	0.079	0.062	0.049	0.038	0.030	0.023	0.018	0.014	0.010	0.007	0.004
70	0.363	0.343	0.320	0.294	0.266	0.235	0.204	0.172	0.143	0.116	0.093	0.073	0.058	0.045	0.036	0.028	0.022	0.016	0.011	0.006
80	0.507	0.479	0.448	0.413	0.374	0.332	0.289	0.245	0.205	0.168	0.136	0.109	0.087	0.069	0.055	0.043	0.034	0.025	0.018	0.010
90	0.699	0.661	0.618	0.571	0.518	0.462	0.403	0.345	0.290	0.240	0.196	0.159	0.128	0.103	0.083	0.066	0.052	0.040	0.028	0.015
100	0.951	0.899	0.843	0.780	0.710	0.634	0.556	0.479	0.405	0.338	0.279	0.228	0.186	0.152	0.123	0.100	0.079	0.061	0.043	0.024
110	1.277	1.209	1.135	1.052	0.960	0.861	0.758	0.656	0.559	0.470	0.392	0.324	0.268	0.220	0.181	0.148	0.119	0.092	0.065	0.036
120	1.695	1.607	1.510	1.402	1.283	1.154	1.021	0.889	0.763	0.647	0.544	0.455	0.380	0.316	0.263	0.217	0.176	0.137	0.099	0.056
130	2.226	2.112	1.988	1.850	1.696	1.532	1.361	1.192	1.030	0.881	0.747	0.632	0.532	0.448	0.376	0.313	0.257	0.202	0.147	0.083
140	2.893	2.748	2.591	2.415	2.221	2.012	1.796	1.582	1.376	1.186	1.016	0.867	0.738	0.628	0.532	0.448	0.371	0.295	0.216	0.124
150	3.723	3.540	3.343	3.122	2.879	2.618	2.347	2.078	1.821	1.582	1.367	1.177	1.013	0.870	0.746	0.634	0.529	0.425	0.314	0.183
160	4.747	4.519	4.273	4.000	3.698	3.374	3.039	2.706	2.387	2.090	1.821	1.584	1.376	1.194	1.033	0.887	0.748	0.607	0.453	0.267
170	6.000	5.717	5.416	5.079	4.709	4.312	3.902	3.493	3.101	2.736	2.405	2.110	1.851	1.622	1.418	1.229	1.047	0.858	0.647	0.386
180	7.520	7.174	6.807	6.397	5.947	5.465	4.968	4.472	3.995	3.551	3.148	2.787	2.468	2.184	1.928	1.688	1.451	1.201	0.917	0.555
190	9.350	8.931	8.488	7.994	7.452	6.873	6.275	5.680	5.107	4.573	4.086	3.650	3.262	2.914	2.598	2.297	1.994	1.669	1.290	0.793
200	11.538	11.035	10.504	9.916	9.270	8.580	7.869	7.160	6.479	5.842	5.262	4.740	4.275	3.856	3.470	3.098	2.718	2.300	1.802	1.129
210	14.136	13.538	12.910	12.213	11.449	10.635	9.796	8.962	8.160	7.410	6.725	6.110	5.559	5.061	4.598	4.146	3.675	3.147	2.502	1.600
220	17.201	16.496	15.758	14.941	14.047	13.095	12.115	11.141	10.205	9.331	8.534	7.817	7.175	6.592	6.045	5.504	4.932	4.277	3.455	2.262
230	20.796	19.971	19.111	18.162	17.124	16.020	14.886	13.760	12.679	11.672	10.754	9.930	9.192	8.522	7.889	7.255	6.573	5.777	4.751	3.196
240	24.986	24.029	23.037	21.943	20.748	19.479	18.179	16.889	15.654	14.506	13.463	12.530	11.696	10.938	10.221	9.496	8.703	7.759	6.508	4.520
250	29.844	28.744	27.607	26.358	24.996	23.549	22.070	20.608	19.212	17.917	16.748	15.708	14.783	13.946	13.153	12.346	11.452	10.369	8.891	6.413

The values in Table 2-16 were generated from the NIST REFPROP software (Lemmon, E. W., McLinden, M. O., and Huber, M. L., NIST Standard Reference Database 23: Reference Fluid Thermodynamic and Transport Properties—REFPROP, Version 7.0, National Institute of Standards and Technology, Standard Reference Data Program, Gaithersburg, Md., 2002). The primary source for the properties of aqueous ammonia mixtures is R. Tillner-Roth and D. G. Friend, "A Helmholtz Free Energy Formulation of the Thermodynamic Properties of the Mixture {Water + Ammonia}," *J. Phys. Chem. Ref. Data* **27:**63–96 (1998).

TABLE 2-17 Partial Pressures of H_2O over Aqueous Solutions of Sodium Carbonate*

t, °C	mmHg %Na_2CO_3 0	5	10	15	20	25	30
0	4.5	4.5					
10	9.2	9.0	8.8				
20	17.5	17.2	16.8	16.3			
30	31.8	31.2	30.4	29.6	28.8	27.8	26.4
40	55.3	54.2	53.0	57.6	50.2	48.4	46.1
50	92.5	90.7	88.7	86.5	84.1	81.2	77.5
60	149.5	146.5	143.5	139.9	136.1	131.6	125.7
70	239.8	235	230.5	225	219	211.5	202.5
80	355.5	348	342	334	325	315	301
90	526.0	516	506	494	482	467	447
100	760.0	746	731	715	697	676	648

**International Critical Tables*, vol. 3, p. 372.

TABLE 2-18 Partial Pressures of H_2O and CH_3OH over Aqueous Solutions of Methyl Alcohol*

Mole fraction CH_3OH	39.9°C P_{H_2O}, mmHg	39.9°C P_{CH_3OH}, mmHg	Mole fraction CH_3OH	59.4°C P_{H_2O}, mmHg	59.4°C P_{CH_3OH}, mmHg
0	54.7	0	0	145.4	0
14.99	39.2	66.1	22.17	106.9	210.1
17.85	38.5	75.5	27.40	102.2	240.2
21.07	37.2	85.2	33.24	96.6	272.1
27.31	35.8	100.6	39.80	91.7	301.9
31.06	34.9	108.8	47.08	84.8	335.6
40.1	32.8	127.7	55.5	76.9	373.7
47.0	31.5	141.6	69.2	57.8	439.4
55.8	27.3	158.4	78.5	43.8	486.6
68.9	20.7	186.6	85.9	30.1	526.9
86.0	10.1	225.2	100.0	0	609.3
100.0	0	260.7			

**International Critical Tables*, vol. 3, p. 290.

TABLE 2-19 Partial Pressures of H_2O over Aqueous Solutions of Sodium Hydroxide*

Conc. g NaOH/100 g H_2O	mmHg Temperature, °C 0	20	40	60	80	100	120	160	200	250	300	350
0	4.6	17.5	55.3	149.5	355.5	760.0	1,489	4,633	11,647	29,771	64,200	123,600
5	4.4	16.9	53.2	143.5	341.5	730.0	1,430	4,450	11,200	28,600	61,800	118,900
10	4.2	16.0	50.6	137.0	325.5	697.0	1,365	4,260	10,750	27,500	59,300	114,100
20	3.6	13.9	44.2	120.5	288.5	621.0	1,225	3,860	9,800	25,300	54,700	105,400
30	2.9	11.3	36.6	101.0	246.0	537.0	1,070	3,460	8,950	23,300	50,800	98,000
40	2.2	8.7	28.7	81.0	202.0	450.0	920	3,090	8,150	21,500	47,200	91,600
50		6.3	20.7	62.5	160.5	368.0	770	2,690	7,400	19,900	44,100	85,800
60		4.4	15.5	47.0	124.0	294.0	635	2,340	6,750	18,400	41,200	80,700
70		3.0	10.9	34.5	94.0	231.0	515	2,030	6,100	17,100	38,700	76,000
80		2.0	7.6	24.5	70.5	179.0	415	1,740	5,500	15,800	36,300	71,900
90		1.3	5.2	17.5	53.0	138.0	330	1,490	5,000	14,700	34,200	68,100
100		0.9	3.6	12.5	38.5	105.0	262	1,300	4,500	13,650	32,200	64,600
120			1.7	6.3	20.5	61.0	164	915	3,650	11,800	28,800	58,600
140				3.0	11.0	35.5	102	765	2,980	10,300	25,900	53,400
160				1.5	6.0	20.5	63	470	2,430	8,960	23,300	49,000
180					3.5	12.0	40	340	1,980	7,830	21,200	45,100
200					2.0	7.0	25	245	1,620	6,870	19,200	41,800
250					0.5	2.0	8	110	985	5,000	15,400	35,000
300					0.1	0.5	2.7	50	610	3,690	12,500	29,800
350							0.9	23	380	2,750	10,300	25,700
400								11	240	2,080	8,600	22,400
500									100	1,210	6,100	17,500
700										440	3,300	11,500
1000											1,470	6,800
2000											150	1,760
4000												120
8000												7

**International Critical Tables*, vol. 3, p. 370.

WATER VAPOR CONTENT IN GASES

The accompanying figure is useful in determining the water vapor content of air at high pressure in contact with liquid water.

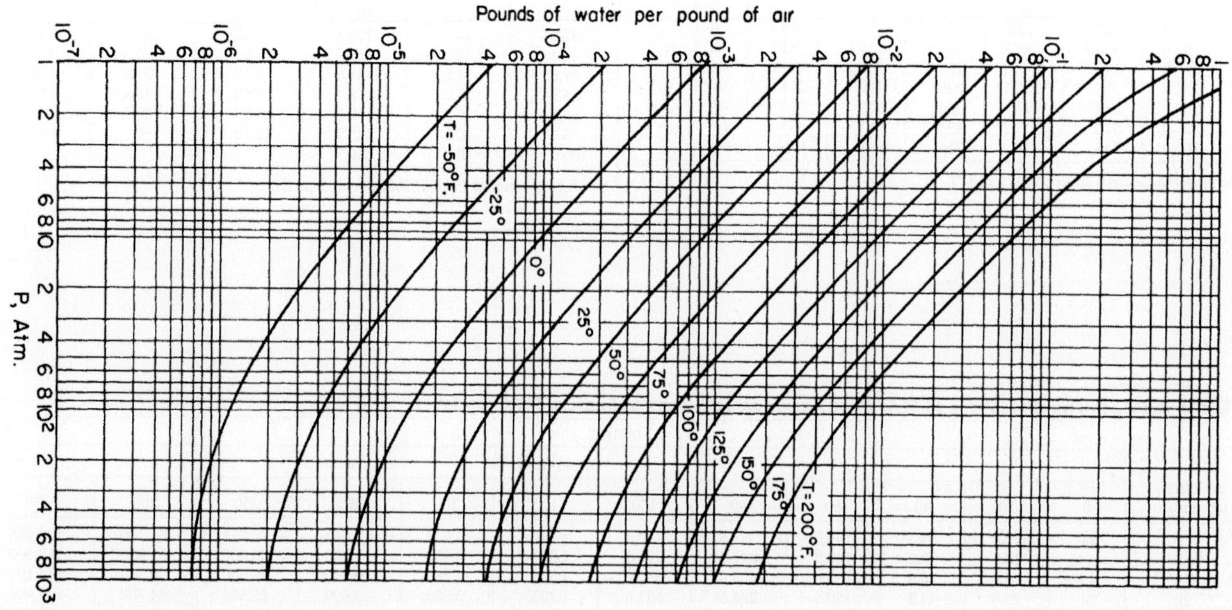

FIG. 2-2 Water content in air at pressures over atmospheric. (Landsbaum, E.M., W.S. Dodds, and L.F. Stutzman. Reprinted from vol. 47, January 1955 issue of *Ind. Eng. Chem.* [p. 192]. Copyright 1955 by the American Chemical Society and reproduced by permission of the copyright owner.) For other water-in-air data, see Table 2-111, Fig. 2-3 and Section 12 figures and tables.

SOLUBILITIES

Unit Conversions For this subsection, the following unit conversions are applicable: °F = $\frac{9}{5}$°C + 32. To convert cubic centimeters to cubic feet, multiply by 3.532×10^{-5}. To convert millimeters of mercury to pounds-force per square inch, multiply by 0.01934. To convert grams per liter to pounds per cubic foot, multiply by 6.243×10^{-2}.

Introduction The database containing solubilities was originally published in the International Union for Pure and Applied Chemistry (IUPAC)-National Institute of Standards and Technology (NIST) Solubility Data Series. It is available at no cost online at http://srdata.nist.gov/solubility.

The H in the following tables is the proportionality constant in Henry's law, $p = Hx$, where x is the mole fraction of the solute in the aqueous liquid phase; p is the partial pressure in atm of the solute in the gas phase; and H is a proportionality constant, generally referred to as Henry's constant. Values of H often have considerable uncertainty and are strong functions of temperature. To convert values of H at 25°C from atm to atm/(mol/m^3), divide by the molar density of water at 25°C, which is 55,342 mol/m^3. Henry's law is valid only for dilute solutions.

Additional values of Henry's constant can be found in "Environmental Simulation Program," OLI Systems, Inc., Morris Plains, N.J.; "Estimated Henry's Law Constant," EPA Online Tools for Site Assessment Calculation (http://www.epa.gov/athens/learn2model/part-two/onsite/esthenry.htm); Rolf Sander, "Compilation of Henry's Law Constants for Inorganic and Organic Species of Potential Importance in Environmental Chemistry," Air Chemistry Department, Max-Planck Institute of Chemistry, Mainz, Germany; Rolf Sander, "Modeling Atmospheric Chemistry: Interactions between Gas-Phase Species and Liquid Cloud/Aerosol Particles," *Surv. Geophys.* **20:** 1–31, 1999 (http://www.henrys-law.org).

TABLE 2-20 Solubilities of Inorganic Compounds in Water at Various Temperatures*

This table shows the grams of anhydrous substance that are soluble in 100 g of water at the temperature in degrees Celsius as indicated; when the name is followed by †, the value is expressed in grams of substance in 100 cm^3 of saturated solution. Solid phase gives the hydrated form in equilibrium with the saturated solution.

	Substance	Formula	Solid phase	0°C	10°C	20°C	30°C	40°C	50°C	60°C	70°C	80°C	90°C	100°C	
1	Aluminum chloride	$AlCl_3$	$6H_2O$			$69.86^{15°}$									1
2	sulfate	$Al_2(SO_4)_3$	$18H_2O$	31.2	33.5	36.4	40.4	46.1	52.2	59.2	66.1	73.0	80.8	89.0	2
3	Ammonium aluminum sulfate	$(NH_4)_2Al_2(SO_4)_4$	$24H_2O$	2.1	4.99	7.74	10.94	14.88	20.10	26.70				$109.7^{96°}$	3
4	bicarbonate	NH_4HCO_3		11.9	15.8	21	27								4
5	bromide	NH_4Br		60.6	68	75.5	83.2	91.1	99.2	107.8	116.8	126	135.6	145.6	5
6	chloride	NH_4Cl		29.4	33.3	37.2	41.4	45.8	50.4	55.2	60.2	65.6	71.3	77.3	6
7	chloroplatinate	$(NH_4)_2PtCl_6$			0.7									1.25	7
8	chromate	$(NH_4)_2CrO_4$					40.4								8
9	chromium sulfate	$(NH_4)_2Cr_2(SO_4)_4$	$24H_2O$			$10.78^{25°}$									9
10	dichromate	$(NH_4)_2Cr_2O_7$					47.17								10
11	dihydrogen phosphite	$NH_4H_2PO_3$		171		$190^{14.5°}$	$260^{31°}$								11
12	hydrogen phosphate	$(NH_4)_2HPO_4$				131^{15}									12
13	iodide	NH_4I		154.2	163.2	172.3	181.4	190.5	199.6	208.9	218.7	228.8		250.3	13
14	magnesium phosphate	NH_4MgPO_4	$6H_2O$	0.023		0.052		0.036	0.030	0.040	0.016	0.019			14
15	manganese phosphate	NH_4MnPO_4	$7H_2O$			0		0		0	0.005	0.007			15
16	nitrate	NH_4NO_3		118.3		192	241.8	297.0	344.0	421.0	499.0	580.0	740.0	871.0	16
17	oxalate	$(NH_4)_2C_2O_4$	$1H_2O$	2.2	3.1	4.4	5.9	8.0	10.3						17
18	perchlorate†	NH_4ClO_4†		11.56		20.85		30.58		39.05		48.19		57.01	18
19	persulfate	$(NH_4)_2S_2O_8$		58.2											19
20	sulfate	$(NH_4)_2SO_4$		70.6	73.0	75.4	78.0	81.0		88.0		95.3		103.3	20
21	thiocyanate	NH_4CNS		119.8	144	170	207.7								21
22	vanadate (meta)	NH_4VO_3				0.48	0.84	1.32	1.78		3.05				22
23	Antimonious fluoride	SbF_3		384.7		444.7	563.6								23
24	sulfide	Sb_2S_3				$0.000175^{18°}$									24
25	Arsenic oxide	As_2O_5		59.5	62.1	65.8	69.5	71.2		73.0		75.1		76.7	25
26	Arsenious sulfide	As_2S_3		5.17×10^{-5} at 18°											26
27	Barium acetate	$Ba(C_2H_3O_2)_2$	$3H_2O$	59	63	71									27
28	acetate	$Ba(C_2H_3O_2)_2$	$1H_2O$				75	79	77	74	74			75	28
29	carbonate	$BaCO_3$			$0.0016^{8°}$	$0.0022^{18°}$	0.0024 at 24.2°								29
30	chlorate	$Ba(ClO_3)_2$	$1H_2O$	20.34	26.95	33.80	41.70	49.61		66.81		84.84		104.9	30
31	chloride	$BaCl_2$	$2H_2O$	31.6	33.3	35.7	38.2	40.7	43.6	46.4	49.4	52.4		58.8	31
32	chromate	$BaCrO_4$		0.0002	0.00028	0.00037	0.00046								32
33	hydroxide	$Ba(OH)_2$	$8H_2O$	1.67	2.48	3.89	5.59	8.22	13.12	20.94		101.4			33
34	iodide	BaI_2	$6H_2O$	170.2	185.7	203.1	219.6								34
35	iodide	BaI_2	$2H_2O$					231.9		247.3		261.0		271.7	35
36	nitrate	$Ba(NO_3)_2$		5.0	7.0	9.2	11.6	14.2	17.1	20.3		27.0		34.2	36
37	nitrite	$Ba(NO_2)_2$	$1H_2O$			67.5						205.8		300	37
38	oxalate	BaC_2O_4			$0.0016^{8°}$	$0.0022^{18°}$	0.0024 at 24.2°								38
39	perchlorate	$Ba(ClO_4)_2$	$3H_2O$	205.8		289.1		358.7	426.3		495.2		562.3		39
40	sulfate	$BaSO_4$		1.15×10^{-4}	2.0×10^{-4}	2.4×10^{-4}	2.85×10^{-4}								40
41	Beryllium sulfate	$BeSO_4$	$6H_2O$				52		60.67						41
42	sulfate	$BeSO_4$	$4H_2O$				43.78	46.74			62		83	100	42
43	sulfate	$BeSO_4$	$2H_2O$									84.76	98	110	43
44	Boric acid	H_3BO_3		2.66	3.57	5.04	6.60	8.72	11.54	14.81	16.73	23.75	30.38	40.25	44
45	Boron oxide	B_2O_3		1.1	1.5	2.2		4.0		6.2		9.5		15.7	45
46	Bromine	Br_2		4.22	3.4	3.20	3.13								46
47	Cadmium chloride	$CdCl_2$	$4H_2O$	97.59	125.1										47
48	chloride	$CdCl_2$	$2\frac{1}{2}H_2O$	90.01			132.1								48
49	chloride	$CdCl_2$	$1H_2O$		135.1	134.5		135.3		136.5		140.4		147.0	49
50	cyanide	$Cd(CN)_2$				$1.7^{15°}$									50
51	hydroxide	$Cd(OH)_2$					2.6×10^{-4} at 25°								51
52	sulfate	$CdSO_4$		76.48	76.00	76.60		78.54		83.68			63.13	60.77	52
53	Calcium acetate	$Ca(C_2H_3O_2)_2$	$2H_2O$	37.4	36.0	34.7	33.8	33.2		32.7		33.5			53
54	acetate	$Ca(C_2H_3O_2)_2$	$1H_2O$										31.1	29.7	54

(*Continued*)

TABLE 2-20 Solubilities of Inorganic Compounds in Water at Various Temperatures* *(Continued)*

This table shows the grams of anhydrous substance that are soluble in 100 g of water at the temperature in degrees Celsius as indicated; when the name is followed by †, the value is expressed in grams of substance in 100 cm^3 of saturated solution. Solid phase gives the hydrated form in equilibrium with the saturated solution.

	Substance	Formula	Solid phase	0°C	10°C	20°C	30°C	40°C	50°C	60°C	70°C	80°C	90°C	100°C	
1	Calcium bicarbonate	$Ca(HCO_3)_2$		16.15		16.60		17.05		17.50		17.95		18.40	1
2	chloride	$CaCl_2$	$6H_2O$	59.5	65.0	74.5	102								2
3	chloride	$CaCl_2$	$2H_2O$							136.8	141.7	147.0	152.7	159	3
4	fluoride	CaF_2				$0.0016^{18°}$	$0.0017^{26°}$								4
5	hydroxide	$Ca(OH)_2$		0.185	0.176	0.165	0.153	0.141	0.128	0.116	0.106	0.094	0.085	0.077	5
6	nitrate	$Ca(NO_3)_2$	$4H_2O$	102.0	115.3	129.3	152.6	195.9							6
7	nitrate	$Ca(NO_3)_2$	$3H_2O$					237.5	281.5						7
8	nitrate	$Ca(NO_3)_2$										358.7		363.6	8
9	nitrite	$Ca(NO_2)_2$	$4H_2O$	62.07		76.68									9
10	nitrite	$Ca(NO_2)_2$	$2H_2O$							132.6	151.9		244.8		10
11	oxalate	CaC_2O_4			6.7×10^{-4} at 13°	6.8×10^{-4} at 25°	9.5×10^{-4} at 50°	14×10^{-4} at 95°							11
12	sulfate	$CaSO_4$	$2H_2O$	0.1759	0.1928		0.2090	0.2097		0.2047	0.1966			0.1619	12
13	Carbon dioxide, 760 mm ‡	CO_2		0.3346	0.2318	0.1688	0.1257	0.0973	0.0761	0.0576				0	13
14	monoxide, 760 mm ‡	CO		0.0044	0.0035	0.0028	0.0024	0.0021	0.0018	0.0015	0.0013	0.0010	0.0006	0	14
15	Cesium chloride	CsCl		161.4	174.7	186.5	197.3	208.0	218.5	229.7	239.5	250.0	260.1	270.5	15
16	nitrate	$CsNO_3$		9.33	14.9	23.0	33.9	47.2	64.4	83.8	107.0	134.0	163.0	197.0	16
17	sulfate	Cs_2SO_4		167.1	173.1	178.7	184.1	189.9	194.9	199.9	205.0	210.3	214.9	220.3	17
18	Chlorine, 760 mm ‡	Cl_2		1.46	0.980	0.716	0.562	0.451	0.386	0.324	0.274	0.219	0.125	0	18
19	Chromic anhydride	CrO_3		164.9				174.0	182.1				217.5	206.8	19
20	Cuprio chloride	$CuCl_2$	$2H_2O$	70.7	73.76	77.0	80.34	83.8	87.44	91.2		99.2		107.9	20
21	nitrate	$Cu(NO_3)_2$	$6H_2O$	81.8	95.28	125.1									21
22	nitrate	$Cu(NO_3)_2$	$3H_2O$					159.8		178.8		207.8			22
23	sulfate	$CuSO_4$	$5H_2O$	14.3	17.4	20.7	25	28.5	33.3	40		55		75.4	23
24	sulfide	CuS				3.3×10^{-5} at 18°									24
25	Cuprous chloride	CuCl				$1.52^{25°}$									25
26	Ferric chloride	$FeCl_3$		74.4	81.9	91.8			315.1			525.8		535.7	26
27	Ferrous chloride	$FeCl_2$	$4H_2O$		64.5		73.0	77.3	82.5	88.7		100			27
28	chloride	$FeCl_2$											105.3	105.8	28
29	nitrate	$Fe(NO_3)_2$	$6H_2O$	71.02		83.8				165.6					29
30	sulfate	$FeSO_4$	$7H_2O$	15.65	20.51	26.5	32.9	40.2	48.6						30
31	sulfate	$FeSO_4$	$1H_2O$								50.9	43.6	37.3		31
32	Hydrobromic acid, 760 mm	HBr		221.2	210.3	198			171.5					130	32
33	Hydrochloric acid, 760 mm	HCl		82.3			67.3	63.3	59.6	56.1					33
34	Iodine	I_2				0.029	0.04	0.056	0.078						34
35	Lead acetate	$Pb(C_2H_3O_2)_2$	$3H_2O$				$55.04^{25°}$								35
36	bromide	$PbBr_2$		0.4554		0.85	1.15	1.53	1.94	2.36		3.34		4.75	36
37	carbonate	$PbCO_3$				0.00011									37
38	chloride	$PbCl_2$		0.6728		0.99	1.20	1.45	1.70	1.98		2.62		3.34	38
39	chromate	$PbCrO_4$				7×10^{-6}									39
40	fluoride	PbF_2			0.060	0.064	0.068								40
41	nitrate	$Pb(NO_3)_2$		38.8	48.3	56.5	66	75	85	95		115		38.8	41
42	sulfate	$PbSO_4$		0.0028	0.0035	0.0041	0.0049	0.0056							42
43	Magnesium bromide	$MgBr_2$	$6H_2O$	91.0	94.5	96.5	99.2	101.6	104.1	107.5		113.7		120.2	43
44	chloride	$MgCl_2$	$6H_2O$	52.8	53.5	54.5		57.5		61.0		66.0		73.0	44
45	hydroxide	$Mg(OH)_2$				$0.0009^{18°}$									45
46	nitrate	$Mg(NO_3)_2$	$6H_2O$	66.55				84.74					137.0		46
47	sulfate	$MgSO_4$	$7H_2O$		30.9	35.5	40.8	45.6							47
48	sulfate	$MgSO_4$	$6H_2O$	40.8	42.2	44.5	45.3		50.4	53.5	59.5	64.2	69.0	74.0	48
49	sulfate	$MgSO_4$	$1H_2O$									62.9		68.3	49
50	Manganous sulfate	$MnSO_4$	$7H_2O$	53.23	60.01										50
51	sulfate	$MnSO_4$	$5H_2O$		59.5	62.9	67.76								51
52	sulfate	$MnSO_4$	$4H_2O$			64.5	66.44	68.8	72.6						52
53	sulfate	$MnSO_4$	$1H_2O$						58.17	55.0	52.0	48.0	42.5	34.0	53
54	Mercurous chloride	HgCl		0.00014		0.0002		0.0007							54
55	Molybdic oxide	MoO_3	$2H_2O$			0.138	0.264	0.476	0.687	1.206	2.055	2.106			55
56	Nickel chloride	$NiCl_2$	$6H_2O$	53.9	59.5	64.2	68.9	73.3	78.3	82.2	85.2			87.6	56
57	nitrate	$Ni(NO_3)_2$	$6H_2O$	79.58		96.31		122.2							57
58	nitrate	$Ni(NO_3)_2$	$3H_2O$							163.1	169.1		235.1		58
59	sulfate	$NiSO_4$	$7H_2O$	27.22	32		42.46								59
60	sulfate	$NiSO_4$	$6H_2O$						50.15	54.80	59.44	63.17		76.7	60
61	Nitric oxide, 760 mm	NO		0.00984	0.00757	0.00618	0.00517	0.00440	0.00376	0.00324	0.00267	0.00199	0.00114	0	61
62	Nitrous oxide	N_2O			0.1705	0.1211									62

1	Potassium acetate	$KC_2H_3O_2$	1½H_2O	216.7	233.9	255.6	283.8	323.3							1
2	acetate	$KC_2H_3O_2$	½H_2O						337.3	350	364.8	380.1	396.3		2
3	alum	$K_2SO_4{\cdot}Al_2(SO_4)_3$	24H_2O	3.0	4.0	5.9	8.39	11.70	17.00	24.75	40.0	71.0	109.0		3
4	bicarbonate	$KHCO_3$		22.4	27.7	33.2	39.1	45.4		60.0					4
5	bisulfate	$KHSO_4$		36.3		51.4		67.3						121.6	5
6	bitartrate	$KHC_4H_4O_6$		0.32	0.40	0.53	0.90	1.32	1.83	2.46		4.6		6.95	6
7	carbonate	K_2CO_3	2H_2O	105.5	108	110.5	113.7	116.9	121.2	126.8	133.1	139.8	147.5	155.7	7
8	chlorate	$KClO_3$		3.3	5	7.4	10.5	14	19.3	24.5		38.5		57	8
9	chloride	KCl		27.6	31.0	34.0	37.0	40.0	42.6	45.5	48.3	51.1	54.0	56.7	9
10	chromate	K_2CrO_4		58.2	60.0	61.7	63.4	65.2	66.8	68.6	70.4	72.1	73.9	75.6	10
11	dichromate	$K_2Cr_2O_7$		5	7	12	20	26	34	43	52	61	70	80	11
12	ferricyanide	$K_3Fe(CN)_6$		31	36	43	50	60		66				82.6^{104}	12
13	hydroxide	KOH	2H_2O	97	103	112	126								13
14	hydroxide	KOH	1H_2O						140					178	14
15	nitrate	KNO_3		13.3	20.9	31.6	45.8	63.9	85.5	110.0	138	169	202	246	15
16	nitrite	KNO_2		278.8		298.4		334.9						412.8	16
17	perchlorate	$KClO_4$		0.75	1.05	1.80	2.6	4.4	6.5	9	11.8	14.8	18	21.8	17
18	permanganate	$KMnO_4$		2.83	4.4	6.4	9.0	12.56	16.89	22.2					18
19	persulfate†	$K_2S_2O_8$†	†	1.62	2.60	4.49	7.19	9.89							19
20	sulfate	K_2SO_4		7.35	9.22	11.11	12.97	14.76	16.50	18.17	19.75	21.4	22.8	24.1	20
21	thiocyanate	$KCNS$		177.0		217.5									21
22	Silver cyanide	$AgCN$				2.2×10^{-5}									22
23	nitrate	$AgNO_3$		122	170	222	300	376	455	525		669		952	23
24	sulfate	Ag_2SO_4		0.573	0.695	0.796	0.888	0.979	1.08	1.15	1.22	1.30	1.36	1.41	24
25	Sodium acetate	$NaC_2H_3O_2$	3H_2O	36.3	40.8	46.5	54.5	65.5	83	139					25
26	acetate	$NaC_2H_3O_2$		119	121	123.5	126	129.5	134	139.5	146	153	161	170	26
27	bicarbonate	$NaHCO_3$		6.9	8.15	9.6	11.1	12.7	14.45	16.4					27
28	carbonate	Na_2CO_3	10H_2O	7	12.5	21.5	38.8								28
29	carbonate	Na_2CO_3	1H_2O				50.5	48.5		46.4		45.8		45.5	29
30	chlorate	$NaClO_8$		79	89	101	113	126	140	155	172	189		230	30
31	chloride	$NaCl$		35.65	35.72	35.89	36.09	36.37	36.69	37.04	37.46	37.93	38.47	38.99	31
32	chromate	Na_2CrO_4	10H_2O	31.70	50.17	88.7									32
33	chromate	Na_2CrO_4	4H_2O				88.7	95.96	104	114.6					33
34	chromate	Na_2CrO_4									123.0	124.8		125.9	34
35	dichromate	$Na_2Cr_2O_7$	2H_2O	163.0		177.8			244.8		316.7	376.2			35
36	dichromate	$Na_2Cr_2O_7$												426.3	36
37	dihydrogen phosphate	NaH_2PO_4	2H_2O	57.9	69.9	85.2	106.5	138.2							37
38	dihydrogen phosphate	NaH_2PO_4	1H_2O						158.6						38
39	dihydrogen phosphate	NaH_2PO_4								179.3	190.3	207.3	225.3	246.6	39
40	hydrogen arsenate	Na_2HAsO_4	12H_2O	7.3	15.5	26.5	37	47		65		85			40
41	hydrogen phosphate	Na_2HPO_4	12H_2O	1.67	3.6	7.7	20.8								41
42	hydrogen phosphate	Na_2HPO_4	7H_2O					51.8							42
43	hydrogen phosphate	Na_2HPO_4	2H_2O						80.2	82.9	88.1	92.4	102.9		43
44	hydrogen phosphate	Na_2HPO_4												102.2	44
45	hydroxide	$NaOH$	4H_2O	42											45
46	hydroxide	$NaOH$	3½H_2O		51.5										46
47	hydroxide	$NaOH$	1H_2O			109	119	129	145	174					47
48	hydroxide	$NaOH$											313	347	48
49	nitrate	$NaNO_3$		73	80	88	96	104	114	124		148		180	49
50	nitrite	$NaNO_2$		72.1	78.0	84.5	91.6	98.4	104.1			132.6		163.2	50
51	oxalate	$Na_2C_2O_4$				3.7								6.33	51
52	phosphate, tri-	Na_3PO_4	12H_2O	1.5	4.1	11	20	31	43	55		81		108	52
53	pyrophosphate	$Na_4P_2O_7$	10H_2O	3.16	3.95	6.23	9.95	13.50	17.45	21.83		30.04		40.26	53
54	sulfate	Na_2SO_4	10H_2O	5.0	9.0	19.4	40.8								54
55	sulfate	Na_2SO_4	7H_2O	19.5	30	44									55
56	sulfate	Na_2SO_4						48.8	46.7	45.3		43.7		42.5	56
57	sulfide	Na_2S	9H_2O		15.42	18.8	22.5	28.5							57
58	sulfide	Na_2S	5½H_2O						39.82	42.69	45.73	51.40	59.23		58
59	sulfide	Na_2S	6H_2O						36.4	39.1	43.31	49.14	57.28		59
60	sulfite	Na_2SO_3	7H_2O	13.9	20	26.9	36								60
61	sulfite	Na_2SO_3						28	28.2	28.8		28.3			61
62	tetraborate	$Na_2B_4O_7$	10H_2O	1.3	1.6	2.7	3.9		10.5	20.3					62
63	tetraborate	$Na_2B_4O_7$	5H_2O								24.4	31.5	41	52.5	63
64	vanadate (meta)	$NaVO_8$	2H_2O			$15.3^{25°}$		30.2		68.4					64

*By N. A. Lange; abridged from "Table of Solubilities of Inorganic Compounds in Water at Various Temperatures" in *Lange's Handbook of Chemistry,* 10th ed., McGraw-Hill, New York, 1961 (except for NaCl, which is from *CRC Handbook of Chemistry and Physics,* 86th ed., CRC Press, 2005). For tables of the solubility of gases in water at various temperatures, Atack (*Handbook of Chemical Data,* Reinhold, New York, 1957) gives values at closer temperature intervals, usually 1 or 5°C, than are tabulated here. For materials marked by ‡, additional data are given in tables subsequent to this one. For the solubility of various hydrocarbons in water at high pressures see *J. Chem. Eng. Data,* **4,** 212 (1959).

TABLE 2-20 Solubilities of Inorganic Compounds in Water at Various Temperatures *(Continued)*

This table shows the grams of anhydrous substance that are soluble in 100 g of water at the temperature in degrees Celsius as indicated; when the name is followed by †, the value is expressed in grams of substance in 100 cm³ of saturated solution. Solid phase gives the hydrated form in equilibrium with the saturated solution.

	Substance	Formula	Solid phase	0°C	10°C	20°C	30°C	40°C	50°C	60°C	70°C	80°C	90°C	100°C	
1	Sodium vanadate (meta)	$NaVO_3$				$21.10^{25°}$		26.23		32.97	36.9	$38.8^{75°}$			1
2	Stannous chloride	$SnCl_2$		83.9		$269.8^{15°}$									2
3	sulfate	$SnSO_4$				19								18	3
4	Strontium acetate	$Sr(C_2H_3O_2)_2$	$4H_2O$	36.9	43.61										4
5	acetate	$Sr(C_2H_3O_2)_2$	$\frac{1}{2}H_2O$		42.95	41.6	39.5		37.35		36.24	36.10		36.4	5
6	chloride	$SrCl_2$	$6H_2O$	43.5	47.7	52.9	58.7	65.3	72.4	81.8					6
7	chloride	$SrCl_2$	$2H_2O$								85.9	90.5		100.8	7
8	nitrate	$Sr(NO_3)_2$	$1H_2O$	52.7		64.0			83.8	97.2			130.4	139	8
9	nitrate	$Sr(NO_3)_2$	$4H_2O$	40.1		70.5									9
10	nitrate	$Sr(NO_3)_2$					88.6	90.1		93.8	96	98	100		10
11	sulfate	$SrSO_4$		0.0113		0.0114	0.0114								11
12	Sulfur dioxide, 760 mm†	SO_2		22.83	16.21	11.29	7.81	5.41	4.5						12
13	Thallium sulfate	Tl_2SO_4		2.70	3.70	4.87	6.16		9.21	10.92	12.74	14.61	16.53	18.45	13
14	Thorium sulfate	$Th(SO_4)_2$	$9H_2O$	0.74	0.98	1.38	1.995	2.998	5.22						14
15	sulfate	$Th(SO_4)_2$	$8H_2O$	1.0	1.25	1.62									15
16	sulfate	$Th(SO_4)_2$	$6H_2O$	1.50		1.90	2.45			6.64					16
17	sulfate	$Th(SO_4)_2$	$4H_2O$					4.04	2.54	1.63	1.09				17
18	Zinc chlorate	$ZnClO_3$	$6H_2O$	145.0	152.5										18
19	chlorate	$ZnClO_3$	$4H_2O$			200.3	209.2	223.2	273.1						19
20	nitrate	$Zn(NO_3)_2$	$6H_2O$	94.78		118.3									20
21	nitrate	$Zn(NO_3)_2$	$3H_2O$					206.9							21
22	sulfate	$ZnSO_4$	$7H_2O$	41.9	47	54.4									22
23	sulfate	$ZnSO_4$	$6H_2O$					70.1	76.8						23
24	sulfate	$ZnSO_4$	$1H_2O$									86.6	83.7	80.8	24

TABLE 2-21 Solubility as a Function of Temperature and Henry's Constant at 25°C for Gases in Water

Name	Formula	*A*	*B*	*C*	*D*	*T* range, K	*H* at 25°C, atm
Acetylene	C_2H_2	−156.51	8,160.2	21.403	0	274–343	1,330
Carbon dioxide	CO_2	−159.854	8,741.68	21.6694	−1.10261E-03	273–353	1,635
Carbon monoxide	CO	−171.764	8,296.9	23.3376	0	273–353	58,000
Ethane	C_2H_6	−250.812	12,695.6	34.7413	0	275–323	29,400
Ethylene	C_2H_4	−153.027	7,965.2	20.5248	0	287–346	11,726
Helium	He	−105.9768	4,259.62	14.0094	0	273–348	142,900
Hydrogen	H_2	−125.939	5,528.45	16.8893	0	273–345	70,800
Methane	CH_4	−338.217	13,282.1	51.9144	−0.0425831	273–523	39,200
Nitrogen	N_2	−181.587	8,632.13	24.7981	0	273–350	84,600
Oxygen	O_2	−171.2542	8,391.24	23.24323	0	273–333	43,400

The constants can be used to calculate solubility by the equation $\ln x = A + B/T + C \ln T + DT$, where T is in K and x is the mole fraction of the solute dissolved in water when the solute partial pressure is 1 atm. With the assumption that Henry's law is valid up to 1 atm, $H = 1/x$. Values of the constants are from P. G. T. Fogg and W. Gerrard, *Solubility of Gases in Liquids,* Wiley, 1991, New York, and *Solubility Data Series*, vol. 1, *Helium and Neon*, IUPAC, Pergamon Press, Oxford, 1979. For higher-temperature behavior and an up-to-date reference list, see R. Fernandez-Prini, J. L. Alvarez, and A. H. Harvey, *J. Phys. Chem. Ref. Data* **32**(2):903, 2003. To find H at temperatures other than 25°C, first find the solubility and then take the reciprocal.

TABLE 2-22 Henry's Constant *H* for Various Compounds in Water at 25°C

Group	Compound	Formula	CAS	*H*, atm†	Rating*
Paraffin hydrocarbons	Methane	CH_4	74-82-8	36,600	4
	Ethane	C_2H_6	74-84-0	26,700	3
	Propane	C_3H_8	74-98-6	37,800	3
	Butane	C_4H_{10}	106-97-8	51,100	3
	Pentane	C_5H_{12}	109-66-0	70,000	3
	Octane	C_8H_{18}	111-65-9	2,74,000	3
	Nonane	C_9H_{20}	111-84-2	3,29,000	3
Olefins	Ethylene	C_2H_4	74-85-1	11,700	3
	Propylene	C_3H_6	115-07-1	11,700	4
Aromatics	Benzene	C_6H_6	71-43-2	299	10
	Toluene	C_7H_8	108-88-3	354	10
	o-Xylene	C_8H_{10}	95-47-6	272	10
	Cumene	C_9H_{12}	98-82-8	724	9
	Phenol	C_6H_6O	108-95-2	0.0394	7
Aldehydes	Acetaldehyde	C_2H_4O	75-07-0	5.56	3
	Propionaldehyde	C_3H_6O	123-38-6	4.36	4
Ketones	Methylethyl ketone	C_4H_8O	78-93-3	2.59	5
Esters	Methyl formate	$C_2H_4O_2$	107-31-3	13.6	3
	Ethyl formate	$C_3H_6O_2$	109-94-4	13.6	3
	Methyl acetate	$C_3H_6O_2$	79-20-9	5.04	3
	Butyl acetate	$C_6H_{12}O_2$	123-86-4	13.6	3
Chlorine containing	Chloromethane	CH_3Cl	74-87-3	556	?
	Chloroethane	C_2H_5Cl	75-00-3	681	10
	Chlorobenzene	C_6H_5Cl	108-90-7	204	10
Alcohols	Methanol	CH_4O	67-56-1	0.272	4
	Ethanol	C_2H_6O	64-17-5	0.272	4
	1-Propanol	C_3H_8O	71-23-8	0.507	3
	1-Butanol	$C_4H_{10}O$	71-36-3	0.482	3
Miscellaneous	Acrylonitrile	C_3H_3N	107-13-1	5.54	3
	Dimethyl sulfide	C_2H_6S	75-18-3	121	3
	Dimethyl disulfide	$C_2H_6S_2$	624-92-0	68.1	3
	Methyl mercaptan	CH_4S	74-93-1	177	3
	Ethyl mercaptan	C_2H_6S	75-08-1	161	3
	Pyridine	C_5H_5N	110-86-1	0.817	3

Values in this table were taken from the Design Institute for Physical Properties (DIPPR) of the American Institute of Chemical Engineers (AIChE), 801 Critically Evaluated Gold Standard™ Database, copyright 2016 AIChE, and reproduced with permission of AIChE and of the DIPPR Evaluated Process Design Data Project Steering Committee. Their source should be cited as R. L. Rowley, W. V. Wilding, J. L. Oscarson, T. A. Knotts, N. F. Giles, DIPPR® Data Compilation of Pure Chemical Properties, Design Institute for Physical Properties, AIChE, New York (2016).

*The ratings reflect DIPPR® ESP's effort to provide a critical evaluation and quality assessment of each data point with 15 being the highest score possible. The rating is not directly correlated with the estimated experimental uncertainty.

†Henry's constant is a strong nonlinear function of temperature. A single value measured at one temperature, if used for calculation at a different temperature, can lead to serious errors. Procedures for extrapolation of single-point values over the ambient temperature range ($4°C < T < 50°C$) are presented in Sec. 22, under "Air Pollution Control" > "Biological APC Technologies" > "Estimating Henry's law constants". Estimation procedures for the larger range ($4°C < T < 200°C$) are presented in F. L. Smith and A. H. Harvey, "Avoid Common Pitfalls When Using Henry's Law," *Chem. Eng. Prog.*, 103(9), 2007. See also Y.-L. Huang, J. D. Olson, and G. E. Keller II, "Steam Stripping for Removal of Organic Pollutants from Water. 2. Vapor-Liquid Equilibrium Data," *Ind. Eng. Chem. Res.*, 31, pp. 1759–1768, 1992. (Also see the Supplementary Material, which contains the databank of 404 compounds of environmental interest and other useful property data.)

TABLE 2-23 Henry's Constant *H* for Various Compounds in Water at 25°C from Infinite Dilution Activity Coefficients

Compound	CAS no.	Formula	$H = \gamma^{\infty} P_{vp}$, atm
Pentane	109660	C_5H_{12}	63700
Hexane	1100543	C_6H_{14}	84600
Heptane	142825	C_7H_{16}	120000
Benzene	71432	C_6H_6	309
Toluene	108883	C_7H_8	344
o-Xylene	95476	C_8H_{10}	267
Cumene	98,828	C_9H_{12}	613
Styrene	100425	C_8H_8	145
Formaldehyde	50000	CH_2O	14.3
Acetaldehyde	75070	C_2H_4O	4.54
Propanal	123386	C_3H_6O	5.45
Acetone	67641	C_3H_6O	2.13
Methyl ethyl ketone	78933	C_4H_8O	3.11
Methyl *n*-propyl ketone	107879	$C_5H_{10}O$	4.60
Formic acid	64186	CH_2O_2	0.0404
Methyl acetate	79209	$C_3H_6O_2$	6.38
Ethyl acetate	141786	$C_4H_8O_2$	8.01
Butyl acetate	123864	$C_6H_{12}O_2$	12.3
Chloroethane	75003	C_2H_5Cl	626
1-Chloropropane	74986	C_3H_7Cl	792
Chlorobenzene	108907	C_6H_5Cl	219
Methanol	67561	CH_4O	0.263
Ethanol	64175	C_2H_6O	0.293
Pyridine	110861	C_5H_5N	0.544
Diethyl ether	60297	$C_4H_{10}O$	48.7
Thiophene	110021	C_4H_4S	160

Henry's constant H at 25°C is the vapor pressure at 25°C times the infinite dilution activity coefficient, also at 25°C. Infinite dilution activity coefficients are from Mitchell and Jurs, *J. Chem. Inf. Comput. Sci.* **38:** 200 (1998). Henry's constant is a strong nonlinear function of temperature. A single value measured at one temperature, if used for calculation at a different temperature, can lead to serious errors. Procedures for extrapolation of single-point values over the ambient temperature range (4°C < T < 50°C) are presented in Sec. 22, pp. 22–49, under "Estimating Henry's law constants." Estimation procedures for the larger range (4°C < T < 200°C) are presented in F. L. Smith and A. H. Harvey, "Avoid Common Pitfalls When Using Henry's Law," *Chem. Eng. Prog.,* **103**(9), 2007. See also Y.-L. Huang, J. D. Olson, and G. E. Keller II, "Steam Stripping for Removal of Organic Pollutants from Water. 2. Vapor-Liquid Equilibrium Data," *Ind. Eng. Chem. Res.*, **31**, pp. 1759–1768, 1992. (Also see the Supplementary Material, which contains the databank of 404 compounds of environmental interest and other useful property data.)

TABLE 2-24 Air*

t, °C	0	5	10	15	20	25	30	35
$10^{-4} \times H^{\dagger}$	4.32	4.88	5.49	6.07	6.64	7.20	7.71	8.23
t, °C	40	45	50	60	70	80	90	100
$10^{-4} \times H^{\dagger}$	8.70	9.11	9.46	10.1	10.5	10.7	10.8	10.7

**International Critical Tables,* vol. 3, p. 257.

†H is calculated from the absorption coefficients of O_2 and N_2, taking into consideration the correction for constant argon content.

TABLE 2-25 Ammonia-Water at 10 and 20°C*

	10°C		20°C	
Mass fraction NH_3 in liquid	P, kPa	Mass fraction NH_3 in vapor	P, kPa	Mass fraction NH_3 in vapor
0.0	1.23	0.0	2.34	0.0
0.00467	1.37	0.1		
0.00495			2.60	0.1
0.1	7.07	0.84164	11.95	0.82096
0.2	20.07	0.95438	32.34	0.94541
0.3	47.37	0.98565	73.85	0.98199
0.4	99.84	0.99544	150.56	0.99393
0.5	184.44	0.99848	269.50	0.99783
0.6	292.15	0.99943	416.63	0.99913
0.7	399.03	0.99975	560.61	0.99960
0.8	486.44	0.99988	678.61	0.99980
0.9	554.33	0.99995	771.87	0.99991
1.0	615.05	1.0	857.48	1.0

*Selected values from R. Tillner-Roth and D. G. Friend, *J. Phys. Chem. Ref. Data* **27:**63 (1998). This reference lists solubilities for temperatures from −70 to 340°C. Densities, enthalpies, and entropies are listed for both the two-phase and single-phase regions for pressures up to 40 MPa.

TABLE 2-26 Carbon Dioxide (CO_2)*

	Liquid mol fraction $CO_2 \times 10^3$								
Total pressure, atm	0°C	10°C	15°C	20°C	25°C	35°C	50°C	75°C	100°C
1	1.445	0.985	0.802	0.692	0.608	0.473	0.342	0.248	0.187
2	2.89	1.946	1.587	1.374	1.207	0.943	0.683	0.495	0.373
10	12.71	8.81	7.32	6.44	5.74	4.54	3.30	2.41	1.841
20	21.23	15.38	13.13	11.84	10.75	8.64	6.34	4.65	3.62
30	25.79	19.80	17.49	16.22	15.05	12.80	9.10	6.78	5.35
36		21.45	19.42	18.30	17.29	14.80	10.63	7.90	6.35

*Values selected from G. Houghton, A. M. McLean, and P. D. Ritchie, *Chem. Eng. Sci.* **6:**132–137, 1957.

TABLE 2-27 Chlorine (Cl_2)

Partial pressure of Cl_2, mmHg	Solubility, g of Cl_2 per liter					
	0°C	10°C	20°C	30°C	40°C	50°C
5	0.488	0.451	0.438	0.424	0.412	0.398
10	0.679	0.603	0.575	0.553	0.532	0.512
30	1.221	1.024	0.937	0.873	0.821	0.781
50	1.717	1.354	1.210	1.106	1.025	0.962
100	2.79	2.08	1.773	1.573	1.424	1.313
150	3.81	2.73	2.27	1.966	1.754	1.599
200	4.78	3.35	2.74	2.34	2.05	1.856
250	5.71	3.95	3.19	2.69	2.34	2.09
300		4.54	3.63	3.03	2.61	2.31
350		5.13	4.06	3.35	2.86	2.53
400		5.71	4.48	3.69	3.11	2.74
450		6.26	4.88	3.98	3.36	2.94
500		6.85	5.29	4.30	3.61	3.14
550		7.39	5.71	4.60	3.84	3.33
600		7.97	6.12	4.91	4.08	3.52
650		8.52	6.52	5.21	4.32	3.71
700		9.09	6.90	5.50	4.54	3.89
750		9.65	7.29	5.80	4.77	4.07
800		10.21	7.69	6.08	4.99	4.27
900			8.46	6.68	5.44	4.62
1000			9.27	7.27	5.89	4.97
1200	$Cl_2.8H_2O_2$ separates		10.84	8.42	6.81	5.67
1500			13.23	10.14	8.05	6.70
2000			17.07	13.02	10.22	8.38
2500			21.0	15.84	12.32	10.03
3000				18.73	14.47	11.70
3500				21.7	16.62	13.38
4000				24.7	18.84	15.04
4500				27.7	20.7	16.75
5000				30.8	23.3	18.46

Partial pressure of Cl_2, mmHg	Solubility, g of Cl_2 per liter					
	60°C	70°C	80°C	90°C	100°C	110°C
5	0.383	0.369	0.351	0.339	0.326	0.316
10	0.492	0.470	0.447	0.431	0.415	0.402
30	0.743	0.704	0.671	0.642	0.627	0.598
50	0.912	0.863	0.815	0.781	0.747	0.722
100	1.228	1.149	1.085	1.034	0.987	0.950
150	1.482	1.382	1.294	1.227	1.174	1.137
200	1.706	1.580	1.479	1.396	1.333	1.276
250	1.914	1.764	1.642	1.553	1.480	1.413
300	2.10	1.932	1.793	1.700	1.610	1.542
350	2.28	2.10	1.940	1.831	1.736	1.661
400	2.47	2.25	2.08	1.965	1.854	1.773
450	2.64	2.41	2.22	2.09	1.972	1.880
500	2.80	2.55	2.35	2.21	2.08	1.986
550	2.97	2.69	2.47	2.32	2.19	2.09
600	3.13	2.83	2.59	2.43	2.29	2.19
650	3.29	2.97	2.72	2.55	2.41	2.28
700	3.44	3.10	2.84	2.66	2.50	2.37
750	3.59	3.23	2.96	2.76	2.60	2.47
800	3.75	3.37	3.08	2.87	2.69	2.56
900	4.04	3.63	3.30	3.08	2.89	2.74
1000	4.36	3.88	3.53	3.28	3.07	2.91
1200	4.92	4.37	3.95	3.67	3.43	3.25
1500	5.76	5.09	4.58	4.23	3.95	3.74
2000	7.14	6.26	5.63	5.17	4.78	4.49
2500	8.48	7.40	6.61	6.05	5.59	5.25
3000	9.83	8.52	7.54	6.92	6.38	5.97
3500	11.22	9.65	8.53	7.79	7.16	6.72
4000	12.54	10.76	9.52	8.65	7.94	7.42
4500	13.88	11.91	10.46	9.49	8.72	8.13
5000	15.26	13.01	11.42	10.35	9.48	8.84

TABLE 2-28 Chlorine Dioxide (ClO_2)

Vol % of ClO_2 in gas phase	Weight of ClO_2, grams per liter of solution						
	0°C	5°C	10°C	15°C	20°C	30°C	40°C
1	2.00	1.50	1.25	1.00	0.90	0.60	0.46
3	6.00	4.7	3.85	3.20	2.70	1.95	1.30
5	10.0	7.8	6.30	5.25	4.30	3.20	2.25
7	14.0	10.9	8.95	7.35	6.15	4.40	3.20
10	20.0	15.5	12.8	10.5	8.80	6.30	4.50
11		17.0	14.0	11.7	9.70	7.00	5.00
12		18.6	15.3	12.8	10.55	7.50	5.45
13		20.3	16.6	13.8	11.5	8.20	5.85
14			18.0	14.9	12.3	8.80	6.35
15			19.2	16.0	13.2	9.50	6.80
16			20.3	17.0	14.2	10.1	7.20

Ishi, *Chem. Eng. (Japan)*, **22**:153 (1958).

TABLE 2-29 Hydrogen Chloride (HCl)

Weights of HCl per 100 weights of H_2O	Partial pressure of HCl, mmHg			
	0°C	10°C	20°C	30°C
78.6	510	840		
66.7	130	233	399	627
56.3	29.0	56.4	105.5	188
47.0	5.7	11.8	23.5	44.5
38.9	1.0	2.27	4.90	9.90
31.6	0.175	0.43	1.00	2.17
25.0	0.0316	0.084	0.205	0.48
19.05	0.0056	0.016	0.0428	0.106
13.64	0.00099	0.00305	0.0088	0.0234
8.70	0.000118	0.000583	0.00178	0.00515
4.17	0.000018	0.000069	0.00024	0.00077
2.04		0.0000117	0.000044	0.000151

Weights of HCl per 100 weights of H_2O	Partial pressure of HCl, mm Hg		
	50°C	80°C	110°C
78.6			
66.7			
56.3	535		
47.0	141	623	
38.9	35.7	188	760
31.6	8.9	54.5	253
25.0	2.21	15.6	83
19.05	0.55	4.66	28
13.64	0.136	1.34	9.3
8.70	0.0344	0.39	3.10
4.17	0.0064	0.095	0.93
2.04	0.00140	0.0245	0.280

Enthalpy and phase-equilibrium data for the binary system HCl-H_2O are given by Van Nuys, *Trans. Am. Inst. Chem. Engrs.*, **39,** 663 (1943).

TABLE 2-30 Hydrogen Sulfide (H_2S)

t, °C	0	5	10	15	20	25	30	35
$10^{-2} \times H$	2.68	3.15	3.67	4.23	4.83	5.45	6.09	6.76
t, °C	40	45	50	60	70	80	90	100
$10^{-2} \times H$	7.45	8.14	8.84	10.3	11.9	13.5	14.4	14.8

International Critical Tables, vol. 3, p. 259.

DENSITIES

Unit Conversions Unless otherwise noted, densities are given in grams per cubic centimeter. To convert to pounds per cubic foot, multiply by 62.43. Temperature conversion: °F = 9/5°C + 32.

Additional References and Comments The aqueous solution data tables are from *International Critical Tables,* vol. 3, pp. 115–129, unless otherwise stated. All compositions are in weight percent in vacuo. All density values are $d_4^t = g/mL$ in vacuo. For more detailed data on densities, see also the *CRC Handbook of Chemistry and Physics,* Chemical Rubber Publishing Co., 97th ed.; or http://hbcponline.com.

DENSITIES OF PURE SUBSTANCES

TABLE 2-31 Density (kg/m³) of Saturated Liquid Water from the Triple Point to the Critical Point

T, K	ρ, kg/m³	*T*, K	ρ, kg/m³	*T*, K	ρ, kg/m³	*T*, K	ρ, kg/m³	*T*, K	ρ, kg/m³
273.160*	999.793	352	972.479	432	908.571	512	814.982	592	669.930
274	999.843	354	971.235	434	906.617	514	812.164	594	664.974
276	999.914	356	969.972	436	904.645	516	809.318	596	659.907
278	999.919	358	968.689	438	902.656	518	806.441	598	654.722
280	999.862	360	967.386	440	900.649	520	803.535	600	649.411
282	999.746	362	966.064	442	898.624	522	800.597	602	643.97
284	999.575	364	964.723	444	896.580	524	797.629	604	638.38
286	999.352	366	963.363	446	894.519	526	794.628	606	632.64
288	999.079	368	961.984	448	892.439	528	791.594	608	626.74
290	998.758	370	960.587	450	890.341	530	788.527	610	620.65
292	998.392	372	959.171	452	888.225	532	785.425	612	614.37
294	997.983	374	957.737	454	886.089	534	782.288	614	607.88
296	997.532	376	956.285	456	883.935	536	779.115	616	601.15
298	997.042	378	954.815	458	881.761	538	775.905	618	594.16
300	996.513	380	953.327	460	879.569	540	772.657	620	586.88
302	995.948	382	951.822	462	877.357	542	769.369	622	579.26
304	995.346	384	950.298	464	875.125	544	766.042	624	571.25
306	994.711	386	948.758	466	872.873	546	762.674	626	562.81
308	994.042	388	947.199	468	870.601	548	759.263	628	553.84
310	993.342	390	945.624	470	868.310	550	755.808	630	544.25
312	992.610	392	944.030	472	865.997	552	752.308	632	533.92
314	991.848	394	942.420	474	863.664	554	748.762	634	522.71
316	991.056	396	940.793	476	861.310	556	745.169	636	510.42
318	990.235	398	939.148	478	858.934	558	741.525	638	496.82
320	989.387	400	937.486	480	856.537	560	737.831	640	481.53
322	988.512	402	935.807	482	854.118	562	734.084	641	473.01
324	987.610	404	934.111	484	851.678	564	730.283	642	463.67
326	986.682	406	932.398	486	849.214	566	726.425	643	453.14
328	985.728	408	930.668	488	846.728	568	722.508	644	440.73
330	984.750	410	928.921	490	844.219	570	718.530	645	425.05
332	983.747	412	927.157	492	841.686	572	714.489	646	402.96
334	982.721	414	925.375	494	839.130	574	710.382	647	357.34
336	981.671	416	923.577	496	836.549	576	706.206	647.096†	322
338	980.599	418	921.761	498	833.944	578	701.959		
340	979.503	420	919.929	500	831.313	580	697.638		
342	978.386	422	918.079	502	828.658	582	693.238		
344	977.247	424	916.212	504	825.976	584	688.757		
346	976.086	426	914.328	506	823.269	586	684.190		
348	974.904	428	912.426	508	820.534	588	679.533		
350	973.702	430	910.507	510	817.772	590	674.781		

*Triple point

†Critical point

From Wagner, W., and Pruss, A., "The IAPWS Formulation 1995 for the Thermodynamic Properties of Ordinary Water Substance for General and Scientific Use," *J. Phys. Chem. Ref. Data* **31**(2):387–535, 2002.

TABLE 2-32 Densities of Inorganic and Organic Liquids (mol/dm³)

Eqn	Cmpd. no.	Name	Formula	CAS	Mol. wt.	C_1	C_2	C_3	C_4	C_5	C_6	C_7	T_{min}, K	Density at T_{min}	T_{max}, K	Density at T_{max}
105	1	Acetaldehyde	C_2H_4O	75-07-0	44.05256	1.711365	0.26355	466	0.28571				149.78	21.423	466.00	6.4935
105	2	Acetamide	C_2H_5NO	60-35-5	59.0672	1.016	0.21845	761	0.26116				353.33	16.936	761.00	4.6509
105	3	Acetic acid	$C_2H_4O_2$	64-19-7	60.052	1.4486	0.25892	591.95	0.2529				289.81	17.492	591.95	5.5948
105	4	Acetic anhydride	$C_4H_6O_3$	108-24-7	102.08864	0.79388	0.24119	606	0.29817				200.15	11.626	606.00	3.2915
105	5	Acetone	C_3H_6O	67-64-1	58.07914	1.2332	0.25886	508.2	0.2913				178.45	15.683	508.20	4.7640
105	6	Acetonitrile	C_2H_3N	75-05-8	41.0519	1.0693	0.20656	545.5	0.24699				229.32	20.544	545.50	5.1767
105	7	Acetylene	C_2H_2	74-86-2	26.03728	2.4507	0.27448	308.3	0.28752				192.40	23.692	308.30	8.9285
105	8	Acrolein	C_3H_4O	107-02-8	56.06326	1.3261	0.26124	506	0.2489				185.45	16.822	506.00	5.0762
105	9	Acrylic acid	$C_3H_4O_2$	79-10-7	72.06266	1.2414	0.25822	615	0.30701				286.15	14.693	615.00	4.8075
105	10	Acrylonitrile	C_3H_3N	107-13-1	53.0626	1.0379	0.22465	540	0.28921				189.63	17.254	540.00	4.6201
105	11	Air	Mixture	132259-10-0	28.96	2.8963	0.26733	132.45	0.27341				59.15	33.279	132.45	10.8340
105	12	Ammonia	H_3N	7664-41-7	17.03052	3.5383	0.25443	405.65	0.2888				195.41	43.141	405.65	13.9070
105	13	Anisole	C_7H_8O	100-66-3	108.13782	0.77488	0.26114	645.6	0.28234				235.65	9.6675	645.60	2.9673
105	14	Argon	Ar	7440-37-1	39.948	3.8469	0.2881	150.86	0.29783				83.78	35.491	150.86	13.3530
105	15	Benzamide	C_7H_7NO	55-21-0	121.13658	0.7371	0.25487	824	0.28571				403.00	8.9381	824.00	2.8921
105	16	Benzene	C_6H_6	71-43-2	78.11184	1.0259	0.26666	562.05	0.28394				278.68	11.422	562.05	3.8472
105	17	Benzenethiol	C_6H_6S	108-98-5	110.17684	0.83573	0.26326	689	0.30798				258.27	10.074	689.00	3.1745
105	18	Benzoic acid	$C_7H_6O_2$	65-85-0	122.12134	0.71587	0.24812	751	0.2857				395.45	8.8935	751.00	2.8852
105	19	Benzonitrile	C_7H_5N	100-47-0	103.1213	0.72184	0.24606	702.3	0.28789				260.28	10.008	702.30	2.9336
105	20	Benzophenone	$C_{13}H_{10}O$	119-61-9	182.2179	0.43743	0.24833	830	0.27555				321.35	5.9496	830.00	1.7615
105	21	Benzyl alcohol	C_7H_8O	100-51-6	108.13782	0.59867	0.22849	720.15	0.23567				257.85	9.9051	720.15	2.6201
105	22	Benzyl ethyl ether	$C_9H_{12}O$	539-30-0	136.19098	0.60917	0.26925	662	0.2632				275.65	7.0651	662.00	2.2625
105	23	Benzyl mercaptan	C_7H_8S	100-53-8	124.20342	0.70797	0.25982	718	0.32144				243.95	8.8623	718.00	2.7248
105	24	Biphenyl	$C_{12}H_{10}$	92-52-4	154.2078	0.52257	0.25833	773	0.27026				342.20	6.4251	773.00	2.0229
105	25	Bromine	Br_2	7726-95-6	159.808	2.1872	0.29527	584.15	0.3295				265.85	20.109	584.15	7.4075
105	26	Bromobenzene	C_6H_5Br	108-86-1	157.0079	0.8226	0.26632	670.15	0.2821				242.43	9.9087	670.15	3.0888
105	27	Bromoethane	C_2H_5Br	74-96-4	108.965	1.3285	0.2708	503.8	0.3012				154.25	15.809	503.80	4.9058
105	28	Bromomethane	CH_3Br	74-83-9	94.93852	1.796	0.27065	464	0.28947				173.00	20.787	464.00	6.6359
105	29	1,2-Butadiene	C_4H_6	590-19-2	54.09044	1.187	0.26114	452	0.3065				136.95	15.123	452.00	4.5455
105	30	1,3-Butadiene	C_4H_6	106-99-0	54.09044	1.2346	0.27216	425	0.28707				164.25	14.058	425.00	4.5363
105	31	Butane	C_4H_{10}	106-97-8	58.1222	1.0677	0.27188	425.12	0.28688				134.86	12.62	425.12	3.9271
105	32	1,2-Butanediol	$C_4H_{10}O_2$	584-03-2	90.121	0.81696	0.24755	680	0.24535				220.00	11.734	680.00	3.3002
105	33	1,3-Butanediol	$C_4H_{10}O_2$	107-88-0	90.121	0.81856	0.24967	676	0.22023				196.15	11.872	676.00	3.2786
105	34	1-Butanol	$C_4H_{10}O$	71-36-3	74.1216	0.98279	0.2683	563.1	0.25488				183.85	12.035	563.10	3.6630
105	35	2-Butanol	$C_4H_{10}O$	78-92-2	74.1216	0.97552	0.26339	535.9	0.26864				158.45	12.473	535.90	3.7037
105	36	1-Butene	C_4H_8	106-98-9	56.10632	1.0877	0.26454	419.5	0.2843				87.80	14.264	419.50	4.1117
105	37	*cis*-2-Butene	C_4H_8	590-18-1	56.10632	1.1591	0.27085	435.5	0.28116				134.26	13.894	435.50	4.2795
105	38	*trans*-2-Butene	C_4H_8	624-64-6	56.10632	1.1448	0.27154	428.6	0.28419				167.62	13.08	428.60	4.2160
105	39	Butyl acetate	$C_6H_{12}O_2$	123-86-4	116.15828	0.67794	0.2637	575.4	0.29318				199.65	8.3365	575.40	2.5709
105	40	Butylbenzene	$C_{10}H_{14}$	104-51-8	134.21816	0.50812	0.25238	660.5	0.29373				185.30	7.0264	660.50	2.0133
105	41	Butyl mercaptan	$C_4H_{10}S$	109-79-5	90.1872	0.89458	0.27463	570.1	0.28512				157.46	10.585	570.10	3.2574
105	42	sec-Butyl mercaptan	$C_4H_{10}S$	513-53-1	90.1872	0.89137	0.27365	554	0.2953				133.02	10.761	554.00	3.2573
105	43	1-Butyne	C_4H_6	107-00-6	54.09044	1.3409	0.27892	440	0.29661				147.43	14.901	440.00	4.8075
105	44	Butyraldehyde	C_4H_8O	123-72-8	72.10572	1.033873	0.266739	537.2	0.28571				176.80	12.602	537.20	3.8760
105	45	Butyric acid	$C_4H_8O_2$	107-92-6	88.1051	0.88443	0.25828	615.7	0.248				267.95	11.087	615.70	3.4243
105	46	Butyronitrile	C_4H_7N	109-74-0	69.1051	0.79716	0.23168	585.4	0.28071				161.30	13.087	585.40	3.4408
105	47	Carbon dioxide	CO_2	124-38-9	44.0095	2.768	0.26212	304.21	0.2908				216.58	26.828	304.21	10.5600
105	48	Carbon disulfide	CS_2	75-15-0	76.1407	1.7968	0.28749	552	0.3226				161.11	19.064	552.00	6.2500
105	49	Carbon monoxide	CO	630-08-0	28.0101	2.897	0.27532	132.92	0.2813				68.15	30.18	132.92	10.5220
105	50	Carbon tetrachloride	CCl_4	56-23-5	153.8227	0.99835	0.274	556.35	0.287				250.33	10.843	556.35	3.6436
105	51	Carbon tetrafluoride	CF_4	75-73-0	88.0043	1.955	0.27884	227.51	0.28571				89.56	21.211	227.51	7.0112
105	52	Chlorine	Cl_2	7782-50-5	70.906	2.23	0.27645	417.15	0.2926				172.12	24.242	417.15	8.0666

(*Continued*)

TABLE 2-32 Densities of Inorganic and Organic Liquids (mol/dm³) *(Continued)*

Eqn	Cmpd. no.	Name	Formula	CAS	Mol. wt.	C_1	C_2	C_3	C_4	C_5	C_6	C_7	T_{min}, K	Density at T_{min}	T_{max}, K	Density at T_{max}
105	53	Chlorobenzene	C_6H_5Cl	108-90-7	112.5569	0.8711	0.26805	632.35	0.2799				227.95	10.385	632.35	3.2498
105	54	Chloroethane	C_2H_5Cl	75-00-3	64.5141	1.39625	0.26867	460.35	0.28571				136.75	17.055	460.35	5.1969
105	55	Chloroform	$CHCl_3$	67-66-3	119.37764	1.0841	0.2581	536.4	0.2741				209.63	13.702	536.40	4.2003
105	56	Chloromethane	CH_3Cl	74-87-3	50.4875	1.8651	0.2627	416.25	0.28571				175.43	22.272	416.25	7.0997
105	57	1-Chloropropane	C_3H_7Cl	540-54-5	78.54068	1.12465	0.2728	503.15	0.28571				150.35	13.333	503.15	4.1226
105	58	2-Chloropropane	C_3H_7Cl	75-29-6	78.54068	1.1202	0.27669	489	0.27646				155.97	12.855	489.00	4.0486
105	59	*m*-Cresol	C_7H_8O	108-39-4	108.13782	0.9061	0.28268	705.85	0.2707				285.39	9.6115	705.85	3.2054
105	60	*o*-Cresol	C_7H_8O	95-48-7	108.13782	0.95937	0.2882	697.55	0.2857				304.19	9.5725	697.55	3.3288
105	61	*p*-Cresol	C_7H_8O	106-44-5	108.13782	1.1503	0.31861	704.65	0.30104				307.93	9.4494	704.65	3.6104
105	62	Cumene	C_9H_{12}	98-82-8	120.19158	0.58711	0.25583	631	0.28498				177.14	7.9387	631.00	2.2949
105	63	Cyanogen	C_2N_2	460-19-5	52.0348	1.7805	0.26846	400.15	0.26079				245.25	18.517	400.15	6.6323
105	64	Cyclobutane	C_4H_8	287-23-0	56.10632	1.3931	0.29255	459.93	0.24913				182.48	14.074	459.93	4.7619
105	65	Cyclohexane	C_6H_{12}	110-82-7	84.15948	0.88998	0.27376	553.8	0.28571				279.69	9.3804	553.80	3.2509
105	66	Cyclohexanol	$C_6H_{12}O$	108-93-0	100.15888	0.8243	0.26545	650.1	0.28495				296.60	9.4693	650.10	3.1053
105	67	Cyclohexanone	$C_6H_{10}O$	108-94-1	98.143	0.86464	0.26888	653	0.29943				242.00	10.09	653.00	3.2157
105	68	Cyclohexene	C_6H_{10}	110-83-8	82.1436	0.92997	0.27056	560.4	0.28943				169.67	11.16	560.40	3.4372
105	69	Cyclopentane	C_5H_{10}	287-92-3	70.1329	1.0897	0.28356	511.7	0.25142				179.28	11.906	511.70	3.8429
105	70	Cyclopentene	C_5H_8	142-29-0	68.11702	1.1035	0.27035	507	0.28699				138.13	13.47	507.00	4.0817
105	71	Cyclopropane	C_3H_6	75-19-4	42.07974	1.7411	0.28205	398	0.29598				145.59	18.658	398.00	6.1730
105	72	Cyclohexyl mercaptan	$C_6H_{12}S$	1569-69-3	116.22448	0.78578	0.27882	664	0.31067				189.64	8.9048	664.00	2.8182
105	73	Decanal	$C_{10}H_{20}O$	112-31-2	156.2652	0.478542	0.275162	674	0.28571				285.00	5.2396	674.00	1.7391
105	74	Decane	$C_{10}H_{22}$	124-18-5	142.28168	0.41084	0.25175	617.7	0.28571				243.51	5.3927	617.70	1.6319
105	75	Decanoic acid	$C_{10}H_{20}O_2$	334-48-5	172.265	0.39348	0.2492	722.1	0.28571				304.55	5.1809	722.10	1.5790
105	76	1-Decanol	$C_{10}H_{22}O$	112-30-1	158.28108	0.38208	0.24645	688	0.26125				280.05	5.2609	688.00	1.5503
105	77	1-Decene	$C_{10}H_{20}$	872-05-9	140.2658	0.43981	0.25661	616.6	0.29148				206.89	5.7328	616.60	1.7139
105	78	Decyl mercaptan	$C_{10}H_{22}S$	143-10-2	174.34668	0.44289	0.27636	696	0.27668				247.56	5.0048	696.00	1.6026
105	79	1-Decyne	$C_{10}H_{18}$	764-93-2	138.24992	0.46877	0.25875	619.85	0.29479				229.15	5.8954	619.85	1.8117
105	80	Deuterium	D_2	7782-39-0	4.0316	5.2115	0.315	38.35	0.28571				18.73	42.945	38.35	16.5440
105	81	1,1-Dibromoethane	$C_2H_4Br_2$	557-91-5	187.86116	0.95523	0.26364	628	0.29825				210.15	11.799	628.00	3.6232
105	82	1,2-Dibromoethane	$C_2H_4Br_2$	106-93-4	187.86116	1.0132	0.26634	650.15	0.28571				282.85	11.704	650.15	3.8042
105	83	Dibromomethane	CH_2Br_2	74-95-3	173.83458	1.1136	0.24834	611	0.27583				220.60	15.358	611.00	4.4842
105	84	Dibutyl ether	$C_8H_{18}O$	142-96-1	130.22792	0.55941	0.27243	584.1	0.29932				175.30	6.6071	584.10	2.0534
105	85	*m*-Dichlorobenzene	$C_6H_4Cl_2$	541-73-1	147.00196	0.74495	0.26147	683.95	0.31526				248.39	9.1207	683.95	2.8491
105	86	*o*-Dichlorobenzene	$C_6H_4Cl_2$	95-50-1	147.00196	0.74404	0.26112	705	0.30815				256.15	9.1658	705.00	2.8494
105	87	*p*-Dichlorobenzene	$C_6H_4Cl_2$	106-46-7	147.00196	0.74858	0.26276	684.75	0.30788				326.14	8.5175	684.75	2.8489
105	88	1,1-Dichloroethane	$C_2H_4Cl_2$	75-34-3	98.95916	1.1055	0.26533	523	0.287				176.19	13.549	523.00	4.1665
105	89	1,2-Dichloroethane	$C_2H_4Cl_2$	107-06-2	98.95916	1.2591	0.27698	561.6	0.30492				237.49	13.462	561.60	4.5458
105	90	Dichloromethane	CH_2Cl_2	75-09-2	84.93258	1.3897	0.25678	510	0.2902				178.01	17.974	510.00	5.4120
105	91	1,1-Dichloropropane	$C_3H_6Cl_2$	78-99-9	112.98574	0.9551	0.27794	560	0.24132				192.50	10.925	560.00	3.4364
105	92	1,2-Dichloropropane	$C_3H_6Cl_2$	78-87-5	112.98574	0.89833	0.26142	572	0.2868				172.71	11.526	572.00	3.4363
105	93	Diethanol amine	$C_4H_{11}NO_2$	111-42-2	105.13564	0.68184	0.23796	736.6	0.2062				301.15	10.39	736.60	2.8654
105	94	Diethyl amine	$C_4H_{11}N$	109-89-7	73.13684	0.85379	0.25675	496.6	0.27027				223.35	10.575	496.60	3.3254
105	95	Diethyl ether	$C_4H_{10}O$	60-29-7	74.1216	0.9554	0.26847	466.7	0.2814				156.85	11.487	466.70	3.5587
105	96	Diethyl sulfide	$C_4H_{10}S$	352-93-2	90.1872	0.82227	0.26314	557.15	0.27369				169.20	10.47	557.15	3.1248
105	97	1,1-Difluoroethane	$C_2H_4F_2$	75-37-6	66.04997	1.4345	0.25774	386.44	0.28178				154.56	18.006	386.44	5.5657
105	98	1,2-Difluoroethane	$C_2H_4F_2$	624-72-6	66.04997	1.173	0.22856	445	0.28571				179.60	18.336	445.00	5.1321
105	99	Difluoromethane	CH_2F_2	75-10-5	52.02339	1.9973	0.24653	351.26	0.28153				136.95	27.399	351.26	8.1017
105	100	Di–sopropyl amine	$C_6H_{15}N$	108-18-9	101.19	0.6181	0.25786	523.1	0.271				176.85	8.0541	523.10	2.3970
105	101	Di–sopropyl ether	$C_6H_{14}O$	108-20-3	102.17476	0.69213	0.26974	500.05	0.28571				187.65	8.0673	500.05	2.5659
105	102	Di–sopropyl ketone	$C_7H_{14}O$	565-80-0	114.18546	0.64619	0.26881	576	0.28036				204.81	7.6796	576.00	2.4039
105	103	1,1-Dimethoxyethane	$C_4H_{10}O_2$	534-15-6	90.121	0.89368	0.26599	507.8	0.28571				159.95	11.029	507.80	3.3598
105	104	1,2-Dimethoxypropane	$C_5H_{12}O_2$	7778-85-0	104.14758	0.76327	0.26742	543	0.28571				226.10	8.8431	543.00	2.8542

105	105	Dimethyl acetylene	C_4H_6	503-17-3	54.09044	1.1717	0.25895	473.2	0.27289				240.91	13.767	473.20	4.5248
105	106	Dimethyl amine	C_2H_7N	124-40-3	45.08368	1.5436	0.27784	437.2	0.2572				180.96	16.964	437.20	5.5557
105	107	2,3-Dimethylbutane	C_6H_{14}	79-29-8	86.17536	0.7565	0.27305	500	0.27408				145.19	9.031	500.00	2.7706
105	108	1,1-Dimethylcyclohexane	C_8H_{16}	590-66-9	112.21264	0.55873	0.25143	591.15	0.27758				239.66	7.3417	591.15	2.2222
105	109	*cis*-1,2-Dimethylcyclohexane	C_8H_{16}	2207-01-4	112.21264	0.52953	0.24358	606.15	0.26809				223.16	7.5783	606.15	2.1739
105	110	*trans*-1,2-Dimethylcyclohexane	C_8H_{16}	6876-23-9	112.21264	0.54405	0.25026	596.15	0.2658				184.99	7.6258	596.15	2.1739
105	111	Dimethyl disulfide	$C_2H_6S_2$	624-92-0	94.19904	1.1058	0.27866	615	0.31082				188.44	12.413	615.00	3.9683
105	112	Dimethyl ether	C_2H_6O	115-10-6	46.06844	1.5693	0.2679	400.1	0.2882				131.65	18.95	400.10	5.8578
105	113	*N,N*-Dimethyl formamide	C_3H_7NO	68-12-2	73.09378	0.89615	0.23478	649.6	0.28091				212.72	13.954	649.60	3.8170
105	114	2,3-Dimethylpentane	C_7H_{16}	565-59-3	100.20194	0.72352	0.28629	537.3	0.27121				141.23	7.9932	537.30	2.5272
105	115	Dimethyl phthalate	$C_{10}H_{10}O_4$	131-11-3	194.184	0.47977	0.25428	766	0.30722				274.18	6.2334	766.00	1.8868
105	116	Dimethylsilane	C_2H_8Si	1111-74-6	60.17042	1.0214	0.26351	402	0.28421				122.93	12.898	402.00	3.8761
105	117	Dimethyl sulfide	C_2H_6S	75-18-3	62.134	1.4029	0.27991	503.04	0.2741				174.88	15.556	503.04	5.0120
105	118	Dimethyl sulfoxide	C_2H_6OS	67-68-5	78.13344	1.1096	0.25189	729	0.3311				291.67	14.111	729.00	4.4051
105	119	Dimethyl terephthalate	$C_{10}H_{10}O_4$	120-61-6	194.184	0.48611	0.25715	777.4	0.28571				413.79	5.6397	777.40	1.8904
105	120	1,4-Dioxane	$C_4H_8O_2$	123-91-1	88.10512	1.1819	0.2813	587	0.3047				284.95	11.838	587.00	4.2016
105	121	Diphenyl ether	$C_{12}H_{10}O$	101-84-8	170.2072	0.52133	0.26218	766.8	0.31033				300.03	6.2648	766.80	1.9884
105	122	Dipropyl amine	$C_6H_{15}N$	142-84-7	101.19	0.659	0.26428	550	0.2766				210.15	7.9929	550.00	2.4936
105	123	Dodecane	$C_{12}H_{26}$	112-40-3	170.33484	0.33267	0.24664	658	0.28571				263.57	4.5205	658.00	1.3488
105	124	Eicosane	$C_{20}H_{42}$	112-95-8	282.54748	0.18166	0.23351	768	0.28571				309.58	2.7293	768.00	0.7780
105	125	Ethane	C_2H_6	74-84-0	30.069	1.9122	0.27937	305.32	0.29187				90.35	21.64	305.32	6.8447
105	126	Ethanol	C_2H_6O	64-17-5	46.06844	1.6288	0.27469	514	0.23178				159.05	19.41	514.00	5.9296
105	127	Ethyl acetate	$C_4H_8O_2$	141-78-6	88.10512	0.8996	0.25856	523.3	0.278				189.60	11.478	523.30	3.4793
105	128	Ethyl amine	C_2H_7N	75-04-7	45.08368	1.0936	0.22636	456.15	0.25522				192.15	17.588	456.15	4.8312
105	129	Ethylbenzene	C_8H_{10}	100-41-4	106.165	0.70041	0.26162	617.15	0.28454				178.20	9.0407	617.15	2.6772
105	130	Ethyl benzoate	$C_9H_{10}O_2$	93-89-0	150.1745	0.48864	0.23894	698	0.28421				238.45	7.2908	698.00	2.0450
105	131	2-Ethyl butanoic acid	$C_6H_{12}O_2$	88-09-5	116.15828	0.66085	0.25707	655	0.31103				258.15	8.2198	655.00	2.5707
105	132	Ethyl butyrate	$C_6H_{12}O_2$	105-54-4	116.15828	0.63566	0.25613	571	0.27829				175.15	8.4912	571.00	2.4818
105	133	Ethylcyclohexane	C_8H_{16}	1678-91-7	112.21264	0.61587	0.26477	609.15	0.28054				161.84	7.8679	609.15	2.3261
105	134	Ethylcyclopentane	C_7H_{14}	1640-89-7	98.18606	0.71751	0.26903	569.5	0.27733				134.71	9.0179	569.50	2.6670
105	135	Ethylene	C_2H_4	74-85-1	28.05316	2.0961	0.27657	282.34	0.29147				104.00	23.326	282.34	7.5789
105	136	Ethylenediamine	$C_2H_8N_2$	107-15-3	60.09832	0.7842	0.20702	593	0.20254				284.29	15.055	593.00	3.7880
105	137	Ethylene glycol	$C_2H_6O_2$	107-21-1	62.06784	1.315	0.25125	720	0.21868				260.15	18.31	720.00	5.2338
105	138	Ethyleneimine	C_2H_5N	151-56-4	43.0678	1.3462	0.23289	537	0.23357				195.20	21.45	537.00	5.7804
105	139	Ethylene oxide	C_2H_4O	75-21-8	44.05256	1.836	0.26024	469.15	0.2696				160.65	23.477	469.15	7.0550
105	140	Ethyl formate	$C_3H_6O_2$	109-94-4	74.07854	1.1343	0.26168	508.4	0.2791				193.55	14.006	508.40	4.3347
105	141	2-Ethyl hexanoic acid	$C_8H_{16}O_2$	149-57-5	144.211	0.47428	0.25028	674.6	0.25442				155.15	6.926	674.60	1.8950
105	142	Ethylhexyl ether	$C_8H_{18}O$	5756-43-4	130.22792	0.55729	0.2714	583	0.29538				180.00	6.612	583.00	2.0534
105	143	Ethylisopropyl ether	$C_5H_{12}O$	625-54-7	88.14818	0.8185	0.26929	489	0.30621				140.00	9.9236	489.00	3.0395
105	144	Ethylisopropyl ketone	$C_6H_{12}O$	565-69-5	100.15888	0.68162	0.25152	567	0.3182				204.15	8.9749	567.00	2.7100
105	145	Ethyl mercaptan	C_2H_6S	75-08-1	62.13404	1.3047	0.2694	499.15	0.27866				125.26	16.242	499.15	4.8430
105	146	Ethyl propionate	$C_5H_{10}O_2$	105-37-3	102.1317	0.7405	0.25563	546	0.2795				199.25	9.6317	546.00	2.8968
105	147	Ethylpropyl ether	$C_5H_{12}O$	628-32-0	88.14818	0.7908	0.266	500.23	0.292				145.65	9.8474	500.23	2.9729
105	148	Ethyltrichlorosilane	$C_2H_5Cl_3Si$	115-21-9	163.506	0.61243	0.24681	559.95	0.30858				167.55	8.6934	559.95	2.4814
105	149	Fluorine	F_2	7782-41-4	37.9968064	4.2895	0.28587	144.12	0.28776				53.48	44.888	144.12	15.0050
105	150	Fluorobenzene	C_6H_5F	462-06-6	96.1023032	1.0146	0.27277	560.09	0.28291				230.94	11.374	560.09	3.7196
105	151	Fluoroethane	C_2H_5F	353-36-6	48.0595	1.693858	0.269323	375.31	0.28571				129.95	20.099	375.31	6.2893
105	152	Fluoromethane	CH_3F	593-53-3	34.03292	2.2261	0.25072	317.42	0.27343				131.35	29.345	317.42	8.8788
105	153	Formaldehyde	CH_2O	50-00-0	30.02598	3.897011	0.331636	420	0.28571				155.15	30.92	420.00	11.7510
105	154	Formamide	CH_3NO	75-12-7	45.04062	1.2486	0.20352	771	0.25178				275.60	25.488	771.00	6.1350
105	155	Formic acid	CH_2O_2	64-18-6	46.0257	1.938	0.24225	588	0.24435				281.45	26.806	588.00	8.0000
105	156	Furan	C_4H_4O	110-00-9	68.07396	1.1339	0.24741	490.15	0.2612				187.55	15.702	490.15	4.5831
105	157	Helium-4	He	7440-59-7	4.0026	7.2475	0.41865	5.2	0.24096				2.20	37.115	5.20	17.3120
105	158	Heptadecane	$C_{17}H_{36}$	629-78-7	240.46774	0.21897	0.23642	736	0.28571				295.13	3.2189	736.00	0.9262
105	159	Heptanal	$C_7H_{14}O$	111-71-7	114.18546	0.577362	0.250575	620	0.28571				229.80	7.7462	620.00	2.3041

(*Continued*)

TABLE 2-32 Densities of Inorganic and Organic Liquids (mol/dm³) *(Continued)*

Eqn	Cmpd. no.	Name	Formula	CAS	Mol. wt.	C_1	C_2	C_3	C_4	C_5	C_6	C_7	T_{min}, K	Density at T_{min}	T_{max}, K	Density at T_{max}
105	160	Heptane	C_7H_{16}	142-82-5	100.20194	0.61259	0.26211	540.2	0.28141				182.57	7.6998	540.20	2.3371
105	161	Heptanoic acid	$C_7H_{14}O_2$	111-14-8	130.185	0.53066	0.24729	677.3	0.28289				265.83	7.2212	677.30	2.1459
105	162	1-Heptanol	$C_7H_{16}O$	111-70-6	116.20134	0.55687	0.24725	632.3	0.31471				239.15	7.5022	632.30	2.2523
105	163	2-Heptanol	$C_7H_{16}O$	543-49-7	116.20134	0.59339	0.2602	608.3	0.26968				220.00	7.5173	608.30	2.2805
105	164	3-Heptanone	$C_7H_{14}O$	106-35-4	114.18546	0.59268	0.25663	606.6	0.27766				234.15	7.5751	606.60	2.3095
105	165	2-Heptanone	$C_7H_{14}O$	110-43-0	114.18546	0.58247	0.25279	611.4	0.29818				238.15	7.5514	611.40	2.3042
105	166	1-Heptene	C_7H_{14}	592-76-7	98.18606	0.66016	0.26657	537.4	0.28571				154.12	8.2257	537.40	2.4765
105	167	Heptyl mercaptan	$C_7H_{16}S$	1639-09-4	132.26694	0.58622	0.2726	645	0.29644				229.92	6.7277	645.00	2.1505
105	168	1-Heptyne	C_7H_{12}	628-71-7	96.17018	0.67304	0.26045	547	0.28388				192.22	8.4922	547.00	2.5841
105	169	Hexadecane	$C_{16}H_{34}$	544-76-3	226.44116	0.23289	0.23659	723	0.28571				291.31	3.415	723.00	0.9844
105	170	Hexanal	$C_6H_{12}O$	66-25-1	100.15888	0.668504	0.252695	594	0.28571				214.93	8.8708	594.00	2.6455
105	171	Hexane	C_6H_{14}	110-54-3	86.17536	0.70824	0.26411	507.6	0.27537				177.83	8.747	507.60	2.6816
105	172	Hexanoic acid	$C_6H_{12}O_2$	142-62-1	116.158	0.62833	0.25598	660.2	0.25304				269.25	8.0964	660.20	2.4546
105	173	1-Hexanol	$C_6H_{14}O$	111-27-3	102.17476	0.70093	0.26776	611.3	0.24919				228.55	8.456	611.30	2.6178
105	174	2-Hexanol	$C_6H_{14}O$	626-93-7	102.175	0.67393	0.25948	585.3	0.26552				223.00	8.5181	585.30	2.5972
105	175	2-Hexanone	$C_6H_{12}O$	591-78-6	100.15888	0.67816	0.25634	587.61	0.28365				217.35	8.7319	587.61	2.6455
105	176	3-Hexanone	$C_6H_{12}O$	589-38-8	100.15888	0.67666	0.25578	582.82	0.27746				217.50	8.7631	582.82	2.6455
105	177	1-Hexene	C_6H_{12}	592-41-6	84.15948	0.76925	0.26809	504	0.28571				133.39	9.5815	504.00	2.8694
105	178	3-Hexyne	C_6H_{10}	928-49-4	82.1436	0.78045	0.26065	544	0.28571				170.05	10.021	544.00	2.9942
105	179	Hexyl mercaptan	$C_6H_{14}S$	111-31-9	118.24036	0.66372	0.27345	623	0.29185				192.62	7.7733	623.00	2.4272
105	180	1-Hexyne	C_6H_{10}	693-02-7	82.1436	0.84427	0.27185	516.2	0.2771				141.25	10.23	516.20	3.1056
105	181	2-Hexyne	C_6H_{10}	764-35-2	82.1436	0.76277	0.25248	549	0.31611				183.65	10.133	549.00	3.0211
105	182	Hydrazine	H_4N_2	302-01-2	32.04516	1.0516	0.16613	653.15	0.1898				274.69	31.934	653.15	6.3300
105	183	Hydrogen	H_2	1333-74-0	2.01588	5.414	0.34893	33.19	0.2706				13.95	38.487	33.19	15.5160
105	184	Hydrogen bromide	BrH	10035-10-6	80.91194	2.832	0.2832	363.15	0.28571				185.15	27.985	363.15	10.0000
105	185	Hydrogen chloride	ClH	7647-01-0	36.46094	3.342	0.2729	324.65	0.3217				158.97	34.854	324.65	12.2460
105	186	Hydrogen cyanide	CHN	74-90-8	27.02534	1.3413	0.18589	456.65	0.28206				259.83	27.202	456.65	7.2156
105	187	Hydrogen fluoride	FH	7664-39-3	20.0063432	2.8061	0.19362	461.15	0.29847				189.79	58.861	461.15	14.4930
105	188	Hydrogen sulfide	H_2S	7783-06-4	34.08088	2.7672	0.27369	373.53	0.29015				187.68	29.13	373.53	10.1110
105	189	Isobutyric acid	$C_4H_8O_2$	79-31-2	88.10512	0.88575	0.25736	605	0.26265				227.15	11.42	605.00	3.4417
105	190	Isopropyl amine	C_3H_9N	75-31-0	59.11026	1.2801	0.2828	471.85	0.2972				177.95	13.561	471.85	4.5265
105	191	Malonic acid	$C_3H_4O_4$	141-82-2	104.06146	0.87969	0.24543	834	0.28571				409.15	11.417	834.00	3.5843
105	192	Methacrylic acid	$C_4H_6O_2$	79-41-4	86.08924	0.87025	0.24383	662	0.28571				288.15	11.834	662.00	3.5691
105	193	Methane	CH_4	74-82-8	16.0425	2.9214	0.28976	190.56	0.28881				90.69	28.18	190.56	10.0820
105	194	Methanol	CH_4O	67-56-1	32.04186	2.3267	0.27073	512.5	0.24713				175.47	27.915	512.50	8.5942
105	195	*N*-Methyl acetamide	C_3H_7NO	79-16-3	73.09378	0.88268	0.23568	718	0.27379				301.15	13.012	718.00	3.7452
105	196	Methyl acetate	$C_3H_6O_2$	79-20-9	74.07854	1.13	0.2593	506.55	0.2764				175.15	14.475	506.55	4.3579
105	197	Methyl acetylene	C_3H_4	74-99-7	40.06386	1.6085	0.26436	402.4	0.27987				170.45	19.031	402.40	6.0845
105	198	Methyl acrylate	$C_4H_6O_2$	96-33-3	86.08924	0.97286	0.26267	536	0.2508				196.32	12.203	536.00	3.7037
105	199	Methyl amine	CH_5N	74-89-5	31.0571	1.39	0.21405	430.05	0.2275				179.69	25.378	430.05	6.4938
105	200	Methyl benzoate	$C_8H_8O_2$	93-58-3	136.14792	0.53382	0.23274	693	0.28147				260.75	8.2202	693.00	2.2936
105	201	3-Methyl-1,2-butadiene	C_5H_8	598-25-4	68.11702	0.84623	0.24625	490	0.29041				159.53	11.994	490.00	3.4365
105	202	2-Methylbutane	C_5H_{12}	78-78-4	72.14878	0.91991	0.27815	460.4	0.28667				113.25	10.764	460.40	3.3072
105	203	2-Methylbutanoic acid	$C_5H_{10}O_2$	116-53-0	102.1317	0.72762	0.25244	643	0.28571				193.00	9.9915	643.00	2.8823
105	204	3-Methyl-1-butanol	$C_5H_{12}O$	123-51-3	88.1482	0.8189	0.26974	577.2	0.23573				155.95	10.248	577.20	3.0359
105	205	2-Methyl-1-butene	C_5H_{10}	563-46-2	70.1329	0.91619	0.26752	465	0.28164				135.58	11.332	465.00	3.4248
105	206	2-Methyl-2-butene	C_5H_{10}	513-35-9	70.1329	0.93391	0.27275	470	0.2578				139.39	11.216	470.00	3.4241
105	207	2-Methyl -1-butene-3-yne	C_5H_6	78-80-8	66.10114	1.1157	0.27671	492	0.30821				160.15	12.581	492.00	4.0320
105	208	Methylbutyl ether	$C_5H_{12}O$	628-28-4	88.14818	0.8363	0.27514	512.74	0.27553				157.48	9.7581	512.74	3.0395
105	209	Methylbutyl sulfide	$C_5H_{12}S$	628-29-5	104.214	0.75509	0.27183	593	0.29127				175.30	9.0056	593.00	2.7778
105	210	3-Methyl-1-butyne	C_5H_8	598-23-2	68.11702	0.94575	0.26008	463.2	0.30807				183.45	11.519	463.20	3.6364
105	211	Methyl butyrate	$C_5H_{10}O_2$	623-42-7	102.1317	0.76983	0.26173	554.5	0.26879				187.35	9.7638	554.50	2.9413

105	212	Methylchlorosilane	CH_5ClSi	993-00-0	80.5889	1.0674	0.26257	442	0.26569				139.05	13.626	442.00	4.0652
105	213	Methylcyclohexane	C_7H_{14}	108-87-2	98.18606	0.73109	0.26971	572.1	0.29185				146.58	9.0173	572.10	2.7107
105	214	1-Methylcyclohexanol	$C_7H_{14}O$	590-67-0	114.18546	0.7013	0.266	686	0.28571				285.15	8.2091	686.00	2.6365
105	215	*cis*-2-Methylcyclohexanol	$C_7H_{14}O$	7443-70-1	114.18546	0.70973	0.26544	614	0.26016				280.15	8.2931	614.00	2.6738
105	216	*trans*-2-Methylcyclohexanol	$C_7H_{14}O$	7443-52-9	114.18546	0.72836	0.27241	617	0.2478				269.15	8.2628	617.00	2.6738
105	217	Methylcyclopentane	C_6H_{12}	96-37-7	84.15948	0.84758	0.27037	532.7	0.28258				130.73	10.491	532.70	3.1349
105	218	1-Methylcyclopentene	C_6H_{10}	693-89-0	82.1436	0.88824	0.26914	542	0.27874				146.62	10.98	542.00	3.3003
105	219	3-Methylcyclopentene	C_6H_{10}	1120-62-3	82.1436	0.9109	0.276	526	0.26756				168.54	10.538	526.00	3.3004
105	220	Methyldichlorosilane	CH_4Cl_2Si	75-54-7	115.03396	0.97608	0.28209	483	0.22529				182.55	10.789	483.00	3.4602
105	221	Methylethyl ether	C_3H_8O	540-67-0	60.09502	1.2635	0.27878	437.8	0.2744				160.00	13.995	437.80	4.5322
105	222	Methylethyl ketone	C_4H_8O	78-93-3	72.10572	0.93767	0.25035	535.5	0.29964				186.48	12.663	535.50	3.7454
105	223	Methylethyl sulfide	C_3H_8S	624-89-5	76.1606	1.067	0.27102	533	0.29364				167.23	12.671	533.00	3.9370
105	224	Methyl formate	$C_2H_4O_2$	107-31-3	60.05196	1.525	0.2634	487.2	0.2806				174.15	18.811	487.20	5.7897
105	225	Methylisobutyl ether	$C_5H_{12}O$	625-44-5	88.14818	0.84005	0.27638	497	0.27645				188.00	9.3871	497.00	3.0395
105	226	Methylisobutyl ketone	$C_6H_{12}O$	108-10-1	100.15888	0.71687	0.26453	574.6	0.28918				189.15	8.8617	574.60	2.7100
105	227	Methyl Isocyanate	C_2H_3NO	624-83-9	57.05132	1.0228	0.20692	488	0.28571				256.15	17.666	488.00	4.9430
105	228	Methylisopropyl ether	$C_4H_{10}O$	598-53-8	74.1216	0.97887	0.27017	464.48	0.28998				127.93	11.933	464.48	3.6232
105	229	Methylisopropyl ketone	$C_5H_{10}O$	563-80-4	86.1323	0.86567	0.26836	553.4	0.28364				180.15	10.46	553.40	3.2258
105	230	Methylisopropyl sulfide	$C_4H_{10}S$	1551-21-9	90.1872	0.78912	0.25915	553.1	0.26512				171.64	10.352	553.10	3.0450
105	231	Methyl mercaptan	CH_4S	74-93-1	48.10746	1.9323	0.28018	469.95	0.28523				150.18	21.564	469.95	6.8966
105	232	Methyl methacrylate	$C_5H_8O_2$	80-62-6	100.11582	0.7761	0.25068	566	0.29773				224.95	10.176	566.00	3.0960
105	233	2-Methyloctanoic acid	$C_9H_{18}O_2$	3004-93-1	158.23802	0.4416	0.2521	694	0.28532				240.00	5.938	694.00	1.7517
105	234	2-Methylpentane	C_6H_{14}	107-83-5	86.17536	0.72701	0.26754	497.7	0.28268				119.55	9.2041	497.70	2.7174
105	235	Methyl pentyl ether	$C_6H_{14}O$	628-80-8	102.17476	0.71004	0.26981	546.49	0.29974				176.00	8.445	546.49	2.6316
105	236	2-Methylpropane	C_4H_{10}	75-28-5	58.1222	1.0631	0.27506	407.8	0.2758				113.54	12.574	407.80	3.8650
105	237	2-Methyl-2-propanol	$C_4H_{10}O$	75-65-0	74.1216	0.92128	0.25442	506.2	0.27586				298.97	10.556	506.20	3.6211
105	238	2-Methyl propene	C_4H_8	115-11-7	56.10632	1.1446	0.2724	417.9	0.28172				132.81	13.507	417.90	4.2019
105	239	Methyl propionate	$C_4H_8O_2$	554-12-1	88.10512	0.9147	0.2594	530.6	0.2774				185.65	11.678	530.60	3.5262
105	240	Methylpropyl ether	$C_4H_{10}O$	557-17-5	74.1216	0.96145	0.26536	476.25	0.30088				133.97	12.043	476.25	3.6232
105	241	Methylpropyl sulfide	$C_4H_{10}S$	3877-15-4	90.1872	0.87496	0.26862	565	0.30259				160.17	10.689	565.00	3.2572
105	242	Methylsilane	CH_6Si	992-94-9	46.14384	1.3052	0.26757	352.5	0.28799				116.34	15.791	352.50	4.8780
105	243	alpha-Methyl styrene	C_9H_{10}	98-83-9	118.1757	0.64856	0.25877	654	0.31444				249.95	8.0099	654.00	2.5063
105	244	Methyl tert-butyl ether	$C_5H_{12}O$	1634-04-4	88.1482	0.817948	0.269105	497.1	0.28571				164.55	9.7955	497.10	3.0395
105	245	Methyl vinyl ether	C_3H_6O	107-25-5	58.07914	1.2587	0.26433	437	0.25819				151.15	15.691	437.00	4.7619
105	246	Naphthalene	$C_{10}H_8$	91-20-3	128.17052	0.6348	0.25838	748.4	0.27727				333.15	7.7545	748.40	2.4568
105	247	Neon	Ne	7440-01-9	20.1797	7.3718	0.3067	44.4	0.2786				24.56	61.796	44.40	24.0360
105	248	Nitroethane	$C_2H_5NO_2$	79-24-3	75.0666	1.0024	0.23655	593	0.278				183.63	15.556	593.00	4.2376
105	249	Nitrogen	N_2	7727-37-9	28.0134	3.2091	0.2861	126.2	0.2966				63.15	31.063	126.20	11.2170
105	250	Nitrogen trifluoride	F_3N	7783-54-2	71.00191	2.3736	0.2817	234	0.29529				66.46	26.555	234.00	8.4260
105	251	Nitromethane	CH_3NO_2	75-52-5	61.04002	1.3728	0.23793	588.15	0.29601				244.60	19.632	588.15	5.7698
105	252	Nitrous oxide	N_2O	10024-97-2	44.0128	2.781	0.27244	309.57	0.2882				182.30	27.928	309.57	10.2080
105	253	Nitric oxide	NO	10102-43-9	30.0061	5.246	0.3044	180.15	0.242				109.50	44.487	180.15	17.2340
105	254	Nonadecane	$C_{19}H_{40}$	629-92-5	268.5209	0.19199	0.23337	758	0.28571				305.04	2.8889	758.00	0.8227
105	255	Nonanal	$C_9H_{18}O$	124-19-6	142.23862	0.473233	0.256918	658.5	0.28571				267.30	5.9415	658.50	1.8420
105	256	Nonane	C_9H_{20}	111-84-2	128.2551	0.46321	0.25444	594.6	0.28571				219.66	6.0427	594.60	1.8205
105	257	Nonanoic acid	$C_9H_{18}O_2$	112-05-0	158.238	0.41582	0.24284	710.7	0.30036				285.55	5.7592	710.70	1.7123
105	258	1-Nonanol	$C_9H_{20}O$	143-08-8	144.2545	0.43682	0.25161	670.9	0.2498				268.15	5.8496	670.90	1.7361
105	259	2-Nonanol	$C_9H_{20}O$	628-99-9	144.255	0.419258	0.241912	649.5	0.28571				238.15	6.0223	649.50	1.7331
105	260	1-Nonene	C_9H_{18}	124-11-8	126.23922	0.48661	0.25722	593.1	0.28571				191.91	6.3717	593.10	1.8918
105	261	Nonyl mercaptan	$C_9H_{20}S$	1455-21-6	160.3201	0.47377	0.27052	681	0.30284				253.05	5.4532	681.00	1.7513
105	262	1-Nonyne	C_9H_{16}	3452-09-3	124.22334	0.52152	0.25918	598.05	0.29177				223.15	6.5369	598.05	2.0122
105	263	Octadecane	$C_{18}H_{38}$	593-45-3	254.49432	0.20448	0.23474	747	0.28571				301.31	3.0418	747.00	0.8711
105	264	Octanal	$C_8H_{16}O$	124-13-0	128.212	0.525901	0.25664	638.9	0.28571				251.65	6.6608	638.90	2.0492
105	265	Octane	C_8H_{18}	111-65-9	114.22852	0.5266	0.25693	568.7	0.28571				216.38	6.7049	568.70	2.0496
105	266	Octanoic acid	$C_8H_{16}O_2$	124-07-2	144.211	0.48251	0.25196	694.26	0.26842				289.65	6.3107	694.26	1.9150

(*Continued*)

TABLE 2-32 Densities of Inorganic and Organic Liquids (mol/dm³) (*Continued*)

Eqn	Cmpd. no.	Name	Formula	CAS	Mol. wt.	C_1	C_2	C_3	C_4	C_5	C_6	C_7	T_{min}, K	Density at T_{min}	T_{max}, K	Density at T_{max}
105	267	1-Octanol	$C_8H_{18}O$	111-87-5	130.22792	0.48979	0.24931	652.3	0.27824				257.65	6.5738	652.30	1.9646
105	268	2-Octanol	$C_8H_{18}O$	123-96-6	130.228	0.52497	0.26186	629.8	0.25257				241.55	6.5625	629.80	2.0048
105	269	2-Octanone	$C_8H_{16}O$	111-13-7	128.21204	0.50006	0.24851	632.7	0.29942				252.85	6.6477	632.70	2.0122
105	270	3-Octanone	$C_8H_{16}O$	106-68-3	128.21204	0.5108	0.25386	627.7	0.26735				255.55	6.6283	627.70	2.0121
105	271	1-Octene	C_8H_{16}	111-66-0	112.21264	0.55449	0.25952	566.9	0.28571				171.45	7.2155	566.90	2.1366
105	272	Octyl mercaptan	$C_8H_{18}S$	111-88-6	146.29352	0.52577	0.27234	667.3	0.30063				223.95	6.0987	667.30	1.9306
105	273	1-Octyne	C_8H_{14}	629-05-0	110.19676	0.58945	0.26052	574	0.28532				193.55	7.4832	574.00	2.2626
105	274	Oxalic acid	$C_2H_2O_4$	144-62-7	90.03488	1.1911	0.27038	828	0.28571				462.65	12.405	828.00	4.4053
105	275	Oxygen	O_2	7782-44-7	31.9988	3.9143	0.28772	154.58	0.2924				54.35	40.77	154.58	13.6050
105	276	Ozone	O_3	10028-15-6	47.9982	3.3592	0.29884	261	0.28523				80.15	33.361	261.00	11.2410
105	277	Pentadecane	$C_{15}H_{32}$	629-62-9	212.41458	0.25142	0.23837	708	0.28571				283.07	3.6423	708.00	1.0547
105	278	Pentanal	$C_5H_{10}O$	110-62-3	86.1323	0.85658	0.26811	566.1	0.27354				191.59	10.353	566.10	3.1949
105	279	Pentane	C_5H_{12}	109-66-0	72.14878	0.84947	0.26726	469.7	0.27789				143.42	10.474	469.70	3.1784
105	280	Pentanoic acid	$C_5H_{10}O_2$	109-52-4	102.132	0.73455	0.25636	639.16	0.25522				239.15	9.5869	639.16	2.8653
105	281	1-Pentanol	$C_5H_{12}O$	71-41-0	88.1482	0.81754	0.26732	588.1	0.25348				195.56	10.061	588.10	3.0583
105	282	2-Pentanol	$C_5H_{12}O$	6032-29-7	88.1482	0.81577	0.26594	561	0.25551				200.00	10.017	561.00	3.0675
105	283	2-Pentanone	$C_5H_{10}O$	107-87-9	86.1323	0.90411	0.27207	561.08	0.30669				196.29	10.398	561.08	3.3231
105	284	3-Pentanone	$C_5H_{10}O$	96-22-0	86.1323	0.71811	0.24129	560.95	0.27996				234.18	10.102	560.95	2.9761
105	285	1-Pentene	C_5H_{10}	109-67-1	70.1329	0.89816	0.26608	464.8	0.28571				108.02	11.521	464.80	3.3755
105	286	2-Pentyl mercaptan	$C_5H_{12}S$	2084-19-7	104.21378	0.65858	0.25367	584.3	0.28571				160.75	9.073	584.30	2.5962
105	287	Pentyl mercaptan	$C_5H_{12}S$	110-66-7	104.21378	0.75345	0.27047	598	0.30583				197.45	8.8575	598.00	2.7857
105	288	1-Pentyne	C_5H_8	627-19-0	68.11702	0.8491	0.2352	481.2	0.353				167.45	12.532	481.20	3.6101
105	289	2-Pentyne	C_5H_8	627-21-4	68.11702	0.92099	0.25419	519	0.31077				163.83	12.24	519.00	3.6232
105	290	Phenanthrene	$C_{14}H_{10}$	85-01-8	178.2292	0.45554	0.2523	869	0.24841				372.38	5.9853	869.00	1.8055
105	291	Phenol	C_6H_6O	108-95-2	94.11124	1.3798	0.31598	694.25	0.32768				314.06	11.244	694.25	4.3667
105	292	Phenyl isocyanate	C_7H_5NO	103-71-9	119.1207	0.63163	0.23373	653	0.28571				243.15	9.6466	653.00	2.7024
105	293	Phthalic anhydride	$C_8H_4O_3$	85-44-9	148.11556	0.5393	0.22704	791	0.248				404.15	8.2218	791.00	2.3754
105	294	Propadiene	C_3H_4	463-49-0	40.06386	1.6087	0.26543	394	0.29895				136.87	19.479	394.00	6.0607
105	295	Propane	C_3H_8	74-98-6	44.09562	1.3757	0.27453	369.83	0.29359				85.47	16.583	369.83	5.0111
105	296	1-Propanol	C_3H_8O	71-23-8	60.09502	1.2457	0.27281	536.8	0.23994				146.95	15.206	536.80	4.5662
105	297	2-Propanol	C_3H_8O	67-63-0	60.095	1.1799	0.2644	508.3	0.24653				185.26	14.663	508.30	4.4626
105	298	Propenylcyclohexene	C_9H_{14}	13511-13-2	122.20746	0.61255	0.26769	636	0.28571				199.00	7.4763	636.00	2.2883
105	299	Propionaldehyde	C_3H_6O	123-38-6	58.07914	1.2861	0.26236	503.6	0.3004				165.00	16.075	503.60	4.9020
105	300	Propionic acid	$C_3H_6O_2$	79-09-4	74.0785	1.0969	0.25568	600.81	0.26857				252.45	13.935	600.81	4.2901
105	301	Propionitrile	C_3H_5N	107-12-0	55.0785	0.91281	0.22125	561.3	0.26811				180.37	16.067	561.30	4.1257
105	302	Propyl acetate	$C_5H_{10}O_2$	109-60-4	102.1317	0.73041	0.25456	549.73	0.27666				178.15	9.7941	549.73	2.8693
105	303	Propyl amine	C_3H_9N	107-10-8	59.11026	0.9195	0.23878	496.95	0.2461				188.36	13.764	496.95	3.8508
105	304	Propylbenzene	C_9H_{12}	103-65-1	120.19158	0.57233	0.25171	638.35	0.29616				173.55	7.9821	638.35	2.2738
105	305	Propylene	C_3H_6	115-07-1	42.07974	1.4403	0.26852	364.85	0.28775				87.89	18.07	364.85	5.3638
105	306	Propyl formate	$C_4H_8O_2$	110-74-7	88.10512	0.915	0.26134	538	0.28				180.25	11.59	538.00	3.5012
105	307	2-Propyl mercaptan	C_3H_8S	75-33-2	76.16062	1.093	0.27762	517	0.29781				142.61	12.61	517.00	3.9370
105	308	Propyl mercaptan	C_3H_8S	107-03-9	76.16062	1.0714	0.27214	536.6	0.29481				159.95	12.716	536.60	3.9369
105	309	1,2-Propylene glycol	$C_3H_8O_2$	57-55-6	76.09442	1.0923	0.26106	626	0.20459				213.15	14.363	626.00	4.1841
105	310	Quinone	$C_6H_4O_2$	106-51-4	108.09476	0.83228	0.25385	683	0.23658				388.85	10.082	683.00	3.2786
105	311	Silicon tetrafluoride	F_4Si	7783-61-1	104.07911	1.1945	0.24128	259	0.16693				186.35	15.635	259.00	4.9507
105	312	Styrene	C_8H_8	100-42-5	104.14912	0.7397	0.2603	636	0.3009				242.54	9.1088	636.00	2.8417

105	313	Succinic acid	$C_4H_6O_4$	110-15-6	118.08804	0.65882	0.21741	838	0.28571				460.85	10.21	838.00	3.0303
105	314	Sulfur dioxide	O_2S	7446-09-5	64.0638	2.106	0.25842	430.75	0.2895				197.67	25.298	430.75	8.1495
105	315	Sulfur hexafluoride	F_6S	2551-62-4	146.0554192	1.3587	0.2701	318.69	0.2921				223.15	12.631	318.69	5.0304
105	316	Sulfur trioxide	O_3S	7446-11-9	80.0632	1.4969	0.19013	490.85	0.4359				289.95	24.241	490.85	7.8730
105	317	Terephthalic acid	$C_8H_6O_4$	100-21-0	166.13084	0.41922	0.17775	883.6	0.28571				700.15	7.102	883.60	2.3585
105	318	*o*-Terphenyl	$C_{18}H_{14}$	84-15-1	230.30376	0.3448	0.25116	857	0.29268				329.35	4.5526	857.00	1.3728
100	318	*o*-Terphenyl	$C_{18}H_{14}$	84-15-1	230.30376	5.7136	−0.003474						288.15	4.7126	313.19	4.6256
105	319	Tetradecane	$C_{14}H_{30}$	629-59-4	198.388	0.27248	0.24007	693	0.28571				279.01	3.889	693.00	1.1350
105	320	Tetrahydrofuran	C_4H_8O	109-99-9	72.10572	1.2543	0.28084	540.15	0.2912				164.65	13.998	540.15	4.4662
105	321	1,2,3,4-Tetrahydronaphthalene	$C_{10}H_{12}$	119-64-2	132.20228	0.67717	0.27772	720	0.2878				237.38	7.638	720.00	2.4383
105	322	Tetrahydrothiophene	C_4H_8S	110-01-0	88.17132	1.1628	0.28954	631.95	0.28674				176.99	12.408	631.95	4.0160
105	323	2,2,3,3-Tetramethylbutane	C_8H_{18}	594-82-1	114.22852	0.58988	0.27201	568	0.27341				373.96	5.7242	568.00	2.1686
105	324	Thiophene	C_4H_4S	110-02-1	84.13956	1.2874	0.28194	579.35	0.30781				234.94	13.43	579.35	4.5662
105	325	Toluene	C_7H_8	108-88-3	92.13842	0.8792	0.27136	591.75	0.29241				178.18	10.487	591.75	3.2400
105	326	1,1,2-Trichloroethane	$C_2H_3Cl_3$	79-00-5	133.40422	0.9062	0.25475	602	0.31				236.50	11.478	602.00	3.5572
105	327	Tridecane	$C_{13}H_{28}$	629-50-5	184.36142	0.29934	0.2433	675	0.28571				267.76	4.1817	675.00	1.2303
105	328	Triethyl amine	$C_6H_{15}N$	121-44-8	101.19	0.7035	0.27386	535.15	0.2872				158.45	8.2843	535.15	2.5688
105	329	Trimethyl amine	C_3H_9N	75-50-3	59.11026	1.0116	0.25683	433.25	0.2696				156.08	13.144	433.25	3.9388
105	330	1,2,3-Trimethylbenzene	C_9H_{12}	526-73-8	120.19158	0.6531	0.27002	664.5	0.26268				243.15	7.7278	664.50	2.4187
105	331	1,2,4-Trimethylbenzene	C_9H_{12}	95-63-6	120.19158	0.60394	0.25956	649.1	0.27713				229.33	7.689	649.10	2.3268
105	332	2,2,4-Trimethylpentane	C_8H_{18}	540-84-1	114.22852	0.59059	0.27424	543.8	0.2847				165.78	6.9146	543.80	2.1536
105	333	2,3,3-Trimethylpentane	C_8H_{18}	560-21-4	114.22852	0.6028	0.27446	573.5	0.2741				172.22	7.0934	573.50	2.1963
105	334	1,3,5-Trinitrobenzene	$C_6H_3N_3O_6$	99-35-4	213.10452	0.48195	0.23093	846	0.28571				398.40	7.0825	846.00	2.0870
105	335	2,4,6-Trinitrotoluene	$C_7H_5N_3O_6$	118-96-7	227.1311	0.37378	0.21379	828	0.29905				354.00	6.4521	828.00	1.7484
105	336	Undecane	$C_{11}H_{24}$	1120-21-4	156.30826	0.36703	0.24876	639	0.28571				247.57	4.9453	639.00	1.4754
105	337	1-Undecanol	$C_{11}H_{24}O$	112-42-5	172.30766	0.33113	0.23676	703.9	0.2762				288.45	4.8594	703.90	1.3986
105	338	Vinyl acetate	$C_4H_6O_2$	108-05-4	86.08924	0.9591	0.2593	519.13	0.27448				180.35	12.287	519.13	3.6988
105	339	Vinyl acetylene	C_4H_4	689-97-4	52.07456	1.2703	0.26041	454	0.297				173.15	15.664	454.00	4.8781
105	340	Vinyl chloride	C_2H_3Cl	75-01-4	62.49822	1.5115	0.2707	432	0.2716				119.36	18.481	432.00	5.5837
105	341	Vinyl trichlorosilane	$C_2H_3Cl_3Si$	75-94-5	161.48972	0.59595	0.24314	543.15	0.24856				178.35	8.8236	543.15	2.4511
100	342	Water	H_2O	7732-18-5	18.01528	−13.851	0.64038	−0.0019124	1.8211E-06				273.16	55.497	353.15	54.0010
119	342	Water	H_2O	7732-18-5	18.01528	17.874	35.618	19.655	−9.1306	−31.367	−813.56	−17421000	273.16	55.487	647.096	17.8740
105	343	*m*-Xylene	C_8H_{10}	108-38-3	106.165	0.68902	0.26086	617	0.27479				225.30	8.648	617.00	2.6413
105	344	*o*-Xylene	C_8H_{10}	95-47-6	106.165	0.69962	0.26143	630.3	0.27365				247.98	8.6229	630.30	2.6761
105	345	*p*-Xylene	C_8H_{10}	106-42-3	106.165	0.67752	0.25887	616.2	0.27596				286.41	8.1614	616.20	2.6172

Except for *o*-terphenyl and water, liquid density ρ is calculated by Eqn 105: $\rho = C_1/(C_2^{[1+(1-T/C_3)^{C_4}]})$ where ρ is in mol/dm³ and *T* is in K. The pressure is equal to the vapor pressure for pressures greater than 1 atm and equal to 1 atm when the vapor pressure is less than 1 atm.

Equation (2-100), used for the limited temperature ranges as noted for *o*-terphenyl and water, is $\rho = C_1 + C_2T + C_3T^2 + C_4T^3$.

Equation (2-119), used for water, is $\rho = C_1 + C_2\tau^{1/3} + C_3\tau^{2/3} + C_4\tau^{5/3} + C_5\tau^{16/3} + C_6\tau^{43/3} + C_7\tau^{110/3}$ where $\tau = 1 - T/TC$, and TC = critical temperature (647.096 K).

Values in this table were taken from the Design Institute for Physical Properties (DIPPR) of the American Institute of Chemical Engineers (AIChE), 801 Critically Evaluated Gold Standard™ Database, copyright 2016 AIChE, and reproduced with permission of AIChE and of the DIPPR Evaluated Process Design Data Project Steering Committee. Their source should be cited as R. L. Rowley, W. V. Wilding, J. L. Oscarson, T. A. Knotts, and N. F. Giles, *DIPPR® Data Compilation of Pure Chemical Properties*, Design Institute for Physical Properties, AIChE, New York, NY (2016).

DENSITIES OF AQUEOUS INORGANIC SOLUTIONS AT 1 ATM

TABLE 2-33 Ammonia (NH_3)*

%	−15°C	−10°C	−5°C	0°C	5°C	10°C	20°C	25°C	%	d_4^{15}
1		0.9943	0.9954	0.9959	0.9958	0.9955	0.9939	0.993	32	0.889
2		.9906	.9915	.9919	.9917	.9913	.9895	.988	36	.877
4		.9834	.9840	.9842	.9837	.9832	.9811	.980	40	.865
8	0.970	.9701	.9701	.9695	.9686	.9677	.9651	.964	45	.849
12	.958	.9576	.9571	.9561	.9548	.9534	.9501	.948	50	.832
16	.947	.9461	.9450	.9435	.9420	.9402	.9362	.934	60	.796
20		.9353	.9335	.9316	.9296	.9275	.9229		70	.755
24		.9249	.9226	.9202	.9179	.9155	.9101		80	.711
28		.9150	.9122	.9094	.9067	.9040	.8980		90	.665
30		.9101	.9070	.9040	.9012	.8983	.8920		100	.618

**International Critical Tables*, vol. 3, p. 59.

TABLE 2-34 Ammonium Chloride (NH_4Cl)*

%	0°C	10°C	20°C	30°C	50°C	80°C	100°C
1	1.0033	1.0029	1.0013	0.9987	0.9910	0.9749	0.9617
2	1.0067	1.0062	1.0045	1.0018	.9940	.9780	.9651
4	1.0135	1.0126	1.0107	1.0077	.9999	.9842	.9718
8	1.0266	1.0251	1.0227	1.0195	1.0116	.9963	.9849
12	1.0391	1.0370	1.0344	1.0310	1.0231	1.0081	.9975
16	1.0510	1.0485	1.0457	1.0422	1.0343	1.0198	1.0096
20	1.0625	1.0596	1.0567	1.0532	1.0454	1.0312	1.0213
24	1.0736	1.0705	1.0674	1.0641	1.0564	1.0426	1.0327

**International Critical Tables*, vol. 3, p. 60.

TABLE 2-35 Calcium Chloride ($CaCl_2$)*

%	−5°C	0°C	20°C	30°C	40°C	60°C	80°C	100°C	120°C†	140°C
2		1.0171	1.0148	1.0120	1.0084	0.9994	0.9881	0.9748	0.9596	0.9428
4		1.0346	1.0316	1.0286	1.0249	1.0158	1.0046	0.9915	0.9765	0.9601
8	1.0708	1.0703	1.0659	1.0626	1.0586	1.0492	1.0382	1.0257	1.0111	0.9954
12	1.1083	1.1072	1.1015	1.0978	1.0937	1.0840	1.0730	1.0610	1.0466	1.0317
16	1.1471	1.1454	1.1386	1.1345	1.1301	1.1202	1.1092	1.0973	1.0835	1.0691
20	1.1874	1.1853	1.1775	1.1730	1.1684	1.1581	1.1471	1.1352	1.1219	1.1080
25		1.2376	1.2284	1.2236	1.2186	1.2079	1.1965	1.1846		
30		1.2922	1.2816	1.2764	1.2709	1.2597	1.2478	1.2359		
35			1.3373	1.3316	1.3255	1.3137	1.3013	1.2893		
40			1.3957	1.3895	1.3826	1.3700	1.3571	1.3450		

**International Critical Tables*, vol. 3, pp. 72–73.
†Corrected to atmospheric pressure.

TABLE 2-36 Ferric Chloride ($FeCl_3$)*

%	0°C	10°C	20°C	30°C
1	1.0086	1.0084	1.0068	1.0040
2	1.0174	1.0168	1.0152	1.0122
4	1.0347	1.0341	1.0324	1.0292
8	1.0703	1.0692	1.0669	1.0636
12	1.1088	1.1071	1.1040	1.1006
16	1.1475	1.1449	1.1418	1.1386
20	1.1870	1.1847	1.1820	1.1786
25	1.2400	1.2380	1.2340	1.2290
30	1.2970	1.2950	1.2910	1.2850
35	1.3605	1.3580	1.3530	1.3475
40	1.4280	1.4235	1.4175	1.4115
45		1.4920	1.4850	
50		1.5610	1.5510	

**International Critical Tables*, vol. 3, p. 68.

TABLE 2-37 Ferric Sulfate [$Fe_2(SO_4)_3$]*

%	$d_4^{17.5}$
1	1.0072
2	1.0157
4	1.0327
8	1.0670
12	1.1028
16	1.1409
20	1.1811
30	1.3073
40	1.4487
50	1.6127
60	1.7983

**International Critical Tables*, vol. 3, p. 68.

TABLE 2-38 Ferric Nitrate [$Fe(NO_3)_3$]*

%	d_4^{18}
1	1.0065
2	1.0144
4	1.0304
8	1.0636
12	1.0989
16	1.1359
20	1.1748
25	1.2281

**International Critical Tables*, vol. 3, p. 68.

TABLE 2-39 Ferrous Sulfate ($FeSO_4$)*

%	15°C	18°C	20°C
0.2		1.00068	1.0002
0.4		1.00275	1.0022
0.8		1.00645	1.0062
1.0	1.0090	1.0085	1.0082
4.0	1.0380	1.0375	
8.0	1.0790	1.0785	
12.0	1.1235	1.1220	
16.0	1.1690	1.1675	
20.0	1.2150	1.2135	

**International Critical Tables*, vol. 3, p. 68.

TABLE 2-40 Hydrogen Cyanide (HCN)*

%	d_4^{15}
1	0.998
2	0.996
4	0.993
8	0.984
12	0.971
16	0.956
82	0.752
90	0.724
100	0.691

**International Critical Tables*, vol. 3, p. 61.

TABLE 2-41 Hydrogen Chloride (HCl)*

%	−5°C	0°C	10°C	20°C	40°C	60°C	80°C	100°C
1	1.0048	1.0052	1.0048	1.0032	0.9970	0.9881	0.9768	0.9636
2	1.0104	1.0106	1.0100	1.0082	1.0019	0.9930	0.9819	0.9688
4	1.0213	1.0213	1.0202	1.0181	1.0116	1.0026	0.9919	0.9791
6	1.0321	1.0319	1.0303	1.0279	1.0211	1.0121	1.0016	0.9892
8	1.0428	1.0423	1.0403	1.0376	1.0305	1.0215	1.0111	0.9992
10	1.0536	1.0528	1.0504	1.0474	1.0400	1.0310	1.0206	1.0090
12	1.0645	1.0634	1.0607	1.0574	1.0497	1.0406	1.0302	1.0188
14	1.0754	1.0741	1.0711	1.0675	1.0594	1.0502	1.0398	1.0286
16	1.0864	1.0849	1.0815	1.0776	1.0692	1.0598	1.0494	1.0383
18	1.0975	1.0958	1.0920	1.0878	1.0790	1.0694	1.0590	1.0479
20	1.1087	1.1067	1.1025	1.0980	1.0888	1.0790	1.0685	1.0574
22	1.1200	1.1177	1.1131	1.1083	1.0986	1.0886	1.0780	1.0668
24	1.1314	1.1287	1.1238	1.1187	1.1085	1.0982	1.0874	1.0761
26	1.1426	1.1396	1.1344	1.1290	1.1183	1.1076	1.0967	1.0853
28	1.1537	1.1505	1.1449	1.1392	1.1280	1.1169	1.1058	1.0942
30	1.1648	1.1613	1.1553	1.1493	1.1376	1.1260	1.1149	1.1030
32				1.1593				
34				1.1691				
36				1.1789				
38				1.1885				
40				1.1980				

**International Critical Tables*, vol. 3, p. 54.

TABLE 2-42 Hydrogen Peroxide (H_2O_2)*

%	d_4^{18}	%	d_4^{18}
1	1.0022	26	1.0959
2	1.0058	28	1.1040
4	1.0131	30	1.1122
6	1.0204	35	1.1327
8	1.0277	40	1.1536
10	1.0351	45	1.1749
12	1.0425	50	1.1966
14	1.0499	55	1.2188
16	1.0574	60	1.2416
18	1.0649	70	1.2897
20	1.0725	80	1.3406
22	1.0802	90	1.3931
24	1.0880	100	1.4465

**International Critical Tables*, vol. 3, p. 54.

TABLE 2-43 Nitric Acid (HNO_3)*

%	0°C	5°C	10°C	15°C	20°C	25°C	30°C	40°C	50°C	60°C	80°C	100°C
1	1.0058	1.00572	1.00534	1.00464	1.00364	1.00241	1.0009	0.9973	0.9931	0.9882	0.9767	0.9632
2	1.0117	1.01149	1.01099	1.01018	1.00909	1.00778	1.0061	1.0025	0.9982	0.9932	0.9816	0.9681
3	1.0176	1.01730	1.01668	1.01576	1.01457	1.01318	1.0114	1.0077	1.0033	0.9982	0.9865	0.9730
4	1.0236	1.02315	1.02240	1.02137	1.02008	1.01861	1.0168	1.0129	1.0084	1.0033	0.9915	0.9779
5	1.0296	1.02904	1.02816	1.02702	1.02563	1.02408	1.0222	1.0182	1.0136	1.0084	0.9965	0.9829
6	1.0357	1.03497	1.03397	1.03272	1.03122	1.02958	1.0277	1.0235	1.0188	1.0136	1.0015	0.9879
7	1.0418	1.0410	1.0399	1.0385	1.0369	1.0352	1.0333	1.0289	1.0241	1.0188	1.0066	0.9929
8	1.0480	1.0471	1.0458	1.0443	1.0427	1.0409	1.0389	1.0344	1.0295	1.0241	1.0117	0.9980
9	1.0543	1.0532	1.0518	1.0502	1.0485	1.0466	1.0446	1.0399	1.0349	1.0294	1.0169	1.0032
10	1.0606	1.0594	1.0578	1.0561	1.0543	1.0523	1.0503	1.0455	1.0403	1.0347	1.0221	1.0083
11	1.0669	1.0656	1.0639	1.0621	1.0602	1.0581	1.0560	1.0511	1.0458	1.0401	1.0273	1.0134
12	1.0733	1.0718	1.0700	1.0681	1.0661	1.0640	1.0618	1.0567	1.0513	1.0455	1.0326	1.0186
13	1.0797	1.0781	1.0762	1.0742	1.0721	1.0699	1.0676	1.0624	1.0568	1.0509	1.0379	1.0238
14	1.0862	1.0845	1.0824	1.0803	1.0781	1.0758	1.0735	1.0681	1.0624	1.0564	1.0432	1.0289
15	1.0927	1.0909	1.0887	1.0865	1.0842	1.0818	1.0794	1.0739	1.0680	1.0619	1.0485	1.0341
16	1.0992	1.0973	1.0950	1.0927	1.0903	1.0879	1.0854	1.0797	1.0737	1.0675	1.0538	1.0393
17	1.1057	1.1038	1.1014	1.0989	1.0964	1.0940	1.0914	1.0855	1.0794	1.0731	1.0592	1.0444
18	1.1123	1.1103	1.1078	1.1052	1.1026	1.1001	1.0974	1.0913	1.0851	1.0787	1.0646	1.0496
19	1.1189	1.1168	1.1142	1.1115	1.1088	1.1062	1.1034	1.0972	1.0908	1.0843	1.0700	1.0547
20	1.1255	1.1234	1.1206	1.1178	1.1150	1.1123	1.1094	1.1031	1.0966	1.0899	1.0754	1.0598
21	1.1322	1.1300	1.1271	1.1242	1.1213	1.1185	1.1155	1.1090	1.1024	1.0956	1.0808	1.0650
22	1.1389	1.1366	1.1336	1.1306	1.1276	1.1247	1.1217	1.1150	1.1083	1.1013	1.0862	1.0701
23	1.1457	1.1433	1.1402	1.1371	1.1340	1.1310	1.1280	1.1210	1.1142	1.1070	1.0917	1.0753
24	1.1525	1.1501	1.1469	1.1437	1.1404	1.1374	1.1343	1.1271	1.1201	1.1127	1.0972	1.0805
25	1.1594	1.1569	1.1536	1.1503	1.1469	1.1438	1.1406	1.1332	1.1260	1.1185	1.1027	1.0857
26	1.1663	1.1638	1.1603	1.1569	1.1534	1.1502	1.1469	1.1394	1.1320	1.1244	1.1083	1.0910
27	1.1733	1.1707	1.1670	1.1635	1.1600	1.1566	1.1533	1.1456	1.1381	1.1303	1.1139	1.0963
28	1.1803	1.1777	1.1738	1.1702	1.1666	1.1631	1.1597	1.1519	1.1442	1.1362	1.1195	1.1016
29	1.1874	1.1847	1.1807	1.1770	1.1733	1.1697	1.1662	1.1582	1.1503	1.1422	1.1251	1.1069
30	1.1945	1.1917	1.1876	1.1838	1.1800	1.1763	1.1727	1.1645	1.1564	1.1482	1.1307	1.1122
31	1.2016	1.1988	1.1945	1.1906	1.1867	1.1829	1.1792	1.1708	1.1625	1.1542	1.1363	1.1175
32	1.2088	1.2059	1.2014	1.1974	1.1934	1.1896	1.1857	1.1772	1.1687	1.1602	1.1419	1.1228
33	1.2160	1.2131	1.2084	1.2043	1.2002	1.1963	1.1922	1.1836	1.1749	1.1662	1.1476	1.1281
34	1.2233	1.2203	1.2155	1.2113	1.2071	1.2030	1.1988	1.1901	1.1812	1.1723	1.1533	1.1335
35	1.2306	1.2275	1.2227	1.2183	1.2140	1.2098	1.2055	1.1966	1.1876	1.1784	1.1591	1.1390
36	1.2375	1.2344	1.2294	1.2249	1.2205	1.2163	1.2119	1.2028	1.1936	1.1842	1.1645	1.1440
37	1.2444	1.2412	1.2361	1.2315	1.2270	1.2227	1.2182	1.2089	1.1995	1.1899	1.1699	1.1490
38	1.2513	1.2479	1.2428	1.2381	1.2335	1.2291	1.2245	1.2150	1.2054	1.1956	1.1752	1.1540
39	1.2581	1.2546	1.2494	1.2446	1.2399	1.2354	1.2308	1.2210	1.2112	1.2013	1.1805	1.1589
40	1.2649	1.2613	1.2560	1.2511	1.2463	1.2417	1.2370	1.2270	1.2170	1.2069	1.1858	1.1638
41	1.2717	1.2680	1.2626	1.2576	1.2527	1.2480	1.2432	1.2330	1.2229	1.2126	1.1911	1.1687
42	1.2786	1.2747	1.2692	1.2641	1.2591	1.2543	1.2494	1.2390	1.2287	1.2182	1.1963	1.1735
43	1.2854	1.2814	1.2758	1.2706	1.2655	1.2606	1.2556	1.2450	1.2345	1.2238	1.2015	1.1783
44	1.2922	1.2880	1.2824	1.2771	1.2719	1.2669	1.2618	1.2510	1.2403	1.2294	1.2067	1.1831
45	1.2990	1.2947	1.2890	1.2836	1.2783	1.2732	1.2680	1.2570	1.2461	1.2350	1.2119	1.1879
46	1.3058	1.3014	1.2955	1.2901	1.2847	1.2795	1.2742	1.2630	1.2519	1.2406	1.2171	1.1927
47	1.3126	1.3080	1.3021	1.2966	1.2911	1.2858	1.2804	1.2690	1.2577	1.2462	1.2223	1.1976
48	1.3194	1.3147	1.3087	1.3031	1.2975	1.2921	1.2867	1.2750	1.2635	1.2518	1.2275	1.2024
49	1.3263	1.3214	1.3153	1.3096	1.3040	1.2984	1.2929	1.2811	1.2693	1.2575	1.2328	1.2073
50	1.3327	1.3277	1.3215	1.3157	1.3100	1.3043	1.2987	1.2867	1.2748	1.2628	1.2377	1.2118
51	1.3391	1.3339	1.3277	1.3218	1.3160	1.3102	1.3045	1.2923	1.2802	1.2680	1.2425	1.2163
52	1.3454	1.3401	1.3338	1.3278	1.3219	1.3160	1.3102	1.2978	1.2856	1.2731	1.2473	1.2208
53	1.3517	1.3462	1.3399	1.3338	1.3278	1.3218	1.3159	1.3033	1.2909	1.2782	1.2521	1.2252
54	1.3579	1.3523	1.3459	1.3397	1.3336	1.3275	1.3215	1.3087	1.2961	1.2833	1.2568	1.2296
55	1.3640	1.3583	1.3518	1.3455	1.3393	1.3331	1.3270	1.3141	1.3013	1.2883	1.2615	1.2339
56	1.3700	1.3642	1.3576	1.3512	1.3449	1.3386	1.3324	1.3194	1.3064	1.2932	1.2661	1.2382
57	1.3759	1.3700	1.3634	1.3569	1.3505	1.3441	1.3377	1.3246	1.3114	1.2981	1.2706	1.2424
58	1.3818	1.3757	1.3691	1.3625	1.3560	1.3495	1.3430	1.3298	1.3164	1.3029	1.2751	1.2466
59	1.3875	1.3813	1.3747	1.3680	1.3614	1.3548	1.3482	1.3348	1.3213	1.3077	1.2795	1.2507
60	1.3931	1.3868	1.3801	1.3734	1.3667	1.3600	1.3533	1.3398	1.3261	1.3124	1.2839	1.2547
61	1.3986	1.3922	1.3855	1.3787	1.3719	1.3651	1.3583	1.3447	1.3308	1.3169	1.2881	1.2587
62	1.4039	1.3975	1.3907	1.3838	1.3769	1.3700	1.3632	1.3494	1.3354	1.3213	1.2922	1.2625
63	1.4091	1.4027	1.3958	1.3888	1.3818	1.3748	1.3679	1.3540	1.3398	1.3255	1.2962	1.2661
64		1.4078	1.4007	1.3936	1.3866	1.3795	1.3725					

(Continued)

TABLE 2-43 Nitric Acid (HNO_3) (*Continued*)

%	0°C	5°C	10°C	15°C	20°C	25°C	30°C	40°C	50°C	60°C	80°C	100°C
65		1.4128	1.4055	1.3984	1.3913	1.3841	1.3770					
66		1.4177	1.4103	1.4031	1.3959	1.3887	1.3814					
67		1.4224	1.4150	1.4077	1.4004	1.3932	1.3857					
68		1.4271	1.4196	1.4122	1.4048	1.3976	1.3900					
69		1.4317	1.4241	1.4166	1.4091	1.4019	1.3942					
70		1.4362	1.4285	1.4210	1.4134	1.4061	1.3983					
71		1.4406	1.4328	1.4252	1.4176	1.4102	1.4023					
72		1.4449	1.4371	1.4294	1.4218	1.4142	1.4063					
73		1.4491	1.4413	1.4335	1.4258	1.4182	1.4103					
74		1.4532	1.4454	1.4376	1.4298	1.4221	1.4142					
75		1.4573	1.4494	1.4415	1.4337	1.4259	1.4180					
76		1.4613	1.4533	1.4454	1.4375	1.4296	1.4217					
77		1.4652	1.4572	1.4492	1.4413	1.4333	1.4253					
78		1.4690	1.4610	1.4529	1.4450	1.4369	1.4288					
79		1.4727	1.4647	1.4565	1.4486	1.4404	1.4323					
80		1.4764	1.4683	1.4601	1.4521	1.4439	1.4357					
81		1.4800	1.4718	1.4636	1.4555	1.4473	1.4391					
82		1.4835	1.4753	1.4670	1.4589	1.4507	1.4424					
83		1.4869	1.4787	1.4704	1.4622	1.4540	1.4456					
84		1.4903	1.4820	1.4737	1.4655	1.4572	1.4487					
85		1.4936	1.4852	1.4769	1.4686	1.4603	1.4518					
86		1.4968	1.4883	1.4799	1.4716	1.4633	1.4548					
87		1.4999	1.4913	1.4829	1.4745	1.4662	1.4577					
88		1.5029	1.4942	1.4858	1.4773	1.4690	1.4605					
89		1.5058	1.4970	1.4885	1.4800	1.4716	1.4631					
90		1.5085	1.4997	1.4911	1.4826	1.4741	1.4656					
91		1.5111	1.5023	1.4936	1.4850	1.4766	1.4681					
92		1.5136	1.5048	1.4960	1.4873	1.4789	1.4704					
93		1.5156	1.5068	1.4979	1.4892	1.4807	1.4722					
94		1.5177	1.5088	1.4999	1.4912	1.4826	1.4741					
95		1.5198	1.5109	1.5019	1.4932	1.4846	1.4761					
96		1.5220	1.5130	1.5040	1.4952	1.4867	1.4781					
97		1.5244	1.5152	1.5062	1.4974	1.4889	1.4802					
98		1.5278	1.5187	1.5096	1.5008	1.4922	1.4835					
99		1.5327	1.5235	1.5144	1.5056	1.4969	1.4881					
100		1.5402	1.5310	1.5217	1.5129	1.5040	1.4952					

**International Critical Tables*, vol. 3, pp. 58–59.

TABLE 2-44 Perchloric Acid ($HClO_4$)*

%	d_4^{15}	d_4^{20}	d_4^{25}	d_4^{50}	%	d_4^{15}	d_4^{20}	d_4^{50}
1	1.0050		1.0020	0.9933	28	1.1900	1.1851	1.1645
2	1.0109		1.0070	0.9986	30	1.2067	1.2013	1.1800
4	1.0228		1.0169	0.9906	32	1.2239	1.2183	1.1960
6	1.0348		1.0270	1.0205	34	1.2418	1.2359	1.2130
8	1.0471		1.0372	1.0320	36	1.2603	1.2542	1.2310
10	1.0597		1.0475	1.0440	38	1.2794	1.2732	1.2490
12	1.0726			1.0560	40	1.2991	1.2927	1.2680
14	1.0589			1.0680	45	1.3521	1.3450	1.3180
16	1.0995			1.0810	50	1.4103	1.4018	1.3730
18	1.1135			1.0940	55	1.4733	1.4636	1.4320
20	1.1279			1.1070	60	1.5389	1.5298	1.4950
22	1.1428			1.1205	65	1.6059	1.5986	1.5620
24	1.1581			1.1345	70	1.6736	1.6680	1.6290
26	1.1738	1.1697		1.1490				

**International Critical Tables*, vol. 3, p. 54.

TABLE 2-45 Phosphoric Acid (H_3PO_4)*

°C	2%	6%	14%	20%	26%	35%	50%	75%	100%
0	1.0113	1.0339	1.0811	1.1192					
10	1.0109	1.0330	1.0792	1.1167	1.1567	1.221	1.341		
20	1.0092	1.0309	1.0764	1.1134	1.1529	1.216	1.335	1.579	1.870
30	1.0065	1.0279	1.0728	1.1094	1.1484	1.211	1.329	1.572	1.862
40	1.0029	1.0241	1.0685	1.1048					

**International Critical Tables*, vol. 3, p. 61.

TABLE 2-46 Potassium Bicarbonate ($KHCO_3$)*

°C	1%	2%	4%	6%	8%	10%
0	1.0066	1.0134	1.0270			
10	1.0064	1.0132	1.0268			
15	1.0058	1.0125	1.0260	1.0396	1.0534	1.0674
20	1.0049	1.0117	1.0252			
30	1.0024	1.0092	1.0228			
40	0.9990	1.0058	1.0195			
50	0.9949	1.0017	1.0154			
60	0.9901	0.9969	1.0106			
80	0.9786	0.9855	0.9993			
100	0.9653	0.9722	0.9860			

**International Critical Tables*, vol. 3, p. 90.

TABLE 2-47 Potassium Carbonate (K_2CO_3)*

%	0°C	10°C	20°C	40°C	60°C	80°C	100°C
1	1.0094	1.0089	1.0072	1.0010	0.9919	0.9803	0.9670
2	1.0189	1.0182	1.0163	1.0098	1.0005	0.9889	0.9756
4	1.0381	1.0369	1.0345	1.0276	1.0180	1.0063	0.9951
8	1.0768	1.0746	1.0715	1.0640	1.0538	1.0418	1.0291
12	1.1160	1.1131	1.1096	1.1013	1.0906	1.0786	1.0663
16	1.1562	1.1530	1.1490	1.1399	1.1290	1.1170	1.1049
20	1.1977	1.1941	1.1898	1.1801	1.1690	1.1570	1.1451
24	1.2405	1.2366	1.2320	1.2219	1.2106	1.1986	1.1869
28	1.2846	1.2804	1.2756	1.2652	1.2538	1.2418	1.2301
30	1.3071	1.3028	1.2979	1.2873	1.2759	1.2640	1.2522
35	1.3646	1.3600	1.3548	1.3440	1.3324	1.3206	1.3089
40	1.4244	1.4195	1.4141	1.4029	1.3913	1.3795	1.3678
45	1.4867	1.4815	1.4759	1.4644	1.4528	1.4408	1.4290
50	1.5517	1.5462	1.5404	1.5285	1.5169	1.5048	1.4928

**International Critical Tables*, vol. 3, p. 90.

TABLE 2-48 Potassium Chloride (KCl)*

%	0°C	20°C	25°C	40°C	60°C	80°C	100°C
1.0	1.00661	1.00462	1.00342	0.99847	0.9894	0.9780	0.9646
2.0	1.01335	1.01103	1.00977	1.00471	0.9956	0.9842	0.9708
4.0	1.02690	1.02391	1.02255	1.01727	1.0080	0.9966	0.9634
8.0	1.05431	1.05003	1.04847	1.04278	1.0333	1.0219	1.0888
12.0	1.08222	1.07679	1.07506	1.06897	1.0592	1.0478	1.0350
16.0	1.11068	1.10434	1.10245	1.09600	1.0861	1.0746	1.0619
20.0	1.13973	1.13280	1.13072	1.12399	1.1138	1.1024	1.0897
24.0		1.16226	1.15995	1.15299	1.1425	1.1311	1.1185
28.0				1.18304	1.1723	1.1609	1.1483

%	110°C	120°C	130°C	140°C
3.79	0.9733	0.9663	0.9583	0.9502
7.45	0.9978	0.9899	0.9827	0.9745
13.62	1.0388	1.0313	1.0238	1.0159

**International Critical Tables*, vol. 3, p. 87.

TABLE 2-49 Potassium Hydroxide (KOH)*

%	d_4^{15}
1.0	1.0083
2.0	1.0175
4.0	1.0359
6.0	1.0544
8.0	1.0730
10.0	1.0918
15.0	1.1396
20.0	1.1884
25.0	1.2387
30.0	1.2905
35.0	1.3440
40.0	1.3991
45.0	1.4558
50.0	1.5143
51.7	1.5355 (sat'd. soln.)

**International Critical Tables*, vol. 3, p. 86.

TABLE 2-50 Potassium Nitrate (KNO_3)*

%	0°C	10°C	20°C	40°C	60°C	80°C	100°C
1	1.00654	1.00615	1.00447	0.99825	0.9890	0.9776	0.9641
2	1.01326	1.01262	1.01075	1.00430	0.9949	0.9834	0.9699
4	1.02677	1.02566	1.02344	1.01652	1.0068	0.9951	0.9816
8	1.05419	1.05226	1.04940	1.04152	1.0313	1.0192	1.0056
12	1.08221	1.07963	1.07620	1.06740	1.0567	1.0442	1.0304
16			1.10392	1.09432	1.0831	1.0703	1.0562
20			1.13261	1.12240	1.1106	1.0974	1.0831
24			1.16233	1.15175	1.1391	1.1256	1.1110

**International Critical Tables*, vol. 3, p. 89.

TABLE 2-51 Sodium Acetate ($NaC_2H_3O_2$)*

%	d_4^{20}
1	1.0033
2	1.0084
4	1.0186
8	1.0392
12	1.0598
18	1.0807
20	1.1021
26	1.1351
28	1.1462

**International Critical Tables*, vol. 3, p. 83.

TABLE 2-52 Sodium Carbonate (Na_2CO_3)*

%	0°C	10°C	20°C	30°C	40°C	60°C	80°C	100°C
1	1.0109	1.0103	1.0086	1.0058	1.0022	0.9929	0.9814	0.9683
2	1.0219	1.0210	1.0190	1.0159	1.0122	1.0027	0.9910	0.9782
4	1.0439	1.0423	1.0398	1.0363	1.0323	1.0223	1.0105	0.9980
8	1.0878	1.0850	1.0816	1.0775	1.0732	1.0625	1.0503	1.0380
12	1.1319	1.1284	1.1244	1.1200	1.1150	1.1039	1.0914	1.0787
14	1.1543	1.1506	1.1463	1.1417	1.1365	1.1251	1.1125	1.0996
16				1.1636				
18				1.1859				
20				1.2086				
24				1.2552				
28				1.3031				
30				1.3274				

**International Critical Tables*, vol. 3, pp. 82–83.

TABLE 2-53 Sodium Chloride (NaCl)*

%	0°C	10°C	25°C	40°C	60°C	80°C	100°C
1	1.00747	1.00707	1.00409	0.99908	0.9900	0.9785	0.9651
2	1.01509	1.01442	1.01112	1.00593	0.9967	0.9852	0.9719
4	1.03038	1.02920	1.02530	1.01977	1.0103	0.9988	0.9855
8	1.06121	1.05907	1.05412	1.04798	1.0381	1.0264	1.0134
12	1.09244	1.08946	1.08365	1.07699	1.0667	1.0549	1.0420
16	1.12419	1.12056	1.11401	1.10688	1.0962	1.0842	1.0713
20	1.15663	1.15254	1.14533	1.13774	1.1268	1.1146	1.1017
24	1.18999	1.18557	1.17776	1.16971	1.1584	1.1463	1.1331
26	1.20709	1.20254	1.19443	1.18614	1.1747	1.1626	1.1492

**International Critical Tables*, vol. 3, p. 79.

TABLE 2-54 Sodium Hydroxide (NaOH)*

%	0°C	15°C	20°C	40°C	60°C	80°C	100°C
1	1.0124	1.01065	1.0095	1.0033	0.9941	0.9824	0.9693
2	1.0244	1.02198	1.0207	1.0139	1.0045	0.9929	0.9797
4	1.0482	1.04441	1.0428	1.0352	1.0254	1.0139	1.0009
8	1.0943	1.08887	1.0869	1.0780	1.0676	1.0560	1.0432
12	1.1399	1.13327	1.1309	1.1210	1.1101	1.0983	1.0855
16	1.1849	1.17761	1.1751	1.1645	1.1531	1.1408	1.1277
20	1.2296	1.22183	1.2191	1.2079	1.1960	1.1833	1.1700
24	1.2741	1.26582	1.2629	1.2512	1.2388	1.2259	1.2124
28	1.3182	1.3094	1.3064	1.2942	1.2814	1.2682	1.2546
32	1.3614	1.3520	1.3490	1.3362	1.3232	1.3097	1.2960
36	1.4030	1.3933	1.3900	1.3768	1.3634	1.3498	1.3360
40	1.4435	1.4334	1.4300	1.4164	1.4027	1.3889	1.3750
44	1.4825	1.4720	1.4685	1.4545	1.4405	1.4266	1.4127
48	1.5210	1.5102	1.5065	1.4922	1.4781	1.4641	1.4503
50	1.5400	1.5290	1.5253	1.5109	1.4967	1.4827	1.4690

**International Critical Tables*, vol. 3, p. 79.

TABLE 2-55 Sulfuric Acid (H_2SO_4)*

%	0°C	10°C	15°C	20°C	25°C	30°C	40°C	50°C	60°C	80°C	100°C
1	1.0074	1.0068	1.0060	1.0051	1.0038	1.0022	0.9986	0.9944	0.9895	0.9779	0.9645
2	1.0147	1.0138	1.0129	1.0118	1.0104	1.0087	1.0050	1.0006	0.9956	0.9839	0.9705
3	1.0219	1.0206	1.0197	1.0184	1.0169	1.0152	1.0113	1.0067	1.0017	0.9900	0.9766
4	1.0291	1.0275	1.0264	1.0250	1.0234	1.0216	1.0176	1.0129	1.0078	0.9961	0.9827
5	1.0364	1.0344	1.0332	1.0317	1.0300	1.0281	1.0240	1.0192	1.0140	1.0022	0.9888
6	1.0437	1.0414	1.0400	1.0385	1.0367	1.0347	1.0305	1.0256	1.0203	1.0084	0.9950
7	1.0511	1.0485	1.0469	1.0453	1.0434	1.0414	1.0371	1.0321	1.0266	1.0146	1.0013
8	1.0585	1.0556	1.0539	1.0522	1.0502	1.0481	1.0437	1.0386	1.0330	1.0209	1.0076
9	1.0660	1.0628	1.0610	1.0591	1.0571	1.0549	1.0503	1.0451	1.0395	1.0273	1.0140
10	1.0735	1.0700	1.0681	1.0661	1.0640	1.0617	1.0570	1.0517	1.0460	1.0338	1.0204
11	1.0810	1.0773	1.0753	1.0731	1.0710	1.0686	1.0637	1.0584	1.0526	1.0403	1.0269
12	1.0886	1.0846	1.0825	1.0802	1.0780	1.0756	1.0705	1.0651	1.0593	1.0469	1.0335
13	1.0962	1.0920	1.0898	1.0874	1.0851	1.0826	1.0774	1.0719	1.0661	1.0536	1.0402
14	1.1039	1.0994	1.0971	1.0947	1.0922	1.0897	1.0844	1.0788	1.0729	1.0603	1.0469
15	1.1116	1.1069	1.1045	1.1020	1.0994	1.0968	1.0914	1.0857	1.0798	1.0671	1.0537
16	1.1194	1.1145	1.1120	1.1094	1.1067	1.1040	1.0985	1.0927	1.0868	1.0740	1.0605
17	1.1272	1.1221	1.1195	1.1168	1.1141	1.1113	1.1057	1.0998	1.0938	1.0809	1.0674
18	1.1351	1.1298	1.1271	1.1243	1.1215	1.1187	1.1129	1.1070	1.1009	1.0879	1.0744
19	1.1430	1.1375	1.1347	1.1318	1.1290	1.1261	1.1202	1.1142	1.1081	1.0950	1.0814
20	1.1510	1.1453	1.1424	1.1394	1.1365	1.1335	1.1275	1.1215	1.1153	1.1021	1.0885
21	1.1590	1.1531	1.1501	1.1471	1.1441	1.1410	1.1349	1.1288	1.1226	1.1093	1.0957
22	1.1670	1.1609	1.1579	1.1548	1.1517	1.1486	1.1424	1.1362	1.1299	1.1166	1.1029
23	1.1751	1.1688	1.1657	1.1626	1.1594	1.1563	1.1500	1.1437	1.1373	1.1239	1.1102
24	1.1832	1.1768	1.1736	1.1704	1.1672	1.1640	1.1576	1.1512	1.1448	1.1313	1.1176
25	1.1914	1.1848	1.1816	1.1783	1.1750	1.1718	1.1653	1.1588	1.1523	1.1388	1.1250
26	1.1996	1.1929	1.1896	1.1862	1.1829	1.1796	1.1730	1.1665	1.1599	1.1463	1.1325
27	1.2078	1.2010	1.1976	1.1942	1.1909	1.1875	1.1808	1.1742	1.1676	1.1539	1.1400
28	1.2160	1.2091	1.2057	1.2023	1.1989	1.1955	1.1887	1.1820	1.1753	1.1616	1.1476
29	1.2243	1.2173	1.2138	1.2104	1.2069	1.2035	1.1966	1.1898	1.1831	1.1693	1.1553
30	1.2326	1.2255	1.2220	1.2185	1.2150	1.2115	1.2046	1.1977	1.1909	1.1771	1.1630
31	1.2409	1.2338	1.2302	1.2267	1.2232	1.2196	1.2126	1.2057	1.1988	1.1849	1.1708
32	1.2493	1.2421	1.2385	1.2349	1.2314	1.2278	1.2207	1.2137	1.2068	1.1928	1.1787
33	1.2577	1.2504	1.2468	1.2432	1.2396	1.2360	1.2289	1.2218	1.2148	1.2008	1.1866
34	1.2661	1.2588	1.2552	1.2515	1.2479	1.2443	1.2371	1.2300	1.2229	1.2088	1.1946
35	1.2746	1.2672	1.2636	1.2599	1.2563	1.2526	1.2454	1.2383	1.2311	1.2169	1.2027
36	1.2831	1.2757	1.2720	1.2684	1.2647	1.2610	1.2538	1.2466	1.2394	1.2251	1.2109
37	1.2917	1.2843	1.2805	1.2769	1.2732	1.2695	1.2622	1.2550	1.2477	1.2334	1.2192
38	1.3004	1.2929	1.2891	1.2855	1.2818	1.2780	1.2707	1.2635	1.2561	1.2418	1.2276
39	1.3091	1.3016	1.2978	1.2941	1.2904	1.2866	1.2793	1.2720	1.2646	1.2503	1.2361
40	1.3179	1.3103	1.3065	1.3028	1.2991	1.2953	1.2880	1.2806	1.2732	1.2589	1.2446
41	1.3268	1.3191	1.3153	1.3116	1.3079	1.3041	1.2967	1.2893	1.2819	1.2675	1.2532
42	1.3357	1.3280	1.3242	1.3205	1.3167	1.3129	1.3055	1.2981	1.2907	1.2762	1.2619
43	1.3447	1.3370	1.3332	1.3294	1.3256	1.3218	1.3144	1.3070	1.2996	1.2850	1.2707
44	1.3538	1.3461	1.3423	1.3384	1.3346	1.3308	1.3234	1.3160	1.3086	1.2939	1.2796
45	1.3630	1.3553	1.3515	1.3476	1.3437	1.3399	1.3325	1.3251	1.3177	1.3029	1.2886
46	1.3724	1.3646	1.3608	1.3569	1.3530	1.3492	1.3417	1.3343	1.3269	1.3120	1.2976
47	1.3819	1.3740	1.3702	1.3663	1.3624	1.3586	1.3510	1.3435	1.3362	1.3212	1.3067
48	1.3915	1.3835	1.3797	1.3758	1.3719	1.3680	1.3604	1.3528	1.3455	1.3305	1.3159
49	1.4012	1.3931	1.3893	1.3854	1.3814	1.3775	1.3699	1.3623	1.3549	1.3399	1.3253
50	1.4110	1.4029	1.3990	1.3951	1.3911	1.3872	1.3795	1.3719	1.3644	1.3494	1.3348
51	1.4209	1.4128	1.4088	1.4049	1.4009	1.3970	1.3893	1.3816	1.3740	1.3590	1.3444
52	1.4310	1.4228	1.4188	1.4148	1.4109	1.4069	1.3991	1.3914	1.3837	1.3687	1.3540
53	1.4412	1.4329	1.4289	1.4248	1.4209	1.4169	1.4091	1.4013	1.3936	1.3785	1.3637
54	1.4515	1.4431	1.4391	1.4350	1.4310	1.4270	1.4191	1.4113	1.4036	1.3884	1.3735
55	1.4619	1.4535	1.4494	1.4453	1.4412	1.4372	1.4293	1.4214	1.4137	1.3984	1.3834
56	1.4724	1.4640	1.4598	1.4557	1.4516	1.4475	1.4396	1.4317	1.4239	1.4085	1.3934
57	1.4830	1.4746	1.4703	1.4662	1.4621	1.4580	1.4500	1.4420	1.4342	1.4187	1.4035
58	1.4937	1.4852	1.4809	1.4768	1.4726	1.4685	1.4604	1.4524	1.4446	1.4290	1.4137
59	1.5045	1.4959	1.4916	1.4875	1.4832	1.4791	1.4709	1.4629	1.4551	1.4393	1.4240
60	1.5154	1.5067	1.5024	1.4983	1.4940	1.4898	1.4816	1.4735	1.4656	1.4497	1.4344
61	1.5264	1.5177	1.5133	1.5091	1.5048	1.5006	1.4923	1.4842	1.4762	1.4602	1.4449
62	1.5375	1.5287	1.5243	1.5200	1.5157	1.5115	1.5031	1.4950	1.4869	1.4708	1.4554
63	1.5487	1.5398	1.5354	1.5310	1.5267	1.5225	1.5140	1.5058	1.4977	1.4815	1.4660
64	1.5600	1.5510	1.5465	1.5421	1.5378	1.5335	1.5250	1.5167	1.5086	1.4923	1.4766

TABLE 2-55 Sulfuric Acid (H_2SO_4) *(Continued)*

%	0°C	10°C	15°C	20°C	25°C	30°C	40°C	50°C	60°C	80°C	100°C
65	1.5714	1.5623	1.5578	1.5533	1.5490	1.5446	1.5361	1.5277	1.5195	1.5031	1.4873
66	1.5828	1.5736	1.5691	1.5646	1.5602	1.5558	1.5472	1.5388	1.5305	1.5140	1.4981
67	1.5943	1.5850	1.5805	1.5760	1.5715	1.5671	1.5584	1.5499	1.5416	1.5249	1.5089
68	1.6059	1.5965	1.5920	1.5874	1.5829	1.5785	1.5697	1.5611	1.5528	1.5359	1.5198
69	1.6176	1.6081	1.6035	1.5989	1.5944	1.5899	1.5811	1.5724	1.5640	1.5470	1.5307
70	1.6293	1.6198	1.6151	1.6105	1.6059	1.6014	1.5925	1.5838	1.5753	1.5582	1.5417
71	1.6411	1.6315	1.6268	1.6221	1.6175	1.6130	1.6040	1.5952	1.5867	1.5694	1.5527
72	1.6529	1.6433	1.6385	1.6338	1.6292	1.6246	1.6155	1.6067	1.5981	1.5806	1.5637
73	1.6648	1.6551	1.6503	1.6456	1.6409	1.6363	1.6271	1.6182	1.6095	1.5919	1.5747
74	1.6768	1.6670	1.6622	1.6574	1.6526	1.6480	1.6387	1.6297	1.6209	1.6031	1.5857
75	1.6888	1.6789	1.6740	1.6692	1.6644	1.6597	1.6503	1.6412	1.6322	1.6142	1.5966
76	1.7008	1.6908	1.6858	1.6810	1.6761	1.6713	1.6619	1.6526	1.6435	1.6252	1.6074
77	1.7128	1.7026	1.6976	1.6927	1.6878	1.6829	1.6734	1.6640	1.6547	1.6361	1.6181
78	1.7247	1.7144	1.7093	1.7043	1.6994	1.6944	1.6847	1.6751	1.6657	1.6469	1.6286
79	1.7365	1.7261	1.7209	1.7158	1.7108	1.7058	1.6959	1.6862	1.6766	1.6575	1.6390
80	1.7482	1.7376	1.7323	1.7272	1.7221	1.7170	1.7069	1.6971	1.6873	1.6680	1.6493
81	1.7597	1.7489	1.7435	1.7383	1.7331	1.7279	1.7177	1.7077	1.6978	1.6782	1.6594
82	1.7709	1.7599	1.7544	1.7491	1.7437	1.7385	1.7281	1.7180	1.7080	1.6882	1.6692
83	1.7815	1.7704	1.7649	1.7594	1.7540	1.7487	1.7382	1.7279	1.7179	1.6979	1.6787
84	1.7916	1.7804	1.7748	1.7693	1.7639	1.7585	1.7479	1.7375	1.7274	1.7072	1.6878
85	1.8009	1.7897	1.7841	1.7786	1.7732	1.7678	1.7571	1.7466	1.7364	1.7161	1.6966
86	1.8095	1.7983	1.7927	1.7872	1.7818	1.7763	1.7657	1.7552	1.7449	1.7245	1.7050
87	1.8173	1.8061	1.8006	1.7951	1.7897	1.7842	1.7736	1.7632	1.7529	1.7324	1.7129
88	1.8243	1.8132	1.8077	1.8022	1.7968	1.7914	1.7809	1.7705	1.7602	1.7397	1.7202
89	1.8306	1.8195	1.8141	1.8087	1.8033	1.7979	1.7874	1.7770	1.7669	1.7464	1.7269
90	1.8361	1.8252	1.8198	1.8144	1.8091	1.8038	1.7933	1.7829	1.7729	1.7525	1.7331
91	1.8410	1.8302	1.8248	1.8195	1.8142	1.8090	1.7986	1.7883	1.7783	1.7581	1.7388
92	1.8453	1.8346	1.8293	1.8240	1.8188	1.8136	1.8033	1.7932	1.7832	1.7633	1.7439
93	1.8490	1.8384	1.8331	1.8279	1.8227	1.8176	1.8074	1.7974	1.7876	1.7681	1.7485
94	1.8520	1.8415	1.8363	1.8312	1.8260	1.8210	1.8109	1.8011	1.7914		
95	1.8544	1.8439	1.8388	1.8337	1.8286	1.8236	1.8137	1.8040	1.7944		
96	1.8560	1.8457	1.8406	1.8355	1.8305	1.8255	1.8157	1.8060	1.7965		
97	1.8569	1.8466	1.8414	1.8364	1.8314	1.8264	1.8166	1.8071	1.7977		
98	1.8567	1.8463	1.8411	1.8361	1.8310	1.8261	1.8163	1.8068	1.7976		
99	1.8551	1.8445	1.8393	1.8342	1.8292	1.8242	1.8145	1.8050	1.7958		
100	1.8517	1.8409	1.8357	1.8305	1.8255	1.8205	1.8107	1.8013	1.7922		

%	$d_4^{5.96}$	%	$d_4^{13.00}$	$d_4^{18.00}$
0.005	1.000 0140	0.05	0.999 810	0.999 028
.01	1.000 0576	0.1	1.000 185	0.999 400
.02	1.000 1434	0.2	1.000 912	1.000 119
.03	1.000 2276	0.3	1.001 623	1.000 820
.04	1.000 3104	0.4	1.002 326	1.001 512
.05	1.000 3920	0.5	1.003 023	1.002 197
.06	1.000 4726	0.6	1.003 716	1.002 877
.07	1.000 5523	0.8	1.005 090	1.004 227
.08	1.000 6313	1.0	1.006 452	1.005 570
.09	1.000 7098	1.2	1.007 807	1.006 909
.10	1.000 7880	1.4	1.009 159	1.008 247
.15	1.001 1732	1.6	1.010 510	1.009 583
.20	1.001 5514	1.8	1.011 860	1.010 918
.25	1.001 9254	2.0	1.013 209	1.012 252
.30	1.002 2961	2.2	1.014 557	1.013 586
.35	1.002 6639	2.4	1.015 904	1.014 919
.40	1.003 0292			
.45	1.003 3923			
.50	1.003 7534			

**International Critical Tables*, vol. 3, pp. 56–57.

DENSITIES OF AQUEOUS ORGANIC SOLUTIONS

TABLE 2-56 Acetic Acid (CH_3COOH)

%	0°C	10°C	15°C	20°C	25°C	30°C	40°C
0	0.9999	0.9997	0.9991	0.9982	0.9971	0.9957	0.9922
1	1.0016	1.0013	1.0006	0.9996	0.9987	0.9971	0.9934
2	1.0033	1.0029	1.0021	1.0012	1.0000	0.9984	0.9946
3	1.0051	1.0044	1.0036	1.0025	1.0013	0.9997	0.9958
4	1.0070	1.0060	1.0051	1.0040	1.0027	1.0011	0.9970
5	1.0088	1.0076	1.0066	1.0055	1.0041	1.0024	0.9982
6	1.0106	1.0092	1.0081	1.0069	1.0055	1.0037	0.9994
7	1.0124	1.0108	1.0096	1.0083	1.0068	1.0050	1.0006
8	1.0142	1.0124	1.0111	1.0097	1.0081	1.0063	1.0018
9	1.0159	1.0140	1.0126	1.0111	1.0094	1.0076	1.0030
10	1.0177	1.0156	1.0141	1.0125	1.0107	1.0089	1.0042
11	1.0194	1.0171	1.0155	1.0139	1.0120	1.0102	1.0054
12	1.0211	1.0187	1.0170	1.0154	1.0133	1.0115	1.0065
13	1.0228	1.0202	1.0184	1.0168	1.0146	1.0127	1.0077
14	1.0245	1.0217	1.0199	1.0182	1.0159	1.0139	1.0088
15	1.0262	1.0232	1.0213	1.0195	1.0172	1.0151	1.0099
16	1.0278	1.0247	1.0227	1.0209	1.0185	1.0163	1.0110
17	1.0295	1.0262	1.0241	1.0223	1.0198	1.0175	1.0121
18	1.0311	1.0276	1.0255	1.0236	1.0210	1.0187	1.0132
19	1.0327	1.0291	1.0269	1.0250	1.0223	1.0198	1.0142
20	1.0343	1.0305	1.0283	1.0263	1.0235	1.0210	1.0153
21	1.0358	1.0319	1.0297	1.0276	1.0248	1.0222	1.0164
22	1.0374	1.0333	1.0310	1.0288	1.0260	1.0233	1.0174
23	1.0389	1.0347	1.0323	1.0301	1.0272	1.0244	1.0185
24	1.0404	1.0361	1.0336	1.0313	1.0283	1.0256	1.0195
25	1.0419	1.0375	1.0349	1.0326	1.0295	1.0267	1.0205
26	1.0434	1.0388	1.0362	1.0338	1.0307	1.0278	1.0215
27	1.0449	1.0401	1.0374	1.0349	1.0318	1.0289	1.0225
28	1.0463	1.0414	1.0386	1.0361	1.0329	1.0299	1.0234
29	1.0477	1.0427	1.0399	1.0372	1.0340	1.0310	1.0244
30	1.0491	1.0440	1.0411	1.0384	1.0350	1.0320	1.0253
31	1.0505	1.0453	1.0423	1.0395	1.0361	1.0330	1.0262
32	1.0519	1.0465	1.0435	1.0406	1.0372	1.0341	1.0272
33	1.0532	1.0477	1.0446	1.0417	1.0382	1.0351	1.0281
34	1.0545	1.0489	1.0458	1.0428	1.0392	1.0361	1.0289
35	1.0558	1.0501	1.0469	1.0438	1.0402	1.0371	1.0298
36	1.0571	1.0513	1.0480	1.0449	1.0412	1.0380	1.0306
37	1.0584	1.0524	1.0491	1.0459	1.0422	1.0390	1.0314
38	1.0596	1.0535	1.0501	1.0469	1.0432	1.0399	1.0322
39	1.0608	1.0546	1.0512	1.0479	1.0441	1.0408	1.0330
40	1.0621	1.0557	1.0522	1.0488	1.0450	1.0416	1.0338
41	1.0633	1.0568	1.0532	1.0498	1.0460	1.0425	1.0346
42	1.0644	1.0578	1.0542	1.0507	1.0469	1.0433	1.0353
43	1.0656	1.0588	1.0551	1.0516	1.0477	1.0441	1.0361
44	1.0667	1.0598	1.0561	1.0525	1.0486	1.0449	1.0368
45	1.0679	1.0608	1.0570	1.0534	1.0495	1.0456	1.0375
46	1.0689	1.0618	1.0579	1.0542	1.0503	1.0464	1.0382
47	1.0699	1.0627	1.0588	1.0551	1.0511	1.0471	1.0389
48	1.0709	1.0636	1.0597	1.0559	1.0518	1.0479	1.0395
49	1.0720	1.0645	1.0605	1.0567	1.0526	1.0486	1.0402
50	1.0729	1.0654	1.0613	1.0575	1.0534	1.0492	1.0408
51	1.0738	1.0663	1.0622	1.0582	1.0542	1.0499	1.0414
52	1.0748	1.0671	1.0629	1.0590	1.0549	1.0506	1.0421
53	1.0757	1.0679	1.0637	1.0597	1.0555	1.0512	1.0427
54	1.0765	1.0687	1.0644	1.0604	1.0562	1.0518	1.0432
55	1.0774	1.0694	1.0651	1.0611	1.0568	1.0525	1.0438
56	1.0782	1.0701	1.0658	1.0618	1.0574	1.0531	1.0443
57	1.0790	1.0708	1.0665	1.0624	1.0580	1.0536	1.0448
58	1.0798	1.0715	1.0672	1.0631	1.0586	1.0542	1.0453
59	1.0805	1.0722	1.0678	1.0637	1.0592	1.0547	1.0458
60	1.0813	1.0728	1.0684	1.0642	1.0597	1.0552	1.0462
61	1.0820	1.0734	1.0690	1.0648	1.0602	1.0557	1.0466
62	1.0826	1.0740	1.0696	1.0653	1.0607	1.0562	1.0470
63	1.0833	1.0746	1.0701	1.0658	1.0612	1.0566	1.0473
64	1.0838	1.0752	1.0706	1.0662	1.0616	1.0571	1.0477
65	1.0844	1.0757	1.0711	1.0666	1.0621	1.0575	1.0480
66	1.0850	1.0762	1.0716	1.0671	1.0624	1.0578	1.0483
67	1.0856	1.0767	1.0720	1.0675	1.0628	1.0582	1.0486
68	1.0860	1.0771	1.0725	1.0678	1.0631	1.0585	1.0489
69	1.0865	1.0775	1.0729	1.0682	1.0634	1.0588	1.0491
70	1.0869	1.0779	1.0732	1.0685	1.0637	1.0590	1.0493
71	1.0874	1.0783	1.0736	1.0687	1.0640	1.0592	1.0495
72	1.0877	1.0786	1.0738	1.0690	1.0642	1.0594	1.0496
73	1.0881	1.0789	1.0741	1.0693	1.0644	1.0595	1.0497
74	1.0884	1.0792	1.0743	1.0694	1.0645	1.0596	1.0498
75	1.0887	1.0794	1.0745	1.0696	1.0647	1.0597	1.0499
76	1.0889	1.0796	1.0746	1.0698	1.0648	1.0598	1.0499
77	1.0891	1.0797	1.0747	1.0699	1.0648	1.0598	1.0499
78	1.0893	1.0798	1.0747	1.0700	1.0648	1.0598	1.0498
79	1.0894	1.0798	1.0747	1.0700	1.0648	1.0597	1.0497
80	1.0895	1.0798	1.0747	1.0700	1.0647	1.0596	1.0495
81	1.0895	1.0797	1.0745	1.0699	1.0646	1.0594	1.0493
82	1.0895	1.0796	1.0743	1.0698	1.0644	1.0592	1.0490
83	1.0895	1.0795	1.0741	1.0696	1.0642	1.0589	1.0487
84	1.0893	1.0793	1.0738	1.0693	1.0638	1.0585	1.0483
85	1.0891	1.0790	1.0735	1.0689	1.0635	1.0582	1.0479
86	1.0887	1.0787	1.0731	1.0685	1.0630	1.0576	1.0473
87	1.0883	1.0783	1.0726	1.0680	1.0626	1.0571	1.0467
88	1.0877	1.0778	1.0721	1.0675	1.0620	1.0564	1.0460
89	1.0872	1.0773	1.0715	1.0668	1.0613	1.0557	1.0453
90	1.0865	1.0766	1.0708	1.0661	1.0605	1.0549	1.0445
91	1.0857	1.0758	1.0700	1.0652	1.0597	1.0541	1.0436
92	1.0848	1.0749	1.0690	1.0643	1.0587	1.0530	1.0426
93	1.0838	1.0739	1.0680	1.0632	1.0577	1.0518	1.0414
94	1.0826	1.0727	1.0667	1.0619	1.0564	1.0506	1.0401
95	1.0813	1.0714	1.0652	1.0605	1.0551	1.0491	1.0386
96	1.0798		1.0632	1.0588	1.0535	1.0473	1.0368
97	1.0780		1.0611	1.0570	1.0516	1.0454	1.0348
98	1.0759		1.0590	1.0549	1.0495	1.0431	1.0325
99	1.0730		1.0567	1.0524	1.0468	1.0407	1.0299
100	1.0697		1.0545	1.0498	1.0440	1.0380	1.0271

TABLE 2-57 Methyl Alcohol (CH_3OH)*

%	0°C	10°C	15.56°C	20°C	15°C	%	0°C	10°C	15.56°C	20°C	15°C	%	0°C	10°C	15.56°C	20°C	15°C
0	0.9999	0.9997	0.9990	0.9982	0.99913	35	0.9534	0.9484	0.9456	0.9433	0.94570	70	0.8869	0.8794	0.8748	0.8715	0.87507
1	0.9981	0.9980	0.9973	0.9965	0.99727	36	0.9520	0.9469	0.9440	0.9416	0.94404	71	0.8847	0.8770	0.8726	0.8690	0.87271
2	0.9963	0.9962	0.9955	0.9948	0.99543	37	0.9505	0.9453	0.9422	0.9398	0.94237	72	0.8824	0.8747	0.8702	0.8665	0.87033
3	0.9946	0.9945	0.9938	0.9931	0.99370	38	0.9490	0.9437	0.9405	0.9381	0.94067	73	0.8801	0.8724	0.8678	0.8641	0.86792
4	0.9930	0.9929	0.9921	0.9914	0.99198	39	0.9475	0.9420	0.9387	0.9363	0.93894	74	0.8778	0.8699	0.8653	0.8616	0.86546
5	0.9914	0.9912	0.9904	0.9896	0.99029	40	0.9459	0.9403	0.9369	0.9345	0.93720	75	0.8754	0.8676	0.8629	0.8592	0.86300
6	0.9899	0.9896	0.9889	0.9880	0.98864	41	0.9443	0.9387	0.9351	0.9327	0.93543	76	0.8729	0.8651	0.8604	0.8567	0.86051
7	0.9884	0.9881	0.9872	0.9863	0.98701	42	0.9427	0.9370	0.9333	0.9309	0.93365	77	0.8705	0.8626	0.8579	0.8542	0.85801
8	0.9870	0.9865	0.9857	0.9847	0.98547	43	0.9411	0.9352	0.9315	0.9290	0.93185	78	0.8680	0.8602	0.8554	0.8518	0.85551
9	0.9856	0.9849	0.9841	0.9831	0.98394	44	0.9395	0.9334	0.9297	0.9272	0.93001	79	0.8657	0.8577	0.8529	0.8494	0.85300
10	0.9842	0.9834	0.9826	0.9815	0.98241	45	0.9377	0.9316	0.9279	0.9252	0.92815	80	0.8634	0.8551	0.8503	0.8469	0.85048
11	0.9829	0.9820	0.9811	0.9799	0.98093	46	0.9360	0.9298	0.9261	0.9234	0.92627	81	0.8610	0.8527	0.8478	0.8446	0.84794
12	0.9816	0.9805	0.9796	0.9784	0.97945	47	0.9342	0.9279	0.9242	0.9214	0.92436	82	0.8585	0.8501	0.8452	0.8420	0.84536
13	0.9804	0.9791	0.9781	0.9768	0.97802	48	0.9324	0.9260	0.9223	0.9196	0.92242	83	0.8560	0.8475	0.8426	0.8394	0.84274
14	0.9792	0.9778	0.9766	0.9754	0.97660	49	0.9306	0.9240	0.9204	0.9176	0.92048	84	0.8535	0.8449	0.8400	0.8366	0.84009
15	0.9780	0.9764	0.9752	0.9740	0.97518	50	0.9287	0.9221	0.9185	0.9156	0.91852	85	0.8510	0.8422	0.8374	0.8340	0.83742
16	0.9769	0.9751	0.9738	0.9725	0.97377	51	0.9269	0.9202	0.9166	0.9135	0.91653	86	0.8483	0.8394	0.8347	0.8314	0.83475
17	0.9758	0.9739	0.9723	0.9710	0.97237	52	0.9250	0.9182	0.9146	0.9114	0.91451	87	0.8456	0.8367	0.8320	0.8286	0.83207
18	0.9747	0.9726	0.9709	0.9696	0.97096	53	0.9230	0.9162	0.9126	0.9094	0.91248	88	0.8428	0.8340	0.8294	0.8258	0.82937
19	0.9736	0.9713	0.9695	0.9681	0.96955	54	0.9211	0.9142	0.9106	0.9073	0.91044	89	0.8400	0.8314	0.8267	0.8230	0.82667
20	0.9725	0.9700	0.9680	0.9666	0.96814	55	0.9191	0.9122	0.9086	0.9052	0.90839	90	0.8374	0.8287	0.8239	0.8202	0.82396
21	0.9714	0.9687	0.9666	0.9651	0.96673	56	0.9172	0.9101	0.9065	0.9032	0.90631	91	0.8347	0.8261	0.8212	0.8174	0.82124
22	0.9702	0.9673	0.9652	0.9636	0.96533	57	0.9151	0.9080	0.9045	0.9010	0.90421	92	0.8320	0.8234	0.8185	0.8146	0.81849
23	0.9690	0.9660	0.9638	0.9622	0.96392	58	0.9131	0.9060	0.9024	0.8988	0.90210	93	0.8293	0.8208	0.8157	0.8118	0.81568
24	0.9678	0.9646	0.9624	0.9607	0.96251	59	0.9111	0.9039	0.9002	0.8968	0.89996	94	0.8266	0.8180	0.8129	0.8090	0.81285
25	0.9666	0.9632	0.9609	0.9592	0.96108	60	0.9090	0.9018	0.8980	0.8946	0.89781	95	0.8240	0.8152	0.8101	0.8062	0.80999
26	0.9654	0.9618	0.9595	0.9576	0.95963	61	0.9068	0.8998	0.8958	0.8924	0.89563	96	0.8212	0.8124	0.8073	0.8034	0.80713
27	0.9642	0.9604	0.9580	0.9562	0.95817	62	0.9046	0.8977	0.8936	0.8902	0.89341	97	0.8186	0.8096	0.8045	0.8005	0.80428
28	0.9629	0.9590	0.9565	0.9546	0.95668	63	0.9024	0.8955	0.8913	0.8879	0.89117	98	0.8158	0.8068	0.8016	0.7976	0.80143
29	0.9616	0.9575	0.9550	0.9531	0.95518	64	0.9002	0.8933	0.8890	0.8856	0.88890	99	0.8130	0.8040	0.7987	0.7948	0.79859
30	0.9604	0.9560	0.9535	0.9515	0.95366	65	0.8980	0.8911	0.8867	0.8834	0.88662	100	0.8102	0.8009	0.7959	0.7917	0.79577
31	0.9590	0.9546	0.9521	0.9499	0.95213	66	0.8958	0.8888	0.8844	0.8811	0.88433						
32	0.9576	0.9531	0.9505	0.9483	0.95056	67	0.8935	0.8865	0.8820	0.8787	0.88203						
33	0.9563	0.9516	0.9489	0.9466	0.94896	68	0.8913	0.8842	0.8797	0.8763	0.87971						
34	0.9549	0.9500	0.9473	0.9450	0.94734	69	0.8891	0.8818	0.8771	0.8738	0.87739						

*It should be noted that the values for 100 percent do not agree with some data available elsewhere, e.g., *American Institute of Physics Handbook*, McGraw-Hill, New York, 1957. Also, see Atack, *Handbook of Chemical Data*, Reinhold, New York, 1957. Also, see Tables 2-120 and 2-135 for pure methanol and water densities.

TABLE 2-58 Ethyl Alcohol (C_2H_5OH)*

%	10°C	15°C	20°C	25°C	30°C	35°C	40°C	%	10°C	15°C	20°C	25°C	30°C	35°C	40°C
0	0.99973	0.99913	0.99823	0.99708	0.99568	0.99406	0.99225	50	0.92126	0.91776	0.91384	0.90985	0.90580	0.90168	0.89750
1	785	725	636	520	379	217	034	51	0.91943	555	160	760	353	0.89940	519
2	602	542	453	336	194	031	0.98846	52	723	333	0.90936	534	125	710	288
3	426	365	275	157	014	0.98849	663	53	502	110	711	307	0.89896	479	056
4	258	195	103	0.98984	0.98839	672	485	54	279	0.90885	485	079	667	248	0.88823
5	098	032	0.98938	817	670	501	311	55	055	659	258	0.89850	437	016	589
6	0.98946	0.98877	780	656	507	335	142	56	0.90831	433	031	621	206	0.88784	356
7	801	729	627	500	347	172	0.97975	57	607	207	0.89803	392	0.88975	552	122
8	660	584	478	346	189	009	808	58	381	0.89980	574	162	744	319	0.87888
9	524	442	331	193	031	0.97846	641	59	154	752	344	0.88931	512	085	653
10	393	304	187	043	0.97875	685	475	60	0.89927	523	113	699	278	0.87851	417
11	267	171	047	0.97897	723	527	312	61	698	293	0.88882	466	044	615	180
12	145	041	0.97910	753	573	371	150	62	468	062	650	233	0.87809	379	0.86943
13	026	0.97914	775	611	424	216	0.96989	63	237	0.88830	417	0.87998	574	142	705
14	0.97911	790	643	472	278	063	829	64	006	597	183	763	337	0.86905	466
15	800	669	514	334	133	0.96911	670	65	0.88774	364	0.87948	527	100	667	227
16	692	552	387	199	0.96990	760	512	66	541	130	713	291	0.86863	429	0.85987
17	583	433	259	062	844	607	352	67	308	0.87895	477	054	625	190	747
18	473	313	129	0.96923	697	452	189	68	074	660	241	0.86817	387	0.85950	507
19	363	191	0.96997	782	547	294	023	69	0.87839	424	004	579	148	710	266
20	252	068	864	639	395	134	0.95856	70	602	187	0.86766	340	0.85908	470	025
21	139	0.96944	729	495	242	0.95973	687	71	365	0.86949	527	100	667	228	0.84783
22	024	818	592	348	087	809	516	72	127	710	287	0.85859	426	0.84986	540
23	0.96907	689	453	199	0.95929	643	343	73	0.86888	470	047	618	184	743	297
24	787	558	312	048	769	476	168	74	648	229	0.85806	376	0.84941	500	053
25	665	424	168	0.95895	607	306	0.94991	75	408	0.85988	564	134	698	257	0.83809
26	539	287	020	738	442	133	810	76	168	747	322	0.84891	455	013	564
27	406	144	0.95867	576	272	0.94955	625	77	0.85927	505	079	647	211	0.83768	319
28	268	0.95996	710	410	098	774	438	78	685	262	0.84835	403	0.83966	523	074
29	125	844	548	241	0.94922	590	248	79	442	018	590	158	720	277	0.82827
30	0.95977	686	382	067	741	403	055	80	197	0.84772	344	0.83911	473	029	578
31	823	524	212	0.94890	557	214	0.93860	81	0.84950	525	096	664	224	0.82780	329
32	665	357	038	709	370	021	662	82	702	277	0.83848	415	0.82974	530	079
33	502	186	0.94860	525	180	0.93825	461	83	453	028	599	164	724	279	0.81828
34	334	011	679	337	0.93986	626	257	84	203	0.83777	348	0.82913	473	027	576
35	162	0.94832	494	146	790	425	051	85	0.83951	525	095	660	220	0.81774	322
36	0.94986	650	306	0.93952	591	221	0.92843	86	697	271	0.82840	405	0.81965	519	067
37	805	464	114	756	390	016	634	87	441	014	583	148	708	262	0.80811
38	620	273	0.93919	556	186	0.92808	422	88	181	0.82754	323	0.81888	448	003	552
39	431	079	720	353	0.92979	597	208	89	0.82919	492	062	626	186	0.80742	291
40	238	0.93882	518	148	770	385	0.91992	90	654	227	0.81797	362	0.80922	478	028
41	042	682	314	0.92940	558	170	774	91	386	0.81959	529	094	655	211	0.79761
42	0.93842	478	107	729	344	0.91952	554	92	114	688	257	0.80823	384	0.79941	491
43	639	271	0.92897	516	128	733	332	93	0.81839	413	0.80983	549	111	669	220
44	433	062	685	301	0.91910	513	108	94	561	134	705	272	0.79835	393	0.78947
45	226	0.92852	472	085	692	291	0.90884	95	278	0.80852	424	0.79991	555	114	670
46	017	640	257	0.91868	472	069	660	96	0.80991	566	138	706	271	0.78831	388
47	0.92806	426	041	649	250	0.90845	434	97	698	274	0.79846	415	0.78981	542	100
48	593	211	0.91823	429	028	621	207	98	399	0.79975	547	117	684	247	0.77806
49	379	0.91995	604	208	0.90805	396	0.89979	99	094	670	243	0.78814	382	0.77946	507
								100	0.79784	360	0.78934	506	075	641	203

*For data from −78° to 78°C, see p. 2-142, Table 2N-5, *American Institute of Physics Handbook*, McGraw-Hill, New York, 1957. See Tables 2-115 and 2-135 for pure ethanol and pure water densities.

TABLE 2-59 *n*-Propyl Alcohol (C_3H_7OH)

%	0°C	15°C	30°C	%	0°C	15°C	30°C	%	0°C	15°C	30°C	%	0°C	15°C	30°C	%	0°C	15°C	30°C
0	0.9999	0.9991	0.9957	20	0.9789	0.9723	0.9643	40	0.9430	0.9331	0.9226	60	0.9033	0.8922	0.8807	80	0.8634	0.8516	0.8394
1	0.9982	0.9974	0.9940	21	0.9776	0.9705	0.9622	41	0.9411	0.9310	0.9205	61	0.9013	0.8902	0.8786	81	0.8614	0.8496	0.8373
2	0.9967	0.9960	0.9924	22	0.9763	0.9688	0.9602	42	0.9391	0.9290	0.9184	62	0.8994	0.8882	0.8766	82	0.8594	0.8475	0.8352
3	0.9952	0.9944	0.9908	23	0.9748	0.9670	0.9583	43	0.9371	0.9269	0.9164	63	0.8974	0.8861	0.8745	83	0.8574	0.8454	0.8332
4	0.9939	0.9929	0.9893	24	0.9733	0.9651	0.9563	44	0.9352	0.9248	0.9143	64	0.8954	0.8841	0.8724	84	0.8554	0.8434	0.8311
5	0.9926	0.9915	0.9877	25	0.9717	0.9633	0.9543	45	0.9332	0.9228	0.9122	65	0.8934	0.8820	0.8703	85	0.8534	0.8413	0.8290
6	0.9914	0.9902	0.9862	26	0.9700	0.9614	0.9522	46	0.9311	0.9207	0.9100	66	0.8913	0.8800	0.8682	86	0.8513	0.8393	0.8269
7	0.9904	0.9890	0.9848	27	0.9682	0.9594	0.9501	47	0.9291	0.9186	0.9079	67	0.8894	0.8779	0.8662	87	0.8492	0.8372	0.8248
8	0.9894	0.9877	0.9834	28	0.9664	0.9576	0.9481	48	0.9272	0.9165	0.9057	68	0.8874	0.8759	0.8641	88	0.8471	0.8351	0.8227
9	0.9883	0.9864	0.9819	29	0.9646	0.9556	0.9460	49	0.9252	0.9145	0.9036	69	0.8854	0.8739	0.8620	89	0.8450	0.8330	0.8206
10	0.9874	0.9852	0.9804	30	0.9627	0.9535	0.9439	50	0.9232	0.9124	0.9015	70	0.8835	0.8719	0.8600	90	0.8429	0.8308	0.8185
11	0.9865	0.9840	0.9790	31	0.9608	0.9516	0.9418	51	0.9213	0.9104	0.8994	71	0.8815	0.8700	0.8580	91	0.8408	0.8287	0.8164
12	0.9857	0.9828	0.9775	32	0.9589	0.9495	0.9396	52	0.9192	0.9084	0.8973	72	0.8795	0.8680	0.8559	92	0.8387	0.8266	0.8142
13	0.9849	0.9817	0.9760	33	0.9570	0.9474	0.9375	53	0.9173	0.9064	0.8952	73	0.8776	0.8659	0.8539	93	0.8364	0.8244	0.8120
14	0.9841	0.9806	0.9746	34	0.9550	0.9454	0.9354	54	0.9153	0.9044	0.8931	74	0.8756	0.8639	0.8518	94	0.8342	0.8221	0.8098
15	0.9833	0.9793	0.9730	35	0.9530	0.9434	0.9333	55	0.9132	0.9023	0.8911	75	0.8736	0.8618	0.8497	95	0.8320	0.8199	0.8077
16	0.9825	0.9780	0.9714	36	0.9511	0.9413	0.9312	56	0.9112	0.9003	0.8890	76	0.8716	0.8598	0.8477	96	0.8296	0.8176	0.8054
17	0.9817	0.9768	0.9698	37	0.9491	0.9392	0.9289	57	0.9093	0.8983	0.8869	77	0.8695	0.8577	0.8456	97	0.8272	0.8153	0.8031
18	0.9808	0.9752	0.9680	38	0.9471	0.9372	0.9269	58	0.9073	0.8963	0.8849	78	0.8675	0.8556	0.8435	98	0.8248	0.8128	0.8008
19	0.9800	0.9739	0.9661	39	0.9450	0.9351	0.9247	59	0.9053	0.8942	0.8828	79	0.8655	0.8536	0.8414	99	0.8222	0.8104	0.7984
																100	0.8194	0.8077	0.7958

TABLE 2-60 Isopropyl Alcohol (C_3H_7OH)

%	0°C	15°C*	15°C*	20°C	30°C	%	0°C	15°C*	15°C*	20°C	30°C	%	0°C	15°C*	15°C*	20°C	30°C
0	0.9999	0.9991	0.99913	0.9982	0.9957	35	0.9557		0.9446	0.9419	0.9338	70	0.8761	0.8639	0.86346	0.8584	0.8511
1	0.9980	0.9973	0.9972	0.9962	0.9939	36	0.9536		0.9424	0.9399	0.9315	71	0.8738	0.8615	0.8611	0.8560	0.8487
2	0.9962	0.9956	0.9954	0.9944	0.9921	37	0.9514		0.9401	0.9377	0.9292	72	0.8714	0.8592	0.8588	0.8537	0.8464
3	0.9946	0.9938	0.9936	0.9926	0.9904	38	0.9493		0.9379	0.9355	0.9269	73	0.8691	0.8568	0.8564	0.8513	0.8440
4	0.9930	0.9922	0.9920	0.9909	0.9887	39	0.9472		0.9356	0.9333	0.9246	74	0.8668	0.8545	0.8541	0.8489	0.8416
5	0.9916	0.9906	0.9904	0.9893	0.9871	40	0.9450		0.93333	0.9310	0.9224	75	0.8644	0.8521	0.8517	0.8464	0.8392
6	0.9902	0.9892	0.9890	0.9877	0.9855	41	0.9428		0.9311	0.9287	0.9201	76	0.8621	0.8497	0.8493	0.8439	0.8368
7	0.9890	0.9878	0.9875	0.9862	0.9839	42	0.9406		0.9288	0.9264	0.9177	77	0.8598	0.8474	0.8470	0.8415	0.8344
8	0.9878	0.9864	0.9862	0.9847	0.9824	43	0.9384		0.9266	0.9239	0.9154	78	0.8575	0.8450	0.8446	0.8391	0.8321
9	0.9866	0.9851	0.9849	0.9833	0.9809	44	0.9361		0.9243	0.9215	0.9130	79	0.8551	0.8426	0.8422	0.8366	0.8297
10	0.9856	0.9838	0.98362	0.9820	0.9794	45	0.9338		0.9220	0.9191	0.9106	80	0.8528	0.8403	0.83979	0.8342	0.8273
11	0.9846	0.9826	0.9824	0.9808	0.9778	46	0.9315		0.9197	0.9165	0.9082	81	0.8503	0.8379	0.8374	0.8317	0.8248
12	0.9838	0.9813	0.9812	0.9797	0.9764	47	0.9292		0.9174	0.9141	0.9059	82	0.8479	0.8355	0.8350	0.8292	0.8224
13	0.9829	0.9802	0.9800	0.9876	0.9750	48	0.9270		0.9150	0.9117	0.9036	83	0.8456	0.8331	0.8326	0.8268	0.8200
14	0.9821	0.9790	0.9788	0.9776	0.9735	49	0.9247		0.9127	0.9093	0.9013	84	0.8432	0.8307	0.8302	0.8243	0.8175
15	0.9814	0.9779	0.9777	0.9765	0.9720	50	0.9224		0.91043	0.9069	0.8990	85	0.8408	0.8282	0.8278	0.8219	0.8151
16	0.9806	0.9768	0.9765	0.9754	0.9705	51	0.9201		0.9081	0.9044	0.8966	86	0.8384	0.8259	0.8254	0.8194	0.8127
17	0.9799	0.9756	0.9753	0.9743	0.9690	52	0.9178		0.9058	0.9020	0.8943	87	0.8360	0.8234	0.8229	0.8169	0.8201
18	0.9792	0.9745	0.9741	0.9731	0.9675	53	0.9155		0.9035	0.8996	0.8919	88	0.8336	0.8209	0.8205	0.8145	0.8078
19	0.9784	0.9730	0.9728	0.9717	0.9658	54	0.9132		0.9011	0.8971	0.8895	89	0.8311	0.8184	0.8180	0.8120	0.8053
20	0.9777	0.9719	0.97158	0.9703	0.9642	55	0.9109		0.8988	0.8946	0.8871	90	0.8287	0.8161	0.81553	0.8096	0.8029
21	0.9768	0.9704	0.9703	0.9688	0.9624	56	0.9086		0.8964	0.8921	0.8847	91	0.8262	0.8136	0.8130	0.8072	0.8004
22	0.9759	0.9690	0.9689	0.9669	0.9606	57	0.9063		0.8940	0.8896	0.8823	92	0.8237	0.8110	0.8104	0.8047	0.7979
23	0.9749	0.9675	0.9674	0.9651	0.9587	58	0.9040		0.8917	0.8874	0.8800	93	0.8212	0.8085	0.8079	0.8023	0.7954
24	0.9739	0.9660	0.9659	0.9634	0.9569	59	0.9017		0.8893	0.8850	0.8777	94	0.8186	0.8060	0.8052	0.7998	0.7929
25	0.9727	0.9643	0.9642	0.9615	0.9549	60	0.8994		0.88690	0.8825	0.8752	95	0.8160	0.8034	0.8026	0.7973	0.7904
26	0.9714	0.9626	0.9624	0.9597	0.9529	61	0.8970		0.8845	0.8800	0.8728	96	0.8133	0.8008	0.7999	0.7949	0.7878
27	0.9699	0.9608	0.9605	0.9577	0.9509	62	0.8947	0.8829	0.8821	0.8776	0.8704	97	0.8106	0.7981	0.7972	0.7925	0.7852
28	0.9684	0.9590	0.9586	0.9558	0.9488	63	0.8924	0.8805	0.8798	0.8751	0.8680	98	0.8078	0.7954	0.7945	0.7901	0.7826
29	0.9669	0.9570	0.9568	0.9540	0.9467	64	0.8901	0.8781	0.8775	0.8727	0.8656	99	0.8048	0.7926	0.7918	0.7877	0.7799
30	0.9652	0.9551	0.95493	0.9520	0.9446	65	0.8878	0.8757	0.8752	0.8702	0.8631	100	0.8016	0.7896	0.78913	0.7854	0.7770
31	0.9634		0.9530	0.9500	0.9426	66	0.8854	0.8733	0.8728	0.8679	0.8607						
32	0.9615		0.9510	0.9481	0.9405	67	0.8831	0.8710	0.8705	0.8656	0.8583						
33	0.9596		0.9489	0.9460	0.9383	68	0.8807	0.8686	0.8682	0.8632	0.8559						
34	0.9577		0.9468	0.9440	0.9361	69	0.8784	0.8662	0.8658	0.8609	0.8535						

*Two different observers; see *International Critical Tables*, vol. 3, p. 120.

TABLE 2-61 Glycerol*

Glycerol, %	Density					Glycerol, %	Density					Glycerol, %	Density				
	15°C	15.5°C	20°C	25°C	30°C		15°C	15.5°C	20°C	25°C	30°C		15°C	15.5°C	20°C	25°C	30°C
100	1.26415	1.26381	1.26108	1.15802	1.25495	65	1.17030	1.17000	1.16750	1.16475	1.16195	30	1.07455	1.07435	1.07270	1.07070	1.06855
99	1.26160	1.26125	1.25850	1.25545	1.25235	64	1.16755	1.16725	1.16475	1.16200	1.15925	29	1.07195	1.07175	1.07010	1.06815	1.06605
98	1.25900	1.25865	1.25590	1.25290	1.24975	63	1.16480	1.16445	1.16205	1.15925	1.15650	28	1.06935	1.06915	1.06755	1.06560	1.06355
97	1.25645	1.25610	1.25335	1.25030	1.24710	62	1.16200	1.16170	1.15930	1.15655	1.15375	27	1.06670	1.06655	1.06495	1.06305	1.06105
96	1.25385	1.25350	1.25080	1.24770	1.24450	61	1.15925	1.15895	1.15655	1.15380	1.15100	26	1.06410	1.06390	1.06240	1.06055	1.05855
95	1.25130	1.25095	1.24825	1.24515	1.24190	60	1.15650	1.15615	1.15380	1.15105	1.14830	25	1.06150	1.06130	1.05980	1.05800	1.05605
94	1.24865	1.24830	1.24560	1.24250	1.23930	59	1.15370	1.15340	1.15105	1.14835	1.14555	24	1.05885	1.05870	1.05720	1.05545	1.05350
93	1.24600	1.24565	1.24300	1.23985	1.23670	58	1.15095	1.15065	1.14830	1.14560	1.14285	23	1.05625	1.05610	1.05465	1.05290	1.05100
92	1.24340	1.24305	1.24035	1.23725	1.23410	57	1.14815	1.14785	1.14555	1.14285	1.14010	22	1.05365	1.05350	1.05205	1.05035	1.04850
91	1.24075	1.24040	1.23770	1.23460	1.23150	56	1.14535	1.14510	1.14280	1.14015	1.13740	21	1.05100	1.05090	1.04950	1.04780	1.04600
90	1.23810	1.23775	1.23510	1.23200	1.22890	55	1.14260	1.14230	1.14005	1.13740	1.13470	20	1.04840	1.04825	1.04690	1.04525	1.04350
89	1.23545	1.23510	1.23245	1.22935	1.22625	54	1.13980	1.13955	1.13730	1.13465	1.13195	19	1.04590	1.04575	1.04440	1.04280	1.04105
88	1.23280	1.23245	1.22975	1.22665	1.22360	53	1.13705	1.13680	1.13455	1.13195	1.12925	18	1.04335	1.04325	1.04195	1.04035	1.03860
87	1.23015	1.22980	1.22710	1.22400	1.22095	52	1.13425	1.13400	1.13180	1.12920	1.12650	17	1.04085	1.04075	1.03945	1.03790	1.03615
86	1.22750	1.22710	1.22445	1.22135	1.21830	51	1.13150	1.13125	1.12905	1.12650	1.12380	16	1.03835	1.03825	1.03695	1.03545	1.03370
85	1.22485	1.22445	1.22180	1.21870	1.21565	50	1.12870	1.12845	1.12630	1.12375	1.12110	15	1.03580	1.03570	1.03450	1.03300	1.03130
84	1.22220	1.22180	1.21915	1.21605	1.21300	49	1.12600	1.12575	1.12360	1.12110	1.11845	14	1.03330	1.03320	1.03200	1.03055	1.02885
83	1.21955	1.21915	1.21650	1.21340	1.21035	48	1.12325	1.12305	1.12090	1.11840	1.11580	13	1.03080	1.03070	1.02955	1.02805	1.02640
82	1.21690	1.21650	1.21380	1.21075	1.20770	47	1.12055	1.12030	1.11820	1.11575	1.11320	12	1.02830	1.02820	1.02705	1.02560	1.02395
81	1.21425	1.21385	1.21115	1.20810	1.20505	46	1.11780	1.11760	1.11550	1.11310	1.11055	11	1.02575	1.02565	1.02455	1.02315	1.02150
80	1.21160	1.21120	1.20850	1.20545	1.20240	45	1.11510	1.11490	1.11280	1.11040	1.10795	10	1.02325	1.02315	1.02210	1.02070	1.01905
79	1.20885	1.20845	1.20575	1.20275	1.19970	44	1.11235	1.11215	1.11010	1.10775	1.10530	9	1.02085	1.02075	1.01970	1.01835	1.01670
78	1.20610	1.20570	1.20305	1.20005	1.19705	43	1.10960	1.10945	1.10740	1.10510	1.10265	8	1.01840	1.01835	1.01730	1.01600	1.01440
77	1.20335	1.20300	1.20030	1.19735	1.19435	42	1.10690	1.10670	1.10470	1.10240	1.10005	7	1.01600	1.01590	1.01495	1.01360	1.01205
76	1.20060	1.20025	1.19760	1.19465	1.19170	41	1.10415	1.10400	1.10200	1.09975	1.09740	6	1.01360	1.01350	1.01255	1.01125	1.00970
75	1.19785	1.19750	1.19485	1.19195	1.18900	40	1.10145	1.10130	1.09930	1.09710	1.09475	5	1.01120	1.01110	1.01015	1.00890	1.00735
74	1.19510	1.19480	1.19215	1.18925	1.18635	39	1.09875	1.09860	1.09665	1.09445	1.09215	4	1.00875	1.00870	1.00780	1.00655	1.00505
73	1.19235	1.19205	1.18940	1.18650	1.18365	38	1.09605	1.09590	1.09400	1.09180	1.08955	3	1.00635	1.00630	1.00540	1.00415	1.00270
72	1.18965	1.18930	1.18670	1.18380	1.18100	37	1.09340	1.09320	1.09135	1.08915	1.08690	2	1.00395	1.00385	1.00300	1.00180	1.00035
71	1.18690	1.18655	1.18395	1.18110	1.17830	36	1.09070	1.09050	1.08865	1.08655	1.08430	1	1.00155	1.00145	1.00060	0.99945	0.99800
70	1.18415	1.18385	1.18125	1.17840	1.17565	35	1.08800	1.08780	1.08600	1.08390	1.08165	0	0.99913	0.99905	0.99823	0.99708	0.99568
69	1.18135	1.18105	1.17850	1.17565	1.17290	34	1.08530	1.08515	1.08335	1.08125	1.07905						
68	1.17860	1.17830	1.17575	1.17295	1.17020	33	1.08265	1.08245	1.08070	1.07860	1.07645						
67	1.17585	1.17555	1.17300	1.17020	1.16745	32	1.07995	1.07975	1.07800	1.07600	1.07380						
66	1.17305	1.17275	1.17025	1.16745	1.16470	31	1.07725	1.07705	1.07535	1.07335	1.07120						

*Bosart and Snoddy, *Ind. Eng. Chem.*, **20**, (1928): 1378.

TABLE 2-62 Hydrazine (N_2H_4)*

%	d_4^{15}	%	d_4^{15}
1	1.0002	30	1.0305
2	1.0013	40	1.038
4	1.0034	50	1.044
8	1.0077	60	1.047
12	1.0121	70	1.046
16	1.0164	80	1.040
20	1.0207	90	1.030
24	1.0248	100	1.011
28	1.0286		

**International Critical Tables*, vol. 3, p. 55.

TABLE 2-63 Densities of Aqueous Solutions of Miscellaneous Organic Compounds*

d, d_w, and d_s are the density of the solution, pure water, and pure liquid solute, respectively, all in g/mL. p_s is the wt % solute. 0.0_3255 means 2.55×10^{-4}.

Section A $d = d_w + Ap_s + Bp_s^2 + Cp_s^3$

Name	Formula	t, °C	Range, p_s	A	B	C
Acetaldehyde	C_2H_4O	18	0–30	$+0.0_3255$	-0.0_516	
Acetamide	C_2H_5NO	15	0–6	$+0.0_3639$	$+0.0_4171$	
Acetone	C_3H_6O	0	0–100	-0.0_3856	-0.0_5449	-0.0_7588
		4	0–100	-0.0_27648	-0.0_41193	$+0.0_8272$
		15	0–100	-0.0_21009	-0.0_59682	-0.0_8624
		20	0–100	-0.0_21233	-0.0_53529	-0.0_75327
		25	0–100	-0.0_21171	-0.0_5904	-0.0_856
Acetonitrile	C_2H_3N	15	0–16	-0.0_21175	-0.0_42024	
Allyl alcohol	C_3H_6O	0	0–89	-0.0_33729	-0.0_41232	$+0.0_72984$
Benzenepentacarboxylic acid	$C_{11}H_6O_{10}$	25	0–0.6	$+0.0_25615$	-0.0_2117	
Butyl alcohol (n-)	$C_4H_{10}O$	20	0–7.9	-0.0_21651	$+0.0_4285$	
Butyric acid (n-)	$C_4H_8O_2$	18	0–10	$+0.0_3414$	$+0.0_4131$	
		25	0–62	$+0.0_35135$	-0.0_4166	$+0.0_611$
Chloral hydrate	$C_2H_3Cl_3O_2$	0	0–70	$+0.0_24489$	$+0.0_42802$	-0.0_71291
		15	0–78	$+0.0_24455$	$+0.0_42198$	$+0.0_74366$
		30	0–90	$+0.0_24401$	$+0.0_41887$	$+0.0_76549$
Chloroacetic acid	$C_2H_3ClO_2$	20	0–32	$+0.0_23648$	$+0.0_5302$	
		25	0–86	$+0.0_23602$	$+0.0_5552$	$+0.0_722$
Citric acid (hydrate)	$C_6H_8O_7 + H_2O$	18	0–50	$+0.0_23824$	$+0.0_41141$	$+0.0_717$
Dichloroacetic acid	$C_2H_2Cl_2O_2$	20	0–30	$+0.0_24427$	$+0.0_5537$	$+0.0_77534$
		25	0–97	$+0.0_24427$	$+0.0_5537$	$+0.0_77534$
Diethylamine hydrochloride	$C_4H_{12}ClN$	21	0–36	$+0.0_334$	$+0.0_676$	
Ethylamine hydrochloride	C_2H_8ClN	21	0–65	$+0.0_21193$	-0.0_5307	-0.0_747
Ethylene glycol	$C_2H_6O_2$	0	0–100	$+0.0_21483$	$+0.0_52992$	-0.0_75248
		15	0–6	$+0.0_2133$	-0.0_5108	
Ethyl ether	$C_4H_{10}O$	20	0–5	-0.0_2221	$+0.0_448$	
		25	0–4.5	-0.0_2221	$+0.0_435$	
tartrate	$C_8H_{14}O_6$	15	0–95	$+0.0_22367$	$+0.0_5358$	-0.0_76005
Formaldehyde	CH_2O	15	0–40	$+0.0_22518$	-0.0_5658	$+0.0_6542$
Formamide	CH_3NO	25	22–96	$+0.0_21217$	$+0.0_53199$	-0.0_72529
Furfural	$C_5H_4O_2$	20	0–8	$+0.0_21827$	$+0.0_5366$	
		25	0–8	$+0.0_21664$	$+0.0_421$	
Isoamyl alcohol	$C_5H_{12}O$	20	0–2.5	$+0.0_2155$	$+0.0_43$	
Isobutyl alcohol	$C_4H_{10}O$	15	0–8	-0.0_2146	$+0.0_56$	
		20	0–8	-0.0_2169	$+0.0_438$	
Isobutyric acid	$C_4H_8O_2$	15	0–9	$+0.0_352$		
		18	0–9	$+0.0_345$		
		25	0–12	$+0.0_337$		
Isovaleric acid	$C_5H_{10}O_2$	25	0–5	$+0.0_3253$	-0.0_4282	
Lactic acid	C_3H_6O	25	0–9	$+0.0_2231$	$+0.0_5186$	
Maleic acid	$C_4H_4O_4$	25	0–40	$+0.0_234$	$+0.0_575$	
Malic acid	$C_4H_6O_5$	20	0–40	$+0.0_23933$	$+0.0_5957$	
		25	0–40	$+0.0_23736$	$+0.0_4175$	
Malonic acid	$C_3H_4O_4$	20	0–40	$+0.0_2389$	$+0.0_41066$	
Methyl acetate	$C_3H_6O_2$	20	0–20	$+0.0_340$	-0.0_574	
glucoside (α-)	$C_7H_{14}O_6$	0	26–51	$+0.0_23336$	$+0.0_5996$	$+0.01544$
		30	26–51	$+0.0_23151$	$+0.0_5975$	$+0.0_8978$
Nicotine	$C_{10}H_{14}N_2$	20	0–60	$+0.0_3642$	$+0.0_5454$	-0.0_7687
Nitrophenol (p-)	$C_6H_5NO_3$	15	0–1.5	$+0.0_23216$	-0.0_455	
Oxalic acid	$C_2H_2O_4$	0	0–4	$+0.0_25898$	-0.0_33185	$+0.0_441$
		15	0–4	$+0.0_2494$	-0.0_58	
		17.5	0–9	$+0.0_2494$	-0.0_58	
		20	0–4	$+0.0_25264$	-0.0_31996	$+0.0_4254$
		25	0–4	$+0.0_25108$	-0.0_31607	$+0.0_4208$
Phenol	C_6H_6O	15	0–5	$+0.0_2111$	-0.0_4283	
		80	0–65	$+0.0_3462$	-0.0_686	
Phenylglycolic acid	$C_8H_8O_3$	25	0–11	$+0.0_2207$	$+0.0_423$	
Picoline (α-)	C_6H_7N	25	0–70	-0.0_4386	-0.0_51405	-0.0_74167
(β-)	C_6H_7N	25	0–60	-0.0_4683	-0.0_513	
Propionic acid	$C_3H_6O_2$	18	0–10	$+0.0_395$	-0.0_4172	
		25	0–40	$+0.0_39245$	-0.0_599	$+0.0_7361$
Pyridine	C_5H_5N	25	0–60	$+0.0_3229$	-0.0_5204	-0.0_828
Resorcinol	$C_6H_6O_2$	18	0–52	$+0.0_2201$	$+0.0_5519$	-0.0_819
Succinic acid	$C_4H_6O_4$	25	0–5.5	$+0.0_2304$		
Tartaric acid (d, l, or dl)	$C_4H_6O_6$	15	0–15	$+0.0_24482$	$+0.0_4185$	
		17.5	0–50	$+0.0_24455$	$+0.0_4185$	
		20	0–50	$+0.0_24432$	$+0.0_41837$	
		30	0–50	$+0.0_24335$	$+0.0_4185$	
		40	0–50	$+0.0_24265$	$+0.0_4185$	
		50	0–50	$+0.0_24205$	$+0.0_4185$	
		60	0–50	$+0.0_24155$	$+0.0_4185$	

*From *International Critical Tables*, vol. 3, pp. 111–114.

(Continued)

TABLE 2-63 Densities of Aqueous Solutions of Miscellaneous Organic Compounds *(Continued)*

Section A $d = d_w + Ap_s + Bp_s^2 + Cp_s^3$ *(Cont.)*

Name	Formula	t, °C	Range, p_s	A	B	C
Tetraethyl ammonium chloride	$C_8H_{20}ClN$	21	0–63	$+0.0_31884$	$+0.0_56$	$+0.0_7122$
Thiourea	CH_4N_2S	15	0–7	$+0.0_22995$	$+0.0_5374$	
Trichloroacetic acid	$C_2HCl_3O_2$	12.5	0–61	$+0.0_2499$	$+0.0_4153$	
		20	10–30	$+0.0_25053$	$+0.0_41387$	
		25	0–94	$+0.0_25051$	$+0.0_56119$	$+0.0_61038$
Triethylamine hydrochloride	$C_6H_{16}ClN$	21	0–54	$+0.0_46$	$+0.0_5558$	-0.0_869
Trimethyl carbinol	$C_4H_{10}O$	20	0–100	-0.0_2117	-0.0_41908	$+0.0_7957$
		25	0–100	-0.0_21286	-0.0_4176	$+0.0_7887$
Urea	CH_4N_2O	14.8	0–12	$+0.0_23213$	-0.0_44802	$+0.0_51216$
		18	0–51	$+0.0_22718$	$+0.0_51552$	$+0.0_72573$
		20	0–35	$+0.0_22702$	$+0.0_53712$	-0.0_72285
		25	0–10	$+0.0_22728$	-0.0_41817	$+0.0_51379$
Urethane	$C_3H_7NO_2$	20	0–56	$+0.0_21278$	-0.0_5245	-0.0_73437
Valeric acid (n-)	$C_5H_{10}O_2$	25	0–3	$+0.0_334$	-0.0_427	

Section B $d = d_s + Ap_w + Bp_w^2 + Cp_w^3$

Name	Formula	d_s	t, °C	Range, p_w	A	B	C
Butyl alcohol (n-)	$C_4H_{10}O$	0.8097	20	0–20	$+0.0_22103$	-0.0_4113	
Butyric acid (n-)	$C_4H_8O_2$	0.9534	25	0–38	$+0.0_21854$	-0.0_42314	
Ethyl ether	$C_4H_{10}O$	0.7077	25	0–1.1	$+0.0_234$	$+0.0_336$	
Isobutyl alcohol	$C_4H_{10}O$	0.8170	0	0–14	$+0.0_22437$	-0.0_4285	
		0.8055	15	0–16	$+0.0_2224$	-0.0_4129	
Isobutyric acid	$C_4H_8O_2$	0.9425	26	0–80	$+0.0_21808$	-0.0_42358	$+0.0_61253$
Nicotine	$C_{10}H_{14}N_2$	1.0093	20	0–40	$+0.0_2199$	-0.0_4331	$+0.0_7315$
Picoline (α-)	C_6H_7N	0.9404	25	0–30	$+0.0_22715$	-0.0_4393	
(β-)	C_6H_7N	0.9515	25	0–40	$+0.0_21925$	-0.0_4352	$+0.0_625$
Pyridine	C_5H_5N	0.9776	25	0–40	$+0.0_21157$	-0.0_5536	-0.0_62
Trimethyl carbinol	$C_4H_{10}O$	0.7856	20	0–20	$+0.0_22287$	$+0.0_5275$	

Section C $d_t = d_o + At + Bt^2$

Name	Formula	p_s	d_o	Range, °C	A	B
Allyl alcohol	C_3H_6O	76.60	0.9122	0–45	-0.0_38	-0.0_527
Butyl alcohol (n-)	$C_4H_{10}O$	80.95	0.8614	0–43	-0.0_37292	-0.0_675
Chloral hydrate	$C_2H_3Cl_3O_2$	2.00	1.0094	7–80	-0.0_42597	-0.0_54313
		10.00	1.0476	7–80	-0.0_47955	-0.0_54253
Ethyl tartrate	$C_7H_{14}O_6$	5.00	1.0150	15–80	-0.0_32103	-0.0_52544
		10.00	1.0270	15–80	-0.0_32116	-0.0_62929
		25.00	1.0665	15–80	-0.0_3401	-0.0_523
Furfural	$C_5H_4O_2$	4.62	1.0125	22–74	-0.0_3232	-0.0_5254
		5.69	1.0140	22–74	-0.0_3221	-0.0_5268
		6.56	1.0155	22–74	-0.0_3211	-0.0_5290
Pyridine	C_5H_5N	9.34	1.0055	11–73	-0.0_3171	-0.0_53615
		21.20	1.0115	14–73	-0.0_3378	-0.0_5248
		29.50	1.0145	12–72	-0.0_3463	-0.0_5235
		40.40	1.0182	9–74	-0.0_3605	-0.0_5167

DENSITIES OF MISCELLANEOUS MATERIALS

TABLE 2-64 Approximate Specific Gravities and Densities of Miscellaneous Solids and Liquids*

Water at 4°C and normal atmospheric pressure taken as unity. For more detailed data on any material, see the section dealing with the properties of that material.

Substance	Sp. gr.	Aver. density lb/ft³
Metals, Alloys, Ores		
Aluminum, cast-hammered	2.55–2.80	165
bronze	7.7	481
Brass, cast-rolled	8.4–8.7	534
Bronze, 7.9 to 14% Sn	7.4–8.9	509
phosphor	8.88	554
Copper, cast-rolled	8.8–8.95	556
ore, pyrites	4.1–4.3	262
German silver	8.58	536
Gold, cast-hammered	19.25–19.35	1205
coin (U.S.)	17.18–17.2	1073
Iridium	21.78–22.42	1383
Iron, gray cast	7.03–7.13	442
cast, pig	7.2	450
wrought	7.6–7.9	485
spiegeleisen	7.5	468
ferro-silicon	6.7–7.3	437
ore, hematite	5.2	325
ore, limonite	3.6–4.0	237
ore, magnetite	4.9–5.2	315
slag	2.5–3.0	172
Lead	11.34	710
ore, galena	7.3–7.6	465
Manganese	7.42	475
ore, pyrolusite	3.7–4.6	259
Mercury	13.6	849
Monel metal, rolled	8.97	555
Nickel	8.9	537
Platinum, cast-hammered	21.5	1330
Silver, cast-hammered	10.4–10.6	656
Steel, cold-drawn	7.83	489
machine	7.80	487
tool	7.70–7.73	481
Tin, cast-hammered	7.2–7.5	459
cassiterite	6.4–7.0	418
Tungsten	19.22	1200
Zinc, cast-rolled	6.9–7.2	440
blende	3.9–4.2	253
Various Solids		
Cereals, oats, bulk	0.51	26
barley, bulk	0.62	39
corn, rye, bulk	0.73	45
wheat, bulk	0.77	48
Cork	0.22–0.26	15
Cotton, flax, hemp	1.47–1.50	93
Fats	0.90–0.97	58
Flour, loose	0.40–0.50	28
pressed	0.70–0.80	47
Glass, common	2.40–2.80	162
plate or crown	2.45–2.72	161
crystal	2.90–3.00	184
dint	3.2–4.7	247
Hay and straw, bales	0.32	20
Leather	0.86–1.02	59
Paper	0.70–1.15	58
Potatoes, piled	0.67	44
Rubber, caoutchouc	0.92–0.96	59
goods	1.0–2.0	94
Salt, granulated, piled	0.77	48
Saltpeter	1.07	67
Starch	1.53	96
Sulfur	1.93–2.07	125
Wool	1.32	82

Substance	Sp. gr.	Aver. density lb/ft³
Timber, Air-dry		
Apple	0.66–0.74	44
Ash, black	0.55	34
white	0.64–0.71	42
Birch, sweet, yellow	0.71–0.72	44
Cedar, white, red	0.35	22
Cherry, wild red	0.43	27
Chestnut	0.48	30
Cypress	0.45–0.48	29
Elm, white	0.56	35
Fir, Douglas	0.48–0.55	32
balsam	0.40	25
Hemlock	0.45–0.50	29
Hickory	0.74–0.80	48
Locust	0.67–0.77	45
Mahogany	0.56–0.85	44
Maple, sugar	0.68	43
white	0.53	33
Oak, chestnut	0.74	46
live	0.87	54
red, black	0.64–0.71	42
white	0.77	48
Pine, Norway	0.55	34
Oregon	0.51	32
red	0.48	30
Southern	0.61–0.67	38–42
white	0.43	27
Poplar	0.43	27
Redwood, California	0.42	26
Spruce, white, red	0.45	28
Teak, African	0.99	62
Indian	0.66–0.88	48
Walnut, black	0.59	37
Willow	0.42–0.50	28
Various Liquids		
Alcohol, ethyl (100%)	0.789	49
methyl (100%)	0.796	50
Acid, muriatic, 40%	1.20	75
nitric, 91%	1.50	94
sulfuric, 87%	1.80	112
Chloroform	1.500	95
Ether	0.736	46
Lye, soda, 66%	1.70	106
Oils, vegetable	0.91–0.94	58
mineral, lubricants	0.88–0.94	57
Turpentine	0.861–0.867	54
Water, 4°C max. density	1.0	62.428
100°C	0.9584	59.830
ice	0.88–0.92	56
snow, fresh fallen	0.125	8
sea water	1.02–1.03	64
Ashlar Masonry		
Bluestone	2.3–2.6	153
Granite, syenite, gneiss	2.4–2.7	159
Limestone	2.1–2.8	153
Marble	2.4–2.8	162
Sandstone	2.0–2.6	143
Rubble Masonry		
Bluestone	2.2–2.5	147
Granite, syenite, gneiss	2.3–2.6	153
Limestone	2.0–2.7	147
Marble	2.3–2.7	156
Sandstone	1.9–2.5	137

Substance	Sp. gr.	Aver. density lb/ft³
Dry Rubble Masonry		
Granite, syenite, gneiss	1.9–2.3	130
Limestone, marble	1.9–2.1	125
Sandstone, bluestone	1.8–1.9	110
Brick Masonry		
Hard brick	1.8–2.3	128
Medium brick	1.6–2.0	112
Soft brick	1.4–1.9	103
Sand-lime brick	1.4–2.2	112
Concrete Masonry		
Cement, stone, sand	2.2–2.4	144
slag, etc.	1.9–2.3	130
cinder, etc.	1.5–1.7	100
Various Building Materials		
Ashes, cinders	0.64–0.72	40–45
Cement, Portland, loose	1.5	94
Lime, gypsum, loose	0.85–1.00	53–64
Mortar, lime, set	1.4–1.9	103
Portland cement	2.08–2.25	94–135
Portland cement	3.1–3.2	196
Slags, bank slag	1.1–1.2	67–72
bank screenings	1.5–1.9	98–117
machine slag	1.5	96
slag sand	0.8–0.9	49–55
Earth, etc., Excavated		
Clay, dry	1.0	63
damp plastic	1.76	110
and gravel, dry	1.6	100
Earth, dry, loose	1.2	76
dry, packed	1.5	95
moist, loose	1.3	78
moist, packed	1.6	96
mud, flowing	1.7	108
mud, packed	1.8	115
Riprap, limestone	1.3–1.4	80–85
Riprap, sandstone	1.4	90
Riprap, shale	1.7	105
Sand, gravel, dry, loose	1.4–1.7	90–105
gravel, dry, packed	1.6–1.9	100–120
gravel, wet	1.89–2.16	126
Excavations in Water		
Clay	1.28	80
River mud	1.44	90
Sand or gravel	0.96	60
and clay	1.00	65
Soil	1.12	70
Stone riprap	1.00	65
Minerals		
Asbestos	2.1–2.8	153
Barytes	4.50	281
Basait	2.7–3.2	184
Bauxite	2.55	159
Bluestone	2.5–2.6	159
Borax	1.7–1.8	109
Chalk	1.8–2.8	143
Clay, marl	1.8–2.6	137
Dolomite	2.9	181
Feldspar, orthoclase	2.5–2.7	162
Gneiss	2.7–2.9	175
Granite	2.6–2.7	165
Greenstone, trap	2.8–3.2	187
Gypsum, alabaster	2.3–2.8	159
Hornblende	3.0	187
Limestone	2.1–2.86	155
Marble	2.6–2.86	170
Magnesite	3.0	187
Phosphate rock, apatite	3.2	200
Porphyry	2.6–2.9	172

*From *Marks' Standard Handbook for Mechanical Engineers*, 10th ed., McGraw-Hill, 1996.

(*Continued*)

TABLE 2-64 Approximate Specific Gravities and Densities of Miscellaneous Solids and Liquids *(Continued)*

Water at 4°C and normal atmospheric pressure taken as unity. For more detailed data on any material, see the section dealing with the properties of that material.

Substance	Sp. gr.	Aver. density lb/ft³	Substance	Sp. gr.	Aver. density lb/ft³	Substance	Sp. gr.	Aver. density lb/ft³
Minerals (*Cont.*)			Bituminous Substances			Bituminous Substances (*Cont.*)		
Pumice, natural	0.37–0.90	40	Asphaltum	1.1–1.5	81	Petroleum	0.87	54
Quartz, flint	2.5–2.8	165	Coal, anthracite	1.4–1.8	97	refined (kerosene)	0.78–0.82	50
Sandstone	2.0–2.6	143	bituminous	1.2–1.5	84	benzine	0.73–0.75	46
Serpentine	2.7–2.8	171	lignite	1.1–1.4	78	gasoline	0.70–0.75	45
Shale, slate	2.6–2.9	172	peat, turf, dry	0.65–0.85	47	Pitch	1.07–1.15	69
						Tar, bituminous	1.20	75
Soapstone, talc	2.6–2.8	169	charcoal, pine	0.28–0.44	23			
Syenite	2.6–2.7	165	charcoal, oak	0.47–0.57	33	Coal and Coke, Piled		
			coke	1.0–1.4	75	Coal, anthracite	0.75–0.93	47–58
Stone, Quarried, Piled			Graphite	1.64–2.7	135	bituminous, lignite	0.64–0.87	40–54
Basalt, granite, gneiss	1.5	96	Paraffin	0.87–0.91	56	peat, turf	0.32–0.42	20–26
Greenstone, hornblende	1.7	107				charcoal	0.16–0.23	10–14
Limestone, marble, quartz	1.5	95				coke	0.37–0.51	23–32
Sandstone	1.3	82						
Shale	1.5	92						

NOTE: To convert pounds per cubic foot to kilograms per cubic meter, multiply by 16.02. °F = 9⁄5°C + 32.

TABLE 2-65 Density (kg/m³) of Selected Elements as a Function of Temperature

Temperature, K*	Element symbol												
	Al	Be†	Cr	Cu	Au	Ir	Fe	Pb	Mo	Ni	Pt	Ag	Zn†
50	2736	3650	7160	9019	19,490	22,600	7910	11,570	10,260	8960	21,570	10,620	7280
100	2732	3640	7155	9009	19,460	22,580	7900	11,520	10,260	8950	21,550	10,600	7260
150	2726	3630	7150	8992	19,420	22,560	7890	11,470	10,250	8940	21,530	10,575	7230
200	2719	3620	7145	8973	19,380	22,540	7880	11,430	10,250	8930	21,500	10,550	7200
250	2710	3610	7140	8951	19,340	22,520	7870	11,380	10,250	8910	21,470	10,520	7170
300	2701	3600	7135	8930	19,300	22,500	7860	11,330	10,240	8900	21,450	10,490	7135
400	2681	3580	7120	8885	19,210	22,450	7830	11,230	10,220	8860	21,380	10,430	7070
500	2661	3555	7110	8837	19,130	22,410	7800	11,130	10,210	8820	21,330	10,360	7000
600	2639	3530	7080	8787	19,040	22,360	7760	11,010	10,190	8780	21,270	10,300	6935
800	2591		7040	8686	18,860	22,250	7690	10,430	10,160	8690	21,140	10,160	6430
1000	2365	—	7000	8568	18,660	22,140	7650	10,190	10,120	8610	21,010	10,010	6260
1200	2305		6945	8458	18,440	22,030	7620	9,940	10,080	8510	20,870	9,850	
1400	2255		6890	7920	17,230	21,920	7520		10,040	8410	20,720	9,170	
1600			6760	7750	16,950	21,790	7420		10,000	8320	20,570	8,980	
1800			6700	7600		21,660	7320		9,950	7690	20,400		
2000				7460		21,510	7030		9,900	7450	20,220		

NOTE: Above the horizontal line the condensed phase is solid; below the line, it is liquid.
*°R = 9⁄5 K.
†Polycrystalline form tabulated. Similar tables for an additional 45 elements appear in the *Handbook of Heat Transfer,* 2d ed., McGraw-Hill, New York, 1984.

LATENT HEATS

Unit Conversions For this subsection, the following unit conversions are applicable: °F = 9⁄5°C + 32.

To convert calories per gram to British thermal units per pound, multiply by 1.799.

To convert millimeters of mercury to pounds-force per square inch, multiply by 1.934×10^{-2}.

TABLE 2-66 Heats of Fusion and Vaporization of the Elements and Inorganic Compounds*

Unless stated otherwise, the values have been taken from the compilations by K. K. Kelley on "Heats of Fusion of Inorganic Compounds," U.S. Bur. Mines Bull. 393 (1936), and "The Free Energies of Vaporization and Vapor Pressures of Inorganic Substances," U.S. Bur. Mines Bull. 383 (1935).

Substance	mp, °C	Heat of fusion,[a,b] cal/mol	bp at 1 atm, °C	Heat of vaporization,[a,b] cal/mol
Aluminum				
Al	660.0	2,550	2057	61,020
Al_2Br_6	97.5	5,420	256.4	10,920
Al_2Cl_6	192.5	16,960	180.2[c]	26,750[c]
$AlF_3 \cdot 3NaF$	1000	16,380		
Al_2I_6	191.0	7,960	385.5	15,360
Al_2O_3	2045	(26,000)	3000	
Antimony				
Sb	630.5	4,770	1440	46,670
$SbBr_3$	97	3,510		
$SbCl_3$	73.4	3,030	219	10,360
$SbCl_5$	4	2,400	172[d]	11,570
Sb_4O_6	655	(27,000)	1425	17,820
Sb_4S_6	546	11,200		
Argon				
A	−189.3	290	−185.8	1,590
Arsenic				
As	814	(6,620)	610[c]	31,000[c]
$AsBr_3$	31	2,810		
$AsCl_3$	−16	2,420	122	7,570
AsF_5	−80.7	2,800	−52.8	4,980
As_4O_6	313	8,000	457.2	14,300
Barium				
Ba	704	(1,400)[e]	1638	35,670
$BaBr_2$	847	6,000		
$BaCl_2$	960	5,370		
BaF_2	1287	3,000		
$Ba(NO_3)_2$	595	(5,980)		
$Ba_3(PO_4)_2$	1730	18,600		
$BaSO_4$	1350	9,700		
Beryllium				
Be	1280	2,500[e]		
Bismuth				
Bi	271.3	2,505	1420	
$BiBr_3$			461	18,020
$BiCl_3$	224	2,600	441	17,350
Bi_2O_3	817	6,800		
Bi_2S_5	747	8,900		
Boron				
BBr_3			91.3	7,300
BCl_3			12.5	5,680
BF_3	−128	480	−100.9	4,620
B_2H_6	−165.5		−92.4	3,685
B_4H_{10}	−119.8		16	6,470
B_5H_9	−46.9		58	7,700
B_5H_{11}			67	8,500
$B_{10}H_{14}$	99.7	7,800	*f*	11,600
B_2H_5Br	−104		16	6,230
$B_3N_3H_6$	−58		50.4	7,670
Bromine				
Br_2	−7.2	2,580	58.0	7,420
BrF_5	−61.3	1,355	40.4	7,470
Cadmium				
Cd	320.9	1,460	765	23,870
$CdBr_2$	568	(5,000)		
$CdCl_2$	568	5,300	967	29,860
CdF_2	1110	(5,400)		
CdI_2	387	3,660	796	25,400
CdO			1559[c]	53,820[c]
$CdSO_4$	1000	4,790		
Calcium				
Ca	851	2,230	1487	36,580
$CaBr_2$	730	4,180		
$CaCO_3$	1282	(12,700)		
$CaCl_2$	782	6,100		
CaF_2	1392	4,100		
$Ca(NO_3)_2$	561	5,120		
CaO	2707	(12,240)		
$CaO \cdot Al_2O_3 \cdot 2SiO_2$	1550	29,400		
$CaO \cdot MgO \cdot 2SiO_2$	1392	(18,200)		
$CaO \cdot SiO_2$	1512	13,400		
$CaSO_4$	1297	6,700		
Carbon				
C (graphite)	3600	11,000[e]		
CBr_4	90	1,050		
CCl_4	−24.0	644	77	7,280
CF_4			−127.9	3,110
CH_4	−182.5	224	−161.4	2,040
C_2N_2	−27.8	1,938[u]	−21.1	5,576[u]
CNBr	52			11,010[c]
CNCl	−5	2,240	13	6,300
Carbon (*Cont.*)				
CNF			−72.8	5,780[c]
CNI			141	13,980[c]
CO	−205.0	200	−191.5	1,444
CO_2	−57.5	1,900	−78.4[c]	6,030[c,r]
COS	−138.8	1,129[k]	−50.2	4,423[k]
$COCl_2$			8.0	5,990
CS_2	−112.0	1,049[l]		
Cerium				
Ce	775	2,120		
Cesium				
Cs	28.4	500	690	16,320
CsBr			1300	35,990
CsCl	642	3,600	1300	35,690
CsF	715	(2,450)	1251	34,330
CsI			1280	35,930
$CsNO_3$	407	3,250		
Chlorine				
Cl_2	−101.0	1,531[m]	−34.1	4,878[m]
ClF			−101	
ClF_3			11.3	5,890
Cl_2O			2.0	6,280
ClO_2			10.9	7,100
Cl_2O_7			79	8,480
Chromium				
Cr	1550	3,930	2475	
CrO_2Cl_2			117	8,250
Cobalt				
Co	1490	3,660		
$CoCl_2$	727	7,390	1050	27,170
Copper				
Cu	1083.0	3,110	2595	72,810
Cu_2Br_2			1355	16,310
Cu_2Cl_2	430	4,890	1490	11,920
CuI			1336	15,940
$Cu_2(CN)_2$	473	(5,400)		
Cu_2O	1230	(13,400)		
CuO	1447	2,820		
Cu_2S	1127	5,500		
Fluorine				
F_2	−223		−188.2	1,640
F_2O			−144.8	2,650
Gallium				
Ga	29.8	1,336	2071	
Germanium				
Ge	959	(8,300)		
GeH_4	−165		−89.1	3,580
Ge_2H_6	−109		31.4	5,900
Ge_3H_8	−105.6		110.6	7,550
$GeHCl_3$	−71		75[g]	8,000
$GeBr_4$	26.1		189	8,560
$GeCl_4$	−49.5		84	7,030
$Ge(CH_3)_4$	−88		44	6,460
Gold				
Au	1063.0	3,030	2966	81,800
Helium				
He	−271.4		−268.4	22
Hydrogen				
H_2	−259.2	28	−252.7	216
HBr	−86.9	575	−66.7	4,210
HCl	−114.2	476	−85.0	3,860
HCN	−13.2	2,009[i]	25.7	6,027[i]
HF	−83.0	1,094	33.3	7,460
$(HF)_6$			51.2	5,020
HI	−50.8	686		
H_2O	0.0	1,436	100.0	9,729[h,q]
$H^2{}_2O$ (= D_2O)	3.8	1,501[s]	101.4	9,945[r,q]
H_2O_2	−2	2,520[c]	158	10,270
HNO_3	−47	600		
H_3PO_2	17.4	2,310		
H_3PO_3	74	3,070		
H_3PO_4	42.4	2,520		
$H_4P_2O_6$	55	8,300		
H_2S	−85.5	568[t]	−60.3	4,463[t]
H_2S_2	−87.6	1,805		
H_2SO_4	10.5	2,360		
H_2Se			−41.3	4,880
H_2SeO_4	58	3,450		
H_2Te	−48.9	1,670	−2.2	5,650
Indium				
In	156.4	781		

*See also subsection "Thermodynamic Properties."

(*Continued*)

TABLE 2-66 Heats of Fusion and Vaporization of the Elements and Inorganic Compounds *(Continued)*

Substance	mp, °C	Heat of fusion,[a,b] cal/mol	bp at 1 atm, °C	Heat of vaporization,[a,b] cal/mol
Iodine				
I_2	113.0	3,650	183	10,390
ICl(α)	17.2	2,660		
ICl(β)	13.9	2,270		
IF_7			4[c]	7,460[c]
Iron				
Fe	1530	3,560	2735	84,600
$FeCl_2$	677	7,800	1026	30,210
Fe_2Cl_6	304	20,590	319	12,040
$Fe(CO)_5$	−21	3,250	105	9,000
FeO	1380	(7,700)		
FeS	1195	5,000		
Krypton				
Kr	−157	360[e]	152.9	2,310[e]
Lead				
Pb	327.4	1,224	1744	42,060
$PbBr_2$	488	4,290	914	27,700
$PbCl_2$	498	5,650	954	29,600
PbF_2	824	1,860	1293	38,300
PbI_2	412	5,970	872	24,850
$PbMoO_4$	1065	(25,800)		
PbO	890	2,820	1472	51,310
PbS	1114	4,150	1281	(50,000)
$PbSO_4$	1087	9,600		
$PbWO_4$	1123	(15,200)		
Lithium				
Li	179	1,100	1372	32,250
$LiBO_2$	845	(5,570)		
LiBr	552	2,900	1310	35,420
LiCl	614	3,200	1382	35,960
LiF	847	(2,360)	1681	50,970
LiI	440	(1,420)	1171	40,770
LiOH	462	2,480		
Li_2MoO_4	705	4,200		
$LiNO_3$				
Li_2SiO_3	1177	7,210		
Li_4SiO_4	1249	7,430		
Li_2SO_4	857	3,040		
Li_2WO_4	742	(6,700)		
Magnesium				
Mg	650	2,160	1107	32,520
$MgBr_2$	711	8,300		
$MgCl_2$	712	8,100	1418	32,690
MgF_2	1221	5,900		
MgO	2642	18,500		
$Mg_3(PO_4)_2$	1184	(11,300)		
$MgSiO_3$	1524	14,700		
$MgSO_4$	1127	3,500		
$MgZn_2$	589	(8,270)		
Manganese				
Mn	1220	3,450	2152	55,150
$MnCl_2$	650	7,340	1190	29,630
$MnSiO_3$	1274	(8,200)		
$MnTiO_3$	1404	(7,960)		
Mercury				
Hg	−38.9	557	361	13,980
$HgBr_2$	241	3,960	319	14,080
$HgCl_2$	277	4,150	304	14,080
HgI_2	250	4,500	354	14,260
$HgSO_4$	850	(1,440)		
Molybdenum				
Mo	2622	(6,660)	(4800)	(128,000)
MoF_6	17	2,500	36	6,000
MoO_3	745	(2,500)	1151	
Neon				
Ne	−248.5	77	−246.0	440[e]
Nickel				
Ni	1455	4,200	2730	87,300
$NiCl_2$			987[c]	48,360[c]
$Ni(CO)_4$			42.5	7,000
Ni_2S	645	(2,980)		
Ni_3S_2	790	5,800		
Nitrogen				
N_2	−210.0	172	−195.8	1,336
NF_3			−129.0	3,000
NH_3	−77.7	1,352[n]	−33.4	5,581[n]
NH_4CNS	146	(4,700)		
NH_4NO_3	169.6	1,460		
N_2O	−90.8	1,563	−88.5	3,950
NO	−163.6	550	−151.7	3,307
N_2O_4	−13	5,540	30	7,040
N_2O_5			32.4	13,800[c]
NOCl			−6.4	6,140
Osmium				
OsF_8			47.4	6,840
OsO_4 (yellow)	56	4,060	130	9,450
OsO_4 (white)	42	2,340		
Oxygen				
O_2	−218.9	106	−183.0	1,629
O_3			−111	2,880

Substance	mp, °C	Heat of fusion,[a,b] cal/mol	bp at 1 atm, °C	Heat of vaporization,[a,b] cal/mol
Palladium				
Pd	1554	4,120		
Phosphorus				
P_4 (yellow)	44.2	615	280	12,520
P_4 (violet)			417[c]	25,600[c]
P_4 (black)			453[c]	33,100
PCl_3			74.2	7,280
PH_3	−133.8	270[o]	−87.7	3,489[o]
P_4O_6	23.8	3,360	174	10,380
P_4O_{10}(α)	569	17,080	591	20,670
P_4O_{10}(β)			358[c]	
$POCl_3$	1.1	3,110	105.1	8,380
P_2S_3			508	
Platinum				
Pt	1773.5	4,700	(4400)	(107,000)
Potassium				
K	63.5	574	776	18,920
KBO_2	947	(5,700)		
KBr	742	5,000	1383	37,060
KCl	770	6,410	1407	38,840
KCN	623	(3,500)		
KCNS	179	2,250		
K_2CO_3	897	7,800		
K_2CrO_4	984	6,920		
$K_2Cr_2O_7$	398	8,770		
KF	857	6,500		
KI	682	4,100	1324	34,690
K_2MoO_4	922	(4,000)		
KNO_3	338	2,840		
KOH	360	(2,000)	1327	30,850
KPO_3	817	2,110		
K_3PO_4	1340	8,900		
$K_4P_2O_7$	1092	14,000		
K_2SO_4	1074	8,100		
K_2TiO_3	810	(10,600)		
K_2WO_4	927	(4,400)		
Praseodymium				
Pr	932	2,700		
Radon				
Rn	−71		−61.8	4,010
Rhenium				
Re	(3000)			
Re_2O_7	296	15,340	362.4	18,060
Re_2O_8	147	3,800		
Rubidium				
Rb	39.1	525	679	18,110
RbBr	677	3,700	1352	37,120
RbCl	717	4,400	1381	36,920
RbF	833	4,130	1408	39,510
RbI	638	2,990	1304	35,960
$RbNO_3$	305	1,340		
Selenium				
Se_2	217	1,220	753	25,490
Se_6			736	20,600
SeF_6			−45.8[c]	6,350[c]
SeO_2			317[c]	20,900
$SeOCl_2$	10	1,010	168	
Silicon				
Si	1427	9,470	2290	
$SiCl_4$	−67.6	1,845	56.8	6,860
Si_2Cl_6	−1		139	
Si_3Cl_8			211.4	12,340
$(SiCl_3)_2O$	−33		135.6	8,820
SiF_4			−94.8[c]	6,130[c]
Si_2F_6	−18.5	3,900	−18.9[c]	10,400[c]
SiF_3Cl	−138		−70.1	4,460
SiF_2Cl_2	−144		−31.5	5,080
SiH_4	−185		−111.6	2,960
Si_2H_6	−132.5		−14.3	5,110
Si_3H_8	−117		53.1	6,780
Si_4H_{10}	−93.5		100	8,890
SiH_3Br	−93.8		2.4	5,650
SiH_2Br_2	−70.0		70.5	6,840
$SiHCl_3$	−126.5		31.8	6,360
$(SiH_3)_3N$	−105.6		48.7	6,850
$(SiH_3)_2O$	−144		−15.4	5,350
SiO_2 (quartz)	1470	3,400	2230	
SiO_2 (cristobalite)	1700	2,100		
Silver				
Ag	960.5	2,700	2212	60,720
AgBr	430	2,180		
AgCl	455	3,155	1564	42,520
AgCN	350	2,750		
AgI	557	2,250	1506	34,450
$AgNO_3$	209	2,755		
Ag_2S	842	3,360		
Ag_2SO_4	657	(4,300)		
Sodium				
Na	97.7	630	914	23,120
$NaBO_2$	966	8,660		

(Continued)

TABLE 2-66 Heats of Fusion and Vaporization of the Elements and Inorganic Compounds *(Continued)*

Substance	mp, °C	Heat of fusion,[a,b] cal/mol	bp at 1 atm, °C	Heat of vaporization,[a,b] cal/mol
Sodium (*Cont.*)				
NaBr	747	6,140	1392	37,950
NaCl	800	7,220	1465	40,810
$NaClO_3$	255	5,290		
NaCN	562	(4,400)	1500	37,280
NaCNS	323	4,450		
Na_2CO_3	854	7,000		
NaF	992	7,000	1704	53,260
NaI	662	5,240		
Na_2MoO_4	687	3,600		
$NaNO_3$	310	3,760		
NaOH	322	2,000	1378	
$\frac{1}{2}Na_2O{\cdot}\frac{1}{2}Al_2O_3{\cdot}3SiO_2$	1107	13,150		
$NaPO_3$	988	(5,000)		
$Na_4P_2O_7$	970	(13,700)		
Na_2S	920	(1,200)		
Na_2SiO_3	1087	10,300		
$Na_2Si_2O_5$	884	8,460		
Na_2SO_4	884	5,830		
Na_2WO_4	702	5,800		
Strontium				
Sr	757	2,190	1384	33,610
$SrBr_2$	643	4,780		
$SrCl_2$	872	4,100		
SrF_2	1400	4,260		
$Sr_3(PO_4)_2$	1770	18,500		
Sulfur				
S (rhombic)	112.8		444.6	2,200
S (monoclinic)	119.2			
S_2Cl_2			138	8,720
SF_6			−63.5[c]	5,600[c]
SO_2	−75.5	1,769[p]	−5.0	5,960[p]
$SO_3(\alpha)$	17	2,060	44.8	10,190
$SO_3(\beta)$	32.4	2,890		
$SO_3(\gamma)$	62.2	6,310		
$SOBr_2$			139.5	9,920
$SOCl_2$			75.4	7,600
SO_2Cl_2			69.2	7,760
Tellurium				
Te	453	3,230	1090	
$TeCl_4$			392	16,830
TeF_6			−38.6[c]	6,700[c]

Substance	mp, °C	Heat of fusion,[a,b] cal/mol	bp at 1 atm, °C	Heat of vaporization,[a,b] cal/mol
Thallium				
Tl	302.5	1,030	1457	38,810
TlBr	460	5,990	819	23,800
TlCl	427	4,260	807	24,420
Tl_2CO_3	273	4,400		
TlI	440	3,125	823	25,030
$TlNO_3$	207	2,290		
Tl_2S	449	3,000		
Tl_2SO_4	632	5,500		
Tin				
Sn_4	231.8	1,720	2270	68,000
$SnBr_2$	232	(1,700)		
$SnBr_4$	30	3,000		
$SnCl_2$	247	3,050	623	20,740
$SnCl_4$	−33.2	2,190	113	8,330
$Sn(CH_3)_4$			78.3	7,320
SnH_4	−149.8		−52.3	4,420
SnI_4	143.5	(4,300)		
Titanium				
$TiBr_4$	38.2	(2,060)		
$TiCl_4$	−23	2,240	136	8,350
TiO_2	1825	(11,400)		
Tungsten				
W	3390	(8,400)	(5900)	(176,000)
WF_6	−0.4	1,800	17.3	6,350
Uranium				
UF_6			55.1[c]	9,990[c]
Xenon				
Xe	−111.5	740	−108.0	3,110
Zinc				
Zn	419.5	1,595	907	27,430
$ZnCl_2$	283	(5,500)	732	28,710
$Zn(C_2H_5)_2$			118	8,960
ZnO	1975	4,470		
ZnS	1645	(9,000)		
Zirconium				
$ZrBr_4$			357[c]	25,800[c]
$ZrCl_4$			311[c]	25,290[c]
ZrI_4			431[c]	29,030[c]
ZrO_2	2715	20,800		

[a] Values in parentheses are uncertain.
[b] For the freezing point or the normal boiling point unless otherwise stated.
[c] Sublimation.
[d] Decomposes at about 75°C; value obtained by extrapolation.
[e] Bichowsky and Rossini, *Thermochemistry of the Chemical Substances*, Reinhold, New York (1936).
[f] Decomposes before the normal boiling point is reached.
[g] Decomposes at about 40°C; value obtained by extrapolation.
[h] See also pp. 2-304 through 2-307 on steam table.
[i] Giauque and Ruehrwein, *J. Am. Chem. Soc.*, **61** (1939): 2626.
[j] Giauque and Egan, *J. Chem. Phys.*, **5** (1937): 45.
[k] Kemp and Giauque, *J. Am. Chem. Soc.*, **59** (1937): 79.
[l] Brown and Manov, *J. Am. Chem. Soc.*, **59** (1937): 500.
[m] Giauque and Powell, *J. Am. Chem. Soc.*, **61** (1939): 1970.
[n] Overstreet and Giauque, *J. Am. Chem. Soc.*, **59** (1937): 254.
[o] Stephenson and Giauque, *J. Chem. Phys.*, **5** (1937): 149.
[p] Giauque and Stephenson, *J. Am. Chem. Soc.*, **60** (1938): 1389.
[q] Osborne, Stimson, and Ginnings, *Bur. Standards J. Research*, **23,** 197 (1939): 261.
[r] Miles and Menzies, *J. Am. Chem. Soc.*, **58** (1936): 1067.
[s] Long and Kemp, *J. Am. Chem. Soc.*, **58** (1936): 1829.
[t] Giauque and Blue, *J. Am. Chem. Soc.*, **58** (1936): 831.
[u] Ruehrwein and Giauque, *J. Am. Chem. Soc.*, **61** (1939): 2940.

TABLE 2-67 Heats of Fusion of Miscellaneous Materials

Material	mp, °C	Heat of fusion, cal/g
Alloys		
30.5 Pb + 69.5 Sn	183	17
36.9 Pb + 63.1 Sn	179	15.5
63.7 Pb + 36.3 Sn	177.5	11.6
77.8 Pb + 22.2 Sn	176.5	9.54
1 Pb + 9 Sn	236	28
24 Pb + 27.3 Sn + 48.7 Bi	98.8	6.85
25.8 Pb + 14.7 Sn + 52.4 Bi + 7 Cd	75.5	8.4
Silicates		
Anorthite ($CaAl_2Si_2O_8$)		100
Orthoclase ($KAlSi_2O_8$)		100
Microcline ($KAlSi_3O_8$)		83
Wollastonite ($CaSiO_8$)		100
Malacolite ($Ca_8MgSi_4O_{12}$)		94
Diopside ($CaMgSi_2O_4$)		100
Olivine (Mg_2SiO_4)		130
Fayalite (Fe_2SiO_4)		85
Spermaceti	43.9	37.0
Wax (bees')	61.8	42.3

TABLE 2-68 Heats of Fusion of Organic Compounds

The values for the hydrocarbons are from the tables of the American Petroleum Institute Research Project 44 at the National Bureau of Standards, with some from Parks and Huffman, *Ind. Eng. Chem.*, **23,** 1138 (1931).

The values for the nonhydrocarbon compounds were recalculated from data in *International Critical Tables,* vol. 5.

Hydrocarbon compounds	Formula	mp, °C	Heat of fusion, cal/g	Hydrocarbon compounds	Formula	mp, °C	Heat of fusion, cal/g
Paraffins				Aromatics—(*Cont.*)			
Methane	CH_4	−182.48	14.03	1-Methyl-3-ethylbenzene	C_9H_{12}	−95.55	15.14
Ethane	C_2H_6	−183.23	22.712	1-Methyl-4-ethylbenzene	C_9H_{12}	−62.350	25.29
Propane	C_3H_8	−187.65	19.100	1,2,3-Trimethylbenzene	C_9H_{12}	−25.375	16.64
n-Butane	C_4H_{10}	−138.33	19.167	1,2,4-Trimethylbenzene	C_9H_{12}	−43.80	24.54
2-Methylpropane	C_4H_{10}	−159.60	18.668	1,3,5-Trimethylbenzene	C_9H_{12}	−44.720	18.97
n-Pentane	C_5H_{12}	−129.723	27.874	Naphthalene	$C_{10}H_8$	+80.0	36.0
2-Methylbutane	C_5H_{12}	−159.890	17.076	Camphene	$C_{10}H_{12}$	+51	57
2,2-Dimethylpropane	C_5H_{12}	−16.6	10.786	Durene	$C_{10}H_{14}$	+79.3	37.4
n-Hexane	C_6H_{14}	−95.320	36.138	Isodurene	$C_{10}H_{14}$	−24.0	23.0
2-Methylpentane	C_6H_{14}	−153.680	17.407	Prehnitene	$C_{10}H_{14}$	−7.7	20.0
2,2-Dimethylbutane	C_6H_{14}	−99.73	1.607	*p*-Cymene	$C_{10}H_{14}$	−68.9	17.1
2,3-Dimethylbutane	C_6H_{14}	−128.41	2.251	*n*-Butyl benzene	$C_{10}H_{14}$	−88.5	19.5
n-Heptane	C_7H_{16}	−90.595	33.513	*tert*-Butyl benzene	$C_{10}H_{14}$	−58.1	14.9
2-Methylhexane	C_7H_{16}	−118.270	21.158	β-Methyl naphthalene	$C_{11}H_{10}$	+34.1	20.1
3-Ethylpentane	C_7H_{16}	−118.593	22.555	Diphenyl	$C_{12}H_{10}$	+68.6	28.8
2,2-Dimethylpentane	C_7H_{16}	−123.790	13.982	Hexamethyl benzene	$C_{12}H_{18}$	+165.5	30.4
2,4-Dimethylpentane	C_7H_{16}	−119.230	15.968	Diphenyl methane	$C_{13}H_{12}$	+25.2	26.4
3,3-Dimethylpentane	C_7H_{16}	−134.46	16.856	Anthracene	$C_{14}H_{10}$	+216.5	38.7
2,2,3-Trimethylbutane	C_7H_{16}	−24.96	5.250	Phenanthrene	$C_{14}H_{10}$	+96.3	25.0
n-Octane	C_8H_{18}	−56.798	43.169	Tolane	$C_{14}H_{10}$	+60	28.7
2-Methylheptane	C_8H_{18}	−109.04	21.458	Stilbene	$C_{14}H_{12}$	+124	40.0
3-Methylpentane	C_8H_{18}	−120.50	23.795	Dibenzil	$C_{14}H_{14}$	+51.4	30.7
4-Methylheptane	C_8H_{18}	−120.955	22.692	Triphenyl methane	$C_{19}H_{16}$	+92.1	21.1
2,2-Dimethylhexane	C_8H_{18}	−121.18	24.226	Alkyl cyclohexanes			
2,5-Dimethylhexane	C_8H_{18}	−91.200	26.903	Cyclohexane	C_6H_{12}	+6.67	7.569
3,3-Dimethylhexane	C_8H_{18}	−126.10	14.9	Methylcyclohexane	C_7H_{14}	−126.58	16.429
2-Methyl-3-ethylpentane	C_8H_{18}	−114.960	23.690	Alkyl cyclopentanes			
3-Methyl-3-ethylpentane	C_8H_{18}	−90.870	22.657	Cyclopentane	C_5H_{10}	−93.80	2.068
2,2,3-Trimethylpentane	C_8H_{18}	−112.27	18.061	Methylcyclopentane	C_6H_{12}	−142.445	19.68
2,2,4-Trimethylpentane	C_8H_{18}	−107.365	19.278	Ethylcyclopentane	C_7H_{14}	−138.435	11.10
2,3,3-Trimethylpentane	C_8H_{18}	−100.70	3.204	1,1-Dimethylcyclopentane	C_7H_{14}	−69.73	3.36
2,3,4-Trimethylpentane	C_8H_{18}	−109.210	19.392	*cis*-1,2-Dimethylcyclopentane	C_7H_{14}	−53.85	3.87
2,2,3,3-Tetramethylbutane	C_8H_{18}	+100.69	14.900	*trans*-1,2-Dimethylcyclopentane	C_7H_{14}	−117.57	15.68
n-Nonane	C_9H_{20}	−53.9	41.2	*trans*-1,3-Dimethylcyclopentane	C_7H_{14}	−133.680	17.93
n-Decane	$C_{10}H_{22}$	−30.0	48.3	Monoolefins			
n-Undecane	$C_{11}H_{24}$	−25.9	34.1	Ethene (Ethylene)	C_2H_4	−169.15	28.547
n-Dodecane	$C_{12}H_{26}$	−9.6	51.3	Propene (Propylene)	C_3H_6	−185.25	17.054
Eicosane	$C_{20}H_{42}$	+36.4	52.0	1-Butene	C_4H_8	−185.35	16.393
Pentacosane	$C_{25}H_{52}$	+53.3	53.6	*cis*-2-Butene	C_4H_8	−138.91	31.135
Tritriacontane	$C_{33}H_{68}$	+71.1	54.0	*trans*-2-Butene	C_4H_8	−105.55	41.564
Aromatics				2-Methylpropene (isobutene)	C_4H_8	−140.35	25.265
Benzene	C_6H_6	+5.533	30.100	1-Pentene	C_5H_{10}	−165.27	16.82
Methylbenzene (Toluene)	C_7H_8	−94.991	17.171	*cis*-2-pentene	C_5H_{10}	−151.363	24.239
Ethylbenzene	C_8H_{10}	−94.950	20.629	*trans*-2-pentene	C_5H_{10}	−140.235	26.536
o-Xylene	C_8H_{10}	−25.187	30.614	2-Methyl-1-butene	C_5H_{10}	−137.560	26.879
m-Xylene	C_8H_{10}	−47.872	26.045	3-Methyl-1-butene	C_5H_{10}	−168.500	18.009
p-Xylene	C_8H_{10}	+13.263	38.526	2-Methyl-2-butene	C_5H_{10}	−133.780	25.738
n-Propylbenzene	C_9H_{12}	−99.500	16.97	Acetylenes			
Isopropylbenzene	C_9H_{12}	−96.028	19.22	Acetylene	C_2H_2	−81.5	23.04
1-Methyl-2-ethylbenzene	C_9H_{12}	−80.833	21.13	2-Butyne (dimethylacetylene)	C_4H_6	−132.23	40.808

Nonhydrocarbon compounds	Formula	mp, °C	Heat of fusion, cal/g	Nonhydrocarbon compounds	Formula	mp, °C	Heat of fusion, cal/g
Acetic acid	$C_2H_4O_2$	16.7	46.68	Butyl alcohol (*n*-)	$C_4H_{10}O$	−89.2	29.93
Acetone	C_3H_6O	−95.5	23.42	(*t*-)	$C_4H_{10}O$	25.4	21.88
Acrylic acid	$C_3H_4O_2$	12.3	37.03	Butyric acid (*n*-)	$C_4H_8O_2$	−5.7	30.04
Allo-cinnamic acid	$C_9H_8O_2$	68	27.35				
Aminobenzoic acid (*o*-)	$C_7H_7NO_2$	145	35.48	Capric acid (*n*-)	$C_{10}H_{20}O_2$	31.99	38.87
(*m*-)	$C_7H_7NO_2$	179.5	38.03	Caprylic acid (*n*-)	$C_8H_{16}O_2$	16.3	35.40
(*p*-)	$C_7H_7NO_2$	188.5	36.46	Carbazole	$C_{12}H_9N$	243	42.05
Amyl alcohol	$C_5H_{12}O$	−78.9	26.65	Carbon tetrachloride	CCl_4	−22.8	41.57
Anethole	$C_{10}H_{12}O$	22.5	25.80	Carvoxime (*d*-)	$C_{10}H_{15}NO$	71.5	23.29
Aniline	$C_6H_5NH_2$	−6.3	27.09	(*l*-)	$C_{10}H_{15}NO$	71	23.41
Anthraquinone	$C_{14}H_8O_2$	284.8	37.48	(*dl*-)	$C_{10}H_{15}NO$	91	24.61
Apiol	$C_{12}H_{14}O_4$	29.5	25.80	Cetyl alcohol	$C_{16}H_{34}O$	49.27	33.80
Azobenzene	$C_{12}H_{10}N_2$	67.1	28.91	Chloracetic acid (α-)	$C_2H_3ClO_2$	61.2	31.06
Azoxybenzene	$C_{12}H_{10}N_2O$	36	21.62	(β-)	$C_2H_3ClO_2$	56	35.12
				Chloral alcoholate	$C_4H_7Cl_3O_2$	9	24.03
Benzil	$C_{14}H_{10}O_2$	95.2	22.15	hydrate	$C_2H_3Cl_3O_2$	47.4	33.18
Benzoic acid	$C_7H_6O_2$	122.45	33.90	Chloroaniline (*p*-)	C_6H_6ClN	71	37.15
Benzophenone	$C_{13}H_{10}O$	47.85	23.53	Chlorobenzoic acid (*o*-)	$C_7H_5ClO_2$	140.2	39.30
Benzylaniline	$C_{13}H_{13}N$	32.37	21.86	(*m*-)	$C_7H_5ClO_2$	154.25	36.41
Bromocamphor	$C_{10}H_{15}BrO$	78	41.57	(*p*-)	$C_7H_5ClO_2$	239.7	49.21
Bromochlorbenzene (*o*-)	C_6H_4BrCl	−12.6	15.41	Chloronitrobenzene (*m*-)	$C_6H_4ClNO_2$	44.4	29.38
(*m*-)	C_6H_4BrCl	−21.2	15.29	(*p*-)	$C_6H_4ClNO_2$	83.5	31.51
(*p*-)	C_6H_4BrCl	64.6	23.41	Cinnamic acid	$C_9H_8O_2$	133	36.50
Bromoiodobenzene (*o*-)	C_6H_4BrI	21	12.18	anhydride	$C_{18}H_{14}O_3$	48	28.14
(*m*-)	C_6H_4BrI	9.3	10.27	Cresol (*p*-)	C_7H_8O	34.6	26.28
(*p*-)	C_6H_4BrI	90.1	16.60	Crotonic acid (α-)	$C_4H_6O_2$	72	25.32
Bromol hydrate	$C_2H_3Br_3O_2$	46	16.90	(*cis*-)	$C_4H_6O_2$	71.2	34.90
Bromophenol (*p*-)	C_6H_5BrO	63.5	20.50	Cyanamide	CH_2N_2	44	49.81
Bromotoluene (*p*-)	C_7H_7Br	28	20.86	Cyclohexanol	$C_6H_{12}O$	25.46	4.19

(*Continued*)

TABLE 2-68 Heats of Fusion of Organic Compounds *(Continued)*

Nonhydrocarbon compounds	Formula	mp, °C	Heat of fusion, cal/g
Dibromobenzene (*o*-)	$C_6H_4Br_2$	1.8	12.78
(*m*-)	$C_6H_4Br_2$	−6.9	13.38
(*p*-)	$C_6H_4Br_2$	86	20.55
Dibromophenol (2, 4-)	$C_6H_4Br_2O$	12	13.97
Dichloroacetic acid	$C_2H_2Cl_2O_2$	−4(?)	14.21
Dichlorobenzene (*o*-)	$C_6H_4Cl_2$	−16.7	21.02
(*m*-)	$C_6H_4Cl_2$	−24.8	20.55
(*p*-)	$C_6H_4Cl_2$	53.13	29.67
Dihydroxybenzene (*o*-)	$C_6H_6O_2$	104.3	49.40
(*m*-)	$C_6H_6O_2$	109.65	46.20
(*p*-)	$C_6H_6O_2$	172.3	58.77
Di-iodobenzene (*o*-)	$C_6H_4I_2$	23.4	10.15
(*m*-)	$C_6H_4I_2$	34.2	11.54
(*p*-)	$C_6H_4I_2$	129	16.20
Dimethyl tartrate (*dl*-)	$C_6H_{10}O_6$	87	35.12
(*d*-)	$C_6H_{10}O_6$	49	21.50
pyrone	$C_7H_8O_2$	132	56.14
Dinitrobenzene (*o*-)	$C_6H_4N_2O_4$	116.93	32.25
(*m*-)	$C_6H_4N_2O_4$	89.7	24.70
(*p*-)	$C_6H_4N_2O_4$	173.5	39.99
Dinitrotoluene (2, 4-)	$C_7H_6N_2O_4$	70.14	26.40
Dioxane	$C_4H_8O_2$	11.0	34.85
Diphenyl amine	$C_{12}H_{11}N$	52.98	25.23
Elaidic acid	$C_{18}H_{34}O_2$	44.4	52.08
Ethyl acetate	$C_4H_8O_2$	83.8	28.43
alcohol	C_2H_6O	−114.4	25.76
Ethylene dibromide	$C_2H_4Br_2$	10.012	13.52
Ethyl ether	$C_4H_{10}O$	−116.3	23.54
Formic acid	CH_2O_2	8.40	58.89
Glutaric acid	$C_6H_8O_4$	97.5	37.39
Glycerol	$C_3H_8O_3$	18.07	47.49
Glycol, ethylene	$C_2H_6O_2$	−11.5	43.26
Hydrazo benzene	$C_{12}H_{12}N_2$	134	22.89
Hydrocinnamic acid	$C_9H_{10}O_2$	48	28.14
Hydroxyacetanilide	$C_8H_9NO_2$	91.3	33.59
Iodotoluene (*p*-)	C_7H_7I	34	18.75
Isopropyl alcohol	C_3H_8O	−88.5	21.08
ether	$C_6H_{14}O$	−86.8	25.79
Lauric acid (*n*-)	$C_{12}H_{24}O_2$	43.22	43.72
Levulinic acid	$C_5H_8O_3$	33	18.97
Menthol (*l*-) (α)	$C_{10}H_{20}O$	43.5	18.63
Methyl alcohol	CH_4O	−97.8	23.7
Myristic acid	$C_{14}H_{28}O_2$	53.86	47.49
Methyl cinnamate	$C_{10}H_{10}O_2$	36	26.53
fumarate	$C_6H_8O_4$	102	57.93
oxalate	$C_4H_6O_4$	54.35	42.64
phenylpropiolate	$C_{10}H_8O_2$	18	22.86
succinate	$C_6H_{10}O_4$	19.5	35.72
Naphthol (α-)	$C_{10}H_8O$	95.0	38.94
(β-)	$C_{10}H_8O$	120.6	31.30
Naphthylamine (α-)	$C_{10}H_9N$	50	22.34
Nitroaniline (*o*-)	$C_6H_6N_2O_2$	71.2	27.88
(*m*-)	$C_6H_6N_2O_2$	114.0	40.97
(*p*-)	$C_6H_6N_2O_2$	147.3	36.46
Nitrobenzene	$C_6H_5NO_2$	5.85	22.52
Nitrobenzoic acid (*o*-)	$C_7H_5NO_4$	145.8	40.06
(*m*-)	$C_7H_5NO_4$	141.1	27.59
(*p*-)	$C_7H_5NO_4$	239.2	52.80
Nitronaphthalene	$C_{10}H_7NO_2$	56.7	25.44
Nitrophenol (*o*-)	$C_6H_5NO_3$	45.13	26.76
Palmitic acid	$C_{16}H_{32}O_2$	61.82	39.18
Paraldehyde	$C_6H_{12}O_3$	10.5	25.02
Pelargic acid (*n*-) (β-)	$C_9H_{18}O_2$		39.04
Pelargonic acid (*n*-) (α-)	$C_9H_{18}O_2$	12.35	30.63
Phenol	C_6H_6O	40.92	29.03
Phenylacetic acid	$C_8H_8O_2$	76.7	25.44
Phenylhydrazine	$C_6H_8N_2$	19.6	36.31
Propyl ether (*n*)	$C_6H_{14}O$	−126.1	20.66
Quinone	$C_6H_4O_2$	115.7	40.85
Stearic acid	$C_{18}H_{30}O_2$	68.82	47.54
Succinic anhydride	$C_4H_4O_3$	119	48.74
Succinonitrile	$C_4H_4N_2$	54.5	11.71
Tetrachloroxylene (*o*-)	$C_8H_6Cl_4$	86	21.02
(*p*-)	$C_8H_6Cl_4$	95	22.10
Thiophene	C_4H_4S	−39.4	14.11
Thiosinamine	$C_4H_8N_2S$	77	33.45
Thymol	$C_{10}H_{14}O$	51.5	27.47
Toluic acid (*o*-)	$C_8H_8O_2$	103.7	35.40
(*m*-)	$C_8H_8O_2$	108.75	27.59
(*p*-)	$C_8H_8O_2$	179.6	39.90
Toluidine (*p*-)	C_7H_9N	43.3	39.90
Tribromophenol (2, 4, 6-)	$C_6H_3Br_3O$	93	13.38
Trichloroacetic acid	$C_2HCl_3O_2$	57.5	8.60
Trinitroglycerol	$C_3H_5N_3O_9$	12.3	23.02
Trinitrotoluene (2, 4, 6-)	$C_7H_5N_3O_6$	80.83	22.34
Tristearin	$C_{57}H_{110}O_6$	70.8, 54.5	45.63
Undecylic acid (α-) (*n*-)	$C_{11}H_{22}O_2$	28.25	32.20
(β-) (*n*-)	$C_{11}H_{22}O_2$		42.91
Urethane	$C_3H_7NO_2$	48.7	40.85
Veratrol	$C_8H_{10}O_2$	22.5	27.45
Xylene dibromide (*o*-)	$C_8H_8Br_2$	95	24.25
(*m*-)	$C_8H_8Br_2$	77	21.45
dichloride (*o*-)	$C_8H_8Cl_2$	55	29.03
(*m*-)	$C_8H_8Cl_2$	34	26.64
(*p*-)	$C_8H_8Cl_2$	100	32.73

TABLE 2-69 Heats of Vaporization of Inorganic and Organic Liquids (J/kmol)

Cmpd. no.*	Name	Formula	CAS	Mol. wt.	$C_1 \times$ 1E-07	C_2	C_3	C_4	T_{min}, K	ΔH_v at $T_{min} \times$ 1E-07	T_{max}, K	ΔH_v at T_{max}
1	Acetaldehyde	C_2H_4O	75-07-0	44.05256	3.4088	0.043317	0.21502	0.23791	149.780	3.23240	466.000	0
2	Acetamide	C_2H_5NO	60-35-5	59.0672	9.9475	0.94835	–0.51011	0.015094	353.150	6.36890	761.000	0
3	Acetic acid	$C_2H_4O_2$	64-19-7	60.052	6.127546	3.683421	–6.193052	2.977694	289.810	2.44660	591.950	0
4	Acetic anhydride	$C_4H_6O_3$	108-24-7	102.08864	5.8564	0.33055	–0.057073	0.083671	200.150	5.14960	606.000	0
5	Acetone	C_3H_6O	67-64-1	58.07914	4.9258	1.0809	–1.3684	0.69723	178.450	3.66050	508.200	0
6	Acetonitrile	C_2H_3N	75-05-8	41.0519	3.8345	0.033941	0.34283	–0.13415	229.315	3.52490	545.500	0
7	Acetylene	C_2H_2	74-86-2	26.03728	1.7059	–0.52025	1.0982	–0.29832	192.400	1.62620	308.300	0
8	Acrolein	C_3H_4O	107-02-8	56.06326	6.6599	2.2443	–2.9192	1.1113	185.450	3.63950	506.000	0
9	Acrylic acid	$C_3H_4O_2$	79-10-7	72.06266	4.3756	2.2571	–4.5116	2.5738	286.150	2.79650	615.000	0
10	Acrylonitrile	C_3H_3N	107-13-1	53.0626	4.3052	0.095188	0.47381	–0.26294	189.630	3.89890	540.000	0
11	Air	Mixture	132259-10-0	28.96	0.74587	0.47571	–0.71131	0.60517	59.150	0.63247	132.450	0
12	Ammonia	H_3N	7664-41-7	17.03052	3.1523	0.3914	–0.2289	0.2309	195.410	2.52980	405.650	0
13	Anisole	C_7H_8O	100-66-3	108.13782	7.6926	1.4255	–1.6901	0.72371	235.650	5.10000	645.600	0
14	Argon	Ar	7440-37-1	39.948	0.84215	0.28333	0.033281	0.030551	83.780	0.65440	150.860	0
15	Benzamide	C_7H_7NO	55-21-0	121.13658	8.7809	0.1933	0.30877	–0.14162	403.000	7.12860	824.000	0
16	Benzene	C_6H_6	71-43-2	78.11184	5.0007	0.65393	–0.27698	0.029569	278.680	3.49320	562.050	0
17	Benzenethiol	C_6H_6S	108-98-5	110.17684	6.081621	0.2724357	0.4430641	–0.3449689	258.270	5.06340	689.000	0
18	Benzoic acid	$C_7H_6O_2$	65-85-0	122.12134	11.374	1.4864	–2.3097	1.4025	395.450	6.94850	751.000	0
19	Benzonitrile	C_7H_5N	100-47-0	103.1213	6.4966	0.54598	–0.42255	0.2597	260.280	5.33600	702.300	0
20	Benzophenone	$C_{13}H_{10}O$	119-61-9	182.2179	10.523	0.87091	–0.45568		321.350	7.48950	830.000	0
21	Benzyl alcohol	C_7H_8O	100-51-6	108.13782	8.4762	0.35251	0.43853	–0.3026	257.850	6.88000	720.150	0
22	Benzyl ethyl ether	$C_9H_{12}O$	539-30-0	136.19098	8.2051	1.4438	–1.8053	0.79682	275.650	5.24700	662.000	0
23	Benzyl mercaptan	C_7H_8S	100-53-8	124.20342	11.544	2.2311	–2.5186	0.83063	243.950	6.26740	718.000	0
24	Biphenyl	$C_{12}H_{10}$	92-52-4	154.2078	7.6737	0.28923	0.34048	–0.26011	342.200	6.11280	773.000	0
25	Bromine	Br_2	7726-95-6	159.808	5.5242	1.5015	–1.7185	0.6614	265.850	3.28440	584.150	0
26	Bromobenzene	C_6H_5Br	108-86-1	157.0079	5.0392	–0.2027	1.2207	–0.70705	242.430	4.71870	670.150	0
27	Bromoethane	C_2H_5Br	74-96-4	108.965	3.9247	0.28886	0.38616	–0.35786	154.250	3.42380	503.800	0
28	Bromomethane	CH_3Br	74-83-9	94.93852	3.1988	0.2896	0.0344	0.0114	179.440	2.75620	464.000	0
29	1,2-Butadiene	C_4H_6	590-19-2	54.09044	3.039582	0.2698591	–0.3789853	0.5165115	136.950	2.82540	452.000	0
30	1,3-Butadiene	C_4H_6	106-99-0	54.09044	3.8018	0.90446	–0.74555	0.24234	164.250	2.76410	425.000	0
31	Butane	C_4H_{10}	106-97-8	58.1222	3.6238	0.8337	–0.82274	0.39613	134.860	2.86840	425.120	0
32	1,2-Butanediol	$C_4H_{10}O_2$	584-03-2	90.121	9.4943	0.64824	–0.24961	0.058188	220.000	7.58750	680.000	0
33	1,3-Butanediol	$C_4H_{10}O_2$	107-88-0	90.121	11.344	1.4414	–1.9412	1.035	196.150	8.14880	676.000	0
34	1-Butanol	$C_4H_{10}O$	71-36-3	74.1216	7.1274	0.0483	0.8966	–0.5116	183.850	6.36430	563.100	0
35	2-Butanol	$C_4H_{10}O$	78-92-2	74.1216	7.5007	0.09616	1.1444	–0.78448	158.450	6.59780	535.900	0
36	1-Butene	C_4H_8	106-98-9	56.10632	3.3774	0.5107	–0.17304	0.05181	87.800	3.01970	419.500	0
37	*cis*-2-Butene	C_4H_8	590-18-1	56.10632	4.3478	1.3196	–1.5096	0.63987	134.260	3.10310	435.500	0
38	*trans*-2-Butene	C_4H_8	624-64-6	56.10632	3.8671	1.0672	–1.2574	0.62539	167.620	2.77200	428.600	0
39	Butyl acetate	$C_6H_{12}O_2$	123-86-4	116.15828	8.8262	1.7772	–1.926	0.63659	199.650	5.32550	575.400	0
40	Butylbenzene	$C_{10}H_{14}$	104-51-8	134.21816	8.0911	1.2599	–1.2911	0.47381	185.300	5.94710	660.500	0
41	Butyl mercaptan	$C_4H_{10}S$	109-79-5	90.1872	5.0883	0.47166	–0.0078998	–0.071247	157.460	4.37960	570.100	0
42	*sec*-Butyl mercaptan	$C_4H_{10}S$	513-53-1	90.1872	4.7563	0.49657	–0.13123	0.027307	133.020	4.18430	554.000	0
43	1-Butyne	C_4H_6	107-00-6	54.09044	4.3143	1.0149	–0.99196	0.40891	147.430	3.20490	440.000	0
44	Butyraldehyde	C_4H_8O	123-72-8	72.10572	4.17	0.23488	0.020947	0.086255	176.800	3.77230	537.200	0
45	Butyric acid	$C_4H_8O_2$	107-92-6	88.1051	6.1947	1.6524	–2.8505	1.6285	250.000	4.16190	615.700	0
46	Butyronitrile	C_4H_7N	109-74-0	69.1051	5.1323	0.32362	0.16979	–0.18921	161.300	4.57590	585.400	0

47	Carbon dioxide	CO_2	124-38-9	44.0095	2.173	0.382	−0.4339	0.42213	216.580	1.52020	304.210	0
48	Carbon disulfide	CS_2	75-15-0	76.1407	4.0359	1.0897	−1.6483	0.9779	161.110	3.17860	552.000	0
49	Carbon monoxide	CO	630-08-0	28.0101	0.8585	0.4921	−0.326	0.2231	68.130	0.65166	132.920	0
50	Carbon tetrachloride	CCl_4	56-23-5	153.8227	4.6113	0.55241	−0.18725	0.022973	250.330	3.47600	556.350	0
51	Carbon tetrafluoride	CF_4	75-73-0	88.0043	1.9311	0.94983	−1.0615	0.51894	89.560	1.42150	227.510	0
52	Chlorine	Cl_2	7782-50-5	70.906	3.068	0.8458	−0.9001	0.453	172.120	2.28780	417.150	0
53	Chlorobenzene	C_6H_5Cl	108-90-7	112.5569	4.6746	0.013055	0.51777	−0.18852	227.950	4.32240	632.350	0
54	Chloroethane	C_2H_5Cl	75-00-3	64.5141	3.253	0.321	−0.252	0.295	136.750	2.95540	460.350	0
55	Chloroform	$CHCl_3$	67-66-3	119.37764	5.3032	1.0366	−0.79572	0.16746	209.630	3.65460	536.400	0
56	Chloromethane	CH_3Cl	74-87-3	50.4875	2.442	−0.298	0.87	−0.271	175.430	2.41470	416.250	0
57	1-Chloropropane	C_3H_7Cl	540-54-5	78.54068	3.93706	0.14297	0.55088	−0.3511	150.350	3.56930	503.150	0
58	2-Chloropropane	C_3H_7Cl	75-29-6	78.54068	3.9033	0.3867	0.008595	−0.016793	155.970	3.36320	489.000	0
59	*m*-Cresol	C_7H_8O	108-39-4	108.13782	6.87	−0.39158	1.7208	−0.97478	285.390	6.37340	705.850	0
60	*o*-Cresol	C_7H_8O	95-48-7	108.13782	13.355	2.3486	−2.5463	0.74218	304.190	6.06020	697.550	0
61	*p*-Cresol	C_7H_8O	106-44-5	108.13782	8.0979	−0.33815	2.3495	−1.7015	307.930	6.57120	704.650	0
62	Cumene	C_9H_{12}	98-82-8	120.19158	7.5255	1.3714	−1.5024	0.59731	177.140	5.41880	631.000	0
63	Cyanogen	C_2N_2	460-19-5	52.0348	2.3558	−0.29499	0.34496	0.24271	245.250	2.33890	400.150	0
64	Cyclobutane	C_4H_8	287-23-0	56.10632	3.6762	0.76666	−0.74793	0.35979	182.480	2.81720	459.930	0
65	Cyclohexane	C_6H_{12}	110-82-7	84.15948	5.193	1.0019	−1.0159	0.46332	279.690	3.38860	553.800	0
66	Cyclohexanol	$C_6H_{12}O$	108-93-0	100.15888	5.5761	−1.7498	4.5168	−2.4034	296.600	6.25790	650.100	0
67	Cyclohexanone	$C_6H_{10}O$	108-94-1	98.143	6.6898	1.0012	−0.96028	0.37622	242.000	4.84470	653.000	0
68	Cyclohexene	C_6H_{10}	110-83-8	82.1436	4.698	0.44894	0.070295	−0.14736	169.670	3.98460	560.400	0
69	Cyclopentane	C_5H_{10}	287-92-3	70.1329	3.4216	−0.21723	1.0245	−0.49752	179.280	3.30460	511.700	0
70	Cyclopentene	C_5H_8	142-29-0	68.11702	3.6524	0.17652	0.2777	−0.10817	138.130	3.37950	507.000	0
71	Cyclopropane	C_3H_6	75-19-4	42.07974	2.7681	0.44645	−0.28756	0.21791	145.590	2.33840	398.000	0
72	Cyclohexyl mercaptan	$C_6H_{12}S$	1569-69-3	116.22448	6.7798	1.1402	−1.1701	0.45855	189.640	5.10540	664.000	0
73	Decanal	$C_{10}H_{20}O$	112-31-2	156.2652	9.0851	1.3026	−1.6803	0.86441	285.000	6.02700	674.000	0
74	Decane	$C_{10}H_{22}$	124-18-5	142.28168	8.7515	1.3204	−1.2441	0.38061	243.510	5.60450	617.700	0
75	Decanoic acid	$C_{10}H_{20}O_2$	334-48-5	172.265	12.531	0.76281	−0.32459	0.054808	304.550	8.84640	722.100	0
76	1-Decanol	$C_{10}H_{22}O$	112-30-1	158.28108	7.9041	−1.36	4.0854	−2.3871	280.050	8.29590	688.000	0
77	1-Decene	$C_{10}H_{20}$	872-05-9	140.2658	6.6985	0.76944	−0.79975	0.42379	206.890	5.35240	616.600	0
78	Decyl mercaptan	$C_{10}H_{22}S$	143-10-2	174.34668	8.4103	0.40556	0.34553	−0.4009	247.560	6.81720	696.000	0
79	1-Decyne	$C_{10}H_{18}$	764-93-2	138.24992	10.603	1.7758	−1.6849	0.38281	229.150	6.07920	619.850	0
80	Deuterium	D_2	7782-39-0	4.0316	0.11867	−0.31087	0.28353	0.34543	18.730	0.12605	38.350	0
81	1,1-Dibromoethane	$C_2H_4Br_2$	557-91-5	187.86116	4.7061	0.098096	0.20134	0.22064	210.150	4.35520	628.000	0
82	1,2-Dibromoethane	$C_2H_4Br_2$	106-93-4	187.86116	6.057225	1.372193	−2.053024	1.161394	282.850	4.06410	650.150	0
83	Dibromomethane	CH_2Br_2	74-95-3	173.83458	6.1207	1.2282	−1.1989	0.40137	220.600	4.18700	611.000	0
84	Dibutyl ether	$C_8H_{18}O$	142-96-1	130.22792	6.4978	0.77464	−0.67379	0.31825	175.300	5.24340	584.100	0
85	*m*-Dichlorobenzene	$C_6H_4Cl_2$	541-73-1	147.00196	5.3065	0.20288	0.039962	0.12466	248.390	4.77510	683.950	0
86	*o*-Dichlorobenzene	$C_6H_4Cl_2$	95-50-1	147.00196	6.4394	0.67955	−0.58058	0.36746	256.150	5.09850	705.000	0
87	*p*-Dichlorobenzene	$C_6H_4Cl_2$	106-46-7	147.00196	7.0416	0.96641	−0.86362	0.32976	326.140	4.68520	684.750	0
88	1,1-Dichloroethane	$C_2H_4Cl_2$	75-34-3	98.95916	4.7631	1.0048	−1.2457	0.67919	176.190	3.62860	523.000	0
89	1,2-Dichloroethane	$C_2H_4Cl_2$	107-06-2	98.95916	5.6489	1.0038	−0.7936	0.17013	237.490	3.84750	561.600	0
90	Dichloromethane	CH_2Cl_2	75-09-2	84.93258	4.8739	0.9583	−0.79374	0.28069	178.010	3.58500	510.000	0
91	1,1-Dichloropropane	$C_3H_6Cl_2$	78-99-9	112.98574	5.6495	1.0359	−0.98747	0.39006	192.500	4.13210	560.000	0
92	1,2-Dichloropropane	$C_3H_6Cl_2$	78-87-5	112.98574	4.2593	−0.0038971	0.58142	−0.23734	172.710	4.03570	572.000	0
93	Diethanol amine	$C_4H_{11}NO_2$	111-42-2	105.13564	12.931	1.2215	−1.3197	0.50585	301.150	8.64260	736.600	0
94	Diethyl amine	$C_4H_{11}N$	109-89-7	73.13684	2.595917	−1.334101	2.366723	−0.7871881	223.350	3.35400	496.600	0
95	Diethyl ether	$C_4H_{10}O$	60-29-7	74.1216	5.947	1.6416	−1.7394	0.5831	156.850	3.75450	466.700	0
96	Diethyl sulfide	$C_4H_{10}S$	352-93-2	90.1872	4.7806	0.39507	−0.028657	0.014929	169.200	4.15460	557.150	0

(*Continued*)

TABLE 2-69 Heats of Vaporization of Inorganic and Organic Liquids (J/kmol) *(Continued)*

Cmpd. no.*	Name	Formula	CAS	Mol. wt.	$C_1 \times$ 1E-07	C_2	C_3	C_4	T_{min}, K	ΔH_v at $T_{min} \times$ 1E-07	T_{max}, K	ΔH_v at T_{max}
97	1,1-Difluoroethane	$C_2H_4F_2$	75-37-6	66.04997	3.663	0.93553	−0.9806	0.46753	154.560	2.67130	386.440	0
98	1,2-Difluoroethane	$C_2H_4F_2$	624-72-6	66.04997	4.2313	0.90591	−0.59583	0.074323	215.000	2.78200	445.000	0
99	Difluoromethane	CH_2F_2	75-10-5	52.02339	3.3907	1.1148	−1.2957	0.58214	136.950	2.40150	351.255	0
100	Diisopropyl amine	$C_6H_{15}N$	108-18-9	101.19	2.8258	−1.5731	2.9709	−1.1073	176.850	3.76470	523.100	0
101	Diisopropyl ether	$C_6H_{14}O$	108-20-3	102.17476	4.630224	1.265631	−2.325122	1.525306	187.650	3.47860	500.050	0
102	Diisopropyl ketone	$C_7H_{14}O$	565-80-0	114.18546	5.2429	0.80535	−1.4147	1.0288	204.810	4.33570	576.000	0
103	1,1-Dimethoxyethane	$C_4H_{10}O_2$	534-15-6	90.121	4.3872	0.56226	−0.60662	0.4202	159.950	3.75280	507.800	0
104	1,2-Dimethoxypropane	$C_5H_{12}O_2$	7778-85-0	104.14758	4.7999	0.30724	−0.024545	0.091361	226.100	4.05570	543.000	0
105	Dimethyl acetylene	C_4H_6	503-17-3	54.09044	3.6881	0.37958	−0.22063	0.21968	240.910	2.92830	473.200	0
106	Dimethyl amine	C_2H_7N	124-40-3	45.08368	3.4422	−0.49774	1.8024	−0.97741	180.960	3.29670	437.200	0
107	2,3-Dimethylbutane	C_6H_{14}	79-29-8	86.17536	4.8054	1.0013	−1.0356	0.4668	145.190	3.72820	500.000	0
108	1,1-Dimethylcyclohexane	C_8H_{16}	590-66-9	112.21264	5.5503	0.7692	−0.56915	0.2328	239.660	4.11250	591.150	0
109	*cis*-1,2-Dimethylcyclohexane	C_8H_{16}	2207-01-4	112.21264	5.4479	0.56826	−0.29095	0.15397	223.160	4.36640	606.150	0
110	*trans*-1,2-Dimethylcyclohexane	C_8H_{16}	6876-23-9	112.21264	5.8702	1.0022	−1.0188	0.46949	184.990	4.47370	596.150	0
111	Dimethyl disulfide	$C_2H_6S_2$	624-92-0	94.19904	5.8328	0.99061	−0.9035	0.34792	188.440	4.43890	615.000	0
112	Dimethyl ether	C_2H_6O	115-10-6	46.06844	2.6377	−0.072806	0.54324	−0.13977	131.650	2.54380	400.100	0
113	*N,N*-Dimethyl formamide	C_3H_7NO	68-12-2	73.09378	5.9186	0.37731	0.0051489	−0.0027682	212.720	5.09300	649.600	0
114	2,3-Dimethylpentane	C_7H_{16}	565-59-3	100.20194	5.3387	0.9509	−0.97007	0.44354	160.000	4.16640	537.300	0
115	Dimethyl phthalate	$C_{10}H_{10}O_4$	131-11-3	194.184	10.263	1.504	−2.441	1.388	274.180	7.17430	766.000	0
116	Dimethylsilane	C_2H_8Si	1111-74-6	60.17042	2.919	0.47315	−0.19035	0.078322	122.930	2.50210	402.000	0
117	Dimethyl sulfide	C_2H_6S	75-18-3	62.134	4.5493	0.81834	−0.47199	0.047802	174.880	3.43160	503.040	0
118	Dimethyl sulfoxide	C_2H_6OS	67-68-5	78.13344	7.0161	0.9938	−1.4767	0.97462	291.670	5.27280	729.000	0
119	Dimethyl terephthalate	$C_{10}H_{10}O_4$	120-61-6	194.184	7.66109	0.36322	−0.28551	0.23966	413.786	6.19680	777.400	0
120	1,4-Dioxane	$C_4H_8O_2$	123-91-1	88.10512	5.0368	0.37438	−0.0004344	0.0050378	284.950	3.92500	587.000	0
121	Diphenyl ether	$C_{12}H_{10}O$	101-84-8	170.2072	6.9745	0.43414	−0.26069	0.15024	300.030	5.84730	766.800	0
122	Dipropyl amine	$C_6H_{15}N$	142-84-7	101.19	7.993218	1.697066	−1.895364	0.6664379	210.150	4.77500	550.000	0
123	Dodecane	$C_{12}H_{26}$	112-40-3	170.33484	10.962	1.5544	−1.5358	0.46286	263.570	6.52590	658.000	0
124	Eicosane	$C_{20}H_{42}$	112-95-8	282.54748	12.86	0.50351	0.32986	−0.42184	309.580	9.59330	768.000	0
125	Ethane	C_2H_6	74-84-0	30.069	2.1091	0.60646	−0.55492	0.32799	90.350	1.78790	305.320	0
126	Ethanol	C_2H_6O	64-17-5	46.06844	6.5831	1.1905	−1.7666	1.0012	159.050	5.00600	514.000	0
127	Ethyl acetate	$C_4H_8O_2$	141-78-6	88.10512	4.8272	0.2372	0.32434	−0.19429	189.600	4.16260	523.300	0
128	Ethyl amine	C_2H_7N	75-04-7	45.08368	4.275	0.5857	−0.332	0.169	192.150	3.29550	456.150	0
129	Ethylbenzene	C_8H_{10}	100-41-4	106.165	7.4288	1.6218	−2.0278	0.906	178.200	5.08620	617.150	0
130	Ethyl benzoate	$C_9H_{10}O_2$	93-89-0	150.1745	6.8245	1.071	−1.943	1.2788	238.450	5.40830	698.000	0
131	2-Ethyl butanoic acid	$C_6H_{12}O_2$	88-09-5	116.15828	8.7212	0.79255	−0.64882	0.28369	258.150	6.51870	655.000	0
132	Ethyl butyrate	$C_6H_{12}O_2$	105-54-4	116.15828	5.7624	0.46881	−0.14511	0.061942	175.150	4.92230	571.000	0
133	Ethylcyclohexane	C_8H_{16}	1678-91-7	112.21264	6.0933	0.96339	−0.94933	0.44931	161.840	4.84420	609.150	0
134	Ethylcyclopentane	C_7H_{14}	1640-89-7	98.18606	5.7997	1.0161	−0.92313	0.33212	134.710	4.65290	569.500	0
135	Ethylene	C_2H_4	74-85-1	28.05316	2.0639	0.80153	−0.8128	0.4179	104.000	1.59660	282.340	0
136	Ethylenediamine	$C_2H_8N_2$	107-15-3	60.09832	5.6091	0.077011	0.66595	−0.43437	284.290	4.62220	593.000	0
137	Ethylene glycol	$C_2H_6O_2$	107-21-1	62.06784	8.9207	0.83021	−0.88126	0.53255	260.150	6.87400	720.000	0
138	Ethyleneimine	C_2H_5N	151-56-4	43.0678	4.7462	0.37327	0.047488	0.045906	195.200	3.96760	537.000	0
139	Ethylene oxide	C_2H_4O	75-21-8	44.05256	4.4514	1.1569	−1.2336	0.50875	160.650	3.19090	469.150	0
140	Ethyl formate	$C_3H_6O_2$	109-94-4	74.07854	4.4151	0.51536	−0.39281	0.28461	193.550	3.63270	508.400	0
141	2-Ethyl hexanoic acid	$C_8H_{16}O_2$	149-57-5	144.211	11.08845	0.7029	−0.10529	−0.17295	155.150	9.30840	674.600	0

142	Ethylhexyl ether	$C_8H_{18}O$	5756-43-4	130.22792	6.6828	0.6664	−0.4545	0.20227	180.000	5.46390	583.000	0
143	Ethylisopropyl ether	$C_5H_{12}O$	625-54-7	88.14818	4.2527	0.42014	−0.17341	0.14204	140.000	3.73840	489.000	0
144	Ethylisopropyl ketone	$C_6H_{12}O$	565-69-5	100.15888	5.6735	0.85864	−1.1249	0.69714	204.150	4.45040	567.000	0
145	Ethyl mercaptan	C_2H_6S	75-08-1	62.13404	4.292	0.93726	−1.0593	0.54636	125.260	3.50010	499.150	0
146	Ethyl propionate	$C_5H_{10}O_2$	105-37-3	102.1317	5.033	−0.023028	0.84791	−0.44199	199.250	4.53900	546.000	0
147	Ethylpropyl ether	$C_5H_{12}O$	628-32-0	88.14818	5.438	0.60624			145.650	4.41400	500.230	0
148	Ethyltrichlorosilane	$C_2H_5Cl_3Si$	115-21-9	163.506	5.0124	0.48381	−0.1946	0.12282	167.550	4.29170	559.950	0
149	Fluorine	F_2	7782-41-4	37.9968064	0.89107	0.48888	−0.44035	0.31792	53.480	0.75083	144.120	0
150	Fluorobenzene	C_6H_5F	462-06-6	96.1023032	3.7517	−0.33542	1.0497	−0.40021	230.940	3.69360	560.090	0
151	Fluoroethane	C_2H_5F	353-36-6	48.0595	2.4749	0.18492	−0.21197	0.36038	129.950	2.31740	375.310	0
152	Fluoromethane	CH_3F	593-53-3	34.03292	1.9302	−0.2029	0.65339	−0.16704	131.350	1.89050	317.420	0
153	Formaldehyde	CH_2O	50-00-0	30.02598	2.9575	0.098296	0.28373		155.150	2.69310	420.000	0
154	Formamide	CH_3NO	75-12-7	45.04062	5.8307	−0.62844	1.6751	−0.77554	275.700	6.17220	771.000	0
155	Formic acid	CH_2O_2	64-18-6	46.0257	2.3195	1.9091	−5.0003	3.2641	250.000	1.88650	588.000	0
156	Furan	C_4H_4O	110-00-9	68.07396	4.4388	0.82914	−0.72757	0.33552	196.290	3.27960	490.150	0
157	Helium-4	He	7440-59-7	4.0026	0.012504	1.3038	−2.6954	1.7098	2.200	0.00966	5.200	0
158	Heptadecane	$C_{17}H_{36}$	629-78-7	240.46774	15.97	1.977	−2.2318	0.78544	295.130	8.59730	736.000	0
159	Heptanal	$C_7H_{14}O$	111-71-7	114.18546	4.7135	−0.27964	0.89761	−0.33523	229.800	4.69820	620.000	0
160	Heptane	C_7H_{16}	142-82-5	100.20194	5.2516	0.51283	−0.10982	−0.01018	182.570	4.31810	540.200	0
161	Heptanoic acid	$C_7H_{14}O_2$	111-14-8	130.185	12.916	1.4923	−1.3795	0.39603	265.830	7.80040	677.300	0
162	1-Heptanol	$C_7H_{16}O$	111-70-6	116.20134	7.0236	−1.3652	3.987	−2.2545	239.150	7.64980	632.300	0
163	2-Heptanol	$C_7H_{16}O$	543-49-7	116.20134	11.119	1.3264	−1.1057	0.36023	220.000	7.18220	608.300	0
164	3-Heptanone	$C_7H_{14}O$	106-35-4	114.18546	6.067	0.18619	0.47762	−0.26967	234.150	5.16400	606.600	0
165	2-Heptanone	$C_7H_{14}O$	110-43-0	114.18546	6.2857	0.3899	0.17742	−0.19455	238.150	5.08510	611.400	0
166	1-Heptene	C_7H_{14}	592-76-7	98.18606	4.9437	0.35428	0.22149	−0.2353	154.120	4.32080	537.400	0
167	Heptyl mercaptan	$C_7H_{16}S$	1639-09-4	132.26694	6.7011	0.38694	0.24973	−0.26228	229.920	5.51330	645.000	0
168	1-Heptyne	C_7H_{12}	628-71-7	96.17018	4.8235	0.35765	−0.060379	0.045749	192.220	4.15950	547.000	0
169	Hexadecane	$C_{16}H_{34}$	544-76-3	226.44116	14.979	1.89	−2.0762	0.71724	291.310	8.19340	723.000	0
170	Hexanal	$C_6H_{12}O$	66-25-1	100.15888	5.3802	0.52771	−0.4757	0.3242	214.930	4.49940	594.000	0
171	Hexane	C_6H_{14}	110-54-3	86.17536	4.3848	0.34057	0.063282	−0.017037	177.830	3.75320	507.600	0
172	Hexanoic acid	$C_6H_{12}O_2$	142-62-1	116.158	9.0746	0.8926	−0.75172	0.34378	269.250	6.47830	660.200	0
173	1-Hexanol	$C_6H_{14}O$	111-27-3	102.17476	7.035	−0.9575	3.1431	−1.8066	228.550	7.15090	611.300	0
174	2-Hexanol	$C_6H_{14}O$	626-93-7	102.175	9.591	1.236	−1.359	0.717	223.000	6.46500	585.300	0
175	2-Hexanone	$C_6H_{12}O$	591-78-6	100.15888	5.5382	0.19854	0.47139	−0.31556	217.350	4.75590	587.610	0
176	3-Hexanone	$C_6H_{12}O$	589-38-8	100.15888	5.8213	0.44196	0.090968	−0.15346	217.500	4.70770	582.820	0
177	1-Hexene	C_6H_{12}	592-41-6	84.15948	4.249938	0.52336	−0.57323	0.45101	133.390	3.75440	504.000	0
178	3-Hexyne	C_6H_{10}	928-49-4	82.1436	4.282053	0.5862582	−0.9710554	0.8523437	170.050	3.73310	544.000	0
179	Hexyl mercaptan	$C_6H_{14}S$	111-31-9	118.24036	5.9346	0.41114	0.043753	−0.081964	192.620	5.08670	623.000	0
180	1-Hexyne	C_6H_{10}	693-02-7	82.1436	6.8856	1.9737	−2.4886	0.99472	141.250	4.44750	516.200	0
181	2-Hexyne	C_6H_{10}	764-35-2	82.1436	6.0629	1.1597	−0.99686	0.32547	183.650	4.26690	549.000	0
182	Hydrazine	H_4N_2	302-01-2	32.04516	5.9794	0.9424	−1.398	0.8862	274.690	4.52380	653.150	0
183	Hydrogen	H_2	1333-74-0	2.01588	0.10127	0.698	−1.817	1.447	13.950	0.09131	33.190	0
184	Hydrogen bromide	BrH	10035-10-6	80.91194	1.5513	−0.80615	1.1788	−0.070978	185.150	1.81940	363.150	0
185	Hydrogen chloride	ClH	7647-01-0	36.46094	3.4872	2.1553	−2.9128	1.2442	158.970	1.74720	324.650	0
186	Hydrogen cyanide	CHN	74-90-8	27.02534	3.3907	0.43574	−0.56984	0.36017	259.830	2.79840	456.650	0
187	Hydrogen fluoride	FH	7664-39-3	20.0063432	13.451	13.36	-23.383	10.785	277.560	0.71043	461.150	0
188	Hydrogen sulfide	H_2S	7783-06-4	34.08088	2.6092	0.47883	−0.2233	0.12903	187.680	1.97460	373.530	0
189	Isobutyric acid	$C_4H_8O_2$	79-31-2	88.10512	4.0385	0.82698	−2.033	1.4769	227.150	3.55340	605.000	0
190	Isopropyl amine	C_3H_9N	75-31-0	59.11026	5.6917	1.2441	−1.0742	0.32331	177.950	3.74360	471.850	0
191	Malonic acid	$C_3H_4O_4$	141-82-2	104.06146	7.7143	−1.0139	2.2898	−0.91517	409.150	8.31300	834.000	0

(*Continued*)

TABLE 2-69 Heats of Vaporization of Inorganic and Organic Liquids (J/kmol) *(Continued)*

Cmpd. no.*	Name	Formula	CAS	Mol. wt.	$C_1 \times$ 1E-07	C_2	C_3	C_4	T_{min}, K	ΔH_v at $T_{min} \times$ 1E-07	T_{max}, K	ΔH_v at T_{max}
192	Methacrylic acid	$C_4H_6O_2$	79-41-4	86.08924	176.7855	16.29674	−28.8053	14.522	288.150	4.28480	662.000	0
193	Methane	CH_4	74-82-8	16.0425	1.0194	0.26087	−0.14694	0.22154	90.690	0.87235	190.564	0
194	Methanol	CH_4O	67-56-1	32.04186	3.2615	−1.0407	1.8695	−0.60801	175.470	3.97480	512.500	0
195	*N*-Methyl acetamide	C_3H_7NO	79-16-3	73.09378	6.8795	0.012343	0.77544	−0.4379	301.150	5.97080	718.000	0
196	Methyl acetate	$C_3H_6O_2$	79-20-9	74.07854	4.329	0.18771	0.33528	−0.17125	175.150	3.83890	506.550	0
197	Methyl acetylene	C_3H_4	74-99-7	40.06386	3.0066	0.25873	0.033435	0.087053	170.450	2.56480	402.400	0
198	Methyl acrylate	$C_4H_6O_2$	96-33-3	86.08924	6.2689	1.6462	−2.2795	1.0975	196.320	4.04870	536.000	0
199	Methyl amine	CH_5N	74-89-5	31.0571	4.2834	0.90615	−0.93138	0.4776	179.690	3.09550	430.050	0
200	Methyl benzoate	$C_8H_8O_2$	93-58-3	136.14792	5.8474	−0.6042	2.1528	−1.2871	260.750	5.78260	693.000	0
201	3-Methyl-1,2-butadiene	C_5H_8	598-25-4	68.11702	4.2709	0.70788	−0.67299	0.43009	159.530	3.46030	490.000	0
202	2-Methylbutane	C_5H_{12}	78-78-4	72.14878	4.233	0.95448	−0.98289	0.45719	113.250	3.43450	460.400	0
203	2-Methylbutanoic acid	$C_5H_{10}O_2$	116-53-0	102.1317	8.223	0.80923	−0.70838	0.32497	193.000	6.57690	643.000	0
204	3-Methyl-1-butanol	$C_5H_{12}O$	123-51-3	88.1482	10.165	1.4422	−1.6123	0.75941	155.950	7.27510	577.200	0
205	2-Methyl-1-butene	C_5H_{10}	563-46-2	70.1329	4.5217	1.0678	−1.1735	0.55525	135.580	3.46420	465.000	0
206	2-Methyl-2-butene	C_5H_{10}	513-35-9	70.1329	4.897	1.1838	−1.2079	0.43353	139.390	3.61390	470.000	0
207	2-Methyl -1-butene-3-yne	C_5H_6	78-80-8	66.10114	4.5822	1.3506	−1.6049	0.71575	160.150	3.20970	492.000	0
208	Methylbutyl ether	$C_5H_{12}O$	628-28-4	88.14818	4.4918	0.32576	0.1124	−0.067377	157.480	3.94480	512.740	0
209	Methylbutyl sulfide	$C_5H_{12}S$	628-29-5	104.214	6.8872	1.2703	−1.2699	0.44562	175.300	4.96500	593.000	0
210	3-Methyl-1-butyne	C_5H_8	598-23-2	68.11702	3.1821	−0.89979	2.8579	−1.7826	183.450	3.25930	463.200	0
211	Methyl butyrate	$C_5H_{10}O_2$	623-42-7	102.1317	5.1299	0.10033	0.64085	−0.38359	187.350	4.58370	554.500	0
212	Methylchlorosilane	CH_5ClSi	993-00-0	80.5889	4.4696	1.1838	−0.87047	0.056694	139.050	3.16280	442.000	0
213	Methylcyclohexane	C_7H_{14}	108-87-2	98.18606	5.3789	0.71218	−0.28902	−0.014989	146.580	4.45440	572.100	0
214	1-Methylcyclohexanol	$C_7H_{14}O$	590-67-0	114.18546	7.7573	0.56959	0.7221	−0.86278	299.150	5.13430	686.000	0
215	*cis*-2-Methylcyclohexanol	$C_7H_{14}O$	7443-70-1	114.18546	9.4404	0.8722	−0.33173	−0.10938	280.150	6.16980	614.000	0
216	*trans*-2-Methylcyclohexanol	$C_7H_{14}O$	7443-52-9	114.18546	9.4625	0.88768	−0.39167	−0.057899	269.150	6.31440	617.000	0
217	Methylcyclopentane	C_6H_{12}	96-37-7	84.15948	5.1137	0.98237	−0.90553	0.34878	130.730	4.10400	532.700	0
218	1-Methylcyclopentene	C_6H_{10}	693-89-0	82.1436	4.2603	0.34248	−0.088074	0.13072	146.620	3.84130	542.000	0
219	3-Methylcyclopentene	C_6H_{10}	1120-62-3	82.1436	4.2081	0.43515	−0.24963	0.20811	168.540	3.63850	526.000	0
220	Methyldichlorosilane	CH_4Cl_2Si	75-54-7	115.03396	4.8242	1.3456	−1.5783	0.61746	182.550	3.24190	483.000	0
221	Methylethyl ether	C_3H_8O	540-67-0	60.09502	3.7592	0.64544	−0.46384	0.21809	160.000	2.98760	437.800	0
222	Methylethyl ketone	C_4H_8O	78-93-3	72.10572	5.2256	0.9427	−1.0868	0.55491	186.480	3.98780	535.500	0
223	Methylethyl sulfide	C_3H_8S	624-89-5	76.1606	4.9455	0.78235	−0.56637	0.22052	167.230	3.90650	533.000	0
224	Methyl formate	$C_2H_4O_2$	107-31-3	60.05196	4.7691	0.98928	−0.98574	0.42695	174.150	3.51240	487.200	0
225	Methylisobutyl ether	$C_5H_{12}O$	625-44-5	88.14818	4.266	0.37791	0.0037827	−0.001928	188.000	3.56270	497.000	0
226	Methylisobutyl ketone	$C_6H_{12}O$	108-10-1	100.15888	8.1495	1.8479	−2.1328	0.76628	189.150	4.98940	574.600	0
227	Methyl Isocyanate	C_2H_3NO	624-83-9	57.05132	3.2575	−0.58542	1.4307	−0.54833	256.150	3.22260	488.000	0
228	Methylisopropyl ether	$C_4H_{10}O$	598-53-8	74.1216	3.8148	0.38959	−0.15805	0.15228	127.930	3.39970	464.480	0
229	Methylisopropyl ketone	$C_5H_{10}O$	563-80-4	86.1323	2.7567	−1.6298	3.0001	−1.1865	180.150	3.74640	553.400	0
230	Methylisopropyl sulfide	$C_4H_{10}S$	1551-21-9	90.1872	4.0063	−0.17489	0.94886	−0.44746	171.640	3.89410	553.100	0
231	Methyl mercaptan	CH_4S	74-93-1	48.10746	3.0851	−0.29985	1.4733	−0.89559	150.180	2.99210	469.950	0
232	Methyl methacrylate	$C_5H_8O_2$	80-62-6	100.11582	5.6613	0.3132	0.57076	−0.46309	224.950	4.46890	566.000	0
233	2-Methyloctanoic acid	$C_9H_{18}O_2$	3004-93-1	158.23802	10.53	0.7454	−0.39297	0.047214	240.000	8.11060	694.000	0
234	2-Methylpentane	C_6H_{14}	107-83-5	86.17536	5.0351	1.1424	−1.3269	0.62481	119.550	3.97590	497.700	0
235	Methyl pentyl ether	$C_6H_{14}O$	628-80-8	102.17476	5.0003	0.42203	−0.14687	0.11507	176.000	4.30250	546.490	0
236	2-Methylpropane	C_4H_{10}	75-28-5	58.1222	3.9654	1.274	−1.4255	0.60708	113.540	2.93300	407.800	0

237	2-Methyl-2-propanol	$C_4H_{10}O$	75-65-0	74.1216	2.2708	−3.8183	6.7137	−2.7247	298.970	4.65420	506.200	0
238	2-Methyl propene	C_4H_8	115-11-7	56.10632	4.3172	1.5334	−1.9	0.83816	132.810	2.92920	417.900	0
239	Methyl propionate	$C_4H_8O_2$	554-12-1	88.10512	4.9563	0.22568	0.45949	−0.31541	185.650	4.26690	530.600	0
240	Methylpropyl ether	$C_4H_{10}O$	557-17-5	74.1216	4.2364	0.25325	0.58114	−0.4757	133.970	3.73780	476.250	0
241	Methylpropyl sulfide	$C_4H_{10}S$	3877-15-4	90.1872	5.7015	1.0015	−0.95589	0.38421	160.170	4.42340	565.000	0
242	Methylsilane	CH_6Si	992-94-9	46.14384	2.0613	0.33885	−0.63279	0.6454	116.340	1.90240	352.500	0
243	alpha-Methyl styrene	C_9H_{10}	98-83-9	118.1757	5.3293	0.15144	0.15411	0.066538	249.950	4.79340	654.000	0
244	Methyl tert-butyl ether	$C_5H_{12}O$	1634-04-4	88.1482	4.0052	0.19309	0.20658	−0.010244	164.550	3.60720	497.100	0
245	Methyl vinyl ether	C_3H_6O	107-25-5	58.07914	3.2566	0.10042	0.26926	−0.0003252	151.150	2.99980	437.000	0
246	Naphthalene	$C_{10}H_8$	91-20-3	128.17052	5.093	−0.44584	1.0348	−0.19528	353.430	5.09530	748.400	0
247	Neon	Ne	7440-01-9	20.1797	0.19063	−0.048268	0.11183	0.25512	24.560	0.17706	44.400	0
248	Nitroethane	$C_2H_5NO_2$	79-24-3	75.0666	3.8821	−1.2495	3.2285	−1.8283	183.630	4.54440	593.000	0
249	Nitrogen	N_2	7727-37-9	28.0134	0.74905	0.40406	−0.317	0.27343	63.150	0.60243	126.200	0
250	Nitrogen trifluoride	F_3N	7783-54-2	71.00191	1.8859	1.0917	−1.4143	0.76165	66.460	1.46720	234.000	0
251	Nitromethane	CH_3NO_2	75-52-5	61.04002	4.7494	0.1535	0.49623	−0.38464	244.600	4.05640	588.150	0
252	Nitrous oxide	N_2O	10024-97-2	44.0128	2.2724	0.22278	0.29352	−0.13493	182.300	1.66660	309.570	0
253	Nitric oxide	NO	10102-43-9	30.0061	0.94287	−2.0627	3.2659	−1.0186	109.500	1.44210	180.150	0
254	Nonadecane	$C_{19}H_{40}$	629-92-5	268.5209	17.161	1.7444	−1.6657	0.43242	305.040	9.52160	758.000	0
255	Nonanal	$C_9H_{18}O$	124-19-6	142.23862	4.5173	−1.1627	2.3227	−0.89716	267.300	5.47060	658.500	0
256	Nonane	C_9H_{20}	111-84-2	128.2551	7.888	1.3126	−1.3571	0.5034	219.660	5.25710	594.600	0
257	Nonanoic acid	$C_9H_{18}O_2$	112-05-0	158.238	12.126	0.82704	−0.42449	0.08636	285.550	8.59240	710.700	0
258	1-Nonanol	$C_9H_{20}O$	143-08-8	144.2545	7.5429	−1.5966	4.6489	−2.7229	268.150	8.24110	670.900	0
259	2-Nonanol	$C_9H_{20}O$	628-99-9	144.255	14.251	1.418	−0.53849	−0.33162	238.150	8.32860	649.500	0
260	1-Nonene	C_9H_{18}	124-11-8	126.23922	5.9054	0.61039	−0.54533	0.30683	191.910	4.92180	593.100	0
261	Nonyl mercaptan	$C_9H_{20}S$	1455-21-6	160.3201	6.6716	−0.70869	2.636	−1.6685	253.050	6.54750	681.000	0
262	1-Nonyne	C_9H_{16}	3452-09-3	124.22334	8.7405	1.5599	−1.7205	0.64325	223.150	5.46000	598.050	0
263	Octadecane	$C_{18}H_{38}$	593-45-3	254.49432	17.264	2.167	−2.6262	1.0161	301.310	8.94580	747.000	0
264	Octanal	$C_8H_{16}O$	124-13-0	128.212	5.7746	0.16524	0.095968	0.10146	251.650	5.17550	638.900	0
265	Octane	C_8H_{18}	111-65-9	114.22852	6.7138	1.0769	−1.0124	0.37075	216.380	4.69860	568.700	0
266	Octanoic acid	$C_8H_{16}O_2$	124-07-2	144.211	12.626	1.1753	−0.835	0.1489	289.650	7.96680	694.260	0
267	1-Octanol	$C_8H_{18}O$	111-87-5	130.22792	7.2468	−1.2464	3.6797	−2.0665	257.650	7.67930	652.300	0
268	2-Octanol	$C_8H_{18}O$	123-96-6	130.228	12.581	1.3269	−0.69134	−0.08027	241.550	7.57060	629.800	0
269	2-Octanone	$C_8H_{16}O$	111-13-7	128.21204	11.048	2.5722	−3.7155	1.7307	252.850	5.50930	632.700	0
270	3-Octanone	$C_8H_{16}O$	106-68-3	128.21204	6.6142	0.58562	−0.40512	0.22144	255.550	5.20760	627.700	0
271	1-Octene	C_8H_{16}	111-66-0	112.21264	5.4859	0.26207	0.50642	−0.43873	171.450	4.79270	566.900	0
272	Octyl mercaptan	$C_8H_{18}S$	111-88-6	146.29352	7.3618	0.63204	−0.29459	0.063444	223.950	5.90250	667.300	0
273	1-Octyne	C_8H_{14}	629-05-0	110.19676	5.367	0.31607	0.073613	−0.040895	193.550	4.67380	574.000	0
274	Oxalic acid	$C_2H_2O_4$	144-62-7	90.03488	7.7236	−0.55914	1.8363	−0.85806	462.650	6.56310	828.000	0
275	Oxygen	O_2	7782-44-7	31.9988	0.9008	0.4542	−0.4096	0.3183	54.360	0.77419	154.580	0
276	Ozone	O_3	10028-15-6	47.9982	1.7289	0.12106	0.088716	0.10749	80.150	1.63130	261.000	0
277	Pentadecane	$C_{15}H_{32}$	629-62-9	212.41458	10.052	0.37778	0.50709	−0.46599	283.070	7.76350	708.000	0
278	Pentanal	$C_5H_{10}O$	110-62-3	86.1323	5.2373	1.0132	−1.6348	1.0473	191.590	4.12150	566.100	0
279	Pentane	C_5H_{12}	109-66-0	72.14878	4.5087	0.95886	−0.92384	0.39393	143.420	3.47660	469.700	0
280	Pentanoic acid	$C_5H_{10}O_2$	109-52-4	102.132	7.3197	1.2093	−1.9114	1.1591	239.150	5.38130	639.160	0
281	1-Pentanol	$C_5H_{12}O$	71-41-0	88.1482	7.39	−0.1464	1.4751	−0.9208	195.560	6.70050	588.100	0
282	2-Pentanol	$C_5H_{12}O$	6032-29-7	88.1482	8.8703	0.90566	−0.67627	0.3485	200.000	6.48970	561.000	0
283	2-Pentanone	$C_5H_{10}O$	107-87-9	86.1323	5.3818	0.35111	0.40264	−0.42577	196.290	4.45330	561.080	0
284	3-Pentanone	$C_5H_{10}O$	96-22-0	86.1323	4.451	−0.5483	2.1051	−1.3486	234.180	4.22720	560.950	0
285	1-Pentene	C_5H_{10}	109-67-1	70.1329	3.5027	0.3481	−0.19672	0.22394	108.016	3.22320	464.800	0
286	2-Pentyl mercaptan	$C_5H_{12}S$	2084-19-7	104.21378	5.0573	0.45827	−0.22568	0.16393	160.750	4.43430	584.300	0

(*Continued*)

TABLE 2-69 Heats of Vaporization of Inorganic and Organic Liquids (J/kmol) *(Continued)*

Cmpd. no.*	Name	Formula	CAS	Mol. wt.	$C_1 \times$ 1E-07	C_2	C_3	C_4	T_{min}, K	ΔH_v at $T_{min} \times$ 1E-07	T_{max}, K	ΔH_v at T_{max}
287	Pentyl mercaptan	$C_5H_{12}S$	110-66-7	104.21378	5.4925	0.38608	0.12415	−0.13245	197.450	4.65540	598.000	0
288	1-Pentyne	C_5H_8	627-19-0	68.11702	5.1346	1.3829	−1.6264	0.67069	167.450	3.49690	481.200	0
289	2-Pentyne	C_5H_8	627-21-4	68.11702	5.4839	0.98943	−0.46159	−0.064298	163.830	3.99170	519.000	0
290	Phenanthrene	$C_{14}H_{10}$	85-01-8	178.2292	10.336	1.0678	−1.0693	0.39121	372.380	7.05940	869.000	0
291	Phenol	C_6H_6O	108-95-2	94.11124	6.283	−0.64878	2.4219	−1.4972	314.060	5.77350	694.250	0
292	Phenyl isocyanate	C_7H_5NO	103-71-9	119.1207	7.3079	1.3522	−1.6409	0.66839	243.150	4.95580	653.000	0
293	Phthalic anhydride	$C_8H_4O_3$	85-44-9	148.11556	18.461	3.6123	−5.1111	1.9668	404.150	6.24970	791.000	0
294	Propadiene	C_3H_4	463-49-0	40.06386	2.8092	0.30398	0.017572	0.10232	136.870	2.44810	394.000	0
295	Propane	C_3H_8	74-98-6	44.09562	2.9209	0.78237	−0.77319	0.39246	85.470	2.47870	369.830	0
296	1-Propanol	C_3H_8O	71-23-8	60.09502	6.8988	0.6458	−0.5384	0.3317	146.950	5.83560	536.800	0
297	2-Propanol	C_3H_8O	67-63-0	60.095	8.502	1.474	−1.878	0.933	185.258	5.61950	508.300	0
298	Propenylcyclohexene	C_9H_{14}	13511-13-2	122.20746	5.9068	0.44605	−0.18075	0.13426	199.000	5.07850	636.000	0
299	Propionaldehyde	C_3H_6O	123-38-6	58.07914	3.3611	−0.27575	0.66467		165.000	3.43940	503.600	0
300	Propionic acid	$C_3H_6O_2$	79-09-4	74.0785	4	1.3936	−2.9465	1.794	252.450	3.09220	600.810	0
301	Propionitrile	C_3H_5N	107-12-0	55.0785	4.6242	0.12029	0.62187	−0.48327	180.370	4.16430	561.300	0
302	Propyl acetate	$C_5H_{10}O_2$	109-60-4	102.1317	6.4745	0.93113	−0.65971	0.17587	178.150	4.85340	549.730	0
303	Propyl amine	C_3H_9N	107-10-8	59.11026	3.4054	−0.29885	0.72173	−0.080173	188.360	3.46570	496.950	0
304	Propylbenzene	C_9H_{12}	103-65-1	120.19158	7.2986	1.2428	−1.361	0.56435	173.550	5.46050	638.350	0
305	Propylene	C_3H_6	115-07-1	42.07974	2.5216	0.33721	−0.18399	0.22377	87.890	2.31770	364.850	0
306	Propyl formate	$C_4H_8O_2$	110-74-7	88.10512	5.7631	0.70122	−0.15754	−0.11477	180.250	4.44670	538.000	0
307	2-Propyl mercaptan	C_3H_8S	75-33-2	76.16062	4.2077	0.33823	0.2503	−0.21085	142.610	3.70860	517.000	0
308	Propyl mercaptan	C_3H_8S	107-03-9	76.16062	4.4542	0.31385	0.30517	−0.24568	159.950	3.88960	536.600	0
309	1,2-Propylene glycol	$C_3H_8O_2$	57-55-6	76.09442	7.097812	−0.5348227	1.770112	−0.9904166	213.150	7.23780	626.000	0
310	Quinone	$C_6H_4O_2$	106-51-4	108.09476	6.2374	0.73316	−1.3874	1.0391	388.850	4.92650	683.000	0
311	Silicon tetrafluoride	F_4Si	7783-61-1	104.07911	2.3637	0.32997	0.055931	−0.011041	186.350	1.48720	259.000	0
312	Styrene	C_8H_8	100-42-5	104.14912	8.6409	1.8893	−2.1943	0.81388	242.540	4.92460	636.000	0
313	Succinic acid	$C_4H_6O_4$	110-15-6	118.08804	11.447	−0.04418	1.1282	−0.67562	460.850	8.50610	838.000	0
314	Sulfur dioxide	O_2S	7446-09-5	64.0638	2.846	−0.24905	0.62158	−0.020421	197.670	2.79080	430.750	0
315	Sulfur hexafluoride	F_6S	2551-62-4	146.0554192	1.3661	−1.1465	1.5442	−0.15766	223.150	1.62200	318.690	0
316	Sulfur trioxide	O_3S	7446-11-9	80.0632	0.8509	−7.1061	11.558	−4.483	289.950	4.41460	490.850	0
317	Terephthalic acid	$C_8H_6O_4$	100-21-0	166.13084	11.928	−0.063031	0.89651	−0.5152	700.150	7.16890	883.600	0
318	*o*-Terphenyl	$C_{18}H_{14}$	84-15-1	230.30376	13.0705	1.329955	−1.300762	0.5044183	329.350	8.42870	857.000	0
319	Tetradecane	$C_{14}H_{30}$	629-59-4	198.388	12.007	1.445	−1.3846	0.42836	279.010	7.33360	693.000	0
320	Tetrahydrofuran	C_4H_8O	109-99-9	72.10572	4.0907	0.12318	0.46123	−0.23807	164.650	3.74660	540.150	0
321	1,2,3,4-Tetrahydronaphthalene	$C_{10}H_{12}$	119-64-2	132.20228	10.07	1.994	−2.5052	1.0593	237.380	6.02700	720.000	0

322	Tetrahydrothiophene	C_4H_8S	110-01-0	88.17132	5.2918	0.57615	−0.32236	0.15218	176.990	4.49330	631.950	0
323	2,2,3,3-Tetramethylbutane	C_8H_{18}	594-82-1	114.22852	3.8116	−0.60048	1.6501	−0.73052	373.960	3.17800	568.000	0
324	Thiophene	C_4H_4S	110-02-1	84.13956	5.2472	0.78829	−0.47503	0.098333	234.940	3.81710	579.350	0
325	Toluene	C_7H_8	108-88-3	92.13842	5.4643	0.76764	−0.62056	0.25935	178.180	4.40060	591.750	0
326	1,1,2-Trichloroethane	$C_2H_3Cl_3$	79-00-5	133.40422	4.1283	−0.34796	1.0118	−0.32712	236.500	4.13030	602.000	0
327	Tridecane	$C_{13}H_{28}$	629-50-5	184.36142	11.72	1.6004	−1.6689	0.56396	267.760	6.97470	675.000	0
328	Triethyl amine	$C_6H_{15}N$	121-44-8	101.19	4.6139	0.41881	−0.23744	0.20257	158.450	4.05710	535.150	0
329	Trimethyl amine	C_3H_9N	75-50-3	59.11026	5.1056	1.6568	−1.6244	0.41985	156.080	3.08740	433.250	0
330	1,2,3-Trimethylbenzene	C_9H_{12}	526-73-8	120.19158	7.0138	1.0377	−1.1841	0.56211	247.790	5.12030	664.500	0
331	1,2,4-Trimethylbenzene	C_9H_{12}	95-63-6	120.19158	7.8955	1.513	−1.9061	0.85016	229.330	5.22830	649.100	0
332	2,2,4-Trimethylpentane	C_8H_{18}	540-84-1	114.22852	5.935	1.1967	−1.2686	0.51652	165.780	4.34440	543.800	0
333	2,3,3-Trimethylpentane	C_8H_{18}	560-21-4	114.22852	6.0778	1.207	−1.3449	0.58	172.220	4.47800	573.500	0
334	1,3,5-Trinitrobenzene	$C_6H_3N_3O_6$	99-35-4	213.10452	10.688	0.38045	−0.00074017	0.0003222	398.400	8.39050	846.000	0
335	2,4,6-Trinitrotoluene	$C_7H_5N_3O_6$	118-96-7	227.1311	1.9497	−8.4859	17.865	−10.196	354.000	8.84860	828.000	0
336	Undecane	$C_{11}H_{24}$	1120-21-4	156.30826	10.136	1.5084	−1.473	0.44521	247.570	6.19520	639.000	0
337	1-Undecanol	$C_{11}H_{24}O$	112-42-5	172.30766	8.7274	−1.5834	5.0913	−3.2171	288.450	8.90070	703.900	0
338	Vinyl acetate	$C_4H_6O_2$	108-05-4	86.08924	4.6643	0.50913	−0.55117	0.45397	180.350	3.97880	519.130	0
339	Vinyl acetylene	C_4H_4	689-97-4	52.07456	3.649	0.4	0.043		173.150	2.98760	454.000	0
340	Vinyl chloride	C_2H_3Cl	75-01-4	62.49822	4.2629	1.0111	−0.48757	−0.045787	119.360	3.21450	432.000	0
341	Vinyl trichlorosilane	$C_2H_3Cl_3Si$	75-94-5	161.48972	4.3817	0.26434	0.034522	0.071549	178.350	3.91430	543.150	0
342	Water	H_2O	7732-18-5	18.01528	5.66	0.612041	−0.625697	0.398804	273.160	4.49810	647.096	0
343	*m*-Xylene	C_8H_{10}	108-38-3	106.165	6.493	1.0653	−1.1205	0.48226	225.300	4.68030	617.000	0
344	*o*-Xylene	C_8H_{10}	95-47-6	106.165	6.5393	0.98813	−0.91617	0.35023	247.980	4.65030	630.300	0
345	*p*-Xylene	C_8H_{10}	106-42-3	106.165	6.6475	1.1739	−1.2812	0.54229	286.410	4.30350	616.200	0

The heat of vaporization ΔH_v is calculated by

$$\Delta H_v = C_1(1 - T_r)^{(C_2 + C_3T_r + C_4T_r^2)}$$

where $T_r = T/T_C$, T_C is the critical temperature from Table 2-106, ΔH_v is in J/kmol, and T is in K.

All substances are listed by chemical family in Table 2-6 and by formula.

Values in this table were taken from the Design Institute for Physical Properties (DIPPR) of the American Institute of Chemical Engineers (AIChE), 801 Critically Evaluated Gold Standard™ Database, copyright 2016 AIChE, and reproduced with permission of AIChE and of the DIPPR Evaluated Process Design Data Project Steering Committee. Their source should be cited as R. L. Rowley, W. V. Wilding, J. L. Oscarson, T. A. Knotts, and N. F. Giles, *DIPPR® Data Compilation of Pure Chemical Properties*, Design Institute for Physical Properties, AIChE, New York NY (2016).

SPECIFIC HEATS

SPECIFIC HEATS OF PURE COMPOUNDS

Unit Conversions For this subsection, the following unit conversions are applicable: °F = 9⁄5°C + 32 and °R = 1.8 K. To convert calories per gram-kelvin to British thermal units (Btu) per pound-degree Rankine, multiply by 1.0. To convert kilojoules per kilogram-kelvin to British thermal units per pound-degree Rankine, multiply by 0.2388.

Additional References Additional data are contained in the subsection "Thermodynamic Properties." Data on water are also contained in that subsection.

TABLE 2-70 Heat Capacities of the Elements and Inorganic Compounds*

Substance	State†	Heat capacity at constant pressure (T = K; 0°C = 273.1 K), cal/(mol·K)	Range of temperature, K	Uncertainty, %
Aluminum[1]				
Al	*c*	$4.80 + 0.00322T$	273–931	1
	l	7.00	931–1273	5
$AlBr_3$	*c*	$18.74 + 0.01866T$	273–370	3
	l	29.5	370–407	5
$AlCl_3$	*c*	$13.25 + 0.02800T$	273–465	3
	l	31.2	465–504	3
$AlCl_3{\cdot}6H_2O$	*c*	76	288–327	?
AlF_3	*c*	19.3	288–326	?
$AlF_3{\cdot}3\frac{1}{2}H_2O$	*c*	50.5	288–326	?
$AlF_3{\cdot}3NaF$	*c*	$38.63 + 0.04760T - 449200/T^2$	273–1273	2
	l	142	1273–1373	?
AlI_3	*c*	$16.88 + 0.02266T$	273–464	3
	l	28.8	464–480	5
Al_2O_3	*c*	$22.08 + 0.008971T - 522500/T^2$	273–1973	3
$Al_2O_3{\cdot}SiO_2$	*c*, sillimanite	$40.79 + 0.004763T - 992800/T^2$	273–1573	3
	c, disthene	$41.81 + 0.005283T - 1211000/T^2$	273–1673	2
	c, andalusite	$43.96 + 0.001923T - 1086000/T^2$	273–1573	3
$3Al_2O_3{\cdot}2SiO_2$	*c*, mullite	$59.65 + 0.0670T$	273–576	5
$4Al_2O_3{\cdot}3SiO_2$	*c*	$113.2 + 0.0652T$	273–575	3
$Al_2(SO_4)_3$	*c*	63.5	273–373	?
$Al_2(SO_4)_3{\cdot}18H_2O$	*c*	235	288–325	?
Antimony				
Sb	*c*	$5.51 + 0.00178T$	273–903	2
	l	7.15	903–1273	5
$SbBr_3$	*c*	$17.2 + 0.0293T$	273–370	?
$SbCl_3$	*c*	$10.3 + 0.0511T$	273–346	?
Sb_2O_3	*c*	$19.1 + 0.0171T$	273–929	?
Sb_2O_4	*c*	$22.6 + 0.0162T$	273–1198	?
Sb_2S_3	*c*	$24.2 + 0.0132T$	273–821	?
Argon[2]				
A	*g*	4.97	All	0
Arsenic				
As	*c*	$5.17 + 0.00234T$	273–1168	5
$AsCl_3$	*l*	31.9	286–371	?
As_2O_3	*c*	$8.37 + 0.0486T$	273–548	?
As_2S_3	*c*	25.8	293–373	?
Barium				
$BaCl_2$	*c*	$17.0 + 0.00334T$	273–1198	?
$BaCl_2{\cdot}H_2O$	*c*	28.2	273–307	?
$BaCl_2{\cdot}2H_2O$	*c*	37.3	273–307	?
$Ba(ClO_3)_2{\cdot}H_2O$	*c*	51	289–320	?
$BaCO_3$	*c*, α	$17.26 + 0.0131T$	273–1083	5
	c, β	30.0	1083–1255	15
$BaMoO_4$	*c*	34	273–297	?
$Ba(NO_3)_2$	*c*	39.8	285–371	?
$BaSO_4$	*c*	$21.35 + 0.0141T$	273–1323	5
Beryllium[3,4]				
Be	*c*	$4.698 + 0.001555T - 121000/T^2$	273–1173	1
BeO	*c*	$8.69 + 0.00365T - 313000/T^2$	273–1175	5
$BeO \cdot Al_2O_3$	*c*	25.4	273–373	?
$BeSO_4$	*c*	20.8	273–373	?

*From Kelley, U.S. Bur. Mines Bull. 371, 1934. For a revision see Kelley, U.S. Bur. Mines Bull. 477, 1948. Data for many elements and compounds are given by Johnson (ed.), WADD-TR-60-56, 1960, for cryogenic temperatures. Tabulated data for gases can be obtained from many of the references cited in the "Thermodynamic Properties" subsection and other tables in this section. Thinh, Duran, et al., *Hydrocarbon Process.*, **50,** 98 (January 1971), review previous equation fits and give newer fits for 408 hydrocarbons and related compounds. Later publications include Duran, Thinh, et al., *Hydrocarbon Process.*, **55,** 153 (August 1976); Thompson, *J. Chem. Eng. Data,* **22**(4), 431 (1977); and Passut and Danner, *Ind. Eng. Chem. Process Des. Dev.*, **11,** 543 (1972); **13,** 193 (1974).

†The symbols in this column have the following meaning; *c*, crystal; *l*, liquid; *g*, gas; *gls*, glass.

TABLE 2-70 Heat Capacities of the Elements and Inorganic Compounds *(Continued)*

Substance	State†	Heat capacity at constant pressure (T = K; 0°C = 273.1 K), cal/(mol·K)	Range of temperature, K	Uncertainty, %
Bismuth[4]				
Bi	*c*	$5.38 + 0.00260T$	273–544	3
	l	7.60	544–1273	3
Bi_2O_3	*c*	$23.27 + 0.01105T$	273–777	2
Bi_2S_3	*c*	30.4	284–372	?
Boron				
B	*c*	$1.54 + 0.00440T$	273–1174	5
B_2O_3	*gls*	$5.14 + 0.0320T$	273–513	3
	gls	30.4	513–623	3
BN	*c*	$1.61 + 0.00400T$	273–1173	5
Bromine				
Br_2	*g*	9.00	300–2000	5
Cadmium				
Cd	*c*	$5.46 + 0.002466T$	273–594	1
	l	7.13	594–973	5
CdO	*c*	$9.65 + 0.00208T$	273–2086	?
CdS	*c*	$12.9 + 0.00090T$	273–1273	?
$CdSO_4{\cdot}8/3H_2O$	*c*	51.3	293	?
Calcium				
Ca	*c*	$5.31 + 0.00333T$	273–673	2
	c	$6.29 + 0.00140T$	673–873	2
$CaCl_2$	*c*	$16.9 + 0.00386T$	273–1055	?
$CaCO_3$	*c*	$19.68 + 0.01189T - 307600/T^2$	273–1033	3
CaF_2	*c*	$14.7 + 0.00380T$	273–1651	?
$CaMg(CO_3)_2$	*c*	40.1	299–372	?
$CaMoO_4$	*c*	33	273–297	?
CaO	*c*	$10.00 + 0.00484T - 108000/T^2$	273–1173	2
$Ca(OH)_2$	*c*	21.4	276–373	?
$CaO{\cdot}Al_2O_3{\cdot}2SiO_2$	*c*, anorthite	$63.13 + 0.01500T - 1537000/T^2$	273–1673	1
	gls	$67.41 + 0.01048T - 1874000/T^2$	273–973	1
$CaO{\cdot}MgO{\cdot}2SiO_2$	*c*, diopside	$54.46 + 0.005746T - 1500000/T^2$	273–1573	1
	gls	$51.68 + 0.009724T - 1308000/T^2$	273–973	1
$CaO{\cdot}SiO_2$	*c*, wollastonite	$27.95 + 0.002056T - 745600/T^2$	273–1573	1
	c, pseudowollastonite	$25.48 + 0.004132T - 488100/T^2$	273–1673	1
	gls	$23.16 + 0.009672T - 487100/T^2$	273–973	1
CaP_2O_6	*c*	39.5	287–371	?
$CaSO_4$	*c*	$18.52 + 0.02197T - 156800/T^2$	273–1373	5
$CaSO_4{\cdot}2H_2O$	*c*	46.8	282–373	?
$CaWO_4$	*c*	27.9	292–322	?
Carbon[5]				
C	*c*, graphite	$2.673 + 0.002617T - 116900/T^2$	273–1373	2
	c, diamond	$2.162 + 0.003059T - 130300/T^2$	273–1313	3
CH_4	*g*	$5.34 + 0.0115T$	273–1200	2
CO[6]	*g*	$6.60 + 0.00120T$	273–2500	1½
CO_2	*g*	$10.34 + 0.00274T - 195500/T^2$	273–1200	1½
CS_2	*l*	18.4	293	?
Cerium				
Ce	*c*	$5.88 + 0.00123T$	273–908	?
CeO_2	*c*	15.1	273–373	?
$Ce_2(MoO_4)_3$	*c*	96	273–297	?
$Ce_2(SO_4)_3$	*c*	66.4	273–373	?
$Ce_2(SO_4)_3{\cdot}5H_2O$	*c*	131.6	273–319	?
Cesium				
Cs	*c*	$1.96 + 0.0182T$	273–301	3
	l	8.00	302	3
	g	4.97	All	0
CsBr	*c*	$12.6 + 0.00259T$	273–909	?
CsCl	*c*	$11.7 + 0.00309T$	273–752	?
CsF	*c*	$11.3 + 0.00285T$	273–957	?
CsI	*c*	$11.6 + 0.00268T$	273–894	?
Chlorine				
Cl_2	*g*	$8.28 + 0.00056T$	273–2000	1½
Chromium[4]				
Cr	*c*	$4.84 + 0.00295T$	273–1823	5
	l	9.70	1823–1923	10
$CrCl_3$	*c*	23	286–319	?
Cr_2O_3	*c*	$26.0 + 0.00400T$	273–2263	?
CrSb	*c*	$12.3 + 0.00120T$	273–1383	?
$CrSb_2$	*c*	$19.2 + 0.00184T$	273–949	?
$Cr_2(SO_4)_3$	*c*	67.4	273–373	?
Cobalt[4]				
Co	*c*	$5.12 + 0.00333T$	273–1763	5
	l	8.40	1763–1873	5
$CoAs_2{\cdot}CoS_2$	*c*	32.9	283–373	?
CoSb	*c*	$11.7 + 0.00156T$	273–1464	?
Co_2Sn	*c*	$15.83 + 0.00950T$	273–903	2
CoS	*c*	$10.6 + 0.00251T$	273–1373	?
$CoSO_4{\cdot}7H_2O$	*c*	96	286–303	?

(Continued)

TABLE 2-70 Heat Capacities of the Elements and Inorganic Compounds *(Continued)*

Substance	State[†]	Heat capacity at constant pressure (T = K; 0°C = 273.1 K), cal/(mol·K)	Range of temperature, K	Uncertainty, %
Copper[7]				
Cu	*c*	$5.44 + 0.001462T$	273–1357	1
	l	7.50	1357–1573	3
CuAl	*c*	$9.88 + 0.00500T$	273–733	2
$CuAl_2$	*c*	$16.78 + 0.00366T$	273–773	2
Cu_3Al	*c*	$19.61 + 0.01054T$	273–775	2
CuI	*c*	$12.1 + 0.00286T$	273–675	?
CuI_2	*c*	20.1	274–328	?
CuO	*c*	$10.87 + 0.003576T - 150600/T^2$	273–810	2
$CuO·SiO_2·H_2O$	*c*	29	293–323	?
CuS	*c*	$10.6 + 0.00264T$	273–1273	?
Cu_2S	*c*, α	$9.38 + 0.0312T$	273–376	3
	c, β	20.9	376–1173	2
CuS·FeS	*c*	24	292–321	?
Cu_2Sb	*c*	$13.73 + 0.01350T$	273–573	2
Cu_3Sb	*c*	$21.79 + 0.00900T$	273–693	2
Cu_2Se	*c*, α	20.85	273–383	5
	c, β	20.35	383–488	5
Cu_3Si	*c*	$20.3 + 0.00587T$	273–1135	?
$CuSO_4$	*c*	24.1	282	?
$CuSO_4·H_2O$	*c*	31.3	282	?
$CuSO_4·3H_2O$	*c*	49.0	282	?
$CuSO_4·5H_2O$	*c*	67.2	282	?
Fluorine[8]				
F_2	*g*	$6.50 + 0.00100T$	300–3000	5
Gallium				
Ga_2O_3	*c*	$18.2 + 0.0252T$	273–923	?
$Ga_2(SO_4)_3$	*c*	62.4	273–373	?
Germanium[4]				
Ge	*c*			
Gold				
Au	*c*	$5.61 + 0.00144T$	273–1336	2
	l	7.00	1336–1573	5
$AuSb_2$	*c*, α	$17.12 + 0.00465T$	273–628	1
	c, βγ	$11.47 + 0.01756T$	628–713	?
Helium[9]				
He	*g*	4.97	All	0
Hydrogen[10]				
H	*g*	4.97	All	0
H_2	*g*	$6.62 + 0.00081T$	273–2500	2
HBr	*g*	$6.80 + 0.00084T$	273–2000	2
HCl	*g*	$6.70 + 0.00084T$	273–2000	1½
HI	*g*	$6.93 + 0.00083T$	273–2000	2
H_2O	*l*	See Tables 2-72 and 2-136		
	g	$8.22 + 0.00015T + 0.00000134T^2$	300–2500	?
H_2S	*g*	$7.20 + 0.00360T$	300–600	8
$H_2S_2O_7$	*c*	27	281	?
	l	58	308	?
Indium				
In	*c*			
Iodine				
I_2	*g*	9.00	300–2000	5
Iridium				
Ir	*c*	$5.50 + 0.00148T$	273–1873	1
Iron[4]				
Fe	*c*, α	$4.13 + 0.00638T$	273–1041	3
	c, β	$6.12 + 0.00336T$	1041–1179	3
	c, γ	8.40	1179–1674	5
	c, δ	10.0	1674–1803	5
	l	8.15	1803–1873	5
$FeAs_2$	*c*	17.8	283–373	?
Fe_3C	*c*	$25.17 + 0.00223T$	273–1173	10
$FeCO_3$	*c*	22.7	293–368	?
FeO	*c*	$12.62 + 0.001492T - 76200/T^2$	273–1173	2
Fe_2O_3	*c*	$24.72 + 0.01604T - 423400/T^2$	273–1097	2
Fe_3O_4	*c*	$41.17 + 0.01882T - 979500/T^2$	273–1065	2
$Fe_2O_3·3H_2O$	*c*	47.8	286–373	?
FeS	*c*, α	$2.03 + 0.0390T$	273–411	5
	c, β	$12.05 + 0.00273T$	411–1468	3
FeS_2	*c*	$10.7 + 0.01336T$	273–773	?
FeSi	*c*	$10.54 + 0.00458T$	273–903	2
Fe_2SiO_4	*c*	$33.57 + 0.01907T - 879700/T^2$	273–1161	2
$FeSO_4$	*c*	22	293–373	?
$Fe_2(SO_4)_3$	*c*	66.2	273–373	?
$FeSO_4·4H_2O$	*c*	63.6	282	?
$FeSO_4·7H_2O$	*c*	96	291–319	?
Krypton				
Kr	*g*	4.97	All	0

TABLE 2-70 Heat Capacities of the Elements and Inorganic Compounds *(Continued)*

Substance	State†	Heat capacity at constant pressure (T = K; 0°C = 273.1 K), cal/(mol·K)	Range of temperature, K	Uncertainty, %
Lanthanum				
La	*c*	$5.91 + 0.00100T$	273–1009	?
La_2O_3	*c*	$22.6 + 0.00544T$	273–2273	?
$La_2(MoO_4)_3$	*c*	86	273–307	?
$La_2(SO_4)_3$	*c*	66.9	273–373	?
$La_2(SO_4)_3{\cdot}9H_2O$	*c*	152	273–319	?
Lead[4]				
Pb	*c*	$5.77 + 0.00202T$	273–600	2
	l	6.8	600–1273	5
$Pb_3(AsO_4)_2$	*c*	65.5	286–370	?
PbB_2O_4	*c*	26.5	288–371	?
PbB_4O_7	*c*	41.4	289–371	?
$PbBr_2$	*c*	$18.13 + 0.00310T$	273–761	2
	l	27.4	761–860	10
$PbCl_2$	*c*	$15.88 + 0.00835T$	273–771	2
	l	27.2	771–851	10
$2PbCl_2{\cdot}NH_4Cl$	*c*	53.1	293	?
$PbCO_3$	*c*	21.1	286–320	?
$PbCrO_4$	*c*	29.1	292–323	?
PbF_2	*c*	$16.5 + 0.00412T$	273–1091	?
PbI_2	*c*	$18.66 + 0.00293T$	273–648	2
	l	32.3	648–776	20
$PbMoO_4$	*c*	30.4	292–322	?
$Pb(NO_3)_2$	*c*	36.4	286–320	?
PbO	*c*	$10.33 + 0.00318T$	273–544	2
PbO_2	*c*	$12.7 + 0.00780T$	273–?	?
$Pb_2P_2O_7$	*c*	48.3	284–371	?
PbS	*c*	$10.63 + 0.00401T$	273–873	3
$PbSO_4$	*c*	26.4	293–372	?
PbS_2O_3	*c*	29	293–373	?
$PbWO_4$	*c*	35	273–297	?
Lithium				
Li	*c*	$0.68 + 0.0180T$	273–459	10
	g	4.97	All	0
LiBr	*c*	$11.5 + 0.00302T$	273–825	?
$LiBr{\cdot}H_2O$	*c*	22.6	278–318	?
LiCl	*c*	$11.0 + 0.00339T$	273–887	?
$LiCl{\cdot}H_2O$	*c*	23.6	279–360	?
LiF	*c*	$8.20 + 0.00520T$	273–1117	?
LiI	*c*	$12.5 + 0.00208T$	273–723	?
$LiI{\cdot}H_2O$	*c*	23.6	277–359	?
$LiI{\cdot}2H_2O$	*c*	32.9	277–345	?
$LiI{\cdot}3H_2O$	*c*	43.2	277–347	?
$LiNO_3$	*c*	$9.17 + 0.0360T$	273–523	5
	l	26.8	523–575	5
Magnesium[4]				
Mg	*c*	$6.20 + 0.00133T - 67800/T^2$	273–923	1
	l	7.4	923–1048	10
MgAg	*c*	$10.58 + 0.00412T$	273–905	2
Mg_4Al_3	*c*	$34.4 + 0.0198T$	273–736	?
MgAu	*c*	$11.3 + 0.00189T$	273–1433	?
Mg_2Au	*c*	$16.2 + 0.00451T$	273–1073	?
Mg_3Au	*c*	$21.2 + 0.00614T$	273–1103	?
$MgCl_2$	*c*	$17.3 + 0.00377T$	273–991	?
$MgCl_2{\cdot}6H_2O$	*c*	77.1	292–342	?
$MgCO_3$	*c*	16.9	290	?
$MgCu_2$	*c*	$14.96 + 0.00776T$	273–903	3
Mg_2Cu	*c*	$15.5 + 0.00652T$	273–843	?
$MgNi_2$	*c*	$15.87 + 0.00692T$	273–903	2
MgO	*c*	$10.86 + 0.001197T - 208700/T^2$	273–2073	2
$MgO{\cdot}Al_2O_3$	*c*	28	288–319	?
$MgO{\cdot}SiO_2$	*c*, amphibole	$25.60 + 0.004380T - 674200/T^2$	273–1373	1
	c, pyroxene	$23.35 + 0.008062T - 558800/T^2$	273–773	1
	gls	$23.30 + 0.007734T - 542000/T^2$	273–973	1
$6MgO{\cdot}MgCl_2{\cdot}8B_2O_3$	*c*, α	$58.7 + 0.408T$	273–538	5
	c, β	$107.2 + 0.2876T$	538–623	5
$Mg(OH)_2$	*c*	18.2	292–323	?
Mg_3Sb_2	*c*	$28.2 + 0.00560T$	273–1234	?
Mg_2Si	*c*	$15.4 + 0.00415T$	273–1343	?
$MgSO_4$	*c*	26.7	296–372	?
$MgSO_4{\cdot}H_2O$	*c*	33	282	?
$MgSO_4{\cdot}6H_2O$	*c*	80	282	?
$MgSO_4{\cdot}7H_2O$	*c*	89	291–319	?

(Continued)

TABLE 2-70 Heat Capacities of the Elements and Inorganic Compounds *(Continued)*

Substance	State†	Heat capacity at constant pressure (T = K; 0°C = 273.1 K), cal/(mol·K)	Range of temperature, K	Uncertainty, %
Manganese				
Mn	c, α	$3.76 + 0.00747T$	273–1108	5
	c, β	$5.06 + 0.00395T$	1108–1317	5
	c, γ	$4.80 + 0.00422T$	1317–1493	5
	l	11.0	1493–1673	10
$MnCl_2$	c	$16.2 + 0.00520T$	273–923	?
$MnCO_3$	c	$7.79 + 0.0421T + 0.0000090T^2$	273–773	?
MnO	c	$7.43 + 0.01038T - 0.000003621T^2$	273–1923	?
Mn_2O_3	c	$10.33 + 0.0530T - 0.0000257T^2$	273–1173	?
Mn_3O_4	c	$19.25 + 0.0538T - 0.0000209T^2$	273–1773	?
MnO_2	c	$1.92 + 0.0471T - 0.0000297T^2$	273–773	?
$Mn_2O_3{\cdot}H_2O$	c	31	291–322	?
MnS	c	$10.21 + 0.00656T - 0.000002422T^2$	273–1883	?
$MnSO_4$	c	27.5	293–373	?
$MnSO_4{\cdot}5H_2O$	c	78	290–319	?
Mercury[11]				
Hg	l	6.61	273–630	1
	g	4.97	All	0
Hg_2	g	9.00	300–2000	5
HgCl	c	$11.05 + 0.00370T$	273–798	?
$HgCl_2$	c	$15.3 + 0.0103T$	273–553	?
$Hg(CN)_2$	c	25	285–319	?
HgI	c	$11.4 + 0.00461T$	273–563	?
HgI_2	c, α	$17.4 + 0.004001T$	273–403	3
	c, β	20.2	403–523	3
HgO	c	11.5	278–371	?
HgS	c	$10.9 + 0.00365T$	273–853	?
Hg_2SO_4	c	31.0	273–307	?
Molybdenum				
Mo	c	$5.69 + 0.00188T - 50300/T^2$	273–1773	5
MoO_3	c	$15.1 + 0.0121T$	273–1068	?
MoS_2	c	$19.7 + 0.00315T$	273–729	?
Neon[12]				
Ne	g	4.97	All	0
Nickel[4]				
Ni	c, α	$4.26 + 0.00640T$	273–626	2
	c, β	$6.99 + 0.000905T$	626–1725	5
	l	8.55	1725–1903	10
NiO	c	$11.3 + 0.00215T$	273–1273	?
NiS	c	$9.25 + 0.00640T$	273–597	3
Ni_2Si	c	$15.8 + 0.00329T$	273–1582	?
NiSi	c	$10.0 + 0.00312T$	273–1273	?
Ni_3Sn	c	$20.78 + 0.0102T$	273–904	2
$NiSO_4$	c	33.4	293–373	?
$NiSO_4{\cdot}6H_2O$	c	82	291–325	?
NiTe	c	$11.00 + 0.00433T$	273–700	2
Nitrogen[13]				
N_2	g	$6.50 + 0.00100T$	300–3000	3
NH_3	g	$6.70 + 0.00630T$	300–800	1½
NH_4Br	c	22.8	274–328	?
NH_4Cl	c, α	$9.80 + 0.0368T$	273–457	5
	c, β	$5.0 + 0.0340T$	457–523	5
NH_4I	c	17.8	273–328	?
NH_4NO_3	c	31.8	273–293	?
$(NH_4)_2SO_4$	c	51.6	275–328	?
NO	g	$8.05 + 0.000233T - 156300/T^2$	300–5000	2
Osmium				
Os	c	$5.686 + 0.000875T$	273–1877	1
Oxygen[14]				
O_2	g	$8.27 + 0.000258T - 187700/T^2$	300–5000	1
Palladium				
Pd	c	$5.41 + 0.00184T$	273–1822	2
Phosphorus				
P	c, yellow	5.50	273–317	5
	c, red	$0.21 + 0.0180T$	273–472	10
	l	6.6	317–373	10
PCl_3	l	28.7	284–371	?
P_4O_{10}	c	$15.72 + 0.1092T$	273–631	2
	g	73.6	631–1371	3
Platinum[4]				
Pt	c	$5.92 + 0.00116T$	273–1873	1
Potassium				
K	c	$5.24 + 0.00555T$	273–336	5
	l	7.7	336–373	5

TABLE 2-70 Heat Capacities of the Elements and Inorganic Compounds *(Continued)*

Substance	State[†]	Heat capacity at constant pressure (T = K; 0°C = 273.1 K), cal/(mol·K)	Range of temperature, K	Uncertainty, %
Potassium—(*Cont.*)				
K	*g*	4.97	All	0
K_2	*g*	9.00	300–2000	5
$KAsO_3$	*c*	25.3	290–372	?
KBO_2	*c*	$12.6 + 0.0126T$	273–1220	?
$K_2B_4O_7$	*c*	51.3	290–372	?
KBr	*c*	$11.49 + 0.00360T$	273–543	2
KCl	*c*	$10.93 + 0.00376T$	273–1043	2
$KClO_3$	*c*	25.7	289–371	?
$KClO_4$	*c*	26.3	287–318	?
$2KCl·CuCl_2·2H_2O$	*c*	63	292–323	?
$2KCl·PtCl_4$	*c*	55	286–319	?
$2KCl·SnCl_4$	*c*	54.5	292–323	?
$2KCl·ZnCl_2$	*c*	43.4	279–319	?
$2KCN·Zn(CN)_2$	*c*	57.4	277–319	?
K_2CO_3	*c*	29.9	296–372	?
K_2CrO_4	*c*	35.9	289–371	?
$K_2Cr_2O_7$	*c*	$42.80 + 0.0410T$	273–671	5
	l	96.9	671–757	5
KF	*c*	$10.8 + 0.00284T$	273–1129	?
$K_4Fe(CN)_6$	*c*	80.1	273–319	?
$K_4Fe(CN)_6·3H_2O$	*c*	114.5	273–310	?
KH_2AsO_4	*c*	32	289–319	?
KH_2PO_4	*c*	28.3	290–320	?
$KHSO_4$	*c*	30	292–324	?
$KMnO_4$	*c*	28	287–318	?
KNO_3	*c*	$6.42 + 0.0530T$	273–401	10
	c	28.8	401–611	5
	l	29.5	611–683	10
$K_2O·Al_2O_3·3SiO_2$	*c*, orthoclase	$69.26 + 0.00821T - 2331000/T^2$	273–1373	1½
	gls, orthoclase	$69.81 + 0.01053 - 2403000/T^2$	273–1373	1½
	c, microcline	$65.65 + 0.01102T - 1748000/T^2$	273–1373	1½
	gls, microcline	$64.83 + 0.01438T - 1641000/T^2$	273–1373	1½
$K_4P_2O_7$	*c*	63.1	290–371	?
K_2SO_4	*c*	33.1	287–371	?
$K_2S_2O_3$	*c*	37	293–373	?
$K_2SO_4·Al_2(SO_4)_3·24H_2O$	*c*	352	292–322	?
$K_2SO_4·Cr_2(SO_4)_3·24H_2O$	*c*	324	292–324	?
$K_2SO_4·MgSO_4·6H_2O$	*c*	106	292–323	?
$K_2SO_4·NiSO_4·6H_2O$	*c*	107	289–319	?
$K_2SO_4·ZnSO_4·6H_2O$	*c*	120	293–317	?
Prometheum				
Pr	*c*			
Radon				
Rn	*g*	4.97	All	0
Rhenium				
Re	*c*	$6.30 + 0.00053T$	273–2273	?
Rhodium				
Rh	*c*	$5.40 + 0.00219T$	273–1877	2
Rubidium				
Rb	*c*	$3.27 + 0.0131T$	273–312	2
	l	7.85	312–373	5
RbBr	*c*	$11.6 + 0.00255T$	273–954	?
RbCl	*c*	$11.5 + 0.00249T$	273–987	?
Rb_2CO_3	*c*	28.4	291–320	?
RbF	*c*	$11.3 + 0.00256T$	273–1048	?
RbI	*c*	$11.6 + 0.00263T$	273–913	?
Scandium				
Sc_2O_3	*c*	21.1	273–373	?
$Sc_2(SO_4)_3$	*c*	62.0	273–373	?
Selenium				
Se	*c*	$4.53 + 0.00550T$	273–490	2
	l	8.35	490–570	3
Silicon				
Si	*c*	$5.74 + 0.000617T - 101000/T^2$	273–1174	2
SiC	*c*	$8.89 + 0.00291T - 284000/T^2$	273–1629	2
$SiCl_4$	*l*	32.4	293–373	?
SiO_2	*c*, quartz, α	$10.87 + 0.008712T - 241200/T^2$	273–848	1
	c, quartz, β	$10.95 + 0.00550T$	848–1873	3½
	c, cristobalite, α	$3.65 + 0.0240T$	273–523	2½
	c, cristobalite, β	$17.09 + 0.000454T - 897200/T^2$	523–1973	2
	gls	$12.80 + 0.00447T - 302000/T^2$	273–1973	3½
Silver[4]				
Ag	*c*	$5.60 + 0.00150T$	273–1234	1
	l	8.2	1234–1573	3

(Continued)

TABLE 2-70 Heat Capacities of the Elements and Inorganic Compounds *(Continued)*

Substance	State†	Heat capacity at constant pressure (T = K; 0°C = 273.1 K), cal/(mol·K)	Range of temperature, K	Uncertainty, %
Silver—(*Cont.*)				
Ag_3Al	*c*	22.56 + 0.00570T	273–902	2
Ag_2Al	*c*	16.85 + 0.00450T	273–903	2
$AgAl_{12}$	*c*	58.62 + 0.0575T	273–768	5
AgBr	*c*	8.58 + 0.0141T	273–703	6
	l	14.9	703–836	5
AgCl	*c*	9.60 + 0.00929T	273–728	2
	l	14.05	728–806	5
AgCNO	*c*	18.7	273–353	?
AgI	*c*, α	8.58 + 0.0141T	273–423	6
$AgNO_3$	*c*, α	18.83 + 0.0160T	273–433	2
	c, β	25.7	433–482	5
	l	30.2	482–541	5
Ag_3PO_4	*c*	37.5	293–325	?
Ag_2S	*c*, α	18.8	273–448	5
	c, β	21.8	448–597	5
Ag_3Sb	*c*	19.53 + 0.0160T	273–694	5
Ag_2Se	*c*, α	20.2	273–406	5
	c, β	20.4	406–460	5
Sodium[15]				
Na	*c*	5.01 + 0.00536T	273–371	1½
	l	7.50	371–451	2
	g	4.97	All	0
$NaBO_2$	*c*	10.4 + 0.0199T	273–1239	?
$Na_2B_4O_7$	*c*	47.9	289–371	?
$Na_2B_4O_7·10H_2O$	*c*	147	292–323	?
NaBr	*c*	11.74 + 0.00233T	273–543	2
NaCl	*c*	10.79 + 0.00420T	273–1074	2
	l	15.9	1073–1205	3
$NaClO_3$	*c*	9.48 + 0.0468T	273–528	3
	l	31.8	528–572	5
NaCNO	*c*	13.1	273–353	?
Na_2CO_3	*c*	28.9	288–371	?
NaF	*c*	10.4 + 0.00289T	273–1261	?
$Na_2HPO_4·7H_2O$	*c*	86.6	275–307	?
$Na_2HPO_4·12H_2O$	*c*	133.4	275–307	?
NaI	*c*	12.5 + 0.00162T	273–936	?
$NaNO_3$	*c*	4.56 + 0.0580T	273–583	5
	l	37.2	583–703	10
$Na_2O·Al_2O_3·3SiO_2$	*c*, albite	63.78 + 0.01171T − 1678000/T^2	273–1373	1
	gls	61.25 + 0.01768T − 1545000/T^2	273–1173	1
$NaPO_3$	*c*	22.1	290–319	?
$Na_4P_2O_7$	*c*	60.7	290–371	?
Na_2SO_4	*c*	32.8	289–371	?
$Na_2S_2O_3$	*c*	34.9	273–307	?
$Na_2S_2O_3·5H_2O$	*c*	86.2	273–307	?
Sodium-potassium alloys[15]	*l*			
Strontium				
$SrBr_2$	*c*	18.1 + 0.00311T	273–923	?
$SrBr_2·H_2O$	*c*	28.9	277–370	?
$SrBr_2·6H_2O$	*c*	82.1	276–327	?
$SrCl_2$	*c*	18.2 + 0.00244T	273–1143	?
$SrCl_2·H_2O$	*c*	28.7	276–365	?
$SrCl_2·2H_2O$	*c*	38.3	277–366	?
$SrCO_3$	*c*	21.8	281–371	?
SrI_2	*c*	18.6 + 0.00304T	273–783	?
$SrI_2·H_2O$	*c*	28.5	276–363	?
$SrI_2·2H_2O$	*c*	39.1	275–336	?
$SrI_2·6H_2O$	*c*	84.9	275–333	?
$SrMoO_4$	*c*	37	273–297	?
$Sr(NO_3)_2$	*c*	38.3	290–320	?
$SrSO_4$	*c*	26.2	293–369	?
Sulfur[16]				
S	*c*, rhombic	3.63 + 0.00640T	273–368	3
	c, monoclinic	4.38 + 0.00440T	368–392	3
S_2	*g*	8.58 + 0.00030T	300–2500	5
S_2Cl_2	*l*	27.5	273–332	?
SO_2	*g*	7.70 + 0.00530T − 0.00000083T^2	300–2500	2½
Tantalum				
Ta	*c*	5.91 + 0.00099T	273–1173	2
Tellurium				
Te	*c*	5.19 + 0.00250T	273–600	3
Thallium				
Tl	*c*, α	5.32 + 0.00385T	273–500	1
	c, β	8.12	500–576	1

TABLE 2-70 Heat Capacities of the Elements and Inorganic Compounds *(Continued)*

Substance	State†	Heat capacity at constant pressure (T = K; 0°C = 273.1 K), cal/(mol·K)	Range of temperature, K	Uncertainty, %
Thallium—(*Cont.*)				
Tl	*l*	7.12	576–773	3
TlBr	*c*	$12.53 + 0.00100T$	273–733	10
	l	16.0	733–800	10
TlCl	*c*	$12.56 + 0.00088T$	273–700	5
	l	14.2	700–803	10
Thorium				
Th	*c*	6.40	273–373	?
ThO_2	*c*	$14.6 + 0.00507T$	273–1273	?
$Th(SO_4)_2$	*c*	41.2	273–373	?
Tin[4]				
Sn	*c*	$5.05 + 0.00480T$	273–504	2
	l	6.6	504–1273	10
SnAu	*c*	$11.79 + 0.00233T$	273–581	1
$SnCl_2$	*c*	$16.2 + 0.00926T$	273–520	?
$SnCl_4$	*l*	38.4	286–371	?
SnO	*c*	$9.40 + 0.00362T$	273–1273	?
SnO_2	*c*	$13.94 + 0.005657T - 252000/T^2$	273–1373	?
SnPt	*c*	$11.49 + 0.00190T$	273–1318	1
SnS	*c*	$12.1 + 0.00165T$	273–1153	?
SnS_2	*c*	$20.5 + 0.00400T$	273–873	?
Titanium				
Ti	*c*	$8.91 + 0.00114T - 433000/T^2$	273–713	3
$TiCl_4$	*l*	35.7	285–372	?
TiO_2	*c*	$11.81 + 0.00754T - 41900/T^2$	273–713	3
Tungsten				
W	*c*	$5.65 + 0.00866$	273–2073	1
WO_3	*c*	$16.0 + 0.00774T$	273–1550	?
Uranium				
U	*c*	6.64	273–372	?
U_3O_8	*c*	59.8	276–314	?
Vanadium				
V	*c*	$5.57 + 0.00097T$	273–1993	?
Xenon				
Xe	*g*	4.97	All	0
Zinc[4]				
Zn	*c*	$5.25 + 0.00270T$	273–692	1
	l	$7.59 + 0.00055T$	692–1122	3
$ZnCl_2$	*c*	$15.9 + 0.00800T$	273–638	?
ZnO	*c*	$11.40 + 0.00145T - 182400/T^2$	273–1573	1
ZnS	*c*	$12.81 + 0.000095T - 194600/T^2$	273–1173	5
ZnSb	*c*	$11.5 + 0.00313T$	273–810	?
$ZnSO_4$	*c*	28	293–373	?
$ZnSO_4·H_2O$	*c*	34.7	282	?
$ZnSO_4·6H_2O$	*c*	80.8	282	?
$ZnSO_4·7H_2O$	*c*	100.2	273–307	?
Zirconium				
ZrO_2	*c*	$11.62 + 0.01046T - 177700/T^2$	273–1673	5
$ZrO_2·SiO_2$	*c*	26.7	297–372	?

[1] See also Table 2-71. Data to 298 K are also given by Scott, *Cryogenic Engineering,* Van Nostrand, Princeton, N.J., 1959.
[2] For liquid and gas data, see Johnson (ed.), WADD-TR-60-56, 1960.
[3] Stalder, NACA Tech. Note 4141, 1957 (Fig. 5), gives data from 400 to 2600°R.
[4] See also Table 2-71.
[5] For data from 400 to 5500°R see Stalder, NACA Tech. Note 4141, 1975 (Fig. 4).
[6] For solid, liquid, and gas data, see Johnson (ed.), WADD-TR-60-56, 1960.
[7] For data from 400 to 2350°R see Stalder, NACA Tech. Note 4141, 1957.
[8] For solid, liquid, and gas data, see Johnson (ed.), WADD-TR-60-56, 1960.
[9] For liquid and gas data, see Johnson (ed.), WADD-TR-60-56, 1960.
[10] For solid, liquid, and gas data, see Johnson (ed.), WADD-TR-60-56, 1960.
[11] See also Table 2-71. Douglas, Ball, et al., *Bur. Stand. J. Res.,* **46** (1951): 334; Busey and Giaque, *J. Am. Chem. Soc.,* **75** (1953): 806; Sheldon, ASME Pap. 49-A-30, 1949.
[12] For solid, liquid, and gas data, see Johnson (ed.), WADD-TR-56-60, 1960.
[13] For solid, liquid, and gas data, see Johnson (ed.), WADD-TR-56-60, 1960.
[14] For solid, liquid, and gas data, see Johnson (ed.), WADD-TR-56-60, 1960. Ozone: For liquid see Brabets and Waterman, *J. Chem. Phys.,* **28** (1958): 1212.
[15] For data on liquid Na-K alloys to 1500°F and for liquid Na to 1460°F, see Lubarsky and Kaufman, NACA Rep. 1270, 1956.
[16] See also Evans and Wagman, *Bur. Stand. J. Res.* **49** (1952): 141; Gratch, OTS PB 124957, 1950; Guthrie, Scott et al., *J. Am. Chem. Soc.,* **76** (1954): 1488.

TABLE 2-71 Specific Heat [kJ/(kg·K)] of Selected Elements

	Temperature, K														
Symbol	4	6	8	10	20	40	60	80	100	200	250	300	400	600	800
Al	0.00026	0.00050	0.00088	0.00140	0.0089	0.0775	0.214	0.357	0.481	0.797	0.859	0.902	0.949	1.042	1.134
Be	0.00008			0.00028	0.0014				0.195	1.109	1.537	1.840	2.191	2.605	2.823
Bi	0.00054	0.00220	0.00541	0.01040	0.0340	0.0729	0.092	0.102	0.109	0.120	0.121	0.122	0.123	0.142	0.136
Cr	0.00016	0.00029	0.00050	0.00081	0.0021	0.0107	0.059	0.127	0.190	0.382	0.424	0.450	0.501	0.565	0.611
Co	0.00036	0.00059	0.00085	0.00121	0.0048	0.0404	0.110	0.184	0.234	0.376	0.406	0.426	0.451	0.509	0.543
Cu	0.00011	0.00024	0.00048	0.00086	0.0076	0.059	0.137	0.203	0.254	0.357	0.377	0.386	0.396	0.431	0.448
Ge			0.00037	0.00081	0.0129	0.0619	0.108	0.153	0.192	0.286	0.305	0.323	0.343	0.364	0.377
Au	0.00018	0.00047	0.00126	0.00255	0.0163	0.0569	0.084	0.100	0.109	0.124	0.127	0.129	0.131	0.136	0.141
Ir				0.00032	0.0021				0.090	0.122	0.128	0.131	0.133	0.140	0.146
Fe	0.00038	0.00061	0.00090	0.00127	0.0039	0.0276	0.086	0.154	0.216	0.384	0.422	0.450	0.491	0.555	0.692
Pb	0.00075	0.00242	0.00747	0.01350	0.0531	0.0944	0.108	0.114	0.118	0.125	0.127	1.129	0.132	0.142	
Mg	0.00034	0.00080	0.00155	0.00172	0.0148	0.138	0.336	0.513	0.648	0.929	0.985	1.005	1.082	1.177	1.263
Hg	0.00417	0.01420	0.01820	0.02250	0.0515	0.0895	0.107	0.116	0.121	0.136	0.141	0.139	0.136	0.135	0.104
Mo	0.00011	0.00019	0.00032	0.00050	0.0029	0.0236	0.061	0.105	0.140	0.223	0.241	0.248	0.261	0.280	0.292
Ni	0.00054	0.00086	0.00121	0.00178	0.0058	0.0380	0.103	0.173	0.232	0.383	0.416	0.444	0.490	0.590	0.530
Pt	0.00019	0.00028	0.00067	0.00112	0.0077	0.0382	0.069	0.088	0.101	0.127	0.132	0.134	0.136	0.140	0.146
Ag	0.00016	0.00035	0.00093	0.00186	0.0159	0.0778	0.133	0.166	0.187	0.225	0.232	0.236	0.240	0.251	0.264
Sn	0.00024	0.00127	0.00423	0.00776	0.0400	0.108	0.149	0.173	0.189	0.214	0.220	0.222	0.245	0.257	0.257
Zn	0.00011	0.00029	0.00096	0.00250	0.0269	0.123	0.205	0.258	0.295	0.366	0.380	0.389	0.404	0.435	0.479

TABLE 2-72 Heat Capacities of Inorganic and Organic Liquids [J/(kmol·K)]

Eqn	Cmpd. no.	Name	Formula	CAS	Mol. wt.	C_1	C_2	C_3	C_4	C_5	T_{min}, K	C_p at T_{min} × 1E-05	T_{max}, K	C_p at T_{max} × 1E-05
100	1	Acetaldehyde	C_2H_4O	75-07-0	44.05256	152.99	598.64	−0.89481			149.78	0.69743	294.15	0.98820
100	2	Acetamide	C_2H_5NO	60-35-5	59.0672	10,2300	128.7				354.15	1.47880	571.00	1.75790
100	3	Acetic acid	$C_2H_4O_2$	64-19-7	60.052	13,9640	−320.8	0.8985			289.81	1.22130	391.05	1.51590
100	4	Acetic anhydride	$C_4H_6O_3$	108-24-7	102.08864	26,0050	−565.43	1.1035			200.15	1.91090	412.70	2.14650
100	5	Acetone	C_3H_6O	67-64-1	58.07914	13,5600	−177	0.2837	0.000689		178.45	1.16960	329.44	1.32710
100	6	Acetonitrile	C_2H_3N	75-05-8	41.0519	73,381	60.042				229.32	0.87150	354.81	0.94685
100	7	Acetylene	C_2H_2	74-86-2	26.03728	−122,020	3082.7	−15.895	0.027732		192.40	0.80208	250.00	0.88530
100	8	Acrolein	C_3H_4O	107-02-8	56.06326	103,090	−247.8	1.0343			253.00	1.06600	379.50	1.58010
100	9	Acrylic acid	$C_3H_4O_2$	79-10-7	72.06266	55,300	300				286.15	1.41150	375.00	1.67800
100	10	Acrylonitrile	C_3H_3N	107-13-1	53.0626	109,750	−108.61	0.35246			189.63	1.01830	400.00	1.22700
100	11	Air	Mixture	132259-10-0	28.96	−214,460	9185.1	−106.12	0.41616		75.00	0.53065	115.00	0.71317
114	12	Ammonia	H_3N	7664-41-7	17.03052	61.289	80925	799.4	−2651		203.15	0.75753	401.15	4.18470
100	13	Anisole	C_7H_8O	100-66-3	108.13782	150,940	93.455	0.23602			298.15	1.99780	484.20	2.51530
100	14	Argon	Ar	7440-37-1	39.948	134,390	−1989.4	11.043			83.78	0.45230	135.00	0.67080
100	15	Benzamide	C_7H_7NO	55-21-0	121.13658	161,440	260.66				403.00	2.66490	563.15	3.08230
100	16	Benzene	C_6H_6	71-43-2	78.11184	129,440	−169.5	0.64781			278.68	1.32510	353.24	1.50400
100	16	Benzene	C_6H_6	71-43-2	78.11184	162,940	−344.94	0.85562			278.68	1.33260	500.00	2.04380
100	17	Benzenethiol	C_6H_6S	108-98-5	110.17684	119,780	180.34				258.27	1.66360	442.29	1.99540
100	18	Benzoic acid	$C_7H_6O_2$	65-85-0	122.12134	−5,480	647.12				395.45	2.50420	450.00	2.85720
100	19	Benzonitrile	C_7H_5N	100-47-0	103.1213	66,950	333.33				260.28	1.53710	464.15	2.21670
100	20	Benzophenone	$C_{13}H_{10}O$	119-61-9	182.2179	156,130	454.49				321.35	3.02180	640.00	4.47000
100	21	Benzyl alcohol	C_7H_8O	100-51-6	108.13782	−334,997	3644.21	−7.77514	0.00591102		257.85	1.89060	478.60	2.76170
100	22	Benzyl ethyl ether	$C_9H_{12}O$	539-30-0	136.19098	87,500	480				275.65	2.19810	458.15	3.07410
100	23	Benzyl mercaptan	C_7H_8S	100-53-8	124.20342	100,320	346.89				243.95	1.84940	472.03	2.64060
100	24	Biphenyl	$C_{12}H_{10}$	92-52-4	154.2078	121,770	429.3				342.20	2.68680	533.37	3.50750
100	25	Bromine	Br_2	7726-95-6	159.808	179,400	−667.11	1.0701			265.90	0.77675	331.90	0.75866
100	26	Bromobenzene	C_6H_5Br	108-86-1	157.0079	121,600	−9.45	0.358			293.15	1.49600	495.08	2.04670
100	27	Bromoethane	C_2H_5Br	74-96-4	108.965	95,588	−110.94	0.41864			154.25	0.88436	311.49	1.01650
100	28	Bromomethane	CH_3Br	74-83-9	94.93852	102,760	−230.08	0.51796			179.44	0.78152	280.15	0.78955
100	29	1,2-Butadiene	C_4H_6	590-19-2	54.09044	135,150	−311.14	0.97007	−0.0001523		136.95	1.10340	290.00	1.22790
100	30	1,3-Butadiene	C_4H_6	106-99-0	54.09044	128,860	−323.1	1.015	0.000032		165.00	1.03330	350.00	1.41480
100	31	Butane	C_4H_{10}	106-97-8	58.1222	191,030	−1675	12.5	−0.03874	0.000046121	134.86	1.12720	400.00	2.22370
114	32	1,2-Butanediol	$C_4H_{10}O_2$	584-03-2	90.121	55.136	314200	280.19	1413.9		220.00	1.55900	670.00	5.20450
114	33	1,3-Butanediol	$C_4H_{10}O_2$	107-88-0	90.121	42.152	324580	517.35	1449.5		196.15	0.62506	670.00	5.24370
100	34	1-Butanol	$C_4H_{10}O$	71-36-3	74.1216	191,200	−730.4	2.2998			183.85	1.34650	391.00	2.57210
100	35	2-Butanol	$C_4H_{10}O$	78-92-2	74.1216	533,390	−4986.2	18.908	−0.02		158.45	1.38480	372.90	2.66210
100	36	1-Butene	C_4H_8	106-98-9	56.10632	182,050	−1611	11.963	−0.037454	0.000045027	87.80	1.10150	380.00	1.81030
100	37	*cis*-2-Butene	C_4H_8	590-18-1	56.10632	126,680	−65.47	−0.64	0.002912		134.26	1.13400	350.00	1.50220
100	38	*trans*-2-Butene	C_4H_8	624-64-6	56.10632	112,760	−104.7	0.5214			167.62	1.09860	274.03	1.23220
100	39	Butyl acetate	$C_6H_{12}O_2$	123-86-4	116.15828	111,850	384.52				298.15	2.26490	399.26	2.65370
100	40	Butylbenzene	$C_{10}H_{14}$	104-51-8	134.21816	182,470	−13.912	0.72897			185.30	2.04920	400.00	2.93540
100	41	Butyl mercaptan	$C_4H_{10}S$	109-79-5	90.1872	232,190	−804.35	2.7063	−0.0023017		157.46	1.63650	390.00	1.93590
100	42	*sec*-Butyl mercaptan	$C_4H_{10}S$	513-53-1	90.1872	197,890	−491.54	1.7219	−0.0012499		133.02	1.60030	370.00	1.88440
100	43	1-Butyne	C_4H_6	107-00-6	54.09044	136,340	−300.4	1.0216			147.43	1.14260	298.15	1.37590
100	44	Butyraldehyde	C_4H_8O	123-72-8	72.10572	194,170	−532.38	1.4286			176.80	1.44700	347.94	1.81880
100	45	Butyric acid	$C_4H_8O_2$	107-92-6	88.1051	237,700	−746.4	1.829			267.95	1.69020	436.42	2.60310
100	46	Butyronitrile	C_4H_7N	109-74-0	69.1051	154,800	−239.75	0.68616			161.30	1.33980	390.74	1.65880
100	47	Carbon dioxide	CO_2	124-38-9	44.0095	−8,304,300	104370	−433.33	0.60052		220.00	0.78265	290.00	1.66030
100	48	Carbon disulfide	CS_2	75-15-0	76.1407	85,600	−122	0.5605	−0.001452	0.000002008	161.11	0.75774	552.00	1.31250
114	49	Carbon monoxide	CO	630-08-0	28.0101	65.429	28723	−847.39	1959.6		68.15	0.59115	132.00	6.47990

(*Continued*)

TABLE 2-72 Heat Capacities of Inorganic and Organic Liquids [J/(kmol·K)] (*Continued*)

Eqn	Cmpd. no.	Name	Formula	CAS	Mol. wt.	C_1	C_2	C_3	C_4	C_5	T_{min}, K	C_p at $T_{min} \times$ 1E-05	T_{max}, K	C_p at $T_{max} \times$ 1E-05
100	50	Carbon tetrachloride	CCl_4	56-23-5	153.8227	−752,700	8966.1	−30.394	0.034455		250.33	1.27630	388.71	1.63740
100	51	Carbon tetrafluoride	CF_4	75-73-0	88.0043	104,600	−500.6	2.2851			89.56	0.78095	145.10	0.80073
100	52	Chlorine	Cl_2	7782-50-5	70.906	63,936	46.35	−0.1623			172.12	0.67106	239.12	0.65739
100	53	Chlorobenzene	C_6H_5Cl	108-90-7	112.5569	−1,307,500	15338	−53.974	0.063483		227.95	1.36170	360.00	1.81010
100	54	Chloroethane	C_2H_5Cl	75-00-3	64.5141	118,380	−248.915	0.68074			136.75	0.97071	298.15	1.04680
100	55	Chloroform	$CHCl_3$	67-66-3	119.37764	124,850	−166.34	0.43209			233.15	1.09560	366.48	1.21920
100	56	Chloromethane	CH_3Cl	74-87-3	50.4875	107,900	−330.13	0.808			175.43	0.74852	303.15	0.82076
100	57	1-Chloropropane	C_3H_7Cl	540-54-5	78.54068	134,733	−176.332	0.55966			150.35	1.20870	319.67	1.35560
100	58	2-Chloropropane	C_3H_7Cl	75-29-6	78.54068	69,362	215.01				200.00	1.12360	308.85	1.35770
100	59	*m*-Cresol	C_7H_8O	108-39-4	108.13782	−246,700	3256.8	−7.4202	0.0060467		285.39	2.18950	400.00	2.55780
100	60	*o*-Cresol	C_7H_8O	95-48-7	108.13782	−185,150	3148	−8.0367	0.007254		304.20	2.32970	400.00	2.52430
100	61	*p*-Cresol	C_7H_8O	106-44-5	108.13782	259,980	−1112.3	4.9427	−0.0054367		307.93	2.27400	400.00	2.57940
100	62	Cumene	C_9H_{12}	98-82-8	120.19158	61,723	494.81				177.14	1.49370	425.56	2.72290
100	63	Cyanogen	C_2N_2	460-19-5	52.0348	77,461	111.51				245.25	1.04810	253.82	1.05760
100	64	Cyclobutane	C_4H_8	287-23-0	56.10632	101,920	−215.81	0.8103			190.00	0.90168	298.15	1.09610
100	65	Cyclohexane	C_6H_{12}	110-82-7	84.15948	−220,600	3118.3	−9.4216	0.010687		279.69	1.48360	400.00	2.03230
100	66	Cyclohexanol	$C_6H_{12}O$	108-93-0	100.15888	−40,000	853				296.60	2.13000	434.00	3.30200
100	67	Cyclohexanone	$C_6H_{10}O$	108-94-1	98.143	6,110.4	600.94				290.00	1.80380	489.75	3.00420
100	68	Cyclohexene	C_6H_{10}	110-83-8	82.1436	105,850	−60	0.68			169.67	1.15250	356.12	1.70720
100	69	Cyclopentane	C_5H_{10}	287-92-3	70.1329	122,530	−403.8	1.7344	−0.0010975		179.28	0.99559	322.40	1.35840
100	70	Cyclopentene	C_5H_8	142-29-0	68.11702	125,380	−349.7	1.143			138.13	0.98884	317.38	1.29530
100	71	Cyclopropane	C_3H_6	75-19-4	42.07974	89,952	−196.63	0.65237			150.00	0.75136	298.15	0.89318
100	72	Cyclohexyl mercaptan	$C_6H_{12}S$	1569-69-3	116.22448	177,560	−179.12	0.76723			189.64	1.71180	431.95	2.43340
100	73	Decanal	$C_{10}H_{20}O$	112-31-2	156.2652	218,480	374.14	0.11851			285.00	3.34740	481.65	4.26180
100	74	Decane	$C_{10}H_{22}$	124-18-5	142.28168	278,620	−197.91	1.0737			243.51	2.94090	460.00	4.14780
100	75	Decanoic acid	$C_{10}H_{20}O_2$	334-48-5	172.265	219,840	140.41	0.9968			304.75	3.55210	543.15	5.90170
100	76	1-Decanol	$C_{10}H_{22}O$	112-30-1	158.28108	4,988,500	−52898	216.35	−0.37538	0.00023674	280.00	3.53690	503.15	5.01740
100	77	1-Decene	$C_{10}H_{20}$	872-05-9	140.2658	417,440	−1616.5	5.3948	−0.004348		206.89	2.75410	494.00	4.11250
100	78	Decyl mercaptan	$C_{10}H_{22}S$	143-10-2	174.34668	314,570	−160.93	0.95561			247.56	3.33300	512.35	4.82970
100	79	1-Decyne	$C_{10}H_{18}$	764-93-2	138.24992	276,900	−371.23	1.5774			229.15	2.74660	447.15	4.26290
	80	Deuterium	D_2	7782-39-0	4.0316									
100	81	1,1-Dibromoethane	$C_2H_4Br_2$	557-91-5	187.86116	149,400	−231.8	0.5946			210.15	1.26950	381.15	1.47430
100	82	1,2-Dibromoethane	$C_2H_4Br_2$	106-93-4	187.86116	200,560	−491.44	0.9187			282.85	1.35060	410.00	1.53500
100	83	Dibromomethane	CH_2Br_2	74-95-3	173.83458	202,580	−726.3	1.3377			240.00	1.05320	370.10	1.17010
100	84	Dibutyl ether	$C_8H_{18}O$	142-96-1	130.22792	270,720	−259.83	0.95427			175.30	2.54500	450.00	3.47040
100	85	*m*-Dichlorobenzene	$C_6H_4Cl_2$	541-73-1	147.00196	114,880	187.25				248.39	1.61390	400.00	1.89780
100	86	*o*-Dichlorobenzene	$C_6H_4Cl_2$	95-50-1	147.00196	93,093	183.97	0.2314			273.15	1.60610	528.75	2.55060
100	87	*p*-Dichlorobenzene	$C_6H_4Cl_2$	106-46-7	147.00196	133,950	−24.84	0.48191			326.14	1.77110	513.56	2.48290
100	88	1,1-Dichloroethane	$C_2H_4Cl_2$	75-34-3	98.95916	126,340	−94.63	0.32			176.19	1.19600	330.45	1.30010
100	89	1,2-Dichloroethane	$C_2H_4Cl_2$	107-06-2	98.95916	179,170	−444.74	0.93009			237.49	1.26010	356.59	1.38850
100	90	Dichloromethane	CH_2Cl_2	75-09-2	84.93258	98,968	−62.941	0.23265			180.00	0.95176	320.00	1.02650
100	91	1,1-Dichloropropane	$C_3H_6Cl_2$	78-99-9	112.98574	144,560	−53.605	0.30617			192.50	1.45590	361.25	1.65150
100	92	1,2-Dichloropropane	$C_3H_6Cl_2$	78-87-5	112.98574	111,560	149.44				275.00	1.52660	369.52	1.66780
100	93	Diethanol amine	$C_4H_{11}NO_2$	111-42-2	105.13564	184,200	286				301.15	2.70330	541.54	3.39080
100	94	Diethyl amine	$C_4H_{11}N$	109-89-7	73.13684	101,330	243.18				223.35	1.55640	328.60	1.81240
100	95	Diethyl ether	$C_4H_{10}O$	60-29-7	74.1216	44,400	1301	−5.5	0.008763		156.92	1.46980	460.00	3.32020
100	96	Diethyl sulfide	$C_4H_{10}S$	352-93-2	90.1872	238,520	−1038.4	4.0587	−0.0044691		181.95	1.57030	322.08	1.75790
114	97	1,1-Difluoroethane	$C_2H_4F_2$	75-37-6	66.04997	67.155	105580	310.21	−490.54		154.56	0.99146	359.98	1.68740
100	98	1,2-Difluoroethane	$C_2H_4F_2$	624-72-6	66.04997	82,577	109.85				179.60	1.02310	283.65	1.13740
100	99	Difluoromethane	CH_2F_2	75-10-5	52.02339	263,980	−1791.1	4.3666			200.00	0.80424	250.00	0.89118

100	100	Diisopropyl amine	$C_6H_{15}N$	108-18-9	101.19	98,434	429.04				275.00	2.16420	357.05	2.51620
100	101	Diisopropyl ether	$C_6H_{14}O$	108-20-3	102.17476	163,000	−4.5	0.62			187.65	1.83990	341.45	2.33750
100	102	Diisopropyl ketone	$C_7H_{14}O$	565-80-0	114.18546	179,270	28.37	0.5375			204.81	2.07630	410.00	2.81260
100	103	1,1-Dimethoxyethane	$C_4H_{10}O_2$	534-15-6	90.121	187,790	−313.41	1.1023			159.95	1.65860	337.45	2.07550
100	104	1,2-Dimethoxypropane	$C_5H_{12}O_2$	7778-85-0	104.14758	199,930	−191.5	0.87664			226.10	2.01450	366.15	2.47340
100	105	Dimethyl acetylene	C_4H_6	503-17-3	54.09044	88,153	124.16				240.91	1.18060	300.13	1.25420
100	106	Dimethyl amine	C_2H_7N	124-40-3	45.08368	−214,870	3787.2	−13.781	0.016924		180.96	1.19470	298.15	1.37790
100	107	2,3-Dimethylbutane	C_6H_{14}	79-29-8	86.17536	129,450	18.5	0.608			145.19	1.44950	331.13	2.02240
100	108	1,1-Dimethylcyclohexane	C_8H_{16}	590-66-9	112.21264	134,500	8.765	0.81151			239.66	1.83210	392.70	2.63090
100	109	*cis*-1,2-Dimethylcyclohexane	C_8H_{16}	2207-01-4	112.21264	150,130	−62.38	0.8851			223.16	1.80290	402.94	2.68700
100	110	*trans*-1,2-Dimethylcyclohexane	C_8H_{16}	6876-23-9	112.21264	155,560	−145.26	1.0932			184.99	1.66100	396.58	2.69890
100	111	Dimethyl disulfide	$C_2H_6S_2$	624-92-0	94.19904	171,580	−256.67	0.5727			188.44	1.43550	360.00	1.53400
100	112	Dimethyl ether	C_2H_6O	115-10-6	46.06844	110,100	−157.47	0.51853			131.65	0.98356	250.00	1.03140
100	113	*N,N*-Dimethyl formamide	C_3H_7NO	68-12-2	73.09378	147,900	−106	0.384			273.82	1.47670	466.44	1.82000
100	114	2,3-Dimethylpentane	C_7H_{16}	565-59-3	100.20194	146,420	59.2	0.604			90.00	1.56640	380.00	2.56130
100	115	Dimethyl phthalate	$C_{10}H_{10}O_4$	131-11-3	194.184	206,560	325.75				274.16	2.95870	360.00	3.23830
100	116	Dimethylsilane	C_2H_8Si	1111-74-6	60.17042	131,810					298.15	1.31810	298.15	1.31810
100	117	Dimethyl sulfide	C_2H_6S	75-18-3	62.134	146,950	−380.06	1.2035	−0.00084787		174.88	1.12760	310.48	1.19590
100	118	Dimethyl sulfoxide	C_2H_6OS	67-68-5	78.13344	240,300	−595	1.013			291.67	1.52930	422.15	1.69650
100	119	Dimethyl terephthalate	$C_{10}H_{10}O_4$	120-61-6	194.184	195,251	419.918				413.79	3.69010	559.20	4.30070
100	120	1,4-Dioxane	$C_4H_8O_2$	123-91-1	88.10512	956,860	−5559.9	9.6124			284.95	1.53060	374.47	2.22770
100	121	Diphenyl ether	$C_{12}H_{10}O$	101-84-8	170.2072	134,160	447.67				300.03	2.68470	570.00	3.89330
100	122	Dipropyl amine	$C_6H_{15}N$	142-84-7	101.19	49,120	562.24				277.90	2.05370	407.90	2.78460
100	123	Dodecane	$C_{12}H_{26}$	112-40-3	170.33484	508,210	−1368.7	3.1015			263.57	3.62920	433.15	4.97260
100	124	Eicosane	$C_{20}H_{42}$	112-95-8	282.54748	352,720	807.32	0.2122			309.58	6.22990	616.93	9.31540
114	125	Ethane	C_2H_6	74-84-0	30.069	44.009	89718	918.77	−1886		92.00	0.68554	290.00	1.24440
100	126	Ethanol	C_2H_6O	64-17-5	46.06844	102,640	−139.63	−0.030341	0.0020386		159.05	0.87867	390.00	1.64500
100	127	Ethyl acetate	$C_4H_8O_2$	141-78-6	88.10512	226,230	−624.8	1.472			189.60	1.60680	350.21	1.87960
100	128	Ethyl amine	C_2H_7N	75-04-7	45.08368	121,700	38.993				192.15	1.29190	289.73	1.33000
100	129	Ethylbenzene	C_8H_{10}	100-41-4	106.165	154,040	−142.29	0.80539			178.20	1.54260	409.35	2.30750
100	130	Ethyl benzoate	$C_9H_{10}O_2$	93-89-0	150.1745	124,500	370.6				238.45	2.12870	486.55	3.04820
100	131	2-Ethyl butanoic acid	$C_6H_{12}O_2$	88-09-5	116.15828	56,359	603.02				258.15	2.12030	466.95	3.37940
100	132	Ethyl butyrate	$C_6H_{12}O_2$	105-54-4	116.15828	82,434	422.45	0.20992			285.50	2.20150	428.25	3.01850
100	133	Ethylcyclohexane	C_8H_{16}	1678-91-7	112.21264	132,360	72.74	0.64738			161.84	1.61090	404.95	2.67980
100	134	Ethylcyclopentane	C_7H_{14}	1640-89-7	98.18606	178,520	−518.35	2.3255	−0.0016818		134.71	1.46780	301.82	1.87670
100	135	Ethylene	C_2H_4	74-85-1	28.05316	247,390	−4428	40.936	−0.1697	0.00026816	104.00	0.70123	252.70	0.97582
100	136	Ethylenediamine	$C_2H_8N_2$	107-15-3	60.09832	184,440	−150.2	0.37044			284.29	1.71680	390.41	1.82260
100	137	Ethylene glycol	$C_2H_6O_2$	107-21-1	62.06784	35,540	436.78	−0.18486			260.15	1.36660	493.15	2.05980
100	138	Ethyleneimine	C_2H_5N	151-56-4	43.0678	46,848	205.35				250.00	0.98186	329.00	1.14410
100	139	Ethylene oxide	C_2H_4O	75-21-8	44.05256	144,710	−758.87	2.8261	−0.003064		160.65	0.83031	283.85	0.86932
100	140	Ethyl formate	$C_3H_6O_2$	109-94-4	74.07854	80,000	223.6				254.20	1.36840	374.20	1.63670
100	141	2-Ethyl hexanoic acid	$C_8H_{16}O_2$	149-57-5	144.211	207,670	−17.907	1.0493			155.15	2.30150	510.10	4.71570
100	142	Ethylhexyl ether	$C_8H_{18}O$	5756-43-4	130.22792	146,040	458.22				298.15	2.82660	417.15	3.37190
100	143	Ethylisopropyl ether	$C_5H_{12}O$	625-54-7	88.14818	106,250	292.15				298.15	1.93350	326.15	2.01530
100	144	Ethylisopropyl ketone	$C_6H_{12}O$	565-69-5	100.15888	229,250	−404.54	1.1382			204.15	1.94100	386.55	2.42950
100	145	Ethyl mercaptan	C_2H_6S	75-08-1	62.13404	134,670	−234.39	0.59656			125.26	1.14670	315.25	1.20070
100	146	Ethyl propionate	$C_5H_{10}O_2$	105-37-3	102.1317	76,330	400.1				298.15	1.95620	410.00	2.40370
100	147	Ethylpropyl ether	$C_5H_{12}O$	628-32-0	88.14818	103,680	726.3	−2.6047	0.0040957		145.65	1.66860	320.00	2.03580
100	148	Ethyltrichlorosilane	$C_2H_5Cl_3Si$	115-21-9	163.506	173,110	−697.18	3.7615	−0.005289	1.6179E-06	167.55	1.38290	371.05	1.92770
100	149	Fluorine	F_2	7782-41-4	37.9968064	−94,585	7529.9	−139.6	1.1301	−0.0033241	58.00	0.55414	98.00	0.59663
100	149	Fluorine	F_2	7782-41-4	37.9968064	1,724,400	−59924	537.85			53.48	0.57975	56.00	0.55354
100	150	Fluorobenzene	C_6H_5F	462-06-6	96.1023032	148,640	−202.58	0.66374			230.94	1.37260	504.08	2.15180
100	151	Fluoroethane	C_2H_5F	353-36-6	48.0595	65,106	103.44	0.67161	−0.0074083	0.000019119	129.95	0.79084	337.78	1.40050
100	152	Fluoromethane	CH_3F	593-53-3	34.03292	141,790	−814.32	2.2673	0		131.35	0.73946	285.70	0.94206

(*Continued*)

TABLE 2-72 Heat Capacities of Inorganic and Organic Liquids [J/(kmol·K)] *(Continued)*

Eqn	Cmpd. no.	Name	Formula	CAS	Mol. wt.	C_1	C_2	C_3	C_4	C_5	T_{min}, K	C_p at $T_{min} \times$ 1E-05	T_{max}, K	C_p at $T_{max} \times$ 1E-05
100	153	Formaldehyde	CH_2O	50-00-0	30.02598	70,077	−661.79	5.9749	−0.01813	0.00001983	155.15	0.55005	253.85	0.72876
100	154	Formamide	CH_3NO	75-12-7	45.04062	63,400	150.6				292.00	1.07380	493.00	1.37650
100	155	Formic acid	CH_2O_2	64-18-6	46.0257	78,060	71.54				281.45	0.98195	380.00	1.05250
100	156	Furan	C_4H_4O	110-00-9	68.07396	114,370	−215.69	0.72691			187.55	0.99486	304.50	1.16090
100	157	Helium-4	He	7440-59-7	4.0026	387,220	−465570	211800	−42494	3212.9	2.20	0.10866	4.60	0.29652
100	157	Helium-4	He	7440-59-7	4.0026	410,430	−464890	135100			1.80	0.11352	2.10	0.29952
100	158	Heptadecane	$C_{17}H_{36}$	629-78-7	240.46774	376,970	347.82	0.57895			295.13	5.30050	575.30	7.68690
100	159	Heptanal	$C_7H_{14}O$	111-71-7	114.18546	176,120	242.92				229.80	2.31940	426.15	2.79640
114	160	Heptane	C_7H_{16}	142-82-5	100.20194	61.26	314410	1824.6	−2547.9		182.57	1.99890	520.00	4.06570
100	161	Heptanoic acid	$C_7H_{14}O_2$	111-14-8	130.185	194,570	−23.206	0.88395			265.83	2.50870	496.15	4.00650
100	162	1-Heptanol	$C_7H_{16}O$	111-70-6	116.20134	2,416,800	−26105	110.03	−0.19172	0.00011968	239.15	2.35900	448.60	3.87660
100	163	2-Heptanol	$C_7H_{16}O$	543-49-7	116.20134	1,070,000	−9470	33.004	−0.0334	0	220.00	2.28350	432.90	4.45840
100	164	3-Heptanone	$C_7H_{14}O$	106-35-4	114.18546	270,730	−399.89	1.0601			234.15	2.35220	480.00	3.23030
100	165	2-Heptanone	$C_7H_{14}O$	110-43-0	114.18546	265,040	−375.68	1.0024			238.15	2.32420	490.00	3.21630
100	166	1-Heptene	C_7H_{14}	592-76-7	98.18606	267,950	−1315.9	6.5242	−0.011994	9.3808E-06	154.12	1.81500	430.00	2.75540
100	167	Heptyl mercaptan	$C_7H_{16}S$	1639-09-4	132.26694	236,870	−158.01	0.78982			229.92	2.42290	460.00	3.31310
100	168	1-Heptyne	C_7H_{12}	628-71-7	96.17018	46,798	761.13	−0.62882			200.00	1.73870	372.93	2.43190
100	169	Hexadecane	$C_{16}H_{34}$	544-76-3	226.44116	370,350	231.47	0.68632			291.31	4.96020	560.01	7.15210
100	170	Hexanal	$C_6H_{12}O$	66-25-1	100.15888	157,820	157.44				214.93	1.91660	401.15	2.20980
100	171	Hexane	C_6H_{14}	110-54-3	86.17536	172,120	−183.78	0.88734			177.83	1.67500	460.00	2.75340
100	172	Hexanoic acid	$C_6H_{12}O_2$	142-62-1	116.158	161,980	44.116	0.709			269.25	2.25260	478.85	3.45680
100	173	1-Hexanol	$C_6H_{14}O$	111-27-3	102.17476	1,638,600	−17261	71.721	−0.12026	0.000071087	228.55	1.98210	460.00	3.51970
100	174	2-Hexanol	$C_6H_{14}O$	626-93-7	102.175	1,409,400	−12553	40.991	−0.04		223.00	2.04940	412.40	3.98500
100	175	2-Hexanone	$C_6H_{12}O$	591-78-6	100.15888	208,250	−107.47	0.2062	0.00070293		217.35	2.01850	460.00	2.70870
100	176	3-Hexanone	$C_6H_{12}O$	589-38-8	100.15888	235,960	−345.94	0.94278			217.50	2.05320	460.00	2.76320
100	177	1-Hexene	C_6H_{12}	592-41-6	84.15948	164,640	−200.37	0.8784			133.39	1.53540	404.00	2.27060
100	178	3-Hexyne	C_6H_{10}	928-49-4	82.1436	82,795	283.4				300.00	1.67820	354.35	1.83220
100	179	Hexyl mercaptan	$C_6H_{14}S$	111-31-9	118.24036	303,320	−1009	3.3885	−0.002762		192.62	2.14950	430.00	2.76390
100	180	1-Hexyne	C_6H_{10}	693-02-7	82.1436	93,000	326				200.00	1.58200	344.48	2.05300
100	181	2-Hexyne	C_6H_{10}	764-35-2	82.1436	94,860	254.15				300.00	1.71110	357.67	1.85760
100	182	Hydrazine	H_4N_2	302-01-2	32.04516	79,815	50.929	0.043379			274.69	0.97078	653.15	1.31580
114	183	Hydrogen	H_2	1333-74-0	2.01588	66.653	6765.9	−123.63	478.27		13.95	0.12622	32.00	1.31220
100	184	Hydrogen bromide	BrH	10035-10-6	80.91194	57,720	9.9				185.15	0.59553	206.45	0.59764
100	185	Hydrogen chloride	ClH	7647-01-0	36.46094	47,300	90				165.00	0.62150	185.00	0.63950
100	186	Hydrogen cyanide	CHN	74-90-8	27.02534	95,398	−197.52	0.3883			259.83	0.70291	298.85	0.71049
100	187	Hydrogen fluoride	FH	7664-39-3	20.0063432	62,520	−223.02	0.6297			189.79	0.42875	292.67	0.51186
114	188	Hydrogen sulfide	H_2S	7783-06-4	34.08088	64.666	49354	22.493	−1623		187.68	0.67327	370.00	4.91830
100	189	Isobutyric acid	$C_4H_8O_2$	79-31-2	88.10512	127,540	−65.35	0.82867			270.00	1.70310	427.65	2.51140
100	190	Isopropyl amine	C_3H_9N	75-31-0	59.11026	−32,469	1977.1	−7.0145	0.0086913		177.95	1.46210	320.00	1.66710
100	191	Malonic acid	$C_3H_4O_4$	141-82-2	104.06146	138,790	121.24				409.15	1.88400	580.00	2.09110
100	192	Methacrylic acid	$C_4H_6O_2$	79-41-4	86.08924	146,290	−58.59	0.3582			288.15	1.59150	434.15	1.88370
114	193	Methane	CH_4	74-82-8	16.0425	65.708	38883	−257.95	614.07		90.69	0.53605	190.00	14.97800
100	194	Methanol	CH_4O	67-56-1	32.04186	256,040	−2741.4	14.777	−0.035078	0.000032719	175.47	0.71489	503.15	2.46460
100	195	*N*-Methyl acetamide	C_3H_7NO	79-16-3	73.09378	62,600	243.4				359.00	1.49980	538.50	1.93670
100	196	Methyl acetate	$C_3H_6O_2$	79-20-9	74.07854	61,260	270.9				253.40	1.29910	373.40	1.62410
100	197	Methyl acetylene	C_3H_4	74-99-7	40.06386	79,791	89.49				200.00	0.97689	249.94	1.02160
100	198	Methyl acrylate	$C_4H_6O_2$	96-33-3	86.08924	275,500	−1147	2.568			196.32	1.49300	353.35	1.90840
100	199	Methyl amine	CH_5N	74-89-5	31.0571	92,520	37.45				179.69	0.99249	266.82	1.02510
100	200	Methyl benzoate	$C_8H_8O_2$	93-58-3	136.14792	125,630	279.75				260.75	1.98570	472.65	2.57850
100	201	3-Methyl-1,2-butadiene	C_5H_8	598-25-4	68.11702	135,370	−133.34	0.63868			159.53	1.30350	314.56	1.56620

100	202	2-Methylbutane	C_5H_{12}	78-78-4	72.14878	108,300	146	−0.292	0.00151		113.25	1.23280	310.00	1.70480
100	203	2-Methylbutanoic acid	$C_5H_{10}O_2$	116-53-0	102.1317	74,200	417.4				321.50	2.08390	481.50	2.75180
100	204	3-Methyl-1-butanol	$C_5H_{12}O$	123-51-3	88.1482	206,600	−761.14	2.5899			155.95	1.50890	404.15	3.22010
100	205	2-Methyl-1-butene	C_5H_{10}	563-46-2	70.1329	149,510	−247.63	0.91849			135.58	1.32820	304.31	1.59210
100	206	2-Methyl-2-butene	C_5H_{10}	513-35-9	70.1329	151,600	−266.72	0.90847			139.39	1.32070	311.71	1.56730
100	207	2-Methyl -1-butene-3-yne	C_5H_6	78-80-8	66.10114	81,919	181.01				298.15	1.35890	305.40	1.37200
100	208	Methylbutyl ether	$C_5H_{12}O$	628-28-4	88.14818	177,850	−171.57	0.74379			157.48	1.69280	343.31	2.06610
100	209	Methylbutyl sulfide	$C_5H_{12}S$	628-29-5	104.214	198,390	−220.35	0.76096			175.30	1.83150	510.00	2.83940
100	210	3-Methyl-1-butyne	C_5H_8	598-23-2	68.11702	105,200	191.1				200.00	1.43420	299.49	1.62430
100	211	Methyl butyrate	$C_5H_{10}O_2$	623-42-7	102.1317	102,930	129.1	0.62516			277.25	1.86780	415.87	2.64740
100	212	Methylchlorosilane	CH_5ClSi	993-00-0	80.5889	47,726	338.4				250.00	1.32330	325.00	1.57710
100	213	Methylcyclohexane	C_7H_{14}	108-87-2	98.18606	131,340	−63.1	0.8125			146.58	1.39550	320.00	1.94350
100	214	1-Methylcyclohexanol	$C_7H_{14}O$	590-67-0	114.18546	50,578	508.59				300.00	2.03160	441.15	2.74940
100	215	*cis*-2-Methylcyclohexanol	$C_7H_{14}O$	7443-70-1	114.18546	118,600	447.07				300.00	2.52720	438.15	3.14480
100	216	*trans*-2-Methylcyclohexanol	$C_7H_{14}O$	7443-52-9	114.18546	118,170	447.99				300.00	2.52570	440.15	3.15350
100	217	Methylcyclopentane	C_6H_{12}	96-37-7	84.15948	155,920	−490	2.1383	−0.0015585		130.73	1.24920	366.48	1.86820
100	218	1-Methylcyclopentene	C_6H_{10}	693-89-0	82.1436	53,271	327.92				200.00	1.18860	348.64	1.67600
100	219	3-Methylcyclopentene	C_6H_{10}	1120-62-3	82.1436	46,457	346.93				200.00	1.15840	338.05	1.63740
100	220	Methyldichlorosilane	CH_4Cl_2Si	75-54-7	115.03396	27,030	413				250.00	1.30280	350.00	1.71580
100	221	Methylethyl ether	C_3H_8O	540-67-0	60.09502	85,383	199.08	−0.061547			160.00	1.15660	280.50	1.36380
100	222	Methylethyl ketone	C_4H_8O	78-93-3	72.10572	132,300	200.87	−0.9597	0.0019533		186.48	1.49050	373.15	1.75110
100	223	Methylethyl sulfide	C_3H_8S	624-89-5	76.1606	161,240	−288.61	0.78179			167.23	1.34840	339.80	1.53440
100	224	Methyl formate	$C_2H_4O_2$	107-31-3	60.05196	130,200	−396	1.21			174.15	0.97934	304.90	1.21950
100	225	Methylisobutyl ether	$C_5H_{12}O$	625-44-5	88.14818	92,919	324.43				298.15	1.89650	350.00	2.06470
100	226	Methylisobutyl ketone	$C_6H_{12}O$	108-10-1	100.15888	183,650	−79.862	0.60769			189.15	1.90290	389.15	2.44600
100	227	Methyl Isocyanate	C_2H_3NO	624-83-9	57.05132	149,770	−529.82	1.3499			256.15	1.02630	366.00	1.36680
100	228	Methylisopropyl ether	$C_4H_{10}O$	598-53-8	74.1216	143,440	−154.07	0.7255			127.93	1.35600	310.00	1.65400
100	229	Methylisopropyl ketone	$C_5H_{10}O$	563-80-4	86.1323	191,170	−331.04	0.98445			180.15	1.63480	440.00	2.36100
100	230	Methylisopropyl sulfide	$C_4H_{10}S$	1551-21-9	90.1872	211,170	−661.97	2.4216	−0.0021383		171.64	1.58080	357.91	1.86410
100	231	Methyl mercaptan	CH_4S	74-93-1	48.10746	115,300	−263.23	0.60412			150.18	0.89393	298.15	0.90520
100	232	Methyl methacrylate	$C_5H_8O_2$	80-62-6	100.11582	255,100	−938.4	2.413			224.95	1.66110	373.45	2.41180
100	233	2-Methyloctanoic acid	$C_9H_{18}O_2$	3004-93-1	158.23802	226,650	15.421	1.0578			240.00	2.91280	518.15	5.18640
100	234	2-Methylpentane	C_6H_{14}	107-83-5	86.17536	142,220	−47.83	0.739			119.55	1.47060	333.41	2.08420
100	235	Methyl pentyl ether	$C_6H_{14}O$	628-80-8	102.17476	251,890	−468.32	1.2209			176.00	2.07280	372.00	2.46630
100	236	2-Methylpropane	C_4H_{10}	75-28-5	58.1222	172,370	−1783.9	14.759	−0.047909	0.00005805	113.54	0.99613	380.00	2.07250
100	237	2-Methyl-2-propanol	$C_4H_{10}O$	75-65-0	74.1216	−925,460	7894.9	−17.661	0.013617		298.96	2.20160	460.00	2.94550
100	238	2-Methyl propene	C_4H_8	115-11-7	56.10632	87,680	217.1	−0.9153	0.002266		132.81	1.05680	343.15	1.45960
100	239	Methyl propionate	$C_4H_8O_2$	554-12-1	88.10512	71,140	335.5				300.00	1.71790	390.00	2.01990
100	240	Methylpropyl ether	$C_4H_{10}O$	557-17-5	74.1216	144,110	−102.09	0.58113			133.97	1.40860	312.20	1.68880
100	241	Methylpropyl sulfide	$C_4H_{10}S$	3877-15-4	90.1872	179,850	−264.1	0.79202			160.17	1.57870	368.69	1.90140
100	242	Methylsilane	CH_6Si	992-94-9	46.14384	113,470					298.15	1.13470	298.15	1.13470
100	243	alpha-Methyl styrene	C_9H_{10}	98-83-9	118.1757	76,822	421.6				249.95	1.82200	438.65	2.61760
100	244	Methyl tert-butyl ether	$C_5H_{12}O$	1634-04-4	88.1482	134,580	90.833	0.011456	0.00095984		164.55	1.54110	328.20	1.99560
100	245	Methyl vinyl ether	C_3H_6O	107-25-5	58.07914	73,600	184.7				151.15	1.01520	278.65	1.25070
100	246	Naphthalene	$C_{10}H_8$	91-20-3	128.17052	29,800	527.5				353.43	2.16230	491.14	2.88880
100	247	Neon	Ne	7440-01-9	20.1797	1,034,100	−138770	7154	−162.55	1.3841	24.56	0.36664	40.00	0.69796
100	248	Nitroethane	$C_2H_5NO_2$	79-24-3	75.0666	187,740	−497.6	1.0691			183.63	1.32420	387.22	1.55360
100	249	Nitrogen	N_2	7727-37-9	28.0134	281,970	−12281	248	−2.2182	0.0074902	63.15	0.55925	112.00	0.79596
100	250	Nitrogen trifluoride	F_3N	7783-54-2	71.00191	101,400	−682.11	3.8912			117.00	0.74860	175.50	1.01540
100	251	Nitromethane	CH_3NO_2	75-52-5	61.04002	116,270	−135.3	0.345			244.60	1.03820	473.15	1.29490
100	252	Nitrous oxide	N_2O	10024-97-2	44.0128	67,556	54.373				182.30	0.77468	200.00	0.78431
100	253	Nitric oxide	NO	10102-43-9	30.0061	−2,979,600	76602	−652.59	1.8879		109.50	0.62287	150.00	1.99090
100	254	Nonadecane	$C_{19}H_{40}$	629-92-5	268.5209	342,570	762.08	0.20481			305.04	5.94090	603.05	8.76630
100	255	Nonanal	$C_9H_{18}O$	124-19-6	142.23862	195,220	378.71	0.029716			267.30	2.98570	465.52	3.77960

(Continued)

TABLE 2-72 Heat Capacities of Inorganic and Organic Liquids [J/(kmol·K)] (*Continued*)

Eqn	Cmpd. no.	Name	Formula	CAS	Mol. wt.	C_1	C_2	C_3	C_4	C_5	T_{min}, K	C_p at $T_{min} \times$ 1E-05	T_{max}, K	C_p at $T_{max} \times$ 1E-05
100	256	Nonane	C_9H_{20}	111-84-2	128.2551	383,080	−1139.8	2.7101			219.66	2.63480	325.00	2.98900
100	257	Nonanoic acid	$C_9H_{18}O_2$	112-05-0	158.238	224,336	49.726	0.9813			285.55	3.18550	528.75	5.24980
100	258	1-Nonanol	$C_9H_{20}O$	143-08-8	144.2545	10,483,000	−115220	476.87	−0.85381	0.00056246	310.00	3.50590	460.00	4.64940
100	259	2-Nonanol	$C_9H_{20}O$	628-99-9	144.255	1,510,000	−12600	40.7	−0.0386		238.15	2.96270	471.70	5.71150
100	260	1-Nonene	C_9H_{18}	124-11-8	126.23922	254,490	−298.06	1.1707			191.91	2.40410	475.00	3.77050
100	261	Nonyl mercaptan	$C_9H_{20}S$	1455-21-6	160.3201	265,350	−46.22	0.79154			253.05	3.04340	492.95	4.34910
100	262	1-Nonyne	C_9H_{16}	3452-09-3	124.22334	253,580	−366.3	1.4881			223.15	2.45940	423.85	3.65660
100	263	Octadecane	$C_{18}H_{38}$	593-45-3	254.49432	399,430	374.64	0.58156			301.31	5.65110	589.86	8.22760
100	264	Octanal	$C_8H_{16}O$	124-13-0	128.212	171,960	383.28	−0.059074			251.65	2.64670	445.15	3.30870
100	265	Octane	C_8H_{18}	111-65-9	114.22852	224,830	−186.63	0.95891			216.38	2.29340	460.00	3.41890
100	266	Octanoic acid	$C_8H_{16}O_2$	124-07-2	144.211	205,260	44.392	0.8956			289.65	2.93260	512.85	4.63580
100	267	1-Octanol	$C_8H_{18}O$	111-87-5	130.22792	571,370	−4849	19.725	−0.021532		250.00	2.55500	467.10	4.15660
100	268	2-Octanol	$C_8H_{18}O$	123-96-6	130.228	1,115,100	−9773.8	34.252	−0.03454		241.55	2.65930	452.90	5.05560
100	269	2-Octanone	$C_8H_{16}O$	111-13-7	128.21204	300,400	−426.2	1.1172			252.86	2.64060	500.00	3.66600
100	270	3-Octanone	$C_8H_{16}O$	106-68-3	128.21204	289,980	−417.27	1.2218			255.55	2.63140	440.65	3.43350
100	271	1-Octene	C_8H_{16}	111-66-0	112.21264	509,420	−4279.1	21.477	−0.044462	0.000035028	171.45	2.13270	454.00	3.20980
100	272	Octyl mercaptan	$C_8H_{18}S$	111-88-6	146.29352	240,040	−33.198	0.67889			240.00	2.71180	472.19	3.75730
100	273	1-Octyne	C_8H_{14}	629-05-0	110.19676	42,642	886.67	−0.69315			200.00	1.92250	399.35	2.86190
100	274	Oxalic acid	$C_2H_2O_4$	144-62-7	90.03488	63,131	199.92				462.65	1.55620	516.00	1.66290
100	275	Oxygen	O_2	7782-44-7	31.9988	175,430	−6152.3	113.92	−0.92382	0.0027963	54.36	0.53646	142.00	0.90662
100	276	Ozone	O_3	10028-15-6	47.9982	60,046	281.16				90.00	0.85350	150.00	1.02220
100	277	Pentadecane	$C_{15}H_{32}$	629-62-9	212.41458	346,910	219.54	0.65632			283.07	4.61650	543.84	6.60420
100	278	Pentanal	$C_5H_{10}O$	110-62-3	86.1323	102,000	389.95	−0.32545			191.59	1.64760	375.15	2.02490
100	279	Pentane	C_5H_{12}	109-66-0	72.14878	159,080	−270.5	0.99537			143.42	1.40760	390.00	2.04980
100	280	Pentanoic acid	$C_5H_{10}O_2$	109-52-4	102.132	145,050	28.344	0.6372			239.15	1.88270	458.95	2.92280
100	281	1-Pentanol	$C_5H_{12}O$	71-41-0	88.1482	201,200	−651.3	2.275			200.14	1.61980	389.15	2.92270
100	282	2-Pentanol	$C_5H_{12}O$	6032-29-7	88.1482	883,630	−8220.5	29.125	−0.02989		200.00	1.65410	392.20	3.36360
100	283	2-Pentanone	$C_5H_{10}O$	107-87-9	86.1323	194,590	−263.86	0.76808			196.29	1.72390	375.46	2.03800
100	284	3-Pentanone	$C_5H_{10}O$	96-22-0	86.1323	193,020	−176.43	0.5669			234.18	1.82790	375.14	2.06610
100	285	1-Pentene	C_5H_{10}	109-67-1	70.1329	156,100	−456.94	2.255	−0.003163	0.00000238	108.02	1.29390	372.00	1.80920
100	286	2-Pentyl mercaptan	$C_5H_{12}S$	2084-19-7	104.21378	188,200	−140.84	0.63581			160.75	1.81990	385.15	2.28270
100	287	Pentyl mercaptan	$C_5H_{12}S$	110-66-7	104.21378	213,760	−324.4	0.9472			197.45	1.86640	399.79	2.35460
100	288	1-Pentyne	C_5H_8	627-19-0	68.11702	86,200	256.6				200.00	1.37520	313.33	1.66600
100	289	2-Pentyne	C_5H_8	627-21-4	68.11702	68,671	246.66				200.00	1.18000	329.27	1.49890
100	290	Phenanthrene	$C_{14}H_{10}$	85-01-8	178.2292	103,370	527.03				372.39	2.99630	500.00	3.66890
100	291	Phenol	C_6H_6O	108-95-2	94.11124	101,720	317.61				314.06	2.01470	425.00	2.36700
100	292	Phenyl isocyanate	C_7H_5NO	103-71-9	119.1207	60,834	215.89	0.29552			243.15	1.30800	489.75	2.37450
100	293	Phthalic anhydride	$C_8H_4O_3$	85-44-9	148.11556	145,400	252.4				404.15	2.47410	557.65	2.86150
100	294	Propadiene	C_3H_4	463-49-0	40.06386	66,230	98.275				200.00	0.85885	238.65	0.89683
114	295	Propane	C_3H_8	74-98-6	44.09562	62.983	113630	633.21	−873.46		85.47	0.84879	360.00	2.60790
100	296	1-Propanol	C_3H_8O	71-23-8	60.09502	158,760	−635	1.969			146.95	1.07970	400.00	2.19800
100	297	2-Propanol	C_3H_8O	67-63-0	60.095	471,710	−4172.1	14.745	−0.014402		185.26	1.13280	463.00	2.71460
100	298	Propenylcyclohexene	C_9H_{14}	13511-13-2	122.20746	201,400	−450.6	1.7053			199.00	1.79260	431.65	3.24630
100	299	Propionaldehyde	C_3H_6O	123-38-6	58.07914	55,679	406.13	−0.50303			165.00	1.09000	322.15	1.34310
100	300	Propionic acid	$C_3H_6O_2$	79-09-4	74.0785	213,660	−702.7	1.6605			252.45	1.42090	414.32	2.07560
100	301	Propionitrile	C_3H_5N	107-12-0	55.0785	121,750	−149.56	0.47759			180.37	1.10310	370.25	1.31850
100	302	Propyl acetate	$C_5H_{10}O_2$	109-60-4	102.1317	83,400	384.1				274.70	1.88910	404.70	2.38850
100	303	Propyl amine	C_3H_9N	107-10-8	59.11026	139,530	78				188.36	1.54220	340.00	1.66050
100	304	Propylbenzene	C_9H_{12}	103-65-1	120.19158	174,380	−101.8	0.79			173.55	1.80510	432.39	2.78060
100	305	Propylene	C_3H_6	115-07-1	42.07974	114,140	−343.72	1.0905			87.89	0.92354	298.15	1.08600
100	306	Propyl formate	$C_4H_8O_2$	110-74-7	88.10512	75,700	326.1				298.15	1.72930	398.15	2.05540

100	307	2-Propyl mercaptan	C_3H_8S	75-33-2	76.16062	138,390	−117.11	0.47059			142.61	1.31260	350.00	1.55050
100	308	Propyl mercaptan	C_3H_8S	107-03-9	76.16062	167,330	−319.1	0.8127			159.95	1.37080	340.87	1.52990
100	309	1,2-Propylene glycol	$C_3H_8O_2$	57-55-6	76.09442	58,080	445.2				213.15	1.52970	460.75	2.63210
100	310	Quinone	$C_6H_4O_2$	106-51-4	108.09476	45,810	368.33				388.85	1.89040	683.00	2.97380
100	311	Silicon tetrafluoride	F_4Si	7783-61-1	104.07911	829,380	−7331.5	19.203			186.35	1.30000	253.15	2.04030
100	312	Styrene	C_8H_8	100-42-5	104.14912	113,340	290.2	−0.6051	0.0013567		242.54	1.67490	418.31	2.28160
100	313	Succinic acid	$C_4H_6O_4$	110-15-6	118.08804	186,250	247.8				460.85	3.00450	591.00	3.32700
100	314	Sulfur dioxide	O_2S	7446-09-5	64.0638	85,743	5.7443				197.67	0.86878	350.00	0.87754
100	315	Sulfur hexafluoride	F_6S	2551-62-4	146.0554192	119,500					230.15	1.19500	230.15	1.19500
100	316	Sulfur trioxide	O_3S	7446-11-9	80.0632	258,090					303.15	2.58090	303.15	2.58090
100	317	Terephthalic acid	$C_8H_6O_4$	100-21-0	166.13084	131,270	345.64				700.15	3.73270	795.28	4.06150
100	318	*o*-Terphenyl	$C_{18}H_{14}$	84-15-1	230.30376	182,900	635.09				329.35	3.92070	609.15	5.69770
100	319	Tetradecane	$C_{14}H_{30}$	629-59-4	198.388	353,140	29.13	0.86116			279.01	4.28310	526.73	6.07410
100	320	Tetrahydrofuran	C_4H_8O	109-99-9	72.10572	171,730	−800.47	2.8934	−0.0025015		164.65	1.07210	339.12	1.35460
100	321	1,2,3,4-Tetrahydronaphthalene	$C_{10}H_{12}$	119-64-2	132.20228	81,760	455.38				237.38	1.89860	480.77	3.00690
100	322	Tetrahydrothiophene	C_4H_8S	110-01-0	88.17132	123,300	−130.1	0.6229			176.98	1.19790	394.27	1.68830
100	323	2,2,3,3-Tetramethylbutane	C_8H_{18}	594-82-1	114.22852	43,326	630.73				375.41	2.80110	426.00	3.12020
100	324	Thiophene	C_4H_4S	110-02-1	84.13956	84,864	91.725	0.13243			234.94	1.13720	357.31	1.34550
100	325	Toluene	C_7H_8	108-88-3	92.13842	140,140	−152.3	0.695			178.18	1.35070	500.00	2.37740
100	326	1,1,2-Trichloroethane	$C_2H_3Cl_3$	79-00-5	133.40422	103,350	159.3				236.50	1.41020	300.00	1.51140
100	327	Tridecane	$C_{13}H_{28}$	629-50-5	184.36142	350,180	−104.7	1.0022			267.76	3.94000	508.62	5.56190
100	328	Triethyl amine	$C_6H_{15}N$	121-44-8	101.19	111,480	368.13				200.00	1.85110	361.92	2.44710
100	329	Trimethyl amine	C_3H_9N	75-50-3	59.11026	136,050	−288	0.9913			156.08	1.15250	276.02	1.32080
100	330	1,2,3-Trimethylbenzene	C_9H_{12}	526-73-8	120.19158	119,450	324.54				247.79	1.99870	449.27	2.65260
100	331	1,2,4-Trimethylbenzene	C_9H_{12}	95-63-6	120.19158	178,800	−128.47	0.83741			229.33	1.93380	350.00	2.36420
100	332	2,2,4-Trimethylpentane	C_8H_{18}	540-84-1	114.22852	95,275	696.7	−1.3765	0.0021734		165.78	1.82850	520.00	3.90950
100	333	2,3,3-Trimethylpentane	C_8H_{18}	560-21-4	114.22852	388,620	−1439.5	3.2187			280.00	2.37910	320.00	2.57570
100	334	1,3,5-Trinitrobenzene	$C_6H_3N_3O_6$	99-35-4	213.10452	40,364	664.46				398.40	3.05080	475.47	3.56290
100	335	2,4,6-Trinitrotoluene	$C_7H_5N_3O_6$	118-96-7	227.1311	133,530	514.64				354.00	3.15710	475.00	3.77980
100	336	Undecane	$C_{11}H_{24}$	1120-21-4	156.30826	293,980	−114.98	0.96936			247.57	3.24930	433.42	4.26240
100	337	1-Undecanol	$C_{11}H_{24}O$	112-42-5	172.30766	−1,360,200	10964	−20.86	0.013055		289.05	3.81370	523.15	5.35730
100	338	Vinyl acetate	$C_4H_6O_2$	108-05-4	86.08924	136,300	−106.17	0.75175			259.56	1.59390	389.35	2.08920
100	339	Vinyl acetylene	C_4H_4	689-97-4	52.07456	68,720	135				200.00	0.95720	278.25	1.06280
100	340	Vinyl chloride	C_2H_3Cl	75-01-4	62.49822	−10,320	322.8				200.00	0.54240	400.00	1.18800
100	341	Vinyl trichlorosilane	$C_2H_3Cl_3Si$	75-94-5	161.48972	49,516	420.35				178.35	1.24490	363.85	2.02460
100	342	Water	H_2O	7732-18-5	18.01528	276,370	−2090.1	8.125	−0.014116	9.3701E-06	273.16	0.76150	533.15	0.89394
100	343	*m*-Xylene	C_8H_{10}	108-38-3	106.165	133,860	7.8754	0.52265			217.00	1.60180	540.15	2.90600
100	344	*o*-Xylene	C_8H_{10}	95-47-6	106.165	36,500	1017.5	−2.63	0.00302		247.98	1.73140	417.58	2.22690
100	345	*p*-Xylene	C_8H_{10}	106-42-3	106.165	−35,500	1287.2	−2.599	0.002426		286.41	1.76970	600.00	3.25200

For the 11 substances: ammonia; 1,2-butanediol; 1,3-butanediol; carbon monoxide; 1,1-difluoroethane; ethane; heptane; hydrogen; hydrogen sulfide; methane; and propane; the liquid heat capacity C_{pL} is calculated with Eq. (2-114): $C_{pL} = C_1^2/\tau + C_2 - 2C_1C_3\tau - C_1C_4\tau^2 - C_3^2\tau^3/3 - C_3C_4\tau^4/2 - C_4^2\tau^5/5$, where $\tau = 1 - T_r$, $T_r = T/T_C$, T_C is the critical temperature from Table 2-106, C_{pL} is in J/(kmol·K) and T is in K.

For all other compounds, Eqn 100 is used. Eqn 100: $C_{pL} = C_1 + C_2T + C_3T^2 + C_4T^3 + C_5T^4$. For benzene, fluorine, and helium, two sets of constants are given for Eqn 100 that cover different temperature ranges, as shown in the table.

Values in this table were taken from the Design Institute for Physical Properties (DIPPR) of the American Institute of Chemical Engineers (AIChE), 801 Critically Evaluated Gold Standard™ Database, copyright 2016 AIChE, and reproduced with permission of AIChE and of the DIPPR Evaluated Process Design Data Project Steering Committee. Their source should be cited as R. L. Rowley, W. V. Wilding, J. L. Oscarson, T. A. Knotts, and N. F. Giles, *DIPPR® Data Compilation of Pure Chemical Properties*, Design Institute for Physical Properties, AIChE, New York, NY (2016).

TABLE 2-73 Specific Heats of Organic Solids

Recalculated from *International Critical Tables,* vol. 5, pp. 101–105

Compound	Formula	Temperature, °C	sp ht, cal/(g·°C)
Acetic acid	$C_2H_4O_2$	−200 to +25	0.330 + 0.00080*t*
Acetone	C_3H_6O	−210 to −80	0.540 + 0.0156*t*
Aminobenzoic acid (*o*-)	$C_7H_7NO_2$	85 to mp	0.254 + 0.00136*t*
(*m*-)	$C_7H_7NO_2$	120 to mp	0.253 + 0.00122*t*
(*p*-)	$C_7H_7NO_2$	128 to mp	0.287 + 0.00088*t*
Aniline	C_6H_7N		0.741
Anthracene	$C_{14}H_{10}$	50	0.308
		100	0.350
		150	0.382
Anthraquinone	$C_{14}H_8O_2$	0 to 270	0.258 + 0.00069*t*
Apiol	$C_{12}H_{14}O_4$	10	0.299
Azobenzene	$C_{12}H_{10}N_2$	28	0.330
Benzene	C_6H_6	−250	0.0399
		−225	0.0908
		−200	0.124
		−150	0.170
		−100	0.227
		−50	0.299
		0	0.375
Benzoic acid	$C_7H_6O_2$	20 to mp	0.287 + 0.00050*t*
Benzophenone	$C_{13}H_{10}O$	−150	0.115
		−100	0.172
		−50	0.220
		0	0.275
		+20	0.303
Betol	$C_{17}H_{12}O_3$	−150	0.129
		−100	0.167
		0	0.248
		+50	0.308
Bromoiodobenzene (*o*-)	C_6H_4BrI	−50 to 0	0.143 + 0.00025*t*
(*m*-)	C_6H_4BrI	−75 to −15	0.143
(*p*-)	C_6H_4BrI	−40 to 50	0.116 + 0.00032*t*
Bromonaphthalene (β-)	$C_{10}H_7Br$	41	0.260
Bromophenol	C_6H_5BrO	32	0.263
Camphene	$C_{10}H_{16}$	35	0.380
Capric acid	$C_{10}H_{20}O_2$	8	0.695
Caprylic acid	$C_8H_{16}O_2$	−2	0.628
Carbon tetrachloride	CCl_4	−240	0.013
		−200	0.081
		−160	0.131
		−120	0.162
		−80	0.182
		−40	0.201
Cerotic acid	$C_{27}H_{54}O_2$	15	0.387
Chloral alcoholate	$C_4H_7Cl_3O_2$	78	0.509
hydrate	$C_2H_3Cl_3O_2$	32	0.213
Chloroacetic acid	$C_2H_3ClO_2$	60	0.363
Chlorobenzoic acid (*o*-)	$C_7H_5ClO_2$	80 to mp	0.228 + 0.00084*t*
(*m*-)	$C_7H_5ClO_2$	94 to mp	0.232 + 0.00073*t*
(*p*-)	$C_7H_5ClO_2$	180 to mp	0.242 + 0.00055*t*
Chlorobromobenzene (*o*-)	C_6H_4BrCl	−34	0.192
(*m*-)	C_6H_4BrCl	−52	0.150
(*p*-)	C_6H_4BrCl	−40	0.150
Crotonic acid	$C_4H_6O_2$	38 to 70	0.520 + 0.00020*t*
Cyamelide	$C_3H_3N_3O_3$	40	0.263
Cyanamide	CH_2N_2	20	0.547
Cyanuric acid	$C_3H_3N_3O_3$	40	0.318
Dextrin	$(C_6H_{10}O_5)x$	0 to 90	0.291 + 0.00096*t*
Dextrose	$C_6H_{12}O_6$	−250	0.016
		−200	0.077
		−100	0.160
		0	0.277
		20	0.300
Dibenzyl	$C_{14}H_{14}$	28	0.363
Dibromobenzene (*o*-)	$C_6H_4Br_2$	−36	0.248
(*m*-)	$C_6H_4Br_2$	−25	0.134
(*p*-)	$C_6H_4Br_2$	−50 to +50	0.139 + 0.00038*t*
Dichloroacetic acid	$C_2H_2Cl_2O_2$		0.406
Dichlorobenzene (*o*-)	$C_6H_4Cl_2$	−48.5	0.185
(*m*-)	$C_6H_4Cl_2$	−52	0.186
(*p*-)	$C_6H_4Cl_2$	−50 to +53	0.219 + 0.0021*t*
Dicyandiamide	$C_2H_4N_4$	0 to 204	0.456

TABLE 2-73 Specific Heats of Organic Solids *(Continued)*

Recalculated from *International Critical Tables,* vol. 5, pp. 101–105

Compound	Formula	Temperature, °C	sp. ht., cal/(g·°C)
Dihydroxybenzene (*o*-)	$C_6H_6O_2$	−163 to mp	0.278 + 0.00098*t*
(*m*-)	$C_6H_6O_2$	−160 to mp	0.269 + 0.00118*t*
(*p*-)	$C_6H_6O_2$	−250	0.025
		−240	0.038
		−220	0.061
		−200	0.081
		−150 to mp	0.268 + 0.00093*t*
Di-iodobenzene (*o*-)	$C_6H_4I_2$	−50 to +15	0.109 + 0.00026*t*
(*m*-)	$C_6H_4I_2$	−52 to −42	0.100 + 0.00026*t*
(*p*-)	$C_6H_4I_2$	−50 to +80	0.101 + 0.00026*t*
Dimethyl oxalate	$C_4H_6O_4$	10 to 50	0.212 + 0.0044*t*
Dimethylpyrene	$C_7H_8O_2$	50	0.368
Dinitrobenzene (*o*-)	$C_6H_4N_2O_4$	−160 to mp	0.252 + 0.00083*t*
(*m*-)	$C_6H_4N_2O_4$	−160 to mp	0.248 + 0.00077*t*
(*p*-)	$C_6H_4N_2O_4$	119 to mp	0.259 + 0.00057*t*
Diphenyl	$C_{12}H_{10}$	40	0.385
Diphenylamine	$C_{12}H_{11}N$	26	0.337
Dulcitol	$C_6H_{14}O_6$	20	0.282
Erythritol	$C_4H_{10}O_4$	60	0.351
Ethyl alcohol	C_2H_6O (crystalline)	−190	0.232
		−180	0.248
		−160	0.282
		−140	0.318
		−130	0.376
	(vitreous)	−190	0.260
		−180	0.296
		−175	0.380
		−170	0.399
Ethylene glycol	$C_2H_6O_2$	−190 to −40	0.366 + 0.00110*t*
Formic acid	CH_2O_2	−22	0.387
		0	0.430
Glutaric acid	$C_5H_8O_4$	20	0.299
Glycerol	$C_3H_8O_3$	−265	0.009
		−260	0.022
		−250	0.047
		−220	0.085
		−200	0.115
		−100	0.217
		0	0.330
Hexachloroethane	C_2Cl_6	25	0.174
Hexadecane	$C_{16}H_{34}$		0.495
Hydroxyacetanilide	$C_8H_9NO_2$	41 to mp	0.249 + 0.00154*t*
Iodobenzene	C_6H_5I	40	0.191
Isopropyl alcohol	C_3H_8O	−200 to −160	0.051 + 0.00165*t*
Lactose	$C_{12}H_{22}O_{11}$	20	0.287
	$C_{12}H_{22}O_{11}\cdot H_2O$	20	0.299
Lauric acid	$C_{12}H_{24}O_2$	−30 to +40	0.430 + 0.000027*t*
Levoglucosane	$C_6H_{10}O_5$	40	0.607
Levulose	$C_6H_{12}O_6$	20	0.275
Malonic acid	$C_3H_4O_4$	20	0.275
Maltose	$C_{12}H_{22}O_{11}$	20	0.320
Mannitol	$C_6H_{14}O_6$	0 to 100	0.313 + 0.00025*t*
Melamine	$C_3H_6N_6$	40	0.351
Myristic acid	$C_{14}H_{28}O_2$	0 to 35	0.381 + 0.00545*t*
Naphthalene	$C_{10}H_8$	−130 to mp	0.281 + 0.00111*t*
Naphthol (α-)	$C_{10}H_8O$	50 to mp	0.240 + 0.00147*t*
(β-)	$C_{10}H_8O$	61 to mp	0.252 + 0.00128*t*
Naphthylamine (α-)	$C_{10}H_9N$	0 to 50	0.270 + 0.0031*t*
Nitroaniline (*o*-)	$C_6H_6N_2O_2$	−160 to mp	0.269 + 0.000920*t*
(*m*-)	$C_6H_6N_2O_2$	−160 to mp	0.275 + 0.000946*t*
(*p*-)	$C_6H_6N_2O_2$	−160 to mp	0.276 + 0.001000*t*
Nitrobenzoic acid (*o*-)	$C_7H_5NO_4$	−163 to mp	0.256 + 0.00085*t*
(*m*-)	$C_7H_5NO_4$	66 to mp	0.258 + 0.00091*t*
(*p*-)	$C_7H_5NO_4$	−160 to mp	0.247 + 0.00077*t*
Nitronaphthalene	$C_{10}H_7NO_2$	0 to 55	0.236 + 0.00215*t*

(Continued)

TABLE 2-73 Specific Heats of Organic Solids *(Continued)*

Recalculated from *International Critical Tables,* vol. 5, pp. 101–105

Compound	Formula	Temperature, °C	sp ht, cal/(g·°C)
Oxalic acid	$C_2H_2O_4$	−200 to +50	0.259 + 0.00076*t*
	$C_2H_2O_4{\cdot}2H_2O$	−200	0.117
		−100	0.239
		0	0.338
		+50	0.385
		100	0.416
Palmitic acid	$C_{16}H_{32}O_2$	−180	0.167
		−140	0.208
		−100	0.251
		−50	0.306
		0	0.382
		+20	0.430
Phenol	C_6H_6O	14 to 26	0.561
Phthalic acid	$C_8H_6O_4$	20	0.232
Picric acid	$C_6H_3N_3O_7$	−100	0.165
		0	0.240
		+50	0.263
		100	0.297
		120	0.332
Propionic acid	$C_3H_6O_2$	−33	0.726
Propyl alcohol (*n*-)	C_3H_8O	−200	0.170
		−175	0.363
		−150	0.471
		−130	0.497
Pyrotartaric acid	$C_6H_8O_4$	20	0.301
Quinhydrone	$C_{12}H_{10}O_4$	−250	0.017
		−225	0.061
		−200	0.098
		−100	0.191
		0	0.256
Quinone	$C_6H_4O_2$	−250	0.031
		−225	0.082
		−200	0.113
		−150 to mp	0.282 + 0.00083*t*
Salol	$C_{13}H_{10}O_3$	32	0.289
Stearic acid	$C_{18}H_{36}O_2$	15	0.399
Succinic acid	$C_4H_6O_4$	0 to 160	0.248 + 0.00153*t*
Sucrose	$C_{12}H_{22}O_{11}$	20	0.299
Sugar (cane)	$C_{12}H_{22}O_{11}$	22 to 51	0.301
Tartaric acid	$C_4H_6O_6$	36	0.287
Tartaric acid	$C_4H_6O_6{\cdot}H_2O$	−150	0.112
		−100	0.170
		−50	0.231
		0	0.308
		+50	0.366
Tetrachloroethylene	C_2Cl_4	−40 to 0	0.198 + 0.00018*t*
Tetryl	$C_7H_5N_5O_8$	−100	0.182
		−50	0.199
		0	0.212
		+100	0.236
1 Tetryl + 1 picric acid	$C_{13}H_8N_8O_{15}$	−100 to +100	0.253 + 0.00072*t*
1 Tetryl + 2 TNT	$C_{21}H_{15}N_{11}O_{20}$	−100	0.172
		0	0.280
		+50	0.325
Thymol	$C_{10}H_{14}O$	0 to 49	0.315 + 0.0031*t*
Toluic acid (*o*-)	$C_8H_8O_2$	54 to mp	0.277 + 0.00120*t*
(*m*-)	$C_8H_8O_2$	54 to mp	0.239 + 0.00195*t*
(*p*-)	$C_8H_8O_2$	130 to mp	0.271 + 0.00106*t*
Toluidine (*p*-)	C_7H_9N	0	0.337
		20	0.387
		40	0.440
Trichloroacetic acid	$C_2HCl_3O_2$	solid	0.459
Trimethyl carbinol	$C_4H_{10}O$	−4	0.559
Trinitrotoluene	$C_7H_5N_3O_6$	−100	0.170
		−50	0.253
		0	0.311
		+100	0.385
Trinitroxylene	$C_8H_7N_3O_6$	−185 to +23	0.241
		20 to 50	0.423
Triphenylmethane	$C_{19}H_{16}$	0 to 91	0.189 + 0.0027*t*
Urea	CH_4N_2O	20	0.320

TABLE 2-74 Heat Capacity at Constant Pressure of Inorganic and Organic Compounds in the Ideal Gas State Fit to a Polynomial C_p [J/(kmol·K)]

Cmpd. no.	Name	Formula	CAS	Mol. wt.	C_1	C_2	C_3	C_4	C_5	T_{min}, K	C_p at T_{min}	T_{max}, K	C_p at T_{max}
1	Acetaldehyde	C_2H_4O	75-07-0	44.05256	29705	127.43	−0.21793			50	3.553E+04	200	4.647E+04
7	Acetylene	C_2H_2	74-86-2	26.03728	30800	−53.08	0.384			50	2.911E+04	200	3.554E+04
8	Acrolein	C_3H_4O	107-02-8	56.06326	30702	80.95	0.191			50	3.523E+04	200	5.453E+04
14	Argon	Ar	7440-37-1	39.948	20786					100	2.079E+04	1500	2.079E+04
16	Benzene	C_6H_6	71-43-2	78.11184	35978	−101.69	0.939			50	3.324E+04	200	5.320E+04
27	Bromoethane	C_2H_5Br	74-96-4	108.965	27112	117.99				100	3.891E+04	200	5.071E+04
29	1,2-Butadiene	C_4H_6	590-19-2	54.09044	27400	177.6				50	3.628E+04	200	6.292E+04
31	Butane	C_4H_{10}	106-97-8	58.1222	17330	458.16	−0.816			50	3.820E+04	200	7.632E+04
34	1-Butanol	$C_4H_{10}O$	71-36-3	74.1216	25300	371.2	−0.461			50	4.271E+04	200	8.110E+04
37	*cis*-2-Butene	C_4H_8	590-18-1	56.10632	39760	108.8				50	4.520E+04	200	6.152E+04
38	*trans*-2-Butene	C_4H_8	624-64-6	56.10632	20908	324.73	−0.411			50	3.612E+04	200	6.941E+04
43	1-Butyne	C_4H_6	107-00-6	54.09044	25300	183.2				50	3.446E+04	200	6.194E+04
59	*m*-Cresol	C_7H_8O	108-39-4	108.13782	29002	158.79	0.635			50	3.853E+04	200	8.616E+04
60	*o*-Cresol	C_7H_8O	95-48-7	108.13782	16192	469.81	−0.479			50	3.849E+04	200	9.099E+04
61	*p*-Cresol	C_7H_8O	106-44-5	108.13782	29090	166	0.616			50	3.893E+04	200	8.693E+04
64	Cyclobutane	C_4H_8	287-23-0	56.10632	31863	37.226	0.23616			50	3.431E+04	200	4.875E+04
67	Cyclohexanone	$C_6H_{10}O$	108-94-1	98.143	32182	116.87	0.547			50	3.939E+04	200	7.744E+04
81	1,1-Dibromoethane	$C_2H_4Br_2$	557-91-5	187.86116	20560	285.2	−0.332			100	4.576E+04	200	6.432E+04
88	1,1-Dichloroethane	$C_2H_4Cl_2$	75-34-3	98.95916	19560	249.01	−0.22187			100	4.224E+04	200	6.049E+04
95	Diethyl ether	$C_4H_{10}O$	60-29-7	74.1216	26040	388	−0.268			50	4.477E+04	200	9.292E+04
97	1,1-Difluoroethane	$C_2H_4F_2$	75-37-6	66.04997	29736	72.364	0.228			50	3.392E+04	200	5.333E+04
98	1,2-Difluoroethane	$C_2H_4F_2$	624-72-6	66.04997	27581	169.88	−0.1581			50	3.568E+04	200	5.523E+04
99	Difluoromethane	CH_2F_2	75-10-5	52.02339	33851	−20.966	0.17584			50	3.324E+04	200	3.669E+04
112	Dimethyl ether	C_2H_6O	115-10-6	46.06844	25940	178.46	−0.186			50	3.440E+04	200	5.419E+04
120	1,4-Dioxane	$C_4H_8O_2$	123-91-1	88.10512	28345	88.3	0.446			50	3.388E+04	200	6.385E+04
125	Ethane	C_2H_6	74-84-0	30.069	31742	26.567	0.12927			50	3.339E+04	200	4.223E+04
126	Ethanol	C_2H_6O	64-17-5	46.06844	32585	87.4	0.05			50	3.708E+04	200	5.207E+04
134	Ethylcyclopentane	C_7H_{14}	1640-89-7	98.18606	34710	304.96	−0.084			50	4.975E+04	200	9.234E+04
145	Ethyl mercaptan	C_2H_6S	75-08-1	62.13404	23014	271.36	−0.4427			50	3.548E+04	200	5.958E+04
151	Fluoroethane	C_2H_5F	353-36-6	48.0595	30358	62.839	0.1067			50	3.377E+04	200	4.719E+04
156	Furan	C_4H_4O	110-00-9	68.07396	40860	−160.3	0.87			100	3.353E+04	200	4.360E+04
157	Helium-4	He	7440-59-7	4.0026	20786					100	2.079E+04	1500	2.079E+04
182	Hydrazine	H_4N_2	302-01-2	32.04516	32998	−5.2147	0.21379			50	3.327E+04	200	4.051E+04
183	Hydrogen	H_2	1333-74-0	2.01588	64979	−788.17	5.8287	−0.018459	2.164E-05	50	3.797E+04	250	2.834E+04
190	Isopropyl amine	C_3H_9N	75-31-0	59.11026	23590	310.42	−0.274			50	3.843E+04	200	7.471E+04
194	Methanol	CH_4O	67-56-1	32.04186	30270	84.64	−0.188			50	3.403E+04	200	3.968E+04
197	Methyl acetylene	C_3H_4	74-99-7	40.06386	30810	35.8	0.27			50	3.328E+04	200	4.877E+04
217	Methylcyclopentane	C_6H_{12}	96-37-7	84.15948	35465	147.38	0.242			50	4.344E+04	200	7.462E+04
221	Methylethyl ether	C_3H_8O	540-67-0	60.09502	23337	309.03	−0.285			50	3.808E+04	200	7.374E+04
231	Methyl mercaptan	CH_4S	74-93-1	48.10746	31520	60.1				50	3.453E+04	200	4.354E+04
236	2-Methylpropane	C_4H_{10}	75-28-5	58.1222	21380	271.2	−0.092			50	3.471E+04	200	7.194E+04
237	2-Methyl-2-propanol	$C_4H_{10}O$	75-65-0	74.1216	17080	381.7	−0.199			50	3.567E+04	200	8.546E+04
238	2-Methyl propene	C_4H_8	115-11-7	56.10632	24970	211.8				50	3.556E+04	200	6.733E+04
243	alpha-Methyl styrene	C_9H_{10}	98-83-9	118.1757	37735	112.94	0.846			50	4.550E+04	200	9.416E+04
246	Naphthalene	$C_{10}H_8$	91-20-3	128.17052	29120	82.88	0.964			50	3.567E+04	200	8.426E+04
247	Neon	Ne	7440-01-9	20.1797	20786					100	2.079E+04	1500	2.079E+04
248	Nitroethane	$C_2H_5NO_2$	79-24-3	75.0666	33055	89.54	0.238			50	3.813E+04	200	6.048E+04
251	Nitromethane	CH_3NO_2	75-52-5	61.04002	38782	−48.39	0.413			50	3.740E+04	200	4.562E+04

(*Continued*)

TABLE 2-74 Heat Capacity at Constant Pressure of Inorganic and Organic Compounds in the Ideal Gas State Fit to a Polynomial C_p [J/(kmol·K)] *(Continued)*

Cmpd. no.	Name	Formula	CAS	Mol. wt.	C_1	C_2	C_3	C_4	C_5	T_{min}, K	C_p at T_{min}	T_{max}, K	C_p at T_{max}
253	Nitric oxide	NO	10102-43-9	30.0061	34980	−35.32	0.07729	−5.7357E-05	1.4526E-08	100	3.216E+04	1500	3.586E+04
289	2-Pentyne	C_5H_8	627-21-4	68.11702	24330	335.7	−0.37			50	4.019E+04	200	7.667E+04
290	Phenanthrene	$C_{14}H_{10}$	85-01-8	178.2292	27700	210	1.24			50	4.130E+04	200	1.193E+05
294	Propadiene	C_3H_4	463-49-0	40.06386	31690	17.1	0.282			50	3.325E+04	200	4.639E+04
295	Propane	C_3H_8	74-98-6	44.09562	26675	147.04				50	3.403E+04	200	5.608E+04
296	1-Propanol	C_3H_8O	71-23-8	60.09502	28800	257	−0.35			50	4.078E+04	200	6.620E+04
304	Propylbenzene	C_9H_{12}	103-65-1	120.19158	22880	538.46	−0.546			50	4.844E+04	200	1.087E+05
310	Quinone	$C_6H_4O_2$	106-51-4	108.09476	29668	129.07	0.53105			50	3.745E+04	200	7.672E+04
320	Tetrahydrofuran	C_4H_8O	109-99-9	72.10572	36970	−12.28	0.444			50	3.747E+04	200	5.227E+04
321	1,2,3,4-Tetrahydronaphthalene	$C_{10}H_{12}$	119-64-2	132.20228	28560	225.1	0.616			50	4.136E+04	200	9.822E+04
322	Tetrahydrothiophene	C_4H_8S	110-01-0	88.17132	41195	−88.3	0.942			50	3.914E+04	200	6.122E+04
324	Thiophene	C_4H_4S	110-02-1	84.13956	36765	−112.82	0.862			50	3.328E+04	200	4.868E+04
331	1,2,4-Trimethylbenzene	C_9H_{12}	95-63-6	120.19158	35652	323.89	0.305			50	5.261E+04	200	1.126E+05

Constants in this table can be used in the following equation to calculate the ideal gas heat capacity C^0_p. $C^0_p = C_1 + C_2T + C_3T^2 + C_4T^3 + C_5T^4$ where C^0_p is in J/(kmol·K) and T is in K.

Values in this table were taken from the Design Institute for Physical Properties (DIPPR) of the American Institute of Chemical Engineers (AIChE), 801 Critically Evaluated Gold Standard™ Database, copyright 2016 AIChE, and reproduced with permission of AIChE and of the DIPPR Evaluated Process Design Data Project Steering Committee. Their source should be cited as "R. L. Rowley, W. V. Wilding, J. L. Oscarson, T. A. Knotts, and N. F. Giles, *DIPPR® Data Compilation of Pure Chemical Properties*, Design Institute for Physical Properties AIChE New York NY (2016)".

TABLE 2-75 Heat Capacity at Constant Pressure of Inorganic and Organic Compounds in the Ideal Gas State Fit to Hyperbolic Functions C_p [J/(kmol·K)]

Cmpd. no.	Name	Formula	CAS	Mol. wt.	$C_1 \times$ 1E-05	$C_2 \times$ 1E-05	$C_3 \times$ 1E-03	$C_4 \times$ 1E-05	C_5	T_{min}, K	C_p at $T_{min} \times$ 1E-05	T_{max}, K	C_p at $T_{max} \times$ 1E-05
1	Acetaldehyde	C_2H_4O	75-07-0	44.05256	0.48251	1.06650	1.99290	0.78851	912.78	298.15	0.54732	1500	1.29930
2	Acetamide	C_2H_5NO	60-35-5	59.0672	0.34200	1.29400	1.07500	0.64000	502	100	0.34481	1500	1.49970
3	Acetic acid	$C_2H_4O_2$	64-19-7	60.052	0.40200	1.36750	1.26200	0.70030	569.7	50	0.40200	1500	1.57560
4	Acetic anhydride	$C_4H_6O_3$	108-24-7	102.08864	0.87998	1.66350	0.80153	0.76076	2310.1	298.15	1.10440	1500	2.69700
5	Acetone	C_3H_6O	67-64-1	58.07914	0.57040	1.63200	1.60700	0.96800	731.5	200	0.60487	1500	1.88200
6	Acetonitrile	C_2H_3N	75-05-8	41.0519	0.44346	0.84650	1.63980	0.49487	761.47	298.15	0.52233	1500	1.11990
7	Acetylene	C_2H_2	74-86-2	26.03728	0.36921	0.31793	0.67805	0.33430	3036.6	298.15	0.44032	1500	0.75868
8	Acrolein	C_3H_4O	107-02-8	56.06326	0.57019	0.91830	0.76747	0.38554	2375.4	298.15	0.71326	1500	1.56240
9	Acrylic acid	$C_3H_4O_2$	79-10-7	72.06266	0.60590	1.37030	1.64750	1.04460	751.49	250	0.69837	1500	1.74240
10	Acrylonitrile	C_3H_3N	107-13-1	53.0626	0.56303	1.09720	0.91248	−0.44070	1178.4	298.15	0.64356	1500	1.37940
11	Air	Mixture	132259-10-0	28.96	0.28958	0.09390	3.01200	0.07580	1484	50	0.28958	1500	0.34956
12	Ammonia	H_3N	7664-41-7	17.03052	0.33427	0.48980	2.03600	0.22560	882	100	0.33427	1500	0.66465
13	Anisole	C_7H_8O	100-66-3	108.13782	0.76370	2.93770	1.60510	2.17000	751.2	300	1.13020	1200	3.02260
14	Argon	Ar	7440-37-1	39.948	See Table 2-155								
15	Benzamide	C_7H_7NO	55-21-0	121.13658	1.95810	1.70190	1.32570	−37.41700	41.232	298.15	1.27450	1500	3.25010
16	Benzene	C_6H_6	71-43-2	78.11184	0.55238	1.73380	0.76425	0.72545	2445.7	298.15	0.82616	1500	2.41800
17	Benzenethiol	C_6H_6S	108-98-5	110.17684	0.68950	2.32750	1.51200	1.75160	697.9	200	0.76894	1500	2.67390
18	Benzoic acid	$C_7H_6O_2$	65-85-0	122.12134	0.77594	2.64550	1.79250	2.23820	835.9	200	0.81258	1500	2.97120
19	Benzonitrile	C_7H_5N	100-47-0	103.1213	0.76820	2.26350	0.74786	−0.67585	896	298	1.09070	1500	2.68100
20	Benzophenone	$C_{13}H_{10}O$	119-61-9	182.2179	1.00990	4.48980	1.31100	2.83950	627.4	300	1.80010	1500	4.93110
21	Benzyl alcohol	C_7H_8O	100-51-6	108.13782	0.84115	3.14280	1.95390	2.57430	850.06	298.15	1.11980	1500	3.28800
22	Benzyl ethyl ether	$C_9H_{12}O$	539-30-0	136.19098	0.95210	2.88680	0.70207	1.63850	2002.6	300	1.55010	1500	4.34450
23	Benzyl mercaptan	C_7H_8S	100-53-8	124.20342	0.99192	2.96330	1.55830	2.21160	719.16	300	1.41560	1200	3.29570
24	Biphenyl	$C_{12}H_{10}$	92-52-4	154.2078	1.07590	4.21050	1.90410	4.17850	828.81	200	1.14810	1500	4.55570
25	Bromine	Br_2	7726-95-6	159.808	0.30113	0.08009	0.75140	0.10780	314.6	100	0.30901	1500	0.37938
26	Bromobenzene	C_6H_5Br	108-86-1	157.0079	0.72100	2.06400	1.65040	1.68700	765.3	200	0.76789	1500	2.46280
27	Bromoethane	C_2H_5Br	74-96-4	108.965	0.52310	0.89110	0.81205	0.67540	2809	298.15	0.63800	1500	1.54570
28	Bromomethane	CH_3Br	74-83-9	94.93852	0.36241	0.69248	1.74540	0.44781	793.32	298.15	0.42454	1500	0.90758
29	1,2-Butadiene	C_4H_6	590-19-2	54.09044	0.66964	1.09950	0.83737	0.68373	2441.1	298.15	0.79668	1500	1.92080
30	1,3-Butadiene	C_4H_6	106-99-0	54.09044	0.50950	1.70500	1.53240	1.33700	685.6	200	0.57563	1500	1.95550
31	Butane	C_4H_{10}	106-97-8	58.1222	0.80154	1.62420	0.84149	1.05750	2476.1	298.15	0.98586	1500	2.66050
32	1,2-Butanediol	$C_4H_{10}O_2$	584-03-2	90.121	1.04780	2.54900	1.87760	1.87500	833	298.15	1.26670	1500.1	3.02890
33	1,3-Butanediol	$C_4H_{10}O_2$	107-88-0	90.121	1.06600	2.57500	1.96700	1.95100	860.5	298.15	1.26790	1500.15	3.03110
34	1-Butanol	$C_4H_{10}O$	71-36-3	74.1216	0.74540	2.59070	1.60730	1.73200	712.4	298.15	1.07860	1500	2.85090
35	2-Butanol	$C_4H_{10}O$	78-92-2	74.1216	0.90878	2.55080	1.89300	1.85200	832.13	298.15	1.12570	1500	2.87300
36	1-Butene	C_4H_8	106-98-9	56.10632	0.64257	2.06180	1.67680	1.33240	757.06	250	0.75708	1500	2.28980
37	*cis*-2-Butene	C_4H_8	590-18-1	56.10632	0.65121	1.43250	0.85796	0.89648	2477.2	298.15	0.80241	1500	2.27180
38	*trans*-2-Butene	C_4H_8	624-64-6	56.10632	0.74296	1.34760	0.87025	0.89116	2463.4	298.15	0.87766	1500	2.28360
39	Butyl acetate	$C_6H_{12}O_2$	123-86-4	116.15828	1.16840	3.76900	1.95600	2.81800	811.2	298.15	1.52810	1200	3.67240
40	Butylbenzene	$C_{10}H_{14}$	104-51-8	134.21816	1.13800	4.45400	1.55070	3.04970	708.86	200	1.26590	1500	4.84350
41	Butyl mercaptan	$C_4H_{10}S$	109-79-5	90.1872	0.92478	2.77950	1.68370	1.59740	758.68	200	0.97140	1500	3.10080
42	*sec*-Butyl mercaptan	$C_4H_{10}S$	513-53-1	90.1872	0.92367	2.51660	1.61090	1.56410	739.2	200	0.97633	1500	2.96150
43	1-Butyne	C_4H_6	107-00-6	54.09044	0.66492	1.07260	0.79390	0.74240	−2458.4	298.15	0.81441	1500	1.92210
44	Butyraldehyde	C_4H_8O	123-72-8	72.10572	0.89240	1.56750	0.90190	1.09840	2566	298.15	1.02830	1500	2.67780
45	Butyric acid	$C_4H_8O_2$	107-92-6	88.1051	1.48800	1.35220	1.14600	−678.00000	6.98	298.15	1.15330	1500	2.59050
46	Butyronitrile	C_4H_7N	109-74-0	69.1051	0.82142	1.32340	0.84021	0.67932	2313.7	298.15	0.97246	1500	2.28510
47	Carbon dioxide	CO_2	124-38-9	44.0095	0.29370	0.34540	1.42800	0.26400	588	50	0.29370	5000	0.63346
48	Carbon disulfide	CS_2	75-15-0	76.1407	0.30100	0.33380	0.89600	0.28930	374.7	100	0.31003	1500	0.61475
49	Carbon monoxide	CO	630-08-0	28.0101	0.29108	0.08773	3.08510	0.08455	1538.2	60	0.29108	1500	0.35208
50	Carbon tetrachloride	CCl_4	56-23-5	153.8227	0.37582	0.70540	0.51210	0.48500	236.1	100	0.47299	1500	1.06620

(*Continued*)

TABLE 2-75 Heat Capacity at Constant Pressure of Inorganic and Organic Compounds in the Ideal Gas State Fit to Hyperbolic Functions C_p [J/(kmol·K)] *(Continued)*

Cmpd. no.	Name	Formula	CAS	Mol. wt.	$C_1 \times$ 1E-05	$C_2 \times$ 1E-05	$C_3 \times$ 1E-03	$C_4 \times$ 1E-05	C_5	T_{min}, K	C_p at $T_{min} \times$ 1E-05	T_{max}, K	C_p at $T_{max} \times$ 1E-05
51	Carbon tetrafluoride	CF_4	75-73-0	88.0043	0.92004	0.16446	1.07640	−5083.80000	2.3486	298	0.61055	1500	1.04650
52	Chlorine	Cl_2	7782-50-5	70.906	0.29142	0.09176	0.94900	0.10030	425	50	0.29142	1500	0.37930
53	Chlorobenzene	C_6H_5Cl	108-90-7	112.5569	0.80110	2.31000	2.15700	2.04600	897.6	200	0.82193	1500	2.53270
54	Chloroethane	C_2H_5Cl	75-00-3	64.5141	0.52590	1.40200	2.03700	0.99820	861.18	298.15	0.62879	1500	1.55080
55	Chloroform	$CHCl_3$	67-66-3	119.37764	0.39420	0.65730	0.92800	0.49300	399.6	100	0.40484	1500	1.00630
56	Chloromethane	CH_3Cl	74-87-3	50.4875	0.36220	0.69810	1.80500	0.44470	844.27	298.15	0.41193	1500	0.90655
57	1-Chloropropane	C_3H_7Cl	540-54-5	78.54068	0.64710	1.79800	1.67600	1.23300	755.78	298.15	0.84674	1500	2.09750
58	2-Chloropropane	C_3H_7Cl	75-29-6	78.54068	0.61809	1.80230	1.54380	1.18930	685.93	200	0.67679	1500	2.10230
59	*m*-Cresol	C_7H_8O	108-39-4	108.13782	0.90974	2.13210	0.76324	0.93355	2474.5	298.15	1.24780	1500	3.21580
60	*o*-Cresol	C_7H_8O	95-48-7	108.13782	0.79880	2.85300	1.47650	2.04200	664.7	200	0.91584	1500	3.21630
61	*p*-Cresol	C_7H_8O	106-44-5	108.13782	0.92021	2.11060	0.76622	0.95073	2464.6	298.15	1.25080	1500	3.21320
62	Cumene	C_9H_{12}	98-82-8	120.19158	1.08100	3.79320	1.75050	3.00270	794.8	200	1.14800	1500	4.18080
63	Cyanogen	C_2N_2	460-19-5	52.0348	0.45894	0.41286	1.38120	0.33023	559.94	273.15	0.54968	1500	0.81268
64	Cyclobutane	C_4H_8	287-23-0	56.10632	0.50835	1.64870	0.82849	0.86658	2472.4	298.15	0.70636	1500	2.32330
65	Cyclohexane	C_6H_{12}	110-82-7	84.15948	0.43200	3.73500	1.19200	1.63500	530.1	100	0.43657	1500	3.65160
66	Cyclohexanol	$C_6H_{12}O$	108-93-0	100.15888	0.90430	2.57710	0.78820	1.30680	1952.2	200	0.96478	1500	3.82510
67	Cyclohexanone	$C_6H_{10}O$	108-94-1	98.143	0.85860	2.57770	0.84895	0.77780	2401.5	298.15	1.14170	1500	3.47740
68	Cyclohexene	C_6H_{10}	110-83-8	82.1436	0.58171	3.17170	1.54350	2.12730	701.62	150	0.59782	1500	3.21320
69	Cyclopentane	C_5H_{10}	287-92-3	70.1329	0.41600	3.01400	1.46170	1.80950	668.8	100	0.41650	1500	2.92980
70	Cyclopentene	C_5H_8	142-29-0	68.11702	0.48074	2.51590	1.58030	1.74540	718.37	150	0.49182	1500	2.56190
71	Cyclopropane	C_3H_6	75-19-4	42.07974	0.33800	1.68940	1.61350	1.17680	722.8	100	0.33813	1500	1.72130
72	Cyclohexyl mercaptan	$C_6H_{12}S$	1569-69-3	116.22448	0.54305	3.99620	1.35750	2.56230	618.54	300	1.26440	1200	3.72360
73	Decanal	$C_{10}H_{20}O$	112-31-2	156.2652	1.94250	5.14030	1.89780	4.17520	859.95	298.15	2.37630	1500	6.04070
74	Decane	$C_{10}H_{22}$	124-18-5	142.28168	1.67200	5.35300	1.61410	3.78200	742	200	1.79670	1500	6.09320
75	Decanoic acid	$C_{10}H_{20}O_2$	334-48-5	172.265	0.24457	6.54600	1.08990	4.86420	424	298.15	2.52320	1500	6.10990
76	1-Decanol	$C_{10}H_{22}O$	112-30-1	158.28108	1.69840	5.39200	1.56800	3.93800	720.5	298.15	2.43540	1500	6.21860
77	1-Decene	$C_{10}H_{20}$	872-05-9	140.2658	1.71010	5.20890	1.72650	3.59350	782.92	298.15	2.23040	1500	5.87450
78	Decyl mercaptan	$C_{10}H_{22}S$	143-10-2	174.34668	1.93100	5.48150	1.60850	3.74000	754.75	200	2.04340	1500	6.46130
79	1-Decyne	$C_{10}H_{18}$	764-93-2	138.24992	1.50450	4.37940	1.32910	2.55570	632.01	298	2.19380	1500	5.27940
80	Deuterium	D_2	7782-39-0	4.0316	0.30290	0.09750	2.51500	−0.02750	368	100	0.30195	1500	0.34251
81	1,1-Dibromoethane	$C_2H_4Br_2$	557-91-5	187.86116	0.66622	0.81703	0.76285	0.40941	2488.3	298.15	0.79599	1500	1.56840
82	1,2-Dibromoethane	$C_2H_4Br_2$	106-93-4	187.86116	0.74906	1.27250	1.98100	0.94370	845.2	200	0.76345	1500	1.70410
83	Dibromomethane	CH_2Br_2	74-95-3	173.83458	0.39100	0.64800	1.19400	0.42000	501	100	0.39288	1500	0.95987
84	Dibutyl ether	$C_8H_{18}O$	142-96-1	130.22792	1.61220	4.47770	1.68310	2.91800	781.6	200	1.68410	1500	5.21450
85	*m*-Dichlorobenzene	$C_6H_4Cl_2$	541-73-1	147.00196	0.70000	2.07460	1.36640	1.59830	620.16	200	0.82450	1500	2.51610
86	*o*-Dichlorobenzene	$C_6H_4Cl_2$	95-50-1	147.00196	0.69480	2.08040	1.36320	1.59400	619.2	200	0.81978	1500	2.51610
87	*p*-Dichlorobenzene	$C_6H_4Cl_2$	106-46-7	147.00196	0.69780	2.07800	1.36350	1.59650	619.37	200	0.82283	1500	2.51750
88	1,1-Dichloroethane	$C_2H_4Cl_2$	75-34-3	98.95916	0.63412	0.83862	0.76898	0.44030	2533.2	298.15	0.76395	1500	1.56330
89	1,2-Dichloroethane	$C_2H_4Cl_2$	107-06-2	98.95916	0.65271	1.12540	1.73760	0.87800	795.45	200	0.67221	1500	1.57430
90	Dichloromethane	CH_2Cl_2	75-09-2	84.93258	0.36280	0.68040	1.25600	0.42750	548	100	0.36369	1500	0.95430
91	1,1-Dichloropropane	$C_3H_6Cl_2$	78-99-9	112.98574	0.71450	1.73440	1.52400	1.22300	674.2	150	0.72683	1500	2.16090
92	1,2-Dichloropropane	$C_3H_6Cl_2$	78-87-5	112.98574	0.78658	1.74290	1.71570	1.26270	765.1	200	0.82172	1500	2.18940
93	Diethanol amine	$C_4H_{11}NO_2$	111-42-2	105.13564	1.20800	3.06600	2.08900	2.34300	891	298.15	1.41970	1500.1	3.46740
94	Diethyl amine	$C_4H_{11}N$	109-89-7	73.13684	0.91020	2.67400	1.71900	1.79260	794.94	200	0.95017	1500	3.05190
95	Diethyl ether	$C_4H_{10}O$	60-29-7	74.1216	0.99953	1.70380	0.87072	1.07460	2471.3	298.15	1.16950	1500	2.92630
96	Diethyl sulfide	$C_4H_{10}S$	352-93-2	90.1872	0.91273	2.41000	1.66860	1.65200	771.08	200	0.95673	1500	2.87240
97	1,1-Difluoroethane	$C_2H_4F_2$	75-37-6	66.04997	0.55477	1.23610	0.83501	−0.40972	1033.4	298.15	0.67988	1500	1.54560
98	1,2-Difluoroethane	$C_2H_4F_2$	624-72-6	66.04997	0.57793	0.89811	0.84727	0.43249	2424.2	298.15	0.67730	1500	1.55140
99	Difluoromethane	CH_2F_2	75-10-5	52.02339	0.37540	0.53510	0.86687	0.22998	2437.2	298.15	0.42969	1500	0.94201
100	Diisopropyl amine	$C_6H_{15}N$	108-18-9	101.19	1.13840	2.57470	0.73840	1.62000	2143	300	1.59950	1500	4.19410
101	Diisopropyl ether	$C_6H_{14}O$	108-20-3	102.17476	1.09300	3.68300	1.60570	2.34200	699	298.15	1.56690	1500	4.05350
102	Diisopropyl ketone	$C_7H_{14}O$	565-80-0	114.18546	1.08690	4.05400	1.78020	2.97860	791.6	300	1.51020	1500	4.30930

103	1,1-Dimethoxyethane	$C_4H_{10}O_2$	534-15-6	90.121	1.15560	1.83050	0.95919	0.99605	2826.3	298.15	1.27770	1500	3.06780
104	1,2-Dimethoxypropane	$C_5H_{12}O_2$	7778-85-0	104.14758	1.01130	3.23930	1.56110	2.15010	689.3	298.15	1.46380	1500	3.66690
105	Dimethyl acetylene	C_4H_6	503-17-3	54.09044	0.65340	1.61790	1.78370	1.02420	821.4	200	0.67211	1500	1.91480
106	Dimethyl amine	C_2H_7N	124-40-3	45.08368	0.55650	1.63840	1.73410	1.08990	793.04	200	0.58115	1500	1.85850
107	2,3-Dimethylbutane	C_6H_{14}	79-29-8	86.17536	0.77720	4.03200	1.54400	2.50800	649.95	200	0.93628	1500	4.03530
108	1,1-Dimethylcyclohexane	C_8H_{16}	590-66-9	112.21264	1.07760	4.67180	1.65400	3.33970	792.5	200	1.15350	1500	4.95430
109	*cis*-1,2-Dimethylcyclohexane	C_8H_{16}	2207-01-4	112.21264	1.10390	4.64450	1.69430	3.39490	798.35	200	1.17770	1500	4.92430
110	*trans*-1,2-Dimethylcyclohexane	C_8H_{16}	6876-23-9	112.21264	1.09910	4.64010	1.66790	3.37360	781.97	200	1.18200	1500	4.92750
111	Dimethyl disulfide	$C_2H_6S_2$	624-92-0	94.19904	0.78430	1.43640	1.58360	0.87100	730.65	200	0.81551	1500	1.95230
112	Dimethyl ether	C_2H_6O	115-10-6	46.06844	0.57431	0.94494	0.89551	0.65065	2467.4	298.15	0.65866	1500	1.65840
113	*N,N*-Dimethyl formamide	C_3H_7NO	68-12-2	73.09378	0.72200	1.78300	1.53200	1.31000	762	200	0.75937	1500	2.25960
114	2,3-Dimethylpentane	C_7H_{16}	565-59-3	100.20194	0.85438	4.57720	1.51810	2.97400	641.01	200	1.05500	1500	4.59830
115	Dimethyl phthalate	$C_{10}H_{10}O_4$	131-11-3	194.184	1.39600	4.78000	2.19000	3.97050	900.6	300	1.74810	1200	4.47400
116	Dimethylsilane	C_2H_8Si	1111-74-6	60.17042	0.61453	1.74380	1.34180	1.01020	592.09	200	0.70950	1500	2.09440
117	Dimethyl sulfide	C_2H_6S	75-18-3	62.134	0.60370	1.37470	1.64100	0.79880	743.5	200	0.62976	1500	1.69490
118	Dimethyl sulfoxide	C_2H_6OS	67-68-5	78.13344	0.69490	1.52400	1.65140	1.06580	722.2	200	0.73547	1500	1.92550
119	Dimethyl terephthalate	$C_{10}H_{10}O_4$	120-61-6	194.184	1.14025	5.36801	2.08860	4.13440	809.837	298.15	1.67000	1500	4.97220
120	1,4-Dioxane	$C_4H_8O_2$	123-91-1	88.10512	0.68444	1.98020	0.82793	0.90830	2447.1	298.15	0.92284	1500	2.81860
121	Diphenyl ether	$C_{12}H_{10}O$	101-84-8	170.2072	1.09850	4.34120	1.62220	3.64550	743.62	300	1.72980	1200	4.51430
122	Dipropyl amine	$C_6H_{15}N$	142-84-7	101.19	1.21140	2.61270	0.78956	1.69030	2394.4	300	1.59000	1500	4.24840
123	Dodecane	$C_{12}H_{26}$	112-40-3	170.33484	2.12950	6.63300	1.71550	4.51610	777.5	200	2.24420	1500	7.43250
124	Eicosane	$C_{20}H_{42}$	112-95-8	282.54748	3.24810	11.09000	1.63600	7.45000	726.27	200	3.52350	1500	12.21100
125	Ethane	C_2H_6	74-84-0	30.069	0.44256	0.84737	0.87224	0.67130	2430.4	298.15	0.52652	1500	1.45610
126	Ethanol	C_2H_6O	64-17-5	46.06844	0.49200	1.45770	1.66280	0.93900	744.7	273.15	0.61172	1500	1.65760
127	Ethyl acetate	$C_4H_8O_2$	141-78-6	88.10512	0.99810	2.09310	2.02260	1.80300	928.05	200	1.01260	1500	2.65940
128	Ethyl amine	C_2H_7N	75-04-7	45.08368	0.59400	1.61800	1.81200	1.07800	820	200	0.61390	1500	1.85280
129	Ethylbenzene	C_8H_{10}	100-41-4	106.165	0.78440	3.39900	1.55900	2.42600	702	200	0.89121	1500	3.61470
130	Ethyl benzoate	$C_9H_{10}O_2$	93-89-0	150.1745	1.09440	4.17940	0.88375	−1.60900	1183.1	300	1.45980	1500	4.25400
131	2-Ethyl butanoic acid	$C_6H_{12}O_2$	88-09-5	116.15828	1.04550	2.31480	0.71000	1.47100	2061.6	300	1.51020	1200.15	3.63300
132	Ethyl butyrate	$C_6H_{12}O_2$	105-54-4	116.15828	1.11500	3.39100	1.67050	2.51800	733.6	298	1.55830	1200	3.62130
133	Ethylcyclohexane	C_8H_{16}	1678-91-7	112.21264	1.10590	4.63060	1.66280	3.29900	781.1	200	1.18750	1500	4.91840
134	Ethylcyclopentane	C_7H_{14}	1640-89-7	98.18606	0.93177	2.79330	0.78650	1.64590	2303.3	298.15	1.33350	1500	4.14000
135	Ethylene	C_2H_4	74-85-1	28.05316	0.33380	0.94790	1.59600	0.55100	740.8	60	0.33380	1500	1.09870
136	Ethylenediamine	$C_2H_8N_2$	107-15-3	60.09832	0.72860	1.84360	1.68800	1.19900	767.3	300	0.91775	1500	2.20160
137	Ethylene glycol	$C_2H_6O_2$	107-21-1	62.06784	0.63012	1.45840	1.67300	0.97296	773.65	300	0.77997	1500	1.80950
138	Ethyleneimine	C_2H_5N	151-56-4	43.0678	0.34300	1.42700	1.63800	1.03700	744.7	150	0.34798	1500	1.51780
139	Ethylene oxide	C_2H_4O	75-21-8	44.05256	0.33460	1.21160	1.60840	0.82410	737.3	50	0.33460	1500	1.32970
140	Ethyl formate	$C_3H_6O_2$	109-94-4	74.07854	0.53700	1.88600	1.20700	0.86400	496	100	0.54118	1500	2.14850
141	2-Ethyl hexanoic acid	$C_8H_{16}O_2$	149-57-5	144.211	1.57770	4.40170	1.74940	3.23780	792.34	298.15	2.02790	1500	5.12010
142	Ethylhexyl ether	$C_8H_{18}O$	5756-43-4	130.22792	1.63400	4.51190	1.75320	3.10320	809.75	298.15	2.03600	1200	4.87440
143	Ethylisopropyl ether	$C_5H_{12}O$	625-54-7	88.14818	1.09530	3.00320	1.79880	2.13110	817.35	298.15	1.36200	1200	3.22890
144	Ethylisopropyl ketone	$C_6H_{12}O$	565-69-5	100.15888	1.24000	3.20000	1.96700	2.34600	896	298.15	1.44790	1200	3.42340
145	Ethyl mercaptan	C_2H_6S	75-08-1	62.13404	0.60436	0.87524	0.78662	0.62622	−2190	298.15	0.73021	1500	1.66280
146	Ethyl propionate	$C_5H_{10}O_2$	105-37-3	102.1317	0.93700	2.82900	1.64800	2.15500	724.7	300	1.33770	1200	3.05690
147	Ethylpropyl ether	$C_5H_{12}O$	628-32-0	88.14818	1.13200	2.94000	1.82700	2.05500	852	298.15	1.35380	1500	3.45350
148	Ethyltrichlorosilane	$C_2H_5Cl_3Si$	115-21-9	163.506	0.96993	1.08780	0.70467	0.55556	2089.7	298.15	1.18910	1500	2.21700
149	Fluorine	F_2	7782-41-4	37.9968064	0.29122	0.10132	1.45300	0.09410	662.91	50	0.29122	1500	0.38122
150	Fluorobenzene	C_6H_5F	462-06-6	96.1023032	0.73393	2.37390	2.30860	2.45890	906.45	200	0.75730	1500	2.50800
151	Fluoroethane	C_2H_5F	353-36-6	48.0595	0.49090	0.88880	0.83107	0.54120	2446	298.15	0.59646	1500	1.49880
152	Fluoromethane	CH_3F	593-53-3	34.03292	0.35193	0.65344	1.13330	0.15240	5316.2	100	0.35193	6000	1.05710
153	Formaldehyde	CH_2O	50-00-0	30.02598	0.33503	0.49394	1.92800	0.29728	965.04	298.15	0.35440	1500	0.71121
154	Formamide	CH_3NO	75-12-7	45.04062	0.38220	0.93000	1.84500	0.69000	850	150	0.38326	1500	1.12030
155	Formic acid	CH_2O_2	64-18-6	46.0257	0.33810	0.75930	1.19250	0.31800	550	50	0.33810	1500	0.99328
156	Furan	C_4H_4O	110-00-9	68.07396	0.43673	1.28390	0.74699	0.47541	2500.6	298.15	0.65450	1500	1.79520
157	Helium-4	He	7440-59-7	4.0026	See Table 2-155								

(*Continued*)

TABLE 2-75 Heat Capacity at Constant Pressure of Inorganic and Organic Compounds in the Ideal Gas State Fit to Hyperbolic Functions C_p [J/(kmol·K)] *(Continued)*

Cmpd. no.	Name	Formula	CAS	Mol. wt.	$C_1 \times$ 1E-05	$C_2 \times$ 1E-05	$C_3 \times$ 1E-03	$C_4 \times$ 1E-05	C_5	T_{min}, K	C_p at $T_{min} \times$ 1E-05	T_{max}, K	C_p at $T_{max} \times$ 1E-05
158	Heptadecane	$C_{17}H_{36}$	629-78-7	240.46774	2.78780	9.52470	1.69350	6.66510	744.57	200	3.00340	1500	10.41600
159	Heptanal	$C_7H_{14}O$	111-71-7	114.18546	1.30930	3.53810	1.52500	2.23950	740.37	298.15	1.70230	1500	4.27590
160	Heptane	C_7H_{16}	142-82-5	100.20194	1.20150	4.00100	1.67660	2.74000	756.4	200	1.28280	1500	4.42830
161	Heptanoic acid	$C_7H_{14}O_2$	111-14-8	130.185	1.31350	2.33170	0.67567	1.82400	1846	300	1.84970	1500	4.29410
162	1-Heptanol	$C_7H_{16}O$	111-70-6	116.20134	1.22150	3.99100	1.58000	2.83500	717.7	298.15	1.75720	1500	4.53460
163	2-Heptanol	$C_7H_{16}O$	543-49-7	116.20134	1.41060	2.88580	0.80394	1.49680	2456.1	298.15	1.79590	1500	4.59900
164	3-Heptanone	$C_7H_{14}O$	106-35-4	114.18546	1.27680	3.38100	1.38310	1.88800	650.3	200	1.39680	1500	4.13860
165	2-Heptanone	$C_7H_{14}O$	110-43-0	114.18546	1.25070	2.14800	0.69120	1.61900	1759.3	150	1.26880	1200	3.84460
166	1-Heptene	C_7H_{14}	592-76-7	98.18606	1.18510	3.63620	1.73590	2.50480	785.73	298.15	1.54340	1500	4.08360
167	Heptyl mercaptan	$C_7H_{16}S$	1639-09-4	132.26694	1.44200	4.16030	1.66030	2.65720	759.39	200	1.51910	1500	4.78310
168	1-Heptyne	C_7H_{12}	628-71-7	96.17018	1.07120	3.02580	1.52730	2.09750	689.62	200	1.17210	1500	3.59850
169	Hexadecane	$C_{16}H_{34}$	544-76-3	226.44116	2.62830	8.97330	1.69120	6.26400	744.41	200	2.83120	1500	9.81820
170	Hexanal	$C_6H_{12}O$	66-25-1	100.15888	1.18400	3.07260	1.70770	2.11740	790.64	298.15	1.48160	1500	3.66440
171	Hexane	C_6H_{14}	110-54-3	86.17536	1.04400	3.52300	1.69460	2.36900	761.6	200	1.11170	1500	3.86200
172	Hexanoic acid	$C_6H_{12}O_2$	142-62-1	116.158	1.16220	2.07080	0.68661	1.53550	1932.5	298.15	1.61070	1500	3.76360
173	1-Hexanol	$C_6H_{14}O$	111-27-3	102.17476	1.06250	3.52100	1.58350	2.46200	715.75	298.15	1.53110	1500	3.97260
174	2-Hexanol	$C_6H_{14}O$	626-93-7	102.175	1.26150	3.59640	1.84450	2.59400	819.17	298.15	1.58290	1500	4.06720
175	2-Hexanone	$C_6H_{12}O$	591-78-6	100.15888	1.09400	1.80700	0.68900	1.47400	1772	200	1.18150	1200	3.32070
176	3-Hexanone	$C_6H_{12}O$	589-38-8	100.15888	1.12370	2.93600	1.40100	1.60100	650.5	150	1.14430	1500	3.58740
177	1-Hexene	C_6H_{12}	592-41-6	84.15948	1.04340	3.07490	1.74590	2.07280	793.53	298	1.33010	1500	3.48190
178	3-Hexyne	C_6H_{10}	928-49-4	82.1436	0.93760	3.01500	1.90570	1.98600	817	300	1.19090	1500	3.18890
179	Hexyl mercaptan	$C_6H_{14}S$	111-31-9	118.24036	1.26620	3.72940	1.65740	2.30800	757.8	200	1.33400	1500	4.24830
180	1-Hexyne	C_6H_{10}	693-02-7	82.1436	0.91290	2.55770	1.52900	1.73700	683	200	1.00040	1500	3.03710
181	2-Hexyne	C_6H_{10}	764-35-2	82.1436	1.03600	3.00900	2.11600	2.10600	902.4	300	1.22150	1500	3.18940
182	Hydrazine	H_4N_2	302-01-2	32.04516	0.41729	0.54686	0.81130	0.41755	2639.2	298.15	0.48803	1500	1.05830
183	Hydrogen	H_2	1333-74-0	2.01588	0.27617	0.09560	2.46600	0.03760	567.6	250	0.28426	1500	0.32248
184	Hydrogen bromide	BrH	10035-10-6	80.91194	0.29120	0.09530	2.14200	0.01570	1400	50	0.29120	1500	0.34786
185	Hydrogen chloride	ClH	7647-01-0	36.46094	0.29157	0.09048	2.09380	−0.00107	120	50	0.29137	1500	0.34063
186	Hydrogen cyanide	CHN	74-90-8	27.02534	0.30125	0.31710	1.61020	0.21790	626	100	0.30137	1500	0.55224
187	Hydrogen fluoride	FH	7664-39-3	20.0063432	0.29134	0.09325	2.90500	0.00195	1326	50	0.29134	1500	0.32243
188	Hydrogen sulfide	H_2S	7783-06-4	34.08088	0.33288	0.26086	0.91340	−0.17979	949.4	100	0.33288	1500	0.51432
189	Isobutyric acid	$C_4H_8O_2$	79-31-2	88.10512	0.74694	2.43560	1.71500	1.84840	757.75	298.15	1.04270	1200	2.53830
190	Isopropyl amine	C_3H_9N	75-31-0	59.11026	0.79534	1.44250	0.81831	0.95493	2499.9	298.15	0.97640	1500	2.45580
191	Malonic acid	$C_3H_4O_4$	141-82-2	104.06146	0.49522	1.87180	1.29580	1.48520	569.96	300	0.97903	1500	2.14970
192	Methacrylic acid	$C_4H_6O_2$	79-41-4	86.08924	0.72510	2.08900	1.85160	1.64830	798.43	298.15	0.94749	1200.1	2.20570
193	Methane	CH_4	74-82-8	16.0425	0.33298	0.79933	2.08690	0.41602	991.96	50	0.33298	1500	0.88904
194	Methanol	CH_4O	67-56-1	32.04186	0.39252	0.87900	1.91650	0.53654	896.7	273.15	0.42513	1500	1.05330
195	*N*-Methyl acetamide	C_3H_7NO	79-16-3	73.09378	0.61160	2.02900	1.76830	1.33020	835.5	300	0.76980	1500	2.22090
196	Methyl acetate	$C_3H_6O_2$	79-20-9	74.07854	0.55500	1.78200	1.26000	0.85300	562	298	0.84891	1500	2.07540
197	Methyl acetylene	C_3H_4	74-99-7	40.06386	0.51734	0.68157	0.80525	0.51402	2463.8	298.15	0.60784	1500	1.33000
198	Methyl acrylate	$C_4H_6O_2$	96-33-3	86.08924	0.12060	2.37660	1.05430	1.81860	418.8	298.15	0.99083	1200.1	2.16630
199	Methyl amine	CH_5N	74-89-5	31.0571	0.41000	1.05780	1.70800	0.68360	735	150	0.41364	1500	1.23880
200	Methyl benzoate	$C_8H_8O_2$	93-58-3	136.14792	0.93960	2.55900	0.82500	1.36000	3000	300	1.25860	1200	3.35690
201	3-Methyl-1,2-butadiene	C_5H_8	598-25-4	68.11702	0.67100	2.22200	1.42100	1.19400	614.7	150	0.69311	1500	2.50280
202	2-Methylbutane	C_5H_{12}	78-78-4	72.14878	0.74600	3.26500	1.54500	1.92300	666.7	200	0.85462	1500	3.37920
203	2-Methylbutanoic acid	$C_5H_{10}O_2$	116-53-0	102.1317	1.84580	1.74300	1.22000	−56.11000	31.2	300	1.27930	1500	3.22620
204	3-Methyl-1-butanol	$C_5H_{12}O$	123-51-3	88.1482	0.92139	3.33710	1.83610	2.46440	757.83	298.15	1.31350	1500	3.48560
205	2-Methyl-1-butene	C_5H_{10}	563-46-2	70.1329	0.87026	2.55560	1.77570	1.76360	807.82	200	0.90596	1500	2.89230
206	2-Methyl-2-butene	C_5H_{10}	513-35-9	70.1329	0.81924	2.60380	1.75930	1.71950	800.93	200	0.85589	1500	2.87090
207	2-Methyl -1-butene-3-yne	C_5H_6	78-80-8	66.10114	0.79060	1.65600	1.69260	1.21670	788.4	298.15	0.96319	1500.15	2.15020
208	Methylbutyl ether	$C_5H_{12}O$	628-28-4	88.14818	0.82051	3.08690	1.38640	1.78860	613.87	300	1.33000	1200	3.19940
209	Methylbutyl sulfide	$C_5H_{12}S$	628-29-5	104.214	1.07850	2.73880	1.58850	1.90670	749.6	273.15	1.31730	1200	3.16870

210	3-Methyl-1-butyne	C_5H_8	598-23-2	68.11702	0.82740	2.13770	1.75500	1.51490	782	200	0.86459	1500	2.52550
211	Methyl butyrate	$C_5H_{10}O_2$	623-42-7	102.1317	0.89400	2.91000	1.57000	2.07300	678.3	298	1.34610	1200	3.07660
212	Methylchlorosilane	CH_5ClSi	993-00-0	80.5889	0.59895	1.16360	1.56500	0.81581	690.39	200	0.63795	1500	1.55930
213	Methylcyclohexane	C_7H_{14}	108-87-2	98.18606	0.92270	4.11500	1.65040	2.90060	779.48	200	0.99530	1500	4.31800
214	1-Methylcyclohexanol	$C_7H_{14}O$	590-67-0	114.18546	0.79590	2.59600	0.62130	2.28800	1698.6	300	1.53020	1200	4.13590
215	*cis*-2-Methylcyclohexanol	$C_7H_{14}O$	7443-70-1	114.18546	0.92279	2.67090	0.68784	1.98470	1732.4	300	1.50990	1200	4.14670
216	*trans*-2-Methylcyclohexanol	$C_7H_{14}O$	7443-52-9	114.18546	0.92279	2.67090	0.68784	1.98470	1732.4	300	1.50990	1200	4.14670
217	Methylcyclopentane	C_6H_{12}	96-37-7	84.15948	0.78439	2.50070	0.81937	1.30010	2416.4	298.15	1.09680	1500	3.54830
218	1-Methylcyclopentene	C_6H_{10}	693-89-0	82.1436	0.69411	3.02090	1.69030	2.12090	781.56	200	0.74637	1500	3.14960
219	3-Methylcyclopentene	C_6H_{10}	1120-62-3	82.1436	0.64220	3.07110	1.63870	2.12980	750.25	200	0.70833	1500	3.15490
220	Methyldichlorosilane	CH_4Cl_2Si	75-54-7	115.03396	0.72830	1.03070	1.54290	0.78110	668.94	200	0.77172	1500	1.58930
221	Methylethyl ether	C_3H_8O	540-67-0	60.09502	0.79188	1.31660	0.87136	0.86597	2468	298.15	0.92283	1500	2.29440
222	Methylethyl ketone	C_4H_8O	78-93-3	72.10572	0.78400	2.10320	1.54880	1.18550	693	200	0.83967	1500	2.48160
223	Methylethyl sulfide	C_3H_8S	624-89-5	76.1606	0.75083	1.95770	1.64240	1.19490	749.19	273.16	0.90040	1500	2.31780
224	Methyl formate	$C_2H_4O_2$	107-31-3	60.05196	0.50600	1.21900	1.63700	0.89400	743	250	0.58880	1500	1.51090
225	Methylisobutyl ether	$C_5H_{12}O$	625-44-5	88.14818	0.72840	3.17130	1.35200	1.89480	585.14	300	1.32000	1200	3.19870
226	Methylisobutyl ketone	$C_6H_{12}O$	108-10-1	100.15888	1.22700	2.19500	0.84200	1.19100	2460	298.15	1.47550	1500.15	3.65320
227	Methyl Isocyanate	C_2H_3NO	624-83-9	57.05132	0.47400	1.22600	2.18800	0.85983	1008.2	298.15	0.51946	1500	1.35950
228	Methylisopropyl ether	$C_4H_{10}O$	598-53-8	74.1216	0.89232	2.47650	1.69600	1.55980	791.4	200	0.92804	1500	2.86960
229	Methylisopropyl ketone	$C_5H_{10}O$	563-80-4	86.1323	1.59140	1.76400	1.20760	−407.40000	10.503	300	1.12910	1500	2.99910
230	Methylisopropyl sulfide	$C_4H_{10}S$	1551-21-9	90.1872	0.99247	2.72750	2.00300	1.89740	849.64	273	1.13770	1500	2.99520
231	Methyl mercaptan	CH_4S	74-93-1	48.10746	0.43697	0.50387	0.80924	0.42223	2192.4	298.15	0.50277	1500	1.06940
232	Methyl methacrylate	$C_5H_8O_2$	80-62-6	100.11582	0.86400	1.81100	0.75430	0.80000	2160	298.15	1.16210	1500	2.86370
233	2-Methyloctanoic acid	$C_9H_{18}O_2$	3004-93-1	158.23802	1.74830	4.92880	1.73840	3.58970	788.01	298.15	2.25670	1500	5.71770
234	2-Methylpentane	C_6H_{14}	107-83-5	86.17536	0.90300	3.80100	1.60200	2.45300	691.6	200	1.01920	1500	3.96170
235	Methyl pentyl ether	$C_6H_{14}O$	628-80-8	102.17476	0.94326	3.59650	1.35330	2.05690	599.92	300	1.56000	1200	3.74090
236	2-Methylpropane	C_4H_{10}	75-28-5	58.1222	0.76394	1.68020	0.82654	1.02850	2483.1	298.15	0.96745	1500	2.66680
237	2-Methyl-2-propanol	$C_4H_{10}O$	75-65-0	74.1216	0.90658	1.71370	0.80201	1.04240	2489.7	298.15	1.13730	1500	2.85290
238	2-Methyl propene	C_4H_8	115-11-7	56.10632	0.73226	1.36060	0.84872	0.88667	2499.8	298.15	0.88184	1500	2.28420
239	Methyl propionate	$C_4H_8O_2$	554-12-1	88.10512	0.77650	2.44200	1.71400	1.81800	716	300	1.12420	1200	2.52760
240	Methylpropyl ether	$C_4H_{10}O$	557-17-5	74.1216	0.92151	2.39430	1.69360	1.48960	797.79	298	1.12510	1200	2.63910
241	Methylpropyl sulfide	$C_4H_{10}S$	3877-15-4	90.1872	0.93775	2.61780	1.72910	1.62360	783.23	298.15	1.17280	1500	2.99040
242	Methylsilane	CH_6Si	992-94-9	46.14384	0.46149	1.27810	1.45650	0.79115	643.23	200	0.51411	1500	1.52530
243	alpha-Methyl styrene	C_9H_{10}	98-83-9	118.1757	1.00010	2.65370	0.77176	1.11620	2405.2	298.15	1.40620	1500	3.86080
244	Methyl tert-butyl ether	$C_5H_{12}O$	1634-04-4	88.1482	0.98059	3.08940	1.64560	2.09850	732.6	298.15	1.35330	1500	3.47810
245	Methyl vinyl ether	C_3H_6O	107-25-5	58.07914	0.60865	1.59650	1.61900	0.93783	739.55	300	0.77480	1500	1.88710
246	Naphthalene	$C_{10}H_8$	91-20-3	128.17052	0.89232	2.67720	0.76122	1.02010	2435.5	298.15	1.32040	1500	3.73860
247	Neon	Ne	7440-01-9	20.1797	See Table 2-155								
248	Nitroethane	$C_2H_5NO_2$	79-24-3	75.0666	0.64084	1.16310	0.80970	0.59591	2425.6	298.15	0.79235	1500	1.92450
249	Nitrogen	N_2	7727-37-9	28.0134	0.29105	0.08615	1.70160	0.00103	909.79	50	0.29105	1500	0.34838
250	Nitrogen trifluoride	F_3N	7783-54-2	71.00191	0.33284	0.49837	0.70930	0.23264	372.91	100	0.34036	1500	0.80919
251	Nitromethane	CH_3NO_2	75-52-5	61.04002	0.47876	0.78357	0.82960	0.37215	2433.8	298.15	0.57242	1500	1.32860
252	Nitrous oxide	N_2O	10024-97-2	44.0128	0.29338	0.32360	1.12380	0.21770	479.4	100	0.29475	1500	0.58278
253	Nitric oxide	NO	10102-43-9	30.0061	See Table 2-155								
254	Nonadecane	$C_{19}H_{40}$	629-92-5	268.5209	3.10620	10.57500	0.76791	−4.56610	912.03	200	3.35330	1500	11.61300
255	Nonanal	$C_9H_{18}O$	124-19-6	142.23862	1.71190	4.50580	1.71000	3.36580	807.38	298.15	2.15310	1500	5.42420
256	Nonane	C_9H_{20}	111-84-2	128.2551	1.51750	4.91500	1.64480	3.47000	749.6	200	1.62570	1500	5.54070
257	Nonanoic acid	$C_9H_{18}O_2$	112-05-0	158.238	0.12660	6.01100	1.08150	4.59460	418.2	298.15	2.29530	1500	5.52670
258	1-Nonanol	$C_9H_{20}O$	143-08-8	144.2545	1.54000	4.93600	1.57800	3.58800	721.11	298.15	2.20920	1500	5.66060
259	2-Nonanol	$C_9H_{20}O$	628-99-9	144.255	1.81180	3.59270	0.81841	2.17920	2550.1	298.15	2.26250	1500	5.85550
260	1-Nonene	C_9H_{18}	124-11-8	126.23922	1.53520	4.68440	1.72880	3.23040	783.67	298.15	2.00140	1500	5.27760
261	Nonyl mercaptan	$C_9H_{20}S$	1455-21-6	160.3201	1.76460	5.04400	1.61820	3.38570	755.48	200	1.86580	1500	5.90820
262	1-Nonyne	C_9H_{16}	3452-09-3	124.22334	1.62890	3.97080	1.89280	3.21360	855.52	298.15	1.96930	1500	4.79240
263	Octadecane	$C_{18}H_{38}$	593-45-3	254.49432	2.95020	10.03400	0.77107	−4.30120	916.73	200	3.18000	1500	11.01600

(Continued)

TABLE 2-75 Heat Capacity at Constant Pressure of Inorganic and Organic Compounds in the Ideal Gas State Fit to Hyperbolic Functions C_p [J/(kmol·K)] *(Continued)*

Cmpd. no.	Name	Formula	CAS	Mol. wt.	$C_1 \times$ 1E-05	$C_2 \times$ 1E-05	$C_3 \times$ 1E-03	$C_4 \times$ 1E-05	C_5	T_{min}, K	C_p at $T_{min} \times$ 1E-05	T_{max}, K	C_p at $T_{max} \times$ 1E-05
264	Octanal	$C_8H_{16}O$	124-13-0	128.212	1.59550	3.14670	0.85788	1.47130	2679.4	298.15	1.92770	1500	4.91940
265	Octane	C_8H_{18}	111-65-9	114.22852	1.35540	4.43100	1.63560	3.05400	746.4	200	1.45290	1500	4.97640
266	Octanoic acid	$C_8H_{16}O_2$	124-07-2	144.211	1.40820	4.34360	1.46620	2.76870	659.38	298.15	2.06520	1500	5.04110
267	1-Octanol	$C_8H_{18}O$	111-87-5	130.22792	1.38050	4.45900	1.57510	3.20160	718.8	298.15	1.98320	1500	5.09650
268	2-Octanol	$C_8H_{18}O$	123-96-6	130.228	1.58030	3.23480	0.79814	1.78820	2434.3	298.15	2.02310	1500	5.20600
269	2-Octanone	$C_8H_{16}O$	111-13-7	128.21204	1.39010	3.80600	1.37170	2.25730	660.96	150	1.41620	1500	4.65470
270	3-Octanone	$C_8H_{16}O$	106-68-3	128.21204	1.49520	4.41030	0.80211	−2.09580	981.95	200	1.57750	1500	4.90670
271	1-Octene	C_8H_{16}	111-66-0	112.21264	1.35990	4.16050	1.73170	2.86750	784.47	298.15	1.77230	1500	4.68070
272	Octyl mercaptan	$C_8H_{18}S$	111-88-6	146.29352	1.59810	4.60630	1.62950	3.03010	756.28	200	1.68810	1500	5.35490
273	1-Octyne	C_8H_{14}	629-05-0	110.19676	1.23070	3.49420	1.52800	2.46170	694.81	200	1.34480	1500	4.16040
274	Oxalic acid	$C_2H_2O_4$	144-62-7	90.03488	0.56777	1.11940	0.62070	−0.38079	676.72	298.15	0.79711	1500	1.56180
275	Oxygen	O_2	7782-44-7	31.9988	0.29103	0.10040	2.52650	0.09356	1153.8	50	0.29103	1500	0.36533
276	Ozone	O_3	10028-15-6	47.9982	0.33483	0.29577	1.52170	0.27151	680.35	100	0.33489	1500	0.59282
277	Pentadecane	$C_{15}H_{32}$	629-62-9	212.41458	2.46790	8.42120	1.68650	5.85370	743.6	200	2.65860	1500	9.22090
278	Pentanal	$C_5H_{10}O$	110-62-3	86.1323	1.06000	2.85000	1.93000	2.01000	879.23	298.15	1.25200	1500	3.24590
279	Pentane	C_5H_{12}	109-66-0	72.14878	0.88050	3.01100	1.65020	1.89200	747.6	200	0.94039	1500	3.29270
280	Pentanoic acid	$C_5H_{10}O_2$	109-52-4	102.132	2.83600	1.08000	2.10700	−3.56000	283	298.15	1.38240	1500	3.29520
281	1-Pentanol	$C_5H_{12}O$	71-41-0	88.1482	0.90600	3.06200	1.60540	2.11500	717.97	298.15	1.30440	1500	3.41330
282	2-Pentanol	$C_5H_{12}O$	6032-29-7	88.1482	1.08530	3.07470	1.86720	2.22710	825.4	298.15	1.35390	1500	3.47010
283	2-Pentanone	$C_5H_{10}O$	107-87-9	86.1323	0.90053	2.70850	1.65920	1.80120	743.96	200	0.95908	1500	3.07970
284	3-Pentanone	$C_5H_{10}O$	96-22-0	86.1323	0.96896	2.49070	1.41770	1.30100	646.7	200	1.05360	1500	3.03580
285	1-Pentene	C_5H_{10}	109-67-1	70.1329	0.82523	2.59430	1.72910	1.76800	778.7	298.15	1.08560	1500	2.88970
286	2-Pentyl mercaptan	$C_5H_{12}S$	2084-19-7	104.21378	1.13270	2.94700	1.74180	2.09870	795.78	298	1.42020	1500	3.49940
287	Pentyl mercaptan	$C_5H_{12}S$	110-66-7	104.21378	1.09740	3.29590	1.67610	1.94860	757.67	200	1.15470	1500	3.69560
288	1-Pentyne	C_5H_8	627-19-0	68.11702	0.75300	2.09050	1.53070	1.37800	672.8	200	0.82759	1500	2.47540
289	2-Pentyne	C_5H_8	627-21-4	68.11702	0.82096	1.46770	0.84463	0.96258	2452.3	298.15	0.98524	1500	2.50600
290	Phenanthrene	$C_{14}H_{10}$	85-01-8	178.2292	1.27200	3.56890	0.75021	1.32990	2409.4	298.15	1.86940	1500	5.06820
291	Phenol	C_6H_6O	108-95-2	94.11124	0.43400	2.44500	1.15200	1.51200	507	100	0.44014	1500	2.60450
292	Phenyl isocyanate	C_7H_5NO	103-71-9	119.1207	0.59683	2.55330	1.23970	1.55190	576.78	298.15	1.10540	1500	2.83900
293	Phthalic anhydride	$C_8H_4O_3$	85-44-9	148.11556	0.73640	2.54400	1.08520	0.80800	573	298.15	1.07450	1000.15	2.67370
294	Propadiene	C_3H_4	463-49-0	40.06386	0.48308	0.73665	0.78152	0.48698	2480	298.15	0.59127	1500	1.33810
295	Propane	C_3H_8	74-98-6	44.09562	0.59474	1.26610	0.84431	0.86165	2482.7	298.15	0.73665	1500	2.05600
296	1-Propanol	C_3H_8O	71-23-8	60.09502	0.61900	2.02130	1.62930	1.29560	727.4	298.15	0.85428	1500	2.24580
297	2-Propanol	C_3H_8O	67-63-0	60.095	0.73145	2.03130	1.93750	1.48150	843.37	298.15	0.89664	1500	2.27600
298	Propenylcyclohexene	C_9H_{14}	13511-13-2	122.20746	1.05630	4.33970	1.60980	3.18100	729.66	300	1.63920	1500	4.65270
299	Propionaldehyde	C_3H_6O	123-38-6	58.07914	0.71306	1.16890	0.92731	1.02100	2512.8	298.15	0.80337	1500	2.11890
300	Propionic acid	$C_3H_6O_2$	79-09-4	74.0785	0.69590	1.77780	1.70980	1.26540	763.78	298.15	0.89382	1500	2.12480
301	Propionitrile	C_3H_5N	107-12-0	55.0785	0.52525	1.46630	1.54760	0.93033	674.15	298.15	0.73244	1500	1.72030
302	Propyl acetate	$C_5H_{10}O_2$	109-60-4	102.1317	1.79940	1.75300	1.19600	−4.12000	108.2	298.15	1.35940	1500	3.20240
303	Propyl amine	C_3H_9N	107-10-8	59.11026	0.76078	2.10490	1.72560	1.39360	789.03	200	0.79326	1500	2.43530
304	Propylbenzene	C_9H_{12}	103-65-1	120.19158	1.13460	2.80980	0.79504	1.23760	2449.5	298.15	1.52430	1500	4.16280
305	Propylene	C_3H_6	115-07-1	42.07974	0.43852	1.50600	1.39880	0.74754	616.46	130	0.44363	1500	1.68170

306	Propyl formate	$C_4H_8O_2$	110-74-7	88.10512	0.87100	2.44700	1.92540	1.88800	821.3	298.15	1.10220	1500	2.74840
307	2-Propyl mercaptan	C_3H_8S	75-33-2	76.16062	0.73815	1.95290	1.59540	1.23560	730.5	200	0.78247	1500	2.32870
308	Propyl mercaptan	C_3H_8S	107-03-9	76.16062	0.74740	1.95230	1.63100	1.21120	750.92	200	0.78483	1500	2.32160
309	1,2-Propylene glycol	$C_3H_8O_2$	57-55-6	76.09442	2.01140	0.80820	1.86560	−2.44040	279.98	298.15	1.02180	1000.15	2.11750
310	Quinone	$C_6H_4O_2$	106-51-4	108.09476	0.80992	1.57510	0.74707	0.60196	2344.9	298.15	1.07700	1500	2.49790
311	Silicon tetrafluoride	F_4Si	7783-61-1	104.07911	0.36810	0.71245	0.65201	0.46721	286.03	100	0.41815	1500	1.05370
312	Styrene	C_8H_8	100-42-5	104.14912	0.89300	2.15030	0.77200	0.99900	2442	100	0.89310	1500	3.24160
313	Succinic acid	$C_4H_6O_4$	110-15-6	118.08804	0.71806	2.26690	1.27390	1.73420	537.65	300	1.33700	1200	2.58230
314	Sulfur dioxide	O_2S	7446-09-5	64.0638	0.33375	0.25864	0.93280	0.10880	423.7	100	0.33538	1500	0.56950
315	Sulfur hexafluoride	F_6S	2551-62-4	146.0554192	0.35256	1.22700	0.67938	0.78407	351.27	100	0.38719	1500	1.53970
316	Sulfur trioxide	O_3S	7446-11-9	80.0632	0.33408	0.49677	0.87322	0.28563	393.74	100	0.34081	1500	0.79673
317	Terephthalic acid	$C_8H_6O_4$	100-21-0	166.13084	1.00130	2.61780	0.87239	1.28310	3521.5	298.15	1.26040	1500	3.59670
318	*o*-Terphenyl	$C_{18}H_{14}$	84-15-1	230.30376	2.07190	6.26680	2.40440	6.34500	967.71	298.15	2.47630	1500	6.69470
319	Tetradecane	$C_{14}H_{30}$	629-59-4	198.388	2.30820	7.86780	1.68230	5.44860	743.1	200	2.48640	1500	8.62250
320	Tetrahydrofuran	C_4H_8O	109-99-9	72.10572	0.54850	1.84910	0.83310	0.89089	2458.5	298.15	0.76617	1500	2.55380
321	1,2,3,4-Tetrahydronaphthalene	$C_{10}H_{12}$	119-64-2	132.20228	1.05550	3.21010	0.78248	1.43950	2433	298.15	1.52510	1500	4.53700
322	Tetrahydrothiophene	C_4H_8S	110-01-0	88.17132	0.65341	1.71150	0.77705	0.91824	2432.6	298.15	0.90956	1500	2.56890
323	2,2,3,3-Tetramethylbutane	C_8H_{18}	594-82-1	114.22852	1.13520	5.63310	1.62110	3.38290	681.9	200	1.30690	1500	5.57840
324	Thiophene	C_4H_4S	110-02-1	84.13956	0.48694	1.23760	0.71271	0.47248	2484.2	298.15	0.72827	1500	1.81130
325	Toluene	C_7H_8	108-88-3	92.13842	0.58140	2.86300	1.44060	1.89800	650.43	200	0.70157	1500	3.00290
326	1,1,2-Trichloroethane	$C_2H_3Cl_3$	79-00-5	133.40422	0.66554	1.12570	1.54540	0.97196	717.04	298.15	0.84963	1500	1.64330
327	Tridecane	$C_{13}H_{28}$	629-50-5	184.36142	2.14960	7.30450	1.66950	4.99980	741.02	200	2.31560	1500	8.02510
328	Triethyl amine	$C_6H_{15}N$	121-44-8	101.19	1.27660	2.55590	0.80937	1.48290	2231.7	200	1.32780	1500	4.20460
329	Trimethyl amine	C_3H_9N	75-50-3	59.11026	0.71070	1.50510	0.79662	0.84537	2187.6	200	0.74387	1500	2.43220
330	1,2,3-Trimethylbenzene	C_9H_{12}	526-73-8	120.19158	1.05200	3.79000	1.48140	2.33100	667.3	200	1.18320	1500	4.19830
331	1,2,4-Trimethylbenzene	C_9H_{12}	95-63-6	120.19158	1.22100	2.68650	0.82886	1.42030	2443	298.15	1.54310	1500	4.18780
332	2,2,4-Trimethylpentane	C_8H_{18}	540-84-1	114.22852	1.13900	5.28600	1.59400	3.35100	677.94	200	1.31390	1500	5.37690
333	2,3,3-Trimethylpentane	C_8H_{18}	560-21-4	114.22852	0.98200	5.40200	1.53100	3.49300	639.9	200	1.21940	1500	5.37540
334	1,3,5-Trinitrobenzene	$C_6H_3N_3O_6$	99-35-4	213.10452	2.03670	1.81810	1.20890	0.79777	1060.8	298.15	2.10540	1500	3.75850
335	2,4,6-Trinitrotoluene	$C_7H_5N_3O_6$	118-96-7	227.1311	2.15400	2.44320	1.11260	0.58651	950.59	298.15	2.27260	1500	4.35600
336	Undecane	$C_{11}H_{24}$	1120-21-4	156.30826	1.95290	6.09980	1.70870	4.13020	775.4	200	2.05940	1500	6.83420
337	1-Undecanol	$C_{11}H_{24}O$	112-42-5	172.30766	1.85900	5.86900	1.57180	4.32600	722.7	298.15	2.66140	1500	6.78340
338	Vinyl acetate	$C_4H_6O_2$	108-05-4	86.08924	0.53600	2.11900	1.19800	1.14700	510	100	0.54044	1500	2.37500
339	Vinyl acetylene	C_4H_4	689-97-4	52.07456	0.55978	1.21410	1.61020	0.89079	710.4	200	0.59670	1500	1.55900
340	Vinyl chloride	C_2H_3Cl	75-01-4	62.49822	0.42364	0.87350	1.64920	0.65560	739.07	200	0.44572	1500	1.14230
341	Vinyl trichlorosilane	$C_2H_3Cl_3Si$	75-94-5	161.48972	0.84894	1.14710	1.38000	0.90000	644.61	298.15	1.07540	1500	1.85950
342	Water	H_2O	7732-18-5	18.01528	0.33363	0.26790	2.61050	0.08896	1169	100	0.33363	2273.15	0.52760
343	*m*-Xylene	C_8H_{10}	108-38-3	106.165	0.75680	3.39240	1.49600	2.24700	675.9	200	0.87588	1500	3.59200
344	*o*-Xylene	C_8H_{10}	95-47-6	106.165	0.85210	3.29540	1.49440	2.11500	675.8	200	0.96428	1500	3.59650
345	*p*-Xylene	C_8H_{10}	106-42-3	106.165	0.75120	3.39700	1.49280	2.24700	675.1	200	0.87096	1500	3.59230

Constants in this table can be used in the following equation to calculate the ideal gas heat capacity C^0_p. $C^0_p = C_1 + C_2[C_3/T/\sinh(C_3/T)]^2 + C_4[C_5/T/\cosh(C_5/T)]^2$ where C^0_p is in J/(kmol·K) and T is in K.

Values in this table were taken from the Design Institute for Physical Properties (DIPPR) of the American Institute of Chemical Engineers (AIChE), 801 Critically Evaluated Gold Standard™ Database, copyright 2016 AIChE, and reproduced with permission of AIChE and of the DIPPR Evaluated Process Design Data Project Steering Committee. Their source should be cited as "R. L. Rowley, W. V. Wilding, J. L. Oscarson, T. A. Knotts, N. F. Giles, *DIPPR® Data Compilation of Pure Chemical Properties*, Design Institute for Physical Properties, AIChE, New York, NY (2016)".

TABLE 2-76 C_p/C_v: Ratios of Specific Heats of Gases at 1 atm Pressure*

Compound	Formula	Temperature, °C	Ratio of specific heats, $(\gamma) = C_p/C_v$	Compound	Formula	Temperature, °C	Ratio of specific heats, $(\gamma) = C_p/C_v$
Acetaldehyde	C_2H_4O	30	1.14	Hydrogen (*Cont.*)			
Acetic acid	$C_2H_4O_2$	136	1.15	iodide	HI	20–100	1.40
Acetylene	C_2H_2	15	1.26	sulfide	H_2S	15	1.332
		−71	1.31			−45	1.350
Air		925	1.36			−57	1.356
		17	1.403				
		−78	1.408	Iodine	I_2	185	1.30
		−118	1.415	Isobutane	C_4H_{10}	15	1.110
Ammonia	NH_3	15	1.320				
Argon	Ar	15	1.670	Krypton	Kr	19	1.672
		−180	1.715				
		0–100	1.67	Mercury	Hg	360	1.67
				Methane	CH_4	600	1.113
Benzene	C_6H_6	90	1.10			300	1.196
Bromine	Br_2	20–350	1.32			15	1.310
						−80	1.339
Carbon dioxide	CO_2	15	1.299			−115	1.347
		−75	1.37	Methyl acetate	$C_3H_6O_2$	15	1.14
disulfide	CS_2	100	1.21	alcohol	CH_4O	77	1.237
monoxide	CO	15	1.402	ether	C_2H_6O	6–30	1.11
		−180	1.433	Methylal	$C_3H_8O_2$	13	1.06
Chlorine	Cl_2	15	1.355			40	1.09
Chloroform	$CHCl_3$	100	1.15				
Cyanogen	$(CN)_2$	15	1.256	Neon	Ne	19	1.667
Cyclohexane	C_6H_{12}	80	1.315	Nitric oxide	NO	15	1.400
						−45	1.39
Dichlorodifluormethane	CCl_2F_2	25	1.139			−80	1.38
				Nitrogen	N_2	15	1.402
Ethane	C_2H_6	100	1.157			−181	1.433
		15	1.200	Nitrous oxide	N_2O	100	1.28
		−82	1.28			15	1.303
Ethyl alcohol	C_2H_6O	90	1.13			−30	1.31
ether	$C_4H_{10}O$	35	1.08			−70	1.34
		80	1.086				
Ethylene	C_2H_4	100	1.201	Oxygen	O_2	15	1.398
		15	1.253			−76	1.405
		−91	1.345			−181	1.439
Helium	He	−180	1.667	Pentane (*n*-)	C_5H_{12}	86	1.071
Hexane (*n*-)	C_6H_{14}	80	1.066	Phosphorus	P	300	1.17
Hydrogen	H_2	15	1.407	Potassium	K	850	1.77
		−76	1.441				
		−181	1.607	Sodium	Na	750–920	1.68
bromide	HBr	20	1.42	Sulfur dioxide	SO_2	15	1.290
chloride	HCl	15	1.41				
		100	1.40	Xenon	Xe	19	1.678
cyanide	HCN	65	1.31				
		140	1.28				
		210	1.24				

*For compounds that appear in Tables 2-109 to 2-122, values are from E. W. Lemmon, M. O. McLinden, and D. G. Friend, "Thermophysical Properties of Fluid Systems" in *NIST Chemistry WebBook*, **NIST Standard Reference Database Number 69**, Eds. P. J. Linstrom and W. G. Mallard, June 2005, National Institute of Standards and Technology, Gaithersburg, Md. (http://webbook.nist.gov). Values for other compounds are from *International Critical Tables*, vol. 5, pp. 80–82.

SPECIFIC HEATS OF AQUEOUS SOLUTIONS

Additional References Most of the tables below are from *International Critical Tables*, vol. 5, pp. 115–116, 122–125. Specific heats for other compounds in aqueous solution can also be found in the same reference.

TABLE 2-77 Acetic Acid (at 38°C)

Mole % acetic acid	0	6.98	30.9	54.5	100
Cal/(g·°C)	1.0	0.911	0.73	0.631	0.535

TABLE 2-78 Ammonia

Mole % NH_3	Specific heat, cal/(g·°C) 2.4°C	20.6°C	41°C	61°C
0	1.01	1.0	0.995	1.0
10.5	0.98	0.995	1.06	1.02
20.9	0.96	0.99	1.03	
31.2	0.956	1.0		
41.4	0.985			

TABLE 2-79 Ethyl Alcohol

Mole % C_2H_5OH	Specific heat, cal/(g·°C) 3°C	23°C	41°C
4.16	1.05	1.02	1.02
11.5	1.02	1.03	1.03
37.0	0.805	0.86	0.875
61.0	0.67	0.727	0.748
100.0	0.54	0.577	0.621

TABLE 2-80 Glycerol

Mole % $C_3H_5(OH)_3$	Specific heat, cal/(g·°C) 15°C	32°C
2.12	0.961	0.960
4.66	0.929	0.924
11.5	0.851	0.841
22.7	0.765	0.758
43.9	0.67	0.672
100.0	0.555	0.576

TABLE 2-81 Hydrochloric Acid

Mole % HCl	Specific heat, cal/(g·°C) 0°C	10°C	20°C	40°C	60°C
0.0	1.00				
9.09	0.72	0.72	0.74	0.75	0.78
16.7	0.61	0.605	0.631	0.645	0.67
20.0	0.58	0.575	0.591	0.615	0.638
25.9	0.55				0.61

TABLE 2-82 Methyl Alcohol

Mole % CH_3OH	Specific heat, cal/(g·°C) 5°C	20°C	40°C
5.88	1.02	1.0	0.995
12.3	0.975	0.982	0.98
27.3	0.877	0.917	0.92
45.8	0.776	0.811	0.83
69.6	0.681	0.708	0.726
100	0.576	0.60	0.617

TABLE 2-83 Nitric Acid

% HNO_3 by Weight	Specific heat at 20°C, cal/(g·°C)
0	1.000
10	0.900
20	0.810
30	0.730
40	0.675
50	0.650
60	0.640
70	0.615
80	0.575
90	0.515

TABLE 2-84 Phosphoric Acid*

%H_3PO_4	C_p at 21.3°C cal/(g·°C)	%H_3PO_4	C_p at 21.3°C cal/(g·°C)
2.50	0.9903	50.00	0.6350
3.80	0.9970	52.19	0.6220
5.33	0.9669	53.72	0.6113
8.81	0.9389	56.04	0.5972
10.27	0.9293	58.06	0.5831
14.39	0.8958	60.23	0.5704
16.23	0.8796	62.10	0.5603
19.99	0.8489	64.14	0.5460
22.10	0.8300	66.13	0.5349
24.56	0.8125	68.14	0.5242
25.98	0.8004	69.97	0.5157
28.15	0.7856	69.50	0.5160
29.96	0.7735	71.88	0.5046
32.09	0.7590	73.71	0.4940
33.95	0.7432	75.79	0.4847
36.26	0.7270	77.69	0.4786
38.10	0.7160	79.54	0.4680
40.10	0.7024	80.00	0.4686
42.08	0.6877	82.00	0.4593
44.11	0.6748	84.00	0.4500
46.22	0.6607	85.98	0.4419
48.16	0.6475	88.01	0.4359
49.79	0.6370	89.72	0.4206

*_Z. Physik. Chem._, A167, **42** (1933).

TABLE 2-85 Potassium Chloride

Mole % KCl	Specific heat, cal/(g·°C) 6°C	20°C	33°C	40°C
0.99	0.945	0.947	0.947	0.947
3.85	0.828	0.831	0.835	0.837
5.66	0.77	0.775	0.778	0.775
7.41		0.727		

TABLE 2-86 Potassium Hydroxide (at 19°C)

Mole % KOH	0	0.497	1.64	4.76	9.09
Cal/(g·°C)	1.0	0.975	0.93	0.814	0.75

TABLE 2-87 Normal Propyl Alcohol

Mole % C_3H_7OH	Specific heat, cal/(g·°C) 5°C	20°C	40°C
1.55	1.03	1.02	1.01
5.03	1.07	1.06	1.03
11.4	1.035	1.032	0.99
23.1	0.877	0.90	0.91
41.2	0.75	0.78	0.815
73.0	0.612	0.645	0.708
100.0	0.534	0.57	0.621

TABLE 2-88 Sodium Carbonate*

% Na_2CO_3 by weight	Temperature, °C 17.6	30.0	76.6	98.0
0.000	0.9992	0.9986	1.0098	1.0084
1.498	0.9807			
2.000		0.9786		
2.901	0.9597			
4.000		0.9594		
5.000	0.9428		0.9761	
6.000		0.9392		
8.000	0.9183			
10.000	0.9086		0.9452	
13.790	0.8924			
13.840		0.8881		
20.000		0.8631	0.8936	
25.000			0.8615	0.8911

*_J. Chem. Soc._ 3062–3079 (1931).

TABLE 2-89 Sodium Chloride

Mole % NaCl	Specific heat, cal/(g·°C) 6°C	20°C	33°C	57°C
0.249		0.99		
0.99	0.96	0.97	0.97	
2.44	0.91	0.915	0.915	0.923
9.09	0.805	0.81	0.81	0.82

TABLE 2-90 Sodium Hydroxide (at 20°C)

Mole % NaOH	0	0.5	1.0	9.09	16.7	28.6	37.5
Cal/(g·°C)	1.0	0.985	0.97	0.835	0.80	0.784	0.782

TABLE 2-91 Sulfuric Acid*

%H_2SO_4	C_p at 20°C, cal/(g·°C)	%H_2SO_4	C_p at 20°C, cal/(g·°C)
0.34	0.9968	35.25	0.7238
0.68	0.9937	37.69	.7023
1.34	0.9877	40.49	.6770
2.65	0.9762	43.75	.6476
3.50	0.9688	47.57	.6153
5.16	0.9549	52.13	.5801
9.82	0.9177	57.65	.5420
15.36	0.8767	64.47	.5012
21.40	0.8339	73.13	.4628
22.27	0.8275	77.91	.4518
23.22	0.8205	81.33	.4481
24.25	0.8127	82.49	.4467
25.39	0.8041	84.48	.4408
26.63	0.7945	85.48	.4346
28.00	0.7837	89.36	.4016
29.52	0.7717	91.81	.3787
30.34	0.7647	94.82	.3554
31.20	0.7579	97.44	.3404
33.11	0.7422	100.00	.3352

*Vinal and Craig, _Bur. Standards J. Research_, **24,** 475 (1940).

SPECIFIC HEATS OF MISCELLANEOUS MATERIALS

TABLE 2-92 Specific Heats of Miscellaneous Liquids and Solids

Material	Specific heat, cal/(g·°C)
Alumina	0.2 (100°C); 0.274 (1500°C)
Alundum	0.186 (100°C)
Asbestos	0.25
Asphalt	0.22
Bakelite	0.3 to 0.4
Brickwork	About 0.2
Carbon	0.168 (26 to 76°C)
	0.314 (40 to 892°C)
	0.387 (56 to 1450°C)
(gas retort)	0.204
(see under Graphite)	
Cellulose	0.32
Cement, Portland Clinker	0.186
Charcoal (wood)	0.242
Chrome brick	0.17
Clay	0.224
Coal	0.26 to 0.37
tar oils	0.34 (15 to 90°C)
Coal tars	0.35 (40°C); 0.45 (200°C)
Coke	0.265 (21 to 400°C)
	0.359 (21 to 800°C)
	0.403 (21 to 1300°C)
Concrete	0.156 (70 to 312°F); 0.219 (72 to 1472°F)
Cryolite	0.253 (16 to 55°C)
Diamond	0.147
Fireclay brick	0.198 (100°C); 0.298 (1500°C)
Fluorspar	0.21 (30°C)
Gasoline	0.53
Glass (crown)	0.16 to 0.20
(flint)	0.117
(pyrex)	0.20
(silicate)	0.188 to 0.204 (0 to 100°C)
	0.24 to 0.26 (0 to 700°C)
wool	0.157
Granite	0.20 (20 to 100°C)
Graphite	0.165 (26 to 76°C); 0.390 (56 to 1450°C)
Gypsum	0.259 (16 to 46°C)
Kerosene	0.47
Limestone	0.217
Litharge	0.055
Magnesia	0.234 (100°C); 0.188 (1500°C)
Magnesite brick	0.222 (100°C); 0.195 (1500°C)
Marble	0.21 (18°C)
Porcelain, fired Berlin	0.189 (60°C)
Porcelain, green Berlin	0.185 (60°C)
Porcelain, fired earthenware	0.186 (60°C)
Porcelain, green earthenware	0.181 (60°C)
Pyrex glass	0.20
Pyrites (copper)	0.131 (30°C)
Pyrites (iron)	0.136 (30°C)
Pyroxylin plastics	0.34 to 0.38
Quartz	0.17 (0°C); 0.28 (350°C)
Rubber (vulcanized)	0.415
Sand	0.191
Silica	0.316
Silica brick	0.202 (100°C); 0.195 (1500°C)
Silicon carbide brick	0.202 (100°C)
Silk	0.33
Steel	0.12
Stone	about 0.2
Stoneware (common)	0.188 (60°C)
Turpentine	0.42 (18°C)
Wood (Oak)	0.570
Woods, miscellaneous	0.45 to 0.65
Wool	0.325
Zirconium oxide	0.11 (100°C); 0.179 (1500°C)

TABLE 2-93 Oils (Animal, Vegetable, Mineral Oils)

$$C_p[\text{cal/(g·°C)}] = A/\sqrt{d_4^{15}} + B(t-15)$$

where d = density, g/cm^3.

°F = 9⁄5°C + 32; to convert calories per gram-degree Celsius to British thermal units per pound-degree Fahrenheit, multiply by 1.0; to convert grams per cubic centimeter to pounds per cubic foot, multiply by 62.43.

Oils	A	B
Castor	0.500	0.0007
Citron	(0.438 at 54°C)	
Fatty drying	0.440	0.0007
nondrying	0.450	0.0007
semidrying	0.445	0.0007
oils (except castor)	0.450	0.0007
Naphthene base	0.405	0.0009
Olive	(0.47 at 7°C)	
Paraffin base	0.425	0.0009
Petroleum oils	0.415	0.0009

PROPERTIES OF FORMATION AND COMBUSTION REACTIONS

Unit Conversions °F = 9⁄5°C + 32; to convert kilocalories per gram-mole to British thermal units per pound-mole, multiply by 1.799×10^{-3}.

TABLE 2-94 Heats and Free Energies of Formation of Inorganic Compounds*

The values given in the following table for the heats and free energies of formation of inorganic compounds are derived from (*a*) Bichowsky and Rossini, "Thermochemistry of the Chemical Substances," Reinhold, New York, 1936; (*b*) Latimer, "Oxidation States of the Elements and Their Potentials in Aqueous Solution," Prentice-Hall, New York, 1938; (*c*) the tables of the American Petroleum Institute Research Project 44 at the National Bureau of Standards; and (*d*) the tables of Selected Values of Chemical Thermodynamic Properties of the National Bureau of Standards. The reader is referred to the preceding books and tables for additional details as to methods of calculation, standard states, and so on.

Compound	State[†]	Heat of formation[‡§] ΔH (formation) at 25°C, kcal/mol	Free energy of formation[‖¶] ΔF (formation) at 25°C, kcal/mol
Aluminum			
Al	c	0.00	0.00
$AlBr_3$	c	−123.4	
	aq	−209.5	−189.2
Al_4C_3	c	−30.8	−29.0
$AlCl_3$	c	−163.8	
	aq, 600	−243.9	−209.5
AlF_3	c	−329	
	aq	−360.8	−312.6
AlI_3	c	−72.8	
	aq	−163.4	−152.5
AlN	c	−57.7	−50.4
$Al(NH_4)(SO_4)_2$	c	−561.19	−486.17
$Al(NH_4)(SO_4)_2 \cdot 12H_2O$	c	−1419.36	−1179.26
$Al(NO_3)_3 \cdot 6H_2O$	c	−680.89	−526.32
$Al(NO_3)_3 \cdot 9H_2O$	c	−897.59	
Al_2O_3	c, corundum	−399.09	−376.87
$Al(OH)_3$	c	−304.8	−272.9
$Al_2O_3 \cdot SiO_2$	c, sillimanite	−648.7	
$Al_2O_3 \cdot SiO_2$	c, disthene	−642.4	
$Al_2O_3 \cdot SiO_2$	c, andalusite	−642.0	
$3Al_2O_3 \cdot 2SiO_2$	c, mullite	−1874	
Al_2S_3	c	−121.6	
$Al_2(SO_4)_3$	c	−820.99	−739.53
	aq	−893.9	−759.3
$Al_2(SO_4)_3 \cdot 6H_2O$	c	−1268.15	−1103.39
$Al_2(SO_4)_3 \cdot 18H_2O$	c	−2120	
Antimony			
Sb	c	0.00	0.00
$SbBr_3$	c	−59.9	
$SbCl_3$	c	−91.3	−77.8
$SbCl_5$	l	−104.8	
SbF_3	c	−216.6	
SbI_3	c	−22.8	
Sb_2O_3	c, I, orthorhombic	−165.4	−146.0
	c, II, octahedral	−166.6	
Sb_2O_4	c	−213.0	−186.6
Sb_2O_5	c	−230.0	−196.1
Sb_2S_3	c, black	−38.2	−36.9
Arsenic			
As	c	0.00	0.00
$AsBr_3$	c	−45.9	
$AsCl_3$	l	−80.2	−70.5
AsF_3	l	−223.76	−212.27
AsH_3	g	43.6	37.7
AsI_3	c	−13.6	
As_2O_3	c	−154.1	−134.8
As_2O_5	c	−217.9	−183.9
As_2S_3	c	−20	−20
	amorphous	−34.76	
Barium			
Ba	c	0.00	0.00
$BaBr_2$	c	−180.38	
	aq, 400	−185.67	−183.0
$BaCl_2$	c	−205.25	
	aq, 300	−207.92	−196.5
$Ba(ClO_3)_2$	c	−176.6	
	aq, 1600	−170.0	−134.4
$Ba(ClO_4)_2$	c	−210.2	
	aq, 800		−155.3
$Ba(CN)_2$	c	−48	
$Ba(CNO)_2$	c	−212.1	
	aq		−180.7
$BaCN_2$	c	−63.6	
$BaCO_3$	c, witherite	−284.2	−271.4
$BaCrO_4$	c	−342.2	
BaF_2	c	−287.9	
	aq, 1600	−284.6	−265.3
BaH_2	c	−40.8	−31.5
$Ba(HCO_3)_2$	aq	−459	−414.4
BaI_2	c	−144.6	
	aq, 400	−155.17	−158.52
Barium (*Cont.*)			
$Ba(IO_3)_2$	c	−264.5	
	aq	−237.50	−198.35
$BaMoO_4$	c	−370	
Ba_3N_2	c	−90.7	
$Ba(NO_2)_2$	c	−184.5	
	aq	−179.05	−150.75
$Ba(NO_3)_2$	c	−236.99	−189.94
	aq, 600	−227.74	
BaO	c	−133.0	
$Ba(OH)_2$	c	−225.9	
	aq, 400	−237.76	−209.02
$BaO \cdot SiO_2$	c	−363	
$Ba_3(PO_4)_2$	c	−992	
$BaPtCl_6$	c	−284.9	
BaS	c	−111.2	
$BaSO_3$	c	−282.5	
$BaSO_4$	c	−340.2	−313.4
$BaWO_4$	c	−402	
Beryllium			
Be	c	0.00	0.00
$BeBr_2$	c	−79.4	
	aq	−142	−127.9
$BeCl_2$	c	−112.6	
	aq	−163.9	−141.4
BeI_2	c	−39.4	
	aq	−112	−103.4
Be_3N_2	c	−134.5	−122.4
BeO	c	−145.3	−138.3
$Be(OH)_2$	c	−215.6	
BeS	c	−56.1	
$BeSO_4$	c	−281	
	aq		−254.8
Bismuth			
Bi	c	0.00	0.00
$BiCl_3$	c	−90.5	−76.4
	aq	−101.6	
BiI_3	c	−24	
	aq	−27	
BiO	c	−49.5	−43.2
Bi_2O_3	c	−137.1	−117.9
$Bi(OH)_3$	c	−171.1	
Bi_2S_3	c	−43.9	−39.1
$Bi_2(SO_4)_3$	c	−607.1	
Boron			
B	c	0.00	0.00
BBr_3	l	−52.7	
	g	−44.6	−50.9
BCl_3	g	−94.5	−90.8
BF_3	g	−265.2	−261.0
B_2H_6	g	7.5	19.9
BN	c	−32.1	−27.2
B_2O_3	c	−302.0	−282.9
	gls	−297.6	−280.3
$B(OH)_3$	c	−260.0	−229.4
B_2S_3	c	−56.6	
Bromine			
Br_2	l	0.00	0.00
	g	7.47	0.931
$BrCl$	g	3.06	−0.63
Cadmium			
Cd	c	0.00	0.00
$CdBr_2$	c	−75.8	−70.7
	aq, 400	−76.6	−67.6
$CdCl_2$	c	−92.149	−81.889
	aq, 400	−96.44	−81.2
$Cd(CN)_2$	c	36.2	
$CdCO_3$	c	−178.2	−163.2
CdI_2	c	−48.40	
	aq, 400	−47.46	−43.22
Cd_3N_2	c	39.8	
$Cd(NO_3)_2$	aq, 400	−115.67	−71.05

*For footnotes see end of table.

(*Continued*)

TABLE 2-94 Heats and Free Energies of Formation of Inorganic Compounds *(Continued)*

Compound	State†	Heat of formation‡§ ΔH (formation) at 25°C, kcal/mol	Free energy of formation‖¶ ΔF (formation) at 25°C, kcal/mol
Cadmium (*Cont.*)			
CdO	c	−62.35	−55.28
$Cd(OH)_2$	c	−135.0	−113.7
CdS	c	−34.5	−33.6
$CdSO_4$	c	−222.23	
	aq, 400	−232.635	−194.65
Calcium			
Ca	c	0.00	0.00
$CaBr_2$	c	−162.20	
	aq, 400	−187.19	−181.86
CaC_2	c	−14.8	−16.0
$CaCl_2$	c	−190.6	−179.8
	aq	−209.15	−195.36
$CaCN_2$	c	−85	
$Ca(CN)_2$	c	−43.3	
	aq		−54.0
$CaCO_3$	c, calcite	−289.5	−270.8
	c, aragonite	−289.54	−270.57
$CaCO_3{\cdot}MgCO_3$	c	−558.8	
CaC_2O_4	c	−332.2	
$Ca(C_2H_3O_2)_2$	c	−356.3	
	aq	−364.1	−311.3
CaF_2	c	−290.2	
	aq	−286.5	−264.1
CaH_2	c	−46	−35.7
CaI_2	c	−128.49	
	aq, 400	−156.63	−157.37
Ca_3N_2	c	−103.2	−88.2
$Ca(NO_3)_2$	c	−224.05	−177.38
	aq, 400	−228.29	
$Ca(NO_3)_2{\cdot}2H_2O$	c	−367.95	−293.57
$Ca(NO_3)_2{\cdot}3H_2O$	c	−439.05	−351.58
$Ca(NO_3)_2{\cdot}4H_2O$	c	−509.43	−409.32
CaO	c	−151.7	−144.3
$Ca(OH)_2$	c	−235.58	−213.9
	aq, 800	−239.2	−207.9
$CaO{\cdot}SiO_2$	c, II, wollastonite	−377.9	−357.5
	c, I, pseudo-wollastonite	−376.6	−356.6
CaS	c	−114.3	−113.1
$CaSO_4$	c, insoluble form	−338.73	−311.9
	c, soluble form α	−336.58	−309.8
	c, soluble form β	−335.52	−308.8
$CaSO_4{\cdot}\frac{1}{2}H_2O$	c	−376.13	
$CaSO_4{\cdot}2H_2O$	c	−479.33	−425.47
$CaWO_4$	c	−387	
Carbon			
C	c, graphite	0.00	0.00
	c, diamond	0.453	0.685
CO	g	−26.416	−32.808
CO_2	g	−94.052	−94.260
Cerium			
Ce	c	0.00	0.00
CeN	c	−78.2	−70.8
Cesium			
Cs	c	0.00	0.00
$CsBr$	c	−97.64	
	aq, 500	−91.39	−94.86
$CsCl$	c	−106.31	
	aq, 400	−102.01	−101.61
Cs_2CO_3	c	−271.88	
CsF	c	−131.67	
	aq, 400	−140.48	−135.98
CsH	c	−12	−7.30
$CsHCO_3$	c	−230.6	
	aq, 2000	−226.6	−210.56
CsI	c	−83.91	
	aq, 400	−75.74	−82.61
$CsNH_2$	c	−28.2	
$CsNO_3$	c	−121.14	
	aq, 400	−111.54	−96.53
Cs_2O	c	−82.1	
$CsOH$	c	−100.2	
	aq, 200	−117.0	−107.87
Cs_2S	c	−87	
Cs_2SO_4	c	−344.86	
	aq	−340.12	−316.66

Compound	State†	Heat of formation‡§ ΔH (formation) at 25°C, kcal/mol	Free energy of formation‖¶ ΔF (formation) at 25°C, kcal/mol
Chlorine			
Cl_2	g	0.00	0.00
ClF	g	−25.7	
ClO	g	33	
ClO_2	g	24.7	29.5
ClO_3	g	37	
Cl_2O	g	18.20	22.40
Cl_2O_7	g	63	
Chromium			
Cr	c	0.00	0.00
$CrBr_3$	aq		−122.7
Cr_3C_2	c	−21.008	−21.20
Cr_4C	c	−16.378	−16.74
$CrCl_2$	c	−103.1	−93.8
	aq		−102.1
CrF_2	c	−152	
CrF_3	c	−231	
CrI_2	c	−63.7	
	aq		−64.1
CrO_3	c	−139.3	
Cr_2O_3	c	−268.8	−249.3
$Cr_2(SO_4)_3$	aq		−626.3
Cobalt			
Co	c	0.00	0.00
$CoBr_2$	c	−55.0	
	aq	−73.61	−61.96
Co_3C	c	9.49	7.08
$CoCl_2$	c	−76.9	−66.6
	aq, 400	−95.58	−75.46
$CoCO_3$	c	−172.39	−155.36
CoF_2	aq	−172.98	−144.2
CoI_2	c	−24.2	
	aq	−43.15	−37.4
$Co(NO_3)_2$	c	−102.8	
	aq	−114.9	−65.3
CoO	c	−57.5	
Co_3O_4	c	−196.5	
$Co(OH)_2$	c	−131.5	−108.9
$Co(OH)_3$	c	−177.0	−142.0
CoS	c	−22.3	−19.8
Co_2S_3	c	−40.0	
$CoSO_4$	c	−216.6	
	aq, 400		−188.9
Columbium			
Cb	c	0.00	0.00
Cb_2O_5	c	−462.96	
Copper			
Cu	c	0.00	0.00
$CuBr$	c	−26.7	−23.8
$CuBr_2$	c	−34.0	
	aq	−42.4	−33.25
$CuCl$	c	−31.4	−24.13
$CuCl_2$	c	−48.83	
	aq, 400	−64.7	
$CuClO_4$	aq	−28.3	1.34
$Cu(ClO_3)_2$	aq, 400		15.4
$Cu(ClO_4)_2$	aq		−5.5
CuI	c	−17.8	−16.66
CuI_2	c	−4.8	
	aq	−11.9	−8.76
Cu_3N	c	17.78	
$Cu(NO_3)_2$	c	−73.1	
	aq, 200	−83.6	−36.6
CuO	c	−38.5	−31.9
Cu_2O	c	−43.00	−38.13
$Cu(OH)_2$	c	−108.9	−85.5
CuS	c	−11.6	−11.69
Cu_2S	c	−18.97	−20.56
$CuSO_4$	c	−184.7	−158.3
	aq, 800	−200.78	−160.19
Cu_2SO_4	c	−179.6	
	aq		−152.0
Erbium			
Er	c	0.00	0.00
$Er(OH)_3$	c	−326.8	
Fluorine			
F_2	g	0.00	0.00
F_2O	g	5.5	9.7

TABLE 2-94 Heats and Free Energies of Formation of Inorganic Compounds *(Continued)*

Compound	State[†]	Heat of formation[‡§] ΔH (formation) at 25°C, kcal/mol	Free energy of formation[‖¶] ΔF (formation) at 25°C, kcal/mol
Gallium			
Ga	c	0.00	0.00
$GaBr_3$	c	−92.4	
$GaCl_3$	c	−125.4	
GaN	c	−26.2	
Ga_2O	c	−84.3	
Ga_2O_3	c	−259.9	
Germanium			
Ge	c	0.00	0.00
Ge_3N_4	c	−15.7	
GeO_2	c	−128.6	
Gold			
Au	c	0.00	0.00
AuBr	c	−3.4	
$AuBr_3$	c	−14.5	
	aq	−11.0	24.47
AuCl	c	−8.3	
$AuCl_3$	c	−28.3	
	aq	−32.96	4.21
AuI	c	0.2	−0.76
Au_2O_3	c	11.0	18.71
$Au(OH)_3$	c	−100.6	
Hafnium			
Hf	c	0.00	0.00
HfO_2	c	−271.1	−258.2
Hydrogen			
H_3AsO_3	aq	−175.6	−153.04
H_3AsO_4	c	−214.9	
	aq	−214.8	−183.93
HBr	g	−8.66	−12.72
	aq, 400	−28.80	−24.58
HBrO	aq	−25.4	−19.90
$HBrO_3$	aq	−11.51	5.00
HCl	g	−22.063	−22.778
	aq, 400	−39.85	−31.330
HCN	g	31.1	27.94
	aq, 100	24.2	26.55
HClO	aq, 400	−28.18	−19.11
$HClO_3$	aq	−23.4	−0.25
$HClO_4$	aq, 660	−31.4	−10.70
$HC_2H_3O_2$	l	−116.2	−93.56
	aq, 400	−116.74	−96.8
$H_2C_2O_4$	c	−196.7	
	aq, 300	−194.6	−165.64
HCOOH	l	−97.8	−82.7
	aq, 200	−98.0	−85.1
H_2CO_3	aq	−167.19	−149.0
HF	g	−64.2	−64.7
	aq, 200	−75.75	
HI	g	6.27	0.365
	aq, 400	−13.47	−12.35
HIO	aq	−38	−23.33
HIO_3	c	−56.77	
	aq	−54.8	−32.25
HN_3	g	70.3	78.50
HNO_3	g	−31.99	−17.57
	l	−41.35	−19.05
	aq, 400	−49.210	
$HNO_3 \cdot H_2O$	l	−112.91	−78.36
$HNO_3 \cdot 3H_2O$	l	−252.15	−193.70
H_2O	g	−57.7979	−54.6351
	l	−68.3174	−56.6899
H_2O_2	l	−45.16	−28.23
	aq, 200	−45.80	−31.47
H_3PO_2	c	−145.5	
	aq	−145.6	−120.0
H_3PO_3	c	−232.2	
	aq	−232.2	−204.0
H_3PO_4	c	−306.2	
	aq, 400	−309.32	−270.0
H_2S	g	−4.77	−7.85
	aq, 2000	−9.38	
H_2S_2	l	−3.6	
H_2SO_3	aq, 200	−146.88	−128.54
H_2SO_4	l	−193.69	
	aq, 400	−212.03	
H_2Se	g	20.5	17.0
	aq	18.1	18.4
Hydrogen (*Cont.*)			
H_2SeO_3	c	−126.5	
	aq	−122.4	−101.36
H_2SeO_4	c	−130.23	
	aq, 400	−143.4	
H_2SiO_3	c	−267.8	−247.9
H_4SiO_4	c	−340.6	
H_2Te	g	36.9	33.1
H_2TeO_3	c	−145.0	−115.7
	aq	−145.0	
H_2TeO_4	aq	−165.6	
Indium			
In	c	0.00	0.00
$InBr_3$	c	−97.2	
	aq	−112.9	−97.2
$InCl_3$	c	−128.5	
	aq	−145.6	−117.5
InI_3	c	−56.5	
	aq	−67.2	−60.5
InN	c	−4.8	
In_2O_3	c	−222.47	
Iodine			
I_2	c	0.00	0.00
	g	14.88	4.63
IBr	g	10.05	1.24
ICl	g	4.20	−1.32
ICl_3	c	−21.8	−6.05
I_2O_5	c	−42.5	
Iridium			
Ir	c	0.00	0.00
IrCl	c	−20.5	−16.9
$IrCl_2$	c	−40.6	−32.0
$IrCl_3$	c	−60.5	−46.5
IrF_6	l	−130	
IrO_2	c	−40.14	
Iron			
Fe	c, α	0.00	0.00
$FeBr_2$	c	−57.15	
	aq, 540	−78.7	−69.47
$FeBr_3$	aq	−95.5	−76.26
Fe_3C	c	5.69	4.24
$Fe(CO)_5$	l	−187.6	
$FeCO_3$	c, siderite	−172.4	−154.8
$FeCl_2$	c	−81.9	−72.6
	aq	−100.0	−83.0
$FeCl_3$	c	−96.4	
	aq, 2000	−128.5	−96.5
FeF_2	aq, 1200	−177.2	−151.7
FeI_2	c	−24.2	
	aq	−47.7	−45
FeI_3	aq	−49.7	−39.5
Fe_4N	c	−2.55	0.862
$Fe(NO_3)_2$	aq	−118.9	−72.8
$Fe(NO_3)_3$	aq, 800	−156.5	−81.3
FeO	c	−64.62	−59.38
Fe_2O_3	c	−198.5	−179.1
Fe_3O_4	c	−266.9	−242.3
$Fe(OH)_2$	c	−135.9	−115.7
$Fe(OH)_3$	c	−197.3	−166.3
$FeO \cdot SiO_2$	c	−273.5	
Fe_2P	c	−13	
FeSi	c	−19.0	
FeS	c	−22.64	−23.23
FeS_2	c, pyrites	−38.62	−35.93
	c, marcasite	−33.0	
$FeSO_4$	c	−221.3	−195.5
	aq, 400	−236.2	−196.4
$Fe_2(SO_4)_3$	aq, 400	−653.3	−533.4
$FeTiO_3$	c, ilmenite	−295.51	−277.06
Lanthanum			
La	c	0.00	0.00
$LaCl_3$	c	−253.1	
	aq	−284.7	
La_3H_8	c	−160	
LaN	c	−72.0	−64.6
La_2O_3	c	−539	
LaS_2	c	−148.3	
La_2S_3	c	−351.4	
$La_2(SO_4)_3$	aq	−972	

(Continued)

TABLE 2-94 Heats and Free Energies of Formation of Inorganic Compounds *(Continued)*

Compound	State[†]	Heat of formation[‡§] ΔH (formation) at 25°C, kcal/mol	Free energy of formation[‖¶] ΔF (formation) at 25°C, kcal/mol
Lead			
Pb	c	0.00	0.00
$PbBr_2$	c	−66.24	−62.06
	aq	−56.4	−54.97
$PbCO_3$	c, cerussite	−167.6	−150.0
$Pb(C_2H_3O_2)_2$	c	−232.6	
	aq, 400	−234.2	−184.40
PbC_2O_4	c	−205.3	
$PbCl_2$	c	−85.68	−75.04
	aq	−82.5	−68.47
PbF_2	c	−159.5	−148.1
PbI_2	c	−41.77	−41.47
$Pb(NO_3)_2$	c	−106.88	
	aq, 400	−99.46	−58.3
PbO	c, red	−51.72	−45.53
	c, yellow	−50.86	−43.88
PbO_2	c	−65.0	−52.0
Pb_3O_4	c	−172.4	−142.2
$Pb(OH)_2$	c	−123.0	−102.2
PbS	c	−22.38	−21.98
$PbSO_4$	c	−218.5	−192.9
Lithium			
Li	c	0.00	0.00
LiBr	c	−83.75	
	aq, 400	−95.40	−95.28
$LiBrO_3$	aq	−77.9	−65.70
Li_2C_2	c	−13.0	
LiCN	aq	−31.4	−31.35
LiCNO	aq	−101.2	−94.12
$LiC_2H_3O_2$	aq	−183.9	−160.00
Li_2CO_3	c	−289.7	−269.8
	aq, 1900	−293.1	−267.58
LiCl	c	−97.63	
	aq, 278	−106.45	−102.03
$LiClO_3$	aq	−87.5	−70.95
$LiClO_4$	aq	−106.3	−81.4
LiF	c	−145.57	
	aq, 400	−144.85	−136.40
LiH	c	−22.9	
$LiHCO_3$	aq, 2000	−231.1	−210.98
LiI	c	−65.07	
	aq, 400	−80.09	−83.03
$LiIO_3$	aq	−121.3	−102.95
Li_3N	c	−47.45	−37.33
$LiNO_3$	c	−115.350	
	aq, 400	−115.88	−96.95
Li_2O	c	−142.3	
Li_2O_2	c	−151.9	−138.0
	aq	−159	
LiOH	c	−116.58	−106.44
	aq, 400	−121.47	−108.29
$LiOH{\cdot}H_2O$	c	−188.92	
$Li_2O{\cdot}SiO_2$	gls	−374	
Li_2Se	c	−84.9	
	aq	−95.5	−105.64
Li_2SO_4	c	−340.23	−314.66
	aq, 400	−347.02	
$Li_2SO_4{\cdot}H_2O$	c	−411.57	−375.07
Magnesium			
Mg	c	0.00	0.00
$Mg(AsO_4)_2$	c	−731.3	
	aq	−749	−630.14
$MgBr_2$	c	−123.9	
	aq, 400	−167.33	−156.94
$Mg(CN)_2$	aq	−39.7	−29.08
$MgCN_2$	c	−61	
$Mg(C_2H_3O_2)_2$	aq	−344.6	−286.38
$MgCO_3$	c	−261.7	−241.7
$MgCl_2$	c	−153.220	−143.77
	aq, 400	−189.76	
Magnesium *(Cont.)*			
$MgCl_2{\cdot}H_2O$	c	−230.970	−205.93
$MgCl_2{\cdot}2H_2O$	c	−305.810	−267.20
$MgCl_2{\cdot}4H_2O$	c	−453.820	−387.98
$MgCl_2{\cdot}6H_2O$	c	−597.240	−505.45
MgF_2	c	−263.8	
MgI_2	c	−86.8	
	aq, 400	−136.79	−132.45
$MgMoO_4$	c	−329.9	
Mg_3N_2	c	−115.2	−100.8
$Mg(NO_3)_2$	c	−188.770	−140.66
	aq, 400	−209.927	−160.28
$Mg(NO_3)_2{\cdot}2H_2O$	c	−336.625	
$Mg(NO_3)_2{\cdot}6H_2O$	c	−624.48	−496.03
MgO	c	−143.84	−136.17
$MgO{\cdot}SiO_2$	c	−347.5	−326.7
$Mg(OH)_2$	c, ppt.	−221.90	−200.17
	c, brucite	−223.9	−193.3
MgS	c	−84.2	
	aq	−108	
$MgSO_4$	c	−304.94	−277.7
	aq, 400	−325.4	−283.88
MgTe	c	−25	
$MgWO_4$	c	−345.2	
Manganese			
Mn	c, α	0.00	0.00
$MnBr_2$	c	−91	
	aq	−106	−97.8
Mn_3C	c	1.1	1.26
$Mn(C_2H_3O_2)_2$	c	−270.3	
	aq	−282.7	−227.2
$MnCO_3$	c	−211	−192.5
MnC_2O_4	c	−240.9	
$MnCl_2$	c	−112.0	−102.2
	aq, 400	−128.9	
MnF_2	aq, 1200	−206.1	−180.0
MnI_2	c	−49.8	
	aq	−76.2	−73.3
Mn_5N_2	c	−57.77	−46.49
$Mn(NO_3)_2$	c	−134.9	
	aq, 400	−148.0	−101.1
$Mn(NO_3)_2.6H_2O$	c	−557.07	−441.2
MnO	c	−92.04	−86.77
MnO_2	c	−124.58	−111.49
Mn_2O_3	c	−229.5	−209.9
Mn_3O_4	c	−331.65	−306.22
$MnO.SiO_2$	c	−301.3	−282.1
$Mn(OH)_2$	c	−163.4	−143.1
$Mn(OH)_3$	c	−221	−190
$Mn_3(PO_4)_2$	c	−736	
MnSe	c	−26.3	−27.5
MnS	c, green	−47.0	−48.0
$MnSO_4$	c	−254.18	−228.41
	aq, 400	−265.2	
$Mn_2(SO_4)_3$	c	−635	
	aq	−657	
Mercury			
Hg	l	0.00	0.00
HgBr	g	23	18
$HgBr_2$	c	−40.68	−38.8
	aq	−38.4	−9.74
$Hg(C_2H_3O_2)_2$	c	−196.3	
	aq	−192.5	−139.2
$HgCl_2$	c	−53.4	−42.2
	aq	−50.3	−23.25
HgCl	g	19	14
Hg_2Cl_2	c	−63.13	
$Hg(CN)_2$	c	62.8	
	aq, 1110	66.25	
HgC_2O_4	c	−159.3	

TABLE 2-94 Heats and Free Energies of Formation of Inorganic Compounds *(Continued)*

Compound	State†	Heat of formation‡§ ΔH (formation) at 25°C, kcal/mol	Free energy of formation‖¶ ΔF (formation) at 25°C, kcal/mol
Mercury (*Cont.*)			
HgH	g	57.1	52.25
HgI_2	c, red	−25.3	−24.0
HgI	g	33	23
Hg_2I_2	c	−28.88	−26.53
$Hg(NO_3)_2$	aq	−56.8	−13.09
$Hg_2(NO_3)_2$	aq	−58.5	−15.65
HgO	c, red	−21.6	−13.94
	c, yellow ppt.	−20.8	
Hg_2O	c	−21.6	−12.80
HgS	c, black	−10.7	−8.80
$HgSO_4$	c	−166.6	
Hg_2SO_4	c	−177.34	−149.12
Molybdenum			
Mo	c	0.00	0.00
Mo_2C	c	4.36	2.91
Mo_2N	c	−8.3	
MoO_2	c	−130	−118.0
MoO_3	c	−180.39	−162.01
MoS_2	c	−56.27	−54.19
MoS_3	c	−61.48	−57.38
Nickel			
Ni	c	0.00	0.00
$NiBr_2$	c	−53.4	
	aq	−72.6	−60.7
Ni_3C	c	9.2	8.88
$Ni(C_2H_3O_2)_2$	aq	−249.6	−190.1
$Ni(CN)_2$	aq	230.9	66.3
$NiCl_2$	c	−75.0	
	aq, 400	−94.34	−74.19
NiF_2	c	−157.5	
	aq	−171.6	−142.9
NiI_2	c	−22.4	
	aq	−42.0	−36.2
$Ni(NO_3)_2$	c	−101.5	
	aq, 200	−113.5	−64.0
NiO	c	−58.4	−51.7
$Ni(OH)_2$	c	−129.8	−105.6
$Ni(OH)_3$	c	−163.2	
NiS	c	−20.4	
$NiSO_4$	c	−216	
	aq, 200	−231.3	−187.6
Nitrogen			
N_2	g	0.00	0.00
NF_3	g	−27	
NH_3	g	−10.96	−3.903
	aq, 200	−19.27	
NH_4Br	c	−64.57	
	aq	−60.27	−43.54
$NH_4C_2H_3O_2$	c	−148.1	
	aq, 400	−148.58	−108.26
NH_4CN	c	−0.7	
	aq	3.6	20.4
NH_4CNS	c	−17.8	
	aq	−12.3	4.4
$(NH_4)_2CO_3$	aq	−223.4	−164.1
$(NH_4)_2C_2O_4$	c	−266.3	
	aq	−260.6	−196.2
NH_4Cl	c	−75.23	−48.59
	aq, 400	−71.20	
NH_4ClO_4	c	−69.4	
	aq	−63.2	−21.1
$(NH_4)_2CrO_4$	c	−276.9	
	aq	−271.3	−209.3
NH_4F	c	−111.6	
	aq	−110.2	−84.7
NH_4I	c	−48.43	
	aq	−44.97	−31.3
NH_4NO_3	c	−87.40	
	aq, 500	−80.89	
Nitrogen (*Cont.*)			
NH_4OH	aq	−87.59	
$(NH_4)_2S$	aq, 400	−55.21	−14.50
$(NH_4)_2SO_4$	c	−281.74	−215.06
	aq, 400	−279.33	−214.02
N_2H_4	l	12.06	
$N_2H_4 \cdot H_2O$	l	−57.96	
$N_2H_4 \cdot H_2SO_4$	c	−232.2	
N_2O	g	19.55	24.82
NO	g	21.600	20.719
NO_2	g	7.96	12.26
N_2O_4	g	2.23	23.41
N_2O_5	c	−10.0	
NOBr	l	11.6	19.26
NOCl	g	12.8	16.1
Osmium			
Os	c	0.00	0.00
OsO_4	c	−93.6	−70.9
	g	−80.1	−68.1
Oxygen			
O_2	g	0.00	0.00
O_3	g	33.88	38.86
Palladium			
Pd	c	0.00	0.00
PdO	c	−20.40	
Phosphorus			
P	c, white ("yellow")	0.00	0.00
	c, red ("violet")	−4.22	−1.80
P	g	150.35	141.88
P_2	g	33.82	24.60
P_4	g	13.2	5.89
PBr_3	l	−45	
PBr_5	c	−60.6	
PCl_3	g	−70.0	−65.2
	l	−76.8	−63.3
PCl_5	g	−91.0	−73.2
PH_3	g	2.21	−1.45
PI_3	c	−10.9	
P_2O_5	c	−360.0	
$POCl_3$	g	−138.4	−127.2
Platinum			
Pt	c	0.00	0.00
$PtBr_4$	c	−40.6	
	aq	−50.7	
$PtCl_2$	c	−34	
$PtCl_4$	c	−62.6	
	aq	−82.3	
PtI_4	c	−18	
$Pt(OH)_2$	c	−87.5	−67.9
PtS	c	−20.18	−18.55
PtS_2	c	−26.64	−24.28
Potassium			
K	c	0.00	0.00
K_3AsO_3	aq	−323.0	
K_3AsO_4	aq	−390.3	−355.7
KH_2AsO_4	c	−271.2	−236.7
KBr	c	−94.06	−90.8
	aq, 400	−89.19	−92.0
$KBrO_3$	c	−81.58	−60.30
	aq, 1667	−71.68	
$KC_2H_3O_2$	c	−173.80	
	aq, 400	−177.38	−156.73
KCl	c	−104.348	−97.76
	aq, 400	−100.164	−98.76
$KClO_3$	c	−93.5	−69.30
	aq, 400	−81.34	
$KClO_4$	c	−103.8	−72.86
	aq, 400	−101.14	
KCN	c	−28.1	
	aq, 400	−25.3	−28.08

(Continued)

TABLE 2-94 Heats and Free Energies of Formation of Inorganic Compounds *(Continued)*

Compound	State[†]	Heat of formation[‡§] ΔH (formation) at 25°C, kcal/mol	Free energy of formation[‖¶] ΔF (formation) at 25°C, kcal/mol
Potassium (*Cont.*)			
KCNO	c	−99.6	
	aq	−94.5	−90.85
KCNS	c	−47.0	
	aq, 400	−41.07	−44.08
K_2CO_3	c	−274.01	
	aq, 400	−280.90	−264.04
$K_2C_2O_4$	c	−319.9	
	aq, 400	−315.5	−293.1
K_2CrO_4	c	−333.4	
	aq, 400	−328.2	−306.3
$K_2Cr_2O_7$	c	−488.5	
	aq, 400	−472.1	−440.9
KF	c	−134.50	
	aq, 180	−138.36	−133.13
$K_3Fe(CN)_6$	c	−48.4	
	aq	−34.5	
$K_4Fe(CN)_6$	c	−131.8	
	aq	−119.9	
KH	c	−10	−5.3
$KHCO_3$	c	−229.8	
	aq, 2000	−224.85	−207.71
KI	c	−78.88	−77.37
	aq, 500	−73.95	−79.76
KIO_3	c	−121.69	−101.87
	aq, 400	−115.18	−99.68
KIO_4	aq	−98.1	
$KMnO_4$	c	−192.9	−169.1
	aq, 400	−182.5	−168.0
K_2MoO_4	aq, 880	−364.2	−342.9
KNH_2	c	−28.25	
KNO_2	aq	−86.0	−75.9
KNO_3	c	−118.08	−94.29
	aq, 400	−109.79	−93.68
K_2O	c	−86.2	
$K_2O \cdot Al_2O_3 \cdot SiO_2$	c, leucite	−1379.6	
	gls	−1368.2	
$K_2O \cdot Al_2O_3 \cdot SiO_2$	c, adularia	−1784.5	
	c, microcline	−1784.5	
	gls	−1747	
KOH	c	−102.02	
	aq, 400	−114.96	−105.0
K_3PO_3	aq	−397.5	
K_3PO_4	aq	−478.7	−443.3
KH_2PO_4	c	−362.7	−326.1
K_2PtCl_4	c	−254.7	
	aq	−242.6	−226.5
K_2PtCl_6	c	−299.5	−263.6
	aq, 9400	−286.1	
K_2Se	c	−74.4	
	aq	−83.4	−99.10
K_2SeO_4	aq	−267.1	−240.0
K_2S	c	−121.5	
	aq, 400	−110.75	−111.44
K_2SO_3	c	−267.7	
	aq	−269.7	−251.3
K_2SO_4	c	−342.65	−314.62
	aq, 400	−336.48	−310.96
$K_2SO_4 \cdot Al_2(SO_4)_3$	c	−1178.38	−1068.48
$K_2SO_4 \cdot Al_2(SO_4)_3 \cdot 24H_2O$	c	−2895.44	−2455.68
$K_2S_2O_6$	c	−418.62	
Rhenium			
Re	c	0.00	0.00
ReF_6	g	−274	
Rhodium			
Rh	c	0.00	0.00
RhO	c	−21.7	
Rh_2O	c	−22.7	
Rh_2O_3	c	−68.3	
Rubidium			
Rb	c	0.00	0.00
RbBr	c	−95.82	
	g	−45.0	−52.50
	aq, 500	−90.54	−93.38
RbCN	aq	−25.9	
Rb_2CO_3	c	−273.22	
	aq, 220	−282.61	−263.78
RbCl	c	−105.06	−98.48
	g	−53.6	−57.9
	aq, ∞	−101.06	−100.13
RbF	c	−133.23	
	aq, 400	−139.31	−134.5
$RbHCO_3$	c	−230.01	
	aq, 2000	−225.59	−209.07
RbI	c	−81.04	
	g	−31.2	−40.5
	aq, 400	−74.57	−81.13
$RbNH_2$	c	−27.74	
$RbNO_3$	c	−119.22	
	aq, 400	−110.52	−95.05
Rb_2O	c	−82.9	
Rb_2O_2	c	−107	
RbOH	c	−101.3	
	aq, 200	−115.8	−106.39
Ruthenium			
Ru	c	0.00	0.00
RuS_2	c	−46.99	−44.11
Selenium			
Se	c, I, hexagonal	0.00	0.00
	c, II, red, monoclinic	0.2	
Se_2Cl_2	l	−22.06	−13.73
SeF_6	g	−246	−222
SeO_2	c	−56.33	
Silicon			
Si	c	0.00	0.00
$SiBr_4$	l	−93.0	
SiC	c	−28	−27.4
$SiCl_4$	l	−150.0	−133.9
	g	−142.5	−133.0
SiF_4	g	−370	−360
SiH_4	g	−14.8	−9.4
SiI_4	c	−29.8	
Si_3N_4	c	−179.25	−154.74
SiO_2	c, cristobalite, 1600° form	−202.62	
	c, cristobalite, 1100° form	−202.46	
	c, quartz	−203.35	−190.4
	c, tridymite	−203.23	
Silver			
Ag	c	0.00	0.00
AgBr	c	−23.90	−23.02
Ag_2C_2	c	84.5	
$AgC_2H_3O_2$	c	−95.9	
	aq	−91.7	−70.86
AgCN	c	33.8	38.70
Ag_2CO_3	c	−119.5	−103.0
$Ag_2C_2O_4$	c	−158.7	
AgCl	c	−30.11	−25.98
AgF	c	−48.7	
	aq, 400	−53.1	−47.26
AgI	c	−15.14	−16.17
$AgIO_3$	c	−42.02	−24.08
$AgNO_2$	c	−11.6	3.76
	aq	−2.9	9.99
$AgNO_3$	c	−29.4	−7.66
	aq, 6500	−24.02	−7.81
Ag_2O	c	−6.95	−2.23

TABLE 2-94 Heats and Free Energies of Formation of Inorganic Compounds *(Continued)*

Compound	State[†]	Heat of formation[‡§] ΔH (formation) at 25°C, kcal/mol	Free energy of formation[‖¶] ΔF (formation) at 25°C, kcal/mol
Silver *(Cont.)*			
Ag_2S	c	−5.5	−7.6
Ag_2SO_4	c	−170.1	−146.8
	aq	−165.8	−139.22
Sodium			
Na	c	0.00	0.00
Na_3AsO_3	aq, 500	−314.61	
Na_3AsO_4	c	−366	
	aq, 500	−381.97	−341.17
NaBr	c	−86.72	
	aq, 400	−86.33	−87.17
NaBrO	aq	−78.9	
$NaBrO_3$	aq, 400	−68.89	−57.59
$NaC_2H_3O_2$	c	−170.45	
	aq, 400	−175.450	−152.31
NaCN	c	−22.47	
	aq, 200	−22.29	−23.24
NaCNO	c	−96.3	
	aq	−91.7	−86.00
NaCNS	c	−39.94	
	aq, 400	−38.23	−39.24
Na_2CO_3	c	−269.46	−249.55
	aq, 1000	−275.13	−251.36
$NaCO_2NH_2$	c	−142.17	
$Na_2C_2O_4$	c	−313.8	
	aq, 600	−309.92	−283.42
NaCl	c	−98.321	−91.894
	aq, 400	−97.324	−93.92
$NaClO_3$	c	−83.59	
	aq, 400	−78.42	−62.84
$NaClO_4$	c	−101.12	
	aq, 476	−97.66	−73.29
Na_2CrO_4	c	−319.8	
	aq, 800	−323.0	−296.58
$Na_2Cr_2O_7$	aq, 1200	−465.9	−431.18
NaF	c	−135.94	−129.0
	aq, 400	−135.711	−128.29
NaH	c	−14	−9.30
$NaHCO_3$	c	−226.0	−202.66
	aq	−222.1	−202.87
NaI	c	−69.28	
	aq, ∞	−71.10	−74.92
$NaIO_3$	aq, 400	−112.300	−94.84
Na_2MoO_4	c	−364	
	aq	−358.7	−333.18
$NaNO_2$	c	−86.6	
	aq	−83.1	−71.04
$NaNO_3$	c	−111.71	−87.62
	aq, 400	−106.880	−88.84
Na_2O	c	−99.45	−90.06
Na_2O_2	c	−119.2	−105.0
$Na_2O \cdot SiO_2$	c	−383.91	−361.49
$Na_2O \cdot Al_2O_3 \cdot 3SiO_2$	c, natrolite	−1180	
$Na_2O \cdot Al_2O_3 \cdot 4SiO_2$	c	−1366	
NaOH	c	−101.96	−90.60
	aq, 400	−112.193	−100.18
Na_3PO_3	aq, 1000	−389.1	
Na_3PO_4	c	−457	
	aq, 400	−471.9	−428.74
Na_2PtCl_4	aq	−237.2	−216.78
Na_2PtCl_6	c	−272.1	
	aq	−280.9	
Na_2Se	c	−59.1	
	aq, 440	−78.1	−89.42
Na_2SeO_4	c	−254	
	aq, 800	−261.5	−230.30
Na_2S	c	−89.8	
	aq, 400	−105.17	−101.76
Na_2SO_3	c	−261.2	−240.14
	aq, 800	−264.1	−241.58
Na_2SO_4	c	−330.50	−302.38
	aq, 1100	−330.82	−301.28
Sodium *(Cont.)*			
$Na_2SO_4 \cdot 10H_2O$	c	−1033.85	−870.52
Na_2WO_4	c	−391	
	aq	−381.5	−345.18
Strontium			
Sr	c	0.00	0.00
$SrBr_2$	c	−171.0	
	aq, 400	−187.24	−182.36
$Sr(C_2H_3O_2)_2$	c	−358.0	
	aq	−364.4	−311.80
$Sr(CN)_2$	aq	−59.5	−54.50
$SrCO_3$	c	−290.9	−271.9
$SrCl_2$	c	−197.84	
	aq, 400	−209.20	−195.86
SrF_2	c	−289.0	
$Sr(HCO_3)_2$	aq	−459.1	−413.76
SrI_2	c	−136.1	
	aq, 400	−156.70	−157.87
Sr_3N_2	c	−91.4	−76.5
$Sr(NO_3)_2$	c	−233.2	
	aq, 400	−228.73	−185.70
SrO	c	−140.8	−133.7
$SrO \cdot SiO_2$	gls	−364	
SrO_2	c	−153.3	−139.0
Sr_2O	c	−153.6	
$Sr(OH)_2$	c	−228.7	
	aq, 800	−239.4	−208.27
$Sr_3(PO_4)_2$	c	−980	
	aq	−985	−881.54
SrS	c	−113.1	
	aq	−120.4	−109.78
$SrSO_4$	c	−345.3	
	aq, 400	−345.0	−309.30
$SrWO_4$	c	−393	
Sulfur			
S	c, rhombic	0.00	0.00
	c, monoclinic	−0.071	0.023
	l, λ	0.257	0.072
	l, λμ equilibrium		0.071
	g	53.25	43.57
S_2	g	31.02	19.36
S_6	g	27.78	13.97
S_8	g	27.090	12.770
S_2Br_2	l	−4	
SCl_4	l	−13.7	
S_2Cl_2	l	−14.2	−5.90
S_2Cl_4	l	−24.1	
SF_6	g	−262	−237
SO	g	19.02	12.75
SO_2	g	−70.94	−71.68
SO_3	g	−94.39	−88.59
	l	−103.03	−88.28
	c, α	−105.09	−88.22
	c, β	−105.92	−88.34
	c, γ	−109.34	−88.98
SO_2Cl_2	g	−82.04	−74.06
	l	−89.80	−75.06
Tantalum			
Ta	c	0.00	0.00
TaN	c	−51.2	−45.11
Ta_2O_5	c	−486.0	−453.7
Tellurium			
Te	c	0.00	0.00
$TeBr_4$	c	−49.3	
$TeCl_4$	c	−77.4	−57.4
TeF_6	g	−315	−292
TeO_2	c	−77.56	−64.66
Thallium			
Tl	c	0.00	0.00
TlBr	c	−41.5	−39.43
	aq	−28.0	−32.34
TlCl	c	−49.37	−44.46
	aq	−38.4	−39.09

(Continued)

TABLE 2-94 Heats and Free Energies of Formation of Inorganic Compounds *(Continued)*

Compound	State†	Heat of formation‡§ ΔH (formation) at 25°C, kcal/mol	Free energy of formation‖¶ ΔF (formation) at 25°C, kcal/mol
Thallium (*Cont.*)			
$TlCl_3$	c	−82.4	
	aq	−91.0	−44.25
TlF	aq	−77.6	−73.46
TlI	c	−31.1	−31.3
	aq	−12.7	−20.09
$TlNO_3$	c	−58.2	−36.32
	aq	−48.4	−34.01
Tl_2O	c	−43.18	
Tl_2O_3	c	−120	
TlOH	c	−57.44	−45.54
	aq	−53.9	−45.35
Tl_2S	c	−22	
Tl_2SO_4	c	−222.8	−197.79
	aq, 800	−214.1	−191.62
Thorium			
Th	c	0.00	0.00
$ThBr_4$	c	−281.5	
	aq	−352.0	−295.31
ThC_2	c	−45.1	
$ThCl_4$	c	−335	
	aq	−392	−322.32
ThI_4	aq	−292.0	−246.33
Th_3N_4	c	−309.0	−282.3
ThO_2	c	−291.6	−280.1
$Th(OH)_4$	c, "soluble"	−336.1	
$Th(SO_4)_2$	c	−632	
	aq	−668.1	−549.2
Tin			
Sn	c, II, tetragonal	0.00	0.00
	c, III, "gray," cubic	0.6	1.1
$SnBr_2$	c	−61.4	
	aq	−60.0	−55.43
$SnBr_4$	c	−94.8	
	aq	−110.6	−97.66
$SnCl_2$	c	−83.6	
	aq	−81.7	−68.94
$SnCl_4$	l	−127.3	−110.4
	aq	−157.6	−124.67
SnI_2	c	−38.9	
	aq	−33.3	−30.95
SnO	c	−67.7	−60.75
SnO_2	c	−138.1	−123.6
$Sn(OH)_2$	c	−136.2	−115.95
$Sn(OH)_4$	c	−268.9	−226.00
SnS	c	−18.61	
Titanium			
Ti	c	0.00	0.00
TiC	c	−110	−109.2
$TiCl_4$	l	−181.4	−165.5
TiN	c	−80.0	−73.17
TiO_2	c, III, rutil	−225.0	−211.9
	amorphous	−214.1	−201.4
Tungsten			
W	c	0.00	0.00
WO_2	c	−130.5	−118.3
WO_3	c	−195.7	−177.3
WS_2	c	−84	
Uranium			
U	c	0.00	0.00
UC_2	c	−29	
UCl_3	c	−213	
UCl_4	c	−251	
U_3N_4	c	−274	−249.6
UO_2	c	−256.6	−242.2
$UO_2(NO_3)_2{\cdot}6H_2O$	c	−756.8	−617.8
UO_3	c	−291.6	
U_3O_8	c	−845.1	
Vanadium			
V	c	0.00	0.00
VCl_2	c	−147	
VCl_3	l	−187	
VCl_4	l	−165	
VN	c	−41.43	−35.08
V_2O_2	c	−195	
V_2O_3	c	−296	−277
V_2O_4	c	−342	−316
V_2O_5	c	−373	−342
Zinc			
Zn	c	0.00	0.00
ZnSb	c	−3.6	−3.88
$ZnBr_2$	c	−77.0	−72.9
	aq, 400	−93.6	
$Zn(C_2H_3O_2)_2$	c	−259.4	
	aq, 400	−269.4	−214.4
$Zn(CN)_2$	c	17.06	
$ZnCO_3$	c	−192.9	−173.5
$ZnCl_2$	c	−99.9	−88.8
	aq, 400	−115.44	
ZnF_2	aq	−192.9	−166.6
ZnI_2	c	−50.50	−49.93
	aq	−61.6	
$Zn(NO_3)_2$	aq, 400	−134.9	−87.7
ZnO	c, hexagonal	−83.36	−76.19
$ZnO{\cdot}SiO_2$	c	−282.6	
$Zn(OH)_2$	c, rhombic	−153.66	
ZnS	c, wurtzite	−45.3	−44.2
$ZnSO_4$	c	−233.4	
	aq, 400	−252.12	−211.28
Zirconium			
Zr	c	0.00	0.00
ZrC	c	−29.8	−34.6
$ZrCl_4$	c	−268.9	
ZrN	c	−82.5	−75.9
ZrO_2	c, monoclinic	−258.5	−244.6
$Zr(OH)_4$	c	−411.0	
$ZrO(OH)_2$	c	−337	−307.6

†The physical state is indicated as follows: *c*, crystal (solid); *l*, liquid; *g*, gas; *gls*, glass or solid supercooled liquid; *aq*, in aqueous solution. A number following the symbol *aq* applies only to the values of the heats of formation (not to those of free energies of formation); and indicates the number of moles of water per mole of solute; when no number is given, the solution is understood to be dilute. For the free energy of formation of a substance in aqueous solution, the concentration is always that of the hypothetical solution of unit molality.

‡The increment in heat content, ΔH, is the reaction of forming the given substance from its elements in their standard states. When ΔH is negative, heat is evolved in the process, and, when positive, heat is absorbed.

§The heat of solution in water of a given solid, liquid, or gaseous compound is given by the difference in the value for the heat of formation of the given compound in the solid, liquid, or gaseous state and its heat of formation in aqueous solution. The following two examples serve as an illustration of the procedure: (1) For NaCl(*c*) and NaCl(*aq*, 400H_2O), the values of ΔH(formation) are, respectively, −98.321 and −97.324 kcal/mol. Subtraction of the first value from the second gives ΔH = 0.998 kcal/mol for the reaction of dissolving crystalline sodium chloride in 400 mol of water. When this process occurs at a constant pressure of 1 atm, 0.998 kg-cal of energy are absorbed. (2) For HCl(*g*) and HCl(*aq*, 400H_2O), the values for ΔH(formation) are, respectively, −22.06 and −39.85 kcal/mol. Subtraction of the first from the second gives ΔH = −17.79 kcal/mol for the reaction of dissolving gaseous hydrogen chloride in 400 mol of water. At a constant pressure of 1 atm, 17.79 kcal of energy are evolved in this process.

‖The increment in the free energy, ΔF, is the reaction of forming the given substance in its standard state from its elements in their standard states. The standard states are: for a gas, fugacity (approximately equal to the pressure) of 1 atm; for a pure liquid or solid, the substance at a pressure of 1 atm; for a substance in aqueous solution, the hypothetical solution of unit molality, which has all the properties of the infinitely dilute solution except the property of concentration.

¶The free energy of solution of a given substance from its normal standard state as a solid, liquid, or gas to the hypothetical one molal state in aqueous solution may be calculated in a manner similar to that described in footnote § for calculating the heat of solution.

TABLE 2-95 Enthalpies and Gibbs Energies of Formation, Entropies, and Net Enthalpies of Combustion of Inorganic and Organic Compounds at 298.15 K

Cmpd. no.	Name	Formula	CAS	Mol. wt.	Ideal gas enthalpy of formation, J/kmol × 1E-07	Ideal gas Gibbs energy of formation, J/kmol × 1E-07	Ideal gas entropy, J/(kmol·K) × 1E-05	Standard net enthalpy of combustion, J/kmol × 1E-09
1	Acetaldehyde	C_2H_4O	75-07-0	44.05256	−17.1	−13.78	2.6384	−1.1046
2	Acetamide	C_2H_5NO	60-35-5	59.0672	−23.83	−15.96	2.722	−1.0741
3	Acetic acid	$C_2H_4O_2$	64-19-7	60.052	−43.28	−37.45	2.825	−0.7866
4	Acetic anhydride	$C_4H_6O_3$	108-24-7	102.08864	−57.55	−47.6	3.899	−1.675
5	Acetone	C_3H_6O	67-64-1	58.07914	−21.57	−15.13	2.954	−1.659
6	Acetonitrile	C_2H_3N	75-05-8	41.0519	6.467	8.241	2.438	−1.18118
7	Acetylene	C_2H_2	74-86-2	26.03728	22.82	21.068	2.0081	−1.257
8	Acrolein	C_3H_4O	107-02-8	56.06326	−8.18	−5.68	2.97	−1.5468
9	Acrylic acid	$C_3H_4O_2$	79-10-7	72.06266	−35.591	−30.6	3.15	−1.32717
10	Acrylonitrile	C_3H_3N	107-13-1	53.0626	17.97	18.92	2.77267	−1.71238
11	Air	Mixture	132259-10-0	28.96	0	0	1.94452	0
12	Ammonia	H_3N	7664-41-7	17.03052	−4.5898	−1.64	1.9266	−0.31683
13	Anisole	C_7H_8O	100-66-3	108.13782	−6.79	2.27	3.61	−3.6072
14	Argon	Ar	7440-37-1	39.948	0	0	1.54845	0
15	Benzamide	C_7H_7NO	55-21-0	121.13658	−10.09	−0.211	3.641	−3.39877
16	Benzene	C_6H_6	71-43-2	78.11184	8.288	12.96	2.693	−3.136
17	Benzenethiol	C_6H_6S	108-98-5	110.17684	11.15	14.76	3.369	−3.4474
18	Benzoic acid	$C_7H_6O_2$	65-85-0	122.12134	−29.41	−21.42	3.69	−3.0951
19	Benzonitrile	C_7H_5N	100-47-0	103.1213	21.57	25.8	3.21	−3.524
20	Benzophenone	$C_{13}H_{10}O$	119-61-9	182.2179	5.68	17.3	4.4	−6.2876
21	Benzyl alcohol	C_7H_8O	100-51-6	108.13782	−9.025	−0.254	3.713	−3.56
22	Benzyl ethyl ether	$C_9H_{12}O$	539-30-0	136.19098	−11.5	3.37	4.39	−4.83
23	Benzyl mercaptan	C_7H_8S	100-53-8	124.20342	9.33	16.3	3.607	−4.06
24	Biphenyl	$C_{12}H_{10}$	92-52-4	154.2078	17.849	27.63	3.9367	−6.248
25	Bromine	Br_2	7726-95-6	159.808	3.091	0.314	2.4535	0
26	Bromobenzene	C_6H_5Br	108-86-1	157.0079	10.5018	13.8532	3.24386	−3.01917
27	Bromoethane	C_2H_5Br	74-96-4	108.965	−6.36	−2.574	2.873	−1.301
28	Bromomethane	CH_3Br	74-83-9	94.93852	−3.77	−2.7037	2.421	−0.7185
29	1,2-Butadiene	C_4H_6	590-19-2	54.09044	16.23	19.86	2.93	−2.4617
30	1,3-Butadiene	C_4H_6	106-99-0	54.09044	10.924	14.972	2.7889	−2.409
31	Butane	C_4H_{10}	106-97-8	58.1222	−12.579	−1.67	3.0991	−2.65732
32	1,2-Butanediol	$C_4H_{10}O_2$	584-03-2	90.121	−44.58	−30.44	4.065	−2.2678
33	1,3-Butanediol	$C_4H_{10}O_2$	107-88-0	90.121	−43.32	−29.18	4.065	−2.2824
34	1-Butanol	$C_4H_{10}O$	71-36-3	74.1216	−27.51	−15.07	3.618	−2.454
35	2-Butanol	$C_4H_{10}O$	78-92-2	74.1216	−29.29	−16.7	3.566	−2.446
36	1-Butene	C_4H_8	106-98-9	56.10632	−0.05	7.041	3.074	−2.5408
37	*cis*-2-Butene	C_4H_8	590-18-1	56.10632	−0.74	6.536	3.012	−2.5339
38	*trans*-2-Butene	C_4H_8	624-64-6	56.10632	−1.1	6.32	2.965	−2.53
39	Butyl acetate	$C_6H_{12}O_2$	123-86-4	116.15828	−48.56	−31.26	4.425	−3.28
40	Butylbenzene	$C_{10}H_{14}$	104-51-8	134.21816	−1.314	14.54	4.3949	−5.5644
41	Butyl mercaptan	$C_4H_{10}S$	109-79-5	90.1872	−8.78	1.139	3.752	−2.9554
42	*sec*-Butyl mercaptan	$C_4H_{10}S$	513-53-1	90.1872	−9.66	0.512	3.667	−2.949
43	1-Butyne	C_4H_6	107-00-6	54.09044	16.52	20.225	2.9039	−2.4647
44	Butyraldehyde	C_4H_8O	123-72-8	72.10572	−20.62	−11.48	3.418	−2.301
45	Butyric acid	$C_4H_8O_2$	107-92-6	88.1051	−47.58	−36	3.601	−2.008
46	Butyronitrile	C_4H_7N	109-74-0	69.1051	3.342	10.57	3.337	−2.4146
47	Carbon dioxide	CO_2	124-38-9	44.0095	−39.351	−39.437	2.13677	
48	Carbon disulfide	CS_2	75-15-0	76.1407	11.69	6.68	2.379	−1.0769
49	Carbon monoxide	CO	630-08-0	28.0101	−11.053	−13.715	1.97556	−0.283

(*Continued*)

TABLE 2-95 Enthalpies and Gibbs Energies of Formation, Entropies, and Net Enthalpies of Combustion of Inorganic and Organic Compounds at 298.15 K *(Continued)*

Cmpd. no.	Name	Formula	CAS	Mol. wt.	Ideal gas enthalpy of formation, J/kmol × 1E-07	Ideal gas Gibbs energy of formation, J/kmol × 1E-07	Ideal gas entropy, J/(kmol·K) × 1E-05	Standard net enthalpy of combustion, J/kmol × 1E-09
50	Carbon tetrachloride	CCl_4	56-23-5	153.8227	−9.581	−5.354	3.0991	−0.2653
51	Carbon tetrafluoride	CF_4	75-73-0	88.0043	−92.21	−87.76	2.62	0.5286
52	Chlorine	Cl_2	7782-50-5	70.906	0	0	2.23079	0
53	Chlorobenzene	C_6H_5Cl	108-90-7	112.5569	5.109	9.829	3.1403	−2.976
54	Chloroethane	C_2H_5Cl	75-00-3	64.5141	−11.23	−6.045	2.758	−1.279
55	Chloroform	$CHCl_3$	67-66-3	119.37764	−10.29	−7.01	2.956	−0.38
56	Chloromethane	CH_3Cl	74-87-3	50.4875	−8.57	−6.209	2.341	−0.6705
57	1-Chloropropane	C_3H_7Cl	540-54-5	78.54068	−13.32	−5.251	3.155	−1.864
58	2-Chloropropane	C_3H_7Cl	75-29-6	78.54068	−14.477	−6.136	3.0594	−1.863
59	*m*-Cresol	C_7H_8O	108-39-4	108.13782	−13.23	−4.019	3.5604	−3.52783
60	*o*-Cresol	C_7H_8O	95-48-7	108.13782	−12.857	−3.543	3.5259	−3.528
61	*p*-Cresol	C_7H_8O	106-44-5	108.13782	−12.535	−3.166	3.5075	−3.52256
62	Cumene	C_9H_{12}	98-82-8	120.19158	0.4	13.79	3.86	−4.951
63	Cyanogen	C_2N_2	460-19-5	52.0348	30.894	29.76	2.4117	−1.096
64	Cyclobutane	C_4H_8	287-23-0	56.10632	2.85	11.22	2.64396	−2.5678
65	Cyclohexane	C_6H_{12}	110-82-7	84.15948	−12.33	3.191	2.97276	−3.656
66	Cyclohexanol	$C_6H_{12}O$	108-93-0	100.15888	−28.62	−10.95	3.277	−3.4639
67	Cyclohexanone	$C_6H_{10}O$	108-94-1	98.143	−22.61	−9.028	3.3426	−3.299
68	Cyclohexene	C_6H_{10}	110-83-8	82.1436	−0.46	10.77	3.10518	−3.532
69	Cyclopentane	C_5H_{10}	287-92-3	70.1329	−7.703	3.885	2.929	−3.0709
70	Cyclopentene	C_5H_8	142-29-0	68.11702	3.23	11.05	2.91267	−2.9393
71	Cyclopropane	C_3H_6	75-19-4	42.07974	5.33	10.44	2.37378	−1.9593
72	Cyclohexyl mercaptan	$C_6H_{12}S$	1569-69-3	116.22448	−9.602	4.886	3.646	−3.968
73	Decanal	$C_{10}H_{20}O$	112-31-2	156.2652	−33.17	−6.349	5.672	−5.958
74	Decane	$C_{10}H_{22}$	124-18-5	142.28168	−24.946	3.318	5.457	−6.29422
75	Decanoic acid	$C_{10}H_{20}O_2$	334-48-5	172.265	−59.43	−30.5	5.99	−5.72
76	1-Decanol	$C_{10}H_{22}O$	112-30-1	158.28108	−39.85	−10.02	5.971	−6.116
77	1-Decene	$C_{10}H_{20}$	872-05-9	140.2658	−12.47	12.27	5.433	−6.1809
78	Decyl mercaptan	$C_{10}H_{22}S$	143-10-2	174.34668	−21.09	6.165	6.116	−6.6161
79	1-Decyne	$C_{10}H_{18}$	764-93-2	138.24992	4.1	25.16	5.263	−6.1037
80	Deuterium	D_2	7782-39-0	4.0316	0	0	1.4486	−0.24625
81	1,1-Dibromoethane	$C_2H_4Br_2$	557-91-5	187.86116	−4.08	−1.181	3.276	−1.16
82	1,2-Dibromoethane	$C_2H_4Br_2$	106-93-4	187.86116	−3.89	−1.054	3.297	−1.1769
83	Dibromomethane	CH_2Br_2	74-95-3	173.83458			2.92964	
84	Dibutyl ether	$C_8H_{18}O$	142-96-1	130.22792	−33.34	−8.827	5.014	−4.94691
85	*m*-Dichlorobenzene	$C_6H_4Cl_2$	541-73-1	147.00196	2.57	7.79	3.4353	−2.825
86	*o*-Dichlorobenzene	$C_6H_4Cl_2$	95-50-1	147.00196	3.02	8.29	3.4185	−2.826
87	*p*-Dichlorobenzene	$C_6H_4Cl_2$	106-46-7	147.00196	2.25	7.67	3.3674	−2.802
88	1,1-Dichloroethane	$C_2H_4Cl_2$	75-34-3	98.95916	−12.941	−7.259	3.0501	−1.1104
89	1,2-Dichloroethane	$C_2H_4Cl_2$	107-06-2	98.95916	−12.979	−7.3945	3.0828	−1.105
90	Dichloromethane	CH_2Cl_2	75-09-2	84.93258	−9.552	−6.896	2.7018	−0.51388
91	1,1-Dichloropropane	$C_3H_6Cl_2$	78-99-9	112.98574	−15.08	−6.52	3.448	−1.72
92	1,2-Dichloropropane	$C_3H_6Cl_2$	78-87-5	112.98574	−16.28	−8.018	3.548	−1.707
93	Diethanol amine	$C_4H_{11}NO_2$	111-42-2	105.13564	−40.847	−22.574	4.29	−2.4105
94	Diethyl amine	$C_4H_{11}N$	109-89-7	73.13684	−7.142	7.308	3.522	−2.8003
95	Diethyl ether	$C_4H_{10}O$	60-29-7	74.1216	−25.21	−12.21	3.423	−2.5035
96	Diethyl sulfide	$C_4H_{10}S$	352-93-2	90.1872	−8.356	1.774	3.681	−2.9607
97	1,1-Difluoroethane	$C_2H_4F_2$	75-37-6	66.04997	−49.7	−43.9485	2.824	−0.773662
98	1,2-Difluoroethane	$C_2H_4F_2$	624-72-6	66.04997	−44.77	−39.19	2.88194	−0.823

99	Difluoromethane	CH_2F_2	75-10-5	52.02339	−45.23	−42.4747	2.4658	−0.183031
100	Diisopropyl amine	$C_6H_{15}N$	108-18-9	101.19	−14.38	6.42	4.12	−3.99
101	Diisopropyl ether	$C_6H_{14}O$	108-20-3	102.17476	−31.92	−12.48	3.989	−3.70261
102	Diisopropyl ketone	$C_7H_{14}O$	565-80-0	114.18546	−31.14	−12.37	4.27	−4.095
103	1,1-Dimethoxyethane	$C_4H_{10}O_2$	534-15-6	90.121	−38.97	−23.8	3.726	−2.394
104	1,2-Dimethoxypropane	$C_5H_{12}O_2$	7778-85-0	104.14758	−38.42	−20.11	4.038	−2.996
105	Dimethyl acetylene	C_4H_6	503-17-3	54.09044	14.57	18.49	2.833	−2.4189
106	Dimethyl amine	C_2H_7N	124-40-3	45.08368	−1.845	6.839	2.7296	−1.6146
107	2,3-Dimethylbutane	C_6H_{14}	79-29-8	86.17536	−17.68	−0.3125	3.6592	−3.84761
108	1,1-Dimethylcyclohexane	C_8H_{16}	590-66-9	112.21264	−18.1	3.52293	3.65012	−4.8639
109	*cis*-1,2-Dimethylcyclohexane	C_8H_{16}	2207-01-4	112.21264	−17.2172	4.12124	3.7451	−4.87084
110	*trans*-1,2-Dimethylcyclohexane	C_8H_{16}	6876-23-9	112.21264	−17.9996	3.44761	3.70912	−4.86435
111	Dimethyl disulfide	$C_2H_6S_2$	624-92-0	94.19904	−2.42	1.516	3.35291	−2.0441
112	Dimethyl ether	C_2H_6O	115-10-6	46.06844	−18.41	−11.28	2.667	−1.3284
113	*N,N*-Dimethyl formamide	C_3H_7NO	68-12-2	73.09378	−19.17	−8.84	3.26	−1.78871
114	2,3-Dimethylpentane	C_7H_{16}	565-59-3	100.20194	−19.41	0.5717	4.1455	−4.46075
115	Dimethyl phthalate	$C_{10}H_{10}O_4$	131-11-3	194.184	−60.5	−46.7749	6.6	−4.4662
116	Dimethylsilane	C_2H_8Si	1111-74-6	60.17042	−9.47	−1.925	2.9953	−2.569
117	Dimethyl sulfide	C_2H_6S	75-18-3	62.134	−3.724	0.7302	2.8585	−1.7443
118	Dimethyl sulfoxide	C_2H_6OS	67-68-5	78.13344	−15.046	−8.1441	3.0627	−1.6054
119	Dimethyl terephthalate	$C_{10}H_{10}O_4$	120-61-6	194.184	−62.742	−41.97	4.245	−4.41057
120	1,4-Dioxane	$C_4H_8O_2$	123-91-1	88.10512	−31.58	−18.16	3.0012	−2.1863
121	Diphenyl ether	$C_{12}H_{10}O$	101-84-8	170.2072	5.2	17.5	4.13	−5.8939
122	Dipropyl amine	$C_6H_{15}N$	142-84-7	101.19	−11.6	11.96	3.2	−4.0189
123	Dodecane	$C_{12}H_{26}$	112-40-3	170.33484	−29.072	4.981	6.2415	−7.51368
124	Eicosane	$C_{20}H_{42}$	112-95-8	282.54748	−45.646	11.57	9.3787	−12.3908
125	Ethane	C_2H_6	74-84-0	30.069	−8.382	−3.192	2.2912	−1.42864
126	Ethanol	C_2H_6O	64-17-5	46.06844	−23.495	−16.785	2.8064	−1.235
127	Ethyl acetate	$C_4H_8O_2$	141-78-6	88.10512	−44.45	−32.8	3.597	−2.061
128	Ethyl amine	C_2H_7N	75-04-7	45.08368	−4.715	3.616	2.848	−1.5874
129	Ethylbenzene	C_8H_{10}	100-41-4	106.165	2.992	13.073	3.6063	−4.3448
130	Ethyl benzoate	$C_9H_{10}O_2$	93-89-0	150.1745	−32.6	−19.05	4.55	−4.41
131	2-Ethyl butanoic acid	$C_6H_{12}O_2$	88-09-5	116.15828	−53.78	−35.9	4.23	−3.21203
132	Ethyl butyrate	$C_6H_{12}O_2$	105-54-4	116.15828	−48.55	−31.22	4.417	−3.284
133	Ethylcyclohexane	C_8H_{16}	1678-91-7	112.21264	−17.15	3.955	3.826	−4.87051
134	Ethylcyclopentane	C_7H_{14}	1640-89-7	98.18606	−12.69	4.48	3.783	−4.2839
135	Ethylene	C_2H_4	74-85-1	28.05316	5.251	6.844	2.192	−1.323
136	Ethylenediamine	$C_2H_8N_2$	107-15-3	60.09832	−1.73	10.3	3.21833	−1.691
137	Ethylene glycol	$C_2H_6O_2$	107-21-1	62.06784	−39.22	−30.18	3.04891	−1.0527
138	Ethyleneimine	C_2H_5N	151-56-4	43.0678	12.3428	17.7987	2.5062	−1.481
139	Ethylene oxide	C_2H_4O	75-21-8	44.05256	−5.263	−1.323	2.4299	−1.218
140	Ethyl formate	$C_3H_6O_2$	109-94-4	74.07854	−38.83	−30.31	3.282	−1.50696
141	2-Ethyl hexanoic acid	$C_8H_{16}O_2$	149-57-5	144.211	−55.95	−32.49	5.097	−4.448
142	Ethylhexyl ether	$C_8H_{18}O$	5756-43-4	130.22792	−33.37	−9.042	5.076	−4.943
143	Ethylisopropyl ether	$C_5H_{12}O$	625-54-7	88.14818	−28.58	−12.64	3.8	−3.103
144	Ethylisopropyl ketone	$C_6H_{12}O$	565-69-5	100.15888	−28.61	−13.3	4.069	−3.4863
145	Ethyl mercaptan	C_2H_6S	75-08-1	62.13404	−4.63	−0.4814	2.961	−1.7366
146	Ethyl propionate	$C_5H_{10}O_2$	105-37-3	102.1317	−46.36	−31.93	4.025	−2.674
147	Ethylpropyl ether	$C_5H_{12}O$	628-32-0	88.14818	−27.22	−11.52	3.881	−3.12
148	Ethyltrichlorosilane	$C_2H_5Cl_3Si$	115-21-9	163.506	−59.15	−50.66	4.07	−1.67471
149	Fluorine	F_2	7782-41-4	37.9968064	0	0	2.02789	
150	Fluorobenzene	C_6H_5F	462-06-6	96.1023032	−11.6566	−6.9036	3.02629	−2.81451

(*Continued*)

TABLE 2-95 Enthalpies and Gibbs Energies of Formation, Entropies, and Net Enthalpies of Combustion of Inorganic and Organic Compounds at 298.15 K *(Continued)*

Cmpd. no.	Name	Formula	CAS	Mol. wt.	Ideal gas enthalpy of formation, J/kmol × 1E-07	Ideal gas Gibbs energy of formation, J/kmol × 1E-07	Ideal gas entropy, J/(kmol·K) × 1E-05	Standard net enthalpy of combustion, J/kmol × 1E-09
151	Fluoroethane	C_2H_5F	353-36-6	48.0595	−26.44	−21.23	2.644	−1.127
152	Fluoromethane	CH_3F	593-53-3	34.03292	−23.43	−21.03	2.22734	−0.5219
153	Formaldehyde	CH_2O	50-00-0	30.02598	−10.86	−10.26	2.19	−0.5268
154	Formamide	CH_3NO	75-12-7	45.04062	−19.22	−14.71	2.4857	−0.5021
155	Formic acid	CH_2O_2	64-18-6	46.0257	−37.88	−35.11	2.487	−0.2115
156	Furan	C_4H_4O	110-00-9	68.07396	−3.48	0.08225	2.6714	−1.9959
157	Helium-4	He	7440-59-7	4.0026	0	0	1.26152	0
158	Heptadecane	$C_{17}H_{36}$	629-78-7	240.46774	−39.445	9.083	8.2023	−10.5618
159	Heptanal	$C_7H_{14}O$	111-71-7	114.18546	−26.48	−8.367	4.5	−4.136
160	Heptane	C_7H_{16}	142-82-5	100.20194	−18.765	0.8165	4.2798	−4.46473
161	Heptanoic acid	$C_7H_{14}O_2$	111-14-8	130.185	−53.62	−33.4	4.8	−3.839
162	1-Heptanol	$C_7H_{16}O$	111-70-6	116.20134	−33.68	−12.55	4.795	−4.285
163	2-Heptanol	$C_7H_{16}O$	543-49-7	116.20134	−35.3	−13.7	4.66	−4.27
164	3-Heptanone	$C_7H_{14}O$	106-35-4	114.18546	−30.1	−12.25	4.58	−4.098
165	2-Heptanone	$C_7H_{14}O$	110-43-0	114.18546	−30.0453	−11.96	4.486	−4.09952
166	1-Heptene	C_7H_{14}	592-76-7	98.18606	−6.289	9.482	4.252	−4.3499
167	Heptyl mercaptan	$C_7H_{16}S$	1639-09-4	132.26694	−14.95	3.622	4.939	−4.7865
168	1-Heptyne	C_7H_{12}	628-71-7	96.17018	10.3	22.7	4.085	−4.2717
169	Hexadecane	$C_{16}H_{34}$	544-76-3	226.44116	−37.417	8.216	7.8102	−9.95145
170	Hexanal	$C_6H_{12}O$	66-25-1	100.15888	−24.8	−9.92	4.22	−3.524
171	Hexane	C_6H_{14}	110-54-3	86.17536	−16.694	−0.006634	3.8874	−3.8551
172	Hexanoic acid	$C_6H_{12}O_2$	142-62-1	116.158	−51.19	−33.8	4.41	−3.23
173	1-Hexanol	$C_6H_{14}O$	111-27-3	102.17476	−31.62	−13.39	4.402	−3.675
174	2-Hexanol	$C_6H_{14}O$	626-93-7	102.175	−33.46	−15.06	4.349	−3.67
175	2-Hexanone	$C_6H_{12}O$	591-78-6	100.15888	−27.9826	−13.0081	4.17856	−3.49
176	3-Hexanone	$C_6H_{12}O$	589-38-8	100.15888	−27.76	−12.6	4.092	−3.492
177	1-Hexene	C_6H_{12}	592-41-6	84.15948	−4.167	8.7	3.863	−3.7397
178	3-Hexyne	C_6H_{10}	928-49-4	82.1436	10.6	19.9	3.76	−3.64
179	Hexyl mercaptan	$C_6H_{14}S$	111-31-9	118.24036	−12.92	2.759	4.546	−4.1762
180	1-Hexyne	C_6H_{10}	693-02-7	82.1436	12.37	21.85	3.694	−3.661
181	2-Hexyne	C_6H_{10}	764-35-2	82.1436	10.5	19.9	3.72	−3.64
182	Hydrazine	H_4N_2	302-01-2	32.04516	9.5353	15.917	2.3861	−0.5342
183	Hydrogen	H_2	1333-74-0	2.01588	0	0	1.30571	−0.24182
184	Hydrogen bromide	BrH	10035-10-6	80.91194	−3.629	−5.334	1.98591	−0.06904
185	Hydrogen chloride	ClH	7647-01-0	36.46094	−9.231	−9.53	1.86786	−0.0286
186	Hydrogen cyanide	CHN	74-90-8	27.02534	13.5143	12.4725	2.01719	−0.62329
187	Hydrogen fluoride	FH	7664-39-3	20.0063432	−27.33	−27.54	1.7367	0.1524
188	Hydrogen sulfide	H_2S	7783-06-4	34.08088	−2.063	−3.344	2.056	−0.518
189	Isobutyric acid	$C_4H_8O_2$	79-31-2	88.10512	−48.41	−36.21	3.412	−2.0004
190	Isopropyl amine	C_3H_9N	75-31-0	59.11026	−8.38	3.192	3.124	−2.1566
191	Malonic acid	$C_3H_4O_4$	141-82-2	104.06146	−77.89	−69.29	4.003	−0.7732
192	Methacrylic acid	$C_4H_6O_2$	79-41-4	86.08924	−36.8	−28.8	3.5	−1.93
193	Methane	CH_4	74-82-8	16.0425	−7.452	−5.049	1.8627	−0.80262
194	Methanol	CH_4O	67-56-1	32.04186	−20.094	−16.232	2.3988	−0.6382
195	*N*-Methyl acetamide	C_3H_7NO	79-16-3	73.09378	−24	−13.5	3.2	−1.71
196	Methyl acetate	$C_3H_6O_2$	79-20-9	74.07854	−41.19	−32.42	3.198	−1.461
197	Methyl acetylene	C_3H_4	74-99-7	40.06386	18.49	19.384	2.4836	−1.8487
198	Methyl acrylate	$C_4H_6O_2$	96-33-3	86.08924	−33.3	−25.7	3.66	−1.9303
199	Methyl amine	CH_5N	74-89-5	31.0571	−2.297	3.207	2.433	−0.97508

200	Methyl benzoate	$C_8H_8O_2$	93-58-3	136.14792	−28.79	−18.1	4.14	−3.772
201	3-Methyl-1,2-butadiene	C_5H_8	598-25-4	68.11702	12.908	19.75	3.2151	−3.032
202	2-Methylbutane	C_5H_{12}	78-78-4	72.14878	−15.37	−1.405	3.4374	−3.23954
203	2-Methylbutanoic acid	$C_5H_{10}O_2$	116-53-0	102.1317	−49.8	−34.99	3.9	−2.622
204	3-Methyl-1-butanol	$C_5H_{12}O$	123-51-3	88.1482	−30.3	−14.54	3.869	−3.062
205	2-Methyl-1-butene	C_5H_{10}	563-46-2	70.1329	−3.53	6.668	3.395	−3.1159
206	2-Methyl-2-butene	C_5H_{10}	513-35-9	70.1329	−4.18	6.045	3.386	−3.1088
207	2-Methyl -1-butene-3-yne	C_5H_6	78-80-8	66.10114	26	30.25	2.78	−2.93
208	Methylbutyl ether	$C_5H_{12}O$	628-28-4	88.14818	−25.81	−10.17	3.901	−3.12818
209	Methylbutyl sulfide	$C_5H_{12}S$	628-29-5	104.214	−10.2	2.691	4.118	−3.5723
210	3-Methyl-1-butyne	C_5H_8	598-23-2	68.11702	13.8	20.72	3.189	−3.046
211	Methyl butyrate	$C_5H_{10}O_2$	623-42-7	102.1317	−45.07	−30.53	3.988	−2.686
212	Methylchlorosilane	CH_5ClSi	993-00-0	80.5889	−21.5	−16.61	2.98277	−1.693
213	Methylcyclohexane	C_7H_{14}	108-87-2	98.18606	−15.48	2.733	3.433	−4.25714
214	1-Methylcyclohexanol	$C_7H_{14}O$	590-67-0	114.18546	−33.2	−12.9	3.75	−4.058
215	*cis*-2-Methylcyclohexanol	$C_7H_{14}O$	7443-70-1	114.18546	−32.7	−12.68	3.853	−4.0574
216	*trans*-2-Methylcyclohexanol	$C_7H_{14}O$	7443-52-9	114.18546	−35.26	−15.24	3.853	−4.0318
217	Methylcyclopentane	C_6H_{12}	96-37-7	84.15948	−10.62	3.63	3.399	−3.6741
218	1-Methylcyclopentene	C_6H_{10}	693-89-0	82.1436	−0.38	10.38	3.264	−3.534
219	3-Methylcyclopentene	C_6H_{10}	1120-62-3	82.1436	0.74	11.38	3.305	−3.5464
220	Methyldichlorosilane	CH_4Cl_2Si	75-54-7	115.03396	−40.2	−34.83	3.287	−1.357
221	Methylethyl ether	C_3H_8O	540-67-0	60.09502	−21.64	−11.71	3.0881	−1.9314
222	Methylethyl ketone	C_4H_8O	78-93-3	72.10572	−23.9	−14.7	3.394	−2.268
223	Methylethyl sulfide	C_3H_8S	624-89-5	76.1606	−5.96	1.147	3.332	−2.354
224	Methyl formate	$C_2H_4O_2$	107-31-3	60.05196	−35.24	−29.5	2.852	−0.8924
225	Methylisobutyl ether	$C_5H_{12}O$	625-44-5	88.14818	−26.6	−10.7	3.81	−3.122
226	Methylisobutyl ketone	$C_6H_{12}O$	108-10-1	100.15888	−28.64	−13.51	4.129	−3.4762
227	Methyl Isocyanate	C_2H_3NO	624-83-9	57.05132	−6.24	0.0244	1.955	−1.06
228	Methylisopropyl ether	$C_4H_{10}O$	598-53-8	74.1216	−25.2	−12.18	3.416	−2.5311
229	Methylisopropyl ketone	$C_5H_{10}O$	563-80-4	86.1323	−26.26	−13.93	3.699	−2.877
230	Methylisopropyl sulfide	$C_4H_{10}S$	1551-21-9	90.1872	−8.96	1.4509	3.59	−2.957
231	Methyl mercaptan	CH_4S	74-93-1	48.10746	−2.29	−0.98	2.55	−1.1517
232	Methyl methacrylate	$C_5H_8O_2$	80-62-6	100.11582	−36	−25.4	4.01	−2.54
233	2-Methyloctanoic acid	$C_9H_{18}O_2$	3004-93-1	158.23802	−57.95	−31.8	5.533	−5.056
234	2-Methylpentane	C_6H_{14}	107-83-5	86.17536	−17.455	−0.5338	3.8089	−3.84915
235	Methyl pentyl ether	$C_6H_{14}O$	628-80-8	102.17476	−27.8	−9.321	4.32	−3.739
236	2-Methylpropane	C_4H_{10}	75-28-5	58.1222	−13.499	−2.144	2.955	−2.64812
237	2-Methyl-2-propanol	$C_4H_{10}O$	75-65-0	74.1216	−31.24	−17.76	3.263	−2.4239
238	2-Methyl propene	C_4H_8	115-11-7	56.10632	−1.71	5.808	2.9309	−2.5242
239	Methyl propionate	$C_4H_8O_2$	554-12-1	88.10512	−42.75	−31.1	3.596	−2.078
240	Methylpropyl ether	$C_4H_{10}O$	557-17-5	74.1216	−23.82	−11.1	3.52	−2.51739
241	Methylpropyl sulfide	$C_4H_{10}S$	3877-15-4	90.1872	−8.23	1.793	3.717	−2.962
242	Methylsilane	CH_6Si	992-94-9	46.14384	−2.91	1.853	2.565	−1.999
243	alpha-Methyl styrene	C_9H_{10}	98-83-9	118.1757	11.83	21.73	3.725	−4.8214
244	Methyl tert-butyl ether	$C_5H_{12}O$	1634-04-4	88.1482	−28.3	−11.7	3.58	−3.11
245	Methyl vinyl ether	C_3H_6O	107-25-5	58.07914	−10.8	−4.73	3.08	−1.77431
246	Naphthalene	$C_{10}H_8$	91-20-3	128.17052	15.058	22.408	3.3315	−4.9809
247	Neon	Ne	7440-01-9	20.1797	0	0	1.46327	0
248	Nitroethane	$C_2H_5NO_2$	79-24-3	75.0666	−10.21	−0.6125	3.168	−1.25
249	Nitrogen	N_2	7727-37-9	28.0134	0	0	1.91609	
250	Nitrogen trifluoride	F_3N	7783-54-2	71.0019096	−13.2089	−9.063	2.60773	

(*Continued*)

TABLE 2-95 Enthalpies and Gibbs Energies of Formation, Entropies, and Net Enthalpies of Combustion of Inorganic and Organic Compounds at 298.15 K *(Continued)*

Cmpd. no.	Name	Formula	CAS	Mol. wt.	Ideal gas enthalpy of formation, J/kmol × 1E-07	Ideal gas Gibbs energy of formation, J/kmol × 1E-07	Ideal gas entropy, J/(kmol·K) × 1E-05	Standard net enthalpy of combustion, J/kmol × 1E-09
251	Nitromethane	CH_3NO_2	75-52-5	61.04002	−7.47	−0.6934	2.751	−0.6432
252	Nitrous oxide	N_2O	10024-97-2	44.0128	8.205	10.416	2.1985	−0.0820482
253	Nitric oxide	NO	10102-43-9	30.0061	9.025	8.657	2.106	−0.0902489
254	Nonadecane	$C_{19}H_{40}$	629-92-5	268.5209	−43.579	10.74	8.9866	−11.7812
255	Nonanal	$C_9H_{18}O$	124-19-6	142.23862	−31.09	−7.136	5.266	−5.35
256	Nonane	C_9H_{20}	111-84-2	128.2551	−22.874	2.498	5.064	−5.68455
257	Nonanoic acid	$C_9H_{18}O_2$	112-05-0	158.238	−57.73	−31.7	5.59	−5.061
258	1-Nonanol	$C_9H_{20}O$	143-08-8	144.2545	−37.79	−10.86	5.579	−5.506
259	2-Nonanol	$C_9H_{20}O$	628-99-9	144.255	−39.71	−12.61	5.523	−5.506
260	1-Nonene	C_9H_{18}	124-11-8	126.23922	−10.35	11.23	5.041	−5.5716
261	Nonyl mercaptan	$C_9H_{20}S$	1455-21-6	160.3201	−19.08	5.28	5.724	−6.006
262	1-Nonyne	C_9H_{16}	3452-09-3	124.22334	6.17	24.34	4.8699	−5.493
263	Octadecane	$C_{18}H_{38}$	593-45-3	254.49432	−41.512	9.91	8.5945	−11.1715
264	Octanal	$C_8H_{16}O$	124-13-0	128.212	−29.02	−8	4.896	−4.74
265	Octane	C_8H_{18}	111-65-9	114.22852	−20.875	1.6	4.6723	−5.07415
266	Octanoic acid	$C_8H_{16}O_2$	124-07-2	144.211	−55.6	−32.5	5.2	−4.448
267	1-Octanol	$C_8H_{18}O$	111-87-5	130.22792	−35.73	−11.7	5.187	−4.895
268	2-Octanol	$C_8H_{18}O$	123-96-6	130.228	−37.62	−13.43	5.132	−4.894
269	2-Octanone	$C_8H_{16}O$	111-13-7	128.21204	−32.16	−11.38	4.962	−4.6984
270	3-Octanone	$C_8H_{16}O$	106-68-3	128.21204	−33.9	−12.81	4.879	−4.711
271	1-Octene	C_8H_{16}	111-66-0	112.21264	−8.194	10.57	4.637	−4.961
272	Octyl mercaptan	$C_8H_{18}S$	111-88-6	146.29352	−17.01	4.457	5.331	−5.3962
273	1-Octyne	C_8H_{14}	629-05-0	110.19676	8.23	23.5	4.478	−4.88145
274	Oxalic acid	$C_2H_2O_4$	144-62-7	90.03488	−71.95	−66.24	3.608	−0.1989
275	Oxygen	O_2	7782-44-7	31.9988	0	0	2.05147	0
276	Ozone	O_3	10028-15-6	47.9982	14.2671	16.3164	2.38823	−0.142671
277	Pentadecane	$C_{15}H_{32}$	629-62-9	212.41458	−35.311	7.426	7.4181	−9.34237
278	Pentanal	$C_5H_{10}O$	110-62-3	86.1323	−22.78	−10.67	3.777	−2.91
279	Pentane	C_5H_{12}	109-66-0	72.14878	−14.676	−0.8813	3.4945	−3.24494
280	Pentanoic acid	$C_5H_{10}O_2$	109-52-4	102.132	−49.13	−34.7	4.02	−2.617
281	1-Pentanol	$C_5H_{12}O$	71-41-0	88.1482	−29.57	−14.23	4.01	−3.064
282	2-Pentanol	$C_5H_{12}O$	6032-29-7	88.1482	−31.37	−15.88	3.958	−3.058
283	2-Pentanone	$C_5H_{10}O$	107-87-9	86.1323	−25.92	−13.83	3.786	−2.87956
284	3-Pentanone	$C_5H_{10}O$	96-22-0	86.1323	−25.79	−13.44	3.7	−2.8804
285	1-Pentene	C_5H_{10}	109-67-1	70.1329	−2.162	7.837	3.462	−3.13037
286	2-Pentyl mercaptan	$C_5H_{12}S$	2084-19-7	104.21378	−11.3	1.814	4.05	−3.564
287	Pentyl mercaptan	$C_5H_{12}S$	110-66-7	104.21378	−10.84	1.94408	4.154	−3.5641
288	1-Pentyne	C_5H_8	627-19-0	68.11702	14.44	21.03	3.298	−3.051
289	2-Pentyne	C_5H_8	627-21-4	68.11702	12.89	19.45	3.3084	−3.0291
290	Phenanthrene	$C_{14}H_{10}$	85-01-8	178.2292	20.12	30.219	3.945	−6.8282
291	Phenol	C_6H_6O	108-95-2	94.11124	−9.6399	−3.2637	3.1481	−2.921
292	Phenyl isocyanate	C_7H_5NO	103-71-9	119.1207	−1.454	4.87212	3.527	−3.298
293	Phthalic anhydride	$C_8H_4O_3$	85-44-9	148.11556	−37.14	−30.7001	3.995	−3.1715
294	Propadiene	C_3H_4	463-49-0	40.06386	19.05	20.08	2.439	−1.8563
295	Propane	C_3H_8	74-98-6	44.09562	−10.468	−2.439	2.702	−2.04311
296	1-Propanol	C_3H_8O	71-23-8	60.09502	−25.46	−15.99	3.226	−1.844
297	2-Propanol	C_3H_8O	67-63-0	60.095	−27.21	−17.52	3.175	−1.834
298	Propenylcyclohexene	C_9H_{14}	13511-13-2	122.20746	4.677	20.85	4.233	−5.232
299	Propionaldehyde	C_3H_6O	123-38-6	58.07914	−18.49	−12.37	3.065	−1.684

300	Propionic acid	$C_3H_6O_2$	79-09-4	74.0785	−45.35	−35.82	2.949	−1.395
301	Propionitrile	C_3H_5N	107-12-0	55.0785	5.155	9.688	2.877	−1.80056
302	Propyl acetate	$C_5H_{10}O_2$	109-60-4	102.1317	−46.48	−32.04	4.023	−2.672
303	Propyl amine	C_3H_9N	107-10-8	59.11026	−7.05	4.17	3.242	−2.165
304	Propylbenzene	C_9H_{12}	103-65-1	120.19158	0.79	13.76	4.0014	−4.95415
305	Propylene	C_3H_6	115-07-1	42.07974	2.023	6.264	2.67	−1.9262
306	Propyl formate	$C_4H_8O_2$	110-74-7	88.10512	−40.76	−29.36	3.678	−2.041
307	2-Propyl mercaptan	C_3H_8S	75-33-2	76.16062	−7.59	−0.218	3.243	−2.3398
308	Propyl mercaptan	C_3H_8S	107-03-9	76.16062	−6.75	0.2583	3.365	−2.3458
309	1,2-Propylene glycol	$C_3H_8O_2$	57-55-6	76.09442	−42.15	−30.4	3.52	−1.6476
310	Quinone	$C_6H_4O_2$	106-51-4	108.09476	−12.29	−6.92	3.205	−2.658
311	Silicon tetrafluoride	F_4Si	7783-61-1	104.0791128	−161.494	−157.27	2.82651	0.7055
312	Styrene	C_8H_8	100-42-5	104.14912	14.74	21.39	3.451	−4.219
313	Succinic acid	$C_4H_6O_4$	110-15-6	118.08804	−81.6	−70.11	4.398	−1.3591
314	Sulfur dioxide	O_2S	7446-09-5	64.0638	−29.684	−30.012	2.481	
315	Sulfur hexafluoride	F_6S	2551-62-4	146.0554192	−122.047	−111.653	2.91625	0.924
316	Sulfur trioxide	O_3S	7446-11-9	80.0632	−39.572	−37.095	2.5651	0.1422
317	Terephthalic acid	$C_8H_6O_4$	100-21-0	166.13084	−66.94	−55.01	4.48	−3.19
318	*o*-Terphenyl	$C_{18}H_{14}$	84-15-1	230.30376	27.66	42.3	5.263	−9.053
319	Tetradecane	$C_{14}H_{30}$	629-59-4	198.388	−33.244	6.599	7.0259	−8.73282
320	Tetrahydrofuran	C_4H_8O	109-99-9	72.10572	−18.418	−7.969	2.9729	−2.325
321	1,2,3,4-Tetrahydronaphthalene	$C_{10}H_{12}$	119-64-2	132.20228	2.661	16.71	3.6964	−5.3575
322	Tetrahydrothiophene	C_4H_8S	110-01-0	88.17132	−3.376	4.59	3.1	−2.76549
323	2,2,3,3-Tetramethylbutane	C_8H_{18}	594-82-1	114.22852	−22.56	2.239	3.893	−5.0639
324	Thiophene	C_4H_4S	110-02-1	84.13956	11.544	12.67	2.784	−2.4352
325	Toluene	C_7H_8	108-88-3	92.13842	5.017	12.22	3.2099	−3.734
326	1,1,2-Trichloroethane	$C_2H_3Cl_3$	79-00-5	133.40422	−14.2	−8.097	3.371	−0.9685
327	Tridecane	$C_{13}H_{28}$	629-50-5	184.36142	−31.177	5.771	6.6337	−8.1229
328	Triethyl amine	$C_6H_{15}N$	121-44-8	101.19	−9.58	11.41	4.054	−4.0405
329	Trimethyl amine	C_3H_9N	75-50-3	59.11026	−2.431	9.899	2.87	−2.2449
330	1,2,3-Trimethylbenzene	C_9H_{12}	526-73-8	120.19158	−0.95	12.61	3.805	−4.934
331	1,2,4-Trimethylbenzene	C_9H_{12}	95-63-6	120.19158	−1.38	11.71	3.961	−4.9307
332	2,2,4-Trimethylpentane	C_8H_{18}	540-84-1	114.22852	−22.401	1.394	4.2296	−5.06528
333	2,3,3-Trimethylpentane	C_8H_{18}	560-21-4	114.22852	−21.845	1.828	4.2702	−5.06876
334	1,3,5-Trinitrobenzene	$C_6H_3N_3O_6$	99-35-4	213.10452	6.24	26.79	4.435	−2.6867
335	2,4,6-Trinitrotoluene	$C_7H_5N_3O_6$	118-96-7	227.1311	4.34	28.44	4.607	−3.2959
336	Undecane	$C_{11}H_{24}$	1120-21-4	156.30826	−27.043	4.116	5.8493	−6.9036
337	1-Undecanol	$C_{11}H_{24}O$	112-42-5	172.30766	−41.9	−9.177	6.363	−6.726
338	Vinyl acetate	$C_4H_6O_2$	108-05-4	86.08924	−31.49	−22.79	3.28	−1.95
339	Vinyl acetylene	C_4H_4	689-97-4	52.07456	30.46	30.6	2.794	−2.362
340	Vinyl chloride	C_2H_3Cl	75-01-4	62.49822	2.845	4.195	2.7354	−1.178
341	Vinyl trichlorosilane	$C_2H_3Cl_3Si$	75-94-5	161.48972	−48.116	−42.5514	3.73966	−1.544
342	Water	H_2O	7732-18-5	18.01528	−24.1818	−22.8572	1.88825	
343	*m*-Xylene	C_8H_{10}	108-38-3	106.165	1.732	11.876	3.5854	−4.3318
344	*o*-Xylene	C_8H_{10}	95-47-6	106.165	1.908	12.2	3.5383	−4.333
345	*p*-Xylene	C_8H_{10}	106-42-3	106.165	1.803	12.14	3.52165	−4.333

The compounds are considered to be formed from the elements in their standard states at 298.15 K and 1 bar. These include C (graphite) and S (rhombic). Enthalpy of combustion is the net value for the compound in its standard state at 298.15 K and 1 bar. Products of combustion are taken to be CO_2 (gas), H_2O (gas), F_2(gas), Cl_2 (gas), Br_2 (gas), I_2 (gas), SO_2 (gas), N_2 (gas), P_4O_{10} (crystalline), SiO_2 (crystobalite), and Al_2O_3 (crystal, alpha).

Values in this table were taken from the Design Institute for Physical Properties (DIPPR) of the American Institute of Chemical Engineers (AIChE), 801 Critically Evaluated Gold Standard™ Database, copyright 2016 AIChE, and reproduced with permission of AIChE and of the DIPPR Evaluated Process Design Data Project Steering Committee. Their source should be cited as "R. L. Rowley, W. V. Wilding, J. L. Oscarson, T. A. Knotts, N. F. Giles, *DIPPR® Data Compilation of Pure Chemical Properties*, Design Institute for Physical Properties, AIChE, New York, NY (2016)".

TABLE 2-96 Ideal Gas Sensible Enthalpies, $h_T - h_{298}$ (kJ/kmol), of Combustion Products

Temperature, K	CO	CO_2	H	OH	H_2	N	NO	NO_2	N_2	N_2O	O	O_2	SO_2	H_2O
200	−2858	−3414	−2040	−2976	−2774	−2040	−2951	−3495	−2857	−3553	−2186	−2868	−3736	−3282
240	−1692	−2079	−1209	−1756	−1656	−1209	−1743	−2104	−1692	−2164	−1285	−1703	−2258	−1948
260	−1110	−1383	−793	−1150	−1091	−793	−1142	−1392	−1110	−1438	−840	−1118	−1496	−1279
280	−529	−665	−377	−546	−522	−378	−543	−672	−528	−692	−398	−533	−718	−609
298.15	0	0	0	0	0	0	0	0	0	0	0	0	0	0
300	54	69	38	55	53	38	55	68	54	72	41	54	74	62
320	638	823	454	654	630	454	652	816	636	854	478	643	881	735
340	1221	1594	870	1251	1209	870	1248	1571	1219	1654	913	1234	1702	1410
360	1805	2382	1285	1847	1791	1286	1845	2347	1802	2470	1346	1828	2538	2088
380	2389	3184	1701	2442	2373	1701	2442	3130	2386	3302	1777	2425	3387	2769
400	2975	4003	2117	3035	2959	2117	3040	3927	2971	4149	2207	3025	4250	3452
420	3563	4835	2532	3627	3544	2533	3638	4735	3557	5010	2635	3629	5126	4139
440	4153	5683	2948	4219	4131	2949	4240	5557	4143	5884	3063	4236	6015	4829
460	4643	6544	3364	4810	4715	3364	4844	6392	4731	6771	3490	4847	6917	5523
480	5335	7416	3779	5401	5298	3780	5450	7239	5320	7670	3918	5463	7831	6222
500	5931	8305	4196	5992	5882	4196	6059	8099	5911	8580	4343	6084	8758	6925
550	7428	10572	5235	7385	6760	5235	7592	10340	7395	10897	5402	7653	11123	8699
600	8942	12907	6274	8943	8811	6274	9144	12555	8894	13295	6462	9244	13544	10501
650	10477	15303	7314	10423	10278	7314	10716	14882	10407	15744	7515	10859	16022	12321
700	12023	17754	8353	11902	11749	8353	12307	17250	11937	18243	8570	12499	18548	14192
750	13592	20260	9392	13391	13223	9329	13919	19671	13481	20791	9620	14158	21117	16082
800	15177	22806	10431	14880	14702	10431	15548	22136	15046	23383	10671	15835	23721	18002
850	16781	25398	11471	16384	16186	11471	17195	24641	16624	26014	11718	17531	26369	19954
900	18401	28030	12510	17888	17676	12510	18858	27179	18223	28681	12767	19241	29023	21938
950	20031	30689	13550	19412	19175	13550	20537	29749	19834	31381	13812	20965	31714	23954
1000	21690	33397	14589	20935	20680	14589	22229	32344	21463	34110	14860	22703	34428	26000
1100	25035	38884	16667	24024	23719	16667	25653	37605	24760	39647	16950	26212	39914	30191
1200	28430	44473	18746	27160	26797	18746	29120	42946	28109	45274	19039	29761	45464	34506
1300	31868	50148	20824	30342	29918	20824	32626	48351	31503	50976	21126	33344	51069	38942
1400	35343	55896	22903	33569	33082	22903	36164	53808	34936	56740	23212	36957	56718	43493
1500	38850	61705	24982	36839	36290	24982	39729	59309	38405	62557	25296	40599	62404	48151
1600	42385	67569	27060	40151	39541	27060	43319	64846	41904	68420	27381	44266	68123	52908
1700	45945	73480	29139	43502	42835	29139	46929	70414	45429	74320	29464	47958	73870	57758
1800	49526	79431	31217	46889	46169	31218	50557	76007	48978	80254	31547	51673	79642	62693
1900	53126	85419	33296	50310	49541	33296	54201	81624	52548	86216	33630	55413	85436	67706
2000	56744	91439	35375	53762	52951	35375	57859	87259	56137	92203	35713	59175	91250	72790
2100	60376	97488	37453	57243	56397	37454	61530	92911	59742	98212	37796	62961	97081	77941
2200	64021	103562	39532	60752	59876	39534	65212	98577	63361	104240	39878	66769	102929	83153
2300	67683	109660	41610	64285	63387	41614	68904	104257	66995	110284	41962	70600	108792	88421
2400	71324	115779	43689	67841	66928	43695	72606	109947	70640	116344	44045	74453	114669	93741
2500	74985	121917	45768	71419	70498	45777	76316	115648	74296	122417	46130	78328	120559	99108
2600	78673	128073	47846	75017	74096	47860	80034	121357	77963	128501	48216	82224	126462	104520
2700	82369	134246	49925	78633	77720	49945	83759	127075	81639	134596	50303	86141	132376	109973
2800	86074	140433	52004	82267	81369	52033	87491	132799	85323	140701	52391	90079	138302	115464
2900	89786	146636	54082	85918	85043	54124	91229	138530	89015	146814	54481	94036	144238	120990
3000	93504	152852	56161	89584	88740	56218	94973	144267	92715	152935	56574	98013	150184	126549
3500	112185	184109	66554	108119	107555	66769	113768	173020	111306	183636	67079	118165	180057	154768
4000	130989	215622	75947	126939	126874	77532	132671	201859	130027	214453	77675	188705	210145	183552
4500	149895	247354	87340	145991	146660	88614	151662	230756	148850	245348	88386	159572	240427	212764
5000	168890	279283	97733	165246	166876	100111	170730	259692	167763	276299	99222	180749	270893	242313

Converted and usually rounded off from JANAF Thermochemical Tables, NSRDS-NBS-37, 1971 (1141 pp.).

TABLE 2-97 Ideal Gas Entropies s°, kJ/(kmol·K), of Combustion Products

Temperature, K	CO	CO_2	H	OH	H_2	N	NO	NO_2	N_2	N_2O	O	O_2	SO_2	H_2O
200	186.0	200.0	106.4	171.6	119.4	145.0	198.7	225.9	180.0	205.6	152.2	193.5	233.0	175.5
240	191.3	206.0	110.1	177.1	124.5	148.7	204.1	232.2	185.2	211.9	156.2	198.7	239.9	181.4
260	193.7	208.8	111.8	179.5	126.8	150.4	206.6	235.0	187.6	214.8	158.0	201.1	242.8	184.1
280	195.3	211.5	113.3	181.8	129.2	151.9	208.8	237.7	189.8	217.5	159.7	203.3	245.8	186.6
298.15	197.7	213.8	114.7	183.7	130.7	153.3	210.8	240.0	191.6	220.0	161.1	205.1	248.2	188.8
300	197.8	214.0	114.8	183.9	130.9	153.4	210.9	240.3	191.8	220.2	161.2	205.3	248.5	189.0
320	199.7	216.5	116.2	185.9	132.8	154.8	212.9	242.7	193.7	222.7	162.6	207.2	251.1	191.2
340	201.5	218.8	117.4	187.7	134.5	156.0	214.7	245.0	195.5	225.2	163.9	209.0	253.6	193.3
360	203.2	221.0	118.6	189.4	136.2	157.2	216.4	247.2	197.2	227.5	165.2	210.7	256.0	195.2
380	204.7	223.2	119.7	191.0	137.7	158.3	218.0	249.3	198.7	229.7	166.3	212.5	258.2	197.1
400	206.2	225.3	120.8	192.5	139.2	159.4	219.5	251.3	200.2	231.9	167.4	213.8	260.4	198.8
420	207.7	227.3	121.8	194.0	140.6	160.4	221.0	253.2	201.5	234.0	168.4	215.3	262.5	200.5
440	209.0	229.3	122.8	195.3	141.9	161.4	222.3	255.1	202.9	236.0	169.4	216.7	264.6	202.0
460	210.4	231.2	123.7	196.6	143.2	162.3	223.7	257.0	204.2	238.0	170.4	218.0	266.6	203.6
480	211.6	233.1	124.6	197.9	144.5	163.1	225.0	258.8	205.5	239.9	171.3	219.4	268.5	205.1
500	212.8	234.9	125.5	199.1	145.7	164.0	226.3	260.6	206.7	241.8	172.2	220.7	270.5	206.5
550	215.7	239.2	127.5	201.8	148.6	166.0	229.1	264.7	209.4	246.2	174.2	223.7	274.9	210.5
600	218.3	243.3	129.3	204.4	151.1	167.8	231.9	268.8	212.2	250.4	176.1	226.5	279.2	213.1
650	220.8	247.1	131.0	206.8	153.4	169.4	234.4	272.6	214.6	254.3	177.7	229.1	283.1	215.9
700	223.1	250.8	132.5	209.0	155.6	171.0	236.8	276.0	216.9	258.0	179.3	231.5	286.9	218.7
750	225.2	255.4	133.9	211.1	157.6	172.5	239.0	279.3	219.0	261.5	180.7	233.7	290.4	221.3
800	227.3	257.5	135.2	213.0	159.5	173.8	241.1	282.5	221.0	264.8	182.1	235.9	293.8	223.8
850	229.2	260.6	136.4	214.8	161.4	175.1	243.0	285.5	223.0	268.0	183.4	237.9	297.0	226.2
900	231.1	263.6	137.7	216.5	163.1	176.3	245.0	288.4	224.8	271.1	184.6	239.9	300.1	228.5
950	232.8	266.5	138.8	218.1	164.7	177.4	246.8	291.3	226.5	274.0	185.7	241.8	303.0	230.6
1000	234.5	269.3	139.9	219.7	166.2	178.5	248.4	293.9	228.2	276.8	186.8	243.6	305.8	232.7
1100	237.7	274.5	141.9	222.7	169.1	180.4	251.8	298.9	231.3	282.1	188.8	246.9	311.0	236.7
1200	240.7	279.4	143.7	225.4	171.8	182.2	254.8	303.6	234.2	287.0	190.6	250.0	315.8	240.5
1300	243.4	283.9	145.3	228.0	174.3	183.9	257.6	307.9	236.9	291.5	192.3	252.9	320.3	244.0
1400	246.0	288.2	146.9	230.3	176.6	185.4	260.2	311.9	239.5	295.8	193.8	255.6	324.5	247.4
1500	248.4	292.2	148.3	232.6	178.8	186.9	262.7	315.7	241.9	299.8	195.3	258.1	328.4	250.6
1600	250.7	296.0	149.6	234.7	180.9	188.2	265.0	319.3	244.1	303.6	196.6	260.4	332.1	253.7
1700	252.9	299.6	150.9	236.8	182.9	189.5	267.2	322.7	246.3	307.2	197.9	262.7	335.6	256.6
1800	254.9	303.0	152.1	238.7	184.8	190.7	269.3	325.9	248.3	310.6	199.1	264.8	338.9	259.5
1900	256.8	306.2	153.2	240.6	186.7	191.8	271.3	328.9	250.2	313.8	200.2	266.8	342.0	262.2
2000	258.7	309.3	154.3	242.3	188.4	192.9	273.1	331.8	252.1	316.9	201.3	268.7	345.0	264.8
2100	260.5	312.2	155.3	244.0	190.1	193.9	274.9	334.5	253.8	319.8	202.3	270.6	347.9	267.3
2200	262.2	315.1	156.3	245.7	191.7	194.8	276.6	337.2	255.5	322.6	203.2	272.4	350.6	269.7
2300	263.8	317.8	157.2	247.2	193.3	195.8	278.3	339.7	257.1	325.3	204.2	274.1	353.2	272.0
2400	265.4	320.4	158.1	248.7	194.8	196.7	279.8	342.1	258.7	327.9	205.0	275.7	355.7	274.3
2500	266.9	322.9	158.9	250.2	196.2	197.5	281.4	344.5	260.2	330.4	205.9	277.3	358.1	276.5
2600	268.3	325.3	159.7	251.6	197.7	198.3	282.8	346.7	261.6	332.7	206.7	278.8	360.4	278.6
2700	269.7	327.6	160.5	253.0	199.0	199.1	284.2	348.9	263.0	335.0	207.5	280.3	362.6	380.7
2800	271.0	329.9	161.3	254.3	200.3	199.9	285.6	350.9	264.3	337.3	208.3	281.7	364.8	282.7
2900	272.3	332.1	162.0	255.6	201.6	200.6	286.9	352.9	265.6	339.4	209.0	283.1	366.9	284.6
3000	273.6	334.2	162.7	256.8	202.9	201.3	288.2	354.9	266.9	341.5	209.7	284.4	368.9	286.5
3500	279.4	343.8	165.9	262.5	208.7	204.6	294.0	363.8	272.6	350.9	212.9	290.7	378.1	295.2
4000	284.4	352.2	168.7	267.6	213.8	207.4	299.0	371.5	277.6	359.2	215.8	296.2	386.1	302.9
4500	288.8	359.7	171.1	272.1	218.5	210.1	303.5	378.3	282.1	366.5	218.3	301.1	393.3	309.8
5000	292.8	366.4	173.3	276.1	222.8	212.5	307.5	384.4	286.0	373.0	220.6	305.5	399.7	316.0

Usually rounded off from JANAF Thermochemical Tables, NSRDS-NBS-37, 1971 (1141 pp.). Equilibrium constants can be calculated by combining Δh_f° values from Table 2-95, $h_T - h_{298}$ from Table 2-96, and s° values from the above, using the formula $\ln k_p = -\Delta G/(RT)$, where $\Delta G = \Delta h_f^\circ + (h_T - h_{298}) - T_s^\circ$.

HEATS OF SOLUTION

TABLE 2-98 Heats of Solution of Inorganic Compounds in Water

Heat evolved, in kilocalories per gram formula weight, on solution in water at 18°C. Computed from data in Bichowsky and Rossini, *Thermochemistry of Chemical Substances,* Reinhold, New York, 1936.

Substance	Dilution*	Formula	Heat, kcal/mol
Aluminum bromide	aq	$AlBr_3$	+85.3
chloride	600	$AlCl_3$	+77.9
	600	$AlCl_3\cdot 6H_2O$	+13.2
fluoride	aq	AlF_3	+31
	aq	$AlF_3\cdot\frac{1}{2}H_2O$	+19.0
	aq	$AlF_3\cdot 3\frac{1}{2}H_2O$	−1.7
iodide	aq	AlI_3	+89.0
sulfate	aq	$Al_2(SO_4)_3$	+126
	aq	$Al_2(SO_4)_3\cdot 6H_2O$	+56.2
	aq	$Al_2(SO_4)_3\cdot 18H_2O$	+6.7
Ammonium bromide	aq	NH_4Br	−4.45
chloride	∞	NH_4Cl	−3.82
chromate	aq	$(NH_4)_2CrO_4$	−5.82
dichromate	600	$(NH_4)_2Cr_2O_7$	−12.9
iodide	aq	NH_4I	−3.56
nitrate	∞	NH_4NO_3	−6.47
perborate	aq	$NH_4BO_3\cdot H_2O$	−9.0
sulfate	∞	$(NH_4)_2SO_4$	−2.75
sulfate, acid	800	NH_4HSO_4	+0.56
sulfite	aq	$(NH_4)_2SO_3$	−1.2
	aq	$(NH_4)_2SO_3\cdot H_2O$	−4.13
Antimony fluoride	aq	SbF_3	−1.7
iodide	aq	SbI_3	−0.8
Arsenic acid	aq	H_3AsO_4	−0.4
Barium bromate	∞	$Ba(BrO_3)_2\cdot H_2O$	−15.9
bromide	∞	$BaBr_2$	+5.3
	∞	$BaBr_2\cdot H_2O$	−0.8
	∞	$BaBr_2\cdot 2H_2O$	−3.87
chlorate	∞	$Ba(ClO_3)_2$	−6.7
	∞	$Ba(ClO_3)_2\cdot H_2O$	−10.6
chloride	∞	$BaCl_2$	+2.4
	∞	$BaCl_2\cdot H_2O$	−2.17
	∞	$BaCl_2.2H_2O$	−4.5
cyanide	aq	$Ba(CN)_2$	+1.5
	aq	$Ba(CN)_2\cdot H_2O$	−2.4
	aq	$Ba(CN)_2\cdot 2H_2O$	−4.9
iodate	∞	$Ba(IO_3)_2$	−9.1
	∞	$Ba(IO_3)_2\cdot H_2O$	−11.3
iodide	∞	BaI_2	+10.5
	∞	$BaI_2\cdot H_2O$	+2.7
	∞	$BaI_2\cdot 2H_2O$	+0.14
	∞	$BaI_2\cdot 2\frac{1}{2}H_2O$	−0.58
	∞	$BaI_2\cdot 7H_2O$	−6.61
nitrate	∞	$Ba(NO_3)_2$	−10.2
perchlorate	∞	$Ba(ClO_4)_2$	−2.8
	∞	$Ba(ClO_4)_2\cdot 3H_2O$	−10.5
sulfide	∞	BaS	+7.2
Beryllium bromide	aq	$BeBr_2$	+62.6
chloride	aq	$BeCl_2$	+51.1
iodide	aq	BeI_2	+72.6
sulfate	aq	$BeSO_4$	+18.1
	aq	$BeSO_4\cdot H_2O$	+13.5
	aq	$BeSO_4\cdot 2H_2O$	+7.9
	aq	$BeSO_4\cdot 4H_2O$	+1.1
Bismuth iodide	aq	BiI_3	+3
Boric acid	aq	H_3BO_3	−5.4
Cadmium bromide	400	$CdBr_2$	+0.4
	400	$CdBr_2\cdot 4H_2O$	−7.3
chloride	400	$CdCl_2$	+3.1
	400	$CdCl_2\cdot H_2O$	+0.6
	400	$CdCl_2\cdot 2\frac{1}{2}H_2O$	−3.00
nitrate	400	$Cd(NO_3)_2\cdot H_2O$	+4.17
	400	$Cd(NO_3)_2\cdot 4H_2O$	−5.08
sulfate	400	$CdSO_4$	+10.69
	400	$CdSO_4\cdot H_2O$	+6.05
	400	$CdSO_4\cdot 2\frac{2}{3}H_2O$	+2.51
Calcium acetate	∞	$Ca(C_2H_3O_2)_2$	+7.6
	∞	$Ca(C_2H_3O_2)_2\cdot H_2O$	+6.5
bromide	∞	$CaBr_2$	+24.86
	∞	$CaBr_2\cdot 6H_2O$	−0.9
chloride	∞	$CaCl_2$	+4.9
	∞	$CaCl_2\cdot H_2O$	+12.3
	∞	$CaCl_2\cdot 2H_2O$	+12.5
	∞	$CaCl_2\cdot 4H_2O$	+2.4
	∞	$CaCl_2\cdot 6H_2O$	−4.11
Calcium—(*Cont.*)			
formate	400	$Ca(CHO_2)_2$	+0.7
iodide	∞	CaI_2	+28.0
	∞	$CaI_2\cdot 8H_2O$	+1.8
nitrate	∞	$Ca(NO_3)_2$	+4.1
	∞	$Ca(NO_3)_2\cdot H_2O$	+0.7
	∞	$Ca(NO_3)_2\cdot 2H_2O$	−3.2
	∞	$Ca(NO_3)_2\cdot 3H_2O$	−4.2
	∞	$Ca(NO_3)_2\cdot 4H_2O$	−7.99
phosphate, mono-	aq	$Ca(H_2PO_4)_2\cdot H_2O$	−0.6
dibasic	aq	$CaHPO_4\cdot 2H_2O$	−1
sulfate	∞	$CaSO_4$	+5.1
	∞	$CaSO_4\cdot\frac{1}{2}H_2O$	+3.6
	∞	$CaSO_4\cdot 2H_2O$	−0.18
Chromous chloride	aq	$CrCl_2$	+18.6
		$CrCl_2\cdot 3H_2O$	+5.3
		$CrCl_2\cdot 4H_2O$	+2.0
iodide	aq	CrI_2	+5.7
Cobaltous bromide	aq	$CoBr_2$	+18.4
	aq	$CoBr_2\cdot 6H_2O$	−1.25
chloride	400	$CoCl_2$	+18.5
	400	$CoCl_2\cdot 2H_2O$	+9.8
	400	$CoCl_2\cdot 6H_2O$	−2.9
iodide	aq	CoI_2	+18.8
sulfate	400	$CoSO_4$	+15.0
	400	$CoSO_4\cdot 6H_2O$	−1.4
	400	$CoSO_4\cdot 7H_2O$	−3.6
Cupric acetate	aq	$Cu(C_2H_3O_2)_2$	+2.4
formate	aq	$Cu(CHO_2)_2$	+0.5
nitrate	200	$Cu(NO_3)_2$	+10.3
	200	$Cu(NO_3)_2\cdot 3H_2O$	−2.6
	200	$Cu(NO_3)_2\cdot 6H_2O$	−10.7
sulfate	800	$CuSO_4$	+15.9
		$CuSO_4\cdot H_2O$	+9.3
		$CuSO_4\cdot 3H_2O$	+3.65
		$CuSO_4\cdot 5H_2O$	−2.85
Cuprous sulfate	aq	Cu_2SO_4	+11.6
Ferric chloride	1000	$FeCl_3$	+31.7
	1000	$FeCl_3\cdot 2\frac{1}{2}H_2O$	+21.0
	1000	$FeCl_3\cdot 6H_2O$	+5.6
nitrate	800	$Fe(NO_3)_3\cdot 9H_2O$	−9.1
Ferrous bromide	aq	$FeBr_2$	+18.0
chloride	400	$FeCl_2$	+17.9
	400	$FeCl_2\cdot 2H_2O$	+8.7
	400	$FeCl_2\cdot 4H_2O$	+2.7
iodide	aq	FeI_2	+23.3
sulfate	400	$FeSO_4$	+14.7
	400	$FeSO_4\cdot H_2O$	+7.35
	400	$FeSO_4\cdot 4H_2O$	+1.4
	400	$FeSO_4\cdot 7H_2O$	−4.4
Lead acetate	400	$Pb(C_2H_3O_2)_2$	+1.4
	400	$Pb(C_2H_3O_2)_2\cdot 3H_2O$	−5.9
bromide	aq	$PbBr_2$	−10.1
chloride	aq	$PbCl_2$	−3.4
formate	aq	$Pb(CHO_2)_2$	−6.9
nitrate	400	$Pb(NO_3)_2$	−7.61
Lithium bromide	∞	$LiBr$	+11.54
	∞	$LiBr\cdot H_2O$	+5.30
	∞	$LiBr\cdot 2H_2O$	+2.05
	∞	$LiBr\cdot 3H_2O$	−1.59
chloride	∞	$LiCl$	+8.66
	∞	$LiCl\cdot H_2O$	+4.45
	∞	$LiCl\cdot 2H_2O$	+1.07
	∞	$LiCl\cdot 3H_2O$	−1.98
fluoride	∞	LiF	−0.74
hydroxide	∞	$LiOH$	+4.74
	∞	$LiOH\cdot\frac{1}{8}H_2O$	+4.39
	∞	$LiOH\cdot H_2O$	+9.6
iodide	∞	LiI	+14.92
	∞	$LiI\cdot\frac{1}{2}H_2O$	+10.08
	∞	$LiI\cdot H_2O$	+6.93
	∞	$LiI\cdot 2H_2O$	+3.43
	∞	$LiI\cdot 3H_2O$	−0.17
nitrate	∞	$LiNO_3$	+0.466
	∞	$LiNO_3\cdot 3H_2O$	−7.87

*The numbers represent moles of water used to dissolve 1 g formula weight of substance; ∞ means "infinite dilution"; and *aq* means "aqueous solution of unspecified dilution."

TABLE 2-98 Heats of Solution of Inorganic Compounds in Water *(Continued)*

Substance	Dilution*	Formula	Heat, kcal/mol
Lithium—(*Cont.*)			
sulfate	∞	Li_2SO_4	+6.71
	∞	$Li_2SO_4 \cdot H_2O$	+3.77
Magnesium bromide	∞	$MgBr_2$	+43.7
	∞	$MgBr_2 \cdot H_2O$	+35.9
	∞	$MgBr_2 \cdot 6H_2O$	+19.8
chloride	∞	$MgCl_2$	+36.3
	∞	$MgCl_2 \cdot 2H_2O$	+20.8
	∞	$MgCl_2 \cdot 4H_2O$	+10.5
	∞	$MgCl_2 \cdot 6H_2O$	+3.4
iodide	∞	MgI_2	+50.2
nitrate	∞	$Mg(NO_3)_2 \cdot 6H_2O$	−3.7
phosphate	aq	$Mg_3(PO_4)_2$	+10.2
sulfate	∞	$MgSO_4$	+21.1
	∞	$MgSO_4 \cdot H_2O$	+14.0
	∞	$MgSO_4 \cdot 2H_2O$	+11.7
	∞	$MgSO_4 \cdot 4H_2O$	+4.9
	∞	$MgSO_4 \cdot 6H_2O$	+0.55
	∞	$MgSO_4 \cdot 7H_2O$	−3.18
sulfide	aq	MgS	+25.8
Manganic nitrate	400	$Mn(NO_3)_2$	+12.9
	400	$Mn(NO_3)_2 \cdot 3H_2O$	−3.9
	400	$Mn(NO_3)_2 \cdot 6H_2O$	−6.2
sulfate	aq	$Mn_2(SO_4)_3$	+22
Manganous acetate	aq	$Mn(C_2H_3O_2)_2$	+12.2
	aq	$Mn(C_2H_3O_2)_2 \cdot 4H_2O$	+1.6
bromide	aq	$MnBr_2$	+15
	aq	$MnBr_2 \cdot H_2O$	+14.4
	aq	$MnBr_2 \cdot 4H_2O$	+16.1
chloride	400	$MnCl_2$	+16.0
	400	$MnCl_2 \cdot 2H_2O$	+8.2
	400	$MnCl_2 \cdot 4H_2O$	+1.5
formate	aq	$Mn(CHO_2)_2$	+4.3
	aq	$Mn(CHO_2)_2 \cdot 2H_2O$	−2.9
iodide	aq	MnI_2	+26.2
	aq	$MnI_2 \cdot H_2O$	+24.1
	aq	$MnI_2 \cdot 2H_2O$	+22.7
	aq	$MnI_2 \cdot 4H_2O$	+19.9
	aq	$MnI_2 \cdot 6H_2O$	+21.2
sulfate	400	$MnSO_4$	+13.8
	400	$MnSO_4 \cdot H_2O$	+11.9
	400	$MnSO_4 \cdot 7H_2O$	−1.7
Mercuric acetate	aq	$Hg(C_2H_3O_2)_2$	−4.0
bromide	aq	$HgBr_2$	−2.4
chloride	aq	$HgCl_2$	−3.3
nitrate	aq	$Hg(NO_3)_2 \cdot ½H_2O$	−0.7
Mercurous nitrate	aq	$Hg_2(NO_3)_2 \cdot 2H_2O$	−11.5
Nickel bromide	aq	$NiBr_2$	+19.0
	aq	$NiBr_2 \cdot 3H_2O$	+0.2
Nickel chloride	800	$NiCl_2$	+19.23
	800	$NiCl_2 \cdot 2H_2O$	+10.4
	800	$NiCl_2 \cdot 4H_2O$	+4.2
	800	$NiCl_2 \cdot 6H_2O$	−1.15
iodide	aq	NiI_2	+19.4
nitrate	200	$Ni(NO_3)_2$	+11.8
	200	$Ni(NO_3)_2 \cdot 6H_2O$	−7.5
sulfate	200	$NiSO_4$	+15.1
	200	$NiSO_4 \cdot 7H_2O$	−4.2
Phosphoric acid, ortho-	400	H_3PO_4	+2.79
	400	$H_3PO_4 \cdot ½H_2O$	−0.1
pyro-	aq	$H_4P_2O_7$	+25.9
	aq	$H_4P_2O_7 \cdot 1½H_2O$	+4.65
Potassium acetate	∞	$KC_2H_3O_2$	+3.55
aluminum sulfate	600	$KAl(SO_4)_2$	+48.5
	600	$KAl(SO_4)_2 \cdot 3H_2O$	+26.6
		$KAl(SO_4)_2 \cdot 12H_2O$	−10.1
bicarbonate	2000	$KHCO_3$	−5.1
bromate	∞	$KBrO_3$	−10.13
bromide	∞	KBr	−5.13
carbonate	∞	K_2CO_3	+6.58
		$K_2CO_3 \cdot ½H_2O$	+4.25
		$K_2CO_3 \cdot 1½H_2O$	−0.43
chlorate	∞	$KClO_3$	−10.31
chloride	∞	KCl	−4.404
chromate	2185	K_2CrO_4	−4.9
chrome sulfate	600	$KCr(SO_4)_2$	+55
		$KCr(SO_4)_2 \cdot H_2O$	+42
		$KCr(SO_4)_2 \cdot 2H_2O$	+33
		$KCr(SO_4)_2 \cdot 6H_2O$	+7
		$KCr(SO_4)_2 \cdot 12H_2O$	−9.5
Potassium—(*Cont.*)			
cyanide	200	KCN	−3.0
dichromate	1600	$K_2Cr_2O_7$	−17.8
fluoride	∞	KF	+3.96
	∞	$KF \cdot 2H_2O$	−1.85
	∞	$KF \cdot 4H_2O$	−6.05
hydrosulfide	∞	KHS	+0.86
	∞	$KHS \cdot ¼H_2O$	+1.21
hydroxide	∞	KOH	+12.91
	∞	$KOH \cdot ¾H_2O$	+4.27
	∞	$KOH \cdot H_2O$	+3.48
	∞	$KOH \cdot 7H_2O$	+0.86
iodate	∞	KIO_3	−6.93
iodide	∞	KI	−5.23
nitrate	∞	KNO_3	−8.633
oxalate	400	$K_2C_2O_4$	−4.6
		$K_2C_2O_4 \cdot H_2O$	−7.5
perchlorate	∞	$KClO_4$	−12.94
permanganate	400	$KMnO_4$	−10.4
phosphate, dihydrogen	aq	KH_2PO_4	+4.7
pyrosulfite	aq	$K_2S_2O_5$	−11.0
	aq	$K_2S_2O_5 \cdot ½H_2O$	−10.22
sulfate	∞	K_2SO_4	−6.32
sulfate, acid	800	$KHSO_4$	−3.10
sulfide	∞	K_2S	−11.0
sulfite	aq	K_2SO_3	+1.8
	aq	$K_2SO_3 \cdot H_2O$	+1.37
thiocyanate	∞	$KCNS$	−6.08
thionate, di-	aq	$K_2S_2O_6$	−13.0
thiosulfate	∞	$K_2S_2O_3$	−4.5
Silver acetate	aq	$AgC_2H_3O_2$	−5.4
nitrate	200	$AgNO_3$	−4.4
Sodium acetate	∞	$NaC_2H_3O_2$	+4.085
	∞	$NaC_2H_3O_2 \cdot 3H_2O$	−4.665
arsenate	500	Na_3AsO_4	+15.6
	500	$Na_3AsO_4 \cdot 12H_2O$	−12.61
bicarbonate	1800	$NaHCO_3$	−4.1
borate, tetra-	900	$Na_2B_4O_7$	+10.0
	900	$Na_2B_4O_7 \cdot 10H_2O$	−16.8
bromide	∞	$NaBr$	−0.58
	∞	$NaBr \cdot 2H_2O$	−4.57
carbonate	∞	Na_2CO_3	+5.57
	∞	$Na_2CO_3 \cdot H_2O$	+2.19
	∞	$Na_2CO_3 \cdot 7H_2O$	−10.81
	∞	$Na_2CO_3 \cdot 10H_2O$	−16.22
chlorate	∞	$NaClO_3$	−5.37
chloride	∞	$NaCl$	−1.164
chromate	800	Na_2CrO_4	+2.50
	800	$Na_2CrO_4 \cdot 4H_2O$	−7.52
	800	$Na_2CrO_4 \cdot 10H_2O$	−16.0
cyanide	200	$NaCN$	−0.37
	200	$NaCN \cdot ½H_2O$	−0.92
	200	$NaCN \cdot 2H_2O$	−4.41
fluoride	∞	NaF	−0.27
hydrosulfide	∞	$NaHS$	+4.62
	∞	$NaHS \cdot 2H_2O$	−1.49
Sodium hydroxide	∞	$NaOH$	+10.18
	∞	$NaOH \cdot ½H_2O$	+8.17
	∞	$NaOH \cdot ⅔H_2O$	+7.08
	∞	$NaOH \cdot ¾H_2O$	+6.48
	∞	$NaOH \cdot H_2O$	+5.17
iodide	∞	NaI	+1.57
	∞	$NaI \cdot 2H_2O$	−3.89
metaphosphate	600	$NaPO_3$	+3.97
nitrate	∞	$NaNO_3$	−5.05
nitrite	aq	$NaNO_2$	−3.6
perchlorate	∞	$NaClO_4$	−4.15
phosphate di	1600	Na_2HPO_4	+5.21
tri-	1600	Na_3PO_4	+13
phosphate di	1600	$Na_3PO_4 \cdot 12H_2O$	−15.3
di-	1600	$Na_2HPO_4 \cdot 2H_2O$	−0.82
	1600	$Na_2HPO_4 \cdot 7H_2O$	−12.04
	1600	$Na_2HPO_4 \cdot 12H_2O$	−23.18
phosphite, mono-	600	NaH_2PO_3	+0.90
	600	$NaH_2PO_3 \cdot 2½H_2O$	−5.29
di-	800	Na_2HPO_3	+9.30
	800	$Na_2HPO_3 \cdot 5H_2O$	−4.54
pyrophosphate	1600	$Na_4P_2O_7$	+11.9
	1600	$Na_4P_2O_7 \cdot 10H_2O$	−11.7
di-	1200	$Na_2H_2P_2O_7$	−2.2
	1200	$Na_2H_2P_2O_7 \cdot 6H_2O$	−14.0

(Continued)

TABLE 2-98 Heats of Solution of Inorganic Compounds in Water *(Continued)*

Substance	Dilution*	Formula	Heat, kcal/mol
Sodium—(*Cont.*)			
sulfate	∞	Na_2SO_4	+0.28
	∞	$Na_2SO_4·10H_2O$	−18.74
sulfate, acid	800	$NaHSO_4$	+1.74
	800	$NaHSO_4·H_2O$	+0.15
sulfide	∞	Na_2S	+15.2
	∞	$Na_2S·4½H_2O$	+0.09
	∞	$Na_2S·5H_2O$	−6.54
	∞	$Na_2S·9H_2O$	−16.65
sulfite	∞	Na_2SO_3	+2.8
	∞	$Na_2SO_3·7H_2O$	−11.1
thiocyanate	∞	NaCNS	−1.83
thionate, di-	aq	$Na_2S_2O_6$	−5.80
	aq	$Na_2S_2O_6·2H_2O$	−11.86
Sodium thiosulfate	aq	$Na_2S_2O_3$	+2.0
	aq	$Na_2S_2O_3·5H_2O$	−11.30
Stannic bromide	aq	$SnBr_4$	+15.5
Stannous bromide	aq	$SnBr_2$	−1.6
iodide	aq	SnI_2	−5.8
Strontium acetate	∞	$Sr(C_2H_3O_2)_2$	+6.2
	∞	$Sr(C_2H_3O_2)_2·½H_2O$	+5.9
bromide	∞	$SrBr_2$	+16.4
	∞	$SrBr_2·H_2O$	+9.25
	∞	$SrBr_2·2H_2O$	+6.5
	∞	$SrBr_2·4H_2O$	+0.4
	∞	$SrBr_2·6H_2O$	−6.1
Strontium—(*Cont.*)			
chloride	∞	$SrCl_2$	+11.54
	∞	$SrCl_2·H_2O$	+6.4
	∞	$SrCl_2·2H_2O$	+2.95
	∞	$SrCl_2·6H_2O$	−7.1
iodide	∞	SrI_2	+20.7
	∞	$SrI_2·H_2O$	+12.65
	∞	$SrI_2·2H_2O$	+10.4
	∞	$SrI_2·6H_2O$	−4.5
nitrate	∞	$Sr(NO_3)_2$	−4.8
	∞	$Sr(NO_3)_2·4H_2O$	−12.4
sulfate	∞	$SrSO_4$	+0.5
Sulfuric acid, pyro-	∞	$H_2S_2O_7$	−18.08
Zinc acetate	400	$Zn(C_2H_3O_2)_2$	+9.8
	400	$Zn(C_2H_3O_2)_2·H_2O$	+7.0
	400	$Zn(C_2H_3O_2)_2·2H_2O$	+3.9
bromide	400	$ZnBr_2$	+15.0
chloride	400	$ZnCl_2$	+15.72
iodide	aq	ZnI_2	+11.6
nitrate	400	$Zn(NO_3)_2·3H_2O$	−5
	400	$Zn(NO_3)_2·6H_2O$	−6.0
sulfate	400	$ZnSO_4$	+18.5
	400	$ZnSO_4·H_2O$	+10.0
	400	$ZnSO_4·6H_2O$	−0.8
	400	$ZnSO_4·7H_2O$	−4.3

NOTE: To convert kilocalories per mole to British thermal units per pound-mole, multiply by 1.799×10^{-3}.

TABLE 2-99 Heats of Solution of Organic Compounds in Water (at Infinite Dilution and Approximately Room Temperature)

Recalculated and rearranged from *International Critical Tables,* vol. 5, pp. 148–150. cal/mol = Btu/(lb·mol) × 1.799.

Solute	Heat of solution, cal/mol solute*
Acetic acid (solid), $C_2H_4O_2$	−2,251
Acetylacetone, $C_5H_8O_2$	−641
Acetylurea, $C_3H_6N_2O_2$	−6,812
Aconitic acid, $C_6H_6O_6$	−4,206
Ammonium benzoate, $C_7H_9NO_2$	−2,700
picrate	−8,700
succinate (*n*-)	−3,489
Aniline, hydrochloride, C_6H_8ClN	−2,732
Barium picrate	−4,708
Benzoic acid, $C_7H_6O_2$	−6,501
Camphoric acid, $C_{10}H_{16}O_4$	−502
Citric acid, $C_6H_8O_7$	−5,401
Dextrin, $C_{12}H_{20}O_{10}$	268
Fumaric acid, $C_4H_4O_4$	−5,903
Hexamethylenetetramine, $C_6H_{12}N_4$	4,780
Hydroxybenzamide (*m*-), $C_7H_7NO_2$	−4,161
(*m*-), (HCl)	−7,003
(*o*-), $C_7H_7NO_2$	−4,340
(*p*-)	−5,392
Hydroxybenzoic acid (*o*-), $C_7H_6O_3$	−6,350
(*p*-), $C_7H_6O_3$	−5,781
Hydroxybenzyl alcohol (*o*-), $C_7H_8O_2$	−3,203
Inulin, $C_{36}H_{62}O_{31}$	−96
Isosuccinic acid, $C_4H_6O_4$	−3,420
Itaconic acid, $C_5H_6O_4$	−5,922
Lactose, $C_{12}H_{22}O_{11}·H_2O$	−3,705
Lead picrate	−7,098
($2H_2O$)	−13,193
Magnesium picrate	14,699
($8H_2O$)	−15,894
Maleic acid, $C_4H_4O_4$	−4,441
Malic acid, $C_4H_6O_5$	−3,150
Malonic acid, $C_3H_4O_4$	−4,493
Mandelic acid, $C_8H_8O_3$	−3,090
Mannitol, $C_6H_{14}O_6$	−5,260
Menthol, $C_{10}H_{20}O$	0
Nicotine dihydrochloride, $C_{10}H_{16}Cl_2N_2$	6,561
Nitrobenzoic acid (*m*-), $C_7H_5NO_4$	−5,593
(*o*-), $C_7H_5NO_4$	−5,306
(*p*-), $C_7H_5NO_4$	−8,891
Nitrophenol (*m*-), $C_6H_5NO_3$	−5,210
(*o*-), $C_6H_5NO_3$	−6,310
(*p*-), $C_6H_5NO_3$	−4,493
Oxalic acid, $C_2H_2O_4$	−2,290
($2H_2O$)	−8,485
Phenol (solid), C_6H_6O	−2,605
Phthalic acid, $C_8H_6O_4$	−4,871
Picric acid, $C_6H_3N_3O_7$	−7,098
Piperic acid, $C_{12}H_{10}O_4$	−10,492
Piperonylic acid, $C_8H_6O_4$	−9,106
Potassium benzoate	−1,506
citrate	2,820
tartrate (*n*-) (0.5 H_2O)	−5,562
Pyrogallol, $C_6H_6O_3$	−3,705
Pyrotartaric acid	−5,019
Quinone	−3,991
Raffinose, $C_{18}H_{32}O_{16}$ ($5H_2O$)	−9,703
Resorcinol, $C_6H_6O_2$	−3,960
Silver malonate (*n*-)	−9,799
Sodium citrate (tri-)	5,270
picrate	−6,441
potassium tartrate	−1,817
($4H_2O$)	−12,342
succinate (*n*-)	2,390
($6H_2O$)	−10,994
tartrate (*n*-)	−1,121
($2H_2O$)	−5,882
Strontium picrate	7,887
($6H_2O$)	−14,412
Succinic acid, $C_4H_6O_4$	−6,405
Succinimide, $C_4H_5NO_2$	−4,302
Sucrose, $C_{12}H_{22}O_{11}$	−1,319
Tartaric acid (*d*-)	−3,451
Thiourea, CH_4N_2S	−5,330
Urea, CH_4N_2O	−3,609
acetate	−8,795
formate	−7,194
nitrate	−10,803
oxalate	−17,806
Vanillic acid	−5,160
Vanillin	−5,210
Zinc picrate	−11,496
($8H_2O$)	−15,894

*+ denotes heat evolved, and − denotes heat absorbed. The data in the *International Critical Tables* were calculated by E. Anderson.

THERMAL EXPANSION AND COMPRESSIBILITY

Unit Conversion For this subsection, the following unit conversion is applicable: °F = 9/5°C + 32.

Additional References Some of the tables given under this subject are reprinted by permission from the *Smithsonian Tables.* For other data on thermal expansion, see *International Critical Tables.* The tabular index is in volume 3, and the data are in volume 2.

Thermal Expansion of Gases No tables of coefficients of thermal expansion of gases are given in this edition. The coefficient at constant pressure, $1/\upsilon\,(\partial u/\partial T)_p$, for an ideal gas is merely the reciprocal of the absolute temperature. For a real gas or liquid, both it and the coefficient at constant volume $1/p\,(\partial p/\partial T)_v$ should be calculated either from the equation of state or from tabulated *PVT* data.

For expansion of liquids and solids, see the following tables.

TABLE 2-100 Linear Expansion of the Solid Elements*

C is the true expansion coefficient at the given temperature; *M* is the mean coefficient between given temperatures; where one temperature is given, the true coefficient at that temperature is indicated; α and β are coefficients in formula $l_t = l_0(1 + \alpha t + \beta t^2)$; l_0 is length at 0°C (unless otherwise indicated, when, if *x* is the reference temperature, $l_t = l_x[1 + \alpha(t - t_x) + \beta(t - t_x)^2]$; l_t is length at *t* °C).

Element	Temp., °C	$C \times 10^4$	Temp. range, °C	$M \times 10^4$	Temp. range, °C	$\alpha \times 10^4$	$\beta \times 10^6$
Aluminum	20	0.224	100	0.235	0, 500	0.22	0.009
Aluminum	300	0.284	500	0.311			
Antimony	20	0.136‖	20	0.080⊥			
Arsenic	20	0.05					
Bismuth	20	0.014‖	20	0.103⊥			
Cadmium	0	0.54‖	−180, −140	0.59‖	20, 100	0.526‖	
Cadmium	0	0.20⊥	−180, −140	0.117⊥	20, 100	0.214⊥	
Carbon, diamond	50	0.012					
graphite	50	0.06					
Chromium			20, 100	0.068	20, 500	0.086	
Cobalt	20	0.123			6, 121	0.121	0.0064
Copper	20	0.162	100	0.166	0, 625	0.161	0.0040
Copper	200	0.170	300	0.175			
Gold	20	0.140	17, 100	0.143	0, 520	0.142	0.0022
Gold			−191, 17	0.132			
Indium	40	0.417					
Iodine			−190, 17	0.837			
Iridium	20	0.065			0, 80	0.0636	0.0032
Iridium					1070, 1720	0.0679	0.0011
Iron, soft	40	0.1210	0, 100	0.11			
cast	20	0.118			0, 750	0.1158	0.0053
wrought	20	0.119			0, 750	0.1170	0.0053
steel	20	0.114			0, 750	0.1118	0.0053
Lead (99.9)			20, 100	0.291	100, 240	0.269	0.011
	100	0.291	20, 200	0.300			
	280	0.343					
Magnesium	20	0.254	−100, +20	0.240	+20, 500	0.2480	0.0096
			20, 100	0.260			
Manganese	20	0.233	0, 100	0.228			
			−190, 0	0.159	20, 300	0.216	0.0121
Molybdenum†	20	0.053	0, 100	0.052	−142, 19	0.0515	0.0057
			25, 100	0.049	19, +305	0.0501	0.0014
			25, 500	0.055			
Nickel	20	0.126	0, 100	0.130	−190, +20	0.1308	0.0166
					+20, +300	0.1236	0.0066
					500, 1000	0.1346	0.0033
Osmium	40	0.066					
Palladium	20	0.1173			−190, +100	0.1152	0.00517
					0, 1000	0.1167	0.0022
Platinum	20	0.0887			−190, −100	0.0875	0.00314
	20	0.0893			0, +80	0.0890	0.00121
					0, 1000	0.0887	0.00132
Potassium			0, 50	0.83			
Rhodium	40	0.0850	6, 21	0.0876	−75, −112	0.0746	
Ruthenium	40	0.0963					
Selenium	0	0.439	0, 100	0.660			
Silicon	40	0.0763	−3, +18	0.0249	−75, −67	0.0182	
Silver	20	0.1846	0, 100	0.197	0, 875	0.1827	0.00479
	20	0.195			20, 500	0.1939	0.00295
Sodium			−190, −17	0.622	0, 50	0.72	
Steel, 36.4Ni			20, 260	0.031	260, 500	0.144	
			20, 340	0.055	340, 500	0.136	
Tantalum†	20	0.065	−78, 0	0.059	20, 400	0.0646	0.0009
			0, 100	0.0655			
Tellurium	20	0.016‖	20	0.272⊥			
Thallium	40	0.302					
Tin	20	0.214			8, 95	0.2033	0.0263
	20	0.305‖	20	0.154⊥			
Tungsten†	27	0.0444	0, 100	0.045	−105, +502	0.0428	0.00058
Zinc	20‡	0.643‖	−140, −100	0.656‖	+0, 400	0.354	0.010
	20‡	0.125⊥	+20, 100	0.639‖			
	20	0.358	+20, 100	0.141⊥			

**Smithsonian Tables.* For more complete tabulations see Table 142, *Smithsonian Physical Tables,* 9th ed., 1954; *Handbook of Chemistry and Physics,* 40th ed., pp. 2239–2245. Chemical Rubber Publishing Co.; Goldsmith, and Waterman, WADC-TR-58-476, 1959; Johnson (ed.), WADD-TR-60-56, 1960, etc.

†Molybdenum, 300 to 2500°C; $l_t = l_{300}[1 + 5.00 \times 10^{-6}(t - 300) + 10.5 \times 10^{-10}(t - 300)^2]$

Tantalum, 300 to 2800°C; $l_t = l_{300}[1 + 6.60 \times 10^{-6}(t - 300) + 5.2 \times 10^{-10}(t - 300)^2]$

Tungsten, 300 to 2700°C; $l_t = l_{300}[1 + 4.44 \times 10^{-6}(t - 300) + 4.5 \times 10^{-10}(t - 300)^2]$

Beryllium, 20 to 100°C; 12.3×10^{-6} per °C.

Columbium, 0 to 100°C; 7.2×10^{-6} per °C.

Tantalum, 20 to 100°C; 6.6×10^{-6} per °C.

‡These values for zinc were taken from Grüneisen and Goens, *Z. Physik.,* **29:**141 (1924).

TABLE 2-101 Linear Expansion of Miscellaneous Substances*

The coefficient of cubical expansion may be taken as three times the linear coefficient. In the following table, *t* is the temperature or range of temperature, and *C*, the coefficient of expansion.

Substance	*t*, °C	$C \times 10^4$
Amber	0–30	0.50
	0–09	0.61
Bakelite, bleached	20–60	0.22
Brass:		
Cast	0–100	0.1875
Wire	0–100	0.1930
Wire	0–100	0.1783–0.193
71.5 Cu + 27.7 Zn + 0.3 Sn + 0.5 Pb	40	0.1859
71 Cu + 29 Zn	0–100	0.1906
Bronze:		
3 Cu + 1 Sn	16.6–100	0.1844
3 Cu + 1 Sn	16.6–350	0.2116
3 Cu + 1 Sn	16.6–957	0.1737
86.3 Cu + 9.7 Sn + 4 Zn	40	0.1782
97.6 Cu + 2.2 Sn + 0.2 P, hard	0–80	0.1713
97.6 Cu + 2.2 Sn + 0.2 P, soft	0–80	0.1708
Caoutchouc		0.657–0.686
Caoutchouc	16.7–25.3	0.770
Celluloid	20–70	1.00
Constantan	4–29	0.1523
Duralumin, 94Al	20–100	0.23
	20–300	0.25
Ebonite	25.3–35.4	0.842
Fluorspar, CaF_2	0–100	0.1950
German silver	0–100	0.1836
Gold-platinum, 2 Au + 1 Pt	0–100	0.1523
Gold-copper, 2 Au + 1 Cu	0–100	0.1552
Glass:		
Tube	0–100	0.0833
Tube	0–100	0.0828
Plate	0–100	0.0891
Crown (mean)	0–100	0.0897
Crown (mean)	50–60	0.0954
Flint	50–60	0.0788
Jena thermometer 16III normal	0–100	0.081

Substance	*t*, °C	$C \times 10^4$
Jena thermometer 59III	0–100	0.058
Jena thermometer 59III	−191–+16	0.424
Gutta percha	20	1.983
Ice	−20–−1	0.51
Iceland spar:		
Parallel to axis	0–80	0.2631
Perpendicular to axis	0–80	0.0544
Lead tin (solder) 2 Pb + 1 Sn	0–100	0.2508
Limestone	25–100	0.09
Magnalium	12–39	0.238
Manganin		0.181
Marble	15–100	0.117
Monel metal	25–100	0.14
	25–600	0.16
Paraffin	0–16	1.0662
Paraffin	16–38	1.3030
Paraffin	38–49	4.7707
Platinum-iridium, 10 Pt + 1 Ir	40	0.0884
Platinum-silver, 1 Pt + 2 Ag	0–100	0.1523
Porcelain	20–790	0.0413
Porcelain Bayeux	1000–1400	0.0553
Quartz:		
Parallel to axis	0–80	0.0797
Parallel to axis	−190 to + 16	0.0521
Perpend. to axis	0–80	0.1337
Quartz glass	−190 to + 16	−0.0026
Quartz glass	16 to 500	0.0057
Quartz glass	16 to 1000	0.0058
Rock salt	40	0.4040
Rubber, hard	0	0.691
Rubber, hard	−160	0.300
Speculum metal	0–100	0.1933
Steel, 0.14 C, 34.5 Ni	25–100	0.037
	25–600	0.136

Substance	*t*, °C	$C \times 10^4$
Topas:		
Parallel to lesser horizontal axis	0–100	0.0832
Parallel to greater horizontal axis	0–100	0.0836
Parallel to vertical axis	0–100	0.0472
Tourmaline:		
Parallel to longitudinal axis	0–100	0.0937
Parallel to horizontal axis	0–100	0.0773
Type metal	16.6–254	0.1952
Vulcanite	0–18	0.6360
Wedgwood ware	0–100	0.0890
Wood:		
Parallel to fiber:		
Ash	0–100	0.0951
Beech	2.34	0.0257
Chestnut	2.34	0.0649
Elm	2.34	0.0565
Mahogany	2.34	0.0361
Maple	2.34	0.0638
Oak	2.34	0.0492
Pine	2.34	0.0541
Walnut	2.34	0.0658
Across the fiber:		
Beech	2.34	0.614
Chestnut	2.34	0.325
Elm	2.34	0.443
Mahogany	2.34	0.404
Maple	2.34	0.484
Oak	2.34	0.544
Pine	2.34	0.341
Walnut	2.34	0.484
Wax white	10–26	2.300
Wax white	26–31	3.120
Wax white	31–43	4.860
Wax white	43–57	15.227

**Smithsonian Tables.* For a more complete tabulation see Tables 143, 144. *Smithsonian Physical Tables.* 9th ed., 1954, also reprinted in *American Institute of Physics Handbook*, McGraw-Hill, New York, 1957; *Handbook of Chemistry and Physics*, 40th ed., pp. 2239–2245, Chemical Rubber Publishing Co. For data on many solids prior to 1926, see Gruneisen, *Handbuch der Physik*, vol. 10, pp. 1–52, 1926, translation available as *N.A.S.A.* RE 2-18-59W, 1959. For eight plastic solids below 300 K, see Scott, *Cryogenic Engineering*, p. 331, Van Nostrand, Princeton, NJ, 1959. For 11 other materials to 300 K, see Scott, *loc. cit.*, p. 333. For quartz and silica, see Cook, *Brit. J. Appl. Phys.*, 7, 285 (1956).

TABLE 2-102 Volume Expansion of Liquids*

If V_0 is the volume at 0°, then at $t°$ the expansion formula is $V_t = V_0(1 + \alpha t + \beta t^2 + \gamma t^3)$. The table gives values of α, β, and γ, and of C, the true coefficient of volume expansion at 20° for some liquids and solutions. The temperature range of the observation is Δt. Values for the coefficient of volume expansion of liquids can be derived from the tables of specific volumes of the saturated liquid given as a function of temperature later in this section. $C = (dV/dt)/V_0$

Liquid	Range	$\alpha \times 10^3$	$\beta \times 10^6$	$\gamma \times 10^8$	$C \times 10^3$ at 20°
Acetic acid	16–107	1.0630	0.12636	1.0876	1.071
Acetone	0–54	1.3240	3.8090	−0.87983	1.487
Alcohol:					
Amyl	−15–80	0.9001	0.6573	1.18458	0.902
Ethyl, 30% by volume	18–39	0.2928	10.790	−11.87	
Ethyl, 50% by volume	0–39	0.7450	1.85	0.730	
Ethyl, 99.3% by volume	27–46	1.012	2.20		1.12
Ethyl, 500 atm pressure	0–40	0.866			
Ethyl, 3000 atm pressure	0–40	0.524			
Methyl	0–61	1.1342	1.3635	0.8741	1.199
Benzene	11–81	1.17626	1.27776	0.80648	1.237
Bromine	0–59	1.06218	1.87714	−0.30854	1.132
Calcium chloride:					
5.8% solution	18–25	0.07878	4.2742		0.250
40.9% solution	17–24	0.42383	0.8571		0.458
Carbon disulfide	−34–60	1.13980	1.37065	1.91225	1.218
500 atm pressure	0–50	0.940			
3000 atm pressure	0–50	0.581			
Carbon tetrachloride	0–76	1.18384	0.89881	1.35135	1.236
Chloroform	0–63	1.10715	4.66473	−1.74328	1.273
Ether	−15–38	1.51324	2.35918	4.00512	1.656
Glycerin		0.4853	0.4895		0.505
Hydrochloric acid, 33.2% solution	0–33	0.4460	0.215		0.455
Mercury	0–100	0.18182	0.0078		0.18186
Olive oil		0.6821	1.1405	−0.539	0.721
Pentane	0–33	1.4646	3.09319	1.6084	1.608
Potassium chloride, 24.3% solution	16–25	0.2695	2.080		0.353
Phenol	36–157	0.8340	0.10732	0.4446	1.090
Petroleum, 0.8467 density	24–120	0.8994	1.396		0.955
Sodium chloride, 20.6% solution	0–29	0.3640	1.237		0.414
Sodium sulfate, 24% solution	11–40	0.3599	1.258		0.410
Sulfuric acid:					
10.9% solution	0–30	0.2835	2.580		0.387
100.0%	0–30	0.5758	−0.432		0.558
Turpentine	−9–106	0.9003	1.9595	−0.44998	0.973
Water	0–33	−0.06427	8.5053	−6.7900	0.207

Smithsonian Tables*, Table 269. For a detailed discussion of mercury data, see Cook, *Brit. J. Appl. Phys.*, **7, 285 (1956). For data on nitrogen and argon, see Johnson (ed.), WADD-TR-60-56, 1960.

Bromoform[1] 7.7 – 50°C.

$V_t = 0.34204[1 + 0.00090411(t - 7.7) + 0.0000006766(t - 7.7)^2]$

0.34204 is the specific volume of bromoform at 7.7°C.

Glycerin[2] −62 to 0°C.

$V_t = V_0(1 + 4.83 \times 10^{-4}t - 0.49 \times 10^{-6}t^2)$

0 – 80°C.

$V_t = V_0(1 + 4.83 \times 10^{-4}t + 0.49 \times 10^{-6}t^2)$

Mercury[3] 0 – 300°C.

$V_t - V_0[1 + 10^{-8}(18{,}153.8t + 0.7548t^2 + 0.001533t^3 + 0.00000536t^4)]$

[1]Sherman and Sherman, *J. Am. Chem. Soc.*, **50**, 1119 (1928). (An obvious error in their equation has been corrected.)

[2]Samsoen, *Ann. phys.*, (10) **9**, 91 (1928).

[3]Harlow, *Phil. Mag.*, (7) **7**, 674 (1929).

TABLE 2-103 Volume Expansion of Solids*

If v_2 and v_1 are the volumes at t_2 and t_1, respectively, then $v_2 = v_1(1 + C\Delta t)$, C being the coefficient of cubical expansion and Δt the temperature interval. Where only a single temperature is stated, C represents the true coefficient of volume expansion at that temperature.

Substance	t or Δt	$C \times 10^4$
Antimony	0–100	0.3167
Beryl	0–100	0.0105
Bismuth	0–100	0.3948
Copper†	0–100	0.4998
Diamond	40	0.0354
Emerald	40	0.0168
Galena	0–100	0.558
Glass, common tube	0–100	0.276
hard	0–100	0.214
Jena, borosilicate 59 III	20–100	0.156
pure silica	0–80	0.0129
Gold	0–100	0.4411
Ice	−20 to −1	1.1250
Iron	0–100	0.3550
Lead†	0–100	0.8399
Paraffin	20	5.88
Platinum	0–100	0.265
Porcelain, Berlin	20	0.0814
chloride	0–100	1.094
nitrate	0–100	1.967
sulfate	20	1.0754
Quartz	0–100	0.3840
Rock salt	50–60	1.2120
Rubber	20	4.87
Silver	0–100	0.5831
Sodium	20	2.13
Stearic acid	33.8–45.4	8.1
Sulfur, native	13.2–50.3	2.23
Tin	0–100	0.6889
Zinc†	0–100	0.8928

**Smithsonian Tables*, Table 268.

†See additional data below.

Aluminum[1] 100 – 530°C.

$V = V_0(1 + 2.16 \times 10^{-5}t + 0.95 \times 10^{-8}t^2)$

Cadmium[1] 130 – 270°C.

$V = V_0(1 + 8.04 \times 10^{-5}t + 5.9 \times 10^{-8}t^2)$

Copper[1] 110 – 300°C.

$V = V_0(1 + 1.62 \times 10^{-5}t + 0.20 \times 10^{-8}t^2)$

Colophony[2] 0 – 34°C.

$V = V_0(1 + 2.21 \times 10^{-4}t + 0.31 \times 10^{-6}t^2)$

34 – 150°C.

$V = V_{34}[1 + 7.40 \times 10^{-4}(t - 34) + 5.91 \times 10^{-6}(t - 34)^2]$

Lead[1] 100 – 280°C.

$V = V_0(1 + 1.60 \times 10^{-5}t + 3.2 \times 10^{-8}t^2)$

Shellac[2] 0 – 46°C.

$V = V_0(1 + 2.73 \times 10^{-4}t + 0.39 \times 10^{-6}t^2)$

46 – 100°C.

$V = V_{46}[1 + 13.10 \times 10^{-4}(t - 46) + 0.62 \times 10^{-6}(t - 46)^2]$

Silica (vitreous)[3] 0 – 300°C.

$V_t = V_0[1 + 10^{-8}(93.6t + 0.7776t^2 - 0.003315t^3 + 0.000005244t^4)$

Sugar (cane, amorphous)[2] 0 – 67°C.

$V_t = V_0(1 + 2.34 \times 10^{-4}t + 0.14 \times 10^{-6}t^2)$

67 – 160°C.

$V_t = V_{67}[1 + 5.02 \times 10^{-4}(t - 67) + 0.43 \times 10^{-6}(t - 67)^2]$

Zinc[1] 120 – 360°C.

$V_t = V_0(1 + 8.50 \times 10^{-5}t + 3.9 \times 10^{-8}t^2)$

[1]Uffelmann, *Phil. Mag.*, (7) **10**, 633 (1930).

[2]Samsoen, *Ann. phys.*, (10) **9**, 83 (1928).

[3]Harlow, *Phil. Mag.*, (7) **7**, 674 (1929).

GAS EXPANSION: JOULE-THOMSON EFFECT

Introduction The Joule-Thomson coefficient, $(\partial T/\partial P)_H$, is the change in gas temperature with pressure during an adiabatic expansion (a throttling process, at constant enthalpy H). The temperature at which the Joule-Thomson coefficient changes sign is called the *Joule-Thomson inversion temperature.*

Joule-Thomson coefficients for substances listed in Table 2-104 are given in tables in the Thermodynamic Properties section.

Unit Conversions To convert the Joule-Thomson coefficient μ, in degrees Celsius per atmosphere to degrees Fahrenheit per atmosphere, multiply by 1.8. Temperature conversion: °F = 9⁄5°C + 32; °R = 9⁄5 K.

To convert bars to pounds-force per square inch, multiply by 14.504; to convert bars to kilopascals, multiply by 100.

TABLE 2-104 Additional References Available for the Joule-Thomson Coefficient

	Pressure range, atm				Temp. range, °C			
Gas	0–10	10–50	50–200	>200	<0	0–300	>300	Other references
Air	12, 15, 16 19, 35	12, 15, 19 35	15, 19, 35		19, 35	12, 15, 16 19, 35		3, 4, 18
Ammonia	28					28		2, 3
Argon	39	39	39		39	39		
Benzene	31	31	31			31	31	
Butane	26	26				26		
Carbon dioxide	7, 8, 28 37	7, 8, 37	7, 8, 37		7, 8, 37	7, 8, 9, 10 37		
Carbon monoxide	17	17			17	17		
Deuterium		22, 24, 25 1*	1,* 22, 24 25		1,* 22, 24, 25			
Dowtherm A	46	46				46	46	
Ethane	45	45				45		
Ethylene						9, 10		
Helium	1, 38	1, 38	38		1, 38	38		48
Hydrogen	24, 30	22, 24, 25 30	24, 30		22, 24, 25 30	24		
Methane		6	6			6		
Mixtures						9, 11		
Natural gas			33	33	33	33		
Nitrogen	13, 28, 40	13, 40	13, 40	13	13, 40	9, 10, 13 28, 40	13	19
Nitrous oxide						9, 10		
Pentane	26, 34, 44	34	34			26, 34, 44		
Propane	41	43				43		
Steam	28, 29, 42	29, 42, 47	42, 47			28, 29, 42 45	29, 42, 47	29, 47

*See also 14 (generalized chart); 18 (review, to 1919); 20–22; 23 (review, to 1948); 27 (review, to 1905); 32, 36, 41, 50.

REFERENCES: 1. Baehr. *Z. Elektrochem.*, **60,** 515 (1956). 2. Beattie, *J. Math. Phys.*, 9, 11 (1930). 3. Beattie, *Phys. Rev.*, **35,** 643 (1930). 4. Bradley and Hale, *Phys. Rev.*, **29,** 258 (1909). 5. Brown and Dean, *Bur. Stand. J. Res.*, **60,** 161 (1958). 6. Budenholzer, Sage, et al., *Ind. Eng. Chem.*, **29,** 658 (1937). 7. Burnett, *Phys. Rev.*, **22,** 590 (1923). 8. Burnett, Univ. Wisconsin Bull. 9(6), 1926. 9. Charnley, Ph.D. thesis. University of Manchester, 1952. 10. Charnley, Isles, et al., *Proc. R. Soc. (London)*, **A217,** 133 (1953). 11. Charnley, Rowlinson, et al., *Proc. R. Soc. (London)*, **A230,** 354 (1955). 12. Dalton, *Commun. Phys. Lab. Univ. Leiden*, no. 109c, 1909. 13. Deming and Deming, *Phys. Rev.*, **48,** 448 (1935). 14. Edmister, *Pet. Refiner*, **28,** 128 (1949). 15. Eucken, Clusius, et al., *Z. Tech. Phys.*, **13,** 267 (1932). 16. Eumorfopoulos and Rai, *Phil. Mag.*, **7,** 961 (1926). 17. Huang, Lin, et al., *Z. Phys.*, **100,** 594 (1936). 18. Hoxton, *Phys. Rev.*, **13,** 438 (1919). 19. Ishkin and Kaganev, *J. Tech. Phys. U.S.S.R.*, **26,** 2323 (1956). 20. Isles, Ph.D. thesis, Leeds University. 21. Jenkin and Pye, *Phil. Trans. R. Soc. (London)*, **A213,** 67 (1914); **A215,** 353 (1915). 22. Johnston, *J. Am. Chem. Soc.*, **68,** 2362 (1946). 23. Johnston, *Trans. Am. Soc. Mech. Eng.*, **70,** 651 (1948). 24. Johnston, Bezman, et al., *J. Am. Chem. Soc.*, **68,** 2367 (1946). 25. Johnston, Swanson, et al., *J. Am. Chem. Soc.*, **68,** 2373 (1946). 26. Kennedy, Sage, et al., *Ind. Eng. Chem.*, **28,** 718 (1936). 27. Kester, *Phys. Rev.*, **21,** 260 (1905). 28. Keyes and Collins, *Proc. Nat. Acad. Sci.*, **18,** 328 (1932). 29. Kleinschmidt, *Mech. Eng.*, **45,** 165 (1923); **48,** 155 (1926). 30. Koeppe, *Kältetechnik*, **8,** 275 (1956). 31. Lindsay and Brown, *Ind. Eng. Chem.*, **27,** *817* (1935). 32. Noell, dissertation, Munich, 1914, *Forschungs-dienst*, 184, p. 1, 1916. 33. Palienko, *Tr. Inst. Ispol' z. Gaza, Akad. Nauk Ukr. SSR*, no. 4, p. 87, 1956. 34. Pattee and Brown, *Ind. Eng. Chem.*, **26,** 511, (1934). 35. Roebuck, *Proc. Am. Acad. Arts Sci.*, **60,** 537 (1925); **64,** 287 (1930). 36. Roebuck, see 49 below, 37. Roebuck and Murrell, *Phys. Rev.*, **55,** 240 (1939). 38. Roebuck and Osterberg, *Phys. Rev.*, **37,** 110 (1931); **43,** 60 (1933). 39. Roebuck and Osterberg, *Phys. Rev.*, **46,** 785 (1934). 40. Roebuck and Osterberg, *Phys. Rev.*, **48,** 450 (1935). 41. Roebuck, Murrell, et al., *J. Am. Chem. Soc.*, **64,** 400 (1942). 42. Sage, unpublished data, California Institute of Technology, 1959. 43. Sage and Lacy, *Ind. Eng. Chem.*, **27,** 1484 (1934). 44. Sage, Kennedy, et al., *Ind. Eng. Chem.*, **28,** 601 (1936). 45. Sage, Webster, et al., *Ind. Eng. Chem.*, **29,** 658 (1937). 46. Ullock, Gaffert, et al., *Trans. Am. Inst. Chem. Eng.*, **32,** 73 (1936). 47. Yang, *Ind. Eng. Chem.*, **45,** 786 (1953). 48. Zelmanov, *J. Phys. U.S.S.R.*, **3,** 43 (1940). 49. Roebuck, recalculated data. 50. Michels et al., van der Waals laboratory publications. Gunn, Cheuh, and Prausnitz, *Cryogenics*, **6,** 324 (1966), review equations relating the inversion temperatures and pressures. The ability of various equations of state to relate these was also discussed by Miller, *Ind. Eng. Chem. Fundam.*, **9,** 585 (1970); and Juris and Wenzel, *Am. Inst. Chem. Eng. J.*, **18,** 684 (1972). Perhaps the most detailed review is that of Hendricks, Peller, and Baron. NASA Tech. Note D 6807, 1972.

TABLE 2-105 Approximate Inversion-Curve Locus in Reduced Coordinates ($T_r = T/T_c$; $P_r = P/P_c$)*

P_r	0	0.5	1	1.5	2	2.5	3	4
T_{rL}	0.782	0.800	0.818	0.838	0.859	0.880	0.903	0.953
T_{rU}	4.984	4.916	4.847	4.777	4.706	4.633	4.550	4.401
P_r	5	6	7	8	9	10	11	11.79
T_{rL}	1.01	1.08	1.16	1.25	1.35	1.50	1.73	2.24
T_{rU}	4.23	4.06	3.88	3.68	3.45	3.18	2.86	2.24

*Calculated from the best three-constant equation recommended by Miller, *Ind. Eng. Chem. Fundam.*, **9,** 585 (1970). T_{rL} refers to the lower curve, and T_{rU}, to the upper curve.

CRITICAL CONSTANTS

Additional References For other inorganic substances see Mathews, *Chem. Rev.*, **72** (1972):71–100. For other organics see Kudchaker, Alani, and Zwolinski, *Chem. Rev.*, **68** (1968): 659–735.

TABLE 2-106 Critical Constants and Acentric Factors of Inorganic and Organic Compounds

Cmpd. no.	Name	Formula	CAS	Mol. wt.	T_C, K	P_C, MPa	V_C, m^3/kmol	Z_C	Acentric factor
1	Acetaldehyde	C_2H_4O	75-07-0	44.05256	466	5.57	0.154	0.221	0.262493
2	Acetamide	C_2H_5NO	60-35-5	59.0672	761	6.6	0.215	0.224	0.421044
3	Acetic acid	$C_2H_4O_2$	64-19-7	60.052	591.95	5.786	0.177	0.208	0.466521
4	Acetic anhydride	$C_4H_6O_3$	108-24-7	102.08864	606	4	0.304	0.241	0.455328
5	Acetone	C_3H_6O	67-64-1	58.07914	508.2	4.701	0.209	0.233	0.306527
6	Acetonitrile	C_2H_3N	75-05-8	41.0519	545.5	4.85	0.193	0.206	0.341926
7	Acetylene	C_2H_2	74-86-2	26.03728	308.3	6.138	0.112	0.268	0.191185
8	Acrolein	C_3H_4O	107-02-8	56.06326	506	5	0.197	0.234	0.319832
9	Acrylic acid	$C_3H_4O_2$	79-10-7	72.06266	615	5.66	0.208	0.23	0.538324
10	Acrylonitrile	C_3H_3N	107-13-1	53.0626	540	4.66	0.216	0.224	0.310664
11	Air	Mixture	132259-10-0	28.96	132.45	3.774	0.09147	0.313	0
12	Ammonia	H_3N	7664-41-7	17.03052	405.65	11.28	0.07247	0.242	0.252608
13	Anisole	C_7H_8O	100-66-3	108.13782	645.6	4.25	0.337	0.267	0.350169
14	Argon	Ar	7440-37-1	39.948	150.86	4.898	0.07459	0.291	0
15	Benzamide	C_7H_7NO	55-21-0	121.13658	824	5.05	0.346	0.255	0.5585
16	Benzene	C_6H_6	71-43-2	78.11184	562.05	4.895	0.256	0.268	0.2103
17	Benzenethiol	C_6H_6S	108-98-5	110.17684	689	4.74	0.315	0.261	0.262789
18	Benzoic acid	$C_7H_6O_2$	65-85-0	122.12134	751	4.47	0.344	0.246	0.602794
19	Benzonitrile	C_7H_5N	100-47-0	103.1213	702.3	4.215	0.3132	0.226	0.343214
20	Benzophenone	$C_{13}H_{10}O$	119-61-9	182.2179	830	3.352	0.5677	0.276	0.501941
21	Benzyl alcohol	C_7H_8O	100-51-6	108.13782	720.15	4.374	0.382	0.279	0.363116
22	Benzyl ethyl ether	$C_9H_{12}O$	539-30-0	136.19098	662	3.11	0.442	0.25	0.433236
23	Benzyl mercaptan	C_7H_8S	100-53-8	124.20342	718	4.06	0.367	0.25	0.312604
24	Biphenyl	$C_{12}H_{10}$	92-52-4	154.2078	773	3.38	0.497	0.261	0.402873
25	Bromine	Br_2	7726-95-6	159.808	584.15	10.3	0.135	0.286	0.128997
26	Bromobenzene	C_6H_5Br	108-86-1	157.0079	670.15	4.5191	0.324	0.263	0.250575
27	Bromoethane	C_2H_5Br	74-96-4	108.965	503.8	5.565	0.204	0.271	0.205275
28	Bromomethane	CH_3Br	74-83-9	94.93852	464	6.929	0.152	0.273	0.153426
29	1,2-Butadiene	C_4H_6	590-19-2	54.09044	452	4.36	0.22	0.255	0.165877
30	1,3-Butadiene	C_4H_6	106-99-0	54.09044	425	4.32	0.221	0.27	0.195032
31	Butane	C_4H_{10}	106-97-8	58.1222	425.12	3.796	0.255	0.274	0.200164
32	1,2-Butanediol	$C_4H_{10}O_2$	584-03-2	90.121	680	5.21	0.303	0.279	0.630463
33	1,3-Butanediol	$C_4H_{10}O_2$	107-88-0	90.121	676	4.02	0.305	0.218	0.704256
34	1-Butanol	$C_4H_{10}O$	71-36-3	74.1216	563.1	4.414	0.273	0.258	0.58828
35	2-Butanol	$C_4H_{10}O$	78-92-2	74.1216	535.9	4.1885	0.27	0.254	0.580832
36	1-Butene	C_4H_8	106-98-9	56.10632	419.5	4.02	0.241	0.278	0.184495
37	*cis*-2-Butene	C_4H_8	590-18-1	56.10632	435.5	4.21	0.234	0.272	0.201877
38	*trans*-2-Butene	C_4H_8	624-64-6	56.10632	428.6	4.1	0.238	0.274	0.217592
39	Butyl acetate	$C_6H_{12}O_2$	123-86-4	116.15828	575.4	3.09	0.389	0.251	0.439393
40	Butylbenzene	$C_{10}H_{14}$	104-51-8	134.21816	660.5	2.89	0.497	0.262	0.394149
41	Butyl mercaptan	$C_4H_{10}S$	109-79-5	90.1872	570.1	3.97	0.307	0.257	0.271361
42	sec-Butyl mercaptan	$C_4H_{10}S$	513-53-1	90.1872	554	4.06	0.307	0.271	0.25059
43	1-Butyne	C_4H_6	107-00-6	54.09044	440	4.6	0.208	0.262	0.246976
44	Butyraldehyde	C_4H_8O	123-72-8	72.10572	537.2	4.41	0.258	0.255	0.282553
45	Butyric acid	$C_4H_8O_2$	107-92-6	88.1051	615.7	4.06	0.293	0.232	0.675003
46	Butyronitrile	C_4H_7N	109-74-0	69.1051	585.4	3.88	0.291	0.232	0.3601
47	Carbon dioxide	CO_2	124-38-9	44.0095	304.21	7.383	0.094	0.274	0.223621
48	Carbon disulfide	CS_2	75-15-0	76.1407	552	7.9	0.16	0.275	0.110697
49	Carbon monoxide	CO	630-08-0	28.0101	132.92	3.499	0.0944	0.299	0.0481621
50	Carbon tetrachloride	CCl_4	56-23-5	153.8227	556.35	4.56	0.276	0.272	0.192552
51	Carbon tetrafluoride	CF_4	75-73-0	88.0043	227.51	3.745	0.143	0.283	0.178981
52	Chlorine	Cl_2	7782-50-5	70.906	417.15	7.71	0.124	0.276	0.0688183

(*Continued*)

TABLE 2-106 Critical Constants and Acentric Factors of Inorganic and Organic Compounds *(Continued)*

Cmpd. no.	Name	Formula	CAS	Mol. wt.	T_C, K	P_C, MPa	V_C, m^3/kmol	Z_C	Acentric factor
53	Chlorobenzene	C_6H_5Cl	108-90-7	112.5569	632.35	4.5191	0.308	0.265	0.249857
54	Chloroethane	C_2H_5Cl	75-00-3	64.5141	460.35	5.27	0.192	0.264	0.188591
55	Chloroform	$CHCl_3$	67-66-3	119.37764	536.4	5.472	0.239	0.293	0.221902
56	Chloromethane	CH_3Cl	74-87-3	50.4875	416.25	6.68	0.141	0.272	0.151
57	1-Chloropropane	C_3H_7Cl	540-54-5	78.54068	503.15	4.425	0.243	0.257	0.215047
58	2-Chloropropane	C_3H_7Cl	75-29-6	78.54068	489	4.54	0.247	0.276	0.198553
59	*m*-Cresol	C_7H_8O	108-39-4	108.13782	705.85	4.56	0.312	0.242	0.448034
60	*o*-Cresol	C_7H_8O	95-48-7	108.13782	697.55	5.01	0.282	0.244	0.43385
61	*p*-Cresol	C_7H_8O	106-44-5	108.13782	704.65	5.15	0.277	0.244	0.50721
62	Cumene	C_9H_{12}	98-82-8	120.19158	631	3.209	0.434	0.265	0.327406
63	Cyanogen	C_2N_2	460-19-5	52.0348	400.15	5.924	0.151	0.269	0.275605
64	Cyclobutane	C_4H_8	287-23-0	56.10632	459.93	4.98	0.21	0.273	0.18474
65	Cyclohexane	C_6H_{12}	110-82-7	84.15948	553.8	4.08	0.308	0.273	0.208054
66	Cyclohexanol	$C_6H_{12}O$	108-93-0	100.15888	650.1	4.26	0.322	0.254	0.369047
67	Cyclohexanone	$C_6H_{10}O$	108-94-1	98.143	653	4	0.311	0.229	0.299006
68	Cyclohexene	C_6H_{10}	110-83-8	82.1436	560.4	4.35	0.291	0.272	0.212302
69	Cyclopentane	C_5H_{10}	287-92-3	70.1329	511.7	4.51	0.26	0.276	0.194874
70	Cyclopentene	C_5H_8	142-29-0	68.11702	507	4.8	0.245	0.279	0.19611
71	Cyclopropane	C_3H_6	75-19-4	42.07974	398	5.54	0.162	0.271	0.127829
72	Cyclohexyl mercaptan	$C_6H_{12}S$	1569-69-3	116.22448	664	3.97	0.355	0.255	0.264134
73	Decanal	$C_{10}H_{20}O$	112-31-2	156.2652	674	2.6	0.575	0.267	0.520066
74	Decane	$C_{10}H_{22}$	124-18-5	142.28168	617.7	2.11	0.617	0.254	0.492328
75	Decanoic acid	$C_{10}H_{20}O_2$	334-48-5	172.265	722.1	2.28	0.639	0.243	0.813724
76	1-Decanol	$C_{10}H_{22}O$	112-30-1	158.28108	688	2.308	0.645	0.26	0.606986
77	1-Decene	$C_{10}H_{20}$	872-05-9	140.2658	616.6	2.223	0.584	0.253	0.480456
78	Decyl mercaptan	$C_{10}H_{22}S$	143-10-2	174.34668	696	2.13	0.624	0.23	0.587421
79	1-Decyne	$C_{10}H_{18}$	764-93-2	138.24992	619.85	2.37	0.552	0.254	0.51783
80	Deuterium	D_2	7782-39-0	4.0316	38.35	1.6617	0.060263	0.314	−0.14486
81	1,1-Dibromoethane	$C_2H_4Br_2$	557-91-5	187.86116	628	6.03	0.276	0.319	0.125025
82	1,2-Dibromoethane	$C_2H_4Br_2$	106-93-4	187.86116	650.15	5.4769	0.2616	0.265	0.206724
83	Dibromomethane	CH_2Br_2	74-95-3	173.83458	611	7.17	0.223	0.315	0.20945
84	Dibutyl ether	$C_8H_{18}O$	142-96-1	130.22792	584.1	2.46	0.487	0.247	0.447646
85	*m*-Dichlorobenzene	$C_6H_4Cl_2$	541-73-1	147.00196	683.95	4.07	0.351	0.251	0.27898
86	*o*-Dichlorobenzene	$C_6H_4Cl_2$	95-50-1	147.00196	705	4.07	0.351	0.244	0.219189
87	*p*-Dichlorobenzene	$C_6H_4Cl_2$	106-46-7	147.00196	684.75	4.07	0.351	0.251	0.284638
88	1,1-Dichloroethane	$C_2H_4Cl_2$	75-34-3	98.95916	523	5.07	0.24	0.28	0.233943
89	1,2-Dichloroethane	$C_2H_4Cl_2$	107-06-2	98.95916	561.6	5.37	0.22	0.253	0.286595
90	Dichloromethane	CH_2Cl_2	75-09-2	84.93258	510	6.08	0.185	0.265	0.198622
91	1,1-Dichloropropane	$C_3H_6Cl_2$	78-99-9	112.98574	560	4.24	0.291	0.265	0.252928
92	1,2-Dichloropropane	$C_3H_6Cl_2$	78-87-5	112.98574	572	4.24	0.291	0.259	0.256391
93	Diethanol amine	$C_4H_{11}NO_2$	111-42-2	105.13564	736.6	4.27	0.349	0.243	0.952882
94	Diethyl amine	$C_4H_{11}N$	109-89-7	73.13684	496.6	3.71	0.301	0.27	0.303856
95	Diethyl ether	$C_4H_{10}O$	60-29-7	74.1216	466.7	3.64	0.28	0.263	0.281065
96	Diethyl sulfide	$C_4H_{10}S$	352-93-2	90.1872	557.15	3.96	0.318	0.272	0.29002
97	1,1-Difluoroethane	$C_2H_4F_2$	75-37-6	66.04997	386.44	4.5198	0.179	0.252	0.275052
98	1,2-Difluoroethane	$C_2H_4F_2$	624-72-6	66.04997	445	4.34	0.195	0.229	0.222428
99	Difluoromethane	CH_2F_2	75-10-5	52.02339	351.255	5.784	0.123	0.244	0.277138
100	Di–isopropyl amine	$C_6H_{15}N$	108-18-9	101.19	523.1	3.2	0.418	0.308	0.388315
101	Di–isopropyl ether	$C_6H_{14}O$	108-20-3	102.17476	500.05	2.88	0.386	0.267	0.338683
102	Di–isopropyl ketone	$C_7H_{14}O$	565-80-0	114.18546	576	3.02	0.416	0.262	0.404427
103	1,1-Dimethoxyethane	$C_4H_{10}O_2$	534-15-6	90.121	507.8	3.773	0.297	0.265	0.32768
104	1,2-Dimethoxypropane	$C_5H_{12}O_2$	7778-85-0	104.14758	543	3.446	0.35	0.267	0.352222
105	Dimethyl acetylene	C_4H_6	503-17-3	54.09044	473.2	4.87	0.221	0.274	0.238542

106	Dimethyl amine	C_2H_7N	124-40-3	45.08368	437.2	5.34	0.18	0.264	0.299885
107	2,3-Dimethylbutane	C_6H_{14}	79-29-8	86.17536	500	3.15	0.361	0.274	0.249251
108	1,1-Dimethylcyclohexane	C_8H_{16}	590-66-9	112.21264	591.15	2.93843	0.45	0.269	0.232569
109	*cis*-1,2-Dimethylcyclohexane	C_8H_{16}	2207-01-4	112.21264	606.15	2.93843	0.46	0.268	0.232443
110	*trans*-1,2-Dimethylcyclohexane	C_8H_{16}	6876-23-9	112.21264	596.15	2.93843	0.46	0.273	0.237864
111	Dimethyl disulfide	$C_2H_6S_2$	624-92-0	94.19904	615	5.36	0.252	0.264	0.205916
112	Dimethyl ether	C_2H_6O	115-10-6	46.06844	400.1	5.37	0.17	0.2744	0.200221
113	*N,N*-Dimethyl formamide	C_3H_7NO	68-12-2	73.09378	649.6	4.42	0.26199	0.214	0.31771
114	2,3-Dimethylpentane	C_7H_{16}	565-59-3	100.20194	537.3	2.91	0.393	0.256	0.296407
115	Dimethyl phthalate	$C_{10}H_{10}O_4$	131-11-3	194.184	766	2.78	0.53	0.231	0.656848
116	Dimethylsilane	C_2H_8Si	1111-74-6	60.17042	402	3.56	0.258	0.275	0.129957
117	Dimethyl sulfide	C_2H_6S	75-18-3	62.134	503.04	5.53	0.201	0.266	0.194256
118	Dimethyl sulfoxide	C_2H_6OS	67-68-5	78.13344	729	5.65	0.227	0.212	0.280551
119	Dimethyl terephthalate	$C_{10}H_{10}O_4$	120-61-6	194.184	777.4	2.76	0.529	0.226	0.580691
120	1,4-Dioxane	$C_4H_8O_2$	123-91-1	88.10512	587	5.2081	0.238	0.254	0.279262
121	Diphenyl ether	$C_{12}H_{10}O$	101-84-8	170.2072	766.8	3.08	0.503	0.243	0.43889
122	Dipropyl amine	$C_6H_{15}N$	142-84-7	101.19	550	3.14	0.402	0.276	0.449684
123	Dodecane	$C_{12}H_{26}$	112-40-3	170.33484	658	1.82	0.755	0.251	0.576385
124	Eicosane	$C_{20}H_{42}$	112-95-8	282.54748	768	1.16	1.34	0.243	0.906878
125	Ethane	C_2H_6	74-84-0	30.069	305.32	4.872	0.1455	0.279	0.099493
126	Ethanol	C_2H_6O	64-17-5	46.06844	514	6.137	0.168	0.241	0.643558
127	Ethyl acetate	$C_4H_8O_2$	141-78-6	88.10512	523.3	3.88	0.286	0.255	0.366409
128	Ethyl amine	C_2H_7N	75-04-7	45.08368	456.15	5.62	0.207	0.307	0.284788
129	Ethylbenzene	C_8H_{10}	100-41-4	106.165	617.15	3.609	0.374	0.263	0.30347
130	Ethyl benzoate	$C_9H_{10}O_2$	93-89-0	150.1745	698	3.18	0.489	0.268	0.477055
131	2-Ethyl butanoic acid	$C_6H_{12}O_2$	88-09-5	116.15828	655	3.41	0.389	0.244	0.632579
132	Ethyl butyrate	$C_6H_{12}O_2$	105-54-4	116.15828	571	2.95	0.403	0.25	0.401075
133	Ethylcyclohexane	C_8H_{16}	1678-91-7	112.21264	609.15	3.04	0.43	0.258	0.245525
134	Ethylcyclopentane	C_7H_{14}	1640-89-7	98.18606	569.5	3.4	0.375	0.269	0.270095
135	Ethylene	C_2H_4	74-85-1	28.05316	282.34	5.041	0.131	0.281	0.0862484
136	Ethylenediamine	$C_2H_8N_2$	107-15-3	60.09832	593	6.29	0.264	0.337	0.472367
137	Ethylene glycol	$C_2H_6O_2$	107-21-1	62.06784	720	8.2	0.191	0.262	0.506776
138	Ethyleneimine	C_2H_5N	151-56-4	43.0678	537	6.85	0.173	0.265	0.200735
139	Ethylene oxide	C_2H_4O	75-21-8	44.05256	469.15	7.19	0.140296	0.25876	0.197447
140	Ethyl formate	$C_3H_6O_2$	109-94-4	74.07854	508.4	4.74	0.229	0.257	0.284736
141	2-Ethyl hexanoic acid	$C_8H_{16}O_2$	149-57-5	144.211	674.6	2.778	0.528	0.262	0.801289
142	Ethylhexyl ether	$C_8H_{18}O$	5756-43-4	130.22792	583	2.46	0.487	0.247	0.494378
143	Ethylisopropyl ether	$C_5H_{12}O$	625-54-7	88.14818	489	3.41	0.329	0.276	0.305629
144	Ethylisopropyl ketone	$C_6H_{12}O$	565-69-5	100.15888	567	3.32	0.369	0.26	0.389061
145	Ethyl mercaptan	C_2H_6S	75-08-1	62.13404	499.15	5.49	0.207	0.274	0.187751
146	Ethyl propionate	$C_5H_{10}O_2$	105-37-3	102.1317	546	3.362	0.345	0.256	0.394373
147	Ethylpropyl ether	$C_5H_{12}O$	628-32-0	88.14818	500.23	3.37007	0.339	0.275	0.347328
148	Ethyltrichlorosilane	$C_2H_5Cl_3Si$	115-21-9	163.506	559.95	3.33	0.403	0.288	0.269778
149	Fluorine	F_2	7782-41-4	37.9968064	144.12	5.1724	0.066547	0.287	0.0530336
150	Fluorobenzene	C_6H_5F	462-06-6	96.1023032	560.09	4.55051	0.269	0.263	0.247183
151	Fluoroethane	C_2H_5F	353-36-6	48.0595	375.31	5.028	0.159	0.256	0.217903
152	Fluoromethane	CH_3F	593-53-3	34.03292	317.42	5.87511	0.113	0.252	0.194721
153	Formaldehyde	CH_2O	50-00-0	30.02598	420	6.59	0.0851	0.161	0.167887
154	Formamide	CH_3NO	75-12-7	45.04062	771	7.8	0.163	0.198	0.412381
155	Formic acid	CH_2O_2	64-18-6	46.0257	588	5.81	0.125	0.149	0.312521
156	Furan	C_4H_4O	110-00-9	68.07396	490.15	5.5	0.218	0.294	0.201538
157	Helium-4	He	7440-59-7	4.0026	5.2	0.2275	0.0573	0.302	−0.390032
158	Heptadecane	$C_{17}H_{36}$	629-78-7	240.46774	736	1.34	1.11	0.244	0.769688
159	Heptanal	$C_7H_{14}O$	111-71-7	114.18546	620	3.16	0.434	0.266	0.405751
160	Heptane	C_7H_{16}	142-82-5	100.20194	540.2	2.74	0.428	0.261	0.349469

(*Continued*)

TABLE 2-106 Critical Constants and Acentric Factors of Inorganic and Organic Compounds *(Continued)*

Cmpd. no.	Name	Formula	CAS	Mol. wt.	T_C, K	P_C, MPa	V_C, m³/kmol	Z_C	Acentric factor
161	Heptanoic acid	$C_7H_{14}O_2$	111-14-8	130.185	677.3	3.043	0.466	0.252	0.759934
162	1-Heptanol	$C_7H_{16}O$	111-70-6	116.20134	632.3	3.085	0.444	0.261	0.562105
163	2-Heptanol	$C_7H_{16}O$	543-49-7	116.20134	608.3	3	0.447	0.265	0.567733
164	3-Heptanone	$C_7H_{14}O$	106-35-4	114.18546	606.6	2.92	0.433	0.251	0.407565
165	2-Heptanone	$C_7H_{14}O$	110-43-0	114.18546	611.4	2.94	0.434	0.251	0.418982
166	1-Heptene	C_7H_{14}	592-76-7	98.18606	537.4	2.92	0.402	0.263	0.343194
167	Heptyl mercaptan	$C_7H_{16}S$	1639-09-4	132.26694	645	2.77	0.465	0.24	0.422568
168	1-Heptyne	C_7H_{12}	628-71-7	96.17018	547	3.21	0.387	0.273	0.377799
169	Hexadecane	$C_{16}H_{34}$	544-76-3	226.44116	723	1.4	1.04	0.243	0.717404
170	Hexanal	$C_6H_{12}O$	66-25-1	100.15888	594	3.46	0.378	0.266	0.361818
171	Hexane	C_6H_{14}	110-54-3	86.17536	507.6	3.025	0.371	0.266	0.301261
172	Hexanoic acid	$C_6H_{12}O_2$	142-62-1	116.158	660.2	3.308	0.408	0.246	0.733019
173	1-Hexanol	$C_6H_{14}O$	111-27-3	102.17476	611.3	3.446	0.382	0.259	0.558598
174	2-Hexanol	$C_6H_{14}O$	626-93-7	102.175	585.3	3.311	0.385	0.262	0.553
175	2-Hexanone	$C_6H_{12}O$	591-78-6	100.15888	587.61	3.287	0.378	0.254	0.384626
176	3-Hexanone	$C_6H_{12}O$	589-38-8	100.15888	582.82	3.32	0.378	0.259	0.380086
177	1-Hexene	C_6H_{12}	592-41-6	84.15948	504	3.21	0.348	0.267	0.285121
178	3-Hexyne	C_6H_{10}	928-49-4	82.1436	544	3.53	0.331	0.258	0.218301
179	Hexyl mercaptan	$C_6H_{14}S$	111-31-9	118.24036	623	3.08	0.412	0.245	0.368101
180	1-Hexyne	C_6H_{10}	693-02-7	82.1436	516.2	3.62	0.322	0.272	0.332699
181	2-Hexyne	C_6H_{10}	764-35-2	82.1436	549	3.53	0.331	0.256	0.221387
182	Hydrazine	H_4N_2	302-01-2	32.04516	653.15	14.7	0.158	0.428	0.314282
183	Hydrogen	H_2	1333-74-0	2.01588	33.19	1.313	0.064147	0.305	−0.215993
184	Hydrogen bromide	BrH	10035-10-6	80.91194	363.15	8.552	0.1	0.283	0.073409
185	Hydrogen chloride	ClH	7647-01-0	36.46094	324.65	8.31	0.081	0.249	0.131544
186	Hydrogen cyanide	CHN	74-90-8	27.02534	456.65	5.39	0.139	0.197	0.409913
187	Hydrogen fluoride	FH	7664-39-3	20.0063432	461.15	6.48	0.069	0.117	0.382283
188	Hydrogen sulfide	H_2S	7783-06-4	34.08088	373.53	8.96291	0.0985	0.284	0.0941677
189	Isobutyric acid	$C_4H_8O_2$	79-31-2	88.10512	605	3.7	0.292	0.215	0.61405
190	Isopropyl amine	C_3H_9N	75-31-0	59.11026	471.85	4.54	0.221	0.256	0.275913
191	Malonic acid	$C_3H_4O_4$	141-82-2	104.06146	834	6.1	0.279	0.245	0.738273
192	Methacrylic acid	$C_4H_6O_2$	79-41-4	86.08924	662	4.79	0.28	0.244	0.331817
193	Methane	CH_4	74-82-8	16.0425	190.564	4.599	0.0986	0.286	0.0115478
194	Methanol	CH_4O	67-56-1	32.04186	512.5	8.084	0.117	0.222	0.565831
195	*N*-Methyl acetamide	C_3H_7NO	79-16-3	73.09378	718	4.98	0.267	0.223	0.435111
196	Methyl acetate	$C_3H_6O_2$	79-20-9	74.07854	506.55	4.75	0.228	0.257	0.331255
197	Methyl acetylene	C_3H_4	74-99-7	40.06386	402.4	5.63	0.164	0.276	0.211537
198	Methyl acrylate	$C_4H_6O_2$	96-33-3	86.08924	536	4.25	0.27	0.258	0.342296
199	Methyl amine	CH_5N	74-89-5	31.0571	430.05	7.46	0.154	0.321	0.281417
200	Methyl benzoate	$C_8H_8O_2$	93-58-3	136.14792	693	3.59	0.436	0.272	0.420541
201	3-Methyl-1,2-butadiene	C_5H_8	598-25-4	68.11702	490	3.83	0.291	0.274	0.187439
202	2-Methylbutane	C_5H_{12}	78-78-4	72.14878	460.4	3.38	0.306	0.27	0.227875
203	2-Methylbutanoic acid	$C_5H_{10}O_2$	116-53-0	102.1317	643	3.89	0.347	0.252	0.589443
204	3-Methyl-1-butanol	$C_5H_{12}O$	123-51-3	88.1482	577.2	3.93	0.329	0.269	0.59002
205	2-Methyl-1-butene	C_5H_{10}	563-46-2	70.1329	465	3.447	0.292	0.26	0.234056
206	2-Methyl-2-butene	C_5H_{10}	513-35-9	70.1329	470	3.42	0.292	0.256	0.28703
207	2-Methyl -1-butene-3-yne	C_5H_6	78-80-8	66.10114	492	4.38	0.248	0.266	0.137046
208	Methylbutyl ether	$C_5H_{12}O$	628-28-4	88.14818	512.74	3.371	0.329	0.26	0.313008
209	Methylbutyl sulfide	$C_5H_{12}S$	628-29-5	104.214	593	3.47	0.36	0.253	0.3229
210	3-Methyl-1-butyne	C_5H_8	598-23-2	68.11702	463.2	4.2	0.275	0.3	0.308085
211	Methyl butyrate	$C_5H_{10}O_2$	623-42-7	102.1317	554.5	3.473	0.34	0.256	0.377519
212	Methylchlorosilane	CH_5ClSi	993-00-0	80.5889	442	4.17	0.246	0.279	0.225204
213	Methylcyclohexane	C_7H_{14}	108-87-2	98.18606	572.1	3.48	0.369	0.27	0.236055

214	1-Methylcyclohexanol	$C_7H_{14}O$	590-67-0	114.18546	686	4	0.374	0.262	0.221299
215	*cis*-2-Methylcyclohexanol	$C_7H_{14}O$	7443-70-1	114.18546	614	3.79	0.374	0.278	0.68049
216	*trans*-2-Methylcyclohexanol	$C_7H_{14}O$	7443-52-9	114.18546	617	3.79	0.374	0.276	0.67904
217	Methylcyclopentane	C_6H_{12}	96-37-7	84.15948	532.7	3.79	0.319	0.273	0.228759
218	1-Methylcyclopentene	C_6H_{10}	693-89-0	82.1436	542	4.13	0.303	0.278	0.23179
219	3-Methylcyclopentene	C_6H_{10}	1120-62-3	82.1436	526	4.13	0.303	0.286	0.229606
220	Methyldichlorosilane	CH_4Cl_2Si	75-54-7	115.03396	483	3.95	0.289	0.284	0.275755
221	Methylethyl ether	C_3H_8O	540-67-0	60.09502	437.8	4.4	0.221	0.267	0.231374
222	Methylethyl ketone	C_4H_8O	78-93-3	72.10572	535.5	4.15	0.267	0.249	0.323369
223	Methylethyl sulfide	C_3H_8S	624-89-5	76.1606	533	4.26	0.254	0.244	0.209108
224	Methyl formate	$C_2H_4O_2$	107-31-3	60.05196	487.2	6	0.172	0.255	0.255551
225	Methylisobutyl ether	$C_5H_{12}O$	625-44-5	88.14818	497	3.41	0.329	0.272	0.307786
226	Methylisobutyl ketone	$C_6H_{12}O$	108-10-1	100.15888	574.6	3.27	0.369	0.253	0.355671
227	Methyl Isocyanate	C_2H_3NO	624-83-9	57.05132	488	5.48	0.202	0.273	0.300694
228	Methylisopropyl ether	$C_4H_{10}O$	598-53-8	74.1216	464.48	3.762	0.276	0.269	0.26555
229	Methylisopropyl ketone	$C_5H_{10}O$	563-80-4	86.1323	553.4	3.8	0.31	0.256	0.320845
230	Methylisopropyl sulfide	$C_4H_{10}S$	1551-21-9	90.1872	553.1	4.021	0.328	0.28718	0.24611
231	Methyl mercaptan	CH_4S	74-93-1	48.10746	469.95	7.23	0.145	0.268	0.158174
232	Methyl methacrylate	$C_5H_8O_2$	80-62-6	100.11582	566	3.68	0.323	0.253	0.280233
233	2-Methyloctanoic acid	$C_9H_{18}O_2$	3004-93-1	158.23802	694	2.54	0.572	0.252	0.791271
234	2-Methylpentane	C_6H_{14}	107-83-5	86.17536	497.7	3.04	0.368	0.27	0.279149
235	Methyl pentyl ether	$C_6H_{14}O$	628-80-8	102.17476	546.49	3.042	0.38	0.254	0.344201
236	2-Methylpropane	C_4H_{10}	75-28-5	58.1222	407.8	3.64	0.259	0.278	0.183521
237	2-Methyl-2-propanol	$C_4H_{10}O$	75-65-0	74.1216	506.2	3.972	0.275	0.26	0.615203
238	2-Methyl propene	C_4H_8	115-11-7	56.10632	417.9	4	0.239	0.275	0.19484
239	Methyl propionate	$C_4H_8O_2$	554-12-1	88.10512	530.6	4.004	0.282	0.256	0.346586
240	Methylpropyl ether	$C_4H_{10}O$	557-17-5	74.1216	476.25	3.801	0.276	0.265	0.276999
241	Methylpropyl sulfide	$C_4H_{10}S$	3877-15-4	90.1872	565	3.97	0.307	0.259	0.273669
242	Methylsilane	CH_6Si	992-94-9	46.14384	352.5	4.7	0.205	0.329	0.131449
243	alpha-Methyl styrene	C_9H_{10}	98-83-9	118.1757	654	3.36	0.399	0.247	0.32297
244	Methyl tert-butyl ether	$C_5H_{12}O$	1634-04-4	88.1482	497.1	3.286	0.329	0.262	0.246542
245	Methyl vinyl ether	C_3H_6O	107-25-5	58.07914	437	4.67	0.21	0.27	0.241564
246	Naphthalene	$C_{10}H_8$	91-20-3	128.17052	748.4	4.05	0.407	0.265	0.302034
247	Neon	Ne	7440-01-9	20.1797	44.4	2.653	0.0417	0.3	−0.0395988
248	Nitroethane	$C_2H_5NO_2$	79-24-3	75.0666	593	5.16	0.236	0.247	0.380324
249	Nitrogen	N_2	7727-37-9	28.0134	126.2	3.4	0.08921	0.289	0.0377215
250	Nitrogen trifluoride	F_3N	7783-54-2	71.00191	234	4.4607	0.11875	0.272	0.119984
251	Nitromethane	CH_3NO_2	75-52-5	61.04002	588.15	6.31	0.173	0.223	0.348026
252	Nitrous oxide	N_2O	10024-97-2	44.0128	309.57	7.245	0.0974	0.274	0.140894
253	Nitric oxide	NO	10102-43-9	30.0061	180.15	6.48	0.058	0.251	0.582944
254	Nonadecane	$C_{19}H_{40}$	629-92-5	268.5209	758	1.21	1.26	0.242	0.852231
255	Nonanal	$C_9H_{18}O$	124-19-6	142.23862	658.5	2.68	0.543	0.266	0.473309
256	Nonane	C_9H_{20}	111-84-2	128.2551	594.6	2.29	0.551	0.255	0.44346
257	Nonanoic acid	$C_9H_{18}O_2$	112-05-0	158.238	710.7	2.514	0.584	0.248	0.778706
258	1-Nonanol	$C_9H_{20}O$	143-08-8	144.2545	670.9	2.527	0.576	0.261	0.584074
259	2-Nonanol	$C_9H_{20}O$	628-99-9	144.255	649.5	2.5408	0.577	0.271	0.6092
260	1-Nonene	C_9H_{18}	124-11-8	126.23922	593.1	2.428	0.524	0.258	0.436736
261	Nonyl mercaptan	$C_9H_{20}S$	1455-21-6	160.3201	681	2.31	0.571	0.233	0.52604
262	1-Nonyne	C_9H_{16}	3452-09-3	124.22334	598.05	2.61	0.497	0.261	0.470974
263	Octadecane	$C_{18}H_{38}$	593-45-3	254.49432	747	1.27	1.19	0.243	0.811359
264	Octanal	$C_8H_{16}O$	124-13-0	128.212	638.9	2.96	0.488	0.272	0.441993
265	Octane	C_8H_{18}	111-65-9	114.22852	568.7	2.49	0.486	0.256	0.399552
266	Octanoic acid	$C_8H_{16}O_2$	124-07-2	144.211	694.26	2.779	0.523	0.252	0.773427
267	1-Octanol	$C_8H_{18}O$	111-87-5	130.22792	652.3	2.783	0.509	0.261	0.569694
268	2-Octanol	$C_8H_{18}O$	123-96-6	130.228	629.8	2.749	0.512	0.269	0.58814

(*Continued*)

TABLE 2-106 Critical Constants and Acentric Factors of Inorganic and Organic Compounds *(Continued)*

Cmpd. no.	Name	Formula	CAS	Mol. wt.	T_C, K	P_C, MPa	V_C, m³/kmol	Z_C	Acentric factor
269	2-Octanone	$C_8H_{16}O$	111-13-7	128.21204	632.7	2.64	0.497	0.249	0.454874
270	3-Octanone	$C_8H_{16}O$	106-68-3	128.21204	627.7	2.704	0.496953	0.257	0.440561
271	1-Octene	C_8H_{16}	111-66-0	112.21264	566.9	2.663	0.464	0.262	0.392149
272	Octyl mercaptan	$C_8H_{18}S$	111-88-6	146.29352	667.3	2.52	0.518	0.235	0.449744
273	1-Octyne	C_8H_{14}	629-05-0	110.19676	574	2.88	0.442	0.267	0.42329
274	Oxalic acid	$C_2H_2O_4$	144-62-7	90.03488	828	8.2	0.227	0.27	0.286278
275	Oxygen	O_2	7782-44-7	31.9988	154.58	5.043	0.0734	0.288	0.0221798
276	Ozone	O_3	10028-15-6	47.9982	261	5.57	0.089	0.228	0.211896
277	Pentadecane	$C_{15}H_{32}$	629-62-9	212.41458	708	1.48	0.969	0.244	0.68632
278	Pentanal	$C_5H_{10}O$	110-62-3	86.1323	566.1	3.845	0.313	0.256	0.313152
279	Pentane	C_5H_{12}	109-66-0	72.14878	469.7	3.37	0.313	0.27	0.251506
280	Pentanoic acid	$C_5H_{10}O_2$	109-52-4	102.132	639.16	3.63	0.35	0.239	0.706632
281	1-Pentanol	$C_5H_{12}O$	71-41-0	88.1482	588.1	3.897	0.326	0.258	0.57483
282	2-Pentanol	$C_5H_{12}O$	6032-29-7	88.1482	561	3.7	0.326	0.259	0.554979
283	2-Pentanone	$C_5H_{10}O$	107-87-9	86.1323	561.08	3.694	0.301	0.238	0.343288
284	3-Pentanone	$C_5H_{10}O$	96-22-0	86.1323	560.95	3.74	0.336	0.269	0.344846
285	1-Pentene	C_5H_{10}	109-67-1	70.1329	464.8	3.56	0.2934	0.27	0.237218
286	2-Pentyl mercaptan	$C_5H_{12}S$	2084-19-7	104.21378	584.3	3.536	0.385	0.28	0.26853
287	Pentyl mercaptan	$C_5H_{12}S$	110-66-7	104.21378	598	3.47	0.359	0.251	0.320705
288	1-Pentyne	C_5H_8	627-19-0	68.11702	481.2	4.17	0.277	0.289	0.289925
289	2-Pentyne	C_5H_8	627-21-4	68.11702	519	4.03	0.276	0.258	0.175199
290	Phenanthrene	$C_{14}H_{10}$	85-01-8	178.2292	869	2.9	0.554	0.222	0.470716
291	Phenol	C_6H_6O	108-95-2	94.11124	694.25	6.13	0.229	0.243	0.44346
292	Phenyl isocyanate	C_7H_5NO	103-71-9	119.1207	653	4.06	0.37	0.277	0.412323
293	Phthalic anhydride	$C_8H_4O_3$	85-44-9	148.11556	791	4.72	0.421	0.302	0.702495
294	Propadiene	C_3H_4	463-49-0	40.06386	394	5.25	0.165	0.264	0.104121
295	Propane	C_3H_8	74-98-6	44.09562	369.83	4.248	0.2	0.276	0.152291
296	1-Propanol	C_3H_8O	71-23-8	60.09502	536.8	5.169	0.219	0.254	0.6209
297	2-Propanol	C_3H_8O	67-63-0	60.095	508.3	4.765	0.222	0.25	0.663
298	Propenylcyclohexene	C_9H_{14}	13511-13-2	122.20746	636	3.12	0.437	0.258	0.341975
299	Propionaldehyde	C_3H_6O	123-38-6	58.07914	503.6	5.038	0.204	0.246	0.281254
300	Propionic acid	$C_3H_6O_2$	79-09-4	74.0785	600.81	4.668	0.235	0.22	0.579579
301	Propionitrile	C_3H_5N	107-12-0	55.0785	561.3	4.26	0.242	0.221	0.350057
302	Propyl acetate	$C_5H_{10}O_2$	109-60-4	102.1317	549.73	3.36	0.345	0.254	0.388902
303	Propyl amine	C_3H_9N	107-10-8	59.11026	496.95	4.74	0.26	0.298	0.279839
304	Propylbenzene	C_9H_{12}	103-65-1	120.19158	638.35	3.2	0.44	0.265	0.344391
305	Propylene	C_3H_6	115-07-1	42.07974	364.85	4.6	0.185	0.281	0.137588
306	Propyl formate	$C_4H_8O_2$	110-74-7	88.10512	538	4.02	0.285	0.256	0.308779
307	2-Propyl mercaptan	C_3H_8S	75-33-2	76.16062	517	4.75	0.254	0.281	0.21381
308	Propyl mercaptan	C_3H_8S	107-03-9	76.16062	536.6	4.63	0.254	0.264	0.231789

309	1,2-Propylene glycol	$C_3H_8O_2$	57-55-6	76.09442	626	6.1	0.239	0.28	1.10651
310	Quinone	$C_6H_4O_2$	106-51-4	108.09476	683	5.96	0.291	0.305	0.494515
311	Silicon tetrafluoride	F_4Si	7783-61-1	104.07911	259	3.72	0.202	0.349	0.38584
312	Styrene	C_8H_8	100-42-5	104.14912	636	3.84	0.352	0.256	0.297097
313	Succinic acid	$C_4H_6O_4$	110-15-6	118.08804	838	5	0.33	0.237	0.743044
314	Sulfur dioxide	O_2S	7446-09-5	64.0638	430.75	7.8841	0.122	0.269	0.245381
315	Sulfur hexafluoride	F_6S	2551-62-4	146.0554192	318.69	3.76	0.19852	0.282	0.215146
316	Sulfur trioxide	O_3S	7446-11-9	80.0632	490.85	8.21	0.127	0.255	0.42396
317	Terephthalic acid	$C_8H_6O_4$	100-21-0	166.13084	883.6	3.486	0.424	0.201	0.94695
318	*o*-Terphenyl	$C_{18}H_{14}$	84-15-1	230.30376	857	2.99	0.731	0.307	0.551265
319	Tetradecane	$C_{14}H_{30}$	629-59-4	198.388	693	1.57	0.897	0.244	0.643017
320	Tetrahydrofuran	C_4H_8O	109-99-9	72.10572	540.15	5.19	0.224	0.259	0.225354
321	1,2,3,4-Tetrahydronaphthalene	$C_{10}H_{12}$	119-64-2	132.20228	720	3.65	0.408	0.249	0.335255
322	Tetrahydrothiophene	C_4H_8S	110-01-0	88.17132	631.95	5.16	0.249	0.245	0.199551
323	2,2,3,3-Tetramethylbutane	C_8H_{18}	594-82-1	114.22852	568	2.87	0.461	0.28	0.244953
324	Thiophene	C_4H_4S	110-02-1	84.13956	579.35	5.69	0.219	0.259	0.196972
325	Toluene	C_7H_8	108-88-3	92.13842	591.75	4.108	0.316	0.264	0.264012
326	1,1,2-Trichloroethane	$C_2H_3Cl_3$	79-00-5	133.40422	602	4.48	0.281	0.252	0.259135
327	Tridecane	$C_{13}H_{28}$	629-50-5	184.36142	675	1.68	0.826	0.247	0.617397
328	Triethyl amine	$C_6H_{15}N$	121-44-8	101.19	535.15	3.04	0.39	0.266	0.316193
329	Trimethyl amine	C_3H_9N	75-50-3	59.11026	433.25	4.07	0.254	0.287	0.206243
330	1,2,3-Trimethylbenzene	C_9H_{12}	526-73-8	120.19158	664.5	3.454	0.414	0.259	0.366553
331	1,2,4-Trimethylbenzene	C_9H_{12}	95-63-6	120.19158	649.1	3.232	0.43	0.258	0.37871
332	2,2,4-Trimethylpentane	C_8H_{18}	540-84-1	114.22852	543.8	2.57	0.468	0.266	0.303455
333	2,3,3-Trimethylpentane	C_8H_{18}	560-21-4	114.22852	573.5	2.82	0.455	0.269	0.2903
334	1,3,5-Trinitrobenzene	$C_6H_3N_3O_6$	99-35-4	213.10452	846	3.39	0.479	0.231	0.862257
335	2,4,6-Trinitrotoluene	$C_7H_5N_3O_6$	118-96-7	227.1311	828	3.04	0.572	0.253	0.897249
336	Undecane	$C_{11}H_{24}$	1120-21-4	156.30826	639	1.95	0.685	0.252	0.530316
337	1-Undecanol	$C_{11}H_{24}O$	112-42-5	172.30766	703.9	2.119	0.715	0.259	0.623622
338	Vinyl acetate	$C_4H_6O_2$	108-05-4	86.08924	519.13	3.958	0.27	0.248	0.351307
339	Vinyl acetylene	C_4H_4	689-97-4	52.07456	454	4.86	0.205	0.264	0.106852
340	Vinyl chloride	C_2H_3Cl	75-01-4	62.49822	432	5.67	0.179	0.283	0.100107
341	Vinyl trichlorosilane	$C_2H_3Cl_3Si$	75-94-5	161.48972	543.15	3.06	0.408	0.276	0.281543
342	Water	H_2O	7732-18-5	18.01528	647.096	22.064	0.0559472	0.229	0.344861
343	*m*-Xylene	C_8H_{10}	108-38-3	106.165	617	3.541	0.375	0.259	0.326485
344	*o*-Xylene	C_8H_{10}	95-47-6	106.165	630.3	3.732	0.37	0.264	0.31013
345	*p*-Xylene	C_8H_{10}	106-42-3	106.165	616.2	3.511	0.378	0.259	0.321839

Values in this table were taken from the Design Institute for Physical Properties (DIPPR) of the American Institute of Chemical Engineers (AIChE), 801 Critically Evaluated Gold Standard™ Database, copyright 2016 AIChE, and reproduced with permission of AIChE and of the DIPPR Evaluated Process Design Data Project Steering Committee. Their source should be cited as "R. L. Rowley, W. V. Wilding, J. L. Oscarson, T. A. Knotts, N. F. Giles, *DIPPR® Data Compilation of Pure Chemical Properties*, Design Institute for Physical Properties, AIChE, New York, NY (2016)".

COMPRESSIBILITIES

Introduction The compressibility factor Z can be calculated by using the defining equation $Z = PV/(RT)$, where P is pressure, V is molar volume, R is the gas constant, and T is absolute temperature. Values of P, V, and T for substances listed in Table 2-109 are given in tables in the Thermodynamic Properties section. For the units used in these tables, R is 0.008314472 MPa·dm^3/(mol·K). Values at temperatures and pressures other than those in the tables can be generated for many of the substances in Table 2-109 by going to http://webbook.nist.gov and selecting NIST Chemistry WebBook, then Thermophysical Properties of Fluid Systems High Accuracy Data. Results can be pasted into a spreadsheet to facilitate calculation of the compressibility factor.

Unit Conversions For this subsection, the following unit conversion is applicable: °R = 9/5 K. To convert bars to pounds-force per cubic inch, multiply by 14.504. To convert bars to kilopascals, multiply by 100.

TABLE 2-107 Compressibilities of Liquids*

At the constant temperature T, the compressibility $\beta = (1/V_0)(dV/dP)$. In general as P increases, β decreases rapidly at first and then slowly; the change of β with T is large at low pressures but very small at pressures above 1000 to 2000 megabars. 1 megabar = 0.987 atm = 10^6 dynes/cm^2 based upon the older usage, 1 bar = 1 dyne/cm^2.

Substance	Temp., °C	Pressure, megabars	Compressibility per megabar $\beta \times 10^6$	Substance	Temp., °C	Pressure, megabars	Compressibility per megabar $\beta \times 10^6$	Substance	Temp., °C	Pressure, megabars	Compressibility per megabar $\beta \times 10^6$
Acetone	14	23	111	Ethyl acetate	20	400	75	Methyl alcohol	15	23	103
Acetone	20	500	61	alcohol	14	23	100	alcohol	20	200	95
Acetone	20	1,000	52	alcohol	20	500	63	alcohol	20	400	80
Acetone	40	12,000	9	alcohol	20	1,000	54	alcohol	20	500	65
Amyl alcohol	14	23	88	alcohol	20	12,000	8	alcohol	20	1,000	54
alcohol, iso.	20	200	84	bromide	20	200	100	alcohol	20	12,000	8
alcohol, iso.	20	400	70	bromide	20	400	82	Nitric acid	0	17	32
alcohol, *n*	20	500	61	bromide	20	500	70	Oils:			
alcohol, *n*	20	1,000	46	bromide	20	1,000	54	Almond	15	5	53
alcohol, *n*	20	12,000	8	bromide	20	12,000	8	Castor	15	5	46
alcohol, *n*	40	12,000	8	chloride	15	23	151	Linseed	15	5	51
Benzene	17	5	89	chloride	20	500	102	Olive	15	5	55
Benzene	20	200	77	chloride	20	1,000	66	Rapeseed	20		59
Benzene	20	400	67	chloride	20	12,000	8	Phosphorus trichloride	10	250	71
Bromine	20	200	56	ether	25	23	188	trichloride	20	500	63
Bromine	20	400	51	ether	20	500	84	trichloride	20	1,000	47
Butyl alcohol, iso	18	8	97	ether	20	1,000	61	trichloride	20	12,000	8
alcohol, iso	20	200	81	ether	20	12,000	10	Propyl alcohol (*n*)	20	200	77
alcohol, iso	20	400	64	iodide	20	200	81	alcohol (*n*)	20	400	67
alcohol, iso	20	500	56	iodide	20	400	69	alcohol (*n*?)	20	500	65
alcohol, iso	20	1,000	46	iodide	20	500	64	alcohol (*n*?)	20	1,000	47
alcohol, iso	20	12,000	8	iodide	20	1,000	50	alcohol (*n*?)	20	12,000	7
Carbon bisulfide	16	21	86	iodide	20	12,000	8	Toluene	20	200	74
bisulfide	20	500	57	Gallium	30	300	3.97	Toluene	20	400	64
bisulfide	20	1,000	48	Glycerol	15	5	22	Turpentine	20		74
bisulfide	20	12,000	6	Hexane	20	200	117	Water	20	13	49
tetrachloride	20	200	86	Hexane	20	400	91	Water	20	200	43
tetrachloride	20	400	73	Kerosene	20	500	55	Water	20	400	41
Chloroform	20	200	83	Kerosene	20	1,000	45	Water	20	500	39
Chloroform	20	400	70	Kerosene	20	12,000	8	Water	40	500	38
Dichloroethylsulfide	32	1,000	34	Mercury	20	300	3.95	Water	40	1,000	33
Dichloroethylsulfide	32	2,000	24	Mercury	22	500	3.97	Water	40	12,000	9
Ethyl acetate	13	23	103	Mercury	22	1,000	3.91	Xylene, meta	20	200	69
acetate	20	200	90	Mercury	22	12,000	2.37	meta	20	400	60

* *Smithsonian Tables*, Table 106.

Scott (*Cryogenic Engineering*, Van Nostrand, Princeton, N.J., 1959) gives data for liquid nitrogen (p. 283), oxygen (p. 276), and hydrogen (p. 303). For a convenient index to the high-pressure work of Bridgman, see *American Institute of Physics Handbook*, p. 2-163, McGraw-Hill, New York, 1957.

TABLE 2-108 Compressibilities of Solids

Many data on the compressibility of solids obtained prior to 1926 are contained in Gruneisen, *Handbuch der Physik*, vol. 10, Springer, Berlin, 1926, pp. 1–52; also available as translation, NASA RE 2-18-59W, 1959. See also Tables 271, 273, 276, 278, and other material in *Smithsonian Physical Tables*, 9th ed., 1954. For a review of high-pressure work to 1946, see Bridgman, *Rev. Mod. Phys.*, **18**, 1 (1946).

THERMODYNAMIC PROPERTIES

Explanation of Tables The following subsection presents thermodynamic properties of a number of fluids. In some cases, transport properties are also included.

Property tables generated from the NIST database (Lemmon, E. W., M. O. McLinden, and M. L. Huber, NIST Standard Reference Database 23) are listed in Table 2-109. The number of digits provided in these tables was chosen for uniformity of appearance and formatting and does not represent the uncertainties of the physical quantities: They are the result of calculations from the standard thermophysical property formulations within a fixed format. They were generated using REFPROP software (Reference Fluid Thermodynamic and Transport Properties—REFPROP, National Institute of Standards and Technology, Standard Reference Data Program, Gaithersburg, Md., 2002, Version 7.1). Megan Friend helped produce these tables initially for Perry's 8th edition.

Because properties for many compounds also can be generated by the user at the NIST website, only more commonly used compounds' properties are given here. For other compounds, go to http://webbook.nist.gov and select NIST Chemistry WebBook > Thermophysical Properties of Fluid Systems High Accuracy Data. After selecting the desired unit system and temperature and/or pressure increments for which properties are to be generated, the resulting table can be copied into a spreadsheet.

Notation

c_p = isobaric specific heat
c_v = isochoric specific heat
e = specific internal energy
h = enthalpy
k = thermal conductivity
p = pressure
s = specific entropy
t = temperature
T = absolute temperature
u = specific internal energy
μ = viscosity
v = specific volume
f = subscript denoting saturated liquid
g = subscript denoting saturated vapor

Unit Conversions For this subsection, the following unit conversions are applicable:

c_p, specific heat: To convert kilojoules per kilogram-kelvin to British thermal units (Btu) per pound–degree Fahrenheit, multiply by 0.23885.

e, internal energy: To convert kilojoules per kilogram to Btu per pound, multiply by 0.42992.

g, gravity acceleration: To convert meters per second squared to feet per second squared, multiply by 3.2808.

h, enthalpy: To convert kilojoules per kilogram to Btu per pound, multiply by 0.42992.

k, thermal conductivity: To convert watts per meter-kelvin to Btu–feet per hour–square foot–degree Fahrenheit, multiply by 0.57779.

p, pressure: To convert bars to kilopascals, multiply by 100; to convert bars to pounds-force per square inch, multiply by 14.504; and to convert millimeters of mercury to pounds-force per square inch, multiply by 0.01934.

s, entropy: To convert kilojoules per kilogram-kelvin to Btu per pound–degree Rankine, multiply by 0.23885.

t, temperature: $°F = \frac{9}{5}°C + 32$.

T, absolute temperature: $°R = \frac{9}{5}$ K.

u, internal energy: To convert kilojoules per kilogram to Btu per pound, multiply by 0.42992.

μ, viscosity: To convert pascal-seconds to pound-force–seconds per square foot, multiply by 0.020885; to convert pascal-seconds to c_p, multiply by 1000.

v, specific volume: To convert cubic meters per kilogram to cubic feet per pound, multiply by 16.018.

ρ, density: To convert kilograms per cubic meter to pounds per cubic foot, multiply by 0.062428.

Additional References Bretsznajder, *Prediction of Transport and Other Physical Properties of Fluids,* Pergamon, New York, 1971. D'Ans and Lax, *Handbook for Chemists and Physicists* (in German), 3 vols., Springer-Verlag, Berlin. *Engineering Data Book,* 12th ed., 2004, Natural Gas Processors Suppliers Association, Tulsa, Okla. Ganic, Hartnett, and Rohsenow, *Handbook of Heat Transfer,* 2nd ed., McGraw-Hill, New York, 1984. Gray, *American Institute of Physics Handbook,* 3d ed., McGraw-Hill, New York, 1972. Kay and Laby, *Tables of Physical and Chemical Constants,* Longman, London, various editions and dates. *Landolt-Börnstein Tables,* many volumes and dates, Springer-Verlag, Berlin. Partington, *Advanced Treatise on Physical Chemistry,* Longman, London, 1950. Raznjevic, *Handbook of Thermodynamic Tables and Charts,* McGraw-Hill, New York, 1976 and other editions. Reynolds, *Thermodynamic Properties in SI,* Department of Mechanical Engineering, Stanford University, 1979. Stephan and Lucas, *Viscosity of Dense Fluids,* Plenum, New York and London, 1979. Vargaftik, *Tables of the Thermophysical Properties of Gases and Liquids,* Wiley, New York, 1975. Vargaftik, Filippov, Tarzimanov, and Totskiy, *Thermal Conductivity of Liquids and Gases* (in Russian), Standartov, Moscow, 1978. Weast, *Handbook of Chemistry and Physics,* Chemical Rubber Co., Boca Raton, FL, 97th print edition (2016) and online.

TABLE 2-109 Thermodynamic Properties of Acetone

Temperature K	Pressure MPa	Density mol/dm^3	Volume dm^3/mol	Int. energy kJ/mol	Enthalpy kJ/mol	Entropy kJ/(mol·K)	C_v kJ/(mol·K)	C_p kJ/(mol·K)	Sound speed m/s	Joule-Thomson K/MPa
Saturated Properties										
178.50	2.3265E-06	15.723	0.063601	0.47366	0.47366	0.0080825	0.082500	0.11544	1765.7	−0.43351
180.00	2.8743E-06	15.695	0.063715	0.64687	0.64687	0.0090488	0.082598	0.11550	1757.0	−0.43308
195.00	1.9454E-05	15.416	0.064868	2.3835	2.3835	0.018316	0.083407	0.11604	1672.3	−0.42849
210.00	9.6588E-05	15.141	0.066048	4.1282	4.1282	0.026935	0.084076	0.11660	1591.8	−0.42274
225.00	0.00037556	14.867	0.067264	5.8823	5.8823	0.035003	0.084758	0.11731	1514.4	−0.41520
240.00	0.0012008	14.593	0.068525	7.6487	7.6488	0.042602	0.085541	0.11825	1439.4	−0.40545
255.00	0.0032765	14.319	0.069840	9.4311	9.4314	0.049806	0.086468	0.11946	1366.3	−0.39322
270.00	0.0078514	14.041	0.071218	11.234	11.234	0.056674	0.087553	0.12094	1294.8	−0.37827
285.00	0.016899	13.760	0.072673	13.060	13.062	0.063259	0.088794	0.12270	1224.5	−0.36033
300.00	0.033259	13.474	0.074217	14.915	14.918	0.069601	0.090180	0.12474	1155.2	−0.33907
315.00	0.060720	13.181	0.075867	16.802	16.807	0.075739	0.091697	0.12704	1086.7	−0.31399
330.00	0.10404	12.880	0.077643	18.725	18.733	0.081702	0.093329	0.12962	1018.8	−0.28437
345.00	0.16891	12.568	0.079569	20.687	20.701	0.087517	0.095063	0.13249	951.24	−0.24915
360.00	0.26188	12.243	0.081677	22.693	22.714	0.093209	0.096886	0.13568	883.84	−0.20678
375.00	0.39033	11.904	0.084008	24.746	24.779	0.098798	0.098794	0.13924	816.36	−0.15495
390.00	0.56235	11.545	0.086616	26.852	26.900	0.10431	0.10078	0.14328	748.57	−0.090162
405.00	0.78681	11.163	0.089578	29.015	29.085	0.10975	0.10286	0.14794	680.21	−0.0069455
420.00	1.0733	10.753	0.093001	31.243	31.343	0.11516	0.10504	0.15350	610.99	0.10371
435.00	1.4324	10.304	0.097051	33.546	33.685	0.12056	0.10736	0.16042	540.51	0.25760
450.00	1.8759	9.8043	0.10200	35.938	36.130	0.12599	0.10986	0.16967	468.19	0.48516
465.00	2.4172	9.2319	0.10832	38.445	38.707	0.13150	0.11265	0.18350	392.99	0.85357
480.00	3.0725	8.5423	0.11706	41.117	41.476	0.13720	0.11600	0.20893	312.66	1.5474
495.00	3.8632	7.6072	0.13145	44.096	44.604	0.14341	0.12077	0.28551	221.66	3.3240
508.10	4.6924	4.7000	0.21277	49.249	50.247	0.15437			0	14.310
178.50	2.3265E-06	1.5677E-06	637,900.	36.689	38.173	0.21928	0.050120	0.058440	172.60	3845.4
180.00	2.8743E-06	1.9207E-06	520,660.	36.764	38.260	0.21801	0.050280	0.058600	173.29	3637.4
195.00	1.9454E-05	1.2001E-05	83,324.	37.528	39.149	0.20686	0.051928	0.060265	179.95	2139.7
210.00	9.6588E-05	5.5355E-05	18,065.	38.314	40.059	0.19803	0.053740	0.062119	186.29	1312.0
225.00	0.00037556	0.00020108	4,973.1	39.121	40.989	0.19103	0.055800	0.064267	192.29	834.10
240.00	0.0012008	0.00060385	1,656.0	39.947	41.936	0.18546	0.058169	0.066795	197.94	547.82
255.00	0.0032765	0.0015555	642.89	40.790	42.897	0.18104	0.060883	0.069763	203.19	370.79
270.00	0.0078514	0.0035368	282.74	41.649	43.869	0.17754	0.063945	0.073198	207.99	258.27
285.00	0.016899	0.0072603	137.74	42.522	44.849	0.17479	0.067329	0.077094	212.26	184.97
300.00	0.033259	0.013699	72.996	43.406	45.834	0.17266	0.070988	0.081429	215.93	136.14
315.00	0.060720	0.024107	41.482	44.302	46.821	0.17102	0.074863	0.086172	218.90	102.93
330.00	0.10404	0.040034	24.979	45.207	47.806	0.16980	0.078895	0.091302	221.08	79.878
345.00	0.16891	0.063362	15.782	46.119	48.784	0.16892	0.083030	0.096822	222.35	63.590
360.00	0.26188	0.096367	10.377	47.033	49.751	0.16831	0.087227	0.10277	222.60	51.884
375.00	0.39033	0.14184	7.0503	47.946	50.698	0.16791	0.091459	0.10927	221.70	43.343
390.00	0.56235	0.20329	4.9192	48.849	51.615	0.16768	0.095718	0.11649	219.53	37.032
405.00	0.78681	0.28530	3.5050	49.733	52.490	0.16754	0.10001	0.12481	215.94	32.325
420.00	1.0733	0.39420	2.5368	50.582	53.305	0.16745	0.10438	0.13483	210.76	28.797
435.00	1.4324	0.53918	1.8547	51.376	54.033	0.16734	0.10887	0.14772	203.80	26.154
450.00	1.8759	0.73472	1.3611	52.083	54.636	0.16711	0.11357	0.16583	194.82	24.184
465.00	2.4172	1.0061	0.99393	52.648	55.050	0.16664	0.11865	0.19480	183.50	22.717
480.00	3.0725	1.4051	0.71168	52.968	55.154	0.16569	0.12436	0.25197	169.39	21.551
495.00	3.8632	2.0767	0.48154	52.771	54.631	0.16367	0.13126	0.42947	151.36	20.240
508.10	4.6924	4.7000	0.21277	49.249	50.247	0.15437			0	14.310
Single-Phase Properties										
200.00	0.10000	15.325	0.065254	2.9626	2.9691	0.021248	0.083638	0.11621	1645.6	−0.42678
250.00	0.10000	14.411	0.069389	8.8328	8.8397	0.047436	0.086143	0.11902	1391.1	−0.39768
300.00	0.10000	13.475	0.074210	14.913	14.921	0.069594	0.090180	0.12473	1155.7	−0.33922
328.84	0.10000	12.903	0.077500	18.575	18.583	0.081247	0.093199	0.12941	1024.0	−0.28685
328.84	0.10000	0.038565	25.930	45.137	47.730	0.16988	0.078579	0.090892	220.94	81.384
350.00	0.10000	0.035712	28.002	46.843	49.643	0.17552	0.079533	0.090386	229.44	58.339
400.00	0.10000	0.030709	32.563	50.998	54.255	0.18783	0.085418	0.094849	246.85	30.192
450.00	0.10000	0.027083	36.923	55.474	59.166	0.19939	0.092823	0.10175	262.23	18.173
500.00	0.10000	0.024272	41.200	60.316	64.436	0.21049	0.10033	0.10903	276.40	12.201
550.00	0.10000	0.022008	45.437	65.522	70.066	0.22122	0.10753	0.11612	289.72	8.8355
200.00	1.0000	15.333	0.065220	2.9486	3.0138	0.021178	0.083649	0.11619	1649.7	−0.42708
250.00	1.0000	14.423	0.069336	8.8130	8.8824	0.047357	0.086152	0.11896	1396.0	−0.39848
300.00	1.0000	13.491	0.074123	14.885	14.959	0.069499	0.090182	0.12460	1162.0	−0.34115
350.00	1.0000	12.483	0.080107	21.312	21.392	0.089316	0.095644	0.13326	936.35	−0.24033
400.00	1.0000	11.308	0.088431	28.263	28.351	0.10788	0.10213	0.14605	707.25	−0.042437
416.48	1.0000	10.852	0.092149	30.714	30.806	0.11389	0.10452	0.15210	627.32	0.074613
416.48	1.0000	0.36582	2.7336	50.387	53.120	0.16747	0.10335	0.13228	212.13	29.536
450.00	1.0000	0.31254	3.1996	54.081	57.281	0.17709	0.10087	0.11921	233.76	20.211
500.00	1.0000	0.26538	3.7681	59.388	63.156	0.18947	0.10402	0.11743	256.99	12.984
550.00	1.0000	0.23391	4.2751	64.832	69.107	0.20081	0.10950	0.12100	275.55	9.1542

TABLE 2-109 Thermodynamic Properties of Acetone *(Continued)*

Temperature K	Pressure MPa	Density mol/dm^3	Volume dm^3/mol	Int. energy kJ/mol	Enthalpy kJ/mol	Entropy kJ/(mol·K)	C_v kJ/(mol·K)	C_p kJ/(mol·K)	Sound speed m/s	Joule-Thomson K/MPa
					Single-Phase Properties (*Cont.*)					
200.00	5.0000	15.367	0.065073	2.8871	3.2125	0.020868	0.083704	0.11609	1667.9	−0.42837
250.00	5.0000	14.471	0.069106	8.7271	9.0726	0.047011	0.086197	0.11871	1417.7	−0.40187
300.00	5.0000	13.560	0.073747	14.762	15.130	0.069085	0.090197	0.12408	1189.0	−0.34909
350.00	5.0000	12.588	0.079439	21.128	21.525	0.088784	0.095584	0.13214	972.15	−0.25988
400.00	5.0000	11.490	0.087035	27.958	28.393	0.10711	0.10186	0.14320	759.27	−0.10136
450.00	5.0000	10.123	0.098782	35.450	35.944	0.12488	0.10898	0.16059	538.79	0.26123
500.00	5.0000	7.8139	0.12798	44.435	45.075	0.14406	0.11961	0.23343	262.33	2.3418
550.00	5.0000	1.7344	0.57657	60.563	63.446	0.17943	0.12191	0.17820	205.69	10.650
200.00	10.000	15.410	0.064894	2.8125	3.4614	0.020488	0.083781	0.11598	1689.9	−0.42983
250.00	10.000	14.528	0.068831	8.6237	9.3120	0.046589	0.086264	0.11843	1443.6	−0.40569
300.00	10.000	13.641	0.073307	14.616	15.349	0.068589	0.090234	0.12351	1220.9	−0.35775
350.00	10.000	12.709	0.078687	20.916	21.703	0.088163	0.095554	0.13100	1013.1	−0.27983
400.00	10.000	11.683	0.085592	27.629	28.485	0.10626	0.10166	0.14066	815.03	−0.15336
450.00	10.000	10.491	0.095320	34.864	35.818	0.12352	0.10827	0.15332	622.74	0.080235
500.00	10.000	8.9733	0.11144	42.815	43.930	0.14060	0.11552	0.17314	433.48	0.63674
550.00	10.000	6.6600	0.15015	52.079	53.581	0.15896	0.12442	0.22174	255.34	2.7218
250.00	100.00	15.320	0.065276	7.2620	13.790	0.040421	0.088285	0.11631	1791.8	−0.43634
300.00	100.00	14.657	0.068228	12.852	19.675	0.061873	0.092127	0.11946	1616.6	−0.42000
350.00	100.00	14.023	0.071312	18.631	25.763	0.080632	0.097243	0.12424	1466.4	−0.39555
400.00	100.00	13.409	0.074574	24.654	32.112	0.097579	0.10299	0.12980	1337.4	−0.36734
450.00	100.00	12.813	0.078044	30.941	38.745	0.11320	0.10892	0.13553	1226.9	−0.33807
500.00	100.00	12.234	0.081739	37.489	45.663	0.12777	0.11478	0.14112	1133.0	−0.30922
550.00	100.00	11.674	0.085664	44.286	52.852	0.14147	0.12045	0.14639	1053.8	−0.28171
450.00	500.00	15.616	0.064037	27.237	59.256	0.097266	0.11562	0.13393	2201.1	−0.39010
500.00	500.00	15.306	0.065335	33.413	66.081	0.11164	0.12123	0.13909	2129.8	−0.37710
550.00	500.00	15.012	0.066615	39.856	73.163	0.12514	0.12669	0.14416	2067.5	−0.36510

The values in this table were generated from the NIST REFPROP software (Lemmon, E. W., McLinden, M. O., and Huber, M. L., NIST Standard Reference Database 23: Reference Fluid Thermodynamic and Transport Properties—REFPROP, National Institute of Standards and Technology, Standard Reference Data Program, Gaithersburg, Md., 2002, Version 7.1). The primary source for the thermodynamic properties is Lemmon, E. W., and Span, R., "Short Fundamental Equations of State for 20 Industrial Fluids," *J. Chem. Eng. Data*, **51**(3):785–850, 2006. Validated equations for the viscosity and thermal conductivity are not currently available for this fluid.

Properties at the triple point temperature and the critical point temperature are given in the first and last entries of the saturation tables, respectively. In the single-phase table, when the temperature range for a given isobar includes a vapor-liquid phase boundary, the temperature of phase equilibrium is noted, and properties for both the saturated liquid and saturated vapor are given (with liquid properties given in the upper line). Lines are omitted from the temperature-pressure grid of the single-phase table, when the system would be in the solid phase or if there are potential problems with the source property surface.

The uncertainties in the equation of state are 0.1% in the saturated liquid density between 280 and 310 K, 0.5% in density in the liquid phase below 380 K, and 1% in density elsewhere, including all states at pressures above 100 MPa. The uncertainties in vapor pressure are 0.5% above 270 K (0.25% between 290 and 390 K), and the uncertainties in heat capacities and speeds of sound are 1%. These uncertainties (in caloric properties and sound speeds) may be higher at pressures above the saturation pressure and at temperatures above 320 K in the liquid phase and at supercritical conditions.

TABLE 2-110 Thermodynamic Properties of Air

Temperature K	Pressure MPa	Density mol/dm^3	Volume dm^3/mol	Int. energy kJ/mol	Enthalpy kJ/mol	Entropy kJ/(mol·K)	C_v kJ/(mol·K)	C_p kJ/(mol·K)	Sound speed m/s	Joule-Thomson K/MPa	Therm. cond. mW/(m·K)	Viscosity μPa·s
						Saturated Properties						
59.75	0.005265	33.067	0.030242	−1.0619	−1.0617	−0.01536	0.034011	0.055064	1030.3	−0.40785	171.43	376.64
60	0.005546	33.031	0.030275	−1.0481	−1.0480	−0.01513	0.033955	0.055062	1028.3	−0.40743	171.02	371.92
61	0.006797	32.888	0.030406	−0.99308	−0.99287	−0.01422	0.033731	0.055060	1020.3	−0.40565	169.40	353.83
62	0.008270	32.745	0.030539	−0.93803	−0.93778	−0.01333	0.033512	0.055062	1012.2	−0.40375	167.78	336.91
63	0.009994	32.601	0.030674	−0.88298	−0.88267	−0.01245	0.033298	0.055069	1004.0	−0.40173	166.16	321.09
64	0.012000	32.457	0.030810	−0.82792	−0.82755	−0.01158	0.033089	0.055081	995.77	−0.39958	164.53	306.27
65	0.014320	32.312	0.030949	−0.77286	−0.77241	−0.01073	0.032884	0.055098	987.48	−0.39729	162.91	292.39
66	0.016988	32.166	0.031089	−0.71777	−0.71725	−0.00989	0.032683	0.055120	979.13	−0.39485	161.28	279.38
67	0.020042	32.020	0.031231	−0.66267	−0.66205	−0.00906	0.032486	0.055148	970.72	−0.39227	159.65	267.17
68	0.023520	31.873	0.031375	−0.60755	−0.60681	−0.00824	0.032294	0.055181	962.24	−0.38952	158.01	255.71
69	0.027461	31.725	0.031521	−0.55239	−0.55152	−0.00744	0.032105	0.055220	953.70	−0.38660	156.37	244.94
70	0.031908	31.576	0.031669	−0.49720	−0.49619	−0.00664	0.031920	0.055266	945.10	−0.38352	154.73	234.81
71	0.036905	31.427	0.031820	−0.44196	−0.44079	−0.00586	0.031739	0.055317	936.43	−0.38024	153.09	225.28
72	0.042498	31.277	0.031972	−0.38669	−0.38533	−0.00508	0.031562	0.055376	927.70	−0.37677	151.44	216.31
73	0.048733	31.126	0.032127	−0.33135	−0.32979	−0.00432	0.031388	0.055441	918.90	−0.37310	149.79	207.85
74	0.055659	30.974	0.032285	−0.27597	−0.27417	−0.00357	0.031217	0.055514	910.04	−0.36922	148.14	199.88
75	0.063326	30.821	0.032445	−0.22051	−0.21846	−0.00282	0.031050	0.055594	901.11	−0.36511	146.49	192.35
76	0.071786	30.668	0.032608	−0.16499	−0.16265	−0.00209	0.030886	0.055682	892.11	−0.36076	144.83	185.23
77	0.081091	30.513	0.032773	−0.10939	−0.10673	−0.00136	0.030725	0.055779	883.05	−0.35616	143.16	178.51
78	0.091294	30.357	0.032941	−0.05371	−0.05070	−0.00064	0.030568	0.055884	873.91	−0.35130	141.50	172.14
79	0.10245	30.200	0.033112	0.002063	0.005456	6.86E-05	0.030413	0.055998	864.71	−0.34616	139.83	166.11
80	0.11462	30.042	0.033287	0.057934	0.061749	0.000772	0.030262	0.056122	855.44	−0.34074	138.15	160.39
81	0.12785	29.883	0.033464	0.11391	0.11819	0.001467	0.030113	0.056256	846.09	−0.33500	136.48	154.96
82	0.14221	29.722	0.033645	0.17000	0.17478	0.002156	0.029968	0.056400	836.67	−0.32894	134.80	149.80
83	0.15775	29.560	0.033829	0.22621	0.23155	0.002838	0.029826	0.056556	827.18	−0.32254	133.11	144.90
84	0.17453	29.397	0.034017	0.28255	0.28849	0.003513	0.029686	0.056723	817.61	−0.31577	131.42	140.23
85	0.19262	29.232	0.034209	0.33903	0.34562	0.004181	0.029550	0.056902	807.96	−0.30862	129.78	135.78
86	0.21207	29.066	0.034404	0.39566	0.40296	0.004844	0.029417	0.057094	798.24	−0.30107	128.11	131.54
87	0.23295	28.898	0.034604	0.45245	0.46051	0.005501	0.029286	0.057300	788.44	−0.29308	126.44	127.50
88	0.25531	28.729	0.034808	0.50940	0.51829	0.006153	0.029158	0.057521	778.56	−0.28464	124.76	123.63
89	0.27922	28.558	0.035017	0.56653	0.57631	0.006799	0.029033	0.057757	768.59	−0.27572	123.07	119.93
90	0.30475	28.385	0.035230	0.62386	0.63459	0.007440	0.028911	0.058009	758.55	−0.26628	121.38	116.38
91	0.33196	28.210	0.035449	0.68138	0.69315	0.008077	0.028792	0.058278	748.42	−0.25629	119.69	112.98
92	0.36091	28.033	0.035672	0.73912	0.75199	0.008708	0.028676	0.058566	738.20	−0.24573	118.00	109.72
93	0.39166	27.854	0.035901	0.79709	0.81115	0.009336	0.028563	0.058874	727.90	−0.23455	116.30	106.59
94	0.42429	27.673	0.036137	0.85529	0.87062	0.009960	0.028453	0.059202	717.51	−0.22270	114.61	103.58
95	0.45886	27.489	0.036378	0.91375	0.93044	0.010579	0.028346	0.059553	707.03	−0.21016	112.91	100.68
96	0.49543	27.304	0.036625	0.97248	0.99063	0.011195	0.028241	0.059928	696.46	−0.19686	111.21	97.879
97	0.53408	27.115	0.036880	1.0315	1.0512	0.011808	0.028140	0.060329	685.80	−0.18275	109.51	95.179
98	0.57486	26.924	0.037142	1.0908	1.1122	0.012418	0.028042	0.060757	675.05	−0.16779	107.81	92.571
99	0.61786	26.730	0.037411	1.1505	1.1736	0.013025	0.027948	0.061216	664.20	−0.15189	106.11	90.048
100	0.66313	26.533	0.037688	1.2104	1.2354	0.013630	0.027856	0.061707	653.26	−0.13501	104.41	87.605
101	0.71074	26.333	0.037975	1.2708	1.2978	0.014232	0.027768	0.062232	642.22	−0.11705	102.71	85.236
102	0.76077	26.130	0.038270	1.3315	1.3606	0.014833	0.027684	0.062796	631.08	−0.09794	101.01	82.937
103	0.81329	25.923	0.038575	1.3926	1.4240	0.015431	0.027603	0.063401	619.84	−0.07758	99.316	80.703
104	0.86836	25.713	0.038891	1.4542	1.4880	0.016029	0.027525	0.064052	608.50	−0.05588	97.623	78.529
105	0.92606	25.499	0.039217	1.5162	1.5525	0.016625	0.027452	0.064753	597.06	−0.03271	95.933	76.412
106	0.98645	25.281	0.039556	1.5787	1.6177	0.017221	0.027383	0.065508	585.51	−0.00795	94.247	74.347
107	1.0496	25.058	0.039908	1.6417	1.6836	0.017816	0.027317	0.066323	573.85	0.018543	92.565	72.331
108	1.1156	24.831	0.040273	1.7053	1.7502	0.018411	0.027256	0.067206	562.09	0.046927	90.888	70.361
109	1.1845	24.598	0.040653	1.7695	1.8176	0.019006	0.027200	0.068163	550.21	0.077386	89.216	68.432
110	1.2564	24.361	0.041050	1.8343	1.8858	0.019602	0.027149	0.069205	538.21	0.11012	87.551	66.542
111	1.3314	24.118	0.041464	1.8997	1.9549	0.020200	0.027103	0.070341	526.10	0.14538	85.893	64.688
112	1.4095	23.868	0.041896	1.9659	2.0250	0.020799	0.027062	0.071585	513.86	0.18342	84.242	62.867
113	1.4908	23.613	0.042350	2.0329	2.0960	0.021400	0.027028	0.072951	501.48	0.22456	82.599	61.075
114	1.5753	23.350	0.042826	2.1007	2.1682	0.022004	0.027000	0.074459	488.97	0.26917	80.965	59.311

115	1.6633	23.080	0.043328	2.1695	2.2415	0.022611	0.026979	0.076131	476.31	0.31767	79.340	57.571
116	1.7546	22.801	0.043857	2.2392	2.3161	0.023223	0.026965	0.077996	463.48	0.37057	77.724	55.852
117	1.8495	22.514	0.044417	2.3100	2.3922	0.023840	0.026961	0.080090	450.49	0.42848	76.119	54.152
118	1.9479	22.217	0.045011	2.3821	2.4697	0.024462	0.026966	0.082459	437.29	0.49214	74.523	52.467
119	2.0499	21.908	0.045645	2.4554	2.5490	0.025092	0.026982	0.085163	423.88	0.56243	72.938	50.794
120	2.1557	21.588	0.046323	2.5303	2.6302	0.025731	0.027010	0.088280	410.23	0.64047	71.363	49.130
121	2.2653	21.253	0.047052	2.6069	2.7135	0.026380	0.027053	0.091919	396.30	0.72765	69.798	47.469
122	2.3787	20.903	0.047841	2.6854	2.7992	0.027041	0.027113	0.096227	382.04	0.82574	68.243	45.809
123	2.4960	20.534	0.048700	2.7662	2.8878	0.027717	0.027194	0.10142	367.40	0.93703	66.700	44.141
124	2.6173	20.144	0.049643	2.8496	2.9796	0.028412	0.027300	0.10781	352.31	1.0646	65.170	42.460
125	2.7427	19.727	0.050691	2.9363	3.0753	0.029131	0.027438	0.11589	336.67	1.2125	63.658	40.755
126	2.8721	19.278	0.051871	3.0269	3.1759	0.029880	0.027618	0.12645	320.36	1.3865	62.176	39.013
127	3.0055	18.788	0.053225	3.1227	3.2827	0.030668	0.027855	0.14089	303.21	1.5951	60.751	37.215
128	3.1431	18.242	0.054818	3.2253	3.3976	0.031512	0.028171	0.16186	285.00	1.8510	59.445	35.332
129	3.2845	17.616	0.056765	3.3379	3.5243	0.032436	0.028607	0.19519	265.37	2.1752	58.409	33.316
130	3.4295	16.863	0.059300	3.4661	3.6695	0.033492	0.029242	0.25624	243.75	2.6058	58.054	31.072
131	3.5770	15.869	0.063015	3.6243	3.8497	0.034804	0.030266	0.40151	219.07	3.2246	59.591	28.384
132	3.7228	14.198	0.070432	3.8680	4.1302	0.036863	0.032343	1.0148	189.12	4.2808	67.802	24.467
132.63	3.7858	10.448	0.095715	4.4004	4.7627	0.041603			0	6.3978		
59.75	0.002432	0.004907	203.80	4.8774	5.3730	0.096708	0.020805	0.029217	154.83	58.283	5.2938	4.2197
60	0.002584	0.005192	192.59	4.8825	5.3800	0.096323	0.020809	0.029225	155.14	57.634	5.3199	4.2382
61	0.003274	0.006475	154.45	4.9025	5.4081	0.094825	0.020825	0.029261	156.38	55.151	5.4244	4.3119
62	0.004111	0.008005	124.93	4.9225	5.4361	0.093392	0.020843	0.029302	157.60	52.832	5.5291	4.3855
63	0.005120	0.009817	101.86	4.9424	5.4639	0.092020	0.020864	0.029348	158.81	50.666	5.6340	4.4590
64	0.006325	0.011948	83.693	4.9621	5.4915	0.090705	0.020886	0.029399	159.99	48.640	5.7391	4.5324
65	0.007756	0.014438	69.263	4.9817	5.5189	0.089445	0.020911	0.029455	161.16	46.742	5.8444	4.6057
66	0.009442	0.017326	57.715	5.0012	5.5461	0.088235	0.020938	0.029518	162.30	44.963	5.9500	4.6788
67	0.011416	0.020659	48.406	5.0205	5.5731	0.087074	0.020968	0.029587	163.42	43.293	6.0559	4.7519
68	0.013713	0.024481	40.849	5.0397	5.5998	0.085959	0.021000	0.029663	164.53	41.724	6.1621	4.8248
69	0.016372	0.028841	34.673	5.0587	5.6263	0.084887	0.021035	0.029746	165.60	40.248	6.2688	4.8976
70	0.019431	0.033789	29.595	5.0774	5.6525	0.083855	0.021072	0.029836	166.66	38.858	6.3759	4.9703
71	0.022933	0.039379	25.394	5.0960	5.6784	0.082862	0.021113	0.029934	167.69	37.548	6.4835	5.0429
72	0.026921	0.045664	21.899	5.1144	5.7040	0.081906	0.021156	0.030040	168.70	36.313	6.5917	5.1154
73	0.031443	0.052702	18.975	5.1326	5.7292	0.080983	0.021201	0.030155	169.69	35.146	6.7005	5.1878
74	0.036547	0.060550	16.515	5.1505	5.7541	0.080094	0.021250	0.030278	170.65	34.043	6.8099	5.2602
75	0.042282	0.069268	14.437	5.1682	5.7786	0.079235	0.021302	0.030410	171.58	32.999	6.9202	5.3325
76	0.048702	0.078918	12.671	5.1856	5.8027	0.078406	0.021356	0.030552	172.49	32.010	7.0312	5.4048
77	0.055859	0.089564	11.165	5.2028	5.8264	0.077604	0.021414	0.030703	173.37	31.072	7.1431	5.4771
78	0.063810	0.10127	9.8746	5.2196	5.8497	0.076828	0.021474	0.030865	174.23	30.183	7.2560	5.5494
79	0.072611	0.11410	8.7639	5.2362	5.8726	0.076076	0.021538	0.031037	175.05	29.337	7.3700	5.6217
80	0.082321	0.12813	7.8043	5.2525	5.8949	0.075348	0.021605	0.031220	175.85	28.534	7.4851	5.6940
81	0.093001	0.14343	6.9721	5.2684	5.9169	0.074643	0.021674	0.031415	176.62	27.769	7.6014	5.7664
82	0.10471	0.16006	6.2475	5.2841	5.9383	0.073957	0.021747	0.031621	177.36	27.041	7.7192	5.8389
83	0.11751	0.17811	5.6145	5.2994	5.9591	0.073292	0.021822	0.031840	178.07	26.346	7.8384	5.9116
84	0.13147	0.19765	5.0595	5.3143	5.9795	0.072645	0.021901	0.032072	178.75	25.684	7.9591	5.9844
85	0.14665	0.21875	4.5715	5.3289	5.9993	0.072016	0.021983	0.032317	179.40	25.051	8.0817	6.0574
86	0.16312	0.24150	4.1408	5.3431	6.0185	0.071403	0.022068	0.032577	180.02	24.447	8.2060	6.1307
87	0.18094	0.26598	3.7597	5.3569	6.0372	0.070806	0.022155	0.032851	180.61	23.869	8.3324	6.2043
88	0.20018	0.29228	3.4214	5.3703	6.0552	0.070224	0.022246	0.033141	181.17	23.316	8.4610	6.2781
89	0.22091	0.32048	3.1203	5.3832	6.0726	0.069655	0.022340	0.033447	181.69	22.786	8.5919	6.3524
90	0.24320	0.35068	2.8516	5.3958	6.0893	0.069099	0.022436	0.033770	182.19	22.278	8.7254	6.4272
91	0.26712	0.38298	2.6111	5.4079	6.1054	0.068556	0.022536	0.034111	182.65	21.791	8.8616	6.5024
92	0.29273	0.41747	2.3954	5.4195	6.1207	0.068024	0.022638	0.034472	183.08	21.324	9.0008	6.5782
93	0.32011	0.45426	2.2014	5.4307	6.1354	0.067503	0.022744	0.034853	183.47	20.876	9.1433	6.6547
94	0.34934	0.49345	2.0265	5.4413	6.1492	0.066991	0.022852	0.035256	183.84	20.445	9.2893	6.7318
95	0.38047	0.53517	1.8686	5.4514	6.1624	0.066489	0.022964	0.035681	184.17	20.031	9.4390	6.8098
96	0.41359	0.57953	1.7255	5.4610	6.1747	0.065995	0.023078	0.036132	184.46	19.632	9.5929	6.8887
97	0.44878	0.62667	1.5957	5.4701	6.1862	0.065510	0.023196	0.036610	184.72	19.249	9.7513	6.9686
98	0.48609	0.67671	1.4777	5.4785	6.1968	0.065031	0.023317	0.037116	184.95	18.879	9.9145	7.0495
99	0.52562	0.72980	1.3702	5.4864	6.2066	0.064560	0.023441	0.037654	185.14	18.523	10.083	7.1317
100	0.56742	0.78609	1.2721	5.4936	6.2154	0.064094	0.023568	0.038225	185.30	18.180	10.257	7.2153
101	0.61159	0.84575	1.1824	5.5002	6.2233	0.063633	0.023698	0.038834	185.42	17.848	10.438	7.3003

(*Continued*)

TABLE 2-110 Thermodynamic Properties of Air *(Continued)*

Temperature K	Pressure MPa	Density mol/dm³	Volume dm³/mol	Int. energy kJ/mol	Enthalpy kJ/mol	Entropy kJ/(mol·K)	C_v kJ/(mol·K)	C_p kJ/(mol·K)	Sound speed m/s	Joule-Thomson K/MPa	Therm. cond. mW/(m·K)	Viscosity μPa·s
102	0.65820	0.90895	1.1002	5.5060	6.2302	0.063177	0.023833	0.039483	185.51	17.528	10.626	7.3870
103	0.70732	0.97587	1.0247	5.5112	6.2360	0.062726	0.023970	0.040176	185.55	17.218	10.821	7.4755
104	0.75903	1.0467	0.95535	5.5156	6.2408	0.062277	0.024112	0.040918	185.57	16.918	11.024	7.5659
105	0.81341	1.1217	0.89147	5.5193	6.2444	0.061832	0.024258	0.041714	185.54	16.628	11.237	7.6586
106	0.87055	1.2011	0.83254	5.5221	6.2469	0.061389	0.024408	0.042570	185.48	16.346	11.459	7.7537
107	0.93052	1.2852	0.77810	5.5240	6.2481	0.060947	0.024563	0.043492	185.38	16.072	11.693	7.8514
108	0.9934	1.3742	0.72772	5.5250	6.2480	0.060506	0.024722	0.044490	185.24	15.805	11.939	7.9521
109	1.0593	1.4684	0.68102	5.5251	6.2465	0.060065	0.024887	0.045573	185.07	15.546	12.198	8.0560
110	1.1282	1.5682	0.63767	5.5241	6.2436	0.059623	0.025058	0.046751	184.85	15.292	12.473	8.1634
111	1.2004	1.6740	0.59737	5.5221	6.2391	0.059180	0.025234	0.048038	184.60	15.044	12.764	8.2749
112	1.2757	1.7862	0.55985	5.5188	6.2330	0.058735	0.025418	0.049450	184.30	14.800	13.074	8.3907
113	1.3545	1.9053	0.52486	5.5143	6.2252	0.058286	0.025608	0.051005	183.97	14.561	13.406	8.5114
114	1.4366	2.0318	0.49217	5.5085	6.2156	0.057833	0.025807	0.052727	183.59	14.324	13.762	8.6375
115	1.5223	2.1664	0.46160	5.5012	6.2039	0.057375	0.026015	0.054644	183.17	14.090	14.145	8.7696
116	1.6115	2.3097	0.43296	5.4924	6.1901	0.056910	0.026232	0.056790	182.71	13.856	14.559	8.9086
117	1.7045	2.4625	0.40608	5.4819	6.1740	0.056437	0.026461	0.059209	182.21	13.623	15.008	9.0552
118	1.8013	2.6259	0.38082	5.4695	6.1554	0.055955	0.026701	0.061956	181.66	13.388	15.499	9.2104
119	1.9020	2.8009	0.35702	5.4550	6.1341	0.055461	0.026956	0.065102	181.08	13.151	16.039	9.3755
120	2.0067	2.9889	0.33457	5.4383	6.1097	0.054954	0.027226	0.068738	180.45	12.909	16.635	9.5518
121	2.1156	3.1913	0.31335	5.4190	6.0819	0.054432	0.027514	0.072988	179.78	12.661	17.298	9.7412
122	2.2287	3.4103	0.29323	5.3969	6.0504	0.053890	0.027823	0.078015	179.06	12.405	18.042	9.9456
123	2.3462	3.6481	0.27412	5.3715	6.0147	0.053326	0.028155	0.084052	178.31	12.137	18.884	10.168
124	2.4682	3.9078	0.25590	5.3424	5.9740	0.052735	0.028516	0.091426	177.51	11.854	19.849	10.411
125	2.5949	4.1934	0.23847	5.3089	5.9277	0.052112	0.028910	0.10063	176.68	11.553	20.968	10.681
126	2.7266	4.5101	0.22173	5.2701	5.8746	0.051448	0.029344	0.11241	175.81	11.229	22.288	10.982
127	2.8633	4.8653	0.20554	5.2248	5.8133	0.050732	0.029827	0.12801	174.91	10.874	23.877	11.324
128	3.0055	5.2697	0.18976	5.1713	5.7417	0.049950	0.030371	0.14959	173.96	10.480	25.841	11.720
129	3.1536	5.7405	0.17420	5.1069	5.6563	0.049076	0.030994	0.18134	172.98	10.033	28.367	12.191
130	3.3084	6.3074	0.15854	5.0268	5.5513	0.048067	0.031726	0.23261	171.93	9.5119	31.807	12.775
131	3.4712	7.0343	0.14216	4.9209	5.4143	0.046830	0.032619	0.32992	170.79	8.8740	37.001	13.553
132	3.6462	8.1273	0.12304	4.7566	5.2053	0.045064	0.033814	0.59804	169.40	7.9854	46.996	14.798
132.63	3.7858	10.448	0.095715	4.4004	4.7627	0.041603			0	6.3978		
					Single-Phase Properties							
100	0.1	0.12283	8.1414	5.6800	6.4941	0.080463	0.021087	0.030116	198.24	17.423	9.4692	7.1068
300	0.1	0.040103	24.936	9.8544	12.348	0.11269	0.020796	0.029149	347.36	2.2510	26.384	18.537
500	0.1	0.024046	41.586	14.072	18.231	0.12770	0.021504	0.029830	446.40	0.50305	39.944	27.090
700	0.1	0.017175	58.223	18.500	24.323	0.13794	0.022817	0.031137	523.89	−0.12430	51.755	34.176
900	0.1	0.013359	74.855	23.201	30.686	0.14593	0.024150	0.032467	589.60	−0.41124	62.543	40.394
1100	0.1	0.010931	91.486	28.145	37.293	0.15255	0.025246	0.033562	648.15	−0.56194	72.680	46.051
1300	0.1	0.009249	108.12	33.282	44.094	0.15823	0.026091	0.034406	701.76	−0.64963	82.381	51.325
1500	0.1	0.008016	124.75	38.568	51.042	0.16320	0.026734	0.035049	751.59	−0.70457	91.781	56.325
1700	0.1	0.007073	141.38	43.966	58.104	0.16762	0.027229	0.035544	798.38	−0.74078	100.97	61.127
1900	0.1	0.006329	158.00	49.453	65.253	0.17160	0.027619	0.035934	842.62	−0.76547	110.01	65.783
100	1	26.593	0.037604	1.2007	1.2383	0.013532	0.027868	0.061355	658.25	−0.14308	104.97	88.326
106.22	1	25.232	0.039632	1.5924	1.6321	0.017351	0.027368	0.065680	582.97	−0.00232	93.879	73.903
108.1	1	1.3836	0.72278	5.5251	6.2479	0.060461	0.024739	0.044597	185.23	15.779	11.965	7.9625
300	1	0.40205	2.4873	9.8022	12.289	0.093372	0.020859	0.029563	348.45	2.1789	26.684	18.672
500	1	0.23974	4.1711	14.046	18.218	0.10851	0.021526	0.029954	448.46	0.47425	40.110	27.179
700	1	0.17119	5.8415	18.485	24.326	0.11877	0.022830	0.031194	525.96	−0.13809	51.868	34.242
900	1	0.13319	7.5079	23.190	30.698	0.12677	0.024159	0.032498	591.54	−0.41899	62.628	40.446
1100	1	0.10902	9.1727	28.138	37.311	0.13340	0.025253	0.033582	649.96	−0.56686	72.748	46.094
1300	1	0.092279	10.837	33.278	44.114	0.13908	0.026096	0.034419	703.44	−0.65304	82.438	51.361
1500	1	0.079999	12.500	38.565	51.065	0.14405	0.026738	0.035057	753.17	−0.70711	91.830	56.357
1700	1	0.070604	14.163	43.964	58.128	0.14847	0.027233	0.035550	799.86	−0.74278	101.01	61.155
1900	1	0.063185	15.827	49.451	65.278	0.15245	0.027622	0.035939	844.02	−0.76711	110.05	65.808
100	5	27.222	0.036735	1.0983	1.2820	0.012483	0.028034	0.058181	710.56	−0.21837	111.13	96.436
300	5	2.0232	0.49426	9.5710	12.042	0.079244	0.021131	0.031423	355.63	1.8817	28.389	19.420

500	5	1.1814	0.84642	13.935	18.167	0.094907	0.021621	0.030478	458.30	0.36370	40.969	27.606
700	5	0.84321	1.1859	18.417	24.347	0.10529	0.022885	0.031434	535.45	−0.19118	52.433	34.545
900	5	0.65711	1.5218	23.146	30.755	0.11334	0.024197	0.032632	600.34	−0.44905	63.045	40.682
1100	5	0.53874	1.8562	28.107	37.388	0.11999	0.025282	0.033664	658.10	−0.58606	73.076	46.287
1300	5	0.45667	2.1898	33.256	44.205	0.12568	0.026119	0.034473	711.01	−0.66646	82.707	51.523
1500	5	0.39636	2.5229	38.550	51.165	0.13066	0.026757	0.035095	760.23	−0.71716	92.057	56.497
1700	5	0.35015	2.8559	43.954	58.234	0.13509	0.027249	0.035577	806.49	−0.75073	101.21	61.278
1900	5	0.31361	3.1887	49.445	65.389	0.13907	0.027636	0.035958	850.28	−0.77366	110.22	65.917
100	10	27.863	0.035889	0.99444	1.3533	0.011382	0.028284	0.055716	763.47	−0.27969	117.77	105.78
300	10	4.0370	0.24771	9.2885	11.766	0.072612	0.021441	0.033664	369.50	1.5212	31.116	20.637
500	10	2.3157	0.43183	13.802	18.120	0.088894	0.021733	0.031078	471.81	0.25100	42.260	28.194
700	10	1.6542	0.60452	18.336	24.382	0.099422	0.022952	0.031710	547.83	−0.24405	53.257	34.944
900	10	1.2922	0.77388	23.092	30.831	0.10752	0.024243	0.032786	611.64	−0.47890	63.641	40.985
1100	10	1.0618	0.94184	28.070	37.489	0.11420	0.025317	0.033760	668.47	−0.60517	73.538	46.531
1300	10	0.90165	1.1091	33.231	44.321	0.11990	0.026146	0.034537	720.60	−0.67990	83.082	51.728
1500	10	0.78374	1.2759	38.532	51.292	0.12489	0.026780	0.035139	769.17	−0.72730	92.372	56.673
1700	10	0.69321	1.4426	43.943	58.368	0.12932	0.027268	0.035608	814.87	−0.75881	101.48	61.432
1900	10	0.62149	1.6090	49.438	65.528	0.13330	0.027653	0.035981	858.18	−0.78039	110.45	66.054
100	100	33.161	0.030156	0.24746	3.2631	0.001378	0.031980	0.048218	1192.4	−0.47290	179.20	252.46
300	100	21.138	0.047309	7.0356	11.767	0.049067	0.023981	0.038366	818.47	−0.49747	86.312	53.642
500	100	15.089	0.066273	12.371	18.999	0.067619	0.023117	0.034686	772.41	−0.55640	71.549	42.159
700	100	11.803	0.084722	17.367	25.840	0.079134	0.023855	0.034011	790.14	−0.62591	73.572	43.339
900	100	9.7481	0.10258	22.408	32.667	0.087711	0.024903	0.034331	821.78	−0.67702	79.057	46.948
1100	100	8.3307	0.12004	27.580	39.584	0.09465	0.025831	0.034845	857.40	−0.71435	85.797	51.158
1300	100	7.2877	0.13722	32.880	46.602	0.10051	0.026565	0.035323	894.00	−0.74281	93.151	55.511
1500	100	6.4847	0.15421	38.287	53.708	0.10559	0.027131	0.035723	930.40	−0.76506	100.84	59.875
1700	100	5.8456	0.17107	43.779	60.886	0.11009	0.027569	0.036049	966.13	−0.78264	108.75	64.208
1900	100	5.3239	0.18783	49.340	68.123	0.11411	0.027915	0.036317	1001.0	−0.79653	116.78	68.504
300	500	34.106	0.029320	6.2145	20.875	0.033155	0.028875	0.039265	1678.8	−0.57656	208.23	181.12
500	500	29.826	0.033528	11.583	28.348	0.052311	0.026614	0.036111	1573.6	−0.65015	178.50	120.62
700	500	26.714	0.037433	16.768	35.484	0.064323	0.026496	0.035494	1514.8	−0.67879	161.67	97.470
900	500	24.283	0.041180	22.008	42.598	0.073261	0.026991	0.035702	1482.8	−0.68796	151.95	86.531
1100	500	22.305	0.044833	27.358	49.775	0.080460	0.027539	0.036073	1468.3	−0.69130	146.88	81.387
1300	500	20.651	0.048423	32.814	57.025	0.086515	0.028000	0.036415	1465.1	−0.69354	144.95	79.411
1500	500	19.243	0.051966	38.354	64.337	0.091746	0.02836	0.036693	1469.3	−0.69594	145.84	79.312
1700	500	18.027	0.055473	43.961	71.698	0.096353	0.02864	0.036911	1478.5	−0.69875	148.48	80.393
1900	500	16.963	0.058952	49.623	79.098	0.10047	0.02886	0.037085	1491.1	−0.70188	152.39	82.251
300	1000	40.130	0.024919	6.8286	31.747	0.024761	0.032271	0.041510	2208.5	−0.50493	274.96	337.76
500	1000	36.567	0.027347	12.271	39.618	0.044944	0.029334	0.037843	2104.7	−0.57316	247.30	219.41
700	1000	33.895	0.029503	17.554	47.057	0.057468	0.028754	0.036801	2033.9	−0.60504	230.60	174.51
900	1000	31.736	0.031510	22.890	54.399	0.066695	0.028917	0.036702	1984.7	−0.61882	219.72	149.43
1100	1000	29.916	0.033427	28.327	61.754	0.074073	0.029215	0.036858	1951.3	−0.62560	212.46	133.76
1300	1000	28.338	0.035288	33.857	69.145	0.080246	0.029476	0.037051	1929.3	−0.62968	207.70	123.58
1500	1000	26.946	0.037111	39.461	76.573	0.085561	0.029675	0.037224	1915.7	−0.63251	204.81	116.94
1700	1000	25.701	0.038909	45.123	84.032	0.090229	0.029821	0.037369	1908.3	−0.63465	203.41	112.74
1900	1000	24.577	0.040688	50.830	91.519	0.094392	0.029928	0.037491	1905.8	−0.63632	203.25	110.27

This table was generated for a standard three-component dry air containing mole fractions 0.7812 nitrogen, 0.2096 oxygen, and 0.0092 argon. The values in this table were generated from the NIST REFPROP software (Lemmon, E. W., McLinden, M. O., and Huber, M. L., NIST Standard Reference Database 23: Reference Fluid Thermodynamic and Transport Properties—REFPROP, National Institute of Standards and Technology, Standard Reference Data Program, Gaithersburg, Md., 2002, Version 7.1). The primary source for the thermodynamic properties is Lemmon, E. W., Jacobsen, R. T, Penoncello, S. G., and Friend, D. G., "Thermodynamic Properties of Air and Mixtures of Nitrogen, Argon, and Oxygen from 60 to 2000 K at Pressures to 2000 MPa," *J. Phys. Chem. Ref. Data* **29**(3):331–385, 2000. The source for viscosity and thermal conductivity is Lemmon, E. W., and Jacobsen, R. T., "Viscosity and Thermal Conductivity Equations for Nitrogen, Oxygen, Argon, and Air," *Int. J. Thermophys.* **25**:21–69, 2004.

Properties at the freezing point temperature and the critical point temperature are given in the first and last entries of the saturation tables, respectively. In the single-phase table, when the temperature range for a given isobar includes a vapor-liquid phase boundary, the temperature of phase equilibrium is noted, and properties for both the saturated liquid and saturated vapor are given (with liquid properties given in the upper line). Lines are omitted from the temperature-pressure grid of the single-phase table, when the system would be in the solid phase or if there are potential problems with the source property surface.

In the range from the solidification point to 873 K at pressures to 70 MPa, the estimated uncertainty of density values calculated with the equation of state is 0.1%. The estimated uncertainty of calculated speed of sound values is 0.2% and that for calculated heat capacities is 1%. At temperatures above 873 K and 70 MPa, the estimated uncertainty of calculated density values is 0.5%, increasing to 1.0% at 2000 K and 2000 MPa. For viscosity, the uncertainty is 1% in the dilute gas. The uncertainty is around 2% between 270 and 300 K and increases to 5% outside of this region. There are very few measurements between 130 and 270 K for air to validate this claim, and the uncertainties may be even higher in this supercritical region. For thermal conductivity, the uncertainty for the dilute gas is 2% with increasing uncertainties near the triple points. The uncertainties range from 3% between 140 and 300 K to 5% at the triple point and at high temperatures. The uncertainties above 100 MPa are not known due to a lack of experimental data.

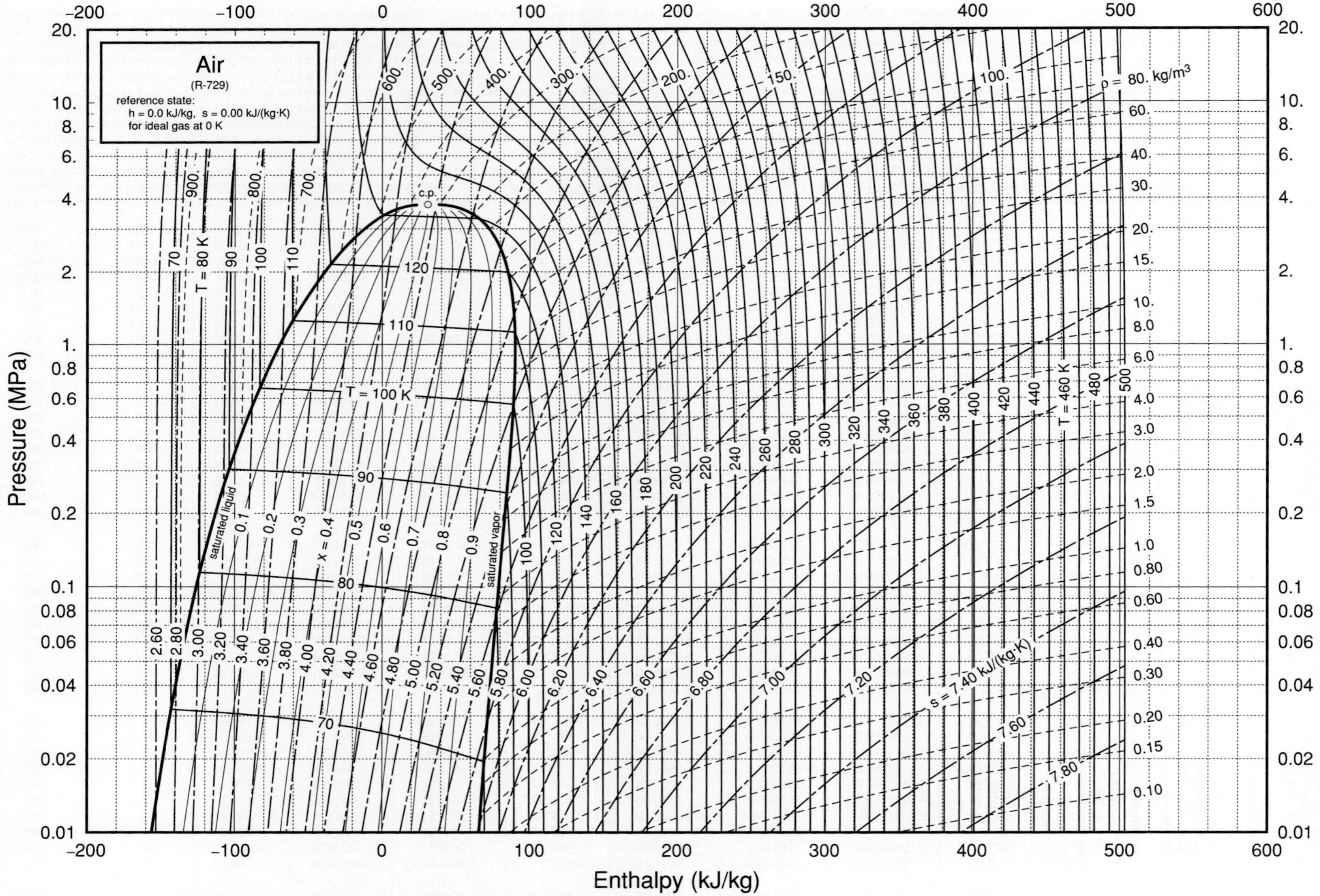

FIG. 2-3 Pressure-enthalpy diagram for dry air. Properties computed with the NIST REFPROP Database, Version 7.0 (Lemmon, E. W., M. O. McLinden, and M. L. Huber, 2002, NIST Standard Reference Database 23, NIST Reference Fluid Thermodynamic and Transport Properties—REFPROP, Version 7.0, Standard Reference Data Program, National Institute of Standards and Technology), based on the equation of state of E. W. Lemmon, R. T. Jacobsen,, S. G. Penoncello, and D. G. Friend.

TABLE 2-111 Air

Other tables include Stewart, R. B., S. G. Penoncello, et al., University of Idaho CATS report, 85-5, 1985 (0.1-700 bar, 85-750 K), and Lemmon, E. W., Jacobsen, R. T., Penoncello, S. G., and Friend, D. G., Thermodynamic Properties of Air and Mixtures of Nitrogen, Argon, and Oxygen from 60 to 2000 K at Pressures to 2000 MPa, *J. Phys. Chem. Ref. Data,* **29**(3): 331-385, 2000. Tables including reactions with hydrocarbons include Gordon, S., NASA Techn. Paper 1907, 4 vols., 1982. See also Gupta, R. N., K-P. Lee, et al., NASA RP 1232, 1990 (89 pp.) and RP 1260, 1991 (75 pp.). Analytic expressions for high temperatures were given by Matsuzaki, R., *Jap. J. Appl. Phys.,* **21,** 7 (1982): 1009-1013 and Japanese National Aerospace Laboratory report NAL TR 671, 1981 (45 pp.). Functions from 1500 to 15,000 K were tabulated by Hilsenrath, J. and M. Klein, AEDC-TR-65-58 = AD 612 301, 1965 (333 pp.). Tables from 10000 to 10,000,000 K were authored by Gilmore, F. R., Lockheed rept. 3-27-67-1, vol 1., 1967 (340 pp.), also published as *Radiative Properties of Air,* IFI/Plenum, New York, 1969 (648 pp.). Saturation and superheat tables and a chart to 7000 psia, 660°R appear in Stewart, R. B., R. T. Jacobsen, et al., *Thermodynamic Properties of Refrigerants,* ASHRAE, Atlanta, Ga, 1986 (521 pp.). For specific heat, thermal conductivity, and viscosity see *Thermophysical Properties of Refrigerants,* ASHRAE, 1993.

Air, Moist

For other data in this handbook, please see Figure 2-2 and the psychrometric tables, figures and descriptions in Section 12.

An ASHRAE publication, *Thermodynamic Properties of Dry Air and Water and S. I. Psychrometric Charts,* 1983 (360 pp.), extensively reviews moist air properties. Gandiduson, P., *Chem. Eng.,* Oct. 29, 1984 gives on page 118 a nomograph from 50 to 120°F, while equations in SI units were given by Nelson, B., *Chem. Eng. Progr.* **76,** 5 (May 1980): 83–85. Liley, P. E., *2000 Solved Problems in M.E. Thermodynamics,* McGraw-Hill, New York, 1989, gives four simple equations with which most calculations can be made. Devres, Y.O., *Appl. Energy* **48** (1994): 1–18 gives equations with which three known properties can be used to determine four others. Klappert, M. T. and G. F. Schilling, Rand RM-4244-PR = AD 604 856, 1984 (40 pp.) gives tables from 100 to 270 K, while programs from −60 to 2°F are given by Sando, F. A., *ASHRAE Trans.,* **96,** 2 (1990): 299–308.

Viscosity references include Kestin, J. and J. H. Whitelaw, *Int. J. Ht. Mass Transf.* **7,** 11 (1964): 1245–1255; Studnokov, E. L., *Inz.-Fiz. Zhur.* **19,** 2 (1970): 338–340; Hochramer, D. and F. Munczak, *Setzb. Ost. Acad. Wiss II* **175,** 10 (1966): 540–550. For thermal conductivity see, for instance, Mason, E. A. and L. Monchick, *Humidity and Moisture Control in Science and Industry,* Reinhold, New York, 1965 (257–272).

TABLE 2-112 Thermodynamic Properties of Ammonia

Temperature K	Pressure MPa	Density mol/dm³	Volume dm³/mol	Int. energy kJ/mol	Enthalpy kJ/mol	Entropy kJ/(mol·K)	C_v kJ/(mol·K)	C_p kJ/(mol·K)	Sound speed m/s	Joule-Thomson K/MPa	Therm. cond. mW/(m·K)	Viscosity μPa·s
						Saturated Properties						
195.50	0.0060912	43.035	0.023237	0.00000	0.00014154	0.00000	0.049972	0.071565	2124.2	−0.23362	818.99	559.57
200.00	0.0086509	42.754	0.023389	0.32333	0.32353	0.0016351	0.049837	0.071988	2080.2	−0.22917	803.14	507.28
210.00	0.017739	42.111	0.023747	1.0480	1.0484	0.0051707	0.049521	0.072971	1992.7	−0.21883	768.02	414.98
220.00	0.033790	41.442	0.024130	1.7825	1.7833	0.0085874	0.049207	0.073950	1913.7	−0.20813	733.17	346.68
230.00	0.060407	40.748	0.024541	2.5265	2.5279	0.011894	0.048906	0.074883	1839.2	−0.19712	698.80	294.94
240.00	0.10223	40.032	0.024980	3.2793	3.2818	0.015098	0.048613	0.075764	1766.9	−0.18561	665.09	254.85
250.00	0.16494	39.293	0.025450	4.0403	4.0445	0.018205	0.048327	0.076608	1695.6	−0.17327	632.16	223.08
260.00	0.25531	38.533	0.025952	4.8093	4.8160	0.021222	0.048047	0.077448	1624.5	−0.15963	600.07	197.34
270.00	0.38107	37.748	0.026491	5.5862	5.5963	0.024154	0.047774	0.078328	1553.1	−0.14414	568.85	176.06
280.00	0.55092	36.939	0.027072	6.3712	6.3861	0.027010	0.047511	0.079296	1481.0	−0.12612	538.50	158.12
290.00	0.77436	36.101	0.027700	7.1651	7.1866	0.029797	0.047266	0.080412	1407.8	−0.10470	508.99	142.74
300.00	1.0617	35.230	0.028385	7.9691	7.9993	0.032525	0.047044	0.081747	1333.2	−0.078790	480.25	129.33
310.00	1.4240	34.320	0.029138	8.7850	8.8265	0.035203	0.046856	0.083390	1256.7	−0.046923	452.23	117.49
320.00	1.8728	33.363	0.029973	9.6153	9.6714	0.037843	0.046715	0.085465	1177.9	−0.0070718	424.83	106.91
330.00	2.4205	32.350	0.030912	10.463	10.538	0.040458	0.046636	0.088145	1096.5	0.043673	397.96	97.325
340.00	3.0802	31.264	0.031986	11.333	11.432	0.043065	0.046642	0.091701	1011.8	0.10967	371.51	88.555
350.00	3.8660	30.087	0.033237	12.232	12.361	0.045682	0.046767	0.096576	923.38	0.19774	345.32	80.430
360.00	4.7929	28.788	0.034737	13.169	13.335	0.048339	0.047064	0.10357	830.62	0.31928	319.25	72.796
370.00	5.8778	27.321	0.036602	14.158	14.373	0.051075	0.047619	0.11435	732.78	0.49497	293.07	65.493
380.00	7.1402	25.606	0.039054	15.224	15.503	0.053961	0.048589	0.13314	628.75	0.76738	266.57	58.315
390.00	8.6045	23.465	0.042616	16.424	16.790	0.057149	0.050319	0.17550	515.88	1.2455	239.65	50.877
400.00	10.305	20.232	0.049426	17.969	18.478	0.061223	0.054109	0.38707	384.58	2.3557	216.00	41.802
405.40	11.339	13.212	0.075690	20.640	21.499	0.068559			0	5.0513		
195.50	0.0060912	0.0037635	265.71	23.661	25.279	0.12931	0.026510	0.035130	354.12	171.13	19.636	6.8396
200.00	0.0086509	0.0052305	191.19	23.770	25.424	0.12714	0.026650	0.035345	357.91	152.55	19.684	6.9515
210.00	0.017739	0.010249	97.573	24.006	25.737	0.12273	0.027053	0.035961	365.94	120.01	19.860	7.2115
220.00	0.033790	0.018721	53.415	24.233	26.038	0.11884	0.027583	0.036783	373.38	96.215	20.132	7.4846
230.00	0.060407	0.032214	31.043	24.450	26.325	0.11536	0.028245	0.037836	380.19	78.430	20.503	7.7679
240.00	0.10223	0.052667	18.987	24.655	26.596	0.11224	0.029043	0.039142	386.30	64.852	20.978	8.0587
250.00	0.16494	0.082417	12.133	24.846	26.847	0.10942	0.029978	0.040728	391.66	54.280	21.560	8.3552
260.00	0.25531	0.12421	8.0506	25.021	27.077	0.10684	0.031050	0.042623	396.20	45.905	22.258	8.6558
270.00	0.38107	0.18126	5.5168	25.179	27.281	0.10447	0.032253	0.044859	399.86	39.175	23.079	8.9595
280.00	0.55092	0.25729	3.8867	25.317	27.459	0.10227	0.033581	0.047476	402.59	33.701	24.034	9.2664
290.00	0.77436	0.35664	2.8040	25.435	27.606	0.10021	0.035028	0.050530	404.30	29.207	25.138	9.5771
300.00	1.0617	0.48448	2.0641	25.528	27.720	0.098259	0.036584	0.054099	404.95	25.489	26.408	9.8938
310.00	1.4240	0.64702	1.5455	25.595	27.796	0.096395	0.038244	0.058302	404.45	22.391	27.872	10.220
320.00	1.8728	0.85202	1.1737	25.632	27.830	0.094589	0.040004	0.063320	402.70	19.794	29.568	10.561
330.00	2.4205	1.1094	0.90139	25.634	27.816	0.092817	0.041868	0.069443	399.61	17.599	31.559	10.927
340.00	3.0802	1.4325	0.69810	25.595	27.746	0.091046	0.043844	0.077150	395.05	15.728	33.945	11.330
350.00	3.8660	1.8399	0.54350	25.505	27.606	0.089242	0.045954	0.087280	388.86	14.112	36.900	11.792
360.00	4.7929	2.3598	0.42377	25.350	27.381	0.087355	0.048233	0.10141	380.83	12.690	40.752	12.346
370.00	5.8778	3.0375	0.32922	25.107	27.042	0.085316	0.050744	0.12286	370.69	11.400	46.149	13.053
380.00	7.1402	3.9558	0.25279	24.734	26.539	0.083003	0.053589	0.16000	357.96	10.172	54.556	14.025
390.00	8.6045	5.2979	0.18875	24.144	25.768	0.080169	0.056957	0.24170	341.67	8.9038	70.114	15.527
400.00	10.305	7.6973	0.12992	23.047	24.386	0.075992	0.061281	0.59477	318.22	7.3513	113.54	18.529
405.40	11.339	13.212	0.075690	20.640	21.499	0.068559			0	5.0513		

200.00	0.10000	42.756	0.023388	0.32270	0.32504	0.0016320	0.049842	0.071983	2080.3	−0.22921	803.24	507.47
239.56	0.10000	40.064	0.024960	3.2461	3.2486	0.014960	0.048626	0.075726	1770.0	−0.18613	666.56	256.42
239.56	0.10000	0.051595	19.382	24.646	26.584	0.11237	0.029005	0.039079	386.05	65.377	20.955	8.0459
300.00	0.10000	0.040502	24.690	26.378	28.847	0.12080	0.028021	0.036849	434.39	27.493	25.100	10.161
400.00	0.10000	0.030171	33.144	29.297	32.612	0.13162	0.030417	0.038883	497.93	10.681	37.215	13.971
500.00	0.10000	0.024091	41.509	32.514	36.665	0.14065	0.033897	0.042280	550.96	5.5276	53.119	17.863
600.00	0.10000	0.020060	49.849	36.096	41.081	0.14869	0.037731	0.046083	597.69	3.2702	68.607	21.682
700.00	0.10000	0.017188	58.179	40.068	45.885	0.15609	0.041678	0.050015	640.16	2.0841	78.312	25.391
200.00	1.0000	42.774	0.023379	0.31651	0.33989	0.0016010	0.049890	0.071938	2081.5	−0.22959	804.23	509.28
298.05	1.0000	35.403	0.028246	7.8111	7.8393	0.031996	0.047085	0.081465	1347.9	−0.084271	485.81	131.82
298.05	1.0000	0.45697	2.1883	25.512	27.700	0.098633	0.036271	0.053356	404.91	26.163	26.145	9.8313
300.00	1.0000	0.45215	2.2117	25.592	27.804	0.098979	0.035866	0.052493	407.16	25.620	26.308	9.9115
400.00	1.0000	0.31157	3.2095	29.019	32.229	0.11177	0.031641	0.041627	488.94	10.494	38.087	13.927
500.00	1.0000	0.24426	4.0940	32.359	36.453	0.12119	0.034312	0.043338	546.79	5.4884	53.750	17.877
600.00	1.0000	0.20197	4.9513	35.994	40.945	0.12937	0.037928	0.046628	595.60	3.2544	69.123	21.717
700.00	1.0000	0.17248	5.7977	39.994	45.792	0.13684	0.041791	0.050341	639.15	2.0746	78.751	25.434
200.00	5.0000	42.852	0.023336	0.28942	0.40611	0.0014649	0.050097	0.071739	2086.8	−0.23126	808.60	517.30
300.00	5.0000	35.450	0.028209	7.8852	8.0263	0.032243	0.047090	0.080899	1361.2	−0.089577	487.57	132.49
362.03	5.0000	28.505	0.035081	13.365	13.540	0.048887	0.047152	0.10538	811.17	0.34968	313.94	71.291
362.03	5.0000	2.4828	0.40277	25.309	27.323	0.086956	0.048722	0.10501	378.95	12.419	41.693	12.475
400.00	5.0000	1.8706	0.53459	27.540	30.213	0.094581	0.038466	0.061581	441.81	9.6373	45.730	14.036
500.00	5.0000	1.3046	0.76650	31.630	35.462	0.10634	0.036193	0.048779	528.14	5.2830	57.294	18.073
600.00	5.0000	1.0412	0.96040	35.527	40.329	0.11521	0.038798	0.049210	586.79	3.1693	71.791	21.941
700.00	5.0000	0.87563	1.1420	39.662	45.373	0.12298	0.042289	0.051836	635.17	2.0254	80.941	25.662
200.00	10.000	42.947	0.023284	0.25644	0.48928	0.0012980	0.050342	0.071495	2093.5	−0.23328	814.02	527.29
300.00	10.000	35.714	0.028000	7.7848	8.0648	0.031903	0.047164	0.079960	1394.2	−0.10159	496.50	136.36
398.32	10.000	20.945	0.047744	17.655	18.132	0.060394	0.053149	0.30653	409.04	2.0704	218.73	43.632
398.32	10.000	7.1390	0.14008	23.303	24.704	0.076892	0.060447	0.46915	323.12	7.6606	101.04	17.793
400.00	10.000	6.5455	0.15278	23.801	25.329	0.078458	0.057611	0.30552	336.28	7.7633	95.455	17.230
500.00	10.000	2.8656	0.34897	30.616	34.106	0.098525	0.038603	0.057806	505.64	4.9335	63.922	18.722
600.00	10.000	2.1650	0.46190	34.920	39.539	0.10844	0.039862	0.052806	577.38	3.0278	76.053	22.393
700.00	10.000	1.7835	0.56069	39.241	44.848	0.11663	0.042896	0.053796	631.50	1.9491	84.235	26.035
300.00	100.00	38.995	0.025644	6.5830	9.1474	0.027511	0.048894	0.072740	1774.7	−0.19551	622.86	193.71
400.00	100.00	33.105	0.030207	13.432	16.453	0.048523	0.046636	0.073557	1378.2	−0.11309	431.98	96.237
500.00	100.00	27.067	0.036945	20.212	23.907	0.065147	0.045999	0.075495	1081.8	0.049919	305.65	60.386
600.00	100.00	21.518	0.046473	26.825	31.472	0.078942	0.046723	0.075193	918.11	0.23722	234.79	46.188
700.00	100.00	17.303	0.057794	33.074	38.854	0.090326	0.048331	0.072317	861.52	0.32753	196.04	41.237
300.00	500.00	45.670	0.021896	4.7114	15.660	0.018023	0.052877	0.067831	2597.1	−0.25055	989.00	376.31
400.00	500.00	42.416	0.023576	10.633	22.421	0.037482	0.051527	0.066802	2353.2	−0.25064	804.05	188.46
500.00	500.00	39.515	0.025307	16.367	29.021	0.052215	0.050431	0.065418	2176.9	−0.25260	674.00	120.77
600.00	500.00	36.909	0.027094	22.007	35.554	0.064127	0.050614	0.065476	2044.6	−0.24722	582.63	91.251
700.00	500.00	34.550	0.028943	27.680	42.152	0.074295	0.051816	0.066615	1943.8	−0.23682	511.57	77.538
300.00	1000.0	49.944	0.020022	4.1818	24.204	0.011750	0.055176	0.065784	3230.2	−0.25989	1324.0	554.62
400.00	1000.0	47.551	0.021030	9.8612	30.891	0.030984	0.054864	0.066677	2997.6	−0.25431	1138.9	274.91
500.00	1000.0	45.362	0.022045	15.432	37.477	0.045686	0.053323	0.065150	2842.8	−0.26084	996.49	174.11
600.00	1000.0	43.378	0.023053	20.911	43.964	0.057514	0.052940	0.064819	2728.6	−0.26235	887.73	129.02
700.00	1000.0	41.556	0.024064	26.418	50.481	0.067559	0.053649	0.065697	2639.0	−0.25820	797.25	107.14

The values in these tables were generated from the NIST REFPROP software (Lemmon, E. W., McLinden, M. O., and Huber, M. L., NIST Standard Reference Database 23: Reference Fluid Thermodynamic and Transport Properties—REFPROP, National Institute of Standards and Technology, Standard Reference Data Program, Gaithersburg, Md., 2002, Version 7.1). The primary source for the thermodynamic properties is Tillner-Roth, R., Harms-Watzenberg, F., and Baehr, H. D., "Eine neue Fundamentalgleichung fuer Ammoniak," *DKV-Tagungsbericht*, **20:**167–181, 1993. The source for viscosity is Fenghour, A., Wakeham, W. A., Vesovic, V., Watson, J. T. R., Millat, J., and Vogel, E., "The Viscosity of Ammonia," *J. Phys. Chem. Ref. Data* **24:**1649–1667, 1995. The source for thermal conductivity is Tufeu, R., Ivanov, D. Y., Garrabos, Y., and Le Neindre, B., "Thermal Conductivity of Ammonia in a Large Temperature and Pressure Range Including the Critical Region," *Ber. Bunsenges. Phys. Chem.* **88:**422–427, 1984.

Properties at the triple point temperature and the critical point temperature are given in the first and last entries of the saturation tables, respectively. In the single-phase table, when the temperature range for a given isobar includes a vapor-liquid phase boundary, the temperature of phase equilibrium is noted, and properties for both the saturated liquid and saturated vapor are given (with liquid properties given in the upper line). Lines are omitted from the temperature-pressure grid of the single-phase table, when the system would be in the solid phase or if there are potential problems with the source property surface.

The uncertainties of the equation of state are 0.2% in density, 2% in heat capacity, and 2% in the speed of sound, except in the critical region. The uncertainty in vapor pressure is 0.2%. The uncertainty varies from 0.5% for the viscosity of the dilute gas phase at moderate temperatures to about 5% for the viscosity at high pressures and temperatures. The uncertainty in thermal conductivity is 2%.

TABLE 2-113 Thermodynamic Properties of Carbon Dioxide

Temperature K	Pressure MPa	Density mol/dm³	Volume dm³/mol	Int. energy kJ/mol	Enthalpy kJ/mol	Entropy kJ/(mol·K)	C_v kJ/(mol·K)	C_p kJ/(mol·K)	Sound speed m/s	Joule-Thomson K/MPa	Therm. cond. mW/(m·K)	Viscosity μPa·s
						Saturated Properties						
216.59	0.51796	26.777	0.037345	3.5030	3.5223	0.022943	0.042895	0.085960	975.85	−0.14430	180.63	256.70
220.00	0.59913	26.497	0.037740	3.7943	3.8169	0.024279	0.042682	0.086338	951.21	−0.13180	176.15	242.01
225.00	0.73509	26.078	0.038347	4.2235	4.2517	0.026209	0.042383	0.087024	915.16	−0.11104	169.67	222.19
230.00	0.89291	25.646	0.038992	4.6554	4.6902	0.028110	0.042103	0.087886	879.09	−0.086994	163.28	204.23
235.00	1.0747	25.201	0.039680	5.0908	5.1334	0.029986	0.041843	0.088954	842.88	−0.059053	156.98	187.88
240.00	1.2825	24.742	0.040418	5.5303	5.5821	0.031840	0.041605	0.090263	806.38	−0.026454	150.75	172.96
245.00	1.5185	24.264	0.041213	5.9749	6.0375	0.033678	0.041393	0.091866	769.44	0.011808	144.58	159.30
250.00	1.7850	23.767	0.042075	6.4256	6.5007	0.035505	0.041212	0.093831	731.78	0.057087	138.47	146.74
255.00	2.0843	23.246	0.043018	6.8836	6.9733	0.037326	0.041079	0.096251	693.01	0.11121	132.40	135.14
260.00	2.4188	22.697	0.044059	7.3505	7.4571	0.039148	0.041029	0.099258	652.58	0.17663	126.35	124.40
265.00	2.7909	22.114	0.045219	7.8282	7.9544	0.040979	0.041109	0.10306	610.07	0.25672	120.31	114.40
270.00	3.2033	21.491	0.046531	8.3190	8.4681	0.042829	0.041351	0.10798	565.46	0.35639	114.25	105.02
275.00	3.6589	20.817	0.048037	8.8266	9.0024	0.044711	0.041750	0.11457	519.14	0.48324	108.17	96.174
280.00	4.1607	20.077	0.049808	9.3560	9.5633	0.046643	0.042270	0.12385	471.54	0.64959	102.03	87.731
285.00	4.7123	19.247	0.051957	9.9154	10.160	0.048657	0.042900	0.13790	422.75	0.87650	95.810	79.548
290.00	5.3177	18.284	0.054693	10.519	10.810	0.050805	0.043734	0.16176	371.95	1.2037	89.546	71.409
295.00	5.9822	17.100	0.058480	11.197	11.547	0.053196	0.045175	0.21098	315.91	1.7218	83.558	62.936
300.00	6.7131	15.434	0.064793	12.036	12.471	0.056151	0.049288	0.38279	245.67	2.7258	80.593	53.107
304.13	7.3773	10.625	0.094118	13.928	14.622	0.063094			0	5.8665		
216.59	0.51796	0.31268	3.1982	17.286	18.943	0.094138	0.027691	0.039992	222.78	26.174	11.014	10.951
220.00	0.59913	0.35941	2.7824	17.329	18.996	0.093276	0.028120	0.040943	223.15	25.084	11.301	11.135
225.00	0.73509	0.43766	2.2849	17.387	19.067	0.092055	0.028782	0.042489	223.49	23.617	11.745	11.409
230.00	0.89291	0.52878	1.8912	17.438	19.127	0.090878	0.029488	0.044244	223.57	22.288	12.221	11.689
235.00	1.0747	0.63442	1.5762	17.481	19.175	0.089736	0.030241	0.046248	223.40	21.077	12.736	11.976
240.00	1.2825	0.75654	1.3218	17.515	19.210	0.088622	0.031042	0.048555	222.96	19.969	13.297	12.272
245.00	1.5185	0.89743	1.1143	17.538	19.230	0.087526	0.031899	0.051242	222.24	18.950	13.917	12.579
250.00	1.7850	1.0599	0.94353	17.550	19.234	0.086439	0.032827	0.054421	221.22	18.005	14.610	12.902
255.00	2.0843	1.2472	0.80180	17.549	19.220	0.085352	0.033844	0.058244	219.87	17.117	15.396	13.245
260.00	2.4188	1.4637	0.68320	17.532	19.185	0.084254	0.034955	0.062912	218.19	16.277	16.306	13.614
265.00	2.7909	1.7149	0.58314	17.498	19.125	0.083133	0.036164	0.068721	216.15	15.476	17.381	14.017
270.00	3.2033	2.0080	0.49800	17.441	19.037	0.081972	0.037482	0.076168	213.75	14.704	18.687	14.469
275.00	3.6589	2.3535	0.42490	17.359	18.913	0.080750	0.038949	0.086123	210.96	13.947	20.325	14.987
280.00	4.1607	2.7663	0.36150	17.241	18.746	0.079437	0.040628	0.10020	207.72	13.185	22.468	15.601
285.00	4.7123	3.2702	0.30579	17.078	18.519	0.077987	0.042629	0.12177	203.94	12.387	25.424	16.361
290.00	5.3177	3.9074	0.25593	16.848	18.209	0.076319	0.045155	0.15906	199.45	11.509	29.821	17.357
295.00	5.9822	4.7654	0.20985	16.509	17.764	0.074270	0.048677	0.23904	193.84	10.459	37.215	18.792
300.00	6.7131	6.1028	0.16386	15.935	17.035	0.071364	0.054908	0.52463	185.33	9.0093	53.689	21.306
304.13	7.3773	10.625	0.094118	13.928	14.622	0.063094			0	5.8665		

Single-Phase Properties

250.00	0.10000	0.048542	20.601	18.448	20.509	0.11415	0.026766	0.035428	247.79	17.399	12.950	12.565
450.00	0.10000	0.026758	37.372	24.664	28.401	0.13712	0.034775	0.043148	324.41	4.0212	29.346	21.901
650.00	0.10000	0.018506	54.037	32.199	37.602	0.15397	0.040192	0.048529	385.01	1.6551	45.466	29.873
850.00	0.10000	0.014148	70.683	40.636	47.705	0.16750	0.043944	0.052271	437.11	0.78058	60.295	36.707
1050.0	0.10000	0.011452	87.321	49.704	58.436	0.17883	0.046573	0.054895	483.65	0.34646	73.843	42.692
250.00	1.0000	0.53250	1.8779	18.023	19.901	0.093263	0.029361	0.042504	235.08	17.606	13.584	12.691
450.00	1.0000	0.27038	3.6985	24.546	28.244	0.11771	0.034954	0.043866	322.89	3.9880	29.620	21.954
650.00	1.0000	0.18527	5.3976	32.133	37.530	0.13473	0.040239	0.048779	385.36	1.6311	45.651	29.907
850.00	1.0000	0.14131	7.0767	40.591	47.668	0.14830	0.043965	0.052397	438.06	0.76632	60.435	36.732
1050.0	1.0000	0.11430	8.7487	49.671	58.419	0.15965	0.046585	0.054970	484.84	0.33777	73.956	42.712
250.00	5.0000	24.060	0.041563	6.2824	6.4902	0.034925	0.041321	0.090937	762.21	0.015208	142.22	153.15
287.43	5.0000	18.798	0.053196	10.202	10.468	0.049681	0.043268	0.14775	398.39	1.0195	92.760	75.598
287.43	5.0000	3.5600	0.28090	16.977	18.381	0.077209	0.043774	0.13705	201.86	11.974	27.323	16.808
450.00	5.0000	1.4155	0.70647	24.000	27.533	0.10313	0.035769	0.047478	317.50	3.8034	31.164	22.429
650.00	5.0000	0.92982	1.0755	31.842	37.219	0.12091	0.040445	0.049898	387.59	1.5263	46.589	30.157
850.00	5.0000	0.70241	1.4237	40.395	47.513	0.13469	0.044055	0.052945	442.64	0.70611	61.117	36.899
1050.0	5.0000	0.56658	1.7650	49.524	58.349	0.14613	0.046637	0.055297	490.31	0.30129	74.494	42.836
250.00	10.000	24.459	0.040885	6.0862	6.4950	0.034120	0.041488	0.087624	804.05	−0.034849	147.52	162.47
450.00	10.000	2.9910	0.33433	23.276	26.619	0.095787	0.036785	0.052935	314.60	3.4705	33.917	23.679
650.00	10.000	1.8632	0.53671	31.482	36.849	0.11461	0.040693	0.051293	391.91	1.3965	48.005	30.687
850.00	10.000	1.3930	0.71790	40.155	47.334	0.12866	0.044164	0.053603	449.04	0.63635	62.093	37.224
1050.0	10.000	1.1205	0.89248	49.347	58.271	0.14021	0.046701	0.055685	497.48	0.25964	75.242	43.066
250.00	100.00	28.075	0.035619	4.3002	7.8621	0.026023	0.043569	0.073521	1227.6	−0.27302	206.28	287.05
450.00	100.00	19.246	0.051959	16.560	21.756	0.067062	0.040841	0.066107	753.30	−0.11128	106.65	83.996
650.00	100.00	13.677	0.073117	27.132	34.444	0.090445	0.043108	0.061252	646.36	−0.054084	86.093	58.868
850.00	100.00	10.636	0.094022	37.076	46.478	0.10660	0.045620	0.059534	646.61	−0.13292	87.259	54.445
1050.0	100.00	8.7929	0.11373	46.995	58.368	0.11916	0.047676	0.059512	668.90	−0.21482	94.022	55.058
450.00	500.00	28.922	0.034576	13.014	30.302	0.050604	0.047702	0.063434	1576.4	−0.38514	239.59	303.64
650.00	500.00	25.661	0.038969	23.302	42.786	0.073576	0.048419	0.061885	1404.7	−0.40369	197.25	191.14
850.00	500.00	23.144	0.043208	33.551	55.155	0.090166	0.049676	0.061912	1320.1	−0.41098	177.31	145.07
1050.0	500.00	21.126	0.047334	43.903	67.570	0.10328	0.050818	0.062247	1278.7	−0.41674	168.50	123.33

The values in these tables were generated from the NIST REFPROP software (Lemmon, E. W., McLinden, M. O., and Huber, M. L., NIST Standard Reference Database 23: Reference Fluid Thermodynamic and Transport Properties—REFPROP, National Institute of Standards and Technology, Standard Reference Data Program, Gaithersburg, Md., 2002, Version 7.1). The primary source for the thermodynamic properties is Span, R., and Wagner, W., "A New Equation of State for Carbon Dioxide Covering the Fluid Region from the Triple-Point Temperature to 1100 K at Pressures up to 800 MPa," *J. Phys. Chem. Ref. Data* **25**(6):1509–1596, 1996. The source for viscosity is Fenghour, A., Wakeham, W. A., and Vesovic, V., "The Viscosity of Carbon Dioxide," *J. Phys. Chem. Ref. Data* **27**:31–44, 1998. The source for thermal conductivity is Vesovic, V., Wakeham, W. A., Olchowy, G. A., Sengers, J. V., Watson, J. T. R., and Millat, J., "The Transport Properties of Carbon Dioxide," *J. Phys. Chem. Ref. Data* **19**:763–808, 1990.

Properties at the triple point temperature and the critical point temperature are given in the first and last entries of the saturation tables, respectively. In the single-phase table, when the temperature range for a given isobar includes a vapor-liquid phase boundary, the temperature of phase equilibrium is noted, and properties for both the saturated liquid and saturated vapor are given (with liquid properties given in the upper line). Lines are omitted from the temperature-pressure grid of the single-phase table, when the system would be in the solid phase or if there are potential problems with the source property surface.

At pressures up to 30 MPa and temperatures up to 523 K, the estimated uncertainty ranges from 0.03% to 0.05% in density, 0.03% (in the vapor) to 1% in the speed of sound (0.5% in the liquid), and 0.15% (in the vapor) to 1.5% (in the liquid) in heat capacity. Special interest has been focused on the description of the critical region and the extrapolation behavior of the formulation (to the limits of chemical stability). The uncertainty in viscosity ranges from 0.3% in the dilute gas near room temperature to 5% at the highest pressures. The uncertainty in thermal conductivity is less than 5%.

TABLE 2-114 Thermodynamic Properties of Carbon Monoxide

Temperature K	Pressure MPa	Density mol/dm^3	Volume dm^3/mol	Int. energy kJ/mol	Enthalpy kJ/mol	Entropy kJ/(mol·K)	C_v kJ/(mol·K)	C_p kJ/(mol·K)	Sound speed m/s	Joule-Thomson K/MPa	Therm. cond. mW/(m·K)	Viscosity μPa·s
						Saturated Properties						
68.160	0.015537	30.330	0.032971	−0.81158	−0.81106	−0.010820	0.035351	0.060430	998.20	−0.36906	180.28	274.18
70.000	0.021053	30.064	0.033262	−0.70065	−0.69995	−0.0092140	0.034805	0.060226	980.50	−0.36553	175.49	252.15
72.000	0.028718	29.773	0.033588	−0.58046	−0.57950	−0.0075210	0.034248	0.060064	961.22	−0.36074	170.45	232.14
74.000	0.038447	29.478	0.033924	−0.46058	−0.45927	−0.0058785	0.033724	0.059961	941.89	−0.35489	165.55	215.32
76.000	0.050599	29.180	0.034270	−0.34088	−0.33915	−0.0042823	0.033232	0.059917	922.49	−0.34794	160.76	201.01
78.000	0.065559	28.878	0.034628	−0.22127	−0.21900	−0.0027285	0.032768	0.059930	903.01	−0.33981	156.06	188.69
80.000	0.083738	28.573	0.034999	−0.10165	−0.098716	−0.0012138	0.032329	0.060002	883.44	−0.33041	151.45	177.96
82.000	0.10556	28.262	0.035383	0.018099	0.021834	0.00026503	0.031915	0.060132	863.76	−0.31966	146.89	168.52
84.000	0.13148	27.947	0.035782	0.13806	0.14277	0.0017110	0.031522	0.060324	843.95	−0.30742	142.40	160.13
86.000	0.16196	27.626	0.036197	0.25835	0.26421	0.0031269	0.031150	0.060578	824.00	−0.29356	137.96	152.60
88.000	0.19748	27.300	0.036630	0.37906	0.38629	0.0045153	0.030798	0.060899	803.89	−0.27794	133.57	145.77
90.000	0.23852	26.967	0.037082	0.50030	0.50915	0.0058787	0.030463	0.061291	783.60	−0.26034	129.23	139.52
92.000	0.28559	26.627	0.037556	0.62218	0.63291	0.0072195	0.030146	0.061760	763.12	−0.24056	124.94	133.75
94.000	0.33919	26.280	0.038052	0.74482	0.75773	0.0085399	0.029846	0.062314	742.41	−0.21834	120.69	128.38
96.000	0.39983	25.924	0.038574	0.86835	0.88377	0.0098422	0.029562	0.062962	721.45	−0.19335	116.51	123.34
98.000	0.46805	25.559	0.039125	0.99289	1.0112	0.011129	0.029294	0.063716	700.22	−0.16523	112.38	118.57
100.00	0.54438	25.184	0.039708	1.1186	1.1402	0.012402	0.029043	0.064590	678.68	−0.13353	108.31	114.02
102.00	0.62934	24.798	0.040326	1.2457	1.2710	0.013663	0.028809	0.065604	656.78	−0.097704	104.30	109.66
104.00	0.72348	24.399	0.040985	1.3742	1.4039	0.014916	0.028592	0.066781	634.50	−0.057078	100.36	105.45
106.00	0.82736	23.987	0.041689	1.5045	1.5390	0.016162	0.028395	0.068153	611.77	−0.010824	96.482	101.35
108.00	0.94154	23.560	0.042446	1.6368	1.6768	0.017404	0.028218	0.069759	588.54	0.042104	92.679	97.342
110.00	1.0666	23.114	0.043263	1.7713	1.8175	0.018646	0.028066	0.071656	564.73	0.10304	88.948	93.404
112.00	1.2031	22.649	0.044151	1.9085	1.9616	0.019891	0.027941	0.073916	540.25	0.17371	85.290	89.510
114.00	1.3517	22.161	0.045124	2.0487	2.1097	0.021142	0.027850	0.076648	515.01	0.25641	81.702	85.641
116.00	1.5130	21.646	0.046197	2.1925	2.2624	0.022406	0.027800	0.080005	488.86	0.35427	78.180	81.774
118.00	1.6877	21.099	0.047395	2.3405	2.4205	0.023688	0.027803	0.084225	461.63	0.47167	74.716	77.888
120.00	1.8765	20.513	0.048749	2.4938	2.5853	0.024996	0.027874	0.089692	433.11	0.61495	71.296	73.954
122.00	2.0802	19.878	0.050307	2.6536	2.7583	0.026343	0.028038	0.097070	403.00	0.79382	67.896	69.940
124.00	2.2997	19.179	0.052141	2.8221	2.9420	0.027744	0.028333	0.10762	370.88	1.0239	64.476	65.797
126.00	2.5360	18.390	0.054377	3.0024	3.1403	0.029230	0.028826	0.12411	336.15	1.3325	60.972	61.448
128.00	2.7904	17.464	0.057259	3.2010	3.3608	0.030854	0.029646	0.15392	297.82	1.7728	57.261	56.748
130.00	3.0647	16.288	0.061393	3.4328	3.6210	0.032745	0.031097	0.22603	254.03	2.4703	53.107	51.348
132.86	3.4982	10.850	0.092166	4.2912	4.6137	0.040039			0	6.1475		
68.160	0.015537	0.027707	36.091	5.1252	5.6859	0.084499	0.021089	0.029785	167.25	40.804	6.6865	4.6366
70.000	0.021053	0.036656	27.281	5.1600	5.7343	0.082704	0.021155	0.029947	169.22	38.426	6.8845	4.7768
72.000	0.028718	0.048780	20.500	5.1971	5.7859	0.080887	0.021238	0.030153	171.27	36.126	7.1009	4.9329
74.000	0.038447	0.063796	15.675	5.2334	5.8361	0.079194	0.021333	0.030394	173.22	34.080	7.3188	5.0934
76.000	0.050599	0.082130	12.176	5.2688	5.8849	0.077613	0.021441	0.030672	175.07	32.250	7.5382	5.2589
78.000	0.065559	0.10424	9.5935	5.3031	5.9320	0.076131	0.021563	0.030993	176.80	30.604	7.7592	5.4300
80.000	0.083738	0.13059	7.6573	5.3363	5.9775	0.074739	0.021699	0.031360	178.42	29.116	7.9820	5.6076
82.000	0.10556	0.16171	6.1841	5.3682	6.0210	0.073426	0.021850	0.031777	179.92	27.763	8.2067	5.7922
84.000	0.13148	0.19810	5.0478	5.3988	6.0625	0.072185	0.022017	0.032250	181.29	26.527	8.4335	5.9847
86.000	0.16196	0.24036	4.1605	5.4280	6.1019	0.071007	0.022199	0.032783	182.54	25.392	8.6627	6.1860
88.000	0.19748	0.28906	3.4595	5.4556	6.1388	0.069885	0.022397	0.033383	183.66	24.345	8.8944	6.3968
90.000	0.23852	0.34486	2.8997	5.4816	6.1733	0.068813	0.022611	0.034057	184.65	23.377	9.1291	6.6182
92.000	0.28559	0.40845	2.4483	5.5058	6.2050	0.067785	0.022842	0.034813	185.51	22.477	9.3672	6.8512
94.000	0.33919	0.48058	2.0808	5.5280	6.2338	0.066796	0.023089	0.035661	186.22	21.638	9.6091	7.0969
96.000	0.39983	0.56209	1.7791	5.5482	6.2595	0.065840	0.023352	0.036615	186.80	20.853	9.8555	7.3566
98.000	0.46805	0.65388	1.5293	5.5661	6.2819	0.064912	0.023633	0.037690	187.23	20.118	10.107	7.6317
100.00	0.54438	0.75700	1.3210	5.5816	6.3007	0.064007	0.023931	0.038906	187.52	19.426	10.366	7.9239
102.00	0.62934	0.87260	1.1460	5.5945	6.3157	0.063120	0.024248	0.040288	187.67	18.773	10.632	8.2350
104.00	0.72348	1.0020	0.99799	5.6044	6.3265	0.062248	0.024586	0.041869	187.66	18.154	10.909	8.5675
106.00	0.82736	1.1468	0.87198	5.6112	6.3327	0.061385	0.024945	0.043694	187.51	17.565	11.198	8.9238
108.00	0.94154	1.3088	0.76404	5.6145	6.3339	0.060526	0.025329	0.045820	187.20	17.001	11.502	9.3073
110.00	1.0666	1.4903	0.67102	5.6138	6.3295	0.059665	0.025741	0.048326	186.73	16.458	11.828	9.7221
112.00	1.2031	1.6938	0.59039	5.6088	6.3191	0.058797	0.026186	0.051322	186.11	15.930	12.181	10.173

114.00	1.3517	1.9228	0.52008	5.5986	6.3016	0.057914	0.026671	0.054966	185.33	15.411	12.569	10.667
116.00	1.5130	2.1815	0.45841	5.5827	6.2762	0.057008	0.027203	0.059493	184.38	14.894	13.005	11.213
118.00	1.6877	2.4754	0.40397	5.5598	6.2416	0.056070	0.027794	0.065263	183.27	14.372	13.507	11.821
120.00	1.8765	2.8123	0.35558	5.5286	6.1959	0.055084	0.028462	0.072864	181.99	13.833	14.101	12.509
122.00	2.0802	3.2027	0.31224	5.4872	6.1367	0.054034	0.029229	0.083320	180.52	13.265	14.826	13.301
124.00	2.2997	3.6629	0.27301	5.4324	6.0602	0.052892	0.030133	0.098585	178.84	12.648	15.747	14.234
126.00	2.5360	4.2194	0.23700	5.3595	5.9605	0.051613	0.031233	0.12291	176.93	11.956	16.981	15.373
128.00	2.7904	4.9212	0.20320	5.2594	5.8264	0.050117	0.032636	0.16759	174.68	11.140	18.777	16.840
130.00	3.0647	5.8832	0.16998	5.1113	5.6322	0.048216	0.034579	0.27599	171.86	10.100	21.845	18.936
132.86	3.4982	10.850	0.092166	4.2912	4.6137	0.040039			0	6.1475		

Single-Phase Properties

100.00	0.10000	0.12298	8.1315	5.7653	6.5785	0.080014	0.021118	0.030153	201.29	17.820	10.075	6.9147
200.00	0.10000	0.060293	16.586	7.8674	9.5259	0.10048	0.020812	0.029239	288.05	5.3111	19.227	12.897
300.00	0.10000	0.040104	24.935	9.9522	12.446	0.11231	0.020833	0.029191	353.12	2.5186	26.605	17.731
400.00	0.10000	0.030062	33.265	12.045	15.371	0.12073	0.021028	0.029364	407.29	1.2653	33.106	21.870
500.00	0.10000	0.024045	41.588	14.169	18.328	0.12733	0.021479	0.029807	454.00	0.56244	39.272	25.540
100.00	1.0000	25.261	0.039586	1.1047	1.1443	0.012262	0.029062	0.064114	685.44	−0.14414	112.87	113.83
108.96	1.0000	23.349	0.042829	1.7009	1.7437	0.017998	0.028142	0.070627	577.22	0.070176	90.884	95.450
108.96	1.0000	1.3931	0.71782	5.6147	6.3325	0.060114	0.025522	0.046966	186.99	16.739	11.655	9.5017
200.00	1.0000	0.61727	1.6200	7.7647	9.3847	0.080819	0.020996	0.030510	286.20	5.1924	19.474	13.192
300.00	1.0000	0.40214	2.4867	9.8936	12.380	0.092976	0.020895	0.029646	354.42	2.4256	26.760	17.918
400.00	1.0000	0.29999	3.3334	12.005	15.338	0.10149	0.021064	0.029598	409.43	1.2088	33.222	22.024
500.00	1.0000	0.23962	4.1732	14.140	18.313	0.10812	0.021505	0.029948	456.39	0.52786	39.364	25.676
100.00	5.0000	25.864	0.038663	0.99666	1.1900	0.011154	0.029263	0.060925	737.92	−0.21740	152.30	112.19
200.00	5.0000	3.4130	0.29299	7.2656	8.7305	0.064994	0.021878	0.038000	285.27	4.3757	22.190	15.094
300.00	5.0000	2.0232	0.49426	9.6364	12.108	0.078767	0.021174	0.031673	362.95	2.0288	27.812	18.716
400.00	5.0000	1.4824	0.67458	11.834	15.207	0.087691	0.021224	0.030585	420.15	0.98413	33.871	22.588
500.00	5.0000	1.1786	0.84845	14.015	18.258	0.094498	0.021618	0.030535	467.56	0.39254	39.837	26.139
100.00	10.000	26.482	0.037761	0.88669	1.2643	0.0099878	0.029539	0.058409	792.04	−0.27800	200.46	110.33
200.00	10.000	7.4298	0.13459	6.5960	7.9419	0.056188	0.022832	0.048831	307.84	2.8854	34.772	19.114
300.00	10.000	4.0263	0.24837	9.3290	11.813	0.072068	0.021511	0.034036	379.01	1.5731	30.972	19.862
400.00	10.000	2.9079	0.34389	11.634	15.073	0.081462	0.021420	0.031689	435.68	0.74980	35.414	23.298
500.00	10.000	2.3052	0.43381	13.870	18.208	0.088461	0.021755	0.031188	482.48	0.25486	40.797	26.682
100.00	50.000	29.422	0.033988	0.39153	2.0910	0.0040257	0.031398	0.052094	1066.8	−0.43831	567.46	99.463
200.00	50.000	20.591	0.048566	4.5424	6.9707	0.038097	0.025036	0.045541	706.01	−0.28689	256.88	50.929
300.00	50.000	14.766	0.067725	7.7949	11.181	0.055259	0.023212	0.039083	609.73	−0.21242	139.18	34.086
400.00	50.000	11.439	0.087418	10.518	14.889	0.065951	0.022620	0.035519	604.51	−0.27153	94.476	31.319
500.00	50.000	9.3865	0.10654	13.024	18.350	0.073681	0.022659	0.033937	622.30	−0.36911	76.899	32.167
100.00	100.00	31.474	0.031772	0.095937	3.2732	−0.00053857	0.033037	0.050530	1282.4	−0.47725	1005.7	90.560
200.00	100.00	24.888	0.040181	3.9022	7.9203	0.031951	0.026437	0.043352	987.69	−0.50923	536.84	73.380
300.00	100.00	20.200	0.049505	7.0625	12.013	0.048608	0.024358	0.038799	866.60	−0.54516	331.69	54.648
400.00	100.00	16.970	0.058928	9.8487	15.741	0.059353	0.023555	0.036051	822.33	−0.59581	229.18	45.748
500.00	100.00	14.662	0.068206	12.449	19.269	0.067230	0.023434	0.034683	808.91	−0.64123	173.58	42.504

The values in these tables were generated from the NIST REFPROP software (Lemmon, E. W., McLinden, M. O., and Huber, M. L., NIST Standard Reference Database 23: Reference Fluid Thermodynamic and Transport Properties—REFPROP, National Institute of Standards and Technology, Standard Reference Data Program, Gaithersburg, Md., 2002, Version 7.1). The primary source for the thermodynamic properties is Lemmon, E. W., and Span, R., "Short Fundamental Equations of State for 20 Industrial Fluids," *J. Chem. Eng. Data*, **51**(3):785–850, 2006. The source for viscosity and thermal conductivity is Version 9.08 of the NIST14 database.

Properties at the triple point temperature and the critical point temperature are given in the first and last entries of the saturation tables, respectively. In the single-phase table, when the temperature range for a given isobar includes a vapor-liquid phase boundary, the temperature of phase equilibrium is noted, and properties for both the saturated liquid and saturated vapor are given (with liquid properties given in the upper line). Lines are omitted from the temperature-pressure grid of the single-phase table, when the system would be in the solid phase or if there are potential problems with the source property surface.

The equation of state is valid from the triple point to 500 K with pressures to 100 MPa. At higher pressures, the deviations from the equation increase rapidly, and it is not recommended to use the equation above 100 MPa. The uncertainties in the equation are 0.3% in density (approaching 1% near the critical point), 0.2% in vapor pressure, and 2% in heat capacities. For viscosity, estimated uncertainty is 2%. For thermal conductivity, estimated uncertainty, except near the critical region, is 4–6%.

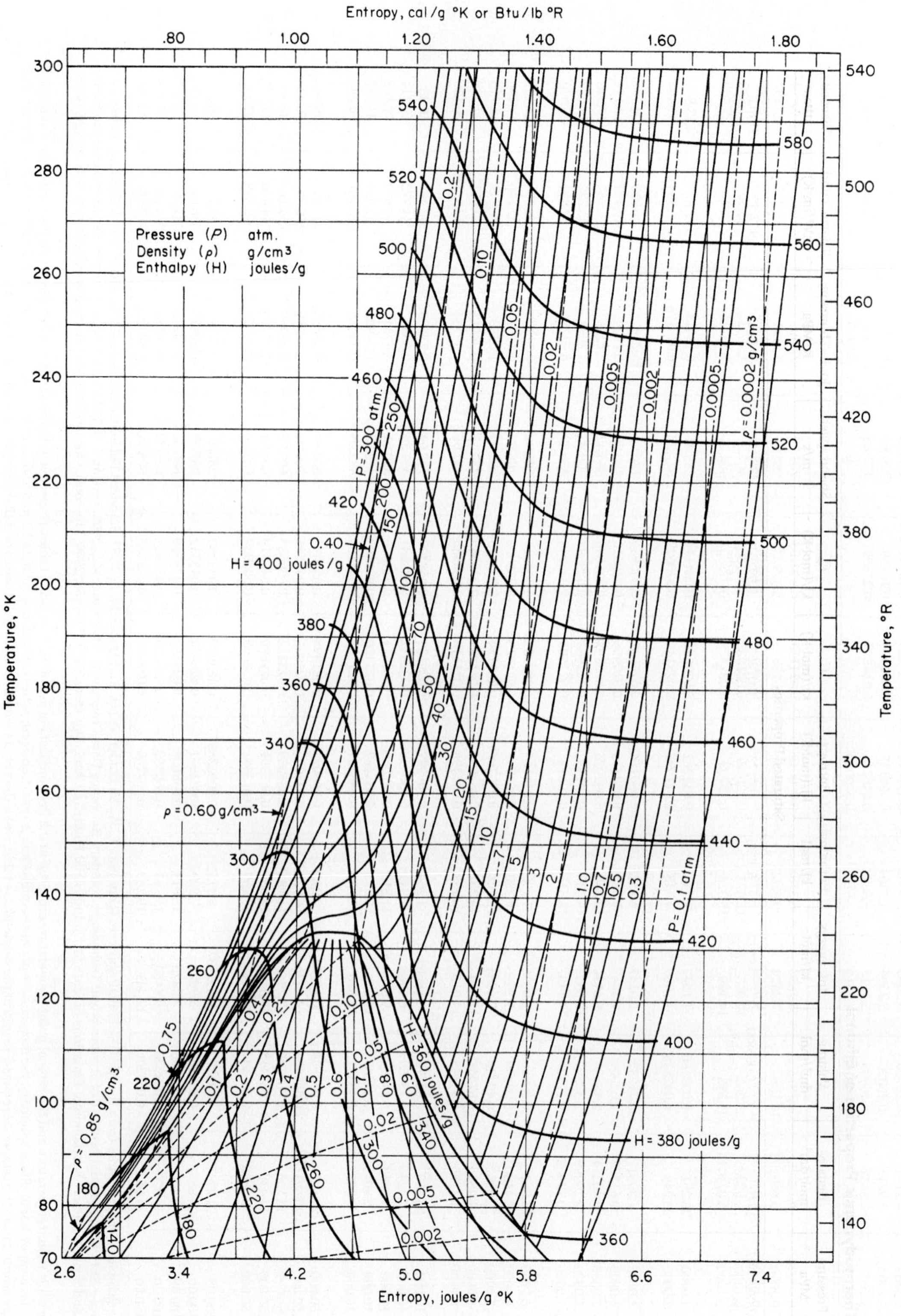

FIG. 2-4 Temperature-entropy diagram for carbon monoxide. Pressure *P*, in atmospheres; density ρ, in grams per cubic centimeter; enthalpy *H*, in joules per gram. (From J.G. Hust and R.B. Stewart, NBS Tech. Note 202, 1963.)

TABLE 2-115 Thermodynamic Properties of Ethanol

Temperature K	Pressure MPa	Density mol/dm^3	Volume dm^3/mol	Int. energy kJ/mol	Enthalpy kJ/mol	Entropy kJ/(mol·K)	C_v kJ/(mol·K)	C_p kJ/(mol·K)	Sound speed m/s	Joule-Thomson K/MPa	Therm. cond. mW/(m·K)	Viscosity μPa·s
Saturated Properties												
250.00	0.00027007	17.911	0.055831	6.9274	6.9275	0.037330	0.076657	0.093612	1325.0	−0.44553	178.12	3140.9
265.00	0.00089527	17.642	0.056681	8.3792	8.3793	0.042968	0.083798	0.10028	1260.8	−0.41423	173.58	2182.0
280.00	0.0025823	17.376	0.057551	9.9424	9.9426	0.048704	0.091653	0.10829	1202.8	−0.37872	169.56	1564.4
295.00	0.0066146	17.106	0.058460	11.630	11.631	0.054574	0.099433	0.11678	1149.2	−0.34323	165.87	1152.5
310.00	0.015298	16.828	0.059426	13.445	13.446	0.060574	0.10670	0.12524	1098.1	−0.30910	162.38	869.40
325.00	0.032394	16.537	0.060469	15.385	15.387	0.066684	0.11322	0.13340	1048.0	−0.27615	159.01	669.49
340.00	0.063544	16.231	0.061610	17.444	17.448	0.072875	0.11893	0.14115	997.94	−0.24356	155.69	524.87
355.00	0.11663	15.905	0.062872	19.615	19.622	0.079123	0.12381	0.14847	947.31	−0.21011	152.39	417.88
370.00	0.20205	15.557	0.064281	21.892	21.905	0.085403	0.12792	0.15543	895.56	−0.17428	149.09	337.04
385.00	0.33279	15.181	0.065871	24.268	24.290	0.091699	0.13130	0.16215	842.31	−0.13410	145.78	274.76
400.00	0.52446	14.774	0.067684	26.740	26.775	0.098000	0.13405	0.16883	787.16	−0.086812	142.47	225.91
415.00	0.79509	14.331	0.069779	29.307	29.362	0.10430	0.13625	0.17576	729.67	−0.028333	139.18	186.93
430.00	1.1649	13.843	0.072241	31.970	32.054	0.11061	0.13798	0.18341	669.25	0.047976	135.93	155.35
445.00	1.6559	13.298	0.075202	34.737	34.862	0.11695	0.13934	0.19262	605.07	0.15384	132.78	129.37
460.00	2.2916	12.676	0.078889	37.629	37.810	0.12335	0.14041	0.20504	535.80	0.31228	129.85	107.62
475.00	3.0963	11.941	0.083745	40.684	40.943	0.12991	0.14134	0.22469	459.19	0.57597	127.41	88.972
490.00	4.0954	11.007	0.090848	44.002	44.374	0.13684	0.14234	0.26508	371.03	1.0976	126.33	72.213
505.00	5.3159	9.5842	0.10434	47.926	48.480	0.14485	0.14382	0.41790	264.74	2.5369	129.43	55.104
513.90	6.1480	5.9910	0.16692	53.880	54.906	0.15723			0	8.6373		
250.00	0.00027007	0.00012998	7693.7	49.039	51.116	0.21409	0.058885	0.067215	226.86	149.30	14.936	7.2715
265.00	0.00089527	0.00040670	2458.8	49.932	52.134	0.20808	0.060795	0.069146	233.03	111.11	15.737	7.7433
280.00	0.0025823	0.0011115	899.69	50.851	53.174	0.20310	0.062753	0.071149	238.89	87.283	16.612	8.2114
295.00	0.0066146	0.0027080	369.28	51.792	54.234	0.19899	0.064753	0.073238	244.41	71.858	17.566	8.6756
310.00	0.015298	0.0059814	167.18	52.749	55.307	0.19561	0.066816	0.075464	249.49	61.180	18.602	9.1353
325.00	0.032394	0.012150	82.305	53.717	56.383	0.19282	0.068988	0.077921	254.02	53.164	19.731	9.5902
340.00	0.063544	0.022975	43.525	54.684	57.450	0.19053	0.071336	0.080736	257.88	46.697	20.969	10.040
355.00	0.11663	0.040873	24.466	55.640	58.494	0.18862	0.073932	0.084059	260.92	41.226	22.341	10.486
370.00	0.20205	0.069039	14.485	56.573	59.500	0.18701	0.076838	0.088058	263.02	36.486	23.886	10.929
385.00	0.33279	0.11160	8.9606	57.469	60.451	0.18562	0.080106	0.092930	264.03	32.353	25.659	11.372
400.00	0.52446	0.17385	5.7521	58.312	61.329	0.18438	0.083774	0.098936	263.82	28.756	27.741	11.820
415.00	0.79509	0.26261	3.8080	59.087	62.115	0.18322	0.087876	0.10646	262.27	25.644	30.251	12.283
430.00	1.1649	0.38683	2.5851	59.774	62.785	0.18208	0.092450	0.11610	259.21	22.967	33.369	12.774
445.00	1.6559	0.55876	1.7897	60.348	63.312	0.18088	0.097544	0.12898	254.44	20.681	37.377	13.318
460.00	2.2916	0.79629	1.2558	60.777	63.654	0.17954	0.10323	0.14727	247.61	18.747	42.735	13.961
475.00	3.0963	1.1286	0.88602	61.004	63.747	0.17792	0.10966	0.17612	238.10	17.136	50.248	14.786
490.00	4.0954	1.6143	0.61945	60.916	63.453	0.17578	0.11709	0.23200	224.59	15.831	61.578	15.982
505.00	5.3159	2.4339	0.41086	60.144	62.328	0.17228	0.12644	0.42053	203.70	14.728	82.512	18.148
513.90	6.1480	5.9910	0.16692	53.880	54.906	0.15723			0	8.6373		

(*Continued*)

TABLE 2-115 Thermodynamic Properties of Ethanol *(Continued)*

Temperature K	Pressure MPa	Density mol/dm³	Volume dm³/mol	Int. energy kJ/mol	Enthalpy kJ/mol	Entropy kJ/(mol·K)	C_v kJ/(mol·K)	C_p kJ/(mol·K)	Sound speed m/s	Joule-Thomson K/MPa	Therm. cond. mW/(m·K)	Viscosity μPa·s
						Single-Phase Properties						
300.00	0.10000	17.016	0.058768	12.219	12.225	0.056554	0.10193	0.11962	1132.5	−0.33179	164.74	1047.2
351.05	0.10000	15.993	0.062527	19.033	19.040	0.077475	0.12261	0.14658	960.72	−0.21908	153.26	443.11
351.05	0.10000	0.035314	28.317	55.390	58.222	0.18909	0.073221	0.083127	260.21	42.587	21.965	10.369
400.00	0.10000	0.030577	32.704	59.207	62.477	0.20043	0.080640	0.089997	279.09	28.685	26.374	11.853
500.00	0.10000	0.024191	41.338	67.925	72.058	0.22176	0.092910	0.10162	312.39	11.830	37.865	14.768
600.00	0.10000	0.020086	49.786	77.796	82.775	0.24127	0.10403	0.11252	341.55	5.6356	52.622	17.543
300.00	1.0000	17.034	0.058707	12.202	12.261	0.056497	0.10191	0.11954	1137.9	−0.33273	165.24	1053.2
400.00	1.0000	14.795	0.067589	26.715	26.783	0.097937	0.13400	0.16857	791.85	−0.089821	142.87	227.32
423.85	1.0000	14.049	0.071181	30.866	30.938	0.10802	0.13732	0.18015	694.41	0.013963	137.25	167.52
423.85	1.0000	0.33095	3.0216	59.504	62.526	0.18255	0.090516	0.11184	260.65	24.014	32.003	12.568
500.00	1.0000	0.25567	3.9114	67.014	70.925	0.20078	0.096953	0.11008	300.64	12.007	39.539	14.859
600.00	1.0000	0.20473	4.8846	77.311	82.195	0.22131	0.10581	0.11605	337.33	5.6301	53.583	17.678
300.00	5.0000	17.111	0.058443	12.129	12.421	0.056249	0.10185	0.11922	1161.4	−0.33665	167.43	1079.6
400.00	5.0000	14.961	0.066842	26.516	26.851	0.097435	0.13359	0.16658	829.44	−0.11211	146.07	238.82
500.00	5.0000	10.220	0.097846	46.419	46.908	0.14179	0.14311	0.32410	308.88	1.7596	127.42	61.882
501.39	5.0000	10.013	0.099875	46.876	47.375	0.14272	0.14340	0.35152	292.31	2.0063	128.00	59.510
501.39	5.0000	2.1809	0.45852	60.445	62.737	0.17336	0.12389	0.34099	209.80	15.000	75.676	17.454
600.00	5.0000	1.1372	0.87939	74.966	79.363	0.20395	0.11419	0.13659	314.06	5.6703	61.725	18.972
300.00	10.000	17.203	0.058131	12.041	12.623	0.055950	0.10179	0.11885	1189.5	−0.34096	170.07	1111.8
400.00	10.000	15.147	0.066020	26.293	26.953	0.096860	0.13313	0.16456	872.36	−0.13414	149.80	252.40
500.00	10.000	11.521	0.086800	44.752	45.620	0.13830	0.14031	0.22204	464.50	0.60618	130.15	80.680
600.00	10.000	2.8001	0.35713	71.266	74.837	0.19172	0.12599	0.18744	273.66	5.6926	84.190	23.411
300.00	100.00	18.389	0.054380	10.984	16.422	0.051802	0.10149	0.11571	1558.1	−0.37198	207.54	1611.3
400.00	100.00	17.030	0.058722	24.075	29.947	0.090466	0.12901	0.15081	1348.2	−0.25352	195.29	435.09
500.00	100.00	15.408	0.064899	39.356	45.846	0.12589	0.13221	0.16433	1166.1	−0.17199	188.35	192.15
600.00	100.00	13.601	0.073523	55.055	62.407	0.15608	0.12575	0.16553	1015.1	−0.082822	187.31	109.49
300.00	200.00	19.244	0.051963	10.349	20.742	0.048505	0.10196	0.11495	1830.4	−0.37578	238.67	2085.4
400.00	200.00	18.138	0.055134	22.905	33.931	0.086238	0.12678	0.14539	1660.3	−0.28090	228.57	591.02
500.00	200.00	16.878	0.059250	37.295	49.145	0.12014	0.12868	0.15623	1525.6	−0.22946	224.49	269.30
600.00	200.00	15.505	0.064494	51.902	64.801	0.14869	0.12066	0.15566	1422.7	−0.19099	226.40	148.43

The values in these tables were generated from the NIST REFPROP software (Lemmon, E. W., McLinden, M. O., and Huber, M. L., NIST Standard Reference Database 23: Reference Fluid Thermodynamic and Transport Properties—REFPROP, National Institute of Standards and Technology, Standard Reference Data Program, Gaithersburg, Md., 2002, Version 7.1). The primary source for the thermodynamic properties is Dillon, H. E., and Penoncello, S. G., "A Fundamental Equation for Calculation of the Thermodynamic Properties of Ethanol," *Int. J. Thermophys.*, **25**(2):321–335, 2004. The source for viscosity is Kiselev, S. B., Ely, J. F., Abdulagatov, I. M., and Huber, M. L., "Generalized SAFT-DFT/DMT Model for the Thermodynamic, Interfacial, and Transport Properties of Associating Fluids: Application for *n*-Alkanols," *Ind. Eng. Chem. Res.*, **44**:6916–6927, 2005. The source for thermal conductivity is unpublished, 2004; however, the fit uses functional form found in Marsh, K., Perkins, R., and Ramires, M. L. V., "Measurement and Correlation of the Thermal Conductivity of Propane from 86 to 600 K at Pressures to 70 MPa," *J. Chem. Eng. Data*, **47**(4):932–940, 2002.

Properties at the critical point temperature are given in the last entry of the saturation tables. In the single-phase table, when the temperature range for a given isobar includes a vapor-liquid phase boundary, the temperature of phase equilibrium is noted, and properties for both the saturated liquid and saturated vapor are given (with liquid properties given in the upper line). Lines are omitted from the temperature-pressure grid of the single-phase table, when the system would be in the solid phase or if there are potential problems with the source property surface.

The uncertainties in the equation of state are 0.2% in density, 3% in heat capacities, 1% in speed of sound, and 0.5% in vapor pressure and saturation densities. The estimated uncertainty in the liquid phase along the saturation boundary is approximately 3%, increasing to 10% at pressures to 100 MPa, and is estimated at 10% in the vapor phase. The estimated uncertainty in the liquid phase is approximately 5% and is estimated as 10% in the vapor phase.

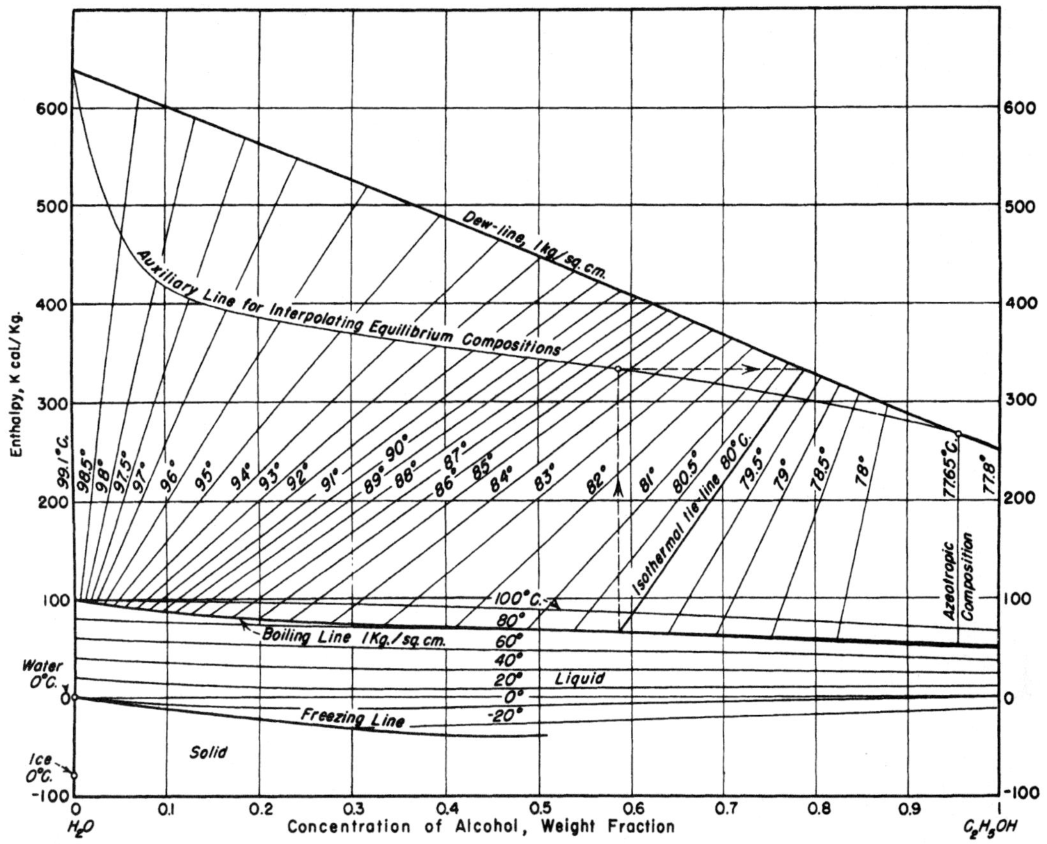

FIG. 2-5 Enthalpy-concentration diagram for aqueous ethyl alcohol. Reference states: Enthalpies of liquid water and ethyl alcohol at 0°C are zero. Note: In order to interpolate equilibrium compositions, a vertical may be erected from any liquid composition on the boiling line and its intersection with the auxiliary line determined. A horizontal from this intersection will establish the equilibrium vapor composition on the dew line. (F. Bosnjakovic, *Technische Thermodynamik*, T. Steinkopff, Leipzig, 1935.)

TABLE 2-116 Thermodynamic Properties of Normal Hydrogen

Temperature K	Pressure MPa	Density mol/dm³	Volume dm³/mol	Int. energy kJ/mol	Enthalpy kJ/mol	Entropy kJ/(mol·K)	C_v kJ/(mol·K)	C_p kJ/(mol·K)	Sound speed m/s	Joule-Thomson K/MPa	Therm. cond. mW/(m·K)	Viscosity μPa·s
						Saturated Properties						
13.957	0.0077031	38.148	0.026214	−0.10434	−0.10414	−0.0059480	0.011064	0.015654	1361.1	−1.4137	76.293	25.463
14.000	0.0078936	38.129	0.026226	−0.10367	−0.10346	−0.0059000	0.010957	0.015547	1359.6	−1.4230	76.650	25.310
15.000	0.013436	37.701	0.026524	−0.088896	−0.088539	−0.0048799	0.0096961	0.014420	1318.5	−1.5204	84.106	22.215
16.000	0.021534	37.261	0.026838	−0.074446	−0.073868	−0.0039471	0.0096482	0.014709	1271.6	−1.4695	90.079	19.784
17.000	0.032848	36.802	0.027172	−0.059414	−0.058521	−0.0030355	0.010003	0.015557	1226.6	−1.3623	94.784	17.815
18.000	0.048078	36.321	0.027533	−0.043440	−0.042116	−0.0021219	0.010462	0.016642	1185.5	−1.2409	98.405	16.182
19.000	0.067960	35.812	0.027923	−0.026375	−0.024477	−0.0011983	0.010915	0.017842	1147.3	−1.1194	101.10	14.799
20.000	0.093249	35.274	0.028350	−0.0081516	−0.0055080	−0.00026211	0.011323	0.019120	1110.7	−1.0003	103.01	13.607
21.000	0.12472	34.702	0.028817	0.011274	0.014868	0.00068790	0.011677	0.020476	1074.7	−0.88232	104.24	12.565
22.000	0.16314	34.092	0.029333	0.031947	0.036732	0.0016528	0.011978	0.021935	1038.2	−0.76268	104.87	11.641
23.000	0.20932	33.439	0.029905	0.053929	0.060188	0.0026344	0.012235	0.023539	1000.5	−0.63795	104.98	10.811
24.000	0.26406	32.738	0.030546	0.077308	0.085375	0.0036357	0.012457	0.025351	960.99	−0.50414	104.60	10.057
25.000	0.32818	31.979	0.031271	0.10222	0.11248	0.0046610	0.012655	0.027465	919.10	−0.35648	103.79	9.3625
26.000	0.40250	31.152	0.032101	0.12884	0.14176	0.0057167	0.012840	0.030024	874.29	−0.18882	102.53	8.7151
27.000	0.48788	30.242	0.033067	0.15744	0.17357	0.0068122	0.013025	0.033265	826.00	0.0073109	100.83	8.1034
28.000	0.58524	29.225	0.034217	0.18843	0.20846	0.0079614	0.013224	0.037610	773.58	0.24446	98.654	7.5160
29.000	0.69554	28.067	0.035629	0.22245	0.24723	0.0091865	0.013460	0.043909	716.22	0.54283	95.935	6.9409
30.000	0.81989	26.706	0.037444	0.26061	0.29132	0.010527	0.013764	0.054194	652.73	0.93827	92.547	6.3620
31.000	0.95964	25.017	0.039973	0.30524	0.34360	0.012063	0.014198	0.074872	581.16	1.5038	88.221	5.7518
32.000	1.1168	22.637	0.044175	0.36302	0.41236	0.014035	0.014926	0.14185	497.24	2.4292	82.176	5.0391
33.190	1.3301	14.940	0.066934	0.53004	0.61907	0.020012			0	5.3208		
13.957	0.0077031	0.067540	14.806	0.68715	0.80120	0.058918	0.013157	0.021964	304.61	31.943	10.375	0.66345
14.000	0.0078936	0.069018	14.489	0.68764	0.80201	0.058777	0.013129	0.021944	305.17	31.808	10.431	0.66695
15.000	0.013436	0.11050	9.0494	0.69864	0.82024	0.055705	0.012872	0.021898	316.15	28.572	11.624	0.74268
16.000	0.021534	0.16764	5.9651	0.70899	0.83745	0.053010	0.012907	0.022199	325.05	25.724	12.681	0.81064
17.000	0.032848	0.24349	4.1069	0.71875	0.85365	0.050622	0.012992	0.022618	333.00	23.407	13.681	0.87421
18.000	0.048078	0.34126	2.9303	0.72783	0.86871	0.048480	0.013083	0.023121	340.22	21.522	14.669	0.93555
19.000	0.067960	0.46437	2.1535	0.73614	0.88249	0.046537	0.013178	0.023724	346.75	19.961	15.675	0.99611
20.000	0.093249	0.61652	1.6220	0.74359	0.89484	0.044755	0.013280	0.024449	352.59	18.642	16.716	1.0569
21.000	0.12472	0.80187	1.2471	0.75005	0.90558	0.043103	0.013392	0.025329	357.75	17.507	17.806	1.1186
22.000	0.16314	1.0251	0.97549	0.75541	0.91455	0.041554	0.013514	0.026401	362.25	16.513	18.956	1.1819
23.000	0.20932	1.2919	0.77406	0.75951	0.92154	0.040085	0.013650	0.027724	366.11	15.629	20.180	1.2472
24.000	0.26406	1.6089	0.62153	0.76218	0.92630	0.038674	0.013802	0.029376	369.34	14.829	21.493	1.3151
25.000	0.32818	1.9848	0.50383	0.76318	0.92853	0.037303	0.013973	0.031482	371.95	14.091	22.916	1.3863
26.000	0.40250	2.4307	0.41141	0.76224	0.92783	0.035950	0.014167	0.034234	373.96	13.396	24.477	1.4619
27.000	0.48788	2.9618	0.33763	0.75895	0.92368	0.034594	0.014392	0.037960	375.38	12.726	26.218	1.5433
28.000	0.58524	3.6003	0.27775	0.75276	0.91531	0.033206	0.014655	0.043253	376.19	12.061	28.202	1.6331
29.000	0.69554	4.3810	0.22826	0.74277	0.90154	0.031749	0.014971	0.051322	376.39	11.374	30.535	1.7362
30.000	0.81989	5.3643	0.18642	0.72747	0.88031	0.030160	0.015358	0.065054	375.97	10.624	33.407	1.8628
31.000	0.95964	6.6763	0.14978	0.70374	0.84748	0.028317	0.015854	0.093486	374.91	9.7362	37.226	2.0375
32.000	1.1168	8.6823	0.11518	0.66274	0.79136	0.025879	0.016535	0.18606	373.31	8.5059	43.200	2.3378
33.190	1.3301	14.940	0.066934	0.53004	0.61907	0.020012			0	5.3208		

Single-Phase Properties

25.000	0.10000	0.50823	1.9676	0.81207	1.0088	0.049309	0.012734	0.022519	403.66	12.894	20.761	1.3142
100.00	0.10000	0.12030	8.3127	1.7949	2.6261	0.079163	0.014263	0.022637	808.92	1.4058	68.334	4.1896
175.00	0.10000	0.068680	14.560	3.0224	4.4785	0.092882	0.018150	0.026480	1026.9	0.13575	117.11	6.1845
250.00	0.10000	0.048077	20.800	4.4642	6.5442	0.10269	0.020003	0.028323	1209.1	−0.22980	160.59	7.9025
325.00	0.10000	0.036986	27.037	5.9947	8.6984	0.11022	0.020681	0.028998	1371.7	−0.38965	197.72	9.4561
400.00	0.10000	0.030054	33.273	7.5545	10.882	0.11626	0.020865	0.029180	1519.7	−0.47650	234.06	10.892
25.000	1.0000	32.746	0.030538	0.089693	0.12023	0.0041410	0.012580	0.025394	985.14	−0.51115	106.80	9.9923
31.268	1.0000	24.474	0.040861	0.31894	0.35980	0.012531	0.014353	0.084709	560.14	1.7031	86.829	5.5759
31.268	1.0000	7.1182	0.14049	0.69511	0.83559	0.027747	0.016014	0.10713	374.52	9.4548	38.524	2.0997
100.00	1.0000	1.2044	0.83027	1.7679	2.5982	0.059751	0.014331	0.023244	817.03	1.3036	70.413	4.2550
175.00	1.0000	0.68243	1.4653	3.0101	4.4754	0.073667	0.018190	0.026659	1035.8	0.11718	118.31	6.2213
250.00	1.0000	0.47788	2.0926	4.4576	6.5501	0.083515	0.020028	0.028401	1217.3	−0.23428	161.46	7.9283
325.00	1.0000	0.36797	2.7176	5.9910	8.7086	0.091062	0.020699	0.029039	1379.1	−0.39039	198.43	9.4759
400.00	1.0000	0.29924	3.3418	7.5525	10.894	0.097113	0.020878	0.029204	1526.4	−0.47605	234.65	10.908
25.000	5.0000	35.661	0.028042	0.046611	0.18682	0.0021443	0.012376	0.020610	1223.4	−0.90198	119.36	13.101
100.00	5.0000	5.9683	0.16755	1.6549	2.4927	0.045314	0.014583	0.025613	865.94	0.86369	80.395	4.5875
175.00	5.0000	3.3132	0.30183	2.9582	4.4673	0.059998	0.018352	0.027370	1077.3	0.032971	123.58	6.3871
250.00	5.0000	2.3268	0.42978	4.4292	6.5781	0.070022	0.020136	0.028723	1254.0	−0.25678	165.19	8.0420
325.00	5.0000	1.7990	0.55587	5.9750	8.7543	0.077631	0.020776	0.029211	1412.0	−0.39563	201.36	9.5631
400.00	5.0000	1.4680	0.68120	7.5440	10.950	0.083710	0.020937	0.029304	1556.2	−0.47537	237.09	10.979
25.000	10.000	37.930	0.026364	0.020221	0.28386	0.00059913	0.012222	0.018499	1402.1	−1.0762	131.12	16.625
100.00	10.000	11.417	0.087588	1.5346	2.4105	0.038585	0.014838	0.027423	955.43	0.39679	94.196	5.1692
175.00	10.000	6.3697	0.15699	2.9000	4.4699	0.053931	0.018532	0.028063	1133.3	−0.068718	130.39	6.6199
250.00	10.000	4.5028	0.22209	4.3966	6.6175	0.064133	0.020260	0.029065	1300.6	−0.28786	169.92	8.1898
325.00	10.000	3.5006	0.28567	5.9563	8.8130	0.071811	0.020867	0.029402	1453.0	−0.40519	205.05	9.6733
400.00	10.000	2.8687	0.34859	7.5339	11.020	0.077921	0.021006	0.029419	1593.1	−0.47687	240.15	11.067
100.00	50.000	31.993	0.031257	1.1768	2.7397	0.023964	0.016349	0.026254	1710.1	−0.57213	192.47	10.534
175.00	50.000	22.700	0.044053	2.6415	4.8442	0.039587	0.019545	0.029321	1632.6	−0.46415	189.52	9.1377
250.00	50.000	17.524	0.057066	4.2297	7.0830	0.050225	0.020999	0.030163	1690.7	−0.46214	211.67	9.7772
325.00	50.000	14.304	0.069911	5.8539	9.3494	0.058153	0.021434	0.030202	1784.8	−0.48443	237.63	10.827
400.00	50.000	12.107	0.082595	7.4773	11.607	0.064404	0.021458	0.029985	1887.2	−0.51016	267.11	11.965
175.00	100.00	33.019	0.030286	2.5589	5.5875	0.033643	0.020316	0.029140	2128.4	−0.52750	282.17	13.079
250.00	100.00	27.257	0.036688	4.1604	7.8292	0.044291	0.021603	0.030346	2125.4	−0.50698	303.19	12.218
325.00	100.00	23.228	0.043051	5.8083	10.113	0.052281	0.021923	0.030469	2170.6	−0.51215	327.83	12.546
400.00	100.00	20.261	0.049356	7.4555	12.391	0.058588	0.021864	0.030246	2235.8	−0.52481	356.83	13.289

The values in these tables were generated from the NIST REFPROP software (Lemmon, E. W., McLinden, M. O., and Huber, M. L., NIST Standard Reference Database 23: Reference Fluid Thermodynamic and Transport Properties—REFPROP, National Institute of Standards and Technology, Standard Reference Data Program, Gaithersburg, Md., 2002, Version 7.1). The primary source for the thermodynamic properties is Younglove, B. A., "Thermophysical Properties of Fluids. I. Argon, Ethylene, Parahydrogen, Nitrogen, Nitrogen Trifluoride, and Oxygen," *J. Phys. Chem. Ref. Data*, Suppl. 1, **11:** 1–11, 1982. The source for viscosity is McCarty, R. D., and Weber, L. A., "Thermophysical Properties of Parahydrogen from the Freezing Liquid Line to 5000 R for Pressures to 10,000 psia," *N.B.S. Tech. Note* 617, 1972. The source for thermal conductivity is McCarty, R. D., and Weber, L. A., "Thermophysical Properties of Parahydrogen from the Freezing Liquid Line to 5000 R for Pressures to 10,000 psia," *N.B.S. Tech. Note* 617, 1972.

Properties at the triple point temperature and the critical point temperature are given in the first and last entries of the saturation tables, respectively. In the single-phase table, when the temperature range for a given isobar includes a vapor-liquid phase boundary, the temperature of phase equilibrium is noted, and properties for both the saturated liquid and saturated vapor are given (with liquid properties given in the upper line). Lines are omitted from the temperature-pressure grid of the single-phase table, when the system would be in the solid phase or if there are potential problems with the source property surface.

The uncertainties in density are 0.1% in the liquid phase, 0.25% in the vapor phase, and 0.2% in the supercritical region. The uncertainty in heat capacity is 3%, and the uncertainty in speed of sound is 2% in the liquid phase and 1% elsewhere. The uncertainty in viscosity ranges from 4% to 15%. The uncertainty in thermal conductivity below 100 K is estimated to be 3% below 150 atm and up to 10% below 700 atm. For temperatures around 100 K at low densities, the uncertainty is about 1%. Above 100 K, the uncertainty is estimated to be on the order of 10%.

TABLE 2-117 Saturated Hydrogen Peroxide*

T, K	P, bar	v_f, m^3/kg	v_g, m^3/kg	h_f, kJ/kg	h_g, kJ/kg	s_f, kJ/(kg·K)	s_g, kJ/(kg·K)	c_{pf}, kJ/(kg·K)	μ_f, 10^{-4} Pa·s	k_f, W/(m·K)
273	0.0004	0.00068	1672	−5577	−4027	2.990	8.662	1.45	18.0	0.483
300	0.0031	0.00069	235	−5510	−3995	3.224	8.269	1.48	11.3	0.481
350	0.0564	0.00072	15.1	−5376	−3933	3.631	7.758	1.54	4.3	0.474
400	0.4521	0.00076	2.12	−5238	−3878	4.032	7.440	1.61	2.2	0.464
450	2.143	0.00081	0.487	−5091	−3820	4.346	7.172	1.68	1.3	0.453
500	7.126	0.00088	0.155	−4945	−3777	4.656	6.992	1.75	0.89	0.443
550	18.56	0.00095	0.0605	−4794	−3745	4.941	6.846	1.82	0.65	0.431
600	40.75	0.00107	0.0268	−4635	−3731	5.209	6.720	1.90	0.50	0.416
650	79.27	0.00125	0.0125	−4463	−3746	5.485	6.582			
700	141.7	0.00171	0.0048	−4195	−3860	5.682	6.339			
708.5^c	155.3	0.00284	0.0028	−4012	−4012	5.732	5.732			

*Values reproduced or converted from a tabulation by Tsykalo and Tabachnikov in V. A. Rabinovich (ed.), *Thermophysical Properties of Gases and Liquids*, Standartov, Moscow, 1968; NBS-NSF transl. TT 69-55091, 1970. The reader may be reminded that very pure hydrogen peroxide is very difficult to obtain owing to its decomposition or instability. c = critical point. The FMC Corp., Philadelphia, PA tech. bull. 67, 1969 (100 pp.) contains an enthalpy-pressure diagram to 3000 psia, 1100 K.

TABLE 2-118 Thermodynamic Properties of Hydrogen Sulfide

Temperature K	Pressure MPa	Density mol/dm³	Volume dm³/mol	Int. energy kJ/mol	Enthalpy kJ/mol	Entropy kJ/(mol·K)	C_v kJ/(mol·K)	C_p kJ/(mol·K)	Sound speed m/s	Joule-Thomson K/MPa	Therm. cond. mW/(m·K)	Viscosity μPa·s
						Saturated Properties						
187.70	0.023259	29.116	0.034345	−1.7210	−1.7202	−0.0085877	0.044390	0.068835	1437.8	−0.34039	254.24	439.13
190.00	0.027106	29.003	0.034479	−1.5628	−1.5619	−0.0077504	0.044124	0.068707	1425.8	−0.33923	251.74	428.67
200.00	0.050340	28.505	0.035082	−0.87841	−0.87664	−0.0042394	0.043042	0.068273	1373.4	−0.33253	240.93	385.68
210.00	0.087474	27.998	0.035717	−0.19759	−0.19446	−0.00091743	0.042067	0.068029	1321.0	−0.32287	230.26	346.75
220.00	0.14366	27.480	0.036390	0.48138	0.48661	0.0022415	0.041188	0.067975	1268.4	−0.30988	219.81	311.74
230.00	0.22485	26.949	0.037107	1.1602	1.1686	0.0052596	0.040393	0.068115	1215.6	−0.29305	209.52	280.37
240.00	0.33767	26.403	0.037875	1.8406	1.8534	0.0081564	0.039677	0.068461	1162.3	−0.27174	199.43	252.29
250.00	0.48934	25.838	0.038702	2.5245	2.5434	0.010949	0.039030	0.069032	1108.5	−0.24509	189.56	227.14
260.00	0.68751	25.253	0.039599	3.2135	3.2408	0.013654	0.038449	0.069859	1053.9	−0.21197	179.91	204.60
270.00	0.94022	24.642	0.040580	3.9100	3.9481	0.016285	0.037930	0.070989	998.54	−0.17084	170.48	184.32
280.00	1.2558	24.002	0.041662	4.6161	4.6685	0.018857	0.037470	0.072490	942.08	−0.11959	161.26	166.02
290.00	1.6429	23.327	0.042868	5.3348	5.4053	0.021385	0.037070	0.074466	884.34	−0.055218	152.24	149.44
300.00	2.1103	22.609	0.044230	6.0696	6.1629	0.023885	0.036732	0.077082	825.04	0.026636	143.40	134.32
310.00	2.6672	21.838	0.045791	6.8248	6.9469	0.026373	0.036462	0.080603	763.84	0.13260	134.71	120.43
320.00	3.3233	21.000	0.047618	7.6068	7.7650	0.028873	0.036273	0.085498	700.23	0.27324	126.16	107.58
330.00	4.0889	20.073	0.049818	8.4246	8.6283	0.031414	0.036191	0.092666	633.51	0.46655	117.71	95.533
340.00	4.9755	19.021	0.052573	9.2932	9.5548	0.034044	0.036265	0.10410	562.59	0.74618	109.36	84.050
350.00	5.9969	17.776	0.056256	10.241	10.578	0.036848	0.036600	0.12534	485.59	1.1841	101.19	72.784
360.00	7.1713	16.172	0.061837	11.335	11.779	0.040035	0.037471	0.17963	398.86	1.9714	93.864	61.060
370.00	8.5294	13.436	0.074429	12.903	13.538	0.044599	0.040079	0.63367	292.76	3.9324	92.754	46.102
373.10	8.9987	10.190	0.098135	14.470	15.353	0.049374			0	6.3885		
187.70	0.023259	0.015024	66.559	16.328	17.876	0.095815	0.025347	0.034000	245.84	55.730	10.628	8.0025
190.00	0.027106	0.017314	57.758	16.382	17.947	0.094930	0.025386	0.034078	247.20	53.868	10.775	8.1053
200.00	0.050340	0.030704	32.569	16.611	18.250	0.091395	0.025586	0.034487	252.82	46.796	11.429	8.5566
210.00	0.087474	0.051165	19.545	16.832	18.541	0.088301	0.025837	0.035021	257.96	41.090	12.107	9.0159
220.00	0.14366	0.080932	12.356	17.043	18.818	0.085567	0.026142	0.035698	262.58	36.435	12.816	9.4844
230.00	0.22485	0.12253	8.1613	17.244	19.079	0.083129	0.026502	0.036537	266.64	32.601	13.566	9.9634
240.00	0.33767	0.17879	5.5933	17.431	19.320	0.080933	0.026917	0.037563	270.10	29.412	14.365	10.455
250.00	0.48934	0.25286	3.9547	17.604	19.539	0.078934	0.027388	0.038807	272.91	26.737	15.227	10.961
260.00	0.68751	0.34834	2.8707	17.761	19.735	0.077092	0.027914	0.040312	275.05	24.476	16.166	11.485
270.00	0.94022	0.46937	2.1305	17.899	19.902	0.075375	0.028496	0.042139	276.47	22.550	17.202	12.031
280.00	1.2558	0.62086	1.6107	18.016	20.039	0.073752	0.029136	0.044378	277.15	20.897	18.360	12.604
290.00	1.6429	0.80887	1.2363	18.108	20.139	0.072193	0.029838	0.047166	277.05	19.466	19.675	13.213
300.00	2.1103	1.0411	0.96050	18.171	20.198	0.070669	0.030608	0.050723	276.12	18.212	21.197	13.867
310.00	2.6672	1.3280	0.75300	18.199	20.207	0.069149	0.031458	0.055410	274.34	17.097	22.997	14.582
320.00	3.3233	1.6843	0.59373	18.183	20.156	0.067594	0.032407	0.061879	271.64	16.081	25.187	15.380
330.00	4.0889	2.1323	0.46898	18.109	20.027	0.065956	0.033485	0.071400	268.00	15.116	27.946	16.300
340.00	4.9755	2.7096	0.36906	17.957	19.793	0.064157	0.034746	0.086837	263.35	14.142	31.600	17.405
350.00	5.9969	3.4881	0.28669	17.684	19.403	0.062064	0.036293	0.11617	257.65	13.053	36.820	18.833
360.00	7.1713	4.6442	0.21532	17.192	18.736	0.059360	0.038364	0.19265	250.84	11.629	45.513	20.940
370.00	8.5294	6.9933	0.14299	16.046	17.266	0.054674	0.041755	0.80649	242.80	9.0701	70.939	25.604
373.10	8.9987	10.190	0.098135	14.470	15.353	0.049374			0	6.3885		

(*Continued*)

TABLE 2-118 Thermodynamic Properties of Hydrogen Sulfide *(Continued)*

Temperature K	Pressure MPa	Density mol/dm³	Volume dm³/mol	Int. energy kJ/mol	Enthalpy kJ/mol	Entropy kJ/(mol·K)	C_v kJ/(mol·K)	C_p kJ/(mol·K)	Sound speed m/s	Joule-Thomson K/MPa	Therm. cond. mW/(m·K)	Viscosity μPa·s
						Single-Phase Properties						
200.00	0.10000	28.506	0.035080	−0.87902	−0.87551	−0.0042425	0.043042	0.068269	1373.6	−0.33258	240.95	385.79
212.60	0.10000	27.865	0.035888	−0.021243	−0.017654	−0.000082766	0.041830	0.067997	1307.4	−0.31984	227.54	337.30
212.60	0.10000	0.057900	17.271	16.888	18.615	0.087559	0.025911	0.035183	259.21	39.791	12.288	9.1366
300.00	0.10000	0.040389	24.759	19.164	21.640	0.099486	0.025979	0.034563	309.73	16.968	17.999	12.954
400.00	0.10000	0.030157	33.160	21.830	25.146	0.10956	0.027268	0.035693	356.35	8.9467	24.990	17.172
500.00	0.10000	0.024088	41.515	24.642	28.794	0.11770	0.028923	0.037297	396.06	5.5432	32.218	21.094
600.00	0.10000	0.020059	49.853	27.626	32.611	0.12465	0.030708	0.039059	431.20	3.7250	39.592	24.714
700.00	0.10000	0.017187	58.182	30.789	36.607	0.13081	0.032534	0.040873	463.05	2.6185	47.091	28.082
200.00	1.0000	28.528	0.035053	−0.89011	−0.85506	−0.0042980	0.043058	0.068210	1377.6	−0.33351	241.31	387.91
272.07	1.0000	24.513	0.040795	4.0550	4.0958	0.016821	0.037830	0.071266	986.97	−0.16116	168.56	180.39
272.07	1.0000	0.49800	2.0080	17.925	19.933	0.075033	0.028623	0.042564	276.67	22.188	17.430	12.147
300.00	1.0000	0.43539	2.2968	18.775	21.072	0.079019	0.027626	0.039427	296.60	17.337	19.015	13.337
400.00	1.0000	0.31003	3.2255	21.626	24.852	0.089907	0.027708	0.037234	351.09	9.0111	25.609	17.465
500.00	1.0000	0.24394	4.0993	24.507	28.606	0.098281	0.029100	0.038036	393.73	5.5366	32.691	21.319
600.00	1.0000	0.20183	4.9547	27.525	32.480	0.10534	0.030799	0.039492	430.30	3.7023	39.980	24.893
700.00	1.0000	0.17237	5.8015	30.710	36.511	0.11155	0.032589	0.041157	462.94	2.5947	47.423	28.227
200.00	5.0000	28.625	0.034935	−0.93837	−0.76369	−0.0045411	0.043127	0.067957	1394.7	−0.33745	242.90	397.26
300.00	5.0000	22.858	0.043749	5.9433	6.1620	0.023458	0.036740	0.075047	855.61	−0.017492	146.15	139.57
340.26	5.0000	18.992	0.052654	9.3164	9.5797	0.034113	0.036269	0.10449	560.70	0.75500	109.14	83.760
340.26	5.0000	2.7266	0.36675	17.952	19.786	0.064108	0.034782	0.087361	263.22	14.116	31.710	17.437
400.00	5.0000	1.8047	0.55412	20.549	23.319	0.073747	0.030043	0.048271	325.86	9.1749	29.786	19.070
500.00	5.0000	1.2939	0.77288	23.863	27.728	0.083609	0.029940	0.041996	384.33	5.4471	35.063	22.459
600.00	5.0000	1.0365	0.96476	27.065	31.889	0.091196	0.031216	0.041590	427.30	3.5816	41.773	25.773
700.00	5.0000	0.87211	1.1466	30.353	36.086	0.097665	0.032838	0.042472	463.26	2.4840	48.924	28.935
200.00	10.000	28.741	0.034793	−0.99643	−0.64850	−0.0048367	0.043212	0.067668	1415.5	−0.34197	244.82	408.90
300.00	10.000	23.238	0.043033	5.7496	6.1800	0.022795	0.036779	0.072377	902.78	−0.077399	150.49	148.08
400.00	10.000	5.0473	0.19812	18.370	20.351	0.062081	0.034651	0.10189	291.29	8.2243	44.719	23.639
500.00	10.000	2.8037	0.35667	22.959	26.526	0.076030	0.031080	0.048875	375.68	5.1487	38.963	24.438
600.00	10.000	2.1399	0.46730	26.466	31.139	0.084452	0.031755	0.044597	426.01	3.3812	44.210	27.165
700.00	10.000	1.7663	0.56617	29.903	35.564	0.091275	0.033155	0.044212	465.45	2.3347	50.878	30.013
300.00	75.000	26.050	0.038388	4.3332	7.2123	0.017506	0.037754	0.061705	1276.9	−0.33612	187.18	232.59
400.00	75.000	21.973	0.045510	9.9713	13.384	0.035260	0.035381	0.061962	983.51	−0.18247	134.39	124.22
500.00	75.000	17.947	0.055720	15.404	19.583	0.049092	0.034762	0.061649	786.35	0.057516	103.90	81.074
600.00	75.000	14.519	0.068874	20.474	25.640	0.060142	0.034994	0.059227	688.11	0.27314	87.712	62.531
700.00	75.000	11.974	0.083511	25.148	31.412	0.069045	0.035721	0.056291	654.55	0.36585	82.684	54.563
300.00	150.00	27.794	0.035979	3.5100	8.9069	0.013888	0.038777	0.058983	1538.4	−0.40376	214.24	311.25
400.00	150.00	24.751	0.040403	8.6429	14.703	0.030575	0.036402	0.057226	1302.6	−0.37006	165.64	174.85
500.00	150.00	21.937	0.045585	13.539	20.377	0.043238	0.035779	0.056273	1132.1	−0.31802	135.75	119.57
600.00	150.00	19.449	0.051416	18.248	25.960	0.053420	0.036044	0.055409	1019.1	−0.26874	118.38	93.323
700.00	150.00	17.335	0.057687	22.811	31.464	0.061906	0.036804	0.054711	949.06	−0.23292	110.03	79.581

The values in these tables were generated from the NIST REFPROP software (Lemmon, E. W., McLinden, M. O., and Huber, M. L., NIST Standard Reference Database 23: Reference Fluid Thermodynamic and Transport Properties—REFPROP, National Institute of Standards and Technology, Standard Reference Data Program, Gaithersburg, Md., 2002, Version 7.1). The primary source for the thermodynamic properties is Lemmon, E. W., and Span, R., "Short Fundamental Equations of State for 20 Industrial Fluids," *J. Chem. Eng. Data* **51**(3): 785–850, 2006. The source for viscosity and thermal conductivity is NIST14, Version 9.08.

Properties at the triple point temperature and the critical point temperature are given in the first and last entries of the saturation tables, respectively. In the single-phase table, when the temperature range for a given isobar includes a vapor-liquid phase boundary, the temperature of phase equilibrium is noted, and properties for both the saturated liquid and saturated vapor are given (with liquid properties given in the upper line). Lines are omitted from the temperature-pressure grid of the single-phase table, when the system would be in the solid phase or if there are potential problems with the source property surface.

The uncertainties in density are 0.1% in the liquid phase below the critical temperature, 0.4% in the vapor phase, 1% at supercritical temperatures up to 500 K, and 2.5% at higher temperatures. Uncertainties will be higher near the critical point, and may be lower than 0.5% between 400 and 500 K. The uncertainty in vapor pressure is 0.25%, and the uncertainty in heat capacities is estimated to be 1%. For viscosity, estimated uncertainty is 2%. For thermal conductivity, estimated uncertainty, except near the critical region, is 4–6%.

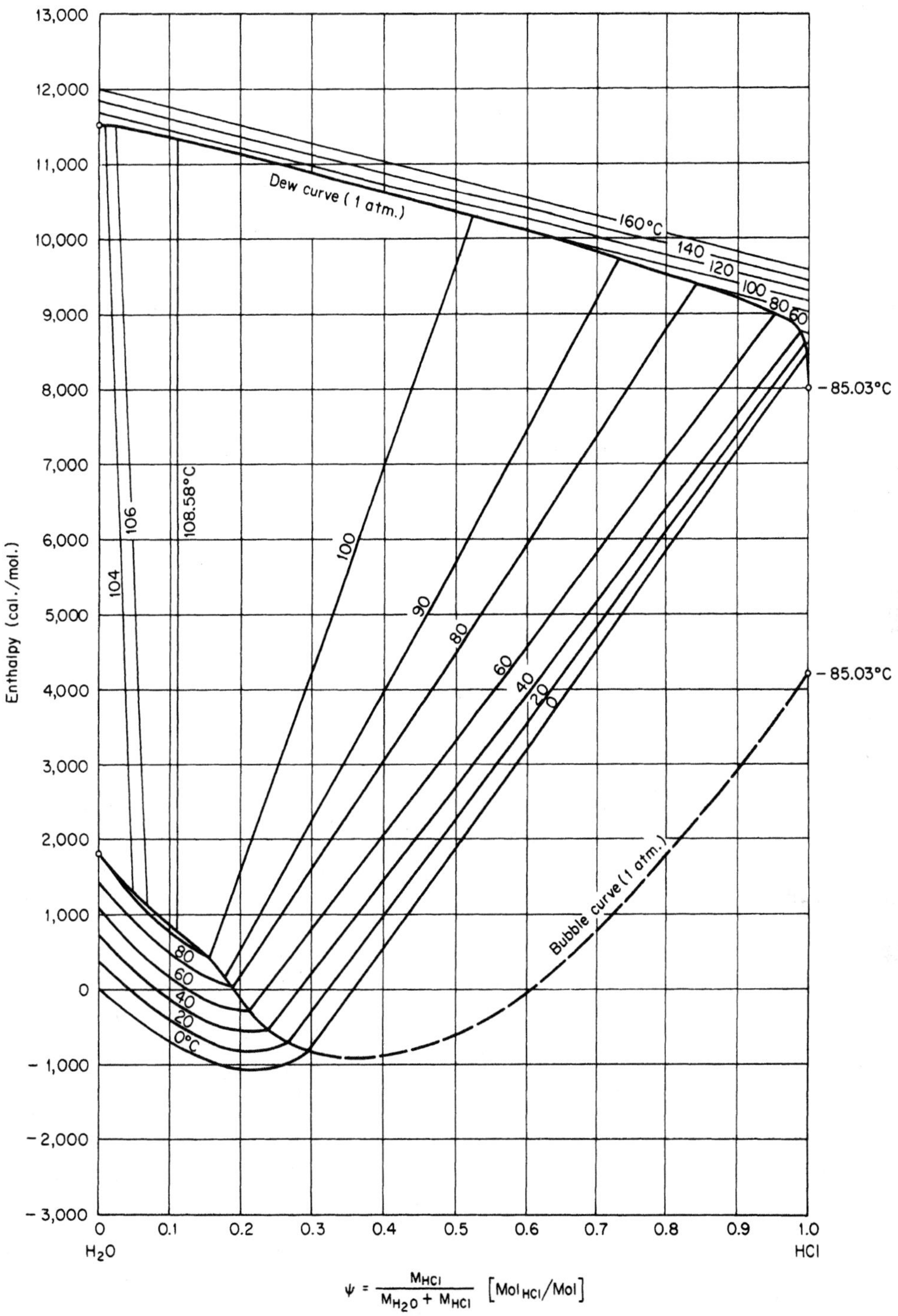

FIG. 2-6 Enthalpy-concentration diagram for aqueous hydrogen chloride at 1 atm. Reference states: enthalpy of liquid water at 0°C is zero; enthalpy of pure saturated HCl vapor at 1 atm (−85.03°C) is 8000 kcal/mol. Note: It should be observed that the weight basis includes the vapor, which is particularly important in the two-phase region. Saturation values may be read at the ends of the tie lines [C.C. Van Nuys, *Trans. Am. Inst. Chem. Eng* **39:** 663 (1943)].

TABLE 2-119 Thermodynamic Properties of Methane

Temperature K	Pressure MPa	Density mol/dm³	Volume dm³/mol	Int. energy kJ/mol	Enthalpy kJ/mol	Entropy kJ/(mol·K)	C_v kJ/(mol·K)	C_p kJ/(mol·K)	Sound speed m/s	Joule-Thomson K/MPa	Therm. cond. mW/(m·K)	Viscosity μPa·s
						Saturated Properties						
90.694	0.011696	28.142	0.035534	−1.1526	−1.1522	−0.011389	0.034776	0.054029	1538.6	−0.48191	211.24	204.52
100.00	0.034376	27.357	0.036554	−0.64728	−0.64602	−0.0060856	0.033908	0.054681	1452.0	−0.45812	199.67	155.78
105.00	0.056377	26.923	0.037143	−0.37306	−0.37097	−0.0034096	0.033500	0.055135	1403.9	−0.44202	193.03	136.86
110.00	0.088130	26.478	0.037768	−0.096585	−0.093257	−0.00083691	0.033115	0.055656	1354.7	−0.42328	186.18	121.34
115.00	0.13221	26.021	0.038431	0.18242	0.18750	0.0016441	0.032749	0.056253	1304.6	−0.40145	179.21	108.39
120.00	0.19143	25.551	0.039138	0.46425	0.47174	0.0040439	0.032400	0.056941	1253.5	−0.37589	172.15	97.432
125.00	0.26876	25.065	0.039896	0.74927	0.75999	0.0063722	0.032069	0.057741	1201.3	−0.34578	165.04	88.031
130.00	0.36732	24.562	0.040714	1.0379	1.0529	0.0086383	0.031757	0.058684	1148.1	−0.31006	157.91	79.868
135.00	0.49035	24.038	0.041600	1.3307	1.3511	0.010851	0.031469	0.059809	1093.6	−0.26735	150.78	72.699
140.00	0.64118	23.491	0.042569	1.6284	1.6557	0.013020	0.031206	0.061169	1037.7	−0.21579	143.65	66.333
145.00	0.82322	22.917	0.043636	1.9317	1.9676	0.015154	0.030974	0.062840	980.17	−0.15286	136.54	60.620
150.00	1.0400	22.309	0.044825	2.2418	2.2884	0.017264	0.030780	0.064932	920.85	−0.075032	129.43	55.437
155.00	1.2950	21.661	0.046165	2.5602	2.6199	0.019362	0.030631	0.067613	859.39	0.022798	122.32	50.682
160.00	1.5921	20.964	0.047702	2.8887	2.9647	0.021462	0.030541	0.071156	795.43	0.14836	115.19	46.266
165.00	1.9351	20.202	0.049500	3.2304	3.3262	0.023584	0.030531	0.076044	728.42	0.31398	108.01	42.105
170.00	2.3283	19.355	0.051667	3.5895	3.7098	0.025755	0.030634	0.083218	657.52	0.54087	100.73	38.115
175.00	2.7765	18.384	0.054394	3.9734	4.1244	0.028021	0.030920	0.094816	581.27	0.86918	93.324	34.196
180.00	3.2852	17.218	0.058078	4.3965	4.5873	0.030467	0.031554	0.11699	497.01	1.3866	85.799	30.193
185.00	3.8617	15.668	0.063825	4.8955	5.1420	0.033313	0.033085	0.17822	398.59	2.3397	78.733	25.773
190.00	4.5186	12.515	0.079902	5.7074	6.0685	0.038000	0.041746	1.5082	250.31	5.2488	96.970	18.982
190.56	4.5992	10.139	0.098628	6.2136	6.6672	0.041109			0	6.8877		
90.694	0.011696	0.015630	63.981	6.8310	7.5793	0.084885	0.025243	0.033851	249.13	47.921	8.8517	3.6388
100.00	0.034376	0.042048	23.782	7.0469	7.8644	0.079019	0.025487	0.034425	260.09	37.826	10.015	3.9976
105.00	0.056377	0.066154	15.116	7.1582	8.0104	0.076413	0.025652	0.034853	265.31	33.883	10.669	4.1951
110.00	0.088130	0.099622	10.038	7.2654	8.1501	0.074103	0.025842	0.035378	270.01	30.662	11.350	4.3964
115.00	0.13221	0.14457	6.9171	7.3680	8.2825	0.072036	0.026056	0.036016	274.17	28.004	12.062	4.6019
120.00	0.19143	0.20332	4.9183	7.4652	8.4067	0.070168	0.026295	0.036786	277.76	25.790	12.811	4.8123
125.00	0.26876	0.27844	3.5915	7.5562	8.5215	0.068464	0.026560	0.037714	280.76	23.928	13.604	5.0285
130.00	0.36732	0.37278	2.6825	7.6403	8.6257	0.066891	0.026854	0.038836	283.13	22.347	14.449	5.2517
135.00	0.49035	0.48962	2.0424	7.7165	8.7180	0.065421	0.027182	0.040203	284.86	20.993	15.355	5.4833
140.00	0.64118	0.63279	1.5803	7.7837	8.7970	0.064029	0.027549	0.041885	285.93	19.819	16.334	5.7254
145.00	0.82322	0.80691	1.2393	7.8406	8.8608	0.062694	0.027965	0.043985	286.31	18.789	17.402	5.9806
150.00	1.0400	1.0177	0.98256	7.8856	8.9074	0.061391	0.028439	0.046657	285.97	17.870	18.581	6.2526
155.00	1.2950	1.2728	0.78568	7.9166	8.9340	0.060098	0.028989	0.050144	284.88	17.035	19.904	6.5462
160.00	1.5921	1.5821	0.63206	7.9306	8.9369	0.058789	0.029636	0.054849	283.01	16.255	21.423	6.8688
165.00	1.9351	1.9603	0.51014	7.9238	8.9109	0.057431	0.030412	0.061496	280.30	15.500	23.225	7.2313
170.00	2.3283	2.4294	0.41163	7.8898	8.8482	0.055982	0.031374	0.071527	276.66	14.732	25.477	7.6515
175.00	2.7765	3.0268	0.33038	7.8184	8.7357	0.054371	0.032615	0.088273	271.99	13.896	28.545	8.1609
180.00	3.2852	3.8257	0.26139	7.6893	8.5480	0.052471	0.034338	0.12151	266.04	12.892	33.392	8.8251
185.00	3.8617	5.0137	0.19945	7.4515	8.2217	0.049961	0.037087	0.21701	258.03	11.492	43.706	9.8238
190.00	4.5186	7.8027	0.12816	6.7850	7.3641	0.044819	0.045796	2.2590	238.55	8.4951	119.40	12.455
190.56	4.5992	10.139	0.098628	6.2136	6.6672	0.041109			0	6.8877		

Single-Phase Properties

100.00	0.10000	27.360	0.036549	−0.64803	−0.64438	−0.0060931	0.033911	0.054672	1452.6	−0.45829	199.74	155.91
111.51	0.10000	26.341	0.037963	−0.012738	−0.0089413	−0.000079677	0.033003	0.055828	1339.7	−0.41705	184.09	117.20
111.51	0.10000	0.11186	8.9395	7.2969	8.1908	0.073456	0.025904	0.035558	271.33	29.808	11.561	4.4579
200.00	0.10000	0.060518	16.524	9.5570	11.209	0.093427	0.025259	0.033784	369.98	9.2893	21.941	7.8096
300.00	0.10000	0.040158	24.901	12.175	14.665	0.10741	0.027479	0.035869	449.74	4.3216	34.552	11.245
400.00	0.10000	0.030082	33.243	15.151	18.475	0.11834	0.032300	0.040652	510.56	2.2395	50.127	14.272
500.00	0.10000	0.024055	41.572	18.673	22.831	0.12803	0.038196	0.046533	561.86	1.2245	68.564	16.976
600.00	0.10000	0.020042	49.895	22.795	27.784	0.13705	0.044179	0.052509	608.04	0.68722	88.921	19.431
100.00	1.0000	27.403	0.036493	−0.65829	−0.62179	−0.0061960	0.033950	0.054562	1459.6	−0.46060	200.62	157.63
149.14	1.0000	22.416	0.044610	2.1878	2.2325	0.016902	0.030810	0.064535	931.21	−0.089695	130.66	56.297
149.14	1.0000	0.97852	1.0220	7.8788	8.9007	0.061614	0.028353	0.046147	286.08	18.022	18.368	6.2043
200.00	1.0000	0.64363	1.5537	9.3582	10.912	0.073276	0.025879	0.036730	357.81	9.5001	23.028	8.0145
300.00	1.0000	0.40776	2.4524	12.072	14.524	0.087922	0.027621	0.036721	447.04	4.2699	35.152	11.367
400.00	1.0000	0.30205	3.3108	15.083	18.393	0.099023	0.032360	0.041056	510.57	2.2001	50.558	14.357
500.00	1.0000	0.24058	4.1567	18.623	22.780	0.10879	0.038227	0.046766	562.99	1.1998	68.902	17.040
600.00	1.0000	0.20012	4.9971	22.755	27.752	0.11784	0.044198	0.052659	609.73	0.67124	89.200	19.483
100.00	5.0000	27.586	0.036250	−0.70190	−0.52065	−0.0066393	0.034116	0.054117	1490.0	−0.46993	204.45	165.28
200.00	5.0000	5.4706	0.18279	7.8197	8.7337	0.051495	0.032029	0.11667	291.29	8.9784	40.612	10.828
300.00	5.0000	2.1799	0.45874	11.590	13.884	0.072954	0.028262	0.041234	439.25	3.9428	38.480	12.194
400.00	5.0000	1.5333	0.65221	14.779	18.040	0.084897	0.032614	0.042903	513.11	2.0089	52.693	14.872
500.00	5.0000	1.2013	0.83240	18.401	22.563	0.094971	0.038361	0.047789	569.49	1.0870	70.509	17.410
600.00	5.0000	0.99281	1.0072	22.581	27.617	0.10417	0.044277	0.053309	618.13	0.60013	90.498	19.768
100.00	10.000	27.802	0.035969	−0.75239	−0.39270	−0.0071652	0.034314	0.053642	1525.7	−0.47979	209.07	174.83
200.00	10.000	16.593	0.060268	5.1551	5.7578	0.034542	0.030129	0.085085	567.92	1.0266	84.234	29.399
300.00	10.000	4.6859	0.21340	10.942	13.077	0.065137	0.028995	0.048165	444.53	3.2606	44.730	13.896
400.00	10.000	3.1002	0.32256	14.401	17.627	0.078246	0.032902	0.045220	522.58	1.7355	55.941	15.766
500.00	10.000	2.3887	0.41863	18.132	22.318	0.088698	0.038516	0.049007	580.99	0.94125	72.781	18.011
600.00	10.000	1.9619	0.50971	22.371	27.468	0.098073	0.044371	0.054070	630.63	0.51200	92.268	20.217
200.00	100.00	25.496	0.039222	3.0510	6.9732	0.020596	0.032058	0.048512	1541.0	−0.51619	188.05	80.392
300.00	100.00	21.266	0.047024	7.0865	11.789	0.040126	0.031823	0.048281	1267.5	−0.44889	137.68	47.835
400.00	100.00	17.881	0.055926	11.121	16.713	0.054276	0.035273	0.050523	1115.8	−0.37484	120.38	37.584
500.00	100.00	15.305	0.065340	15.405	21.939	0.065922	0.040312	0.054139	1044.8	−0.32811	120.87	33.590
600.00	100.00	13.357	0.074869	20.074	27.561	0.076160	0.045724	0.058364	1018.4	−0.30439	130.36	32.111
200.00	500.00	33.003	0.030301	2.3322	17.482	0.0061671	0.037832	0.047821	2664.2	−0.53926	429.60	205.24
300.00	500.00	30.786	0.032482	5.9505	22.192	0.025271	0.037006	0.047114	2500.0	−0.55416	358.93	106.90
400.00	500.00	28.929	0.034567	9.7401	27.024	0.039152	0.039890	0.049933	2360.3	−0.52806	312.36	78.768
500.00	500.00	27.331	0.036588	13.934	32.228	0.050747	0.044407	0.054280	2250.3	−0.49035	285.41	66.669
600.00	500.00	25.934	0.038559	18.612	37.892	0.061061	0.049344	0.059017	2168.1	−0.45514	272.14	60.413

The values in these tables were generated from the NIST REFPROP software (Lemmon, E. W., McLinden, M. O., and Huber, M. L., NIST Standard Reference Database 23: Reference Fluid Thermodynamic and Transport Properties—REFPROP, National Institute of Standards and Technology, Standard Reference Data Program, Gaithersburg, Md., 2002, Version 7.1). The primary source for the thermodynamic properties is Setzmann, U., and Wagner, W., "A New Equation of State and Tables of Thermodynamic Properties for Methane Covering the Range from the Melting Line to 625 K at Pressures up to 1000 MPa," *J. Phys. Chem. Ref. Data* **20**(6):1061–1151, 1991. The source for viscosity is Younglove, B. A., and Ely, J. F., "Thermophysical Properties of Fluids. II. Methane, Ethane, Propane, Isobutane and Normal Butane," *J. Phys. Chem. Ref. Data* **16**:577–798, 1987. The source for thermal conductivity is Friend, D. G., Ely, J. F., and Ingham, H., "Tables for the Thermophysical Properties of Methane," *NIST Tech. Note* 1325, 1989.

Properties at the triple point temperature and the critical point temperature are given in the first and last entries of the saturation tables, respectively. In the single-phase table, when the temperature range for a given isobar includes a vapor-liquid phase boundary, the temperature of phase equilibrium is noted, and properties for both the saturated liquid and saturated vapor are given (with liquid properties given in the upper line). Lines are omitted from the temperature-pressure grid of the single-phase table, when the system would be in the solid phase or if there are potential problems with the source property surface.

The uncertainties in density are 0.03% for pressures below 12 MPa and temperatures below 350 K and up to 0.07% for pressures less than 50 MPa. For the speed of sound, the uncertainty ranges from 0.03% (in the vapor phase) to 0.3% depending on temperature and pressure. Heat capacities may be generally calculated within an uncertainty of 1%. The uncertainty in viscosity is 2%, except in the critical region which is 5%. The uncertainty in thermal conductivity of the dilute gas between 130 and 625 K is 2.5%. For temperatures below 130 K, the uncertainty is less than 10%. Excluding the dilute gas, the uncertainty is 2% between 110 and 725 K at pressures up to 70 MPa, except near the critical point which has an uncertainty of 5% or greater. For the vapor at lower temperatures and the dense liquid near the triple point, an uncertainty of 10% is possible.

TABLE 2-120 Thermodynamic Properties of Methanol

Temperature K	Pressure MPa	Density mol/dm³	Volume dm³/mol	Int. energy kJ/mol	Enthalpy kJ/mol	Entropy kJ/(mol·K)	C_v kJ/(mol·K)	C_p kJ/(mol·K)	Sound speed m/s	Joule-Thomson K/MPa
Saturated Properties										
175.61	1.8635E-07	28.230	0.035423	−12.440	−12.440	−0.049524	0.056728	0.070390	1625.1	−0.40884
180.00	3.7619E-07	28.096	0.035592	−12.130	−12.130	−0.047781	0.056689	0.070750	1590.2	−0.40373
195.00	3.2175E-06	27.629	0.036194	−11.067	−11.067	−0.042108	0.056604	0.070897	1496.4	−0.39791
210.00	1.9841E-05	27.163	0.036815	−10.001	−10.001	−0.036846	0.057072	0.071215	1425.3	−0.39361
225.00	9.4330E-05	26.703	0.037449	−8.9277	−8.9277	−0.031908	0.057992	0.072004	1363.2	−0.38674
240.00	0.00036348	26.250	0.038096	−7.8395	−7.8395	−0.027226	0.059275	0.073141	1304.6	−0.37793
255.00	0.0011791	25.802	0.038756	−6.7318	−6.7318	−0.022750	0.060916	0.074617	1248.2	−0.36733
270.00	0.0033166	25.360	0.039432	−5.5991	−5.5990	−0.018434	0.062917	0.076487	1194.1	−0.35457
285.00	0.0082787	24.922	0.040125	−4.4351	−4.4347	−0.014239	0.065250	0.078803	1142.6	−0.33915
300.00	0.018682	24.484	0.040844	−3.2329	−3.2322	−0.010129	0.067864	0.081584	1093.5	−0.32073
315.00	0.038692	24.041	0.041595	−1.9858	−1.9842	−0.0060725	0.070693	0.084823	1046.3	−0.29904
330.00	0.074453	23.590	0.042390	−0.68700	−0.68385	−0.0020451	0.073674	0.088505	1000.2	−0.27385
345.00	0.13447	23.124	0.043244	0.66961	0.67543	0.0019747	0.076752	0.092616	954.32	−0.24479
360.00	0.22992	22.638	0.044174	2.0901	2.1003	0.0060049	0.079881	0.097164	907.64	−0.21121
375.00	0.37483	22.123	0.045203	3.5806	3.5975	0.010061	0.083033	0.10219	859.31	−0.17199
390.00	0.58617	21.571	0.046358	5.1475	5.1746	0.014159	0.086189	0.10776	808.48	−0.12529
405.00	0.88399	20.973	0.047681	6.7983	6.8404	0.018314	0.089346	0.11405	754.41	−0.068163
420.00	1.2914	20.315	0.049226	8.5423	8.6058	0.022546	0.092514	0.12133	696.47	0.0042669
435.00	1.8349	19.579	0.051075	10.392	10.485	0.026879	0.095722	0.13007	634.17	0.10021
450.00	2.5433	18.741	0.053360	12.364	12.500	0.031347	0.099032	0.14120	567.09	0.23444
465.00	3.4456	17.759	0.056310	14.488	14.682	0.036009	0.10257	0.15679	494.36	0.43762
480.00	4.5713	16.553	0.060411	16.820	17.096	0.040977	0.10666	0.18345	412.12	0.79465
495.00	5.9794	14.880	0.067203	19.521	19.923	0.046590	0.11250	0.25717	308.94	1.6506
510.00	7.7496	11.689	0.085547	23.297	23.960	0.054351	0.12653	1.1088	192.83	4.6061
513.38	8.2159	8.7852	0.11383	25.917	26.852	0.059911			0	6.7425
175.61	1.8635E-07	1.2764E-07	7,834,400.	28.219	29.679	0.19032	0.031874	0.040287	239.95	1187400.
180.00	3.7619E-07	2.5140E-07	3,977,700.	28.353	29.850	0.18544	0.032397	0.040854	242.62	857090.
195.00	3.2175E-06	1.9855E-06	503,660.	28.810	30.430	0.17069	0.035224	0.043954	251.06	293110.
210.00	1.9841E-05	1.1378E-05	87,892.	29.259	31.003	0.15841	0.040104	0.049389	258.49	105090.
225.00	9.4330E-05	5.0556E-05	19,780.	29.698	31.564	0.14806	0.047248	0.057480	265.30	39363.
240.00	0.00036348	0.00018304	5,463.4	30.123	32.109	0.13923	0.056324	0.067973	271.89	15552.
255.00	0.0011791	0.00056065	1,783.7	30.534	32.637	0.13164	0.066572	0.080135	278.43	6557.5
270.00	0.0033166	0.0014959	668.48	30.932	33.149	0.12508	0.077055	0.093000	284.90	2971.8
285.00	0.0082787	0.0035581	281.05	31.321	33.648	0.11938	0.086920	0.10564	291.19	1449.9
300.00	0.018682	0.0076845	130.13	31.703	34.134	0.11442	0.095581	0.11740	297.15	759.96
315.00	0.038692	0.015300	65.359	32.077	34.606	0.11009	0.10279	0.12798	302.63	426.34
330.00	0.074453	0.028438	35.164	32.442	35.060	0.10627	0.10860	0.13749	307.47	254.84
345.00	0.13447	0.049870	20.052	32.789	35.485	0.10287	0.11331	0.14644	311.48	161.53
360.00	0.22992	0.083267	12.009	33.108	35.869	0.099808	0.11736	0.15559	314.48	108.02
375.00	0.37483	0.13344	7.4940	33.385	36.194	0.096984	0.12125	0.16605	316.20	75.802
390.00	0.58617	0.20674	4.8370	33.601	36.436	0.094317	0.12546	0.17917	316.34	55.467
405.00	0.88399	0.31179	3.2073	33.736	36.571	0.091723	0.13033	0.19663	314.53	42.002
420.00	1.2914	0.46071	2.1706	33.767	36.570	0.089128	0.13587	0.21986	310.36	32.587
435.00	1.8349	0.67055	1.4913	33.687	36.423	0.086505	0.14101	0.24709	303.71	25.532
450.00	2.5433	0.96219	1.0393	33.541	36.184	0.083978	0.14238	0.26502	295.26	19.860
465.00	3.4456	1.3555	0.73775	33.439	35.981	0.081813	0.13589	0.25879	285.25	15.568
480.00	4.5713	1.9102	0.52352	33.258	35.652	0.079634	0.12618	0.27959	267.83	13.904
495.00	5.9794	2.9050	0.34423	32.267	34.325	0.075685	0.12608	0.42448	247.46	12.099
510.00	7.7496	5.1706	0.19340	29.688	31.187	0.068520	0.13259	1.9096	212.65	9.5115
513.38	8.2159	8.7852	0.11383	25.917	26.852	0.059911			0	6.7425

Single-Phase Properties

200.00	0.10000	27.474	0.036398	−10.713	−10.709	−0.040317	0.056702	0.070943	1471.5	−0.39677
300.00	0.10000	24.486	0.040839	−3.2341	−3.2300	−0.010133	0.067862	0.081580	1094.1	−0.32081
337.30	0.10000	23.366	0.042798	−0.034546	−0.030266	−0.000089518	0.075163	0.090451	977.93	−0.26023
337.30	0.10000	0.037626	26.577	32.613	35.271	0.10457	0.11100	0.14187	309.54	202.71
400.00	0.10000	0.030452	32.839	36.075	39.359	0.11581	0.044972	0.054208	349.19	40.941
500.00	0.10000	0.024157	41.396	40.921	45.060	0.12851	0.051823	0.060380	387.15	12.933
600.00	0.10000	0.020089	49.779	46.476	51.454	0.14014	0.059065	0.067441	420.71	4.3382
200.00	1.0000	27.491	0.036376	−10.720	−10.684	−0.040354	0.056724	0.070932	1475.1	−0.39705
300.00	1.0000	24.514	0.040793	−3.2472	−3.2064	−0.010176	0.067848	0.081541	1100.0	−0.32177
400.00	1.0000	21.193	0.047185	6.2298	6.2770	0.016901	0.088257	0.11177	775.46	−0.090229
409.75	1.0000	20.772	0.048143	7.3401	7.3883	0.019645	0.090347	0.11623	736.51	−0.047179
409.75	1.0000	0.35352	2.8287	33.758	36.586	0.090904	0.13203	0.20333	313.48	38.678
500.00	1.0000	0.25202	3.9680	40.335	44.303	0.10818	0.056676	0.068069	376.08	13.330
600.00	1.0000	0.20501	4.8778	46.300	51.178	0.12070	0.061344	0.070369	413.09	4.5635
200.00	5.0000	27.561	0.036283	−10.752	−10.571	−0.040517	0.056820	0.070883	1490.9	−0.39825
300.00	5.0000	24.635	0.040592	−3.3039	−3.1010	−0.010367	0.067795	0.081377	1125.5	−0.32568
400.00	5.0000	21.441	0.046640	6.0896	6.3228	0.016546	0.087676	0.11029	818.58	−0.11504
484.95	5.0000	16.076	0.062203	17.655	17.966	0.042725	0.10826	0.19836	381.15	0.98684
484.95	5.0000	2.1711	0.46060	33.047	35.350	0.078574	0.12499	0.32263	260.06	13.826
500.00	5.0000	1.7679	0.56566	35.907	38.735	0.085457	0.098975	0.17315	301.43	12.155
600.00	5.0000	1.1389	0.87808	45.247	49.638	0.10553	0.072927	0.087489	379.21	5.1009
200.00	10.000	27.648	0.036169	−10.791	−10.430	−0.040716	0.056935	0.070820	1509.9	−0.39966
300.00	10.000	24.779	0.040357	−3.3713	−2.9677	−0.010598	0.067746	0.081196	1155.3	−0.32990
400.00	10.000	21.717	0.046048	5.9321	6.3925	0.016141	0.087087	0.10880	865.91	−0.13884
500.00	10.000	15.932	0.062765	19.374	20.002	0.046226	0.10760	0.18959	424.68	0.83939
600.00	10.000	2.6640	0.37537	43.262	47.015	0.096406	0.088868	0.12122	343.60	5.0965
200.00	100.00	28.911	0.034588	−11.305	−7.8460	−0.043691	0.057827	0.068992	1772.1	−0.41976
300.00	100.00	26.630	0.037552	−4.2043	−0.44914	−0.013840	0.067889	0.079818	1515.8	−0.35799
400.00	100.00	24.493	0.040827	4.3449	8.4277	0.011565	0.082823	0.098951	1334.7	−0.26099
500.00	100.00	22.020	0.045413	14.917	19.458	0.036085	0.095694	0.12152	1164.1	−0.14407
600.00	100.00	19.139	0.052250	27.300	32.525	0.059862	0.10406	0.13787	977.50	−0.023905
300.00	500.00	30.547	0.032736	−5.4195	10.949	−0.022106	0.070293	0.080627	2316.0	−0.34762
400.00	500.00	29.154	0.034300	2.2795	19.430	0.0022308	0.077541	0.089761	2194.8	−0.30897
500.00	500.00	27.670	0.036140	11.020	29.089	0.023726	0.084883	0.10419	2123.2	−0.24751
600.00	500.00	26.003	0.038457	21.094	40.322	0.044161	0.092281	0.12017	2074.1	−0.19279

The values in these tables were generated from the NIST REFPROP software (Lemmon, E. W., McLinden, M. O., and Huber, M. L., NIST Standard Reference Database 23: Reference Fluid Thermodynamic and Transport Properties—REFPROP, National Institute of Standards and Technology, Standard Reference Data Program, Gaithersburg, Md., 2002, Version 7.1). The primary source for the thermodynamic properties is de Reuck, K. M., and Craven, R. J. B., "Methanol, International Thermodynamic Tables of the Fluid State—12," *IUPAC*, Blackwell Scientific Publications, London, 1993. Validated equations for the viscosity and thermal conductivity are not currently available for this fluid.

Properties at the triple point temperature and the critical point temperature are given in the first and last entries of the saturation tables, respectively. In the single-phase table, when the temperature range for a given isobar includes a vapor-liquid phase boundary, the temperature of phase equilibrium is noted, and properties for both the saturated liquid and saturated vapor are given (with liquid properties given in the upper line). Lines are omitted from the temperature-pressure grid of the single-phase table, when the system would be in the solid phase or if there are potential problems with the source property surface.

The uncertainties of the equation of state are generally 0.1% in density and 2% in the speed of sound, except in the critical region and high pressures.

TABLE 2-121 Thermodynamic Properties of Nitrogen

Temperature K	Pressure MPa	Density mol/dm³	Volume dm³/mol	Int. energy kJ/mol	Enthalpy kJ/mol	Entropy kJ/(mol·K)	C_v kJ/(mol·K)	C_p kJ/(mol·K)	Sound speed m/s	Joule-Thomson K/MPa	Therm. cond. mW/(m·K)	Viscosity μPa·s
						Saturated Properties						
63.151	0.012520	30.957	0.032303	−4.2230	−4.2226	0.067951	0.032951	0.056033	995.28	−0.40419	173.24	311.59
65.000	0.017404	30.685	0.032589	−4.1194	−4.1188	0.069569	0.032591	0.056121	976.36	−0.39833	169.51	282.07
67.000	0.024300	30.387	0.032909	−4.0071	−4.0063	0.071270	0.032207	0.056231	956.04	−0.39135	165.49	254.55
69.000	0.033213	30.085	0.033239	−3.8946	−3.8935	0.072924	0.031831	0.056360	935.83	−0.38364	161.47	230.85
71.000	0.044527	29.779	0.033581	−3.7819	−3.7804	0.074535	0.031463	0.056512	915.66	−0.37508	157.47	210.32
73.000	0.058656	29.468	0.033935	−3.6689	−3.6669	0.076105	0.031106	0.056690	895.49	−0.36560	153.46	192.43
75.000	0.076043	29.153	0.034302	−3.5556	−3.5530	0.077637	0.030760	0.056899	875.28	−0.35506	149.47	176.75
77.000	0.097152	28.832	0.034683	−3.4419	−3.4385	0.079133	0.030427	0.057142	855.00	−0.34334	145.48	162.94
79.000	0.12247	28.506	0.035080	−3.3278	−3.3235	0.080597	0.030105	0.057425	834.61	−0.33029	141.50	150.71
81.000	0.15251	28.175	0.035493	−3.2132	−3.2078	0.082030	0.029795	0.057752	814.07	−0.31574	137.55	139.82
83.000	0.18780	27.837	0.035924	−3.0980	−3.0913	0.083436	0.029499	0.058130	793.36	−0.29951	133.61	130.07
85.000	0.22886	27.492	0.036375	−2.9822	−2.9739	0.084815	0.029215	0.058566	772.44	−0.28135	129.66	121.31
87.000	0.27626	27.139	0.036847	−2.8657	−2.8555	0.086172	0.028944	0.059068	751.28	−0.26099	125.72	113.38
89.000	0.33055	26.779	0.037343	−2.7483	−2.7360	0.087507	0.028687	0.059647	729.84	−0.23813	121.77	106.18
91.000	0.39230	26.409	0.037865	−2.6301	−2.6152	0.088823	0.028444	0.060315	708.09	−0.21237	117.83	99.602
93.000	0.46210	26.030	0.038417	−2.5107	−2.4930	0.090123	0.028215	0.061088	685.99	−0.18326	113.89	93.568
95.000	0.54052	25.640	0.039002	−2.3902	−2.3691	0.091408	0.028001	0.061983	663.50	−0.15025	109.95	88.004
97.000	0.62817	25.238	0.039623	−2.2683	−2.2434	0.092682	0.027804	0.063026	640.57	−0.11264	106.02	82.847
99.000	0.72566	24.822	0.040288	−2.1449	−2.1156	0.093946	0.027624	0.064246	617.14	−0.069613	102.08	78.042
101.00	0.83358	24.390	0.041000	−2.0196	−1.9854	0.095204	0.027464	0.065684	593.17	−0.020100	98.144	73.543
103.00	0.95259	23.941	0.041769	−1.8923	−1.8525	0.096459	0.027327	0.067392	568.58	0.037239	94.208	69.306
105.00	1.0833	23.471	0.042605	−1.7625	−1.7163	0.097715	0.027214	0.069443	543.30	0.10414	90.272	65.292
107.00	1.2264	22.978	0.043520	−1.6298	−1.5765	0.098977	0.027133	0.071937	517.24	0.18288	86.337	61.464
109.00	1.3826	22.457	0.044530	−1.4938	−1.4323	0.10025	0.027088	0.075021	490.29	0.27654	82.404	57.786
111.00	1.5526	21.902	0.045658	−1.3537	−1.2828	0.10154	0.027089	0.078914	462.32	0.38936	78.472	54.224
113.00	1.7371	21.306	0.046935	−1.2086	−1.1271	0.10285	0.027149	0.083966	433.19	0.52741	74.544	50.740
115.00	1.9370	20.658	0.048407	−1.0571	−0.96336	0.10420	0.027290	0.090771	402.67	0.69974	70.626	47.290
117.00	2.1533	19.943	0.050144	−0.89741	−0.78944	0.10561	0.027545	0.10044	370.43	0.92076	66.728	43.824
119.00	2.3869	19.134	0.052262	−0.72635	−0.60161	0.10710	0.027981	0.11531	335.85	1.2154	62.883	40.270
121.00	2.6391	18.187	0.054985	−0.53833	−0.39322	0.10873	0.028755	0.14140	297.68	1.6317	59.196	36.509
123.00	2.9116	16.997	0.058834	−0.32093	−0.14962	0.11059	0.030317	0.20028	253.32	2.2811	56.121	32.310
125.00	3.2069	15.210	0.065747	−0.031475	0.17937	0.11310	0.034680	0.46831	195.48	3.5308	56.435	26.935
126.19	3.3958	11.184	0.089414	0.51527	0.81891	0.11807			0	6.0831		
63.151	0.012520	0.024070	41.546	1.2945	1.8147	0.16355	0.021007	0.029647	161.11	40.718	5.6209	4.3763
65.000	0.017404	0.032594	30.680	1.3299	1.8639	0.16161	0.021059	0.029788	163.20	38.268	5.8164	4.5123
67.000	0.024300	0.044300	22.573	1.3675	1.9160	0.15966	0.021123	0.029969	165.37	35.907	6.0298	4.6601
69.000	0.033213	0.059031	16.940	1.4042	1.9668	0.15786	0.021196	0.030180	167.43	33.803	6.2457	4.8088
71.000	0.044527	0.077273	12.941	1.4400	2.0162	0.15618	0.021278	0.030427	169.39	31.922	6.4645	4.9585
73.000	0.058656	0.099542	10.046	1.4747	2.0639	0.15461	0.021370	0.030712	171.23	30.231	6.6870	5.1096
75.000	0.076043	0.12638	7.9124	1.5082	2.1099	0.15314	0.021472	0.031039	172.95	28.707	6.9138	5.2621
77.000	0.097152	0.15838	6.3140	1.5404	2.1539	0.15176	0.021585	0.031413	174.55	27.328	7.1458	5.4164
79.000	0.12247	0.19613	5.0986	1.5713	2.1957	0.15046	0.021709	0.031839	176.03	26.074	7.3839	5.5727
81.000	0.15251	0.24030	4.1614	1.6007	2.2353	0.14923	0.021845	0.032323	177.38	24.931	7.6295	5.7313
83.000	0.18780	0.29157	3.4297	1.6284	2.2725	0.14806	0.021994	0.032873	178.60	23.884	7.8837	5.8924
85.000	0.22886	0.35069	2.8515	1.6544	2.3070	0.14694	0.022157	0.033496	179.68	22.923	8.1483	6.0565
87.000	0.27626	0.41846	2.3897	1.6784	2.3386	0.14587	0.022334	0.034204	180.63	22.035	8.4251	6.2238
89.000	0.33055	0.49576	2.0171	1.7005	2.3672	0.14485	0.022528	0.035008	181.43	21.212	8.7163	6.3948
91.000	0.39230	0.58355	1.7137	1.7203	2.3925	0.14385	0.022738	0.035925	182.10	20.446	9.0247	6.5700
93.000	0.46210	0.68291	1.4643	1.7377	2.4143	0.14289	0.022967	0.036973	182.62	19.730	9.3533	6.7499
95.000	0.54052	0.79504	1.2578	1.7525	2.4324	0.14195	0.023217	0.038177	182.99	19.057	9.7060	6.9353
97.000	0.62817	0.92134	1.0854	1.7645	2.4463	0.14103	0.023489	0.039568	183.21	18.421	10.087	7.1270
99.000	0.72566	1.0634	0.94038	1.7733	2.4557	0.14012	0.023787	0.041185	183.28	17.815	10.503	7.3260
101.00	0.83358	1.2231	0.81759	1.7788	2.4603	0.13922	0.024113	0.043081	183.18	17.236	10.960	7.5334
103.00	0.95259	1.4027	0.71291	1.7804	2.4595	0.13832	0.024471	0.045326	182.93	16.676	11.467	7.7509
105.00	1.0833	1.6049	0.62309	1.7778	2.4528	0.13742	0.024860	0.048012	182.51	16.132	12.035	7.9804
107.00	1.2264	1.8331	0.54553	1.7703	2.4394	0.13651	0.025284	0.051276	181.93	15.600	12.679	8.2245

109.00	1.3826	2.0916	0.47811	1.7573	2.4183	0.13557	0.025750	0.055332	181.19	15.075	13.419	8.4867
111.00	1.5526	2.3860	0.41911	1.7377	2.3884	0.13461	0.026284	0.060528	180.28	14.546	14.284	8.7716
113.00	1.7371	2.7240	0.36711	1.7102	2.3479	0.13360	0.026924	0.067435	179.15	13.996	15.315	9.0860
115.00	1.9370	3.1162	0.32091	1.6730	2.2946	0.13253	0.027721	0.077010	177.75	13.409	16.580	9.4395
117.00	2.1533	3.5786	0.27944	1.6234	2.2251	0.13138	0.028723	0.091003	176.01	12.767	18.186	9.8474
119.00	2.3869	4.1370	0.24172	1.5572	2.1341	0.13009	0.029997	0.11312	173.87	12.045	20.329	10.336
121.00	2.6391	4.8380	0.20670	1.4665	2.0119	0.12860	0.031683	0.15295	171.17	11.203	23.424	10.953
123.00	2.9116	5.7846	0.17287	1.3343	1.8376	0.12675	0.034185	0.24490	167.43	10.148	28.604	11.813
125.00	3.2069	7.3244	0.13653	1.1039	1.5417	0.12400	0.039278	0.66512	160.26	8.6030	41.535	13.326
126.19	3.3958	11.184	0.089414	0.51527	0.81891	0.11807			0	6.0831		
Single-Phase Properties												
100.00	0.10000	0.12268	8.1514	2.0396	2.8547	0.15950	0.021049	0.030012	201.64	16.082	9.3806	6.9581
600.00	0.10000	0.020037	49.908	12.573	17.564	0.21217	0.021796	0.030118	496.27	0.021483	44.840	29.577
1100.0	0.10000	0.010930	91.489	24.284	33.433	0.23131	0.024932	0.033248	660.05	−0.65654	70.075	44.199
1600.0	0.10000	0.0075152	133.06	37.272	50.579	0.24414	0.026815	0.035130	788.94	−0.81543	92.344	56.398
100.00	1.0000	24.658	0.040554	−2.0907	−2.0501	0.094493	0.027546	0.064564	609.42	−0.054514	100.58	76.255
103.75	1.0000	23.768	0.042073	−1.8441	−1.8020	0.096928	0.027281	0.068113	559.22	0.060996	92.738	67.783
103.75	1.0000	1.4754	0.67778	1.7800	2.4577	0.13799	0.024612	0.046272	182.79	16.471	11.671	7.8351
600.00	1.0000	0.19960	5.0099	12.554	17.564	0.19300	0.021812	0.030198	498.66	0.0061465	44.992	29.626
1100.0	1.0000	0.10899	9.1755	24.277	33.452	0.21216	0.024938	0.033267	662.07	−0.65940	70.155	44.221
1600.0	1.0000	0.074993	13.335	37.270	50.605	0.22499	0.026820	0.035138	790.64	−0.81612	92.399	56.411
100.00	5.0000	25.436	0.039314	−2.2176	−2.0210	0.093188	0.027713	0.059868	673.24	−0.17096	108.13	84.510
600.00	5.0000	0.98084	1.0195	12.469	17.567	0.17948	0.021881	0.030539	509.60	−0.057679	45.797	29.882
1100.0	5.0000	0.53797	1.8588	24.247	33.541	0.19875	0.024969	0.033350	671.08	−0.67112	70.555	44.330
1600.0	5.0000	0.37146	2.6921	37.259	50.720	0.21161	0.026839	0.035170	798.18	−0.81886	92.663	56.476
100.00	10.000	26.188	0.038186	−2.3398	−1.9580	0.091882	0.028004	0.056646	734.22	−0.25658	115.90	93.648
600.00	10.000	1.9183	0.52130	12.368	17.581	0.17355	0.021965	0.030926	523.87	−0.12928	46.995	30.284
1100.0	10.000	1.0590	0.94433	24.211	33.654	0.19296	0.025006	0.033447	682.37	−0.68394	71.130	44.493
1600.0	10.000	0.73435	1.3618	37.246	50.864	0.20583	0.026863	0.035209	807.57	−0.82170	93.033	56.570
600.00	500.00	27.434	0.036451	10.778	29.003	0.13791	0.026493	0.035336	1574.4	−0.70223	177.40	103.10
1100.0	500.00	21.868	0.045729	23.840	46.705	0.15935	0.027586	0.035848	1501.4	−0.72394	149.37	79.801
1600.0	500.00	18.335	0.054541	37.584	64.855	0.17295	0.028647	0.036665	1506.6	−0.73166	147.47	79.226
600.00	1000.0	34.270	0.029180	11.714	40.894	0.13093	0.029169	0.036905	2107.2	−0.61888	278.97	208.00
1100.0	1000.0	29.362	0.034057	25.065	59.122	0.15303	0.029373	0.036577	1985.1	−0.65271	232.49	129.86
1600.0	1000.0	25.920	0.038580	38.999	77.579	0.16685	0.029977	0.037212	1942.0	−0.65439	214.36	110.35

The values in these tables were generated from the NIST REFPROP software (Lemmon, E. W., McLinden, M. O., and Huber, M. L., NIST Standard Reference Database 23: Reference Fluid Thermodynamic and Transport Properties—REFPROP, National Institute of Standards and Technology, Standard Reference Data Program, Gaithersburg, Md., 2002, Version 7.1). The primary source for the thermodynamic properties is Span, R., Lemmon, E. W., Jacobsen, R. T., Wagner, W., and Yokozeki, A., "A Reference Quality Thermodynamic Property Formulation for Nitrogen," *J. Phys. Chem. Ref. Data* **29**(6):1361–1433, 2000. See also *Int. J. Thermophys.* **14**(4):1121–1132, 1998. The source for viscosity is Lemmon, E. W., and Jacobsen, R. T., "Viscosity and Thermal Conductivity Equations for Nitrogen, Oxygen, Argon, and Air," *Int. J. Thermophys.* **25**:21–69, 2004. The source for thermal conductivity is Lemmon, E. W., and Jacobsen, R. T., "Viscosity and Thermal Conductivity Equations for Nitrogen, Oxygen, Argon, and Air," *Int. J. Thermophys.* **25**:21–69, 2004.

Properties at the triple point temperature and the critical point temperature are given in the first and last entries of the saturation tables, respectively. In the single-phase table, when the temperature range for a given isobar includes a vapor-liquid phase boundary, the temperature of phase equilibrium is noted, and properties for both the saturated liquid and saturated vapor are given (with liquid properties given in the upper line). Lines are omitted from the temperature-pressure grid of the single-phase table, when the system would be in the solid phase or if there are potential problems with the source property surface.

The uncertainty in density of the equation of state is 0.02% from the triple point up to temperatures of 523 K and pressures up to 12 MPa and from temperatures of 240 to 523 K at pressures less than 30 MPa. In the range from 270 to 350 K at pressures less than 12 MPa, the uncertainty in density is 0.01%. The uncertainty at very high pressures (>1 GPa) is 0.6% in density. The uncertainty in pressure in the critical region is estimated to be 0.02%. In the gaseous and supercritical region, the speed of sound can be calculated with a typical uncertainty of 0.005% to 0.1%. At liquid states and at high pressures, the uncertainty increases to 0.5% to 1.5%. For pressures up to 30 MPa, the estimated uncertainty for heat capacities ranges from 0.3% at gaseous and gaslike supercritical states up to 0.8% at liquid states and at certain gaseous and supercritical states at low temperatures. The uncertainty is 2% for pressures up to 200 MPa and larger at higher pressures. The estimated uncertainties of vapor pressure, saturated-liquid density, and saturated-vapor density are in general 0.02% for each property. The formulation yields a reasonable extrapolation behavior up to the limits of chemical stability of nitrogen.

For viscosity, the uncertainty is 0.5% in dilute gas. Away from the dilute gas (pressures greater than 1 MPa and in the liquid), the uncertainties are as low as 1% between 270 and 300 K at pressures less than 100 MPa, and increase outside that range. The uncertainties are around 2% at temperatures of 180 K and higher. Below this and away from the critical region, the uncertainties steadily increase to around 5% at the triple points of the fluids. The uncertainties in the critical region are higher.

For thermal conductivity, the uncertainty for the dilute gas is 2% with increasing uncertainties near the triple point. For the nondilute gas, the uncertainty is 2% for temperatures greater than 150 K. The uncertainty is 3% at temperatures less than the critical point and 5% in the critical region, except for states very near the critical point.

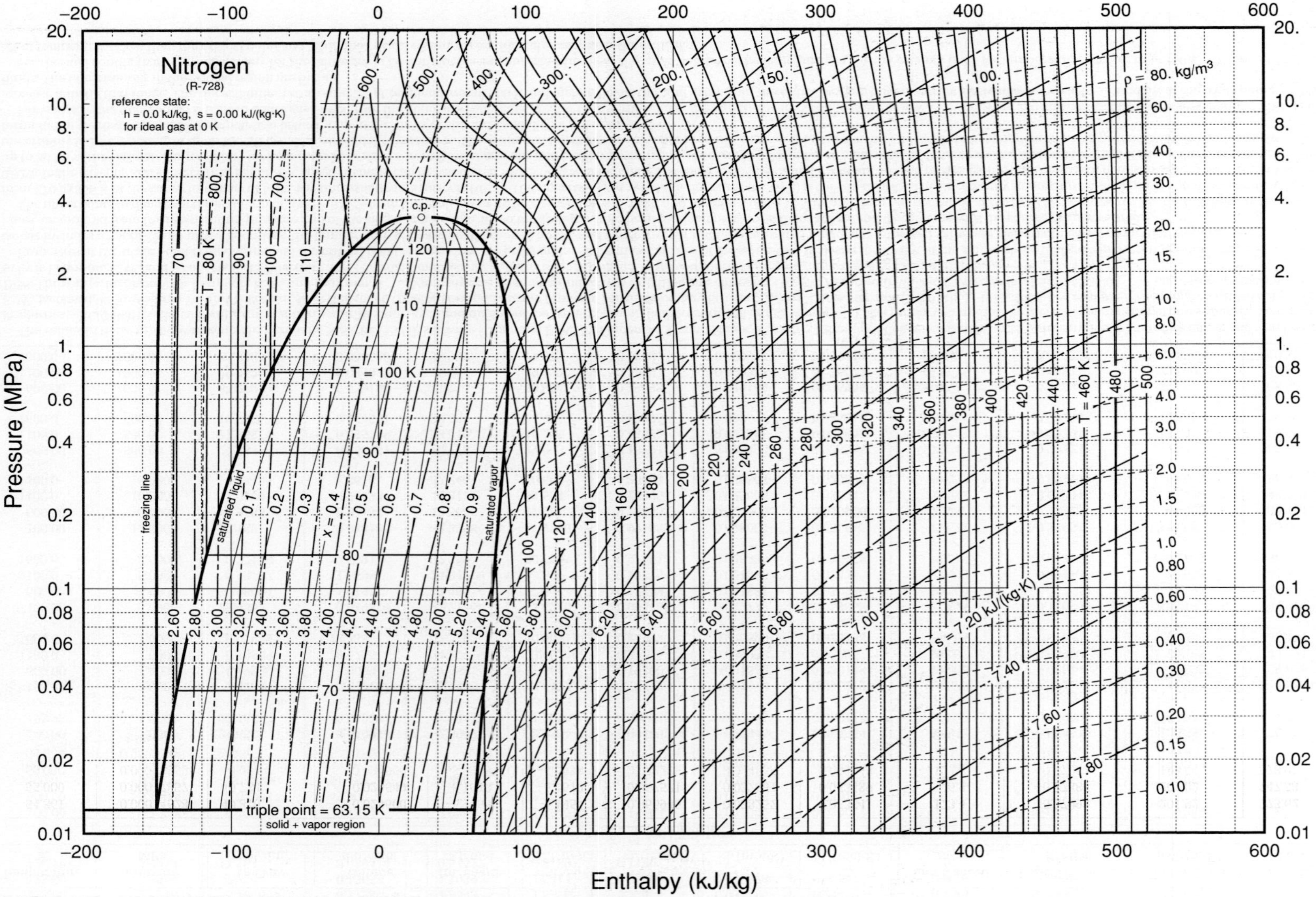

FIG. 2-7 Pressure-enthalpy diagram for nitrogen. Properties computed with the NIST REFPROP Database, Version 7.0 (Lemmon, E. W., M. O. McLinden, and M. L. Huber, 2002, NIST Standard Reference Database 23, NIST Reference Fluid Thermodynamic and Transport Properties—REFPROP, Version 7.0, Standard Reference Data Program, National Institute of Standards and Technology), based on the equation of state of Span, R., E. W. Lemmon, R. T. Jacobsen, W. Wagner, and A. Yokozeki, "A Reference Equation of State for the Thermodynamic Properties of Nitrogen for Temperatures from 63.151 to 1000 K and Pressures to 2200 MPa.," *J. Phys. Chem. Ref. Data* **29**:1361–1433, 2000.

TABLE 2-122 Thermodynamic Properties of Oxygen

Temperature K	Pressure MPa	Density mol/dm³	Volume dm³/mol	Int. energy kJ/mol	Enthalpy kJ/mol	Entropy kJ/(mol·K)	C_v kJ/(mol·K)	C_p kJ/(mol·K)	Sound speed m/s	Joule-Thomson K/MPa	Therm. cond. mW/(m·K)	Viscosity μPa·s
Saturated Properties												
54.361	0.00014628	40.816	0.024500	−6.1954	−6.1954	0.066946	0.038252	0.053541	1123.4	−0.37992	201.92	773.62
55.000	0.00017857	40.734	0.024549	−6.1613	−6.1612	0.067571	0.037651	0.053489	1126.9	−0.37886	201.02	747.53
60.000	0.00072582	40.064	0.024960	−5.8938	−5.8938	0.072225	0.034835	0.053548	1127.4	−0.37011	193.94	578.07
65.000	0.0023349	39.367	0.025402	−5.6258	−5.6257	0.076516	0.033469	0.053668	1101.7	−0.36312	186.82	457.94
70.000	0.0062623	38.656	0.025869	−5.3573	−5.3572	0.080495	0.032532	0.053697	1066.3	−0.35686	179.70	371.79
75.000	0.014547	37.936	0.026360	−5.0889	−5.0885	0.084199	0.031745	0.053719	1027.5	−0.34972	172.58	308.66
80.000	0.030123	37.203	0.026879	−4.8202	−4.8194	0.087667	0.031030	0.053808	987.43	−0.34056	165.44	261.22
85.000	0.056831	36.457	0.027430	−4.5510	−4.5495	0.090931	0.030365	0.054012	946.87	−0.32856	158.27	224.62
90.000	0.099350	35.692	0.028017	−4.2806	−4.2778	0.094023	0.029745	0.054361	905.90	−0.31302	151.05	195.64
95.000	0.16308	34.905	0.028649	−4.0084	−4.0038	0.096967	0.029169	0.054880	864.40	−0.29316	143.81	172.12
100.00	0.25400	34.092	0.029333	−3.7337	−3.7263	0.099787	0.028636	0.055599	822.19	−0.26804	136.55	152.56
105.00	0.37853	33.245	0.030079	−3.4556	−3.4442	0.10250	0.028146	0.056557	779.06	−0.23637	129.25	135.93
110.00	0.54340	32.360	0.030903	−3.1732	−3.1564	0.10513	0.027703	0.057816	734.77	−0.19639	121.92	121.52
115.00	0.75559	31.426	0.031820	−2.8853	−2.8612	0.10770	0.027311	0.059469	689.03	−0.14551	114.57	108.81
120.00	1.0223	30.434	0.032858	−2.5904	−2.5568	0.11022	0.026976	0.061666	641.52	−0.079899	107.23	97.426
125.00	1.3509	29.367	0.034051	−2.2867	−2.2407	0.11271	0.026712	0.064659	591.86	0.0063780	99.912	87.086
130.00	1.7491	28.203	0.035457	−1.9711	−1.9091	0.11520	0.026536	0.068905	539.50	0.12309	92.634	77.571
135.00	2.2250	26.907	0.037165	−1.6394	−1.5567	0.11773	0.026485	0.075327	483.69	0.28750	85.404	68.687
140.00	2.7878	25.415	0.039347	−1.2839	−1.1742	0.12035	0.026634	0.086099	423.10	0.53357	78.217	60.223
145.00	3.4477	23.599	0.042375	−0.88908	−0.74298	0.12319	0.027189	0.10778	355.20	0.93865	71.056	51.869
150.00	4.2186	21.110	0.047372	−0.41330	−0.21346	0.12654	0.028982	0.17484	273.80	1.7389	64.190	42.900
154.58	5.0428	13.630	0.073368	0.66752	1.0375	0.13442			0	5.0628		
54.361	0.00014628	0.00032370	3089.2	1.1195	1.5714	0.20982	0.021241	0.029631	140.32	507.90	4.4204	4.0962
55.000	0.00017857	0.00039060	2560.2	1.1327	1.5898	0.20850	0.021297	0.029698	141.11	480.26	4.4842	4.1481
60.000	0.00072582	0.0014561	686.75	1.2355	1.7339	0.19935	0.021815	0.030320	147.03	284.62	4.9840	4.5528
65.000	0.0023349	0.0043291	230.99	1.3377	1.8770	0.19194	0.022310	0.030934	152.65	156.71	5.4863	4.9555
70.000	0.0062623	0.010804	92.556	1.4393	2.0189	0.18587	0.022565	0.031294	158.07	87.254	5.9925	5.3557
75.000	0.014547	0.023509	42.536	1.5397	2.1584	0.18083	0.022513	0.031336	163.33	52.570	6.5051	5.7533
80.000	0.030123	0.045891	21.791	1.6377	2.2941	0.17659	0.022239	0.031177	168.36	35.817	7.0277	6.1486
85.000	0.056831	0.082138	12.175	1.7320	2.4239	0.17297	0.021896	0.031019	173.06	27.728	7.5654	6.5423
90.000	0.099350	0.13710	7.2938	1.8209	2.5455	0.16984	0.021624	0.031053	177.30	23.649	8.1241	6.9355
95.000	0.16308	0.21627	4.6239	1.9031	2.6571	0.16708	0.021515	0.031420	180.99	21.338	8.7113	7.3301
100.00	0.25400	0.32579	3.0695	1.9772	2.7569	0.16462	0.021605	0.032204	184.06	19.753	9.3362	7.7281
105.00	0.37853	0.47267	2.1156	2.0421	2.8430	0.16238	0.021894	0.033461	186.44	18.446	10.010	8.1324
110.00	0.54340	0.66506	1.5036	2.0966	2.9136	0.16032	0.022361	0.035245	188.14	17.250	10.748	8.5467
115.00	0.75559	0.91283	1.0955	2.1391	2.9668	0.15838	0.022978	0.037647	189.13	16.118	11.571	8.9760
120.00	1.0223	1.2284	0.81405	2.1678	3.0000	0.15652	0.023726	0.040839	189.41	15.045	12.509	9.4273
125.00	1.3509	1.6285	0.61407	2.1801	3.0097	0.15471	0.024597	0.045146	188.96	14.029	13.607	9.9112
130.00	1.7491	2.1366	0.46803	2.1722	2.9908	0.15289	0.025604	0.051204	187.75	13.062	14.940	10.445
135.00	2.2250	2.7893	0.35852	2.1380	2.9357	0.15100	0.026794	0.060349	185.74	12.120	16.641	11.061
140.00	2.7878	3.6487	0.27407	2.0670	2.8311	0.14896	0.028269	0.075824	182.82	11.155	18.977	11.823
145.00	3.4477	4.8412	0.20656	1.9383	2.6505	0.14659	0.030276	0.10781	178.78	10.071	22.582	12.881
150.00	4.2186	6.7170	0.14888	1.6938	2.3219	0.14345	0.033574	0.21201	172.82	8.6358	29.666	14.721
154.58	5.0428	13.630	0.073368	0.66752	1.0375	0.13442			0	5.0628		

(Continued)

TABLE 2-122 Thermodynamic Properties of Oxygen *(Continued)*

Temperature K	Pressure MPa	Density mol/dm³	Volume dm³/mol	Int. energy kJ/mol	Enthalpy kJ/mol	Entropy kJ/(mol·K)	C_v kJ/(mol·K)	C_p kJ/(mol·K)	Sound speed m/s	Joule-Thomson K/MPa	Therm. cond. mW/(m·K)	Viscosity μPa·s
Single-Phase Properties												
100.00	0.10000	0.12316	8.1192	2.0355	2.8474	0.17297	0.020885	0.029925	188.37	18.479	9.0852	7.7121
300.00	0.10000	0.040116	24.928	6.2338	8.7265	0.20531	0.021078	0.029435	329.72	2.6530	26.485	20.652
500.00	0.10000	0.024050	41.579	10.604	14.762	0.22069	0.022781	0.031108	421.27	0.75388	41.046	30.486
700.00	0.10000	0.017177	58.216	15.357	21.179	0.23147	0.024672	0.032992	493.31	0.10517	53.966	38.653
900.00	0.10000	0.013360	74.849	20.438	27.923	0.23994	0.026045	0.034363	555.60	−0.18735	65.867	45.806
100.00	1.0000	34.158	0.029276	−3.7444	−3.7151	0.099680	0.028683	0.055399	826.85	−0.27181	137.23	153.89
119.62	1.0000	30.512	0.032774	−2.6131	−2.5803	0.11003	0.027000	0.061476	645.19	−0.085501	107.79	98.249
119.62	1.0000	1.2018	0.83209	2.1662	2.9983	0.15666	0.023665	0.040564	189.41	15.124	12.433	9.3921
300.00	1.0000	0.40337	2.4791	6.1772	8.6563	0.18598	0.021148	0.029887	329.90	2.6066	26.894	20.846
500.00	1.0000	0.24010	4.1649	10.576	14.741	0.20149	0.022802	0.031240	422.68	0.73726	41.288	30.630
700.00	1.0000	0.17135	5.8360	15.340	21.176	0.21230	0.024682	0.033052	494.87	0.098376	54.139	38.766
900.00	1.0000	0.13328	7.5029	20.426	27.929	0.22078	0.026051	0.034395	557.14	−0.19062	66.001	45.899
100.00	5.0000	34.497	0.028988	−3.7983	−3.6533	0.099132	0.028935	0.054458	850.39	−0.28978	140.71	160.92
154.36	5.0000	16.011	0.062457	0.35374	0.66602	0.13204	0.038878	3.5718	163.89	4.2044	75.954	29.668
154.36	5.0000	11.160	0.089610	1.0294	1.4774	0.13729	0.041906	4.2513	158.85	6.0016	72.313	20.574
300.00	5.0000	2.0616	0.48505	5.9227	8.3480	0.17177	0.021448	0.032003	332.25	2.3730	28.797	21.766
500.00	5.0000	1.1908	0.83975	10.454	14.653	0.18787	0.022894	0.031815	429.36	0.66261	42.362	31.267
700.00	5.0000	0.84728	1.1802	15.264	21.165	0.19881	0.024726	0.033309	501.98	0.068114	54.901	39.261
900.00	5.0000	0.65931	1.5167	20.373	27.956	0.20734	0.026076	0.034537	564.07	−0.20519	66.593	46.305
100.00	10.000	34.885	0.028665	−3.8593	−3.5726	0.098498	0.029235	0.053516	877.07	−0.30803	144.82	169.49
300.00	10.000	4.2056	0.23778	5.6024	7.9802	0.16499	0.021790	0.034749	339.35	2.0332	31.466	23.153
500.00	10.000	2.3538	0.42484	10.306	14.554	0.18182	0.022999	0.032491	438.67	0.56900	43.708	32.074
700.00	10.000	1.6705	0.59861	15.171	21.157	0.19292	0.024776	0.033613	511.24	0.030534	55.839	39.873
900.00	10.000	1.3010	0.76866	20.307	27.993	0.20150	0.026104	0.034706	572.92	−0.22339	67.321	46.804
100.00	25.000	35.884	0.027867	−4.0109	−3.3142	0.096845	0.030037	0.051627	945.24	−0.34532	155.97	194.38
300.00	25.000	10.393	0.096215	4.7194	7.1247	0.15490	0.022521	0.040917	390.80	1.0167	41.851	29.605
500.00	25.000	5.6243	0.17780	9.8920	14.337	0.17346	0.023256	0.034167	472.62	0.30658	47.943	34.705
700.00	25.000	3.9923	0.25048	14.907	21.169	0.18495	0.024901	0.034397	541.32	−0.076019	58.651	41.714
900.00	25.000	3.1222	0.32028	20.117	28.124	0.19369	0.026174	0.035155	600.66	−0.27597	69.464	48.271
100.00	75.000	38.263	0.026135	−4.3340	−2.3739	0.092788	0.031906	0.049123	1115.1	−0.39472	184.96	274.96
300.00	75.000	21.603	0.046289	3.1884	6.6601	0.14315	0.023601	0.041272	645.54	−0.18640	75.261	53.378
500.00	75.000	13.760	0.072675	8.8798	14.330	0.16284	0.023725	0.036534	619.75	−0.20732	64.149	45.084
700.00	75.000	10.201	0.098029	14.192	21.544	0.17498	0.025126	0.035903	657.04	−0.31840	68.835	48.269
900.00	75.000	8.1749	0.12233	19.571	28.745	0.18403	0.026293	0.036153	701.72	−0.40609	76.863	53.163

The values in these tables were generated from the NIST REFPROP software (Lemmon, E. W., McLinden, M. O., and Huber, M. L., NIST Standard Reference Database 23: Reference Fluid Thermodynamic and Transport Properties—REFPROP, National Institute of Standards and Technology, Standard Reference Data Program, Gaithersburg, Md., 2002, Version 7.1). The primary source for the thermodynamic properties is Schmidt, R., and Wagner, W., "A New Form of the Equation of State for Pure Substances and Its Application to Oxygen," *Fluid Phase Equilibria*, **19:**175–200, 1985. The source for viscosity and thermal conductivity is Lemmon, E. W., and Jacobsen, R. T., "Viscosity and Thermal Conductivity Equations for Nitrogen, Oxygen, Argon, and Air," *Int. J. Thermophys.* **25:**21–69, 2004.

Properties at the triple point temperature and the critical point temperature are given in the first and last entries of the saturation tables, respectively. In the single-phase table, when the temperature range for a given isobar includes a vapor-liquid phase boundary, the temperature of phase equilibrium is noted, and properties for both the saturated liquid and saturated vapor are given (with liquid properties given in the upper line). Lines are omitted from the temperature-pressure grid of the single-phase table, when the system would be in the solid phase or if there are potential problems with the source property surface.

The uncertainties of the equation of state are 0.1% in density, 2% in heat capacity, and 1% in the speed of sound, except in the critical region. For viscosity, the uncertainty is 1% in the dilute gas at temperatures above 200 K, and 5% in the dilute gas at lower temperatures. The uncertainty is around 2% between 270 and 300 K, and increases to 5% outside of this region. The uncertainty may be higher in the liquid near the triple point. The uncertainty for the dilute gas is 2% with increasing uncertainties near the triple point. For thermal conductivity, the uncertainties range from 3% between 270 and 300 K to 5% elsewhere. The uncertainties above 100 MPa are not known due to a lack of experimental data.

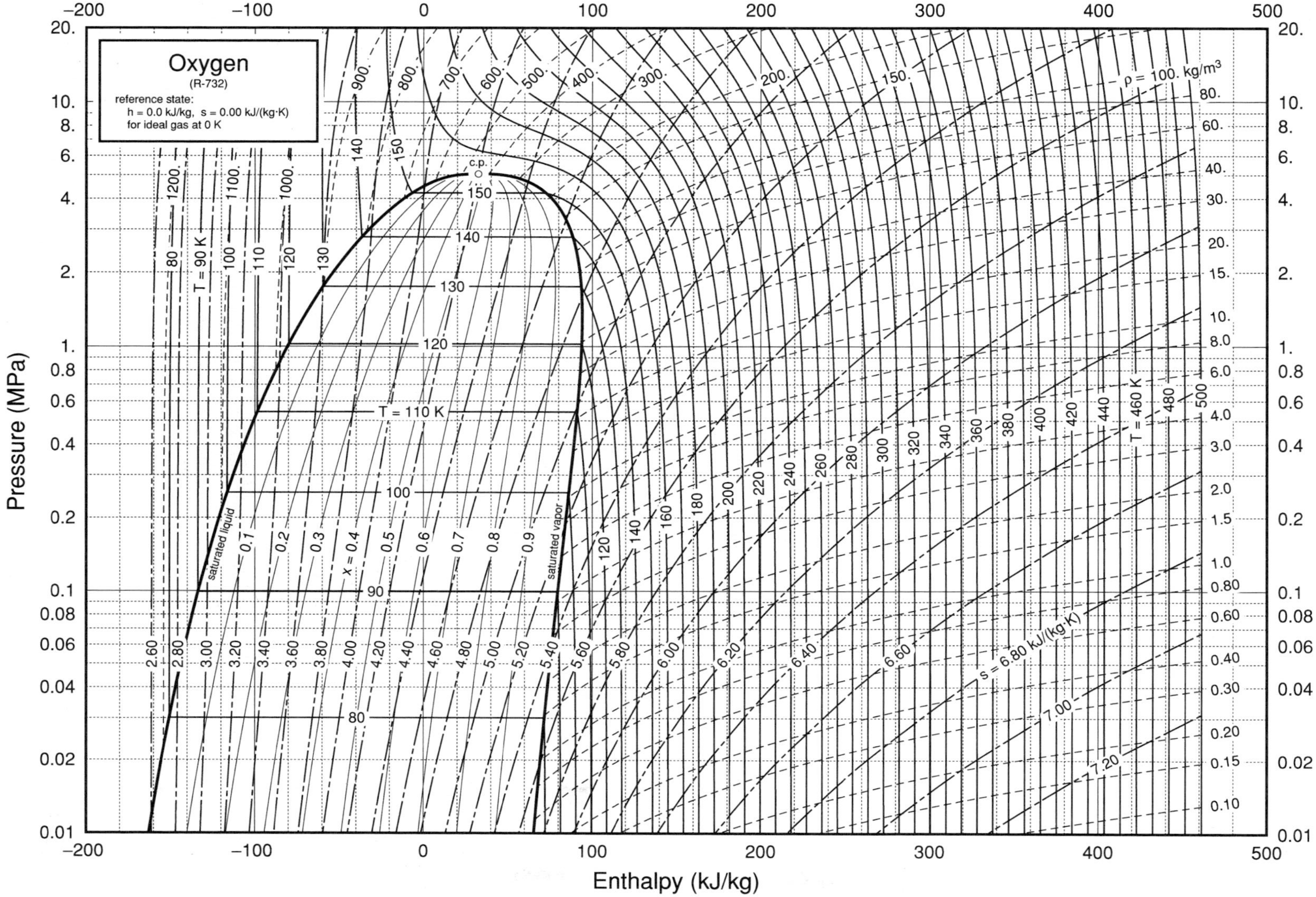

FIG. 2-8 Pressure-enthalpy diagram for oxygen. Properties computed with the NIST REFPROP Database, Version 7.0 (Lemmon, E. W., M. O. McLinden, and M. L. Huber, 2002, NIST Standard Reference Database 23, NIST Reference Fluid Thermodynamic and Transport Properties—REFPROP, Version 7.0, Standard Reference Data Program, National Institute of Standards and Technology), based on the equation of state of Schmidt, R., and W. Wagner, "A New Form of the Equation of State for Pure Substances and Its Application to Oxygen," *Fluid Phase Equilibria* **19:** 175–200, 1985.

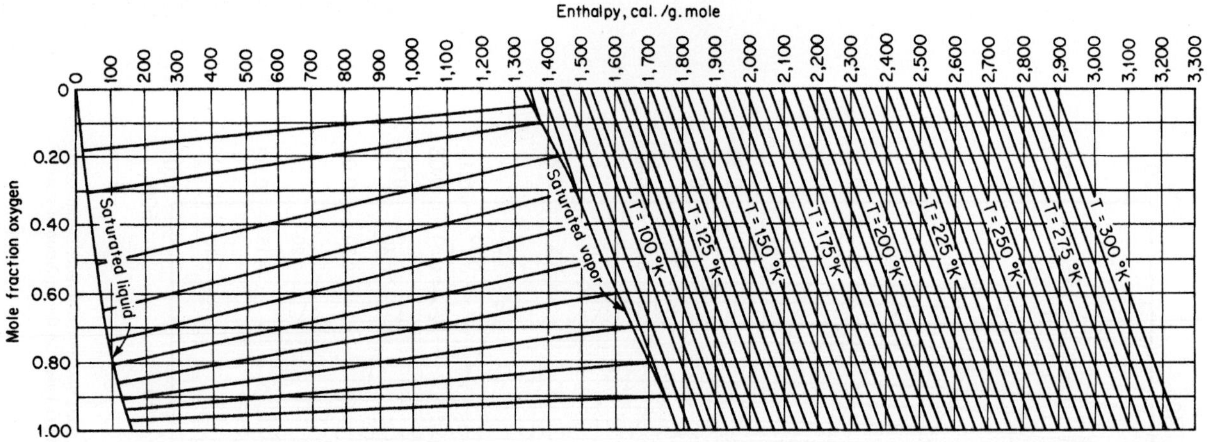

FIG. 2-9 Enthalpy-concentration diagram for oxygen-nitrogen mixture at 1 atm. Reference states: Enthalpies of liquid oxygen and liquid nitrogen at the normal boiling point of nitrogen are zero. (Dodge, B.F. *Chemical Engineering Thermodynamics*, McGraw-Hill, New York, 1944.) Wilson, G.M., P.M. Silverberg, and M.G. Zellner, AFAPL TDR 64-64 (AD 603151), 1964, p. 314, present extensive vapor-liquid equilibrium data for the three-component system argon-nitrogen-oxygen as well as for binary systems including oxygen-nitrogen. Calculations for this mixture are also available with the NIST REFPROP software.

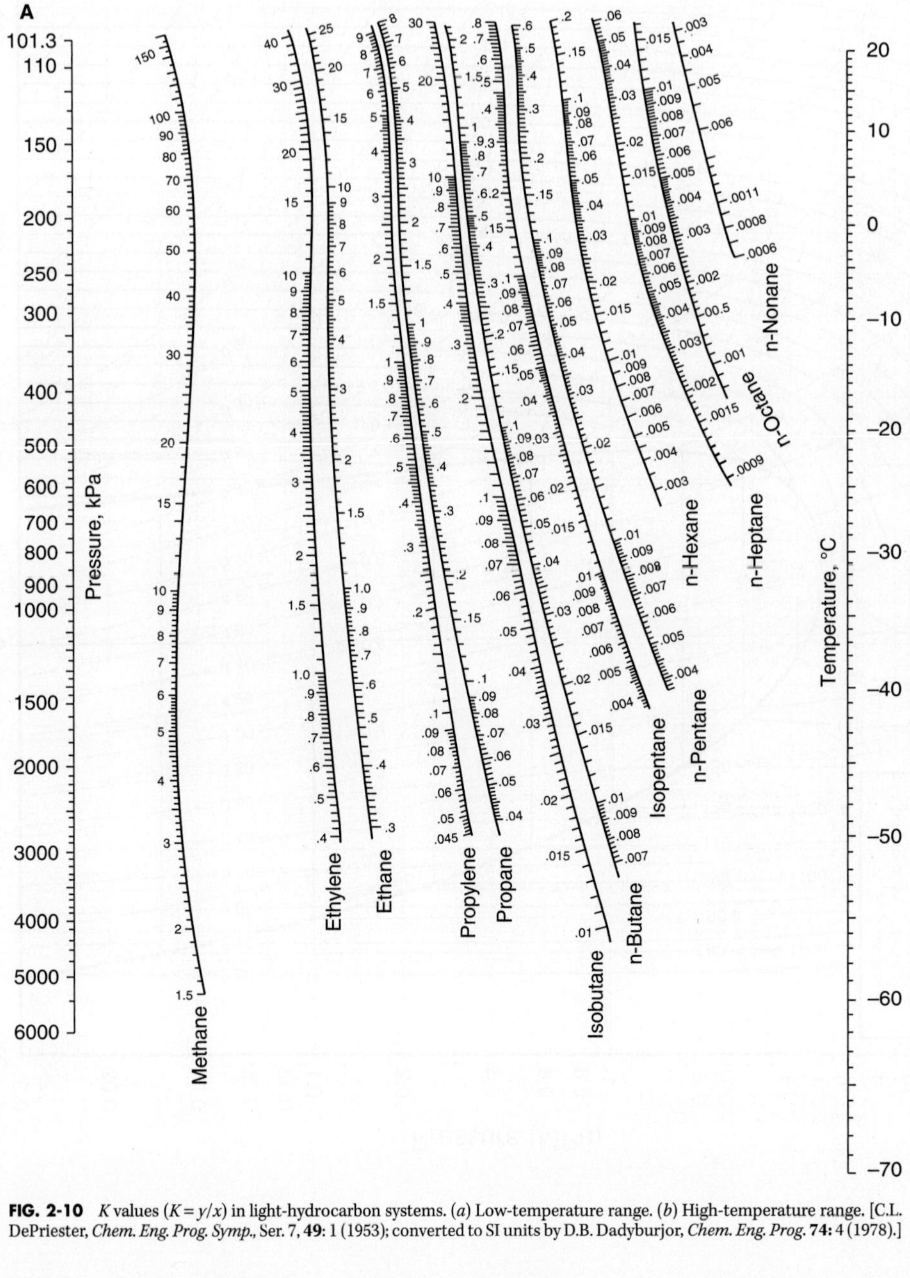

FIG. 2-10 *K* values ($K = y/x$) in light-hydrocarbon systems. (*a*) Low-temperature range. (*b*) High-temperature range. [C.L. DePriester, *Chem. Eng. Prog. Symp.*, Ser. 7, **49**: 1 (1953); converted to SI units by D.B. Dadyburjor, *Chem. Eng. Prog.* **74**: 4 (1978).]

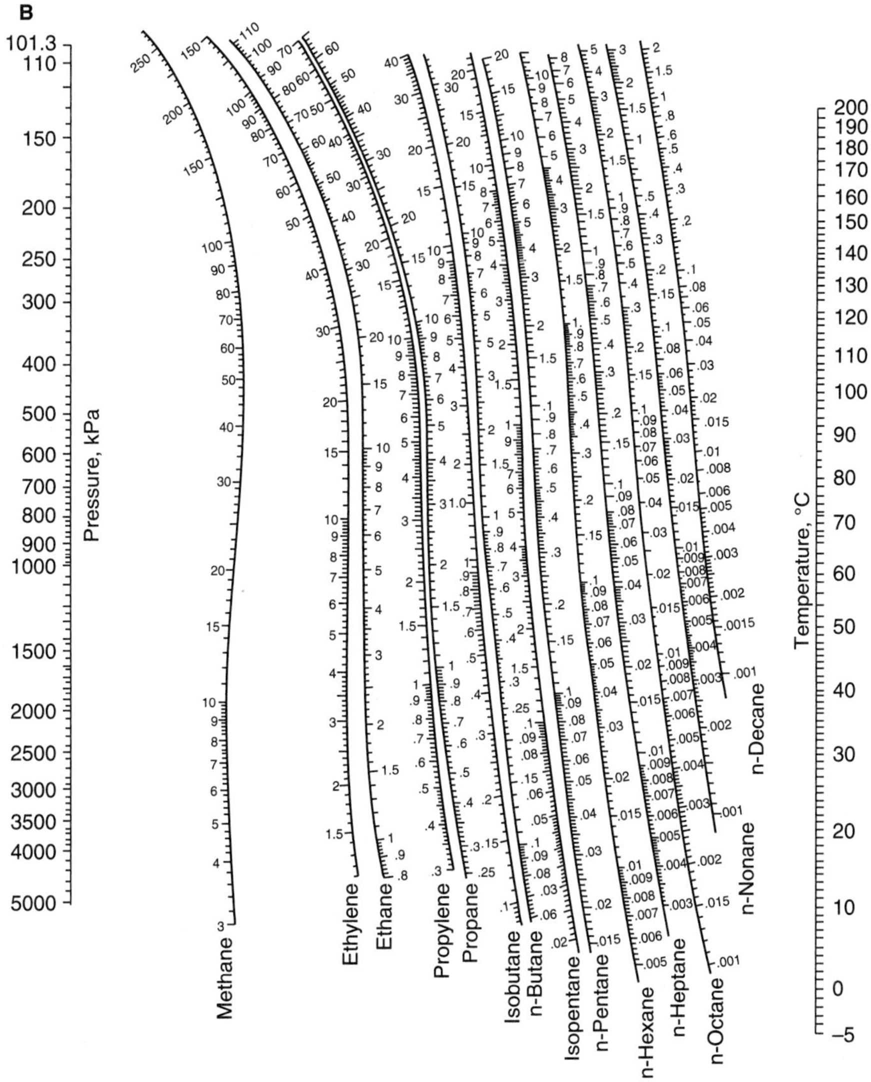

FIG. 2-10 *(Continued)*

TABLE 2-123 Composition of Selected Refrigerant Mixtures*

Mixed Product Name	Property Table/Figure	Composition, mass%				Ozone Depletion Potential (ODP)†	Global Warming Potential (GWP)‡ (100 year)
		R-32	R-125	R-134a	R-143a		
R-32	2-126, Fig. 2-12	100				0	650
R-125	2-127, Fig. 2-13		100			0	3400
R-134a	2-128, Fig. 2-14			100		0	1300
R-143a	2-129				100	0	4300
R-404A	2-130		44	4	52	0	3300
R-407C	2-131, Fig. 2-15	23	25	52		0	1600
R-410A	2-132	50	50			0	2088

*All products listed here are HFCs (hydrofluorocarbons), the primary replacement for hydrochlorofluorocarbons (HCFCs) like R-22.
†The ODP of the old CFC refrigerants R-11 and R-12 is 1.
‡CO_2 is the GWP reference: GWP of CO_2 = 1.

TABLE 2-124 Thermodynamic Properties of R-22, Chlorodifluoromethane

Temperature K	Pressure MPa	Density mol/dm^3	Volume dm^3/mol	Int. energy kJ/mol	Enthalpy kJ/mol	Entropy kJ/(mol·K)	C_v kJ/(mol·K)	C_p kJ/(mol·K)	Sound speed m/s	Joule-Thomson K/MPa
					Saturated Properties					
115.73	3.7947E-07	19.907	0.050235	2.5595	2.5595	0.0065813	0.061918	0.092976	1410.9	−0.44463
120.00	9.9588E-07	19.777	0.050564	2.9559	2.9559	0.0099451	0.061567	0.092700	1388.4	−0.44526
135.00	1.7187E-05	19.325	0.051747	4.3400	4.3400	0.020814	0.060123	0.091960	1312.3	−0.44448
150.00	0.00015627	18.873	0.052985	5.7179	5.7179	0.030493	0.059099	0.091824	1239.0	−0.43872
165.00	0.00089946	18.420	0.054289	7.0951	7.0951	0.039243	0.058356	0.091789	1166.5	−0.43084
180.00	0.0037009	17.963	0.055670	8.4715	8.4717	0.047227	0.057679	0.091751	1095.0	−0.42063
195.00	0.011835	17.500	0.057141	9.8483	9.8490	0.054574	0.057097	0.091898	1024.5	−0.40608
210.00	0.031218	17.028	0.058726	11.230	11.232	0.061399	0.056707	0.092431	954.55	−0.38505
225.00	0.070909	16.542	0.060453	12.622	12.627	0.067804	0.056579	0.093482	884.79	−0.35555
240.00	0.14319	16.036	0.062359	14.034	14.043	0.073877	0.056737	0.095120	814.96	−0.31561
255.00	0.26329	15.506	0.064493	15.472	15.489	0.079692	0.057171	0.097391	744.97	−0.26263
270.00	0.44888	14.944	0.066919	16.945	16.975	0.085308	0.057856	0.10037	674.69	−0.19248
285.00	0.71966	14.341	0.069730	18.462	18.513	0.090782	0.058767	0.10424	603.91	−0.097827
300.00	1.0970	13.686	0.073070	20.034	20.114	0.096166	0.059887	0.10941	532.11	0.035333
315.00	1.6039	12.956	0.077181	21.676	21.800	0.10152	0.061218	0.11688	458.37	0.23608
330.00	2.2661	12.114	0.082547	23.418	23.605	0.10696	0.062814	0.12929	381.15	0.57205
345.00	3.1130	11.069	0.090340	25.325	25.606	0.11267	0.064858	0.15550	298.16	1.2334
360.00	4.1837	9.5229	0.10501	27.613	28.053	0.11931	0.068488	0.25950	201.90	3.0745
369.30	4.9900	6.0582	0.16506	30.901	31.725	0.12907			0	10.366
115.73	3.7947E-07	3.9436E-07	2,535,700.	27.807	28.769	0.23305	0.028465	0.036779	119.91	398.80
120.00	9.9588E-07	9.9813E-07	1,001,900.	27.929	28.927	0.22637	0.028872	0.037186	121.91	367.18
135.00	1.7187E-05	1.5313E-05	65,305.	28.373	29.495	0.20715	0.030386	0.038703	128.58	269.83
150.00	0.00015627	0.00012533	7,979.0	28.840	30.087	0.19295	0.031990	0.040316	134.79	197.57
165.00	0.00089946	0.00065634	1,523.6	29.329	30.699	0.18230	0.033655	0.042018	140.59	146.30
180.00	0.0037009	0.0024808	403.09	29.836	31.328	0.17421	0.035388	0.043849	145.98	110.28
195.00	0.011835	0.0073561	135.94	30.357	31.966	0.16800	0.037218	0.045881	150.87	84.902
210.00	0.031218	0.018161	55.062	30.885	32.603	0.16317	0.039177	0.048204	155.15	66.865
225.00	0.070909	0.038991	25.647	31.411	33.229	0.15937	0.041300	0.050920	158.70	53.872
240.00	0.14319	0.075182	13.301	31.928	33.833	0.15634	0.043615	0.054144	161.35	44.348
255.00	0.26329	0.13342	7.4950	32.428	34.401	0.15386	0.046138	0.058023	162.96	37.240
270.00	0.44888	0.22208	4.5029	32.900	34.922	0.15178	0.048881	0.062763	163.38	31.852
285.00	0.71966	0.35201	2.8409	33.335	35.380	0.14996	0.051851	0.068713	162.45	27.725
300.00	1.0970	0.53822	1.8580	33.717	35.755	0.14830	0.055064	0.076534	159.98	24.549
315.00	1.6039	0.80363	1.2443	34.021	36.017	0.14666	0.058574	0.087659	155.73	22.102
330.00	2.2661	1.1882	0.84159	34.207	36.114	0.14486	0.062516	0.10579	149.36	20.201
345.00	3.1130	1.7777	0.56253	34.184	35.935	0.14261	0.067254	0.14389	140.39	18.613
360.00	4.1837	2.8529	0.35052	33.661	35.127	0.13896	0.074178	0.29995	127.92	16.641
369.30	4.9900	6.0582	0.16506	30.901	31.725	0.12907			0	10.366

Single-Phase Properties

150.00	0.10000	18.874	0.052982	5.7167	5.7220	0.030484	0.059102	0.091820	1239.3	−0.43876
232.06	0.10000	16.306	0.061325	13.284	13.290	0.070699	0.056618	0.094177	851.94	−0.33819
232.06	0.10000	0.053734	18.610	31.656	33.517	0.15786	0.042364	0.052366	160.06	49.032
250.00	0.10000	0.049441	20.226	32.442	34.465	0.16180	0.043860	0.053391	166.39	38.154
350.00	0.10000	0.034652	28.859	37.325	40.211	0.18105	0.053191	0.061845	196.16	13.439
450.00	0.10000	0.026819	37.287	43.093	46.822	0.19762	0.061672	0.070141	221.08	6.7638
550.00	0.10000	0.021901	45.660	49.620	54.186	0.21238	0.068475	0.076877	243.30	4.0604
150.00	1.0000	18.885	0.052952	5.7053	5.7582	0.030408	0.059134	0.091780	1241.8	−0.43921
250.00	1.0000	15.711	0.063648	14.961	15.024	0.077665	0.057020	0.096318	773.25	−0.28642
296.57	1.0000	13.841	0.072248	19.669	19.741	0.094938	0.059612	0.10807	548.68	0.00029448
296.57	1.0000	0.48960	2.0425	33.635	35.677	0.14867	0.054305	0.074523	160.70	25.205
350.00	1.0000	0.37703	2.6523	36.754	39.407	0.16025	0.055369	0.068138	184.85	14.183
450.00	1.0000	0.27693	3.6110	42.774	46.385	0.17776	0.062289	0.072300	216.22	6.8582
550.00	1.0000	0.22209	4.5027	49.400	53.903	0.19283	0.068737	0.077971	241.16	4.0624
150.00	5.0000	18.931	0.052823	5.6555	5.9197	0.030074	0.059277	0.091614	1253.0	−0.44109
250.00	5.0000	15.837	0.063142	14.822	15.138	0.077105	0.057135	0.095206	797.37	−0.30700
350.00	5.0000	11.141	0.089759	25.585	26.034	0.11341	0.064765	0.14356	317.33	1.0314
450.00	5.0000	1.6422	0.60893	41.131	44.175	0.16067	0.065284	0.087278	194.56	7.0507
550.00	5.0000	1.1832	0.84520	48.377	52.603	0.17760	0.069855	0.083655	232.93	3.9600
150.00	10.000	18.987	0.052667	5.5953	6.1220	0.029665	0.059460	0.091420	1266.7	−0.44331
250.00	10.000	15.982	0.062569	14.663	15.289	0.076450	0.057274	0.094054	825.41	−0.32844
350.00	10.000	12.008	0.083275	24.782	25.615	0.11098	0.063647	0.11843	412.24	0.39255
450.00	10.000	4.2433	0.23566	38.423	40.780	0.14893	0.068870	0.12461	184.82	5.5220
550.00	10.000	2.5432	0.39321	47.019	50.951	0.16944	0.071068	0.092204	229.21	3.5059
150.00	30.000	19.198	0.052089	5.3741	6.9367	0.028113	0.060251	0.090769	1318.0	−0.45090
250.00	30.000	16.469	0.060719	14.132	15.953	0.074181	0.057799	0.091018	921.00	−0.38542
350.00	30.000	13.518	0.073976	23.279	25.499	0.10621	0.063170	0.10053	603.47	−0.13764
450.00	30.000	10.241	0.097648	33.086	36.016	0.13259	0.069234	0.10864	392.35	0.40072
550.00	30.000	7.3810	0.13548	42.738	46.802	0.15425	0.073417	0.10533	317.51	0.89994
150.00	60.000	19.480	0.051336	5.0916	8.1717	0.026007	0.061560	0.089983	1384.9	−0.45992
250.00	60.000	17.029	0.058724	13.533	17.056	0.071434	0.058519	0.088686	1034.6	−0.42947
350.00	60.000	14.650	0.068258	22.111	26.206	0.10216	0.063912	0.094730	764.59	−0.31532
450.00	60.000	12.381	0.080770	31.083	35.929	0.12657	0.070220	0.099099	589.21	−0.17799
550.00	60.000	10.405	0.096108	40.152	45.919	0.14662	0.075020	0.10030	492.11	−0.055349

The values in these tables were generated from the NIST REFPROP software (Lemmon, E. W., McLinden, M. O., and Huber, M. L., NIST Standard Reference Database 23: Reference Fluid Thermodynamic and Transport Properties—REFPROP, National Institute of Standards and Technology, Standard Reference Data Program, Gaithersburg, Md., 2002, Version 7.1). The primary source for the thermodynamic properties is Kamei, A., Beyerlein, S. W., and Jacobsen, R. T., "Application of Nonlinear Regression in the Development of a Wide Range Formulation for HCFC-22," *Int. J. Thermophys.* **16:**1155–1164, 1995. Validated equations for the viscosity and thermal conductivity are not currently available for this fluid.

Properties at the triple point temperature and the critical point temperature are given in the first and last entries of the saturation tables, respectively. In the single-phase table, when the temperature range for a given isobar includes a vapor-liquid phase boundary, the temperature of phase equilibrium is noted, and properties for both the saturated liquid and saturated vapor are given (with liquid properties given in the upper line). Lines are omitted from the temperature-pressure grid of the single-phase table, when the system would be in the solid phase or if there are potential problems with the source property surface.

The uncertainties of the equation of state are 0.1% in density, 1% in heat capacity, and 0.3% in the speed of sound, except in the critical region. The uncertainty in vapor pressure is 0.2%.

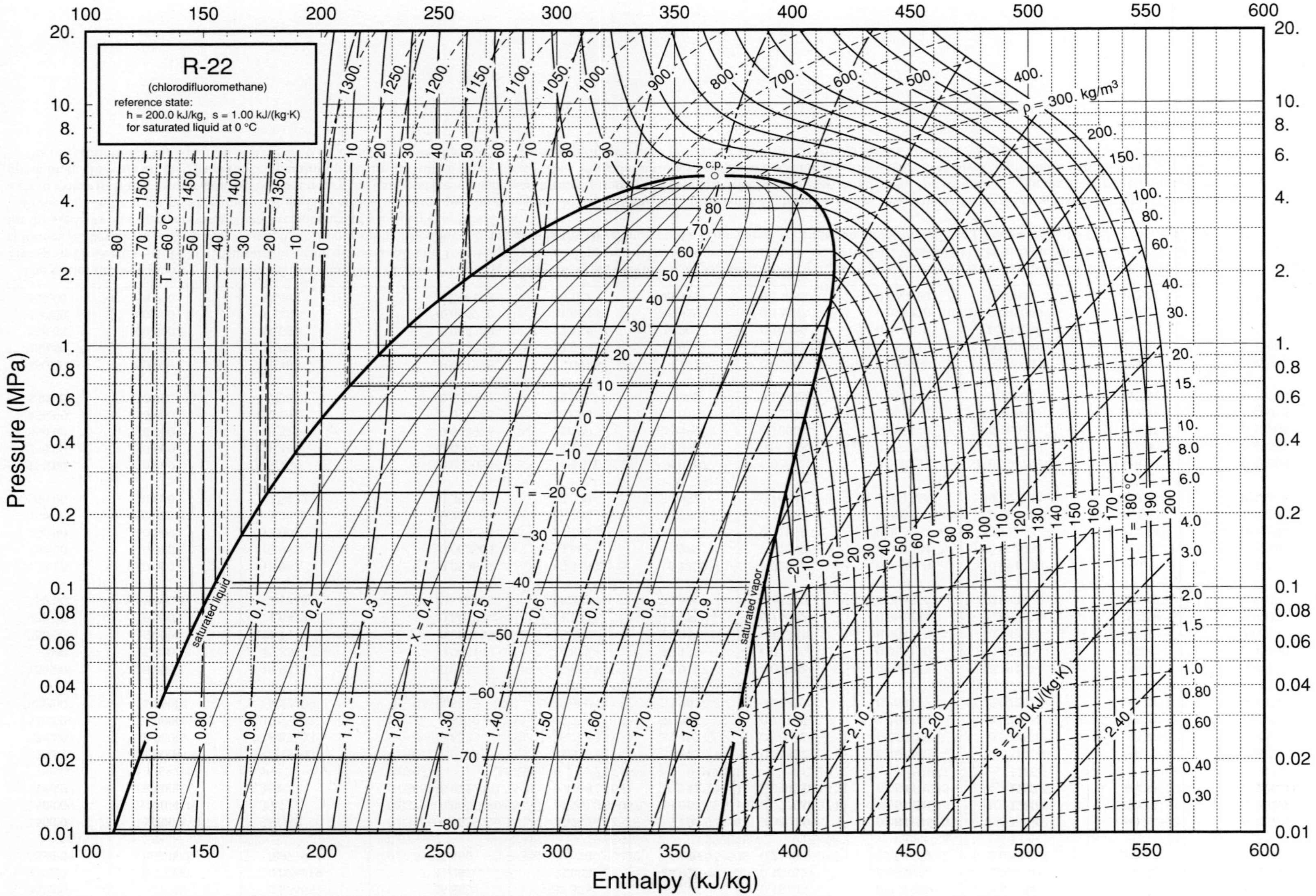

FIG. 2-11 Pressure-enthalpy diagram for Refrigerant 22. Properties computed with the NIST REFPROP Database, Version 7.0 (Lemmon, E. W., M. O. McLinden, and M. L. Huber, 2002, NIST Standard Reference Database 23, NIST Reference Fluid Thermodynamic and Transport Properties—REFPROP, Version 7.0, Standard Reference Data Program, National Institute of Standards and Technology), based on the equation of state of Kamei, A., S. W. Beyerlein, and R. T. Jacobsen, "Application of Nonlinear Regression in the Development of a Wide Range Formulation for HCFC-22," *Int. J. Thermophysics* **16:**1155–1164, 1995.

TABLE 2-125 Thermodynamic Properties of R-32, Difluoromethane

Temperature K	Pressure MPa	Density mol/dm³	Volume dm³/mol	Int. energy kJ/mol	Enthalpy kJ/mol	Entropy kJ/(mol·K)	C_v kJ/(mol·K)	C_p kJ/(mol·K)	Sound speed m/s	Joule-Thomson K/MPa	Therm. cond. mW/(m·K)
Saturated Properties											
136.34	4.8000E-05	27.473	0.036399	−0.99220	−0.99220	−0.0054608	0.055447	0.082847	1414.4	−0.33760	242.91
140.00	8.3535E-05	27.302	0.036627	−0.68946	−0.68946	−0.0032696	0.054980	0.082588	1395.1	−0.33728	241.74
150.00	0.00032474	26.835	0.037265	0.13324	0.13325	0.0024067	0.053793	0.081975	1342.3	−0.33542	237.64
160.00	0.0010410	26.364	0.037930	0.95053	0.95057	0.0076815	0.052740	0.081513	1289.9	−0.33191	232.45
170.00	0.0028536	25.889	0.038626	1.7640	1.7641	0.012613	0.051818	0.081215	1237.6	−0.32650	226.39
180.00	0.0068782	25.409	0.039357	2.5753	2.5756	0.017251	0.051021	0.081087	1185.7	−0.31891	219.64
190.00	0.014904	24.921	0.040127	3.3862	3.3868	0.021635	0.050345	0.081137	1133.8	−0.30886	212.37
200.00	0.029545	24.424	0.040944	4.1983	4.1995	0.025800	0.049783	0.081373	1082.1	−0.29600	204.70
210.00	0.054344	23.916	0.041814	5.0135	5.0158	0.029778	0.049333	0.081803	1030.4	−0.27988	196.75
220.00	0.093819	23.394	0.042745	5.8337	5.8377	0.033594	0.048988	0.082443	978.59	−0.25996	188.62
230.00	0.15345	22.858	0.043749	6.6608	6.6675	0.037271	0.048747	0.083313	926.62	−0.23550	180.39
240.00	0.23965	22.303	0.044838	7.4969	7.5077	0.040830	0.048604	0.084442	874.35	−0.20551	172.14
250.00	0.35967	21.726	0.046028	8.3443	8.3609	0.044291	0.048559	0.085874	821.65	−0.16864	163.92
260.00	0.52157	21.124	0.047340	9.2056	9.2303	0.047671	0.048610	0.087676	768.33	−0.12292	155.78
270.00	0.73415	20.491	0.048802	10.084	10.120	0.050989	0.048761	0.089947	714.18	−0.065518	147.75
280.00	1.0069	19.820	0.050454	10.983	11.034	0.054264	0.049019	0.092852	658.88	0.0079177	139.86
290.00	1.3501	19.102	0.052350	11.908	11.979	0.057517	0.049399	0.096659	602.05	0.10428	132.11
300.00	1.7749	18.323	0.054577	12.866	12.963	0.060775	0.049934	0.10185	543.11	0.23517	124.48
310.00	2.2934	17.460	0.057273	13.867	13.998	0.064076	0.050685	0.10938	481.27	0.42163	116.94
320.00	2.9194	16.477	0.060691	14.930	15.107	0.067477	0.051776	0.12140	415.41	0.70602	109.42
330.00	3.6686	15.299	0.065364	16.088	16.328	0.071088	0.053487	0.14404	343.84	1.1876	101.80
340.00	4.5614	13.740	0.072779	17.428	17.760	0.075179	0.056594	0.20457	263.77	2.1660	94.166
350.00	5.6311	10.732	0.093180	19.453	19.977	0.081343	0.066340	1.2085	163.70	5.4955	97.067
351.26	5.7826	8.1501	0.12270	20.836	21.546	0.085769			0	8.0731	
136.34	4.8000E-05	4.2353E-05	23,611.	21.981	23.115	0.17135	0.025987	0.034319	169.60	881.12	6.9492
140.00	8.3535E-05	7.1788E-05	13,930.	22.076	23.239	0.16765	0.026110	0.034451	171.76	769.01	6.9554
150.00	0.00032474	0.00026061	3,837.2	22.335	23.581	0.15872	0.026507	0.034889	177.47	541.12	7.0006
160.00	0.0010410	0.00078411	1,275.3	22.593	23.921	0.15125	0.027014	0.035477	182.88	391.73	7.0875
170.00	0.0028536	0.0020270	493.35	22.850	24.258	0.14493	0.027667	0.036272	187.97	291.02	7.2166
180.00	0.0068782	0.0046295	216.01	23.103	24.589	0.13955	0.028505	0.037336	192.69	221.02	7.3887
190.00	0.014904	0.0095503	104.71	23.350	24.910	0.13492	0.029560	0.038728	197.00	170.81	7.6049
200.00	0.029545	0.018112	55.213	23.588	25.219	0.13090	0.030843	0.040483	200.85	133.79	7.8668
210.00	0.054344	0.032028	31.223	23.816	25.513	0.12738	0.032341	0.042613	204.20	106.00	8.1765
220.00	0.093819	0.053428	18.717	24.032	25.788	0.12428	0.034016	0.045105	206.99	84.924	8.5374
230.00	0.15345	0.084890	11.780	24.234	26.042	0.12151	0.035821	0.047943	209.19	68.870	8.9546
240.00	0.23965	0.12949	7.7224	24.421	26.272	0.11901	0.037709	0.051127	210.73	56.605	9.4365
250.00	0.35967	0.19093	5.2375	24.590	26.474	0.11674	0.039648	0.054696	211.57	47.194	9.9965
260.00	0.52157	0.27370	3.6537	24.738	26.643	0.11464	0.041621	0.058741	211.65	39.922	10.656
270.00	0.73415	0.38340	2.6083	24.860	26.775	0.11268	0.043631	0.063434	210.90	34.246	11.449
280.00	1.0069	0.52726	1.8966	24.952	26.862	0.11079	0.045693	0.069063	209.26	29.758	12.431
290.00	1.3501	0.71503	1.3985	25.006	26.894	0.10895	0.047840	0.076110	206.64	26.148	13.691
300.00	1.7749	0.96054	1.0411	25.011	26.858	0.10709	0.050119	0.085424	202.95	23.187	15.376
310.00	2.2934	1.2848	0.77830	24.950	26.735	0.10516	0.052598	0.098649	198.02	20.693	17.748
320.00	2.9194	1.7233	0.58029	24.797	26.491	0.10305	0.055390	0.11948	191.66	18.510	21.309
330.00	3.6686	2.3442	0.42658	24.503	26.068	0.10060	0.058707	0.15836	183.49	16.477	27.173
340.00	4.5614	3.3211	0.30111	23.943	25.316	0.097403	0.063103	0.26199	172.68	14.312	38.601
350.00	5.6311	5.7166	0.17493	22.380	23.365	0.091024	0.071998	1.9028	154.59	10.637	87.141
351.26	5.7826	8.1501	0.12270	20.836	21.546	0.085769			0	8.0731	

(*Continued*)

TABLE 2-125 Thermodynamic Properties of R-32, Difluoromethane *(Continued)*

Temperature K	Pressure MPa	Density mol/dm^3	Volume dm^3/mol	Int. energy kJ/mol	Enthalpy kJ/mol	Entropy kJ/(mol·K)	C_v kJ/(mol·K)	C_p kJ/(mol·K)	Sound speed m/s	Joule-Thomson K/MPa	Therm. cond. mW/(m·K)
					Single-Phase Properties						
150.00	0.10000	26.836	0.037263	0.13226	0.13599	0.0024002	0.053795	0.081971	1342.7	−0.33547	237.67
221.24	0.10000	23.329	0.042866	5.9359	5.9402	0.034057	0.048953	0.082538	972.15	−0.25719	187.60
221.24	0.10000	0.056727	17.628	24.058	25.821	0.12392	0.034234	0.045439	207.30	82.688	8.5859
225.00	0.10000	0.055592	17.988	24.191	25.989	0.12467	0.033681	0.044513	209.39	76.114	8.7199
300.00	0.10000	0.040576	24.645	26.760	29.224	0.13708	0.035327	0.044188	241.95	25.024	12.643
375.00	0.10000	0.032244	31.013	29.631	32.733	0.14749	0.041082	0.049630	267.64	12.117	18.907
150.00	1.0000	26.851	0.037243	0.12350	0.16075	0.0023417	0.053807	0.081935	1345.7	−0.33584	237.89
225.00	1.0000	23.159	0.043179	6.2265	6.2697	0.035360	0.048867	0.082690	957.47	−0.25084	185.10
279.77	1.0000	19.836	0.050414	10.962	11.013	0.054190	0.049012	0.092777	660.15	0.0060372	140.04
279.77	1.0000	0.52357	1.9100	24.951	26.861	0.11084	0.045646	0.068922	209.31	29.848	12.406
300.00	1.0000	0.46100	2.1692	25.950	28.120	0.11518	0.041298	0.057916	223.83	23.889	13.334
375.00	1.0000	0.33917	2.9484	29.227	32.175	0.12727	0.042549	0.053538	259.50	11.929	19.059
150.00	5.0000	26.915	0.037154	0.085198	0.27097	0.0020846	0.053863	0.081784	1358.9	−0.33741	238.83
225.00	5.0000	23.298	0.042923	6.1394	6.3541	0.034970	0.048922	0.082027	978.81	−0.26135	187.71
300.00	5.0000	18.710	0.053448	12.637	12.905	0.060001	0.049679	0.096911	586.17	0.13431	128.72
344.33	5.0000	12.808	0.078078	18.130	18.520	0.077304	0.059015	0.28240	224.84	2.9944	91.412
344.33	5.0000	3.9887	0.25071	23.524	24.777	0.095475	0.065786	0.39574	166.61	13.148	48.235
375.00	5.0000	2.3228	0.43051	26.928	29.080	0.10755	0.050835	0.090625	218.44	10.800	26.239
150.00	10.000	26.993	0.037047	0.038702	0.40917	0.0017692	0.053932	0.081608	1375.0	−0.33924	239.97
225.00	10.000	23.461	0.042625	6.0369	6.4632	0.034504	0.048996	0.081303	1004.1	−0.27287	190.81
300.00	10.000	19.196	0.052094	12.346	12.867	0.058997	0.049538	0.092105	639.69	0.031716	134.49
375.00	10.000	10.448	0.095714	21.112	22.069	0.085980	0.057265	0.21222	231.14	3.5190	77.284
150.00	30.000	27.285	0.036650	−0.13357	0.96593	0.00056834	0.054209	0.081027	1436.1	−0.34524	244.02
225.00	30.000	24.030	0.041614	5.6813	6.9297	0.032836	0.049307	0.079197	1094.1	−0.30661	201.89
300.00	30.000	20.517	0.048739	11.542	13.004	0.056105	0.049670	0.083697	789.57	−0.15824	152.78
375.00	30.000	16.472	0.060708	17.786	19.607	0.075712	0.052941	0.093069	536.98	0.21392	112.94
225.00	70.000	24.916	0.040135	5.1400	7.9495	0.030109	0.049916	0.076915	1240.7	−0.34341	219.54
300.00	70.000	22.090	0.045270	10.583	13.752	0.052355	0.050281	0.078370	986.17	−0.28333	179.29
375.00	70.000	19.309	0.051788	16.127	19.752	0.070195	0.053544	0.081767	788.81	−0.18583	147.00

The values in these tables were generated from the NIST REFPROP software (Lemmon, E. W., McLinden, M. O., and Huber, M. L., NIST Standard Reference Database 23: Reference Fluid Thermodynamic and Transport Properties—REFPROP, National Institute of Standards and Technology, Standard Reference Data Program, Gaithersburg, Md., 2002, Version 7.1). The primary source for the thermodynamic properties is Tillner-Roth, R., and Yokozeki, A., "An International Standard Equation of State for Difluoromethane (R-32) for Temperatures from the Triple Point at 136.34 K to 435 K and Pressures up to 70 MPa," *J. Phys. Chem. Ref. Data* **26**(6):1273–1328, 1997. Validated equations for the viscosity are not currently available for this fluid. The source for thermal conductivity is unpublished; however, the fit uses the functional form found in Marsh, K., Perkins, R., and Ramires, M. L. V., "Measurement and Correlation of the Thermal Conductivity of Propane from 86 to 600 K at Pressures to 70 MPa," *J. Chem. Eng. Data* **47**(4):932–940, 2002.

Properties at the triple point temperature and the critical point temperature are given in the first and last entries of the saturation tables, respectively. In the single-phase table, when the temperature range for a given isobar includes a vapor-liquid phase boundary, the temperature of phase equilibrium is noted, and properties for both the saturated liquid and saturated vapor are given (with liquid properties given in the upper line). Lines are omitted from the temperature-pressure grid of the single-phase table, when the system would be in the solid phase or if there are potential problems with the source property surface.

For the equation of state, typical uncertainties are 0.05% for density, 0.02% for the vapor pressure, and 0.5% to 1% for the heat capacity and speed of sound in the liquid phase. In the vapor phase, the uncertainty in the speed of sound is 0.02%. For thermal conductivity, the estimated uncertainty of the correlation is 5%, except for the dilute gas and points approaching critical where the uncertainty rises to 10%.

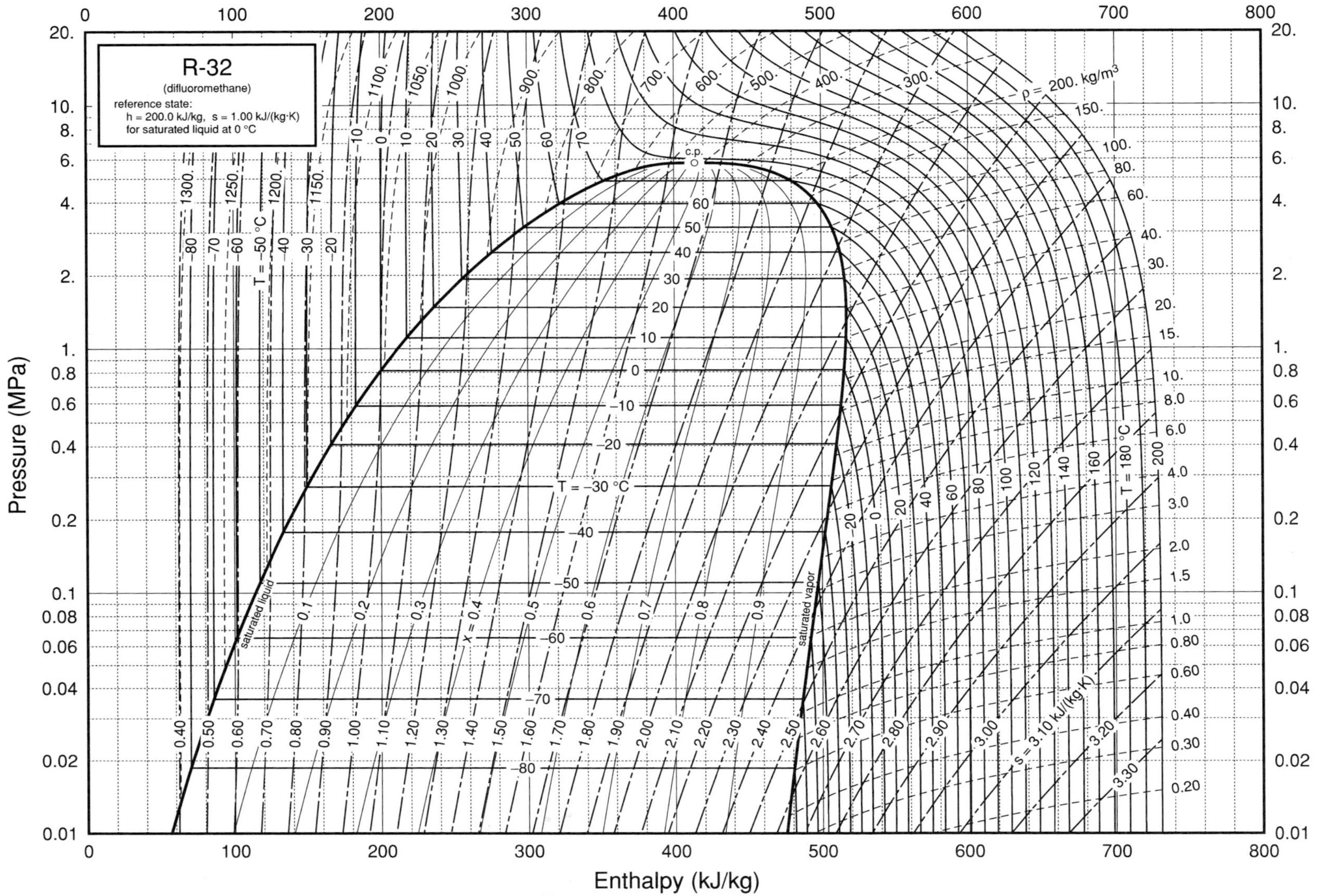

FIG. 2-12 Pressure-enthalpy diagram for Refrigerant 32. Properties computed with the NIST REFPROP Database, Version 7.0 (Lemmon, E. W., M. O. McLinden, and M. L. Huber, 2002, NIST Standard Reference Database 23, NIST Reference Fluid Thermodynamic and Transport Properties—REFPROP, Version 7.0, Standard Reference Data Program, National Institute of Standards and Technology), based on the equation of state of Tillner-Roth, R., and A. Yokozeki, "An International Standard Equation of State for Difluoromethane (R-32) for Temperatures from the Triple Point at 136.34 K to 435 K and Pressures up to 70 MPa," *J. Phys. Chem. Ref. Data* **26**(6): 1273–1328, 1997.

TABLE 2-126 Thermodynamic Properties of R-125, Pentafluoroethane

Temperature K	Pressure MPa	Density mol/dm³	Volume dm³/mol	Int. energy kJ/mol	Enthalpy kJ/mol	Entropy kJ/(mol·K)	C_v kJ/(mol·K)	C_p kJ/(mol·K)	Sound speed m/s	Joule-Thomson K/MPa	Therm. cond. mW/(m·K)	Viscosity μPa·s
						Saturated Properties						
172.52	0.0029140	14.086	0.070990	10.457	10.457	0.058837	0.081329	0.12417	932.57	−0.38374	116.02	1152.4
180.00	0.0056285	13.885	0.072020	11.389	11.389	0.064124	0.082012	0.12500	893.63	−0.37406	112.52	957.54
190.00	0.012328	13.613	0.073461	12.646	12.647	0.070919	0.083102	0.12647	843.11	−0.35818	107.79	768.40
200.00	0.024602	13.336	0.074988	13.919	13.921	0.077448	0.084327	0.12825	793.91	−0.33901	103.06	631.00
210.00	0.045417	13.052	0.076615	15.210	15.214	0.083750	0.085644	0.13028	745.67	−0.31627	98.331	527.00
220.00	0.078505	12.762	0.078360	16.523	16.529	0.089856	0.087029	0.13254	698.07	−0.28935	93.653	445.76
230.00	0.12833	12.461	0.080247	17.858	17.869	0.095792	0.088472	0.13505	650.89	−0.25723	89.019	380.67
240.00	0.20004	12.150	0.082305	19.219	19.235	0.10158	0.089971	0.13785	603.92	−0.21839	84.443	327.41
250.00	0.29934	11.824	0.084572	20.607	20.632	0.10725	0.091529	0.14102	556.99	−0.17062	79.940	283.01
260.00	0.43250	11.481	0.087098	22.025	22.063	0.11282	0.093153	0.14468	510.02	−0.11056	75.520	245.39
270.00	0.60624	11.117	0.089954	23.478	23.532	0.11830	0.094843	0.14903	462.91	−0.032921	71.187	213.02
280.00	0.82782	10.724	0.093245	24.971	25.048	0.12374	0.096594	0.15440	415.46	0.070972	66.940	184.73
290.00	1.1050	10.295	0.097130	26.511	26.619	0.12916	0.098430	0.16135	367.36	0.21606	62.772	159.60
300.00	1.4463	9.8162	0.10187	28.112	28.259	0.13461	0.10043	0.17099	318.17	0.43036	58.667	136.86
310.00	1.8610	9.2637	0.10795	29.793	29.994	0.14015	0.10274	0.18593	267.31	0.77370	54.597	115.81
320.00	2.3600	8.5923	0.11638	31.595	31.869	0.14593	0.10571	0.21395	213.55	1.4029	50.534	95.602
330.00	2.9579	7.6744	0.13030	33.632	34.017	0.15231	0.11043	0.29625	153.34	2.9184	46.661	74.602
339.17	3.6179	4.7790	0.20925	37.417	38.174	0.16438			0	12.361		
172.52	0.0029140	0.0020381	490.65	31.863	33.293	0.19120	0.059815	0.068285	116.43	90.257	5.2349	7.4339
180.00	0.0056285	0.0037809	264.49	32.307	33.795	0.18860	0.061648	0.070217	118.54	77.516	5.7185	7.7624
190.00	0.012328	0.0078784	126.93	32.913	34.477	0.18582	0.064126	0.072893	121.15	64.018	6.3724	8.1999
200.00	0.024602	0.015031	66.529	33.530	35.167	0.18368	0.066646	0.075712	123.47	53.589	7.0353	8.6344
210.00	0.045417	0.026661	37.508	34.157	35.860	0.18207	0.069223	0.078713	125.44	45.456	7.7081	9.0657
220.00	0.078505	0.044514	22.465	34.788	36.552	0.18087	0.071864	0.081939	127.01	39.066	8.3929	9.4944
230.00	0.12833	0.070679	14.148	35.421	37.237	0.18000	0.074575	0.085437	128.11	34.014	9.0929	9.9221
240.00	0.20004	0.10763	9.2907	36.052	37.911	0.17940	0.077362	0.089271	128.68	29.998	9.8136	10.353
250.00	0.29934	0.15835	6.3153	36.678	38.568	0.17900	0.080230	0.093526	128.64	26.787	10.563	10.791
260.00	0.43250	0.22645	4.4159	37.292	39.202	0.17874	0.083141	0.098283	127.93	24.232	11.356	11.246
270.00	0.60624	0.31661	3.1585	37.890	39.805	0.17857	0.086003	0.10368	126.44	22.293	12.213	11.732
280.00	0.82782	0.43510	2.2983	38.460	40.363	0.17844	0.088869	0.11025	124.10	20.938	13.169	12.266
290.00	1.1050	0.59084	1.6925	38.988	40.858	0.17826	0.092050	0.11918	120.81	20.043	14.286	12.884
300.00	1.4463	0.79742	1.2540	39.453	41.267	0.17797	0.095933	0.13255	116.42	19.438	15.680	13.638
310.00	1.8610	1.0777	0.92787	39.828	41.554	0.17744	0.10077	0.15449	110.71	19.016	17.586	14.635
320.00	2.3600	1.4778	0.67670	40.054	41.651	0.17649	0.10679	0.19725	103.29	18.738	20.574	16.104
330.00	2.9579	2.1269	0.47016	39.964	41.355	0.17454	0.11481	0.32843	93.550	18.404	26.607	18.766
339.17	3.6179	4.7790	0.20925	37.417	38.174	0.16438			0	12.361		

Single-Phase Properties

200.00	0.10000	13.337	0.074979	13.916	13.924	0.077436	0.084327	0.12823	794.34	−0.33920	103.09	631.60
224.79	0.10000	12.619	0.079245	17.160	17.168	0.092721	0.087714	0.13371	675.42	−0.27468	91.425	412.85
224.79	0.10000	0.055877	17.897	35.092	36.881	0.18042	0.073155	0.083578	127.60	36.498	8.7263	9.6994
300.00	0.10000	0.040689	24.576	41.155	43.613	0.20616	0.086960	0.095910	149.16	13.603	14.156	13.041
400.00	0.10000	0.030228	33.082	50.746	54.054	0.23608	0.10398	0.11252	172.25	5.6807	22.115	17.070
500.00	0.10000	0.024108	41.479	61.864	66.012	0.26271	0.11764	0.12607	192.24	3.1093	30.917	20.691
200.00	1.0000	13.355	0.074878	13.888	13.963	0.077295	0.084329	0.12805	799.42	−0.34145	103.50	638.78
286.46	1.0000	10.452	0.095673	25.960	26.055	0.12724	0.097767	0.15865	384.49	0.15862	64.240	168.19
286.46	1.0000	0.53072	1.8842	38.807	40.691	0.17833	0.090862	0.11564	122.09	20.318	13.866	12.653
300.00	1.0000	0.48261	2.0720	40.159	42.231	0.18359	0.092286	0.11224	129.90	16.539	14.732	13.270
400.00	1.0000	0.31780	3.1466	50.310	53.456	0.21585	0.10520	0.11626	165.47	5.9354	22.513	17.404
500.00	1.0000	0.24593	4.0661	61.591	65.657	0.24302	0.11806	0.12764	189.33	3.1109	31.336	21.014
200.00	5.0000	13.432	0.074450	13.768	14.140	0.076686	0.084364	0.12732	820.94	−0.35061	105.29	671.00
300.00	5.0000	10.214	0.097901	27.606	28.095	0.13288	0.099404	0.15790	379.39	0.16755	62.727	155.81
400.00	5.0000	2.1222	0.47120	47.739	50.095	0.19593	0.11164	0.15240	136.55	6.6581	27.340	22.244
500.00	5.0000	1.3333	0.75004	60.288	64.038	0.22710	0.12001	0.13643	180.53	2.9952	33.551	23.530
200.00	10.000	13.522	0.073953	13.626	14.366	0.075959	0.084459	0.12655	845.71	−0.36040	107.42	712.07
300.00	10.000	10.597	0.094369	27.096	28.039	0.13109	0.098728	0.14971	441.85	−0.0032571	67.164	177.37
400.00	10.000	5.5436	0.18039	43.724	45.528	0.18103	0.11417	0.19389	165.42	2.8474	40.583	46.568
500.00	10.000	2.8438	0.35165	58.555	62.072	0.21813	0.12198	0.14925	183.47	2.4316	37.529	29.745
200.00	30.000	13.835	0.072281	13.138	15.306	0.073355	0.085197	0.12439	927.73	−0.38765	115.19	889.29
300.00	30.000	11.489	0.087039	25.838	28.449	0.12645	0.098551	0.13883	597.35	−0.23171	79.855	245.90
400.00	30.000	9.0272	0.11078	39.705	43.028	0.16830	0.11217	0.15193	391.05	0.060756	59.824	112.78
500.00	30.000	6.8239	0.14654	54.141	58.538	0.20288	0.12339	0.15662	307.14	0.34144	53.649	67.694
300.00	60.000	12.259	0.081575	24.732	29.626	0.12196	0.10001	0.13426	735.32	−0.32748	93.938	338.04
400.00	60.000	10.465	0.095559	37.854	43.587	0.16206	0.11339	0.14453	563.68	−0.23451	75.022	173.26
500.00	60.000	8.9121	0.11221	51.703	58.436	0.19516	0.12470	0.15193	471.23	−0.15582	67.810	113.08

The values in these tables were generated from the NIST REFPROP software (Lemmon, E. W., McLinden, M. O., and Huber, M. L., NIST Standard Reference Database 23: Reference Fluid Thermodynamic and Transport Properties—REFPROP, National Institute of Standards and Technology, Standard Reference Data Program, Gaithersburg, Md., 2002, Version 7.1). The primary source for the thermodynamic properties is Lemmon, E. W., and Jacobsen, R. T., "A New Functional Form and New Fitting Techniques for Equations of State with Application to Pentafluoroethane (HFC-125)," *J. Phys. Chem. Ref. Data* **34**(1):69–108, 2005. The source for viscosity is Huber, M. L., and Laesecke, A., "Correlation for the Viscosity of Pentafluoroethane (R125) from the Triple Point to 500 K at Pressures up to 60 MPa," *Ind. Eng. Chem. Res.*, **45**(12):4447–4453, 2006. The source for thermal conductivity is Perkins, R., and Huber, M. L., "Measurement and Correlation of the Thermal Conductivity of Pentafluoroethane (R125) from 190 K to 512 K at Pressures to 70 MPa," *J. Chem. Eng. Data* **51**(3):898–904, 2006.

Properties at the triple point temperature and the critical point temperature are given in the first and last entries of the saturation tables, respectively. In the single-phase table, when the temperature range for a given isobar includes a vapor-liquid phase boundary, the temperature of phase equilibrium is noted, and properties for both the saturated liquid and saturated vapor are given (with liquid properties given in the upper line). Lines are omitted from the temperature-pressure grid of the single-phase table, when the system would be in the solid phase or if there are potential problems with the source property surface.

The uncertainty in density is 0.1% at temperatures from the triple point to 400 K at pressures up to 60 MPa, except in the critical region, where an uncertainty of 0.2% in pressure is generally attained. In the limited region between 340 and 400 K and at pressures from 4 to 10 MPa, as well as for all states above 400 K, the uncertainty in density increases to 0.5%. At temperatures below 330 K and pressures below 30 MPa, the uncertainty in density in the liquid phase may be as low as 0.04%. In the vapor and supercritical region, speed of sound data are represented within 0.05% at pressures below 1 MPa. The estimated uncertainty for heat capacities is 0.5%, and the estimated uncertainty for the speed of sound in the liquid phase is 0.5% for $T > 250$ K. The estimated uncertainties of vapor pressures and saturated liquid densities calculated using the Maxwell criterion are 0.1% for each property, and the estimated uncertainty for saturated vapor densities is 0.2%. The uncertainty in density increases as the critical point is approached, while the accompanying uncertainty in calculated pressures is 0.2%. The viscosity correlation has an estimated uncertainty of 3.0% along the saturation boundary in the liquid phase, and 0.8% in the vapor. For thermal conductivity, the estimated uncertainty of the correlation is 3%, except for the dilute gas and points approaching critical, where the uncertainty rises to 5%.

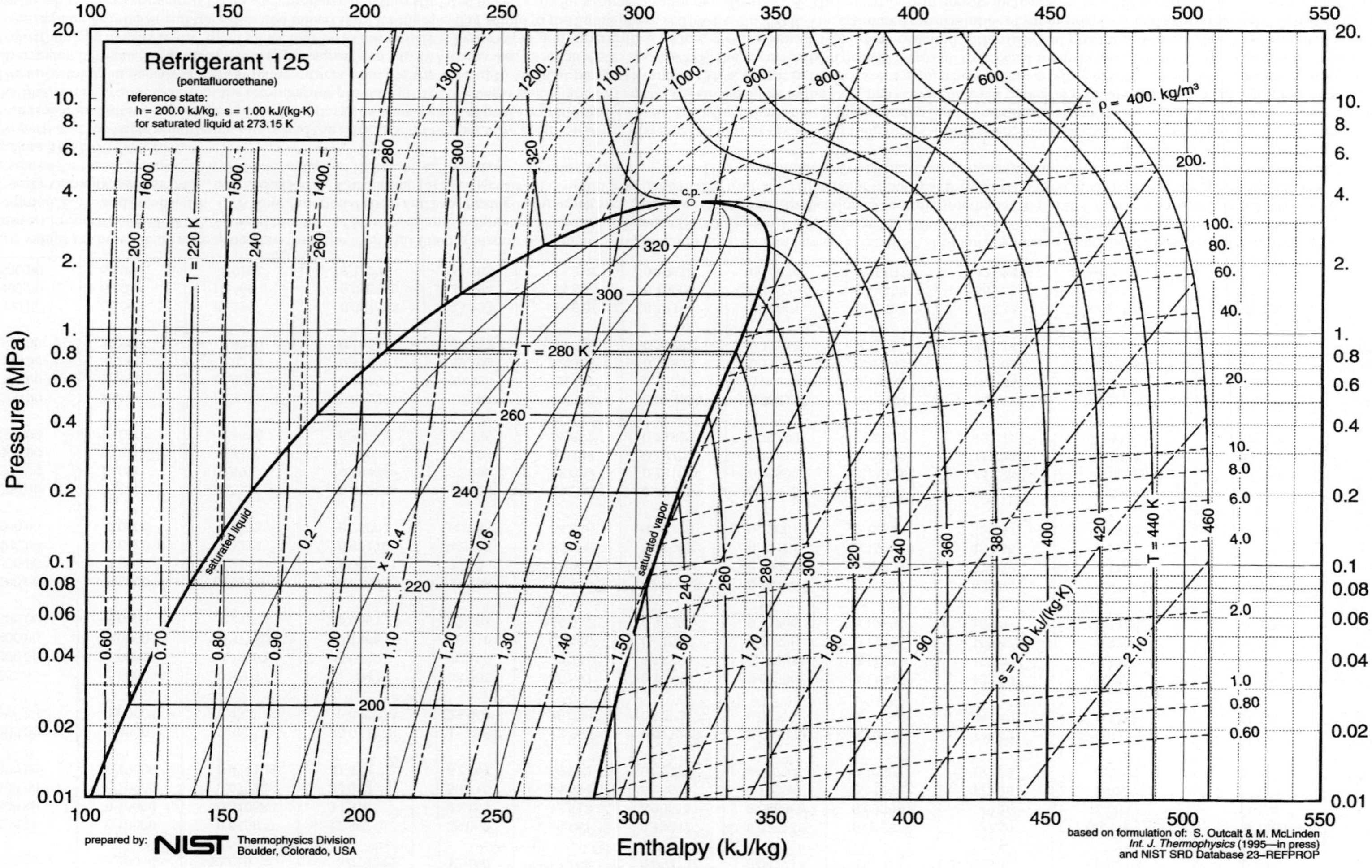

FIG. 2-13 Pressure-enthalpy diagram for Refrigerant 125.

TABLE 2-127 Thermodynamic Properties of R-134a, 1,1,1,2-Tetrafluoroethane

Temperature K	Pressure MPa	Density mol/dm³	Volume dm³/mol	Int. energy kJ/mol	Enthalpy kJ/mol	Entropy kJ/(mol·K)	C_v kJ/(mol·K)	C_p kJ/(mol·K)	Sound speed m/s	Joule-Thomson K/MPa	Therm. cond. mW/(m·K)	Viscosity μPa·s
						Saturated Properties						
169.85	0.00038956	15.594	0.064126	7.2907	7.2907	0.042100	0.080831	0.12079	1120.0	−0.38145	145.24	2153.6
170.00	0.00039617	15.590	0.064142	7.3088	7.3088	0.042207	0.080824	0.12079	1119.2	−0.38136	145.15	2139.7
180.00	0.0011275	15.331	0.065228	8.5179	8.5179	0.049117	0.080732	0.12112	1068.3	−0.37370	139.12	1479.1
190.00	0.0028170	15.069	0.066362	9.7328	9.7330	0.055686	0.081114	0.12193	1017.7	−0.36352	133.32	1106.2
200.00	0.0063130	14.804	0.067550	10.957	10.958	0.061966	0.081784	0.12303	967.61	−0.35119	127.74	867.31
210.00	0.012910	14.535	0.068798	12.194	12.195	0.067999	0.082633	0.12434	918.33	−0.33678	122.36	702.27
220.00	0.024433	14.262	0.070116	13.444	13.446	0.073815	0.083595	0.12582	869.85	−0.32011	117.17	582.15
230.00	0.043287	13.984	0.071512	14.710	14.713	0.079441	0.084636	0.12746	822.11	−0.30082	112.14	491.22
240.00	0.072481	13.699	0.072999	15.992	15.997	0.084899	0.085734	0.12927	775.00	−0.27839	107.27	420.20
250.00	0.11561	13.406	0.074593	17.293	17.301	0.090209	0.086879	0.13126	728.39	−0.25204	102.53	363.25
260.00	0.17684	13.104	0.076311	18.613	18.627	0.095389	0.088067	0.13348	682.14	−0.22073	97.922	316.57
270.00	0.26082	12.791	0.078179	19.956	19.976	0.10046	0.089298	0.13597	636.12	−0.18299	93.414	277.54
280.00	0.37271	12.465	0.080227	21.322	21.352	0.10543	0.090576	0.13883	590.17	−0.13675	88.995	244.34
290.00	0.51805	12.121	0.082499	22.716	22.759	0.11032	0.091908	0.14216	544.15	−0.079015	84.644	215.64
300.00	0.70282	11.758	0.085050	24.141	24.201	0.11516	0.093303	0.14615	497.89	−0.0052732	80.341	190.46
310.00	0.93340	11.368	0.087965	25.603	25.685	0.11996	0.094777	0.15108	451.23	0.091533	76.063	168.04
320.00	1.2166	10.945	0.091364	27.108	27.219	0.12475	0.096352	0.15740	404.00	0.22306	71.781	147.78
330.00	1.5599	10.478	0.095439	28.667	28.816	0.12956	0.098067	0.16598	355.90	0.41006	67.464	129.20
340.00	1.9715	9.9483	0.10052	30.297	30.495	0.13446	0.10001	0.17863	306.37	0.69376	63.075	111.81
350.00	2.4611	9.3237	0.10725	32.029	32.293	0.13952	0.10241	0.20012	254.06	1.1714	58.581	95.095
360.00	3.0405	8.5279	0.11726	33.932	34.289	0.14496	0.10601	0.24863	196.05	2.1419	54.062	78.146
370.00	3.7278	7.2558	0.13782	36.283	36.797	0.15159	0.11372	0.52085	127.23	5.1434	51.767	57.956
374.21	4.0591	5.0171	0.19932	38.947	39.756	0.15938			0	11.931		
169.85	0.00038956	0.00027611	3621.7	32.764	34.175	0.20038	0.051318	0.059719	126.79	373.57	3.0801	6.8294
170.00	0.00039617	0.00028055	3564.4	32.772	34.184	0.20029	0.051354	0.059756	126.84	370.78	3.0921	6.8353
180.00	0.0011275	0.00075481	1324.8	33.287	34.781	0.19502	0.053742	0.062208	130.05	234.43	3.8934	7.2319
190.00	0.0028170	0.0017896	558.79	33.821	35.395	0.19075	0.056118	0.064682	133.11	160.10	4.6952	7.6253
200.00	0.0063130	0.0038201	261.77	34.371	36.023	0.18729	0.058489	0.067201	135.98	116.94	5.4978	8.0147
210.00	0.012910	0.0074704	133.86	34.934	36.662	0.18451	0.060874	0.069802	138.63	90.215	6.3018	8.3993
220.00	0.024433	0.013574	73.669	35.508	37.308	0.18228	0.063296	0.072534	141.01	72.584	7.1080	8.7786
230.00	0.043287	0.023188	43.125	36.090	37.956	0.18050	0.065783	0.075455	143.06	60.236	7.9176	9.1524
240.00	0.072481	0.037603	26.593	36.675	38.602	0.17909	0.068357	0.078618	144.73	51.130	8.7324	9.5209
250.00	0.11561	0.058360	17.135	37.261	39.242	0.17797	0.071031	0.082078	145.98	44.137	9.5551	9.8853
260.00	0.17684	0.087278	11.458	37.844	39.870	0.17709	0.073812	0.085888	146.75	38.613	10.389	10.247
270.00	0.26082	0.12651	7.9043	38.420	40.482	0.17640	0.076698	0.090115	146.99	34.169	11.241	10.611
280.00	0.37271	0.17865	5.5976	38.986	41.073	0.17586	0.079686	0.094850	146.63	30.561	12.118	10.980
290.00	0.51805	0.24685	4.0511	39.538	41.636	0.17542	0.082776	0.10023	145.61	27.621	13.035	11.363
300.00	0.70282	0.33512	2.9840	40.069	42.166	0.17504	0.085974	0.10650	143.88	25.230	14.011	11.771
310.00	0.93340	0.44874	2.2285	40.573	42.653	0.17469	0.089297	0.11404	141.33	23.301	15.081	12.219
320.00	1.2166	0.59505	1.6805	41.038	43.083	0.17432	0.092780	0.12355	137.86	21.768	16.303	12.735
330.00	1.5599	0.78498	1.2739	41.451	43.438	0.17387	0.096484	0.13638	133.33	20.578	17.780	13.358
340.00	1.9715	1.0363	0.96498	41.785	43.687	0.17326	0.10052	0.15548	127.57	19.687	19.711	14.164
350.00	2.4611	1.3818	0.72368	41.994	43.775	0.17232	0.10510	0.18870	120.33	19.033	22.525	15.300
360.00	3.0405	1.8973	0.52707	41.973	43.576	0.17075	0.11074	0.26594	111.25	18.448	27.365	17.140
370.00	3.7278	2.8805	0.34717	41.323	42.617	0.16731	0.11928	0.70016	99.370	17.050	40.137	21.336
374.21	4.0591	5.0171	0.19932	38.947	39.756	0.15938			0	11.931		

(Continued)

TABLE 2-127 Thermodynamic Properties of R-134a, 1,1,1,2-Tetrafluoroethane *(Continued)*

Temperature K	Pressure MPa	Density mol/dm³	Volume dm³/mol	Int. energy kJ/mol	Enthalpy kJ/mol	Entropy kJ/(mol·K)	C_v kJ/(mol·K)	C_p kJ/(mol·K)	Sound speed m/s	Joule-Thomson K/MPa	Therm. cond. mW/(m·K)	Viscosity μPa·s
						Single-Phase Properties						
200.00	0.10000	14.805	0.067543	10.955	10.962	0.061955	0.081787	0.12301	968.03	−0.35132	127.78	868.18
246.79	0.10000	13.501	0.074068	16.873	16.880	0.088519	0.086506	0.13060	743.31	−0.26099	104.04	380.27
246.79	0.10000	0.050898	19.647	37.073	39.037	0.17830	0.070161	0.080931	145.63	46.198	9.2899	9.7687
275.00	0.10000	0.044972	22.236	39.124	41.348	0.18716	0.073781	0.083445	154.76	29.047	11.540	10.906
350.00	0.10000	0.034753	28.775	45.154	48.032	0.20860	0.086248	0.095065	175.31	12.552	17.537	13.823
425.00	0.10000	0.028455	35.143	52.096	55.610	0.22818	0.098386	0.10695	192.97	6.7852	23.539	16.650
200.00	1.0000	14.819	0.067479	10.933	11.001	0.061846	0.081812	0.12291	972.08	−0.35256	128.11	876.60
275.00	1.0000	12.657	0.079009	20.597	20.676	0.10281	0.089915	0.13695	619.10	−0.16746	91.627	262.84
312.54	1.0000	11.264	0.088775	25.980	26.069	0.12117	0.095165	0.15252	439.31	0.12101	74.978	162.71
312.54	1.0000	0.48242	2.0729	40.695	42.768	0.17460	0.090164	0.11623	140.54	22.877	15.374	12.343
350.00	1.0000	0.39132	2.5555	44.290	46.846	0.18694	0.090315	0.10603	159.63	13.885	17.989	13.936
425.00	1.0000	0.29983	3.3352	51.597	54.933	0.20785	0.099891	0.11116	185.14	7.0297	23.806	16.917
200.00	5.0000	14.880	0.067202	10.839	11.175	0.061371	0.081929	0.12246	989.55	−0.35768	129.56	915.11
275.00	5.0000	12.804	0.078103	20.385	20.776	0.10203	0.089864	0.13495	651.41	−0.19924	94.015	277.35
350.00	5.0000	9.8674	0.10134	31.397	31.904	0.13765	0.10066	0.17178	320.01	0.59952	63.012	109.32
425.00	5.0000	2.0736	0.48225	48.647	51.058	0.18734	0.10791	0.15381	148.25	7.9048	28.574	20.974
200.00	10.000	14.954	0.066874	10.727	11.395	0.060796	0.082085	0.12196	1010.3	−0.36339	131.31	966.05
275.00	10.000	12.967	0.077121	20.149	20.920	0.10115	0.089868	0.13304	687.36	−0.22964	96.744	295.02
350.00	10.000	10.478	0.095440	30.642	31.597	0.13537	0.099573	0.15486	400.60	0.21924	68.919	128.79
425.00	10.000	6.1370	0.16295	43.563	45.193	0.17038	0.11141	0.20870	177.89	3.0434	44.888	46.711
200.00	30.000	15.216	0.065720	10.326	12.298	0.058683	0.082769	0.12047	1084.1	−0.38053	137.79	1211.0
275.00	30.000	13.479	0.074190	19.398	21.624	0.098210	0.090220	0.12865	801.47	−0.30014	105.87	364.87
350.00	30.000	11.662	0.085750	29.071	31.644	0.13038	0.098885	0.13885	582.52	−0.15240	82.955	183.26
425.00	30.000	9.7202	0.10288	39.385	42.471	0.15838	0.10808	0.14967	425.63	0.10364	67.154	107.03
275.00	70.000	14.181	0.070517	18.373	23.310	0.093839	0.091314	0.12519	962.41	−0.35619	119.84	521.91
350.00	70.000	12.797	0.078141	27.492	32.961	0.12484	0.099542	0.13226	787.10	−0.30093	99.868	277.09
425.00	70.000	11.494	0.087004	37.066	43.157	0.15121	0.10829	0.13963	661.39	−0.23655	86.640	181.77

The values in these tables were generated from the NIST REFPROP software (Lemmon, E. W., McLinden, M. O., and Huber, M. L., NIST Standard Reference Database 23: Reference Fluid Thermodynamic and Transport Properties—REFPROP, National Institute of Standards and Technology, Standard Reference Data Program, Gaithersburg, Md., 2002, Version 7.1). The primary source for the thermodynamic properties is Tillner-Roth, R., and Baehr, H. D., "An International Standard Formulation of the Thermodynamic Properties of 1,1,1,2-Tetrafluoroethane (HFC-134a) for Temperatures from 170 K to 455 K at Pressures up to 70 MPa," *J. Phys. Chem. Ref. Data* **23**:657–729, 1994. The source for viscosity is Huber, M. L., Laesecke, A., and Perkins, R. A., "Model for the Viscosity and Thermal Conductivity of Refrigerants, Including a New Correlation for the Viscosity of R134a," *Ind. Eng. Chem. Res.* **42**:3163–3178, 2003. The source for thermal conductivity is Perkins, R. A., Laesecke, A., Howley, J., Ramires, M. L. V., Gurova, A. N., and Cusco, L., "Experimental Thermal Conductivity Values for the IUPAC Round-Robin Sample of 1,1,1,2-Tetrafluoroethane (R134a)," NISTIR, 2000.

Properties at the triple point temperature and the critical point temperature are given in the first and last entries of the saturation tables, respectively. In the single-phase table, when the temperature range for a given isobar includes a vapor-liquid phase boundary, the temperature of phase equilibrium is noted, and properties for both the saturated liquid and saturated vapor are given (with liquid properties given in the upper line). Lines are omitted from the temperature-pressure grid of the single-phase table, when the system would be in the solid phase or if there are potential problems with the source property surface.

Typical uncertainties are 0.05% for density, 0.02% for vapor pressure, 0.5% to 1% for heat capacity, 0.05% for vapor speed of sound, and 1% for liquid speed of sound, except in the critical region. The uncertainty in viscosity is 1.5% along the saturated-liquid line, 3% in the liquid phase, 0.5% in the dilute gas, 3% to 5% in the vapor phase, and 5% in the supercritical region, rising to 8% at pressures above 40 MPa. Below 200 K, the uncertainty is 8%. The uncertainty in thermal conductivity is 5%.

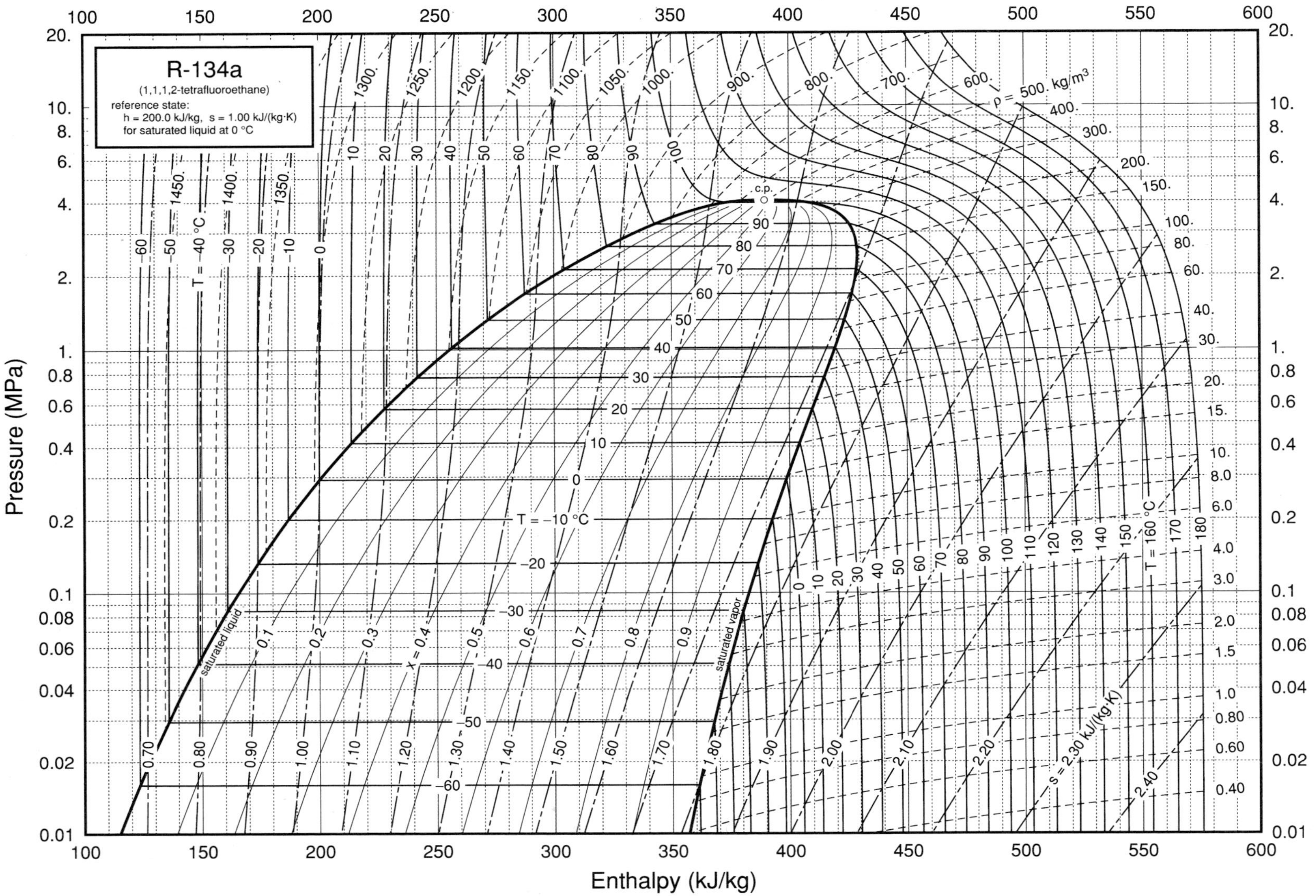

FIG. 2-14 Pressure-enthalpy diagram for Refrigerant 134a. Properties computed with the NIST REFPROP Database, Version 7.0 (Lemmon, E. W., M. O. McLinden, and M. L. Huber, 2002, NIST Standard Reference Database 23, NIST Reference Fluid Thermodynamic and Transport Properties—REFPROP, Version 7.0, Standard Reference Data Program, National Institute of Standards and Technology), based on the equation of state of Tillner-Roth, R., and H. D. Baehr, "An International Standard Formulation of the Thermodynamic Properties of 1,1,1,2-Tetrafluoroethane (HFC-134a) Covering Temperatures from 170 K to 455 K at Pressures up to 70 MPa," *J. Phys. Chem. Ref. Data* **23**(5): 657–729, 1994.

TABLE 2-128 Thermodynamic Properties of R-143a, 1,1,1-Trifluoroethane

Temperature K	Pressure MPa	Density mol/dm³	Volume dm³/mol	Int. energy kJ/mol	Enthalpy kJ/mol	Entropy kJ/(mol·K)	C_v kJ/(mol·K)	C_p kJ/(mol·K)	Sound speed m/s	Joule-Thomson K/MPa
					Saturated Properties					
161.34	0.0010749	15.832	0.063163	4.4138	4.4138	0.026403	0.068393	0.10179	1058.1	−0.43936
170.00	0.0025084	15.583	0.064174	5.2969	5.2971	0.031735	0.068179	0.10225	1016.7	−0.42914
180.00	0.0059324	15.291	0.065399	6.3240	6.3244	0.037606	0.068405	0.10325	969.14	−0.41402
190.00	0.012629	14.994	0.066693	7.3629	7.3637	0.043223	0.068990	0.10460	921.61	−0.39585
200.00	0.024624	14.692	0.068062	8.4164	8.4181	0.048626	0.069825	0.10621	874.04	−0.37472
210.00	0.044602	14.384	0.069519	9.4869	9.4900	0.053849	0.070836	0.10803	826.46	−0.35034
220.00	0.075908	14.069	0.071077	10.576	10.581	0.058915	0.071969	0.11005	778.88	−0.32211
230.00	0.12252	13.745	0.072753	11.685	11.694	0.063848	0.073190	0.11227	731.28	−0.28906
240.00	0.18902	13.410	0.074570	12.817	12.831	0.068665	0.074475	0.11474	683.65	−0.24979
250.00	0.28049	13.062	0.076556	13.973	13.995	0.073385	0.075809	0.11750	635.90	−0.20231
260.00	0.40251	12.698	0.078751	15.156	15.188	0.078027	0.077186	0.12066	587.94	−0.14368
270.00	0.56112	12.314	0.081208	16.368	16.414	0.082607	0.078605	0.12435	539.61	−0.069548
280.00	0.76276	11.904	0.084004	17.615	17.680	0.087149	0.080078	0.12879	490.70	0.026895
290.00	1.0144	11.461	0.087249	18.903	18.991	0.091675	0.081625	0.13438	440.95	0.15683
300.00	1.3234	10.975	0.091119	20.239	20.360	0.096221	0.083293	0.14180	389.93	0.34002
310.00	1.6983	10.426	0.095914	21.638	21.801	0.10083	0.085172	0.15244	337.04	0.61491
320.00	2.1483	9.7846	0.10220	23.125	23.344	0.10559	0.087455	0.16980	281.21	1.0681
330.00	2.6850	8.9829	0.11132	24.750	25.048	0.11065	0.090641	0.20591	220.39	1.9469
340.00	3.3250	7.7913	0.12835	26.688	27.114	0.11659	0.096654	0.35704	149.32	4.3897
345.86	3.7618	5.1285	0.19499	29.429	30.163	0.12527			0	12.397
161.34	0.0010749	0.00080362	1244.4	25.521	26.859	0.16552	0.044397	0.052938	137.57	385.09
170.00	0.0025084	0.0017832	560.78	25.895	27.302	0.16118	0.046371	0.055037	140.62	262.77
180.00	0.0059324	0.0039967	250.20	26.340	27.824	0.15705	0.048691	0.057550	143.91	176.43
190.00	0.012629	0.0081006	123.45	26.796	28.355	0.15370	0.051040	0.060156	146.92	124.70
200.00	0.024624	0.015110	66.180	27.262	28.892	0.15100	0.053424	0.062886	149.60	92.835
210.00	0.044602	0.026311	38.007	27.736	29.431	0.14881	0.055867	0.065796	151.89	72.442
220.00	0.075908	0.043269	23.111	28.213	29.968	0.14704	0.058395	0.068954	153.71	58.764
230.00	0.12252	0.067850	14.738	28.693	30.498	0.14560	0.061026	0.072428	155.02	49.133
240.00	0.18902	0.10227	9.7783	29.170	31.018	0.14444	0.063766	0.076287	155.74	42.049
250.00	0.28049	0.14916	6.7041	29.641	31.521	0.14349	0.066612	0.080611	155.81	36.657
260.00	0.40251	0.21175	4.7225	30.102	32.003	0.14270	0.069560	0.085515	155.15	32.456
270.00	0.56112	0.29409	3.4004	30.548	32.456	0.14202	0.072606	0.091184	153.70	29.136
280.00	0.76276	0.40144	2.4910	30.972	32.872	0.14141	0.075756	0.097932	151.36	26.498
290.00	1.0144	0.54109	1.8481	31.366	33.240	0.14081	0.079035	0.10632	148.04	24.406
300.00	1.3234	0.72367	1.3818	31.715	33.544	0.14017	0.082489	0.11743	143.59	22.765
310.00	1.6983	0.96617	1.0350	32.000	33.758	0.13940	0.086214	0.13359	137.85	21.499
320.00	2.1483	1.2991	0.76974	32.183	33.837	0.13838	0.090400	0.16076	130.56	20.526
330.00	2.6850	1.7898	0.55871	32.184	33.684	0.13682	0.095488	0.22018	121.34	19.682
340.00	3.3250	2.6696	0.37458	31.732	32.977	0.13384	0.10298	0.47999	109.29	18.259
345.86	3.7618	5.1285	0.19499	29.429	30.163	0.12527			0	12.397

Single-Phase Properties

200.00	0.10000	14.694	0.068054	8.4143	8.4211	0.048616	0.069829	0.10620	874.45	−0.37493
225.63	0.10000	13.888	0.072006	11.198	11.205	0.061709	0.072648	0.11128	752.07	−0.30416
225.63	0.10000	0.056043	17.843	28.483	30.268	0.14619	0.059863	0.070868	154.52	52.958
300.00	0.10000	0.040759	24.534	33.380	35.833	0.16745	0.070758	0.079793	179.93	18.414
400.00	0.10000	0.030247	33.061	41.260	44.566	0.19247	0.086121	0.094695	207.36	7.5341
500.00	0.10000	0.024114	41.469	50.551	54.698	0.21503	0.099065	0.10751	231.13	4.1010
600.00	0.10000	0.020066	49.836	61.004	65.987	0.23558	0.10950	0.11790	252.55	2.5191
200.00	1.0000	14.715	0.067956	8.3891	8.4570	0.048489	0.069866	0.10604	879.27	−0.37736
289.48	1.0000	11.485	0.087067	18.835	18.922	0.091441	0.081543	0.13406	443.55	0.14902
289.48	1.0000	0.53292	1.8765	31.346	33.223	0.14084	0.078861	0.10583	148.24	24.503
300.00	1.0000	0.49298	2.0285	32.269	34.298	0.14449	0.077834	0.099455	155.53	20.787
400.00	1.0000	0.32001	3.1249	40.778	43.903	0.17212	0.087343	0.098681	198.53	7.7199
500.00	1.0000	0.24665	4.0544	50.244	54.298	0.19527	0.099519	0.10927	227.30	4.0976
600.00	1.0000	0.20246	4.9391	60.781	65.720	0.21607	0.10975	0.11891	251.11	2.4913
200.00	5.0000	14.806	0.067539	8.2811	8.6188	0.047943	0.070021	0.10541	900.01	−0.38724
300.00	5.0000	11.380	0.087876	19.812	20.251	0.094764	0.082741	0.13222	452.11	0.11636
400.00	5.0000	2.2504	0.44436	38.004	40.225	0.15165	0.093496	0.13815	159.25	8.1026
500.00	5.0000	1.3639	0.73318	48.802	52.468	0.17903	0.10135	0.11872	213.49	3.8903
600.00	5.0000	1.0491	0.95318	59.789	64.554	0.20106	0.11075	0.12364	246.86	2.2996
200.00	10.000	14.913	0.067054	8.1545	8.8250	0.047292	0.070195	0.10473	924.61	−0.39776
300.00	10.000	11.776	0.084916	19.382	20.231	0.093258	0.082583	0.12585	514.06	−0.037998
400.00	10.000	6.3531	0.15740	33.679	35.253	0.13608	0.095653	0.17697	196.98	2.9315
500.00	10.000	3.0122	0.33199	46.933	50.252	0.16976	0.10303	0.13206	211.16	3.0876
600.00	10.000	2.1596	0.46305	58.580	63.211	0.19339	0.11178	0.12945	248.28	1.9304
200.00	50.000	15.598	0.064113	7.3673	10.573	0.042937	0.070934	0.10170	1089.7	−0.44295
300.00	50.000	13.358	0.074862	17.612	21.355	0.086486	0.083360	0.11389	794.12	−0.33796
400.00	50.000	11.268	0.088747	28.859	33.296	0.12077	0.096003	0.12454	590.09	−0.20571
500.00	50.000	9.4018	0.10636	40.839	46.157	0.14943	0.10668	0.13213	478.93	−0.077157
600.00	50.000	7.8978	0.12662	53.306	59.637	0.17399	0.11548	0.13720	431.69	−0.0052131
300.00	100.00	14.343	0.069721	16.500	23.472	0.081539	0.084059	0.11166	1008.9	−0.40055
400.00	100.00	12.767	0.078330	27.298	35.131	0.11502	0.097186	0.12126	833.69	−0.35302
500.00	100.00	11.435	0.087453	38.923	47.668	0.14296	0.10821	0.12921	723.22	−0.31534
600.00	100.00	10.314	0.096952	51.227	60.923	0.16711	0.11716	0.13566	656.57	−0.28902

The values in these tables were generated from the NIST REFPROP software (Lemmon, E. W., McLinden, M. O., and Huber, M. L., NIST Standard Reference Database 23: Reference Fluid Thermodynamic and Transport Properties—REFPROP, National Institute of Standards and Technology, Standard Reference Data Program, Gaithersburg, Md., 2002, Version 7.1). The primary source for the thermodynamic properties is Lemmon, E. W., and Jacobsen, R. T., "An International Standard Formulation for the Thermodynamic Properties of 1,1,1-Trifluoroethane (HFC-143a) for Temperatures from 161 to 450 K and Pressures to 50 MPa," *J. Phys. Chem. Ref. Data* **29**(4):521–552, 2000. Validated equations for the viscosity and thermal conductivity are not currently available for this fluid.

Properties at the triple point temperature and the critical point temperature are given in the first and last entries of the saturation tables, respectively. In the single-phase table, when the temperature range for a given isobar includes a vapor-liquid phase boundary, the temperature of phase equilibrium is noted, and properties for both the saturated liquid and saturated vapor are given (with liquid properties given in the upper line). Lines are omitted from the temperature-pressure grid of the single-phase table, when the system would be in the solid phase or if there are potential problems with the source property surface.

The estimated uncertainties of properties calculated using the equation of state are 0.1% in density, 0.5% in heat capacities, 0.02% in the speed of sound for the vapor at pressures less than 1 MPa, 0.5% in speed of sound elsewhere, and 0.1% in vapor pressure, except in the critical region.

TABLE 2-129 Thermodynamic Properties of R-404A

Temperature K	Pressure MPa	Density mol/dm^3	Volume dm^3/mol	Int. energy kJ/mol	Enthalpy kJ/mol	Entropy kJ/(mol·K)	C_v kJ/(mol·K)	C_p kJ/(mol·K)	Sound speed m/s	Joule-Thomson K/MPa
					Saturated Properties					
200.00	0.022649	14.209	0.070377	10.353	10.355	0.058867	0.076939	0.11881	859.89	−0.34161
205.00	0.030989	14.059	0.071131	10.948	10.950	0.061803	0.077522	0.11915	831.56	−0.33384
210.00	0.041658	13.907	0.071905	11.544	11.547	0.064678	0.078116	0.11965	804.42	−0.32460
215.00	0.055101	13.755	0.072703	12.144	12.148	0.067498	0.078719	0.12028	778.20	−0.31394
220.00	0.071804	13.600	0.073527	12.746	12.751	0.070269	0.079326	0.12103	752.69	−0.30185
225.00	0.092293	13.444	0.074380	13.353	13.359	0.072995	0.079940	0.12188	727.72	−0.28830
230.00	0.11713	13.286	0.075265	13.963	13.972	0.075679	0.080558	0.12282	703.16	−0.27318
235.00	0.14693	13.126	0.076185	14.578	14.590	0.078326	0.081183	0.12386	678.91	−0.25636
240.00	0.18232	12.963	0.077145	15.199	15.213	0.080939	0.081815	0.12499	654.88	−0.23766
245.00	0.22397	12.796	0.078147	15.824	15.842	0.083520	0.082456	0.12621	631.01	−0.21683
250.00	0.27258	12.627	0.079197	16.456	16.478	0.086073	0.083107	0.12754	607.23	−0.19356
255.00	0.32888	12.453	0.080301	17.094	17.120	0.088600	0.083770	0.12899	583.50	−0.16749
260.00	0.39363	12.275	0.081465	17.738	17.770	0.091104	0.084446	0.13057	559.77	−0.13814
265.00	0.46763	12.092	0.082697	18.390	18.429	0.093589	0.085137	0.13229	535.99	−0.10493
270.00	0.55168	11.904	0.084006	19.049	19.096	0.096056	0.085846	0.13419	512.13	−0.067104
275.00	0.64664	11.709	0.085402	19.717	19.772	0.098510	0.086574	0.13630	488.15	−0.023728
280.00	0.75338	11.508	0.086899	20.394	20.460	0.10095	0.087326	0.13866	464.02	0.026418
285.00	0.87280	11.298	0.088513	21.081	21.159	0.10339	0.088104	0.14133	439.69	0.084928
290.00	1.0059	11.078	0.090266	21.780	21.870	0.10583	0.088914	0.14438	415.13	0.15392
295.00	1.1536	10.848	0.092182	22.490	22.597	0.10826	0.089761	0.14793	390.28	0.23626
300.00	1.3169	10.605	0.094295	23.215	23.339	0.11071	0.090653	0.15211	365.11	0.33594
305.00	1.4970	10.346	0.096652	23.956	24.101	0.11317	0.091603	0.15715	339.56	0.45865
310.00	1.6950	10.069	0.099314	24.717	24.885	0.11566	0.092625	0.16339	313.56	0.61280
315.00	1.9122	9.7686	0.10237	25.500	25.695	0.11818	0.093744	0.17139	287.00	0.81141
320.00	2.1499	9.4384	0.10595	26.310	26.538	0.12075	0.094998	0.18212	259.73	1.0757
325.00	2.4096	9.0688	0.11027	27.157	27.423	0.12341	0.096455	0.19751	231.53	1.4433
330.00	2.6932	8.6431	0.11570	28.054	28.365	0.12619	0.098241	0.22197	201.95	1.9881
335.00	3.0027	8.1285	0.12302	29.026	29.396	0.12918	0.10064	0.26871	170.20	2.8807
340.00	3.3414	7.4362	0.13448	30.147	30.597	0.13261	0.10445	0.40392	134.59	4.6353
345.00	3.7150	5.7429	0.17413	32.108	32.755	0.13874	0.11650	8.2559	89.976	10.564
345.27	3.7348	4.9400	0.20243	32.875	33.631	0.14126			0	12.409
200.00	0.021264	0.013010	76.866	29.920	31.555	0.16521	0.058696	0.068032	138.13	88.073
205.00	0.029285	0.017550	56.979	30.185	31.853	0.16408	0.059968	0.069509	139.28	77.215
210.00	0.039592	0.023271	42.971	30.451	32.152	0.16307	0.061245	0.071021	140.34	68.305
215.00	0.052629	0.030378	32.919	30.718	32.451	0.16217	0.062526	0.072573	141.29	60.948
220.00	0.068883	0.039095	25.579	30.987	32.749	0.16138	0.063815	0.074172	142.12	54.835
225.00	0.088879	0.049667	20.134	31.256	33.046	0.16068	0.065113	0.075826	142.84	49.721
230.00	0.11318	0.062359	16.036	31.525	33.340	0.16006	0.066424	0.077546	143.43	45.413
235.00	0.14240	0.077463	12.909	31.795	33.633	0.15952	0.067750	0.079344	143.88	41.761
240.00	0.17718	0.095292	10.494	32.063	33.922	0.15903	0.069095	0.081232	144.18	38.642
245.00	0.21817	0.11619	8.6063	32.330	34.208	0.15861	0.070463	0.083226	144.33	35.962
250.00	0.26610	0.14055	7.1149	32.596	34.489	0.15823	0.071855	0.085343	144.32	33.645
255.00	0.32169	0.16879	5.9247	32.859	34.765	0.15789	0.073276	0.087604	144.14	31.630
260.00	0.38571	0.20137	4.9659	33.119	35.034	0.15759	0.074728	0.090033	143.77	29.871
265.00	0.45896	0.23885	4.1867	33.375	35.296	0.15732	0.076214	0.092660	143.22	28.327
270.00	0.54225	0.28183	3.5483	33.626	35.550	0.15707	0.077737	0.095524	142.46	26.970
275.00	0.63645	0.33102	3.0210	33.872	35.794	0.15684	0.079300	0.098671	141.49	25.773
280.00	0.74245	0.38725	2.5823	34.111	36.028	0.15661	0.080909	0.10217	140.29	24.718
285.00	0.86115	0.45152	2.2148	34.342	36.249	0.15639	0.082567	0.10609	138.85	23.789
290.00	0.99353	0.52501	1.9047	34.563	36.455	0.15616	0.084282	0.11056	137.16	22.972
295.00	1.1406	0.60922	1.6414	34.773	36.645	0.15592	0.086062	0.11574	135.19	22.256
300.00	1.3034	0.70599	1.4165	34.968	36.814	0.15566	0.087917	0.12186	132.92	21.633
305.00	1.4830	0.81772	1.2229	35.147	36.960	0.15536	0.089863	0.12929	130.34	21.094

310.00	1.6806	0.94761	1.0553	35.304	37.078	0.15501	0.091922	0.13856	127.41	20.630
315.00	1.8975	1.1001	0.90903	35.435	37.159	0.15459	0.094123	0.15062	124.10	20.234
320.00	2.1351	1.2815	0.78032	35.530	37.196	0.15408	0.096513	0.16713	120.38	19.889
325.00	2.3950	1.5019	0.66583	35.578	37.173	0.15343	0.099162	0.19141	116.21	19.571
330.00	2.6789	1.7781	0.56239	35.558	37.065	0.15257	0.10220	0.23111	111.51	19.233
335.00	2.9893	2.1438	0.46645	35.429	36.824	0.15136	0.10585	0.30867	106.19	18.763
340.00	3.3299	2.6882	0.37199	35.084	36.323	0.14946	0.11074	0.53035	100.03	17.851
345.00	3.7109	4.2113	0.23746	33.615	34.496	0.14379	0.12022	8.6291	90.307	14.130
345.27	3.7348	4.9400	0.20243	32.875	33.631	0.14126			0	12.409
Single-Phase Properties										
226.65	0.10000	13.392	0.074669	13.554	13.561	0.073887	0.080143	0.12218	719.55	−0.28348
227.41	0.10000	0.055492	18.021	31.386	33.188	0.16038	0.065742	0.076645	143.14	47.558
300.00	0.10000	0.040750	24.540	36.596	39.050	0.18269	0.076875	0.085907	166.24	16.998
400.00	0.10000	0.030243	33.066	45.121	48.428	0.20956	0.092844	0.10141	191.81	6.9229
500.00	0.10000	0.024113	41.471	55.103	59.250	0.23365	0.10612	0.11456	213.92	3.7455
289.79	1.0000	11.088	0.090189	21.750	21.840	0.10572	0.088879	0.14425	416.16	0.15078
290.23	1.0000	0.52866	1.8916	34.573	36.464	0.15615	0.084363	0.11078	137.07	22.936
300.00	1.0000	0.49205	2.0323	35.486	37.518	0.15972	0.083854	0.10554	143.42	19.634
400.00	1.0000	0.31957	3.1292	44.644	47.773	0.18922	0.094142	0.10545	183.69	7.1401
500.00	1.0000	0.24648	4.0571	54.804	58.861	0.21391	0.10659	0.11631	210.41	3.7496
300.00	5.0000	10.994	0.090955	22.770	23.225	0.10919	0.089725	0.14193	427.56	0.11428
400.00	5.0000	2.2256	0.44932	41.867	44.113	0.16875	0.10069	0.14547	148.00	7.6582
500.00	5.0000	1.3561	0.73741	53.389	57.076	0.19774	0.10859	0.12579	198.35	3.5826
300.00	10.000	11.371	0.087944	22.323	23.203	0.10763	0.089307	0.13525	489.01	−0.035312
400.00	10.000	6.1241	0.16329	37.557	39.190	0.15323	0.10301	0.18569	184.17	2.8522
500.00	10.000	2.9622	0.33758	51.539	54.915	0.18852	0.11046	0.13925	198.11	2.8527
300.00	25.000	12.107	0.082594	21.419	23.484	0.10432	0.089393	0.12713	614.12	−0.22002
400.00	25.000	9.2730	0.10784	34.286	36.982	0.14304	0.10166	0.14227	381.56	0.20064
500.00	25.000	6.6326	0.15077	47.758	51.527	0.17548	0.11215	0.14633	292.53	0.65319
300.00	50.000	12.867	0.077719	20.472	24.358	0.10057	0.090168	0.12262	753.19	−0.32391
400.00	50.000	10.834	0.092300	32.538	37.153	0.13731	0.10251	0.13302	559.24	−0.19415
500.00	50.000	9.0225	0.11083	45.317	50.859	0.16786	0.11339	0.14051	455.37	−0.070101

The values in these tables were generated from the NIST REFPROP software (Lemmon, E. W., McLinden, M. O., and Huber, M. L., NIST Standard Reference Database 23: Reference Fluid Thermodynamic and Transport Properties—REFPROP, National Institute of Standards and Technology, Standard Reference Data Program, Gaithersburg, Md., 2002, Version 7.1). The primary source for the thermodynamic properties is Lemmon, E. W., "Pseudo Pure-Fluid Equations of State for the Refrigerant Blends R-410A, R-404A, R-507A, and R-407C," *Int. J. Thermophys.* **24**(4):991–1006, 2003. Validated equations for the viscosity and thermal conductivity are not currently available for this fluid.

Properties at the critical point temperature are given in the last entry of the saturation tables. In the single-phase table, when the temperature range for a given isobar includes a vapor-liquid phase boundary, the temperature of phase equilibrium is noted, and properties for both the saturated liquid and saturated vapor are given (with liquid properties given in the upper line). Lines are omitted from the temperature-pressure grid of the single-phase table, when the system would be in the solid phase or if there are potential problems with the source property surface.

The estimated uncertainty of density values calculated with the equation of state is 0.1%. The estimated uncertainty of calculated heat capacities and speed of sound values is 0.5%. Uncertainties of bubble and dew point pressures are 0.5%.

TABLE 2-130 Thermodynamic Properties of R-407C

Temperature K	Pressure MPa	Density mol/dm³	Volume dm³/mol	Int. energy kJ/mol	Enthalpy kJ/mol	Entropy kJ/(mol·K)	C_v kJ/(mol·K)	C_p kJ/(mol·K)	Sound speed m/s	Joule-Thomson K/MPa
Saturated Properties										
200.00	0.019158	17.036	0.058698	8.8272	8.8283	0.050593	0.070988	0.11073	956.60	−0.31996
210.00	0.035795	16.697	0.059892	9.9359	9.9380	0.056002	0.071320	0.11118	903.06	−0.30662
220.00	0.062640	16.352	0.061156	11.051	11.055	0.061189	0.071817	0.11203	851.40	−0.28934
230.00	0.10366	15.999	0.062503	12.175	12.182	0.066188	0.072410	0.11319	801.12	−0.26790
240.00	0.16353	15.637	0.063949	13.312	13.323	0.071026	0.073074	0.11465	751.77	−0.24161
250.00	0.24755	15.264	0.065512	14.464	14.480	0.075728	0.073803	0.11641	703.01	−0.20937
260.00	0.36157	14.877	0.067218	15.632	15.657	0.080314	0.074597	0.11853	654.53	−0.16951
270.00	0.51193	14.472	0.069099	16.822	16.857	0.084805	0.075463	0.12111	606.09	−0.11964
280.00	0.70540	14.045	0.071198	18.035	18.085	0.089223	0.076412	0.12427	557.44	−0.056172
290.00	0.94916	13.591	0.073576	19.278	19.348	0.093590	0.077458	0.12822	508.32	0.026372
300.00	1.2507	13.102	0.076322	20.555	20.651	0.097931	0.078624	0.13331	458.46	0.13683
310.00	1.6182	12.567	0.079573	21.877	22.006	0.10228	0.079950	0.14013	407.51	0.29038
320.00	2.0599	11.969	0.083552	23.255	23.427	0.10668	0.081509	0.14989	354.98	0.51547
330.00	2.5851	11.278	0.088671	24.711	24.940	0.11119	0.083453	0.16541	300.07	0.87274
340.00	3.2038	10.435	0.095832	26.287	26.594	0.11596	0.086157	0.19551	241.20	1.5203
350.00	3.9255	9.2661	0.10792	28.110	28.534	0.12137	0.090943	0.28993	174.57	3.0499
359.35	4.6317	5.2600	0.19011	32.145	33.025	0.13372			0	10.947
200.00	0.011312	0.0068643	145.68	30.051	31.699	0.16726	0.048920	0.057805	149.59	109.12
210.00	0.022624	0.013151	76.041	30.504	32.224	0.16412	0.050967	0.060200	152.36	88.122
220.00	0.041929	0.023450	42.644	30.957	32.745	0.16151	0.053143	0.062854	154.78	72.006
230.00	0.072846	0.039384	25.391	31.409	33.259	0.15933	0.055439	0.065784	156.79	59.560
240.00	0.11979	0.062913	15.895	31.857	33.761	0.15749	0.057839	0.069010	158.33	49.904
250.00	0.18793	0.096374	10.376	32.298	34.248	0.15593	0.060328	0.072563	159.36	42.378
260.00	0.28317	0.14256	7.0147	32.728	34.715	0.15460	0.062897	0.076500	159.80	36.483
270.00	0.41203	0.20484	4.8820	33.145	35.157	0.15343	0.065542	0.080917	159.60	31.840
280.00	0.58173	0.28739	3.4796	33.544	35.568	0.15240	0.068266	0.085971	158.69	28.163
290.00	0.80008	0.39560	2.5278	33.918	35.940	0.15144	0.071085	0.091920	156.99	25.236
300.00	1.0757	0.53670	1.8632	34.259	36.263	0.15051	0.074027	0.099203	154.41	22.898
310.00	1.4179	0.72101	1.3869	34.556	36.523	0.14956	0.077140	0.10861	150.83	21.024
320.00	1.8375	0.96439	1.0369	34.790	36.696	0.14852	0.080507	0.12170	146.11	19.516
330.00	2.3470	1.2939	0.77283	34.931	36.745	0.14727	0.084283	0.14208	140.03	18.284
340.00	2.9627	1.7642	0.56682	34.916	36.595	0.14560	0.088801	0.18030	132.24	17.206
350.00	3.7100	2.5260	0.39588	34.578	36.047	0.14299	0.095065	0.28700	122.09	15.951
359.35	4.6317	5.2600	0.19011	32.145	33.025	0.13372			0	10.947

Single-Phase Properties

200.00	0.10000	17.038	0.058692	8.8253	8.8312	0.050583	0.070990	0.11072	956.99	−0.32010
229.25	0.10000	16.026	0.062399	12.091	12.097	0.065819	0.072363	0.11310	804.85	−0.26966
236.25	0.10000	0.053062	18.846	31.690	33.574	0.15814	0.056928	0.067764	157.81	53.242
300.00	0.10000	0.040722	24.557	35.535	37.991	0.17467	0.063341	0.072378	179.00	20.041
400.00	0.10000	0.030231	33.079	42.554	45.862	0.19722	0.076588	0.085147	205.99	7.7925
500.00	0.10000	0.024109	41.479	50.849	54.997	0.21756	0.088895	0.097330	229.27	4.0471
291.84	1.0000	13.504	0.074050	19.510	19.584	0.094388	0.077662	0.12906	499.23	0.044233
297.47	1.0000	0.49738	2.0105	34.177	36.187	0.15075	0.073268	0.097199	155.15	23.442
300.00	1.0000	0.48865	2.0465	34.384	36.431	0.15156	0.072744	0.095419	156.77	22.566
400.00	1.0000	0.31821	3.1425	42.101	45.244	0.17694	0.078067	0.089213	198.26	7.9701
500.00	1.0000	0.24608	4.0637	50.576	54.639	0.19786	0.089416	0.099001	225.88	4.0390
300.00	5.0000	13.412	0.074559	20.240	20.613	0.096862	0.078093	0.12762	507.10	0.027202
400.00	5.0000	2.1880	0.45703	39.458	41.743	0.15675	0.086188	0.12964	161.10	8.5257
500.00	5.0000	1.3504	0.74050	49.289	52.992	0.18193	0.091753	0.10811	213.00	3.9036
300.00	10.000	13.740	0.072780	19.898	20.626	0.095679	0.077756	0.12301	558.94	−0.063246
400.00	10.000	7.1029	0.14079	34.426	35.834	0.13888	0.090433	0.20031	184.71	3.3408
500.00	10.000	2.9957	0.33381	47.547	50.885	0.17282	0.094263	0.12254	207.67	3.3327
300.00	25.000	14.443	0.069238	19.146	20.877	0.092972	0.077634	0.11624	672.43	−0.19834
400.00	25.000	10.899	0.091752	30.990	33.284	0.12855	0.087056	0.13223	399.92	0.28513
500.00	25.000	7.3363	0.13631	43.479	46.886	0.15889	0.096319	0.13592	289.22	0.97116
300.00	50.000	15.220	0.065703	18.302	21.587	0.089730	0.078160	0.11176	802.78	−0.28843
400.00	50.000	12.648	0.079064	29.255	33.209	0.12310	0.087179	0.12077	579.57	−0.12975
500.00	50.000	10.260	0.097468	40.787	45.660	0.15086	0.096753	0.12761	457.54	0.047330

The values in these tables were generated from the NIST REFPROP software (Lemmon, E. W., McLinden, M. O., and Huber, M. L., NIST Standard Reference Database 23: Reference Fluid Thermodynamic and Transport Properties—REFPROP, National Institute of Standards and Technology, Standard Reference Data Program, Gaithersburg, Md., 2002, Version 7.1). The primary source for the thermodynamic properties is Lemmon, E. W., "Pseudo Pure-Fluid Equations of State for the Refrigerant Blends R-410A, R-404A, R-507A, and R-407C," *Int. J. Thermophys.* **24**(4):991–1006, 2003. Validated equations for the viscosity and thermal conductivity are not currently available for this fluid.

Properties at the critical point temperature are given in the last entry of the saturation tables. In the single-phase table, when the temperature range for a given isobar includes a vapor-liquid phase boundary, the temperature of phase equilibrium is noted, and properties for both the saturated liquid and saturated vapor are given (with liquid properties given in the upper line). Lines are omitted from the temperature-pressure grid of the single-phase table, when the system would be in the solid phase or if there are potential problems with the source property surface.

The estimated uncertainty of density values calculated with the equation of state is 0.1%. The estimated uncertainty of calculated heat capacities and speed of sound values is 0.5%. Uncertainties of bubble and dew point pressures are 0.5%.

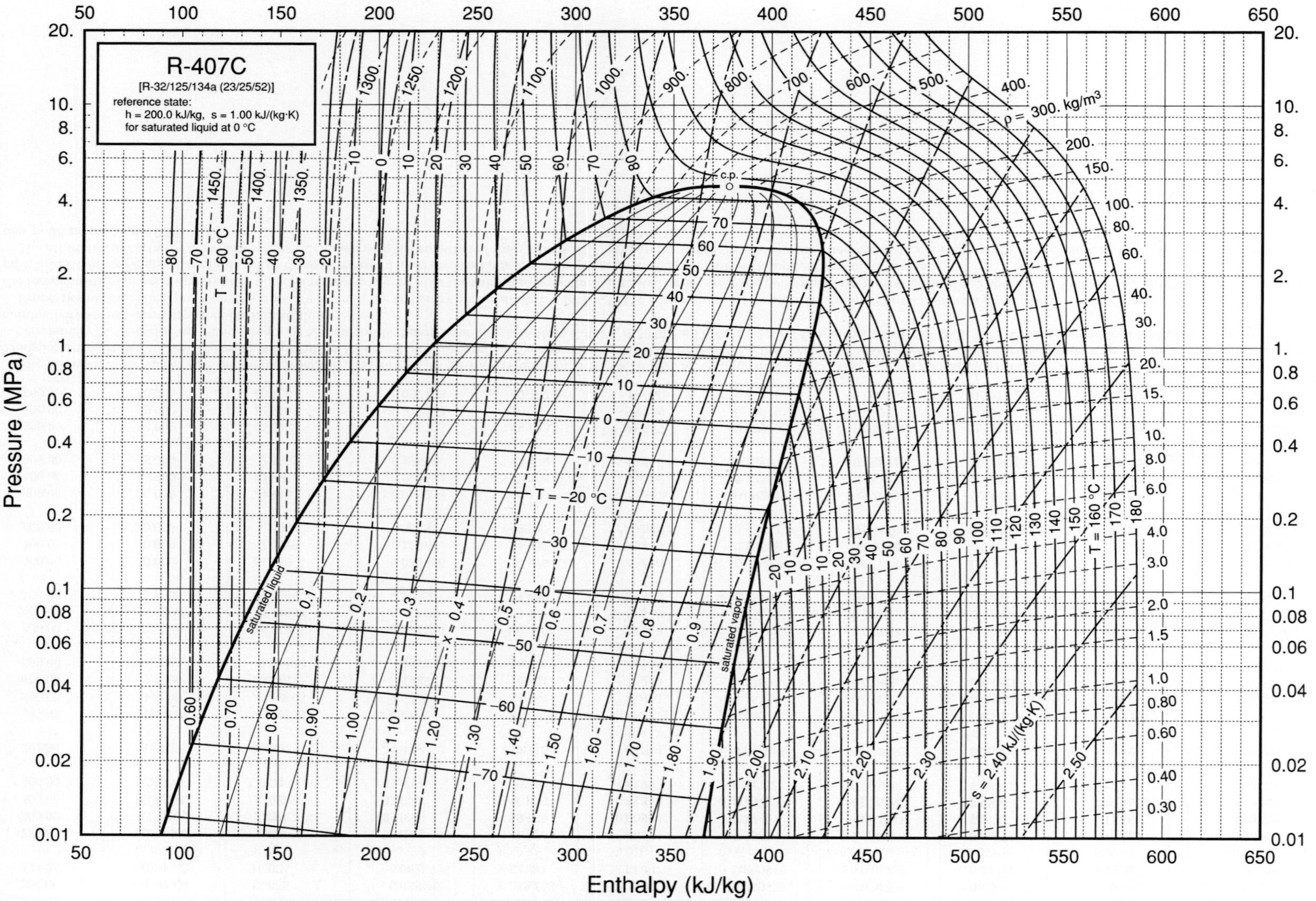

FIG. 2-15 Pressure-enthalpy diagram for Refrigerant 407C. Properties computed with the NIST REFPROP Database, Version 7.0 (Lemmon, E. W., M. O. McLinden, and M. L. Huber, 2002, NIST Standard Reference Database 23, NIST Reference Fluid Thermodynamic and Transport Properties—REFPROP, Version 7.0, Standard Reference Data Program, National Institute of Standards and Technology), based on the mixture model of Lemmon, E. W., and R. T. Jacobsen, "Equations of State for Mixtures of R-32, R-125, R-134a, R-143a, and R-152a," *J. Phys. Chem. Ref. Data* **33**: 593–620, 2004.

TABLE 2-131 Thermodynamic Properties of R-410A

Temperature K	Pressure MPa	Density mol/dm³	Volume dm³/mol	Int. energy kJ/mol	Enthalpy kJ/mol	Entropy kJ/(mol·K)	C_v kJ/(mol·K)	C_p kJ/(mol·K)	Sound speed m/s	Joule-Thomson K/MPa
					Saturated Properties					
200.00	0.029160	19.510	0.051256	7.0380	7.0395	0.040995	0.062260	0.097942	929.01	−0.30179
210.00	0.053727	19.093	0.052375	8.0188	8.0217	0.045781	0.062050	0.098396	879.84	−0.28524
215.00	0.071143	18.881	0.052962	8.5112	8.5149	0.048098	0.062014	0.098729	855.20	−0.27544
220.00	0.092819	18.667	0.053571	9.0052	9.0101	0.050370	0.062020	0.099138	830.52	−0.26446
225.00	0.11946	18.449	0.054202	9.5012	9.5077	0.052600	0.062066	0.099628	805.81	−0.25217
230.00	0.15182	18.229	0.054858	9.9997	10.008	0.054791	0.062151	0.10020	781.06	−0.23841
235.00	0.19070	18.005	0.055542	10.501	10.512	0.056948	0.062271	0.10088	756.26	−0.22300
240.00	0.23697	17.776	0.056255	11.006	11.019	0.059073	0.062426	0.10165	731.41	−0.20574
245.00	0.29152	17.543	0.057002	11.514	11.530	0.061169	0.062615	0.10253	706.48	−0.18637
250.00	0.35531	17.305	0.057786	12.026	12.047	0.063240	0.062837	0.10353	681.45	−0.16459
255.00	0.42933	17.062	0.058611	12.543	12.568	0.065289	0.063092	0.10466	656.31	−0.14006
260.00	0.51461	16.812	0.059482	13.065	13.096	0.067318	0.063380	0.10594	631.02	−0.11232
265.00	0.61223	16.555	0.060406	13.593	13.630	0.069331	0.063701	0.10738	605.55	−0.080861
270.00	0.72330	16.290	0.061388	14.127	14.172	0.071331	0.064057	0.10902	579.88	−0.045000
275.00	0.84899	16.016	0.062439	14.669	14.722	0.073321	0.064451	0.11088	553.95	−0.0039006
280.00	0.99048	15.732	0.063567	15.218	15.281	0.075304	0.064884	0.11300	527.72	0.043515
285.00	1.1490	15.436	0.064785	15.776	15.851	0.077284	0.065363	0.11543	501.14	0.098651
290.00	1.3260	15.127	0.066109	16.344	16.432	0.079266	0.065893	0.11825	474.14	0.16337
295.00	1.5226	14.802	0.067559	16.924	17.026	0.081254	0.066483	0.12156	446.66	0.24022
300.00	1.7404	14.459	0.069160	17.516	17.636	0.083253	0.067147	0.12550	418.60	0.33275
305.00	1.9809	14.095	0.070948	18.123	18.263	0.085270	0.067901	0.13029	389.87	0.44607
310.00	2.2456	13.704	0.072969	18.747	18.911	0.087314	0.068773	0.13630	360.33	0.58788
315.00	2.5364	13.282	0.075293	19.392	19.583	0.089398	0.069800	0.14413	329.82	0.77028
320.00	2.8550	12.816	0.078025	20.064	20.287	0.091537	0.071046	0.15493	298.10	1.0135
325.00	3.2037	12.294	0.081343	20.772	21.032	0.093762	0.072616	0.17109	264.83	1.3544
330.00	3.5848	11.685	0.085578	21.531	21.837	0.096123	0.074717	0.19853	229.46	1.8665
335.00	4.0009	10.930	0.091491	22.376	22.742	0.098732	0.077843	0.25685	190.98	2.7232
340.00	4.4556	9.8413	0.10161	23.414	23.867	0.10194	0.083650	0.46832	147.49	4.4554
344.49	4.9012	6.3240	0.15813	25.988	26.763	0.11022			0	9.7623
200.00	0.029010	0.017797	56.190	26.495	28.125	0.14644	0.042482	0.052236	164.41	113.67
210.00	0.053489	0.031567	31.678	26.835	28.530	0.14345	0.044604	0.055055	167.03	90.100
215.00	0.070844	0.041089	24.338	27.002	28.726	0.14212	0.045719	0.056590	168.16	80.508
220.00	0.092447	0.052763	18.953	27.167	28.919	0.14087	0.046862	0.058205	169.16	72.155
225.00	0.11900	0.066925	14.942	27.329	29.107	0.13972	0.048026	0.059899	170.03	64.889
230.00	0.15125	0.083936	11.914	27.488	29.290	0.13864	0.049206	0.061674	170.75	58.571
235.00	0.19000	0.10420	9.5972	27.645	29.468	0.13762	0.050400	0.063533	171.32	53.077
240.00	0.23611	0.12814	7.8039	27.798	29.640	0.13667	0.051603	0.065483	171.73	48.299
245.00	0.29049	0.15625	6.4000	27.947	29.806	0.13577	0.052814	0.067535	171.97	44.137
250.00	0.35407	0.18905	5.2895	28.092	29.965	0.13492	0.054033	0.069705	172.04	40.508
255.00	0.42786	0.22714	4.4026	28.232	30.116	0.13411	0.055260	0.072011	171.93	37.336
260.00	0.51287	0.27117	3.6877	28.367	30.258	0.13333	0.056497	0.074481	171.62	34.560
265.00	0.61019	0.32190	3.1066	28.496	30.392	0.13259	0.057747	0.077148	171.11	32.123
270.00	0.72092	0.38018	2.6303	28.619	30.515	0.13187	0.059014	0.080057	170.39	29.979
275.00	0.84622	0.44702	2.2371	28.733	30.626	0.13116	0.060302	0.083265	169.45	28.087
280.00	0.98729	0.52357	1.9100	28.839	30.725	0.13047	0.061618	0.086846	168.27	26.412
285.00	1.1454	0.61123	1.6360	28.935	30.809	0.12978	0.062969	0.090901	166.84	24.924
290.00	1.3218	0.71170	1.4051	29.019	30.876	0.12908	0.064364	0.095568	165.16	23.598
295.00	1.5179	0.82707	1.2091	29.090	30.925	0.12837	0.065814	0.10104	163.20	22.410
300.00	1.7351	0.95997	1.0417	29.144	30.951	0.12764	0.067335	0.10760	160.94	21.338
305.00	1.9749	1.1138	0.89779	29.178	30.951	0.12688	0.068945	0.11566	158.36	20.365
310.00	2.2390	1.2933	0.77322	29.189	30.920	0.12606	0.070671	0.12589	155.44	19.470
315.00	2.5291	1.5048	0.66456	29.170	30.850	0.12517	0.072551	0.13943	152.13	18.632
320.00	2.8472	1.7576	0.56894	29.112	30.732	0.12418	0.074643	0.15831	148.40	17.826

(*Continued*)

TABLE 2-131 Thermodynamic Properties of R-410A *(Continued)*

Temperature K	Pressure MPa	Density mol/dm³	Volume dm³/mol	Int. energy kJ/mol	Enthalpy kJ/mol	Entropy kJ/(mol·K)	C_v kJ/(mol·K)	C_p kJ/(mol·K)	Sound speed m/s	Joule-Thomson K/MPa
Saturated Properties (*Continued*)										
325.00	3.1955	2.0668	0.48384	29.002	30.548	0.12305	0.077041	0.18670	144.16	17.018
330.00	3.5766	2.4582	0.40681	28.817	30.272	0.12169	0.079915	0.23464	139.30	16.153
335.00	3.9935	2.9848	0.33503	28.510	29.848	0.11995	0.083629	0.33370	133.59	15.123
340.00	4.4504	3.7974	0.26334	27.951	29.123	0.11740	0.089197	0.65947	126.39	13.641
344.49	4.9012	6.3240	0.15813	25.988	26.763	0.11022			0	9.7623
Single-Phase Properties										
221.45	0.10000	18.604	0.053751	9.1488	9.1541	0.051020	0.062030	0.099271	823.36	−0.26104
221.53	0.10000	0.056810	17.603	27.217	28.977	0.14051	0.047215	0.058714	169.44	69.827
300.00	0.10000	0.040605	24.628	31.028	33.491	0.15794	0.050980	0.059877	198.35	19.643
400.00	0.10000	0.030202	33.111	36.670	39.981	0.17654	0.061554	0.070067	227.37	7.7467
500.00	0.10000	0.024099	41.495	43.331	47.480	0.19323	0.071249	0.079663	252.59	4.0758
280.32	1.0000	15.713	0.063641	15.253	15.317	0.075429	0.064913	0.11314	526.05	0.046760
280.42	1.0000	0.53054	1.8849	28.848	30.733	0.13041	0.061731	0.087169	168.16	26.279
300.00	1.0000	0.46599	2.1460	30.151	32.297	0.13580	0.057548	0.075210	180.65	20.254
400.00	1.0000	0.31478	3.1769	36.304	39.481	0.15648	0.062713	0.073258	220.92	7.7768
500.00	1.0000	0.24505	4.0808	43.106	47.187	0.17364	0.071665	0.081012	249.79	4.0343
300.00	5.0000	14.870	0.067248	17.202	17.539	0.082188	0.066139	0.11773	472.56	0.17344
400.00	5.0000	1.9755	0.50621	34.349	36.880	0.13813	0.068570	0.097588	192.34	7.6786
500.00	5.0000	1.3185	0.75845	42.072	45.864	0.15821	0.073521	0.087959	239.51	3.7957
300.00	10.000	15.342	0.065180	16.830	17.482	0.080897	0.065435	0.11125	533.86	0.036775
400.00	10.000	5.7949	0.17257	30.845	32.570	0.12363	0.074099	0.16518	182.45	5.0121
500.00	10.000	2.8642	0.34914	40.710	44.202	0.14982	0.075667	0.098492	233.93	3.3106
300.00	25.000	16.289	0.061392	16.058	17.592	0.078110	0.065000	0.10273	658.09	−0.14678
400.00	25.000	11.685	0.085582	26.530	28.670	0.10987	0.072157	0.11830	379.59	0.50709
500.00	25.000	7.4115	0.13493	37.197	40.570	0.13644	0.078574	0.11515	291.31	1.2724
300.00	50.000	17.287	0.057845	15.231	18.123	0.074923	0.065499	0.097459	792.80	−0.26163
400.00	50.000	14.049	0.071182	24.722	28.281	0.10409	0.072480	0.10533	566.13	−0.063564
500.00	50.000	11.128	0.089864	34.526	39.019	0.12804	0.079657	0.10880	449.76	0.13566

The values in these tables were generated from the NIST REFPROP software (Lemmon, E. W., McLinden, M. O., and Huber, M. L., NIST Standard Reference Database 23: Reference Fluid Thermodynamic and Transport Properties—REFPROP, National Institute of Standards and Technology, Standard Reference Data Program, Gaithersburg, Md., 2002, Version 7.1). The primary source for the thermodynamic properties is Lemmon, E. W., "Pseudo Pure-Fluid Equations of State for the Refrigerant Blends R-410A, R-404A, R-507A, and R-407C," *Int. J. Thermophys.* **24**(4):991–1006, 2003. Validated equations for the viscosity and thermal conductivity are not currently available for this fluid.

Properties at the critical point temperature are given in the last entry of the saturation tables. In the single-phase table, when the temperature range for a given isobar includes a vapor-liquid phase boundary, the temperature of phase equilibrium is noted, and properties for both the saturated liquid and saturated vapor are given (with liquid properties given in the upper line). Lines are omitted from the temperature-pressure grid of the single-phase table, when the system would be in the solid phase or if there are potential problems with the source property surface.

The estimated uncertainty of density values calculated with the equation of state is 0.1%. The estimated uncertainty of calculated heat capacities and speed of sound values is 0.5%. Uncertainties of bubble and dew point pressures are 0.5%.

TABLE 2-132 Opteon™ YF (R-1234yf)

Saturation Properties—Temperature Table

Temp [°C]	Pressure [kPa]	Volume [m³/kg]		Density [kg/m³]		Enthalpy [kJ/kg]			Entropy [kJ/kg·K]		Temp [°C]
		Liquid v_f	Vapor v_g	Liquid d_f	Vapor d_g	Liquid h_f	Latent h_{fg}	Vapor h_g	Liquid s_f	Vapor s_g	
–40	62.367	0.000774	0.2635	1291.9	3.795	151.1	185.5	336.6	0.807	1.603	–40
–39	65.454	0.000776	0.2519	1289.2	3.970	152.2	185.0	337.3	0.812	1.603	–39
–38	68.661	0.000777	0.2409	1286.5	4.152	153.4	184.5	337.9	0.817	1.602	–38
–37	71.992	0.000779	0.2304	1283.8	4.340	154.6	184.0	338.6	0.822	1.602	–37
–36	75.450	0.000781	0.2205	1281.0	4.535	155.7	183.5	339.3	0.827	1.601	–36
–35	79.039	0.000782	0.2111	1278.3	4.737	156.9	183.0	339.9	0.832	1.601	–35
–34	82.761	0.000784	0.2022	1275.6	4.946	158.1	182.5	340.6	0.837	1.600	–34
–33	86.620	0.000786	0.1937	1272.8	5.162	159.3	182.0	341.3	0.842	1.600	–33
–32	90.620	0.000787	0.1857	1270.1	5.386	160.4	181.5	342.0	0.847	1.600	–32
–31	94.764	0.000789	0.1780	1267.3	5.617	161.6	181.0	342.6	0.852	1.599	–31
–30	99.056	0.000791	0.1708	1264.5	5.855	162.8	180.5	343.3	0.857	1.599	–30
–29	103.500	0.000793	0.1639	1261.8	6.102	164.0	180.0	344.0	0.861	1.599	–29
–28	108.098	0.000794	0.1573	1259.0	6.357	165.2	179.5	344.7	0.866	1.598	–28
–27	112.856	0.000796	0.1511	1256.2	6.620	166.4	178.9	345.3	0.871	1.598	–27
–26	117.775	0.000798	0.1451	1253.4	6.891	167.6	178.4	346.0	0.876	1.598	–26
–25	122.861	0.000800	0.1394	1250.5	7.171	168.8	177.9	346.7	0.881	1.598	–25
–24	128.117	0.000801	0.1340	1247.7	7.460	170.0	177.4	347.4	0.886	1.598	–24
–23	133.548	0.000803	0.1289	1244.9	7.758	171.2	176.8	348.0	0.891	1.597	–23
–22	139.155	0.000805	0.1240	1242.0	8.066	172.4	176.3	348.7	0.895	1.597	–22
–21	144.945	0.000807	0.1193	1239.2	8.383	173.7	175.7	349.4	0.900	1.597	–21
–20	150.921	0.000809	0.1148	1236.3	8.709	174.9	175.2	350.1	0.905	1.597	–20
–19	157.086	0.000811	0.1105	1233.4	9.046	176.1	174.6	350.7	0.910	1.597	–19
–18	163.444	0.000813	0.1065	1230.5	9.392	177.3	174.1	351.4	0.915	1.597	–18
–17	170.001	0.000815	0.1026	1227.6	9.750	178.6	173.5	352.1	0.919	1.597	–17
–16	176.759	0.000817	0.0988	1224.7	10.117	179.8	172.9	352.7	0.924	1.597	–16
–15	183.724	0.000818	0.0953	1221.8	10.496	181.0	172.4	353.4	0.929	1.597	–15
–14	190.898	0.000820	0.0919	1218.8	10.885	182.3	171.8	354.1	0.934	1.597	–14
–13	198.287	0.000822	0.0886	1215.9	11.286	183.5	171.2	354.7	0.939	1.597	–13
–12	205.895	0.000824	0.0855	1212.9	11.699	184.8	170.6	355.4	0.943	1.597	–12
–11	213.726	0.000826	0.0825	1209.9	12.123	186.0	170.0	356.1	0.948	1.597	–11
–10	221.783	0.000829	0.0796	1207.0	12.559	187.3	169.5	356.7	0.953	1.597	–10
–9	230.072	0.000831	0.0769	1203.9	13.008	188.5	168.9	357.4	0.958	1.597	–9
–8	238.597	0.000833	0.0742	1200.9	13.469	189.8	168.3	358.0	0.962	1.597	–8
–7	247.363	0.000835	0.0717	1197.9	13.943	191.0	167.7	358.7	0.967	1.597	–7
–6	256.373	0.000837	0.0693	1194.9	14.431	192.3	167.0	359.4	0.972	1.597	–6
–5	265.632	0.000839	0.0670	1191.8	14.931	193.6	166.4	360.0	0.976	1.597	–5
–4	275.144	0.000841	0.0647	1188.7	15.446	194.9	165.8	360.7	0.981	1.597	–4
–3	284.915	0.000843	0.0626	1185.6	15.974	196.1	165.2	361.3	0.986	1.597	–3
–2	294.948	0.000846	0.0605	1182.5	16.517	197.4	164.6	362.0	0.991	1.598	–2
–1	305.249	0.000848	0.0586	1179.4	17.074	198.7	163.9	362.6	0.995	1.598	–1
0	315.821	0.000850	0.0567	1176.3	17.647	200.0	163.3	363.3	1.000	1.598	0
1	326.670	0.000852	0.0548	1173.1	18.234	201.3	162.6	363.9	1.005	1.598	1
2	337.800	0.000855	0.0531	1170.0	18.837	202.6	162.0	364.6	1.009	1.598	2
3	349.216	0.000857	0.0514	1166.8	19.457	203.9	161.3	365.2	1.014	1.598	3

(*Continued*)

TABLE 2-132 Opteon™ YF (R-1234yf) *(Continued)*

Saturation Properties—Temperature Table

Temp [°C]	Pressure [kPa]	Volume [m³/kg]		Density [kg/m³]		Enthalpy [kJ/kg]			Entropy [kJ/kg·K]		Temp [°C]
		Liquid v_f	Vapor v_g	Liquid d_f	Vapor d_g	Liquid h_f	Latent h_{fg}	Vapor h_g	Liquid s_f	Vapor s_g	
4	360.923	0.000859	0.0498	1163.6	20.092	205.2	160.7	365.9	1.019	1.599	4
5	372.925	0.000862	0.0482	1160.4	20.744	206.5	160.0	366.5	1.023	1.599	5
6	385.227	0.000864	0.0467	1157.2	21.413	207.8	159.3	367.2	1.028	1.599	6
7	397.833	0.000867	0.0452	1153.9	22.100	209.1	158.7	367.8	1.033	1.599	7
8	410.750	0.000869	0.0439	1150.6	22.804	210.5	158.0	368.4	1.037	1.599	8
9	423.981	0.000872	0.0425	1147.3	23.526	211.8	157.3	369.1	1.042	1.600	9
10	437.532	0.000874	0.0412	1144.0	24.267	213.1	156.6	369.7	1.047	1.600	10
11	451.408	0.000877	0.0400	1140.7	25.027	214.4	155.9	370.3	1.051	1.600	11
12	465.613	0.000879	0.0387	1137.4	25.807	215.8	155.2	371.0	1.056	1.600	12
13	480.152	0.000882	0.0376	1134.0	26.606	217.1	154.5	371.6	1.061	1.601	13
14	495.031	0.000884	0.0365	1130.6	27.425	218.5	153.8	372.2	1.065	1.601	14
15	510.255	0.000887	0.0354	1127.2	28.266	219.8	153.0	372.8	1.070	1.601	15
16	525.828	0.000890	0.0343	1123.8	29.127	221.2	152.3	373.4	1.075	1.601	16
17	541.756	0.000893	0.0333	1120.3	30.011	222.5	151.6	374.1	1.079	1.602	17
18	558.044	0.000895	0.0323	1116.9	30.916	223.9	150.8	374.7	1.084	1.602	18
19	574.697	0.000898	0.0314	1113.4	31.845	225.2	150.1	375.3	1.088	1.602	19
20	591.721	0.000901	0.0305	1109.9	32.796	226.6	149.3	375.9	1.093	1.602	20
21	609.120	0.000904	0.0296	1106.3	33.772	228.0	148.5	376.5	1.098	1.603	21
22	626.901	0.000907	0.0288	1102.8	34.772	229.3	147.7	377.1	1.102	1.603	22
23	645.068	0.000910	0.0279	1099.2	35.797	230.7	147.0	377.7	1.107	1.603	23
24	663.626	0.000913	0.0271	1095.5	36.848	232.1	146.2	378.3	1.112	1.603	24
25	682.582	0.000916	0.0264	1091.9	37.925	233.5	145.4	378.9	1.116	1.604	25
26	701.940	0.000919	0.0256	1088.2	39.029	234.9	144.6	379.5	1.121	1.604	26
27	721.707	0.000922	0.0249	1084.5	40.161	236.3	143.7	380.0	1.125	1.604	27
28	741.887	0.000925	0.0242	1080.8	41.321	237.7	142.9	380.6	1.130	1.605	28
29	762.487	0.000928	0.0235	1077.1	42.510	239.1	142.1	381.2	1.135	1.605	29
30	783.511	0.000932	0.0229	1073.3	43.729	240.5	141.2	381.8	1.139	1.605	30
31	804.966	0.000935	0.0222	1069.5	44.979	241.9	140.4	382.3	1.144	1.605	31
32	826.857	0.000938	0.0216	1065.7	46.260	243.4	139.5	382.9	1.148	1.606	32
33	849.190	0.000942	0.0210	1061.8	47.573	244.8	138.7	383.4	1.153	1.606	33
34	871.971	0.000945	0.0204	1057.9	48.920	246.2	137.8	384.0	1.158	1.606	34
35	895.206	0.000949	0.0199	1054.0	50.301	247.6	136.9	384.5	1.162	1.606	35
36	918.900	0.000952	0.0193	1050.0	51.717	249.1	136.0	385.1	1.167	1.607	36
37	943.060	0.000956	0.0188	1046.0	53.169	250.5	135.1	385.6	1.171	1.607	37
38	967.691	0.000960	0.0183	1042.0	54.658	252.0	134.1	386.1	1.176	1.607	38
39	992.800	0.000963	0.0178	1037.9	56.186	253.4	133.2	386.7	1.181	1.607	39
40	1018.393	0.000967	0.0173	1033.8	57.753	254.9	132.3	387.2	1.185	1.608	40
41	1044.476	0.000971	0.0168	1029.6	59.360	256.4	131.3	387.7	1.190	1.608	41
42	1071.055	0.000975	0.0164	1025.5	61.010	257.8	130.3	388.2	1.194	1.608	42
43	1098.137	0.000979	0.0159	1021.2	62.702	259.3	129.4	388.7	1.199	1.608	43
44	1125.728	0.000983	0.0155	1017.0	64.440	260.8	128.4	389.2	1.204	1.608	44
45	1153.834	0.000988	0.0151	1012.6	66.223	262.3	127.4	389.7	1.208	1.608	45
46	1182.462	0.000992	0.0147	1008.3	68.053	263.8	126.3	390.1	1.213	1.609	46
47	1211.618	0.000996	0.0143	1003.9	69.933	265.3	125.3	390.6	1.217	1.609	47

48	1241.310	0.001001	0.0139	999.4	71.863	266.8	124.3	391.1	1.222	1.609	48
49	1271.543	0.001005	0.0135	994.9	73.846	268.3	123.2	391.5	1.227	1.609	49
50	1302.325	0.001010	0.0132	990.4	75.884	269.9	122.1	392.0	1.231	1.609	50
51	1333.663	0.001014	0.0128	985.8	77.978	271.4	121.0	392.4	1.236	1.609	51
52	1365.563	0.001019	0.0125	981.1	80.130	272.9	119.9	392.8	1.241	1.609	52
53	1398.032	0.001024	0.0121	976.4	82.343	274.5	118.8	393.3	1.245	1.609	53
54	1431.079	0.001029	0.0118	971.6	84.619	276.0	117.7	393.7	1.250	1.609	54
55	1464.709	0.001034	0.0115	966.7	86.961	277.6	116.5	394.1	1.254	1.610	55
56	1498.931	0.001040	0.0112	961.8	89.371	279.2	115.3	394.5	1.259	1.610	56
57	1533.751	0.001045	0.0109	956.8	91.852	280.7	114.1	394.9	1.264	1.610	57
58	1569.178	0.001051	0.0106	951.7	94.407	282.3	112.9	395.2	1.269	1.609	58
59	1605.219	0.001056	0.0103	946.6	97.040	283.9	111.7	395.6	1.273	1.609	59
60	1641.882	0.001062	0.0100	941.3	99.754	285.5	110.4	395.9	1.278	1.609	60
61	1679.174	0.001068	0.0098	936.0	102.552	287.1	109.1	396.3	1.283	1.609	61
62	1717.104	0.001075	0.0095	930.6	105.438	288.8	107.8	396.6	1.287	1.609	62
63	1755.680	0.001081	0.0092	925.1	108.418	290.4	106.5	396.9	1.292	1.609	63
64	1794.911	0.001088	0.0090	919.5	111.496	292.1	105.1	397.2	1.297	1.609	64
65	1834.805	0.001094	0.0087	913.7	114.676	293.7	103.7	397.5	1.302	1.609	65
66	1875.370	0.001101	0.0085	907.9	117.964	295.4	102.3	397.7	1.307	1.608	66
67	1916.617	0.001109	0.0082	901.9	121.367	297.1	100.9	398.0	1.311	1.608	67
68	1958.553	0.001116	0.0080	895.8	124.891	298.8	99.4	398.2	1.316	1.608	68
69	2001.189	0.001124	0.0078	889.6	128.544	300.5	97.9	398.4	1.321	1.607	69
70	2044.535	0.001132	0.0076	883.2	132.332	302.2	96.3	398.6	1.326	1.607	70
71	2088.600	0.001141	0.0073	876.7	136.266	304.0	94.8	398.7	1.331	1.606	71
72	2133.395	0.001149	0.0071	870.0	140.355	305.7	93.1	398.9	1.336	1.606	72
73	2178.931	0.001159	0.0069	863.1	144.611	307.5	91.5	399.0	1.341	1.605	73
74	2225.219	0.001168	0.0067	856.1	149.044	309.3	89.8	399.1	1.346	1.605	74
75	2272.271	0.001178	0.0065	848.8	153.671	311.1	88.0	399.1	1.351	1.604	75
76	2320.100	0.001189	0.0063	841.4	158.505	313.0	86.2	399.2	1.356	1.603	76
77	2368.717	0.001199	0.0061	833.7	163.566	314.8	84.3	399.2	1.361	1.602	77
78	2418.137	0.001211	0.0059	825.7	168.874	316.7	82.4	399.1	1.366	1.601	78
79	2468.375	0.001223	0.0057	817.5	174.454	318.6	80.4	399.0	1.372	1.600	79
80	2519.445	0.001236	0.0055	809.0	180.333	320.5	78.4	398.9	1.377	1.599	80

(*Continued*)

TABLE 2-132 Opteon™ YF (R-1234yf) *(Continued)*

Superheated Vapor—Constant Pressure Tables

V = Volume in m^3/kg | H = Enthalpy in kJ/kg | S = Entropy in kJ/kg·K | Saturation Properties in Light Gray

	Absolute Pressure, kPa												
	90			100			101.325			110			
	−32.15°C			−29.78°C			−29.49°C			−27.60°C			
	V	H	S	V	H	S	V	H	S	V	H	S	
Temp [°C]	0.1869	341.9	1.600	0.1693	343.5	1.599	0.1672	343.7	1.599	0.1548	344.9	1.598	Temp [°C]
−30	0.1888	343.6	1.607										−30
−25	0.1933	347.7	1.623	0.1732	347.4	1.615	0.1708	347.3	1.614	0.1567	347.1	1.607	−25
−20	0.1978	351.8	1.640	0.1773	351.5	1.631	0.1748	351.5	1.630	0.1604	351.2	1.623	−20
−15	0.2022	355.9	1.656	0.1813	355.6	1.647	0.1788	355.6	1.646	0.1642	355.4	1.640	−15
−10	0.2066	360.1	1.672	0.1853	359.9	1.664	0.1828	359.8	1.663	0.1678	359.6	1.656	−10
−5	0.2110	364.3	1.688	0.1893	364.1	1.680	0.1867	364.1	1.679	0.1715	363.9	1.672	−5
0	0.2153	368.6	1.704	0.1932	368.4	1.695	0.1906	368.4	1.694	0.1751	368.2	1.688	0
5	0.2197	372.9	1.719	0.1971	372.7	1.711	0.1945	372.7	1.710	0.1787	372.5	1.704	5
10	0.2240	377.3	1.735	0.2010	377.1	1.727	0.1983	377.1	1.726	0.1822	376.9	1.719	10
15	0.2283	381.7	1.750	0.2049	381.5	1.742	0.2022	381.5	1.741	0.1858	381.3	1.735	15
20	0.2325	386.1	1.766	0.2088	386.0	1.758	0.2060	385.9	1.757	0.1893	385.8	1.750	20
25	0.2368	390.6	1.781	0.2126	390.5	1.773	0.2098	390.4	1.772	0.1929	390.3	1.765	25
30	0.2410	395.2	1.796	0.2165	395.0	1.788	0.2136	395.0	1.787	0.1964	394.9	1.781	30
35	0.2453	399.8	1.811	0.2203	399.6	1.803	0.2174	399.6	1.802	0.1999	399.5	1.796	35
40	0.2495	404.4	1.826	0.2241	404.2	1.818	0.2211	404.2	1.817	0.2034	404.1	1.811	40
45	0.2537	409.1	1.841	0.2279	408.9	1.833	0.2249	408.9	1.832	0.2068	408.8	1.825	45
50	0.2579	413.8	1.855	0.2317	413.6	1.847	0.2286	413.6	1.846	0.2103	413.5	1.840	50
55	0.2621	418.5	1.870	0.2355	418.4	1.862	0.2324	418.4	1.861	0.2138	418.3	1.855	55
60	0.2663	423.3	1.884	0.2393	423.2	1.877	0.2361	423.2	1.876	0.2172	423.1	1.869	60
65	0.2705	428.2	1.899	0.2431	428.0	1.891	0.2399	428.0	1.890	0.2207	427.9	1.884	65
70	0.2746	433.0	1.913	0.2468	432.9	1.905	0.2436	432.9	1.904	0.2241	432.8	1.898	70
75	0.2788	438.0	1.927	0.2506	437.9	1.920	0.2473	437.8	1.919	0.2275	437.8	1.912	75
80	0.2830	442.9	1.942	0.2544	442.8	1.934	0.2510	442.8	1.933	0.2310	442.7	1.927	80
85	0.2871	447.9	1.956	0.2581	447.8	1.948	0.2547	447.8	1.947	0.2344	447.7	1.941	85
90	0.2913	453.0	1.970	0.2619	452.9	1.962	0.2584	452.9	1.961	0.2378	452.8	1.955	90
95	0.2954	458.1	1.984	0.2656	458.0	1.976	0.2621	458.0	1.975	0.2412	457.9	1.969	95
100	0.2996	463.2	1.997	0.2693	463.1	1.990	0.2658	463.1	1.989	0.2446	463.0	1.982	100
105	0.3037	468.4	2.011	0.2731	468.3	2.003	0.2695	468.3	2.002	0.2480	468.2	1.996	105
110	0.3078	473.6	2.025	0.2768	473.5	2.017	0.2731	473.5	2.016	0.2514	473.4	2.010	110
115	0.3120	478.8	2.038	0.2805	478.7	2.031	0.2768	478.7	2.030	0.2548	478.7	2.024	115

	Absolute Pressure, kPa												
	120			130			140			150			
	−25.56°C			−23.65°C			−21.85°C			−20.15°C			
	V	*H*	*S*	*V*	*H*	*S*	*V*	*H*	*S*	*V*	*H*	*S*	
Temp [°C]	0.1426	346.3	1.598	0.1322	347.6	1.598	0.1233	348.8	1.597	0.1155	349.9	1.597	Temp [°C]
−25	0.1430	346.8	1.600										−25
−20	0.1464	350.9	1.616	0.1346	350.7	1.610	0.1244	350.4	1.603	0.1156	350.1	1.598	−20
−15	0.1499	355.1	1.633	0.1378	354.9	1.626	0.1274	354.6	1.620	0.1184	354.3	1.614	−15
−10	0.1533	359.4	1.649	0.1409	359.1	1.642	0.1304	358.9	1.636	0.1212	358.6	1.631	−10
−5	0.1566	363.6	1.665	0.1441	363.4	1.659	0.1333	363.2	1.653	0.1240	362.9	1.647	−5
0	0.1600	367.9	1.681	0.1472	367.7	1.675	0.1362	367.5	1.669	0.1267	367.3	1.663	0
5	0.1633	372.3	1.697	0.1503	372.1	1.690	0.1391	371.9	1.684	0.1294	371.7	1.679	5
10	0.1666	376.7	1.712	0.1533	376.5	1.706	0.1420	376.3	1.700	0.1321	376.1	1.695	10
15	0.1699	381.1	1.728	0.1564	380.9	1.722	0.1448	380.8	1.716	0.1348	380.6	1.710	15
20	0.1731	385.6	1.743	0.1594	385.4	1.737	0.1477	385.3	1.731	0.1375	385.1	1.726	20
25	0.1764	390.1	1.759	0.1624	390.0	1.752	0.1505	389.8	1.747	0.1401	389.6	1.741	25
30	0.1796	394.7	1.774	0.1655	394.5	1.768	0.1533	394.4	1.762	0.1428	394.2	1.756	30
35	0.1828	399.3	1.789	0.1684	399.1	1.783	0.1561	399.0	1.777	0.1454	398.8	1.772	35
40	0.1861	403.9	1.804	0.1714	403.8	1.798	0.1589	403.6	1.792	0.1480	403.5	1.787	40
45	0.1893	408.6	1.819	0.1744	408.5	1.813	0.1616	408.3	1.807	0.1506	408.2	1.801	45
50	0.1925	413.4	1.833	0.1773	413.2	1.827	0.1644	413.1	1.822	0.1532	413.0	1.816	50
55	0.1956	418.1	1.848	0.1803	418.0	1.842	0.1671	417.9	1.836	0.1557	417.7	1.831	55
60	0.1988	422.9	1.863	0.1832	422.8	1.857	0.1699	422.7	1.851	0.1583	422.6	1.846	60
65	0.2020	427.8	1.877	0.1862	427.7	1.871	0.1726	427.6	1.865	0.1609	427.4	1.860	65
70	0.2051	432.7	1.892	0.1891	432.6	1.885	0.1753	432.5	1.880	0.1634	432.4	1.875	70
75	0.2083	437.6	1.906	0.1920	437.5	1.900	0.1781	437.4	1.894	0.1660	437.3	1.889	75
80	0.2114	442.6	1.920	0.1949	442.5	1.914	0.1808	442.4	1.908	0.1685	442.3	1.903	80
85	0.2146	447.6	1.934	0.1978	447.5	1.928	0.1835	447.4	1.922	0.1711	447.3	1.917	85
90	0.2177	452.7	1.948	0.2007	452.6	1.942	0.1862	452.5	1.937	0.1736	452.4	1.931	90
95	0.2209	457.8	1.962	0.2037	457.7	1.956	0.1889	457.6	1.950	0.1761	457.5	1.945	95
100	0.2240	462.9	1.976	0.2065	462.8	1.970	0.1916	462.7	1.964	0.1786	462.6	1.959	100
105	0.2271	468.1	1.990	0.2094	468.0	1.984	0.1943	467.9	1.978	0.1812	467.8	1.973	105
110	0.2302	473.3	2.003	0.2123	473.2	1.997	0.1970	473.1	1.992	0.1837	473.1	1.987	110
115	0.2334	478.6	2.017	0.2152	478.5	2.011	0.1997	478.4	2.005	0.1862	478.3	2.000	115
120	0.2365	483.9	2.031	0.2181	483.8	2.025	0.2023	483.7	2.019	0.1887	483.6	2.014	120

(*Continued*)

TABLE 2-132 Opteon™ YF (R-1234yf) *(Continued)*

Superheated Vapor—Constant Pressure Tables

V = Volume in m³/kg *H* = Enthalpy in kJ/kg *S* = Entropy in kJ/kg·K Saturation Properties in Light Gray

	Absolute Pressure, kPa												
	160			170			180			190			
	−18.54°C			−17.00°C			−15.53°C			−14.12°C			
	V	*H*	*S*	*V*	*H*	*S*	*V*	*H*	*S*	*V*	*H*	*S*	
Temp [°C]	0.1086	351.0	1.597	0.1026	352.1	1.597	0.0972	353.0	1.597	0.0923	354.0	1.597	Temp [°C]
−15	0.1105	354.1	1.609	0.1036	353.8	1.603	0.0974	353.5	1.598				−15
−10	0.1132	358.4	1.625	0.1061	358.1	1.620	0.0998	357.8	1.615	0.0942	357.6	1.610	−10
−5	0.1158	362.7	1.641	0.1086	362.4	1.636	0.1022	362.2	1.632	0.0965	362.0	1.627	−5
0	0.1184	367.1	1.658	0.1111	366.8	1.653	0.1045	366.6	1.648	0.0987	366.4	1.643	0
5	0.1210	371.5	1.674	0.1135	371.2	1.669	0.1069	371.0	1.664	0.1009	370.8	1.659	5
10	0.1235	375.9	1.689	0.1159	375.7	1.684	0.1092	375.5	1.680	0.1031	375.3	1.675	10
15	0.1261	380.4	1.705	0.1183	380.2	1.700	0.1114	380.0	1.696	0.1053	379.8	1.691	15
20	0.1286	384.9	1.721	0.1207	384.7	1.716	0.1137	384.5	1.711	0.1074	384.3	1.707	20
25	0.1311	389.4	1.736	0.1231	389.3	1.731	0.1160	389.1	1.727	0.1096	388.9	1.722	25
30	0.1335	394.0	1.751	0.1254	393.9	1.747	0.1182	393.7	1.742	0.1117	393.5	1.738	30
35	0.1360	398.7	1.766	0.1277	398.5	1.762	0.1204	398.4	1.757	0.1138	398.2	1.753	35
40	0.1385	403.3	1.782	0.1301	403.2	1.777	0.1226	403.0	1.772	0.1159	402.9	1.768	40
45	0.1409	408.1	1.796	0.1324	407.9	1.792	0.1248	407.8	1.787	0.1180	407.6	1.783	45
50	0.1433	412.8	1.811	0.1347	412.7	1.807	0.1270	412.5	1.802	0.1201	412.4	1.798	50
55	0.1458	417.6	1.826	0.1370	417.5	1.821	0.1292	417.3	1.817	0.1222	417.2	1.813	55
60	0.1482	422.4	1.841	0.1393	422.3	1.836	0.1313	422.2	1.832	0.1242	422.1	1.827	60
65	0.1506	427.3	1.855	0.1415	427.2	1.850	0.1335	427.1	1.846	0.1263	427.0	1.842	65
70	0.1530	432.2	1.870	0.1438	432.1	1.865	0.1356	432.0	1.861	0.1283	431.9	1.856	70
75	0.1554	437.2	1.884	0.1461	437.1	1.879	0.1378	437.0	1.875	0.1303	436.9	1.871	75
80	0.1578	442.2	1.898	0.1483	442.1	1.894	0.1399	442.0	1.889	0.1324	441.9	1.885	80
85	0.1602	447.2	1.912	0.1506	447.1	1.908	0.1420	447.0	1.903	0.1344	446.9	1.899	85
90	0.1626	452.3	1.926	0.1528	452.2	1.922	0.1442	452.1	1.917	0.1364	452.0	1.913	90
95	0.1649	457.4	1.940	0.1551	457.3	1.936	0.1463	457.2	1.931	0.1384	457.1	1.927	95
100	0.1673	462.6	1.954	0.1573	462.5	1.950	0.1484	462.4	1.945	0.1405	462.3	1.941	100
105	0.1697	467.7	1.968	0.1595	467.6	1.963	0.1505	467.6	1.959	0.1425	467.5	1.955	105
110	0.1720	473.0	1.982	0.1618	472.9	1.977	0.1526	472.8	1.973	0.1445	472.7	1.969	110
115	0.1744	478.2	1.995	0.1640	478.1	1.991	0.1547	478.1	1.987	0.1465	478.0	1.982	115
120	0.1767	483.5	2.009	0.1662	483.4	2.004	0.1568	483.4	2.000	0.1485	483.3	1.996	120
125	0.1791	488.9	2.023	0.1684	488.8	2.018	0.1589	488.7	2.014	0.1505	488.6	2.010	125
130	0.1815	494.2	2.036	0.1706	494.2	2.031	0.1610	494.1	2.027	0.1524	494.0	2.023	130

	Absolute Pressure, kPa												
	200			210			220			230			
	−12.77°C			−11.47°C			−10.22°C			−9.01°C			
	V	*H*	*S*	*V*	*H*	*S*	*V*	*H*	*S*	*V*	*H*	*S*	
Temp [°C]	0.0879	354.9	1.597	0.0839	355.7	1.597	0.0802	356.6	1.597	0.0769	357.4	1.597	Temp [°C]
−10	0.0891	357.3	1.606	0.0845	357.0	1.602	0.0803	356.8	1.597				−10
−5	0.0913	361.7	1.623	0.0866	361.5	1.618	0.0824	361.2	1.614	0.0785	360.9	1.610	−5
0	0.0934	366.1	1.639	0.0887	365.9	1.635	0.0843	365.7	1.631	0.0804	365.4	1.627	0
5	0.0956	370.6	1.655	0.0907	370.4	1.651	0.0863	370.1	1.647	0.0823	369.9	1.643	5
10	0.0977	375.1	1.671	0.0927	374.9	1.667	0.0883	374.7	1.663	0.0842	374.4	1.659	10
15	0.0997	379.6	1.687	0.0947	379.4	1.683	0.0902	379.2	1.679	0.0860	379.0	1.675	15
20	0.1018	384.2	1.703	0.0967	384.0	1.699	0.0921	383.8	1.695	0.0878	383.6	1.691	20
25	0.1039	388.7	1.718	0.0987	388.6	1.714	0.0940	388.4	1.710	0.0896	388.2	1.707	25
30	0.1059	393.4	1.733	0.1006	393.2	1.730	0.0958	393.0	1.726	0.0914	392.9	1.722	30
35	0.1079	398.0	1.749	0.1025	397.9	1.745	0.0977	397.7	1.741	0.0932	397.6	1.737	35
40	0.1099	402.7	1.764	0.1045	402.6	1.760	0.0995	402.4	1.756	0.0950	402.3	1.753	40
45	0.1119	407.5	1.779	0.1064	407.3	1.775	0.1013	407.2	1.771	0.0967	407.0	1.768	45
50	0.1139	412.3	1.794	0.1083	412.1	1.790	0.1032	412.0	1.786	0.0985	411.8	1.783	50
55	0.1159	417.1	1.809	0.1101	417.0	1.805	0.1050	416.8	1.801	0.1002	416.7	1.798	55
60	0.1178	421.9	1.823	0.1120	421.8	1.819	0.1068	421.7	1.816	0.1020	421.6	1.812	60
65	0.1198	426.8	1.838	0.1139	426.7	1.834	0.1086	426.6	1.830	0.1037	426.5	1.827	65
70	0.1217	431.8	1.852	0.1158	431.6	1.849	0.1103	431.5	1.845	0.1054	431.4	1.841	70
75	0.1237	436.7	1.867	0.1176	436.6	1.863	0.1121	436.5	1.859	0.1071	436.4	1.856	75
80	0.1256	441.7	1.881	0.1195	441.6	1.877	0.1139	441.5	1.874	0.1088	441.4	1.870	80
85	0.1275	446.8	1.895	0.1213	446.7	1.891	0.1157	446.6	1.888	0.1105	446.5	1.884	85
90	0.1295	451.9	1.909	0.1231	451.8	1.906	0.1174	451.7	1.902	0.1122	451.6	1.899	90
95	0.1314	457.0	1.923	0.1250	456.9	1.920	0.1192	456.8	1.916	0.1139	456.7	1.913	95
100	0.1333	462.2	1.937	0.1268	462.1	1.934	0.1209	462.0	1.930	0.1155	461.9	1.927	100
105	0.1352	467.4	1.951	0.1286	467.3	1.947	0.1227	467.2	1.944	0.1172	467.1	1.940	105
110	0.1371	472.6	1.965	0.1305	472.5	1.961	0.1244	472.4	1.958	0.1189	472.3	1.954	110
115	0.1390	477.9	1.979	0.1323	477.8	1.975	0.1261	477.7	1.971	0.1206	477.6	1.968	115
120	0.1409	483.2	1.992	0.1341	483.1	1.988	0.1279	483.0	1.985	0.1222	482.9	1.981	120
125	0.1428	488.5	2.006	0.1359	488.5	2.002	0.1296	488.4	1.998	0.1239	488.3	1.995	125
130	0.1447	493.9	2.019	0.1377	493.8	2.015	0.1313	493.8	2.012	0.1255	493.7	2.008	130
135	0.1466	499.3	2.032	0.1395	499.3	2.029	0.1331	499.2	2.025	0.1272	499.1	2.022	135

(Continued)

TABLE 2-132 Opteon™ YF (R-1234yf) *(Continued)*

Superheated Vapor—Constant Pressure Tables

V = Volume in m³/kg H = Enthalpy in kJ/kg S = Entropy in kJ/kg·K   Saturation Properties in Light Gray

Absolute Pressure, kPa

	240			250			260			270			
	−7.84°C			−6.70°C			−5.60°C			−4.54°C			
	V	H	S	V	H	S	V	H	S	V	H	S	
Temp [°C]	0.0738	358.2	1.597	0.0710	358.9	1.597	0.0684	359.6	1.597	0.0659	360.3	1.597	Temp [°C]
−5	0.0749	360.7	1.606	0.0716	360.4	1.603	0.0686	360.2	1.599				−5
0	0.0768	365.2	1.623	0.0734	364.9	1.619	0.0703	364.7	1.616	0.0675	364.4	1.612	0
5	0.0786	369.7	1.639	0.0752	369.5	1.636	0.0721	369.2	1.632	0.0691	369.0	1.629	5
10	0.0804	374.2	1.656	0.0769	374.0	1.652	0.0738	373.8	1.649	0.0708	373.6	1.645	10
15	0.0822	378.8	1.672	0.0787	378.6	1.668	0.0754	378.4	1.665	0.0724	378.2	1.661	15
20	0.0840	383.4	1.687	0.0804	383.2	1.684	0.0771	383.0	1.681	0.0740	382.8	1.677	20
25	0.0857	388.0	1.703	0.0821	387.9	1.700	0.0787	387.7	1.696	0.0756	387.5	1.693	25
30	0.0874	392.7	1.719	0.0837	392.5	1.715	0.0803	392.4	1.712	0.0772	392.2	1.709	30
35	0.0891	397.4	1.734	0.0854	397.2	1.731	0.0819	397.1	1.727	0.0787	396.9	1.724	35
40	0.0908	402.1	1.749	0.0870	402.0	1.746	0.0835	401.8	1.743	0.0803	401.7	1.739	40
45	0.0925	406.9	1.764	0.0887	406.8	1.761	0.0851	406.6	1.758	0.0818	406.5	1.755	45
50	0.0942	411.7	1.779	0.0903	411.6	1.776	0.0867	411.4	1.773	0.0833	411.3	1.770	50
55	0.0959	416.5	1.794	0.0919	416.4	1.791	0.0882	416.3	1.788	0.0848	416.1	1.785	55
60	0.0976	421.4	1.809	0.0935	421.3	1.806	0.0898	421.2	1.802	0.0863	421.0	1.799	60
65	0.0992	426.3	1.824	0.0951	426.2	1.820	0.0913	426.1	1.817	0.0878	426.0	1.814	65
70	0.1009	431.3	1.838	0.0967	431.2	1.835	0.0928	431.1	1.832	0.0893	430.9	1.829	70
75	0.1025	436.3	1.852	0.0983	436.2	1.849	0.0944	436.1	1.846	0.0907	435.9	1.843	75
80	0.1041	441.3	1.867	0.0998	441.2	1.864	0.0959	441.1	1.861	0.0922	441.0	1.858	80
85	0.1058	446.4	1.881	0.1014	446.3	1.878	0.0974	446.2	1.875	0.0937	446.1	1.872	85
90	0.1074	451.5	1.895	0.1030	451.4	1.892	0.0989	451.3	1.889	0.0951	451.2	1.886	90
95	0.1090	456.6	1.909	0.1045	456.5	1.906	0.1004	456.4	1.903	0.0966	456.3	1.900	95
100	0.1106	461.8	1.923	0.1061	461.7	1.920	0.1019	461.6	1.917	0.0980	461.5	1.914	100
105	0.1122	467.0	1.937	0.1076	466.9	1.934	0.1034	466.8	1.931	0.0995	466.7	1.928	105
110	0.1138	472.3	1.951	0.1092	472.2	1.948	0.1049	472.1	1.945	0.1009	472.0	1.942	110
115	0.1154	477.5	1.965	0.1107	477.4	1.961	0.1064	477.4	1.958	0.1023	477.3	1.956	115
120	0.1170	482.9	1.978	0.1122	482.8	1.975	0.1078	482.7	1.972	0.1038	482.6	1.969	120
125	0.1186	488.2	1.992	0.1138	488.1	1.989	0.1093	488.0	1.986	0.1052	488.0	1.983	125
130	0.1202	493.6	2.005	0.1153	493.5	2.002	0.1108	493.4	1.999	0.1066	493.4	1.996	130
135	0.1218	499.0	2.019	0.1168	499.0	2.015	0.1123	498.9	2.012	0.1080	498.8	2.010	135
140	0.1234	504.5	2.032	0.1184	504.4	2.029	0.1137	504.3	2.026	0.1094	504.3	2.023	140

	Absolute Pressure, kPa												
	280			290			300			310			
	−3.50°C			−2.49°C			−1.51°C			−0.55°C			
	V	*H*	*S*	*V*	*H*	*S*	*V*	*H*	*S*	*V*	*H*	*S*	
Temp [°C]	0.0637	361.0	1.597	0.0615	361.7	1.597	0.0596	362.3	1.598	0.0577	362.9	1.598	Temp [°C]
0	0.0648	364.2	1.609	0.0623	363.9	1.606	0.0600	363.7	1.603	0.0579	363.4	1.600	0
5	0.0664	368.8	1.626	0.0639	368.5	1.622	0.0616	368.3	1.619	0.0594	368.1	1.616	5
10	0.0680	373.4	1.642	0.0655	373.2	1.639	0.0631	372.9	1.636	0.0609	372.7	1.633	10
15	0.0696	378.0	1.658	0.0670	377.8	1.655	0.0646	377.6	1.652	0.0623	377.4	1.649	15
20	0.0712	382.6	1.674	0.0685	382.4	1.671	0.0661	382.2	1.668	0.0638	382.0	1.665	20
25	0.0727	387.3	1.690	0.0700	387.1	1.687	0.0675	386.9	1.684	0.0652	386.8	1.681	25
30	0.0742	392.0	1.706	0.0715	391.8	1.703	0.0690	391.7	1.700	0.0666	391.5	1.697	30
35	0.0757	396.7	1.721	0.0730	396.6	1.718	0.0704	396.4	1.715	0.0680	396.2	1.713	35
40	0.0772	401.5	1.736	0.0744	401.4	1.734	0.0718	401.2	1.731	0.0693	401.0	1.728	40
45	0.0787	406.3	1.752	0.0758	406.2	1.749	0.0732	406.0	1.746	0.0707	405.9	1.743	45
50	0.0802	411.1	1.767	0.0773	411.0	1.764	0.0746	410.9	1.761	0.0720	410.7	1.758	50
55	0.0816	416.0	1.782	0.0787	415.9	1.779	0.0759	415.7	1.776	0.0733	415.6	1.773	55
60	0.0831	420.9	1.797	0.0801	420.8	1.794	0.0773	420.6	1.791	0.0747	420.5	1.788	60
65	0.0845	425.8	1.811	0.0815	425.7	1.808	0.0786	425.6	1.806	0.0760	425.5	1.803	65
70	0.0860	430.8	1.826	0.0829	430.7	1.823	0.0800	430.6	1.820	0.0773	430.5	1.818	70
75	0.0874	435.8	1.840	0.0842	435.7	1.838	0.0813	435.6	1.835	0.0786	435.5	1.832	75
80	0.0888	440.9	1.855	0.0856	440.8	1.852	0.0827	440.6	1.849	0.0799	440.5	1.847	80
85	0.0902	445.9	1.869	0.0870	445.8	1.866	0.0840	445.7	1.864	0.0812	445.6	1.861	85
90	0.0916	451.1	1.883	0.0884	451.0	1.880	0.0853	450.9	1.878	0.0825	450.7	1.875	90
95	0.0930	456.2	1.897	0.0897	456.1	1.894	0.0866	456.0	1.892	0.0837	455.9	1.889	95
100	0.0944	461.4	1.911	0.0911	461.3	1.909	0.0879	461.2	1.906	0.0850	461.1	1.903	100
105	0.0958	466.6	1.925	0.0924	466.5	1.922	0.0892	466.4	1.920	0.0863	466.3	1.917	105
110	0.0972	471.9	1.939	0.0938	471.8	1.936	0.0905	471.7	1.934	0.0875	471.6	1.931	110
115	0.0986	477.2	1.953	0.0951	477.1	1.950	0.0918	477.0	1.947	0.0888	476.9	1.945	115
120	0.1000	482.5	1.966	0.0964	482.4	1.964	0.0931	482.3	1.961	0.0901	482.3	1.958	120
125	0.1013	487.9	1.980	0.0978	487.8	1.977	0.0944	487.7	1.975	0.0913	487.6	1.972	125
130	0.1027	493.3	1.993	0.0991	493.2	1.991	0.0957	493.1	1.988	0.0926	493.0	1.986	130
135	0.1041	498.7	2.007	0.1004	498.6	2.004	0.0970	498.6	2.002	0.0938	498.5	1.999	135
140	0.1055	504.2	2.020	0.1017	504.1	2.017	0.0983	504.0	2.015	0.0950	504.0	2.012	140
145	0.1068	509.7	2.033	0.1031	509.6	2.031	0.0996	509.6	2.028	0.0963	509.5	2.026	145

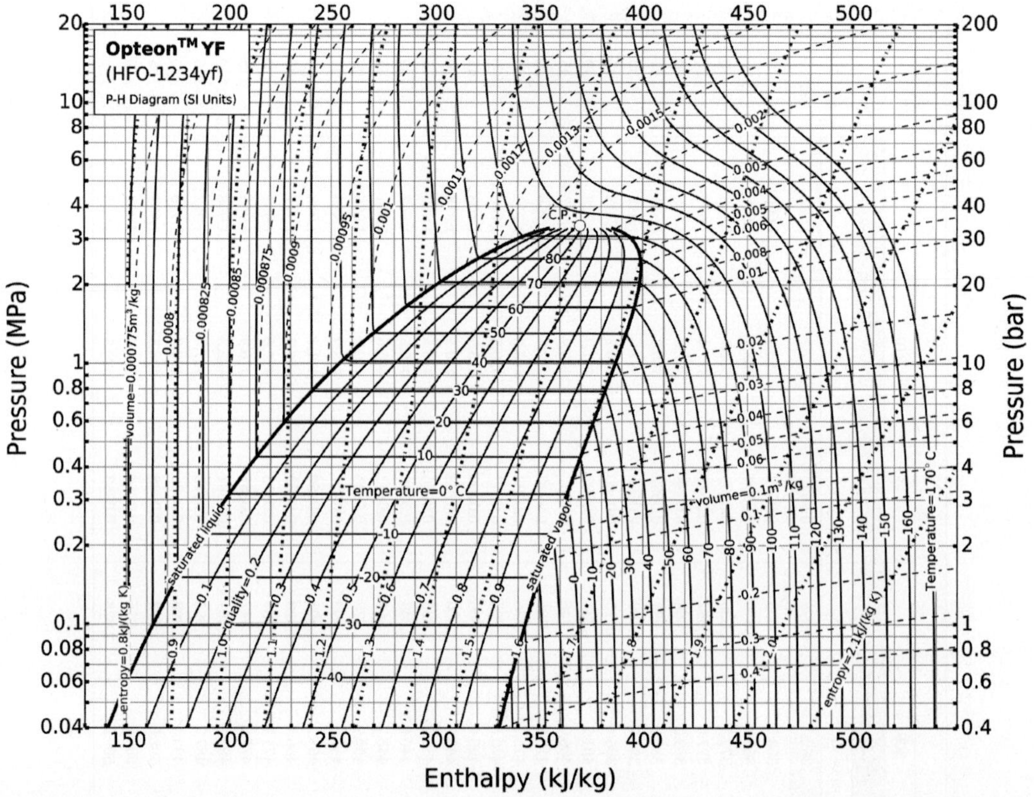

FIG. 2-16 Pressure-enthalpy diagram for Refrigerant 1234yf. Properties computed with the NIST REFPROP Database, Version 7.0 (Lemmon, E. W., M.O. McLinden, and M. L. Huber, 2002, NIST Standard Reference Database 23, NIST Reference Fluid Thermodynamic and Transport Properties—REFPROP, Version 7.0, Standard Reference Data Program, National Institute of Standards and Technology). Provided by Chemours.

TABLE 2-133 Thermophysical Properties of Saturated Seawater

Temp., °C	Pressure, bar	v, (m^3/kg)10^3	c_p, kJ/(kg·K)	μ, Ns/m^2	k, W/(m·K)	N_{Pr}	$10^5\kappa$, 1/bar
0	0.005993	1.000158	4.000	0.001884	0.560	13.46	5.06
1	0.006438	1.000099	4.000	0.001827	0.563	12.98	5.02
2	0.006916	1.000057	4.000	0.001772	0.565	12.55	4.98
3	0.007427	1.000033	4.000	0.001720	0.567	12.13	4.95
4	0.007970	1.000025	4.001	0.001669	0.569	11.74	4.92
5	0.008548	1.000033	4.001	0.001620	0.571	11.35	4.89
6	0.009163	1.000057	4.001	0.001574	0.574	10.97	4.86
7	0.009816	1.000096	4.002	0.001529	0.576	10.62	4.83
8	0.010511	1.000149	4.002	0.001486	0.578	10.29	4.80
9	0.011248	1.000261	4.002	0.001445	0.580	9.97	4.78
10	0.01203	1.000298	4.003	0.001405	0.582	9.70	4.76
11	0.01286	1.000392	4.003	0.001367	0.584	9.37	4.74
12	0.01374	1.000500	4.003	0.001330	0.586	9.09	4.72
13	0.01467	1.000620	4.004	0.001294	0.588	8.81	4.70
14	0.01566	1.000727	4.004	0.001259	0.590	8.54	4.68
15	0.01671	1.000899	4.005	0.001226	0.592	8.29	4.66
16	0.01781	1.001055	4.005	0.001195	0.594	8.06	4.65
17	0.01898	1.001224	4.006	0.001165	0.595	7.82	4.63
18	0.02022	1.001404	4.006	0.001136	0.597	7.62	4.62
19	0.02153	1.001595	4.007	0.001107	0.599	7.41	4.60
20	0.02291	1.001796	4.007	0.001080	0.600	7.21	4.59
21	0.02437	1.002009	4.007	0.001054	0.602	7.02	4.57
22	0.02591	1.002232	4.008	0.001029	0.604	6.82	4.56
23	0.02753	1.002465	4.008	0.001005	0.605	6.66	4.55
24	0.02924	1.002708	4.009	0.000981	0.607	6.48	4.54
25	0.03104	1.002961	4.009	0.000958	0.608	6.31	4.53
26	0.03294	1.003224	4.009	0.000936	0.609	6.16	4.52
27	0.03494	1.003496	4.010	0.000915	0.611	6.01	4.51
28	0.03705	1.003778	4.010	0.000895	0.612	5.86	4.50
29	0.03926	1.004069	4.011	0.000875	0.614	5.72	4.49
30	0.04159	1.004369	4.011	0.000855	0.615	5.58	4.48

$\kappa = (-1/V)(\partial v/\partial p)_T \cdot 10^5$. Thus, at 0°C, the compressibility is 5.06×10^{-5}/bar.

For further information see, for instance, Bromley, LeR. A., *J. Chem. Eng. Data,* **12,** 2 (1967): 202–206; **13,** 1 (1968): 60–62 and **13,** 3: 399–402; **15,** 2 (1970): 246–253; and *A.I.Ch.E.J.,* **20,** 2 (1974): 326–335.

Thermal conductivity data sources include Castelli, V. J., E. M. Stanley, et al., *Deep Sea Res.,* **211** (1974): 311–318; Levy, F. L., *Int. J. Refrig.,* **5,** 3 (1982): 155–159.

For velocity of sound, see, for instance, U.S. Naval Oceanographic Office SP 58, 1962 (50 pp.). More recent information is contained in UNESCO technical papers. See *Marine Science* No. 38, 1981 (6 pp.) and No. 44, 1983 (53 pp.).

For sea ice properties, see Fukusako, S., *Int. J. Thermophys.,* **11,** 2 (1990): 353–372.

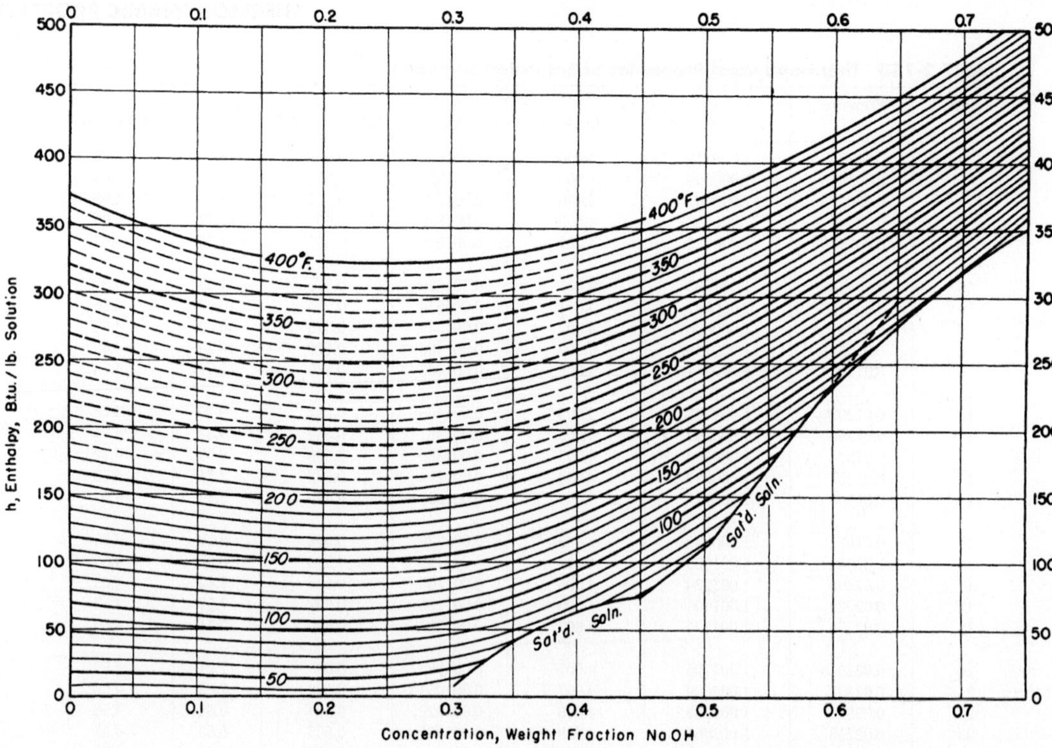

FIG. 2-17 Enthalpy-concentration diagram for aqueous sodium hydroxide at 1 atm. Reference states: enthalpy of liquid water at 32°F and vapor pressure is zero; partial molal enthalpy of infinitely dilute NaOH solution at 64°F and 1 atm is zero. [W.L. McCabe, *Trans. Am. Inst. Chem. Eng.*, **31:** 129 (1935).]

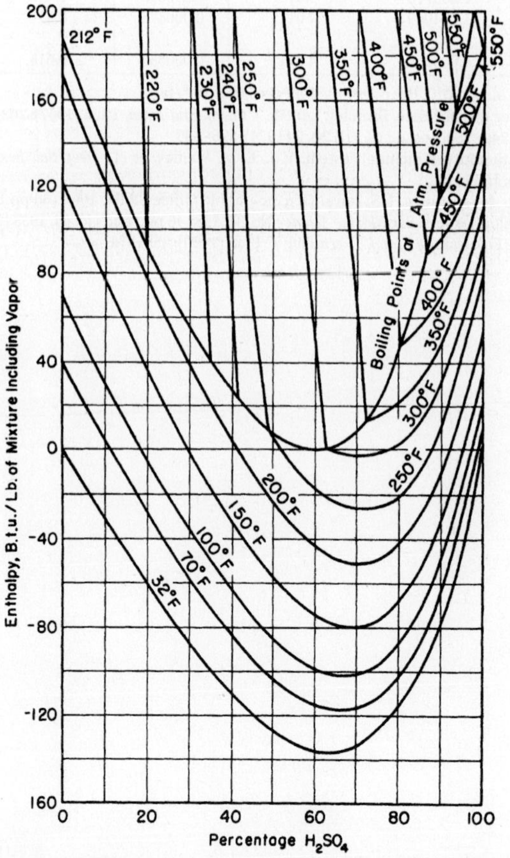

FIG. 2-18 Enthalpy-concentration diagram for aqueous sulfuric acid at 1 atm. Reference states: enthalpies of pure-liquid components at 32°F and vapor pressures are zero. Note: It should be observed that the weight basis includes the vapor, which is particularly important in the two-phase region. The upper ends of the tie lines in this region are assumed to be pure water. (O.A. Hougen and K.M. Watson, *Chemical Process Principles*, part I, Wiley, New York, 1943.)

TABLE 2-134 Saturated Solid/Vapor Water*

Temp., °F	Pressure, lb/in² abs.	Volume, ft³/lb		Enthalpy, Btu/lb		Entropy, Btu/(lb)(°F)	
		Solid	Vapor	Solid	Vapor	Solid	Vapor
−160	4.949.−8	0.01722	3.607.+9	−222.05	990.38	−0.4907	3.5549
−150	1.620.−7	0.01723	1.139.+9	−218.82	994.80	−0.4801	3.4387
−140	4.928.−7	0.01724	3.864.+8	−215.49	999.21	−0.4695	3.3301
−130	1.403.−6	0.01725	1.400.+8	−212.08	1003.63	−0.4590	3.2284
−120	3.757.−6	0.01726	5.386.+7	−208.58	1008.05	−0.4485	3.1330
−110	9.517.−6	0.01728	2.189.+7	−204.98	1012.47	−0.4381	3.0434
−100	2.291.−5	0.01729	9.352.+6	−201.28	1016.89	−0.4277	2.9591
−90	5.260.−5	0.01730	4.186.+6	−197.49	1021.31	−0.4173	2.8796
−80	1.157.−4	0.01731	1.955.+6	−193.60	1025.73	−0.4069	2.8045
−70	2.443.−4	0.01732	9.501.+5	−189.61	1030.15	−0.3965	2.7336
−60	4.972.−4	0.01734	4.788.+5	−185.52	1034.58	−0.3862	2.6664
−50	9.776.−4	0.01735	2.496.+5	−181.34	1039.00	−0.3758	2.6028
−45	1.354.−3	0.01736	1.824.+5	−179.21	1041.21	−0.3707	2.5723
−40	1.861.−3	0.01737	1.343.+5	−177.06	1043.42	−0.3655	2.5425
−35	2.540.−3	0.01737	9.961.+4	−174.88	1045.63	−0.3604	2.5135
−30	3.440.−3	0.01738	7.441.+4	−172.68	1047.84	−0.3552	2.4853
−25	4.627.−3	0.01739	5.596.+4	−170.46	1050.05	−0.3501	2.4577
−20	6.181.−3	0.01739	4.237.+4	−168.21	1052.26	−0.3449	2.4308
−15	8.204.−3	0.01740	3.228.+4	−165.94	1054.47	−0.3398	2.4046
−10	1.082.−2	0.01741	2.475.+4	−163.65	1056.67	−0.3347	2.3791
−5	1.419.−2	0.01741	1.909.+4	−161.33	1058.88	−0.3295	2.3541
0	1.849.−2	0.01742	1.481.+4	−158.98	1061.09	−0.3244	2.3297
5	2.396.−2	0.01743	1.155.+4	−156.61	1063.29	−0.3193	2.3039
10	3.087.−2	0.01744	9.060.+3	−154.22	1065.50	−0.3142	2.2827
15	3.957.−2	0.01744	7.144.+3	−151.80	1067.70	−0.3090	2.2600
16	4.156.−2	0.01745	6.817.+3	−151.32	1068.14	−0.3080	2.2555
18	4.581.−2	0.01745	6.210.+3	−150.34	1069.02	−0.3060	2.2466
20	5.045.−2	0.01745	5.662.+3	−149.36	1069.90	−0.3039	2.2378
22	5.552.−2	0.01746	5.166.+3	−148.38	1070.38	−0.3019	2.2291
24	6.105.−2	0.01746	4.717.+3	−147.39	1071.66	−0.2998	2.2205
26	6.708.−2	0.01746	4.311.+3	−146.40	1072.53	−0.2978	2.2119
28	7.365.−2	0.01746	3.943.+3	−145.40	1073.41	−0.2957	2.2034
30	8.080.−2	0.01747	3.608.+3	−144.40	1074.29	−0.2937	2.1950
31	8.461.−2	0.01747	3.453.+3	−143.90	1074.73	−0.2927	2.1908
32	8.858.−2	0.01747	3.305.+3	−143.40	1075.16	−0.2916	2.1867

*Condensed from *Fundamentals,* American Society of Heating, Refrigerating and Air-Conditioning Engineers, 1967 and 1972. Reproduced by permission. The validity of many standard reference tables has been critically reviewed by Jancso, Pupezin, and van Hook, *J. Phys. Chem.,* **74** (1970):2984. Current information on the properties of solid, vapor, and liquid water properties can be found at http://www.iapws.org. The notation 4.949.−8, 3.607.+9, etc., means 4.949×10^{-8}, 3.607×10^{9}, etc.

TABLE 2-135 Thermodynamic Properties of Water

Temperature K	Pressure MPa	Density mol/dm³	Volume dm³/mol	Int. energy kJ/mol	Enthalpy kJ/mol	Entropy kJ/(mol·K)	C_v kJ/(mol·K)	C_p kJ/(mol·K)	Sound speed m/s	Joule-Thomson K/MPa	Therm. cond. mW/(m·K)	Viscosity μPa·s
Saturated Properties												
273.16	0.000612	55.497	0.018019	0	1.1E-05	0	0.075978	0.076023	1402.3	−0.24142	561.04	1791.2
280	0.000992	55.501	0.018018	0.51875	0.51877	0.001876	0.075669	0.075688	1434.1	−0.23515	574.04	1433.7
290	0.001920	55.440	0.018038	1.2742	1.2742	0.004527	0.075095	0.075429	1472.1	−0.22720	592.73	1084.0
300	0.003537	55.315	0.018078	2.0278	2.0279	0.007082	0.074412	0.075320	1501.4	−0.22024	610.28	853.84
310	0.006231	55.139	0.018136	2.7808	2.7810	0.009551	0.073645	0.075294	1523.2	−0.21393	626.05	693.54
320	0.010546	54.919	0.018209	3.5339	3.5340	0.011941	0.072811	0.075317	1538.7	−0.20804	639.71	577.02
330	0.017213	54.662	0.018294	4.2873	4.2876	0.014260	0.071927	0.075373	1548.7	−0.20241	651.18	489.49
340	0.027188	54.371	0.018392	5.0414	5.0419	0.016511	0.071008	0.075456	1553.9	−0.19690	660.55	421.97
350	0.041682	54.049	0.018502	5.7964	5.7972	0.018700	0.070070	0.075567	1554.8	−0.19140	668.00	368.77
360	0.062194	53.698	0.018623	6.5526	6.5538	0.020830	0.069124	0.075708	1552.0	−0.18581	673.76	326.10
370	0.090535	53.321	0.018754	7.3104	7.3121	0.022906	0.068180	0.075883	1545.8	−0.18005	678.02	291.36
380	0.12885	52.918	0.018897	8.0701	8.0725	0.024932	0.067247	0.076098	1536.5	−0.17404	681.00	262.69
390	0.17964	52.490	0.019051	8.8320	8.8354	0.026911	0.066331	0.076357	1524.3	−0.16769	682.83	238.77
400	0.24577	52.038	0.019217	9.5966	9.6013	0.028847	0.065438	0.076664	1509.5	−0.16092	683.64	218.60
410	0.33045	51.563	0.019394	10.364	10.371	0.030743	0.064570	0.077026	1492.2	−0.15366	683.52	201.43
420	0.43730	51.064	0.019583	11.136	11.144	0.032602	0.063731	0.077447	1472.5	−0.14581	682.53	186.68
430	0.57026	50.541	0.019786	11.911	11.923	0.034427	0.062920	0.077934	1450.6	−0.13728	680.70	173.91
440	0.73367	49.994	0.020003	12.692	12.706	0.036222	0.062140	0.078495	1426.5	−0.12794	678.05	162.77
450	0.9322	49.421	0.020234	13.477	13.496	0.037988	0.061390	0.079136	1400.4	−0.11767	674.59	152.98
460	1.1709	48.824	0.020482	14.269	14.293	0.039729	0.060671	0.079869	1372.2	−0.10631	670.28	144.31
470	1.4551	48.199	0.020748	15.068	15.098	0.041448	0.059984	0.080706	1342.0	−0.09369	665.12	136.58
480	1.7905	47.545	0.021033	15.875	15.913	0.043147	0.059327	0.081662	1309.8	−0.07959	659.07	129.64
490	2.1831	46.861	0.021340	16.690	16.737	0.044830	0.058702	0.082757	1275.7	−0.06372	652.06	123.37
500	2.6392	46.145	0.021671	17.515	17.573	0.046498	0.058109	0.084013	1239.6	−0.04578	644.05	117.66
510	3.1655	45.393	0.022030	18.352	18.421	0.048156	0.057548	0.085464	1201.5	−0.02534	634.95	112.42
520	3.7690	44.603	0.022420	19.200	19.285	0.049807	0.057023	0.087149	1161.3	−0.00189	624.68	107.57
530	4.4569	43.770	0.022847	20.064	20.165	0.051454	0.056536	0.089124	1119.1	0.025264	613.15	103.05
540	5.2369	42.889	0.023316	20.943	21.065	0.053102	0.056089	0.091464	1074.6	0.057002	600.26	98.792
550	6.1172	41.954	0.023836	21.841	21.987	0.054756	0.055690	0.094275	1027.9	0.094527	585.95	94.746
560	7.1062	40.956	0.024417	22.762	22.935	0.056422	0.055347	0.097713	978.54	0.13949	570.21	90.857
570	8.2132	39.885	0.025072	23.709	23.915	0.058106	0.055071	0.10201	926.44	0.19425	553.08	87.074
580	9.448	38.725	0.025823	24.688	24.932	0.059821	0.054881	0.10754	871.23	0.26220	534.74	83.342
590	10.821	37.456	0.026698	25.707	25.996	0.061577	0.054808	0.11491	812.49	0.34857	515.43	79.600
600	12.345	36.048	0.027741	26.777	27.119	0.063396	0.054902	0.12526	749.57	0.46172	495.46	75.773
610	14.033	34.451	0.029026	27.917	28.324	0.065309	0.055258	0.14100	681.27	0.61660	475.03	71.759
620	15.901	32.577	0.030697	29.160	29.648	0.067371	0.056100	0.16852	604.73	0.84473	454.10	67.382
630	17.969	30.210	0.033101	30.585	31.180	0.069715	0.058152	0.23108	513.19	1.2251	432.51	62.244
640	20.265	26.729	0.037413	32.422	33.180	0.072737	0.064521	0.46736	400.66	1.9542	414.93	55.247
647.1	22.064	17.874	0.055948	36.314	37.548	0.079393			0	3.7410		
273.16	0.000612	0.000269	3711.0	42.785	45.055	0.16494	0.025553	0.033947	409.00	592.65	17.071	9.2163
280	0.000992	0.000426	2345.4	42.954	45.280	0.16174	0.025657	0.034073	413.92	477.26	17.442	9.3815
290	0.001920	0.000797	1254.3	43.201	45.609	0.15741	0.025816	0.034270	420.99	351.65	18.031	9.6414
300	0.003537	0.001420	704.01	43.446	45.936	0.15344	0.025982	0.034483	427.89	264.35	18.673	9.9195
310	0.006231	0.002424	412.60	43.690	46.261	0.14981	0.026158	0.034716	434.63	203.74	19.369	10.213
320	0.010546	0.003978	251.39	43.931	46.582	0.14647	0.026350	0.034980	441.18	161.25	20.117	10.518
330	0.017213	0.006304	158.62	44.169	46.900	0.14339	0.026568	0.035287	447.54	130.92	20.922	10.833
340	0.027188	0.009681	103.30	44.404	47.212	0.14054	0.026821	0.035653	453.68	108.77	21.784	11.157
350	0.041682	0.014448	69.213	44.634	47.519	0.13791	0.027118	0.036091	459.58	92.178	22.707	11.487
360	0.062194	0.021014	47.586	44.860	47.819	0.13546	0.027469	0.036617	465.22	79.440	23.695	11.823
370	0.090535	0.029859	33.491	45.079	48.111	0.13317	0.027883	0.037249	470.57	69.427	24.750	12.162
380	0.12885	0.041537	24.075	45.291	48.393	0.13104	0.028372	0.038004	475.61	61.373	25.875	12.504
390	0.17964	0.056683	17.642	45.496	48.665	0.12904	0.028944	0.038903	480.32	54.749	27.074	12.848
400	0.24577	0.076014	13.156	45.691	48.924	0.12715	0.029608	0.039963	484.67	49.181	28.347	13.192
410	0.33045	0.10034	9.9666	45.876	49.170	0.12537	0.030369	0.041203	488.65	44.405	29.699	13.538
420	0.43730	0.13055	7.6601	46.050	49.400	0.12369	0.031230	0.042634	492.22	40.237	31.128	13.883
430	0.57026	0.16765	5.9649	46.211	49.613	0.12208	0.032187	0.044269	495.39	36.550	32.638	14.228
440	0.73367	0.21276	4.7002	46.359	49.807	0.12054	0.033234	0.046114	498.12	33.259	34.230	14.573

450	0.93220	0.26711	3.7438	46.492	49.982	0.11907	0.034362	0.048177	500.41	30.307	35.904	14.917
460	1.1709	0.33209	3.0113	46.609	50.134	0.11764	0.035561	0.050469	502.24	27.653	37.663	15.261
470	1.4551	0.40925	2.4435	46.708	50.263	0.11627	0.036821	0.053005	503.60	25.265	39.512	15.606
480	1.7905	0.50035	1.9986	46.788	50.367	0.11493	0.038137	0.055809	504.45	23.118	41.455	15.952
490	2.1831	0.60738	1.6464	46.848	50.442	0.11362	0.039503	0.058919	504.78	21.187	43.502	16.300
500	2.6392	0.73265	1.3649	46.885	50.487	0.11233	0.040920	0.062388	504.55	19.450	45.666	16.653
510	3.1655	0.87884	1.1379	46.898	50.500	0.11105	0.042391	0.066289	503.71	17.886	47.969	17.011
520	3.7690	1.0491	0.95318	46.883	50.475	0.10979	0.043920	0.070723	502.23	16.475	50.442	17.377
530	4.4569	1.2473	0.80174	46.838	50.411	0.10852	0.045519	0.075827	500.05	15.197	53.130	17.755
540	5.2369	1.4780	0.67659	46.758	50.302	0.10724	0.047197	0.081789	497.10	14.035	56.102	18.149
550	6.1172	1.7471	0.57238	46.641	50.142	0.10595	0.048968	0.088873	493.31	12.973	59.456	18.563
560	7.1062	2.0620	0.48497	46.478	49.925	0.10462	0.050848	0.097461	488.58	11.997	63.341	19.007
570	8.2132	2.4325	0.41110	46.264	49.641	0.10324	0.052856	0.10813	482.79	11.093	67.981	19.489
580	9.4480	2.8720	0.34819	45.988	49.278	0.10180	0.055017	0.12178	475.80	10.248	73.721	20.024
590	10.821	3.3994	0.29417	45.636	48.819	0.10026	0.057361	0.13994	467.41	9.4499	81.108	20.634
600	12.345	4.0434	0.24732	45.188	48.242	0.098600	0.059939	0.16540	457.33	8.6837	91.052	21.350
610	14.033	4.8497	0.20620	44.613	47.506	0.096755	0.062831	0.20384	445.11	7.9329	105.17	22.229
620	15.901	5.9009	0.16946	43.855	46.550	0.094631	0.066197	0.26923	429.99	7.1743	126.66	23.374
630	17.969	7.3737	0.13562	42.801	45.238	0.092029	0.070465	0.40819	410.21	6.3669	163.44	25.018
640	20.265	9.8331	0.10170	41.095	43.156	0.088324	0.077576	0.94736	379.64	5.3854	250.01	27.938
647.1	22.064	17.874	0.055948	36.314	37.548	0.079393			0	3.7410		
Single-Phase Properties												
300	0.1	55.317	0.018078	2.0277	2.0295	0.007081	0.074406	0.075315	1501.5	−0.22024	610.32	853.83
372.76	0.1	53.212	0.018793	7.5196	7.5214	0.02347	0.067921	0.075938	1543.5	−0.17843	678.97	282.91
372.76	0.1	0.032769	30.517	45.138	48.190	0.13257	0.02801	0.037444	471.99	67.038	25.053	12.256
400	0.1	0.030397	32.898	45.900	49.189	0.13516	0.02717	0.036170	490.31	47.254	27.008	13.285
500	0.1	0.024154	41.401	48.619	52.759	0.14313	0.02717	0.035693	548.31	19.298	35.861	17.270
600	0.1	0.020086	49.786	51.387	56.365	0.14970	0.028103	0.036513	598.61	10.567	46.367	21.407
700	0.1	0.017201	58.136	54.256	60.069	0.15541	0.029225	0.037592	643.92	6.6444	57.964	25.564
800	0.1	0.015044	66.471	57.240	63.887	0.16050	0.030431	0.038778	685.47	4.5167	70.385	29.669
900	0.1	0.013369	74.799	60.347	67.827	0.16514	0.031687	0.040024	724.03	3.2280	83.466	33.685
1000	0.1	0.012030	83.123	63.581	71.893	0.16943	0.032963	0.041293	760.17	2.3885	97.085	37.592
1100	0.1	0.010936	91.444	66.941	76.085	0.17342	0.034228	0.042554	794.33	1.8122	111.15	41.382
1200	0.1	0.010024	99.763	70.426	80.402	0.17718	0.035458	0.043781	826.85	1.4006	125.58	45.054
300	1	55.340	0.018070	2.0263	2.0444	0.007077	0.074353	0.075270	1503.0	−0.22022	610.73	853.67
400	1	52.060	0.019209	9.5914	9.6106	0.028834	0.065422	0.076628	1511.3	−0.16113	684.10	218.80
453.03	1	49.243	0.020307	13.717	13.737	0.038518	0.061169	0.079348	1392.0	−0.11435	673.37	150.24
453.03	1	0.28559	3.5015	46.529	50.030	0.11863	0.034718	0.048846	501.02	29.473	36.427	15.021
500	1	0.25158	3.9749	48.111	52.086	0.12295	0.030084	0.041065	535.74	19.741	38.799	17.051
600	1	0.20466	4.8861	51.123	56.009	0.13011	0.029002	0.038358	592.58	10.615	47.636	21.329
700	1	0.17377	5.7547	54.087	59.842	0.13602	0.029629	0.038495	640.55	6.6387	58.735	25.550
800	1	0.15134	6.6074	57.121	63.729	0.14121	0.030651	0.039301	683.48	4.5077	70.983	29.687
900	1	0.13418	7.4524	60.258	67.710	0.14590	0.031821	0.040358	722.85	3.2212	84.000	33.718
1000	1	0.12058	8.2932	63.511	71.804	0.15021	0.033051	0.041522	759.50	2.3837	97.573	37.630
1100	1	0.10951	9.1313	66.885	76.016	0.15422	0.034290	0.042719	794.01	1.8089	111.57	41.420
1200	1	0.10032	9.9677	70.380	80.347	0.15799	0.035504	0.043905	826.77	1.3982	125.89	45.088
300	5	55.439	0.018038	2.0204	2.1106	0.007057	0.074119	0.075070	1509.8	−0.22012	612.54	853.00
400	5	52.173	0.019167	9.5643	9.6601	0.028766	0.065337	0.076438	1520.9	−0.16222	686.54	219.84
500	5	46.267	0.021614	17.474	17.582	0.046415	0.058082	0.083643	1250.0	−0.04945	646.52	118.27
537.09	5	43.151	0.023175	20.685	20.801	0.052622	0.056215	0.090740	1087.8	0.047232	604.15	100.01
537.09	5	1.4072	0.71063	46.785	50.338	0.10762	0.046699	0.079952	498.04	14.362	55.203	18.032
600	5	1.1320	0.88340	49.734	54.151	0.11436	0.034611	0.051045	561.07	10.407	54.653	21.062
700	5	0.91269	1.0957	53.286	58.765	0.12148	0.031678	0.043318	624.59	6.5536	62.680	25.547
800	5	0.77805	1.2853	56.576	63.002	0.12714	0.031683	0.041848	674.39	4.4532	73.950	29.806
900	5	0.68224	1.4658	59.855	67.183	0.13207	0.032430	0.041922	717.57	3.1856	86.626	33.891
1000	5	0.60918	1.6416	63.197	71.405	0.13652	0.033447	0.042571	756.57	2.3599	99.971	37.821
1100	5	0.55109	1.8146	66.632	75.705	0.14061	0.034565	0.043465	792.63	1.7924	113.64	41.606
1200	5	0.50355	1.9859	70.172	80.101	0.14444	0.035704	0.044458	826.45	1.3865	127.51	45.257

(*Continued*)

TABLE 2-135 Thermodynamic Properties of Water *(Continued)*

Temperature K	Pressure MPa	Density mol/dm³	Volume dm³/mol	Int. energy kJ/mol	Enthalpy kJ/mol	Entropy kJ/(mol·K)	C_v kJ/(mol·K)	C_p kJ/(mol·K)	Sound speed m/s	Joule-Thomson K/MPa	Therm. cond. mW/(m·K)	Viscosity μPa·s
Single-Phase Properties (*Continued*)												
300	10	55.561	0.017998	2.0131	2.1931	0.007031	0.073834	0.074829	1518.2	−0.21999	614.81	852.28
400	10	52.312	0.019116	9.5311	9.7222	0.028682	0.065233	0.076208	1532.7	−0.16351	689.57	221.13
500	10	46.517	0.021497	17.389	17.604	0.046244	0.058028	0.082910	1271.3	−0.05669	651.64	119.55
584.15	10	38.213	0.026169	25.105	25.367	0.060543	0.054835	0.11032	847.33	0.29540	526.83	81.795
584.15	10	3.0787	0.32482	45.852	49.100	0.10117	0.055964	0.128640	472.51	9.9124	76.543	20.267
600	10	2.7628	0.36195	47.183	50.802	0.10405	0.047271	0.092535	503.34	9.4382	71.110	21.036
700	10	1.9625	0.50956	52.145	57.241	0.11405	0.034838	0.051779	602.20	6.3228	69.301	25.704
800	10	1.6157	0.61893	55.851	62.040	0.12046	0.033089	0.045603	662.61	4.3529	78.476	30.054
900	10	1.3945	0.71709	59.334	66.505	0.12572	0.033219	0.044062	710.98	3.1289	90.516	34.176
1000	10	1.2345	0.81002	62.798	70.898	0.13035	0.033947	0.043952	753.03	2.3241	103.50	38.111
1100	10	1.1111	0.90002	66.314	75.314	0.13456	0.034908	0.044427	791.02	1.7683	116.73	41.882
1200	10	1.0119	0.98820	69.910	79.792	0.13846	0.035954	0.045164	826.16	1.3695	130.00	45.506
300	100	57.573	0.017369	1.8921	3.6290	0.006516	0.069812	0.071696	1667.9	−0.21618	654.50	856.88
400	100	54.500	0.018349	9.0423	10.877	0.027360	0.063582	0.073086	1717.3	−0.17905	741.80	243.50
500	100	49.914	0.020034	16.289	18.292	0.043895	0.057324	0.075607	1555.7	−0.12564	730.42	138.92
600	100	43.935	0.022761	23.820	26.097	0.058109	0.052776	0.081104	1300.4	−0.02079	645.83	101.51
700	100	36.179	0.027640	31.916	34.680	0.071320	0.049610	0.091576	1020.0	0.21155	510.14	79.363
800	100	26.768	0.037359	40.700	44.435	0.084331	0.047143	0.10108	813.97	0.65939	351.46	62.042
900	100	19.073	0.052429	48.805	54.048	0.095669	0.043932	0.088057	765.30	1.0399	257.03	53.250
1000	100	14.734	0.067868	55.188	61.975	0.10404	0.041345	0.071678	792.50	1.0944	232.07	51.518
1100	100	12.246	0.081656	60.470	68.635	0.11039	0.040131	0.062539	832.67	0.98401	223.70	52.497
1200	100	10.631	0.094062	65.222	74.628	0.11561	0.039810	0.057826	872.28	0.83544	219.07	54.415
300	500	63.750	0.015686	1.5247	9.3678	0.003746	0.063403	0.068296	2228.6	−0.19915	763.82	1089.4
400	500	60.862	0.016431	7.9635	16.179	0.023347	0.059634	0.067603	2258.7	−0.19486	929.09	320.18
500	500	57.695	0.017332	14.264	22.930	0.038412	0.055769	0.067522	2200.7	−0.18339	1096.6	189.08
600	500	54.316	0.018411	20.481	29.687	0.050731	0.052734	0.067584	2093.8	−0.16883	1097.9	141.83
700	500	50.847	0.019667	26.606	36.439	0.061141	0.050315	0.067436	1970.5	−0.15188	935.15	118.47
800	500	47.385	0.021104	32.615	43.167	0.070124	0.048442	0.067080	1850.1	−0.13256	738.72	104.70
900	500	44.018	0.022718	38.492	49.851	0.077998	0.047068	0.066596	1743.4	−0.11124	572.49	95.388
1000	500	40.814	0.024501	44.233	56.484	0.084987	0.046126	0.066041	1655.7	−0.08910	445.17	88.418
1100	500	37.834	0.026432	49.839	63.055	0.091251	0.045537	0.065356	1589.3	−0.06907	350.97	83.021
1200	500	35.124	0.028470	55.312	69.547	0.096900	0.045218	0.064451	1543.9	−0.05511	282.78	78.952
400	1000	65.942	0.015165	7.4792	22.644	0.019833	0.057934	0.065743	2718.6	−0.19303	1172.7	329.93
500	1000	63.253	0.015810	13.357	29.167	0.034391	0.055063	0.064967	2677.2	−0.19158	2199.5	190.55
600	1000	60.572	0.016509	19.141	35.650	0.046212	0.053055	0.064676	2602.3	−0.18789	3250.5	137.73
700	1000	57.937	0.017260	24.836	42.096	0.056150	0.051393	0.064219	2513.7	−0.18439	3202.2	108.98
800	1000	55.384	0.018056	30.435	48.491	0.064689	0.050059	0.063663	2423.7	−0.18105	2408.7	91.430
900	1000	52.937	0.018890	35.938	54.828	0.072155	0.049062	0.063101	2338.7	−0.17779	1610.7	80.198
1000	1000	50.611	0.019759	41.354	61.113	0.078776	0.048373	0.062594	2261.5	−0.17459	1052.9	72.716
1100	1000	48.415	0.020655	46.695	67.350	0.084722	0.047942	0.062176	2193.2	−0.17139	703.41	67.520
1200	1000	46.349	0.021575	51.976	73.551	0.090117	0.047713	0.061861	2133.6	−0.16808	487.61	63.774

The values in these tables were generated from the NIST REFPROP software (Lemmon, E. W., McLinden, M. O., and Huber, M. L., NIST Standard Reference Database 23: Reference Fluid Thermodynamic and Transport Properties—REFPROP, National Institute of Standards and Technology, Standard Reference Data Program, Gaithersburg, Md., 2002, Version 7.1). The primary source for the thermodynamic properties is Wagner, W., and Pruss, A., "The IAPWS Formulation 1995 for the Thermodynamic Properties of Ordinary Water Substance for General and Scientific Use," *J. Phys. Chem. Ref. Data* **31**(2):387–535, 2002. The source for viscosity is International Association for the Properties of Water and Steam, *Revised Release on the IAPS Formulation 1985 for the Viscosity of Ordinary Water Substance,* IAPWS, 1997. The source for thermal conductivity is the International Association for the Properties of Water and Steam, *Revised Release on the IAPS Formulation 1985 for the Thermal Conductivity of Ordinary Water Substance,* IAPWS, 1998.

Properties at the triple point temperature and the critical point temperature are given in the first and last entries of the saturation tables, respectively. In the single-phase table, when the temperature range for a given isobar includes a vapor-liquid phase boundary, the temperature of phase equilibrium is noted, and properties for both the saturated liquid and saturated vapor are given (with liquid properties given in the upper line). Lines are omitted from the temperature-pressure grid of the single-phase table, when the system would be in the solid phase or if there are potential problems with the source property surface.

The uncertainty in density of the equation of state is 0.0001% at 1 atm in the liquid phase, and 0.001% at other liquid states at pressures up to 10 MPa and temperatures to 423 K. In the vapor phase, the uncertainty is 0.05% or less. The uncertainties rise at higher temperatures and/or pressures, but are generally less than 0.1% in density except at extreme conditions. The uncertainty in pressure in the critical region is 0.1%. The uncertainty of the speed of sound is 0.15% in the vapor and 0.1% or less in the liquid, and increases near the critical region and at high temperatures and pressures. The uncertainty in isobaric heat capacity is 0.2% in the vapor and 0.1% in the liquid, with increasing values in the critical region and at high pressures. The uncertainties of saturation conditions are 0.025% in vapor pressure, 0.0025% in saturated-liquid density, and 0.1% in saturated-vapor density. The uncertainties in the saturated densities increase substantially as the critical region is approached. For the uncertainties in the viscosity and thermal conductivity, see the IAPWS Release.

TABLE 2-136 Thermodynamic Properties of Water Substance along the Melting Line

P, bar	T, °C	$10^3\ v_f$, m³/kg	h_f, kJ/kg	s_f, kJ/kg·K	c_{pf}, kJ/kg·K	c_{melt}, kJ/kg·K	$10^6\alpha_f$, K⁻¹	$10^6K_{f,T}$ bar⁻¹
6.117×10^{-3t}	0.0100	1.00021	0	0	4.219	3.969	−67.42	50.90
1.01325	0.0026	1.00016	0.0719	−0.0001	4.218	3.970	−67.17	50.88
50	−0.3618	0.99770	3.5140	−0.0054	4.196	3.997	−54.92	50.30
100	−0.7410	0.99523	6.9794	−0.0110	4.174	4.023	−42.52	49.73
150	−1.1249	0.99278	10.3964	−0.0167	4.152	4.047	−30.24	49.17
200	−1.5166	0.99037	13.7648	−0.0225	4.132	4.070	−18.05	48.63
250	−1.9151	0.98798	17.0843	−0.0285	4.112	4.092	−5.93	48.11
300	−2.3206	0.98562	20.3547	−0.0347	4.092	4.113	6.12	47.59
400	−3.1532	0.98098	26.7472	−0.0474	4.056	4.150	30.09	46.61
500	−4.0156	0.97643	32.9403	−0.0607	4.022	4.184	53.97	45.68
600	−4.909	0.97196	38.932	−0.0747	3.992	4.215	77.87	44.80
800	−6.790	0.96326	50.300	−0.1046	3.937	4.270	126.18	43.19
1000	−8.803	0.95493	60.836	−0.1371	3.893	4.320	175.98	41.74

Condensed from U. Grigull, Private communication, January 18, 1995.

Materials prepared at Technical University München, Germany by U. Grigull and S. Marek. For a table as a function of temperature, see Grigull, U. and S. Marek, *Warme u. Stoff.*, **30** (1994): 1–8.

t = the triple point (at 6.117×10^{-3} bar, 0.01°C); v_f = 0.0010021 m³/kg: $\alpha_f = -67.42 \times 10^{-6}$/K.

Other equations for properties are given by Jones, F. E. and G. L. Harris, *J. Res. N.I.S.T.*, **97,** 3 (1992): 335–340, and by Wagner, W. and A. Pruss, *J. Phys. Chem. Ref. Data,* **22,** 3 (1993): 783–787. Steam tables include Walker, W. A., U.S. Naval Ordn. Lab. rept. NOLTR NOLTR-66-217 = AD 651105 (0–1000 bar, 0–150°C), 1967 (72 pp.); Grigull, U., J. Straub, et al., *Steam Tables in S.I. Units* (0.01–1000 bar, 0–1000°C), Springer-Verlag, Berlin, 1990 (133 pp.); Tseng, C. M., T. A. Hamp, et al., Atomic Energy of Canada rept. (30 props, sat liq & vap., 1–220 bar), AECL-5910 1977 (90 pp.). For dissociation, see e.g., Knonicek, V., *Rozpr. Cesko Acad Ved., Rada techn ved* (0.01–100 bar, 1000–5000 K). **77,** 1 (1967). The proceedings of the 10th international conference on the properties of steam were edited by Sytchev, V. V. and A. A. Aleksandrov, Plenum, NY, 1984; and for the 11th conference by Pichal, M. and O. Sifner, Hemisphere, 1989 (550 pp.). Current information on the properties of solid, vapor, and liquid water properties can be found at http://www.iapws.org.

For electrical conductivity, see e.g., Marshall, W. L., *J. Chem. Eng. Data,* **32** (1987): 221–226.

TRANSPORT PROPERTIES

Introduction The tables and nomographs in this subsection are organized roughly with mass transport properties first (surface tension, viscosity, diffusion coefficient) followed by thermal transport properties.

Unit Conversions For this subsection, the following unit conversions are applicable:

Diffusivity: to convert square centimeters per second to square feet per hour, multiply by 3.8750; to convert square meters per second to square feet per hour, multiply by 38,750.

Pressure: to convert bars to pounds-force per square inch, multiply by 14.504.

Temperature: °F = ⁹⁄₅°C + 32; °R = ⁹⁄₅ K.

Thermal conductivity: to convert watts per meter-kelvin to British thermal unit–feet per hour–square foot–degree Fahrenheit, multiply by 0.57779; and to convert British thermal unit–feet per hour–square foot–degree Fahrenheit to watts per meter-kelvin, multiply by 1.7307.

Viscosity: to convert pascal-seconds to centipoise, multiply by 1000.

Additional References An extensive coverage of the general pressure and temperature variation of thermal conductivity is given in the monograph by Vargaftik, N. B., L. P. Filippov, A. A. Tarzimanov and E. E. Totskiy, *Thermal Conductivity of Liquids and Gases* (in Russian), Standards Press, Moscow, 1978, now published in English translation by CRC Press, Miami, Fla.

For a similar work on viscosity, see Stephan and Lucas, *Viscosity of Dense Fluids,* Plenum, New York and London, 1979. Tables and polynomial fits for refrigerants in both the gaseous and the liquid states are contained in *ASHRAE Handbook—Fundamentals,* SI ed., ASHRAE, Atlanta, 2005. Other sources for viscosity include Fischer & Porter Co. catalog 10-A-94, "Fluid Densities and Viscosities," 1953 (200 industrial fluids in 48 pp.) and D. van Velzen, R. L. Cardozo et al., EURATOM Ispra, Italy rept. 4735 e, 1972 (160 pp.). Liquid viscosity, 314 cpds, is summarized in *I&EC Fundtls.,* 11 (1972): 20–26. Five hundred forty-nine binary and ternary systems are discussed in Skubla, P., *Coll. Czech. Chem. Commun.,* **46** (1981): 303–339.

See also Duhne, C. R., *Chem. Eng.* (NY), **86:** 15 (July 16, 1979): 83–91 (equations and 326 liquids); and Rao, K. V. K., *Chem. Eng.* (NY), **90,** 11 (May 30, 1983): 90–91 (nomograph, 87 liquids). For rheology, non-Newtonian behavior, see, for instance, Barnes, H., *The Chem. Engr.* (UK), (June 24, 1993): 17–23; Hyman, W. A., *I&EC Fundtls.,* **16** (1976): 215–218; and Ferguson, J., and Z. Kemblowski, *Applied Fluid Rheology,* Elsevier, 1991 (325 pp.). Other sources for thermal conductivity include Ho, C. Y., R. W. Powell et al., *J. Phys. Chem. Ref. Data,* **1** (1972) and **3,** suppl. 1 (1974); Childs, Ericks et al., *N.B.S. Monogr.* 131, 1973; Jamieson, D. T., J. B. Irving et al., *Liquid Thermal Conductivity,* H.M.S.O., Edinburgh, Scotland, 1975 (220 pp.).

Other references include B. Poling, J. Prausnitz, and J. O'Connell, *The Properties of Gases and Liquids*, 5th ed., McGraw-Hill, New York, 2000; N.B. Vargaftik, Y.K. Vinogradov, and V.S. Yargin, *Handbook of Physical Properties of Liquids and Gases*, Begell House, New York, 1996; Carl Yaws, *Chemical Properties Handbook: Physical, Thermodynamics, Environmental Transport, Safety & Health Related Properties for Organic & Inorganic Chemicals*, McGraw-Hill, New York, 1998; and M.R. Riazi, *Characterization and Properties of Petroleum Fractions*, ASTM, West Conshohocken, Pa., 2005. Free web resources include the NIST Webbook at http://webbook.nist.gov and the KDB (Korea thermophysical properties) database at http://www.cheric.org/research/kdb/.

MASS TRANSPORT PROPERTIES

TABLE 2-137 Surface Tension σ (dyn/cm) of Various Liquids

Compound	*T*, K	σ	Compound	*T*, K	σ	Compound	*T*, K	σ
Acetic acid	293	27.59	*p*-Cresol	313	34.88	Isobutyric acid	293	25.04
	333	23.62		373	29.32		313	23.2
Acetone	298	24.02	Cyclohexane	293	25.24		333	21.36
	308	22.34		313	22.87		363	18.6
	318	21.22		333	20.49	Methyl formate	293	24.62
Aniline	293	42.67	Cyclopentane	293	22.61		323	20.05
	313	40.5		313	19.68		373	12.9
	333	38.33	Diethyl ether	288	17.56		423	6.3
	353	36.15		303	16.2		473	0.87
Benzene	293	28.88	2,3-Dimethylbutane	293	17.38	Methyl alcohol	293	22.56
	313	26.25		313	15.38		313	20.96
	333	23.67	Ethyl acetate	293	23.97		333	19.41
	353	21.2		313	21.65	Phenol	313	39.27
Benzonitrile	293	39.37		333	19.32		333	37.13
	323	35.89		353	17		373	32.96
	363	31.26		373	14.68	*n*-Propyl alcohol	293	23.71
Bromobenzene	293	35.82	Ethyl benzoate	293	35.04		313	22.15
	323	32.34		313	32.92		333	20.6
	373	26.54		333	30.81		363	18.27
n-Butane	203	23.31	Ethyl bromide	283	25.36	*n*-Propyl benzene	293	29.98
	233	19.69		303	23.04		313	26.83
	293	12.46	Ethyl mercaptan	288	23.87		333	24.68
Carbon disulfide	293	32.32		303	22.68		353	22.53
	313	29.35	Formamide	298	57.02		373	20.38
Carbon tetrachloride	288	27.65		338	53.66	Pyridine	293	37.21
	308	25.21		373	50.71		313	34.6
	328	22.76	*n*-Heptane	293	20.14		333	31.98
	348	20.31		313	18.18			
	368	17.86		333	16.22			
Chlorobenzene	293	33.59		353	14.26			
	323	30.01						
	373	24.06						

Methyl formate values from D. B. Macleod, *Trans. Faradaay Soc.* **19**:38, 1923. All others from J. J. Jasper, *J. Phys. Chem. Ref. Data* **1**:841, 1972.

TABLE 2-138 Vapor Viscosity of Inorganic and Organic Substances (Pa·s)

Cmpd. no.	Name	Formula	CAS	Mol. wt.	C_1	C_2	C_3	C_4	T_{min}, K	Viscosity at T_{min}	T_{max}, K	Viscosity at T_{max}
1	Acetaldehyde	C_2H_4O	75-07-0	44.05256	1.9703E-05	0.17646	1564.6		149.78	4.166E-06	1000	2.600E-05
2	Acetamide	C_2H_5NO	60-35-5	59.0672	1.4230E-07	0.7574	272.14		353.33	6.842E-06	1000	2.093E-05
3	Acetic acid	$C_2H_4O_2$	64-19-7	60.052	1.5640E-08	1.078			289.81	7.053E-06	1000	2.681E-05
4	Acetic anhydride	$C_4H_6O_3$	108-24-7	102.08864	1.0939E-05	0.23466	1209.5		200.15	5.386E-06	1000	2.504E-05
5	Acetone	C_3H_6O	67-64-1	58.07914	3.1005E-08	0.9762	23.139		178.45	4.329E-06	1000	2.571E-05
6	Acetonitrile	C_2H_3N	75-05-8	41.0519	4.7754E-07	0.60273	327.16		229.32	5.208E-06	1000	2.314E-05
7	Acetylene	C_2H_2	74-86-2	26.03728	1.2025E-06	0.4952	291.4		192.40	6.468E-06	600	1.923E-05
8	Acrolein	C_3H_4O	107-02-8	56.06326	6.5230E-07	0.579	410.8		185.45	4.174E-06	1000	2.523E-05
9	Acrylic acid	$C_3H_4O_2$	79-10-7	72.06266	1.7154E-07	0.7418	138.4		286.15	7.679E-06	1000	2.532E-05
10	Acrylonitrile	C_3H_3N	107-13-1	53.0626	2.4910E-08	0.98882			189.63	4.455E-06	1000	2.306E-05
11	Air	Mixture	132259-10-0	28.96	1.4250E-06	0.5039	108.3		80.00	5.508E-06	2000	6.227E-05
12	Ammonia	H_3N	7664-41-7	17.03052	4.1855E-08	0.9806	30.8		195.41	6.378E-06	1000	3.551E-05
13	Anisole	C_7H_8O	100-66-3	108.13782	1.7531E-07	0.72	176.17		235.65	5.122E-06	1000	2.154E-05
14	Argon	Ar	7440-37-1	39.948	9.2121E-07	0.60529	83.24		83.78	6.742E-06	3273.1	1.205E-04
15	Benzamide	C_7H_7NO	55-21-0	121.13658	2.5082E-08	0.96663			403.00	8.274E-06	1000	1.992E-05
16	Benzene	C_6H_6	71-43-2	78.11184	3.1340E-08	0.9676	7.9		278.68	7.077E-06	1000	2.486E-05
17	Benzenethiol	C_6H_6S	108-98-5	110.17684	1.1184E-07	0.8002	152.43		442.29	1.089E-05	1000	2.441E-05
18	Benzoic acid	$C_7H_6O_2$	65-85-0	122.12134	7.4266E-08	0.8289	91.197		395.45	8.578E-06	1000	2.087E-05
19	Benzonitrile	C_7H_5N	100-47-0	103.1213	3.4647E-05	0.12396	3260.2		260.28	5.104E-06	1000	1.915E-05
20	Benzophenone	$C_{13}H_{10}O$	119-61-9	182.2179	3.7790E-07	0.6005	409		321.35	5.324E-06	1000	1.698E-05
21	Benzyl alcohol	C_7H_8O	100-51-6	108.13782	6.9022E-08	0.84014	74.746		257.85	5.680E-06	1000	2.129E-05
22	Benzyl ethyl ether	$C_9H_{12}O$	539-30-0	136.19098	1.5600E-07	0.7181	180		458.15	9.122E-06	1000	1.886E-05
23	Benzyl mercaptan	C_7H_8S	100-53-8	124.20342	4.0138E-08	0.90735	34.714		243.95	5.151E-06	1000	2.045E-05
24	Biphenyl	$C_{12}H_{10}$	92-52-4	154.2078	1.3874E-06	0.4434	678.22		342.20	6.186E-06	1000	1.768E-05
25	Bromine	Br_2	7726-95-6	159.808	7.3534E-08	0.93798			265.85	1.383E-05	600	2.967E-05
26	Bromobenzene	C_6H_5Br	108-86-1	157.0079	2.2320E-07	0.7146	184.9		429.24	1.187E-05	1000	2.623E-05
27	Bromoethane	C_2H_5Br	74-96-4	108.965	6.2597E-08	0.9115			154.25	6.182E-06	1000	3.397E-05
28	Bromomethane	CH_3Br	74-83-9	94.93852	6.5411E-08	0.92914			179.44	8.126E-06	1000	4.009E-05
29	1,2-Butadiene	C_4H_6	590-19-2	54.09044	6.0259E-07	0.5309	199.64		136.95	3.340E-06	1000	1.966E-05
30	1,3-Butadiene	C_4H_6	106-99-0	54.09044	2.6960E-07	0.6715	134.7		164.25	4.553E-06	1000	2.457E-05
31	Butane	C_4H_{10}	106-97-8	58.1222	3.4387E-08	0.94604			134.86	3.559E-06	1000	2.369E-05
32	1,2-Butanediol	$C_4H_{10}O_2$	584-03-2	90.121	7.5626E-08	0.83521	71.798		220.00	5.157E-06	1000	2.260E-05
33	1,3-Butanediol	$C_4H_{10}O_2$	107-88-0	90.121	7.0728E-08	0.84383	64.391		196.15	4.580E-06	1000	2.259E-05
34	1-Butanol	$C_4H_{10}O$	71-36-3	74.1216	1.4031E-06	0.4611	537		183.85	3.961E-06	1000	2.207E-05
35	2-Butanol	$C_4H_{10}O$	78-92-2	74.1216	1.2114E-07	0.76972	92.661		158.45	3.772E-06	1000	2.259E-05
36	1-Butene	C_4H_8	106-98-9	56.10632	6.9744E-07	0.5462	305.25		87.80	1.795E-06	1000	2.325E-05
37	*cis*-2-Butene	C_4H_8	590-18-1	56.10632	4.2898E-08	0.91349			134.26	3.770E-06	1000	2.360E-05
38	*trans*-2-Butene	C_4H_8	624-64-6	56.10632	1.0500E-06	0.4867	358.7		167.62	4.044E-06	1000	2.229E-05
39	Butyl acetate	$C_6H_{12}O_2$	123-86-4	116.15828	1.0060E-07	0.77881	95.108		199.65	4.216E-06	1000	1.993E-05
40	Butylbenzene	$C_{10}H_{14}$	104-51-8	134.21816	3.4205E-07	0.59764	234.21		185.30	3.424E-06	1000	1.720E-05
41	Butyl mercaptan	$C_4H_{10}S$	109-79-5	90.1872	5.4539E-08	0.88896	43.687		157.46	3.833E-06	1000	2.427E-05
42	*sec*-Butyl mercaptan	$C_4H_{10}S$	513-53-1	90.1872	3.1378E-08	0.96513			133.02	3.520E-06	1000	2.466E-05
43	1-Butyne	C_4H_6	107-00-6	54.09044	2.7856E-06	0.377	663.14		147.43	3.329E-06	800	1.893E-05
44	Butyraldehyde	C_4H_8O	123-72-8	72.10572	4.2200E-05	0.10118	2840		176.80	4.175E-06	1000	2.211E-05
45	Butyric acid	$C_4H_8O_2$	107-92-6	88.1051	1.2566E-08	1.0939			267.95	5.692E-06	1000	2.404E-05
46	Butyronitrile	C_4H_7N	109-74-0	69.1051	1.8178E-05	0.17513	2110.6		161.30	3.144E-06	1000	1.959E-05
47	Carbon dioxide	CO_2	124-38-9	44.0095	2.1480E-06	0.46	290		194.67	9.749E-06	1500	5.203E-05
48	Carbon disulfide	CS_2	75-15-0	76.1407	5.8204E-08	0.9262	44.581		161.11	5.048E-06	800	2.693E-05
49	Carbon monoxide	CO	630-08-0	28.0101	1.1127E-06	0.5338	94.7		68.15	4.434E-06	1250	4.654E-05
50	Carbon tetrachloride	CCl_4	56-23-5	153.8227	3.1370E-06	0.3742	491.5		250.33	8.361E-06	1000	2.789E-05
51	Carbon tetrafluoride	CF_4	75-73-0	88.0043	2.1709E-06	0.45853	208		89.56	5.132E-06	1000	4.267E-05

(*Continued*)

TABLE 2-138 Vapor Viscosity of Inorganic and Organic Substances (Pa·s) *(Continued)*

Cmpd. no.	Name	Formula	CAS	Mol. wt.	C_1	C_2	C_3	C_4	T_{min}, K	Viscosity at T_{min}	T_{max}, K	Viscosity at T_{max}
52	Chlorine	Cl_2	7782-50-5	70.906	2.6000E-07	0.7423	98.3		200.00	8.900E-06	1000	3.992E-05
53	Chlorobenzene	C_6H_5Cl	108-90-7	112.5569	1.0650E-07	0.7942	94.7		227.95	5.611E-06	1000	2.348E-05
54	Chloroethane	C_2H_5Cl	75-00-3	64.5141	3.5554E-08	0.98455			136.75	4.506E-06	1000	3.195E-05
55	Chloroform	$CHCl_3$	67-66-3	119.37764	1.6960E-07	0.7693	96.6		209.63	7.091E-06	1000	3.143E-05
56	Chloromethane	CH_3Cl	74-87-3	50.4875	6.2860E-08	0.907			175.43	6.820E-06	1000	3.307E-05
57	1-Chloropropane	C_3H_7Cl	540-54-5	78.54068	4.7100E-08	0.911			150.35	4.533E-06	1000	2.547E-05
58	2-Chloropropane	C_3H_7Cl	75-29-6	78.54068	3.8802E-07	0.6367	205.08		155.97	4.175E-06	1000	2.618E-05
59	*m*-Cresol	C_7H_8O	108-39-4	108.13782	1.4427E-07	0.7438	166.15		285.39	6.113E-06	1000	2.108E-05
60	*o*-Cresol	C_7H_8O	95-48-7	108.13782	8.7371E-08	0.80775	98.538		304.19	6.687E-06	1000	2.108E-05
61	*p*-Cresol	C_7H_8O	106-44-5	108.13782	1.4305E-07	0.7451	159.8		307.93	6.731E-06	1000	2.120E-05
62	Cumene	C_9H_{12}	98-82-8	120.19158	3.3699E-07	0.60751	221.17		177.14	3.480E-06	1000	1.834E-05
63	Cyanogen	C_2N_2	460-19-5	52.0348	3.7385E-08	0.98433			245.25	8.411E-06	1000	3.355E-05
64	Cyclobutane	C_4H_8	287-23-0	56.10632	1.0881E-06	0.48359	330.86		182.48	4.797E-06	1000	2.308E-05
65	Cyclohexane	C_6H_{12}	110-82-7	84.15948	6.7700E-08	0.8367	36.7		279.69	6.671E-06	900	1.928E-05
66	Cyclohexanol	$C_6H_{12}O$	108-93-0	100.15888	7.9581E-08	0.8376	104.97		296.60	6.917E-06	1000	2.346E-05
67	Cyclohexanone	$C_6H_{10}O$	108-94-1	98.143	5.2312E-08	0.89422	58.008		242.00	5.714E-06	1000	2.381E-05
68	Cyclohexene	C_6H_{10}	110-83-8	82.1436	1.3326E-06	0.4537	445		169.67	3.778E-06	1000	2.118E-05
69	Cyclopentane	C_5H_{10}	287-92-3	70.1329	2.3619E-07	0.67465	139		179.28	4.409E-06	1000	2.191E-05
70	Cyclopentene	C_5H_8	142-29-0	68.11702	3.0260E-07	0.64991	167.14		138.13	3.369E-06	1000	2.309E-05
71	Cyclopropane	C_3H_6	75-19-4	42.07974	1.7578E-06	0.4265	370.34		145.59	4.150E-06	1000	2.441E-05
72	Cyclohexyl mercaptan	$C_6H_{12}S$	1569-69-3	116.22448	3.9150E-08	0.91427	22.264		189.64	4.238E-06	1000	2.118E-05
73	Decanal	$C_{10}H_{20}O$	112-31-2	156.2652	3.5018E-05	0.11725	3394.6		285.00	5.262E-06	1000	1.791E-05
74	Decane	$C_{10}H_{22}$	124-18-5	142.28168	2.6400E-08	0.9487	71		243.51	3.755E-06	1000	1.729E-05
75	Decanoic acid	$C_{10}H_{20}O_2$	334-48-5	172.265	7.1748E-08	0.7982	109.38		304.55	5.070E-06	1000	1.604E-05
76	1-Decanol	$C_{10}H_{22}O$	112-30-1	158.28108	5.5065E-08	0.8341	79.56		280.05	4.715E-06	1000	1.622E-05
77	1-Decene	$C_{10}H_{20}$	872-05-9	140.2658	6.1192E-08	0.82546	77.434		206.89	3.632E-06	1000	1.701E-05
78	Decyl mercaptan	$C_{10}H_{22}S$	143-10-2	174.34668	3.2720E-08	0.9302	39.13		247.56	4.761E-06	1000	1.944E-05
79	1-Decyne	$C_{10}H_{18}$	764-93-2	138.24992	5.6914E-07	0.50744	273.3		229.15	4.091E-06	1000	1.488E-05
80	Deuterium	D_2	7782-39-0	4.0316	2.4999E-07	0.6878	0.5962		60.00	4.137E-06	480	1.744E-05
81	1,1-Dibromoethane	$C_2H_4Br_2$	557-91-5	187.86116	1.4125E-07	0.8097	83.243		210.15	7.685E-06	1000	3.502E-05
82	1,2-Dibromoethane	$C_2H_4Br_2$	106-93-4	187.86116	1.1379E-07	0.8502	93.816		282.85	1.038E-05	1000	3.696E-05
83	Dibromomethane	CH_2Br_2	74-95-3	173.83458	2.9444E-07	0.728	154.74		370.10	1.538E-05	1000	3.895E-05
84	Dibutyl ether	$C_8H_{18}O$	142-96-1	130.22792	7.7147E-08	0.79906	80.765		175.30	3.278E-06	1000	1.781E-05
85	*m*-Dichlorobenzene	$C_6H_4Cl_2$	541-73-1	147.00196	2.3340E-07	0.714	260		248.39	5.850E-06	1000	2.569E-05
86	*o*-Dichlorobenzene	$C_6H_4Cl_2$	95-50-1	147.00196	1.6030E-07	0.763	205		256.15	6.127E-06	1000	2.588E-05
87	*p*-Dichlorobenzene	$C_6H_4Cl_2$	106-46-7	147.00196	1.5913E-07	0.7639	193.14		326.14	8.313E-06	1000	2.611E-05
88	1,1-Dichloroethane	$C_2H_4Cl_2$	75-34-3	98.95916	2.0135E-07	0.73421	111.98		176.19	5.487E-06	1000	2.887E-05
89	1,2-Dichloroethane	$C_2H_4Cl_2$	107-06-2	98.95916	1.4321E-07	0.7785	98.159		237.49	7.164E-06	1000	2.824E-05
90	Dichloromethane	CH_2Cl_2	75-09-2	84.93258	7.6787E-07	0.5741	276.16		178.01	5.895E-06	1000	3.175E-05
91	1,1-Dichloropropane	$C_3H_6Cl_2$	78-99-9	112.98574	1.4906E-07	0.7617	105.9		200.00	5.515E-06	1000	2.599E-05
92	1,2-Dichloropropane	$C_3H_6Cl_2$	78-87-5	112.98574	1.1989E-07	0.79108	84.37		172.71	4.742E-06	1000	2.611E-05
93	Diethanol amine	$C_4H_{11}NO_2$	111-42-2	105.13564	3.3628E-08	0.9426	39.587		301.15	6.450E-06	1000	2.176E-05
94	Diethyl amine	$C_4H_{11}N$	109-89-7	73.13684	4.3184E-07	0.6035	247		223.35	5.364E-06	1000	2.239E-05
95	Diethyl ether	$C_4H_{10}O$	60-29-7	74.1216	1.9480E-06	0.41	495.8		156.85	3.720E-06	1000	2.212E-05
96	Diethyl sulfide	$C_4H_{10}S$	352-93-2	90.1872	6.5492E-08	0.86232	59.455		169.20	4.046E-06	1000	2.388E-05
97	1,1-Difluoroethane	$C_2H_4F_2$	75-37-6	66.04997	2.7228E-06	0.39531	445.07		154.56	5.148E-06	1000	2.891E-05
98	1,2-Difluoroethane	$C_2H_4F_2$	624-72-6	66.04997	4.3934E-07	0.64867	169.64		215.00	8.001E-06	1000	3.317E-05
99	Difluoromethane	CH_2F_2	75-10-5	52.02339	7.7484E-07	0.57978	198.7		136.95	5.478E-06	1000	3.547E-05
100	Diisopropyl amine	$C_6H_{15}N$	108-18-9	101.19	4.1380E-07	0.5999	269.5		357.05	8.016E-06	1000	2.055E-05
101	Diisopropyl ether	$C_6H_{14}O$	108-20-3	102.17476	1.6910E-07	0.7114	124		187.65	4.218E-06	1000	2.049E-05
102	Diisopropyl ketone	$C_7H_{14}O$	565-80-0	114.18546	9.2797E-08	0.7819	93.399		204.81	4.089E-06	1000	1.881E-05

103	1,1-Dimethoxyethane	$C_4H_{10}O_2$	534-15-6	90.121	4.4172E-08	0.91098			159.95	4.497E-06	1000	2.388E-05
104	1,2-Dimethoxypropane	$C_5H_{12}O_2$	7778-85-0	104.14758	3.9833E-08	0.91566			226.10	5.701E-06	1000	2.224E-05
105	Dimethyl acetylene	C_4H_6	503-17-3	54.09044	1.9377E-06	0.4093	492.69		240.91	6.006E-06	1000	2.194E-05
106	Dimethyl amine	C_2H_7N	124-40-3	45.08368	2.7570E-07	0.6841	133.2		180.96	5.563E-06	1000	2.744E-05
107	2,3-Dimethylbutane	C_6H_{14}	79-29-8	86.17536	6.8567E-07	0.52542	278.82		145.19	3.211E-06	1000	2.021E-05
108	1,1-Dimethylcyclohexane	C_8H_{16}	590-66-9	112.21264	7.8220E-07	0.4994	371.6		392.70	7.936E-06	1000	1.796E-05
109	*cis*-1,2-Dimethylcyclohexane	C_8H_{16}	2207-01-4	112.21264	8.4576E-07	0.487	398		402.94	7.900E-06	1000	1.749E-05
110	*trans*-1,2-Dimethylcyclohexane	C_8H_{16}	6876-23-9	112.21264	9.9104E-07	0.4723	436.89		396.58	7.957E-06	1000	1.801E-05
111	Dimethyl disulfide	$C_2H_6S_2$	624-92-0	94.19904	3.2282E-08	0.97742			188.44	5.405E-06	1000	2.762E-05
112	Dimethyl ether	C_2H_6O	115-10-6	46.06844	2.6800E-06	0.3975	534		131.65	3.688E-06	1000	2.722E-05
113	*N,N*-Dimethyl formamide	C_3H_7NO	68-12-2	73.09378	3.5538E-06	0.3766	1176.1		212.72	4.097E-06	1000	2.202E-05
114	2,3-Dimethylpentane	C_7H_{16}	565-59-3	100.20194	5.0372E-07	0.54462	227.44		160.00	3.300E-06	1000	1.766E-05
115	Dimethyl phthalate	$C_{10}H_{10}O_4$	131-11-3	194.184	5.2195E-08	0.85584	69.036		274.18	5.089E-06	1000	1.804E-05
116	Dimethylsilane	C_2H_8Si	1111-74-6	60.17042	4.7238E-08	0.90849			122.93	3.739E-06	1000	2.511E-05
117	Dimethyl sulfide	C_2H_6S	75-18-3	62.134	5.2854E-07	0.6112	302.85		174.88	4.544E-06	1000	2.766E-05
118	Dimethyl sulfoxide	C_2H_6OS	67-68-5	78.13344	8.6101E-08	0.8345	167.86		291.67	6.231E-06	1000	2.350E-05
119	Dimethyl terephthalate	$C_{10}H_{10}O_4$	120-61-6	194.184	3.9554E-08	0.892597			413.79	8.569E-06	1000	1.884E-05
120	1,4-Dioxane	$C_4H_8O_2$	123-91-1	88.10512	2.7334E-07	0.7393	129.93		284.95	1.226E-05	1000	3.995E-05
121	Diphenyl ether	$C_{12}H_{10}O$	101-84-8	170.2072	2.8451E-08	0.93622			300.03	5.933E-06	1000	1.831E-05
122	Dipropyl amine	$C_6H_{15}N$	142-84-7	101.19	1.2900E-07	0.744	117.03		210.15	4.429E-06	1000	1.970E-05
123	Dodecane	$C_{12}H_{26}$	112-40-3	170.33484	6.3440E-08	0.8287	219.5		263.57	3.511E-06	1000	1.593E-05
124	Eicosane	$C_{20}H_{42}$	112-95-8	282.54748	2.9236E-07	0.62458	702.84		309.58	3.214E-06	1000	1.284E-05
125	Ethane	C_2H_6	74-84-0	30.069	2.5906E-07	0.67988	98.902		90.35	2.643E-06	1000	2.583E-05
126	Ethanol	C_2H_6O	64-17-5	46.06844	1.0613E-07	0.8066	52.7		200.00	6.029E-06	1000	2.651E-05
127	Ethyl acetate	$C_4H_8O_2$	141-78-6	88.10512	3.2140E-06	0.3572	667		189.60	4.632E-06	1000	2.274E-05
128	Ethyl amine	C_2H_7N	75-04-7	45.08368	4.9340E-07	0.5924	239.17		192.15	4.953E-06	1000	2.384E-05
129	Ethylbenzene	C_8H_{10}	100-41-4	106.165	4.2231E-07	0.58154	239.21		178.20	3.673E-06	1000	1.893E-05
130	Ethyl benzoate	$C_9H_{10}O_2$	93-89-0	150.1745	6.3441E-08	0.8369	73.63		238.45	4.733E-06	1000	1.915E-05
131	2-Ethyl butanoic acid	$C_6H_{12}O_2$	88-09-5	116.15828	9.2371E-08	0.7908	102.32		258.15	5.344E-06	1000	1.975E-05
132	Ethyl butyrate	$C_6H_{12}O_2$	105-54-4	116.15828	1.6175E-07	0.7163	142.27	3590	175.15	3.392E-06	1000	1.989E-05
133	Ethylcyclohexane	C_8H_{16}	1678-91-7	112.21264	4.1070E-07	0.57143	230.06		161.84	3.103E-06	1000	1.729E-05
134	Ethylcyclopentane	C_7H_{14}	1640-89-7	98.18606	2.1696E-06	0.3812	577.77		134.71	2.659E-06	1000	1.914E-05
135	Ethylene	C_2H_4	74-85-1	28.05316	2.0789E-06	0.4163	352.7		169.41	5.714E-06	1000	2.726E-05
136	Ethylenediamine	$C_2H_8N_2$	107-15-3	60.09832	1.3744E-07	0.7557	122.8		284.29	6.863E-06	1000	2.264E-05
137	Ethylene glycol	$C_2H_6O_2$	107-21-1	62.06784	8.6706E-08	0.83923	75.512		260.15	7.150E-06	1000	2.655E-05
138	Ethyleneimine	C_2H_5N	151-56-4	43.0678	2.8132E-07	0.6792	238.46		329.00	8.359E-06	1000	2.477E-05
139	Ethylene oxide	C_2H_4O	75-21-8	44.05256	4.3403E-08	0.94806			160.65	5.356E-06	1000	3.032E-05
140	Ethyl formate	$C_3H_6O_2$	109-94-4	74.07854	6.7610E-07	0.5804	354.9		193.55	5.069E-06	1000	2.750E-05
141	2-Ethyl hexanoic acid	$C_8H_{16}O_2$	149-57-5	144.211	2.5704E-08	0.94738			155.15	3.058E-06	1000	1.787E-05
142	Ethylhexyl ether	$C_8H_{18}O$	5756-43-4	130.22792	7.9129E-08	0.79565	83.193		180.00	3.371E-06	1000	1.781E-05
143	Ethylisopropyl ether	$C_5H_{12}O$	625-54-7	88.14818	1.3974E-07	0.74266	98.58		140.00	3.219E-06	1000	2.150E-05
144	Ethylisopropyl ketone	$C_6H_{12}O$	565-69-5	100.15888	1.0498E-07	0.76988	100.41		204.15	4.224E-06	1000	1.946E-05
145	Ethyl mercaptan	C_2H_6S	75-08-1	62.13404	8.5992E-08	0.8427	58.148		125.26	3.441E-06	1000	2.742E-05
146	Ethyl propionate	$C_5H_{10}O_2$	105-37-3	102.1317	5.5300E-07	0.6061	273.66		199.25	5.768E-06	1000	2.857E-05
147	Ethylpropyl ether	$C_5H_{12}O$	628-32-0	88.14818	5.1539E-07	0.5726	288.76		145.65	2.994E-06	1000	2.088E-05
148	Ethyltrichlorosilane	$C_2H_5Cl_3Si$	115-21-9	163.506	2.6635E-05	0.15779	2173.5		167.55	4.277E-06	1000	2.496E-05
149	Fluorine	F_2	7782-41-4	37.9968064	6.3600E-07	0.6638	61.6		53.48	4.148E-06	1000	5.873E-05
150	Fluorobenzene	C_6H_5F	462-06-6	96.1023032	2.1174E-07	0.7087	157.42		357.88	9.491E-06	1000	2.446E-05
151	Fluoroethane	C_2H_5F	353-36-6	48.0595	4.0868E-06	0.35526	651.07		129.95	3.832E-06	1000	2.880E-05
152	Fluoromethane	CH_3F	593-53-3	34.03292	3.9346E-08	1.0027			131.35	5.237E-06	1000	4.009E-05
153	Formaldehyde	CH_2O	50-00-0	30.02598	1.5948E-05	0.21516	1151.1		155.15	5.608E-06	1000	3.277E-05
154	Formamide	CH_3NO	75-12-7	45.04062	6.8290E-08	0.8774	54.864		275.60	7.882E-06	1000	2.776E-05
155	Formic acid	CH_2O_2	64-18-6	46.0257	5.0702E-08	0.9114			281.45	8.658E-06	1000	2.749E-05
156	Furan	C_4H_4O	110-00-9	68.07396	6.4320E-07	0.5854	325.3		187.55	5.037E-06	1000	2.768E-05

(*Continued*)

TABLE 2-138 Vapor Viscosity of Inorganic and Organic Substances (Pa·s) *(Continued)*

Cmpd. no.	Name	Formula	CAS	Mol. wt.	C_1	C_2	C_3	C_4	T_{min}, K	Viscosity at T_{min}	T_{max}, K	Viscosity at T_{max}
157	Helium-4	He	7440-59-7	4.0026	3.2530E-07	0.7162	−9.6	107	20.00	3.530E-06	2000	7.561E-05
158	Heptadecane	$C_{17}H_{36}$	629-78-7	240.46774	3.1338E-07	0.6238	692.2		295.13	3.254E-06	1000	1.377E-05
159	Heptanal	$C_7H_{14}O$	111-71-7	114.18546	4.2392E-05	0.1011	3420		229.80	4.625E-06	1000	1.928E-05
160	Heptane	C_7H_{16}	142-82-5	100.20194	6.6720E-08	0.82837	85.752		182.57	3.391E-06	1000	1.878E-05
161	Heptanoic acid	$C_7H_{14}O_2$	111-14-8	130.185	1.3633E-08	1.0595			265.83	5.052E-06	1000	2.056E-05
162	1-Heptanol	$C_7H_{16}O$	111-70-6	116.20134	2.5720E-07	0.6502	248.6		239.15	4.440E-06	1000	1.838E-05
163	2-Heptanol	$C_7H_{16}O$	543-49-7	116.20134	3.4649E-05	0.10705	2900.7		220.00	4.351E-06	1000	1.861E-05
164	3-Heptanone	$C_7H_{14}O$	106-35-4	114.18546	8.9656E-08	0.78236	100.14		234.15	4.485E-06	1000	1.812E-05
165	2-Heptanone	$C_7H_{14}O$	110-43-0	114.18546	8.8629E-08	0.78376	100.18		238.15	4.550E-06	1000	1.809E-05
166	1-Heptene	C_7H_{14}	592-76-7	98.18606	7.7509E-08	0.81089	69.927		154.12	3.169E-06	1000	1.962E-05
167	Heptyl mercaptan	$C_7H_{16}S$	1639-09-4	132.26694	4.6970E-08	0.8932	57.6		229.92	4.832E-06	1000	2.124E-05
168	1-Heptyne	C_7H_{12}	628-71-7	96.17018	5.9501E-07	0.52758	274.02		192.22	3.932E-06	1000	1.787E-05
169	Hexadecane	$C_{16}H_{34}$	544-76-3	226.44116	1.2463E-07	0.7322	395	6000	291.31	3.274E-06	1000	1.399E-05
170	Hexanal	$C_6H_{12}O$	66-25-1	100.15888	4.0986E-05	0.10349	3180.6		214.93	4.523E-06	1000	2.004E-05
171	Hexane	C_6H_{14}	110-54-3	86.17536	1.7514E-07	0.70737	157.14		177.83	3.631E-06	1000	2.005E-05
172	Hexanoic acid	$C_6H_{12}O_2$	142-62-1	116.158	1.2145E-08	1.0861			269.25	5.294E-06	1000	2.201E-05
173	1-Hexanol	$C_6H_{14}O$	111-27-3	102.17476	1.5773E-07	0.7189	163.3		228.55	4.567E-06	1000	1.945E-05
174	2-Hexanol	$C_6H_{14}O$	626-93-7	102.175	1.0652E-07	0.77022	105.85		223.00	4.650E-06	1000	1.970E-05
175	2-Hexanone	$C_6H_{12}O$	591-78-6	100.15888	9.7820E-08	0.7772	99.53		217.35	4.397E-06	1000	1.909E-05
176	3-Hexanone	$C_6H_{12}O$	589-38-8	100.15888	9.8882E-08	0.7755	99.825		217.50	4.403E-06	1000	1.907E-05
177	1-Hexene	C_6H_{12}	592-41-6	84.15948	8.0060E-08	0.81293	65.274		133.39	2.871E-06	1000	2.064E-05
178	3-Hexyne	C_6H_{10}	928-49-4	82.1436	5.2127E-07	0.5444	237.01		170.05	3.567E-06	1000	1.811E-05
179	Hexyl mercaptan	$C_6H_{14}S$	111-31-9	118.24036	4.3636E-08	0.90747	42.32		192.62	4.235E-06	1000	2.209E-05
180	1-Hexyne	C_6H_{10}	693-02-7	82.1436	2.9986E-07	0.62647	178.17		141.25	2.947E-06	1000	1.928E-05
181	2-Hexyne	C_6H_{10}	764-35-2	82.1436	5.5562E-07	0.5337	244.38		183.65	3.851E-06	1000	1.782E-05
182	Hydrazine	H_4N_2	302-01-2	32.04516	2.3489E-07	0.7151	205.05		274.69	7.460E-06	1673.15	4.225E-05
183	Hydrogen	H_2	1333-74-0	2.01588	1.7970E-07	0.685	−0.59	140	13.95	6.517E-07	3000	4.330E-05
184	Hydrogen bromide	BrH	10035-10-6	80.91194	9.1700E-08	0.9273			206.45	1.285E-05	800	4.512E-05
185	Hydrogen chloride	ClH	7647-01-0	36.46094	4.9240E-07	0.6702	157.7		200.00	9.594E-06	1000	4.358E-05
186	Hydrogen cyanide	CHN	74-90-8	27.02534	1.2780E-08	1.0631	340		300.00	2.576E-06	425	4.421E-06
187	Hydrogen fluoride	FH	7664-39-3	20.0063432	4.5101E-14	3.0005	−521.83	76,111	285.50	9.931E-06	472.68	2.019E-05
188	Hydrogen sulfide	H_2S	7783-06-4	34.08088	3.9314E-08	1.0134			250.00	1.058E-05	480	2.050E-05
189	Isobutyric acid	$C_4H_8O_2$	79-31-2	88.10512	1.1202E-07	0.7822	100.3		227.15	5.415E-06	1000	2.261E-05
190	Isopropyl amine	C_3H_9N	75-31-0	59.11026	5.2542E-08	0.88063			177.95	5.037E-06	1000	2.304E-05
191	Malonic acid	$C_3H_4O_4$	141-82-2	104.06146	6.7978E-05	0.092766	4637.3		409.15	9.629E-06	1000	2.289E-05
192	Methacrylic acid	$C_4H_6O_2$	79-41-4	86.08924	9.1130E-08	0.8222	93.57		288.15	7.242E-06	1000	2.440E-05
193	Methane	CH_4	74-82-8	16.0425	5.2546E-07	0.59006	105.67		90.69	3.470E-06	1000	2.800E-05
194	Methanol	CH_4O	67-56-1	32.04186	3.0663E-07	0.69655	205		240.00	7.523E-06	1000	3.128E-05
195	*N*-Methyl acetamide	C_3H_7NO	79-16-3	73.09378	8.0599E-08	0.8392	77.332		301.15	7.714E-06	1000	2.464E-05
196	Methyl acetate	$C_3H_6O_2$	79-20-9	74.07854	1.3226E-06	0.4885	504.3		250.00	6.505E-06	800	2.125E-05
197	Methyl acetylene	C_3H_4	74-99-7	40.06386	1.1630E-06	0.4787	316		170.45	4.769E-06	800	2.045E-05
198	Methyl acrylate	$C_4H_6O_2$	96-33-3	86.08924	1.6480E-06	0.4444	510.66		196.32	4.781E-06	1000	2.350E-05
199	Methyl amine	CH_5N	74-89-5	31.0571	5.6409E-07	0.5863	231.9		179.69	5.167E-06	1000	2.628E-05
200	Methyl benzoate	$C_8H_8O_2$	93-58-3	136.14792	7.4106E-08	0.82436	83.086		260.75	5.515E-06	1000	2.034E-05
201	3-Methyl-1,2-butadiene	C_5H_8	598-25-4	68.11702	4.0824E-07	0.5923	208.22		159.53	3.572E-06	1000	2.021E-05
202	2-Methylbutane	C_5H_{12}	78-78-4	72.14878	2.4344E-08	0.97376	−91.597	18,720	150.00	2.621E-06	1000	2.190E-05
203	2-Methylbutanoic acid	$C_5H_{10}O_2$	116-53-0	102.1317	1.8690E-07	0.7096	192		450.15	1.000E-05	1000	2.109E-05
204	3-Methyl-1-butanol	$C_5H_{12}O$	123-51-3	88.1482	8.9348E-08	0.80197	77.653		155.95	3.422E-06	1000	2.111E-05
205	2-Methyl-1-butene	C_5H_{10}	563-46-2	70.1329	5.0602E-07	0.55258	199.82		135.58	3.083E-06	1000	1.918E-05
206	2-Methyl-2-butene	C_5H_{10}	513-35-9	70.1329	8.5423E-07	0.47389	239.34		139.39	3.263E-06	1000	1.820E-05

207	2-Methyl -1-butene-3-yne	C_5H_6	78-80-8	66.10114	5.6844E-07	0.553	227.18		160.15	3.893E-06	1000	2.112E-05
208	Methylbutyl ether	$C_5H_{12}O$	628-28-4	88.14818	3.9342E-08	0.91086			157.48	3.947E-06	1000	2.125E-05
209	Methylbutyl sulfide	$C_5H_{12}S$	628-29-5	104.214	4.9950E-08	0.89479	44.662		175.30	4.052E-06	1000	2.312E-05
210	3-Methyl-1-butyne	C_5H_8	598-23-2	68.11702	4.0748E-08	0.92709			183.45	5.112E-06	1000	2.463E-05
211	Methyl butyrate	$C_5H_{10}O_2$	623-42-7	102.1317	3.7330E-07	0.6177	256.5		187.35	3.993E-06	1000	2.118E-05
212	Methylchlorosilane	CH_5ClSi	993-00-0	80.5889	4.8806E-08	0.92549			139.05	4.698E-06	1000	2.917E-05
213	Methylcyclohexane	C_7H_{14}	108-87-2	98.18606	6.5281E-07	0.5294	310.59		146.58	2.934E-06	1000	1.930E-05
214	1-Methylcyclohexanol	$C_7H_{14}O$	590-67-0	114.18546	8.5736E-08	0.80277	100.77		299.15	6.232E-06	1000	1.994E-05
215	*cis*-2-Methylcyclohexanol	$C_7H_{14}O$	7443-70-1	114.18546	2.4000E-07	0.68	210		280.15	6.331E-06	1000	2.175E-05
216	*trans*-2-Methylcyclohexanol	$C_7H_{14}O$	7443-52-9	114.18546	2.0000E-07	0.704	187		269.15	6.062E-06	1000	2.181E-05
217	Methylcyclopentane	C_6H_{12}	96-37-7	84.15948	9.0798E-07	0.495	355.89		130.73	2.722E-06	1000	2.046E-05
218	1-Methylcyclopentene	C_6H_{10}	693-89-0	82.1436	3.7026E-08	0.92849			146.62	3.800E-06	1000	2.259E-05
219	3-Methylcyclopentene	C_6H_{10}	1120-62-3	82.1436	3.9771E-08	0.92242			115.00	3.165E-06	1000	2.327E-05
220	Methyldichlorosilane	CH_4Cl_2Si	75-54-7	115.03396	1.9770E-07	0.7453	131.22		182.55	5.574E-06	1000	3.009E-05
221	Methylethyl ether	C_3H_8O	540-67-0	60.09502	2.6098E-07	0.68276	133.4		160.00	4.551E-06	1000	2.573E-05
222	Methylethyl ketone	C_4H_8O	78-93-3	72.10572	2.6552E-08	0.98316			186.48	4.534E-06	1000	2.364E-05
223	Methylethyl sulfide	C_3H_8S	624-89-5	76.1606	8.6219E-08	0.83591	72.564		167.23	4.341E-06	1000	2.588E-05
224	Methyl formate	$C_2H_4O_2$	107-31-3	60.05196	6.9755E-06	0.3154	1034.5		174.15	5.117E-06	1000	3.029E-05
225	Methylisobutyl ether	$C_5H_{12}O$	625-44-5	88.14818	1.5035E-07	0.7338	108.5		150.00	3.448E-06	1000	2.157E-05
226	Methylisobutyl ketone	$C_6H_{12}O$	108-10-1	100.15888	9.4257E-08	0.7845	90.183		189.15	3.901E-06	1000	1.951E-05
227	Methyl Isocyanate	C_2H_3NO	624-83-9	57.05132	3.1573E-07	0.66404	173.59		256.15	7.481E-06	1000	2.642E-05
228	Methylisopropyl ether	$C_4H_{10}O$	598-53-8	74.1216	1.9250E-07	0.7091	109		127.93	3.242E-06	1000	2.327E-05
229	Methylisopropyl ketone	$C_5H_{10}O$	563-80-4	86.1323	1.0826E-07	0.77382	93.349		180.15	3.968E-06	1000	2.076E-05
230	Methylisopropyl sulfide	$C_4H_{10}S$	1551-21-9	90.1872	8.6077E-08	0.81669	71.294		171.64	4.065E-06	1000	2.265E-05
231	Methyl mercaptan	CH_4S	74-93-1	48.10746	1.6370E-07	0.76706	107.97		150.18	4.450E-06	1000	2.956E-05
232	Methyl methacrylate	$C_5H_8O_2$	80-62-6	100.11582	4.8890E-07	0.6096	342.23		224.95	5.265E-06	1000	2.456E-05
233	2-Methyloctanoic acid	$C_9H_{18}O_2$	3004-93-1	158.23802	7.2131E-08	0.80319	99.437		240.00	4.162E-06	1000	1.685E-05
234	2-Methylpentane	C_6H_{14}	107-83-5	86.17536	1.1164E-06	0.4537	374.74		119.55	2.366E-06	1000	1.865E-05
235	Methyl pentyl ether	$C_6H_{14}O$	628-80-8	102.17476	1.0546E-07	0.77106	93.745		176.00	3.707E-06	1000	1.983E-05
236	2-Methylpropane	C_4H_{10}	75-28-5	58.1222	1.0871E-07	0.78135	70.639		150.00	3.707E-06	1000	2.242E-05
237	2-Methyl-2-propanol	$C_4H_{10}O$	75-65-0	74.1216	9.6050E-07	0.4856	381		298.97	6.727E-06	600	1.312E-05
238	2-Methyl propene	C_4H_8	115-11-7	56.10632	9.0981E-07	0.49288	260.08		132.81	3.423E-06	1000	2.174E-05
239	Methyl propionate	$C_4H_8O_2$	554-12-1	88.10512	3.5642E-07	0.6327	232.2		185.65	4.316E-06	1000	2.288E-05
240	Methylpropyl ether	$C_4H_{10}O$	557-17-5	74.1216	4.4941E-08	0.90199			133.97	3.725E-06	1000	2.284E-05
241	Methylpropyl sulfide	$C_4H_{10}S$	3877-15-4	90.1872	5.8223E-08	0.88057	48.298		160.17	3.908E-06	1000	2.434E-05
242	Methylsilane	CH_6Si	992-94-9	46.14384	3.8926E-07	0.63159	169.45		116.34	3.196E-06	1000	2.612E-05
243	alpha-Methyl styrene	C_9H_{10}	98-83-9	118.1757	7.1455E-07	0.49832	303.31		249.95	5.057E-06	1000	1.714E-05
244	Methyl tert-butyl ether	$C_5H_{12}O$	1634-04-4	88.1482	1.5779E-07	0.73224	112.15		164.55	3.938E-06	1000	2.232E-05
245	Methyl vinyl ether	C_3H_6O	107-25-5	58.07914	7.6460E-07	0.5476	284		278.65	8.264E-06	1000	2.616E-05
246	Naphthalene	$C_{10}H_8$	91-20-3	128.17052	6.4318E-07	0.5389	400.16		353.43	7.125E-06	1000	1.900E-05
247	Neon	Ne	7440-01-9	20.1797	7.1900E-07	0.6659	5.3		30.00	5.884E-06	3273.1	1.573E-04
248	Nitroethane	$C_2H_5NO_2$	79-24-3	75.0666	2.4391E-07	0.702	280		183.63	3.752E-06	1000	2.432E-05
249	Nitrogen	N_2	7727-37-9	28.0134	6.5592E-07	0.6081	54.714		63.15	4.372E-06	1970	6.432E-05
250	Nitrogen trifluoride	F_3N	7783-54-2	71.00191	8.2005E-07	0.61423	114.58		66.46	3.964E-06	1000	5.122E-05
251	Nitromethane	CH_3NO_2	75-52-5	61.04002	4.0700E-07	0.6485	367.5		244.60	5.756E-06	1000	2.625E-05
252	Nitrous oxide	N_2O	10024-97-2	44.0128	2.1150E-06	0.4642	305.7		182.30	8.854E-06	1000	4.000E-05
253	Nitric oxide	NO	10102-43-9	30.0061	1.4670E-06	0.5123	125.4		110.00	7.618E-06	1500	5.737E-05
254	Nonadecane	$C_{19}H_{40}$	629-92-5	268.5209	3.0465E-07	0.62218	705.34		305.04	3.231E-06	1000	1.314E-05
255	Nonanal	$C_9H_{18}O$	124-19-6	142.23862	3.8518E-05	0.10867	3502.7		267.30	5.013E-06	1000	1.812E-05
256	Nonane	C_9H_{20}	111-84-2	128.2551	1.0344E-07	0.77301	220.47		219.66	3.335E-06	1000	1.767E-05
257	Nonanoic acid	$C_9H_{18}O_2$	112-05-0	158.238	1.8105E-08	0.99668			285.55	5.074E-06	1000	1.769E-05
258	1-Nonanol	$C_9H_{20}O$	143-08-8	144.2545	1.2000E-07	0.74	180		268.15	4.499E-06	1000	1.688E-05
259	2-Nonanol	$C_9H_{20}O$	628-99-9	144.255	3.5879E-05	0.10109	3258.2		238.15	4.250E-06	1000	1.694E-05

(Continued)

TABLE 2-138 Vapor Viscosity of Inorganic and Organic Substances (Pa·s) (*Continued*)

Cmpd. no.	Name	Formula	CAS	Mol. wt.	C_1	C_2	C_3	C_4	T_{min}, K	Viscosity at T_{min}	T_{max}, K	Viscosity at T_{max}
260	1-Nonene	C_9H_{18}	124-11-8	126.23922	6.6329E-08	0.82027	76.204		191.91	3.542E-06	1000	1.781E-05
261	Nonyl mercaptan	$C_9H_{20}S$	1455-21-6	160.3201	3.8673E-08	0.91142	50.646		253.05	4.995E-06	1000	1.996E-05
262	1-Nonyne	C_9H_{16}	3452-09-3	124.22334	6.1447E-07	0.50705	287.19		223.15	4.170E-06	1000	1.585E-05
263	Octadecane	$C_{18}H_{38}$	593-45-3	254.49432	3.2095E-07	0.61839	709.09		301.31	3.266E-06	1000	1.345E-05
264	Octanal	$C_8H_{16}O$	124-13-0	128.212	3.9500E-05	0.10787	3390		251.65	4.955E-06	1000	1.896E-05
265	Octane	C_8H_{18}	111-65-9	114.22852	3.1191E-08	0.92925	55.092		216.38	3.677E-06	1000	1.813E-05
266	Octanoic acid	$C_8H_{16}O_2$	124-07-2	144.211	1.5557E-08	1.0299			289.65	5.338E-06	1000	1.913E-05
267	1-Octanol	$C_8H_{18}O$	111-87-5	130.22792	1.7520E-07	0.6941	206.8		257.65	4.583E-06	1000	1.755E-05
268	2-Octanol	$C_8H_{18}O$	123-96-6	130.228	3.4163E-05	0.10661	3028		241.55	4.530E-06	1000	1.771E-05
269	2-Octanone	$C_8H_{16}O$	111-13-7	128.21204	8.0901E-08	0.79062	99.338		252.85	4.611E-06	1000	1.733E-05
270	3-Octanone	$C_8H_{16}O$	106-68-3	128.21204	6.1515E-11	1.8808			255.55	2.075E-06	1000	2.700E-05
271	1-Octene	C_8H_{16}	111-66-0	112.21264	5.0324E-05	0.077611	3604.6		171.45	3.406E-06	1000	1.868E-05
272	Octyl mercaptan	$C_8H_{18}S$	111-88-6	146.29352	3.3253E-08	0.9351	32.426		223.95	4.579E-06	1000	2.057E-05
273	1-Octyne	C_8H_{14}	629-05-0	110.19676	5.7084E-07	0.52446	271.76		193.55	3.757E-06	1000	1.681E-05
274	Oxalic acid	$C_2H_2O_4$	144-62-7	90.03488	6.3032E-05	0.10487	4210.1		462.65	1.188E-05	1000	2.496E-05
275	Oxygen	O_2	7782-44-7	31.9988	1.1010E-06	0.5634	96.3		54.35	3.773E-06	1500	6.371E-05
276	Ozone	O_3	10028-15-6	47.9982	1.1960E-07	0.84797			80.15	4.922E-06	1000	4.184E-05
277	Pentadecane	$C_{15}H_{32}$	629-62-9	212.41458	4.0828E-08	0.8766	212.68		283.07	3.288E-06	1000	1.436E-05
278	Pentanal	$C_5H_{10}O$	110-62-3	86.1323	4.3300E-05	0.098676	3090		191.59	4.246E-06	1000	2.093E-05
279	Pentane	C_5H_{12}	109-66-0	72.14878	6.3412E-08	0.84758	41.718		143.42	3.305E-06	1000	2.124E-05
280	Pentanoic acid	$C_5H_{10}O_2$	109-52-4	102.132	1.0971E-08	1.11			239.15	4.793E-06	1000	2.346E-05
281	1-Pentanol	$C_5H_{12}O$	71-41-0	88.1482	1.8903E-07	0.7031	175.9		410.95	9.111E-06	1000	2.068E-05
282	2-Pentanol	$C_5H_{12}O$	6032-29-7	88.1482	1.1749E-07	0.7649	103.78		200.00	4.452E-06	1000	2.098E-05
283	2-Pentanone	$C_5H_{10}O$	107-87-9	86.1323	2.4630E-07	0.6653	208.7		196.29	4.003E-06	1000	2.019E-05
284	3-Pentanone	$C_5H_{10}O$	96-22-0	86.1323	1.1640E-07	0.7615	107.94		234.18	5.079E-06	1000	2.023E-05
285	1-Pentene	C_5H_{10}	109-67-1	70.1329	1.6378E-06	0.44337	636.11	−26,218	108.02	2.813E-06	1000	2.176E-05
286	2-Pentyl mercaptan	$C_5H_{12}S$	2084-19-7	104.21378	8.8646E-08	0.81492	85.198		160.75	3.638E-06	1000	2.275E-05
287	Pentyl mercaptan	$C_5H_{12}S$	110-66-7	104.21378	2.7467E-08	0.97555			197.45	4.766E-06	1000	2.320E-05
288	1-Pentyne	C_5H_8	627-19-0	68.11702	4.1022E-08	0.90585			167.45	4.242E-06	1000	2.141E-05
289	2-Pentyne	C_5H_8	627-21-4	68.11702	5.7650E-07	0.53498	235.2		163.83	3.621E-06	1000	1.879E-05
290	Phenanthrene	$C_{14}H_{10}$	85-01-8	178.2292	4.3478E-07	0.5272	238.27		372.38	6.010E-06	1000	1.340E-05
291	Phenol	C_6H_6O	108-95-2	94.11124	1.0094E-07	0.799	103.1		314.06	7.514E-06	1000	2.283E-05
292	Phenyl isocyanate	C_7H_5NO	103-71-9	119.1207	8.5360E-08	0.80872	88.273		243.15	5.324E-06	1000	2.093E-05
293	Phthalic anhydride	$C_8H_4O_3$	85-44-9	148.11556	4.3511E-08	0.908	102.73		404.15	8.072E-06	1000	2.090E-05
294	Propadiene	C_3H_4	463-49-0	40.06386	6.0758E-07	0.53845	173.45		136.87	3.788E-06	1000	2.135E-05
295	Propane	C_3H_8	74-98-6	44.09562	4.9054E-08	0.90125			85.47	2.702E-06	1000	2.480E-05
296	1-Propanol	C_3H_8O	71-23-8	60.09502	7.9420E-07	0.5491	415.8		200.00	4.732E-06	1000	2.490E-05
297	2-Propanol	C_3H_8O	67-63-0	60.095	1.2003E-06	0.494	479.78		187.35	4.471E-06	1000	2.461E-05
298	Propenylcyclohexene	C_9H_{14}	13511-13-2	122.20746	5.4749E-07	0.53893	283.52		199.00	3.914E-06	1000	1.765E-05
299	Propionaldehyde	C_3H_6O	123-38-6	58.07914	3.8397E-05	0.10821	2510.9		165.00	4.114E-06	1000	2.309E-05
300	Propionic acid	$C_3H_6O_2$	79-09-4	74.0785	1.4807E-08	1.0733			252.45	5.607E-06	1000	2.457E-05
301	Propionitrile	C_3H_5N	107-12-0	55.0785	9.6891E-06	0.24601	1537.6		180.37	3.652E-06	1000	2.089E-05
302	Propyl acetate	$C_5H_{10}O_2$	109-60-4	102.1317	2.1372E-07	0.6894	178.57		178.15	3.802E-06	1000	2.122E-05
303	Propyl amine	C_3H_9N	107-10-8	59.11026	1.6200E-07	0.7285	117		188.36	4.540E-06	1000	2.223E-05
304	Propylbenzene	C_9H_{12}	103-65-1	120.19158	3.0387E-07	0.61945	210.35		173.55	3.350E-06	1000	1.812E-05
305	Propylene	C_3H_6	115-07-1	42.07974	7.3919E-07	0.5423	263.73		87.89	2.093E-06	1000	2.477E-05
306	Propyl formate	$C_4H_8O_2$	110-74-7	88.10512	6.0741E-07	0.5863	367.29		180.25	4.203E-06	1000	2.550E-05
307	2-Propyl mercaptan	C_3H_8S	75-33-2	76.16062	3.5532E-08	0.95654			142.61	4.085E-06	1000	2.632E-05
308	Propyl mercaptan	C_3H_8S	107-03-9	76.16062	7.9457E-08	0.84656	65.878		159.95	4.132E-06	1000	2.583E-05
309	1,2-Propylene glycol	$C_3H_8O_2$	57-55-6	76.09442	4.5430E-08	0.9173	61		213.15	4.832E-06	1000	2.418E-05

310	Quinone	$C_6H_4O_2$	106-51-4	108.09476	1.1085E-07	0.8008	152.51		388.85	9.439E-06	1000	2.429E-05
311	Silicon tetrafluoride	F_4Si	7783-61-1	104.07911	2.1671E-07	0.76757	16.28		250.00	1.410E-05	500	2.475E-05
312	Styrene	C_8H_8	100-42-5	104.14912	6.3863E-07	0.5254	295.1		242.54	5.158E-06	1000	1.858E-05
313	Succinic acid	$C_4H_6O_4$	110-15-6	118.08804	5.7821E-05	0.099467	4409.6		460.85	1.007E-05	1000	2.125E-05
314	Sulfur dioxide	O_2S	7446-09-5	64.0638	6.8630E-07	0.6112	217		197.67	8.280E-06	1000	3.844E-05
315	Sulfur hexafluoride	F_6S	2551-62-4	146.0554192	5.3986E-07	0.6349	34.5	19,000	205.15	9.790E-06	5000	1.195E-04
316	Sulfur trioxide	O_3S	7446-11-9	80.0632	3.9067E-06	0.3845	470.1		297.93	1.355E-05	694.19	2.883E-05
317	Terephthalic acid	$C_8H_6O_4$	100-21-0	166.13084	3.9218E-05	0.12589	3861.1		700.15	1.373E-05	1000	1.925E-05
318	*o*-Terphenyl	$C_{18}H_{14}$	84-15-1	230.30376	7.0859E-07	0.51971	652.24		329.35	4.837E-06	1000	1.554E-05
319	Tetradecane	$C_{14}H_{30}$	629-59-4	198.388	5.1567E-09	1.1561			279.01	3.465E-06	1000	1.516E-05
320	Tetrahydrofuran	C_4H_8O	109-99-9	72.10572	3.7780E-07	0.6533	271.01		164.65	4.006E-06	1000	2.710E-05
321	1,2,3,4-Tetrahydronaphthalene	$C_{10}H_{12}$	119-64-2	132.20228	5.0784E-07	0.5614	328.55		237.38	4.592E-06	1000	1.847E-05
322	Tetrahydrothiophene	C_4H_8S	110-01-0	88.17132	8.5988E-08	0.82841	68.172		176.99	4.520E-06	1000	2.461E-05
323	2,2,3,3-Tetramethylbutane	C_8H_{18}	594-82-1	114.22852	8.1458E-07	0.50257	380.29		373.96	7.930E-06	1000	1.900E-05
324	Thiophene	C_4H_4S	110-02-1	84.13956	1.0300E-06	0.5497	569.4		234.94	6.049E-06	1000	2.926E-05
325	Toluene	C_7H_8	108-88-3	92.13842	8.7268E-07	0.49397	323.79		178.18	4.008E-06	1000	2.000E-05
326	1,1,2-Trichloroethane	$C_2H_3Cl_3$	79-00-5	133.40422	2.7081E-07	0.6955	187.93		236.50	6.756E-06	1000	2.782E-05
327	Tridecane	$C_{13}H_{28}$	629-50-5	184.36142	3.5585E-08	0.8987	165.3		267.76	3.344E-06	1000	1.517E-05
328	Triethyl amine	$C_6H_{15}N$	121-44-8	101.19	2.4110E-07	0.6845	223		158.45	3.210E-06	1000	2.230E-05
329	Trimethyl amine	C_3H_9N	75-50-3	59.11026	1.2434E-06	0.4832	447.7		156.08	3.689E-06	1000	2.418E-05
330	1,2,3-Trimethylbenzene	C_9H_{12}	526-73-8	120.19158	7.8498E-07	0.49855	362.79		247.79	4.975E-06	1000	1.803E-05
331	1,2,4-Trimethylbenzene	C_9H_{12}	95-63-6	120.19158	6.8812E-07	0.51063	330.88		229.33	4.520E-06	1000	1.760E-05
332	2,2,4-Trimethylpentane	C_8H_{18}	540-84-1	114.22852	1.1070E-07	0.746	72.4		165.78	3.488E-06	1000	1.786E-05
333	2,3,3-Trimethylpentane	C_8H_{18}	560-21-4	114.22852	8.2418E-07	0.4931	371.44		387.91	7.958E-06	1000	1.812E-05
334	1,3,5-Trinitrobenzene	$C_6H_3N_3O_6$	99-35-4	213.10452	3.4066E-08	0.95252	43.528		398.40	9.208E-06	1000	2.352E-05
335	2,4,6-Trinitrotoluene	$C_7H_5N_3O_6$	118-96-7	227.1311	2.8471E-08	0.96571	30.83		354.00	7.581E-06	1000	2.179E-05
336	Undecane	$C_{11}H_{24}$	1120-21-4	156.30826	3.5940E-08	0.9052	125		247.57	3.506E-06	1000	1.660E-05
337	1-Undecanol	$C_{11}H_{24}O$	112-42-5	172.30766	5.9537E-08	0.81842	90.245		288.45	4.677E-06	1000	1.558E-05
338	Vinyl acetate	$C_4H_6O_2$	108-05-4	86.08924	1.3880E-07	0.7599	98		180.35	4.659E-06	1000	2.407E-05
339	Vinyl acetylene	C_4H_4	689-97-4	52.07456	6.7484E-07	0.5304	230.17		173.15	4.459E-06	1000	2.140E-05
340	Vinyl chloride	C_2H_3Cl	75-01-4	62.49822	2.3790E-07	0.71517	102.84		119.36	3.907E-06	1000	3.016E-05
341	Vinyl trichlorosilane	$C_2H_3Cl_3Si$	75-94-5	161.48972	3.6429E-08	0.95924			178.35	5.260E-06	1000	2.749E-05
342	Water	H_2O	7732-18-5	18.01528	1.7096E-08	1.1146			273.16	8.882E-06	1073.15	4.082E-05
343	*m*-Xylene	C_8H_{10}	108-38-3	106.165	6.8293E-07	0.52199	324.17		225.30	4.735E-06	1000	1.898E-05
344	*o*-Xylene	C_8H_{10}	95-47-6	106.165	8.3436E-07	0.49713	365.86		247.98	5.225E-06	1000	1.894E-05
345	*p*-Xylene	C_8H_{10}	106-42-3	106.165	9.3485E-07	0.47683	371.96		286.41	6.037E-06	1000	1.836E-05

The vapor viscosity is calculated by

$$\mu = C_1 T^{C_2}/(1 + C_3/T + C_4/T^2)$$

where μ is the viscosity in Pa·s and T is the temperature in K. Viscosities are at either 1 atm or the vapor pressure, whichever is lower.

Values in this table were taken from the Design Institute for Physical Properties (DIPPR) of the American Institute of Chemical Engineers (AIChE), 801 Critically Evaluated Gold Standard™ Database, copyright 2016 AIChE, and reproduced with permission of AIChE and of the DIPPR Evaluated Process Design Data Project Steering Committee. Their source should be cited as "R. L. Rowley, W. V. Wilding, J. L. Oscarson, T. A. Knotts, and N. F. Giles, *DIPPR® Data Compilation of Pure Chemical Properties*, Design Institute for Physical Properties, AIChE, New York, NY (2016)".

TABLE 2-139 Viscosity of Inorganic and Organic Liquids (Pa·s)

Eqn	Cmpd. no.	Name	Formula	CAS	Mol. wt.	C_1	C_2	C_3	C_4	C_5	T_{min}, K	Viscosity at T_{min}	T_{max}, K	Viscosity at T_{max}
101	1	Acetaldehyde	C_2H_4O	75-07-0	44.05256	−10.976	755.12				149.78	2.647E-03	294.15	2.229E-04
101	2	Acetamide	C_2H_5NO	60-35-5	59.0672	1.5525	1376.4	−2.0126			353.33	1.728E-03	494.3	2.895E-04
101	3	Acetic acid	$C_2H_4O_2$	64-19-7	60.052	−9.03	1212.3	−0.322			289.81	1.265E-03	391.05	3.890E-04
101	4	Acetic anhydride	$C_4H_6O_3$	108-24-7	102.08864	−20.457	1638.6	1.3834			200.15	7.159E-03	412.7	2.874E-04
101	5	Acetone	C_3H_6O	67-64-1	58.07914	−14.918	1023.4	0.5961			190	1.655E-03	329.44	2.351E-04
101	6	Acetonitrile	C_2H_3N	75-05-8	41.0519	5.4711	143.99	−2.4432			229.32	7.616E-04	354.81	2.100E-04
101	7	Acetylene	C_2H_2	74-86-2	26.03728	6.224	−151.8	−2.6554			193.15	1.958E-04	273.15	9.819E-05
101	8	Acrolein	C_3H_4O	107-02-8	56.06326	−12.032	867.34	0.19534			185.45	1.773E-03	353.22	2.181E-04
101	9	Acrylic acid	$C_3H_4O_2$	79-10-7	72.06266	−28.12	2280.2	2.3956			286.15	1.359E-03	460	2.086E-04
101	10	Acrylonitrile	C_3H_3N	107-13-1	53.0626	−0.24126	350.57	−1.5676			189.63	1.340E-03	350.45	2.191E-04
101	11	Air	Mixture	132259-10-0	28.96	−20.077	285.15	1.784	−6.238E-22	10	59.15	3.430E-04	130	4.276E-05
101	12	Ammonia	H_3N	7664-41-7	17.03052	−6.743	598.3	−0.7341	−3.690E-27	10	195.41	5.240E-04	393.15	4.858E-05
101	13	Anisole	C_7H_8O	100-66-3	108.13782	−15.407	1518.7	0.60172			235.65	3.429E-03	426.73	2.736E-04
101	14	Argon	Ar	7440-37-1	39.948	−8.8685	204.29	−0.38305	−1.294E-22	10	83.78	2.950E-04	150	3.823E-05
101	15	Benzamide	C_7H_7NO	55-21-0	121.13658	−12.632	2668.2				403	2.451E-03	563.15	3.730E-04
101	16	Benzene	C_6H_6	71-43-2	78.11184	7.5117	294.68	−2.794			278.68	7.761E-04	545	7.106E-05
101	17	Benzenethiol	C_6H_6S	108-98-5	110.17684	−8.4562	1024.4	−0.30635			258.27	2.047E-03	442.29	3.333E-04
101	18	Benzoic acid	$C_7H_6O_2$	65-85-0	122.12134	−12.947	2557.9				395.52	1.534E-03	600.8	1.683E-04
101	19	Benzonitrile	C_7H_5N	100-47-0	103.1213	−23.268	1880.5	1.7994			260.28	2.393E-03	464.15	2.836E-04
101	20	Benzophenone	$C_{13}H_{10}O$	119-61-9	182.2179	−148.6	8377.2	20.559	−0.0000133	2	321.35	5.369E-03	664	2.614E-04
101	21	Benzyl alcohol	C_7H_8O	100-51-6	108.13782	−14.152	2652				257.85	2.092E-02	478.6	1.821E-04
101	22	Benzyl ethyl ether	$C_9H_{12}O$	539-30-0	136.19098	−11.46	1497	−0.043397			275.65	1.886E-03	458.15	2.121E-04
101	23	Benzyl mercaptan	C_7H_8S	100-53-8	124.20342	−11.459	1334.4	0.00049694			243.95	2.513E-03	472.03	1.788E-04
101	24	Biphenyl	$C_{12}H_{10}$	92-52-4	154.2078	−9.9265	1576.3	−0.21119			342.2	1.427E-03	723.15	1.076E-04
101	25	Bromine	Br_2	7726-95-6	159.808	16.775	-314	−3.9763			265.85	1.353E-03	350	6.021E-04
101	26	Bromobenzene	C_6H_5Br	108-86-1	157.0079	−20.611	1656.5	1.4415			242.43	2.842E-03	429.24	3.310E-04
101	27	Bromoethane	C_2H_5Br	74-96-4	108.965	−5.0539	645.8	−0.87689			154.25	5.065E-03	393.15	1.751E-04
101	28	Bromomethane	CH_3Br	74-83-9	94.93852	−16.615	931.44	0.94366			179.44	1.464E-03	363.15	2.060E-04
101	29	1,2-Butadiene	C_4H_6	590-19-2	54.09044	−10.143	472.79	−0.028241			136.95	1.081E-03	284	1.773E-04
101	30	1,3-Butadiene	C_4H_6	106-99-0	54.09044	17.844	−310.2	−4.5058			250	2.547E-04	400	4.880E-05
101	31	Butane	C_4H_{10}	106-97-8	58.1222	−7.2471	534.82	−0.57469	−4.6625E-27	10	134.86	2.243E-03	420	3.566E-05
101	32	1,2-Butanediol	$C_4H_{10}O_2$	584-03-2	90.121	−393.86	19,042	59.978	−0.049479	1	220	2.020E+02	544	3.441E-04
101	33	1,3-Butanediol	$C_4H_{10}O_2$	107-88-0	90.121	−390.03	18,609	60.014	−0.055844	1	196.15	4.410E+04	540.8	2.890E-04
101	34	1-Butanol	$C_4H_{10}O$	71-36-3	74.1216	−82.851	4481.8	11.182	−0.000020943	2	190	2.602E-01	391.9	3.845E-04
101	35	2-Butanol	$C_4H_{10}O$	78-92-2	74.1216	−16.323	3141.7				238	4.404E-02	372.9	3.715E-04
101	36	1-Butene	C_4H_8	106-98-9	56.10632	−10.773	591.61				87.8	1.769E-02	335.6	1.222E-04
101	37	*cis*-2-Butene	C_4H_8	590-18-1	56.10632	−10.346	522.3	−0.011847			134.26	1.483E-03	276.87	1.982E-04
101	38	*trans*-2-Butene	C_4H_8	624-64-6	56.10632	−10.335	521.39	−0.013184			167.62	6.810E-04	274.03	2.022E-04
101	39	Butyl acetate	$C_6H_{12}O_2$	123-86-4	116.15828	−17.488	1478.2	0.91828			250	1.496E-03	399.26	2.521E-04
101	40	Butylbenzene	$C_{10}H_{14}$	104-51-8	134.21816	−23.802	1887.2	1.8479			200	1.030E-02	456.46	2.359E-04
101	41	Butyl mercaptan	$C_4H_{10}S$	109-79-5	90.1872	−10.807	966.74	−0.014851			157.46	8.716E-03	373.15	2.475E-04
101	42	*sec*-Butyl mercaptan	$C_4H_{10}S$	513-53-1	90.1872	−10.903	932.82	0.023034			133.02	2.287E-02	358.13	2.851E-04
101	43	1-Butyne	C_4H_6	107-00-6	54.09044	−3.4644	334.5	−1.0811			147.43	1.369E-03	373.15	1.271E-04
101	44	Butyraldehyde	C_4H_8O	123-72-8	72.10572	−6.4551	744.7	−0.67524			176.8	3.223E-03	347.94	2.570E-04
101	45	Butyric acid	$C_4H_8O_2$	107-92-6	88.1051	−9.817	1388	−0.238			267.95	2.561E-03	436.42	3.087E-04
101	46	Butyronitrile	C_4H_7N	109-74-0	69.1051	−11.13	1084.1				161.3	1.217E-02	390.74	2.351E-04
101	47	Carbon dioxide	CO_2	124-38-9	44.0095	18.775	−402.92	−4.6854	−6.9171E-26	10	216.58	2.488E-04	303.15	5.652E-05
101	48	Carbon disulfide	CS_2	75-15-0	76.1407	−10.306	703.01				161.58	2.592E-03	441.6	1.643E-04
101	49	Carbon monoxide	CO	630-08-0	28.0101	−4.9735	97.67	−1.1088			68.15	2.688E-04	131.37	6.515E-05

101	50	Carbon tetrachloride	CCl_4	56-23-5	153.8227	−8.0738	1121.1	−0.4726			250	2.032E-03	455	2.030E-04
101	51	Carbon tetrafluoride	CF_4	75-73-0	88.0043	−9.9212	300.5				89.56	1.408E-03	145.1	3.897E-04
101	52	Chlorine	Cl_2	7782-50-5	70.906	−9.5412	456.62				172.12	1.020E-03	333.72	2.822E-04
101	53	Chlorobenzene	C_6H_5Cl	108-90-7	112.5569	0.15772	540.5	−1.6075			250	1.422E-03	540	1.291E-04
101	54	Chloroethane	C_2H_5Cl	75-00-3	64.5141	10.9222	−118.895	−3.305			136.75	2.026E-03	423.15	8.727E-05
101	55	Chloroform	$CHCl_3$	67-66-3	119.37764	−14.109	1049.2	0.5377			209.63	1.970E-03	353.2	3.410E-04
101	56	Chloromethane	CH_3Cl	74-87-3	50.4875	10.39	−134.38	−3.262			175.43	7.234E-04	416.25	6.726E-05
101	57	1-Chloropropane	C_3H_7Cl	540-54-5	78.54068	10.27183	−67.2235	−3.1664			150.35	2.362E-03	423.15	1.190E-04
101	58	2-Chloropropane	C_3H_7Cl	75-29-6	78.54068	−15.458	1086	0.654			250	5.514E-04	308.85	2.767E-04
101	59	*m*-Cresol	C_7H_8O	108-39-4	108.13782	−914.12	38,855	139.11	−0.00014757	2	273.15	8.438E-02	564.68	1.793E-05
101	60	*o*-Cresol	C_7H_8O	95-48-7	108.13782	−377.23	17,909	55.565	−0.00004841	2	293.15	9.548E-03	558.04	1.514E-04
101	61	*p*-Cresol	C_7H_8O	106-44-5	108.13782	−851.12	36,686	129.13	−0.00013329	2	273.15	9.674E-02	563.72	2.992E-05
101	62	Cumene	C_9H_{12}	98-82-8	120.19158	−24.988	1807.9	2.0556			200	6.363E-03	400	2.881E-04
101	63	Cyanogen	C_2N_2	460-19-5	52.0348	−11.794	992.33				245.25	4.317E-04	320.12	1.676E-04
101	64	Cyclobutane	C_4H_8	287-23-0	56.10632	−3.4968	397.94	−1.1087			182.48	8.345E-04	367.94	1.278E-04
101	65	Cyclohexane	C_6H_{12}	110-82-7	84.15948	−33.763	2497.2	3.2236			279.69	1.264E-03	443.04	2.070E-04
101	66	Cyclohexanol	$C_6H_{12}O$	108-93-0	100.15888	280.87	−31,869	−38.837	3,994,500	−2.002	296.6	6.328E-02	520.08	1.652E-04
101	67	Cyclohexanone	$C_6H_{10}O$	108-94-1	98.143	−44.877	3227.7	4.887			242	8.960E-03	428.58	4.402E-04
101	68	Cyclohexene	C_6H_{10}	110-83-8	82.1436	−11.641	1154.3	0.066511			200	4.017E-03	373.15	2.877E-04
101	69	Cyclopentane	C_5H_{10}	287-92-3	70.1329	−3.2612	614.16	−1.156			225	1.122E-03	325	3.167E-04
101	70	Cyclopentene	C_5H_8	142-29-0	68.11702	−4.1508	599.77	−1.0308			138.13	7.531E-03	405.6	1.416E-04
101	71	Cyclopropane	C_3H_6	75-19-4	42.07974	−3.524	342.54	−1.1599			145.59	9.601E-04	318.4	1.080E-04
101	72	Cyclohexyl mercaptan	$C_6H_{12}S$	1569-69-3	116.22448	−11.338	1304.1	0.000092396			189.64	1.155E-02	431.95	2.440E-04
101	73	Decanal	$C_{10}H_{20}O$	112-31-2	156.2652	4.1184	629.98	−2.2076			285	2.134E-03	481.65	2.718E-04
101	74	Decane	$C_{10}H_{22}$	124-18-5	142.28168	−97.663	4342.7	13.645	−0.000019319	2	240.05	2.741E-03	494.16	1.292E-04
101	75	Decanoic acid	$C_{10}H_{20}O_2$	334-48-5	172.265	−12.305	2324.1	−0.055494			304.55	6.798E-03	543.15	2.304E-04
101	76	1-Decanol	$C_{10}H_{22}O$	112-30-1	158.28108	−69.985	5818.8	8.0715			285	1.937E-02	503	2.727E-04
101	77	1-Decene	$C_{10}H_{20}$	872-05-9	140.2658	−15.868	1434.8	0.68071			206.89	4.975E-03	443.75	2.064E-04
101	78	Decyl mercaptan	$C_{10}H_{22}S$	143-10-2	174.34668	−11.464	1510.1	−0.012754			247.56	4.364E-03	512.35	1.848E-04
101	79	1-Decyne	$C_{10}H_{18}$	764-93-2	138.24992	−2.3633	791.93	−1.2272			229.15	3.786E-03	505.6	2.167E-04
100	80	Deuterium	D_2	7782-39-0	4.0316	0.000001348					20.35	1.348E-06	20.35	1.348E-06
101	81	1,1-Dibromoethane	$C_2H_4Br_2$	557-91-5	187.86116	−10.457	1101.1	−0.0031354			210.15	5.331E-03	381.15	5.071E-04
101	82	1,2-Dibromoethane	$C_2H_4Br_2$	106-93-4	187.86116	−17.582	1635.4	0.9932			282.85	2.042E-03	404.51	5.120E-04
101	83	Dibromomethane	CH_2Br_2	74-95-3	173.83458	−10.013	921.31				220.6	2.919E-03	488.8	2.951E-04
101	84	Dibutyl ether	$C_8H_{18}O$	142-96-1	130.22792	10.027	206	−3.1607			175.3	5.931E-03	414.15	1.989E-04
101	85	*m*-Dichlorobenzene	$C_6H_4Cl_2$	541-73-1	147.00196	−114.7	4905.4	16.358	−0.000020577	2	248.39	2.463E-03	547.16	1.565E-04
101	86	*o*-Dichlorobenzene	$C_6H_4Cl_2$	95-50-1	147.00196	−30.6	2153.4	2.9371			256.15	2.726E-03	453.57	3.761E-04
101	87	*p*-Dichlorobenzene	$C_6H_4Cl_2$	106-46-7	147.00196	31.63	−1080	−6.114			326.14	8.543E-04	447.21	3.039E-04
101	88	1,1-Dichloroethane	$C_2H_4Cl_2$	75-34-3	98.95916	−8.991	870.2	−0.2805			176.19	4.076E-03	330.45	3.407E-04
101	89	1,2-Dichloroethane	$C_2H_4Cl_2$	107-06-2	98.95916	15.312	−41.12	−3.919			237.49	1.839E-03	400	2.557E-04
101	90	Dichloromethane	CH_2Cl_2	75-09-2	84.93258	−13.071	940.03	0.3733			208.38	1.406E-03	373.93	2.374E-04
101	91	1,1-Dichloropropane	$C_3H_6Cl_2$	78-99-9	112.98574	−10.872	1033.1	−0.00067435			192.5	4.051E-03	361.25	3.301E-04
101	92	1,2-Dichloropropane	$C_3H_6Cl_2$	78-87-5	112.98574	−11.269	1195.3	0.012736			172.71	1.381E-02	369.52	3.495E-04
101	93	Diethanol amine	$C_4H_{11}NO_2$	111-42-2	105.13564	−375.21	17,177	66.66	−3.6367	0.5	293.15	8.128E-01	589.28	1.090E-04
101	94	Diethyl amine	$C_4H_{11}N$	109-89-7	73.13684	−17.57	1385.7	0.85647			223.35	1.190E-03	329.1	2.260E-04
101	95	Diethyl ether	$C_4H_{10}O$	60-29-7	74.1216	10.197	−63.8	−3.226			200	7.359E-04	373.15	1.141E-04
101	96	Diethyl sulfide	$C_4H_{10}S$	352-93-2	90.1872	−5.135	667.5	−0.8553			225	1.113E-03	365.25	2.354E-04
101	97	1,1-Difluoroethane	$C_2H_4F_2$	75-37-6	66.04997	10.501	−52.181	−3.3459			154.56	1.229E-03	343.15	1.026E-04
101	98	1,2-Difluoroethane	$C_2H_4F_2$	624-72-6	66.04997	−10.072	710.48	−0.14677			179.6	1.030E-03	283.65	2.257E-04
101	99	Difluoromethane	CH_2F_2	75-10-5	52.02339	−17.723	850.2	1.0601	−1.1719E-18	7	137	1.832E-03	343.15	6.050E-05
101	100	Diisopropyl amine	$C_6H_{15}N$	108-18-9	101.19	−1.7366	599.8	−1.4237			250	7.479E-04	357.05	2.193E-04

(*Continued*)

TABLE 2-139 Viscosity of Inorganic and Organic Liquids (Pa·s) *(Continued)*

Eqn	Cmpd. no.	Name	Formula	CAS	Mol. wt.	C_1	C_2	C_3	C_4	C_5	T_{min}, K	Viscosity at T_{min}	T_{max}, K	Viscosity at T_{max}
101	101	Diisopropyl ether	$C_6H_{14}O$	108-20-3	102.17476	−11.5	993	0.022			187.65	2.258E-03	341.45	2.110E-04
101	102	Diisopropyl ketone	$C_7H_{14}O$	565-80-0	114.18546	−15.097	1426.9	0.51512			204.81	4.569E-03	397.55	2.194E-04
101	103	1,1-Dimethoxyethane	$C_4H_{10}O_2$	534-15-6	90.121	−10.968	885.49				159.95	4.375E-03	337.45	2.378E-04
101	104	1,2-Dimethoxypropane	$C_5H_{12}O_2$	7778-85-0	104.14758	−10.631	1086.4				226.1	2.950E-03	366.15	4.695E-04
101	105	Dimethyl acetylene	C_4H_6	503-17-3	54.09044	0.10842	300.2	−1.6831			240.91	3.796E-04	371	1.186E-04
101	106	Dimethyl amine	C_2H_7N	124-40-3	45.08368	−10.93	699.5				200	5.917E-04	308.15	1.734E-04
101	107	2,3-Dimethylbutane	C_6H_{14}	79-29-8	86.17536	7.2565	221.4	−2.7946			220	1.103E-03	331.13	2.509E-04
101	108	1,1-Dimethylcyclohexane	C_8H_{16}	590-66-9	112.21264	−10.716	1140.5	−0.047736			239.66	1.992E-03	392.7	3.045E-04
101	109	*cis*-1,2-Dimethylcyclohexane	C_8H_{16}	2207-01-4	112.21264	−11.796	1463.5				223.16	5.311E-03	484.92	1.541E-04
101	110	*trans*-1,2-Dimethylcyclohexane	C_8H_{16}	6876-23-9	112.21264	−11.344	1168.9	0.04513			184.99	8.315E-03	396.58	2.956E-04
101	111	Dimethyl disulfide	$C_2H_6S_2$	624-92-0	94.19904	−10.577	1172.6	−0.14244			188.44	6.093E-03	382.9	2.336E-04
101	112	Dimethyl ether	C_2H_6O	115-10-6	46.06844	−10.62	448.99	0.000083967			131.65	7.398E-04	248.31	1.490E-04
101	113	*N,N*-Dimethyl formamide	C_3H_7NO	68-12-2	73.09378	−20.425	1515.5	1.4444			240	2.041E-03	425.15	2.981E-04
101	114	2,3-Dimethylpentane	C_7H_{16}	565-59-3	100.20194	−12.08	1112.2	0.09654			160	9.669E-03	362.93	2.147E-04
101	115	Dimethyl phthalate	$C_{10}H_{10}O_4$	131-11-3	194.184	152.9	−10,183	−22.709	50,373,000,000	−4	274.18	6.023E-02	612.8	1.109E-04
	116	Dimethylsilane	C_2H_8Si	1111-74-6	60.17042									
101	117	Dimethyl sulfide	C_2H_6S	75-18-3	62.134	−17.641	1067.5	1.0317			225	6.696E-04	310.48	2.528E-04
101	118	Dimethyl sulfoxide	C_2H_6OS	67-68-5	78.13344	−37.347	2835	3.7937			291.67	2.253E-03	464	3.547E-04
101	119	Dimethyl terephthalate	$C_{10}H_{10}O_4$	120-61-6	194.184	−16.0542	2221.79	0.63829			413.79	1.071E-03	559.2	3.214E-04
101	120	1,4-Dioxane	$C_4H_8O_2$	123-91-1	88.10512	−46.166	3086.2	5.104			284.95	1.525E-03	374.65	4.610E-04
101	121	Diphenyl ether	$C_{12}H_{10}O$	101-84-8	170.2072	−12.373	2017.5				293.15	4.124E-03	613.44	1.134E-04
101	122	Dipropyl amine	$C_6H_{15}N$	142-84-7	101.19	−15.404	1390	0.5564			260	9.454E-04	382.35	2.118E-04
101	123	Dodecane	$C_{12}H_{26}$	112-40-3	170.33484	−134.91	6054.2	19.337	−0.00002443	2	262.15	3.002E-03	526.4	1.220E-04
101	124	Eicosane	$C_{20}H_{42}$	112-95-8	282.54748	−18.315	2283.5	0.95485			309.58	4.242E-03	616.93	2.078E-04
101	125	Ethane	C_2H_6	74-84-0	30.069	−7.0046	276.38	−0.6087	−3.11E-18	7	90.35	1.247E-03	300	3.587E-05
101	126	Ethanol	C_2H_6O	64-17-5	46.06844	7.875	781.98	−3.0418			200	1.315E-02	440	1.416E-04
101	127	Ethyl acetate	$C_4H_8O_2$	141-78-6	88.10512	14.354	−154.6	−3.7887			220	1.132E-03	473.15	9.061E-05
101	128	Ethyl amine	C_2H_7N	75-04-7	45.08368	19.822	−0.12598	−4.9793			192.15	1.727E-03	289.73	2.236E-04
101	129	Ethylbenzene	C_8H_{10}	100-41-4	106.165	−13.563	1208.6	0.377			178.2	8.012E-03	413.1	2.326E-04
101	130	Ethyl benzoate	$C_9H_{10}O_2$	93-89-0	150.1745	−40.706	3035	4.2655			250	6.643E-03	486.55	3.109E-04
101	131	2-Ethyl butanoic acid	$C_6H_{12}O_2$	88-09-5	116.15828	−12.24	1836.4	0.021868			258.15	6.705E-03	466.95	2.822E-04
101	132	Ethyl butyrate	$C_6H_{12}O_2$	105-54-4	116.15828	−15.485	1325.6	0.6432			250	1.319E-03	394.65	2.533E-04
101	133	Ethylcyclohexane	C_8H_{16}	1678-91-7	112.21264	−22.11	1673	1.641			200	6.406E-03	404.94	2.956E-04
101	134	Ethylcyclopentane	C_7H_{14}	1640-89-7	98.18606	−6.894	818.6	−0.5941			253.15	9.605E-04	378.15	2.599E-04
101	135	Ethylene	C_2H_4	74-85-1	28.05316	1.8878	78.865	−2.1554			104	6.334E-04	250	6.142E-05
101	136	Ethylenediamine	$C_2H_8N_2$	107-15-3	60.09832	−53.908	4030.8	5.9704			284.29	2.487E-03	483.15	1.723E-04
101	137	Ethylene glycol	$C_2H_6O_2$	107-21-1	62.06784	−290.36	14,251	42.486	−0.000040369	2	260.15	1.305E-01	576	1.276E-04
101	138	Ethyleneimine	C_2H_5N	151-56-4	43.0678	−11.012	967.4				250	7.909E-04	329	3.123E-04
101	139	Ethylene oxide	C_2H_4O	75-21-8	44.05256	−8.521	634.2	−0.3314			160.65	1.918E-03	283.85	2.863E-04
101	140	Ethyl formate	$C_3H_6O_2$	109-94-4	74.07854	−9.8417	876.4	−0.1708			245	7.435E-04	345	2.486E-04
101	141	2-Ethyl hexanoic acid	$C_8H_{16}O_2$	149-57-5	144.211	−13.037	2346				155.15	8.035E+00	510.1	2.165E-04
101	142	Ethylhexyl ether	$C_8H_{18}O$	5756-43-4	130.22792	−11.311	1337.2	−0.02982			180	1.765E-02	417.15	2.522E-04
101	143	Ethylisopropyl ether	$C_5H_{12}O$	625-54-7	88.14818	−11.331	908.46	0.00042478			140	7.908E-03	326.15	1.949E-04
101	144	Ethylisopropyl ketone	$C_6H_{12}O$	565-69-5	100.15888	−11.452	1172.7	−0.00010095			204.15	3.319E-03	386.55	2.207E-04
101	145	Ethyl mercaptan	C_2H_6S	75-08-1	62.13404	−9.7574	729.43	−0.14912			125.26	9.520E-03	308.15	2.626E-04
101	146	Ethyl propionate	$C_5H_{10}O_2$	105-37-3	102.1317	−8.9215	950.8	−0.32687			250	9.848E-04	372.25	2.480E-04
101	147	Ethylpropyl ether	$C_5H_{12}O$	628-32-0	88.14818	0.7109	386.51	−1.7754			200	1.156E-03	337.01	2.086E-04
101	148	Ethyltrichlorosilane	$C_2H_5Cl_3Si$	115-21-9	163.506	−11.499	1122.6				167.55	8.239E-03	371.05	2.089E-04

101	149	Fluorine	F_2	7782-41-4	37.9968064	8.18	−75.6	−3.5148			53.48	7.317E-04	140	5.954E-05
101	150	Fluorobenzene	C_6H_5F	462-06-6	96.1023032	−10.064	1058.7	−0.17162			232.15	1.599E-03	453.15	1.542E-04
101	151	Fluoroethane	C_2H_5F	353-36-6	48.0595	−10.118	464.42				129.95	1.438E-03	235.45	2.900E-04
101	152	Fluoromethane	CH_3F	593-53-3	34.03292	−10.501	427.78	0.0086309			131.35	7.450E-04	194.82	2.587E-04
101	153	Formaldehyde	CH_2O	50-00-0	30.02598	−7.6591	603.36	−0.53378			155.15	1.560E-03	253.85	2.645E-04
101	154	Formamide	CH_3NO	75-12-7	45.04062	−74.521	5081.5	9.0873			273.15	7.171E-03	493	3.829E-04
101	155	Formic acid	CH_2O_2	64-18-6	46.0257	−48.529	3394.7	5.3903			281.45	2.319E-03	373.71	5.444E-04
101	156	Furan	C_4H_4O	110-00-9	68.07396	−10.923	894.63	−0.00068418			200	1.575E-03	304.5	3.392E-04
101	157	Helium-4	He	7440-59-7	4.0026	−9.6312	−3.841	−1.458	−1.065E-08	10	2.2	3.628E-06	5.1	2.532E-06
101	158	Heptadecane	$C_{17}H_{36}$	629-78-7	240.46774	−19.991	2245.1	1.1982			295.13	3.814E-03	575.3	2.088E-04
101	159	Heptanal	$C_7H_{14}O$	111-71-7	114.18546	−9.5468	1147.2	−0.23251			229.8	2.971E-03	426.15	2.580E-04
101	160	Heptane	C_7H_{16}	142-82-5	100.20194	−98.159	3592.6	14.197	−0.000029555	2	180.15	4.341E-03	432.16	1.003E-04
101	161	Heptanoic acid	$C_7H_{14}O_2$	111-14-8	130.185	−40.543	3328.3	4.1804			265.83	9.242E-03	496.15	3.754E-04
101	162	1-Heptanol	$C_7H_{16}O$	111-70-6	116.20134	−66.654	5325.8	7.66	−2.2512E-28	9.9041	239.15	8.805E-02	448.6	3.190E-04
101	163	2-Heptanol	$C_7H_{16}O$	543-49-7	116.20134	−125.81	7996	16.412	−7.6643E-17	6	220	3.856E-01	432.9	2.707E-04
101	164	3-Heptanone	$C_7H_{14}O$	106-35-4	114.18546	−9.3874	1204.9	−0.32618			234.15	2.427E-03	421.15	2.040E-04
101	165	2-Heptanone	$C_7H_{14}O$	110-43-0	114.18546	−13.929	1321.9	0.40382			250	1.642E-03	424.18	2.318E-04
101	166	1-Heptene	C_7H_{14}	592-76-7	98.18606	−10.819	841.33				154.12	4.701E-03	429.92	1.417E-04
101	167	Heptyl mercaptan	$C_7H_{16}S$	1639-09-4	132.26694	−11.812	1291.9	0.076469			229.92	3.097E-03	450.09	2.087E-04
101	168	1-Heptyne	C_7H_{12}	628-71-7	96.17018	−2.7947	563.86	−1.1636			192.22	2.528E-03	447.2	1.777E-04
101	169	Hexadecane	$C_{16}H_{34}$	544-76-3	226.44116	−20.182	2203.5	1.2289			291.31	3.536E-03	564.15	2.054E-04
101	170	Hexanal	$C_6H_{12}O$	66-25-1	100.15888	0.1369	633.77	−1.6659			214.93	2.849E-03	401.15	2.563E-04
101	171	Hexane	C_6H_{14}	110-54-3	86.17536	−56.569	2140.5	7.5175	−0.000017676	2	174.65	2.379E-03	406.08	1.164E-04
101	172	Hexanoic acid	$C_6H_{12}O_2$	142-62-1	116.158	−46.402	3448.6	5.0849			269.25	5.854E-03	478.85	4.019E-04
101	173	1-Hexanol	$C_6H_{14}O$	111-27-3	102.17476	−39.324	3841	3.6933	−2.12E-30	10.485	228.55	8.570E-02	429.9	3.343E-04
101	174	2-Hexanol	$C_6H_{14}O$	626-93-7	102.175	−82.705	7404.9	6.4721	1.5016	0.41014	223	4.919E-01	412.4	3.274E-04
101	175	2-Hexanone	$C_6H_{12}O$	591-78-6	100.15888	−11.445	1187.2	0.0029076			217.35	2.561E-03	400.7	2.108E-04
101	176	3-Hexanone	$C_6H_{12}O$	589-38-8	100.15888	−13.684	1283.4	0.33755			217.5	2.563E-03	396.65	2.185E-04
101	177	1-Hexene	C_6H_{12}	592-41-6	84.15948	−10.903	796.19				133.39	7.197E-03	336.63	1.959E-04
101	178	3-Hexyne	C_6H_{10}	928-49-4	82.1436	−4.2684	647.6	−1.0087			170.05	3.550E-03	432	1.377E-04
101	179	Hexyl mercaptan	$C_6H_{14}S$	111-31-9	118.24036	−10.073	1123.3	−0.16515			192.62	6.035E-03	425.81	2.172E-04
101	180	1-Hexyne	C_6H_{10}	693-02-7	82.1436	−4.7263	594.43	−0.86247			141.25	8.332E-03	412	2.083E-04
101	181	2-Hexyne	C_6H_{10}	764-35-2	82.1436	−3.7464	624.2	−1.084			183.65	2.483E-03	435	1.368E-04
101	182	Hydrazine	H_4N_2	302-01-2	32.04516	−75.781	4175.4	9.6508	−7.27E-09	3	274.69	1.451E-03	522.52	2.191E-04
101	183	Hydrogen	H_2	1333-74-0	2.01588	−11.661	24.7	−0.261	−4.10E-16	10	13.95	2.546E-05	33	3.906E-06
101	184	Hydrogen bromide	BrH	10035-10-6	80.91194	−11.633	316.38	0.56191			185.15	9.207E-04	206.45	8.206E-04
101	185	Hydrogen chloride	ClH	7647-01-0	36.46094	−116.34	3834.6	16.864	−2.5875E-10	4	158.97	1.003E-03	318.15	5.777E-05
101	186	Hydrogen cyanide	CHN	74-90-8	27.02534	−21.927	1266.5	1.5927			259.83	2.754E-04	298.85	1.821E-04
101	187	Hydrogen fluoride	FH	7664-39-3	20.0063432	353.99	13,928	−41.717	−2962	−0.5	189.79	1.545E-03	368.92	1.185E-04
101	188	Hydrogen sulfide	H_2S	7783-06-4	34.08088	−10.905	762.11	−0.11863			187.68	5.726E-04	350	8.089E-05
101	189	Isobutyric acid	$C_4H_8O_2$	79-31-2	88.10512	−11.497	1365.7	0.036966			250	2.938E-03	450	2.649E-04
101	190	Isopropyl amine	C_3H_9N	75-31-0	59.11026	−31.157	1926	2.925			250	6.737E-04	453.15	1.214E-04
101	191	Malonic acid	$C_3H_4O_4$	141-82-2	104.06146	−117.73	9943.3	14.589			409.15	3.386E-03	580	4.281E-04
101	192	Methacrylic acid	$C_4H_6O_2$	79-41-4	86.08924	−14.527	1497.7	0.51747			288.15	1.664E-03	434.15	3.582E-04
101	193	Methane	CH_4	74-82-8	16.0425	−6.1572	178.15	−0.95239	−9.0606E-24	10	90.69	2.063E-04	188	2.262E-05
101	194	Methanol	CH_4O	67-56-1	32.04186	−25.317	1789.2	2.069			175.47	1.193E-02	337.85	3.442E-04
101	195	*N*-Methyl acetamide	C_3H_7NO	79-16-3	73.09378	−4.648	1832	−1.2191			301.15	3.995E-03	478.15	2.392E-04
101	196	Methyl acetate	$C_3H_6O_2$	79-20-9	74.07854	13.557	−187.3	−3.6592			250	6.135E-04	425	1.198E-04
101	197	Methyl acetylene	C_3H_4	74-99-7	40.06386	−2.8737	301.35	−1.2271			170.45	6.045E-04	373.15	8.846E-05
101	198	Methyl acrylate	$C_4H_6O_2$	96-33-3	86.08924	10.848	75	−3.297			275	6.126E-04	400	1.636E-04

(*Continued*)

TABLE 2-139 Viscosity of Inorganic and Organic Liquids (Pa·s) *(Continued)*

Eqn	Cmpd. no.	Name	Formula	CAS	Mol. wt.	C_1	C_2	C_3	C_4	C_5	T_{min}, K	Viscosity at T_{min}	T_{max}, K	Viscosity at T_{max}
101	199	Methyl amine	CH_5N	74-89-5	31.0571	−17.044	1074	0.84203			179.69	1.236E-03	273.15	2.275E-04
101	200	Methyl benzoate	$C_8H_8O_2$	93-58-3	136.14792	−21.971	2267.4	1.4173			288.15	2.299E-03	472.65	2.149E-04
101	201	3-Methyl-1,2-butadiene	C_5H_8	598-25-4	68.11702	−10.481	648.37	−0.041947			159.53	1.321E-03	314	1.739E-04
101	202	2-Methylbutane	C_5H_{12}	78-78-4	72.14878	−12.596	889.11	0.20469			150	3.542E-03	310	1.928E-04
101	203	2-Methylbutanoic acid	$C_5H_{10}O_2$	116-53-0	102.1317	−1.035	1048.5	−1.5474			298.15	1.774E-03	450.15	2.859E-04
101	204	3-Methyl-1-butanol	$C_5H_{12}O$	123-51-3	88.1482	−46.377	4169.6	4.7			155.95	5.989E+01	404.15	3.891E-04
101	205	2-Methyl-1-butene	C_5H_{10}	563-46-2	70.1329	−10.755	705.48	−0.011113			135.58	3.675E-03	304.3	2.034E-04
101	206	2-Methyl-2-butene	C_5H_{10}	513-35-9	70.1329	−8.4453	639.21	−0.38409			139.39	3.164E-03	311.7	1.841E-04
101	207	2-Methyl -1-butene-3-yne	C_5H_6	78-80-8	66.10114	−3.6585	441.1	−1.0547			160.15	1.915E-03	390.15	1.476E-04
101	208	Methylbutyl ether	$C_5H_{12}O$	628-28-4	88.14818	−11.278	949.12	−0.00012343			157.48	5.239E-03	343.31	2.006E-04
101	209	Methylbutyl sulfide	$C_5H_{12}S$	628-29-5	104.214	−10.97	1067.3	−0.017484			175.3	6.930E-03	396.58	2.286E-04
101	210	3-Methyl-1-butyne	C_5H_8	598-23-2	68.11702	−1.8842	433.58	−1.3238			183.45	1.628E-03	364	2.035E-04
101	211	Methyl butyrate	$C_5H_{10}O_2$	623-42-7	102.1317	−12.206	1141.7	0.15014			200	3.339E-03	375.9	2.539E-04
101	212	Methylchlorosilane	CH_5ClSi	993-00-0	80.5889	−12.002	1009.7				139.05	8.734E-03	353.6	1.066E-04
101	213	Methylcyclohexane	C_7H_{14}	108-87-2	98.18606	−11.358	1213.1				146.58	4.587E-02	457.68	1.653E-04
101	214	1-Methylcyclohexanol	$C_7H_{14}O$	590-67-0	114.18546	−6.1534	3219	−1.4494			299.15	2.584E-02	548.8	8.025E-05
101	215	*cis*-2-Methylcyclohexanol	$C_7H_{14}O$	7443-70-1	114.18546	−6.6904	3150.5	−1.392			280.15	3.729E-02	491.2	1.360E-04
101	216	*trans*-2-Methylcyclohexanol	$C_7H_{14}O$	7443-52-9	114.18546	−6.6915	3173.2	−1.3046			269.15	1.107E-01	493.6	2.356E-04
101	217	Methylcyclopentane	C_6H_{12}	96-37-7	84.15948	−1.8553	612.62	−1.3774			248.15	9.288E-04	353.15	2.742E-04
101	218	1-Methylcyclopentene	C_6H_{10}	693-89-0	82.1436	−4.8515	679.07	−0.93238			146.62	7.669E-03	433.6	1.301E-04
101	219	3-Methylcyclopentene	C_6H_{10}	1120-62-3	82.1436	−6.7424	788.86	−0.69862			168.54	3.539E-03	420.8	1.129E-04
101	220	Methyldichlorosilane	CH_4Cl_2Si	75-54-7	115.03396	−10.517	745.32				275	4.070E-04	314.7	2.891E-04
101	221	Methylethyl ether	C_3H_8O	540-67-0	60.09502	−11.104	627.18	0.036581			160	9.133E-04	280.5	1.731E-04
101	222	Methylethyl ketone	C_4H_8O	78-93-3	72.10572	−1.0598	520.68	−1.4961			186.48	2.266E-03	535.5	7.577E-05
101	223	Methylethyl sulfide	C_3H_8S	624-89-5	76.1606	−10.842	863.65	−0.00074603			167.23	3.409E-03	339.8	2.474E-04
101	224	Methyl formate	$C_2H_4O_2$	107-31-3	60.05196	−39.641	2113.3	4.308			250	6.104E-04	304.9	3.134E-04
101	225	Methylisobutyl ether	$C_5H_{12}O$	625-44-5	88.14818	−11.27	888.42	0.024736			188	1.637E-03	331.7	2.143E-04
101	226	Methylisobutyl ketone	$C_6H_{12}O$	108-10-1	100.15888	−11.394	1168.7	−0.007539			189.15	5.222E-03	389.15	2.170E-04
	227	Methyl isocyanate	C_2H_3NO	624-83-9	57.05132									
101	228	Methylisopropyl ether	$C_4H_{10}O$	598-53-8	74.1216	−11.216	737.75	0.019308			127.93	4.722E-03	303.92	1.703E-04
101	229	Methylisopropyl ketone	$C_5H_{10}O$	563-80-4	86.1323	−11.272	1048.9	0.00030493			180.15	4.305E-03	367.55	2.212E-04
101	230	Methylisopropyl sulfide	$C_4H_{10}S$	1551-21-9	90.1872	−11.075	990.72				171.64	4.977E-03	553.1	9.292E-05
101	231	Methyl mercaptan	CH_4S	74-93-1	48.10746	−10.628	645	0.025885			150.18	2.022E-03	279.11	2.826E-04
101	232	Methyl methacrylate	$C_5H_8O_2$	80-62-6	100.11582	−0.099	496	−1.5939			260	8.635E-04	400	2.229E-04
101	233	2-Methyloctanoic acid	$C_9H_{18}O_2$	3004-93-1	158.23802	−12.579	2224.2				240	3.646E-02	518.15	2.519E-04
101	234	2-Methylpentane	C_6H_{14}	107-83-5	86.17536	−12.86	946.91	0.26191			119.55	2.506E-02	333.41	2.038E-04
101	235	Methyl pentyl ether	$C_6H_{14}O$	628-80-8	102.17476	−11.391	1090.8	1.0752E-07			176	5.554E-03	372	2.120E-04
101	236	2-Methylpropane	C_4H_{10}	75-28-5	58.1222	−13.912	797.09	0.45308			110	1.072E-02	310.95	1.588E-04
101	237	2-Methyl-2-propanol	$C_4H_{10}O$	75-65-0	74.1216	400.35	−30,387	−56.971	550,680,000	−3	295.56	5.334E-03	451.21	1.006E-04
101	238	2-Methyl propene	C_4H_8	115-11-7	56.10632	−10.385	599.59	−0.046088			132.81	2.253E-03	266.25	2.270E-04
101	239	Methyl propionate	$C_4H_8O_2$	554-12-1	88.10512	−4.841	696.7	−0.9194			250	8.002E-04	352.6	2.593E-04
101	240	Methylpropyl ether	$C_4H_{10}O$	557-17-5	74.1216	−10.705	788.94	−0.048383			133.97	6.390E-03	312.2	2.127E-04
101	241	Methylpropyl sulfide	$C_4H_{10}S$	3877-15-4	90.1872	−10.569	952.38	−0.063873			160.17	7.103E-03	368.69	2.333E-04
	242	Methylsilane	CH_6Si	992-94-9	46.14384									
101	243	alpha-Methyl styrene	C_9H_{10}	98-83-9	118.1757	−11.632	1251.6	0.071692			249.95	1.972E-03	438.65	2.382E-04
101	244	Methyl tert-butyl ether	$C_5H_{12}O$	1634-04-4	88.1482	−13.415	1050.5	0.33157	0	0	164.55	4.801E-03	328.2	2.502E-04
101	245	Methyl vinyl ether	C_3H_6O	107-25-5	58.07914	−10.34	519.61	−0.013899			151.15	9.377E-04	278.65	1.929E-04
101	246	Naphthalene	$C_{10}H_8$	91-20-3	128.17052	−19.308	1822.5	1.218			353.43	9.077E-04	633.15	1.892E-04
101	247	Neon	Ne	7440-01-9	20.1797	−17.945	115.57	1.428	−2.14E-17	10	25.09	1.602E-04	44.13	2.706E-05

101	248	Nitroethane	$C_2H_5NO_2$	79-24-3	75.0666	−4.438	746.5	−0.9385			200	3.420E-03	387.22	3.027E-04
101	249	Nitrogen	N_2	7727-37-9	28.0134	16.004	−181.61	−5.1551			63.15	2.633E-04	124	3.331E-05
	250	Nitrogen trifluoride	F_3N	7783-54-2	71.00191									
101	251	Nitromethane	CH_3NO_2	75-52-5	61.04002	−9.5556	981.64	−0.19453			244.6	1.344E-03	374.35	3.078E-04
101	252	Nitrous oxide	N_2O	10024-97-2	44.0128	19.329	−381.68	−4.8618			210	2.065E-04	283.09	7.730E-05
101	253	Nitric oxide	NO	10102-43-9	30.0061	−246.65	3150.3	49.98	−0.22541	1	109.5	3.858E-04	180.05	3.791E-05
101	254	Nonadecane	$C_{19}H_{40}$	629-92-5	268.5209	−16.403	2119.5	0.6881			305.04	4.012E-03	603.15	2.068E-04
101	255	Nonanal	$C_9H_{18}O$	124-19-6	142.23862	−4.3492	1052.7	−1.0035			267.3	2.432E-03	465.52	2.606E-04
101	256	Nonane	C_9H_{20}	111-84-2	128.2551	−68.54	3165.3	9.0919	−0.000013519	2	218.15	3.306E-03	593.15	4.997E-05
101	257	Nonanoic acid	$C_9H_{18}O_2$	112-05-0	158.238	−48.851	4095	5.294			285.55	1.030E-02	528.75	3.670E-04
101	258	1-Nonanol	$C_9H_{20}O$	143-08-8	144.2545	−39.863	4089	3.7631			280	1.733E-02	486.25	2.823E-04
101	259	2-Nonanol	$C_9H_{20}O$	628-99-9	144.255	−98.854	7183.8	12.283			238.15	2.310E-01	471.7	3.334E-04
101	260	1-Nonene	C_9H_{18}	124-11-8	126.23922	−11.069	1081.7				191.91	4.372E-03	420.02	2.048E-04
101	261	Nonyl mercaptan	$C_9H_{20}S$	1455-21-6	160.3201	−11.319	1428	−0.022545			253.05	3.026E-03	492.95	1.912E-04
101	262	1-Nonyne	C_9H_{16}	3452-09-3	124.22334	−2.3409	715.52	−1.222			223.15	3.206E-03	487.2	2.172E-04
101	263	Octadecane	$C_{18}H_{38}$	593-45-3	254.49432	−22.688	2466	1.5703			301.31	3.926E-03	589.86	2.057E-04
101	264	Octanal	$C_8H_{16}O$	124-13-0	128.212	−2.5373	900.91	−1.2685			251.65	2.555E-03	445.15	2.614E-04
101	265	Octane	C_8H_{18}	111-65-9	114.22852	−98.805	3905.5	14.103	−0.000025112	2	211.15	2.629E-03	454.96	1.111E-04
101	266	Octanoic acid	$C_8H_{16}O_2$	124-07-2	144.211	−60.795	4617.8	7.028			289.65	6.652E-03	512.85	3.576E-04
101	267	1-Octanol	$C_8H_{18}O$	111-87-5	130.22792	−0.22128	3018.4	−2.8054	0.000013141	2	280	1.472E-02	468.35	2.902E-04
101	268	2-Octanol	$C_8H_{18}O$	123-96-6	130.228	−145.99	9296.7	19.285			241.55	1.856E-01	452.9	5.409E-04
101	269	2-Octanone	$C_8H_{16}O$	111-13-7	128.21204	−11.736	1415.2	0.0003618			252.85	2.161E-03	446.15	1.913E-04
101	270	3-Octanone	$C_8H_{16}O$	106-68-3	128.21204	−20.804	1834.6	1.3403			255.55	2.039E-03	440.65	2.075E-04
101	271	1-Octene	C_8H_{16}	111-66-0	112.21264	−11.19	1057.4				171.45	6.587E-03	453.52	1.422E-04
101	272	Octyl mercaptan	$C_8H_{18}S$	111-88-6	146.29352	−11.498	1362.1	0.015575			223.95	4.837E-03	472.19	1.999E-04
101	273	1-Octyne	C_8H_{14}	629-05-0	110.19676	−3.8552	684.22	−1.0071			193.55	3.614E-03	468	1.868E-04
101	274	Oxalic acid	$C_2H_2O_4$	144-62-7	90.03488	−27.978	2915.1	2.3374			462.65	6.539E-04	516	4.399E-04
101	275	Oxygen	O_2	7782-44-7	31.9988	−4.1476	94.04	−1.207			54.36	7.170E-04	150	6.990E-05
101	276	Ozone	O_3	10028-15-6	47.9982	−10.94	415.96				77.55	3.787E-03	208.8	1.300E-04
101	277	Pentadecane	$C_{15}H_{32}$	629-62-9	212.41458	−19.299	2088.6	1.1091			283.07	3.486E-03	543.84	2.091E-04
101	278	Pentanal	$C_5H_{10}O$	110-62-3	86.1323	−8.2185	919.43	−0.42363			191.59	3.532E-03	375.15	2.539E-04
101	279	Pentane	C_5H_{12}	109-66-0	72.14878	−53.509	1836.6	7.1409	−0.000019627	2	143.42	3.529E-03	465.15	4.796E-05
101	280	Pentanoic acid	$C_5H_{10}O_2$	109-52-4	102.132	−37.067	2856.7	3.7344			270	3.773E-03	458.95	3.510E-04
101	281	1-Pentanol	$C_5H_{12}O$	71-41-0	88.1482	−36.561	3542.2	3.3364	−8.0487E-37	12.84	253.15	1.649E-02	410.95	3.842E-04
101	282	2-Pentanol	$C_5H_{12}O$	6032-29-7	88.1482	−16.456	3209.9				200	6.660E-01	392.2	2.557E-04
101	283	2-Pentanone	$C_5H_{10}O$	107-87-9	86.1323	−11.055	1005.3	0.0039301			250	9.009E-04	375.46	2.354E-04
101	284	3-Pentanone	$C_5H_{10}O$	96-22-0	86.1323	−2.8695	596.32	−1.2025			234.18	1.024E-03	375.14	2.232E-04
101	285	1-Pentene	C_5H_{10}	109-67-1	70.1329	−10.667	659.56				108.02	1.045E-02	303.22	2.051E-04
101	286	2-Pentyl mercaptan	$C_5H_{12}S$	2084-19-7	104.21378	−6.9168	818.76	−0.59628			220	1.643E-03	385.15	2.385E-04
101	287	Pentyl mercaptan	$C_5H_{12}S$	110-66-7	104.21378	−11.677	1091.2	0.10658			197.45	3.745E-03	399.79	2.463E-04
101	288	1-Pentyne	C_5H_8	627-19-0	68.11702	−1.7273	424.34	−1.342			167.45	2.322E-03	378	1.898E-04
101	289	2-Pentyne	C_5H_8	627-21-4	68.11702	−3.7241	516.54	−1.1167			163.83	1.902E-03	415.2	9.980E-05
101	290	Phenanthrene	$C_{14}H_{10}$	85-01-8	178.2292	−22.472	2566.9	1.5749			372.38	1.920E-03	610.03	2.849E-04
101	291	Phenol	C_6H_6O	108-95-2	94.11124	−15.822	3301.8				291.45	1.119E-02	555.4	5.134E-05
101	292	Phenyl isocyanate	C_7H_5NO	103-71-9	119.1207	−11.31	1280				243.15	2.368E-03	522.4	1.420E-04
101	293	Phthalic anhydride	$C_8H_4O_3$	85-44-9	148.11556	195.25	−11,072	−29.084			404.15	1.229E-03	557.65	1.986E-04
101	294	Propadiene	C_3H_4	463-49-0	40.06386	−6.3528	240.85	−0.58229			136.87	5.772E-04	298.15	1.416E-04
101	295	Propane	C_3H_8	74-98-6	44.09562	−17.156	646.25	1.1101	−7.3439E-11	4	85.47	9.458E-03	360	4.275E-05
101	296	1-Propanol	C_3H_8O	71-23-8	60.09502	23.467	116.07	−5.3372	2,880,100,000	−4.0267	146.95	2.069E+01	370.35	4.735E-04
101	297	2-Propanol	C_3H_8O	67-63-0	60.095	−8.8918	2357.6	−0.91376			185.26	3.917E-01	355.3	4.892E-04
101	298	Propenylcyclohexene	C_9H_{14}	13511-13-2	122.20746	−11.208	1079.8				199	3.083E-03	508.8	1.133E-04
101	299	Propionaldehyde	C_3H_6O	123-38-6	58.07914	−5.9402	617.95	−0.74183			165	2.522E-03	322.15	2.470E-04

(Continued)

TABLE 2-139 Viscosity of Inorganic and Organic Liquids (Pa·s) *(Continued)*

Eqn	Cmpd. no.	Name	Formula	CAS	Mol. wt.	C_1	C_2	C_3	C_4	C_5	T_{min}, K	Viscosity at T_{min}	T_{max}, K	Viscosity at T_{max}
101	300	Propionic acid	$C_3H_6O_2$	79-09-4	74.0785	−23.931	1834.6	1.9124			252.45	2.275E-03	414.32	3.430E-04
101	301	Propionitrile	C_3H_5N	107-12-0	55.0785	−6.698	753.58	−0.63783			180.37	2.928E-03	370.25	2.172E-04
101	302	Propyl acetate	$C_5H_{10}O_2$	109-60-4	102.1317	17.797	−252.43	−4.291			250	1.002E-03	473.15	1.045E-04
101	303	Propyl amine	C_3H_9N	107-10-8	59.11026	−9.8074	1010.4	−0.25697			188.36	3.060E-03	321	2.908E-04
101	304	Propylbenzene	C_9H_{12}	103-65-1	120.19158	−18.282	1549.7	1.0454			200	6.774E-03	432.39	2.357E-04
101	305	Propylene	C_3H_6	115-07-1	42.07974	−92.082	1907.3	15.639	−0.043098	1	87.89	1.549E-02	333.15	5.147E-05
101	306	Propyl formate	$C_4H_8O_2$	110-74-7	88.10512	−73.735	2668.2	10.993	−0.018364	1	180.25	5.852E-03	353.97	2.810E-04
101	307	2-Propyl mercaptan	C_3H_8S	75-33-2	76.16062	−5.7244	638.2	−0.76415			142.61	6.477E-03	325.71	2.784E-04
101	308	Propyl mercaptan	C_3H_8S	107-03-9	76.16062	−10.153	840.71	−0.093763			159.95	4.641E-03	340.87	2.656E-04
101	309	1,2-Propylene glycol	$C_3H_8O_2$	57-55-6	76.09442	−804.54	30,487	130.79	−0.15449	1	213.15	9.502E+02	500.8	3.307E-04
101	310	Quinone	$C_6H_4O_2$	106-51-4	108.09476	−14.846	1829.4	0.3729			388.85	3.642E-04	454	1.965E-04
	311	Silicon tetrafluoride	F_4Si	7783-61-1	104.07911									
101	312	Styrene	C_8H_8	100-42-5	104.14912	−22.675	1758	1.6701			242.54	1.919E-03	418.31	2.268E-04
101	313	Succinic acid	$C_4H_6O_4$	110-15-6	118.08804	−104.32	9615.1	12.587			460.85	1.913E-03	591	4.426E-04
101	314	Sulfur dioxide	O_2S	7446-09-5	64.0638	46.223	−1378	−8.7475			225	6.900E-04	400	6.557E-05
101	315	Sulfur hexafluoride	F_6S	2551-62-4	146.0554192	3.8305	41.21	−2.1342			223.15	5.388E-04	318.69	2.383E-04
101	316	Sulfur trioxide	O_3S	7446-11-9	80.0632	−88.793	6400.7	10.709			289.95	2.477E-03	318.15	9.456E-04
101	317	Terephthalic acid	$C_8H_6O_4$	100-21-0	166.13084	−11.566	2843.2				700.15	5.502E-04	795.28	3.385E-04
101	318	*o*-Terphenyl	$C_{18}H_{14}$	84-15-1	230.30376	−215.09	11,612	31.849	−0.026882	1	329.35	1.736E-02	723.15	1.522E-04
101	319	Tetradecane	$C_{14}H_{30}$	629-59-4	198.388	−136.73	6421.3	19.493	−0.00002297	2	277.65	3.350E-03	554.4	1.170E-04
101	320	Tetrahydrofuran	C_4H_8O	109-99-9	72.10572	−10.321	900.92	−0.069128			164.65	5.505E-03	373.15	2.446E-04
101	321	1,2,3,4-Tetrahydronaphthalene	$C_{10}H_{12}$	119-64-2	132.20228	−118.86	5829.5	16.605	−0.000016991	2	237.4	1.183E-02	576	1.458E-04
101	322	Tetrahydrothiophene	C_4H_8S	110-01-0	88.17132	−10.843	1165.2				293.15	1.040E-03	303.15	9.125E-04
101	323	2,2,3,3-Tetramethylbutane	C_8H_{18}	594-82-1	114.22852	5.5351	632.38	−2.6576			373.96	1.999E-04	454	8.859E-05
101	324	Thiophene	C_4H_4S	110-02-1	84.13956	−16.671	1342.5	0.8388			250	1.269E-03	393.15	2.625E-04
101	325	Toluene	C_7H_8	108-88-3	92.13842	−226.08	6805.7	37.542	−0.060853	1	178.18	1.569E-02	383.78	2.428E-04
101	326	1,1,2-Trichloroethane	$C_2H_3Cl_3$	79-00-5	133.40422	0.388	736.5	−1.7063			236.5	2.955E-03	387	3.798E-04
101	327	Tridecane	$C_{13}H_{28}$	629-50-5	184.36142	−111.98	5468.6	15.579	−0.000016992	2	267.67	3.399E-03	540	1.520E-04
101	328	Triethyl amine	$C_6H_{15}N$	121-44-8	101.19	−3.7067	585.78	−1.0926			250	6.135E-04	359.05	2.028E-04
101	329	Trimethyl amine	C_3H_9N	75-50-3	59.11026	10.142	−130.41	−3.2199			200	5.156E-04	308.15	1.612E-04
101	330	1,2,3-Trimethylbenzene	C_9H_{12}	526-73-8	120.19158	−11.756	1483.1	−0.040387			247.79	2.495E-03	449.27	1.663E-04
101	331	1,2,4-Trimethylbenzene	C_9H_{12}	95-63-6	120.19158	−9.6461	1281.2	−0.29478			229.33	3.477E-03	442.53	1.942E-04
101	332	2,2,4-Trimethylpentane	C_8H_{18}	540-84-1	114.22852	−12.928	1137.5	0.25725	−3.6929E-28	10	165.78	8.636E-03	541.15	4.530E-05
101	333	2,3,3-Trimethylpentane	C_8H_{18}	560-21-4	114.22852	−4.0309	990.76	−1.1771			172.22	1.305E-02	387.91	2.049E-04
101	334	1,3,5-Trinitrobenzene	$C_6H_3N_3O_6$	99-35-4	213.10452	−10.707	1818.5				398.4	2.150E-03	676.8	3.288E-04
101	335	2,4,6-Trinitrotoluene	$C_7H_5N_3O_6$	118-96-7	227.1311	−11.504	3301	−0.39102			353.15	1.167E-02	625	1.601E-04
101	336	Undecane	$C_{11}H_{24}$	1120-21-4	156.30826	52.176	−4951.9	−8.5676	570,980	−2	247.57	3.240E-03	511.2	1.569E-04
101	337	1-Undecanol	$C_{11}H_{24}O$	112-42-5	172.30766	−69.778	5905.2	8.0214			288.45	2.089E-02	590.15	1.856E-04
101	338	Vinyl acetate	$C_4H_6O_2$	108-05-4	86.08924	−22.407	1462.8	1.7006			225	1.237E-03	345.65	2.654E-04
101	339	Vinyl acetylene	C_4H_4	689-97-4	52.07456	−2.2333	320.37	−1.2915			173.15	8.764E-04	364	1.273E-04
101	340	Vinyl chloride	C_2H_3Cl	75-01-4	62.49822	0.26297	276.55	−1.7282			130	2.425E-03	400	8.272E-05
101	341	Vinyl trichlorosilane	$C_2H_3Cl_3Si$	75-94-5	161.48972	−10.37	823.31				178.35	3.171E-03	434.52	2.086E-04
101	342	Water	H_2O	7732-18-5	18.01528	−52.843	3703.6	5.866	−5.879E-29	10	273.16	1.702E-03	646.15	5.028E-05
101	343	*m*-Xylene	C_8H_{10}	108-38-3	106.165	−11.91	1094.9	0.13825			225.3	1.834E-03	413.1	2.189E-04
101	344	*o*-Xylene	C_8H_{10}	95-47-6	106.165	−15.489	1393.5	0.63711			247.98	1.735E-03	418.1	2.459E-04
101	345	*p*-Xylene	C_8H_{10}	106-42-3	106.165	−7.381	911.7	−0.54152			286.41	7.021E-04	413.1	2.169E-04

Except for deuterium, the liquid viscosity is calculated by Eqn 101: $\mu = \exp(C_1 + C_2/T + C_3 \ln T + C_4 T^{C_5})$ where μ is the viscosity in Pa·s and T is the temperature in K. Viscosity is either 1 atm or the vapor pressure, whichever is higher. For deuterium, liquid viscosity is calculated by Eqn 100: $\mu = C_1 + C_2 T + C_3 T^2 + C_4 T^3 + C_5 T^4$ where μ is the viscosity in Pa·s and T is the temperature in K.

Values in this table were taken from the Design Institute for Physical Properties (DIPPR) of the American Institute of Chemical Engineers (AIChE), 801 Critically Evaluated Gold Standard™ Database, copyright 2016 AIChE, and reproduced with permission of AIChE and of the DIPPR Evaluated Process Design Data Project Steering Committee. Their source should be cited as "R. L. Rowley, W. V. Wilding, J. L. Oscarson, T. A. Knotts, N. F. Giles, *DIPPR® Data Compilation of Pure Chemical Properties*, Design Institute for Physical Properties, AIChE, New York, NY (2016)".

TABLE 2-140 Viscosities of Liquids: Coordinates for Use with Fig. 2-19

Liquid	*X*	*Y*	Liquid	*X*	*Y*
Acetaldehyde	15.2	4.8	Glycerol, 100%	2.0	30.0
Acetic acid, 100%	12.1	14.2	Glycerol, 50%	6.9	19.6
Acetic acid, 70%	9.5	17.0	Heptane	14.1	8.4
Acetic anhydride	12.7	12.8	Hexane	14.7	7.0
Acetone, 100%	14.5	7.2	Hydrochloric acid, 31.5%	13.0	16.6
Acetone, 35%	7.9	15.0	Iodobenzene	12.8	15.9
Acetonitrile	14.4	7.4	Isobutyl alcohol	7.1	18.0
Acrylic acid	12.3	13.9	Isobutyric acid	12.2	14.4
Allyl alcohol	10.2	14.3	Isopropyl iodide	13.7	11.2
Allyl bromide	14.4	9.6	Kerosene	10.2	16.9
Allyl iodide	14.0	11.7	Linseed oil, raw	7.5	27.2
Ammonia, 100%	12.6	2.0	Mercury	18.4	16.4
Ammonia, 26%	10.1	13.9	Methanol, 100%	12.4	10.5
Amyl acetate	11.8	12.5	Methanol, 90%	12.3	11.8
Amyl alcohol	7.5	18.4	Methanol, 40%	7.8	15.5
Aniline	8.1	18.7	Methyl acetate	14.2	8.2
Anisole	12.3	13.5	Methyl acrylate	13.0	9.5
Arsenic trichloride	13.9	14.5	Methyl *i*-butyrate	12.3	9.7
Benzene	12.5	10.9	Methyl *n*-butyrate	13.2	10.3
Brine, $CaCl_2$, 25%	6.6	15.9	Methyl chloride	15.0	3.8
Brine, NaCl, 25%	10.2	16.6	Methyl ethyl ketone	13.9	8.6
Bromine	14.2	13.2	Methyl formate	14.2	7.5
Bromotoluene	20.0	15.9	Methyl iodide	14.3	9.3
Butyl acetate	12.3	11.0	Methyl propionate	13.5	9.0
Butyl acrylate	11.5	12.6	Methyl propyl ketone	14.3	9.5
Butyl alcohol	8.6	17.2	Methyl sulfide	15.3	6.4
Butyric acid	12.1	15.3	Naphthalene	7.9	18.1
Carbon dioxide	11.6	0.3	Nitric acid, 95%	12.8	13.8
Carbon disulfide	16.1	7.5	Nitric acid, 60%	10.8	17.0
Carbon tetrachloride	12.7	13.1	Nitrobenzene	10.6	16.2
Chlorobenzene	12.3	12.4	Nitrogen dioxide	12.9	8.6
Chloroform	14.4	10.2	Nitrotoluene	11.0	17.0
Chlorosulfonic acid	11.2	18.1	Octane	13.7	10.0
Chlorotoluene, *ortho*	13.0	13.3	Octyl alcohol	6.6	21.1
Chlorotoluene, *meta*	13.3	12.5	Pentachloroethane	10.9	17.3
Chlorotoluene, *para*	13.3	12.5	Pentane	14.9	5.2
Cresol, *meta*	2.5	20.8	Phenol	6.9	20.8
Cyclohexanol	2.9	24.3	Phosphorus tribromide	13.8	16.7
Cyclohexane	9.8	12.9	Phosphorus trichloride	16.2	10.9
Dibromomethane	12.7	15.8	Propionic acid	12.8	13.8
Dichloroethane	13.2	12.2	Propyl acetate	13.1	10.3
Dichloromethane	14.6	8.9	Propyl alcohol	9.1	16.5
Diethyl ketone	13.5	9.2	Propyl bromide	14.5	9.6
Diethyl oxalate	11.0	16.4	Propyl chloride	14.4	7.5
Diethylene glycol	5.0	24.7	Propyl formate	13.1	9.7
Diphenyl	12.0	18.3	Propyl iodide	14.1	11.6
Dipropyl ether	13.2	8.6	Refrigerant R-22	17.2	4.7
Dipropyl oxalate	10.3	17.7	Sodium	16.4	13.9
Ethyl acetate	13.7	9.1	Sodium hydroxide, 50%	3.2	25.8
Ethyl acrylate	12.7	10.4	Stannic chloride	13.5	12.8
Ethyl alcohol, 100%	10.5	13.8	Succinonitrile	10.1	20.8
Ethyl alcohol, 95%	9.8	14.3	Sulfur dioxide	15.2	7.1
Ethyl alcohol, 40%	6.5	16.6	Sulfuric acid, 110%	7.2	27.4
Ethyl benzene	13.2	11.5	Sulfuric acid, 100%	8.0	25.1
Ethyl bromide	14.5	8.1	Sulfuric acid, 98%	7.0	24.8
2-Ethyl butyl acrylate	11.2	14.0	Sulfuric acid, 60%	10.2	21.3
Ethyl chloride	14.8	6.0	Sulfuryl chloride	15.2	12.4
Ethyl ether	14.5	5.3	Tetrachloroethane	11.9	15.7
Ethyl formate	14.2	8.4	Thiophene	13.2	11.0
2-Ethyl hexyl acrylate	9.0	15.0	Titanium tetrachloride	14.4	12.3
Ethyl iodide	14.7	10.3	Toluene	13.7	10.4
Ethyl propionate	13.2	9.9	Trichloroethylene	14.8	10.5
Ethyl propyl ether	14.0	7.0	Triethylene glycol	4.7	24.8
Ethyl sulfide	13.8	8.9	Turpentine	11.5	14.9
Ethylene bromide	11.9	15.7	Vinyl acetate	14.0	8.8
Ethylene chloride	12.7	12.2	Vinyl toluene	13.4	12.0
Ethylene glycol	6.0	23.6	Water	10.2	13.0
Ethylidene chloride	14.1	8.7	Xylene, *ortho*	13.5	12.1
Fluorobenzene	13.7	10.4	Xylene, *meta*	13.9	10.6
Formic acid	10.7	15.8	Xylene, *para*	13.9	10.9

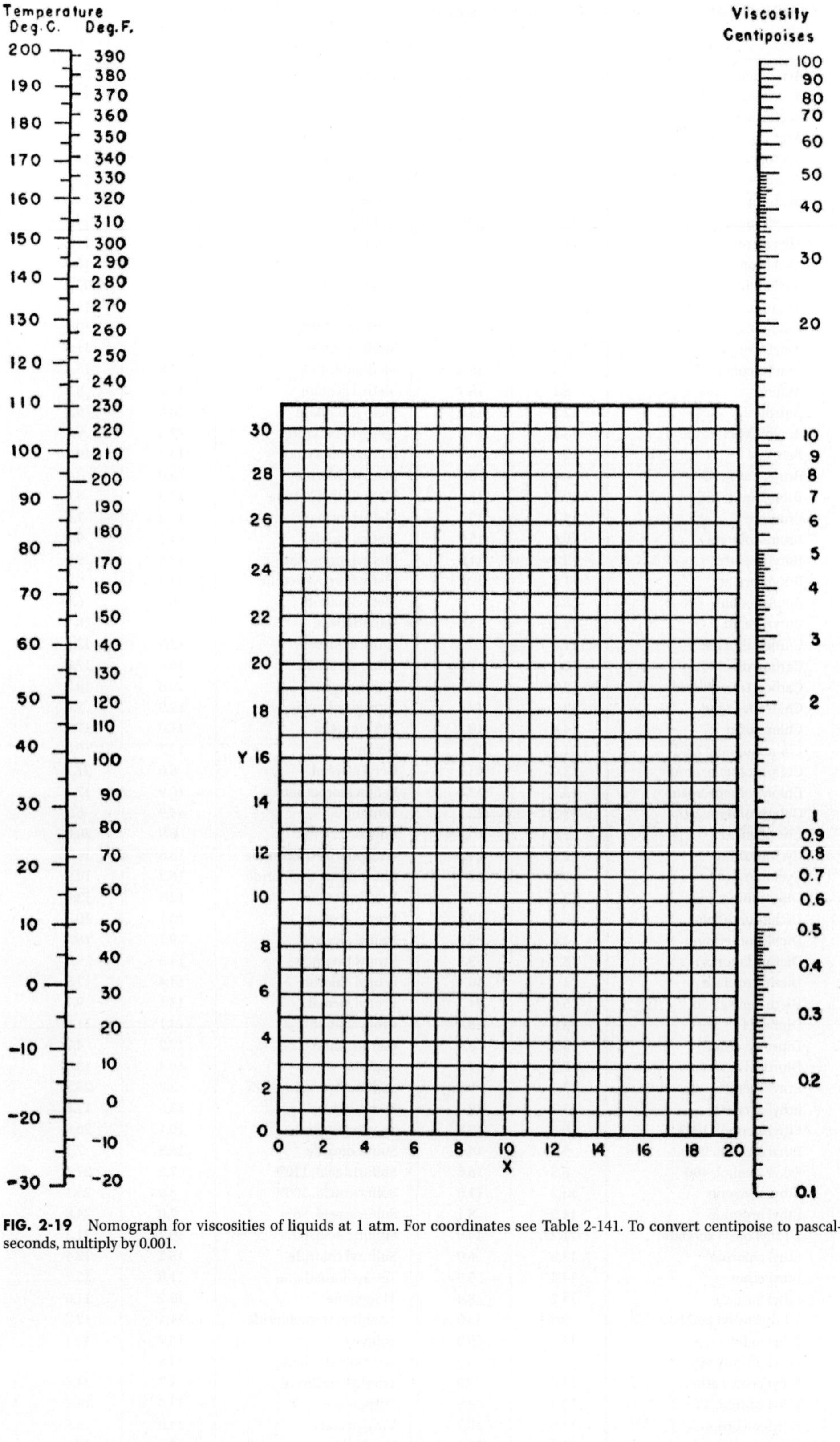

FIG. 2-19 Nomograph for viscosities of liquids at 1 atm. For coordinates see Table 2-141. To convert centipoise to pascal-seconds, multiply by 0.001.

TABLE 2-141 Diffusivities of Pairs of Gases and Vapors (1 atm)

		D_v in cm^2/s												
Substance	Temp., °C	Air	A	H_2	O_2	N_2	CO_2	N_2O	CH_4	C_2H_6	C_2H_4	n-C_4H_{10}	i-C_4H_{10}	Ref.
Acetic acid	0	0.1064		0.416			0.0716							8
Acetone	0	.109		.361										6, 16
n-Amyl alcohol	0	.0589		.235			.0422							8
sec-Amyl alcohol	30	.072												5
Amyl butyrate	0	.040												8
Amyl formate	0	.0543												8
i-Amyl formate	0	.058												8
Amyl isobutyrate	0	.0419		.171										8
Amyl propionate	0	.046		.1914			.0347							8
Aniline	0	.0610												8
	30	.075												5
Anthracene	0	.0421												8
Argon	20					0.194								18
Benzene	0	.077		.306	0.0797		.0528							8, 15
Benzidine	0	.0298												8
Benzyl chloride	0	.066												8
n-Butyl acetate	0	.058												8
i-Butyl acetate	0	.0612		.2364			.0425							8
n-Butyl alcohol	0	.0703		.2716			.0476							8
	30	.088												5
i-Butyl alcohol	0	.0727		.2771			.0483							8
Butyl amine	0	.0821												8
i-Butyl amine	0	.0853												8
i-Butyl butyrate	0	.0468		.185			.0327							8
i-Butyl formate	0	.0705												8
i-Butyl isobutyrate	0	.0457		.191			.0364							8
i-Butyl proprionate	0	.0529		.203			.0366							8
i-Butyl valerate	0	.0424		.173			.0308							8
Butyric acid	0	.067		.264			.0476							8
i-Butyric acid	0	.0679		.271			.0471							8
Cadmium	0					.17								13
Caproic acid	0	.050												8
i-Caproic acid	0	.0513												8
Carbon dioxide	0	.138		.550	.139			0.096	0.153					8
	20					.163								19
	25							.0996*	.00215†					1, 9
	500‡				.9									18
Carbon disulfide	0	.0892		.369			.063							8
Carbon monoxide	0			.651	.185		.137				0.116			8
	450‡				1.0									18
Carbon tetrachloride	0			.293	0.0636									16, 17
Chlorobenzene	30	.075												5
Chloroform	0	.091												6
Chloropicrin	25	.088												10
m-Chlorotoluene	0	.054												8
o-Chlorotoluene	0	.059												8
p-Chlorotoluene	0	.051												8
Cyanogen chloride	0	.111												10
Cyclohexane	15		0.0719	.319	.0744	.0760								3
	45	.086												6
n-Decane	90			.306		.0841								3
Diethylamine	0	.0884												8
2,3-Dimethyl butane	15		.0657	.301	.0753	.0751								3
Diphenyl	0	.0610												8
n-Dodecane	126			.308		.0813								3
Ethane	0			.459										8
Ethanol	0			.377			.0686							20
Ether (diethyl)	0	.0778		.298			.0546							7, 8
Ethyl acetate	0	.0715		.273			.0487							8
	30	.089												5
Ethyl alcohol	0	.102		.375			.0685							8
Ethyl benzene	0	.0658												8
Ethyl *n*-butyrate	0	.0579		.224			.0407							8
Ethyl *i*-butyrate	0	.0591		.229			.0413							8
Ethylene	0			.486										8
Ethyl formate	0	.0840		.337			.0573							8
Ethyl propionate	0	.068		.236			.0450							4, 8
Ethyl valerate	0	.0512		.205			.0367							8
Eugenol	0	.0377												8
Formic acid	0	.1308		.510			.0874							8
Helium	0		.641											8
	20					.705								19
n-Heptane	38								.066§					
n-Hexane	15		.0663	.290	.0753	.0757								3
Hexyl alcohol	0	.0499		.200			.0351							8
Hydrogen	0	.611			.697	.674	.550	.535	.625	0.459	0.486	0.272	0.277	8
	25						.646			.537	.726			2
	500				4.2									18

(*Continued*)

TABLE 2-141 Diffusivities of Pairs of Gases and Vapors (1 atm) *(Continued)*

		D_v in cm^2/s												
Substance	Temp., °C	Air	A	H_2	O_2	N_2	CO_2	N_2O	CH_4	C_2H_6	C_2H_4	n-C_4H_{10}	i-C_4H_{10}	Ref.
Hydrogen cyanide	0	0.173												10
Hydrogen peroxide	60	.188												11
Iodine	0	.07				0.070								8, 12, 14
Mercury	0	.112		0.53		.13								8, 12, 13
Mesitylene	0	.056												8
Methane	500				1.1								18	
Methyl acetate	0	.084		.333			0.0567							8
Methyl alcohol	0	.132		.506			.0879							8
Methyl butyrate	0	.0633		.242			.0446							8
Methyl *i*-butyrate	0	.0639		.257			.0451							8
Methyl cyclopentane	15		0.0731	.318	0.0742	0.0758								3
Methyl formate	0	.0872												8
Methyl propionate	0	.0735		.295			.0528							8
Methyl valerate	0	0.0569												8
Naphthalene	0	.0513												8
Nitrogen	0				0.181									8
	25						0.165			0.148	0.163	0.0960	0.0908	2
Nitrous oxide	0			0.535			.096							8
n-Octane	0	.0505												8
	30		0.0642	.271	0.0705	0.0710								3
Oxygen	0	.178		.697		0.181	.139							8
Phosgene	0	.095												10
Propionic acid	0	.0829		.330			.0588							8
Propyl acetate	0	.067												8
n-Propyl alcohol	0	.085		.315			.0577							8
i-Propyl alcohol	0	.0818												8
	30	.101												5
n-Propyl benzene	0	.0481												8
i-Propyl benzene	0	.0489												8
n-Propyl bromide	0	.085												8
i-Propyl bromide	0	.0902												8
Propyl butyrate	0	.0530		.206			.0364							8
Propyl formate	0	.0712		.281			.0490							8
n-Propyl iodide	0	.079												8
i-Propyl iodide	0	.0802												8
n-Propyl isobutyrate	0	.0549		.212			.0388							8
i-Propyl isobutyrate	0	.059												8
Propyl propionate	0	.057		.212			.0395							8
Propyl valerate	0	.0466		.189			.0341							8
Safrol	0	.0434												8
i-Safrol	0	.0455												8
Sulfur hexafluoride	25			.418										2
Toluene	0	.076	0.071											4, 8
	30	.088												5
Trimethyl carbinol	0	.087												8
2,2,4-Trimethyl pentane	30		0.0618	.288	0.0688	0.0705								3
2,2,3-Trimethyl heptane	90			.270		0.0684								3
n-Valeric acid	0	0.050												8
i-Valeric acid	0	0.0544		.212			.0376							8
Water	0	0.220		.75			.138							8, 20
	450				1.3									18

*320 mmHg.
†40 atm.
‡Also at other temperatures.
§Strong function of concentration.

References

[1]Amdur, Irvine, Mason, and Ross, *J. Chem. Phys.*, **20,** 436 (1952).
[2]Boyd, Stein, Steingrimsson, and Rumpel, *J. Chem. Phys.*, **19,** 548 (1951).
[3]Cummings and Ubbelohde, *J. Chem. Soc. (London)*, 1953, p. 3751.
[4]Fairbanks and Wilke, *Ind. Eng. Chem.*, **42,** 471 (1950).
[5]Gilliland, *Ind. Eng. Chem.*, **26,** 681 (1934).
[6]Gorynnova and Kuvskinskii, *Zhur. Tekh. Fiz.*, **18,** 1421 (1948).
[7]Hansen, Dissertation, Jena, 1907.
[8]*International Critical Tables*, vol. 5, p. 62.
[9]Jeffries and Drickamer, *J. Chem. Phys.*, **22,** 436 (1954).
[10]Klotz and Miller, *J. Am. Chem. Soc.*, **69,** 2557 (1947).
[11]McMurtrie and Keyes, *J. Am. Chem. Soc.*, **70,** 3755 (1948).
[12]Mullaly and Jacques, *Phil. Mag.*, **48,** 6, 1105 (1924).
[13]Spier, *Physica*, 6 (1939): 453; **7,** 381 (1940).
[14]Topley and Whytlaw-Gray, *Phil. Mag.*, **4,** 873 (1927).
[15]Trautz and Ludwig, *Ann. Physik*, **5,** 5, 887 (1930).
[16]Trautz and Muller, *Ann. Physik*, **22,** 353 (1935).
[17]Trautz and Ries, *Ann. Physik*, **8,** 163 (1931).
[18]Walker and Westenberg, *J. Chem. Phys.*, **32,** 136 (1960).
[19]Westenberg and Walker, *J. Chem. Phys.*, **26,** 1753 (1957).
[20]Winkelmann, *Wied. Ann.*, **22,** 152 (1884); **23,** 203 (1884); **26,** 105 (1885); **33,** 445 (1888); **36,** 92 (1889).

Table 2-143 has a representative selection of diffusion coefficients. The subsection "Prediction and Correlation of Physical Properties" should be consulted for estimation techniques.

TABLE 2-142 Diffusivities in Liquids (25°C)

Dilute solutions and 1 atm unless otherwise noted; use $D_L\mu/T$ = constant to estimate effect of temperature; * indicates that reference gives effect of concentration.

Solute	Solvent	$D_L \times 10^5$, sq cm/sec	Estimated possible, error, ± %1	Ref.
Acetal*	Ethanol	1.25	5	11
Acetamide*	Ethanol	0.68	5	11
Acetamide*	Water	1.19	3	11
Acetic acid	Acetone	3.31		4
Acetic acid	Benzene	2.11		1, 4
Acetic acid	Carbon tetrachloride	1.49		4
Acetic acid	Ethylene glycol	0.13		4
Acetic acid	Toluene	2.26		4
Acetic acid*	Water	1.24	3	11
Acetonitrile	Water	1.66	5	11
Acetylene	Water	1.78, 2.11		1, 24
Allyl alcohol*	Ethanol	1.06	5	11
Allyl alcohol	Water	1.19	6	11
Ammonia*	Water	1.7, 2.0, 2.3		1, 11
i-Amyl alcohol*	Ethanol	0.87	5	11
i-Amyl alcohol	Water	1.0	8	11, 25
Benzene	Carbon tetrachloride	1.53		7
Benzene (50 mole %)	*n*-Decane	1.72		26
Benzene (50 mole %)	2,4-Dimethyl pentane	2.49		26
Benzene (50 mole %)	*n*-Dodecane	1.40		26
Benzene (50 mole %)	*n*-Heptane	2.47		26
Benzene (50 mole %)	*n*-Hexadecane	0.96		26
Benzene (50 mole %)	*n*-Octadecane	0.86		26
Benzoic acid	Acetone	2.62		4
Benzoic acid	Benzene	1.38		4
Benzoic acid	Carbon tetrachloride	0.91		4
Benzoic acid	Ethylene glycol	0.043		4
Benzoic acid	Toluene	1.49		4
Bromine	Benzene	2.7		11
Bromine	Carbon disulfide	4.1		11
Bromine	Water	1.3		11
Bromobenzene	Benzene	2.30		25
Bromoform*	Acetone	2.90		11
Bromoform	*i*-Amyl alcohol	0.53		11
Bromoform	Ethanol	1.08	5	11
Bromoform*	Ethyl ether	3.62		11
Bromoform	Methanol	2.20		23
Bromoform	*n*-Propanol	0.94		11
n-Butanol	Water	0.96	5	1, 11, 18, 25
Caffeine	Water	0.63	6	11
Carbon dioxide	Ethanol	4.0	6	11
Carbon dioxide	Water	1.96	1	1, 3, 5, 20, 24, 28
Carbon disulfide (50 mole %, 200 atm.)	*n*-Butanol	3.57		14
Carbon disulfide (50 mole %, 200 atm.)	*i*-Butanol	2.42		14
Carbon disulfide (50 mole %, 218 atm.)	Chlorobenzene	3.00		14
Carbon disulfide (50 mole %, 200 atm.)	2,4-Dimethyl pentane	3.63		14
Carbon disulfide (50 mole %, 100 atm.)	*n*-Heptane	3.0		14
Carbon disulfide (50 mole %, 50 atm.)	Methyl cyclohexane	3.5		14
Carbon disulfide (50 mole %, 200 atm.)	*n*-Octane	3.10		14
Carbon disulfide (50 mole %)	Toluene	2.06		14
Carbon tetrachloride	Benzene	2.04	3	7, 9
Carbon tetrachloride*	Cyclohexane	1.49	2	9, 10*
Carbon tetrachloride	Decalin	0.776	2	9
Carbon tetrachloride	Dioxane	1.02	2	9
Carbon tetrachloride*	Ethanol	1.50	2	9, 10*
Carbon tetrachloride	*n*-Heptane	3.17	2	9
Carbon tetrachloride	Kerosene	0.961	2	9
Carbon tetrachloride	Methanol	2.30	2	9
Carbon tetrachloride	*i*-Octane	2.57	2	9
Carbon tetrachloride	Tetralin	0.735	2	9
Chloral*	Ethanol	0.68	5	11
Chloral hydrate	Water	0.77	7	11

(*Continued*)

TABLE 2-142 Diffusivities in Liquids (25°C) *(Continued)*

Dilute solutions and 1 atm unless otherwise noted; use $D_L\mu / T$ = constant to estimate effect of temperature; * indicates that reference gives effect of concentration.

Solute	Solvent	$D_L \times 10^5$, sq cm/sec	Estimated possible, error, ± %1	Ref.
Chlorine	Water	1.44	4	1, 28
Chlorobenzene	Benzene	2.66		25
Chloroform	Benzene	2.50	6	1, 25
Chloroform	Ethanol	1.38	3	11
Cinnamic acid	Acetone	2.41		4
Cinnamic acid	Benzene	1.12		4
Cinnamic acid	Carbon tetrachloride	0.76		4
Cinnamic acid	Toluene	2.41		4
1,1′-Dichloropropanol	Water	1.0	6	11
Dicyanodiamide*	Water	1.18	4	11
Diethyl ether	Benzene	2.73		25
Diethyl ether	Water	0.85		2
2,4-Dimethyl pentane (50 mole %)	*n*-Dodecane	1.44		26
2,4-Dimethyl pentane (50 mole %)	*n*-Hexadecane	0.88		26
Ethanol*	Water	1.28	4	1, 7, 9,* 11,* 22
Ethyl acetate	Ethyl benzoate	0.94		6
Ethylene dichloride	Benzene	2.8		1, 25
Formic acid	Acetone	3.77		4
Formic acid	Benzene	2.28		4
Formic acid	Carbon tetrachloride	1.89		4
Formic acid	Ethylene glycol	0.094		4
Formic acid	Toluene	2.65		
Formic acid	Water	1.37	10	11
Glucose	Water	0.69	6	11
Glycerol	*i*-Amyl alcohol	0.12		11
Glycerol	Ethanol	0.56		11
Glycerol*	Water	0.94	6	1, 11*
n-Heptane (50 mole %)	*n*-Dodecane	1.58		26
n-Heptane (50 mole %)	*n*-Hexadecane	1.00		26
n-Heptane (50 mole %)	*n*-Octadecane	0.92		26
n-Heptane (50 mole %)	*n*-Tetradecane	1.29		26
Hexamethylene tetramine	Water	0.67		11
Hydrogen chloride*	Water	3.10	3	4, 11,* 12*
Hydrogen	Water	5.85 (4.4)		1, 11, 24(?)
Hydrogen sulfide	Water	1.61		1
Hydroquinone*	Ethanol	0.53	5	11
Hydroquinone*	Water	0.88, 1.12		2, 11*
Iodine	Acetic acid	1.13		11
Iodine	Anisole	1.25		11
Iodine	Benzene	1.98		9, 19, 23
Iodine	Bromobenzene	1.25	10	4, 11, 19
Iodine	Carbon disulfide	3.2		11, 19, 23
Iodine	Carbon tetrachloride	1.45	8	9, 11, 19
Iodine	Chloroform	2.30	3	11, 23
Iodine	Cyclohexane	1.80		4
Iodine	Dioxane	1.07		9
Iodine*	Ethanol	1.30		4, 11*
Iodine	Ethyl acetate	2.2		11, 19
Iodine	Ethyl ether	3.61		11
Iodine	Ethylene bromide	0.93		11
Iodine	*n*-Heptane	3.4, 2.5		9, 11, 19
Iodine	*n*-Hexane	4.15		4, 9
Iodine	Mesitylene	1.49		9
Iodine	Methanol	1.74		19
Iodine	Methyl cyclohexane	2.1		4
Iodine	*n*-Octane	2.76		4
Iodine	Tetrabromoethane	2.0		11
Iodine	*n*-Tetradecane	0.96		4
Iodine	Toluene	2.1		11
Iodine	*m*-Xylene	1.82		9, 11
Iodobenzene	Ethanol	1.09	3	11
Lactose*	Water	0.49	5	11
Maltose*	Water	0.48	5	11
Mannitol*	Water	0.65	5	11
Methanol	Water	1.6		1, 7, 11
Nicotine*	Water	0.60	8	11
Nitric acid*	Water	2.98	2	11
Nitrobenzene	Carbon tetrachloride	1.00		7
Nitrogen	Water	1.9		1, 24
Nitrous oxide	Water	1.8		1, 11
Oxalic acid*	Water	1.61	2	11

TABLE 2-142 Diffusivities in Liquids (25°C) *(Continued)*

Dilute solutions and 1 atm unless otherwise noted; use $D_L\mu/T$ = constant to estimate effect of temperature; * indicates that reference gives effect of concentration.

Solute	Solvent	$D_L \times 10^5$, sq cm/sec	Estimated possible, error, ± %1	Ref.
Oxygen	Glycerol*-water (106 poise)	0.24		13
Oxygen	Sucrose*-water (125 poise)	0.25		13
Oxygen	Water	2.5	20	1, 3, 15, 21, 24
Pentaerythritol*	Water	0.77	4	11
Phenol	*i*-Amyl alcohol	0.2		11
Phenol	Benzene	1.68		1
Phenol	Carbon disulfide	3.7		11
Phenol	Chloroform	2.0		11
Phenol	Ethanol	0.89		11
Phenol	Ethyl ether	3.9		11
n-Propanol	Water	1.1		1, 7, 11
Pyridine*	Ethanol	1.24	3	11
Pyridine	Water	0.76	7	11
Pyrogallol	Water	0.74	7	11
Raffinose*	Water	0.41	4	11
Resorcinol*	Ethanol	0.46	5	11
Resorcinol*	Water	0.87	4	11
Saccharose*	Water	0.49	4	11
Stearic acid*	Ethanol	0.65	5	11
Succinic acid*	Water	0.94		11
Sucrose	Water	0.56	6	2, 27
Sulfur dioxide	Water	1.7		15, 17
Sulfuric acid*	Water	1.97	3	11
Tartaric acid*	Water	0.80	10	11
1,1,2,2-Tetrabromoethane	1,1,2,2-Tetra-chloroethane	0.61	4	11
Toluene	*n*-Decane	2.09		4
Toluene	*n*-Dodecane	1.38		4
Toluene	*n*-Heptane	3.72		4
Toluene	*n*-Hexane	4.21		4
Toluene	*n*-Tetradecane	1.02		4
Urea	Ethanol	0.73		11
Urea	Water	1.37	2	8, 11
Urethane	Water	1.06		11, 25
Water	Glycerol	0.021		16

References

[1]Arnold, *J. Am. Chem. Soc.*, **52,** 3937 (1930).
[2]Calvet, *J. Chim. Phys.*, **44,** 47 (1947).
[3]Carlson, *J. Am. Chem. Soc.*, **33,** 1027 (1911).
[4]Chang and Wilke, *J. Phys. Chem.*, **59,** 592 (1955).
[5]Davidson and Cullen, *Trans. Inst. Chem. Eng.*, **35,** 51 (1957).
[6]Dummer, Z. *Anorg. Chem.*, **109,** 31 (1949).
[7]Gerlach, *Ann. Phys. (Leipzig)*, **10,** 437 (1931).
[8]Gosting and Akeley, *J. Am. Chem. Soc.*, **74,** 2058 (1952).
[9]Hammond and Stokes, *Trans. Faraday Soc.*, **49,** 890 (1953); **49,** 886 (1953).
[10]Hammond and Stokes, *Trans. Faraday Soc.*, **52,** 781 (1956).
[11]*International Critical Tables*, vol. 5, p. 63.
[12]James, Hollingshead, and Gordon, *J. Chem. Phys.*, **7,** 89 (1939); **7,** 836 (1939).
[13]Jordon, Ackermann, and Berger, *J. Am. Chem. Soc.*, **78,** 2979 (1956).
[14]Koeller and Drickamer, *J. Chem. Phys.*, **21,** 575 (1953).
[15]Kolthoff and Miller, *J. Am. Chem. Soc.*, **63,** 1013 (1941).

THERMAL TRANSPORT PROPERTIES

TABLE 2-143 Transport Properties of Selected Gases at Atmospheric Pressure*

	Thermal conductivity, W/(m · K) Temperature, K					Viscosity, 10^{-4} Pa · s Temperature, K					Prandtl number, dimensionless Temperature, K			
Substance	250	300	400	500	600	250	300	400	500	600	250	300	400	500
Acetone	0.0080	0.0115	0.0201	0.0310			0.077	0.101	0.128	0.156				
Acetylene	0.0162	0.0213	0.0332	0.0452	0.0561		0.104	0.135	0.164					
Benzene	0.0077	0.0104	0.0195	0.0335	0.0524		0.076	0.101	0.127	0.154				
Bromine	0.0038	0.0048	0.0067					0.203	0.260	0.291				
CCl_4	0.0053	0.0067	0.0099	0.0126			0.101	0.131	0.162	0.191				
Chlorine	0.0071	0.0089	0.0124	0.0156	0.0190		0.136	0.178	0.218	0.259				
Deuterium	0.122	0.141	0.176			0.111	0.126	0.153	0.178	0.201				
Propylene	0.0114	0.0168	0.0226	0.0430	0.0580	0.073	0.087	0.115	0.141		0.860	0.797	0.762	
R 22	0.0080	0.0109	0.0170	0.0230	0.0290	0.109	0.129	0.168			0.820	0.771	0.760	
SO_2	0.0078	0.0096	0.0143	0.0200	0.0256		0.129	0.175	0.217	0.256				

*An approximate interpolation scheme is to plot the logarithm of the viscosity or the thermal conductivity versus the logarithm of the absolute temperature. At 250 K the viscosity of gaseous argon is to be read as 1.95×10^{-5} Pa · s = 0.0000195 N · s/m^2.

TABLE 2-144 Prandtl Number of Air*

	Pressure, bar											
Temperature, K	1	5	10	20	30	40	50	60	70	80	90	100
80	mix	2.31	2.32	2.35	2.37	2.40	2.42	2.45	2.48	2.51	2.54	2.57
90	0.796	1.76	1.77	1.78	1.79	1.81	1.82	1.83	1.85	1.87	1.89	1.91
100	0.786	0.872	1.54	1.53	1.53	1.53	1.53	1.53	1.53	1.54	1.54	1.55
120	0.773	0.813	0.89	1.44	1.65	1.54	1.48	1.43	1.40	1.38	1.36	1.34
140	0.763	0.782	0.82	0.94	1.20	1.59	2.14	2.43	2.07	1.78	1.62	1.52
160	0.754	0.765	0.78	0.84	0.92	1.03	1.13	1.25	1.37	1.65	1.83	1.72
180	0.745	0.754	0.763	0.792	0.830	0.876	0.932	1.00	1.07	1.14	1.20	1.25
200	0.738	0.743	0.749	0.766	0.788	0.812	0.841	0.87	0.90	0.95	0.97	1.00
240	0.724	0.727	0.729	0.737	0.746	0.756	0.767	0.78	0.80	0.81	0.81	0.82
280	0.710	0.711	0.713	0.717	0.721	0.726	0.731	0.737	0.742	0.75	0.75	0.76
300	0.705	0.707	0.708	0.712	0.715	0.717	0.721	0.725	0.728	0.732	0.737	0.742
350	0.699	0.699	0.699	0.701	0.703	0.705	0.707	0.709	0.711	0.712	0.714	0.716
400	0.694	0.694	0.694	0.695	0.696	0.697	0.698	0.699	0.700	0.701	0.703	0.704
450	0.691	0.691	0.691	0.691	0.692	0.692	0.693	0.693	0.694	0.695	0.695	0.696
500	0.689	0.689	0.689	0.689	0.689	0.690	0.690	0.690	0.690	0.691	0.691	0.691
600	0.690	0.690	0.690	0.689	0.689	0.689	0.689	0.689	0.689	0.690	0.690	0.690
700	0.696	0.696	0.695	0.695	0.695	0.695	0.695	0.695	0.695	0.695	0.695	0.695
800	0.705	0.704	0.704	0.704	0.704	0.703	0.703	0.703	0.703	0.702	0.702	0.702
900	0.709	0.709	0.708	0.708	0.708	0.708	0.708	0.708	0.708	0.708	0.708	0.708
1000	0.711	0.711	0.711	0.711	0.711	0.710	0.710	0.710	0.710	0.709	0.709	0.709

*Compiled by P. E. Liley from tables of specific heat at constant pressure, thermal conductivity, and viscosity given in SI units for integral kelvin temperatures and pressures in bars by Vasserman. *Thermophysical Properties of Air and Its Components* and *Thermophysical Properties of Liquid Air and Its Components.* Nauka, Moscow, and in translated form by the National Bureau of Standards, Washington. The number of significant figures given above reflects the similar numbers appearing for the constituent properties in the source references. While reasonable agreement occurs for atmospheric pressure with some other works, the fragmentary data available for the saturated, etc., states show large deviations.

TABLE 2-145 Vapor Thermal Conductivity of Inorganic and Organic Substances [W/(m·K)]

Eqn	Cmpd. no.	Name	Formula	CAS	Mol. wt.	C_1	C_2	C_3	C_4	T_{min}, K	Thermal cond. at T_{min}	T_{max}, K	Thermal cond. at T_{max}
102	1	Acetaldehyde	C_2H_4O	75-07-0	44.05256	1.0943E-07	2.0279			294.15	0.01110	1000	0.13269
102	2	Acetamide	C_2H_5NO	60-35-5	59.0672	0.00013195	0.97	728.3		494.3	0.02189	1000	0.06206
100	3	Acetic acid	$C_2H_4O_2$	64-19-7	60.052	2.4148	−0.020867	0.000059409	−5.4718E-08	391.05	0.06749	458.15	0.06259
100	3	Acetic acid	$C_2H_4O_2$	64-19-7	60.052	1.0879	−0.0038977	3.6227E-06		458.15	0.06258	541.5	0.03955
102	3	Acetic acid	$C_2H_4O_2$	64-19-7	60.052	3.3901E-06	1.9588	36053	14,086,000	541.5	0.03925	1000	0.11105
102	4	Acetic anhydride	$C_4H_6O_3$	108-24-7	102.08864	3.1289E-06	1.4618			412.7	0.02084	1000	0.07600
102	5	Acetone	C_3H_6O	67-64-1	58.07914	−26.8	0.9098	126,500,000		329.44	0.01363	1000	0.11362
102	6	Acetonitrile	C_2H_3N	75-05-8	41.0519	8.3653E-07	1.6481			339.09	0.01238	1000	0.07358
102	7	Acetylene	C_2H_2	74-86-2	26.03728	0.000075782	1.0327	−36.227	31,432	189.35	0.01011	1000	0.09545
102	8	Acrolein	C_3H_4O	107-02-8	56.06326	0.024098	0.3285	1325.3	577,830	325.84	0.01534	1000	0.08028
102	9	Acrylic acid	$C_3H_4O_2$	79-10-7	72.06266	0.0009265	0.7035	627.58	112,460	414.15	0.02027	1000	0.06867
102	10	Acrylonitrile	C_3H_3N	107-13-1	53.0626	−0.000861	0.77281	−2555.2		298.15	0.00929	1000	0.11525
102	11	Air	Mixture	132259-10-0	28.96	0.00031417	0.7786	−0.7116	2121.7	70	0.00603	2000	0.11675
102	12	Ammonia	H_3N	7664-41-7	17.03052	9.6608E-06	1.3799			200	0.01446	900	0.11523
102	13	Anisole	C_7H_8O	100-66-3	108.13782	0.00059858	0.7527	354.04	241,830	426.73	0.01809	1000	0.06796
102	14	Argon	Ar	7440-37-1	39.948	0.000633	0.6221	70		90	0.00585	3273.1	0.09525
102	15	Benzamide	C_7H_7NO	55-21-0	121.13658	0.025389	0.28547	1018.3	1,228,600	563.15	0.02317	1000	0.05618
102	16	Benzene	C_6H_6	71-43-2	78.11184	0.00001652	1.3117	491		339.15	0.01407	1000	0.09542
102	17	Benzenethiol	C_6H_6S	108-98-5	110.17684	0.00047951	0.7818	463.4	189,410	442.29	0.01861	1000	0.06427
102	18	Benzoic acid	$C_7H_6O_2$	65-85-0	122.12134	0.0001163	0.9705	740		522.4	0.02090	1000	0.05452
102	19	Benzonitrile	C_7H_5N	100-47-0	103.1213	1.3917E-06	1.5389			464.15	0.01767	1000	0.05758
102	20	Benzophenone	$C_{13}H_{10}O$	119-61-9	182.2179	0.0001235	0.9495	778.7		579.24	0.02213	1000	0.04899
102	21	Benzyl alcohol	C_7H_8O	100-51-6	108.13782	0.00023476	0.8639	187.8	193,840	478.6	0.02167	1000	0.06636
102	22	Benzyl ethyl ether	$C_9H_{12}O$	539-30-0	136.19098	0.00096451	0.69225	519.99	278,930	458.15	0.01936	1000	0.06398
102	23	Benzyl mercaptan	C_7H_8S	100-53-8	124.20342	0.00015525	0.9446	715.78		472.03	0.02071	1000	0.06171
102	24	Biphenyl	$C_{12}H_{10}$	92-52-4	154.2078	2.8646E-06	1.4098	−391.35	156,820	373.15	0.01123	1000	0.06347
102	25	Bromine	Br_2	7726-95-6	159.808	1.0404E-06	1.4685			300	0.00452	500	0.00956
102	26	Bromobenzene	C_6H_5Br	108-86-1	157.0079	0.00027085	0.7932	278.33	165,880	429.24	0.01302	1000	0.04495
102	27	Bromoethane	C_2H_5Br	74-96-4	108.965	0.00099879	0.71894	2358.4		311.49	0.00723	1000	0.04267
102	28	Bromomethane	CH_3Br	74-83-9	94.93852	5.7816E-07	1.6666			273	0.00664	1000	0.05779
102	29	1,2-Butadiene	C_4H_6	590-19-2	54.09044	0.000088221	1.0273	75.316	99,063	284	0.01172	1000	0.09071
102	30	1,3-Butadiene	C_4H_6	106-99-0	54.09044	−20890	0.9593	−93,820,000,000		268.74	0.01281	1000	0.16809
102	31	Butane	C_4H_{10}	106-97-8	58.1222	0.051094	0.45253	5455.5	1,979,800	272.65	0.01357	1000	0.13799
102	32	1,2-Butanediol	$C_4H_{10}O_2$	584-03-2	90.121	0.00014035	1.0032	711.66		469.57	0.02672	1000	0.08383
102	33	1,3-Butanediol	$C_4H_{10}O_2$	107-88-0	90.121	−918.39	−0.21199	334420	−2,884,200,000	481.38	0.02110	1000	0.08332
102	34	1-Butanol	$C_4H_{10}O$	71-36-3	74.1216	0.0011484	0.87647	3253.7		370.7	0.02097	712.94	0.06536
102	35	2-Butanol	$C_4H_{10}O$	78-92-2	74.1216	4.5894E-06	1.4484			372.9	0.02435	1000	0.10161
102	36	1-Butene	C_4H_8	106-98-9	56.10632	0.000096809	1.1153	781.82		266.91	0.01252	1000	0.12049
102	37	*cis*-2-Butene	C_4H_8	590-18-1	56.10632	0.000067737	1.0709	−65.881	129,390	273.15	0.01105	1273.15	0.13926
102	38	*trans*-2-Butene	C_4H_8	624-64-6	56.10632	0.000078576	1.0565	14.63	105,920	274.03	0.01200	1257	0.13704
102	39	Butyl acetate	$C_6H_{12}O_2$	123-86-4	116.15828	5.86E-09	2.376	−401.32	69,280	273	0.00783	800	0.07634
102	40	Butylbenzene	$C_{10}H_{14}$	104-51-8	134.21816	0.1807	0.0082225	−129.42	1,691,500	456.46	0.02151	1000	0.07465
102	41	Butyl mercaptan	$C_4H_{10}S$	109-79-5	90.1872	0.00097826	0.78643	1531.5	67,115	371.61	0.01832	1000	0.08610
102	42	*sec*-Butyl mercaptan	$C_4H_{10}S$	513-53-1	90.1872	0.9719	−0.111	1167.2	3,163,200	358.13	0.01749	1000	0.08470
102	43	1-Butyne	C_4H_6	107-00-6	54.09044	0.000037269	1.1427	−43.844	79,421	281.22	0.01268	1000	0.09644
102	44	Butyraldehyde	C_4H_8O	123-72-8	72.10572	9.9652E-07	1.6558			347.94	0.01610	1000	0.09245
100	45	Butyric acid	$C_4H_8O_2$	107-92-6	88.1051	0.7873	−0.0036161	5.6641E-06	−2.8451E-09	436.42	0.05147	706.95	0.05647
102	45	Butyric acid	$C_4H_8O_2$	107-92-6	88.1051	9.2069E-08	2.0312			706.95	0.05647	1000	0.11421
102	46	Butyronitrile	C_4H_7N	109-74-0	69.1051	1.3751E-06	1.5786			390.74	0.01698	1000	0.07484
102	47	Carbon dioxide	CO_2	124-38-9	44.0095	3.69	−0.3838	964	1,860,000	194.67	0.00887	1500	0.09025

(Continued)

TABLE 2-145 Vapor Thermal Conductivity of Inorganic and Organic Substances [W/(m·K)] (Continued)

Eqn	Cmpd. no.	Name	Formula	CAS	Mol. wt.	C_1	C_2	C_3	C_4	T_{min}, K	Thermal cond. at T_{min}	T_{max}, K	Thermal cond. at T_{max}
102	48	Carbon disulfide	CS_2	75-15-0	76.1407	0.0003467	0.7345	479		273.15	0.00776	1000	0.03745
102	49	Carbon monoxide	CO	630-08-0	28.0101	0.00059882	0.6863	57.13	501.92	70	0.00576	1500	0.08724
102	50	Carbon tetrachloride	CCl_4	56-23-5	153.8227	0.00016599	0.94375	1449.6		349.79	0.00812	1000	0.04595
102	51	Carbon tetrafluoride	CF_4	75-73-0	88.0043	0.000092004	1.0164	270.83		145.1	0.00505	1000	0.08108
102	52	Chlorine	Cl_2	7782-50-5	70.906	0.0009993	0.5472	458.6		200	0.00551	1000	0.03002
102	53	Chlorobenzene	C_6H_5Cl	108-90-7	112.5569	0.0004783	0.8994	1845.5	163,000	400	0.01579	1000	0.07935
102	54	Chloroethane	C_2H_5Cl	75-00-3	64.5141	4.91778E-07	1.70639	−232.008	46603.4	285.45	0.01004	1000	0.07943
102	55	Chloroform	$CHCl_3$	67-66-3	119.37764	0.00043073	0.83878	1874.5		334.33	0.00854	1000	0.04920
102	56	Chloromethane	CH_3Cl	74-87-3	50.4875	−3263.77	0.0675	−46,803,200	−25,000,700,000	248.95	0.00801	1000	0.07246
102	57	1-Chloropropane	C_3H_7Cl	540-54-5	78.54068	0.01652	0.44154	2444.42	793,392	319.67	0.01285	1000	0.08232
102	58	2-Chloropropane	C_3H_7Cl	75-29-6	78.54068	0.00009154	1.0681	746.6		308.85	0.01222	1000	0.08389
102	59	*m*-Cresol	C_7H_8O	108-39-4	108.13782	0.00019307	0.9248	710		475.43	0.02316	1000	0.06716
102	60	*o*-Cresol	C_7H_8O	95-48-7	108.13782	0.00018648	0.9302	709.37		464.15	0.02230	1000	0.06736
102	61	*p*-Cresol	C_7H_8O	106-44-5	108.13782	0.00019063	0.9282	716.91		475.13	0.02319	1000	0.06762
102	62	Cumene	C_9H_{12}	98-82-8	120.19158	1.6743E-07	1.8369	−449.46	112,760	380	0.01534	1000	0.08181
102	63	Cyanogen	C_2N_2	460-19-5	52.0348	0.000014433	1.2104			251.9	0.01164	1000	0.06174
102	64	Cyclobutane	C_4H_8	287-23-0	56.10632	−449910	0.27364	−10,001,000,000	−9.8654E+12	285.66	0.01356	1000	0.14994
102	65	Cyclohexane	C_6H_{12}	110-82-7	84.15948	0.000000859	1.7709	243		325	0.01380	1000	0.14198
102	66	Cyclohexanol	$C_6H_{12}O$	108-93-0	100.15888	0.0032207	0.5991	608.69	509,290	434	0.02399	1000	0.09535
102	67	Cyclohexanone	$C_6H_{10}O$	108-94-1	98.143	−1095.5	−0.023408	498, 780	−7,835,500,000	428.58	0.02291	1000	0.12704
102	68	Cyclohexene	C_6H_{10}	110-83-8	82.1436	0.0000901	1.0897	655		356.12	0.01914	1000	0.10116
102	69	Cyclopentane	C_5H_{10}	287-92-3	70.1329	9.5461E-06	1.4641	632.62		273	0.01061	1000	0.14429
102	70	Cyclopentene	C_5H_8	142-29-0	68.11702	0.0010949	0.71644	175.55	346,040	317.38	0.01360	1000	0.10148
102	71	Cyclopropane	C_3H_6	75-19-4	42.07974	−91.383	0.89718	−283,310,000		240.37	0.01061	1000	0.15854
102	72	Cyclohexyl mercaptan	$C_6H_{12}S$	1569-69-3	116.22448	0.0000813	1.0674	697.6		431.95	0.02022	1000	0.07629
102	73	Decanal	$C_{10}H_{20}O$	112-31-2	156.2652	1.9749E-06	1.5349			481.65	0.02590	1000	0.07948
102	74	Decane	$C_{10}H_{22}$	124-18-5	142.28168	−668.4	0.9323	−4,071,000,000		447.3	0.02173	1000	0.10286
102	75	Decanoic acid	$C_{10}H_{20}O_2$	334-48-5	172.265	3.3251E-09	2.4876	−124.9		543.15	0.02746	1000	0.11029
102	76	1-Decanol	$C_{10}H_{22}O$	112-30-1	158.28108	−0.3072	0.489	−67,500	−29,400,000	504	0.02590	1000	0.09389
102	77	1-Decene	$C_{10}H_{20}$	872-05-9	140.2658	0.000027232	1.257	751.7		443.75	0.02149	1000	0.09175
102	78	Decyl mercaptan	$C_{10}H_{22}S$	143-10-2	174.34668	0.00012058	1.0111	740		512.35	0.02709	1000	0.07482
102	79	1-Decyne	$C_{10}H_{18}$	764-93-2	138.24992	0.000016707	1.2128	−206.08	153,850	447.15	0.02092	1000	0.07667
102	80	Deuterium	D_2	7782-39-0	4.0316	0.00028527	0.9874	−200.51	21,807	233.15	0.11474	1500	0.44547
102	81	1,1-Dibromoethane	$C_2H_4Br_2$	557-91-5	187.86116	0.00021231	0.8052	649.51		381.15	0.00940	1000	0.03351
102	82	1,2-Dibromoethane	$C_2H_4Br_2$	106-93-4	187.86116	0.00015878	0.8636	659.5		404.51	0.01077	1000	0.03729
102	83	Dibromomethane	CH_2Br_2	74-95-3	173.83458	0.00021302	0.8719	1620		370.1	0.00687	1000	0.03356
102	84	Dibutyl ether	$C_8H_{18}O$	142-96-1	130.22792	0.0032694	0.58633	1259.9	300,890	323.15	0.01244	1000	0.07330
102	85	*m*-Dichlorobenzene	$C_6H_4Cl_2$	541-73-1	147.00196	−1067.8	0.754	−3,036,100,000		446.23	0.01561	1000	0.06430
102	86	*o*-Dichlorobenzene	$C_6H_4Cl_2$	95-50-1	147.00196	−1420	0.7614	−4,504,000,000		453.57	0.01507	1000	0.06066
102	87	*p*-Dichlorobenzene	$C_6H_4Cl_2$	106-46-7	147.00196	−1520.8	0.754	−433,2800,000		447.21	0.01564	1000	0.06417
102	88	1,1-Dichloroethane	$C_2H_4Cl_2$	75-34-3	98.95916	0.0001315	1.0113	1023.8		330.45	0.01132	1000	0.07025
102	89	1,2-Dichloroethane	$C_2H_4Cl_2$	107-06-2	98.95916	0.00021054	0.9574	1414		356.59	0.01177	1000	0.06498
102	90	Dichloromethane	CH_2Cl_2	75-09-2	84.93258	0.0014796	0.69531	2657.4		312.9	0.00847	1000	0.04931
102	91	1,1-Dichloropropane	$C_3H_6Cl_2$	78-99-9	112.98574	0.000057603	1.1148	849.98		361.25	0.01220	1000	0.06881
102	92	1,2-Dichloropropane	$C_3H_6Cl_2$	78-87-5	112.98574	0.000062435	1.103	913.43		369.52	0.01222	1000	0.06647
102	93	Diethanol amine	$C_4H_{11}NO_2$	111-42-2	105.13564	−11,633	0.4621	−3,793,900,000		541.54	0.03044	1000	0.07463
102	94	Diethyl amine	$C_4H_{11}N$	109-89-7	73.13684	0.00001706	1.248	−112.8	77,960	273.15	0.01148	1000	0.09804
102	95	Diethyl ether	$C_4H_{10}O$	60-29-7	74.1216	−0.0044894	0.6155	−3266.3		200	0.00764	600	0.05181
102	96	Diethyl sulfide	$C_4H_{10}S$	352-93-2	90.1872	0.0018097	0.67406	1179.7	174,850	365.25	0.01743	1000	0.08089
102	97	1,1-Difluoroethane	$C_2H_4F_2$	75-37-6	66.04997	0.000059249	1.0713	101.84	45,974	248.95	0.01016	1000	0.08447

102	98	1,2-Difluoroethane	$C_2H_4F_2$	624-72-6	66.04997	2.4194E-06	1.4456			303.65	0.00938	993.65	0.05206
102	99	Difluoromethane	CH_2F_2	75-10-5	52.02339	0.000013015	1.1897			221.5	0.00803	1000	0.04826
102	100	Diisopropyl amine	$C_6H_{15}N$	108-18-9	101.19	0.00051305	0.8076	360.19	154,510	357.05	0.01836	1000	0.08967
102	101	Diisopropyl ether	$C_6H_{14}O$	108-20-3	102.17476	0.00019879	0.9423	306.8	106,230	328.05	0.01598	1000	0.09444
102	102	Diisopropyl ketone	$C_7H_{14}O$	565-80-0	114.18546	−8.5357	−0.0056423	1882.1	−65,622,000	397.55	0.02015	1000	0.13085
102	103	1,1-Dimethoxyethane	$C_4H_{10}O_2$	534-15-6	90.121	0.00046265	0.81968	539.34	104,530	337.45	0.01554	1000	0.08099
102	104	1,2-Dimethoxypropane	$C_5H_{12}O_2$	7778-85-0	104.14758	3.7962E-06	1.4462			366.15	0.01936	1000	0.08279
102	105	Dimethyl acetylene	C_4H_6	503-17-3	54.09044	0.00021761	0.9187	217	132,070	300.13	0.01288	1000	0.09199
102	106	Dimethyl amine	C_2H_7N	124-40-3	45.08368	1.6085	−0.1103	2160.3	2,989,300	280.03	0.01845	1000	0.12209
102	107	2,3-Dimethylbutane	C_6H_{14}	79-29-8	86.17536	0.000034741	1.1646	−99.956	130,820	331.13	0.01581	1000	0.10506
102	108	1,1-Dimethylcyclohexane	C_8H_{16}	590-66-9	112.21264	0.008856	0.4215	−50.645	764,580	392.7	0.01884	1000	0.09500
102	109	*cis*-1,2-Dimethylcyclohexane	C_8H_{16}	2207-01-4	112.21264	0.013298	0.3692	0.1027	852,540	402.94	0.01948	1000	0.09196
102	110	*trans*-1,2-Dimethylcyclohexane	C_8H_{16}	6876-23-9	112.21264	0.012144	0.3854	52.191	803,590	396.58	0.01952	1000	0.09376
102	111	Dimethyl disulfide	$C_2H_6S_2$	624-92-0	94.19904	0.00022578	0.892	697		382.9	0.01613	1000	0.06310
102	112	Dimethyl ether	C_2H_6O	115-10-6	46.06844	0.059975	0.2667	1018.6	1,098,800	248.31	0.01139	1500	0.19458
102	113	*N,N*-Dimethyl formamide	C_3H_7NO	68-12-2	73.09378	0.014449	0.3612	595.22	728,130	425.15	0.02001	1000	0.07539
102	114	2,3-Dimethylpentane	C_7H_{16}	565-59-3	100.20194	0.000022421	1.2137	−146.91	131,830	362.93	0.01797	1000	0.09962
102	115	Dimethyl phthalate	$C_{10}H_{10}O_4$	131-11-3	194.184	0.00012822	0.9324	752.5		556.85	0.01981	1000	0.04587
102	116	Dimethylsilane	C_2H_8Si	1111-74-6	60.17042	0.0011808	0.742	1131	6400	253.55	0.01291	1000	0.09296
102	117	Dimethyl sulfide	C_2H_6S	75-18-3	62.134	0.00023614	0.9204	638		310.48	0.01520	1000	0.08319
102	118	Dimethyl sulfoxide	C_2H_6OS	67-68-5	78.13344	0.00064761	0.7716	1013.3	82,563	462.15	0.02059	1000	0.06379
102	119	Dimethyl terephthalate	$C_{10}H_{10}O_4$	120-61-6	194.184	0.00402358	0.57548	3598.32		559.2	0.02063	1000	0.04661
102	120	1,4-Dioxane	$C_4H_8O_2$	123-91-1	88.10512	6.4032E-07	1.7194			337.85	0.01427	768.01	0.05855
102	121	Diphenyl ether	$C_{12}H_{10}O$	101-84-8	170.2072	0.00014629	0.9377	745.89		531.46	0.02188	1000	0.05449
102	122	Dipropyl amine	$C_6H_{15}N$	142-84-7	101.19	0.0001123	0.9958	183.2	98,000	279.65	0.01055	1000	0.08515
102	123	Dodecane	$C_{12}H_{26}$	112-40-3	170.33484	0.000005719	1.4699	579.4		489.47	0.02354	1000	0.09301
102	124	Eicosane	$C_{20}H_{42}$	112-95-8	282.54748	−375.32	1.0708	−8,783,600,000		616.93	0.02563	1000	0.06968
102	125	Ethane	C_2H_6	74-84-0	30.069	0.000073869	1.1689	500.73		184.55	0.00886	1000	0.15807
102	126	Ethanol	C_2H_6O	64-17-5	46.06844	−0.010109	0.6475	−7332	−268,000	293.15	0.01475	1000	0.13417
102	127	Ethyl acetate	$C_4H_8O_2$	141-78-6	88.10512	1.3575E-07	1.9681			273.15	0.00847	990.21	0.10681
102	128	Ethyl amine	C_2H_7N	75-04-7	45.08368	0.3935	0.0131	1380	1,710,000	289.73	0.01622	1000	0.10532
102	129	Ethylbenzene	C_8H_{10}	100-41-4	106.165	0.000017537	1.3144	560.65		409.35	0.02007	1000	0.09859
102	130	Ethyl benzoate	$C_9H_{10}O_2$	93-89-0	150.1745	0.00002012	1.1513	−89.583	125,410	486.55	0.01855	1000	0.05524
102	131	2-Ethyl butanoic acid	$C_6H_{12}O_2$	88-09-5	116.15828	0.00017727	0.9428	712.4		466.95	0.02306	1000	0.06973
102	132	Ethyl butyrate	$C_6H_{12}O_2$	105-54-4	116.15828	829.29	1.0156	8,955,300,000		394.65	0.01583	1000	0.10314
102	133	Ethylcyclohexane	C_8H_{16}	1678-91-7	112.21264	0.0000748	1.1103	686		404.95	0.02180	1000	0.09505
102	134	Ethylcyclopentane	C_7H_{14}	1640-89-7	98.18606	0.0043244	0.5429	333.67	570,470	376.62	0.01832	1000	0.09659
102	135	Ethylene	C_2H_4	74-85-1	28.05316	8.6806E-06	1.4559	299.72	−29,403	170	0.00879	590.92	0.06613
102	136	Ethylenediamine	$C_2H_8N_2$	107-15-3	60.09832	0.1655	0.1798	3827.9	1,600,000	390.41	0.02272	1000	0.08915
102	137	Ethylene glycol	$C_2H_6O_2$	107-21-1	62.06784	−8145800	−0.30502	1,832,500,000	−1.1842E+13	470.45	0.02513	1000	0.09896
102	138	Ethyleneimine	C_2H_5N	151-56-4	43.0678	0.00077079	0.7713	446.16	197,930	329	0.01610	1000	0.09659
102	139	Ethylene oxide	C_2H_4O	75-21-8	44.05256	−0.0003788	1.115	−5641		273.15	0.01004	1000	0.18063
102	140	Ethyl formate	$C_3H_6O_2$	109-94-4	74.07854	508	0.9023	2,170,000,000		327.46	0.01426	1000	0.11921
102	141	2-Ethyl hexanoic acid	$C_8H_{16}O_2$	149-57-5	144.211	2.5804E-06	1.4669			500.66	0.02353	1000	0.06492
102	142	Ethylhexyl ether	$C_8H_{18}O$	5756-43-4	130.22792	0.0052833	0.52982	1415.7	378,180	417.15	0.01967	1000	0.07348
102	143	Ethylisopropyl ether	$C_5H_{12}O$	625-54-7	88.14818	0.00021652	0.94192	632.16		326.15	0.01717	1000	0.08882
102	144	Ethylisopropyl ketone	$C_6H_{12}O$	565-69-5	100.15888	−152400	−0.049106	80,955,000	−9.3122E+11	386.55	0.01889	1000	0.12768
102	145	Ethyl mercaptan	C_2H_6S	75-08-1	62.13404	0.0015251	0.70243	1347.5	35,085	308.15	0.01487	1000	0.08195
102	146	Ethyl propionate	$C_5H_{10}O_2$	105-37-3	102.1317	1.0507E-07	1.9854			400	0.01540	1000	0.09499
102	147	Ethylpropyl ether	$C_5H_{12}O$	628-32-0	88.14818	5.8174E-08	2.0116	−372.68	57,690	273.15	0.01133	550	0.03690
102	148	Ethyltrichlorosilane	$C_2H_5Cl_3Si$	115-21-9	163.506	2.7142E-06	1.4281			371.05	0.01268	1000	0.05223
102	149	Fluorine	F_2	7782-41-4	37.9968064	0.00012144	0.93831			70	0.00654	700	0.05675
102	150	Fluorobenzene	C_6H_5F	462-06-6	96.1023032	0.000053432	1.1576	760.75		357.88	0.01546	600	0.03874

(*Continued*)

TABLE 2-145 Vapor Thermal Conductivity of Inorganic and Organic Substances [W/(m·K)] (*Continued*)

Eqn	Cmpd. no.	Name	Formula	CAS	Mol. wt.	C_1	C_2	C_3	C_4	T_{min}, K	Thermal cond. at T_{min}	T_{max}, K	Thermal cond. at T_{max}
102	151	Fluoroethane	C_2H_5F	353-36-6	48.0595	6.3522E-06	1.346			235.45	0.00990	1000	0.06933
102	152	Fluoromethane	CH_3F	593-53-3	34.03292	0.000048998	1.0175			194.82	0.01047	1000	0.05529
102	153	Formaldehyde	CH_2O	50-00-0	30.02598	5.2201E-06	1.417			253.85	0.01333	1000	0.09304
102	154	Formamide	CH_3NO	75-12-7	45.04062	0.00025893	0.9083	723.6		493	0.02930	1000	0.07973
100	155	Formic acid	CH_2O_2	64-18-6	46.0257	−0.8303	0.0046141	−5.7466E-06		420	0.09392	470	0.06890
100	155	Formic acid	CH_2O_2	64-18-6	46.0257	1.8897	−0.006901	6.4407E-06		470	0.06898	537.9	0.04118
102	155	Formic acid	CH_2O_2	64-18-6	46.0257	0.00072291	1.8898	4,877,600	−1,889,300,000	537.9	0.04120	1000	0.11296
102	156	Furan	C_4H_4O	110-00-9	68.07396	−644950	0.2862	−16,794,000,000	−1.7372E+13	304.5	0.01367	1000	0.13631
102	157	Helium-4	He	7440-59-7	4.0026	0.00226	0.7305	−18.63	440	30	0.03124	2000	0.58820
102	158	Heptadecane	$C_{17}H_{36}$	629-78-7	240.46774	−114.41	1.0566	−2,211,400,000		575.3	0.02454	1000	0.07649
102	159	Heptanal	$C_7H_{14}O$	111-71-7	114.18546	1.4326E-06	1.5896			426.15	0.02168	1000	0.08413
102	160	Heptane	C_7H_{16}	142-82-5	100.20194	−0.070028	0.38068	−7049.9	−2,400,500	339.15	0.01583	1000	0.11493
100	161	Heptanoic acid	$C_7H_{14}O_2$	111-14-8	130.185	−0.088162	0.00065022	−1.2803E-06	9.1349E-10	496.15	0.03085	643.11	0.04346
102	161	Heptanoic acid	$C_7H_{14}O_2$	111-14-8	130.185	4.449E-08	2.133			643.11	0.04349	1000	0.11150
102	162	1-Heptanol	$C_7H_{16}O$	111-70-6	116.20134	−0.061993	0.2792	−3336	−1,642,000	449.45	0.02345	1000	0.10722
102	163	2-Heptanol	$C_7H_{16}O$	543-49-7	116.20134	0.00018818	0.96338	696.02		432.9	0.02501	1000	0.08616
102	164	3-Heptanone	$C_7H_{14}O$	106-35-4	114.18546	1348.6	1.0313	14,832,000,000		420.55	0.01943	1000	0.11287
102	165	2-Heptanone	$C_7H_{14}O$	110-43-0	114.18546	2049.3	1.0323	22,983,000,000		424.18	0.01951	1000	0.11145
102	166	1-Heptene	C_7H_{14}	592-76-7	98.18606	0.00002133	1.2885	487.8		366.79	0.01845	1000	0.10518
102	167	Heptyl mercaptan	$C_7H_{16}S$	1639-09-4	132.26694	0.0083145	0.51862	2253	532,590	450.09	0.02289	1000	0.07899
102	168	1-Heptyne	C_7H_{12}	628-71-7	96.17018	0.000060732	1.0586	−102.79	143,140	372.93	0.01827	1000	0.08751
102	169	Hexadecane	$C_{16}H_{34}$	544-76-3	226.44116	0.000004438	1.4949	682		560.01	0.02568	1000	0.08055
102	170	Hexanal	$C_6H_{12}O$	66-25-1	100.15888	1.5427E-06	1.5824			401.15	0.02031	1000	0.08620
102	171	Hexane	C_6H_{14}	110-54-3	86.17536	−650.5	0.8053	−1,412,100,000		339.09	0.01704	1000	0.12003
102	172	Hexanoic acid	$C_6H_{12}O_2$	142-62-1	116.158	12,049,00,000	−4.0059	−1668.8	722,550	478.85	0.03317	641.42	0.04435
102	172	Hexanoic acid	$C_6H_{12}O_2$	142-62-1	116.158	6.1268E-08	2.0874			641.42	0.04435	1000	0.11206
102	173	1-Hexanol	$C_6H_{14}O$	111-27-3	102.17476	−4935500	−0.1653	1,563,100,000	−1.5752E+13	429.9	0.02220	1000	0.11104
102	174	2-Hexanol	$C_6H_{14}O$	626-93-7	102.175	0.00018361	0.97199	677.05		412.4	0.02421	1000	0.09022
102	175	2-Hexanone	$C_6H_{12}O$	591-78-6	100.15888	−1.2158	0.026637	−1711.6	−13,176,000	273	0.00775	1000	0.10523
102	176	3-Hexanone	$C_6H_{12}O$	589-38-8	100.15888	−0.33262	0.12054	−2472.6	−5,493,400	273	0.00800	1000	0.10980
102	177	1-Hexene	C_6H_{12}	592-41-6	84.15948	0.000064256	1.1355	445.15	64,810	336.63	0.01644	1000	0.10850
102	178	3-Hexyne	C_6H_{10}	928-49-4	82.1436	6.9682E-06	1.347	−214.35	110,480	354.35	0.01485	1000	0.08546
102	179	Hexyl mercaptan	$C_6H_{14}S$	111-31-9	118.24036	0.074318	0.30035	4470.1	1,775,800	425.81	0.02151	1000	0.08167
102	180	1-Hexyne	C_6H_{10}	693-02-7	82.1436	0.000058116	1.0724	−77.165	123,900	344.48	0.01679	1000	0.09155
102	181	2-Hexyne	C_6H_{10}	764-35-2	82.1436	0.000011631	1.2753	−202.84	122,990	357.67	0.01506	1000	0.08466
102	182	Hydrazine	H_4N_2	302-01-2	32.04516	0.00043196	0.86603	641.48		386.65	0.02828	1000	0.10430
102	183	Hydrogen	H_2	1333-74-0	2.01588	0.002653	0.7452	12		22	0.01718	1600	0.64299
102	184	Hydrogen bromide	BrH	10035-10-6	80.91194	0.00049725	0.63088	331.62		206.45	0.00551	600	0.01812
102	185	Hydrogen chloride	ClH	7647-01-0	36.46094	0.001865	0.49755	358		190	0.00880	700	0.03213
102	186	Hydrogen cyanide	CHN	74-90-8	27.02534	4.6496E-06	1.3669	−210.76	58,295	273.15	0.00985	673.15	0.04185
102	187	Hydrogen fluoride	FH	7664-39-3	20.0063432	0.000034629	1.1224	18.744		350	0.02356	450	0.03160
102	188	Hydrogen sulfide	H_2S	7783-06-4	34.08088	1.381E-07	1.8379	−352.09	46,041	212.8	0.00724	600	0.03258
102	189	Isobutyric acid	$C_4H_8O_2$	79-31-2	88.10512	0.000214	0.9248	698		427.85	0.02206	1000	0.07497
102	190	Isopropyl amine	C_3H_9N	75-31-0	59.11026	0.00028183	0.92094	619.17		304.92	0.01804	1000	0.10081
102	191	Malonic acid	$C_3H_4O_4$	141-82-2	104.06146	4.8284E-06	1.3599			580	0.02766	1000	0.05801
102	192	Methacrylic acid	$C_4H_6O_2$	79-41-4	86.08924	0.00019847	0.9284	678.69		434.15	0.02176	1000	0.07210
102	193	Methane	CH_4	74-82-8	16.0425	8.3983E-06	1.4268	−49.654		111.63	0.01263	600	0.08425
102	194	Methanol	CH_4O	67-56-1	32.04186	5.7992E-07	1.7862			273	0.01303	684.37	0.06726
102	195	*N*-Methyl acetamide	C_3H_7NO	79-16-3	73.09378	0.034177	0.3312	2070	1,195,600	478.15	0.02498	1000	0.07895
102	196	Methyl acetate	$C_3H_6O_2$	79-20-9	74.07854	−25343	−0.1934	11,164,000	−67,259,000,000	330.09	0.01415	1000	0.11878

102	197	Methyl acetylene	C_3H_4	74-99-7	40.06386	0.00026544	0.8921	222.19	79,869	249.94	0.01154	1000	0.09675
102	198	Methyl acrylate	$C_4H_6O_2$	96-33-3	86.08924	0.4734	−0.1111	533.57	1,649,600	353.35	0.01569	1000	0.06904
102	199	Methyl amine	CH_5N	74-89-5	31.0571	−55.13	1.065	−448,200,000		266.82	0.01259	650	0.07917
102	200	Methyl benzoate	$C_8H_8O_2$	93-58-3	136.14792	0.000023963	1.1308	−67.272	125,720	472.65	0.01784	1000	0.05588
102	201	3-Methyl-1,2-butadiene	C_5H_8	598-25-4	68.11702	0.0002509	0.899	253.4	149,500	314	0.01326	1000	0.08902
102	202	2-Methylbutane	C_5H_{12}	78-78-4	72.14878	0.0008968	0.7742	456	230,640	273.15	0.01198	1000	0.11176
102	203	2-Methylbutanoic acid	$C_5H_{10}O_2$	116-53-0	102.1317	0.0001799	0.9457	704.6		450.15	0.02266	1000	0.07253
102	204	3-Methyl-1-butanol	$C_5H_{12}O$	123-51-3	88.1482	2054.5	0.90109	8,760,500,000		404.15	0.02116	1000	0.11843
102	205	2-Methyl-1-butene	C_5H_{10}	563-46-2	70.1329	0.00019098	0.9341	84.07	155,720	304.3	0.01348	1000	0.09771
102	206	2-Methyl-2-butene	C_5H_{10}	513-35-9	70.1329	0.00021736	0.9171	112.3	177,690	311.71	0.01320	1000	0.09504
102	207	2-Methyl -1-butene-3-yne	C_5H_6	78-80-8	66.10114	0.00015498	0.9364	15.366	137,400	305.4	0.01304	1000	0.08664
102	208	Methylbutyl ether	$C_5H_{12}O$	628-28-4	88.14818	0.000023993	1.1976	58.59	35,667	273.15	0.01173	1000	0.08586
102	209	Methylbutyl sulfide	$C_5H_{12}S$	628-29-5	104.214	0.079414	0.23442	2671.9	1,366,100	396.58	0.01966	1000	0.07960
102	210	3-Methyl-1-butyne	C_5H_8	598-23-2	68.11702	0.000065855	1.072	−36.369	106,430	302.15	0.01468	1000	0.10120
102	211	Methyl butyrate	$C_5H_{10}O_2$	623-42-7	102.1317	1333.1	0.9962	12,317,000,000		375.9	0.01495	1000	0.10543
102	212	Methylchlorosilane	CH_5ClSi	993-00-0	80.5889	0.00037057	0.81367	609.17		281.85	0.01155	1000	0.06357
102	213	Methylcyclohexane	C_7H_{14}	108-87-2	98.18606	0.0000719	1.1274	667		374.08	0.02056	1000	0.10399
102	214	1-Methylcyclohexanol	$C_7H_{14}O$	590-67-0	114.18546	0.00011359	1.0311	709.27		441.15	0.02322	1000	0.08238
102	215	*cis*-2-Methylcyclohexanol	$C_7H_{14}O$	7443-70-1	114.18546	0.069565	0.1633	208.7	1,209,500	438.15	0.02415	1000	0.08888
102	216	*trans*-2-Methylcyclohexanol	$C_7H_{14}O$	7443-52-9	114.18546	0.075448	0.155	218.44	1,252,500	440.15	0.02435	1000	0.08908
102	217	Methylcyclopentane	C_6H_{12}	96-37-7	84.15948	0.0024385	0.61774	223.01	477,570	344.96	0.01592	1000	0.10227
102	218	1-Methylcyclopentene	C_6H_{10}	693-89-0	82.1436	0.0040082	0.54462	242.12	559,040	348.64	0.01544	1000	0.09578
102	219	3-Methylcyclopentene	C_6H_{10}	1120-62-3	82.1436	0.0019845	0.6393	227.11	434,120	338.05	0.01501	1000	0.09888
102	220	Methyldichlorosilane	CH_4Cl_2Si	75-54-7	115.03396	0.00041077	0.75688	591.5		314.7	0.01109	1000	0.04813
102	221	Methylethyl ether	C_3H_8O	540-67-0	60.09502	0.00024036	0.93177	588.14		273	0.01419	1000	0.09447
102	222	Methylethyl ketone	C_4H_8O	78-93-3	72.10572	−4202700	−0.1524	2,084,600,000	−1.4577E+13	352.79	0.01546	1000	0.11740
102	223	Methylethyl sulfide	C_3H_8S	624-89-5	76.1606	0.0034805	0.61906	1810.8	166,290	339.8	0.01653	1000	0.08415
102	224	Methyl formate	$C_2H_4O_2$	107-31-3	60.05196	−800040	−0.2285	248,100,000	−1.5034E+12	300	0.01369	1000	0.13148
102	225	Methylisobutyl ether	$C_5H_{12}O$	625-44-5	88.14818	0.00020053	0.95381	644.42		331.7	0.01729	1000	0.08863
102	226	Methylisobutyl ketone	$C_6H_{12}O$	108-10-1	100.15888	−2483300	−0.046517	1,313,100,000	−1.5798E+13	389.65	0.01869	1000	0.12433
102	227	Methyl Isocyanate	C_2H_3NO	624-83-9	57.05132	0.0026136	0.62	1631.7	126,720	312	0.01221	1000	0.06864
102	228	Methylisopropyl ether	$C_4H_{10}O$	598-53-8	74.1216	2.1191	−0.19015	1453.4	3,575,500	303.92	0.01606	1000	0.09451
102	229	Methylisopropyl ketone	$C_5H_{10}O$	563-80-4	86.1323	−5935000	−0.089497	3,098,800,000	−2.7994E+13	367.55	0.01760	1000	0.12847
102	230	Methylisopropyl sulfide	$C_4H_{10}S$	1551-21-9	90.1872	0.0071536	0.53907	2700.7	241,730	171.64	0.00459	1000	0.07516
102	231	Methyl mercaptan	CH_4S	74-93-1	48.10746	0.00002653	1.1631	29.996	32,519	273.15	0.01171	1000	0.07704
102	232	Methyl methacrylate	$C_5H_8O_2$	80-62-6	100.11582	0.00072502	0.7395	365.68	204,360	373.45	0.01680	1000	0.07637
102	233	2-Methyloctanoic acid	$C_9H_{18}O_2$	3004-93-1	158.23802	0.0001813	0.92912	793.45		518.15	0.02383	1000	0.06195
102	234	2-Methylpentane	C_6H_{14}	107-83-5	86.17536	0.000061119	1.0861	−59.592	141,260	333.41	0.01606	1000	0.10242
102	235	Methyl pentyl ether	$C_6H_{14}O$	628-80-8	102.17476	0.93312	−0.1172	1154.3	2,961,700	372	0.01828	1000	0.08117
102	236	2-Methylpropane	C_4H_{10}	75-28-5	58.1222	0.089772	0.18501	639.23	1,114,700	261.43	0.01273	1000	0.11701
102	237	2-Methyl-2-propanol	$C_4H_{10}O$	75-65-0	74.1216	1.1776E-06	1.6618			333.82	0.01839	766.87	0.07325
102	238	2-Methyl propene	C_4H_8	115-11-7	56.10632	−488.1	0.8877	−1,448,500,000		266.25	0.01276	1000	0.15513
102	239	Methyl propionate	$C_4H_8O_2$	554-12-1	88.10512	−200.9	−0.1321	104,000	−846,000,000	350	0.01402	1000	0.10886
102	240	Methylpropyl ether	$C_4H_{10}O$	557-17-5	74.1216	0.011136	0.4831	21,70.3	281,220	312.2	0.01648	1000	0.09079
102	241	Methylpropyl sulfide	$C_4H_{10}S$	3877-15-4	90.1872	0.0023574	0.67434	1804.1	155,660	368.69	0.01802	1000	0.08398
102	242	Methylsilane	CH_6Si	992-94-9	46.14384	12.248	−0.5611	−1067	2,715,200	216.25	0.01108	1000	0.09590
102	243	alpha-Methyl styrene	C_9H_{10}	98-83-9	118.1757	0.21276	−0.022299	−194.68	1,708,700	438.65	0.01969	1000	0.07255
102	244	Methyl tert-butyl ether	$C_5H_{12}O$	1634-04-4	88.1482	0.0002084	0.93034	364.832	73,041	328.2	0.01638	1000	0.08958
102	245	Methyl vinyl ether	C_3H_6O	107-25-5	58.07914	0.00032359	0.8892	623.22		278.65	0.01493	1000	0.09273
102	246	Naphthalene	$C_{10}H_8$	91-20-3	128.17052	0.000091828	1.0345	731.78		491.14	0.02243	1000	0.06730
102	247	Neon	Ne	7440-01-9	20.1797	0.0011385	0.6646	8.7		30	0.00846	3273.1	0.24616
102	248	Nitroethane	$C_2H_5NO_2$	79-24-3	75.0666	0.0011282	0.6895	679.11	238,800	387.22	0.01580	1000	0.06887
102	249	Nitrogen	N_2	7727-37-9	28.0134	0.00033143	0.7722	16.323	373.72	63.15	0.00602	2000	0.11638

(*Continued*)

TABLE 2-145 Vapor Thermal Conductivity of Inorganic and Organic Substances [W/(m·K)] *(Continued)*

Eqn	Cmpd. no.	Name	Formula	CAS	Mol. wt.	C_1	C_2	C_3	C_4	T_{min}, K	Thermal cond. at T_{min}	T_{max}, K	Thermal cond. at T_{max}
102	250	Nitrogen trifluoride	F_3N	7783-54-2	71.00191	2.1443	−0.30545	1860.3	1,216,700	144.09	0.00648	1000	0.06377
102	251	Nitromethane	CH_3NO_2	75-52-5	61.04002	0.00003135	1.1119	−91.6	128,000	374.35	0.01365	1000	0.06553
102	252	Nitrous oxide	N_2O	10024-97-2	44.0128	0.001096	0.667	540		182.3	0.00891	1000	0.07133
102	253	Nitric oxide	NO	10102-43-9	30.0061	0.0004096	0.7509	45.6		121.38	0.01094	750	0.05567
102	254	Nonadecane	$C_{19}H_{40}$	629-92-5	268.5209	0.000049571	1.2652	3332.3		603.05	0.02502	1000	0.07147
102	255	Nonanal	$C_9H_{18}O$	124-19-6	142.23862	0.00000175	1.5534			465.52	0.02440	1000	0.08003
102	256	Nonane	C_9H_{20}	111-84-2	128.2551	−0.065771	0.27198	−3482.3	−1,580,300	423.97	0.02130	1000	0.10597
102	257	Nonanoic acid	$C_9H_{18}O_2$	112-05-0	158.238	46.08	−1.0037	−2460.2	1,867,000	528.75	0.02815	1000	0.11042
102	258	1-Nonanol	$C_9H_{20}O$	143-08-8	144.2545	−30.715	−0.1075	8107	−156,830,000	485.2	0.02436	1000	0.09895
102	259	2-Nonanol	$C_9H_{20}O$	628-99-9	144.255	0.00016806	0.96876	713.67		471.7	0.02603	1000	0.07904
102	260	1-Nonene	C_9H_{18}	124-11-8	126.23922	0.000021269	1.2943	662.21		420.02	0.02051	1000	0.09772
102	261	Nonyl mercaptan	$C_9H_{20}S$	1455-21-6	160.3201	0.047041	0.29733	2460.6	1,367,200	492.95	0.02559	1000	0.07598
102	262	1-Nonyne	C_9H_{16}	3452-09-3	124.22334	0.000016681	1.218	−199.41	144,580	423.85	0.01981	1000	0.07956
102	263	Octadecane	$C_{18}H_{38}$	593-45-3	254.49432	−291.08	1.0615	−6,019,900,000		589.86	0.02491	1000	0.07395
102	264	Octanal	$C_8H_{16}O$	124-13-0	128.212	0.00000166	1.5669			445.15	0.02345	1000	0.08333
102	265	Octane	C_8H_{18}	111-65-9	114.22852	−8758	0.8448	−27,121,000,000		339	0.01503	1000	0.11053
100	266	Octanoic acid	$C_8H_{16}O_2$	124-07-2	144.211	−0.20973	0.0012201	−2.1843E-06	1.3942E-09	512.85	0.02955	637.35	0.04157
102	266	Octanoic acid	$C_8H_{16}O_2$	124-07-2	144.211	3.2003E-08	2.18			637.35	0.04157	1000	0.11097
102	267	1-Octanol	$C_8H_{18}O$	111-87-5	130.22792	−0.0030238	0.8745	−13352		468.35	0.02380	1000	0.10288
102	268	2-Octanol	$C_8H_{18}O$	123-96-6	130.228	0.00016915	0.97238	698.55		452.9	0.02545	1000	0.08229
102	269	2-Octanone	$C_8H_{16}O$	111-13-7	128.21204	−0.0020184	1.0027	−20406		446.15	0.02046	1000	0.10597
102	270	3-Octanone	$C_8H_{16}O$	106-68-3	128.21204	8.1833E-08	2.0418			440.65	0.02050	1000	0.10923
102	271	1-Octene	C_8H_{16}	111-66-0	112.21264	0.0000133	1.3554	504.59		394.41	0.01926	1000	0.10295
102	272	Octyl mercaptan	$C_8H_{18}S$	111-88-6	146.29352	−3965.5	0.5213	−1,851,900,000		472.19	0.02505	1000	0.07845
102	273	1-Octyne	C_8H_{14}	629-05-0	110.19676	0.000060734	1.0516	−124.91	158,300	399.35	0.01967	1000	0.08394
102	274	Oxalic acid	$C_2H_2O_4$	144-62-7	90.03488	2.7969E-06	1.3164			516	0.01041	1000	0.02488
102	275	Oxygen	O_2	7782-44-7	31.9988	0.00044994	0.7456	56.699		80	0.00691	2000	0.12655
102	276	Ozone	O_3	10028-15-6	47.9982	0.0043147	0.47999	700.09		161.85	0.00931	1000	0.06990
102	277	Pentadecane	$C_{15}H_{32}$	629-62-9	212.41458	4.7796E-06	1.4851	643.13		543.84	0.02529	1000	0.08299
102	278	Pentanal	$C_5H_{10}O$	110-62-3	86.1323	0.00000113	1.6323			375.15	0.01799	1000	0.08912
102	279	Pentane	C_5H_{12}	109-66-0	72.14878	−684.4	0.764	−1,055,000,000		273.15	0.01288	1000	0.12707
100	280	Pentanoic acid	$C_5H_{10}O_2$	109-52-4	102.132	0.44736	−0.0019667	2.9973E-06	−1.4141E-09	458.95	0.03938	706.95	0.05536
102	280	Pentanoic acid	$C_5H_{10}O_2$	109-52-4	102.132	7.5284E-08	2.0589			706.95	0.05537	1000	0.11308
102	281	1-Pentanol	$C_5H_{12}O$	71-41-0	88.1482	2896	0.8985	12,735,000,000		410.9	0.02084	990.95	0.11087
102	282	2-Pentanol	$C_5H_{12}O$	6032-29-7	88.1482	0.00019575	0.9692	664.04		392.2	0.02372	1000	0.09509
102	283	2-Pentanone	$C_5H_{10}O$	107-87-9	86.1323	−0.01719	0.4832	−3798	−1,235,000	273	0.00877	1000	0.12002
102	284	3-Pentanone	$C_5H_{10}O$	96-22-0	86.1323	22.775	1.0019	191,000,000		273	0.00898	1000	0.12082
102	285	1-Pentene	C_5H_{10}	109-67-1	70.1329	2.7081E-06	1.5493	41.075	8301.3	303.22	0.01546	1000	0.11472
102	286	2-Pentyl mercaptan	$C_5H_{12}S$	2084-19-7	104.21378	0.00022307	0.93358	794.16		385.15	0.01890	1000	0.07858
102	287	Pentyl mercaptan	$C_5H_{12}S$	110-66-7	104.21378	0.00011261	1.034	693.05		399.79	0.02019	1000	0.08412
102	288	1-Pentyne	C_5H_8	627-19-0	68.11702	0.000052415	1.0948	−51.09	101,160	313.33	0.01517	1000	0.09608
102	289	2-Pentyne	C_5H_8	627-21-4	68.11702	0.00025623	1.0073	1423.7		329.27	0.01653	1000	0.11119
102	290	Phenanthrene	$C_{14}H_{10}$	85-01-8	178.2292	0.00010167	0.988	797		610.03	0.02490	1000	0.05208
102	291	Phenol	C_6H_6O	108-95-2	94.11124	0.038846	0.2392	985.81	937,170	454.99	0.02183	1000	0.06936
102	292	Phenyl isocyanate	C_7H_5NO	103-71-9	119.1207	0.00016675	0.91777	730.1		439.43	0.01669	1000	0.05461
102	293	Phthalic anhydride	$C_8H_4O_3$	85-44-9	148.11556	0.0000593	1.046	765.5		557.65	0.01864	1000	0.04615
102	294	Propadiene	C_3H_4	463-49-0	40.06386	0.000061629	1.0731	1.8579	70,128	238.65	0.00980	1000	0.09526
102	295	Propane	C_3H_8	74-98-6	44.09562	−1.12	0.10972	−9834.6	−7,535,800	231.11	0.01114	1000	0.14599
102	296	1-Propanol	C_3H_8O	71-23-8	60.09502	−613.84	0.7927	−1,157,400,000		370.35	0.02135	720.25	0.07034
102	297	2-Propanol	C_3H_8O	67-63-0	60.095	7.3907E-07	1.7419			355.3	0.02049	1000	0.12428

102	298	Propenylcyclohexene	C_9H_{14}	13511-13-2	122.20746	0.00010242	1.0486	701.56		431.65	0.02262	1000	0.08421
102	299	Propionaldehyde	C_3H_6O	123-38-6	58.07914	9.0711E-07	1.6709			322.15	0.01407	1000	0.09340
100	300	Propionic acid	$C_3H_6O_2$	79-09-4	74.0785	1.0014	−0.0045954	7.1517E-06	−3.5878E-09	414.32	0.06993	616.15	0.04578
102	300	Propionic acid	$C_3H_6O_2$	79-09-4	74.0785	1.8905E-07	1.93			616.15	0.04578	1000	0.11657
102	301	Propionitrile	C_3H_5N	107-12-0	55.0785	1.1671E-06	1.6033			370.25	0.01532	1000	0.07534
102	302	Propyl acetate	$C_5H_{10}O_2$	109-60-4	102.1317	1325.3	1	12,235,000,000		374.65	0.01520	1000	0.10832
102	303	Propyl amine	C_3H_9N	107-10-8	59.11026	0.2833	0.055046	1325.9	1,817,600	321	0.01709	1000	0.10000
102	304	Propylbenzene	C_9H_{12}	103-65-1	120.19158	0.16992	0.021288	−54.484	1,624,800	432.39	0.02022	1000	0.07658
102	305	Propylene	C_3H_6	115-07-1	42.07974	0.0000449	1.2018	421		225.45	0.01054	1000	0.12737
102	306	Propyl formate	$C_4H_8O_2$	110-74-7	88.10512	740.1	0.9732	5,646,000,000		353.97	0.01403	1000	0.10893
102	307	2-Propyl mercaptan	C_3H_8S	75-33-2	76.16062	0.00018367	0.9627	646.01		325.71	0.01616	1000	0.08624
102	308	Propyl mercaptan	C_3H_8S	107-03-9	76.16062	0.0087425	0.51733	2358.1	334,590	340.87	0.01654	1000	0.08439
102	309	1,2-Propylene glycol	$C_3H_8O_2$	57-55-6	76.09442	0.0001666	0.9765	706		460.75	0.02624	1000	0.08302
102	310	Quinone	$C_6H_4O_2$	106-51-4	108.09476	−5678600	−0.045252	2,615,700,000	−3.5415E+13	454	0.02593	1000	0.12665
102	311	Silicon tetrafluoride	F_4Si	7783-61-1	104.07911	0.0000955	0.928	63.6		333.55	0.01761	702.45	0.03837
102	312	Styrene	C_8H_8	100-42-5	104.14912	0.010048	0.4033	553.74	685,570	418.31	0.01837	1000	0.07276
102	313	Succinic acid	$C_4H_6O_4$	110-15-6	118.08804	5.5263E-06	1.344			591	0.02934	1000	0.05949
102	314	Sulfur dioxide	O_2S	7446-09-5	64.0638	10.527	−0.7732	−1333	1,506,400	250	0.00745	900	0.03969
102	315	Sulfur hexafluoride	F_6S	2551-62-4	146.0554192	0.00048883	0.6518	−117.08	78,863	273.15	0.01163	1000	0.04587
102	316	Sulfur trioxide	O_3S	7446-11-9	80.0632	1.0702	−0.2348	2010.4	1,277,000	317.9	0.01386	1000	0.04930
102	317	Terephthalic acid	$C_8H_6O_4$	100-21-0	166.13084	3.4082E-06	1.3647			795.28	0.03097	1000	0.04233
102	318	*o*-Terphenyl	$C_{18}H_{14}$	84-15-1	230.30376	0.000078652	0.95174	−282.82	289,490	373.15	0.00950	1000	0.05598
102	319	Tetradecane	$C_{14}H_{30}$	629-59-4	198.388	−163.62	0.9193	−1,087,600,000		526.73	0.02517	1000	0.08615
102	320	Tetrahydrofuran	C_4H_8O	109-99-9	72.10572	9.5521E-06	1.4561	662.22		339.12	0.01564	1000	0.13419
102	321	1,2,3,4-Tetrahydronaphthalene	$C_{10}H_{12}$	119-64-2	132.20228	0.00007754	1.0778	729		480.77	0.02395	1000	0.07676
102	322	Tetrahydrothiophene	C_4H_8S	110-01-0	88.17132	0.00085604	0.7297	531.99	213,840	394.27	0.01801	1000	0.07579
102	323	2,2,3,3-Tetramethylbutane	C_8H_{18}	594-82-1	114.22852	0.000015235	1.2816	−111.88	124,120	379.44	0.01964	1000	0.10528
102	324	Thiophene	C_4H_4S	110-02-1	84.13956	0.00013384	0.98115	645.95		357.31	0.01525	1000	0.07139
102	325	Toluene	C_7H_8	108-88-3	92.13842	0.00002392	1.2694	537		383.78	0.01901	1000	0.10007
102	326	1,1,2-Trichloroethane	$C_2H_3Cl_3$	79-00-5	133.40422	0.0000952	1.0423	1243.3		387	0.01125	1000	0.05684
102	327	Tridecane	$C_{13}H_{28}$	629-50-5	184.36142	5.3701E-06	1.4751	599.09		508.62	0.02422	1000	0.08942
102	328	Triethyl amine	$C_6H_{15}N$	121-44-8	101.19	0.000106	1.0161	91	132,900	273.15	0.01018	1000	0.09680
102	329	Trimethyl amine	C_3H_9N	75-50-3	59.11026	0.00027648	0.901	167.68	132,200	273.15	0.01280	1000	0.10734
102	330	1,2,3-Trimethylbenzene	C_9H_{12}	526-73-8	120.19158	0.000098408	1.0452	720.49		449.27	0.02238	1000	0.07816
102	331	1,2,4-Trimethylbenzene	C_9H_{12}	95-63-6	120.19158	0.00008498	1.061	708		442.53	0.02098	1000	0.07583
102	332	2,2,4-Trimethylpentane	C_8H_{18}	540-84-1	114.22852	0.00001758	1.3114	392.9		355.15	0.01846	1000	0.10847
102	333	2,3,3-Trimethylpentane	C_8H_{18}	560-21-4	114.22852	0.000020248	1.2284	−174.72	147,800	387.91	0.02001	1000	0.10079
102	334	1,3,5-Trinitrobenzene	$C_6H_3N_3O_6$	99-35-4	213.10452	0.00020544	0.87137	807.3		629.6	0.02474	1000	0.04675
102	335	2,4,6-Trinitrotoluene	$C_7H_5N_3O_6$	118-96-7	227.1311	0.00018189	0.88744	803.39		625	0.02410	1000	0.04635
102	336	Undecane	$C_{11}H_{24}$	1120-21-4	156.30826	0.038012	0.68615	34,663	8,721,900	469.08	0.02259	1000	0.09798
102	337	1-Undecanol	$C_{11}H_{24}O$	112-42-5	172.30766	2498.8	0.95209	20,167,000,000		520.3	0.02486	1000	0.08899
102	338	Vinyl acetate	$C_4H_6O_2$	108-05-4	86.08924	−3279500	−0.12941	1,710,400,000	−1.2727E+13	345.65	0.01515	1000	0.12177
102	339	Vinyl acetylene	C_4H_4	689-97-4	52.07456	0.000054197	1.0632	−70.589	90,617	278.25	0.01123	1000	0.08222
102	340	Vinyl chloride	C_2H_3Cl	75-01-4	62.49822	−229.41	0.59582	−169,430,000		259.25	0.00963	1000	0.08300
102	341	Vinyl trichlorosilane	$C_2H_3Cl_3Si$	75-94-5	161.48972	3510.8	0.225	401,720,000		363.85	0.01198	1000	0.04135
102	342	Water	H_2O	7732-18-5	18.01528	6.2041E-06	1.3973			273.16	0.01574	1073.15	0.10652
102	343	*m*-Xylene	C_8H_{10}	108-38-3	106.165	3.0593E-09	2.4182	−569.28	121,060	320	0.00867	1000	0.09965
102	344	*o*-Xylene	C_8H_{10}	95-47-6	106.165	4.9707E-06	1.3787	−225.64	66,786	320	0.01492	1000	0.08084
102	345	*p*-Xylene	C_8H_{10}	106-42-3	106.165	9.9305E-08	1.9229	−469.93	113,460	320	0.01019	1000	0.09060

Except for acetic acid, butyric acid, formic acid, heptanoic acid, octanoic acid, pentanoic acid, propionic acid, the vapor thermal conductivity is calculated by Eqn 102: $k = C_1T^{C_2}/(1 + C_3/T + C_4/T^2)$ where k is the thermal conductivity in W/(m·K) and T is the temperature in K. Thermal conductivities are at either 1 atm or the vapor pressure, whichever is lower.

Eqn 100, used for the limited temperature ranges as noted for the associating compounds above, $k = C_1 + C_2T + C_3T^2 + C_4T^3$

Values in this table were taken from the Design Institute for Physical Properties (DIPPR) of the American Institute of Chemical Engineers (AIChE), 801 Critically Evaluated Gold Standard™ Database, copyright 2016 AIChE, and reproduced with permission of AIChE and of the DIPPR Evaluated Process Design Data Project Steering Committee. Their source should be cited as "R. L. Rowley, W. V. Wilding, J. L. Oscarson, T. A. Knotts, N. F. Giles, *DIPPR® Data Compilation of Pure Chemical Properties*, Design Institute for Physical Properties, AIChE, New York, NY (2016)".

TABLE 2-146 Thermophysical Properties of Miscellaneous Saturated Liquids

		Temperature, °C															
Substance	Property	−50	−40	−30	−20	−10	0	10	20	30	40	50	60	70	80	90	100
Acetaldehyde	ρ (kg/m^3)	863	852	840	828	816	804	794	783								
	c_p (kJ/kg·K)	2.05	2.08	2.11	2.14	2.17	2.20	2.24	2.28								
	μ (10^{-6}Pa·s)	460	404	358	321	290	263	241	222								
	k (W/m·K)	0.211	0.206	0.200	0.195	0.189	0.184	0.182	0.180								
	Pr	4.47	4.08	3.78	3.52	3.33	3.14	2.97	2.81								
Acetic acid	ρ (kg/m^3)								1049	1039	1028	1018	1006	995	984	972	960
	c_p (kJ/kg·K)								2.031								
	μ (10^{-6}Pa·s)								1210	1102	1010	795	600				
	k (W/m·K)								0.173	0.170	0.168	0.167	0.165	0.163	0.161		
	Pr								14.2								
Aniline	ρ (kg/m^3)	—	—	—	—	—	1039	1030	1022	1013	1005	996	987	978	969	960	951
	c_p (kJ/kg·K)	—	—	—	—	—	2.024	2.047	2.071	2.093	2.113	2.132	2.17	2.20	2.23	2.27	2.32
	μ (10^{-6}Pa·s)	—	—	—	—	—	10200	6500	4400	3160	2370	1850	1510	1270	1090	935	825
	k (W/m·K)	—	—	—	—	—	0.186	0.184	0.182	0.180	0.177	0.174	0.171	0.169	0.168	0.167	0.167
	Pr	—	—	—	—	—	111	72	50	36.7	28.3	22.7	19.2	16.5	14.5	12.7	11.5
Butanol	ρ (kg/m^3)	845	841	837	833	829	825	817	810	803	797	791	784	776	768	760	753
	c_p (kJ/kg·K)	1.947	1.996	2.046	2.100	2.153	2.202	2.262	2.345	2.437	2.524	2.621					
	μ (10^{-6}Pa·s)	34700	22400	14700	10300	7400	5190	3870	2950	2300	1780	1410	1140	930	760	630	535
	k (W/m·K)	0.175	0.174	0.173	0.172	0.171	0.170	0.168	0.167	0.166	0.165	0.164	0.163	0.162	0.161	0.160	0.159
	Pr	3860	2570	1740	1260	930	670	120	41	33.8	27.2	22.5					
Carbon disulfide	ρ (kg/m^3)	1362	1348	1334	1320	1306	1292	1278	1263								
	c_p (kJ/kg·K)	0.988	0.989	0.990	0.991	0.993	0.996	1.004	1.017								
	μ (10^{-6}Pa·s)	630	580	535	496	463	435	405	375	350	330						
	k (W/m·K)	0.194	0.190	0.186	0.182	0.178	0.174	0.170	0.166	0.161	0.158	0.156	0.154	0.152	0.150		
	Pr	3.21	3.02	2.85	2.70	2.58	2.49	2.39	2.30								
Cyclohexane	ρ (kg/m^3)	—	—	—	—	—	—	789	779	769	759	750	740	731	721		
	c_p (kJ/kg·K)	—	—	—	—	—	—	2.068	2.081	2.094	2.106	2.119					
	μ (10^{-6}Pa·s)	—	—	—	—	—	—	1175	980	820	710	605	540				
	k (W/m·K)	—	—	—	—	—	—	0.122	0.120	0.119	0.118	0.117	0.116	0.114	0.112		
	Pr	—	—	—	—	—	—	19.9	17.0	14.4	12.7	11.0					
Ethanol	ρ (kg/m^3)						806	798	789	781	776	763	754	745	735	725	716
	c_p (kJ/kg·K)	2.01	2.04	2.08	2.13	2.19	2.27	2.35	2.43	2.52	2.62	2.73	2.83	2.93	3.03	3.19	3.30
	μ (10^{-6}Pa·s)	6400	4790	3650	2825	2220	1770	1470	1200	1000	835	700	590	500	435	370	314
	k (W/m·K)	0.188	0.186	0.184	0.181	0.179	0.177	0.175	0.173	0.171	0.168	0.165	0.162	0.159	0.156	0.153	0.151
	Pr	68.4	52.5	41.3	33.2	27.2	22.7	19.7	16.9	14.7	13.0	11.6	10.3	9.2	8.4	7.7	6.9
Ethyl acetate	ρ (kg/m^3)				947	935	924	912	901	888	876	863	851	838	825	811	797
	c_p (kJ/kg·K)								2.01								
	μ (10^{-6}Pa·s)	1090					580	510	455	400	370	345	310	280	250	230	220
	k (W/m·K)								0.145	0.142	0.139	0.136	0.133	0.130	0.127	0.123	0.119
	Pr								6.3								
Ethylamine	ρ (kg/m^3)	761	750	739	729	718	707	695	683	671	658	646	633	620	607		
	c_p (kJ/kg·K)	2.95	2.97	2.98	3.00	3.01	3.03										
	μ (10^{-6}Pa·s)	580	500	435	390	350	320										
	k (W/m·K)	0.204	0.201	0.199	0.196	0.194	0.191										
	Pr	8.39	7.39	6.51	5.97	5.43	5.08										
Ethyl ether	ρ (kg/m^3)	790	780	769	758	747	736	725	714	702	689	676	666	653	640	625	611
	c_p (kJ/kg·K)	2.135	2.156	2.179	2.205	2.233	2.265	2.299	2.332	2.36	2.39	2.43	2.47	2.51			
	μ (10^{-6}Pa·s)	550	470	410	365	330	290	265	233	214	197	181	166	153	140	129	118
	k (W/m·K)	0.159	0.155	0.151	0.147	0.144	0.140	0.139	0.134	0.129	0.125	0.120	0.116	0.112			
	Pr	7.39	6.54	5.92	5.48	5.12	4.69	4.38	4.05	3.92	3.77	3.67	3.54	3.43			
Ethyl iodide	ρ (kg/m^3)																
	c_p (kJ/kg·K)			0.656	0.663	0.670	0.677	0.684	0.691	0.698	0.705	0.712	0.718	0.724			
	μ (10^{-6}Pa·s)						730	655	590	539	495	455	420	390			
	k (W/m·K)						0.092	0.090	0.088	0.086	0.085	0.083	0.081	0.080			
	Pr						5.37	4.98	4.63	4.30	4.11	3.90	3.72	3.53			
Ethylene glycol	ρ (kg/m^3)						1127	1120	1113	1106	1099	1092	1085	1077	1070	1063	1056
	c_p (kJ/kg·K)						2.272	2.327	2.381	2.431	2.484	2.536	2.586	2.636	2.685	2.734	2.779
	μ (10^{-6}Pa·s)						57000	33300	20200	13400	9100	7070	4000	3450	3000	2440	2000
	k (W/m·K)						0.254	0.255	0.256	0.258	0.259	0.260					
	Pr						510	305	190	126	87.3	69.0					
Formic acid	ρ (kg/m^3)						1241	1231	1220	1209	1196	1184	1170	1156	1140	1124	1108
	c_p (kJ/kg·K)																
	μ (10^{-6}Pa·s)							2260	1800	1470	1220	1030	890	780	680	615	550
	k (W/m·K)						0.265	0.261	0.257	0.257	0.253	0.250	0.246	0.243	0.240	0.236	0.232
	Pr																

TABLE 2-146 Thermophysical Properties of Miscellaneous Saturated Liquids *(Continued)*

		Temperature, °C															
Substance	Property	−50	−40	−30	−20	−10	0	10	20	30	40	50	60	70	80	90	100
Gasoline	ρ (kg/m³)				784	775	767	759	751	743	735	721	717	708	699	690	681
	c_p (kJ/kg·K)				1.88	1.92	1.97	2.02	2.06	2.11	2.15	2.20	2.25	2.30	2.35	2.41	2.46
	μ (10^{-6}Pa·s)	1710	1400	1170	990	850	735	645	530	464	410	367	330	298	270	246	225
	k (W/m·K)	0.131	0.128	0.125	0.123	0.121	0.120	0.118	0.116	0.114	0.112	0.110	0.108	0.106	0.104	0.102	0.100
	Pr				15.1	13.5	12.1	11.0	9.41	8.59	7.87	7.34	6.88	6.47	6.10	5.81	5.54
Glycerol	ρ (kg/m³)	—	—	—	—	—	1276	1270	1260	1254	1248	1242					
	c_p (kJ/kg·K)								2.393	2.406	2.457	2.504	2.548	2.588	2.625	2.657	2.686
	μ (10^{-6}Pa·s)						1.2.+7	4.0.+6	1.5.+6								
	k (W/m·K)								0.284	0.285	0.287	0.288	0.289	0.291	0.293	0.294	0.295
	Pr								12650								
Kerosene	ρ (kg/m³)						781	774	767	760	754	748	742				
	c_p (kJ/kg·K)						1.91	1.96	2.02	2.07	2.13	2.18	2.23	2.28	2.32	2.35	2.38
	μ (10^{-6}Pa·s)	1150	725	500	360	275	215	173	149	126	108	95	83	73	66	60	55
	k (W/m·K)						0.140	0.139	0.139	0.138	0.138	0.137	0.137				
	Pr						2.93	2.44	2.17	1.89	1.67	1.51	1.35				
Methanol	ρ (kg/m³)									783	774	766	756	746	736	725	711
	c_p (kJ/kg·K)	2.30	2.32	2.35	2.37	2.40	2.42	2.45	2.47	2.49	2.52	2.55	2.65	2.78	2.94	3.13	3.30
	μ (10^{-6}Pa·s)	2305	1800	1410	1170	975	820	692	590	510	455	400	355	315	271	240	218
	k(W/m·K)	0.225	0.222	0.219	0.216	0.212	0.209	0.206	0.203	0.199	0.195	0.192	0.189	0.187	0.184	0.182	0.180
	Pr	23.6	18.8	15.1	12.9	11.0	9.53	8.23	7.18	6.38	5.88	5.31	4.98	4.68	4.34	4.13	3.99
Methyl formate	ρ (kg/m³)	1069	1056	1043	1030	1017	1003	989	975	960	944	929	913	897	880	863	845
	c_p (kJ/kg·K)	1.84	1.86	1.88	1.90	1.92	1.95	1.99	2.03	2.08							
	μ (10^{-6}Pa·s)	830	711	618	544	481	430	380	345	315							
	k (W/m·K)	0.217	0.213	0.209	0.205	0.200	0.195	0.191	0.186	0.180							
	Pr	7.04	6.21	5.56	5.04	4.62	4.30	3.96	3.77	3.64							
Oil, castor	ρ (kg/m³)																
	c_p (kJ/kg·K)																
	μ (10^{-6}Pa·s)							2,420,000	986,000	451,000	231,000	125,000	74,000	43,000			
	k (W/m·K)							0.182	0.181	0.180	0.179	0.178	0.177	0.176	0.175	0.174	0.17
	Pr																
Oil, olive	ρ (kg/m³)								914								
	c_p (kJ/kg·K)								1.633								
	μ (10^{-6}Pa·s)							138,000	84,000	52,000	36,300	24,500	17,000	12,400			
	k (W/m·K)							0.170	0.169	0.168	0.167	0.166	0.166	0.165	0.165	0.164	0.164
	Pr								810								
Pentane	ρ (kg/m³)	693	684	674	665	656	646	636	626	616	606	596	585	574	562	550	538
	c_p (kJ/kg·K)	2.060	2.084	2.110	2.137	2.167	2.206	2.239	2.273								
	μ (10^{-6}Pa·s)	489	428	379	339	307	279	254	234	209	190	175	161	148	137	124	113
	k (W/m·K)	0.142	0.139	0.136	0.132	0.128	0.125	0.122	0.119	0.115	0.112	0.108	0.105	0.101	0.098	0.095	0.091
	Pr	7.14	6.42	5.88	5.49	5.20	4.92	4.66	4.47								
Propanol	ρ (kg/m³)	849					819	811	814	796	788	779	770	761	752	747	743
	c_p (kJ/kg·K)	1.955					2.219										
	μ (10^{-6}Pa·s)	20,200	13,500	9500	6900	5110	3900	2900	2245	1720	1400	1130	921	760	630	508	447
	k (W/m·K)	0.167	0.166	0.165						0.171	0.169	0.168	0.167	0.165	0.164	0.163	0.162
	Pr	236															
Sulfuric acid	ρ (kg/m³)								1834								
	c_p (kJ/kg·K)								1.382								
	μ (10^{-6}Pa·s)						48,400	35,200	25,400	15,700	11,500	8820	7220	6090	5190		
	k (W/m·K)						0.314										
	Pr																
Toluene	ρ (kg/m³)	932	923	913	904	895	886	876	867	858	848	839	829	820	810	800	790
	c_p (kJ/kg·K)	1.514	1.535	1.556	1.579	1.602	1.633	1.652	1.675	1.701	1.73	1.76	1.80	1.83	1.87	1.92	1.97
	μ (10^{-6}Pa·s)	2120	1670	1345	1100	915	770	670	590	520	470	420	380	355	325	295	270
	k (W/m·K)	0.152	0.149	0.147	0.144	0.142	0.139	0.137	0.134	0.132	0.129	0.126	0.124	0.122	0.119	0.117	0.114
	Pr	21.1	17.8	14.2	12.1	10.3	9.0	8.1	7.4	6.7	6.3	5.9	5.5	5.3	5.1	4.8	4.7
Turpentine	ρ (kg/m³)																
	c_p (kJ/kg·K)						1.72	1.76	1.80			1.93					
	μ (10^{-6}Pa·s)						2250	1780	1490	1270	1070	925	820	730	675		
	k (W/m·K)						0.130	0.129	0.128	0.127	0.126	0.125					
	Pr						29.8	24.3	20.9	18.4	16.1	14.3					

TABLE 2-147 Thermal Conductivity of Inorganic and Organic Liquids [W/(m·K)]

Cmpd. no.	Name	Formula	CAS	Mol. wt.	C_1	C_2	C_3	C_4	C_5	T_{min}, K	Thermal cond. at T_{min}	T_{max}, K	Thermal cond. at T_{max}
1	Acetaldehyde	C_2H_4O	75-07-0	44.05256	0.33515	−0.00055227				149.78	0.2524	294.15	0.1727
2	Acetamide	C_2H_5NO	60-35-5	59.0672	0.39363	−0.00037053				353.33	0.2627	494.3	0.2105
3	Acetic acid	$C_2H_4O_2$	64-19-7	60.052	0.214	−0.0001834				289.81	0.1608	391.05	0.1423
4	Acetic anhydride	$C_4H_6O_3$	108-24-7	102.08864	0.23638	−0.00024263				200.15	0.1878	412.7	0.1362
5	Acetone	C_3H_6O	67-64-1	58.07914	0.2878	−0.000427				178.45	0.2116	343.15	0.1413
6	Acetonitrile	C_2H_3N	75-05-8	41.0519	0.30755	−0.000402				229.32	0.2154	354.81	0.1649
7	Acetylene	C_2H_2	74-86-2	26.03728	0.33363	−0.00083655				192.4	0.1727	250	0.1245
8	Acrolein	C_3H_4O	107-02-8	56.06326	0.2703	−0.0003764				185.45	0.2005	325.84	0.1477
9	Acrylic acid	$C_3H_4O_2$	79-10-7	72.06266	0.2441	−0.0002904				286.15	0.1610	484.5	0.1034
10	Acrylonitrile	C_3H_3N	107-13-1	53.0626	0.30751	−0.000487				189.63	0.2152	350.45	0.1368
11	Air	Mixture	132259-10-0	28.96	0.28472	−0.0017393				75	0.1543	125	0.0673
12	Ammonia	H_3N	7664-41-7	17.03052	1.169	−0.002314				195.41	0.7168	400.05	0.2433
13	Anisole	C_7H_8O	100-66-3	108.13782	0.23494	−0.00026477				235.65	0.1725	512.5	0.0993
14	Argon	Ar	7440-37-1	39.948	0.1819	−0.0003176	−0.00000411			83.78	0.1264	150	0.0418
15	Benzamide	C_7H_7NO	55-21-0	121.13658	0.28485	−0.00025225				403	0.1832	563.15	0.1428
16	Benzene	C_6H_6	71-43-2	78.11184	0.23444	−0.00030572				278.68	0.1492	413.1	0.1081
17	Benzenethiol	C_6H_6S	108-98-5	110.17684	0.20996	−0.0002146				258.27	0.1545	442.29	0.1150
18	Benzoic acid	$C_7H_6O_2$	65-85-0	122.12134	0.2391	−0.0002325				395.45	0.1472	596	0.1005
19	Benzonitrile	C_7H_5N	100-47-0	103.1213	0.20603	−0.00021023				260.28	0.1513	464.15	0.1085
20	Benzophenone	$C_{13}H_{10}O$	119-61-9	182.2179	0.25867	−0.00022516				321.35	0.1863	664	0.1092
21	Benzyl alcohol	C_7H_8O	100-51-6	108.13782	0.17847	−0.000065843				257.85	0.1615	478.6	0.1470
22	Benzyl ethyl ether	$C_9H_{12}O$	539-30-0	136.19098	0.2029	−0.0002226				275.65	0.1415	528.6	0.0852
23	Benzyl mercaptan	C_7H_8S	100-53-8	124.20342	0.20316	−0.00019912				243.95	0.1546	472.03	0.1092
24	Biphenyl	$C_{12}H_{10}$	92-52-4	154.2078	0.19053	−0.00015145				342.2	0.1387	723.15	0.0810
25	Bromine	Br_2	7726-95-6	159.808	−0.2185	0.0042143	−0.000017753	3.1041E-08	−2.0108E-11	266	0.1299	584	0.0316
26	Bromobenzene	C_6H_5Br	108-86-1	157.0079	0.16983	−0.0001981				242.43	0.1218	429.24	0.0848
27	Bromoethane	C_2H_5Br	74-96-4	108.965	0.1629	−0.00021198				154.25	0.1302	327	0.0936
28	Bromomethane	CH_3Br	74-83-9	94.93852	0.16143	−0.00021287				179.44	0.1232	413.15	0.0735
29	1,2-Butadiene	C_4H_6	590-19-2	54.09044	0.21966	−0.0003436				136.95	0.1726	284	0.1221
30	1,3-Butadiene	C_4H_6	106-99-0	54.09044	0.22231	−0.0003664				164.25	0.1621	268.74	0.1238
31	Butane	C_4H_{10}	106-97-8	58.1222	0.27349	−0.00071267	5.1555E-07			134.86	0.1868	400	0.0709
32	1,2-Butanediol	$C_4H_{10}O_2$	584-03-2	90.121	0.064621	0.00067625	−1.0491E-06			220	0.1626	469.57	0.1508
33	1,3-Butanediol	$C_4H_{10}O_2$	107-88-0	90.121	−0.0032865	0.0011463	−1.5525E-06			196.15	0.1618	481.38	0.1888
34	1-Butanol	$C_4H_{10}O$	71-36-3	74.1216	0.22888	−0.00025				183.85	0.1829	391	0.1311
35	2-Butanol	$C_4H_{10}O$	78-92-2	74.1216	0.18599	−0.00017227				158.45	0.1587	372.9	0.1218
36	1-Butene	C_4H_8	106-98-9	56.10632	0.22153	−0.00035023				87.8	0.1908	266.91	0.1281
37	*cis*-2-Butene	C_4H_8	590-18-1	56.10632	0.21378	−0.00035445				134.26	0.1662	276.87	0.1156
38	*trans*-2-Butene	C_4H_8	624-64-6	56.10632	0.21153	−0.00035056				167.62	0.1528	274.03	0.1155
39	Butyl acetate	$C_6H_{12}O_2$	123-86-4	116.15828	0.21721	−0.00026563				199.65	0.1642	453.75	0.0967
40	Butylbenzene	$C_{10}H_{14}$	104-51-8	134.21816	0.18707	−0.00020037				185.3	0.1499	473.15	0.0923
41	Butyl mercaptan	$C_4H_{10}S$	109-79-5	90.1872	0.21143	−0.000258				157.46	0.1708	371.61	0.1156
42	*sec*-Butyl mercaptan	$C_4H_{10}S$	513-53-1	90.1872	0.2069	−0.0002568				133.02	0.1727	358.13	0.1149
43	1-Butyne	C_4H_6	107-00-6	54.09044	0.22334	−0.0003515				147.43	0.1715	281.22	0.1245
44	Butyraldehyde	C_4H_8O	123-72-8	72.10572	0.24962	−0.000325				176.8	0.1922	347.94	0.1365
45	Butyric acid	$C_4H_8O_2$	107-92-6	88.1051	0.1967	−0.000168				267.95	0.1517	573.15	0.1004
46	Butyronitrile	C_4H_7N	109-74-0	69.1051	0.24077	−0.00028665				161.3	0.1945	390.74	0.1288
47	Carbon dioxide	CO_2	124-38-9	44.0095	0.4406	−0.0012175				216.58	0.1769	300	0.0754
48	Carbon disulfide	CS_2	75-15-0	76.1407	0.2333	−0.000275				161.11	0.1890	319.37	0.1455
49	Carbon monoxide	CO	630-08-0	28.0101	0.2855	−0.001784				68.15	0.1639	125	0.0625
50	Carbon tetrachloride	CCl_4	56-23-5	153.8227	0.1589	−0.0001987				250.33	0.1092	349.79	0.0894
51	Carbon tetrafluoride	CF_4	75-73-0	88.0043	0.20771	−0.00078883				89.56	0.1371	145.1	0.0933

52	Chlorine	Cl_2	7782-50-5	70.906	0.2246	−0.000064	−0.000000788			172.12	0.1902	410	0.0659
53	Chlorobenzene	C_6H_5Cl	108-90-7	112.5569	0.1841	−0.0001917				227.95	0.1404	404.87	0.1065
54	Chloroethane	C_2H_5Cl	75-00-3	64.5141	0.23779	−0.000395209				136.75	0.1837	348.15	0.1002
55	Chloroform	$CHCl_3$	67-66-3	119.37764	0.1778	−0.0002023				209.63	0.1354	400	0.0969
56	Chloromethane	CH_3Cl	74-87-3	50.4875	0.25381	−0.000431803				175.43	0.1781	333	0.1100
57	1-Chloropropane	C_3H_7Cl	540-54-5	78.54068	0.21851	−0.00033762				150.35	0.1677	393.15	0.0858
58	2-Chloropropane	C_3H_7Cl	75-29-6	78.54068	0.21232	−0.0003149				155.97	0.1632	386.7	0.0906
59	*m*-Cresol	C_7H_8O	108-39-4	108.13782	0.18241	−0.00011109				285.39	0.1507	475.43	0.1296
60	*o*-Cresol	C_7H_8O	95-48-7	108.13782	0.19186	−0.0001303				304.19	0.1522	464.15	0.1314
61	*p*-Cresol	C_7H_8O	106-44-5	108.13782	0.17971	−0.00012037				307.93	0.1426	475.13	0.1225
62	Cumene	C_9H_{12}	98-82-8	120.19158	0.1855	−0.00020895				177.14	0.1485	413.15	0.0992
63	Cyanogen	C_2N_2	460-19-5	52.0348	0.37845	−0.00069945				245.25	0.2069	251.9	0.2023
64	Cyclobutane	C_4H_8	287-23-0	56.10632	0.22262	−0.00034082				182.48	0.1604	285.66	0.1253
65	Cyclohexane	C_6H_{12}	110-82-7	84.15948	0.19813	−0.0002505				279.69	0.1281	353.87	0.1095
66	Cyclohexanol	$C_6H_{12}O$	108-93-0	100.15888	0.1715	−0.0001255				296.6	0.1343	563.15	0.1008
67	Cyclohexanone	$C_6H_{10}O$	108-94-1	98.143	0.17557	−0.00012392				242	0.1456	428.58	0.1225
68	Cyclohexene	C_6H_{10}	110-83-8	82.1436	0.20926	−0.00026037				169.67	0.1651	356.12	0.1165
69	Cyclopentane	C_5H_{10}	287-92-3	70.1329	0.2066	−0.0002696				179.28	0.1583	322.4	0.1197
70	Cyclopentene	C_5H_8	142-29-0	68.11702	0.21776	−0.00027783				138.13	0.1794	333.15	0.1252
71	Cyclopropane	C_3H_6	75-19-4	42.07974	0.24348	−0.00042568				145.59	0.1815	240.37	0.1412
72	Cyclohexyl mercaptan	$C_6H_{12}S$	1569-69-3	116.22448	0.18374	−0.0001925				189.64	0.1472	431.95	0.1006
73	Decanal	$C_{10}H_{20}O$	112-31-2	156.2652	0.21363	−0.00023004				285	0.1481	481.65	0.1028
74	Decane	$C_{10}H_{22}$	124-18-5	142.28168	0.2063	−0.00025				243.51	0.1454	447.3	0.0945
75	Decanoic acid	$C_{10}H_{20}O_2$	334-48-5	172.265	0.206	−0.0002				304.75	0.1451	543.15	0.0974
76	1-Decanol	$C_{10}H_{22}O$	112-30-1	158.28108	0.236171	−0.00025				280.05	0.1662	503	0.1104
77	1-Decene	$C_{10}H_{20}$	872-05-9	140.2658	0.20237	−0.00024187				206.89	0.1523	443.75	0.0950
78	Decyl mercaptan	$C_{10}H_{22}S$	143-10-2	174.34668	0.20134	−0.00020826				247.56	0.1498	512.35	0.0946
79	1-Decyne	$C_{10}H_{18}$	764-93-2	138.24992	0.20839	−0.00023622				229.15	0.1543	447.15	0.1028
80	Deuterium	D_2	7782-39-0	4.0316	1.264					20.4	1.2640	20.4	1.2640
81	1,1-Dibromoethane	$C_2H_4Br_2$	557-91-5	187.86116	0.1426	−0.00016402				210.15	0.1081	498.4	0.0609
82	1,2-Dibromoethane	$C_2H_4Br_2$	106-93-4	187.86116	0.13622	−0.0001179				282.85	0.1029	404.51	0.0885
83	Dibromomethane	CH_2Br_2	74-95-3	173.83458	0.17558	−0.00022499				220.6	0.1259	370.1	0.0923
84	Dibutyl ether	$C_8H_{18}O$	142-96-1	130.22792	0.19418	−0.00022246				175.3	0.1552	523.15	0.0778
85	*m*-Dichlorobenzene	$C_6H_4Cl_2$	541-73-1	147.00196	0.16694	−0.0001667				248.39	0.1255	446.23	0.0926
86	*o*-Dichlorobenzene	$C_6H_4Cl_2$	95-50-1	147.00196	0.16994	−0.0001637				262.87	0.1269	351.71	0.1124
87	*p*-Dichlorobenzene	$C_6H_4Cl_2$	106-46-7	147.00196	0.16977	−0.0001799				326.14	0.1111	548	0.0712
88	1,1-Dichloroethane	$C_2H_4Cl_2$	75-34-3	98.95916	0.18881	−0.00026083				176.19	0.1429	416.9	0.0801
89	1,2-Dichloroethane	$C_2H_4Cl_2$	107-06-2	98.95916	0.214	−0.000266				253.15	0.1467	356.59	0.1191
90	Dichloromethane	CH_2Cl_2	75-09-2	84.93258	0.23847	−0.00033366				178.01	0.1791	325	0.1300
91	1,1-Dichloropropane	$C_3H_6Cl_2$	78-99-9	112.98574	0.18	−0.00023144				192.5	0.1354	438	0.0786
92	1,2-Dichloropropane	$C_3H_6Cl_2$	78-87-5	112.98574	0.19653	−0.00025012				172.71	0.1533	457.6	0.0821
93	Diethanol amine	$C_4H_{11}NO_2$	111-42-2	105.13564	0.0218	0.0010315	−0.000001355			301.15	0.2095	673.15	0.1022
94	Diethyl amine	$C_4H_{11}N$	109-89-7	73.13684	0.2587	−0.00054343	4.2097E-07			223.35	0.1583	453.15	0.0989
95	Diethyl ether	$C_4H_{10}O$	60-29-7	74.1216	0.2495	−0.000407				156.85	0.1857	433.15	0.0732
96	Diethyl sulfide	$C_4H_{10}S$	352-93-2	90.1872	0.21065	−0.0002623				169.2	0.1663	365.25	0.1148
97	1,1-Difluoroethane	$C_2H_4F_2$	75-37-6	66.04997	0.27019	−0.000661	3.443E-07			154.56	0.1763	363.15	0.0756
98	1,2-Difluoroethane	$C_2H_4F_2$	624-72-6	66.04997	0.23171	−0.00038503				179.6	0.1626	372.8	0.0882
99	Difluoromethane	CH_2F_2	75-10-5	52.02339	0.37296	−0.00088707	2.5762E-07			136.95	0.2563	302.56	0.1282
100	Di-isopropyl amine	$C_6H_{15}N$	108-18-9	101.19	0.1844	−0.000239				176.85	0.1421	357.05	0.0991
101	Di-isopropyl ether	$C_6H_{14}O$	108-20-3	102.17476	0.19162	−0.0002762				187.65	0.1398	400.1	0.0811
102	Di-isopropyl ketone	$C_7H_{14}O$	565-80-0	114.18546	0.22076	−0.00027624				204.81	0.1642	460	0.0937
103	1,1-Dimethoxyethane	$C_4H_{10}O_2$	534-15-6	90.121	0.22078	−0.00031271				159.95	0.1708	337.45	0.1153
104	1,2-Dimethoxypropane	$C_5H_{12}O_2$	7778-85-0	104.14758	0.22998	−0.00030372				226.1	0.1613	366.15	0.1188
105	Dimethyl acetylene	C_4H_6	503-17-3	54.09044	0.22773	−0.00034804				240.91	0.1439	300.13	0.1233
106	Dimethyl amine	C_2H_7N	124-40-3	45.08368	0.2454	−0.000338				180.96	0.1842	403.15	0.1091

(*Continued*)

TABLE 2-147 Thermal Conductivity of Inorganic and Organic Liquids [W/(m·K)] *(Continued)*

Cmpd. no.	Name	Formula	CAS	Mol. wt.	C_1	C_2	C_3	C_4	C_5	T_{min}, K	Thermal cond. at T_{min}	T_{max}, K	Thermal cond. at T_{max}
107	2,3-Dimethylbutane	C_6H_{14}	79-29-8	86.17536	0.1774	−0.0002436				145.19	0.1420	331.15	0.0967
108	1,1-Dimethylcyclohexane	C_8H_{16}	590-66-9	112.21264	0.1807	−0.0002177				239.66	0.1285	392.7	0.0952
109	*cis*-1,2-Dimethylcyclohexane	C_8H_{16}	2207-01-4	112.21264	0.18092	−0.0002108				223.16	0.1339	402.94	0.0960
110	*trans*-1,2-Dimethylcyclohexane	C_8H_{16}	6876-23-9	112.21264	0.17675	−0.0002077				184.99	0.1383	596.15	0.0529
111	Dimethyl disulfide	$C_2H_6S_2$	624-92-0	94.19904	0.21373	−0.0002447				188.44	0.1676	382.9	0.1200
112	Dimethyl ether	C_2H_6O	115-10-6	46.06844	0.31174	−0.0005638				131.65	0.2375	320.03	0.1313
113	*N,N*-Dimethyl formamide	C_3H_7NO	68-12-2	73.09378	0.26	−0.000255				250	0.1963	425.15	0.1516
114	2,3-Dimethylpentane	C_7H_{16}	565-59-3	100.20194	0.17964	−0.000246				160	0.1403	362.93	0.0904
115	Dimethyl phthalate	$C_{10}H_{10}O_4$	131-11-3	194.184	0.13905	0.0001509	−3.978E-07			273.15	0.1506	556.85	0.0997
116	Dimethylsilane	C_2H_8Si	1111-74-6	60.17042	0.25547	−0.0004411				122.93	0.2012	253.55	0.1436
117	Dimethyl sulfide	C_2H_6S	75-18-3	62.134	0.23942	−0.0003311				174.88	0.1815	310.48	0.1366
118	Dimethyl sulfoxide	C_2H_6OS	67-68-5	78.13344	0.3142	−0.00030809				291.67	0.2243	464	0.1712
119	Dimethyl terephthalate	$C_{10}H_{10}O_4$	120-61-6	194.184	0.21956	−0.000209955				413.79	0.1327	559.2	0.1022
120	1,4-Dioxane	$C_4H_8O_2$	123-91-1	88.10512	0.3027	−0.0004827				284.95	0.1652	374.47	0.1219
121	Diphenyl ether	$C_{12}H_{10}O$	101-84-8	170.2072	0.18686	−0.00014953				300.03	0.1420	531.46	0.1074
122	Dipropyl amine	$C_6H_{15}N$	142-84-7	101.19	0.2224	−0.000314				210.15	0.1564	382	0.1025
123	Dodecane	$C_{12}H_{26}$	112-40-3	170.33484	0.2047	−0.0002326				263.57	0.1434	489.47	0.0909
124	Eicosane	$C_{20}H_{42}$	112-95-8	282.54748	0.2178	−0.0002233				309.58	0.1487	616.93	0.0800
125	Ethane	C_2H_6	74-84-0	30.069	0.35758	−0.0011458	6.1866E-07			90.35	0.2591	300	0.0695
126	Ethanol	C_2H_6O	64-17-5	46.06844	0.2468	−0.000264				159.05	0.2048	353.15	0.1536
127	Ethyl acetate	$C_4H_8O_2$	141-78-6	88.10512	0.2501	−0.0003563				189.6	0.1825	350.21	0.1253
128	Ethyl amine	C_2H_7N	75-04-7	45.08368	0.30059	−0.000581	6.602E-07			192.15	0.2133	293.15	0.1870
129	Ethylbenzene	C_8H_{10}	100-41-4	106.165	0.1999	−0.00023823				178.2	0.1574	413.1	0.1015
130	Ethyl benzoate	$C_9H_{10}O_2$	93-89-0	150.1745	0.20771	−0.00021265				238.45	0.1570	549.4	0.0909
131	2-Ethyl butanoic acid	$C_6H_{12}O_2$	88-09-5	116.15828	0.2175	−0.0002407				258.15	0.1554	516.5	0.0932
132	Ethyl butyrate	$C_6H_{12}O_2$	105-54-4	116.15828	0.21043	−0.00024903				175.15	0.1668	453.15	0.0976
133	Ethylcyclohexane	C_8H_{16}	1678-91-7	112.21264	0.17662	−0.0002014				161.84	0.1440	404.94	0.0951
134	Ethylcyclopentane	C_7H_{14}	1640-89-7	98.18606	0.18334	−0.0002228				134.71	0.1533	376.62	0.0994
135	Ethylene	C_2H_4	74-85-1	28.05316	0.4194	−0.001591	0.000001306			104	0.2681	280	0.0763
136	Ethylenediamine	$C_2H_8N_2$	107-15-3	60.09832	0.36434	−0.0004433				284.29	0.2383	390.41	0.1913
137	Ethylene glycol	$C_2H_6O_2$	107-21-1	62.06784	0.088067	0.00094712	−1.3114E-06			260.15	0.2457	470.45	0.2434
138	Ethyleneimine	C_2H_5N	151-56-4	43.0678	0.3097	−0.0004023				195.2	0.2312	329	0.1773
139	Ethylene oxide	C_2H_4O	75-21-8	44.05256	0.26957	−0.0003984				160.65	0.2056	283.85	0.1565
140	Ethyl formate	$C_3H_6O_2$	109-94-4	74.07854	0.2587	−0.00033				193.55	0.1948	433.15	0.1158
141	2-Ethyl hexanoic acid	$C_8H_{16}O_2$	149-57-5	144.211	0.20954	−0.00022251				155.15	0.1750	500.66	0.0981
142	Ethylhexyl ether	$C_8H_{18}O$	5756-43-4	130.22792	0.19356	−0.00024102				180	0.1502	466.4	0.0812
143	Ethylisopropyl ether	$C_5H_{12}O$	625-54-7	88.14818	0.21928	−0.00032568				140	0.1737	391.2	0.0919
144	Ethylisopropyl ketone	$C_6H_{12}O$	565-69-5	100.15888	0.22873	−0.0002913				204.15	0.1693	450.1	0.0976
145	Ethyl mercaptan	C_2H_6S	75-08-1	62.13404	0.23392	−0.0003206				125.26	0.1938	308.15	0.1351
146	Ethyl propionate	$C_5H_{10}O_2$	105-37-3	102.1317	0.2137	−0.0002515				199.25	0.1636	495	0.0892
147	Ethylpropyl ether	$C_5H_{12}O$	628-32-0	88.14818	0.22717	−0.0003298				145.65	0.1791	400.07	0.0952
148	Ethyltrichlorosilane	$C_2H_5Cl_3Si$	115-21-9	163.506	0.19653	−0.00016907	−1.6698E-07			167.55	0.1635	371.05	0.1108
149	Fluorine	F_2	7782-41-4	37.9968064	0.2758	−0.0016297				53.48	0.1886	130	0.0639
150	Fluorobenzene	C_6H_5F	462-06-6	96.1023032	0.20962	−0.00028034				238.15	0.1429	353.15	0.1106
151	Fluoroethane	C_2H_5F	353-36-6	48.0595	0.25866	−0.000498				129.95	0.1939	235.45	0.1414
152	Fluoromethane	CH_3F	593-53-3	34.03292	0.48162	−0.0010709	0	0		131.35	0.3410	194.82	0.2730
153	Formaldehyde	CH_2O	50-00-0	30.02598	0.336003243	−0.00054				155.15	0.2522	253.85	0.1989
154	Formamide	CH_3NO	75-12-7	45.04062	0.3847	−0.0001065				275.7	0.3553	493	0.3322
155	Formic acid	CH_2O_2	64-18-6	46.0257	0.302	−0.000108				281.45	0.2716	373.71	0.2616
156	Furan	C_4H_4O	110-00-9	68.07396	0.2198	−0.00031405				187.55	0.1609	304.5	0.1242

157	Helium-4	He	7440-59-7	4.0026	−0.013833	0.022913	−0.0054872	0.0004585		2.2	0.0149	4.8	0.0204
158	Heptadecane	$C_{17}H_{36}$	629-78-7	240.46774	0.20926	−0.0002215				295.13	0.1439	575.3	0.0818
159	Heptanal	$C_7H_{14}O$	111-71-7	114.18546	0.22841	−0.00026273				229.8	0.1680	426.15	0.1164
160	Heptane	C_7H_{16}	142-82-5	100.20194	0.215	−0.000303				182.57	0.1597	371.58	0.1024
161	Heptanoic acid	$C_7H_{14}O_2$	111-14-8	130.185	0.202	−0.0002				265.83	0.1488	496.15	0.1028
162	1-Heptanol	$C_7H_{16}O$	111-70-6	116.20134	0.234063	−0.00025				239.15	0.1743	573.15	0.0908
163	2-Heptanol	$C_7H_{16}O$	543-49-7	116.20134	0.21142	−0.00024793				220	0.1569	432.9	0.1041
164	3-Heptanone	$C_7H_{14}O$	106-35-4	114.18546	0.2026	−0.0002234				234.15	0.1503	553.15	0.0790
165	2-Heptanone	$C_7H_{14}O$	110-43-0	114.18546	0.2108	−0.000246				238.15	0.1522	424.05	0.1065
166	1-Heptene	C_7H_{14}	592-76-7	98.18606	0.19664	−0.00016623	−2.5241E-07			154.12	0.1650	366.79	0.1017
167	Heptyl mercaptan	$C_7H_{16}S$	1639-09-4	132.26694	0.2037	−0.0002252				229.92	0.1519	450.09	0.1023
168	1-Heptyne	C_7H_{12}	628-71-7	96.17018	0.21098	−0.00026652				192.22	0.1597	372.93	0.1116
169	Hexadecane	$C_{16}H_{34}$	544-76-3	226.44116	0.20749	−0.00021917				291.31	0.1436	560.01	0.0848
170	Hexanal	$C_6H_{12}O$	66-25-1	100.15888	0.22832	−0.00026482				214.93	0.1714	401.15	0.1221
171	Hexane	C_6H_{14}	110-54-3	86.17536	0.22492	−0.0003533				177.83	0.1621	370	0.0942
172	Hexanoic acid	$C_6H_{12}O_2$	142-62-1	116.158	0.1855	−0.000146				269.25	0.1462	603.15	0.0974
173	1-Hexanol	$C_6H_{14}O$	111-27-3	102.17476	0.230656	−0.00025				228.55	0.1735	575	0.0869
174	2-Hexanol	$C_6H_{14}O$	626-93-7	102.175	0.21391	−0.00026042				223	0.1558	412.4	0.1065
175	2-Hexanone	$C_6H_{12}O$	591-78-6	100.15888	0.21076	−0.00024				217.35	0.1586	400.85	0.1146
176	3-Hexanone	$C_6H_{12}O$	589-38-8	100.15888	0.23493	−0.0002912				217.5	0.1716	466	0.0992
177	1-Hexene	C_6H_{12}	592-41-6	84.15948	0.19112	−0.000083519	−5.1407E-07			133.39	0.1708	336.63	0.1048
178	3-Hexyne	C_6H_{10}	928-49-4	82.1436	0.20996	−0.0002692				170.05	0.1642	354.35	0.1146
179	Hexyl mercaptan	$C_6H_{14}S$	111-31-9	118.24036	0.2058	−0.0002324				192.62	0.1610	425.81	0.1068
180	1-Hexyne	C_6H_{10}	693-02-7	82.1436	0.21492	−0.0002899				141.25	0.1740	344.48	0.1151
181	2-Hexyne	C_6H_{10}	764-35-2	82.1436	0.2119	−0.00027048				183.65	0.1622	357.67	0.1152
182	Hydrazine	H_4N_2	302-01-2	32.04516	1.3675	−0.0015895				274.69	0.9309	623.15	0.3770
183	Hydrogen	H_2	1333-74-0	2.01588	−0.0917	0.017678	−0.000382	−3.3324E-06	1.0266E-07	13.95	0.0754	31	0.0848
184	Hydrogen bromide	BrH	10035-10-6	80.91194	0.234	−0.0004636				185.15	0.1482	290.62	0.0993
185	Hydrogen chloride	ClH	7647-01-0	36.46094	0.8045	−0.002102				273.15	0.2303	323.15	0.1252
186	Hydrogen cyanide	CHN	74-90-8	27.02534	0.43454	−0.0007008				259.83	0.2525	298.85	0.2251
187	Hydrogen fluoride	FH	7664-39-3	20.0063432	0.7516	−0.0010874				189.79	0.5452	394.45	0.3227
188	Hydrogen sulfide	H_2S	7783-06-4	34.08088	0.4842	−0.001184				193.15	0.2555	292.42	0.1380
189	Isobutyric acid	$C_4H_8O_2$	79-31-2	88.10512	0.21668	−0.0002556				227.15	0.1586	482.75	0.0933
190	Isopropyl amine	C_3H_9N	75-31-0	59.11026	0.237	−0.000332				177.95	0.1779	305.55	0.1356
191	Malonic acid	$C_3H_4O_4$	141-82-2	104.06146	0.28231	−0.00024019				409.15	0.1840	580	0.1430
192	Methacrylic acid	$C_4H_6O_2$	79-41-4	86.08924	0.2306	−0.00025201				288.15	0.1580	530	0.0970
193	Methane	CH_4	74-82-8	16.0425	0.41768	−0.0024528	3.5588E-06			90.69	0.2245	180	0.0915
194	Methanol	CH_4O	67-56-1	32.04186	0.2837	−0.000281				175.47	0.2344	337.85	0.1888
195	*N*-Methyl acetamide	C_3H_7NO	79-16-3	73.09378	0.23743	−0.0002362				301.15	0.1663	478.15	0.1245
196	Methyl acetate	$C_3H_6O_2$	79-20-9	74.07854	0.2777	−0.000417				175.15	0.2047	386.15	0.1167
197	Methyl acetylene	C_3H_4	74-99-7	40.06386	0.23648	−0.00041639				170.45	0.1655	249.94	0.1324
198	Methyl acrylate	$C_4H_6O_2$	96-33-3	86.08924	0.26082	−0.0003506				196.32	0.1920	421	0.1132
199	Methyl amine	CH_5N	74-89-5	31.0571	0.33446	−0.00067427	8.033E-07			179.69	0.2392	283.15	0.2079
200	Methyl benzoate	$C_8H_8O_2$	93-58-3	136.14792	0.22142	−0.00022759				260.75	0.1621	547.9	0.0967
201	3-Methyl-1,2-butadiene	C_5H_8	598-25-4	68.11702	0.1983	−0.0002822				159.53	0.1533	314	0.1097
202	2-Methylbutane	C_5H_{12}	78-78-4	72.14878	0.21246	−0.00033581				113.25	0.1744	368.13	0.0888
203	2-Methylbutanoic acid	$C_5H_{10}O_2$	116-53-0	102.1317	0.22284	−0.0002516				357.15	0.1330	480.9	0.1018
204	3-Methyl-1-butanol	$C_5H_{12}O$	123-51-3	88.1482	0.17471	−0.0001256				155.95	0.1551	404.15	0.1239
205	2-Methyl-1-butene	C_5H_{10}	563-46-2	70.1329	0.19447	−0.0002901				135.58	0.1551	304.3	0.1062
206	2-Methyl-2-butene	C_5H_{10}	513-35-9	70.1329	0.19636	−0.000291				139.39	0.1558	311.7	0.1057
207	2-Methyl -1-butene-3-yne	C_5H_6	78-80-8	66.10114	0.20385	−0.0002874				160.15	0.1578	305.4	0.1161
208	Methylbutyl ether	$C_5H_{12}O$	628-28-4	88.14818	0.22235	−0.0003044				157.48	0.1744	463.15	0.0814
209	Methylbutyl sulfide	$C_5H_{12}S$	628-29-5	104.214	0.20698	−0.00024439				175.3	0.1641	396.58	0.1101
210	3-Methyl-1-butyne	C_5H_8	598-23-2	68.11702	0.20348	−0.0003106				183.45	0.1465	302.15	0.1096
211	Methyl butyrate	$C_5H_{10}O_2$	623-42-7	102.1317	0.21748	−0.00025913				187.35	0.1689	493.15	0.0897

(*Continued*)

TABLE 2-147 Thermal Conductivity of Inorganic and Organic Liquids [W/(m·K)] (*Continued*)

Cmpd. no.	Name	Formula	CAS	Mol. wt.	C_1	C_2	C_3	C_4	C_5	T_{min}, K	Thermal cond. at T_{min}	T_{max}, K	Thermal cond. at T_{max}
212	Methylchlorosilane	CH_5ClSi	993-00-0	80.5889	0.24683	−0.00038854				139.05	0.1928	281.85	0.1373
213	Methylcyclohexane	C_7H_{14}	108-87-2	98.18606	0.1791	−0.0002291				273.15	0.1165	374.08	0.0934
214	1-Methylcyclohexanol	$C_7H_{14}O$	590-67-0	114.18546	0.21558	−0.00022728				299.15	0.1476	548.8	0.0909
215	*cis*-2-Methylcyclohexanol	$C_7H_{14}O$	7443-70-1	114.18546	0.21839	−0.00025776				280.15	0.1462	484.2	0.0936
216	*trans*-2-Methylcyclohexanol	$C_7H_{14}O$	7443-52-9	114.18546	0.21828	−0.0002557				269.15	0.1495	484.8	0.0943
217	Methylcyclopentane	C_6H_{12}	96-37-7	84.15948	0.1929	−0.0002492				130.73	0.1603	344.95	0.1069
218	1-Methylcyclopentene	C_6H_{10}	693-89-0	82.1436	0.20023	−0.00025581				146.62	0.1627	348.64	0.1110
219	3-Methylcyclopentene	C_6H_{10}	1120-62-3	82.1436	0.1994	−0.00026149				168.54	0.1553	338.05	0.1110
220	Methyldichlorosilane	CH_4Cl_2Si	75-54-7	115.03396	0.21956	−0.00032153				182.55	0.1609	314.7	0.1184
221	Methylethyl ether	C_3H_8O	540-67-0	60.09502	0.27304	−0.0004518				160	0.2008	341.34	0.1188
222	Methylethyl ketone	C_4H_8O	78-93-3	72.10572	0.2197	−0.0002505				186.48	0.1730	352.79	0.1313
223	Methylethyl sulfide	C_3H_8S	624-89-5	76.1606	0.22136	−0.00028938				167.23	0.1730	339.8	0.1230
224	Methyl formate	$C_2H_4O_2$	107-31-3	60.05196	0.3246	−0.000468				174.15	0.2431	373.15	0.1500
225	Methylisobutyl ether	$C_5H_{12}O$	625-44-5	88.14818	0.222	−0.00032217				188	0.1614	390	0.0964
226	Methylisobutyl ketone	$C_6H_{12}O$	108-10-1	100.15888	0.2301	−0.00028899				189.15	0.1754	451.42	0.0996
227	Methyl Isocyanate	C_2H_3NO	624-83-9	57.05132	0.2822	−0.00042037				256.15	0.1745	312	0.1510
228	Methylisopropyl ether	$C_4H_{10}O$	598-53-8	74.1216	0.24154	−0.0003774				127.93	0.1933	370	0.1019
229	Methylisopropyl ketone	$C_5H_{10}O$	563-80-4	86.1323	0.2332	−0.0003044				180.15	0.1784	435.9	0.1005
230	Methylisopropyl sulfide	$C_4H_{10}S$	1551-21-9	90.1872	0.20978	−0.00026468				171.64	0.1644	357.91	0.1150
231	Methyl mercaptan	CH_4S	74-93-1	48.10746	0.26119	−0.00038345				150.18	0.2036	279.11	0.1542
232	Methyl methacrylate	$C_5H_8O_2$	80-62-6	100.11582	0.2583	−0.000379				290.15	0.1483	363.45	0.1206
233	2-Methyloctanoic acid	$C_9H_{18}O_2$	3004-93-1	158.23802	0.20911	−0.00021852				208.2	0.1636	555.2	0.0878
234	2-Methylpentane	C_6H_{14}	107-83-5	86.17536	0.19334	−0.00028038				119.55	0.1598	389.25	0.0842
235	Methyl pentyl ether	$C_6H_{14}O$	628-80-8	102.17476	0.21698	−0.00028998				176	0.1659	432.3	0.0916
236	2-Methylpropane	C_4H_{10}	75-28-5	58.1222	0.20455	−0.00036589				113.54	0.1630	400	0.0582
237	2-Methyl-2-propanol	$C_4H_{10}O$	75-65-0	74.1216	0.21258	−0.00029864				298.97	0.1233	404.96	0.0916
238	2-Methyl propene	C_4H_8	115-11-7	56.10632	0.2802	−0.000786	6.516E-07			132.81	0.1873	395.2	0.0713
239	Methyl propionate	$C_4H_8O_2$	554-12-1	88.10512	0.22534	−0.0002683				185.65	0.1755	475	0.0979
240	Methylpropyl ether	$C_4H_{10}O$	557-17-5	74.1216	0.24817	−0.0003774				133.97	0.1976	373	0.1074
241	Methylpropyl sulfide	$C_4H_{10}S$	3877-15-4	90.1872	0.21103	−0.00025985				160.17	0.1694	368.69	0.1152
242	Methylsilane	CH_6Si	992-94-9	46.14384	0.2774	−0.00054608				116.34	0.2139	216.25	0.1593
243	alpha-Methyl styrene	C_9H_{10}	98-83-9	118.1757	0.19657	−0.0002118				249.95	0.1436	438.65	0.1037
244	Methyl tert-butyl ether	$C_5H_{12}O$	1634-04-4	88.1482	0.22526	−0.00037235	1.1689E-07	0	0	164.55	0.1672	328.2	0.1156
245	Methyl vinyl ether	C_3H_6O	107-25-5	58.07914	0.28035	−0.0004646				151.15	0.2101	341.1	0.1219
246	Naphthalene	$C_{10}H_8$	91-20-3	128.17052	0.17096	−0.00010059				353.43	0.1354	646.97	0.1059
247	Neon	Ne	7440-01-9	20.1797	0.2971	−0.017356	0.0005911	−0.000007421		25	0.1167	44	0.0457
248	Nitroethane	$C_2H_5NO_2$	79-24-3	75.0666	0.247	−0.0002814				183.63	0.1953	387.22	0.1380
249	Nitrogen	N_2	7727-37-9	28.0134	0.2654	−0.001677				63.15	0.1595	124	0.0575
250	Nitrogen trifluoride	F_3N	7783-54-2	71.00191									
251	Nitromethane	CH_3NO_2	75-52-5	61.04002	0.3276	−0.000405				244.6	0.2285	374.35	0.1760
252	Nitrous oxide	N_2O	10024-97-2	44.0128	0.10112					277.59	0.1011	277.59	0.1011
253	Nitric oxide	NO	10102-43-9	30.0061	0.1878	0.0010293	−0.00000943			110	0.1869	176.4	0.0759
254	Nonadecane	$C_{19}H_{40}$	629-92-5	268.5209	0.21229	−0.00022				305.04	0.1452	603.05	0.0796
255	Nonanal	$C_9H_{18}O$	124-19-6	142.23862	0.21905	−0.00024013				267.3	0.1549	465.52	0.1073
256	Nonane	C_9H_{20}	111-84-2	128.2551	0.209	−0.000264				219.66	0.1510	423.97	0.0971
257	Nonanoic acid	$C_9H_{18}O_2$	112-05-0	158.238	0.204	−0.0002				285.55	0.1469	528.75	0.0983
258	1-Nonanol	$C_9H_{20}O$	143-08-8	144.2545	0.240538	−0.00025				268.15	0.1735	578.65	0.0959
259	2-Nonanol	$C_9H_{20}O$	628-99-9	144.255	0.2081	−0.00022869				238.15	0.1536	471.7	0.1002
260	1-Nonene	C_9H_{18}	124-11-8	126.23922	0.20468	−0.00025738				191.91	0.1553	420.02	0.0966
261	Nonyl mercaptan	$C_9H_{20}S$	1455-21-6	160.3201	0.20244	−0.00021343				253.05	0.1484	492.95	0.0972
262	1-Nonyne	C_9H_{16}	3452-09-3	124.22334	0.20954	−0.00024588				223.15	0.1547	423.85	0.1053

263	Octadecane	$C_{18}H_{38}$	593-45-3	254.49432	0.2137	−0.0002252				301.31	0.1458	589.86	0.0809
264	Octanal	$C_8H_{16}O$	124-13-0	128.212	0.22273	−0.00025037				251.65	0.1597	445.15	0.1113
265	Octane	C_8H_{18}	111-65-9	114.22852	0.2156	−0.00029483				216.38	0.1518	398.83	0.0980
266	Octanoic acid	$C_8H_{16}O_2$	124-07-2	144.211	0.203	−0.0002				289.65	0.1451	512.85	0.1004
267	1-Octanol	$C_8H_{18}O$	111-87-5	130.22792	0.235281	−0.00025				257.65	0.1709	570.15	0.0927
268	2-Octanol	$C_8H_{18}O$	123-96-6	130.228	0.20955	−0.00023733				241.55	0.1522	452.9	0.1021
269	2-Octanone	$C_8H_{16}O$	111-13-7	128.21204	0.2132	−0.0002494				252.85	0.1501	499	0.0888
270	3-Octanone	$C_8H_{16}O$	106-68-3	128.21204	0.21732	−0.00024969				255.55	0.1535	440.65	0.1073
271	1-Octene	C_8H_{16}	111-66-0	112.21264	0.20467	−0.0002675				171.45	0.1588	394.41	0.0992
272	Octyl mercaptan	$C_8H_{18}S$	111-88-6	146.29352	0.2012	−0.0002142				223.95	0.1532	472.19	0.1001
273	1-Octyne	C_8H_{14}	629-05-0	110.19676	0.2095	−0.00025334				193.55	0.1605	399.35	0.1083
274	Oxalic acid	$C_2H_2O_4$	144-62-7	90.03488	0.26335	−0.00022461				462.65	0.1594	516	0.1475
275	Oxygen	O_2	7782-44-7	31.9988	0.2741	−0.00138				60	0.1913	150	0.0671
276	Ozone	O_3	10028-15-6	47.9982	0.17483	0.00075288	−2.5228E-06			77.35	0.2180	161.85	0.2306
277	Pentadecane	$C_{15}H_{32}$	629-62-9	212.41458	0.20649	−0.00021911				283.07	0.1445	543.84	0.0873
278	Pentanal	$C_5H_{10}O$	110-62-3	86.1323	0.23894	−0.00029724				191.59	0.1820	375.15	0.1274
279	Pentane	C_5H_{12}	109-66-0	72.14878	0.2537	−0.000576	0.000000344			143.42	0.1782	445	0.0655
280	Pentanoic acid	$C_5H_{10}O_2$	109-52-4	102.132	0.1848	−0.0001434				239.15	0.1505	458.65	0.1190
281	1-Pentanol	$C_5H_{12}O$	71-41-0	88.1482	0.223042	−0.00025				273.15	0.1548	353.15	0.1348
282	2-Pentanol	$C_5H_{12}O$	6032-29-7	88.1482	0.21875	−0.00027849				200	0.1631	392.2	0.1095
283	2-Pentanone	$C_5H_{10}O$	107-87-9	86.1323	0.2161	−0.00024866				196.29	0.1673	375.46	0.1227
284	3-Pentanone	$C_5H_{10}O$	96-22-0	86.1323	0.21569	−0.00024081				234.18	0.1593	375.14	0.1254
285	1-Pentene	C_5H_{10}	109-67-1	70.1329	0.21361	−0.00030777				108.02	0.1804	303.22	0.1203
286	2-Pentyl mercaptan	$C_5H_{12}S$	2084-19-7	104.21378	0.20597	−0.00024518				160.75	0.1666	385.15	0.1115
287	Pentyl mercaptan	$C_5H_{12}S$	110-66-7	104.21378	0.2086	−0.00024536				197.45	0.1602	399.79	0.1105
288	1-Pentyne	C_5H_8	627-19-0	68.11702	0.22102	−0.000322				167.45	0.1671	313.33	0.1201
289	2-Pentyne	C_5H_8	627-21-4	68.11702	0.21282	−0.0002856				163.83	0.1660	329.27	0.1188
290	Phenanthrene	$C_{14}H_{10}$	85-01-8	178.2292	0.13753	−0.000025247				372.38	0.1281	610.03	0.1221
291	Phenol	C_6H_6O	108-95-2	94.11124	0.18831	−0.0001				314.06	0.1569	454.99	0.1428
292	Phenyl isocyanate	C_7H_5NO	103-71-9	119.1207	0.16326	−0.00017777				243.15	0.1200	439.43	0.0851
293	Phthalic anhydride	$C_8H_4O_3$	85-44-9	148.11556	0.22946	−0.00021345				404.15	0.1432	557.65	0.1104
294	Propadiene	C_3H_4	463-49-0	40.06386	0.23081	−0.0004078				136.87	0.1750	238.65	0.1335
295	Propane	C_3H_8	74-98-6	44.09562	0.26755	−0.00066457	2.774E-07			85.47	0.2128	350	0.0689
296	1-Propanol	C_3H_8O	71-23-8	60.09502	0.23144	−0.00025				200	0.1814	370.35	0.1389
297	2-Propanol	C_3H_8O	67-63-0	60.095	0.20161	−0.00021529				185.26	0.1617	425	0.1101
298	Propenylcyclohexene	C_9H_{14}	13511-13-2	122.20746	0.1831	−0.00020275				199	0.1428	431.65	0.0956
299	Propionaldehyde	C_3H_6O	123-38-6	58.07914	0.31721	−0.000528				165	0.2301	322.15	0.1471
300	Propionic acid	$C_3H_6O_2$	79-09-4	74.0785	0.1954	−0.000164				252.45	0.1540	543.15	0.1063
301	Propionitrile	C_3H_5N	107-12-0	55.0785	0.26743	−0.00033418				180.37	0.2072	370.25	0.1437
302	Propyl acetate	$C_5H_{10}O_2$	109-60-4	102.1317	0.2332	−0.0003096				178.15	0.1780	434.82	0.0986
303	Propyl amine	C_3H_9N	107-10-8	59.11026	0.2632	−0.0004278	0.000000412			188.36	0.1972	333.15	0.1664
304	Propylbenzene	C_9H_{12}	103-65-1	120.19158	0.18707	−0.00019846				173.55	0.1526	583.15	0.0713
305	Propylene	C_3H_6	115-07-1	42.07974	0.24719	−0.00048824				87.89	0.2043	340.49	0.0810
306	Propyl formate	$C_4H_8O_2$	110-74-7	88.10512	0.2247	−0.000264				180.25	0.1771	483.15	0.0972
307	2-Propyl mercaptan	C_3H_8S	75-33-2	76.16062	0.21706	−0.00028952				142.61	0.1758	325.71	0.1228
308	Propyl mercaptan	C_3H_8S	107-03-9	76.16062	0.2202	−0.00028535				159.95	0.1746	340.87	0.1229
309	1,2-Propylene glycol	$C_3H_8O_2$	57-55-6	76.09442	0.2152	−0.0000497				213.15	0.2046	460.75	0.1923
310	Quinone	$C_6H_4O_2$	106-51-4	108.09476	0.26524	−0.00028676				388.85	0.1537	545	0.1090
311	Silicon tetrafluoride	F_4Si	7783-61-1	104.07911									
312	Styrene	C_8H_8	100-42-5	104.14912	0.20215	−0.0002201				242.54	0.1488	418.31	0.1101
313	Succinic acid	$C_4H_6O_4$	110-15-6	118.08804	0.27216	−0.00023183				460.85	0.1653	591	0.1351
314	Sulfur dioxide	O_2S	7446-09-5	64.0638	0.38218	−0.0006254				197.67	0.2586	400	0.1320
315	Sulfur hexafluoride	F_6S	2551-62-4	146.0554192	0.2544	−0.0006595				223.15	0.1072	318.69	0.0442
316	Sulfur trioxide	O_3S	7446-11-9	80.0632	0.92882	−0.0030803	0.00000266			289.95	0.2593	481.47	0.0624
317	Terephthalic acid	$C_8H_6O_4$	100-21-0	166.13084	0.3063	−0.00028541				700.15	0.1065	795.28	0.0793

(*Continued*)

TABLE 2-147 Thermal Conductivity of Inorganic and Organic Liquids [W/(m·K)] *(Continued)*

Cmpd. no.	Name	Formula	CAS	Mol. wt.	C_1	C_2	C_3	C_4	C_5	T_{min}, K	Thermal cond. at T_{min}	T_{max}, K	Thermal cond. at T_{max}
318	*o*-Terphenyl	$C_{18}H_{14}$	84-15-1	230.30376	0.16853	−0.00010817				329.35	0.1329	723.15	0.0903
319	Tetradecane	$C_{14}H_{30}$	629-59-4	198.388	0.20293	−0.00021798				279.01	0.1421	526.73	0.0881
320	Tetrahydrofuran	C_4H_8O	109-99-9	72.10572	0.19428	−0.000249				164.65	0.1533	339.12	0.1098
321	1,2,3,4-Tetrahydronaphthalene	$C_{10}H_{12}$	119-64-2	132.20228	0.14563	−0.0000536				237.38	0.1329	480.77	0.1199
322	Tetrahydrothiophene	C_4H_8S	110-01-0	88.17132	0.20414	−0.00021217				176.98	0.1666	394.27	0.1205
323	2,2,3,3-Tetramethylbutane	C_8H_{18}	594-82-1	114.22852	0.17835	−0.00023704				373.96	0.0897	426	0.0774
324	Thiophene	C_4H_4S	110-02-1	84.13956	0.20571	−0.00020028				234.94	0.1587	357.31	0.1341
325	Toluene	C_7H_8	108-88-3	92.13842	0.20463	−0.00024252				178.18	0.1614	474.85	0.0895
326	1,1,2-Trichloroethane	$C_2H_3Cl_3$	79-00-5	133.40422	0.20731	−0.00024997				236.5	0.1482	482	0.0868
327	Tridecane	$C_{13}H_{28}$	629-50-5	184.36142	0.20447	−0.00022612				267.76	0.1439	508.62	0.0895
328	Triethyl amine	$C_6H_{15}N$	121-44-8	101.19	0.1918	−0.0002453				158.45	0.1529	483.15	0.0733
329	Trimethyl amine	C_3H_9N	75-50-3	59.11026	0.23813	−0.00038397				156.08	0.1782	276.02	0.1321
330	1,2,3-Trimethylbenzene	C_9H_{12}	526-73-8	120.19158	0.18854	−0.0001963				247.79	0.1399	449.27	0.1003
331	1,2,4-Trimethylbenzene	C_9H_{12}	95-63-6	120.19158	0.19216	−0.0002105				229.33	0.1439	442.53	0.0990
332	2,2,4-Trimethylpentane	C_8H_{18}	540-84-1	114.22852	0.1659	−0.00022686				165.78	0.1283	372.39	0.0814
333	2,3,3-Trimethylpentane	C_8H_{18}	560-21-4	114.22852	0.16815	−0.00020535				172.22	0.1328	387.91	0.0885
334	1,3,5-Trinitrobenzene	$C_6H_3N_3O_6$	99-35-4	213.10452	0.18421	−0.00016097				398.4	0.1201	629.6	0.0829
335	2,4,6-Trinitrotoluene	$C_7H_5N_3O_6$	118-96-7	227.1311	0.19898	−0.00017659				354	0.1365	625	0.0886
336	Undecane	$C_{11}H_{24}$	1120-21-4	156.30826	0.20515	−0.00023933				247.57	0.1459	469.08	0.0929
337	1-Undecanol	$C_{11}H_{24}O$	112-42-5	172.30766	0.218744	−0.00025				281	0.1485	561.2	0.0784
338	Vinyl acetate	$C_4H_6O_2$	108-05-4	86.08924	0.256	−0.0003542				180.35	0.1921	410	0.1108
339	Vinyl acetylene	C_4H_4	689-97-4	52.07456	0.22838	−0.00035173				173.15	0.1675	278.25	0.1305
340	Vinyl chloride	C_2H_3Cl	75-01-4	62.49822	0.2333	−0.00039223				119.36	0.1865	345.6	0.0978
341	Vinyl trichlorosilane	$C_2H_3Cl_3Si$	75-94-5	161.48972	0.21831	−0.00029122				178.35	0.1664	434.52	0.0918
342	Water	H_2O	7732-18-5	18.01528	−0.432	0.0057255	−0.000008078	1.861E-09		273.16	0.5672	633.15	0.4272
343	*m*-Xylene	C_8H_{10}	108-38-3	106.165	0.20044	−0.00023544				225.3	0.1474	413.1	0.1032
344	*o*-Xylene	C_8H_{10}	95-47-6	106.165	0.19989	−0.0002299				247.98	0.1429	417.58	0.1039
345	*p*-Xylene	C_8H_{10}	106-42-3	106.165	0.20003	−0.00023573				286.41	0.1325	413.1	0.1026

The liquid thermal conductivity is calculated by $k = C_1 + C_2T + C_3T_2 + C_4T_3 + C_5T_4$ where k is the thermal conductivity in W/(m·K) and T is the temperature in K. Thermal conductivities are at either 1 atm or the vapor pressure, whichever is higher.

Values in this table were taken from the Design Institute for Physical Properties (DIPPR) of the American Institute of Chemical Engineers (AIChE), 801 Critically Evaluated Gold Standard™ Database, copyright 2016 AIChE, and reproduced with permission of AIChE and of the DIPPR Evaluated Process Design Data Project Steering Committee. Their source should be cited as "R. L. Rowley, W. V. Wilding, J. L. Oscarson, T. A. Knotts, N. F. Giles, *DIPPR® Data Compilation of Pure Chemical Properties*, Design Institute for Physical Properties, AIChE, New York, NY (2016)".

TABLE 2-148

No.	Compound	Range, °C	Exp pt	% Avg abs devi
31	Acetaldehyde	0 – 30	2	0.4
40	Acetic acid	18 – 79	5	0.1
29	Acetone	0 – 40	3	0.4
20	Aniline	0 – 93	6	1.3
2	Benzaldehyde	16 – 68	4	0.7
13	Benzene	20 – 116	5	0.6
47	n-Butane	−21 – 0	2	0.2
16	n-Butanol	0 – 102	7	0.4
48	i-Butanol	0 – 20	2	0.6
49	s-Butanol	0 – 65	3	0.2
18	t-Butanol	20 – 77	3	2.5
28	Butyl acetate	0 – 38	4	1.1
23	Carbon tetrachloride	−20 – 25	4	1.1
25	Chlorobenzene	−40 – 80	6	0.5
42	Chloroform	0 – 40	3	1.3
19	m-Cresol	20 – 80	2	0.2
11	Cyclohexane	20 – 38	2	0.4
7	n-Decane	20 – 76	5	1.0
30	Diethyl ether	0 – 25	3	0.7
9	2,3-Dimethylbutane	32 – 49	2	0.3
38	n-Dodecane	0 – 63	6	1.1
15	Ethanol	17 – 77	7	0.5
27	Ethyl acetate	6 – 160	9	1.6
14	Ethylbenzene	0 – 80	3	0.7
24	Ethyl bromide	0 – 30	3	0.3
4	n-Heptane	0 – 60	3	0.3
3	n-Hexane	17 – 40	5	1.4
26	Iodobenzene	−20 – 80	4	0.7
41	Methanol	20 – 62	6	0.2
39	Methyl acetate	4 – 21	2	0.4
12	Methylcyclopentane	20 – 38	2	0.1
22	Methylene chloride	−20 – 20	3	0.9
8	2-Methylpentane	32 – 49	2	0.5
6	n-Nonane	16 – 77	5	2.0
5	n-Octane	0 – 77	3	0.3
10	i-Octane	20 – 77	4	1.6
17	n-Octanol	20 – 212	3	1.3
43	n-Pentanol	0 – 38	3	0.4
1	Propane	−60 – 50	4	1.6
32	n-Propanol	16 – 66	6	0.6
21	Propionic acid	12 – 30	3	1.7
44	n-Propyl acetate	0 – 38	2	0.1
33	i-Propylbenzene	0 – 38	4	0.2
34	Toluene	−70 – 200	5	1.8
55	n-Tridecane	20 – 226	5	2.4
36	m-Xylene	0 – 208	5	1.0
35	o-Xylene	0 – 208	5	0.7
37	p-Xylene	0 – 251	5	1.4

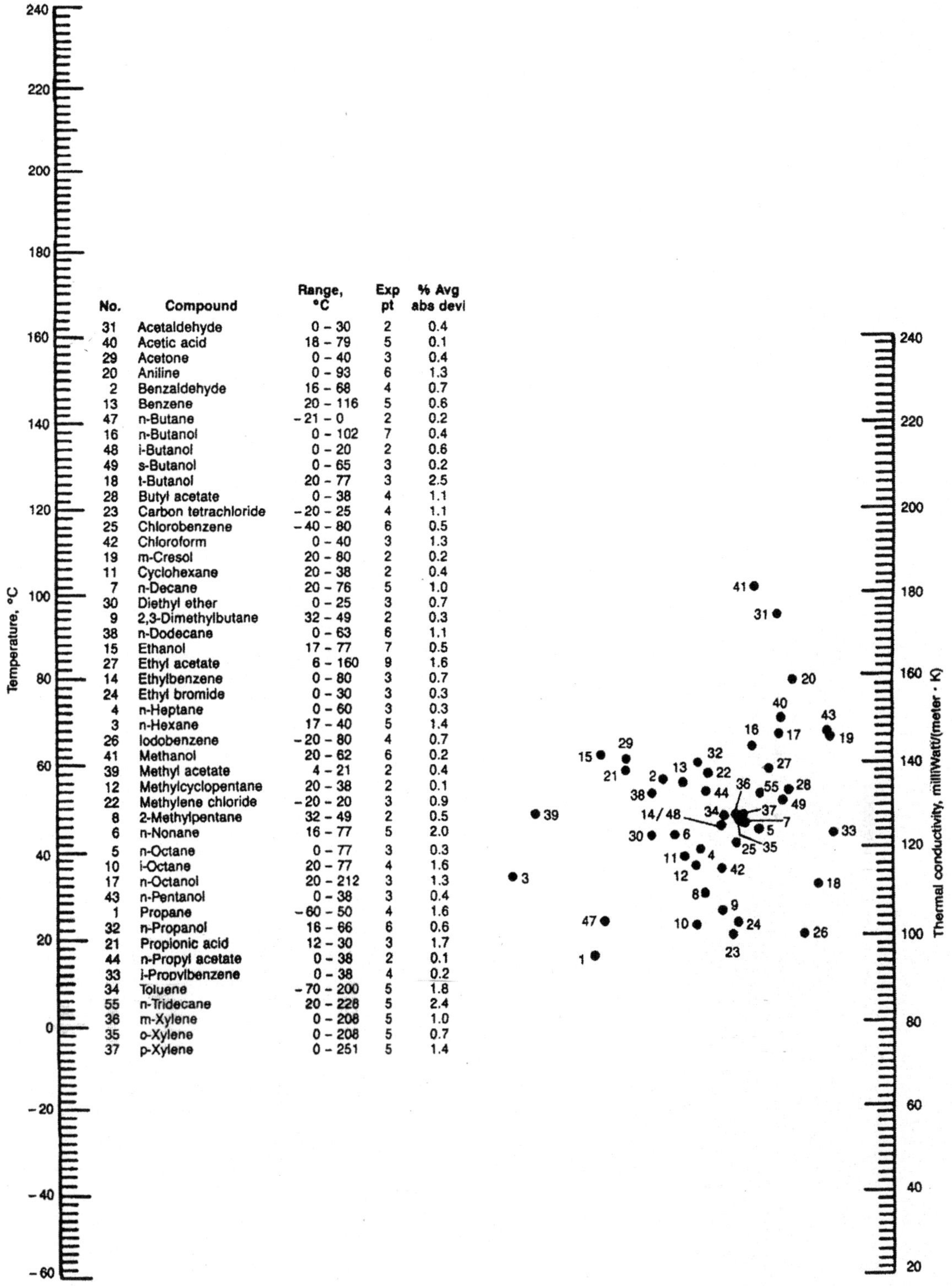

FIG. 2-20 and TABLE 2-148 Nomograph (*right*) for thermal conductivity of organic liquids. (From B.V. Mallu and Y.J. Rao, *Hydroc. Proc.* 78, 1988.)

TABLE 2-149 Thermal-Conductivity-Temperature Table for Metals and Nonmetals*

Thermal conductivities tabulated in watts per meter-kelvin

Substance	Temperature, K														
	10	20	40	60	80	100	200	300	400	500	600	800	1000	1200	1400
Alumina	7	32	121	174	160	125	55	36	26	20	16	10	8	7	6
Aluminum	38,000	13,500	2,300	850	380	300	237	273	240	237	232	220	93	99	105
Antimony	470	230	110	80	60	48	32	26	22	20					
Beryllium oxide	47	196	810	1,400	1,650	1,490	480	272	196	146	111	70	47	33	25
Bismuth	240	100	45	31	24	22	18	16	14	12					
Boron	165	305	400	327	230	170	45	25	15	12					
Cadmium	900	250	150	120	110	110	105	104	101	99					
Chromium	400	570	450	250	180	158	111	90	87	85	81	71	65	62	61
Cobalt	250	450	380	250	190	160	120	100	85	70					
Constantan	4	9	16	18	19	20	23	25	27	30					
Copper	19,000	10,700	2,100	850	570	483	413	398	392	388	383	371	357	342	
Gallium	2,200	640	250	200	170	140	100	85							
Gold	2,800	1,500	520	380	350	345	327	315	312	309	304	292	278	262	
Graphite†	27	108	135	81	54	39	15	10	7	5	4	3	3	2	2
Graphite‡	81	420	1,630	2,980	4,290	4,980	3,250	2,000	1,460	1,140	930	680	530	440	370
Hastelloy	1	3	4	5	6	7	9	10	11	13					
Inconel	2	4	8	10	11	11	14	15							
Iridium	1,300	1,900	750	360	230	172	147	145	143	140					
Iron	710	1,000	560	270	170	132	94	80	69	61	55	43	33	28	31
Lead	175	57	43	42	41	40	37	35	34	33	31	19	22	24	26
Magnesium	1,200	1,300	620	290	190	169	159	156	153	151	149	146	84	98	112
Magnesium oxide	1,100	3,100	2,200	950	460	260	75	48	36	27	21	13	10	8	7
Manganese	2	2	4	5	5	6	7	8	9	9					
Manganin	2	4	9	11	13	13	17	22	28	34	40				
Mercury	54	40	35	33	33	32	32	8	10	11	12	13	14		
Molybdenum	150	280	350	250	210	179	143	138	134	130	126	118	112	105	100
Nickel	2,600	1,700	570	290	200	158	106	91	80	72	66	67	72	76	80
Nylon	0.04	0.10	0.17	0.20	0.23	0.25	0.28	0.30							
Palladium	1,200	610	160	100	88	80	78	78	78	80					
Platinum	1,200	490	130	92	82	79	75	73	72	72	72	73	78	78	81
PTFE§	0.94	1.43	1.94	2.1	2.15	2.16	2.20	2.25	2.3	2.5					
Pyrex	0.12	0.20	0.33	0.42	0.51	0.57	0.88	1.1	1.6	2.1					
Quartz	1,200	480	82	40	30										
Rhodium	2,900	3,900	1,000	370	250	190	160	150	145	140					
Rubber			0.13	0.15	0.16	0.17	0.20	0.22	0.24	0.25					
Selenium (axis)	140	57	25	15	10	8	6	4	3	2					
Silica								1.34	1.52	1.70	1.87	2.22	2.60		
Silver	16,500	5,200	1,100	630	500	430	425	424	420	413	405	389	374	358	
Tantalum	108	146	88	68	62	59	58	57	58	58	59	59	60	61	62
Tellurium	300	93	29	17	13	11	6	4	3	3					
Tin		320	130	101	90	84	72	67	62	60					
Titanium	14	28	39	37	33	31	26	21	20	20	19				
Tungsten			880	330	310	280	190	180	170	150	140				
Uranium				20	22	23	26	28	30	32					
Zinc				150	135	130	123	120	116	110	110				
Zirconium	100	110	59	42	38	34	25	23	22	21	21				

*Especially at low temperatures, the thermal conductivity can often be markedly reduced by even small traces of impurities. This table, for the highest-purity specimens available, should thus be used with caution in applications with commercial materials. From Perry, *Engineering Manual,* 3d ed., McGraw-Hill, New York, 1976. A more detailed table appears as Section 5.5.6 in the *Heat Exchanger Design Handbook,* Hemisphere Pub. Corp., Washington, DC, 1983.

†Parallel to basal plane.

‡Perpendicular to basal plane.

§Also known as Teflon, etc.

TABLE 2-150 Thermal Conductivity of Chromium Alloys*

k = Btu/(h·ft²)(°F/ft)

American iron and steel institute type no.	k at 212°F	k at 932°F
301, 302, 302B, 303, 304, 316†	9.4	12.4
308	8.8	12.5
309, 310	8.0	10.8
321, 347	9.3	12.8
403, 406, 410, 414, 416†	14.4	16.6
430, 430F†	15.1	15.2
442	12.5	14.2
501, 502†	21.2	19.5

*Table 2-150 is based on information from manufacturers.
†Shelton and Swanger (National Bureau of Standards), *Trans. Am. Soc. Steel Treat.*, **21**, 1061–1078 (1933).

TABLE 2-151 Thermal Conductivity of Some Alloys at High Temperature*

°R	Thermal conductivity, Btu/(ft)(hr)(°R)					
	Kovar	Advance	Monel	Hastelloy A	Inconel	Nichrome V
500	7.8		9.0	5.6	6.0	5.5
600	8.3	11.4	10.2	6.2	6.5	6.1
700	8.6	12.6	11.2	6.8	7.0	6.7
800	8.7	13.9	12.3	7.3	7.6	7.3
900	8.7	15.1	13.4	7.8	8.1	7.8
1000	8.9	16.4	14.4	8.4	8.6	8.4
1100	9.2	17.6	15.4	9.0	9.1	9.0
1200	9.5	18.8	16.5	9.5	9.7	9.5
1300	9.8	20.0	17.6	10.1	10.2	10.1
1400	10.2	21.2	18.7	10.7	10.8	10.7
1500	10.5	22.5	19.8	11.3	11.3	11.3
1600	10.8	23.8	20.8	11.8	11.8	11.9
1700	11.1	25.0	21.9	12.3	12.4	12.4
1800	11.3	26.2	23.0	12.9	13.0	13.0
1900	11.5	27.4	24.0	13.4	13.6	13.5
2000	11.8	28.7	25.1	14.0	14.0	14.1
2100	12.1	30.0	26.1	14.6	14.5	14.7
2200	12.3		27.2	15.1	15.0	15.3

*Silverman, *J. Metals*, **5**, 631 (1953). Copyright American Institute of Mining, Metallurgical and Petroleum Engineers, Inc.

TABLE 2-152 Thermophysical Properties of Selected Nonmetallic Solid Substances

Material	Density, kg/m³	Emissivity	Specific heat, kJ/(kg·K)	Thermal conductivity, W/(m·K)	Thermal diffusivity, m²/s × 10⁶
Alumina	3975		0.765	36	11.9
Asphalt	2110		0.920	0.06	0.03
Bakelite	1300		1.465	1.4	0.74
Beryllia	3000	0.82	1.030	270	88
Brick	1925	0.93	0.835	0.72	0.45
Brick, fireclay	2640	0.93	0.960	1.0	0.39
Carbon, amorphous	1950	0.86	0.724	1.6	1.13
Clay	1460	0.91	0.880	1.3	1.01
Coal	1350	0.80	1.26	0.26	0.15
Cotton	80		1.30	0.06	0.58
Diamond	3500		0.509	2300	1290
Granite	2630		0.775	2.79	1.37
Hardboard	1000		1.38	0.15	0.11
Magnesite	3025	0.38	1.13	4.0	1.2
Magnesia	3635	0.72	0.943	48	14
Oak	770	0.90	2.38	0.18	0.10
Paper	930	0.83	1.34	0.011	0.01
Pine	525	0.84	2.75	0.12	0.54
Plaster board	800	0.91		0.17	
Plywood	540		1.22	0.12	0.18
Pyrex	2250	0.92	0.835	1.4	0.74
Rubber	1150	0.92	2.00	0.2	0.09
Rubber, foam	70	0.90		0.03	
Salt		0.34	0.854	7.1	
Sandstone	2150	0.59	0.745	2.9	1.8
Silica		0.79	0.743	1.3	
Sapphire	3975	0.48	0.765	46	15
Silicon carbide	3160	0.86	0.675	110	230
Soil	2050	0.38	1.84	0.52	0.14
Teflon	2200	0.92	0.35	0.26	0.34
Thoria	4160	0.28	0.71	14	4.7
Urethane foam	70		1.05	0.03	0.36
Vermiculite	120		0.84	0.06	0.60

NOTE: Difficulties of accurately characterizing many of the specimens mean that many of the values presented here must be regarded as being of order of magnitude only. For some materials, actual measurement may be the only way to obtain data of the required accuracy. To convert kilograms per cubic meter to pounds per cubic foot, multiply by 0.062428; to convert kilojoules per kilogram-kelvin to British thermal units per pound-degree Fahrenheit, multiply by 0.23885.

TABLE 2-153 Lower and Upper Flammability Limits, Flash Points, and Autoignition Temperatures for Selected Hydrocarbons

Group	Compound	CAS	Formula	LFL	UFL	Flash point (K)	Autoignition T (K)
Paraffin hydrocarbons	Methane	74-82-8	CH_4	5.00	15.00	87.12	810.00
Paraffin hydrocarbons	Ethane	74-84-0	C_2H_6	3.00	12.40	139.00	745.00
Paraffin hydrocarbons	Propane	74-98-6	C_3H_8	2.10	9.50	171.00	723.00
Paraffin hydrocarbons	*n*-Butane	106-97-8	C_4H_{10}	1.60	8.40	199.15	561.00
Paraffin hydrocarbons	Isobutane	75-28-5	C_4H_{10}	1.80	8.40	191.00	733.15
Paraffin hydrocarbons	*n*-Pentane	109-66-0	C_5H_{12}	1.40	7.80	224.15	516.00
Paraffin hydrocarbons	Isopentane	78-78-4	C_5H_{12}	1.40	7.60	218.00	693.15
Paraffin hydrocarbons	Neopentane	463-82-1	C_5H_{12}	1.40	7.50	205.00	723.15
Paraffin hydrocarbons	*n*-Hexane	110-54-3	C_6H_{14}	1.20	7.20	250.15	498.00
Paraffin hydrocarbons	*n*-Heptane	142-82-5	C_7H_{16}	1.05	6.70	269.00	477.00
Paraffin hydrocarbons	2,3-Dimethylpentane	565-59-3	C_7H_{16}	1.10	6.70	261.00	608.15
Paraffin hydrocarbons	*n*-Octane	111-65-9	C_8H_{18}	0.96	6.50	287.15	479.00
Paraffin hydrocarbons	2,2,4-Trimethylpentane	540-84-1	C_8H_{18}	0.95	6.00	265.00	684.15
Paraffin hydrocarbons	*n*-Nonane	111-84-2	C_9H_{20}	0.85	5.60	304.15	478.00
Paraffin hydrocarbons	*n*-Decane	124-18-5	$C_{10}H_{22}$	0.75	5.40	322.85	474.00
Olefins	Ethylene	74-85-1	C_2H_4	2.70	36.00	129.00	723.15
Olefins	Propylene	115-07-1	C_3H_6	2.15	11.20	169.00	728.15
Olefins	1-Butene	106-98-9	C_4H_8	1.60	10.00	198.00	657.00
Olefins	*cis*-2-Butene	590-18-1	C_4H_8	1.70	9.70	205.00	598.00
Olefins	*trans*-2-Butene	624-64-6	C_4H_8	1.70	9.70	203.00	597.00
Olefins	1-Pentene	109-67-1	C_5H_{10}	1.40	8.70	222.00	546.00
Acetylenes	Acetylene	74-86-2	C_2H_2	2.50	80.00	151.00	578.15
Acetylenes	Vinylacetylene	689-97-4	C_4H_4	2.20	31.70	211.00	Decomposes violently on heating. Forms explosive peroxides with air or oxygen.
Acetylenes	Methylacetylene	74-99-7	C_3H_4	1.70	57.30	192.00	613.15
Aromatics	Benzene	71-43-2	C_6H_6	1.20	8.00	262.00	833.15
Aromatics	Toluene	108-88-3	C_7H_8	1.10	7.10	279.15	753.15
Aromatics	*o*-Xylene	95-47-6	C_8H_{10}	1.10	6.40	305.15	736.15
Aromatics	Ethylbenzene	100-41-4	C_8H_{10}	1.00	6.70	296.15	703.15
Aromatics	Cumene	98-82-8	C_9H_{12}	0.88	6.50	309.15	697.00
Aromatics	Anthracene	120-12-7	$C_{14}H_{10}$	0.60	5.20	458.15	813.15
Cyclic hydrocarbons	Cyclopropane	75-19-4	C_3H_6	2.40	10.40	180.00	771.00
Cyclic hydrocarbons	Furan	110-00-9	C_4H_4O	2.00	23.00	237.00	663.15
Cyclic hydrocarbons	Cyclopentadiene	542-92-7	C_5H_6	1.70	14.60	227.00	913.15
Cyclic hydrocarbons	Cyclohexane	110-82-7	C_6H_{12}	1.30	7.80	255.93	518.15
Cyclic hydrocarbons	Methylcyclohexane	108-87-2	C_7H_{14}	1.15	6.70	269.15	523.15
Cyclic hydrocarbons	Phenol	108-95-2	C_6H_6O	1.70	8.60	352.15	988.00
Cyclic hydrocarbons	Dicyclopentadiene	77-73-6	$C_{10}H_{12}$	0.80	6.30	318.15	783.15
Alcohols	Methanol	67-56-1	CH_4O	7.18	36.50	284.15	737.00
Alcohols	Ethanol	64-17-5	C_2H_6O	3.30	19.00	286.15	696.00
Alcohols	Allyl Alcohol	107-18-6	C_3H_6O	2.50	18.00	294.00	651.00
Alcohols	1-Propanol	71-23-8	C_3H_8O	2.10	14.00	297.59	644.00
Alcohols	Isopropanol	67-63-0	C_3H_8O	2.00	12.70	285.15	728.75
Alcohols	1-Butanol	71-36-3	$C_4H_{10}O$	1.70	11.30	310.50	616.00
Alcohols	2-Butanol	78-92-2	$C_4H_{10}O$	1.70	9.80	296.15	663.15
Alcohols	2-Methyl-1-propanol	78-83-1	$C_4H_{10}O$	1.70	11.00	302.32	681.15
Alcohols	2-Methyl-2-propanol	75-65-0	$C_4H_{10}O$	1.84	9.00	284.26	751.00
Alcohols	Cyclohexanol	108-93-0	$C_6H_{12}O$	1.20	11.10	334.15	573.15
Aldehydes	Formaldehyde	50-00-0	CH_2O	7.00	73.00	219.80	697.15
Aldehydes	Acetaldehyde	75-07-0	C_2H_4O	4.00	30.00	232.00	449.15
Aldehydes	Acrolein	107-02-8	C_3H_4O	2.80	31.00	247.15	507.00

Aldehydes	Propanal	123-38-6	C_3H_6O	2.60	17.00	243.15	500.15
Aldehydes	*trans*-Crotonaldehyde	123-73-9	C_4H_6O	2.10	15.50	286.15	505.00
Aldehydes	*cis*-Crotonaldehyde	15798-64-8	C_4H_6O	2.10	15.50	285.93	505.00
Aldehydes	2-Methylpropanal	78-84-2	C_4H_8O	1.60	11.00	254.15	478.00
Aldehydes	Butanal	123-72-8	C_4H_8O	1.90	12.50	262.15	503.15
Aldehydes	Furfural	98-01-1	$C_5H_4O_2$	2.10	19.30	333.15	589.00
Ethers	Dimethyl ether	115-10-6	C_2H_6O	3.30	26.20	193.00	499.15
Ethers	Methyl vinyl ether	107-25-5	C_3H_6O	2.60	39.00	217.15	560.15
Ethers	Diethyl ether	60-29-7	$C_4H_{10}O$	1.70	46.00	228.15	433.15
Ethers	Diphenyl ether	101-84-8	$C_{12}H_{10}O$	0.80	6.00	388.15	891.15
Ketones	Acetone	67-64-1	C_3H_6O	2.60	13.00	253.15	738.15
Ketones	Methyl ethyl ketone	78-93-3	C_4H_8O	1.80	11.00	264.15	789.00
Ketones	Acetophenone	98-86-2	C_8H_8O	1.10	6.70	350.15	843.15
Acids	Acetic acid	64-19-7	$C_2H_4O_2$	4.00	19.90	312.04	700.00
Acids	Hydrogen cyanide	74-90-8	CHN	5.60	40.00	255.00	811.00
Acids	Formic acid	64-18-6	CH_2O_2	12.00	38.00	323.15	753.00
Esters	Methyl formate	107-31-3	$C_2H_4O_2$	5.20	23.00	247.00	729.00
Esters	Ethyl formate	109-94-4	$C_3H_6O_2$	2.76	15.70	254.15	728.15
Esters	Methyl acetate	79-20-9	$C_3H_6O_2$	3.13	14.00	260.15	775.00
Esters	Vinyl acetate	108-05-4	$C_4H_6O_2$	2.60	13.40	265.37	700.00
Esters	Ethyl acetate	141-78-6	$C_4H_8O_2$	2.18	11.50	269.00	700.00
Esters	*n*-Propyl acetate	109-60-4	$C_5H_{10}O_2$	1.80	8.00	283.71	723.00
Esters	Isopropyl acetate	108-21-4	$C_5H_{10}O_2$	1.76	7.20	274.82	733.15
Esters	*n*-Butyl acetate	123-86-4	$C_6H_{12}O_2$	1.40	7.60	298.15	694.00
Esters	Isobutyl acetate	110-19-0	$C_6H_{12}O_2$	1.42	8.00	291.00	696.00
Esters	*n*-Pentyl acetate	628-63-7	$C_7H_{14}O_2$	1.10	7.10	310.15	633.15
Inorganic	Hydrogen	1333-74-0	H_2	4.00	75.00	14.00	793.15
Inorganic	Ammonia	7664-41-7	H_3N	15.00	28.00	209.00	924.00
Inorganic	Cyanogen	460-19-5	C_2N_2	6.60	32.00	214.00	984.00
Oxides	Carbon monoxide	630-08-0	CO	12.50	74.20	71.00	882.00
Oxides	Ethylene oxide	75-21-8	C_2H_4O	3.00	100.00	225.00	702.00
Oxides	1,2-Propylene oxide	75-56-9	C_3H_6O	2.20	35.50	236.00	703.15
Oxides	1,4-Dioxane	123-91-1	$C_4H_8O_2$	2.00	22.00	284.15	453.15
Oxides	Mesityl oxide	141-79-7	$C_6H_{10}O$	1.30	8.80	301.00	618.00
Peroxides	Di-*t*-Butyl peroxide	110-05-4	$C_8H_{18}O_2$	0.74	8.20	277.15	Organic peroxides can ignite easily
Sulfur containing	Carbon disulfide	75-15-0	CS_2	1.30	50.00	243.15	363.15
Sulfur containing	Hydrogen sulfide	7783-06-4	H_2S	4.00	44.00	167.00	533.15
Sulfur containing	Carbonyl sulfide	463-58-1	COS	12.00	29.00	186.00	477.00
Sulfur containing	Dimethyl sulfide	75-18-3	C_2H_6S	2.20	19.70	237.15	478.15
Chlorine containing	Methyl chloride	74-87-3	CH_3Cl	8.10	17.20	203.00	905.00
Chlorine containing	Ethyl chloride	75-00-3	C_2H_5Cl	3.80	15.40	223.15	802.00
Chlorine containing	Isopropyl chloride	75-29-6	C_3H_7Cl	2.80	10.70	238.15	866.00
Chlorine containing	1,2-Dichloroethane	107-06-2	$C_2H_4Cl_2$	4.50	16.00	286.00	686.00
Chlorine containing	1,2-Dichloropropane	78-87-5	$C_3H_6Cl_2$	3.30	14.50	286.15	830.00
Chlorine containing	Dichloromethane	75-09-2	CH_2Cl_2	14.00	22.00	265.00	888.15
Chlorine containing	2-Chloroethanol	107-07-3	C_2H_5ClO	4.90	15.90	328.15	698.15
Chlorine containing	Trichloroethylene	79-01-6	C_2HCl_3	12.00	29.00	305.15	683.15
Chlorine containing	Hexachloro-1,3-Butadiene	87-68-3	C_4Cl_6	2.90	15.70	389.00	883.15
Chlorine containing	Vinyl chloride	75-01-4	C_2H_3Cl	3.60	33.00	205.00	745.00
Chlorine containing	Monochlorobenzene	108-90-7	C_6H_5Cl	1.30	9.60	301.15	911.00
Chlorine containing	Benzyl chloride	100-44-7	C_7H_7Cl	1.10	7.10	333.15	858.15
Bromides	Bromomethane	74-83-9	CH_3Br	10.10	16.00	230.00	800.00
Glycols	Ethylene glycol	107-21-1	$C_2H_6O_2$	3.10	42.00	384.15	669.00
Glycols	Diethylene glycol	111-46-6	$C_4H_{10}O_3$	1.70	37.00	413.15	636.15

(*Continued*)

TABLE 2-153 Lower and Upper Flammability Limits, Flash Points, and Autoignition Temperatures for Selected Hydrocarbons *(Continued)*

Group	Compound	CAS	Formula	LFL	UFL	Flash point (K)	Autoignition T (K)
Glycols	Triethylene glycol	112-27-6	$C_6H_{14}O_4$	0.90	9.20	429.15	644.00
Amines	Methylamine	74-89-5	CH_5N	4.90	20.70	217.00	703.15
Amines	Ethylamine	75-04-7	C_2H_7N	2.70	14.00	227.00	657.00
Amines	Dimethylamine	124-40-3	C_2H_7N	2.80	14.40	223.15	595.00
Amines	Isopropylamine	75-31-0	C_3H_9N	2.00	10.40	236.15	673.15
Amines	Trimethylamine	75-50-3	C_3H_9N	2.00	11.60	207.00	463.15
Amines	Allylamine	107-11-9	C_3H_7N	2.03	24.30	252.00	647.039
Amines	Diethylamine	109-89-7	$C_4H_{11}N$	1.70	10.10	245.15	583.15
Amines	Tert-Butylamine	75-64-9	$C_4H_{11}N$	1.70	8.90	236.00	648.15
Amines	Triethylamine	121-44-8	$C_6H_{15}N$	1.20	8.00	262.15	522.15
Amines	Cyclohexylamine	108-91-8	$C_6H_{13}N$	0.66	9.40	299.65	566.15
Amines	Monoethanolamine	141-43-5	C_2H_7NO	3.00	13.10	366.55	683.15
Amines	Diethanolamine	111-42-2	$C_4H_{11}NO_2$	1.70	9.80	445.15	935.00
Amines	Dimethylethanolamine	108-01-0	$C_4H_{11}NO$	1.40	12.20	312.15	568.15
Miscellaneous	Acrylonitrile	107-13-1	C_3H_3N	3.05	17.00	268.15	754.00
Miscellaneous	Aniline	62-53-3	C_6H_7N	1.30	11.00	344.15	890.00
Miscellaneous	Diborane	19287-45-7	B_2H_6	0.80	88.00	142.00	325.00
Miscellaneous	Methyl methacrylate	80-62-6	$C_5H_8O_2$	1.70	12.50	284.15	708.15
Miscellaneous	Styrene	100-42-5	C_8H_8	1.10	6.10	305.00	763.15
Miscellaneous	Biphenyl	92-52-4	$C_{12}H_{10}$	0.70	5.80	383.15	813.15
Miscellaneous	Methyl acrylate	96-33-3	$C_4H_6O_2$	2.18	14.40	270.00	741.15
Miscellaneous	Phthalic anhydride	85-44-9	$C_8H_4O_3$	1.20	9.20	425.00	857.00

Values in this table were taken from the Design Institute for Physical Properties (DIPPR) of the American Institute of Chemical Engineers (AIChE), 801 Critically Evaluated Gold Standard™ Database, copyright 2016 AIChE, and reproduced with permission of AIChE and of the DIPPR Evaluated Process Design Data Project Steering Committee. Their source should be cited as "R. L. Rowley, W. V. Wilding, J. L. Oscarson, T. A. Knotts, N. F. Giles, *DIPPR® Data Compilation of Pure Chemical Properties*, Design Institute for Physical Properties, AIChE, New York, NY (2016)".

PREDICTION AND CORRELATION OF PHYSICAL PROPERTIES*

INTRODUCTION

Physical property values, sufficiently accurate for many engineering applications, can be estimated in the absence of reliable experimental data. The purpose of this section is to provide a set of recommended prediction methods for general engineering use. It is not intended to be a comprehensive review, and many additional methods are available in the literature. Methods recommended in this section were selected on the basis of accuracy, generality, and, in most cases, simplicity or ease of use. They generally correspond to the methods tested and given priority in the DIPPR 801 database project.*

Properties included in this subsection are divided into 10 categories: (1) physical constants including critical properties, normal melting and boiling points, acentric factor, radius of gyration, dipole moment, refractive index, and dielectric constant; (2) liquid and solid vapor pressure; (3) thermal properties including enthalpy and Gibbs energy of formation and ideal gas entropy; (4) latent enthalpies of vaporization, fusion, and sublimation; (5) heat capacities for ideal and real gases, liquids, and solids; (6) densities of gas, liquid, and solid phases; (7) gas and liquid viscosity; (8) gas and liquid thermal conductivity; (9) surface tension; and (10) flammability properties including flash point, flammability limits, and autoignition temperature. Each of the 10 subsections gives a definition of the properties and a description of one or more recommended prediction methods. Each description lists the type of method, its uncertainty, its limitations, and the expected uncertainty of the predicted value. A numerical example is also given to illustrate use of the method. For brevity, symbols used for physical properties and for variables and constants in the equations are defined under Nomenclature and are not necessarily defined after their first use except where doing so clarifies usage. A list of equation and table numbers in which variables appear is included in the Nomenclature section for quick cross-referencing. Although emphasis is on pure-component properties, some mixture estimation techniques have been included for physical constants, density, viscosity, thermal conductivity, surface tension, and flammability. Correlation and estimation of properties that are inherently multicomponent (e.g., diffusion coefficients, mixture excess properties, activity coefficients) are treated elsewhere in this handbook.

*The Design Institute for Physical Properties (DIPPR) is an industrial consortium under the auspices of AIChE; Project 801, Evaluated Process Design Data, is a pure-component database of industrially important compounds. Values and procedures used with permission of the DIPPR 801 Technical Committee.

UNITS

The International System (SI) of metric units has been used throughout this section. Where possible, the estimation equations are set up in dimensionless groups to eliminate the need to specify units of variables and to facilitate unit conversions. For example, rather than use P_c as an equation variable, the dimensionless group (P_c/Pa) is used. When a value for P_c expressed in any units (say, P_c = 6.53 MPa) is inserted into this group, the result is dimensionless with an explicit indication of conversion factors that must be included, such as

$$\frac{P_c}{\text{Pa}} = \frac{6.53\ \text{MPa}}{\text{Pa}} = \left(\frac{6.53\ \text{MPa}}{\text{Pa}}\right)\left(\frac{10^6\ \text{Pa}}{\text{MPa}}\right) = 6.53 \times 10^6$$

Appropriate unit conversion factors are found in Sec. 1 of this handbook.

Nomenclature

Physical constants	Definition	Value
h	Planck's constant	6.626×10^{-34} J·s
k	Boltzmann's constant	1.3806×10^{-23} J/(molecule·K)
N_A	Avogadro's number	6.022×10^{26} molecule/kmol
R	Gas constant	8314.3 Pa·m^3/(kmol·K)

Properties	Definition	Typical units
A, B, C	Molecular principal moments of inertia	kg·m^2
AIT	Autoignition temperature	K
A_{vdw}	Van der Waals area	m^2/kmol
$B, B(T)$	Second virial coefficient	m^3/kmol
C_P	Isobaric molar heat capacity	J/(kmol·K)
C_p^o	Ideal gas isobaric molar heat capacity	J/(kmol·K)
C_v	Constant-volume molar heat capacity	J/(kmol·K)
H_i	Enthalpy of compound i	J/kmol
k	Thermal conductivity	W/(m·K)
LFL	Lower flammability limit	%
M	Molecular weight	kg/kmol
n	Refractive index	unitless
P	Pressure	Pa
P	Parachor	unitless
P_c	Critical pressure	Pa
P_r	Reduced pressure; $P_r = P/P_c$	unitless
P^*	Vapor pressure	Pa
P^*_{meas}	Measured vapor pressure value	Pa
P_r^*	Reduced vapor pressure; $P_r^* = P^*/P_c$	unitless
P_t^*	Vapor pressure at triple point	Pa
R_D	Molar refraction	cm^3/mol
R_g	Radius of gyration	m
S^o	Ideal gas entropy	J/(kmol·K)
S^s	Standard state entropy	J/(kmol·K)
S_r	Rotational contribution to entropy	J/(kmol·K)
S_{vib}	Vibrational contribution to entropy	J/(kmol·K)

Nomenclature *(Continued)*

Properties	Definition	Typical units
T	Temperature	K
T_{ad}	Adiabatic flame temperature	K
T_b	Normal boiling point temperature	K
T_{br}	Reduced temperature at T_b; $T_{br} = T_b/T_c$	unitless
T_c	Critical temperature	K
T_{FP}	Flash point temperature	K
T_m	Melting temperature	K
T_{meas}	T at which a dependent property was measured	K
T_r	Reduced temperature; $T_r = T/T_c$	unitless
UFL	Upper flammability limit	%
V	Molar volume	m^3/kmol
V_c	Critical volume	m^3/kmol
V_r	Reduced volume; $V_r = ZT_r/P_r$	unitless
w_i	Mass fraction of component i	unitless
x_i	Mole fraction of component i	unitless
y_i	Mole fraction of component i in vapor phase	unitless
Z	Compressibility factor; $Z = PV/RT$	unitless
Z_c	Critical compressibility factor; $Z_c = P_cV_c/RT_c$	unitless
Z_i	Compressibility factor of reference fluid i	unitless
ΔG_f^o	Ideal gas standard Gibbs energy of formation	J/kmol
ΔG_f^s	Standard state Gibbs energy of formation	J/kmol
ΔH_f^o	Ideal gas standard enthalpy of formation	J/kmol
ΔH_f^s	Standard state enthalpy of formation	J/kmol
ΔH_{fus}	Enthalpy of fusion	J/kmol
ΔH_{rxn}	Enthalpy change per mole of reaction as written	J/kmol
ΔH_{sub}	Enthalpy of sublimation	J/kmol
ΔH_v	Enthalpy of vaporization	J/kmol
ΔS_f^s	Standard state entropy of formation	J/(kmol · K)
ΔS_f^o	Ideal gas entropy of formation	J/(kmol · K)
ΔS_{fus}	Latent entropy of fusion	J/(kmol · K)
ΔZ_v	Change in compressibility factor upon vaporization	unitless
δ	Solubility parameter	$J^{1/2} \cdot m^{-3/2}$
ε	Dielectric constant	unitless
η	Viscosity	Pa · s
η^o	Viscosity at low pressure	Pa · s
μ	Dipole moment	D
μ_r	Reduced dipole moment [defined in Eq. (2-66)]	unitless
ρ	Molar density; $\rho = V^{-1}$	kmol/m^3
ρ_c	Critical molar density; $\rho_c = V_c^{-1}$	kmol/m^3
ρ_r	Reduced molar density; $\rho_r = \rho/\rho_c$	unitless
ρ_S, ρ_L, ρ_V	Density of solid, liquid, vapor, respectively	kmol/m^3
σ	Surface tension	mN/m
σ_m	Surface tension of mixture	mN/m
τ	Complementary reduced temperature $(1 - T_r)$	unitless
τ_b	τ at the normal boiling point $(1 - T_{br})$	unitless
ϕ_i	Volume fraction of component i	unitless
ω	Acentric factor	unitless

Equation variables	Definition	Appears in (Eq. 2-?) or [Table 2-?]
a	EoS constant	(70), [172]
$a, b, c, \ldots$	GC values for C_p and η	(54), (57), (96), [174]
a, b, c	Correlation coefficients	(25), (27), (42), (43), (44), (69)
a_i	GC values	(46), (96), [164], [174]
a, b	Terms in second virial correlation	(65)
a, b	Chickos correlation parameters	(42), (43), (44)
a_i, b_i, d_i	GC values for liquid C_p	(54), [166]
$a\alpha$	EoS constant for mixture	(78)
$A, B, C, \ldots$	Correlation constants/parameters	(2), (23), (24), (26), (28), (28a), (38), (40), (53), (54), (56), (69), (71), (82), (84), (86), (87), (94), (95), (100), (101), (102)
A	Factor in liquid k correlation	(110), [176]
A_i	Constants in C_p^o correlation	(48), (49), (70)
b	EoS constant	(70), [172]
$b_i, c_i, \ldots$	Reference EoS constants	(69), [171]
b_i	GC value for AIT	(129), [180]

Nomenclature *(Continued)*

Equation variables	Definition	Appears in (Eq. 2-?) or [Table 2-?]
$\bar{b}$	EoS constant for mixture	(77)
$B^{(i)}$	Second virial expansion term	(62), (63), (64), (65)
C	Number of components in mixture	(74), (75), (76), (77), (78), (79), (80), (81), (98)
C	Parameter in modified Pachaiyappan method	(109)
C_i	GC values for some methods	(9), (10), (11), (18), (86), [175], [173]
C_{ij}	Group-group intramolecular interaction pair	(12), [156]
$(C_p^o)_i$	GC values for ideal gas heat capacity	(52), [165]
Cs_j	Chickos GC value for C—H group	(44), [162]
C_{st}	Fuel concentration for stoichiometric combustion	(127), (128)
Ct_j	Chickos GC value for functional group	(44), [163]
f_i	Halogen correction for ΔH_{sub} correlation	(46), [164]
$F^{(i)}$	Vapor pressure deviation function	(29)
F	Factor in surface tension equation	(118), (119)
GI	Group-group interaction correction term	(9), (10), (11), (12)
G_{ij}	Adjustable mixture viscosity parameter	(97)
g_c^E	UNIFAC combinatorial excess Gibbs energy	(98)
g_r^E	UNIFAC residual excess Gibbs energy	(98)
h	Parameter in Riedel vapor pressure equation	(28a)
K	Parameter in Riedel vapor pressure equation	(28a)
m	Parameter in modified Pachaiyappan method	(109)
n_{hvy}	Number of non-hydrogen atoms	(11), (12), (18)
n_A	Number of atoms in the molecule	(1), (34), (35), (51)
n_E	Number of occurrences of element E in compound	(58)
NG	Number of interacting groups	(12)
n_i	Number of occurrences of group i	(9), (10), (11), (13), (15), (16), (31), (46), (52), (54), (55), (86), (117), (124), (127), (129)
n_f	Number of different functional groups	(44)
n_s	Number of C—H groups bonded to functional groups	(44)
n_x	Number of halogen and H atoms	(46)
N	Total number of groups in molecule	(18), (31) (46), (54), (57), (58), (86), (96), (117), (124), (127), (129)
N_C	Number of C atoms	(123)
Nf_i	Number of functional groups of type i	(44)
Ng_i	Number of C—H groups of type i bonded to C	(44)
N_H	Number of H atoms	(123)
N_{CR}	Number of CH_2 groups forming cyclic paraffin	(43)
N_O	Number of O atoms	(123)
N_R	Number of nonaromatic rings	(43)
N_S	Number of S atoms	(123)
Ns_i	Number of C—H groups bonded to functional group	(44)
N_{Si}	Number of Si atoms	(123)
N_X	Number of halogen atoms	(123)
$\bar{P}_c$	Pseudocritical pressure for mixture	(75)
q	Rackett equation power for Z_c	(72), (80)
q_i	UNIFAC molecular surface area	following (99)
Q_k	UNIFAC group surface area	following (99)
r_i	UNIFAC molecular volume	following (99)
r^*	Dimensionless separation distance	(4)
R_k	UNIFAC group volume	following (99)
$(S^o)_i$	GC value for entropy	(31), [161]
t	Total number of functional groups	(44)
$t_{m1,i}$	First-order GC contribution for T_m	(16), [158]
$t_{m2,i}$	Second-order GC contribution for T_m	(16), [159]
$\bar{T}_c$	Pseudocritical temperature for mixture	(74), (75), (79)
$T_{c,ij}$	Cross term in mixing rule	(79)
x_P	Term in the Pailhes method [= log(1 atm/P)]	(17)
U^*	Dimensionless intermolecular potential	(4)
UFL_i	GC contribution	(127), [178]
$Z^{(0)}$	Compressibility factor of simple fluid	(68), [169]
$Z^{(1)}$	Acentric deviation term for Z	(68), [170]
$Z_{c,ij}$	Cross term in mixing rule	(79)
Z_{RA}	Modified Rackett correlation parameter	following (72)
$\bar{Z}_{RA}$	Modified Rackett parameter for mixture	(80), (81)
$\alpha, \beta, \gamma, \ldots$	Correlation parameters for k	(107), (108), (110), [176]

Nomenclature *(Continued)*

Equation variables	Definition	Appears in (Eq. 2-?) or [Table 2-?]
$\alpha(T_r)$	EoS temperature-dependent function	(70), [172]
α_c	Parameter in Riedel vapor pressure equation	(28a)
α_{mn}	Viscosity group-group interactions	(99), [175]
β	Reference EoS constant	(69), [171]
β	Stoichiometric coefficient for combustion	(122), (123), (128)
β_i	Nonlinear correction term in correlation	(46), (57), [164], [167]
γ	Reference EoS constant	(69), [171]
δ	= 0 for nonlinear molecules; = 1 for linear	(1), (35), before (50), (51)
δ	EoS parameter	(70), [172]
Δ_E	Contribution of element *E* to heat capacity	(58), [168]
Δ_P	GC contribution to P_c	(7), [154]
Δ_T	GC contribution to T_c	(6), [154]
Δ_V	GC contribution to V_c	(8), [154]
$(\Delta H_f^o)_i$	GC value for enthalpy of formation	(31), [161]
$\Delta \mathbf{P}_i$	GC for Parachor	(117), [177]
Δpc_i	Group *i* contribution to critical pressure	(15)
Δs_i	GC value for group *i*	(44), [162, 163]
$\Delta T_{ad,i}$	Group *i* contribution to adiabatic flame temperature	(124)
Δtc_i	Group *i* contribution to critical temperature	(13)
ε	Lennard-Jones well depth parameter	following (4)
ε	EoS parameter	(70), [172]
ϕ	UNIFAC molecular volume fraction	following (99)
ν	LFL enthalpic term	(126)
ν_i	Stoichiometric coefficient (+ for product and − for reactant) for compound *i* in reaction	(32), (33), (34)
ν_j	Frequency of vibrational mode *j*	(50)
θ	UNIFAC molecular surface fraction	following (99)
Θ	UNIFAC group surface fraction	following (99)
$\Theta_A, \Theta_B, \Theta_C$	Characteristic rotational *T* of molecule	before and following (35)
Θ_j	Characteristic vibrational *T* of mode *j*	(1), (35)
σ	Lennard-Jones size parameter	following (4)
σ	Rotational external symmetry number	following (35)
μ_r^*	Modified reduced dipole moment	(84), (85)
ψ	Parameter in Riedel vapor pressure equation	(28a)
ψ	Parameter in correlation of *k* for gases	(106), (107)
ψ_{mn}	UNIFAC interaction factor	(99)
ξ	Viscosity de-dimensionalizing factor	(88), (89), (90), (91), (92), (93)
ϖ	Pseudo-acentric factor for mixture	(76)

Acronyms and abbreviations	Definition
CC	Computational chemistry
CS	Corresponding states
DIPPR	Design Institute for Physical Properties
EoS	Equation of state
GC	Group contributions
LJ	Lennard-Jones
MC	Monte Carlo
MD	Molecular dynamics
QSPR	Quantitative structure-property relationships

GENERAL REFERENCES

Prediction Methods

[PGL4] Reid, R. C., J. M. Prausnitz, and B. E. Poling, *The Properties of Gases and Liquids,* 4th ed., McGraw-Hill, New York, 1987.

[PGL5] Poling, B. E., J. M. Prausnitz, and J. P. O'Connell, *The Properties of Gases and Liquids,* 5th ed., McGraw-Hill, New York, 2001.

Property Databases

[DIPPR] Rowley, R. L., et al., *DIPPR Data Compilation of Pure Chemicals Properties,* Design Institute for Physical Properties, AIChE, New York, 2007.

[TRC] *TRC Thermodynamic Tables—Non-Hydrocarbons,* Thermodynamics Research Center, The Texas A&M University System, College Station, Tex., extant 2004; *TRC Thermodynamic Tables—Hydrocarbons,* Thermodynamics Research Center, The Texas A&M University System, College Station, Tex., extant 2004.

[JANAF] Chase, M. W., Jr., et al., "JANAF Thermochemical Tables," *J. Phys. Chem. Ref. Data,* **14,** suppl. 1, 1985.

[SWS] Stull, D. R., F. F. Westrum, Jr., and G. C. Sinke, *The Chemical Thermodynamics of Organic Compounds,* John Wiley & Sons, New York, 1969.

[TDS] Daubert, T. E., and R. P. Danner, *Technical Data Book—Petroleum Refining,* 5th ed., American Petroleum Institute, Washington, extant 1994.

CLASSIFICATION OF ESTIMATION METHODS

Physical property estimation methods may be classified into six general areas: (1) theory and empirical extension of theory, (2) corresponding states, (3) group contributions, (4) computational chemistry, (5) empirical and quantitative structure-property relations (QSPR) correlations, and (6) molecular simulation. A quick overview of each class is given below to provide context for the methods and to define the general assumptions, accuracies, and limitations inherent in each.

Theory and Empirical Extension of Theory Methods based on theory generally provide better extrapolation capability than empirical fits of experimental data. Assumptions required to simplify the theory to a manageable equation suggest accuracy limitations and possible improvements, if necessary. For example, the ideal gas isobaric heat capacity, rigorously obtained from statistical mechanics under the assumption of independent harmonic vibrational modes, is (Rowley, R. L., *Statistical Mechanics for Thermophysical Property Calculations*, Prentice-Hall, Englewood Cliffs, N.J., 1994)

$$\frac{C_p^o}{R} = \frac{8-\delta}{2} + \sum_{j=1}^{3n_A-6+\delta} \left(\frac{\Theta_j}{T}\right)^2 \frac{e^{\Theta_j/T}}{(e^{\Theta_j/T}-1)^2} \tag{2-1}$$

$$\delta = \begin{cases} 0 & \text{nonlinear molecules} \\ 1 & \text{linear molecules} \end{cases}$$

where Θ_j is the characteristic temperature for the *j*th vibrational frequency in a molecule of n_A atoms. The temperature dependence of this equation is exact to the extent that the frequencies are harmonic.

Extension of theory often requires introduction of empirical models and parameters in lieu of terms that cannot be rigorously calculated. Good accuracy is expected in the region where the model parameters were fitted to experimental data, but only limited accuracy when an empirical model is extrapolated to other conditions. For example, a simplified theory suggests that vapor pressure should have the form

$$\ln P^* = A - \frac{B}{T} \tag{2-2}$$

where the empirical parameter B is given by

$$B = \frac{\Delta H_\upsilon}{R\Delta Z_\upsilon} \tag{2-3}$$

and ΔH_υ and ΔZ_υ are differences between the vapor and liquid enthalpies and compressibility factors, respectively. Equation (2-2) can be used to correlate vapor pressures over a moderate temperature range, but it is inadequate to represent vapor pressures over the whole liquid temperature range because ΔH_υ also varies with temperature.

Corresponding States (CS) The principle of CS applies to conformal fluids [Leland, T. L., Jr., and P. S. Chappelear, *Ind. Eng. Chem.*, **60** (1968): 15]. Two fluids are conformal if their intermolecular interactions are equivalent when scaled in dimensionless form. For example, the Lennard-Jones (LJ) intermolecular pair potential energy U can be written in dimensionless form as

$$U^* = 4(r^{*-12} - r^{*-6}) \tag{2-4}$$

where $r^* = r/\sigma$, $U^* = U/\varepsilon$, σ is the LJ size parameter, and ε is the LJ attractive well depth parameter. At equivalent scaled temperatures kT/ε (k is Boltzmann's constant) and pressures $P\sigma^3/\varepsilon$, all LJ fluids will have identical dimensionless properties because the molecules interact through the identical scaled intermolecular potential given by Eq. (2-4). Generalization of this scaling principle is commonly done using critical temperature T_c and critical pressure P_c as scaling factors. At the same reduced coordinates ($T_r = T/T_c$ and $P_r = P/P_c$) conformal fluids will have the same dimensionless properties. For example, $Z = Z(T_r, P_r)$ where the compressibility factor is defined as $Z = PV/RT$. A correlation of experimental data for one fluid can then be used as the reference for the properties of all conformal fluids. Nonconformality is the main accuracy limitation. For instance, interactions between nonspherical or polar molecules are not adequately represented by Eq. (2-4), and so the scaled properties of these fluids will not conform to those of a fluid with interactions well represented by Eq. (2-4). A correction for nonconformality is usually made by the addition of one or more reference fluids whose deviations from the first reference fluid are used to characterize the effect of nonconformality. For example, in the Lee-Kesler method [Lee, B. I., and M. G. Kesler, *AIChE J.*, **21** (1975): 510] *n*-octane is used as a second, nonspherical reference fluid, and deviations of *n*-octane scaled properties from those of the spherical reference fluid at equivalent reduced conditions are assumed to be a linear function of the acentric factor.

Group Contributions (GCs) Physical properties generally correlate well with molecular structure. GC methods assume a summative behavior of the structural groups of the constituent molecules. For example, ethanol (CH_3—CH_2—OH) properties would be obtained as the sum of contributions from the —CH_3, —CH_2, and —OH groups. The contribution of each group is obtained by regression of experimental data that include as many different compounds containing that group as possible. Structural groups must be used exactly as defined in the original correlation of the groups. A general principle when parsing a structure into constituent groups is that the more specific the group, the higher its priority. For example, the structural piece —$COOCH_3$ in a methyl ester could be divided in more than one way, but if the —COO— and —CH_3 groups are available in the method, then they should be used rather than the combination of the two less specific groups —(C=O)— and —O—. These latter group values were most likely regressed only from ketone and ether data, respectively. Excellent accuracy can usually be expected from GC methods in which the group values were regressed from large quantities of experimental data. However, if the ratio of the number of groups to regressed experimental data is large, significant errors can result when the method is applied to new compounds (extrapolation). Such excessive specificity in the group definitions leads to poor extrapolation capabilities even though the fit of the regressed data may have been excellent.

First-order GC methods assume simple summations of the group values are adequate to represent the molecular value. Second-order effects, caused by steric and electron induction effects from neighboring groups, can alter group values. Second-order GC methods require considerably more experimental data to tune the method, and large tables of group values are required because differences in bonded neighbors require separate groups.

Computational Chemistry (CC) Commercial software is available that solves the Schrödinger equation by using approximate forms of the wave function. Various levels of sophistication (termed *model chemistry*) for the wave function can be chosen at the expense of computational time. Results include structural information (bond lengths, bond angles, dihedral angles, etc.), electron/charge distribution information, internal vibrational modes (for ideal gas properties), and energy of the molecule, valid for the chosen model chemistry. Because calculations are usually performed on individual molecules, the results are best suited for ideal gas properties. Relative energies for the same model chemistry are more accurately obtained than absolute energies, so enthalpies and entropies of reaction are also common industrial uses of CC predictions.

Empirical QSPR Correlations Quantitative structure-property relationship (QSPR) methods correlate physical properties with molecular descriptors that characterize the structural and electronic character of the molecule. Large amounts of experimental data are used to statistically determine the most significant descriptors to be used in the correlation and their contributions. The resultant correlations are simple to apply if the descriptors are available. Descriptors must be generated by the user with computational chemistry software or obtained from some tabulation. QSPR methods are often very accurate for specific families of compounds for which the correlation was developed, but extrapolation to other families generally results in considerable loss of accuracy.

Molecular Simulations Molecular simulations are useful for predicting properties of bulk fluids and solids. Molecular dynamics (MD) simulations solve Newton's equations of motion for a small number (on the order of 10^3) of molecules to obtain the time evolution of the system. MD methods can be used for equilibrium and transport properties. Monte Carlo (MC) simulations use a model for the potential energy between molecules to simulate configurations of the molecules in proportion to their probability of occurrence. Statistical averages of MC configurations are useful for equilibrium properties, particularly for saturated densities, vapor pressures, etc. Property estimations using molecular simulation techniques are not illustrated in the remainder of this section as commercial software implementations are not commonly available.

PHYSICAL CONSTANTS

Critical Properties The critical temperature T_c, pressure P_c, and volume V_c of a compound are important, widely used constants. They are important in determining the phase boundaries of a compound and (particularly T_c and P_c) are required input parameters for many property estimation methods, particularly CS methods.

The critical temperature of a compound is the temperature above which a liquid phase cannot be formed, regardless of the system pressure. The critical pressure is the vapor pressure of the compound at the critical temperature. The molar critical volume is the volume occupied by 1 mol of a chemical at its critical temperature and pressure. The critical compressibility factor Z_c is determined from the experimental or predicted values of the critical properties by its definition

$$Z_c = \frac{P_c V_c}{RT_c} \tag{2-5}$$

Recommended Methods The Ambrose method is recommended for all three critical properties of hydrocarbons and *n*-alcohols. The Nannoolal method is recommended for all three critical properties of all other organic molecules. The Wilson-Jasperson method is a simple method also recommended for estimating T_c and P_c for organic and some inorganic chemicals.

The first-order Wilson-Jasperson method often gives better results than the second-order method except strongly polar, hydrogen-bonding, and associating fluids.

Method: Ambrose method.

Reference: Ambrose, D., *Natl. Phys. Lab. Report Chem.* **92** (1978); *Natl. Phys. Lab Report Chem.* **98** (1979).

Classification: Group contributions.

Expected uncertainty: ~6 K for T_c (about 1 percent), ~2 bar for P_c (about 5 percent), ~8 cm³/mol for V_c (about 3 percent).

Applicability: Organic compounds.

Input data: T_b, M, group contributions Δ_T, Δ_P, and Δ_V from Table 2-154.

Description: A GC method with first-order contributions and corrections (delta Platt number) for branched alkanes. Variables T_c, P_c, and V_c are given by the following relations:

$$T_c = T_b\left[1+\left(1.242+\sum\Delta_T\right)^{-1}\right] \tag{2-6}$$

$$\frac{P_c}{\text{bar}} = \frac{M}{\text{kg/kmol}}\left(0.339+\sum\Delta_P\right)^{-2} \tag{2-7}$$

$$\frac{V_c}{\text{cm}^3/\text{mol}} = 40+\sum\Delta_V \tag{2-8}$$

Example Use the Ambrose method to estimate the critical constants of 2,2,4-trimethylpentane.

Required data: From the DIPPR 801 database, T_b = 372.39 K and M = 114.229 kg/kmol.

Structure:

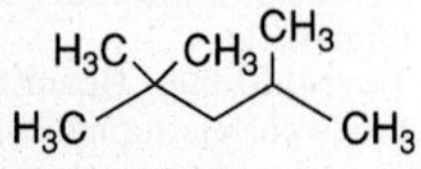

Group contributions from Table 2-154:

Group	n_i	Δ_T	Δ_P	Δ_V
Alkyl carbons	8	0.138	0.226	55.1
>CH— (correction)	1	−0.043	−0.006	−8
>C< (correction)	1	−0.120	−0.030	−17
Delta Platt no.	0	−0.023	−0.026	—

Calculations using Eqs. (2-6), (2-7), and (2-8):

$$\sum\Delta_T = (8)(0.138) + (1)(-0.043) + (1)(-0.120) = 0.941$$

$$T_c = T_b(1.4581) = (372.39\text{ K})(1.4581) = 543.0\text{ K}$$

$$\sum\Delta_P = (8)(0.226) + (1)(-0.006) + (1)(-0.030) = 1.772$$

$$\frac{P_c}{\text{bar}} = \frac{M}{\text{kg/kmol}}\left(0.339+\sum\Delta_P\right)^{-2} = \frac{114.229}{(0.339+1.772)^2} = 25.63 \qquad P_c = 25.63\text{ bar}$$

$$\sum\Delta_V = (8)(55.1) + (1)(-8) + (1)(-17) = 415.8$$

$$V_c = (40 + 415.8)\text{ cm}^3/\text{mol} = 455.8\text{ cm}^3/\text{mol}$$

Results:

Property	DIPPR recommended value	Ambrose estimation	% Difference
T_c/K	543.8	543.0	−0.15
P_c/bar	25.70	25.63	0.27
V_c/(cm³/mol)	468.0	455.8	−2.6

Method: Nannoolal method.

Reference: Nannoolal, Y., J. Rarey, and D. Ramjugernath, *Fluid Phase Equilib.* **252** (2007): 1.

Classification: Group contributions.

Expected uncertainty: ~6 K or 1 percent for T_c; ~2 bar or 5 percent for P_c; ~8 cm³/mol or 3 percent for V_c.

Applicability: Organic compounds.

Input data: T_b, group contributions C_i from Table 2-155, intramolecular group-group interactions C_{ij}, from Table 2-156, and the number of nonhydrogen atoms in the molecule n_{hvy}.

Description: A GC method with first-order contributions. Variables T_c, P_c, and V_c are given by the following relations:

$$T_c = T_b\left[0.6990 + \frac{1}{0.9889+\left(\sum_i n_iC_i+\text{GI}\right)^{0.8607}}\right] \tag{2-9}$$

$$\frac{P_c}{\text{kPa}} = \frac{\left(\dfrac{M}{\text{kg/kmol}}\right)^{-0.14041}}{\left(0.00939+\sum_i n_iC_i+\text{GI}\right)^2} \tag{2-10}$$

$$\frac{V_c}{10^{-6}\text{m}^3/\text{mol}} = \frac{\sum_i n_iC_i+\text{GI}}{n_{\text{hvy}}^{-0.2266}} + 86.1539 \tag{2-11}$$

where n_i is the number of groups of type i; C_i are group contributions from Table 2-155; M is molecular weight; and GI is the total correction for group-group interactions calculated using

$$\text{GI} = \frac{1}{n_{\text{hvy}}}\sum_{i=1}^{\text{NG}}\sum_{j=1}^{\text{NG}}\frac{C_{ij}}{\text{NG}-1} \tag{2-12}$$

where $C_{ji} = C_{ij}$. The values for the interactions are shown in this format in Table 2-156. The sum of all group pairs within the molecule is divided by the number of nonhydrogen atoms, n_{hvy}, and by 1 less than the number of interacting groups NG. In the example below, there are no group-group interactions. The calculation of GI using Eq. (2-12) is illustrated later in an example calculation for the normal boiling point.

Example Estimate the critical constants of *o*-xylene using the Nannoolal method.

Structure:

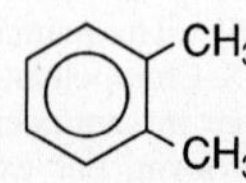

Required input data: From the DIPPR 801 database, T_b = 417.58 K. From Table 2-155:

Group	n_i	$C_i(T_C)$	$C_i(P_C)$	$C_i(V_C)$
=C(a)—	4	0.0161154	0.00021064	19.402
CH₃–(a)	2	−0.001071	0.0004166	26.7237
=C(a)<(ne)	2	0.0682045	0.00041826	25.0434
ortho	1	0.0012823	0.00007061	−3.5964
GI	—	0	0	0

From Eqs. (2-9), (2-10), and (2-11):

$$\sum C_i(T_c) = (4)(0.0161154) + (2)(-0.001071) + (2)(0.0682045)$$
$$+ (1)(0.0012823) = 0.20001$$

$$T_c = T_b\left[0.6990 + \frac{1}{0.9889+(0.20001)^{0.8607}}\right] = 1.5060T_b$$

$$T_c = (1.5060)(417.58\text{ K}) = 628.87\text{ K}$$

$$\sum C_i(P_c) = (4)(0.00021064) + (2)(0.0004166) + (2)(0.00041826)$$
$$+(1)(0.00007061) = 0.0025829$$

$$\frac{P_c}{\text{kPa}} = \frac{(106.165)^{-0.14041}}{(0.00939+0.00258289)^2} = 3623.55$$

$$\sum C_i(V_c) = (4)(19.402) + (2)(26.7237) + (2)(25.0434)$$
$$+ (1)(-3.5964) = 177.5458$$

$$n_{\text{hvy}} = 8 \qquad \frac{V_c}{10^{-6}\text{m}^3/\text{mol}} = \frac{177.5458}{n_{\text{hvy}}^{-0.2266}} + 86.1539 = \frac{177.5458}{(8)^{-0.2266}} + 86.1539 = 370.57$$

TABLE 2-154 Ambrose Group[a] Contributions for Critical Constants

Group	Δ_T	Δ_P	Δ_V
Carbon atoms in alkyl groups	0.138	0.226	55.1
Corrections			
>CH— (each)	−0.043	−0.006	−8
>C< (each)	−0.120	−0.030	−17
Double bonds (nonaromatic)	−0.050	−0.065	−20
Triple bonds	−0.200	−0.170	−40
Delta Platt number,[b] multiply by	−0.023	−0.026	—
Aliphatic functional groups:			
—O—	0.138	0.160	20
>CO	0.220	0.282	60
—CHO	0.220	0.220	55
—COOH	0.578	0.450	80
—CO—O—OC—	1.156	0.900	160
—CO—O—	0.330	0.470	80
—NO_2—	0.370	0.420	78
—NH_2	0.208	0.095	30
—NH—	0.208	0.135	30
>N—	0.088	0.170	30
—CN	0.423	0.360	80
—S—	0.105	0.270	55
—SH	0.090	0.270	55
—SiH_3	0.200	0.460	119
—O—$Si(CH_3)_2$	0.496	—	—
—F	0.055	0.223	14
—Cl	0.055	0.318	45
—Br	0.055	0.500	67
—I	0.055	—	90
Halogen correction in aliphatic compounds:			
F is present	0.125		
F is absent, but Cl, Br, I are present	0.055		
Aliphatic alcohols[c]	[d]	[e]	15
Ring compound increments (listed only when different from aliphatic values):			
—CH_2—, >CH—, >C<	0.090	0.182	44.5
>CH— in fused ring	0.030	0.182	44.5
Double bond	−0.030	—	−15
—O—	0.090	—	10
—NH—	0.090	—	—
—S—	0.090	—	30
Aromatic compounds:			
Benzene	0.448	0.924	[f]
Pyridine	0.448	0.850	
C_4H_4 (fused as in naphthalene)	0.220	0.515	
—F	0.080	0.183	
—Cl	0.080	0.318	
—Br	0.080	0.600	
—I	0.080	0.850	
—OH	0.198	−0.025	
Corrections for nonhalogenated substitutions:			
First	0.010	0	
Each subsequent	0.030	0.020	
Ortho pairs containing —OH	−0.080	−0.050	
Ortho pairs with no —OH	−0.040	−0.050	
Highly fluorinated aliphatic compounds:			
—CF_3, —CF_2—, >CF—	0.200	0.550	
—CF_2—, >CF— (ring)	0.140	0.420	
>CF— (in fused ring)	0.030	—	
—H (monosubstitution)	−0.050	−0.350	
Double bond (nonring)	−0.150	−0.500	
Double bond (ring)	−0.030	—	
(other increments as in nonfluorina ted compounds)			

[a]Ambrose, D., Correlation and Estimation of Vapour-Liquid Critical Properties. I. Critical Temperatures of Organic Compounds, *Natl. Phys. Lab Report Chem.* **92** (1978); Correlation and Estimation of Vapour-Liquid Critical Properties. II. Critical Pressures and Volumes of Organic Compounds, *Natl. Phys. Lab Report Chem.* **98** (1979).

[b]The delta Platt number is defined as the Platt number of the isomer minus the Platt number of the corresponding alkane. (For n-alkanes the Platt number is $n - 3$.) The Platt number is the total number of groups of four carbon atoms three bonds apart [Platt, J. R., *J. Chem. Phys.*, **15**(1947): 419; **56**(1952): 328]. This correction is used only for branched alkanes.

[c]Includes naphthenic alcohols and glycols but not aromatic alcohols such as xylenol.

[d]First determine the hydrocarbon homomorph, i.e., substitute —CH_3 for each —OH and calculate $\Sigma\Delta_T$ for this compound. Subtract 0.138 from $\Sigma\Delta_T$ for each —OH substituted. Next, add $0.87 - 0.11n + 0.003n^2$ where $n = [T_b/\text{K (alcohol)} - 314]/19.2$. Exceptions include methanol ($\Sigma\Delta_T = 0$), ethanol ($\Sigma\Delta_T = 0.939$), and any alcohol whose value of n exceeds 10.

[e]Determine the hydrocarbon homomorph as in footnote *d*. Calculate $\Sigma\Delta_p$ and subtract 0.226 for each —OH substituted. Add $0.100 - 0.013n$, where n is computed as in footnote *d*.

[f]When estimating the critical volumes of aromatic substances, use ring compound values, if available, and correct for double bonds.

TABLE 2-155 Group Contributions for the Nannoolal et al. Method for Critical Constants[a] and Normal Boiling Point[b]

Table-specific nomenclature: (e) = connected to N, O, F, Cl; (ne) = not connected to N, O, F, Cl; (r) = in a ring; (c) = in a chain; (a) = aromatic, not necessarily carbon; (Ca) = aromatic carbon; b = any nonhydrogen atom

ID	Group	Description	TC × 10^3	PC × 10^4	VC	NBP
1	CH_3—(ne)	CH_3— not connected to N, O, F, or Cl	41.8682	8.1620	28.7855	177.3066
2	CH_3—(e)	CH_3— connected to N, O, F, or Cl	33.1371	5.5262	28.8811	251.8338
3	CH_3—(a)	CH_3— connected to an aromatic atom (not necessarily C)	−1.0710	4.1660	26.7237	157.9527
4	—C(c)H_2—	—CH_2— in a chain	40.0977	5.2623	32.0493	239.4531
5	>C(c)H—	>CH— in a chain	30.2069	2.3009	32.1108	240.6785
6	>C(c)<	>C< in a chain	−3.8778	−2.9925	28.0534	249.5809
7	>C(c)<(e)	>C< in a chain connected to at least one F, Cl, N, or O	52.8003	3.4310	33.7577	266.8769
8	>C(c)<(Ca)	>C< in a chain connected to at least one aromatic carbon	9.4422	2.3665	28.8792	201.0115
9	—C(r)H_2—	—CH_2— in a ring	21.2898	3.4027	24.8517	239.4957
10	>C(r)H—	>CH— in a ring	26.3513	3.6162	30.9323	222.1163
11	>C(r)<	>C< in a ring	−17.0459	−5.1299	5.9550	209.9749
12	>C(r)<(e, c)	>C< in a ring; connected to at least one N, O, Cl, or F not in the ring	51.7974	4.1421	29.5901	250.9584
13	>C(r)<(e, r)	>C< in a ring connected to at least one N or O which is part of the ring	18.9549	0.8765	20.2325	492.0707
14	>C(r)<(Ca)	>C< in a ring connected to at least one aromatic carbon	−29.1568	−0.1320	10.5669	244.3581
15	=C(a)H—	aromatic =CH—	16.1154	2.1064	19.4020	235.3462
16	=C(a)<(ne)	aromatic =C< not connected to O, N, Cl, or F	68.2045	4.1826	25.0434	315.4128
17	=C(a)<(e)	aromatic =C< connected to O, N, Cl, or F	68.1923	3.5500	5.6704	348.2779
18	(a)= C(a)<2(a)	aromatic =C< with three aromatic neighbors	29.8039	1.0997	16.4118	367.9649
19	F—(C, Si)	F— connected to C or Si	15.6068	0.7328	−5.0331	106.5492
20	—CF= C<	F— on a C=C (vinyl fluoride)	11.0757	4.3757	1.5646	49.2701
21	F—(C,Si)(F)(2b)	F— connected to C or Si substituted with at least one F and two other atoms	18.1302	3.4933	3.3646	53.1871
22	F—(C,Si)([F, Cl])(b)	F— connected to a C or Si substituted with one F or Cl and one other atom	19.1772	2.6558	1.0897	78.7578
23	F—(C,Si)([F, Cl]2)	F— connected to C or Si already substituted with two F or Cl atoms	20.8519	1.6547	1.1084	103.5672
24	F—(Ca)	F— connected to an aromatic carbon	−24.0220	0.5236	19.3190	−19.5575
25	Cl—(C,Si)	Cl— connected to C or Si not already substituted with F or Cl	−1.3329	−2.2611	22.0457	330.9117
26	Cl—(C,Si)([F, Cl])	Cl— connected to C or Si already substituted with one F or Cl	2.6113	−1.4992	23.9279	287.1863
27	Cl—(C, Si)([F, Cl]2)	Cl— connected to C or Si already substituted with at least two F or Cl	15.5010	0.4883	26.2582	267.4170
28	Cl—(Ca)	Cl— connected to aromatic C	−16.1905	−0.9280	36.7624	205.7363
29	—CCl=C<	Cl— on a C=C (vinyl chloride)	60.1907	11.8687	34.4110	292.5816
30	Br—(C,Si)	Br— connected to a nonaromatic C or Si	5.2621	−4.3170	36.0223	419.4959
31	Br—(Ca)	Br— connected to an aromatic C	−21.5199	−2.2409	30.7004	377.6775
32	I—(C, Si)	I— connected to C or Si	−8.6881	−4.7841	48.2989	556.3944
33	—OH *tert*	—OH connected to tertiary carbon	84.8567	−7.4244	10.6790	349.9409
34	HO—(C,Si) *sec*	—OH connected to secondary C or Si	79.3047	−4.4735	5.6645	390.2446
35	HO—(C,Si) long	—OH connected to primary C or Si; chain >4 C or Si	49.5968	−1.8153	2.0869	443.8712
36	HO—(C,Si) short	—OH connected to primary C or Si; chain <5 C or Si	130.1320	−6.8991	3.7778	488.0819
37	—OH (Ca)	—OH connected to an aromatic C (phenols)	14.0159	−12.1664	25.6584	361.4775
38	(C,Si)—O—(C,Si)	ether —O— connected to two C or Si	12.5082	2.0592	11.6284	146.4836
39	>(OC_2)<	>(OC_2)< (epoxide)	41.3490	0.1759	46.7680	820.7118
40	NH_2—(C, Si)	NH_2— connected to either C or Si	18.3404	−4.4164	13.2571	321.1759
41	NH_2—(Ca)	NH_2— connected to an aromatic C	−50.6419	−9.0065	73.7444	441.4388
42	(C,Si)—NH—(C,Si)	—NH— connected to two C or Si (secondary amine)	17.1780	−0.4086	20.5722	223.0992
43	(C,Si)2>N—(C,Si)	>N— connected to three C or Si (tertiary amine)	−0.5820	2.3625	6.0178	126.2952
44	COOH—(C)	—COOH connected to C	199.9042	3.9873	40.3909	1080.3139
45	(C)—COO—(C)	—COO— connected to two C (ester)	75.7089	4.3592	42.6733	636.2020
46	HCOO—(C)	HCOO— connected to C (formic acid ester)	58.0782	1.0266	36.1286	642.0427
47	—C(r)OO—	—COO— in ring, C is connected to C (lactone)	109.1930	0.4329		1142.6119
48	—CON<	—CON< disubstituted amide	102.1024	0.5172	64.3506	1052.6072
49	—CONH—	—CONH— (monosubstituted amide)				1364.5333
50	—$CONH_2$	—$CONH_2$ (amide)				1487.4109
51	O=C<(Can)2	—CO— connected to two nonaromatic C (ketones)	56.1572	0.1190	30.9229	618.9782
52	CHO—(Can)	CHO— connected to nonaromatic C (aldehydes)	44.2000	−2.3615	25.5034	553.8090
53	SH—(C)	—SH connected to C (thiols)	−7.1070	−9.4154	34.7699	434.0811
54	(C)—S—(C)	—S— connected to two C	0.5887	−8.2595	38.0185	461.5784
55	(C)—S—S—(C)	—S—S— (disulfide) connected to two C				864.5074
56	—S(a)—	—S— in an aromatic ring	−7.7181	−4.9259	20.3127	304.3321
57	(C)—C≡N	—C≡N (cyanide) connected to C	117.1330	5.1666	43.7983	719.2462
58	>C(c)=C(c)<	>C=C< (both C have at least one non-H neighbor)	45.1531	7.1581		475.7958
59	>C(c)=C(c)<(Ca)	noncyclic >C=C< connected to at least one aromatic C				586.1413
60	—(e)C(c)=C(c)<	noncyclic >C=C< with at least one F, Cl, N, or O	67.9821	−6.2791	51.0710	500.2434
61	H_2C(c)=C<	H_2C=C< (1-ene)	45.4406	9.6413	48.1957	412.6276
62	>C(r)=C(r)<	cyclic >C=C<	56.4059	3.4731	34.1240	475.9623
63	—C≡C—	—C≡C—	−19.9737	−2.2718	40.9263	512.2893
64	HC≡C—	HC≡C— (1-yne)	36.0883	2.4489	29.8612	422.2307
65	(Ca)—O(a)—(Ca)	—O—in an aromatic ring with aromatic C neighbors	10.4146	−0.5403	4.7476	37.1936
66	=N(a)—(r5)	aromatic —N— in a five-member ring, free electron pair	18.9903	8.3052	−25.3680	453.3397
67	=N(a)—(r6)	aromatic =N— in a six-member ring	10.9495	−4.7101	23.6094	306.7139
68	NO_2—(C)	NO_2— connected to aliphatic C	82.6239	−5.0929	34.8472	866.5843
69	NO_2—(Ca)	NO_2— connected to aromatic C				821.4141
70	>Si<	>Si<	25.4209	5.7270	75.7193	282.0181
71	>Si<(O)	>Si< connected to at least one O	72.5587	2.7602	69.5645	207.9312
72	NO_3—	nitrate (esters of nitric acid)				920.3617
74	O=N—O—(C)	nitrites (esters of nitrous acid)				494.2668

TABLE 2-155 Group Contributions for the Nannoolal Method for Critical Constants[a] and Normal Boiling Point[b] *(Continued)*

Table-specific nomenclature: (e) = connected to N, O, F, Cl; (ne) = not connected to N, O, F, Cl; (r) = in a ring; (c) = in a chain; (a) = aromatic, not necessarily carbon; (Ca) = aromatic carbon; b = any nonhydrogen atom

ID	Group	Description	TC × 10^3	PC × 10^4	VC	NBP
76	—C=O—O—C=O—	anhydride connected to two C	164.3355	4.0458		1251.2675
77	COCl—	COCl— connected to C (acid chloride)				778.9151
78	>Si<(F,Cl)	>Si< connected to at least one F or Cl	157.3401	12.6786		540.0895
79	O=C(—O—)2	noncyclic carbonate	97.2830	0.2822	52.8789	879.7062
80	OCN—	OCN— connected to C or Si (cyanate)	153.7225		27.1026	660.4645
81	SCN—(C)	SCN— (thiocyanate) connected to C				1018.4865
82	(C)—SO_2—(C)	noncyclic sulfone connected to two C (sulfones)	90.9726	−23.9221	68.0701	1559.9840
83	(C)2>Sn<(C)2	>Sn< connected to four carbons	62.3642	0.7043		510.4223
87	>C=C=C<	cumulated double bond	53.6350	12.6128		664.0903
88	>C=C—C=C<(r)	conjugated double bond in a ring	24.7302	−10.2451	64.4616	957.6388
89	>C=C—C=C<(c)	conjugated double bond in a chain				928.9954
90	CHO—(Ca)	CHO— connected to aromatic C (aldehydes)	38.4681	−4.0133	20.0440	560.1024
91	(C,Si) =N—	double-bonded amine connected to at least one C or Si				229.2288
92	(O= C< (C)2)a	—CO— connected to two C with at least one aromatic C (ketones)	63.6504	−5.0403	28.7127	606.1797
93	>Si<(C,H)2	>Si< attached to two carbon or hydrogen	34.2058	3.2023	55.3822	
94	—O—O—	peroxide				273.1755
97	(C,Si)a—NH—(Ca,Si)a	—NH— connected to two C or Si, at least one aromatic (secondary amines)	27.3441	−4.3834	29.3068	201.3224
99	—OCON<	—CO connected to O and N (carbamate)				886.7613
100	>N—(C=O)—N<	—CO connected to two N (urea)				1045.0343
101	(C,Si)2>N< (C,Si)2	Quaternary amine connected to four C or Si				−109.6269
102	F—(C,Si)(Cl)(b)2	F— connected to C or Si with at least one Cl and two other atoms	1.3231	3.3971	1.3597	111.0590
103	—OCOO—	—CO connected to two O (carbonates)	764.9595	58.9190		1573.3769
104	>SO_4	S(= O)$_2$ connected to two O (sulfates)				1483.1289
107	>S=O	sulfoxide				1379.4485
109	>N(C=O)	—CO connected to N				492.0707
111	(N)—C≡N	—C≡N (cyanide) connected to N				971.0365
113	>P<	phosphorus connected to at least 1 C or S (phosphine)				428.8911
115	—ON=(C,Si)	—ON= connected to C or Si (isoazole)	36.0361	−5.1116	16.2688	612.9506
		Corrections				
119	(C=O)—C([F,Cl]2,3)	carbonyl connected to C with two or more halogens	32.1829	7.3149	−3.8033	−82.2328
120	(C=O)—C([F,Cl]2,3)2	carbonyl connected to two C, each with at least two halogens	11.4437	4.1439	27.5326	−247.8893
121	C—[F,Cl]3	carbon with three halogens	−1.3023	0.4387	1.5807	−20.3996
122	(C)2—C—[F,Cl]2	secondary carbon with two halogens	−34.3037	−4.2678	−2.6235	15.4720
123	No hydrogen	component has no hydrogen	−1.3798	4.8944	−5.3091	−172.4201
124	One hydrogen	component has one hydrogen	−2.7180	2.8103	−6.1909	−99.8035
125	(3,4) ring	a three- or four-member nonaromatic ring	11.3251	−0.3035	3.2219	−62.3740
126	5-ring	a five-member nonaromatic ring	−4.7516	0.0930	−6.3900	−40.0058
127	*Ortho* pair(s)	*ortho*- position counted only once and only if no *meta* or *para* pairs	1.2823	0.7061	−3.5964	−27.2705
128	*Meta* pair(s)	*meta*- position counted only once and only if no *para* or *ortho* pairs	6.7099	−0.7246	1.5196	−3.5075
129	*Para* pair(s)	*para*- position counted only once and only if no *meta* or *ortho* pairs				16.1061
130	((C=)(C)C—CC3)	carbon with four carbon neighbors and one double-bonded carbon neighbor	−33.8201	−8.8457	−4.6483	25.8348
131	C2C—CC2	carbon with four carbon neighbors, two on each side	−18.4815	−2.2542	−5.0563	35.8330
132	C3C—CC2	carbon with five carbon neighbors	−23.6024	−3.2460	−6.3267	51.9098
133	C3C—CC3	carbon with six carbon neighbors	−24.5802	−5.3113	4.9392	111.8372
134	C=C—C=O	—C=O connected to sp^3 carbon	−35.6113	1.0934	2.8889	40.205

[a]Nannoolal, Y., et al., *Fluid Phase Equilib.* **252** (2007): 1.
[b]Nannoolal, Y., et al., *Fluid Phase Equilib.* **226** (2004): 45.

Results:

Property	DIPPR 801 recommendation	Nannoolal estimation	% Difference
T_c/K	630.3	628.9	−0.2
P_c/bar	37.32	36.24	−2.9
V_c/(cm^3/mol)	370	370.6	0.2

Method: Wilson-Jasperson method.

Reference: Wilson, G. M., and L. V. Jasperson, "Critical Constants T_c, P_c, Estimation Based on Zero, First and Second Order Methods," AIChE Spring Meeting, New Orleans, La., 1996.

Classification: Group contributions.

Expected uncertainty: ~6 K or 1 percent for T_c; ~2 bar or 5 percent for P_c.

Applicability: Organic and some inorganic compounds.

Input data: M, T_b, group contributions C_i from Table 2-157, and molecular structure.

Description: A GC method with first- and some second-order contributions. Variables T_c, P_c, and V_c are given by the following relations:

$$T_c = \frac{T_b}{\left(0.048271 - 0.019846 n_r + \sum_k n_k \Delta tc_k + \sum_j n_j \Delta tc_j\right)^{0.2}} \tag{2-13}$$

$$\frac{P_c}{\text{bar}} = \frac{0.0186233(T_c/\text{K})}{\exp(Y) - 0.96601} \tag{2-14}$$

$$Y = -0.00922295 - 0.0290403 n_r + 0.041\left(\sum_k n_k \Delta pc_k + \sum_j n_j \Delta pc_j\right) \tag{2-15}$$

where n_r is the number of rings in the molecule; Δtc_k and Δpc_k are the first-order group contributions tabulated in Table 2-157 with n_k the number of such occurrences in the molecule; and Δtc_j and Δpc_j are the second-order

TABLE 2-156 Intermolecular Interaction Corrections for the Nannoolal et al. Method for Critical Constants[a] and Normal Boiling Point[b]

	TC × 10^3	PC × 10^4	VC	NBP
—OH :: —OH	−434.8568	−5.6023		291.7985
—OH :: —COOH				146.7286
—OH :: —O—	−146.7881	7.3373	19.7707	135.3991
—OH :: >(OC$_2$)<				226.4980
—OH :: —COOC—				211.6814
—OH :: —CO—				46.3754
—OH :: —O(a)—				435.0923
—OH :: —S(na)—				-74.0193
—OH :: —SH				38.6974
—OH :: —NH$_2$	120.9166	69.8200		314.6126
—OH :: >NH	−30.4354	6.1331	−8.0423	286.9698
—OH :: —CN				306.3979
—OH :: ═N(a)-(r6)				1334.6747
—OH(a) :: —OH(a)	144.4697	57.8350	97.5425	288.6155
—OH(a) :: —COOH				-1477.9671
—OH(a) :: —O—				130.3742
—OH(a) :: —COOC—				−1184.9784
—OH(a) :: —CHO				43.9722
—OH(a) :: —NH$_2$				797.4327
—OH(a) :: Nitrate				−1048.124
—OH(a) :: ═N(a)-(r6)				−614.3624
—COOH :: —COOH				117.2044
—COOH :: —O—				612.8821
—COOH :: —COOC—				−183.2986
—COOH :: —CO—				−55.9871
—O— :: —O—	162.6878	2.6751	−23.6366	91.4997
—O— :: >(OC$_2$)<	707.4116	88.8752	−329.5074	178.7845
—O— :: —COOC—	128.2740	−1.0295	−55.5112	322.5671
—O— :: —CO—				15.6980
—O— :: —CHO				17.0400
—O— :: —O(a)—				329.0050
—O— :: —S(na)—	−654.1363	25.8246	−37.2468	394.5505
—O— :: —NH$_2$	−738.0515	−125.5983		124.3549
—O— :: >NH				101.8475
—O— :: —CN	741.8565			293.5974
—O— :: Nitrate				963.6518
>(OC$_2$)< :: >(OC$_2$)<				1006.388
>(OC$_2$)< :: —CO—				22.5208
>(OC$_2$)< :: —CHO				163.5475
—COOC— :: —COOC—	366.2663	0.5195	−74.8680	431.0990
—COOC— :: —CO—				22.5208
—COOC— :: —O(a)—				707.9404
—COOC— :: —NH$_2$				182.6291
—COOC— :: >NH				317.0200
—COOC— :: —CN				517.0677
—COOC— :: Nitrate				−205.6165
—CO— :: —CO—	1605.564	−78.2743	−413.3976	−303.9653
—CO— :: —CHO				−391.3690
—CO— :: —O(a)—				176.5481
—CO— :: —S(a)—				381.0107
—CO— :: >NH				−215.3532
—CO— :: —CN				−574.2230
—CO— :: Nitrate				−3628.903
—CO— :: ═N(a)-(r6)				124.1943
—CHO— :: —CHO—				562.1763
—CHO— :: —O(a)—				674.6858
—CHO— :: —S(a)—				397.575
—CHO— :: Nitrate				140.9644
—O(a)— :: —NH$_2$				395.4093
—O(a)— :: ═N(a)-(r5)	24.0243	−35.1998	217.9243	-888.612
—S(na)— :: —S(na)—	-861.1528	43.9001	−403.1196	-11.9406
—S(na)— :: —NH$_2$				-562.306
—S(a)— :: —CN				-101.232
—S(a)— :: ═N(a)-(r5)	131.7924	-19.7033	164.2930	-348.740
—SH :: —SH				217.6360
—NH$_2$ :: —NH$_2$	-60.9217	-0.6754		174.0258
—NH$_2$ :: >NH				510.3473
—NH$_2$ :: Nitrate				663.8009
—NH$_2$ :: ═N(a)-(r6)				27.2735
>NH :: >NH	-49.7641	22.1871	-57.1233	239.8076
>NH :: ═N(a)-(r6)				758.9855
—OCN :: —OCN	-1866.097		44.1062	-356.5017
—OCN :: Nitrate				-263.0807
—CN :: ═N(a)-(r6)				-370.9729
Nitrate :: Nitrate				65.1432
═N(a)-(r6) :: ═N(a)-(r6)	-32.3208	12.5371	-26.4556	-271.9449

[a]Nannoolal, Y., et al., *Fluid Phase Equilib.* **252** (2007): 1.
[b]Nannoolal, Y., et al., *Fluid Phase Equilib.* **226** (2004): 45.

group contributions, also tabulated in Table 2-157, with n_j occurrences of these second-order groups in the molecule.

Example Estimate T_c and P_c of *sec*-butanol by using the Wilson-Jasperson method.
Required input data: From DIPPR 801 database, $T_b = 372.9$ K.
Structure:

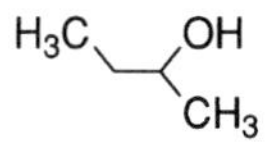

Group contributions from Table 2-157:

Group	n_k	Δtc_k	Δpc_k	n_j	Δtc_j	Δpc_j
H	10	0.002793	0.12660	—	—	—
O	1	0.020341	0.43360	—	—	—
C	4	0.008532	0.72983	—	—	—
–OH, C4 or less	—	—	—	1	0.0350	0

From Eqs. (2-13), (2-14), and (2-15):

$$\sum n_k \Delta tc_k = (10)(0.002793) + (1)(0.020341) + (4)(0.008532) = 0.082399$$

$$T_c = \frac{T_b}{(0.048271 + 0.082399 + 0.0350)^{0.2}} = \frac{372.9\text{ K}}{0.6980} = 534.25\text{ K}$$

$$\sum n_k \Delta pc_k = (10)(0.12660) + (1)(0.43360) + (4)(0.72983) = 4.61892$$

$$Y = -0.00922295 + 0.041(4.61892) = 0.18015$$

$$\frac{P_c}{\text{bar}} = \frac{0.0186233(T_c/\text{K})}{\exp(Y) - 0.96601} = \frac{(0.0186233)(534.25)}{\exp(0.18015) - 0.96601} = 43.00$$

Results:

Property	DIPPR 801 recommendation	Wilson-Jasperson estimation	% Difference
T_c/K	536.2	534.25	-0.4
P_c/bar	42.02	43.00	2.3

Normal Melting Point The *normal melting point* is defined as the temperature at which melting occurs at atmospheric pressure. Methods to estimate the melting point are not particularly effective because the melting point depends strongly on solid crystal structure and that structure is not effectively correlated with standard GC or CS methods.

Recommended Method The method of Constantinou and Gani is recommended with caution.
Reference: Constantinou, L., and R. Gani, *AIChE J.*, **40** (1994): 1697.
Classification: Group contributions.
Expected uncertainty: 25 percent.
Applicability: Organic compounds.
Input data: First- and second-order group contributions from molecular structure.
Description: A group contribution method given by

$$T_m = (102.425\text{ K}) \cdot \ln\left(\sum_i n_i t_{m1,i} + \sum_j n_j t_{m2,j}\right) \qquad (2\text{-}16)$$

where n_i, n_j = number of first- and second-order groups, respectively
$t_{m1,i}$ = first-order group contributions from Table 2-158
$t_{m2,i}$ = second-order group contributions from Table 2-159

Example Estimate the melting point of 2,6-dimethylpyridine.
Structure and group contributions:

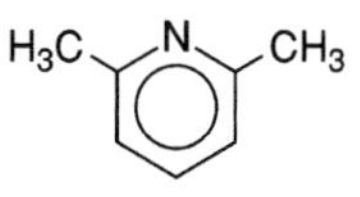

Group	n_i	$t_{m1,i}$	$t_{m2,i}$
$-CH_3$	2	0.4640	
$-C_5H_3(N)-$	1	12.6275	
Six-member ring	1		1.5656

TABLE 2-157 Wilson-Jasperson First- and Second-Order Contributions for Critical Temperature and Pressure[a]

First-order atom	Δtc_k	Δpc_k
H, D, T	0.002793	0.12660
He	0.320000	0.43400
B	0.019000	0.91000
C	0.008532	0.72983
N	0.019181	0.44805
O	0.020341	0.43360
F	0.008810	0.32868
Ne	0.036400	0.12600
Al	0.088000	6.05000
Si	0.020000	1.34000
P	0.012000	1.22000
S	0.007271	1.04713
Cl	0.011151	0.97711
Ar	0.016800	0.79600
Ti	0.014000	1.19000
V	0.018600	—
Ga	0.059000	—
Ge	0.031000	1.42000
As	0.007000	2.68000
Se	0.010300	1.20000
Br	0.012447	0.97151
Kr	0.013300	1.11000
Rb	−0.027000	—
Zr	0.175000	1.11000
Nb	0.017600	2.71000
Mo	0.007000	1.69000
Sn	0.020000	1.95000
Sb	0.010000	—
Te	0.000000	0.43000
I	0.005900	1.31593
Xe	0.017000	1.66000
Cs	−0.027500	6.33000
Hf	0.219000	1.07000
Ta	0.013000	—
W	0.011000	1.08000
Re	0.014000	—
Os	−0.050000	—
Hg	0.000000	−0.08000
Bi	0.000000	0.69000
Rn	0.007000	2.05000
U	0.015000	2.04000
Second-order group	Δtc_j	Δpc_j
—OH, C_4 or less	0.0350	0.00
—OH, C_5 or more	0.0100	0.00
—O—	−0.0075	0.00
$—NH_2$, >NH, >N—	−0.0040	0.00
—CHO	0.0000	0.50
>CO	−0.0550	0.00
—COOH	0.0170	0.50
—COO—	−0.0150	0.00
—CN	0.0170	1.50
$—NO_2$	−0.0200	1.00
Organic halides (once per molecule)	0.0020	0.00
—SH, —S—, —SS—	0.0000	0.00
Siloxane bond	−0.0250	−0.50

[a]As cited in PGL5.

Calculation using Eq. (2-16):

$$T_m = (102.425\text{ K}) \ln [(2)(0.4640) + 12.6275 + 1.5656] = 278\text{ K}$$

The predicted value is 4 percent higher than the recommended experimental value of 267 K in the DIPPR 801 database.

Normal Boiling Point The normal boiling temperature T_b is the temperature at which the vapor pressure of the liquid equals 101.325 kPa (1.0 atm). If there are sufficient vapor pressure data available, then T_b may be found from a regression of the data using an appropriate vapor pressure equation [e.g., Eqs. (2-24) to (2-28)]. If two or more vapor pressure values are available in the approximate temperature range of T_b, they can be used to obtain T_b by using Eq. (2-2) to linearly interpolate ln P^* versus $1/T$ values. When one or more low-temperature vapor pressure points are available, a common occurrence, then the method of Pailhes can be used to estimate T_b.

TABLE 2-158 First-Order Groups and Their Contributions for Melting Point*

Group	$t_{m1,i}$	Group	$t_{m1,i}$	Group	$t_{m1,i}$
$—CH_3$	0.4640	$—COOCH_2—$	3.5572	$—CCl_3$	10.2337
$>CH_2$	0.9246	—OOCH	4.2250	>ACCl	2.7336
>CH—	0.3557	$—OCH_3$	2.9248	$—CH_2NO_2$	5.5424
>C<	1.6479	$—OCH_2—$	2.0695	$>CHNO_2$	4.9738
$—CH{=}CH_2$	1.6472	—OCH<	4.0352	$>ACNO_2$	8.4724
—CH=CH—	1.6322	$—OCH_2F$	4.5047	$—CH_2SH$	3.0044
$>C{=}CH_2$	1.7899	$—CH_2NH_2$	6.7684	—I	4.6089
>C=CH—	2.0018	$>CHNH_2$	4.1187	—Br	3.7442
>C=C<	5.1175	$—NHCH_3$	4.5341	—C≡CH	3.9106
$—CH{=}C{=}CH_2$	3.3439	$—CH_2NH—$	6.0609	—C≡C—	9.5793
>ACH	1.4669	>CHNH—	3.4100	>C=CCl—	1.5598
>AC—	0.2098	$>NCH_3$	4.0580	>ACF	2.5015
$>ACCH_3$	1.8635	$—NCH_2—$	0.9544	$—CF_3$	3.2411
$>ACCH_2—$	0.4177	$>ACNH_2$	10.1031	—COO—	3.4448
>ACCH<	−1.7567	$—C_5H_3(N)—$	12.6275	$—CCl_2F$	7.4756
—OH	3.5979	$—CH_2CN$	4.1859	$—CClF_2$	2.7523
>ACOH	13.7349	—COOH	11.5630	—F (other)	1.9623
$—COCH_3$	4.8776	$—CH_2Cl$	3.3376	$—CONH_2$	31.2786
$—COCH_2—$	5.6622	>CHCl	2.9933	$—CON(CH_3)_2$	11.3770
—CHO	4.2927	>CCl—	9.8409	$—CH_3S$	5.0506
$—COOCH_3$	4.0823	$—CHCl_2$	5.1638	$>CH_2S$	3.1468

*Constantinou, L., and R. Gani, *AIChE J.*, **40** (1994): 1697.

TABLE 2-159 Second-Order Groups and Their Contributions for Melting Point*

Group	$t_{m2I,i}$	Group	$t_{m2I,i}$
$—CH(CH_3)_2$	0.0381	CHCOOH; CCOOH	−3.1034
$—C(CH_3)_3$	−0.2355	ACCOOH	28.4324
$—CH(CH_3)CH(CH_3)—$	0.4401	CH_3COOCH; CH_3COOC	0.4838
$—CH(CH_3)C(CH_3)_2—$	−0.4923	$COCH_2COO$ or COCHCOO or COCCOO	0.0127
$—C(CH_3)_2C(CH_3)_2—$	6.0650	CO—O—CO	−2.3598
Three-member ring	1.3772	ACCOO	−2.0198
Five-member ring	0.6824	CHOH	−0.5480
Six-member ring	1.5656	COH	0.3189
Seven-member ring	6.9709	$CH_m(OH)CH_n(OH)$ [$m, n = 0, 1, 2$]	0.9124
$CH_n{=}CH_m—CH_p{=}CH_k$ [$k, n, m, p = 0, 1, 2$]	1.9913	$CH_{m\,cyclic}—OH$ [$m = 0, 1$]	9.5209
$CH_3CH_m{=}CH_n$ [$m, n = 0, 1, 2$]	0.2476	$CH_m(OH)CH_n(NH_p)$ [$m, n, p = 0, 1, 2, 3$]	2.7826
$CH_2CH_m{=}CH_n$ [$m, n = 0, 1, 2$]	−0.5870	$CH_m(NH_2)CH_n(NH_2)$ [$m, n = 0, 1, 2$]	2.5114
$CHCH_m{=}CH_n$ or $CCH_m{=}CH_n$ [$m, n = 0, 1, 2$]	−0.2361	$CH_{m\,cyclic}—NH_p—CH_{n\,cyclic}$ [$m, n, p = 0, 1, 2$]	1.0729
Alicyclic side chain: $C_{cyclic}C_m$ [$m > 1$]	−2.8298	$CH_m—O—CH_n{=}CH_p$ [$m, n, p = 0, 1, 2$]	0.2476
CH_3CH_3	1.4880	$AC—O—CH_m$ [$m = 0, 1, 2, 3$]	0.1175
CHCHO; CCHO	2.0547	$CH_{m\,cyclic}—S—CH_{n\,cyclic}$ [$m, n = 0, 1, 2$]	−0.2914
CH_3COCH_2	−0.2951	$CH_m{=}CH_n—F$ [$m, n = 0, 1, 2$]	−0.0514
CH_3COCH; CH_3COC	−0.2986	$CH_m{=}CH_n—Br$ [$m, n = 0, 1, 2$]	−1.6425
$C_{cyclic}(=O)$	0.7143	ACBr	2.5832
ACHO	−0.6697	ACl	−1.5511

*Constantinou, L., and R. Gani, *AIChE J.*, **40** (1994): 1697.

The most accurate method for prediction of normal boiling temperatures without experimental data is the Nannoolal method.

Recommended Method Pailhes method.

Reference: Pailhes, F., *Fluid Phase Equilib.*, **41** (1988): 97.

Classification: Group contributions.

Expected uncertainty: ~3 K (1 to 2 percent).

Applicability: Organic compounds.

Input data: Molecular structure and one measured vapor pressure value P^*_{meas} (often at a low pressure). The method requires estimation of the reduced normal boiling point, T_{br}, and P_c, which in the example below are obtained using the Wilson-Jasperson first-order method and the Ambrose method, respectively.

Description: A simple group contribution method is given by

$$T_b = T_{meas}\left[\frac{\log(P_c/\text{bar}) + (1 - T_{br})\,x_P}{\log(P_c/\text{bar})}\right] - 3x_p - 1.49x_p^2 \qquad (2\text{-}17)$$

where T_b = estimated normal boiling point

P_c = critical pressure estimated from group contributions

$x_P = \log(1\ \text{atm}/P^*_{meas})$

T_{meas} = temperature at which experimental vapor pressure P^*_{meas} is known

Example The vapor pressure of *n*-decylacetate (M = 200.32 kg/kmol) at 348.65 K is 106.66 Pa. Estimate the normal boiling point of this compound, using the Pailhes method.

Structure and group contributions from Tables 2-154 and 2-157:

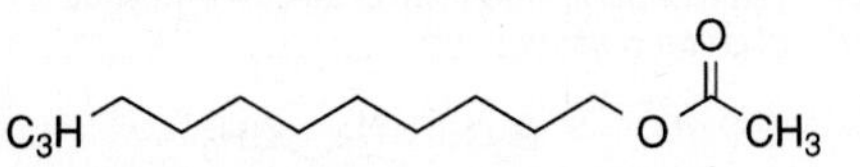

Wilson-Jasperson Groups	n_i	Δtc_i
H	24	0.002793
O	2	0.020341
C	12	0.008532

Ambrose Groups	n_i	$\Delta_{P,i}$
–COO–	1	0.470
C (alkyl)	11	0.226

Group contribution calculations using Eq. (2-13) for T_{br} and Eq. (2-7) for P_c:

$$\sum n_i \Delta tc_i = (24)(0.002793) + (2)(0.020341) + (12)(0.008532) = 0.210098$$

$$T_{br} = (0.048271 + 0.210098)^{0.2} = 0.7629$$

$$\sum n_i \Delta_{P,i} = (1)(0.470) + (11)(0.226) = 2.956$$

$$P_c = \frac{200.32}{(0.339 + 2.956)^2}\text{bar} = 18.450 \text{ bar}$$

Calculation of auxiliary quantities:

$$x_P = \log\left(\frac{1 \text{ atm}}{P^*_{\text{meas}}}\right) = \log\left(\frac{101{,}325 \text{ Pa}}{106.66 \text{ Pa}}\right) = 2.9777$$

Calculation of normal boiling point using Eq. (2-17):

$$\frac{T_b}{\text{K}} = (348.65)\left[\frac{\log(18.450) + (1 - 0.7629)(2.9777)}{\log(18.450)}\right] - 3(2.9777) - 1.49(2.9777)^2$$

$$T_b = 520.94 \text{ K}$$

The estimated value is 0.7 percent higher than the DIPPR 801 recommended value of 517.15 K.

Recommended Method: Nannoolal method.

Reference: Nannoolal, Y., J. Rarey, D. Ramjugernath, and W. Cordes, *Fluid Phase Equilib.*, **226** (2004): 45.

Classification: Group contributions.

Expected uncertainty: ~7 K (about 2 percent).

Applicability: Organic compounds.

Input data: C_i values in Table 2-155; intramolecular group-group interactions C_{ij} in Table 2-156; and the number of nonhydrogen atoms in the molecule.

Description: A GC method that includes second-order corrections for steric effects and intramolecular interactions. T_b is calculated from

$$\frac{T_b}{K} = \frac{\sum_{i=1}^{N} n_i \cdot C_i + \text{GI}}{n_{\text{hvy}}^{0.6583} + 1.6868} + 84.3395 \qquad (2\text{-}18)$$

where n_{hvy} = number of nonhydrogen (heavy) atoms
n_i = number of occurrences of group i
C_i = group contribution from Table 2-155
GI = total group-group interaction as calculated using Eq. (2-12) and Table 2-156

Example Estimate the normal boiling point of di-isopropanolamine by using the Nannoolal method.

Structure:

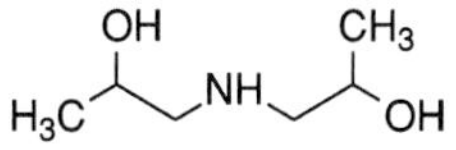

Group contributions and values:

Group	n_i	C_i	Group total
—CH$_3$	2	177.3066	354.6132
>C(c)<(e)	4	266.8769	1067.508
—OH sec	2	390.2446	780.4892
—NH—	1	223.0992	223.0992
GI			
—OH:: —OH	2/(9 × 2)	291.7985	32.42206
—OH:: —NH—	4/(9 × 2)	286.9698	63.77107
		Total	2521.902

Note that the frequencies of the interaction correction terms are calculated in the following manner: There are three interacting groups (—OH, —OH, —NH—) in the molecule, so NG − 1 = 2. The four —OH:: —NH— interactions and two —OH:: —OH interactions are each divided by 2 and by the number of nonhydrogen atoms $n_{\text{hvy}} = 9$, according to Eq. (2-12).

Calculation using Eq. (2-18):

$$\frac{T_b}{\text{K}} = \frac{2521.902}{9^{0.6583} + 1.6868} + 84.3395 = 509.3 \qquad T_b = 509.3 \text{ K}$$

This value differs by −2.4 percent from the DIPPR 801 recommended value of 521.9 K.

CHARACTERIZING AND CORRELATING CONSTANTS

Acentric Factor The acentric factor of a compound ω is defined in terms of the reduced vapor pressure evaluated at a reduced temperature of 0.7 as

$$\omega = -\log P_r^*\Big|_{T_r=0.7} - 1.0000 \qquad (2\text{-}19)$$

It is primarily used as a third parameter (in addition to T_c and P_c) in CS predictions as a measure of deviations from nonspherical molecular shape, hence the name, suggesting molecular interactions that are not between centers of molecules. However, as defined in Eq. (2-19), ω also contains polarity information, and it increases with increasing polarity for molecules of similar size and shape. The value of ω is close to zero for small, spherically shaped, nonpolar molecules (argon, methane, etc.). It increases in value with larger deviations of molecular shape from spherical (longer chain lengths, less chain branching, etc.) and with increasing molecular polarity. When possible, ω should be obtained from experimental vapor pressure correlations by using Eq. (2-19), but an accurate estimation of ω can be made by using the critical constants and a single vapor pressure point by application of CS vapor pressure equations.

Recommended Method 1 Definition.

Classification: Theory and empirical extension of theory.

Expected uncertainty: Within 3 percent if an experimental vapor pressure correlation is available; within 10 percent from a predicted vapor pressure correlation.

Applicability: Most organic compounds.

Input data: Vapor pressure correlation or T_c, P_c, and T_b if an experimental vapor pressure correlation is unavailable.

Description: Equation (2-19) is applied directly to the appropriate vapor pressure equation. A predictive vapor pressure equation can also be used as in the second example.

Example Calculate the acentric factor of chlorobenzene with a known value for T_b.

Input information: From the DIPPR 801 database, T_b = 404.87 K, T_c = 632.35 K, and P_c = 45.1911 bar.

Calculation of auxiliary quantities (see Eq. (2-28a) for these equations):

$$T_{br} = \frac{T_b}{T_c} = \frac{404.87}{632.35} = 0.64 \qquad K = 0.0838$$

$$\psi = -35 + \frac{36}{T_{br}} + 42 \cdot \ln(T_{br}) - T_{br}^6 = -35 + \frac{36}{0.64} + 42 \cdot \ln(0.64) - (0.64)^6 = 2.4312$$

$$\alpha_c = \frac{(3.758)K\psi + \ln(P_c/1.01325\text{bar})}{K\psi - \ln(T_{br})}$$

$$= \frac{(3.758)(0.0838)(2.4312) + \ln\left(\frac{45.1911}{1.01325}\right)}{(0.0838)(2.4312) - \ln(0.64)} = 7.025$$

$$D = K(\alpha_c - 3.758) = (0.0838)(7.025 - 3.758) = 0.2738$$

$$A = 35D = 9.581 \quad B = -36D = -9.855 \quad C = \alpha_c - 42D = -4.473$$

Calculation using Eq. (2-28) at T_r = 0.7:

$$\ln(P_r) = 9.581 - \frac{9.855}{0.7} - 4.473 \cdot \ln(0.7) + 0.2738 \cdot (0.7)^6 = -2.870$$

Calculation using Eq. (2-19):

$$\omega = -\frac{\ln(P_r)}{2.303} - 1.0000 = \frac{2.870}{2.303} - 1.0000 = 0.246$$

This value differs by −1.5 percent from DIPPR 801 recommended value of 0.2499.

Recommended Method 2 Corresponding states.

Reference: [PGL5].

Classification: Corresponding states.

Expected uncertainty: Generally within 5 percent, worse for strongly polar fluids.

Applicability: Most organic compounds.

Input data: T_c, P_c, and a single vapor pressure point (e.g., the normal boiling point T_b).

Description: See Eq. (2-29) for the equations used in this method. The vapor pressure equation is inverted to obtain the acentric factor from a single vapor pressure point.

Example Repeat the above calculation of the acentric factor of chlorobenzene, using the Walton-Ambrose modification of the Lee-Kesler vapor pressure equation, Eq. (2-29).

Input information: From the DIPPR 801 database, T_b = 404.87 K, T_c = 632.35 K, and P_c = 45.1911 bar.

Calculation of auxiliary quantities:

$$T_{br} = \frac{T_b}{T_c} = \frac{404.87}{632.35} = 0.64 \qquad \tau = 1 - 0.64 = 0.36$$

$$f^{(0)} = \frac{(-5.97616)(0.36) + (1.29874)(0.36)^{1.5} - (0.60394)(0.36)^{2.5} - (1.06841)(0.36)^5}{0.64}$$

$$= -3.0034$$

$$f^{(1)} = \frac{(-5.03365)(0.36) + (1.11505)(0.36)^{1.5} - (5.41217)(0.36)^{2.5} - (7.46628)(0.36)^5}{0.64}$$

$$= -3.1788$$

$$f^{(2)} = \frac{(-0.64771)(0.36) + (2.41539)(0.36)^{1.5} - (4.26979)(0.36)^{2.5} - (3.25259)(0.36)^5}{0.64}$$

$$= -0.037$$

Calculation using Eq. (2-29) at the normal boiling point:

$$\ln \frac{1.01325}{45.1911} = -3.798 = f^{(0)} + \omega f^{(1)} + \omega^2 f^{(2)} = -3.0034 - 3.1788\omega - 0.037\omega^2$$

Back solution of the quadratic equation for ω: $\omega = 0.249$

Radius of Gyration The radius of gyration R_g is a measure of the mass distribution about the center of mass of a molecule. Radius R_g increases with molecular size. It is useful in CS applications to separate molecular size and shape effects from polar effects. It is defined in terms of the principal moments of inertia of a molecule (*A, B,* and *C*) as

$$R_g = \sqrt{\frac{(AB)^{1/2} N_A}{M}} \tag{2-20}$$

for planar molecules and as

$$R_g = \sqrt{\frac{2\pi (ABC)^{1/3} N_A}{M}} \tag{2-21}$$

for nonplanar molecules. Radii of gyration can be calculated from these defining equations using principal moments of inertia obtained from spectral data or from computational chemistry software.

Recommended Method Principal moments of inertia.

Classification: Computational chemistry.

Expected uncertainty: Less than 5 percent.

Applicability: All molecules.

Input data: *M* and molecular structure.

Description: Computational chemistry software is used to optimize the geometry of the molecule and obtain the principal moments of inertia to be used in Eqs. (2-20) and (2-21).

Example Calculate the radius of gyration for hydrazine.

Input information: From the DIPPR 801 database, $M = 32.0452$ kg/kmol. The structure of hydrazine is

$$H_2N—NH_2$$

Calculation of the principal moments of inertia: Optimizing hydrazine with HF/6-31G model chemistry gives the following principal moments of inertia:

$$A = 12.24050 \text{ amu} \cdot \text{Bohr}^2$$

$$B = 72.41081 \text{ amu} \cdot \text{Bohr}^2$$

$$C = 79.16893 \text{ amu} \cdot \text{Bohr}^2$$

Conversion from atomic units to SI gives

$$A = (12.24050 \text{ amu} \cdot \text{Bohr}^2)\left(\frac{5.29177 \times 10^{-11}\text{m}}{\text{Bohr}}\right)^{-2}\left(\frac{1.66054 \times 10^{-27}\text{kg}}{\text{amu}}\right)$$

$$= 5.692 \times 10^{-47} \text{kg} \cdot \text{m}^2$$

$$B = (72.41081 \text{ amu} \cdot \text{Bohr}^2)\left(\frac{4.65010^{-48} \text{kg} \cdot \text{m}^2}{\text{amu} \cdot \text{Bohr}^2}\right) = 3.367 \times 10^{-46} \text{kg} \cdot \text{m}^2$$

$$C = (79.16893 \text{ amu} \cdot \text{Bohr}^2)\left(\frac{4.65010^{-48} \text{kg} \cdot \text{m}^2}{\text{amu} \cdot \text{Bohr}^2}\right) = 3.681 \times 10^{-46} \text{kg} \cdot \text{m}^2$$

Calculation using Eq. (2-21):

$$(ABC)^{1/3} = [(5.692 \times 10^{-47})(3.367 \times 10^{-46})(3.681 \times 10^{-46})]^{1/3} \text{kg} \cdot \text{m}^2$$

$$= 1.918 \times 10^{-46} \text{kg} \cdot \text{m}^2$$

$$R_g = \sqrt{\frac{2\pi\,(1.918 \times 10^{-46} \text{kg} \cdot \text{m}^2)(6.022 \cdot 10^{26} \text{kmol}^{-1})}{32.0452 \text{ kg/kmol}}} = 1.505 \times 10^{-10} \text{m}$$

This is 3.8 percent below the DIPPR 801 database value of 1.564×10^{-10} m which was obtained from spectral principal moments of inertia.

Dipole Moment The dipole moment of a molecule is the first moment of the electric charge density expansion. All normal paraffins have a value of zero. Charge separation within the molecule due to electronegativity differences between bonded atoms increases the dipole moment. Computational chemistry software uses the electron density distribution of the optimized molecule to calculate dipole moments.

Recommended Method Electron density distribution.

Classification: Computational chemistry.

Expected uncertainty: Uncertainty varies depending upon the model chemistry chosen, but it can be as large as 60 percent.

Applicability: All molecules.

Input data: Molecular structure.

Example Calculate the dipole moment for methanol.

Draw structure and optimize molecule by using computational chemistry software: The dipole moment obtained from a geometry optimized with the HF/6-31G model chemistry for methanol is 2.288 D. This value is 35 percent larger than the experimental gas-phase value of 1.700 D in the DIPPR 801 database.

Refractive Index Refractive index is the ratio of the speed of light in a vacuum to the speed of light in the medium. The incident light is the sodium D line (5.896×10^{-7} m). Refractive index is dimensionless and generally ranges between 1.3 and 1.5 for organic liquids.

Recommended Method Wildman-Crippen method.

Reference: Wildman, S. A., and G. M. Crippen, *J. Chem. Inf. Comput. Sci.* **39** (1999): 868.

Classification: Theory and group contribution.

Expected uncertainty: Generally less than 3 percent for liquids.

Applicability: Most organic molecules (currently not applicable to organic acids).

Input data: Molecular structure, molecular weight, and density at the desired temperature.

Description: This method is based on the Lorentz-Lorenz relation between the molar refraction R_D and the refractive index, which can be written in the form

$$n = \sqrt{\frac{M + 2\left(\frac{\rho}{\text{gm} \cdot \text{cm}^{-3}}\right)R_D}{M - \left(\frac{\rho}{\text{gm} \cdot \text{cm}^{-3}}\right)R_D}} \tag{2-22}$$

where *n* is refractive index at the same temperature as the density ρ. Wildman and Crippen developed a GC method for R_D with the atomic contributions shown in Table 2-160 for each type of atom with its bonded neighbors.

Example Calculate the refractive index of *m*-ethylphenol at 298.15 K. The various types of atoms corresponding to the descriptions in Table 2-160 are identified in the 2-D structural diagram shown here.

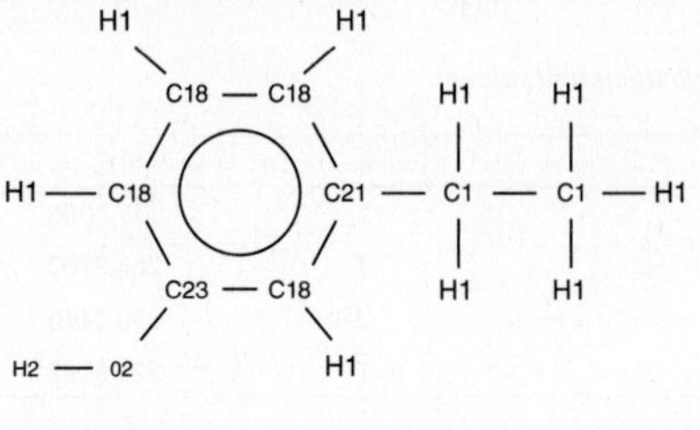

The molecular weight of *m*-ethylphenol is 122.16 kg/kmol, and its liquid density at 298.15 K is given in the DIPPR database as 1.00651 g/cm³. The group contributions are summed up as shown in this table:

Type	Description	Number	Value	Contribution
C1	1° & 2° aliphatic	2	2.503	5.006
C18	aromatic	4	3.350	13.40
C21	4° aromatic –aliphatic C	1	3.509	3.509
C23	4° aromatic –O attached	1	3.853	3.853
02	alcohol	1	0.8238	0.8238
H1	hydrocarbon	9	1.057	9.513
H2	alcohol	1	1.395	1.395
	Group Sum	19	R_D	37.4998

TABLE 2-160 Wildman-Crippen Contributions for Refractive Index[a]

Table-specific nomenclature: e = N, O, P, S, F, Cl, Br, or I; ! = not (e.g., !e = not any of the e elements); c = aromatic carbon; n = aromatic nitrogen; o = aromatic oxygen; A = any nonhydrogen atom; a = aromatic, not necessarily carbon; bond types are single (—), double (═), triple (#), and aromatic (:)

Type	Description	MR	Type	Description	MR
C1 1°, 2° aliphatic	C(4H), C(3H)(C), C(2H)(2C)	2.503	N1 1° amine	N(2H)(A)	2.262
C2 3°, 4° aliphatic	C(H)(3C), C(4C)	2.433	N2 2° amine	N(H)(2A)	2.173
C3 1°, 2° aliphatic e	C(3H)(e), C(2H)(2e)	2.753	N3 1° aromatic amine	N(2H)(a)	2.827
C4 3°, 4° aliphatic e	C(H)(3e), C(4e)	2.731	N4 2° aromatic amine	N(H)(a)(A, a)	3.000
C5 olefin e	C(═e)	5.007	N5 imine	N(H)(═A, ═a)	1.757
C6 olefin	C(2H)(═A), C(H)(═A), C(═A)	3.513	N6 substituted imine	N[2(═A, ═a)]	2.428
C7 acetylene	C(#A)	3.888	N7 3° amine	N(3A)	1.839
C8 1° aromatic c	C(3H)(c)	2.464	N8 3° aromatic amine	N(a)[2(a, A)]	2.819
C9 1° aromatic e	C(3H)(ae)	2.412	N9 nitrile	N(#A)	1.725
C10 2° aromatic	C(2H)(a)	2.488	N11 aromatic N	n	2.202
C11 3° aromatic	C(H)(a)	2.582	N13 4° amine	N(4A,), N(═A)[2(A, a)]	0.2604
C12 4° aromatic	C(a)	2.576	NS supplemental N	any other N	2.134
C13 aromatic e	c(!e)	4.041	O1 aromatic	o	1.080
C14 aromatic F	c(F)	3.257	O2 alcohol	O(H), O(2H)	0.8238
C15 aromatic Cl	c(Cl)	3.564	O3 aliphatic ether	O[2(C, A)]	1.085
C16 aromatic Br	c(Br)	3.180	O4 aromatic ether	O(a)(A, a)	1.182
C17 aromatic I	c(I)	3.104	O5 oxide	O(═O, ═N), O(A)(N)	3.367
C18 aromatic	c(H)	3.350	O6 oxide	O(A)(S)	0.7774
C19 bridgehead	c(3:a)	4.346	O7 oxide	O(A)(!N, !S)	0.000
C20 4° aromatic	c(2:a)(a)	3.904	O8 aromatic carbonyl	O(═c)	3.135
C21 4° aromatic	c(2:a)(C)	3.509	O9 aliphatic carbonyl	O(═C)[2(C, H, N, A]	0.000
C22 4° aromatic	c(2:a)(N)	4.067	O10 aromatic carbonyl	O(═C)(c)(C, H, A, a)	0.2215
C23 4° aromatic	c(2:a)(O)	3.853	O11 carbonyl (e)	O(═C)(A, a)	0.3890
C24 4° aromatic	c(2:a)(S)	2.673	O12 acid	O(C═O)	—
C25 4° aromatic	c(2:a)(═C), c(2:a)(═N, ═O)	3.135	OS supplemental O	any other O	0.6865
C26 C═C aromatic	C(═C)(a)	4.305	F fluorine	F(A)	1.108
C27 aliphatic e	C(4!e)	2.693	Cl chlorine	Cl(A)	5.853
CS supplemental C	any other C	3.243	Br bromine	Br(A)	8.927
H1 hydrocarbon	H(H), H(C)	1.057	I iodine	I(A)	14.02
H2 alcohol	H(O)	1.395	P phosphorous	P(A)	6.920
H3 amine/amide	H(N), H(O)(N)	0.9627	S1 aliphatic S	S(A)	7.591
H4 acid	H(COO), H(COS), H(OO)	1.805	S3 aromatic S	s(A)	6.691
HS supplemental H	any other H	1.112	pblk all remaining p-block elements		5.754

[a]Wildman, S. A., and G. M. Crippen, *J. Chem. Inf. Comput. Sci.* **39** (1999): 868.

This value for R_D is used in Eq. (2.22) to obtain

$$n=\sqrt{\frac{122.16+2(1.00651)(37.4998)}{122.16-(1.00651)(37.4998)}}=1.530$$

The predicted value differs by 0.3 percent from the experimental value of 1.535 given in the DIPPR database.

Dielectric Constant The dielectric constant is the ratio of the electric field strength in vacuum to that in the material for the same charge distribution. Equivalently, it is the ratio of the capacitance between two parallel charged plates when filled with the material to that of a vacuum with identical charges on the plates.

Recommended Method Liu method.

Reference: Liu, J-P, W. V. Wilding, N. F. Giles, and R. L. Rowley, *J. Chem. Eng. Data* **55** (2010): 41–45.

Classification: QSPR.

Expected uncertainty: Generally less than 1 percent for nonpolar organic liquids and less than 20 percent for polar organic liquids.

Applicability: Organic liquids. Not valid if the predicted dielectric constant is greater than 50.

Input data: For hydrocarbons and nonpolar molecules, the dipole moment μ, solubility parameter δ, and refractive index n are required. For polar and nonhydrocarbon molecules, the van der Waals area A_{vdw} and number of oxygen-containing groups are additionally required.

Description: The general correlation for the dielectric constant ε is

$$\ln\varepsilon=C_0+C_1\left(\frac{\mu}{\mathrm{D}}\right)+C_2\left(\frac{A_{vdw}}{\mathrm{m}^2\cdot\mathrm{kmol}^{-1}}\right)^{-1}+C_3\left(\frac{\delta}{\mathrm{J}^{1/2}\cdot\mathrm{m}^{-3/2}}\right)+C_4n^2+\sum_i^{\mathrm{O\ Groups}}\frac{G_i}{k_i} \tag{2-23}$$

with the coefficients given by

Applicability	C_0	C_1	C_2	C_3	C_4
Hydrocarbons and nonpolar	−0.1694	0.1283	0	2.8251×10^{-5}	0.2150
	−0.3416	0.5239	4.072×10^{8}	7.408×10^{-5}	−0.3248

The summation term shown in Eq. (2.23) is only for oxygen-containing groups in the molecule in which G_i is the contribution shown below and k_i ($k_i > 1$) is the number of occurrences of that group in the molecule.

Group	Example	G_i	Group	Example	G_i
[S, N, P] = O	thionyl chloride	0.2879	–OH(na)	alcohol	0.2230
>C=O	ketone	0.3615	–OH(a)	phenol	0.0990
>C=O ring	2-pyrrolidone	0.0075	–OH(C < 5)*	ethanol	0.3348
–COO–	ester	−0.0650	–CHO	aldehyde	0.1617
–COOH	acid	−0.5900			

*Applied in addition to regular –OH group for molecules with fewer than 5 C atoms.

Example Calculate the dielectric constant of salicylaldehyde at 303 K. The structure of salicylaldehyde is shown below with the two different oxygen-containing groups and their contributions that are to be used in Eq. (2.23).

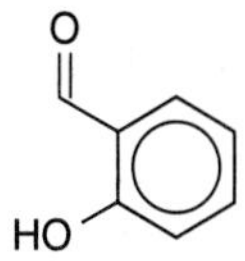

Group	G_i	k_i
—CHO	0.1617	1
—OH(a)	0.0990	1

Values of the input properties for Eq. (2.23) obtained from the DIPPR database are $\mu = 3.08794$ D, $A_{vdw} = 8.43 \times 10^8$ m^2/kmol, $\delta = 21330$ J$^{1/2}$·m$^{-3/2}$, $n = 1.57017$. Equation (2.23) is then used to obtain the dielectric constant:

$$\ln \varepsilon = -0.3416 + (0.5239)(3.08794) + \frac{4.072}{8.43}$$
$$+ (7.408 \times 10^{-5})(21330) - (0.3248)(1.57017)^2 + 0.1617 + 0.0990$$
$$\ln \varepsilon = 2.799 \text{ and } \varepsilon = 16.43$$

A few reported experimental values are 13.9 at 293 K, 17.1 at 303 K, and 18.35 at 293.15 K.

VAPOR PRESSURE

Liquids Vapor pressure is the equilibrium pressure at a given temperature of pure, coexisting liquid and vapor phases. The vapor pressure curve is a monotonic function of temperature from its minimum value (the triple point pressure) at the triple point temperature T_t, to its maximum value, P_c, at T_c.

Liquid vapor pressure data over a limited temperature range can be correlated with the Antoine equation [Antoine, C., *C.R.*, **107** (1888): 681, 836]

$$\ln \frac{P^*}{\text{Pa}} = A - \frac{B}{T/\text{K} + C} \qquad (2\text{-}24)$$

Data from the triple point to the critical point can be correlated with either a modified form of the Wagner equation [Wagner, W., *A New Correlation Method for Thermodynamic Data Applied to the Vapor-Pressure Curve of Argon, Nitrogen, and Water*, J. T. R. Watson (trans. and ed.), IUPAC Thermodynamic Tables Project Centre, London, 1977; Ambrose, D., *J. Chem. Thermodyn.*, **18** (1986): 45; Ambrose, D., and N. B. Ghiassee, *J. Chem. Thermodyn.*, **19** (1987): 903, 911]

$$\ln P_r^* = \frac{a\tau + b\tau^{1.5} + c\tau^{2.5} + d\tau^5}{1-\tau} \quad \text{where } \tau \equiv 1 - T_r \qquad (2\text{-}25)$$

or the Riedel equation [Riedel, L., *Chem. Ing. Tech.*, **26** (1954): 679]

$$\ln \frac{P^*}{\text{Pa}} = A + \frac{B}{T/\text{K}} + C \ln \frac{T}{\text{K}} + D\left(\frac{T}{\text{K}}\right)^E \qquad (2\text{-}26)$$

In its original form, E in Eq. (2-26) was assigned a value of 6, but other integer values of E from 1 to 6 have been found to be more effective for different families of chemicals in representing the vapor pressure over the whole liquid range. With the best value of E, either the Riedel or the Wagner equation can be used to correlate most fluids over the whole liquid range, but a fifth term is used in the Wagner equation for alcohols [Poling, B. E., *Fluid Phase Equilib.*, **116** (1996): 102]:

$$\ln P_r^* = \frac{a\tau + b\tau^{1.5} + c\tau^{2.5} + d\tau^5 + e\tau^6}{1-\tau} \quad \text{(for alcohols)} \qquad (2\text{-}27)$$

Correlation of experimental data within a few tenths of a percent over the entire fluid range can usually be obtained with either the Wagner or Riedel equations.

Two prediction methods are recommended for liquid vapor pressure. The first method is based on the Riedel equation; the second is a CS method. Both methods require T_c and P_c as input, but these can be estimated by the methods shown earlier if experimental values are unavailable.

Recommended Method 1 Riedel method.

Reference: Riedel, L., *Chem. Ing. Tech.*, **26** (1954): 679.

Classification: Empirical extension of theory and corresponding states.

Expected uncertainty: Varies strongly depending upon relative T, but 1 percent or less above T_b is typical with uncertainties of 5 to 30 percent near the triple point.

Applicability: Most organic compounds.

Input data: T_b, T_c, P_c.

Description: Equation (2-26) in reduced form

$$\ln P_r = A + \frac{B}{T_r} + C \ln T_r + DT_r^6 \qquad (2\text{-}28)$$

is used with the constants for this equation determined from the following set of relationships:

$$\psi = -35 + \frac{36}{T_{br}} + 42 \ln T_{br} - T_{br}^6 \qquad \alpha_c = \frac{3.758K\psi + \ln(P_c/1.01325 \text{ bar})}{K\psi - \ln T_{br}}$$
$$h = T_{br}\frac{\ln(P_c/1.01325 \text{ bar})}{1 - T_{br}} \qquad D = K(\alpha_c - 3.758)$$
$$C = \alpha_c - 42D \qquad B = -36D \qquad A = 35D \qquad (2\text{-}28a)$$

Values of the constant K [Vetere, A., *Ind. Eng. Chem. Res.*, **30** (1991): 2487] are as follows:

Class	Value
Acids	$K = -0.120 + 0.025h$
Alcohols	$K = 0.373 - 0.030h$
All other organic compounds	$K = 0.0838$

Example Estimate the vapor pressure of chlorobenzene at 50 K intervals from 300 to 600 K.

Input information: From the DIPPR 801 database, $T_b = 404.87$ K, $T_c = 632.35$ K, and $P_c = 45.1911$ bar.

Auxiliary Quantities:

$$K = 0.0838 \qquad T_{br} = 404.87/632.35 = 0.640$$
$$\psi = -35 + \frac{36}{0.640} + 42\ln(0.640) - (0.640)^6 = 2.431$$
$$\alpha_c = \frac{(3.758)(0.0838)(2.431) + \ln(45.191/1.01325)}{(0.0838)(2.431) - \ln(0.640)} = 7.0248$$
$$D = (0.0838)(7.0248 - 3.758) = 0.2738 \qquad C = 7.0248 - (42)(0.2738) = -4.4729$$
$$B = -(36)(0.2738) = -9.8552 \qquad A = -(35)(0.2738) = 9.5814$$

Calculation using Eq. (2-28) at each T (detailed calculation shown for T = 500 K):

$$T_r = 500/632.35 = 0.7907$$
$$\ln P_r = 9.5814 - \frac{9.8552}{0.7907} - 4.4729 \ln 0.7907 + (0.2738)(0.7907)^6 = -1.7651$$
$$P_r = \exp(-1.7651) = 0.1712 \qquad P = P_rP_c = (0.1712)(45.1911 \text{ bar}) = 7.74 \text{ bar}$$

T/K	T_r	$\ln P_r$	P/bar	P_{DIPPR}/bar	% Error
300	0.4744	−7.8532	0.0176	0.0175	0.3
350	0.5535	−5.5704	0.172	0.172	0.1
400	0.6326	−3.9323	0.886	0.880	0.6
450	0.7116	−2.7101	3.01	2.98	0.9
500	0.7907	−1.7651	7.74	7.67	0.9
550	0.8698	−1.0067	16.51	16.39	0.8
600	0.9488	−0.3705	31.20	31.11	0.3

Recommended Method 2 Ambrose-Walton method.

References: Ambrose, D., and J. Walton, *Pure & Appl. Chem.*, **61** (1989): 1395; Lee, B. I., and M. G. Kesler, *AIChE J.*, **21** (1975): 510.

Classification: Corresponding states.

Expected uncertainty: Varies strongly with relative T, but less than 1 percent is typical above T_b if the acentric factor is known.

Applicability: Most organic compounds.

Input data: T_b, T_c, P_c, and ω.

Description: The acentric factor is used to interpolate within the simple-fluid and deviation terms for ln P^*. The $f^{(i)}$ terms have been obtained from correlations of the reference fluid vapor pressures with the Wagner vapor pressure equation

$$\ln P_r^* = f^{(0)} + \omega f^{(1)} + \omega^2 f^{(2)}$$
$$f^{(0)} = \frac{-5.97616\tau + 1.29874\tau^{1.5} - 0.60394\tau^{2.5} - 1.06841\tau^5}{1-\tau}$$
$$f^{(1)} = \frac{-5.03365\tau + 1.11505\tau^{1.5} - 5.41217\tau^{2.5} - 7.46628\tau^5}{1-\tau} \qquad (2\text{-}29)$$
$$f^{(2)} = \frac{-0.64771\tau + 2.41539\tau^{1.5} - 4.26979\tau^{2.5} + 3.25259\tau^5}{1-\tau}$$

where $\tau = 1 - T_r$.

Example Repeat the calculation of the liquid vapor pressure of chlorobenzene at 50-K intervals from 300 to 600 K using the Ambrose-Walton method.

Input information: From the DIPPR 801 database, $T_c = 632.35$ K, $P_c = 45.1911$ bar, and $\omega = 0.249857$.

Auxiliary quantities:

$$T_r = 500/632.35 = 0.7907 \qquad \tau = 1 - 0.7907 = 0.2093$$

Simple-fluid and deviation vapor pressure terms at each T (shown for T = 500 K):

$$f^{(0)} = \frac{(-5.97616)(0.2093) + (1.29874)(0.2093)^{1.5} - (0.60394)(0.2093)^{2.5} - (1.06841)(0.2093)^{5}}{0.7907}$$
$$= -1.4405$$

$$f^{(1)} = \frac{(-5.03365)(0.2093) + (1.11505)(0.2093)^{1.5} - (5.41217)(0.2093)^{2.5} - (7.46628)(0.2093)^{5}}{0.7907}$$
$$= -1.3383$$

$$f^{(2)} = \frac{(-0.64771)(0.2093) + (2.41539)(0.2093)^{1.5} - (4.26979)(0.2093)^{2.5} + (3.25259)(0.2093)^{5}}{0.7907}$$
$$= 0.0145$$

Calculation using Eq. (2-29):

$$\ln P_r^* = -1.4405 + (0.249857)(-1.3383) + (0.249857)^2(0.0145) = -1.774$$

$$P^* = (45.1911 \text{ bar})[\exp(-1.774)] = 7.667 \text{ bar}$$

T	τ	$f^{(0)}$	$f^{(1)}$	$f^{(2)}$	$\ln P^*_r$	P^*/bar	P^*_{DIPPR}/bar	% Error
300	0.5256	−5.9228	−7.5966	−0.3050	−7.840	0.0178	0.0175	1.4
350	0.4465	−4.3006	−5.0017	−0.1439	−5.559	0.174	0.172	1.5
400	0.3674	−3.1036	−3.3106	−0.0437	−3.933	0.885	0.880	0.5
450	0.2884	−2.1800	−2.1576	0.0043	−2.719	2.98	2.98	0.0
500	0.2093	−1.4405	−1.3383	0.0145	−1.774	7.67	7.67	0.0
550	0.1302	−0.8289	−0.7318	0.0036	−1.012	16.43	16.39	0.3
600	0.0512	−0.3068	−0.2612	−0.0081	−0.373	31.14	31.11	0.1

Solids Below the triple point, the pressure at which the solid and vapor phases of a pure component are in equilibrium at any given temperature is the vapor pressure of the solid. It is a monotonic function of temperature with a maximum at the triple point. Solid vapor pressures can be correlated with the same equations used for liquids. Estimation of solid vapor pressure can be made from the integrated form of the Clausius-Clapeyron equation

$$\ln \frac{P^*}{P_t^*} = \frac{\Delta H_{\text{sub}}}{RT_t}\left(1 - \frac{T_t}{T}\right) \tag{2-30}$$

where T_t = triple point temperature
P_t^* = triple point pressure
ΔH_{sub} = enthalpy of sublimation

The liquid and solid vapor pressures are identical at the triple point. A good vapor pressure correlation that is valid at the triple point may be used to obtain the triple point pressure. Estimating solid vapor pressures by using Eq. (2-30) generally requires an estimation of ΔH_{sub}, and so the illustrative example is combined with the example on enthalpy of sublimation in the section on latent enthalpy.

THERMAL PROPERTIES

Enthalpy of Formation The standard enthalpy (heat) of formation is the enthalpy change upon formation of 1 mole of the compound in its standard state from its constituent elements in their standard states. Two different standard enthalpies of formation are commonly defined based on the chosen standard state. The standard enthalpy of formation ΔH_f^s uses the naturally occurring phase at 298.15 K and 1 bar as the standard state while the ideal gas enthalpy (heat) of formation ΔH_f^o uses the compound in the ideal gas state at 298.15 K and 1 bar as the standard state. In both cases, the standard state for the elements is their naturally occurring state of aggregation at 298.15 K and 1 atm. Sources for data include DIPPR, TRC, SWS, JANAF, and TDB. The Domalski-Hearing method is the most accurate general method for estimating either ΔH_f^s or ΔH_f^o if the appropriate GC values are available, but a CC method is also as accurate for estimating ΔH_f^o if an isodesmic reaction can be formulated and used. The Domalski-Hearing method also applies to entropies, and the entropy predictive equations are listed in this section for convenience because they are equivalent in form to the enthalpy equations. However, discussion and illustration of the estimation methods for entropy are delayed to the next subsection.

Recommended Method Domalski-Hearing method.

Reference: Domalski, E. S., and E. D. Hearing, *J. Phys. Chem. Ref. Data*, **22** (1993): 805.

Classification: Group contributions.

Expected uncertainty: 3 percent.

Applicability: Organic compounds for which group contributions have been regressed.

Input data: Molecular structure.

Description: GC values from Table 2-161 are directly additive for both enthalpy of formation and absolute third-law entropies:

$$\frac{\Delta H_f^o}{\text{kJ/mol}} = \sum_{i=1}^{N} n_i (\Delta H_f^o)_i \qquad \frac{S^o}{\text{J}\cdot\text{mol}^{-1}\text{K}^{-1}} = \sum_{i=1}^{N} n_i (S^o)_i \tag{2-31}$$

where $(\Delta H_f^o)_i$ = enthalpy of formation GC value and $(S^o)_i$ = entropy GC value, both obtained from Table 2-161.

Group values in Table 2-161 are defined by the central, nonhydrogen group and the atoms bonded to that group. Thus, C—(2H)(2C) represents a C atom to which 2 H and 2 C atoms are bonded. For example, propane (CH_3—CH_2—CH_3) is composed of three groups: two C—(3H)(C) and one C—(2H)(2C).

Example Estimate the standard and ideal gas enthalpies of formation of *o*-toluidine.

Input information: Because the melting point (256.8 K) and boiling point (473.49 K) for *o*-toluidine bracket 298.15 K, the standard state phase at 298.15 K and 1 bar is liquid.

Structure:

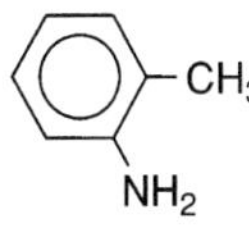

Group contributions:

Group	n_i	ΔH_f^o gas	ΔH_f^o liq.	S^o gas	S^s liq.
Cb—(H)(2Cb)	4	13.81	8.16	48.31	28.87
Cb—(C)(2Cb)	1	23.64	19.16	−35.61	−19.50
Cb—(N)(2Cb)	1	−1.30	1.50	−43.53	−24.43
C—(3H)(C)	1	−42.26	−47.61	127.32	83.30
N—(2H)(Cb)	1	19.25	−11.00	126.90	71.71
	Total	54.57	−5.31	368.32	226.56

Calculation from Eq. (2-31):

$$\frac{\Delta H_f^o}{\text{kJ/mol}} = 54.57 \qquad \frac{\Delta H_f^s}{\text{kJ/mol}} = -5.31$$

$$\frac{S^o}{\text{J/(mol}\cdot\text{K)}} = 368.32 \qquad \frac{S^s}{\text{J/(mol}\cdot\text{K)}} = 226.56$$

The recommended DIPPR 801 standard enthalpies of formation are ΔH_f^o = 53.20 kJ/mol and ΔH_f^s = −4.72 kJ/mol. The estimated values are higher than the recommended values by 2.6 and 12.5 percent, respectively. The recommended DIPPR 801 standard entropies are S^o = 355.8 J/(mol · K) and S^s = 231.2 J/(mol · K). The estimated values differ from these by 3.5 and −2.0 percent, respectively.

Recommended Method Isodesmic reaction.

Reference: Foresman, J. B., and A. Frisch, *Exploring Chemistry with Electronic Structure Methods*, 2d ed., Gaussian Inc., Pittsburgh, Pa., 1996.

Classification: Computational chemistry.

Expected uncertainty: 5 to 10 percent depending upon the level of theory and basis set size used.

Applicability: Compounds for which an isodesmic reaction can be formulated.

Input data: Experimental ΔH_f^o values for all *other* participants in the isodesmic reaction.

Description: While *ab initio* calculations of *absolute* enthalpies are not currently as accurate as GC methods, *relative* enthalpies of molecules calculated with the same level of theory and basis set can be very accurate, as in the case of isodesmic reactions. An isodesmic reaction is one in which the number and type of bonds are preserved during the reaction. For example, the reaction of acetaldehyde with ethane to form acetone and methane is

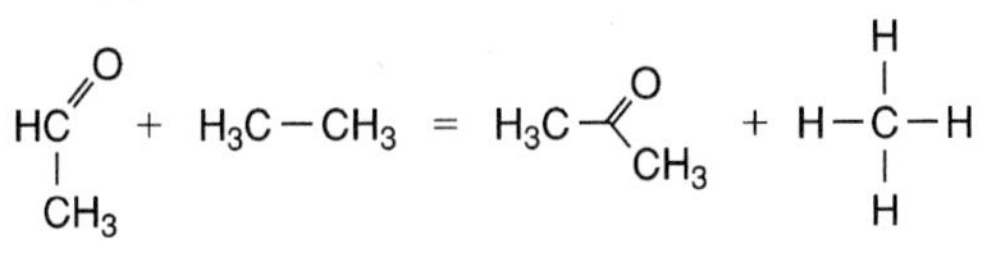

TABLE 2-161 Domalski-Hearing* Group Contribution Values for Standard State Thermal Properties

This table is a partial listing of GC values available from the original Domalski-Hearing tables. Table-specific nomenclature: Cd = carbon with double bond; Ct = carbon with triple bond; Cb = carbon in benzene ring; Ca = allenic carbon; corr = correction term; Cbf = fused benzene ring; N_A = azo nitrogen; N_I = imino nitrogen.

Group	ΔH_f^o	S^o	ΔH_f^s liq.	S^s liq.	ΔH_f^s solid	S^s solid
			CH Groups			
C—(3H)(C)	−42.26	127.32	−47.61	83.30	−46.74	56.69
C—(2H)(2C)	−20.63	39.16	−25.73	32.38	−29.41	23.01
C—(H)(3C)	−1.17	−53.60	−4.77	−23.89	−5.98	−16.89
—$CH_{3\,corr}$ (tertiary)	−2.26	0.00	−2.18	0.00	−2.34	0.00
C—(4C)	19.20	−149.49	17.99	−98.65	12.47	−33.19
—$CH_{3\,corr}$ (quaternary)	−4.56	0.00	−4.39	0.00	−4.35	0.00
—$CH_{3\,corr}$ (tert/quat)	−1.80	0.00	−1.77	0.00	−2.70	0.00
—$CH_{3\,corr}$ (quat/quat)	−0.64	0.00	−0.64	0.00	−2.24	0.00
Cd—(2H)	26.32	115.52	21.75	86.19	22.43	
Cd—(H)(C)	36.32	33.05	31.05	28.58	25.48	
Cd—(2C)	44.14	−50.84	39.16	−29.83	32.97	
Cd—(H)(Cd)	28.28	27.74	22.18	13.30	17.53	21.75
Cd—(C)(Cd)	36.78	−61.33	30.42	−41.92	27.91	
Cd—(Cd)(Cb)					56.07	
Cd—(H)(Cb)	28.28	27.74	22.18	13.30	17.53	21.75
Cd—(C)(Cb)	37.95	−51.97	38.58			
Cd—(H)(Ct)	28.28	27.74	22.18	13.30	17.53	21.75
C—(4H), Methane	−74.48	206.92				
Cd—(2Cb)	32.88		30.83		49.91	
C—(2H)(C)(Cd)	−20.88	38.20	−25.73	31.67	−24.35	
C—(H)(2C)(Cd)	−1.63	−50.38	−5.02	−28.07	−6.49	
—$CH_{3\,corr}$ (tertiary)	−2.26	0.00	−2.18	0.00	−2.34	0.00
C—(3C)(Cd)	22.13	−150.23	20.79	−108.20	12.51	
—$CH_{3\,corr}$ (quaternary)	−4.56	0.00	−4.39	0.00	−4.35	0.00
C—(H)(C)(2Cd)	−1.17	−53.60	−4.77	−23.89	−5.98	−16.89
C—(2H)(2Cd)	−18.92	42.08	−24.43	19.32	−21.60	
C—(2H)(Cd)(Cb)			−24.73			
C—(H)(C)(Cd)(Cb)			−6.90			
cis (unsat) $_{corr}$	4.85	5.06	5.27	0.00	5.73	0.00
tert—Butyl *cis* $_{corr}$	17.24	0.00	17.48	0.00	17.57	0.00
Ct—(H)	113.50	101.96	104.47	67.57	110.34	
Ct—(C)	115.10	26.32	107.15	14.25	101.66	
Ct—(Cd)	121.42	39.92	114.77			
Ct—(Cb)	120.76	17.77	119.00		103.28	
Ct—(Ct)	120.76	25.94	104.80		103.28	
C—(2H)(C)(Ct)	−19.70	42.80	−22.13	32.36	−29.41	
C—(H)(2C)(Ct)	−3.16	−45.69				
—$CH_{3\,corr}$ (tertiary)	−2.26	0.00	−2.18	0.00	−2.34	0.00
C—(3C)(Ct)			22.83		26.38	
—$CH_{3\,corr}$ (quaternary)	−4.56	0.00	−4.39	0.00	−4.35	0.00
C—(2H)(2Ct)	−41.14		−39.08			
C—(2C)(2Ct)			20.67			
Ca	142.67	26.28	134.68	14.39	131.08	
Cb—(H)(2Cb)	13.81	48.31	8.16	28.87	6.53	22.75
Cb—(C)(2Cb)	23.64	−35.61	19.16	−19.50	13.90	−5.50
Cb—(Cd)(2Cb)	24.17	−33.85	19.12	−9.04	20.27	−10.00
Cb—(Ct)(2Cb)	24.17	−33.85	19.12	−9.04	20.07	−10.00
Cb—(3Cb)	21.66	−36.57	17.21		17.03	−6.00
C—(2C)(2Cb)					52.81	
C—(2H)(C)(Cb)	−21.34	42.59	−24.81	47.40	−22.10	26.90
C—(H)(2C)(Cb)	−4.52	−48.00	−5.82	−13.90	−3.50	
C—(Cb)(3C)	18.28	−147.19	18.70	−96.10	21.57	
C—(2H)(2Cb)	−46.43		−26.50	51.97	−21.44	22.85
C—(H)(C)(2Cb)			−21.47	28.12	16.40	
C—(H)(3Cb)	−6.86				34.48	−12.62
C—(3Cb)(C)					116.25	
C—(4Cb)	27.04				64.89	
Cbf—(Cbf)(2Cb)	20.10	0.00	15.83	−5.54	14.10	−6.00
Cbf—(Cb)(2Cbf)	16.00		11.50		12.00	2.00
Cbf—(3Cbf)	3.59		−0.90		1.94	7.00
Cb—(2Cb)(Cbf)					−8.77	
Cb—(Cb)(2Cbf)	22.46				47.93	
ortho $_{corr}$ hydrocarbons	1.26	−2.50	3.26	0.00	5.00	0.00
meta $_{corr}$ hydrocarbons	−0.63	0.00	0.00	0.00	2.00	0.00
Cyclopropane rsc (unsub)	115.15	134.86	111.58			
Cyclobutane rsc	110.89	126.04	106.64	51.48	114.43	
Cyclopentane rsc (unsub)	26.75	116.22	22.84	42.24	34.00	
Cyclohexane rsc (unsub)	0.68	78.18	−1.77	10.07	10.94	
Cycloheptane rsc	26.34	73.97	23.50	15.89		
Cyclooctane rsc	40.65	70.78	38.10	2.96		
Cyclononane rsc	52.91		50.40			
Cyclodecane rsc	51.99		50.61			

TABLE 2-161 Domalski-Hearing* Group Contribution Values for Standard State Thermal Properties *(Continued)*

This table is a partial listing of GC values available from the original Domalski-Hearing tables. Table-specific nomenclature: Cd = carbon with double bond; Ct = carbon with triple bond; Cb = carbon in benzene ring; Ca = allenic carbon; corr = correction term; Cbf = fused benzene ring; N_A = azo nitrogen; N_I = imino nitrogen.

Group	ΔH_f^o	S^o	ΔH_f^s liq.	S^s liq.	ΔH_f^s solid	S^s solid
			CHO Groups			
CO—(2H), formaldehyde	−108.60	224.54				
CO—(C)(CO)	−121.29		−135.04		−140.75	
CO—(H)(CO)	−105.98					
CO—(CO)(Cb)	−112.30				−117.75	
CO—(O)(CO)	−123.75		−123.30		−120.81	
CO—(Cd)(O)	−136.73	62.59	−155.56		−134.10	32.90
CO—(C)(O)	−137.24	62.59	−149.37	32.72	−153.60	32.13
CO—(H)(O)	−124.39	147.03	−142.42	94.68		
CO—(2O)	−111.88		−122.00		−123.00	−42.92
CO—(H)(Cd)	−126.96		−153.05			
CO—(2Cb)	−110.00		−119.00		−116.00	
CO—(C)(Cb)	−148.82		−145.22		−143.70	23.72
CO—(H)(Cb)	−121.35		−138.12		−160.18	
CO—(O)(Cb)	−125.00		−140.00		−145.00	32.13
CO—(2C)	−132.67	64.31	−152.76	33.81	−157.95	
CO—(H)(C)	−124.39	147.03	−142.42	93.55		
CO—(C)(Cd)						
O—(2CO), aliphatic	−214.50	34.16	−230.50		−235.00	
O—(2CO), aromatic	−238.30		−220.90		−207.00	
O—(Cd)(CO)	−198.03		−201.42			
O—(C)(CO)	−188.87	36.03	−196.02	38.28	−210.60	12.09
O—(H)(CO)	−254.30	101.71	−285.64	38.28	−282.15	21.78
O—(Cb)(CO)	−167.00		−165.50		−170.00	45.32
O—(C)(O)	−20.75		−23.50		−30.20	
O—(H)(O)	−72.26		−101.75		−105.30	
O—(2Cd)	−139.29		−137.32			
O—(H)(Cd)						
O—(C)(Cd)	−129.33		−133.72			
O—(2Cb)	−77.66		−85.27	23.31	−96.20	3.14
O—(C)(Cb)	−92.55		−104.85		−122.87	
O—(H)(Cb)	−160.30	121.50	−191.75	43.89	−199.25	28.62
O—(2C)	−101.42	29.33	−110.83	26.78	−119.00	
O—(H)(C)	−159.33	121.50	−191.50	43.89	−199.66	28.62
Cd—(H)(CO)	32.30	35.19	26.61		7.82	27.53
Cd—(C)(CO)						
Cd—(O)(Cd)	36.78	−61.34	30.42	−41.92	27.91	
Cd—(O)(C)	44.14	−50.84	39.08	−29.83	32.97	
Cd—(O)(H)	36.32	33.05	31.05	28.58	25.48	
Ct—(CO)					144.52	
Cb—(CO)(2Cb)	15.50		10.50		8.15	0.08
Cb—(O)(2Cb)	−4.75	−43.72	−5.61	−10.59	1.00	1.59
C—(2H)(2CO)	−30.74		−23.06		−19.10	
C—(CO)(3C)	23.93		26.15	−85.98	24.02	
C—(H)(CO)(2C)	−0.25		−3.89	−24.52	−9.83	
C—(2H)(CO)(C)	−21.84	39.58	−24.14	39.87	−27.90	24.73
C—(3H)(CO)	−42.26	127.32	−47.61	83.30	−46.74	56.69
C—(2H)(CO)(Cd)	−16.95		−19.62			
C—(2H)(CO)(Ct)	−25.48		−26.61			
C—(2H)(CO)(Cb)	−16.20		−11.67			
C—(H)(CO)(C)(Cb)					14.81	
C—(H)(O)(CO)(C)	126.63		123.43	−46.71	−14.39	8.08
C—(4O)	−152.46		−133.34			
C—(H)(3O)	−113.97		−107.74			
C—(3O)(C)	−114.39		−99.54			
C—(2O)(2C)	−53.56		−41.30			
C—(H)(2O)(C)	−57.78		−51.42			
C—(2H)(2O)	−62.22		−62.89	23.85		
C—(2H)(O)(Cb)	−33.76		−29.17			
C—(2H)(O)(Cd)	−27.49	37.49	−28.62			
C—(H)(CO)(C)(Cb)					−14.39	
C—(H)(CO)(2Cb)					3.72	
C—(O)(3Cb)					60.46	
C—(O)(3C) (ethers, esters)	9.50	−141.92	0.79	−94.68	−0.50	
C—(H)(O)(2C) (ethers, esters)	−19.46	−52.80	−21.00	−25.31	−20.08	
C—(O)(3C) (alcohols, peroxides)	−13.50	−144.60	−11.13	−122.48	−12.25	−14.77
C—(H)(O)(2C) (alcohols, peroxides)	−26.10	−43.05	−27.60	−29.83	−29.08	6.95
C—(2H)(O)(C)	−32.90	43.43	−35.80	32.59	−33.00	24.73
C—(3H)(O)	−42.26	127.32	−47.61	83.30	−46.74	56.69
O—(CO)(O)	−88.00		−90.00		−80.50	
C—(2C)(O)(Cb)	15.30		25.80		29.30	
C—(H)(C)(2O)					−52.50	

(Continued)

TABLE 2-161 Domalski-Hearing* Group Contribution Values for Standard State Thermal Properties *(Continued)*

This table is a partial listing of GC values available from the original Domalski-Hearing tables. Table-specific nomenclature: Cd = carbon with double bond; Ct = carbon with triple bond; Cb = carbon in benzene ring; Ca = allenic carbon; corr = correction term; Cbf = fused benzene ring; N_A = azo nitrogen; N_I = imino nitrogen.

Group	ΔH_f^o	S^o	ΔH_f^s liq.	S^s liq.	ΔH_f^s solid	S^s solid
			CHN and CHNO Groups			
C—(3H)(N)	−42.26	127.32	−47.61	83.30	−46.74	56.69
C—(2H)(C)(N)	−28.30	42.26	−30.80	32.38	−34.00	23.01
C—(H)(2C)(N)	−16.70	−63.55	−14.65	−20.00	−13.90	
—$CH_{3\ corr}$ (tertiary)	−2.26	0.00	−2.18	0.00	−2.34	0.00
C—(3C)(N)	0.29	−152.59	5.10	−87.99	1.00	
—$CH_{3\ corr}$ (quaternary)	−4.56	0.00	−4.39	0.00	−4.35	0.00
C—(2H)(2N)	−30.00				−26.00	
C—(2H)(Cb)(N)	−24.14		−26.09		−33.31	
N—(2H)(C) (first, amino acids)	19.25	124.40	0.33	71.71	−6.30	39.00
N—(2H)(C) (second, amino acids)	19.25	126.90	0.33	71.71	−46.00	48.75
N—(H)(2C)	67.55	33.96	51.50	32.09	47.80	
N—(3C)	116.50	−61.71	112.00	−38.62	101.00	
N—(2H)(N)	47.70	122.18	25.30	60.58	18.97	
N—(H)(C)(N)	89.16		75.00	22.05		
N—(2C)(N)	120.71		119.00	−26.94		
N—(2Cb)(N)					137.35	
N—(H)(Cb)(N)	87.50		73.40		66.90	
N—(2CO)(N)					73.62	
N—(H)(2Cd)	83.55		50.50		45.40	
N—(C)(2Cd)	120.64		97.38		88.92	
N—(2H)(Cb)	19.25	126.90	−11.00	71.71	−21.60	70.00
N—(H)(C)(Cb)	59.00		26.25		36.55	
N—(2C)(Cb)	126.40		109.40		96.50	
N—(C)(2Cb)	120.44		97.38		89.30	
N—(H)(2Cb)	83.55		50.50		45.40	
N—(3Cb)	123.15		121.80		107.50	
N_I—(C)	81.46		73.68			
N_I—(Cb)	69.00	47.01	54.50	36.40	57.00	
N_A—(C)	109.50		104.85		103.00	
N_A—(Cb)	109.50		104.85		103.00	
N_A—(oxide)(C)	40.80		22.65			
C—(2H)(C)(N_A)	−20.70		−25.70		−29.41	
C—(H)(2C)(N_A)	−2.66		−5.42			
C—(3C)(N_A)	11.50		15.50		10.50	
Cd—(H)(N)	−16.00		−15.50		−13.00	
Cd—(C)(N)	−5.74		−5.62		−3.95	
Cb—(N)(2Cb)	−1.30	−43.53	1.50	−24.43	9.75	−37.57
Cb—(NO)(2Cb)	21.50				23.00	
Cb—(NO_2)(2Cb)	−1.45		−28.30	79.95	−32.50	110.46
Cb—(CNO)(2Cb)	−177.63				155.69	
Cb—(CN)(2Cb)	151.00	85.25	122.38	64.75	121.20	50.45
Cb—(N_A)(2Cb)	22.55		20.08		18.65	
Cb—(H)(2N_I)	6.30				0.25	
CO—(H)(N)	−124.39	147.03	−188.00	93.55		
CO—(C)(N)	−133.26	56.70	−185.00		−194.60	40.00
CO—(Cb)(N) (amides)					−177.75	
CO—(Cb)(N) (amino acids)					−177.75	
CO—(Cd)(N)	−171.80					
CO—(2N)	−111.00	96.00	−190.50		−203.10	69.00
N—(2H)(CO) (amides, ureas)	−63.00	88.25	−63.90		−65.25	18.00
N—(2H)(CO) (amino acids)	−63.00		−63.90		−59.75	33.03
N—(H)(C)(CO) (amides, ureas)	−16.28		−17.10		−9.80	
N—(H)(C)(CO) (amino acids)	−16.28		−17.10		5.50	
N—(2C)(CO)	45.00		62.00		55.00	
N—(H)(Cb)(CO)	−20.84				−3.50	
N—(H)(2CO)	−91.00				−30.80	
N—(C)(2CO)	−11.64		56.20		64.00	
N—(Cb)(2CO)	9.12					
N-(2Cb)(CO)					60.85	
N—(C)(Cb)(CO)					72.00	
C—(3H)(CN), acetonitrile	74.04	252.60	40.56	149.62		
C—(2H)(C)(CN)	94.52	167.25	66.07	106.02	69.85	96.15
C—(H)(2C)(CN)	113.50	67.86	81.50		69.00	
C—(3C)(CN)	137.96		116.20	−17.91	102.07	
C—(2C)(2CN)						74.57
C—(2H)(Cd)(CN)	95.31		66.40			
Cd—(H)(CN)	146.65	158.41	117.28	92.72		
Ct—(CN)	264.60		250.20			
C—(3H)(NO_2), nitromethane	−74.86	284.14	−112.60	171.75		
C—(2H)(2NO_2), dinitromethane	−58.90		−104.90			
C—(H)(3NO_2), trinitromethane	−0.30		−32.80		−48.00	
C—(4NO_2), tetranitromethane	82.30		38.30			
C—(2H)(C)(NO_2)	−60.50	203.60	−93.50		−99.00	

TABLE 2-161 Domalski-Hearing* Group Contribution Values for Standard State Thermal Properties *(Continued)*

This table is a partial listing of GC values available from the original Domalski-Hearing tables. Table-specific nomenclature: Cd = carbon with double bond; Ct = carbon with triple bond; Cb = carbon in benzene ring; Ca = allenic carbon; corr = correction term; Cbf = fused benzene ring; N_A = azo nitrogen; N_I = imino nitrogen.

Group	ΔH_f^o	S^o	ΔH_f^s liq.	S^s liq.	ΔH_f^s solid	S^s solid
CHN and CHNO Groups						
C—(H)(2C)(NO_2)	−53.00	115.32	−82.50		−89.00	
C—(3C)(NO_2)	−36.65		−61.20		−76.55	
C—(2H)(Cb)(NO_2)	−62.00		−82.76		−81.00	
C—(H)(C)(2NO_2)	−36.80		−88.80		−91.50	
C—(2C)(2NO_2)	−28.50		−77.20		−90.30	
C—(H)(C)(CO)(N)	−18.70				−11.65	−4.00
C—(2H)(CO)(N)	−3.10				−30.95	24.00
C—(H)(Cb)(CO)(N)						
O—(C)(NO)	−24.23	166.11	−46.50			
O—(C)(NO_2)	−79.71	191.92	−108.96	127.50	−124.00	
N—(H)(C)(NO_2)					16.50	
N—(H)(Cb)(NO_2)						
N—(H)(CO)(NO_2)					−14.00	
N—(C)(2NO_2)	100.30		53.50			
N—(C)(Cb)(NO_2)	183.00		167.00		150.50	
N—(2C)(NO)	90.00		59.00		55.00	
N—(2C)(NO_2)	88.00		50.00		40.00	
C—(2H)(C)(N_3)			321.70			
C—(H)(2C)(N_3)	274.00		255.00			
C—(2H)(Cb)(N_3)	347.00		327.40			
C—(3Cb)(N_3)	328.60				346.50	
Cb—(N_3)(2Cb)	320.00		303.50			
CHS and CHSO Groups						
C—(3H)(S)	−42.26	127.32	−47.61	83.30	−46.74	56.69
C—(2H)(C)(S)	−23.17	41.87	−26.77	41.09		
C—(H)(2C)(S)	−5.88	−47.36	−6.07	−16.61		
—$CH_{3\ corr}$ (tertiary)	−2.26	0.00	−2.18	0.00	−2.34	0.00
C—(3C)(S)	13.52	−145.38	16.69	−86.86		
—$CH_{3\ corr}$ (quaternary)	−4.56	0.00	−4.39	0.00	−4.35	0.00
—$CH_{3\ corr}$ (tert/quat)	−1.80	0.00	−1.77	0.00	−2.70	0.00
—$CH_{3\ corr}$ (quat/quat)	−0.64	0.00	−0.64	0.00	−2.24	0.00
C—(2H)(Cb)(S)	−18.53		−23.82			
C—(2H)(Cd)(S)	−25.93		−32.44			
C—(2H)(2S)	−25.10					
Cb—(S)(2Cb)	−4.75	43.72	−5.61	−10.59	1.00	1.59
Cd—(H)(S)	36.32	33.05	31.05	28.58	25.48	
Cd—(C)(S)	45.73	−51.92				
S—(C)(H)	18.64	137.67	0.06	85.95		
S—(Cb)(H)	48.10	57.34	28.51	89.04		
S—(2C)	46.99	55.19	29.82	29.80		
S—(H)(Cd)	25.52					
S—(C)(Cd)	54.39					
S—(2Cd)	102.60	68.59				
S—(Cb)(C)	76.21		58.20	35.44	42.00	
S—(C)(S)	27.62	50.50	14.36	30.84		
S—(Cb)(S)	57.45				40.60	
S—(2S)	12.59	56.07				
S—(2Cb)	102.60	68.59	93.02			
S—(H)(S)	7.95					
S—(H)(CO)	−5.90	130.54				
CO—(C)(S)	−132.67	64.31	−152.76	33.81		
C—(3H)(SO)	−42.26	127.32	−47.61	83.30	−46.74	56.69
C—(2H)(C)(SO)	−29.16		−36.88			
C—(H)(2C)(SO)						
—$CH_{3\ corr}$ (tertiary)	−2.26	0.00	−2.18	0.00	−2.34	0.00
C—(3C)(SO)	4.56		0.97			
—$CH_{3\ corr}$ (quaternary)	−4.56	0.00	−4.39	0.00	−4.35	0.00
C—(2H)(Cd)(SO)	−27.56		−32.63			
cis correction	4.11	5.06	5.27	0.00	5.73	0.00
Cb—(SO)(2Cb)	15.48		25.44		7.55	0.08
O—(SO)(H)	−158.60					
O—(C)(SO)	−92.60					
SO—(2C)	−66.78	75.73	−108.98	22.18		
SO—(2Cb)	−62.26					
SO—(2O)	−213.00					
SO—(C)(Cb)	−72.00					
C—(3H)(SO_2)	−42.26	127.32	−47.61	83.30	−46.74	56.69
C—(2H)(C)(SO_2)	−27.03		−33.76		−35.96	
C—(H)(2C)(SO_2)	−14.00					
—$CH_{3\ corr}$ (tertiary)	−2.26	0.00	−2.18	0.00	−2.34	0.00
C—(3C)(SO_2)	1.52		2.00		3.78	
—$CH_{3\ corr}$ (quaternary)	−4.56	0.00	−4.39	0.00	−4.35	0.00

(Continued)

TABLE 2-161 Domalski-Hearing* Group Contribution Values for Standard State Thermal Properties *(Continued)*

This table is a partial listing of GC values available from the original Domalski-Hearing tables. Table-specific nomenclature: Cd = carbon with double bond; Ct = carbon with triple bond; Cb = carbon in benzene ring; Ca = allenic carbon; corr = correction term; Cbf = fused benzene ring; N_A = azo nitrogen; N_I = imino nitrogen.

Group	ΔH_f^o	S^o	ΔH_f^s liq.	S^s liq.	ΔH_f^s solid	S^s solid
CHN and CHNO Groups						
—$CH_{3\ corr}$ (quat/quat)	−0.64		−0.64		−2.24	
C—(2H)(Cd)(SO_2)	−29.49		−49.05			
C—(H)(C)(Cd)(SO_2)	−71.99					
C—(2H)(Cb)(SO_2)	−29.80					
C—(2H)(Ct)(SO_2)	16.36					
Cb—(SO_2)(2Cb)	15.48		25.44		7.55	0.08
Cd—(H)(SO_2)	51.58					
Cd—(C)(SO_2)	64.01					
Ct—(SO_2)	177.10					
SO_2—(Cd)(Cb)	−291.55					
SO_2—(2Cd)	−306.70					
SO_2—(2C)	−288.58	87.37	−341.14		−356.62	32.10
SO_2—(C)(Cb)	−289.10					
SO_2—(2Cb)	−287.76				−305.40	
SO_2—(SO_2)(Cb)	−325.18				−361.75	
SO_2—(2O)	−417.30					
SO_2—(C)(Cd)	−316.80					
SO_2—(Ct)(Cb)	−296.30					
O—(SO_2)(H)	−158.60					
O—(C)(SO_2)	−91.40					
CHX and CHXO Groups						
C—(3H)(F), methyl fluoride	−247.00	231.93				
C—(3H)(Cl), methyl chloride	−81.90	243.60				
C—(3H)(Br), methyl bromide	−37.66	254.94	−61.10			
C—(3H)(I), methyl iodide	14.30	263.14	−11.70			
C—(C)(3F)	−673.81	178.22	−709.07	135.56		
C—(2H)(C)(F)	−221.12	146.80				
C—(H)(2C)(F)	−204.46	55.76				
C—(3C)(F)	−202.92					
C—(H)(C)(2F)	−454.74	164.32	−487.23			
C—(2C)(2F)	−411.39	74.48	−400.37		−428.77	
C—(C)(Cl)(2F)	−462.70	169.45	−466.00	138.31		
C—(H)(C)(Cl)(F)	−271.14					
C—(C)(3Cl)	−81.98	202.14	−112.93	145.91		
C—(H)(C)(2Cl)	−79.10	183.28	−102.60	128.45		
C—(2H)(C)(Cl)	−69.45	159.24	−86.90	104.27	−85.65	
C—(2C)(2Cl)	−79.56	95.41	−101.80			
C—(H)(2C)(Cl)	−55.61	71.34	−71.17			
C—(3C)(Cl)	−43.70	−24.26	−56.78			
C—(C)(3Br)		233.05				
C—(H)(C)(2Br)						
C—(2H)(C)(Br)	−21.78	173.31	−42.65	113.00		
C—(2C)(2Br)						
C—(H)(2C)(Br)	−10.75	84.69	−27.31			
C—(3C)(Br)	7.26	−13.46	−7.40			
C—(C)(3I)						
C—(H)(C)(2I)	108.78	228.45				
C—(2H)(C)(I)	33.54	177.78	4.14		3.65	
C—(2C)(2I)						
C—(H)(2C)(I)	48.74	88.10	24.78			
C—(3C)(I)	68.46	−3.21	48.60			
C—(H)(C)(Br)(Cl)	−18.45	191.21				
N—(C)(2F)	−32.64					
C—(H)(C)(Cl)(O)	−90.37	66.53				
C—(2H)(I)(O)	15.90	170.29				
C—(C)(2Cl)(F)	−322.54		−343.87	141.71		
C—(C)(Br)(2F)	−394.55			149.70		
C—(C)(2Br)(F)						
C—(Br)(Cl)(F)						
Cd—(H)(F)	−165.12	137.24				
Cd—(H)(Cl)	4.37	147.85	−12.67			
Cd—(H)(Br)	50.94	159.91				
Cd—(H)(I)	102.36	169.45				
Cd—(C)(Cl)	−5.06	62.76	−2.23			
Cd—(2F)	−329.90	155.63				
Cd—(2Cl)	−11.51	175.41	−32.08	115.35		
Cd—(2Br)		199.16				
Cd—(2I)						
Cd—(Cl)(F)	−235.10	175.61				
Cd—(Br)(F)		177.82				
Cd—(Cl)(Br)		188.70				
Ct—(F)						

TABLE 2-161 Domalski-Hearing* Group Contribution Values for Standard State Thermal Properties *(Continued)*

This table is a partial listing of GC values available from the original Domalski-Hearing tables. Table-specific nomenclature: Cd = carbon with double bond; Ct = carbon with triple bond; Cb = carbon in benzene ring; Ca = allenic carbon; corr = correction term; Cbf = fused benzene ring; N_A = azo nitrogen; N_I = imino nitrogen.

Group	ΔH_f^o	S^o	ΔH_f^s liq.	S^s liq.	ΔH_f^s solid	S^s solid
		CHX and CHXO Groups				
Ct—(Cl)		140.00				
Ct—(Br)		151.30				
Ct—(I)	35.53					
Cb—(F)(2Cb)	−181.26	67.52	−191.20	54.19	−194.00	39.79
Cb—(Cl)(2Cb)	−17.03	77.08	−32.20	55.47	−32.00	43.37
Cb—(Br)(2Cb)	36.35	88.60	19.90	74.85	13.50	54.45
Cb—(I)(2Cb)	94.50	98.26	73.70	61.08	70.40	
cis $_{corr}$—(I)(I)	3.00	0.00	0.00	0.00	0.00	0.00
C—(2H)(CO)(Cl)	−44.26		−58.41		−74.75	
C—(H)(CO)(2Cl)	−40.40		−55.11			
CO—(C)(F)	−379.84		−419.59			
C—(Cb)(3F)	−691.79	179.08	−696.66			
C—(2H)(Cb)(Br)	−29.49		−44.06			
C—(2H)(Cb)(I)	7.31		−7.24			
C—(2H)(Cb)(Cl)	−73.79		−92.56			
CO—(C)(Cl)	−200.54	176.66	−225.29			
CO—(Cb)(Cl)			−216.67		−212.99	
CO—(C)(Br)	−148.54		−175.49			
CO—(C)(I)	−83.94		−117.09			
C—(H)(C)(CO)(Cl)	−39.88		−35.46			
C—(C)(CO)(2Cl)						
ortho $_{corr}$—(I)(I)	7.56	0.00	6.96	0.00	5.50	0.00
ortho $_{corr}$—(F)(F)	20.90	0.00	25.00	0.00	25.50	0.00
ortho $_{corr}$—(Cl)(Cl)	9.50	0.00	14.00	0.00	8.50	0.00
ortho $_{corr}$—(alkyl)(X)	2.51	0.00	6.30	0.00	0.00	0.00
cis $_{corr}$—(Cl)(Cl)	−4.00	0.00	0.00	0.00	0.00	0.00
cis $_{corr}$—(CH_3)(Br)	−4.00	0.00	0.00	0.00	0.00	0.00
ortho $_{corr}$—(F)(Cl)	13.50	0.00	18.50	0.00	19.50	0.00
ortho $_{corr}$—(F)(Br)	37.25	0.00	40.60	0.00	42.50	0.00
ortho $_{corr}$—(F)(I)	85.40	0.00	83.55	0.00	85.20	0.00
meta $_{corr}$—(I)(I)	0.00	0.00	0.00	0.00	20.08	0.00
meta $_{corr}$—(COCl)(COCl)	0.00	0.00	0.00	0.00	16.06	0.00
ortho $_{corr}$—(COCl)(COCl)	0.00	0.00	0.00	0.00		0.00
ortho $_{corr}$—(F)(CF_3)	111.00	0.00	112.00	0.00	0.00	0.00
meta $_{corr}$—(F)(CF_3)	2.00	0.00	6.00	0.00	0.00	0.00
ortho $_{corr}$—(F)(CH_3)	−3.30	0.00	−6.00	0.00	0.00	0.00
ortho $_{corr}$—(F)(F')	8.00	0.00	8.00	0.00	8.00	0.00
ortho $_{corr}$—(Cl)(Cl')	8.00	0.00	8.00	0.00	8.00	0.00
meta $_{corr}$—(F)(F)	0.00	0.00	6.00	0.00	8.50	0.00
meta $_{corr}$—(Cl)(Cl)	−5.00	0.00	10.00	0.00	4.00	0.00
ortho $_{corr}$—(Cl)(CHO)	−6.75	0.00	8.50	0.00	0.00	0.00
ortho $_{corr}$—(F)(COOH)	20.00	0.00	0.00	0.00	20.00	0.00
ortho $_{corr}$—(Cl)(COCl)	0.00	0.00	34.43	0.00	0.00	0.00
ortho $_{corr}$—(F)(OH)	25.50	0.00	23.00	0.00	20.00	0.00
ortho $_{corr}$—(Cl)(COOH)	0.00	0.00	0.00	0.00	20.00	0.00
ortho $_{corr}$—(Br)(COOH)	0.00	0.00	0.00	0.00	20.00	0.00
ortho $_{corr}$—(I)(COOH)	0.00	0.00	0.00	0.00	20.00	0.00
ortho $_{corr}$—(NH_2)(NH_2)	−10.00	0.00	0.00	0.00	0.00	0.00
meta $_{corr}$—(NH_2)(NH_2)	0.00	0.00	0.00	0.00	14.00	0.00
ortho $_{corr}$—(OH)(Cl)	7.50	0.00	0.00	0.00	11.00	0.00
cis $_{corr}$—(CH_3)(I)	−4.00	0.00	0.00	0.00	0.00	0.00

*Domalski, E. S., and E. D. Hearing, *J. Phys. Chem. Ref. Data*, **22** (1993): 805.

isodesmic with 12 single bonds and 1 double bond in both reactants and products. To use this method, one devises an isodesmic reaction involving the compound for which ΔH_f° is to be determined with other compounds for which experimental ΔH_f° values are available. *Ab initio* calculations are performed on all the participating compounds, all at the same level of theory and basis set size, to obtain the enthalpy for each at 298.15 K. The enthalpy of reaction is then calculated from

$$\Delta H_{rxn} = \sum \nu_i H_i \tag{2-32}$$

where ν_i = stoichiometric coefficient of i (+ for products, − for reactants). The enthalpy of reaction is also related to ΔH_f^o by

$$\Delta H_{rxn} = \sum \nu_i (\Delta H_f^\circ)_i \tag{2-33}$$

With experimental values available for all ΔH_f° except the desired compound, its value can be back-calculated from Eq. (2-33).

Example Estimate the standard ideal gas enthalpy of formation of acetaldehyde.

Input information: The isodesmic reaction shown above will be used. The recommended ΔH_f^o values from DIPPR 801 for the other three compounds are as follows:

Acetone	Methane	Ethane
−215.70 kJ/mol	−74.52 kJ/mol	−83.82 kJ/mol

Ab initio calculations of enthalpy: With structures optimized using HF/6-31G(d) model chemistry and energies calculated with B3LYP/6-311+G(3df,2p), the following enthalpies are obtained (including the zero-point energy):

Acetone	Methane	Ethane	Acetaldehyde
-5.071×10^5 kJ/mol	-1.063×10^5 kJ/mol	-2.095×10^5 kJ/mol	-4.039×10^5 kJ/mol

Calculation using Eq. (2-32):

$$\Delta H_{rxn} = (-1.063 - 5.071 + 2.095 + 4.039) \times 10^5 \text{ kJ/mol} = -41.67 \text{ kJ/mol}$$

Calculation using Eq. (2-33):

$$\Delta H^\circ_{f,\text{acetaldehyde}} = \Delta H^\circ_{f,\text{acetone}} + \Delta H^\circ_{f,\text{methane}} - \Delta H^\circ_{f,\text{ethane}} - \Delta H_{rxn}$$

$$\Delta H^\circ_{f,\text{acetaldehyde}} = (-215.70 - 74.52 + 83.82 + 41.67)\frac{\text{kJ}}{\text{mol}} = -164.73\frac{\text{kJ}}{\text{mol}}$$

The estimated value is 1.0 percent above the DIPPR 801 recommended value of −166.40 kJ/mol.

Entropy Absolute or third-law entropies (relative to a perfectly ordered crystal at 0 K) of a compound in its standard state S^s or of an ideal gas S^o at 298.15 K and 1 bar can be found in various literature sources (DIPPR, JANAF, TRC, SWS, and TDB). Very good estimates for S^s or S^o can be obtained by using the Domalski-Hearing method. Excellent S^o values can also be obtained from statistical mechanics by using experimental vibrational frequencies or values of the frequencies generated from computational chemistry.

The standard ΔS_f^s and ideal gas ΔS_f^o entropies of formation at 298.15 K and 1 bar are related to the standard entropies by

$$\Delta S_f^s = S^s_{\text{compound}} - \sum_{i=1}^{n_A} \nu_i S^s_{\text{element},i} \qquad \Delta S_f^o = S^o_{\text{compound}} - \sum_{i=1}^{n_A} \nu_i S^s_{\text{element},i} \tag{2-34}$$

where $S^s_{\text{element},i}$ is the absolute entropy of element i in its standard state at 298.15 K and 1 bar.

Recommended Method Domalski-Hearing method.

Reference: Domalski, E. S., and E. D. Hearing, *J. Phys. Chem. Ref. Data*, **22** (1993): 805.

Classification: Group contributions.

Expected uncertainty: 3 percent.

Applicability: Organic compounds for which group contributions have been regressed.

Input data: Molecular structure.

Description: See description given under Enthalpy of Formation above.

Example Estimate the standard and ideal gas entropies of formation of *o*-toluidine.

Standard state entropies: Estimation of S^s and S^o using the Domalski-Hearing method was illustrated above in the Enthalpy of Formation section. The standard entropies of formation can be obtained from the values determined in that example.

Formula: C_7H_9N. The standard entropies of the elements from the DIPPR 801 database are as follows:

Compound	N_2	H_2	C, graphite
ν_i	1/2	9/2	7
S_i^s/[J(kmol·K)]	1.9151×10^5	1.3057×10^5	5740

Entropies of formation can be calculated from these values by using Eq. (2-34):

$$\Delta S_f^s = \left[2.2656 - \left(\frac{1}{2}\right)(1.9151) - \left(\frac{9}{2}\right)(1.3057) - (7)(0.0574)\right]\frac{10^5\,\text{J}}{\text{kmol}\cdot\text{K}}$$

$$= -4.969 \cdot 10^5 \frac{\text{J}}{\text{kmol}\cdot\text{K}}$$

$$\Delta S_f^o = \left[3.6832 - \left(\frac{1}{2}\right)(1.9151) - \left(\frac{9}{2}\right)(1.3057) - (7)(0.0574)\right]\frac{10^5\,\text{J}}{\text{kmol}\cdot\text{K}}$$

$$= -3.552 \cdot 10^5 \frac{\text{J}}{\text{kmol}\cdot\text{K}}$$

Recommended Method Statistical mechanics.

Classification: Theory and computational chemistry.

Expected uncertainty: 0.2 percent if vibrational frequencies (or their characteristic temperatures) are experimentally available; uncertainty depends upon model chemistry if frequencies are determined from computational chemistry, but generally within about 5 percent.

Applicability: Ideal gases.

Input data: M; σ (external symmetry number); characteristic rotational temperature(s) (Θ_A for linear molecules; Θ_A, Θ_B, and Θ_C for nonlinear molecules); and $3n_A - 6 + \delta$ characteristic vibrational temperatures Θ_j.

Description: For harmonic frequencies, the rigorous temperature dependence of S^o is given by

$$\frac{S^o}{R} = \frac{3}{2}\ln\left(6175\frac{M}{\text{kg/kmol}}\right) + \frac{S_r}{R} + \sum_{j=1}^{3n_A-6+\delta}\left[\left(\frac{\Theta_j}{T}\right)(e^{\Theta_j/T}-1)^{-1} - \ln(1-e^{-\Theta_j/T})\right] \tag{2-35}$$

$$\text{where } \delta = \begin{cases} 0 & \text{nonlinear} \\ 1 & \text{linear} \end{cases}$$

$$\text{and } \frac{S_r}{R} = \begin{cases} \ln\left[\dfrac{1}{\sigma}\left(\dfrac{\pi T^3 e^3}{\Theta_A\Theta_B\Theta_C}\right)^{1/2}\right] & \text{nonlinear} \\ \ln\left[\left(\dfrac{Te}{\sigma\Theta_A}\right)\right] & \text{linear} \end{cases}$$

Example Calculate S^o for ammonia.

Structure: NH_3.

Input data: M = 17 kg/kmol. McQuarrie [McQuarrie, D. A., *Statistical Mechanics*, Harper & Row, New York, 1976] gives the following $3n_A - 6 + \delta = 12 - 6 + 0 = 6$ characteristic vibrational temperatures (in K): 1360, 2330, 2330, 4800, 4880, 4880. The characteristic rotational temperatures given by McQuarrie are Θ_A = 13.6 K, Θ_B = 13.6 K, and Θ_C = 8.92 K. For NH_3, σ = 3.

Vibrational contribution: The table below shows a spreadsheet calculation of the vibrational terms inside the summation sign in Eq. (2-35).

Θ_j/K	Θ_j/(298.15 K)	S_{vib}
1207.91	4.051	0.08929
1850.16	6.205	0.01457
1850.16	6.205	0.01457
3688.19	12.370	0.00006
3821.36	12.817	0.00004
3821.36	12.817	0.00004
		Sum 0.1186

Rotational contribution:

$$\frac{S_r}{R} = \ln\left\{\frac{1}{3}\cdot\left[\frac{(298.15\text{ K})^3\pi e^3}{(13.6\text{ K})(13.6\text{ K})(8.92\text{ K})}\right]^{1/2}\right\} = 5.81593$$

Calculation using Eq. (2-35):

$$\frac{S^o_{298}}{R} = \frac{3}{2}\ln(6175.17) + 5.81593 + 0.1186 = 23.277$$

$$S^o_{298} = 1.935 \times 10^5 \frac{\text{J}}{\text{kmol}\cdot\text{K}}$$

The calculated value differs from the DIPPR 801 recommended value of 1.927×10^5 J/(kmol·K) by 0.5 percent.

Gibbs Energy of Formation The standard Gibbs energy of formation is the Gibbs energy change upon formation of 1 mole of the compound in its standard state from its constituent elements in their standard states. The standard Gibbs energy of formation ΔG_f^s uses the naturally occurring phase at 298.15 K and 1 bar as the standard state, while the ideal gas Gibbs energy of formation ΔG_f^o uses the compound in the ideal gas state at 298.15 K and 1 bar as the standard state. In both cases, the standard state for the elements is their naturally occurring state of aggregation at 298.15 K and 1 bar. Sources for data include DIPPR, TRC, JANAF, and TDB. The Gibbs energies of formation are related to the corresponding enthalpies and entropies of formation by

$$\Delta G_f^o = \Delta H_f^o - T\Delta S_f^o \quad \text{and} \quad \Delta G_f^s = \Delta H_f^s - T\Delta S_f^s \tag{2-36}$$

and predicted values of ΔG_f^s and ΔG_f^o are obtained from Eq. (2-36) by estimating the enthalpies and entropies of formation as shown above.

LATENT ENTHALPY

Enthalpy of Vaporization The enthalpy (heat) of vaporization ΔH_υ is the difference between the molar enthalpies of the saturated vapor and saturated liquid at a temperature between the triple point and critical point (at the corresponding vapor pressure). Variable ΔH_υ is related to the vapor pressure P^* by the thermodynamically exact Clapeyron equation

$$\Delta H_\upsilon = -R\,\Delta Z_\upsilon \frac{d\ln P^*}{d(1/T)} = RT^2\Delta Z_\upsilon \frac{d\ln P^*}{dT} \tag{2-37}$$

where $\Delta Z_\upsilon = Z_G - Z_L$, $Z_G = Z$ of saturated vapor, and $Z_L = Z$ of saturated liquid. Experimental heats of vaporization can be effectively correlated with

$$\Delta H_\upsilon = A(1-T_r)^{B+CT_r+DT_r^2+ET_r^3} \tag{2-38}$$

A simple method for obtaining ΔH_υ at one temperature from a known value at a reference temperature, say at the normal boiling point, is to truncate Eq. (2-38) after the B term, set $B = 0.38$, and take a ratio of the ΔH_υ values at the two conditions to give the Watson [Thek, R. E., and L. I. Stiel, *AIChE J.*, **12** (1966): 599; **13** (1967): 626] correlation

$$\Delta H_\upsilon = \Delta H_{\upsilon,\text{ref}}\left(\frac{1-T_r}{1-T_{r,\text{ref}}}\right)^{0.38} \tag{2-39}$$

If an accurate correlation for P^* and accurate values for Z_G and Z_L are available, Eq. (2-37) is the preferred method for obtaining enthalpies of vaporization. Otherwise, the CS methods shown below should be used.

Recommended Method 1 Vapor pressure correlation.

Classification: Extension of theory.

Expected uncertainty: The uncertainty varies significantly with temperature and with the quality and temperature range of the vapor pressure data used in the correlation.

Applicability: Organic compounds for which group contributions have been regressed.

Input data: Correlations for P^*, Z_G, and Z_L.

Description: An expression for ΔH_υ can be obtained from Eq. (2-37) by using an appropriate vapor pressure correlation. If one differentiates the Riedel vapor pressure correlation, Eq. (2-26), in accordance with Eq. (2-37), one obtains the heat of vaporization as

$$\Delta H_\upsilon = R\,\Delta Z_\upsilon\,(-B + CT + DET^{E+1}) \tag{2-40}$$

The Z_G and Z_L values can be evaluated using the methods given in the section on densities below.

Example Calculate ΔH_υ for anisole at 452 K.

Input data: The vapor pressure coefficients in the DIPPR 801 database for Eq. (2-26) are

$$A = 128.06 \qquad B = -9307.7 \qquad C = -16.693 \qquad D = 0.014919 \qquad E = 1$$

The vapor pressure at 452 K is therefore

$$\ln\left(\frac{P^*}{\text{Pa}}\right) = 128.06 - \frac{9307.7}{452} - 16.693\ln(452) + 0.014919\,(452)^1 = 12.155$$

$$P^* = \exp\,(12.155)\cdot \text{Pa} = 1.901\times 10^5\ \text{Pa}$$

Determine ΔZ: Required data from the DIPPR 801 database for this calculation are T_c = 645.6 K, P_c = 4.25 MPa, and ω = 0.35017. These values are used to determine the reduced conditions,

$$T_r = \frac{452}{645.6} = 0.7 \qquad P_r = \frac{0.1901}{4.25} = 0.045$$

and the values of Z_G and Z_L from the Lee-Kesler corresponding states method as discussed in the section on density. Interpolation of the P_r values in Tables 2-169 and 2-170 at a T_r of 0.7 gives

$$Z_G^{(0)} = 0.9904 + \frac{0.045 - 0.010}{0.050 - 0.010}(0.9504 - 0.9904) = 0.9554$$

$$Z_G^{(1)} = -0.0064 + \frac{0.045 - 0.010}{0.050 - 0.010}(-0.0507 + 0.0064) = -0.0452$$

$$Z_G = Z_G^{(0)} + \omega Z_G^{(1)} = 0.9554 + (0.35017)(-0.0452) = 0.94$$

At this low pressure, Z_L is very small compared to Z_G and may be neglected; so

$$\Delta Z_V = Z_G - Z_L = 0.94$$

Calculation using Eq. (2-40):

$$\Delta H_\upsilon = \left(8.314\frac{\text{J}}{\text{mol}\cdot\text{K}}\right)(0.94)[9307.7 - (16.693)(452) + (0.014919)(1)(452)^2]$$

$$= 37.59\frac{\text{kJ}}{\text{mol}\cdot\text{K}}$$

This value is 0.2 percent higher than the value of 37.51 kJ/(mol·K) obtained from the DIPPR 801 database.

Recommended Method 2 Corresponding states correlation.

Reference: [PGL5], p. 7.18.

Classification: Corresponding states.

Expected uncertainty: Less than about 6 percent.

Applicability: Organic compounds.

Input data: T_c, P_c, and ω.

Description: The following correlation is used:

$$\frac{\Delta H_\upsilon}{RT_c} = 7.08\tau^{0.354} + 10.95\omega\tau^{0.456} \qquad \text{where } \tau = 1 - T_r \tag{2-41}$$

Example Repeat the above calculation for anisole's ΔH_υ at 452 K.

Input data: T_c = 645.6 K, P_c = 4.25 MPa, and ω = 0.35017.

Auxiliary quantities: From the previous example, the reduced temperature variables are

$$T_r = 0.7 \qquad \tau = 1 - 0.7 = 0.3$$

Calculation using Eq. (2-41):

$$\frac{\Delta H_\upsilon}{RT_c} = 7.08(0.3)^{0.354} + 10.95\,(0.35017)(0.3)^{0.456} = 6.838$$

$$\Delta H_\upsilon = (6.838)\left(8.314\frac{\text{J}}{\text{mol}\cdot\text{K}}\right)(645.6\ \text{K}) = 36.70\frac{\text{kJ}}{\text{mol}\cdot\text{K}}$$

This value is 2.2 percent below the DIPPR 801 recommended value of 37.51 kJ/(mol·K).

Enthalpy of Fusion The enthalpy (heat) of fusion ΔH_{fus} is the difference between the molar enthalpies of the equilibrium liquid and solid at the melting temperature and 1.0 atm pressure. There is no generally applicable, high-accuracy estimation method for ΔH_{fus}, but the GC method of Chickos can be used to obtain approximate results if the melting temperature is known.

Recommended Method Chickos method.

Reference: Chickos, J. S., C. M. Braton, D. G. Hesse, and J. R. Liebman, *J. Org. Chem.*, **56** (1991): 927.

Classification: QSPR and group contributions.

Expected uncertainty: Considerable variation but generally less than 50 percent.

Applicability: Only valid at the melting temperature. The method is based on the ΔS_{fus} between a solid at 0 K and the liquid at the T_m so no solid-solid transitions are taken into account. Values of ΔH_{fus} will be overestimated if there are solid-solid transitions for the actual material.

Input data: T_m and molecular structure.

Description:

$$\frac{\Delta H_{\text{fus}}}{\text{J/mol}} = \frac{\Delta S_{\text{fus}}}{\text{J/(mol}\cdot\text{K)}}\left(\frac{T_m}{\text{K}}\right) = (T_m/\text{K})\,(a+b) \tag{2-42}$$

$$a = \begin{cases} 0 & \text{no nonaromatic rings} \\ 35.19N_R + 4.289(N_{CR} - 3N_R) & \text{nonaromatic rings} \end{cases} \tag{2-43}$$

$$b = \sum_{i=1}^{ng} Ng_i\Delta s_i + \sum_{j=1}^{ns} Ns_j Cs_j \Delta s_j + \sum_{k=1}^{nf} Nf_k Ct_k\,\Delta s_k \tag{2-44}$$

where Ng_i = number of C—H groups of type i bonded to other carbon atoms
n_g = number of different nonring or aromatic C—H groups bonded to other carbon atoms
Ns_j = number of C—H groups of type j bonded to at least one functional group or atom
n_s = number of different nonring or aromatic C—H groups bonded to at least one functional group or atom
Nf_k = number of functional groups of type k
n_f = number of different functional groups or atoms
t = total number of functional groups or atoms with the exception that F atoms count as one regardless of number of occurrences
Cs_j = value from Table 2-162 for C—H group j bonded to at least one functional group or atom
Ct_k = value from Table 2-163 for functional group k
N_R = number of nonaromatic rings
N_{CR} = number of $-CH_2-$ groups in nonaromatic ring(s) required to form cyclic paraffin of same ring size(s)
Δs_i = contribution from Table 2-162 for group i
Δs_k = contribution from Table 2-163 for group k

Note that nonaromatic ring $-CH_2$ groups are accounted for in the a term and are *not* included in the b term.

Example Calculate ΔH_{fus} at the melting point for (*a*) benzothiophene, (*b*) furfuryl alcohol, and (*c*) *cis*-crotonaldehyde.

Structures:

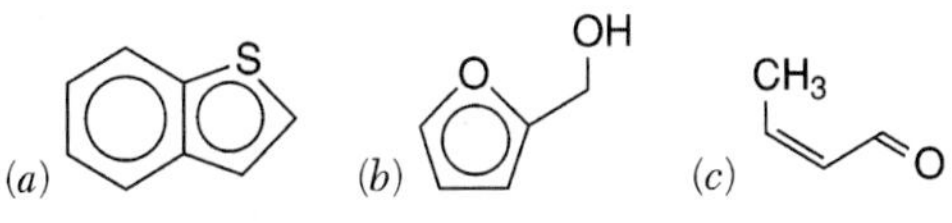

(*a*) (*b*) (*c*)

TABLE 2-162 Cs (C—H) Group Values for Chickos Estimation* of ΔH_{fus}

Group	Description	*Cs*	Δ*s*	Group	Description	*Cs*	Δ*s*
$—CH_3$	methyl	1.0	18.33	$—CH_{Ar}$	aromatic C	1.0	6.44
$>CH_2$	methylene	1.0	9.41	$—C_{Ar}—$	ar. C bonded to paraffinic C	1.0	−10.33
>CH—	secondary C	0.69	−16.91	$—C_{Ar}—$	ar. C bonded to olefinic C or non-C group	1.0	−4.27
>C<	tertiary C	0.67	−38.70	$—C_{Ar}—$	ar. C bonded to acetylinic C	1.0	−2.51
CH_2=	terminal alkene	1.0	14.56	$>C_rH—$	ring structure	0.76	−15.98
—CH=	alkene	3.23	4.85	$>C_r<$	ring structure	1.0	−32.97
>C=	subst. alkene	1.0	−11.38	$—C_rH=$	ring structure	0.62	−4.35
≡CH	term. alkyne	1.0	10.88	$>C_r=$	ring structure	0.86	−11.72
≡C—	alkyne	1.0	2.18	$≡C_r—$ or $=C_r=$	ring structure	1.0	−5.36

*Chickos, J. S., et al., *J. Org. Chem.*, **56** (1991): 927.

(*a*) $t = 1$ (1 total "functional group"), so the C_1 column in Table 2-163 is used.

$N_R = 1 \quad N_{CR} = 5 \quad a = 35.19 + (5 - 3)(4.289) = 43.77$

Group	Description	*N*	*C*	Δ*s*	Total
=CH—	aromatic (*Ng* type)	4	1	6.44	25.76
=C—	ring (*Ng* type)	1	1	−11.72	−11.72
=C—	ring (*Ns* type)	1	0.86	−11.72	−10.08
=CH—	ring (*Ng* type)	1	1	−4.35	−4.35
=CH—	ring (*Ns* type)	1	0.62	−4.35	−2.70
—S—	ring	1	1	2.18	2.18
				Total	−0.91

$T_m = 304.5$ K from DIPPR 801 database

$\Delta H_{fus} = (T_m/\text{K})(a + b)$ J/mol = (304.5)(43.77 − 0.91) J/mol = 13.05 kJ/mol

This value is 10 percent higher than the DIPPR 801 recommended value of 11.83 kJ/mol.

(*b*) $t = 2$ (2 total "functional groups"), so the C_2 column in Table 2-163 is used.

$N_R = 1 \quad N_{CR} = 5 \quad a = 35.19 + (5 - 3)(4.289) = 43.77$

Group	Description	*N*	*C*	Δ*s*	Total
=CH—	ring (*Ng* type)	2	1	−4.35	−8.70
=CH—	ring (*Ns* type)	1	0.62	−4.35	−2.70
=C<	ring (*Ns* type)	1	0.86	−11.72	−10.08
=O—	ring ether	1	1	1.34	1.34
$—CH_2—$	*Ns* type	1	1	9.41	9.41
—OH	alcohol	1	12.6	1.13	14.24
				Total	3.51

TABLE 2-163 Ct (Functional) Group Values for Chickos Estimation* of ΔH_{fus}

Group	Description	C_1	C_2	C_3	C_4	Δs
—OH	alcohol	1.0	12.6	18.9	26.4	1.13
—OH	phenol	1.0	1.0	1.0	1.0	16.57
—O—	nonring ether	1.0	1.0	1.0	1.0	1.09
—O—	ring ether	1.0	1.0	1.0	1.0	1.34
>C=O	nonring ketone	1.0	1.0			3.14
>C=O	ring ketone	1.0	1.0			−1.88
—CHO	aldehyde	1.0	1.0			19.66
—COOH	acid	1.0	1.83	1.88	1.72	14.90
—COO—	ester	1.0	1.0	1.0	1.0	3.68
$—NH_2$	aliphatic	1.0	1.0			16.23
$—NH_2$	aromatic	1.0	1.0			15.48
>NH	nonring	1.0	1.0			−2.18
>NH	ring	1.0	1.0			1.84
>N—	nonring	1.0	1.0			−15.90
>N—	ring	1.0	1.0			−17.07
=N—	ring	1.0	1.0			1.67
=N—	aromatic	1.0	1.0	1.0		7.32
—CN	nitrile	1.0	1.4			9.62
$—NO_2$	nitro	1.0	1.0	1.0		17.36
$—CONH_2$	primary amide	1.0	1.0			26.19
—CONH—	secondary amide	1.0	1.0			−0.42
—SH		1.0	1.0			17.99
—S—	nonring	1.0	1.0		0.36	7.20
—S—	ring	1.0	1.0			2.18
$—SO_2$	nonring	1.0	1.0			3.26
—F	on aliph. C	1.0	1.0	1.0	1.0	14.73
—F	on olefinic C	1.0	1.0	1.0	1.0	13.01
—F	on ring C	1.0	1.0	1.0	1.0	15.90
—Cl		1.0	2.0	2.0	1.93	8.37
—Br		1.0	1.0	1.0	0.82	17.95
—I		1.0	1.0			16.95

*Chickos, J. S., et al., *J. Org. Chem.*, **56** (1991): 927.

$T_m = 258.52$ K from DIPPR 801 database

$\Delta H_{fus} = (T_m/\text{K})(a + b)$ J/mol = (258.52)(43.77 + 3.51) J/mol = 12.22 kJ/mol

This value is 7 percent lower than the DIPPR 801 recommended value of 13.13 kJ/mol.

(*c*) $t = 1 \quad N_R = 0 \quad a = 0$

Group	Description	*N*	*C*	Δ*s*	Total
$—CH_3$	nonring (*Ng* type)	1	1	18.33	18.33
=CH—	nonring (*Ng* type)	1	1	4.85	4.85
=CH—	nonring (*Ns* type)	1	3.23	4.85	15.67
—CHO	aldehyde	1	1	19.66	19.66
				Total	58.51

$T_m = 158.38$ K from DIPPR 801 database

$\Delta H_{fus} = (T_m/\text{K})(a + b)$ J/mol = (158.38)(0 + 58.51) J/mol = 9.27 kJ/mol

This value is 5 percent higher than the DIPPR 801 recommended value of 8.86 kJ/mol.

Enthalpy of Sublimation The enthalpy (heat) of sublimation ΔH_{sub} is the difference between the molar enthalpies of the equilibrium vapor and solid along the sublimation curve below the triple point. The effects of pressure on ΔH_{sub} and melting temperature are very small so that T_t and the normal melting point are nearly equal and

$$\Delta H_{sub}(T_t) = \Delta H_v(T_t) + \Delta H_{fus}(T_t) \quad (2\text{-}45)$$

Equation (2-45) can be used to estimate ΔH_{sub} at the triple point if ΔH_v is accurately known at T_t. Because ΔH_v is usually obtained from Eq. (2-37), $\Delta H_v(T)$ correlations may be less accurate near T_t where $P^*(T_t)$ is very small and difficult to measure. In this case, it is better to estimate ΔH_{sub} directly by using the following recommended method. ΔH_{sub} is only a weak function of temperature and can generally be treated as a constant from the triple point temperature down to the first solid-solid phase transition.

Recommended Method Goodman method.

Reference: Goodman, B. T., W. V. Wilding, J. L. Oscarson, and R. L. Rowley, *Int. J. Thermophys.* **25** (2004): 337.

Classification: QSPR and group contributions.

Expected uncertainty: 6 percent.

Applicability: Organic compounds for which group contributions have been regressed.

Input data: Molecular structure and radius of gyration R_G.

Description:

$$\frac{\Delta H_{sub}(T_t)}{R\,\text{K}} = 698.04 + 3.83798\times10^{12}\left(\frac{R_G}{\text{m}}\right) + \sum_{i=1}^{N} n_i a_i + \sum_{i=1}^{N} n_i^2 \beta_i + \sum_{i=1}^{N} \frac{n_i}{n_x} f_i \quad (2\text{-}46)$$

where a_i = GC values from Table 2-164

β_i = nonlinear corrections for $>CH_2$ and Ar—CH = groups

f_i = halogen corrections

n_x = total number of all halogen and hydrogen atoms attached to C and Si atoms

Example Calculate ΔH_{sub} and the solid vapor pressure for 1,2,3-trichlorobenzene at 301.15 K.

Structure:

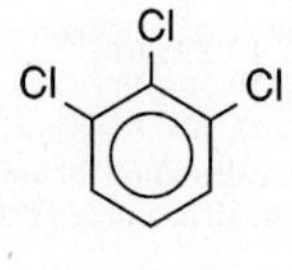

TABLE 2-164 Group Contributions and Corrections* for ΔH_{sub}

Group	Description	a_i	Group	Description	a_i
—CH_3	methyl	736.5889	>C=O	ketone	1816.093
>CH_2	methylene	561.3543	—COO—	ester	2674.525
>CH—	secondary C	111.0344	—COOH	acid	5006.188
>C<	tertiary C	−800.517	—NH_2	primary amine	2219.148
CH_2=	terminal alkene	572.6245	—NH—	sec. amine	1561.222
—CH=	alkene	541.2918	>N—	tertiary amine	325.9442
>C=	substituted alkene	117.9504	—NO_2	nitro	3661.233
Ar—CH=	aromatic C	626.7621	—SH	thiol/mercaptan	1921.097
Ar >C=	subst. aromatic C	348.8092	—S—	sulfide	1930.84
Ar—O—	furan O	763.284	—SS—	disulfide	2782.054
Ar—N=	pyridine N	1317.056	—F	fluoride	626.4494
Ar—S—	thiophene S	911.2903	—Cl	chloride	1243.445
—O—	ether	970.4474	—Br	bromide	669.9302
—OH	alcohol	3278.446	>Si<	silane	−83.7034
—COH	aldehyde	2402.093	>Si(O—)—	siloxane	−16.0597
Nonlinear terms		β_i	Halogen correction terms		f_i
>CH_2	methylene	9.5553	—F	F fraction	−1397.4
Ar—CH=	aromatic C	−2.21614	—Cl	Cl fraction	−1543.66
			—Br	Br fraction	5812.49

*Goodman, B., et al., *Int. J. Thermophys.*, **25** (2004): 337.

Group contributions:

Linear groups			Nonlinear and correction terms			
Group	n_i	a_i	Group	n_i	β_i	f_i
Ar—CH=	3	626.7621	Ar—CH=	3	−2.21614	
Ar >C=	3	348.8092	—Cl	3		−1543.66
—Cl	3	1243.445	n_x	6		
	$\sum_i n_i a_i = 6657.049$					

Input data: The value of R_G from the DIPPR 801 database is 4.455×10^{-10} m.
Calculation using Eq. (2-46):

$$\frac{\Delta H_{sub}(T_t)}{RK} = 698.04 + (3.838\times10^{12})(4.455\times10^{-10}) + 6657.05 + (3^2)(-2.21614) + \left(\frac{3}{6}\right)(-1543.66)$$

$$\Delta H_{sub}(T_t) = (8273\ \text{K})\left(0.008314\frac{\text{kJ}}{\text{mol}\cdot\text{K}}\right) = 68.78\frac{\text{kJ}}{\text{mol}}$$

The estimated value is 5.6 percent above the DIPPR 801 recommended value of 65.11 kJ/mol.

Estimate the solid vapor pressure at 301.15 K: The solid vapor pressure can be calculated from Eq. (2-30) by using the estimated ΔH_{sub} and one additional solid vapor pressure point. In this example the triple point temperature and vapor pressure ($T_t = 325.65$ K; $P_t^* = 182.957$ Pa) from the DIPPR 801 database are used in Eq. (2-30):

$$\ln\frac{P^*}{182.957\ \text{Pa}} = \frac{68.78\ \text{kJ/mol}}{[0.008314\ \text{kJ/(mol}\cdot\text{K)}](325.65\ \text{K})}\left(1-\frac{325.65}{301.15}\right) = -2.067$$

$$P^* = (182.957\ \text{Pa})\,[\exp(-2.067)] = 23.16\ \text{Pa}$$

The estimated value is 0.3 percent above the DIPPR 801 recommended value of 23.09 Pa.

HEAT CAPACITY

The isobaric heat capacity C_P is defined as the energy required to change the temperature of a unit mass (specific heat) or mole (molar heat capacity) of the material by one degree at constant pressure. Typical units are J/(kg·K).

Gases The isobaric heat capacity of a gas is related rigorously to the ideal gas value C_P^o by

$$C_P = C_P^o - T\int_0^P \left(\frac{\partial^2 V}{\partial T^2}\right)_P dP \tag{2-47}$$

The second term, giving the deviation of the real fluid heat capacity from the ideal gas value, can be neglected at low to moderate pressures, or it can be calculated directly from an appropriate EoS.

Ideal gas heat capacities are available from several sources (DIPPR, JANAF, TRC, and SWS). Two common correlating equations for C_P^o are the Aly-Lee equation [Aly, F. A., and L. L. Lee, *Fluid Phase Equilib.*, **6** (1981): 169]

$$C_P^o = A_0 + A_1\left[\frac{A_2/T}{\sinh(A_2/T)}\right]^2 + A_3\left[\frac{A_4/T}{\cosh(A_4/T)}\right]^2 \tag{2-48}$$

and a polynomial form (generally fourth-order)

$$C_P^o = \sum_{i=0}^{4} A_i T^i \tag{2-49}$$

Ideal gas heat capacities may also be estimated from several techniques, of which two of the most accurate and commonly used are recommended here.

Recommended Method 1 Statistical mechanics.

Reference: Rowley, R. L., *Statistical Mechanics for Thermophysical Property Calculations*, Prentice-Hall, Englewood Cliffs, N.J., 1994.

Classification: Theory and computational chemistry.

Expected uncertainty: 0.2 percent if vibrational frequencies (or their characteristic temperatures) are experimentally available; accuracy depends upon model chemistry if frequencies are determined from computational chemistry, but generally within 3 percent.

Applicability: Ideal gases.

Input data: $3n_A - 6 + \delta$ vibrational frequencies ν_j, or the corresponding characteristic vibrational temperatures Θ_j. The two are related by

$$\Theta_j = h\nu_j/k \tag{2-50}$$

Description: For harmonic frequencies, the rigorous temperature dependence of C_P^o is given by

$$\frac{C_P^o}{R} = \frac{8-\delta}{2} + \sum_{j=1}^{3n_A-6+\delta}\left(\frac{\Theta_j}{T}\right)^2\left[\frac{e^{\Theta_j/T}}{(e^{\Theta_j/T}-1)^2}\right] \qquad \delta = \begin{cases}0 & \text{nonlinear}\\ 1 & \text{linear}\end{cases} \tag{2-51}$$

Example Calculate the ideal gas heat capacity of ammonia at 300 K.
Structure:

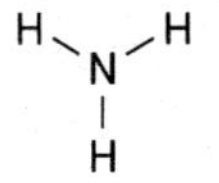

Input data: McQuarrie (McQuarrie, D. A., *Statistical Mechanics*, Harper & Row, New York, 1976) gives the following $3n_A - 6 + \delta = 12 - 6 + 0 = 6$ characteristic vibrational temperatures (in K): 1360, 2330, 2330, 4800, 4880, and 4880. Alternatively, a computational chemistry package gives the following *scaled* frequencies for HF/6-31G+ model chemistry (10^{13} Hz): 3.24, 4.97, 4.97, 9.90, 10.26, and 10.26.

Calculation: The table on the left uses the experimental Θ values to determine the individual terms in the summation of Eq. (2-51). The table on the right uses the scaled frequencies from computational chemistry software and Eq. (2-50) to obtain Θ values and the individual terms in Eq. (2-51).

Experimental frequencies			HF/6-31G+ scaled frequencies*			
Θ/K	Θ/(300 K)	Term	$\nu_{scaled}/10^{13}$ Hz	Θ/K	Θ/(300 K)	Term
1360	4.533	0.2256	3.24	1555.0	5.183	0.1524
2330	7.767	0.0256	4.97	2385.3	7.951	0.0223
2330	7.767	0.0256	4.97	2385.3	7.951	0.0223
4800	16.000	0.0000	9.90	4751.4	15.838	0.0000
4880	16.267	0.0000	10.26	4924.2	16.414	0.0000
4880	16.267	0.0000	10.26	4924.2	16.414	0.0000
	Sum:	0.2768			Sum:	0.1970

*Empirical scaling factors have been developed for each model chemistry to help correct theoretical frequencies for anharmonic effects [Scott, A. P., and L. Radom, *J. Phys. Chem.*, **100** (1996): 16502].

From experimental frequencies:

$$C_p^o = \left(\frac{8}{2} + 0.2768\right) R = (4.2768)\left(8.3143 \frac{\text{J}}{\text{mol}\cdot\text{K}}\right) = 35.56 \frac{\text{J}}{\text{mol}\cdot\text{K}}$$

From computational chemistry frequencies:

$$C_p^o = \left(\frac{8}{2} + 0.197\right) R = (4.197)\left(8.3143 \frac{\text{J}}{\text{mol}\cdot\text{K}}\right) = 34.90 \frac{\text{J}}{\text{mol}\cdot\text{K}}$$

The value calculated from experimental frequencies is 0.1 percent lower than the DIPPR 801 recommended value of 35.61 J/(mol · K); the value calculated from frequencies generated from computational chemistry software is 2.0 percent lower than the DIPPR 801 value.

Recommended Method 2 Benson method as implemented in CHETAH program.

References: Benson, S. W., et al., *Chem. Rev.*, **69** (1969): 279; CHETAH Version 8.0: The ASTM Computer Program for Chemical Thermodynamic and Energy Release Evaluation (NIST Special Database 16).

Classification: Group contributions.

Expected uncertainty: 4 percent.

Applicability: Ideal gases of organic compounds.

Input data: Table 2-165 group values at the seven specified temperatures.

Description: Groups are summed at each individual temperature:

$$C_p^o = \sum_{i=1}^{N} n_i \cdot (C_p^o)_i \qquad (2\text{-}52)$$

where n_i = number of occurrences of group i and $(C_p^o)_i$ = individual group contribution. Either Eq. (2-48) or Eq. (2-49) can be used to interpolate between the discrete temperatures.

Example Calculate the ideal gas heat capacity of isoprene (2-methyl-1,3-butadiene) at 400 K.

Structure:

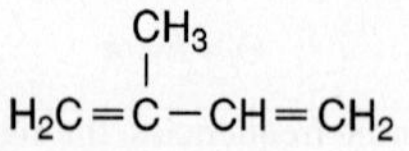

Group identification and values:

Group	No.	Value, J/(mol · K)	Contribution, J/(mol · K)
$=CH_2$	2	26.62	53.24
=C—(2C)	1	19.3	19.3
$—CH_3—(=C)$	1	32.82	32.82
=CH—(C)	1	21.05	21.05
		Total	126.41

The value of 126.4 J/(mol · K) is 3.1 percent below the DIPPR 801 recommended value of 130.4 J/(mol · K).

Liquids Liquid isobaric heat capacity increases with increasing temperature, although a minimum occurs near the triple point for many compounds. Usually liquid heat capacity is correlated as a function of temperature with a polynomial equation; a third-order polynomial is usually adequate.

Estimation of liquid heat capacity can be done by using a number of methods [Ruzicka, V., and E. S. Domalski, *J. Phys. Chem. Ref. Data*, **22** (1993): 597, 619; Chueh, C. F., and A. C. Swanson, *Chem. Eng. Prog.*, **69**, 7 (1973): 83; Lee, B. I., and M. G. Kesler, *AIChE J.*, **21** (1975): 510; Tarakad, R. R., and R. P. Danner, *AIChE J.*, **23** (1977): 944] and thermodynamic differentiation. The Ruzicka-Domalski method is generally accurate at low temperature, but the cubic behavior can overestimate the temperature rise at higher temperatures. The Lee-Kesler method is accurate for nonpolar and slightly polar fluids, but has less accuracy for strongly polar or associating fluids.

Recommended Method 1 Ruzicka-Domalski.

References: Ruzicka, V., and E. S. Domalski, *J. Phys. Chem. Ref. Data*, **22** (1993): 597, 619.

Classification: Group contributions.

Expected uncertainty: 4 percent.

Applicability: Organic compounds for which group values are available.

Input data: Molecular structure and Table 2-166 values.

Description: Groups are summed to find the temperature coefficients for a cubic polynomial correlation:

$$\frac{C_p}{R} = A + B\left(\frac{T}{100\text{ K}}\right) + D\left(\frac{T}{100\text{ K}}\right)^2 \qquad (2\text{-}53)$$

$$A = \sum_{i=1}^{N} n_i a_i \quad B = \sum_{i=1}^{N} n_i b_i \quad D = \sum_{i=1}^{N} n_i d_i \qquad (2\text{-}54)$$

where n_i = number of occurrences of group i and a_i, b_i, d_i = individual group contributions.

Example Estimate the liquid heat capacity for 2-methyl-2-propanol at 340 K.

Structure:

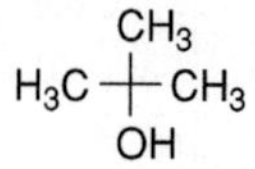

Group contributions:

Group	n_i	a_i	b_i	d_i
C—(3C, O) (alcohol)	1	−44.690	31.769	−4.8791
O—(H)(C)	1	12.952	−10.145	2.6261
C—(3H)(C)	3	3.8452	−0.33997	0.19489
	Sum	−20.202	20.604	−1.668

$$C_p = \left(8.3143 \frac{\text{J}}{\text{mol}\cdot\text{K}}\right)\left[-20.202 + 20.604\left(\frac{304}{100}\right) - 1.668\left(\frac{340}{100}\right)^2\right]$$

$$= 254.16 \frac{\text{J}}{\text{mol}\cdot\text{K}}$$

This value is 0.7 percent higher than the DIPPR 801 recommended value of 252.40 J/(mol · K).

Recommended Method 2 Lee-Kesler.

References: [PGL5]

Classification: Corresponding states.

Expected uncertainty: 4 percent.

Applicability: Organic compounds other than those that are strongly polar or associate.

Input data: T_c, ω, and the ideal gas heat capacity at the same temperature.

Description: The isobaric liquid heat capacity is calculated at the reduced temperature T_r using

$$\frac{C_p}{R} = \frac{C_p^o}{R} + 1.586 + \frac{0.49}{1-T_r} + \omega\left[4.2775 + \frac{6.3(1-T_r)^{1/3}}{T_r} + \frac{0.4355}{1-T_r}\right] \qquad (2\text{-}55)$$

Example Calculate the isobaric liquid heat capacity for 1,4-dioxane at 320 K.

Auxiliary data: From the DIPPR 801 database: T_c = 597.0 K, ω = 0.2793, and C_p^o/R = 11.94. The reduced temperature is therefore T_r = (320 K)/(597.0 K) = 0.536.

From Eq. (2.55),

$$\frac{C_p}{R} = 11.94 + 1.586 + \frac{0.49}{1-0.536} + (0.2793)\left[4.2775 + \frac{6.3(1-0.536)^{1/3}}{0.536} + \frac{0.4355}{1-0.536}\right] = 18.58$$

and C_p = 154.5 J/(mol · K). This is 4.6 percent below the DIPPR recommended value of 162.0 J/(mol · K).

TABLE 2-165 Benson* and CHETAH† Group Contributions for Ideal Gas Heat Capacity

Table-specific nomenclature: Cb = carbon in benzene ring; Ct = carbon with a triple bond, (=C) = carbon with a double bond; Cp = carbon in fused ring; Naz = azide; Nim = imino.

Group	298 K	400 K	500 K	600 K	800 K	1000 K	1500 K
			CH_3 Groups				
CH_3—(Cb)	25.91	32.82	39.35	45.17	54.5	61.83	73.59
CH_3—(CO)	25.91	32.82	39.35	45.17	54.5	61.83	73.59
CH_3—(Ct)	25.91	32.82	39.35	45.17	54.5	61.83	73.59
CH_3—(C)	25.91	32.82	39.35	45.17	54.5	61.83	73.59
CH_3—(N)	25.95	32.65	39.35	45.21	54.42	61.95	73.67
CH_3—(O)	25.91	32.82	39.35	45.17	54.54	61.83	73.59
CH_3—(PO)	25.91	32.82	39.35	45.17	54.54	61.83	73.59
CH_3—(P)	25.91	32.82	39.35	45.17	54.54	61.83	73.59
CH_3—(P=N)	25.91	32.82	39.35	45.17	54.54	61.83	73.59
CH_3—(Si)	25.91	32.82	39.35	45.17	54.5	61.83	73.59
CH_3—(SO_2)	25.91	32.82	39.35	45.17	54.5	61.83	
CH_3—(SO)	25.91	32.82	39.35	45.17	54.5	61.83	
CH_3—(S)	25.91	32.82	39.35	45.17	54.5	61.83	
CH_3—(=C)	25.91	32.82	39.35	45.17	54.5	61.83	73.59
			Ct Groups				
Ct—(Cb)	10.76	14.82	14.65	20.59	22.35	23.02	24.28
Ct—(Ct)	14.82	16.99	18.42	19.42	20.93	21.89	23.32
Ct—(C)	13.1	14.57	15.95	17.12	19.25	20.59	26.58
Ct—(=C)	10.76	14.82	14.65	20.59	22.35	23.02	24.28
CtBr	34.74	36.42	37.67	38.51	39.77	40.6	
CtCl	33.07	35.16	36.42	37.67	39.35	40.18	
CtF	28.55	31.65	33.99	35.79	38.3	39.85	41.77
CtH	22.06	25.07	27.17	28.76	31.27	33.32	37.04
CtI	35.16	36.84	38.09	38.93	40.18	41.02	
Ct(CN)	43.11	47.3	50.65	53.16	56.93	59.86	64.04
			CH_2 Groups				
CH_2—(2CO)	16.03	26.66	32.15	37.8	45.46	51.74	
CH_2—(2C)	23.02	29.09	34.53	39.14	46.34	51.65	59.65
CH_2—(2O)	11.85	21.18	31.48	38.17	43.2	47.26	
CH_2—(2=C)	19.67	28.46	35.16	40.18	47.3	52.74	60.28
CH_2—(Cb,O)	15.53	26.26	34.66	40.98	49.35	55.25	
CH_2—(Cb,SO_2)	15.53	27.5	34.66	40.98	49.77	55.25	
CH_2—(Cb,S)	38.09	49.02	57.43	63.71	72.58	78.82	
CH_2—(Cb,=C)	19.67	28.46	35.16	40.18	47.3	52.74	60.28
CH_2—(C,Cb)	24.45	31.85	37.59	41.9	48.1	52.49	57.6
CH_2—(C,CO)	25.95	32.23	36.42	39.77	46.46	51.07	
CH_2—(C,Ct)	20.72	27.46	33.19	38.01	45.46	51.03	59.44
CH_2—(C,N)	21.77	28.88	34.74	39.35	46.46	51.49	
CH_2—(C,O)	20.89	28.67	34.74	39.47	46.5	51.61	61.11
CH_2—(C,SO_2)	17.12	24.99	31.56	36.84	44.58	49.94	
CH_2—(C,SO)	19.05	26.87	33.28	38.34	45.84	51.15	
CH_2—(C,S)	22.52	29.64	36	41.73	51.32	59.23	
CH_2—(C,=C)	21.43	28.71	34.83	39.72	46.97	52.24	60.11
CH_2—(=C,O)	19.51	29.18	36.21	41.36	48.3	53.29	
CH_2—(=C,SO_2)	20.34	28.51	34.95	40.1	47.17	52.49	
CH_2—(=C,SO)	18.42	26.62	29.05	38.72	45.92	51.28	
CH_2—(=C,S)	22.23	28.59	34.45	40.85	50.98	59.48	
			CH Groups				
CH—(2C,Cb)	20.43	27.88	33.07	36.63	40.73	42.9	44.7
CH—(2C,CO)	18.96	25.87	30.89	35.12	41.11	43.99	
CH—(2C,Ct)	16.7	23.48	28.67	32.57	38.09	41.44	46.55
CH—(2C,N)	19.67	26.37	31.81	35.16	40.18	42.7	
CH—(2C,O)	20.09	27.79	33.91	36.54	41.06	43.53	
CH—(2C,SO_2)	18.5	26.16	31.65	35.5	40.35	43.11	
CH—(2C,S)	20.3	27.25	32.57	36.38	41.44	44.24	
CH—(2C,=C)	17.41	24.74	30.72	34.28	39.6	42.65	47.22
CH—(3C)	19	25.12	30.01	33.7	38.97	42.07	46.76
CH—(C,2O)	22.02	23.06	27.67	31.77	35.41	38.97	
			C Groups				
C—(2C,2O)	19.25	19.25	23.02	25.53	27.63	28.46	
C—(3C,Cb)	19.72	28.42	33.86	36.75	38.47	37.51	31.94
C—(3C,CO)	9.71	18.33	23.86	27.17	30.43	31.69	
C—(3C,Ct)	0.33	7.33	14.36	19.97	25.2	26.71	
C—(3C,N)	18.42	25.95	30.56	33.07	35.58	35.58	
C—(3C,O)	18.12	25.91	30.35	32.23	34.32	34.49	
C—(3C,SO_2)	9.71	18.33	23.86	27.17	30.43	31.23	
C—(3C,SO)	12.81	19.17	20.26	27.63	31.56	33.32	
C—(3C,S)	19.13	26.25	31.18	34.11	36.5	33.91	
C—(3C,=C)	16.7	25.28	31.1	34.58	37.34	37.51	34.45
C—(4C)	18.29	25.66	30.81	33.99	36.71	36.67	33.99

(Continued)

TABLE 2-165 Benson* and CHETAH† Group Contributions for Ideal Gas Heat Capacity *(Continued)*

Table-specific nomenclature: Cb = carbon in benzene ring; Ct = carbon with a triple bond, (=C) = carbon with a double bond; Cp = carbon in fused ring; Naz = azide; Nim = imino.

Group	298 K	400 K	500 K	600 K	800 K	1000 K	1500 K
Aromatic (Cb and Cp Groups)							
Cb—(Cb)	13.94	17.66	20.47	22.06	24.11	24.91	25.32
Cb—(CO)	11.18	13.14	15.4	17.37	20.76	22.77	
Cb—(Ct)	15.03	16.62	18.33	19.76	22.1	23.48	24.07
Cb—(C)	11.18	13.14	15.4	17.37	20.76	22.77	25.03
Cb—(N)	16.53	21.81	24.86	26.45	27.33	27.46	
Cb—(O)	16.32	22.19	25.95	27.63	28.88	28.88	
Cb—(Si)	11.18	13.14	15.4	17.37	20.76	22.77	25.03
Cb—(SO$_2$)	11.18	13.14	15.4	17.37	20.76	22.77	
Cb—(SO)	11.18	13.14	15.4	17.37	20.76	22.77	
Cb—(S)	16.32	22.19	25.95	27.63	28.88	28.88	
Cb—(=C)	15.03	16.62	18.33	19.76	22.1	23.48	24.07
Cb—(=Nim)	16.53	21.81	24.86	26.45	27.33	27.46	
CbBr	32.65	36.42	39.35	41.44	43.11	43.95	
CbCl	30.98	35.16	38.51	40.6	42.7	43.53	
CbF	26.37	31.81	35.58	38.09	41.02	42.7	
CbH	13.56	18.59	22.85	26.37	31.56	35.2	40.73
CbI	33.49	37.25	40.18	41.44	43.11	43.95	
Cb(CHN$_2$)	47.3						
Cb(CN)	41.86	48.14	52.74	55.67	59.86	62.79	
Cb(N$_3$)	34.74						
Cb(NCO)	55.25	64.04	70.32	74.51	79.95	82.88	85.81
Cb(NCS)	32.23						
Cb(NO$_2$)	38.93	50.23	59.44	66.56	76.18	80.37	
Cb(SO$_2$OH)	65.42	79.49	84.51	97.61	109.25	113.31	
Cp—(2Cb,Cp)	12.56	15.49	17.58	19.25	21.77	23.02	
Cp—(3Cp)	8.37	12.14	14.65	16.74	19.67	21.35	
Cp—(Cb,2Cp)	12.56	15.49	17.58	19.25	21.77	23.02	
=C=,=C—,=CH—Groups							
=C—(2C)	17.16	19.3	20.89	22.02	24.28	25.45	26.62
=C—(CO,O)	23.4	29.3	31.31	32.44	33.57	34.03	
=C—(C,Cb)	18.42	22.48	24.82	25.87	27.21	27.71	28.13
=C—(C,CO)	22.94	29.22	31.02	31.98	33.53	34.32	
=C—(C,O)	17.16	19.3	20.89	22.02	24.28	25.45	
=C—(C,SO$_2$)	15.49	26.04	33.32	38.51	44.62	47.47	
=C—(C,S)	14.65	14.94	16.03	17.12	18.46	20.93	
=C—(C,=C)	18.42	22.48	24.82	25.87	27.21	27.71	28.13
=CC—(=C,O)	18.42	22.9	24.82	26.29	27.21	27.71	
=CH—(Cb)	18.67	24.24	28.25	31.06	34.95	37.63	41.77
=CH—(CO)	31.73	37.04	38.8	40.31	43.45	46.21	
=CH—(Ct)	18.67	24.24	28.25	31.06	34.95	37.63	41.77
=CH—(C)	17.41	21.05	24.32	27.21	32.02	35.37	40.27
=CH—(O)	17.41	21.05	24.32	27.21	32.02	35.37	40.27
=CH—(SO$_2$)	12.72	19.55	24.82	28.63	32.94	36.29	
=CH—(S)	17.41	21.05	24.32	27.21	32.02	35.37	
=CH—(=C)	18.67	24.24	28.25	31.06	34.95	37.63	41.77
=CH$_2$	21.35	26.62	31.44	35.58	42.15	47.17	55.21
=C=	16.32	18.42	19.67	20.93	22.19	23.02	23.86
Oxygen Groups							
O—(2C)	14.23	15.49	15.49	15.91	18.42	19.25	
O—(2O)	15.49	15.49	15.49	15.49	17.58	17.58	20.09
O—(2=C)	14.02	16.32	17.58	18.84	21.35	22.6	
O—(Cb,CO)	8.62	11.3	13.02	14.32	16.24	17.5	
O—(CO,O)	1.51	6.28	9.63	11.89	15.28	17.33	
O—(C,Cb)	2.6	3.01	4.94	7.45	11.89	14.99	
O—(C,CO)	11.64	15.86	18.33	19.8	20.55	21.05	
O—(C,O)	15.49	15.49	15.49	15.49	17.58	17.58	20.09
O—(C,=C)	12.72	13.9	14.65	15.49	17.54	18.96	
O—(=C,CO)	6.03	12.47	16.66	18.79	20.8	21.77	
OH—(Cb)	18	18.84	20.09	21.77	25.12	27.63	
OH—(CO)	15.95	20.85	24.28	26.54	30.01	32.44	37.34
OH—(C)	18.12	18.63	20.18	21.89	25.2	27.67	33.65
OH—(O)	21.64	24.24	26.29	27.88	29.93	31.44	34.2
O(CN)—(Cb)	34.74						
O(CN)—(C)	41.86						
O(CN)—(=C)	54.42						
O(NO$_2$)—(C)	39.93	48.3	55.5	65.3	68.61	72.75	
O(NO)—(C)	38.09	43.11	46.88	50.23	55.67	58.18	60.69
(CO)Cl—(C)	42.28	46.04	49.39	51.9	55.67	57.76	
(CO)H—(Cb)	33.53	44.2	48.77	59.48	68.56	74.01	
(CO)H—(CO)	28.13	32.78	37.25	41.4	47.84	50.73	

TABLE 2-165 Benson* and CHETAH† Group Contributions for Ideal Gas Heat Capacity *(Continued)*

Table-specific nomenclature: Cb = carbon in benzene ring; Ct = carbon with a triple bond, (=C) = carbon with a double bond; Cp = carbon in fused ring; Naz = azide; Nim = imino.

Group	298 K	400 K	500 K	600 K	800 K	1000 K	1500 K
			Oxygen Groups				
(CO)H—(C)	29.43	32.94	36.92	40.52	46.71	51.07	
(CO)H—(N)	29.43	32.94	36.92	40.52	46.71	51.07	
(CO)H—(O)	29.43	32.94	36.92	40.52	46.71	51.07	
(CO)H—(=C)	24.32	30.22	39.77	48.77	63.12	74.68	
CO—(Cb)(O)	9.12	11.51	16.65	21.05	26.32	29.54	
			Halide Groups				
CBr—(3C)	39.35	47.72	52.74	55.25	56.93	56.09	
CBr_3—(C)	72.12	78.65	82.92	85.64	88.66	89.66	
CCl—(3C)	36.96	43.87	47.72	49.52	52.07	53.12	
CCl_2—(2C)	51.07	62.29	66.76	68.98	70.99	71.24	
CCl_3—(C)	68.23	75.35	79.95	82.88	86.23	87.9	
$CClF_2$—(C)	57.35	67.39	73.25	77.86	82.88	85.39	
CF—(3C)	28.46	37.09	42.7	46.71	52.03	53.24	
CF_2—(2C)	39.01	46.97	53.24	57.85	63.46	65.84	
CF_3—(Cb)	52.32	64.04	72	77.44	84.14	87.9	
CF_3—(C)	53.16	62.79	68.65	74.93	80.79	83.72	
CF_3—(S)	41.36	54.46	62.08	68.52	76.06	79.99	
CH_2Br—(Cb)	30.51	46.46	52.2	57.3	65.26	69.95	
CH_2Br—(C)	38.09	46.04	52.74	57.35	64.88	70.32	
CH_2Br—(=C)	40.6	47.72	54.42	59.86	67.81	73.67	
CH_2Cl—(C)	37.25	44.79	51.49	56.09	64.04	69.9	
CH_2F—(C)	33.91	41.86	50.23	54.42	63.62	69.49	
CH_2I—(Cb)	33.91	45.17	53.7	59.9	68.15	73.8	
CH_2I—(C)	38.51	46.04	54	58.18	66.14	72	
CH_2I—(O)	34.41	43.91	51.19	56.72	64.25	69.36	
CHBr—(2C)	37.38	44.62	50.06	53.75	58.81	61.62	
CHBrCl—(C)	51.9	58.6	63.3	68.23	74.93	79.53	
CHCl—(2C)	35.45	42.7	48.89	53.41	59.82	64.38	
CHCl—(C,O)	37.67	41.44	43.95	46.88			
$CHCl_2$—(C)	50.65	58.6	64.46	69.07	74.93	78.28	
CHF—(2C)	30.56	37.84	43.83	48.39	54.83	58.64	
CHF_2—(C)	41.44	50.23	57.35	63.21	69.9	74.51	
CHI—(2C)	38.64	45.67	50.9	54.42	59.31	61.95	
CHI_2—(C)	56.93	63.42	69.61	74.17	79.7	81.58	
CI—(3C)	41.15	49.18	54.08	56.3	57.72	56.93	
$=CBr_2$	51.49	55.25	58.18	59.86	62.37	63.62	
=CBrCl	50.65	53.16	56.51	59.02	61.53	62.79	
=CBrF	45.21	50.23	53.58	56.51	59.86	61.53	
$=CCl_2$	47.72	52.32	55.67	58.18	61.11	62.79	
=CClF	43.11	48.97	52.74	55.67	59.44	61.53	
$=CF_2$	40.6	46.04	50.23	53.16	57.76	60.69	
=CHBr	33.91	39.77	44.37	47.72	51.9	55.25	
=CHCl	33.07	38.51	43.11	46.88	51.49	54.83	
=CHF	28.46	35.16	39.77	43.95	49.39	53.16	
=CHI	36.84	41.86	45.63	48.56	52.74	55.67	
			Nitrogen Groups				
$CH_2(N_3)$—(C)	64.46						
$=CH(N_3)$	54.42						
N—(2C,Cb)	2.6	8.46	13.69	17.29	21.89	23.4	
N—(2C,CO)	13.02	19.17	23.52	26.16	28.42	28.76	
N—(2C,SO_2)	25.2	26.58	31.56	34.45	37.8	38.47	
N—(2C,SO)	17.58	24.61	25.62	27.33	28.59	34.91	
N—(2C,S)	15.99	21.64	25.99	29.05	30.93	38.68	
N—(3C)	14.57	19.09	22.73	24.99	27.46	27.92	27.21
N—(Cb,2CO)	4.1	12.81	17.71	20.3	22.1	22.14	
N—(C,2CO)	4.48	12.99	18.04	20.93	22.94	27.08	
Nb pyrid—N	10.88	13.48	15.95	17.66	20.05	21.43	
NF_2—(C)	26.5	34.58	40.9	45.63	50.9	53.54	
NH—(2Cb)	9.04	13.06	17.29	21.35	28.3	32.98	
NH—(2CO)	15.03	23.19	28.05	30.93	33.28	34.28	
NH—(2C)	17.58	21.81	25.66	28.59	33.07	36.21	39.97
NH—(Cb,CO)	2.39	6.32	9.96	13.94	16.91	18.21	
NH—(C,Cb)	15.99	20.47	23.9	26.29	30.1	32.36	
NH—(C,CO)	2.76	6.49	10.3	14.57	17.75	18.96	
NH—(C,N)	20.09	24.28	27.21	29.3	32.65	34.74	37.67
NH_2—(Cb)	23.94	27.25	30.64	33.78	39.39	43.83	51.4
NH_2—(CO)	17.04	24.03	29.85	34.7	41.69	46.97	
NH_2—(C)	23.94	27.25	30.64	33.78	39.39	43.83	51.4
NH_2—N	25.53	30.98	35.16	38.93	43.95	48.14	55.25
=Naz—(C)	11.3	17.16	20.59	22.35	23.82	23.9	
=Naz—(N)	8.87	17.5	23.06	28.34	28.71	29.51	

(Continued)

TABLE 2-165 Benson* and CHETAH† Group Contributions for Ideal Gas Heat Capacity *(Continued)*

Table-specific nomenclature: Cb = carbon in benzene ring; Ct = carbon with a triple bond, (═C) = carbon with a double bond; Cp = carbon in fused ring; Naz = azide; Nim = imino.

Group	298 K	400 K	500 K	600 K	800 K	1000 K	1500 K
Nitrogen Groups							
═NazH	18.33	20.47	22.77	24.86	28.34	31.06	35.33
═Nim—(Cb)	12.56						
═Nim—(C)	10.38	13.98	16.53	17.96	19.21	19.25	
═NimH	12.35	19.17	27	32.27	38.22	41.52	
Sulfur Groups							
S—(2Cb)	8.37	8.41	9.38	11.47	15.91	19.72	
S—(2C)	20.89	20.76	21.01	21.22	22.65	23.98	
S—(2S)	19.67	20.93	21.35	21.77	22.19	22.6	
S—(2═C)	20.05	23.36	23.15	26.33	33.24	40.73	
S—(Cb,S)	12.1	14.19	15.57	17.37	20.01	21.35	
S—(C,Cb)	12.64	14.19	15.53	16.91	19.34	20.93	
S—(C,S)	21.89	22.69	23.06	23.06	22.52	21.43	
S—(C,═C)	17.66	21.26	23.27	24.15	24.57	24.57	
SH—(Cb)	21.43	22.02	23.32	25.24	29.26	32.82	
SH—(CO)	31.94	33.86	33.99	34.2	35.58	34.49	
SH—(C)	24.53	25.95	27.25	28.38	30.56	32.27	
SO—(2Cb)	23.94	38.05	40.6	47.93	47.97	47.09	
SO—(2C)	37.17	41.98	43.95	45.17	45.96	46.76	
SO_2—(2Cb)	34.99	46.17	56.72	62.54	66.39	66.81	
SO_2—(2C)	48.22	50.1	55.88	59.77	64.38	66.47	
SO_2—(2═C)	48.22	50.1	55.88	59.77	64.38	66.47	
SO_2—(Cb,SO_2)	41.06	48.14	56.59	61.66	65.76	67.1	
SO_2—(Cb,═C)	41.4	48.14	55.88	61.16	65.8	66.64	
SO_2—(C,Cb)	41.61	48.14	56.3	60.74	65.38	66.64	
S(CN)—(Cb)	39.77						
S(CN)—(C)	46.88						
S(CN)—(═C)	59.44						
Boron and Silicon Groups							
Si—(4C)	113.23	134.95	154.5	171.2	198.62	219.72	252.91
SiH_3—(C)	−39.64						
Monovalent Ligands							
CH_2(CN)—(C)	47.72	56.93	64.04	70.74	80.79	85.81	
CH_2(NCS)—(C)	61.95						
$CH_2(NO_2)$—(C)	52.7	66.22	77.52	86.48	99.58	108.41	
CH(CN)—(2C)	45.21	54	60.69	66.14	72	79.11	
$CH(NO_2)$—(2C)	50.19	63.67	74.17	82.08	92.84	99.2	
$CH(NO_2)_2$—(C)	80.79	101.3	117.2	129.76	146.09	156.13	
C(CN)—(3C)	36.21	46.71	53.96	58.81	64.92	67.77	
$C(CN)_2$—(2C)	61.62	74.47	83.72	90.46	99.54	104.48	
$C(NO_2)$—(3C)	41.4	55.84	66.39	73.75	82.92	87.32	
═$CH(CHN_2)$	72.42						
═CH(CN)	43.11	50.23	56.09	61.11	68.65	73.67	
═CH(NCS)	51.90						
═$CH(NO_2)$	51.49	63.21	72.83	80.37	90.41	97.11	105.9
═$C(CN)_2$	56.93	69.28	78.19	84.76	93.51	98.74	
3,4 Member Ring Corrections							
cyclobutane ring	−19.3	−16.28	−13.14	−11.05	−7.87	−5.78	−2.8
cyclobutene ring	−10.59	−9.17	−7.91	−7.03	−6.2	−5.57	−5.11
cyclopropane ring	−12.77	−10.59	−8.79	−7.95	−7.41	−6.78	−6.36
ethylene oxide ring	−8.37	−11.72	−12.56	−10.88	−9.63	−8.63	
ethylene sulfide ring	−11.93	−10.84	−11.13	−12.64	18.09	24.35	
thietane ring	−19.21	−17.5	−16.37	−16.37	−19.25	−23.86	
trimethylene oxide ring	−19.25	−20.93	−17.58	−14.56	−10.88	0.84	
5,6 Member Ring Corrections							
1,4 dioxane ring	−19.21	−20.8	−15.91	−10.97	−6.4	−1.8	
cyclohexane ring	−24.28	−17.16	−12.14	−5.44	4.6	9.21	13.81
cyclohexene ring	−17.92	−12.72	−8.29	−5.99	−1.21	0.33	3.39
cyclopentadiene ring	−14.44	−11.85	−8.96	−6.91	−5.36	−4.35	
cyclopentane ring	−27.21	−23.02	−18.84	−15.91	−11.72	−8.08	−1.55
cylopentene ring	−25.03	−22.39	−20.47	−17.33	−12.26	−9.46	−4.52
furan ring	−20.51	−18	−15.07	−12.56	−10.88	−10.05	
piperidine ring	−24.7	−19.67	−12.14	−3.77	9.21	17.58	
pyrrolidine ring	−25.83	−23.36	−20.09	−16.74	−12.01	−9.08	
tetrahydrofuran ring	−25.12	−24.28	−20.09	−15.91	−11.3	−7.53	
thiacyclohexane ring	−26.04	−17.83	−9.38	−2.89	3.6	5.4	
thiolane ring	−20.51	−19.55	−15.4	−15.32	−18.46	−23.32	
thiophene ring	−20.51	−19.55	−15.4	−15.32	−18.46	−23.32	

TABLE 2-165 Benson* and CHETAH† Group Contributions for Ideal Gas Heat Capacity *(Continued)*

Table-specific nomenclature: Cb = carbon in benzene ring; Ct = carbon with a triple bond, (=C) = carbon with a double bond; Cp = carbon in fused ring; Naz = azide; Nim = imino.

Group	298 K	400 K	500 K	600 K	800 K	1000 K	1500 K
7 and 8 Member Ring Corrections							
cycloheptane ring	−38.01						
cyclooctane ring	−44.16						
Gauche and 1,5 Repulsion Corrections							
but-2-ene structure C—C═C—C	−5.61	−4.56	−3.39	−2.55	−1.63	−1.09	
but-3-ene structure C—C—C═C	−5.61	−4.56	−3.39	−2.55	−1.63	−1.09	
cis- between 2 t-butyl groups	−5.61	−4.56	−3.39	−2.55	−1.63	−1.09	
cis- involving 1 t-butyl group	−5.61	−4.56	−3.39	−2.55	−1.63	−1.09	
cis-(not with t-butyl group)	−5.61	−4.56	−3.39	−2.55	−1.63	−1.09	
ortho- between Cl atoms	−2.09	5.02	2.09	−2.51	−1.26		
ortho- between F atoms		−0.84	−0.42	1.26	2.93		
other *ortho*- (nonpolar-nonpolar)	4.69	5.65	5.44	4.9	3.68	2.76	−0.21

*Benson, S. W., et al., *Chem. Rev.*, **69** (1969): 279.
†CHETAH Version 8.0: *The ASTM Computer Program for Chemical Thermodynamic and Energy Release Evaluation (NIST Special Database 16).*

TABLE 2-166 Liquid Heat Capacity Group Parameters for Ruzicka-Domalski Method*

Table-specific nomenclature: Ct refers to a carbon atom with a triple bond; Cb refers to a carbon atom in benzene ring; ═C refers to a carbon atom with a double bond; Cp refers to a carbon atom in a fused benzene ring; ═C═ refers to an allenic carbon atom.

Group Definition	*a*	*b*	*d*	*T* range (K)
Hydrocarbon Groups				
C—(3H,C)	3.8452	−0.33997	0.19489	80–490
C—(2H,2C)	2.7972	−0.054967	0.10679	80–490
C—(H,3C)	−0.42867	0.93805	0.0029498	85–385
C—(4C)	−2.9353	1.4255	−0.085271	145–395
═C—(2H)	4.1763	−0.47392	0.099928	90–355
═C—(H,C)	4.0749	−1.0735	0.21413	90–355
═C—(2C)	1.9570	−0.31938	0.11911	140–315
═C—(H,═C)	3.6968	−1.6037	0.55022	130–305
═C—(C,═C)	1.0679	−0.50952	0.33607	130–305
C—(3H,═C)	3.8452	−0.33997	0.19489	80–490
C—(2H,C,═C)	2.0268	−0.20137	0.11624	90–355
C—(H,2C,═C)	−0.87558	0.82109	0.18415	110–300
C—(3C,═C)	−4.8006	2.6004	−0.040688	165–295
C—(2H,2═C)	1.4973	−0.46017	0.52861	130–300
Ct—(H)	9.1633	−4.6695	1.1400	150–275
Ct—(C)	1.4822	1.0770	−0.19489	150–285
═C═	3.0880	−0.62917	0.25779	140–315
Ct—(Cb)	12.377	−7.5742	1.3760	230–550
Cb—(H)	2.2609	−0.2500	0.12592	180–670
Cb—(C)	1.5070	−0.13366	0.011799	180–670
Cb—(═C)	−5.7020	5.8271	−1.2013	230–550
Cb—(Cb)	5.8685	−0.86054	−0.063611	295–670
C—(2H,C,Ct)	2.0268	−0.20137	0.11624	90–355
C—(3H,Ct)	3.8452	−0.33997	0.19489	80–490
C—(3H,Cb)	3.8452	−0.33997	0.19489	80–490
C—(2H,C,Cb)	1.4142	0.56919	0.0053465	180–470
C—(H,2C,Cb)	−0.10495	1.0141	−0.071918	180–670
C—(3C,Cb)	1.2367	−1.3997	0.41385	220–295
C—(2H,2Cb)	−18.583	11.344	−1.4108	300–420
C—(H,3Cb)	−46.611	24.987	−3.0249	375–595
═C—(H,Cb)	3.6968	−1.6037	0.55022	130–305
═C—(C,Cb)	1.0679	−0.50952	0.33607	130–305
Cp—(Cp,2Cb)	−3.5572	2.8308	−0.39125	250–510
Cp—(2Cp,Cb)	−11.635	6.4068	−0.78182	370–510
Cp—(3Cp)	26.164	−11.353	1.2756	385–480
Halogen Groups				
C—(C,3F)	15.423	−9.2464	2.8647	125–345
C—(2C,2F)	−8.9527	10.550	−1.9986	125–345
C—(C,3Cl)	8.5430	2.6966	−0.42564	245–310
C—(H,C,2Cl)	10.880	−0.35391	0.08488	180–355
C—(2H,C,Cl)	9.6663	−1.8601	0.41360	140–360
C—(2H,═C,Cl)	9.6663	−1.8601	0.41360	140–360
C—(H,2C,Cl)	−2.0600	5.3281	−0.82721	275–360
C—(2H,C,Br)	6.3944	−0.10298	0.19403	168–360
C—(H,2C,Br)	10.784	−2.4754	0.33288	190–420
C—(2H,C,I)	0.037620	5.6204	−0.92054	245–340
C—(C,2Cl,F)	13.532	−3.2794	0.80145	240–420
C—(C,Cl,2F)	7.2295	0.41759	0.15892	180–420
C—(C,Br,2F)	8.7956	−0.19165	0.24596	165–415
═C—(H,Cl)	7.1564	−0.84442	0.27199	120–300
═C—(2F)	7.6646	−2.0750	0.82003	120–240
═C—(2Cl)	9.3249	−1.2478	0.44241	155–300
═C—(Cl,F)	7.8204	−0.69005	0.19165	120–240
Cb—(F)	3.0794	0.46959	−0.0055745	210–365
Cb—(Cl)	4.5479	0.22250	−0.0097873	230–460
Cb—(Br)	2.2857	2.2573	−0.40942	245–370
Cb—(I)	2.9033	2.9763	−0.62960	250–320
C—(Cb,3F)	7.4477	−0.92230	0.39346	210–365
C—(2H,Cb,Cl)	16.752	−6.7938	1.2520	245–345
Nitrogen Groups				
C—(3H,N)	3.8452	−0.33997	0.19489	80–490
C-(2H,C,N)	2.4555	1.0431	−0.24054	190–375
C—(2H,Cb,N)	2.4555	1.0431	−0.24054	190–375
C—(H,2C,N)	2.6322	−2.0135	0.45109	240–370
C—(3C,N)	1.9630	−1.7235	0.31086	255–375
N—(2H,C)	8.2758	−0.18365	0.035272	185–455
N—(2H,Cb)	8.2758	−0.18365	0.035272	185–455
N—(H,2C)	−0.10987	0.73024	0.89325	170–400
N—(3C)	4.5942	−2.2134	0.55316	160–360
N—(H,C,Cb)	0.49631	3.4617	−0.57161	240–380
N—(2C,Cb)	−0.23640	16.260	−2.5258	285–390
N—(C,2Cb)	4.5942	−2.2134	0.55316	160–360
Cb—(N)	−0.78169	1.5059	−0.25287	240–455
N—(2H,N)	6.8050	−0.72563	0.15634	215–465
N—(H,C,N)	1.1411	3.5981	−0.69350	205–300
N—(2C,N)	−1.0570	4.0038	−0.71494	205–300
N—(H,Cb,N)	−0.74531	3.6258	−0.53306	295–385
C—(2H,C,CN)	11.976	−2.4886	0.52358	185–345
C—(3C,CN)	2.5774	3.5218	−0.58466	295–345
═C—(H,CN)	9.0789	−0.86929	0.32986	195–345
Cb—(CN)	1.9389	3.0269	−0.47276	265–480
C—(2H,C,NO_2)	18.520	−5.4568	1.05080	190–300
O—(C,NO_2)	−2.0181	10.505	−1.83980	180–350
Cb—(NO_2)	15.277	−4.4049	0.71161	280–415
N—(H,2Cb) (pyrrole)	−7.3662	6.3622	−0.68137	255–450
Nb—(2Cb)	0.84237	1.25560	−0.20336	210–395
Oxygen Groups				
O—(H,C)	12.952	−10.145	2.6261	155–505
O—(H,C) (diol)	5.2302	−1.5124	0.54075	195–475
O—(H,Cb) (diol)	5.2302	−1.5124	0.54075	195–475
O—(H,Cb)	−7.9768	8.10450	−0.87263	285–400
C-(3H,O)	3.8452	−0.33997	0.19489	80–490
C-(2H,C,O)	1.4596	1.4657	−0.27140	135–505
C-(2H,Cb,O)	−35.127	28.409	−4.9593	260–460
C-(2H,═C,O)	−35.127	28.409	−4.9593	260–460
C—(H,2C,O) (alcohol)	2.2209	−1.4350	0.69508	185–460
C—(H,2C,O) (ether, ester)	0.98790	0.39403	−0.016124	130–170
C—(3C,O) (alcohol)	−44.690	31.769	−4.8791	200–355
C—(3C,O) (ether, ester)	−3.3182	2.6317	−0.44354	170–310
O—(2C)	5.0312	−1.5718	0.37860	130–350
O—(C,Cb)	−22.5240	13.1150	−1.44210	320–350
O—(2Cb)	−4.5788	0.94150	0.31655	300–535
C—(2H,2O)	1.0852	1.5402	−0.31693	170–310

(Continued)

TABLE 2-166 Liquid Heat Capacity Group Parameters for Ruzicka-Domalski Method* *(Continued)*

Table-specific nomenclature: Ct refers to a carbon atom with a triple bond; Cb refers to a carbon atom in benzene ring; =C refers to a carbon atom with a double bond; Cp refers to a carbon atom in a fused benzene ring; =C= refers to an allenic carbon atom.

Group Definition	a	b	d	T range (K)
Oxygen Groups				
C—(2C,2O)	−12.955	9.10270	−1.53670	275–335
Cb—(O)	−1.0686	3.52210	−0.79259	285–530
C—(3H,CO)	3.8452	−0.33997	0.19489	80–490
C—(2H,C,CO)	6.6782	−2.44730	0.47121	180–465
C—(H,2C,CO)	3.92380	−2.12100	0.49646	185–375
C—(3C,CO)	−2.2681	1.75580	−0.25674	225–360
CO—(H,C)	−3.82680	7.67190	−1.27110	180–430
CO—(H,=C)	−8.00240	3.63790	−0.15377	220–430
CO—(H,Cb)	−8.00240	3.63790	−0.15377	220–430
CO—(2C)	5.4375	0.72091	−0.18312	185–380
CO—(C,=C)	41.507	−32.632	6.0326	275–355
CO—(C,Cb)	−47.21100	24.36800	−2.82740	300–465
CO—(H,O)	13.11800	16.12000	−5.12730	280–340
CO—(C,O)	29.24600	3.42610	−2.89620	180–445
CO—(=C,O)	41.61500	−12.78900	0.53631	195–350
CO—(O,CO)	23.99000	6.25730	−3.24270	320–345
O—(C,CO)	−21.43400	−4.01640	3.05310	175–440
O—(H,CO)	−27.58700	−0.16485	2.74830	230–500
=C—(H,CO)	−9.01080	15.14800	−3.04360	195–355
=C—(C,CO)	−12.81800	15.99700	−3.05670	195–430
Cb—(CO)	12.15100	−1.67050	−0.12758	175–500
CO—(Cb,O)	16.58600	5.44910	−2.68490	175–500
Sulfur Groups				
C—(3H,S)	3.84520	−0.33997	0.19489	80–490
C—(2H,C,S)	1.54560	0.88228	−0.08349	130–390
C—(H,2C,S)	−1.64300	2.30700	−0.31234	150–390
C—(3C,S)	−5.38250	4.50230	−0.72356	190–365
Cb—(S)	−4.45070	4.43240	−0.75674	260–375
S—(H,C)	10.99400	−3.21130	0.47368	130–380
S—(H,Cb)	10.99400	−3.21130	0.47368	130–380
S—(2C)	9.23060	−3.00870	0.45625	165–390
S—(2Cb)	9.23060	−3.00870	0.45625	165–390
S—(C,S)	6.65900	−1.35570	0.17938	170–350
S—(Cb,S)	9.23060	−3.00870	0.45625	165–390
S—(2Cb) (thiophene)	3.84610	0.36718	−0.06131	205–345

Group Definition	a	b	d	T range (K)
Ring Strain Contributions				
Hydrocarbons (ring strain)				
cyclopropane	4.4297	−4.3392	1.0222	155–240
cyclobutane	1.2313	−2.8988	0.75099	140–300
cyclopentane (unsub)	−0.33642	−2.8663	0.70123	180–300
cyclopentane (sub)	0.21983	−1.5118	0.28172	135–365
cyclohexane	−2.0097	−0.72656	0.14758	145–485
cycloheptane	−11.460	4.9507	−0.74754	270–300
cyclooctane	−4.1696	0.52991	−0.018423	295–320
spiropentane	5.9700	−3.7965	0.74612	175–310
cyclopentene	0.21433	−2.5214	0.63136	140–300
cyclohexene	−1.2086	−1.5041	0.42863	160–320
cycloheptene	−5.6817	1.5073	−0.19810	220–300
cyclooctene	−14.885	7.4878	−1.0879	260–330
cyclohexadiene	−8.9683	6.4959	−1.5272	170–300
cyclooctadiene	−7.2890	3.1119	−0.43040	205–320
cycloheptatriene	−8.7885	8.2530	−2.4573	200–310
cyclooctatetraene	−12.914	13.583	−4.0230	275–330
indan	−6.1414	3.5709	−0.48620	170–395
1H-indene	−3.6501	2.4707	−0.60531	280–375
tetrahydronaphthalene	−6.3861	2.6257	−0.19578	250–320
decahydronaphthalene	−6.8984	0.66846	−0.070012	235–485
hexahydroindan	−3.9271	−0.29239	0.048561	210–425
dodecahydrofluorene	−19.687	8.8265	−1.4031	315–485
tetradecahydrophenanthrene	−0.67632	−1.4753	−0.13087	315–485
hexadecahydropyrene	61.213	−30.927	3.2269	310–485
Nitrogen compounds				
ethyleneimine	15.281	−2.3360	−0.13720	195–330
pyrrolidine	12.703	1.3109	−1.18130	170–400
piperidine	25.681	−7.0966	0.14304	265–370
Oxygen compounds				
ethylene oxide	6.8459	−5.8759	1.2408	135–325
trimethylene oxide	−7.0148	7.3764	−2.1901	185–300
1,3-dioxolane	−2.3985	−0.48585	0.10253	175–300
furan	9.6704	−2.8138	0.11376	190–305
tetrahydrofuran	3.2842	−5.8260	1.2681	160–320
tetrahydropyran	−13.017	3.7416	−0.15622	295–325
Sulfur compounds				
thiacyclobutane	−0.73127	−1.3426	0.40114	200–320
thiacyclopentane	−3.2899	0.38399	0.089358	170–390
thiacyclohexane	−12.766	5.2886	−0.59558	295–340

*Ruzicka, V., and E. S. Domalski, *J. Phys. Chem. Ref. Data*, **22** (1993): 597, 619.

Solids Solid heat capacity increases with increasing temperature and is proportional to T^3 near absolute zero. The heat capacity at a solid-solid phase transition becomes large, and there can be a substantial difference in the heat capacity of the two equilibrium solid phases that exist on either side of the transition temperature. The heat capacity generally rises steeply with increasing temperature near the triple point.

For a quick estimation of solid heat capacity specifically at 298.15 K, the very simple modification of Kopp's rule [Kopp, H., *Ann. Chem. Pharm. (Liebig)*, **126** (1863): 362] by Hurst and Harrison [Hurst, J. E., and B. K. Harrison, *Chem. Eng. Comm.*, **112** (1992): 21] can be used. At other temperatures and to obtain the temperature dependence of the solid heat capacity, the method given below by Goodman et al. should be used.

Recommended Method 1 Goodman method.

Reference: Goodman, B. T., W. V. Wilding, J. L. Oscarson, and R. L. Rowley, *J. Chem. Eng. Data*, **49** (2004): 24.

Classification: Group contributions.

Expected uncertainty: 10 percent.

Applicability: Organic compounds for which group values are available.

Input data: Molecular structure and Table 2-167 group values.

Description:

$$\frac{C_p}{\text{J/(mol·K)}} = \frac{A}{1000}\left(\frac{T}{\text{K}}\right)^{0.79267} \quad (2\text{-}56)$$

$$A = \exp\left(6.7796 + \sum_{i=1}^{N} n_i a_i + \sum_{i=1}^{N} n_i^2 \beta_i\right) \quad (2\text{-}57)$$

where n_i = number of occurrences of group i

a_i = individual group i contribution

β_i = nonlinear correction terms for chain and aromatic carbons

Example Estimate the solid heat capacity for *p*-cresol at 307.93 K.

Structure:

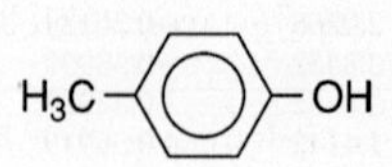

Group contributions:

Group	n_i	a_i	β_i
—CH_3	1	0.20184	0
Ar—CH=	4	0.082478	−0.00033
Ar >C=	2	0.012958	0
—OH	1	0.10341	0

From Eq. (2-57):

$$A = \exp\,[6.7796 + 0.20184 + (4)\,(0.082478) + (2)(0.012958) + 0.10341 + (4)^2(-0.00033)] = 1694.9$$

From Eq. (2-56):

$$C_p = \frac{1694.9}{1000}(307.93)^{0.79267}\frac{\text{J}}{\text{mol·K}} = 159.1\frac{\text{J}}{\text{mol·K}}$$

This value is 2.5 percent higher than the DIPPR 801 recommended value of 155.2 J/(mol · K).

Recommended Method 2 Modified Kopp's rule.

Reference: Kopp, H., *Ann. Chem. Pharm. (Liebig)*, **126** (1863): 362; Hurst, J. E., and B. K. Harrison, *Chem. Eng. Comm.*, **112** (1992): 21.

TABLE 2-167 Group Values and Nonlinear Correction Terms for Estimation of Solid Heat Capacity with the Goodman et al.* Method

Group	Description	a_i	Group	Description	a_i
—CH₃	methyl	0.20184	—CO₃—	carbonate	0.2517
>CH₂	methylene	0.11644	—NH₂	primary amine	0.056138
>CH—	secondary C	0.030492	>NH	secondary amine	−0.00717
>C<	tertiary C	−0.04064	>N—	tertiary amine	−0.01661
CH₂═	terminal alkene	0.18511	═NH	double-bond NH	0.17689
—CH═	alkene	0.11224	#N	nitrile	0.015355
>C═	subst. alkene	0.028794	—N═N—	diazide	0.3687
═C═	allene	0.053464	—NO₂	nitro	0.23327
#CH	terminal alkyne	−0.02914	—N═C═O	isocyanate	0.2698
#C—	alkyne	0.13298	—SH	thiol/mercaptan	0.21123
Ar —CH═	arom. C	0.082478	—S—	sulfide	0.14232
Ar >C═	subst. arom. C	0.012958	—SS—	disulfide	0.31457
Ar —O—	furan O	0.066027	═S	sulfur double bond	0.13753
Ar —N═	pyridine N	0.056641	>S═O	sulfoxide	0.040002
Ar >N—	subst. pyrrole N	0.008938	—F	fluoride	0.15511
Ar —NH—	pyrrole N	−0.05246	—Cl	chloride	0.16995
Ar —S—	thiophene S	0.090926	—Br	bromide	0.19112
—O—	ether	0.064068	—I	iodide	0.11318
—OH	alcohol	0.10341	>Si<	silane	0.12213
—COH	aldehyde	0.15699	>Si(O—)—	linear siloxane	0.10125
>C═O	ketone	0.12939	cyc >Si(O—)—	cyclic siloxane	0.063438
—COO—	ester	0.13686	P(═O)(O—)₃	phosphate	0.15016
—COOH	acid	0.21019	>P—	phosphine	0.069602
—COOCO—	anhydride	0.33091	>P(═O)—	phosphine oxide	0.21875

Nonlinear Terms

Groups	β_i	Usage
>CH₂	−0.00188	Methylene
Ar═CH—	−0.00033	Aromatic carbon

*Goodman, B. T., W. V. Wilding, J. L. Oscarson, and R. L. Rowley, *J. Chem. Eng. Data*, **49** (2004): 24.

Classification: Group contributions.
Expected uncertainty: 10 percent.
Applicability: At 298.15 K; organic compounds that are solids at 298.15 K.
Input data: Compound chemical formula and element contributions of Table 2-168.
Description:

$$\frac{C_P}{\text{J/(mol}\cdot\text{K)}} = \sum_{E=1}^{N} n_E \Delta_E \tag{2-58}$$

where N = number of different elements in compound
n_E = number of occurrences of element E in compound
Δ_E = contribution of element E from Table 2-168

Example Estimate the solid heat capacity at 298.15 K for dibenzothiophene.
Structure: $C_{12}H_8S$.
Group values from Table 2-168:

$$\Delta_C = 10.89 \qquad \Delta_H = 7.56 \qquad \Delta_S = 12.36$$

Calculation using Eq. (2-54):

$$C_P = (12)(10.89) + (8)(7.56) + (1)(12.36) = 203.52 \text{ J/(mol}\cdot\text{K)}$$

This value is 2.5 percent higher than the DIPPR 801 recommended value of 198.45 J/(mol · K).

TABLE 2-168 Element Contributions to Solid Heat Capacity for the Modified Kopp's Rule*†

Element	Δ_E	Element	Δ_E	Element	Δ_E
C	10.89	Ba	32.37	Mo	29.44
H	7.56	Be	12.47	Na	26.19
O	13.42	Ca	28.25	Ni	25.46
N	18.74	Co	25.71	Pb	31.60
S	12.36	Cu	26.92	Si	17.00
F	26.16	Fe	29.08	Sr	28.41
Cl	24.69	Hg	27.87	Ti	27.24
Br	25.36	K	28.78	V	29.36
I	25.29	Li	23.25	W	30.87
Al	18.07	Mg	22.69	Zr	26.82
B	10.10	Mn	28.06	All others	26.63

*Kopp, H., *Ann. Chem. Pharm. (Liebig)*, **126** (1863): 362.
†Hurst, J. E., and B. K. Harrison, *Chem. Eng. Comm.*, **112** (1992): 21.

Mixtures The molar heat capacity of liquid and vapor mixtures can be estimated as a mole fraction average of the pure-component values

$$C_{P,m} = \sum_{i=1}^{C} x_i C_{P,i} \tag{2-59}$$

This neglects the excess heat capacity, which, if available, can be added to the mole fraction average to improve the estimated value.

DENSITY

Density is defined as the mass of a substance per unit volume. Density is given in kg/m³ in SI units, but lb_m/ft^3 and g/cm³ are common AES and cgs units, respectively. Other commonly used forms of density include *molar density* (density divided by molecular weight) in kmol/m³, *relative density* (density relative to water at 15°C), and the older term *specific gravity* (density relative to water at 60°F). Often the inverse of density, *specific volume*, and the inverse of molar density, *molar volume*, are correlated and used to convey equivalent information.

Gases Gases/vapors are compressible and their densities are strong functions of both temperature and pressure. Equations of state (EoS) are commonly used to correlate molar densities or molar volumes. The most accurate EoS are those developed for specific fluids with parameters regressed from all available data for that fluid. Super EoS are available for some of the most industrially important gases and may contain 50 or more constants specific to that chemical. Different predictive methods may be used for gas densities depending upon the conditions:

1. At *very low densities* (high temperatures, generally above the critical, and very low pressures, generally below a few bar), the ideal gas EoS

$$Z \equiv \frac{PV}{RT} = 1 \tag{2-60}$$

may be applied.

2. At *moderate densities* (below 40 percent of the critical density), the virial equation truncated after the second virial coefficient

$$Z = 1 + \frac{B(T)}{V} \tag{2-61}$$

may be used. Second virial coefficients $B(T)$ are available in the DIPPR 801 database for many chemicals and can be estimated using the Tsonopoulos method.

Recommended Method Tsonopoulos method.

Reference: Tsonopoulos, C., *AIChE J.*, **20** (1974): 263; **21** (1975): 827; **24** (1978): 1112.

Classification: Corresponding states.

Expected uncertainty: 8 percent for $B(T)$.

Applicability: Nonpolar organic compounds and some classes of polar compounds.

Input data: Class of fluid, ω, P_c, T_c, and μ.

Description:

$$\frac{BP_c}{RT_c} = B^{(0)} + \omega B^{(1)} + B^{(2)} \tag{2-62}$$

where

$$B^{(0)} = 0.1445 - \frac{0.330}{T_r} - \frac{0.1385}{T_r^2} - \frac{0.0121}{T_r^3} - \frac{0.000607}{T_r^8} \tag{2-63}$$

$$B^{(1)} = 0.0637 + \frac{0.331}{T_r^2} - \frac{0.423}{T_r^3} - \frac{0.008}{T_r^8} \tag{2-64}$$

$$B^{(2)} = \frac{a}{T_r^6} - \frac{b}{T_r^8} \tag{2-65}$$

$$\mu_r = \left(\frac{\mu}{\text{D}}\right)^2 \left(\frac{P_c}{\text{bar}}\right)\left(\frac{T_c}{\text{K}}\right)^{-2} \tag{2-66}$$

and μ = dipole moment. The values of a and b used in Eq. (2-65) depend upon the class of fluid, as given in the table below:

Class	a	b
Nonpolar fluids	0	0
Ketones, aldehydes, nitriles, ethers, esters, NH_3, H_2S, HCN	$-21.4\mu_r - 4.308 \times 10^{19}\mu_r^8$	0
Monoalkylhalides, mercaptans, sulfides	$-2.188 \times 10^{16}\mu_r^4 - 7.831 \times 10^{19}\mu_r^8$	0
1-Alcohols except methanol	0.0878	$0.00908 + 69.57\mu_r$
Methanol	0.0878	0.0525

Example Estimate the molar volume of ammonia at 430 K and 2.82 MPa.

Input properties: Recommended values from the DIPPR 801 database are T_c = 405.65 K, P_c = 11.28 MPa, μ = 1.469 D, and ω = 0.252608.

Reduced conditions:

$$T_r = (430\text{ K})/(405.65\text{ K}) = 1.06$$

$$P_r = (2.82\text{ MPa})/(11.28\text{ MPa}) = 0.25$$

$$\mu_r = (1.469)^2(112.8)/(405.65)^2 = 0.0014793$$

Second virial coefficient from Eqs. (2-63) to (2-66):

$$B^{(0)} = 0.1445 - 0.330/1.06 - 0.1385/(1.06)^2 - 0.0121/(1.06)^3 - 0.000607/(1.06)^8 = -0.301$$

$$B^{(1)} = 0.0637 + 0.331/(1.06)^2 - 0.423/(1.06)^3 - 0.008/(1.06)^8 = -0.00189$$

$$a = (-21.4)(0.0014793) - (4.308 \times 10^{19})(0.0014793)^8 = -0.033$$

$$b = 0$$

$$B^{(2)} = (-0.033)/(1.06)^6 = -0.023$$

From Eq. (2-62):

$$BP_c/(RT_c) = -0.301 - (0.252608)(0.00189) - 0.023 = -0.324$$

$$B = (-0.324)[0.008314\text{ m}^3\cdot\text{MPa}/(\text{kmol}\cdot\text{K})](405.65\text{ K})/(11.28\text{ MPa}) = -0.097\text{ m}^3/\text{kmol}$$

Molar volume from Eq. (2-61):

$$V = \frac{RT}{P}\left(1+\frac{B}{V}\right) = \frac{\left(0.0083143\frac{\text{m}^3\cdot\text{MPa}}{\text{kmol}\cdot\text{K}}\right)(430\text{ K})}{2.82\text{ MPa}}\left(1+\frac{-0.097\frac{\text{m}^3}{\text{kmol}}}{V}\right) = 1.162\frac{\text{m}^3}{\text{kmol}}$$

Note that the ideal gas value, 1.268 m^3/kmol, deviates by 9.1 percent from this more accurate value. The truncated virial EoS should be valid for this density since $\rho = V^{-1}$ = 0.86 kmol/m^3 is much less than 40 percent of the critical density (the DIPPR 801 recommended value for the critical density is 13.8 kmol/m^3).

3. For *higher gas densities*, the Lee-Kesler method described below provides excellent predictions for nonpolar and slightly polar fluids. Extended four-parameter corresponding-states methods are available for polar and slightly associating compounds.

Recommended Method Lee-Kesler method.

Reference: Lee, B. I., and M. G. Kesler, *AIChE J.*, **21** (1975): 510.

Classification: Corresponding states.

Expected uncertainty: 1 percent except near the critical point where errors can be up to 30 percent.

Applicability: Nonpolar and moderately polar compounds. An extended Lee-Kesler method, not described here, may be used for polar and slightly associating compounds [Wilding, W. V., and R. L. Rowley, *Int. J. Thermophys.*, **8** (1986): 525].

Input data: T_c, P_c, ω, $Z^{(0)}$, $Z^{(1)}$.

Description:

$$Z = Z^{(0)} + \omega Z^{(1)} \tag{2-67}$$

where Z = compressibility factor
$Z^{(0)}$ = compressibility factor of simple fluid obtained from Table 2-169
$Z^{(1)}$ = deviation from simple fluid obtained from Table 2-170

Analytical expressions for $Z^{(0)}$ and $Z^{(1)}$ can also be generated by using

$$Z^{(0)} = Z_0 \qquad Z^{(1)} = \frac{Z_1 - Z_0}{0.3978} \tag{2-68}$$

where Z_0 and Z_1 are determined from

$$\begin{aligned} Z_i &= \frac{P_rV_r}{T_r} = 1 + \frac{B}{V_r} + \frac{C}{V_r^2} + \frac{D}{V_r^5} + \frac{c_4}{T_r^3V_r^2}\left(\beta + \frac{\gamma}{V_r^2}\right)\exp\left(\frac{-\gamma}{V_r^2}\right) \\ B &= b_1 - \frac{b_2}{T_r} - \frac{b_3}{T_r^2} - \frac{b_4}{T_r^3} \\ C &= c_1 - \frac{c_2}{T_r} + \frac{c_3}{T_r^2} \\ D &= d_1 + \frac{d_2}{T_r} \end{aligned} \tag{2-69}$$

as applied to the simple reference fluid and to the acentric reference fluid (*n*-octane), respectively. The constants for Eq. (2-69) for the two reference fluids are given in Table 2-171.

Example Estimate the molar volume of saturated *n*-decane vapor at 540.5 K.

Input properties: Recommended values from the DIPPR 801 database are T_c = 617.7 K, P_c = 2.11 MPa, P^*(540.5 K) = 0.6799 MPa, and ω = 0.492328.

Reduced conditions:

$$T_r = (540.5\text{ K})/(617.7\text{ K}) = 0.875 \quad\text{and}\quad P_r = (0.6799\text{ MPa})/(2.11\text{ MPa}) = 0.322$$

LK compressiblity factor: Since vapor phase values are needed, the appropriate values from Tables 2-169 and 2-170 that can be used to double-interpolate are as follows:

$Z^{(0)}$		
$T_r\backslash P_r$	0.2	0.4
0.85	0.8810	(0.7222)
0.90	0.9015	0.7800

$Z^{(1)}$		
$T_r\backslash P_r$	0.2	0.4
0.85	−0.0715	(−0.1503)
0.90	−0.0442	−0.1118

Double linear interpolation within these values gives $Z^{(0)}$ = 0.8058 and $Z^{(1)}$ = −0.1025.

From Eq. (2-67):

$$Z = 0.8058 + (0.492328)(-0.1025) = 0.7553$$

Note: If the analytical form available in Eq. (2-69) is used, the following more accurate values are obtained: $Z^{(0)}$ = 0.8131, $Z^{(1)}$ = − 0.1067, and Z = 0.7606.

Molar volume:

$$V = \frac{ZRT}{P} = \frac{(0.7553)\left(0.0083143\frac{\text{m}^3\cdot\text{MPa}}{\text{kmol}\cdot\text{K}}\right)(540.5\text{ K})}{0.6799\text{ M Pa}} = 4.992\frac{\text{m}^3}{\text{kmol}}$$

4. *Cubic EoS* can be used to obtain both vapor and liquid densities as an alternative method to those mentioned above.

TABLE 2-169 Simple Fluid Compressibility Factors $Z^{(0)}$

Values in parentheses are for the opposite phase and may be used to interpolate to or near the phase boundary [PGL4; Wilding, W. V., J. K. Johnson, and R. L. Rowley, *Int. J. Thermophys.*, **8**(1987):717].

*Tr**Pr*	0.010	0.050	0.100	0.200	0.400	0.600	0.800	1.000	1.200	1.500	2.000	3.000	5.000	7.000	10.000
0.30	0.0029	0.0145	0.0290	0.0579	0.1158	0.1737	0.2315	0.2892	0.3470	0.4335	0.5775	0.8648	1.4366	2.0048	2.8507
0.35	0.0026	0.0130	0.0261	0.0522	0.1043	0.1564	0.2084	0.2604	0.3123	0.3901	0.5195	0.7775	1.2902	1.7987	2.5539
0.40	0.0024	0.0119	0.0239	0.0477	0.0953	0.1429	0.1904	0.2379	0.2853	0.3563	0.4744	0.7095	1.1758	1.6373	2.3211
0.45	0.0022 (0.9648)	0.0110	0.0221	0.0442	0.0882	0.1322	0.1762	0.2200	0.2638	0.3294	0.4384	0.6551	1.0841	1.5077	2.1338
0.50	0.0021 (0.9741)	0.0103 (0.8699)	0.0207	0.0413	0.0825	0.1236	0.1647	0.2056	0.2465	0.3077	0.4092	0.6110	1.0094	1.4017	1.9801
0.55	0.9804 (0.0020)	0.0098 (0.9000)	0.0195 (0.7995)	0.0390	0.0778	0.1166	0.1553	0.1939	0.2323	0.2899	0.3853	0.5747	0.9475	1.3137	1.8520
0.60	0.9849 (0.0019)	0.0093 (0.9211)	0.0186 (0.8405)	0.0371	0.0741	0.1109	0.1476	0.1842	0.2207	0.2753	0.3657	0.5446	0.8959	1.2398	1.7440
0.65	0.9881 (0.0018)	0.9377 (0.0089)	0.0178 (0.8707)	0.0356 (0.7367)	0.0710	0.1063	0.1415	0.1765	0.2113	0.2634	0.3495	0.5197	0.8526	1.1773	1.6519
0.70	0.9904	0.9504 (0.0086)	0.8958 (0.0172)	0.0344 (0.7805)	0.0687	0.1027	0.1366	0.1703	0.2038	0.2538	0.3364	0.4991	0.8161	1.1241	1.5729
0.75	0.9922	0.9598 (0.0085)	0.9165 (0.0169)	0.0336 (0.8181)	0.0670 (0.6122)	0.1001	0.1330	0.1656	0.1981	0.2464	0.3260	0.4823	0.7854	1.0787	1.5047
0.80	0.9935	0.9669	0.9319 (0.0168)	0.8539 (0.0332)	0.0661 (0.6659)	0.0985 (0.4746)	0.1307	0.1626	0.1942	0.2411	0.3182	0.4690	0.7598	1.0400	1.4456
0.85	0.9946	0.9725	0.9436	0.8810 (0.0336)	0.0661 (0.7222)	0.0983 (0.5346)	0.1301	0.1614	0.1924	0.2382	0.3132	0.4591	0.7388	1.0071	1.3943
0.90	0.9954	0.9768	0.9528	0.9015 (0.0364)	0.7800 (0.0685)	0.1006 (0.6040)	0.1321 (0.4034)	0.1630	0.1935	0.2383	0.3114	0.4527	0.7220	0.9793	1.3496
0.93	0.9959	0.9790	0.9573	0.9115	0.8059 (0.7350)	0.6635 (0.1047)	0.1359 (0.4499)	0.1664	0.1963	0.2405	0.3122	0.4507	0.7138	0.9648	1.3257
0.95	0.9961	0.9803	0.9600	0.9174	0.8206 (0.0822)	0.6967 (0.1116)	0.1410 0.4853)	0.1705	0.1998	0.2432	0.3138	0.4501	0.7092	0.9561	1.3108
0.97	0.9963	0.9815	0.9625	0.9227	0.8338	0.7240 (0.1312)	0.5580 (0.1532)	0.1779	0.2055	0.2474	0.3164	0.4504	0.7052	0.9480	1.2968
0.98	0.9965	0.9821	0.9637	0.9253	0.8398	0.7360	0.5887 (0.1703)	0.1844	0.2097	0.2503	0.3182	0.4508	0.7035	0.9442	1.2901
0.99	0.9966	0.9826	0.9648	0.9277	0.8455	0.7471	0.6138 (0.2324)	0.1959	0.2154	0.2538	0.3204	0.4514	0.7018	0.9406	1.2835
1.00	0.9967	0.9832	0.9659	0.9300	0.8509	0.7574	0.6353	0.2901	0.2237	0.2583	0.3229	0.4522	0.7004	0.9372	1.2772
1.01	0.9968	0.9837	0.9669	0.9322	0.8561	0.7671	0.6542	0.4648	0.2370	0.2640	0.3260	0.4533	0.6991	0.9339	1.2710
1.02	0.9969	0.9842	0.9679	0.9343	0.8610	0.7761	0.6710	0.5146	0.2629	0.2715	0.3297	0.4547	0.6980	0.9307	1.2650
1.05	0.9971	0.9855	0.9707	0.9401	0.8743	0.8002	0.7130	0.6026	0.4437	0.3131	0.3452	0.4604	0.6956	0.9222	1.2481
1.10	0.9975	0.9874	0.9747	0.9485	0.8930	0.8323	0.7649	0.6880	0.5984	0.4580	0.3953	0.4770	0.6950	0.9110	1.2232
1.15	0.9978	0.9891	0.9780	0.9554	0.9081	0.8576	0.8032	0.7443	0.6803	0.5798	0.4760	0.5042	0.6987	0.9033	1.2021
1.20	0.9981	0.9904	0.9808	0.9611	0.9205	0.8779	0.8330	0.7858	0.7363	0.6605	0.5605	0.5425	0.7069	0.8990	1.1844
1.30	0.9985	0.9926	0.9852	0.9702	0.9396	0.9083	0.8764	0.8438	0.8111	0.7624	0.6908	0.6344	0.7358	0.8998	1.1580
1.40	0.9988	0.9942	0.9884	0.9768	0.9534	0.9298	0.9062	0.8827	0.8595	0.8256	0.7753	0.7202	0.7761	0.9112	1.1419
1.50	0.9991	0.9954	0.9909	0.9818	0.9636	0.9456	0.9278	0.9103	0.8933	0.8689	0.8328	0.7887	0.8200	0.9297	1.1339
1.60	0.9993	0.9964	0.9928	0.9856	0.9714	0.9575	0.9439	0.9308	0.9180	0.9000	0.8738	0.8410	0.8617	0.9518	1.1320
1.70	0.9994	0.9971	0.9943	0.9886	0.9775	0.9667	0.9563	0.9463	0.9367	0.9234	0.9043	0.8809	0.8984	0.9745	1.1343
1.80	0.9995	0.9977	0.9955	0.9910	0.9823	0.9739	0.9659	0.9583	0.9511	0.9413	0.9275	0.9118	0.9297	0.9961	1.1391
1.90	0.9996	0.9982	0.9964	0.9929	0.9861	0.9796	0.9735	0.9678	0.9624	0.9552	0.9456	0.9359	0.9557	1.0157	1.1452
2.00	0.9997	0.9986	0.9972	0.9944	0.9892	0.9842	0.9796	0.9754	0.9715	0.9664	0.9599	0.9550	0.9772	1.0328	1.1516
2.20	0.9998	0.9992	0.9983	0.9967	0.9937	0.9910	0.9886	0.9865	0.9847	0.9826	0.9806	0.9827	1.0094	1.0600	1.1635
2.40	0.9999	0.9996	0.9991	0.9983	0.9969	0.9957	0.9948	0.9941	0.9936	0.9935	0.9945	1.0011	1.0313	1.0793	1.1728
2.60	1.0000	0.9998	0.9997	0.9994	0.9991	0.9990	0.9990	0.9993	0.9998	1.0010	1.0040	1.0137	1.0463	1.0926	1.1792
2.80	1.0000	1.0000	1.0001	1.0002	1.0007	1.0013	1.0021	1.0031	1.0042	1.0063	1.0106	1.0223	1.0565	1.1016	1.1830
3.00	1.0000	1.0002	1.0004	1.0008	1.0018	1.0030	1.0043	1.0057	1.0074	1.0101	1.0153	1.0284	1.0635	1.1075	1.1848
3.50	1.0001	1.0004	1.0008	1.0017	1.0035	1.0055	1.0075	1.0097	1.0120	1.0156	1.0221	1.0368	1.0723	1.1138	1.1834
4.00	1.0001	1.0005	1.0010	1.0021	1.0043	1.0066	1.0090	1.0115	1.0140	1.0179	1.0249	1.0401	1.0741	1.1136	1.1773

TABLE 2–170 Acentric Deviations $Z^{(1)}$ from the Simple Fluid Compressibility Factor

Values in parentheses are for the opposite phase and may be used to interpolate to or near the phase boundary [PGL4; Wilding, W. V., J. K. Johnson, and R. L. Rowley, *Int. J. Thermophys.*, 8(1987):717].

*Tr**Pr*	0.010	0.050	0.100	0.200	0.400	0.600	0.800	1.000	1.200	1.500	2.000	3.000	5.000	7.000	10.000
0.30	−0.0008	−0.0040	−0.0081	−0.0161	−0.0323	−0.0484	−0.0645	−0.0806	−0.0966	−0.1207	−0.1608	−0.2407	−0.3996	−0.5572	−0.7915
0.35	−0.0009	−0.0046	−0.0093	−0.0185	−0.0370	−0.0554	−0.0738	−0.0921	−0.1105	−0.1379	−0.1834	−0.2738	−0.4523	−0.6279	−0.8863
0.40	−0.0010	−0.0048	−0.0095	−0.0190	−0.0380	−0.0570	−0.0758	−0.0946	−0.1134	−0.1414	−0.1879	−0.2799	−0.4603	−0.6365	−0.8936
0.45	−0.0009	−0.0047	−0.0094	−0.0187	−0.0374	−0.0560	−0.0745	−0.0929	−0.1113	−0.1387	−0.1840	−0.2734	−0.4475	−0.6162	−0.8606
	(−0.0740)														
0.50	−0.0009	−0.0045	−0.0090	−0.0181	−0.0360	−0.0539	−0.0716	−0.0893	−0.1069	−0.1330	−0.1762	−0.2611	−0.4253	−0.5831	−0.8099
	(−0.0457)	(−0.2270)													
0.55	−0.0314	−0.0043	−0.0086	−0.0172	−0.0343	−0.0513	−0.0682	−0.0849	−0.1015	−0.1263	−0.1669	−0.2465	−0.3991	−0.5446	−0.7521
	(−0.0009)	(−0.1438)	(−0.2864)												
0.60	−0.0205	−0.0041	−0.0082	−0.0164	−0.0326	−0.0487	−0.0646	−0.0803	−0.0960	−0.1192	−0.1572	−0.2312	−0.3718	−0.5047	−0.6928
	(0.0008)	(0.0949)	(−0.1857)												
0.65	−0.0137	−0.0772	−0.0078	−0.0156	−0.0309	−0.0461	−0.0611	−0.0759	−0.0906	−0.1122	−0.1476	−0.2160	−0.3447	−0.4653	−0.6346
	(−0.0008)	(0.0039)	(−0.1262)	(−0.2424)											
0.70	−0.0093	−0.0507	−0.1161	−0.0148	−0.0294	−0.0438	−0.0579	−0.0718	−0.0855	−0.1057	−0.1385	−0.2013	−0.3184	−0.4270	−0.5785
		(−0.0038)	(−0.0075)	(−0.1685)											
0.75	−0.0064	−0.0339	−0.0744	−0.0143	−0.0282	−0.0417	−0.0550	−0.0681	−0.0808	−0.0996	−0.1298	−0.1872	−0.2929	−0.3901	−0.5250
		(−0.0037)	(−0.0072)	(−0.1298)	(−0.2203)										
0.80	−0.0044	−0.0228	−0.0487	−0.1160	−0.0272	−0.0401	−0.0526	−0.0648	−0.0767	−0.0940	−0.1217	−0.1736	−0.2682	−0.3545	−0.4740
			(−0.0073)	(−0.0139)	(−0.1682)	(−0.2185)									
0.85	−0.0029	−0.0152	−0.0319	−0.0715	−0.0268	−0.0391	−0.0509	−0.0622	−0.0731	−0.0888	−0.1138	−0.1602	−0.2439	−0.3201	−0.4254
				(−0.0144)	(−0.1503)	(−0.1692)									
0.90	−0.0019	−0.0099	−0.0205	−0.0442	−0.1118	−0.0396	−0.0503	−0.0604	−0.0701	−0.0840	−0.1059	−0.1463	−0.2195	−0.2862	−0.3788
				(−0.0179)	(−0.0286)	(−0.1580)	(−0.1464)								
0.93	−0.0015	−0.0075	−0.0154	−0.0326	−0.0763	−0.1662	−0.0514	−0.0602	−0.0687	−0.0810	−0.1007	−0.1374	−0.2045	−0.2661	−0.3516
					(−0.0340)	(−0.0424)	(−0.1418)								
0.95	−0.0012	−0.0062	−0.0126	−0.0262	−0.0589	−0.1110	−0.0540	−0.0607	−0.0678	−0.0788	−0.0967	−0.1310	−0.1943	−0.2526	−0.3339
					(−0.0444)	(−0.0490)	(−0.1532)								
0.97	−0.0010	−0.0050	−0.0101	−0.0208	−0.0450	−0.0770	−0.1647	−0.0623	−0.0669	−0.0759	−0.0921	−0.1240	−0.1837	−0.2391	−0.3163
						(−0.0714)	(−0.0643)								
0.98	−0.0009	−0.0044	−0.0090	−0.0184	−0.0390	−0.0641	−0.1100	−0.0641	−0.0661	−0.0740	−0.0893	−0.1202	−0.1783	−0.2322	−0.3075
							(−0.0828)								
0.99	−0.0008	−0.0039	−0.0079	−0.0161	−0.0335	−0.0531	−0.0796	−0.0680	−0.0646	−0.0715	−0.0861	−0.1162	−0.1728	−0.2254	−0.2989
							(−0.1621)								
1.00	−0.0007	−0.0034	−0.0069	−0.0140	−0.0285	−0.0435	−0.0588	−0.0879	−0.0609	−0.0678	−0.0824	−0.1118	−0.1672	−0.2185	−0.2902
1.01	−0.0006	−0.0030	−0.0060	−0.0120	−0.0240	−0.0351	−0.0429	−0.0223	−0.0473	−0.0621	−0.0778	−0.1072	−0.1615	−0.2116	−0.2816
1.02	−0.0005	−0.0026	−0.0051	−0.0102	−0.0198	−0.0277	−0.0303	−0.0062	0.0227	−0.0524	−0.0722	−0.1021	−0.1556	−0.2047	−0.2731
1.05	−0.0003	−0.0015	−0.0029	−0.0054	−0.0092	−0.0097	−0.0032	0.0220	0.1059	0.0451	−0.0432	−0.0838	−0.1370	−0.1835	−0.2476
1.10	0.0000	0.0000	0.0001	0.0007	0.0038	0.0106	0.0236	0.0476	0.0897	0.1630	0.0698	−0.0373	−0.1021	−0.1469	−0.2056
1.15	0.0002	0.0011	0.0023	0.0052	0.0127	0.0237	0.0396	0.0625	0.0943	0.1548	0.1667	0.0332	−0.0611	−0.1084	−0.1642
1.20	0.0004	0.0019	0.0039	0.0084	0.0190	0.0326	0.0499	0.0719	0.0991	0.1477	0.1990	0.1095	−0.0141	−0.0678	−0.1231
1.30	0.0006	0.0030	0.0061	0.0125	0.0267	0.0429	0.0612	0.0819	0.1048	0.1420	0.1991	0.2079	0.0875	0.0176	−0.0423
1.40	0.0007	0.0036	0.0072	0.0147	0.0306	0.0477	0.0661	0.0857	0.1063	0.1383	0.1894	0.2397	0.1737	0.1008	0.0350
1.50	0.0008	0.0039	0.0078	0.0158	0.0323	0.0497	0.0677	0.0864	0.1055	0.1345	0.1806	0.2433	0.2309	0.1717	0.1058
1.60	0.0008	0.0040	0.0080	0.0162	0.0330	0.0501	0.0677	0.0855	0.1035	0.1303	0.1729	0.2381	0.2631	0.2255	0.1673
1.70	0.0008	0.0040	0.0081	0.0163	0.0329	0.0497	0.0667	0.0838	0.1008	0.1259	0.1658	0.2305	0.2788	0.2628	0.2179
1.80	0.0008	0.0040	0.0081	0.0162	0.0325	0.0488	0.0652	0.0816	0.0978	0.1216	0.1593	0.2224	0.2846	0.2871	0.2576
1.90	0.0008	0.0040	0.0079	0.0159	0.0318	0.0477	0.0635	0.0792	0.0947	0.1173	0.1532	0.2144	0.2848	0.3017	0.2876
2.00	0.0008	0.0039	0.0078	0.0155	0.0310	0.0464	0.0617	0.0767	0.0916	0.1133	0.1476	0.2069	0.2819	0.3097	0.3096
2.20	0.0007	0.0037	0.0074	0.0147	0.0293	0.0437	0.0579	0.0719	0.0857	0.1057	0.1374	0.1932	0.2720	0.3135	0.3355
2.40	0.0007	0.0035	0.0070	0.0139	0.0276	0.0411	0.0544	0.0675	0.0803	0.0989	0.1285	0.1812	0.2602	0.3089	0.3459
2.60	0.0007	0.0033	0.0066	0.0131	0.0260	0.0387	0.0512	0.0634	0.0754	0.0929	0.1207	0.1706	0.2484	0.3009	0.3475
2.80	0.0006	0.0031	0.0062	0.0124	0.0245	0.0365	0.0483	0.0598	0.0711	0.0876	0.1138	0.1613	0.2372	0.2915	0.3443
3.00	0.0006	0.0029	0.0059	0.0117	0.0232	0.0345	0.0456	0.0565	0.0672	0.0828	0.1076	0.1529	0.2268	0.2817	0.3385
3.50	0.0005	0.0026	0.0052	0.0103	0.0204	0.0303	0.0401	0.0497	0.0591	0.0728	0.0949	0.1356	0.2042	0.2584	0.3194
4.00	0.0005	0.0023	0.0046	0.0091	0.0182	0.0270	0.0357	0.0443	0.0527	0.0651	0.0849	0.1219	0.1857	0.2378	0.2994

TABLE 2-171 Constants for the Two Reference Fluids Used in Lee-Kesler Method*

Constant	Simple reference fluid	Acentric reference fluid
b_1	0.1181193	0.2026579
b_2	0.265728	0.331511
b_3	0.154790	0.027655
b_4	0.030323	0.203488
c_1	0.0236744	0.0313385
c_2	0.0186984	0.0503618
c_3	0.0	0.016901
c_4	0.042724	0.041577
$d_1 \times 10^4$	0.155488	0.48736
$d_2 \times 10^4$	0.623689	0.0740336
β	0.65392	1.226
γ	0.060167	0.03754

*Lee, B. I., and M. G. Kesler, *AIChE J.*, **21** (1975): 510.

Recommended Method Cubic EoS.

Classification: Empirical extension of theory.

Expected uncertainty: Varies depending upon compound and conditions, but a general expectation is 10 to 20 percent.

Applicability: Nonpolar and moderately polar compounds.

Input data: T_c, P_c, ω.

Description: The more common cubic EoS can be written in the form

$$Z = \frac{V}{V-b} - \frac{V}{V^2+\delta V+\varepsilon}\,\frac{a\alpha(T_r)}{RT} \tag{2-70}$$

where a, b, δ, and ε are constants that depend upon the model EoS chosen, as does the temperature dependence of the function $\alpha(T_r)$. Definitions of these constants and $\alpha(T_r)$ for some of the more commonly used EoS models are shown in Table 2-172. The corresponding relations for many other EoS models in this same form are available [Soave, G., *Chem. Eng. Sci.*, **27** (1972): 1197]. The independent parameters a and b in these models can be regressed from experimental data to correlate densities or can be obtained from known critical constants to predict density data.

Of the cubic EoS given in Table 2-172, the Soave and Peng-Robinson are the most accurate, but there is no general rule for which EoS produces the best estimated volumes for specific fluids or conditions. The Peng-Robinson equation has been better tuned to liquid densities, while the Soave equation has been better tuned to vapor-liquid equilibrium and vapor densities. In solving the cubic equation for volume, a convenient initial guess to find the vapor root is the ideal gas value, while an initial value of $1.05b$ is convenient to locate the liquid root.

Example Estimate the molar density of liquid and vapor saturated ammonia at 353.15 K, using the Soave and Peng-Robinson EoS.

Required properties: Recommended values in the DIPPR 801 database are

$$T_c = 405.65\text{ K} \qquad P_c = 112.8\text{ bar} \qquad \omega = 0.252608$$

$$P^*(353.15\text{ K}) = 41.352\text{ bar (vapor pressure at 353.15 K)}$$

EoS parameters (shown for Soave EoS):

$$a = \frac{0.42748\,(RT_c)^2}{P_c} = \frac{0.42748\left[\left(83.145\,\dfrac{\text{bar}\cdot\text{cm}^3}{\text{mol}\cdot\text{K}}\right)(405.65\text{ K})\right]^2}{112.8\text{ bar}} = 4.311\times10^6\,\frac{\text{cm}^6\cdot\text{bar}}{\text{mol}^2}$$

$$b = \frac{0.08664\,(RT_c)}{P_c} = \frac{0.08664\left(83.145\,\dfrac{\text{bar}\cdot\text{cm}^3}{\text{mol}\cdot\text{K}}\right)(405.65\text{ K})}{112.8\text{ bar}} = 25.906\,\frac{\text{cm}^3}{\text{mol}}$$

$$T_r = (353.15\text{ K})/(405.65\text{ K}) = 0.871$$

$$\alpha = \{1 + [0.48 + (1.574)(0.252608) - (0.176)(0.252608)^2]\,[1 - (0.871)^{0.5}]\}^2 = 1.119$$

Rearrange and solve Eq. (2-70) for V:

$$P = \frac{RT}{V-b} - \frac{a\alpha}{V(V+b)} \quad \text{or} \quad PV^3 - RTV^2 + (a\alpha - bRT - Pb^2)V - ab\alpha = 0$$

$$41.352\left(\frac{V}{\text{m}^3/\text{mol}}\right)^3 - \left(0.029\frac{\text{m}^3}{\text{mol}}\right)\left(\frac{V}{\text{m}^3/\text{mol}}\right)^2 + \left(4.037\times10^{-6}\frac{\text{m}^6}{\text{mol}^2}\right)\left(\frac{V}{\text{m}^3/\text{mol}}\right) - 1.25\times10^{-10} = 0$$

Vapor root (initial guess of $V = 7.1 \times 10^{-7}$ m³/mol from ideal gas equation):

$$V_{\text{vap}} = 5.395 \times 10^{-4}\text{ m}^3/\text{mol} \quad \text{and} \quad \rho_{\text{vap}} = 1/V_{\text{vap}} = 1.854\text{ kmol/m}^3$$

Liquid root (initial guess of $V = 2.72 \times 10^{-5}$ m³/mol from 1.05b):

$$V_{\text{liq}} = 4.441 \times 10^{-5}\text{ m}^3/\text{mol} \quad \text{and} \quad \rho_{\text{liq}} = 1/V_{\text{liq}} = 22.516\text{ kmol/m}^3$$

The corresponding values and equation for the Peng-Robinson EoS are

$$a = 4.611 \times 10^6\text{ cm}^6\cdot\text{bar/mol}^2 \qquad b = 23.262\text{ cm}^3/\text{mol} \qquad \alpha = 1.103$$

$$P = \frac{RT}{V-b} - \frac{a\alpha}{V^2+2bV-b^2}$$

or

$$PV^3 + (bP - RT)V^2 + (a\alpha - 2bRT - 3Pb^2)V + (bP^3 + RTb^2 - ab\alpha) = 0$$

$$41.352\left(\frac{V}{\text{m}^3/\text{mol}}\right)^3 - \left(0.0284\frac{\text{m}^3}{\text{mol}}\right)\left(\frac{V}{\text{m}^3/\text{mol}}\right)^2 + \left(3.651\times10^{-6}\frac{\text{m}^6}{\text{mol}^2}\right)\left(\frac{V}{\text{m}^3/\text{mol}}\right) - 1.018\times10^{-10} = 0$$

Solve for the two physical roots of this equation:

$$V_{\text{vap}} = 5.286 \times 10^{-4}\text{ m}^3/\text{mol and } \rho_{\text{vap}} = 1.892\text{ kmol/m}^3$$

$$V_{\text{liq}} = 3.914 \times 10^{-5}\text{ m}^3/\text{mol and } \rho_{\text{liq}} = 25.55\text{ kmol/m}^3$$

The liquid density calculated from the Soave EoS is 24.2 percent below the DIPPR 801 recommended value of 29.69 kmol/m³; that calculated from the Peng-Robinson EoS is 13.9 percent below the recommended value.

Liquids For most liquids, the saturated molar liquid density ρ can be effectively correlated with

$$\rho = \frac{A}{B^{[1+(1-T/C)^D]}} \tag{2-71}$$

adapted from the Rackett prediction equation [Rackett, H. G., *J. Chem. Eng. Data*, **15** (1970): 514]. The regression constants A, B, and D are determined from the nonlinear regression of available data, while C is usually taken as the critical temperature. The liquid density decreases approximately linearly from the triple point to the normal boiling point and then nonlinearly to the critical density (the reciprocal of the critical volume). A few compounds such as water cannot be fit with this equation over the entire range of temperature.

The recommended method for estimation of saturated liquid density for pure organic compounds is the Rackett prediction method.

Recommended Method Rackett method.

Reference: Rackett, H. G., *J. Chem. Eng. Data*, **15** (1970): 514.

Classification: Corresponding states.

Expected uncertainty: 15 percent as purely predictive equation; 2 percent if a liquid density value is available.

TABLE 2-172 Relationships for Eq. (2-70) for Common Cubic EoS

EoS	δ	ε	$\alpha(T_r)$	$aP_c/(RT_c)^2$	$bP_c/(RT_c)$
van der Waals*	0	0	1	0.42188	0.125
Relich-Kwong†	0	0	$T_r^{-0.5}$	0.42748	0.08664
Soave‡	b	0	$[1 + (0.48 + 1.574\omega - 0.176\omega^2)(1 - T_r^{0.5})]^2$	0.42748	0.08664
Peng-Robinson§	$2b$	$-b^2$	$[1 + (0.37464 + 1.54226\omega - 0.2699\omega^2)(1 - T_r^{0.5})]^2$	0.45724	0.0778

*van der Waal, J. H., *Z. Phys. Chem.*, **5** (1890): 133.
†Redlich, O., and J. N. S. Kwong, *Chem. Rev.*, **44** (1949): 233.
‡Soave, G., *Chem. Eng. Sci.*, **27** (1972): 1197.
§Peng, D. Y., and D. B. Robinson, *Ind. Eng. Chem. Fundam.*, **15** (1976): 59.

Applicability: Saturated liquid densities of organic compounds.
Input data: T_c, P_c, and Z_c (or, equivalently, V_c).
Description: A predictive form of the equation is given by

$$\frac{1}{\rho}=V=\left(\frac{RT_c}{P_c}\right)Z_c^q \quad \text{where } q=1+(1-T_r)^{2/7} \tag{2-72}$$

When one or more liquid density data points are available, Z_c in Eq. (2-72) can be replaced with an adjustable parameter fitted from the data (Z_{RA} in the notation of Spencer and Danner [Spencer, C. F., and R. P. Danner, *J. Chem. Eng. Data* **17** (1972): 236]). This produces densities in good agreement with experiment and permits accurate interpolation of the densities over most of the liquid temperature range, but it does not give the correct critical density unless $Z_{RA} = Z_c$.

Example Estimate the saturated liquid density of acetonitrile at 376.69 K.
Required properties: The recommended values from the DIPPR 801 database are

$$T_c = 545.5 \text{ K} \qquad P_c = 4.83 \text{ MPa} \qquad Z_c = 0.184$$

Calculate supporting quantities:

$$T_r = (376.69 \text{ K})/(545.5 \text{ K}) = 0.691$$
$$q = 1 + (1 - 0.691)^{2/7} = 1.715$$

Calculate saturated liquid density from Eq. (2-72):

$$\rho=\left[\frac{4.83\times10^6 \text{ Pa}}{\left(8.314\frac{\text{Pa}\cdot\text{m}^3}{\text{mol}\cdot\text{K}}\right)(545.5 \text{ K})}\right](0.184)^{-1.715}=19.42\frac{\text{kmol}}{\text{m}^3}$$

The estimated density is 16 percent above the DIPPR 801 value of 16.73 kmol/m^3.

Calculate ρ_{sat} from Eq. (2-72) with a known liquid density: Kratzke and Muller [Kratzke, H., and S. Muller, *J. Chem. Thermo.*, **17** (1985): 151] reported an experimental density of 18.919 kmol/m^3 at 298.08 K. Use of this experimental value in Eq. (2-72) to calculate Z_{RA} gives

$$T_r = (298.08 \text{ K})/(545.5 \text{ K}) = 0.546 \qquad q = 1 + (1 - 0.546)^{2/7} = 1.798$$

$$Z_{RA}=\left[\frac{4.83\times10^6 \text{ Pa}}{\left(8.314\frac{\text{Pa}\cdot\text{m}^3}{\text{kmol}\cdot\text{K}}\right)(545.5 \text{ K})\left(18.919\frac{\text{kmol}}{\text{m}^3}\right)}\right]^{1/1.798}=0.202$$

$$\rho=\left[\frac{4.83\times10^6 \text{ Pa}}{\left(8.314\frac{\text{Pa}\cdot\text{m}^3}{\text{mol}\cdot\text{K}}\right)(545.5 \text{ K})}\right](0.202)^{-1.715}=16.577\frac{\text{kmol}}{\text{m}^3}$$

The value obtained by the modified Rackett method is 0.9 percent below the DIPPR 801 recommended value. Note, however, that with $Z_{RA} = 0.202$ instead of Z_c, Eq. (2-72) gives $\rho_c = 5.28$ kmol/m^3 instead of $\rho_c = P_c/(Z_cRT_c) = 5.79$ kmol/m^3.

Solids Solid density data are sparse and usually available only within a narrow temperature range. For most solids, density decreases approximately linearly with increasing temperature. No accurate method for prediction of solid densities is available, but an approximate correlation has been found between the density of the liquid phase at the triple point and the solid that is stable at the triple point conditions.

Recommended Method Goodman method.

Reference: Goodman, B. T., W. V. Wilding, J. L. Oscarson, and R. L. Rowley, *J. Chem. Eng. Data*, **49** (2004): 1512.

Classification: Empirical correlation.

Expected uncertainty: 6 percent.

Applicability: Organic compounds; applicable to the stable solid phase at the triple point temperature T_t; applicable T range is from T_t down to either the first solid-phase transition temperature or to approximately $0.3T_t$.

Input data: Liquid density at the triple point.

Description: The density for the solid phase that is stable at the triple point has been correlated as a function of temperature and the liquid density at T_t as

$$\rho_s=\left(1.28-0.16\frac{T}{T_t}\right)\rho_L(T_t) \tag{2-73}$$

Example Estimate the density of solid naphthalene at 281.46 K.
Required properties: The recommended values from the DIPPR 801 database for T_t and the liquid density at T_t are

$$T_t = 353.43 \text{ K} \qquad \rho_L(T_t) = 7.6326 \text{ kmol/m}^3$$

From Eq. (2-73):

$$\rho_s=\left(1.28-0.16\frac{281.46 \text{ K}}{353.43 \text{ K}}\right)\left(7.6326\frac{\text{kmol}}{\text{m}^3}\right)=8.797\frac{\text{kmol}}{\text{m}^3}$$

The estimated value is 4.3 percent lower than the DIPPR 801 recommended value of 9.1905 kmol/m^3.

Mixtures Both liquid and vapor densities can be estimated using pure-component CS and EoS methods by treating the fluid as a pseudo-pure component with effective parameters calculated from the pure-component parameters using ad hoc mixing rules.

To apply the Lee-Kesler CS method to mixtures, pseudo-pure fluid constants are required. One of the simplest set of mixing rules for these quantities is [Prausnitz, J. M., and R. D. Gunn, *AIChE J.*, **4** (1958): 430, 494; Joffe, J., *Ind. Eng. Chem. Fundam.*, **10** (1971): 532]:

$$\overline{T}_c=\sum_{i=1}^{C}x_iT_{c,i} \tag{2-74}$$

$$\overline{P}_c=\frac{\sum_{i=1}^{C}x_iZ_{c,i}}{\sum_{i=1}^{C}x_iV_{c,i}}R\overline{T}_c \tag{2-75}$$

$$\overline{\omega}=\sum_{i=1}^{C}x_i\omega_i \tag{2-76}$$

The procedures are identical to those for pure components with the replacement of T_c, P_c, and ω with the effective mixture values obtained from the above equations.

To use a cubic EoS for a mixture, mixing rules are used to calculate effective mixture parameters in terms of the pure-component values. Although more complex mixing rules may improve prediction accuracy, the simple forms recommended here provide reasonable accuracy without adjustable parameters:

$$\overline{b}=\sum_{i=1}^{C}x_ib_i \tag{2-77}$$

$$\overline{a\alpha}=\left[\sum_{i=1}^{C}x_i(a_i\alpha_i)^{1/2}\right]^2 \tag{2-78}$$

Mixture calculations are then identical to the pure-component calculations using these effective mixture parameters for the pure-component $a\alpha$ and b values.

The modified Rackett method has also been extended to liquid mixtures [Spencer, C. F., and R. P. Danner, *J. Chem. Eng. Data*, **17** (1972): 236] using the following combining and mixing rules as modified by Li [Li, C. C., *Can. J. Chem. Eng.*, **19** (1971): 709]:

$$T_{c,ij}=\sqrt{T_{c,i}T_{c,j}} \qquad \phi_i=\frac{x_iV_{c,i}}{\sum_{j=1}^{C}x_jV_{c,j}} \qquad \overline{T}_c=\sum_{i=1}^{C}\sum_{j=1}^{C}\phi_i\phi_jT_{c,ij} \tag{2-79}$$

Recommended Method Spencer-Danner-Li mixing rules with Rackett equation.

References: Spencer, C. F., and R. P. Danner, *J. Chem. Eng. Data*, **17** (1972): 236; Li, C. C., *Can. J. Chem. Eng.*, **19** (1971): 709.

Classification: Corresponding states.

Expected uncertainty: About 7 percent on average; higher near the T_c of any of the components.

Applicability: Saturated (at the bubble point) liquid mixtures.

Input data: T_c, V_c, and x_i.

Description: The predictive form of the equation is given by

$$\frac{1}{\rho}=V=R\left(\sum_{i=1}^{C}\frac{x_iT_{c,i}}{P_{c,i}}\right)\overline{Z}_{RA}^q \qquad q=1.0+(1.0-T_r)^{2/7} \tag{2-80}$$

where

$$\bar{Z}_{RA} = 0.29056 - 0.08775 \sum_{i=1}^{C} x_i \omega_i \quad \text{and} \quad T_r = \frac{T}{\bar{T}_c} \tag{2-81}$$

Example Estimate the saturated liquid density of a liquid mixture of 50 mol% ethane(1) and 50 mol% *n*-decane(2) at 377.6 K.

Required properties: The recommended values from the DIPPR 801 database for the required properties are as follows:

	T_c/K	V_c/(m³ · kmol⁻¹)	P_c/bar	ω
Ethane	305.32	0.1455	48.72	0.0995
Decane	617.7	0.617	21.1	0.4923

Auxiliary quantities from Eq. (2-79):

$$\phi_1 = \frac{(0.5)(0.1455)}{(0.5)(0.1455)+(0.5)(0.617)} = 0.191; \quad \phi_2 = 0.809$$

$$T_{c,12} = \sqrt{(305.32\ \text{K})(617.7\ \text{K})} = 434.3\ \text{K}$$

$$\frac{\bar{T}_c}{\text{K}} = \phi_1^2 T_{c,1} + 2\phi_1\phi_2 T_{c,12} + \phi_2^2 T_{c,2}$$

$$= (0.191)^2(305.32) + (2)(0.191)(0.809)(434.3) + (0.809)^2(617.7)$$

$$\bar{T}_c = 549.68\ \text{K}$$

Calculations from Eqs. (2-80) and (2-81):

$$T_r = (377.6\ \text{K})/(549.63\ \text{K}) = 0.687 \quad q = 1 + (1 - 0.687)^{2/7} = 1.718$$

$$\bar{Z}_{RA} = 0.29056 - 0.08775[(0.5)(0.0995) + (0.5)(0.4923)] = 0.2646$$

$$V = \left(0.08314 \frac{\text{m}^3 \cdot \text{bar}}{\text{K} \cdot \text{kmol}}\right)\left[\frac{(0.5)(305.32\ \text{K})}{48.72\ \text{bar}} + \frac{(0.5)(617.7\ \text{K})}{21.1\ \text{bar}}\right](0.2646)^{1.718} = 0.151 \frac{\text{m}^3}{\text{kmol}}$$

The experimental value [Reamer, H. H., and B. H. Sage, *J. Chem. Eng. Data*, **7** (1962): 161] is 0.149 m³/kmol, and the error in the estimated value is 1.3 percent.

VISCOSITY

Viscosity is defined as the shear stress per unit area at any point in a confined fluid, divided by the velocity gradient in the direction perpendicular to the direction of flow. The *absolute viscosity* η is the shear stress at a point, divided by the velocity gradient at that point. The SI unit of viscosity is Pa · s [1 kg/(m · s)], but the cgs units of poise (P) [1 g/(cm · s)] and centipoise (cP = 0.01 P) are also frequently used (1 cP = 1 mPa · s). The *kinematic viscosity* ν is defined as the ratio of the absolute viscosity to density at the same temperature and pressure. The SI unit for ν is m^2/s, but again cgs units are very common and ν is often given in stokes (1 St = 1 cm^2/s) or centistokes (1 cSt = 0.01 cm^2/s).

Gases Experimental data for gases and vapors at low density are often correlated with

$$\eta^o = \frac{AT^B}{1 + C/T + D/T^2} \tag{2-82}$$

Over smaller temperature ranges, parameters C and D may not be necessary as ln(η) is often reasonably linear with ln(T). Care should be taken in extrapolating using Eq. (2-82) as there can be unintended mathematical poles where the denominator approaches zero.

Numerous methods have been developed for estimation of vapor viscosity. For nonpolar vapors, the Yoon-Thodos CS method works well, but for polar fluids the Reichenberg method is preferred. Both methods are illustrated below.

Recommended Method 1 Yoon-Thodos method.
Reference: Yoon, P., and G. Thodos, *AIChE J.*, **16** (1970): 300.
Classification: Corresponding states.
Expected uncertainty: 5 percent.
Applicability: Nonpolar and slightly polar organic vapors.
Input data: T_c, P_c, and M.
Description: The correlation for viscosity as a function of reduced temperature is

$$\frac{\eta^o}{\text{Pa}\cdot\text{s}} = \frac{46.1T_r^{0.618} - 20.4\exp(-0.449T_r) + 19.4\exp(-4.058\,T_r) + 1}{2.173424\times10^{11}(T_c/\text{K})^{1/6}(M/\text{g}\cdot\text{mol}^{-1})^{-1/2}(P_c/\text{Pa})^{-2/3}} \tag{2-83}$$

Example Estimate the low-pressure vapor viscosity of propane at 353 K.
Required constants: The DIPPR 801 database recommends the following values:

$$T_c = 369.83\ \text{K} \qquad P_c = 4.248\ \text{MPa} \qquad M = 44.0956\ \text{g/mol}$$

Reduced temperature:

$$T_r = (353\ \text{K})/(369.83\ \text{K}) = 0.9545$$

Calculation using Eq. (2-83):

$$\frac{\eta^o}{\text{Pa}\cdot\text{s}} = \frac{(46.1)(0.9545)^{0.618} - 20.4\exp[-(0.449)(0.9545)] + 19.4\exp[-4.058(0.9545)] + 1}{(2.173424\times10^{11})(369.83)^{-1/6}(44.0956)^{-1/2}(4.248\times10^6)^{-2/3}}$$

$$= 9.84\times10^{-6}$$

This value is 1.5 percent higher than the DIPPR 801 recommended value of 9.70 × 10^{-6} Pa · s.

Recommended Method 2 Reichenberg method.
Reference: Reichenberg, D., *AIChE J.*, **21** (1975): 181.
Classification: Group contributions and corresponding states.
Expected uncertainty: 5 percent.
Applicability: Nonpolar and polar organic and inorganic vapors.
Input data: T_c, P_c, M, μ, and molecular structure.
Description: The temperature dependence of the viscosity is given by

$$\frac{\eta^o}{\text{Pa}\cdot\text{s}} = \frac{AT_r^2}{[1 + 0.36\,T_r(T_r - 1)]^{1/6}}\left[\frac{1 + 270(\mu_r^*)^4}{T_r + 270(\mu_r^*)^4}\right] \tag{2-84}$$

where the parameter A is determined from group contributions and the modified reduced dipole μ_r^* is found from

$$\mu_r^* = 52.46\mu_r \tag{2-85}$$

and Eq. (2-66).

For organic compounds, A is found from the group values C_i, listed in Table 2-173, using

$$A = 10^{-7}\frac{\left(\dfrac{M}{\text{kg/kmol}}\right)^{1/2}(T_c/\text{K})}{\sum_{i=1}^{N} n_i C_i} \tag{2-86}$$

For inorganic gases, A is obtained from

$$A = 1.6104\times10^{-10}\left[\left(\frac{M}{\text{g/mol}}\right)^{1/2}\left(\frac{P_c}{\text{Pa}}\right)^{2/3}\left(\frac{T_c}{\text{K}}\right)^{-1/6}\right] \tag{2-87}$$

TABLE 2-173 Reichenberg* Group Contribution Values

Group	C_i	Group	C_i
—CH₃	9.04	—F	4.46
>CH₂	6.47	—Cl	10.06
>CH—	2.67	—Br	12.83
>C<	−1.53	—OH alcohol	7.96
=CH₂	7.68	>O	3.59
=CH—	5.53	>C=O	12.02
>C=	1.78	—CHO	14.02
≡CH	7.41	—COOH	18.65
≡C—	5.24	—COO— or HCOO—	13.41
>CH₂ ring	6.91	—NH₂	9.71
>CH— ring	1.16	>NH	3.68
>C< ring	0.23	=N— ring	4.97
=CH— ring	5.90	—CN	18.13
>C= ring	3.59	>S ring	8.86

*Reichenberg, D., *AIChE J.*, **21** (1975): 181.

Example Estimate the low-pressure vapor viscosity of ethyl acetate at 401.25 K.
Required constants: The DIPPR 801 database recommends the following values:

$M = 88.1051$ g/mol $T_c = 523.3$ K $P_c = 3.88$ MPa $\mu = 1.78\ D$

Supporting quantities:
Structural groups:

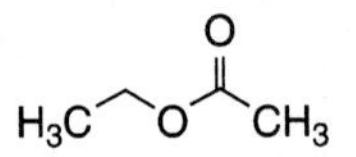

Group	n_i	C_i	Contribution
$-CH_3$	2	9.04	18.08
$>CH_2$	1	6.47	6.47
—COO—	1	13.41	13.41
		Total	37.96

$T_r = (401.25\ \text{K})/(523.3\ \text{K}) = 0.767$

From Eqs. (2-66) and (2-85):

$$\mu_r^* = 52.46\frac{(1.78)^2(38.8)}{(523.3)^2} = 0.024$$

From Eq. (2-86):

$$A = 10^{-7}\frac{(88.1051)^{1/2}(523.3)}{37.96} = 1.294\times10^{-5}$$

Calculation using Eq. (2-84):

$$\frac{\eta^o}{\text{Pa}\cdot\text{s}} = \frac{(1.294\times10^{-5})(0.767)^2}{[1+(0.36)(0.767)(0.767-1)]^{1/6}}\frac{1+(270)(0.024)^4}{0.767+(270)(0.024)^4} = 1.003\times10^{-5}$$

The estimated value is 1.5 percent lower than the DIPPR 801 recommended value of 1.018×10^{-5} Pa · s.

The dependence of viscosity upon pressure is principally a density effect. Estimation of vapor viscosity at elevated pressures is commonly done by correlating density deviations from the low-pressure values estimated. Several methods are available, but the method developed by Jossi et al. and extended to polar fluids by Stiel and Thodos is relatively accurate and easy to apply.

Recommended Method Jossi-Stiel-Thodos method.
References: Stiel, L. I., and G. Thodos, *AIChE J.*, **10** (1964): 26; Jossi, J. A., L. I. Stiel, and G. Thodos, *AIChE J.*, **8** (1962): 59.
Classification: Empirical correlation and corresponding states.
Expected uncertainty: 9 percent—often less for nonpolar gases, larger for polar gases.
Applicability: Nonassociating gases; $\rho_r < 2.6$.
Input data: M, T_c, P_c, Z_c, μ, η^o (low-pressure viscosity at same T may be estimated by using methods given above), and ρ (may be calculated from T and P by using density methods given above).
Description: Deviation of η from the low-pressure value η^o is given by one of the following correlations depending upon its polarity and reduced density range:
For nonpolar gases, $0.1 < \rho_r < 3.0$:

$$\left[\left(\frac{\eta-\eta^o}{\text{mPa}\cdot\text{s}}\right)\xi+1\right]^{1/4} = 1.0230+0.23364\rho_r+0.58533\rho_r^2-0.40758\rho_r^3+0.093324\rho_r^4 \quad (2\text{-}88)$$

For polar gases, $\rho_r \le 0.1$:

$$\left(\frac{\eta-\eta^o}{\text{mPa}\cdot\text{s}}\right)\xi = 1.656\rho_r^{1.111} \quad (2\text{-}89)$$

For polar gases, $0.1 < \rho_r \le 0.9$:

$$\left(\frac{\eta-\eta^o}{\text{mPa}\cdot\text{s}}\right)\xi = 0.0607\,(9.045\rho_r+0.63)^{1.739} \quad (2\text{-}90)$$

For polar gases, $0.9 < \rho_r \le 2.2$:

$$\log\left\{4-\log\left[\left(\frac{\eta-\eta^o}{\text{mPa}\cdot\text{s}}\right)\xi\right]\right\} = 0.6439-0.1005\rho_r \quad (2\text{-}91)$$

For polar gases, $2.2 < \rho_r \le 2.6$:

$$\log\left\{4-\log\left[\left(\frac{\eta-\eta^o}{\text{mPa}\cdot\text{s}}\right)\xi\right]\right\} = 0.6439-0.1005\rho_r-0.000475(\rho_r^3-10.65)^2 \quad (2\text{-}92)$$

where $\rho_c = P_c/(Z_cRT_c)$ and

$$\xi = 2173.4\left(\frac{T_c}{\text{K}}\right)^{1/6}\left(\frac{M}{\text{kg/kmol}}\right)^{-1/2}\left(\frac{P_c}{\text{MPa}}\right)^{-2/3} \quad (2\text{-}93)$$

Example Estimate the vapor viscosity of CO_2 at 350 K and 20 MPa if $\eta^o = 0.0174$ mPa · s and $Z = 0.4983$ (estimated from Lee-Kesler method, see section on density).
Required properties: From the DIPPR 801 database,

$M = 44.01$ kg/kmol $T_c = 304.21$ K $P_c = 7.383$ MPa
$Z_c = 0.274$ $\mu = 0$ D (nonpolar)

Auxiliary quantities:

$$\xi = (2173.4)(304.21)^{1/6}(44.01)^{-1/2}(7.383)^{-2/3} = 224.1$$

$$\rho_c = \frac{7.383\ \text{MPa}}{0.274\,[0.008314\ \text{m}^3\text{MPa}/(\text{K}\cdot\text{kmol})](304.21\ \text{K})} = 10.654\frac{\text{kmol}}{\text{m}^3}$$

$$\rho_r = \frac{\rho}{\rho_c} = \frac{P}{ZRT\rho_c} = \frac{20\ \text{MPa}}{0.4983[0.008314\text{m}^3\cdot\text{MPa}/(\text{K}\cdot\text{kmol})](350\ \text{K})(10.654\ \text{m}^3\cdot\text{kmol})} = 1.295$$

Calculation using Eq. (2-88) for nonpolar fluids:

$$\left[224.1\left(\frac{\eta-\eta^o}{\text{mPa}\cdot\text{s}}\right)+1\right]^{1/4} = 1.0230+0.23364(1.295)+0.58533(1.295)^2$$

$$-0.40758(1.295)^3+0.093324(1.295)^4 = 1.684$$

$$\eta = \frac{1.684^4-1}{224.1}\text{mPa}\cdot\text{s}+0.0174\ \text{mPa}\cdot\text{s} = 0.0489\ \text{mPa}\cdot\text{s}$$

This differs from the experimental value of 0.0473 mPa · s by 3.4 percent.

Liquids Liquid viscosity can be correlated as a function of temperature for low pressures. Usually the correlation is based on the Andrade equation [Andrade, E. N. da C., *Nature*, **125** (1930): 309]

$$\ln(\eta) = A+\frac{B}{T} \quad (2\text{-}94)$$

or an extension of it. For example, the DIPPR 801 database uses the equation

$$\ln(\eta) = A+\frac{B}{T}+C\ln T+DT^E \quad (2\text{-}95)$$

which is analogous to the Riedel [Riedel, L., *Chem. Ing. Tech.*, **26** (1954): 83] vapor pressure equation.

Currently the most accurate method for predicting pure liquid viscosity is the GC method by Hsu et al. It has been found that most liquids have a viscosity between 0.15 mPa · s (or cP) and 0.55 mPa · s at the normal boiling point, and this "rule" can be used as a valuable criterion to validate estimated viscosities as a function of temperature.

Recommended Method Hsu method.
Reference: Hsu, H.-C., Y.-W. Sheu, and C.-H. Tu, *Chem. Eng. J.*, **88** (2002): 27.
Classification: Group contributions.
Expected uncertainty: 20 percent.
Applicability: Organic liquids; $T_r < 0.75$.
Input data: P_c and molecular structure.
Description: The temperature dependence of the liquid viscosity is given by

$$\ln\left(\frac{\eta}{\text{mPa}\cdot\text{s}}\right) = \sum_{i=1}^{N}a_i+T\sum_{i=1}^{N}b_i+\frac{\sum_{i=1}^{N}c_i}{T^2}+\left(\sum_{i=1}^{N}d_i\right)\ln\left(\frac{P_c}{\text{bar}}\right) \quad (2\text{-}96)$$

where P_c is critical pressure and a_i, b_i, c_i, and d_i are the group contributions obtained from Table 2-174.

Example Estimate the liquid viscosity of benzotrifluoride at 303.15 K.
Structural information:

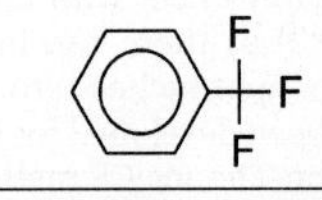

Group	Number	a	$100b$	$0.0001c$	d
>C<	1	1.0031	−0.3677	−6.0316	1.1972
$(=CH-)_A$	5	−0.8570	−0.0098	2.4376	0.1311
$(=C<)_A$	1	0.7896	−0.0231	−0.9222	0.1928
$(-F)_3$	1	1.5394	0.8465	17.8121	−2.9915
	Total	−0.9529	0.4067	23.0463	−0.9460

TABLE 2-174 Group Contributions for the Hsu et al. Method*

Table-specific nomenclature: R = in nonaromatic ring, A = in aromatic ring, RC = attached to nonaromatic ring, AC = attached to aromatic ring, X = halogen, $(-X)_n$ = n X atoms attached to same C atom

Group	*a*	100*b*	0.0001*c*	*d*
		C, H Groups		
CH_4	−1.7296	−1.0563	0.8928	−0.0019
—CH_3	0.0570	−0.2382	0.7556	−0.1765
—CH_2—	−0.1497	0.0060	1.4157	0.0751
>CH—	−2.2942	0.4028	4.5094	0.6679
>C<	1.0031	−0.3677	−6.0316	1.1972
=CH_2	0.9256	−0.2656	0.9860	−0.4417
=CH—	1.3365	0.1612	1.9408	0.2507
=C<	−3.5020	0.4305	3.1287	1.0465
≡CH	87.6040	−0.1106	4.4245	−24.1836
≡C—	−91.6154	−0.0111	0.3265	25.0542
$(-CH_2-)_R$	6.0416	−0.1778	0.8437	−1.5184
$(>CH-)_R$	−33.8745	0.7637	7.2433	8.5951
$(=CH-)_R$ cycloalkene	1.2028	−0.0120	2.0143	−0.3677
$(>C<)_R$ spirocyclane	−56.2158	1.7694	19.0452	13.3885
$(=CH-)_A$	−0.8570	−0.0098	2.4376	0.1311
$(=C<)_A$ cycloalkene	0.7896	−0.0231	−0.9222	0.1928
$(=C<)_A$ bi/terphenyl	2.0973	0.0444	8.1690	−0.4351
$(=C<)_A$ naphthalene	0.4392	0.0683	8.8426	−0.1685
$(=C<)_A$ turpentine	27.3350	1.2165	34.2857	−11.6500
$(=C<)_A$ tetralin	14.2586	−0.8665	−14.7474	−2.7574
		O, S Groups		
—OH primary for C<3	5.7852	−0.5310	9.5499	−1.0300
—OH primary for C>2	1.4351	−1.0010	13.8366	0.3418
—OH secondary	−2.6895	−0.3645	29.8404	0.4246
—OH tertiary	−18.5630	2.4275	78.5417	0.9650
$(-OH)_{RC}$	16.7808	0.8509	77.1759	−6.9285
—OH polyhydric	−0.0125	−0.3634	23.2329	−0.0172
$(-OH)_{AC}$	−2.0856	0.6362	50.0840	−1.0539
—OH alkoxyalcohol	−2.6991	−0.4377	17.2243	0.7139
—O—	−0.7185	0.0985	2.9405	0.1149
$(-O-)_R$	−29.8045	−0.2847	−4.3145	8.3131
$(-O-)_{AC}$	−2.3454	0.0872	6.4296	0.5389
—CHO	−0.8288	−0.2612	3.7241	0.2386
>CO	−2.6622	0.1142	6.7008	0.7348
$(>CO)_R$	45.9143	−0.2405	3.8828	−12.4994
HCOOH	−2.7291	0.0413	27.4079	0.0002
—COOH for C<7	−4.0451	−0.1841	12.6878	1.1139
—COOH for C>6	−0.6721	−0.1693	20.0309	0.0279
HCOO—	−3.3731	−0.0113	9.4694	0.6071
—COO— for C<8	−0.0635	−0.2162	1.9325	0.4686
—COO— for C>7	−2.5390	0.0006	5.4231	0.8717
>CHO—	−5.4872	1.5834	34.5474	−0.4244
—(CO)—O—(CO)— anhydride	−11.8236	0.0111	7.2831	3.6587
—O—(CO)—O— carbonate	−8.0314	0.2848	9.3746	2.1486
$(>NO)_R$	−16.9531	1.0614	49.1049	2.8583
—NO_2	−13.0333	0.1801	12.9392	2.8987
=$CHNO_2$	−1.9653	0.1322	15.8672	−0.0701
$(-NO_2)_{AC}$	−1.2954	0.0427	12.1837	−0.0948
—S—	−3.2767	0.0779	4.4123	0.9549
—SH primary	−2.1030	−0.0965	6.0066	0.3464
—SH secondary	−0.2481	−0.3285	1.9387	0.1148
—SH tertiary	−12.3498	1.2621	23.1473	1.3950
—CSO— for C<13	−15.2678	0.5248	14.2694	3.7646
—CSO— for C>12	3.7475	−1.2592	−23.9353	0.8329
>SO	−32.8607	0.6232	27.5184	7.7525
		N, X Groups		
—NH_2	−1.1345	−0.2126	7.0544	0.1336
—NH—	−6.9489	−0.1723	5.7804	1.6467
—N<	−2.1403	0.4842	6.1893	0.4718
$(-NH_2)_{AC}$	−6.3646	−0.0180	23.2752	1.0653
$(-NH-)_{AC}$	−1.7592	0.2208	14.9707	0.1171
$(-N<)_{AC}$	−1.2982	0.5975	14.0415	−0.0031
$HCONH_2$	−1.5435	−0.2774	31.8007	0.0001
HCONH—	−8.1097	0.0432	20.9135	1.8795
HCON<	−122.3280	26.4615	394.1670	0.3530
—$CONH_2$	−6.7363	0.1316	45.5193	1.2172
—CONH—	8.9977	1.5664	60.8742	−4.6399
—$COONH_2$	17.8400	−4.5188	−62.0987	1.2353
—COONH—	−10.1316	0.6712	37.9465	1.9199
$(>NH)_R$	−0.1589	0.1910	12.0578	−0.0276

(Continued)

TABLE 2-174 Group Contributions for the Hsu et al. Method* *(Continued)*

Table-specific nomenclature: R = in nonaromatic ring, A = in aromatic ring, RC = attached to nonaromatic ring, AC = attached to aromatic ring, X = halogen, $(\text{—X})_n$ = n X atoms attached to same C atom

Group	a	$100b$	$0.0001c$	d
		N, X Groups		
$(\text{=N—})_R$	−4.7601	0.1120	6.98437	0.9719
—C≡N	−2.7194	−0.1324	7.7955	0.6293
$(\text{—C≡N})_{AC}$	0.9435	−0.0086	8.6310	−0.6443
—Cl primary	−1.7997	−0.3851	3.0118	0.5524
=CHCl	1.5851	−0.1934	3.7798	−0.4748
$(\text{—Cl})_2$	−3.0561	−1.0770	0.1882	1.2223
$(\text{—Cl})_3$	−1.3357	−0.3220	8.8683	0.1702
$(\text{—Cl})_4$	4.2070	−0.4130	13.3194	−1.1972
$(\text{—Cl})_{AC}$	−0.3083	−0.0623	4.1382	−0.2644
—F primary	−9.4982	0.2607	11.3406	1.8461
$(\text{—F})_2$	−10.3980	−1.1189	1.3134	2.6681
$(\text{—F})_3$	1.5394	0.8465	17.8121	−2.9915
$(\text{—F})_{AC}$	0.4079	−0.2352	−0.1505	−0.2893
(—F)(—Cl)	−0.8565	−0.3682	4.6451	−0.0751
$(\text{—F})(\text{—Cl})_2$	−3.4552	−0.5629	3.6831	0.3613
$(\text{—F})_2(\text{—Cl})$	54.2824	0.0109	5.9474	−14.5771
$(\text{—F})_2(\text{—Cl})_2$	−2.1710	0.1403	10.3743	−1.1972
—Br primary	−0.7586	−0.6623	−2.4228	0.7385
—Br secondary	−279.0030	−0.3420	1.4253	73.6293
$(\text{—Br})_{AC}$	−8.1919	−0.1635	3.0150	0.0621
—I primary	−1.4672	−0.2787	4.3362	0.5635
$(\text{—I})_{AC}$	70.9918	−0.0245	7.2061	−18.9106
—(CO)—Cl	−2.3300	−0.0470	8.2815	0.4485

*Hsu, H.-C., Y.-W. Sheu, and C.-H. Tu, *Chem. Eng. J.*, **88** (2002): 27

Supporting values:

$$P_c = 32.1 \text{ MPa}$$

Calculation using Eq. (2-96):

$$\frac{\eta}{\text{mPa}\cdot\text{s}} = \exp\left[-0.9529 + (0.004067)(303.15) + \frac{230{,}463}{(303.15)^2} - 0.9460\ln(32.1)\right] = 0.610$$

The estimated value is 20 percent higher than the DIPPR 801 value of 0.509 mPa · s. Note that when the calculation is repeated at the normal boiling point (375.2 K), one obtains 0.343 mPa · s which is within the range of the aforementioned empirical rule.

Liquid Mixtures Most methods for estimating liquid mixture viscosity interpolate between the pure-component values at the same temperature. The Grunberg-Nissan equation [Grunberg, L., and A. H. Nissan, *Nature*, **164** (1949): 799]

$$\ln \eta = \sum_i^C x_i \ln \eta_i + \frac{1}{2}\sum_{i=1}^C \sum_{j=1}^C x_i x_j G_{ij} \tag{2-97}$$

is commonly used for nonaqueous mixtures. The parameter G_{ij} generally must be regressed from an experimental mixture viscosity. However, G_{ij} can be set to zero for hydrocarbon mixtures with expected errors in the mixture viscosity of about 15 percent.

Estimation of liquid mixture viscosity without any mixture data is difficult because the viscosity is strongly affected by large molecular size differences and strong cross-interactions between different types of molecules. The UNIFAC-VISCO method described below can be used to predict liquid viscosity of organic mixtures without any mixture data. It can estimate mixture viscosity to a limited accuracy, but it is limited in scope by the small number of group contributions currently available.

Recommended Method UNIFAC-VISCO method.

Reference: Chevalier, J. L., P. Petrino, and Y. Gaston-Bonhomme, *Chem. Eng. Sci.*, **43** (1988): 1303; Gaston-Bonhomme, Y., P. Petrino, and J. L. Chevalier, *Chem. Eng. Sci.*, **49** (1994): 1799.

Classification: Group contributions.

Expected uncertainty: 20 percent.

Applicability: Organic liquids.

Input data: Molecular structure; pure-component molar volumes and viscosities at the mixture temperature.

Description: Liquid mixture viscosity can be estimated in a manner similar to the UNIFAC method employed for mixture excess Gibbs energy and activity coefficients. The primary equation is

$$\ln\left(\frac{\eta}{\text{mPa}\cdot\text{s}}\right) = \sum_{i=1}^C x_i \ln\left(\frac{\eta_i}{\text{mPa}\cdot\text{s}}\cdot\frac{V_i}{V_m}\right) + \frac{g_c^E}{RT} - \frac{g_r^E}{RT} \tag{2-98}$$

where V_m is the mixture molar volume and V_i is the pure-component molar volume of component i. The combinatorial and residual excess Gibbs energies are calculated as in the standard UNIFAC method for activity coefficients (see [PGL5]) and for brevity is not shown here. However, the group interactions ψ_{mn} are calculated using the interaction parameters α_{mn} obtained from Table 2-175 in the equation

$$\psi_{mn} = \exp\left(-\frac{\alpha_{mn}}{298.15}\right) \tag{2-99}$$

TABLE 2-175 UNIFAC-VISCO* Group Interaction Parameters α_{mn}

m/n	CH_2	CH_3	CH_{2cy}	CH_{ar}	Cl	CO	COO	OH	CH_3OH
CH_2	0	66.53	224.9	406.7	60.30	859.5	1172.0	498.6	−219.7
CH_3	−709.5	0	−130.7	−119.5	82.41	11.86	−172.4	594.4	−228.7
CH_{2cy}	−538.1	187.3	0	8.958	251.4	−125.4	−165.7	694.4	−381.53
CH_{ar}	−623.7	237.2	50.89	0	177.2	128.4	−49.85	419.3	−88.81
Cl	−710.3	375.3	−163.3	−139.8	0	−404.3	−525.4	960.2	−165.4
CO	586.2	−21.56	740.6	−117.9	−4.145	0	29.20	221.5	55.52
COO	541.6	−44.25	416.2	−36.17	240.5	22.92	0	186.8	69.62
OH	−634.5	1209.0	−138	197.7	195.7	664.1	68.35	0	416.4
CH_3OH	−526.1	653.1	751.3	51.31	−140.9	−22.59	−286.2	−23.91	0

*Chevalier, J. L., P. Petrino, and Y. Gaston-Bonhomme, *Chem. Eng. Sci.*, **43** (1988): 1303; Gaston-Bonhomme, Y., P. Petrino, and J. L. Chevalier, *Chem. Eng. Sci.*, **49** (1994): 1799.

Example Estimate the viscosity of a mixture of 51.13 mol% ethanol(1) and 48.87 mol% benzene(2) at 298.15 K.

Required input: Values from the DIPPR 801 database for the pure components at 298.15 K are η_1 = 1.0774 mPa · s, η_2 = 0.5997 mPa · s, V_1 = 0.05862 m³/kmol, and V_2 = 0.08948 m³/kmol.

Groups, area fractions, and volume fractions:

Group	R	Q	N_1	N_2
CH_3	0.9011	0.8480	1	0
CH_2	0.6744	0.5400	1	0
CH_{ar}	0.5313	0.4000	0	6
OH	1.0000	1.2000	1	0
Group	r_1	q_1	r_2	q_2
CH_3	0.9011	0.848	0	0
CH_2	0.6744	0.54	0	0
CH_{ar}	0	0	3.1878	2.4
OH	1	1.2	0	0
Total	2.5755	2.588	3.1878	2.4

where in the above table

$$q_i = \sum_{k=1}^{N} N_{i,k} Q_k \quad \text{and} \quad r_i = \sum_{k=1}^{N} N_{i,k} R_k$$

$$\theta_1 = \frac{x_1 q_1}{\sum_{i=1}^{4} x_i q_i} = \frac{(0.5113)(2.588)}{(0.5113)(2.588) + (0.4887)(2.4)} = 0.53 \qquad \theta_2 = 0.47$$

$$\phi_1 = \frac{x_1 r_1}{\sum_{i=1}^{4} x_i r_i} = \frac{(0.5113)(2.5755)}{(0.5113)(2.5755) + (0.4887)(3.1878)} = 0.458 \qquad \phi_2 = 0.542$$

UNIFAC combinatorial term:

$$\frac{g_C^E}{RT} = \sum_{i=1}^{2} x_i \ln\frac{\phi_i}{x_i} + 5\sum_{i=1}^{2} x_i q_i \ln\frac{\theta_i}{\phi_i} = 0.124$$

Group interactions:

	α_{mn}			
m/n group	CH_3	CH_2	CH_{ar}	OH
CH_3	0	−709.5	−119.5	594.4
CH_2	66.53	0	406.7	498.6
CH_{ar}	237.2	−623.7	0	419.3
OH	1209	−634.5	197.7	0
	ψ_{mn}			
m/n group	CH_3	CH_2	CH_{ar}	OH
CH_3	1.000	10.801	1.493	0.136
CH_2	0.800	1.000	0.256	0.188
CH_{ar}	0.451	8.100	1.000	0.245
OH	0.017	8.399	0.515	1.000

The α_{mn} values were obtained from Table 2-175, and ψ_{mn} values were calculated from Eq. (2-99).

Group fractions in the mixture:

Group	N	X	XQ	Θ	$\ln\gamma$
CH_3	0.5113	0.1145	0.097083	0.17370	0.293
CH_2	0.5113	0.1145	0.061822	0.11061	−0.873
CH_{ar}	2.9322	0.6565	0.262618	0.46988	0.066
OH	0.5113	0.1145	0.137382	0.24581	1.077
Sum	4.4661		0.558905		

Here $\Theta_m = \frac{X_m Q_m}{\sum_{i=1}^{4} X_i Q_i}$ and $\ln\gamma_m = Q_m\left[1 - \ln\left(\sum_{i=1}^{4}\Theta_i\psi_{i,m}\right) - \sum_{i=1}^{4}\left(\frac{\Theta_i\psi_{m,i}}{\sum_{j=1}^{4}\Theta_j\psi_{j,i}}\right)\right]$

Group fractions in pure components:

Ethanol					
Group	N	X	XQ	Θ	$\ln\gamma$
CH_3	1	0.3333	0.283	0.3277	0.5306
CH_2	1	0.3333	0.180	0.2087	−0.9405
CH_{ar}	0	0.0000	0.000	0.0000	0.2095
OH	1	0.3333	0.400	0.4637	0.6179
Sum	3		0.863		

Benzene					
Group	N	X	XQ	Θ	$\ln\gamma$
CH_3	0	0	0	0	0.257
CH_2	0	0	0	0	−0.728
CH_{ar}	6	1	0.4	1	0.000
OH	0	0	0	0	2.270
Sum	6		0.4		

The pure-component Θ and ln γ equations are the same as shown above for the mixture groups.

UNIFAC residual term:

$$\frac{g_r^E}{RT} = \sum_{i=1}^{2} x_i\left[\sum_{m=1}^{4} N_{m,i}(\ln\gamma_m - \ln\gamma_{m,i})\right] = 0.3425$$

where N_m and ln γ_m refer to the mixture and $N_{m,i}$ and ln $\gamma_{m,i}$ refer to the pure-component values.

Mixture volume:

$$V_m = \sum_{i=1}^{2} x_i V_i = 0.5113\left(0.05862\frac{\text{m}^3}{\text{kmol}}\right) + 0.4887\left(0.08948\frac{\text{m}^3}{\text{kmol}}\right) = 0.07370\frac{\text{m}^3}{\text{kmol}}$$

Using Eq. (2-98):

$$\ln\left(\frac{\eta}{\text{mPa}\cdot\text{s}}\right) = 0.5113\ \ln\left[1.0774\left(\frac{0.05862}{0.07370}\right)\right] + 0.4887\ \ln\left[0.5997\left(\frac{0.08948}{0.07370}\right)\right] + 0.124 - 0.3425 = -0.4523$$

$$\eta = \exp(-0.4523)\ \text{mPa}\cdot\text{s} = 0.636\ \text{mPa}\cdot\text{s}$$

The estimated value is 6.6 percent below the reported experimental value of 0.681 mPa · s [Kouris, S., and C. Panayiotou, *J. Chem. Eng. Data*, **34** (1989): 200].

THERMAL CONDUCTIVITY

Thermal conductivity, *k*, is a measure of the rate at which heat conducts through the material and is defined as the proportionality constant in Fourier's law of heat conduction that relates the gradient of temperature to the heat flux or flow per unit area. In SI, it has the units of W/(m · K). The conduction mechanism in gases is primarily via molecular collisions, and *k* increases with increasing temperature (increasing molecular velocity). The temperature dependence of low-pressure, gas-phase thermal conductivity is adequately correlated with

$$k = \frac{AT^B}{1 + \frac{C}{T}} \tag{2-100}$$

In dense media such as liquids, energy transfers more efficiently through the intermolecular force fields than through collisions. As a result, liquid thermal conductivity generally decreases with increasing temperature (except for water, aqueous solutions, and a few multihydroxy and multiamine compounds), corresponding to the decrease in density with increased temperature. The temperature dependence of liquid thermal conductivity at low to moderate pressures has been found to be well correlated by [Jamieson, D. T., *J. Chem. Eng. Data* **24** (1979): 244]

$$k = A(1 + B\tau^{1/3} + C\tau^{2/3} + D\tau) \tag{2-101}$$

where $\tau = 1 - T/T_C$. For nonassociating liquids, this equation can be simplified to two parameters by setting $C = 1 - 3B$ and $D = 3B$, generally without much loss in accuracy. Below or near the normal boiling point, the temperature dependence of liquid thermal conductivity is nearly linear for modest temperature ranges and can be represented by

$$k = A - BT \tag{2-102}$$

where B is generally in the range of 1×10^{-4} to 3×10^{-4} W/(m · K²).

Gases Methods for estimating low-pressure gas thermal conductivities are based on kinetic theory and generally correlate the dimensionless group $kM/\eta C_\upsilon$ (M = molecular weight, η = viscosity, C_υ = isochoric heat capacity), known as the Eucken factor. The method of Stiel and Thodos is recommended for pure nonpolar compounds, and the method of Chung is recommended for pure polar compounds.

Recommended Method Stiel-Thodos method.

Reference: Stiel, L. I., and G. Thodos, *AIChE J.*, **10** (1964): 26.

Classification: Empirical extension of theory.

Expected uncertainty: 15 percent.

Applicability: Pure nonpolar gases at low pressure.

Input data: M, T_c, η, and C_υ.

Description: The following equations may be used depending upon the molecular shape:

$$\frac{kM}{\eta C_\upsilon} = 2.5 \qquad \text{monatomic} \tag{2-103}$$

$$\frac{kM}{\eta C_\upsilon} = 1.30 + \left(\frac{R}{C_\upsilon}\right)\left(1.7614 - \frac{0.3523}{T_r}\right) \qquad \text{linear molecules} \tag{2-104}$$

$$\frac{kM}{\eta C_\upsilon} = 1.15 + 2.033\left(\frac{R}{C_\upsilon}\right) \qquad \text{nonlinear molecules} \tag{2-105}$$

where η = viscosity at same conditions as desired for k. Because this method is only applicable at low pressures, C_υ may usually be calculated as $C_p^o - R$, where C_p^o is the ideal gas isobaric heat capacity.

Example Estimate the low-pressure thermal conductivity of toluene vapor at 500 K.

Required properties from the DIPPR 801 database:

$$T_c = 591.75 \text{ K} \qquad M = 92.138 \text{ g/mol} \qquad \eta(500 \text{ K}) = 1.1408 \times 10^{-5} \text{ Pa} \cdot \text{s}$$

$$C_\upsilon = C_p^o - R = (170.78 - 8.314) \text{ J/(mol} \cdot \text{K)} = 162.47 \text{ J/(mol} \cdot \text{K)}$$

Auxiliary quantities:

$$T_r = 500/591.75 = 0.845 \qquad R/C_\upsilon = (8.314)/(162.47) = 0.0512$$

From Eq. (2-105):

$$k = [1.15 + (2.033)(0.0512)]\left[\frac{(1.1408 \times 10^{-5} \text{ Pa}\cdot\text{s})\left(162.47 \frac{\text{J}}{\text{mol}\cdot\text{K}}\right)}{92.138 \frac{\text{g}}{\text{mol}}}\right] = 25.2 \frac{\text{mW}}{\text{m}\cdot\text{K}}$$

The estimated value is 18 percent below the DIPPR 801 value of 30.76 mW/(m · K).

Recommended Method Chung-Lee-Starling method.

Reference: Chung, T.-H., L. L. Lee, and K. E. Starling, *Ind. Eng. Chem. Fundam.*, **23** (1984): 8.

Classification: Corresponding states.

Expected uncertainty: 15 percent.

Applicability: Pure organic gases at low pressure.

Input data: C_υ, ω, T_c, M, and η.

Description: The following equations apply:

$$\frac{kM}{\eta C_\upsilon} = 3.75\, \Psi\left(\frac{R}{C_\upsilon}\right) \tag{2-106}$$

$$\Psi = 1 + \alpha\left(\frac{0.215 + 0.28288\alpha - 1.061\beta + 0.26665\gamma}{0.6366 + \beta\gamma + 1.061\alpha\beta}\right)$$

$$\alpha = \frac{C_\upsilon}{R} - 1.5 \qquad \beta = 0.7862 - 0.7109\omega + 1.3168\omega^2 \tag{2-107}$$

$$\gamma = 2.0 + 10.5\, T_r^2 \tag{2-108}$$

Example Estimate the low-pressure thermal conductivity of naphthalene vapor at 500 K.

Required properties from the DIPPR 801 database:

$$T_c = 748.4 \text{ K} \qquad M = 128.17 \text{ g/mol} \qquad \omega = 0.30203$$

$$\eta(500 \text{ K}) = 1.0173 \times 10^{-5} \text{ Pa} \cdot \text{s}$$

$$C_\upsilon = C_p^o - R = (219.82 - 8.314) \text{ J/(mol} \cdot \text{K)} = 211.51 \text{ J/(mol} \cdot \text{K)}$$

Auxiliary quantities [Eqs. (2-107) and (2-108)]:

$$T_r = 500/748.4 = 0.6681 \qquad R/C_\upsilon = (8.314)/(211.51) = 0.0393$$

$$\gamma = 2.0 + (10.5)(0.6681)^2 = 6.6866 \qquad \alpha = (0.0393)^{-1} - 1.5 = 23.9388$$

$$\beta = 0.7862 - (0.7109)(0.30203) + (1.3168)(0.30203)^2 = 0.6916$$

$$\Psi = 1 + (23.9388)\left[\frac{0.215 + 0.28288\,(23.9388) - 1.061(0.6916) + 0.26665(6.6866)}{0.6366 + 0.6916\,(6.6866) + 1.061(23.9388)\,(0.6916)}\right]$$

$$= 9.4273$$

From Eq. (2-106):

$$k = (3.75)\,(9.4273)\left[\frac{(1.1408 \times 10^{-5} \text{ Pa}\cdot\text{s})\left(8.314 \frac{\text{J}}{\text{mol}\cdot\text{K}}\right)}{128.17 \frac{\text{g}}{\text{mol}}}\right] = 23.33 \frac{\text{mW}}{\text{m}\cdot\text{K}}$$

The estimated value is 1.0 percent above the DIPPR 801 value of 23.09 mW/(m·K).

Liquids For hydrocarbons at low to moderate pressures, a modification of the Pachaiyappan method should be used. For nonhydrocarbons, the Baroncini method provides accurate liquid thermal conductivity estimates for compounds clearly belonging to one of the chemical families specified below. Otherwise, the Missenard method is recommended as a general method for estimating thermal conductivity of pure liquids at ambient pressure.

Recommended Method Modified Pachaiyappan.

Reference: Pachaiyappan, V., S. H. Ibrahim, and N. R. Kuloor, *Chem. Eng.* **74**(4) (1967): 140; *API Technical Databook*, 10th ed., chap. 12, 2017.

Classification: Empirical correlation.

Expected uncertainty: 10 percent.

Applicability: Hydrocarbons only; low to moderate pressures.

Input data: M, T_b, and T_c.

Description:

$$\frac{k}{\text{W}\cdot\text{m}^{-1}\text{K}^{-1}} = \frac{C\left(\dfrac{M}{\text{g}\cdot\text{mol}^{-1}}\right)^m}{\left(\dfrac{V_{293}}{\text{cm}^3\cdot\text{mol}^{-1}}\right)}\left[\frac{3 + 20(1 - T_r)^{2/3}}{3 + 20(1 - T_{r,293})^{2/3}}\right] \tag{2-109}$$

where M is molecular weight, V_{293} is the molar volume at 293.15 K, T_r is the reduced temperature, $T_{r,293}$ = (293.15 K)/(T_c) and the correlation parameters C and m are obtained from the table below:

Classification	C	m
Unbranched, straight-chain hydrocarbon	0.1811	1.001
All branched, cyclic and aromatic hydrocarbons	0.4407	0.7717

Example Estimate the thermal conductivity of liquid *n*-butylbenzene at low pressure and 333.15 K.

Required properties from DIPPR 801 database:

$$M = 134.218 \text{ g/mol} \qquad T_c = 660.5 \text{ K} \qquad V_{293} = 162.01 \text{ cm}^3\text{/mol}$$

Auxiliary properties:

$$T_r = (333.15 \text{ K})/(660.5 \text{ K}) = 0.5044 \qquad T_{r,293} = (293.15 \text{ K})/(660.5 \text{ K}) = 0.4438$$

Since this is an aromatic hydrocarbon,

$$C = 0.4407 \text{ and } m = 0.7717 \text{ (from the above table)}$$

From Eq. (2-109):

$$\frac{k}{(\text{W}\cdot\text{m}^{-1}\text{K}^{-1})} = \frac{(0.4407)(134.218)^{0.7717}}{162.01}\left[\frac{3 + 20(1 - 0.5044)^{2/3}}{3 + 20(1 - 0.4438)^{2/3}}\right] = 0.112$$

The estimated value is 5 percent below the experimental value of 0.118 W/(m·K) reported by Rastorguev and Pugach [Rastorguev, Yu. L., and V. V. Pugach, *Izv. Vyssh. Uchebn. Zaved., Neft Gaz*, **13** (1970): 69].

Recommended Method 1 Baroncini method.
Reference: Baroncini, C., F. DiFilippo, G. Latini, and M. Pacetti, *Int. J. Thermophys.*, **2** (1981): 21.
Classification: Empirical correlation.
Expected uncertainty: 10 percent.
Applicability: Particularly accurate for the following families: acetates, aliphatic ethers, halogenated compounds, dicarboxylic acids, ketones, aliphatic alcohols, aliphatic acids, propionates and butyrates, and unsaturated aliphatic esters.
Input data: M, T_b, and T_c.
Description:

$$\frac{k}{\text{W/(m}\cdot\text{K)}} = A\left(\frac{T_b}{\text{K}}\right)^{\alpha}\left(\frac{M}{\text{g/mol}}\right)^{-\beta}\left(\frac{T_c}{\text{K}}\right)^{-\gamma}\frac{(1-T_r)^{0.38}}{T_r^{1/6}} \tag{2-110}$$

where A, α, β, and γ are obtained from Table 2-176.

Example Estimate the thermal conductivity of liquid *p*-cresol at 400 K.
Required properties from DIPPR 801 database:

$$M = 108.1378 \text{ g/mol} \qquad T_c = 704.65 \text{ K} \qquad T_b = 475.133 \text{ K}$$

Auxiliary properties:

$$T_r = T/T_c = (400 \text{ K})/(704.65 \text{ K}) = 0.5677$$

From Table 2-176 for alcohols:

$$A = 0.00339 \qquad \alpha = 1.2 \qquad \beta = \frac{1}{2} \qquad \gamma = 0.167$$

From Eq. (2-110):

$$\frac{k}{\text{W/(m}\cdot\text{K)}} = (0.00339)(475.13)^{1.2}(108.1378)^{-1/2}(704.65)^{-0.167}\frac{(1-0.5677)^{0.38}}{0.5677^{1/6}} = 0.142$$

The estimated value is 7.6 percent higher than the DIPPR 801 value of 0.132 W/(m·K).

Recommended Method 2 Missenard method.
Reference: Missenard, A., *Comptes Rendus*, **260** (1965): 5521.
Classification: Corresponding states.
Expected uncertainty: 20 percent.
Applicability: Organic compounds; nonassociating.
Input data: T_c, n_A (number of atoms in molecule), ρ_{273} (liquid density at 273.15 K), T_b, M, $C_{p,273}$ (liquid heat capacity at 273.15 K).
Description:

$$\frac{k_{273}}{\text{mW/(m}\cdot\text{K)}} = \left(\frac{8.4}{N_A^{1/4}}\right)\left(\frac{T_b}{\text{K}}\right)^{1/2}\left(\frac{\rho_{273}}{\text{g/m}^3}\right)^{1/2}\left(\frac{M}{\text{g/mol}}\right)^{-1/2}\left[\frac{C_{p,273}}{\text{J/(mol}\cdot\text{K)}}\right] \tag{2-111}$$

$$k = \frac{k_{273}[3+20(1-T_r)^{2/3}]}{3+20(1-T_{r,273})^{2/3}} \tag{2-112}$$

where $T_{r,273} = (273 \text{ K})/T_c$.

TABLE 2-176 Correlation Parameters for Baroncini et al. Method* for Estimation of Thermal Conductivity

Family	A	α	β	γ
Saturated hydrocarbons	0.00350	1.2	0.5	0.167
Olefins	0.0361	1.2	1	0.167
Cycloparaffins	0.0310	1.2	1	0.167
Aromatics	0.0346	1.2	1	0.167
Alcohols	0.00339	1.2	0.5	0.167
Organic acids	0.00319	1.2	0.5	0.167
Ketones	0.00383	1.2	0.5	0.167
Esters	0.0415	1.2	1	0.167
Ethers	0.0385	1.2	1	0.167
Refrigerants				
R20, R21, R22, R23	0.562	0	0.5	−0.167
Others	0.494	0	0.5	−0.167

*Baroncini, C., et al., *Int. J. Thermophys.*, **2** (1981): 21.

Example Estimate the thermal conductivity of *m*-xylene at 350 K.
Required properties from DIPPR 801 database:

$$T_c = 617 \text{ K} \qquad n_A = 18 \qquad \rho_{273} = 7.6812 \text{ kmol/m}^3$$

$$T_b = 412.27 \text{ K} \qquad M = 106.165 \text{ kg/kmol} \qquad C_{p,273} = 200.64 \text{ kJ/(kmol}\cdot\text{K)}$$

Auxiliary properties:

$$T_r = 350/617 = 0.5673 \qquad T_{br} = 412.27/617 = 0.6682$$

$$T_{r273} = 273/617 = 0.4425$$

From Eq. (2-111):

$$\frac{k_{273}}{\text{mW/(m}\cdot\text{K)}} = (8.4)(412.27)^{1/2}(0.007681)^{1/2}(106.165)^{-1/2}(200.64)(18)^{-0.25} = 141.3$$

From Eq. (2-112):

$$k = \frac{k_{273}[3+20(1-T_r)^{2/3}]}{3+20(1-T_{r,273})^{2/3}} = \frac{\left(141.3\frac{\text{mW}}{\text{m}\cdot\text{K}}\right)[3+20(1-0.5673)^{2/3}]}{3+20(1-0.4425)^{2/3}} = 123.3\frac{\text{mW}}{\text{m}\cdot\text{K}}$$

The estimated value is 4.5 percent above the DIPPR 801 value of 118.0 mW/(m·K).

Liquid Mixtures The thermal conductivity of liquid mixtures generally shows a modest negative deviation from a linear mass-fraction average of the pure-component values. Although more complex methods with some improved accuracy are available, two simple methods are recommended here that require very little additional information. The first method applies only to binary mixtures while the second can be used for multiple components.
Recommended Method Filippov correlation.
References: Filippov, L. P., *Vest. Mosk. Univ., Ser. Fiz. Mat. Estestv. Nauk*, **10** (1955): 67; Filippov, L. P., and N. S. Novoselova, Sugden, *Vest. Mosk. Univ., Ser. Fiz. Mat. Estestv. Nauk*, **10** (1955): 37.
Classification: Empirical correlation.
Expected uncertainty: 4 to 8 percent.
Applicability: Binary liquid mixtures.
Input data: Pure-component thermal conductivities k_i at mixture conditions; w_i.
Description: The mixture thermal conductivity is calculated from the pure-component values using

$$k = w_1k_1 + w_2k_2 - 0.72w_1w_2|k_2 - k_1| \tag{2-113}$$

where w_i is the mass fraction of pure fluid i and k_i is the thermal conductivity of pure component i at the mixture temperature.
Recommended Method Li correlation.
References: Li, C. C., *AIChE J.*, **22** (1976): 927.
Classification: Empirical correlation.
Expected uncertainty: 4 to 8 percent.
Applicability: Liquid mixtures.
Input data: Pure-component thermal conductivities k_i at mixture conditions; $\rho_{L,i}$
Description: The mixture thermal conductivity is correlated as a function of the mixture volume fractions ϕ_i:

$$k = \sum_{i=1}^{C}\sum_{j=1}^{C}\phi_i\phi_j\frac{2k_ik_j}{k_i+k_j} \tag{2-114}$$

$$\text{where } \phi_i = \frac{x_i\rho_{L,i}^{-1}}{\sum_{j=1}^{C}x_j\rho_{L,j}^{-1}}$$

Example Estimate the thermal conductivity of a mixture containing 30.2 mol% diethyl ether(1) and 69.8 mol% methanol(2) at 273.15 K and 0.1 MPa, using the Filippov and Li correlations.
Auxiliary data: The pure-component thermal conductivities and molar densities at 273.15 K recommended in the DIPPR 801 database are

$$k_1 = 0.1383 \text{ W/(m}\cdot\text{K)} \qquad \rho_1 = 9.9335 \text{ kmol/m}^3 \qquad M_1 = 74.1216 \text{ kg/kmol}$$

$$k_2 = 0.2069 \text{ W/(m}\cdot\text{K)} \qquad \rho_2 = 25.371 \text{ kmol/m}^3 \qquad M_2 = 32.0419 \text{ kg/kmol}$$

The mass fractions corresponding to the mole fractions given above are

$$w_1 = 0.5 \qquad w_2 = 0.5$$

The volume fractions are

$$\phi_1 = \frac{(0.302)(9.9335)^{-1}}{(0.302)(9.9335)^{-1} + (0.698)(25.371)^{-1}} = 0.525 \qquad \phi_2 = 0.475$$

Calculation using Eq. (2-113):

$$k = [(0.5)(0.1383) + (0.5)(0.2069) - (0.72)(0.5)(0.5)|0.2069 - 0.1383|]\frac{\text{W}}{\text{m}\cdot\text{K}}$$
$$= 0.160\ \text{W/(m}\cdot\text{K)}$$

Calculation using Eq. (2-114):

$$k = \left[(0.525)^2(0.1383) + 2\cdot\frac{(0.525)(0.475)(2)(0.1383)(0.2069)}{0.1383 + 0.2069} + (0.475)^2(0.2069)\right]\frac{\text{W}}{\text{m}\cdot\text{K}}$$
$$= 0.167\ \text{W/(m}\cdot\text{K)}$$

The Filippov value is 7.5 percent lower than the experimental value of 0.173 W/(m · K) [Jamieson, D. T., and B. K. Hastings, *Thermal Conductivity, Proceedings of the Eighth Conference*, C. Y. Ho and R. E. Taylor, eds., Plenum Press, New York, 1969]; the Li value is 3.5 percent lower than the experimental value.

SURFACE TENSION

The surface at a vapor-liquid interface is in tension due to the difference in attractive forces experienced by molecules at the interface between the dense liquid phase and the low-density gas phase. This causes the liquid to contract to minimize the surface area. *Surface tension* is defined as the force in the surface plane per unit length. Jasper [Jasper, J. J., *J. Phys. Chem. Ref. Data*, **1** (1972): 841] has made a critical evaluation of experimental surface tension data for approximately 2200 pure chemicals and correlated surface tension σ (mN/m = dyn/cm) with temperature as

$$\sigma = A - BT \qquad (2\text{-}115)$$

Jasper's evaluation also includes values of A and B for most of the tabulated chemicals. Surface tension decreases with increasing temperature and increasing pressure.

Pure Liquids An approach suggested by Macleod [Macleod, D. B., *Trans. Faraday Soc.*, **19** (1923): 38] and modified by Sugden [Sugden, S. J., *Chem. Soc.*, **125** (1924): 32] relates σ to the liquid and vapor *molar* densities and a temperature-independent parameter called the Parachor **P**

$$\frac{\sigma}{\text{mN/m}} = \left[\mathbf{P}\cdot\left(\frac{\rho_L - \rho_V}{10^3\text{kmol/m}^3}\right)\right]^4 \qquad (2\text{-}116)$$

where ρ_L and ρ_V are the saturated molar liquid and vapor densities, respectively. At low temperatures, where $\rho_L >> \rho_V$, the vapor density can be neglected, but at higher temperatures the density of both phases must be calculated. The surface tension is zero at the critical point where $\rho_L = \rho_V$. Quayle [Quayle, O. R., *Chem. Rev.*, **53** (1953): 439] proposed a group contribution method for estimating **P** that has been improved in recent years by Knotts et al. [Knotts, T. A., et. al., *J. Chem. Eng. Data*, **46** (2001): 1007]. This method using **P** is recommended when groups are available; otherwise, the Brock-Bird [Brock, J. R., and R. B. Bird, *AIChE J.*, **1** (1955): 174] corresponding-states method as modified by Miller [Miller, D. G., *Ind. Eng. Chem. Fundam.*, **2** (1963): 78] may be used to estimate surface tension for compounds that are not strongly polar or associating.

Recommended Method 1 Parachor method.

References: Macleod, D. B., *Trans. Faraday Soc.*, **19** (1923): 38; Sugden, S. J., *Chem. Soc.*, **125** (1924): 32; Knotts, T. A., W. V. Wilding, J. L. Oscarson, and R. L. Rowley, *J. Chem. Eng. Data*, **46** (2001): 1007.

Classification: Group contributions and QSPR.

Expected uncertainty: 4 percent.

Applicability: Organic compounds for which group values are available.

Input data: ρ_L, molecular structure, and Table 2-177.

Description: Equation (2-116) is used with **P** calculated from

$$\mathbf{P} = \sum_{i=1}^{N} n_i\,\Delta\mathbf{P}_i \qquad (2\text{-}117)$$

Group values for the Parachor are given in Table 2-177.

Example Estimate the surface tension of ethylacetylene at 237.45 K.
Structure:

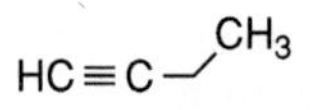

Group	n_i	ΔP_i	$n_i\,\Delta P_i$
≡CH	1	43.64	43.64
≡C—	1	28.64	28.64
>CH_2 (n = 1–11)	1	39.92	39.92
CH_3	1	55.25	55.25
		Total	167.45

Required properties: The DIPPR 801 database gives ρ_L = 13.2573 kmol/m³ at 237.45 K.
Calculation using Eq. (2-116):

$$\sigma = \left[(167.45)\left(\frac{13.2573}{1000}\right)\right]^4 \frac{\text{mN}}{\text{m}} = 0.02429\frac{\text{N}}{\text{m}}$$

The estimated value is 0.9 percent above the DIPPR 801 recommended value of 0.02407 N/m.

Recommended Method 2 Brock-Bird method.

Reference: Brock, J. R., and R. B. Bird, *AIChE J.*, **1** (1955): 174; Miller, D. G., *Ind. Eng. Chem. Fundam.*, **2** (1963): 78.

Classification: Corresponding states.

Expected uncertainty: 5 percent.

Applicability: Nonpolar and moderately polar organic compounds.

Input data: T_c, P_c, and T_b.

Description:

$$\frac{\sigma}{\text{mN/m}} = (5.553\times10^{-5})\left(\frac{P_c}{\text{Pa}}\right)^{2/3}\left(\frac{T_c}{\text{K}}\right)^{1/3} F(1-T_r)^{11/9} \qquad (2\text{-}118)$$

where

$$F = \frac{T_{br}[\ln(P_c/\text{Pa}) - 11.5261]}{1 - T_{br}} - 1.3281 \qquad (2\text{-}119)$$

Example Estimate the surface tension for ethyl mercaptan at 303.15 K.
Required properties from DIPPR 801:

T_c = 499.15 K P_c = 5.49 × 10⁶ Pa T_b = 308.15 K

Supporting quantities:

T_r = (303.15 K)/(499.15 K) = 0.6073
T_{br} = (308.15 K)/(499.15 K) = 0.6173
F = {0.6173[ln (5.49 × 10⁶) – 11.5261]/(1 – 0.6173)} – 1.3281 = 5.113 [from Eq. (2-119)]

From Eq. (2-118):

$\sigma = (5.553 \times 10^{-5})(5.49 \times 10^6)^{2/3}(499.15)^{1/3}(5.113)(1 - 0.6073)^{11/9}$ mN/m
= 22.36 mN/m

The estimated value is 1.4 percent lower than the DIPPR 801 value of 22.68 mN/m.

Liquid Mixtures Compositions at the liquid-vapor interface are not the same as in the bulk liquid, and so simple (bulk) composition-weighted averages of the pure-fluid values do not provide quantitative estimates of the surface tension at the vapor-liquid interface of a mixture. The behavior of aqueous mixtures is more difficult to correlate and estimate than that of nonpolar mixtures because small amounts of organic material can have a pronounced effect upon the surface concentrations and the resultant surface tension. These effects are usually modeled with thermodynamic methods that account for the activity coefficients. For example, a UNIFAC method [Suarez, J. T., C. Torres-Marchal, and P. Rasmussen, *Chem. Eng. Sci.*, **44** (1989): 782] is recommended and illustrated in [PGL5]. For nonaqueous systems the extension of the Parachor method, used above for pure fluids, is a simple and reasonably effective method for estimating σ for mixtures.

Recommended Method Parachor correlation.

Reference: Hugill, J. A., and A. J. van Welsenes, *Fluid Phase Equilib.*, **29** (1986): 383; Macleod, D. B., *Trans. Faraday Soc.*, **19** (1923): 38; Sugden, S. J., *Chem. Soc.*, **125** (1924).

Classification: Corresponding states.

Expected uncertainty: 3 to 10 percent.

Applicability: Nonaqueous mixtures.

TABLE 2-177 Knotts* Group Contributions for the Parachor in Estimating Surface Tension

Group	ΔP_i
(a) Nonring C	
—CH_3	55.25
>CH_2 (n = 1–11)	39.92
>CH_2 (n = 12–20)	40.11
>CH_2 (n > 20)	40.51
>CH—	28.90
>C<	15.76
=CH_2	49.76
=CH—	34.57
=C<	24.50
=C=	24.76
≡CH	43.64
≡C—	28.64
Branch corrections	
Per branch	−6.02
sec-sec adjacency	−2.73
sec-tert adjacency	−3.61
tert-tert adjacency	−6.10
(b) Nonaromatic ring C	
—CH_2—	39.21
>CH—	23.94
>C<	7.19
=CH—	34.07
=C<	18.85
>CH— (fused ring)	22.05
Ring corrections	
Three-member ring	12.67
Four-member ring	15.76
Five-member ring	7.04
Six-member ring	5.19
Seven-member ring	3.00
(c) Aromatic ring C	
>CH	34.36
>C—	16.07
—C— (fused arom/arom)	19.73
—C— (fused arom/aliph)	14.41
Arom ring corr	
ortho	−0.60
para	3.40
meta	2.24
subst. naphthalene corr	−7.07
(d) Oxygen groups	
—OH (alc, primary)	31.42
—OH (alc, sec)	22.68
—OH (alc, tertiary)	20.66
—OH (phenol)	30.32
—O— (nonring)	20.61
—O— (ring)	21.67
—O— (aromatic)	23.54
>C=O (nonring)	47.02
>C=O (ring)	50.04
O=CH— (aldehyde)	66.06
CHOOH (formic)	94.01
—COOH (acid)	74.57
—OCHO (formate)	82.29
—COO— (ester)	64.97
—COOCO— (acid anhyd)	115.07
—OC(=O)O— (ring)	84.05
(e) Nitrogen groups	
R—NH_2 (primary R)	44.98
R—NH_2 (sec R)	44.63
R—NH_2 (tert R)	46.44
A—NH_2 (attached to arom ring)	46.53
>NH (nonring)	29.04
>NH (ring)	31.97
>NH (in arom ring)	33.92
>N- (nonring)	10.77
>N- (ring)	15.71
—N= (nonring)	23.24
>N (aromatic)	26.49
HC≡N (hyd cyanide)	80.94
—C≡N	65.23
—C≡N (aromatic)	67.54
(f) Nitrogen and oxygen groups	
—C=ONH_2 (amides)	93.43
—C=ONH- (amides)	73.64
—C=ON< (amides)	57.05
—NHCHO	91.69
>NCHO	77.12
—N=O	64.32
—NO_2	73.86
—NO_2 (aromatic)	75.05
(g) Sulfur groups	
R-SH (primary R)	66.89
R-SH (sec R)	63.34
R-SH (tert R)	65.33
—SH (aromatic)	68.30
—S— (nonring)	51.37
—S— (ring)	51.75
—S— (aromatic)	51.47
>S=O (nonring)	72.21
>SO_2 (nonring)	93.20
>SO_2 (ring)	90.13
(h) Halogen groups	
—F	21.81
—Cl	26.24
—Br	51.16
—I	54.56
—F (aromatic)	66.30
—Cl (aromatic)	70.39
—Br (aromatic)	90.84
—I (aromatic)	92.04
(i) Si groups	
SiH_4	105.11
>SiH—	54.50
>Si<	44.93
>Si< (ring)	28.64
(j) Other inorganic groups	
—PO_4	115.59
>P—	48.84
>B—	22.65
>Al—	25.06
—ClO_3	106.03

*Knotts, T. A., et al., *J. Chem. Eng. Data*, **46** (2001): 1007.

Input data: Liquid and vapor ρ at mixture T; Parachors of pure components; x_i.

Description:

$$\frac{\sigma_m}{\text{mN/m}} = \left(\mathbf{P}_{L,m}\frac{\rho_{L,m}}{10^3\,\text{kmol/m}^3} - \mathbf{P}_{V,m}\frac{\rho_{V,m}}{10^3\,\text{kmol/m}^3}\right)^4 \qquad (2\text{-}120)$$

where σ_m = surface tension of the mixture
$\mathbf{P}_{L,m}$, $\mathbf{P}_{V,m}$ = Parachor of liquid and vapor mixtures, respectively
$\rho_{L,m}$, $\rho_{V,m}$ = mixture molar density of liquid and vapor, respectively

The following definitions are used for the liquid and vapor mixture Parachors:

$$\mathbf{P}_{L,m} = \frac{1}{2}\sum_{i=1}^{C}\sum_{j=1}^{C} x_i x_j(\mathbf{P}_i + \mathbf{P}_j) \qquad \mathbf{P}_{V,m} = \frac{1}{2}\sum_{i=1}^{C}\sum_{j=1}^{C} y_i y_j(\mathbf{P}_i + \mathbf{P}_j) \qquad (2\text{-}121)$$

where x_i is the mole fraction of component i in the liquid and y_i is the mole fraction of component i in the vapor.

Note that ρ_V is generally very small compared to ρ_L at temperatures substantially lower than T_c and can often be neglected.

Example Estimate the surface tension for a 16.06 mol% n-pentane(1) + 83.94 mol% dichloromethane(2) mixture at 298.15 K.

Required properties from DIPPR 801:

	P	ρ_L/(kmol·m^{-3}) at 298.15 K
n-Pentane	231.1	8.6173
Dichloromethane	146.6	15.5211

Mixture Parachor from Eq. (2-121) and mixture density:

$$\mathbf{P}_{L,m} = (0.1606)^2(231.1) + (0.1606)(0.8394)(231.1 + 146.6) + (0.8394)^2(146.6) = 160.17$$

$$\rho_{L,m} = \left(\sum_{i=1}^{C}\frac{x_i}{\rho_i}\right)^{-1} = \left(\frac{0.1606}{8.6173} + \frac{0.8394}{15.5211}\right)^{-1}\frac{\text{kmol}}{\text{m}^3} = 13.752\frac{\text{kmol}}{\text{m}^3}$$

Calculation using Eq. (2-120): Because the temperature is low, the density of the vapor can be neglected, and

$$\frac{\sigma_m}{\text{mN/m}} = [(160.17)(0.013752)]^4 = 23.54\,\frac{\text{mN}}{\text{m}}$$

The estimated value is 2.9 percent below the experimental value of 24.24 mN/m reported by De Soria [De Soria, M. L. G., et al., *J. Colloid Interface Sci.*, **103** (1985): 354].

FLAMMABILITY PROPERTIES

Flash Point The flash point is the lowest temperature at which a liquid gives off sufficient vapor to form an ignitable mixture with air near the surface of the liquid or within the vessel used. ASTM test methods include procedures using a closed-cup apparatus (ASTM D 56, ASTM D 93, and ASTM D 3828), which is preferred, and an open-cup apparatus (ASTM D 92 and ASTM D 1310). Closed-cup values are typically lower than open-cup values. Estimation methods cannot take into account the apparatus and procedural influences on the observed flash point.

Recommended Method Leslie-Geniesse method.

Reference: Leslie, E. H., and J. C. Geniesse, *International Critical Tables*, vol. 2, McGraw-Hill, New York, 1927, p. 161.

Classification: GC (element contributions).

Expected uncertainty: ~4 K or about 1.5 percent.

Applicability: Organic compounds.

Input data: Chemical structure and vapor pressure correlation.

Description: The flash point T_{FP} is obtained from the moles of oxygen required for stoichiometric combustion β, by back-solving from the vapor pressure correlation using

$$\frac{P^*(T_{FP})}{\text{atm}} = \frac{1}{8\beta} \tag{2-122}$$

where P^* = vapor pressure at the flash point

$$\beta = N_C + N_{Si} + N_S + \frac{N_H - N_X - 2N_O}{4} \tag{2-123}$$

N_C, N_{Si}, N_S, N_H, N_X, N_O = number of carbon, silicon, sulfur, hydrogen, halogen, and oxygen atoms in the molecule, respectively

Example Estimate the flash point of phenol.
Structure:

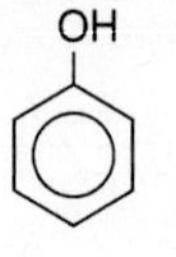

Atomic contributions:

Atom type	Number
C	6
H	6
O	1

From Eq. (2-123), $\beta = 6 + (6 - 2\cdot1)/4 = 7$
The DIPPR 801 correlation for the vapor pressure of phenol is

$$\frac{P^*}{\text{Pa}} = \exp\left[95.444 - \frac{10{,}113\text{ K}}{T} - 10.09\ln\left(\frac{T}{\text{K}}\right) + 6.7603\times10^{-18}\left(\frac{T}{\text{K}}\right)^6\right]$$

When this expression is used in Eq. (2-122) and solved for temperature, one obtains T_{FP} = 350.84 K, which is 0.4 percent below the DIPRR recommended value of 352.15 K.

Flammability Limits The lower flammability limit (LFL) is the equilibrium-mixture boundary-line volume percent of vapor or gas in air which if ignited will just propagate a flame away from the ignition source. Similarly, the upper flammability limit (UFL) is the upper volume percent boundary at which a flame can propagate in an ignited fuel/air equilibrium mixture. Each of these limits has a temperature at which the corresponding volumetric percent is reached. The lower flammability limit temperature corresponds approximately to the flash point, but since the flash point is determined with downward flame propagation and nonuniform mixtures and the lower flammability temperature is determined with upward flame propagation and uniform vapor mixtures, the measured lower flammability temperature is generally slightly lower than the flash point.

Recommended Method Rowley method.

Reference: Rowley, J. R., R. L. Rowley, and W. V. Wilding, *J. Hazard. Materials*, **186** (2011): 551; Rowley, J. R., "Flammability Limits, Flash Points, and Their Consanguinity: Critical Analysis, Experimental Exploration, and Prediction," Ph.D. Dissertation, Brigham Young University, 2010.

Classification: GC and extended theory.

Expected uncertainty: 10 percent for the lower limit; 25 percent for the upper limit.

Applicability: Organic compounds.

Input data: Group contributions from Tables 2-178, ΔH_f^o, and the thermal properties (ideal gas heat of formation and average isobaric heat capacity) of the combustion products. These latter quantities are given in Table 2-179. A vapor pressure correlation is also required to obtain the corresponding flammability limit temperature.

Description: A GC method is used to obtain the adiabatic flame temperature (T_{ad}) of a lower-limit fuel-air mixture using the $\Delta T_{ad,j}$ contributions shown in Table 2-178:

$$T_{ad} = \frac{\sum_j n_j \cdot \Delta T_{ad,j}}{N} \tag{2-124}$$

where N is the total number of groups in the molecule. The ideal gas enthalpies H_i of the combustion products and oxygen at T_{ad} are then calculated from the ideal gas enthalpies of formation at 298 K and the average isobaric heat capacities (given in Table 2-179) with Eq. (2-125):

$$\frac{H_i(T_{ad})}{\text{kJ/mol}} = \frac{\Delta H^o_{f,i}}{\text{kJ/mol}} + \frac{C_{p,i}}{(\text{kJ/mol}\cdot\text{K})}(T_{ad} - 298)\text{K} \tag{2-125}$$

The lower flammability limit in volume percent is then calculated from

$$\text{LFL} = \frac{100\%}{1+\nu} \quad \nu = \frac{\Delta H^o_{f,\text{fuel}} - \sum_{\text{products}} n_i H_i(T_{ad}) + \beta H_{O_2}(T_{ad})}{C_{p,\text{air}}(T_{ad} - 298)\text{ K}} \tag{2-126}$$

where β is defined in Eq. (2-123).

The upper flammability limit in volume percent is obtained from the UFL group values given in Table 2-178 and

$$\frac{\text{UFL}}{\%} = \left[4.30C_{st}^{0.72} + \frac{\sum_j n_j \cdot \text{UFL}_j}{N}\right] \tag{2-127}$$

where C_{st} is the fuel concentration required for stoichiometric combustion given by

$$C_{st} = \frac{100}{1 + 4.773\beta} \tag{2-128}$$

Example Estimate the lower and upper flammability limits of toluene.
Structure:

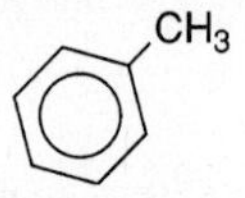

Group contributions:

Group	n_j	ΔT_{ad}	UFL_j
CH_3—c	1	1862.04	−4.49
c—	1	1719.69	5.50
c—H	5	1731.92	−1.25

Auxiliary calculations:

T_{ad} = [1862.04 + 1719.69 + (5)(1731.92)]/7 = 1748.8

$\beta = 7 + 8/4 = 9$

TABLE 2-178 Group Contributions for Quantities Used to Estimate Flammability Limits by Rowley et al.* Method for Organic Compounds (special notation: lower case indicates aromatic atom; # = triple bond; R = ring)

Group	Example	$\Delta T_{ad,i}$	UFL_i	Group	Example	$\Delta T_{ad,i}$	UFL_i
#C—	vinyl acetate	991.44	−8.65	n	pyridine	2622.13	4.46
#CH	acetylene	1237.85	61.25	n	piperazine	2124.88	13.32
=C<	isobutene	1834.42	−7.15	>NH	*n*-pentylamine	1566.76	−0.78
=CH	*trans*-2-butene	1751.82	0.30	>N—(c)	N-ethylaniline	2695.31	−7.25
$=CH_2$	1-hexene	1558.49	3.06	N#C	benzonitrile	939.73	−9.72
=CH—(c)	styrene	−76.72	−11.24	N=C=O	methyl isocyanate	1147.48	4.95
=C—(c)	α-methylstyrene	2091.10	−5.13	$—NO_2$	nitroethane	1777.58	−11.46
>C<	neopentane	1957.78	−0.23	—S—	thiophene	1056.05	23.55
—CH	isopropanol	1558.73	0.62	—SH	ethyl mercaptan	1727.5	12.67
$—CH_2$	propane	1705.21	−0.30	S=	carbon disulfide	272.36	53.67
$—CH_3$	butane	1856.30	−1.12	Si	trimethylsilane	−55.66	78.90
CH_3—c	toluene	1862.04	−4.49	$Si(O_3)$	tetraethoxysilane	2095.22	120.24
c—	toluene	1719.69	5.50	(Si)—O—	octamethyltrisiloxane	2347.17	−67.75
cH	benzene	1731.92	−1.25	Si-(Cl)	monochlorosilane	1062.27	−13.93
OH—(C)	1-methylcyclohexanol	786.14	4.90	$Si—(Cl_2)$	dichlorosilane	554.54	62.48
OH—(CH)	isopropanol	1508.33	0.12	$Si—(Cl_3)$	methyl trichlorosilane	−34.35	−18.52
$OH—(CH_2)$	butanol	1397.73	5.32	$F_2—(C)$	1,1-difluoroethane	2556.15	−4.95
OH—(c)	phenol	1337.25	9.15	$F_2—(C{=}C)$	1,1-difluoroethylene	2088.23	3.43
OH—(CC#C)	propargyl alcohol	2209.35	15.57	$F_3—(C)$	3,3,3-trifluoropropene	2451.95	−12.81
O=C	3-pentanone	1532.45	2.50	F—(C)	methyl fluoride	1841.54	0.80
$O{=}C_R$	cyclohexanone	954.03	−11.84	F—(C=C)	vinyl fluoride	1477.04	15.38
O=C—C=C	methacrolein	1761.66	6.00	$Cl_2—(C)$	dichloromethane	2882.45	−22.73
O=COC	hexyl formate	1492.23	0.47	$Cl_2—(C{=}C)$	1,1-dichloroethylene	2956.55	−15.50
(C)—O—(C)	diethyl ether	1325.57	13.38	$Cl_3—(C)$	1,1,1-trichloroethane	3046.39	−26.31
—COOH	formic acid	1252.38	−5.12	Cl—(C)	isopropyl chloride	1948.51	−5.20
$—O_R—$	furan	1402.11	26.05	Cl—(C=C)	chloropropene	2294.79	0.16
—O—O—	ethyl peroxide	−728.23	0.76	Cl—(cc—Cl)	*o*-dichlorobenzene	3257.79	−13.14
>N—	triethylamine	1442.71	8.85	Br—	methylbromide	3389.83	−24.38

*Rowley, J. R., R. L. Rowley, and W. V. Wilding, *J. Hazard. Materials*, **186** (2011): 551; Rowley, J. R., "Flammability Limits, Flash Points, and Their Consanguinity: Critical Analysis, Experimental Exploration, and Prediction," Ph.D. Dissertation, Brigham Young University, 2010.

Calculation of $H(T_{ad})$ from Eq. (2-125) and Table 2-179:

Species	$H°$(298 K)/(kJ/mol)	C_p/[kJ/(mol·K)]	$H(T_{ad})$/(kJ/mol)
Toluene	50.17	—	—
CO_2	−393.51	0.0372433	−339.48
H_2O	−241.81	0.0335780	−193.10
O_2	0	0.0293468	42.58
Air	0	0.0289937	—

From Eq. (2-126) and the stoichiometry of the combustion reaction, $C_7H_8 + 9O_2 = 7CO_2 + 4H_2O$:

$$v = \left[\frac{50.17-[(7)(-339.48)+(4)(-193.10)]+(9)(42.58)}{(0.0289937)(1749-298)}\right] = 85.148$$

TABLE 2-179 Ideal Gas Enthalpies of Formation and Average Heat Capacities of Combustion Gases for Use in Eq. (2-125)

Species	$H°$/(kJ/mol)	C_p/[J/(mol·K)]
Air	0	28.9937
O_2	0	29.3468
N_2	0	29.1260
CO_2	−393.51	37.2433
H_2O	−241.81	33.5780
SO_2	−296.84	39.8980
SiO_2	−305.43	44.0254
HF	−273.30	29.1361
HCl	−92.31	29.1436
HBr	−36.29	29.1327
HI	−26.50	29.1583

$$\text{LFL} = \frac{100\%}{1+85.148} = 1.16\%$$

The UFL is found from Eqs. (2-127) and (2-128):

$$\text{UFL} = (4.30)\left[\frac{100}{1+(4.773)(9)}\right]^{0.72} + \frac{-4.49+5.50+(5)(-1.25)}{7} = 7.02\%$$

These values agree well with the DIPPR 801 recommended values of 1.2 and 7.1 percent, respectively.

Flammability limit temperatures are found by determining the temperature at which the vapor pressure equals the partial pressure corresponding to the LFL or UFL. The vapor pressure correlation for toluene from DIPPR 801 is

$$\frac{P^*}{\text{Pa}} = \exp\left[76.945 - \frac{6729.8\text{ K}}{T} - 8.179\ln\left(\frac{T}{\text{K}}\right) + 5.3017\times10^{-6}\left(\frac{T}{\text{K}}\right)^2\right]$$

Back-solving for T using the partial pressures of 0.0116 atm for LFL and 0.0702 atm for UFL gives

$$T_{LFL} = 277\text{ K and } T_{UFL} = 311\text{ K}$$

Autoignition Temperature The autoignition temperature (AIT) is the minimum temperature for a substance to initiate self-combustion in air in the absence of an ignition source. Methods to estimate AIT are in general rather approximate. The method illustrated here may provide reasonable estimates, but significant errors can also result. Estimated values should not be assumed to be reliable for design and safety purposes.

TABLE 2-180 Group Contributions for Pintar* Autoignition Temperature Method for Organic Compounds

Group	b_i	Group	b_i	Group	b_i
—CH_3	301.91	—Cl_3	1073.47	—SO_3—	—
>CH_2	−10.86	—F	360.60	—SO_4—	−31.71
>CH—	−275.17	—F_2	755.54	—CO_3—	442.26
>C<	−570.43	—F_3	1082.00	—P=	−334.91
—H	391.48	—Br	420.96	—PO—	−549.59
—OH	324.10	—Br_2	607.69	—OPO_2—	—
—O—	−18.60	—Br_3	1260.00	—PO_4=	−329.45
—O—O—	−397.61	—I	310.53	Si—C†	−147.69
=C=O	57.65	—I_2	—	Si—O†	−136.99
—CHO	195.20	—I_3	—	Si—H†	−310.52
—COOH	370.75	—NH_2	354.11	Si—Cl†	−200.88
—COO—	43.90	>NH	9.88	Si—N†	—
—CO—O—CO—	46.11	—N=	−249.91	Si—Si	—
—C_6H_5	380.27	—CN	469.67	Al	—
m—C_6H_4	153.15	=C=N—	−273.70	B	—
o—C_6H_4	77.48	=N—NH_2	378.27	Cr	—
p—C_6H_4	99.87	>N—NH_2	−215.02	Na	534.29
Aromatic ring	−1339.65	—NO_2	292.57	*cis*	−29.19
=	578.72	—SH	273.84	*trans*	−38.31
≡	1116.50	—S—	−60.75	Nonarom.ring	605.97
—Cl	347.39	—SO—	−91.10	Add'l.ring	565.11
—Cl_2	726.03	—SO_2—	—	Zn	349.02

*Pintar, A. J., *Estimation of Autoignition Temperature,* Technical Support Document DIPPR Project 912, Michigan Technological University, Houghton, 1996.
†Does not include contribution of atoms attached to silicon.

Recommended Method Pintar method.

Reference: Pintar, A. J., *Estimation of Autoignition Temperature,* Technical Support Document DIPPR Project 912, Michigan Technological University, Houghton, 1996.

Classification: Group contributions.

Expected uncertainty: 25 percent.

Applicability: Organic compounds.

Input data: Group contributions from Table 2-180.

Description: A simple GC method with first-order contributions is given by

$$\text{AIT} = \sum_{i=1}^{N} n_i b_i \qquad (2\text{-}129)$$

where n_i is the number of groups of type i in the molecule and b_i is the contribution of group i to the autoignition temperature. A more accurate but somewhat more complicated logarithmic GC method was also developed by Pintar in the same reference cited here.

Example Estimate the autoignition temperature of 2,3-dimethylpentane.

Structure and group information:

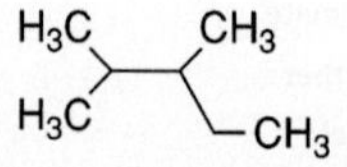

Group	n_i	b_i
—CH_3	4	301.91
>CH_2	1	−10.86
>CH—	2	−275.17

Calculation using Eq. (2-129):

$$\text{AIT} = 4(301.91) - 10.86 + 2(-275.17) = 646.4 \text{ K}$$

The estimated value is 6.3 percent above the DIPPR 801 recommended value of 608.15 K.

Section 3

Mathematics

Bruce A. Finlayson, Ph.D. *Rehnberg Professor Emeritus, Department of Chemical Engineering, University of Washington; Member, National Academy of Engineering (Section Editor, numerical methods and all general material)*

Lorenz T. Biegler, Ph.D. *Bayer Professor of Chemical Engineering, Carnegie Mellon University; Member, National Academy of Engineering (Optimization)*

PLANE TRIGONOMETRY

DIFFERENTIAL AND INTEGRAL CALCULUS

INFINITE SERIES

COMPLEX VARIABLES

DIFFERENTIAL EQUATIONS

DIFFERENCE EQUATIONS

INTEGRAL EQUATIONS

INTEGRAL TRANSFORMS (OPERATIONAL METHODS)

MATRIX ALGEBRA AND MATRIX COMPUTATIONS

NUMERICAL APPROXIMATIONS TO SOME EXPRESSIONS

NUMERICAL ANALYSIS AND APPROXIMATE METHODS

NUMERICAL SOLUTION OF ORDINARY DIFFERENTIAL EQUATIONS AS INITIAL-VALUE PROBLEMS

OPTIMIZATION

STATISTICS

DIMENSIONAL ANALYSIS

PROCESS SIMULATION

GENERAL REFERENCES

Courant, R., and D. Hilbert, *Methods of Mathematical Physics*, vol. I, Interscience, New York, 1953; Finlayson, B. A., *Nonlinear Analysis in Chemical Engineering*, McGraw-Hill, New York, 1980; Finlayson, B. A., L. T. Biegler, and I. E. Grossmann, *Mathematics in Chemical Engineering*, Ullmann's Encyclopedia of Industrial Chemistry, Published Online: 15 DEC 2006, DOI: 10.1002/14356007.b01_01.pub2, Wiley, New York, 2006; Jeffrey, A., *Mathematics for Engineers and Scientists*, 6th ed., Chapman & Hall/CRC, New York, 2004; Kaplan, W., *Advanced Calculus*, 5th ed., Addison-Wesley, Redwood City, Calif., 2003; Lipschultz, S., M. Spiegel, and J. Liu, *Schaum's Outline of Mathematical Handbook of Formulas and Tables*, 4th ed., McGraw-Hill Education, New York, 2012; Logan, J. D., and W. R. Wolesensky, *Mathematical Methods in Biology*, Wiley, New York, 2009; Olver, F. W. J., D. W. Lozier, R. F. Boisvert, and C. W. Clark, eds., *NIST Handbook of Mathematical Functions*, Cambridge University Press, London, 2010; see also http://dlmf.nist.gov; Press, W. H., S. A. Teukolsky, W. T. Vetterling, and B. P. Plannery, *Numerical Recipes*, 3d ed., Cambridge University Press, London, 2007; Rice, R. G., and D. D. Do, *Applied Mathematics and Modeling for Chemical Engineers*, 2d ed., Wiley, New York, 2012; Stroud, K. A., and D. J. Booth, *Engineering Mathematics*, 7th ed., Industrial Press, South Norwick, Conn., 2013; Thompson, W. J., *Atlas for Computing Mathematical Functions*, Wiley, New York, 1997; Varma, A., and M. Morbidelli, *Mathematical Methods in Chemical Engineering*, Oxford Press, New York, 1997; Weisstein, E. W., *CRC Concise Encyclopedia of Mathematics*, 3d ed., CRC Press, New York, 2009; Wrede, R. C., and M. R. Spiegel, *Schaum's Outline of Theory and Problems of Advanced Calculus*, 3d ed., McGraw-Hill, New York, 2010.

MATHEMATICS

GENERAL

The basic problems of the sciences and engineering fall broadly into three categories:

1. *Steady-state problems.* In such problems the configuration of the system is to be determined. This solution does not change with time but continues indefinitely in the same pattern, hence the name *steady state.* Typical chemical engineering examples include steady temperature distributions in heat conduction, equilibrium in chemical reactions, and steady diffusion problems.

2. *Eigenvalue problems.* These are extensions of equilibrium problems in which critical values of certain parameters are to be determined in addition to the corresponding steady-state configurations. The determination of eigenvalues may also arise in propagation problems and stability problems. Typical chemical engineering problems include those in heat transfer and resonance in which certain boundary conditions are prescribed.

3. *Propagation problems.* These problems are concerned with predicting the subsequent behavior of a system from a knowledge of the initial state. For this reason they are often called the *transient* (time-varying) or *unsteady-state* phenomena. Chemical engineering examples include the transient state of chemical reactions (kinetics), the propagation of pressure waves in a fluid, transient behavior of an adsorption column, and the rate of approach to equilibrium of a packed distillation column.

The mathematical treatment of engineering problems involves four basic steps:

1. *Formulation.* This involves the expression of the problem in mathematical language. That translation is based on the appropriate physical laws governing the process.

2. *Solution.* Appropriate mathematical and numerical operations are carried out so that logical deductions may be drawn from the mathematical model.

3. *Interpretation.* This process develops relations between the mathematical results and their meaning in the physical world.

4. *Refinement.* The procedure is recycled to obtain better predictions, as indicated by experimental checks.

Steps 1 and 2 are of primary interest here. The actual details are left to the various subsections, and only general approaches will be discussed.

The formulation step may result in algebraic equations, difference equations, differential equations, integral equations, or combinations of these. In any event these mathematical models usually arise from statements of physical laws such as the laws of mass and energy conservation in the form

$$\text{Input of } x - \text{output of } x + \text{production of } x = \text{accumulation of } x$$

or

$$\text{Rate of input of } x - \text{rate of output of } x + \text{rate of production of } x = \text{rate of accumulation of } x$$

where x = mass, energy, etc. These statements may be abbreviated by the statement

$$\text{Input} - \text{output} + \text{production} = \text{accumulation}$$

Many general laws of the physical universe are expressible by differential equations. Specific phenomena are then singled out from the infinity of solutions of these equations by assigning the individual initial or boundary conditions which characterize the given problem. For steady-state or boundary-value problems (Fig. 3-1), the solution must satisfy the differential equation inside the region and the prescribed conditions on the boundary.

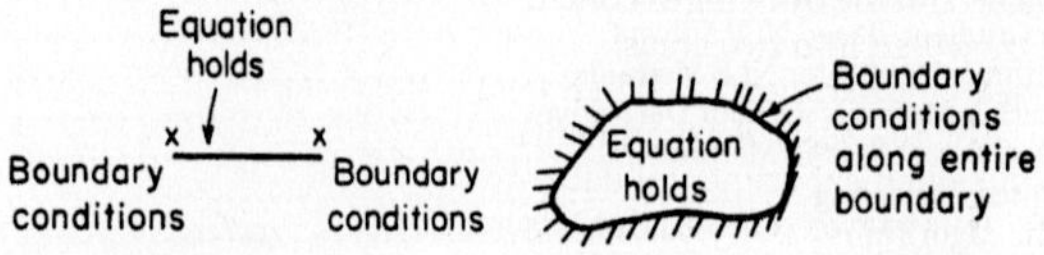

FIG. 3-1 Boundary conditions.

In mathematical language, the propagation problem is known as an *initial-value problem* (Fig. 3-2). Schematically, the problem is characterized by a differential equation plus an open region in which the equation holds. The solution of the differential equation must satisfy the initial conditions plus any "side" boundary conditions.

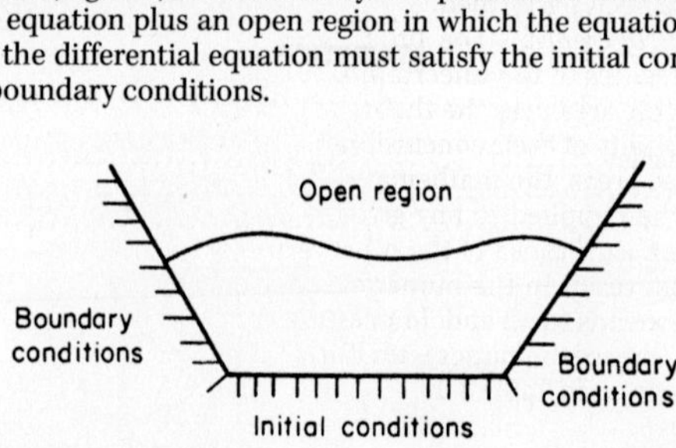

FIG. 3-2 Propagation problem.

The description of phenomena in a continuous medium such as a gas or a fluid often leads to partial differential equations. In particular, phenomena of "wave" propagation are described by a class of partial differential equations called *hyperbolic,* and these are essentially different in their properties from other classes such as those that describe equilibrium (*elliptic*) or diffusion and heat transfer (*parabolic*). Prototypes are as follows:

1. *Elliptic.* Laplace's equation

$$\frac{\partial^2 u}{\partial x^2}+\frac{\partial^2 u}{\partial y^2}=0$$

Poisson's equation

$$\frac{\partial^2 u}{\partial x^2}+\frac{\partial^2 u}{\partial y^2}=g(x,y)$$

These do not contain the variable t (time) explicitly; accordingly, their solutions represent equilibrium configurations. Laplace's equation corresponds to a "natural" equilibrium, while Poisson's equation corresponds to an equilibrium under the influence of $g(x, y)$. Steady heat-transfer and mass-transfer problems are elliptic.

2. *Parabolic.* The heat equation

$$\frac{\partial u}{\partial t}=\frac{\partial^2 u}{\partial x^2}+\frac{\partial^2 u}{\partial y^2}$$

describes unsteady or propagation states of diffusion as well as heat transfer.

3. *Hyperbolic.* The wave equation

$$\frac{\partial^2 u}{\partial t^2}=\frac{\partial^2 u}{\partial x^2}+\frac{\partial^2 u}{\partial y^2}$$

describes wave propagation of all types when the assumption is made that the wave amplitude is small and that interactions are linear.

The solution phase has been characterized in the past by a concentration on methods to obtain analytic solutions to the mathematical equations. These efforts have been most fruitful in the area of linear equations such as those just given. However, many natural phenomena are nonlinear. While there are a few nonlinear problems that can be solved analytically, most cannot. In those cases, numerical methods are used. Due to the widespread availability of software for computers, the engineer has quite good tools available. Numerical methods almost never fail to provide an answer to any particular situation, but they can never furnish a general solution of any problem. The mathematical details outlined here include both analytic and numeric techniques useful in obtaining solutions to problems.

Our discussion to this point has been confined to those areas in which the governing laws are well known. However, in many areas, information on the governing laws is lacking and statistical methods are used. Broadly speaking, statistical methods may be of use whenever conclusions are to be drawn or decisions made on the basis of experimental evidence. Since statistics could be defined as the technology of the scientific method, it is primarily concerned with the first two aspects of the method, namely, the performance of experiments and the drawing of conclusions from experiments. Traditionally the field is divided into two areas:

1. *Design of experiments.* When conclusions are to be drawn or decisions made on the basis of experimental evidence, statistical techniques are most useful when experimental data are subject to errors. First, the design of experiments may be carried out in such a fashion as to avoid some of the sources of experimental error and make the necessary allowances for that portion which is unavoidable. Second, the results can be presented in terms of probability statements which express the reliability of the results. Third, a statistical approach frequently forces a more thorough evaluation of the experimental aims and leads to a more definitive experiment than would otherwise have been performed.

2. *Statistical inference.* The broad problem of statistical inference is to provide measures of the uncertainty of conclusions drawn from experimental data. This area uses the theory of probability, enabling scientists to assess the reliability of their conclusions in terms of probability statements.

Both of these areas, the mathematical and the statistical, are intimately intertwined when applied to any given situation. The methods of one are often combined with those of the other. And both, in order to be successfully used, must result in the numerical answer to a problem, that is, they constitute the means to an end. Increasingly the numerical answer is being obtained from the mathematics with the aid of computers. The mathematical notation is given in Table 3-1.

MISCELLANEOUS MATHEMATICAL CONSTANTS AND FORMULAS

Numerical values of the constants that follow are approximate to the number of significant digits given.

$\pi = 3.1415926536$	Pi
$e = 2.7182818285$	Napierian (natural) logarithm base
$\gamma = 0.5772156649$	Euler's constant
Radian = 57.2957795131°	

Integral Exponents (Powers and Roots) If m and n are positive integers and a and b are numbers or functions, then the following properties hold:

$$a^{-n} = 1/a^n \qquad a \neq 0$$

$$(ab)^n = a^n b^n$$

$$(a^n)^m = a^{nm} \qquad a^n a^m = a^{n+m}$$

$$\sqrt[n]{a} = a^{1/n} \quad \text{if} \quad a > 0$$

$$a^0 = 1\,(a \neq 0)$$

$$0^a = 0\,(a \neq 0)$$

TABLE 3-1 Mathematical Signs, Symbols, and Abbreviations

$\mp$ ($\pm$)	plus or minus (minus or plus)
:	divided by, ratio sign
::	proportional sign
$<$	less than
$\not<$	not less than
$>$	greater than
$\not>$	not greater than
$\cong$	approximately equals, congruent to
$\sim$	similar to
$\Leftrightarrow$	equivalent to
$\neq$	not equal to
$\doteq$	approaches, is approximately equal to
$\propto$	varies as
∞	infinity
$\therefore$	therefore
$\cap$	intersection
$\sqrt{\ }$	square root
$\sqrt[3]{\ }$	cube root
$\sqrt[n]{\ }$	*n*th root
$\angle$	angle
$\perp$	perpendicular to
$\parallel$	parallel to
$\lvert x \rvert$	numerical value of x
log (or $\log_{10}$)	common logarithm or Briggsian logarithm
ln (or $\log_e$)	natural logarithm or hyperbolic logarithm or Napierian logarithm
e	base (2.718) of natural system of logarithms
a°	an angle a degrees
a'	a prime, an angle a minutes
a''	a double prime, an angle a seconds, a second
sin	sine
cos	cosine
tan	tangent
cot (or ctn)	cotangent
sec	secant
csc	cosecant
$\sin^{-1}$	inverse sin, anti-sine, or angle whose sine is
sinh	hyperbolic sine
cosh	hyperbolic cosine
tanh	hyperbolic tangent
$\sinh^{-1}$	anti-hyperbolic sine or angle whose hyperbolic sine is
$f(x)$ or $\phi(x)$	function of x
Δx	increment of x, delta x
Σ	summation of
dx	differential of x
dy/dx or y'	derivative of y with respect to x
d^2y/dx^2 or y''	second derivative of y with respect to x
d^ny/dx^n	*n*th derivative of y with respect to x
$\partial y/\partial x$	partial derivative of y with respect to x
$\partial^n y/\partial x^n$	*n*th partial derivative of y with respect to x
$\frac{\partial^n z}{\partial x \partial y}$	*n*th partial derivative with respect to x and y
$\int$	integral of
$\int_a^b$	integral between the limits a and b
$\dot{y}$	first derivative of y with respect to time
$\ddot{y}$	second derivative of y with respect to time
Δ or ∇^2	laplacian $\left(\frac{\partial^2}{\partial x^2}+\frac{\partial^2}{\partial y^2}+\frac{\partial^2}{\partial z^2}\right)$
δ	sign of a variation
$\oint$	sign for integration around a closed path

Logarithms

$$\log ab = \log a + \log b,\ a>0, b>0$$
$$\log a^n = n \log a$$
$$\log (a/b) = \log a - \log b$$
$$\log \sqrt[n]{a} = (1/n) \log a$$

The common logarithm (base 10) is denoted log a (or $\log_{10} a$ in some texts). The natural logarithm (base e) is denoted ln a (or in some texts $\log_e a$). If the text is ambiguous (perhaps using log x for ln x), test the formula by evaluating it.

ALGEBRAIC INEQUALITIES

Arithmetic-Geometric Inequality Let A_n and G_n denote, respectively, the arithmetic and the geometric means of a set of positive numbers $a_1, a_2, \ldots, a_n$. Then $A_n \ge G_n$, that is,

$$\frac{a_1 + a_2 + \cdots + a_n}{n} \ge (a_1 a_2 \cdots a_n)^{1/n}$$

The equality holds only if all the numbers a_i are equal.

Carleman's Inequality The arithmetic and geometric means just defined satisfy the inequality

$$\sum_{r=1}^{n} (a_1 a_2 \cdots a_r)^{1/r} \le neA_n$$

where e is the best possible constant in this inequality.

Cauchy-Schwarz Inequality Let $a = (a_1, a_2, \ldots, a_n)$ and $b = (b_1, b_2, \ldots, b_n)$, where the a_i and b_i are real or complex numbers. Then

$$\left|\sum_{k=1}^{n} (a_k \bar{b}_k)\right|^2 \le \left(\sum_{k=1}^{n} |a_k|^2\right)\left(\sum_{k=1}^{n} |b_k|^2\right)$$

The equality holds if, and only if, the vectors a and b are linearly dependent (i.e., one vector is a scalar times the other vector).

Minkowski's Inequality Let $a_1, a_2, \ldots, a_n$ and $b_1, b_2, \ldots, b_n$ be any two sets of complex numbers. Then for any real number $p > 1$,

$$\left(\sum_{k=1}^{n} |a_k + b_k|^p\right)^{1/p} \le \left(\sum_{k=1}^{n} |a_k|^p\right)^{1/p} + \left(\sum_{k=1}^{n} |b_k|^p\right)^{1/p}$$

Hölder's Inequality Let $a_1, a_2, \ldots, a_n$ and $b_1, b_2, \ldots, b_n$ be any two sets of complex numbers, and let p and q be positive numbers with $1/p + 1/q = 1$. Then

$$\left|\sum_{k=1}^{n} a_k \bar{b}_k\right| \le \left(\sum_{k=1}^{n} |a_k|^p\right)^{1/p} \left(\sum_{k=1}^{n} |b_k|^q\right)^{1/q}$$

The equality holds if, and only if, the sequences $|a_1|^p, |a_2|^p, \ldots, |a_n|^p$ and $|b_1|^q, |b_2|^q, \ldots, |b_n|^q$ are proportional and the argument (angle) of the complex numbers $a_k \bar{b}_k$ is independent of k. This last condition is of course automatically satisfied if $a_1, \ldots, a_n$ and $b_1, \ldots, b_n$ are positive numbers.

Lagrange's Inequality Let $a_1, a_2, \ldots, a_n$ and $b_1, b_2, \ldots, b_n$ be real numbers. Then

$$\left(\sum_{k=1}^{n} a_k b_k\right)^2 = \left(\sum_{k=1}^{n} a_k^2\right)\left(\sum_{k=1}^{n} b_k^2\right) - \sum_{1\le k \le j \le n} (a_k b_j - a_j b_k)^2$$

Example Two chemical engineers, Mary and John, purchase stock in the same company at times $t_1, t_2, \ldots, t_n$, when the price per share is, respectively, $p_1, p_2, \ldots, p_n$. Their methods of investment are different, however: John purchases x shares each time, whereas Mary invests P dollars each time (fractional shares can be purchased). Who is doing better?

While one can argue intuitively that the average cost per share for Mary does not exceed that for John, we illustrate a mathematical proof using inequalities. The average cost per share for John is equal to

$$\frac{\text{Total money invested}}{\text{Number of shares purchased}} = \frac{x\sum_{i=1}^{n} p_i}{nx} = \frac{1}{n}\sum_{i=1}^{n} p_i$$

The average cost per share for Mary is

$$\frac{nP}{\sum_{i=1}^{n} \frac{P}{p_i}} = \frac{n}{\sum_{i=1}^{n} \frac{1}{p_i}}$$

Thus the average cost per share for John is the arithmetic mean of $p_1, p_2, \ldots, p_n$, whereas that for Mary is the harmonic mean of these n numbers. Since the harmonic mean is less than or equal to the arithmetic mean for any set of positive numbers and the two means are equal only if $p_1 = p_2 = \ldots = p_n$, we conclude that the average cost per share for Mary is less than that for John if two of the prices p_i are distinct. One can also give a proof based on the Cauchy-Schwarz inequality. To this end, define the vectors

$$a = (p_1^{-1/2}, p_2^{-1/2}, \ldots, p_n^{-1/2}) \quad b = (p_1^{1/2}, p_2^{1/2}, \ldots, p_n^{1/2})$$

Then $a \cdot b = 1 + \ldots + 1 = n$, and so by the Cauchy-Schwarz inequality

$$(a \cdot b)^2 = n^2 \le \sum_{i=1}^{n} \frac{1}{p_i} \sum_{j=1}^{n} p_j$$

with the equality holding only if $p_1 = p_2 = \ldots = p_n$. Therefore

$$\frac{n}{\sum_{i=1}^{n} \frac{1}{p_i}} \le \frac{\sum_{i=1}^{n} p_i}{n}$$

MENSURATION FORMULAS

REFERENCE: http://mathworld.wolfram.com/SphericalSector.html, etc.

PLANE GEOMETRIC FIGURES WITH STRAIGHT BOUNDARIES

Let A denote area and V volume in the following.

Triangles (see also "Plane Trigonometry") $A = \frac{1}{2}bh$ where b = base, h = altitude.

Rectangle $A = ab$ where a and b are the lengths of the sides.

Parallelogram (opposite sides parallel) $A = ah = ab \sin \alpha$ where a and b are the lengths of the sides, h is the height, and α is the angle between the sides. See Fig. 3-3.

Rhombus (equilateral parallelogram) $A = \frac{1}{2}ab$ where a and b are the lengths of the diagonals.

Trapezoid (four sides, two parallel) $A = \frac{1}{2}(a + b)h$ where the lengths of the parallel sides are a and b and h = height.

Quadrilateral (four-sided) $A = \frac{1}{2}ab \sin \theta$ where a and b are the lengths of the diagonals and the acute angle between them is θ.

Regular Polygon of n Sides See Fig. 3-4.

$$A = \frac{1}{4} nl^2 \cot \frac{180°}{n} \quad \text{where } l = \text{length of each side}$$

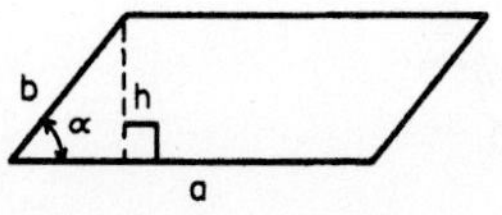

FIG. 3-3 Parallelogram.

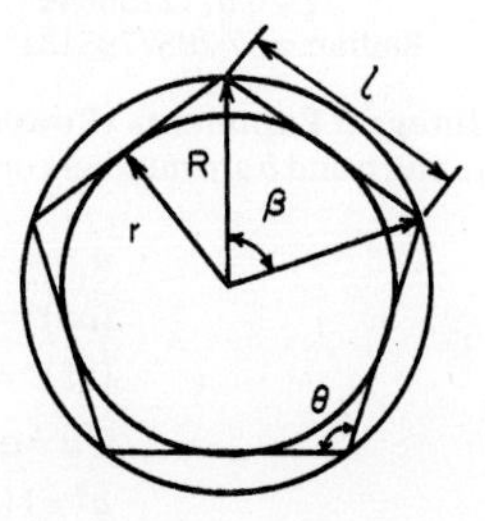

FIG. 3-4 Regular polygon.

$$R=\frac{l}{2}\csc\frac{180°}{n}$$ where R is the radius of the circumscribed circle

$$r=\frac{l}{2}\cot\frac{180°}{n}$$ where r is the radius of the inscribed circle

***Radius* r *of Circle Inscribed in Triangle with Sides* a, b, c**

$$r=\sqrt{\frac{(s-a)(s-b)(s-c)}{s}} \quad \text{where } s=\tfrac{1}{2}(a+b+c)$$

Radius* R *of Circumscribed Circle

$$R=\frac{abc}{4\sqrt{s(s-a)(s-b)(s-c)}}$$

***Area of Regular Polygon of* n *Sides Inscribed in a Circle of Radius* r**

$$A=(nr^2/2)\sin(360°/n)$$

Perimeter of Inscribed Regular Polygon

$$P=2nr\sin(180°/n)$$

***Area of Regular Polygon Circumscribed about a Circle of Radius* r**

$$A=nr^2\tan(180°/n)$$

Perimeter of Circumscribed Regular Polygon

$$P=2nr\tan\frac{180°}{n}$$

PLANE GEOMETRIC FIGURES WITH CURVED BOUNDARIES

Circle (see Fig. 3-5). Let
C = circumference
r = radius
D = diameter
A = area
S = arc length subtended by θ
l = chord length subtended by θ
H = maximum rise of arc above chord, $r - H = d$
θ = central angle (rad) subtended by arc S
$C = 2\pi r = \pi D$ $\quad(\pi = 3.14159\ldots)$
$S = r\theta = \tfrac{1}{2} D\theta$

$$l=2\sqrt{r^2-d^2}=2r\sin(\theta/2)=2d\tan(\theta/2)$$

$$d=\frac{1}{2}\sqrt{4r^2-l^2}=\frac{1}{2}l\cot\frac{\theta}{2}$$

$$\theta=\frac{S}{r}=2\cos^{-1}\frac{d}{r}=2\sin^{-1}\frac{l}{D}$$

A (circle) $= \pi r^2 = \tfrac{1}{4}\pi D^2$
A (sector) $= \tfrac{1}{2} rS = \tfrac{1}{2} r^2\theta$
A (segment) $= A$ (sector) $- A$ (triangle) $= \tfrac{1}{2} r^2(\theta - \sin\theta)$

Ring (area between two circles of radii r_1 and r_2) The circles need not be concentric, but one of the circles must enclose the other.

$$A=\pi(r_1+r_2)(r_1-r_2) \qquad r_1>r_2$$

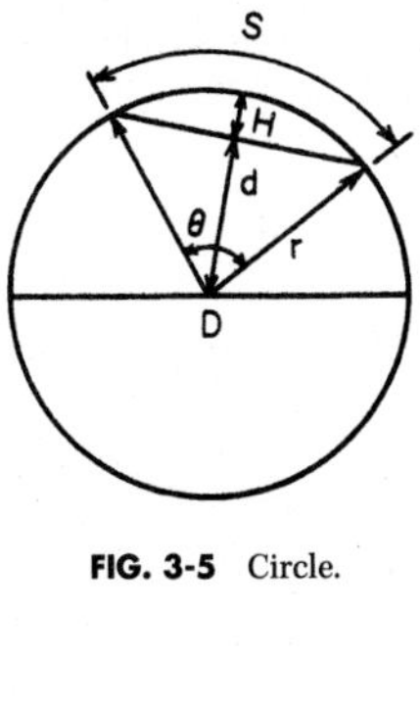

FIG. 3-5 Circle.

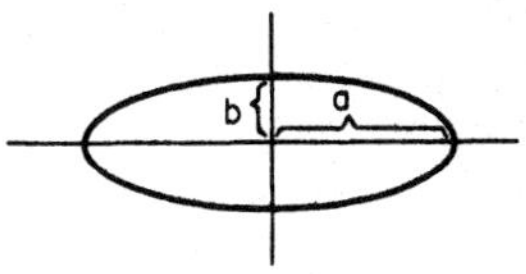

FIG. 3-6 Ellipse.

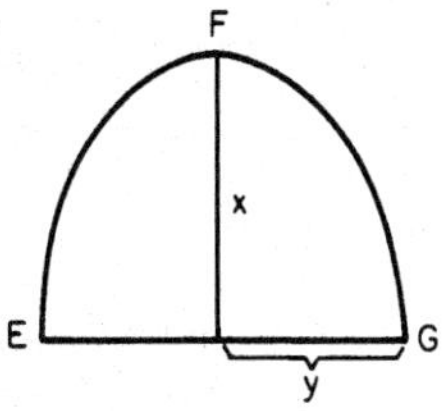

FIG. 3-7 Parabola.

Ellipse (Fig. 3-6) Let the semiaxes of the ellipse be a and b.

$$A=\pi ab$$
$$C=4aE(e)$$

where $e^2 = 1 - b^2/a^2$ and $E(e)$ is the complete elliptic integral of the second kind

$$E(e)=\frac{\pi}{2}\left[1-\left(\frac{1}{2}\right)^2 e^2+\cdots\right]$$

[an approximation for the circumference $C=2\pi\sqrt{(a^2+b^2)/2}$].

Parabola (Fig. 3-7)

$$\text{Length of arc } EFG=\sqrt{4x^2+y^2}+\frac{y^2}{2x}\ln\frac{2x+\sqrt{4x^2+y^2}}{y}$$

$$\text{Area of section } EFG=\frac{4}{3}xy$$

Catenary (the curve formed by a cord of uniform weight suspended freely between two points A and B; Fig. 3-8)

$$y=a\cosh(x/a)$$

The length of arc between points A and B is equal to $2a\sinh(L/a)$. The sag of the cord is $D = a\cosh(L/a) - a$.

SOLID GEOMETRIC FIGURES WITH PLANE BOUNDARIES

Cube Volume = a^3; total surface area = $6a^2$; diagonal = $a\sqrt{3}$, where a = length of one side of the cube.

Rectangular Parallelepiped Volume = abc; surface area = $2(ab + ac + bc)$; diagonal = $\sqrt{a^2+b^2+c^2}$, where a, b, and c are the lengths of the sides.

Prism Volume = (area of base) × (altitude); lateral surface area = (perimeter of right section) × (lateral edge).

Pyramid Volume = ⅓ (area of base) × (altitude); lateral area of regular pyramid = ½ (perimeter of base) × (slant height) = ½ (number of sides) (length of one side) (slant height).

Frustum of Pyramid It is formed from the pyramid by cutting off the top with a plane

$$V=\tfrac{1}{3}(A_1+A_2+\sqrt{A_1\cdot A_2})h$$

where h = altitude and A_1 and A_2 are the areas of the base; lateral area of a regular figure = ½ (sum of the perimeters of base) × (slant height).

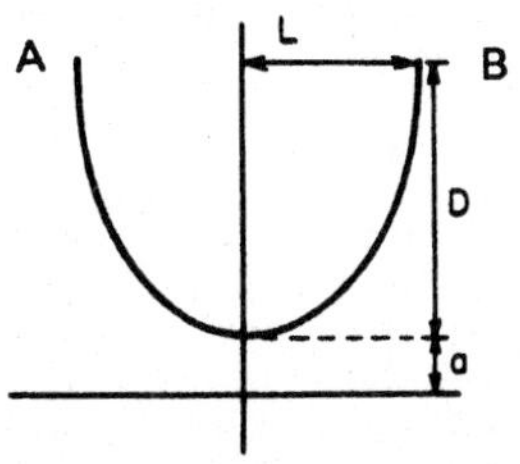

FIG. 3-8 Catenary.

Volume and Surface Area of Regular Polyhedra with Edge *l*

Type of surface	Name	Volume	Surface area
4 equilateral triangles	Tetrahedron	$0.1179l^3$	$1.7321l^2$
6 squares	Hexahedron (cube)	$1.0000l^3$	$6.0000l^2$
8 equilateral triangles	Octahedron	$0.4714l^3$	$3.4641l^2$
12 pentagons	Dodecahedron	$7.6631l^3$	$20.6458l^2$
20 equilateral triangles	Icosahedron	$2.1817l^3$	$8.6603l^2$

SOLIDS BOUNDED BY CURVED SURFACES

Cylinders (Fig. 3-9) V = (area of base) × (altitude); lateral surface area = (perimeter of right section) × (lateral edge).

Right Circular Cylinder $V = \pi\,(\text{radius})^2 \times (\text{altitude})$; lateral surface area = 2π (radius) × (altitude).

Truncated Right Circular Cylinder

$$V = \pi r^2 h \qquad \text{lateral area} = 2\pi r h$$

$$h = \tfrac{1}{2}(h_1 + h_2)$$

Hollow Cylinders Volume = $\pi h(R^2 - r^2)$, where r and R are the internal and external radii, respectively, and h is the height of the cylinder.

Sphere See Fig. 3-10.

$V(\text{sphere}) = \tfrac{4}{3}\pi R^3 = \tfrac{1}{6}\pi D^3$

$V(\text{spherical sector}) = \tfrac{2}{3}\pi R^2 h_1$

$V(\text{spherical segment of one base}) = \tfrac{1}{6}\pi h_1(3r_2^2 + h_1^2)$

$V(\text{spherical segment of two bases}) = \tfrac{1}{6}\pi h_2(3r_1^2 + 3r_2^2 + h_2^2)$

$A\,(\text{sphere}) = 4\pi R^2 = \pi D^2$

$A\,(\text{zone}) = 2\pi R h = \pi D h$

A (lune on surface included between two great circles, with inclination of θ radians) = $2R^2\theta$.

Cone $V = \tfrac{1}{3}$ (area of base) × (altitude).

Right Circular Cone $V = (\pi/3)r^2h$, where h is the altitude and r is the radius of the base; curved surface area $= \pi r\sqrt{r^2 + h^2}$, curved surface of the frustum of a right cone $= \pi(r_1 + r_2)\sqrt{h^2 + (r_1 - r_2)^2}$, where r_1 and r_2 are the radii of the base and top, respectively, and h is the altitude; volume of the the frustum of a right cone $= \pi(h/3)(r_1^2 + r_1r_2 + r_2^2) = h/3\,(A_1 + A_2 + \sqrt{A_1A_2})$, where A_1 = area of base and A_2 = area of top.

Ellipsoid $V = (\tfrac{4}{3})\pi abc$, where a, b, and c are the lengths of the semiaxes.

Torus (obtained by rotating a circle of radius r about a line whose distance is $R > r$ from the center of the circle)

$$V = 2\pi^2 R r^2 \qquad \text{Surface area} = 4\pi^2 R r$$

Prolate Spheroid (formed by rotating an ellipse about its major axis $2a$)

$$\text{Surface area} = 2\pi b^2 + 2\pi(ab/e)\sin^{-1} e \qquad V = \tfrac{4}{3}\pi a b^2$$

where a and b are the major and minor axes and e = eccentricity ($e < 1$).

Oblate Spheroid (formed by the rotation of an ellipse about its minor axis $2b$)

$$\text{Surface area} = 2\pi a^2 + \pi\frac{b^2}{e}\ln\frac{1+e}{1-e} \qquad V = \tfrac{4}{3}\pi a^2 b$$

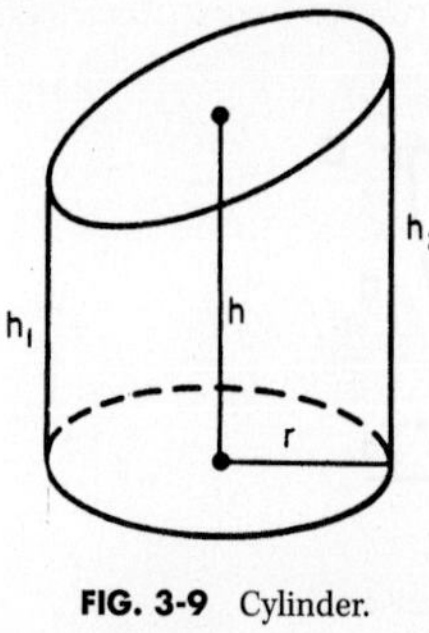

FIG. 3-9 Cylinder.

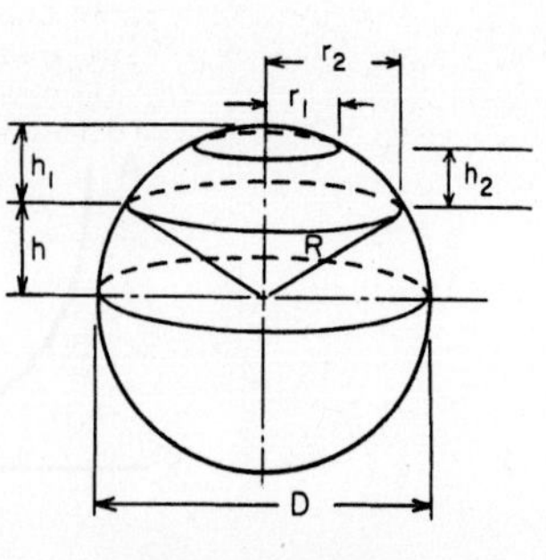

FIG. 3-10 Sphere.

For process vessels, the formulas reduce to the following:

Hemisphere $V = \dfrac{\pi}{12}D^3 \qquad A = \dfrac{\pi}{2}D^2$

For a hemisphere (concave up) partially filled to a depth h_1, use the formulas for spherical segment with one base, which simplify to

$$V = \pi h_1^2(R - h_1/3) = \pi h_1^2(D/2 - h_1/3)$$

$$A = 2\pi R h_1 = \pi D h_1$$

For a hemisphere (concave down) partially filled from the bottom, use the formulas for a spherical segment of two bases, one of which is a plane through the center, where h = distance from the center plane to the surface of the partially filled hemisphere.

$$V = \pi h(R^2 - h^2/3) = \pi h(D^2/4 - h^2/3)$$

$$A = 2\pi R h = \pi D h$$

Cone For a cone partially filled, use the same formulas as for right circular cones, but use r and h for the region filled.

Ellipsoid If the base of a vessel is one-half of an oblate spheroid (the cross section fitting to a cylinder is a circle with radius of $D/2$ and the minor axis is smaller), then use the formulas for one-half of an oblate spheroid.

$$V = 0.1745D^3 \qquad S = 1.236D^2 \qquad \text{minor axis} = D/3$$

$$V = 0.1309D^3 \qquad S = 1.084D^2 \qquad \text{minor axis} = D/4$$

MISCELLANEOUS FORMULAS

See also "Differential and Integral Calculus."

Volume of a Solid Revolution (the solid generated by rotating a plane area about the x axis)

$$V = \pi\int_a^b [f(x)]^2\,dx$$

where $y = f(x)$ is the equation of the plane curve and $a \le x \le b$.

Area of a Surface of Revolution

$$S = 2\pi\int_a^b y\,ds$$

where $ds = \sqrt{1 + (dy/dx)^2}\,dx$ and $y = f(x)$ is the equation of the plane curve rotated about the x axis to generate the surface.

Area Bounded by $f(x)$, the x Axis, and the Lines $x = a$, $x = b$

$$A = \int_a^b f(x)\,dx \qquad [f(x) \ge 0]$$

Length of Arc of a Plane Curve

If $y = f(x)$,

$$\text{Length of arc } s = \int_a^b \sqrt{1 + \left(\frac{dy}{dx}\right)^2}\,dx$$

If $x = f(t)$, $y = g(t)$,

$$\text{Length of arc } s = \int_{t_0}^{t_1} \sqrt{\left(\frac{dx}{dt}\right)^2 + \left(\frac{dy}{dt}\right)^2}\,dt$$

In general, $(ds)^2 = (dx)^2 + (dy)^2$.

IRREGULAR AREAS AND VOLUMES

Irregular Areas Let $y_0, y_1, \ldots, y_n$ be the lengths of a series of equally spaced parallel chords and h be their distance apart (Fig. 3-11). The area of the figure is given approximately by any of the following:

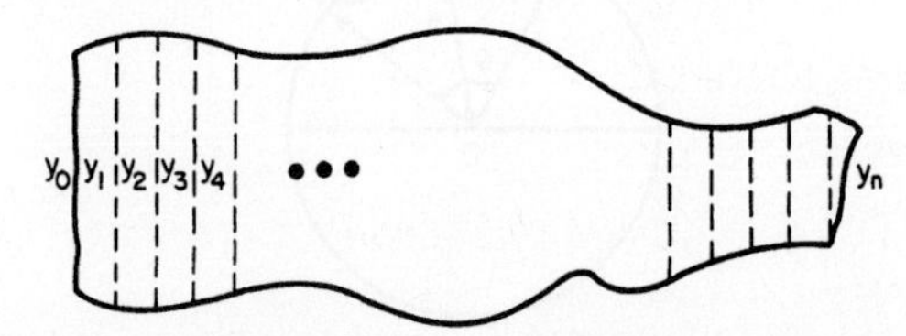

FIG. 3-11 Irregular area.

$$A_T = (h/2)[(y_0 + y_n) + 2(y_1 + y_2 + \cdots + y_{n-1})] \quad \text{(trapezoidal rule)}$$

$$A_s = (h/3)[(y_0 + y_n) + 4(y_1 + y_3 + y_5 + \cdots + y_{n-1}) + 2(y_2 + y_4 + \cdots + y_{n-2})] \quad (n \text{ even, Simpson's rule})$$

The greater the value of n, the greater the accuracy of the approximation.

Irregular Volumes To find the volume, replace the y's by cross-sectional areas A_j and use the results in the preceding equations.

ELEMENTARY ALGEBRA

REFERENCES: Stillwell, J., *Elements of Algebra*, Springer-Verlag, New York, 2010; Rich, B., and P. Schmidt, *Schaum's Outline of Elementary Algebra*, 3d ed., McGraw-Hill Education, New York, 2009.

OPERATIONS ON ALGEBRAIC EXPRESSIONS

An algebraic expression will be denoted here as a combination of letters and numbers such as

$$3ax - 3xy + 7x^2 + 7x^{3/2} - 2.8xy$$

Addition and Subtraction Only like terms can be added or subtracted in two algebraic expressions.

Example $(3x + 4xy - x^2) + (3x^2 + 2x - 8xy) = 5x - 4xy + 2x^2$.

Multiplication Multiplication of algebraic expressions is term by term, and corresponding terms are combined.

Example $(2x + 3y - 2xy)(3 + 3y) = 6x + 9y + 9y^2 - 6xy^2$.

Division This operation is analogous to that in arithmetic.

Example Divide $3e^{2x} + e^x + 1$ by $e^x + 1$.

	Dividend	
Divisor $e^x + 1$	$3e^{2x} + e^x + 1$	$3e^x - 2$ quotient
	$3e^{2x} + 3e^x$	
	$-2e^x + 1$	
	$-2e^x - 2$	
	$+3$ (remainder)	

Therefore, $3e^{2x} + e^x + 1 = (e^x + 1)(3e^x - 2) + 3$.

Operations with Zero All numerical computations (except division) can be done with zero. Both $a/0$ and $0/0$ have no meaning.

Fractional Operations

$$-\frac{x}{y} = -\left(\frac{-x}{-y}\right) = \frac{x}{-y} = \frac{-x}{y} \qquad \frac{x}{y} = \frac{-x}{-y} \qquad \frac{x}{y} = \frac{ax}{ay} \quad \text{if } a \neq 0$$

$$\frac{x}{y} \pm \frac{z}{y} = \frac{x \pm z}{y} \qquad \left(\frac{x}{y}\right)\left(\frac{z}{t}\right) = \frac{xz}{yt} \qquad \frac{x/y}{z/t} = \left(\frac{x}{y}\right)\left(\frac{t}{z}\right) = \frac{xt}{yz}$$

Factoring It is that process of analysis consisting of reducing a given expression to the product of two or more simpler expressions, called *factors*. Some of the more common expressions are factored here:

(1) $x^2 - y^2 = (x - y)(x + y)$

(2) $x^2 + 2xy + y^2 = (x + y)^2$

(3) $x^3 - y^3 = (x - y)(x^2 + xy + y^2)$

(4) $x^3 + y^3 = (x + y)(x^2 - xy + y^2)$

(5) $x^4 - y^4 = (x - y)(x + y)(x^2 + y^2)$

(6) $x^5 + y^5 = (x + y)(x^4 - x^3y + x^2y^2 - xy^3 + y^4)$

(7) $x^n - y^n = (x - y)(x^{n-1} + x^{n-2}y + x^{n-3}y^2 + \cdots + y^{n-1})$

Laws of Exponents

$$(a^n)^m = a^{nm};\ a^{n+m} = a^n \cdot a^m;\ a^{n/m} = (a^n)^{1/m};\ a^{n-m} = a^n/a^m;\ a^{1/m} = \sqrt[m]{a};$$

$$a^{1/2} = \sqrt{a};\ \sqrt{x^2} = |x| \text{ (absolute value of } x\text{). For } x > 0, y > 0, \sqrt{xy} = \sqrt{x}\sqrt{y};$$

$$\text{for } x > 0\ \sqrt[n]{x^m} = x^{m/n};\ \sqrt[n]{1/x} = 1/\sqrt[n]{x}$$

BINOMIAL THEOREM

If n is a positive integer, then

$$(a+b)^n = a^n + na^{n-1}b + \frac{n(n-1)}{2!}a^{n-2}b^2 + \frac{n(n-1)(n-2)}{3!}a^{n-3}b^3 + \cdots + b^n = \sum_{j=0}^{n}\binom{n}{j}a^{n-j}b^j$$

where $\binom{n}{j} = \dfrac{n!}{j!(n-j)!}$ = number of combination of n things taken j at a time and $n! = 1 \cdot 2 \cdot 3 \cdot 4 \cdots n$, $0! = 1$.

Example $(x+y)^4 = x^4 + 4x^3y + 6x^2y^2 + 4xy^3 + y^4$.

If n is not a positive integer, the sum formula no longer applies and an infinite series results for $(a + b)^n$.

Example $(1 + x)^{1/2} = 1 + \tfrac{1}{2}x - \tfrac{1}{2} \cdot \tfrac{1}{4}x^2 + \tfrac{1}{2} \cdot \tfrac{1}{4} \cdot \tfrac{3}{6}x^3 \cdots$ (convergent for $x^2 < 1$). Additional discussion can be found under "Infinite Series."

PROGRESSIONS

An *arithmetic progression* is a succession of terms such that each term, except the first, is derivable from the preceding by the addition of a quantity d, called the *common difference*. All arithmetic progressions have the form a, $a + d$, $a + 2d$, $a + 3d$, ... With a = first term, l = last term, d = common difference, n = number of terms, and s = sum of the terms, the following relations hold:

$$l = a + (n-1)d = \frac{s}{n} + \frac{n-1}{2}d$$

$$s = \frac{n}{2}[2a + (n-1)d] = \frac{n}{2}(a + l) = \frac{n}{2}[2l - (n-1)d]$$

$$a = l - (n-1)d = \frac{s}{n} - \frac{(n-1)d}{2} = \frac{2s}{n} - l$$

$$d = \frac{l-a}{n-1} = \frac{2(s-an)}{n(n-1)} = \frac{2(nl-s)}{n(n-1)}$$

$$n = \frac{l-a}{d} + 1 = \frac{2s}{l+a}$$

The *arithmetic mean* or *average* of two numbers a and b is $(a + b)/2$ and of n numbers $a_1, \ldots, a_n$ is $(a_1 + a_2 + \cdots + a_n)/n$.

A *geometric progression* is a succession of terms such that each term, except the first, is derivable from the preceding by the multiplication of a quantity r called the *common ratio*. All such progressions have the form $a, ar, ar^2, \ldots, ar^{n-1}$. With a = first term, l = last term, r = ratio, n = number of terms, and s = sum of the terms, the following relations hold:

$$l = ar^{n-1} = \frac{a + (r-1)s}{r} = \frac{(r-1)sr^{n-1}}{r^n - 1}$$

$$s = \frac{a(r^n - 1)}{r - 1} = \frac{a(1 - r^n)}{1 - r} = \frac{rl - a}{r - 1} = \frac{lr^n - l}{r^n - r^{n-1}}$$

$$a = \frac{l}{r^{n-l}} = \frac{(r-1)s}{r^n - 1},\ r = \frac{s-a}{s-l},\ \log r = \frac{\log l - \log a}{n-1}$$

$$n = \frac{\log l - \log a}{\log r} + 1 = \frac{\log[a + (r-1)s] - \log a}{\log r}$$

The geometric mean of two nonnegative numbers a and b is $\sqrt{ab}$; of n numbers is $(a_1a_2 \ldots a_n)^{1/n}$. The geometric mean of a set of positive numbers is less than or equal to the arithmetic mean.

Example Find the sum of $1 + \tfrac{1}{2} + \tfrac{1}{4} + \cdots + \tfrac{1}{64}$. Here $a = 1$, $r = \tfrac{1}{2}$, $n = 7$. Thus

$$s = \frac{\tfrac{1}{2}(\tfrac{1}{64}) - 1}{\tfrac{1}{2} - 1} = 127/64$$

$$s = a + ar + ar^2 + \cdots + ar^{n-1} = \frac{a}{1-r} - \frac{ar^n}{1-r}$$

If $|r| < 1$, then $\displaystyle\lim_{n\to\infty} s = \frac{a}{1-r}$

which is called the *sum of the infinite geometric progression*.

Example The present worth (PW) of a series of cash flows C_k at the end of year k is

$$\text{PW} = \sum_{k=1}^{n} \frac{C_k}{(1+i)^k}$$

where i is an assumed interest rate. (Thus the present worth always requires specification of an interest rate.) If all the payments are the same, $C_k = R$, then the present worth is

$$\text{PW} = R \sum_{k=1}^{n} \frac{1}{(1+i)^k}$$

This can be rewritten as

$$\text{PW} = \frac{R}{1+i} \sum_{k=1}^{n} \frac{1}{(1+i)^{k-1}} = \frac{R}{1+i} \sum_{j=0}^{n-1} \frac{1}{(1+i)^j}$$

This is a geometric series with $r = 1/(1+i)$ and $a = R/(1+i)$. The formulas above give

$$\text{PW}\,(=s) = \frac{R}{i}\,\frac{(1+i)^n - 1}{(1+i)^n}$$

The same formula applies to the value of an annuity (PW) now, to provide for equal payments R at the end of each of n years, with interest rate i.

A progression of the form a, $(a+d)r$, $(a+2d)r^2$, $(a+3d)r^3$, etc., is a combined arithmetic and geometric progression. The sum of n such terms is

$$s = \frac{a - [a+(n-1)d]r^n}{1-r} + \frac{rd(1-r^{n-1})}{(1-r)^2}$$

If $|r|<1$, $\lim_{n\to\infty} s = \dfrac{a}{1-r} + rd/(1-r)^2$.

The nonzero numbers a, b, c, etc., form a *harmonic progression* if their reciprocals $1/a$, $1/b$, $1/c$, etc., form an arithmetic progression.

Example The progression 1, ⅓, ⅕, ⅐, …, $\frac{1}{31}$ is harmonic since 1, 3, 5, 7, …, 31 form an arithmetic progression.

The *harmonic mean* of two numbers a and b is $2ab/(a+b)$.

PERMUTATIONS, COMBINATIONS, AND PROBABILITY

Each separate arrangement of all or a part of a set of things is called a *permutation*. The number of permutations of n things taken r at a time is written

$$P(n,r) = \frac{n!}{(n-r)!} = n(n-1)(n-2)\cdots(n-r+1)$$

Each separate selection of objects that is possible irrespective of the order in which they are arranged is called a *combination*. The number of combinations of n things taken r at a time is written $C(n, r) = n!/[r!(n-r)!]$.

An important relation is $r!C(n, r) = P(n, r)$.

If an event can occur in p ways and can fail to occur in q ways, with all ways being equally likely, the *probability* of its occurrence is $p/(p+q)$, and that of its failure is $q/(p+q)$.

Example Two dice may be thrown in 36 separate ways. What is the probability of throwing such that their sum is 7? The number 7 may arise in 6 ways: 1 and 6, 2 and 5, 3 and 4, 4 and 3, 5 and 2, 6 and 1. The probability of shooting a 7 is ⅙.

THEORY OF EQUATIONS

Linear Equations A linear equation is one of the first degree (i.e., only the first powers of the variables are involved), and the process of obtaining definite values for the unknown is called *solving the equation*. Every linear equation in one variable is written $Ax + B = 0$ or $x = -B/A$. Linear equations in n variables have the form

$$\begin{aligned} a_{11}x_1 + a_{12}x_2 + \cdots + a_{1n}x_n &= b_1 \\ a_{21}x_1 + a_{22}x_2 + \cdots + a_{2n}x_n &= b_2 \\ &\vdots \\ a_{m1}x_1 + a_{m2}x_2 + \cdots + a_{mn}x_n &= b_m \end{aligned}$$

The solution of the system may then be found by elimination or matrix methods if a solution exists (see Matrix Algebra and Matrix Computations).

Quadratic Equations Every quadratic equation in one variable is expressible in the form $ax^2 + bx + c = 0$, $a \neq 0$. This equation has two solutions, say, x_1 and x_2, given by

$$\left.\begin{matrix} x_1 \\ x_2 \end{matrix}\right\} = \frac{-b \pm \sqrt{b^2 - 4ac}}{2a}$$

If a, b, and c are real, the discriminant $b^2 - 4ac$ gives the character of the roots. If $b^2 - 4ac > 0$, the roots are real and unequal. If $b^2 - 4ac < 0$, the roots are complex conjugates. If $b^2 - 4ac = 0$, the roots are *real* and *equal*. Two quadratic equations in two variables in general can be solved only by numerical methods (see Numerical Analysis and Approximate Methods).

Cubic Equations A cubic equation in one variable has the form $x^3 + bx^2 + cx + d = 0$. Every cubic equation having complex coefficients has three complex roots. If the coefficients are real numbers, then at least one of the roots must be real. The cubic equation $x^3 + bx^2 + cx + d = 0$ may be reduced by the substitution $x = y - b/3$ to the form $y^3 + py + q = 0$, where $p = \frac{1}{3}(3c - b^2)$ and $q = \frac{1}{27}(27d - 9bc + 2b^3)$.

This reduced equation has the solutions

$$y_1 = A + B,\ y_2 = -\tfrac{1}{2}(A+B) + (i\sqrt{3}/2)(A-B),$$

$$y_3 = -\tfrac{1}{2}(A+B) - (i\sqrt{3}/2)(A-B),\ \text{where } i^2 = -1,\ A = \sqrt[3]{-q/2 + \sqrt{R}},$$

$$B = \sqrt[3]{-q/2 - \sqrt{R}},\ \text{and } R = (p/3)^3 + (q/2)^2.$$

If b, c, and d are all real and if $R > 0$, there are one real root and two conjugate complex roots; if $R = 0$, there are three real roots, of which at least two are equal; if $R < 0$, there are three real unequal roots. If $R < 0$, which requires $p < 0$, these formulas are impractical. In this case, the roots are given by $y_k = \mp 2\sqrt{-p/3}\cos[(\varphi/3) + 120k]$, $k = 0, 1, 2$, where

$$\phi = \cos^{-1}\sqrt{\frac{q^2/4}{-p^3/27}}$$

and the negative sign applies if $q > 0$, and the positive sign applies if $q < 0$.

Example Many equations of state involve solving cubic equations for the compressibility factor Z. For example, the Soave-Redlich-Kwong equation of state requires solving

$$Z^3 - Z^2 + cZ + d = 0 \qquad d < 0$$

where c and d depend on critical constants of the chemical species and temperature and pressure. In this case, only positive solutions, $Z > 0$, are desired.

Quartic Equations See Olver et al. (2010) in General References.

General Polynomials of the *n*th Degree If $n > 4$, there is no formula that gives the roots of the general equation. The roots can be found numerically (see "Numerical Analysis and Approximate Methods").

Fundamental Theorem of Algebra Every polynomial of degree n has exactly n real or complex roots, counting multiplicities.

Determinants Consider the system of two linear equations

$$a_{11}x_1 + a_{12}x_2 = b_1$$

$$a_{21}x_1 + a_{22}x_2 = b_2$$

If the first equation is multiplied by a_{22} and the second by $-a_{12}$ and the results are added, we obtain

$$(a_{11}a_{22} - a_{21}a_{12})x_1 = b_1a_{22} - b_2a_{12}$$

The expression $a_{11}a_{22} - a_{21}a_{12}$ may be represented by the symbol

$$\begin{vmatrix} a_{11} & a_{12} \\ a_{21} & a_{22} \end{vmatrix} = a_{11}a_{22} - a_{21}a_{12}$$

This symbol is called a *determinant of second order*. The value of the square array of n^2 quantities a_{ij}, where $i = 1, \ldots, n$, is the row index, $j = 1, \ldots, n$. The column index, written in the form

$$|A| = \begin{vmatrix} a_{11} & a_{12} & a_{13}\cdots a_{1n} \\ a_{21} & a_{22} & \cdots\cdots a_{2n} \\ \vdots & & \\ a_{n1} & a_{n2} & a_{n3}\cdots a_{nn} \end{vmatrix}$$

is called a *determinant.* The n^2 quantities a_{ij} are called the *elements* of the determinant. In the determinant $|A|$, let the ith row and jth column be deleted and a new determinant be formed having $n-1$ rows and columns. This new determinant is called the *minor* of a_{ij}, denoted M_{ij}.

Example

$$\begin{vmatrix} a_{11} & a_{12} & a_{13} \\ a_{21} & a_{22} & a_{23} \\ a_{31} & a_{32} & a_{33} \end{vmatrix} \text{ The minor of } a_{23} \text{ is } M_{23} = \begin{vmatrix} a_{11} & a_{12} \\ a_{31} & a_{32} \end{vmatrix}$$

The cofactor A_{ij} of the element a_{ij} is the signed minor of a_{ij} determined by the rule $A_{ij} = (-1)^{i+j}M_{ij}$. The *value* of $|A|$ is obtained by forming any of the equivalent expressions $\sum_{j=1}^{n} a_{ij}A_{ij}$, $\sum_{i=1}^{n} a_{ij}A_{ij}$, where the elements a_{ij} must be taken from a single row or a single column of A.

Example

$$\begin{vmatrix} a_{11} & a_{12} & a_{13} \\ a_{21} & a_{22} & a_{23} \\ a_{31} & a_{32} & a_{33} \end{vmatrix} = a_{31}A_{31} + a_{32}A_{32} + a_{33}A_{33}$$

$$= a_{31}\begin{vmatrix} a_{12} & a_{13} \\ a_{22} & a_{23} \end{vmatrix} - a_{32}\begin{vmatrix} a_{11} & a_{13} \\ a_{21} & a_{23} \end{vmatrix} + a_{33}\begin{vmatrix} a_{11} & a_{12} \\ a_{21} & a_{22} \end{vmatrix}$$

In general, A_{ij} will be determinants of order $n-1$, but they may in turn be expanded by the rule. Also,

$$\sum_{j=1}^{n} a_{ji}A_{jk} = \sum_{j=1}^{n} a_{ij}A_{jk} = \begin{cases} |A| & i=k \\ 0 & i \neq k \end{cases}$$

Fundamental Properties of Determinants

1. The value of a determinant $|A|$ is not changed if the rows and columns are interchanged.
2. If the elements of one row (or one column) of a determinant are all zero, the value of $|A|$ is zero.
3. If the elements of one row (or column) of a determinant are multiplied by the same constant factor, the value of the determinant is multiplied by this factor.
4. If one determinant is obtained from another by interchanging any two rows (or columns), the value of either is the negative of the value of the other.
5. If two rows (or columns) of a determinant are identical, the value of the determinant is zero.
6. If two determinants are identical except for one row (or column), the sum of their values is given by a single determinant obtained by adding corresponding elements of dissimilar rows (or columns) and leaving unchanged the remaining elements.
7. The value of a determinant is not changed if one row (or column) is multiplied by a constant and added to another row (or column).

ANALYTIC GEOMETRY

REFERENCES: Gersting, J. L., *Technical Calculus with Analytic Geometry*, Dover, Mineola, N.Y., 2010.

Analytic geometry uses algebraic equations and methods to study geometric problems. It also permits one to visualize algebraic equations in terms of geometric curves, which frequently clarifies abstract concepts.

PLANE ANALYTIC GEOMETRY

Coordinate Systems The basic concept of analytic geometry is the establishment of a one-to-one correspondence between the points of the plane and number pairs (x, y). This correspondence may be done in a number of ways. The rectangular or cartesian coordinate system consists of two straight lines intersecting at right angles (Fig. 3-12). A point is designated by (x, y). Another common coordinate system is the polar coordinate system (Fig. 3-13). In this system the position of a point is designated by the pair (r, θ), with $r=\sqrt{x^2+y^2}$ being the distance to the origin $O(0, 0)$ and θ being the angle the line r makes with the positive x axis (polar axis). To change from polar to rectangular coordinates, use $x = r\cos\theta$ and $y = r\sin\theta$. To change from rectangular to polar coordinates, use $r=\sqrt{x^2+y^2}$ and $\theta = \tan^{-1}(y/x)$ if $x \neq 0$; $\theta = \pi/2$ if $x = 0$. The distance between two points (x_1, y_1) and (x_2, y_2) is defined by $d=\sqrt{(x_1-x_2)^2+(y_1-y_2)^2}$ in rectangular coordinates or by $d=\sqrt{r_1^2+r_2^2-2r_1r_2\cos(\theta_1-\theta_2)}$ in polar coordinates. Other coordinate systems are sometimes used. For example, on the surface of a sphere, latitude and longitude prove useful.

Straight Line See Fig. 3-14. The slope m of a straight line is the tangent of the inclination angle θ made with the positive x axis. If (x_1, y_1) and (x_2, y_2) are any two points on the line, then slope $= m = (y_2 - y_1)/(x_2 - x_1)$. The slope of a line parallel to the x axis is zero; the slope of a line parallel to the y axis is undefined. Two lines are parallel if and only if they have the same slope. Two lines are perpendicular if and only if the product of their slopes is -1 (the exception being that case when the lines are parallel to the coordinate axes). Every equation of the type $Ax + By + C = 0$ represents a straight line, and every straight line has an equation of this form. A straight line is determined by a variety of conditions:

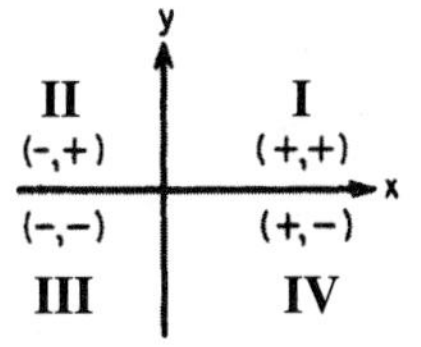

FIG. 3-12 Rectangular coordinates.

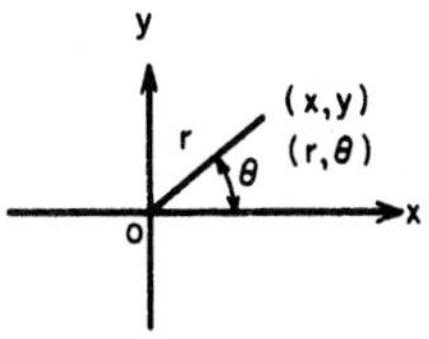

FIG. 3-13 Polar coordinates.

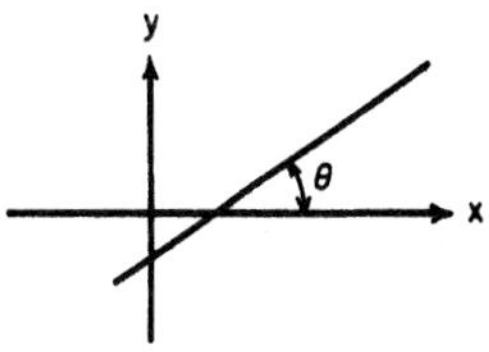

FIG. 3-14 Straight line.

Given conditions	Equation of line
1. Parallel to x axis	y = constant
2. Parallel to y axis	x = constant
3. Point (x_1, y_1) and slope m	$y - y_1 = m(x - x_1)$
4. Intercept on y axis $(0, b)$, m	$y = mx + b$
5. Intercept on x axis $(a, 0)$, m	$y = m(x - a)$
6. Two points (x_1, y_1), (x_2, y_2)	$y - y_1 = \dfrac{y_2 - y_1}{x_2 - x_1}(x - x_1)$
7. Two intercepts $(a, 0)$, $(0, b)$	$x/a + y/b = 1$

The angle β that a line with slope m_1 makes with a line having slope m_2 is given by $\tan\beta = (m_2 - m_1)/(m_1m_2 + 1)$. The distance from a point (x_1, y_1) to a line with equation $Ax + By + C = 0$ is

$$d = \frac{|Ax_1 + By_1 + C|}{\sqrt{A^2 + B^2}}$$

Occasionally some nonlinear algebraic equations can be reduced to linear equations under suitable substitutions or changes of variables.

Example Consider $y = bx^n$ and $B = \log b$. Taking logarithms gives $\log y = n\log x + \log b$. Let $Y = \log y$, $X = \log x$, and $B = \log b$. The equation then has the form $Y = nX + B$, which is a linear equation. Consider $k = k_0 \exp(-E/RT)$; taking logarithms gives $\ln k = \ln k_0 - E/(RT)$. Let

$Y = \ln k$, $B = \ln k_0$, $m = -E/R$, and $X = 1/T$, and the result is $Y = mX + B$.

Asymptotes The limiting position of the tangent to a curve, as the point of contact tends to an infinite distance from the origin, is called an *asymptote.*

Conic Sections The curves included in this group are obtained from plane sections of the cone. They include the circle, ellipse, parabola, hyperbola, and degeneratively the point and straight line. A *conic* is the locus of a point whose distance from a fixed point called the *focus* is in a constant ratio to its distance from a fixed line, called the *directrix*. This ratio is the eccentricity e. If $e = 0$, the conic is a circle; if $0 < e < 1$, the conic is an ellipse; if $e = 1$, the conic is a parabola; if $e > 1$, the conic is a hyperbola. Every conic section is representable by an equation of second degree. Conversely, every equation of second degree in two variables represents a conic. The general equation of the second degree is $Ax^2 + Bxy + Cy^2 + Dx + Ey + F = 0$. Let Δ be defined as the determinant

$$\Delta = \begin{vmatrix} 2A & B & D \\ B & 2C & E \\ D & E & 2F \end{vmatrix}$$

The table characterizes the curve represented by the equation.

	$B^2 - 4AC < 0$	$B^2 - 4AC = 0$	$B^2 - 4AC > 0$
$\Delta \neq 0$	$A\Delta < 0, A \neq C$, an ellipse $A\Delta < 0, A = C$, a circle $A\Delta > 0$, no locus	Parabola	Hyperbola
$\Delta = 0$	Point	Two parallel lines if $Q = D^2 + E^2 - 4(A + C)F > 0$ One straight line if $Q = 0$ no locus if $Q < 0$	Two intersecting straight lines

(1) $(x-h)^2+(y-k)^2-a^2$	$x=h+a\cos\theta$ $y=k+a\sin\theta$	Circle (Fig. 3-15) parameter is angle θ Ellipse (Fig. 3-16) parameter is angle θ
(2) $\frac{(x-h)^2}{a^2}+\frac{(y-k)^2}{b^2}=1$	$x=h+a\cos\phi$ $y=k+a\sin\phi$	
(3) $x^2+y^2=a^2$	$\begin{cases} x=\frac{-at}{\sqrt{t^2+1}} \\ y=\frac{a}{\sqrt{t^2+1}} \end{cases}$	Circle parameter is $t=\frac{dy}{dx}=$ slope of tangent at (x, y)
(4) $x^2=y+k$		Parabola (Fig. 3-17)
(5) $\frac{x^2}{a^2}-\frac{y^2}{b^2}=1$		Hyperbola with the origin at the center (Fig. 3-18)
(6) $y=a\cosh\frac{x}{a}$	$\begin{cases} x=a\sinh^{-1}\frac{s}{a} \\ y^2=a^2+s^2 \end{cases}$	Catenary (such as hanging cable under gravity) Parameter s = arc length from $(0, a)$ to (x, y)
(7) Cycloid	$\begin{cases} x=a(\phi-\sin\phi) \\ y=a(1-\cos\phi) \end{cases}$	(Fig. 3-19)

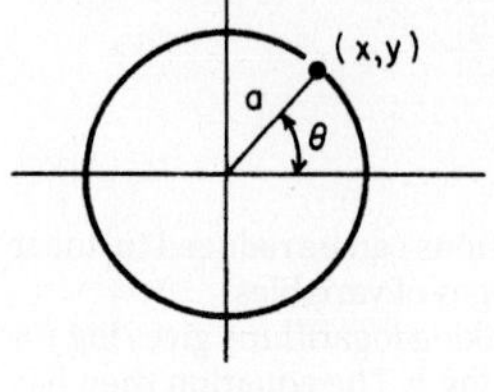

FIG. 3-15 Circle.

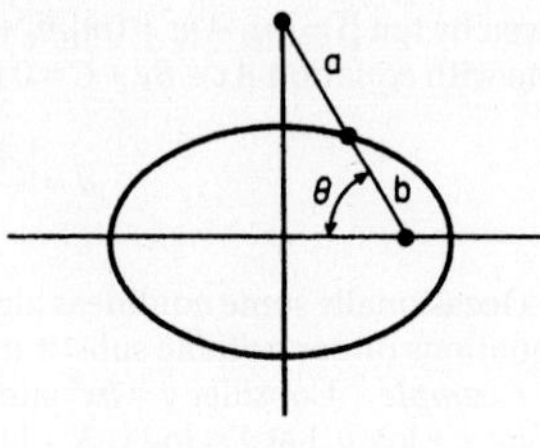

FIG. 3-16 Ellipse, $0 < e < 1$.

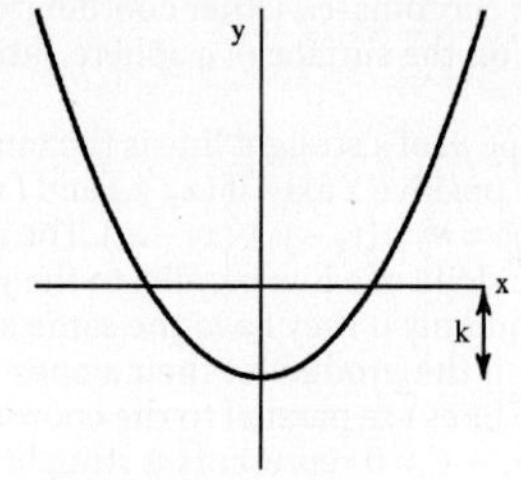

FIG. 3-17 Parabola, $e = 1$.

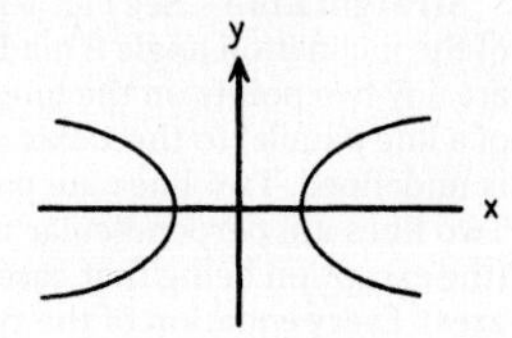

FIG. 3-18 Hyperbola, $e > 1$.

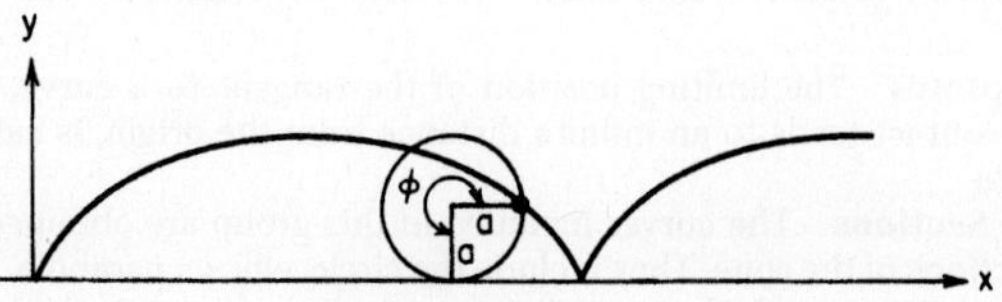

FIG. 3-19 Cycloid.

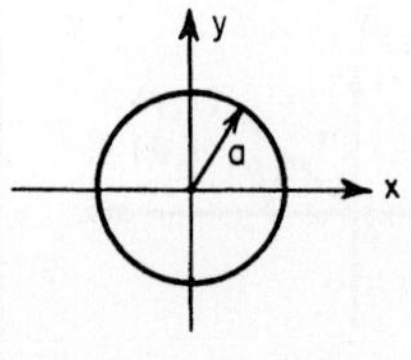

FIG. 3-20 Circle center (0, 0), $r = a$.

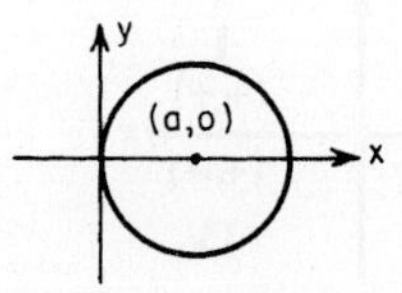

FIG. 3-21 Circle center $(a, 0)$, $r = 2a \cos \theta$.

Example $3x^2 + 4xy - 2y^2 + 3x - 2y + 7 = 0$.

$$\Delta = \begin{vmatrix} 6 & 4 & 3 \\ 4 & -4 & -2 \\ 3 & -2 & 14 \end{vmatrix} = -596 \neq 0, B^2 - 4AC = 40 > 0$$

The curve is therefore a hyperbola.

Polar equation	Type of curve
(1) $r = a$	Circle, Fig. 3-20
(2) $r = 2a \cos \theta$	Circle, Fig. 3-21
(3) $r = 2a \sin \theta$	Circle, Fig. 3-22
(4) $r^2 - 2br \cos (\theta - \beta) + b^2 - a^2 = 0$	Circle at (b, β), radius a
(5) $r = \frac{ke}{1 - e\cos\theta}$	$e = 1$ parabola, Fig. 3-17 $0 < e < 1$ ellipse, Fig. 3-16 $e > 1$ hyperbola, Fig. 3-18

Parametric Equations It is frequently useful to write the equations of a curve in terms of a parameter. For example, a circle of radius a, center at (0, 0), can be written in the equivalent form $x = a \cos \phi$, $y = a \sin \phi$, where ϕ is the parameter. Similarly, $x = a \cos \phi$ and $y = b \sin \phi$ are the parametric equations of the ellipse $x^2/a^2 + y^2/b^2 = 1$ with parameter ϕ.

SOLID ANALYTIC GEOMETRY

Coordinate Systems There are three commonly used coordinate systems. Others may be used in specific problems (see Morse, P. M., and H. Feshbach, *Methods of Theoretical Physics*, vols. 1 and I2, McGraw-Hill, New York, 1953). The *rectangular* (cartesian) system (Fig. 3-23) consists of mutually orthogonal axes x, y, and z. A triple of numbers (x, y, z) is used to represent each point. The *cylindrical* coordinate system (r, θ, z; Fig. 3-24) is frequently used to locate a point in space. These are essentially the polar coordinates (r, θ) coupled with the z coordinate. As before, $x = r \cos \theta$, $y = r \sin \theta$, $z = z$ and $r^2 = x^2 + y^2$, $y/x = \tan \theta$. If r is held constant and θ and z are allowed to vary, the locus of (r, θ, z) is a right circular cylinder of radius r along the z axis. The locus of $r = C$ is a circle, and θ = constant is a plane containing the z axis and making an angle θ with the xz plane. Cylindrical coordinates are convenient to use when the problem has an axis of symmetry.

The **spherical** coordinate system is convenient if there is a point of symmetry in the system. This point is taken as the origin and the coordinates (ρ, ϕ, θ) are illustrated in Fig. 3-25. The relations are $x = \rho \sin \phi \cos \theta$, $y = \rho \sin \phi \sin \theta$, $z = \rho \cos \phi$, and $r = \rho \sin \phi$. Also θ = constant is a plane containing the z axis and making an angle θ with the xz plane; ϕ = constant is a cone with vertex at 0; ρ = constant is the surface of a sphere of radius ρ, center at the origin 0. Every point in the space may be given spherical coordinates restricted to the ranges $0 \leq \phi \leq \pi$, $\rho \geq 0$, $0 \leq \theta < 2\pi$.

Lines and Planes The distance between two points (x_1, y_1, z_1), (x_2, y_2, z_2) is $d = \sqrt{(x_1 - x_2)^2 + (y_1 - y_2)^2 + (z_1 - z_2)^2}$. There is nothing in the geometry of three dimensions quite analogous to the slope of a line in

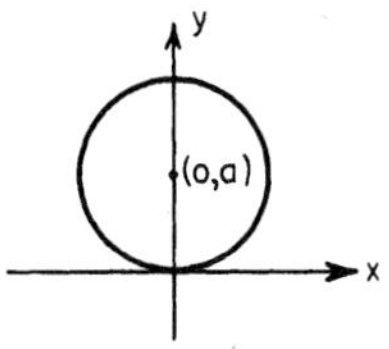

FIG. 3-22 Circle center $(0, a)$, $r = 2a \sin \theta$.

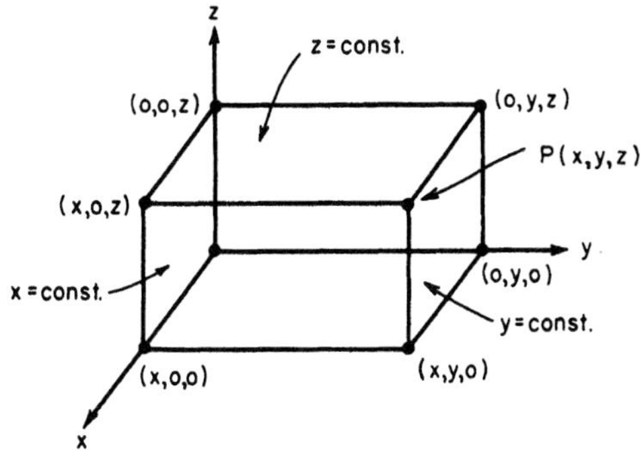

FIG. 3-23 Cartesian coordinates.

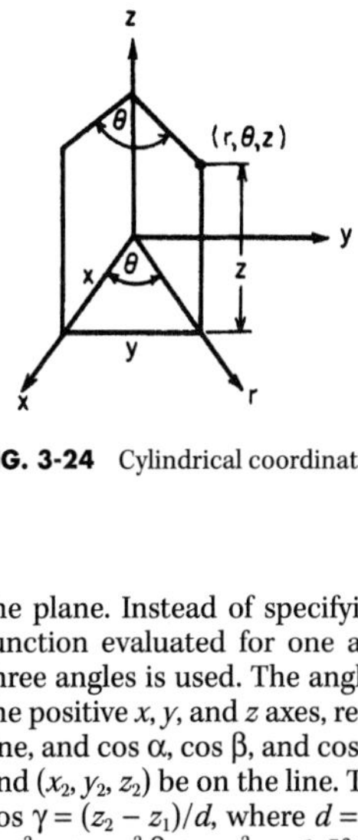

FIG. 3-24 Cylindrical coordinates.

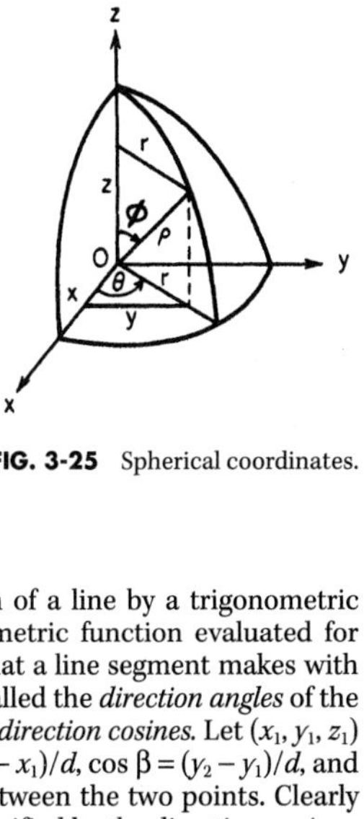

FIG. 3-25 Spherical coordinates.

the plane. Instead of specifying the direction of a line by a trigonometric function evaluated for one angle, a trigonometric function evaluated for three angles is used. The angles α, β, and γ that a line segment makes with the positive x, y, and z axes, respectively, are called the *direction angles* of the line, and $\cos \alpha$, $\cos \beta$, and $\cos \gamma$ are called the *direction cosines.* Let (x_1, y_1, z_1) and (x_2, y_2, z_2) be on the line. Then $\cos \alpha = (x_2 - x_1)/d$, $\cos \beta = (y_2 - y_1)/d$, and $\cos \gamma = (z_2 - z_1)/d$, where d = the distance between the two points. Clearly $\cos^2 \alpha + \cos^2 \beta + \cos^2 \gamma = 1$. If two lines are specified by the direction cosines $(\cos \alpha_1, \cos \beta_1, \cos \gamma_1)$ and $(\cos \alpha_2, \cos \beta_2, \cos \gamma_2)$, then the angle θ between the lines is $\cos \theta = \cos \alpha_1 \cos \alpha_2 + \cos \beta_1 \cos \beta_2 + \cos \gamma_1 \cos \gamma_2$. Thus the lines are perpendicular if and only if $\theta = 90°$ or $\cos \alpha_1 \cos \alpha_2 + \cos \beta_1 \cos \beta_2 + \cos \gamma_1 \cos \gamma_2 = 0$. The equation of a line with direction cosines $(\cos \alpha, \cos \beta, \cos \gamma)$ passing through (x_1, y_1, z_1) is $(x - x_1)/\cos \alpha = (y - y_1)/\cos \beta = (z - z_1)/\cos \gamma$.

The equation of every plane is of the form $Ax + By + Cz + D = 0$. The numbers

$$\frac{A}{\sqrt{A^2+B^2+C^2}}, \frac{B}{\sqrt{A^2+B^2+C^2}}, \frac{C}{\sqrt{A^2+B^2+C^2}}$$

are direction cosines of the normal lines to the plane. The plane through the point (x_1, y_1, z_1) whose normals have these as direction cosines is $A(x - x_1) + B(y - y_1) + C(z - z_1) = 0$.

Example Find the equation of the plane through (1, 5, −2) perpendicular to the line $(x + 9)/7 = (y - 3)/(-1) = z/8$. The numbers (7, −1, 8) are called *direction numbers.* They are a constant multiple of the direction cosines $\cos \alpha = 7/114$, $\cos \beta = -1/114$, and $\cos \gamma = 8/114$. The plane has the equation $7(x - 1) - 1(y - 5) + 8(z + 2) = 0$ or $7x - y + 8z + 14 = 0$.

The distance from the point (x_1, y_1, z_1) to the plane $Ax + By + Cz + D = 0$ is

$$d = \frac{|Ax_1 + By_1 + Cz_1 + D|}{\sqrt{A^2+B^2+C^2}}$$

Space Curves Space curves are usually specified as the set of points whose coordinates are given parametrically by a system of equations $x = f(t)$, $y = g(t)$, $z = h(t)$ in the parameter t.

Example The equation of a straight line in space is $(x - x_1)/a = (y - y_1)/b = (z - z_1)/c$. Since all these quantities must be equal (say, to t), we may write $x = x_1 + at$, $y = y_1 + bt$, and $z = z_1 + ct$, which represent the parametric equations of the line.

Example The equations $z = a \cos \beta t$, $y = a \sin \beta t$, and $z = bt$, with a, β, and b positive constants, represent a circular helix.

Surfaces The locus of points (x, y, z) satisfying $f(x, y, z) = 0$, broadly speaking, may be interpreted as a surface. The simplest surface is the *plane.* The next simplest is a *cylinder.*

Example The parabolic cylinder $y = x^2$ (Fig. 3-26) is generated by a straight line parallel to the z axis passing through $y = x^2$ in the plane $z = 0$. A surface whose equation is a quadratic in the variables x, y, and z is called a *quadric surface.* Some of the more common such surfaces are tabulated and pictured in Figs. 3-26 to 3-34.

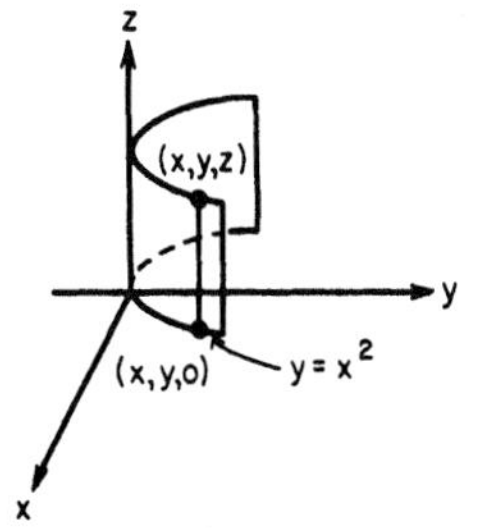

FIG. 3-26 Parabolic cylinder.

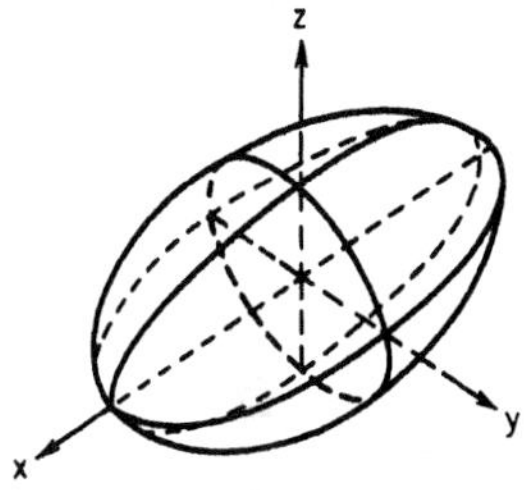

FIG. 3-27 Ellipsoid.

$$\frac{x^2}{a^2}+\frac{y^2}{b^2}+\frac{z^2}{c^2}=1 \text{ (sphere if } a = b = c)$$

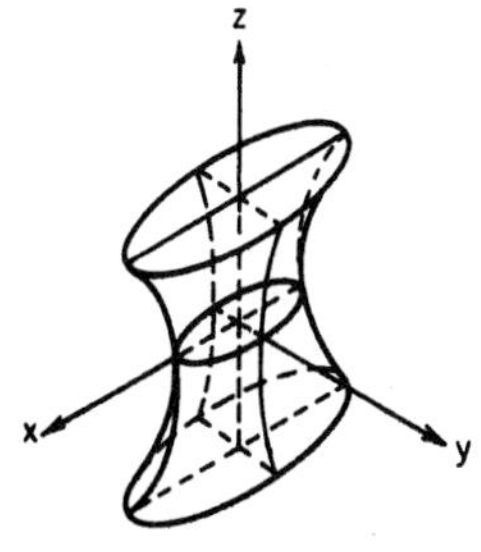

FIG. 3-28 Hyperboloid of one sheet.

$$\frac{x^2}{a^2}+\frac{y^2}{b^2}-\frac{z^2}{c^2}=1$$

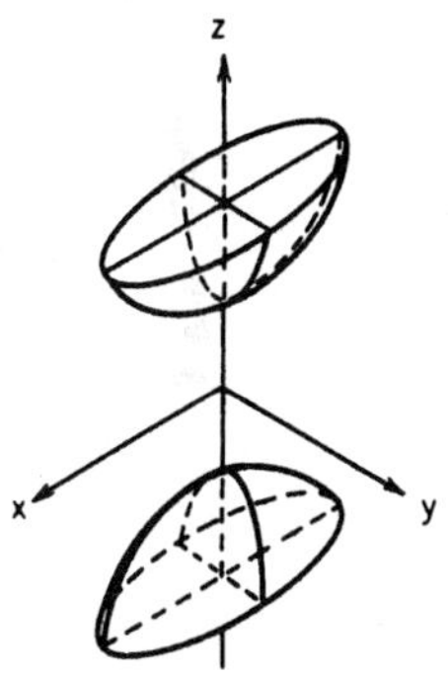

FIG. 3-29 Hyperboloid of two sheets.

$$\frac{x^2}{a^2}+\frac{y^2}{b^2}-\frac{z^2}{c^2}=-1$$

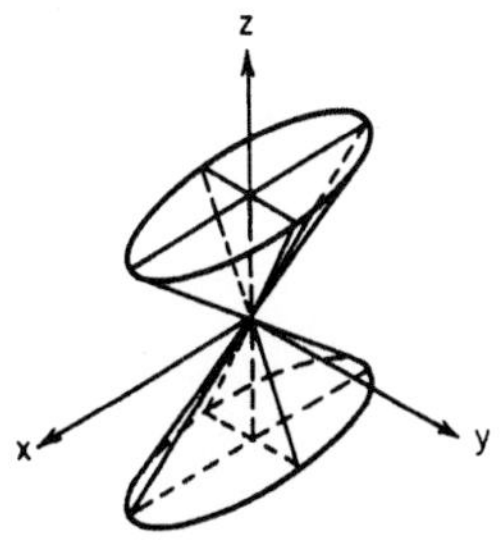

FIG. 3-30 Cone. $\frac{x^2}{a^2}+\frac{y^2}{b^2}+\frac{z^2}{c^2}=0$

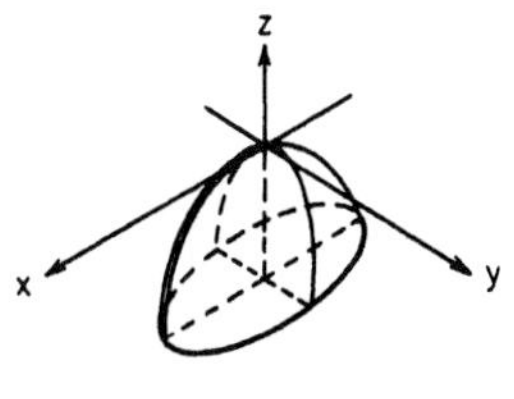

FIG. 3-31 Elliptic paraboloid.

$$\frac{x^2}{a^2}+\frac{y^2}{b^2}+cz=0$$

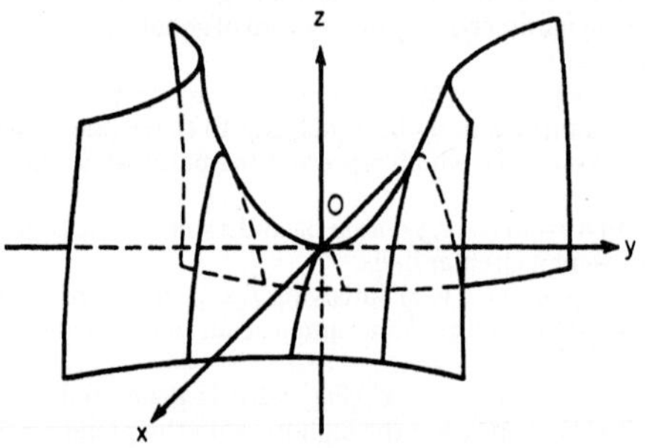

FIG. 3-32 Hyperbolic paraboloid. $\frac{x^2}{a^2}-\frac{y^2}{b^2}+cz=0$

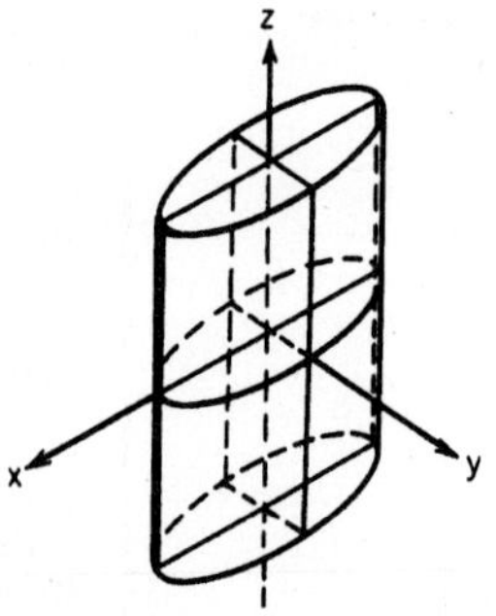

FIG. 3-33 Elliptic cylinder. $\frac{x^2}{a^2}+\frac{y^2}{b^2}=1$

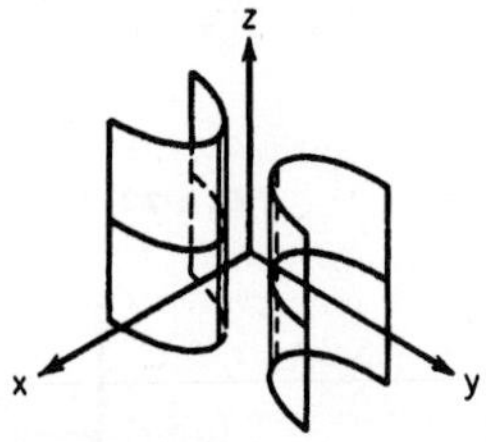

FIG. 3-34 Hyperbolic cylinder. $\frac{x^2}{a^2}-\frac{y^2}{b^2}=1$

PLANE TRIGONOMETRY

References: Gelfand, I. M., and M. Saul, *Trigonometry*, Birkhäuser, Boston, 2001; Heineman, E. Richard, and J. Dalton Tarwater, *Plane Trigonometry*, 7th ed., McGraw-Hill, New York, 1993.

ANGLES

An angle is generated by the rotation of a line about a fixed center from some initial position to some terminal position. If the rotation is clockwise, the angle is negative; if it is counterclockwise, the angle is positive. Angle size is unlimited. If α and β are two angles such that $\alpha + \beta = 90°$, they are complementary; they are supplementary if $\alpha + \beta = 180°$. Angles are most commonly measured in the sexagesimal system or by radian measure. In the first system there are 360° in 1 complete revolution (1 r); 1° = 1/90 of a right angle. The degree is subdivided into 60 minutes; the minute is subdivided into 60 seconds. In the radian system, 1 radian (1 rad) is the angle at the center of a circle subtended by an arc whose length is equal to the radius of the circle. Thus 2π rad = 360°; 1 rad = 57.29578°; 1° = 0.01745 rad; 1 min = 0.00029089 rad. The advantage of radian measure is that it is *dimensionless*. The quadrants are conventionally labeled, as Fig. 3-35 shows.

FUNCTIONS OF CIRCULAR TRIGONOMETRY

The trigonometric functions of angles are the ratios between the various sides of the reference triangles shown in Fig. 3-36 for the various quadrants. Clearly $r=\sqrt{x^2+y^2} \geq 0$. The fundamental functions (see Figs. 3-37, 3-38, 3-39) are as follows:

Plane Trigonometry

Sine of $\theta = \sin\theta = y/r$	Secant of $\theta = \sec\theta = r/x$
Cosine of $\theta = \cos\theta = x/r$	Cosecant of $\theta = \csc\theta = r/y$
Tangent of $\theta = \tan\theta = y/x$	Cotangent of $\theta = \cot\theta = x/y$

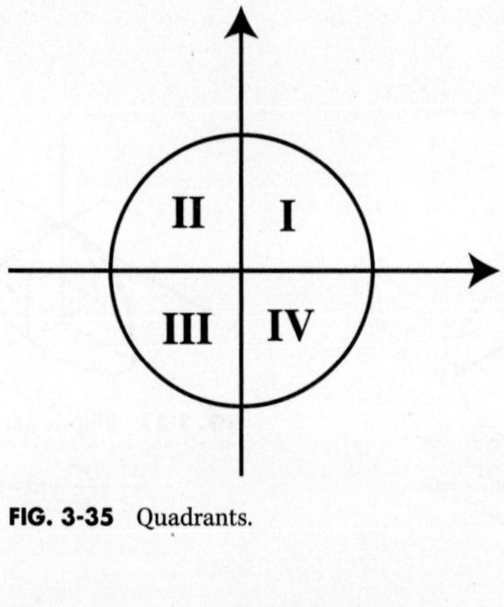

FIG. 3-35 Quadrants.

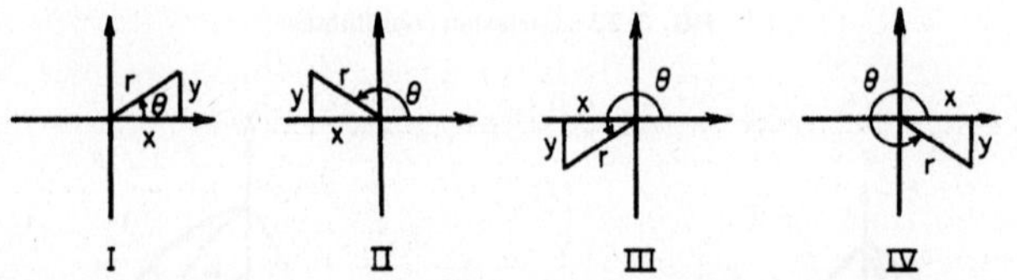

FIG. 3-36 Triangles.

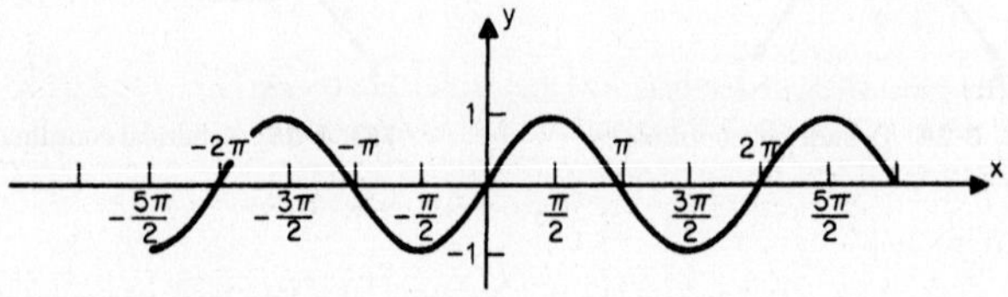

FIG. 3-37 Graph of $y = \sin x$.

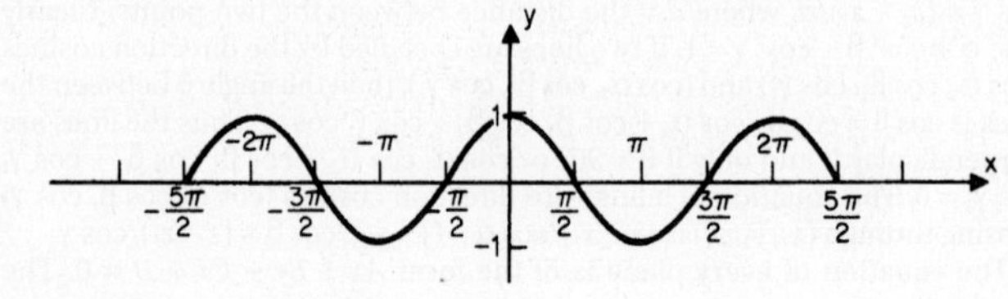

FIG. 3-38 Graph of $y = \cos x$.

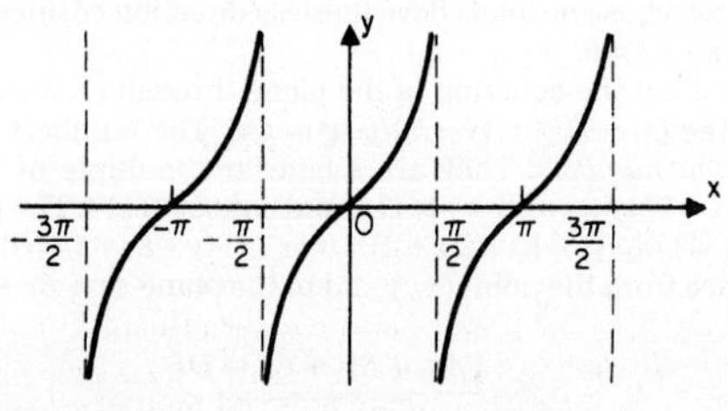

FIG. 3-39 Graph of $y = \tan x$.

Values of the Trigonometric Functions for Common Angles

θ°	θ, rad	sin θ	cos θ	tan θ
0	0	0	1	0
30	$\pi/6$	1/2	$\sqrt{3}/2$	$\sqrt{3}/3$
45	$\pi/4$	$\sqrt{2}/2$	$\sqrt{2}/2$	1
60	$\pi/3$	$\sqrt{3}/2$	1/2	$\sqrt{3}$
90	$\pi/2$	1	0	$+\infty$

If $90° \le \theta \le 180°$, $\sin\theta = \sin(180° - \theta)$; $\cos\theta = -\cos(180° - \theta)$; $\tan\theta = -\tan(180° - \theta)$. If $180° \le \theta \le 270°$, $\sin\theta = -\sin(270° - \theta)$; $\cos\theta = -\cos(270° - \theta)$; $\tan\theta = \tan(270° - \theta)$. If $270° \le \theta \le 360°$, $\sin\theta = -\sin(360° - \theta)$; $\cos\theta = \cos(360° - \theta)$; $\tan\theta = -\tan(360° - \theta)$. The reciprocal properties may be used to find the values of the other functions.

If it is desired to find the angle when a function of it is given, the procedure is as follows: There will in general be two angles between 0° and 360° corresponding to the given value of the function.

Given $a > 0$	Find an acute angle θ_0 such that	Required angles are
$\sin\theta = +a$	$\sin\theta_0 = a$	θ_0 and $180° - \theta_0$
$\cos\theta = +a$	$\cos\theta_0 = a$	θ_0 and $360° - \theta_0$
$\tan\theta = +a$	$\tan\theta_0 = a$	θ_0 and $180° + \theta_0$
$\sin\theta = -a$	$\sin\theta_0 = a$	$180° + \theta_0$ and $360° - \theta_0$
$\cos\theta = -a$	$\cos\theta_0 = a$	$180° - \theta_0$ and $180° + \theta_0$
$\tan\theta = -a$	$\tan\theta_0 = a$	$180° - \theta_0$ and $360° - \theta_0$

Relations between Functions of a Single Angle $\sec\theta = 1/\cos\theta$; $\csc\theta = 1/\sin\theta$, $\tan\theta = \sin\theta/\cos\theta = \sec\theta/\csc\theta = 1/\cot\theta$; $\sin^2\theta + \cos^2\theta = 1$; $1 + \tan^2\theta = \sec^2\theta$; $1 + \cot^2\theta = \csc^2\theta$. For $0 \le \theta \le 90°$ the following results hold:

$$\sin\theta = 2\sin\left(\frac{\theta}{2}\right)\cos\left(\frac{\theta}{2}\right)$$

and

$$\cos\theta = \cos^2\left(\frac{\theta}{2}\right) - \sin^2\left(\frac{\theta}{2}\right)$$

The cofunction property is very important. $\cos\theta = \sin(90° - \theta)$, $\sin\theta = \cos(90° - \theta)$, $\tan\theta = \cot(90° - \theta)$, $\cot\theta = \tan(90° - \theta)$, etc.

Functions of Negative Angles $\sin(-\theta) = -\sin\theta$, $\cos(-\theta) = \cos\theta$, $\tan(-\theta) = -\tan\theta$, $\sec(-\theta) = \sec\theta$, $\csc(-\theta) = -\csc\theta$, $\cot(-\theta) = -\cot\theta$.

Identities

Sum and Difference Formulas Let x, y be two angles. $\sin(x \pm y) = \sin x \cos y \pm \cos x \sin y$; $\cos(x \pm y) = \cos x \cos y \mp \sin x \sin y$; $\tan(x \pm y) = (\tan x \pm \tan y)/(1 \mp \tan x \tan y)$; $\sin x \pm \sin y = 2 \sin \tfrac{1}{2}(x \pm y) \cos \tfrac{1}{2}(x \mp y)$; $\cos x + \cos y = 2\cos\tfrac{1}{2}(x + y)\cos\tfrac{1}{2}(x - y)$; $\cos x - \cos y = -2\sin\tfrac{1}{2}(x + y)\sin\tfrac{1}{2}(x - y)$; $\tan x \pm \tan y = [\sin(x \pm y)]/(\cos x \cos y)$; $\sin^2 x - \sin^2 y = \cos^2 y - \cos^2 x = \sin(x + y)\sin(x - y)$; $\cos^2 x - \sin^2 y = \cos^2 y - \sin^2 x = \cos(x + y) \times \cos(x - y)$; $\sin(45° + x) = \cos(45° - x)$; $\sin(45° - x) = \cos(45° + x)$; $\tan(45° \pm x) = \cot(45° \mp x)$.

Multiple and Half-Angle Identities Let x = angle, $\sin 2x = 2\sin x\cos x$; $\sin x = 2\sin\tfrac{1}{2}x \times \cos\tfrac{1}{2}x$; $\cos 2x = \cos^2 x - \sin^2 x = 1 - 2\sin^2 x = 2\cos^2 x - 1$. $\tan 2x = (2\tan x)/(1 - \tan^2 x)$; $\sin 3x = 3\sin x - 4\sin^3 x$; $\cos 3x = 4\cos^3 x - 3\cos x$. $\tan 3x = (3\tan x - \tan^3 x)/(1 - 3\tan^2 x)$; $\sin 4x = 4\sin x\cos x - 8\sin^3 x\cos x$; $\cos 4x = 8\cos^4 x - 8\cos^2 x + 1$.

$$\sin\left(\frac{x}{2}\right) = \sqrt{\tfrac{1}{2}(1 - \cos x)}$$

$$\cos\left(\frac{x}{2}\right) = \sqrt{\tfrac{1}{2}(1 + \cos x)}$$

$$\tan\left(\frac{x}{2}\right) = \sqrt{\frac{1 - \cos x}{1 + \cos x}} = \frac{\sin x}{1 + \cos x} = \frac{1 - \cos x}{\sin x}$$

INVERSE TRIGONOMETRIC FUNCTIONS

Note that $y = \sin^{-1} x = \arcsin x$ is the angle y whose sine is x.

Example $y = \sin^{-1}(\tfrac{1}{2})$, y is 30°.

The complete solution of the equation $x = \sin y$ is $y = (-1)^n \sin^{-1} x + n(180°)$, $-\pi/2 \le \sin^{-1} x \le \pi/2$ where $\sin^{-1} x$ is the principal value of the angle whose sine is x. The range of principal values of $\cos^{-1} x$ is $0 \le \cos^{-1} x \le \pi$ and $-\pi/2 \le \tan^{-1} x \le \pi/2$. If these restrictions are allowed to hold, the following formulas result:

$$\sin^{-1} x = \cos^{-1}\sqrt{1 - x^2} = \tan^{-1}\frac{x}{\sqrt{1 - x^2}} = \cot^{-1}\frac{\sqrt{1 - x^2}}{x}$$

$$= \sec^{-1}\frac{1}{\sqrt{1 - x^2}} = \csc^{-1}\frac{1}{x} = \frac{\pi}{2} - \cos^{-1} x$$

$$\cos^{-1} x = \sin^{-1}\sqrt{1 - x^2} = \tan^{-1}\frac{\sqrt{1 - x^2}}{x}$$

$$= \cot^{-1}\frac{x}{\sqrt{1 - x^2}} = \sec^{-1}\frac{1}{x}$$

$$= \csc^{-1}\frac{1}{\sqrt{1 - x^2}} = \frac{\pi}{2} - \sin^{-1} x$$

$$\tan^{-1} x = \sin^{-1}\frac{x}{\sqrt{1 - x^2}} = \cos^{-1}\frac{1}{\sqrt{1 + x^2}}$$

$$= \cot^{-1}\frac{1}{x} = \sec^{-1}\sqrt{1 + x^2} = \csc^{-1}\frac{\sqrt{1 + x^2}}{x}$$

RELATIONS BETWEEN ANGLES AND SIDES OF TRIANGLES

Solutions of Triangles (Fig. 3-40) Let a, b, and c denote the sides and α, β, and γ the angles opposite the sides in the triangle. Let $2s = a + b + c$, A = area, r = radius of the inscribed circle, R = radius of the circumscribed circle, and h = altitude. In any triangle $\alpha + \beta + \gamma = 180°$.

Law of Sines $\sin\alpha/a = \sin\beta/b = \sin\gamma/c = 1/(2R)$.

Law of Tangents

$$\frac{a + b}{a - b} = \frac{\tan\tfrac{1}{2}(\alpha + \beta)}{\tan\tfrac{1}{2}(\alpha - \beta)}; \quad \frac{b + c}{b - c} = \frac{\tan\tfrac{1}{2}(\beta + \gamma)}{\tan\tfrac{1}{2}(\beta - \gamma)}; \quad \frac{a + c}{a - c} = \frac{\tan\tfrac{1}{2}(\alpha + \gamma)}{\tan\tfrac{1}{2}(\alpha - \gamma)}$$

Law of Cosines $a^2 = b^2 + c^2 - 2bc\cos\alpha$; $b^2 = a^2 + c^2 - 2ac\cos\beta$; $c^2 = a^2 + b^2 - 2ab\cos\gamma$.

More formulas can be generated by replacing a by b, b by c, c by a, α by β, β by γ, and γ by α.

$$A = \frac{1}{2}bh = \frac{1}{2}ab\sin\gamma = \sqrt{s(s - a)(s - b)(s - c)} = rs$$

where $$r = \sqrt{\frac{(s - a)(s - b)(s - c)}{s}}$$

$$R = a/(2\sin\alpha) = abc/4A \qquad h = c\sin\alpha = a\sin\gamma = 2rs/b$$

Right Triangle (Fig. 3-41) Given one side and any acute angle α or any two sides, the remaining parts can be obtained from the following formulas:

$$a = \sqrt{(c + b)(c - b)} = c\sin\alpha = b\tan\alpha$$

$$b = \sqrt{(c + a)(c - a)} = c\cos\alpha = a\cot\alpha$$

$$c = \sqrt{a^2 + b^2} \quad \sin\alpha = \frac{a}{c} \quad \cos\alpha = \frac{b}{c} \quad \tan\alpha = \frac{a}{b} \quad \beta = 90° - \alpha$$

$$A = \frac{1}{2}ab = \frac{a^2}{2\tan\alpha} = \frac{b^2\tan\alpha}{2} = \frac{c^2\sin 2\alpha}{4}$$

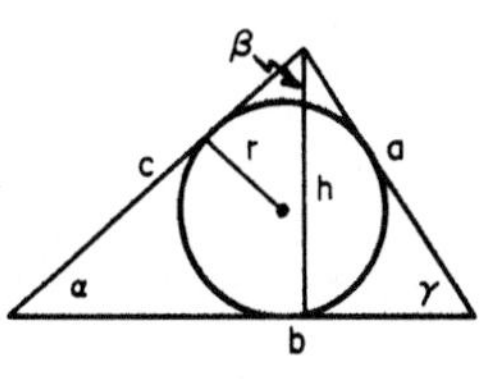

FIG. 3-40 Triangle.

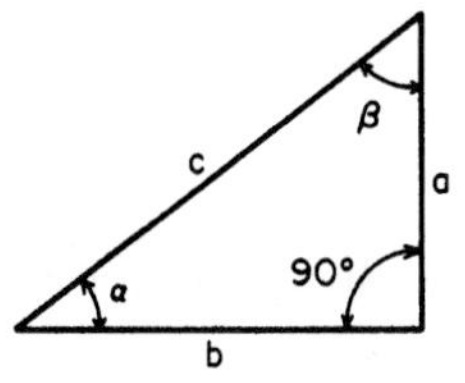

FIG. 3-41 Right triangle.

HYPERBOLIC TRIGONOMETRY

The hyperbolic functions are certain combinations of exponentials e^x and e^{-x}.

$$\cosh x=\frac{e^x+e^{-x}}{2};\ \sinh x=\frac{e^x-e^{-x}}{2};\ \tanh x=\frac{\sinh x}{\cosh x}=\frac{e^x-e^{-x}}{e^x+e^{-x}}$$

$$\coth x=\frac{e^x+e^x}{e^x-e^{-x}}=\frac{1}{\tanh x}=\frac{\cosh x}{\sinh x};\ \operatorname{sech} x=\frac{1}{\cosh x}=\frac{2}{e^x+e^{-x}};$$

$$\operatorname{csch} x=\frac{1}{\sinh x}=\frac{2}{e^x-e^{-x}}$$

Fundamental Relationships $\sinh x+\cosh x=e^x$; $\cosh x-\sinh x=e^{-x}$; $\cosh^2 x-\sinh^2 x=1$; $\operatorname{sech}^2 x+\tanh^2 x=1$; $\coth^2 x-\operatorname{csch}^2 x=1$; $\sinh 2x=2\sinh x\cosh x$; $\cosh 2x=\cosh^2 x+\sinh^2 x=1+2\sinh^2 x=2\cosh^2 x-1$. $\tanh 2x=(2\tanh x)/(1+\tanh^2 x)$; $\sinh(x\pm y)=\sinh x\cosh y\pm\cosh x\sinh y$; $\cosh(x\pm y)=\cosh x\cosh y\pm\sinh x\sinh y$; $2\sinh^2 x/2=\cosh x-1$; $2\cosh^2 x/2=\cosh x+1$; $\sinh(-x)=-\sinh x$; $\cosh(-x)=\cosh x$; $\tanh(-x)=-\tanh x$.

When $u=a\cosh x$ and $\upsilon=a\sinh x$, then $u^2-\upsilon^2=a^2$, which is the equation for a hyperbola. In other words, the hyperbolic functions in the parametric equations $u=a\cosh x$ and $\upsilon=a\sinh x$ have the same relation to the hyperbola $u^2-\upsilon^2=a^2$ that the equations $u=a\cos\theta$ and $\upsilon=a\sin\theta$ have to the circle $u^2+\upsilon^2=a^2$.

Inverse Hyperbolic Functions If $x=\sinh y$, then y is the inverse hyperbolic sine of x, written as $y=\sinh^{-1}x$ or arcsinh x. $\sinh^{-1}x=\ln_e(x+\sqrt{x^2+1})$

$$\cosh^{-1}x=\ln_e(x+\sqrt{x^2-1});\ \tanh^{-1}x=\frac{1}{2}\ln_e\frac{1+x}{1-x};$$

$$\coth^{-1}x=\frac{1}{2}\ln_e\frac{x+1}{x-1};\ \operatorname{sech}^{-1}x=\ln_e\left(\frac{1+\sqrt{1-x^2}}{x}\right);$$

$$\operatorname{csch}^{-1}=\ln_e\left(\frac{1+\sqrt{1+x^2}}{x}\right)$$

Magnitude of the Hyperbolic Functions $\cosh x\geq 1$ with equality only for $x=0$; $-\infty<\sinh x<\infty$; $-1<\tanh x<1$. $\cosh x\sim e^x/2$ as $x\to\infty$; $\sinh x\to e^x/2$ as $x\to\infty$.

APPROXIMATIONS FOR TRIGONOMETRIC FUNCTIONS

For small values of θ (θ measured in radians) $\sin\theta\approx\theta$, $\tan\theta\approx\theta$; $\cos\theta\approx 1-\theta^2/2$.

DIFFERENTIAL AND INTEGRAL CALCULUS

REFERENCES: Larson, R., and B. H. Edwards, *Calculus*, 10th ed., Brooks/Cole, Pacific Grove, Calif., 2013.

DIFFERENTIAL CALCULUS

Limits The limit of function $f(x)$ as x approaches a (a is finite or else x is said to increase without bound) is the number N.

$$\lim_{x\to a} f(x)=N$$

This states that $f(x)$ can be calculated as close to N as desirable by making x sufficiently close to a. This does not put any restriction on $f(x)$ when $x=a$. Alternatively, for any given positive number ε, a number δ can be found such that $0<|a-x|<\delta$ implies that $|N-f(x)|<\varepsilon$.

The following operations with limits (when they exist) are valid:

$$\lim_{x\to a} bf(x)=b\lim_{x\to a} f(x)$$

$$\lim_{x\to a}[f(x)+g(x)]=\lim_{x\to a} f(x)+\lim_{x\to a} g(x)$$

$$\lim_{x\to a}[f(x)\,g(x)]=\lim_{x\to a} f(x)\cdot\lim_{x\to a} g(x)$$

$$\lim_{x\to a}\frac{f(x)}{g(x)}=\frac{\lim_{x\to a} f(x)}{\lim_{x\to a} g(x)}\quad\text{if }\lim_{x\to a} g(x)\neq 0$$

See "Indeterminant Forms" below when $g(a)=0$.

Continuity A function $f(x)$ is continuous at the point $x=a$ if

$$\lim_{h\to 0}[f(a+h)-f(a)]=0$$

Rigorously, it is stated that $f(x)$ is continuous at $x=a$ if for any positive ε there exists a $\delta>0$ such that $|f(a+h)-f(a)|<\varepsilon$ for all x with $|x-a|<\delta$. For example, the function $(\sin x)/x$ is not continuous at $x=0$ and therefore is said to be *discontinuous*. Discontinuities are classified into three types:

1. Removable $y=(\sin x)/x$ at $x=0$
2. Infinite $y=1/x$ at $x=0$
3. Jump $y=10/(1+e^{1/x})$ at $x=0^+$ $y=0^+$; $x=0$ $y=0$; $x=0^-$ $y=10$

Derivative The function $f(x)$ has a derivative at $x=a$, denoted as $f'(a)$, if

$$\lim_{h\to 0}\frac{f(a+h)-f(a)}{h}$$

exists. This implies continuity at $x=a$. However, a function may be continuous but not have a derivative. The derivative function is

$$f'(x)=\frac{df}{dx}=\lim_{h\to 0}\frac{f(x+h)-f(x)}{h}$$

Differentiation Define $\Delta y=f(x+\Delta x)-f(x)$. Then dividing by Δx gives

$$\frac{\Delta y}{\Delta x}=\frac{f(x+\Delta x)-f(x)}{\Delta x}$$

Call
$$\lim_{\Delta x\to 0}\frac{\Delta y}{\Delta x}=\frac{dy}{dx}$$

Then
$$\frac{dy}{dx}=\lim_{\Delta x\to 0}\frac{f(x+\Delta x)-f(x)}{\Delta x}$$

Differential Operations The following differential operations are valid: $f, g, \ldots$ are differentiable functions of x; c and n are constants; e is the base of the natural logarithms.

$$\frac{dc}{dx}=0 \tag{3-1}$$

$$\frac{dx}{dx}=1 \tag{3-2}$$

$$\frac{d}{dx}(f+g)=\frac{df}{dx}+\frac{dg}{dx} \tag{3-3}$$

$$\frac{d}{dx}(f\times g)=f\frac{dg}{dx}+g\frac{df}{dx} \tag{3-4}$$

$$\frac{dy}{dx}=\frac{1}{dx/dy}\quad\text{if}\quad\frac{dx}{dy}\neq 0 \tag{3-5}$$

$$\frac{d}{dx}f^n=nf^{n-1}\frac{df}{dx} \tag{3-6}$$

$$\frac{d}{dx}\left(\frac{f}{g}\right)=\frac{g(df/dx)-f(dg/dx)}{g^2} \tag{3-7}$$

$$\frac{df}{dx}=\frac{df}{d\upsilon}\times\frac{d\upsilon}{dx}\ \text{(chain rule)} \tag{3-8}$$

$$\frac{df^g}{dx} = gf^{g-1}\frac{df}{dx} + f^g \ln f \frac{dg}{dx} \quad (3\text{-}9)$$

$$\frac{da^x}{dx} = (\ln a)\,a^x \quad (3\text{-}10)$$

Example Derive dy/dx for $x^2 + y^3 = x + xy + A$.

Here $$\frac{d}{dx}x^2 + \frac{d}{dx}y^3 = \frac{d}{dx}x + \frac{d}{dx}xy + \frac{d}{dx}A$$

$$2x + 3y^2\frac{dy}{dx} = 1 + y + x\frac{dy}{dx} + 0$$

by the rules in Eqs. (3-6), (3-6), (3-2), (3-4), and (3-1), respectively.

Thus $$\frac{dy}{dx} = \frac{2x-1-y}{x-3y^2}$$

Differentials

$$de^x = e^x\,dx \quad (3\text{-}11)$$

$$d(a^x) = a^x \ln a\,dx \quad (3\text{-}12)$$

$$d\ln x = (1/x)\,dx \quad (3\text{-}13)$$

$$d\log x = (\log e/x)\,dx \quad (3\text{-}14)$$

$$d\sin x = \cos x\,dx \quad (3\text{-}15)$$

$$d\cos x = -\sin x\,dx \quad (3\text{-}16)$$

$$d\tan x = \sec^2 x\,dx \quad (3\text{-}17)$$

$$d\cot x = -\csc^2 x\,dx \quad (3\text{-}18)$$

$$d\sec x = \tan x \sec x\,dx \quad (3\text{-}19)$$

$$d\csc x = -\cot x \csc x\,dx \quad (3\text{-}20)$$

$$d\sin^{-1} x = (1-x^2)^{-1/2}\,dx \quad (3\text{-}21)$$

$$d\cos^{-1} x = -(1-x^2)^{-1/2}\,dx \quad (3\text{-}22)$$

$$d\tan^{-1} x = (1+x^2)^{-1}\,dx \quad (3\text{-}23)$$

$$d\cot^{-1} x = -(1+x^2)^{-1}\,dx \quad (3\text{-}24)$$

$$d\sec^{-1} x = x^{-1}(x^2-1)^{-1/2}\,dx \quad (3\text{-}25)$$

$$d\csc^{-1} x = -x^{-1}(x^2-1)^{-1/2}\,dx \quad (3\text{-}26)$$

$$d\sinh x = \cosh x\,dx \quad (3\text{-}27)$$

$$d\cosh x = \sinh x\,dx \quad (3\text{-}28)$$

$$d\tanh x = \text{sech}^2 x\,dx \quad (3\text{-}29)$$

$$d\coth x = -\text{csch}^2 x\,dx \quad (3\text{-}30)$$

$$d\,\text{sech}\,x = -\text{sech}\,x \tanh x\,dx \quad (3\text{-}31)$$

$$d\,\text{csch}\,x = -\text{csch}\,x \coth x\,dx \quad (3\text{-}32)$$

$$d\sinh^{-1} x = (x^2+1)^{-1/2}\,dx \quad (3\text{-}33)$$

$$d\cosh^{-1} = (x^2-1)^{-1/2}\,dx \quad (3\text{-}34)$$

$$d\tanh^{-1} x = (1-x^2)^{-1}\,dx \quad (3\text{-}35)$$

$$d\coth^{-1} x = -(x^2-1)^{-1}\,dx \quad (3\text{-}36)$$

$$d\,\text{sech}^{-1} x = -(1/x)(1-x^2)^{-1/2}\,dx \quad (3\text{-}37)$$

$$d\,\text{csch}^{-1} x = -x^{-1}(x^2+1)^{-1/2}\,dx \quad (3\text{-}38)$$

Example Find dy/dx for $y = \sqrt{x}\cos(1-x^2)$. Using

$$\frac{dy}{dx} = \sqrt{x}\frac{d}{dx}\cos(1-x^2) + \cos(1-x^2)\frac{d}{dx}\sqrt{x} \quad (3\text{-}4)$$

$$\frac{d}{dx}\cos(1-x^2) = -\sin(1-x^2)\frac{d}{dx}(1-x^2) \quad (3\text{-}16)$$

$$= -\sin(1-x^2)(0-2x) \quad (3\text{-}1), (3\text{-}6)$$

$$\frac{d\sqrt{x}}{dx} = \frac{1}{2}x^{-1/2}$$

$$\frac{dy}{dx} = 2x^{3/2}\sin(1-x^2) + \frac{1}{2}x^{-1/2}\cos(1-x^2) \quad (3\text{-}6)$$

Example Find the derivative of $\tan x$ with respect to $\sin x$.
Let

$$\upsilon = \sin x$$
$$y = \tan x$$

Then,

Using

$$\frac{d\tan x}{d\sin x} = \frac{dy}{d\upsilon} = \frac{dy}{dx}\frac{dx}{d\upsilon} \quad (3\text{-}8)$$

$$= \frac{d\tan x}{dx}\frac{1}{\dfrac{d\sin x}{dx}} \quad (3\text{-}5)$$

$$= \sec^2 x/\cos x \quad (3\text{-}17), (3\text{-}15)$$

If the functions and derivatives are known only numerically at some point, the same formulas may be used.

Higher Differentials The first derivative of $f(x)$ with respect to x is denoted by f' or df/dx. The derivative of the first derivative is called the *second derivative* of $f(x)$ with respect to x and is denoted by f'', $f^{(2)}$, or d^2f/dx^2; and similarly for the higher-order derivatives.

Example Given $f(x) = 3x^3 + 2x + 1$, calculate all derivative values.

$$\frac{df(x)}{dx} = 9x^2 + 2, \frac{d^2f(x)}{dx^2} = 18x, \frac{d^3f(x)}{dx^3} = 18, \frac{d^nf(x)}{dx^n} = 0 \quad \text{for } n \geq 4$$

If $f'(x) > 0$ on (a, b), then f is increasing on (a, b). If $f'(x) < 0$ on (a, b), then f is decreasing on (a, b). The graph of a function $y = f(x)$ is concave up if f' is increasing on (a, b); it is concave down if f' is decreasing on (a, b). If $f''(x)$ exists on (a, b) and if $f''(x) > 0$, then f is concave up on (a, b). If $f''(x) < 0$, then f is concave down on (a, b).

An *inflection point* is a point at which a function changes the direction of its concavity.

Indeterminate Forms: L'Hôpital's Theorem Forms of the type 0/0, ∞/∞, $0 \times \infty$, etc., are called *indeterminates.* To find the limiting values that the corresponding functions approach, L'Hôpital's theorem is useful: If two functions $f(x)$ and $g(x)$ both become zero at $x = a$, then the limit of their quotient is equal to the limit of the quotient of their separate derivatives, if the limit exists, or is $+\infty$ or $-\infty$.

Example Find $\lim\limits_{n\to 0}\dfrac{\sin x}{x}$.

Here $$\lim_{x\to 0}\frac{\sin x}{x} = \lim_{x\to 0}\frac{d\sin x}{dx}\Big/\lim_{x\to 0}\frac{dx}{dx} = \lim_{x\to 0}\frac{\cos x}{1} = 1$$

Example Find $\lim\limits_{x\to\infty} x^3e^{-x}$.

$$\lim_{x\to\infty} x^3e^{-x} = \lim_{x\to\infty}\frac{x^3}{e^x} = \lim_{x\to\infty}\frac{6}{e^x} = 0$$

Example Find $\lim\limits_{x\to 0}(1-x)^{1/x}$.

Let $y = (1-x)^{1/x}$ $\quad \ln y = (1/x)\ln(1-x)$

Then

$$\lim_{x\to 0}(\ln y) = \lim_{x\to 0}\frac{\ln(1-x)}{x} = \frac{\lim_{x\to 0}\ln(1-x)}{\lim_{x\to 0}x}$$

$$= \frac{d[\ln(1-x)]/dx|_{x=0}}{dx/dx|_{x=0}} = \frac{1}{1-x}(-1)\bigg|_{x=0} = -1$$

Therefore, $$\lim_{x\to 0} y = e^{-1}$$

Partial Derivative The abbreviation $z = f(x, y)$ means that z is a function of the two variables x and y. The derivative of z with respect to x, treating y as a constant, is called the *partial derivative* with respect to x and is usually denoted as $\partial z/\partial x$ or $\partial f(x, y)/\partial x$ or simply f_x. Partial differentiation, like full differentiation, is quite simple to apply. Conversely, the solution of partial differential equations is appreciably more difficult than that of differential equations.

Example Find $\partial z/\partial x$ and $\partial z/\partial y$ for $z = ye^{x^2} + xe^y$.

$$\frac{\partial z}{\partial x} = y\frac{\partial e^{x^2}}{\partial x} + e^y\frac{\partial x}{\partial x} = 2xye^{x^2} + e^y$$

$$\frac{\partial z}{\partial y} = e^{x^2}\frac{\partial y}{\partial y} + x\frac{\partial e^y}{\partial y} = e^{x^2} + xe^y$$

Order of Differentiation It is generally true that the order of differentiation is immaterial for any number of differentiations or variables, provided the function and the appropriate derivatives are continuous. For $z = f(x, y)$ it follows that

$$\frac{\partial^3 f}{\partial y^2\,\partial x} = \frac{\partial^3 f}{\partial y\,\partial x\,\partial y} = \frac{\partial^3 f}{\partial x\,\partial y^2}$$

General Form for Partial Differentiation

1. Given $f(x, y) = 0$ and $x = g(t)$, $y = h(t)$.

Then
$$\frac{df}{dt} = \frac{\partial f}{\partial x}\frac{dx}{dt} + \frac{\partial f}{\partial y}\frac{dy}{dt}$$

$$\frac{d^2 f}{dt^2} = \frac{\partial^2 f}{\partial x^2}\left(\frac{dx}{dt}\right)^2 + 2\frac{\partial^2 f}{\partial x\,\partial y}\frac{dx}{dt}\frac{dy}{dt} + \frac{\partial^2 f}{\partial y^2}\left(\frac{dy}{dt}\right)^2 + \frac{\partial f}{\partial x}\frac{d^2 x}{dt^2} + \frac{\partial f}{\partial y}\frac{d^2 y}{dt^2}$$

Example Find df/dt for $f = xy$, $x = \rho \sin t$, and $y = \rho \cos t$.

$$\frac{df}{dt} = \frac{\partial(xy)}{\partial x}\left(\frac{d\rho\sin t}{dt}\right) + \frac{\partial(xy)}{\partial y}\left(\frac{d\rho\cos t}{dt}\right)$$

$$= y(\rho\cos t) + x(-\rho\sin t)$$

$$= \rho^2\cos^2 t - \rho^2\sin^2 t$$

2. Given $f(x, y) = 0$ and $x = g(t, s)$, $y = h(t, s)$.

Then
$$\frac{\partial f}{\partial t} = \frac{\partial f}{\partial x}\frac{\partial x}{\partial t} + \frac{\partial f}{\partial y}\frac{\partial y}{\partial t}$$

$$\frac{\partial f}{\partial s} = \frac{\partial f}{\partial x}\frac{\partial x}{\partial s} + \frac{\partial f}{\partial y}\frac{\partial y}{\partial s}$$

Differentiation of Composite Function

Rule 1. Given $f(x, y) = 0$, then $\dfrac{dy}{dx} = -\dfrac{\partial f/\partial x}{\partial f/\partial y}\left(\dfrac{\partial f}{\partial y} \neq 0\right)$.

Rule 2. Given $f(u) = 0$ where $u = g(x)$, then

$$\frac{df}{dx} = f'(u)\frac{du}{dx}$$

$$\frac{d^2 f}{dx^2} = f''(u)\left(\frac{du}{dx}\right)^2 + f'(u)\frac{d^2 u}{dx^2}$$

Rule 3. Given $f(u) = 0$ where $u = g(x, y)$, then

$$\frac{\partial^2 f}{\partial x^2} = f''\left(\frac{\partial u}{\partial x}\right)^2 + f'\frac{\partial^2 u}{\partial x^2}$$

$$\frac{\partial^2 f}{\partial x\,\partial y} = f''\frac{\partial u}{\partial x}\frac{\partial u}{\partial y} + f'\frac{\partial^2 u}{\partial x\,\partial y}$$

$$\frac{\partial^2 f}{\partial y^2} = f''\left(\frac{\partial u}{\partial y}\right)^2 + f'\frac{\partial^2 u}{\partial y^2}$$

MULTIVARIABLE CALCULUS APPLIED TO THERMODYNAMICS

Many of the functional relationships needed in thermodynamics are direct applications of the rules of multivariable calculus. This section reviews those rules in the context of the needs of thermodynamics. These ideas were expounded in one of the classic books on chemical engineering thermodynamics (see Hougen, O. A., et al., Part II, "Thermodynamics," in *Chemical Process Principles*, 2d ed., Wiley, New York, 1959).

State Functions State functions depend only on the state of the system, not on history or how one got there. If z is a function of two variables x and y, then $z(x, y)$ is a state function, since z is known once x and y are specified. The differential of z is

$$dz = M\,dx + N\,dy$$

The line integral

$$\int_c (M\,dx + N\,dy)$$

is independent of the path in xy space if and only if

$$\frac{\partial M}{\partial y} = \frac{\partial N}{\partial x} \qquad (3\text{-}39)$$

and dz is called an exact differential. The total differential can be written as

$$dz = \left(\frac{\partial z}{\partial x}\right)_y dx + \left(\frac{\partial z}{\partial y}\right)_x dy \qquad (3\text{-}40)$$

and thus the following application of Eq. (3-39) guarantees path independence.

$$\frac{\partial}{\partial y}\left(\frac{\partial z}{\partial x}\right)_y = \frac{\partial}{\partial x}\left(\frac{\partial z}{\partial y}\right)_x$$

or
$$\frac{\partial^2 z}{\partial y\,\partial x} = \frac{\partial^2 z}{\partial x\,\partial y} \qquad (3\text{-}41)$$

Example Suppose z is constant and apply Eq. (3-40).

$$\left[0 = \left(\frac{\partial z}{\partial x}\right)_y dx + \left(\frac{\partial z}{\partial y}\right)_x dy\right]_z$$

Rearrangement gives the triple product rule

$$\left(\frac{\partial z}{\partial x}\right)_y = -\left(\frac{\partial y}{\partial x}\right)_z\left(\frac{\partial z}{\partial y}\right)_x = -\frac{(\partial y/\partial x)_z}{(\partial y/\partial z)_x} \text{ or } \left(\frac{\partial z}{\partial x}\right)_y\left(\frac{\partial x}{\partial y}\right)_z\left(\frac{\partial y}{\partial z}\right)_x = -1 \qquad (3\text{-}42)$$

Alternatively, divide Eq. (3-40) by dy when holding some other variable w constant to obtain

$$\left(\frac{\partial z}{\partial y}\right)_w = \left(\frac{\partial z}{\partial x}\right)_y\left(\frac{\partial x}{\partial y}\right)_w + \left(\frac{\partial z}{\partial y}\right)_x \qquad (3\text{-}43)$$

Also divide both numerator and denominator of a partial derivative by dw while holding a variable y constant to get the chain rule.

$$\left(\frac{\partial z}{\partial x}\right)_y = \frac{(\partial z/\partial w)_y}{(\partial x/\partial w)_y} = \left(\frac{\partial z}{\partial w}\right)_y\left(\frac{\partial w}{\partial x}\right)_y \qquad (3\text{-}44)$$

Thermodynamic State Functions In thermodynamics, the state functions include the internal energy U, enthalpy H, and Helmholtz and Gibbs free energies A and G, respectively, defined as follows:

$$H = U + PV$$

$$A = U - TS$$

$$G = H - TS = U + PV - TS = A + PV$$

where S is the entropy, T the absolute temperature, P the pressure, and V the volume. These are also state functions, in that the entropy is specified once two variables (such as T and P) are specified, for example. Likewise, V is specified once T and P are specified; it is therefore a state function.

In an open system, extensive properties, such as the total internal energy, are functions of two thermodynamic variables plus the mass or moles of each component. The mathematical derivations below are for a single-component system of constant mass. They are applicable when the mass stays constant, i.e., in an intensive system (or else an additional variable for moles N must be added). However, the relations between the thermodynamic variables can be regarded as internal energy per moles in a closed system, or at a point in an open system. The formulas illustrate the use of calculus in thermodynamics.

If a process is reversible and only P-V work is done, the first law and differentials can be expressed as follows:

$$dU = T\,dS - P\,dV \tag{3-45}$$

$$dH = T\,dS + V\,dP \tag{3-46}$$

$$dA = -S\,dT - P\,dV \tag{3-47}$$

$$dG = -S\,dT + V\,dP \tag{3-48}$$

Alternatively, if the internal energy is considered a function of S and V, then the differential is

$$dU = \left(\frac{\partial U}{\partial S}\right)_V dS + \left(\frac{\partial U}{\partial V}\right)_S dV$$

This is the equivalent of Eq. (3-43) and gives the following definitions:

$$T = \left(\frac{\partial U}{\partial S}\right)_V, \quad P = -\left(\frac{\partial U}{\partial V}\right)_S$$

Since the internal energy is a state function, Eq. (3-44) must be satisfied.

$$\frac{\partial^2 U}{\partial V\,\partial S} = \frac{\partial^2 U}{\partial S\,\partial V}$$

This is

$$\left(\frac{\partial T}{\partial V}\right)_S = -\left(\frac{\partial P}{\partial S}\right)_V$$

This is one of the Maxwell relations, and the other Maxwell relations can be derived in a similar fashion by applying Eq. (3-41). See Sec. 4, Thermodynamics, "Constant-Composition Systems."

Partial Derivatives of Intensive Thermodynamic Functions The various partial derivatives of the thermodynamic functions can be classified into six groups. In the general formulas below, the variables U, H, A, G, and S are denoted by Greek letters (these can be extensive properties), while the variables V, T, and P are denoted by Latin letters (T and P can only be intensive properties).

Type I (3 possibilities plus reciprocals)

$$\text{General: } \left(\frac{\partial a}{\partial b}\right)_c \qquad \text{specific: } \left(\frac{\partial P}{\partial T}\right)_V$$

Equation (3-42) gives

$$\left(\frac{\partial P}{\partial T}\right)_V = -\left(\frac{\partial V}{\partial T}\right)_P \left(\frac{\partial P}{\partial V}\right)_T = -\frac{(\partial V/\partial T)_P}{(\partial V/\partial P)_T}$$

Type II (30 possibilities plus reciprocals)

$$\text{General: } \left(\frac{\partial \alpha}{\partial b}\right)_c \qquad \text{specific: } \left(\frac{\partial G}{\partial T}\right)_V$$

The differential for G is from Eq. (3-48) or Eq. (3-43) with $x \rightarrow P$:

$$\left(\frac{\partial G}{\partial T}\right)_V = -S + V\left(\frac{\partial P}{\partial T}\right)_V$$

Using the other equations for U, H, A, or S gives the other possibilities.

Type III (15 possibilities plus reciprocals)

$$\text{General: } \left(\frac{\partial a}{\partial b}\right)_\alpha \qquad \text{specific: } \left(\frac{\partial V}{\partial T}\right)_S$$

First evaluate the derivative, using Eq. (3-45).

$$\left(\frac{\partial V}{\partial T}\right)_S = -\left(\frac{\partial S}{\partial T}\right)_V \left(\frac{\partial V}{\partial S}\right)_T = -\frac{(\partial S/\partial T)_V}{(\partial S/\partial V)_T}$$

Then evaluate the numerator and denominator as type II derivatives. Use Eq. (3-45) and Eq. (3-41) to get $(\partial S/\partial T)_V = C_v/T$. Use Eqs. (3-47) and (3-41) to get the Maxwell relation $(\partial P/\partial T)_V = (\partial S/\partial V)_T$. Finally use Eq. (3-42).

$$\left(\frac{\partial V}{\partial T}\right)_S = -\frac{\frac{C_v}{T}}{-\left(\frac{\partial V}{\partial T}\right)_P \left(\frac{\partial P}{\partial V}\right)_T} = \frac{C_v}{T}\frac{\left(\frac{\partial V}{\partial P}\right)_T}{\left(\frac{\partial V}{\partial T}\right)_P}$$

These derivatives are of importance for reversible, adiabatic processes (such as in an ideal turbine or compressor), since then the entropy is constant. An example is the Joule-Thomson coefficient for constant H.

$$\left(\frac{\partial T}{\partial P}\right)_H = \frac{1}{C_p}\left[-V + T\left(\frac{\partial V}{\partial T}\right)_P\right]$$

Type IV (30 possibilities plus reciprocals)

$$\text{General: } \left(\frac{\partial \alpha}{\partial \beta}\right)_c \qquad \text{specific: } \left(\frac{\partial G}{\partial A}\right)_P$$

Use Eq. (3-47) to introduce a new variable T.

$$\left(\frac{\partial G}{\partial A}\right)_P = \left(\frac{\partial G}{\partial T}\right)_P \left(\frac{\partial T}{\partial A}\right)_P = \frac{(\partial G/\partial T)_P}{(\partial A/\partial T)_P}$$

This operation has created two type II derivatives; using the differential Eqs. (3-47) and (3-48), we obtain

$$\left(\frac{\partial G}{\partial A}\right)_P = \frac{S}{S + P(\partial V/\partial T)_P}$$

Type V (60 possibilities plus reciprocals)

$$\text{General: } \left(\frac{\partial \alpha}{\partial b}\right)_\beta \qquad \text{specific: } \left(\frac{\partial G}{\partial P}\right)_A$$

Start from the differential for dG. Then we get

$$\left(\frac{\partial G}{\partial P}\right)_A = -S\left(\frac{\partial T}{\partial P}\right)_A + V$$

The derivative is type III and can be evaluated by using Eq. (3-42).

$$\left(\frac{\partial G}{\partial P}\right)_A = S\frac{(\partial A/\partial P)_T}{(\partial A/\partial T)_P} + V$$

The two type II derivatives are then evaluated using the differential Eq. (3-47).

$$\left(\frac{\partial G}{\partial P}\right)_A = \frac{SP(\partial V/\partial P)_T}{S + P(\partial V/\partial T)_P} + V$$

These derivatives are also of interest for free expansions or isentropic changes.

Type VI (30 possibilities plus reciprocals)

$$\text{General: } \left(\frac{\partial \alpha}{\partial \beta}\right)_\gamma \qquad \text{specific: } \left(\frac{\partial G}{\partial A}\right)_H$$

We use Eq. (3-44) to obtain two type V derivatives.

$$\left(\frac{\partial G}{\partial A}\right)_H = \frac{(\partial G/\partial T)_H}{(\partial A/\partial T)_H}$$

These can then be evaluated using the procedures for type V derivatives.

INTEGRAL CALCULUS

Indefinite Integral If $f'(x)$ is the derivative of $f(x)$, an antiderivative of $f'(x)$ is $f(x)$. Symbolically, the indefinite integral of $f'(x)$ is

$$\int f'(x)\,dx = f(x) + c$$

where c is an arbitrary constant to be determined by the problem. By virtue of the known formulas for differentiation, the following relationships hold (a is a constant):

$$\int (du + d\upsilon + dw) = \int du + \int d\upsilon + \int dw \quad (3\text{-}49)$$

$$\int a\,d\upsilon = a \int d\upsilon \quad (3\text{-}50)$$

$$\int \upsilon^n\,d\upsilon = \frac{\upsilon^{n+1}}{n+1} + c \quad (n \neq -1) \quad (3\text{-}51)$$

$$\int \frac{d\upsilon}{\upsilon} = \ln|\upsilon| + c \quad (3\text{-}52)$$

$$\int a^\upsilon\,d\upsilon = \frac{a^\upsilon}{\ln a} + c \quad (3\text{-}53)$$

$$\int e^\upsilon\,d\upsilon = e^\upsilon + c \quad (3\text{-}54)$$

$$\int \sin\upsilon\,d\upsilon = -\cos\upsilon + c \quad (3\text{-}55)$$

$$\int \cos\upsilon\,d\upsilon = \sin\upsilon + c \quad (3\text{-}56)$$

Other integrals can be found at en.wikipedia.org/wiki/Lists_of_integrals.

Example Find $\int (3x^2 + e^x - 10)\,dx$ using Eq. (3-49).

$$\int (3x^2 + e^x - 10)dx = 3\int x^2\,dx + \int e^x\,dx - 10\int dx = x^3 + e^x - 10x + c$$

Example: Constant of Integration By definition the derivative of x^3 is $3x^2$, and x^3 is therefore the integral of $3x^2$. However, if $f = x^3 + 10$, it follows that $f' = 3x^2$, and $x^3 + 10$ is therefore also the integral of $3x^2$. For this reason the constant c in $\int 3x^2\,dx = x^3 + c$ must be determined by the problem conditions, i.e., the value of f for a specified x.

Methods of Integration In practice it is rare when generally encountered functions can be directly integrated. For example, the integrand in $\int \sqrt{\sin x}\,dx$ which appears quite simple has no elementary function whose derivative is $\sqrt{\sin x}$. In general, there is no explicit way of determining whether a particular function can be integrated into an elementary form. When they do not exist or cannot be found either from tabled integration formulas or directly, the only recourse is series expansion, as illustrated later. Indefinite integrals cannot be solved numerically unless they are redefined as definite integrals (see "Definite Integral"), that is, $F(x) = \int f(x)\,dx$ is indefinite, whereas $F(x) = \int_a^x f(t)dt$ is definite.

Direct Formula Many integrals can be solved by transformation in the integrand to one of the forms given previously.

Example Find $\int x^2\sqrt{3x^3+10}\,dx$. Let $\upsilon = 3x^3 + 10$ for which $d\upsilon = 9x^2\,dx$. Thus

$$\int x^2\sqrt{3x^3+10}\,dx = \int (3x^3+10)^{1/2}(x^2\,dx)$$
$$= \frac{1}{9}\int (3x^3+10)^{1/2}(9x^2\,dx) = \frac{1}{9}\int \upsilon^{1/2}\,d\upsilon$$
$$= \frac{1}{9}\frac{\upsilon^{3/2}}{3/2} + c \quad \text{[by Eq. (3-51)]}$$
$$= \frac{2}{27}(3x^3+10)^{3/2} + c$$

Trigonometric Substitution This technique is particularly well adapted to integrands in the form of radicals. For these the function is transformed to a trigonometric form. In the latter form they may be more easily recognizable relative to the identity formulas. These functions and their transformations are as follows:

$\sqrt{x^2 - a^2}$ Let $x = a\sec\theta$

$\sqrt{x^2 + a^2}$ Let $x = a\tan\theta$

$\sqrt{a^2 - x^2}$ Let $x = a\sin\theta$

Example Find $\int \frac{\sqrt{4-9x^2}}{x^2}\,dx$. Let $x = \frac{2}{3}\sin\theta$; then $dx = \frac{2}{3}\cos\theta\,d\theta$.

$$3\int \frac{\sqrt{(2/3)^2 - x^2}}{x^2}\,dx = 3\int \frac{2/3\sqrt{1-\sin^2\theta}}{(2/3)^2\sin^2\theta}\left(\frac{2}{3}\cos\theta\,d\theta\right)$$
$$= 3\int \frac{\cos^2\theta}{\sin^2\theta}\,d\theta = 3\int \cot^2\theta\,d\theta$$
$$= -3\cot\theta - 3\theta + c \text{ by trigonometric transform}$$
$$= -\frac{\sqrt{4-9x^2}}{x} - 3\sin^{-1}\frac{3}{2}x + c \text{ in terms of } x$$

Algebraic Substitution Functions containing elements of the type $(a + bx)^{1/n}$ are best handled by the algebraic transformation $y^n = a + bx$.

Example Find $\int \frac{x\,dx}{(3+4x)^{1/4}}$. Let $3 + 4x = y^4$; then $4\,dx = 4y^3\,dy$ and

$$\int \frac{x\,dx}{(3+4x)^{1/4}} = \int \frac{\frac{y^4-3}{4}y^3\,dy}{y} = \frac{1}{4}\int y^2(y^4-3)\,dy$$
$$= \frac{1}{4}\frac{y^7}{7} - \frac{3}{4}\frac{y^3}{3} + c = \frac{1}{28}(3+4x)^{7/4} - \frac{1}{4}(3+4x)^{3/4} + c$$

Partial Fractions Rational functions are of the type $f(x)/g(x)$ where $f(x)$ and $g(x)$ are polynomial expressions of degrees m and n, respectively. If the degree of f is higher than the degree of g, perform the algebraic division—the remainder will then be at least one degree less than the denominator. Consider the following types:

Example Reducible denominator to linear unequal factors.

$$\frac{1}{x^3 - x^2 - 4x + 4} = \frac{1}{(x+2)(x-2)(x-1)}$$
$$= \frac{A}{x+2} + \frac{B}{x-2} + \frac{C}{x-1}$$
$$= \frac{A(x-2)(x-1) + B(x+2)(x-1) + C(x+2)(x-2)}{(x+2)(x-2)(x-1)}$$
$$= \frac{x^2(A+B+C) + x(-3A+B) + (2A-2B-4C)}{(x+2)(x-2)(x-1)}$$

Equate coefficients and solve for A, B, and C.

$$A + B + C = 0 \qquad -3A + B = 0 \qquad 2A - 2B - 4C = 1$$

$$A = 1/12 \qquad B = 1/4 \qquad C = -1/3$$

$$\frac{1}{x^2 - x^2 - 4x + 4} = \frac{1}{12(x+2)} + \frac{1}{4(x-2)} - \frac{1}{3(x-1)}$$

Hence

$$\int \frac{dx}{x^3 - x^2 - 4x + 4} = \int \frac{dx}{12(x+2)} + \int \frac{dx}{4(x-2)} - \int \frac{dx}{3(x-1)}$$

Integration by Parts An extremely useful formula for integration is the relation $d(u\upsilon) = u\,d\upsilon + \upsilon\,du$

and $\quad u\upsilon = \int u\,d\upsilon + \int \upsilon\,du$

or $\quad \int u\,d\upsilon = u\upsilon - \int \upsilon\,du$

It is particularly useful for trigonometric and exponential functions.

Example Find $\int xe^x\,dx$. Let

$$u = x \quad \text{and} \quad dv = e^x\,dx$$
$$du = dx \qquad \upsilon = e^x$$

Therefore $\int xe^x\,dx = xe^x - \int e^x\,dx = xe^x - e^x + c$

Example Find $\int e^x \sin x\,dx$. Let

$$u = e^x \qquad dv = \sin x\,dx$$
$$du = e^x\,dx \qquad v = -\cos x$$
$$\int e^x \sin x\,dx = -e^x \cos x + \int e^x \cos x\,dx$$

Again

$$u = e^x \qquad dv = \cos x\,dx$$
$$du = e^x\,dx \qquad v = \sin x$$
$$\int e^x \sin x\,dx = -e^x \cos x + e^x \sin x - \int e^x \sin x\,dx + c$$
$$= (e^x/2)(\sin x - \cos x) + \frac{c}{2}$$

Series Expansion When an explicit function cannot be found, the integration can sometimes be carried out by a series expansion.

Example Find $\int e^{-x^2}\,dx$. Since

$$e^{-x^2} = 1 - x^2 + \frac{x^4}{2!} - \frac{x^6}{3!} + \cdots$$

$$\int e^{-x^2}\,dx = \int dx - \int x^2\,dx + \int \frac{x^4}{2!}\,dx - \int \frac{x^6}{3!}\,dx + \cdots$$

$$= x - \frac{x^3}{3} + \frac{x^5}{5(2!)} - \frac{x^7}{7(3!)} + \cdots \quad \text{for all } x$$

Definite Integral The value of a definite integral depends on the limits a and b and any selected variable coefficients in the function but not on the dummy variable of integration x. Symbolically

$$F(x) = \int f(x)\,dx \qquad \text{indefinite integral where } dF/dx = f(x)$$

$$F(a,b) = \int_a^b f(x)dx \qquad \text{definite integral}$$

$$F(\alpha) = \int_a^b f(x,\alpha)dxF$$

There are certain restrictions of the integration definition: The function $f(x)$ must be continuous in the finite interval (a, b) with at most a finite number of finite discontinuities. Relaxing two of these restrictions gives rise to so-called *improper integrals* and requires special handling. These occur when

1. The limits of integration are not both finite, i.e., $\int_0^\infty e^{-x}\,dx$.
2. The function becomes infinite within the interval of integration, i.e.,

$$\int_0^1 \frac{1}{\sqrt{x}}\,dx$$

Techniques for determining when integration is valid under these conditions are available in the references.

Properties The fundamental theorem of calculus states

$$\int_a^b f(x)\,dx = F(b) - F(a)$$

where $\qquad dF(x)/dx = f(x)$

Other properties of the definite integral are as follows:

$$\int_a^b c[f(x)\,dx] = c\int_a^b f(x)\,dx$$

$$\int_a^b [f_1(x) + f_2(x)]\,dx = \int_a^b f_1(x)\,dx + \int_a^b f_2(x)\,dx$$

$$\int_a^b f(x)\,dx = -\int_b^a f(x)\,dx$$

$$\int_a^b f(x)\,dx = \int_a^c f(x)\,dx + \int_c^b f(x)\,dx$$

$$\int_a^b f(x)\,dx = (b-a)f(\xi) \qquad \text{for some } \xi \text{ in } (a, b)$$

$$\frac{\partial}{\partial b}\int_a^b f(x)\,dx = f(b)$$

$$\frac{\partial}{\partial b}\int_a^b f(x)\,dx = -f(a)$$

$$\frac{dF(\alpha)}{d\alpha} = \int_a^b \frac{\partial f(x,\alpha)}{\partial \alpha}\,dx \qquad \text{if} \quad a \text{ and } b \text{ are constant} \tag{3-57}$$
$$\int_a^b dx \int_c^d f(x,\alpha)\,d\alpha = \int_c^d d\alpha \int_a^b f(x,\alpha)\,dx$$

When $F(x) = \int_{a(x)}^{b(x)} f(x,y)\,dy$ the Leibniz rule gives

$$\frac{dF}{dx} = \frac{db}{dx} f[x, b(x)] - \frac{da}{dx} f[x, a(x)] + \int_{a(x)}^{b(x)} \frac{\partial f}{\partial x}\,dy$$

Example Find $\int_0^2 \frac{dx}{(x-1)^2}$. Direct application of the formula would yield the incorrect value

$$\int_0^2 \frac{dx}{(x-1)^2} = \left[-\frac{1}{x-1}\right]_0^2 = -2$$

Note that $f(x) = 1/(x-1)^2$ becomes unbounded as $x \to 1$ and by rule 2 the integral diverges and hence is said not to exist.

Methods of Integration All the methods of integration available for the indefinite integral can be used for definite integrals. In addition, several others are available for the latter integrals and are indicated below.

Change of Variable This substitution is basically the same as previously indicated for indefinite integrals. However, for definite integrals, the limits of integration must also be changed: i.e., for $x = \phi(t)$,

$$\int_a^b f(x)\,dx = \int_{t_0}^{t_1} f[\phi(t)]\,\phi'(t)\,dt$$

where

$$t = t_0 \text{ when } x = a$$
$$t = t_1 \text{ when } x = b$$

Example Find $\int_0^4 \sqrt{16 - x^2}\,dx$. Let

$$x = 4\sin\theta \qquad (x = 0, \theta = 0)$$
$$dx = 4\cos\theta\,d\theta \qquad (x = 4, \theta = \pi/2)$$

Then $\quad \int_0^4 \sqrt{16-x^2}\,dx = 16\int_0^{\pi/2} \cos^2\theta\,d\theta = 16[\tfrac{1}{2}\theta + \tfrac{1}{4}\sin 2\theta]_0^{\pi/2} = 4\pi$

Integration It is sometimes useful to generate a double integral to solve a problem. By this approach, the fundamental theorem indicated in Eq. (3-57) can be used.

Example Find $\int_0^1 \frac{x^b - x^a}{\ln x}\,dx$.

Consider $\quad \int_0^1 x^\alpha\,dx = \frac{1}{\alpha+1} \quad (\alpha > -1)$

Multiply both sides by $d\alpha$ and integrate between a and b.

$$\int_a^b d\alpha \int_0^1 x^\alpha\,dx = \int_a^b \frac{d\alpha}{\alpha+1} = \ln\left|\frac{b+1}{a+1}\right|$$

But also

$$\int_a^b d\alpha \int_0^1 x^\alpha\,dx = \int_0^1 dx \int_a^b x^\alpha\,d\alpha = \int_0^1 \frac{x^b - x^a}{\ln x}\,dx$$

Therefore

$$\int_0^1 \frac{x^b - x^a}{\ln x}\,dx = \ln\left|\frac{b+1}{a+1}\right|$$

INFINITE SERIES

References: de Brujin, N. G., *Asymptotic Methods in Analysis*, Dover, New York, 2010; Zwillinger, D., *Table of Integrals, Series, and Products*, 8th ed., Academic, New York, 2014.

DEFINITIONS

A succession of numbers or terms formed according to some definite rule is called a *sequence.* The indicated sum of the terms of a sequence is called a *series.* A series of the form $a_0 + a_1(x - c) + a_2(x - c)^2 + \cdots + a_n(x - c)^n + \cdots$ is called a *power series.*

Consider the sum of a finite number of terms in the geometric series (a special case of a power series).

$$S_n = a + ar + ar^2 + ar^3 + \cdots + ar^{n-1} \tag{3-58}$$

For any number of terms n, the sum equals

$$S_n = a\frac{1-r^n}{1-r}$$

In this form, the geometric series is assumed finite.

In the form of Eq. (3-58), it can further be defined that the terms in the series be nonending and therefore an infinite series.

$$S = a + ar + ar^2 + \cdots + ar^n + \cdots \tag{3-59}$$

However, the defined sum of the terms [Eq. (3-59)]

$$S_n = a\frac{1-r^n}{1-r} \qquad r \neq 1$$

while valid for any finite value of r and n, now takes on a different interpretation. In this sense it is necessary to consider the limit of S_n as n increases indefinitely:

$$S = \lim_{n\to\infty} S_n$$
$$= a \lim_{n\to\infty} \frac{1-r^n}{1-r}$$

The infinite series converges if the limit of S_n approaches a fixed finite value as n approaches infinity. Otherwise, the series is *divergent.* If r is less than 1 but greater than −1, the infinite series is convergent. For values outside of the range $-1 < r < 1$, the series is divergent because the sum is not defined. The range $-1 < r < 1$ is called the *region of convergence.* (We assume $a \neq 0$.)

There are also two types of convergent series. Consider the new series

$$S = 1 - \frac{1}{2} + \frac{1}{3} - \frac{1}{4} + \cdots + (-1)^{n+1}\frac{1}{n} + \cdots \tag{3-60}$$

It can be shown that series (3-60) does converge to the value $S = \ln 2$. However, if each term is replaced by its absolute value, the series becomes unbounded and therefore divergent (unbounded divergent):

$$S = 1 + \frac{1}{2} + \frac{1}{3} + \frac{1}{4} + \frac{1}{5} + \cdots \tag{3-61}$$

In this case series (3-60) is defined as a conditionally convergent series. If the replacement series of absolute values also converges, the series is defined to converge absolutely. Series (3-60) is further defined as an alternating series, while series (3-61) is referred to as a positive series.

OPERATIONS WITH INFINITE SERIES

1. The convergence or divergence of an infinite series is unaffected by the removal of a finite number of finite terms. This is a trivial theorem but useful to remember, especially when using the comparison test to be described in the subsection "Tests for Convergence and Divergence."

2. A power series can be inverted, provided the first-degree term is not zero. Given

$$y = b_1x + b_2x^2 + b_3x^3 + b_4x^4 + b_5x^5 + b_6x^6 + b_7x^7 + \cdots$$

then
$$x = B_1y + B_2y^2 + B_3y^3 + B_4y^4 + B_5y^5 + B_6y^6 + B_7y^7 + \cdots$$

where
$$B_1 = 1/b_1$$
$$B_2 = -b_2/b_1^3$$
$$B_3 = (1/b_1^5)(2b_2^2 - b_1b_3)$$
$$B_4 = (1/b_1^7)(5b_1b_2b_3 - b_1^2b_4 - 5b_2^3)$$

Additional coefficients are available in the references.

3. Two series may be added or subtracted term by term provided each is a convergent series. The joint sum is equal to the sum (or difference) of the individuals.

4. The sum of two divergent series can be convergent. Similarly, the sum of a convergent series and a divergent series must be divergent.

5. A power series may be integrated term by term to represent the integral of the function within an interval of the region of convergence. If $f(x) = a_0 + a_1x + a_2x^2 + \cdots$, then

$$\int_{x_1}^{x_2} f(x)\,dx = \int_{x_1}^{x_2} a_0\,dx + \int_{x_1}^{x_2} a_1x\,dx + \int_{x_1}^{x_2} a_2x^2\,dx + \cdots$$

6. A power series may be differentiated term by term and represents the function $df(x)/dx$ within the same region of convergence as $f(x)$.

TESTS FOR CONVERGENCE AND DIVERGENCE

In general, the problem of determining whether a given series will converge can require a great deal of ingenuity and resourcefulness. It is necessary to apply one or more of the developed theorems in an attempt to ascertain the convergence or divergence of the series under study. The following defined tests are given in relative order of effectiveness. For examples, see references on advanced calculus.

1. *Comparison test.* A series will converge if the absolute value of each term (with or without a finite number of terms) is less than the corresponding term of a known convergent series. Similarly, a positive series is divergent if it is termwise larger than a known divergent series of positive terms.

2. *nth-Term test.* A series is divergent if the nth term of the series does not approach zero as n becomes increasingly large.

3. *Ratio test.* If the absolute ratio of the $n + 1$ term divided by the nth term as n becomes unbounded approaches
   a. A number less than 1, the series is absolutely convergent.
   b. A number greater than 1, the series is divergent.
   c. A number equal to 1, the test is inconclusive.

Example For the power series

$$a_0 + a_1(x - x_0) + a_2(x - x_0)^2 + \cdots$$

the absolute ratio gives

$$\varepsilon = \lim_{n\to\infty}\left|\frac{a_{n+1}}{a_n}\right||x - x_0| = \frac{1}{R}|x - x_0|$$

where R is the inverse of the limit. For convergence $\varepsilon < 1$; therefore the series converges for $|x - x_0| < R$.

4. *Alternating-series Leibniz test.* If the terms of a series are alternately positive and negative and never increase in value, the absolute series will converge, provided that the terms tend to zero as a limit.

5. *Cauchy's root test.* If the nth root of the absolute value of the nth term, as n becomes unbounded, approaches
   a. A number less than 1, the series is absolutely convergent.
   b. A number greater than 1, the series is divergent.
   c. A number equal to 1, the test is inconclusive.

6. *Maclaurin's integral test.* Suppose Σa_n is a series of positive terms and f is a continuous decreasing function such that $f(x) \geq 0$ for $1 \leq x < \infty$ and $f(n) = a_n$. Then the series and the improper integral $\int_1^\infty f(x)\,dx$ either both converge or both diverge.

SERIES SUMMATION AND IDENTITIES

Sums for the First *n* Numbers to Integer Powers

$$\sum_{j=1}^{n} j = \frac{n(n+1)}{2} = 1 + 2 + 3 + 4 + \cdots + n$$

$$\sum_{j=1}^{n} j^2 = \frac{n(n+1)(2n+1)}{6} = 1^2 + 2^2 + 3^2 + 4^2 + \cdots + n^2$$

$$\sum_{j=1}^{n} j^3 = \frac{n^2(n+1)^2}{4} = 1^3 + 2^3 + 3^3 + \cdots + n^3$$

Arithmetic Progression

$$\sum_{k=1}^{n}[a+(k-1)d]=a+(a+d)+(a+2d)$$
$$+(a+3d)+\cdots+[a+(n-1)]d$$
$$=na+\frac{1}{2}n(n-1)d$$

Geometric Progression

$$\sum_{j=1}^{n}ar^{j-1}=a+ar+ar^2+ar^3+\cdots+ar^{n-1}=a\frac{1-r^n}{1-r}\qquad r\neq 1$$

Harmonic Progression

$$\sum_{k=0}^{n}\frac{1}{a+kd}=\frac{1}{a}+\frac{1}{a+d}+\frac{1}{a+2d}+\frac{1}{a+3d}+\frac{1}{a+4d}+\cdots+\frac{1}{a+nd}$$

The reciprocals of the terms of the arithmetic progression series are called a *harmonic progression.* No general summation formulas are available for this series.

Binomial Series (See Also Elementary Algebra)

$$(1\pm x)^n=1\pm nx+\frac{n(n-1)}{2!}x^2\pm\frac{n(n-1)(n-2)}{3!}x^3+\cdots\qquad (x^2<1)$$

Taylor's Series

$$f(h+x)=f(h)+xf'(h)+\frac{x^2}{2!}f''(h)+\frac{x^3}{3!}f'''(h)+\cdots$$

or $$f(x)=f(x_0)+f'(x_0)(x-x_0)+\frac{f'''(x_0)}{2!}(x-x_0)^2+\frac{f'''(x_0)}{3!}(x-x_0)^3+\cdots$$

Example Find a series expansion for $f(x)=\ln(1+x)$ about $x_0=0$.

$$f'(x)=(1+x)^{-1},\ f''(x)=-(1+x)^{-2},\ f'''(x)=2(1+x)^{-3},\ \text{etc.}$$

Thus $$f(0)=0,\ f'(0)=1,\ f''(0)=-1,\ f'''(1)=2,\ \text{etc.}$$

$$\ln(x+1)=x-\frac{x^2}{2}+\frac{x^3}{3}-\frac{x^4}{4}+\cdots+(-1)^{n+1}\frac{x^n}{n}+\cdots$$

which converges for $-1<x\leq 1$.

Maclaurin's Series

$$f(x)=f(0)+xf'(0)+\frac{x^2}{2!}f''(0)+\frac{x^3}{3!}f'''(0)+\cdots$$

This is simply a special case of Taylor's series when h is set to zero.

Exponential Series

$$e^x=1+x+\frac{x^2}{2!}+\frac{x^3}{3!}+\cdots+\frac{x^n}{n!}+\cdots\qquad -\infty<x<\infty$$

Logarithmic Series

$$\ln x=\frac{x-1}{x}+\frac{1}{2}\left(\frac{x-1}{x}\right)^2+\frac{1}{3}\left(\frac{x-1}{x}\right)^3+\cdots\qquad (x>\tfrac{1}{2})$$

$$\ln x=2\left[\left(\frac{x-1}{x+1}\right)+\frac{1}{3}\left(\frac{x-1}{x+1}\right)^3\right]+\cdots\qquad (x>0)$$

Trigonometric Series*

$$\sin x=x-\frac{x^3}{3!}+\frac{x^5}{5!}-\frac{x^7}{7!}+\cdots\qquad -\infty<x<\infty$$

$$\cos x=1-\frac{x^2}{2!}+\frac{x^4}{4!}-\frac{x^6}{6!}+\cdots\qquad -\infty<x<\infty$$

$$\sin^{-1}x=x+\frac{x^3}{6}+\frac{1}{2}\cdot\frac{3}{4}\cdot\frac{x^5}{5}+\frac{1}{2}\cdot\frac{3}{4}\cdot\frac{5}{6}\cdot\frac{x^7}{7}+\cdots\qquad (x^2<1)$$

$$\tan^{-1}x=x-\frac{1}{3}x^3+\frac{1}{5}x^5-\frac{1}{7}x^7+\cdots\qquad (x^2<1)$$

Taylor Series The Taylor series for a function of two variables, expanded about the point (x_0, y_0), is

$$f(x,y)=f(x_0,y_0)+\left.\frac{\partial f}{\partial x}\right|_{x_0,y_0}(x-x_0)+\left.\frac{\partial f}{\partial y}\right|_{x_0,y_0}(y-y_0)$$

$$+\frac{1}{2!}\left[\left.\frac{\partial^2 f}{\partial x^2}\right|_{x_0,y_0}(x-x_0)^2+2\left.\frac{\partial^2 f}{\partial x\partial y}\right|_{x_0,y_0}(x-x_0)(y-y_0)+\left.\frac{\partial^2 f}{\partial y^2}\right|_{x_0,y_0}(y-y_0)^2\right]+\cdots$$

Partial Sums of Infinite Series, and How They Grow Calculus textbooks devote much space to tests for convergence and divergence of series that are of little practical value, since a convergent series either converges rapidly, in which case almost any test (among those presented in the preceding subsections) will do, or it converges slowly, in which case it is not going to be of much use unless there is some way to get at its sum without adding an unreasonable number of terms. To find out, as accurately as possible, how fast a convergent series converges and how fast a divergent series diverges, see Boas, R. P., Jr., *Am. Math. Mon.* **84:** 237–258 (1977).

*The tan x series has awkward coefficients and should be computed as $\left[(\text{sign})\dfrac{\sin x}{\sqrt{1-\sin^2 x}}\right]$

COMPLEX VARIABLES

References: Ablowitz, M. J., and A. S. Fokas, *Complex Variables: Introduction and Applications,* 2d ed., Cambridge University Press, New York, 2012; Asmar, N., and G. C. Jones, *Applied Complex Analysis with Partial Differential Equations,* Prentice-Hall, Upper Saddle River, N.J., 2002; Brown, J. W., and R. V. Churchill, *Complex Variables and Applications,* 9th ed., McGraw-Hill, New York, 2013; Kwok, Y. K., *Applied Complex Variables for Scientists and Engineers,* 2d ed., Cambridge University Press, New York, 2010.

Numbers of the form $z=x+iy$, where x and y are real, $i^2=-1$, are called *complex numbers.* The numbers $z=x+iy$ are representable in the plane, as shown in Fig. 3-42. The following definitions and terminology are used:

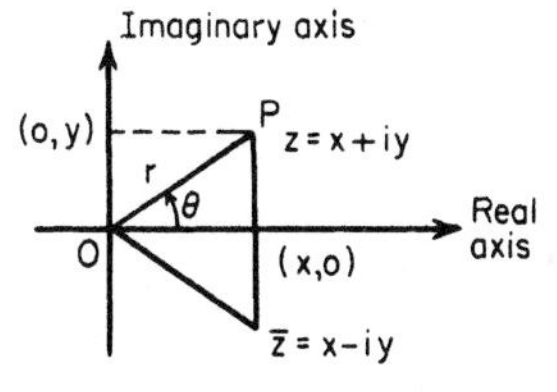

FIG. 3-42 Complex plane.

1. Distance $OP=r=$ modulus of z written $|z|$. $|z|=\sqrt{x^2+y^2}$
2. x is the real part of z.
3. y is the imaginary part of z.
4. The angle θ, $0\leq\theta<2\pi$, measured counterclockwise from the positive x axis to OP, is the argument of z. $\theta=\arctan y/x=\arcsin y/r=\arccos x/r$ if $x\neq 0$, $\theta=\pi/2$ if $x=0$ and $y>0$.
5. The numbers r, θ are the polar coordinates of z.
6. $\bar{z}=x-iy$ is the complex conjugate of z.

ALGEBRA

Let $z_1=x_1+iy_1$ and $z_2=x_2+iy_2$.

Equality $z_1=z_2$ if and only if $x_1=x_2$ and $y_1=y_2$.

Addition $z_1+z_2=(x_1+x_2)+i(y_1+y_2)$.

Subtraction $z_1-z_2=(x_1-x_2)+i(y_1-y_2)$.

Multiplication $z_1z_2=(x_1x_2-y_1y_2)+i(x_1y_2+x_2y_1)$.

Division $z_1/z_2=\dfrac{x_1x_2+y_1y_2}{x_2^2+y_2^2}+i\dfrac{x_2y_1-x_1y_2}{x_2^2+y_2^2},\ z_2\neq 0.$

SPECIAL OPERATIONS

$z\bar{z} = x^2 + y^2 = |z|^2$; $\overline{z_1 \pm z_2} = \bar{z}_1 \pm \bar{z}_2$; $\bar{\bar{z}}_1 = z_1$; $\overline{z_1 z_2} = \bar{z}_1\,\bar{z}_2$; $|z_1 \cdot z_2| = |z_1| \cdot |z_2|$; arg $(z_1 \cdot z_2)$ = arg z_1 + arg z_2; arg (z_1/z_2) = arg z_1 − arg z_2; $i^{4n} = 1$ for n any integer; $i^{2n} = -1$ where n is any odd integer; $z + \bar{z} = 2x$; $z - \bar{z} = 2iy$.

Every complex quantity can be expressed in the form $x + iy$.

TRIGONOMETRIC REPRESENTATION

By referring to Fig. 3-42, there results $x = r\cos\theta$ and $y = r\sin\theta$ so that $z = x + iy = r(\cos\theta + i\sin\theta)$, which is called the *polar form* of the complex number. $\cos\theta + i\sin\theta = e^{i\theta}$. Hence $z = x + iy = re^{i\theta}$. $\bar{z} = x - iy = re^{-i\theta}$. Two important results from this are $\cos\theta = (e^{i\theta} + e^{-i\theta})/2$ and $\sin\theta = (e^{i\theta} - e^{-i\theta})/2i$. Let $z_1 = r_1 e^{i\theta_1}$ and $z_2 = r_2 e^{i\theta_2}$. This form is convenient for multiplication for $z_1 z_2 = r_1 r_2 e^{i(\theta_1+\theta_2)}$ and for division for $z_1/z_2 = (r_1/r_2)e^{i(\theta_1-\theta_2)}$, $z_2 \neq 0$.

POWERS AND ROOTS

If n is a positive integer, $z^n = (re^{i\theta})^n = r^n e^{in\theta} = r^n(\cos n\theta + i\sin n\theta)$.

If n is a positive integer,

$$z^{1/n} = r^{1/n} e^{i[(\theta+2k\pi)/n]} = r^{1/n}\left[\cos\left(\frac{\theta+2k\pi}{n}\right) + i\sin\left(\frac{\theta+2k\pi}{n}\right)\right]$$

and selecting values of $k = 0, 1, 2, 3, \ldots, n-1$ gives the n distinct values of $z^{1/n}$. The n roots of a complex quantity are uniformly spaced around a circle with radius $r^{1/n}$ in the complex plane in a symmetric fashion.

Example Find the three cube roots of −8. Here $r = 8$, $\theta = \pi$. The roots are $z_0 = 2(\cos\pi/3 + i\sin\pi/3) = 1 + i\sqrt{3}$, $z_1 = 2(\cos\pi + i\sin\pi) = -2$, and $z_2 = 2(\cos 5\pi/3 + i\sin 5\pi/3) = 1 - i\sqrt{3}$.

ELEMENTARY COMPLEX FUNCTIONS

Polynomials A polynomial in z, $a_n z^n + a_{n-1}z^{n-1} + \cdots + a_0$, where n is a positive integer, is simply a sum of complex numbers times integral powers of z which have already been defined. Every polynomial of degree n has precisely n complex roots provided each multiple root of multiplicity m is counted m times.

Exponential Functions The exponential function e^z is defined by the equation $e^z = e^{x+iy} = e^x \cdot e^{iy} = e^x(\cos y + i\sin y)$. Properties: $e^0 = 1$; $e^{z_1}e^{z_2} = e^{z_1+z_2}$; $e^{z_1}/e^{z_2} = e^{z_1-z_2}$; $e^{z+2k\pi i} = e^z$, k an integer.

Trigonometric Functions $\sin z = (e^{iz} - e^{-iz})/2i$; $\cos z = (e^{iz} + e^{-iz})/2$; $\tan z = \sin z/\cos z$; $\cot z = \cos z/\sin z$; $\sec z = 1/\cos z$; $\csc z = 1/\sin z$. Fundamental identities for these functions are the same as their real counterparts. Thus $\cos^2 z + \sin^2 z = 1$, $\cos(z_1 \pm z_2) = \cos z_1 \cos z_2 \mp \sin z_1 \sin z_2$, $\sin(z_1 \pm z_2) = \sin z_1 \cos z_2 \pm \cos z_1 \sin z_2$. The sine and cosine of z are periodic functions of period 2π; thus $\sin(z + 2\pi) = \sin z$. For computation purposes $\sin z = \sin(x + iy) = \sin x \cosh y + i\cos x \sinh y$, where $\sin x$, $\cosh y$, etc., are the real trigonometric and hyperbolic functions. Similarly, $\cos z = \cos x \cosh y - i\sin x \sinh y$. If $x = 0$ in the results given, $\cos iy = \cosh y$ and $\sin iy = i\sinh y$.

Example Find all solutions of $\sin z = 3$. From previous data $\sin z = \sin x \cosh y + i\cos x \sinh y = 3$. Equating real and imaginary parts gives $\sin x \cosh y = 3$ and $\cos x \sinh y = 0$. The second equation can hold for $y = 0$ or for $x = \pi/2, 3\pi/2, \ldots$. If $y = 0$, $\cosh 0 = 1$ and $\sin x = 3$ is impossible for real x. Therefore, $x = \pm\pi/2, \pm 3\pi/2, \ldots, \pm(2n+1)\pi/2$, $n = 0, \pm 1, \pm 2, \ldots$. However, $\sin 3\pi/2 = -1$ and $\cosh y \geq 1$. Hence $x = \pi/2, 5\pi/2, \ldots$. The solution is $z = [(4n+1)\pi]/2 + i\cosh^{-1}3$, $n = 0, 1, 2, 3, \ldots$.

Example Find all solutions of $e^z = -i$. $e^z = e^x(\cos y + i\sin y) = -i$. Equating real and imaginary parts gives $e^x \cos y = 0$, $e^x \sin y = -1$ from the first $y = \pm\pi/2$, $\pm 3\pi/2, \ldots$. But $e^x > 0$. Therefore, $y = 3\pi/2, 7\pi/2, -\pi/2, \ldots$. Then $x = 0$. The solution is $z = i[(4n+3)\pi]/2$.

Two important facets of these functions should be recognized. First, $\sin z$ is *unbounded*; second, e^z takes *all* complex values *except* 0.

Hyperbolic Functions $\sinh z = (e^z - e^{-z})/2$; $\cosh z = (e^z + e^{-z})/2$; $\tanh z = \sinh z/\cosh z$; $\coth z = \cosh z/\sinh z$; $\operatorname{csch} z = 1/\sinh z$; $\operatorname{sech} z = 1/\cosh z$. Identities are $\cosh^2 z - \sinh^2 z = 1$; $\sinh(z_1 + z_2) = \sinh z_1 \cosh z_2 + \cosh z_1 \sinh z_2$; $\cosh(z_1 + z_2) = \cosh z_1 \cosh z_2 + \sinh z_1 \sinh z_2$; $\cosh z + \sinh z = e^z$; $\cosh z - \sinh z = e^{-z}$. The hyperbolic sine and hyperbolic cosine are periodic functions with the imaginary period $2\pi i$. That is, $\sinh(z + 2\pi i) = \sinh z$.

Logarithms The logarithm of z, $\log z = \log|z| + i(\theta + 2n\pi)$, where $\log|z|$ is taken to the base e and θ is the principal argument of z, that is, the particular argument lying in the interval $0 \leq \theta < 2\pi$. The logarithm of z is infinitely many valued. If $n = 0$, the resulting logarithm is called the *principal value*. The familiar laws $\log z_1 z_2 = \log z_1 + \log z_2$, $\log z_1/z_2 = \log z_1 - \log z_2$, and $\log z^n = n\log z$ hold for the principal value.

General powers of z are defined by $z^\alpha = e^{\alpha\log z}$. Since $\log z$ is infinitely many valued, so too is z^α unless α is a rational number.

DeMoivre's formula can be derived from properties of e^z.

$$z^n = r^n(\cos\theta + i\sin\theta)^n = r^n(\cos n\theta + i\sin n\theta)$$

Thus $$(\cos\theta + i\sin\theta)^n = \cos n\theta + i\sin n\theta$$

COMPLEX FUNCTIONS (ANALYTIC)

In the real-number system a greater than b ($a > b$) and b less than c ($b < c$) define an order relation. These relations have no meaning for complex numbers. The absolute value is used for ordering. Some important relations follow: $|z| \geq x$; $|z| \geq y$; $|z_1 \pm z_2| \leq |z_1| + |z_2|$; $|z_1 - z_2| \geq ||z_1| - |z_2||$; $|z| \geq (|x| + |y|)/\sqrt{2}$. Parts of the complex plane, commonly called *regions* or *domains*, are described by using inequalities.

Example $|z - 3| \leq 5$. This is equivalent to $\sqrt{(x-3)^2 + y^2} \leq 5$, which is the set of all points within and on the circle, centered at $x = 3$, $y = 0$ of radius 5.

Example $|z - 1| \leq x$ represents the set of all points inside and on the parabola $2x = y^2 + 1$ or, equivalently, $2x \geq y^2 + 1$.

Functions of a Complex Variable If $z = x + iy$, $w = u + iv$ and if for each value of z in some region of the complex plane one or more values of w are defined, then w is said to be a function of z, $w = f(z)$. Some of these functions have already been discussed, such as $\sin z$ and $\log z$. All functions are reducible to the form $w = u(x, y) + iv(x, y)$, where u and v are real functions of the real variables x and y.

Example $z^3 = (x + iy)^3 = x^3 + 3x^2(iy) + 3x(iy)^2 + (iy)^3 = (x^3 - 3xy^2) + i(3x^2y - y^3)$.

Differentiation The *derivative* of $w = f(z)$ is

$$\frac{dw}{dz} = \lim_{\Delta z \to 0} \frac{f(z + \Delta z) - f(z)}{\Delta z}$$

and for the derivative to exist, the limit must be the same no matter how Δz approaches zero. If w_1 and w_2 are differentiable functions of z, the following rules apply:

$$\frac{d(w_1 \pm w_2)}{dz} = \frac{dw_1}{dz} \pm \frac{dw_2}{dz} \qquad \frac{d(w_1 w_2)}{dz} = w_2\frac{dw_1}{dz} + w_1\frac{dw_2}{dz}$$

$$\frac{d(w_1/w_2)}{dz} = \frac{w_2(dw_1/dz) - w_1(dw_2/dz)}{w_2^2}$$

and $$\frac{dw_1^n}{dz} = nw_1^{n-1}\frac{dw_1}{dz}$$

For $w = f(z)$ to be differentiable, it is necessary that $\partial u/\partial x = \partial v/\partial y$ and $\partial v/\partial x = -\partial u/\partial y$. The last two equations are called the *Cauchy-Riemann equations*. The derivative

$$\frac{dw}{dz} = \frac{\partial u}{\partial x} + i\frac{\partial v}{\partial x} = \frac{\partial v}{\partial y} - i\frac{\partial u}{\partial y}$$

If $f(z)$ possesses a derivative at z_0 and at every point in some neighborhood of z_0, then $f(z)$ is said to be *analytic* or *homomorphic* at z_0. If the Cauchy-Riemann equations are satisfied and

$$u, v, \frac{\partial u}{\partial x}, \frac{\partial u}{\partial y}, \frac{\partial v}{\partial x}, \frac{\partial v}{\partial y}$$

are continuous in a region of the complex plane, then $f(z)$ is analytic in that region.

Example $w = z\bar{z} = x^2 + y^2$. Here $u = x^2 + y^2$, $v = 0$. $\partial u/\partial x = 2x$, $\partial u/\partial y = 2y$, $\partial v/\partial x = \partial v/\partial y = 0$. These are continuous everywhere, but the Cauchy-Riemann equations hold only at the origin. Therefore, w is nowhere analytic, but it is differentiable at $z = 0$ only.

Example $w = e^z = e^x\cos y + ie^x\sin y$. $u = e^x\cos y$ and $v = e^x\sin y$. $\partial u/\partial x = e^x\cos y$, $\partial u/\partial y = -e^x\sin y$, $\partial v/\partial x = e^x\sin y$, $\partial v/\partial y = e^x\cos y$. The continuity and Cauchy-Riemann requirements are satisfied for all finite z. Hence e^z is analytic (except at ∞) and $dw/dz = \partial u/\partial x + i(\partial v/\partial x) = e^z$.

Example $$w = \frac{1}{z} = \frac{x - iy}{x^2 + y^2} = \frac{x}{x^2 + y^2} - i\frac{y}{x^2 + y^2}$$

It is easy to see that dw/dz exists except at $z = 0$. Thus $1/z$ is analytic except at $z = 0$.

Singular Points If $f(z)$ is analytic in a region except at certain points, those points are called *singular points.*

Example $1/z$ has a singular point at zero.

Example $\tan z$ has singular points at $z = \pm(2n+1)(\pi/2)$, $n = 0, 1, 2, \ldots$. The derivatives of the common functions, given earlier, are the same as their real counterparts.

Example $(d/dz)(\ln z) = 1/z$, $(d/dz)(\sin z) = \cos z$.

Harmonic Functions Both the *real* and the *imaginary* parts of any analytic function $f = u + iv$ satisfy Laplace's equation $\partial^2\phi/\partial x^2 + \partial^2\phi/\partial y^2 = 0$. A function which possesses continuous second partial derivatives and satisfies Laplace's equation is called a *harmonic function.*

Example $e^z = e^x \cos y + ie^x \sin y$. $u = e^x \cos y$, $\partial u/\partial x = e^x \cos y$, $\partial^2 u/\partial x^2 = e^x \cos y$, $\partial u/\partial y = -e^x \sin y$, $\partial^2 u/\partial y^2 = -e^x \cos y$. Clearly $\partial^2 u/\partial x^2 + \partial^2 u/\partial y^2 = 0$. Similarly, $v = e^x \sin y$ is also harmonic.

If $w = u + iv$ is analytic, the curves $u(x, y) = c$ and $v(x, y) = k$ intersect at right angles, if $w'(z) \neq 0$.

Integration In much of the work with complex variables a simple extension of integration called *line* or *curvilinear integration* is of fundamental importance. Since any complex line integral can be expressed in terms of real line integrals, we define only real line integrals. Let $F(x, y)$ be a real, continuous function of x and y, and let c be any continuous curve of finite length joining points A and B (Fig. 3-43). $F(x, y)$ is not related to the curve c. Divide c into n segments, Δs_i, whose projection on the x axis is Δx_i and on the y axis is Δy_i. Let (ε_i, η_i) be the coordinates of an arbitrary point on Δs_i. The limits of the sums

$$\lim_{\Delta s_i \to 0} \sum_{i=1}^{n} F(\varepsilon_i, \eta_i)\,\Delta s_i = \int_c F(x, y)\,ds$$

$$\lim_{\Delta s_i \to 0} \sum_{i=1}^{n} F(\varepsilon_i, \eta_i)\,\Delta x_i = \int_c F(x, y)\,dx$$

$$\lim_{\Delta s_i \to 0} \sum_{i=1}^{n} F(\varepsilon_i, \eta_i)\,\Delta y_i = \int_c F(x, y)\,dy$$

are known as *line integrals.* Much of the initial strangeness of these integrals will vanish if it is observed that the ordinary definite integral $\int_a^b f(x)\,dx$ is just a line integral in which the curve c is a line segment on the x axis and $F(x, y)$ is a function of x alone. The evaluation of line integrals can be reduced to evaluation of ordinary integrals.

Example $\int_c y(1+x)\,dy$, where c: $y = 1 - x^2$ from $(-1, 0)$ to $(1, 0)$. Clearly $y = 1 - x^2$, $dy = -2x\,dx$. Thus $\int_c y(1+x)\,dy = -2\int_{-1}^{1}(1-x^2)(1+x)x\,dx = -\frac{8}{15}$.

Let $f(z)$ be any function of z, analytic or not, and c any curve as above. The complex integral is calculated as $\int_c f(z)\,dz = \int_c (u\,dx - v\,dy) + i\int_c (v\,dx + u\,dy)$, where $f(z) = u(x, y) + iv(x, y)$. Properties of line integrals are the same as for ordinary integrals. That is, $\int_c [f(z) \pm g(z)]\,dz = \int_c f(z)\,dz \pm \int_c g(z)\,dz$; $\int_c kf(z)\,dz = k\int_c f(z)\,dz$ for any constant k, etc.

Example $\int_c (x^2 + iy)\,dz$ along c: $y = x$, 0 to $1 + i$. This becomes

$$\int_c (x^2 + iy)\,dz = \int_c (x^2\,dx - y\,dy)$$

$$+ i\int_c (y\,dx + x^2\,dy) = \int_0^1 x^2\,dx - \int_0^1 x\,dx + i\int_0^1 x\,dx + i\int_0^1 x^2\,dx = -\tfrac{1}{6} + 5i/6$$

Conformal Mapping Every function of a complex variable $w = f(z) = u(x, y) + iv(x, y)$ transforms the x, y plane into the u, v plane in some manner. A conformal transformation is one in which angles between curves are preserved in *magnitude* and *sense.* Every analytic function, except at those points where $f'(z) = 0$, is a conformal transformation. See Fig. 3-44.

Example $w = z^2$. $u + iv = (x^2 - y^2) + 2ixy$ or $u = x^2 - y^2$, $v = 2xy$. These are the transformation equations between the (x, y) and (u, v) planes. Lines parallel to the x axis, $y = c_1$ map into curves in the u, v plane with parametric equations $u = x^2 - c_1^2$, $v = 2c_1x$. Eliminating x, $u = (v^2/4c_1^2) - c_1^2$, which represents a family of parabolas with the origin of the w plane as focus, the line $v = 0$ as axis and opening to the right. Similar arguments apply to $x = c_2$.

The principles of complex variables are useful in the solution of a variety of applied problems, including Laplace transforms (see Integral Transforms) and process control (Sec. 8).

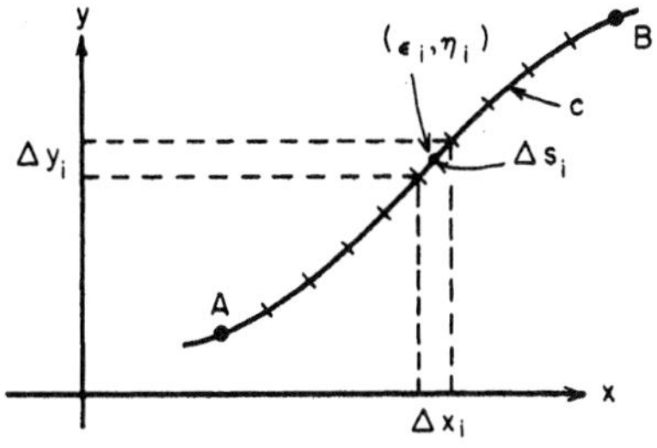

FIG. 3-43 Line integral.

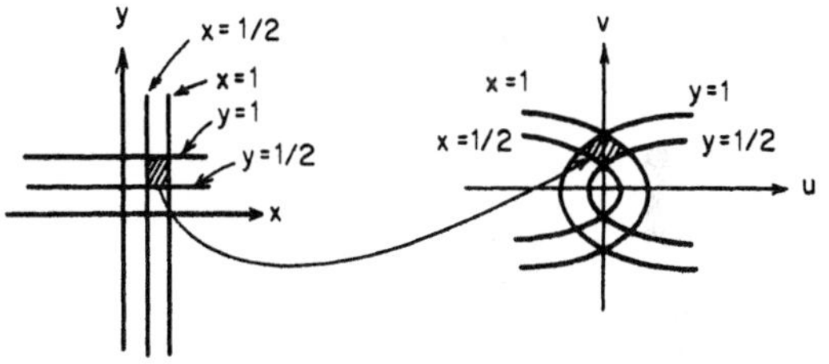

FIG. 3-44 Conformal transformation.

DIFFERENTIAL EQUATIONS

References: Ames, W. F., *Nonlinear Partial Differential Equations in Engineering,* Academic Press, New York, 1965; Aris, R., and N. R. Amundson, *Mathematical Methods in Chemical Engineering,* vol. 2, *First-Order Partial Differential Equations with Applications,* Prentice-Hall, Englewood Cliffs, N.J., 1973; Asmar, N. H., *Partial Differential Equations with Fourier Series and Boundary Value Problems,* 3rd ed., Pearson, New York, 2016. Asmar, N., *Applied Complex Analysis with Partial Differential Equations,* Prentice-Hall, Upper Saddle River, N.J., 2002; Bronson, R., and G. Costa, *Schaum's Outline of Differential Equations,* 4th ed., McGraw-Hill, New York, 2014; Brown, J. W., and R. V. Churchill, *Fourier Series and Boundary Value Problems,* 8th ed., McGraw-Hill Education, New York, 2011; Duffy, D., *Green's Functions with Applications,* 2d ed., Chapman and Hall/CRC, New York, 2015; Kreyszig, E., *Advanced Engineering Mathematics,* 10th ed., Wiley, New York, 2011; Ramkrishna, D., and N. R. Amundson, *Linear Operator Methods in Chemical Engineering with Applications to Transport and Chemical Reaction Systems,* Prentice-Hall, Englewood Cliffs, N.J., 1985.

The natural laws in any scientific or technological field are not regarded as precise and definitive until they have been expressed in mathematical form. Such a form, often an equation, is a relation between the quantity of interest, say, product yield, and independent variables such as time and temperature upon which yield depends. When it happens that this equation involves, besides the function itself, one or more of its derivatives it is called a differential equation.

Example The rate of the homogeneous bimolecular reaction $A + B \xrightarrow{k} C$ is characterized by the differential equation $dx/dt = k(a - x)(b - x)$, where a = initial concentration of A, b = initial concentration of B, and $x = x(t)$ = concentration of C as a function of time t.

Example The differential equation of heat conduction in a moving fluid with velocity components v_x, v_y is

$$\frac{\partial T}{\partial t} + v_x \frac{\partial T}{\partial x} + v_y \frac{\partial T}{\partial y} = \frac{k}{\rho c_p}\left(\frac{\partial^2 T}{\partial x^2} + \frac{\partial^2 T}{\partial y^2}\right)$$

where $T = T(x, y, t)$ = temperature, k = thermal conductivity, ρ = density, and c_p = specific heat at constant pressure.

ORDINARY DIFFERENTIAL EQUATIONS

When the function involved in the equation depends upon only one variable, its derivatives are ordinary derivatives and the differential equation is called an ordinary differential equation. When the function depends upon several independent variables, then the equation is called a partial differential equation. The theories of ordinary and partial differential equations are quite different. In almost every respect the latter is more difficult.

Whichever the type, a differential equation is said to be of nth order if it involves derivatives of order n but no higher. The equation in the first example is of first order and that in the second example of second order. The degree of a differential equation is the power to which the derivative of the highest order is raised after the equation has been cleared of fractions and radicals in the dependent variable and its derivatives.

A relation between the variables, involving no derivatives, is called a solution of the differential equation if this relation, when substituted in the equation, satisfies the equation. A solution of an ordinary differential equation which includes the maximum possible number of "arbitrary" constants is called the *general solution.* The maximum number of "arbitrary" constants is exactly equal to the order of the differential equation. If any set of specific values of the constants is chosen, the result is called a *particular solution.*

Example The general solution of $(d^2x/dt^2) + k^2x = 0$ is $x = A \cos kt + B \sin kt$, where A and B are arbitrary constants. A particular solution is $x = \frac{1}{2} \cos kt + 3 \sin kt$.

In the case of some equations still other solutions exist called singular solutions. A *singular solution* is any solution of the differential equation which is not included in the general solution.

Example $y = x(dy/dx) - \frac{1}{4}(dy/dx)^2$ has the general solution $y = cx - \frac{1}{4}c^2$, where c is an arbitrary constant; $y = x^2$ is a singular solution, as is easily verified.

ORDINARY DIFFERENTIAL EQUATIONS OF THE FIRST ORDER

Equations with Separable Variables Every differential equation of the first order and of the first degree can be written in the form $M(x, y)\,dx + N(x, y)dy = 0$. If the equation can be transformed so that M does not involve y and N does not involve x, then the variables are said to be separated. The solution can then be obtained by quadrature, which means that $y = \int f(x)dx + c$, which may or may not be expressible in simpler form.

Exact Equations The equation $M(x, y)\,dx + N(x, y)\,dy = 0$ is exact if and only if $\partial M/\partial y = \partial N/\partial x$. In this case there exists a function $w = f(x, y)$ such that $\partial f/\partial x = M$, $\partial f/\partial y = N$, and $f(x, y) = C$ is the required solution. $f(x, y)$ is found as follows: treat y as though it were constant and evaluate $\int M(x, y)\,dx$. Then treat x as though it were constant and evaluate $\int N(x, y)\,dy$. The sum of all unlike terms in these two integrals (including no repetitions) is $f(x, y)$.

Example $(2xy - \cos x)\,dx + (x^2 - 1)\,dy = 0$ is exact for $\partial M/\partial y = 2x$, $\partial N/\partial x = 2x$. $\int M\,dx = \int(2xy - \cos x)\,dx = x^2y - \sin x$, $\int N\,dy = \int(x^2 - 1)\,dy = x^2y - y$. The solution is $x^2y - \sin x - y = C$, as may easily be verified.

Linear Equations A differential equation is said to be linear when it is of first degree in the dependent variable and its derivatives. The general linear first-order differential equation has the form $dy/dx + P(x)y = Q(x)$. Its general solution is

$$y = e^{-\int P\,dx}\left[\int Q e^{\int P\,dx}\,dx + C\right]$$

Example A tank initially holds 200 gal of a salt solution in which 100 lb is dissolved. Six gallons of brine containing 4 lb of salt run into the tank per minute. If mixing is perfect and the output rate is 4 gal/min, what is the amount A of salt in the tank at time t? The differential equation of A is $dA/dt = 4 - 2A/[100 + t]$. Its general solution is $A = (4/3)(100 + t) + C/(100 + t)^2$. At $t = 0, A = 100$; so the particular solution is $A = (4/3)(100 + t) - (1/3) \times 10^6/(100 + t)^2$.

ORDINARY DIFFERENTIAL EQUATIONS OF HIGHER ORDER

The higher-order differential equations, especially those of order 2, are of great importance because of physical situations describable by them.

Equation $y^{(n)} = f(x)$. The superscript (n) means n derivatives. Such a differential equation can be solved by n integrations. The solution will contain n arbitrary constants.

Linear Differential Equations with Constant Coefficients and Right-Hand Member of Zero (Homogeneous) The solution of $y'' + ay' + by = 0$ depends upon the nature of the roots of the characteristic equation $m^2 + am + b = 0$ obtained by substituting the trial solution $y = e^{mx}$ in the equation.

Distinct Real Roots If the roots of the characteristic equation are distinct real roots, r_1 and r_2, say, the solution is $y = Ae^{r_1x} + Be^{r_2x}$, where A and B are arbitrary constants.

Example $y'' + 4y' + 3 = 0$. The characteristic equation is $m^2 + 4m + 3 = 0$. The roots are -3 and -1, and the general solution is $y = Ae^{-3x} + Be^{-x}$.

Multiple Real Roots If $r_1 = r_2$, the solution of the differential equation is $y = e^{r_1x}(A + Bx)$.

Example $y'' + 4y + 4 = 0$. The characteristic equation is $m^2 + 4m + 4 = 0$ with roots -2 and -2. The solution is $y = e^{-2x}(A + Bx)$.

Complex Roots If the characteristic roots are $p \pm iq$, then the solution is $y = e^{px} \times (A \cos qx + B \sin qx)$.

Example The differential equation $My'' + Ay' + ky = 0$ represents the vibration of a linear system of mass M, spring constant k, and damping constant A. If $A < 2\sqrt{kM}$, the roots of the characteristic equation

$$Mm^2 + Am + k = 0 \text{ are complex: } -\frac{A}{2M} \pm i\sqrt{\frac{k}{M} - \left(\frac{A}{2M}\right)^2}$$

and the solution is

$$y = e^{-(At/2M)}\left[c_1 \cos\left(\sqrt{\frac{k}{M} - \left(\frac{A}{2M}\right)^2}\right)t + ic_2 \sin\left(\sqrt{\frac{k}{M} - \left(\frac{A}{2M}\right)^2}\right)t\right]$$

This solution is oscillatory, representing undercritical damping.

All these results generalize to homogeneous linear differential equations with constant coefficients of order higher than 2. These equations (especially of order 2) have been much used because of the ease of solution. Oscillations, electric circuits, diffusion processes, and heat flow problems are a few examples for which such equations are useful.

Second-Order Equations: Dependent Variable Missing Such an equation is of the form

$$F\left(x, \frac{dy}{dx}, \frac{d^2y}{dx^2}\right) = 0$$

It can be reduced to a first-order equation by substituting $p = dy/dx$ and $dp/dx = d^2y/dx^2$.

Second-Order Equations: Independent Variable Missing Such an equation is of the form

$$F\left(y, \frac{dy}{dx}, \frac{d^2y}{dx^2}\right) = 0$$

Set $$\frac{du}{dx} = p, \quad \frac{d^2y}{dx^2} = p\frac{dp}{dy}$$

The result is a first-order equation in p

$$F\left(y, p, p\frac{dp}{dy}\right) = 0$$

Example The capillary curve for one vertical plate is given by

$$\frac{d^2y}{dx^2} = \frac{4y}{c^2}\left[1 + \left(\frac{dy}{dx}\right)^2\right]^{3/2}$$

Its solution by this technique is

$$x + \sqrt{c^2 - y^2} - \sqrt{c^2 - h_0^2} = \frac{c}{2}\left(\cosh^{-1}\frac{c}{y} - \cosh^{-1}\frac{c}{h_0}\right)$$

where c and h_0 are physical constants.

Example The equation governing chemical reaction in a porous catalyst in plane geometry of thickness L is

$$D\frac{d^2c}{dx^2} = k\,f(c), \quad \frac{dc}{dx}(0) = 0, \quad c(L) = c_0$$

where D is a diffusion coefficient, k is a reaction rate parameter, c is the concentration, $kf(c)$ is the rate of reaction, and c_0 is the concentration at the boundary. Making the substitution $p = \dfrac{dc}{ds}$ gives (Finlayson, 1980, p. 92)

$$p\frac{dp}{dc} = \frac{k}{D}f(c)$$

Integrating gives $$\frac{p^2}{2} = \frac{k}{D}\int_{c(0)}^{c} f(c)\,dc$$

If the reaction is very fast, $c(0) \approx 0$ and the average reaction rate is related to $p(L)$. This variable is given by

$$p(L)=\left[\frac{2k}{D}\int_0^{c(0)} f(c)\,dc\right]^{1/2}$$

Thus, the average reaction rate can be calculated without solving the complete problem.

Linear Nonhomogeneous Differential Equations

Linear Differential Equations Right-Hand Member $f(x) \neq 0$ Again the specific remarks for $y''+ay'+by=f(x)$ apply to differential equations of similar type but higher order. We shall discuss two general methods.

Method of Undetermined Coefficients Use of this method is limited to equations exhibiting both constant coefficients and particular forms of the function $f(x)$. In most cases $f(x)$ will be a sum or product of functions of the type constant, x^n (n a positive integer), e^{mx}, cos kx, sin kx. When this is the case, the solution of the equation is $y = H(x) + P(x)$, where $H(x)$ is a solution of the homogeneous equations found by the method of the preceding subsection and $P(x)$ is a particular integral found by using the following table subject to these conditions: (1) When $f(x)$ consists of the sum of several terms, the appropriate form of $P(x)$ is the sum of the particular integrals corresponding to these terms individually. (2) When a term in any of the trial integrals listed is already a part of the homogeneous solution, the indicated form of the particular integral is multiplied by x.

Form of Particular Integral

If $f(x)$ is	Then $P(x)$ is
a (constant)	A (constant)
ax^n	$A_nx^n + A_{n-1}x^{n-1} + \cdots + A_1x + A_0$
ae^{rx}	Be^{rx}
$c \cos kx$, $d \sin kx$	$A \cos kx + B \sin kx$
$g\,x^ne^{rx} \cos kx$, $h\,x^ne^{rx} \sin kx$	$(A_nx^n + \cdots + A_0)e^{rx} \cos kx + (B_nx^n + \cdots + B_0)e^{rx} \sin kx$

Since the form of the particular integral is known, the constants may be evaluated by substitution in the differential equation.

Example $y''+2y'+y=3e^{2x} - \cos x + x^3$. The characteristic equation is $(m + 1)^2 = 0$ so that the homogeneous solution is $y = (c_1 + c_2x)e^{-x}$. To find a particular solution we use the trial solution from the table, $y = a_1e^{2x} + a_2 \cos x + a_3 \sin x + a_4x^3 + a_5x^2 + a_6x + a_7$. By substituting this in the differential equation and collecting and equating like terms, there results $a_1 = 1/3$, $a_2 = 0$, $a_3 = -1/2$, $a_4 = 1$, $a_5 = -6$, $a_6 = 18$, and $a_7 = -24$. The solution is $y = (c_1 + c_2x)e^{-x} + \frac{1}{3}e^{2x} - \frac{1}{2}\sin x + x^3 - 6x^2 + 18x - 24$.

Method of Variation of Parameters This method is applicable to any linear equation. The technique is developed for a second-order equation but immediately extends to higher order. Let the equation be $y''+a(x)y'+b(x)y=R(x)$, and let the solution of the homogeneous equation, found by some method, be $y = c_1f_1(x) + c_2f_2(x)$. It is now assumed that a particular integral of the differential equation is of the form $P(x) = uf_1 + vf_2$, where u and v are functions of x to be determined by two equations. One equation results from the requirement that $uf_1 + vf_2$ satisfy the differential equation, and the other is a degree of freedom open to the analyst. The best choice proves to be

$$u'f_1+v'f_2=0 \quad \text{and} \quad u'f_1'+vf_2'=0$$

Then

$$u'=\frac{du}{dx}=-\frac{f_2}{f_1f_2'-f_2f_1'}R(x)$$

$$v'=\frac{dv}{dx}=\frac{f_1}{f_1f_2'-f_2f_1'}R(x)$$

and since f_1, f_2, and R are known, u, v may be found by direct integration.

Perturbation Methods If the ordinary differential equation has a parameter that is small and is not multiplying the highest derivative, perturbation methods can give solutions for small values of the parameter.

Example Consider the differential equation for reaction and diffusion in a catalyst; the reaction is second-order: $c'' = ac^2$, $c'(0) = 0$, $c(1) = 1$. The solution is expanded in the following Taylor series in a.

$$c(x, a) = c_0(x) + ac_1(x) + a^2c_2(x) + \cdots$$

The goal is to find equations governing the functions $\{c_i(x)\}$ and solve them. Substitution into the equations gives the following equations:

$$c_0''(x)+ac_1''(x)+a^2c_2''(x)+\cdots=a[c_0(x)+ac_1(x)+a^2c_2(x)+\cdots]^2$$
$$c_0'(0)+ac_1'(0)+a^2c_2'(0)+\cdots=0$$
$$c_0(1)+ac_1(1)+a^2c_2(1)+\cdots=1$$

Like terms in powers of a are collected to form the individual problems.

$$c_0''=0, c_0'(0)=0,\quad c_0(1)=1$$
$$c_1''=c_0^2, c_1'(0)=0,\quad c_1(1)=0$$
$$c_2''=2c_0c_1, c_2'(0)=0,\quad c_2(1)=0$$

The solution proceeds in turn.

$$c_0(x)=1,\ c_1(x)=\frac{(x^2-1)}{2},\ c_2(x)=\frac{5-6x^2+x^4}{12}$$

SPECIAL DIFFERENTIAL EQUATIONS

See Olver et al. (2010) in General References.

Euler's Equation The linear equation $x^ny^{(n)} + a_1x^{n-1}y^{n-1} + \cdots + a_{n-1}xy' + a_ny = R(x)$ can be reduced to a linear equation with constant coefficients by the change of variable $x = e^t$. To solve the homogeneous equation substitute $y = x^r$ into it, cancel the powers of x, which are the same for all terms, and solve the resulting polynomial for r. In case of multiple or complex roots there results the form $y = x^r(\log x)^r$ and $y = x^\alpha[\cos(\beta \log x) + i \sin(\beta \log x)]$.

Bessel's Equation The linear equation $x^2(d^2y/dx^2) + x(dy/dx) + (x^2 - p^2)y = 0$ is the Bessel equation of integer order. By series methods, not to be discussed here, this equation can be shown to have the solution

$$J_p(x)=\left(\frac{x}{2}\right)^p\sum_{k=0}^{\infty}\frac{(-1)^k(x/2)^{2k}}{k!(p+k)!}$$

(Bessel function of the first kind of order p) and

$$Y_p(x)=\frac{J_p(x)\cos(p\pi)-J_{-p}(x)}{\sin(p\pi)}$$

(Bessel function of the second kind) (replace right-hand side by limiting value if p is an integer or zero).

The series converges for all x. Much of the importance of Bessel's equation and Bessel functions lies in the fact that the solutions of numerous linear differential equations can be expressed in terms of them.

Legendre's Equation The Legendre equation $(1 - x^2)y'' - 2xy' + n(n + 1)y = 0$, $n \geq 0$, has the solution P_n for n an integer.

The polynomials P_n are the so-called Legendre polynomials, $P_0(x) = 1$, $P_1(x) = x$, $P_2(x) = \frac{1}{2}(3x^2 - 1)$, $P_3(x) = \frac{1}{2}(5x^3 - 3x)$, … For n positive and not an integer, see Olver et al. (2010) in General References.

Laguerre's Equation The Laguerre equation $x(d^2y/dx^2) + (c - x)(dy/dx) - ay = 0$ is satisfied by the confluent hypergeometric function. See Olver et al. (2010) in General References.

Hermite's Equation The Hermite equation $y''-2xy'+2ny=0$ is satisfied by the Hermite polynomial of degree n, $y = AH_n(x)$, if n is a positive integer or zero. $H_0(x) = 1$, $H_1(x) = 2x$, $H_2(x) = 4x^2 - 2$, $H_3(x) = 8x^3 - 12x$, $H_4(x) = 16x^4 - 48x^2 + 12$, $H_{r+1}(x) = 2xH_r(x) - 2rH_{r-1}(x)$.

Chebyshev's Equation The equation $(1-x^2)y''-xy'+n^2y=0$ for n a positive integer or zero is satisfied by the nth Chebyshev polynomial $y = AT_n(x)$. $T_0(x) = 1$, $T_1(x) = x$, $T_2(x) = 2x^2 - 1$, $T_3(x) = 4x^3 - 3x$, $T_4(x) = 8x^4 - 8x^2 + 1$; $T_{r+1}(x) = 2xT_r(x) - T_{r-1}(x)$.

PARTIAL DIFFERENTIAL EQUATIONS

The analysis of situations involving two or more independent variables frequently results in a partial differential equation.

Example The equation $\partial T/\partial t = k(\partial^2T/\partial x^2)$ represents the unsteady one-dimensional conduction of heat.

Example The equation for the unsteady transverse motion of a uniform beam clamped at the ends is

$$\frac{\partial^4 y}{\partial x^4}+\frac{\rho}{EI}\frac{\partial^2 y}{\partial t^2}=0$$

Example The expansion of a gas behind a piston is characterized by the simultaneous equations

$$\frac{\partial u}{\partial t}+u\frac{\partial u}{\partial x}+\frac{c^2}{\rho}\frac{\partial \rho}{\partial x}=0 \quad \text{and} \quad \frac{\partial \rho}{\partial t}+u\frac{\partial u}{\partial x}+\rho\frac{\partial u}{\partial x}=0$$

The partial differential equation $\partial^2 f/(\partial x\,\partial y)=0$ can be solved by two integrations yielding the solution $f=g(x)+h(y)$, where $g(x)$ and $h(y)$ are arbitrary differentiable functions. This result is an example of the fact that the general solution of partial differential equations involves arbitrary functions in contrast to the solution of ordinary differential equations, which involve only arbitrary constants. A number of methods are available for finding the general solution of a partial differential equation. In most applications of partial differential equations, the general solution is of limited use. In such applications the solution of a partial differential equation must satisfy both the equation and certain auxiliary conditions called *initial* and/or *boundary* conditions, which are dictated by the problem. Examples of these include those in which the wall temperature is a fixed constant $T(x_0)=T_0$, there is no diffusion across a nonpermeable wall, and the like. In ordinary differential equations, these auxiliary conditions allow definite numbers to be assigned to the constants of integration.

Partial Differential Equations of Second and Higher Order Many of the applications to scientific problems fall naturally into partial differential equations of second order, although there are important exceptions in elasticity, vibration theory, and elsewhere. A second-order differential equation can be written as

$$a\frac{\partial^2 u}{\partial x^2}+b\frac{\partial^2 u}{\partial x\partial y}+c\frac{\partial^2 u}{\partial y^2}=f$$

where *a*, *b*, *c*, and *f* depend upon *x*, *y*, *u*, $\partial u/\partial x$, and $\partial u/\partial y$. This equation is hyperbolic, parabolic, or elliptic, depending on whether the discriminant $b^2-4ac>0$, $=0$, or <0, respectively. Since *a*, *b*, *c*, and *f* depend on the solution, the type of equation can be different at different *x* and *y* locations. If the equation is hyperbolic, discontinuities can be propagated. See Courant and Hilbert (1953, 1962) and LeVeque, R. J., *Numerical Methods for Conservation Laws*, Birkhäuser, Basel, Switzerland, 1992.

Phenomena of *propagation* such as vibrations are characterized by equations of "hyperbolic" type which are essentially different in their properties from other classes such as those which describe equilibrium (elliptic) or unsteady diffusion and heat transfer (parabolic). Prototypes are as follows:

Elliptic Laplace's equation $\partial^2u/\partial x^2+\partial^2u/\partial y^2=0$ and Poisson's equation $\partial^2u/\partial x^2+\partial^2u/\partial y^2=g(x,y)$ do not contain the variable time explicitly and consequently represent equilibrium configurations. Laplace's equation is satisfied by static electric or magnetic potential at points free from electric charges or magnetic poles. Other important functions satisfying Laplace's equation are the velocity potential of the irrotational motion of an incompressible fluid, used in hydrodynamics; the steady temperature at points in a homogeneous solid; and the steady state of diffusion through a homogeneous body.

Parabolic The heat equation $\partial T/\partial t=\partial^2T/\partial x^2+\partial^2T/\partial y^2$ represents nonequilibrium or unsteady states of heat conduction and diffusion.

Hyperbolic The wave equation $\partial^2u/\partial t^2=c^2(\partial^2u/\partial x^2+\partial^2u/\partial y^2)$ represents wave propagation of many varied types.

Quasilinear first-order differential equations are like

$$a\frac{\partial u}{\partial x}+b\frac{\partial u}{\partial y}=f$$

where *a*, *b*, and *f* depend on *x*, *y*, and *u*, with $a^2+b^2\neq 0$. This equation can be solved using the method of characteristics, which writes the solution in terms of a parameter *s*, which defines a path for the characteristic.

$$\frac{dx}{ds}=a,\quad \frac{dy}{ds}=b,\quad \frac{du}{ds}=f$$

These equations are integrated from some initial conditions. For a specified value of *s*, the value of *x* and *y* shows the location where the solution is *u*. The equation is semilinear if *a* and *b* depend just on *x* and *y* (and not *u*), and the equation is linear if *a*, *b*, and *f* all depend on *x* and *y*, but not *u*. Such equations give rise to shock propagation, and conditions have been derived to deduce the presence of shocks. Courant and Hilbert (1953, 1962); Rhee, H. K., R. Aris, and N. R. Amundson, *First-Order Partial Differential Equations*, vol. 1, *Theory and Applications of Single Equations*, Prentice-Hall, Englewood Cliffs, N.J., 1986; and LeVeque (1992), ibid.

An example of a linear hyperbolic equation is the advection equation for flow of contaminants when the *x* and *y* velocity components are *u* and *v*, respectively.

$$\frac{\partial c}{\partial t}+u\frac{\partial c}{\partial x}+v\frac{\partial c}{\partial y}=0$$

The equations for flow and adsorption in a packed bed or chromatography column give a quasilinear equation.

$$\phi\frac{\partial c}{\partial t}+\phi u\frac{\partial c}{\partial x}+(1-\phi)\frac{df}{dc}\frac{\partial c}{\partial t}=0$$

Here $n=f(c)$ is the relation between concentration on the adsorbent and fluid concentration.

The solution of problems involving partial differential equations often revolves about an attempt to reduce the partial differential equation to one or more ordinary differential equations. The solutions of the ordinary differential equations are then combined (if possible) so that the boundary conditions and the original partial differential equation are simultaneously satisfied. Three of these techniques are illustrated.

Similarity Variables The physical meaning of the term "similarity" relates to internal similitude, or self-similitude. Thus, similar solutions in boundary-layer flow over a horizontal flat plate are those for which the horizontal component of velocity *u* has the property that two velocity profiles located at different coordinates *x* differ only by a scale factor. The mathematical interpretation of the term similarity is a transformation of variables carried out so that a reduction in the number of independent variables is achieved. There are essentially two methods for finding similarity variables, "separation of variables" (not the classical concept) and the use of "continuous transformation groups." The basic theory is available in Ames (1965).

Example The equation $\partial\theta/\partial x=(A/y)(\partial^2\theta/\partial y^2)$ with the boundary conditions $\theta=0$ at $x=0, y>0$; $\theta=0$ at $y=\infty, x>0$; $\theta=1$ at $y=0, x>0$ represents the nondimensional temperature θ of a fluid moving past an infinitely wide flat plate immersed in the fluid. Turbulent transfer is neglected, as is molecular transport except in the *y* direction. It is now assumed that the equation and the boundary conditions can be satisfied by a solution of the form $\theta=f(y/x^n)=f(u)$, where $\theta=0$ at $u=\infty$ and $\theta=1$ at $u=0$. The purpose here is to replace the independent variables *x* and *y* by the single variable *u* when it is hoped that a value of *n* exists which will allow *x* and *y* to be completely eliminated in the equation. In this case since $u=y/x^n$, there results after some calculation $\partial\theta/\partial x=-(nu/x)(d\theta/du)$, $\partial^2\theta/\partial y^2=(1/x^{2n})(d^2\theta/du^2)$, and when these are substituted in the equation, $-(1/x)nu\,(d\theta/du)=(1/x^{3n})(A/u)(d^2\theta/du^2)$. For this to be a function of *u* only, choose $n=\frac{1}{3}$. There results $(d^2\theta/du^2)+(u^2/3A)(d\theta/du)=0$. Two integrations and use of the boundary conditions for this ordinary differential equation give the solution

$$\theta=\int_u^\infty \exp(-u^3/9A)\,du\Big/\int_0^\infty \exp(-u^3/9A)\,du$$

Group Method The type of transformation can be deduced using group theory. For a complete exposition, see Ames (1965) and Hill, J. M., *Differential Equations and Group Methods for Scientists and Engineers*, CRC Press, New York, 1992; a shortened version can be found in Finlayson (1980). Basically, a similarity transformation should be considered when one of the independent variables has no physical scale (perhaps it goes to infinity). The boundary conditions must also simplify (and combine) since each transformation leads to a differential equation with one fewer independent variable.

Example A similarity variable is found for the problem

$$\frac{\partial c}{\partial t}=\frac{\partial}{\partial x}\left(\frac{D(c)}{D_0}\frac{\partial c}{\partial x}\right),\quad c(0,t)=1,\quad c(\infty,t)=0,\quad c(x,0)=0$$

Note that the length dimension goes to infinity, so there is no length scale in the problem statement; this is a clue to try a similarity transformation. The transformation examined here is

$$\bar{t}=a^\alpha t,\quad \bar{x}=a^\beta x,\quad \bar{c}=a^\gamma c$$

With this substitution, the equation becomes

$$a^{\alpha-\gamma}\frac{\partial\bar{c}}{\partial\bar{t}}=a^{2\beta-\gamma}\frac{\partial}{\partial\bar{x}}\left[\frac{D(a^{-\gamma}\bar{c})}{D_0}\frac{\partial\bar{c}}{\partial\bar{x}}\right]$$

Group theory says a system is conformally invariant if it has the same form in the new variables; here, that is

$$\gamma=0 \qquad \alpha-\gamma=2\beta-\gamma \qquad \text{or } \alpha=2\beta$$

The invariants are

$$\eta=\frac{x}{t^\delta},\quad \delta=\frac{\beta}{\alpha}$$

and the solution is

$$c(x, t) = f(\eta)t^{\gamma/\alpha}$$

We can take $\gamma = 0$ and $\delta = \beta/\alpha = 1/2$. Note that the boundary conditions combine because the point $x = \infty$ and $t = 0$ gives the same value of η and the conditions on c at $x = \infty$ and $t = 0$ are the same. We thus make the transformation

$$\eta = \frac{x}{\sqrt{4D_0 t}}, \quad c(x,t) = f(\eta)$$

The use of the 4 and D_0 makes the analysis below simpler. The result is

$$\frac{d}{d\eta}\left[\frac{D(c)}{D_0}\frac{df}{d\eta}\right] + 2\eta\frac{df}{d\eta} = 0, \quad f(0) = 1, \quad f(\infty) = 0$$

Thus, we solve a two-point boundary-value problem instead of a partial differential equation. When the diffusivity is constant, the solution is the error function, a tabulated function.

$$c(x,t) = 1 - \text{erf}\,\eta = \text{erfc}\,\eta$$

$$\text{erf}\,\eta = \int_0^\eta e^{-\xi^2}\,d\xi \Big/ \int_0^\infty e^{-\xi^2}\,d\xi$$

Separation of Variables This powerful, well-utilized method is applicable in certain circumstances. It consists of assuming that the solution for a partial differential equation has the form $U = f(x)g(y)$. If it is then possible to obtain an ordinary differential equation on one side of the equation depending on only x and on the other side on only y, the partial differential equation is said to be *separable* in the variables x and y. If this is the case, one side of the equation is a function of x alone and the other of y alone. The two can be equal only if each is a constant, say, λ. Thus the problem has again been reduced to the solution of ordinary differential equations.

Example Laplace's equation $\partial^2V/\partial x^2 + \partial^2V/\partial y^2 = 0$ plus the boundary conditions $V(0, y) = 0$, $V(l, y) = 0$, $V(x, \infty) = 0$, $V(x, 0) = f(x)$ represents the steady-state potential in a thin plate (in the z direction) of infinite extent in the y direction and of width l in the x direction. A potential $f(x)$ is impressed (at $y = 0$) from $x = 0$ to $x = 1$, and the sides are grounded. To obtain a solution of this boundary-value problem, assume $V(x, y) = f(x)g(y)$. Substitution in the differential equation yields $f''(x)g(y) + f(x)g''(y) = 0$ or $g''(y)/g(y) = -f''(x)/f(x) = \lambda^2$ (say). This system becomes $g''(y) - \lambda^2 g(y) = 0$ and $f''(y) + \lambda^2 f(y) = 0$. The solutions of these ordinary differential equations are, respectively, $g(y) = Ae^{\lambda y} + Be^{-\lambda y}$ and $f(x) = C\sin\lambda x + D\cos\lambda x$. Then $f(x)g(y) = (Ae^{\lambda y} + Be^{-\lambda y})(C\sin\lambda x + D\cos\lambda x)$. Now $V(0, y) = 0$ so that $f(0)g(y) = (Ae^{\lambda y} + Be^{-\lambda y})D \equiv 0$ for all y. Hence $D = 0$. The solution then has the form $\sin\lambda x\,(Ae^{\lambda y} + Be^{-\lambda y})$ where the multiplicative constant C has been eliminated. Since $V(l, y) = 0$, $\sin\lambda l(Ae^{\lambda y} + Be^{-\lambda y}) \equiv 0$. Clearly the bracketed function of y is not zero, for the solution would then be the identically zero solution. Hence $\sin\lambda l = 0$ or $\lambda_n = n\pi/l$, $n = 1, 2, \ldots$, where $\lambda_n = n$th eigenvalue.

The solution now has the form $\sin(n\pi x/l)(Ae^{n\pi y/l} + Be^{-n\pi y/l})$. Since $V(x, \infty) = 0$, A must be taken to be zero because e^y becomes arbitrarily large as $y \to \infty$. The solution then reads $B_n \sin(n\pi x/l)e^{-n\pi y/l}$, where B_n is the multiplicative constant. The differential equation is linear and homogeneous so that $\sum_{n=1}^{\infty} B_n e^{-n\pi y/l}\sin(n\pi x/l)$ is also a solution. Satisfaction of the last boundary condition is ensured by taking

$$B_n = \frac{2}{l}\int_0^l f(x)\sin(n\pi x/l)\,dx = \text{Fourier sine coefficients of } f(x)$$

Further, convergence and differentiability of this series are established quite easily. Thus the solution is

$$V(x, y) = \sum_{n=1}^{\infty} B_n e^{-n\pi y/l}\sin\frac{n\pi x}{l}$$

Example The diffusion problem in a slab of thickness L

$$\frac{\partial c}{\partial t} = D\frac{\partial^2 c}{\partial x^2}, \quad c(0,t) = 1, \quad c(L,t) = 0, \quad c(x,0) = 0$$

can be solved by separation of variables. First transform the problem so that the boundary conditions are homogeneous (having zeros on the right-hand side). Let

$$c(x,t) = 1 - \frac{x}{L} + u(x,t)$$

Then $u(x, t)$ satisfies

$$\frac{\partial u}{\partial t} = D\frac{\partial^2 u}{\partial x^2}, \quad u(x,0) = \frac{x}{L} - 1, \quad u(0,t) = 0, \quad u(L,t) = 0$$

Assume a solution of the form $u(x, t) = X(x)T(t)$, which gives

$$\frac{1}{DT}\frac{dT}{dt} = \frac{1}{X}\frac{d^2X}{dx^2}$$

Since both sides are constant, this gives the following ordinary differential equations to solve:

$$\frac{1}{DT}\frac{dT}{dt} = -\lambda, \quad \frac{1}{X}\frac{d^2X}{dx^2} = -\lambda$$

The solution of these is

$$T = Ae^{-\lambda Dt} \qquad X = B\cos\sqrt{\lambda}\,x + E\sin\sqrt{\lambda}\,x$$

The combined solution for $u(x, t)$ is

$$u = A(B\cos\sqrt{\lambda}\,x + E\sin\sqrt{\lambda}\,x)e^{-\lambda Dt}$$

Apply the boundary condition that $u(0, t) = 0$ to give $B = 0$. Then the solution is

$$u = A(\sin\sqrt{\lambda}\,x)e^{-\lambda Dt}$$

where the multiplicative constant E has been eliminated. Apply the boundary condition at $x = L$.

$$0 = A(\sin\sqrt{\lambda}\,L)e^{-\lambda Dt}$$

This can be satisfied by choosing $A = 0$, which gives no solution. However, it can also be satisfied by choosing λ such that

$$\sin\sqrt{\lambda}\,L = 0, \quad \sqrt{\lambda}\,L = n\pi$$

Thus

$$\lambda = \frac{n^2\pi^2}{L^2}$$

The combined solution can now be written as

$$u = A\left(\frac{\sin n\pi x}{L}\right)e^{-n^2\pi^2 Dt/L^2}$$

Since the initial condition must be satisfied, we use an infinite series of these functions.

$$u = \sum_{n=1}^{\infty} A_n\left(\frac{\sin n\pi x}{L}\right)e^{-n^2\pi^2 Dt/L^2}$$

At $t = 0$, we satisfy the initial condition.

$$\frac{x}{L} - 1 = \sum_{n=1}^{\infty} A_n\left(\frac{\sin n\pi x}{L}\right)$$

This is done by multiplying the equation by

$$\frac{\sin m\pi x}{L}$$

and integrating over x: $0 \to L$. (This is the same as minimizing the mean square error of the initial condition.) This gives

$$\frac{A_m L}{2} = \int_0^L \left(\frac{x}{L} - 1\right)\sin\frac{m\pi x}{L}\,dx$$

which completes the solution.

Integral-Transform Method A number of integral transforms are used in the solution of differential equations. Only one, the Laplace transform, is discussed here [for others, see Integral Transforms (Operational Methods)].

The one-sided Laplace transform indicated by $L[f(t)]$ is defined by the equation $L[f(t)]\ \int_0^\infty f(t)e^{-st}dt$. It has numerous important properties. The ones of interest here are $L[f'(t)] = sL[f(t)] - f(0)$ $L[f''(t)] = s^2L[f(t)] - sf(0) - f'(0)$; $L[f^{(n)}(t)] = s^nL[f(t)] - s^{n-1}f(0) - s^{n-2}f'(0) - \cdots - f^{(n-1)}(0)$ for ordinary derivatives. For partial derivatives an indication of which variable is being transformed avoids confusion. Thus, if

$$y = y(x,t),\ L_t\left[\frac{\partial y}{\partial t}\right] = sL[y(x,t)] - y(x,0)$$

whereas

$$L_t\left[\frac{\partial y}{\partial x}\right] = \frac{dL_t[y(x,t)]}{dx}$$

since $L[y(x, t)]$ is "really" only a function of x. Otherwise the results are similar. These facts coupled with the linearity of the transform, i.e., $L[af(t) + bg(t)] = aL[f(t)] + bL[g(t)]$, make it a useful device in solving some linear differential equations. Its use reduces the solution of ordinary differential equations to the solution of algebraic equations for $L[y]$. the inverse transform must be obtained either from tables or by use of complex inversion methods.

Example The equation $\partial c/\partial t = D(\partial^2 c/\partial x^2)$ represents the diffusion in a semi-infinite medium, $x \geq 0$. Under the boundary conditions $c(0, t) = c_0$ and $c(x, 0) = 0$, find a solution of the diffusion equation. By taking the Laplace transform of both sides with respect to t,

$$\int_0^\infty e^{-st}\frac{\partial^2 c}{\partial x^2}dt = \frac{1}{D}\int_0^\infty e^{-st}\frac{\partial c}{\partial t}dt$$

or

$$\frac{d^2F}{dx^2} = (1/D)sF - c(x,0) = \frac{sF}{D}$$

where $F(x, s) = L_t[c(x, t)]$. Hence

$$\frac{d^2F}{dx^2} - \left(\frac{s}{D}\right)F = 0$$

The other boundary condition transforms into $F(0, s) = c_0/s$. Finally the solution of the ordinary differential equation for F subject to $F(0, s) = c_0/s$ and F remains finite as $x \to \infty$ is $F(x,s) = (c_0/s)e^{-\sqrt{s/D}x}$. Reference to a table shows that the function having this as its Laplace transform is

$$c(x,t) = c_0\left[1 - \frac{2}{\sqrt{\pi}}\int_0^{x/2\sqrt{Dt}} e^{-u^2}du\right] = C_0 \operatorname{erfc}\left(\frac{x}{\sqrt{4Dt}}\right)$$

This is the same solution obtained above by the group method.

Matched-Asymptotic Expansions Sometimes the coefficient in front of the highest derivative is a small number. Special perturbation techniques can then be used, provided the proper scaling laws are found. See Holmes, M. H., *Introduction to Perturbation Methods*, 2d ed., Springer, New York, 2013.

DIFFERENCE EQUATIONS

References: Elaydi, Saber, *An Introduction to Difference Equations*, 3d ed., Springer-Verlag, New York, 2005; Kelley, W. G., and A. C. Peterson, *Difference Equations: An Introduction with Applications*, 2d ed., Harcourt/Academic, San Diego, Calif., 2001.

Some models have independent variables that do not vary continuously, but have meaning only for discrete values. Stagewise processes such as distillation, staged extraction systems, absorption columns, and continuous stirred tank reactors (CSTRs) are such processes. The dependent variable varies between stages, and the independent variable is the integral number of the stage. Difference equations arise in discrete models of environmental problems (see Logan and Wolesensky). Difference equations also arise in the solution of partial differential equations using the finite difference method, and those are treated below (Numerical Analysis and Approximate Methods). Examined here are solution methods applicable to the chemical engineering problems; for more detailed information see the references. The methods for difference equations mirror those for differential equations. In particular, find complementary solution and then a particular solution. The order of the difference equation is the difference between the largest and smallest arguments.

Consider the countercurrent cascade shown in Fig. 3-45. We let y_i be the ratio of the mass of solute to mass of solvent in the ith cell; x_i is the ratio of mass of solute to mass of carrier solvent in the ith cell. For illustration we take the equilibrium relation as linear

$$y_i = Kx_i$$

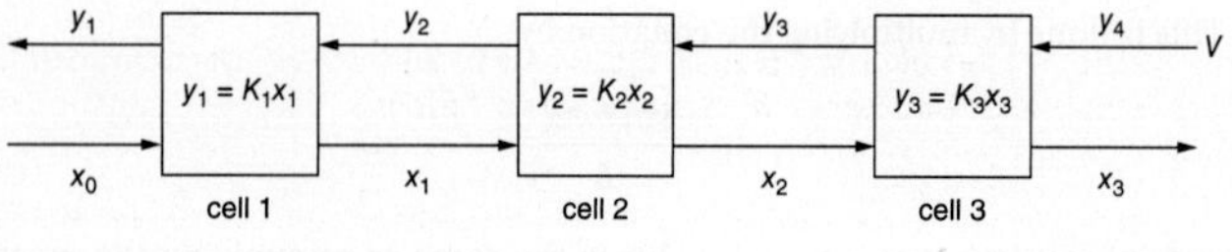

FIG. 3-45 Countercurrent cascade.

A material balance on the ith stage gives

$$Lx_{i-1} + Vy_{i+1} - Lx_i - Vy_i = 0$$

Using the equilibrium relation transforms this equation to the form

$$(L/K)y_{i-1} + Vy_{i+1} - (L/K)y_i - Vy_i = 0$$

or

$$y_{i+1} - [(L/VK) + 1]y_i + (L/VK)y_{i-1} = 0$$

With $\alpha = L/VK$ the final form of the difference equation is $y_{i+1} - (\alpha + 1)y_i + \alpha y_{i-1} = 0$. The solution is obtained by trying the general form $y_i = r^i$. This gives the characteristic equation $r^2 - (\alpha + 1)r + \alpha = 0$. One root is $r = 1$, and call the other root β. The solution is then $y_i = A + B\beta^i$. This completes the complementary solution. The number of units is taken as N. The particular solution is found by choosing A and B to fit boundary conditions. Here they are taken as the inlet feed composition x_0 and the inlet solvent composition y_{N+1}. Using $y_0 = Kx_0$, we obtain two equations for A and B. The solutions are $A = Kx_0 - B$ and $B = (Kx_0 - y_{N+1})/(1 - \beta^{N+1})$. The exit concentration is $y_1 = A + B\beta$.

Nonlinear Difference Equations: Riccati Difference Equation The Riccati equation $y_{i+1}y_i + ay_{i+1} + by_i + c = 0$ is a nonlinear difference equation which can be solved by reduction to linear form. Set $y = z + h$. The equation becomes $z_{i+1}z_i + (h + a)z_{i+1} + (h + b)z_i + h^2 + (a + b)h + c = 0$. If h is selected as a root of $h^2 + (a + b)h + c = 0$ and the equation is divided by $z_{i+1}z_i$, there results $(h + b)/z_{i+1} + (h + a)/z_i + 1 = 0$. This is a linear equation with constant coefficients for $w_i = 1/z_i$. The solution is

$$\frac{1}{y_i - h} = K\left[-\frac{a+h}{b+h}\right]^i - \frac{1}{(a+h)+(b+h)}$$

where K is a constant chosen to fit conditions at one point. This equation is obtained in distillation problems, among others, in which the number of theoretical plates is required. If the relative volatility is assumed to be constant, the plates are theoretically perfect, and the molal liquid and vapor rates are constant, then a material balance around the nth plate of the enriching section yields a Riccati difference equation.

INTEGRAL EQUATIONS

References: Davis, H. T., *Introduction to Nonlinear Differential and Integral Equations*, Dover, New York, 2010; Statgold, I., and M. J. Holst, *Green's Functions and Boundary Value Problems*, 3d ed., Interscience, New York, 2011.

An integral equation is any equation in which the unknown function appears under the sign of integration and possibly outside the sign of integration. If derivatives of the dependent variable appear elsewhere in the equation, the equation is said to be *integrodifferential*.

CLASSIFICATION OF INTEGRAL EQUATIONS

Volterra's integral equations have an integral with a variable limit, whereas Fredholm's integral equations have a fixed limit. The Volterra equation of the second kind is

$$u(x)=f(x)+\lambda\int_a^x K(x,t)u(t)dt$$

whereas a Volterra equation of the first kind is

$$u(x)=\lambda\int_a^x K(x,t)u(t)dt$$

Equations of the first kind are very sensitive to solution errors so that they present severe numerical problems. Volterra equations are similar to initial-value problems.

A Fredholm equation of the second kind is

$$u(x)=f(x)+\lambda\int_a^b K(x,t)u(t)dt$$

whereas a Fredholm equation of the first kind is

$$u(x)=\int_a^b K(x,t)u(t)dt$$

The limits of integration are fixed, and these problems are analogous to boundary value problems.

An eigenvalue problem is a homogeneous equation of the second kind, and solutions exist only for certain λ.

$$u(x)=\lambda\int_a^b K(x,t)u(t)dt$$

An example of a Volterra equation is the heat conduction problem in a semi-infinite domain.

$$\rho C_p\frac{\partial T}{\partial t}=k\frac{\partial^2 T}{\partial x^2}\qquad 0\le x<\infty, t>0$$

$$T(x,0)=0\qquad \frac{\partial T}{\partial x}(0,t)=-g(t)$$

$$\lim_{x\to 0}T(x,t)=0\qquad \lim_{x\to\infty}\frac{\partial T}{\partial x}(x,t)=0$$

If this is solved by using Fourier transforms [see Integral Transforms (Operational Methods)], the solution is

$$T(x)=\frac{1}{k}\int_0^1 G(x,y)Q(y)dy$$

$$T(x,t)=\frac{1}{\sqrt{\pi}}\int_0^1 g(s)\frac{1}{\sqrt{t-s}}e^{-x^2/4(t-s)}ds$$

Integral equations can arise from the formulation of a problem by using Green's function. The equation governing heat conduction with a variable heat generation rate is represented in differential form as

$$\frac{d^2T}{dx^2}=\frac{Q(x)}{k}\qquad T(0)=T(1)=0$$

In integral form the same problem is

$$T(x)=\frac{1}{k}\int_0^1 G(x,y)Q(y)dy$$

$$G(x,y)=\begin{cases}-x(1-y) & x\le y\\ -y(1-x) & y\le x\end{cases}$$

The Poisson equation governs electric charges

$$\nabla^2\Psi=-4\pi\rho$$

and the formulation as an integral equation is

$$\Psi(\mathbf{r})=\int_V \rho(r_0)G(r,r_0)\,dV_0$$

where Green's function in three dimensions is

$$G(\mathbf{r},\mathbf{r}_0)=\frac{1}{r}, r=\sqrt{(x-x_0)^2+(y-y_0)^2+(z-z_0)^2}$$

and in two dimensions is

$$G(\mathbf{r},\mathbf{r}_0)=-2\ln r, r=\sqrt{(x-x_0)^2+(y-y_0)^2}$$

See the references for other examples.

Integral equations can be solved numerically, too. The methods are analogous to the usual methods for integrating differential equations (Runge-Kutta, predictor-corrector, Adams methods, etc.). Explicit methods are fast and efficient until the time step is very small, to meet the stability requirements. Then implicit methods are used, even though sets of simultaneous algebraic equations must be solved. The major part of the calculation is the evaluation of integrals, however, so that the added time to solve the algebraic equations is not excessive. Thus, implicit methods tend to be preferred. Volterra equations of the first kind are not well posed, and small errors in the solution can have disastrous consequences. The boundary element method uses Green's functions and integral equations to solve differential equations. See Brebbia, C. A., and J. Dominguez, *Boundary Elements—An Introductory Course*, 2d ed., Computational Mechanics Publications, Southhampton, UK, 1992; and Mackerle, J., and C. A. Brebbia, eds., *Boundary Element Reference Book*, Springer Verlag, Berlin, 1988.

INTEGRAL TRANSFORMS (OPERATIONAL METHODS)

References: Davies, B., *Integral Transforms and Their Applications*, 3d ed., Springer, New York, 2002; Debnath, L., and D. Bhatta, *Integral Transforms and Their Applications*, 3d ed., Chapman and Hall/CRC, New York, 2014; Duffy, D. G., *Transform Methods for Solving Partial Differential Equations*, Chapman & Hall/CRC, New York, 2nd ed., 2004; see also references for Differential Equations.

The term *operational method* implies a procedure of solving differential and difference equations by which the boundary or initial conditions are automatically satisfied in the course of the solution. The technique offers a very powerful tool in the applications of mathematics, but it is limited to linear problems.

Most integral transforms are special cases of the equation $g(s) = \int_a^b f(t)K(s,t)dt$ in which $g(s)$ is said to be the *transform* of $f(t)$ and $K(s, t)$ is called the *kernel* of the transform. A tabulation of the more important kernels and the interval (a, b) of applicability follows.

Name of transform	(a, b)	$K(s, t)$
Laplace	$(0, \infty)$	e^{-st}
Fourier	$(-\infty, \infty)$	$\frac{1}{\sqrt{2\pi}}e^{-ist}$
Fourier cosine	$(0, \infty)$	$\sqrt{\frac{2}{\pi}}\cos st$
Fourier sine	$(0, \infty)$	$\sqrt{\frac{2}{\pi}}\sin st$

LAPLACE TRANSFORM

The Laplace transform of a function $f(t)$ is defined by $F(s) = L\{f(t)\} = \int_0^\infty e^{-st} f(t)dt$, where s is a complex variable. Note that the transform is an improper integral and therefore may not exist for all continuous functions and all values of s. We restrict consideration to those values of s and those functions f for which this improper integral converges. The Laplace transform is used in process control (see Sec. 8).

The function $L[f(t)] = g(s)$ is called the *direct transform*, and $L^{-1}[g(s)] = f(t)$ is called the *inverse transform*. Both the direct and the inverse transforms are tabulated for many often recurring functions. In general,

$$L^{-1}[g(s)] = \frac{1}{2\pi i}\int_{\alpha-i\infty}^{\alpha+i\infty} e^{st} g(s)ds$$

and to evaluate this integral requires a knowledge of complex variables, the theory of residues, and contour integration.

A function is said to be *piecewise continuous* on an interval if it has only a finite number of finite (or jump) discontinuities. A function f on $0 < t < \infty$ is said to be of *exponential growth at infinity* if there exist constants M and α such that $|f(t)| \le Me^{\alpha t}$ for sufficiently large t.

Sufficient Conditions for the Existence of the Laplace Transform Suppose f is a function which is (1) piecewise continuous on every finite interval $0 < t < T$, (2) of exponential growth at infinity, and (3) for which $\int_0^\delta |f(t)|dt$ exists (finite) for every finite $\delta > 0$. Then the Laplace transform of f exists for all complex numbers s with a sufficiently large real part.

Note that condition 3 is automatically satisfied if f is assumed to be piecewise continuous on every finite interval $0 \le t < T$. The function $f(t) = t^{-1/2}$ is not piecewise continuous on $0 \le t < T$ but satisfies conditions 1 to 3.

Let Λ denote the class of all functions on $0 < t < \infty$ which satisfy conditions 1 to 3.

Example Let $f(t)$ be the Heaviside step function at $t = t_0$; that is, $f(t) = 0$ for $t \le t_0$ and $f(t) = 1$ for $t > t_0$. Then

$$L\{f(t)\} = \int_{t_0}^{\infty} e^{-st}\,dt = \lim_{T\to\infty}\int_{t_0}^{T} e^{-st}\,dt = \lim_{T\to\infty}\frac{1}{s}(e^{-st_0} - e^{-sT}) = \frac{e^{-st_0}}{s} \quad \text{provided } s > 0$$

Example Let $f(t) = e^{at}$, $t \ge 0$, where a is a real number. Then $L\{e^{at}\} = \int_0^\infty e^{-(s-a)t}\,dt = 1/(s-a)$ provided Re $s > a$.

Properties of the Laplace Transform

1. The Laplace transform is a linear operator: $L\{af(t) + bg(t)\} = aL\{f(t)\} + bL\{g(t)\}$ for any constants a and b and any two functions f and g whose Laplace transforms exist.

2. The Laplace transform of a real-valued function is real for real s. If $f(t)$ is a complex-valued function $f(t) = u(t) + iv(t)$, where u and v are real, then $L\{f(t)\} = L\{u(t)\} + iL\{v(t)\}$. Thus $L\{u(t)\}$ is the real part of $L\{f(t)\}$, and $L\{v(t)\}$ is the imaginary part of $L\{f(t)\}$.

3. The Laplace transform of a function in the class Λ has derivatives of all orders, and $L\{t^k f(t)\} = (-1)^k d^kF(s)/ds^k$, $k = 1, 2, 3, \ldots$, where $F(s)$ is the Laplace transform of $f(t)$.

$$\textbf{\textit{Example}} \quad \int_0^\infty e^{st} \sin at\,dt = \frac{a}{s^2+a^2}, s > 0.$$

By property 3, $L\{t \sin at\} = \int_0^\infty e^{-st} t \sin at\,dt = \dfrac{2as}{(s^2+a^2)^2}$

Example By applying property 3 with $f(t) = 1$ and using the preceding results, we obtain

$$L\{t^k\} = (-1)^k \frac{d^k}{ds^k}\left(\frac{1}{s}\right) = \frac{k!}{s^{k+1}}$$

provided Re $s > 0$ for $k = 1, 2, \ldots$. Similarly, we obtain

$$L\{t^k e^{at}\} = (-1)^k \frac{d^k}{ds^k}\left(\frac{1}{s-a}\right) = \frac{k!}{(s-a)^{k+1}}$$

4. Frequency-shift property (or, equivalently, the transform of an exponentially modulated function). If $F(s)$ is the Laplace transform of a function $f(t)$ in class Λ, then for any constant a, $L\{e^{at}f(t)\} = F(s - a)$.

$$\textbf{\textit{Example}} \quad L\{te^{-at}\} = \frac{1}{(s+a)^2} \quad (s > 0).$$

5. Time-shift property. Let $u(t - a)$ be the unit step function at $t = a$. Then $L\{f(t - a)u(t - a)\} = e^{-as}F(s)$.

6. Transform of a derivative. Let f be a differentiable function such that both f and f' belong to the class Λ. Then $L\{f'(t)\} = sF(s) - f(0)$.

7. Transform of a higher-order derivative. Let f be a function which has continuous derivatives up to order n on $(0, \infty)$, and suppose that f and its derivatives up to order n belong to the class Λ. Then $L\{f^{(j)}(t)\} = s^jF(s) - s^{j-1}f(0) - s^{j-2}f'(0) - \ldots - sf^{(j-2)}(0) - f^{(j-1)}(0)$ for $j = 1, 2, \ldots, k$.

Example $L\{f''(t)\} = s^2L\{f(t)\} - sf(0) - f'(0)$

Example Solve $y'' + y = 2e^t$, $y(0) = y'(0) = 2$. $L[y''] = -y'(0) - sy(0) + s^2L[y] = -2 - 2s + s^2L[y]$. Thus

$$-2 - 2s + s^2L[y] + L[y] = 2L[e^t] = \frac{2}{s-1}$$

$$L[y] = \frac{2s^2}{(s-1)(s^2+1)} = \frac{1}{s-1} + \frac{s}{s^2+1} + \frac{1}{s^2+1}$$

Hence $y = e^t + \cos t + \sin t$.

A short table (Table 3-2) of very common Laplace transforms and inverse transforms follows. The references and computer programs include more detailed tables. In Mathematica, the command "LaplaceTransform[cosh[a*t],t,s]" returns $s/(s^2-a^2)$. **NOTE:** $\Gamma(n+1) = \int_0^\infty x^n e^{-x}\,dx$ (gamma function); $J_n(t)$ = Bessel function of the first kind of order n.

8. $L\left[\int_a^t f(t)dt\right] = \dfrac{1}{s}L[f(t)] + \dfrac{1}{s}\int_a^0 f(t)dt$

TABLE 3-2 Laplace Transforms

$f(t)$	$g(s)$	$f(t)$	$g(s)$
1	$1/s$	$e^{-at}(1-at)$	$\dfrac{s}{(s+a)^2}$
t^n, (n = + integer)	$\dfrac{n!}{s^{n+1}}$	$\dfrac{t \sin at}{2a}$	$\dfrac{s}{(s^2+a^2)^2}$
t^n, ($n \ne$ + integer)	$\dfrac{\Gamma(n+1)}{s^{n+1}}$	$\dfrac{1}{2a^2}\sin at \sinh at$	$\dfrac{s}{s^4+4a^4}$
$\cos at$	$\dfrac{s}{s^2+a^2}$	$\cos at \cosh at$	$\dfrac{s^3}{s^2+4a^4}$
$\sin at$	$\dfrac{a}{s^2+a^2}$	$\dfrac{1}{2a}(\sinh at + \sin at)$	$\dfrac{s^2}{s^4-a^4}$
$\cosh at$	$\dfrac{s}{s^2-a^2}$	½$(\cosh at + \cos at)$	$\dfrac{s^3}{s^4-a^4}$
$\sinh at$	$\dfrac{a}{s^2-a^2}$	$\dfrac{\sin at}{t}$	$\tan^{-1}\dfrac{a}{s}$
e^{-at}	$\dfrac{1}{s+a}$	$J_0(at)$	$\dfrac{1}{\sqrt{s^2+a^2}}$
$e^{-bt}\cos at$	$\dfrac{s+b}{(s+b)^2+a^2}$	$na^n\dfrac{J_n(at)}{t}$	$(\sqrt{s^2+a^2}-s)^n (n > 0)$
$e^{-bt}\sin at$	$\dfrac{a}{(s+b)^2+a^2}$	$J_0(2\sqrt{at})$	$\dfrac{1}{s}e^{-a/s}$
$\operatorname{erfc}\dfrac{k}{2\sqrt{t}}$	$\dfrac{1}{s}e^{-k\sqrt{s}}$	$t^{n-1}e^{at}$	$\dfrac{\Gamma(n)}{(s-a)^n}(n > 0)$

Example Find $f(t)$ if $L[f(t)] = \dfrac{1}{s^2}\left[\dfrac{1}{s^2-a^2}\right]$. $L\left[\dfrac{1}{a}\sinh at\right] = \dfrac{1}{s^2-a^2}$.

$$\text{Therefore } f(t) = \int_0^t\left[\int_0^t \frac{1}{a}\sinh at\,dt\right]dt = \frac{1}{a^2}\left[\frac{\sinh at}{a} - t\right].$$

9. $L\left[\dfrac{f(t)}{t}\right] = \int_s^\infty g(s)\,ds \qquad L\left[\dfrac{f(t)}{t^k}\right] = \underbrace{\int_s^\infty \cdots \int_s^\infty}_{k \text{ integrals}} g(s)(ds)^k$

$$\textbf{\textit{Example}} \quad L\left[\frac{\sin at}{t}\right] = \int_s^\infty L[\sin at]\,ds = \int_s^\infty \frac{a\,ds}{s^2+a^2} = \cot^{-1}\frac{s}{a}$$

10. The unit step function

$$u(t-a) = \begin{cases} 0 & t < a \\ 1 & t > a \end{cases} \qquad L[u(t-a)] = \frac{e^{-as}}{s}$$

11. The unit impulse function is

$$\delta(a) = u'(t-a) = \begin{cases} \infty & \text{at } t = a \\ 0 & \text{elsewhere} \end{cases} \qquad L[u'(t-a)] = e^{-as}$$

12. $L^{-1}[e^{-as}g(s)] = f(t-a)u(t-a)$ (second shift theorem).
13. If $f(t)$ is periodic of period b, that is, $f(t+b) = f(t)$, then

$$L[f(t)] = \left[\frac{1}{1-e^{-bs}}\right]\int_0^b e^{-st} f(t)dt$$

Example The partial differential equations relating gas composition to position and time in a gas chromatograph are $\partial y/\partial n + \partial x/\partial \theta = 0$ and $\partial y/\partial n = x - y$, where $x = mx'$, $n = (k_G aP/G_m)h$, $\theta = (mk_G aP/\rho_B)t$ and G_M = molar velocity, y = mole fraction of the component in the gas phase, ρ_B = bulk density, h = distance from entrance, P = pressure, k_G = mass-transfer coefficient, and m = slope of the equilibrium line. These equations are equivalent to $\partial^2 y/\partial n\, \partial\theta + \partial y/\partial n + \partial y/\partial\theta = 0$, where the boundary conditions considered here are $y(0, \theta) = 0$ and $x(n, 0) = y(n, 0) + (\partial y/\partial n)(n, 0) = \delta(0)$ (see property 11). The problem is conveniently solved by using the Laplace transform of y with respect to n; write $g(s,\theta) = \int_0^\infty e^{-ns} y(n,\theta)\, dn$. Operating on the partial differential equation gives $s(dg/d\theta) - (\partial y/\partial\theta)(0, \theta) + sg - y(0, \theta) + dg/d\theta = 0$ or $(s+1)(dg/d\theta) + sg = (\partial y/\partial\theta)(0, \theta) + y(0, \theta) = 0$. The second boundary condition gives $g(s, 0) + sg(s, 0) - y(0, 0) = 1$ or $g(s, 0) + sg(s, 0) = 1$ ($L[\delta(0)] = 1$). A solution of the ordinary differential equation for g consistent with this second condition is

$$g(s,\theta) = \frac{1}{s+1} e^{-s\theta/(s+1)}$$

Inversion of this transform gives the solution $y(n,\theta) = e^{-(n+\theta)} I_0(2\sqrt{n\theta})$ where I_0 = zero-order Bessel function of an imaginary argument. For large u, $I_n(u) \sim e^u/\sqrt{2\pi u}$. For large n,

$$y(n,\theta) \sim \frac{\exp[-(\sqrt{\theta} - \sqrt{n})^2]}{2\pi^{1/2}(n\theta)^{1/4}}$$

or for sufficiently large n, the peak concentration occurs near $\theta = n$.

Other applications of Laplace transforms are given under Differential Equations.

CONVOLUTION INTEGRAL

The convolution integral of two functions $f(t)$ and $r(t)$ is $x(t) = f(t)*r(t) = \int_0^t f(\tau)r(t-\tau)d\tau$.

Example $t * \sin t = \int_0^t \tau \sin(t-\tau)\, d\tau = t - \sin t$.

$$L[f(t)]L[h(t)] = L[f(t)*h(t)]$$

FOURIER TRANSFORM

References: https://en.wikipedia.org/wiki/Fourier_transform#Tables_of_important_Fourier_transforms; Varma and Morbidelli (1997), see General References.

The Fourier transform is given by

$$F[f(t)] = \frac{1}{\sqrt{2\pi}}\int_{-\infty}^{\infty} f(t)^{-ist}\, dt = g(s)$$

and its inverse by

$$F^{-1}[g(s)] = \frac{1}{\sqrt{2\pi}}\int_{-\infty}^{\infty} g(s)e^{ist}\, dt = f(t)$$

In brief, the condition for the Fourier transform to exist is that $\int_{-\infty}^{\infty} |f(t)|\, dt < \infty$, although certain functions may have a Fourier transform even if this is violated.

Example The function $f(t) = \begin{cases} 1 & -a \le t \le a \\ 0 & \text{elsewhere} \end{cases}$ has $F[f(t)] = \int_{-a}^{a} e^{-ist}dt$

$$= \int_0^a e^{ist}\, dt + \int_0^a e^{-ist} dt = 2\int_0^a \cos st\, dt = \frac{2 \sin sa}{s}$$

Properties of the Fourier Transform Let $F[f(t)] = g(s)$; $F^{-1}[g(s)] = f(t)$.

1. $F[f^{(n)}(t)] = (is)^n F[f(t)]$.
2. $F[af(t) + bh(t)] = aF[f(t)] + bF[h(t)]$.
3. $F[f(-t)] = g(-s)$.
4. $F[f(at)] = \frac{1}{a} g\left(\frac{s}{a}\right)$, $a > 0$.
5. $F[e^{-iwt}f(t)] = g(s+w)$.
6. $F[f(t+t_1)] = e^{ist_1}g(s)$.
7. $F[f(t)] = G(is) + G(-is)$ if $f(t) = f(-t)$ (f even)
   $F[f(t)] = G(is) - G(-is)$ if $f(t) = -f(-t)$ (f odd)

where $G(s) = L[f(t)]$. This result allows the use of the Laplace transform tables to obtain the Fourier transforms.

Example Find $F[e^{-a|t|}]$ by property 7. Now $e^{-a|t|}$ is even. So $L[e^{-at}] = 1/(s+a)$. Therefore, $F[e^{-a|t|}] = 1/(is+a) + 1/(-is+a) = 2a/(s^2+a^2)$.

FOURIER COSINE TRANSFORM

The Fourier cosine transform is given by

$$F_c[f(t)] = g(s) = \sqrt{\frac{2}{\pi}}\int_0^\infty f(t)\cos st\, dt$$

and its inverse by

$$F_c^{-1}[g(s)] = f(t) = \sqrt{\frac{2}{\pi}}\int_0^\infty g(s)\cos st\, ds$$

The Fourier sine transform F_s is obtainable by replacing the cosine by the sine in these integrals. They can be used to solve linear differential equations; see the transform references.

MATRIX ALGEBRA AND MATRIX COMPUTATIONS

References: Anton, H., and C. Rorres, *Elementary Linear Algebra with Applications*, 9th ed., Wiley, New York, 2004; Bernstein, D. S., *Matrix Mathematics: Theory, Facts, and Formulas with Application to Linear Systems Theory*, 2d ed., Princeton University Press, Princeton, N.J., 2009.

MATRIX ALGEBRA

Matrices A rectangular array of mn quantities, arranged in m rows and n columns,

$$A = (a_{ij}) = \begin{bmatrix} a_{11} & \cdots & a_{1n} \\ a_{21} & \cdots & a_{2n} \\ \vdots & & \\ a_{m1} & & a_{mn} \end{bmatrix}$$

is called a *matrix*. The elements a_{ij} may be real or complex. The notation a_{ij} means the element in the ith row and jth column; i is called the *row index* and j the *column index*. If $m = n$, the matrix is said to be *square* and of order n. A matrix, even if it is square, does not have a numerical value, as a determinant does. However, if the matrix A is square, a determinant can be formed which has the same elements as matrix A. This is called the *determinant* of the matrix and is written det (A) or $|A|$. If A is square and det (A) $\neq 0$, then A is said to be *nonsingular*; if det (A) = 0, then A is said to be *singular*. A matrix A has rank r if and only if it has a nonvanishing determinant of order r and no nonvanishing determinant of order > r.

Equality of Matrices Let $A = (a_{ij})$, $B = (b_{ij})$. Two matrices A and B are *equal* (=) if and only if they are identical; that is, they have the same number of rows and the same number of columns and equal corresponding elements ($a_{ij} = b_{ij}$ for all i and j).

Addition and Subtraction The operations of addition (+) and subtraction (−) of two or more matrices are possible if and only if the matrices have the same number of rows and columns. Thus $A \pm B = (a_{ij} \pm b_{ij})$; i.e., addition and subtraction are of corresponding elements.

Transposition The matrix obtained from A by interchanging the rows and columns of A is called the *transpose* of A, written A' or A^T.

Example $A = \begin{bmatrix} 1 & 3 & 4 \\ 2 & 1 & 6 \end{bmatrix} \qquad A^T = \begin{bmatrix} 1 & 2 \\ 3 & 1 \\ 4 & 6 \end{bmatrix}$

Note that $(A^T)^T = A$.

Multiplication Let $A = (a_{ij})$, $i = 1, \ldots, m_1$; $j = 1, \ldots, m_2$, and $B = (b_{ij})$, $i = 1, \ldots, n_1, j = 1, \ldots, n_2$. The product AB is defined if and only if the number of columns of A (m_2) equals the number of rows of B (n_1), that is, $n_1 = m_2$. For two such matrices the product $P = AB$ is defined by summing the element-by-element products of a row of A by a column of B.

This is the row-by-column rule. Thus

$$P_{ij} = \sum_{k=1}^{n_1} a_{ik} b_{kj}$$

The resulting matrix has m_1 rows and n_2 columns.

Example
$$\begin{bmatrix} 3 & 2 \\ 1 & 1 \\ 5 & 4 \end{bmatrix} \begin{bmatrix} 0 & 1 & 5 & 6 \\ -2 & 0 & 1 & 3 \end{bmatrix} = \begin{bmatrix} -4 & 3 & 17 & 24 \\ -2 & 1 & 6 & 9 \\ -8 & 5 & 29 & 42 \end{bmatrix}$$

It is helpful to remember that the element P_{ij} is formed from the ith row of the first matrix and the jth column of the second matrix. The matrix product is not commutative. That is, $AB \neq BA$ in general.

Inverse of a Matrix A square matrix A is said to have an inverse if there exists a matrix B such that $AB = BA = I$, where I is the identity matrix of order n.

$$\begin{bmatrix} 1 & 0 & \cdots & 0 \\ 0 & 1 & \cdot\cdot & \\ \vdots & & 1 & 0 \\ 0 & \cdots & 0 & 1 \end{bmatrix}$$

The inverse B is a square matrix of the order of A, designated by A^{-1}. Thus $AA^{-1} = A^{-1}A = I$. A square matrix A has an inverse if and only if A is nonsingular.

Certain relations are important:

(1) $(AB)^{-1} = B^{-1}A^{-1}$
(2) $(AB)^T = B^TA^T$
(3) $(A^{-1})^T = (A^T)^{-1}$
(4) $(ABC)^{-1} = C^{-1}B^{-1}A^{-1}$

Scalar Multiplication Let c be any real or complex number. Then $cA = (ca_{ij})$.

Linear Equations in Matrix Form Every set of n nonhomogeneous linear equations in n unknowns

$$\begin{aligned} a_{11}x_1 + a_{12}x_2 + \cdots + a_{1n}x_n &= b_1 \\ a_{21}x_1 + a_{22}x_2 + \cdots + a_{2n}x_n &= b_2 \\ \vdots \qquad\qquad & \quad \vdots \\ a_{n1}x_1 + a_{n2}x_2 + \cdots + a_{nn}x_n &= b_n \end{aligned}$$

can be written in matrix form as $AX = B$, where $A = (a_{ij})$, $X^T = [x_1 \cdots x_n]$, and $B^T = [b_1 \cdots b_n]$. The solution for the unknowns is $X = A^{-1}B$.

Special Square Matrices

1. A *triangular matrix* is a matrix all of whose elements above or below the main diagonal (set of elements $a_{11}, \ldots, a_{nn}$) are zero.

If A is triangular, $\det(A) = a_{11}a_{22}\cdots a_{nn}$.

2. A *diagonal matrix* is one such that all elements both above and below the main diagonal are zero (that is, $a_{ij} = 0$ for all $i \neq j$). If all diagonal elements are equal, the matrix is called *scalar.* If A is diagonal, $A = (a_{ij})$, $A^{-1} = (1/a_{ij})$.
3. If $a_{ij} = a_{ji}$ for all i and j (that is, $A = A^T$), the matrix is *symmetric.*
4. If $a_{ij} = -a_{ji}$ for $i \neq j$ but not all the a_{ij} are zero, the matrix is *skew.*
5. If $a_{ij} = -a_{ji}$ for all i and j (that is, $a_{ii} = 0$), the matrix is *skew symmetric.*
6. If $A^T = A^{-1}$, the matrix A is *orthogonal.*
7. If the matrix $A^* = (\bar{a}_{ij})^T$ and $\bar{a}_{ij}$ = complex conjugate of a_{ij}, then A^* is the *hermitian transpose* of A.
8. If $A = A^{-1}$, then A is *involutory.*
9. If $A = A^*$, then A is *hermitian.*
10. If $A = -A^*$, then A is *skew hermitian.*
11. If $A^{-1} = A^*$, then A is *unitary.*

If A is any matrix, then AA^T and A^TA are square symmetric matrices, usually of different order.

By using a program such as MATLAB, these are easily calculated.

Matrix Calculus

Differentiation Let the elements of $A = [a_{ij}(t)]$ be differentiable functions of t. Then $\dfrac{dA}{dt} = \left[\dfrac{da_{ij}(t)}{dt}\right]$.

Example
$$A = \begin{bmatrix} \sin t & \cos t \\ -\cos t & \sin t \end{bmatrix} \qquad \frac{dA}{dt} = \begin{bmatrix} \cos t & -\sin t \\ \sin t & \cos t \end{bmatrix}$$

Integration The integral $\int A\,dt = [\int a_{ij}(t)\,dt]$.

Example
$$A = \begin{bmatrix} t & 2 \\ t^2 & e^t \end{bmatrix} \qquad \int A\,dt = \begin{bmatrix} t^2/2 & 2t \\ t^3/3 & e^t \end{bmatrix}.$$

The matrix $B = A - \lambda I$ is called the *characteristic matrix* or *eigenmatrix* of A. Here A is square of order n, λ is a scalar parameter, and I is the $n \times n$ identity matrix. So $\det B = \det(A - \lambda I) = 0$ is the characteristic equation (or eigenequation) for A. The characteristic equation is always of the same degree as the order of A. The roots of the characteristic equation are called the *eigenvalues* of A or *characteristic values* of A.

Example
$$A = \begin{bmatrix} 1 & 2 \\ 3 & 8 \end{bmatrix} \qquad B = \begin{bmatrix} 1 & 2 \\ 3 & 8 \end{bmatrix} - \begin{bmatrix} \lambda & 0 \\ 0 & \lambda \end{bmatrix} = \begin{bmatrix} 1-\lambda & 2 \\ 3 & 8-\lambda \end{bmatrix}.$$

Above is the characteristic matrix and $f(\lambda) = \det(B) = \det(A - \lambda I) = (1 - \lambda)(8 - \lambda) - 6 = 2 - 9\lambda + \lambda^2 = 0$ is the characteristic equation. The eigenvalues of A are the roots of $\lambda^2 - 9\lambda + 2 = 0$, which are $(9 \pm \sqrt{73})/2$.

A nonzero matrix X_i, which has one column and n rows, a column vector, satisfying the equation

$$(A - \lambda I)X_i = 0$$

and associated with the ith characteristic root λ_i is called an *eigenvector.*

Vector and Matrix Norms To carry out error analysis for approximate and iterative methods for the solutions of linear systems, one needs notions for vectors in R^n and for matrices that are analogous to the notion of length of a geometric vector. Let R^n denote the set of all vectors with n components, $x = (x_1, \ldots, x_n)$. In dealing with matrices it is convenient to treat vectors in R^n as columns, and so $x = (x_1, \ldots, x_n)^T$; however, here we shall write them simply as row vectors. A *norm* on R^n is a real-valued function f defined on R^n with the following properties:

1. $f(x) \geq 0$ for all $x \in R^n$.
2. $f(x) = 0$ if and only if $x = (0, 0, \ldots, 0)$.
3. $f(ax) = |a| f(x)$ for all real numbers a and $x \in R^n$.
4. $f(x + y) \leqq f(x) + f(y)$ for all $x, y \in R^n$.

The usual notation for a norm is $f(x) = \|x\|$.

The norm of a matrix is $\kappa(A) \equiv \|A\|\,\|A^{-1}\|$

where $\quad \|A\| \sup_{x \neq 0} = \dfrac{\|Ax\|}{\|x\|} = \max_k \sum_{j=1}^{n} |a_{jk}|$

The norm is useful when doing numerical calculations. If the computer's floating-point precision is 10^{-6}, then $\kappa = 10^6$ indicates an ill-conditioned matrix. If the floating-point precision is 10^{-12} (double precision), then a matrix with $\kappa = 10^{12}$ may be ill-conditioned. Two other measures are useful and are more easily calculated:

$$\text{Ratio} = \frac{\max_k |a_{kk}^{(k)}|}{\min_k |a_{kk}^{(k)}|} \qquad V = \frac{|\det A|}{\alpha_1 \alpha_2 \cdots \alpha_n} \qquad \alpha_i = (\alpha_{i1}^2 + \alpha_{i2}^2 + \cdots + \alpha_{in}^2)^{1/2}$$

where $a_{kk}^{(k)}$ are the diagonal elements of the LU decomposition.

MATRIX COMPUTATIONS

The principal topics in linear algebra involve systems of linear equations, matrices, vector spaces, linear transformations, eigenvalues and eigenvectors, and least-squares problems. The calculations are routinely done on a computer.

***LU* Factorization of a Matrix** Let L be an $n \times n$ lower triangular matrix with unit diagonal elements. Let U be an $n \times n$ upper triangular matrix. If all the principal submatrices of an $n \times n$ matrix A are nonsingular, then it is possible to represent $A = LU$. The Gauss elimination method is in essence an algorithm to determine L and U.

Solution of $Ax = b$ by Using *LU* Factorization Suppose that the indicated system is compatible and that $A = LU$. Let $z = Ux$. Then $Ax = LUx = b$ implies that $Lz = b$. Thus to solve $Ax = b$ we first solve $Lz = b$ for z and then solve $Ux = z$ for x. This procedure does not require that A be invertible and can be used to determine all solutions of a compatible system $Ax = b$. Note that the systems $Lz = b$ and $Ux = z$ are both in triangular form and thus can be easily solved.

The LU decomposition is essentially a gaussian elimination, arranged for maximum efficiency. The chief reason for doing an LU decomposition is that it takes fewer multiplications than would be needed to find an inverse. Also, once the LU decomposition has been found, it is possible to solve for multiple right-hand sides with little increase in work. The multiplication count for an $n \times n$ matrix and m right-hand sides is

$$\text{Operation count} = \frac{1}{3}n^3 - \frac{1}{3}n + mn^2$$

If an inverse is desired, it can be calculated by solving for the LU decomposition and then solving n problems with right-hand sides consisting of all zeros except one entry. Thus $4n^2/3 - n/3$ multiplications are required for the inverse. The determinant is given by

$$\text{Det}\,A = \prod_{i=1}^{n} a_{ii}^{(i)}$$

where $a_{ii}^{(i)}$ are the diagonal elements obtained in the LU decomposition.

A *tridiagonal matrix* is one in which the only nonzero entries lie on the main diagonal and on the diagonal just above and just below the main diagonal. The set of equations can be written as

$$a_i x_{i-1} + b_i x_i + c_i x_{i+1} = d_i$$

The LU decomposition is

$$
\begin{aligned}
& b_1 = b_1 \\
& \text{for } k = 2, n \text{ do} \\
& \quad a_k' = \frac{a_k}{b_{k-1}'}, \quad b_k' = b_k - \frac{a_k}{b_{k-1}'} c_{k-1} \\
& \text{enddo} \\
& \quad d_1' = d_1 \\
& \text{for } k = 2, n \text{ do} \\
& \quad d_k' = d_k - a_k' d_{k-1}' \\
& \text{enddo} \\
& \quad x_n = d_n'/b_n' \\
& \text{for } k = n-1, 1 \text{ do} \\
& \quad x_k = \frac{d_k' - c_k x_{k+1}}{d_k'} \\
& \text{enddo}
\end{aligned}
$$

The operation count for an $n \times n$ matrix with m right-hand sides is

$$2(n-1) + m(3n-2)$$

If $|b_i| > |a_i| + |c_i|$, no pivoting is necessary, and this is true for many boundary-value problems and partial differential equations.

Sparse matrices are ones in which the majority of the elements are zero. If the structure of the matrix is exploited, the solution time on a computer is greatly reduced. See Duff, I. S., A. M. Erisman, and J. K. Reid, *Direct Methods for Sparse Matrices*, Clarendon Press, Oxford, UK, 1986; Davis, T. A., *Direct Methods for Sparse Linear Systems*, Society for Industrial and Applied Mathematics, Philadelphia, Penn., 2006. The conjugate gradient method is one method for solving sparse matrix problems, since it only involves multiplication of a matrix times a vector. Thus the sparseness of the matrix is easy to exploit. The conjugate gradient method is an iterative method that converges for sure in n iterations where the matrix is an $n \times n$ matrix.

Matrix methods, in particular finding the rank of the matrix, can be used to find the number of independent reactions in a reaction set. If the stoichiometric numbers for the reactions and molecules are put in the form of a matrix, the rank of the matrix gives the number of independent reactions. See Amundson, N. R., *Mathematical Methods in Chemical Engineering*, Prentice-Hall, Englewood Cliffs, N.J., 1966, p. 50. See also Dimensional Analysis.

QR Factorization of a Matrix If A is an $m \times n$ matrix with $m \geq n$, there exists an $m \times m$ unitary matrix $Q = [q_1, q_2, ..., q_m]$ and an $m \times n$ right triangular matrix R such that $A = QR$. The QR factorization is frequently used in the actual computations when the other transformations are unstable.

Singular-Value Decomposition If A is an $m \times n$ matrix with $m \geq n$ and rank $k \leq n$, consider the two following matrices.

$$AA^* \qquad \text{and} \qquad A^*A$$

An $m \times m$ unitary matrix U is formed from the eigenvectors $\mathbf{u}_i$ of the first matrix.

$$U = [\mathbf{u}_1, \mathbf{u}_2, \ldots, \mathbf{u}_m]$$

An $n \times n$ unitary matrix V is formed from the eigenvectors $\mathbf{v}_i$ of the second matrix.

$$V = [\mathbf{v}_1, \mathbf{v}_2, \ldots, \mathbf{v}_n]$$

Then matrix A can be decomposed into

$$A = U \Sigma V^*$$

where Σ is a $k \times k$ diagonal matrix with diagonal elements $d_{ii} = \sigma_i > 0$ for $1 \leq i \leq k$. The eigenvalues of $\Sigma^*\Sigma$ are σ_i^2. The vectors $\mathbf{u}_i$ for $k+1 \leq i \leq m$ and $\mathbf{v}_i$ for $k+1 \leq i \leq n$ are eigenvectors associated with the eigenvalue zero; the eigenvalues for $1 \leq i \leq k$ are σ_i^2. The values of σ_i are called the *singular values* of matrix A. If A is real, then U and V are real and hence orthogonal matrices. The value of the singular-value decomposition comes when a process is represented by a linear transformation and the elements of A and a_{ij} are the contribution to an output i for a particular variable as input variable j. The input may be the size of a disturbance, and the output is the gain (Seborg, D. E., T. F. Edgar, and D. A. Mellichamp, *Process Dynamics and Control*, 2d ed., Wiley, New York, 2004). If the rank is less than n, not all the variables are independent and they cannot all be controlled. Furthermore, if the singular values are widely separated, the process is sensitive to small changes in the elements of the matrix, and the process will be difficult to control.

Example Consider the following example from Noble and Daniel (*Applied Linear Algebra*, Prentice-Hall, Upper Saddle River, N.J., 1987) with the MATLAB commands to do the analysis. Define the following real matrix with $m = 3$ and $n = 2$ (whose rank $k = 1$).

```
>>a=[1 1
2 2
2 2]
```

The following MATLAB commands are used.

```
a1=a*a
a2=a*a'
[v,d1]=eig(a1)
[u,d2]=eig(a2)
```

The results are

```
v=[-0.7071 0.7071
    0.7071 0.7071]
d1=[0 0
    0 18]

u=[ 0.8944  0.2981  0.3333
   -0.4472  0.5963  0.6667
    0      -0.7454  0.6667]

d2=0 0 0
   0 0 0
   0 0 18
```

Thus, $\sigma_1^2 = 18$ and the eigenfunctions are the rows of $\mathbf{v}$ and $\mathbf{u}$. The second column of $\mathbf{v}$ is associated with the eigenvalue $\sigma_1^2 = 18$, and the third column of $\mathbf{u}$ is associated with the eigenvalue $\sigma_1^2 = 18$.

If A is square and nonsingular, the vector $\mathbf{x}$ that minimizes

$$||A\mathbf{x} - \mathbf{b}|| \tag{3-62}$$

is obtained by solving the linear equation

$$\mathbf{x} = A^{-1}\mathbf{b}$$

When A is not square, the solution to

$$A\mathbf{x} = \mathbf{b}$$

is

$$\mathbf{x} = V\mathbf{y}$$

where $y_i = b_i'/\sigma_i$ for $i = 1, \ldots, k$, $\mathbf{b}' = U^T\mathbf{b}$, and $y_{k+1}, y_{k+2}, \ldots, y_m$ are arbitrary. The matrices U and V are those obtained in the singular-value decomposition. The solution which minimizes the norm, Eq. (3-62), is $\mathbf{x}$ with $y_{k+1}, y_{k+2}, \ldots, y_m$ zero. These techniques can be used to monitor process variables. See Montgomery, D. C., *Introduction to Statistical Quality Control*, 6th ed., Wiley, New York, 2008; Piovos, M. J., and K. A. Hoo, "Multivariate Statistics for Process Control," *IEEE Control Systems* **22**(5):8 (2002).

Principal Component Analysis (PCA) PCA is used to recognize patterns in data and reduce the dimensionality of the problem. Let the matrix A now represent data with the columns of A representing different samples and the rows representing different variables. The *covariance matrix* is defined as

$$\operatorname{cov}(A) = \frac{A^T A}{m-1}$$

This is just the same matrix discussed with singular-value decomposition. For data analysis, however, it is necessary to adjust the columns to have zero mean by subtracting from each entry in the column the average of the column entries. Once this is done, the *loadings* are the $\mathbf{v}_i$ and satisfy

$$\operatorname{cov}(A)\,\mathbf{v}_i = \sigma_i^2 v_i$$

and the *score* vector $\mathbf{u}_i$ is given by

$$A\mathbf{v}_i = \sigma_i \mathbf{u}_i$$

In process analysis, the columns of A represent different measurement techniques (temperatures, pressures, etc.), and the rows represent the measurement output at different times. In that case the columns of A are adjusted to have a zero mean and a variance of 1.0 (by dividing each entry in the column by the variance of the column). The goal is to represent the essential variation of the process with as few variables as possible. The $\mathbf{u}_i$, $\mathbf{v}_i$ pairs are arranged in descending order according to the associated σ_i. The σ_i can be thought of as the variance, and the $\mathbf{u}_i$, $\mathbf{v}_i$ pair captures the greatest amount of variation in the data. Instead of having to deal with n variables, one can capture most of the variation of the data by using only the first few pairs. An excellent example of this is given by Wise, B. M., and B. R. Kowalski, "Process Chemometrics," Chap. 8 in *Process Analytical Chemistry*, eds. F. McLennan and B. Kowalski, Blackie Academic & Professional, London, 1995. When modeling a slurry-fed ceramic melter, they were able to capture 97 percent of the variation by using only four eigenvalues and eigenvectors, even though there were 16 variables (columns) measured.

NUMERICAL APPROXIMATIONS TO SOME EXPRESSIONS

APPROXIMATION IDENTITIES

For the following relationships the sign $\cong$ means approximately equal to, when X is small. These equations are derived by using a Taylor's series (see Series Summation and Identities).

Approximation	Approximation
$\frac{1}{1 \pm X} \cong 1 \mp X$	$\sqrt{1 \pm X} \cong 1 \pm \frac{X}{2}$

Approximation	Approximation
$(1 \pm X)^n \cong 1 \pm nX$	$(1 \pm X)^{-n} \cong 1 \mp nX$
$(a \pm X)^2 = a^2 \pm 2aX$	$e^x \cong 1 + X$
$\sin X \cong X$ (X rad)	$\tan X \cong X$
$\sqrt{Y(Y+X)} \cong \frac{2Y+X}{2}$	$\sqrt{Y^2+X^2} \cong Y + \frac{X^2}{2Y}\left(\frac{X}{Y} \text{ small}\right)$
Stirling's approximation	$\ln N! \cong N \ln N - N$

NUMERICAL ANALYSIS AND APPROXIMATE METHODS

References: Ascher, U. M., and C. Greif, *A First Course in Numerical Methods*, SIAM-Soc. Ind. Appl. Math., 2011; Atkinson, K., W. Han, and D. E. Stewart, *Numerical Solution of Ordinary Differential Equations*, Wiley, New York, 2009; Burden, R. L., J. D. Faires, A. C. Reynolds, and A. M. Burden, *Numerical Analysis*, 10th ed., Brookes/Cole, Pacific Grove, Calif., 2015; Chapra, S. C., and R. P. Canal, *Numerical Methods for Engineers*, 5th ed., McGraw-Hill, New York, 2006; Heys, Jeffrey, J., *Chemical and Biomedical Engineering Calculations Using Python*, Wiley, New York (2017); Johnson, C., *Numerical Solution of Partial Differential Equations by the Finite Element Method*, Dover, New York, 2009; Lau, H. T., *A Numerical Library in C for Scientists and Engineers*, CRC Press, Boca Raton, Fla., 3rd ed. 2007; LeVeque, R. J., *Finite Volume Methods for Hyperbolic Problems*, Cambridge University Press, Cambridge 2002; Morton, K. W., and D. F. Mayers, *Numerical Solution of Partial Differential Equations: An Introduction*, 2d ed., Cambridge University Press, Cambridge, 2005; Quarteroni, A., and A. Valli, *Numerical Approximation of Partial Differential Equations*, 2d ed., Springer, New York, 2008; Reddy, J. N., and D. K. Gartling, *The Finite Element Method in Heat Transfer and Fluid Dynamics*, 3d ed., CRC Press, Boca Raton, Fla., 2010; Zienkiewicz, O. C., R. L. Taylor, and J. Z. Zhu, *The Finite Element Method: Its Basis and Fundamentals*, 7th ed., Butterworth-Heinemann Elsevier, Oxford, UK, 2013.

INTRODUCTION

The goal of approximate and numerical methods is to provide convenient techniques for obtaining useful information from mathematical formulations of physical problems. Often this mathematical statement is not solvable by analytical means. Or perhaps analytic solutions are available but in a form that is inconvenient for direct interpretation. In the first case, it is necessary either to attempt to approximate the problem satisfactorily by one that will be amenable to analysis, to obtain an approximate solution to the original problem by numerical means, or to use the two techniques in combination. Numerical methods have been used to model polymerization, yeast fermentation, chemical vapor deposition, catalytic converters, pressure swing adsorption, insulin purification, ion exchange, and affinity chromatography, plus many other chemical engineering applications.

Numerical techniques therefore do not yield exact results in the sense of the mathematician. Since most numerical calculations are inexact, the concept of error is an important feature.

The four sources of error are as follows:

1. *Gross errors.* These result from unpredictable human, mechanical, or electrical mistakes.

2. *Rounding errors.* These are the consequence of using a number specified by m correct digits to approximate a number which requires more than m digits for its exact specification. For example, approximate the irrational number $\sqrt{2}$ by 1.414. Such errors are often present in experimental data, in which case they may be called *inherent errors*, due either to empiricism or to the fact that the computer dictates the number of digits. Such errors may be especially damaging in areas such as matrix inversion or the numerical solution of partial differential equations when the number of algebraic operations is extremely large.

3. *Truncation errors.* These errors arise from the substitution of a finite number of steps for an infinite sequence of steps which would yield the exact result. To illustrate this error, consider the infinite series for e^{-x}: $e^{-x} = 1 - x + x^2/2 - x^3/6 + E_T(x)$, where E_T is the truncation error, $E_T = (1/24)e^{-\varepsilon}x^4$, for $0 < \varepsilon < x$. If x is positive, ε is also positive. Hence $e^{-\varepsilon} < 1$. The approximation $e^{-x} \approx 1 - x + x^2/2 - x^3/6$ is in error by a positive amount smaller than $(1/24)x^4$.

A variety of general-purpose computer programs are available commercially. Mathematica (http://www.wolfram.com/), Maple (http://www.maplesoft.com/), and Mathcad (https://www.ptc.com/en/engineering-math-software/mathcad) and MATLAB (http://www.mathworks.com/

product/symbolic) all have the capability of doing symbolic manipulation so that algebraic solutions can be obtained. Different packages can solve some ordinary and partial differential equations analytically, solve nonlinear algebraic equations, make simple graphs and do linear algebra, and combine the symbolic manipulation with numerical techniques. In this section, examples are given for the use of MATLAB (http://www.mathworks.com/), a package of numerical analysis tools, some of which are accessed by simple commands and others of which are accessed by writing programs in C. Spreadsheets can also be used to solve certain problems, and these are described below too. A popular program used in chemical engineering education is Polymath (http://www.polymath-software.com/), which can numerically solve sets of linear or nonlinear equations, ordinary differential equations as initial-value problems, and perform data analysis and regression.

NUMERICAL SOLUTION OF LINEAR EQUATIONS

See the section Matrix Algebra and Matrix Computation.

NUMERICAL SOLUTION OF NONLINEAR EQUATIONS IN ONE VARIABLE

Methods for Nonlinear Equations in One Variable

Successive Substitutions Let $f(x) = 0$ be the nonlinear equation to be solved. If this is rewritten as $x = F(x)$, then an iterative scheme can be set up in the form $x_{k+1} = F(x_k)$. To start the iteration, an initial guess must be obtained graphically or otherwise. The convergence or divergence of the procedure depends upon the method of writing $x = F(x)$, of which there will usually be several forms. However, if a is a root of $f(x) = 0$, and if $|F'(a)| < 1$, then for any initial approximation sufficiently close to a, the method converges to a. This process is called *first-order* because the error in x_{k+1} is proportional to the first power of the error in x_k for large k.

One way of writing the equation is $x_{k+1} = x_k + \beta f(x_k)$. The choice of β is made such that $|1 + \beta\, df/dx(a)| < 1$. Convergence is guaranteed by the theorem given for simultaneous equations.

Methods of Perturbation Let $f(x) = 0$ be the equation. In general, the iterative relation is

$$x_{k+1} = x_k - [f(x_k)/\alpha_k]$$

where the iteration begins with x_0 as an initial approximation and α_k as some functional, derived below.

Newton-Raphson Procedure This variant chooses $\alpha_k = f'(x_k)$ where $f' = df/dx$ and geometrically consists of replacing the graph of $f(x)$ by the tangent line at $x = x_k$ in each successive step. If $f'(x)$ and $f'' \leq (x)$ have the same sign throughout an interval $a \leq x \leq b$ containing the solution, with $f(a)$ and $f(b)$ of opposite signs, then the process converges starting from any x_0 in the interval $a \leq x \leq b$. The process is second-order.

Method of False Position This variant is commenced by finding x_0 and x_1 such that $f(x_0)$ and $f(x_1)$ are of opposite sign. Then α_1 = slope of secant line joining $[x_0, f(x_0)]$ and $[x_1, f(x_1)]$ so that

$$x_2 = x_1 - \frac{x_1 - x_0}{f(x_1) - f(x_0)} f(x_1)$$

In each of the following steps α_k is the slope of the line joining $[x_k, f(x_k)]$ to the most recently determined point where $f(x_j)$ has the opposite sign from that of $f(x_k)$. This method is first-order. If one uses the most recently determined point (regardless of sign), the method is a *secant method.*

Method of Wegstein This is a variant of the method of successive substitutions which forces and/or accelerates convergence. The iterative procedure $x_{k+1} = F(x_k)$ is revised by setting $\hat{x}_{k+1} = F(x_k)$ and then taking $x_{k+1} = qx_k + (1-q)\hat{x}_{k+1}$, where q is a suitably chosen number which may be taken as constant throughout or may be adjusted at each step. Wegstein found that suitable q's are as follows:

Behavior of successive substitution process		Range of optimum q
Oscillatory convergence		$0 < q < ½$
Oscillatory divergence	without Wegstein	$½ < q < 1$
Monotonic convergence		$q < 0$
Monotonic divergence	without Wegstein	$1 < q$

At each step q may be calculated to give a locally optimum value by setting

$$q = \frac{\hat{x}_{k+1} - \hat{x}_k}{\hat{x}_{k+1} - 2\hat{x}_k + \hat{x}_{k-1}}$$

The Wegstein method is a secant method applied to $g(x) \equiv x - F(x)$. In Microsoft Excel, roots are found by using Goal Seek or Solver (an Add-In). Assign one cell to be x, put the equation for $f(x)$ in another cell, and let Goal Seek or Solver find the value of x that makes the equation cell zero. In MATLAB, the process is similar except that a function (m–file) is defined and the command fzero ('f', x0) provides the solution x, starting from the initial guess $x0$. The Wegstein method is sometimes used to promote convergence when solving a mass and energy balance problem for a chemical process with recycle streams.

METHODS FOR MULTIPLE NONLINEAR EQUATIONS

Method of Successive Substitutions Write a system of equations as

$$\alpha_i = f_i(\alpha) \qquad \text{or} \qquad \alpha = \mathbf{f}(\alpha)$$

The following theorem guarantees convergence. Let α be the solution to $\alpha_i = f_i(\alpha)$. Assume that given $h > 0$, there exists a number $0 < \mu < 1$ such that

$$\sum_{j=1}^{n} \left| \frac{\partial f_i}{\partial x_j} \right| \leq \mu \qquad \text{for} \qquad |x_i - \alpha_i| < h, \; i = 1, \ldots, n$$

$$x_i^{k+1} = f_i(x_i^k)$$

Then $\qquad x_i^k \to \alpha_i$

as k increases [see Finlayson (1980)].

Newton-Raphson Method To solve the set of equations

$$F_i(x_1, x_2, \ldots, x_n) = 0 \qquad \text{or} \qquad F_i(\{x_j\}) = 0 \qquad \text{or} \qquad F_i(\mathbf{x}) = 0$$

one uses a truncated Taylor series to get

$$0 = F_i(\{x^k\}) + \sum_{j=1}^{n} \left. \frac{\partial F_i}{\partial x_j} \right|_{x^k} (x_j^{k+1} + x_j^k)$$

Thus one solves iteratively from one point to another.

$$\sum_{j=1}^{n} A_{ij}^k (x_j^{k+1} - x_j^k) = -F_i(\{x^k\})$$

where $\qquad A_{ij}^k = \left. \frac{\partial F_i}{\partial x_j} \right|_{x^k}$

This method requires solution of sets of linear equations until either the functions are zero to some tolerance or the changes of the solution between iterations are small enough. Convergence is guaranteed provided the norm of matrix **A** is bounded, **F**(**x**) is bounded for the initial guess, and the second derivative of **F**(**x**) with respect to all variables is bounded. See Finlayson (1980) in General References. Homotopy methods are also possible; see Finlayson et al. (2006) in General References.

INTERPOLATION

When a function is known at several points, it is sometimes useful to have a means to interpolate and assign a value between those points. The interpolation can be a global approximation, i.e., a function defined using all the points, or piecewise approximation, i.e., a collection of functions, each defined over several different subsets of the points.

Lagrange Interpolation Formulas A global polynomial is defined over the entire region of space

$$P_m(x) = \sum_{j=0}^{m} c_j x^j$$

This polynomial is of degree m (highest power is x^m) and order $m + 1$ ($m + 1$ parameters $\{c_j\}$). If we are given a set of $m + 1$ points

$$y_1 = f(x_1), y_2 = f(x_2), \ldots, y_{m+1} = f(x_{m+1})$$

then Lagrange's formula gives a polynomial of degree m that goes through the $m+1$ points:

$$P_m(x)=\frac{(x-x_2)(x-x_3)\cdots(x-x_{m+1})}{(x_1-x_2)(x_1-x_3)\cdots(x_1-x_{m+1})}y_1+\frac{(x-x_1)(x-x_3)\cdots(x-x_{m+1})}{(x_2-x_1)(x_2-x_3)\cdots(x_2-x_{m+1})}y_2+\cdots+\frac{(x-x_1)(x-x_2)\cdots(x-x_{m+1})}{(x_{m+1}-x_1)(x_{m+1}-x_2)\cdots(x_{m+1}-x_m)}y_{m+1}$$

Note that each coefficient of y_j is a polynomial of degree m that vanishes at the points $\{x_j\}$ (except for one value of j) and takes the value of 1.0 at that point:

$$P_m(x_j)=y_j \qquad j=1,2,\ldots,m+1$$

If the function $f(x)$ is known, the error in the approximation is [www.netliborg/lapack]

$$|\text{error}(x)|\le\frac{|x_{m+1}-x_1|^{m+1}}{(n+2)!}\max_{x_1\le x\le x_{m+1}}|f^{(n+2)}(x)|$$

The evaluation of $P_m(x)$ at a point other than at the defining points can be made with Neville's algorithm [Press et al. (2007) in General References].

Orthogonal Polynomials Another form of polynomials is obtained by defining them so that they are orthogonal. It is required that $P_m(x)$ be orthogonal to $P_k(x)$ for $k=0,\ldots,m-1$.

$$\int_a^b W(x)P_k(x)P_m(x)dx=0 \qquad k=0,1,\ldots,m-1$$

The orthogonality includes a nonnegative weight function $W(x)\ge 0$ for all $a\le x\le b$. This procedure specifies the set of polynomials to within multiplicative constants, which are set by requiring the leading coefficient to be 1.0 or by requiring the norm to be 1.0.

$$\int_a^b W(x)P_m^2(x)dx=1$$

The polynomial $P_m(x)$ has m roots in the closed interval a to b.

The polynomial

$$p(x)=c_0P_0(x)+c_1P_1(x)+\cdots c_mP(x)$$

minimizes

$$I=\int_a^b W(x)[f(x)-p(x)]^2\,dx$$

for a function $f(x)$ when

$$c_j=\int_a^b W(x)f(x)P_j(x)dx\,/\,W_j \qquad W_j=\int_a^b W(x)P_j^2(x)dx$$

Note that each c_j is independent of m, the number of terms retained in the series. The minimum value of I is

$$I_{\min}=\int_a^b W(x)f^2(x)dx-\sum_{j=0}^n W_jc_j^2$$

Such functions are useful for continuous data, i.e., when $f(x)$ is known for all x.

The types of orthogonal polynomials include Chebyshev ($a=-1$, $b=1$, $W(x)=1$, used in spectral methods), Legendre ($a=-1$, $b=1$, $W(x)=1/\sqrt{1-x^2}$), shifted Legendre ($a=0$, $b=1$, $W(x)=1$), used in the orthogonal collocation method), Jacobi, Hermite ($a=-\infty$, $b=\infty$, $W(x)=e^{-x^2}$), and Laguerre polynomials.

Linear Interpolation The simplest piecewise continuous interpolation is a straight line between the points. If a function $f(x)$ is approximately linear in a certain range, then the ratio

$$\frac{f(x_1)-f(x_0)}{x_1-x_0}=f[x_0,x_1]$$

is approximately independent of x_0 and x_1 in the range. The linear approximation to the function $f(x)$, $x_0<x<x_1$ then leads to the interpolation formula

$$f(x)\approx f(x_0)+(x-x_0)f[x_0-x_1]$$
$$\approx f(x_0)+\frac{x-x_0}{x_1-x_0}[f(x_1)-f(x_0)]$$
$$\approx\frac{1}{x_1-x_0}[(x_1-x)f(x_0)-(x_0-x)f(x_1)]$$

Higher-order interpolation is also possible.

Equally Spaced Forward Differences If the ordinates are *equally spaced*, that is, $x_j-x_{j-1}=\Delta x$ for all j, then the first differences are denoted by $\Delta f(x_0)=f(x_1)-f(x_0)$ or $\Delta y_0=y_1-y_0$, where $y=f(x)$. The differences of these first differences, called *second differences*, are denoted by $\Delta^2y_0, \Delta^2y_1,\ldots,\Delta^2y_n$. Thus

$$\Delta^2y_0=\Delta y_1-\Delta y_0=y_2-y_1-y_1+y_0=y_2-2y_1+y_0$$

and in general

$$\Delta^j y_0=\sum_{n=0}^{j}(-1)^n\binom{j}{n}y_{j-n}$$

where $\binom{j}{n}=\frac{j!}{n!(j-n)!}$ = binomial coefficients

If the ordinates are equally spaced,

$$x_{n+1}-x_n=\Delta x$$
$$y_n=y(x_n)$$

then the first and second differences are denoted by

$$\Delta y_n=y_{n+1}-y_n$$
$$\Delta^2y_n=\Delta y_{n+1}-\Delta y_n=y_{n+2}-2y_{n+1}+y_n$$

A new variable is defined

$$\alpha=\frac{x_a-x_0}{\Delta x}$$

and the finite interpolation formula through the points $y_0, y_1,\ldots,y_n$ is written as follows:

$$y_\alpha=y_0+\alpha\,\Delta\,y_0+\frac{\alpha(\alpha-1)}{2!}\Delta^2y_0+\cdots+\frac{\alpha(\alpha+1)\cdots(\alpha-n+1)}{n!}\Delta^ny_0 \tag{3-63}$$

Keeping only the first two terms gives a straight line through (x_0, y_0) and (x_1, y_1); keeping the first three terms gives a quadratic function of position going through those points plus (x_2, y_2). The value $\alpha=0$ gives $x=x_0$; $\alpha=1$ gives $x=x_1$; and so on.

Equally Spaced Backward Differences *Backward differences* are defined by

$$\nabla y_n=y_n-y_{n-1}$$
$$\nabla^2y_n=\nabla y_n-\nabla y_{n-1}=y_n-2y_{n-1}+y_{n-2}$$

The interpolation polynomial of order n through the points $y_0, y_{-1},\ldots,y_{-n}$ is

$$y_\alpha=y_0+\alpha\,\nabla\,y_0+\frac{\alpha(\alpha+1)}{2!}\nabla^2y_0+\cdots+\frac{\alpha(\alpha+1)\cdots(\alpha+n-1)}{n!}\nabla^ny_0$$

The value of $\alpha=0$ gives $x=x_0$; $\alpha=-1$ gives $x=x_{-1}$, and so on. Alternatively, the interpolation polynomial of order n through the points $y_1, y_0, y_{-1},\ldots,y_{-n}$ is

$$y_\alpha=y_1+(\alpha-1)\nabla y_1+\frac{\alpha(\alpha-1)}{2!}\nabla^2y_1+\cdots+\frac{(\alpha-1)\alpha(\alpha+1)\cdots(\alpha+n-2)}{n!}\nabla^ny_1 \tag{3-64}$$

Now $\alpha=1$ gives $x=x_1$; $\alpha=0$ gives $x=x_0$.

Central Differences The central difference denoted by

$$\delta f(x)=f\left(x+\frac{h}{2}\right)-f\left(x-\frac{h}{2}\right)$$
$$\delta^2f(x)=\delta f\left(x+\frac{h}{2}\right)-\delta f\left(x-\frac{h}{2}\right)=f(x+h)-2f(x)+f(x-h)$$
$$\delta^nf(x)=\delta^{n-1}f\left(x+\frac{h}{2}\right)-\delta^{n-1}f\left(x-\frac{h}{2}\right)$$

is useful for calculating at the interior points of tabulated data.

Finite Element Method In the finite element method (see Ordinary Differential Equations—Boundary Value Problems) the independent variable x is divided into regions called *elements.* The simplest approximation is to use linear interpolation on each element, as described above. More useful is to use a quadratic interpolation between the two endpoints of the element and its midpoint. The points of an element are shown in Fig. 3-46.

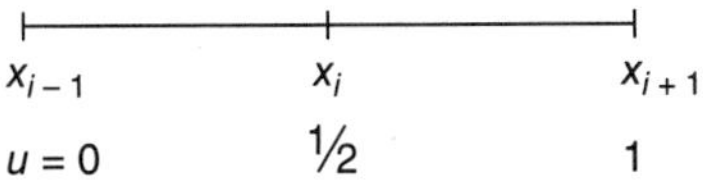

FIG. 3-46 Quadratic finite element.

The element extends from x_{i-1} to x_{i+1}. Define a new variable which takes the values $u = 0, 0.5$, and 1 at the three points, respectively. The interpolation is then

$$y = 2(u-1)\left(u-\frac{1}{2}\right)y_{i-1} + 4u(1-u)y_i + 2u\left(u-\frac{1}{2}\right)y_{i+1}$$

The interpolation clearly takes the correct values at $u = 0$, 0.5, and 1. Over the whole domain in x the interpolated function is continuous, but the first derivative is only piecewise continuous. Other types of finite elements include cubic functions, which are also continuous but the derivatives are only piecewise continuous. When Hermite cubic functions are used, however, the function and its first derivative are continuous throughout the domain in x.

Spline Functions Splines are functions that match given values at the points $x_1, \ldots, x_{NT}$ and have continuous derivatives up to some order at the knots, or the points $x_2, \ldots, x_{NT-1}$. Cubic splines are most common. The function is represented by a cubic polynomial within each interval (x_i, x_{i+1}) and has continuous first and second derivatives at the knots. Two more conditions can be specified arbitrarily. These are usually the second derivatives at the two endpoints, which are commonly taken as zero; this gives the natural cubic splines. Spline functions are useful because the interpolation error can be made small even with low-order polynomials. Some of the other methods may oscillate wildly between the quadrature points. See Schumaker, L. L., *Spline Functions: Computational Methods*, Soc. Ind. Appl. Math. (SIAM), 2015.

NUMERICAL DIFFERENTIATION

Numerical differentiation should be avoided whenever possible, particularly when data are empirical and subject to appreciable observation errors. Errors in data can affect numerical derivatives quite strongly; i.e., differentiation is a roughening process. When such a calculation must be made, it is usually desirable first to smooth the data to a certain extent.

Use of Interpolation Formula If the data are given over equidistant values of the independent variable x, an interpolation formula such as the Newton formula [Eq. (3-63) or (3-64)] may be used and the resulting formula differentiated analytically. If the independent variable is not at equidistant values, then Lagrange's formulas must be used. By differentiating three-point Lagrange interpolation formulas the following differentiation formulas result for equally spaced tabular points:

Three-Point Formulas Let x_0, x_1, and x_2 be the three points.

$$f'(x_0) = \frac{1}{2h}[-3f(x_0) + 4f(x_1) - f(x_2)] + \frac{h^2}{3}f'''(\varepsilon)$$

$$f'(x_1) = \frac{1}{2h}[-f(x_0) + f(x_2)] - \frac{h^2}{6}f'''(\varepsilon)$$

$$f'(x_2) = \frac{1}{2h}[f(x_0) - 4f(x_1) + 3f(x_2)] + \frac{h^2}{3}f'''(\varepsilon)$$

where the last term is an error term $\min_j x_j < \varepsilon < \max_j x_j$.

Smoothing Techniques These techniques involve the approximation of the tabular data by a least-squares fit of the data by using some known functional form, usually a polynomial (for the concept of least squares see Statistics). In place of approximating $f(x)$ by a single least-squares polynomial of degree n over the entire range of the tabulation, it is often desirable to replace each tabulated value by the value taken on by a least-squares polynomial of degree n relevant to a subrange of $2M + 1$ points centered, when possible, at the point for which the entry is to be modified. Thus each smoothed value replaces a tabulated value. Let $f_j = f(x_j)$ be the tabular points and y_j = smoothed values.

First-Degree Least Squares with Three Points

$$y_0 = \tfrac{1}{6}[5f_0 + 2f_1 - f_2]$$
$$y_1 = \tfrac{1}{3}[f_0 + f_1 + f_2]$$
$$y_2 = \tfrac{1}{6}[-f_0 + 2f_1 + 5f_2]$$

The derivatives at all the points are

$$f_0' = f_1' = f_2' = \frac{1}{2}h[y_2 - y_1]$$

Second-Degree Least Squares with Five Points For five evenly spaced points x_{-2}, x_{-1}, x_0, x_1, and x_2 (separated by distance h) and their ordinates f_{-2}, f_{-1}, f_0, f_1, and f_2, assume a parabola is fit by least squares. Then the derivative at the center point is

$$f_0' = 1/10h\,[-2f_{-2} - f_{-1} + f_1 + 2f_2]$$

The derivatives at the other points are

$$f_{-2}' = 1/70h[-54f_{-2} + 13f_{-1} + 40f_0 + 27f_1 - 26f_2]$$
$$f_{-1}' = 1/70h[-34f_{-2} + 3f_{-1} + 20f_0 + 17f_1 - 6f_2]$$
$$f_1' = 1/70h[6f_{-2} - 17f_{-1} + 20f_0 - 3f_1 + 34f_2]$$
$$f_2' = 1/70h[26f_{-2} - 27f_{-1} - 40f_0 - 13f_1 + 54f_2]$$

Numerical Derivatives The results given above can be used to obtain numerical derivatives when solving problems on the computer, in particular for the Newton-Raphson method and homotopy methods. Suppose one has a program, subroutine, or other function evaluation device that will calculate f, given x. One can estimate the value of the first derivative at x_0 using

$$\left.\frac{df}{dx}\right|_{x_0} \approx \frac{f[x_0(1+\varepsilon)] - f[x_0]}{\varepsilon x_0}$$

(a first-order formula) or

$$\left.\frac{df}{dx}\right|_{x_0} \approx \frac{f[x_0(1+\varepsilon)] - f[x_0(1-\varepsilon)]}{2\varepsilon x_0}$$

(a second-order formula). The value of ε is important; a value of 10^{-6} is typical, but smaller or larger values may be necessary depending on the computer precision and the application. One must also be sure that the value of x_0 is not zero and use a different increment in that case.

NUMERICAL INTEGRATION (QUADRATURE)

A multitude of formulas have been developed to accomplish numerical integration, which consists of computing the value of a definite integral from a set of numerical values of the integrand.

Newton-Cotes Integration Formulas (Equally Spaced Ordinates) for Functions of One Variable The definite integral $\int_a^b f(x)\,dx$ is to be evaluated.

Trapezoidal Rule This formula consists of subdividing the interval $a \le x \le b$ into n subintervals a to $a + h$, $a + h$ to $a + 2h$, … and replacing the graph of $f(x)$ by the result of joining the ends of adjacent ordinates by line segments. If $f_j = f(x_j) = f(a + jh)$, $f_0 = f(a)$, and $f_n = f(b)$, the integration formula is

$$\int_a^b f(x)\,dx = \frac{h}{2}[f_0 + 2f_1 + 2f_2 + \cdots + 2f_{n-1} + f_n] + E_n$$

where

$$|E_n| = \frac{nh^3}{12}|f''(\varepsilon)| = \frac{(b-a)^3}{12n^2}|f''(\varepsilon)| \qquad a < \varepsilon < b$$

This procedure is not of high accuracy. However, if $f'' \le (x)$ is continuous in $a < x < b$, the error goes to zero as $1/n^2$, $n \to \infty$. When the finite element method is used with linear trial functions and equal-size elements, quadrature is the same as the trapezoid rule.

Parabolic Rule (Simpson's Rule) This procedure consists of subdividing the interval $a < x < b$ into $n/2$ subintervals, each of length $2h$, where n is an *even* integer. By using the notation as above the integration formula is

$$\int_a^b f(x)\,dx = \frac{h}{3}[f_0 + 4f_1 + 2f_2 + 4f_3 + 2f_4 + \cdots + 4f_{n-3} + 2f_{n-2} + 4f_{n-1} + f_n] + E_n$$

where

$$|E_n| = \frac{nh^5}{180}|f^{(\mathrm{IV})}(\varepsilon)| = \frac{(b-a)^5}{180n^4}|f^{(\mathrm{IV})}(\varepsilon)| \qquad a < \varepsilon < b$$

This method approximates $f(x)$ by a parabola on each subinterval. This rule is generally more accurate than the trapezoidal rule. It is the most widely used integration formula. When the finite element method is used with quadratic trial functions and equal-size elements, quadrature is the same as Simpson's rule.

Gaussian Quadrature Gaussian quadrature provides a highly accurate formula based on irregularly spaced points, but the integral needs to be transformed onto the interval from 0 to 1.

$$x = a + (b-a)u \qquad dx = (b-a)du$$

$$\int_a^b f(x)dx = (b-a)\int_0^1 f(u)du$$

$$\int_a^b f(u)du = \sum_{i=1}^{m} W_i f(u_i)$$

The quadrature is exact when f is a polynomial of degree $2m-1$ in x. Because there are m weights and m Gauss points, we have $2m$ parameters that are chosen to exactly represent a polynomial of degree $2m-1$, which has $2m$ parameters. The Gauss points and weights are given in the table.

Gaussian Quadrature Points and Weights

m	u_i	W_i
2	0.21132 48654	0.50000 00000
	0.78867 51346	0.50000 00000
3	0.11270 16654	0.27777 77778
	0.50000 00000	0.44444 44445
	0.88729 83346	0.27777 77778
4	0.06943 18442	0.17392 74226
	0.33000 94783	0.32607 25774
	0.66999 05218	0.32607 25774
	0.93056 81558	0.17392 74226
5	0.04691 00771	0.11846 34425
	0.23076 53450	0.23931 43353
	0.50000 00000	0.28444 44444
	0.76923 46551	0.23931 43353
	0.95308 99230	0.11846 34425

Example Calculate the value of the following integral.

$$I = \int_0^1 e^{-x} \sin x \, dx$$

Using the gaussian quadrature formulas gives the following values for various values of m. Clearly, three internal points, requiring evaluation of the integrand at only three points, give excellent results.

m	I
2	0.24609 64306
3	0.24583 48774
4	0.24583 70044
5	0.24583 70070

Romberg's Method Romberg's method uses extrapolation techniques to improve the answer [Press et al. (2007)]. If we let I_1 be the value of the integral obtained using interval size $h = \Delta x$, I_2 be the value of I obtained when using interval size $h/2$, I_3 be the value obtained when using an interval of size $h/4$, etc., and I_0 is the true value of I, then the error in a method is approximately h^m, or

$$I \approx I_0 + ch^m$$

$$I_2 \approx I_0 + c\left(\frac{h}{2}\right)^m$$

Replacing the ≈ by an equality (an approximation) and solving for c and I_0 give

$$I_0 = \frac{2^m I_2 - I_1}{2^m - 1}$$

To obtain the most accurate value, first calculate $I_1, I_2, \ldots$, by halving h each time. Then calculate new estimates from each pair, calling them $J_1, J_2, \ldots$; that is, in the formula above, replace I_0 with J_1. The formulas are reapplied for each pair of J to obtain $K_1, K_2, \ldots$. The process continues until the required tolerance is obtained.

$$\begin{array}{cccc} I_1 & I_2 & I_3 & I_4 \\ & J_1 & J_2 & J_3 \\ & & K_1 & K_2 \\ & & & L_1 \end{array}$$

Romberg's method is most useful for a low-order method (small m) because significant improvement is then possible.

Example Evaluate the same integral by using the trapezoid rule and then apply the Romberg method. Use 11, 21, 41, and 81 points with $m = 2$. To achieve six-digit accuracy, any result from J_2 through L_1 is suitable, even though the base results (I_1 through I_4) are not that accurate.

$I_1 = 0.24491\ 14823$	$I_2 = 0.24560\ 56002$	$I_3 = 0.24577\ 91537$	$I_4 = 0.24582\ 25436$
	$J_1 = 0.24583\ 69728$	$J_2 = 0.24583\ 70049$	$J_3 = 0.24583\ 70069$
		$K_1 = 0.24583\ 70156$	$K_2 = 0.24583\ 70075$
			$L_1 = 0.24583\ 70049$

Orthogonal Polynomials The quadrature formulas for orthogonal polynomials are the same as for gaussian quadrature above, with different points and different weights.

Cubic Splines The quadrature formula is

$$\int_{x_1}^{x_{NT}} y(x)dx = \frac{1}{2}\sum_{i=1}^{NT-1} \Delta x_i (y_i + y_{i+1}) - \frac{1}{24}\sum_{i=1}^{NT-1} \Delta x_i^3 (y_i'' + y_{i+1}'')$$

with $y_1' = 0$, $y_{NT}'' = 0$ for natural cubic splines.

Computer Methods These methods are easily programmed in a spreadsheet program such as Microsoft Excel. In MATLAB, the trapezoid rule can be calculated by using the command trapz(x,y), where x is a vector of x values x_i and y is a vector of values $y(x_i)$. Alternatively, use the commands

```
F = @(x) exp(−x).*sin(x)
Q = quad(F,0,1)
```

Monte Carlo methods can be used, too (see Monte Carlo Simulations).

Singularities When the integrand has singularities, a variety of techniques can be tried. The integral may be divided into one part that can be integrated analytically near the singularity and another part that is integrated numerically. Sometimes a change of argument allows analytical integration. Series expansion might be helpful, too. When the domain is infinite, it is possible to use Gauss-Legendre or Gauss-Hermite quadrature. Also a transformation can be made. For example, let $u = 1/x$ and then

$$\int_a^b f(x)\,dx = \int_{1/b}^{1/a} \frac{1}{u^2} f\left(\frac{1}{u}\right) du \qquad ab > 0$$

Two-Dimensional Formula Two-dimensional integrals can be calculated by breaking down the integral into one-dimensional integrals.

$$\int_a^b \int_{g_1(x)}^{g_2(x)} f(x,y)\,dx\,dy = \int_a^b G(x)\,dx$$

$$G(x) = \int_{g_1(x)}^{g_2(x)} f(x,y)\,dy$$

Gaussian quadrature can also be used in two dimensions, provided the integration is on a square or can be transformed to one. (Domain transformations might be used to convert the domain to a square.)

$$\int_0^1 \int_0^1 f(x,y)dx\,dy = \sum_{i=1}^{mx} W_{xi} \sum_{j=1}^{my} W_{yj} f(x_i, y_j)$$

NUMERICAL SOLUTION OF ORDINARY DIFFERENTIAL EQUATIONS AS INITIAL-VALUE PROBLEMS

A differential equation for a function that depends on only one variable, often the variable time, is called an *ordinary differential equation*. The general solution to the differential equation includes many possibilities; the boundary or initial conditions are needed to specify which of those are desired. If all conditions are at one point, then the problem is an initial-value problem and can be integrated from that point on. If some of the conditions are available at one point and others at another point, then the ordinary differential equations become two-point boundary-value problems, which are treated in the next section. Initial-value problems as ordinary differential equations arise in control of lumped-parameter models, transient models of stirred tank reactors, and in all models where there are no spatial gradients in the unknowns. Many computer packages exist to solve initial-value problems, but it is important to understand the choices one must make and how to interpret the output (and change the choices) when the results are anomalous. Furthermore, many problems can be solved using spreadsheets (universally available) provided one understands the methods. It is important to know, too, when simple methods in spreadsheets won't work.

A higher-order differential equation

$$z^{(n)}+F(z^{(n-1)},z^{(n-2)},\ldots,z)=0$$

with initial conditions for z, and its first $n-1$ derivatives can be converted into a set of first-order equations using

$$y_i \equiv z^{(i-1)}=\frac{d^{(i-1)}z}{dt^{(i-1)}}=\frac{d}{dt}z^{(i-2)}=\frac{dy_{i-1}}{dt}$$

The higher-order equation can be written as a set of first-order equations.

$$\frac{dy_1}{dt}=y_2,\frac{dy_2}{dt}=y_3,\frac{dy_3}{dt}=y_4,\ldots,\frac{dy_n}{dt}=-F(y_{n-1},y_{n-2},\ldots,y_2,y_1)$$

The set of equations is then written as

$$\frac{d\mathbf{y}}{dt}=\mathbf{f}(\mathbf{y},t),\mathbf{y}(0)=\mathbf{y}_0$$

The methods in this section are described for a single equation, but they all apply to multiple equations.

The simplest method is Euler's method, which is first-order.

$$y^{n+1}=y^n+\Delta t\, f(y^n)$$

and errors are proportional to Δt. The second-order Adams-Bashforth method is

$$y^{n+1}=y^n+\frac{\Delta t}{2}[3f(y^n)-f(y^{n-1})]$$

Errors are proportional to Δt^2, and high-order methods are available. Notice that the higher-order explicit methods require knowing the solution (or the right-hand side) evaluated at times in the past. Since these were calculated to get to the current time, this presents no problem except for starting the problem. Then it may be necessary to use Euler's method with a very small step size for several steps in order to generate starting values at a succession of time points. The methods, error terms, order of the method, function evaluations per step, and stability limitations are listed in Finlayson (1980) in General References. The advantage of the high-order Adams-Bashforth method is that it uses only one function evaluation per step, yet achieves high-order accuracy. The disadvantage is the necessity of using another method to start. In MATLAB the function ode113 uses a version of the Adams-Bashforth method.

These methods can be used for simple problems when all the variables change on the same time scale and precise results are not needed. Euler's method is easily done in a spreadsheet. Figure 3-47 shows the commands in a spreadsheet for two differential equations, columns 1 to 3.

$$dy_1/dt=y_2-y_1,\quad dy_2/dt=-y_2,\quad y_1(0)=1,\quad y_2(0)=1$$

Once columns 4 to 6 are created, the formulas for the additional time steps are created by copying down. The Richardson extrapolation (see below) can be used to improve the accuracy.

Runge-Kutta methods are explicit methods that use several function evaluations for each time step. Runge-Kutta methods are traditionally written

Equations	Equations	Equations		delta t	0.1
Eq. time	Eq. 1	Eq. 2	time	Results 1	Results 2
0	1	1	0	1	1
=D3+$F$1	=E3+$F$1*(F3–E3)	=F3–$F$1*F3	0.1	1	0.9
=D4+$F$1	=E4+$F$1*(F4–E4)	=F4–$F$1*F4	0.2	0.99	0.81
=D5+$F$1	=E5+$F$1*(F5–E5)	=F5–$F$1*F5	0.3	0.972	0.729
=D6+$F$1	=E6+$F$1*(F6–E6)	=F6–$F$1*F6	0.4	0.9477	0.6561
=D7+$F$1	=E7+$F$1*(F7–E7)	=F7–$F$1*F7	0.5	0.91854	0.59049

FIG. 3-47 Spreadsheet for Euler's method.

for $f(t, y)$. The first-order Runge-Kutta method is Euler's method. A second-order Runge-Kutta method is

$$y^{n+1}=y^n+\frac{\Delta t}{2}[f^n+f(t^n+\Delta t,y^n+\Delta t\, f^n)]$$

while the midpoint scheme is also a second-order Runge-Kutta method

$$y^{n+1}=y^n+\Delta t\, f\left(t^n+\frac{\Delta t}{2},y^n+\frac{\Delta t}{2}f^n\right)$$

A popular fourth-order Runge-Kutta method uses the Runge-Kutta-Fehlberg formulas, which have the property that the method is fourth-order but achieves fifth-order accuracy. The coefficients are available at en.wikipedia.org/wiki/Runge-Kutta-Fehlberg_method. An extension of this method is ode45 in MATLAB.

Usually one would use a high-order method to achieve high accuracy. The Runge-Kutta-Fehlberg method is popular because it is high-order and does not require a starting method (as does an Adams-Bashforth method). However, it does require four function evaluations per time step, or four times as many as a fourth-order Adams-Bashforth method. For problems in which the function evaluations are a significant portion of the calculation time, this might be important. Given the speed and availability of desktop computers, the efficiency of the methods is most important only for very large problems that are going to be solved many times or for problems in which some variables change rapidly while others change slowly. For other problems, the most important criterion for choosing a method is probably the time the user spends setting up the problem.

The stability limits for the explicit methods are based on the largest eigenvalue of the linearized system of equations

$$\frac{dy_i}{dt}=\sum_{j=1}^{n}A_{ij}y_j,\ A_{ij}=\left.\frac{\delta f_i}{\delta y_j}\right|_y$$

For linear problems, the eigenvalues do not change, so that the stability and oscillation limits must be satisfied for *every* eigenvalue of matrix **A**. In solving nonlinear problems, the equations are linearized about the solution at the local time, and the analysis applies for small changes in time, after which a new analysis about the new solution must be made. Thus, for nonlinear problems, the eigenvalues keep changing, and the largest stable time step changes, too. The stability limits are as follows:

Euler method, $\lambda\,\Delta t \le 2$

Runge-Kutta, second-order, $\lambda\,\Delta t < 2$

Runge-Kutta-Fehlberg, $\lambda\,\Delta t < 3.0$

Richardson extrapolation can be used to improve the accuracy of a method. Suppose we step forward one step Δt with a pth-order method. Then redo the problem, this time stepping forward from the same initial point, but in two steps of length $\Delta t/2$, thus ending at the same point. Call the solution of the one-step calculation y_1 and the solution of the two-step calculation y_2. Then an improved solution at the new time is given by

$$y=\frac{2^p y_2-y_1}{2^p-1}$$

This gives a good estimate provided Δt is small enough that the method is truly convergent with order p. This process can also be repeated in the same way Romberg's method was used for quadrature.

The error term in the various methods can be used to deduce a step size that will give a user-specified accuracy. Most packages today are based on a user-specified tolerance; the step size is changed during the calculation to achieve that accuracy. The accuracy itself is not guaranteed, but it improves as the tolerance is decreased.

Implicit Methods When some dependent variables change rapidly while others change slowly, we say the problem is *stiff* and implicit methods are needed. Implicit methods use different interpolation formulas involving y^{n+1} and result in nonlinear equations to be solved for y^{n+1}. Then iterative methods must be used to solve the equations.

The backward Euler method is a first-order method:

$$y^{n+1} = y^n + \Delta t f(y^{n+1})$$

Errors are proportional to Δt for small Δt. The trapezoid rule is a second-order method.

$$y^{n+1} = y^n + \frac{\Delta t}{2}[f(y^n) + f(y^{n+1})]$$

Errors are proportional to Δt^2 for small Δt. When the trapezoid rule is used with the finite difference method for solving partial differential equations, it is called the *Crank-Nicolson method*. The implicit methods are stable for any step size but do require the solution of a set of nonlinear equations, which must be solved iteratively. The set of equations can be solved using the successive substitution method or Newton-Raphson method. See Bogacki, M. B., K. Alejski, and J. Szymanewski, *Comp. Chem. Eng.* **13:** 1081–1085 (1989) for an application to dynamic distillation problems.

The best packages for stiff equations (see below) use backward-difference formulas. Gear first developed these, and the first two orders are given below (Gear, G. W., *Numerical Initial Value Problems in Ordinary Differential Equations*, Prentice-Hall, Englewood Cliffs, N.J., 1971).

1. $y^{n+1} = y^n + \Delta t f(y^{n+1})$
2. $y^{n+1} = \frac{4}{3}y^n + \frac{1}{3}y^{n-1} + \frac{2}{3}\Delta t\, f(y^{n+1})$

These methods require solving sets of nonlinear equations. By adroit manipulation and estimation, a package will change the order to achieve a required accuracy with a minimum number of time steps and iterations. The programs ode15s and ode23s in MATLAB use these techniques.

Stiffness The concept of stiffness is described for a system of linear equations.

$$\frac{d\mathbf{y}}{dt} = \mathbf{Ay}$$

Let λ_i be the eigenvalues of matrix **A**. The *stiffness ratio* SR is defined as

$$\text{SR} = \frac{\max_i |\text{Re}(\lambda_i)|}{\max_i |\text{Re}(\lambda_i)|} \qquad (3\text{-}65)$$

SR = 20 is not stiff, SR = 10^3 is stiff, and SR = 10^6 is very stiff. If the problem is nonlinear, then the solution is expanded about the current state.

$$\frac{dy_i}{dt} = f_i[y(t^n)] + \sum_{j=1}^{n} \frac{\partial f_i}{\partial y_j}[y_j - y_j(t^n)]$$

The question of stiffness then depends on the solution at the current time. Consequently nonlinear problems can be stiff during one time period and not stiff during another. While the chemical engineer may not actually calculate the eigenvalues, it is useful to know that they determine the stability and accuracy of the numerical scheme and the step size used.

Problems are stiff when the time constants for different phenomena have very different magnitudes. Consider flow through a packed bed reactor. The time constants for different phenomena are as follows:

1. Time for device flow-through

$$t_{\text{flow}} = \frac{L}{u} = \frac{\phi AL}{Q}$$

where Q is the volumetric flow rate, A is the cross-sectional area, L is the length of the packed bed, and ϕ is the void fraction.

2. Time for reaction

$$t_{rxn} = \frac{1}{k}$$

where k is a rate constant (time^{-1}).

3. Time for diffusion inside the catalyst

$$t_{\text{internal diffusion}} = \frac{\varepsilon R^2}{D_e}$$

where ε is the porosity of the catalyst, R is the catalyst radius, and D_e is the effective diffusion coefficient inside the catalyst.

4. Time for heat transfer is

$$t_{\text{internal heat transfer}} = \frac{R^2}{\alpha} = \frac{\rho_s C_s R^2}{k_e}$$

where ρ_s is the catalyst density, C_s is the catalyst heat capacity per unit mass, k_e is the effective thermal conductivity of the catalyst, and α is the thermal diffusivity. For example, in the model of a catalytic converter for an automobile [Ferguson, N. B., and B. A. Finlayson, *AIChE J.* **20:** 539–550 (1974)], the time constant for internal diffusion was 0.3 s; internal heat transfer, 21 s; and device flow-through, 0.003 s. The device flow-through is so fast that it might as well be instantaneous. The stiffness is approximately 7000, and implicit methods must be used to integrate the equations. Alternatively, a quasistatic model can be developed. In this case the time derivative is deleted for the variables that change rapidly on the grounds that those variables are essentially in steady state with respect to the rest of the problem, even if the steady state changes slowly.

Differential-Algebraic Systems Sometimes models involve ordinary differential equations subject to some algebraic constraints. For example, the equations governing one equilibrium stage (as in a distillation column) are

$$M\frac{dx^n}{dt} = V^{n+1}y^{n+1} - L^n x^n - V^n y^n + L^{n-1}x^{x-1}$$

$$x^{n-1} - x^n = E^n(x^{n-1} - x^{*,n})$$

$$\sum_{i=1}^{N} x_i = 1$$

where x and y are the mole fraction in the liquid and vapor, respectively; L and V are liquid and vapor flow rates, respectively; M is the holdup; and the superscript is the stage number. The efficiency is E, and the concentration in equilibrium with the vapor is x^*. The first equation is an ordinary differential equation for the mass of one component on the stage, while the third equation represents a constraint that the mass fractions add to 1. This is a *differential-algebraic* system of equations.

Differential-algebraic equations can be written in the general notation

$$F\left(t, y, \frac{dy}{dt}\right) = 0$$

To solve the general problem by using the backward Euler method, replace the nonlinear differential equation with the nonlinear algebraic equation for one step.

$$F\left(t, y^{n+1}, \frac{y^{n+1} - y^n}{\Delta t}\right) = 0$$

This equation must be solved for y^{n+1}. The Newton-Raphson method can be used, and if convergence is not achieved within a few iterations, the time step can be reduced and the step repeated. In actuality, the higher-order backward-difference Gear methods are used in DASSL (Ascher, U. M., and L. R. Petzold, *Computer Methods for Ordinary Differential Equations and Differential-Algebraic Equations*, SIAM, Philadelphia, Penn., 1998). The program ode15s in MATLAB can be used to solve differential-algebraic equations.

Differential-algebraic systems are more complicated than differential systems because the solution may not always be defined. See Pontelides et al. [*Comp. Chem. Eng.* **12:** 449–454 (1988)] for a model of a distillation column in which the column pressure strongly affects the possible solutions and initial conditions. Byrne and Ponzi [*Comp. Chem. Eng.* **12:** 377–382 (1988)] and Chan, T. F. C., and H. B. Keller [*SIAM J. Sci. Stat. Comput.* **3:** 173–194 (1982)] also list several chemical engineering examples of differential-algebraic systems and solve one involving two-phase flow.

Computer Software Efficient computer packages are available for solving ordinary differential equations as initial-value problems. The packages are widely available and good enough that most chemical engineers use them and do not write their own. On the NIST web page http://gams.nist.gov/Problem.html insert "ordinary differential equations" to find packages that can be downloaded. On the Netlib website http://www.netlib.org/, search the Netlib repository, and choose "ode" to find packages that can be downloaded. Using Microsoft Excel to solve ordinary differential equations is cumbersome, except for the simplest problems.

Stability, Bifurcations, and Limit Cycles Some aspects of this subject involve the solution of nonlinear equations; other aspects involve the integration of ordinary differential equations; applications include chaos and fractals as well as the unusual operation of some chemical engineering equipment. Kubicek, M., and M. Marek, *Computational Methods in Bifurcation Theory and Dissipative Structures*, Springer-Verlag, Berlin (1983, 2012), give an excellent introduction to the subject and the details needed to apply the methods. A concise survey with some chemical engineering examples is given in Doherty, M. F., and J. M. Ottino, *Chem. Eng. Sci.* **43:** 139–183 (1988). Bifurcation results are closely connected with the stability of the steady states, which is essentially a transient phenomenon.

Sensitivity Analysis When one is solving differential equations, it is frequently necessary to know the solution as well as the sensitivity of the solution to the value of a parameter. Such information is useful when doing parameter estimation (to find the best set of parameters for a model) and for deciding if a parameter needs to be measured accurately. An added equation is created by differentiating the ordinary differential equation with respect to the parameter and solving that equation concurrently. See Finlayson et al. (2006) in General References.

Molecular Dynamics Special integration methods have been developed for molecular dynamics calculations owing to the structure of the equations. A very large number of equations are to be integrated, with the following form based on molecular interactions between molecules.

$$m_i \frac{d^2\mathbf{r}_i}{dt^2} = \mathbf{F}_i(\{\mathbf{r}\}) \quad \mathbf{F}_i(\{\mathbf{r}\}) = -\nabla V$$

The symbol m_i is the mass of the ith particle, $\mathbf{r}_i$ is the position of the ith particle, $\mathbf{F}_i$ is the force acting on the ith particle, and V is the potential energy that depends upon the location of all the particles (but not their velocities). Since the major part of the calculation lies in the evaluation of the forces, or potentials, a method must be used that minimizes the number of times the forces are calculated to move from one time to another time. Rewrite this equation in the form of an acceleration as

$$\frac{d^2\mathbf{r}_i}{dt^2} = \frac{1}{m}\mathbf{F}_i(\{\mathbf{r}\}) \equiv \mathbf{a}_i$$

In the Verlet method, this equation is written by using central finite differences (see Interpolation and Finite Differences). Note that the accelerations do not depend upon the velocities.

$$\mathbf{r}_i(t+\Delta t) = 2\mathbf{r}_i(t) - \mathbf{r}_i(t-\Delta t) + \mathbf{a}_i(t)\Delta t^2$$

The calculations are straightforward, and no explicit velocity is needed. The storage requirement is modest, and the precision is modest (it is a second-order method). Note that one must start the calculation with values of $\{\mathbf{r}\}$ at times t and $t - \Delta t$.

In the Verlet velocity method, an equation is written for the velocity, too.

$$\frac{d\mathbf{v}_i}{dt} = \mathbf{a}_i$$

The trapezoid rule [see Numerical Integration (Quadrature)] is applied to obtain

$$\mathbf{v}_i(t+\Delta t) = \mathbf{v}_i(t) + \frac{1}{2}[\mathbf{a}_i(t) + \mathbf{a}_i(t+\Delta t)]\Delta t$$

The position of the particles is expanded in a Taylor series.

$$\mathbf{r}_i(t+\Delta t) = \mathbf{r}_i(t) + \mathbf{v}_i\Delta t + \frac{1}{2}\mathbf{a}_i(t)\Delta t^2$$

Beginning with values of $\{\mathbf{r}\}$ and $\{\mathbf{v}\}$ at time 0, one calculates the new positions and then the new velocities. This method is second-order in Δt too. Molecular dynamics is used in chemical engineering for a variety of applications, including drug design, protein folding, nucleation and growth processes, and the phase behavior of polymeric, colloidal, and self-assembled systems [see Pamer, J. C., and P. G. Debenedettii, Recent Advances in Molecular Simulation: A Chemical Engineering Perspective, *AIChE J.* **61**, 370–383 (2015)]. For additional details about the method, see Hinchliffe, A., *Molecular Modelling for Beginners*, 2d ed., Wiley, New York, 2008; Jensen, J. H., *Molecular Modeling Basics*, CRC Press, Boca Raton, Fla., 2010; Leach, A. R., *Molecular Modelling: Principles and Applications*, 2d ed., Prentice Hall, Upper Saddle River, N.J., 2001; Schlick, T., *Molecular Modeling and Simulations*, 2d ed., Springer, New York, 2010. See https://en.wikipedia.org/wiki/List_of_software_for_molecular_mechanics_modeling for computer packages, especially the free programs LAMMPS (lammps.sandia.gov) and GROMACS (www.gromacs.org, especially for biological molecules).

See also Calvetti, D. E., and E. Somersalo, *Computational Mathematical Modeling: An Integrated Approach Across Scales*, SIAM, 2012, for methods to include phenomena that occur on different physical scales.

ORDINARY DIFFERENTIAL EQUATIONS—BOUNDARY-VALUE PROBLEMS

Diffusion problems in one dimension lead to boundary-value problems. The boundary conditions are applied at two different spatial locations: at one side the concentration may be fixed and at the other side the flux may be fixed. Because the conditions are specified at two different locations, the problems are not initial-value in character. It is not possible to begin at one position and integrate directly because at least one of the conditions is specified somewhere else and there are not enough conditions to begin the calculation. Thus, methods have been developed especially for boundary-value problems.

Boundary-value methods provide a description of the solution either by providing values at specific locations or by an expansion in a series of functions. Thus, the key issues are the method of representing the solution, the number of points (i.e., the mesh) or the number of terms in the series, and how the approximation converges to the exact answer, i.e., how the error changes with the number of points or number of terms in the series. In addition, boundary conditions and nonlinear transport coefficients are handled differently in the various methods. These issues are discussed for each of the methods: finite difference, orthogonal collocation, and Galerkin finite element methods. Sometimes the solution has singularities or the domain is semi-infinite, and these situations require special treatment.

The first approach is to try to find an analytical solution. Flow in a pipe is governed by the equation

$$\frac{1}{r}\frac{d}{dr}\left(\mu r \frac{du}{dr}\right) = -\frac{\Delta P}{L}$$

where u is the velocity, r is the radial position, μ is the viscosity, and $\Delta P/L$ is the pressure drop per length. The solution is finite at the origin, $r = 0$, and takes the value zero at the radius of the pipe R. For a newtonian fluid, the viscosity is constant. This equation can be integrated once to obtain

$$r\frac{du}{dr} = -\frac{\Delta P}{\mu L}\frac{r^2}{2} + c_1 \quad \text{or} \quad \frac{du}{dr} = -\frac{\Delta P}{\mu L}\frac{r}{2} + \frac{c_1}{r}$$

and integrated again to get

$$u = -\frac{\Delta P}{\mu L}\frac{r^2}{4} + c_1 \ln r + c_2$$

Since the velocity is finite at the origin, c_1 is taken as zero; c_2 is taken as

$$c_2 = \frac{\Delta P}{\mu L}\frac{R^2}{4}$$

so that the velocity is zero at $r = R$. The solution is then

$$u = \frac{\Delta P}{4\mu L}(R^2 - r^2)$$

This problem requires no numerical methods. But if the viscosity were appropriate to a non-newtonian fluid and depended upon the shear rate, e.g., for a Bird-Carreau fluid

$$\mu = \frac{\eta_0}{\left[1 + \lambda\left(\frac{du}{dr}\right)^2\right]^{(1-n)/2}}$$

then numerical methods would be required, as described in this subsection.

Finite Difference Method To apply the finite difference method, we first spread grid points through the domain. Figure 3-48 shows a uniform mesh of n points (nonuniform meshes are possible too). The unknown, here $c(x)$, at a grid point x_i is assigned the symbol $c_i = c(x_i)$. The finite difference

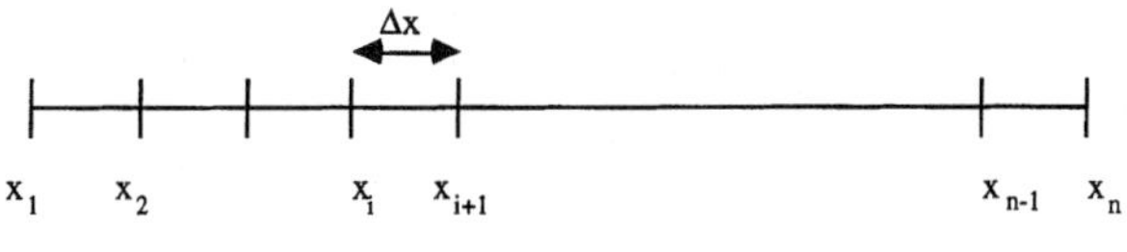

FIG. 3-48 Finite difference mesh; Δx uniform.

method can be derived easily by using a Taylor expansion of the solution about this point. Expressions for the derivatives are

$$\left.\frac{dc}{dx}\right|_i = \frac{c_{i+1}-c_i}{\Delta x} - \left.\frac{d^2c}{dx^2}\right|_i \frac{\Delta x}{2} + \cdots, \quad \left.\frac{dc}{dx}\right|_i = \frac{c_i - c_{i-1}}{\Delta x} + \left.\frac{d^2c}{dx^2}\right|_i \frac{\Delta x}{2} + \cdots$$

$$\left.\frac{dc}{dx}\right|_i = \frac{c_{i+1}-c_{i-1}}{2\Delta x} - \left.\frac{d^3c}{dx^3}\right|_i \frac{\Delta x^2}{3!} + \cdots$$

The truncation error in the first two expressions is proportional to Δx, and the methods are said to be first-order. The truncation error in the third expression is proportional to Δx^2, and the method is said to be second-order. Usually the last equation is used to ensure the best accuracy. The finite difference representation of the second derivative is

$$\left.\frac{d^2c}{dx^2}\right|_i = \frac{c_{i+1} - 2c_i + c_{i-1}}{\Delta x^2} - \left.\frac{d^4c}{dx^4}\right|_i \frac{2\Delta x^2}{4!} + \cdots$$

The truncation error is proportional to Δx^2. To solve a differential equation, it is evaluated at a point i and then these expressions are inserted for the derivatives.

Example Consider the equation for convection, diffusion, and reaction in a tubular reactor.

$$\frac{1}{\text{Pe}}\frac{d^2c}{dx^2} - \frac{dc}{dx} = \text{Da}\, R(c)$$

Pe is the Peclet number and Da is the Damköhler number. The finite difference representation is

$$\frac{1}{\text{Pe}}\frac{c_{i+1} - 2c_i + c_{i-1}}{\Delta x^2} - \frac{c_{i+1} - c_{i-1}}{2\Delta x} = \text{Da}\, R(c_i)$$

This equation is written for $i = 2$ to $n - 1$, or the internal points. The equations would then be coupled but would also involve the values of c_1 and c_n as well. These are determined from the boundary conditions.

If the boundary condition involves a derivative, it is important that the derivatives be evaluated using points that exist. Three possibilities exist; the first two are

$$\left.\frac{dc}{dx}\right|_1 = \frac{c_2 - c_1}{\Delta x}$$

$$\left.\frac{dc}{dx}\right|_1 = \frac{-3c_1 + 4c_2 - c_3}{2\Delta x}$$

The third alternative is to add a false point, outside the domain, as $c_0 = c(x = -\Delta x)$.

$$\left.\frac{dc}{dx}\right|_1 = \frac{c_2 - c_0}{2\Delta x}$$

Since this equation introduces a new variable c_0, another equation is needed and is obtained by writing the finite difference equation for $i = 1$ too. The sets of equations can be solved by using the Newton-Raphson method. The first form of the derivative gives a tridiagonal system of equations, and the standard routines for solving tridiagonal equations suffice. For the other two options, some manipulation is necessary to put them into a tridiagonal form.

Frequently, the transport coefficients, such as the diffusion coefficient or thermal conductivity, depend on the dependent variable, concentration, or temperature, respectively. Then the differential equation might look like

$$\frac{d}{dx}\left(D(c)\frac{dc}{dx}\right) = 0$$

This could be written as two equations.

$$-\frac{dJ}{dx} = 0 \qquad J = -D(c)\frac{dc}{dx}$$

Because the coefficient depends on c, the equations are more complicated. A finite difference method can be written in terms of the fluxes at the midpoints $i + 1/2$.

$$-\frac{J_{i+1/2} - J_{i-1/2}}{\Delta x} = 0 \qquad J_{i+1/2} = -D(c_{i+1/2})\frac{c_{i+1} - c_i}{\Delta x}$$

These are combined to give the complete equation.

$$\frac{D(c_{i+1/2})(c_{i+1} - c_i) - D(c_{i-1/2})(c_i - c_{i-1})}{\Delta x^2} = 0$$

This represents a set of nonlinear algebraic equations that can be solved with the Newton-Raphson method. However, in this case, a viable iterative strategy is to evaluate the transport coefficients at the last value and then solve

$$\frac{D(c^k_{i+1/2})(c^{k+1}_{i+1} - c^{k+1}_i) - D(c^k_{i-1/2})(c^{k+1}_i - c^{k+1}_{i-1})}{\Delta x^2} = 0$$

The advantage of this approach is that it is easier to program than a full Newton-Raphson method. If the transport coefficients do not vary radically, then the method converges. If the method does not converge, then it may be necessary to use the full Newton-Raphson method.

There are two common ways to evaluate the transport coefficient at the midpoint: Use the average value of the solution on each side to evaluate the diffusivity, or use the average value of the diffusivity on each side. Both methods have truncation error Δx^2. The spacing of the grid points need not be uniform. See Finlayson (1980) and Finlayson et al. (2006) in General References.

Example A reaction diffusion problem is solved with the finite difference method.

$$\frac{d^2c}{dx^2} = \phi^2 c, \quad \frac{dc}{dx}(0) = 0 \quad c(1) = 1$$

The solution is derived for $\phi = 2$. It is solved several times, first with two intervals and three points (at $x = 0$, 0.5, 1), then with four intervals, then with eight intervals. The reason is that when an exact solution is not known, one must use several Δx values and see that the solution converges as Δx approaches zero. With two intervals, the equations are as follows. The points are $x_1 = 0$, $x_2 = 0.5$, and $x_3 = 1.0$; and the solutions at those points are c_1, c_2, and c_3, respectively. A false boundary is used at $x_0 = -0.5$.

$$\frac{c_0 - c_2}{2\Delta x} = 0, \quad \frac{c_0 - 2c_1 + c_2}{\Delta x^2} - \phi^2 c_1 = 0, \quad \frac{c_1 - 2c_2 + c_3}{\Delta x^2} - \phi^2 c_2 = 0, \quad c_3 = 1$$

The solution is $c_1 = 0.2857$, $c_2 = 0.4286$, and $c_3 = 1.0$. The problem is solved again with four and then eight intervals. The value of concentration at $x = 0$ takes the following values for different Δx values. These values are extrapolated using the Richardson extrapolation technique to get $c(0) = 0.265718$. Using this value as the best estimate of the exact solution, the errors in the solution are tabulated versus Δx. Clearly the errors go as Δx^2 (decreasing by a factor of 4 when Δx decreases by a factor of 2), thus validating the solution. The exact solution is 0.265802.

$n - 1$	Δx	$c(0)$
2	0.5	0.285714
4	0.25	0.271043
8	0.125	0.267131

$n - 1$	Δx	Error in $c(0)$
2	0.5	0.02000
4	0.25	0.00532
8	0.125	0.00141

Finite Difference Methods Solved with Spreadsheets A convenient way to solve the finite difference equations for simple problems is to use a computer spreadsheet. The equations for the problem solved in the example can be cast into the following form:

$$c_1 = \frac{2c_2}{2 + \phi^2 \Delta x^2}$$

$$c_i = \frac{c_{i+1} + c_{i-1}}{2 + \phi^2 \Delta x^2}$$

$$c_{n+1} = 1$$

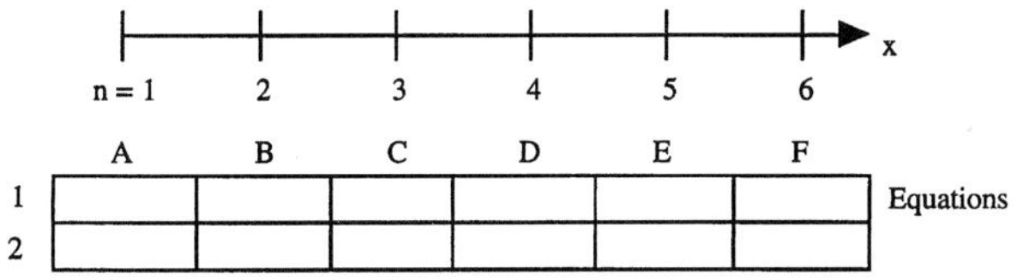

FIG. 3-49 Finite difference method using spreadsheets.

Let us solve the problem using 6 nodes, or 5 intervals. Then the connection between the cell in the spreadsheet and the nodal value is shown in Fig. 3-49. The following equations are placed into the various cells.

A1: = 2*B1/(2.+(phi*dx)**2)
B1: = (A1 + C1)/(2.+(phi*dx)**2)
F1: = 1.

The equation in cell B1 is copied into cells C1 through E1. Then turn on the iteration scheme in the spreadsheet and watch the solution converge. Whether convergence is achieved can depend on how you write the equations, so some experimentation may be necessary. Theorems for convergence of the successive substitution method are useful in this regard.

Orthogonal Collocation The orthogonal collocation method has found widespread application in chemical engineering, particularly for chemical reaction engineering. In the collocation method, the dependent variable is expanded in a series of orthogonal polynomials. See Interpolation: Lagrange Interpolation Formulas.

$$c(x)=\sum_{m=0}^{N} a_m P_m(x)$$

The differential equation is evaluated at certain collocation points. The collocation points are the roots to an orthogonal polynomial, as first used by Lanczos [Lanczos, C., *J. Math. Phys.* **17**:123–199 (1938); and Lanczos, C., *Applied Analysis*, Prentice Hall, Upper Saddle River, N.J., 1956]. A major improvement was proposed by Villadsen and Stewart [Villadsen, J. V., and W. E. Stewart, *Chem. Eng. Sci.* **22**:1483–1501 (1967)], who proposed that the entire solution process be done in terms of the solution at the collocation points rather than the coefficients in the expansion. This method is especially useful for reaction diffusion problems that frequently arise when modeling chemical reactors. It is highly efficient when the solution is smooth, but the finite difference method is preferred when the solution changes steeply in some region of space. The error decreases very rapidly as N is increased since it is proportional to $[1/(1-N)]^{N-1}$. See Finlayson (1980) in General References.

Galerkin Finite Element Method In the finite element method, the domain is divided into elements, and an expansion is made for the solution on each finite element (see Interpolation: Finite Element Method). In the Galerkin finite element method, an additional idea is introduced: the Galerkin method is used to solve the equation. The Galerkin method is explained using the equations for reaction and diffusion in a porous catalyst pellet.

$$\frac{d^2c}{dx^2}=\phi^2 R(c)$$

$$\frac{dc}{dx}(0)=0,\quad c(1)=1$$

The unknown solution is expanded in a series of known functions $\{b_i(x)\}$ with unknown coefficients $\{a_i\}$.

$$c(x)=\sum_{i=1}^{NT} a_i b_i(x)$$

The trial solution is substituted into the differential equation to obtain the residual.

$$\text{Residual}=\sum_{i=1}^{NT} a_i \frac{d^2 b_i}{dx^2}-\phi^2 R\left[\sum_{i=1}^{NT} a_i b_i(x)\right]$$

The residual is then made orthogonal to the set of basis functions.

$$\int_0^1 b_j(x)\left\{\sum_{i=1}^{NT} a_i \frac{d^2 b_i}{dx^2}-\phi^2 R\left[\sum_{i=1}^{NT} a_i b_i(x)\right]\right\}dx=0 \quad j=1,\ldots,NT$$

This is the process that makes the method a Galerkin method. The basis for the orthogonality condition is that any function that is orthogonal to each member of a complete set is zero. The residual is being made orthogonal; and if the basis functions are complete and you use infinitely many of them, then the residual is zero. Once the residual is zero, the problem is solved.

This equation is integrated by parts to give the following equation:

$$-\sum_{i=1}^{NT}\int_0^1 \frac{db_j}{dx}\frac{db_i}{dx}dx a_i=\phi^2\int_0^1 b_j(x) R\left[\sum_{i=1}^{NT} a_i b_i(x)\right]dx \quad j=1,\ldots,NT-1 \qquad (3\text{-}66)$$

This equation defines the Galerkin method, and a solution that satisfies this equation (for all $j = 1, \ldots, \infty$) is called a *weak solution*. For an approximate solution, the equation is written once for each member of the trial function, $j = 1, \ldots, NT - 1$, and the boundary condition is applied.

$$\sum_{i=1}^{NT} a_i b_i(1)=c_B$$

The Galerkin finite element method results when the Galerkin method is combined with a finite element trial function. The domain is divided into elements separated by nodes, as in the finite difference method. The solution is approximated by a linear (or sometimes quadratic) function to provide the Galerkin finite element equations. For example, with the grid shown in Fig. 3-48, a linear interpolation would be used between points x_i and x_{i+1}.

$$c(x)=c_i(1-u)+c_{i+1}u \qquad u\equiv\frac{x-x_i}{x_{i+1}-x_i}$$

A finite element method based on these functions would have an error proportional to Δx^2. The finite element representations for the first derivative and second derivative are the same as in the finite difference method, but this is not true for other functions or derivatives. With quadratic finite elements, take the region from x_{i-1} and x_{i+1} as one element with $x = x_{i-1}$ at $u = 0$, $x = x_i$ at $u = ½$, and $x = x_{i+1}$ at $u = 1$. Then the interpolation would be

$$c(x)=c_{i-1}N_1(u)+c_i N_2(u)+c_{i+1}N_3(u)$$

$$N_1(u)=2(u-1)\left(u-\frac{1}{2}\right) \quad N_2(u)=4u(1-u)$$

$$N_3(u)=2u\left(u-\frac{1}{2}\right)$$

A finite element method based on these functions would have an error proportional to Δx^3. Thus, it would converge faster than one based on linear interpolation.

Adaptive Meshes In many two-point boundary-value problems, the difficulty in the problem lies in the formation of a boundary-layer region, or a region in which the solution changes very dramatically. In such cases, it is prudent to use small mesh spacing there, with either the finite difference method or the finite element method. If the region is known a priori, small mesh spacings can be assumed at the boundary layer. If the region is not known, however, other techniques must be used. These techniques are known as adaptive mesh techniques. The mesh size is made small where some property of the solution is large. For example, if the truncation error of the method is nth-order, then the nth-order derivative of the solution is evaluated and a small mesh is used where it is large. Alternatively, the residual (the differential equation with the numerical solution substituted into it) can be used as a criterion. It is also possible to define the error that is expected from a method one order higher and one order lower. Then a decision about whether to increase or decrease the order of the method can be made, taking into account the relative work of the different orders. This provides a method of adjusting both the mesh spacing (Δx, or sometimes called h) and the degree of polynomial (p). Such methods are called *h-p* methods. Many finite element programs have the capability to do this mesh refinement automatically.

Singular Problems and Infinite Domains If the solution being sought has a singularity, it may be difficult to find a good numerical solution. Sometimes even the location of the singularity may not be known. One method of solving such problems is to refine the mesh near the singularity, relying on the better approximation due to a smaller Δx. Another approach is to incorporate the singular trial function into the approximation. Thus, if the solution approaches $f(x)$ as x goes to zero and $f(x)$ becomes infinite, one may define a new variable $u(x) = y(x) - f(x)$ and derive an equation for u. The differential equation is more complicated, but the solution is better near the singularity. See Press et al. (2007) in General References.

Sometimes the domain is semi-infinite, as in boundary-layer flow. The domain can be transformed from the x domain $(0 - \infty)$ to the η domain $(1 - 0)$ using the transformation $\eta = \exp(-x)$. Another approach is to use a variable mesh, perhaps with the same transformation. For example, use $\eta = \exp(-\beta x)$ and a constant mesh size in η; the value of β is found experimentally. Still another approach is to solve on a finite mesh in which the last point is far enough away that its location does not influence the solution. A location that is far enough away must be found by trial and error.

Packages to solve boundary-value problems are available on the Internet. On the NIST web page http://gams.nist.gov/Problem.html insert "ordinary differential equations" to find packages for boundary-value problems. On the Netlib website http://www.netlib.org/ search on "boundary-value problem." Any spreadsheet that has an iteration capability can be used with the finite difference method. Some packages for partial differential equations also have a capability for solving one-dimensional boundary-value problems (e.g., Comsol Multiphysics).

NUMERICAL SOLUTION OF INTEGRAL EQUATIONS

This subsection considers a method of solving numerically the Fredholm integral equation of the second kind:

$$u(x) = f(x) + \lambda \int_a^b k(x,t)\,u(t)\,dt \quad \text{for } u(x) \tag{3-67}$$

The method discussed arises because a definite integral can be closely approximated by any of several numerical integration formulas (each of which arises by approximating the function by some polynomial over an interval). Thus the definite integral in Eq. (3-67) can be replaced by an integration formula, and Eq. (3-67) may be written

$$u(x) = f(x) + \lambda(b-a)\left[\sum_{i=1}^{n} c_i k(x,t_i)u(t_i)\right] \tag{3-68}$$

where $t_1, \ldots, t_n$ are points of subdivision of the t axis, $a \le t \le b$, and the c's are coefficients whose values depend upon the type of numerical integration formula used. Now Eq. (3-68) must hold for all values of x, $a \le x \le b$; so it must hold for $x = t_1, x = t_2, \ldots, x = t_n$. Substituting for x successively $t_1, t_2, \ldots, t_n$ and setting $u(t_i) = u_i$ and $f(t_i) = f_i$, we get n linear algebraic equations for the n unknowns $u_1, \ldots, u_n$. That is,

$$u_i = f_i + \lambda(b-a)[c_1k(t_i,t_1)u_1 + c_2k(t_i,t_2)u_2 + \cdots + c_nk(t_i,t_n)u_n] \qquad i = 1, 2, \ldots, n$$

These u_j may be solved for by the methods under Numerical Solution of Linear Equations and Associated Problems and substituted into Eq. (3-68) to yield an approximate solution for Eq. (3-67).

Because of the work involved in solving large systems of simultaneous linear equations it is desirable that only a small number of u values be computed. Thus the gaussian integration formulas are useful because of the economy they offer.

Solutions for Volterra equations are done in a similar fashion, except that the solution can proceed point by point or in small groups of points depending on the quadrature scheme. See Linz, P., *Analytical and Numerical Methods for Volterra Equations*, SIAM, Philadelphia, Penn., 1985. There are methods that are analogous to the usual methods for integrating differential equations (Runge-Kutta, predictor-corrector, Adams methods, etc.). Explicit methods are fast and efficient until the time step is very small to meet the stability requirements. Then implicit methods are used, even though sets of simultaneous algebraic equations must be solved. The major part of the calculation is the evaluation of integrals, however, so that the added time to solve the algebraic equations is not excessive. Thus, implicit methods tend to be preferred. Volterra equations of the first kind are not well posed, and small errors in the solution can have disastrous consequences. The boundary element method uses Green's functions and integral equations to solve differential equations. See Brebbia, C. A., and J. Dominguez, *Boundary Elements—An Introductory Course*, 2d ed., Computational Mechanics Publications, Southhampton, UK, 1992; Poljak, D., and C. A. Brebbia, *Boundary Element Methods for Electrical Engineers*, WIT Press, Ashurst, UK, 2005.

MONTE CARLO SIMULATIONS

Some physical problems, such as those involving the interaction of molecules, are usually formulated as integral equations. Monte Carlo methods are especially well suited to their solution. This section cannot give a comprehensive treatment of such methods, but their use in calculating the value of an integral will be illustrated. Suppose we wish to calculate the integral

$$G = \int_{\Omega_0} g(x) f(x)\,dx$$

where the distribution function $f(x)$ satisfies

$$f(x) \ge 0, \quad \int_{\Omega_0} f(x)\,dx = 1$$

The distribution function $f(x)$ can be taken as constant, for example, $1/\Omega_0$. We choose variables $x_1, x_2, \ldots, x_N$ randomly from $f(x)$ and form the arithmetic mean

$$G_N = \frac{1}{N}\sum_i g(x_i)$$

The quantity G_N is an estimation of G, and the fundamental theorem of Monte Carlo guarantees that the expected value of G_N is G, if G exists (Kalos, M. H., and P. A. Whitlock, *Monte Carlo Methods*, vol. 1, Wiley, New York, 1986). The error in the calculation is given by

$$\varepsilon = \frac{\sigma_1}{N^{1/2}}$$

where σ_1^2 is calculated from

$$\sigma_1^2 = \int_{\Omega_0} g^2(x) f(x)\,dx - G^2$$

Thus the number of terms needed to achieve a specified accuracy ε can be calculated once an estimate of σ_1^2 is known.

$$N = \frac{\sigma_1^2}{\varepsilon^2}$$

Various methods, such as influence sampling, can be used to reduce the number of calculations needed. See also Lapeyre, B., *Introduction to Monte-Carlo Methods for Transport and Diffusion Equations*, Oxford University Press, London, 2003; Liu, J. S., *Monte Carlo Strategies in Scientific Computing*, Springer, New York, 2008; and Thomopoulos, N. T., *Essentials of Monte Carlo Simulation: Statistical Methods for Building Simulation Models*, Springer, New York, 2013. Some computer programs are available that perform simple Monte Carlo calculations using Microsoft Excel. Monte Carlo methods for molecular simulation lead to an equilibrium configuration of the molecules. Thus, the approach to that equilibrium is not modeled, and this is an advantage over molecular dynamics (see below) when the equilibrium configuration is the desired result, since the Monte Carlo method is faster. A good open-source Monte Carlo program is CASSANDRA at the University of Notre Dame.

NUMERICAL SOLUTION OF PARTIAL DIFFERENTIAL EQUATIONS

The numerical methods for partial differential equations can be classified according to the type of equation (see Partial Differential Equations): parabolic, elliptic, and hyperbolic. This section uses the finite difference method to illustrate the ideas, and these results can be programmed for simple problems. For more complicated problems, however, it is common to rely on computer packages. Thus, some discussion is given to the issues that arise in using computer packages. These methods are used in modeling microfluidics (with small Reynolds numbers) and turbulence (with large Reynolds numbers).

Parabolic Equations in One Dimension By combining the techniques applied to initial-value problems and boundary-value problems, it is possible to easily solve parabolic equations in one dimension. The method is often called the *method of lines*. It is illustrated here using the finite difference method, but the Galerkin finite element method and the orthogonal collocation method can also be combined with initial-value methods in similar ways. The analysis is done by example. The finite volume method is described under Hyperbolic Equations.

Example Consider the diffusion equation, with boundary- and initial-value conditions.

$$\frac{\partial c}{\partial t} = D\frac{\partial^2 c}{\partial x^2}$$

$$c(x, 0) = 0,$$

$$c(0, t) = 1, \qquad c(1, t) = 0$$

We denote by c_i the value of $c(x_i, t)$ at any time. Thus, c_i is a function of time, and differential equations in c_i are ordinary differential equations.

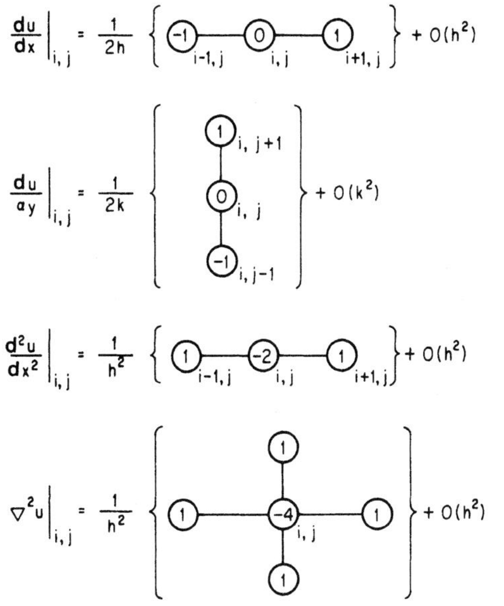

FIG. 3-50 Computational molecules. $h = \Delta x = \Delta y$.

By evaluating the diffusion equation at the ith node and replacing the derivative with a finite difference equation, the following working equation is derived for each node i, $i = 2, \ldots, n$ (see Fig. 3-50).

$$\frac{dc_i}{dt} = D\frac{c_{i+1} - 2c_i + c_{i-1}}{\Delta x^2}$$

This can be written in the general form of a set of ordinary differential equations by defining matrix **AA**.

$$\frac{d\mathbf{c}}{dt} = \mathbf{AAc}$$

This set of ordinary differential equations can be solved using any of the standard methods, and the stability of the integration of these equations is governed by the largest eigenvalue of **AA**. When Euler's method is used to integrate in time, the equations become

$$\frac{c_i^{n+1} - c_i^n}{\Delta t} = D\frac{c_{i+1}^n - 2c_i^n + c_{i-1}^n}{\Delta x^2}$$

where $c_i^n = c(x_i, t^n)$. Notice that if the solution is known at every point at one time n, then it is a straightforward calculation to find the solution at every point at the new time $n + 1$.

If Euler's method is used for integration, the time step is limited by

$$\Delta t \le \frac{2}{|\lambda|_{\max}}$$

whereas if the Runge-Kutta-Fehlberg method is used, the 2 in the numerator is replaced by 3.0. The largest eigenvalue of **AA** is bounded by Gerschgorin's theorem.

$$|\lambda|_{\max} \le \max_{2<j<n} \sum_{i=2}^{n} |\mathbf{AA}_{ji}| = \frac{4D}{\Delta x^2}$$

This gives the well-known stability limit

$$\Delta t \frac{D}{\Delta x^2} \le \frac{1}{2}$$

The smallest eigenvalue is independent of Δx (it is $D\pi^2/L^2$) so that the ratio of largest to smallest eigenvalue is proportional to $1/\Delta x^2$. Thus, the problem becomes stiff as Δx approaches zero. See Eq. (3-65).

The effect of the increased stiffness is that a smaller and smaller time step (Δt) must be taken as the mesh is refined ($\Delta x^2 \to 0$). At the same time, the number of points is increasing, so the computation becomes very lengthy. Implicit methods are used to overcome this problem.

Write a finite difference form for the time derivative and average the right-hand sides, evaluated at the old and new times.

$$\frac{c_i^{n+1} - c_i^n}{\Delta t} = D(1-\theta)\frac{c_{i+1}^n - 2c_i^n + c_{i-1}^n}{\Delta x^2} + D\theta\frac{c_{i+1}^{n+1} - 2c_i^{n+1} + c_{i-1}^{n+1}}{\Delta x^2}$$

Now the equations are of the form

$$-\frac{D\Delta t\theta}{\Delta x^2}c_{i+1}^{n+1} + \left(1 + 2\frac{D\,\Delta t\theta}{\Delta x^2}\right)c_i^{n+1} - \frac{D\Delta t\theta}{\Delta x^2}c_{i-1}^{n+1}$$
$$= c_i^n + \frac{D\Delta t(1-\theta)}{\Delta x^2}(c_{i+1}^n - 2c_i^n + c_{i-1}^n)$$

and require solving a set of simultaneous equations, which have a tridiagonal structure. Using $\theta = 0$ gives the Euler method (as above), $\theta = 0.5$ gives the Crank-Nicolson method, and $\theta = 1$ gives the backward Euler method. The Crank-Nicolson method is also the same as applying the trapezoid rule to do the integration. The stability limit is given by

$$\Delta t \frac{D}{\Delta x^2} \le \frac{0.25}{1 - 2\theta}$$

The price of using implicit methods is that one now has a system of equations to solve at each time step, and the solution methods are more complicated (particularly for nonlinear problems) than the straightforward explicit methods. Phenomena that happen quickly can also be obliterated or smoothed over by using a large time step, so implicit methods are not suitable in all cases. The engineer must decide if she or he wants to track those fast phenomena, and choose an appropriate method that handles the time scales that are important in the problem.

Other methods can be used in space, such as the finite element method, the orthogonal collocation method, or the method of orthogonal collocation on finite elements. One simply combines the methods for ordinary differential equations (see Ordinary Differential Equations—Boundary-Value Problems) with the methods for initial-value problems (see Numerical Solution of Ordinary Differential Equations as Initial-Value Problems). Fast Fourier transforms can also be used on regular grids (see Fast Fourier Transform).

Elliptic Equations Elliptic equations can be solved with both finite difference and finite element methods. One-dimensional elliptic problems are two-point boundary-value problems. Two- and three-dimensional elliptic problems are often solved with iterative methods when the finite difference method is used and with direct methods when the finite element method is used. So there are two aspects to consider: how the equations are discretized to form sets of algebraic equations and how the algebraic equations are then solved.

The prototype elliptic problem is steady-state heat conduction or diffusion

$$k\left(\frac{\partial^2 T}{\partial x^2} + \frac{\partial^2 T}{\partial y^2}\right) = Q$$

possibly with a heat generation term per unit volume Q. The boundary conditions taken here are $T = f(x, y)$ on the boundary (S) with f a known function. Illustrations are given for constant thermal conductivity k while Q is a known function of position. The finite difference formulation is given using the following nomenclature:

$$T_{i,j} = T(i\,\Delta x, j\,\Delta y)$$

The finite difference formulation is then (see Fig. 3-50)

$$\frac{T_{i+1,j} - 2T_{i,j} + T_{i-1,j}}{\Delta x^2} + \frac{T_{i,j+1} - 2T_{i,j} + T_{i,j-1}}{\Delta y^2} = Q_{i,j} \qquad (3\text{-}69)$$
$$T_{i,j} = f(x_i, y_j) \text{ on } S$$

If the boundary is parallel to a coordinate axis, any derivative is evaluated as in the section on boundary-value problems, using either a one-sided, centered difference or a false boundary. If the boundary is more irregular and not parallel to a coordinate line, then more complicated expressions are needed and the finite element method may be the better method.

Equation (3-69) provides a set of linear equations that must be solved. These equations and their boundary conditions may be written in matrix form as

$$\mathbf{At} = \mathbf{f}$$

where **t** is the set of temperatures at all the points, **f** is the set of heat generation terms at all points, and **A** is formed from the coefficients of T_{ij} in Eq. (3-69).

The solution can be obtained simply by solving the set of linear equations. For three-dimensional problems, the matrix **A** is sparse, and iterative methods are used. These include Gauss-Seidel, alternating direction, overrelaxation methods, conjugate gradient, and multigrid methods. In Gauss-Seidel methods, one writes the equation for T_{ij} in terms of the other temperatures and cycles through all the points over and over. In the alternating direction method, one solves along one line (that is, x = constant), keeping the side values fixed, and then repeats this for all lines, and then repeats the process. Multigrid methods solve the problem on successively refined grids, which has advantages for both convergence and error estimation. Conjugate gradient methods frequently use a preconditioned matrix. The equation is multiplied by another matrix, which is chosen so that the resulting problem is easier to solve than the original one. Finding such matrices is an art, but it can speed convergence. The generalized minimal residual method is described in http://mathworld.wolfram.com/ GeneralizedMinimalResidualMethod.html. Additional resources can be found at http://www.netlib.org/linalg/html_templates/Templates.html. When the problem is nonlinear, the iterative methods may not converge, or the mesh may have to be refined before they converge, so some experimentation is sometimes necessary.

Spreadsheets can be used to solve two-dimensional problems on rectangular grids. The equation for T_{ij} is obtained by rearranging Eq. (3-69).

$$2\left(1+\frac{\Delta x^2}{\Delta y^2}\right)T_{i,j}=T_{i+1,j}+T_{i-1,j}+\frac{\Delta x^2}{\Delta y^2}(T_{i,j+1}+T_{1,j-1})-\Delta x^2\frac{Q_{i,j}}{k}$$

This equation is inserted into a cell and copied throughout the space represented by all the cells; when the iteration feature is turned on, the solution is obtained.

The Galerkin finite element method (FEM) is useful for solving elliptic problems and is particularly effective when the domain or geometry is irregular. As an example, cover the domain with triangles and define a trial function on each triangle. The trial function takes the value 1.0 at one corner and the value 0.0 at the other corners and is linear in between. For a triangle with corners at $(x, y) = (0, 0.58)$, $(0.66, 0)$, and $(1, 0.66)$ one of three trial functions is shown in Fig. 3-51. These trial functions on each triangle are pieced together to give a trial function on the whole domain. General treatments of the finite element method are available (see references). The steps in the solution method are similar to those described for boundary-value problems, except now the problems are much bigger so that the numerical analysis must be done very carefully to be efficient. Most engineers, however, just use a finite element program without generating it. There are three major caveats that must be addressed. First, the solution is dependent on the mesh laid down, and the only way to assess the accuracy of the solution is to solve the problem with a more refined mesh. Second, the solution obeys the shape of the trial function inside the element. Thus, if linear functions are used on triangles, a three-dimensional view of the solution, plotting the solution versus x and y, consists of a series of triangular planes joined together at the edges, as in a geodesic dome. Third, the Galerkin finite element method is applied to both the differential equations and the boundary conditions. Computer programs are usually quite general and may allow the user to specify boundary conditions that are not realistic. Also, natural boundary conditions are satisfied if no other boundary condition (ones involving derivatives) is set at a node. Thus, the user of finite element codes must be very clear what boundary conditions and differential equations are built into the computer code. When the problem is nonlinear, the Newton-Raphson method is used to iterate from an initial guess. Nonlinear problems lead to complicated integrals to evaluate, and they are usually evaluated using gaussian quadrature.

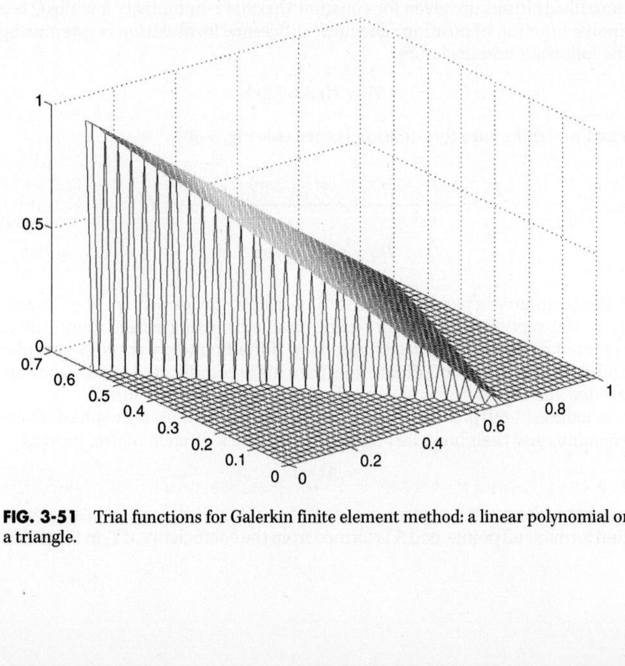

FIG. 3-51 Trial functions for Galerkin finite element method: a linear polynomial on a triangle.

One nice feature of the finite element method is the use of natural boundary conditions. It may be possible to solve the problem on a domain that is shorter than needed to reach some limiting condition (such as at an outflow boundary). The externally applied flux is still applied at the shorter domain, and the solution *inside* the truncated domain is still valid. Examples are given in Chang, M. W., and B. A. Finlayson, *Int. J. Num. Methods Eng.* **15,** 935–942 (1980), and Finlayson, B. A., *Numerical Methods for Problems with Moving Fronts*, Ravenna Park Publishing, Seattle, Wash. (1992). The effect of this is to allow solutions in domains that are smaller, thus saving computation time and permitting the solution in semi-infinite domains.

The trial functions in the finite element method are not limited to linear ones. Quadratic functions and even higher-order functions are frequently used. The same considerations hold as for boundary-value problems: The higher-order trial functions converge faster, but require more work. It is possible to refine both the mesh h and the power of polynomial in the trial function p in an h-p method. Some problems have constraints on some of the variables. For flow problems, the pressure must usually be approximated by using a trial function that is one order lower than the polynomial used to approximate the velocity.

Hyperbolic Equations The most common situation yielding hyperbolic equations involves unsteady phenomena with convection. Two typical equations are the convective diffusive equation

$$\frac{\partial c}{\partial t}+u\frac{\partial c}{\partial x}=D\frac{\partial^2 c}{\partial x^2}$$

and the chromatography equation. (See Partial Differential Equations.) If the diffusion coefficient is zero, the convective diffusion equation is hyperbolic. If D is small, the phenomenon may be essentially hyperbolic, even though the equations are parabolic. Thus the numerical methods for hyperbolic equations may be useful even for special parabolic equations.

Equations for several methods are given here. If the convective term is treated with a centered difference expression, the solution exhibits oscillations from node to node, and these go away only if a very fine grid is used. The simplest way to avoid the oscillations with a hyperbolic equation is to use upstream derivatives. If the flow is from left to right, this would give

$$\frac{dc_i}{dt}+u\frac{c_i-c_{i-1}}{\Delta x}=D\frac{c_{i+1}-2c_i+c_{i-1}}{\Delta x^2}$$

The effect of using upstream derivatives is to add artificial or numerical diffusion to the model. This can be ascertained by rearranging the finite difference form of the convective diffusion equation

$$\frac{dc_i}{dt}+u\frac{c_{i+1}-c_{i-1}}{2\Delta x}=\left(D+\frac{u\Delta x}{2}\right)\frac{c_{i+1}-2c_i+c_{i-1}}{\Delta x^2}$$

Thus the diffusion coefficient has been changed from

$$D \text{ to } D+\frac{u\Delta x}{2}$$

Alternatively, the diffusion coefficient has been multiplied by the factor

$$D'=D\left(1+\frac{u\Delta x}{2D}\right)=D\left(1+\frac{\text{Pe}_{\text{cell}}}{2}\right)$$

where $\text{Pe}_{\text{cell}}=\frac{u\Delta x}{D}=\frac{uL}{D}\frac{\Delta x}{L}=\text{Pe}\frac{\Delta x}{L}$ is called the *cell Peclet number.* When the diffusion coefficient is very small (or diffusion is slow compared with convection), the Peclet number will be large. In that case, extraneous diffusion will be included in the solution unless the mesh size (denoted by Δx) is small compared with the characteristic length of the problem. To avoid this problem (by keeping the factor small), very fine meshes must be used, and the smaller the diffusion coefficient, the smaller the required mesh size.

A variety of other methods are used to obtain a good solution without using extremely fine meshes. The flux correction methods keep track of the flux of material into and out of a cell (from one node to another) and put limits on the flux to make sure that no more material leaves the cell than is there originally plus the input amount. See Finlayson, B. A., *Numerical*

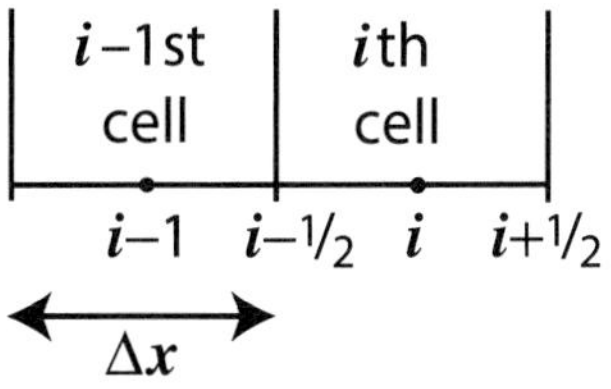

FIG. 3-52 Nomenclature for finite volume method.

Methods for Problems with Moving Fronts, Ravenna Park Publishing, Seattle, Wash., 1992, for many examples.

All the methods have a limit to the time step that is set by the convection term. Essentially, the time step should not be so big as to take the material farther than it can go at its velocity. This is usually expressed as a *Courant number limitation.*

$$\mathrm{Co} = \frac{u\Delta t}{\Delta x} \le 1$$

Some methods require a smaller limit, depending upon the amount of diffusion present (see Finlayson, 1992, Appendix).

In the finite element method, Petrov-Galerkin methods are used to minimize the unphysical oscillations. The Petrov-Galerkin method essentially adds a small amount of diffusion in the flow direction to smooth the unphysical oscillations. The amount of diffusion is usually proportional to Δx so that it becomes negligible as the mesh size is reduced. The value of the Petrov-Galerkin method lies in being able to obtain a smooth solution when the mesh size is large, so that the computation is feasible. This is not so crucial in one-dimensional problems, but it is essential in two- and three-dimensional problems and purely hyperbolic problems.

Finite Volume Methods Finite volume methods are utilized extensively in computational fluid dynamics. An excellent presentation is by LeVeque (2002). In this method, a mass balance is made over a cell, accounting for the change in what is in the cell, and the flow in and out. Figure 3-52 illustrates the geometry of the ith cell. A mass balance made on this cell (with area A perpendicular to the paper) is

$$A\,\Delta x\,(c_i^{n+1} - c_i^n) = \Delta t\,A\,(J_{i-1/2} - J_{i+1/2})$$

where J is the flux due to convection and diffusion, positive in the $+x$ direction.

$$J = uc - D\frac{\partial c}{\partial x},\quad J_{i-1/2} = u_{i-1/2}c_{i-1/2} - D\frac{c_i - c_{i-1}}{\Delta x}$$

The concentration at the edge of the cell is taken as

$$c_{i-1/2} = \frac{1}{2}(c_i + c_{i-1})$$

Rearrangement for the case when the velocity u is the same for all nodes gives

$$\frac{c_i^{n+1} - c^n{}_i}{\Delta t} + \frac{u(c_{i+1} - c_{i-1})}{2\Delta x} = \frac{D}{\Delta x^2}(c_{i+1} - 2c_i + c_{i+1})$$

This is the same equation obtained by using the finite difference method. This coincidence occurs only when the velocity is constant, which isn't usually true. In two and three dimensions, the mesh need not be rectangular, as long as it is possible to compute the velocity normal to an edge of the cell. The finite volume method is useful for applications involving filling, such as injection molding, when only part of the cell is filled with fluid. Such applications do involve some approximations, since the interface is not tracked precisely, but they are useful engineering approximations.

Parabolic Equations in Two or Three Dimensions Computations become much more lengthy when there are two or more spatial dimensions. For example, we may have the unsteady heat conduction equation

$$\rho C_p \frac{\partial T}{\partial t} = k\left(\frac{\partial^2 T}{\partial x^2} + \frac{\partial^2 T}{\partial y^2}\right) - Q$$

Most engineers use computer packages to solve such problems. If there is both convection and diffusion in the problem, the same considerations apply: A fine mesh is needed when the Peclet number is large. The upstream weighting and Petrov-Galerkin methods can be used, but it is important to apply the smoothing only in the direction of flow, since smoothing in the direction transverse to the flow direction would be incorrect. Some transverse smoothing is unavoidable, but the engineer needs to be sure that the smoothing is just enough to allow a good solution without creating large errors. See Finlayson (1980) in General References; Kuzmin, D., and J. Hämäläinen, *Finite Element Methods for Computational Fluid Dynamics: A Practical Guide*, SIAM-Soc. Ind. Appl. Math., 2014; and Layton, W., *Introduction to the Numerical Analysis of Incompressible Viscous Flows*, SIAM, 2008.

Computer Software When one is choosing computer software to solve a problem, there are a number of important considerations. The first decision is whether to use an approximate, engineering flow model, developed from correlations, or to solve the partial differential equations that govern the problem. Correlations are quick and easy to apply, but they may not be appropriate to the problem or give the needed detail. When one is using a computer package to solve partial differential equations, the first task is always to generate a mesh covering the problem domain. This is not a trivial task, and special methods have been developed to permit importation of a geometry from a computer-aided design (CAD) program. Then the mesh must be created automatically. If the boundary is irregular, the finite element method is especially well suited, although special embedding techniques can be used in finite difference methods (which are designed to be solved on rectangular meshes). Another capability to consider is the ability to track free surfaces that move during the computation. This phenomenon introduces the same complexity that occurs in problems with a large Peclet number, with the added difficulty that the free surface moves between mesh points and improper representation can lead to unphysical oscillations. The method used to solve the equations is important, and both explicit and implicit methods (as described above) can be used. Implicit methods may introduce unacceptable extra diffusion, so the engineer needs to examine the solution carefully. The methods used to smooth unphysical oscillations from node to node are also important, and the engineer needs to verify that the added diffusion or smoothing does not give inaccurate solutions. Since current-day problems are mostly nonlinear, convergence is always an issue since the problems are solved iteratively. Robust programs provide several methods for convergence, each of which is best in some circumstance or other. It is wise to have a program that includes many iterative methods. If the iterative solver is not very robust, the only recourse to solving a steady-state problem may be to integrate the time-dependent problem to steady state. The solution time may be long, and the final result may be further from convergence than would be the case if a robust iterative solver were used.

A variety of computer programs are available on the Internet, some free. First consider general-purpose programs. The website http://www.netlib.org/pdes/index.html lists programs for 2D elliptic partial differential equations as well as Clawpack for hyperbolic systems of equations from LeVeque (2002). On the NIST website http://gams.nist.gov/ search on "partial differential equations." Lau (2007) provides many programs in C++ (also see http://numerical.recipes /). The multiphysics program Comsol Multiphysics also solves many standard equations arising in mathematical physics.

Computational fluid dynamics (CFD) programs are more specialized, and most have been designed to solve sets of equations that are appropriate to specific industries. They can then include approximations and correlations for some features that would be difficult to solve for directly. ANSYS (http://www.ansys.com) is a major program having incorporated both Fluent and CFX. Comsol Multiphysics (http://www.comsol.com) is particularly useful because it incorporates many different types of physics (and equations), has a convenient graphical-user interface, permits easy mesh generation and refinement (including adaptive mesh refinement), allows the user to add phenomena and equations easily, permits solution by continuation methods (thus enhancing convergence), and has extensive graphical output capabilities. Other packages are also available (see http://cfd-online.com/), and these may contain features and correlations specific to the engineer's industry. One important point to note is that for turbulent flow, all the programs contain approximations, using the k-epsilon models of turbulence, or large eddy simulations; the direct numerical simulation of turbulence is too slow to apply to very big problems, although it does give insight (independent of any approximations) that is useful for interpreting turbulent phenomena. Thus, the method used to include those turbulent correlations is important, and the method also may affect convergence or accuracy.

FAST FOURIER TRANSFORM

The discrete Fourier transform can be used to differentiate a function, and this is used in the spectral method for solving differential equations as well as in modeling turbulent flow. Gottlieb, D., and S. A. Orszag, *Numerical Analysis of Spectral Methods: Theory and Applications*, SIAM, Philadelphia, Penn., 1977, discusses why they work; Trefethen, L. N., *Spectral Methods in Matlab*, SIAM, Philadelphia, Penn., 2000, shows how to use them in MATLAB. Suppose we have a grid of equidistant points

$$x_n = n\Delta x,\ \ n = 0, 1, 2, \ldots, 2N-1,\ \ \Delta x = \frac{L}{2N}$$

The solution is known at each of these grid points $\{y(x_n)\}$. First the discrete Fourier transform is taken:

$$Y_k = \frac{1}{2N}\sum_{n=0}^{2N-1} y(x_n)e^{-2ik\pi x_n/L} \qquad k = -N, -N+1, \ldots, 0, \ldots, N-1, N$$

The inverse transformation is

$$y(x) = \frac{1}{L}\sum_{k=-N}^{N} Y_k e^{2ik\pi x/L}$$

Differentiate this to get

$$\frac{dy}{dx} = \frac{1}{L}\sum_{k=-N}^{N} Y_k \frac{2\pi ik}{L} e^{2ik\pi x/L}$$

Thus at the grid points

$$\left.\frac{dy}{dx}\right|_n = \sum_{k=-N}^{N} Y_k \frac{2\pi ik}{L} e^{2ik\pi x_n/L}$$

The process works as follows. From the solution at all grid points the Fourier transform is obtained by using the fast Fourier transform (FFT), $\{Y_k\}$. Then this is multiplied by $2\pi ik/L$ to obtain the Fourier transform of the derivative.

$$Y_k' = Y_k \frac{2\pi ik}{L}$$

Then the inverse Fourier transform is taken using FFT, giving the value of the derivative at each of the grid points.

$$\left.\frac{dy}{dx}\right|_n = \frac{1}{L}\sum_{k=-N}^{N} Y_k' e^{2ik\pi x_n/L}$$

The spectral method is used for direct numerical simulation (DNS) of turbulence. The Fourier transform is taken of the differential equation, and the resulting equation is solved. Then the inverse transformation gives the solution. When there are nonlinear terms, as in turbulent flow, they are calculated at each node in physical space, and the Fourier transform is taken of the result. This technique is especially suited to time-dependent problems, and the major computational effort is in the fast Fourier transform.

OPTIMIZATION

References: General references include the following textbooks. For nonlinear programming, see Nocedal, J., and S. J. Wright, *Numerical Optimization*, Springer, New York, 2006; Conn, A. R., N. Gould, and P. Toint, *Trust Region Methods*, SIAM, Philadelphia, Penn., 2000; Biegler, L. T., *Nonlinear Programming: Concepts, Algorithms and Applications to Chemical Engineering*, SIAM, Philadelphia, Penn., 2010; Edgar, T. F., D. M. Himmelblau, and L. S. Lasdon, *Optimization of Chemical Processes*, McGraw-Hill, New York, 2002. For operations research and linear programming, Hillier, F. S., and G. J. Lieberman, *Introduction to Operations Research*, McGraw-Hill, New York, 2015. For mixed integer programming, Nemhauser, G. L., and L. A. Wolsey, *Integer and Combinatorial Optimization*, Wiley, New York, 1999. For global optimization and mixed integer nonlinear programming, Floudas, C. A., *Deterministic Global Optimization: Theory, Algorithms and Applications*, Kluwer, Norwell, Mass., 2000; Tawarmalani, M., and N. Sahinidis, *Convexification and Global Optimization in Continuous and Mixed-Integer Nonlinear Programming: Theory, Algorithms, Software, and Applications*, Kluwer, 2002. Many useful resources including descriptions, trial software, and examples can be found on the NEOS server maintained at Argonne National Laboratory. Background material for this section includes the two previous sections on matrix algebra and numerical analysis.

INTRODUCTION

Optimization is a key enabling tool for decision making in chemical engineering. It has evolved from a methodology of academic interest into a technology that continues to have a significant impact on engineering research and practice. Optimization algorithms form the core tools for (1) experimental design, parameter estimation, model development, and statistical analysis; (2) process synthesis analysis, design, and retrofit; (3) model predictive control and real-time optimization; and (4) planning, scheduling, and the integration of process operations into the supply chain.

As shown in Fig. 3-53, optimization problems that arise in chemical engineering can be classified in terms of continuous and discrete variables. For the former, nonlinear programming (NLP) problems form the most general case, and widely applied specializations include linear programming (LP) and quadratic programming (QP). An important distinction for NLP is whether the optimization problem is convex or nonconvex. The latter NLP problem may have multiple local optima, and an important question is whether a global solution is required for the NLP. Another important distinction is whether the problem is assumed to be differentiable or not.

Mixed integer problems also include discrete variables. These can be written as mixed integer nonlinear programs (MINLPs), or as mixed integer linear programs (MILP), if all variables appear linearly in the constraint and objective functions. For the latter an important case occurs when all the variables are integer; this gives rise to an integer programming (IP) problem. IP problems can be further classified into many special problems (e.g., assignment, traveling salesperson, etc.), which are not shown in Fig. 3-53. Similarly, the MINLP problem also gives rise to special problem classes, although here the main distinction is whether its relaxation is convex or nonconvex.

The ingredients of formulating optimization problems include a mathematical model of the system, an objective function that quantifies a criterion to be extremized, variables that can serve as decisions, and, optionally, inequality constraints on the system. When represented in algebraic form, the general formulation of discrete and continuous optimization problems can be written as the following mixed integer optimization problem:

$$\text{Min}\, f(x, y) \quad \text{subject to} \quad h(x, y) = 0 \;\; g(x, y) \le 0 \;\; x \in \mathcal{R}^n \;\; y \in \{0, 1\}^t \qquad (3\text{-}70)$$

where $f(x, y)$ is the objective function (e.g., cost, energy consumption, etc.), $h(x, y) = 0$ are the equations that describe the performance of the system (e.g., material balances, production rates), and the inequality constraints $g(x, y) \le 0$ may define process specifications or constraints for feasible plans and schedules. Note that the operator max $f(x, y)$ is equivalent to Min$[-f(x, y)]$. We define the real n vector x to represent the continuous variables while the t vector y represents the discrete variables, which, without loss of generality, are often restricted to take values of 0 or 1 to define logical or discrete decisions, such as assignment of equipment and sequencing of tasks. (These variables can also be formulated to take on other integer values as well.) Problem (3-70) corresponds to a mixed integer nonlinear program

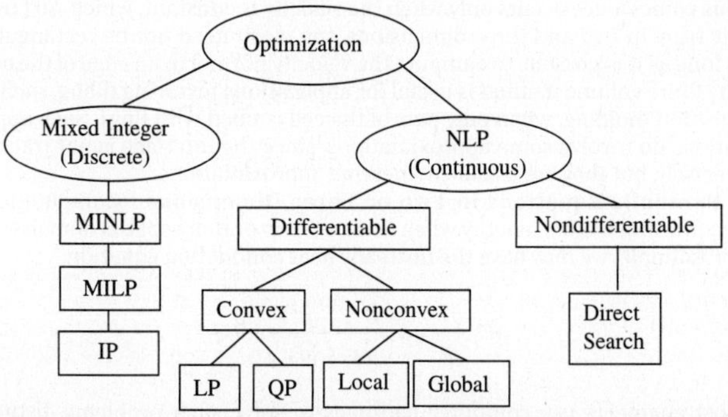

FIG. 3-53 Classes of optimization problems and algorithms.

when any of the functions involved are nonlinear. If all functions are linear, it corresponds to a mixed integer linear program (3-89). If there are no 0–1 variables, then problem (3-70) reduces to a nonlinear program (3-71) or linear program (3-78) depending on whether the functions are linear.

Following the road map in Fig. 3-53, we start with continuous variable optimization and consider in the next section the solution of NLP problems with differentiable objective and constraint functions. If only local solutions are required for the NLP problem, then very efficient large-scale methods can be considered. This is followed by methods that are not based on local optimality criteria; we consider direct search optimization methods that do not require derivatives as well as deterministic global optimization methods. Following this, we consider the solution of mixed integer problems and outline the main characteristics of algorithms for their solution. Finally, we conclude with a discussion of optimization modeling software and its implementation on engineering models.

GRADIENT-BASED NONLINEAR PROGRAMMING

For continuous variable optimization, we consider Eq. (3-70) without discrete variable y. The general NLP problem (3-71) is presented here:

$$\text{Min} f(x) \quad \text{subject to} \quad h(x) = 0, g(x) \leq 0 \tag{3-71}$$

and we assume that the functions $f(x)$, $h(x)$, and $g(x)$ have continuous first and second derivatives. A key characteristic of Eq. (3-71) is whether the problem is convex or not, i.e., whether it has a convex objective function and a convex feasible region. A function $\phi(x)$ of x in some domain X is convex if and only if, for all points $x_1, x_2 \in X$,

$$\phi[\alpha x_1 + (1-\alpha)x_2] \leq \alpha\phi[x_1 + (1-\alpha)x_2] + (1-\alpha)\phi(x_2) \tag{3-72}$$

holds for all $\alpha \in (0, 1)$. [Strict convexity requires that the inequality Eq. (3-72) be strict.] Convex feasible regions require $g(x)$ to be a convex function and $h(x)$ to be linear. Referring to Fig. 3-53, problem (3-71) is convex if and only if $f(x)$ and $g(x)$ are convex functions and $h(x)$ is a linear function. Otherwise, the problem is nonconvex. If Eq. (3-71) is a convex problem, then any local solution is guaranteed to be a global solution to Eq. (3-71). Moreover, if the objective function is strictly convex, then this solution x^* is unique. On the other hand, nonconvex problems may have multiple local solutions, i.e., feasible solutions x^* only within some nonvanishing neighborhood.

We first consider methods that find only local solutions to nonconvex problems, as more difficult (and expensive) search procedures are required to find a global solution. Local methods are currently very efficient and have been developed to deal with very large NLP problems. Moreover, by considering the structure of convex NLP problems (including LP and QP problems), even more powerful methods can be applied. To study these methods, we first consider conditions for local optimality.

Local Optimality Conditions: A Kinematic Interpretation Instead of a formal development of conditions that define a local optimum, we present a more intuitive kinematic illustration. Consider the contour plot of the objective function $f(x)$, given in Fig. 3-54, as a smooth valley in space of variables x_1 and x_2. For the contour plot of this unconstrained problem, Min $f(x)$, consider a ball rolling in this valley to the lowest point of $f(x)$, denoted by x^*. This point is at least a local minimum and is defined by a point with a zero gradient and at least nonnegative curvature in all (nonzero) directions p. We use the first-derivative (*gradient*) vector $\nabla f(x)$ and second-derivative (*hessian*) matrix $\nabla_{xx} f(x)$ to state the necessary first- and second-order conditions for unconstrained optimality:

$$\nabla_x f(x^*) = 0 \quad p^T \nabla_{xx} f(x^*) p \geq 0 \quad \text{for all } p \neq 0 \tag{3-73}$$

These necessary conditions for local optimality can be strengthened to sufficient conditions by making the inequality in Eq. (3-73) strict (i.e., positive curvature in all directions). Equivalently, the sufficient (necessary) curvature conditions can be stated as follows: $\nabla_{xx} f(x^*)$ has all positive (nonnegative) eigenvalues and is therefore defined as a positive (semidefinite) definite matrix.

Now consider the imposition of inequality $g(x) \leq 0$ and equality constraints $h(x) = 0$ in Fig. 3-55. Continuing the kinematic interpretation, the inequality constraints $g(x) \leq 0$ act as "fences" in the valley, and equality constraints $h(x) = 0$ act as "rails." Consider now a ball, constrained on a rail and within fences, to roll to its lowest point. This stationary point occurs when the normal forces exerted by the fences $[-\nabla g(x^*)]$ and rails $[-\nabla h(x^*)]$ on the ball are balanced by the force of gravity $[-\nabla f(x^*)]$. This condition can be stated by the following *Karush-Kuhn-Tucker (KKT) necessary conditions* for constrained optimality.

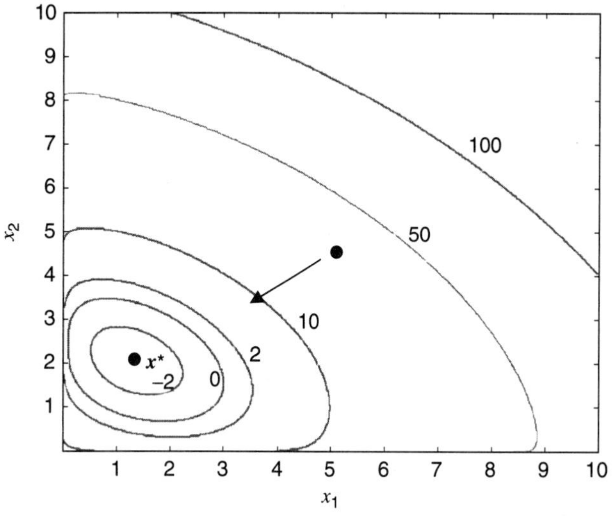

FIG. 3-54 Unconstrained minimum.

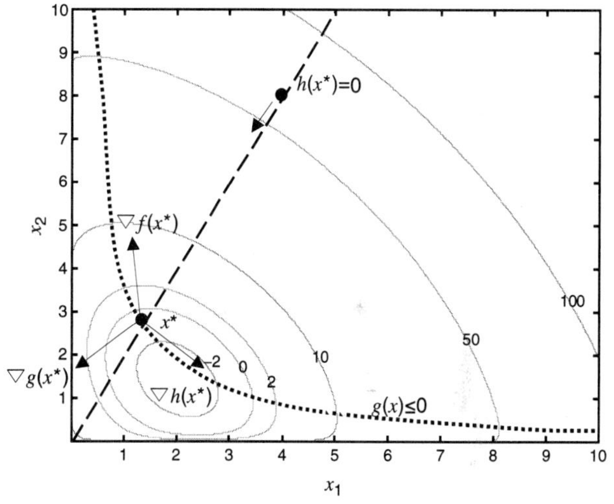

FIG. 3-55 Constrained minimum.

Balance of Forces It is convenient to define the Lagrange function $L(x, \lambda, \nu) = f(x) + g(x)^T\lambda + h(x)^T\nu$, along with "weights" or multipliers λ and ν for the constraints. The stationarity condition (balance of forces acting on the ball) is then given by

$$\nabla L(x, \lambda, \nu) = \nabla f(x) + \nabla h(x)\,\lambda + \nabla g(x)\,\nu = 0 \tag{3-74}$$

Feasibility Both inequality and equality constraints must be satisfied (the ball must lie on the rail and within the fences):

$$h(x) = 0 \qquad g(x) \leq 0 \tag{3-75}$$

Complementarity Inequality constraints are either strictly satisfied (active) or inactive, in which case they are irrelevant to the solution. In the latter case the corresponding KKT multiplier must be zero. This is written as

$$\nu^T g(x) = 0 \qquad \nu \geq 0 \tag{3-76}$$

Constraint Qualification For a local optimum to satisfy the KKT conditions, an additional regularity condition is required on the constraints. This can be defined in several ways. A typical condition is that the active constraints at x^* be linearly independent; i.e., the matrix $[\nabla h(x^*) | \nabla g_A(x^*)]$ is full column rank, where g_A is the vector of inequality constraints with elements that satisfy $g_{A,i}(x^*) = 0$. With this constraint qualification, the KKT multipliers (λ, ν) are guaranteed to be unique at the optimal solution.

Second-Order Conditions As with unconstrained optimization, nonnegative (positive) curvature is necessary (sufficient) in all the allowable (i.e., constrained) nonzero directions p. The necessary second-order conditions can be stated as

$$p^T \nabla_{xx} L(x^*) p \geq 0$$

$$\text{for all } p \neq 0 \text{ with } \nabla h(x^*)^T p = 0, \nabla g_A(x^*)^T p \leq 0, \nabla g_{A,i}(x^*)^T p = 0 \text{ for } \nu_i > 0 \qquad (3\text{-}77)$$

and the corresponding sufficient conditions require the first inequality in Eq. (3-77) to be strict. Note that in Fig. 3-54, the allowable directions p span the entire space for x while in Fig. 3-55 there are *no* allowable directions p.

Convex Cases of NLP Problems Linear programs and quadratic programs are special cases of Eq. (3-71) that allow for more efficient solution, based on application of KKT conditions Eq. (3-74) through Eq. (3-77). Because these are convex problems, any locally optimal solution is a global solution. In particular, if the objective and constraint functions in Eq. (3-71) are linear, then the following linear program (LP)

$$\text{Min } c^T x \quad \text{subject to} \quad Ax = b \text{ and } Cx \leq d \qquad (3\text{-}78)$$

can be solved in a finite number of steps, and the optimal solution lies at a vertex of the polyhedron described by the linear constraints. This is shown in Fig. 3-56, and in so-called primal degenerate cases, multiple vertices can be alternate optimal solutions, with the same values of the objective function. The standard method to solve Eq. (3-78) is the simplex method, developed in the late 1940s, although since Karmarkar's discovery in 1984 interior point methods have also become quite advanced and competitive for highly constrained problems. The simplex method proceeds by moving successively from vertex to vertex with improved objective function values. Methods to solve Eq. (3-78) are well implemented and widely used, especially in planning and logistical applications. They also form the basis for MILP methods discussed later. Currently, state-of-the-art LP solvers can handle millions of variables and constraints, and the application of further decomposition methods leads to the solution of problems that are two or three orders of magnitude larger than this. See the general references of Hillier and Lieberman (2015) and Nocedal and Wright (2006) for more details. Also, the interior point method is described below from the perspective of more general NLP problems.

Quadratic programs (QPs) represent a slight modification of Eq. (3-78) and can be stated as

$$\text{Min } c^T x + 1/2\, x^T Q x \qquad \text{subject to} \qquad Ax = b \quad Cx \leq d \qquad (3\text{-}79)$$

If the matrix Q is positive semidefinite (positive definite) when projected into the null space of the active constraints, then Eq. (3-79) is (strictly) convex and the QP is a global (and unique) minimum. Otherwise, local solutions may exist for Eq. (3-79), and more extensive global optimization methods are needed to obtain the global solution. Like LPs, convex QPs can be solved in a finite number of steps. However, as seen in Fig. 3-57, these optimal solutions may lie on a vertex, on a constraint boundary, or in the interior. A number of active set strategies have been created that solve the KKT conditions of the QP and incorporate efficient updates of active constraints. Popular methods include null space algorithms, range space methods, and Schur complement methods. As with LPs, QP problems can also be solved with interior point methods.

Solving the General NLP Problem Solution techniques for Eq. (3-71) deal with satisfaction of the KKT conditions, Eq. (3-74) through Eq. (3-77). Many NLP solvers are based on successive quadratic programming (SQP) as it allows the construction of a number of NLP algorithms based on the Newton-Raphson method for equation solving (see the Numerical Analysis section). SQP solvers have been shown to require the fewest function evaluations to solve NLP problems, and they can be tailored to a broad range of process engineering problems with different structure.

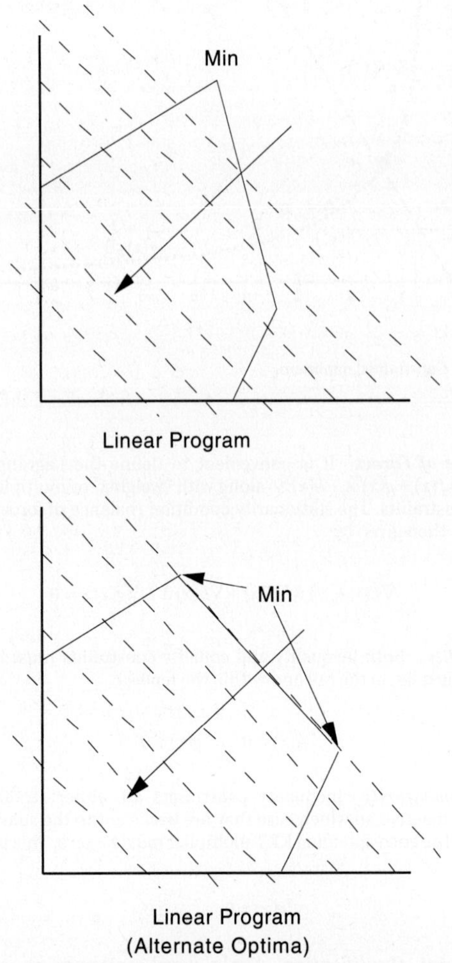

FIG. 3-56 Contour plots of linear programs.

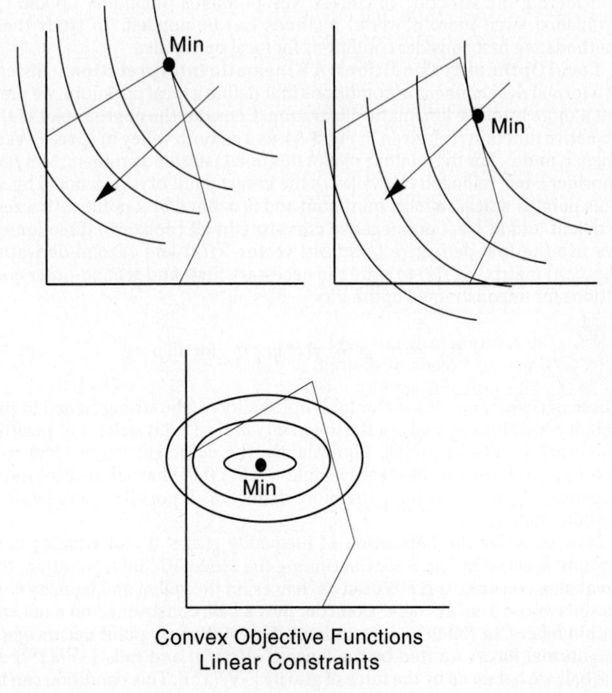

FIG. 3-57 Contour plots of convex quadratic programs.

The SQP strategy applies the equivalent of a Newton step to the KKT conditions of the nonlinear programming problem, and this leads to a fast rate of convergence. By adding slack variables s, the first-order KKT conditions can be rewritten as

$$\nabla f(x) + \nabla h(x)\,\lambda + \nabla g(x)\,\nu = 0 \tag{3-80a}$$
$$h(x) = 0 \tag{3-80b}$$
$$g(x) + s = 0 \tag{3-80c}$$
$$SVe = 0 \tag{3-80d}$$
$$(s, \nu) \geq 0 \tag{3-80e}$$

where $e = [1, 1, \ldots, 1]^T$, $S = \text{diag}\{s\}$, and $V = \text{diag}\{\nu\}$. SQP methods find solutions that satisfy Eq. (3-80) by generating Newton-like search directions at iteration k. However, Eq. (3-80d) and active bounds Eq. (3-80e) are dependent at the solution and serve to make the KKT system ill conditioned near the solution. SQP algorithms treat these conditions in two ways. In the *active set strategy*, discrete decisions are made regarding the active constraint set $i \in I = \{i \mid g_i(x^*) = 0\}$, and Eq. (3-80$d$) is replaced by $s_i = 0$, $i \in I$, and $\nu_i = 0$, $i \notin I$. Determining the active set is a combinatorial problem, and a straightforward way to determine an estimate of the active set [and to satisfy Eq. (3-80e)] is to formulate and solve, at a point x^k, the following QP at iteration k:

$$\text{Min } \nabla f(x^k)^T p + 1/2\, p^T \nabla_{xx} L(x^k, \lambda^k, \nu^k) p$$
$$\text{subject to:}\quad h(x^k) + \nabla h(x^k)^T p = 0 \quad g(x^k) + \nabla g(x^k)^T p + s = 0 \quad s \geq 0 \tag{3-81}$$

The KKT conditions of Eq. (3-81) are given by

$$\nabla f(x^k) + \nabla^2 L(x^k, \lambda^k, \nu^k) p + \nabla h(x^k)\,\lambda + \nabla g(x^k)\,\nu = 0 \tag{3-82a}$$
$$h(x^k) + \nabla h(x^k)^T p = 0 \tag{3-82b}$$
$$g(x^k) + \nabla g(x^k)^T p + s = 0 \tag{3-82c}$$
$$SVe = 0 \tag{3-82d}$$
$$(s, \nu) \geq 0 \tag{3-82e}$$

where the hessian of the Lagrange function $\nabla_{xx} L(x, \lambda, \nu) = \nabla_{xx}[f(x) + h(x)^T\lambda + g(x)^T\nu]$ is calculated directly or through a quasi-Newton approximation (created by differences of gradient vectors). If Eq. (3-81) is strictly convex, it is easy to show that Eqs. (3-82a) through (3-82c) correspond to a Newton-Raphson step for Eqs. (3-80a) through (3-80c) applied at iteration k. Also, selection of the active set is now handled at the QP level by satisfying the conditions of Eqs. (3-82d) and (3-82e). To evaluate and change candidate active sets, QP algorithms apply inexpensive matrix updating strategies to the KKT matrix associated with Eq. (3-82). Details of this approach can be found in Nocedal and Wright (2006).

As alternatives that avoid the combinatorial problem of selecting the active set, interior point (or barrier) methods modify the NLP problem Eq. (3-71) to form

$$\text{Min} f(x^k) - \mu \Sigma_i \ln s_i \quad \text{subject to} \quad h(x^k) = 0 \quad g(x^k) + s = 0 \tag{3-83}$$

where the solution to Eq. (3-84) has $s > 0$ for the penalty parameter $\mu > 0$. Decreasing μ to 0 leads to solution of problem Eq. (3-71). The KKT conditions for this problem can be written as

$$\begin{aligned} &\nabla f(x^*) + \nabla h(x^*)\,\lambda + \nabla g(x^*)\,\nu = 0 \\ &h(x^*) = 0 \\ &g(x^*) + s = 0 \\ &SVe = \mu e \end{aligned} \tag{3-84}$$

and for $\mu > 0$, $s > 0$, and $\nu > 0$, Newton steps generated to solve Eqs. (3-84) are well behaved and analogous to Eq. (3-82), with a modification on the right-hand side of Eq. (3-82d). A detailed description of a particular interior point algorithm, called IPOPT, can be found in Wächter and Biegler [*Math. Prog.* **106**(1): 25–57 (2006)].

Both active set and interior point methods possess clear trade-offs. Interior point methods may require more iterations to solve Eqs. (3-84) for various values of μ, while active set methods require the solution of the more expensive QP subproblem Eq. (3-81). Thus, if there are few inequality constraints or an active set is known (say from a good starting guess, or a known QP solution from a previous iteration), then solving Eq. (3-81) is not expensive and the active set method is favored. However, for problems with many inequality constraints, interior point methods are often faster, as they avoid the combinatorial problem of selecting the active set. This is especially true for large-scale problems where a large number of bounds are active. Examples that demonstrate the application of these approaches include the solution of model predictive control (MPC) problems and the solution of large optimal control problems using barrier NLP solvers. For instance, IPOPT allows the solution of problems with more than 1,000,000 variables and up to 50,000 degrees of freedom [see Biegler et al., *Chem. Eng. Sci.* **57**(4): 575–593 (2002); Laird et al., *ASCE J. Water Resource Management and Planning* **131**(2):125 (2005)].

Other Gradient-Based NLP Solvers In addition to SQP methods, a number of NLP solvers have been developed and adapted for large-scale problems. Generally these methods require more function evaluations than for SQP methods, but they perform very well when interfaced to optimization modeling platforms, where function evaluations are cheap. All these can be derived from the perspective of applying Newton steps to portions of the KKT conditions.

LANCELOT (Conn et al., 2000) is based on the solution of bound-constrained subproblems. Here an augmented lagrangian is formed from Eq. (3-71), and the following subproblem is solved:

$$\text{Min} f(x) + \lambda^T h(x) + \nu^T[g(x) + s] + 1/2\,\rho\|h(x), g(x) + s\|^2 \quad \text{subject to} \quad s \geq 0 \tag{3-85}$$

The above subproblem can be solved very efficiently for fixed values of the multipliers λ and ν and penalty parameter ρ. Here a gradient projection trust region method is applied. Once subproblem Eq. (3-85) is solved, the multipliers and penalty parameter are updated in an outer loop, and the cycle repeats until the KKT conditions for Eq. (3-71) are satisfied. LANCELOT works best when exact second derivatives are available. This promotes a fast convergence rate in solving each subproblem and allows a bound-constrained trust region method to exploit directions of negative curvature in the hessian matrix.

Reduced gradient methods are active set strategies that rely on partitioning the variables and solving Eq. (3-80) in a nested manner. Without loss of generality, problem Eq. (3-71) can be rewritten as Min $f(z)$ subject to $c(z) = 0$ and $a \leq z \leq b$. Variables are partitioned as nonbasic variables (those fixed to their bounds), basic variables (those that can be solved from the equality constraints), and superbasic variables (those remaining variables between bounds that serve to drive the optimization); this leads to $z^T = [z_N^T, z_B^T, z_S^T]$. This partition is derived from local information and may change over the course of the optimization iterations. The corresponding KKT conditions can be written as

$$\nabla_N f(z) + \nabla_N c(z)\gamma = \beta_a - \beta_b \tag{3-86a}$$
$$\nabla_B f(z) + \nabla_B c(z)\gamma = 0 \tag{3-86b}$$
$$\nabla_S f(z) + \nabla_S c(z)\gamma = 0 \tag{3-86c}$$
$$c(z) = 0 \tag{3-86d}$$
$$z_{N,j} = a_j \text{ or } b_j \quad \beta_{a,j} \geq 0 \quad \beta_{b,j} = 0 \quad \text{or} \quad \beta_{b,j} \geq 0 \quad \beta_{a,j} = 0 \tag{3-86e}$$

where γ and β are the KKT multipliers for the equality and bound constraints, respectively, and Eq. (3-86e) replaces the complementarity conditions in Eq. (3-76). Reduced gradient methods work by nesting equations Eqs. (3-86b and d) within Eqs. (3-86a and c). At iteration k, for fixed values of z_N^k and z_S^k, we can solve for z_B by using Eq. (3-86d) and for γ by using Eq. (3-86b). Moreover, linearization of these equations leads to sensitivity information (i.e., constrained derivatives or reduced gradients) that indicates how z_B changes with respect to z_S and z_N. The algorithm then proceeds by updating z_S by using reduced gradients derived from Eq. (3-86b) and given by

$$df(z)/dz_S = \nabla_S f(z) + \nabla_S c(z)\,\gamma = \nabla_S f(z) - \nabla_S c(z)\,\nabla_B c(z)^{-1}\,\nabla_B f(z) \tag{3-87}$$

Driving df/dz_S to zero, with quasi-Newton or Newton iterations, solves Eq. (3-86c). Following this, bound multipliers β are calculated from Eq. (3-86a). Over the course of the iterations, if the variable z_B or z_S exceeds its bounds or if some bound multipliers β become negative, then the variable partition needs to be changed and Eqs. (3-86) are reconstructed. These reduced gradient methods are embodied in the popular GRG2, CONOPT, and SOLVER codes (Edgar et al., 2002). The SOLVER code has been incorporated into Microsoft Excel.

Algorithmic Details for NLP Methods All the above NLP methods incorporate concepts from the Newton-Raphson method for equation solving. Essential features of these methods are that they provide (1) accurate derivative information to solve for the KKT conditions, (2) stabilization strategies to promote convergence of the Newton-like method from poor starting points, and (3) regularization of the jacobian matrix in Newton's method (the so-called KKT matrix) if it becomes singular or ill conditioned.

1. *NLP methods that use first and second derivatives.* The KKT conditions require first derivatives to define stationary points, so accurate first derivatives are essential to determine locally optimal solutions for differentiable NLPs. Moreover, Newton-Raphson methods that are applied to the KKT conditions, as well as the task of checking second-order KKT conditions, necessarily require second-derivative information. (Note that second-order conditions are not checked by methods that do not use second derivatives.) With the recent development of automatic differentiation tools, many modeling and simulation platforms can provide exact first and second derivatives for optimization. When second derivatives are available for the objective or constraint functions, they can be used directly in LANCELOT as well as SQP and reduced gradient methods. Otherwise, on problems with few superbasic variables, both reduced gradient methods and SQP methods [with reduced gradient methods applied to the QP subproblem Eq. (3-81)] can benefit from positive definite quasi-Newton approximations (Nocedal and Wright, 2006) applied to reduced second-derivative quantities (the so-called reduced hessian). Finally, for problems with least squares functions (see Statistics subsection), as in data reconciliation, parameter estimation, and model predictive control, one often assumes that the values of the objective function and its gradient at the solution are vanishingly small. Under these conditions, one can show that the multipliers (λ, ν) also vanish and $\nabla_{xx}L(x, \lambda, \nu)$ can be substituted by $\nabla_{xx}f(x^*)$. This *Gauss-Newton approximation* has been shown to be very efficient for the solution of least squares problems (see Nocedal and Wright, 2006).

2. *Line search and trust region methods promote convergence from poor starting points.* These are commonly used with the search directions calculated from NLP subproblems such as Eq. (3-81). In a *trust region approach,* the constraint $||p|| \le \Delta$ is added, and the iteration step is taken if there is sufficient reduction of some merit function (e.g., the objective function weighted with some measure of the constraint violations). The size of the trust region Δ is adjusted based on the agreement of the reduction of the actual merit function compared to its predicted reduction from the subproblem (see Conn et al., 2000). Such methods have strong global convergence properties and are especially appropriate for ill-conditioned NLPs. This approach has been applied in the KNITRO code (see Nocedal and Wright, 2006). *Line search methods* can be more efficient on problems with reasonably good starting points and well-conditioned subproblems, as in real-time optimization. Typically, once a search direction is calculated from Eq. (3-81), or other related subproblem, a step size $\alpha \in (0, 1)$ is chosen so that $x^k + \alpha p$ leads to a sufficient decrease of a merit function. As a recent alternative, a novel *filter* stabilization strategy (for both line search and trust region approaches) has been developed based on a bicriterion minimization, with the objective function and constraint infeasibility as competing objectives [Fletcher et al., *SIAM J. Optim.* **13**(3):635 (2002)]. This method often leads to better performance than that based on merit functions.

3. *Regularization of the KKT matrix for the NLP subproblem is essential for good performance of general-purpose algorithms.* For instance, to obtain a unique solution to Eq. (3-81), active constraint gradients must be full rank and the hessian matrix, when projected into the null space of the active constraint gradients, must be positive definite. These properties may not hold far from the solution, and corrections to the hessian in SQP may be necessary. Regularization methods ensure that subproblems such as Eq. (3-81) remain well conditioned; they include addition of positive constants to the diagonal of the hessian matrix to ensure its positive definiteness, judicious selection of active constraint gradients to ensure that they are linearly independent, and scaling the subproblem to reduce the propagation of numerical errors. Often these strategies are heuristics built into particular NLP codes. While quite effective, most of these heuristics do not provide convergence guarantees for general NLPs.

From the conceptual descriptions as well as algorithmic details given above, it is clear that NLP solvers are complex algorithms that have required considerable research and development to turn them into reliable and efficient software tools. Practitioners who are confronted with engineering optimization problems should therefore leverage these efforts, rather than write their own codes. Table 3-3 presents a sampling of available NLP codes that represent the above classifications.

OPTIMIZATION METHODS WITHOUT DERIVATIVES

A broad class of optimization strategies does not require derivative information. These methods have the advantage of easy implementation and little prior knowledge of the optimization problem. In particular, such methods are well suited for "quick and dirty" optimization studies that explore the scope of optimization for new problems, prior to investing effort for more sophisticated modeling and solution strategies. Most of these methods are derived from heuristics that naturally spawn numerous variations. As a result, a very broad literature describes these methods. Here we discuss only a few important trends in this area.

TABLE 3-3 Representative NLP Solvers

Method	Algorithm type	Stabilization	Second-order information
CONOPT (Drud, 1994)	Reduced gradient	Line search	Exact and quasi-Newton
GRG2 (Edgar et al., 2002)	Reduced gradient	Line search	Quasi-Newton
IPOPT	SQP, barrier	Line search	Exact
KNITRO (Byrd et al., 1997)	SQP, barrier	Trust region	Exact and quasi-Newton
LANCELOT	Augmented Lagrangian, bound constrained	Trust region	Exact and quasi-Newton
LOQO	SQP, barrier	Line search	Exact
MINOS	Reduced gradient, augmented Lagrangian	Line search	Quasi-Newton
NPSOL	SQP, active set	Line search	Quasi-Newton
SNOPT	Reduced space SQP, active set	Line search	Quasi-Newton
SOCS	SQP, active set	Line search	Exact
SOLVER	Reduced gradient	Line search	Quasi-Newton
SRQP	Reduced space SQP, active set	Line search	Quasi-Newton

Classical Direct Search Methods Developed in the 1960s and 1970s, these methods include one-at-a-time search and methods based on experimental designs (EVOP). At that time, direct search methods were the most popular optimization methods in chemical engineering. Methods that fall into this class include the pattern search of Hooke and Jeeves [*J. ACM* **8:** 212 (1961)], the conjugate direction method of Powell (1964), the simplex search of Nelder-Mead [*Comput. J.* **7:** 308 (1965)], and the adaptive random search methods of Luus-Jaakola [*AIChE J.* **19:** 760 (1973)], Goulcher and Cesares Long [*Comp. Chem. Engr.* **2:** 23 (1978)], and Banga et al. [in *State of the Art in Global Optimization*, C. Floudas and P. Pardalos, eds., Kluwer, Dordrecht, 1996, p. 563]. All these methods require only objective function values for unconstrained minimization. Associated with these methods are numerous studies on a wide range of process problems. Moreover, many of these methods include heuristics that prevent premature termination (e.g., directional flexibility in the complex search as well as random restarts and direction generation).

Simulated Annealing This strategy is related to random search methods and derives from a class of heuristics with analogies to the motion of molecules in the cooling and solidification of metals (Laarhoven and Aarts, *Simulated Annealing: Theory and Applications*, Reidel Publishing, Dordrecht, 1987). Here a temperature parameter θ can be raised or lowered to influence the probability of accepting points that do not improve the objective function. The method starts with a base point x and objective value $f(x)$. The next point x' is chosen at random from a distribution. If $f(x') < f(x)$, the move is accepted with x' as the new point. Otherwise, x' is accepted with probability $p(\theta, x', x)$. Options include the Metropolis distribution $p(\theta, x, x') = \exp\{-[f(x') - f(x)]/\theta\}$ and the Glauber distribution, $p(\theta, x, x') = \exp\{-[f(x') - f(x)]/\theta\}/(1 + \exp\{-[f(x') - f(x)]/\theta\})$. The θ parameter is then reduced, and the method continues until no further progress is made.

Genetic Algorithms This approach, described in Holland, J. H., *Adaptations in Natural and Artificial Systems* (University of Michigan Press, Ann Arbor, 1975), is based on the analogy of improving a population of solutions through modifying their gene pool. It also has similar performance characteristics as random search methods and simulated annealing. Two forms of genetic modification, crossover or mutation, are used, and the elements of the optimization vector x are represented as binary strings. Crossover deals with random swapping of vector elements (among parents with highest objective function values or other rankings of population) or any linear combinations of two parents. Mutation deals with the addition of a random variable to elements of the vector. Genetic algorithms (GAs) have seen widespread use in process engineering, and a number of codes are available. Edgar et al. (2002) describe a related GA that is available in MS Excel.

Derivative-Free Optimization (DFO) Over the past two decades, the availability of parallel computers and faster computing hardware and the need to incorporate complex simulation models within optimization studies have led a number of optimization researchers to reconsider classical direct search approaches. In particular, Dennis and Torczon [*SIAM J. Optim.* **1:** 448 (1991)] developed a multidimensional search algorithm that extends the simplex approach of Nelder and Mead (1965). They note that the Nelder-Mead algorithm fails as the number of variables increases, even for very simple problems. To overcome this, their multidimensional pattern search approach combines reflection, expansion, and contraction steps that act as line search algorithms for a number of linearly independent search directions. This approach is easily adapted to parallel computation, and the method can be tailored to the number of processors available. Moreover, this approach converges to locally optimal solutions for unconstrained problems and observes an unexpected performance synergy when multiple processors are used. The work of Dennis and Torczon (1991) has spawned considerable research on the analysis and code development for DFO methods. In addition, Conn et al. (*Introduction to Derivative Free Optimization*, SIAM, Philadelphia, Penn., 2009) constructed a multivariable DFO algorithm that uses a surrogate model for the objective function within a trust region method. Here points are sampled to obtain a well-defined quadratic interpolation model, and descent conditions from trust region methods enforce convergence properties. A comprehensive overview and convergence analysis of pattern search, surrogate, and trust region DFO methods is presented in Conn, Scheinberg, and Vicente (2009). Moreover, several DFO codes have been developed that lead to black box optimization implementations for large, complex simulation models [see Audet and Dennis, *SIAM J. Optim.* **13:** 889 (2003); Kolda et al., *SIAM Rev.* **45**(3): 385 (2003)].

Direct search methods are easy to apply to a wide variety of problem types and optimization models. Moreover, because their termination criteria are not based on gradient information and stationary points, they are more likely to favor the search for globally optimal rather than locally optimal solutions. These methods can also be adapted easily to include integer variables. However, no rigorous convergence properties to globally optimal solutions have yet been discovered. Also, these methods are best suited for unconstrained problems or for problems with simple bounds. Otherwise, they may have difficulties with constraints, as the only options open for handling constraints are equality constraint elimination and addition of penalty functions for inequality constraints. Both approaches can be unreliable and may lead to failure of the optimization algorithm. Finally, the performance of direct search methods scales poorly (and often exponentially) with the number of decision variables. While performance can be improved with the use of parallel computing, these methods are rarely applied to problems with more than a few dozen decision variables.

GLOBAL OPTIMIZATION

Deterministic optimization methods are available for nonconvex nonlinear programming problems of the form of Eq. (3-71) that guarantee convergence to the global optimum. More specifically, one can show under mild conditions that they converge to an ε distance to the global optimum in a finite number of steps. These methods are generally more expensive than local NLP methods, and they require the exploitation of the structure of the nonlinear program.

Because global optima cannot be characterized by properties analogous to KKT conditions for local optima, global optimization methods work by partitioning the problem domain (i.e., containing the feasible region) into subregions. Upper bounds on the objective function are computed over all subregions of the problem. In addition, lower bounds can be derived from convex relaxations of the objective function and constraints for each subregion. The algorithm then proceeds to eliminate all subregions that have infeasible constraint relaxations or lower bounds that are greater than the least upper bound. After this, the remaining regions are further partitioned to create new subregions, and the cycle continues until the upper and lower bounds converge.

This basic concept leads to a wide variety of global algorithms, with the following features that can exploit different problem classes. *Bounding* strategies relate to the calculation of upper and lower bounds. For the former, any feasible point or, preferably, a locally optimal point in the subregion can be used. For the lower bound, convex relaxations of the objective and constraint functions are derived. The *refining* step deals with the construction of partitions in the domain and further partitioning them during the search process. Finally, the *selection* step decides on the order of exploring the open subregions.

For simplicity, consider the problem Min $f(x)$ subject to $g(x) \le 0$ where each function can be defined by additive terms. Convex relaxations for $f(x)$ and $g(x)$ can be derived in the following ways:

- Convex additive terms remain unmodified in these functions.
- Concave additive unary terms are replaced by linear underestimating functions that match the terms at the boundaries of their subregions.
- Nonconvex polynomial terms can be replaced by a set of scalar bilinear terms, with new variables introduced to define the higher-order polynomials.
- The scalar bilinear terms can be relaxed by using the McCormick underestimator; e.g., the bilinear term xz is replaced by a new variable w and linear inequality constraints

$$\begin{aligned} w &\ge x_l z + z_l x - x_l z_l \\ w &\ge x_u z + z_u x - x_u z_u \\ w &\le x_u z + z_l x - x_u z_l \\ w &\le x_l z + z_u x - x_l z_u \end{aligned} \tag{3-88}$$

where the subregions are defined by $x_l \le x \le x_u$ and $z_l \le z \le z_u$. Thus the feasible region and the objective function are replaced by convex envelopes to form relaxed problems.

Solving these convex relaxed problems leads to global solutions that are lower bounds to the NLP in the particular subregion. Finally, we see that gradient-based NLP solvers play an important role in global optimization algorithms, as they often yield the lower and upper bounds for the subregions. The *spatial branch and bound* global optimization algorithm can therefore be given by the following steps:

0. *Initialize algorithm.* Calculate upper and lower bounds over the entire (relaxed) feasible region.

For iteration k with a set of partitions $M_{k,j}$ and bounds in each subregion f_{Lj} and f_{Uj}:

1. *Bound.* Define the best upper bound $f_U = \text{Min}_j f_{Uj}$ and delete (fathom) all subregions j with lower bounds $f_{Lj} \ge f_U$. If the remaining subregions satisfy $f_{Lj} \ge f_U - \varepsilon$, stop.
2. *Refine.* Divide the remaining active subregions into partitions $M_{k,j1}$ and $M_{k,j2}$. (Many branching rules are available for this step.)
3. *Select.* Solve the convex relaxed NLP in the new partitions to obtain f_{Lj1} and f_{Lj2}. Delete the partition if there is no feasible solution.
4. *Update.* Obtain upper bounds f_{Uj1} and f_{Uj2} to new partitions, if present. Set $k = k + 1$, update partition sets, and go to step 1.

Note that a number of improvements can be made to the bounding, refinement, and selection strategies in the algorithm that accelerate the convergence of this method. A comprehensive discussion of all these options can be found in Floudas (2000) and Tawarlamani and Sahinidis (2002). Also, a number of efficient global optimization codes have recently been developed, including αBB, BARON, LGO, and OQNLP. An interesting numerical comparison of these and other codes can be found in Neumaier et al., *Math. Prog. B* **103**(2): 335 (2005).

MIXED INTEGER PROGRAMMING

Mixed integer programming deals with both discrete and continuous decision variables. For this presentation we consider discrete decisions as binary variables, that is, $y_i = 0$ or 1, and we consider the mixed integer problem (3-70). Unlike in local optimization methods, there are no optimality conditions, such as the KKT conditions, that can be applied directly. Instead, as in global optimization methods, a systematic search of the solution space, coupled with upper and lower bounding information, is applied. As with global optimization problems, large mixed integer programs can be expensive to solve, and some care is needed in problem formulation.

Mixed Integer Linear Programming If the objective and constraint functions are all linear, then Eq. (3-70) becomes a mixed integer linear programming problem given by

$$\text{Min } a^T x + c^T y \quad \text{subject to} \quad Ax + By \le b \quad x \ge 0 \quad y \in \{0, 1\}^t \tag{3-89}$$

Note that if we relax the t binary variables by the inequalities $0 \le y \le 1$, then Eq. (3-89) becomes a linear program with a (global) solution that is a lower bound to the MILP Eq. (3-89). There are specific MILP classes where the LP relaxation of Eq. (3-89) has the same solution as the MILP. Among these problems is the well-known *assignment problem.* Other MILPs that can be solved with efficient special-purpose methods are the *knapsack* problem, the *set covering* and *set partitioning* problems, and the *traveling salesperson* problem. See Nemhauser and Wolsey (1999) for a detailed treatment of these problems.

More generally, MILPs are solved with branch and bound algorithms, similar to the spatial branch and bound method of the previous section, that explore the search space. As seen in Fig. 3-58, binary variables are used to define the search tree, and a number of bounding properties can be noted from the structure of Eq. (3-89).

Upper bounds on the objective function can be found from any feasible solution to Eq. (3-89), with y set to integer values. These can be found at the bottom or "leaf" nodes of a branch and bound tree (and sometimes at intermediate nodes as well). The top, or root, node in the tree is the solution to the linear programming relaxation of Eq. (3-89); this is a lower bound to Eq. (3-89). On the other hand, as one proceeds down the tree with a partial assignment of the binary variables, a lower bound for any leaf node in that branch can be found from solution of the linear program at this intermediate node with the remaining binary variables relaxed. This leads to the following properties:

- Any intermediate node with an infeasible LP relaxation has infeasible leaf nodes and can be fathomed (i.e., all remaining children of this node can be eliminated).
- If the LP solution at an intermediate node is not less than an existing integer solution, then the node can be fathomed.

These properties lead to pruning of the search tree. Branching then continues in the tree until the upper and lower bounds converge.

This basic concept leads to a wide variety of MILP algorithms with the following features. *LP solutions* at intermediate nodes are relatively easy to calculate with the simplex method. If the solution of the parent node is known, multiplier information from this solution can be used to calculate (via efficient pivoting operations) the LP solution at the child node. *Branching strategies* to navigate the tree take a number of forms. More common *depth-first strategies* expand the most recent node to a leaf node or infeasible node and then backtrack to other branches in the tree. These strategies are simple to program and require little storage of past nodes. On the other hand, *breadth-first strategies* expand all the nodes at each level of the tree, select the node with the lowest objective function, and then proceed until the leaf nodes are reached. Here more storage is required, but generally fewer nodes are evaluated than in depth-first search. In addition, *selection of binary variable for branching* is based on a number of criteria, including choosing the variable with the relaxed value closest to 0 or 1, or the one leading to the largest change in the objective. A number of improved branching rules can accelerate the convergence of this method, and a number of efficient, large-scale MILP codes are widely used, including CPLEX, OSL, XPRESS, and ZOOM. Additional description of these strategies can be found in Nemhauser and Wolsey (1999).

Example To illustrate the branch and bound approach, we consider the MILP:

$$\begin{aligned}&\text{Min } Z = x + y_1 + 2y_2 + 3y_3\\ &\text{subject to } -x + 3y_1 + y_2 + 2y_3 \le 0\\ &\qquad -4y_1 - 8y_2 - 3y_3 \le -10\\ &\qquad x \ge 0, y_1, y_2, y_3 \in \{0,1\}\end{aligned}$$

The solution to this problem is given by $x = 4$, $y_1 = 1$, $y_2 = 1$, $y_3 = 0$, and $Z = 7$. Here we use a depth-first strategy and branch on the variables closest to 0 or 1. Figure 3-58 shows the progress of the branch and bound algorithm as the binary variables are selected and the bounds are updated. The sequence numbers for each node in Fig. 3-58 show the order in which they are processed. The grayed partitions correspond to the deleted nodes, and at termination of the algorithm we see that $Z = 7$ and an integer solution is obtained at an intermediate node where coincidentally $y_3 = 0$.

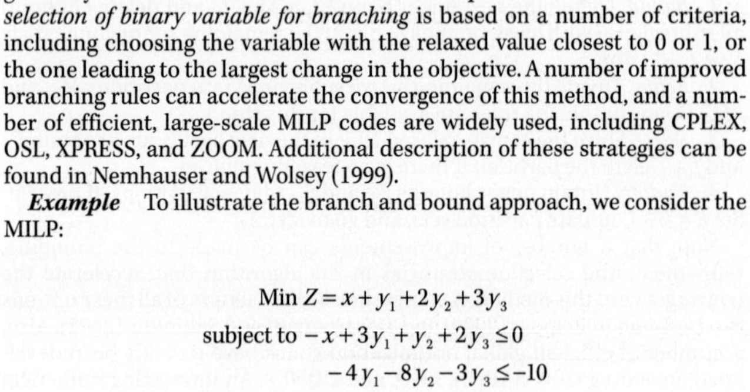

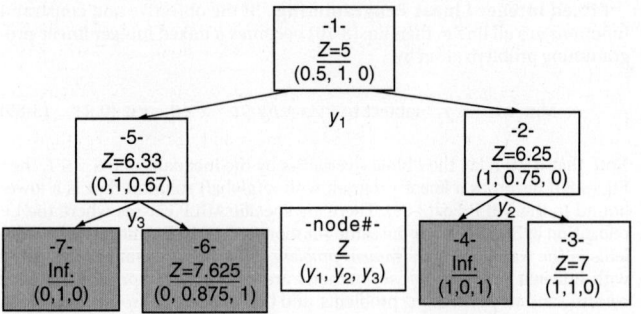

FIG. 3-58 Branch and bound sequence for MILP example.

Mixed Integer Nonlinear Programming Without loss of generality, we can rewrite the MINLP in Eq. (3-71) as

$$\text{Min } f(x) + c^T y \quad \text{subject to} \quad g(x) + By \le b \quad x \ge 0 \quad y \in \{0, 1\}^t \tag{3-90}$$

where the binary variables are kept as separate linear terms. MINLP strategies can be classified into two types. The first deals with nonlinear extensions of the branch and bound method discussed above for MILPs. The second deals with outer approximation decomposition strategies that provide lower and upper bounding information for convergence.

Nonlinear Branch and Bound The MINLP Eq. (3-90) can be solved in a similar manner to Eq. (3-89). If the functions $f(x)$ and $g(x)$ in Eq. (3-90) are convex, then direct extensions to the branch and bound method can be made. A relaxed NLP can be solved at the root node, upper bounds to the solution of Eq. (3-90) can be found at the leaf nodes, and the bounding properties due to NLP solutions at intermediate nodes still hold. However, this approach is more expensive than the corresponding MILP method. First, NLPs are more expensive than LPs to solve. Second, unlike with relaxed LP solutions, NLP solutions at child nodes cannot be updated directly from solutions at parent nodes. Instead, the NLP needs to be solved again (but one hopes with a better starting guess). The NLP branch and bound method is used in the SBB code interfaced to GAMS. In addition, Leyffer [*Comput. Optim. Appl.* **18:** 295 (2001)] proposed a hybrid MINLP strategy nested within an SQP algorithm. At each iteration, a mixed integer quadratic program is formed, and a branch and bound algorithm is executed to solve it.

If $f(x)$ and $g(x)$ are nonconvex, additional difficulties can occur. In this case, nonunique, local solutions can be obtained at intermediate nodes, and consequently lower bounding properties would be lost. In addition, the nonconvexity in $g(x)$ can lead to locally infeasible problems at intermediate nodes, even if feasible solutions can be found in the corresponding leaf node. To overcome problems with nonconvexities, global solutions to relaxed NLPs can be solved at the intermediate nodes. This preserves the lower bounding information and allows nonlinear branch and bound to inherit the convergence properties from the linear case. However, as noted above, this leads to much more expensive solution strategies.

Outer Approximation Decomposition Methods Again, we consider the MINLP Eq. (3-90) with convex $f(x)$ and $g(x)$. Note that the NLP with binary variables fixed at $\bar{y}$

$$\text{Min } f(x) + c^T\bar{y} \quad \text{subject to} \quad g(x) + B\bar{y} \le b \quad x \ge 0 \tag{3-91}$$

if feasible, leads to a solution that is an upper bound on the MINLP solution. In addition, linearizations of a convex function $\phi(x)$ leads to underestimation of the function itself, i.e.,

$$\phi(x) \ge \phi(x^k) + \nabla\phi(x^k)^T(x - x^k) \tag{3-92}$$

Consequently, linearization of Eq. (3-90) at a point x^k, to form the problem

$$\begin{aligned}&\text{Min} \quad f(x^k) + \nabla f(x^k)^T(x - x^k) + c^T y\\ &\text{subject to} \quad g(x^k) + \nabla g(x^k)^T(x - x^k) + By \le b\\ &\qquad x \ge 0 \qquad y \in \{0, 1\}^t\end{aligned} \tag{3-93}$$

leads to overapproximation of the feasible region and underapproximation of the objective function in Eq. (3-90). Consequently, solution of Eq. (3-93) is a lower bound to the solution of Eq. (3-90). Adding more linearizations from other points does not change the bounding property, so for a set of points x^l, $l = 1, \ldots, k$, the problem

$$\begin{aligned}&\text{Min } \alpha\\ &\text{subject to} \quad \left.\begin{aligned}&\alpha \ge f(x^l) + \nabla f(x^l)^T(x - x^l) + c^T y\\ &g(x^l) + \nabla g(x^l)^T(x - x^l) + By \le b\end{aligned}\right\} \quad l = 1, \ldots k\\ &\qquad x \ge 0 \quad y \in \{0, 1\}^t\end{aligned} \tag{3-94}$$

where α is a scalar variable, still has a solution that is a lower bound to Eq. (3-90). The outer approximation strategy is depicted in Fig. 3-59.

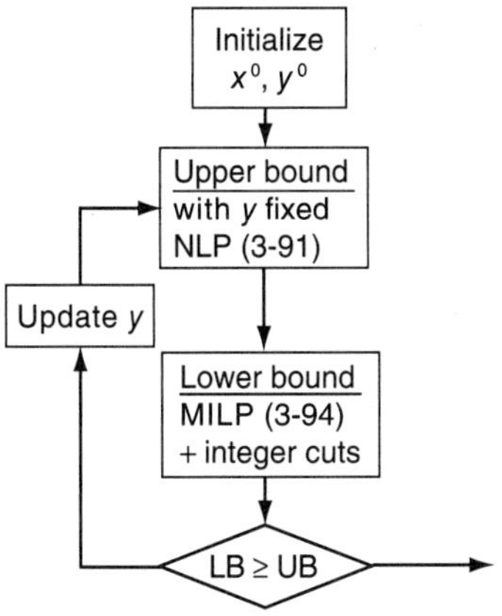

FIG. 3-59 Outer approximation MINLP algorithm.

The outer approximation algorithm first initializes the process, either with a predetermined starting guess or by solving a relaxed NLP based on Eq. (3-90). An upper bound to the solution is then generated by fixing the binary variables to their current values y^k and solving the NLP Eq. (3-91). This solution determines the continuous variable values x^k for the MILP Eq. (3-94). [If Eq. (3-94) is an infeasible problem, any point may be chosen for x^k, or the linearizations could be omitted.] Note that this MILP also contains linearizations from previous solutions of Eq. (3-91). Finally, the integer cut $\sum |y_i - y_i^k| \geq 1$ is added to Eq. (3-94) to avoid revisiting previously encountered values of binary variables. Solution of Eq. (3-94) yields new values of y and (without the integer cut) must lead to a lower bound to the solution of Eq. (3-90). Consequently, if the objective function of the lower bounding MILP is greater than the least upper bound determined in solutions of Eq. (3-91), then the algorithm terminates. Otherwise, the new values of y are used to solve the next NLP Eq. (3-91).

Compared to nonlinear branch and bound, the outer approximation algorithm usually requires very few solutions of the MILP and NLP subproblems. This is especially advantageous on problems where the NLPs are large and expensive to solve. Moreover, there are three variations of outer approximation that may be suitable for particular problem types:

In *generalized benders decomposition* (GBD) the lower bounding problem Eq. (3-94) is replaced by the MILP

$$\text{Min } \alpha \text{ subject to} \quad x \geq 0, y \in \{0, 1\}^t$$
$$\left.\begin{aligned} \alpha \geq f(x^l) + c^T y + [g(x^l) + By]^T v^l \\ \sum_i |y_i - y_i^l| \geq 1 \end{aligned}\right\} \quad l = 1, \ldots, k \tag{3-95}$$

where v^l is the vector of KKT multipliers from the solution of Eq. (3-91) at iteration l. This MILP can be derived through a reformulation of the MILP used in Fig. 3-59 with the *inactive constraints* from Eq. (3-91) dropped. Solution of Eq. (3-95) leads to a weaker lower bound than Eq. (3-94), and consequently, more solutions of the NLP and MILP subproblems are needed to converge to the solution. However, Eq. (3-95) contains only a single continuous variable and far fewer inequality constraints and is much less expensive to solve than Eq. (3-94). Thus, GBD is favored over outer approximation if Eq. (3-91) is relatively inexpensive to solve or solution of Eq. (3-94) is too expensive.

The *extended cutting plane* (ECP) algorithm is complementary to GBD. While the lower bounding problem in Fig. 3-59 remains essentially the same, the continuous variables x^k are chosen from the MILP solution and the NLP Eq. (3-91) is replaced by a simple evaluation of the objective and constraint functions. As a result, only MILP problems [Eq. (3-94) plus integer cuts] need to be solved. Consequently, the ECP approach has weaker upper bounds than outer approximation and requires more MILP solutions. It has advantages over outer approximation when the NLP Eq. (3-91) is expensive to solve.

The third extension to the outer approximation approach is based on a *branch-and-cut algorithm*, which solves a continuous linear program at each node of the search tree, and therefore improves the lower bounds while branching on integer variables. BONMIN, a comprehensive MINLP code described in Bonami et al. ["An Algorithmic Framework for Convex Mixed Integer Nonlinear Programs," *Discrete Optimization* **5**(2): 186–204 (2008)] incorporates NLP branch and bound, branch and cut, and outer approximation as options, along with hybrids of these strategies.

Additional difficulties arise for the outer approximation algorithm and its GBD, ECP, and branch and cut extensions when either $f(x)$ or $g(x)$ is nonconvex. Under these circumstances, the lower bounding properties resulting from the linearization and formulation of the MILP subproblem are lost, and the MILP solution may actually exclude the solution of Eq. (3-90). Hence, these algorithms need to be applied with care to nonconvex problems. To deal with nonconvexities, one can relax the linearizations in Eq. (3-94) through the introduction of additional *deviation variables* that can be penalized in the objective function. Alternately, the linearizations in Eq. (3-94) can be replaced by valid underestimating functions, such as those derived for global optimization [e.g., Eq. (3-86)]. However, this requires specific structural knowledge of Eq. (3-90) and may lead to weak lower bounds for the resulting MILP.

Finally, the performance of both MILP and MINLP algorithms is strongly dependent on the problem formulations Eq. (3-89) and Eq. (3-90). In particular, the efficiency of the approach is impacted by the lower bounds produced by the relaxation of the binary variables and subsequent solution of the linear program in the branch and bound tree. A number of approaches have been proposed to improve the quality of the lower bounds, including these:

- Logic-based methods such as *generalized disjunctive programming* (GDP) can be used to formulate MINLPs with fewer discrete variables that have tighter relaxations. The imposition of logic-based constraints prevents the generation of unsuitable alternatives, leading to less expensive searches. In addition, constrained logic programming (CLP) methods offer efficient alternatives to MILP solvers for highly combinatorial problems. See Jain and Grossmann, *INFORMS Journal of Computing*, **13:** 258–276 (2001) for more details.
- Convex hull formulations of MILPs and MINLPs lead to relaxed problems that have much tighter lower bounds. This leads to the examination of far fewer nodes in the branch and bound tree. See Grossmann and Lee, *Comput. Optim. Applic.* **26:** 83 (2003) for more details.
- *Reformulation* and *preprocessing* strategies including bound tightening of the variables, coefficient reduction, lifting facets, and special ordered set constraints frequently lead to improved lower bounds and significant performance improvements in mixed integer programming algorithms. See Bixby, R., and E. Rothberg, *Annals of Operations Research*, **49**(1): 37–41 (2007) for more details.

A number of efficient codes are available for the solution of MINLPs, including AlphaECP, BARON, BONMIN, DICOPT, MINLP, and SBB. All are available within the GAMS modeling platform.

DEVELOPMENT OF OPTIMIZATION MODELS

The most important aspect to a successful optimization study is the formulation of the optimization model. These models must reflect the real-world problem so that meaningful optimization results are obtained; they also must satisfy the properties of the problem classes in Fig. 3-53. For instance, NLPs addressed by gradient-based methods need to have functions that are defined in the variable domain and have bounded and continuous first and second derivatives. In mixed integer problems, proper formulations are also needed to yield good lower bounds for efficient search. With increased understanding of optimization methods and the development of efficient and reliable optimization codes, optimization practitioners now focus on the *formulation* of optimization models that are realistic, well posed, and inexpensive to solve. Finally, convergence properties of NLP, MILP, and MINLP solvers require accurate first (and often second) derivatives from the optimization model. If these contain numerical errors (say, through finite difference approximations), then the performance of these solvers can deteriorate considerably. As a result of these characteristics, modeling platforms are essential for the formulation task. These are classified into two broad areas: optimization modeling platforms and simulation platforms with optimization.

Optimization modeling platforms provide general-purpose interfaces for optimization algorithms and remove the need for the user to interface to the solver directly. These platforms allow the general formulation for all problem classes discussed above with direct interfaces to state-of-the-art optimization codes. Three representative platforms are GAMS (General Algebraic Modeling Systems), AMPL (A Mathematical Programming Language), and AIMMS (Advanced Integrated Multidimensional Modeling Software). All three require problem model input via a declarative modeling language and provide exact gradient and hessian information through automatic differentiation strategies. Although it is possible, these platforms were not designed to handle externally added procedural models. As a result, these platforms are best applied on optimization models that can be developed entirely within their modeling framework. Nevertheless, these platforms are widely used for large-scale research and industrial applications. In addition, the MATLAB platform allows for flexible formulation of optimization models as well, although it currently has only limited capabilities for automatic differentiation and limited optimization solvers.

Simulation platforms with optimization are often dedicated, application-specific modeling tools to which optimization solvers have been interfaced. These lead to very useful optimization studies, but because they were not originally designed for optimization models, they need to be used with some caution. In particular, most of these platforms do not provide exact derivatives to the optimization solver; often they are approximated through finite differences. In addition, the models themselves are constructed and calculated through numerical procedures, instead of through an open declarative language. Examples of these include widely used process simulators such as Aspen/Plus, PRO/II, and Hysys. Also note that more recent platforms such as Aspen Custom Modeler, GPROMS, and MOSAIC include declarative models and exact first derivatives.

Finally, for optimization tools that must be linked to procedural models, reliable and efficient automatic differentiation (AD) tools that provide exact first (often second) derivatives are available that link to models written in C, C++, FORTRAN, Python, and other modeling platforms. Example AD tools include ADIC, ADOL-C, CasADi, CppAD, and TAPENADE. When used with care, these can be applied to existing procedural models and, when linked to modern NLP and MINLP algorithms, can lead to powerful optimization capabilities.

STATISTICS

References: Box, G. P., J. S. Hunter, and W. G. Hunter, *Statistics for Experimenters: Design, Innovation, and Discovery*, 2d ed., Wiley, New York, 2005; Cropley, J. B., "Heuristic Approach to Complex Kinetics," pp. 292–302 in *Chemical Reaction Engineering—Houston*, ACS Symposium Series 65, American Chemical Society, Washington, D.C., 1978; Schiller, Jr., J. J., R. A. Srinivasan, and M. Spiegel, *Schaum's Outline of Probability and Statistics*, 4th ed., McGraw-Hill, New York, 2012; Mendenhall, W., and T. Sincich, *Statistics for Engineering and the Sciences*, 5th ed., Pearson, Boston, 2006; Moore, D. S., G. P. McCabe, and B. Craig, *Introduction to the Practice of Statistics*, 8th ed., Freeman, San Francisco, 2014; Montgomery, D. C., and G. C. Runger, *Applied Statistics and Probability for Engineers*, 6th ed., Wiley, New York, 2013; see also Logan and Wolesensky (2009) in General References and https://cloud.r-project.org/ for Statistics in R.

INTRODUCTION

Statistics represents a body of knowledge that enables one to deal with quantitative data reflecting any degree of uncertainty. There are six basic aspects of applied statistics:

1. Type of data
2. Random variables
3. Models
4. Parameters
5. Sample statistics
6. Characterization of chance occurrences

From these can be developed strategies and procedures for dealing with (1) estimation and (2) inferential statistics. The following has been directed more toward inferential statistics because of its broader utility.

Detailed illustrations and examples are used throughout to develop basic statistical methodology for dealing with a broad area of applications. If you are new to statistics, look first at the examples and find one that is appropriate to your application. In addition to this material, there are many specialized topics as well as some very subtle areas that have not been discussed. The references should be used for more detailed information. Section 8 discusses the use of statistics in statistical process control (SPC).

Type of Data In general, statistics deals with two types of data: counts and measurements. Counts represent the number of discrete outcomes, such as the number of defective parts in a shipment, the number of lost-time accidents, and so forth. Measurement data are treated as a continuum. For example, the tensile strength of a synthetic yarn theoretically could be measured to any degree of precision. A subtle aspect associated with count and measurement data is that some types of count data can be dealt with through the application of techniques that have been developed for measurement data alone. This ability is due to the fact that some simplified measurement statistics serve as an excellent approximation for the more tedious count statistics.

Random Variables Applied statistics deals with quantitative data. In tossing a fair coin the successive outcomes would tend to be different, with heads and tails occurring randomly over time. Given a long strand of synthetic fiber, the tensile strength of successive samples would tend to vary significantly from sample to sample. Counts and measurements are characterized as random variables, that is, observations which are susceptible to chance. Virtually all quantitative data are susceptible to chance in one way or another.

Models Part of the foundation of statistics consists of the mathematical models that characterize an experiment. The models themselves are mathematical ways of describing the probability, or relative likelihood, of observing specified values of random variables. For example, in tossing a coin once, a random variable x could be defined by assigning to x the value 1 for a head and 0 for a tail. Given a fair coin, the probability of observing a head on a toss would be .5, and similarly for a tail. Therefore, the mathematical model governing this experiment can be written as

x	$P(x)$
0	.5
1	.5

where $P(x)$ stands for what is called a *probability function.* This term is reserved for count data, in that probabilities can be defined for particular outcomes. The probability function that has been displayed is a very special case of the more general case, which is called the *binomial probability distribution.*

For measurement data which are considered continuous, the term *probability density* is used. For example, consider a spinner wheel which conceptually can be thought of as being marked off on the circumference infinitely precisely from 0 up to, but not including, 1. In spinning the wheel, the probability of the wheel's stopping at a specified marking point at any particular x value, where $0 \le x < 1$, is 0, for example, stopping at the value $x = \sqrt{.5}$. For the spinning wheel, the probability density function would be defined by $f(x) = 1$ for $0 \le x < 1$. Graphically, this is shown in Fig. 3-60. The relative probability concept refers to the fact that density reflects the relative likelihood of occurrence; in this case, each number between 0 and 1 is equally likely. For measurement data, probability is defined by the area under the curve between specified limits. A density function always must have a total area of 1.

Example For the density of Fig. 3-60

$$P[0 \le x \le .4] = .4 \qquad P[.2 \le x \le .9] = .7 \qquad P[.6 \le x < 1] = .4$$

and so forth. Since the probability associated with any particular point value is zero, it makes no difference whether the limit point is defined by a closed interval (≤ or ≥) or an open interval (< or >).

Many different types of models are used as the foundation for statistical analysis. These models are also referred to as *populations.*

Parameters As a way of characterizing probability functions and densities, certain types of quantities called *parameters* can be defined. For example, the center of gravity of the distribution is defined to be the population mean, which is designated as μ. For the coin toss $\mu = .5$, which corresponds to the average value of x; i.e., for one-half of the time x will take on a value 0 and for the other half a value 1. The average would be .5. For the spinning wheel, the average value would also be .5.

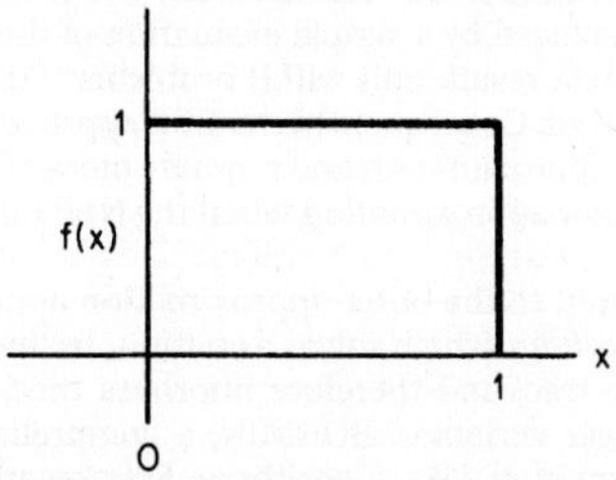

FIG. 3-60 Density function.

Another parameter is called the *standard deviation*, which is designated as σ. The square of the standard deviation is used frequently and is called the *variance* σ^2. Basically, the standard deviation is a quantity which measures the spread or dispersion of the distribution from its mean μ. If the spread is broad, then the standard deviation will be larger than if it were more constrained.

For specified probability and density functions, the respective *mean*, or *expected value* $E(x)$, *variance* $\mathrm{Var}(x)$, and *standard deviation* σ are defined by the following:

Probability functions (discrete variables and counts)	Probability density functions (continuous variables)
$E(x)=\mu=\sum_x xp(x)$	$E(x)=\mu=\int_x xf(x)\,dx$
$\mathrm{Var}(x)=\sigma^2=\sum_x (x-\mu)^2 P(x)$	$\mathrm{Var}(x)=\sigma^2=\int_x (x-\mu)^2 f(x)\,dx$

Sample Statistics Many types of sample statistics will be defined. Two very special types are the *sample mean*, designated as $\bar{x}$, and the *sample standard deviation*, designated as s. These are, by definition, random variables. Parameters such as μ and σ are not random variables; they are fixed constants corresponding to a probability function or distribution.

Example In an experiment, six random numbers (rounded to four decimal places) were observed from the uniform distribution $f(x) = 1$ for $0 \le x < 1$:

0.1009, 0.3754, 0.0842, 0.9901, 0.1280, 0.6606

The *sample mean* corresponds to the arithmetic average of the observations, which will be designated as x_1 through x_6, where

$$\bar{x}=\frac{1}{n}\sum_{i=1}^{n}x_i \quad \text{with } n=6 \quad \bar{x}=0.3899 \tag{3-96}$$

The *sample standard deviation s* is defined by the computation

$$s=\sqrt{\frac{\sum(x_i-\bar{x})^2}{n-1}}=\sqrt{\frac{n\sum x_i^2-(\sum x_i)^2}{n(n-1)}} \tag{3-97}$$

In effect, this represents the root of a statistical average of the squares. The divisor quantity $n - 1$ will be referred to as the *degrees of freedom*. The sample value of the standard deviation for the data given is .3686.

The value of $n - 1$ is used in the denominator because the deviations from the sample average must total zero, or

$$\sum(x_i-\bar{x})=0$$

Thus knowing $n - 1$ values of $x_i - \bar{x}$ permits calculation of the nth value of $x_i - \bar{x}$.

The sample mean and sample standard deviation are obtained by using Microsoft Excel with the commands AVERAGE(B2:B7) and STDEV(B2:B7) when the observations are in cells B2 to B7.

In effect, the standard deviation quantifies the relative magnitude of the deviation numbers, i.e., a special type of "average" of the distance of points from their center. In statistical theory, it turns out that the corresponding variance quantities s^2 have remarkable properties which make possible broad generalities for sample statistics and therefore also their counterparts, the standard deviations.

For the corresponding population, the parameter values are $\mu = .50$ and $\sigma = .2887$, which are obtained by calculating the integrals defined above with $f(x) = 1$ and integrating x from 0 to 1. If, instead of using individual observations only, averages of 6 were reported, then the corresponding population parameter values would be $\mu = .50$ and $\sigma_{\bar{x}} = \sigma/\sqrt{6} = .1179$. The corresponding variance for an average will be written occasionally as $\mathrm{Var}(\bar{x}) = \mathrm{var}(x)/n$. In effect, the variance of an average is inversely proportional to the sample size n, which reflects the fact that sample averages will tend to cluster much more closely than individual observations. This is illustrated in greater detail under Measurement Data and Sampling Densities.

Characterization of Chance Occurrences To deal with a broad area of statistical applications, it is necessary to characterize the way in which random variables will vary by chance alone. The basic foundation for this characteristic is laid through a density called the *gaussian*, or *normal, distribution*.

Determining the area under the normal curve is a very tedious procedure. However, by standardizing a random variable that is normally distributed, it is possible to relate all normally distributed random variables to one table. The standardization is defined by the identity $z = (x - \mu)/\sigma$, where z is called the *unit normal*. Further, it is possible to standardize the sampling distribution of averages $\bar{x}$ by the identity $z = (\bar{x} - \mu)/(\sigma/\sqrt{n})$.

A remarkable property of the normal distribution is that, almost regardless of the distribution of x, sample averages $\bar{x}$ will approach the gaussian distribution as n gets large. Even for relatively small values of n, of about 10, the approximation in most cases is quite close. For example, sample averages of size 10 from the uniform distribution will have essentially a gaussian distribution. Also, in many applications involving count data, the normal distribution can be used as a close approximation. In particular, the approximation is quite close for the binomial distribution within certain guidelines.

The normal probability distribution function can be obtained in Microsoft Excel by using the NORM.DIST function and supplying the desired mean and standard deviation. The cumulative value can also be determined. In the MATLAB Statistics Toolbox the corresponding command is normcdf(x, μ, σ).

ENUMERATION DATA AND PROBABILITY DISTRIBUTIONS

Introduction Many types of statistical applications are characterized by enumeration data in the form of counts. Examples are the number of lost-time accidents in a plant, the number of defective items in a sample, and the number of items in a sample that fall within several specified categories.

The sampling distribution of count data can be characterized through probability distributions. In many cases, count data are appropriately interpreted through their corresponding distributions. However, in other situations analysis is greatly facilitated through distributions which have been developed for measurement data. Examples of each will be illustrated in the following subsections.

Binomial Probability Distribution

Nature Consider an experiment in which each outcome is classified into one of two categories, one of which will be defined as a success and the other as a failure. Given that the probability of success p is constant from trial to trial, then the probability of observing a specified number of successes x in n trials is defined by the binomial distribution. The sequence of outcomes is called a *Bernoulli process*.

Nomenclature Let $\hat{p} = x/n$ be the proportion of successes in n trials.

Probability Law

$$p(x)=p\left(\frac{x}{n}\right)=\binom{n}{x}p^x(1-p)^{n-x}\quad x=0,1,2,\ldots,n$$

where $\displaystyle\binom{n}{x}=\frac{n!}{n!(n-x)!}$

Properties

$$E(x)=np \qquad \mathrm{Var}(x)=np(1-p)$$
$$E(\hat{p})=p \qquad \mathrm{Var}(\hat{p})=p(1-p)/n$$

Example In three tosses of a coin, what is the probability of seeing three heads? This problem uses the binomial probability distribution because each toss is independent of the previous ones. Assuming the coins are "fair" and the probability of heads is ½, then the probability of 3 heads in 3 tosses is

$$P=\frac{3!}{3!0!}\left(\frac{1}{2}\right)^3\left(\frac{1}{2}\right)^0=\frac{1}{8}$$

Likewise, the probability of 2 heads and 1 tail in 3 tosses is

$$P=\frac{3!}{2!1!}\left(\frac{1}{2}\right)^2\left(\frac{1}{2}\right)^1=\frac{3}{8}$$

Geometric Probability Distribution

Nature Consider an experiment in which each outcome is classified into one of two categories, one of which will be defined as a success and the other as a failure. Given that the probability of success p is constant from trial to trial, then the probability of observing the first success on the xth trial is defined by the geometric distribution.

Probability Law

$$P(x) = p(1-p)^{x-1} \qquad x = 1, 2, 3, \ldots$$

Properties

$$E(x) = 1/p \quad \text{Var}(x) = (1-p)/p^2$$

If the event is described as $y = x - 1$, that is, y is the number of failures before the first success, then

$$E(y) = (1-p)/p \qquad \text{Var}(x) = (1-p)/p^2$$

Example Let y be the number of tosses of a die prior to the toss in which a 2 or 3 first appears. Since the probability of a 2 or 3 in a single toss is 1/3, the probability function of x and the expected value is

$$E(y) = \frac{1-p}{p} = \frac{1-1/3}{1/3} = 2$$

That is, you will (on average) do two tosses *before* you get a 2 or 3. Likewise, the expected value of getting your first 2 or 3 on the xth toss is

$$E(x) = \frac{1}{p} = \frac{1}{1/3} = 3$$

The difference between these is simply in the first case you are counting the tosses before the "success" and in the second case you are including the toss giving the "success."

Poisson Probability Distribution

Nature The Poisson probability distribution is used to assess the number of events that will occur in a span of time, regardless of when the event occurred last, provided you know the average rate of events and the events are independent of the time since the last event. For example, in monitoring a moving thread line, one criterion of quality would be the frequency of broken filaments. These can be identified as they occur through the thread line by a broken filament detector mounted adjacent to the thread line. In this context, the random occurrences of broken filaments can be modeled by the Poisson distribution. This is called a *Poisson process* and corresponds to a probabilistic description of the frequency of defects or, in general, what are called arrivals at points on a continuous line or in time. Other examples include these:

1. The number of cars (arrivals) that pass a point on a high-speed highway between 10:00 and 11:00 A.M. on Wednesdays
2. The number of customers arriving at a bank between 10:00 and 10:10 A.M.
3. The number of telephone calls received through a switchboard between 9:00 and 10:00 A.M.
4. The number of insurance claims that are filed each week
5. The number of spinning machines that break down during 1 day at a large plant.

Nomenclature

x = total number of arrivals in a total length L or total period T
a = average rate of arrivals for a unit length or unit time
$\lambda = aL$ = expected or average number of arrivals for the total length L
$\lambda = aT$ = expected or average number of arrivals for the total time T

Probability Law Given that a is constant for the total length L or period T, the probability of observing x arrivals in some period L or T is given by

$$P(x) = \frac{\lambda^x}{x!} e^{-\lambda} \qquad x = 0, 1, 2, \ldots$$

Properties

$$E(x) = \lambda \qquad \text{Var}(x) = \lambda$$

Example The number of broken filaments in a thread line has been averaging .015 per yard. What is the probability of observing exactly two broken filaments in the next 100 yd? In this example, a = .015/yd and L = 100 yd; therefore $\lambda = (.015)(100) = 1.5$:

$$P(x=2) = \frac{(1.5)^2}{2!} e^{-1.5} = .2510$$

Example A commercial item is sold in a retail outlet as a unit product. In the past, sales have averaged 10 units per month with no seasonal variation. The retail outlet must order replacement items 2 months in advance. If the outlet starts the next 2-month period with 25 items on hand, what is the probability that it will run out of stock before the end of the second month?

Given a = 10/month, then $\lambda = 10 \times 2 = 20$ for the total period of 2 months:

$$P(x \ge 26) = \sum_{26}^{\infty} p(x) = 1 - \sum_{0}^{25} p(x)$$

$$\sum_{0}^{25} \frac{20^x}{x!} e^{-20} = e^{-20}\left[1 + \frac{20}{1} + \frac{20^2}{2!} + \cdots + \frac{20^{25}}{25!}\right]$$

$$= 0.888$$

Therefore $P(x \ge 26) = .112$ or roughly an 11 percent chance of a stockout.

Hypergeometric Probability Distribution

Nature In an experiment in which one samples from a relatively small group of items, each of which is classified in one of two categories, A or B, the hypergeometric distribution can be defined. One example is the probability of drawing two red and two black cards from a deck of cards. The hypergeometric distribution is the analog of the binomial distribution when successive trials are not independent, i.e., when the total group of items is not infinite. This happens when the drawn items are not replaced.

	Population	Sample
Category A	X	x
Category B	$N-X$	$n-x$
Total	N	n

Probability Law

$$P(x) = \binom{N-X}{n-x}\binom{X}{x} \Big/ \binom{N}{n}$$

$$E(x) = \frac{nX}{N}$$

$$\text{var}(x) = n\frac{X}{N}\frac{N-X}{N}\frac{N-n}{N-1}$$

Example What is the probability that an appointed special committee of 4 has no female members when the members are randomly selected from a candidate group of 10 males and 7 females? Here $N = 17$, $X = 7$, $n = 4$, $x = 0$, and

$$P(x=0) = \frac{\binom{10}{4}\binom{7}{0}}{\binom{17}{4}} = .0882$$

This distribution can be used in Texas Hold'em poker. See http: en.wikipedia.org/wiki/Hypergeometric-distribution.

To compute these probabilities in Microsoft Excel, put the value of x in cell B2, say, and use the functions

Binomial distribution: = BINOM.DIST(B2, n, p, 0)

Poisson distribution: = POISSON.DIST(B2, λ, 0)

Hypergeometric distribution: = HYPGEOM.DIST(B2, n, X, N, 1)

The factorial function is FACT(n) in Microsoft Excel and factorial(n) in MATLAB. Be sure that x is an integer.

Conditional Probability It is useful to predict the probability of one event, given that a second event has already occurred. For example, suppose you draw a card from a deck of 52 cards, half red and half black (event A). The probability of the first card being red is 26/52 = 1/2. But suppose the question is: What is the probability that the first two cards drawn are red (without replacing the first card)? In the second draw (event B) there are only 25 red cards of 51 total cards, so that the probability of drawing a red card is 25/51. Then the probability of drawing two red cards is

$$P(\mathrm{AB})=\left(\frac{1}{2}\right)\left(\frac{25}{51}\right)=\frac{25}{102}$$

This is sometimes written as

$$P(\mathrm{A}\cap\mathrm{B})$$

i.e., the probability of both events A and B occurring.

The $P(\mathrm{B}|\mathrm{A})$ is called the *conditional probability* of event B, given that event A has occurred. Conditional probabilities satisfy the general equation

$$P(\mathrm{B}|\mathrm{A})=\frac{P(\mathrm{AB})}{P(\mathrm{A})}\quad \text{if}\quad P(\mathrm{A})\neq 0$$

which is a restatement of the numerical equation in a different form, $P(\mathrm{B}|\mathrm{A}) = 25/51$, $P(\mathrm{A}) = 1/2$, and $P(\mathrm{AB}) = 25/102$. Likewise

$$P(\mathrm{A}|\mathrm{B})=\frac{P(\mathrm{AB})}{P(\mathrm{B})}\quad \text{if}\quad P(\mathrm{B})\neq 0$$

The probabilities satisfy Bayes' theorem

$$P(\mathrm{A}|\mathrm{B})=\frac{P(\mathrm{B}|\mathrm{A})P(\mathrm{A})}{P(\mathrm{B}|\mathrm{A})P(\mathrm{A})+P(\mathrm{B}|\mathrm{A}^c)P(\mathrm{A}^c)}$$

where A^c is the complement of A, that is, A did not occur.

Example The table below gives the numbers of Bachelor's degrees in engineering in the United States in 2013–2014. Given that a graduate is a woman, what is the probability that she obtained a degree in chemical engineering, biological engineering, or biomedical engineering?

	Total	Women	% Women
Aerospace	3695	510	13.8
Biological/Biomedical	6150	2573	41.8
Chemical	8110	2944	36.3
Civil	12333	2393	19.4
Civil/Environ.	1893	573	30.3
Electrical	14088	1927	13.7
Mechanical	23675	3196	13.5
Materials	1440	379	26.3
Industrial	4877	1541	31.6
subtotal	76261	16036	21.0
Other	22912	3303	14.4
Total	99173	19339	19.5

The data are from the American Society of Engineering Education, ASEE Report 11-47.pdf, Brian L. Yoder, accessed July 14, 2015. http://www.asee.org/papers-and-publications/publications/college-profiles/14EngineeringbytheNumbersPart1.pdf

Define event A as being a female graduate, B as graduating in chemical engineering (male or female), C as graduating in biological/biomedical engineering (male or female). We thus have

$$P(\mathrm{A}) = 19339/99173 = 0.1950,\ P(\mathrm{B}) = 8110/99173 = 0.0818$$

$$P(\mathrm{C}) = 6150/99173 = 0.0620$$

Also $P(\mathrm{AB}) = 2944/99173 = 0.02968$.

We want $P(\mathrm{B}|\mathrm{A})$ which is $P(\mathrm{AB})/P(\mathrm{A}) = 0.02968/0.19500 = 0.152$. This should be the same as 2944/19339, which it is. Likewise $P(\mathrm{C}|\mathrm{A}) = 0.133$. Since events B and C are independent, the probability of a woman graduate being in one of these fields is the sum of these, or 0.285. This is considerably higher than the probability of a graduate being a woman, .195; or if one looks just at the other engineering fields, the probability of a woman graduate being in them is only .163.

MEASUREMENT DATA AND SAMPLING DENSITIES

This section describes the probability of measurement data. If the number of samples is large, the data often form a normal distribution, so that is discussed first. If the sample size is smaller (somewhat less than 30), the data may be described by the t distribution. The chi-square test allows us to find whether an observed frequency of observation differs significantly from those expected from a model. Finally, the F test is used to compare variances and their properties. While tables exist to compute the various functions, here the commands will be given to compute them using Microsoft Excel. Similar commands are available in MATLAB and Mathematica.

Microsoft Excel command	Designated symbol	Variable	Sampling distribution of
NORM.DIST(X, μ, σ, 0)	z	$\frac{x-\mu}{\sigma}$	Observations*
NORM.DIST(X, μ, σ, 0)	z	$\frac{\bar{x}-\mu}{\sigma\sqrt{n}}$	Averages
TINV(α, df)	t	$\frac{\bar{x}-\mu}{s\sqrt{n}}$	Averages when σ is unknown*
CHIINV(α, df)	χ^2	$(s^2/\sigma^2)(\mathrm{df})$	Variances*
FINV(α, df$_1$, df$_2$)	F	s_1^2/s_2^2	Ratio of two independent sample variances*

*When sampling from a gaussian distribution.

Normal Distribution The most common probability density function is the gaussian or normal probability function. This function describes a bell-shaped curve that indicates the probability of a measurement deviating from the average of many measurements. The formula is

$$f(x,\mu,\sigma)=\frac{1}{\sqrt{2\pi}\,\sigma}\exp\left[-\frac{(x-\mu)^2}{2\sigma^2}\right]$$

The curve is typically scaled so that the mean μ is 0; the symbol σ is the standard deviation, see Fig. 3-61. The area under the curve is 1.0. The Microsoft Excel function NORM.DIST(x, μ, σ, 1) gives the probability that a sample measurement is less than x when the measurements have a mean of μ and a standard deviation of σ, that is, the integral from negative infinity to x. For example, NORM.DIST(1, 0, 1, 1) (mean 0 and standard deviation of 1) gives the value .8413. Thus, the probability of a measurement x being less than 1 (i.e., the mean plus 1 standard deviation) is .8413. NORM.DIST(−1, 0, 1, 1) gives the value .1587 for the probability of a measurement being less than −1, or less than the mean minus 1 standard deviation. Thus, the probability that x is between +1 standard deviation and −1 standard deviation is .8413 − .1587 = .6827. The probability of a measurement falling within 2 standard

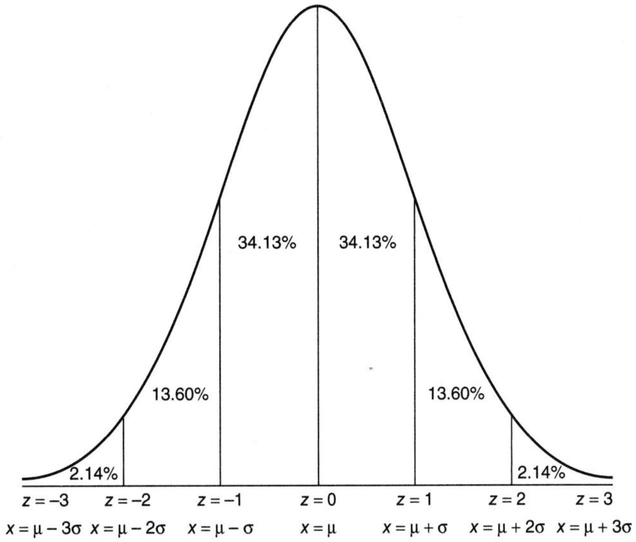

FIG. 3-61 Normal probability distribution.

deviations of the mean is .9545, and the probability of a measurement falling with 3 standard deviations of the mean is 0.9975. In the Excel formulas, the last 1 is a logical variable and can be replaced by TRUE (for 1).

The *standard normal variable* is defined as

$$z = \frac{x - \mu}{\sigma}$$

Then z is a normal random variable with a mean of 0 and a standard deviation of 1. Likewise, $x = \mu + \sigma z$.

Example Suppose one measures the concentration of some product coming off a production line. After 25 measurements one computes the average of 5.2 (in some units) with a variance of .15, using Eqs. (3-96) and (3-97). One believes the variation is randomly distributed so that it would be modeled by the normal distribution. What is the probability that a product will have the concentration between 5.0 and 5.4? The first method uses the data as they come. Calculate NORM.DIST(5.4, 5.2, .15, 1) = .90879 to get the probability that the concentration is below 5.4 and NORM.DIST(5.0, 5.2, .15, 1) = .09121 to get the probability that the concentration is below 5.0. The probability that the concentration is between 5.0 and 5.4 is then .90879 − .09121 = .81758, or 82 percent.

An alternative solution is to calculate the standard normal variables.

$$z = \frac{5.4 - 5.2}{0.15} = 1.33333 \quad \text{and} \quad z = \frac{5.0 - 5.2}{0.15} = -1.33333$$

Then NORM.DIST(1.33333, 0, 1, 1) − NORM.DIST(−1.33333, 0, 1, 1) = .90879 − .09121 = .81758.

The *central limit theorem* says that a set of random variables approaches a normal distribution as n, the number of measurements, goes to infinity. In addition, averages turn out to vary less than individual measurements. Suppose one calculates the average of n concentrations this week, next week, etc. The corresponding relationship for the Z scale is

$$z = \frac{\bar{x} - \mu}{\sigma}\sqrt{n} \quad \text{or} \quad \bar{x} = \mu + \frac{\sigma}{\sqrt{n}} z$$

The Microsoft Excel command CONFIDENCE(α, σ, n) gives the confidence interval about the mean for a sample size n, where σ is the standard deviation and α is the confidence level.

Example Suppose one makes measurements of the concentration several times a week, and the average and variance of individual measurements are as given above. For a variance of 0.15, at a 95 percent confidence level what is the probable range of measurements?

The formula CONFIDENCE(0.05, 0.15, 100) = 0.0293. Thus, with 95 percent confidence the weekly averages will be 5.2 ± 0.029 or between 4.91 and 5.49.

***t* Distribution of Averages** The normal curve relies on a knowledge of σ, or in special cases, when it is unknown, s can be used with the normal curve as an approximation when $n > 30$. For example, with $n > 30$ the intervals $\pm s$ and $\pm 2s$ will include roughly 68 and 95 percent of the sample values, respectively, when the distribution is normal.

In applications, sample sizes are usually small and σ is unknown. In these cases, the t distribution can be used where

$$t = (\bar{x} - \mu)(s/\sqrt{n}) \quad \text{or} \quad \bar{x} = \mu + ts\sqrt{n}$$

See Fig. 3-62. The t distribution is also symmetric and centered at zero. It is said to be robust in the sense that even when the individual observations x are not normally distributed, sample averages of x have distributions that tend toward normality as n gets large. Even for small n of 5 through 10, the approximation is usually relatively accurate. It is sometimes called the Student's t distribution.

Since the t distribution relies on the sample standard deviation s, the resultant distribution will differ according to the sample size n. To designate this difference, the respective distributions are classified according to what are called the *degrees of freedom* and abbreviated as df. In simple problems, the df are just the sample size minus 1. In general, degrees of freedom are the number of quantities minus the number of constraints. The mathematical definition of the t distribution is

$$A(t, \text{df}) = \frac{1}{\text{df}^{1/2} B\left(\frac{1}{2}, \frac{\text{df}}{2}\right)} \int_{-t}^{t} \left(1 + \frac{x^2}{\text{df}}\right)^{-\frac{\text{df}+1}{2}} dx$$

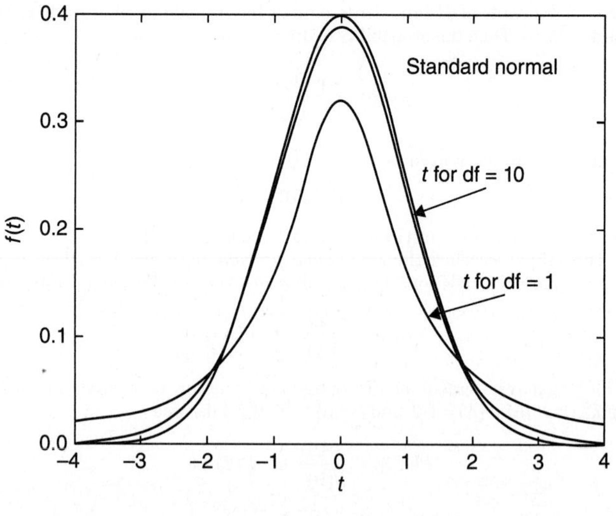

FIG. 3-62 The t distribution function.

where B is the incomplete beta function. $A(t, \text{df})$ is the probability, for degrees of freedom df, that a certain statistic t (measuring the observed difference of means) would be smaller than the observed value if the means were in fact the same. Limiting values are $A(0, \text{df}) = 0$ and $A(\infty, \text{df}) = 1$. The Microsoft Excel function TDIST(X, df,1) gives the right-tail probability, and TDIST(X, df, 2) gives twice that. The probability that $t \le X$ is 1 − TDIST(X, df, 1) when $X \ge 0$ and TDIST(abs(X), df, 1) when $X < 0$. The probability that $-X \le t \le +X$ is 1 − TDIST(X, df, 2). To find the limits for a given confidence level, one uses the Microsoft Excel function TINV(α, df). For a two-tailed distribution, to achieve 95 percent confidence, the two tails represent 2.5 percent each, and one uses $\alpha = 0.05$.

Example For a sample size $n = 5$, what values of t define a midarea of 90 percent? For 4 df using Microsoft Excel, TINV(.1, 4) = 2.132. Thus, $P[-2.132 \le t \le 2.132] = .90$. Also, TDIST(2.132, 4, 2) = 0.10 and 1 − TDIST(2.132, 4, 2) = 0.90.

***t* Distribution for the Difference in Two Sample Means with Equal Variances.** The t distribution can be readily extended to the difference in two sample means when the respective populations have the same variance. Calculate the sample means $\bar{x}_1$ and $\bar{x}_2$ and sample variances s_1^2 and s_2^2, with sample sizes n_1 and n_2, respectively. The variance is estimated by pooling the sum of variances of both samples.

$$s_p^2 = \frac{(n_1 - 1)s_1^2 + (n_2 - 1)s_2^2}{(n_1 - 1) + (n_2 - 1)} \tag{3-98}$$

The significance of the difference is measured by the ratio of the difference to its standard deviation, and it is denoted by t.

$$t = \frac{\bar{x}_1 - \bar{x}_2}{s_p\sqrt{1/n_1 + 1/n_2}} = \frac{\bar{x}_1 - \bar{x}_2}{s_p\sqrt{\dfrac{n_1 + n_2}{n_1 n_2}}} \tag{3-99}$$

Example Suppose we have two sets of data, each with 10 degrees of freedom. The means are 5.23 and 4.95, and the pooled sample variance is 0.3. Are these significantly different at the 10 percent level? Equation (3-99) gives

$$t = \frac{5.23 - 4.95}{\left|0.3\sqrt{\dfrac{10 + 10}{10 * 10}}\right|} = 2.09$$

There are 20 − 2 = 18 degrees of freedom, and TINV(0.1,18) = 1.07. Thus, the probability that t is between +1.07 and −1.07 is $P[-1.07 \le t \le 1.07] = .90$. These data are significantly different, and the hypothesis that they are from the same distribution is disproved.

***t* Distribution for the Difference in Two Sample Means with Unequal Variances** When population variances are unequal, an approximate t quantity can be used:

with

$$t = \frac{\bar{x}_1 - \bar{x}_2 - (\mu_1 - \mu_2)}{\sqrt{a+b}}$$

$$a = s_1^2/n_1 \quad b = s_2^2/n_2$$

and

$$\text{df} = \frac{(a+b)^2}{a^2/(n_1-1)+b^2(n_2-1)} = 2.09$$

Chi-Square Distribution For some industrial applications, product uniformity is of primary importance. The sample standard deviation s is most often used to characterize uniformity. In dealing with this problem, the chi-square distribution can be used where $\chi^2 = (s^2/\sigma^2)$ (df). The chi-square distribution is a family of distributions which are defined by the degrees of freedom associated with the sample variance. For most applications, df is equal to the sample size minus 1. See Fig. 3-63.

The probability density function is

$$C(\chi^2, \text{df}) = \frac{(\chi^2)^{\text{df}/2-1}}{2^{\text{df}/2}\Gamma(\text{df}/2)} e^{-\chi^2/2}$$

and the integral with respect to χ^2 from 0 to infinity is 1.0 and Γ is the gamma function. The cumulative probability function $P(\chi^2, \text{df})$ is the integral of C from 0 to χ^2; different functions are obtained for different degrees of freedom. A plot of C is shown in Fig. 3-63 for 3 and 5 df. The shaded area gives the probability that the chi-squared for a correct model with df degrees of freedom is more than χ^2, here 0.0373. Thus the probability that the χ^2 is larger than 9.2 is .0373. Since the number is small, it means that the probability of getting a χ^2 that large or larger is 3.7 percent. The null hypothesis is that the two samples are from the same normal distribution. If the probability is very small, then there is evidence to reject the hypothesis. If the probability is larger, one can only say that the hypothesis is accepted. That is so because one example cannot prove a statement, but it can disprove a statement.

Also

$$P[\chi_1^2 \le (s^2/\sigma^2)(\text{df}) \le \chi_2^2] = 1 - \alpha$$

where χ_1^2 corresponds to a lower-tail area of $\alpha/2$ and χ_2^2 to an upper-tail area of $\alpha/2$. The basic underlying assumption for the mathematical derivation of chi squared is that a random sample was selected from a normal distribution with variance σ^2. When the population is not normal but skewed, chi-squared probabilities could be substantially in error.

Example On the basis of a sample size $n = 5$, what midrange of values will include the sample ratio s/σ with a probability of 95 percent?

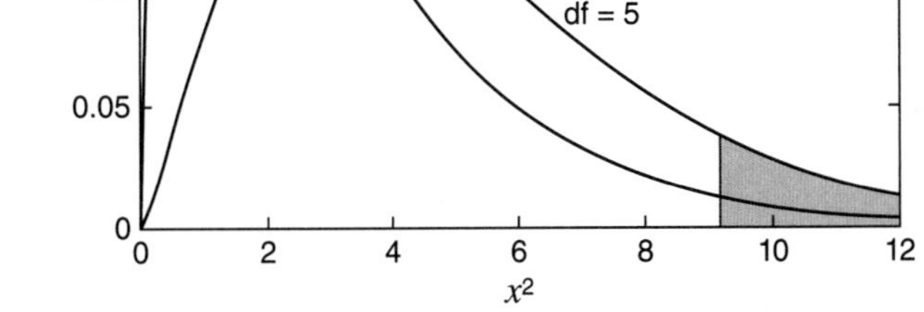

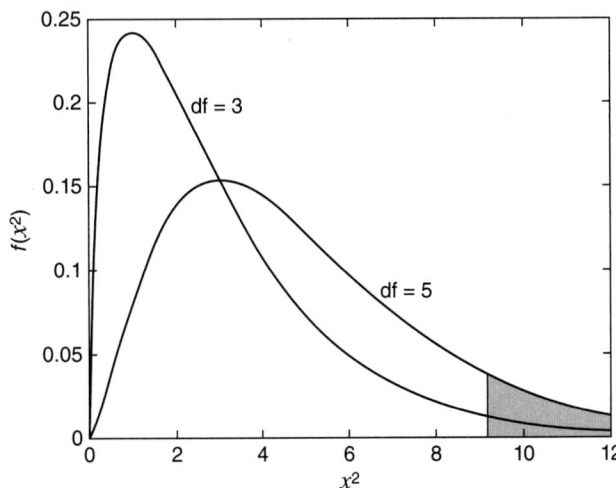

FIG. 3-63 The χ^2 distribution function.

We use the Microsoft Excel function CHIINV to answer this. CHIINV (0.025, 4) = 11.1 and CHIINV(0.975, 4) = 0.484. Then

$$P[.484 \le (s^2/\sigma^2)(4) \le 11.1] = .95$$

or

$$P[.35 \le s/\sigma \le 1.66] = .95$$

This states that the sample standard deviation will be at least 35 percent and not more than 166 percent of the population variance 95 percent of the time. Conversely, 5 percent of the time the standard deviation will underestimate or overestimate the population standard deviation by the corresponding amount.

The chi-squared distribution can be applied to other types of application which are of an entirely different nature. These include applications discussed under Goodness-of-Fit Test and Two-Way Test for Independence of Count Data. In these applications, the mathematical formulation and context are entirely different, but they do result in the same table of values.

***F* Distribution** To test if two samples are from the same population, it is necessary to test whether the means are the same and variances are the same, within some confidence level. The test for variances is done using the F test. The null hypothesis is that the two samples are from the same population. The *F ratio* is defined by the ratio of sample variances.

$$F(\text{df}_1, \text{df}_2) = s_1^2/s_2^2$$

Here df_1 and df_2 correspond to the respective df's for the sample variances. The F distribution, similar to the chi-squared distribution, is sensitive to the basic assumption that sample values were selected randomly from a normal distribution. The Microsoft Excel function FINV(percent, df_1, df_2) gives the largest ratio (and the reciprocal gives the smallest ratio) that agrees with the null hypothesis at that level of significance.

Example For two sample variances with 4 df each, what limits will bracket their ratio with a midarea probability of 90 percent?

Use FINV with 4 df and Percent = 0.05 (to get both sides totaling 10 percent). FINV(.05, 4, 4) gives 6.39. Thus

$$P[1/6.39 \le s_1^2/s_2^2 \le 6.39] = .90$$

or

$$P[.40 \le s_1/s_2 \le 2.53] = .90$$

FDIST(X, df_1, df_2) gives the upper percentage points of the F distribution. FDIST(6.39, 4, 4) = 0.05.

Confidence Interval for a Mean Suppose a change has been made in a process and one wishes to assess the confidence of the population mean (unknown). Thus, one wants

$$P\left[\bar{x} - t\frac{s}{\sqrt{n}} \le \mu \le \bar{x} + t\frac{s}{\sqrt{n}}\right] = 1 - \alpha$$

where t is defined for an upper-tail area of $\alpha/2$ with $n - 1$ df. In this application, the interval $\bar{x} - ts/\sqrt{n}$ limits are random variables which will cover the unknown parameter μ with probability $1 - \alpha$. The converse—that we are $100(1 - \alpha)$ percent sure that the parameter value is within the interval—is not correct. This statement defines a probability for the parameter rather than the probability for the interval. The probability density function is the t distribution and uses the Microsoft Excel commands TDIST and TINV.

Example For the example in Normal Distribution, the sample variance of 25 measurements was 0.15. What is the 90 percent confidence interval for μ?

Using $\text{df} = n - 1 = 24$, $\bar{x} = 5.2$, $s^2 = 0.15$, and a 10 percent value in Microsoft Excel

TINV(0.10, 24) gives $t = 1.71$. Thus

$$P\left[5.2 - 1.71\sqrt{\frac{0.15}{24}} < \mu < 5.2 + 1.71\sqrt{\frac{0.15}{24}}\right] = .90$$

or

$$P(5.065 \le \mu \le 5.335) = 0.90$$

Confidence Interval for the Difference in Two Population Means The confidence interval for a mean can be extended to include the difference between two population means. This interval is based on the assumption that the respective populations have the same variance σ^2:

$$\bar{x}_1 - \bar{x}_2 - ts_p\sqrt{1/n_1 + 1/n_2} \le \mu_1 - \mu_2 \le \bar{x}_1 - \bar{x}_2 + ts_p\sqrt{1/n_1 + 1/n_2}$$

The value of t is obtained from the t distribution, Microsoft Excel function TINV(α, df). To achieve 95 percent confidence, use $\alpha = 0.05$.

Confidence Interval for a Variance The chi-square distribution can be used to derive a confidence interval for a population variance σ^2 when the parent population is normally distributed. For a 100(1 − α) percent confidence interval

$$\frac{(\text{df})s^2}{\chi_2^2} \le \sigma^2 \le \frac{(\text{df})s^2}{\chi_1^2}$$

where χ_1^2 corresponds to a lower-tail area of $\alpha/2$ and χ_2^2 to an upper-tail area of $\alpha/2$.

Example For the example in Normal Distribution, the sample variance of 25 measurements was 0.15. What is range of the population variance σ^2 for a 90 percent confidence interval?

Using $\text{df} = n - 1 = 24$, $s^2 = 0.15$, and 5 percent and 95 percent values in Microsoft Excel function CHIINV gives CHIINV(0.05, 24) = 13.85 and CHIINV(0.95, 24) = 36.42. Thus

$$\frac{24(0.15)}{36.4} \le \sigma^2 \le \frac{24(0.15)}{13.8}$$

or

$$0.0989 \le \sigma^2 \le 0.261$$

$$0.31 \le \sigma \le 0.51$$

Thus, the population variance will be between 0.0989 and 0.261 with 90 percent confidence.

TESTS OF HYPOTHESIS

General Nature of Tests The general nature of tests can be illustrated with a simple example. In a court of law, when a defendant is charged with a crime, the judge instructs the jury initially to presume that the defendant is innocent of the crime. The jurors are then presented with evidence and counterargument as to the defendant's guilt or innocence. If the evidence suggests beyond a reasonable doubt that the defendant did, in fact, commit the crime, the jurors have been instructed to find the defendant guilty; otherwise, not guilty. The burden of proof is on the prosecution. Jury trials represent a form of decision making. In statistics, an analogous procedure for making decisions falls into an area of statistical inference called *hypothesis testing.*

Suppose that a company has been using a certain supplier of raw materials in one of its chemical processes. A new supplier approaches the company and states that its material, at the same cost, will increase the process yield. If the new supplier has a good reputation, the company might be willing to run a limited test. On the basis of the test results it would then make a decision to change suppliers or not. Good management would dictate that an improvement must be demonstrated (beyond a reasonable doubt) for the new material. That is, the burden of proof is tied to the new material. In setting up a test of hypothesis for this application, the initial assumption would be defined as a null hypothesis and symbolized as H_0. The null hypothesis would state that yield for the new material is no greater than for the conventional material. The symbol μ_0 would be used to designate the known current level of yield for the standard material and μ for the unknown population yield for the new material. Thus, the *null hypothesis* can be symbolized as H_0: $\mu \le \mu_0$. The alternative to H_0 is called the *alternative hypothesis* and is symbolized as H_1: $\mu > \mu_0$. To prove the alternative hypothesis, we must show that the null hypothesis is not valid within some probability.

Given a series of tests with the new material, the average yield $\bar{x}$ would be compared with μ_0. If $\bar{x} < \mu_0$, the new supplier would be dismissed. If $\bar{x} > \mu_0$, the question would be: Is it sufficiently greater in the light of its corresponding reliability, i.e., beyond a reasonable doubt? If the confidence interval for μ included μ_0, the answer would be no; but if it did not include μ_0, the answer would be yes. In this simple application, the formal test of hypothesis would result in the same conclusion as that derived from the confidence interval. However, the utility of tests of hypothesis lies in their generality, whereas confidence intervals are restricted to a few special cases.

Nomenclature for All Examples

μ = mean of the population or differences from which the sample has been drawn

σ = standard deviation of the population or differences from which the sample has been drawn

μ_0 = base or reference level

H_0 = null hypothesis

H_1 = alternative hypothesis

α = significance level, usually set at .10, .05, or .01

n = number of observations

t = t value corresponding to the significance level α. For a two-tailed test, each corresponding tail would have an area of $\alpha/2$, and for a one-tailed test, one-tail area would be equal to α. If σ^2 is known, then z would be used rather than t.

Test of Hypothesis for a Mean Procedure

$$t = (\bar{x} - \mu_0)/(s/\sqrt{n}) \quad \text{sample value of the test statistic}$$

Assumptions

1. The n observations $x_1, x_2, \ldots, x_n$ have been selected randomly.

2. The population from which the observations were obtained is normally distributed with an unknown mean μ and standard deviation σ. In actual practice, this is a robust test, in the sense that in most types of problems it is not sensitive to the normality assumption when the sample size is 10 or greater.

Test of Hypothesis

1. The null hypothesis is that the sample came from a population whose mean μ is equivalent to some base or reference designated by μ_0. This can take one of the three forms shown in Table 3-4.

2. If the null hypothesis is assumed to be true, say, in form 1, then the distribution of the test statistic t is known. Given a random sample in a two-sided test, one can predict how far its sample value of t might be expected to deviate from zero (the midvalue of t) by chance alone. If the sample value of t does, in fact, deviate too far from zero, then this is defined to be sufficient evidence to refute the assumption of the null hypothesis. It is consequently rejected, and the converse or alternative hypothesis is accepted.

3. The rule for accepting H_0 is specified by selection of the α level, as indicated in Fig. 3-64. For forms 2 and 3 the α area is defined to be in the upper and the lower tail, respectively. The parameter α is the probability of rejecting the null hypothesis when it is actually true.

4. The decision rules for each of the three forms are defined as follows: If the sample t falls within the acceptance region, accept H_0 for lack of contrary evidence. If the sample t falls in the critical region, reject H_0 at a significance level of 100α percent.

TABLE 3-4 Options for Null Hypothesis and Alternate Hypothesis

Form 1	Form 2	Form 3
H_0: $\mu = \mu_0$	H_0: $\mu \le \mu_0$	H_0: $\mu \ge \mu_0$
H_1: $\mu \ne \mu_0$	H_1: $\mu > \mu_0$	H_1: $\mu < \mu_0$
Two-tailed test	Upper-tailed test	Lower-tailed test

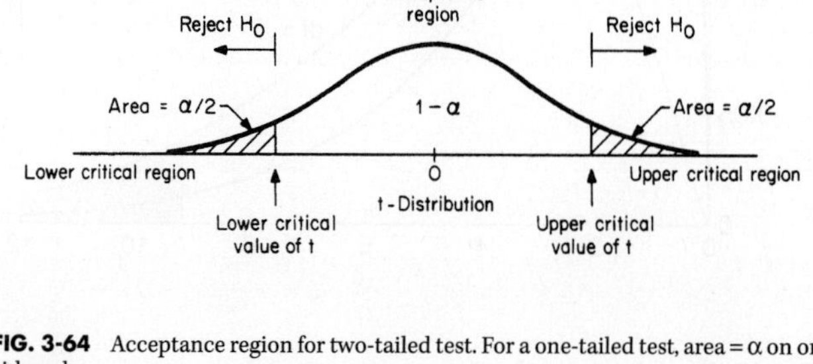

FIG. 3-64 Acceptance region for two-tailed test. For a one-tailed test, area = α on one side only.

Example

Application. In the past, the yield for a chemical process has been established at 89.6 percent with a standard deviation of 3.4 percent. A new supplier of raw materials will be used and tested for 7 days. [Note: many problems with chemical processes arise because of bad raw materials.]

Procedure

1. The standard of reference is $\mu_0 = 89.6$ with a known $\sigma = 3.4$.
2. It is of interest to demonstrate whether an increase in yield is achieved with the new material; H_0 says it has not; therefore,

$$H_0: \mu \le 89.6 \qquad H_1: \mu > 89.6$$

3. Select $\alpha = .05$, and since σ is known (assuming the new material would not affect the day-to-day variability in yield), we use z. The corresponding critical value is TINV(0.10, ∞) = 1.645. Remember TINV is the value for the two-tailed distribution, so the one-tailed distribution is TINV(2α, df).
4. The decision rule is

$$\text{Accept } H_0 \text{ if sample } z < 1.645.$$
$$\text{Reject } H_0 \text{ if sample } z > 1.645.$$

5. A 7-day test was carried out, and daily yields averaged 91.6 percent with a sample standard deviation $s = 3.6$ (this is not needed for the test of hypothesis).
6. For the data sample $z = (91.6 - 89.6)/(3.4/\sqrt{7}) = 1.56$.
7. Since this is less than 1.645, accept the null hypothesis for lack of contrary evidence; i.e., an improvement has not been demonstrated beyond a reasonable doubt.

Example

Application. In the past, the break strength of a synthetic yarn has averaged 34.6 lb. The first-stage draw ratio of the spinning machines has been increased. Production management wants to determine whether the break strength has changed under the new condition.

Procedure

1. The standard of reference is $\mu_0 = 34.6$.
2. It is of interest to demonstrate whether a change has occurred; therefore,

$$H_0: \mu = 34.6 \qquad H_1: \mu \ne 34.6$$

3. Select $\alpha = .05$, and since with the change in draw ratio the uniformity might change, the sample standard deviation would be used, and therefore t would be the appropriate test statistic.
4. A sample of 21 ends was selected randomly and tested on an Instron with the results $\bar{x} = 35.55$ and $s = 2.041$.
5. For 20 df and a two-tailed α level of 5 percent, TINV(0.05, 20) = ±2.086.

$$\text{Accept } H_0 \text{ if } -2.086 < \text{sample } t < 2.086.$$
$$\text{Reject } H_0 \text{ if sample } t < -2.086 \text{ or } t > 2.086.$$

6. For the data sample $t = (35.55 - 34.6)/(2.041/\sqrt{21}) = 2.133$.
7. Since 2.133 > 2.086, reject H_0 and accept H_1. It has been demonstrated that an improvement in break strength has been achieved.

Two-Population Test of Hypothesis for Means

Nature Two samples were selected from different locations in a plastic-film sheet and measured for thickness. The thickness of the respective samples was measured at 10 close but equally spaced points in each of the samples. It was of interest to compare the average thickness of the respective samples to detect whether they were significantly different. That is, was there a significant variation in thickness between locations?

From a modeling standpoint, statisticians would define this problem as a two-population test of hypothesis. They would define the respective sample sheets as two populations from which 10 sample thickness determinations were measured for each.

To compare populations based on their respective samples, it is necessary to have some basis of comparison. This basis is predicated on the distribution of the t statistic. In effect, the t statistic characterizes the way in which two sample means from two separate populations will tend to vary by chance alone when the population means and variances are equal. Consider the following:

Population 1		Population 2	
Normal	Sample 1	Normal	Sample 2
μ_1	n_1	μ_2	n_2
	$\bar{x}_1$		$\bar{x}_2$
σ_1^2	s_1^2	σ_2^2	s_2^2

Consider the hypothesis $\mu_1 = \mu_2$. If, in fact, the hypothesis is correct, that is, $\mu_1 = \mu_2$ (under the condition $\sigma_1^2 = \sigma_2^2$), then the sampling distribution of $\bar{x}_1 - \bar{x}_2$ is predictable through the t distribution. (We use t rather than z because the variance is unknown.) The observed sample values then can be compared with the corresponding t distribution.

Example

Application. Two samples were selected from different locations in a plastic-film sheet. The thickness of the respective samples was measured at 10 close but equally spaced points.

Procedure

1. Demonstrate whether the thicknesses of the respective sample locations are significantly different from each other; therefore,

$$H_0: \mu_1 = \mu_2 \qquad H_1: \mu_1 \ne \mu_2$$

2. Select $\alpha = .05$.
3. Summarize the statistics for the respective samples:

Sample 1		Sample 2	
1.473	1.367	1.474	1.417
1.484	1.276	1.501	1.448
1.484	1.485	1.485	1.469
1.425	1.462	1.435	1.474
1.448	1.439	1.348	1.452
$\bar{x}_1 = 1.434$	$s_1 = .0664$	$\bar{x}_2 = 1.450$	$s_2 = .0435$

4. The first step is to use the F test on the ratio of sample variances. The null hypothesis is $H_0: \sigma_1^2 = \sigma_2^2$, and it would be tested against $H_1: \sigma_1^2 \ne \sigma_2^2$. Since this is a two-tailed test, the procedure is to use the largest ratio and the corresponding ordered degrees of freedom. However, since the largest ratio is arbitrary, it is necessary to define the true α level as twice the desired value. Therefore, using FINV(0.05, 9, 9) = 3.18 would be for a true $\alpha = .10$. For the sample,

$$\text{Sample } F = (.0664/.0435)^2 = 2.33$$

Therefore, the ratio of sample variances is no larger than one might expect to observe when in fact $\sigma_1^2 = \sigma_2^2$. There is not sufficient evidence to reject the null hypothesis that $\sigma_1^2 = \sigma_2^2$.

5. Turn next to the t test. For 18 df and a two-tailed α level of 5 percent, the critical values of t are given by TINV(0.05, 18) = ±2.101.
6. The decision rule is

$$\text{Accept } H_0 \text{ if } -2.101 \le \text{sample } t \le 2.101.$$
$$\text{Reject } H_0 \text{ otherwise.}$$

7. For the sample the pooled variance estimate is given by Eq. (3-98).

$$s_p^2 = \frac{9(.0664)^2 + 9(.0435)^2}{9+9} = .00315, s_p = .056$$

8. The sample statistic value of t is given by Eq. (3-99).

$$\text{Sample } t = \frac{1.434 - 1.450}{.056\sqrt{1/10 + 1/10}} = -.64$$

9. Since the sample value of t falls within the acceptance region, accept H_0 for lack of contrary evidence; i.e., there is insufficient evidence to demonstrate that thickness differs between the two selected locations.

Test of Hypothesis for Paired Observations

Nature In some types of applications, associated pairs of observations are defined. For example, (1) pairs of samples from two populations are treated in the same way, or (2) two types of measurements are made on the same unit. For applications of this type, it is not only more effective but also necessary to define the random variable as the difference between the pairs of observations. The difference numbers can then be tested by the standard t distribution.

Test of Hypothesis for Matched Pairs: Procedure

Nomenclature

d_i = sample difference between the ith pair of observations

s = sample standard deviation of differences

μ = population mean of differences

σ = population standard deviation of differences

t = value with $(n-1)$ df

$t=(\bar{d}-\mu_0)/(s/\sqrt{n})$, the sample value of t

Assumptions

1. The n pairs of samples have been selected and assigned for testing in a random way.

2. The population of differences is normally distributed with a mean μ and variance σ^2. As in the previous application of the t distribution, this is a robust procedure, i.e., not sensitive to the normality assumption if the sample size is 10 or greater in most situations.

Test of Hypothesis

1. Under the null hypothesis, it is assumed that the sample came from a population whose mean μ is equivalent to some base or reference level, designated by μ_0. For most applications of this type, the value of μ_0 is defined to be zero; that is, it is of interest generally to demonstrate a difference not equal to zero. The hypothesis can take one of three forms shown in Table 3-4.

2. If the null hypothesis is assumed to be true, say, in the case of a lower-tailed test, form 3, then the distribution of the test statistic t is known under the null hypothesis that limits $\mu=\mu_0$. Given a random sample, one can predict how far its sample value of t might be expected to deviate from zero by chance alone when $\mu=\mu_0$. If the sample value of t is too small, as in the case of a negative value, then this would be defined as sufficient evidence to reject the null hypothesis.

3. Select α.

4. The critical values or value of t would be defined by the value of t with $n-1$ df corresponding to a tail area of α. For a two-tailed test use TINV(α, df), and for a one-tailed test use TINV(2α, df).

5. The decision rule for each of the three forms would be to reject the null hypothesis if the sample value of t fell in that area of the t distribution defined by α, which is called the *critical region.* Otherwise, the alternative hypothesis would be accepted for lack of contrary evidence.

Example, Two-Sided Test

Application. Pairs of pipes have been buried in 11 different locations to determine corrosion on nonbituminous pipe coatings for underground use. One type includes a lead-coated steel pipe and the other a bare steel pipe.

Procedure

1. The standard of reference is taken as $\mu_0=0$, corresponding to no difference in the two types.

2. It is of interest to demonstrate whether either type of pipe has a greater corrosion resistance than the other. Therefore,

$$H_0: \mu=0 \qquad H_1: \mu\neq 0$$

3. Select $\alpha=.05$. TINV(0.05, 10) = 2.228.

4. The decision rule is then

Accept H_0 if $-2.228\le$ sample $t\le 2.228$

Reject H_0 otherwise

5. The sample of 11 pairs of corrosion determinations and their differences is as follows:

Soil type	Lead-coated steel pipe	Bare steel pipe	d = difference
A	27.3	41.4	−14.1
B	18.4	18.9	−0.5
C	11.9	21.7	−9.8
D	11.3	16.8	−5.5
E	14.8	9.0	5.8
F	20.8	19.3	1.5
G	17.9	32.1	−14.2
H	7.8	7.4	0.4
I	14.7	20.7	−6.0
J	19.0	34.4	−15.4
K	65.3	76.2	−10.9

6. The sample statistics, Eq. (3-97)

$$\bar{d}=-6.245 \quad s^2\frac{11\sum d^2-(\sum d)^2}{11\times 10}=52.59$$

or

$$s=7.25$$

$$\text{Sample } t=(-6.245-0)/(7.25/\sqrt{11})$$
$$=-2.86$$

7. Since the sample t of $-2.86<$ tabled t of -2.228, reject H_0 and accept H_1; that is, it has been demonstrated that, on the basis of the evidence, lead-coated steel pipe has a greater corrosion resistance than bare steel pipe.

Example, One-Sided Test

Application. A stimulus was tested for its effect on blood pressure. Ten men were selected randomly, and their blood pressure was measured before and after the stimulus was administered. It was of interest to determine whether the stimulus had caused a significant increase in the blood pressure.

Procedure

1. The standard of reference was taken as $\mu_0\le 0$, corresponding to no increase.

2. It was of interest to demonstrate an increase in blood pressure if in fact an increase did occur. Therefore,

$$H_0: \mu_0\le 0 \qquad H_1: \mu_0>0$$

3. Select $\alpha=.05$. Since only increases are of interest, use a one-sided value TINV(0.05*2, 9) = 1.833.

4. The decision rule is

Accept H_0 if sample $t<1.833$.

Reject H_0 if sample $t>1.833$.

5. The sample of 10 pairs of blood pressure and their differences was as follows:

Individual	Before	After	d = difference
1	138	146	8
2	116	118	2
3	124	120	−4
4	128	136	8
5	155	174	19
6	129	133	4
7	130	129	−1
8	148	155	7
9	143	148	5
10	159	155	−4

6. The sample statistics:

$$\bar{d}=4.4 \quad s=6.85$$

$$\text{Sample } t=(44-0)/(6.85/\sqrt{10})=2.03$$

7. Since the sample t = 2.03 > critical t = 1.833, reject the null hypothesis. It has been demonstrated that the population of men from whom the sample was drawn tend, as a whole, to have an increase in blood pressure after the stimulus has been given. The distribution of differences d seems to indicate that the degree of response varies by individuals.

Test of Hypothesis for a Proportion

Nature Some types of statistical applications deal with counts and proportions rather than measurements. Examples are (1) the proportion of workers in a plant who are out sick, (2) lost-time worker accidents per month, (3) defective items in a shipment lot, and (4) preference in consumer surveys.

The procedure for testing the significance of a sample proportion follows that for a sample mean. In this case, however, owing to the nature of the problem the appropriate test statistic is Z. This follows from the fact that the null hypothesis requires the specification of the goal or reference quantity p_0, and since the distribution is a binomial proportion, the associated variance under the null hypothesis is $[p_0(1-p_0)]n$. The primary requirement is that the sample size n satisfy normal approximation criteria for a binomial proportion, roughly $np > 5$ and $n(1-p) > 5$.

Test of Hypothesis for a Proportion: Procedure

Nomenclature

p = mean proportion of the population from which the sample has been drawn

p_0 = base or reference proportion

$[p_0(1-p_0)]/n$ = base or reference variance

$\hat{p} = x/n$ = sample proportion, where x refers to the number of observations out of n which have the specified attribute

z = Z value corresponding to the significance level α

$z = (\hat{p} - p_0)/\sqrt{p_0(1-p_0)/n}$, the sample value of the test statistic

Assumptions

1. The n observations have been selected randomly.
2. The sample size n is sufficiently large to meet the requirement for the Z approximation.

Test of Hypothesis

1. Under the null hypothesis, it is assumed that the sample came from a population with a proportion p_0 of items having the specified attribute. For example, in tossing a coin the population could be thought of as having an unbounded number of potential tosses. If it is assumed that the coin is fair, this would dictate $p_0 = 1/2$ for the proportional number of heads in the population. The null hypothesis can take one of three forms:

Form 1	Form 2	Form 3
H_0: $p = p_0$	H_0: $p \le p_0$	H_0: $p \ge p_0$
H_1: $p \ne p_0$	H_1: $p > p_0$	H_1: $p < p_0$
Two-tailed test	Upper-tailed test	Lower-tailed test

2. If the null hypothesis is assumed to be true, then the sampling distribution of the test statistic Z is known. Given a random sample, it is possible to predict how far the sample proportion x/n might deviate from its assumed population proportion p_0 through the Z distribution. When the sample proportion deviates too far, as defined by the significance level α, this serves as the justification for rejecting the assumption, that is, rejecting the null hypothesis.
3. The decision rule is given by

Form 1: Accept H_0 if lower critical z < sample z < upper critical z.
Reject H_0 otherwise.
Form 2: Accept H_0 if sample z < upper critical z.
Reject H_0 otherwise.
Form 3: Accept H_0 if lower critical z < sample z.
Reject H_0 otherwise.

Example

Application. A company has received a very large shipment of rivets. One product specification required that no more than 2 percent of the rivets have diameters greater than 14.28 mm. Any rivet with a diameter greater than this would be classified as defective. A random sample of 600 was selected and tested with a go–no go gauge. Of these, 16 rivets were found to be defective. Is this sufficient evidence to conclude that the shipment contains more than 2 percent defective rivets?

Procedure

1. The quality goal is $p \le .02$. It would be assumed initially that the shipment meets this standard; that is, H_0: $p \le .02$.
2. The assumption in step 1 would first be tested by obtaining a random sample. Under the assumption that $p \le .02$, the distribution for a sample proportion would be defined by the z distribution. This distribution would define an upper bound corresponding to the upper critical value for the sample proportion. It would be unlikely that the sample proportion would rise above that value if, in fact, $p \le .02$. If the observed sample proportion exceeds that limit, corresponding to what would be a very unlikely chance outcome, this would lead one to question the assumption that $p \le .02$. That is, one would conclude that the null hypothesis is false. To test, set

$$H_0: p \le .02 \qquad H_1: p > .02$$

3. Select $\alpha = .05$.
4. With $\alpha = .05$, the upper critical value of Z = TINV(0.05*2, ∞) = 1.645 for a one-sided test.
5. The decision rule:

Accept H_0 if sample $z < 1.645$.
Reject H_0 if sample $z > 1.645$.

6. The sample z is given by

$$\text{Sample } z = \frac{(16/600) - .02}{\sqrt{(.02)(.98)/600}} = 1.17$$

7. Since the sample $z < 1.645$, accept H_0 for lack of contrary evidence; there is not sufficient evidence to demonstrate that the defect proportion in the shipment is greater than 2 percent.

Test of Hypothesis for Two Proportions

Nature In some types of engineering and management science problems, we may be concerned with a random variable that represents a proportion, for example, the proportional number of defective items per day. The method described previously relates to a single proportion. In this subsection two proportions will be considered.

A certain change in a manufacturing procedure for producing component parts is being considered. Samples are taken by using both the existing and the new procedures to determine whether the new procedure results in an improvement. In this application, it is of interest to demonstrate statistically whether the population proportion p_2 for the new procedure is less than the population proportion p_1 for the old procedure on the basis of a sample of data.

Test of Hypothesis for Two Proportions: Procedure

Nomenclature

p_i = population i = 1 or 2 proportion

n_i = sample size from population 1 or 2

x_i = number of observations out of that have the designated attribute

$\hat{p}_i = x_i/n_i$, the sample proportion from population 1 or 2

z = Z value corresponding to the stated significance level α

$$z = \frac{\hat{p}_1 - \hat{p}_2}{\sqrt{\hat{p}_1(1-\hat{p}_1)/n_1 + \hat{p}_2(1-\hat{p}_2)/n_2}} \text{ the sample value of } Z$$

Assumptions

1. The respective two samples of n_1 and n_2 observations have been selected randomly.
2. The sample sizes n_1 and n_2 are sufficiently large to meet the requirement for the Z approximation; that is, $x_1 > 5$ and $x_2 > 5$.

Test of Hypothesis

1. Under the null hypothesis, it is assumed that the respective two samples have come from populations with equal proportions $p_1 = p_2$. Under this hypothesis, the sampling distribution of the corresponding Z statistic is known. On the basis of the observed data, if the resultant sample value of Z represents an unusual outcome, that is, if it falls within the critical region, this would cast doubt on the assumption of equal proportions. Therefore, it will have been demonstrated statistically that the population proportions are in fact not equal. The various hypotheses can be stated:

Form 1	Form 2	Form 3
H_0: $p_1 = p_2$	H_0: $p_1 \le p_2$	H_0: $p_1 \ge p_2$
H_1: $p_1 \ne p_2$	H_1: $p_1 > p_2$	H_1: $p_1 < p_2$
Two-tailed test	Upper-tailed test	Lower-tailed test

2. The decision rule for form 1 is given by

Accept H_0 if lower critical $z <$ sample $z <$ upper critical z.

Reject H_0 otherwise.

Example

Application. A change was made in a manufacturing procedure for component parts. Samples were taken during the last week of operations with the old procedure and during the first week of operations with the new procedure. Determine whether the proportional numbers of defects for the respective populations differ on the basis of the sample information.

Procedure

1. The hypotheses are

$$H_0: p_1 = p_2 \qquad H_1: p_1 \neq p_2$$

2. Select $\alpha = .05$. Therefore, the critical values of z are ± 1.96 since TINV(0.05, ∞) = 1.96.

3. For the samples, 75 out of 1720 parts from the previous procedure and 80 out of 2780 parts under the new procedure were found to be defective; therefore,

$$\hat{p}_1 = 75/1720 = .0436 \qquad \hat{p}_2 = 80/2780 = .0288$$

4. The decision rule:

Accept H_0 if $-1.96 \leq$ sample $z \leq 1.96$.

Reject H_0 otherwise.

5. The sample statistic:

$$\text{Sample } z = \frac{.0436 - .0288}{\sqrt{(.0436)(.9564)/1720 + (.0288)(.9712)/2780}} = 2.53$$

6. Since the sample z of $2.53 > z = 1.96$, reject H_0 and conclude that the new procedure has resulted in a reduced defect rate.

Goodness-of-Fit Test

Nature A standard die has six sides numbered from 1 to 6. If one were really interested in determining whether a particular die was well balanced, one would have to carry out an experiment. To do this, it might be decided to count the frequencies of outcomes, 1 through 6, in tossing the die N times. On the assumption that the die is perfectly balanced, one would expect to observe $N/6$ occurrences each for 1, 2, 3, 4, 5, and 6. However, chance dictates that exactly $N/6$ occurrences each will not be observed. For example, given a perfectly balanced die, the probability is only 1 chance in 65 that one will observe 1 outcome each, for 1 through 6, in tossing the die 6 times. Therefore, an outcome different from 1 occurrence each can be expected. Conversely, an outcome of six 3s would seem to be too unusual to have occurred by chance alone.

Some industrial applications involve the concept outlined here. The basic idea is to test whether a group of observations follows a preconceived distribution. In the case cited, the distribution is uniform; i.e., each face value should *tend* to occur with the same frequency.

Goodness-of-Fit Test: Procedure

Nomenclature Each experimental observation can be classified into one of r possible categories or cells.

r = total number of cells

O_j = number of observations occurring in cell j

E_j = expected number of observations for cell j based on the preconceived distribution

N = total number of observations = $\sum_{j=1}^{r} O_j$

df = degrees of freedom for the test. In general, this will be equal to $r - 1$ minus the number of statistical quantities on which the E_j's are based (see the examples that follow for details).

Assumptions

1. The observations represent a sample selected randomly from a population that has been specified.

2. The number of expectation counts E_j within each category should be roughly 5 or more. If an E_j count is significantly less than 5, that cell should be pooled with an adjacent cell.

Computation for E_j On the basis of the specified population, the probability of observing a count in cell j is defined by p_j. For a sample of size N, corresponding to N total counts, the expected frequency is given by $E_j = Np_j$.

Test Statistics: Chi Square

$$\chi^2 = \sum_{j=1}^{r} \frac{(O_j - E_j)^2}{E_j} \quad \text{with df degrees of freedom}$$

Test of Hypothesis

1. H_0: The sample came from the specified theoretical distribution.
   H_1: The sample did not come from the specified theoretical distribution.
2. For a stated level of α,
   Reject H_0 if sample $\chi^2 >$ CHIINV χ^2.
   Accept H_0 if sample $\chi^2 <$ CHIINV χ^2.

Example

Application. A production-line product is rejected if one of its characteristics does not fall within specified limits. The standard goal is that no more than 2 percent of the production should be rejected.

Computation

1. Of 950 units produced during the day, 28 units were rejected; there are two cells, so $r = 2$.

2. The hypotheses:

H_0: process is in control

H_1: process is not in control

3. Assume that $\alpha = .05$; therefore, the critical value of $\chi^2(1)$ is CHIINV (0.05, 1) = 3.84.

4. The decision rule:

Reject H_0 if sample $\chi^2 > 3.84$.

Accept H_0 otherwise.

5. Since it is assumed that $p = .02$, this would dictate that in a sample of 950 there would be on average (.02)(950) = 19 defective items and 931 acceptable items:

Category	Observed O_j	Expectation $E_j = 950p_j$
Acceptable	922	931
Not acceptable	28	19
Total	950	950

$$\text{Sample } \chi^2 = \frac{(922-931)^2}{931} + \frac{(28-19)^2}{19}$$
$$= 4.35 \text{ with critical } \chi^2 = 3.84$$

6. Conclusion. Since the sample value exceeds the critical value, the process is not in control.

Example

Application. A frequency count of 52 workers was tabulated according to the number of defective items that they produced. An unresolved question is whether the observed distribution is a Poisson distribution. That is, do observed and expected frequencies agree within chance variation?

Computation

1. The hypotheses:
   H_0: there are no significant differences, in number of defective units, between workers.
   H_1: there are significant differences.
2. Assume that $\alpha = .05$.
3. Test statistic:

No. of defective units = n_j	O_j		$O_j n_j$	E_j	
0	3	10	0	2.06	8.70 pool
1	7		7	6.64	
2	9		18	10.73	
3	12		36	11.55	
4	9		36	9.33	
5	6		30	6.03	
6	3	6	18	3.24	5.66 pool
7	2		14	1.50	
8	0		0	0.60	
9	1		9	0.22	
>10	0		0	0.10	
Sum	52		168	52	

The expectation numbers E_j were computed as follows: For the Poisson distribution, $\lambda = E(x)$; therefore, an estimate of λ is the average number of defective units per worker, that is, $\lambda = (1/52)(0\times3 + 1\times7 + \cdots + 9\times1) = 3.23$. Given this approximation, the probability of no defective units for a worker would be $(3.23)^0/0!)e^{-3.23} = .0396$. For the 52 workers, the number of workers producing no defective units would have an expectation $E = 52(0.0396) = 2.06$, and so forth.

The sample chi-square value is computed from

$$\chi^2 = \frac{(10-8.70)^2}{8.70} + \frac{(9-10.73)^2}{10.73} + \cdots + \frac{(6-5.66)^2}{5.66}$$
$$= .522$$

4. The critical value of χ^2 would be based on 4 degrees of freedom. This corresponds to $(r-1)-1=4$, since one statistical quantity λ was computed from the sample and used to derive the expectation numbers.

5. The critical value of χ^2 is CHIINV(0.05, 4) = 9.49; therefore, accept H_0.

Two-Way Test for Independence for Count Data

Nature When individuals or items are observed and classified according to two different criteria, the resultant counts can be statistically analyzed. For example, a market survey may examine whether a new product is preferred and if it is preferred due to a particular characteristic.

Count data, based on a random selection of individuals or items which are classified according to two different criteria, can be statistically analyzed through the χ^2 distribution. The purpose of this analysis is to determine whether the respective criteria are dependent. That is, is the product preferred because of a particular characteristic?

Two-Way Test for Independence for Count Data: Procedure

Nomenclature

1. Each observation is classified according to two categories:
   a. The first one into 2, 3, ..., or r categories
   b. The second one into 2, 3, ..., or c categories
2. O_{ij} = number of observations (observed counts) in cell (i, j) with

$$i = 1, 2, \ldots, r$$
$$j = 1, 2, \ldots, c$$

3. N = total number of observations
4. E_{ij} = computed number for cell (i, j) which is an expectation based on the assumption that two characteristics are independent
5. R_i = subtotal of counts in row i
6. C_j = subtotal of counts in column j
7. χ^2 = critical value of χ^2 corresponding to the significance level α and $(r-1)(c-1)$ df
8. $$\text{Sample } \chi^2 = \frac{\sum_{i,j}^{c,r}(O_{ij}-E_{ij})^2}{E_{ij}}$$

Assumptions

1. The observations represent a sample selected randomly from a large total population.

2. The number of expectation counts E_{ij} within each cell should be approximately 2 or more for arrays 3×3 or larger. If any cell contains a number smaller than 2, appropriate rows or columns should be combined to increase the magnitude of the expectation count. For arrays 2×2, approximately 4 or more are required; if the number is less than 4, the exact Fisher test should be used.

Test of Hypothesis Under the null hypothesis, the classification criteria are assumed to be independent, i.e.,

H_0: criteria are independent
H_1: criteria are not independent

For the stated level of α,

Reject H_0 if sample χ^2 > CHIINV χ^2.
Accept H_0 otherwise.

Computation for E_{ij} Compute E_{ij} across rows or down columns by using either of the following identities:

$$E_{ij} = C_j\left(\frac{R_i}{N}\right) \quad \text{across rows}$$

$$E_{ij} = R_j\left(\frac{C_i}{N}\right) \quad \text{down columns}$$

Sample χ^2 Value

$$\chi^2 = \sum_{i,j}\frac{(O_{ij}-E_{ij})^2}{E_{ij}}$$

In the special case of $r = 2$ and $c = 2$, a more accurate and simplified formula that does not require the direct computation of E_{ij} can be used:

$$\chi^2 = \frac{\left[|O_{11}O_{22} - O_{12}O_{21}| - \frac{1}{2}N\right]^2 N}{R_1R_2C_1C_1}$$

Example

Application. A market research study was carried out to relate the subjective "feel" of a consumer product to consumer preference. In other words, is the consumer's preference for the product associated with the feel of the product, or is the preference independent of the product feel?

Procedure

1. It was of interest to demonstrate whether an association exists between feel and preference; therefore, assume

H_0: feel and preference are independent
H_1: they are not independent

2. A sample of 200 people was asked to classify the product according to two criteria:

a. Liking for this product
b. Liking for the feel of the product

		Like feel		
		Yes	No	R_i
Like product	Yes	114	13	127
	No	55	18	73
	C_j	169	31	200

3. Select $\alpha = .05$; therefore, with $(r-1)(c-1) = 1$ df, the critical value of χ^2 is CHIINV(0.05, 1) =3.84.

4. The decision rule:

Accept H_0 if sample $\chi^2 < 3.84$.
Reject H_0 otherwise.

5. The sample value of χ^2 by using the special formula is

$$\text{Sample } \chi^2 = \frac{\left[|114\times18 - 13\times55| - 100\right]^2 200}{(169)(31)(127)(73)} = 6.30$$

6. Since the sample χ^2 of 6.30 > CHIINV χ^2 of 3.84, reject H_0 and accept H_1. The relative proportionality of $E_{11} = 169(127/200) = 107.3$ to the observed 114 compared with $E_{22} = 31(73/200) = 11.3$ to the observed 18 suggests that when the consumer likes the feel, the consumer tends to like the product, and conversely for not liking the feel. The proportions 169/200 = 84.5 percent and 127/200 = 63.5 percent suggest further that there are other attributes of the product which tend to nullify the beneficial feel of the product.

LEAST SQUARES

When experimental data are to be fit with a mathematical model, it is necessary to allow for the fact that the data have errors. The engineer is interested in finding the parameters in the model as well as the uncertainty in their determination. In the simplest case, the model is a linear equation with only two parameters, and they are found by a least-squares minimization of the errors in fitting the data. Multiple regression is just linear least squares applied with more terms. Nonlinear regression allows the parameters of the model to enter in a nonlinear fashion. See Press et al. (2007) in General References for a description of maximum likelihood as it applies to both linear and nonlinear least squares. Since many calculators include least-squares calculations, the emphasis here is on the estimates and their uncertainty.

In a least-squares parameter estimation, it is desired to find parameters that minimize the sum of squares of the deviation between the experimental data and the theoretical equation.

$$\chi^2 = \sum_{i=1}^{N}\left[\frac{y_i - y(x_i; a_1, a_2, \cdots, a_M)}{\sigma_i}\right]^2$$

where y_i is the ith experimental data point for the value x_i, σ_i is the standard deviation for the ith point, y_i, $y(x_i; a_1, a_2, \ldots, a_M)$ is the theoretical equation at x_i, and the parameters $\{a_1, a_2, \ldots, a_M\}$ are to be determined to minimize χ^2. If the uncertainties in y_i are not known, then assume a constant $\sigma = \sigma_i$ for all i. After calculation the variance will be minimized, giving σ, and χ^2 can be calculated.

$$\sigma^2 = \sum_{i=1}^{N} \frac{[y_i - y(x_i; a_1, a_2, \ldots, a_M)]^2}{N}$$

Linear Least Squares When the model is a straight line $y_i - a - bx_i$, one is minimizing

$$\chi^2 = \sum_{i=1}^{N} (y_i - a - bx_i)^2$$

The linear correlation coefficient r is defined by

$$r = \frac{\sum_{i=1}^{N} (x_i - \bar{x})(y_i - \bar{y})}{\sqrt{\sum_{i=1}^{N} (x_i - \bar{x})^2} \sqrt{\sum_{i=1}^{N} (y_i - \bar{y})^2}}$$

and

$$\chi^2 = (1 - r^2) \sum_{i=1}^{N} (y_i - \bar{y})^2$$

where $\bar{y}$ is the average of the y_i values. Values of r near 1 indicate a positive correlation; r near −1 means a negative correlation, and r near 0 means no correlation. These parameters are easily found by using standard programs.

The solution for $\hat{a}$ and $\hat{b}$ is

$$\hat{a} = \left(\sum_{i=1}^{N} x_i^2 \sum_{i=1}^{N} y_i - \sum_{i=1}^{N} x_i \sum_{i=1}^{N} x_i y_i \right) / \text{dem}$$

$$\hat{b} = \left(N \sum_{i=1}^{N} x_i y_i - \sum_{i=1}^{N} x_i \sum_{i=1}^{N} y_i \right) / \text{dem}$$

$$\text{dem} \equiv N \sum_{i=1}^{N} x_i^2 - \left(\sum_{i=1}^{N} x_i \right)^2$$

The variance of the estimate is the χ^2 given above with the N replaced by $N - 2$ since the line has two constraints.

$$s^2 = \sum_{i=1}^{N} (y_i - \hat{a} - \hat{b}x_i)^2 / (N-2)$$

The variances of $\hat{a}$ and $\hat{b}$ are

$$s_{\hat{a}}^2 = \left(\frac{1}{N} + \frac{\bar{x}^2}{\sum_{i=1}^{N} (x_i - \bar{x})^2} \right) s^2 \quad \text{and} \quad s_{\hat{b}}^2 = \frac{s^2}{\sum_{i=1}^{N} (x_i - \bar{x})^2}$$

The solution is found in Microsoft Excel by putting the values for x and y in two columns (for example, A1:A10, B1:B10). The commands = SLOPE(A1:A10,B1-B10), INTERCEPT(A1:A10,B1-B10), and RSQ(A1:A10,B1-B10) give the slope, intercept, and residual squared. You can also use the LINEST function. See Microsoft Excel Help menu for the function LINEST which can give the statistics for multiple linear regression.

A t test can give the significance. For example, using

$$t = \frac{\hat{b}}{s_{\hat{b}}}$$

with a two-sided test, the value would be rejected if outside the range given by $t(\alpha/2)$ = TINV(α, $N - 2$) in Microsoft Excel.

When there are more terms, i.e., multiple linear regression, similar formulas can be found, usually using the computer. Estimates for the variances and t tests are available, e.g., in Mendenhall and Sincich (2006). In Microsoft Excel, one simply adds columns to the spreadsheet for the additional independent variables.

Polynomial Regression In polynomial regression, one expands the function in a polynomial in x and the same considerations apply.

$$y(x) = \sum_{j=1}^{M} a_j x^{j-1}$$

For N measurements write this as

$$y_i = \sum_{j=1}^{M} a_j x_i^{j-1}$$

In Microsoft Excel, the instructions above hold with the columns $x_i^0, x_i^1, x_i^2, x_i^3$, etc.

Multiple Nonlinear Regression In multiple nonlinear regression, any set of functions can be used, not just polynomials, such as

$$y(x) = \sum_{j=1}^{M} = a_j f_j(x)$$

where the set of functions $\{f_j(x)\}$ is known and specified. Note that the unknown parameters $\{a_j\}$ enter the equation linearly. In this case, the spreadsheet can be expanded to have a column for x and then successive columns for $f_j(x)$. Then this works in the same way as for linear multiple regression.

Nonlinear Least Squares There are no analytic methods for determining the most appropriate model for a particular set of data. In many cases, however, the engineer has some basis for a model. If the parameters occur in a nonlinear fashion, then the analysis becomes more difficult. For example, in relating the temperature to the elapsed time of a fluid cooling in the atmosphere, a model that has an asymptotic property would be the appropriate model (temp = $a + b$ exp($-c$ time), where a represents the asymptotic temperature corresponding to $t \to \infty$). In this case, the parameter c appears nonlinearly. The usual practice is to concentrate on model development and computation rather than on statistical aspects. In general, nonlinear regression should be applied only to problems in which there is a well-defined, clear association between the two variables; therefore, a test of hypothesis on the significance of the fit would be somewhat ludicrous. In addition, the generalization of the theory for the associate confidence intervals for nonlinear coefficients is not well developed.

Example

Application. Data were collected on the cooling of water in the atmosphere as a function of time.

Sample data

Time x	Temperature y
0	92.0
1	85.5
2	79.5
3	74.5
5	67.0
7	60.5
10	53.5
15	45.0
20	39.5

Model. The data are fit to the formula $y = a + be^{cx}$ using optimization techniques in MATLAB, giving $a = 33.54$, $b = 57.89$, $c = 0.11$. The value of χ^2 is 1.83. Using an alternative form, $y = a + b/(c + x)$, gives $a = 9.872$, $b = 925.7$, $c = 11.27$, and $\chi = 0.19$. Since this model had a smaller value of χ^2, it might be the chosen one, but it is only a fit of the specified data and may not be generalized beyond that. Both forms give equivalent plots.

ERROR ANALYSIS OF EXPERIMENTS

Consider the problem of assessing the accuracy of a series of measurements. If measurements are for independent, identically distributed observations, then the errors are independent and uncorrelated. Then $\bar{y}$, the experimentally determined mean, varies about $E(y)$, the true mean, with variance σ^2/n, where n is the number of observations in $\bar{y}$. Thus, if one measures something several times today, and each day, and the measurements have the same distribution, then the variance of the means decreases with the number of samples in each day's measurement n. Of course, other factors (weather, weekends) may make the observations on different days *not* distributed identically.

Consider next the problem of estimating the error in a variable that cannot be measured directly but must be calculated based on results of other measurements. Suppose the computed value Y is a linear combination of the measured variables $\{y_i\}$, $Y = \alpha_1 y_1 + \alpha_2 y_2 + \ldots$. Let the random variables $y_1, y_2, \ldots$ have means $E(y_1), E(y_2), \ldots$ and variances $\sigma^2(y_1), \sigma^2(y_2), \ldots$. The variable Y has mean

$$E(Y) = \alpha_1 E(y_1) + \alpha_2 E(y_2) + \cdots$$

and variance (Cropley, 1978)

$$\sigma^2(Y) = \sum_{i=1}^{n} \alpha_i^2 \sigma_i^2(y_i) + 2\sum_{i=1}^{n}\sum_{j=i+1}^{n} \alpha_i \alpha_j \operatorname{Cov}(y_i, y_j)$$

If the variables are uncorrelated and have the same variance, then

$$\sigma^2(Y) = \left(\sum_{i=1}^{n} \alpha_i^2\right)\sigma^2$$

Next suppose the model relating Y to $\{y_i\}$ is nonlinear, but the errors are small and independent of one another. Then a change in Y is related to changes in y_i by

$$dY = \frac{\partial Y}{\partial y_1} dy_1 + \frac{\partial Y}{\partial y_2} dy_2 + \cdots$$

If the changes are indeed small, then the partial derivatives are constant among all the samples. Then the expected value of the change $E(dY)$ is zero. The variances are related by the following equation (Box et al., 2005):

$$\sigma^2(dY) = \sum_{i=1}^{N} \left(\frac{\partial Y}{\partial y_i}\right)^2 \sigma_i^2$$

Thus, the variance of the desired quantity Y can be found. This gives an independent estimate of the errors in measuring the quantity Y from the errors in measuring each variable it depends upon.

Example Suppose one wants to measure the thermal conductivity of a solid k. To do this, one needs to measure the heat flux q, the thickness of the sample d, and the temperature difference across the sample ΔT. Each measurement has some error. The heat flux q may be the rate of electric heat input $\dot{Q}$ divided by the area A, and both quantities are measured to some tolerance. The thickness of the sample is measured with some accuracy, and the temperatures are probably measured with a thermocouple to some accuracy. These measurements are combined, however, to obtain the thermal conductivity, and it is desired to know the error in the thermal conductivity. The formula is

$$k = \frac{d}{A\Delta T}\dot{Q}$$

The variance in the thermal conductivity is then

$$\sigma_k^2 = \left(\frac{k}{d}\right)^2 \sigma_d^2 + \left(\frac{k}{\dot{Q}}\right)^2 \sigma_Q^2 + \left(\frac{k}{A}\right)^2 \sigma_A^2 + \left(\frac{k}{\Delta T}\right)^2 \sigma_{\Delta T}^2$$

ANALYSIS OF VARIANCE (ANOVA) AND FACTORIAL DESIGN OF EXPERIMENTS

Statistically designed experiments consider the effect of primary variables, but they also consider the effect of extraneous variables and the interactions between variables, and they include a measure of the random error. Primary variables are those whose effect you wish to determine. These variables can be quantitative or qualitative. The quantitative variables are ones you may fit to a model in order to determine the model parameters (see the section Least Squares). Qualitative variables are ones whose effect you wish to know, but you do not try to quantify that effect other than to assign possible errors or magnitudes. Qualitative variables can be further subdivided into Type I variables, whose effect you wish to determine directly, and Type II variables, which contribute to the performance variability and whose effect you wish to average out. For example, if you are studying the effect of several catalysts on yield in a chemical reactor, each different type of catalyst would be a Type I variable because you would like to know the effect of each. However, each time the catalyst is prepared, the results are slightly different due to random variations; thus, you may have several batches of what purports to be the same catalyst. The variability between batches is a Type II variable. Since the ultimate use will require using different batches, you would like to know the overall effect including that variation, since knowing precisely the results from one batch of one catalyst might not be representative of the results obtained from all batches of the same catalyst. A randomized block design, incomplete block design, or Latin square design (Box et al., 2005), for example, all keep the effect of experimental error in the blocked variables from influencing the effect of the primary variables. Other uncontrolled variables are accounted for by introducing randomization in parts of the experimental design. To study all variables and their interaction requires a factorial design, involving all possible combinations of each variable, or a fractional factorial design, involving only a selected set. Statistical techniques are then used to determine which are the important variables, what are the important interactions, and what the error is in estimating these effects. The discussion here is only a brief overview of the excellent book by Box et al. (2005).

ANOVA Suppose we have two methods of preparing some product and we wish to see which treatment is better. When there are only two treatments, then the sampling analysis discussed in the section Two-Population Test of Hypothesis for Means can be used to deduce if the means of the two treatments differ significantly. When there are more treatments, the analysis is more detailed. The goal is to see if the treatments differ significantly from each other; that is, whether their means are different when the samples have the same variance. The hypothesis is that the treatments are all the same, and the null hypothesis is that they are different. The statistical validity of the hypothesis is determined by an analysis of variance.

Example Suppose the experimental results of the four treatments are arranged as shown in the table: several measurements for each treatment. Are the treatments significantly different from each other? The data are a modified table from Box et al. (2005).

Analysis of Variance: Estimating the Variance of Four Treatments

	Treatment			
	1	2	3	4
	63	65	71	62
	59	63	68	61
	63	71	68	63
		64	68	56
		67		64
				56
Treatment average	61.67	66.00	68.75	60.33
Grand average		64.00		
Sample variance	5.333	10.000	2.250	12.267

The data for $k = 4$ treatments is arranged in the table. For each treatment, there are n_t experiments, and the outcome of the ith experiment with treatment t is called y_{ti}. Compute the treatment average

$$\bar{y}_t = \frac{\sum_{i=1}^{n_t} y_{ti}}{n_t}$$

Also compute the grand average

$$\bar{y} = \frac{\sum_{t=1}^{k} n_t \bar{y}_t}{N}, N = \sum_{t=1}^{k} n_t$$

Next compute the sum of squares of deviations from the average within the tth treatment

$$S_t = \sum_{i=1}^{n_t} (y_{ti} - \bar{y}_t)^2$$

Since each treatment has n_t experiments, the number of degrees of freedom is $n_t - 1$. Then the sample variances are

$$s_t^2 = \frac{S_t}{n_t - 1}$$

The within-treatment sum of squares is

$$S_R = \sum_{t=1}^{k} S_t$$

and the within-treatment sample variance is

$$s_R^2 = \frac{S_R}{N-k} = 8.482$$

Now, if there is no difference between treatments, a second estimate of σ^2 could be obtained by calculating the variation of the treatment averages about the grand average. Thus compute the between-treatment mean square

$$S_T = \sum_{t=1}^{k} n_t (\bar{y}_t - \bar{y})^2, \quad s_T^2 = \frac{S_T}{k-1} = 69.08$$

Basically the test for whether the hypothesis is true hinges on a comparison of the within-treatment estimate s_R^2(with $\nu_R = N - k$ degrees of freedom) with the between-treatment estimate s_T^2 (with $\nu_T = k - 1$ degrees of freedom). The ratio of variances $s_T^2/s_R^2 = 8.145$. The test is made based on the F distribution for ν_T and ν_R degrees of freedom, FINV($\alpha/2$, ν_T, ν_R) = f(the order of the degrees of freedom is important) where

$$\text{Probability}\,[1/f \le s_T^2/s_R^2 \le f] = \alpha$$

Here FINV(0.05, 3, 14) = 3.344; the rejection region is $F > 3.344$. Since the ratio of variances is 8.145 and larger than 3.344, the hypothesis is rejected; the four treatments are not statistically the same at the 10 percent level. Alternatively, $F = 8.145$ at a p value of about 0.002, and the null hypothesis is rejected.

Randomized blocking can be used to eliminate the effect of some variable whose effect is of no interest, such as the batch-to-batch variation of the catalysts in the chemical reactor example. See Box et al., 2005 for details.

Factorial Design To measure the effects of variables on a single outcome, a factorial design is appropriate. In a two-level factorial design, each variable is considered at two levels only, a high and low value, often designated as a + and −. The two-level factorial design is useful for indicating trends and showing interactions, and it is also the basis for a fractional factorial design. As an example, consider a 2^3 factorial design with 3 variables and 2 levels for each. The experiments are indicated in the factorial design table.

Two-Level Factorial Design with Three Variables

	Variable		
Run	1	2	3
1	−	−	−
2	+	−	−
3	−	+	−
4	+	+	−
5	−	−	+
6	+	−	+
7	−	+	+
8	+	+	+

The main effects are calculated by calculating the difference between results from all high values of a variable and all low values of a variable; the result is divided by the number of experiments at each level. For example, for the first variable

$$\text{Effect of variable } 1 = \frac{(y_2 + y_4 + y_6 + y_8) - (y_1 + y_3 + y_5 + y_7)}{4}$$

Note that all observations are being used to supply information on each of the main effects, and each effect is determined with the precision of a fourfold replicated difference. The advantage of a one-at-a-time experiment is the gain in precision if the variables are additive and the measure of nonadditivity if it occurs (Box et al., 2005).

Interaction effects between variables 1 and 2 are obtained by calculating the difference between the results obtained with the high and low value of 1 at the low value of 2 compared with the results obtained with the high and low value of 1 at the high value of 2. The 12 interaction is

$$12\text{-interaction} = \frac{(y_4 - y_3 + y_8 - y_7) - (y_2 - y_1 + y_6 - y_5)}{2}$$

The key step is to determine the errors associated with the effect of each variable and each interaction so that the significance can be determined. Thus, standard errors need to be assigned. This can be done by repeating the experiments, but it can also be done by using higher-order interactions (such as 123 interactions in a 2^4 factorial design). These are assumed negligible in their effect on the mean but can be used to estimate the standard error. Then calculated effects that are large compared with the standard error are considered important, while those that are small compared with the standard error are considered to be due to random variations and are unimportant.

In a fractional factorial design, one does only part of the possible experiments. When there are k variables, a factorial design requires 2^k experiments. When k is large, the number of experiments can be large; for $k = 5$, $2^5 = 32$. For a k this large, Box et al. (2005) do a fractional factorial design. In the fractional factorial design with $k = 5$, only 16 experiments are done. Cropley (1978) gives an example of how to combine heuristics and statistical arguments in application to kinetics mechanisms in chemical engineering.

DIMENSIONAL ANALYSIS

Dimensional analysis allows the engineer to reduce the number of variables that must be considered to model experiments or correlate data. Consider a simple example in which two variables F_1 and F_2 have the units of force, and two additional variables L_1 and L_2 have the units of length. Rather than having to deduce the relation of one variable on the other three, $F_1 = \text{fn}(F_2, L_1, L_2)$, dimensional analysis can be used to show that the relation must be of the form $F_1/F_2 = \text{fn}(L_1/L_2)$. Thus considerable experimentation is saved. Historically, dimensional analysis can be done using the Rayleigh method or the Buckingham pi method. This brief discussion is equivalent to the Buckingham pi method but uses concepts from linear algebra; see Amundson, N. R., *Mathematical Methods in Chemical Engineering*, Prentice-Hall, Englewood Cliffs, N.J., 1966, p. 54, for further information.

The general problem is posed as finding the minimum number of variables necessary to define the relationship between n variables. Let $\{Q_i\}$ represent a set of fundamental units, such as length, time, force, and so on.

Let $[P_i]$ represent the dimensions of a physical quantity P_i; there are n physical quantities. Then form the matrix α_{ij}

	$[P_1]$	$[P_2]$	…	$[P_n]$
Q_1	α_{11}	α_{12}	…	α_{1n}
Q_2	α_{21}	α_{22}	…	α_{2n}
…				
Q_m	α_{m1}	α_{m2}	…	α_{mn}

in which the entries are the number of times each fundamental unit appears in the dimensions $[P_i]$. The dimensions can then be expressed as follows:

$$[P_i] = Q_1^{\alpha_{1i}} Q_2^{\alpha_{2i}} \cdots Q_m^{\alpha_{mi}}$$

Let m be the rank of the α matrix. Then $p = n - m$ is the number of dimensionless groups that can be formed. One can choose m variables $\{P_i\}$ to be the basis and express the other p variables in terms of them, giving p dimensionless quantities.

Example: Buckingham Pi Method—Heat-Transfer Film Coefficient It is desired to determine a complete set of dimensionless groups with which to correlate experimental data on the film coefficient of heat transfer between the walls of a straight conduit with circular cross section and a fluid flowing in that conduit. The variables and the dimensional constant believed to be involved and their dimensions in the engineering system are given below:

Film coefficient $= h = F/L\theta T$

Conduit internal diameter $= D = L$

Fluid linear velocity $= V = L/\theta$

Fluid density $= \rho = M/L^3$

Fluid absolute viscosity $= \mu = M/L\theta$

Fluid thermal conductivity $= k = F/\theta T$

Fluid specific heat $= c_p = FL/MT$

Dimensional constant $= g_c = ML/F\theta^2$

The matrix α in this case is as follows:

		$[P_i]$							
		h	D	V	ρ	μ	k	C_p	g_c
	F	1	0	0	0	0	1	1	−1
	M	0	0	0	1	1	0	−1	1
Q_j	L	−1	1	1	−3	−1	0	1	1
	θ	−1	0	−1	0	−1	−1	0	−2
	T	−1	0	0	0	0	−1	−1	0

Here $m \le 5$, $n = 8$, and $p \ge 3$. Choose D, V, μ, k, and g_c as the primary variables. By examining the 5×5 matrix associated with those variables, we can see that its determinant is not zero, so the rank of the matrix is $m = 5$; thus, $p = 3$. These variables are thus a possible basis set. The dimensions of the other three variables h, ρ, and C_p must be defined in terms of the primary variables. This can be done by inspection, although linear algebra can be used, too.

$$[h] = D^{-1}k^{+1};\text{ thus } \frac{h}{D^{-1}k} = \frac{hD}{k} \text{ is a dimensionless group}$$

$$[\rho] = \mu^1 V^{-1} D^{-1};\text{ thus } \frac{\rho}{\mu^1 V^{-1} D^{-1}} = \frac{\rho VD}{\mu} \text{ is a dimensionless group}$$

$$[C_p] = k^1 \mu^{-1};\text{ thus } \frac{C_p}{k^{+1}\mu^{-1}} = \frac{C_p\mu}{k} \text{ is a dimensionless group}$$

Thus, the dimensionless groups are

$$\frac{[P_i]}{Q_1^{\alpha 1i} Q_1^{\alpha 2i} \cdots Q_m^{\alpha mi}} : \frac{hD}{k}, \frac{\rho VD}{\mu}, \frac{C_p\mu}{k}$$

The dimensionless group hD/k is called the Nusselt number, N_{Nu}, and the group $C_p\mu/k$ is the Prandtl number, N_{Pr}. The group $DV\rho/\mu$ is the familiar Reynolds number, N_{Re}, encountered in fluid-friction problems. These three dimensionless groups are frequently used in heat-transfer-film-coefficient correlations. Functionally, their relation may be expressed as

$$\phi(N_{Nu}, N_{Pr}, N_{Re}) = 0 \tag{3-100}$$

or as

$$N_{Nu} = \phi_1(N_{Pr}, N_{Re})$$

It has been found that these dimensionless groups may be correlated well by an equation of the type

$$hD/k = K(c_p\mu/k)^a(DV\rho/\mu)^b$$

in which K, a, and b are experimentally determined dimensionless constants. However, any other type of algebraic expression or perhaps simply a graphical relation among these three groups that accurately fits the experimental data would be an equally valid manner of expressing Eq. (3-100).

Naturally, other dimensionless groups might have been obtained in the example by employing a different set of five repeating quantities that would not form a dimensionless group among themselves. Some of these groups may be found among those presented in Table 3-5. Such a complete set of three dimensionless groups might consist of Stanton, Reynolds, and Prandtl numbers or of Stanton, Peclet, and Prandtl numbers. Also such a complete set different from that obtained in the preceding example will result from a multiplication of appropriate powers of the Nusselt, Prandtl, and Reynolds numbers. For such a set to be complete, however, it must satisfy the condition that each of the three dimensionless groups is independent of the other two.

TABLE 3-5 Dimensionless Groups in the Engineering System of Dimensions

Biot number	N_{Bi}	hL/k
Condensation number	N_{Co}	$(h/k)(\mu^2/\rho^2 g)^{1/3}$
Number used in condensation of vapors	N_{Cv}	$L^3\rho^2 g\lambda/(k\mu\,\Delta t)$
Euler number	N_{Eu}	$g_c(-dp)/\rho V^2$
Fourier number	N_{Fo}	$k\,\theta/\rho c L^2$
Froude number	N_{Fr}	V^2/Lg
Graetz number	N_{Gz}	wc/kL
Grashof number	N_{Gr}	$L^3\rho^2\beta g\,\Delta T/\mu^2$
Mach number	N_{Ma}	V/V_a
Nusselt number	N_{Nu}	hD/k
Peclet number	N_{Pe}	$DV\rho c/k$
Prandtl number	N_{Pr}	$c\mu/k$
Reynolds number	N_{Re}	$DV\rho/\mu$
Schmidt number	N_{Sc}	$\mu/\rho D_u$
Stanton number	N_{St}	$h/cV\rho$
Weber number	Nw_e	$LV^2\rho/\sigma g_c$

PROCESS SIMULATION

References: Jana, A. K., *Chemical Process Modelling and Computer Simulation*, PHI Learning Pvt. Ltd., New Delhi, India, 2011; Jana, A. K., *Process Simulation and Control Using Aspen*, PHI Learning Pvt. Ltd., New Delhi, India, 2012; Mah, R. S. H., *Chemical Process Structure and Information Flows*, Butterworths-Heinemann, Oxford, 1990; Sandler, S. I., *Using Aspen Plus in Thermodynamics Instruction*, Wiley, New York, 2015; Schefflan, R., *Teach Yourself the Basics of Aspen Plus*, Wiley, New York, 2011; Seader, J. D., *Computer Modeling of Chemical Processes*, AIChE Monograph Series no. 15, American Institute of Chemical Engineers, New York, 1985; Seider, W. D., J. D. Seader, D. R. Lewin, and S. Widagdo, *Product and Process Design Principles: Synthesis, Analysis, and Evaluation*, 3d ed., Wiley, New York, 2009.

CLASSIFICATION

Process simulation refers to the activity in which mathematical systems of chemical processes and refineries are modeled with equations, usually on the computer. The usual distinction must be made between steady-state models and transient models, following the ideas presented in the introduction to this section. In a chemical process, of course, the process is nearly always in a transient mode, at some level of precision, but when the time-dependent fluctuations are below some value, a steady-state model can be formulated. This subsection presents briefly the ideas behind steady-state process simulation (also called *flowsheeting*), which are embodied in commercial codes. The transient simulations are important for designing the start-up of plants and are especially useful for the operation of chemical plants.

THERMODYNAMICS

The most important aspect of the simulation is that the thermodynamic data of the chemicals be modeled correctly. It is necessary to decide what equation of state to use for the vapor phase (ideal gas, Soave-Redlich-Kwong, Peng-Robinson, etc.) and what model to use for liquid activity coefficients [ideal solutions, solubility parameters, Wilson equation, nonrandom two liquid (NRTL), UNIFAC, etc.]. See Sec. 4, Thermodynamics, and Sandler (2015). It is necessary to consider mixtures of chemicals, and the interaction parameters must be predictable. The best case is to determine them from data, and the next-best case is to use correlations based on the molecular weight, structure, and normal boiling point. To validate the model, the computer results of vapor-liquid equilibria could be checked against experimental data to ensure their validity before the data are used in more complicated computer calculations.

PROCESS MODULES OR BLOCKS

At the first level of detail, it is not necessary to know the internal parameters for all the units, since what is desired is just the overall performance. For example, in a heat exchanger design, it suffices to know the heat duty, total area, and temperatures of the output streams; the details such as the percentage baffle cut, tube layout, or baffle spacing can be specified later when the details of the proposed plant are better defined. It is important to realize the level of detail modeled by a commercial computer program. For example, a chemical reactor could be modeled as an equilibrium reactor, in which the input stream is brought to a new temperature and pressure and the output stream is in chemical equilibrium at those new conditions. Or, it may suffice to simply specify the conversion, and the computer program will calculate the outlet compositions. In these cases, the model equations are algebraic ones, and you do not learn the volume of the reactor. A more complicated reactor might be a stirred tank reactor, and then you would have to specify kinetic information so that the simulation can be made, and one output would be either the volume of the reactor or the conversion possible in a volume you specify. Such models are also composed of sets of algebraic equations. A plug flow reactor is modeled as a set of ordinary differential equations as initial-value problems, and the computer program must use numerical methods to integrate them. See Numerical Solution of Ordinary Differential Equations as Initial-Value Problems. Kinetic information must be specified, and one learns the conversion possible in a given reactor volume, or, in some cases, the volume reactor that will achieve a given conversion. The simulation engineer determines what a reactor of a given volume will do for the specified kinetics and reactor volume. The design engineer, however, wants to achieve a certain result and wants to know the volume necessary. Simulation packages are best suited for the simulation engineer, and the design engineer must vary specifications to achieve the desired output.

Distillation simulations can be based on shortcut methods, using correlations based on experience, but more rigorous methods involve solving for the vapor-liquid equilibrium on each tray. The shortcut method uses relatively simple equations, and the rigorous method requires solution of huge sets of nonlinear equations. The computation time of the latter is significant, but the rigorous method may be necessary when the chemicals you wish to distill are not well represented in the correlations. Then the designer must specify the number of trays and determine the separation that is possible. This, of course, is not what she or he wants: the number of trays needed to achieve a specified objective. Thus, again, some adjustment of parameters is necessary in a design situation.

Absorption columns can be modeled in a plate-to-plate fashion (even if it is a packed bed) or as a packed bed. The former model is a set of nonlinear algebraic equations, and the latter model is an ordinary differential equation. Since streams enter at both ends, the differential equation is a two-point boundary-value problem, and numerical methods are used (see Numerical Solution of Ordinary Differential Equations as Initial-Value Problems).

If one wants to model a process unit that has significant flow variation, and possibly some concentration distributions as well, one can consider using computational fluid dynamics (CFD) to do so. These calculations are very time-consuming, however, so that they are often left until the mechanical design of the unit. The exception would occur when the flow variation and concentration distribution had a significant effect on the output of the unit so that mass and energy balances couldn't be made without it.

The process units are described in greater detail in other sections of the Handbook. In each case, parameters of the unit are specified (size, temperature, pressure, area, and so forth). In addition, in a computer simulation, the computer program must be able to take any input to the unit and calculate the output for those parameters. Since the entire calculation is done iteratively, there is no assurance that the input stream is a "reasonable" one, so that the computer codes must be written to give some sort of output even when the input stream is unreasonable. This difficulty makes the iterative process even more complicated.

PROCESS TOPOLOGY

A chemical process usually consists of a series of units, such as distillation towers, reactors, and so forth (see Fig. 3-65). If the feed to the process is known and the operating parameters of the units are specified by the user, then one can begin with the first unit, take the process input, calculate the unit output, carry that output to the input of the next unit, and continue the process. However, if the process involves a recycle stream, as nearly all chemical processes do, then when the calculation is begun, it is discovered that the recycle stream is unknown. This situation leads to an iterative process: the flow rates, temperature, and pressure of the unknown recycle stream are guessed, and the calculations proceed as before. When one reaches the end of the process, where the recycle stream is formed to return to the first unit, it is necessary to check to see if the recycle stream is the same as assumed. If not, an iterative procedure must be used to cause convergence. Possible techniques are described in Numerical Solutions of Nonlinear Equations in One Variable and Numerical Solution of Simultaneous Equations. The direct method (or successive substitution method) just involves calculating around the process over and over. The Wegstein method accelerates convergence for a single variable, and Broyden's method does the same for multiple variables. The Newton method can be used provided there is some way to calculate the derivatives (possibly by using a numerical derivative). Optimization methods can also be used (see Optimization in this section). In the description given here, the recycle stream is called the *tear stream*: this is the stream that must be guessed to begin the calculation. When there are multiple recycle streams, convergence is even more difficult, since more guesses are necessary, and what happens in one recycle stream may cause difficulties for the guesses in other recycle streams. See Seader (1985) and Mah (1990).

It is sometimes desired to control some stream by varying an operating parameter. For example, in a reaction/separation system, if there is an impurity that must be purged, a common objective is to set the purge fraction so that the impurity concentration into the reactor is kept at some moderate value. Commercial packages contain procedures for doing this, using what are often called *control blocks*. However, this can also make the solution more difficult to find.

An alternative method of solving the equations is to solve them as simultaneous equations. In that case, one can specify the design variables and the desired specifications and let the computer figure out the process parameters that will achieve those objectives. It is possible to overspecify the system or to give impossible conditions. However, the biggest drawback to

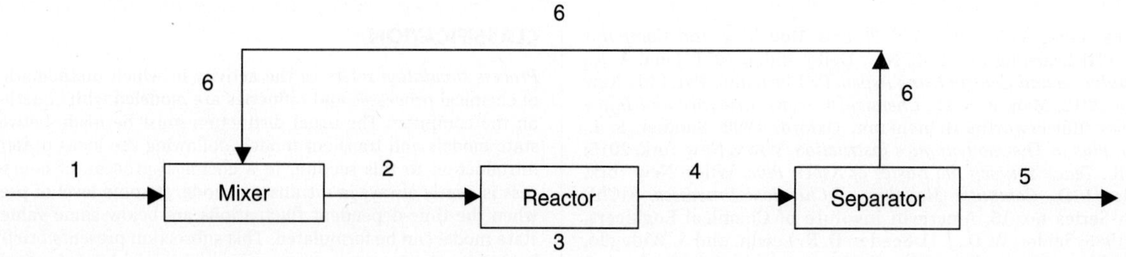

FIG. 3-65 Prototype flowsheet.

this method of simulation is that large sets (tens of thousands) of nonlinear algebraic equations must be solved simultaneously. As computers become faster, this is less of an impediment, provided efficient software is available.

Dynamic simulations are also possible, and these require solving differential equations, sometimes with algebraic constraints. If some parts of the process change extremely quickly when there is a disturbance, that part of the process may be modeled in the steady state for the disturbance at any instant. Such situations are called *stiff,* and the methods for them are discussed in Numerical Solution of Ordinary Differential Equations as Initial-Value Problems. It must be realized, though, that a dynamic calculation can also be time-consuming, and sometimes the allowable units are lumped-parameter models that are simplifications of the equations used for the steady-state analysis. Thus, as always, the assumptions need to be examined critically before accepting the computer results. The dynamic simulators can also be used to simulate operations with the objective to maintain purity and standards of the product.

COMMERCIAL PACKAGES

Computer programs are provided by many companies, and they range from empirical models to deterministic models. For example, if one wanted to know the pressure drop in a piping network, one would normally use a correlation for friction factor as a function of Reynolds number to calculate the pressure drop in each segment. A sophisticated turbulence model of fluid flow is not needed in that case. As computers become faster, however, more and more models are deterministic. Since the commercial codes have been used by many customers, the data in them have been verified, but possibly not for the case you want to solve. Thus, you must test the thermodynamics correlations carefully. In 2015, there were a number of computer codes, but the company names change constantly. Here are a few of them for process simulation: Aspen Tech (Aspen Plus), Chemstations (CHEMCAD), Honeywell (UniSim Design), ProSim (ProSimPlus), and Pro II. The CAPE-OPEN project is working to make details as transferable as possible.

Section 4

Thermodynamics

J. Richard Elliott, Ph.D. *Professor, Department of Chemical and Biomolecular Engineering, University of Akron; Member, American Institute of Chemical Engineers; Member, American Chemical Society; Member, American Society of Engineering Educators (Section Coeditor)*

Carl T. Lira, Ph.D. *Associate Professor, Department of Chemical and Materials Engineering, Michigan State University; Member, American Institute of Chemical Engineers; Member, American Chemical Society; Member, American Society of Engineering Educators (Section Coeditor)*

Timothy C. Frank, Ph.D. *Fellow, The Dow Chemical Company; Fellow, American Institute of Chemical Engineers (Section Coeditor)*

Paul M. Mathias, Ph.D. *Senior Fellow and Technical Director, Fluor Corporation; Fellow, American Institute of Chemical Engineers (Section Coeditor)*

INTRODUCTION

GENERAL BALANCES

PROPERTY CALCULATIONS FROM EQUATIONS OF STATE

SYSTEMS OF VARIABLE COMPOSITION

The contributions of Profs. Hendrick C. van Ness and Michael A. Abbott, Section Editors, 8th ed., are acknowledged.

TRENDS IN PHASE BEHAVIOR

TEMPERATURE DEPENDENCE OF INFINITE-DILUTION ACTIVITY COEFFICIENTS

THERMODYNAMICS FOR CONCEPTUAL DESIGN

REACTING SYSTEMS

Nomenclature and Units

Correlation- and application-specific symbols are not shown.

Symbol	Definition	SI units	U.S. Customary System units
A	Molar (or unit-mass) Helmholtz energy† or dimensionless equation of state parameter	J/mol [J/kg]	Btu/lb·mol [Btu/lbm]
A_x	Cross-sectional area in flow	m^2	ft^2
a_i	Activity of species i in solution	Dimensionless	Dimensionless
B	2d virial coefficient or dimensionless equation of state parameter	cm^3/mol	cm^3/mol
B_{ij}	Interaction 2d virial coefficient	cm^3/mol	cm^3/mol
C_P	Heat capacity at constant pressure	J/(mol·K)	Btu/(lb·mol·°R)
C_V	Heat capacity at constant volume	J/(mol·K)	Btu/(lb·mol·°R)
f_i	Fugacity of pure species i	kPa	psi
$\hat{f}_i$	Fugacity of species i in solution	kPa	psi
G	Molar (or unit-mass) Gibbs energy†	J/mol [J/kg]	Btu/(lb·mol) [Btu/lbm]
g	Acceleration of gravity	m/s^2	ft/s^2
H	Molar (or unit-mass) enthalpy†	J/mol [J/kg]	Btu/(lb·mol) [Btu/lbm]
h_i	Henry's volatility constant for solute species at T and $P°$	kPa	psi
K_i	Equilibrium K value, y_i/x_i	Dimensionless	Dimensionless
$K_{a,j}$	Equilibrium constant for chemical reaction j	Dimensionless	Dimensionless
$k_{H,i}$	Henry's solubility constant for solute species i in molal units	mol/kg·kPa	mol/kg·psi
M	Molar or unit-mass solution property† (A, G, H, S, U, V)		
M_i	Molar or unit-mass pure-species property† ($A_i, G_i, H_i, S_i, U_i, V_i$)		
$\bar{M}_i$	Partial property of species i in solution ($\bar{A}_i, \bar{G}_i, \bar{H}_i, \bar{S}_i, \bar{U}_i, \bar{V}_i$)		
M^{dep}	Departure thermodynamic property† ($A^{dep}, G^{dep}, H^{dep}, S^{dep}, U^{dep}, V^{dep}$)		
M^E	Excess thermodynamic property† ($A^E, G^E, H^E, S^E, U^E, V^E$)		
$\bar{M}_i^E$	Partial molar excess thermodynamic property		
ΔM	Property change of mixing† ($\Delta A, \Delta G, \Delta H, \Delta S, \Delta U, \Delta V$)		
$\Delta M_j°$	Standard property change of reaction j ($\Delta G_j°, \Delta H_j°, \Delta C_{Pj}°$)		
m	Mass	kg	lbm
$\dot{m}$	Mass flow rate	kg/s	lbm/s
n	Number of moles		
$\dot{n}$	Molar flow rate		
n_i	Number of moles of species i		
P	Pressure	kPa	psi
P_i^{sat}	Saturation or vapor pressure of species i	kPa	psi
q	Quality (vapor fraction)	Dimensionless	Dimensionless
Q	Heat per mole or mass†	J/mol [J/kg]	Btu/lb·mol [Btu/lbm]
$\dot{Q}$	Rate of heat transfer per mole or mass†	J/(mol·s) [J/kg·s]	Btu/(lb·mol·s) [Btu/(lbm·s)]
R	Universal gas constant	J/(mol·K)	Btu/(lb·mol·°R)
S	Molar (or unit-mass) entropy†	J/(mol·K) [J/(kg·K)]	Btu/(lb·mol·°R) [Btu/(lbm·°R)]
$\dot{S}_{gen}$	Rate of entropy generation per mole or unit mass†	J/(K·mol·s) [J/(K·kg·s)]	Btu/(lb·mol·°R·s) [Btu/(lbm·°R·s)]
T	Temperature	K	°R
T_c	Critical temperature	K	°R
U	Molar (or unit-mass) internal energy†	J/mol [J/kg]	Btu/(lb·mol) [Btu/lbm]
v	Fluid velocity	m/s	ft/s
V	Molar (or unit-mass) volume†	m^3/mol [m^3/kg]	ft^3/(lb·mol) [ft^3/lbm]
W	Work per mole or unit mass†	J/mol [J/kg]	Btu/lb·mol [Btu/lbm]
$\dot{W}_s$	Shaft work for flow process†	J/mol [J/kg]	Btu/lb·mol [Btu/lbm]
$\dot{W}_S$	Shaft power for flow process†	J/(mol·s) [J/(kg·s]	Btu/(lb·mol·s) [Btu/(lbm·s)]
x_i	Mole fraction in general		
x_i	Mole fraction of species i in liquid phase		
y_i	Mole fraction of species i in vapor phase		
Z	Compressibility factor	Dimensionless	Dimensionless
z	Elevation above a datum level	m	ft

Superscripts

Symbol	Definition
cond	Condensed phase, e.g. liquid or solid
dep	Departure thermodynamic property
E	Excess thermodynamic property
id	Value for an ideal solution
ig	Value for an ideal gas
L	Liquid phase
V	Vapor phase
vap	Phase transition, liquid to vapor
∞	Value at infinite dilution

Subscripts

Symbol	Definition
c	Value for the critical state
cv	Control volume
fs	Flowing streams
n	Normal boiling point
r	Reduced value
R	Reference state

Greek Letters

Symbol	Definition	SI units	U.S. Customary System units
α, β	As superscripts, identify phases		
β	Volume expansivity	K^{-1}	°R^{-1}
ε_j	Reaction coordinate for reaction j	mol	lb·mol
γ	Heat capacity ratio C_P/C_V	Dimensionless	Dimensionless
γ_i	Activity coefficient of species i in solution for the Lewis-Randall rule, a superscript * denotes Henry's law activity coefficient	Dimensionless	Dimensionless
κ	Isothermal compressibility	kPa^{-1}	psi^{-1}
μ_i	Chemical potential of species i	J/mol	Btu/(lb·mol)
$\nu_{i,j}$	Stoichiometric number of species i in reaction j	Dimensionless	Dimensionless
ρ	Molar or mass density	mol/m^3 [kg/m^3]	lb·mol/ft^3 [lbm/ft^3]
ϕ_i	Fugacity coefficient of pure species i	Dimensionless	Dimensionless
$\hat{\phi}_i$	Fugacity coefficient of species i in solution	Dimensionless	Dimensionless
ω	Acentric factor	Dimensionless	Dimensionless

†When underlined, denotes extensive thermodynamic property [e.g., energy] or total work or heat [e.g., energy].

General References: 1. Abbott, M. M., and H. C. Van Ness, *Schaum's Outline of Theory and Problems of Thermodynamics*, 2d ed., McGraw-Hill, New York, 1989. 2. Chen, C.-C., and P. M. Mathias, "Applied Thermodynamics for Process Modeling," *AIChE J.* **48**(2): 194–200 (2002). 3. Elliott, J. R., and C. T. Lira, *Introductory Chemical Engineering Thermodynamics*, 2d ed., Prentice Hall PTR, Upper Saddle River, N.J., 2012. 4. O'Connell, J. P., and J. M. Haile, *Thermodynamics. Fundamentals for Applications*, Cambridge University Press, London, 2005. 5. Poling, B. E., J. M. Prausnitz, and J. P. O'Connell, *The Properties of Gases and Liquids*, 5th ed., McGraw-Hill, New York, 2001. 6. Prausnitz, J. M., R. N. Lichtenthaler, and E. G. de Azevedo, *Molecular Thermodynamics of Fluid-Phase Equilibria*, 3d ed., Prentice-Hall PTR, Upper Saddle River, N.J., 1999. 7. Rafal, M., J. E. Berthold, N. C. Scrivner, and S. L. Grise, "Models for Electrolyte Solutions," in *Models for Thermodynamic and Phase Equilibria Calculations*, ed. S. I. Sandler, Marcel Dekker, New York, 1994. 8. Sandler, S. I., *Chemical, Biochemical, and Engineering Thermodynamics*, 4th ed., Wiley, Hoboken, N.J., 2006. 9. Smith, J. M., H. C. Van Ness, and M. M. Abbott, *Introduction to Chemical Engineering Thermodynamics*, 7th ed., McGraw-Hill, New York, 2005. 10. Tester, J. W., and M. Modell, *Thermodynamics and Its Applications*, 3d ed., Prentice-Hall PTR, Upper Saddle River, N.J., 1997. 11. Walas, S. M., *Phase Equilibria in Chemical Engineering*, Butterworth-Heinemann, Boston, 1985. 12. Van Ness, H. C., and M. M. Abbott, *Classical Thermodynamics of Nonelectrolyte Solutions: With Applications to Phase Equilibria*, McGraw-Hill, New York, 1982.

INTRODUCTION

Thermodynamics is the branch of science that deals with energy transformation and the state of equilibrium in macroscopic systems. The laws of thermodynamics are shown by experience to apply to all such transformations. The first law states that energy can take many forms, but it cannot be created or destroyed (except that nuclear reactions may contribute components outside the norm). The second law concerns the distribution of the energy and of the material components comprising a system, traditionally described in terms of the order or disorder of the system. It states that maintaining a nonequilibrium or unnaturally ordered state requires work. The systematic analysis of these two laws leads to profound insights pervading chemistry, physics, and biology, especially when combined with molecular insights through statistical thermodynamics. In the context of chemical engineering, it is important to include one additional conservation law, the material balance. Similar to energy, mass is neither created nor destroyed in (nonnuclear) systems. The material balance is not technically a law of thermodynamics, but it is necessary to fully characterize the equilibrium systems that are central to thermodynamics.

While the first law of thermodynamics is the basis for the energy balance, the second law is the basis for the concept of entropy. The second law states that the entropy of the universe (defined in terms of reversible heat flow divided by absolute temperature) must increase through the conduct of any practical process, meaning that the entropy of individual subsystems may increase or decrease but the sum of entropy changes across all subsystems and surroundings must increase. As a consequence, thermal energy spontaneously flows from a hotter body to a cooler one, and statistical mechanics indicates a general tendency for a system to move toward spatial homogeneity.

Energy, on the other hand, has a tendency to pull things together. Molecules are attracted to one another as evidenced by the energy of vaporization required to increase the intermolecular distances when converting liquid to vapor. Thus, nature exhibits a competition between energetic and entropic driving forces when temperature and pressure are fixed. When these competing driving forces are perfectly balanced, the situation is described as *equilibrium*. A simple form of equilibrium is evidenced by a vapor in equilibrium with a liquid at its boiling point. Entropy is driving the molecules toward the vapor while energy is pulling molecules into the liquid. At a given temperature for a pure fluid, the rate of evaporation equals the rate of condensation at one specific pressure, comprising equilibrium, and referred to as the *saturation pressure* or *vapor pressure*.

Remarkably, the same equations and concepts of energy, entropy, and equilibrium describe all the phenomena of phase equilibria in mixtures. Tracing the energy transformations through a process is relatively straightforward. However, mixing and separation become quite complicated in the presence of aqueous streams mixed with organic compounds and possibly electrolytes, the mixing of which may form multiple solid, liquid, or vapor phases. The application of thermodynamic theory in chemical engineering practice yields models describing all these different phases at equilibrium. A deviation from the equilibrium composition is the driving force for many chemical separation processes, and many are modeled as equilibrium staged processes even when perfect equilibrium is not achieved at any particular point in the process. As a real process can only approach equilibrium (the thermodynamic limit), such an analysis allows the process designer to characterize separation difficulty and the magnitude of the opportunity for further improvement.

Another form of equilibrium in mixtures occurs when one considers that individual atoms can be rearranged within and among molecules, also known as *chemical reaction equilibrium*. By controlling the components in a mixture and through the use of catalysts that favor selected pathways, chemical engineers can synthesize desirable products from crude raw materials on a very large scale. Each step in the synthesis process is constrained by reaction thermodynamics. The desired products can be formed only if the equilibrium constant is favorable.

The mass, energy, and entropy balances of multicomponent, multiphase, reacting systems at thermodynamic equilibrium comprise significant coverage of the chemical engineering discipline. The rates at which systems move toward equilibrium comprise another fundamental field of study, and often an analysis of process performance requires an assessment as to which phenomenon is the dominant factor controlling performance—the equilibrium state or the rate of mass transfer or chemical reaction exhibited by a system in moving toward that state (for fundamentals, see Sections 5-7). Identifying and addressing key equilibrium limitations and/or rate-limiting resistances is a fundamental approach to improving process designs.

For most chemical engineers, a process simulator is the primary interface for engaging thermodynamics. The intent of this section is to expose the thermodynamics while simplifying the computational rigor, with the emphasis placed on nonelectrolyte systems.

ELEMENTARY VARIABLES AND DEFINITIONS

It is necessary to define several common quantities before developing the key equations to be applied in further analysis.

Mass m *Mass* is the magnitude of the interaction of a physical body in response to an external force. (We ignore relativistic influences in this discussion.) Commonly, the external force is gravity, and the mass is given by the weight at sea level. The mass also describes the resistance of the body to acceleration in the presence of any force, as in $F = ma$.

System or Control Volume A region in space that identifies the portion of the universe under consideration at a particular juncture.

Density ρ The mass or moles per unit volume. We use the same symbol ρ for both mass and molar density where the units are inferred by the particular context.

Pressure P The force per unit area of molecules on the surface of their container.

Internal Energy U Energy can be transformed into many forms, such as work, heat, or kinetic energy. To be clear, it is necessary to define each form of energy distinctly. To begin, internal energy is energy inherent to a system as determined by the kinetic and intermolecular potential energy of its constituent molecules. The kinetic energy of the molecules is described below in terms of temperature. The intermolecular potential energy arises from the tendency of molecules to attract and repel one another. Attractions are responsible for the heat of vaporization. Repulsions explain why "you can't put two things in the same place at the same time." Depending on the reference state, U may also include the energy of forming the molecule from the elements.

Heat Capacity at Constant Volume C_V $C_V \equiv (\partial U/\partial T)_V$ The translational and vibrational molecular energies are largely unaffected by changes in density and can be represented by the ideal gas heat capacity C_V^{ig}. The departures of internal energy and heat capacity from ideal gas behavior are discussed in the subsection Departure Functions from PVT Correlations.

Enthalpy H Enthalpy is a combination of internal energy, pressure, and molar volume ($H \equiv U + PV$) that is convenient for computations involving systems that are classified as "open," as defined shortly after Eq. (4-3). Note that molar volume V is the reciprocal of molar density.

Heat Capacity at Constant Pressure C_P $C_P \equiv (\partial H/\partial T)_P$ Similar to the heat capacity at constant volume, the enthalpy of a fluid varies with temperature. Empirical equations relating C_p^{ig} to T are available for many pure gases; a common form is either a polynomial like Eq. (4-1) or the form used by DIPPR, Eq. (4-2) (R. L. Rowley et al., *DIPPR Data Compilation of Pure Chemical Properties,* Design Institute for Physical Properties, AIChE, New York, 2006)

$$C_p^{ig} = A + BT + CT^2 + DT^3 + ET^4 \tag{4-1}$$

$$C_p^{ig} = A + B\left[\frac{C/T}{\sinh(C/T)}\right]^2 + D\left[\frac{E/T}{\cosh(E/T)}\right]^2 \tag{4-2}$$

where $A, B, C, D,$ and E are constants characteristic of the particular gas, and for Eq. (4-2) either C or D is zero. The DIPPR form derives from the plateaus inherent in heat capacity due to quantum energy levels [Aly, F. A., and L. L. Lee, *Fluid Phase Equilibr.* **6**: 169 (1981)].

Expansion/Contraction Work W_{EC} Work interaction of the system with the surroundings due to force at the surface of interaction through a distance is given by

$$W_{EC} = -\int P\,dV \tag{4-3}$$

Mass, Energy, and Entropy Balances Mass, energy, and entropy balances for any system are written with respect to a region of space known as a *system* or *control volume,* bounded by an imaginary *control surface* that separates it from the surroundings, forming a system or subsystem. This surface may follow fixed walls or be arbitrarily placed; it may be rigid or flexible. A primary step in any chemical process analysis or design is the mass balance. A system is defined as *open* if any mass crosses the system boundary. If the mass flowing into the system equals the mass flowing out, and all intensive (state) variables are invariant with time at all positions within the system, then the system is said to be at *steady state.*

GENERAL BALANCES

THE MASS BALANCE

Because mass is conserved, the time rate of change of mass within the control volume equals the net rate of flow of mass into the control volume (cv). The flow is positive when directed into the control volume and negative when directed out. The mass balance is expressed mathematically by

$$\sum_{\text{streams}} \dot{m}_i\, dt = d(m_{cv}) \tag{4-4}$$

The mass flow rate $\dot{m}$ can be expressed in terms of the stream velocity as

$$\dot{m}_i = (\rho v A_x)_i \,[=]\, \text{kg/s} \tag{4-5}$$

Substitution gives

$$\Delta(\rho v A_x)_{fs} = \frac{dm_{cv}}{dt} \tag{4-6}$$

The operator Δ signifies the difference between exit and entrance flows, and the subscript fs indicates that the term encompasses all flowing streams. This form of the mass balance equation is often called the *continuity equation,* an important equation in the analysis of transport processes, such as fluid flow or absorption. For the special case of *steady-state* flow, the control volume contains a constant mass of fluid, and the right-hand side of Eq. (4-6) is zero. Additional constraints for steady-state systems are discussed in the subsection The General Energy Balance.

Mass Balances for Chemical Manufacturing Processes Mass balances can be especially useful for multicomponent processes of multiple-unit operations. A spreadsheet calculation can suffice for many applications. A process flow diagram (PFD) is required, representing the unit operations and streams connecting them. Figure 4-1 shows a sample PFD for the dimethyl ether (DME) synthesis process ($2CH_3OH \rightarrow CH_3OCH_3 + H_2O$). Streams are numbered to provide unique identifiers. The masses in each stream are computed sequentially, depending on the unit operation.

For example, the mass of each component in stream 2 is the sum of the component mass entering from stream 1 and stream 11. Note that the flow of stream 11 may not be known at the outset, requiring an iterative process to determine its value. As another example, the mass of methanol in stream 5 is determined from the fractional conversion specification for methanol (X), and the masses of the other components are determined by the reaction stoichiometry. The flow rates of components in streams 8 and 9 are determined from the *split* specifications on the distillation column. The *split* is defined as the fraction of the component that exits the column as distillate. The light key component is the least volatile component that has a split fraction greater than 0.5. In consequence, any components more volatile than the light key are often assumed to exit completely in the distillate with a split of 100 percent. Similarly, the heavy key is the most volatile component that has a split fraction less than 0.5, and components less volatile than the heavy key are often assumed to exit the column with a split of 0 percent. The requisite computations to complete the mass balance are illustrated in Example 4-1.

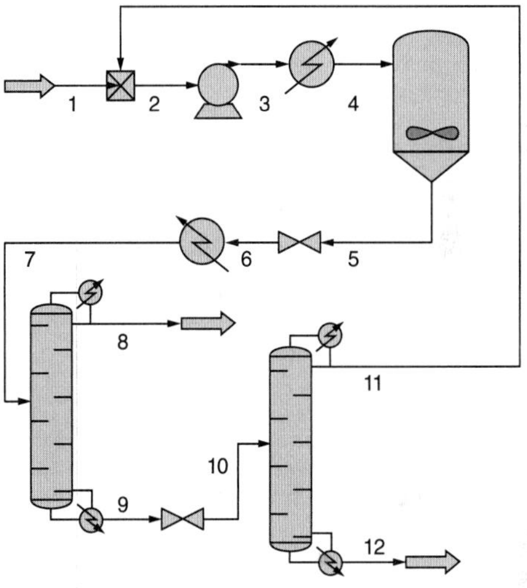

FIG. 4-1 PFD for DME synthesis.

Example 4-1 Mass Balances for the DME Process Dimethyl ether (DME) synthesis provides a simple prototype of many petrochemical processes. Ten tonnes (10,000 kg) per hour of methanol are fed at 25°C. The entire process operates at roughly 10 bar. The feed stream is mixed with the recycle stream (11), compressed and passed through a heat exchange to form stream 4 at 75°C. The methanol is 50 percent converted to DME and water at 250°C. The reactor effluent is cooled to 75°C and sent to a distillation column where 99 percent of the entering DME exits the top with 1 percent of the entering methanol and no water. This DME product stream (8) exits as liquid at 44°C while the bottom stream (9) exits at 152°C. The bottoms of the first column (9) are sent to a second column where 99 percent of the entering methanol exits the top as liquid at 136°C, along with all DME and 1 percent of the entering water, and is recycled. The bottoms of the second column exit at 179°C and are sent for wastewater treatment. Determine the masses of each component in each stream.

Solution A preliminary step is to write the reaction stoichiometry: $2CH_3OH \rightarrow H_2O + DME$. Since the reaction requires mass balances in terms of molar stoichiometry, we convert 10,000 kg/h to 312.11 kmol/h for stream 1. The solution for stream 2 depends on the recycle stream, for which the flow is not known at the outset. Modern spreadsheets facilitate an iterative solution for the recycle stream. Making an initial guess that 50 percent of the feed methanol (156.05 kmol/h) is being recycled in stream 11 with zero DME or H_2O gives a flow of 468.16 for stream 2, as tabulated below. With 50 percent conversion of the methanol in stream 2, the tabulated flows of methanol, DME, and H_2O in stream 5 are obtained. The masses are unaffected by the heat exchanger, resulting in stream 7. Applying the 99 and 1 percent splits gives the flow of streams 8 and 9. Similar computations give the flow of streams 11 and 12, at which point we note that the initial guess for stream 11 was substantially in error. At this point, the correct component masses of stream 11 could be added to stream 1 and the next iteration could proceed. Alternatively, Microsoft Excel offers an "iteration" feature that can be enabled through the calculation options. Implementing this feature leads to the mass flows (kmol/h) in the second table.

Initial guess assuming 156.05 kmol/h MeOH recycle.

Stream	1	2	5	7	8	9	11	12
T (°C)	25.0	73.3	250.0	75.0	44.0	152.0	136.0	179.0
Methanol	312.11	468.16	234.08	234.08	2.34	231.74	229.42	2.32
DME	0.00	0.00	117.04	117.04	115.87	1.17	1.17	0.00
H_2O	0.00	0.00	117.04	117.04	0.00	117.04	1.17	115.87

Final result after iterating stream 11 to convergence.

Stream	1	2	5	7	8	9	11	12
T (°C)	25.0	80.9	250.0	75.0	45.0	152.0	139.0	179.0
Methanol	312.11	612.04	306.02	306.02	3.06	302.96	299.93	3.03
DME	0.00	1.55	154.56	154.56	153.01	1.55	1.55	0.00
H_2O	0.00	1.55	154.56	154.56	0.00	154.56	1.55	153.01

INTRODUCTORY STATE CALCULATIONS

Values of energy require calculation of fluid properties for each component relative to the reference state, plus the property state changes involved in mixing components. For changes in state properties in nonreactive systems, the reference state drops out. However, for reactive systems, the reference states must be included. The energy balances are generalized most easily by including a reference state of the elements that comprise each molecule in the stream. Most process simulators default to use a reference state of the elements in their naturally occurring state at 1 bar and 298.15 K. For expediency, we provide ideal gas properties here to introduce the energy balances and in subsection Departure Functions for PVT Relations we discuss the contributions due to nonideal gas behavior.

For an ideal gas,

$$PV = RT \quad \text{or} \quad P\underline{V} = nRT \tag{4-7}$$

Using a reference state of the elements at 298.15 K and 1 bar,

$$U^{ig} = \int_{298.15}^{T} C_V\, dT + \Delta U^f_{298.15} \tag{4-8}$$

where $\Delta U^f_{298.15}$ is the energy of formation. More typically for flow problems, the enthalpy is used. Enthalpy is defined for convenience because the combination $U + PV$ appears often:

$$H \equiv U + PV \tag{4-9}$$

The enthalpy change of an ideal gas is

$$\Delta H^{ig} = \int_{T_1}^{T_2} C_P\, dT \tag{4-10}$$

The enthalpy of an ideal gas is

$$H^{ig} = \int_{298.15}^{T} C_P\, dT + \Delta H^f_{298.15} \tag{4-11}$$

where $\Delta H^f_{298.15}$ is the enthalpy of formation. Enthalpies of formation are commonly available (c.f. Sec. 2, Tables 2-94 and 2-95). The energy of formation can be calculated from the enthalpy of formation by adapting Eq. (4-9), $\Delta U^f_{298.15} = \Delta H^f_{298.15} - \Delta(PV)$, where the PV term is on the basis of 1 mol of the substance being formed. The PV term is typically negligible for condensed phases (e.g., solid carbon), and the value is RT for each mole of ideal gas, so the correction term can typically be written $\Delta n^{ig}RT$ when the formation reaction stoichiometry is balanced for 1 mol of the substance being formed.

Gas-phase nonidealities are calculated with the departure function at the same temperature and pressure, denoted for enthalpy as $(H - H^{ig})$ as discussed in the subsection Departure Functions from PVT Calculations resulting in $H^V = \Delta H^V_{\text{mix},T} + \sum_i x_i[(H^V - H^{ig})_T + \int_{298.15}^{T} C_P\, dT + \Delta H^f_{298.15}]$. For condensed phases the enthalpy of phase transformations are added (subsection *Phase Changes*), and any state changes of mixing (subsection *Property Changes of Mixing*) and a pressure correction. Mixing changes are large for acids, bases, and salts with water, but are generally a small contribution for organic mixture streams. When mixing changes are included, the enthalpy of a liquid stream with conventional liquid components would be

$$H^L = \Delta H^L_{\text{mix},T} + \sum_i x_i[V_i^L(P - P^{\text{sat}}_{i,T}) - \Delta H^{\text{vap}}_T + (H^V - H^{ig})_T + \int_{298.15}^{T} C_P\, dT + \Delta H^f_{298.15}] \tag{4-12}$$

For equation of state modeling of vapor phases, mixing process are usually included in the mixture departure function, resulting in $H^V = (H^V - H^{ig})_T + \sum_i x_i[\int_{298.15}^{T} C_P\, dT + \Delta H^f_{298.15}]$. For equation of state modeling of liquid phases, the mixing, pressure correction, and vaporization terms are included in the departure function, $H^L = (H^L - H^{ig})_T + \sum_i x_i[\int_{298.15}^{T} C_P\, dT + \Delta H^f_{298.15}]$.

Phase Changes Vaporization of a pure fluid occurs at constant temperature at the species vapor pressure $P^{\text{sat}}(T)$. The heat of vaporization is directly related to the slope of the vapor-pressure curve.

$$\Delta H^{\text{vap}} = T\Delta V^{\text{vap}} \frac{dP^{\text{sat}}}{dT} \tag{4-13}$$

Known as the *Clapeyron equation*, this exact thermodynamic relation provides the connection between the properties of the liquid and vapor phases at saturation. An empirical parameter frequently used in characterization of fluid properties is the *acentric factor*, defined by

$$\omega \equiv -1 - \log(P^{\text{sat}}/P_c)\big|_{T_r=0.7} \tag{4-14}$$

where $T_r \equiv T/T_c$ is the *reduced temperature*. In application, an empirical vapor pressure versus temperature relation is commonly used such as the Antoine equation

$$\log(P^{\text{sat}}) = A - \frac{B}{T + C} \tag{4-15}$$

Experimentally, $\log P^{\text{sat}}$ is not quite linear with $1/T$ (in K^{-1}); however, use of Eq. (4-15) with $C = 0$ and T in K is also sufficient for interpolation between reasonably spaced values of T. The acentric factor can be used for crude estimates of vapor pressure by neglecting the slight curvature

$$\log(P^{\text{sat}}/P_c) = (7/3)(1+\omega)(1 - T_c/T) \tag{4-16}$$

where T is in Kelvin. This approximation is sometimes referred to as *Wilson's vapor pressure equation*, but the exact attribution has been lost. More accurate equations are listed in the correlations of Sec. 2. Accurate correlations for ΔH^{vap} are available in Sec. 2, but in this section we apply a simple approximation

$$\frac{\Delta H^{\text{vap}}}{RT_c} = 7(1 - T_r)^{0.354} + 11\omega(1 - T_r)^{0.456} \quad \text{when} \quad P > P^{\text{sat}} \tag{4-17}$$

Equations (4-16) and (4-17) are accurate to roughly 15 percent for hydrocarbons when $T_r > 0.5$.

THE GENERAL ENERGY BALANCE

Because energy, like mass, is conserved, the time rate of change of energy within the control volume equals the net rate of energy transfer into the control volume. Streams flowing into and out of the control volume have energy associated with them in the internal, potential, and kinetic forms, and all contribute to the energy change of the system. Energy may also flow across the control surface as heat and work. General References 1, 3 through 6, and 8 through 12 show that the general energy balance for flow processes is

$$\frac{d\left[m\left(U + \dfrac{v^2}{2g_c} + \dfrac{zg}{g_c}\right)\right]_{cv}}{dt} = \dot{Q} + \dot{\underline{W}} + \sum_{\text{streams}}\left[\left(H + \frac{v^2}{2g_c} + \frac{zg}{g_c}\right)\dot{m}\right]_i \tag{4-18}$$

The work rate $\dot{W}$ may be of several forms. Most commonly there is shaft work $\dot{W}_s$. Work may be associated with expansion or contraction of the control volume, and there may be stirring work. Note that when a gas expands from inlet to outlet across a pressure drop, flow work is inherently included in the definition of enthalpy and not the work term. The velocity v in the kinetic energy term is the bulk mean velocity as defined by the equation $v = \dot{m}/(\rho A_x)$; z is elevation above a datum level, g is the local acceleration of gravity, and g_c is the gravitational units conversion constant.

Energy Balances for Closed Systems In closed systems, all mass flows across system boundaries are zero, so the last term of Eq. (4-18) is zero. The simplified energy balance then becomes

$$d\left[m\left(U + \frac{v^2}{2g_c} + \frac{zg}{g_c}\right)\right]_{cv} = \dot{\underline{Q}}\, dt + \dot{\underline{W}}\, dt \tag{4-19}$$

The most common form of energy balance is obtained by noting that changes in the velocity and altitude of most systems are usually negligible when temperature changes are present. Noting that $\underline{Q} \equiv \int \dot{\underline{Q}} dt$ and $\underline{W} \equiv \int \dot{\underline{W}} dt$, we find that

$$\Delta U = Q + W \tag{4-20}$$

Example 4-2 Adiabatic Reversible Compression of Air One stroke of a positive displacement compressor is analogous to the piston-cylinder arrangement of a bicycle pump. If the stroke is fast enough, heat transfer can be neglected for the purpose of a single stroke. Suppose a pump is 35 cm long and has a 3-cm diameter with ambient air initial pressure $P_1 = 0.1$ MPa. Estimate the pressure and temperature achieved at the end of the stroke using the ideal gas law and the work done (J/mol), assuming air enters at 25°C and a weight of 80 kg is applied.

Solution $dU = C_V\, dT = dQ + dW = dW_{EC} = -P\, dV = -RT\, dV/V$; let $L \equiv$ length (cm) of the cylinder after compression. Rearranging gives $C_V\, dT/T = R\, dV/V$ => $T_2/T_1 = (V_1/V_2)^{R/C_V} = (P_2/P_1)^{R/C_P}$. For the given conditions, $P_{2,\text{weight}} = 80$ kg/$(0.015^2\ \pi) = 113{,}177$ kg/m^2 · 9.8066 N/kg = 1.1099 MPa = 161 psig, to which we add 0.1 MPa for absolute pressure, $P_2 = 1.21$ MPa. By the ideal gas law, $P_2/P_1 = (T_2/T_1)(35/L) = (P_2/P_1)^{R/C_P}(35/L)$. Rearranging gives $L = 35(P_1/P_2)^{C_V/C_P} = 5.89$ cm; $T_2 = 608$ K = 335°C. The work done is $W_{EC} = (5/2)(8.314)(607.9 - 273.15) = 6960$ J/mol.

This example includes the ideal gas law approximation. Air near room temperature and pressure can be approximated as an ideal gas composed of nitrogen and oxygen, both of which have roughly constant C_P values of 3.5R, and $C_V = C_P - R$ for ideal gases. Guidelines for ideal gas behavior are:

$$T/T_c \equiv T_r > 0.5 + 2(P/P_c) \quad \text{or} \quad V/V_c > 4 \quad \text{Conditions applicable to ideal gases} \tag{4-21}$$

The definition of C_V leads to the substitution for the dU term. The temperature rise during adiabatic compression can be quite large. We have implicitly assumed that the pressure inside the cylinder is uniformly equal to the pressure applied externally, signifying a reversible process. Hence the result for T_2/T_1 can be applied to any reversible, adiabatic, ideal gas process. Through the ideal gas law, this result becomes,

$$T_2/T_1 = (V_1/V_2)^{(R/C_V)} = (P_2/P_1)^{(R/C_P)} \Rightarrow PV^{\gamma} = \text{constant, ad. rev. ideal gas, } \gamma \equiv C_P/C_V \tag{4-22}$$

Energy Balances for Steady-State Flow Processes Flow processes for which the left-hand side of Eq. (4-18) is zero are said to occur at *steady state*. As discussed with respect to the mass balance, this means that the mass of the system within the control volume is constant; it also means that no changes occur with time in the properties of the fluid within the control volume or at its entrances and exits. The only work of the process commonly present is shaft work, and the simplified form of the general energy balance for a single inlet and single outlet, becomes

$$\Delta\left[\left(H + \frac{1}{2}\frac{v^2}{g_c} + \frac{zg}{g_c}\right)\dot{m}\right] = \dot{\underline{Q}} + \dot{\underline{W}}_s \tag{4-23}$$

Note that the sign appears to change relative to Eq. (4-18) because Δ is defined as outlet – inlet and here the absolute values of the mass flows should be used, whereas outlet mass flows inherently have negative signs in Eq. (4-18). Simplifying further, the most common energy balance for open steady-state systems neglects changes in kinetic energy and altitude.

$$\Delta H = Q + W_s \tag{4-24}$$

Example 4-3 Continuous Adiabatic Reversible Compression of Air The energy balance changes when the system is viewed from a steady-state perspective. The individual strokes of the compressor, or even whether it is a positive displacement or centrifugal compressor, are irrelevant. Only the continuous flows of energy into and out of the system matter. To illustrate, consider the following example where air enters a continuous reversible compressor at 25°C and 1 bar and is adiabatically compressed to 6 bar. Compute the outlet temperature and work requirement (J/mol).

Solution Due to the adiabatic assumption, the heat term drops out of Eq. (4-24) and we seek ΔH to find the work. The air is treated as an ideal gas. Noting the words *adiabatic, reversible,* and *ideal gas*, we can immediately apply Eq. (4-22). The definition of C_P leads to a subtle but significant distinction relative to the previous example. The energy balance simplifies as $\Delta H = \int C_P dT = Q + W = W_s$

$$W_s = C_P(T_2 - T_1) = C_P T_1\left[\left(\frac{P_2}{P_1}\right)^{R/C_P} - 1\right] \tag{4-25}$$

Substituting numerical values gives

$$T_2 = T_1(6/1)^{(1/3.5)} = 497.5\text{ K} = 224.3°\text{C} \quad W_s = 3.5 \times 8.314(497.5 - 298.15) = 5801\text{ J/mol}$$

Energy Balances for Chemical Manufacturing Processes Similar to mass balances, energy balances are applicable to composite systems as well as individual subsystems. Thus it is valuable to extend PFD computations to include stream enthalpies as well as component mass flows. Highly accurate estimation of stream enthalpies is quite complicated because it involves accurate estimation of the enthalpy of compressed gases and nonideal liquids that may exhibit substantial heats of mixing. Such an approach would be very computationally intensive, hence the necessity of a process simulator. We convey the general concepts by presenting the general equations, and then we illustrate the connection between the general equations and the pathway to properties by using computationally expedient models. The general equations can be revisited at various stages to show improvements in accuracy with increasingly sophisticated computational models.

The essential relation for estimating stream enthalpies is given by

$$H(T,P,z) = \sum_i z_i\left(\Delta H^{o,\text{ig}}_{f,i,T_R}(T_R) + \int_{T_R}^{T} C_P^{\text{ig}}\, dT\right) + q(H^V(T,y) - H^{\text{ig}}) + (1-q)(H^L(T,x) - H^{\text{ig}}) \tag{4-26}$$

where z_i is the overall mole fraction, and q is the stream's molar vapor fraction. Here we have assumed that the reference state is defined relative to the elements at 25°C and 1 bar as in Eq. (4-12). To apply this rigorously, equations of state are used (see the subsection *Departure Functions from PVT Correlations*). For shortcut calculations, the contributions can be written,

$$(H^L(T,x) - H^{\text{ig}}) = \Delta H^L_{\text{mix}} + \sum_i x_i\left(\left(H_i^V(T, P_i^{\text{sat}}) - H_i^{\text{ig}}\right) - \Delta H_T^{\text{vap}} + V_i^L\left(P - P_{i,T}^{\text{sat}}\right)\right) \tag{4-27}$$

$$(H^V(T,y) - H^{\text{ig}}) = \left(\sum_i x_i\left(H_i^V - H_i^{\text{ig}}\right)\right) + \Delta H^V_{\text{mix}} \tag{4-28}$$

where H^{ig} is calculated using Eq. (4-11). For the purpose of illustrating the pathway to computing stream enthalpies, we can make the following shortcut approximations: (1) $C_P^{\text{ig}} = C_P^{\text{ig}}(25°\text{C}) =$ constant. This is reasonably accurate for $T < 100$°C. (2) $(H^V - H^{\text{ig}}) = 0$. It neglects departures for compressed vapors and heats of mixing for vapors, but is reasonable when $P < 5$ bar. (3) $(H^L - H^{\text{ig}}) = -\sum x_i \Delta H_i^{\text{vap}}$. This neglects heats of mixing and enthalpy increases due to increased pressure.

Example 4-4 Energy Balances for the DME Process The formulas above make it possible to compute an enthalpy flow (MJ/h) for each stream in Example 4-1. Heats of formation for methanol, DME, and H_2O are −200.94, −184.1, and −241.835 kJ/mol. The ideal gas heat capacities at 25°C are 43.9, 65.7, and 33.6 J/mol·K. All the streams are liquid except stream 5. Tabulate these enthalpies and compute (1) the net heat flow (kW) of the heat exchanger after the reactor and (2) the net power flow (kW) for the overall process.

Solution Illustrating the procedure for one liquid stream and one vapor stream should suffice. Stream 5 is all vapor. From Example 4-1, the stream species flows are 306.0, 154.6, and 154.6 kmol/h. Applying Eq. (4-28),

$$\dot{\underline{H}}_5 = [306.0(-200940 + 43.9(250 - 25)) + 155.6(-184100 + 65.7(250 - 25)) + 155.6(-241835 + 33.6(250 - 25))]/1000 = -120846\text{ MJ/h}$$

Note that we retain a larger number of significant figures than would normally be warranted for such imprecise estimates. This is so because the heats of formation play a significant role in each stream enthalpy. When we take differences, the large heat of formation terms cancel for control volumes without reactions, but are necessary for control volumes that include reactions. Stream 7 is interesting as a sample stream that is liquid and relates to the heat exchanger. The ideal gas contribution can be computed as for stream 5.

$$\dot{\underline{H}}_7\,(\text{ig}) = [306.0(-200940 + 43.9\,(75 - 25)) + 155.6\,(-184100 + 65.7(75 - 25)) + 155.6(-241835 + 33.6(75 - 25))]/1000 = -125883\text{ MJ/h}$$

Applying Eq. (4-17) at 75°C for stream 7 gives 35.75, 14.19, and 42.99 kJ/mol for the heats of vaporization. Adding this to the ideal gas contribution gives

$$\dot{\underline{H}}_7 = -125883 + [306.0(-35750) + 155.6(-14190) + 155.6(-42990)]/1000 = -145661\text{ MJ/h}$$

Repeating the procedure for the other streams gives the enthalpy flows $\dot{\underline{H}} = \dot{n}H$,

Stream	1	2	5	7	8	9	11	12
$\dot{\underline{H}}$(GJ/h)	−75.1	−143.8	−120.8	−145.7	−31.3	−110.5	−68.4	−42.4

The energy balance for heat exchanger: No work is accomplished so $\Delta\underline{\dot{H}} = \underline{\dot{Q}} + \underline{\dot{W}} = \underline{\dot{Q}}$.

$$\underline{\dot{Q}} = (-145.7 + 120.8)\,(1{,}000{,}000)/(3600\text{ s/h}) = -6900\text{ kW}$$

The net energy balance for process involves streams 1, 8, and 12: No pumps or turbines appear in this process, so $\Delta\underline{\dot{H}} = \underline{\dot{Q}}$.

$$\underline{\dot{Q}} = (-42.4 - 31.3 + 75.1)(1{,}000{,}000)/(3600\text{ s/h}) = 390\text{ kW}$$

These results show that net heat addition is required even though the heat of reaction is negative. Note that the outlet streams are hotter than the inlet streams, and despite the exothermic heat of reaction, the heat exchangers, reboilers, and condensers must balance this heat duty.

THE GENERAL ENTROPY BALANCE

The primary engineering purpose of entropy is to evaluate process thermodynamic reversibility. Entropy is defined in a closed system as

$$\Delta\underline{S} = \int_{T_1}^{T_2} \frac{Q_{\text{rev}}}{T_{\text{cv}}}\,dT \tag{4-29}$$

where Q_{rev} is reversible heat transfer, and T_{cv} is the control volume temperature at the surface where heat is transferred. Entropy changes for an ideal gas using T and P as independent variables can be calculated using a formula derived from Eq. (4-29) by combining an isothermal step and an isobaric step:

$$\Delta S^{\text{ig}} = \int_{T_1}^{T_2} \frac{C_P}{T}\,dT - R\ln\frac{P_2}{P_1} \quad\text{or}\quad S^{\text{ig}} = \int_{T_{\text{ref}}}^{T_2} \frac{C_P}{T}\,dT - R\ln\frac{P_2}{P_{\text{ref}}} + S_{\text{ref}} \tag{4-30}$$

Commonly $S_{\text{ref}} = 0$. If the reference state uses the elements in the naturally occurring state of aggregation, then $\Delta S^f_{T_{\text{ref}},P_{\text{ref}}} = (\Delta H^f_{T_{\text{ref}},P_{\text{ref}}} - \Delta G^f_{T_{\text{ref}},P_{\text{ref}}})/T_{\text{ref}}$ is added to the last expression. The entropy balance differs from an energy balance in a very important way—entropy is not conserved and is generated by irreversibilities. Entropy generation is always $\dot{\underline{S}}_{\text{gen}} \ge 0$, where the equality applies for (hypothetical) reversible processes. The statement of balance for a control volume, expressed as rates, is therefore

$$\left\{\begin{matrix}\text{Rate of entropy}\\ \text{change of}\\ \text{control volume}\end{matrix}\right\} = \left\{\begin{matrix}\text{net rate of}\\ \text{entropy transport}\\ \text{into control volume}\end{matrix}\right\} + \left\{\begin{matrix}\text{rate of entropy}\\ \text{change due to heat}\\ \text{transfer at surfaces}\end{matrix}\right\} + \left\{\begin{matrix}\text{rate of}\\ \text{entropy}\\ \text{generation}\end{matrix}\right\}$$

The equivalent *entropy balance* is

$$\frac{d(mS)_{\text{cv}}}{dt} = -\Delta(\dot{m}S)_{\text{fs}} + \sum_{\text{surfaces}} \frac{\dot{Q}}{T_{\text{cv}}} + \dot{\underline{S}}_{\text{gen}} \tag{4-31}$$

where $\dot{\underline{S}}_{\text{gen}}$ is the entropy generation term. This equation is the general *rate* form of the entropy balance, applicable at any instant. In general application, the contribution of flowing streams is most easily incorporated by adding the sum of entropy flow $\sum(\dot{m}S)$ of the incoming streams and subtracting the sum of entropy flow of outgoing streams. The entropy calculations for streams extend the process set forth above for enthalpy. However, the entropy of mixing (see subsections *Property Changes on Mixing* and *Ideal Solution Model and Henry's Law*) cannot and should not be neglected.

For any process, the two kinds of irreversibility are (1) those *internal* to the control volume and (2) those resulting from heat transfer across finite temperature differences that may exist between the system and surroundings. When a temperature gradient exists at a boundary, the entropy balance for the boundary itself must be included when determining the entropy change of the universe. In the limiting case where for the universe (system + boundary + surroundings) $\dot{\underline{S}}_{\text{gen}} = 0$, the process is *completely reversible*, implying that

- The process is internally reversible within the control volume.
- Heat transfer between the control volume and its surroundings is reversible.

A sample application of the entropy balance is given below under the heading Turbines.

Entropy Balances for Composite Systems As for the energy balance, entropy balances can be useful in analyzing processes from an overall perspective. The most common applications involve idealized 100 percent reversible processes such as the Carnot engine. However, it can be meaningful to consider irreversible processes using entropy as a measure of overall thermodynamic efficiency, as in the case of availability or exergy analysis.

Example 4-5 Carnot Efficiency A heat engine is to run between 340 and 260 K. As an approximation, we can assume that the engine follows the Carnot process of adiabatic reversible compression to 340 K, isothermal heat addition, adiabatic reversible expansion to 260 K, and isothermal heat removal. For the purposes of this illustration, assume the process is working on propane, where heat addition or removal could be accomplished isothermally by boiling and condensing. The entropies of the saturated vapor and saturated liquid at 340 K would define the entropy range of operation.

The equations that apply to reversible Carnot engines are as follows:

By the first law: $\quad |\underline{W}_{\text{net}}| = |Q_H| - |Q_C|$

By the second law (for a reversible process): $\quad \dfrac{Q_H}{T_H} + \dfrac{Q_C}{T_C} = 0 \Rightarrow \dfrac{|Q_H|}{|Q_C|} = \dfrac{T_H}{T_C}$

The initial implementation of the second law recognizes that heat flows are of opposite sign for heating and cooling. The second form helps to minimize sign confusion during application.

In combination:

$$|\underline{W}_{\text{net}}| = |Q_C|\left(\frac{T_H}{T_C} - 1\right) = |Q_H|\left(\frac{T_H - T_C}{T_H}\right) \tag{4-32}$$

Here $|\underline{W}_{\text{net}}|$ is the net work produced by the Carnot engine after accounting for both compression and expansion; $|Q_H|$ is the heat transferred at the hot temperature, i.e., to vaporize the propane; T_H and T_C are the hot and cold temperatures of the heat reservoirs between which the heat engine operates, or 340 and 260 K, respectively.

FUNDAMENTAL RELATIONS OF CLASSICAL THERMODYNAMICS

Multivariable calculus provides a number of relations between thermodynamic variables that are quite useful for estimating stream properties. The key point is that specification of two independent variables suffices to define the state of a pure system. For example, if T and ρ are known, the other properties (such as P, U, H, S) and their changes from state are implied. Through the equations of classical thermodynamics, we find that all the properties can be derived from an equation of state $P = P(\rho, T)$ by characterizing departures from ideal gas behavior.

The Fundamental Property Relation for Pure Fluids Energy and entropy balances can be combined to eliminate references to heat and work in favor of state variables. For a single-component, reversible, closed system, Eq. (4-31) becomes

$$d(S)_{\text{cv}} = \frac{\dot{Q}dt}{T_{\text{cv}}} \tag{4-33}$$

Similarly, Eqs. (4-33) and (4-19) combine to give for a simple system with uniform T,

$$d(U)_{\text{cv}} = \dot{Q}dt + \dot{W}dt = T\,dS + \dot{W}dt \tag{4-34}$$

Noting that only W_{EC} is relevant for a reversible, closed system, Eqs. (4-3) and (4-34) give

$$dU = T\,dS - P\,dV \tag{4-35}$$

Equation (4-35) is the fundamental property relation. After substituting the definition of $H \equiv U + PV$, then $dH = dU + d(PV) = dU + P\,dV + V\,dP$, and

$$dH = T\,dS + V\,dP \tag{4-36}$$

The transformation from Eq. (4-35) to Eq. (4-36) suggests two additional relations, $A \equiv U - TS$ and $G \equiv U + PV - TS$, resulting in

$$dA = -S\,dT - P\,dV \tag{4-37}$$

$$dG = -S\,dT + V\,dP \tag{4-38}$$

Relations Using Desired Independent Variables For practical application, it is useful to select easily measured properties as desired independent variables for use in calculation of U, H, A, and G. Because the differentials of these state functions are *exact differential expressions*, application of the reciprocity relation for such expressions produces the common Maxwell relations as described in Sec. 3 in the subsection *Multivariable Calculus Applied to Thermodynamics*, and the four most frequently used Maxwell relations are developed in texts [see Elliott and Lira (2012)]. Combining Maxwell's relations with Eqs. (4-35) through (4-38) and the chain rule provides a number of useful relations.

$$dU(T,V) = \left(\frac{\partial U}{\partial T}\right)_V dT + \left(\frac{\partial U}{\partial V}\right)_T dV = C_V\,dT + \left[T\left(\frac{\partial P}{\partial T}\right)_V - P\right]dV \quad (4\text{-}39)$$

$$dH(T,P) = C_P\,dT + \left[V - T\left(\frac{\partial V}{\partial T}\right)_P\right]dP \quad (4\text{-}40)$$

$$dS(T,P) = \frac{C_P}{T}dT - \left(\frac{\partial V}{\partial T}\right)_P dP \quad (4\text{-}41)$$

$$dS(T,V) = \frac{C_V}{T}dT + \left(\frac{\partial P}{\partial T}\right)_V dV \quad (4\text{-}42)$$

As an example application of these differentials, consider the pressure correction for enthalpy due to pressure at constant temperature. If we neglect the usually small contribution of $T(\partial V/\partial T)_P$ in Eq. (4-40), then for a liquid where the fluid is approximately incompressible, the effect of pressure gives $V^L(P_2 - P_1)$ as implemented in Eq. (4-27) relative to the species vapor pressure.

BALANCE APPLICATIONS TO FLOW PROCESSES

Duct Flow of Compressible Fluids Thermodynamics provides equations interrelating pressure changes, velocity, duct cross-sectional area, enthalpy, entropy, and specific volume within a flowing stream. Consider the adiabatic, steady-state, one-dimensional flow of a compressible fluid in the absence of shaft work and changes in potential energy. The appropriate energy balance Eq. (4-23) with Q, W_s, and Δz all set equal to zero is

$$\Delta H + \frac{\Delta v^2}{2g_c} = 0$$

In differential form,

$$dH = -v\,dv/g_c \quad (4\text{-}43)$$

The mass balance (continuity) Eq. (4-6) here becomes $d(\rho v A_x) = d(vA_x/V) = 0$, which gives

$$\frac{dV}{V} - \frac{dv}{vg_c} - \frac{dA_x}{A_x} = 0 \quad (4\text{-}44)$$

Most common chemical engineering processes occur at fluid velocities substantially less than sonic. Therefore, we confine further discussion to subsonic flow. Discussion relating to near sonic or supersonic flow is available in Smith, Van Ness, and Abbott (2005). Flow rates should always be checked and recourse taken to account for supersonic effects if high flow rates are experienced.

Nozzles Nozzle flow is quite specialized in that a properly designed nozzle varies its cross-sectional area with length in such a way as to make the flow nearly frictionless. The limit is adiabatic, reversible flow, for which the rate of entropy increase is zero. An analytical expression relating velocity to pressure in an isentropic nozzle is readily derived for an ideal gas with constant heat capacities. Combination of Eqs. (4-36) and (4-43) for isentropic flow gives

$$v\,dv/g_c = -V\,dP$$

Integration, with nozzle entrance and exit conditions denoted by 1 and 2, yields for an ideal gas with constant $\gamma \equiv C_P/C_V$

$$\frac{v_2^2 - v_1^2}{g_c} = -2\int_{P_1}^{P_2} V\,dP = \frac{2\gamma P_1 V_1}{\gamma - 1}\left[1 - \left(\frac{P_2}{P_1}\right)^{(\gamma-1)/\gamma}\right] \quad (4\text{-}45)$$

where the final term is obtained upon elimination of V by PV^γ = constant, following Eq. (4-22).

Throttling Processes Fluid flowing through a restriction, such as an orifice, without appreciable change in kinetic or potential energy undergoes a finite pressure drop. This *throttling process* produces no shaft work, and in the absence of heat transfer, Eq. (4-24) reduces to $\Delta H = 0$ or $H_2 = H_1$. The process therefore occurs at constant enthalpy.

The temperature of an ideal gas is not changed by a throttling process because $dH^{ig} = C_P\,dT$. For most real gases at moderate conditions of T and P, a reduction in pressure at constant enthalpy results in a decrease in temperature, although the effect is usually small.

Throttling of a wet vapor causes the liquid to evaporate, resulting in a considerable temperature drop because of the evaporation of liquid, and depending on the pressure drop, the evaporation may be complete.

If a saturated liquid is throttled to a lower pressure, some of the liquid vaporizes or *flashes*, producing a mixture of saturated liquid and vapor at the lower pressure. For a pure fluid, the outlet temperature is the saturation temperature at the outlet pressure, which may be very cold.

Turbines (Expanders) High-velocity streams from nozzles impinging on blades attached to a rotating shaft from a turbine (or expander) through which vapor or gas flows in a steady-state expansion process that converts the internal energy of a high-pressure stream into shaft work. The motive force is usually provided by a (steam) turbine or a high-pressure (gas) expander.

In any properly designed adiabatic turbine, heat transfer and changes in potential and kinetic energy are negligible. Equation (4-24) therefore reduces to

$$W_s = \Delta H = H_2 - H_1 \quad (4\text{-}46)$$

The rate form of this equation is

$$\dot{W}_s = \dot{m}\Delta H = \dot{m}(H_2 - H_1) \quad (4\text{-}47)$$

When inlet conditions T_1 and P_1 and discharge pressure P_2 are known, the value of H_1 is fixed. In Eq. (4-46) both H_2 and W_s are unknown, and the energy balance alone does not allow their calculation. However, if the fluid expands *reversibly and adiabatically* in the turbine, then the process is isentropic ($S_2 = S_1$). This entropy balance establishes the final state of the fluid and allows calculation of H_2. Equation (4-47) then gives the isentropic (reversible) work, and the prime denotes the reversible process:

$$W_s' = \Delta H' \quad (4\text{-}48)$$

The absolute value $|W_s'|$ is the *maximum* work that can be produced by an adiabatic turbine with given inlet conditions and given discharge pressure. Because the actual expansion process is irreversible, *expander efficiency* is defined as

$$\eta_E \equiv \frac{W_s}{W_s'}$$

where W_s is the actual shaft work. By Eqs. (4-47) and (4-48),

$$\eta_E = \frac{\Delta H}{\Delta H'} \quad (4\text{-}49)$$

Values of η_E usually range from 0.7 to 0.8.

Example 4-6 Turbine Process Design Steam is expanded in a turbine from 500°C and 1.4 MPa to an outlet of 0.6 MPa. If the turbine is 75 percent efficient, how much work can be obtained per kilogram of steam (kJ/kg)? Use the steam tables from Elliott and Lira (2012).

Solution The inlet conditions are H_1 = 3474.8 kJ/kg and S_1 = 7.6047 kJ/kg·K. To apply Eq. (4-49) the reversible calculation is performed first. Interpolating at the outlet state pressure using S_2' = 7.6047 kJ/kg·K gives H_2' = 3202.8 kJ/kg. Then $\Delta H = W_S = \Delta H'\eta_E$ = (3202.8 − 3474.8)(0.75) = −204 kJ/kg.

Compressors Compressors, pumps, fans, blowers, and vacuum pumps are all devices designed to produce pressure increases. The energy Eqs. (4-43) through (4-48) are the same for adiabatic compression, based on the same assumptions, as for turbines or expanders. A specialized equation of state (EOS) would be applied for steam or polar fluids, whereas a generalized EOS typically would be applied for the fluids involved in compressors.

The isentropic work of compression, as given by Eq. (4-48), is the *minimum* shaft work required for compression of a gas from a given initial state to a given discharge pressure. *Compressor efficiency* is defined as (again using the prime to denote the reversible process)

$$\eta_C \equiv \frac{W_s'}{W_s}$$

The relation between the reversible and actual process is inverted relative to a turbine. In view of Eqs. (4-46) and (4-48), this becomes

$$\eta_C \equiv \frac{\Delta H'}{\Delta H} \tag{4-50}$$

Compressor efficiencies are usually in the range of 0.7 to 0.8.

Pumps Liquids are moved by pumps, and the same equations apply to adiabatic pumps as to adiabatic compressors. Thus, Eqs. (4-46) to (4-48) and (4-50) are valid. However, application of Eq. (4-46) requires values of the enthalpy of compressed (subcooled) liquids, and these are seldom available. The enthalpy relation, Eq. (4-36), provides an alternative. For an isentropic process,

$$dH = V\,dP \;(\text{constant } S)$$

Combining this with Eq. (4-48) yields

$$W_s' = \Delta H' = \int_{P_1}^{P_2} V\,dP$$

The usual assumption for liquids (at conditions well removed from the critical point) is that V is independent of P. Integration then gives

$$W_s' = \Delta H' = V(P_2 - P_1) = \Delta P/\rho \tag{4-51}$$

PROPERTY CALCULATIONS FROM EQUATIONS OF STATE

The most satisfactory calculation procedure for the thermodynamic properties of gases and vapors is based on ideal gas state heat capacities and quantification of the nonidealities using departure functions. Of primary interest are the enthalpy and entropy departures, defined as the difference between the state properties of the real fluid and an ideal gas at the same *pressure* and *temperature*:

$$(H - H^{ig}) \equiv H^{dep} \quad \text{and} \quad (S - S^{ig}) \equiv S^{dep} \tag{4-52}$$

These departures are integrated into the process calculations, e.g., see Eq. (4-12). The reader is cautioned that departure functions are sometimes called *residual properties*, and sign conventions for the definitions differ in literature.

DEPARTURE FUNCTIONS FROM *PVT* CORRELATIONS

The departure functions of gases and vapors depend on their *PVT* behavior. This is often expressed through correlations for the compressibility factor Z, defined by

$$Z \equiv PV/(RT) \tag{4-53}$$

Analytical expressions for Z as functions of T and P or T and V are known as *equations of state* (EOSs). Since most EOSs are in terms of T and V, the most useful relations are Eqs. (4-37) and (4-39). Elliott and Lira (2012) show how these can be rearranged in the most convenient form:

$$\left(\frac{\partial U}{\partial V}\right)_T = T\left(\frac{\partial P}{\partial T}\right)_V - P \Rightarrow \frac{U^{dep}}{RT} = \int_0^{\rho} -T\left(\frac{\partial Z}{\partial T}\right)_{\rho} \frac{d\rho}{\rho} \tag{4-54}$$

$$\left(\frac{\partial A}{\partial V}\right)_T V = -PV \Rightarrow \frac{A_{T,V}^{dep}}{RT} = \int_0^{\rho} \frac{Z-1}{\rho}\,d\rho \tag{4-55}$$

The subscript T, V in Eq. (4-55) indicates that this departure is evaluated at the same T, V for the real fluid and ideal gas: $A_{T,V}^{dep} = A(T,V) - A^{ig}(T,V)$. Most applications require residuals at given T, P. The properties of the real fluid imply unique values of $T, P,$ and V, but the pressure obtained from the ideal gas equation given T, V is not equal to the real fluid's P. A translation in the ideal gas state of ln Z is used to obtain $A^{dep} = A(T,P) - A^{ig}(T,P)$.

$$\frac{A^{dep}}{RT} = \frac{A_{T,V}^{dep}}{RT} - \ln Z = \int_0^{\rho} \frac{Z-1}{\rho}\,d\rho - \ln Z \tag{4-56}$$

Other departure functions can be derived from the definitions of $U, H, A, G,$ and S.

$$\frac{S^{dep}}{R} = \frac{U^{dep} - A^{dep}}{RT} - \ln Z = -\int_0^{\rho}\left[T\left(\frac{\partial Z}{\partial T}\right)_{\rho} + (Z-1)\right]\frac{d\rho}{\rho} - \ln Z \tag{4-57}$$

$$\frac{H^{dep}}{RT} = \frac{U^{dep} + PV - (PV)^{ig}}{RT} = -\int_0^{\rho} T\left(\frac{\partial Z}{\partial T}\right)_{\rho}\frac{d\rho}{\rho} + Z - 1 \tag{4-58}$$

$$\frac{G^{dep}}{RT} = \frac{A^{dep}}{RT} + Z - 1 = \int_0^{\rho} \frac{Z-1}{\rho}\,d\rho + Z - 1 - \ln Z \tag{4-59}$$

A few EOSs may be reformulated to give P as a function of T and V or V as a function of T and P, in which case Eqs. (4-38) and (4-40) are more convenient.

$$\left(\frac{\partial H}{\partial P}\right)_T = V - T\left(\frac{\partial V}{\partial T}\right)_P \Rightarrow \frac{H^{dep}}{RT} = -\int_0^{P} T\left(\frac{\partial Z}{\partial T}\right)_P \frac{dP}{P} \tag{4-60}$$

$$P\left(\frac{\partial G}{\partial P}\right)_T = PV \Rightarrow \frac{G^{dep}}{RT} = \int_0^{P} \frac{Z-1}{P}\,dP \tag{4-61}$$

$$\frac{S^{dep}}{R} = -\int_0^{P}\left[T\left(\frac{\partial Z}{\partial T}\right)_P + Z - 1\right]\frac{dP}{P} \tag{4-62}$$

CHEMICAL POTENTIAL, FUGACITY, AND FUGACITY COEFFICIENT

The chemical potential μ plays a vital role in both phase and chemical reaction equilibria. However, the chemical potential exhibits certain unfortunate characteristics that discourage its use in the solution of practical problems. For pure fluids, $\mu = G \equiv H - TS$ defines μ in terms of the internal energy and entropy, both primitive quantities for which absolute values are unknown. Moreover, μ approaches negative infinity when P approaches zero. While these characteristics do not preclude the use of chemical potentials, the application of equilibrium criteria is facilitated by introduction of the *fugacity*, a quantity that takes the place of μ and overcomes its less desirable characteristics. The Gibbs energy departure of a real fluid is related to the fugacity by

$$G^{dep} = G - G^{ig} = RT\ln\frac{f}{P} \tag{4-63}$$

The dimensionless ratio f/P is another new property called the *fugacity coefficient* ϕ. Thus,

$$G^{dep} = RT\ln\phi \tag{4-64}$$

where

$$\phi \equiv \frac{f}{P} \quad \text{or} \quad f = \phi P \tag{4-65}$$

The definition of fugacity is completed by setting the ideal gas state fugacity of pure species i equal to its pressure, $f_i^{ig} = P$. Thus for the special case of an ideal gas, $G_i^{dep} = 0$, $\phi_i = 1$. From the phase equilibrium criterion, $\mu^{\alpha} = \mu^{\beta}$ when phases α and β are in equilibrium. Substitution into Eq. (4-63) shows that $f^{\alpha} = f^{\beta}$ is equivalent.

$$\mu^{\alpha} = \mu^{\beta} \qquad f^{\alpha} = f^{\beta} \tag{4-66}$$

For condensed phases, Eq. (4-65) is used to calculate the saturation fugacity (at the vapor pressure or sublimation pressure), and then Eq. (4-38) is used to add a pressure correction,

$$f^{cond} = \phi^{sat}P^{sat}\exp[V^{cond}(P - P^{sat})/(RT)] \tag{4-67}$$

where the exponential term is known as the *Poynting correction*, and V^{cond} is the molar volume of the condensed phase. As written, the Poynting correction assumes the condensed phase is incompressible.

APPLICATIONS OF DEPARTURE FUNCTIONS

Virial Equations of State The virial equation in *density* is an infinite series expansion of the compressibility factor Z in powers of molar density ρ (or reciprocal molar volume V^{-1}) about the real gas state at zero density (zero pressure):

$$Z=1+B\rho+C\rho^2+D\rho^3+\cdots \tag{4-68}$$

The density series virial coefficients $B, C, D, \ldots$ depend on temperature and composition only. In practice, truncation is to two terms. For engineering purposes, P is more convenient than density, and the pressure through mathematical reversion of the series is

$$Z=1+BP/(RT)+\cdots \tag{4-69}$$

For a pure fluid, Eq. (4-61) gives

$$\phi=\exp(BP/(RT)) \tag{4-70}$$

The composition dependency of B is given by the exact *mixing rule*

$$B=\sum_i\sum_j y_i y_j B_{ij} \tag{4-71}$$

where y_i and y_j are mole fractions for a gas mixture and i and j identify species. The coefficient B_{ij} characterizes a bimolecular interaction between molecules i and j, and therefore $B_{ij}=B_{ji}$. Two kinds of second virial coefficient arise: B_{ii} and B_{jj} (the subscripts are the same), and B_{ij} (they are different). The first is a virial coefficient for a pure species; the second is a mixture property, called a *cross-coefficient.*

An extensive set of three-parameter corresponding-states correlations has been developed by Pitzer and coworkers [Pitzer, *Thermodynamics*, 3d ed., App. 3, McGraw-Hill, New York, 1995]:

$$\frac{BP_c}{RT_c}=B^0+\omega B^1 \tag{4-72}$$

with the *acentric factor* defined by Eq. (4-14). For pure chemical species B^0 and B^1 are functions of reduced temperature only. Substitution for B in Eq. (4-69) by this expression gives

$$Z=1+(B^0+\omega B^1)\frac{P_r}{T_r} \tag{4-73}$$

where $T_r=T/T_c$ and $P_r=P/P_c$ are the *reduced temperature* and *reduced pressure.* Detailed discussion of B^0 and B^1 and their derivatives is given in Elliott and Lira (2012, p. 259):

$$B^0=0.083-\frac{0.422}{T_r^{1.6}} \tag{4-74}$$

$$B^1=0.139-\frac{0.172}{T_r^{4.2}} \tag{4-75}$$

Substituting into Eqs. (4-60) and (4-62) and integrating give

$$\frac{H^{\text{dep}}}{RT}=-P_r\left(\frac{1.0972}{T_r^{2.6}}-\frac{0.083}{T_r}+\omega\left(\frac{0.8944}{T_r^{5.2}}-\frac{0.139}{T_r}\right)\right) \tag{4-76}$$

$$\frac{S^{\text{dep}}}{R}=-P_r\left(\frac{0.675}{T_r^{2.6}}+\omega\frac{0.722}{T_r^{5.2}}\right) \tag{4-77}$$

$$\frac{G^{\text{dep}}}{RT}=\frac{H^{\text{dep}}}{RT}-\frac{S^{\text{dep}}}{R} \tag{4-78}$$

Although limited to pressures where the two-term virial equation in pressure has approximate validity, these correlations are applicable for many chemical processing conditions. The second virial equation is reliable at higher pressures when the temperature is also higher in accordance with the following guideline:

$$T_r>0.686+0.439P_r \quad \text{or} \quad V_r>2.0 \tag{4-79}$$

Values for the cross coefficients B_{ij}, with $i \neq j$, and their derivatives are provided by Eq. (4-72) written in extended form:

$$B_{ij}=\frac{RT_{cij}}{P_{cij}}(B^0+\omega_{ij}B^1) \tag{4-80}$$

where B^0 and B^1 are the same functions of T_r as given by Eqs. (4-74) and (4-75), and $T_{rij}=T/T_{cij}$. The combining rules for ω_{ij}, T_{cij}, and P_{cij} are given by Elliott and Lira (2012, p. 580).

A primary merit of Eqs. (4-74) and (4-75) for second virial coefficients is simplicity. Generalized correlations for B are given by Meng, Duan, and Li [*Fluid Phase Equilibr.* **226:**109–120 (2004)]. More complex correlations of somewhat wider applicability include those by Tsonopoulos [*AIChE J.* **20:** 263–272 (1974); **21:** 827–829 (1975); **24:** 1112–1115 (1978); *Adv. in Chemistry Series* 182, pp. 143–162 (1979)]. For polar and associating molecules, the correlation of Hayden and O'Connell [*Ind. Eng. Chem. Proc. Des. Dev.* **14:** 209–216 (1975)] is generally preferred. For aqueous systems, see Bishop and O'Connell [*Ind. Eng. Chem. Res.* **44:** 630–633 (2005)].

Extended Virial and Multiparameter Equations Another class of equations, known as an *extended* virial equation, was introduced by Benedict, Webb, and Rubin [*J. Chem. Phys.* **8:** 334–345 (1940); **10:** 747–758 (1942)]. This equation contains eight parameters, all functions of composition. It and its modifications, despite their complexity, find application in the petroleum and natural gas industries for light hydrocarbons and a few other commonly encountered gases [Lee and Kesler, *AIChE J.* **21:** 510–527 (1975)].

Similar in spirit to the Benedict, Webb, and Rubin (BWR) model, highly accurate equations can be developed when extensive experimental data are available. These equations are generally written as density expansions of the Helmholtz energy that may involve up to 54 parameters. For example, the IAPWS equation [Wagner, W., and A. Pruss, *J. Phys. Chem. Ref. Data* **31:** 387–535 (2002)] for the properties of steam applies this approach. Solution for the compressibility factor and internal energy can be obtained by differentiating the Helmholtz energy according to

$$Z-1=\frac{\rho}{RT}\left(\frac{\partial(A-A^{\text{ig}})_{T,V}}{\partial\rho}\right)_T \tag{4-81}$$

$$\frac{U-U^{\text{ig}}}{RT}=\frac{1}{RT}\left(\frac{\partial((A-A^{\text{ig}})_{T,V}/T)}{\partial(1/T)}\right)_\rho \tag{4-82}$$

The NIST Chemistry Webbook [E. W. Lemmon, M. O. McLinden, and D. G. Friend, "Thermophysical Properties of Fluid Systems," in NIST Standard Reference Database 69, eds. W. G. Mallard and P. J. Linstrom, http://webbook.nist.gov, Gaithersburg, MD, 2016 (retrieved Nov. 8, 2016)] implements the IAPWS equation and similar equations for roughly 100 compounds common in natural gas and refrigeration industries. The relatively small list of compounds for which multiparameter equations exist has been somewhat limiting for these types of models. Recent progress has expanded this list considerably, however, with the promise of greater expansions in the near future. Another traditional limitation has been the extension to mixtures, but similar recent progress has established the GERG-2008 model as a viable method for high-accuracy treatment of streams related to the natural gas industry [O. Kunz and W. Wagner, *J. Chem. Eng. Data* **57:** 3032 (2012)]. It is likely that highly accurate equations will become available for 200 to 300 pure compounds and nonpolar mixtures within the next 5 to 10 years.

Cubic Equations of State The modern development of cubic equations of state started in 1949 with publication of the Redlich-Kwong (RK) equation [*Chem. Rev.* **44:** 233–244 (1949)], and many others have since been proposed. An extensive review is given by Valderrama [*Ind. Eng. Chem. Res.* **42:** 1603–1618 (2003)]. Of the equations published more recently, the two most popular are the Soave-modified RK (SRK) equation [*Chem. Eng. Sci.* **27:** 1197–1203 (1972)] and the Peng-Robinson (PR) equation [*Ind. Eng. Chem. Fundam.* **15:** 59–64 (1976)]. Since these two are functionally equivalent, the present discussion focuses arbitrarily on the PR model

$$P=\frac{RT}{V-b}-\frac{a(T)}{(V^2+2Vb-b^2)},\quad Z=\frac{1}{1-b\rho}-\frac{a}{bRT}\frac{b\rho}{[1+2b\rho-(b\rho)^2]} \tag{4-83}$$

where parameters $a(T)$ and b are substance-dependent.

$$a(T)=0.45723553\frac{\alpha(T_r)R^2T_c^2}{P_c} \tag{4-84}$$

$$b=0.0777960\frac{RT_c}{P_c} \tag{4-85}$$

Function $\alpha(T_r)$ is an empirical expression specific to a particular form of the equation of state.

$$\alpha \equiv \left[1+\kappa\left(1-\sqrt{T_r}\right)\right]^2 \qquad \kappa \equiv 0.37464+1.54226\omega-0.26992\omega^2$$

As an equation cubic in V or ρ, Eq. (4-83) has three roots, of which two may be complex numbers. Physically meaningful values of V are always real numbers, positive and greater than parameter b. The quantity $b\rho$ is effectively the packing fraction and must range between 0 and 1.0. When $T > T_c$, solution at any positive value of P yields only one real positive root. When $T = T_c$, this is also true, except at the critical pressure, where three roots exist, all equal. For $T < T_c$, only one real positive (liquid-like) root exists at high pressures, but for a range of lower pressures there are three. Here, the middle root is of no significance; the smallest root is a liquid or liquid-like volume, and the largest root is a vapor or vapor-like volume. In principle, cubic equations have the advantage that they can be solved analytically. This may be convenient for some calculations, but most process simulators apply iterative solution. Reasons are that the analytical solution may have round-off errors for the liquid root at low temperatures [R. Monroy-Loperena, *Ind. Eng. Chem. Res.* **51:** 6972 (2012)], and also because iterative Newton-like methods enable avoiding trivial solutions through the use of the pseudo-root technique, as described on page 4-25.

Cubic equations of state may be applied to mixtures through expressions that give the parameters as functions of composition. No established theory strictly prescribes the form of this dependence, and empirical *mixing rules* are often used to relate mixture parameters to pure-species parameters. The simplest realistic expressions (known as van der Waal's mixing rules) are a linear mixing rule for parameter b and a quadratic mixing rule for parameter a

$$b = \sum_i x_i b_i \tag{4-86}$$

$$a = \sum_i \sum_j x_i x_j a_{ij} \tag{4-87}$$

with $a_{ij} = a_{ji}$. The a_{ij} are of two types: pure-species parameters (identical subscripts) and interaction parameters (unlike subscripts). Parameter b_i is for pure species i. The interaction parameter a_{ij} is often evaluated from pure-species parameters by a geometric mean *combining rule* known as the *Lorentz-Berthelot rule*

$$a_{ij} = (a_i a_j)^{1/2}(1 - k_{ij}) \tag{4-88}$$

where k_{ij} is an empirical binary parameter that should be fit to experimental data. These traditional equations yield mixture parameters solely from parameters for the pure constituent species. They are most likely to be satisfactory for mixtures composed of simple and chemically similar molecules. Because cubic equations provide reasonable results for nonpolar mixtures, and they have been available since the mid-1970s, they have become the workhorses for chemical process modeling. In cases where their deficiencies are unacceptable, customized empirical adaptations are generally developed. This leads to some fracturing of modeling efforts as specialists in different companies make the adaptations and they become private. Over the long term, however, there is a tendency for the more accurate adaptations to find their way into process simulators. In particular, the modified Huron-Vidal [cf. M. Michelsen, *Fluid Phase Equilibr.* **60:** 213 (1990)] and Wong-Sandler [*AIChE J.* **38:** 671 (1992)] mixing rules are similar in flexibility to the activity models discussed below, while maintaining the applicability of equations of state to dense, near-critical fluids.

One desirable feature of the cubic EOSs is the simplicity of their working equations for departure functions. As an example, the departure functions for the Peng-Robinson EOS are

$$\frac{H^{\text{dep}}}{RT} = Z - 1 - \frac{A}{B\sqrt{8}}\left(1+\frac{\kappa\sqrt{T_r}}{\sqrt{\alpha}}\right)\ln\left[\frac{Z+(1+\sqrt{2})B}{Z+(1-\sqrt{2})B}\right] \tag{4-89}$$

$$\frac{S^{\text{dep}}}{R} = \ln(Z-B) - \frac{A}{B\sqrt{8}}\frac{\kappa\sqrt{T_r}}{\sqrt{\alpha}}\ln\left[\frac{Z+(1+\sqrt{2})B}{Z+(1-\sqrt{2})B}\right] \tag{4-90}$$

$$\frac{G^{\text{dep}}}{RT} \equiv \ln\phi = Z - 1 - \ln(Z-B) - \frac{A}{B\sqrt{8}}\ln\left[\frac{Z+(1+\sqrt{2})B}{Z+(1-\sqrt{2})B}\right] \tag{4-91}$$

where the parameters are made dimensionless

$$A \equiv aP/(RT)^2 \qquad B \equiv bP/(RT) \tag{4-92}$$

and for a mixture, the parameters of Eqs. (4-86) to (4-88) are similarly made dimensionless. Do not confuse A with Helmholtz energy, nor confuse B with the virial coefficient.

Example 4-7 Estimating Enthalpy Using the PR EOS Compute the enthalpies (kJ/kg) of saturated vapor and liquid methane and the enthalpy of vaporization, using the PR EOS at 115 K, and compare to the values given in the NIST Webbook. Also compare the heat of vaporization computed by the shortcut equation. Use the ideal gas elements at 25°C and 1 bar as the reference state for the PR EOS.

Solution The density is most easily solved by rearranging the PR EOS in terms of Z:

$$Z = \frac{1}{1-b\rho} - \frac{a}{bRT}\frac{b\rho}{[1+2b\rho-(b\rho)^2]} = \frac{1}{1-B/Z} - \frac{A}{B}\frac{B/Z}{[1+2B/Z-(B/Z)^2]} \tag{4-93}$$

Cross-multiplying and collecting terms give

$$Z^3 - Z^2(1-B) + Z(A - 3B^2 - 2B) - (AB - B^2 - B^3) = 0 \tag{4-94}$$

In principle, solving for vapor pressure using an EOS involves trial and error. We can take Eq. (4-16) as an initial guess, giving $P^{\text{sat}} \approx 0.130$ MPa. This leads to $A = 0.04178$ and $B = 0.003627$. Solving the cubic equation with an initial guess of $Z = 1$ gives $Z^V = 0.9606$ and $f^V = 0.1246$ MPa. Solving with an initial guess of $Z = 0$ gives $Z^L = 0.004627$ and $f^L = 0.1280$ MPa. Iterating on P^{sat} to obtain $f^V = f^L$ gives $P^{\text{sat}} = 0.1332$, with $\rho^V = 0.002329$ g/cm^3, $(H^{\text{dep}})^V = -93.63$ J/mol; $f^V = 0.1280$ MPa; $\rho^L = 0.4695$ g/cm^3, $(H^{\text{dep}})^L = -8200.51$ J/mol; $f^L = 0.1280$ MPa. A polynomial form for methane is $C_P^{ig} = 19.25 + 0.05213T + 1.197(10^{-5})T^2 - 1.132(10^{-8})T^3$. Noting that $\Delta H_f^\circ = -74{,}893.6$ J/mol and applying Eq. (4-26), we have $H^V = -5022.61$ and $H^L = -5528.05$ kJ/kg. Taking the difference gives $\Delta H^{\text{vap}} = 505.4$ kJ/kg. This compares to 506.4 from Eq. (4-17). The NIST Webbook gives $P = 0.1322$ MPa, $H^L = 11.687$, and $H^V = 516.28$, leading to $\Delta H^{\text{vap}} = 504.6$ kJ/kg. Using the Webbook as a basis for comparison, the PR EOS gives a 0.76 percent deviation in P^{sat} and 0.17 percent in H^{vap}. Equations (4-16) and (4-17) give 2.0 percent and 0.34 percent deviations. Example 4-7 gives a 0.040 percent deviation in ΔH^{vap}. Several notes can be made about these results:

1. These deviations pertain to comparisons at only a single point. Similar comparisons at 100 K give 1.2 percent and 0.087 percent deviations in P^{sat} and ΔH^{vap} for the PR EOS relative to 2.4 percent and 1.8 percent for Eqs. (4-16) and (4-17), respectively. More comparisons at 175 K give 0.81 percent and 3.6 percent deviations in P^{sat} and ΔH^{vap} for the PR EOS relative to 2.2 percent and 0.53 percent for Eqs. (4-16) and (4-17).

2. For whatever model an engineer may be using, the model estimates should be validated against the experiment. When a multiparameter EOS is available, the NIST results can be relied upon as accurate characterizations of experimental data. In general, NIST's ThermoLit resource [http://trc.nist.gov/thermolit/main/home.html#home] provides a reliable summary of the available experimental literature. The above example illustrates the procedure for validating models for P^{sat} and ΔH^{vap}. It is simply a summary of deviations over the conditions' range of interest.

3. The PR EOS generally provides superior accuracy relative to Eqs. (4-16) and (4-17). This might be more apparent if our comparison were based on a component other than methane, which played a substantial role in the development of Eqs. (4-16) and (4-17).

4. Equations (4-16) and (4-17) provide reasonable estimates, especially for methane. Equation (4-16) is exact at $T_r = 0.7$ and 1.0 because it is a linear interpolation between these two points, so comparisons at these conditions (133 K and 191 K for methane) would make the model look uncharacteristically accurate.

5. When using Eq. (4-17) with (4-27), the accuracy of H^L depends on inclusion of the $(H^{\text{dep}})^V$ even when Eq. (4-17) is accurate. While it might be possible to provide a shortcut estimate of $(H^{\text{dep}})^V$, the PR EOS is reliable and readily available in process simulators. Shortcut estimates should only be used as checks that can be performed with hand calculations.

6. The values of enthalpy from the various sources cannot be compared directly. For example, the values of H^V at 115 K are −5022.61 kJ/kg by the PR EOS, and 516.28 on the Webbook. These large discrepancies are due to different reference states. If interest is limited to a single component, then the reference state can be chosen arbitrarily, but that would be a poor practice in the general case of multicomponent process simulations.

7. Solving for vapor pressure of an EOS requires iteration until the fugacities of vapor and liquid are equal.

8. Incorporation of the PR EOS into manual calculations is facilitated by software available at websites such as CheThermo.net. The PREOS.xls workbook was used for the computations illustrated here.

Pitzer (Lee-Kesler) Correlations In addition to the corresponding-states correlation for the second virial coefficient, Pitzer and coworkers [*Thermodynamics*, 3d ed., App. 3, McGraw-Hill, New York, 1995] developed a full set of generalized correlations. They have as their basis an equation for the compressibility factor, given by

$$Z = Z^0 + \omega Z^1 \tag{4-95}$$

where Z^0 and Z^1 are each functions of reduced temperature T_r and reduced pressure P_r. The acentric factor ω is defined by Eq. (4-14).

Pitzer's original correlations for Z and the derived quantities were determined graphically and presented in tabular form. Since then, analytical refinements to the tables have been developed, with extended range and accuracy. The most popular Pitzer-type correlation is that of Lee and Kesler [*AIChE J.* **21:** 510–527 (1975)]; the advent of computers has made the original tabular and graphical implementations obsolete.

Although the Pitzer correlations are based on data for pure materials, they may also be used for the calculation of mixture properties. A set of recipes is required relating the parameters T_c, P_c, and ω for a mixture to the pure-species values and to composition. One such set is given by Eqs. (2-80) through (2-82) in the Seventh Edition of *Perry's Chemical Engineers' Handbook* (1997). These equations define *pseudoparameters*, so called because the defined values of T_{pc}, P_{pc}, and ω_{pc} have no physical significance for the mixture. The Lee-Kesler correlations provide reliable data for nonpolar and slightly polar fluids; errors of less than 3 percent are likely. Larger errors can be expected in applications to highly polar and associating fluids.

Wertheim's Theory and SAFT Equations of State The reader may have noticed caveats pertaining to polar molecules for all the models mentioned so far. To clarify, the term *polar* is generally applied to molecules that may be either mildly polar, such as CO_2, or associating, such as H_2O. A particularly common form of association is hydrogen bonding, which occurs in H_2O, alcohols, aldehydes, some amides, and some amines. The primary distinction between association and mild polarity is that association leads to specific orientations between molecules where interactions are quite strong, while polarity leads to a broader distribution of orientations that are generally favored. For example, inaccuracies in the PR EOS may be small for mildly polar pure fluids, but larger for associating fluids. The inaccuracies might be much larger when mixing mildly polar fluids with associating fluids because polarity without association is generally correlated with strong asymmetry in either acid or base character. The statistical associating fluid theory (SAFT) family of equations was developed to address limitations related to molecular polarity [W. G. Chapman, K. E. Gubbins, G. Jackson, and M. Radosz, *Fluid Phase Equilibr.* **52:** 31 (1989)]. Many implementations of this theory have been developed since the 1990s. A recent review [S. P. Tan, H. Adidharma, and M. Radosz, *Ind. Eng. Chem. Res.* **47:** 8063 (2008)] concluded that PC-SAFT [J. Gross and G. Sadowski, *Ind. Eng. Chem. Res.* **40:** 1244 (2001)] provided a reasonable representation of what can generally be achieved. The SAFT EOSs are best expressed in terms of the Helmholtz energy:

$$\frac{(A-A^{ig})_{T,V}}{RT}=\frac{(A_0-A^{ig})_{T,V}}{RT}+\frac{A_1}{T}+\frac{A_2}{T^2}+A^{assoc} \tag{4-96}$$

The equations for A_0, A_1, A_2, and A^{assoc} are semiempirical in the sense that their qualitative behavior has been validated with comparison to molecular simulation data.

The association term was specifically developed based on Wertheim's theory, a rigorous theory for associating molecules [M. S. Wertheim, *J. Stat. Phys.* **35:** 19 (1984)]. Wertheim's theory is equivalent to modeling association as weak chemical reactions under certain conditions [J. R. Elliott, S. J. Suresh, M. D. Donohue, *Ind. Eng. Chem. Res.* **29:** 1476, (1990), A. M. Bala, C. T. Lira, *Fluid Phase Equilibr,* **430:** 47 (2016)], but Wertheim's use of site balances rather than species balances facilitates more general application. Molecules that are polar but not strictly associating can also be approximated with this theory. The significance of having a close relationship between the equation of state and the rigors of molecular simulation is that the firm theoretical basis provides insights into the proper mixing rules. Another advantage of Wertheim's theory is realized when taking the limit of the association energy to infinity. A reasonable model of a covalently bonded chain is obtained, mimicking polymeric species, again with validated behavior relative to molecular simulation.

Wertheim's theory has also been implemented to achieve a smooth transition between cubic equations and an association model. The Cubic Plus Association (CPA) model applies the SRK model for nonassociating species and an adaptation of Wertheim's theory for associating species [G. M. Kontogeorgis, M. L. Michelsen, G. K. Folas, S. Derawi, N. von Solms, and E. H. Stenby, *Ind. Eng. Chem. Res.* **45:** 4869 (2006)]. This has the advantage of carrying over accumulated expertise based on the SRK model while gaining the benefits of Wertheim's theory when necessary. A related alternative is the ESD model [S. J. Suresh, J. R. Elliott, *Ind. Eng. Chem. Res.* **31:** 2783 (1992)], which naturally reduces to a cubic equation in the absence of association. When molecules in a mixture have similar site types such that the geometric mean of the association constants can be used for cross-interactions, computational efficiency can be improved [J. R. Elliott, *Ind. Eng. Chem. Res.* **35:** p. 1624, (1996)], which sacrifices slightly on generality but is roughly 3 times faster for binary mixtures, and much faster for multicomponent mixtures.

At present, SAFT models are superior to cubic equations for mixtures involving molecules with high molecular weights, above about 1200 g/mol. Readers should be careful to validate the model as implemented in their software of choice. Check that parameters are available for the components of interest, especially the association parameters. As always, include comparison of the model to experimental data for as many systems as are available.

LIQUID-PHASE PROPERTIES

The simplicity and generality of Eq. (4-26) recommend it when properties need to be computed consistently for streams that may contain mixed phases and reactive compositions, and modern equations of state can provide accurate characterizations of vapor-liquid transitions. However, calculation of property changes from one liquid state to another can be based on Eqs. (4-40) and (4-41) where the pressure-dependent contributions are either ignored or treated as small corrections. The main challenge for mixtures is to estimate the properties with greater accuracy than can be obtained from the pathway of Eq. (4-26).

SYSTEMS OF VARIABLE COMPOSITION

The composition of a system may vary because the system is open or because of chemical reactions even in a closed system. The equations developed here apply regardless of the cause of composition changes. The objectives of this analysis are twofold: (1) to enable more accurate estimation of mixed stream properties with Eq. (4-26) through a more detailed treatment of departure functions and heats of mixing and (2) to articulate the necessary relations for evaluating the activities of components in mixtures.

While computations of thermodynamic properties such as U, H, and S dominate in the analysis of processes involving pure compounds, processes involving mixtures tend to focus foremost on computing the equilibrium phase behavior. State conditions such as T, P that lead to a vapor phase at one composition may yield a liquid at another composition, so determining the state(s) of the phase(s) in question is not as straightforward as for single-component systems. Occasionally, paradoxical quantities are encountered, such as the liquid solubility of a "noncondensable" gas. Computation of bulk phase thermodynamic properties is straightforward once the phase behavior has been resolved.

Coverage of mixtures begins with a review of fundamentals. Briefly, the Gibbs energy is minimized at equilibrium, suggesting the importance of derivative properties. This leads to the formulation of phase and reaction equilibrium criteria. These criteria are general, but they require models of the Gibbs energy for implementation. The two classes of models most commonly used are equations of state and activity models. Activity models are quite successful for modeling in most common industrial situations. Therefore, we cover first activity models and then EOSs. We return to Henry's law after EOSs to facilitate discussion of how gaseous species are treated in the two approaches.

CHEMICAL POTENTIAL

For an *open* single-phase system, we add the dependence of energy on composition, $nU = \underline{U}(nS, nV, n_1, n_2, n_3, \ldots)$. In consequence,

$$d(nU)=T\,d(nS)-P\,d(nV)+\sum_i\left[\frac{\partial(nU)}{\partial n_i}\right]_{nS,nV,n_{j\neq i}}dn_i \tag{4-97}$$

where the summation is over all species present in the system and subscript $n_{j\neq i}$ indicates that all mole numbers are held constant except the ith. Equation (4-97) is the *fundamental property relation* for mixed single-phase *PVT* systems, from which all other equations connecting properties of such systems are derived. The partial derivative in Eq. (4-97) has special significance for phase equilibria in mixtures, and it is called the *chemical potential.* Treatment of the other basic properties H, A, and G results in similar relations, the most important of which is

$$d(nG)=-nS\,dT+nV\,dP+\sum_i\left(\frac{\partial(nG)}{\partial n_i}\right)_{T,P,n_{j\neq i}}dn_i \tag{4-98}$$

where chemical potential is given equivalently by

$$\mu_i\equiv\left(\frac{\partial(nU)}{\partial n_i}\right)_{nS,nV,n_{j\neq i}}=\left(\frac{\partial(nG)}{\partial n_i}\right)_{T,P,n_{j\neq i}} \tag{4-99}$$

PARTIAL MOLAR PROPERTIES

For a homogeneous *PVT* system composed of any number of chemical species, let symbol *M* represent the molar value of an extensive thermodynamic property, say, *U, H, S, A,* or *G*. The extensive quantity can be expressed as

$$nM = \underline{M}(T, P, n_1, n_2, n_3, \ldots)$$

Their derivatives at constant *T, P* and all *n* except *i* are given the generic symbol $\bar{M}_i$ and are defined as a *partial molar property* by

$$\bar{M}_i \equiv \left[\frac{\partial(nM)}{\partial n_i}\right]_{T,P,n_{j\neq i}} \tag{4-100}$$

where the derivative constraints are part of the definition. A result of this definition is that the molar property can be obtained by

$$M = \sum_i x_i \bar{M}_i \quad \text{and} \quad nM = \sum_i n_i \bar{M}_i \tag{4-101}$$

The definition of a partial molar quantity can be applied to all intensive properties yielding the partial-property relations

$$\bar{H}_i = \bar{U}_i + P\bar{V}_i \qquad \bar{A}_i = \bar{U}_i - T\bar{S}_i \qquad \bar{G}_i = \bar{H}_i - T\bar{S}_i$$

These equations illustrate the parallelism that exists between the equations for a constant-composition solution and those for the corresponding partial properties. This parallelism exists whenever the solution properties in the parent equation are related linearly (in the algebraic sense). The partial molar Gibbs energy should be recognized as the *chemical potential*, μ.

The Gibbs-Duhem Equation Partial molar quantities must satisfy the Gibbs-Duhem relation [cf. Tester and Modell (1997)]

$$\left(\frac{\partial M}{\partial T}\right)_{P,x} dT + \left(\frac{\partial M}{\partial P}\right)_{T,x} dP - \sum_i x_i \, d\bar{M}_i = 0 \tag{4-102}$$

Frequently, the first two terms are small, resulting in the approximate relation

$$\sum_i x_i \, d\bar{M}_i = 0 \qquad (\text{constant } T, P) \tag{4-103}$$

PROPERTIES OF IDEAL GAS MIXTURES

The ideal gas mixture model is useful because it is molecularly based, is analytically simple, is realistic in the limit of zero pressure, and provides a conceptual basis for solution thermodynamics. A simple molar average applies for internal energy, *U*, and volume, *V*, and thus enthalpy, $H = U + PV$,

$$M^{\text{ig}} = \sum_i y_i \, M_i^{\text{ig}} \tag{4-104}$$

where M^{ig} can represent *U, V,* or *H*. For the entropy an additional term is required to account for the distinguishability of species in a mixture:

$$S^{\text{ig}} = \sum_i y_i \, S_i^{\text{ig}} - R \sum_i y_i \ln y_i \tag{4-105}$$

For the Gibbs energy, $G^{\text{ig}} = H^{\text{ig}} - TS^{\text{ig}}$ whence by Eqs. (4-104) and (4-105)

$$G^{\text{ig}} = \sum_i y_i \, G_i^{\text{ig}} + RT \sum_i y_i \ln y_i \tag{4-106}$$

The ideal gas model may serve as a reasonable approximation to reality under conditions indicated by Eq. (4-21), where molar averages are applied to the critical properties.

Chemical potential for an ideal gas is obtained by applying Eq. (4-100)

$$\mu_i^{\text{ig}} \equiv \bar{G}_i^{\text{ig}} = G_i^{\text{ig}} + RT \ln y_i \tag{4-107}$$

Elimination of G_i^{ig} from this equation is accomplished through Eq. (4-38), written for pure species *i* as an ideal gas at the temperature of the system:

$$dG_i^{\text{ig}} = V_i^{\text{ig}} \, dP = \frac{RT}{P} dP = RT \, d \ln P \qquad (\text{constant } T)$$

Integration at constant temperature and standard state pressure $P°$ gives

$$G_i^{\text{ig}} = \Gamma_i(T) + RT \ln P \tag{4-108}$$

where the integration constant $\Gamma_i(T)$ includes $RT \ln P°$ and is a function of temperature and standard state pressure only. Equation (4-107) now becomes

$$\mu_i^{\text{ig}} = \bar{G}_i^{\text{ig}} = \Gamma_i(T) + RT \ln(y_i P) \tag{4-109}$$

leading to

$$G^{\text{ig}} = \sum_i y_i \Gamma_i(T) + RT \sum_i \ln(y_i P) \tag{4-110}$$

A dimensional ambiguity is implied with Eqs. (4-108) through (4-110) in that *P* has units, whereas ln *P* must be dimensionless. Although the units cancel with the standard state pressure, in practice this is of no consequence, because only *differences* in Gibbs energy appear, along with *ratios* of the quantities with units of pressure in the arguments of the logarithm. Consistency in the units of pressure is, of course, required; if the standard state is 1 bar, use bars for all computations involving reactions.

COMPONENT FUGACITY

The definition of the fugacity of a species in solution is parallel to the definition of the pure-species fugacity. An equation analogous to the ideal gas expression, Eq. (4-109), is written for species *i* in a fluid mixture

$$\mu_i \equiv \Gamma_i(T) + RT \ln \hat{f}_i \tag{4-111}$$

where the partial pressure $y_i P$ is replaced by $\hat{f}_i$, the fugacity of species *i* in solution. Because it is not a partial property, it is identified by a circumflex rather than an overbar. The fugacity of an ideal gas component is apparent by comparing Eqs. (4-111) and (4-109):

$$\hat{f}_i^{\text{ig}} = y_i P \tag{4-112}$$

Ideal Solution Model and Henry's Law The ideal gas model is useful as a standard of comparison for real gas behavior. This is formalized through departure functions. The *ideal solution* is similarly useful as a standard to which real solution behavior may be compared and is common for liquid solutions.

The partial molar Gibbs energy or chemical potential of species *i* in an ideal gas mixture is given by Eq. (4-107), written as

$$\mu_i^{\text{ig}} = \bar{G}_i^{\text{ig}} = G_i^{\text{ig}}(T, P) + RT \ln y_i$$

This equation takes on new meaning when $G_i^{\text{ig}}(T, P)$ is replaced by $G_i(T, P)$, the Gibbs energy of pure species *i* in its *real physical state* of gas, liquid, or solid at the mixture *T* and *P*. The ideal solution is therefore *defined* as one for which

$$\mu_i^{\text{id}} = \bar{G}_i^{\text{id}} \equiv G_i(T, P) + RT \ln x_i \tag{4-113}$$

where superscript id denotes an ideal solution property and x_i represents the mole fraction because application is usually to liquids.

This relation requires

$$\bar{V}_i^{\text{id}} = V_i; \qquad \bar{U}_i^{\text{id}} = U_i \tag{4-114}$$

and

$$\bar{S}_i^{\text{id}} = S_i - R \ln x_i \tag{4-115}$$

Because $\bar{H}_i^{\text{id}} = \bar{G}_i^{\text{id}} + T\bar{S}_i^{\text{id}}$, substitutions by Eqs. (4-113) and (4-115) yield

$$\bar{H}_i^{\text{id}} = H_i \tag{4-116}$$

The mixture property can be calculated by the mole-fraction-weighted sum of partial molar properties, Eq. (4-101). For the special case of an ideal solution, incorporating Eqs. (4-113) through (4-116) gives:

$$G^{\text{id}} = \sum_i x_i G_i + RT \sum_i x_i \ln x_i \tag{4-117}$$

$$V^{\text{id}} = \sum_i x_i V_i \tag{4-118}$$

$$S^{\text{id}} = \sum_i x_i S_i - R \sum_i x_i \ln x_i \tag{4-119}$$

$$H^{\text{id}} = \sum_i x_i H_i \tag{4-120}$$

A simple equation for the fugacity of a species in an ideal solution follows. For the special case of species i in an ideal solution, Eq. (4-113) becomes

$$\mu_i^{id} = \bar{G}_i^{id} = \Gamma_i(T) + RT \ln \hat{f}_i^{id} \tag{4-121}$$

When this equation and Eq. (4-111) are combined with Eq. (4-117), $\Gamma_i(T)$ is eliminated, and the resulting expression reduces to

$$\hat{f}_i^{id} = x_i f_i \tag{4-122}$$

This equation, known as the *Lewis-Randall rule*, shows that the fugacity of each species in an ideal solution is proportional to its mole fraction; the proportionality constant is the fugacity of *pure* species i in the same physical state as the solution and at the same T and P. Division of both sides of Eq. (4-122) by $x_i P$ and substitution of $\hat{\phi}_i^{id}$ for $\hat{f}_i^{id}/x_i P$ [Eq. (4-130)] and of ϕ_i for f_i/P [Eq. (4-64)] give the alternative form for equations of state

$$\hat{\phi}_i^{id} = \phi_i \tag{4-123}$$

Thus the fugacity coefficient of species i in an ideal solution equals the fugacity coefficient of *pure* species i in the same physical state as the solution and at the same T and P. Ideal solution behavior is often approximated by solutions composed of molecules not too different in size and of the same chemical nature. Thus, a mixture of isomers conforms very closely to ideal solution behavior. So do mixtures of adjacent members of a homologous series.

An alternative ideal solution results when the standard state is at infinite dilution rather than at purity, which results in the Henry law ideal solution, here written using the Henry volatility constant common in chemical engineering literature:

$$\hat{f}_i^{id} = x_i h_i \tag{4-124}$$

While Eqs. (4-122) and (4-124) hint that h_i and f_i might be the same, they are equal only when a solution follows ideal solution behavior at all compositions, which is rare. Henry's law is named after the English chemist who examined the solubilities of gases in water in the early 19th century [*Phil. Trans. R. Soc. Lond.* **93:** pp. 29 and 274 (1803)].

In casual terms we would state that the concentration of a dissolved solute in a liquid is proportional to its partial pressure in the vapor phase, and the proportionality constant, at a given temperature, is referred to as Henry's "constant" (although it varies with temperature). Formally, Henry's law can be expressed as a limiting fugacity

$$h_i = \lim_{x_i \to 0} \frac{\hat{f}_i}{x_i} \tag{4-125}$$

where $\hat{f}_i$ is the fugacity of component i and x_i is its liquid mole fraction. Application of the rigorous limit is often a problem in practice, and hence other considerations need to be adopted. For example, a reference fluid of pure sodium ions would be impractical when concerned with salt solutions.

Over the past two centuries, scientists and engineers have built upon Henry's seminal discovery to develop a comprehensive theoretical framework and extensive databases for the correlation of solute solubilities over a wide range of temperature, pressure, and liquid- and vapor-phase concentrations. This background is presented in the section following discussion of activity models.

PHASE EQUILIBRIA CRITERIA

The criteria for internal thermal and mechanical equilibrium simply require uniformity of temperature and pressure throughout the system. The criteria for phase equilibria at constant T and P require that the Gibbs energy for the overall system be minimized. For a two-phase system, each phase taken separately is an *open* system, capable of exchanging mass with the other. The criteria for phase equilibria are derived in textbooks. The general result is

$$\mu_i' = \mu_i'' = \mu''' = \cdots \tag{4-126}$$

Substitution for each μ_i by Eq. (4-111) produces the equivalent result:

$$\hat{f}_i' = \hat{f}_i'' = \hat{f}_i''' = \cdots \tag{4-127}$$

These are the criteria of phase equilibrium applied in the solution of practical problems.

For the case of equilibrium with respect to chemical reaction within a single-phase closed system, at constant T and P, Eq. (4-98) simplifies to

$$\sum_i \mu_i \, dn_i = 0 \tag{4-128}$$

For a system in which both phase and chemical reaction equilibrium prevail, the criteria of Eqs. (4-127) and (4-128) are superimposed.

PHASE RULE

The *intensive* state of a *PVT* system is established when its temperature and pressure and the compositions of all phases are fixed. However, for equilibrium states not all these variables are independent, and fixing a limited number of them automatically establishes the others. This number of independent intensive variables is given by the phase rule, and it is called the *number of degrees of freedom* of the system. It is the number of variables that may be arbitrarily specified and that must be so specified in order to fix the *intensive* state of a system at equilibrium. This number is the difference between the number of variables needed to characterize the system and the number of equations that may be written connecting these variables.

For a system containing N chemical species distributed at equilibrium among π phases, the phase rule variables are T and P, presumed uniform throughout the system, and $N - 1$ mole fractions in each phase. The number of these variables is $2 + (N - 1)\pi$. The masses of the phases are not phase rule variables, because they have nothing to do with the intensive state of the system.

The equilibrium equations that may be written to express chemical potentials or fugacities as functions of T, P, the phase compositions, and the phase rule variables:

1. Equation (4-127) for each species, giving $(\pi - 1)N$ phase equilibrium equations
2. Equation (4-128) for each independent chemical reaction, giving r equations

The total number of independent equations is therefore $(\pi - 1)N + r$. Because the degrees of freedom F is the difference between the number of variables and the number of equations,

$$F = 2 + (N - 1)\pi - (\pi - 1)N - r$$

or

$$F = 2 - \pi + N - r \tag{4-129}$$

The number of independent chemical reactions r can be determined as follows:

1. Write *formation* reactions from the elements for each chemical compound present.
2. Combine these reaction equations so as to eliminate from the set all elements not present as elements in the system. A systematic procedure is to select one equation and combine it with each of the other equations of the set so as to eliminate a particular element. This usually reduces the set by one equation for each element eliminated, although two or more elements may be simultaneously eliminated.

The resulting set of r equations is a complete set of independent reactions. More than one such set is often possible, but all sets number r and are equivalent.

Example 4-8 Application of the Phase Rule Consider the following cases.

a. For a system of two miscible nonreacting species in vapor/liquid equilibrium,

$$F = 2 - \pi + N - r = 2 - 2 + 2 - 0 = 2$$

The 2 degrees of freedom for this system may be satisfied by setting T and P, or T and y_1, or P and x_1, or x_1 and y_1, etc., at fixed values. Thus for equilibrium at a particular T and P, this state (if possible at all) exists only at one liquid and one vapor composition. Once the 2 degrees of freedom are used up, no further specification is possible that would restrict the phase rule variables. For example, one cannot *in addition* require that the system form an azeotrope (assuming this is possible), for this requires $x_1 = y_1$, an equation not taken into account in the derivation of the phase rule. Thus the requirement that the system form an azeotrope imposes a special constraint, making $F = 1$.

b. For a gaseous system consisting of CO, CO_2, H_2, H_2O, and CH_4 in chemical reaction equilibrium,

$$F = 2 - \pi + N - r = 2 - 1 + 5 - 2 = 4$$

The value of $r = 2$ is found from the formation reactions:

$$C + \tfrac{1}{2}O_2 \rightarrow CO \qquad C + O_2 \rightarrow CO_2$$

$$H_2 + \tfrac{1}{2}O_2 \rightarrow H_2O \qquad C + 2H_2 \rightarrow CH_4$$

Systematic elimination of C and O_2 from this set of chemical equations reduces the set to 2. Three possible pairs of equations may result, depending on how the combination of equations is effected. Any *pair* of the following three equations represents a complete set of independent reactions, and all pairs are equivalent.

$$CH_4 + H_2O \rightarrow CO + 3H_2$$

$$CO + H_2O \rightarrow CO_2 + H_2$$

$$CH_4 + 2H_2O \rightarrow CO_2 + 4H_2$$

The result, $F = 4$, means that one is free to specify, for example, T, P, and two mole fractions in an equilibrium mixture of these five chemical species, provided nothing else is arbitrarily set. Thus it cannot simultaneously be required that the system be prepared from specified amounts of particular constituent species.

APPROACHES FOR PHASE AND REACTION EQUILIBRIA MODELING

Component Fugacity and Activity Coefficients While Eqs. (4-126) to (4-128) form the basis of phase and chemical equilibria, they do not dictate the methods to be used to calculate the properties of the phases. Acknowledging that equilibria can be expressed in terms of chemical potential or fugacity, chemical engineering practice has evolved to use fugacity. We express the fugacity of the real gas or liquid relative to one of the idealized states (ideal gas, Lewis-Randall ideal solution, or Henry's law ideal solution). When the fugacity is calculated relative to the ideal gas, the departure function is used, resulting in the component *fugacity coefficient.* Subtracting Eq. (4-107) from Eq. (4-111), both written for the same temperature, pressure, and composition, yields after including (4-108)

$$\mu_i - \mu_i^{ig} = RT \ln \frac{\hat{f}_i}{y_i P} \equiv RT \ln \hat{\phi}_i \tag{4-130}$$

where by definition $\hat{\phi}_i \equiv \hat{f}_i / (y_i P)$, or

$$\hat{f}_i = y_i \hat{\phi}_i P \tag{4-131}$$

The dimensionless ratio $\hat{\phi}_i$ is called the *fugacity coefficient of species i in solution.*

Using the Lewis-Randall rule, typically for condensed phases and thus the use of x gives

$$\hat{f}_i^{\text{cond}} = x_i \gamma_i f_i^{\text{cond}} = x_i \gamma_i \phi_i^{\text{sat}} P_i^{\text{sat}} \exp\left(\frac{V_i^{\text{cond}}(P - P_i^{\text{sat}})}{RT}\right) \tag{4-132}$$

where the activity coefficient γ_i characterizes the deviations from an ideal solution and f_i^{cond} is given by Eq. (4-67). For Henry's law volatility constant, typically used for liquid phases,

$$\hat{f}_i^L = h_i(T, P_i^o) x_i \gamma_i^* \exp\left(\int_{P_i^o}^{P} \frac{\overline{V}_i^\infty}{RT} dP\right) \tag{4-133}$$

The activity coefficient γ_i^* of Eq. (4-133) is related, but not equal to the activity coefficient γ_i in Eq. (4-132) as discussed later in Eq. (4-199).

For the EOS approach, we use Eq. (4-131) for both vapor and liquid phases, resulting in

$$\hat{\phi}_i^V y_i = \hat{\phi}_i^L x_i \tag{4-134}$$

This introduces compositions x_i and y_i into the equilibrium equations, but neither is explicit, because the $\hat{\phi}_i$ are functions of composition as well as T and P. Thus, Eq. (4-134) represents N complex relationships connecting T, P, $\{x_i\}$, and $\{y_i\}$. The EOS approach is typically successful for nonpolar substances and must be used when fluids are near critical points. Polar substances can be modeled by including association effects or by use of sophisticated mixing rules.

For polar substances the gas phase may be modeled with the EOS approach, while the liquid phase is modeled with deviations from the Lewis-Randall rule. The fugacity of species i in the liquid phase is given by Eq. (4-132), and the vapor-phase fugacity is given by Eq. (4-131). By Eqs. (4-127) and (4-132) the relation becomes

$$\gamma_i x_i \phi_i^{\text{sat}} P_i^{\text{sat}} \exp(V^L(P - P_i^{\text{sat}})/(RT)) = \hat{\phi}_i y_i P \qquad i = 1, 2, \ldots, N \tag{4-135}$$

Identifying superscripts L and V are omitted here with the understanding that γ_i is a liquid-phase property, whereas $\hat{\phi}_i$ is a vapor-phase property. Applications of Eq. (4-135) represent the *gamma/phi* approach to VLE calculations, generally applicable below 10 bar.

For Henry's law, we use Eq. (4-131) for the vapor phase and Eq. (4-133) for the liquid phase,

$$\hat{\phi}_i y_i P = h_i(T, P_i^o) x_i \gamma_i^* \exp\left(\int_{P_i^o}^{P_i} \frac{\overline{V}_i^\infty}{RT} dP\right) \tag{4-136}$$

When Henry's law is applied, it is common to use it for some of the components (normally noncondensable components) while using the Lewis-Randall approach, Eq. (4-135), for the remainder of components.

Excess Properties An *excess property* M^E is defined as the difference between the actual property value of a solution and the value it would have as an ideal solution at the same T, P, and composition. Thus,

$$M^E \equiv M - M^{\text{id}} \tag{4-137}$$

where M represents the molar (or unit-mass) value of any extensive thermodynamic property (say, V, U, H, S, G). This definition is similar to the definition of a departure function as given by Eq. (4-52). However, excess properties have no meaning for pure species, whereas departure functions exist for pure species as well as for mixtures. Partial molar excess properties $\overline{M}_i^E$ are defined analogously:

$$\overline{M}_i^E = \overline{M}_i - \overline{M}_i^{\text{id}} \tag{4-138}$$

Of particular interest is the partial molar excess Gibbs energy. Rewriting Eq. (4-111) as

$$\overline{G}_i = \Gamma_i(T) + RT \ln \hat{f}_i$$

in accord with Eq. (4-122) for an ideal solution, this becomes

$$\overline{G}_i^{\text{id}} = \Gamma_i(T) + RT \ln x_i f_i$$

By differences
$$\overline{G}_i - \overline{G}_i^{\text{id}} = RT \ln \frac{\hat{f}_i}{x_i f_i}$$

The left side is the partial excess Gibbs energy $\overline{G}_i^E$; the dimensionless ratio $\hat{f}_i / x_i f_i$ on the right is the *activity coefficient of species i in solution,* given the symbol γ_i, and *by definition,*

$$\gamma_i \equiv \frac{\hat{f}_i}{x_i f_i} \tag{4-139}$$

Thus,
$$\overline{G}_i^E = RT \ln \gamma_i \tag{4-140}$$

Comparison with Eq. (4-130) shows that Eq. (4-140) relates γ_i to $\overline{G}_i^E$ exactly as Eq. (4-130) relates $\hat{\phi}_i$ to $\overline{G}_i^{\text{dep}}$. For an ideal solution, $\overline{G}_i^E = 0$, and therefore $\gamma_i^{\text{id}} = 1$.

Activity coefficients are key descriptors in the design of chemical separation and reaction operations, liquid product formulations, and other technologies where liquid composition is a key factor affecting performance. In practice, the value of γ_i serves as a correction factor for the solute mole fraction concentration x_i to better account for the solute's true chemical potential–driven activity which determines phase equilibrium behavior and reactivity,

$$\mu_i - G_i^\circ = RT \ln a_i \tag{4-141}$$

where *activity* is given by $a_i \equiv \hat{f}_i / (f_i^\circ)$ and for the Lewis-Randall rule, $a_i = \gamma_i x_i$. For a Lewis-Randall ideal solution, γ_i is unity; each component acts as if surrounded by its own kind, so phase equilibrium properties are determined according to molar concentration or mole fraction ($a_i = x_i$). For other mixtures, the solute's activity coefficient is usually greater than unity, and the solute behaves as if there is more of it present in the mixture than its mole fraction would indicate ($a_i > x_i$, a *positive* deviation from ideality). This may indicate that solute-solvent interactions are repulsive relative to solvent-solvent interactions or may lead to segregation of the mixture (negative entropic effects). For a vapor-liquid system, such a component has an enhanced tendency to escape from the liquid into the vapor. The solute activity coefficient can also be less than unity such that its effective mole fraction in the mixture is reduced ($a_i < x_i$, a *negative* deviation). This behavior may result from mixing molecules that differ greatly in molecular size (an entropic effect) or by exothermic formation of multicomponent molecular structures or complexes in solution. The activity coefficient is used to quantify a deviation from ideal mixture behavior, although often the molecular mechanisms responsible for the observed deviation are not fully understood.

The *infinite-dilution* (or limiting) activity coefficient $\gamma_{ij}^\infty = \lim_{x_i \to 0} \gamma_i$ is a particularly useful quantity because it represents nonideal interactions for a solute i completely surrounded by solvent j. Its value normally is the most extreme γ_i value for a given binary, and so it serves to characterize the nonideality of the mixture. The limiting activity coefficient is related to the partial molar excess Gibbs energy involved in moving a molecule of solute i from its pure liquid reference state into a pool of solvent j molecules:

$$RT \ln \gamma_{ij}^\infty = \bar{G}_{ij}^{E,\infty} = \bar{H}_{ij}^{E,\infty} - T\bar{S}_{ij}^{E,\infty} \tag{4-142}$$

Once a value for γ_{ij}^∞ has been determined, by either experiment or prediction, values at other compositions can be estimated by extrapolation using a suitable correlation equation such as those discussed below. In many cases, knowledge of γ_{ij}^∞ for all binary pairs allows reliable extrapolation to higher concentrations in multicomponent mixed solution—with results suitable for many applications or at least for initial screening studies. In special cases, data for ternary or higher numbers of components in solution may be needed to improve the correlation for final design purposes, especially for systems with unusually strong multicomponent intermolecular interactions or strong association of molecules of the same kind. A database of over 4000 values of γ_{ij}^∞ has been published by Lazzaroni et al. [*Ind. Chem. Eng. Res.* **44:** 4075–4083 (2005)].

Property Changes of Mixing A *property change of mixing* is defined by

$$\Delta M \equiv M - \sum_i x_i M_i \tag{4-143}$$

where M represents a molar thermodynamic property of a homogeneous solution and M_i is the molar property of pure species i at the T and P of the solution and in the same physical state. In addition, ΔG, ΔV, ΔS, and ΔH are the Gibbs energy change of mixing, the volume change of mixing, the entropy change of mixing, and the enthalpy change of mixing, respectively. Applications are usually to liquids.

Each of Eqs. (4-117) through (4-120) is an expression for an ideal solution property, and each may be combined with the defining equation for an excess property. For an ideal solution, each excess property is zero.

Property changes of mixing and excess properties are easily calculated one from the other. The most common property changes of mixing are the volume change of mixing ΔV and the enthalpy change of mixing ΔH, commonly called the *heat of mixing*. These properties are identical to the corresponding excess properties. Moreover, they are directly measurable, providing an experimental entry into the network of equations of solution thermodynamics.

Excess and Departure Property Relations Equations for excess properties are developed in much the same way as those for departure properties. The following equations are in complete analogy to those for departure properties:

$$\frac{V^E}{RT} = \left[\frac{\partial(G^E/RT)}{\partial P}\right]_{T,x} \tag{4-144}$$

$$\frac{H^E}{RT} = -T\left[\frac{\partial(G^E/RT)}{\partial T}\right]_{P,x} \tag{4-145}$$

$$\ln \gamma_i = \left[\frac{\partial n(G^E/RT)}{\partial n_i}\right]_{T,P,n_{j\neq i}} \tag{4-146}$$

This last equation demonstrates that $\ln \gamma_i$ is a partial property with respect to G^E/RT, implying also the sum of mole-fraction-weighted partial molar properties to give the excess Gibbs energy can be written using activity coefficients

$$\frac{G^E}{RT} = \sum_i x_i \ln \gamma_i \tag{4-147}$$

$$\left(\frac{\partial \ln \gamma_i}{\partial P}\right)_{T,x} = \frac{\bar{V}_i^E}{RT} \tag{4-148}$$

$$\left(\frac{\partial \ln \gamma_i}{\partial T}\right)_{P,x} = -\frac{\bar{H}_i^E}{RT^2} \tag{4-149}$$

Equation (4-149) is a version of the Gibbs-Helmholtz equation. For a detailed discussion of the origins of this equation, see Mathias [*Ind. Eng. Chem. Res.* **55:** 1076–1087 (2016)]. Analogous to Eqs. (4-144) and (4-145), we can write

$$\frac{V^{\text{dep}}}{RT} = \left[\frac{\partial(G^{\text{dep}}/RT)}{\partial P}\right]_{T,x} \tag{4-150}$$

$$\frac{H^{\text{dep}}}{RT} = -T\left[\frac{\partial(G^{\text{dep}}/RT)}{\partial T}\right]_{P,x} \tag{4-151}$$

Also implicit in Eq. (4-130) is the relation

$$\ln \hat{\phi}_i = \left[\frac{\partial n(G^{\text{dep}}/RT)}{\partial n_i}\right]_{T,P,n_{j\neq i}} \tag{4-152}$$

This equation demonstrates that $\ln \hat{\phi}_i$ is a partial property with respect to G^{dep}/RT. The sum of the mole-fraction-weighted partial molar properties to give the mixture property, relation (4-101), therefore applies, and

$$\frac{G^{\text{dep}}}{RT} = \sum_i x_i \ln \hat{\phi}_i \tag{4-153}$$

Recognizing $\ln \hat{\phi}_i$ as a partial property leads to

$$\left(\frac{\partial \ln \hat{\phi}_i}{\partial P}\right)_{T,x} = \frac{\bar{V}_i^{\text{dep}}}{RT} \tag{4-154}$$

$$\left(\frac{\partial \ln \hat{\phi}_i}{\partial T}\right)_{P,x} = -\frac{\bar{H}_i^{\text{dep}}}{RT^2} \tag{4-155}$$

Component Fugacity Coefficients from an EOS Equation (4-152) can be applied readily only to departure functions explicit in T and P. For departure functions explicit in T and V, such as a cubic equation of state, an alternative method is used:

$$\ln \hat{\phi}_i = \left[\frac{\partial n(A^{\text{dep}}/RT)}{\partial n_i}\right]_{T,\underline{V},n_{j\neq i}} - \ln Z \tag{4-156}$$

Example 4-9 Derivation of Fugacity Coefficient Expressions Application of Eq. (4-152) to an expression giving G^{dep} as a function of composition yields an equation for $\ln \hat{\phi}_i$. In the simplest case of a gas mixture for which the virial equation [Eqs. (4-69) and (4-71)] is appropriate, Eq. (4-61) provides the relation

$$\frac{nG^{\text{dep}}}{RT} = \frac{P}{RT}(nB)$$

Differentiation in accord with Eqs. (4-152) yields

$$\frac{RT \ln \hat{\phi}_i}{P} = \left[\frac{\partial(nB)}{\partial n_i}\right]_{T,n_{j\neq i}} = \left[\frac{\partial\left(\frac{1}{n}\sum\sum n_k n_j B_{kj}\right)}{\partial n_i}\right]_{T,n_{j\neq i}} = 2\sum_1^n y_j B_{ij} - B \tag{4-157}$$

For EOSs involving $Z(T, V)$ such as cubic EOSs, the differentiation follows Eq. (4-156). For example, when the mixing and combining rules follow Eqs. (4-86) to (4-88) as nondimensionalized by Eq. (4-92), the component fugacity coefficient for the PR EOS is given by

$$\ln \hat{\phi}_i^\alpha = \ln\left(\frac{f_i^\alpha}{y_i P}\right) = \frac{B_i}{B^\alpha}(Z^\alpha - 1) - \ln(Z^\alpha - B^\alpha) - \frac{A^\alpha}{B^\alpha\sqrt{8}}\left(\frac{2\Sigma z_j A_{ij}}{A^\alpha} - \frac{B_i}{B^\alpha}\right)\ln\left[\frac{Z^\alpha + (1+\sqrt{2})B^\alpha}{Z^\alpha + (1-\sqrt{2})B^\alpha}\right] \tag{4-158}$$

where α indicates the phase (that is, V or L) and z_j indicates the mole fraction in that phase (typically y_j for vapor and x_j for liquid). The A here is not Helmholtz energy and B here is not the virial coefficient. Although the formula is the same for all phases, the values of Z^V and Z^L are naturally quite distinct. Similarly, A^V, A^L and B^V, and B^L differ owing to compositions. Standard texts describe this derivation in detail.

CORRELATIVE MODELS FOR THE EXCESS GIBBS ENERGY

Excess properties find application in the treatment of liquid solutions. The excess volume for liquid mixtures is usually small, and in accord with Eq. (4-144) the pressure dependence of G^E is usually ignored. Thus, engineering efforts to model G^E center on representing its composition and temperature dependence. For educational purposes, G^E models such as the Redlich-Kister expansion, Margules models, and the van Laar model are typically covered, but these are simple empirical or semiempirical relations that are best applied to binary systems. Since most realistic applications focus on multicomponent systems, we focus our discussion here on multicomponent G^E models. All of these are typically available in chemical process simulators.

Margules, Wilson, NRTL, UNIQUAC When experimental data are available for a system of interest, correlative models are preferred over predictive models if the quality of the data is high. We provide in later sections some recommendations for assessing data by evaluating trends in homologous series of compounds. Methods used to assess thermodynamic consistency are discussed by Kang et al. [*J. Chem. Eng. Data* **55:** 3631 (2010)]. If pure component vapor pressure data are in error, the thermodynamic consistency tests and reliability of the resulting model for multicomponent mixtures likely will be poor. Two subsections immediately following this one address sources of data and methods of reducing the data to relevant model parameters. Four G^E models are applied to correlation most often, most based on the concepts of local compositions and Lewis-Randall activity coefficients: the Margules model, the Wilson model, the NRTL model, and the UNIQUAC model. To illustrate the general form of each model, in the discussion that follows formulas are listed for the activity coefficients of a binary mixture and evaluated for 1-propanol and water at 120°C as an example.

Margules Equation The *Margules equation* is empirical. It is typically equivalent to the Redlich-Kister expansion in its one- to three-parameter form. The two-parameter Gibbs excess energy and activity coefficients can be written as

$$\frac{G^E}{RT} = x_1 x_2 (A_{21} x_1 + x_2 A_{12}) \tag{4-159}$$

$$\ln \gamma_1 = x_2^2 (A_{12} + 2(A_{21} - A_{12}) x_1) \tag{4-160}$$

$$\ln \gamma_2 = x_1^2 (A_{21} + 2(A_{12} - A_{21}) x_2) \tag{4-161}$$

For a mixture of 20 mol% propanol in water: A_{12} = 1.164; A_{21} = 2.244; γ_1 = 2.777; and γ_2 = 1.021. The Margules two-parameter model reduces to the one-parameter model when $A_{12} = A_{21}$. The most common extension of the Margules model to multicomponent mixtures is Wohl's expansion [cf. Prausnitz, Lichtenthaler, and de Azevedo (1999, Sec. 6.14)]. Extension of polynomial models for G^E to multicomponent mixtures must be done carefully because the expression for G^E must be invariant to division into identical subcomponents [Michelsen and Kistenmacher, *Fluid Phase Equilib.* **58:** 229 (1990); Mathias, Klotz, and Prausnitz, *Fluid Phase Equilib.* **67:** 31 (1991)]—the so-called invariance criterion. The Wohl expansion shown in the reference does not violate the invariance criterion. However, when models with higher-order binary summations are used, they should be evaluated to ensure that they do not violate this criterion.

Theoretical developments in the molecular thermodynamics of liquid solution behavior are often based on the concept of *local composition*, presumed to account for the short-range order and nonrandom molecular orientations resulting from differences in molecular size and intermolecular forces. Introduced by G. M. Wilson [*J. Am. Chem. Soc.* **86:** 127–130 (1964)] with the publication of a model for G^E, this concept prompted the development of alternative local composition models, most notably the NRTL (non-random two-liquid) equation of Renon and Prausnitz [*AIChE J.* **14:** 135–144 (1968)] and the UNIQUAC (UNIversal QUAsi-Chemical) equation of Abrams and Prausnitz [*AIChE J.* **21:** 116–128 (1975)].

Wilson Equation The *Wilson equation* contains just two parameters per binary system (a_{ij} and a_{ji}),

$$\frac{G^E}{RT} = -\sum_i x_i \ln\left(\sum_j x_j \Lambda_{ij}\right) \tag{4-162}$$

$$\ln \gamma_i = 1 - \ln\left(\sum_j x_j \Lambda_{ij}\right) - \sum_k \frac{x_k \Lambda_{ki}}{\sum_j x_j \Lambda_{kj}} \tag{4-163}$$

The temperature dependence of the parameters is estimated by

$$\Lambda_{ij} = \frac{V_j}{V_i} \exp\frac{-a_{ij}}{RT} \qquad i \neq j \tag{4-164}$$

where V_j and V_i are the molar volumes of pure liquids j and i, respectively, and a_{ij} is a constant independent of composition and temperature. Molar volumes V_j and V_i, themselves weak functions of temperature, form ratios that in practice may be taken as independent of T and are usually evaluated at or near 25°C, $\Lambda_{ij} = 1$ for $i = j$, etc. All indices in these equations refer to the same species, and all summations are over *all* species. For each i, j pair there are two parameters, because $\Lambda_{ij} \neq \Lambda_{ji}$. For example, in a ternary system the three possible i, j pairs are associated with the parameters Λ_{12}, Λ_{21}; Λ_{13}, Λ_{31}; and Λ_{23}, Λ_{32}. At infinite dilution for a binary mixture,

$$\ln\gamma_{1,2}^\infty = -\ln\Lambda_{12} + 1 - \Lambda_{21}; \quad \ln\gamma_{2,1}^\infty = -\ln\Lambda_{21} + 1 - \Lambda_{12}$$

By Eq. 4-164, both Λ_{12} and Λ_{21} must be positive numbers. These binary relations may be helpful when inferring values of the parameters from experimental data. For a mixture of 20 percent propanol in water, a_{12} = 3793 J/mol; a_{21} = 5844; V_1 = 18.76; V_2 = 85.71; γ_1 = 2.561; and γ_2 = 1.168. The Wilson equation has a well-known limitation owing to the positive nature of Eq. (4-164); it cannot correlate liquid-liquid equilibria (LLE). This can be an advantage for systems that exhibit large positive deviations from ideality, but do not exhibit LLE. More often it is a disadvantage because most systems with such nonideality do phase-separate at some set conditions.

NRTL Equation The *NRTL equation* contains three parameters for a binary system and is written in multicomponent form as [C. Cohen and H. Renon, *Canadian J. Chem. Eng.* **48:** 291–296 (1970)]

$$\frac{G^E}{RT} = \sum_i x_i \frac{\sum_j x_j \tau_{ji} G_{ji}}{\sum_k x_k G_{ki}} \tag{4-165}$$

$$\ln\gamma_i = \frac{\sum_j x_j \tau_{ji} G_{ji}}{\sum_k x_k G_{ki}} + \sum_j \frac{x_i G_{ij}}{\sum_k x_k G_{kj}}\left(\tau_{ij} - \frac{\sum_m x_m \tau_{mj} G_{mj}}{\sum_k x_k G_{kj}}\right) \tag{4-166}$$

where G and τ are intermediate variables. Here i identifies the species, and j, k, m are dummy variables.

$$G_{ij} = \exp(-\alpha_{ij}\tau_{ij}) \tag{4-167}$$

and $\qquad \alpha_{12} = \alpha_{21} \equiv \alpha$ for a single binary; $\tau_{12} = \dfrac{b_{12}}{RT}$; $\tau_{21} = \dfrac{b_{21}}{RT}$

where α, b_{12}, and b_{21}, parameters specific to a particular pair of species, are independent of composition and temperature. The infinite-dilution values of binary activity coefficients are

$$\ln\gamma_{1,2}^\infty = \tau_{21} + \tau_{12}\exp(-\alpha\tau_{12}) \qquad \ln\gamma_{2,1}^\infty = \tau_{12} + \tau_{21}\exp(-\alpha\tau_{21})$$

For 20 percent propanol in water with α = 0.3: b_{12} = 75.3 J/mol; b_{21} = 7259; γ_1 = 2.772; and γ_2 = 1.131.

UNIQUAC Equation The *UNIQUAC equation* treats G^E/RT as made up of two additive parts, a *combinatorial* term G^{comb}, accounting for molecular size and shape differences, and a *residual* term G^{res} (which is not the same as "residual property," i.e., departure function), accounting for molecular interactions:

$$G^E = G^{\text{comb}} + G^{\text{res}} \tag{4-168}$$

Function G^{comb} contains pure-species parameters only, whereas function G^{res} incorporates two binary parameters for each pair of molecules. For a multicomponent system,

$$\frac{G^{\text{comb}}}{RT} = \sum_i x_i \ln\frac{\Phi_i}{x_i} + 5\sum_i q_i x_i \ln\frac{\theta_i}{\Phi_i} \tag{4-169}$$

$$\frac{G^{\text{res}}}{RT} = -\sum_i q_i x_i \ln\left(\sum_j \theta_j \tau_{ji}\right) \tag{4-170}$$

where q is a relative surface area of the molecule and r is the relative molecular volume.

$$\Phi_i \equiv \frac{x_i r_i}{\sum_j x_j r_j}; \quad \theta_i \equiv \frac{x_i q_i}{\sum_j x_j q_j} \tag{4-171}$$

Subscript i identifies species, and j is a dummy index; all summations are over all species. Note that $\tau_{ji} \neq \tau_{ij}$; nevertheless, when $i = j$, then $\tau_{ii} = \tau_{jj} = 1$. In these equations r_i (a relative molecular volume) and q_i (a relative molecular surface area) are pure-species constants. The influence of temperature on G^E/RT enters through the interaction parameters τ_{ji} of Eq. (4-170), which are temperature-dependent:

$$\tau_{ji} = \exp\frac{-a_{ji}}{RT} \tag{4-172}$$

Parameters for the UNIQUAC equation are therefore values of a_{ji}.

An expression for ln γ_i is found by application of Eq. (4-146) to the UNIQUAC model for G^E/RT [Eq. (4-168)]. The result is given by the following equations:

$$\ln\gamma_i = \ln\gamma_i^{comb} + \ln\gamma_i^{res} \tag{4-173}$$

$$\ln\gamma_i^{comb} = 1 - \frac{\Phi_i}{x_i} + \ln\frac{\Phi_i}{x_i} - 5q_i\left(1 - \frac{\Phi_i}{\theta_i} + \ln\frac{\Phi_i}{\theta_i}\right) \tag{4-174}$$

$$\ln\gamma_i^{res} = q_i\left(1 - \ln\sum_k \theta_k\tau_{ki} - \sum_j \theta_j \frac{\tau_{ij}}{\sum_k \theta_k\tau_{kj}}\right) \tag{4-175}$$

Again subscript *i* identifies species, and *j* and *k* are dummy indices.

For a mixture of 20 percent propanol in water: $a_{21} = 20.4$ J/mol; $a_{12} = 2551$; $r_1 = 3.25$; $q_1 = 3.13$; $r_2 = 0.94$; $q_2 = 1.40$; $\gamma_1 = 2.721$; and $\gamma_2 = 1.140$.

The NRTL equation is the most flexible for fitting experimental data because it has three parameters per binary system, compared to two parameters for the Wilson or UNIQUAC models. For most applications, the default value of $\alpha = 0.3$ suffices for the NRTL model, making it similar to the others when limited data are available. For multicomponent systems, the subscripted form of α should be used to distinguish, say, α_{12} (= α_{21}) from α_{13} (= α_{31}). The Wilson parameters Λ_{ij}, NRTL parameters G_{ij}, and UNIQUAC parameters τ_{ij} inherit a Boltzmann-type *T* dependence from the origins of the expressions for G^E, but it is only approximate. Computations of properties sensitive to this dependence (e.g., heats of mixing and liquid/liquid solubility) are in general only qualitatively correct. All parameters can be characterized from data for binary systems (in contrast to multicomponent), and this makes parameter determination for the local composition models a manageable task.

PHASE EQUILIBRIUM DATA SOURCES

The literature on phase equilibrium measurements is vast and continually increasing. Keeping track of all the data generated throughout time and across the globe is part of the mission of the Thermodynamics Research Center (TRC) in Boulder, Colorado, a part of the Physical and Chemical Properties Division of the National Institute of Standards and Technology (NIST). The NIST TRC group has developed a website (ThermoLit [ibid]) that compiles literature sources ostensibly covering all known physical property data pertaining to one-, two-, or three-component systems. The resource compiles citations of data for vapor-liquid, liquid-liquid, and solid-fluid equilibria. The Korean Database (KDB) (http://www.cheric.org/research/kdb/hcvle/hcvle.php) provides online tabulation of some experimental data. Another database available in many university libraries is the DECHEMA database of Gmehling, Onken, and Arlt [*Vapor-Liquid Equilibrium Data Collection,* Chemistry Data Series, vol. 1, parts 1–8, DECHEMA, Frankfurt/Main, 1974–1990]. An older but still useful data collection is that of Stephens and Stephens [*Solubilities of Inorganic and Organic Compounds,* vol. 1, pts. 1 and 2, Pergamon, Oxford, England, 1960]. A database of infinite-dilution activity coefficients is included in the supporting information submitted with the article by Lazzaroni et al. [*Ind. Eng. Chem. Res.* **44**(11): 4075–4083 (2005)].

A number of other sources have compiled data from the literature into a single volume or series that may be more convenient than referring to the original literature. Comprehensive collections of phase equilibrium data (including vapor-liquid, liquid-liquid, and solid-liquid data) and infinite-dilution activity coefficients are maintained by the TRC and by DDBST, GmbH. Another database called Infotherm is available from Wiley. Other sources of thermodynamic data include the IUPAC Solubility Data Series published by Oxford University Press. Additional sources of data are discussed by Skrzecz [*Pure Appl. Chem.* (IUPAC), **69**(5): 943–950 (1997)].

Data Reduction Correlations for G^E and the activity coefficients are based on VLE data taken at low to moderate pressures. The process of finding a suitable analytic relation for G^E/RT as a function of its independent variables *T* and x_1, thus producing a correlation of VLE data, is known as *data reduction.* Although in principle G^E/RT is also a function of *P*, the dependence is so weak as to be usually neglected. The adjustable parameters of the models are regressed by minimizing the residuals. The maximum-likelihood method [T. F. Anderson and J. M. Prausnitz, *Ind. Eng. Chem. Proc. Res. Dev.* **17:** 552 (1978)] provides consideration that every measurement may include experimental error, but sometimes the method is difficult to reliably converge and thus a least-squares approach on bubble pressure, bubble temperature, or liquid-liquid phase behavior is typical. See also Van Ness [*J. Chem. Thermodyn.* **27:** 113–134 (1995); *Pure & Appl. Chem.* **67:** 859–872 (1995)]. Although the discussion focuses on fitting experimental data, recognize that the predictive methods discussed next may be used to generate excess Gibbs energy or activity coefficient information, and then the model parameters can be regressed against the predictions to provide a tractable multicomponent engineering process model.

PREDICTIVE AND ADAPTIVE MODELS FOR THE EXCESS GIBBS ENERGY

Predictive Models: UNIFAC, Solubility Parameter Models, COSMO For the design of processes that often involve synthesis of new compounds in new combinations or at new conditions, the need to predict mixture behavior is inevitable. Some models, such as UNIFAC, represent extensive correlations with large databases and scores of parameters based on regression of group contributions. Because UNIFAC is correlated by fitting group parameters to experimental data, it might be viewed as interpolations with molecular structure as the independent variable. The group contribution approach makes UNIFAC work well computationally, but provides little intuitive insight. Other models rely on leveraging insights from the analysis of the chemical nature of the molecular structure, such as hydrogen bonding tendencies or localized electron density. These models may be less accurate when compared to a large database, but can be helpful during the conceptual stages of process or product design.

The UNIFAC Model Perhaps the most widely used activity model is the UNIFAC family of group contribution methods. These methods are based on the UNIQUAC equation, such that UNIFAC stands for UNIQUAC functional-group activity coefficients, proposed by Fredenslund, Jones, and Prausnitz [*AIChE J.* **21:** 1086–1099 (1975)] and given detailed treatment by Fredenslund, Gmehling, and Rasmussen [*Vapor-Liquid Equilibrium Using UNIFAC,* Elsevier, Amsterdam, 1977], Fredenslund et al. [*Ind. Eng. Chem. Proc. Des. Dev.* **16**(4): 450–462 (1977)]; and Wittig et al. [*Ind. Eng. Chem. Res.* **42**(1): 183–188 (2003)]. Also see Jakob et al. [*Ind. Eng. Chem. Res.* **45:** 7924–7933 (2006)].

Subsequent development has led to a variety of separate correlations, each focused on specific applications, including liquid/liquid equilibria [Magnussen, Rasmussen, and Fredenslund, *Ind. Eng. Chem. Process Des. Dev.* **20:** 331–339 (1981)], solid/liquid equilibria [Anderson and Prausnitz, *Ind. Eng. Chem. Fundam.* **17:** 269–273 (1978)], solvent activities in polymer solutions [Oishi and Prausnitz, *Ind. Eng. Chem. Process Des. Dev.* **17:** 333–339 (1978)], vapor pressures of pure species [Jensen, Fredenslund, and Rasmussen, *Ind. Eng. Chem. Fundam.* **20:** 239–246 (1981)], gas solubilities [Sander, Skjold-Jørgensen, and Rasmussen, *Fluid Phase Equilibr.* **11:** 105–126 (1983)], and excess enthalpies [Dang and Tassios, *Ind. Eng. Chem. Process Des. Dev.* **25:** 22–31 (1986)].

The range of applicability of the original UNIFAC model has been greatly extended and its reliability enhanced. Its most recent revision and extension is treated by Wittig et al. (2003), wherein are cited earlier pertinent papers. Because it is based on temperature-independent parameters, its application is largely restricted to 0 to 150°C.

Two modified versions of the UNIFAC model, based on temperature-dependent parameters, have come into use. Not only do they provide a wide temperature range of applicability, but also they allow correlation of various kinds of property data, including phase equilibria, infinite-dilution activity coefficients, and excess properties. The most recent revision and extension of the modified UNIFAC (Dortmund) model is provided by Gmehling et al. [*Ind. Eng. Chem. Res.* **41:** 1678–1688 (2002)]. An extended UNIFAC model called KT-UNIFAC is described in detail by Kang et al. [*Ind. Eng. Chem. Res.* **41:** 3260–3273 (2003)], and updated [*Fluid Ph. Equilibr.* **309:** 68–75 (2011)].

The use of UNIFAC for estimating LLE is discussed by Gupte and Danner [*Ind. Eng. Chem. Res.* **26**(10): 2036–2042 (1987)] and by Hooper, Michel, and Prausnitz [*Ind. Eng. Chem. Res.* **27**(11): 2182–2187 (1988)]. Vakili-Nezhand, Modarress, and Mansoori [*Chem. Eng. Technol.* **22**(10): 847–852 (1999)] discuss its use for representing a complex stream containing a large number of components for which available LLE data are incomplete. Similar to UNIQUAC, UNIFAC calculates activity coefficients in two parts:

$$\ln\gamma_i = \ln\gamma_i^{comb} + \ln\gamma_i^{res} \tag{4-176}$$

The combinatorial part ln γ_i^{comb} is calculated from pure-component properties. The residual part ln γ_i^{res} is calculated by using binary interaction parameters for solute-solvent group pairs determined by regressing the group parameters against a large set of phase equilibrium data. Thus, the predictions are most reliable when the method is applied to monofunctional molecules similar to those used in the regression. With this approach, a molecule is treated as a mixture of various functional groups. The proximity of the groups to one another in the molecule is not taken into account.

Solubility Parameter Models A number of methods based on regular solution theory are also available. Only pure-component parameters are needed to make estimates, so they may be applied when UNIFAC group-interaction parameters are not available. These methods are also sufficiently simple that they provide intuitive guides as to what compounds might blend well or contribute to desirable solution behavior, such as increasing solution ideality.

Scatchard-Hildebrand Theory Scatchard-Hildebrand solution theory defines G^E in terms of

$$\frac{G^E}{RT}=\frac{U^E}{RT}=\sum x_i V_i(\delta_i - <\delta>)^2 + V << k_{mm} >> \quad (4\text{-}177)$$

$$RT \ln \gamma_i = V_i[(\delta_i - <\delta>)^2 + 2<k_{im}> - << k_{mm} >>] \quad (4\text{-}178)$$

where $<\delta> = \sum\Phi_j\delta_j$, $<<k_{mm}>> = \sum\Phi_i\delta_i<k_{im}>$, $<k_{im}> = \sum\Phi_j\delta_j k_{ij}$, $\Phi_i = x_i V_i/(\sum x_j V_j)$, and $k_{ij} = k_{ji}$ = a single binary parameter per binary system.

The parameter δ is known as the *Hildebrand solubility parameter* and defined in terms of pure-component properties at 25°C.

$$\delta \equiv \sqrt{\frac{\Delta U^{\text{vap}}}{V^L}} \approx \sqrt{\frac{\Delta H^{\text{vap}} - 298.15R}{V^L}} \quad (4\text{-}179)$$

Assuming $k_{ij} = 0$ for all i, j, the theory is predictive, but always predicts positive deviations from ideal solution behavior. This theory is generally reasonable for hydrocarbons and slightly polar substances, but not for complexing or hydrogen bonding systems. The theory can be derived from the van der Waals EOS based on the assumption of a constant packing fraction for all liquids, so its pedigree is similar in quality to that of most EOS methods. There are two guidelines that are apparent from the defining equations: (1) Systems are more nearly ideal ($G^E \approx 0$) when all the solubility parameters are equal. (2) Larger molecules tend to amplify the nonideality. The first guideline may be more familiar in the form "like dissolves like," although the mathematical model provides a more quantitative suggestion. The second guideline may sound reasonable if you are familiar with the poor mutual solubility of polymers in one another; even polyethylene and polypropylene blend poorly.

Flory-Huggins Model For polymer solutions and blends, the primary workhorse continues to be the Flory-Huggins model. This model is very similar to regular solution theory, but adds a term to recognize that excess entropy (S^E) is significant for polymers as well as excess energy (U^E).

$$\ln \gamma_i = V_i[(\delta_i - <\delta>)^2 + 2<k_{im}> - <<k_{mm}>>]/(RT) + \left(1 - \Phi_i/x_i\right) + \ln\left(\Phi_i/x_i\right) \quad (4\text{-}180)$$

Hansen Solubility Parameters The Hansen solubility parameter model divides the Hildebrand solubility parameter into three parts to obtain parameters δ^d, δ^p, and δ^h accounting for nonpolar (dispersion), polar, and hydrogen-bonding effects [Hansen, *J. Paint Technol.* **39:** 104–117 (1967)]. An activity coefficient may be estimated by using an equation of the form

$$RT \ln \gamma_i = V_i\{(\delta_i^d - <\delta^d>)^2 + 0.25[(\delta_i^p - <\delta^p>)^2 + (\delta_i^h - <\delta^h>)^2]\} \quad (4\text{-}181)$$

where $\delta^2 = (\delta^d)^2 + 0.25[(\delta^p)^2 + (\delta^h)^2]$ [Frank, Downey, and Gupta, *Chem. Eng. Prog.* **95**(12): 41–61 (1999)]. Equation (4-181) is equivalent to Eq. (4-178) for nonpolar mixtures with zero binary interaction parameters. The Hansen model has been used for many years to screen solvents and facilitate development of product formulations. Hansen parameters have been determined for more than 500 solvents [Hansen, *Hansen Solubility Parameters: A User's Handbook,* CRC, Boca Raton, Fla., 2000); and *CRC Handbook of Solubility Parameters and Other Cohesion Parameters,* 2d ed., ed. Barton (CRC, Boca Raton, Fla., 1991)].

MOSCED and SPACE Models MOSCED (Modified Separation of Cohesive Energy Density) is another modified Scatchard-Hildebrand solution model. MOSCED utilizes two parameters to represent hydrogen bonding: one for proton donor capability (acidity) and one for proton acceptor capability (basicity) [Thomas and Eckert, *Ind. Eng. Chem. Proc. Des. Dev.* **23**(2): 194–209 (1984)]. This provides a more realistic representation of hydrogen bonding that allows more accurate modeling of a wider range of solvents, and unlike the Hansen model, MOSCED can predict negative deviations from ideal solution (activity coefficients less than 1.0). MOSCED calculates infinite-dilution activity coefficients by using

$$RT \ln \gamma_{2,1}^{\infty} = V_2\left[(\delta_2^d - \delta_1^d)^2 + \frac{q_1^2 q_2^2 (\tau_2^T - \tau_1^T)^2}{\psi_1} + \frac{(\alpha_2^T - \alpha_1^T)(\beta_2^T - \beta_1^T)}{\xi_1}\right] + \left(1 - V_2/V_1\right)^{aa} + aa \ln\left(V_2/V_1\right) \quad (4\text{-}182)$$

where $\alpha_i^T = \alpha_i\left(\frac{293}{T}\right)^{0.8}$, $\beta_i^T = \beta_i\left(\frac{293}{T}\right)^{0.8}$, $\tau_i^T = \tau_i\left(\frac{293}{T}\right)^{0.4}$

There are five adjustable parameters per molecule: the dispersion parameter δ^d originally represented as λ by Thomas and Eckert; the induction parameter q; the polarity parameter τ; the hydrogen-bond acidity parameter α; and the hydrogen-bond basicity parameter β. The induction parameter q often is set to a value of 0.9 or 1.0, yielding a four-parameter model. The terms aa, ψ, and ξ are asymmetry factors calculated from α, β, and τ as a function of temperature. The complete model equations and a database of parameter values for approximately 150 compounds are given by Lazzaroni et al. [*Ind. Eng. Chem. Res.* **44**(11): 4075–4083 (2005)]. An application of MOSCED in the study of liquid-liquid extraction is described by Escudero, Cabezas, and Coca [*Chem. Eng. Comm.* **173:** 135–146 (1999)]. Also see Frank et al. [*Ind. Eng. Chem. Res.* **46:** 4621–4625 (2007)]. Methods for predicting unavailable MOSCED parameters have been discussed by Gnap and Elliott, [*Fluid Phase Eq.*, in press (2018)].

Another method closely related to the MOSCED model is the SPACE model for estimating infinite-dilution activity coefficients [Hait et al., *Ind. Eng. Chem. Res.* **32**(11): 2905–2914 (1993)]. The SPACE model utilizes refractive indices and solvatochromic parameters. The solvatochromic parameters are α (acidity), β (basicity), π (polarity), and δ (polarizability). These have been measured independently of phase equilibria data using spectroscopic techniques such as NMR and UV. The number of parameters is fewer in the SPACE model, in principle, but there are several generalized correlations required to implement the method. The more recent paper by Lazzaroni et al. offers the simplest and most reliable method between SPACE and MOSCED.

Table 4-1 shows typical values for MOSCED parameters over a range of compounds. In the absence of hydrogen bonding, as in the case of acetone + *n*-octane, mixing follows the formula based on differences in δ^d and τ; positive deviations are predicted. The acidity and basicity provide the strongest indication of solution nonideality. When both α and β are significant for a given compound, mixing with a compound that has small α and β leads to large positive deviations from ideality, as in the case of phenol + *n*-decane. An azeotrope or liquid-liquid equilibrium should be suspected for such a system. When one compound is relatively acidic and the other relatively basic, as for phenol + pyridine, negative (exothermic) deviations from ideality should be expected. Finally, when both compounds have similar acidity and basicity, the influences of hydrogen bonding may cancel and the mixture behavior returns to being ideal, as in the case of phenol + benzyl alcohol.

TABLE 4-1 Sampling of MOSCED Parameters

	ρ_{298}	δ(J/cm^3)$^{1/2}$	α	β	δ^d	τ
Acetic Acid	1.04	19	24.03	7.5	14.96	3.23
Acetone	0.79	19.64	0	11.14	13.71	8.3
Aniline	1.02	24.12	6.51	6.34	16.51	9.41
Benzene	0.87	18.73	0.63	2.24	16.71	3.95
Benzyl Alcohol	1.04	24.7	15.01	6.69	16.56	5.03
Chloroform	1.48	18.92	5.8	0.12	15.61	4.5
n-Decane	0.73	15.7	0	0	15.7	0
Ethanol	0.79	26.13	12.58	13.29	14.37	2.53
Iso-octane	0.7	14.11	0	0	14.11	0
Methanol	0.79	29.59	17.43	14.49	14.43	3.77
MTBE	0.74	15.17	0	7.4	15.17	2.48
n-Octane	0.7	15.5	0	0	15.5	0
Phenol	1.06	24.63	25.14	5.35	16.66	4.5
Pyridine	0.98	21.56	1.61	14.93	16.39	6.13
Water	1	47.86	52.78	15.86	10.58	10.48
p-xylene	0.86	17.9	0.27	1.87	16.06	2.7

In general, perusing Table 4-1 shows that most alcohols exhibit balanced acidity and basicity, although the magnitudes of α and β decrease as the molecular volume increases. Ketones, ethers, aldehydes, amines, and esters tend to be relatively basic. Distinctly acidic behavior is less common, except for aromatic alcohols and, of course, carboxylic acids. These elementary insights can go a long way toward making solution behavior seem less mysterious. We revisit these insights when we consider the guidelines of phase diagrams and Robbins' table.

COSMO Models: COSMO-RS and COSMO-SAC The thermodynamic methods described above glean information from available data to make estimates for other systems. As an alternative approach, quantum chemistry calculations and molecular simulation methods are finding greater use in engineering applications [Gupta and Olson, *Ind. Eng. Chem. Res.* **42**(25): 6359–6374 (2003); and Chen and Mathias, *AIChE J.* **48**(2): 194–200 (2002)]. These methods minimize the need for data; however, the computational effort and specialized expertise required to use them are generally higher, and the accuracy of the results may not be known. An important method gaining increasing application in the chemical industry is the conductor-like screening model (COSMO) introduced by Klamt and colleagues [Klamt, *J. Phys. Chem.* **99:** 2224 (1995); Klamt and Eckert, *Fluid Phase Equilibr.* **172:** 43–72 (2000); Eckert and Klamt, *AIChE J.* **48**(2): 369–385 (2002); and Klamt, *From Quantum Chemistry to Fluid Phase Thermodynamics and Drug Design*, Elsevier, Amsterdam, 2005]. Also see Grensemann and Gmehling, *Ind. Eng. Chem. Res.* **44**(5): 1610–1624 (2005). This method utilizes computational quantum mechanics to calculate a two-dimensional electron density profile to characterize a given molecule. This profile is then used to estimate phase equilibrium through application of statistical mechanics and solvation theory. The Klamt model is called COSMO-RS (for realistic solvation). A similar model is COSMO-SAC (for segment activity coefficient) published by Lin and Sandler [*Ind. Eng. Chem. Res.* **41**(5): 899–913, 2332 (2002)]. Databases of electron density profiles (sigma profiles) are available from a number of vendors and universities. A sigma-profile database of more than 1000 molecules is available from the Virginia Polytechnic Institute and State University [Mullins et al., *Ind. Eng. Chem. Res.* **45**(12): 4389–4415 (2006)]. An application of COSMOS-RS to predict liquid-liquid equilibria is discussed by Banerjee et al. [*Ind. Eng. Chem. Res.* **46**(4): 1292–1304 (2007)].

Adaptive Models LSER, NRTL-SAC When data are available for a homologous series of compounds, but not the specific compound of interest, linear solvation energy relationships (LSERs) may be useful. A method developed by Meyer and Maurer [*Ind. Eng. Chem. Res.* **34**(1): 373–381 (1995)] uses the LSER model [Taft et al., *Nature* **313:** 384 (1985); and Taft et al., *J. Pharma Sci.* **74:** 807–814 (1985)] to estimate infinite-dilution partition ratios for solutes distributed between water and an organic solvent. The model uses 36 generalized parameters and 4 solvatochromic parameters to characterize a given solute. Also see Abraham, Ibrahim, and Zissimos, *J. Chromatography* **1037:** 29–47 (2004).

Other Estimation Methods Another method for estimating activity coefficients is described by Chen and Song [*Ind. Eng. Chem. Res.* **43**(26): 8354–8362 (2004); **44**(23): 8909–8921 (2005)]. This method involves regression of a small data set in a manner similar to the way the Hansen and MOSCED models typically are used. The model is based on a modified NRTL framework called NRTL-SAC (for segment activity coefficient) that utilizes only pure-component parameters to represent polar, hydrophobic, and hydrophilic segments of a molecule. An electrolyte parameter may be added to characterize ion-ion and ion-molecule interactions attributed to ionized segments of species in solution. The resulting model may be used to estimate activity coefficients and related properties for nonionic organics plus electrolytes in aqueous and nonaqueous solvents.

Another approach involves use of molecular simulation or electron density calculations to predict values of parameters for phase equilibrium models. An example involves prediction of MOSCED parameters [R. Ley, G. Fuerst, B. Redeker, and A. Paluch, *Ind. Eng. Chem. Res.* **55**(18): 5415–5430 (2016); J. Phifer, K. Soloman, K. Young, and A. Paluch, *AIChE J.* **63:** 781–791 (2017)]. This approach combines molecular modeling and phase equilibrium theory to obtain a predictive tool well suited to early-stage process development.

MODEL SELECTION

Model selection can seem overwhelming due to the large number of possible models. Use of correlative methods fitted to experimental data is preferred over predictive methods, and sometimes a local fit of parameters is necessary [P. M. Mathias, *J. Chem. Eng. Data.* **61:** 4077–4084 (2016)]. Predictive methods should be used cautiously when the compounds of interest differ from those used in the model development or when multifunctional molecules are present.

For subcritical systems (say, $P < 15$ bar), an activity model is likely to suffice. The NRTL model has the advantage of a third parameter when needed, so try it first with $\alpha = 0.3$, then adjust α if necessary. If the comparison indicates liquid-liquid equilibrium where there is none, try the Wilson model. LLE is indicated by a minimax in the predicted y-x curve. If a substantial gap exists in the experimental data along the x axis, it may be that the system exhibits LLE.

Careful model validation against experiment is especially critical for the heavy and light key components in distillation. Multiple columns would have multiple keys so each pair would need to be checked. For the components of secondary importance, experimental data should be sought but predictive methods (such as UNIFAC) can be applied if necessary. Generally, different model parameters are needed for VLE and LLE, even with the same model. If a different model or parameter set is best for different unit operations, customize each operation. If multicomponent data are available, a y-x comparison to experiment is possible by applying a pseudocomponent basis, $y' = y_L/(y_L + y_H)$, for example. Of course, the experimental data should be as close to process conditions as possible, but data within 50°C of process conditions should suffice. Processes involving components with significant association in the vapor phase (e.g., carboxylic acids) should include an association model such as that of Hayden-O'Connell.

For processes with fluids near a critical point or with retrograde condensation, it is advisable to try an EOS method. The General References include discussions of phase diagrams in the critical region [e.g., Elliott and Lira, chapter 16 (2012)]. It may be necessary to compare several models in these cases. For predictions, the predictive SRK method provides a reasonable start [Horstmann et al., *Fluid Phase Equilibr.* **167:** 173–186 (2000)]. For correlation, the modified Huron-Vidal method prevails in most comparisons. If systems involve heavy components or polymers in important roles, the PC-SAFT model should be considered. The PC-SAFT model shows promise as a basis for both prediction and correlation, but it has not been fully implemented in all process simulators. After EOS methods are tried, activity models should be considered also as long as one of the key components is not supercritical. The model that agrees best with experiment is preferred.

PRELIMINARY ESTIMATES

It may be advisable to consider some relatively quick guidelines before delving deeply into computer calculations. Often, considering the kinds of phase behavior to be encountered before seeing the computational output may facilitate a critical evaluation of the output. In other cases, applications of interest may require formulations that involve multiple components designed to achieve a certain process objective such as moderating the solution nonideality for an extractive distillation or finding a liquid solvent that extracts the solute of interest from a diluent while maintaining minimal mutual solubility between solvent and diluent. A purely computational approach might involve many random trials and errors, while phenomenological consideration of a model such as MOSCED could help to guide the search. Two useful approaches are outlined briefly below.

Robbins' Table The interactions of polar and hydrogen bonding forces evident in the MOSCED, Hansen, and COSMO models lead to intuitive insights about how combinations of chemicals may behave in solution. This insight can be helpful when choosing an entrainer for an azeotropic system or a solvent for liquid-liquid extraction, for example. A well-known guide is Robbins' table (Table 4-2) of solute-solvent interactions [*Chem. Eng. Prog.* **76**(10): 58–61 (1980).] This table indicates whether interactions between compounds are likely to yield positive, negative, or near-zero deviations from ideal solution behavior. Similar tables for anticipating solvent-solute interactions are often cited in discussions of distillation and liquid extraction. These rely largely on classifications of hydrogen bonding and polarity that are similar to the intent to those of Robbins' table, with similar results.

TABLE 4-2 Robbins' Table (Modified by Gnap and Elliott) of Solute–Solvent Interactions*

	α′(J)½	β′(J)½	Class	1	2	3	4	5	6	7	8	9	10	11
Acids, etc.	157	11	1	0	0	0	–	–	–	0	–	–	+	+
Thiol	11	55	2	0	0	+	–	–	0	0	0	0	+	+
Alcohol, water	200	70	3	0	+	0	+	0	–	–	+	+	+	+
Ketones, etc.	0.1	26	4	–	–	+	0	0	+	+	0	0	0	+
3′amine	0.1	33	5	–	–	0	0	0	+	+	0	+	0	0
2′amine	13	211	6	–	0	–	+	+	0	0	0	0	0	+
1′amines, etc.	65	755	7	0	0	–	+	+	0	0	+	+	+	+
Ethers, etc.	0.1	33	8	–	0	+	0	0	0	+	0	0	0	+
Esters, etc.	0.1	99	9	–	0	+	0	+	0	+	0	0	0	+
Aromatics, etc.	0.1	4	10	+	+	+	0	0	0	+	0	0	0	0
Nonpolar	0.1	0	11	+	+	+	+	0	+	+	+	+	0	0

*Detailed class descriptions are given in the text.

The listings of acidity and basicity may be viewed in terms of average acidity or basicity as characterized by the MOSCED model. The entries in the classic Robbins' table can be predicted about 70 percent of the time with the formula

$$\Delta^{is} = \frac{(\alpha_2' - \alpha_1')(\beta_2' - \beta_1')}{(\alpha_2' + \alpha_1')(\beta_2' + \beta_1')} \tag{4-183}$$

where α_i' and β_i' are the generalized values. When $\Delta^{is} < -0.2$, a negative deviation is indicated. When $\Delta^{is} > 0.4$, a positive deviation is indicated. For $-0.2 < \Delta^{is} < 0.4$, relatively ideal solution behavior can be expected. The primary source of discrepancies between Eq. (4-183) and Robbins' table involves differences in the assessment of polarity. For example, mixtures of esters with paraffins normally give positive deviations from ideality, but calculated Δ^{is} values can be close to zero. This discrepancy might be anticipated by including the MOSCED term for polarity in the assessment, but this level of detail would undermine the simplicity of Robbins' table.

In this modified version, the classifications of phenols, acids, and halogenated acids have been adjusted somewhat to take polarity effects into account. The detailed groupings for each classification are as follows: (1) Acids, phenols, active H on multihalogen paraffin; (2) thiols; (3) alcohol, water; (4) ketones, tertiary amide, sulfone, phosphine; (5) tertiary amine; (6) secondary amine; (7) primary amine, primary amide, NH_3; (8) ether, oxide, sulfoxide; (9) ester, aldehyde, carbonate, phosphate, nitrate, nitrite, nitrile, intramolecular H bonding (e.g., *o*-nitrophenol); (10) aromatic, olefin, halogenated aromatic, multihalogen paraffin without active H, and monohalogen paraffin; (11) paraffin and carbon disulfide.

Example 4-10 Entrainer Selection for Extractive Distillation A common problem in gasohol production is overcoming the ethanol + water azeotrope. Extractive distillation involves the addition of a relatively nonvolatile entrainer that is miscible in both components. interacts favorably with the less volatile component (water in this case), and moderates the solution nonideality. Candidates for the entrainer are glycerol triacetate, monoethanolamine, and ethylene glycol. Which candidate is most promising from the perspective of Robbins' table?

Solution Glycerol triacetate is an ester with a boiling point near 258°C, monoethanolamine is an alcohol/primary amine with boiling point near 170°C, and ethylene glycol is a diol with a boiling point of 198°C (chemspider.com is convenient for this kind of search). Although the acid/base combination seems favorable for the triacetate, Robbins' table indicates a positive deviation from ideality. The glycol is simply another alcohol, so it indicates zero deviation. The ethanolamine is similar to glycol except that one hydroxyl has been replaced by a primary amine. The primary amine indicates a negative deviation from ideality, suggesting that it should provide more powerful suppression of the water activity, requiring less entrainer. Therefore, monoethanolamine would be recommended by Robbins' perspective. Other considerations such as cost, toxicity, reactivity (the acetate would likely hydrolyze and monoethanolamine may react with trace impurities), or ease of entrainer regeneration must also be considered.

Phase Diagrams Either solutions can be ideal, or they can exhibit positive ($\gamma >$ unity) or negative ($\gamma <$ unity) deviations from ideality. Ideal solutions cannot exhibit liquid-liquid equilibrium (LLE) or azeotropes. Negative deviations from ideality cannot result in LLE, but they can result in azeotrope formation. Positive deviations from ideality can result in azeotropes and LLE if the deviations are large enough. For a binary solution, if the geometric mean of the infinite-dilution activity coefficients exceeds 10, the prospect of LLE should be checked. In the case of vapor-liquid equilibria (VLE), the relative volatility is important to consider.

$$\alpha_{ij} \equiv \frac{K_i}{K_j} = \frac{\gamma_i P_i^{\text{sat}}}{\gamma_j P_j^{\text{sat}}} \tag{4-184}$$

where $K_i = y_i/x_i$ is the ratio of vapor mole fraction to liquid mole fraction and the order of i and j is chosen such that $P_i^{\text{sat}} > P_j^{\text{sat}}$, indicating that $\alpha_{ij} > 1$ for an ideal solution. If $\alpha_{ij} < 1$ for some range of compositions, an azeotrope occurs. Azeotropes are important because distillation fails when the vapor and liquid phases have the same composition. Checking the relative volatility of key components at both top and bottom of the column is recommended, deliberately verifying whether it crosses unity.

These cases can be illustrated by Fig. 4-2. *T-xy* diagrams have two advantages: (1) the onset of LLE is easier to show than in a *P-xy* diagram and (2) most

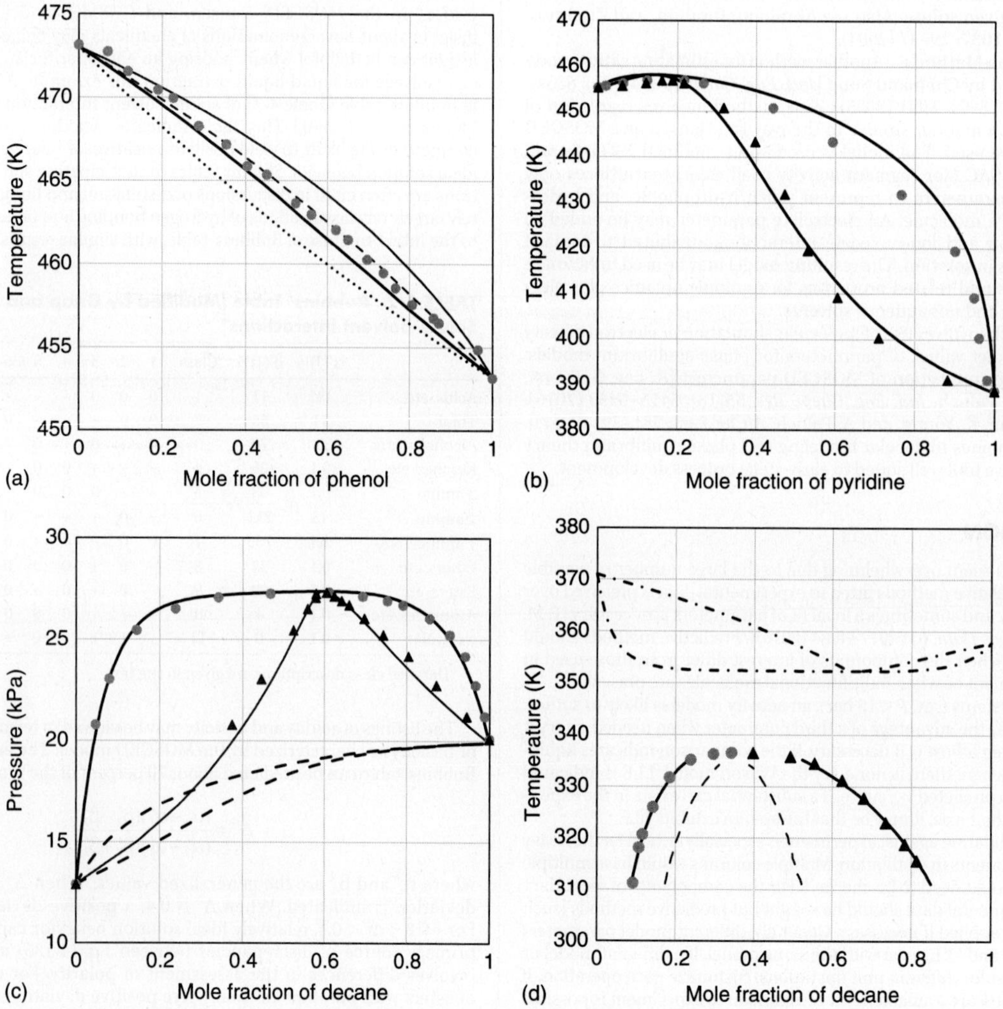

FIG. 4-2 Phase diagrams fit using the NRTL model. (*a*) *T-xy* for phenol(1) + *p*-cresol(2) at 96.4 kPa. b_{12} = 1567.2 J/mol, b_{21} = −2010.3, α = 0.3. Data of Selvam et al. [*Fluid Phase Equilibr.* **78:** 261–267 (1992). γ_1^∞(NRTL) = 0.85, γ_2^∞ = 0.81. Dotted lines show the ideal solution model. (*b*) *T-xy* for pyridine(1) + phenol(2) at 101.32 kPa. b_{12} = −7235.8 J/mol, b_{21} = 1886.5, α = 0.3. Data of F. A. Assal [*Bull. Acad. Pol. Sci. Ser. Sci. Chim.* **14:** 603 (1966)]. γ_1^∞(NRTL) = 0.055, γ_2^∞ = 0.17. (*c*) *P-xy* for *n*-decane(1) + phenol(2) at 119.8°C. Solid lines use b_{12} = 4259.6 J/mol, b_{21} = 5729.5, α = 0.43, γ_1^∞(NRTL) = 12.2, γ_2^∞ = 8.4. Dotted lines use parameters from figure (*d*), γ_1^∞(NRTL) = 2.5, γ_2^∞ = 1.5. (*d*) *T-xy* for *n*-decane(1) + phenol(2) at 5 kPa. Solid lines use b_{12} = −71.382*T*(K) + 26038 J/mol, b_{21} = 1.9217*T*(K) + 4704.8, α = 0.3. Dotted lines using parameters from (*c*). Figures (*c*) and (*d*) use data of Gmehling [*J. Chem. Eng. Data* **27:** 371 (1982)].

phase separation processes are conducted at nearly constant pressure. Figure 4-2*a* shows the case of a nearly ideal solution. Figure 4-2*b* shows a maximum boiling azeotrope. When phenol mixes with *p*-cresol, the solution is nearly ideal, but when it mixes with pyridine, an exothermic acid-base interaction ensues. Since the vapor pressures of phenol and pyridine are similar, the ratio of activity coefficients overwhelms the ratio of vapor pressures in the relative volatility, and a maximum boiling azeotrope results. A similar phenomenon occurs for *n*-decane + phenol, but with large positive deviations from ideality overwhelming the vapor pressure ratio and causing a minimum boiling azeotrope, as shown in Fig. 4-2*c* and *d*. The rationale for associating the system with a minimum boiling azeotrope is most evident in Fig. 4-2*d*, where it is evident that *boiling* refers to the bubble temperature, not bubble pressure. Figure 4-2*c* shows that the *P-xy* diagram is the flipped top-to-bottom image of the *T-xy* diagram for the same system. Also evident in Fig. 4-2*d* is the onset of LLE at temperatures below the bubble point. At higher pressures, the bubble curve would increase to higher temperatures and the LLE would be relatively unaffected. In some mixtures, where the nonideality is larger than for the system illustrated, the VLE curve may intersect with the LLE curve, resulting in VLLE. Finally, Fig. 4-2*c* and *d* illustrates a challenge in representing VLE and LLE simultaneously with universal parameters. The parameters fitted to VLE in Fig. 4-2*c* give a poor representation of LLE in Fig. 4-2*d*. Similarly, the parameters fitted to LLE in Fig. 4-2*d* give a poor representation of VLE in Fig. 4-2*c*. Therefore, the "optimal" assessment depends on the job at hand.

VAPOR/LIQUID EQUILIBRIUM

Vapor/liquid equilibrium (VLE) relationships (as well as other interphase equilibrium relationships) are needed in the solution of many engineering problems. The general VLE problem treats a multicomponent system of N constituent species for which the independent variables are T, P, $N-1$ liquid-phase mole fractions and $N-1$ vapor-phase mole fractions. (Note that $\sum_i x_i = 1$ and $\sum_i y_i = 1$, where x_i and y_i represent liquid and vapor mole fractions, respectively.) Thus there are $2N$ independent variables, and application of the phase rule shows that exactly N of these variables must be fixed to establish the intensive state of the system. This means that once N variables have been specified, the remaining N variables can be determined by simultaneous solution of the N equilibrium relations [Eq. (4-127)]. In practice, either T or P *and* either the liquid-phase or vapor-phase composition are specified, thus fixing $1 + N - 1 = N$ independent variables.

***K* Values, VLE, and Flash Calculations** A measure of the distribution of a chemical species between liquid and vapor phases is the K value, defined as the *equilibrium ratio*:

$$K_i \equiv \frac{y_i}{x_i} \tag{4-185}$$

It has no thermodynamic content, but may make for computational convenience through elimination of one set of mole fractions in favor of the other. It does characterize "lightness" of a constituent species. A "light" species, with $K > 1$, tends to concentrate in the vapor phase, whereas a "heavy" species, with $K < 1$, tends to concentrate in the liquid phase. In practice, at least one $K_i \geq 1$, and at least one $K_i \leq 1$.

The defining equation for K can be rearranged as $y_i = K_i x_i$. The sum $\sum_i y_i = 1$ then yields

$$\sum_i K_i x_i = 1 \tag{4-186}$$

With the alternative rearrangement $x_i = y_i/K_i$, the sum $\sum_i x_i = 1$ yields

$$\sum_i \frac{y_i}{K_i} = 1 \tag{4-187}$$

Thus for bubble point calculations, where the x_i are known, the problem is to find the set of K values that satisfies Eq. (4-186), whereas for dew point calculations, where the y_i are known, the problem is to find the set of K values that satisfies Eq. (4-187).

The *flash calculation* is a very common application of VLE. Considered here is the P, T flash, in which are calculated the quantities and compositions of the vapor and liquid phases in equilibrium at known T, P, and *overall* composition. This problem is determinate on the basis of Duhem's theorem: *For any closed system formed initially from given masses of prescribed chemical species, the equilibrium state is completely determined when any two independent variables are fixed.* The independent variables are here T and P, and systems are formed from given masses of nonreacting chemical species.

For F moles fed of a system with *overall* composition represented by the set of mole fractions $\{z_i\}$, let L represent the moles of the system that are liquid (mole fractions $\{x_i\}$) and let V represent the moles that are vapor (mole fractions $\{y_i\}$). The material balance equations are

$$L + V = F \qquad \text{and} \qquad z_i F = x_i L + y_i V \qquad i = 1, 2, \ldots, N \tag{4-188}$$

Rearranging for x_i and y_i yields

$$x_i = z_i/(1+(V/F)(K_i-1)), \quad y_i = z_i K_i/(1+(V/F)(K_i-1)) \tag{4-189}$$

Taking the difference in the sums results in the Rachford-Rice flash method [Elliott and Lira (2012, Sec. 10.3)]

$$\sum_i \frac{z_i(1-K_i)}{1+(V/F)(K_i-1)} = 0 \tag{4-190}$$

The initial step in solving a P, T flash problem is to find the value of V/F which satisfies Eq. (4-190), and then mole fractions are determined by Eq. (4-189).

Gamma/Phi Approach For many VLE systems of interest, the pressure is low enough that evaluation of $\hat{\phi}_i$ is usually by Eq. (4-157), based on the two-term virial equation of state. Liquid-phase behavior, on the other hand, uses activity coefficients γ_i, based on Eq. (4-146) applied to an expression for G^E/RT, as described in the section Models for the Excess Gibbs Energy.

Equation (4-135) may now be written as

$$y_i P \Phi_i = x_i \gamma_i P_i^{\text{sat}} \qquad i = 1, 2, \ldots, N \tag{4-191}$$

where

$$\Phi_i = \frac{\hat{\phi}_i}{\phi_i^{\text{sat}}} \exp\left[\frac{-V_i^L(P-P_i^{\text{sat}})}{RT}\right] \tag{4-192}$$

If evaluation of ϕ_i^{sat} is based on Eq. (4-70) evaluated at P^{sat}, and $\hat{\phi}_i$ by Eq. (4-157), this reduces to

$$\Phi_i = \exp\left[\frac{P(2\sum_k y_k B_{ki} - B) - P_i^{\text{sat}} B_{ii} - V_i^l (P - P_i^{\text{sat}})}{RT}\right] \tag{4-193}$$

The N equations represented by Eq. (4-191) in conjunction with Eq. (4-193) may be solved for N unknown phase equilibrium variables. For a multicomponent system the calculation is best done by computer.

Raoult's Law When Eq. (4-191) is applied to VLE for which the vapor phase is an ideal gas and the liquid phase is an ideal solution, it reduces to a very simple expression. For ideal gases, fugacity coefficients $\hat{\phi}_i$ and ϕ_i^{sat} are unity, and the right side of Eq. (4-192) reduces to the Poynting factor. For the systems of interest here, this factor is always very close to unity, and for practical purposes $\Phi_i = 1$. For ideal solutions, the activity coefficients γ_i are also unity, and Eq. (4-191) reduces to

$$y_i P = x_i P_i^{\text{sat}}, \qquad K_i = P_i^{\text{sat}}/P \qquad i = 1, 2, \ldots, N \tag{4-194}$$

an equation which expresses *Raoult's law*. It is the simplest possible equation for VLE and as such fails to provide a realistic representation of real behavior for most systems. Nevertheless, it is useful as a standard of comparison.

Modified Raoult's Law Of the qualifications that lead to Raoult's law, the one least often reasonable is the supposition of solution ideality for the liquid phase. Real solution behavior is reflected by values of activity coefficients that differ from unity. When γ_i of Eq. (4-191) is retained in the equilibrium equation, the result is the *modified Raoult's law*:

$$y_i P = x_i \gamma_i P_i^{\text{sat}}, \quad K_i = \gamma_i P_i^{\text{sat}}/P \quad i = 1, 2, \ldots, N \tag{4-195}$$

This equation is often adequate when applied to systems at low to moderate pressures and is therefore widely used. Bubble point and dew point calculations are only a bit more complex than the same calculations with Raoult's law.

For a bubble calculation, because $\sum_i y_i = 1$, Eq. (4-195) may be summed over all species to yield

$$P = \sum_i x_i \gamma_i P_i^{sat} \tag{4-196}$$

As discussed in relation to Eq. (4-139), the value of γ_i serves as a correction factor for the solute mole fraction concentration x_i to better account for the solute's true chemical potential–driven activity which determines phase equilibrium behavior and reactivity, where activity is given by $a_i = \gamma_i x_i$.

For dew calculation, Eq. (4-195) may be solved for x_i, in which case summing over all species yields

$$P = \frac{1}{\sum_i y_i / \gamma_i P_i^{sat}} \tag{4-197}$$

The application of this equation requires iteration because the values of γ_i cannot be determined without an estimate of $\{x_i\}$.

Example 4-11 Bubble, Dew, Azeotrope, and Flash Calculations As indicated by Example 4-8, a binary mixture in vapor/liquid equilibrium has 2 degrees of freedom. Thus of the four phase rule variables T, P, x_1, and y_1, two must be fixed to allow calculation of the other two, regardless of the formulation of the equilibrium equations. Modified Raoult's law [Eq. (4-195)] may therefore be applied to the calculation of any pair of phase rule variables, given the other two.

The necessary vapor pressures and activity coefficients are supplied by data correlations. For the system acetone(1)/n-hexane(2), vapor pressures are given by Eq. (4-15), with parameters for P_i^{sat} (kPa) and T (K),

i	A_i	B_i	C_i
1	14.3145	2756.22	−45.090
2	13.8193	2696.04	−48.833

Activity coefficients are calculated by Eq. (4-163), the Wilson equation, here adapted for a binary system in Eqs. (A) and (B) which will be referenced below:

$$\ln \gamma_1 = -\ln(x_1 + x_2\Lambda_{12}) + x_2\lambda \tag{A}$$

$$\ln \gamma_2 = -\ln(x_2 + x_1\Lambda_{21}) - x_1\lambda \tag{B}$$

where

$$\lambda \equiv \frac{\Lambda_{12}}{x_1 + x_2\Lambda_{12}} - \frac{\Lambda_{21}}{x_2 + x_1\Lambda_{21}}$$

By Eq. (4-164)

$$\Lambda_{ij} = \frac{V_j}{V_i} \exp\frac{-a_{ij}}{RT} \qquad i \neq j$$

with parameters [Gmehling et al., *Vapor-Liquid Data Collection*, Chemistry Data Series, vol. 1, part 3, DECHEMA, Frankfurt/Main, 1983]

a_{12} cal mol^{-1}	a_{21} cal mol^{-1}	V_1 cm^3 mol^{-1}	V_2 cm^3 mol^{-1}
985.05	453.57	74.05	131.61

When T and x_1 are given, the calculation is direct, with final values for vapor pressures by Eq. (4-15) and activity coefficients from Eqs. (A) and (B) above. In all other cases either T or x_1 or both are initially unknown, and calculations require iteration. For each part of this example, results are tabulated in the table at the end where given values are in italic; calculated values are in boldface.

a. BUBL P calculation: Find y_1 and P, given $x_1 = 0.40$ and $T = 325.15$ K (52°C). Noting that T and x_1 are given and following the procedure above yields the values listed in the summary table in the following column. Equations (4-196) and (4-195) then become

$$P = x_1\gamma_1 P_1^{sat} + x_2\gamma_2 P_2^{sat} = (0.40)(1.8053)(87.616) + (0.60)(1.2869)(58.105)$$

$$= 108.134 \text{ kPa}$$

$$y_i = \frac{x_i\gamma_i P_i^{sat}}{P} = \frac{(0.40)(1.8053)(87.616)}{108.134} = 0.5851$$

b. DEW P calculation: Find x_1 and P, for $y_1 = 0.4$ and $T = 325.15$ K (52°C). With x_1 an unknown, the activity coefficients cannot be immediately calculated. However, an iteration scheme based on Eq. (4-197) is easily developed. Starting values result from setting each $\gamma_i = 1$ and refining by using Eqs. (A) and (B) after finding $\{x\}$; results of successive substitution of $\{x\}$ to refine γ_i values are listed in the accompanying table.

c. BUBL T calculation: Find y_1 and T, given $x_1 = 0.32$ and $P = 80$ kPa. With T unknown, neither the vapor pressures nor the activity coefficients can be initially calculated. An iteration scheme based on Eq. (4-196) matches P and results in values listed in the accompanying table.

d. DEW T calculation: Find x_1 and T for $y_1 = 0.60$ and $P = 101.33$ kPa. Start with $\gamma_i = 1$. Iterate on $\sum y_i P/(\gamma_i P_i^{sat}) = 1$. Find $x_i = y_i P/(\gamma_i P_i^{sat})$ and new values of γ_i using Eqs. (A) and (B). Iterate. Results are listed in the accompanying table.

e., f. Azeotrope calculations: Find the azeotrope composition and (*e*) P at 46°C and (*f*) T at 101.33 kPa. As noted in Example 4-8, only a single degree of freedom exists for this special case. The most sensitive quantity for identifying the azeotropic state is the relative volatility defined in Eq. (4-184). Because $y_i = x_i$ for the azeotropic state, $\alpha_{12} = 1$. Substitution for the K ratios by Eq. (4-195) provides an equation for calculation of $\alpha_{12} = \gamma_1 P_1^{sat}/(\gamma_2 P_2^{sat})$. Because α_{12} is a monotonic function of x_1, the test of whether an azeotrope exists at a given T or P is provided by values of α_{12} in the limits of $x_1 = 0$ and $x_1 = 1$. If *both* values are either >1 or <1, no azeotrope exists. But if one value is <1 and the other >1, an azeotrope necessarily exists at the given T or P. Given T, the azeotropic composition and pressure is found by seeking the value of x that makes $\alpha_{12} = 1$ using bubble P calculations at each guess. At 46°C, the limiting values of α_{12} are 8.289 at $x_1 = 0$ and 0.223 at $x_1 = 1$. Similarly, given P, one finds the azeotropic composition and temperature, using bubble T calculations at each guess. The results are tabulated below.

g. Isothermal flash: $z_1 = 0.4000$ at $T = 325.15$ K and $P = 101.33$ kPa. Determine V/F, x_1, and y_1. Looking at the summary table for parts (*a*) and (*b*) reveals that the pressure is between the bubble and dew conditions; thus there will be two phases. For a binary system, a trial and error method can be formulated by starting with x and using Eq. (4-189) to find V/F and checking if the y values sum to 1. However, this method cannot be generalized to a multicomponent system. A more general method is to start with Raoult's law and use Eq. (4-190) to find V/F, generate $\{x\}$ from Eq. (4-189), then use modified Raoult's law (4-195) for K_i and loop back to Eq. (4-190), repeating the loop until it converges. The converged value is $V/F = 0.578$.

	T (K)	P_1^{sat} (kPa)	P_2^{sat} (kPa)	γ_1	γ_2	x_1	y_1	P (kPa)
a.	*325.15*	87.616	58.105	1.8053	1.2869	*0.4000*	**0.5851**	**108.134**
b.	*325.15*	87.616	58.105	3.5535	1.0237	**0.1130**	*0.4000*	**87.939**
c.	**317.24**	65.830	43.591	2.1286	1.1861	*0.3200*	**0.5605**	*80.000*
d.	**322.98**	81.125	53.779	1.6473	1.3828	**0.4550**	*0.6000*	*101.330*
e.	*319.15*	70.634	46.790	1.2700	1.9172	**0.6445** =	**0.6445**	**89.707**
f.	**322.58**	79.986	53.021	1.2669	1.9111	**0.6454** =	**0.6454**	*101.330*
g.	*325.15*	87.616	58.105	2.5297	1.0997	**0.2373**	**0.5190**	*101.330*

Equation-of-State Approach The gamma/phi method is generally applicable to systems away from critical points. Equations of state can treat near-critical conditions and the transition to supercritical conditions seamlessly. Of course, there is a cost in terms of robustness of convergence and computational complexity. By Eq. (4-134),

$$x_i\hat{\phi}_i^l = y_i\hat{\phi}_i^v \Rightarrow K_i = \hat{\phi}_i^l/\hat{\phi}_i^v, \quad i = 1, 2, \ldots, N \tag{4-198}$$

This introduces compositions x_i and y_i into the equilibrium equations, but neither is explicit, because the $\hat{\phi}_i$ are functions not only of T and P, but of composition. Thus, Eq. (4-198) represents N complex relationships connecting T, P, $\{x_i\}$, and $\{y_i\}$ where the fugacity coefficients are given by an equation of state, for example Eq. (4-158).

For mixtures the presumption is that the equation of state has exactly the same form as when written for pure species. Liquid and vapor mixtures in equilibrium in general have different compositions, and thus the B and $A(T)$ parameters differ for each phase, resulting in a different isotherm for each phase. The PV isotherms generated by an equation of state for these different compositions are represented in Fig. 4-3 by two similar lines: the solid line for the liquid-phase composition and the dashed line for the vapor-phase composition. They are displaced from each other because the equation-of-state parameters are different for the two compositions.

Each isotherm in the illustration includes a liquid root segment (steep segment to the left of the minimum) and vapor root segment (segment to the right of the maximum). Each liquid segment contains a bubble point (saturated liquid), and each right segment contains a dew point (saturated vapor). Because these points for a given line are for the same composition, they do not represent phases in equilibrium and do not lie at the same pressure. Shown in the figure is a bubble point B on the solid line and a dew point D on the dashed line. Because they lie at the same P, they represent phases in equilibrium, and the lines are characterized by the liquid and vapor compositions.

As a consequence of the isotherm shape depending on composition, a particular iteration may result in a trial composition such that the root pertaining to the phase of interest is imaginary. For example, consider a

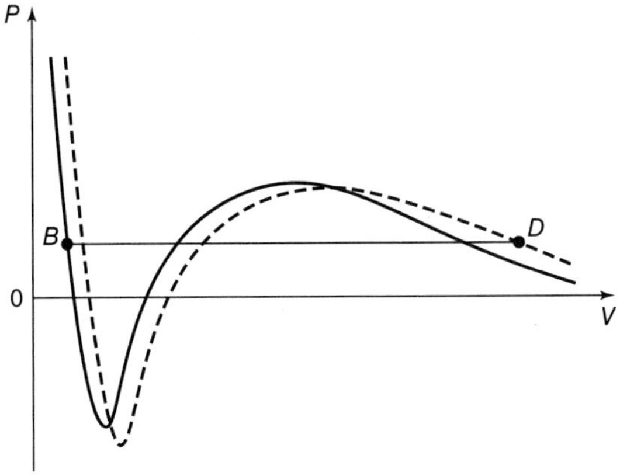

FIG. 4-3 Two *PV* isotherms at the same *T* for mixtures. The solid line is for a liquid-phase composition; the dashed line is for a vapor-phase composition. Point *B* represents a bubble point with the liquid-phase composition; point *D* represents a dew point with the vapor-phase composition. When these points lie at the same *P* (as shown), they represent phases in equilibrium [Smith, Van Ness, and Abbott, *Introduction to Chemical Engineering Thermodynamics*, 7th ed., McGraw-Hill, New York, 2005, p. 560].

mixture of methane and nitrogen at 150 K and 10 bar. The nitrogen is above its critical temperature, so the pressure vs. density can have only one real root, while pure methane at these conditions has three roots, with a local minimum between the liquid root and the middle root. As the composition varies, the pressure vs. density evolves such that the liquid root rises to the local minimum in the cubic function, then becomes imaginary as the local minimum detaches from the *x* axis. As the composition approaches pure nitrogen, the local minimum disappears and there is only an inflection. A similar phenomenon occurs when the pressure is such that the vapor root disappears over a range of compositions. The problem is exacerbated near the critical point, where the distinctions between the phases are small and very sensitive to the trial pressure and composition. Gundersen has shown that convergence can be enhanced by taking the local extremum as the trial value for the missing density root [T. Gundersen, *Comp. Chem. Eng.* **6:** 245 (1982)]. This works as long as the local extremum exists. For computations near the critical region, it may be necessary to first locate the critical point pertaining to a particular *T* and *P*, then work backward to the conditions of interest. A pseudo-root technique can also be used [P. M. Mathias, J. F. Boston, and S. Watanasiri, *AIChE J.* **30:** 182–186 (1984)].

For a *BUBL P* calculation, the temperature and the liquid composition are known, and this fixes the location of the *PV* isotherm for the composition of the liquid phase (solid line). The problem then is to locate a second (dashed) line for a vapor composition such that the line contains a dew point *D* on its vapor segment that lies at the pressure of the bubble point *B* on the liquid segment of the solid line. This pressure is the phase equilibrium pressure, and the composition for the dashed line is that of the equilibrium vapor.

Because of limitations inherent in quadratic mixing and combining rules, such as those given by Eqs. (4-86) to (4-88), the equation-of-state approach has found primary application to systems exhibiting modest deviations from ideal solution behavior in the liquid phase, e.g., to systems containing hydrocarbons and cryogenic fluids. However, since 1990, extensive research has been devoted to developing mixing rules that incorporate the excess Gibbs energy or activity coefficient data available for many systems. The extensive literature on this subject is reviewed by Valderrama [*Ind. Eng. Chem. Res.* **42:** 1603–1618 (2003)] and by Twu, Sim, and Tassone [*Chem. Eng. Progr.* **98**(11): 58–65 (Nov. 2002)]. For example, the UNIFAC model has also been combined with the predictive Soave-Redlich-Kwong (PSRK) equation of state. The procedure is most completely described (with background literature citations) by Horstmann et al. [*Fluid Phase Equilib.* **227:** 157–164 (2005)]. The modified Huron-Vidal and Wong-Sandler methods are the primary examples of this approach. Using these methods, it is possible to describe complex phase behavior with similar accuracy to that of activity models. This may seem like a desirable "universal" theory. Unfortunately, solving the phase equilibrium problem is slower and less reliable than the gamma/phi approach because of the coupling between root selection, density iteration, and composition dependence when applying the equation of state approach.

Solute/Solvent Systems—Henry's Law Henry's law often provides a useful framework to correlate and understand solubilities, especially when the component concentrations are low or when solutes (e.g., ions or supercritical components) do not have a valid pure-component reference state. A prevalent example is the case of liquid solubility of "noncondensable" gases, i.e., components above their critical temperature such as the nitrogen in the methane + nitrogen discussed above. These gases are common to the vast majority of chemical processes because catalytic conditions are often at sufficiently high temperatures that gas by-products are formed, even in cases when gaseous species are not explicitly involved in the desired reactions. Consideration of gas solubilities is critical for safety, for example, when a total condenser is used in a downstream distillation column and the (small but finite) gases could accumulate overhead in the long term.

These considerations are best understood by examining the rigorous vapor-liquid phase-equilibrium relationship for the solute given by Eq. (4-136). Note the superscript in γ^*. This means that the liquid fugacity approaches h_i at Henry's reference state, where γ^* approaches unity. Since most systems exhibit positive deviations from ideal solution behavior, it is common for γ^* to be less than unity while γ (based on the Lewis-Randall perspective) is greater than unity, even for the same system. For nonreacting systems, Eq. (4-136) provides a powerful general framework to analyze and understand even subtle variations in solute solubility, and in many cases the simple, original Henry law may provide an adequate correlation. When a component follows Henry's law, the component activity coefficient must be normalized to approach unity at infinite dilution. The relation with the Lewis-Randall activity coefficient is

$$\ln \gamma_i^* = \ln \gamma_i - \ln \gamma_i^\infty \qquad (4\text{-}199)$$

Because the Gibbs-Duhem equation is still valid, any activity model can be used as long as Eq. (4-199) is applied.

Types of Henry Constant Data or Units of Measure Henry's notion that solute solubility is proportional to partial pressure is widely useful, and it has been successfully applied in many branches of science and engineering. Data for Henry constants are available from a wide variety of sources. Unfortunately, these data can be confusing and error-prone because of variations that arise from the culture of the disciplines, and the constants can be inverted relative to one another. Because Henry's constant varies over many orders of magnitude, errors of units are difficult to detect simply by inspection. This section presents some of the ways in which the equivalents of Henry's constant may be converted to the form of Eq. (4-136). To begin, we should note the form implied by Henry's original definition, which predominates in the older literature. These Henry's constants have been defined as *absorption coefficients*. For example, the Ostwald coefficient is defined as the volume of gas absorbed at a temperature and a reference pressure (typically 101 kPa) divided by the volume of the solvent at the same temperature and pressure. The Ostwald coefficient L_i is quantitatively related to the Henry solubility constant by

$$L_i(T, P) = RT/(h_i V_S) \qquad (4\text{-}200)$$

Note that the Ostwald coefficient varies inversely with Henry's constant. Calculation of the Ostwald coefficient from Henry's constant is illustrated in Table 4-3, which presents calculations of L_i by Rettich et al. [*J. Chem. Thermo.* **32:** 1145 (2000)] for the solubility of oxygen in water. Note that the values decrease with increasing temperature, whereas the analog of vapor pressure would increase. The values of the saturated molar volume of water have been taken from the DIPPR 801 database. The Bunsen coefficient is a special case of the Ostwald coefficient where the temperatures are corrected to a reference value, usually 273.15 K. Other subtle variations in the definition of absorption coefficients are presented by Gamsjäger et al. [*Pure Appl. Chem.* **82:** 1137 (2010)].

TABLE 4-3 Illustrative Example* of Using Eq. (4-200) to Calculate L_i from h_i

Temperature (K)	$h_{O_2} \times 10^{-9}$ Pa	L_{O_2}	V_w (m^3/kmol)	L_{O_2} from Eq. (4-200)
283.15	3.2966	0.039627	0.018024	0.039622
298.15	4.4038	0.031153	0.018071	0.031151
313.15	5.3985	0.026561	0.018158	0.026561
328.15	6.1917	0.024104	0.018278	0.024108

V_w = molar volume of water

*Data (columns 1 to 3) have been taken from Rettich et al. for the solubility of oxygen in water, and the molar volumes of water are from the DIPPR 801 database. The fifth column shows calculations of L_{O_2} by Eq. (4-200).

The most common modern variation of Eq. (4-136) is to change the liquid composition from mole fraction to molarity or molality, and here the units of Henry's constant change from pressure to pressure divided by concentration. In the many cases, the solvent is water, so commonly that it may be (questionably) taken for granted that the solvent is water when referring to Henry's law. In this common form, concentrations are assumed dilute such that $\gamma_i^* \approx 1$ and pressures are low. Then in molarity units

$$y_i P = h_i^C C_i \tag{4-201}$$

where C_i is the concentration (mol/L) of the solute in the aqueous liquid phase. This is the form of Henry's law encountered in environmental applications (Sec. 22) and distillation involving gases (Sec. 13).

Another common form of Henry's law is in the form of the Henry's solubility constant, here written using concentration units of molality

$$k_H = m_i/(y_i P) \quad \text{or} \quad y_i P = (1/k_H) m_i \tag{4-202}$$

This form is tabulated in the NIST Chemistry Webbook [R. Sander, "Henry's Law Constants" in NIST Chemistry Webbook, NIST Standard Reference Database Number 69, eds. P. J. Linstrom and W. G. Mallard, National Institute of Standards and Technology, Gaithersburg, Md., 20899, http://webbook.nist.gov (retrieved November 8, 2016)]. Readers are cautioned that the symbols for Henry's constants are inconsistent in the literature. Here we use h_i as the volatility constant, and we use k_H for the solubility constant. In the atmospheric literature, H_i is used for the solubility constant and K_H for the volatility constant [Sander, *Atmos. Chem. Phys.* **15:** 4399–4981 (2015)]. Units must be included and consulted to avoid misapplication.

In general, specific applications can be expected to have dedicated computerized databases that serve them best, with coverage of compounds peculiar to those applications. In cases where LLE data are available but not Henry's constants, Henry's constant can be estimated by

$$h_i = [(1 - x_w^O) P_i^{\text{sat}}]/x_i \tag{4-203}$$

where x_w^O is the equilibrium mole fraction of water in the organic phase (often negligible), P_i^{sat} is the vapor pressure of the organic solute, and x_i is the equilibrium mole fraction of the solute in the water phase. Any single tabulated Henry constant is published with its measurement temperature, typically 20 or 25°C.

Table 4-4 provides guidance on the conversion factors to use when the Henry's constant is provided in units including the liquid-phase concentration. Note that the conversion factor includes the molecular weight of the solvent when molality units are used, and it becomes temperature-dependent when molarity units are used since the solvent density at the temperature of interest is needed.

TABLE 4-4 Factors to Convert Henry's Constant to the One Defined by Eq. (4-125) When Unit of Liquid Concentration Changes

Type of liquid concentration	Units of Henry constant	Conversion factor	Specific conversion factor with water as solvent	Comment
Mole fraction [Eq. (4-125)]	Pa	1	1	—
Molality	Pa·kg/mol	$1000/MW_{\text{solvent}}$	55.508	Water MW is 18.015
Molarity	Pa·m³/mol	$1000 \times \rho_{\text{solvent}}$	55,339	Water as solvent at 298.15 K, with density of 55.339 kmol/m³

Example 4-12 Solubility of Oxygen in Water by Henry's Law The NIST Webbook lists Henry's solubility constant for oxygen as 0.0013 mol/(kg·bar) at 25°C and the derivative $d(\ln k_H)/d(1/T)$ as 1700 K. Estimate the solubility of oxygen in water at 1 bar and 35°C in ppm by weight (ppmw), assuming air is 21 mol% oxygen.

Solution $k_H(35°C) = 0.0013 \exp[1700(1/308.15 - 1/298.15)] = 0.00108$ mol/kg·bar. For this instance of Henry's law, $\hat{f}_i = y_i P = m_i/k_H$, where m_i is the molality of O_2. Solving gives $m_i = 0.21(1)0.00108 = 0.00023$ mol/kg. Converting to a mass basis, 0.00023·0.032 kg/mol = 7.3 ppmw, where we have neglected the contribution of the mass of oxygen to the total mass of the solution. Note that k_H decreased as the temperature increased for oxygen in water, causing the molality, and hence the ppmw, to decrease. For some solutes, e.g., hydrogen, the solubility may actually increase with increases in temperature.

Temperature Dependence of Henry Constants Henry constants have a significant nonlinear temperature dependence, and ignoring the temperature dependence usually leads to large inaccuracies. Smith and Harvey [*Chem. Eng. Progr.* **103:** 33 (2007)] provide guidance on ways to estimate the temperature dependence of Henry constants. They noted that for subcritical solutes, Henry constants may be approximated as the product of the vapor pressure and the infinite-dilution activity coefficient

$$h_i = \gamma_i^\infty P_i^{\text{sat}} \tag{4-204}$$

If the change in activity coefficient with temperature is small compared to the change in vapor pressure with temperature, then

$$h_i(T_2) = h_i(T_1)(P_2^{\text{sat}}/P_1^{\text{sat}}) \tag{4-205}$$

The simple extrapolation method is quite good up to 50°C and reasonable up to 80°C, though improved results are given by including θ discussed in the section *Temperature Dependence of Infinite-Dilution Activity Coefficients*. However, the vapor pressure–based extrapolation method is poor when Henry's constant goes through a maximum, as often occurs, even for subcritical components. For gaseous components, Henry constants generally go through a maximum, as shown for CO_2 [Smith and Harvey, *Chem. Eng. Progr.* **103:** 33 (2007)]. Figure 4-4, adapted from Smith and Harvey, shows the behavior of Henry's constant for toluene in water using Eq. (4-205) and also by Eq. (4-204) calculating the infinite-dilution activity coefficients as a function of temperature. Activity coefficients are difficult to predict accurately at higher temperature (as depicted by the dashed lines), and hence the uncertainty of the Henry's constant estimate also rises. It is thus better if an estimate of the temperature dependence of the activity coefficient is also used. Hwang et al. [*Ind. Eng. Chem. Res.* **31:** 1759, 1992] studied steam stripping of organic pollutants from water and showed that using the temperature dependence of the activity coefficient estimated by the UNIFAC group-contribution method often enables an improved way to estimate the temperature dependence of h_i. An improved method is referenced in the section *Temperature Dependence of Infinite-Dilution Activity Coefficients.* If data are available over a wide range of temperatures or if the temperature dependence of Henry's constant is being fit to data, an extended equation for the temperature dependence may be used. For example, Henry's constant for H_2S in H_2O is given by

$$h_i(\text{Pa}, T \text{ in K}) = 475.175 - 16{,}752.5/T - 74.904 \ln T + 0.086193T \tag{4-206}$$

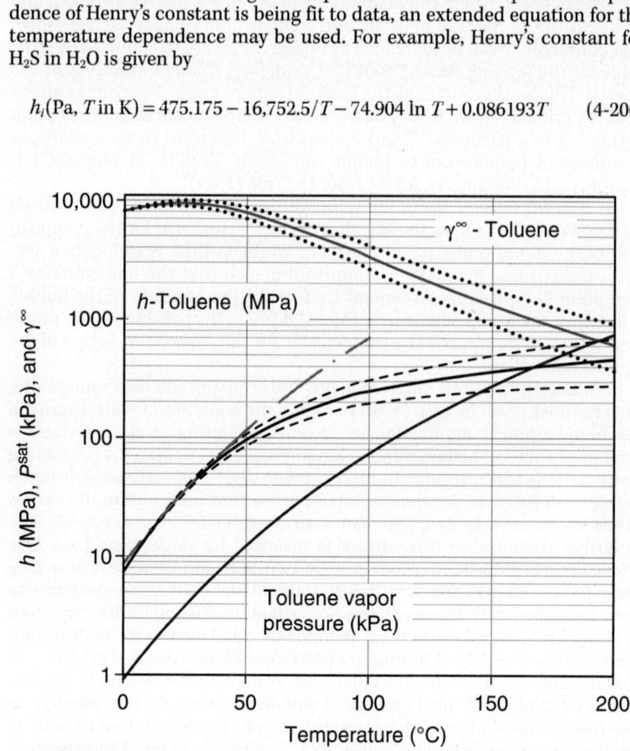

FIG. 4-4 Temperature dependence of Henry's constant for toluene in water, showing the performance of the vapor-pressure extrapolation method starting at 25°C compared to the full calculation including the temperature dependence of the activity coefficient. The dash-dot line is extrapolated as explained in the text. The dotted and dashed lines show uncertainty in γ^∞ and h.

Extension of Henry's Law to Multicomponent Systems Broadly, Henry's constant is an infinite-dilution property that changes when other solutes (e.g., salts) or solvents are added. The Henry constant data may be available in pure solvents when Henry's constant in the mixed solvent is needed. The solubility of gaseous solutes in aqueous solutions was first studied over a century ago by Setchenow [*Ann. Chim. Phys.* **25:** 226 (1892)]. He developed the following expression:

$$h_i = h_i^\circ \exp(kC_S) \tag{4-207}$$

where h_i° is Henry's constant in the pure solvent, C_s is the concentration (e.g., mol/L) of the additive (usually a salt), and k is an empirical parameter. If k is positive, increased concentration of the additive increases Henry's constant, and this is referred to as "salting out." This is the common case since a negative k or "salting in" is rare.

Carroll [*Chem. Eng. Progr.* **88:** 53 (1992)] has discussed in detail the extension of Henry's law to multicomponent systems. Carroll illustrated salting out by using the data of Gamsjäger and Schindler [*Helv. Chim. Acta* **52:** 1395 (1969)] for the effect of dissolved NaCl on Henry's constant of H_2S in H_2O. Setchenow's simple correlation with $k = 0.1742$ provides a good fit of the data. Dankwerts [*Gas-Liquid Reactions*, McGraw-Hill, New York, 1970, p. 18] provided a correlation to estimate k, and in this particular case Carroll (1992) noted that the predicted value of $k = 0.182$ enables an accurate estimate. Finally, note that Henry's constant of H_2S in pure H_2O is 543 bar, and this value agrees within 10 percent with the value of 600 bar calculated by Eq. (4-206), and this is typical of the uncertainty of Henry constant data.

O'Connell and Prausnitz [*Ind. Eng. Chem. Fundam.* **4:** 347 (1964)] derived an equation for Henry's constant of solute i in a binary solvent consisting of components 1 and 2:

$$\ln h_i = x_1 \ln h_i^{(1)} + x_2 \ln h_i^{(2)} + a_{12} x_1 x_2 \tag{4-208}$$

where $h_i^{(1)}$ and $h_i^{(2)}$ are Henry constants of solute i in solvents 1 and 2, respectively, and a_{12} is a semiempirical parameter representing the interactions between the solvent species. This approximation has been evaluated by analyzing Henry constants at 20°C of methane in methanol-water and ethanol-water mixtures reported by Tokunaga and Kawai [*J. Chem. Eng. Japan* **8:** 326 (1975)]. Experience indicates that Eq. (4-208) enables a good fit, and Carroll (1992) has shown how the equation can be extended to multicomponent solvents, but, in general, some mixed solvent data are needed to establish the value of a_{12}.

Sources of Henry Constant Data There are many useful sources of Henry constant data. An excellent source is Sec. 2 of this handbook, and the references provided there. Water is a very important solvent due to its environmental importance, and many sources of Henry's constants in water are available, for example, the extensive compilation by Sander [*Atmos. Chem. Phys.* **15:** 4399 (2015)]. The technical literature is vast, and many search engines are available, e.g., Google Scholar, but the ThermoLit resource generally provides the most reliable place to start. Henry's constants are also available through commercial process simulators, but these values need to be validated using literature or proprietary sources.

Predictions of Henry Constants for Gaseous Species Another problem lies in predicting Henry constants in the absence of experimental data. A convenient equation for estimating gas fugacities in the Lewis-Randall perspective is

$$\log(f_i^L/P_c) = 7(1+\omega)(1-1/T_r)/3 - 3\exp(-E^H/T_r) \tag{4-209}$$

where $E^H = 3$ for most compounds. For $\omega = 0.21$, this equation reproduces the correlation of Prausnitz and Shair [*AIChE J.* **7:** 682 (1961)], effectively extrapolating the shortcut vapor pressure equation. Note that this approach applies the Lewis-Randall perspective, so $h_i = f_i^L \gamma_i^\infty$. Similar correlations have been suggested by Grayson and Streed [Paper 20-PO7, 6th World Petroleum Conference, Frankfurt, June 1963], and Chao and Seader [*AIChE J.* **7:** 598 (1961)]. A graph of this correlation shows that the component fugacity actually decreases with increasing temperature for $T_r > \sim 2$ (see Fig. 4-8). This behavior is counterintuitive, but it is also borne out in EOS models of gas solubility. For a component such as hydrogen, the gas solubility increases with increasing temperature at all common conditions, although it is generally quite a small solubility nevertheless. To obtain a general estimate of Henry's constant, it is necessary to include an estimate for the activity coefficient at infinite dilution. Readers should experiment with their process simulators to infer how gaseous species are treated by comparing multiple solvents and models such as UNIFAC and NRTL. As an example, Henry's constants for H_2 in paraffins can be estimated by Eq. (4-209) with $E^H = 14.1$, $T_c^{H_2} = 42$ K, $P_c^{H_2} = 19$ bar, $\omega^{H_2} = 0$, and an expression developed by regressing infinite-dilution activity coefficients:

$$\text{For } H_2\text{: } \ln(\gamma^\infty) = [3.15 - \ln(M_w^{\text{solv}})]\,(1 - T_c^{\text{solv}}/T) \tag{4-210}$$

where M_w^{solv} and T_c^{solv} are the molecular weight and critical temperature of the solvent. These pseudocritical constants for H_2 were adapted from Prausnitz, Lichtenhaler, and Azevedo (1999, pp. 172–173).

Example 4-13 Solubility of Hydrogen in Hydrocarbons Estimate the mass fraction of hydrogen in *n*-hexadecane and *n*-triacontane at 50°C and 100 bar.

Solution At these conditions, we can ignore the composition of the solvent in the vapor phase. Applying the pseudocritical constants in Eq. (4-209), $1/T_r = 42/323.15 = 0.447$, $f_i^L = (19)10^{[7(1-0.447)/3-3\exp(-14.1(0.447))]} = 674$ bar. By Eq. (4-210), $\gamma^\infty = \exp[(3.15 - \ln 226)(1-723/323.15)] = 16.6$; $f_i = 100 = 674(16.6)x_i$. Solving gives $x_i = 100/(674 \cdot 16.6) = 0.0089$. Converting to mass fraction, we obtain 82 ppmw. Repeating for triacontane, we obtain $\gamma^\infty = 106$ and 6.6 ppmw. The value of 82 for hexadecane compares to 54 ppmw estimated by the method of Trinh et al, *J. Chem. Eng. Data* **61:** 19 (2016).

An alternative to using a hypothetical liquid fugacity such as the Prausnitz and Shair approach is to examine the bubble pressure in the dilute limit using an EOS approach. An estimate of Henry's constant can be inferred by effectively simulating the experimental conditions using an EOS model. Generally, this approach would be most amenable to the Lewis-Randall perspective. It has been observed that EOS models can reproduce the observed maximum in Henry's constant for gaseous components, at least qualitatively.

LIQUID/LIQUID AND VAPOR/LIQUID/LIQUID EQUILIBRIA

Equation (4-127) is the basis for both liquid/liquid equilibria (LLE) and vapor/liquid/liquid equilibria (VLLE). Thus for LLE with superscripts α and β denoting the two phases, Eq. (4-127) is written as

$$\hat{f}_i^\alpha = \hat{f}_i^\beta \qquad i = 1, 2, \ldots, N \tag{4-211}$$

Using modified Raoult's law for fugacities and canceling the P_i^{sat} values gives

$$x_i^\alpha \gamma_i^\alpha = x_i^\beta \gamma_i^\beta \qquad i = 1, 2, \ldots, N \tag{4-212}$$

For most LLE applications, the effect of pressure on γ_i can be ignored, and Eq. (4-212) then constitutes a set of N equations relating equilibrium compositions to one another and to temperature. For a given temperature, solution of these equations requires a single expression for the composition dependence of G^E suitable for both liquid phases. Not all expressions for G^E suffice, even in principle, because some cannot represent liquid/liquid phase splitting. The UNIQUAC equation is suitable, and therefore prediction is possible by UNIFAC models. A special table of parameters for LLE calculations is given by Magnussen et al. [*Ind. Eng. Chem. Process Des. Dev.* **20:** 331–339 (1981)].

A comprehensive treatment of LLE is given by Sorensen et al. [*Fluid Phase Equilibr.* **2:** 297–309 (1979); **3:** 47–82 (1979); **4:** 151–163 (1980)]. Data for LLE are collected in a three-part set compiled by Sorensen and Arlt [*Liquid-Liquid Equilibrium Data Collection*, Chemistry Data Series, vol. 5, parts 1–3, DECHEMA, Frankfurt am Main, 1979–1980]. For vapor/liquid/liquid equilibria, Eq. (4-127) becomes

$$\hat{f}_i^\alpha = \hat{f}_i^\beta = \hat{f}_i^v \qquad i = 1, 2, \ldots, N \tag{4-213}$$

where α and β designate the two liquid phases. With activity coefficients applied to the liquid phases and fugacity coefficients to the vapor phase, the $2N$ equilibrium equations for subcritical VLLE are

$$\left.\begin{aligned} x_i^\alpha \gamma_i^\alpha f_i^\alpha &= y_i \hat{\phi}_i P \\ x_i^\beta \gamma_i^\beta f_i^\beta &= y_i \hat{\phi}_i P \end{aligned}\right\} \quad \text{all } i \tag{4-214}$$

As for LLE, an expression for G^E capable of representing liquid/liquid phase splitting is required; as for VLE, a vapor-phase equation of state for computing the ϕ_i is also needed.

TRENDS IN PHASE BEHAVIOR

Thermophysical properties usually and fortunately fall into regular patterns within and among families of compounds, and these patterns are useful to fill gaps in measurements, to identify outliers that are likely in error, and to educate chemical engineers to anticipate expected behaviors. This idea is especially attractive today because large databases of thermophysical properties are widely available. These databases can easily generate patterns, but also should be tested to identify errors and outliers. The patterns of behavior are more evident for fluid properties than for solid properties. This section provides representative examples of property patterns for the phase behavior of pure fluids and mixtures.

PURE FLUIDS

Figure 4-5 presents a plot of vapor pressures for the 1-alcohols from methanol to 1-docosanol ($C_{22}H_{46}O$), where the data have been taken from the DIPPR 801 database [R. L. Rowley et al., *DIPPR Data Compilation of Pure Chemical Properties*, Design Institute for Physical Properties, AIChE, New York, 2006]. Korsten [*Ind. Eng. Chem. Res.* **39:** 813 (2000)] relates the slopes to the molecular weight within each functional class, and suggests that $\log P^{sat}$ versus $1/T^{1.3}$ is linear, though deviation is evident in Fig. 4-5. Although patterns are evident for fluid properties like the critical points, a pattern is usually not evident for solid properties like the triple point. Extrapolation of the vapor pressures above the critical temperature suggests that they meet at the "infinite point," which was first suggested by Cox [*Ind. Eng. Chem.* **15:** 592 (1923)]. The infinite point remains useful as a visualization tool for the vapor pressures of a family of compounds since it illustrates and explains the rule of thumb that "experimental vapor pressures of a family of compounds do not cross." The extrapolated vapor pressures cross at a hypothetical temperature higher than all the critical temperatures of the members of the family. The pattern is useful to interpolate for missing family members.

Another useful pattern is illustrated in the Othmer plot [*Ind. Eng. Chem.* **32:** 841 (1940)] of vapor-pressure ratios. Figure 4-6 presents the ratio of the vapor pressure of the 1-alcohols to that of water plotted against the vapor pressure of water. For each point on the curve, the same temperature is used to calculate the vapor pressure of the 1-alcohol and water. The chart provides useful education about relative vapor pressures and enthalpies of vaporization (related to the slope of the vapor-pressure curve by the Clausius-Clapeyron equation). Methanol and ethanol have higher vapor pressures than water, and the slopes of their Othmer curves are negative, which means that their enthalpies of vaporization are lower than that of water. The vapor pressure of 1-propanol is very close to that of water, and so is its enthalpy of vaporization. The higher alcohols exhibit successively decreasing vapor pressures and also successively increasing enthalpies of vaporization. The Othmer chart also provides the relative volatility under the ideal solution assumption.

MIXTURES

Solvents such as dimethyl formamide (DMF) and acetonitrile (ACN) are used in extractive-distillation processes that recover 1,3-butadiene from steam-cracker hydrocarbons. Extractive distillation requires accurate correlation of the activity coefficients because the vapor pressures of the hydrocarbons in the feed mixture are very close, which facilitates azeotrope formation, and the basis of extractive distillation is the varying activity coefficients of the different hydrocarbons in the polar solvents. Table 4-5 presents experimental data for the infinite-dilution activity coefficients of various hydrocarbons in DMF and ACN at 313 K, which has been chosen as a representative value since it is typical of temperatures encountered in butadiene extractive-distillation columns. The infinite-dilution activity coefficients have been reported here since at plant conditions the solubility of the hydrocarbons in the liquid phase is relatively low. Table 4-5 demonstrates that the data fall into the pattern expected from "thermodynamic intuition." The infinite-dilution activity coefficients progressively decrease as the hydrocarbon class goes from paraffin to olefin to diolefins to triple bond to olefin plus triple bond. This, of course, is the reason why DMF and ACN are effective as extractive solvents, but the fact that the activity coefficients fall into the expected pattern provides confidence in the experimental data and offers a method to fill in data gaps. Figure 4-7 graphically illustrates

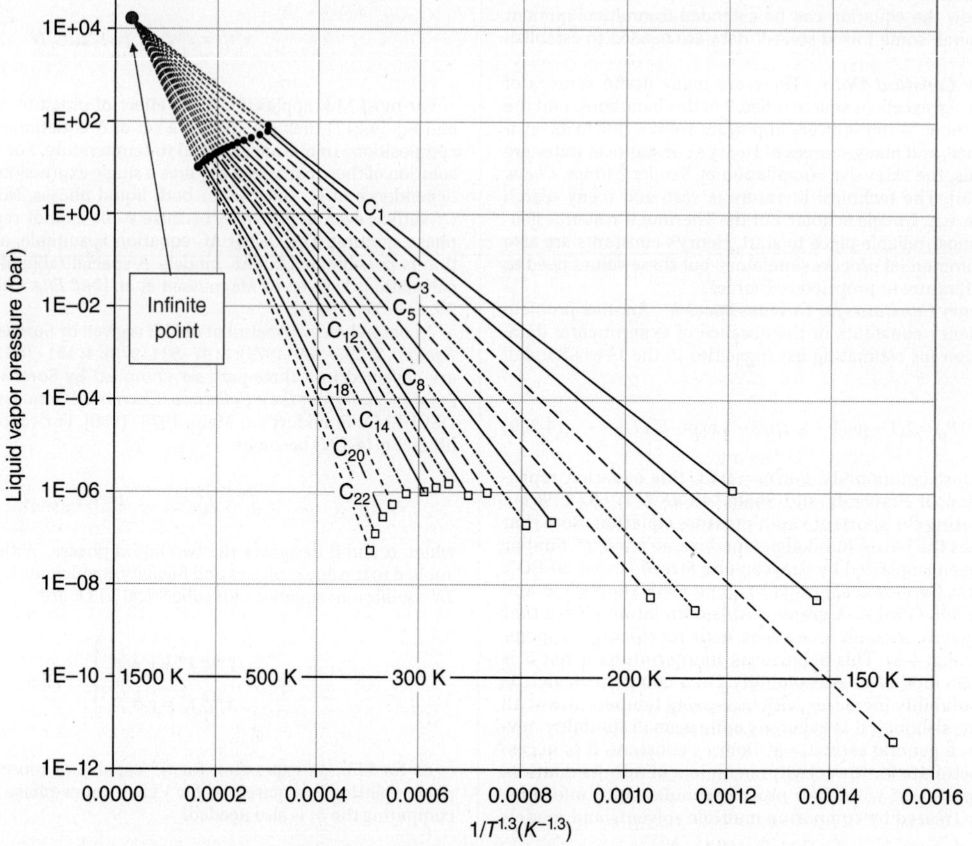

FIG. 4-5 Vapor pressures of the 1-alcohols, including the critical points (solid circles •) and triple points (open squares □).

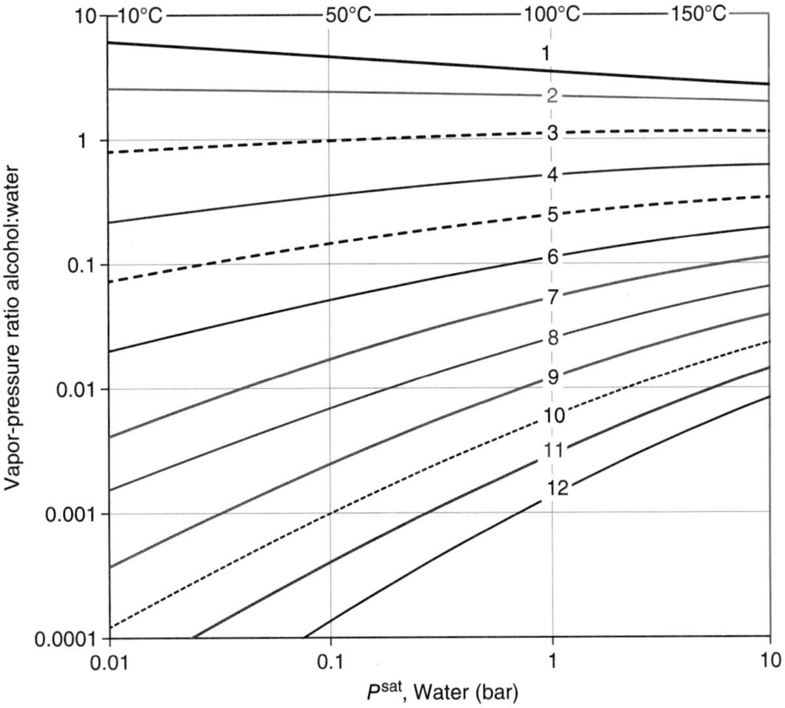

FIG. 4-6 Ratio of vapor pressure of the 1-alcohols to that of water plotted versus the vapor pressure of water. The numbers on each curve show the carbon number of the alcohol. The temperatures at the top of the chart correspond to the water saturation temperature.

TABLE 4-5 Infinite-Dilation Activity Coefficients at 313 K of Hydrocarbons in DMF and ACN

Compound	γ^∞ in DMF	γ^∞ in ACN	Class of hydrocarbon
Vinyl acetylene	0.84	1.78	Olefin and triple bond
1-Propyne	1.09	1.95	Triple bond
1-Butyne		2.98	
1,3-Butadiene	2.38	3.73	Diolefin
1,2-Butadiene	2.8		
cis-2-Butene	4.51	7.44	Olefin
1-Butene	4.97	7.22	
Isobutene	5.21		
trans-2-butene	6.46		
Butane	10.4	10.4	Paraffin
Isobutane	13.5	12.2	

the relationship among the infinite-dilution activity coefficients of hydrocarbons in DMF, ethanol, and water compared to that in ACN. Clearly, the regularity of the data itself—without any theoretical model—reveals a simple, clear pattern that helps to evaluate the consistency of the various data sources and also to fill in data gaps as needed. As an example, it is easy to estimate the infinite-dilution activity coefficient of 1-butyne in DMF, and of 1,2-butadiene, isobutene, and *trans*-2-butene in ACN. For 1-butyne in DMF, taking the value of ln(2.98) = 1.1 from Table 4-5, the y-axis indicates a value of 0.7 for the log value, or 2.0 for the activity coefficient.

Figure 4-8 graphically presents Harvey's correlation [*AIChE J.* **42:** 1491 (1996)] for Henry constants of 12 solutes in hexadecane. Solutes with large Henry constants (low solubility) typically have a negative slope with temperature, which means that the heat of solution is endothermic. As Henry's constant of a solute decreases, the slope becomes increasingly positive, which relates to exothermic absorption. At a sufficiently high temperature, close to the critical temperature of the solvent, most curves have negative slopes, and hence the solutes with low Henry's constants go through a maximum. Figure 4-9 studies the pattern of Henry's constants of various solutes in hexadecane (a variation on Fig. 4-8) by plotting the Henry constant, at a representative temperature of 350 K, versus the solute normal boiling point. (Note that

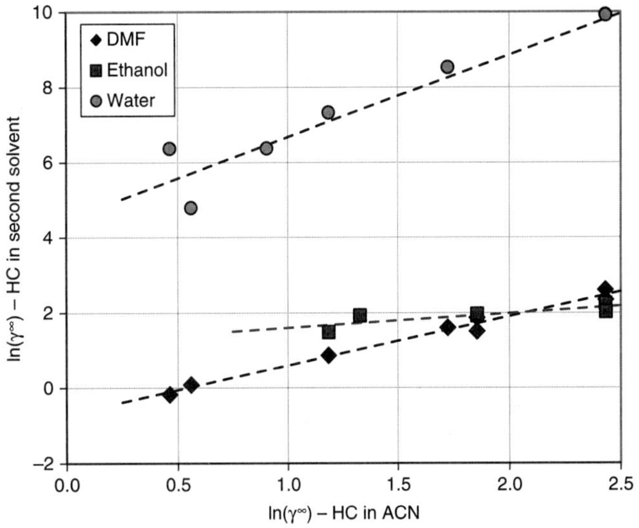

FIG. 4-7 Correlation between hydrocarbon infinite-dilution activity coefficients in dimethyl formamide (DMF), ethanol, and water compared to those in acetonitrile (ACN). All activity coefficients are at 313 K. The dashed lines are best-fit straight lines.

CO_2 is a solid when its vapor pressure is atmospheric, hence an "effective" liquid normal boiling point of 183.7 K was estimated by extrapolating the liquid vapor pressure down to a vapor pressure of 1 atm.) Figure 4-9 indicates that for nonpolar compounds the logarithm of Henry constants at 350 K varies approximately linearly with the normal boiling point. Henry constants of the polar compounds (CO_2, HCl, H_2S, NH_3, and SO_2) are higher than the value based upon the nonpolar estimate, which indicates that they have relatively higher positive deviations from ideality in the nonpolar solvent, hexadecane. Hydrogen is another exception to the simple pattern, and this is likely because of quantum effects in its boiling behavior.

Figure 4-10 presents the aqueous infinite-dilution activity coefficients in water of substances from several families of organic compounds plotted versus their respective normal boiling temperatures. A temperature of 348 K

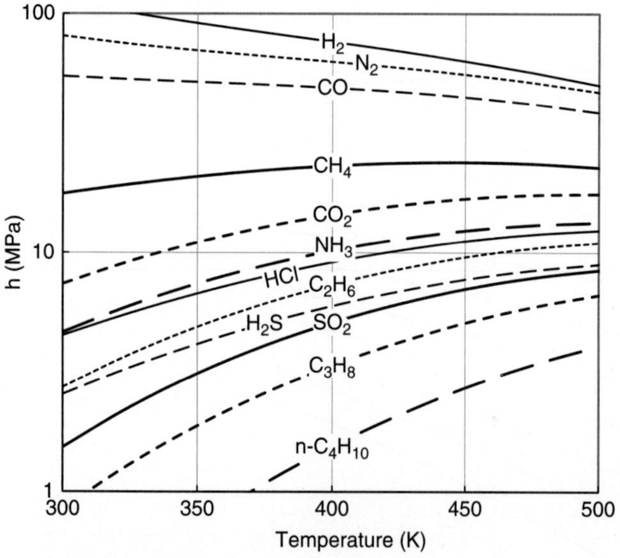

FIG. 4-8 Henry's constants for various solutes in hexadecane.

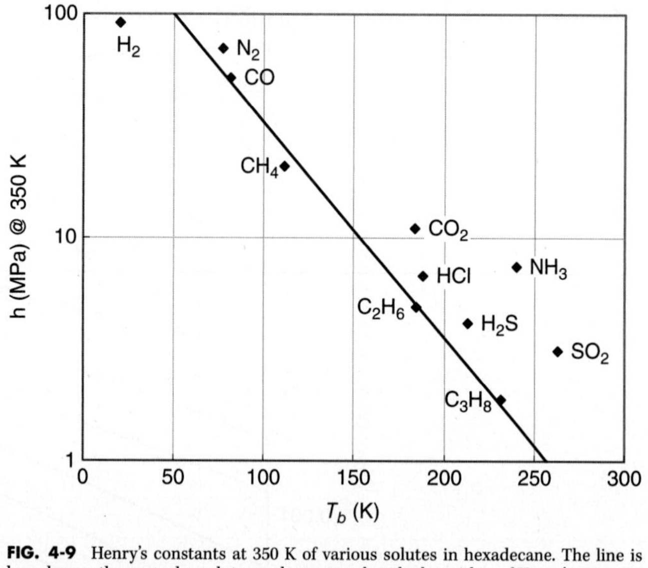

FIG. 4-9 Henry's constants at 350 K of various solutes in hexadecane. The line is based upon the nonpolar solutes and assumes that the logarithm of Henry's constant at 350 K varies linearly with the normal boiling temperature.

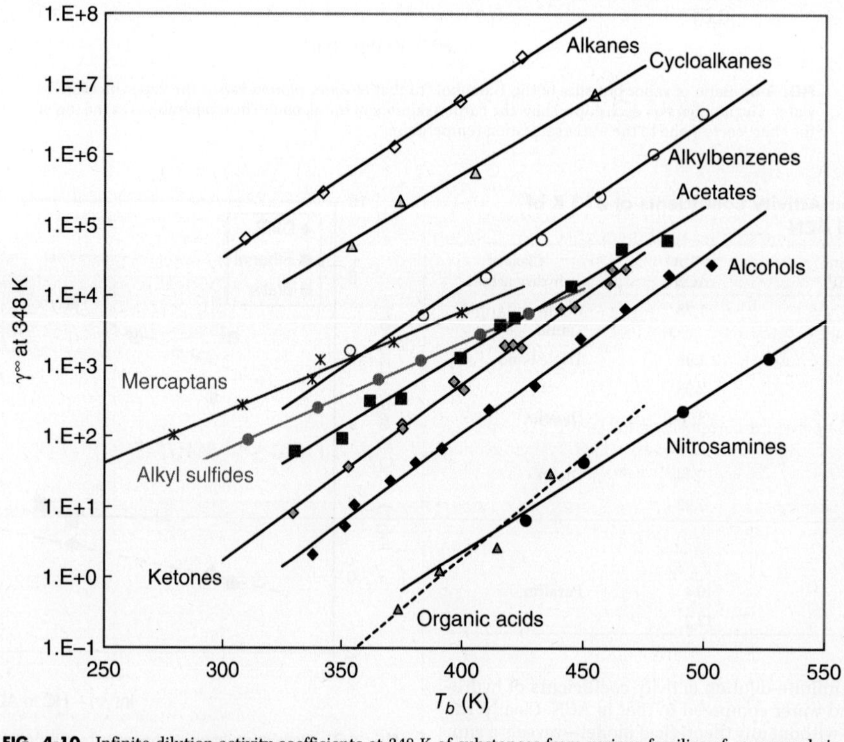

FIG. 4-10 Infinite-dilution activity coefficients at 348 K of substances from various families of compounds in water. The lines are best fits for each compound, assuming that ln γ^∞ varies linearly with the solute normal boiling point.

has been chosen to represent separations where the water concentration is fairly high and the pressure is close to atmospheric; however, the patterns are relatively insensitive to temperature. These patterns help chemical engineers estimate the nonideality of the organic-water pair of particular interest and also identify data that may be in error. The variation of infinite-dilution activity coefficients is extremely large, ranging from organic acids that form nearly ideal mixtures with water to alkane-water mixtures where the activity coefficients are more than 7 orders of magnitude higher. The patterns accentuated by Fig. 4-10 are useful to develop correlations for aqueous separations in emerging biotechnology processes and to evaluate predictive techniques.

The results in Fig. 4-10 may be used to calculate relative volatilities in water at infinite dilution, which is approximately equal to the activity coefficient multiplied by the ratio of the solute and water vapor pressures. These results are very useful in biofuels and biochemical processes since the initial separations are typically performed where the liquid concentrations are dominated by water (i.e., fermentation processes usually occur in relatively dilute aqueous solutions). Figure 4-11 presents relative volatilities at infinite dilution of various families of compounds in water. These results may be surprising and unexpected even to experienced chemical engineers. The relative volatility of most compounds at infinite dilution in water is greater than unity, and the relative volatility increases with the boiling point of the solute. The latter result occurs because as the size or molecular weight of the member of a particular family increases (recall that vapor pressures within a family rarely cross), its vapor pressure at the reference temperature (348 K) decreases, but the activity coefficient increases to a greater extent such that the relative volatility rises.

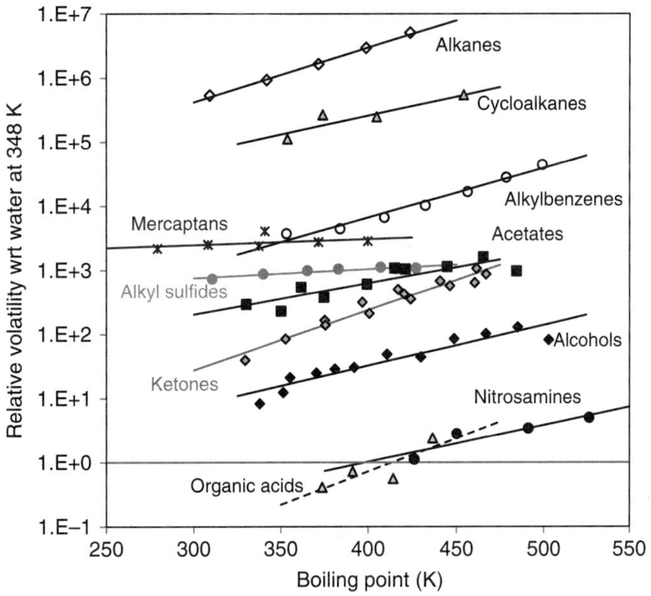

FIG. 4-11 Relative volatilities with respect to (wrt) water for various families of compounds at infinite dilution in water. The relative volatilities have been calculated using the infinite-dilution activity coefficients from Fig. 4-10.

TEMPERATURE DEPENDENCE OF INFINITE-DILUTION ACTIVITY COEFFICIENTS

FUNDAMENTAL RELATIONSHIPS

Most activity coefficient models are concerned primarily with the effect of composition, giving only a rough approximation of the temperature dependence. However, the effect of temperature often is a critical factor in conceptual and process design. Fundamentally, the temperature dependence of γ_{ij}^{∞} is given by a version of the Gibbs-Helmholtz equation:

$$\left[\frac{\partial \ln \gamma_{ij}^{\infty}}{\partial(1/T)}\right]_{P,x} = \frac{\overline{H}_{ij}^{E,\infty}}{R} \tag{4-215}$$

where γ_{ij}^{∞} is evaluated at some constant composition for solute i in solvent j. Equation (4-215) is a version of Eq. (4-149). In cases where $\overline{H}_{ij}^{E,\infty}$ is known and its value is fairly constant over the temperature range of interest, integration allows convenient calculation:

$$\gamma_{ij}^{\infty}\Big|_{\text{at } T_2} \approx \gamma_{ij}^{\infty}\Big|_{\text{at } T_1} \exp\left[\frac{\overline{H}_{ij}^{E,\infty}}{R}\left(\frac{1}{T_2} - \frac{1}{T_1}\right)\right] \tag{4-216}$$

Compilations of available enthalpy of mixing data are given elsewhere (Onken, Rarey-Nies, and Gmehling, *Int. J. Thermophys.* **10**(3): 739–747 (1989); DDBST GmbH, http://www.ddbst.com; Christensen, Hanks, and Izatt, *Handbook of Heats of Mixing,* Wiley, New York, 1982); and Christensen, Rowley, and Izatt, *Handbook of Heats of Mixing, Supplemental Volume,* Wiley, New York, 1988)].

Equations (4-215) and (4-216) are the basis for many correlations of activity coefficient temperature dependence. Methods involving correlation of partial molar excess enthalpy are available for specific classes of compounds [Sherman et al., *J. Phys. Chem.* **99**(28): 11239–11247 (1995)]. Another method involves combining an excess Gibbs energy expression with an equation of state [Kontogeorgis and Coutsikos, *Ind. Eng. Chem. Res.* **51:** 4119–4142 (2012)].

An alternative approach to estimating γ_{ij}^{∞} as a function of temperature involves use of a power-law expression (or stretched exponential), given by

$$\ln \gamma_{ij}^{\infty} = a_{ij}(T_{ref}/T)^{\theta_{ij}} \tag{4-217}$$

where T is temperature in Kelvin and a_{ij} is a constant given by a known reference point, $a_{ij} = \ln \gamma_{ij}^{\infty}$ at $T = T_{\text{ref}}$ [Frank, Arturo, and Holden, *AIChE J.* **60:** 3675–3690 (2014)]. The exponent θ_{ij} is related to the partial molar excess enthalpy and entropy of mixing, such that

$$\theta_{ij} = \frac{1}{1-(T\overline{S}_{ij}^{E,\infty}/\overline{H}_{ij}^{E,\infty})} \tag{4-218}$$

In certain cases, Eq. (4-217) may be used to correlate data with a constant value of θ_{ij}. In doing so, one assumes that the ratio of entropic to enthalpic terms is constant over the temperature range of interest. According to Frank et al. [*AIChE J.* **60:** 3675–3690 (2014)], a constant value of θ_{ij} is able to correlate γ_{ij}^{∞} for many binary types over a reasonably wide temperature span of 50°C to 80°C or more—at normal process conditions far from the critical point. Exceptions (in addition to near-critical mixtures) include a number of hydrogen bonding organic + water binaries such as C_4 to C_7 alcohols dissolved in water, 2-butanone in water, and acetonitrile in water. Water is included in the classification scheme as a solvent but not as a solute because of the many varied and difficult-to-predict ways water can form hydrogen bonds.

CLASSIFICATION SCHEME

It is apparent from Eq. (4-218) that very different types of temperature dependence are possible depending on the signs and relative magnitudes of partial molar excess enthalpy and entropy. With this in mind, Frank et al. [*AIChE J.* **60:** 3675–3690 (2014)] have classified solute-solvent binary pairs into seven types corresponding to distinct domains of $\overline{H}_{ij}^{E,\infty}$, $\overline{S}_{ij}^{E,\infty}$, γ_{ij}^{∞}, and θ_{ij} shown in Table 4-6. Specific interactions that can affect $\overline{H}_{ij}^{E,\infty}$ include static dipole-dipole (polarity effects), induced dipole-dipole, hydrogen bonding (proton donor and proton acceptor interactions), and electron donor/acceptor interactions. Factors affecting $\overline{S}_{ij}^{E,\infty}$ include segregation resulting from these interactions, molecular size differences, and the hydrophobic effect for organic + water mixtures.

In modeling phase equilibrium using the standard activity coefficient correlation equations, it is common practice to represent the effect of temperature for a given binary interaction parameter by using empirical expressions with two or more correlation constants. Typical expressions have the form $\ln A$ or $A = a + b/T + c \ln T$, where A is a model parameter and a, b, and c are correlation constants determined by fitting data, an expression derived from Eq. (4-215) assuming $\overline{H}_{ij}^{E,\infty}$ is a linear function of temperature. As an alternative, Frank et al. have proposed incorporating the parameter θ_{ij} directly into an excess Gibbs energy expression as shown in Table 4-7. In principle, suitable θ_{ij} values may be estimated via Eq. (4-218) by using molecular modeling methods to estimate the dimensionless ratio of $T\overline{S}_{ij}^{E,\infty}/\overline{H}_{ij}^{E,\infty}$, or θ_{ij} may be treated as adjustable model parameters in fitting data. The range of possible θ_{ij} values is bounded by the range of values given in Table 4-6 (at normal conditions). For nonaqueous binaries containing specific classes of compounds, estimates may be obtained from molecular structure using characteristic θ_{ij} values for various classes of compounds, as summarized in Table 4-8. Though developed

TABLE 4-6 Classification of Activity Coefficient Temperature Dependence

Mixture type	Typical characteristics	Excess enthalpy and entropy $\bar{h}_{ij}^{E,\infty}$	$\bar{s}_{ij}^{E,\infty}$	Typical γ_{ij}^{∞} values* and change with temperature	Typical θ_{ij} values†
$\gamma_{ij}^{\infty} \approx 1$					
I	Chemically similar, small molecules (nearly ideal)	≈ 0	≈ 0	$\gamma_{ij}^{\infty} \approx 1$	$\theta_{ij} \approx 0$
$\gamma_{ij}^{\infty} > 1$					
II	Regular-solution-like	Pos.	Pos.	$1.2 < \gamma_{ij}^{\infty} < 1000$ $\frac{\partial \gamma_{ij}^{\infty}}{\partial T} < 0$	$1 < \theta_{ij} < 5$
III	Endothermic with negative excess entropy	Pos.	Neg.	$\gamma_{ij}^{\infty} > 2$ $\frac{\partial \gamma_{ij}^{\infty}}{\partial T} < 0$	$0.15 < \theta_{ij} < 1$
IV	Exothermic with negative excess entropy	Neg.	Neg.	$\gamma_{ij}^{\infty} > 2$ $\frac{\partial \gamma_{ij}^{\infty}}{\partial T} > 0$	$-3 < \theta_{ij} < -0.3$
$\gamma_{ij}^{\infty} < 1$					
V	Net attractive interactions with negative excess entropy	Neg.	Neg.	$0.2 < \gamma_{ij}^{\infty} < 1$ $\frac{\partial \gamma_{ij}^{\infty}}{\partial T} > 0$	$1 < \theta_{ij} < 5$
VI	Net attractive interactions with positive excess entropy	Neg.	Pos.	$0.2 < \gamma_{ij}^{\infty} < 1$ $\frac{\partial \gamma_{ij}^{\infty}}{\partial T} > 0$	$0.3 < \theta_{ij} < 1$
VII	Chemically similar, wide molecular size distribution	Pos.	Pos.	$0.6 < \gamma_{ij}^{\infty} < 1$ $\frac{\partial \gamma_{ij}^{\infty}}{\partial T} < 0$	$-5 < \theta_{ij} < -0.5$

*As a general rule, phase instability may occur at roughly $\gamma_{ij}^{\infty} \approx 7$ to 20 or $\ln \gamma_{ij}^{\infty} \approx 2$ to 3.
†Values determined by analysis of data within the normal temperature range of 0°C to 100°C.
SOURCE: Reprinted from Frank, Arturo, and Holden, *AIChE J.*, **60**(10), pp. 3675–3690 (2014), with permission.

TABLE 4-7 Standard Activity Coefficient Correlation Equations* Modified to Incorporate θ_{ij}

Extended Hansen	$\ln \gamma_{ij}^{\infty} = a_{ij}(T_{ref}/T)^{\theta_{ij}} = \frac{V_i^L (T_{ref}/T)^{\theta_{ij}}}{RT_{ref}} \left\{ (\delta_i^d - \delta_j^d)^2 + 0.25 \left[(\delta_i^p - \delta_j^p)^2 + (\delta_i^h - \delta_j^h)^2 \right] \right\}$
MOSCED*	$\ln \gamma_{ij}^{\infty} = \frac{V_i^L (T_{ref}/T)^{\theta_{ij}}}{RT} \left[(\lambda_j - \lambda_i)^2 + \frac{q_1^2 q_j^2 (\tau_j - \tau_i)^2}{\psi_j} + \frac{(\alpha_j - \alpha_i)(\beta_j - \beta_i)}{\xi_j} \right]$
Wilson†	$\ln \gamma_{ij}^{\infty} = a_{ij}(T_{ref}/T)^{\theta_{ij}} = 1 - \ln \Lambda_{ij} - \Lambda_{ji}$ $\Lambda_{ij} = 1 - a_{ji}(T_{ref}/T)^{\theta_{ji}} - \ln \Lambda_{ji}$ $\Lambda_{ji} = 1 - a_{ij}(T_{ref}/T)^{\theta_{ij}} - \ln \Lambda_{ij}$
NRTL†	$\ln \gamma_{ij}^{\infty} = a_{ij}(T_{ref}/T)^{\theta_{ij}} = \tau_{ij} \exp(-\alpha_{ij}\tau_{ij}) + \tau_{ji}$ $\tau_{ij} = a_{ji}(T_{ref}/T)^{\theta_{ji}} - \tau_{ji} \exp(-\alpha_{ji}\tau_{ji})$ $\tau_{ji} = a_{ij}(T_{ref}/T)^{\theta_{ij}} - \tau_{ij} \exp(-\alpha_{ij}\tau_{ij})$

*A truncated form is shown here for illustrative purposes.
†Model parameters are determined as a function of temperature by simultaneous solution of the given equations.

TABLE 4-8 Average θ for Nonaqueous Solute-Solvent Pairings (Types II and III)

Solute class	Solvent class*	θ	No. of data sets	Relative error (%)
Active hydrogen	Active hydrogen	2.8	11	25.6
Active hydrogen	Aromatic	2.1	6	15.6
Active hydrogen	Nonpolar	2.8	33	24.2
Active hydrogen	Polar	1.8	35	16.8
Aromatic	Active hydrogen	0.6	22	19
Aromatic	Aromatic	1.4	10	12.5
Aromatic	Nonpolar	2.4	8	8.4
Aromatic	Polar	1.5	32	15.3
Nonpolar	Active hydrogen	0.7	53	12.4
Nonpolar	Aromatic	1.6	34	14.7
Nonpolar	Polar	1.4	106	18.8
Polar	Active hydrogen	2.1	63	20.7
Polar	Aromatic	2.8	15	21.7
Polar	Nonpolar	2.3	64	15.4
Polar	Polar	1.4	76	15

*Definitions: *Active hydrogen* species have one or more active hydrogens that may participate in hydrogen-bonding, including halogenated organics containing an active hydrogen. *Aromatic* species involve aromatic rings (with no active hydrogen) with the potential for π-bond interactions. *Nonpolar* species have neither static dipole moment nor aromatic rings. *Polar* species have a static dipole moment, no aromatic rings, and no active hydrogen.
SOURCE: Reprinted from Frank, Arturo, and Holden, *AIChE J.*, **60**(10), pp. 3675–3690 (2014), with permission.

for γ^{∞}_{ij}, applications at more concentrated conditions may be addressed using the standard correlation equations modified to incorporate θ_{ij} as indicated in Table 4-7. For the Wilson or NRTL equation, the original model temperature dependence is not used, and the temperature dependence of the parameters is determined by solving simultaneously the equations provided in the table. The resulting framework is intended for application-directed screening and modeling purposes, focusing on a limited temperature range of up to 80°C or so for a given application of interest.

THERMODYNAMICS FOR CONCEPTUAL DESIGN

PREDICTION OF SPECIES PARTITIONING

Several key quantities are useful in screening separation processes and potential use of extra solvents. The K factor provides the distribution of a single component, but the ratio of K factors provides superior insight. We have previously used the relative volatility, Eq. (4-184). Infinite-dilution activity coefficients can be leveraged for insight. In each of the relations below, the effect of temperature may be estimated by using the θ correction of the previous section [Frank, Arturo, and Holden, *AIChE J.* **60:** 3675–3690 (2014)]. For solvent selection for extractive or azeotropic distillation of components i and j, the relative volatility of the components in the solvent can be evaluated

$$\alpha^{\infty}_{ij} = \gamma^{\infty}_{i,\text{solv}} P^{\text{sat}}_i(T)/(\gamma^{\infty}_{j,\text{solv}} P^{\text{sat}}_j(T)) \tag{4-219}$$

When stripping a component from a solvent, the solvent is nearly pure, so the important quantity is

$$\alpha^{\infty}_{ij} = \gamma^{\infty}_{i,\text{solv}} P^{\text{sat}}_i(T)/(P^{\text{sat}}_j(T)) \tag{4-220}$$

When screening solvents for liquid-liquid extraction under dilute conditions, the LLE K ratio is used

$$K^{\infty}_{ij} = \frac{x_{i,\text{extract}}}{x_{i,\text{raffinate}}} = \frac{\gamma^{\infty}_{i,\text{raffinate}}}{\gamma^{\infty}_{i,\text{extract}}} \tag{4-221}$$

The concept of using ratios of K values as in the relative volatility can be generalized and is called the *separation factor*, representing the relative enrichment of a given component after one theoretical stage of contacting. For cosolutes i and j, and using x and y as generic compositions,

$$\alpha_{ij} = \frac{K_i}{K_j} = \frac{y^{\text{II}}_i/x^{\text{I}}_i}{y^{\text{II}}_j/x^{\text{I}}_j} = \frac{(y_i/y_j)^{\text{II}}}{(x_i/x_j)^{\text{I}}} \tag{4-222}$$

where I and II indicate the phase. For example, I is raffinate and II is extract in liquid extraction.

The enrichment of solute i with respect to solute j can be further increased with the use of multiple contacting stages. Additional discussion of distillation and extraction process fundamentals is given in Secs. 13 and 15.

REACTING SYSTEMS

CHEMICAL REACTION STOICHIOMETRY

For a phase in which a chemical reaction occurs according to the equation

$$|\nu_1|A_1 + |\nu_2|A_2 + \cdots \rightarrow |\nu_3|A_3 + |\nu_4|A_4 + \cdots$$

the $|\nu_i|$ are stoichiometric coefficients and the A_i stand for chemical formulas. The ν_i themselves are called *stoichiometric numbers*, and associated with them is a sign convention such that the value is positive for a product and negative for a reactant. More generally, for a system containing N chemical species, any or all of which can participate in r chemical reactions, the reactions are represented by the equations

$$0 = \sum_i \nu_{i,j} A_i \qquad j = 1, 2, \ldots, r \tag{4-223}$$

where $$\text{sign}(\nu_{i,j}) = \begin{cases} - \text{for a reactant species} \\ + \text{for a product species} \end{cases}$$

If species i does not participate in reaction j, then $\nu_{i,j} = 0$.

The stoichiometric numbers provide relations among the changes in mole numbers of chemical species which occur as the result of chemical reaction. The change in moles for reaction j can be related to the change in a single quantity ε_j, called the *reaction coordinate* for reaction j.

$$\frac{dn_{1,j}}{\nu_{1,j}} = \frac{dn_{2,j}}{\nu_{2,j}} = \ldots = \frac{dn_{N,j}}{\nu_{N,j}} = d\varepsilon_j \tag{4-224}$$

If the initial number of moles of species i is n_{i0} and if the convention is adopted that $\varepsilon_j = 0$ for each reaction in this initial state, then

$$n_i = n_{i_0} + \sum_j \nu_{i,j}\varepsilon_j \qquad i = 1, 2, \ldots, N \tag{4-225}$$

Equation (4-225) is the basic expression of material balance for a closed system in which r chemical reactions occur. It shows for a reacting system that at most r mole-number-related quantities ε_j are capable of independent variation. It is not an equilibrium relation, but merely an accounting scheme, valid for tracking the progress of the reactions to arbitrary levels of conversion. The reaction coordinate has units of moles. A change in ε_j of 1 mol signifies a *mole of reaction*, meaning that reaction j has proceeded to such an extent that the change in mole number of each reactant and product is equal to its stoichiometric number.

CHEMICAL REACTION EQUILIBRIA

The general criterion of chemical reaction equilibria is given by Eq. (4-128). For a system in which just a single reaction occurs, incorporation of Eq. (4-224) leads to

$$\sum_i \nu_i \mu_i = 0 \tag{4-226}$$

Generalization of this result to multiple reactions produces

$$\sum_i \nu_{i,j}\mu_i = 0 \qquad j = \text{I}, \text{II}, \ldots, r \tag{4-227}$$

Standard Property Changes of Reaction For the reaction

$$aA + bB \rightarrow lL + mM$$

a *standard* property change is defined as the property change resulting when a mol of A and b mol of B in their standard states at temperature T react to form l mol of L and m mol of M in their standard states also at temperature T. A *standard state* of species i is its real or hypothetical state as a pure species at temperature T and at a standard state pressure $P°$. The standard property change of reaction j is given by the symbol ΔM°_j, and its general mathematical definition is

$$\Delta M^{\circ}_j \equiv \sum_i \nu_{i,j} M^{\circ}_i \tag{4-228}$$

For species present as gases in the actual reactive system, the standard state is the pure *ideal gas* at pressure $P°$. For liquids and solids, it is usually the state of pure real liquid or solid at $P°$. The standard state pressure $P°$ is fixed at 1 bar. Note that the standard states may represent different physical states for different species; any or all the species may be gases, liquids, or solids.

The most commonly used standard property changes of reaction are

$$\Delta G_j^\circ \equiv \sum_i \nu_{i,j} G_i^\circ = \sum_i \nu_{i,j} \mu_i^\circ \tag{4-229}$$

$$\Delta H_j^\circ \equiv \sum_i \nu_{i,j} H_i^\circ \tag{4-230}$$

$$\Delta C_{P_j}^\circ \equiv \sum_i \nu_{i,j} C_{P_i}^\circ \tag{4-231}$$

The standard Gibbs energy change of reaction ΔG_j° is used in the calculation of equilibrium compositions. The standard heat of reaction ΔH_j° is used in the calculation of the heat effects of chemical reaction, and the standard heat capacity change of reaction is used for extrapolating ΔH_j° and ΔG_j° with T. Numerical values for ΔH_j° and ΔG_j° are computed from tabulated formation data, and $\Delta C_{P_j}^\circ$ is determined from empirical expressions for the T dependence of the $C_{P_j}^\circ$ [see, e.g., Eq. (4-1)].

Equilibrium Constants For practical application, Eq. (4-227) must be reformulated. The initial step is elimination of the μ_i in favor of activities or fugacities. Equation (4-141) gives, upon rearrangement,

$$\sum_i [\nu_{i,j}(G_i^\circ + RT \ln a_i)] = 0$$

or

$$\sum_i \nu_{i,j} G_i^\circ + RT \sum_i \ln a_i^{\nu_{i,j}} = 0$$

or

$$\ln \prod_i a_i^{\nu_{i,j}} = \frac{-\sum_i \nu_{i,j} G_i^\circ}{RT}$$

where

$$a_i \equiv \frac{\hat{f}_i}{f_i^\circ} \tag{4-232}$$

The right side of this equation is a function of temperature only for given reactions and given standard states. Convenience suggests setting it equal to $\ln K_{a,j}$, whence

$$\prod_i a_i^{\nu_{i,j}} = K_{a,j} \quad \text{all } j \tag{4-233}$$

where, by definition, we have the activity-based equilibrium constant

$$K_{a,j} \equiv \exp\left(\frac{-\Delta G_j^\circ}{RT}\right) \tag{4-234}$$

Quantity $K_{a,j}$ is the chemical reaction equilibrium constant for reaction j, and ΔG_j° is the corresponding standard Gibbs energy change of reaction [see Eq. (4-229)]. Although called a "constant," $K_{a,j}$ is a function of T, but only of T.

The activities in Eq. (4-233) provide the connection between the *equilibrium* states of interest and the *standard* states of the constituent species, for which data are presumed available. The standard states are always at the equilibrium temperature. Although the standard state need not be the same for all species, for a *particular* species it must be the state represented by both G_i° and the f_i° upon which activity a_i is based.

The application of Eq. (4-233) requires explicit introduction of composition variables. For gas-phase reactions this is accomplished through the fugacity coefficient

$$a_i \equiv \frac{\hat{f}_i}{f_i^\circ} = \frac{y_i \hat{\phi}_i P}{f_i^\circ}$$

However, the standard state for gases is the ideal gas state at the standard state pressure, for which $f_i^\circ = P^\circ$. Therefore,

$$a_i = \frac{y_i \hat{\phi}_i P}{P^\circ}$$

and Eq. (4-233) becomes

$$\prod_i (y_i \hat{\phi}_i)^{\nu_{i,j}} \left(\frac{P}{P^\circ}\right)^{\nu_j} = K_{a,j} \quad \text{all } j \tag{4-235}$$

where $\nu_j \equiv \sum_i \nu_{i,j}$ and P° is the standard state pressure of 1 bar, expressed in the same units used for P. The y_i may be eliminated in favor of equilibrium values of the reaction coordinates ε_j (see Example 4-14). Then, for fixed temperature Eq. (4-235) relates the ε_j to P. In principle, specification of the pressure allows solution for the ε_j. However, the problem may be complicated by the dependence of $\hat{\phi}_i$ on composition, i.e., on the ε_j. If the equilibrium mixture is assumed an ideal solution, then [Eq. (4-123)] each $\hat{\phi}_i$ becomes ϕ_i, the fugacity coefficient of pure species i at the mixture T and P. An important special case of Eq. (4-235) results for gas-phase reactions when the phase is assumed an ideal gas. In this event $\hat{\phi}_i = 1$, and

$$\prod_i (y_i)^{\nu_{i,j}} \left(\frac{P}{P^\circ}\right)^{\nu_j} = K_{a,j} \quad \text{all } j \tag{4-236}$$

For liquid-phase reactions, Eq. (4-233) is modified by introduction of the activity coefficient $\gamma_i = \hat{f}_i / x_i f_i$, where x_i is the liquid-phase mole fraction. The activity is then

$$a_i \equiv \frac{\hat{f}_i}{f_i^\circ} = \gamma_i x_i \left(\frac{f_i}{f_i^\circ}\right)$$

Both f_i and f_i° represent fugacity of pure liquid i at temperature T, but at pressures P and P°, respectively. Except in the critical region, pressure has little effect on the properties of liquids, and the ratio f_i/f_i° is often taken as unity. When this is not acceptable, this ratio is evaluated by the Poynting equation, Eq. (4-67).

$$\ln \frac{f_i}{f_i^\circ} = \frac{1}{RT} \int_{P^\circ}^{P} V_i \, dP \cong \frac{V_i (P - P^\circ)}{RT}$$

When the ratio f_i/f_i° is taken as unity, $a_i = \gamma_i x_i$, and Eq. (4-233) becomes

$$\prod_i (\gamma_i x_i)^{\nu_{i,j}} = K_{a,j} \quad \text{all } j \tag{4-237}$$

Here the difficulty is to determine the γ_i, which depend on the x_i. In the case of an ideal Lewis-Randall solution, $\gamma_i = 1$, and Eq. (4-237) reduces to

$$\prod_i (x_i)^{\nu_{i,j}} = K_{a,j} \quad \text{all } j \tag{4-238}$$

The significant feature of Eqs. (4-236) and (4-238), the simplest expressions for gas- and liquid-phase reaction equilibrium, respectively, is that the temperature-, pressure-, and composition-dependent terms are distinct and separate.

The effect of temperature on the equilibrium constant is

$$\frac{d \ln K_{a,j}}{dT} = \frac{\Delta H_j^\circ}{RT^2} \tag{4-239}$$

For an endothermic reaction, ΔH_j° is positive and $K_{a,j}$ increases with increasing T; for an exothermic reaction, it is negative and $K_{a,j}$ decreases with increasing T. After integration, the relation is

$$\ln K_a = \ln K_{a,T_0} - \frac{\Delta H_{T_0}^\circ}{R}\left(\frac{1}{T} - \frac{1}{T_0}\right) - \frac{1}{T}\int_{T_0}^{T} \frac{\Delta C_P^\circ}{R} dT + \int_{T_0}^{T} \frac{\Delta C_P^\circ}{R} \frac{dT}{T} \tag{4-240}$$

In the more extensive compilations of data, values of ΔG° and ΔH° for formation reactions are given for a wide range of temperatures, rather than just at the reference temperature $T_0 = 298.15$ K. [See in particular *TRC Thermodynamic Tables—Hydrocarbons* and *TRC Thermodynamic Tables—Non-hydrocarbons*, serial publications of the Thermodynamics Research Center, Texas A & M Univ. System, College Station, Tex.; "The NBS Tables of Chemical Thermodynamic Properties," *J. Phys. Chem. Ref. Data* **11**, supp. 2 (1982).] Where data are lacking, methods of estimation are available; these are reviewed by Poling, Prausnitz, and O'Connell, *The Properties of Gases and Liquids*, 5th ed., McGraw-Hill, New York, 2000, chap. 6. For an estimation procedure based on molecular structure, see Constantinou and Gani, *Fluid Phase Equilibr.* **103:** 11–22 (1995). See also Sec. 2.

Example 4-14 Single-Reaction Equilibrium The hydrogenation of benzene to produce cyclohexane by the reaction

$$C_6H_6 + 3H_2 \rightarrow C_6H_{12}$$

is carried out over a catalyst formulated to favor this reaction. Operating conditions cover a pressure range from 10 to 35 bar and a temperature range from 450 to 670 K. The reaction rate increases with increasing T, but because the reaction is exothermic the equilibrium conversion decreases with increasing T. A comprehensive study of the effect of operating variables on the chemical equilibrium of this reaction has been published by J. Carrero-Mantilla and M. Llano-Restrepo [*Fluid Phase Equilibr.* **219:** 181–193 (2004)]. Presented here are calculations for a single set of operating conditions, namely, $T = 600$ K, $P = 15$ bar, and a molar feed ratio $H_2/C_6H_6 = 3$, the stoichiometric value. For these conditions we determine the fractional conversion of benzene to cyclohexane. Carrero-Mantilla and Llano-Restrepo express ln K as a function of T by an equation which for 600 K yields the value $K = 0.02874$.

A feed stream containing 3 mol H_2 for each 1 mol C_6H_6 is the *basis* of calculation, and for this single reaction, Eq. (4-225) becomes $n_i = n_{i0} + \nu_i\varepsilon$, yielding

$$\begin{aligned} n_B &= 1-\varepsilon && \text{benzene} \\ n_H &= 3-3\varepsilon && \text{hydrogen} \\ n_C &= \varepsilon && \text{cyclohexane} \\ \sum_i n_i &= 4-3\varepsilon \end{aligned}$$

Each mole fraction is therefore given by $y_i = n_i/(4 - 3\varepsilon)$.

Assume first that the equilibrium mixture is an ideal gas, and apply Eq. (4-236), written for a single reaction, with subscript j omitted and $\nu = -3$:

$$\prod_i y_i^{\nu_i}\left(\frac{P}{P^\circ}\right)^{\nu} = \frac{\dfrac{\varepsilon}{4-3\varepsilon}}{\left(\dfrac{1-\varepsilon}{4-3\varepsilon}\right)\left(\dfrac{3-3\varepsilon}{4-3\varepsilon}\right)^3}\left(\frac{15}{1}\right)^{-3} = K = 0.02874$$

whence
$$\frac{\varepsilon}{1-\varepsilon}\left(\frac{4-3\varepsilon}{3-3\varepsilon}\right)^3(15)^{-3} = 0.02874 \quad \text{and} \quad \varepsilon = 0.815$$

Thus, the assumption of ideal gases leads to a calculated conversion of 81.5 percent. Carrero-Mantilla and Llano-Restrep present results for a wide range of conditions, both for the ideal gas assumption and for calculations wherein $\hat{\phi}_i$ values are determined from the Soave-Redlich-Kwong equation of state. In no case are these calculated conversions significantly divergent from ideal gas results.

Complex Chemical Reaction Equilibria When the composition of an equilibrium mixture is determined by a number of simultaneous reactions, calculations based on equilibrium constants become complex and tedious. A more direct procedure (and one suitable for general computer solution) is based on minimization of the total Gibbs energy G^t in accord with Eq. (4-128). The treatment here is limited to gas-phase reactions for which the problem is to find the equilibrium composition for given T and P and for a given initial feed. The method requires constraints on the distribution of elements among the various species proposed to be present in the system.

1. Propose all species that are expected to be present at equilibrium.
2. Formulate the atom balance equations, based on conservation of the total number of atoms of each *element* in a system composed of w elements. Let subscript k identify a particular element, and define A_k as the total number of atomic masses of the kth element in the feed. Further, let a_{ik} be the number of atoms of the kth element present in each molecule of chemical species i. The material balance for element k is then

$$\sum_i n_i a_{ik} = A_k \qquad k = 1, 2, \ldots, w \tag{4-241}$$

3. Eliminate any atom balance constraints that are not unique.

The Gibbs energy of the system is calculated via

$$\underline{G} = \sum_{i=1}^{w} \mu_i n_i$$

The chemical potential is given by Eq. (4-141). For gas-phase reactions and standard states as the pure ideal gases at P°, this equation becomes

$$\mu_i = G_i^\circ + RT \ln \frac{\hat{f}_i}{P^\circ}$$

If G_i° is arbitrarily set equal to zero for all *elements* in their standard states, then for compounds $G_i^\circ = \Delta G_{f_i}^\circ$, the standard Gibbs energy change of formation of species i at the temperature of the system (*not* the reference temperature). In addition, the fugacity is expressed using the fugacity coefficient by Eq. (4-131), $\hat{f}_i = y_i\hat{\phi}_i P$. With these substitutions, the equation for μ_i becomes

$$\mu_i = \Delta G_{f_i}^\circ + RT \ln \frac{y_i \hat{\phi}_i P}{P^\circ} \tag{4-242}$$

The unknowns in these equations are the n_i (note that $y_i = n_i/\Sigma_i n_i$), subject to the atom balance constraints. The problem is most readily solved using a computer. Modern spreadsheets provide constrained optimization. The minimization problem presented here using nonideal gases is best solved with a process simulator.

In this procedure, the question of what chemical reactions are involved never enters directly into any of the equations. However, the choice of a set of species is entirely equivalent to the choice of a set of independent reactions among the species. In any event, a set of species or an equivalent set of independent reactions must always be assumed, and different assumptions produce different results for each reacting system. For example, assuming carbon monoxide as a possible byproduct of a combustion reaction would result in a different equilibrium concentration than ignoring it. Another caveat is that the equilibrium calculation does not necessarily represent the reality of an actual process because there is no consideration of reaction kinetics. For example, though carbon formation is thermodynamically favorable when processing hydrocarbons at high temperatures, industrial processes are successfully run with limited carbon deposition or the catalysts are regenerated. The equilibrium calculations remain an important guideline to explore potential products and/or conversion limitations.

A detailed example of a complex gas-phase equilibrium calculation is given by Smith, Van Ness, and Abbott (2005, pp. 527–528). General application of the method to multicomponent, multiphase systems is treated by Iglesias-Silva et al. [*Fluid Phase Equilibr.* **210:** 229–245 (2003)] and by Sotyan, Ghajar, and Gasem [*Ind. Eng. Chem. Res.* **42:** 3786–3801 (2003)].

HENRY'S LAW FOR REACTING SYSTEMS

Dissociation of weak electrolytes is best modeled as a reversible reaction involving simultaneous reaction and phase equilibrium. These kinds of problems arise in the treatment of acid gases and involve many of the most complex aspects of thermodynamic analysis. Speciation is an important issue that must be handled on a case-by-case basis. To illustrate, we consider a specific example, H_2S absorption.

When the weak acid hydrogen sulfide (H_2S) is dissolved in water, it tends to dissociate into HS^- and S^{2-} anions, which are described by chemical equilibrium equations

$$H_2S + H_2O \leftrightarrow HS^- + H_3O^+ \qquad pK_a = 6.9 \tag{4-243}$$

$$HS^- + H_2O \leftrightarrow S^{2-} + H_3O^+ \qquad pK_a \sim 14 \tag{4-244}$$

The thermodynamic framework and electrolyte activity coefficient models through which chemical equilibrium equations are incorporated into the calculations are beyond the scope of the discussion here; interested readers should consult papers by experts such as Chen and coworkers [*Chem. Eng. Progr.* **111:** 65–75 (2015) and **112:** 3442 (2016)]. However, it should be understood that equations such as Eq. (4-136) relate to only the molecular form of species such as H_2S. At a given partial pressure of H_2S in the gas phase, any dissociation will increase the "apparent concentration" in the liquid phase. At a given overall liquid concentration, the dissociation decreases the molecular (undissociated) concentration and fugacity, and thus the related vapor-phase concentration. For practical purposes, it is useful to define the apparent liquid mole fraction of dissociative species, denoted by the subscript 0, which includes all forms of a given solute. Because the moles are conserved in Eqs. (4-243) and (4-244), we may write

$$x_0(H_2S) \equiv x(H_2S) + x(HS^-) + x(S^{2-}) \tag{4-245}$$

Another useful practical definition is the *apparent Henry's constant*, which is the ratio of the partial pressure to the apparent liquid mole fraction at any equilibrium condition, related to Henry's constant in the following way

$$h_{0i} \equiv \frac{y_i P_i}{x_{0i}} = \frac{\gamma_i^*}{\hat{\phi}_i^V} h_i \exp\left[\frac{1}{RT}\int_{P_i^\circ}^{P} \bar{V}_i^\infty dP\right]\left\{\frac{x_i}{x_{0i}}\right\} \tag{4-246}$$

where x_i is the mole fraction of the molecular undissociated species, H_2S in this discussion. The significance of the definition of h_{0i} in Eq. (4-246) is that experimentally the vapor mole fraction and the overall liquid mole fractions appearing in the first equality are the most important quantities. For a practitioner, knowledge of the apparent Henry's constant facilitates rapid calculations. However, to determine the apparent Henry's constant, detailed calculations are required using the last equality. After the detailed calculations, if h_{0i} is independent of x_{0i} at a given temperature, then the first equality can be used. If not, then the more rigorous model must be used, and the results are conveniently expressed in terms of the apparent Henry's constant, even though the value is not always constant. The discussion here illustrates use of the apparent Henry's constant by first focusing on the H_2S-H_2O binary system and then considers the effect of adding monoethanolamine (MEA) as a chemical solvent. The computations are performed by a process simulator and due to space limitations, the models are discussed in general terms only.

Figure 4-12 presents the apparent Henry's constant of H_2S in H_2O plotted versus its apparent liquid mole fraction; the data are from Lee and Mather [*Ber. Bunsenges. Phys. Chem.* **81:** 1020 (1977)]. The model uses the Redlich-Kwong equation of state for $\hat{\phi}_i^V$, the electrolyte-NRTL model for γ_i^*, and the Brelvi-O'Connell correlation [*AIChE J.* **18:** 1239 (1972)] for $\bar{V}_i^\infty$. The model also includes the dissociation of H_2S into ionic species, as described by Eqs. (4-243) and (4-244), where the dissociation constants and their temperature dependence are determined by the methods described by Chen and co-workers in references at the beginning of this subsection, and Henry's law constant is given by Eq. (4-206).

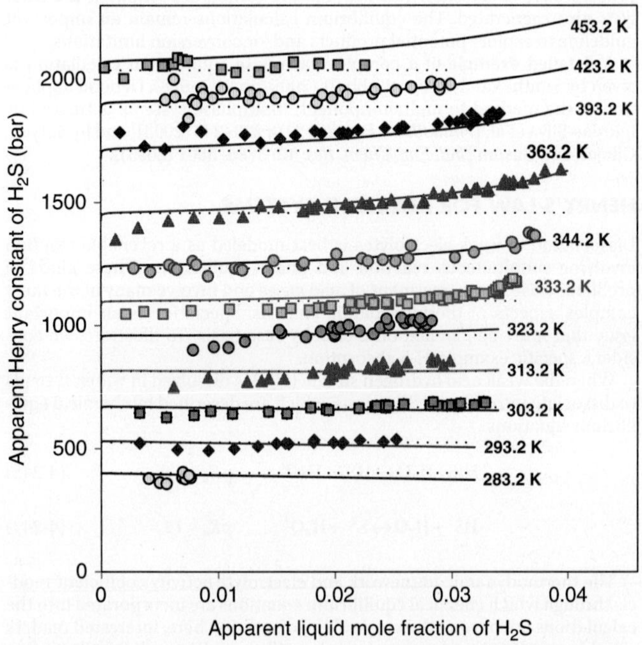

FIG. 4-12 Apparent Henry constants for H_2S in H_2O. The data points are from Lee and Mather [*Ber. Bunsenges. Phys. Chem.* **81:** 1020 (1977)], the solid line is calculated by the model, and the horizontal dashed lines show Henry constants at each temperature.

Figure 4-12 demonstrates that taking h_{0i} as independent of composition in Eq. (4-246) provides a good correlation of the data of Lee and Mather. It is not shown in Fig. 4-12 to avoid cluttering the chart, but note that the model also provides quantitative agreement with the data of Clarke and Glew [*Can. J. Chem.* **49:** 691 (1971)], and Gillespie and Wilson [*Gas. Proc. Assoc.*, RR-48, Tulsa, OK (1982)]. The apparent Henry's constant decreases weakly with increasing H_2S liquid mole fraction at the lowest temperature (283.2 K), and then gradually it has an increasing slope with H_2S composition as the temperature increases. The reason is that both γ_i^* and $\hat{\phi}_i^V$ decrease in this particular case as the composition of H_2S increases, and the ratio remains approximately equal to unity. The full calculations show the ratio x_i/x_{0i} is nearly unity at the concentrations of the data due to the pK_a for the first dissociation. Hence the total variation of h_{0i} with x_{0i} at low temperature is largely controlled by the Poynting correction. The Poynting correction increases with pressure, and the pressure increase for a given change in H_2S composition increases with temperature. It may be expected that the use of the complete model will provide better accuracy for general application; however, note that the simple approximation of assuming that the apparent Henry constant is independent of concentration at a given temperature (horizontal dotted lines in Fig. 4-12) is surprisingly accurate in this case, as the errors do not exceed 10 percent.

Close inspection of Fig. 4-12 shows a sharp decrease in the apparent Henry's constant at very low concentrations, and this is so because H_2S dissociates into ions according to Eqs. (4-243) and (4-244). Only the first dissociation is important in the $H_2S + H_2O$ system, and only at very dilute concentrations. For example, at 333.2 K, the fraction of H_2S dissociated into ions when $x_{0i} = 10^{-5}$ is about 2 percent. Hence, the dissociation of H_2S is effectively zero at the measured data points in Fig. 4-12.

Addition of a base dramatically affects the equilibrium because of the large concentration of neutralized ions. Figure 4-13 presents the apparent Henry's constant of H_2S in a 5*N* solution of aqueous MEA (30.2 wt% MEA or an MEA mole fraction of 0.113 in the H_2S-free solvent). The curves come from a fitted model. Figure 4-13 is dramatically different from Fig. 4-12 because the values of h_{0i} are smaller by 1 to 5 orders of magnitude, indicating that the apparent solubility at a given partial pressure is correspondingly 1 to 5 orders of magnitude higher. Also, for the MEA mixture, the apparent Henry's constant varies significantly with H_2S composition (note the log scale for h_{0i}).

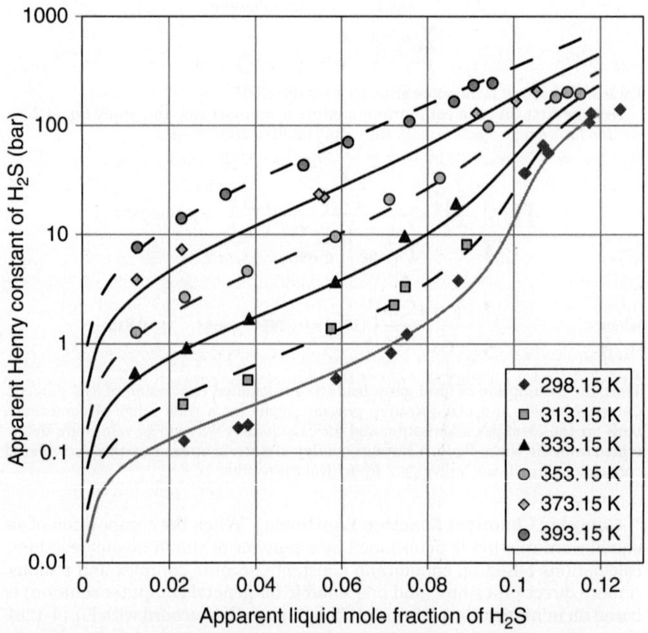

FIG. 4-13 Apparent Henry constants for H_2S in a 5*N* aqueous solution of monoethanolamine (MEA). The data points are from Lee, Otto, and Mather [*J. Chem. Eng. Data* **21:** 207 (1976)], and the curves are from a model that may be considered to be an interpolation of the data.

An issue that becomes clear when one is faced with the strong variation of the apparent Henry's constant with solute concentration is estimation of the true Henry's constant since the limit of Eq. (4-246) cannot be taken in practice. CO_2 is an acid gas similar to H_2S, and a chart analogous to Fig. 4-13 may be constructed, for example, using the data of Jou, Mather, and Otto [*Can. J. Chem. Eng.* **73:** 140 (1995)]. Clarke [*Ind. Eng. Chem. Fundam.* **3:** 239 (1964)] suggested N_2O as a *homomorph* for CO_2 since the two molecules have similar molecular weights, volumes, and structures, and proposed the "N_2O analogy":

$$\left[\frac{h_{CO_2}}{h_{N_2O}}\right]_{\text{Water}} \approx \left[\frac{h_{CO_2}}{h_{N_2O}}\right]_{\text{Reacting Solution}} \tag{4-247}$$

There is an equivalent equation for the diffusion coefficient. The N_2O analogy has been used in equilibrium and mass-transfer correlations for CO_2 in reactive systems, and it has even been tested through molecular modeling by Chen et al. [*Ind. Eng. Chem. Res.* **53:** 18081 (2014)], but there is no equivalent "analogy" for other reactive solutes like H_2S. In the case of H_2S it is generally assumed that Henry's constant in water is equal to that in the reacting solution, but it is hoped that better approximations will be invented in the future.

Section 5

Heat and Mass Transfer

Geoffrey D. Silcox, Ph.D. *Professor of Chemical Engineering, University of Utah; Member, American Institute of Chemical Engineers, American Chemical Society (Conduction, Convection, Heat Transfer with Phase Change, Section Coeditor)*

James J. Noble, Ph.D., P.E., C.Eng. [U.K.] *Research Affiliate, Department of Chemical Engineering, Massachusetts Institute of Technology; Fellow, American Institute of Chemical Engineers; Member, New York Academy of Sciences (Radiation Section Coeditor)*

Adel F. Sarofim, Sc.D. *Deceased; Presidential Professor of Chemical Engineering, Combustion, and Reactors, University of Utah; Member, American Institute of Chemical Engineers, American Chemical Society, Combustion Institute (Radiation)*

Phillip C. Wankat, Ph.D. *Clifton L. Lovell Distinguished Professor of Chemical Engineering Emeritus, Purdue University; Member, American Institute of Chemical Engineers (Mass Transfer Section Coeditor)*

Kent S. Knaebel, Ph.D. *President, Adsorption Research, Inc.; Member, American Institute of Chemical Engineers, International Adsorption Society; Professional Engineer (Ohio) (Mass Transfer Section Coeditor)*

HEAT TRANSFER

HEAT TRANSFER BY CONDUCTION

HEAT TRANSFER BY CONVECTION

HEAT TRANSFER WITH CHANGE OF PHASE

HEAT TRANSFER BY RADIATION

MASS TRANSFER

HEAT TRANSFER

General References Arpaci, V., *Conduction Heat Transfer*, Addison-Wesley, Boston, 1966; Arpaci, V., *Convection Heat Transfer*, Prentice Hall, Upper Saddle River, N.J., 1984; Arpaci, V., *Introduction to Heat Transfer*, Prentice Hall, Upper Saddle River, N.J., 1999; Baehr, H., and K. Stephan, *Heat and Mass Transfer*, Springer, Berlin, 1998; Bejan, A., *Convection Heat Transfer*, Wiley, Hoboken, N.J., 1995; Carslaw, H., and J. Jaeger, *Conduction of Heat in Solids*, Oxford University Press, London, 1959; Edwards, D., *Radiation Heat Transfer Notes*, Hemisphere Publishing, New York, 1981; Hottel, H. C., and A. F. Sarofim, *Radiative Transfer*, McGraw-Hill, New York, 1967; Bergman, T., A. Lavine, F. Incropera, and D. DeWitt, *Fundamentals of Heat and Mass Transfer*, 7th ed., Wiley, Hoboken, N.J., 2011; Kays, W., and M. Crawford, *Convective Heat and Mass Transfer*, 3d ed., McGraw-Hill, New York, 1993; Mills, A., *Heat Transfer*, 2d ed., Prentice Hall, Upper Saddle River, N.J., 1999; Modest, M., *Radiative Heat Transfer*, McGraw-Hill, New York, 1993; Patankar, S., *Numerical Heat Transfer and Fluid Flow*, Taylor and Francis, London, 1980; Pletcher, R., D. Anderson, and J. Tannehill, *Computational Fluid Mechanics and Heat Transfer*, 2d ed., Taylor and Francis, London, 1997; Rohsenow, W., J. Hartnett, and Y. Cho, *Handbook of Heat Transfer*, 3d ed., McGraw-Hill, New York, 1998; Siegel, R., and J. Howell, *Thermal Radiation Heat Transfer*, 4th ed., Taylor and Francis, London, 2001.

MODES OF HEAT TRANSFER

Heat is energy transferred due to a difference in temperature. There are three modes of heat transfer: conduction, convection, and radiation. All three may act at the same time. *Conduction* is the transfer of energy between adjacent particles of matter. It is a local phenomenon and can only occur through matter. *Radiation* is the transfer of energy from a point of higher temperature to a point of lower energy by electromagnetic radiation. Radiation can act at a distance through transparent media and vacuum. *Convection* is the transfer of energy by conduction and radiation in moving, fluid media. The motion of the fluid is an essential part of convective heat transfer.

Nomenclature and Units—Heat Transfer by Conduction, by Convection, and with Phase Change

Symbol	Definition	SI units
A	Area for heat transfer	m^2
A_c	Cross-sectional area	m^2
A_f	Area for heat transfer for finned portion of surface	m^2
A_1	First Fourier coefficient	
b	Geometry: $b = 1$, plane; $b = 2$, cylinder; $b = 3$, sphere	
B_1	First Fourier coefficient	
Bi	Biot number, hR/k	
c	Specific heat	J/(kg·K)
c_p	Specific heat, constant pressure	J/(kg·K)
D	Diameter	m
D_i	Inner diameter	m
D_o	Outer diameter	m
f	Fanning friction factor	
Fo	Dimensionless time or Fourier number, $\alpha t/R^2$	
g	Acceleration of gravity, 9.81 m^2/s	m^2/s
G	Mass velocity, $\dot{m}/A_c$	$kg/(m^2 \cdot s)$
Gz	Graetz number = Re Pr	
h	Heat-transfer coefficient	$W/(m^2 \cdot K)$
$\bar{h}$	Average heat-transfer coefficient	$W/(m^2 \cdot K)$
h_i	Heat-transfer coefficient at inside tube surface	$W/(m^2 \cdot K)$
h_o	Heat-transfer coefficient at outside tube surface	$W/(m^2 \cdot K)$
h_{am}	Heat-transfer coefficient for use with ΔT_{am}, see Eq. (5.40)	$W/(m^2 \cdot K)$
h_{lm}	Heat-transfer coefficient for use with ΔT_{lm}; see Eq. (5-39)	$W/(m^2 \cdot K)$
h_{fg}	Enthalpy of vaporization	kJ/kg
k	Thermal conductivity	W/(m·K)
$\bar{k}$	Average thermal conductivity	W/(m·K)
L	Length of cylinder or length of flat plate in direction of flow or downstream distance	m
	Length of heat-transfer surface	m
$\dot{m}$	Mass flow rate	kg/s
Nu_D	Nusselt number hD/k based on diameter D	
$\overline{Nu}_D$	Average Nusselt number $\bar{h}\,D/k$ based on diameter D	
p	Perimeter	m
P	Perimeter of fin	m
Pr	Prandtl number, ν/α	
q	Flux of heat	W/m^2
Q	Amount of heat transfer	J
$\dot{Q}$	Rate of heat transfer	W
Q/Q_i	Heat loss fraction, $Q/[\rho c V(T_i - T_\infty)]$	
r	Distance from center in plate, cylinder, or sphere	m
R	Thermal resistance or radius	K/W or m
Ra_x	Rayleigh number, $\beta\,\Delta T g x^3/\nu\alpha$	
Re_D	Reynolds number, GD/μ	
S	Volumetric source term	W/m^3
S_1	Fourier spatial function	
t	Time	s
T	Temperature	K or °C
T_b	Bulk or mean temperature at a given x	K
$\bar{T}_b$	Bulk mean temperature, $(T_{b,in} + T_{b,out})/2$	K
T_C	Temperature of cold surface in enclosure	K
T_f	Film temperature, $(T_s + T_e)/2$	K
T_H	Temperature of hot surface in enclosure	K
T_i	Initial temperature	K
T_e	Temperature of free stream	K
T_s	Temperature of surface	K
T_{sat}	Saturation temperature of vapor	K
T_∞	Temperature of fluid in contact with a solid surface	K
U	Overall heat-transfer coefficient	$W/(m^2 \cdot K)$
V	Volume	m^3
V_∞	Velocity upstream of tube bank	m/s
x	Cartesian coordinate direction, characteristic dimension of a surface, or distance from entrance	m
	Greek Symbols	
α	Thermal diffusivity, $k/(\rho c)$	m^2/s
β	Fin parameter defined by Eq. (5-17)	m^{-1}
β	Volumetric coefficient of expansion	K^{-1}
Γ	Mass flow rate per unit length perpendicular to flow	kg/(m·s)
ΔT	Temperature difference	K
ΔT_{am}	Arithmetic mean temperature difference, see Eq. (5-40)	K
ΔT_{lm}	Logarithmic mean temperature difference, see Eq. (5-39)	K
Δx	Thickness of plane wall for conduction	m
δ_1	First dimensionless eigenvalue	
$\delta_{1,0}$	First dimensionless eigenvalue as Bi approaches 0	
$\delta_{1,\infty}$	First dimensionless eigenvalue as Bi approaches ∞	
ε	Emissivity of a surface	
ζ	Dimensionless distance, r/R	
η_f	Fin efficiency	
θ/θ_i	Dimensionless temperature, $(T - T_\infty)/(T_i - T_\infty)$	
μ	Viscosity	kg/(m·s)
ν	Kinematic viscosity, μ/ρ	m^2/s
ρ	Density	kg/m^3
σ	Stefan-Boltzmann constant, 5.67×10^{-8}	$W/(m^2 \cdot K^4)$
σ	Surface tension between liquid and its vapor	N/m
τ	Time constant, time scale	s

HEAT TRANSFER BY CONDUCTION

FOURIER'S LAW

The heat flux due to conduction in the x direction is given by Fourier's law

$$\dot{Q} = -kA\frac{dT}{dx} \tag{5-1}$$

where $\dot{Q}$ is the rate of heat transfer (W), k is the thermal conductivity [W/(m·K)], A is the area perpendicular to the x direction, and T is temperature (K). For the homogeneous wall shown in Fig. 5-1*a*, with constant k, the integrated form of Eq. (5-1) is

$$\dot{Q} = kA\frac{T_1 - T_2}{\Delta x} \tag{5-2}$$

where Δx is the thickness of the wall. Using the thermal circuit shown in Fig. 5-1*b*, Eq. (5-2) can be written in the form

$$\dot{Q} = \frac{T_1 - T_2}{\Delta x / kA} = \frac{T_1 - T_2}{R} \tag{5-3}$$

where R is the thermal resistance (K/W).

THERMAL CONDUCTIVITY

The thermal conductivity k is a transport property whose value for a variety of gases, liquids, and solids is tabulated in Sec. 2. That section also provides methods for predicting and correlating vapor and liquid thermal conductivities. The thermal conductivity is a function of temperature, but the use of constant or averaged values is frequently sufficient. Room temperature values for air, water, concrete, and copper are, respectively, 0.026, 0.61, 1.4, and 400 W/(m·K). Methods for estimating contact resistances and the thermal conductivities of composites and insulation are summarized by Gebhart, *Heat Conduction and Mass Diffusion*, McGraw-Hill, New York, 1993, p. 399.

STEADY CONDUCTION

One-Dimensional Conduction In the absence of energy source terms, $\dot{Q}$ is constant with distance, as shown in Fig. 5-1*a*. For steady conduction, the integrated form of (5-1) for a planar system with constant k and A is Eq. (5-2) or (5-3). For the general case of variables k (k is a function of temperature) and A (cylindrical and spherical systems with radial coordinate r, as sketched in Fig. 5-2), the average heat-transfer area and thermal conductivity are defined such that

$$\dot{Q} = \overline{kA}\frac{T_1 - T_2}{\Delta x} = \frac{T_1 - T_2}{R} \tag{5-4}$$

For a thermal conductivity that depends linearly on T,

$$k = k_0(1 + \gamma T) \tag{5-5}$$

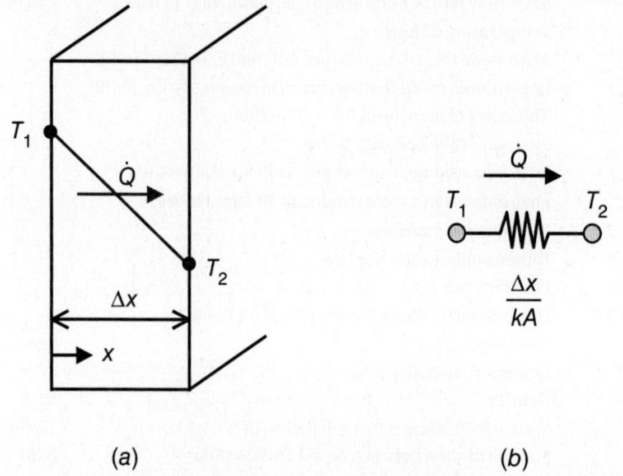

FIG. 5-1 Steady, one-dimensional conduction in a homogeneous planar wall with constant k. The thermal circuit is shown in (*b*) with thermal resistance $\Delta x/(kA)$.

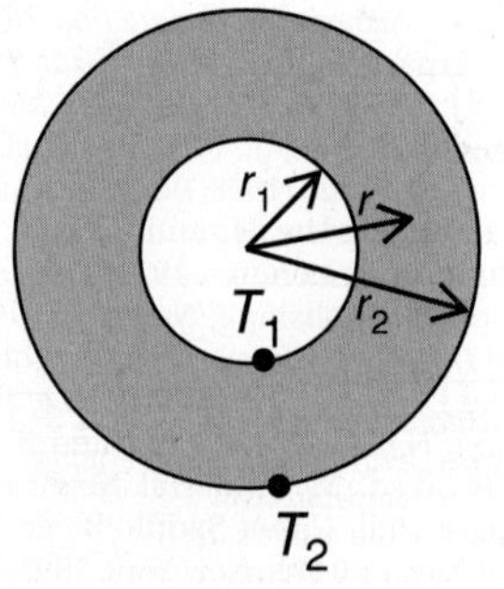

FIG. 5-2 The hollow sphere or cylinder.

and the average heat thermal conductivity is

$$\bar{k} = k_0(1 + \gamma \bar{T}) \tag{5-6}$$

where $\bar{T} = 0.5(T_1 + T_2)$ and γ is a constant.

For cylinders and spheres, A is a function of radial position (see Fig. 5-2): $2\pi rL$ and $4\pi r^2$, where L is the length of the cylinder. For constant k, Eq. (5-4) becomes

$$\dot{Q} = \frac{T_1 - T_2}{[\ln(r_2/r_1)]/(2\pi kL)} \qquad \text{cylinder} \tag{5-7}$$

and

$$\dot{Q} = \frac{T_1 - T_2}{(r_2 - r_1)/(4\pi k r_1 r_2)} \qquad \text{sphere} \tag{5-8}$$

Conduction with Resistances in Series A steady temperature profile in a planar composite wall, with three constant thermal conductivities and no source terms, is shown in Fig. 5-3*a*. The corresponding thermal circuit is given in Fig. 5-3*b*. The rate of heat transfer through each of the layers is the same. The total resistance is the sum of the individual resistances shown in Fig. 5-3*b*:

$$\dot{Q} = \frac{T_1 - T_2}{\dfrac{\Delta x_A}{k_A A} + \dfrac{\Delta x_B}{k_B A} + \dfrac{\Delta x_C}{k_C A}} = \frac{T_1 - T_2}{R_A + R_B + R_C} \tag{5-9}$$

Additional resistances in the series may occur at the surfaces of the solid if they are in contact with a fluid. The rate of convective heat transfer, between a surface of area A and a fluid, is given by Newton's law of cooling as

$$\dot{Q} = hA(T_{\text{surface}} - T_{\text{fluid}}) = \frac{T_{\text{surface}} - T_{\text{fluid}}}{1/(hA)} \tag{5-10}$$

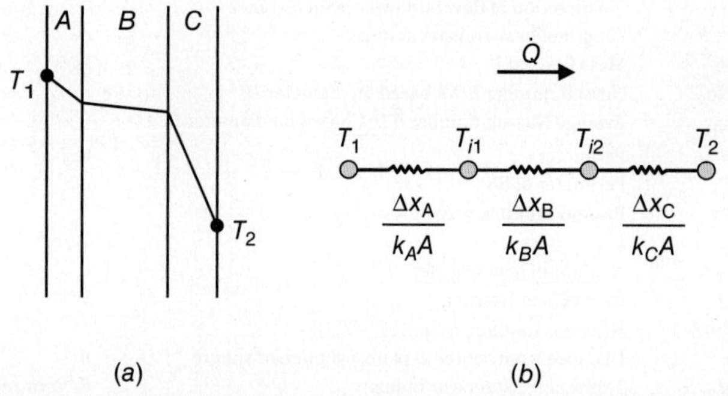

FIG. 5-3 Steady temperature profile in a composite wall with constant thermal conductivities k_A, k_B, and k_C and no energy sources in the wall. The thermal circuit is shown in (*b*). The total resistance is the sum of the three resistances shown.

where $1/(hA)$ is the resistance due to convection (K/W) and the heat-transfer coefficient is h[W/(m^2·K)]. For the cylindrical geometry shown in Fig. 5-2, with convection to inner and outer fluids at temperatures T_i and T_o, with heat-transfer coefficients h_i and h_o, the steady rate of heat transfer is

$$\dot{Q}=\frac{T_i-T_o}{\dfrac{1}{2\pi r_1 L h_i}+\dfrac{\ln(r_2/r_1)}{2\pi k L}+\dfrac{1}{2\pi r_2 L h_o}}=\frac{T_i-T_o}{R_i+R_1+R_o} \tag{5-11}$$

where resistances R_i and R_o are the convective resistances at the inner and outer surfaces. The total resistance is again the sum of the resistances in series.

Example 5-1 Conduction with Resistances in Series and Parallel Figure 5-4 shows the thermal circuit for a furnace wall. The outside surface has a known temperature T_2 = 625 K. The temperature of the surroundings T_{sur} is 290 K. We want to estimate the temperature of the inside surface T_1. The wall consists of three layers: deposit [k_D = 1.6 W/(m·K), Δx_D = 0.080 m], brick [k_B = 1.7 W/(m·K), Δx_B = 0.15 m], and steel [k_S = 45 W/(m·K), Δx_S = 0.00245 m]. The outside surface loses heat by two parallel mechanisms—convection and radiation. The convective heat transfer coefficient h_C = 5.0 W/(m^2·K). The radiative heat-transfer coefficient h_R = 16.3 W/(m^2·K). The latter is calculated from

$$h_R=\varepsilon_2\sigma(T_2^2+T_{sur}^2)(T_2+T_{sur}) \tag{5-12}$$

where the emissivity of surface 2 is ε_2 = 0.76 and the Stefan-Boltzmann constant $\sigma = 5.67\times10^{-8}$ W/(m^2·K^4).

Referring to Fig. 5-4, the steady heat flux q (W/m^2) through the wall is

$$q=\frac{\dot{Q}}{A}=\frac{T_1-T_2}{\dfrac{\Delta x_D}{k_D}+\dfrac{\Delta x_B}{k_B}+\dfrac{\Delta x_S}{k_S}}=(h_C+h_R)(T_2-T_{sur})$$

Solving for T_1 gives

$$T_1=T_2+\left(\frac{\Delta x_D}{k_D}+\frac{\Delta x_B}{k_B}+\frac{\Delta x_S}{k_S}\right)(h_C+h_R)(T_2-T_{sur})$$

and

$$T_1=625+\left(\frac{0.080}{1.6}+\frac{0.15}{1.7}+\frac{0.00254}{45}\right)(5.0+16.3)(625-290)=1610\text{ K}$$

Conduction with Heat Source Application of the law of conservation of energy to a one-dimensional solid, with the heat flux given by Eq. (5-1) and volumetric source term S (W/m^3), results in the following equations for steady conduction in a flat plate of thickness $2R$ (b = 1), a cylinder of diameter $2R$ (b = 2), and a sphere of diameter $2R$ (b = 3). The parameter b is a measure of the curvature. The thermal conductivity is constant, and there is convection at the surface, with heat-transfer coefficient h and fluid temperature T_∞.

$$\frac{d}{dr}\left(r^{b-1}\frac{dT}{dr}\right)+\frac{S}{k}r^{b-1}=0$$

$$\frac{dT(0)}{dr}=0 \quad \text{(symmetry condition)} \tag{5-13}$$

$$-k\frac{dT}{dr}=h[T(R)-T_\infty]$$

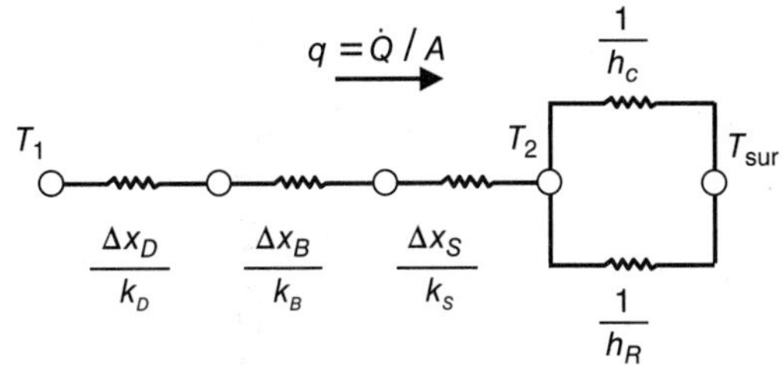

FIG. 5-4 Thermal circuit for Example 5-1. Steady conduction in a furnace wall with heat losses from the outside surface by convection (h_C) and radiation (h_R) to the surroundings at temperature T_{sur}. The thermal conductivities k_D, k_B, and k_S are constant. The heat flux q has units of W/m^2.

The solutions to Eq. (5-13), for uniform S, are

$$\frac{T(r)-T_\infty}{SR^2/k}=\frac{1}{2b}\left[1-\left(\frac{r}{R}\right)^2\right]+\frac{1}{b\text{Bi}} \quad \begin{cases} b=1,\text{ plate, thickness }2R\\ b=2,\text{ cylinder, diameter }2R\\ b=3,\text{ sphere, diameter }2R\end{cases} \tag{5-14}$$

where Bi = hR/k is the Biot number. For Bi << 1, the temperature in the solid is uniform. For Bi >> 1, the surface temperature $T(R) = T_\infty$.

Two- and Three-Dimensional Conduction Application of the law of conservation of energy to a three-dimensional solid, with the heat flux given by Eq. (5-1) and volumetric source term S (W/m^3), results in the following equation for steady conduction in rectangular coordinates.

$$\frac{\partial}{\partial x}\left(k\frac{\partial T}{\partial x}\right)+\frac{\partial}{\partial y}\left(k\frac{\partial T}{\partial y}\right)+\frac{\partial}{\partial z}\left(k\frac{\partial T}{\partial z}\right)+S=0 \tag{5-15}$$

Similar equations apply to cylindrical and spherical coordinate systems. Finite difference, finite volume, or finite element methods are generally necessary to solve Eq. (5-15). Useful introductions to these numerical techniques are given in General References and Sec. 3. Simple forms of Eq. (5-15) (constant k, uniform S) can be solved analytically. See Arpaci, *Conduction Heat Transfer*, Addison-Wesley, Boston, 1966, p. 180, and Carslaw and Jaeger, *Conduction of Heat in Solids*, Oxford University Press, London, 1959. For problems involving heat flow between two surfaces, each isothermal, with all other surfaces being adiabatic, the shape factor approach is useful (Mills, *Heat Transfer*, 2d ed., Prentice-Hall, Upper Saddle River, N.J., 1999, p. 164).

Fins The rate of heat transfer from a surface can be increased by adding fins to increase its area (Mills, *Heat Transfer*, 2nd ed., Prentice-Hall, Upper Saddle River, N.J., 1999, p. 86). Adding fins increases the area, but not the entire added surface is at the temperature of the original surface and it becomes necessary to calculate the efficiency of the fin as follows.

The governing equation and boundary conditions for a pin fin are

$$kA_c\frac{d^2T}{dx^2}-hP(T-T_\infty)=0$$

$$T\big|_{x=0}=T_B \tag{5-16}$$

$$\frac{dT}{dx}\bigg|_{x=L}=0$$

where the cross-sectional area of the fin A_c is constant and heat loss from the tip of the fin is assumed negligible. The temperature distribution, heat loss, and fin efficiency with these assumptions are

$$\frac{T-T_\infty}{T_B-T_\infty}=\frac{\cosh\beta(L-x)}{\cosh\beta L} \quad \text{where} \quad \beta=\left(\frac{hP}{kA_c}\right)^{1/2} \tag{5-17}$$

$$\dot{Q}=kA_c\beta(T_B-T_\infty)\tanh\beta L \tag{5-18}$$

$$\eta_f=\frac{\tanh\beta L}{\beta L} \tag{5-19}$$

For a surface that is covered with fins of efficiency η_f, the total surface efficiency is given by

$$A\eta_t=(A-A_f)+\eta_f A_f \tag{5-20}$$

where A is the total area for heat transfer and A_f is the surface area of the fins. The total efficiency becomes

$$\eta_t=1-\frac{A_f}{A}(1-\eta_f) \tag{5-21}$$

The thermal resistance, based on the total area for heat transfer, becomes

$$R=\frac{1}{h_c A\eta_t} \tag{5-22}$$

Mills (*Heat Transfer*, 2d ed., Prentice-Hall, Upper Saddle River, N.J., 1999, p. 104) provides fin efficiencies for a variety of fin shapes.

UNSTEADY CONDUCTION

Application of the law of conservation of energy to a three-dimensional solid, with the heat flux given by Eq. (5-1) and volumetric source term

S (W/m^3), results in the following equation for unsteady conduction in rectangular coordinates.

$$\rho c \frac{\partial T}{\partial t} = \frac{\partial}{\partial x}\left(k\frac{\partial T}{\partial x}\right) + \frac{\partial}{\partial y}\left(k\frac{\partial T}{\partial y}\right) + \frac{\partial}{\partial z}\left(k\frac{\partial T}{\partial z}\right) + S \qquad (5\text{-}23)$$

The energy storage term is on the left-hand side, and ρ and c are, respectively, the density (kg/m^3) and specific heat [J/(kg·K)]. Solutions to Eq. (5-23) are generally obtained numerically (see General References and Sec. 3). The one-dimensional form of Eq. (5-23), with constant k and no source term, is

$$\frac{\partial T}{\partial t} = \alpha \frac{\partial^2 T}{\partial x^2} \qquad (5\text{-}24)$$

where $\alpha = k/(\rho c)$ is the thermal diffusivity (m^2/s).

One-Dimensional Conduction: Lumped and Distributed Analysis The one-dimensional transient conduction equations in rectangular ($b = 1$), cylindrical ($b = 2$), and spherical ($b = 3$) coordinates, with constant k, initial uniform temperature T_i, $S = 0$, and convection at the surface with heat-transfer coefficient h and fluid temperature T_∞, are

$$\frac{\partial T}{\partial t} = \frac{\alpha}{r^{b-1}}\frac{\partial}{\partial r}\left(r^{b-1}\frac{\partial T}{\partial r}\right) \quad \begin{cases} b=1, \text{plate, thickness } 2R \\ b=2, \text{cylinder, diameter } 2R \\ b=3, \text{sphere, diameter } 2R \end{cases}$$

$$\text{For } t<0, \quad T = T_i \quad \text{(initial temperature)} \qquad (5\text{-}25)$$

$$\text{At } r=0, \quad \frac{\partial T}{\partial r} = 0 \quad \text{(symmetry condition)}$$

$$\text{At } r=R, \quad -k\frac{\partial T}{\partial r} = h(T - T_\infty)$$

The solutions to Eq. (5-25) can be compactly expressed by using dimensionless variables: (1) temperature $\theta/\theta_i = [T(r, t) - T_\infty]/(T_i - T_\infty)$; (2) heat loss fraction $Q/Q_i = Q/[\rho c V(T_i - T_\infty)]$, where V is volume; (3) distance from center $\zeta = r/R$; (4) time $\text{Fo} = \alpha t/R^2$; and (5) Biot number $\text{Bi} = hR/k$. The temperature and heat loss are functions of ζ, Fo, and Bi.

When the Biot number is small, Bi < 0.1, the temperature of the solid is nearly uniform and a lumped analysis is acceptable. The solution to the lumped analysis of Eq. (5-25) is

$$\frac{\theta}{\theta_i} = \exp\left(-\frac{hA}{\rho c V}t\right) \quad \text{and} \quad \frac{Q}{Q_i} = 1 - \exp\left(-\frac{hA}{\rho c V}t\right) \qquad (5\text{-}26)$$

where A is the active surface area and V is the volume. The time scale for the lumped problem is

$$\tau = \frac{\rho c V}{hA} \qquad (5\text{-}27)$$

The time scale is the time required for most of the change in Q/Q_i or θ/θ_i to occur. When $t = \tau$, $\theta/\theta_i = \exp(-1) = 0.368$ and roughly two-thirds of the possible change has occurred.

When a lumped analysis is not valid (Bi > 0.1), the single-term solutions to Eq. (5-25) are convenient:

$$\frac{\theta}{\theta_i} = A_1 \exp(-\delta_1^2 \text{Fo}) S_1(\delta_1 \zeta) \quad \text{and} \quad \frac{Q}{Q_i} = 1 - B_1 \exp(-\delta_1^2 \text{Fo}) \qquad (5\text{-}28)$$

where the first Fourier coefficients A_1 and B_1 and the spatial functions S_1 are given in Table 5-1. The first eigenvalue δ_1 is given by Eq. (5-29) in conjunction with Table 5-2. The one-term solutions are accurate to within 2 percent when Fo > Fo$_c$. The values of the critical Fourier number Fo$_c$ are given in Table 5-2.

TABLE 5-1 Fourier Coefficients and Spatial Functions for Use in Eq. (5-28)

Geometry	A_1	B_1	S_1
Plate	$\dfrac{2\sin\delta_1}{\delta_1 + \sin\delta_1 \cos\delta_1}$	$\dfrac{2\text{Bi}^2}{\delta_1^2(\text{Bi}^2 + \text{Bi} + \delta_1^2)}$	$\cos(\delta_1\zeta)$
Cylinder	$\dfrac{2J_1(\delta_1)}{\delta_1[J_0^2(\delta_1) + J_1^2(\delta_1)]}$	$\dfrac{4\text{Bi}^2}{\delta_1^2(\delta_1^2 + \text{Bi}^2)}$	$J_0(\delta_1\zeta)$
Sphere	$\dfrac{2\text{Bi}[\delta_1^2 + (\text{Bi}-1)^2]^{1/2}}{(\delta_1^2 + \text{Bi}^2 - \text{Bi})}$	$\dfrac{6\text{Bi}^2}{\delta_1^2(\delta_1^2 + \text{Bi}^2 - \text{Bi})}$	$\dfrac{\sin(\delta_1\zeta)}{\delta_1\zeta}$

TABLE 5-2 First Eigenvalues for Bi → 0, Bi → ∞, Correlation Parameter *n*, where the Single-Term Approximations Apply Only If Fo ≥ Fo$_c$

Geometry	Bi → 0	Bi → ∞	n	Fo$_c$
Plate	$\delta_1 \to \sqrt{\text{Bi}}$	$\delta_1 \to \pi/2$	2.139	0.24
Cylinder	$\delta_1 \to \sqrt{2\text{Bi}}$	$\delta_1 \to 2.4048255$	2.238	0.21
Sphere	$\delta_1 \to \sqrt{3\text{Bi}}$	$\delta_1 \to \pi$	2.314	0.18

The first eigenvalue is accurately correlated by Yovanovich (Chap. 3 of Rohsenow, Hartnett, and Cho, *Handbook of Heat Transfer*, 3d ed., McGraw-Hill, New York, 1998, p. 3.25)

$$\delta_1 = \frac{\delta_{1,\infty}}{[1 + (\delta_{1,\infty}/\delta_{1,0})^n]^{1/n}} \qquad (5\text{-}29)$$

Equation (5-29) gives values of δ_1 that differ from the exact values by less than 0.4 percent, and it is valid for all values of Bi. The values of $\delta_{1,\infty}$, $\delta_{1,0}$, n, and Fo$_c$ are given in Table 5-2.

Example 5-2 Correlation of First Eigenvalues by Eq. (5-29) As an example of the use of Eq. (5-29), suppose that we want δ_1 for the flat plate with Bi = 5. From Table 5-2, $\delta_{1,\infty} = \pi/2$, $\delta_{1,0} = \sqrt{\text{Bi}} = \sqrt{5}$, and $n = 2.139$. Equation (5-29) gives

$$\delta_1 = \frac{\pi/2}{[1 + \{(\pi/2)/(\sqrt{5})\}^{2.139}]^{1/2.139}} = 1.312$$

The tabulated value is 1.3138.

Example 5-3 One-Dimensional, Unsteady Conduction Calculation As an example of the use of Eq. (5-28), Table 5-1, and Table 5-2, consider the cooking time required to raise the center of a spherical, 8-cm-diameter dumpling from 20 to 80°C. The initial temperature is uniform. The dumpling is heated with saturated steam at 95°C. The heat capacity, density, and thermal conductivity are estimated to be $c = 3500$ J/(kg·K), $\rho = 1000$ kg/m^3, and $k = 0.5$ W/(m·K), respectively.

Because the heat-transfer coefficient for condensing steam is of order 10^4, the Bi → ∞ limit in Table 5-2 is a good choice and $\delta_1 = \pi$. Because we know the desired temperature at the center, we can calculate θ/θ_i and then solve Eq. (5-28) for the time.

$$\frac{\theta}{\theta_i} = \frac{T(0,t) - T_\infty}{T_i - T_\infty} = \frac{80-95}{20-95} = 0.200$$

For Bi → ∞, A_1 in Table 5-1 is 2 and for $\zeta = 0$, S_1 in Table 5-1 is 1. Equation (5-28) becomes

$$\frac{\theta}{\theta_i} = 2\exp(-\pi^2 \text{Fo}) = 2\exp\left(-\pi^2 \frac{\alpha t}{R^2}\right)$$

Solving for t gives the desired cooking time.

$$t = -\frac{R^2}{\alpha\pi^2}\ln\frac{\theta}{2\theta_i} = -\frac{(0.04\text{ m})^2}{1.43\times 10^{-7}(\text{m}^2/\text{s})\pi^2}\ln\frac{0.2}{2} = 43.5 \text{ min}$$

The one-term approximation is applicable in this case because calculation of Fo gives 0.23, which is greater than Fo$_c$ = 0.18 from Table 5-2.

Example 5-4 Rule of Thumb for Time Required to Diffuse a Distance R A general rule of thumb for estimating the time required to diffuse a distance R is obtained from the one-term approximations. Consider the equation for the temperature of a flat plate of thickness $2R$ in the limit as Bi → ∞. From Table 5-2, the first eigenvalue is $\delta_1 = \pi/2$, and from Table 5-1,

$$\frac{\theta}{\theta_i} = A_1 \exp\left[-\left(\frac{\pi}{2}\right)^2 \frac{\alpha t}{R^2}\right] \cos\delta_1\zeta$$

When $t = R^2/\alpha$, the temperature ratio at the center of the plate ($\zeta = 0$) has decayed to $\exp(-\pi^2/4)$, or 8 percent of its initial value. We conclude that *diffusion through a distance R takes roughly* R^2/α *units of time,* or alternatively, *the distance diffused in time t is about* $(\alpha t)^{1/2}$. More generally, the time scale for Eq. (5-25) for any Bi is approximately

$$\tau \approx \frac{R^2}{b\alpha}\left(\frac{1}{2}+\frac{1}{\text{Bi}}\right)$$

One-Dimensional Conduction: Semi-infinite Plate Consider a semi-infinite plate with an initial uniform temperature T_i. Suppose that the temperature of the surface is suddenly raised to T_∞; that is, the heat-transfer coefficient is infinite. The unsteady temperature of the plate is

$$\frac{T(x,t)-T_\infty}{T_i-T_\infty}=\text{erf}\left(\frac{x}{2\sqrt{\alpha t}}\right) \tag{5-30}$$

where erf(z) is the error function. The depth to which the heat penetrates in time t is approximately $(12\alpha t)^{1/2}$.

If the heat-transfer coefficient is finite,

$$\frac{T(x,t)-T_\infty}{T_i-T_\infty}=\text{erf}\left(\frac{x}{2\sqrt{\alpha t}}\right)+\exp\left(\frac{hx}{k}+\frac{h^2\alpha t}{k^2}\right)\text{erfc}\left(\frac{x}{2\sqrt{\alpha t}}+\frac{h\sqrt{\alpha t}}{k}\right) \tag{5-31}$$

Equations (5-30) and (5-31) are applicable to finite plates provided that their half-thickness is greater than $(12\alpha t)^{1/2}$.

Two- and Three-Dimensional Conduction The one-dimensional solutions discussed above can be used to construct solutions to multidimensional problems. The unsteady temperature of a rectangular, solid box of height, length, and width $2H$, $2L$, and $2W$, respectively, with governing equations in each direction as in Eq. (5-25), is

$$\left(\frac{\theta}{\theta_i}\right)_{2H\times 2L\times 2W}=\left(\frac{\theta}{\theta_i}\right)_{2H}\left(\frac{\theta}{\theta_i}\right)_{2L}\left(\frac{\theta}{\theta_i}\right)_{2W} \tag{5-32}$$

Similar products apply for solids with other geometries, e.g., semi-infinite, cylindrical rods.

HEAT TRANSFER BY CONVECTION

CONVECTIVE HEAT-TRANSFER COEFFICIENT

Convection is the transfer of energy by conduction and radiation in moving, fluid media. The motion of the fluid is an essential part of convective heat transfer. A key step in calculating the rate of heat transfer by convection is the calculation of the heat-transfer coefficient. This section focuses on the estimation of heat-transfer coefficients for natural and forced convection. The conservation equations for mass, momentum, and energy, as presented in Sec. 6, can be used to calculate the rate of convective heat transfer. Our approach in this section is to rely on correlations.

In many cases of industrial importance, heat is transferred from one fluid, through a solid wall, to another fluid. The transfer occurs in a heat exchanger. Section 11 introduces several types of heat exchangers, design procedures, overall heat-transfer coefficients, and mean temperature differences. Section 3 introduces dimensional analysis and the dimensionless groups associated with the heat-transfer coefficient.

Individual Heat-Transfer Coefficient The local rate of convective heat transfer between a surface and a fluid is given by Newton's law of cooling

$$q=h(T_{\text{surface}}-T_{\text{fluid}}) \tag{5-33}$$

where h [W/(m^2·K)] is the local heat-transfer coefficient and q is the energy flux (W/m^2). The definition of h is arbitrary, depending on whether the bulk fluid, centerline, free stream, or some other temperature is used for T_{fluid}. The heat-transfer coefficient may be defined on an average basis as noted below.

Consider a fluid with bulk temperature T, flowing in a cylindrical tube of diameter D, with constant wall temperature T_s. An energy balance on a short section of the tube yields

$$c_p\dot{m}\frac{dT}{dx}=\pi Dh(T_s-T) \tag{5-34}$$

where c_p is the specific heat at constant pressure [J/(kg·K)], $\dot{m}$ is the mass flow rate (kg/s), and x is the distance from the inlet. If the temperature of the fluid at the inlet is T_{in}, the temperature of the fluid at a downstream distance L is

$$\frac{T(L)-T_s}{T_{\text{in}}-T_s}=\exp\left(-\frac{\bar{h}\pi DL}{\dot{m}c_p}\right) \tag{5-35}$$

The average heat-transfer coefficient $\bar{h}$ is defined by

$$\bar{h}=\frac{1}{L}\int_0^L h\,dx \tag{5-36}$$

Overall Heat-Transfer Coefficient and Heat Exchangers A local, overall heat-transfer coefficient U for the cylindrical geometry shown in Fig. 5-2 is defined by using Eq. (5-11):

$$\frac{\dot{Q}}{\Delta x}=\frac{T_i-T_o}{\dfrac{1}{2\pi r_1h_i}+\dfrac{\ln(r_2/r_1)}{2\pi k}+\dfrac{1}{2\pi r_2h_o}}=2\pi r_2U(T_i-T_o) \tag{5-37}$$

where Δx is a short length of tube in the axial direction. Equation (5-37) defines U by using the outside perimeter $2\pi r_2$. The inner perimeter can also be used. Equation (5-37) applies to clean tubes. Additional resistances are present in the denominator for dirty tubes (see Sec. 11).

For counterflow and parallel flow heat exchanges, with high- and low-temperature fluids (T_H and T_C) and flow directions as defined in Fig. 5-5, the total heat transfer for the exchanger is given by

$$\dot{Q}=UA\,\Delta T_{\text{lm}} \tag{5-38}$$

where A is the area for heat exchange and the log mean temperature difference ΔT_{lm} is defined as

$$\Delta T_{\text{lm}}=\frac{(T_H-T_C)_{x=0}-(T_H-T_C)_{x=L}}{\ln\left[(T_H-T_C)_{x=0}/(T_H-T_C)_{x=L}\right]} \tag{5-39}$$

Equation (5-39) applies to both counterflow and parallel-flow exchangers with the nomenclature defined in Fig. 5-5. Correction factors to ΔT_{lm} for various heat exchanger configurations are given in Sec. 11.

In certain applications, the log mean temperature difference is replaced with an arithmetic mean difference:

$$\Delta T_{\text{am}}=\frac{(T_H-T_C)_{x=L}+(T_H-T_C)_{x=0}}{2} \tag{5-40}$$

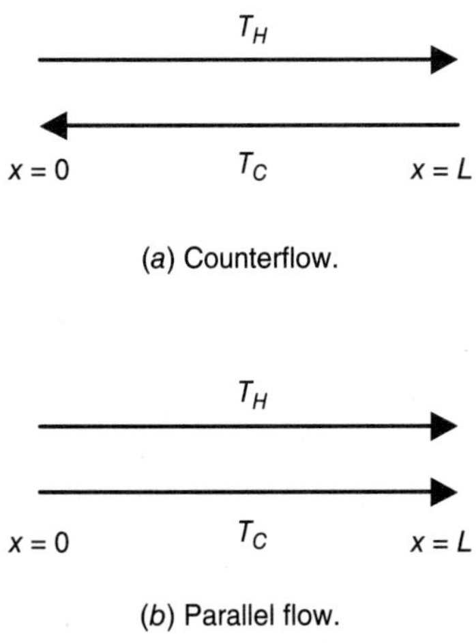

FIG. 5-5 Nomenclature for counterflow and parallel-flow heat exchangers for use with Eqs. (5-38) and (5-39).

Average heat-transfer coefficients are occasionally reported based on Eqs. (5-39) and (5-40) and are written as h_{lm} and h_{am}.

Representation of Heat-Transfer Coefficients Heat-transfer coefficients are usually expressed in two ways: (1) dimensionless equations and (2) dimensional equations. Only the dimensionless approach is used here. The dimensionless form of the heat-transfer coefficient is the Nusselt number. For example, with a cylinder of diameter D in cross flow, the local Nusselt number is defined as $Nu_D = hD/k$, where k is the thermal conductivity of the fluid. The subscript D is important because different characteristic lengths can be used to define Nu. The average Nusselt number is written $\overline{Nu}_D = \bar{h}D/k$.

NATURAL CONVECTION

Natural convection occurs when a fluid is in contact with a solid surface and their temperatures differ. Temperature differences create the density gradients that drive natural or free convection. In addition to the Nusselt number mentioned above, the key dimensionless parameters for natural convection include the Rayleigh number $Ra_x = \beta \Delta T g x^3/\nu\alpha$ and the Prandtl number $Pr = \nu/\alpha$. The properties appearing in Ra and Pr include the volumetric coefficient of expansion β (K^{-1}); the difference ΔT between the surface (T_s) and free stream (T_e) temperatures (K or °C); the acceleration of gravity g (m/s²); a characteristic dimension x of the surface (m); the kinematic viscosity ν (m²/s); and the thermal diffusivity α (m²/s). The volumetric coefficient of expansion for an ideal gas is $\beta = 1/T$, where T is absolute temperature. For a given geometry,

$$\overline{Nu}_x = f(Ra_x, Pr) \tag{5-41}$$

External Natural Flow for Various Geometries For vertical walls, Churchill and Chu [*Int. J. Heat Mass Transfer*, **18:** 1323 (1975)] recommend, for laminar and turbulent flow on isothermal, vertical walls with height L,

$$\overline{Nu}_L = \left\{0.825 + \frac{0.387 Ra_L^{1/6}}{[1+(0.492/Pr)^{9/16}]^{8/27}}\right\}^2 \tag{5-42}$$

where the fluid properties for Eq. (5-42) and $\overline{Nu}_L = \bar{h}L/k$ are evaluated at the film temperature $T_f = (T_s + T_e)/2$. This correlation is valid for all Pr and Ra_L. For vertical cylinders with boundary layer thickness much less than their diameter, Eq. (5-42) is applicable. An expression for uniform heating is available from the same reference.

For laminar and turbulent flow on isothermal, horizontal cylinders of diameter D, Churchill and Chu [*Int. J. Heat Mass Transfer*, **18:** 1049 (1975)] recommend

$$\overline{Nu}_L = \left\{0.60 + \frac{0.387 Ra_D^{1/6}}{[1+(0.559/Pr)^{9/16}]^{8/27}}\right\}^2 \tag{5-43}$$

Fluid properties for Eq. (5-43) should be evaluated at the film temperature $T_f = (T_s + T_e)/2$. This correlation is valid for all Pr and Ra_D.

For long, horizontal, flat plates, the characteristic dimension for the correlations is the width L. With constant surface temperature and hot surfaces facing upward, or cold surfaces facing downward, Lloyd and Moran recommend [ASME Paper 74-WA/HT-66 (1974)]

$$\overline{Nu}_L = 0.54 Ra_L^{1/4} \quad \text{(laminar)} \quad 10^5 < Ra_L < 10^7 \tag{5-44}$$

$$\overline{Nu}_L = 0.15 Ra_L^{1/3} \quad \text{(turbulent)} \quad 10^7 < Ra_L < 10^9 \tag{5-45}$$

and for hot surfaces facing downward, or cold surfaces facing upward, for laminar and turbulent flow,

$$\overline{Nu}_L = 0.27 Ra_L^{1/4} \quad 10^5 < Ra_L < 10^{10} \tag{5-46}$$

Fluid properties for Eqs. (5-44) to (5-46) should be evaluated at the film temperature,

$$T_f = (T_s + T_e)/2 \tag{5-47}$$

Simultaneous Heat Transfer by Radiation and Convection Simultaneous heat transfer by radiation and convection is treated per the procedure outlined in Examples 5-1 and 5-5. A radiative heat-transfer coefficient h_R is defined by Eq. (5-12).

Mixed Forced and Natural Convection Natural convection is commonly assisted or opposed by forced flow. These situations are discussed, e.g., by Mills [*Heat Transfer*, 2d ed., Prentice Hall, Upper Saddle River, N.J., 1999, p. 340] and Raithby and Hollands [Chap. 4 of Rohsenow, Hartnett, and Cho, *Handbook of Heat Transfer*, 3d ed., McGraw-Hill, New York, 1998, p. 4.73].

Enclosed Spaces The rate of heat transfer across an enclosed space is described in terms of a heat-transfer coefficient based on the temperature difference between two surfaces:

$$\bar{h} = \frac{\dot{Q}/A}{T_H - T_C} \tag{5-48}$$

For rectangular cavities, the plate spacing between the two surfaces L is the characteristic dimension that defines the Nusselt and Rayleigh numbers. The temperature difference in the Rayleigh number $Ra_L = \beta \Delta T g L^3/\nu\alpha$ is $\Delta T = T_H - T_C$.

For a horizontal rectangular cavity heated from below, the onset of advection requires $Ra_L > 1708$. Globe and Dropkin [*J. Heat Transfer*, **81:** 24–28 (1959)] propose the correlation

$$\overline{Nu}_L = 0.069 Ra_L^{1/3} Pr^{0.074} \quad 3\times 10^5 < Ra_L < 7\times 10^9 \tag{5-49}$$

All properties in Eq. (5-49) are calculated at the average temperature $(T_H + T_C)/2$.

For vertical rectangular cavities of height H and spacing L, with $Pr \approx 0.7$ (gases) and $40 < H/L < 110$, the equation of Shewen et al. [*J. Heat Transfer*, **118:** 993–995 (1996)] is recommended:

$$\overline{Nu}_L = \left\{1 + \left[\frac{0.0665 Ra_L^{1/3}}{1+(9000/Ra_L)^{1.4}}\right]^2\right\}^{1/2} \quad Ra_L < 10^6 \tag{5-50}$$

All properties in Eq. (5-50) are calculated at the average temperature $(T_H + T_C)/2$.

Example 5-5 Comparison of the Relative Importance of Natural Convection and Radiation at Room Temperature Estimate the heat losses by natural convection and radiation for an undraped person standing in still air. The temperatures of the air, surrounding surfaces, and skin are 19, 15, and 35°C, respectively. The height and surface area of the person are, respectively, 1.8 m and 1.8 m². The emissivity of the skin is 0.95.

We can estimate the Nusselt number by using Eq. (5-42) for a vertical, flat plate of height $L = 1.8$ m. The film temperature is (19 + 35)/2 = 27°C. The Rayleigh number, evaluated at the film temperature, is

$$Ra_L = \frac{\beta \Delta T g L^3}{\nu\alpha} = \frac{(1/300)(35-19)9.81(1.8)^3}{1.589\times 10^{-5}(2.25\times 10^{-5})} = 8.53\times 10^9$$

From Eq. (5-42) with Pr = 0.707, the Nusselt number is 240 and the average heat-transfer coefficient due to natural convection is

$$\bar{h} = \frac{k}{L}\overline{Nu}_L = \frac{0.0263}{1.8}(240) = 3.50 \frac{W}{m^2\cdot K}$$

The radiative heat-transfer coefficient is given by Eq. (5-12):

$$h_R = \varepsilon_{skin}\sigma(T_{skin}^2 + T_{sur}^2)(T_{skin} + T_{sur})$$
$$= 0.95(5.67\times 10^{-8})(308^2 + 288^2)(308 + 288) = 5.71 \frac{W}{m^2\cdot K}$$

The total rate of heat loss is

$$\dot{Q} = \bar{h}A(T_{skin} - T_{air}) + \bar{h}_R A(T_{skin} - T_{sur})$$
$$= 3.50(1.8)(35 - 19) + 5.71(1.8)(35 - 15) = 306 \text{ W}$$

At these conditions, radiation is nearly twice as important as natural convection.

FORCED CONVECTION

Forced convection heat transfer is probably the most common mode in the process industries. Forced flows may be internal or external. This subsection briefly introduces correlations for estimating heat-transfer coefficients for flows in tubes and ducts; flows across plates, cylinders, and spheres; flows through tube banks; and heat transfer to nonevaporating falling films.

Section 11 introduces several types of heat exchangers, design procedures, overall heat-transfer coefficients, and mean temperature differences.

Flow in Round Tubes In addition to the Nusselt ($Nu_D = hD/k$) and Prandtl ($Pr = \nu/\alpha$) numbers introduced above, the key dimensionless parameter for forced convection in round tubes of inside diameter D is the Reynolds number $Re_D = 4\dot{m}/\pi D\mu = \rho VD/\mu$. For internal flow in a tube or duct, the heat-transfer coefficient is defined as

$$q = h(T_s - T_b) \tag{5-51}$$

where T_b is the bulk or mean temperature at a given cross section and T_s is the corresponding surface temperature.

For laminar flow ($Re_D < 2100$) that is fully developed, both hydrodynamically and thermally, the Nusselt number has a constant value for a uniform wall temperature, $Nu_D = 3.66$. For a uniform heat flux through the tube wall, $Nu_D = 4.36$. In both cases, the thermal conductivity of the fluid in Nu_D is evaluated at T_b. The distance x required for a fully developed laminar velocity profile is given by $(x/D)/Re_D \approx 0.05$. The distance x required for fully developed laminar thermal profiles is obtained from $[(x/D)/(Re_D\, Pr)] \approx 0.05$.

For a constant wall temperature, a fully developed laminar velocity profile, and a developing thermal profile, the average Nusselt number is estimated by [Hausen, *Allg. Waermetech.* **9:** 75 (1959)]

$$\overline{Nu}_D = 3.66 + \frac{0.0668(D/L)Re_D Pr}{1 + 0.04[(D/L)Re_D Pr]^{2/3}} \tag{5-52}$$

For large values of L, Eq. (5-52) approaches $Nu_D = 3.66$. Equation (5-52) also applies to developing velocity and thermal profile conditions if $Pr >> 1$. The properties in Eq. (5-52) are evaluated at the bulk mean temperature

$$\overline{T}_b = (T_{b,\text{in}} + T_{b,\text{out}})/2 \tag{5-53}$$

For a constant wall temperature with developing laminar velocity and thermal profiles, the average Nusselt number is approximated by [Sieder and Tate, *Ind. Eng. Chem.* **28:** 1429 (1936)]

$$\overline{Nu}_D = 1.86\left(\frac{D}{L}Re_D Pr\right)^{1/3}\left(\frac{\mu_b}{\mu_s}\right)^{0.14} \tag{5-54}$$

The properties, except for μ_s, are evaluated at the bulk mean temperature per Eq. (5-53) and $0.48 < Pr < 16{,}700$ and $0.0044 < \mu_b/\mu_s < 9.75$.

For fully developed flow in the transition region between laminar and turbulent flow, and for fully developed turbulent flow, Gnielinski's [*Int. Chem. Eng.* **16:** 359 (1976)] equation is recommended:

$$Nu_D = \frac{(f/8)(Re_D - 1000)Pr}{1 + 12.7(f/8)^{1/2}(Pr^{2/3} - 1)}K \tag{5-55}$$

where $0.5 < Pr < 10^5$, $3000 < Re_D < 10^6$, $K = (Pr_b/Pr_s)^{0.11}$ for liquids ($0.05 < Pr_b/Pr_s < 20$), and $K = (T_b/T_s)^{0.45}$ for gases ($0.5 < T_b/T_s < 1.5$). The factor K corrects for variable property effects. For smooth tubes, the Fanning friction factor f for use with Eq. (5-55) is given by

$$f = (0.790 \ln Re_D - 1.64)^{-2} \qquad 3000 < Re_D < 10^6 \tag{5-56}$$

For rough pipes, approximate values of Nu_D are obtained if f is estimated by the Moody diagram of Sec. 6. Equation (5-55) is corrected for entrance effects per Eq. (5-60) and Table 5-3. Sieder and Tate [*Ind. Eng. Chem.* **28:** 1429 (1936)] recommend a simpler but less accurate equation for fully developed turbulent flow

$$Nu_D = 0.027 Re_D^{4/5} Pr^{1/3}\left(\frac{\mu_b}{\mu_s}\right)^{0.14} \tag{5-57}$$

TABLE 5-3 Effect of Entrance Configuration on the Values C and n in Eq. (5-60) for $Pr \approx 1$ (Gases and Other Fluids with Pr about 1)

Entrance configuration	C	n
Long calming section	0.9756	0.760
Open end, 90° edge	2.4254	0.676
180° return bend	0.9759	0.700
90° round bend	1.0517	0.629
90° elbow	2.0152	0.614

where $0.7 < Pr < 16{,}700$, $Re_D < 10{,}000$, and $L/D > 10$. Equations (5-55) and (5-57) apply to both constant temperature and uniform heat flux along the tube. The properties are evaluated at the bulk temperature T_b, except for μ_s, which is at the temperature of the tube. For L/D greater than about 10, Eqs. (5-55) and (5-57) provide an estimate of $\overline{Nu}_D$. In this case, the properties are evaluated at the bulk mean temperature per Eq. (5-53). More complicated and comprehensive predictions of fully developed turbulent convection are available in Churchill and Zajic [*AIChE J.* **48:** 927 (2002)] and Yu, Ozoe, and Churchill [*Chem. Eng. Science,* **56:** 1781 (2001)].

For fully developed turbulent flow of liquid metals, the Nusselt number depends on the wall boundary condition. For a constant wall temperature [Notter and Sleicher, *Chem. Eng. Science,* **27:** 2073 (1972)],

$$Nu_D = 4.8 + 0.0156\, Re_D^{0.85} Pr^{0.93} \tag{5-58}$$

while for a uniform wall heat flux

$$Nu_D = 6.3 + 0.0167\, Re_D^{0.85} Pr^{0.93} \tag{5-59}$$

In both cases the properties are evaluated at T_b and $0.004 < Pr < 0.01$ and $10^4 < Re_D < 10^6$.

Entrance effects for turbulent flow with simultaneously developing velocity and thermal profiles can be significant when $L/D < 10$. Shah and Bhatti correlated entrance effects for gases ($Pr \approx 1$) to give an equation for the average Nusselt number in the entrance region (in Kaka, Shah, and Aung, eds., *Handbook of Single-Phase Convective Heat Transfer,* Chap. 3, Wiley-Interscience, Hoboken, N.J., 1987).

$$\frac{\overline{Nu}_D}{Nu_D} = 1 + \frac{C}{(x/D)^n} \tag{5-60}$$

where Nu_D is the fully developed Nusselt number and the constants C and n are given in Table 5-3 (Ebadian and Dong, Chap. 5 of Rohsenow, Hartnett, and Cho, *Handbook of Heat Transfer,* 3d ed., McGraw-Hill, New York, 1998, p. 5.31). The tube entrance configuration determines the values of C and n as shown in Table 5-3.

Flow in Noncircular Ducts The length scale in the Nusselt and Reynolds numbers for noncircular ducts is the hydraulic diameter $D_h = 4A_c/p$, where A_c is the cross-sectional area for flow and p is the wetted perimeter. For a circular annulus, $D_h = D_o - D_i$, where D_i and D_o are the inner and outer diameters. Nusselt numbers for fully developed laminar flow in a variety of noncircular ducts are given by Mills [*Heat Transfer,* 2d ed., Prentice Hall, Upper Saddle River, N.J., 1999, p. 307]. For turbulent flows, correlations for round tubes can be used with D replaced by D_h.

For annular ducts, the accuracy of the Nusselt number given by Eq. (5-55) is improved by the following multiplicative factors [Petukhov and Roizen, *High Temp.* **2:** 65 (1964)].

$$\text{Inner tube heated} \qquad 0.86\left(\frac{D_i}{D_o}\right)^{-0.16}$$

$$\text{Outer tube heated} \qquad 1 - 0.14\left(\frac{D_i}{D_o}\right)^{0.6}$$

where D_i and D_o are the inner and outer diameters, respectively.

Example 5-6 Turbulent Internal Flow Air at 300 K, 1 bar, and 0.05 kg/s enters a channel of a plate-type heat exchanger (Mills, *Heat Transfer,* 2d ed., Prentice Hall, Upper Saddle River, N.J., 1999) that measures 1 cm wide, 0.5 m high, and 0.8 m long. The walls are at 600 K, and the mass flow rate is 0.05 kg/s. The entrance has a 90° edge. We want to estimate the exit temperature of the air.

Our approach will use Eq. (5-55) to estimate the average heat-transfer coefficient, followed by application of Eq. (5-35) to calculate the exit temperature. We assume ideal gas behavior and an exit temperature of 500 K. The estimated bulk mean temperature of the air is, by Eq. (5-53), 400 K. At this temperature, the properties of the air are $Pr = 0.690$, $\mu = 2.301 \times 10^{-5}$ kg/(m · s), $k = 0.0338$ W/(m · K), and $c_p = 1014$ J/(kg · K).

We start by calculating the hydraulic diameter $D_h = 4A_c/p$. The cross-sectional area for flow A_c is 0.005 m², and the wetted perimeter p is 1.02 m. The hydraulic diameter $D_h = 0.01961$ m. The Reynolds number is

$$Re_{D_h} = \frac{\dot{m}D_h}{A_c\mu} = \frac{0.05(0.01961)}{0.005(2.301 \times 10^{-5})} = 8521$$

The flow is in the transition region, and Eqs. (5-56) and (5-55) apply:

$$f = 0.25(0.790 \ln \mathrm{Re}_{D_h} - 1.64)^{-2} = 0.25(0.790 \ln 8521 - 1.64)^{-2} = 0.008235$$

$$\mathrm{Nu}_D = \frac{(f/2)(\mathrm{Re}_D - 1000)(\mathrm{Pr})}{1 + 12.7(f/2)^{1/2}(\mathrm{Pr}^{2/3} - 1)} K$$

$$= \frac{(0.008235/2)(8521-1000)(0.690)}{1+12.7(0.008235/2)^{1/2}(0.690^{2/3}-1)}\left(\frac{400}{600}\right)^{0.45} = 21.68$$

Entrance effects are included by using Eq. (5-60) for an open-end, 90° edge:

$$\overline{\mathrm{Nu}}_D = \left[1 + \frac{C}{(x/D)^n}\right]\mathrm{Nu}_D = \left[1 + \frac{2.4254}{(0.8/0.01961)^{0.676}}\right](21.68) = 25.96$$

The average heat-transfer coefficient becomes

$$\bar{h} = \frac{k}{D_h}\overline{\mathrm{Nu}}_D = \frac{0.0338}{0.01961}(25.96) = 44.75\ \frac{\mathrm{W}}{\mathrm{m}^2 \cdot \mathrm{K}}$$

The exit temperature is calculated from Eq. (5-35):

$$T(L) = T_s - (T_s - T_{\mathrm{in}})\exp\left(-\frac{\bar{h}pL}{\dot{m}c_p}\right)$$

$$= 600 - (600-300)\exp\left[-\frac{44.75(1.02)0.8}{0.05(1014)}\right] = 450\ \mathrm{K}$$

We conclude that our estimated exit temperature of 500 K is too high. We could repeat the calculations, using fluid properties evaluated at a revised bulk mean temperature of 375 K.

Coiled Tubes For turbulent flow inside helical coils, with tube inside radius a and coil radius R, the Nusselt number for a straight tube Nu_s is related to that for a coiled tube Nu_c by [Rohsenow, Hartnett, and Cho, *Handbook of Heat Transfer*, 3d ed., McGraw-Hill, New York, 1998, p. 5.90]

$$\frac{\mathrm{Nu}_C}{\mathrm{Nu}_S} = 1.0 + 3.6\left(1 - \frac{a}{R}\right)\left(\frac{a}{R}\right)^{0.8} \tag{5-61}$$

where $2 \times 10^4 < \mathrm{Re}_D < 1.5 \times 10^5$ and $5 < R/a < 84$. For lower Reynolds numbers ($1.5 \times 10^3 < \mathrm{Re}_D < 2 \times 10^4$), the same source recommends

$$\frac{\mathrm{Nu}_c}{\mathrm{Nu}_s} = 1.0 + 3.4\frac{a}{R} \tag{5-62}$$

External Flows For a single cylinder in cross flow, Churchill and Bernstein [*J. Heat Transfer*, **99:** 300 (1977)] recommend

$$\overline{\mathrm{Nu}}_D = 0.3 + \frac{0.62\mathrm{Re}_D^{1/2}\mathrm{Pr}^{1/3}}{[1+(0.4/\mathrm{Pr})^{2/3}]^{1/4}}\left[1 + \left(\frac{\mathrm{Re}_D}{282{,}000}\right)^{5/8}\right]^{4/5} \tag{5-63}$$

where $\overline{\mathrm{Nu}}_D = \bar{h}D/k$. Equation (5-63) is for all values of Re_D and Pr, provided that $\mathrm{Re}_D\mathrm{Pr} > 0.4$. The fluid properties are evaluated at the film temperature $(T_e + T_s)/2$, where T_e is the free-stream temperature and T_s is the surface temperature. Equation (5-63) also applies to the uniform heat flux boundary condition provided $\bar{h}$ is based on the perimeter-averaged temperature difference between T_s and T_e.

For an isothermal spherical surface, Whitaker [*AIChE*, **18:** 361 (1972)] recommends

$$\overline{\mathrm{Nu}}_D = 2 + \left(0.4\mathrm{Re}_D^{1/2} + 0.06\mathrm{Re}_D^{2/3}\right)\mathrm{Pr}^{0.4}\left(\frac{\mu_e}{\mu_s}\right)^{1/4} \tag{5-64}$$

This equation is based on data for $0.7 < \mathrm{Pr} < 380$, $3.5 < \mathrm{Re}_D < 8 \times 10^4$, and $1 < \mu_e/\mu_s < 3.2$. The properties are evaluated at the free stream temperature T_e, with the exception of μ_s, which is evaluated at the surface temperature T_s.

The average Nusselt number for laminar flow over an isothermal flat plate of length x is estimated from [Churchill and Ozoe, *J. Heat Transfer*, **95:** 416 (1973)]

$$\overline{\mathrm{Nu}}_x = \frac{1.128\mathrm{Pr}^{1/2}\mathrm{Re}_x^{1/2}}{[1+(0.0468/\mathrm{Pr})^{2/3}]^{1/4}} \tag{5-65}$$

This equation is valid for all values of Pr as long as $\mathrm{Re}_x\ \mathrm{Pr} > 100$ and $\mathrm{Re}_x < 5 \times 10^5$. The fluid properties are evaluated at the film temperature $(T_e + T_s)/2$, where T_e is the free stream temperature and T_s is the surface temperature. For a uniformly heated flat plate, the local Nusselt number is given by [Churchill and Ozoe, *J. Heat Transfer*, **95:** 78 (1973)]

$$\mathrm{Nu}_x = \frac{0.886\mathrm{Pr}^{1/2}\mathrm{Re}_x^{1/2}}{\left[1+(0.0207/\mathrm{Pr})^{2/3}\right]^{1/4}} \tag{5-66}$$

where again the properties are evaluated at the film temperature.

The average Nusselt number for turbulent flow over a smooth, isothermal flat plate of length x is given by [Mills, *Heat Transfer*, 2d ed., Prentice Hall, Upper Saddle River, N.J., 1999, p. 315]

$$\overline{\mathrm{Nu}}_x = 0.664\mathrm{Re}_{\mathrm{cr}}^{1/2}\mathrm{Pr}^{1/3} + 0.036\mathrm{Re}_x^{0.8}\mathrm{Pr}^{0.43}\left[1 - \left(\frac{\mathrm{Re}_{\mathrm{cr}}}{\mathrm{Re}_x}\right)^{0.8}\right] \tag{5-67}$$

The critical Reynolds number $\mathrm{Re}_{\mathrm{cr}}$ is typically taken as 5×10^5, $\mathrm{Re}_{\mathrm{cr}} < \mathrm{Re}_x < 3 \times 10^7$, and $0.7 < \mathrm{Pr} < 400$. The fluid properties are evaluated at the film temperature $(T_e + T_s)/2$, where T_e is the free-stream temperature and T_s is the surface temperature. Equation (5-67) also applies to the uniform heat flux boundary condition provided $\bar{h}$ is based on the average temperature difference between T_s and T_e.

Flow-through Tube Banks Aligned and staggered tube banks are sketched in Fig. 5-6. The tube diameter is D, and the transverse and longitudinal pitches are S_T and S_L. The fluid velocity upstream of the tubes is V_∞. To estimate the overall heat-transfer coefficient for the tube bank, Mills [*Heat Transfer*, 2d ed., Prentice-Hall, 1999, p. 348] proceeds as follows. The Reynolds number for use in Eq. (5-63) is recalculated with an effective average velocity in the space between adjacent tubes:

$$\frac{\bar{V}}{V_\infty} = \frac{S_T}{S_T - (\pi/4)D} \tag{5-68}$$

The heat-transfer coefficient increases from row 1 to about row 5 of the tube bank. The average Nusselt number for a tube bank with 10 or more rows is

$$\overline{\mathrm{Nu}}_D^{10+} = \Phi\overline{\mathrm{Nu}}_D^{1} \tag{5-69}$$

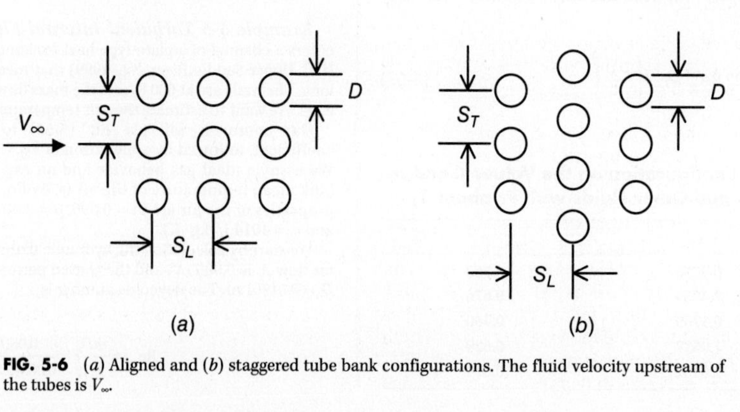

FIG. 5-6 (*a*) Aligned and (*b*) staggered tube bank configurations. The fluid velocity upstream of the tubes is V_∞.

where Φ is an arrangement factor and $\overline{\text{Nu}}_D^1$ is the Nusselt number for the first row, calculated by using the velocity in Eq. (5-68). The arrangement factor is calculated as follows. Define dimensionless pitches as $P_T = S_T/D$ and P_L/D and calculate a factor ψ as follows.

$$\Psi = \begin{cases} 1 - \dfrac{\pi}{4P_t} & \text{if } P_L \geq 1 \\ 1 - \dfrac{\pi}{4P_T P_L} & \text{if } P_L < 1 \end{cases} \tag{5-70}$$

The arrangement factors are

$$\Phi_{\text{aligned}} = 1 + \frac{0.7}{\Psi^{1.5}} \frac{S_L/S_T - 0.3}{(S_L/S_T + 0.7)^2} \tag{5-71}$$

$$\Phi_{\text{staggered}} = 1 + \frac{2}{3P_L} \tag{5-72}$$

If there are fewer than 10 rows,

$$\overline{\text{Nu}}_D = \frac{1 + (N-1)\Phi}{N} \overline{\text{Nu}}_D^1 \tag{5-73}$$

where N is the number of rows.

The fluid properties for gases are evaluated at the average mean film temperature $[(T_{in} + T_{out})/2 + T_s]/2$. For liquids, properties are evaluated at the bulk mean temperature $(T_{in} + T_{out})/2$, with $\overline{\text{Nu}}_D$ from Eq. (5-73) being multiplied by a Prandtl number correction $(\text{Pr}_s/\text{Pr}_b)^{-0.11}$ for cooling and $(\text{Pr}_s/\text{Pr}_b)^{-0.25}$ for heating.

Heat Transfer to Nonevaporating Falling Films When a subcooled liquid flows in a thin layer down a vertical surface, there is little or no evaporation and the heat-transfer coefficient is defined by $q/(T_s - T_b)$ where T_s is the surface temperature and T_b is the bulk fluid temperature.

For laminar flow ($\text{Re}_\delta < 20$–30) the heat-transfer coefficient is given by the equation of Hewitt [Rohsenow, Hartnett, and Cho, *Handbook of Heat Transfer*, 3d ed., McGraw-Hill, New York, 1998, chap. 15]:

$$h = \frac{q}{T_s - T_b} = \frac{280}{141} \left(\frac{4\rho_l^2 g k_l^3}{3\mu_l^2} \right)^{1/3} \text{Re}_\delta^{-1/3} \tag{5-74}$$

where the Reynolds number of the falling film is defined as $\text{Re}_\delta = 4\Gamma/\mu_l$ and Γ is the mass rate of flow of liquid per unit length normal to the direction of flow [kg/(s · m)]. To account for wavy laminar (30–$50 < \text{Re}_\delta < 1600$) and turbulent ($\text{Re}_\delta > 1600$) flow, Wilkie [Rohsenow, Hartnett, and Cho, *Handbook of Heat Transfer*, 3d ed., McGraw-Hill, New York, 1998, chap. 15] recommends

$$\frac{h\delta}{k_l} = C_0 \text{Re}_\delta^m \text{Pr}_l^{0.344} \tag{5-75}$$

where $C_0 = 0.029$ and $m = 0.533$ for $\text{Re}_\delta > 1600$, $C_0 = 0.212 \times 10^{-3}$ and $m = 1.2$ for $1600 < \text{Re}_\delta < 3200$, and $C_0 = 0.181 \times 10^{-2}$ and $m = 0.933$ for $\text{Re}_\delta > 3200$. Equation (5-75) provides an average heat-transfer coefficient, and the value of the film thickness δ for $\text{Re}_\delta < 1600$ is given by

$$\delta = \left(\frac{3\mu_l \Gamma}{\rho_l^2 g} \right)^{1/3} \tag{5-76}$$

and for $\text{Re}_\delta > 1600$ by

$$\left(\frac{\delta^3 \rho_l^2 g}{\mu_l^2} \right)^{0.5} = 0.137 \text{Re}_\delta^{0.75} \tag{5-77}$$

JACKETS AND COILS OF AGITATED VESSELS

See Secs. 11 and 18.

NONNEWTONIAN FLUIDS

Many real fluids are nonnewtonian. Section 6 introduces the dynamics of nonnewtonian fluids in laminar and turbulent regimes. Heat transfer is reviewed by Hartnett and Cho [Rohsenow, Hartnett, and Cho, *Handbook of Heat Transfer*, 3d ed., McGraw-Hill, New York, 1998, chap. 13]. They provide equations, tables, and charts for estimating the Nusselt number in laminar and turbulent internal flow and refer to the literature for external convection, free convection, boiling, suspensions and surfactants, and flow of food products.

HEAT TRANSFER WITH CHANGE OF PHASE

In any process in which a material changes phase, the addition or removal of heat is required to balance the latent heat of the change of phase plus any other sensible heating or cooling that occurs. Heat may be transferred by any one of or a combination of conduction, convection, and radiation. Change of phase involves simultaneous mass and heat transfer.

CONDENSATION

Condensation Mechanisms Condensation occurs when a saturated vapor comes in contact with a surface whose temperature is below the saturation temperature. A film of condensate forms on the surface, and the thickness of the film increases as the liquid flows down the surface. This is called *film-type condensation.*

Another type of condensation, called *dropwise,* occurs when the wall is not uniformly wetted by the condensate, with the result that the condensate appears in many small droplets on the surface. The individual droplets grow, coalesce, and finally form a rivulet.

Film-type condensation is more common and more dependable. Dropwise condensation normally needs to be promoted by introducing an impurity into the vapor stream. Substantially higher (6 to 18 times) coefficients are obtained for dropwise condensation of steam, but it is difficult to maintain. The equations below are for the film type only. For additional details, see Rohsenow, Hartnett, and Cho, *Handbook of Heat Transfer*, 3d ed., McGraw-Hill, New York, 1998 and Bergman, Lavine, Incropera, and DeWitt, *Fundamentals of Heat and Mass Transfer*, 7th ed., Wiley, Hoboken, N.J., 2011.

The *Reynolds number* of the condensate film (falling film) is defined as $\text{Re}_\delta = 4\Gamma/\mu_l$, where Γ is the mass rate of flow of condensate per unit length normal to the direction of flow [kg/(s · m)] and μ_l is the liquid viscosity. For $\text{Re}_\delta < 30$ the flow is laminar and free of waves. When $30 < \text{Re}_\delta < 1800$, the flow is wavy and rippled. At $\text{Re}_\delta > 1800$ the flow is turbulent. The Reynolds number can also be written as

$$\text{Re}_\delta = \frac{4\dot{m}}{\mu_l} = \frac{4\rho_l u_m \delta}{\mu_l} \tag{5-78}$$

where δ is the film thickness.

Condensation Coefficients

Vertical Tubes and Plates For a Reynolds number < 30, the average Nusselt number for laminar condensate films is

$$\overline{\text{Nu}}_L = \frac{\bar{h}_L L}{k_l} = 0.943 \left[\frac{\rho_l g(\rho_l - \rho_v) h'_{fg} L^3}{\mu_l k_l (T_{sat} - T_s)} \right]^{1/4} \tag{5-79}$$

where L is the length of the cooled surface and

$$h'_{fg} = h_{fg} + 0.68 c_{p,l}(T_{sat} - T_s) \tag{5-80}$$

The liquid properties in Eqs. (5-79) and (5-80) are evaluated at the film temperature, $T_f = (T_{sat} + T_s)/2$, and ρ_v and h_{fg} are at T_{sat}. The vapor density in Eq. (5-79) is frequently neglected relative to the liquid density.

The total rate of heat transfer to the surface at temperature T_s is given by

$$\dot{Q} = \bar{h}_L A(T_{sat} - T_s) \tag{5-81}$$

and the rate of condensation is

$$\dot{m} = \frac{\dot{Q}}{h'_{fg}} \tag{5-82}$$

To estimate average Nusselt numbers for laminar, wavy, and turbulent flow, Bergman et al. (2011) recommend the following procedure. A dimensionless parameter P is defined by combining Eqs. (5-82) and (5-78) to give an average modified Nusselt number with characteristic length $\left(\nu_l^2/g\right)^{1/3}$:

$$\begin{aligned} \text{Re}_\delta &= 4P\frac{\bar{h}_L\left(\nu_l^2/g\right)^{1/3}}{k_l} = 4P\overline{\text{Nu}}_L \\ P &= \frac{k_l L(T_{\text{sat}} - T_s)}{\mu_l h'_{fg}\left(\nu_l^2/g\right)^{1/3}} \end{aligned} \tag{5-83}$$

The modified average Nusselt numbers for $\text{Re}_\delta < 30$, $30 < \text{Re}_\delta < 1800$, and $\text{Re}_\delta > 1800$ become

$$\overline{\text{Nu}}_L = \frac{\bar{h}_L\left(\nu_l^2/g\right)^{1/3}}{k_l} = 0.943P^{-1/4} \quad (P \le 15.8) \tag{5-84}$$

$$\overline{\text{Nu}}_L = \frac{\bar{h}_L\left(\nu_l^2/g\right)^{1/3}}{k_l} = \frac{1}{P}(0.68P + 0.89)^{0.82} \quad (15.8 \le P \le 2530) \tag{5-85}$$

$$\overline{\text{Nu}}_L = \frac{\bar{h}_L\left(\nu_l^2/g\right)^{1/3}}{k_l} = \frac{1}{P}\left[(0.024P - 53)\text{Pr}_l^{1/2} + 89\right]^{4/3} \quad (P \ge 2530, \text{Pr}_l \ge 1) \tag{5-86}$$

Equations (5-84) and (5-79) are identical if $\rho_l >> \rho_v$. The fluid properties for Eqs. (5-84), (5-85), and (5-86) are evaluated as described below Eq. (5-80).

Horizontal Smooth Tubes For laminar film condensation on horizontal smooth tubes the average Nusselt number is given by

$$\overline{\text{Nu}}_D = \frac{\bar{h}_D D}{k_l} = 0.729\left[\frac{\rho_l g(\rho_l - \rho_v)h'_{fg}D^3}{\mu_l k_l(T_{\text{sat}} - T_s)}\right]^{1/4} \tag{5-87}$$

The fluid properties for Eq. (5-87) are evaluated as described below Eq. (5-80).

Banks of Horizontal Tubes In the idealized case of N tubes in a vertical row where the total condensate flows smoothly from one tube to the one beneath it, without splashing, and still in laminar flow on the tube, the mean condensing coefficient h_N for the entire row of N tubes is related to the condensing coefficient for the top tube h_1 by

$$h_N = h_1 N^{-s} \tag{5-88}$$

The standard Nusselt theory gives $s = 1/4$ but others recommend $s = 1/6$.

EVAPORATING LIQUID FILMS ON VERTICAL WALLS

Mills' presentation of heat transfer to evaporating falling films [*Heat Transfer*, 2d ed., Prentice Hall, Upper Saddle River, N.J., 1999, pp. 681–685] is given here, with minor modifications. The Reynolds number of the evaporating falling film is defined as $\text{Re}_\delta = 4\Gamma/\mu_l$ and the Nusselt number for evaporation is

$$\text{Nu}_L = \frac{h_L\left(\nu_l^2/g\right)^{1/3}}{k_l} \tag{5-89}$$

For laminar flow the local Nusselt number is

$$\text{Nu} = \left(\frac{3}{4}\text{Re}_\delta\right)^{-1/3} \quad 0 < \text{Re}_\delta < 30, \text{ laminar} \tag{5-90}$$

For wavy laminar and turbulent flows, the correlations of experimental data for water by Chun and Seban [*J. Heat Transfer*, 93: 391–396 (1971)] give

$$\text{Nu} = 0.822\text{Re}_\delta^{-0.22} \quad 30 < \text{Re}_\delta < \text{Re}_{\text{tr}}, \text{ wavy laminar} \tag{5-91}$$

$$\text{Nu} = 0.0038\text{Re}_\delta^{-0.4}\text{Pr}_l^{0.65} \quad \text{Re}_{\text{tr}} < \text{Re}_\delta, \text{ turbulent} \tag{5-92}$$

where

$$\text{Re}_{\text{tr}} = 5800\text{Pr}_l^{0.65} \tag{5-93}$$

Liquid properties in Eqs. (5-88) to (5-92) can be approximated by evaluation at the film temperature: $T_f = (T_{\text{sat}} + T_s)/2$ with h_{fg} at T_{sat}. An energy balance for a vertical surface of length L is

$$\frac{L}{\left(\nu_l^2/g\right)^{1/3}} = -\frac{\text{Pr}_l}{4\text{Ja}_l}\int_{\text{Re}_0}^{\text{Re}_L}\frac{d\text{Re}}{\text{Nu}} \tag{5-94}$$

where L is the length of the surface in the direction of flow and the Jakob number for the liquid is defined as

$$Ja_l = \frac{c_{pl}(T_s - T_{\text{sat}})}{h_{fg}} \tag{5-95}$$

The use of Eqs. (5-88) to (5-95) to estimate an evaporation rate is illustrated by Example 5-7 which is based on an example in Mills [*Heat Transfer*, 2d ed., Prentice Hall, Upper Saddle River, N.J., 1999, pp. 684–685].

Example 5-7 Evaporating Falling Film Water is fed to the outer surface of a single, vertical, 5-cm outer-diameter (OD) tube at the rate of 0.01 kg/s. The tube is 5 m long, and its surface is kept at 311 K by condensing steam on the inside. The saturation temperature at the pressure outside the tube is 308 K. Estimate the evaporation rate. We start by recalling that the liquid properties are approximated at the film temperature [$T_f = (T_s + T_{\text{sat}})/2 = (311 + 308)/2 = 310$ K] and that the enthalpy of vaporization h_{fg} is evaluated at the saturation temperature. At 308 K, $h_{fg} = 2.418 \times 10^6$ J/kg. At the film temperature, $k_l = 0.628$ W/(m · K), $\rho_l = 993$ kg/m^3, $c_{pl} = 4174$ J/(kg · K), $\mu_l = 6.95 \times 10^{-4}$ kg/(m · s), $\nu_l = 0.700 \times 10^{-6}$ m^2/s, and $\text{Pr}_l = 4.6$.

The film Reynolds number at the top of the tube is

$$\text{Re}_0 = \frac{4\Gamma_0}{\mu_l} = \frac{4\dot{m}_0}{\pi D\mu_l} = \frac{4(0.01)}{\pi(0.05)6.95\times10^{-4}} = 366$$

The Reynolds number for transition from wavy laminar to turbulent flow is

$$\text{Re}_{\text{tr}} = 5800\text{Pr}_l^{-1.06} = 5800(4.6)^{-1.06} = 1151$$

Because this is greater than 366 we will assume that we remain in the wavy laminar regime and will check Re_L by using Eq. (5-94).

$$\frac{L}{\left(\nu_l^2/g\right)^{1/3}} = -\frac{\text{Pr}_l}{4\text{Ja}_l}\int_{\text{Re}_0}^{\text{Re}_l}\frac{d\text{Re}}{\text{Nu}} = -\frac{\text{Pr}_l}{4\text{Ja}_l}\int_{\text{Re}_0}^{\text{Re}_l}\frac{d\text{Re}}{0.822\text{Re}^{-0.22}}$$

Integrating and solving for Re_L give

$$\text{Re}_L^{1.22} = -\frac{4.01L}{(\nu_l^2/g)^{1/3}}\frac{\text{Ja}_l}{\text{Pr}_l} + \text{Re}_0^{1.22} = 730$$

where

$$\text{Ja}_l = \frac{c_{pl}(T_s - T_{\text{sat}})}{h_{fg}} = \frac{4174(311-308)}{2.418\times10^6} = 5.18\times10^{-3}$$

$$\left(\frac{\nu_l^2}{g}\right)^{1/3} = \left[\frac{(0.700\times10^{-6})^2}{9.81}\right]^{1/3} = 3.68\times10^{-5}\text{ m}$$

We conclude that the film is entirely in the wavy laminar regime. Solving the film Reynolds number for the mass flow rate at L gives

$$\dot{m}_L = \frac{\pi D\mu_l\text{Re}_L}{4} = \frac{\pi(0.05)6.95\times10^{-4}(222)}{4} = 6.07\times10^{-3}\text{ kg/s}$$

The evaporation rate is

$$\dot{m}_{\text{vap}} = \dot{m}_0 - \dot{m}_L = 0.01 - 0.00607 = 0.0039\text{ kg/s}$$

POOL BOILING

Pool boiling refers to the type of boiling experienced when the heating surface is surrounded by a large body of fluid which is not flowing at any appreciable velocity and is agitated only by the motion of the bubbles and by natural convection currents. Two types of pool boiling are possible: *subcooled* pool boiling, in which the bulk fluid temperature is below the saturation temperature, resulting in collapse of the bubbles before they reach the surface, and *saturated* pool boiling, with the bulk temperature equal to the saturation temperature, resulting in net vapor generation. The following presentation draws heavily from Mills [*Heat Transfer*, 2d ed. (1999)].

In the general shape of the curve relating the heat-transfer coefficient to $\Delta T = T_s - T_{sat}$, the difference between the surface temperature and the saturation temperature is reasonably well understood. The familiar boiling curve was originally demonstrated experimentally by Nukiyama [*J. Soc. Mech. Eng. (Japan)*, **37:** 367 (1934)]. This curve points out one of the great dilemmas for boiling-equipment designers. They are faced with at least four heat-transfer regimes in pool boiling: natural convection (+), nucleate boiling (+), transition to film boiling (−), and film boiling (+). The signs indicate the sign of the derivative $dq/(d\,\Delta T)$. In the transition to film boiling, the heat-transfer rate *decreases* with ΔT. Here we consider nucleate boiling, the peak heat flux, and film boiling.

Nucleate boiling occurs in kettle-type and natural-circulation reboilers commonly used in the process industries. High rates of heat transfer are obtained as a result of bubble formation at the liquid-solid interface. The heat-transfer coefficient is defined by

$$q = h(T_s - T_{sat}) \tag{5-96}$$

where T_{sat} is at the system pressure. The characteristic length used to define the Nusselt number is

$$L_c = \left[\frac{\sigma}{(\rho_l - \rho_v)g}\right]^{1/2} \tag{5-97a}$$

and the Nusselt number is given by Rohsenow [*Trans. ASME*, **74:** 969 (1952)] as

$$\text{Nu} = \frac{hL_c}{k_l} = \frac{\text{Ja}_l^2}{C_{nb}^3 \Pr_l^m} \tag{5-97b}$$

The *Jakob number* is defined as

$$\text{Ja}_l = \frac{c_{pl}(T_s - T_{sat})}{h_{fg}} \tag{5-98}$$

All properties in Eq. (5-97b), including the vapor density, are evaluated at T_{sat}. Typical values for the constant C_{nb} and the exponent m are given in Table 5-4. Equations (5-96) and (5-97b) imply that the rate of heat transfer is proportional to ΔT^3. Errors of 100 percent in q and 25 percent in ΔT are possible with Eq. (5-97b). The designer of heat-transfer equipment is usually more concerned with not exceeding the peak heat flux q_{max} rather than in knowing accurate values of q and ΔT.

TABLE 5-4 The Constant C_{nb} and Exponent m for Use with Rohsenow Eq. (5-97)

Liquid	Surface	C_{nb}	m
Water	Copper, scored	0.0068	2.0
Water	Copper, polished	0.013	2.0
Water	Stainless steel, mechanically polished	0.013	2.0
Ethanol	Chromium	0.0027	4.1
n-Pentane	Chromium	0.015	4.1

The *peak heat flux* may be predicted by the Kutateladse-Zuber [*Trans. ASME*, **80:** 711 (1958)] relationship:

$$q_{max} = C_{max} h_{fg} \left[\sigma \rho_v^2 (\rho_l - \rho_v) g\right]^{1/4} \tag{5-99}$$

where C_{max} is approximately 0.15. All properties in Eq. (5-99) are evaluated at T_{sat}.

For laminar *film boiling*, Bromley's [*Chem. Eng. Prog.* **46:** 221 (1950)] correlation may be used:

$$\bar{h} = C_{fb}\left[\frac{(\rho_l - \rho_v) g h_{fg}^l k_v^3}{\nu_v L(T_s - T_{sat})}\right]^{1/4} \tag{5-100a}$$

where L is a characteristic length. For spheres and horizontal cylinders it is the diameter D. The constant C_{fb} is 0.62 for a horizontal cylinder, 0.67 for a sphere, and 0.71 for a planar vertical surface. The modified latent heat is

$$h_{fg}^l = h_{fg} + 0.35 c_{pv}(T_s - T_{sat}) \tag{5-100b}$$

In Eqs. (5-99) and (5-100a), h_{fg}, ρ_l, and σ are evaluated at T_{sat}; all other properties are at the mean film temperature.

HEAT TRANSFER BY RADIATION

General References: Baukal, C. E., ed., *The John Zink Combustion Handbook,* CRC Press, Boca Raton, Fla., 2001. Blokh, A. G., *Heat Transfer in Steam Boiler Furnaces*, 3d ed., Taylor & Francis, New York, 1987. Brewster, M. Quinn, *Thermal Radiation Heat Transfer and Properties*, Wiley, New York, 1992. Goody, R. M., and Y. L. Yung, *Atmospheric Radiation—Theoretical Basis*, 2d ed., Oxford University Press, London, 1995. Hottel, H. C., and A. F. Sarofim, *Radiative Transfer*, McGraw-Hill, New York, 1967. Howell, John, M. Pinar Mengüç, and Robert Siegel, *Thermal Radiative Heat Transfer*, 6th ed., CRC Press, Boca Raton, Fla., 2015. Modest, Michael F., *Radiative Heat Transfer*, 3d ed., Academic Press, New York, 2013. Noble, James J., "The Zone Method: Explicit Matrix Relations for Total Exchange Areas," *Int. J. Heat Mass Transfer*, **18:** 261–269 (1975). Rhine, J. M., and R. J. Tucker, *Modeling of Gas-Fired Furnaces and Boilers*, British Gas Association with McGraw-Hill, New York, 1991. Sparrow, E. M., and R. D. Cess, *Radiation Heat Transfer*, 3d ed., Taylor & Francis, New York, 1988. Stultz, S. C., and J. B. Kitto, *Steam: Its Generation and Use*, 40th ed., Babcock and Wilcox, Barkerton, Ohio, 1992.

INTRODUCTION

Heat transfer by thermal radiation involves the transport of electromagnetic (EM) energy from a source to a sink. In contrast to other modes of heat transfer, radiation does not require the presence of an intervening medium, e.g., as in the irradiation of the earth by the sun. Most industrially important applications of radiative heat transfer occur in the *near infrared* portion of the EM spectrum (0.7 through 25 μm) and may extend into the *far infrared* region (25 to 1000 μm). For very high temperature sources, such as solar radiation, relevant wavelengths encompass the entire *visible region* (0.4 to 0.7 μm) and may extend down to 0.2 μm in the *ultraviolet* (0.01- to 0.4-μm) portion of the EM spectrum. Radiative transfer can also exhibit unique *action-at-a-distance* phenomena that do not occur in other modes of heat transfer. Radiation differs from conduction and convection with regard to not only mathematical characterization but also its fourth power dependence on temperature. Thus it is usually dominant in high-temperature combustion applications. The temperature at which radiative transfer accounts for roughly one-half of the total heat loss from a surface in air depends on such factors as surface *emissivity* and the convection coefficient. For pipes in free convection, radiation is important at ambient temperatures. For fine wires of low emissivity, it becomes important at temperatures associated with bright red heat (1300 K). Combustion gases at furnace temperatures typically lose more than 90 percent of their energy through radiative emission from constituent carbon dioxide, water vapor, and particulate matter. Radiative transfer methodologies are important in myriad engineering applications. These include semiconductor processing, illumination theory, and gas turbines and rocket nozzles, as well as furnace design.

THERMAL RADIATION FUNDAMENTALS

In a vacuum, the wavelength λ, frequency ν, and wave number η for electromagnetic radiation are interrelated by $\lambda = c/\nu = 1/\eta$, where c is the speed of light. Frequency is independent of the index of refraction of a medium n, but both the speed of light and the wavelength in the medium vary according to $c_m = c/n$ and $\lambda_m = \lambda/n$. When a radiation beam passes into a medium of different refractive index, not only does its wavelength change but also its direction does (Snell's law) as well as the magnitude of its intensity. In most engineering heat-transfer calculations, wavelength is usually employed to characterize radiation while wave number is often used in gas spectroscopy. For a vacuum, air at ambient conditions, and most gases, $n \approx 1.0$. For this reason this presentation sometimes does not distinguish between λ and λ_m. *Dielectric materials* exhibit $1.4 < n < 4$, and the speed of light decreases considerably in such media.

In radiation heat transfer, the *monochromatic intensity* $I_\lambda \equiv I_\lambda(\vec{\mathbf{r}}, \vec{\Omega}, \lambda)$, is a *fundamental* (scalar) *field variable* which characterizes EM energy transport. Intensity defines the radiant energy flux passing through an infinitesimal area dA, oriented *normal* to a radiation beam of arbitrary direction $\vec{\Omega}$. At steady state, the monochromatic intensity is a function of position $\vec{\mathbf{r}}$, direction $\vec{\Omega}$, and wavelength and has units of $W/(m^2 \cdot sr \cdot \mu m)$. In the general case of an *absorbing-emitting* and *scattering* medium, characterized by some absorption coefficient $K(m^{-1})$, intensity in the direction $\vec{\Omega}$ will be modified by attenuation and by *scattering* of radiation into and out of the beam. For the special case of a nonabsorbing (transparent), nonscattering medium of constant refractive index, the radiation intensity *is constant and independent of position in a given direction* $\vec{\Omega}$. This circumstance arises in *illumination theory* where the light intensity in a room is constant in a *given direction* but may vary with respect to *all other directions*. The basic conservation law for radiation intensity is termed the *equation of transfer* or *radiative transfer equation*. The equation of transfer is a *directional* energy balance and mathematically is an *integrodifferential equation*. The relevance of the transport equation to radiation heat transfer is discussed in many sources; see, e.g., Modest, Michael F., *Radiative Heat Transfer*, 3d ed., Academic Press, New York, 2013, or Howell, John, M. Pinar Mengüç, and Robert Siegel, *Thermal Radiative Heat Transfer*, 6th ed., CRC Press, Boca Raton, Fla., 2015.

Introduction to Radiation Geometry Consider a homogeneous medium of constant refractive index n. A pencil of radiation originates at differential area element dA_i and is incident on differential area element dA_j. Designate $\vec{\mathbf{n}}_i$ and $\vec{\mathbf{n}}_j$ as the unit vectors normal to dA_i and dA_j, and let r, with unit direction vector $\vec{\Omega}$, define the distance of separation between the area elements. Moreover, ϕ_i and ϕ_j denote the *confined angles* between $\vec{\Omega}$ and $\vec{\mathbf{n}}_i$ and $\vec{\mathbf{n}}_j$, respectively [i.e., $\cos\phi_i \equiv \cos(\vec{\Omega}, \vec{r}_i)$ and $\cos\phi_j \equiv \cos(\vec{\Omega}, \vec{r}_j)$]. As the beam travels toward dA_j, it will *diverge* and subtend a solid angle

$$d\Omega_j = \frac{\cos\phi_j}{r^2} dA_j \quad \text{in steradian (sr) units}$$

at dA_j. Moreover, the projected area of dA_i in the direction of $\vec{\Omega}$ is given by $\cos(\vec{\Omega}, \vec{r}_j)\, dA_i = \cos\phi_i\, dA_i$. Multiplication of the intensity $I_\lambda \equiv I_\lambda(\vec{r}, \vec{\Omega}, \lambda)$ by $d\Omega_j$ and the apparent area of dA_i then yields an expression for the (*differential*) net *monochromatic radiant energy flux* $dQ_{i,j}$ originating at dA_i and intercepted by dA_j.

$$dQ_{i,j} \equiv I_\lambda(\vec{\Omega}, \lambda)\cos\phi_i \cos\phi_j\, dA_i\, dA_j / r^2 \tag{5-101}$$

The *hemispherical emissive power** E is defined as the radiant flux density (W/m^2) associated with emission from an element of surface area dA into a surrounding unit hemisphere whose base is coplanar with dA. If the monochromatic intensity $I_\lambda(\vec{\Omega}, \lambda)$ of emission from the surface is *isotropic* (independent of the angle of emission $\vec{\Omega}$), then Eq. (5-101) may be integrated over the 2π sr of the surrounding unit hemisphere to yield the simple relation $E_\lambda = \pi I_\lambda$, where $E_\lambda \equiv E_\lambda(\lambda)$ is defined as the monochromatic or *spectral* hemispherical emissive power.

Blackbody Radiation Engineering calculations involving thermal radiation normally employ the *hemispherical blackbody emissive power* as the thermal driving force analogous to temperature in the cases of conduction and convection. A *blackbody* is a theoretical idealization for a *perfect theoretical radiator*; i.e., it absorbs all incident radiation without reflection and emits isotropically. In practice, soot-covered surfaces sometimes approximate blackbody behavior. Let $E_{b,\lambda} = E_{b,\lambda}(T, \lambda)$ denote the monochromatic blackbody hemispherical emissive power *frequency function* defined such that $E_{b,\lambda}(T, \lambda)d\lambda$ represents the fraction of blackbody energy lying in the wavelength region from λ to $\lambda + d\lambda$. The function $E_{b,\lambda} = E_{b,\lambda}(T, \lambda)$ is given by *Planck's law*

$$\frac{E_{b,\lambda}(T,\lambda)}{n^2T^5} = \frac{c_1(\lambda T)^{-5}}{e^{c_2/\lambda T} - 1} \tag{5-102}$$

where $c_1 = 2\pi hc^2$ and $c_2 = hc/k$ are defined as Planck's first and second constants, respectively.

Integration of Eq. (5-102) over all wavelengths yields the *Stefan-Boltzmann law* for the hemispherical blackbody emissive power

$$E_b(T) = \int_{\lambda=0}^{\infty} E_{b,\lambda}(T,\lambda)d\lambda = n^2\sigma T^4 \tag{5-103}$$

*In the literature the emissive power is variously called the *emittance, total hemispherical intensity*, or *radiant flux density*.

where $\sigma = c_1(\pi/c_2)^4/15$ is the Stephan-Boltzmann constant. Since a blackbody is an *isotropic emitter*, it follows that the intensity of blackbody emission is given by the simple formula $I_b = E_b/\pi = n^2\sigma T^4/\pi$. The intensity of radiation emitted over all wavelengths by a blackbody is thus uniquely determined by its temperature. In this presentation, all references to hemispherical emissive power shall be to the *blackbody* emissive power, and the subscript b may be suppressed for expediency.

For short wavelengths $\lambda T \to 0$, the asymptotic form of Eq. (5-102) is known as the *Wien equation*

$$\frac{E_{b,\lambda}(T,\lambda)}{n^2T^5} \cong c_1(\lambda T)^{-5} e^{-c_2/\lambda T} \tag{5-104}$$

The error introduced by use of the Wien equation is less than 1 percent when $\lambda T < 3000\ \mu m \cdot K$. The Wien equation has significant practical value in optical pyrometry for $T < 4600$ K when a red filter ($\lambda = 0.65\ \mu m$) is employed. The long-wavelength asymptotic approximation for Eq. (5-102) is known as the *Rayleigh-Jeans formula*, which is accurate to within 1 percent for $\lambda T > 778{,}000\ \mu m \cdot K$. The Raleigh-Jeans formula is of limited engineering utility since a blackbody emits over 99.9 percent of its total energy below the value of $\lambda T = 53{,}000\ \mu m \cdot K$.

The blackbody fractional energy *distribution function* is defined by

$$F_b(\lambda T) = \frac{\int_{\lambda=0}^{\lambda} E_{b,\lambda}(T,\lambda)d\lambda}{\int_{\lambda=0}^{\infty} E_{b,\lambda}(T,\lambda)d\lambda} \tag{5-105}$$

The function $F_b(\lambda T)$ defines the fraction of total energy in the blackbody spectrum which lies below λT and is a unique function of λT. For purposes of digital computation, the following series expansion for $F_b(\lambda T)$ proves especially useful.

$$F_b(\lambda T) = \frac{15}{\pi^4}\sum_{k=1}^{\infty}\frac{e^{-k\xi}}{k}\left(\xi^3 + \frac{3\xi^2}{k} + \frac{6\xi}{k^2} + \frac{6}{k^3}\right) \quad \text{where} \quad \xi = \frac{c_2}{\lambda T} \tag{5-106}$$

Equation (5-106) converges rapidly and is due to Lowan (1941) as referenced in Chang and Rhee [*Int. Comm. Heat Mass Transfer*, **11:** 451–455 (1984)].

Numerically, in the preceding, $h = 6.6260693 \times 10^{-34}$ J·s is the Planck constant; $c = 2.99792458 \times 10^8$ m/s is the velocity of light in vacuum; and $k = 1.3806505 \times 10^{-23}$ J/K is the Boltzmann constant. These data lead to the following values of Planck's first and second constants: $c_1 = 3.741771 \times 10^{-16}$ $W \cdot m^2$ and $c_2 = 1.438775 \times 10^{-2}$ m·K, respectively. Numerical values of the Stephan-Boltzmann constant σ in several systems of units are as follows: $5.67040 \times 10^{-8}\ W/(m^2 \cdot K^4)$; $1.3544 \times 10^{-12}\ cal/(cm^2 \cdot s \cdot K^4)$; $4.8757 \times 10^{-8}\ kcal/(m^2 \cdot h \cdot K^4)$; $9.9862 \times 10^{-9}\ CHU/(ft^2 \cdot h \cdot K^4)$; and $0.17123 \times 10^{-8}\ Btu/(ft^2 \cdot h \cdot {}^\circ R^4)$ (CHU = Celsius heat unit; 1.0 CHU = 1.8 Btu).

Blackbody Displacement Laws The blackbody energy spectrum is plotted *logarithmically* in Fig. 5-7 as

$$\frac{E_{b,\lambda}(\lambda T)}{n^2T^5} \times 10^{13} \frac{W}{m^2 \cdot \mu m \cdot K^5}$$

versus λT μm·K. For comparison, a companion inset is provided in cartesian coordinates. The upper abscissa of Fig. 5-7 also shows the blackbody energy distribution function $F_b(\lambda T)$. Figure 5-7 indicates that the wavelength-temperature product for which the maximum intensity occurs is $\lambda_{max}T = 2898\ \mu m \cdot K$. This relationship is known as *Wien's displacement law*, which indicates that the wavelength for maximum intensity is *inversely* proportional to the absolute temperature. Blackbody displacement laws are useful in engineering practice to estimate wavelength intervals appropriate to relevant system temperatures. The Wien displacement law can be misleading, however, because the wavelength for maximum intensity depends on whether the intensity is defined in terms of frequency or wavelength interval. Two additional useful displacement laws are defined in terms of either the value of λT corresponding to the maximum energy per unit *fractional change* in wavelength or frequency, that is, $\lambda T = 3670\ \mu m \cdot K$, or to the value of λT corresponding to one-half of the blackbody energy, that is, $\lambda T = 4107\ \mu m \cdot K$. Approximately one-half of the blackbody energy lies within the twofold λT range *geometrically centered* on $\lambda T = 3670\ \mu m \cdot K$, that is, $3670/\sqrt{2} < \lambda T < 3670\sqrt{2}\ \mu m \cdot K$. Some 95 percent of the blackbody energy lies in the interval $1662.6 < \lambda T < 16{,}295\ \mu m \cdot K$. It thus follows that for the temperature range between ambient (300 K) and flame temperatures (2000 K or 3140°F), wavelengths of engineering heat-transfer importance are *bounded* between 0.83 and 54.3 μm.

Nomenclature and Units—Radiative Transfer

a, a_g, a_{g1}	WSGG spectral model clear plus gray weighting constants
$\bar{C}_p, \bar{C}_{p,\text{Prod}}$	Heat capacity per unit mass, $J \cdot kg^{-1} \cdot K^{-1}$
$\overline{ij} = \overline{s_i s_j}$	Shorthand notation for direct exchange area
A, A_i	Area of enclosure or zone i, m^2
c	Speed of light in vacuum, m/s
c_1, c_2	Planck's first and second constants, $W \cdot m^2$ and $m \cdot K$
d_p, r_p	Particle diameter and radius, μm
$E_{b\lambda} = E_{b\lambda}(T, \lambda)$	Monochromatic, blackbody emissive power, $W/(m^2 \cdot \mu m)$
$E_n(x)$	Exponential integral of order n, where $n = 1, 2, 3, \ldots$
E	Hemispherical emissive power, W/m^2
$E_b = n^2 \sigma T^4$	Hemispherical blackbody emissive power, W/m^2
f_v	Volumetric fraction of soot
$f_b(\lambda T)$	Blackbody fractional energy distribution
$F_{i,j}$	Direct view factor from surface zone i to surface zone j
$\bar{F}_{i,j}$	Refractory augmented black view factor; F-bar
$\mathfrak{F}_{i,j}$	Total view factor from surface zone i to surface zone j
h	Planck's constant, $J \cdot s$
h_i	Heat-transfer coefficient, $W/(m^2 \cdot K)$
H_i	Incident flux density for surface zone i, W/m^2
H	Enthalpy rate, W
$\dot{H}_F$	Enthalpy feed rate, W
$I_\lambda \equiv I_\lambda(\vec{r}, \vec{\Omega}, \lambda)$	Monochromatic radiation intensity, $W/(m^2 \cdot \mu m \cdot sr)$
k	Boltzmann's constant, J/K
$k_{\lambda,p}$	Monochromatic line absorption coefficient, $(atm \cdot m)^{-1}$
K	Gas absorption coefficient, m^{-1}
L_M, L_{M0}	Average and optically thin mean beam lengths, m
$\dot{m}$	Mass flow rate, kg/h^{-1}
n	Index of refraction
M, N	Number of surface and volume zones in enclosure
MS, MF	Number of source/sink and flux zones in enclosure where $M = MS + MF$
p_k	Partial pressure of species k, atm
P	Number of WSGG gray gas spectral windows
Q_i	Total radiative flux originating at surface zone i, W
$Q_{i,j}$	Net radiative flux between zone i and zone j, W
T	Temperature, K
U	Overall heat-transfer coefficient in WSCC model
V	Enclosure volume, m^3
W	Leaving flux density (radiosity), W/m^2

Greek characters

$\alpha, \alpha_{1,2}$	Surface absorptivity or absorptance; subscript 1 refers to the surface temperature while subscript 2 refers to the radiation source
$\alpha_{g,1}, \varepsilon_g, \tau_{g,1}$	Gas absorptivity, emissivity, and transmissivity
β	Dimensionless constant in mean beam length equation, $L_M = \beta \cdot L_{M0}$
$\Delta T_e \equiv T_g - T_e$	Adjustable temperature fitting parameter for WSCC model, K
ε	Gray diffuse surface emissivity
$\varepsilon_g(T, r)$	Gas emissivity with path length r
$\varepsilon_\lambda(T, \Omega, \lambda)$	Monochromatic, unidirectional surface emissivity
$\eta = 1/\lambda$	Wave number in vacuum, cm^{-1}
$\lambda = c/\nu$	Wavelength in vacuum, μm
ν	Frequency, Hz
$\rho = 1 - \varepsilon$	Diffuse reflectivity
σ	Stefan-Boltzmann constant, $W/(m^2 \cdot K^4)$
Σ	Number of unique direct surface-to-surface direct exchange areas
$\tau_g = 1 - \varepsilon_g$	Gas transmissivity
Ω	Solid angle, sr (steradians)
Φ	Equivalence ratio of fuel and oxidant
$\Psi^{(3)}(x)$	Pentagamma function of x
ω	Albedo for single scatter

Dimensionless quantities

$D_{\text{eff}} = \dfrac{N_{FD}}{(\overline{S_1 G}_R/A_1) + N_{CR}}$	Effective firing density
$N_{CR} = \dfrac{h}{4\sigma \bar{T}^3_{g,1}}$	Convection-radiation number
$N_{FD} = \dot{H}_f / \sigma T^4_{\text{Ref}} \cdot A_1$	Dimensionless firing density
η_g	Gas-side furnace efficiency
$\eta'_g = \eta_g(1 - \Theta_o)$	Reduced furnace efficiency
$\Theta_i = T_i / T_{\text{Ref}}$	Dimensionless temperature

Vector notation

$\vec{n}_i$ and $\vec{n}_j$	Unit vectors normal to differential area elements dA_i and dA_j
$\vec{r}$	Position vector
$\vec{\Omega}$	Arbitrary unit direction vector

Matrix notation

$\mathbf{I}_M$	Column vector, all of whose elements are unity $[M \times 1]$
$\mathbf{I} = [\delta_{i,j}]$	Identity matrix, where $\delta_{i,j}$ is the Kronecker delta; that is, $\delta_{i,j} = 1$ for $i = j$ and $\delta_{i,j} = 0$ for $i \neq j$
$\mathbf{aI} = [a_i \cdot \delta_{i,j}]$	Diagonal matrix of WSGG gray gas surface zone a-weighting factors $[M \times M]$
$\mathbf{a_g I} = [a_{g,i} \cdot \delta_{i,j}]$	Diagonal matrix of gray gas WSGG volume zone a-weighting factors $[N \times N]$
$\mathbf{A} = [A_{i,j}]$	Arbitrary nonsingular square matrix
$\mathbf{A}^T = [A_{j,i}]$	Transpose of **A**
$\mathbf{A}^{-1} = [A_{i,j}]^{-1}$	Inverse of **A**
$\mathbf{DI} = [D_i \cdot \delta_{i,j}]$	Arbitrary diagonal matrix
$\mathbf{DI}^{-1} = [\delta_{i,j}/D_i]$	Inverse of diagonal matrix
$\mathbf{CDI} = \mathbf{CI} \cdot \mathbf{DI} = [C_i \cdot D_i \cdot \delta_{i,j}]$	Product of two diagonal matrices
$\mathbf{AI} = [A_i \cdot \delta_{i,j}]$	Diagonal matrix of surface zone areas, m^2 $[M \times M]$
$\boldsymbol{\varepsilon}\mathbf{I} = [\varepsilon_i \cdot \delta_{i,j}]$	Diagonal matrix of diffuse zone emissivities $[M \times M]$
$\boldsymbol{\rho}\mathbf{I} = [\rho_i \cdot \delta_{i,j}]$	Diagonal matrix of diffuse zone reflectivities $[M \times M]$
$\mathbf{E} = [E_i] = [\sigma T_i^4]$	Column vector of surface blackbody hemispherical emissive powers, W/m^2 $[M \times 1]$
$\mathbf{EI} = [E_i \cdot \delta_{i,j}] = [\sigma T_i^4 \cdot \delta_{i,j}]$	Diagonal matrix of surface blackbody emissive powers, W/m^2 $[M \times M]$
$\mathbf{E}_g = [E_{g,i}] = [\sigma T_{g,i}^4]$	Column vector of gas blackbody hemispherical emissive powers, W/m^2 $[N \times 1]$
$\mathbf{E_g I} = [E_{g,i} \cdot \delta_{i,j}] = [\sigma T_{g,i}^4 \cdot \delta_{i,j}]$	Diagonal matrix of gas blackbody emissive powers, W/m^2 $[N \times N]$
$\mathbf{H} = [H_i]$	Column vector of surface zone incident flux densities, W/m^2 $[M \times 1]$
$\mathbf{W} = [W_i]$	Column vector of surface zone leaving flux densities, W/m^2 $[M \times 1]$
$\mathbf{Q} = [Q_i]$	Column vector of surface zone fluxes, W $[M \times 1]$
$\mathbf{R} = [\mathbf{AI} - \overline{\mathbf{ss}} \cdot \boldsymbol{\rho}\mathbf{I}]^{-1}$	Inverse multiple-reflection matrix, m^{-2} $[M \times M]$
$\overleftarrow{\mathbf{KI}}_p = [\delta_{i,j} \cdot K_{p,i}]$	Diagonal matrix of WSGG $K_{p,i}$ values for the ith zone and pth gray gas component, m^{-1} $[N \times N]$
$\overleftarrow{\mathbf{KI}}$	Diagonal matrix of WSGG-weighted gray gas absorption coefficients, m^{-1} $[N \times N]$
$\mathbf{S}'$	Column vector for net volume absorption, W $[N \times 1]$
$\overline{\mathbf{ss}} = [\overline{s_i s_j}]$	Array of direct surface-to-surface exchange areas, m^2 $[M \times M]$
$\overline{\mathbf{sg}} = [\overline{s_i g_j}] = \overline{\mathbf{gs}}^T$	Array of direct gas-to-surface exchange areas, m^2 $[M \times N]$
$\overline{\mathbf{gg}} = [\overline{g_i g_j}]$	Array of direct gas-to-gas exchange areas, m^2 $[N \times N]$
$\overline{\mathbf{SS}} = [\overline{S_i S_j}]$	Array of total surface-to-surface exchange areas, m^2 $[M \times M]$
$\overline{\mathbf{SG}} = [\overline{S_i G_j}]$	Array of total gas-to-surface exchange areas, m^2 $[M \times N]$
$\overline{\mathbf{GS}} = \overline{\mathbf{GS}}^T$	Array of total surface-to-gas exchange areas, m^2 $[N \times M]$
$\overline{\mathbf{GG}} = [\overline{G_i G_j}]$	Array of total gas-to-gas exchange areas, m^2 $[N \times N]$
$\overleftarrow{\mathbf{SS}} = [\overleftarrow{S_i S_j}]$	Array of directed surface-to-surface exchange areas, m^2 $[M \times M]$
$\overleftarrow{\mathbf{SG}} = [\overleftarrow{S_i G_j}]$	Array of directed gas-to-surface exchange areas, m^2 $[M \times N]$
$\overleftarrow{\mathbf{GS}} \neq \overleftarrow{\mathbf{SG}}^T$	Array of directed surface-to-gas exchange areas, m^2 $[N \times M]$
$\overleftarrow{\mathbf{GG}} = [\overleftarrow{G_i G_j}]$	Array of directed gas-to-gas exchange areas, m^2 $[N \times N]$
$\mathbf{VI} = [V_i \cdot \delta_{i,j}]$	Diagonal matrix of zone volumes, m^3 $[N \times N]$
$\mathbf{FSQ}$	Array of flux fractions in SSR model, dimensionless $MS \times MS$
$\mathbf{SSb} = [\text{SSb}_{i,j}]$	Array of total refractory-aided exchange areas for black source/sink zones in SSR model, m^2 $MS \times MS$
$\mathbf{SSR} = [\text{SSR}_{i,j}]$	Array of total refractory-aided exchange areas for nonblack source/sink zones in SSR model, m^2 $MS \times MS$

Subscripts

b	Blackbody or denotes a black surface zone
f	Flux surface zone
h	Hemispherical surface emissivity
i, j	Zone number indices
n	Normal component of surface emissivity
p	Index for pth gray gas window
r	Refractory surface zone
s	Source-sink surface zone
λ	Monochromatic variable
Ref	Reference quantity

Abbreviations

CFD	Computational fluid dynamics
DEA, TEA	Direct exchange area and total exchange area
DO, FV	Discrete ordinate and finite volume methods
EM	Electromagnetic
RTE	Radiative transfer equation; equation of transfer
LPFF	Long plug flow furnace
SSR	Source-sink refractory
WSCC	Well-stirred combustion chamber
WSGG	Weighted sum of gray gases

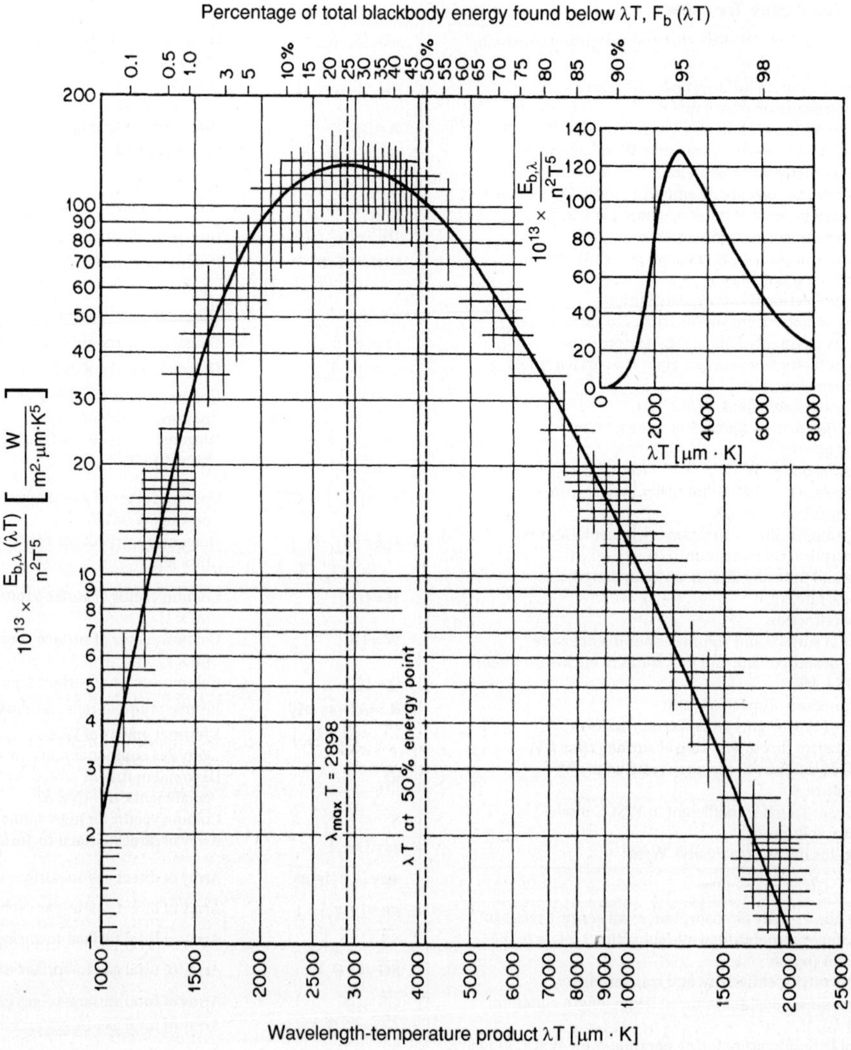

FIG. 5-7 Spectral dependence of monochromatic blackbody hemispherical emissive power.

RADIATIVE PROPERTIES OF OPAQUE SURFACES

Emittance and Absorptance The ratio of the total radiating power of any surface to that of a black surface at the same temperature is called the *emittance* or *emissivity* ε of the surface.* In general, the monochromatic emissivity is a function of temperature, direction, and wavelength, that is, $\varepsilon_\lambda = \varepsilon_\lambda(T, \bar{\Omega}, \lambda)$. The subscripts *n* and *h* are sometimes used to denote the normal and hemispherical values, respectively, of the emittance or emissivity. If radiation is incident on a surface, the fraction absorbed is called the *absorptance* (*absorptivity*). Two subscripts are usually appended to the absorptance $\alpha_{1,2}$ to distinguish between the temperature of the absorbing surface T_1 and the spectral energy distribution of the emitting surface T_2. According to *Kirchhoff's law*, the emissivity and absorptivity of a surface exposed to surroundings at its own temperature are the same for both monochromatic and total radiation. When the temperatures of the surface and its surroundings differ, the total emissivity and absorptivity of the surface are often found to be unequal; but because the absorptivity is substantially independent of irradiation density, the monochromatic emissivity and absorptivity of surfaces are equal for all practical purposes. The difference between *total* emissivity and absorptivity depends on the variation of ε_λ with wavelength and on the difference between the temperature of the surface and the effective temperature of the surroundings.

Consider radiative exchange between a real surface of area A_1 at temperature T_1 with black surroundings at temperature T_2. The *net* radiant interchange is given by

$$Q_{1,2} = A_1 \int_{\lambda=0}^{\infty} [\varepsilon_\lambda(T_1,\lambda)\cdot E_{b,\lambda}(T_1,\lambda) - \alpha_\lambda(T_1,\lambda)\cdot E_{b,\lambda}(T_2,\lambda)]\,d\lambda \tag{5-107a}$$

or

$$Q_{1,2} = A_1\left(\varepsilon_1 \sigma T_1^4 - \alpha_{1,2}\sigma T_2^4\right) \tag{5-107b}$$

where

$$\varepsilon_1(T_1) = \int_{\lambda=0}^{\infty} \varepsilon_\lambda(T_1,\lambda)\cdot\frac{E_{b,\lambda}(T_1,\lambda)}{E_b(T_1)}\,d\lambda \tag{5-108}$$

and since

$$\alpha_\lambda(T,\lambda) = \varepsilon_\lambda(T,\lambda)$$

$$\alpha_{1,2}(T_1,T_2) = \int_{\lambda=0}^{\infty} \varepsilon_\lambda(T_1,\lambda)\cdot\frac{E_{b,\lambda}(T_2,\lambda)}{E_b(T_2)}\,d\lambda \tag{5-109}$$

*In the literature, *emittance* and *emissivity* are often used interchangeably. NIST (the National Institute of Standards and Technology) recommends use of the suffix *-ivity* for pure materials with optically smooth surfaces, and *-ance* for rough and contaminated surfaces. Most real engineering materials fall into the latter category.

For a *gray surface* $\varepsilon_1 = \alpha_{1,2} = \varepsilon_\lambda$. A *selective surface* is one for which $\varepsilon_\lambda(T, \lambda)$ exhibits a strong dependence on wavelength. If the wavelength dependence is *monotonic*, it follows from Eqs. (5-107) to (5-109) that ε_1 and $\alpha_{1,2}$ can differ markedly when T_1 and T_2 are widely separated. For example, in solar energy applications, the nominal temperature of the earth is $T_1 = 294$ K, and the sun may be represented as a blackbody with radiation temperature $T_2 = 5800$ K. For these temperature conditions, a *white paint* can exhibit $\varepsilon_1 = 0.9$ and $\alpha_{1,2} = 0.1$ to 0.2. In contrast, a *thin layer of copper oxide on bright aluminum* can exhibit ε_1 as low as 0.12 and $\alpha_{1,2}$ greater than 0.9.

The effect of radiation source temperature on low-temperature absorptivity for a number of representative materials is shown in Fig. 5-8. Polished aluminum (curve 15) and anodized (surface-oxidized) aluminum (curve 13) are representative of metals and nonmetals, respectively. Figure 5-8 thus demonstrates the generalization that metals and nonmetals respond *in opposite directions* with regard to changes in the radiation source temperature. Since the effective solar temperature is 5800 K (10,440°R), the extreme right-hand side of Fig. 5-8 provides surface absorptivity data relevant to solar energy applications. The dependence of emittance and absorptance on the real and imaginary components of the refractive index and on the geometric structure of the surface layer is quite complex. However, a number of generalizations concerning the radiative properties of opaque surfaces are possible. These are summarized in the following discussion.

Polished Metals

1. In the infrared region, the magnitude of the monochromatic emissivity ε_λ is small and is dependent on free-electron contributions. Emissivity is also a function of the ratio of resistivity to wavelength r/λ, as depicted in Fig. 5-9. At shorter wavelengths, bound-electron contributions become significant, ε_λ is larger in magnitude, and it sometimes exhibits a maximum value.

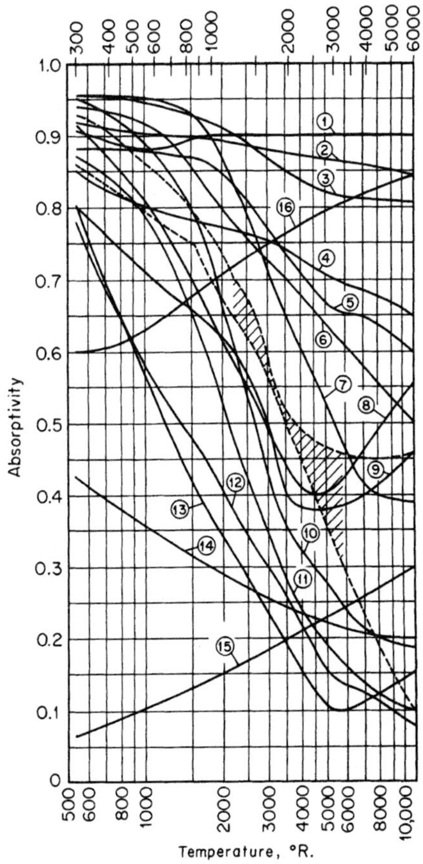

FIG. 5-8 Variation of absorptivity with temperature of radiation source. (1) Slate composition roofing. (2) Linoleum, red brown. (3) Asbestos slate. (4) Soft rubber, gray. (5) Concrete. (6) Porcelain. (7) Vitreous enamel, white. (8) Red brick. (9) Cork. (10) White Dutch tile. (11) White chamotte. (12) MgO, evaporated. (13) Anodized aluminum. (14) Aluminum paint. (15) Polished aluminum. (16) Graphite. The two dashed lines bound the limits of data on gray paving brick, asbestos paper, wood, various cloths, plaster of Paris, lithopone, and paper. To convert degrees Rankine to kelvins, multiply by (5.556)(10^{-1}).

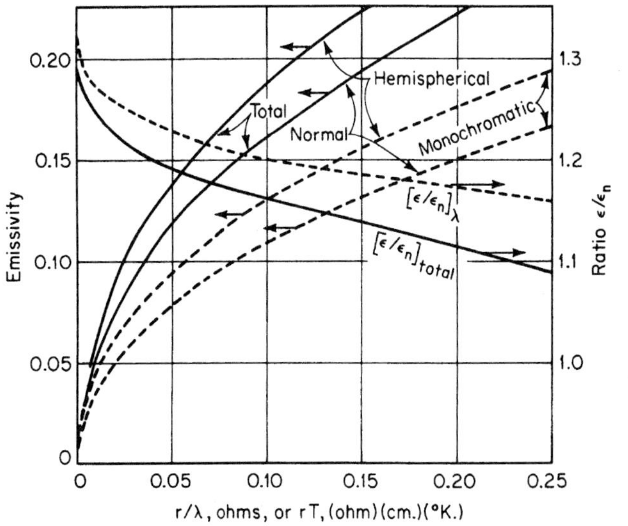

FIG. 5-9 Hemispherical and normal emissivities of metals and their ratio. Dashed lines: monochromatic (spectral) values versus r/λ. Solid lines: total values versus rT. To convert ohmcentimeter-kelvins to ohm-meter-kelvins, multiply by 10^{-2}.

In the visible spectrum, common values for ε_λ are 0.4 to 0.8 and ε_λ decreases slightly as temperature increases. For $0.7 < \lambda < 1.5$ μm, ε_λ is approximately independent of temperature. For $\lambda > 8$ μm, ε_λ is approximately proportional to the square root of temperature since $\varepsilon_\lambda \propto \sqrt{r}$ and $r \propto T$. Here the Drude or Hagen-Rubens relation applies, that is, $\varepsilon_{\lambda,n} \approx 0.0365\sqrt{r}/\lambda$, where r has units of ohm-meters and λ is measured in micrometers.

2. Total emittance is substantially proportional to absolute temperature, and at moderate temperatures $\varepsilon_n = 0.058T\sqrt{rT}$, where T is measured in kelvins.

3. The total absorptance of a metal at temperature T_1 with respect to radiation from a black or gray source at temperature T_2 is equal to the emissivity evaluated at the geometric mean of T_1 and T_2. Figure 5-9 gives values of ε_λ and $\varepsilon_{\lambda,n}$, and their ratio, as a function of the product rT (solid lines). Although Fig. 5-9 is based on free-electron contributions to emissivity in the far infrared, the relations for total emissivity are remarkably good even at high temperatures. Unless extraordinary efforts are taken to prevent oxidation, a metallic surface may exhibit an emittance or absorptance which may be several times that of a polished specimen. For example, the emittance of iron and steel depends strongly on the degree of oxidation and roughness. Clean iron and steel surfaces have an emittance from 0.05 to 0.45 at ambient temperatures and from 0.4 to 0.7 at high temperatures. Oxidized and/or roughened iron and steel surfaces have values of emittance ranging from 0.6 to 0.95 at low temperatures to 0.9 to 0.95 at high temperatures.

Refractory Materials For refractory materials, the dependence of emittance and absorptance on grain size and impurity concentrations is quite important.

1. Most refractory materials are characterized by $0.8 < \varepsilon_\lambda < 1.0$ for the wavelength region $2 < \lambda < 4$ μm. The monochromatic emissivity ε_λ decreases rapidly toward shorter wavelengths for materials that are white in the visible range but demonstrates high values for black materials such as FeO and Cr_2O_3. Small concentrations of FeO and Cr_2O_3, or other colored oxides, can cause marked increases in the emittance of materials that are normally white. The sensitivity of the emittance of refractory oxides to small additions of absorbing materials is demonstrated by the results of calculations presented in Fig. 5-10. Figure 5-10 shows the emittance of a semi-infinite absorbing-scattering medium as a function of its *albedo* $\omega \equiv K_S/(K_a + K_S)$, where K_a and K_S are the scatter and absorption coefficients, respectively. These results are relevant to the radiative properties of fibrous materials, paints, oxide coatings, refractory materials, and other *particulate* media. They demonstrate that over the relatively small range $1 - \omega = 0.005$ to 0.1, the hemispherical emittance ε_h increases from approximately 0.15 to 1.0. For refractory materials, ε_λ varies little with temperature, with the exception of some white oxides which at high temperatures become good emitters in the visible spectrum as a consequence of the induced electronic transitions.

2. For refractory materials at ambient temperatures, the total emittance ε is generally high (0.7 to 1.0). Total refractory emittance decreases

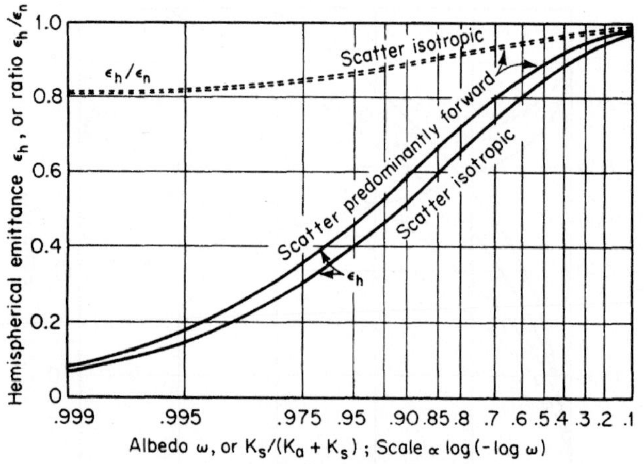

FIG. 5-10 Hemispherical emittance ε_h and the ratio of hemispherical to normal emittance $\varepsilon_h/\varepsilon_n$ for a semi-infinite absorbing-scattering medium.

with increasing temperature, such that a temperature increase from 1000 to 1570°C may result in a 20 to 30 percent reduction in ε.

3. Emittance and absorptance increase with increase in grain size over a grain size range of 1 to 200 μm.

4. The ratio $\varepsilon_h/\varepsilon_n$ of hemispherical to normal emissivity of polished surfaces varies with refractive index n; e.g., the ratio decreases from a value of 1.0 when $n = 1.0$ to a value of 0.93 when $n = 1.5$ (common glass) and increases back to 0.96 at $n = 3.0$.

5. As shown in Fig. 5-10, for a surface composed of particulate matter which scatters isotropically, the ratio $\varepsilon_h/\varepsilon_n$ varies from 1.0 when $\omega < 0.1$ to about 0.8 when $\omega = 0.999$.

6. The total absorptance exhibits a decrease with an increase in temperature of the radiation source similar to the decrease in emittance with an increase in the emitter temperature.

Figure 5-8 shows a regular variation of $\alpha_{1,2}$ with T_2. When T_2 is not very different from T_1, $\alpha_{1,2} = \varepsilon_1(T_2/T_1)^m$. It may be shown that Eq. (5-107*b*) is then approximated by

$$Q_{1,2} = (1 + m/4)\varepsilon_{av} A_1 \sigma (T_1^4 - T_2^4) \tag{5-110}$$

where ε_{av} is evaluated at the arithmetic mean of T_1 and T_2. For metals $m \approx 0.5$ while for nonmetals m is small and negative.

Table 5-5 illustrates values of emittance for materials encountered in engineering practice. It is based on a critical evaluation of early emissivity data. Table 5-5 demonstrates the wide variation possible in the emissivity of a particular material due to variations in surface roughness and thermal pretreatment. With few exceptions the data in Table 5-5 refer to emittances ε_n normal to the surface. The hemispherical emittance ε_h is usually slightly smaller, as demonstrated by the ratio $\varepsilon_h/\varepsilon_n$ depicted in Fig. 5-10. More recent data support the range of emittance values given in Table 5-5 and their dependence on surface conditions. An extensive compilation is provided by Goldsmith, Waterman, and Hirschorn (*Thermophysical Properties of Matter*, Purdue University, Touloukian, ed., Plenum, New York, 1970–1979).

For opaque materials the reflectance ρ is the complement of the absorptance. The directional distribution of the reflected radiation depends on the material, its degree of roughness or grain size, and, if a metal, its state of oxidation. Polished surfaces of homogeneous materials are specular reflectors. In contrast, the intensity of the radiation reflected from a perfectly diffuse or *Lambert surface* is independent of direction. The directional distribution of reflectance of many oxidized metals, refractory materials, and natural products approximates that of a perfectly diffuse reflector. A better model, adequate for many calculation purposes, is achieved by assuming that the total reflectance is the sum of diffuse and specular components ρ_D and ρ_S, as discussed in a subsequent section.

VIEW FACTORS AND DIRECT EXCHANGE AREAS

Consider radiative interchange between two *finite* black surface area elements A_1 and A_2 separated by a transparent medium. Since they are black, the surfaces emit isotropically and totally absorb all incident radiant energy. It is desired to compute the fraction of radiant energy, per unit emissive power E_1, leaving A_1 in all directions which is intercepted and absorbed by A_2. The required quantity is defined as the *direct view factor* and is assigned the notation $F_{1,2}$. Since the *net radiant energy interchange* $Q_{1,2} = A_1F_{1,2}E_1 - A_2F_{2,1}E_2$ *between* surfaces A_1 and A_2 must be zero when their temperatures are equal, it follows thermodynamically that $A_1F_{1,2} = A_2F_{2,1}$. The product of area and view factor $\overline{s_1s_2} \equiv A_1F_{1,2}$ which has the dimensions of area is termed the *direct surface-to-surface exchange area* [DEA] for finite black surfaces. Clearly, direct exchange areas are symmetric with respect to their subscripts, that is, $\overline{s_is_j} = \overline{s_js_i}$, but view factors are not symmetric unless the associated surface areas are equal. This property is referred to as the *symmetry* or *reciprocity relation* for direct exchange areas. The shorthand notation $\overline{s_1s_2} \equiv \overline{12} = \overline{21}$ for direct exchange areas is often found useful in mathematical developments.

Equation (5-101) may also be restated as

$$\frac{\partial^2 \overline{s_is_j}}{\partial A_i \partial A_j} = \frac{\cos\varphi_i \cos\varphi_j}{\pi r^2} \tag{5-111}$$

which leads directly to the required definition of the direct exchange area as a double surface integral

$$\overline{s_is_j} = \oint_{A_i}\oint_{A_j} \frac{\cos\varphi_i \cos\varphi_j}{\pi r^2} dA_j\, dA_i \tag{5-112}$$

All terms in Eq. (5-112) have been previously defined.

Suppose now that Eq. (5-112) is integrated over the entire confining surface of an *enclosure* which has been subdivided into M finite area elements. Each of the M surface *zones* must then satisfy certain *conservation relations* involving *all* the direct exchange areas in the enclosure

$$\sum_{j=1}^{M} \overline{s_is_j} = A_i \quad \text{for } 1 \le i \le M \tag{5-113a}$$

or in terms of view factors

$$\sum_{j=1}^{M} F_{i,j} = 1 \quad \text{for } 1 \le i \le M \tag{5-113b}$$

Contour integration is commonly used to simplify the evaluation of Eq. (5-112) for specific geometries; see Modest (*Radiative Heat Transfer*, 3d ed., Academic Press, New York, 2013, chap. 4) or Siegel and Howell (*Thermal Radiation Heat Transfer*, 4th ed., Taylor and Francis, London, 2001, chap. 5). Two particularly useful view factors are those for perpendicular rectangles of area XZ and YZ with common edge Z and equal parallel rectangles of area XY and distance of separation Z. The formulae for these quantities are given as follows. For *perpendicular rectangles* with common dimension Z

$$(\pi \cdot X)\cdot F_{X,Y} = X\tan^{-1}\frac{1}{X} + Y\tan^{-1}\frac{1}{Y} - \sqrt{X^2+Y^2}\tan^{-1}\sqrt{\frac{1}{(X^2+Y^2)}}$$
$$+\frac{1}{4}\ln\left\{\frac{(1+X^2)(1+Y^2)}{1+X^2+Y^2}\left[\frac{X^2(1+X^2+Y^2)}{(1+X^2)(X^2+Y^2)}\right]^{X^2}\left[\frac{Y^2(1+X^2+Y^2)}{(1+Y^2)(X^2+Y^2)}\right]^{Y^2}\right\} \tag{5-114a}$$

and for *parallel rectangles*, separated by distance Z,

$$\left(\frac{\pi \cdot X \cdot Y}{2}\right)\cdot F_{X,Y} = \frac{1}{2}\cdot\ln\left[\frac{(1+X^2)(1+Y^2)}{1+X^2+Y^2}\right] + X\sqrt{1+Y^2}\tan^{-1}\frac{X}{\sqrt{1+Y^2}}$$
$$+Y\sqrt{1+X^2}\tan^{-1}\frac{Y}{\sqrt{1+X^2}} - X\tan^{-1}X - Y\tan^{-1}Y \tag{5-114b}$$

In Eqs. (5-114) X and Y are normalized whereby $X = x/z$ and $Y = y/z$, and the corresponding *dimensional* direct surface areas are given by $\overline{s_xs_y} = x \cdot z \cdot F_{X,Y}$ and $\overline{s_xs_y} = x\,y\,F_{X,Y}$, respectively.

The exchange area between any two area elements of a sphere is independent of their relative shape and position and is simply the product of the areas, divided by the area of the entire sphere; i.e., any spot on a sphere has equal views of all other spots.

TABLE 5-5 Normal Total Emissivity of Various Surfaces

A. Metals and Their Oxides

Surface	*t*, °F*	Emissivity*
Aluminum		
Highly polished plate, 98.3% pure	440–1070	0.039–0.057
Polished plate	73	0.040
Rough plate	78	0.055
Oxidized at 1110°F	390–1110	0.11–0.19
Aluminum-surfaced roofing	100	0.216
Calorized surfaces, heated at 1110°F		
Copper	390–1110	0.18–0.19
Steel	390–1110	0.52–0.57
Brass		
Highly polished:		
73.2% Cu, 26.7% Zn	476–674	0.028–0.031
62.4% Cu, 36.8% Zn, 0.4% Pb, 0.3% Al	494–710	0.033–0.037
82.9% Cu, 17.0% Zn	530	0.030
Hard rolled, polished:		
But direction of polishing visible	70	0.038
But somewhat attacked	73	0.043
But traces of stearin from polish left on	75	0.053
Polished	100–600	0.096
Rolled plate, natural surface	72	0.06
Rubbed with coarse emery	72	0.20
Dull plate	120–660	0.22
Oxidized by heating at 1110°F	390–1110	0.61–0.59
Chromium; see Nickel Alloys for Ni-Cr steels	100–1000	0.08–0.26
Copper		
Carefully polished electrolytic copper	176	0.018
Commercial, emeried, polished, but pits remaining	66	0.030
Commercial, scraped shiny but not mirror-like	72	0.072
Polished	242	0.023
Plate, heated long time, covered with thick oxide layer	77	0.78
Plate heated at 1110°F	390–1110	0.57
Cuprous oxide	1470–2010	0.66–0.54
Molten copper	1970–2330	0.16–0.13
Gold		
Pure, highly polished	440–1160	0.018–0.035
Iron and steel		
Metallic surfaces (or very thin oxide layer):		
Electrolytic iron, highly polished	350–440	0.052–0.064
Polished iron	800–1880	0.144–0.377
Iron freshly emeried	68	0.242
Cast iron, polished	392	0.21
Wrought iron, highly polished	100–480	0.28
Cast iron, newly turned	72	0.435
Polished steel casting	1420–1900	0.52–0.56
Ground sheet steel	1720–2010	0.55–0.61
Smooth sheet iron	1650–1900	0.55–0.60
Cast iron, turned on lathe	1620–1810	0.60–0.70
Oxidized surfaces:		
Iron plate, pickled, then rusted red	68	0.612
Completely rusted	67	0.685
Rolled sheet steel	70	0.657
Oxidized iron	212	0.736
Cast iron, oxidized at 1100°F	390–1110	0.64–0.78
Steel, oxidized at 1100°F	390–1110	0.79
Smooth oxidized electrolytic iron	260–980	0.78–0.82
Iron oxide	930–2190	0.85–0.89
Rough ingot iron	1700–2040	0.87–0.95
Sheet steel, strong rough oxide layer	75	0.80
Dense shiny oxide layer	75	0.82
Cast plate:		
Smooth	73	0.80
Rough	73	0.82
Cast iron, rough, strongly oxidized	100–480	0.95
Wrought iron, dull oxidized	70–680	0.94
Steel plate, rough	100–700	0.94–0.97
High temperature alloy steels (see Nickel Alloys)		
Molten metal		
Cast iron	2370–2550	0.29
Mild steel	2910–3270	0.28
Lead		
Pure (99.96%), unoxidized	260–440	0.057–0.075
Gray oxidized	75	0.281
Oxidized at 390°F	390	0.63
Mercury	32–212	0.09–0.12
Molybdenum filament	1340–4700	0.096–0.292
Monel metal, oxidized at 1110°F	390–1110	0.41–0.46
Nickel		
Electroplated on polished iron, then polished	74	0.045
Technically pure (98.9% Ni, + Mn), polished	440–710	0.07–0.087
Electroplated on pickled iron, not polished	68	0.11
Wire	368–1844	0.096–0.186
Plate, oxidized by heating at 1110°F	390–1110	0.37–0.48
Nickel oxide	1200–2290	0.59–0.86
Nickel alloys		
Chromnickel	125–1894	0.64–0.76
Nickelin (18–32 Ni; 55–68 Cu; 20 Zn), gray oxidized	70	0.262
KA-2S alloy steel (8% Ni; 18% Cr), light silvery, rough, brown, after heating	420–914	0.44–0.36
After 42 hr. heating at 980°F	420–980	0.62–0.73
NCT-3 alloy (20% Ni; 25% Cr.), brown, splotched, oxidized from service	420–980	0.90–0.97
NCT-6 alloy (60% Ni; 12% Cr), smooth, black, firm adhesive oxide coat from service	520–1045	0.89–0.82
Platinum		
Pure, polished plate	440–1160	0.054–0.104
Strip	1700–2960	0.12–0.17
Filament	80–2240	0.036–0.192
Wire	440–2510	0.073–0.182
Silver		
Polished, pure	440–1160	0.0198–0.0324
Polished	100–700	0.0221–0.0312
Steel, see Iron.		
Tantalum filament	2420–5430	0.194–0.31
Tin—bright tinned iron sheet	76	0.043 and 0.064
Tungsten		
Filament, aged	80–6000	0.032–0.35
Filament	6000	0.39
Zinc		
Commercial, 99.1% pure, polished	440–620	0.045–0.053
Oxidized by heating at 750°F	750	0.11
Galvanized sheet iron, fairly bright	82	0.228
Galvanized sheet iron, gray oxidized	75	0.276

B. Refractories, Building Materials, Paints, and Miscellaneous

Surface	*t*, °F*	Emissivity*
Asbestos		
Board	74	0.96
Paper	100–700	0.93–0.945
Brick		
Red, rough, but no gross irregularities	70	0.93
Silica, unglazed, rough	1832	0.80
Silica, glazed, rough	2012	0.85
Grog brick, glazed	2012	0.75
See Refractory Materials below.		
Carbon		
T-carbon (Gebr. Siemens) 0.9% ash (this started with emissivity at 260°F of 0.72, but on heating changed to values given)	260–1160	0.81–0.79
Carbon filament	1900–2560	0.526
Candle soot	206–520	0.952
Lampblack-waterglass coating	209–362	0.959–0.947

TABLE 5-5 Normal Total Emissivity of Various Surfaces *(Continued)*

B. Refractories, Building Materials, Paints, and Miscellaneous (*Continued*)

Surface	t, °F*	Emissivity*
Same	260–440	0.957–0.952
Thin layer on iron plate	69	0.927
Thick coat	68	0.967
Lampblack, 0.003 in. or thicker	100–700	0.945
Enamel, white fused, on iron	66	0.897
Glass, smooth	72	0.937
Gypsum, 0.02 in. thick on smooth or blackened plate	70	0.903
Marble, light gray, polished	72	0.931
Oak, planed	70	0.895
Oil layers on polished nickel (lube oil)	68	
Polished surface, alone		0.045
+0.001-in. oil		0.27
+0.002-in. oil		0.46
+0.005-in. oil		0.72
Infinitely thick oil layer		0.82
Oil layers on aluminum foil (linseed oil)		
Al foil	212	0.087†
+1 coat oil	212	0.561
+2 coats oil	212	0.574
Paints, lacquers, varnishes		
Snowhite enamel varnish or rough iron plate	73	0.906
Black shiny lacquer, sprayed on iron	76	0.875
Black shiny shellac on tinned iron sheet	70	0.821
Black matte shellac	170–295	0.91
Black lacquer	100–200	0.80–0.95
Flat black lacquer	100–200	0.96–0.98
White lacquer	100–200	0.80–0.95

Surface	t, °F*	Emissivity*
Oil paints, sixteen different, all colors	212	0.92–0.96
Aluminum paints and lacquers		
10% Al, 22% lacquer body, on rough or smooth surface	212	0.52
26% Al, 27% lacquer body, on rough or smooth surface	212	0.3
Other Al paints, varying age and Al content	212	0.27–0.67
Al lacquer, varnish binder, on rough plate	70	0.39
Al paint, after heating to 620°F	300–600	0.35
Paper, thin		
Pasted on tinned iron plate	66	0.924
On rough iron plate	66	0.929
On black lacquered plate	66	0.944
Plaster, rough lime	50–190	0.91
Porcelain, glazed	72	0.924
Quartz, rough, fused	70	0.932
Refractory materials, 40 different	1110–1830	
poor radiators		{0.65, 0.70} – 0.75
good radiators		{0.80, 0.85} – {0.85, 0.90}
Roofing paper	69	0.91
Rubber		
Hard, glossy plate	74	0.945
Soft, gray, rough (reclaimed)	76	0.859
Serpentine, polished	74	0.900
Water	32–212	0.95–0.963

*When two temperatures and two emissivities are given, they correspond, first to first and second to second, and linear interpolation is permissible. °C = (°F − 32)/1.8.
†Although this value is probably high, it is given for comparison with the data by the same investigator to show the effect of oil layers. See Aluminum, Part A of this table.

Figure 5-11, curves 1 through 4, shows view factors for selected parallel opposed disks, squares, and 2:1 rectangles and parallel rectangles with one infinite dimension as a function of the ratio of the smaller diameter or side to the distance of separation. Curves 2 through 4 of Fig. 5-11, for opposed rectangles, can be computed with Eq. (5-114*b*). The view factors for two finite coaxial coextensive cylinders of radii $r \le R$ and height L are shown in Fig. 5-12. The direct view factors for an infinite plane parallel to a system of rows of parallel tubes are given as curves 1 and 3 of Fig. 5-13. The view factors for this two-dimensional geometry can be readily calculated by using the *crossed-strings method.*

The crossed-strings method, due to Hottel (*Radiative Transfer*, McGraw-Hill, New York, 1967), is stated as follows: "The exchange area for two-dimensional surfaces, A_1 and A_2, per unit length (in the infinite dimension) is given by the sum of the lengths of crossed strings from the ends of A_1 to the ends of A_2 less the sum of the uncrossed strings from and to the same points all divided by 2." The strings must be drawn so that all the flux from one surface to the other must cross each of a pair of crossed strings and neither of the pair of uncrossed strings. If one surface can see the other around both sides of an obstruction, two more pairs of strings are involved. The calculation procedure is demonstrated by evaluation of the tube-to-tube view factor for one row of a tube bank, as illustrated in Example 5-8.

Example 5-8 The Crossed-Strings Method Figure 5-14 depicts the transverse cross section of two infinitely long, parallel circular tubes of diameter D and center-to-center distance of separation C. Use the crossed-strings method to formulate the tube-to-tube direct exchange area and view factor, $\overline{s_t s_t}$ and $F_{t,t}$, respectively.

Solution The circumferential area of each tube is $A_t = \pi D$ per unit length in the infinite dimension for this two-dimensional geometry. Application of the crossed-strings procedure then yields simply

$$\overline{s_t s_t} = \frac{2(EFGH - HJ)}{2} = D\cdot[\sin^{-1}(1/R) + \sqrt{R^2-1} - R]$$

and

$$F_{t,t} = \overline{s_t s_t}/A_t = [\sin^{-1}(1/R) + \sqrt{R^2-1} - R]/\pi$$

where $EFGH$ and $HJ = C$ are the indicated line segments and $R \equiv C/D \ge 1$. Curves 1 and 5, respectively, of Fig. 5-13 can be calculated from $F_{t,t}$ with the relations $F_{p,t} = \pi\cdot(0.5 - F_{t,t})/R$ and $\bar{F}_{p,t} = F_{p,t}(2 - F_{p,t})$. Here $\bar{F}_{p,t}$ is defined as the *refractory augmented view factor* from the black plane to the black tubes, as explained in the following section on the zone method.

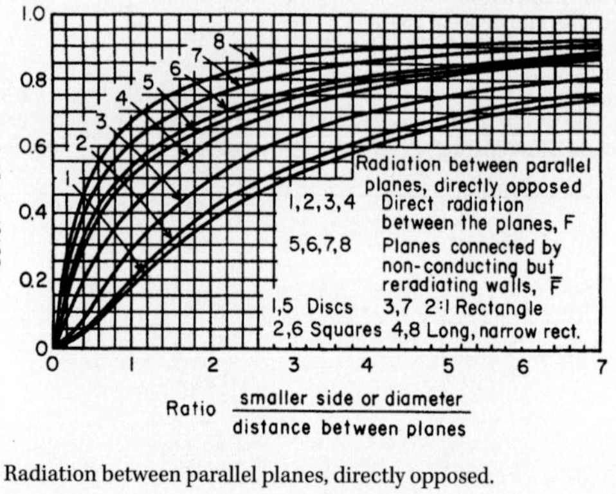

FIG. 5-11 Radiation between parallel planes, directly opposed.

The *Yamauti principle* [Yamauti, *Res. Electrotech. Lab. (Tokyo)*, **148** (1924); **194** (1927); **250** (1929)] is stated as follows: *The exchange areas between two pairs of surfaces are equal when there is a one-to-one correspondence for all sets of symmetrically positioned pairs of differential elements in the two surface combinations.* Figure 5-15 illustrates the Yamauti principle applied to surfaces in perpendicular planes having a common edge. With reference to Fig. 5-15, the Yamauti principle states that the diagonally opposed exchange areas are equal, that is, $\overline{(1)(4)} = \overline{(2)(3)}$. Figure 5-15 also shows a more complex geometric construction for displaced cylinders for which the Yamauti principle also applies. Collectively the three terms *reciprocity* or *symmetry principle, conservation principle,* and *Yamauti principle* are referred to as *view factor or exchange area algebra.*

Example 5-9 Illustration of Exchange Area Algebra Figure 5-15 shows a graphical construction depicting four perpendicular opposed rectangles with a common edge. *Numerically* evaluate the direct exchange areas and view factors for the diagonally opposed (shaded) rectangles A_1 and A_4, that is, $\overline{(1)(4)}$ as well as $\overline{(1)(3+4)}$. The dimensions of the rectangular construction are shown in Fig. 5-15 as $x = 3$, $y = 2$, and $z = 1$.

Solution Using shorthand notation for direct exchange areas, the conservation principle yields

$$\overline{(1+2)(3+4)} = \overline{(1+2)(3)} + \overline{(1+2)(4)} = \overline{(1)(3)} + \overline{(2)(3)} + \overline{(1)(4)} + \overline{(2)(4)}$$

Now by the Yamauti principle we have $\overline{(1)(4)} \equiv \overline{(2)(3)}$. The combination of these two relations yields the first result $\overline{(1)(4)} = \left[\overline{(1+2)(3+4)} - \overline{(1)(3)} - \overline{(2)(4)}\right]/2$. For $\overline{(1)(3+4)}$, again conservation yields $\overline{(1)(3+4)} = \overline{(1)(3)} + \overline{(1)(4)}$, and substitution of the expression

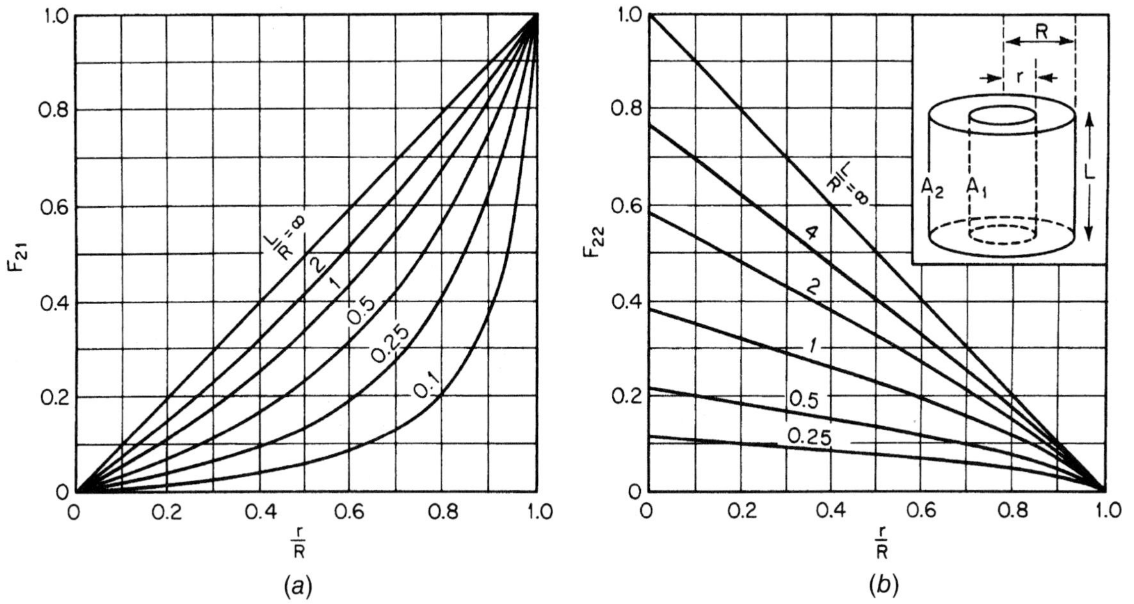

FIG. 5-12 View factors for a system of two concentric coaxial cylinders of equal length. (*a*) Inner surface of outer cylinder to inner cylinder. (*b*) Inner surface of outer cylinder to itself.

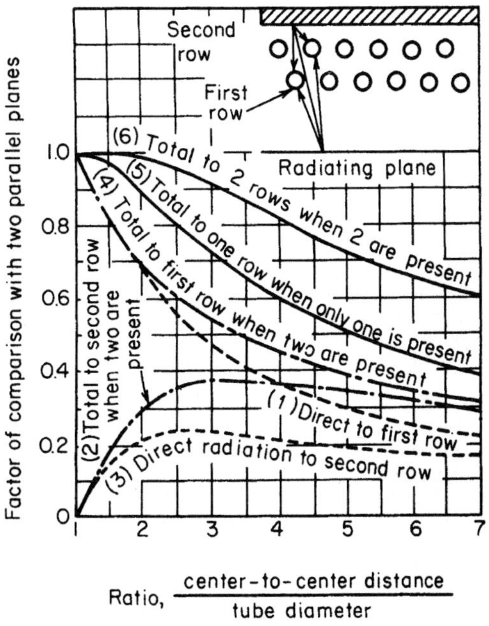

FIG. 5-13 Distribution of radiation to rows of tubes irradiated from one side. Dashed lines: direct view factor F from plane to tubes. Solid lines: total view factor $\bar{F}$ for black tubes backed by a refractory surface.

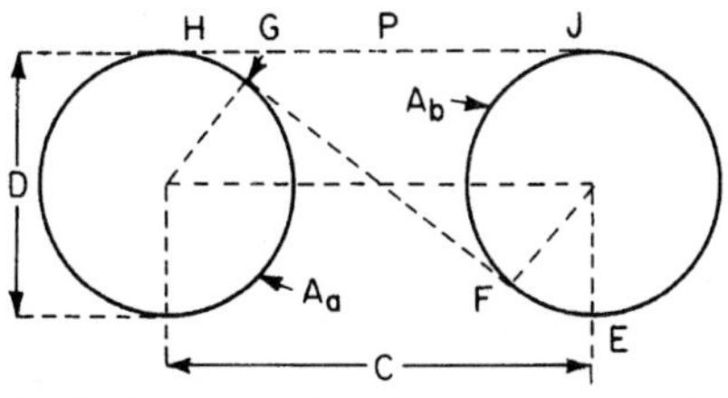

FIG. 5-14 Direct exchange between parallel circular tubes.

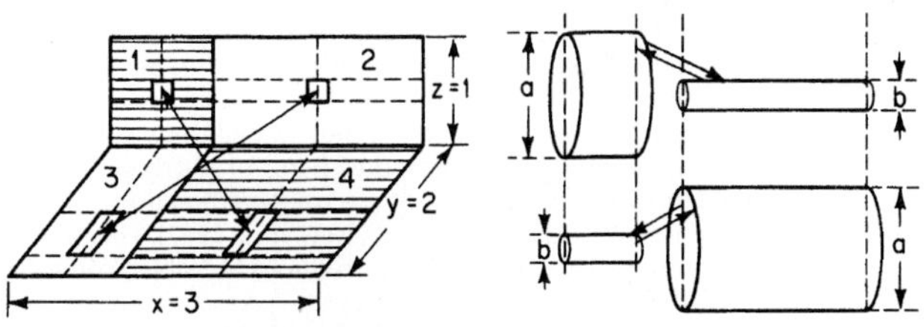

FIG. 5-15 Illustration of the Yamauti principle.

for $\overline{(1)(4)}$ just obtained yields the second result, that is, $\overline{(1)(3+4)} = [\overline{(1+2)(3+4)} + \overline{(1)(3)} - \overline{(2)(4)}]/2.0$. All three required direct exchange areas in these two relations are readily evaluated from Eq. (5-114*a*). Moreover, these equations apply to opposed *parallel* rectangles as well as rectangles with a common edge oriented at *any angle*. Numerically it follows from Eq. (5-114*a*) for $X = 1/3$, $Y = 2/3$, and $z = 3$ that $\overline{(1+2)(3+4)} = 0.95990$; for $X = 1$, $Y = 2$, and $z = 1$ that $\overline{(1)(3)} = 0.23285$; and for $X = 1/2$, $Y = 1$, and $z = 2$ that $\overline{(2)(4)} = 0.585747$. Since $A_1 = 1.0$, this leads to

$$\overline{s_1s_4} = F_{1,4} = (0.95990 - 0.23285 - 0.584747)/2.0 = 0.07115$$

and

$$\overline{s_1s_{3+4}} = F_{1,3+4} = (0.95990 + 0.23285 - 0.584747)/2.0 = 0.30400$$

Many literature sources document closed-form algebraic expressions for view factors. Particularly comprehensive references include the compendia by Modest (*Radiative Heat Transfer*, 3d ed., Academic Press, New York, 2013, app. D) and Howell, Mengüç, and Siegel (*Thermal Radiative Heat Transfer*, 6th ed., CRC Press, Boca Raton, Fla., 2015, app. C). The appendices for both of these textbooks also provide a wealth of resource information for radiative transfer. Appendix F of Modest, for example, references an extensive listing of Fortran computer codes for a variety of radiation calculations which include view factors. These codes are archived in the dedicated Internet website maintained by the publisher. The textbook by Howell, Mengüç, and Siegel has also included an extensive database of view factors archived on a CD-ROM and includes a reference to an author-maintained Internet website. Other historical sources for view factors include Hottel and Sarofim (*Radiative Transfer*, McGraw-Hill, New York, 1967, chap. 2) and Hamilton and Morgan (NACA-TN 2836, December 1952).

RADIATIVE EXCHANGE IN ENCLOSURES—THE ZONE METHOD

Total Exchange Areas

When an enclosure contains *reflective* surface zones, allowance must be made for not only the radiant energy transferred *directly* between any two zones but also the additional transfer attendant to *however many multiple reflections* which occur among the intervening reflective surfaces. Under such circumstances, it can be shown that the *net* radiative flux $Q_{i,j}$ between all such surface zone pairs A_i and A_j, making full allowance for all multiple reflections, may be computed from

$$Q_{i,j} = A_i \mathcal{F}_{i,j} \sigma T_j^4 - A_j \mathcal{F}_{j,i} \sigma T_i^4 \qquad (5\text{-}115a)$$

or

$$Q_{i,j} = \overline{S_iS_j} \cdot E_i - \overline{S_jS_i} \cdot E_j \qquad (5\text{-}115b)$$

Here $\mathcal{F}_{i,j}$ is defined as the *total surface-to-surface view factor* from A_i to A_j, and the quantity $\overline{S_iS_j} \equiv A_i \cdot \mathcal{F}_{i,j}$ is defined as the corresponding *total (surface-to-surface) exchange area* [TEA]. In analogy with the direct exchange areas, the total surface-to-surface exchange areas are also symmetric and thus obey reciprocity, that is, $A_i \cdot \mathcal{F}_{i,j} = A_j \cdot \mathcal{F}_{j,i}$ or $\overline{S_iS_j} = \overline{S_jS_i}$. When applied to an enclosure, *total* exchange areas and *total* view factors also must satisfy appropriate conservation relations. Total exchange areas are functions of the *geometry and radiative properties of the entire enclosure*. They are also independent of temperature if all surfaces and any radiatively participating media are gray. The following subsection presents a general *matrix method* for the *explicit* evaluation of total exchange areas from direct exchange areas and other enclosure parameters.

In what follows, conventional matrix notation is strictly employed as in $\mathbf{A} = [A_{i,j}]$ wherein the scalar subscripts always denote the row and column indices, respectively, and all *matrix entities* defined here are denoted by **boldface** notation. Section 3 of this handbook, "Mathematics," provides an especially convenient reference for introductory matrix algebra and matrix computations.

General Matrix Formulation The *zone method* is perhaps the simplest numerical quadrature of the governing *integral equations* for radiative transfer. It may be derived from first principles by starting with the *equation of transfer* for radiation intensity. The zone method *always conserves radiant energy* since the spatial discretization utilizes macroscopic energy balances involving *spatially averaged* radiative flux quantities. Because large sets of linear algebraic equations can arise in this process, *matrix algebra* provides the most compact notation and the most expeditious methods of solution. The mathematical approach presented here is a matrix generalization of the original (*scalar*) development of the zone method due to Hottel and Sarofim (*Radiative Transfer*, McGraw-Hill, New York, 1967). The present matrix development is abstracted from that introduced by Noble [Noble, J. J., *Int. J. Heat Mass Transfer*, **18:** 261–269 (1975)].

Consider an arbitrary three-dimensional enclosure of total volume V and surface area A which confines an absorbing-emitting medium (gas). Let the enclosure be subdivided (zoned) into M finite surface area A_i and N finite volume elements V_i, each small enough that all such *zones* are substantially isothermal. The mathematical development in this section is then restricted by the following conditions and/or assumptions:

1. The gas temperatures are given a priori.
2. Allowance is made for gas-to-surface radiative transfer.
3. Radiative transfer with respect to the confined gas is either monochromatic or gray. The gray gas absorption coefficient is denoted here by K[m^{-1}]. In subsequent sections the monochromatic absorption coefficient is denoted by $K_\lambda(\lambda)$.
4. All surface emissivities are assumed to be gray and thus independent of temperature.
5. Surface emission and reflection are isotropic or diffuse.
6. The gas does not scatter.

Noble [Noble, J. J., *Int. J. Heat Mass Transfer*, **18:** 261–269 (1975)] has extended the present matrix methodology to the case where the gaseous absorbing-emitting medium also *scatters isotropically*.

In matrix notation the blackbody emissive powers for all surface and volume zones comprising the zoned enclosure are designated as $\mathbf{E} = [E_i] = [\sigma T_i^4]$, an $M \times 1$ vector, and $\mathbf{E}_g = [E_{g,i}] = [\sigma T_{g,i}^4]$, an $N \times 1$ vector, respectively. Moreover, all surface zones are characterized by three $M \times M$ diagonal matrices for surface zone areas $\mathbf{AI} = [A_i \cdot \delta_{i,j}]$, diffuse emissivity $\boldsymbol{\varepsilon}\mathbf{I} = [\varepsilon_i \cdot \delta_{i,j}]$, and diffuse reflectivity, $\boldsymbol{\rho}\mathbf{I} = [(1-\varepsilon_i) \cdot \delta_{i,j}]$, respectively. Here $\delta_{i,j}$ is the *Kronecker delta* (that is, $\delta_{i,j} = 1$ for $i = j$ and $\delta_{i,j} = 0$ for $i \neq j$).

Two arrays of direct exchange areas are now defined; i.e., the matrix $\overline{\mathbf{ss}} = [\overline{s_is_j}]$ is the $M \times M$ array of *direct surface-to-surface exchange areas*, and the matrix $\overline{\mathbf{sg}} = [\overline{s_ig_j}]$ is the $M \times N$ array of *direct gas-to-surface exchange areas*. Here the scalar elements of $\overline{\mathbf{ss}}$ and $\overline{\mathbf{sg}}$ are computed from the integrals

$$\overline{s_is_j} = \oint_{A_i}\oint_{A_j} \frac{e^{-Kr}}{\pi r^2} \cos\phi_i \cos\phi_j \, dA_j \, dA_i \qquad (5\text{-}116a)$$

and

$$\overline{s_ig_j} = \oint_{A_i}\oiint_{V_j} K \frac{e^{-Kr}}{\pi r^2} \cos\phi_i \, dV_j \, dA_i \qquad (5\text{-}116b)$$

Equation (5-116*a*) is a generalization of Eq. (5-112) for the case $K \neq 0$ while $\overline{s_ig_j}$ is a new quantity, which arises *only* for the case $K \neq 0$.

Matrix characterization of the radiative energy balance at each surface zone is then facilitated via definition of three additional M vectors, namely the radiative surface flux $\mathbf{Q} = [Q_i]$, with units of watts; and the vectors $\mathbf{H} = [H_i]$ and $\mathbf{W} = [W_i]$ both having units of W/m^2. The arrays $\mathbf{H}$ and $\mathbf{W}$ define the *incident* and *leaving* flux *densities*, respectively, at each surface zone. The variable $\mathbf{W}$ is also referred to in the literature as the *radiosity* or *exitance*. Subject to the above assumptions the zone method can be stated in three matrix equations in terms of the five vector variables $\mathbf{Q}$, $\mathbf{W}$, $\mathbf{H}$, $\mathbf{E}$, and $\mathbf{E_g}$:

$$\mathbf{W} = \boldsymbol{\varepsilon}\mathbf{I} \cdot \mathbf{E} + \boldsymbol{\rho}\mathbf{I} \cdot \mathbf{H} \qquad (5\text{-}117a)$$

$$\mathbf{AI} \cdot \mathbf{H} = \overline{\mathbf{ss}} \cdot \mathbf{W} + \overline{\mathbf{sg}} \cdot \mathbf{E_g} \qquad (5\text{-}117b)$$

$$\mathbf{Q} = \mathbf{AI} \cdot [\mathbf{W} - \mathbf{H}] \qquad (5\text{-}117c)$$

Implementation of Eqs. (5-117) requires a priori specification of the gas (temperatures) $\mathbf{E_g}$, and M other pieces of information for the surface zones. Elimination of $\mathbf{W}$ between Eqs. (5-117*a*) and (5-117*c*) followed by elimination of $\mathbf{H}$ then leads to two alternative forms for the surface flux vector $\mathbf{Q}$:

$$\mathbf{Q} = \boldsymbol{\varepsilon}\mathbf{AI} \cdot [\mathbf{E} - \mathbf{H}] = \boldsymbol{\rho}\mathbf{I}^{-1} \cdot \boldsymbol{\varepsilon}\mathbf{AI}[\mathbf{E} - \mathbf{W}] \quad \text{for } \rho_i \neq 0 \qquad (5\text{-}117d,e)$$

Explicit Matrix Solution for Total Exchange Areas For gray or monochromatic transfer, the *primary working relation* for zoning calculations via the matrix method is

$$\mathbf{Q} = \boldsymbol{\varepsilon}\mathbf{AI} \cdot \mathbf{E} - \overline{\mathbf{SS}} \cdot \mathbf{E} - \overline{\mathbf{SG}} \cdot \mathbf{E_g} \quad [M \times 1] \qquad (5\text{-}118)$$

Equation (5-118) makes full allowance for multiple reflections in an enclosure of any degree of complexity. To apply Eq. (5-118) for design or simulation purposes, the gas temperatures must be known and surface boundary conditions must be specified for each and every surface zone in the form of *either* E_i or Q_i. In application of Eq. (5-118), *physically impossible values* of E_i may well result if *physically unrealistic* values of Q_i are specified.

In Eq. (5-118), $\overline{\mathbf{SS}}$ and $\overline{\mathbf{SG}}$ are defined as the *required* arrays of *total surface-to-surface exchange areas* and *total gas-to-surface exchange areas*, respectively. The matrices for total exchange areas are calculated *explicitly* from the corresponding arrays of direct exchange areas and the other enclosure parameters by the following matrix formulas:

Surface-to-surface exchange $\overline{\mathbf{SS}} = \varepsilon\mathbf{I}\cdot\mathbf{AI}\cdot\mathbf{R}\cdot\overline{\mathbf{ss}}\cdot\varepsilon\mathbf{I}$ $[M \times M]$ (5-118*a*)

Gas-to-surface exchange $\overline{\mathbf{SG}} = \varepsilon\mathbf{I}\cdot\mathbf{AI}\cdot\mathbf{R}\cdot\overline{\mathbf{sg}}$ $[M \times N]$ (5-118*b*)

where in Eqs. (5-118*a* and *b*), **R** is the explicit *inverse reflectivity* or *multiple reflection matrix*, defined as

$$\mathbf{R} = \left[\mathbf{AI} - \overline{\mathbf{ss}}\cdot\rho\mathbf{I}\right]^{-1} \quad [M \times M] \tag{5-118c}$$

While the **R** matrix is generally not symmetric, the matrix product $\rho\mathbf{I}\cdot\mathbf{R}$ is *always* symmetric. This fact proves useful for error checking.

The most computationally significant aspect of the matrix method is that the inverse reflectivity matrix **R** *always exists* for any physically meaningful enclosure problem. More precisely, **R** always exists provided that $K \neq 0$. Moreover, for a transparent medium, **R** exists provided that there *formally* exists at least one surface zone A_i such that $\varepsilon_i \neq 0$. An important computational corollary of this statement for *transparent media* is that the matrix $[\mathbf{AI} - \overline{\mathbf{ss}}]$ is *always singular* and demonstrates *matrix rank* $M - 1$ [Noble, J. J., *Int. J. Heat Mass Transfer*, **18:** 261–269 (1975)].

Finally, the four matrix arrays $\overline{\mathbf{ss}}$, $\overline{\mathbf{sg}}$, $\overline{\mathbf{SS}}$, and $\overline{\mathbf{SG}}$ of direct and total exchange areas must satisfy *matrix conservation relations* (row sums), i.e.,

Direct exchange areas $\mathbf{AI}\cdot\mathbf{1}_M = \overline{\mathbf{ss}}\cdot\mathbf{1}_M + \overline{\mathbf{sg}}\cdot\mathbf{1}_N$ (5-119*a*)

Total exchange areas $\varepsilon\mathbf{I}\cdot\mathbf{AI}\cdot\mathbf{1}_M = \overline{\mathbf{SS}}\cdot\mathbf{1}_M + \overline{\mathbf{SG}}\cdot\mathbf{1}_N$ (5-119*b*)

Here $\mathbf{1}_M$ is an $M \times 1$ column vector all of whose elements are unity. If $\varepsilon\mathbf{I} = \mathbf{I}$ or equivalently, $\rho\mathbf{I} = 0$, then Eq. (5-118*c*) reduces to $\mathbf{R} = \mathbf{AI}^{-1}$ with the result that Eqs. (5-118*a*) and (5-118*b*) degenerate to simply $\overline{\mathbf{SS}} = \overline{\mathbf{ss}}$ and $\overline{\mathbf{SS}} = \overline{\mathbf{sg}}$, respectively. Further, while the array $\overline{\mathbf{SS}}$ is always symmetric, the array $\overline{\mathbf{SG}}$ is generally not square.

For purposes of digital computation, it is good practice to enter all data for direct exchange surface-to-surface areas $\overline{\mathbf{ss}}$ with a precision of *at least* five significant figures. This need arises because all the scalar elements of $\overline{\mathbf{sg}}$ can be calculated *arithmetically* from appropriate direct surface-to-surface exchange areas by using view factor algebra rather than via the definition of the defining integral, Eq. (5-116*b*). This process often involves small arithmetic differences between two numbers of nearly equal magnitudes, and numerical significance is easily lost.

Computer implementation of matrix methods proves straightforward, given the availability of modern software applications. In particular, several especially user-friendly GUI mathematical utilities are available that perform matrix computations using *essentially algebraic notation.* Many simple zoning problems may be solved with spreadsheets. For large *M* and *N*, matrix methodology can involve management of a large amount of data. Error checks based on symmetry and conservation by calculation of the row sums of the four arrays of direct and total exchange areas then prove indispensable.

Zone Methodology and Conventions For a transparent medium, no more than $\Sigma = M(M-1)/2$ of the M^2 elements of the $\overline{\mathbf{ss}}$ array are *unique.* Further, surface zones are characterized into two generic types. *Source-sink zones* are defined as those for which temperature is specified and whose radiative flux Q_i is to be determined. For *flux zones,* conversely, these conditions are reversed. When both types of zone are present in an enclosure, Eq. (5-118) may be *partitioned* to produce a more efficient computational algorithm. Let *M* = *MS* + *MF* represent the total number of surface zones where *MS* is the number of source-sink zones and *MF* is the number of flux zones. The flux zones are the *last to be numbered.* Equation (5-118) is then partitioned as follows:

$$\begin{bmatrix}\mathbf{Q}_1\\ \mathbf{Q}_2\end{bmatrix} = \begin{bmatrix}\varepsilon\mathbf{AI}_{1,1} & \mathbf{0}\\ \mathbf{0} & \varepsilon\mathbf{AI}_{2,2}\end{bmatrix}\begin{bmatrix}\mathbf{E}_1\\ \mathbf{E}_2\end{bmatrix} - \begin{bmatrix}\overline{\mathbf{SS}}_{1,1} & \overline{\mathbf{SS}}_{1,2}\\ \overline{\mathbf{SS}}_{2,1} & \overline{\mathbf{SS}}_{2,2}\end{bmatrix}\begin{bmatrix}\mathbf{E}_1\\ \mathbf{E}_2\end{bmatrix} - \begin{bmatrix}\overline{\mathbf{SG}}_1\\ \overline{\mathbf{SG}}_2\end{bmatrix}\cdot\mathbf{E}_g \tag{5-120}$$

Here the dimensions of the submatrices $\varepsilon\mathbf{AI}_{1,1}$ and $\overline{\mathbf{SS}}_{1,1}$ are both *MS* × *MS* and $\overline{\mathbf{SG}}_1$ has dimensions *MS* × *N*, where *N* is the number of volume zones. Partition algebra then yields the following two matrix equations for $\mathbf{Q}_1$, the *MS* × 1 vector of unknown source-sink fluxes, and $\mathbf{E}_2$, the *MF* × 1 vector of unknown emissive powers for the flux zones, i.e.,

$$\mathbf{E}_2 = \left[\varepsilon\mathbf{AI}_{2,2} - \overline{\mathbf{SS}}_{2,2}\right]^{-1}\cdot\left[\overline{\mathbf{SS}}_{2,1}\cdot\mathbf{E}_1 + \overline{\mathbf{SG}}_2\cdot\mathbf{E}_g + \mathbf{Q}_2\right] \tag{5-120a}$$

$$\mathbf{Q}_1 = \varepsilon\mathbf{AI}_{1,1}\cdot\mathbf{E}_1 - \overline{\mathbf{SS}}_{1,1}\cdot\mathbf{E}_1 - \overline{\mathbf{SS}}_{1,2}\cdot\mathbf{E}_2 - \overline{\mathbf{SG}}_1\cdot\mathbf{E}_g \tag{5-120b}$$

The inverse matrix in Eq. (5-120*a*) formally does not exist if there is at least one flux zone such that $\varepsilon_i = 0$. However, well-behaved results are usually obtained with Eq. (5-120*a*) by utilizing a *notional zero*, say, $\varepsilon_i \approx 10^{-5}$, to simulate $\varepsilon_i = 0$. Computationally, $\mathbf{E}_2$ is first obtained from Eq. (5-120*a*) and then substituted into either Eq. (5-120*b*) or Eq. (5-118).

Surface zones need not be *contiguous.* For example, in a symmetric enclosure, zones on opposite sides of the plane of symmetry may be "lumped" into a single zone for computational purposes. Lumping nonsymmetrical zones is also possible as long as the zone temperatures and emissivities are equal.

An *adiabatic refractory surface* with surface area A_R and emissivity ε_R, for which $Q_R = 0$, proves quite important in practice. A nearly radiatively adiabatic refractory surface occurs when differences between internal conduction and convection and external heat losses through the refractory wall are small compared with the *magnitude* of the *incident* and *leaving* radiation fluxes. Mathematically, sufficient conditions to model an adiabatic refractory surface are the a priori requirements $Q_R = 0$ for $0 < \varepsilon_R \le 1$ or simply $\varepsilon_R = 0$. Formally, these two conditions imply somewhat different mathematical consequences. First, from Eqs. (5-117) we may write $Q_R = A_R(W_R - H_R) = \varepsilon_R A_R(E_R - H_R)$ such that the requirement $Q_R = 0$ directly implies $E_R = W_R = H_R$. Alternatively, if one specifies $\varepsilon_R = 0$, it follows from the definition of radiosity that $W_R = H_R$ and thus $Q_R \equiv 0$. In this case the value of E_R is not defined and is found to be *entirely immaterial* to the enclosure calculations. Indeed all the total exchange areas for an adiabatic refractory vanish, to wit $\overline{S_iS_R} = 0$ for all $1 \le i \le M$. Thus the value of E_R never even enters the zoning equations. Nonetheless when $\varepsilon_R = 0$, it is customary to use $E_R = W_R = H_R$ to estimate refractory temperatures. A surface zone for which $\varepsilon_R = 0$ is termed a *perfect diffuse mirror.* An adiabatic surface zone is thus also a perfect diffuse mirror. As will be shown, matrix methods automatically deal with *all* options for *flux* and adiabatic refractory surfaces.

Consider an enclosure with a single (lumped) refractory where $\varepsilon_R \neq 0$ and $M_R = 1$ and any number of source/sink and volume zones. The (scalar) refractory emissive power may be calculated from Eq. (5-120*a*) as a weighted sum of all *other known blackbody emissive powers* which characterize the enclosure, i.e.,

$$E_R = \frac{\sum_{j=1}^{M}\overline{S_RS_j}\cdot E_j + \sum_{k=1}^{N}\overline{S_RG_k}\cdot E_{g,k} + Q_R}{\sum_{j=1}^{M}\overline{S_RS_j} + \sum_{k=1}^{N}\overline{S_RG_k}} \quad \text{with } j \neq R \text{ and } \varepsilon_R \neq 0 \tag{5-121}$$

Equation (5-121) specifically *includes* those zones which may not have a *direct* view of the refractory. When $Q_R = 0$, the refractory surface is said to be in *radiative equilibrium* with the entire enclosure. Again, note that Eq. (5-121) is indeterminate if $\varepsilon_R = 0$.

The Limiting Case of a Transparent Medium For the special case of a transparent medium for which $K = 0$, many practical engineering applications can be modeled with the zone method. These include combustion-fired muffle furnaces and electrical resistance furnaces. When $K \to 0$, $\overline{\mathbf{sg}} \to \mathbf{0}$ and $\overline{\mathbf{SG}} \to \mathbf{0}$. Equations (5-118) through (5-119) then reduce to three simple matrix relations

$$\mathbf{Q} = \varepsilon\mathbf{I}\cdot\mathbf{AI}\cdot\mathbf{E} - \overline{\mathbf{SS}}\cdot\mathbf{E} \tag{5-122a}$$

$$\overline{\mathbf{SS}} = \varepsilon\mathbf{I}\cdot\mathbf{AI}\cdot\mathbf{R}\cdot\overline{\mathbf{ss}}\cdot\varepsilon\mathbf{I} \tag{5-122b}$$

with again

$$\mathbf{R} = [\mathbf{AI} - \overline{\mathbf{ss}}\cdot\varepsilon\mathbf{I}]^{-1} \tag{5-122c}$$

The radiant surface flux vector **Q**, as computed from Eq. (5-122*a*), *always* satisfies the (scalar) conservation condition $\sum_{i=1}^{M} Q_i = 0$ [or $\mathbf{1}_M^T\cdot\mathbf{Q} = \mathbf{0}$] which is a statement of the overall *radiant* energy balance. The matrix conservation relations also simplify to

$$\mathbf{AI}\cdot\mathbf{1}_M = \overline{\mathbf{ss}}\cdot\mathbf{I}_M \tag{5-123a}$$

and

$$\varepsilon\mathbf{I}\cdot\mathbf{AI}\cdot\mathbf{1}_M = \overline{\mathbf{SS}}\cdot\mathbf{1}_M \tag{5-123b}$$

And the $M \times M$ arrays for *all* the direct and total view factors can be readily computed from

$$\mathbf{F} = \mathbf{AI}^{-1}\cdot\overline{\mathbf{ss}} \tag{5-124a}$$

and

$$\mathcal{F} = \mathbf{AI}^{-1}\cdot\overline{\mathbf{SS}} \tag{5-124b}$$

where the following matrix conservation relations must also be satisfied

$$\mathbf{F}\cdot\mathbf{1}_M = \mathbf{1}_M \tag{5-125a}$$

and

$$\mathcal{F}\cdot\mathbf{1}_M = \varepsilon\mathbf{I}\cdot\mathbf{1}_M \tag{5-125b}$$

The Two-Zone Enclosure Figure 5-16 depicts four simple enclosure geometries which are particularly useful for engineering calculations characterized by only *two* surface zones. For $M = 2$, the reflectivity matrix $\mathbf{R}$ is readily evaluated in *closed form* since an explicit algebraic inversion formula is available for a 2×2 matrix. In this case knowledge of only $\Sigma = 1$ direct exchange area is required. Direct evaluation of Eqs. (5-122) then leads to

$$\overline{\mathbf{SS}} = \begin{bmatrix} \varepsilon_1 A_1 - \overline{S_1S_2} & \overline{S_1S_2} \\ \overline{S_1S_2} & \varepsilon_2 A_2 - \overline{S_1S_2} \end{bmatrix} \tag{5-126}$$

where

$$\overline{S_1S_2} = \frac{1}{\left(\dfrac{\rho_1}{\varepsilon_1 A_1} + \dfrac{\rho_2}{\varepsilon_2 A_2} + \dfrac{1}{\overline{s_1s_2}}\right)} \tag{5-127}$$

Equation (5-127) is of general utility for any two-zone system for which $\varepsilon_i \neq 0$.

The total exchange areas for the four geometries shown in Fig. 5-16 follow directly from Eqs. (5-126) and (5-127).

1. A planar surface A_1 is completely surrounded by a second surface $A_2 > A_1$. Here $F_{1,1} = 0$, $F_{1,2} = 1$, and $\overline{s_1s_2} = A_1$ result in

$$\overline{\mathbf{SS}} = \begin{bmatrix} \varepsilon_1\rho A_1^2 + \varepsilon_2\rho_1 A_1A_2 & \varepsilon_1\varepsilon_2 A_1A_2 \\ \varepsilon_1\varepsilon_2 A_1A_2 & \varepsilon_2 A_2^2 + \varepsilon_1(\rho_2 - \varepsilon_2)A_1A_2 \end{bmatrix} \Big/ [\varepsilon_1\rho_2 A_1 + \varepsilon_2 A_2]$$

and in particular

$$\overline{S_1S_2} = \frac{A_1}{1/\varepsilon_1 + (A_1/A_2)\cdot(\rho_2/\varepsilon_2)} \tag{5-127a}$$

In the limiting case, where A_1 has no negative curvature and is completely surrounded by a very much larger surface A_2 such that $A_1 << A_2$, Eq. (5-127*a*) leads to the even simpler result that $\overline{S_1S_2} = \varepsilon_1 A_1$. This simple result has widespread engineering utility.

2. Two parallel plates of equal area are large compared to their distance of separation (infinite parallel plates). Case 2 is a limiting form of case 1 with $A_1 = A_2$. Algebraic manipulation then results in

$$\overline{\mathbf{SS}} = \begin{bmatrix} (\varepsilon_1 + \varepsilon_2 - 2\varepsilon_1\varepsilon_2)\cdot A_1 & \varepsilon_1\varepsilon_2 A_1 \\ \varepsilon_1\varepsilon_2 A_1 & (\varepsilon_1 + \varepsilon_2 - 2\varepsilon_1\varepsilon_2)A_1 \end{bmatrix} \Big/ [\varepsilon_1 + \varepsilon_2 + \varepsilon_1\varepsilon_2]$$

and in particular

$$\overline{S_1S_2} = \frac{A_1}{1/\varepsilon_1 + 1/\varepsilon_2 - 1} \tag{5-127b}$$

3. Concentric spheres or cylinders where $A_2 > A_1$. Case 3 is mathematically identical to case 1.

4. A *speckled enclosure* has two surface zones. Here $\mathbf{F} = \dfrac{1}{A_1 + A_2}\begin{bmatrix} A_1 & A_2 \\ A_1 & A_2 \end{bmatrix}$ such that $\overline{\mathbf{ss}} = \dfrac{1}{A_1 + A_2}\begin{bmatrix} A_1^2 & A_1A_2 \\ A_1A_2 & A_2^2 \end{bmatrix}$ and Eqs. (5-126) and (5-127) then produce

$$\overline{\mathbf{SS}} = \begin{bmatrix} \varepsilon_1^2 A_1^2 & \varepsilon_1\varepsilon_2 A_1A_2 \\ \varepsilon_1\varepsilon_2 A_1A_2 & \varepsilon_2^2 A_2^2 \end{bmatrix} \Big/ [\varepsilon_1 A_1 + \varepsilon_2 A_2]$$

with the particular result

$$\overline{S_1S_2} = \frac{1}{1/(\varepsilon_1 A_1) + 1/(\varepsilon_2 A_2)} \tag{5-127c}$$

Physically, a two-zone speckled enclosure is characterized by the fact that the view factor from any point on the enclosure surface to the sink zone is identical to that from any other point on the bounding surface. This is only possible when the two zones are "intimately mixed." The seemingly simplistic concept of a speckled enclosure provides a surprisingly useful *default option* in engineering calculations when the *actual* enclosure geometries are quite complex.

The Generalized Source/Sink Refractory (SSR) Model $M \geq 3$ The major numerical effort involved in implementation of the zone method is the evaluation of the inverse multiple reflection matrix $\mathbf{R}$. For $M = 3$, explicit closed-form algebraic formulas *do indeed exist* for the nine scalar elements of the inverse of any arbitrary nonsingular matrix. These formulas are so algebraically complex, however, that it generally proves impractical to present universal closed-form expressions for the total exchange areas, as has been done for the case $M = 2$. For $M = 3$, a notable exception, which is amenable to hand calculation, is an enclosure comprised of two source/sink zones and one flux zone. Here this method is called the *classical SSR model* and requires inversion of one 2×2 matrix. The generalization of this method to multizone enclosures with $M \geq 3$ follows.

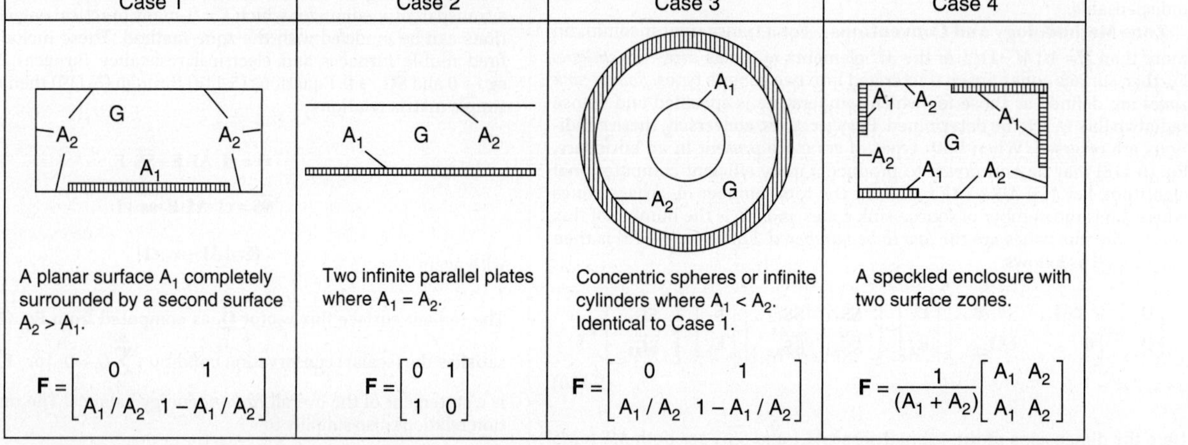

FIG. 5-16 Four enclosure geometries characterized by two surface zones and one volume zone. (*Marks' Standard Handbook for Mechanical Engineers*, McGraw-Hill, New York, 1999, p. 4-73, Table 4.3.5).

Consider an arbitrary multizone enclosure that confines a transparent medium. The bounding surface is comprised of *MS* source/sink and *MF* flux zones such that $M = MS + MF$. The *MS* source/sink zones are numbered first, and *MF* flux zones are numbered last. We then partition the zoning equations exactly per the conventions employed in Eqs. (5-120).

Assume now that all the surface source/sink zones are black and all the flux zones are adiabatic. The partitioned flux equations for this simple black enclosure are then given as

$$\begin{bmatrix}\mathbf{Q}_1\\ \mathbf{0}\end{bmatrix} = \begin{bmatrix}\mathbf{AI}_{1,1} & \mathbf{0}\\ \mathbf{0} & \mathbf{AI}_{2,2}\end{bmatrix} \cdot \begin{bmatrix}\mathbf{E}_1\\ \mathbf{E}_2\end{bmatrix} - \begin{bmatrix}\overline{\mathbf{ss}}_{1,1} & \overline{\mathbf{ss}}_{1,2}\\ \overline{\mathbf{ss}}_{2,1} & \overline{\mathbf{ss}}_{2,2}\end{bmatrix} \cdot \begin{bmatrix}\mathbf{E}_1\\ \mathbf{E}_2\end{bmatrix} \qquad (5\text{-}128a)$$

and the solution to Eq. (5-128*a*) then readily follows as

$$\mathbf{Q}_1 = \left[\mathbf{AI}_{1,1} - \overline{\mathbf{SSb}}\right] \cdot \mathbf{E}_1 \qquad (5\text{-}128b)$$

where
$$\overline{\mathbf{SSb}} = \overline{\mathbf{ss}}_{1,1} + \overline{\mathbf{ss}}_{1,2} \cdot \mathbf{R22} \cdot \overline{\mathbf{ss}}_{2,1} \qquad (5\text{-}128c)$$

and
$$\mathbf{R22} = \left[\mathbf{AI}_{2,2} - \overline{\mathbf{ss}}_{2,2}\right]^{-1} \qquad (5\text{-}128d)$$

Equation (5-128*c*) states that $\overline{\mathbf{SSb}}$ is the sum of the direct radiation between all black source/sink zones plus radiation absorbed and reemitted (*reflected*) from all the adiabatic flux zones.

It then may be shown that if the source/sink zones are nonblack

$$\mathbf{Q}_1 = \varepsilon\mathbf{AI}_1 \cdot \mathbf{E}_1 - \overline{\mathbf{SSR}} \cdot \mathbf{E}_1 - \overline{\mathbf{FSQ}} \cdot \mathbf{Q}_2 \qquad (5\text{-}129a)$$

where $\overline{\mathbf{SSR}}$ and $\overline{\mathbf{FSQ}}$ are specialized total exchange areas defined as follows

$$\overline{\mathbf{SSR}} = \varepsilon\mathbf{AI}_{1,1} \cdot \mathbf{R11b} \cdot \overline{\mathbf{SSb}} \cdot \varepsilon\mathbf{I}_{1,1} \qquad (5\text{-}129b)$$

$$\overline{\mathbf{FSQ}} = \varepsilon\mathbf{AI}_{1,1} \cdot \mathbf{R11b} \cdot \overline{\mathbf{ss}}_{1,2} \cdot \mathbf{R22} \qquad (5\text{-}129c)$$

with
$$\mathbf{R11b} = \left[\mathbf{AI}_{1,1} - \overline{\mathbf{SSb}} \cdot \rho\mathbf{I}_{1,1}\right]^{-1} \qquad (5\text{-}129d)$$

which satisfy the following conservation relations

$$\overline{\mathbf{SSb}} \cdot \mathbf{1}_{MS} = \mathbf{AI}_{1,1} \cdot \mathbf{1}_{MS} \qquad (5\text{-}130e)$$

$$\overline{\mathbf{SSR}} \cdot \mathbf{1}_{MS} = \varepsilon\mathbf{AI}_{1,1} \cdot \mathbf{1}_{MS} \qquad (5\text{-}130f)$$

$$\overline{\mathbf{FSQ}}^{\mathrm{T}} \cdot \mathbf{1}_{MS} = \mathbf{1}_{MF} \qquad (5\text{-}130g)$$

The solution sequence for all the enclosure surface vectors then follows as

$$\begin{aligned}
&\mathbf{W}_1 = \mathbf{E}_1 - \rho\mathbf{I}_{1,1} \cdot \varepsilon\mathbf{AI}_{1,1}^{-1} \cdot \mathbf{Q}_1\\
&\mathbf{W}_2 = \mathbf{R22} \cdot \left[\overline{\mathbf{ss}}_{2,1} \cdot \mathbf{W}_1 + \mathbf{Q}_2\right]\\
&\mathbf{W} = \begin{bmatrix}\mathbf{W}_1\\ \mathbf{W}_1\end{bmatrix} \quad \mathbf{Q} = \begin{bmatrix}\mathbf{Q}_1\\ \mathbf{Q}_2\end{bmatrix}\\
&\mathbf{H} = \mathbf{W} - \mathbf{AI}^{-1} \cdot \mathbf{Q}\\
&\mathbf{E}_2 = \mathbf{W}_2 + \rho\mathbf{I}_{2,2} \cdot \varepsilon\mathbf{AI}_{2,2}^{-1} \cdot \mathbf{Q}_2 \qquad \varepsilon_i \neq 0
\end{aligned} \qquad (5\text{-}131)$$

For the special case $\mathbf{Q}_2 = \mathbf{0}$ it may also be shown that $\mathbf{E}_2 = \overline{\mathbf{FSQ}}^{\mathrm{T}} \cdot \mathbf{E}_1$.

The terminology in Hottel and Sarofim defines $\overline{\mathbf{SSb}}$ as the *refractory augmented exchange area for black source/sink zones*, and $\overline{\mathbf{SSR}}$ is termed the *refractory augmented exchange area for nonblack source/sink zones.* It is of paramount importance here to notice that Eqs. (5-131) lead to the fact that the zoning solution for **W**, **H**, and **Q** *is always independent of the emissivities of all the flux zones.* Moreover, $\mathbf{E}_2$ is also independent of $\varepsilon\mathbf{I}_{2,2}$, provided that $\mathbf{Q}_2 = \mathbf{0}$. In other words, if all the flux zones are adiabatic refractories, then the refractory emissivity is *entirely immaterial* to the zoning calculations. One might first find this consequence counterintuitive since solution of the same problem via the conventional TEA route [Eqs. (5-120)] does indeed require a priori specification of $\varepsilon\mathbf{I}_{2,2}$ even if $\mathbf{Q}_2 = \mathbf{0}$. Note further that in contrast to Eqs. (5-120) this procedure permits $\varepsilon\mathbf{I}_{2,2} = \mathbf{0}$.

The Classical Three-Zone SSR Model Set $M = 3$ and let zones 1 and 2 be the source/sink zones and zone 3 the flux zone with $Q_3 = 0$. Then the general expressions for $\overline{\mathbf{SSR}}$ reduce to

$$\overline{\mathbf{SSR}} = \begin{bmatrix}\varepsilon_1 A_1 - \overline{S_1 S_2} & \overline{S_1 S_2}\\ \overline{S_1 S_2} & \varepsilon_2 A_2 - \overline{S_1 S_2}\end{bmatrix} \qquad (5\text{-}132)$$

where
$$\overline{S_1 S_2} = \cfrac{1}{\cfrac{\rho_1}{\varepsilon_1 A_1} + \cfrac{\rho_2}{\varepsilon_2 A_2} + \cfrac{1}{\overline{SSb}_{1,2}}} \qquad (5\text{-}132a)$$

or
$$\mathcal{F}_{1,2} = \cfrac{1}{\cfrac{\rho_1}{\varepsilon_1} + \left(\cfrac{A_1}{A_2}\right)\cfrac{\rho_2}{\varepsilon_2} + \cfrac{1}{\overline{F}_{1,2}}} \qquad (5\text{-}132b)$$

and
$$\overline{SSb}_{1,2} = \overline{s_1 s_2} + \frac{\overline{s_1 s_3} \cdot \overline{s_3 s_2}}{\overline{s_1 s_3} + \overline{s_2 s_3}} \equiv A_1 \cdot \overline{F}_{1,2} \qquad (5\text{-}132c)$$

where $\varepsilon_i \neq 0$. Notice that Eq. (5-132*a*) appears *deceptively similar* to Eq. (5-127). Moreover Eq. (5-132*c*) is the scalar analog of Eq. (5-128*c*). Further, if $\overline{s_1 s_1} = \overline{s_2 s_2} = 0$, Eq. (5-132*c*) then reduces to

$$\overline{SSb}_{1,2} = \frac{A_1 A_2 - (\overline{s_1 s_2})^2}{A_1 + A_2 - 2 \cdot \overline{s_1 s_2}} \qquad (5\text{-}132d)$$

which necessitates the evaluation of only one direct exchange area. A consequence of Eq. (5-132*d*) is that the classic SSR model with $M = 3$ *cannot distinguish the shape of the refractory.*

Collectively, Eqs. (5-132) along with formulas to compute $\overline{F}_{1,2}$ (F-bar) are sometimes called the three-zone source/sink refractory model.

Refractory Augmented Black View Factors $\overline{F}_{i,j}$ In the older zoning literature, the following definition is employed: $\overline{SSb}_{i,j} = A_i \overline{F}_{i,j}$, where $\overline{F}_{i,j}$ is called the *refractory augmented black view factor,* or $\overline{F}$-bar. This quantity is especially convenient to archive results for a particular enclosure geometry when the enclosure contains only one source/sink pair and any number of refractory zones.

The refractory augmented view factor $\overline{F}_{i,j}$ is documented for a few geometrically simple cases and can be calculated or approximated for others. If A_1 and A_2 are equal parallel disks, squares, or rectangles, connected by nonconducting but reradiating refractory surfaces, then $\overline{F}_{i,j}$ is given by Fig. 5-11 in curves 5 to 8. Let A_1 represent an infinite plane and A_2 represent one or two rows of infinite parallel tubes. If the only other surface is an adiabatic refractory surface located behind the tubes, then $\overline{F}_{2,1}$ is given by curve 5 or 6 of Fig. 5-13. The classic zoning literature thus contains a hierarchy of three distinct surface-to-surface view factors, denoted by $F_{i,j}$, $\overline{F}_{i,j}$, and $\mathcal{F}_{i,j}$.

Accuracy of the Zone Method Experience has shown that despite its limitations even the simple SSR model with $M = 3$ can yield quite useful results for a host of practical engineering applications without resorting to digital computation. The error due to representation of the source and sink by single zones is often small, even if the views of the enclosure from different parts of the same zone are dissimilar, provided the surface emissivities are near unity. The error is also small if the temperature variation of the refractory is small. Any degree of accuracy can, of course, be obtained via matrix methodologies for arbitrarily large M by using a digital computer. From a computational viewpoint, when $M \geq 4$, matrix methods *must be used.* Matrix methods must also be used for finer-scale calculations such as more detailed wall *temperature and flux density profiles.*

The Electrical Network Analog At each surface zone, the *total* radiant flux is proportional to the difference between E_i and W_i, as indicated by the equation $Q_i = \frac{\varepsilon_i A_i}{\rho_i}(E_i - W_i)$. The *net* flux between zones *i* and *j* is also given by $Q_{i,j} = \overline{s_i s_j}(W_i - W_j)$ where $Q_i = \sum_{i=1}^{M} Q_{i,j}$ for all $1 \leq i \leq M$ is the *total* heat flux leaving each zone. These relations suggest a *visual* electrical analog in which E_i and W_i are analogous to voltage potentials. The quantities $\varepsilon_i A_i/\rho_i$ and $\overline{s_i s_j}$ are analogous to conductances (reciprocal impedances), and Q_i or $Q_{i,j}$ is analogous to electric currents. Such an electrical analog has been developed by Oppenheim [Oppenheim, A. K., *Trans. ASME,* **78:** 725–735 (1956)].

Figure 5-17 illustrates a generalized electrical network analogy for a three-zone enclosure consisting of one refractory zone and two gray zones A_1 and A_2. The potential points E_i and W_i are separated by conductances $\varepsilon_i A_i/\rho_i$. The emissive powers E_1 and E_2 represent potential sources or sinks, while W_1, W_2, and W_r are internal node points. In this construction, the nodal point representing each surface is connected to that of every other surface it can see directly. Figure 5-17 can be used to formulate the total exchange area $\overline{S_1 S_2}$ for the SSR model virtually by inspection. The refractory zone is first characterized by a floating potential such that $E_r = W_r$. Next, the resistance for the parallel "current paths" between the internal nodes W_1 and W_2

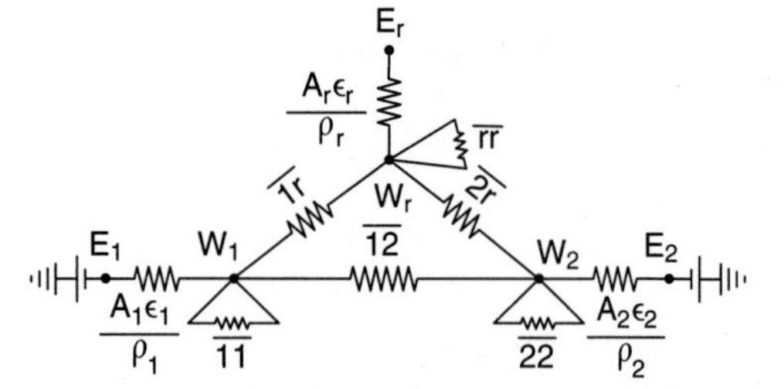

FIG. 5-17 Generalized electrical network analog for a three-zone enclosure. Here A_1 and A_2 are gray surfaces and A_r is a radiatively adiabatic surface (Hottel, H. C., and A. F. Sarofim, *Radiative Transfer*, McGraw-Hill, New York, 1967, p. 91).

is defined by $\dfrac{1}{A_1\bar{F}_{1,2}} \equiv \dfrac{1}{\overline{s_1s_2} + 1\Big/\left(\dfrac{1}{\overline{s_1s_r}} + \dfrac{1}{\overline{s_2s_r}}\right)}$ which is identical to Eq. (5-132*c*).

Finally, the overall impedance between the source E_1 and the sink E_2 is represented simply by three resistors in series and is thus given by

$$\frac{1}{\overline{S_1S_2}} = \frac{\rho_1}{\varepsilon_1 A_1} + \frac{1}{A_1\bar{F}_{1,2}} + \frac{\rho_2}{\varepsilon_2 A_2}$$

or

$$\overline{S_1S_2} = \frac{1}{\dfrac{\rho_1}{\varepsilon_1 A_1} + \dfrac{\rho_2}{\varepsilon_2 A_2} + \dfrac{1}{A_1\bar{F}_{1,2}}} \tag{5-133}$$

This result is identically the same as for the SSR model obtained previously in Eq. (5-132*a*). This equation is also valid for $M_r \geq 1$ as long as $M_b = 2$. The electrical network analog methodology can be generalized for enclosures having $M > 3$.

Some Examples from Furnace Design The theory of the past several subsections is best understood in the context of two engineering examples involving furnace modeling. The engineering idealization of the *equivalent gray plane concept* is introduced first. Figure 5-18 depicts a common furnace configuration in which the heating source is two refractory-backed, internally fired tube banks. Clearly the overall geometry for even this common furnace configuration is too complex to be modeled in an expeditious manner by anything other than *a simple engineering idealization.* Thus the furnace shown in Fig. 5-18 is modeled in Example 5-10, by *partitioning* the entire enclosure into two subordinate furnace compartments. First, the approach defines an imaginary gray plane A_2, located on the inward-facing side of the tube assemblies. Second, the total exchange area between the tubes to this *equivalent* gray plane is calculated, making full allowance for the reflection from the refractory tube backing. This plane-to-tube view factor is then *defined* to be the emissivity of the required equivalent gray plane whose temperature is further assumed to be that of the tubes. This procedure guarantees continuity of the radiant flux *into* the interior radiant portion of the furnace arising from a moderately complicated *external* source.

Example 5-10 demonstrates classical zoning calculations for radiation pyrometry in furnace applications. Example 5-11 is a classical furnace design calculation via zoning an enclosure with a *diathermanous* atmosphere and $M = 5$. The latter calculation can only be addressed with matrix methods. The results of Example 5-11 demonstrate the relative insensitivity of zoning to $M > 3$ and the engineering utility of the generalized SSR model.

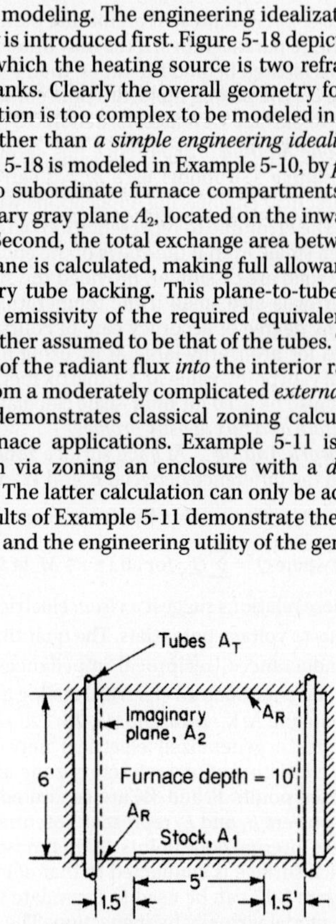

FIG. 5-18 Furnace chamber cross section. To convert feet to meters, multiply by 0.3048.

Example 5-10 Radiation Pyrometry A long tunnel furnace is heated by electrical resistance coils embedded in the ceiling. The stock travels on a floor-mounted conveyer belt and has an estimated emissivity of 0.7. The sidewalls are unheated refractories with emissivity 0.55, and the ceiling emissivity is 0.8. The furnace cross section is rectangular with height 1 m and width 2 m. A total radiation pyrometer is sighted on the walls and indicates the following apparent temperatures: ceiling 1340°C, sidewall readings average about 1145°C and the load indicates about 900°C. (*a*) What are the true temperatures of the furnace walls and stock? (*b*) What is the net heat flux at each surface and each zone pair? (*c*) Compare the adiabatic SSR and TEA matrix models.

$M = 3$ Zones
Zone 1 Source (top)
Zone 2 Sink (bottom)
Zone 3 Refractory (lumped sides)
Physical constants:

$$T_0 \equiv 273.15°\text{C} \qquad \sigma \equiv 5.6704\times10^{-8}\,\frac{\text{W}}{\text{m}^2\cdot\text{K}}$$

Enclosure input parameters:

$$H = 1.0\text{ m} \qquad W = 2.0\text{ m} \qquad L = 1.0\text{ m}$$

Compute direct exchange areas using crossed strings method $\Sigma = 3$

$$\mathbf{ss_4} = \begin{pmatrix} 0.0000 & 1.2361 & 0.3820 & 0.3820 \\ 1.2361 & 0.0000 & 0.3820 & 0.3820 \\ 0.3820 & 0.3820 & 0.0000 & 0.2361 \\ 0.3820 & 0.3820 & 0.2361 & 0.0000 \end{pmatrix} \text{m}^2$$

Lump the four-zone enclosure into a three-zone enclosure by combining rows and columns 3 and 4. Then

$$\overline{\mathbf{ss}} = \begin{pmatrix} \overline{s_1s_1} & \overline{s_1s_2} & \overline{s_1s_3} \\ \overline{s_2s_1} & \overline{s_2s_2} & \overline{s_2s_3} \\ \overline{s_3s_1} & \overline{s_3s_2} & \overline{s_3s_3} \end{pmatrix} = \begin{pmatrix} 0 & 1.2361 & 0.7639 \\ 1.2361 & 0 & 0.7639 \\ 0.7639 & 0.7639 & 0.4721 \end{pmatrix} \text{m}^2$$

$$(\mathbf{AI} - \overline{\mathbf{ss}})\cdot\begin{pmatrix}1\\1\\1\end{pmatrix} = \begin{pmatrix}0\\0\\0\end{pmatrix}$$

$$\mathbf{AI} = \begin{bmatrix} A_1 & 0 & 0 \\ 0 & A_2 & 0 \\ 0 & 0 & A_3 \end{bmatrix} = \begin{bmatrix} W\cdot L & 0 & 0 \\ 0 & W\cdot L & 0 \\ 0 & 0 & 2\cdot H\cdot L \end{bmatrix} = \begin{bmatrix} 2.0 & 0 & 0 \\ 0 & 2.0 & 0 \\ 0 & 0 & 2.0 \end{bmatrix} \text{m}^2$$

$$\boldsymbol{\varepsilon}\mathbf{I} = \begin{bmatrix} \varepsilon_1 & 0 & 0 \\ 0 & \varepsilon_2 & 0 \\ 0 & 0 & \varepsilon_3 \end{bmatrix} = \begin{bmatrix} 0.8 & 0 & 0 \\ 0 & 0.7 & 0 \\ 0 & 0 & 0.55 \end{bmatrix} \qquad \boldsymbol{\rho}\mathbf{I} = \begin{bmatrix} \rho_1 & 0 & 0 \\ 0 & \rho_2 & 0 \\ 0 & 0 & \rho_3 \end{bmatrix} = \begin{bmatrix} 0.2 & 0 & 0 \\ 0 & 0.3 & 0 \\ 0 & 0 & 0.45 \end{bmatrix}$$

Compute Radiosities **W** *from Pyrometer Temperature Readings*

$$\mathbf{T_W} = \begin{bmatrix} 1340.0 \\ 900.0 \\ 1145.0 \end{bmatrix} °\text{C} \qquad \mathbf{W} = \sigma\cdot[\mathbf{T_W} + 273.15]^4 = \begin{bmatrix} 384.0 \\ 107.4 \\ 229.4 \end{bmatrix} \text{kW/m}^2$$

All matrix wall flux density quantities and heat fluxes can then be directly calculated from the radiosity (leaving flux density) vector **W**.

$$\mathbf{H} = \mathbf{AI}^{-1}\cdot\overline{\mathbf{ss}}\cdot\mathbf{W} \qquad \mathbf{Q} = [\mathbf{AI} - \overline{\mathbf{ss}}]\cdot\mathbf{W} \qquad \mathbf{q} = \mathbf{AI}^{-1}\cdot\mathbf{Q} \qquad \mathbf{E} = \boldsymbol{\varepsilon}\mathbf{I}^{-1}\cdot[\mathbf{W} - \boldsymbol{\rho}\mathbf{I}\cdot\mathbf{H}]$$

leading to the final results

ZONE	W	H	q	E	
1	384.0	154.0	230.0	441.5	kW/m²
2	107.4	324.9	−217.5	14.2	
3	229.4	241.8	−12.5	219.1	

and

$$\mathbf{Q} = \begin{pmatrix} 460.0 \\ -435.0 \\ -25.0 \end{pmatrix} \text{kW}$$

Here the sidewalls act as near-adiabatic surfaces since the heat loss through each sidewall is only about 2.7 percent of the total heat flux originating at the source.

Part (a): Actual Wall Temperatures versus Pyrometer Readings

$$\mathbf{T} = \left[\left(\frac{\mathbf{E}}{\sigma}\right)^{0.25} - 273.15\right] = \begin{pmatrix} 1397.3 \\ 434.1 \\ 1128.9 \end{pmatrix} °\text{C} \qquad \mathbf{T_W} = \begin{pmatrix} 1340.0 \\ 900.0 \\ 1145.0 \end{pmatrix} °\text{C}$$

$[k = 1, 3]$

$$\left[\varepsilon\min_k = 1 - \frac{W_k}{H_k}\right] \quad \%\text{Ref}_k = \rho_k \cdot \frac{H_k}{W_k} \quad \varepsilon_{\min} = \begin{pmatrix} \text{NA} \\ 0.669 \\ 0.052 \end{pmatrix} \quad \%\mathbf{Ref} = \begin{pmatrix} 8.0 \\ 90.8 \\ 47.4 \end{pmatrix}$$

The (low) estimated sink emissivity, $\varepsilon_2 = 0.7$, is the dominant parameter in this example. First, the marked disparity in actual and measured sink temperature arises because the sink radiosity is comprised of 90.8 percent reflected energy, that is, $\rho_2 \cdot H_2/W_2 = 0.9075$. Moreover, when a surface flux is negative, $W < H$ and a minimum allowable surface emissivity is defined by $\varepsilon_{\min} = 1 - W/H$. Thus if the sink emissivity is less than $\varepsilon_2 = 1 - 107.4/324.9 = 0.6694$, the sink temperature becomes imaginary. Lastly if the sink were black, $E_2 = W_2 = 107.4\ \text{kW/m}^2$, the pyrometer and sink temperatures would be equal with $T_2 = 900°\text{C}$.

Part (b): Radiant Heat Flux at Each Surface

Define
$$\mathbf{WI}_4 = \begin{pmatrix} 384 & 0 & 0 & 0 \\ 0 & 107.4 & 0 & 0 \\ 0 & 0 & 229.4 & 0 \\ 0 & 0 & 0 & 229.4 \end{pmatrix} \frac{\text{kW}}{\text{m}^2}$$

and
$$\mathbf{Q_P} = \mathbf{WI}_4 \cdot \mathbf{ss}_4 - \mathbf{ss}_4 \cdot \mathbf{WI}_4 = \begin{pmatrix} 0 & 341.9 & 59.1 & 59.1 \\ -341.9 & 0 & -46.6 & -46.6 \\ -59.1 & 46.6 & 0 & 0 \\ -59.1 & 46.6 & 0 & 0 \end{pmatrix} \text{kW}$$

leading to
$$\mathbf{Q} = \mathbf{Q_P} \cdot \begin{pmatrix} 1 \\ 1 \\ 1 \\ 1 \end{pmatrix} = \begin{pmatrix} 460.0 \\ -435.0 \\ -12.5 \\ -12.5 \end{pmatrix} \text{kW}$$

where $\mathbf{Q_p}$ defines the heat flux between all *zone pairs* and $\mathbf{Q}$ is the total heat flux *at* each surface.

Part (c): Compare the Adiabatic SSR and TEA Matrix Approaches

Both approaches require the a prioi specification of three unknowns. Here we shall assume as in part (a) that

$$E_1 = 441.5 \frac{\text{kW}}{\text{m}^2} \qquad E_2 = 14.2 \frac{\text{kW}}{\text{m}^2} \qquad Q_3 = 25.0\ \text{kW}$$

Compute total exchange areas

$$\mathbf{R} \equiv [\mathbf{AI} - \overline{\mathbf{ss}} \cdot \boldsymbol{\rho}\mathbf{I}]^{-1}$$

$$\overline{\mathbf{SS}} = \varepsilon\mathbf{I} \cdot \mathbf{AI} \cdot \mathbf{R} \cdot \overline{\mathbf{ss}} \cdot \varepsilon\mathbf{I} = \begin{bmatrix} 0.2948 & 0.8284 & 0.4769 \\ 0.8284 & 0.1761 & 0.3955 \\ 0.4769 & 0.3955 & 0.2277 \end{bmatrix} \qquad [\varepsilon\mathbf{I} \cdot \mathbf{AI} - \overline{\mathbf{SS}}] \cdot \begin{bmatrix} 1 \\ 1 \\ 1 \end{bmatrix} = \begin{bmatrix} 0 \\ 0 \\ 0 \end{bmatrix}$$

$$E_3 = \frac{(SS_{3,1} \cdot E_1 + SS_{3,2} \cdot E_2 + Q_3)}{SS_{3,1} + SS_{3,2}} = 219.10 \frac{\text{kW}}{\text{m}^2} \quad \text{see Eq. (5-121)}$$

$$\mathbf{Q3} = (\varepsilon\mathbf{I} \cdot \mathbf{AI} - \mathbf{SS}) \cdot \begin{pmatrix} \mathbf{E}_1 \\ \mathbf{E}_2 \\ \mathbf{E}_3 \end{pmatrix} = \begin{pmatrix} 460.0 \\ -435.0 \\ -25.0 \end{pmatrix} \text{kW}$$

Compute SSR matrix for M = 3
$$\mathbf{AI1} = \begin{pmatrix} A_1 & 0 \\ 0 & A_2 \end{pmatrix} \qquad \varepsilon\mathbf{I1} = \begin{pmatrix} \varepsilon_1 & 0 \\ 0 & \varepsilon_2 \end{pmatrix}$$

[Use Eqs. (5-128*a*) and (5-130).]

$$\text{ssb}_{1,2} = \text{ss}_{1,2} + \frac{\text{ss}_{1,3} \cdot \text{ss}_{3,2}}{\text{ss}_{1,3} + \text{ss}_{2,3}} = 1.618\ \text{m}^2$$

$$\text{SSR}_{1,2} = \frac{1}{\dfrac{\rho 1}{\varepsilon_1 \cdot A_1} + \dfrac{\rho 2}{\varepsilon_2 \cdot A_2} + \dfrac{1}{\text{ssb}_{1,2}}} = 1.0446\ \text{m}^2$$

$$\mathbf{SSR} = \begin{pmatrix} \varepsilon_1 \cdot A_1 - \text{SSR}_{1,2} & \text{SSR}_{1,2} \\ \text{SSR}_{1,2} & \varepsilon_2 \cdot A_2 - \text{SSR}_{1,2} \end{pmatrix} = \begin{pmatrix} 0.5554 & 1.0446 \\ 1.0446 & 0.3554 \end{pmatrix} \text{m}^2$$

$$\begin{pmatrix} Q_1 \\ Q_2 \end{pmatrix} = (\varepsilon\mathbf{I1} \cdot \mathbf{AI1} - \mathbf{SSR}) \cdot \begin{pmatrix} \mathbf{E}_1 \\ \mathbf{E}_2 \end{pmatrix} = \begin{pmatrix} 446.3 \\ -446.3 \end{pmatrix} \text{kW}$$

Thus the adiabatic SSR model produces $Q_1 = 446.3$ kW versus the measured value of $Q_1 = 460.0$ with a discrepancy of about 3.0 percent. Mathematically the adiabatic SSR model assumes a value of $Q_3 = 0$ which precludes the sidewall heat loss of $Q_3 = -25.0\ \text{kW/m}^2$. This assumption accounts for *all* the difference between the two values.

Demonstrate relationships between SSR and TEA models

Assume $Q_3 = 0$; then from Eq. (5-121) we have

$$E_{30} = \frac{(\text{SS}_{3,1} \cdot E_1 + \text{SS}_{3,2} \cdot E_2)}{\text{SS}_{3,1} + \text{SS}_{3,2}} = 247.8 \frac{\text{kW}}{\text{m}^2} \qquad \Theta_{30} = \frac{E_{30} - E_2}{E_1 - E_2} = 0.5466$$

and we define
$$\Theta_3 = \frac{\text{SS}_{3,1}}{\text{SS}_{3,1} + \text{SS}_{3,2}} = 0.5466$$

with the result that $\Theta_3 = \Theta_{30}$ and

$$\text{SS12} = \text{SS}_{1,2} + \text{SS}_{2,3} \cdot \Theta_3 = 1.0446\ \text{m}^2$$

Thus $\text{SSR}_{1,2} = \text{SS12}$ is the *refractory-aided* total exchange area between *nonblack* zones 1 and 2. Perhaps counterintuitive, this result is independent of ε_3. Note further that if we recompute the total exchange areas with $\varepsilon_3 = 0$ to obtain $\mathbf{SS_0}$, then **SSR** can be directly evaluated from the (upper) source/sink portion of $\mathbf{SS_0}$ as shown below.

$$\mathbf{SS_0} = \begin{pmatrix} 0.5554 & 1.0446 & 0 \\ 1.0446 & 0.3554 & 0 \\ 0 & 0 & 0 \end{pmatrix}$$

(This example was developed as a MATHCAD 15 worksheet. MATHCAD is a registered trademark of Parametric Technology Corporation.)

Example 5-11 Furnace Simulation via Zoning The furnace chamber depicted in Fig. 5-18 is heated by combustion gases passing through 20 vertical radiant tubes which are backed by refractory sidewalls. The tubes have an outside diameter of $D = 5$ in (12.7 cm) mounted on $C = 12$-in (4.72-cm) centers and a gray body emissivity of 0.8. The interior (radiant) portion of the furnace is a $6 \times 8 \times 10$ ft rectangular parallelepiped with a total surface area of 376 ft^2 (34.932 m^2). A 50-ft^2 (4.645-m^2) sink with emissivity 0.9 is positioned centrally on the floor of the furnace. The tube and sink temperatures are measured with embedded thermocouples as 1500 and 1200°F, respectively. The gray refractory emissivity may be taken as 0.5. While all *other* refractories are assumed to be radiatively adiabatic, the roof of the furnace is estimated to lose heat to the surroundings with a *flux density* (W/m^2) equal to 5 percent of the source and sink emissive power difference. An estimate of the radiant flux arriving at the sink is required, as well as estimates for the roof and average refractory temperatures in consideration of refractory service life.

Part (a): Equivalent Gray Plane Emissivity Algebraically compute the equivalent gray plane emissivity for the refractory-backed tube bank idealized by the imaginary plane A_2, depicted in Fig. 5-18.

Solution Let zone 1 represent *one* tube and zone 2 represent the *effective* plane, that is, the *unit cell* for the tube bank. Then $A_1 = \pi D$ and $A_2 = C$ are the corresponding zone areas, respectively (per unit vertical dimension). Also set $\varepsilon_1 = 0.8$ with $\varepsilon_2 = 1.0$ and define $R = C/D = 12/5 = 2.4$.

For $R = 2.4$, curves 1 and 5 of Fig. 5-13 yield, respectively, $F_{p,t} = 0.57$ and $\overline{F}_{p,t} = 0.81$. Using notation consistent with Example 5-8, more accurate values are calculated as follows:

$$F_{t,t} = [\sin^{-1}(1/R) + \sqrt{R^2 - 1} - R]/\pi = 0.06733$$
$$F_{p,t} = (0.5 - F_{t,t}).\pi/R = 0.56637$$

and
$$\overline{F}_{2,1} = \overline{F}_{p,t} = F_{p,t}(1 - F_{p,t}) = 0.81196$$

To make allowance for nonblack tubes, application of Eq. (5-132*b*) with $\varepsilon_2 = 0.8$ then yields

$$\mathcal{F}_{2,1} = \frac{1}{0/1 + (2.4/\pi) \cdot (0.2/0.8) + 1/\overline{F}_{2,1}} = 0.70295$$

Part (b): Radiant Furnace Chamber with Roof Heat Loss

$M = 5$ zones
Zone 1 = Sink (bottom)
Zone 2 = Source (lumped sides)
Zone 3 = Refractory (roof)
Zone 4 = Refractory (lumped ends)
Zone 5 = Refractory (lumped floor strips)

Boundary conditions: Five pieces of information are required for $M = 5$.

$$\mathbf{T_{IF}} = \begin{pmatrix} 1200.0 \\ 1500.0 \end{pmatrix} °\text{F} \qquad \mathbf{T_{IC}} = \frac{5}{9}(\mathbf{T_{IF}} - 32) = \begin{pmatrix} 648.9 \\ 815.6 \end{pmatrix} °\text{C}$$

$$\mathbf{E_I} = \begin{pmatrix} E_1 \\ E_2 \end{pmatrix} = \sigma \cdot [\mathbf{T_{IC}} + 273.15]^4 = \begin{pmatrix} 40.98 \\ 79.66 \end{pmatrix} \text{kW/m}^2$$

$$Q_3 = -0.5 \cdot [E_2 - E_1] \cdot A_3 / 10.76391 = -14.374\ \text{kW} \qquad Q_4 = Q_5 = 0$$

$$\mathbf{Q_2} = \begin{pmatrix} Q_3 \\ Q_4 \\ Q_5 \end{pmatrix} = \begin{pmatrix} -14.374 \\ 0 \\ 0 \end{pmatrix} \text{kW}$$

Enclosure input parameters:

$$\boldsymbol{\varepsilon}\mathbf{I}=\begin{pmatrix}\boldsymbol{\varepsilon}\mathbf{I}_{11} & \mathbf{O}\\ \mathbf{O}^T & \boldsymbol{\varepsilon}\mathbf{I}_{22}\end{pmatrix}=\begin{bmatrix}\begin{pmatrix}0.9 & 0\\ 0 & 0.70295\end{pmatrix} & \begin{pmatrix}0 & 0 & 0\\ 0 & 0 & 0\end{pmatrix}\\ \begin{pmatrix}0 & 0\\ 0 & 0\\ 0 & 0\end{pmatrix} & \begin{pmatrix}0.5 & 0 & 0\\ 0 & 0.5 & 0\\ 0 & 0 & 0.5\end{pmatrix}\end{bmatrix}$$

$$\boldsymbol{\rho}\mathbf{I}=\begin{pmatrix}\boldsymbol{\rho}\mathbf{I}_{11} & \mathbf{O}\\ \mathbf{O}^T & \boldsymbol{\rho}\mathbf{I}_{22}\end{pmatrix}=\begin{bmatrix}\begin{pmatrix}0.1 & 0\\ 0 & 0.29705\end{pmatrix} & \begin{pmatrix}0 & 0 & 0\\ 0 & 0 & 0\end{pmatrix}\\ \begin{pmatrix}0 & 0\\ 0 & 0\\ 0 & 0\end{pmatrix} & \begin{pmatrix}0.5 & 0 & 0\\ 0 & 0.5 & 0\\ 0 & 0 & 0.5\end{pmatrix}\end{bmatrix}$$

$A_1=50\text{ ft}^2\quad A_2=120\text{ ft}^2\quad A_3=80\text{ ft}^2\quad A_4=96\text{ ft}^2\quad A_5=30\text{ ft}^2$

$$\mathbf{AI}=\begin{pmatrix}\mathbf{AI}_{11} & \mathbf{O}\\ \mathbf{O}^T & \mathbf{AI}_{22}\end{pmatrix}=\begin{bmatrix}\begin{pmatrix}4.6452 & 0\\ 0 & 11.1484\end{pmatrix} & \begin{pmatrix}0 & 0 & 0\\ 0 & 0 & 0\end{pmatrix}\\ \begin{pmatrix}0 & 0\\ 0 & 0\\ 0 & 0\end{pmatrix} & \begin{pmatrix}7.4322 & 0 & 0\\ 0 & 8.9187 & 0\\ 0 & 0 & 2.7871\end{pmatrix}\end{bmatrix}\text{m}^2$$

Compute direct exchange areas: There are $\Sigma = 10$ nonzero direct exchange areas. These are obtained from Eqs. (5-114) and view factor algebra. The final (partitioned) array of direct exchange areas is

$$\mathbf{ss}=\begin{pmatrix}\mathbf{ss}_{11} & \mathbf{ss}_{12}\\ \mathbf{ss}_{21} & \mathbf{ss}_{22}\end{pmatrix}=\begin{bmatrix}\begin{pmatrix}0 & 1.559\\ 1.559 & 2.0777\end{pmatrix} & \begin{pmatrix}1.5747 & 1.5115 & 0\\ 2.8391 & 3.3925 & 1.2802\end{pmatrix}\\ \begin{pmatrix}1.5747 & 2.8391\\ 1.5115 & 3.3925\\ 0 & 1.2802\end{pmatrix} & \begin{pmatrix}0 & 2.2421 & 0.7763\\ 2.2421 & 1.0419 & 0.7306\\ 0.7763 & 0.7306 & 0\end{pmatrix}\end{bmatrix}\text{m}^2$$

$$(\mathbf{AI}-\overline{\mathbf{ss}})\cdot\begin{pmatrix}1\\1\\1\\1\\1\end{pmatrix}=\begin{pmatrix}0\\0\\0\\0\\0\end{pmatrix}\qquad \mathbf{R22}=(\mathbf{AI}_{22}-\mathbf{ss}_{22})^{-1}=\begin{pmatrix}0.1553 & 0.0494 & 0.0562\\ 0.0494 & 0.1459 & 0.052\\ 0.0562 & 0.052 & 0.3881\end{pmatrix}\frac{1}{\text{m}^2}$$

Compute SSR total exchange areas:

$$\mathbf{SSb}=\mathbf{ss}_{11}+\mathbf{ss}_{12}\cdot\mathbf{R22}\cdot\mathbf{ss}_{21}=\begin{pmatrix}0.9537 & 3.6915\\ 3.6915 & 7.4569\end{pmatrix}\text{m}^2$$

$$\mathbf{R11b}=(\mathbf{AI}_{11}-\mathbf{SSb}\cdot\boldsymbol{\rho}\mathbf{I}_{11})^{-1}=\begin{pmatrix}0.222 & 0.0273\\ 0.0092 & 0.1131\end{pmatrix}\frac{1}{\text{m}^2}$$

$$\mathbf{SSR}=\boldsymbol{\varepsilon}\mathbf{I}_{11}\cdot\mathbf{AI}_{11}\cdot\mathbf{R11b}\cdot\mathbf{SSb}\cdot\boldsymbol{\varepsilon}\mathbf{I}_{11}=\begin{pmatrix}1.1751 & 3.0055\\ 3.0055 & 4.8312\end{pmatrix}\text{m}^2$$

$$\mathbf{FSQ}=\boldsymbol{\varepsilon}\mathbf{I}_{11}\cdot\mathbf{AI}_{11}\cdot\mathbf{R11b}\cdot\mathbf{SS}_{12}\cdot\mathbf{R22}=\begin{pmatrix}0.3739 & 0.3568 & 0.2500\\ 0.6261 & 0.6432 & 0.7500\end{pmatrix}\text{m}^2$$

Check row-sum conservation:

$$(\mathbf{AI}_{11}-\mathbf{SSb})\cdot\begin{pmatrix}1\\1\end{pmatrix}=\begin{pmatrix}0\\0\end{pmatrix}\qquad \boldsymbol{\rho}\mathbf{I}_{11}\cdot\mathbf{R11b}-\mathbf{R11b}^T\cdot\boldsymbol{\rho}\mathbf{I}_{11}=\begin{pmatrix}0 & 0\\ 0 & 0\end{pmatrix}\frac{1}{\text{m}^2}$$

$$(\boldsymbol{\varepsilon}\mathbf{I}_{11}\cdot\mathbf{AI}_{11}-\mathbf{SSR})\cdot\begin{pmatrix}1\\1\\1\\1\\1\end{pmatrix}=\begin{pmatrix}0\\0\\0\\0\\0\end{pmatrix}\qquad \mathbf{FSQ}^T\cdot\begin{pmatrix}1\\1\end{pmatrix}=\begin{pmatrix}1\\1\\1\end{pmatrix}$$

Compute wall fluxes and radiosity for source/sink zones:

$$\mathbf{Q}_1=\boldsymbol{\varepsilon}\mathbf{I}_{11}\cdot\mathbf{AI}_{11}\cdot\mathbf{E}_1-\mathbf{SSR}\cdot\mathbf{E}_1-\mathbf{FSQ}\cdot\mathbf{Q}_2=\begin{pmatrix}-110.88\\ 125.25\end{pmatrix}\text{kW}$$

$$\mathbf{Q}=\begin{pmatrix}\mathbf{Q}_1\\ \mathbf{Q}_2\end{pmatrix}=\begin{bmatrix}\begin{pmatrix}-110.88\\ 125.25\end{pmatrix}\\ \begin{pmatrix}-14.37\\ 0\\ 0\end{pmatrix}\end{bmatrix}\text{kW}\qquad \mathbf{q}=\mathbf{AI}^{-1}\cdot\mathbf{Q}=\begin{pmatrix}-23.87\\ 11.23\\ -1.93\\ 0\\ 0\end{pmatrix}\frac{\text{kW}}{\text{m}^2}$$

$$\mathbf{W}_1=\mathbf{E}_1-\boldsymbol{\rho}\mathbf{I}_{11}\cdot(\boldsymbol{\varepsilon}\mathbf{I}_{11}\cdot\mathbf{AI}_{11})^{-1}\cdot\mathbf{Q}_1=\begin{pmatrix}43.636\\ 74.915\end{pmatrix}\frac{\text{kW}}{\text{m}^2}$$

$$\mathbf{W}_2=\mathbf{R22}(\mathbf{ss}_{21}\cdot\mathbf{W}_1+\mathbf{Q}_2)=\begin{pmatrix}62.694\\ 64\cdot874\\ 68\cdot879\end{pmatrix}\frac{\text{kW}}{\text{m}^2}\qquad \mathbf{W}=\begin{pmatrix}\mathbf{W}_1\\ \mathbf{W}_2\end{pmatrix}=\begin{bmatrix}\begin{pmatrix}43\cdot64\\ 74.91\end{pmatrix}\\ \begin{pmatrix}62\cdot69\\ 64\cdot87\\ 68\cdot88\end{pmatrix}\end{bmatrix}\frac{\text{kW}}{\text{m}^2}$$

$$\mathbf{E}_2=\mathbf{W}_2+\boldsymbol{\rho}\mathbf{I}_{22}\cdot(\boldsymbol{\varepsilon}\mathbf{I}_{22}\cdot\mathbf{AI}_{22})^{-1}.\mathbf{Q}_2=\begin{pmatrix}60\cdot76\\ 64\cdot87\\ 68\cdot88\end{pmatrix}\frac{\text{kW}}{\text{m}^2}\qquad \mathbf{E}=\begin{pmatrix}\mathbf{E}_1\\ \mathbf{E}_2\end{pmatrix}=\begin{bmatrix}\begin{pmatrix}40\cdot98\\ 79.66\end{pmatrix}\\ \begin{pmatrix}60\cdot76\\ 64\cdot87\\ 68\cdot88\end{pmatrix}\end{bmatrix}\frac{\text{kW}}{\text{m}^2}$$

$$\mathbf{H}=\mathbf{AI}^{-1}\cdot\mathbf{ss}\cdot\mathbf{W}\qquad T_C=(\mathbf{E}/\sigma)^{0.25}-273.15=\begin{bmatrix}648.9\\ 815.6\\ 744.3\\ 761.1\\ 776.7\end{bmatrix}{}^\circ\text{C}$$

Summary of computed results:
A 5 percent roof heat loss is consistent with practical measurement errors. Sensitivity testing was also performed with $M = 3$, 4, and 5 with and without heat loss. The classic adiabatic SSR model corresponds to $M = 3$ with no roof heat loss. For $M = 4$, the ends (zone 4) and floor strips (zone 5) were lumped together into one noncontiguous refractory zone. The results are summarized in the following tables. With the exception of the temperature of the floor strips, the computed results for **Q** are seen to be remarkably insensitive to M.

5% ROOF HEAT LOSS
Computed results for **W**, **H**, **Q**, and **E** are wholly independent of refractory emissivity except the roof emissive power, E_3, because $Q_3 \neq 0$.

ZONE	W	H	Q	E
1	43.64	67.50	−23.87	40.98
2	74.91	63.68	11.23	79.66
3	62.69	64.63	−1.93	60.76
4	64.87	64.87	0	64.87
5	68.88	68.88	0	68.88

kW/m² and $\mathbf{Q}=\begin{pmatrix}-110.88\\ 125.25\\ -14.37\\ 0\\ 0\end{pmatrix}$ kW

ADIABATIC ROOF
Computed results for **W**, **H**, **Q**, and **E** are wholly independent of the refractory emissivity.

ZONE	W	H	Q	E
1	43.76	68.79	−25.03	40.98
2	75.26	64.83	10.43	79.66
3	65.20	65.20	0	65.20
4	65.86	65.86	0	65.86
5	69.99	69.99	0	69.99

kW/m² and $\mathbf{Q}=\begin{pmatrix}-116.25\\ 116.25\\ 0\\ 0\\ 0\end{pmatrix}$ kW

Effect of zone numbers:

Heat Flux Computations [kW]
5% Roof Heat Loss

ZONE	$M=3$	$M=4$	$M=5$
1	−112.600	−111.869	−110.876
2	126.974	126.242	125.250
3	−14.374	−14.374	−14.374

Adiabatic Roof

[*SINK* ZI (−117.657) (−117.275) (−116.251)]

Computed Temperatures [°C]

5% Roof Heat Loss

$$\begin{bmatrix} \text{ZONE} & M=3 & M=4 & M=5 \\ \begin{pmatrix} 3 \\ 4 \\ 5 \end{pmatrix} & \begin{pmatrix} 756.2 \\ \text{NA} \\ \text{NA} \end{pmatrix} & \begin{pmatrix} 743.9 \\ 764.5 \\ \text{NA} \end{pmatrix} & \begin{pmatrix} 744.3 \\ 761.1 \\ 776.7 \end{pmatrix} \end{bmatrix}$$

Adiabatic Roof

$$\begin{matrix} ZONE & M=3 & M=4 & M=5 \\ \begin{pmatrix} 3 \\ 4 \\ 5 \end{pmatrix} & \begin{pmatrix} 765.8 \\ \text{NA} \\ \text{NA} \end{pmatrix} & \begin{pmatrix} 762.0 \\ 768.4 \\ \text{NA} \end{pmatrix} & \begin{pmatrix} 762.4 \\ 765.0 \\ 780.9 \end{pmatrix} \end{matrix}$$

Part (c): Auxiliary Calculations for Tube Area and Effective Tube Emissivity
Suppose the heating tubes were totally surrounded by an enclosure at the temperature of the sink and the emissivity of the refractory. Calculate the effective emissivity of the tubes for this idealization. Make reference to Eq. (5-127*a*).

Solution: With $D = 5$ in and $H = 6$ ft, the total surface area of the tubes is calculated as $A_{\text{Tubes}} = 20 \cdot \pi \cdot D \cdot H = 157.1 \text{ ft}^2 = 14.59 \text{ m}^2$. Equation (5-127*a*) may be employed to yield

$$\varepsilon_{\text{Tubes}} = \frac{Q_{\text{Tubes}}}{A_{\text{Tubes}}(E_2 - E_1) - \left[\dfrac{1}{R}\right] \cdot \left[\dfrac{1-\varepsilon_{\text{Enc}}}{\varepsilon_{\text{Enc}}}\right] \cdot Q_{\text{Tubes}}}$$

where $Q_{\text{Tubes}} = Q_2 = 125.25$ km and $R = A_{\text{Enc}}/A_{\text{Tubes}} = 376/157.1 = 2.394$.

For $\varepsilon_{\text{Enc}} = 0.50$ there results $\varepsilon_{\text{Tubes}} = 0.2446$ while for a black surround

$$\varepsilon_{\text{Tubes}} = \frac{Q_2}{A_{\text{Tubes}}(E_2 - E_1)} = 0.2219$$

The insensitivity of Eq. (5-127*a*) for $R > 1$ thus demonstrates its significant engineering utility.

[This example was developed as a MATHCAD 15 worksheet. MATHCAD is a registered trademark of Parametric Technology Corporation.]

Allowance for Specular Reflection If the assumption that all surface zones are diffuse emitters and reflectors is relaxed, the zoning equations become much more complex. Here, all surface parameters become functions of the *angles of incidence and reflection* of the radiation beams at each surface. In practice, such details of reflectance and emission are seldom known. When they are, the Monte Carlo method of tracing a large number of beams emitted from random positions and in random initial directions is probably the best way of obtaining a solution. Howell, Mengüç, and Siegel, *Thermal Radiative Heat Transfer*, 6th ed., CRC Press, Boca Raton, Fla., 2015, chap. 7) and Modest (*Radiative Heat Transfer,* 3d ed., Academic Press, New York, 2013, chap. 21) review the utilization of the Monte Carlo approach to a variety of radiant transfer applications. Among these is the Monte Carlo calculation of direct exchange areas for very complex geometries. Monte Carlo techniques are generally *not used* in practice for simpler engineering applications.

A simple *engineering* approach to specular reflection is the *diffuse plus specular reflection model.* Here the total reflectivity $\rho_i = 1 - \varepsilon_i = \rho_{Si} + \rho_{Di}$ is represented as the sum of a *diffuse component* ρ_{Si} and a *specular component* ρ_{Di}. The method yields analytical results for a number of two-surface zone geometries. In particular, the following equation is obtained for exchange between concentric spheres or infinitely long coaxial cylinders for which $A_1 < A_2$:

$$\overline{S_1S_2} = \frac{A_1}{\dfrac{1}{\varepsilon_1} + \dfrac{\rho_2}{\varepsilon_2} \cdot \dfrac{A_1}{A_2} + \dfrac{\rho_{S2}}{(1-\rho_{S2})}(1 - A_1/A_2)} \tag{5-134}$$

For $\rho_{D1} = \rho_{D2} = 0$ (or equivalently $\rho_1 = \rho_{S1}$ with $\rho_2 = \rho_{S2}$), Eq. (5-134) yields the limiting case for wholly specular reflection

Specular limit
$$\overline{S_1S_2} = \frac{A_1}{\dfrac{1}{\varepsilon_1} + \dfrac{1}{\varepsilon_2} - 1} \tag{5-134a}$$

which is independent of the area ratio A_1/A_2. It is important to notice that Eq. (5-134*a*) is *similar* to Eq. (5-127*b*), but the emissivities here are defined as $\varepsilon_1 \equiv 1 - \rho_{s1}$ and $\varepsilon_2 = 1 - \rho_{s2}$. When surface reflection is wholly diffuse [$\rho_{S1} = \rho_{S2} = 0$ or $\rho_1 = \rho_{D1}$ with $\rho_2 = \rho_{D2}$], Eq. (5-134) results in a formula *identical* to Eq. (5-127*a*),

Diffuse limit
$$\overline{S_1S_2} = \frac{A_1}{\left[\dfrac{1}{\varepsilon_1} + \left(\dfrac{A_1}{A_2}\right)\dfrac{\rho_2}{\varepsilon_2}\right]} \tag{5-134b}$$

For the case of (infinite) parallel flat plates where $A_1 = A_2$, Eq. (5-134) leads to a general formula *similar* to Eq. (5-134*a*) but with the stipulation here that $\varepsilon_1 \equiv 1 - \rho_{D1} - \rho_{S1}$ and $\varepsilon_2 \equiv 1 - \rho_{D2} - \rho_{S2}$.

Another particularly interesting limit of Eq. (5-134) occurs when $A_2 >> A_1$, which might represent a small sphere irradiated by infinite surroundings which can reflect radiation originating at A_1 back to A_1. That is, even though $A_2 \to \infty$, the "self" total exchange area does not necessarily vanish, to wit

$$\overline{S_1S_2} = \frac{\varepsilon_1^2 \cdot \rho_{s2} \cdot A_1}{1 - \rho_1 \cdot \rho_{s2}} \quad \text{and} \quad \overline{S_1S_2} = \frac{\varepsilon_1(1-\rho_{s2}) \cdot A_1}{1 - \rho_1 \cdot \rho_{s2}} \tag{5-134c,d}$$

which again exhibits diffuse and specular limits. The diffuse plus specular reflection model becomes significantly more complex for geometries with $M \geq 3$ where digital computation is usually required.

An Exact Solution to the Integral Equations—The Hohlraum Exact solutions of the fundamental *integral equations* for radiative transfer are available for only a few simple cases. One of these is the evaluation of the emittance from a small aperture, of area A_1, in the surface of an isothermal spherical cavity of radius R. In German, this geometry is termed a *hohlraum* for hollow space. For this special case the radiosity W is constant over the inner surface of the cavity. It then follows that the ratio W/E is given by

$$W/E = \frac{\varepsilon}{1 - \rho \cdot [1 - A_1/(4\pi R^2)]} \tag{5-135}$$

where ε and $\rho = 1 - \varepsilon$ are the diffuse emissivity and reflectivity of the interior cavity surface, respectively. The ratio W/E is the effective emittance of the aperture as sensed by an external narrow-angle receiver (radiometer) viewing the cavity interior. Assume that the cavity is constructed of a rough material whose (diffuse) emissivity is $\varepsilon = 0.5$. As a point of reference, if the cavity is to simulate a blackbody emitter to better than 98 percent of an ideal theoretical blackbody, Eq. (5-135) then predicts that the ratio of the aperture to sphere areas $A_1/(4\pi R^2)$ must be less than 2 percent. Equation (5-135) has practical utility in the experimental design of calibration standards for laboratory radiometers.

RADIATION FROM GASES AND SUSPENDED PARTICULATE MATTER

Introduction Flame radiation originates as a result of emission from water vapor and carbon dioxide in the hot gaseous combustion products and from the presence of particulate matter. The latter includes emission both from burning of microscopic and submicroscopic soot particles and from large suspended particles of coal, coke, or ash. Thermal radiation owing to the presence of water vapor and carbon dioxide is not visible. The characteristic blue color of clean natural gas flames is due to chemiluminescence of the excited intermediates in the flame which contribute negligibly to the radiation from combustion products.

Gas Emissivities Radiant transfer in a gaseous medium is characterized by three quantities: the gas emissivity, gas absorptivity, and gas transmissivity. *Gas emissivity* refers to radiation originating within a gas volume that is incident on some reference surface. *Gas absorptivity* and *gas transmissivity,* however, refer to the absorption and transmission of radiation from some *external* surface radiation source characterized by some radiation temperature T_1. The sum of the gas absorptivity and transmissivity must, by definition, be unity. Gas absorptivity may be calculated from an appropriate gas emissivity. The gas emissivity is a function of only the gas temperature T_g while the absorptivity and transmissivity are functions of *both* T_g and T_1.

The *standard hemispherical monochromatic gas emissivity* is defined as the direct volume-to-surface exchange area for a hemispherical gas volume to an infinitesimal area element located at the center of the planar base. Consider monochromatic transfer in a *black* hemispherical enclosure of radius R that confines an isothermal volume of gas at temperature T_g. The temperature of the bounding surfaces is T_1. Let A_2 denote the area of the finite hemispherical surface and dA_1 denote an infinitesimal element of area located at the center of the planar base. The (dimensionless) monochromatic direct exchange area for exchange between the finite hemispherical surface A_2 and dA_1 then follows from direct integration of Eq. (5-116*a*) as

$$\frac{\partial(\overline{s_1 s_2})_\lambda}{\partial A_1} = \int_{\varphi_1=0}^{\pi/2} \frac{e^{-K_\lambda R}}{\pi R^2} \cos\varphi_1\, 2\pi R^2 \sin\varphi_1\, d\varphi_1 = e^{-K_\lambda R} \qquad (5\text{-}136a)$$

and from conservation there results

$$\frac{\partial(\overline{s_1 g})_\lambda}{\partial A_1} = 1 - e^{-K_\lambda R} \qquad (5\text{-}136b)$$

Note that Eq. (5-136*b*) is identical to the expression for the gas emissivity for a column of path length R. In Eqs. (5-136) the gas absorption coefficient is a function of gas temperature, composition, and wavelength, that is, $K_\lambda = K_\lambda(T,\lambda)$. The net monochromatic radiant flux density at dA_1 due to irradiation from the gas volume is then given by

$$q_{1g,\lambda} = \frac{\partial(\overline{s_1 g})_\lambda}{\partial A_1}(E_{1\lambda} - E_{g\lambda}) \equiv \alpha_{g,1\lambda} E_{1\lambda} - \varepsilon_{g\lambda} E_{g\lambda} \qquad (5\text{-}137)$$

In Eq. (5-137), $\varepsilon_{g,\lambda}(T,\lambda) = 1 - \exp(-K_\lambda R)$ is defined as the monochromatic or *spectral gas emissivity* and $\alpha_{g,\lambda}(T,\lambda) = \varepsilon_{g,\lambda}(T,\lambda)$.

If Eq. (5-137) is integrated with respect to wavelength over the entire EM spectrum, an expression for the *total* flux density is obtained

$$q_{1,g} = \alpha_{g,1} E_1 - \varepsilon_g E_g \qquad (5\text{-}138)$$

where

$$\varepsilon_g(T_g) = \int_{\lambda=0}^{\infty} \varepsilon_\lambda(T_{g,\lambda}) \cdot \frac{E_{b\lambda}(T_{g,\lambda})}{E_b(T_g)} d\lambda \qquad (5\text{-}138a)$$

and

$$\alpha_{g,1}(T_1, T_g) = \int_{\lambda=0}^{\infty} \alpha_{g,\lambda}(T_g, \lambda) \cdot \frac{E_{b,\lambda}(T_1, \lambda)}{E_b(T_1)} d\lambda \qquad (5\text{-}138b)$$

define the *total* gas emissivity and absorptivity, respectively. The notation used here is analogous to that used for surface emissivity and absorptivity as previously defined. For a *real* gas $\varepsilon_g = \alpha_{g,1}$ only if $T_1 = T_g$, while for a *gray* gas mass of arbitrarily shaped volume $\varepsilon_g = \alpha_{g,1} = \partial(\overline{s_1 g})/\partial A_1$ is independent of temperature. Because $K_\lambda(T,\lambda)$ is also a function of the composition of the radiating species, it is necessary in what follows to define a second absorption coefficient $k_{p,\lambda}$, where $K_\lambda = k_{p,\lambda} p$. Here p is the partial pressure of the radiating species, and $k_{p,\lambda}$, with units of $(\text{atm}\cdot\text{m})^{-1}$, is referred to as the *monochromatic line absorption coefficient*.

Mean Beam Lengths It is always possible to represent the emissivity of an arbitrarily shaped volume of gray gas (and thus the corresponding direct gas-to-surface exchange area) with an equivalent sphere of radius $R = L_M$. In this context the hemispherical radius $R = L_M$ is referred to as the *mean beam length* of the arbitrary gas volume. Consider, e.g., an isothermal gas layer at temperature T_g confined by two infinite parallel plates separated by distance L. Direct integration of Eq. (5-116*a*) and use of conservation yield a closed-form expression for the requisite surface-gas direct exchange area

$$\frac{\partial(\overline{s_1 g})}{\partial A_1} = [1 - 2E_3(KL)] \qquad (5\text{-}139a)$$

where $E_n(z) = \int_{t=1}^{\infty} \frac{e^{-zt}}{t^n} dt$ is defined as the *nth-order exponential integral* which is readily available. Employing the definition of gas emissivity, the *mean beam length* between the plates L_M is then defined by the expression

$$\varepsilon_g = [1 - 2\cdot E_3(KL)] \equiv 1 - e^{-KL_M} \qquad (5\text{-}139b)$$

Solution of Eq. (5-139*b*) yields $KL_M = -\ln[2E_3(KL)]$, and it is apparent that KL_M is a function of KL. Since $E_n(0) = 1/(n-1)$ for $n > 1$, the mean beam length approximation also correctly predicts the gas emissivity as zero when $K = 0$ and $K \to \infty$.

In the limit $K \to 0$, power series expansion of both sides of Eq. (5-139*b*) leads to $KL_M \to 2KL \equiv KL_{M0}$, where $L_M \equiv L_{M0} = 2L$. Here L_{M0} is defined as the *optically thin mean beam length* for radiant transfer from the entire *infinite* planar gas layer to a differential element of surface area on one of the plates. The optically thin mean beam length for two infinite parallel plates is thus simply *twice* the plate spacing L. In a similar manner it may be shown that for a sphere of diameter D, $L_{M0} = \frac{2}{3}D$, and for an infinitely long cylinder $L_{M0} = D$. A useful default formula for an arbitrary enclosure of volume V and area A is given by $L_{M0} = 4V/A$. This expression predicts $L_{M0} = \frac{8}{3}R$ for the standard hemisphere of radius R because the optically thin mean beam length is averaged over the *entire* hemispherical enclosure.

Use of the optically thin value of the mean beam length yields values of gas emissivities or exchange areas that are too high. It is thus necessary to introduce a dimensionless constant $\beta \le 1$ and define some new *average mean beam length* such that $KL_M \equiv \beta \cdot KL_{M0}$. For the case of parallel plates, we now require that the mean beam length exactly predict the gas emissivity for a *third* value of KL. In this example we find $\beta = -\ln[2E_3(KL)]/2KL$ and for $KL = 0.193095$ there results $\beta = 0.880$. The value $\beta = 0.880$ is not wholly arbitrary. It also happens to minimize the error defined by the so-called shape correction factor $\phi = [\partial(\overline{s_i g})/\partial A_1]/(1 - e^{-KL_M})$ for all $KL > 0$. The required *average* mean beam length for all $KL > 0$ is then taken simply as $L_M = 0.88L_{M0} = 1.76L$. The error in this approximation is less than 5 percent.

For an arbitrary geometry, the *average mean beam length* is defined as the radius of a hemisphere of gas which predicts values of the direct exchange area $\overline{s_1 g}/A_1 = [1 - \exp(-KL_M)]$, subject to the optimization condition indicated above. It has been found that the error introduced by using average mean beam lengths to approximate direct exchange areas is sufficiently small to be appropriate for many engineering calculations. When it is evaluated for a large number of geometries, it is found that $0.8 < \beta < 0.95$. It is recommended here that $\beta = 0.88$ be employed in lieu of any further geometric information. For a single-gas zone, all the requisite direct exchange areas can be approximated for engineering purposes in terms of a *single* appropriately defined average mean beam length.

Emissivities of Combustion Products Absorption or emission of radiation by the constituents of gaseous combustion products is determined primarily by vibrational and rotational transitions between the energy levels of the gaseous molecules. Changes in both vibrational and rotational energy states give rise to *discrete* spectral lines. Rotational lines accompanying vibrational transitions usually overlap, forming a so-called vibration-rotation band. These bands are thus associated with the major vibrational frequencies of the molecules. Each spectral line is characterized by an absorption coefficient $k_{p,\lambda}$ which exhibits a maximum at some central characteristic wavelength or wave number $\eta_0 = 1/\lambda_0$ and is described by a Lorentz* probability distribution. Since the widths of spectral lines are dependent on collisions with other molecules, the absorption coefficient will also depend upon the composition of the combustion gases and the total system pressure. This brief discussion of gas spectroscopy is intended as an introduction to the factors controlling absorption coefficients and thus the factors which govern the empirical correlations to be presented for gas emissivities and absorptivities.

Figure 5-19 shows computed values of the spectral emissivity $\varepsilon_{g,\lambda} \equiv \varepsilon_{g,\lambda}(T, pL, \lambda)$ as a function of wavelength for an equimolar mixture of carbon dioxide and water vapor for a gas temperature of 1500 K, partial pressure of 0.18 atm, and a path length $L = 2$ m. Three principal absorption-emission bands for CO_2 are seen to be centered on 2.7, 4.3, and 15 μm. Two weaker bands at 2 and 9.7 μm are also evident. Three principal absorption-emission bands for water vapor are also identified near 2.7, 6.6, and 20 μm with lesser bands at 1.17, 1.36, and 1.87 μm. The *total* emissivity ε_g and total absorptivity $\alpha_{g,1}$ are calculated by integration with respect to the wavelength of the spectral emissivities, using Eqs. (5-138) in a manner similar to the development of total surface properties.

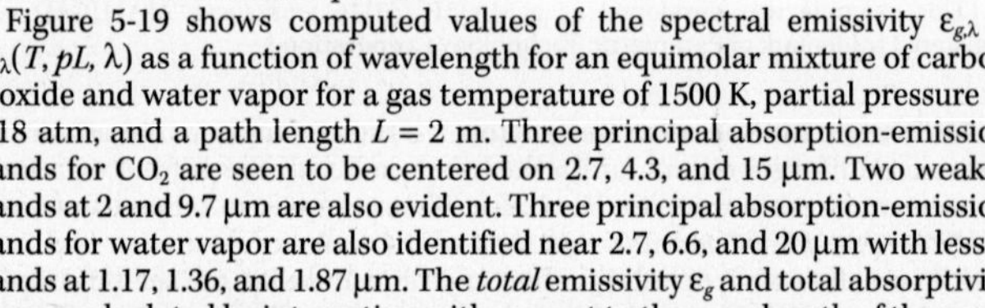

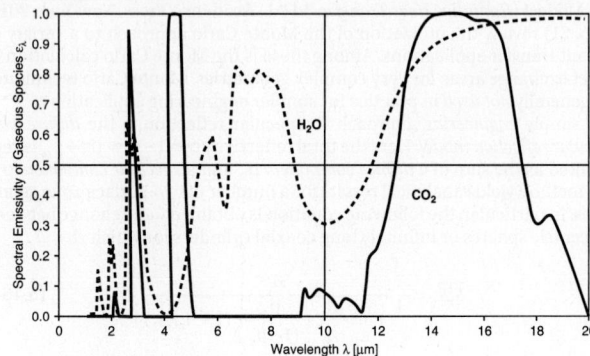

FIG. 5-19 Spectral emittances for carbon dioxide and water vapor after RADCAL. $p_c L = p_w L = 0.36$ atm · m, $T_g = 1500$ K.

*Spectral lines are conventionally described in terms of wave number $\eta = 1/\lambda$, with each line having a peak absorption at wave number η_0. The *Lorentz distribution* is defined as $k_\eta / S = \frac{b_c}{\pi[b_c^2 + (\eta - \eta_0)^2]}$ where S is the integral of k_η over all wave numbers. The parameter S is known as the integrated line intensity, and b_c is defined as the collision line half-width, i.e., the half-width of the line is one-half of its peak centerline value. The units of k_η are $\text{m}^{-1}\,\text{atm}^{-1}$.

Spectral Emissivities Highly resolved spectral emissivities can be generated at ambient temperatures from the HITRAN database (*high*-resolution *tran*smission molecular absorption) that has been developed for atmospheric models [Rothman, L. S., K. Chance, and A. Goldman, eds., *J. Quant. Spectroscopy & Radiative Trans.* **82**(1–4): 2003]. This database includes the chemical species H_2O, CO_2, O_3, N_2O, CO, CH_4, O_2, NO, SO_2, NO_2, NH_3, HNO_3, OH, HF, HCl, HBr, ClO, OCS, H_2CO, HOCl, N_2, HCN, CH_3C, HCl, H_2O_2, C_2H_2, C_2H_6, PH_3, COF_2, SF_6, H_2S, and HCO_2H. These data have been extended to high temperature for CO_2 and H_2O, allowing for the changes in the population of different energy levels and in the line half width [Denison, M. K., and B. W. Webb, *Heat Transfer*, **2**: 19–24 (1994)]. The resolution in the single-line models of emissivities is far greater than that needed in engineering calculations. A number of models are available that average the emissivities over narrow-wavelength regimes or over the entire band. An extensive set of measurements of narrowband parameters performed at NASA (Ludwig, C., et al., *Handbook of Infrared Radiation from Combustion Gases*, NASA SP-3080, 1973) has been used to develop the RADCAL computer code to obtain spectral emissivities for CO_2, H_2O, CH_4, CO, and soot (Grosshandler, W. L., "RADCAL," NIST Technical Note 1402, 1993). The exponential wideband model is available for emissions averaged over a band for H_2O, CO_2, CO, CH_4, NO, SO_2, N_2O, NH_3, and C_2H_2 [Edwards, D. K., and Menard, W. A., *Appl. Optics*, **3**: 621–625 (1964)]. The line and band models have the advantages of being able to account for complexities in determining emissivities of line broadening due to changes in composition and pressure, exchange with spectrally selective walls, and greater accuracy in formulating fluxes in gases with temperature gradients. These models can be used to generate the total emissivities and absorptivies that will be used in this section. RADCAL is a command-line FORTRAN code which is available in the public domain on the Internet.

Total Emissivities and Absorptivities Total emissivities and absorptivities for water vapor and carbon dioxide at present are still based on data embodied in the classical *Hottel emissivity charts.* These data have been adjusted with the more recent measurements in RADCAL and used to develop the correlations of emissivities given in Table 5-6. Two empirical correlations which permit hand calculation of emissivities for water vapor, carbon dioxide, and *four* mixtures of the two gases are presented in Table 5-6. The first section of Table 5-6 provides data for the two constants b and n in the empirical relation

$$\overline{\varepsilon_g T_g} = b[pL - 0.015]^n \qquad (5\text{-}140a)$$

while the second section of Table 5-6 utilizes the *four* constants in the empirical correlation

$$\log(\overline{\varepsilon_g T_g}) = a_0 + a_1 \log(pL) + a_2 \log^2(pL) + a_3 \log^3(pL) \qquad (5\text{-}140b)$$

In both cases the empirical constants are given for the three temperatures of 1000, 1500, and 2000 K. Table 5-6 also includes six values for the partial pressure ratios p_W/p_C of water vapor to carbon dioxide, namely, 0, 0.5, 1.0, 2.0, 3.0, and ∞. These ratios correspond to composition values of $p_C/(p_C + p_W) = 1/(1 + p_W/p_C)$ of 0, 1/3, 1/2, 2/3, 3/4, and unity. For emissivity calculations at other temperatures and mixture compositions, linear interpolation of the constants is recommended.

TABLE 5-6 Emissivity-Temperature Product for CO_2-H_2O Mixtures, $\overline{\varepsilon_g T_g}$

Limited range for furnaces, valid over 25-fold range of $p_{w+c}L$, 0.046–1.15 m · atm (0.15–3.75 ft · atm)

p_w/p_c	0	½	1	2	3	∞
$\frac{p_w}{p_w + p_c}$	0	⅓(0.3–0.42)	½(0.42–0.5)	⅔(0.6–0.7)	¾(0.7–0.8)	1
	CO_2 only	Corresponding to $(CH)_x$, covering coal, heavy oils, pitch	Corresponding to $(CH_2)_x$, covering distillate oils, paraffins, olefines	Corresponding to CH_4, covering natural gas and refinery gas	Corresponding to $(CH_6)_x$, covering future high H_2 fuels	H_2O only

Section 1

Constants b and n of $\overline{\varepsilon_g T_g} = b(pL - 0.015)^n$, pL = m · atm, T = K

T, K	b	n	b	n	b	n	b	n	b	n	b	n
1000	188	0.209	384	0.33	416	0.34	444	0.34	455	0.35	416	0.400
1500	252	0.256	448	0.38	495	0.40	540	0.42	548	0.42	548	0.523
2000	267	0.316	451	0.45	509	0.48	572	0.51	594	0.52	632	0.640

Constants b and n of $\overline{\varepsilon_g T_g} = b(pL - 0.05)^n$, pL = ft · atm, T = °R

T, °R	b	n	b	n	b	n	b	n	b	n	b	n
1800	264	0.209	467	0.33	501	0.34	534	0.34	541	0.35	466	0.400
2700	335	0.256	514	0.38	555	0.40	591	0.42	600	0.42	530	0.523
3600	330	0.316	476	0.45	519	0.48	563	0.51	577	0.52	532	0.640

Section 2

Full range, valid over 2000-fold range of $p_{w+c}L$, 0.005–10.0 m · atm (0.016–32.0 ft · atm)
Constants of $\log_{10} \overline{\varepsilon_g T_g} = a_0 + a_1 \log pL + a_2 \log^2 pL + a_3 \log^3 pL$

		pL = m · atm, T = K					pL = ft · atm, T = °R				
$\frac{p_w}{p_c}$	$\frac{p_w}{p_w + p_c}$	T, K	a_0	a_1	a_2	a_3	T, °R	a_0	a_1	a_2	a_3
		1000	2.2661	0.1742	−0.0390	0.0040	1800	2.4206	0.2176	−0.0452	0.0040
0	0	1500	2.3954	0.2203	−0.0433	0.00562	2700	2.5248	0.2695	−0.0521	0.00562
		2000	2.4104	0.2602	−0.0651	−0.00155	3600	2.5143	0.3621	−0.0627	−0.00155
		1000	2.5754	0.2792	−0.0648	0.0017	1800	2.6691	0.3474	−0.0674	0.0017
½	⅓	1500	2.6451	0.3418	−0.0685	−0.0043	2700	2.7074	0.4091	−0.0618	−0.0043
		2000	2.6504	0.4279	−0.0674	−0.0120	3600	2.6686	0.4879	−0.0489	−0.0120
		1000	2.6090	0.2799	−0.0745	−0.0006	1800	2.7001	0.3563	−0.0736	−0.0006
1	½	1500	2.6862	0.3450	−0.0816	−0.0039	2700	2.7423	0.4561	−0.0756	−0.0039
		2000	2.7029	0.4440	−0.0859	−0.0135	3600	2.7081	0.5210	−0.0650	−0.0135
		1000	2.6367	0.2723	−0.0804	0.0030	1800	2.7296	0.3577	−0.0850	0.0030
2	⅔	1500	2.7178	0.3386	−0.0990	−0.0030	2700	2.7724	0.4384	−0.0944	−0.0030
		2000	2.7482	0.4464	−0.1086	−0.0139	3600	2.7461	0.5474	−0.0871	−0.0139
		1000	2.6432	0.2715	−0.0816	0.0052	1800	2.7359	0.3599	−0.0896	0.0052
3	¾	1500	2.7257	0.3355	−0.0981	0.0045	2700	2.7811	0.4403	−0.1051	0.0045
		2000	2.7592	0.4372	−0.1122	−0.0065	3600	2.7599	0.5478	−0.1021	−0.0065
		1000	2.5995	0.3015	−0.0961	0.0119	1800	2.6720	0.4102	−0.1145	0.0119
∞	1	1500	2.7083	0.3969	−0.1309	0.00123	2700	2.7238	0.5330	−0.1328	0.00123
		2000	2.7709	0.5099	−0.1646	−0.0165	3600	2.7215	0.6666	−0.1391	−0.0165

NOTE: $p_w/(p_w + p_c)$ of ⅓, ½, ⅔, and ¾ may be used to cover the ranges 0.2–0.4, 0.4–0.6, 0.6–0.7, and 0.7–0.8, respectively, with a maximum error in ε_g of 5 percent at pL = 6.5 m · atm, less at lower pL's. Linear interpolation reduces the error generally to less than 1 percent. Linear interpolation or extrapolation on T introduces an error generally below 2 percent, less than the accuracy of the original data.

The absorptivity can be obtained from the emissivity with aid of Table 5-6 by using the following functional equivalence.

$$\overline{\alpha_{g,1}T_1} = \left[\overline{\varepsilon_g T_1(pL\cdot T_1/T_g)}\right]\left(\frac{T_g}{T_1}\right)^{0.5} \qquad (5\text{-}141)$$

Verbally, the absorptivity computed from Eq. (5-141) by using the correlations in Table 5-6 is based on a value for gas emissivity ε_g calculated at a temperature T_1 and at a partial-pressure path length product of $(p_C + p_W)LT_1/T_g$. The absorptivity is then equal to this value of gas emissivity multiplied by $(T_g/T_1)^{0.5}$. It is recommended that spectrally based models such as RADCAL be used particularly when extrapolating beyond the temperature, pressure, or partial-pressure-length product ranges presented in Table 5-6.

A comparison of the results of the predictions of Table 5-6 with values obtained via the integration of the spectral results calculated from the narrowband model in RADCAL is provided in Fig. 5-20. Here calculations are shown for $p_C L = p_W L = 0.12$ atm·m and a gas temperature of 1500 K. The RADCAL predictions are 20 percent higher than the measurements at low values of pL and are 5 percent higher at the large values of pL. An extensive comparison of different sources of emissivity data shows that disparities up to 20 percent are to be expected at the current time [Lallemant, N., Sayre, A., and Weber, R., *Prog. Energy Combust. Sci.* **22:** 543–574 (1996)]. However, smaller errors result for the range of the total emissivity measurements presented in the Hottel emissivity tables. This is demonstrated in Example 5-12.

Example 5-12 Calculations of Gas Emissivity and Absorptivity Consider a slab of gas confined between two infinite parallel plates with a distance of separation of $L = 1$ m. The gas pressure is 101.325 kPa (1 atm), and the gas temperature is 1500 K (2240°F). The gas is an equimolar mixture of CO_2 and H_2O, each with a partial pressure of 12 kPa ($p_C = p_W = 0.12$ atm). The radiative flux to one of its bounding surfaces has been calculated by using RADCAL for two cases. For case (*a*) the flux to the bounding surface is 68.3 kW/m² when the emitting gas is backed by a black surface at an ambient temperature of 300 K (80°F). This (cold) back surface contributes less than 1 percent to the flux. In case (*b*), the flux is calculated as 106.2 kW/m² when the gas is backed by a black surface at a temperature of 1000 K (1340°F). In this example, gas emissivity and absorptivity are to be computed from these flux values and compared with values obtained by using Table 5-6.

Case (a): The flux incident on the surface is equal to $\varepsilon_g \cdot \sigma \cdot T_g^4 = 68.3$ kW/m²; therefore, $\varepsilon_g = 68{,}300/(5.6704 \times 10^{-8} \cdot 1500^4) = 0.238$. To utilize Table 5-6, the mean beam length for the gas is calculated from the relation $L_M = 0.88L_{M0} = 0.88 \cdot 2L = 1.76$ m. For $T_g = 1500$ K and $(p_C + p_W)L_M = 0.24(1.76) = 0.422$ atm·m, the two-constant correlation in Table 5-6 yields $\varepsilon_g = 0.230$ and the four-constant correlation yields $\varepsilon_g = 0.234$. These results are clearly in excellent agreement with the predicted value of $\varepsilon_g = 0.238$ obtained from RADCAL.

Case (b): The flux incident on the surface (106.2 kW/m²) is the sum of that contributed by (1) gas emission $\varepsilon_g \cdot \sigma \cdot T_g^4 = 68.3$ kW/m² and (2) emission from the opposing surface corrected for absorption by the intervening gas using the gas transmissivity, that is, $\tau_{g,1}\sigma \cdot T_1^4$ where $\tau_{g,1} = 1 - \alpha_{g,1}$. Therefore $\alpha_{g,1} = [1 - (106{,}200 - 68{,}300)/(5.6704 \times 10^{-8} \cdot 1000^4)] = 0.332$. Using Table 5-6, the two-constant and four-constant gas emissivities evaluated at $T_1 = 1000$ K and $pL = 0.4224 \times (1000/1500) = 0.282$ atm·m are $\varepsilon_g = 0.2654$ and $\varepsilon_g = 0.2707$, respectively. Multiplication by the factor $(T_g/T_1)^{0.5} = (1500/1000)^{0.5} = 1.225$ produces the final values of the two corresponding gas absorptivities $\alpha_{g,1} = 0.325$ and $\alpha_{g,1} = 0.332$, respectively. Again the agreement with RADCAL is excellent.

Other Gases The most extensive available data for gas emissivity are those for carbon dioxide and water vapor because of their importance in the radiation from the products of fossil fuel combustion. Selected data for other species present in combustion gases are provided in Table 5-7.

Flames and Particle Clouds

Luminous Flames Luminosity conventionally refers to soot radiation. At atmospheric pressure, soot is formed in locally fuel-rich portions of flames in amounts that usually correspond to less than 1 percent of the carbon in the fuel. Because soot particles are small relative to the wavelength of the radiation of interest in flames (primary particle diameters of soot are of the order of 20 nm compared to wavelengths of interest of 500 to 8000 nm), the incident radiation permeates the particles, and the absorption is proportional to the volume of the particles. In the limit of $r_p/\lambda << 1$, the Rayleigh limit, the monochromatic emissivity ε_λ is given by

$$\varepsilon_\lambda = 1 - \exp(-K \cdot f_\upsilon \cdot L/\lambda) \qquad (5\text{-}142)$$

where f_υ is the volumetric soot concentration, L is the path length in the same units as the wavelength λ, and K is dimensionless. The value K will vary with fuel type, experimental conditions, and temperature history of the soot. The values of K for a wide range of systems are within a factor of about 2 of one another. The single most important variable governing the value of K is the hydrogen/carbon ratio of the soot, and the value of K increases as the H/C ratio decreases. A value of $K = 9.9$ is recommended on the basis of seven studies involving 29 fuels [Mulholland, G. W., and Croarkin, C., *Fire and Materials*, **24:** 227–230 (2000)].

The total emissivity of soot ε_S can be obtained by substituting ε_λ from Eq. (5-142) for ε_λ in Eq. (5-138*a*) to yield

$$\varepsilon_S = \int_{\lambda=0}^{\infty} \varepsilon_\lambda \frac{E_{b\lambda}(T_g,\lambda)}{E_b(T_g)} d\lambda = 1 - \frac{15}{4}[\Psi^{(3)}(1 + K \cdot f_\upsilon \cdot L \cdot T/c_2)] \cong (1 + K \cdot f_\upsilon \cdot L \cdot T/c_2)^{-4} \qquad (5\text{-}143)$$

Here $\Psi^{(3)}(x)$ is defined as the pentagamma function of x, and c_2 (m·K) is again Planck's second constant. The approximate relation in Eq. (5-143) is

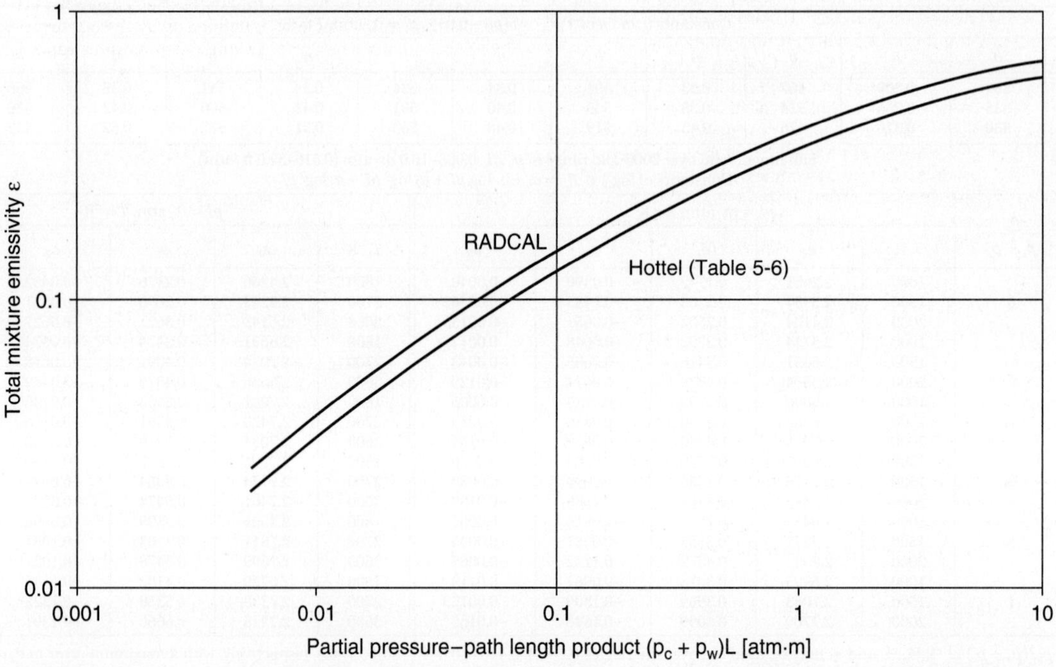

FIG. 5-20 Comparison of Hottel and RADCAL total gas emissivities. Equimolal gas mixture of CO_2 and H_2O with $p_c = p_w = 0.12$ atm and $T_g = 1500$ K.

TABLE 5-7 Total Emissivities of Some Gases

Temperature	1000°R			1600°R			2200°R			2800°R		
P_xL, atm · ft	0.01	0.1	1.0	0.01	0.1	1.0	0.01	0.1	1.0	0.01	0.1	1.0
NH_3^a	0.047	0.20	0.61	0.020	0.120	0.44	0.0057	0.051	0.25	(0.001)	(0.015)	(0.14)
SO_2^b	0.020	0.13	0.28	0.013	0.090	0.32	0.0085	0.051	0.27	0.0058	0.043	0.20
CH_4^c	0.0116	0.0518	0.1296	0.0111	0.0615	0.1880	0.0087	0.0608	0.2004	0.00622	0.04702	0.1525
CO^d	0.0052	0.0167	0.0403	0.0055	0.0196	0.0517	0.0036	0.0145	0.0418	0.00224	0.00986	0.02855
NO^d	0.0046	0.018	0.060	0.0046	0.021	0.070	0.0019	0.010	0.040	0.0078	0.004	0.025
HCl^e	0.00022	0.00079	0.0020	0.00036	0.0013	0.0033	0.00037	0.0014	0.0036	0.00029	0.0010	0.0027

NOTE: Figures in this table are taken from plots in Hottel and Sarofim, *Radiative Transfer,* McGraw-Hill, New York, 1967, chap. 6. Values in parentheses are extrapolated. To convert degrees Rankine to kelvins, multiply by $(5.556)(10^{-1})$. To convert atmosphere-feet to kilopascal-meters, multiply by 30.89.

[a]Total-radiation measurements of Port (Sc.D. thesis in chemical engineering, MIT, 1940) at 1-atm total pressure, $L = 1.68$ ft, T to 2000°R.

[b]Calculations of Guerrieri (S.M. thesis in chemical engineering, MIT, 1932) from room-temperature absorption measurements of Coblentz (*Investigations of Infrared Spectra,* Carnegie Institution, Washington, 1905) with poor allowance for temperature.

[c]Estimated using Grosshandler, W.L., "RADCAL: A Narrow-Band Model for Radial Calculations in a Combustion Environment," NIST Technical Note 1402, 1993.

[d]Calculations of Malkmus and Thompson [*J. Quant. Spectros. Radiat. Transfer,* **2:** 16 (1962)], to $T = 5400$°R and $PL = 30$ atm · ft.

[e]Calculations of Malkmus and Thompson [*J. Quant. Spectros. Radiat. Transfer,* **2:** 16 (1962)], to $T = 5400$°R and $PL = 300$ atm · ft.

accurate to better than 1 percent for arguments yielding values of $\varepsilon_S < 0.7$. At present, the largest uncertainty in estimating total soot emissivities lies in the estimation of the soot volume fraction f_υ. Soot forms in the fuel-rich zones of flames. Soot formation rates are a function of fuel type, mixing rate, local equivalence ratio Φ, temperature, and pressure. The *equivalence ratio* is defined as the quotient of the actual to stoichiometric fuel-to-oxidant ratio $\Phi = [F/O]_{act}/[F/O]_{stoich}$. Soot formation increases with the aromaticity or C/H ratio of fuels with benzene, α-methyl naphthalene, and acetylene having a high propensity to form soot and methane having a low soot formation propensity. Oxygenated fuels, such as alcohols, emit little soot. In practical turbulent diffusion flames, soot forms on the fuel side of the flame front. In premixed flames, at a given temperature, the rate of soot formation increases rapidly for $\Phi > 2$. For temperatures above 1500 K, soot burns out rapidly (in less than 0.1 s) under fuel-lean conditions, $\Phi < 1$. Because of this rapid soot burnout, soot is usually localized in a relatively small fraction of a furnace or combustor volume. Long, poorly mixed diffusion flames promote soot formation while highly back-mixed combustors can burn soot-free. In a typical flame at atmospheric pressure, maximum volumetric soot concentrations are found to be in the range of $10^{-7} < f_\upsilon < 10^{-6}$. This corresponds to a soot formation of about 1.5 to 15 percent of the carbon in the fuel. When f_υ is to be calculated at high pressures, allowance must be made for the significant increase in soot formation with pressure and for the inverse proportionality of f_υ with respect to pressure. Great progress is being made in the ability to calculate soot in premixed flames. For example, predicted and measured soot concentration has been compared in a well-stirred reactor operated over a wide range of temperatures and equivalence ratios [Brown, N. J., Revzan, K. L., and Frenklach, M., *Twenty-seventh Symposium (International) on Combustion,* pp. 1573–1580, 1998]. Moreover, computational fluid dynamics (CFD) and population dynamics modeling have been used to simulate soot formation in a turbulent non-premixed ethylene-air flame [Zucca et al., *Chem. Eng. Sci.* **61:** 87–95 (2006)]. The importance of soot radiation varies widely between combustors. In large boilers the soot is confined to small volumes and is of only local importance. In gas turbines, cooling the combustor liner is of primary importance so that only small incremental soot radiation is of concern. In high-temperature glass tanks, the presence of soot adds 0.1 to 0.2 to emissivities of oil-fired flames. In natural gas-fired flames, efforts to augment flame emissivities with soot generation generally have been unsuccessful. The contributions of soot to the radiation from pool fires often dominates, and thus the presence of soot in such flames directly impacts the safe separation distances from dikes around oil tanks and the location of flares with respect to oil rigs.

Clouds of Large Black Particles The emissivity ε_M of a cloud of black particles with a large perimeter-to-wavelength ratio is

$$\varepsilon_M = 1 - \exp[-(a/\upsilon)L] \tag{5-144}$$

where a/υ is the projected area of the particles per unit volume of space. If the particles have no negative curvature (the particle does not "see" any of itself) and are randomly oriented, $a = a'/4$, where a' is the actual surface area. If the particles are uniform, $a/\upsilon = cA = cA'/4$, where A and A' are the projected and total areas of each particle, respectively, and c is the number concentration of particles. For spherical particles this leads to

$$\varepsilon_M = 1 - \exp[-(\pi/4)cd_p^2L] = 1 - \exp(-1.5 f_\upsilon L/d_p) \tag{5-145}$$

As an example, consider a heavy fuel oil ($CH_{1.5}$, specific gravity of 0.95) atomized to a mean surface particle diameter of d_p burned with 20 percent excess air to produce coke-residue particles having the original drop diameter and suspended in combustion products at 1204°C (2200°F). The flame emissivity due to the particles along a path of L m, with d_p measured in micrometers, is

$$\varepsilon_M = 1 - \exp(-24.3L/d_p) \tag{5-146}$$

For 200-μm particles and $L = 3.05$ m, the particle contribution to emissivity is calculated as 0.31.

Clouds of Nonblack Particles For nonblack particles, emissivity calculations are complicated by multiple scatter of the radiation reflected by each particle. The emissivity ε_M of a cloud of gray particles of individual emissivity ε_1 can be estimated by the use of a simple modification Eq. (5-144), i.e.,

$$\varepsilon_M = 1 - \exp[-\varepsilon_1(a/\upsilon)L] \tag{5-147}$$

Equation (5-147) predicts that $\varepsilon_M \to 1$ as $L \to \infty$. This is *impossible* in a scattering system, and use of Eq. (5-147) is restricted to values of the optical thickness $(a/\upsilon)L < 2$. Instead, the asymptotic value of ε_M is obtained from Fig. 5-12 as $\varepsilon_M = \varepsilon_h$ ($\lim L \to \infty$), where the albedo w is replaced by the particle-surface reflectance $\omega = 1 - \varepsilon_1$. Particles with perimeter-to-wavelength ratios of 0.5 to 5.0 can be analyzed, with significant mathematical complexity, by use of the the *Mie equations* (Bohren, C. F., and Huffman, D. R., *Absorption and Scattering of Light by Small Particles,* Wiley, Hoboken, N.J., 1998).

Combined Gas, Soot, and Particulate Emission In a mixture of emitting species, the emission of each constituent is attenuated on its way to the system boundary by absorption by all other constituents. The transmissivity of a mixture is the product of the transmissivities of its component parts. This statement is a corollary of Beer's law. For present purposes, the transmissivity of "species k" is defined as $\tau_k = 1 - \varepsilon_k$. For a mixture of combustion products consisting of carbon dioxide, water vapor, soot, and oil coke or char particles, the total emissivity ε_T at any wavelength can therefore be obtained from

$$(1-\varepsilon_T)_\lambda = (1-\varepsilon_C)_\lambda(1-\varepsilon_W)_\lambda(1-\varepsilon_S)_\lambda(1-\varepsilon_M)_\lambda \tag{5-148}$$

where the subscripts denote the four flame species. The total emissivity is then obtained by integrating Eq. (5-148) over the entire EM energy spectrum, taking into account the variability of ε_C, ε_W, and ε_S with respect to wavelength. In Eq. (5-148), ε_M is independent of wavelength because absorbing char or coke particles are effectively blackbody absorbers. Computer programs for spectral emissivity, such as RADCAL, perform the integration with respect to wavelength for obtaining total emissivity. Corrections for the overlap of vibration-rotation bands of CO_2 and H_2O are automatically included in the correlations for ε_g for mixtures of these gases. The monochromatic soot emissivity is higher at shorter wavelengths, resulting in higher attenuations of the bands at 2.7 μm for CO_2 and H_2O than at longer wavelengths. The following equation is recommended for calculating the emissivity ε_{g+S} of a mixture of CO_2, H_2O, and soot

$$\varepsilon_{g+S} = \varepsilon_g + \varepsilon_S - M \cdot \varepsilon_g \varepsilon_S \tag{5-149}$$

where M can be represented with acceptable error by the dimensionless function

$$M = 1.12 - 0.27 \cdot (T/1000) + 2.7 \times 10^5 f_\upsilon \cdot L \tag{5-150}$$

In Eq. (5-150), T has units of kelvins and L is measured in meters. Since coke or char emissivities are gray, their addition to those of the CO_2, H_2O, and soot follows simply from Eq. (5-148) as

$$\varepsilon_T = \varepsilon_{g+s} + \varepsilon_M - \varepsilon_{g+s}\varepsilon_M \tag{5-151}$$

with the definition $1-\varepsilon_{g+s} \equiv (1-\varepsilon_C)(1-\varepsilon_W)(1-\varepsilon_S)$.

RADIATIVE EXCHANGE WITH PARTICIPATING MEDIA

Energy Balances for Volume Zones—The Radiation Source Term Reconsider a generalized enclosure with N volume zones confining a *gray* gas. When the N gas temperatures are *unknown*, an *additional* set of N equations is required in the form of radiant energy balances for each volume zone. These N equations are given by the *definition* of the N-vector for the *net radiant volume absorption* $\mathbf{S}' = [S'_j]$ for each volume zone

$$\mathbf{S}' = \overline{\mathbf{GS}}\cdot\mathbf{E} + \overline{\mathbf{GG}}\cdot\mathbf{E}_g - 4K\cdot\mathbf{VI}\cdot\mathbf{E}_g \qquad [N\times 1] \tag{5-152}$$

The radiative source term is a discretized formulation of the *net radiant absorption* for each volume zone which may be incorporated as a *source term* into numerical approximations for the generalized *energy equation.* As such, it permits formulation of energy balances on each zone that may include conductive and convective heat transfer. For $K\to 0$, $\overline{\mathbf{GS}}\to 0$, and $\overline{\mathbf{GG}}\to 0$ leading to $\mathbf{S}'\to\mathbf{0}_N$. When $K\neq 0$ and $\mathbf{S}'=\mathbf{0}_N$, the gas is said to be in a state of *radiative equilibrium.* In the notation usually associated with the discrete ordinate (DO) and finite volume (FV) methods, see Modest (*Radiative Heat Transfer,* 3d ed., Academic Press, New York, 2013, chap. 17), one would write $S'_i/V_i = K[G - 4\cdot E_g] = -\nabla\cdot\mathbf{q}_r$. Here $H_g = G/4$ is the average flux density incident on a given volume zone from all other surface and volume zones. The DO and FV methods are currently available options as "RTE-solvers" in complex simulations of combustion systems using computational fluid dynamics (CFD).*

Implementation of Eq. (5-152) necessitates the definition of two additional *symmetric* $N\times N$ arrays of exchange areas, namely, $\overline{\mathbf{gg}} = [\overline{g_i g_j}]$ and $\overline{\mathbf{GG}} = [\overline{G_i G_j}]$. In Eq. (5-152) $\mathbf{VI} = [V_i\cdot\delta_{i,j}]$ is an $N\times N$ diagonal matrix of zone volumes. The total exchange areas in Eq. (5-151) are *explicit* functions of the direct exchange areas as follows:

Surface-to-gas exchange

$$\overline{\mathbf{GS}} = \overline{\mathbf{SG}}^T \qquad [N\times M] \tag{5-153a}$$

Gas-to-gas exchange

$$\overline{\mathbf{GG}} = \overline{\mathbf{gg}} + \overline{\mathbf{sg}}^T\cdot\rho\mathbf{I}\cdot\mathbf{R}\cdot\overline{\mathbf{sg}} \qquad [N\times M] \tag{5-153b}$$

The matrices $\overline{\mathbf{gg}} = [\overline{g_i g_j}]$ and $\overline{\mathbf{GG}} = [\overline{G_i G_j}]$ must also satisfy the following *matrix conservation relations:*

$$\text{Direct exchange areas:}\quad 4K\cdot\mathbf{VI}\cdot\mathbf{1}_N = \overline{\mathbf{gs}}\cdot\mathbf{1}_M + \overline{\mathbf{gg}}\cdot\mathbf{1}_N \tag{5-154a}$$

$$\text{Total exchange areas:}\quad 4K\cdot\mathbf{VI}\cdot\mathbf{1}_N = \overline{\mathbf{GS}}\cdot\mathbf{1}_M + \overline{\mathbf{GG}}\cdot\mathbf{1}_N \tag{5-154b}$$

The formal integral definition of the direct gas-gas exchange area is

$$\overline{g_i g_j} = \oiiint_{V_i}\oiiint_{V_j} K^2 \frac{e^{-Kr}}{\pi r^2}\, dV_j\, dV_i \tag{5-155}$$

Clearly, when $K = 0$, the two direct exchange areas involving a gas zone $\overline{g_i s_j}$ and $\overline{g_i g_j}$ vanish. Computationally it is *never* necessary to make resort to Eq. (5-155) for calculation of $\overline{g_i g_j}$. This is so because, $\overline{s_i g_j}$, $\overline{s_i g_j}$, and $\overline{g_i g_j}$ may *all* be calculated *arithmetically* from appropriate values of $\overline{s_i s_j}$ by using associated conservation relations and view factor algebra.

Weighted Sum of Gray Gas (WSGG) Spectral Model Even in simple engineering calculations, the assumption of a gray gas is almost never a good one. The zone method is now further generalized to make allowance for nongray radiative transfer via incorporation of the *weighted sum of gray gas* (WSGG) spectral model. Hottel has shown that the emissivity $\varepsilon_g(T,L)$ of an absorbing-emitting gas mixture containing CO_2 and H_2O of known composition can be approximated by a *weighted sum of P gray gases*

$$\varepsilon_g(T,L) \approx \sum_{p=1}^{P} a_p(T)[1-e^{-K_p L}] \tag{5-156a}$$

where

$$\sum_{p=1}^{P} a_p(T) = 1.0 \tag{5-156b}$$

In Eqs. (5-156), K_p is some gray gas absorption coefficient and L is some appropriate path length. In practice, Eqs. (5-156) usually yield acceptable accuracy for $P\le 3$. For $P = 1$, Eqs. (5-156) degenerate to the case of a single gray gas.

The Clear Plus Gray Gas WSGG Spectral Model In principle, the emissivity of all gases approaches unity for infinite path length L. In practice, however, the gas emissivity may fall considerably short of unity for representative values of pL. This behavior results because of the *band* nature of real gas spectral absorption and emission whereby there is usually no significant overlap between dominant absorption bands. Mathematically, this physical phenomenon is modeled by defining one of the gray gas components in the WSGG spectral model to be *transparent.*

For $P = 2$ and path length L_M, Eqs. (5-156) yield the following expression for the gas emissivity

$$\varepsilon_g = a_1(1-e^{-K_1 L_M}) + a_2(1-e^{-K_2 L_M}) \tag{5-157}$$

In Eq. (5-157) if $K_1 = 0$ and $a_2\neq 0$, the limiting value of gas emissivity is $\varepsilon_g(T,\infty)\to a_2$. Put $K_1 = 0$ in Eq. (5-157), $a_g = a_2$, and define $\tau_g = e^{-K_2 L_M}$ as the gray gas transmissivity. Equation (5-157) then simplifies to

$$\varepsilon_g = a_g(1-\tau_g) \tag{5-158}$$

It is important to note in Eq. (5-158) that $0\le a_g, \tau_g\le 1.0$ while $0\le\varepsilon_g\le a_g$.

Equation (5-158) constitutes a two-parameter model which may be fitted with only *two* empirical emissivity data points. To obtain the constants a_g and τ_g in Eq. (5-158) at fixed composition and temperature, denote the two emissivity data points as $\varepsilon_{g,2} = \varepsilon_g(2pL) > \varepsilon_{g,1} = \varepsilon_g(pL)$ and recognize that $\varepsilon_{g,1} = a_g(1-\tau_g)$ and $\varepsilon_{g,2} = a_g(1-\tau_g^2) = a_g(1-\tau_g)(1+\tau_g) = \varepsilon_{g,1}(1+\tau_g)$. These relations lead directly to the final emissivity fitting equations

$$\tau_g = \frac{\varepsilon_{g,2}}{\varepsilon_{g,1}} - 1 \tag{5-159a}$$

and

$$a_g = \frac{\varepsilon_{g,1}}{2-\varepsilon_{g,2}/\varepsilon_{g,1}} \tag{5-159b}$$

The clear plus gray WSGG spectral model also readily leads to values for gas *absorptivity* and *transmissivity*, with respect to some appropriate surface radiation source at temperature T_1, for example,

$$\alpha_{g,1} = a_{g,1}(1-\tau_g) \tag{5-160a}$$

and

$$\tau_{g,1} = a_{g,1}\cdot\tau_g \tag{5-160b}$$

In Eqs. (5-160) the *gray* gas transmissivity τ_g is taken to be identical to that obtained for the gas emissivity ε_g. The constant $a_{g,1}$ in Eq. (5-160*a*) is then obtained with knowledge of *one additional empirical value* for $\alpha_{g,1}$ which may also be obtained from the correlations in Table 5-6. Notice further in the definitions of the three parameters ε_g, $\alpha_{g,1}$, and $\tau_{g,1}$ that *all* the temperature dependence is forced into the *two* WSGG constants a_g and $a_{g,1}$.

The *three* clear plus gray WSGG constants a_g, $a_{g,1}$, and τ_g are functions of total pressure, temperature, and mixture composition. It is not necessary to ascribe any particular physical significance to them. Rather, they may simply be visualized as three constants that happen to fit the gas emissivity data. It is noteworthy that *three* constants are far fewer than the number required to calculate gas emissivity data from fundamental spectroscopic data.

*To further clarify the mathematical differences between zoning and the DO and FV methods, recognize that (neglecting scatter) the matrix expressions $\mathbf{H} = \mathbf{AI}^{-1}\cdot\overline{\mathbf{ss}}\cdot\mathbf{W} + \mathbf{AI}^{-1}\cdot\overline{\mathbf{sg}}\cdot\mathbf{E}_g$ and $4K\cdot\mathbf{H}_g = \mathbf{VI}^{-1}\cdot\overline{\mathbf{gs}}\cdot\mathbf{W} + \mathbf{VI}^{-1}\cdot\overline{\mathbf{gg}}\cdot\mathbf{E}_g$ represent spatial discretizations of the integral form(s) of the RTE applied at any point (zone) on the boundary or interior of an enclosure, respectively, for a gray gas.

The two constants a_g and $a_{g,1}$ defined in Eqs. (5-158) and (5-160) can, however, be interpreted *physically* in a particularly simple manner. Suppose the gas absorption spectrum is idealized by *many* absorption bands (boxes), *all* of which are characterized by the *identical* absorption coefficient K. The a's might then be calculated from the total blackbody energy fraction $F_b\ (\lambda T)$ defined in Eqs. (5-105) and (5-106). That is, a_g simply represents the total *energy fraction* of the blackbody energy distribution in which the gas *absorbs*. This concept may be further generalized to real gas absorption spectra via the *wideband stepwise gray spectral box model* (Modest, *Radiative Heat Transfer,* 3d ed., Academic Press, New York, 2013, chap. 14).

When $P \geq 3$, exponential curve-fitting procedures for the WSGG spectral model become significantly more difficult for hand computation but are quite routine with the aid of a variety of readily available mathematical software utilities. The clear plus gray WSGG fitting procedure is demonstrated in Example 5-13.

The Zone Method and Directed Exchange Areas Spectral dependence of real gas spectral properties is now introduced into the zone method via the WSGG spectral model. It is still assumed, however, that all surface zones are gray isotropic emitters and absorbers.

General Matrix Representation We first define a new set of four *directed exchange areas* $\overleftarrow{\mathbf{SS}}, \overleftarrow{\mathbf{SG}}, \overleftarrow{\mathbf{GS}}$, and $\overleftarrow{\mathbf{GG}}$, which are denoted by an overarrow. The directed exchange areas are obtained from the total exchange areas for gray gases by simple matrix multiplication using weighting factors derived from the WSGG spectral model. The directed exchange areas are denoted by an overarrow to indicate the "sending" and "receiving" zone. The a-weighting factors for transfer originating at a gas zone $a_{g,i}$ are derived from WSGG gas *emissivity* calculations, while those for transfers originating at a surface zone a_i are derived from appropriate WSGG gas *absorptivity* calculations. Let $\mathbf{a}_g\mathbf{I}_p = [a_{p,g,i} \cdot \delta_{i,j}]$ and $\mathbf{aI}_p = [a_{p,i} \cdot \delta_{i,j}]$ represent the $P\ [M \times M]$ and $[N \times N]$ *diagonal* matrices comprised of the appropriate WSGG a constants. The *directed* exchange areas are then computed from the associated *total* gray gas exchange areas via simple *diagonal* matrix multiplication.

$$\overleftarrow{\mathbf{SS}} = \sum_{p=1}^{P} \overline{\mathbf{SS}}_p \cdot \mathbf{aI}_p \qquad [M \times M] \tag{5-161a}$$

$$\overleftarrow{\mathbf{SG}} = \sum_{p=1}^{P} \overline{\mathbf{SG}}_p \cdot \mathbf{a}_g\mathbf{I}_p \qquad [M \times N] \tag{5-161b}$$

$$\overleftarrow{\mathbf{GS}} = \sum_{p=1}^{P} \overline{\mathbf{GS}}_p \cdot \mathbf{aI}_p \qquad [M \times N] \tag{5-161c}$$

$$\overleftarrow{\mathbf{GG}} = \sum_{p=1}^{P} \overline{\mathbf{GG}}_p \cdot \mathbf{a}_g\mathbf{I}_p \qquad [N \times N] \tag{5-161d}$$

with

$$\overleftarrow{\mathbf{KI}} = \sum_{p=1}^{P} \mathbf{KI}_p \cdot \mathbf{a}_g\mathbf{I}_p \qquad [N \times N] \tag{5-161e}$$

In contrast to the *total* exchange areas which are always independent of temperature, the four *directed* arrays $\overleftarrow{\mathbf{SS}}, \overleftarrow{\mathbf{SG}}, \overleftarrow{\mathbf{GS}}$, and $\overleftarrow{\mathbf{GG}}$ are dependent on the temperatures of each and every zone, i.e., as in $a_{p,i} = a_p(T_i)$. Moreover, in contrast to *total* exchange areas, the *directed* arrays $\overleftarrow{\mathbf{SS}}$ and $\overleftarrow{\mathbf{GG}}$ are generally not symmetric and $\overleftarrow{\mathbf{GS}} \neq \overleftarrow{\mathbf{SG}}^{\mathrm{T}}$. Finally, since the directed exchange areas are temperature-*dependent*, iteration may be required to update the $\mathbf{aI}_p$ and $\mathbf{a}_g\mathbf{I}_p$ arrays during the course of a calculation. There is a great deal of latitude with regard to fitting the WSGG a constants in these matrix equations, especially if $N > 1$ and composition variations are to be allowed for in the gas. An extensive discussion of a fitting for $N > 1$ is beyond the scope of this presentation. Details of the fitting procedure, however, are presented in Example 5-13 in the context of a single-gas zone.

Once the directed exchange areas are formulated, the governing matrix equations for the radiative flux equations at each surface zone and the radiant source term are then given as follows:

$$\mathbf{Q} = \varepsilon\mathbf{AI} \cdot \mathbf{E} - \overleftarrow{\mathbf{SS}} \cdot \mathbf{E} - \overleftarrow{\mathbf{SG}} \cdot \mathbf{E}_g \tag{5-162a}$$

$$\mathbf{S}' = \overleftarrow{\mathbf{GG}} \cdot \mathbf{E}_g + \overleftarrow{\mathbf{GS}} \cdot \mathbf{E} - 4 \cdot \overleftarrow{\mathbf{KI}} \cdot \mathbf{VI} \cdot \mathbf{E}_g \tag{5-162b}$$

or the alternative forms

$$\mathbf{Q} = [\mathbf{EI} \cdot \overrightarrow{\mathbf{SS}} - \overleftarrow{\mathbf{SS}} \cdot \mathbf{EI}] \cdot \mathbf{1}_M + [\mathbf{EI} \cdot \overrightarrow{\mathbf{SG}} - \overleftarrow{\mathbf{SG}} \cdot \mathbf{E}_g\mathbf{I}] \cdot \mathbf{1}_N \tag{5-163a}$$

$$\mathbf{S}' = -[\mathbf{E}_g\mathbf{I} \cdot \overrightarrow{\mathbf{GS}} - \overleftarrow{\mathbf{GS}} \cdot \mathbf{EI}] \cdot \mathbf{1}_M - [\mathbf{E}_g\mathbf{I} \cdot \overrightarrow{\mathbf{GG}} - \overleftarrow{\mathbf{GG}} \cdot \mathbf{E}_g\mathbf{I}] \cdot \mathbf{1}_N \tag{5-163b}$$

It may be *proved* that the $\mathbf{Q}$ and $\mathbf{S}'$ vectors computed from Eqs. (5-162) and (5-163) always *exactly* satisfy the overall (scalar) *radiant* energy balance $\mathbf{1}_M^T \cdot \mathbf{Q} = \mathbf{1}_N^T \cdot \mathbf{S}'$. In words, the total radiant gas emission for all gas zones in the enclosure must always exactly equal the total radiant energy received at all surface zones which comprise the enclosure. In Eqs. (5-162) and (5-163), the following definitions are employed for the *four forward-directed* exchange areas

$$\overrightarrow{\mathbf{SS}} = \overleftarrow{\mathbf{SS}} \quad \overrightarrow{\mathbf{SG}} = \overleftarrow{\mathbf{GS}}^{\mathrm{T}} \quad \overrightarrow{\mathbf{GS}} = \overleftarrow{\mathbf{SG}}^{\mathrm{T}} \quad \overrightarrow{\mathbf{GG}} = \overleftarrow{\mathbf{GG}}^{\mathrm{T}} \tag{5-164a,b,c,d}$$

such that formally there are some *eight* matrices of directed exchange areas. The four *backward-directed* arrays of directed exchange areas must satisfy the following conservation relations:

$$\overleftarrow{\mathbf{SS}} \cdot \mathbf{1}_M + \overleftarrow{\mathbf{SG}} \cdot \mathbf{1}_N = \varepsilon\mathbf{I} \cdot \mathbf{AI} \cdot \mathbf{1}_M \tag{5-165a}$$

$$4\overleftarrow{\mathbf{KI}} \cdot \mathbf{VI} \cdot \mathbf{1}_N = \overleftarrow{\mathbf{GS}} \cdot \mathbf{1}_M + \overleftarrow{\mathbf{GG}} \cdot \mathbf{1}_N \tag{5-165b}$$

Subject to the restrictions of no scatter and diffuse surface emission and reflection, the above equations are the most general matrix statement possible for the zone method. When $P = 1$, the *directed* exchange areas all reduce to the *total* exchange areas for a single gray gas. If, in addition, $K = 0$, the much simpler case of radiative transfer in a transparent medium results. If, in addition, all surface zones are *black*, the direct, total, and directed exchange areas are all identical.

Allowance for Flux Zones As in the case of a transparent medium, we now distinguish between source and flux surface zones. Let $M = M_s + M_f$ represent the total number of surface zones where M_s is the number of source-sink zones and M_f is the number of flux zones. The flux zones are the *last to be numbered.* To accomplish this, partition the surface emissive power and flux vectors as $\mathbf{E} = \begin{bmatrix} \mathbf{E}_1 \\ \mathbf{E}_2 \end{bmatrix}$ and $\mathbf{Q} = \begin{bmatrix} \mathbf{Q}_1 \\ \mathbf{Q}_2 \end{bmatrix}$, where subscript 1 denotes surface source/sink zones whose emissive power $\mathbf{E}_1$ is specified a priori and subscript 2 denotes surface flux zones of unknown emissive power vector $\mathbf{E}_2$ and known radiative flux vector $\mathbf{Q}_2$. Suppose the radiative source vector $\mathbf{S}'$ is *known.* Appropriate partitioning of Eqs. (5-162) then produces

$$\begin{bmatrix} \mathbf{Q}_1 \\ \mathbf{Q}_2 \end{bmatrix} = \begin{bmatrix} \varepsilon\mathbf{AI}_{1,1} & \mathbf{0} \\ \mathbf{0} & \varepsilon\mathbf{AI}_{2,2} \end{bmatrix} \cdot \begin{bmatrix} \mathbf{E}_1 \\ \mathbf{E}_2 \end{bmatrix} \begin{bmatrix} \overleftarrow{\mathbf{SS}}_{1,1} & \overleftarrow{\mathbf{SS}}_{1,2} \\ \overleftarrow{\mathbf{SS}}_{2,1} & \overleftarrow{\mathbf{SS}}_{2,2} \end{bmatrix} \cdot \begin{bmatrix} \mathbf{E}_1 \\ \mathbf{E}_2 \end{bmatrix} - \begin{bmatrix} \overleftarrow{\mathbf{SG}}_1 \\ \overleftarrow{\mathbf{SG}}_2 \end{bmatrix} \cdot \mathbf{E}_g \tag{5-166a}$$

and

$$\mathbf{S}' = \overleftarrow{\mathbf{GG}} \cdot \mathbf{E}_g + [\overleftarrow{\mathbf{GS}}_1 \quad \overleftarrow{\mathbf{GS}}_2] \cdot \begin{bmatrix} \mathbf{E}_1 \\ \mathbf{E}_2 \end{bmatrix} - 4 \cdot \overleftarrow{\mathbf{KI}} \cdot \mathbf{VI} \cdot \mathbf{E}_g \tag{5-166b}$$

where the definitions of the matrix partitions follow the conventions with respect to Eq. (5-120). Simultaneous solution of the *two* unknown vectors in Eqs. (5-166) then yields

$$\mathbf{E}_2 = \mathbf{RP} \cdot [\overleftarrow{\mathbf{SS}}_{2,1} + \overleftarrow{\mathbf{SG}}_2 \cdot \mathbf{PP} \cdot \overleftarrow{\mathbf{GS}}_1] \cdot \mathbf{E}_1 + \mathbf{RP} \cdot [\mathbf{Q}_2 - \overleftarrow{\mathbf{SG}}_2 \cdot \mathbf{PP} \cdot \mathbf{S}'] \tag{5-167a}$$

and

$$\mathbf{E}_g = \mathbf{PP} \cdot [\overleftarrow{\mathbf{GS}}_1 \quad \overleftarrow{\mathbf{GS}}_2] \cdot \begin{bmatrix} \mathbf{E}_1 \\ \mathbf{E}_2 \end{bmatrix} - \mathbf{PP} \cdot \mathbf{S}' \tag{167b}$$

where two auxiliary inverse matrices **RP** and **PP** are defined as

$$\mathbf{PP} = [4 \cdot \overleftarrow{\mathbf{KI}} \cdot \mathbf{VI} - \overleftarrow{\mathbf{GG}}]^{-1} \tag{5-168a}$$

$$\mathbf{RP} = [\varepsilon\mathbf{AI}_{2,2} - \overleftarrow{\mathbf{SS}}_{2,2} - \overleftarrow{\mathbf{SG}}_2 \cdot \mathbf{PP} \cdot \overleftarrow{\mathbf{GS}}_2]^{-1} \tag{5-168b}$$

The emissive power vectors $\mathbf{E}$ and $\mathbf{E}_g$ are then both *known* quantities for purposes of subsequent calculation.

Algebraic Formulas for a Single-Gas Zone As shown in Fig. 5-16, the three-zone system with $M = 2$ and $N = 1$ can be employed to simulate a surprisingly large number of useful engineering geometries. These include two infinite parallel plates confining an absorbing-emitting medium, *any* two-surface zone system where a nonconvex surface zone is completely surrounded by a second zone (this includes concentric spheres and cylinders), and the speckled two-surface enclosure. As in the case of a transparent medium, the inverse reflectivity matrix **R** is capable of explicit matrix inversion for $M = 2$. This allows derivation of *explicit* algebraic equations for all the required *directed* exchange areas for the clear plus gray WSGG spectral model with $M = 1$ and 2 and $N = 1$.

The Limiting Case M = 1 and N = 1 The directed exchange areas for this special case correspond to a single well-mixed gas zone completely surrounded by a single-surface zone A_1. Here the reflectivity matrix is a 1×1 scalar quantity which follows directly from the general matrix equations as $\mathbf{R} = 1/(A_1 - \overline{s_1s_1} \cdot \rho_1)$. There are *two* WSCC clear plus gray constants a_1 and a_g and only *one* unique direct exchange area which satisfies the conservation relation $\overline{s_1s_1} + \overline{s_1g} = A_1$. The only two physically meaningful directed exchange areas are those between the surface zone A_1 and the gas zone

$$\overleftarrow{S_1G} = \frac{a_g \cdot \varepsilon_1 A_1 \cdot \overline{s_1g}}{\varepsilon_1 \cdot A_1 + \rho_1 \cdot \overline{s_1g}} \tag{5-169a}$$

$$\overleftarrow{GS}_1 = \frac{a_1 \cdot \varepsilon_1 A_1 \cdot \overline{s_1g}}{\varepsilon_1 \cdot A_1 + \rho_1 \cdot \overline{s_1g}} \tag{5-169b}$$

The total radiative flux Q_1 at surface A_1 and the radiative source term $Q_1 = S$ are given by

$$Q_1 = \overleftarrow{GS}_1 \cdot E_1 - \overleftarrow{S_1G} \cdot E_g \tag{5-169}$$

Directed Exchange Areas for M = 2 and N = 1 For this case there are *four* WSGG constants, a_1, a_2, a_g, and τ_g. There is one required value of K that is readily obtained from the equation $K = -\ln(\tau_g)/L_m$, where $\tau_g = \exp(-KL_m)$. For an enclosure with $M = 2$, $N = 1$, and $K \neq 0$, only *three unique* direct exchange areas are required because conservation stipulates $A_1 = \overline{s_1s_1} + \overline{s_1s_2} + \overline{s_1g}$ and $A_2 = \overline{s_2s_1} + \overline{s_2s_2} + \overline{s_2g}$. For $M = 2$ and $N = 1$, the matrix Eqs. (5-118) readily lead to the *general* gray gas matrix solution for $\overline{\mathbf{SS}}$ and $\overline{\mathbf{SG}}$ with $K \neq 0$ as

$$\overline{\mathbf{SS}} = \begin{bmatrix} \varepsilon_1 A_1 - \overline{S_1S_2} - \overline{S_1G} & \overline{S_1S_2} \\ \overline{S_1S_1} & \varepsilon_2 A_2 - \overline{S_1S_2} - \overline{S_2G} \end{bmatrix} \tag{5-170a}$$

where

$$\overline{S_1S_2} = \varepsilon_1 \varepsilon_2 A_1 A_2 \overline{s_1s_2} / \det \mathbf{R}^{-1} \tag{5-170b}$$

and
$$\overline{\mathbf{SG}} = \begin{bmatrix} \varepsilon_1 A_1 \left[(A_2 - \rho_2 \cdot \overline{s_2s_2}) \cdot \overline{s_1g} + \rho_2 \cdot \overline{s_1s_2} \cdot \overline{s_2g} \right] \\ \varepsilon_2 A_2 \left[(A_2 - \rho_2 \cdot \overline{s_2s_2}) \cdot \overline{s_2g} + \rho_1 \cdot \overline{s_1s_2} \cdot \overline{s_1g} \right] \end{bmatrix} / \det \mathbf{R}^{-1} \tag{5-170c}$$

with $\overline{\mathbf{GS}} = \overline{\mathbf{SG}}^{\mathrm{T}}$ and the indicated determinate of $\mathbf{R}^{-1}$ is evaluated algebraically as

$$\det \mathbf{R}^{-1} = (A_1 - \overline{s_1s_1} \cdot \rho_1) \cdot (A_2 - \overline{s_2s_2} \cdot \rho_2) - \rho_1 \cdot \rho_2 \cdot \overline{s_1s_2}^2 \tag{5-170d}$$

For the WSGG clear gas components we denote $\overline{\mathbf{SS}}]_{K=0} \equiv \overline{\mathbf{SS}}_0$ and $\overline{\mathbf{SG}}]_{K=0} \equiv \overline{\mathbf{SG}}_0 = \mathbf{0}$. Finally the WSGG arrays of directed exchange areas are computed simply from a-weighted sums of the gray gas total exchange areas as

$$\overleftarrow{\mathbf{SS}} = \overline{\mathbf{SS}}_0 \cdot \begin{bmatrix} 1-a_1 & 0 \\ 0 & 1-a_2 \end{bmatrix} + \overline{\mathbf{SS}} \cdot \begin{bmatrix} a_1 & 0 \\ 0 & a_2 \end{bmatrix}$$

$$\overleftarrow{\mathbf{SG}} = \overline{\mathbf{SG}} \cdot a_g \tag{5-171a,b,c}$$

$$\overleftarrow{\mathbf{GS}} = \overline{\mathbf{GS}} \cdot \begin{bmatrix} a_1 & 0 \\ 0 & a_2 \end{bmatrix} \neq \overleftarrow{\mathbf{SG}}^{\mathrm{T}}$$

and finally

$$\overleftarrow{\mathbf{GG}} = a_g \cdot 4KV - \overleftarrow{\mathbf{GS}} \cdot \begin{bmatrix} 1 \\ 1 \end{bmatrix} \tag{5-171d}$$

The results of this development may be further expanded into algebraic form with the aid of Eq. (5-127) to yield

$$\overleftarrow{S_2S_1} = \frac{\varepsilon_1 \varepsilon_2 A_1 A_2 \cdot \overline{s_2s_1}]_0 (1-a_1)}{\varepsilon_1 \varepsilon_2 A_1 A_2 + (\varepsilon_1 A_1 \rho_2 + \varepsilon_2 A_2 \rho_1) \cdot \overline{s_2s_1}]_0} + \frac{\varepsilon_1 \varepsilon_2 A_1 A_2 \overline{s_2s_1} \cdot a_1}{\det \mathbf{R}^{-1}} \tag{5-171e}$$

$$\overleftarrow{\mathbf{SG}} = \begin{bmatrix} \varepsilon_1 A_1 [(A_2 - \rho_2 \cdot \overline{s_2s_2}) \cdot \overline{s_1g} + \rho_2 \cdot \overline{s_1s_2} \cdot \overline{s_2g}] \\ \varepsilon_2 A_2 \left[(A_1 - \rho_1 \cdot \overline{s_1s_1}) \cdot \overline{s_2g} + \rho_1 \cdot \overline{s_1s_2} \cdot \overline{s_1g} \right] \end{bmatrix} \cdot a_g / \det \mathbf{R}^{-1} \tag{5-171f}$$

and
$$\overleftarrow{\mathbf{GS}} = [\overleftarrow{GS}_1 \quad \overleftarrow{GS}_1] \tag{5-171g}$$

whose matrix elements are given by $\overleftarrow{GS}_1 \equiv \varepsilon_1 A_1 \left[(A_2 - \rho_2 \cdot \overline{s_1s_1}) \cdot \overline{s_1g} + \rho_2 \cdot \overline{s_1s_2} \cdot \overline{s_2g} \right] a_1 / \det \mathbf{R}^{-1}$ and $\overleftarrow{GS}_2 \equiv \varepsilon_2 A_2 \left[(A_1 - \rho_1 \cdot \overline{s_1s_1}) \cdot \overline{s_2g} + \rho_1 \cdot \overline{s_1s_2} \cdot \overline{s_1g} \right] a_2 / \det \mathbf{R}^{-1}$. Derivation of the scalar (algebraic) forms for the directed exchange areas here is done primarily for illustrative purposes. Computationally, the only advantage is to obviate the need for a digital computer to evaluate a $[2 \times 2]$ matrix inverse.

Allowance for an Adiabatic Refractory with N = 1 and M = 2 Set $N = 1$ and $M = 2$, and let zone 2 represent the refractory surface. Let $Q_2 = 0$ and $\varepsilon_2 \neq 0$; it then follows that we may define a *refractory-aided* directed exchange area $\overleftarrow{S_1G}_R$ by

$$\overleftarrow{S_1G}_R = \overleftarrow{S_1G} + \frac{\overleftarrow{S_1S_2} \cdot \overleftarrow{S_2G}}{\overleftarrow{S_1S_2} + \overleftarrow{S_2G}} \tag{5-172a}$$

Assuming radiative equilibrium, the emissive power of the refractory may also be calculated from the companion equation

$$E_2 = \frac{\overleftarrow{S_2S_1} \cdot E_1 + \overleftarrow{S_2G} \cdot E_g}{\overleftarrow{S_2S_1} + \overleftarrow{S_2G}} \tag{5-172b}$$

In this circumstance, *all* the radiant energy originating in the gas volume is transferred to the *sole* sink zone A_1. Equation (5-172*a*) is thus tantamount to the statement that $Q_1 = S'$ or that the net emission from the source ultimately must arrive at the sink. Notice that if $\varepsilon_1 = 0$, Eq. (5-172*a*) leads to a physically incongruous statement since *all* the directed exchange areas would vanish and *no sink* would exist. Even for the simple case of $M = 2$ and $N = 1$, the algebraic complexity of Eqs. (5-171) suggests that *numerical matrix manipulation* of directed exchange areas is preferred rather than calculations using algebraic formulas.

Engineering Approximations for Directed Exchange Areas Use of the preceding equations for directed exchange areas with $M = 2$ and $N = 1$ and the WSGG clear plus gray gas spectral approximation requires knowledge of three independent direct exchange areas. It also formally requires evaluation of three WSGG weighting constants a_1, a_2, and a_g with respect to the three temperatures T_1, T_2, and T_g. Further simplifications may be made by assuming that radiant transfer for the *entire* enclosure is characterized by the *single* mean beam length $L_M = 0.88 \cdot 4 \cdot V/A$. The requisite direct exchange areas are then approximated by

$$\overline{\mathbf{ss}} = \tau_g \begin{bmatrix} A_1 \cdot F_{1,1} & A_1 \cdot F_{1,2} \\ A_2 \cdot F_{2,1} & A_2 \cdot F_{2,2} \end{bmatrix} \tag{5-173a}$$

with
$$\overline{\mathbf{sg}} = (1 - \tau_g) \begin{bmatrix} A_1 \\ A_2 \end{bmatrix} \tag{5-173b}$$

and for the particular case of a speckled enclosure

$$\overline{\mathbf{ss}} = \frac{\tau_g}{A_1 + A_2} \begin{bmatrix} A_1^2 & A_1 \cdot A_2 \\ A_1 \cdot A_2 & A_2^2 \end{bmatrix} \tag{5-174a}$$

also with
$$\overline{\mathbf{sg}} = (1 - \tau_g) \begin{bmatrix} A_1 \\ A_2 \end{bmatrix} \tag{5-174b}$$

where again τ_g is obtained from the WSGG fit of gas emissivity. These approximate formulas clearly obviate the need for *exact* values of the direct exchange areas and may be used in conjunction with Eqs. (5-171).

For engineering calculations, an additional simplification is sometimes warranted. Again characterize the system by a single mean beam length $L_M = 0.88 \cdot 4 \cdot V/A$ and employ the *identical* value of $\tau_g = KL_M$ for all surface-gas transfers. The three a constants *might* then be obtained by a WSGG data-fitting procedure for gas emissivity and gas absorptivity which utilizes the three different temperatures T_g, T_1, and T_2. For engineering purposes we choose a simpler method, however. First calculate values of ε_g and α_{g1} for gas temperature T_g with respect to the *dominant* (sink) temperature T_1. The *net* radiative flux between an isothermal gas mass at temperature T_g and a black isothermal bounding surface A_1 at temperature T_1 (the sink) is given by Eq. (5-138) as

$$Q_{1,g} = A_1\sigma\left(\alpha_{g,1}T_1^4 - \varepsilon_g T_g^4\right) \tag{5-175}$$

It is clear that transfer from the gas to the surface and transfer from the surface into the gas are characterized by two different constants of proportionality, ε_g and $\alpha_{g,1}$, respectively. To allow for the difference between gas emissivity and absorptivity, it proves convenient to introduce a single *mean gas emissivity* defined by

$$\sigma\left[\varepsilon_g T_g^4 - \alpha_{g,1}T_1^4\right] = \varepsilon_m\sigma\left(T_g^4 - T_1^4\right) \tag{5-176a}$$

or

$$\varepsilon_m = \frac{\varepsilon_g - \alpha_{g,1}(T_1/T_g)^4}{1-(T_1/T_g)^4} \tag{5-176b}$$

The calculation then proceeds by computing *two* values of ε_m at the given T_g and T_1 temperature pair and the two values of pL_M and $2pL_M$. We thereby obtain the expression $\varepsilon_m = \alpha_m(1-\tau_m)$. It is then assumed that $a_1 = a_2 = a_g = a_m$ for use in Eqs. (5-171). This simplification may be used for $M > 2$ as long as $N = 1$. This simplification is illustrated in Example 5-13.

Example 5-13 WSGG Clear Plus Gray Gas Emissivity Calculations Methane is burned to completion with 20 percent excess air (50 percent relative humidity at 298 K or 0.0088 mol water/mol dry air) in a furnace chamber of floor dimensions 3 × 10 m and height 5 m. The entire surface area of the enclosure is a gray sink with emissivity of 0.8 at 1000 K. The confined gas is well stirred at 1500 K. Evaluate the clear plus gray WSGG constants and the mean effective gas emissivity, and calculate the average radiative flux density to the enclosure surface.

Two-zone model, M = 1, N = 1: A single volume zone completely surrounded by a single sink surface zone.

Function definitions:

Gas emissivity: $\varepsilon_g F(T_g, pL, b, n) = b\cdot(pL - 0.015)^n \div T_g$ Eq. (5-140a)

Gas absorptivity: $$\alpha_{g1}F(T_g, T_1, pL, b, n) = \frac{\varepsilon_g F(T_1, pL\cdot T_1/T_g, b, n)\cdot T_1\cdot(T_g/T_1)^{0.5}}{T_1}$$ Eq. (5-141)

Mean effective gas emissivity: $$\varepsilon_{gm}(\varepsilon_g, \alpha_g, T_g, T_1) = \frac{\varepsilon_g - \alpha_g\cdot(T_1/T_g)^4}{1-(T_1/T_g)^4}$$ Eq. (5-176a)

Physical constants: $\sigma \equiv 5.670400\times10^{-8}\dfrac{\text{W}}{\text{m}^2\cdot\text{K}^4}$

Enclosure input parameters:

$T_g = 1500$ K $T_1 = 1000$ K $A_1 = 190\ \text{m}^2$ $V = 150\ \text{m}^3$

$\varepsilon_1 = 0.8$ $\rho_1 = 1 - \varepsilon_1$ $\rho_1 = 0.2$

$E_1 = \sigma\cdot T_1^4$ $E_g = \sigma\cdot T_g^4$ $E_1 = 56.70\ \dfrac{\text{kW}}{\text{m}^2}$ $E_g = 287.06\ \dfrac{\text{kW}}{\text{m}^2}$

Stoichiometry yields the following mole table:

Mole Table: Basis 1.0 mol Methane

Species	MW	Moles in	Mass in	Moles out	*Y* out
CH_4	16.04	1.00000	16.04	0.00000	0.00000
O_2	32.00	2.40000	76.80	0.40000	0.03193
N_2	28.01	9.02857	252.93	9.02857	0.72061
CO_2	44.01	0.00000	0.00	1.00000	0.07981
H_2O	18.02	0.10057	1.81	2.10057	0.16765
Totals	138.08	12.52914	347.58	12.52914	1.00000

$p_W = 0.16765$ atm $p_C = 0.07981$ atm $p = p_W + p_C$ $p = 0.2475$ atm

$p_W \div p_C = 2.101$

The mean beam length is approximated by

$L_M = 0.88\cdot4\cdot V \div A_1$ $L_M = 2.7789$ m

and $pL_M = p\cdot L_M$ $pL_M = 0.6877$ atm · m $pL_M = 0.6877$

The gas emissivities and absorptivities are then calculated from the two-constant correlation in Table 5-6 (column 5 with $p_W/p_C = 2.0$) as follows:

$\varepsilon_{g1} = \varepsilon_g F(1500, pL_M, 540, 0.42)$ $\varepsilon_{g1} = 0.3048$

$\varepsilon_{g2} = \varepsilon_g F(1500, 2pL_M, 540, 0.42)$ $\varepsilon_{g2} = 0.4097$

$\alpha_{g11} = \alpha_{g1}F(1500, 1000, pL_M, 444, 0.34)$ $\alpha_{g11} = 0.4124$

$\alpha_{g12} = \alpha_{g1}F(1500, 1000, 2pL_M, 444, 0.34)$ $\alpha_{g12} = 0.5250$

Case (a): Compute Flux Density Using Exact Values of the WSGG Constants

$$\tau_g = \frac{\varepsilon_{g2}}{\varepsilon_{g1}} - 1 \qquad a_g = \frac{\varepsilon_{g1}}{1-\tau_g} \qquad \varepsilon_g = a_g\cdot(1-\tau_g) \qquad a_{g1} = \frac{\alpha_{g11}}{1-\tau_g}$$

$\tau_g = 0.3442$ $a_g = 0.4647$ $\varepsilon_g = 0.3048$ $a_{g1} = 0.6289$

and the WSGG gas absorption coefficient (which is necessary for calculation of direct exchange areas) is calculated as

$$K_1 = \frac{-(\ln\tau_g)}{L_M} \quad \text{or} \quad K_1 = 0.3838\,\frac{1}{\text{m}}$$

Compute directed exchange areas:

Given $s_{1g} = (1-\tau_g)\cdot A_1$ $= 124.61\ \text{m}^2$

Eqs. (5-169) yield

$$\text{DS1G} = \frac{a_g\cdot\varepsilon_1\cdot A_1\cdot s_1 g}{\varepsilon_1\cdot A_1 + \rho_1\cdot s_1 g} \qquad \text{DGS1} = \frac{a_{g1}\cdot\varepsilon_1\cdot A_1\cdot s_1 g}{\varepsilon_1\cdot A_1 + \rho_1\cdot s_1 g}$$

$\text{DS1G} = 49.75\ \text{m}^2$ $\text{DGS1} = 67.32\ \text{m}^2$

And finally the gas to sink flux density is computed as

$Q_1 = \text{DGS1}\cdot E_1 - \text{DS1G}\cdot E_g$ $Q_1 = -10{,}464.0$ kW $\dfrac{Q_1}{A_1} = -55.07\ \dfrac{\text{kW}}{\text{m}^2}$

Case (b): Compute the Flux Density Using Mean Effective Gas Emissivity Approximation

$\varepsilon_{gm1} = \varepsilon_{gm}(\varepsilon_{g1}, \alpha_{g11}, T_g, T_1)$ $\varepsilon_{gm1} = 0.2783$

$\varepsilon_{gm2} = \varepsilon_{gm}(\varepsilon_{g2}, \alpha_{g12}, T_g, T_1)$ $\varepsilon_{gm2} = 0.3813$

$$\tau_m = \frac{\varepsilon_{gm2}}{\varepsilon_{gm1}} - 1 \qquad a_m = \frac{\varepsilon_{gm1}}{1-\tau_m} \qquad \varepsilon_{gm} = a_m\cdot(1-\tau_m)$$

$\tau_m = 0.3701$ $a_m = 0.4418$ $\varepsilon_{gm} = 0.2783$ $s_1g_m = (1-\tau_m)\cdot A_1$

$$\text{S1G}_m = \frac{\varepsilon_1\cdot a_m\cdot s_1 g_m\cdot A_1}{\varepsilon_1\cdot A_1 + \rho_1\cdot s_1 g_m}$$ $\text{S1G}_m = 45.68\ \text{m}^2$ $s_1g_m = 119.67\ \text{m}^2$

$$q_{1m} = \frac{\text{SIG}_m\cdot(E_1 - E_g)}{A_1}$$ $q_{1m} = -55.38\ \dfrac{\text{kW}}{\text{m}^2}$

compared with $\dfrac{Q_1}{A_1} = -55.07\ \dfrac{\text{kW}}{\text{m}^2}$

The computed flux densities are nearly equal because there is a *single* sink zone A_1.

(This example was developed as a MATHCAD 15 worksheet. MATHCAD is a registered trademark of Parametric Technology Corporation.)

ENGINEERING MODELS FOR FUEL-FIRED FURNACES

Modern digital computation has evolved methodologies for the design and simulation of fuel-fired combustion chambers and furnaces which incorporate virtually *all* the transport phenomena, chemical kinetics, and thermodynamics studied by chemical engineers. Nonetheless, there still exist many furnace design circumstances where such computational sophistication is not always appropriate. Indeed, a practical need still exists for simple engineering models for purposes of conceptual process design, cost estimation, and the correlation of test performance data. In this section, the zone method is used to develop perhaps the simplest *computational template* available to address some of these practical engineering needs.

Input/Output Performance Parameters for Furnace Operation The term *firing density* is typically used to define the basic *operational input parameter* for fuel-fired furnaces. In practice, firing density is often defined as the input fuel feed rate per unit area (or volume) of furnace heat-transfer surface. Thus defined, the firing density is a *dimensional quantity*. Since the feed enthalpy rate $\dot{H}_f$ is proportional to the feed rate, we employ the sink area A_1 to define a *dimensionless firing density* as $N_{FD} = \dot{H}_f/\sigma T_{\text{Ref}}^4\cdot A_1$ where T_{Ref} is some characteristic reference temperature. In practice, gross furnace *output performance* is often described by using one of several *furnace efficiencies*. The most common is the *gas* or *gas-side furnace efficiency* η_g, defined as the total enthalpy transferred to furnace internals divided by the total available feed enthalpy. Here the total available feed enthalpy is defined to include the lower heating value (LHV) of the fuel plus any air preheat above an arbitrary ambient datum temperature. Under certain conditions the definition of furnace efficiency reduces to some variant of the simple equation $\eta_g = (T_{\text{Ref}} - T_{\text{out}})/(T_{\text{Ref}} - T_0)$ where again T_{Ref} is some reference temperature appropriate to the system in question.

The Long Plug Flow Furnace (LPFF) Model If a combustion chamber of cross-sectional area A_{duct} and perimeter P_{duct} is sufficiently long in the direction of flow, compared to its mean hydraulic radius $L >> R_h = A_{\text{duct}}/P_{\text{duct}}$, then the radiative flux from the gas to the bounding surfaces can sometimes be adequately characterized by the *local* gas temperature. The physical rationale for this is that the *magnitudes* of the opposed upstream and downstream radiative fluxes through a cross section transverse to the direction of flow are sufficiently large as to substantially balance each other. Such a situation is not unusual in engineering practice and is referred to as the *long furnace approximation*. As a result, the radiative flux from the gas to the bounding surface may then be approximated using *two-dimensional* directed exchange areas $\overline{S_1G}/A_1 \equiv \dfrac{\partial(\overline{S_1G})}{\partial A_1}$, calculated using methods as described previously.

Consider a duct of length L and perimeter P, and assume plug flow in the direction of flow z. Further assume high-intensity mixing at the entrance end of the chamber such that combustion is *complete* as the combustion products enter the duct. The duct then acts as a long heat exchanger in which heat is transferred to the walls at constant temperature T_1 by the combined effects of radiation and convection. Subject to the long furnace approximation, a differential energy balance on the duct then yields

$$\dot{m}\overline{C}_p \frac{dT_g}{dz} = P\left[\frac{\overline{S_1G}}{A_1}\sigma\left(T_g^4 - T_1^4\right) + h(T_g - T_1)\right] \tag{5-177}$$

where $\dot{m}$ is the mass flow rate and $\overline{C}_p$ is the heat capacity per unit mass. Equation (5-177) is nonlinear with respect to temperature. To solve Eq. (5-177), first linearize the convective heat-transfer term in the right-hand side with the approximation $\Delta T = T_2 - T_1 \approx \left(T_2^4 - T_1^4\right)/4\overline{T}_{1,2}^3$ where $\overline{T}_{1,2} = (T_1 + T_2)/2$. This linearization underestimates ΔT by no more than 5 percent when $T_2/T_1 < 1.59$. Integration of Eq. (5-177) then leads to the solution

$$\ln\left[\frac{(T_{g,\text{out}} - T_1)(T_{g,\text{in}} + T_1)}{(T_{g,\text{out}} + T_1)(T_{g,\text{in}} - T_1)}\right] + 2.0\tan^{-1}\left[\frac{(T_{g,\text{in}} - T_{g,\text{out}})\cdot T_1}{\left(T_1^2 + T_{g,\text{in}}\cdot T_{g,\text{out}}\right)}\right] = -\frac{4}{D_{\text{eff}}} \tag{5-178}$$

The long plug flow furnace (LPFF) model is described by only two dimensionless parameters, namely, an *effective* firing density and a dimensionless sink temperature

$$D_{\text{eff}} = \frac{N_{\text{FD}}}{\left(\dfrac{\overline{S_1G}}{A_1}\right) + N_{\text{CR}}} \quad \text{and} \quad \Theta_1 = T_1/T_{g,\text{in}} \tag{5-178a,b}$$

Here the dimensionless firing density N_{FD} and a dimensionless convection-radiation (CR) number N_{CR} are defined as

$$N_{\text{FD}} = \frac{\dot{m}\overline{C}_p}{\sigma\cdot T_1^3 A_1} \quad \text{and} \quad N_{\text{CR}} = \frac{h}{4\sigma\overline{T}_{g,1}^3} \tag{5-178c,d}$$

where $A_1 = PL$ is the duct surface area (the sink area), and $\overline{T}_{g1} = (\overline{T}_g + T_1)/2$ is treated as a constant. This definition of the effective dimensionless firing density D_{eff} clearly delineates the relative roles of radiation and convective heat transfer since radiation and convection are identified as *parallel (electrical) conductances*.

In analogy with a conventional heat exchanger, Eq. (5-178) displays two asymptotic limits. First define

$$\eta_f = \frac{T_{g,\text{in}} - T_{g,\text{out}}}{T_{g,\text{in}} - T_1} = 1 - \frac{T_{g,\text{out}} - T_1}{T_{g,\text{in}} - T_1} \tag{5-179}$$

as the efficiency of the long furnace. The two asymptotic limits with respect to firing density are then given by

$$D_{\text{eff}} << 1 \qquad T_{g,\text{out}} \to T_1 \qquad \eta_f \to 1 \tag{5-179a}$$

and

$$D_{\text{eff}} >> 1 \qquad T_{g,\text{out}} \to T_{g,\text{in}}$$

$$\eta_f \to \frac{4}{D_{\text{eff}}\left[1 - \dfrac{R-1}{R+1} - 2\dfrac{R-1}{R^2+1}\right]} \tag{5-179b}$$

where $R \equiv T_{g,\text{in}}/T_1 = 1/\Theta_1$.

For low firing rates, the exit temperature of the furnace gases approaches that of the sink; i.e., sufficient residence time is provided for nearly complete heat removal from the gases. When the combustion chamber is overfired, only a small fraction of the available feed enthalpy heat is removed within the furnace. The exit gas temperature then remains essentially that of the inlet temperature, and the furnace efficiency tends asymptotically to zero.

It is important to recognize that the two-dimensional exchange area $\dfrac{\overline{S_1G}}{A_1} \equiv \dfrac{\partial(\overline{S_1G})}{\partial A_1}$ in the definition of D_{eff} can represent a *lumped* two-dimensional exchange area of somewhat arbitrary complexity. This quantity also contains *all* the information concerning furnace geometry and gas and surface emissivities. To compare the relative importance of radiation with respect to convection, suppose h = 10 Btu/(h·ft²·°R) = 0.057 kW/(K·m²) and $\overline{T}_{g,1}$ = 1250 K, which leads to the numerical value N_{CR} = 0.128; or, in general, N_{CR} is of order 0.1 or less. The importance of the radiation contribution is estimated by bounding the magnitude of the dimensionless directed exchange area. For the case of a single-gas zone completely surrounded by a black enclosure, Eq. (5-169) reduces to simply $\overline{S_1G}/A_1 = \varepsilon_g \le 1.0$, and it is evident that the magnitude of the radiation contribution never exceeds unity. At high temperatures, radiative effects can easily dominate other modes of heat transfer by an order of magnitude or more. When mean beam length calculations are employed, use $L_M/D = 0.94$ for a cylindrical cross section of diameter D, and

$$L_{M0} = \frac{2H\cdot W}{H+W}$$

for a rectangular duct of height H and width W.

The Well-Stirred Combustion Chamber (WSCC) Model Many combustion chambers utilize high-momentum feed conditions with associated high-intensity mixing. The *well-stirred combustion chamber (WSCC) model* assumes a single-gas zone and high-intensity mixing. Moreover, combustion and heat transfer are visualized to occur *simultaneously* within the combustion chamber. The WSCC model is characterized by some *six* temperatures which are listed in rank order as T_0, T_{air}, T_1, T_e, T_g, and T_f. Even though the combustion chamber is well mixed, it is *arbitrarily* assumed that the gas temperature within the enclosure T_g is *not necessarily equal* to the gas exit temperature T_e. Rather the two temperatures are related by the simple relation $\Delta T_{ge} \equiv T_g - T_e$, where $\Delta T_{ge} \approx 170$ K (as a representative value) is introduced as an *adjustable parameter* for the purposes of data fitting and to make allowance for nonideal mixing. In addition, T_0 is the ambient temperature, T_{air} is the air preheat temperature, and T_f is a *pseudoadiabatic flame temperature*, as will be explained in the following development. The condition $\Delta T_{ge} \equiv 0$ is intended to simulate a perfect *continuous well-stirred reactor* (CSTR).

Dimensional WSCC Approach A macroscopic enthalpy balance on the well-stirred combustion chamber is written as

$$-\Delta H = H_{\text{in}} - H_{\text{out}} = Q_{\text{rad}} + Q_{\text{con}} + Q_{\text{ref}} \tag{5-180}$$

Here $Q_{\text{rad}} = \overline{S_1G}_R\sigma\left(T_g^4 - T_1^4\right)$ represents radiative heat transfer to the sink (with due allowance for the presence of any refractory surfaces). And the two terms $Q_{\text{con}} = h_1A_1(T_g - T_1)$ and $Q_{\text{ref}} = UA_R(T_g - T_0)$ formulate the convective heat transfer to the sink and *through* the refractory, respectively.

Formulation of the left-hand side of Eq. (5-180) requires representative thermodynamic data and information on the combustion stoichiometry. In particular, the former includes the lower heating value of the fuel, the temperature-dependent molal heat capacity of the inlet and outlet streams, and the air preheat temperature T_{air}. It proves especially convenient now to introduce the definition of a pseudoadiabatic flame temperature T_f, which is *not* the true adiabatic flame temperature, but rather is an adiabatic flame temperature based on the average heat capacity of the combustion products over the temperature interval $T_0 < T < T_e$. The calculation of T_f does not allow for dissociation of chemical species and is a surrogate for the total enthalpy content of the input fuel-air mixture. It also proves to be an especially convenient system *reference temperature*. Details for the calculation of T_f are illustrated in Example 5-14.

In terms of this particular definition of the pseudoadiabatic flame temperature T_f, the total enthalpy change and gas efficiency are given simply as

$$\Delta H = \dot{H}_f - \dot{m}\cdot\overline{C}_{P,\text{Prod}}(T_e - T_0) = \dot{m}\overline{C}_{P,\text{Prod}}(T_f - T_e) \tag{5-181a,b}$$

where $\dot{H}_f \equiv \dot{m}\cdot\bar{C}_{P,\text{Prod}}(T_f - T_0)$ and $T_e = T_g - \Delta T_{ge}$. This particular definition of T_f leads to an especially convenient formulation of furnace efficiency:

$$\eta_g = Q/\dot{H}_f = \frac{\dot{m}\cdot\bar{C}_{P,\text{Prod}}(T_f - T_e)}{\dot{m}\cdot\bar{C}_{P,\text{Prod}}(T_f - T_0)} = \frac{T_f - T_e}{T_f - T_0} \tag{5-182}$$

In Eq. (5-182), $\dot{m}$ is the total mass flow rate and $\bar{C}_{P,\text{Prod}}$ [J/(kg·K)] is defined as the *average heat capacity* of the product stream over the temperature interval $T_0 < T < T_e$.

The final overall enthalpy balance is then written as

$$\dot{m}\cdot\bar{C}_{P,\text{Prod}}(T_f - T_e) = \overline{S_1G_R}\sigma\left(T_g^4 - T_1^4\right) + h_1A_1(T_g - T_1) + UA_R(T_g - T_0) \tag{5-183}$$

with $T_e = T_g - \Delta T_{ge}$.

Equation (5-183) is a nonlinear algebraic equation which may be solved by a variety of iterative methods. The *sole* unknown quantity, however, in Eq. (5-183) is the gas temperature T_g. It should be recognized, in particular, that T_f, T_e, $\bar{C}_{P,\text{Prod}}$, and the directed exchange area are all explicit functions of T_g. The method of solution of Eq. (5-183) is demonstrated in detail in Example 5-14.

Dimensionless WSCC Approach In Eq. (5-183), assume the convective heat loss through the refractory is negligible, and linearize the convective heat transfer to the sink. These approximations lead to the result

$$\dot{m}\cdot\bar{C}_{P,\text{Prod}}(T_f - T_g + \Delta T_{ge}) = \overline{S_1G_R}\sigma\left(T_g^4 - T_1^4\right) + h_1A_1\left(T_g^4 - T_1^4\right)/4\bar{T}_{g,1}^3 \tag{5-184}$$

where $\bar{T}_{g,1} = (T_g + T_1)/2$ is some characteristic average temperature which is taken as constant. Now normalize all temperatures based on the pseudoadiabatic temperature as in $\Theta_i = T_i/T_f$. Equation (5-184) then leads to the dimensionless equation

$$D_{\text{eff}}(1 - \Theta_g + \Delta^*) = (\Theta_g^4 - \Theta_1^4) \tag{5-185}$$

where again $D_{\text{eff}} = N_{\text{FD}}/(\overline{S_1G_R}/A_1 + N_{\text{CR}})$ is defined *exactly* as in the case of the LPFF model, with the proviso that the WSCC dimensionless firing density is defined here as $N_{\text{FD}} = \dot{m}\bar{C}_{P,\text{Prod}}/\left(\sigma T_f^3\cdot A_1\right)$. The dimensionless furnace efficiency follows directly from Eq. (5-182) as

$$\eta_g = \frac{1-\Theta_e}{1-\Theta_0} = \frac{1-\Theta_g+\Delta^*}{1-\Theta_0} \tag{5-186a}$$

We also define a *reduced furnace efficiency* η_g' as

$$\eta_g' \equiv (1-\Theta_0)\cdot\eta_g = 1 - \Theta_g + \Delta^* \tag{5-186b}$$

Since Eq. (5-186*b*) may be rewritten as $\Theta_g = (1+\Delta^* - \eta_g')$ combination of Eqs. (5-185) and (5-186*b*) yields the final result

$$D_{\text{eff}}\cdot\eta_g' = (1+\Delta^* - \eta_g')^4 - \Theta_1^4 \tag{5-187}$$

Equation (5-187) provides an *explicit* relation between the modified furnace efficiency and the effective firing density directly in which the gas temperature is eliminated.

Equation (5-187) has two asymptotic limits

$$D_{\text{eff}} \ll 1 \qquad \Theta_g \to \Theta_1 \qquad \eta_g' \to 1 - \Theta_1 + \Delta^* \tag{5-188a}$$

and

$$D_{\text{eff}} \gg 1 \quad \Theta_g \to 1+\Delta^* \quad \text{and} \quad \Theta_e \to 1 \qquad \eta_g' \to \frac{(1+\Delta^*)^4 - \Theta_1^4}{D_{\text{eff}} + 4(1+\Delta^*)^3} \tag{5-188b}$$

Figure 5-21 is a plot of η_g' versus D_{eff} computed from Eq. (5-187) for the case $\Delta^* = 0$.

The asymptotic behavior of Eq. (5-189) mirrors that of the LPFF model. Here, however, for low firing densities, the exit temperature of the furnace exit gases approaches $\Theta_e = \Theta_1 - \Delta^*$ rather than the sink temperature. Moreover, for $D_{\text{eff}} \ll 1$ the reduced furnace efficiency adopts the constant value $\eta_g' = 1 - \Theta_e = 1 + \Delta^* - \Theta_1$. Again at very high firing rates, only a very small fraction of the available feed enthalpy heat is recovered within the furnace. Thus the exit gas temperature remains nearly unchanged from the pseudoadiabatic flame temperature [$T_e \approx T_f$], and the gas-side efficiency necessarily approaches zero.

Example 5-14 WSCC Furnace Model Calculations Consider the furnace geometry and combustion stoichiometry described in Example 5-13. The end-fired furnace is 3 m wide, 5 m tall, and 10 m long. Methane at a firing rate of 2500 kg/h is burned to completion with 20 percent excess air which is preheated to 600°C. The speckled furnace model is to be used. The sink (zone 1) occupies 60 percent of the total interior furnace area and is covered with two rows of 5-in (0.127-m) tubes mounted on equilateral centers with a center-to-center distance of twice the tube diameter. The sink temperature is 1000 K, and the tube emissivity is 0.7. Combustion products discharge from a 10-m² duct in the roof which is also covered with tube screen and is

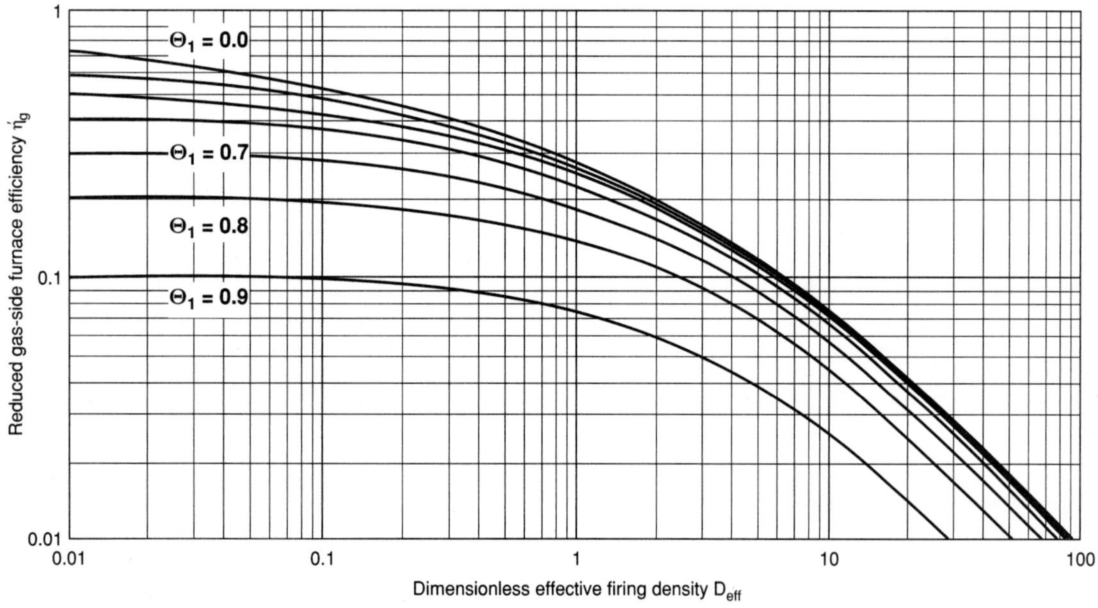

FIG. 5-21 Reduced gas-side furnace efficiency versus effective firing density for well-stirred combustion chamber model. $\Delta^* = 0$, $\Theta_1 = 0.0, 0.4, 0.5, 0.6, 0.7, 0.8, 0.9$.

to be considered part of the sink. The refractory (zone 2) with emissivity 0.6 is radiatively adiabatic but demonstrates a small convective heat loss to be calculated with an overall heat-transfer coefficient U. Compute all unknown furnace temperatures, the gas-side furnace efficiency, and the mean heat flux density through the tube surface. Use the *dimensional* solution approach for the well-stirred combustor model, and compare computed results with the *dimensionless* WSCC and LPFF models. Computed values for mean equivalent gas emissivity obtained from Eq. (5-174*b*) and Table 5-6 for $T_g = 2000$ K for $L_M = 2.7789$ m and $T_1 = 1000$ K are found to be

$$T_g = 1500\text{ K} \quad a_m = 0.44181 \quad \tau_m = 0.37014 \quad \varepsilon_m = 0.27828$$

$$T_g = 2000\text{ K} \quad a_m = 0.38435 \quad \tau_m = 0.41844 \quad \varepsilon_m = 0.22352$$

Over this temperature range the gas emissivity may be calculated by linear interpolation. Additional heat-transfer and thermodynamic data are supplied in context.

Three-zone speckled furnace model, $M = 2$ and $N = 1$:
Zone 1: Sink (60 percent of total furnace area)
Zone 2: Refractory surface (40 percent of total furnace area)
Physical constants:

$$\sigma \equiv 5.670400 \times 10^{-8}\ \frac{\text{W}}{\text{m}^2\cdot\text{K}^4}$$

Linear interpolation function for mean effective gas emissivity constants:

$$\text{LINTF}(T, Y2, Y1) = \frac{Y1\cdot(2000\text{ K} - T) + Y2\cdot(T - 1500\text{ K})}{500\text{ K}} \quad (1500\text{ K} < T < 2000\text{ K})$$

Enclosure input parameters:

$$V_{\text{tot}} = 150\text{ m}^3 \quad A_{\text{tot}} = 190\text{ m}^2 \quad C_1 = 6 \quad C_2 = 1 - C_1$$

$$A_1 = C_1\cdot A_{\text{tot}} \quad A_2 = C_2\cdot A_{\text{tot}} \quad D_{\text{tube}} = 0.127\text{ m}$$

Direct exchange areas for WSGG clear gas component (temperature-independent):

$$F = \begin{pmatrix} C_1 & C_2 \\ C_1 & C_2 \end{pmatrix} \quad AI = \begin{pmatrix} A_1 & 0 \\ 0 & A_2 \end{pmatrix} \quad ss_1 = AI\cdot F \quad ss_1 = \begin{pmatrix} 68.40 & 45.60 \\ 45.60 & 30.40 \end{pmatrix}\text{m}^2$$

Equivalent gray plane emissivity calculations for sink:

$$\varepsilon_{\text{tube}} = 0.7 \quad F_{\text{bar}} = 0.987 \quad (\text{from Fig. 5-11, curve 6 with ratio} = 2.0)$$

$$\varepsilon_{1\text{eq}} = \frac{1}{\left(\frac{1}{1} - 1\right) + 2\cdot\frac{D_{\text{tube}}}{2\cdot\pi\cdot D_{\text{tube}}}\cdot\left(\frac{1}{\varepsilon_{\text{tube}}} - 1\right) + \frac{1}{F_{\text{bar}}}} = 0.86988$$

$$\varepsilon_1 = \varepsilon_{1\text{eq}} \quad \varepsilon_2 = 0.6 \quad \varepsilon I = \begin{pmatrix} \varepsilon_1 & 0 \\ 0 & \varepsilon_2 \end{pmatrix} \quad \rho I = \text{identity}(2) - \varepsilon I$$

Total exchange areas for WSGG clear gas component:

$$R_1 = (AI - ss_1\cdot\rho I)^{-1} \quad SS_1 = \varepsilon I\cdot AI\cdot R_1\cdot ss_1\cdot \varepsilon I \quad SS_1 = \begin{pmatrix} 67.93 & 31.24 \\ 31.24 & 14.36 \end{pmatrix}\text{m}^2$$

Temperature and emissive power input data:

$$T_1 = 1000.0\text{ K} \quad T_{\text{air}} = 873.15\text{ K}$$

$$T_0 = 298.15\text{ K} \quad \Delta T_{ge} = 170\text{ K} \quad E_1 = \sigma\cdot T_1^4 \quad E_1 = 56.704\ \frac{\text{kW}}{\text{m}^2}$$

Mean beam length calculations:

$$L_{M0} = 4\cdot\frac{V_{\text{tot}}}{A_{\text{tot}}} \quad L_{M0} = 3.1579\text{ m} \quad L_M = 0.88L_{M0} \quad L_M = 2.7789\text{ m}$$

Stoichiometric and thermodynamic input data:

$$\Sigma\,\text{Mol} = 12.52914 \quad \text{LHV} = 191{,}760\ \frac{\text{cal}}{\text{mol}} \quad \text{MCH4}_{\text{air}} = 7.31\ \frac{\text{cal}}{\text{mol}\cdot\text{K}}$$

$$(800\text{ K} < T < 1200\text{ K})$$

$\dot{M}$ is the total mass flow rate and $\dot{N}_{CH4}$ is the molal flow rate of CH_4.

$$\dot{M} = \frac{\dot{M}_{\text{CH4}}\cdot \dot{M}W_{\text{in}}\cdot\Sigma\text{Mols}}{\dot{M}W_{\text{CH4}}} \quad \dot{N}_{\text{CH4}} = \frac{\dot{M}_{\text{CH4}}}{\dot{M}W_{\text{CH4}}}$$

$$\dot{M} = 54{,}174.5\ \frac{\text{kg}}{\text{h}} \quad \dot{N}_{\text{CH4}} = 155{,}860.3\ \frac{\text{mol}}{\text{h}}$$

Overall refractory heat-transfer coefficient:

$$D_r = 0.343\text{ m} \quad k_r = 0.00050\ \frac{\text{kW}}{\text{m}\cdot\text{K}}$$

$$h_0 = 0.0114\ \frac{\text{kW}}{\text{m}^2\cdot\text{K}} \quad h_0 = 2.0077\ \frac{\text{Btu}}{\text{h}\cdot\text{ft}^2\cdot{}^\circ\text{R}} \quad h_1 = 0.0170\ \frac{\text{kW}}{\text{m}^2\cdot\text{K}}$$

$$h_1 = 2.9939\ \frac{\text{Btu}}{\text{h}\cdot\text{ft}^2\cdot\text{K}} \quad U = \frac{1}{\frac{1}{h_0} + \frac{1}{h_1} + \frac{D_r}{k_r}}$$

$$U = 0.001201\ \frac{\text{kW}}{\text{m}^2\cdot\text{K}} \quad U = 0.2115\ \frac{\text{Btu}}{\text{h}\cdot\text{ft}^2\cdot{}^\circ\text{R}}$$

START OF ITERATION LOOP: Successive Substitution with T_g as the Trial Variable

Assume $T_g = 1{,}759.1633222447$ K $\quad T_e = T_g - \Delta T_{ge} \quad T_e = 1589.2$ K

$$E_g = \sigma\cdot T_g^4 \quad E_g = 543.05\ \frac{\text{kW}}{\text{m}^2}$$

Compute temperature-dependent mean effective gas emissivity via linear interpolation:

$$\tau_m = \text{LINTF}(T_g, 0.37014, 0.418442) \quad a_m = \text{LINTF}(T_g, 0.44181, 0.38435)$$

$$\tau_m = 0.3934 \quad a_m = 0.4141 \quad \varepsilon_m = a_m(1 - \tau_m) \quad \varepsilon_m = 0.2512$$

Compare interpolated value:

$$\varepsilon_{\text{com}} = \text{LINTF}(T_g, 0.27828, 0.22352) \quad \varepsilon_{\text{com}} = 0.2519$$

Direct and total exchange areas for WSGG gray gas component:

$$ss_2 = \tau_m\cdot ss_1 \quad sg_2 = (AI - ss_2)\begin{bmatrix} 1 \\ 1 \end{bmatrix} \quad R_2 = (AI - ss_2\cdot\rho I)^{-1}$$

$$ss_2 = \begin{pmatrix} 26.91 & 17.94 \\ 17.94 & 11.96 \end{pmatrix}\text{m}^2 \quad sg_2 = \begin{pmatrix} 69.15 \\ 46.10 \end{pmatrix}\text{m}^2 \quad SS_2 = \varepsilon I\cdot AI\cdot R_2\cdot ss_2\cdot\varepsilon I$$

Compute directed exchange areas:

$$\text{DSS} = (1 - a_m)\cdot SS_1 + a_m\cdot SS_2 \quad \text{DSG} = a_m\cdot\varepsilon\text{I}\cdot\text{AI}\cdot R_2\cdot sg_2$$

Refractory augmented directed gas-sink exchange area:

$$\text{DS1GR} = \text{DSG}_1 + \frac{\text{DSS}_{1,2}\cdot\text{DSG}_2}{\text{DSS}_{1,2} + \text{DSG}_2} \quad \text{DS1GR} = 35.59\text{ m}^2 \quad \text{Eq. (5-172}a)$$

Compute refractory temperature (T_2); assume radiative equilibrium:

$$\text{Equation (5-172}b\text{):}\quad E_2 = \frac{\text{DSS}_{2,1}\cdot E_1 + \text{DSG}_2\cdot E_g}{\text{DSS}_{2,1} + \text{DSG}_2}$$

$$E_2 = 231.24\ \frac{\text{kW}}{\text{m}^2}$$

$$T_2 = \left(\frac{E_2}{\sigma}\right)^{0.25} \quad T_2 = 1421.1\text{ K}$$

Enthalpy balance: Basis: 1 mol CH_4:

$$h_{\text{in}} = \text{LHV} + (\Sigma\,\text{Mol} - 1)\cdot\text{MC}p_{\text{air}}\cdot(T_{\text{air}} - T_0) \quad h_{\text{in}} = 240{,}219.9\ \frac{\text{cal}}{\text{mol}}$$

Compute pseudoadiabatic flame temperature T_f:

$$T_f = T_0 + \frac{h_{\text{in}}}{\Sigma\,\text{Mols MC}p(T_e)} \quad h_{\text{out}} = \Sigma\,\text{mols}\cdot\text{MC}p(T_e)\cdot(T_e - T_0)$$

$$T_f = 2580.5\text{ K} \quad h_{\text{out}} = 135{,}880.8\ \frac{\text{cal}}{\text{mol}}$$

$$H_{\text{in}} = \dot{N}_{\text{CH4}}\cdot h_{\text{in}} \quad H_{\text{out}} = \dot{N}_{\text{CH4}}\cdot h_{\text{out}} \quad \Delta H = H_{\text{out}} - H_{\text{in}}$$

$$H_{\text{in}} = 43{,}543.59\text{ kW} \quad H_{\text{out}} = 24{,}630.51\text{ kW} \quad \Delta H = -18{,}913.08\text{ kW}$$

Overall enthalpy balance:

$$Q_{1g} = \text{DS1GR}\cdot(E_g - E_1) \quad Q_{1g} = 17{,}308.45\text{ kW}$$

$$Q_{1\text{con}} = h_1\cdot A_1\cdot(T_g - T_1) \quad Q_{1\text{con}} = 1471.26\text{ kW}$$

$$Q_{2\text{con}} = U\cdot A_2\cdot(T_g - T_0) \quad Q_{2\text{con}} = 133.37\text{ kW}$$

$$\text{ERROR} = Q_{1g} + Q_{1\text{con}} + Q_{2\text{con}} + \Delta H \quad \%\text{ERROR1} = 100\ \frac{\text{ERROR}}{\Delta H}$$

$$\text{ERROR} = -0.0000\text{ kW} \quad \%\text{ERROR1} = 0.00000$$

$$T_{g\text{calc}} = \left[\frac{\text{DS1GR}\cdot E_1 - (\Delta H + Q_{1\text{Con}} + Q_{2\text{Con}})}{\text{DS1GR}\cdot\sigma}\right]^{0.25} \quad T_{g\text{calc}} = 1759.16332\text{ K}$$

Average assumed and calculated temperatures for next iteration:

$$T_{g,\text{new}} = \frac{T_g + T_{g\text{calc}}}{2} \qquad T_{g\text{new}} = 1759.1633222935\ \text{K} \qquad \text{Go to Start}$$

END OF ITERATION LOOP: Final Gas Temperature $T_g = 1759.16$ K

$$\eta_g = \frac{H_{\text{in}} - H_{\text{out}}}{H_{\text{in}}} \ \text{or}\ \eta_{1g} = \frac{T_f - T_e}{T_f - T_0} \qquad \eta_g = 0.43435 \qquad \eta_{1g} = 0.43435$$

Heat flux density calculations:

$$q_1 = \frac{Q_{1g} + Q_{1\text{Con}}}{A_1} \qquad q_{\text{tube}} = q_1 \cdot \left(2 \cdot \frac{D_{\text{tube}}}{2 \cdot \pi \cdot D_{\text{tube}}}\right)$$

$$q_1 = 164.7\ \frac{\text{kW}}{\text{m}^2} \qquad q_{\text{tube}} = 52.44\ \frac{\text{kW}}{\text{m}^2}$$

Note: This example was also solved with $\Delta T_{ge} = 0$. The results were as follows: $T_f = 2552.8$ K, $T_g = T_e = 1707.1$ K, $T_2 = 1381.1$ K, $\eta_g = 37.51$ percent, $D_{\text{eff}} = 0.53371$, and $\Delta H = 16{,}332.7$ kW. The WSCC model with $\Delta T_{ge} = 0$ predicts a lower performance bound.

Compare dimensionless WSCC model:

$$\Theta_0 = \frac{T_0}{T_f} \quad \Theta_1 = \frac{T_1}{T_f} \quad \Delta T_{\text{star}} = \frac{\Delta T_{ge}}{T_f}$$

$$T_{g1} = \frac{T_1 + T_g}{2}$$

$$\Theta_0 = 0.1155 \qquad \Theta_1 = 0.3875 \qquad \Delta T_{\text{star}} = 0.06588 \qquad T_{g1} = 1379.58\ \text{K}$$

$$\text{C}_{P,\text{Prod}} = \frac{\text{MC}p(T_e)}{\text{MW}_{\text{in}}} \qquad \text{C}_{P,\text{Prod}} = 0.000352\ \text{kW} \cdot \frac{\text{h}}{\text{kg} \cdot \text{K}}$$

$$N_{\text{FD}} = \dot{M}\frac{\text{C}_{P,\text{Prod}}}{\sigma \cdot T_f^3 \cdot A_1} \qquad N_{\text{CR}}\frac{h_1}{4 \cdot \sigma \cdot T_{g1}^3} \qquad N_{\text{eff}}\frac{N_{\text{FD}}}{\dfrac{\text{DS1GR}}{A_1} + N_{\text{CR}}}$$

$$N_{\text{FD}} = 0.17176 \qquad N_{\text{CR}} = 0.02855 \qquad D_{\text{eff}} = 0.50409 \qquad \frac{\text{DS1GR}}{A_1} = 0.31218$$

$$\eta' = \eta_g \cdot (1 - \Theta_0) \qquad D_{1\,\text{eff}} = \frac{(1 + \Delta T_{\text{star}} - \eta_{\text{prime}})^4 - \Theta_1^4}{\eta_{\text{prime}}}$$

$$D_{1\,\text{eff}} = 0.50350 \qquad \text{versus} \qquad D_{\text{eff}} = 0.50409$$

This small discrepancy is due to linearization and neglect of convective refractory heat losses in the dimensionless WSCC model.

Compare dimensionless LPFF model:

$$R_{\text{in}} = \frac{1}{\Theta_1} \qquad R_{\text{in}} = 2.58050$$

Trial-and-error calculation to match effective firing densities:

Assume

$$T_{\text{out}} = 1000.13763\ \text{K} \qquad R_{\text{out}} = \frac{T_{\text{out}}}{T_1} \qquad R_{\text{out}} = 1.00014 \qquad \frac{4}{D_{\text{eff}}} = 7.93513$$

$$\text{CLong} = -\ln\left[\frac{(R_{\text{out}} - 1) \cdot (R_{\text{in}} + 1)}{(R_{\text{in}} - 1) \cdot (R_{\text{out}} + 1)}\right] - 2 \cdot \text{atan}\left(\frac{R_{\text{in}} - R_{\text{out}}}{1 + R_{\text{in}} \cdot R_{\text{out}}}\right) \qquad \text{CLong} = 7.93514$$

$$\eta L_f = \frac{R_{\text{in}} - R_{\text{out}}}{R_{\text{in}} - 1} \qquad \eta L_f = 0.99991292 \ \text{versus}\ \eta_g = 0.43435$$

Note: The long plug flow furnace model is so efficient that it would be grossly underfired using the computed WSCC effective firing density. Of the two models, the LPFF model always predicts an upper theoretical performance limit. (This example was developed as a MATHCAD 15 worksheet. MATHCAD is registered trademark of Parametric Technology Corporation.)

WSCC Model Utility and More Complex Zoning Models Despite its simplicity, the WSCC construct has a wide variety of practical uses and is of significant pedagogical value. Here an engineering situation of inordinate complexity is described by the definition of only eight dimensionless quantities D_{eff}, N_{FD}, $\overleftarrow{\text{S}_1\text{G}_\text{R}}/A_1$, N_{CR}, η_g, Δ^*, Θ_0, and Θ_1. The first three are related by the simple algebraic definition $D_{\text{eff}} = N_{\text{FD}} / (\overline{S_1G_R}/A_1 + N_{\text{CR}})$. These dimensionless quantities contain all the physical input information for the model, namely, furnace parameters and geometry, radiative properties of the combustion products, and the stoichiometry and thermodynamics of the combustion process. The WSCC model leads to a dimensionless two-dimensional plot of *reduced* effective furnace efficiency versus dimensionless effective firing density (Fig. 5-21), which is characterized by only two additional parameters, namely, Δ^* and Θ_1.

Of the models presented here, the WSCC model with $\Delta T_{ge} = 0$ produces the lowest furnace efficiencies. The long furnace model usually produces the highest furnace efficiency. This is really not a fair statement because two distinctly different pieces of process equipment are compared. In this regard, a more appropriate definition of the dimensionless firing density for the LPFF model might be $N'_{\text{FD}} = \dot{m}\bar{C}_p / \left(\sigma \cdot T_{g,\text{in}}^3 \cdot A_1\right)$. It may be counterintuitive, but the WSCC and LPFF models *generally do not* characterize the extreme conditions for the performance of combustors as in the case of chemical reactors.

Figure 5-21 has been used to correlate furnace performance data for a multitude of industrial furnaces and combustors. Typical operational domains for a variety of fuel-fired industrial furnaces are summarized in Table 5-8. The WSCC approach (or "speckled" furnace model) is a classic contribution to furnace design methodology which was first due to Hottel [Hottel, H. C., and A. F. Sarofim, *Radiative Transfer*, McGraw-Hill, New York, 1967]. The WSCC model provides a simple *furnace design template* which leads to a host of more complex furnace models. These models include an obvious extension to a tanks-in-series model as well as multizone models utilizing empirical cold-flow velocity patterns. For more information on practical furnace design models, see Hottel and Sarofim (*Radiative Transfer*, McGraw-Hill, New York, 1967, chap. 14). Qualitative aspects of process equipment have been treated in some detail elsewhere (Baukal, C. E., ed., *The John Zink Combustion Handbook*, CRC Press, Boca Raton, Fla., 2001).

TABLE 5-8 Operational Domains for Representative Process Furnaces and Combustors

Domain	Furnace or combustor type	Dimensionless sink temperature	Dimensionless firing density
A	Oil processing furnaces; radiant section of oil tube stills and cracking coils	$\Theta_1 \approx 0.4$	$0.1 < D_{\text{eff}} < 1.0$
B	Domestic boiler combustion chambers	$\Theta_1 \approx 0.2$	$0.5 < D_{\text{eff}} < 1.1$
C	Glass furnaces	$0.7 < \Theta_1 < 0.8$	$0.035 < D_{\text{eff}} < 0.8$
D	Soaking pits	$\Theta_1 \approx 0.6$	$0.7 < D_{\text{eff}} < 1.1$
E	Gas-turbine combustors	$0.4 < \Theta_1 < 0.7$	$4.0 < D_{\text{eff}} < 25.0$

MASS TRANSFER

General References: Benitez, J., *Principles and Modern Applications of Mass Transfer Operations,* 3d ed., Wiley, New York, 2016; Bird, R. B., W. E. Stewart, and E. N. Lightfoot, *Transport Phenomena,* rev. 2d ed., Wiley, New York, 2006; Cussler, E. L., *Diffusion: Mass Transfer in Fluid Systems,* 3d ed., Cambridge University Press, London, 2009; Danner, R. P. and T. E. Daubert, *Manual for Predicting Chemical Process Design Data,* AIChE, New York, 1983; Daubert, T. E. and R. P. Danner, *Physical and Thermodynamic Properties of Pure Chemicals,* Taylor and Francis, Bristol, Pa., 1989–1995; Gammon, B. E., K. N. Marsh, and A. K. R. Dewan, *Transport Properties and Related Thermodynamic Data of Binary Mixtures,* AIChE, New York. Part 1, 1993: Part 2, 1994; Geankoplis, C. J., *Transport Processes and Separation Process Principles,* 4th ed., Prentice-Hall PTR, Upper Saddle River, N.J., 2003; Kirwan, D. J., "Mass Transfer Principles," Chap. 2 in Rousseau, R. W. (ed.), *Handbook of Separation Process Technology,* Wiley, New York, 1987; McCabe, W. L., J. C. Smith, and P. Harriott, *Unit Operations of Chemical Engineering,* 7th ed., McGraw-Hill, New York, 2005; Poling, B. E., J. M. Prausnitz, and J. P. O'Connell, *The Properties of Gases and Liquids,* 5th ed., McGraw-Hill, New York, 2001; Schwartzberg, H. G. and R. Y. Chao, *Food Technol.* **36**(2): 73 (1982); Sherwood, T. K., R. L. Pigford, and C. R. Wilke, *Mass Transfer,* McGraw-Hill, New York, 1975; Skelland, A. H. P., *Diffusional Mass Transfer,* 2d ed., Kreiger Publishing, Malabar, Fla., 1985;

Taylor, R. and R. Krishna, *Multicomponent Mass Transfer,* Wiley, New York, 1993; Treybal, R. E., *Mass-Transfer Operations,* 3d ed., McGraw-Hill, New York, 1980; Welty, J., G. L. Rorrer, and D. G. Foster, *Fundamentals of Momentum, Heat, and Mass Transfer,* 6th ed., Wiley, New York, 2014; Wesselingh, J. A. and R. Krishna, *Mass Transfer in Multicomponent Mixtures,* Delft University Press, Delft, Netherlands, 2000.

References for Diffusivities, Tables 5-11, 5-14, and 5-15

1. Asfour, A. F. A., and F. A. L. Dullien, *Chem. Eng. Sci.* **41:** 1891 (1986).
2. Bosse, D., and H-J. Bart, *Ind. Eng. Chem. Res.* **45:** 1822 (2006).
3. Brokaw, R. S., *Ind. Eng. Chem. Process Des. and Dev.* **8** (2): 240 (1969).
4. Caldwell, C. S., and A. L. Babb, *J. Phys. Chem.* **60:** 51 (1956).
5. Catchpole, O. J., and M. B. King, *Ind. Eng. Chem. Res.* **33:** 1828 (1994).
6. Chen, B. H. C., and S. H. Chen, *Chem. Eng. Sci.* **40:** 1735 (1985).
7. Cullinan, H. T., *AIChE J.* **31:** 1740–1741 (1985).
8. Cussler, E. L., *AIChE J.* **26:** 1 (1980).
9. Fuller, E. N., P. D. Schettler, and J. C. Giddings, *Ind. Eng. Chem.* **58:** 18 (1966).
10. Hayduk, W., and B. S. Minhas, *Can. J. Chem. Eng.* **6:** 195 (1982).
11. He, C. H., and Y-S.Yu, *Ind. Eng. Chem. Res.* **36:** 4430 (1997).
12. Lee, H., and G. Thodos, *Ind. Eng. Chem. Fundam.* **22:** 17–26 (1983).
13. Leffler, J., and H. T. Cullinan, *Ind. Eng. Chem. Fundam.* **9:** 84, 88 (1970).
14. Mathur, G. P., and G. Thodos, *AIChE J.* **11:** 613 (1965).
15. Matthews, M. A., and A. Akgerman, *AIChE J.* **33:** 881 (1987).
16. Rathbun, R. E., and A. L. Babb, *Ind. Eng. Chem. Proc. Des. Dev.* **5:** 273 (1966).
17. Siddiqi, M. A., and K. Lucas, *Can. J. Chem. Eng.* **64:** 839 (1986).
18. Sun, C. K. J., and S. H. Chen, *Ind. Eng. Chem. Res.* **26:** 815 (1987).
19. Tyn, M. T., and W. F. Calus, *J. Chem. Eng. Data* **20:** 310 (1975).
20. Vignes, A., *Ind. Eng. Chem. Fundam.* **5:** 184 (1966).
21. Wilke, C. R., and P. Chang, *AIChE J.* **1:** 164 (1955).
22. Wilke, C. R., and C. Y. Lee, *Ind. Eng. Chem.* **47:** 1253 (1955).

References for Diffusivities in Porous Solids, Table 5-16

23. Ruthven, D. M., *Principles of Adsorption and Adsorption Processes,* Wiley, New York, 1984.
24. Satterfield, C. N., *Mass Transfer in Heterogeneous Catalysis,* MIT Press, Cambridge, Mass., 1970.
25. Suzuki, M., *Adsorption Engineering,* Kodansha—Elsevier, Amsterdam, Netherlands, 1990.
26. Yang, R. T., *Gas Separation by Adsorption Processes,* Butterworths, Oxford, UK, 1987.

References for Tables 5-17 to 5-23

27. Blatt, W. F., et al., *Membrane Science and Technology,* p. 47, Plenum, 1970.
28. Bocquet, S., et al., *AIChE J.* **51:** 1067 (2005).
29. Bolles, W. L., and J. R. Fair, *Institution Chem. Eng. Symp. Ser.* **56**(2): 35 (1979).
30. Bolles, W. L., and J. R. Fair, *Chem. Eng.* **89**(14): 109 (July 12, 1982).
31. Bravo, J. L., and J. R. Fair, *Ind. Eng. Chem. Process Des. Dev.* **21:** 162 (1982).
32. Bravo, J. L., J. A. Rocha, and J. R. Fair, *Hydrocarbon Processing* **91** (Jan. 1985).
33. Brian, P. L. T., and H. B. Hales, *AIChE J.* **15:** 419 (1969).
34. Calderbank, P. H., and M. B. Moo-Young, *Chem. Eng. Sci.* **16:** 39 (1961).
35. Cavatorta, O. N., U. Bohm, and A. Chiappori de del Giorgio, *AIChE J.* **45:** 938 (1999).
36. Chaumat, H., et al., *Chem. Eng. Sci.* **60:** 5930 (2005).
37. Chen, Y. S., C-C. Lin, and H-S. Liu, *Ind. Eng. Chem. Res.* **44:** 7868 (2005).
38. Chilton, T. H., and A. P. Colburn, *Ind. Eng. Chem.* **26:** 1183 (1934).
39. Chowdiah, V. N., G. L Foutch, G. L. and G-C. Lee, *Ind. Eng. Chem. Res.* **42:** 1485 (2003).
40. Colburn, A. P., *Trans. AIChE* **29:** 174 (1933).
41. Cornell, D., W. G. Knapp, and J. R. Fair, *Chem. Engr. Prog.* **56**(7): 68 (1960).
42. Cornet, I., and U. Kaloo, *Proc. 3rd Int'l. Congr. Metallic Corrosion—Moscow,* **3:** 83 (1966).
43. Crause, J. C., and I. Nieuwoudt, *Ind. Eng. Chem. Res.* **38:** 4928 (1999).
44. Dudukovic, A., V. Milosevic, and R. Pjanovic, *AIChE J.* **42:** 269 (1996).
45. Dwivedi, P. N., and S. N. Upadhyay, *Ind. Eng. Chem. Process Des. Develop.* **16:** 1657 (1977).
46. Eisenberg, M., C. W. Tobias, and C. R. Wilke, *Chem. Engr. Prog. Symp. Ser.* **51**(16): 1 (1955).
47. Fair, J. R., "Distillation" *in* Rousseau, R. W. (ed.), *Handbook of Separation Process Technology,* Wiley, New York, 1987.
48. Fan, L. T., Y. C. Yang, and C. Y. Wen, *AIChE J.* **6:** 482 (1960).
49. Garner, F. H., and R. D. Suckling, *AIChE J.* **4:** 114 (1958).
50. Gibilaro, L. G., et al., *Chem. Engr. Sci.* **40:** 1811 (1985).
51. Gilliland. E. R., and T. K. Sherwood, *Ind. Engr. Chem.* **26:** 516 (1934).
52. Gostick, J., et al., *Ind. Eng. Chem. Res.* **42:** 3626 (2003).
53. Griffith, R. M., *Chem. Engr. Sci.* **12:** 198 (1960).
54. Gupta, A. S., and G. Thodos, *AIChE J.* **9:** 751 (1963).
55. Gupta, A. S., and G. Thodos, *Ind. Eng. Chem. Fundam.* **3:** 218 (1964).
56. Harriott, P., *AIChE J.* **8:** 93 (1962).
57. Hines, A. L., and R. N. Maddox, *Mass Transfer: Fundamentals and Applications,* Prentice-Hall, Upper Saddle River, N.J., 1985.
58. Houwing, J., H. A. H. Billiet, and L. A. M. van der Wielin, *AIChE J.* **49:** 1158 (2003).
59. Hsiung, T. H., and G. Thodos, *Int. J. Heat Mass Transfer* **20:** 331 (1977).
60. Hsu, N. T., K. Sato, and B. H. Sage, *Ind. Engr. Chem.* **46:** 870 (1954).
61. Hughmark, G. A., *Ind. Eng. Chem. Fundam.* **6:** 408 (1967).
62. Johnson, A. I., F. Besic, and A. E. Hamielec, *Can. J. Chem. Engr.* **47:** 559 (1969).
63. Johnstone, H. F., and R. L. Pigford, *Trans. AIChE* **38:** 25 (1942).
64. Kafesjian, R., C. A. Plank, and E. R. Gerhard, *AIChE J.* **7:** 463 (1961).
65. Kelly, N. W., and L. K. Swenson, *Chem. Eng. Prog.* **52:** 263 (1956).
66. King, C. J., *Separation Processes,* 2d ed., McGraw-Hill, New York, 1980.
67. Klein, E., R. A. Ward, and R. E. Lacey, "Membrane Processes—Dialysis and Electro-Dialysis" in Rousseau, *Handbook of Separation Process Technology,* Wiley, New York, 1987.
68. Kohl, A. L., "Absorption and Stripping" in Rousseau, R. W. (ed.), *Handbook of Separation Process Technology,* Wiley, New York, 1987.
69. Kojima, M., et al., *J. Chem. Engng. Japan* **20:** 104 (1987).
70. Kreutzer, M. T., et al., *Ind. Eng. Chem. Res.* **44:** 9646 (2005).
71. Larachi, F., et al., *Ind. Eng. Chem. Res.* **42:** 222 (2003).
72. Lee, J. M., *Biochemical Engineering,* Prentice-Hall, Upper Saddle River, N.J., 1992.
73. Lee, J. H., and N. R. Foster, *Appl. Catal.* **63:** 1 (1990).
74. Lee, C. H., and G. D. Holder, *Ind. Engr. Chem. Res.* **34:** 906 (1995).
75. Lee, S., and R. M. Lueptow, *Separ. Sci. Technol.* **39:** 539 (2004).
76. Levich, V. G., *Physicochemical Hydrodynamics,* Prentice-Hall, Upper Saddle River, N.J., 1962.
77. Levins, D. M., and J. R. Gastonbury, *Trans. Inst. Chem. Engr.* **50:** 32, 132 (1972).
78. Linton, W. H., and T. K. Sherwood, *Chem. Engr. Prog.* **46:** 258 (1950).
79. Ludwig, E. E., *Applied Process Design for Chemical and Petrochemical Plants,* 2d ed., vol. 2, Gulf Pub. Co., 1977.
80. Notter, R. H., and C. A. Sleicher, *Chem. Eng. Sci.* **26:** 161 (1971).
81. Ohashi, H., et al., *J. Chem. Engr. Japan* **14:** 433 (1981).
82. Onda, K., H. Takeuchi, and Y. Okumoto, *J. Chem. Engr. Japan* **1:** 56 (1968).
83. Pangarkar, V. G., et al., *Ind. Eng. Chem. Res.* **41:** 4141 (2002).
84. Pasternak, I. S., and W. H. Gauvin, *AIChE J.* **7:** 254 (1961).
85. Pasternak, I. S., and W. H. Gauvin, *Can. J. Chem. Engr.* **38:** 35 (April 1960).
86. Patil, S. S., N. A. Deshmukh, and J. B. Joshi, *Ind. Eng. Chem. Res.* **43:** 2765 (2004).
87. Prasad, R., and K. K. Sirkar, *AIChE J.* **34:** 177 (1988).
88. Rahman, K., and M. Streat, *Chem. Engr. Sci.* **36:** 293 (1981).
89. Ramirez, J. A., and R. H. Davis, *AIChE J.* **45:** 1355 (1999).
90. Ranz, W. E., and W. R. Marshall, *Chem. Engr. Prog.* **48:** 141, 173 (1952).
91. Reiss, L. P., *Ind. Eng. Chem. Process Des. Develop.* **6:** 486 (1967).
92. Rocha, J. A., J. L. Bravo, and J. R. Fair, *Ind. Eng. Chem. Res.* **35:** 1660 (1996).
93. Rowe, P. N., K. T. Claxton, and J. B. Lewis, *Trans. Inst. Chem. Engr. London* **43:** 14 (1965).
94. Ruckenstein, E., and R. Rajagopolan, *Chem. Engr. Commun.* **4:** 15 (1980).
95. Satterfield, C. N., *AIChE J.* **21:** 209 (1975).
96. Schluter, V., and W. D. Deckwer, *Chem. Engr. Sci.* **47:** 2357 (1992).
97. Schugerl, K., et al., *Adv. Biochem. Eng.* **8:** 63 (1978).
98. Scott, K., and J. Lobato, *Ind. Eng. Chem. Res.* **42:** 5697 (2003).
99. Shah, Y. T., et al., *AIChE J.* **28:** 353 (1982).
100. Sherwood, T. K.. et al., *Ind. Eng. Chem. Fundam.* **4:** 113 (1965).
101. Sherwood, T. K., and E. A. L. Holloway, *Trans. Am. Inst. Chem. Eng.* **36:** 39 (1940).
102. Siegel, R., E. M. Sparrow, and T. M. Hallman, *Appl. Sci. Res. Sec. A.,* **7:** 386 (1958).
103. Sirkar, K. K., *Separation of Molecules, Macromolecules and Particles,* Cambridge University Press, London, 2014.
104. Skelland, A. H. P., and A. R. H. Cornish, *AIChE J.* **9:** 73 (1963).
105. Skelland, A. H. P., and D. W. Tedder, "Extraction—Organic Chemicals Processing" in Rousseau, *Handbook of Separation Process Technology,* Wiley, New York, 1987, pp. 405–466.
106. Skelland, A. H. P., and R. M. Wellek, *AIChE J.* **10:** 491, 789 (1964).
107. Slater, M. J., "Rate Coefficients in Liquid-Liquid Extraction Systems," in Godfrey, J. C. and Slater, M. J. (eds.), *Liquid-Liquid Extraction Equipment,* Wiley, New York, 1994, pp. 45–94.
108. Taniguchi, M., and S. Kimura, *AIChE J.* **47:** 1967 (2000).
109. Tournie, P., C. Laguerie, and J. P. Couderc, *Chem. Engr. Sci.* **34:** 1247 (1979).
110. Vandu, C. O., H. Liu, and R. Krishna, *Chem. Eng. Sci.* **60:** 6430 (2005).
111. Von Karman, T., *Trans. ASME* **61:** 705 (1939).
112. Wakao, N., and T. Funazkri, *Chem. Engr. Sci.* **33:** 1375 (1978).
113. Wang, G. Q., X. G. Yuan, and K. T. Yu, *Ind. Eng. Chem. Res.* **44:** 8715 (2005).
114. Wankat, P. C., *Separation Process Engineering,* 4th ed., chap.15, Prentice-Hall, Upper Saddle River, N.J., 2016.
115. Wilson, E. J., and C. J. Geankoplis, *Ind. Eng. Chem. Fundam.* **5:** 9 (1966).
116. Wright, P., and B. J. Glasser, *AIChE J.* **47:** 474 (2001).
117. Yagi, H., and F. Yoshida, *Ind. Eng. Chem. Process Des. Dev.* **14:** 488 (1975).

INTRODUCTION

This part of Sec. 5 provides a concise guide to solving problems in situations commonly encountered by chemical engineers. It deals with diffusivity and mass-transfer coefficient estimation and common flux equations, although material balances are also presented in typical coordinate systems to permit a wide range of problems to be formulated and solved.

Mass-transfer calculations involve transport properties, such as diffusivities, and other empirical factors that have been found to relate mass-transfer rates to measured "driving forces" in myriad geometries and conditions. The context of the problem dictates whether the fundamental or more applied coefficient should be used. One key distinction is that whenever there is flow parallel to an interface through which mass transfer occurs, the relevant coefficient is an empirical combination of properties and conditions. Conversely, when diffusion occurs in stagnant media or in creeping flow without transverse velocity gradients, ordinary diffusivities may be suitable for solving the problem. In either case, it is strongly suggested to employ data, whenever available, instead of relying on correlations.

Units employed in diffusivity correlations commonly followed the cgs system. Similarly, correlations for mass-transfer correlations used the cgs or English system. In both cases, only the most recent correlations employ SI units. Since most correlations involve other properties and physical parameters, often with mixed units, they are repeated here as originally stated. Common conversion factors are listed in Table 1-4.

Fick's First Law This equation (which, as noted, is frequently called Fick's first law, though to call it a law is a misnomer) relates flux of a component to its composition gradient, employing a constant of proportionality called *diffusivity.* It is reasonably accurate for binary mixtures in which the diffusing component (A) is dilute and when no conveyance (explained below) exists; hence, its applicability is very limited. It can be written in several forms, depending on the units and frame of reference. Three that are related but not identical are

$${}_V J_A = -D_{AB}\frac{dc_A}{dz} \approx {}_M J_A = -cD_{AB}\frac{dx_A}{dz} \propto {}_m J_A = -\rho D_{AB}\frac{dw_A}{dz} \qquad (5\text{-}189)$$

The first equality (on the left-hand side) corresponds to the molar flux with respect to the volume average velocity, while the equality in the center represents the molar flux with respect to the molar average velocity and the one on the right is the mass flux with respect to the mass average velocity. These fluxes must be used with consistent flux expressions for fixed coordinates and for N_C components, such as

$$N_A = {}_V J_A + c_A\sum_{i=1}^{N_C} N_i\bar{V}_i = {}_M J_A + x_A\sum_{i=1}^{N_C} N_i = \frac{{}_m J_A + w_A\sum_{i=1}^{N_C} n_i}{M_A} \qquad (5\text{-}190)$$

The summations account for *conveyance,* which is the amount of component A carried by the net flow in the direction of diffusion. Those terms may account for as much as 10 percent of the flux, though in most cases it is much less, and it is frequently ignored. Some people refer to this as the "convective" term, but that usage conflicts with the other sense of convection which is promoted by flow perpendicular to the direction of flux.

Mutual Diffusivity, Mass Diffusivity, and Interdiffusion Coefficient Diffusivity is denoted by D_{AB} and is defined by Fick's first law as the ratio of the flux to the concentration gradient, as in Eq. (5-189). It is analogous to the thermal diffusivity in Fourier's law and to the kinematic viscosity in Newton's law. These analogies are flawed because both heat and momentum are conveniently defined with respect to fixed coordinates, irrespective of the direction of transfer or its magnitude, while mass diffusivity most commonly requires information about bulk motion of the medium in which diffusion occurs. For liquids, D_{AB} usually depends on the concentration of A and only becomes constant as the concentration of A approaches zero, where $D_{AB} = D^{\circ}_{AB}$, the infinite dilution limit diffusivity.

When the flux expressions are consistent, as in Eq. (5-190), the diffusivities in Eq. (5-189) are identical. As a result, experimental diffusivities are often measured under constant-volume conditions but may be used for applications involving open systems. It turns out that the two versions are very nearly equivalent for gas systems because there is negligible volume change on mixing. That is not usually true for liquids, however.

Self-Diffusivity and Tracer Diffusivity *Self-diffusivity* is denoted by $D_{A'A}$ and is the measure of mobility of a species in itself. For instance, A' may represent a small concentration of molecules tagged with a radioactive isotope so they can be detected, but they do not have significantly different properties. Hence, the solution is ideal, and there are practically no gradients to "force" or "drive" diffusion. This kind of diffusion is presumed to be purely stochastic in nature. To cite a specific example, when A is less mobile than B, their self-diffusion coefficients can be used as rough lower and upper bounds of the mutual diffusion coefficient; that is, $D_{A'A} \le D_{AB} \le D_{B'B}$. Obviously, when A is more mobile than B, the inequalities are reversed. Similarly, *tracer diffusivity,* denoted by $D_{A'B}$, is related to both mutual and self-diffusivity. It is evaluated in the presence of a second component B, again using a tagged isotope of the first component. In the dilute range, tagging A merely provides a convenient method for indirect composition analysis. Tracer diffusivities approach mutual diffusivities at the dilute limit, and they approach self-diffusivities at the pure component limit. That is, at the limit of dilute A in B, $D_{A'B} \to D^{\circ}_{AB}$ and $D_{B'A} \to D_{B'B}$; likewise at the limit of dilute B in A, $D_{B'A} \to D^{\circ}_{BA}$ and $D_{A'B} \to D_{A'A}$.

The tracer diffusivity and the self-diffusivity provide a means to understand ordinary diffusion and as order-of-magnitude estimates of mutual diffusivities. Darken's equation [Eq. (5-222)] was derived for tracer diffusivities but is often used to relate mutual diffusivities at moderate concentrations as opposed to infinite dilution. Zhu et al. [*Chem. Eng. Sci.* **132:** 250 (2015)] recently published a model for the prediction of mutual diffusion coefficients in binary liquid mixtures from tracer diffusion coefficients. Data and correlations for self-diffusivity and tracer diffusivity are covered later in this section.

Mass-Transfer Coefficient Denoted by k_c, k_x, K_x, and so on, the mass-transfer coefficient is the ratio of the flux to a concentration (or composition) difference. These coefficients generally represent rates of transfer that are much greater than those that occur by diffusion alone, as a result of convection or turbulence at the interface where mass transfer occurs. There exist several principles that relate that coefficient to the diffusivity and other fluid properties and to the intensity of motion and geometry. Examples that are outlined later are the film theory, the surface renewal theory, and the penetration theory, all of which pertain to idealized cases. For many situations of practical interest such as investigating the flow inside tubes and over flat surfaces as well as measuring external flow through banks of tubes, in fixed beds of particles, and the like, correlations have been developed that follow the same forms as the above theories. Examples of these are provided in the tables of mass-transfer coefficient correlations.

Problem-Solving Methods Most, if not all, problems or applications that involve mass transfer can be approached by a systematic course of action. In the simplest cases, the unknown quantities are obvious. In more complex (e.g., multicomponent, multiphase, multidimensional, nonisothermal, and/or transient) systems, it is more subtle to resolve the known and unknown quantities. For example, in multicomponent systems, one must know the fluxes of the components before predicting their effective diffusivities and vice versa. More will be said about that dilemma later. Once the known and unknown quantities are resolved, however, a combination of conservation equations, definitions, empirical relations, and properties is applied to arrive at an answer. Figure 5-22 is a flowchart that illustrates the primary types of information and their relationships, and it applies to many mass-transfer problems.

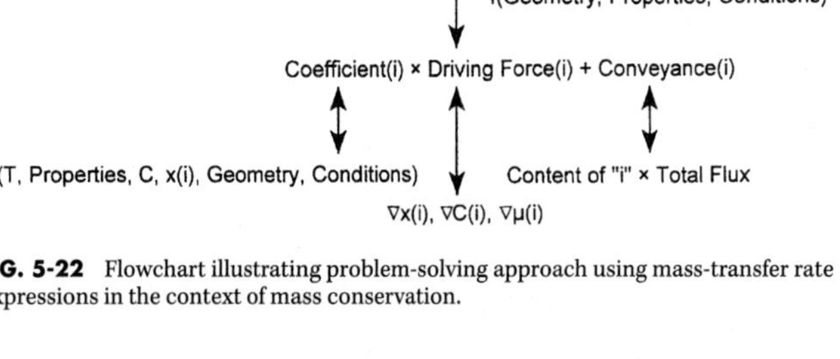

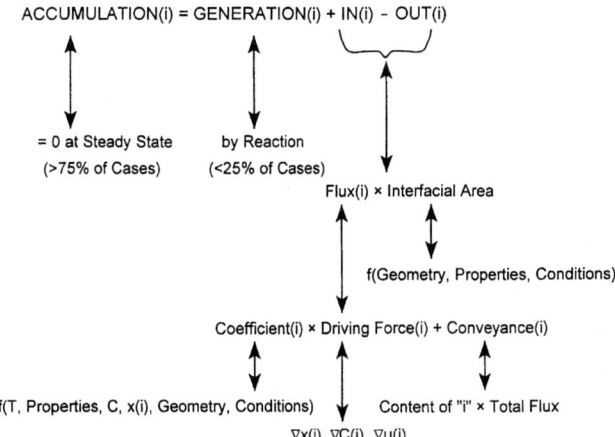

FIG. 5-22 Flowchart illustrating problem-solving approach using mass-transfer rate expressions in the context of mass conservation.

Nomenclature and Units—Mass Transfer

Symbols	Definition	SI units	U.S. Customary System (USCS) units
a, a_p	Effective interfacial mass-transfer area per unit volume	m^2/m^3	ft^2/ft^3
A_{cs}	Cross-sectional area of vessel	m^2 or cm^2	ft^2
A'	Constant (see Table 5-23K)		
C	Concentration = P/RT for ideal gas	mol/m^3, mol/L, or gequiv/L	$lbmol/ft^3$
c_i	Concentration of component $i = x_i c$ at gas-liquid interface	mol/m^3, mol/L, or gequiv/L	$lbmol/ft^3$
c_P	Specific heat	kJ/(kg · K)	Btu/(lb · °F)
d	Characteristic length	m or cm	ft
d_b	Bubble diameter	m	ft
d_c	Column diameter	m or cm	ft
d_{drop}	Sauter mean diameter	m	ft
d_{imp}	Impeller diameter	m	ft
d_{pore}	Pore diameter	m, cm, or μm	ft
$D_{A,A}$	Self-diffusivity (= D_A at $x_A = 1$)	m^2/s or cm^2/s	ft^2/h
D_{AB}	Mutual diffusivity	m^2/s or cm^2/s	ft^2/h
D°_{AB}	Mutual diffusivity at infinite dilution of A in B	m^2/s or cm^2/s	ft^2/h
D_{eff}	Effective diffusivity within a porous solid = $\varepsilon_p D/\tau$	m^2/s	ft^2/h
D_K	Knudson diffusivity for gases in small pores	m^2/s or cm^2/s	ft^2/h
D_L	Liquid-phase diffusion coefficient	m^2/s	ft^2/h
D_S	Surface diffusivity	m^2/s or cm^2/s	ft^2/h
D_{AB}	Maxwell-Stefan mutual diffusivity	m^2/s or cm^2/s	ft^2/h
E	Energy dissipation rate/mass	L^2/t^3	W/kg or HP/lb
E_S	Activation energy for surface diffusion	J/mol or cal/mol	
f	Friction factor for fluid flow	Dimensionless	Dimensionless
F	Faraday's constant	96,487 C/eq	
g	Acceleration due to gravity	m/s^2	ft/h^2
g_c	Conversion factor	1.0	4.17×10^8 lb · ft/(lbf · h^2)
G	Gas-phase mass flux	kg/(s · m^2)	lb/(h · ft^2)
G_a	Dry air flux	kg/(s · m^2)	lb/(h · ft^2)
G_M	Molar gas-phase mass flux	kmol/(s · m^2)	lbmol/(h · ft^2)
h'	Heat-transfer coefficient	W/(m^2 · K) = J/(s · m^2 · K)	Btu/(h · ft^2 · °F)
h_T	Total height of tower packing	m	ft
H	Compartment height	m	ft
H	Henry's law constant	kPa/(mole-fraction solute in liquid phase)	(lbf/in^2)/(mole-fraction solute in liquid phase)
H'	Henry's law constant	kPa/[kmol/(m^3 solute in liquid phase)]	(lbf/in^2)/(lbmol/ft^3 solute in liquid phase) or atm/[lbmol/(ft^3 solute in liquid phase)]
H_G	Height of 1 transfer unit based on gas-phase resistance	m	ft
H_{OG}	Height of 1 overall gas-phase mass-transfer unit	m	ft
H_L	Height of 1 transfer unit based on liquid-phase resistance	m	ft
H_{OL}	Height of 1 overall liquid-phase mass-transfer unit	m	ft
HTU	Height of 1 transfer unit (general)	m	ft
j_D, j_M	Chilton-Colburn factor for mass transfer, Eq. (5-313)	Dimensionless	Dimensionless
j_H	Chilton-Colburn factor for heat transfer	Dimensionless	Dimensionless
${}_mJ_A$	Mass flux of A by diffusion with respect to mean mass velocity	kmol/(m^2 · s) or mol/(cm^2 · s)	lbmol/(ft^2 · h)
${}_MJ_A$	Molar flux of A by diffusion with respect to mean molar velocity	kmol/(m^2 · s) or mol/(cm^2 · s)	lbmol/(ft^2 · h)
${}_VJ_A$	Molar flux of A with respect to mean volume velocity	kmol/(m^2 · s)	lbmol/(ft^2 · h)
J_{Si}	Molar flux by surface diffusion	kmol/(m^2 · s) or gmol/(cm^2 · s)	lbmol/(ft^2 · h)
k	Boltzmann's constant	8.9308×10^{-10} gequiv Ω/s	
k	Film mass-transfer coefficient	m/s or cm/s	ft/h
k	Thermal conductivity	(J · m)/(s · m^2 · K)	Btu/(h · ft · °F)
k'	Mass-transfer coefficient for dilute systems	kmol/[(s · m^2)(kmol/m^3)] or m/s	lbmol/[(h · ft^2) × (lbmol/ft^3)] or ft/h
k_G	Gas-phase mass-transfer coefficient dilute systems	kmol/[(s · m^2)(kPa solute partial pressure)]	lbmol/[(h · ft^2)lbf/in^2 solute partial pressure)]
k'_G	Gas-phase mass-transfer coefficient dilute systems	kmol/[(s · m^2)(mole fraction in gas)]	lbmol/[(h · ft^2)(mole fraction in gas)]
$k_G a$	Volumetric gas-phase mass transfer	kmol/[(s · m^3)(mole fraction)]	(lbmol)/[(h · ft^3) × (mole fraction)]
$\hat{k}_G a$	Overall volumetric gas-phase mass-transfer coefficient concentrated systems	kmol/(s · m^3)	lbmol/(h · ft^3)
$\hat{k}^o_L$	Liquid-phase mass-transfer coefficient for pure absorption (no reaction)	kmol/(s · m^2)	lbmol/(h · ft^2)
k_L	Liquid-phase mass-transfer coefficient dilute systems	kmol/[(s · m^2)(mole-fraction solution in liquid)]	(lbmol)/[(h · ft^2) × (mole-fraction solute in liquid)]
k'_L	Liquid-phase mass-transfer coefficient dilute systems	kmol/[(s · m^2) × (kmol/m^3)] or m/s	(lbmol)/[(h · ft^2) × (lbmol/ft^3)] or ft/h
$\hat{k}_L$	Liquid-phase mass-transfer coefficient concentrated systems	kmol/(s · m^2)	lbmol/(h · ft^2)
$k_L a$	Volumetric liquid-phase mass-transfer coefficient dilute systems	kmol/[(s · m^3) × (mole fraction)]	lbmol/[(h · ft^3) × (mole fraction)]
K	Overall mass-transfer coefficient	m/s or cm/s	ft/h
K	α/R = specific conductance	S/m	
K_G	Overall gas-phase mass-transfer coefficient dilute systems	kmol/[(s · m^2)(mole fraction)]	(lbmol)/[(h · ft^2) × (mole fraction)]
$\hat{K}_G$	Overall gas-phase mass-transfer coefficient concentrated systems	kmol/(s · m^2)	lbmol/(h · ft^2)
$K_G a$	Overall volumetric gas-phase mass-transfer dilute systems	kmol/[(s · m^3) × (mole-fraction solute in gas)]	lbmol/[(h · ft^3(mole-fraction solute in gas)]
$K'_G a$	Overall volumetric gas-phase mass-transfer dilute systems	kmol/[(s · m^3)(kPa solute partial pressure)]	lbmol/[(h · ft^3) × (lbf/in^2 solute partial pressure)]

Nomenclature and Units—Mass Transfer *(Continued)*

Symbols	Definition	SI units	U.S. Customary System (USCS) units
$(Ka)_H$	Overall enthalpy mass-transfer coefficient	kmol/[(s · m^2)(mole fraction)]	lb/[(h · ft^3)(lb water/lb dry air)]
K_L	Overall liquid-phase mass-transfer coefficient	kmol/[(s · m^2)(mole fraction)]	lbmol/[(h · ft^2)(mole fraction)]
$\hat{K}_L$	Liquid-phase mass-transfer coefficient concentrated systems	kmol/(s · m^2)	lbmol/(h · ft^2)
K_La	Overall volumetric liquid-phase mass-transfer coefficient dilute systems	kmol/[(s · m^3)(mole-fraction solute in liquid)]	lbmol/[(h · ft^3)(mole-fraction solute in liquid)]
$\hat{K}_La$	Overall volumetric liquid-phase mass-transfer coefficient concentrated systems	kmol/(s · m^3)	lbmol/(h · ft^3)
l	Mean free path, Eq. (5-261)	m	ft
L	Liquid-phase mass flux	kg/(s · m^2)	lb/(h · ft^2)
L_M	Molar liquid-phase mass flux	kmol/(s · m^2)	lbmol/(h · ft^2)
m	Slope of equilibrium curve = dy/dx (mole-fraction solute in gas)/(mole-fraction solute in liquid)	Dimensionless	Dimensionless
m	Molality of solute	mol/1000 g solvent	
M_i	Molecular weight of species i	kg/kmol or g/mol	lb/lbmol
M	Mass in a control volume V	kg or g	lb
$\lvert n_+\rvert \lvert n_-\rvert$	Valences of cationic and anionic species	Dimensionless	Dimensionless
n'	See Table 5-23K	Dimensionless	Dimensionless
n_A	Mass flux of A with respect to fixed coordinates	kg/(s · m^2)	lb/(h · ft^2)
N	Impeller speed	r/s	r/min
N'	Number of deck levels	Dimensionless	Dimensionless
N_A	Interphase mass-transfer rate of solute A per interfacial area with respect to fixed coordinates	kmol/(s · m^2)	lbmol/(h · ft^2)
N_c	Number of components	Dimensionless	Dimensionless
N_{Fr}	Froude number ($d_{imp} N^2/g$)	Dimensionless	Dimensionless
N_{Gr}	Grashof number $\dfrac{gx^3}{(\mu/\rho)^2}\left(\dfrac{\rho_\infty}{\rho_o}-1\right)$	Dimensionless	Dimensionless
N_{OG}	Number of overall gas-phase mass-transfer units	Dimensionless	Dimensionless
N_{OL}	Number of overall liquid-phase mass-transfer units	Dimensionless	Dimensionless
NTU	Number of transfer units (general)	Dimensionless	Dimensionless
N_{Kn}	Knudson number = l/d_{pore}	Dimensionless	Dimensionless
N_{Pr}	Prandtl number ($c_p\mu/k$)	Dimensionless	Dimensionless
N_{Re}	Reynolds number (Gd/μ_G)	Dimensionless	Dimensionless
N_{Sc}	Schmidt number ($\mu_G/\rho_G D_{AB}$) or ($\mu_L/\rho_L D_L$)	Dimensionless	Dimensionless
N_{Sh}	Sherwood number ($\hat{k}_G RTd/D_{AB}p_T$), see also Tables 5-16 to 5-23	Dimensionless	Dimensionless
N_{St}	Stanton number $\hat{k}_G/G_M$ or $\hat{k}_L/L_M$	Dimensionless	Dimensionless
N_{We}	Weber number ($\rho_c N^2 d^3_{imp}/\sigma$)	Dimensionless	Dimensionless
p	Solute partial pressure in bulk gas	kPa	lbf/in^2
$p_{B,M}$	Log mean partial pressure difference of stagnant gas B	Dimensionless	Dimensionless
p_i	Solute partial pressure at gas-liquid interface	kPa	lbf/in^2
p_T	Total system pressure	kPa	lbf/in^2
P	Pressure	Pa	lbf/in^2 or atm
P	Power	W	
P_c	Critical pressure	Pa	lbf/in^2 or atm
Per	Perimeter/area	m^{-1}	ft^{-1}
Q	Volumetric flow rate	m^3/s	ft^3/h
r_A	Radius of dilute spherical solute	Å	
R	Gas constant	8.314 J/(mol K) = 8.314 Pa · m^3/(mol K) = 82.057 atm · cm^3/(mol K)	10.73 (ft^3 · psia)/(lbmol · h)
$\boldsymbol{R}$	Solution electrical resistance	Ω	Ω
R_i	Radius of gyration of the component i molecule	Å	
s	Fractional surface-renewal rate	s^{-1}	h^{-1}
S	Tower cross-sectional area = $\pi d^2/4$	m^2	ft^2
t	Contact time	s	h
t_f	Formation time of drop	s	h
T	Temperature	K	°R
T_b	Normal boiling point	K	°R
T_c	Critical temperature	K	°R
T_r	Reduced temperature = T/T_c	Dimensionless	Dimensionless
u, υ	Fluid velocity	m/s or cm/s	ft/h
u_∞	Velocity away from object	m/s	ft/h
u_L	Superficial liquid velocity in vertical direction	m/s	ft/h
υ_s	Slip velocity	m/s	ft/h
υ_T	Terminal velocity	m/s	ft/h
υ_{TS}	Stokes law terminal velocity	m/s	ft/h
V	Packed volume in tower	m^3	ft^3
V	Control volume	m^3 or cm^3	ft^3
V_b	Volume at normal boiling point	m^3/kmol or cm^3/mol	ft^3/lbmol
V_i	Molar volume of i at its normal boiling point	m^3/kmol or cm^3/mol	ft^3/lbmol

Nomenclature and Units—Mass Transfer *(Continued)*

Symbols	Definition	SI units	U.S. Customary System (USCS) units
$\bar{\upsilon}_i$	Partial molar volume of i	m³/kmol or cm³/mol	ft³/lbmol
V_{mi}	Molar volume of the liquid-phase component i at melting point	m³/kmol or cm³/mol	ft³/lbmol
V_{tower}	Tower volume per area	m³/m²	ft³/ft²
w	Width of film	m	ft
x	Length along plate	m	ft
x	Mole-fraction solute in bulk-liquid phase	(kmol solute)/(kmol liquid)	(lbmol solute)/(lbmol liquid)
x_A	Mole fraction of component A	kmol A/kmol fluid	lbmol A/lbmol fluid
x^o	Mole-fraction solute in bulk liquid in equilibrium with bulk-gas solute concentration y	(kmol solute)/(kmol liquid)	(lbmol solute)/(lbmol liquid)
x_{BM}	Logarithmic-mean solvent concentration between bulk liquid and interface values	(kmol solvent)/(kmol liquid)	(lbmol solvent)/(lbmol liquid)
x^o_{BM}	Logarithmic-mean inert-solvent concentration between bulk-liquid value and value in equilibrium with bulk gas	(kmol solvent)/(kmol liquid)	(lbmol solvent)/(lbmol liquid)
x_i	Mole-fraction solute in liquid at gas-liquid interface	(kmol solute)/(kmol liquid)	(lbmol solute)/(lbmol liquid)
y	Mole-fraction solute in bulk-gas phase	(kmol solute)/(kmol gas)	(lbmol solute)/(lbmol gas)
y_{BM}	Logarithmic-mean inert-gas concentration [Eq. (5-274a)]	(kmol inert gas)/(kmol gas)	(lbmol inert gas)/(lbmol gas)
y^o_{BM}	Logarithmic-mean inert-gas concentration	(kmol inert gas)/(kmol gas)	(lbmol inert gas)/(lbmol gas)
y_i	Mole fraction solute in gas at interface	(kmole solute)/(kmol gas)	(lbmol solute)/(lbmol gas)
y_i^o	Mole-fraction solute in gas at interface in equilibrium with liquid-phase interfacial solute concentration x_i	(kmol solute)/(kmol gas)	(lbmol solute)/(lbmol gas)
z	Direction of unidimensional diffusion	m	ft
	Greek symbols		
α	$1 + N_B/N_A$	Dimensionless	Dimensionless
α	Conductance cell constant (measured)	cm⁻¹	
β	$M_A^{1/2} P_c^{1/3}/T_c^{5/6}$	Dimensionless	Dimensionless
δ	Effective thickness of stagnant-film layer	m	ft
ε	Fraction of discontinuous phase in continuous phase for two-phase flow	Dimensionless	Dimensionless
ε	Void fraction available for gas flow or fractional gas holdup	m³/m³	ft³/ft³
ε_A	Characteristic Lennard-Jones energy	Dimensionless	Dimensionless
ε_{AB}	$(\varepsilon_A \varepsilon_B)^{1/2}$	Dimensionless	Dimensionless
γ_i	Activity coefficient of solute i	Dimensionless	Dimensionless
$\gamma_\pm$	Solute mean ionic activity coefficient	Dimensionless	Dimensionless
$\lambda_+ \lambda_-$	Infinite dilution conductance of cation and anion	cm²/(gequiv · Ω)	
Λ	$1000\ K/C = \lambda_+ + \lambda_- = \Lambda_o + f(C)$	cm²/Ω gequiv	
Λ_o	Infinite dilution conductance	cm²/gequiv Ω	
μ_i	Dipole moment of i	Debeyes	
μ_i	Viscosity of pure i	cP or Pa s	lb/(h · ft)
μ_G	Gas-phase viscosity	kg/(s · m)	lb/(h · ft)
μ_L	Liquid-phase viscosity	kg/(s · m)	lb/(h · ft)
ν	Kinematic viscosity = ρ/μ	m²/s	ft²/h
ρ	Density of A	kg/m³ or g/cm³	lb/ft³
ρ_c	Critical density of A	kg/m³ or g/cm³	lb/ft³
ρ_c	Density continuous phase	kg/m³	lb/ft³
ρ_G	Gas-phase density	kg/m³	lb/ft³
$\bar{\rho}_L$	Average molar density of liquid phase	kmol/m³	lbmol/ft³
ρ_p	Particle density	kg/m³ or g/cm³	lb/ft³
ρ_r	Reduced density = ρ/ρ_c	Dimensionless	Dimensionless
ψ_i	Parachor of component $i = V_i\sigma^{1/4}$		
ψ	Parameter, Table 5-23G	Dimensionless	Dimensionless
σ	Interfacial tension	dyn/cm	lbf/ft
σ_i	Characteristic length	Å	
σ_i	Surface tension of component i	dyn/cm	
σ_{AB}	Binary pair characteristic length = $(\sigma_A + \sigma_B)/2$	Å	
τ	Intraparticle tortuosity	Dimensionless	Dimensionless
ω	Pitzer's acentric factor = $-[1.0 + \log (P^*/P_c)]$	Dimensionless	Dimensionless
ω	Rotational velocity	rad/s	
Ω	Diffusion collision integral = $f(kT/\varepsilon_{AB})$	Dimensionless	Dimensionless
	Subscripts		
A	Solute component in liquid or gas phase		
B	Inert-gas or inert-solvent component		
G	Gas phase		
m	Mean value		
L	Liquid phase		
super	Superficial velocity		
	Superscript		
*	At equilibrium		

CONTINUITY AND FLUX EXPRESSIONS

Material Balances Whenever mass-transfer applications involve equipment of specific dimensions, flux equations alone are inadequate to assess results. A material balance or continuity equation must also be used. When the geometry is simple, macroscopic balances suffice. The following equation is an overall mass balance for a unit having N_m bulk-flow ports and N_n ports or interfaces through which diffusive flux can occur:

$$\frac{dM}{dt} = \sum_{i=1}^{N_m} m_i + \sum_{i=1}^{N_n} n_i A_{cs_i} \tag{5-191}$$

where M represents the mass in the unit volume V at any time t; m_i is the mass flow rate through the ith port; and n_i is the mass flux through the ith port, which has a cross-sectional area of A_{csi}. The corresponding balance equation for individual components includes a reaction term:

$$\frac{dM_j}{dt} = \sum_{i=1}^{N_m} m_{ij} + \sum_{i=1}^{N_n} n_{ij} A_{cs_i} + r_j V \tag{5-192}$$

For the jth component, $m_{ij} = m_i w_{ij}$ is the component mass flow rate in stream i; w_{ij} is the mass fraction of component j in stream i; and r_j is the net reaction rate (mass generation minus consumption) per unit volume V that contains mass M. If it is inconvenient to measure mass flow rates, the product of density and the volumetric flow rate is used instead.

In addition, most situations that involve mass transfer require material balances, but the pertinent area is ambiguous. Examples are packed columns for absorption, distillation, or extraction. In such cases, flow rates through the discrete ports (nozzles) must be related to the mass-transfer rate in the packing. As a result, the mass-transfer rate is determined via flux equations, and the overall material balance incorporates the stream flow rates m_i and integrated fluxes. In such instances, it is common to begin with the most general, differential material balance equations. Then, by eliminating terms that are negligible, the simplest applicable set of equations remains to be solved. The generic form applies over a unit cross-sectional area and constant volume:

$$\frac{\partial \rho_j}{\partial t} = -\nabla \cdot n_j + r_j \tag{5-193a}$$

where $n_j = \rho \upsilon_j$. Applying Fick's law and expressing composition as concentration give

$$\frac{\partial c_j}{\partial t} = -\upsilon \cdot \nabla c_j + D_j \nabla^2 c_j + r_j \tag{5-193b}$$

Table 5-9 provides material balances for cartesian, cylindrical, and spherical coordinates.

TABLE 5-9 Continuity Equation in Various Coordinate Systems

Coordinate system	Equation	
Cartesian	$\frac{\partial \rho_j}{\partial t} = -\left(\frac{\partial n_{x_j}}{\partial x} + \frac{\partial n_{y_j}}{\partial y} + \frac{\partial n_{z_j}}{\partial z}\right) + r_j$	(5-194)
Cylindrical	$\frac{\partial \rho_j}{\partial t} = -\left(\frac{1}{r}\frac{\partial r n_{r_j}}{\partial r} + \frac{1}{r}\frac{\partial n_{\theta_j}}{\partial \theta} + \frac{\partial n_{z_j}}{\partial z}\right) + r_j$	(5-195)
Spherical	$\frac{\partial \rho_j}{\partial t} = -\left(\frac{1}{r^2}\frac{\partial r^2 n_{r_j}}{\partial r} + \frac{1}{r \sin\theta}\frac{\partial n_{\theta_j} \sin\theta}{\partial \theta} + \frac{1}{r \sin\theta}\frac{\partial n_{\phi_j}}{\partial \phi}\right) + r_j$	(5-196)

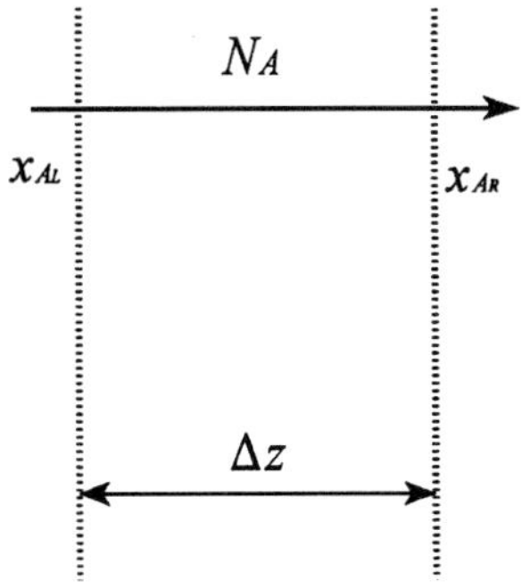

FIG. 5-23 Hypothetical film and boundary conditions.

Flux Expressions: Simple Integrated Forms of Fick's First Law Simplified flux equations that arise from Eqs. (5-189) and (5-190) can be used for one-dimensional, steady-state problems with binary mixtures. The boundary conditions represent the compositions x_{A_L} and x_{A_R} at the left-hand and right-hand sides, respectively, of a hypothetical layer having thickness Δz. The principal restriction of the following equations is that the concentration and diffusivity are assumed to be constant. As written, the flux is positive from left to right, as depicted in Fig. 5-23.

1. Equimolar counterdiffusion ($N_A = -N_B$)

$$N_A = {}_M J_A = -D_{AB}\, c \frac{dx_A}{dz} = \frac{D_{AB}}{\Delta z} c\,(x_{A_L} - x_{A_R}) \tag{5-197}$$

2. Unimolar diffusion ($N_A \neq 0$, $N_B = 0$)

$$N_A = {}_M J_A + x_A N_A = \frac{D_{AB}}{\Delta z} c \ln \frac{1 - x_{A_R}}{1 - x_{A_L}} \tag{5-198}$$

3. Steady-state diffusion ($N_A \neq -N_B \neq 0$)

$$N_A = {}_M J_A + x_A (N_A + N_B) = \frac{N_A}{N_A + N_B} \frac{D_{AB}}{\Delta z} c \ln \left(\frac{\frac{N_A}{N_A + N_B} - x_{A_R}}{\frac{N_A}{N_A + N_B} - x_{A_L}} \right) \tag{5-199}$$

The unfortunate aspect of the last relationship is that one must know a priori the ratio of the fluxes to determine the magnitudes. It is not possible to solve simultaneously the pair of equations that apply for components A and B because the equations are not independent.

DIFFUSIVITY ESTIMATION—GASES

Whenever measured values of diffusivities are available, they should be used. Typically, measurement errors are less than those associated with predictions by empirical or even semitheoretical equations. A few general sources of data can be found in Sec. 2 of this handbook; e.g., experimental values for gas mixtures are listed in Table 2-141. Other pertinent references are Schwartzberg and Chao (1982); Poling et al. (2001); Gammon et al. (1994); and Daubert and Danner (1989–1995). Many other more restricted sources are listed under specific topics later in this subsection.

Before diffusivities from either data or correlations are used, it is a good idea to check their reasonableness with respect to values that have been commonly observed in similar situations. Table 5-10 is a compilation of several rules of thumb. These values are not authoritative; they simply represent guidelines based on experience.

Diffusivity correlations for gases are outlined in Table 5-11. Specific parameters for individual equations are defined in the specific text regarding each equation. References are given at the beginning of the Mass Transfer subsection. The errors reported for Eqs. (5-200) through (5-203) were compiled by Poling et al. (2001), who compared the predictions with 68 experimental values of D_{AB}. Errors cited for Eqs. (5-204) to (5-209) were reported by the authors.

Binary Mixtures at Low Pressure with Nonpolar Components Many evaluations of correlations are available [Elliott and Watts, *Can. J. Chem.* **50:** 31 (1972); Lugg, *Anal. Chem.* **40:** 1072 (1968); Marrero and Mason, *AIChE J.* **19:** 498 (1973)]. The differences in accuracy of the correlations are minor, and thus the major concern is ease of calculation. The Fuller-Schettler-Giddings equation is usually the simplest correlation to use and

TABLE 5-10 Rules of Thumb for Diffusivities (See Cussler, Poling et al., Schwartzberg and Chao)

Continuous phase	D_i magnitude, m²/s	D_i magnitude, cm²/s	D_i range, m²/s	D_i range, cm²/s	Comments
Gas at atmospheric pressure	10^{-5}	0.1	10^{-4}–10^{-6}	1–10^{-2}	Accurate theories exist, generally within ±10%; $D_iP \cong$ constant; $D_i \propto T^{1.66 \text{ to } 2.0}$
Liquid	10^{-9}	10^{-5}	10^{-8}–10^{-10}	10^{-4}–10^{-6}	Approximate correlations exist, generally within ±25%
Liquid occluded in solid matrix	10^{-10}	10^{-6}	10^{-8}–10^{-12}	10^{-4}–10^{-8}	Hard cell walls: D_{eff}/D_i = 0.1 to 0.2. Soft cell walls: D_{eff}/D_i = 0.3 to 0.9
Polymers and glasses	10^{-12}	10^{-8}	10^{-10}–10^{-14}	10^{-6}–10^{-10}	Approximate theories exist for dilute and concentrated limits; strong composition dependence
Solid	10^{-14}	10^{-10}	10^{-10}–10^{-34}	10^{-6}–10^{-30}	Approximate theories exist; strong temperature dependence

TABLE 5-11 Correlations of Diffusivities for Gases

Authors*	Equation		Error, %
1. Binary Mixtures—Low Pressure—Nonpolar			
Chapman-Enskog	$D_{AB} = \dfrac{0.001858 T^{3/2} M_{AB}^{1/2}}{P\sigma_{AB}^2 \Omega_D}$	(5-200)	7.3
Wilke-Lee [22]	$D_{AB} = \dfrac{(0.00217 - 0.0005 M_{AB}^{1/2}) T^{3/2} M_{AB}^{1/2}}{P\sigma_{AB}^2 \Omega_D}$	(5-201)	7.0
Fuller-Schettler-Giddings [9]	$D_{AB} = \dfrac{0.001 T^{1.75} M_{AB}^{1/2}}{P\left[(\Sigma\upsilon)_A^{1/3} + (\Sigma\upsilon)_B^{1/3}\right]^2}$	(5-202)	5.4
2. Binary Mixtures—Low Pressure—Polar			
Brokaw [3]	$D_{AB} = \dfrac{0.001858 T^{3/2} M_{AB}^{1/2}}{P\sigma_{AB}^2 \Omega_D}$	(5-203)	9.0
3. Self-Diffusivity			
Mathur-Thodos [14]	$D_{AA} = \dfrac{10.77 \times 10^{-5} T_r}{\beta\rho_r}\{\rho_r \le 1.5\}$	(5-204)	5
Lee-Thodos [12—#1]	$D_{AA} = \dfrac{0.77 \times 10^{-5} T_r}{\rho_r \delta}\{\rho_r \le 1\}$	(5-205)	0.5
Lee-Thodos [12—#2]	$D_{AA} = \dfrac{(0.007094G + 0.001916)^{2.5} T_r}{\delta}[\rho_r > 1, G < 1]$	(5-206)	17
4. Supercritical Mixtures			
Sun and Chen [18]	$D_{AB} = \dfrac{1.23 \times 10^{-10} T}{\mu^{0.799} V_{C_A}^{0.49}}$	(5-207)	5
Catchpole and King [5]	$D_{AB} = 5.152\, D_c T_r \dfrac{(\rho^{-0.667} - 0.4510)(1 + M_A/M_B)R}{(1 + (V_{C_B}/V_{C_A})^{0.333})^2}$	(5-208)	10
He and Yu [11]	$D_{AB} = \alpha(V_A^k - \beta)\sqrt{\dfrac{T}{M}},\quad \alpha = 10^{-5}\left[0.56392 + 2.1417 \exp\left(\dfrac{-0.95088\sqrt{M_A V_{C_A}}}{P_{c_A}}\right)\right]$ $\beta = 8.9061 + 0.93858\dfrac{\sqrt{M_A V_{C_A}}}{P_{C_A}},\quad k = \dfrac{2}{3}[1 - 0.28 \exp(-0.3\sqrt{M_A}\rho\iota_A)]$	(5-209)	6.9

*References are listed at the beginning of the Mass Transfer subsection.

TABLE 5-12 Estimates for ε_i and σ_i (K, Å, atm, cm³, mol)

Critical point	$\varepsilon/k = 0.75\ T_c$	$\sigma = 0.841\ V_c^{1/3}$ or $2.44\ (T_c/P_c)^{1/3}$
Critical point	$\varepsilon/k = 65.3\ T_c z_c^{3.6}$	$\sigma = \dfrac{1.866\ V_c^{1/3}}{z_c^{1.2}}$
Normal boiling point	$\varepsilon/k = 1.15\ T_b$	$\sigma = 1.18\ V_b^{1/3}$
Melting point	$\varepsilon/k = 1.92\ T_m$	$\sigma = 1.222\ V_m^{1/3}$
Acentric factor	$\varepsilon/k = (0.7915 + 0.1693\ \omega)\ T_c$	$\sigma = (2.3551 - 0.087\ \omega)\left(\dfrac{T_c}{P_c}\right)^{1/3}$

NOTE: These values may not agree closely, so usage of a consistent basis is suggested (e.g., data at the normal boiling point).

is recommended by Poling et al. In several of the correlations, the average molecular weight M_{AB} is defined as

$$M_{AB} = 1/[(1/M_A) + (1/M_B)] \tag{5-210}$$

Chapman-Enskog (Bird et al.) and Wilke and Lee [22] The inherent assumptions of these equations are quite restrictive (i.e., low density, spherical atoms), and the intrinsic potential function is empirical. Despite that, they provide good estimates of D_{AB} for many polyatomic gases and gas mixtures, up to about 1000 K and a maximum of 70 atm. The latter constraint exists because observations for many gases indicate that $D_{AB}P$ is constant up to 70 atm.

The characteristic length is $\sigma_{AB} = (\sigma_A + \sigma_B)/2$ in angstroms. To estimate the collision integral Ω_D for Eq. (5-202) or (5-203), two empirical equations are available. The first is

$$\Omega_D = (44.54T^{*-4.909} + 1.911T^{*-1.575})^{0.10} \tag{5-211a}$$

where $T^* = kT/\varepsilon_{AB}$ and $\varepsilon_{AB} = (\varepsilon_A\varepsilon_B)^{1/2}$. Estimates for σ_i and ε_i are given in Table 5-12. This expression shows that Ω_D is proportional to temperature roughly to the −0.49 power at low temperatures and to the −0.16 power at high temperature. Thus, gas diffusivities are proportional to temperatures to the 2.0 power and 1.66 power, respectively, at low and high temperatures. The second is

$$\Omega_D = \frac{A}{T^{*B}} + \frac{C}{\exp(DT^*)} + \frac{E}{\exp(FT^*)} + \frac{G}{\exp(HT^*)} \tag{5-211b}$$

where $A = 1.06036$, $B = 0.15610$, $C = 0.1930$, $D = 0.47635$, $E = 1.03587$, $F = 1.52996$, $G = 1.76474$, and $H = 3.89411$.

Fuller, Schettler, and Giddings [9] The parameters and constants for this correlation were determined by regression analysis of 340 experimental diffusion coefficient values of 153 binary systems. Values of $\Sigma\upsilon_i$ used in this equation are found in Table 5-13.

Binary Mixtures at Low Pressure with Polar Components The *Brokaw* [3] correlation was based on the Chapman-Enskog equation, but σ_{AB^*} and Ω_{D^*} were evaluated with a modified Stockmayer potential for polar molecules. Hence, slightly different symbols are used. That potential model reduces to the Lennard-Jones 6-12 potential for interactions between nonpolar molecules. As a result, the method should yield accurate predictions for polar as well as nonpolar gas mixtures. Brokaw presented data for 9 relatively polar pairs along with the prediction. The agreement was good: an average absolute error of 6.4 percent, considering the complexity of some of the gas pairs [e.g., $(CH_3)_2O$ and CH_3Cl]. Despite that, Poling et al. (2001) found the average error was 9.0 percent for combinations of mixtures (including several polar-nonpolar gas pairs), temperatures, and pressures. In this equation, Ω_D is calculated as described previously, and other terms are

$$\Omega_{D^*} = \Omega_D + 0.19\delta^2{}_{AB}/T^* \qquad T^* = kT/\varepsilon_{AB^*}$$

$$\sigma_{AB^*} = (\sigma_{A^*}\sigma_{B^*})^{1/2} \qquad \sigma_{i^*} = [1.585V_{bi}/(1 + 1.3\delta_i^2)]^{1/3}$$

$$\delta_{AB} = (\delta_A\delta_B)^{1/2} \qquad \delta_i = 1.94 \times 10^3\ \mu_i^2/V_{bi}T_{bi}$$

$$\varepsilon_{AB^*} = (\varepsilon_{A^*}\varepsilon_{B^*})^{1/2} \qquad \varepsilon_{i^*}/k = 1.18\ (1 + 1.3\delta_i^2)T_{bi}$$

TABLE 5-13 Atomic Diffusion Volumes for Use in Estimating D_{AB} by the Method of Fuller, Schettler, and Giddings [9]

Atomic and Structural Diffusion–Volume Increments, υ_i (cm³/mol)			
C	16.5	(Cl)	19.5
H	1.98	(S)	17.0
O	5.48	Aromatic ring	−20.2
(N)	5.69	Heterocyclic ring	−20.2
Diffusion Volumes for Simple Molecules, $\Sigma\upsilon_i$ (cm³/mol)			
H_2	7.07	CO	18.9
D_2	6.70	CO_2	26.9
He	2.88	N_2O	35.9
N_2	17.9	NH_3	14.9
O_2	16.6	H_2O	12.7
Air	20.1	(CCl_2F_2)	114.8
Ar	16.1	(SF_5)	69.7
Kr	22.8	(Cl_2)	37.7
(Xe)	37.9	(Br_2)	67.2
Ne	5.59	(SO_2)	41.1

Parentheses indicate that the value listed is based on only a few data points.

Binary Mixtures at High Pressure Of the various categories of gas-phase diffusion, this is the least studied. This is so because the effects of diffusion are easily distorted by even a slight pressure gradient, which is difficult to avoid at high pressure. Harstad and Bellan [*Ind. Eng. Chem. Res.* **43:** 645 (2004)] developed a corresponding-states expression that extends the Chapman-Enskog method, covered earlier. They express the diffusivity at high pressure by accounting for the reduced temperature, and they suggest employing an equation of state and shifting from

$$D^o_{AB} = f(T, P) \text{ to } D_{AB} = g(T, V).$$

Self-Diffusivity Self-diffusivity is rarely used for solving separation problems. Despite that, it has been studied extensively under high pressures, e.g., greater than 70 atm. There are few accurate estimation methods for mutual diffusivities at such high pressures, because composition measurements are difficult.

The general observation for gas-phase diffusion $D_{AB}P$ = constant, which holds at low pressure, is not valid at high pressure. Rather, $D_{AB}P$ decreases as pressure increases. In addition, composition effects, which frequently are negligible at low pressure, are very significant at high pressure. Suárez-Iglesias et al. [*J. Chem. Engg. Data* **60:** 2757 (2015)] published a thorough review of experimental self-diffusivities of gases, vapors, and liquids, ranging from noble gases and simple diatomics to complex organic molecules. The methods included both tracer techniques and nuclear magnetic resonance.

Liu and Ruckenstein [*Ind. Eng. Chem. Res.* **36:** 3937 (1997)] studied self-diffusion for both liquids and gases. They compared their estimates to 26 pairs, with a total of 1822 data points, and achieved a relative deviation of 7.3 percent. Zielinski and Hanley [*AIChE J.* **45:** 1 (1999)] developed a model to predict multicomponent diffusivities from self-diffusion coefficients and thermodynamic information. Mathur and Thodos [14] showed that for reduced densities less than unity, the product $D_{AA}\rho$ is approximately constant at a given temperature. In their correlation, $\beta = M_A^{1/2}P_C^{1/3}/T_C^{5/6}$. Lee and Thodos [12] presented generalized correlations of self-diffusivity for gases (and liquids), which have been tested for more than 500 data points each, with an average deviation of the first of 0.51 percent, and that of the second is 17.2 percent. $\delta = M_A^{1/2}/P_C^{1/2}V_C^{5/6}$, s/cm², and where $G = (X^* - X)/(X^* - 1)$, $X = \rho_r/T_r^{0.1}$, and $X^* = \rho_r/T_r^{0.1}$ evaluated at the solid melting point. Lee and Thodos [*Ind. Eng. Chem. Res.* **27:** 17 (1988)] expanded their earlier treatment of self-diffusivity to cover 58 substances and 975 data points, with an average absolute deviation of 5.26 percent. Liu, Silva, and Macedo [*Chem. Eng. Sci.* **53:** 2403 (1998)] and Silva, Liu, and Macedo [*Chem. Eng. Sci.* **53:** 2423 (1998)] present a theoretical approach. For 2047 data points with nonpolar species, their best model yielded 4.5 percent average deviation, while the Lee-Thodos equation yielded 5.2 percent. The new model was also much better than all the other models for over 424 data points with polar species, yielding 4.3 percent deviation, while the Lee-Thodos equation yielded 34 percent.

Supercritical Mixtures Debenedetti and Reid [*AIChE J.* **32:** 2034 (1986) and **33:** 496 (1987)] showed that conventional correlations based on the Stokes-Einstein relation (for liquid phase) tend to overpredict diffusivities in the supercritical state. Nevertheless, they observed that the Stokes-Einstein group $D_{AB}\mu/T$ was constant. Thus, although no general correlation applies, only one data point is necessary to examine variations of fluid viscosity and/or temperature effects. Sun and Chen [18] examined tracer diffusion data of aromatic solutes in alcohols up to the supercritical range and found their data correlated with average deviations of 5 percent and a maximum deviation of 17 percent for their rather limited set of data. Catchpole and King [5] examined binary diffusion data of near-critical fluids in the reduced density range of 1 to 2.5 and found average deviations of 10 percent and a maximum deviation of 60 percent. Liu and Ruckenstein [*Ind. Eng. Chem. Res.* **36:** 888 (1997)] presented a semiempirical equation to estimate diffusivities under supercritical conditions that is based on the Stokes-Einstein relation and the long-range correlation, respectively. They compared their estimates to 33 pairs, 598 data points, and achieved lower deviations (5.7 percent) than the Sun-Chen correlation (13.3 percent) and the Catchpole-King equation (11.0 percent). He and Yu [11] presented a semiempirical equation to estimate diffusivities under supercritical conditions that is based on hard-sphere theory. It is limited to $\rho_r \geq 0.21$, where the reduced density is $\rho_r = \rho_A(T, P)/\rho_{C A}$. They compared their estimates to 107 pairs, 1167 data points, and achieved lower deviations (7.8 percent) than the Catchpole-King equation (9.7 percent), which was restricted to $\rho_r \geq 1$. Silva and Macedo [*Ind. Eng. Chem. Res.* **37:** 1490 (1998)] measured diffusivities of ethers in CO_2 under supercritical conditions and compared them to the Wilke-Chang [Eq. (5-214)], Tyn-Calus [Eq. (5-215)], Catchpole-King [Eq. (5-208)], and their own equations. They found that the Wilke-Chang equation provided the best fit. Gonzalez, Bueno, and Medina [*Ind. Eng. Chem. Res.* **40:** 3711 (2001)] measured diffusivities of aromatic compounds in CO_2 under supercritical conditions and compared them to the Wilke-Chang [Eq. (5-215)], Hayduk-Minhas [Eq. (5-223)], and other equations. They recommended the Wilke-Chang equation (which yielded a relative error of 10.1 percent) but noted that the He and Yu equation provided the best fit (5.5 percent).

Low-Pressure/Multicomponent Mixtures Smith and Taylor [*Ind. Eng. Chem. Fundam.* **22:** 97 (1983)] compared various methods for predicting multicomponent diffusion rates and found that the Maxwell-Stefan approach (see later section) was superior.

Blanc [*J. Phys.* **7:** 825 (1908)] provided a simple limiting case for dilute component *i* diffusing in a stagnant medium (that is, $N \approx 0$), and the result is known as *Blanc's law.*

$$D_{im} = \left(\sum_{j=1}^{N_C} \frac{x_j}{D_{ij}} \right)^{-1} \tag{5-212}$$

The restriction basically means that the compositions of all the components, besides component *i*, are relatively large and uniform.

DIFFUSIVITY ESTIMATION—LIQUIDS

Many more correlations are available for diffusion coefficients in the liquid phase than for the gas phase. Most, however, are restricted to binary diffusion at infinite dilution D°_{AB} or to self-diffusivity $D_{A'A}$. This reflects the much greater complexity of liquids on a molecular level. For example, gas-phase diffusion exhibits negligible composition effects and deviations from thermodynamic ideality. Conversely, liquid-phase diffusion usually involves volumetric and thermodynamic effects due to composition variations. For concentrations greater than a few mole percent of *A* and *B*, corrections are needed to obtain the true diffusivity. Furthermore, many conditions do not fit any of the correlations presented here. Thus, careful consideration is needed to produce a reasonable estimate. Again, if diffusivity data are available at the conditions of interest, then they are strongly preferred over the predictions of any correlations. Experimental values for liquid mixtures are listed in Table 2-142.

Stokes-Einstein and Free-Volume Theories The starting point for many correlations is the Stokes-Einstein equation. This equation is derived from continuum fluid mechanics and classical thermodynamics for the motion of large spherical particles in a liquid. For this case, the need for a molecular theory is cleverly avoided. The Stokes-Einstein equation is (Bird et al.)

$$D_{AB} = \frac{kT}{6\pi r_A \mu_B} \tag{5-213}$$

where *A* refers to the solute and *B* refers to the solvent. This equation is applicable to very large unhydrated molecules ($M > 1000$) in low-molecular-weight solvents or where the molar volume of the solute is greater than 500 cm^3/mol (Reddy and Doraiswamy, *Ind. Eng. Chem. Fundam.* **6:** 77 (1967); Wilke and Chang [21]). Despite its intellectual appeal, this equation is seldom used "as is." Rather, the following principles have been identified: (1) The diffusion coefficient is inversely proportional to the size $r_A \simeq V_A^{1/3}$ of the solute molecules. Experimental observations, however, generally indicate that the exponent of the solute molar volume is larger than one-third. (2) The term $D_{AB}\mu_B/T$ is approximately constant only over a 10 to 15 K interval. Thus, the dependence of liquid diffusivity on properties and conditions does not generally obey the interactions implied by that grouping. For example, Robinson, Edmister, and Dullien [*Ind. Eng. Chem. Fundam.* **5:** 75 (1966)] found that $\ln D_{AB} \propto -1/T$. (3) Finally, pressure does not affect liquid-phase diffusivity much, since μ_B and V_A are only weakly pressure-dependent. Pressure does have an impact at very high levels.

Another advance in the concepts of liquid-phase diffusion was provided by Hildebrand [*Science* **174:** 490 (1971)] who adapted a theory of viscosity to self-diffusivity. He postulated that $D_{A'A} = B(V - V_{ms})/V_{ms}$, where $D_{A'A}$ is the self-diffusion coefficient, *V* is the molar volume, and V_{ms} is the molar volume at which fluidity is zero (i.e., the molar volume of the solid phase at the melting temperature). The difference $V - V_{ms}$ can be thought of as the free volume, which increases with temperature; and *B* is a proportionality constant.

Ertl and Dullien (*AIChE J.* **19:** 1215 (1973)) found that Hildebrand's equation could not fit their data with *B* as a constant. They modified it by applying an empirical exponent *n* (a constant greater than unity) to the volumetric ratio. The new equation is not generally useful, however, since there is no means for predicting *n*. The theory does identify the free volume as an important physical variable, since $n > 1$ for most liquids implies that diffusion is more strongly dependent on free volume than is viscosity.

Dilute Binary Nonelectrolytes: General Mixtures These correlations are outlined in Table 5-14.

Wilke-Chang [22] This correlation for D°_{AB}, which is an empirical modification of the Stokes-Einstein equation, is one of the most widely used. It is not very accurate, however, for water as the solute. Otherwise, it applies to diffusion of very dilute *A* in *B*. The average absolute error for 251 different systems is about 10 percent; ϕ_B is an association factor of solvent *B* that accounts for hydrogen bonding.

Component *B*	ϕ_B
Water	2.26
Methanol	1.9
Ethanol	1.5
Propanol	1.2
Others	1.0

The value of ϕ_B for water was originally stated as 2.6, although when the original data were reanalyzed, the empirical best fit was 2.26. Random comparisons of predictions with 2.26 versus 2.6 show no consistent advantage for either value, however. Kooijman [*Ind. Eng. Chem. Res.* **41:** 3326 (2002)] suggests replacing V_A with $\theta_A V_A$, in which $\theta_A = 1$ except when A = water, $\theta_A = 4.5$. This modification leads to an overall error of 8.7 percent for 41 cases he compared. He suggests retaining $\phi_B = 2.6$ when B = water. It has been suggested to replace the exponent of 0.6 with 0.7 and to use an association factor of 0.7 for systems containing aromatic hydrocarbons. These modifications, however, are not recommended by Umesi and Danner [*Ind. Eng. Chem. Process Des. Dev.* **20:** 662 (1981)], who developed an equation for nonaqueous solvents with nonpolar and polar solutes. The average absolute deviation was 16 percent, compared with 26 percent for the Wilke-Chang equation. Lees and Sarram [*J. Chem. Eng. Data* **16:** 41 (1971)] compare the association parameters. The average absolute error for 87 different solutes in water is 5.9 percent.

Tyn-Calus [19] This correlation requires data in the form of molar volumes and parachors $\psi_i = V_i\sigma_i^{1/4}$ (a property which, over moderate temperature ranges, is nearly constant), measured at the same temperature (not necessarily the temperature of interest). The parachors for the components may also be evaluated at different temperatures from one another. Quale [*Chem. Rev.* **53:** 439 (1953)] compiled values of ψ_i for many chemicals. Group contribution methods are available for estimation purposes (Poling et al.). The following suggestions were made by Poling et al.: The correlation is constrained to cases in which $\mu_B < 30$ cP. If the solute is water or if the solute is an organic acid and the solvent is not water or a short-chain alcohol, then dimerization of solute *A* should be assumed for purposes of estimating its volume and parachor. For example, the appropriate values for water as solute at 25°C are $V_W = 37.4$ cm^3/mol and $\psi_W = 105.2$ $cm^3 g^{1/4}/s^{1/2}$ mol. Finally, if the solute is nonpolar, the solvent volume and parachor should be multiplied by $8\mu_B$. According to Kooijman [*Ind. Eng. Chem. Res.* **41:** 3326 (2002)], if the Brock-Bird method (described in Poling et al.) is used to estimate the surface tension, the error is only increased by about 2 percent, relative to employing experimentally measured values.

Siddiqi-Lucas [17] In an impressive empirical study, these authors examined 1275 organic liquid mixtures. Their equation yielded an average absolute deviation of 13.1 percent, which was less than that for the Wilke-Chang equation (17.8 percent). Note that Eq. (5-216) does not encompass aqueous solutions, which are correlated in Eq. (5-218).

TABLE 5-14 Correlations for Diffusivities of Dilute, Binary Mixtures of Nonelectrolytes in Liquids

Authors*	Equation		Error
1. General Mixtures			
Wilke-Chang [21]	$D^o_{AB} = \dfrac{7.4\times10^{-8}(\phi_B M_B)^{1/2}T}{\mu_B V_A^{0.6}}$	(5-214)	20%
Tyn-Calus [19]	$D^o_{AB} = \dfrac{8.93\times10^{-8}(V_A/V_B^2)^{1/6}(\psi_B/\psi_A)^{0.6}T}{\mu_B}$	(5-215)	10%
Siddiqi-Lucas [17]	$D^o_{AB} = \dfrac{9.89\times10^{-8}(V_B^{0.265})T}{V_A^{0.45}\mu_B^{0.907}}$	(5-216)	13%
2. Gases in Low-Viscosity Liquids			
Chen-Chen [6]	$D^o_{AB} = 2.018\times10^{-9}\dfrac{(\beta V_{C_B})^{2/3}(RT_{C_B})^{1/2}}{M_A^{1/6}(M_B V_{C_A})^{1/3}}(V_r-1)\left(\dfrac{T}{T_{C_B}}\right)^{1/2}$	(5-217)	6%
3. Aqueous Solutions			
Siddiqi-Lucas [17]	$D^o_{AW} = 2.98\times10^{-7}V_A^{-0.5473}\mu_w^{-1.026}T$	(5-218)	13%
4. Hydrocarbon Mixtures			
Hayduk-Minhas [10]	$D^o_{AB} = 13.3\times10^{-8}T^{1.47}\mu_B^{(10.2/V_A-0.791)}V_A^{-0.71}$	(5-219)	5%
Matthews-Akgerman [15]	$D^o_{AB} = 32.88M_A^{-0.61}V_D^{-1.04}T^{0.5}(V_B-V_D)$	(5-220)	5%

*References are listed at the beginning of the Mass Transfer subsection.

Binary Mixtures of Gases in Low-Viscosity, Nonelectrolyte Liquids Sridhar and Potter [*AIChE J.* **23:** 4, 590 (1977)] derived an equation for predicting gas diffusion through liquid by combining existing correlations. Hildebrand had postulated the following dependence of the diffusivity for a gas in a liquid: $D^\circ_{AB} = D_{B'B}(V_{cB}/V_{cA})^{2/3}$, where $D_{B'B}$ is the solvent self-diffusion coefficient and V_{ci} is the critical volume of component *i*, respectively. To correct for minor changes in volumetric expansion, Sridhar and Potter multiplied the resulting equation by V_B/V_{mlB}, where V_{mlB} is the molar volume of the liquid *B* at its melting point. They compared experimentally measured diffusion coefficients for 27 data points of 11 binary mixtures. Their average absolute error was 13.5 percent. This correlation demonstrates the usefulness of self-diffusion as a means to assess mutual diffusivities and the value of observable physical property changes, such as molar expansion, to account for changes in conditions.

Chen-Chen [6] Their correlation was based on diffusion measurements of 50 combinations of conditions with 3 to 4 replicates each and exhibited an average error of 6 percent. In this correlation, $V_r = V_B/[0.9724(V_{mlB} + 0.04765)]$ and V_{mlB} = the liquid molar volume at the melting point, as discussed previously. Their association parameter β [which is different from the definition of that symbol in Eq. (5-221)] accounts for hydrogen bonding of the solvent. Values for acetonitrile and methanol are $\beta = 1.58$ and 2.31, respectively.

Dilute Binary Mixtures of a Nonelectrolyte in Water The correlations that were suggested previously for general mixtures, unless specified otherwise, may also be applied to diffusion of miscellaneous solutes in water. The following correlations are restricted to aqueous systems.

Hayduk and Laudie [*AIChE J.* **20:** 3, 611 (1974)] presented a simple correlation for the infinite dilution diffusion coefficients of nonelectrolytes in water. It has about the same accuracy as the Wilke-Chang equation (about 5.9 percent). There is no explicit temperature dependence, but the variation in water viscosity compensates for the absence of temperature.

Siddiqi and Lucas [17] These authors examined 658 aqueous liquid mixtures in an empirical study. They found an average absolute deviation of 19.7 percent. In contrast, the Wilke-Chang equation gave 35.0 percent and the Hayduk-Laudie correlation gave 30.4 percent.

Dilute Binary Hydrocarbon Mixtures Hayduk and Minhas [10] presented an accurate correlation for normal paraffin mixtures that was developed from 58 data points consisting of solutes from C_5 to C_{32} and solvents from C_5 to C_{16}. The average error was 3.4 percent for the 58 mixtures.

Matthews and Akgerman [15] The free-volume approach of Hildebrand was shown to be valid for binary, dilute liquid paraffin mixtures (as well as self-diffusion), consisting of solutes from C_8 to C_{16} and solvents of C_6 and C_{12}. The term they referred to as the "diffusion volume" was simply correlated with the critical volume, as $V_D = 0.308V_c$. We can infer from Table 5-11 that this is approximately related to the volume at the melting point as $V_D = 0.945V_m$. Their correlation was valid for diffusion of linear alkanes at temperatures up to 300°C and pressures up to 3.45 MPa. Matthews, Rodden, and Akgerman [*J. Chem. Eng. Data* **32:** 317 (1987)] and Erkey and Akgerman [*AIChE J.* **35:** 443 (1989)] completed similar studies of diffusion of alkanes, restricted to *n*-hexadecane and *n*-octane, respectively, as the solvents.

Riazi and Whitson [*Ind. Eng. Chem. Res.* **32:** 3081 (1993)] presented a generalized correlation in terms of viscosity and molar density that was applicable to both gases and liquids. The average absolute deviation for gases was only about 8 percent, while for liquids it was 15 percent. Their expression relies on the Chapman-Enskog correlation [Eq. (5-200)] for the low-pressure diffusivity and correlations for low-pressure viscosity.

Dilute Binary Mixtures of Nonelectrolytes with Water as the Solute Olander [*AIChE J.* **7:** 175 (1961)] modified the Wilke-Chang equation to adapt it to the infinite dilution diffusivity of water as the solute. The modification he recommended is simply the division of the right-hand side of the Wilke-Chang equation by 2.3. Unfortunately, neither the Wilke-Chang equation nor that equation divided by 2.3 fit the data very well. A reasonably valid generalization is that the Wilke-Chang equation is accurate if water is very insoluble in the solvent, such as pure hydrocarbons, halogenated hydrocarbons, and nitro-hydrocarbons. On the other hand, the Wilke-Chang equation divided by 2.3 is accurate for solvents in which water is very soluble, as well as those that have low viscosities. Such solvents include alcohols, ketones, carboxylic acids, and aldehydes. Neither equation is accurate for higher-viscosity liquids, especially diols.

Dilute Dispersions of Macromolecules in Nonelectrolytes The Stokes-Einstein equation predicts accurately the diffusion coefficient of spherical latex particles and globular proteins. Corrections to Stokes-Einstein for molecules approximating spheroids are given by Tanford [*Physical Chemistry of Macromolecules,* Wiley, New York, 1961]. Since solute-solute interactions are ignored in this theory, it applies in the dilute range only.

Hiss and Cussler [AIChE J. 19: 698 (1973)] Their basis is the diffusion of a small solute in a fairly viscous solvent of relatively large molecules, which is the opposite of the Stokes-Einstein assumptions. The large solvent molecules investigated were not polymers or gels but were of moderate molecular weight so that the macroscopic and microscopic viscosities were the same. The major conclusion is that $D^\circ_{AB}\mu^{2/3}$ = constant at a given temperature and for a solvent viscosity from 5×10^{-3} to 5 Pa · s or greater (5 to 5×10^3 cP). This observation is useful if D°_{AB} is known in a given high-viscosity liquid (oils, tars, etc.). Use of the usual relation of $D^\circ_{AB} \propto 1/\mu$ for such an estimate could lead to large errors.

Concentrated Binary Mixtures of Nonelectrolytes Since the infinite dilution values D°_{AB} and D°_{BA} are generally unequal, even a thermodynamically ideal solution like $\gamma_A = \gamma_B = 1$ will exhibit concentration dependence of the diffusivity. In addition, nonideal solutions require the following thermodynamic correction factor because the true "driving force" is the chemical potential gradient, not the composition gradient.

$$\beta_A = 1 + \frac{\partial \ln \gamma_A}{\partial \ln x_A} \tag{5-221}$$

Darken [*Trans. Am. Inst. Mining Met. Eng.* **175:** 184 (1948)] observed that solid-state diffusion in metallurgical applications followed a simple relation. His equation related the tracer diffusivities and mole fractions to the mutual diffusivity:

$$D_{AB} = (x_A D_B + x_B D_A)\beta_A \tag{5-222}$$

Several correlations that predict the composition dependence of D_{AB} are summarized in Table 5-15. Most are based on known values of D°_{AB} and D°_{BA}. In fact, a rule of thumb states that, for many binary systems, D°_{AB} and D°_{BA} bound the D_{AB} vs. x_A curve.

Caldwell and Babb [4] used virtually Darken's equation to evaluate the mutual diffusivity for concentrated mixtures of common liquids.

Van Geet and Adamson [*J. Phys. Chem.* **68:** 238 (1964)] tested that equation for the *n*-dodecane (*A*) and *n*-octane (*B*) system and found the average deviation of D_{AB} from experimental values to be –0.68 percent. Additional tests showed Eq. (5-223) can be expected to be fairly accurate for nonpolar hydrocarbons of similar molecular weight. For systems that depart significantly from thermodynamic ideality, such as polar-polar mixtures, it breaks down, sometimes by a factor of 8. Siddiqi, Krahn, and Lucas [*J. Chem. Eng. Data* **32:** 48 (1987)] found that this relation was superior to those of Vignes and Leffler and Cullinan for a variety of mixtures. Umesi and Danner [*Ind. Eng. Chem. Process Des. Dev.* **20:** 662 (1981)] found an average absolute deviation of 13.9 percent for 198 data points.

Rathbun and Babb [16] suggested that Darken's equation could be improved by raising the thermodynamic correction factor β_A to a power n less than unity. They looked at systems exhibiting negative deviations from Raoult's law and found $n = 0.3$. Furthermore, for polar-nonpolar mixtures, they found $n = 0.6$. Siddiqi and Lucas [17] followed those suggestions and found an average absolute error of 3.3 percent for nonpolar-nonpolar mixtures, 11.0 percent for polar-nonpolar mixtures, and 14.6 percent for polar-polar mixtures. Siddiqi, Krahn, and Lucas [*J. Chem. Eng. Data* **32:** 48 (1987)] examined a few other mixtures and found that $n = 1$ was probably best. Thus, this approach is, at best, highly dependent on the type of components.

Vignes [20] empirically correlated mixture diffusivity data for 12 binary mixtures. Later Ertl, Ghai, and Dollon [*AIChE J.* **20:** 1 (1974)] evaluated 122 binary systems, which showed an average absolute deviation of only 7 percent. None of the latter systems, however, was very nonideal.

TABLE 5-15 Correlations of Diffusivities for Concentrated, Binary Mixtures of Nonelectrolyte Liquids

Authors*	Equation	
Caldwell-Babb [4]	$D_{AB} = (x_A D^\circ_{BA} + x_B D^\circ_{AB})\beta_A$	(5-223)
Rathbun-Babb [16]	$D_{AB} = (x_A D^\circ_{BA} + x_B D^\circ_{AB})\beta^n_A$	(5-224)
Vignes [20]	$D_{AB} = D^{\circ x_B}_{AB}\, D^{\circ x_A}_{BA}\beta_A$	(5-225)
Leffler-Cullinan [13]	$D_{AB}\mu_{mix} = (D^\circ_{AB}\mu_B)^{x_B}(D^\circ_{BA}\mu_A)^{x_A}\beta_A$	(5-226)
Cussler [8]	$D_{AB} = D_0\left[1 + \frac{K}{x_A x_B}\left(\frac{\partial \ln x_A}{\partial \ln a_A} - 1\right)\right]^{-1/2}$	(5-227)
Cullinan [7]	$D_{AB} = \frac{kT}{2\pi\mu_{mix}(V/A)^{1/3}}\left[\frac{2\pi x_A x_B \beta_A}{1 + \beta_A(2\pi x_A x_B - 1)}\right]^{1/2}$	(5-228)
Asfour-Dullien [1]	$D_{AB} = \left(\frac{D^o_{AB}}{\mu_B}\right)^{x_B}\left(\frac{D^o_{BA}}{\mu_A}\right)^{x_A}\zeta\mu\beta_A$	(5-229)
Siddiqi-Lucas [17]	$D_{AB} = (C_B \overline{V}_B D^o_{AB} + C_A \overline{V}_A D^o_{BA})\beta_A$	(5-230)
Bosse and Bart no. 1 [2]	$D_{AB} = (D^\infty_{AB})^{X_B}(D^\infty_{AB})^{X_A}\exp\left(-\frac{g^E}{RT}\right)$	(5-231)
Bosse and Bart no. 2 [2]	$\mu D_{AB} = (\mu\beta D^\infty_{AB})^{X_B}(\mu_A D^\infty_{BA})^{X_A}\exp\left(-\frac{g^E}{RT}\right)$	(5-232)

Relative errors for the correlations in this table are very dependent on the components of interest and are cited in the text.

*See the beginning of the Mass Transfer subsection for references.

Leffler and Cullinan [13] modified Vignes' equation using theoretical arguments to arrive at Eq. (5-226), which the authors compared to Eq. (5-225) for the 12 systems mentioned above. The average absolute maximum deviation was only 6 percent. Umesi and Danner [*Ind. Eng. Chem. Process Des. Dev.* **20:** 662 (1981)], however, found an average absolute deviation of 11.4 percent for 198 data points. For normal paraffins, it is not very accurate. In general, the accuracies of the two equations are not much different, and since Vignes' equation is simpler to use, it is suggested. The application of either should be limited to nonassociating systems that do not deviate much from ideality ($0.95 < \beta_A < 1.05$).

Cussler [8] showed that in concentrated associating systems it is the size of diffusing clusters rather than diffusing solutes that controls diffusion. D_o is a reference diffusion coefficient discussed hereafter; a_A is the activity of component *A*; and *K* is a constant. By assuming that D_o could be predicted by Eq. (5-225) with $\beta = 1$, *K* was found to be equal to 0.5 based on five binary systems and validated with a sixth binary mixture. The limitations of Eq. (5-227) using D_o and *K* defined previously have not been explored, so caution is warranted. Gurkan [*AIChE J.* **33:** 175 (1987)] showed that *K* should actually be closer to 0.3 (rather than 0.5) and discussed the overall results.

Cullinan [7] presented an extension of Cussler's cluster diffusion theory that contains no adjustable constants, does not use diffusivity values at infinite dilution, and relates transport properties and solution thermodynamics. His method accurately accounts for composition and temperature dependence of diffusivity; however, it requires accurate density, viscosity, and activity coefficient data. This equation has been tested for six very different mixtures by Rollins and Knaebel [*AIChE J.* **37:** 470 (1991)], and it was found to agree remarkably well with data for most conditions, considering the absence of empirical parameters (including diffusivity values). In the dilute region (of either *A* or *B*), there are systematic errors probably caused by the breakdown of certain implicit assumptions (that nevertheless appear to be generally valid at higher concentrations).

Asfour and Dullien [1] developed a relation for predicting alkane diffusivities at moderate concentrations that employs

$$\zeta = \left(\frac{V_{fm}}{V_{fx_A} V_{fx_B}}\right)^{2/3} \frac{M_{x_A} M_{x_B}}{M_m} \tag{5-233a}$$

where $V_{fx_i} = V_{fi}^{x_i}$; the fluid free volume is $V_{fi} = V_i - V_{mli}$ for $i = A, B$, and m, in which V_{mli} is the molar volume of the liquid at the melting point and

$$V_{ml_m} = \left(\frac{x_A^2}{V_{ml_A}} + \frac{2x_A x_B}{V_{ml_{AB}}} + \frac{x_B^2}{V_{ml_B}}\right)^{-1} \tag{5-233b}$$

and

$$V_{ml_{AB}} = \left[\frac{V_{ml_A}^{1/3} + V_{ml_B}^{1/3}}{2}\right]^3 \tag{5-233c}$$

and m is the mixture viscosity, M_m is the mixture mean molecular weight, and β_A is defined by Eq. (5-221). The average absolute error of this equation is 1.4 percent, while the Vignes equation and the Leffler-Cullinan equation give 3.3 percent and 6.2 percent, respectively.

Siddiqi and Lucas [17] suggested that component volume fractions might be used to correlate the effects of concentration dependence. They found an average absolute deviation of 4.5 percent for nonpolar-nonpolar mixtures, 16.5 percent for polar-nonpolar mixtures, and 10.8 percent for polar-polar mixtures.

Bosse and Bart [2] added a term to account for excess Gibbs free energy, involved in the activation energy for diffusion, which was previously omitted. Doing so yielded minor modifications of the Vignes and Leffler-Cullinan equations [Eqs. (5-225) and (5-226), respectively]. The UNIFAC method was used to assess the excess Gibbs free energy. Comparing predictions of the new equations with data for 36 pairs and 326 data points yielded relative deviations of 7.8 percent and 8.9 percent, respectively, but which were better than the closely related Vignes (12.8 percent) and Leffler-Cullinan (10.4 percent) equations.

Binary Electrolyte Mixtures When electrolytes are added to a solvent, they dissociate to a certain degree. It would appear that the solution contains at least three components: solvent, anions, and cations. If the solution is to remain neutral in charge at each point (assuming the absence of any applied electric potential field), the anions and cations diffuse effectively as a single component, and diffusion can thus be treated as a binary mixture.

Nernst-Haskell The theory of dilute diffusion of salts is well developed and has been experimentally verified. For dilute solutions of a single salt, the well-known Nernst-Haskell equation (Poling et al.) is applicable:

$$D^{\circ}_{AB}\frac{RT}{F^2}\frac{\left|\frac{1}{n_+}\right|+\left|\frac{1}{n_-}\right|}{\frac{1}{\lambda^0_+}+\frac{1}{\lambda^0_-}}=8.9304\times10^{-10}T\frac{\left|\frac{1}{n_+}\right|+\left|\frac{1}{n_-}\right|}{\frac{1}{\lambda^0_+}+\frac{1}{\lambda^0_-}} \tag{5-234}$$

where D°_{AB} = diffusivity based on molarity rather than normality of dilute salt A in solvent B, cm²/s.

The previous equations can be interpreted in terms of ionic-species diffusivities and conductivities. The latter are easily measured and depend on temperature and composition. The resulting equation of the electrolyte diffusivity is

$$D_{AB}=\frac{|z_+|+|z_-|}{(|z_-|/D_+)+(|z_+|/D_-)} \tag{5-235}$$

where $|z_\pm|$ represents the magnitude of the ionic charge and where the cationic or anionic diffusivities are $D_\pm = 8.9304 \times 10^{-10}\, T\lambda_\pm/|z_\pm|$ cm²/s and $\lambda_\pm$ are the infinite dilution conductances of cation and anion. In practice, the equivalent conductance of the ion pair of interest would be obtained and supplemented with conductances of permutations of those ions and one independent cation and anion. This would allow determination of all the ionic conductances and hence the diffusivity of the electrolyte solution.

According to Gordon [*J. Phys. Chem.* **5:** 522 (1937)] typically, as the concentration of a salt increases from infinite dilution, the diffusion coefficient decreases rapidly from D°_{AB}. As concentration is increased further, however, D_{AB} rises steadily, often becoming greater than D°_{AB}. Gordon proposed the following empirical equation, which is applicable up to concentrations of 2N:

$$D_{AB}=D^{\circ}_{AB}\frac{1}{C_B\bar{V}_B}\frac{\mu_B}{\mu}\left(1+\frac{\ln\gamma_\pm}{\ln m}\right) \tag{5-236}$$

where D°_{AB} is given by the Nernst-Haskell equation. References that tabulate $\gamma_\pm$ as a function of m, as well as other equations for D_{AB}, are given by Poling et al.

Morgan, Ferguson, and Scovazzo [*Ind. Eng. Chem. Res.* **44:** 4815 (2005)] studied diffusion of gases in ionic liquids having moderate to high viscosity (up to about 1000 cP) at 30°C. Their range was limited, and the empirical equation they found was

$$D_{AB}=3.7\times10^{-3}\left(\frac{1}{\mu_B^{0.59}V_A\rho_B^2}\right) \tag{5-237}$$

which yielded a correlation coefficient of 0.975. Of the estimated diffusivities 90 percent were within ±20 percent of the experimental values. The exponent for viscosity approximately confirmed the observation of Hiss and Cussler [*AIChE J.* **19:** 698 (1973)].

Example 5-15 Diffusivity Estimation

a. Estimate the diffusivity of naphthalene (A) in nitrogen (B) at 30°C and 1 atm (abs).

1. Chapman-Enskog equation (refer to Table 5-11, Eqs. (5-200), (5-210), and (5-211*a*), and Table 5-12).
We will use properties of naphthalene (M_A = 128.17) at its melting point (353.5 K, ρ_{LA} = 0.973 g/cm³), and nitrogen (M_B = 28.01) at its boiling point (77.4 K, ρ_{LB} = 0.804 g/cm³) for estimation of parameters:

$$\varepsilon_A/k = 1.92\times353.5 = 678.72\text{ K},\ \sigma_A = 1.222\,(128.17/0.973)^{1/3} = 6.22\text{ Å}$$

$$\varepsilon_B/k = 1.15\times77.4 = 89.01\text{ K},\ \sigma_B = 1.18\,(28.01/0.804)^{1/3} = 3.85\text{ Å}$$

Note that Svehla's (Svehla, R. A., NASA Tech. Rep. R-132, Lewis Research Ctr., Cleveland, Ohio, 1962) values for air are ε_B/k = 78.6 K and σ_B = 3.711 Å.

$$\varepsilon_{AB}/k = (\varepsilon_A\varepsilon_B)^{1/2}/k = (678.72\times89.01)^{1/2} = 245.79\text{ K}$$

$$\sigma_{AB} = (6.22+3.85)/2 = 5.04\text{ Å}$$

$$T^* = kT/\varepsilon_{AB} = 303.2/245.79 = 1.234$$

$$\Omega_D = (44.54T^{*-4.909} + 1.911T^{*-1.575})^{0.10} = 1.056$$

$$D_{AB}=\frac{0.001858\times303.2^{3/2}\times[(1/128.17)+(1/29.0)]^{1/2}}{1\times\sigma_{AB}^2\times\Omega_D}=0.06066\text{ cm}^2/\text{s}$$

b. Estimate the diffusivities of dimethylformamide (DMF = C_3H_7NO = A) in water (B) at 40°C (313.15 K) using the Wilke and Chang method (Table 5-14, Eq. (5-214)) and the following property data.

x_{DMF}	μ (cP)	ρ (kg/m³)	x_{DMF}	D_{AB} (10⁵ cm²/s)
0.0000	0.6529	992.214	0.05013	1.653
0.1960	1.5578	984.062	0.5064	1.482
0.4919	1.3509	962.324	0.9581	2.751
0.7991	0.8768	940.222		
1.0000	0.7143	929.435		

1. ξ_B = 2.26, M_A = 73.09 g/mol, M_B = 18.016 g/mol, μ_B = 0.6529 cP
V_A = 73.09/0.9294 = 78.64 cm³/gmol
D_{AB} = 1.17 × 10⁻¹³ $(\xi_B M_B)^{1/2}\,T/(\mu_B(\varphi V_A)^{0.6})$ = 1.651 × 10⁻⁵ cm²/s
The experimental value is 1.65 × 10⁻⁵ cm²/s.
2. H_2O in DMF (following Kooijman's suggestions):
ξ_B = 2.6 M_A = 18.016 g/mol M_B = 73.09 g/mol μ_B = 0.7143 cP
φV_A = 4.5 × 18.016/0.9922 = 78.64 cm³/gmol
D_{AB} = 1.17 × 10⁻¹³ $(\xi_B M_B)^{1/2}\,T/(\mu_B V_A^{0.6})$ = 3.185 × 10⁻⁵ cm²/s
The experimental value is 2.75 × 10⁻⁵ cm²/s.

MAXWELL-STEFAN ANALYSIS

Fick's law was originally developed for dilute binary diffusion based on analogy to Fourier's law, and then it was successfully extended to concentrated solutions. Although Fick's law has been extended to ternary mixtures, the resulting equations, except for a few special cases, are complex and require additional diffusivity values, some of which may be negative to fit experimental data. The Maxwell-Stefan (M-S) and Fickian analyses give identical results for binary systems, and the choice of which to use becomes one of personal preference. For multicomponent gas systems, the M-S method has clear advantages that are outlined in this section. Curtis and Bird [*Ind. Eng. Chem. Res.* **38:** 2515 (1999)] reconciled the multicomponent Fick's law approach with the more elegant M-S theory.

In the late 1800s, the development of the kinetic theory of gases led to a method developed by Maxwell and Stefan for calculating multicomponent gas diffusion (e.g., the flux of each species in a mixture). The M-S diffusion equations for N_c components in a reference system moving at the average molar velocity are simpler in principle than extensions of Fick's law since they employ binary diffusivities:

$$\nabla x_i=\sum_{j=1}^{N_c}\frac{1}{c\mathcal{D}_{ij}}(x_iN_j-x_jN_i) \tag{5-238}$$

For ideal gases, the values $\mathcal{D}_{ij}$ of this equation are equal to the binary diffusivities for the ij pairs, which are identical to the Fickian diffusivities,

Equation (5-238) can be solved for N_c − 1 independent fluxes. An additional equation (called a "bootstrap" equation) based on the movement of the reference system or the reaction stoichiometry is needed to determine all N_c fluxes. For example, for equimolar counterdiffusion this expression is $\sum_{j=1}^{N_c}(N_j)=0$.

A study of ternary gas diffusion showed that the M-S equations predicted the experimental results within the experimental error [Duncan and Toor, *AIChE J.* **8:** 38 (1962)]. These predictions include a zero component flux despite the presence of that component's concentration gradient, a finite component flux with no component concentration gradient, and flux of a component in the direction opposite the component's concentration gradient. Simplified solutions for ternary diffusion of ideal gases with equimolal counterdiffusion are shown later in Examples 5-15 and 5-16.

For nonideal systems the generalized form of the driving forces is based on the derivative of the chemical potential μ_i (see Taylor and Krishna, Wesselingh and Krishna, Datta and Vilekar [*Chem. Eng. Sci.* **65:** 5976 (2010)] and Krishna and Wesselingh [*Chem. Eng. Sci.* **52:** 861 (1997)]):

$$\frac{x_i}{RT}\nabla\mu_i\big|_{T,p}=\sum_{j=1}^{N_c}\frac{1}{c\mathcal{D}_{ij}}(x_iN_j-x_jN_i) \tag{5-239}$$

For liquids activity coefficients are included in the M-S equations

$$\left(1+\frac{x_i}{\gamma_i}\frac{\partial\gamma_i}{\partial x_i}\right)\nabla x_i=\sum_{j=1}^{N_c}\frac{1}{c\mathcal{D}_{ij}}(x_iN_j-x_jN_i) \tag{5-240}$$

Although in principle extension of the M-S equations to nonideal liquid systems is straightforward, in practice the need for extensive activity coefficient data and the variability of the $\mathcal{D}_{ij}$ can prove daunting.

Almost all reported diffusivities are based on the Fickian model. The relationship between binary Fickian and M-S diffusivities is

$$Đ_{ij} = \frac{D_{ij}}{1+\dfrac{x_i}{\gamma_i}\dfrac{\partial \gamma_i}{\partial x_i}} \tag{5-241}$$

The Fickian and M-S binary diffusivities are equal at the infinite dilution limit and for ideal systems.

The generalized form of the M-S equations in terms of the gradient of the chemical potential can include electromagnetic effects and thermal and pressure diffusion. Since electroneutrality can be included, the M-S method may be applied to electrolyte diffusion [Kraaijeveld, Wesselingh, and Kuiken, *Ind. Eng. Chem. Res.* **33:** 750 (1994)]. Ordinary molecular diffusion and pressure and thermal diffusion in multicomponent mixtures have been studied [Ghorayeb and Firoozabadi, *AIChE J.* **46:** 883 (2000)].

Approximate solutions have been developed by linearization [Toor, H. L., *AIChE J.* **10:** 448 and 460 (1964); Stewart and Prober, *Ind. Eng. Chem. Fundam.* **3:** 224 (1964)]. Those differ in details but yield about the same accuracy as Smith and Taylor [*Ind. Eng. Chem. Fundam.* **22:** 97 (1983)]. More recently, efficient algorithms for solving the equations exactly have been developed; see Benitez, Taylor and Krishna, Krishnamurthy and Taylor [*Chem. Eng. J.* **25:** 47 (1982)] and Taylor and Webb [*Comput. Chem. Eng.* **5:** 61 (1981)]. Amundson, Pan, and Paulson [*AIChE J.* **48:** 813 (2003)] presented numerical methods for coping with mixtures having four or more components, which are nearly intractable via the analytical M-S method.

An even simpler approach than solving the M-S differential equations is to use difference equations (see Wesselingh and Krishna, and Wankat [114]). For a ternary system with mass transfer in the z direction the difference equation is

$$\left(1+\frac{\bar{x}_A}{\bar{\gamma}_A}\frac{\partial \gamma_A}{\partial x_A}\right)\frac{\Delta x_A}{\Delta z} = -\frac{\bar{x}_B\bar{N}_{A,z}-\bar{x}_A\bar{N}_{B,z}}{\bar{c}Đ_{AB}} - \frac{\bar{x}_C\bar{N}_{A,z}-\bar{x}_A\bar{N}_{C,z}}{\bar{c}Đ_{AC}} \tag{5-242}$$

with equivalent equations for the other components. The bars indicate evaluation of the terms at the average conditions. For ideal gases this equation simplifies to

$$\bar{c}\frac{\Delta y_A}{\Delta z} = -\frac{\bar{y}_B\bar{N}_{A,z}-\bar{y}_A\bar{N}_{B,z}}{Đ_{AB}} - \frac{\bar{y}_C\bar{N}_{A,z}-\bar{y}_A\bar{N}_{C,z}}{Đ_{AC}} \tag{5-243}$$

Although difference equation solutions are approximate, they show typical multicomponent behavior. The solutions can be made more exact by including additional Δz segments.

The M-S equations are often used for multicomponent gas mixtures because each $Đ_{ij}$ is practically independent of composition by itself and in a multicomponent mixture. This procedure is illustrated with the difference equation formulation in Example 5-16.

Example 5-16 Maxwell-Stefan Diffusion Without a Gradient Two identical large glass bulbs are filled with gases and connected by a capillary tube that is δ = 0.0090 m long. Bulb 1 at $z = 0$ contains the following mole fractions: $y_{air} = 0.620$, $y_{H2} = 0.380$, and $y_{NH3} = 0.000$. Bulb 2 at $z = \delta$ contains $y_{air} = 0.620$, $y_{H2} = 0.000$, and $y_{NH3} = 0.380$. Operation is assumed to be at pseudo- (or quasi-) steady state. The pressure and temperature are uniform at 1.5 atm and 273 K, respectively. Diffusivity values at 1.0 atm and 273 K are $D_{air\text{-}H2} = 0.611$, $D_{air\text{-}NH3} = 0.198$, and $D_{H2\text{-}NH3} = 0.748$ cm²/s. Assume the gases are ideal. Estimate the fluxes of the three components using the M-S difference equation formulation.

Solution Let A = air, B = hydrogen, and C = ammonia. Since Dp = constant, $D(1.5\text{ atm}) = D(1\text{ atm})/(1.5\text{ atm})$. At pseudo-steady state, mole fractions at the boundaries are constant (e.g., $y_{air} = 0.620$ at $z = 0$ and $y_{air} = 0.620$ at $z = \delta$ so that $\Delta y_{air} = 0.00$). Since temperature and pressure are constant and the molar densities of ideal gases are equal, the system has equimolar counterdiffusion and the total flux is zero, $N_C = -N_A - N_B$. Substitute this expression into Eq. (5-243) and the equivalent equation for B.

$$\bar{\rho}_m\frac{\Delta y_A}{\Delta z} = -\left(\frac{\bar{y}_B}{Đ_{AB}}+\frac{\bar{y}_C}{Đ_{AC}}+\frac{\bar{y}_A}{Đ_{AC}}\right)\bar{N}_A + \left(\frac{\bar{y}_A}{Đ_{AB}}-\frac{\bar{y}_A}{Đ_{AC}}\right)\bar{N}_B \tag{5-244a}$$

$$\bar{\rho}_m\frac{\Delta y_B}{\Delta z} = \left(\frac{\bar{y}_B}{Đ_{BA}}-\frac{\bar{y}_B}{Đ_{BC}}\right)\bar{N}_A - \left(\frac{\bar{y}_A}{Đ_{BA}}+\frac{\bar{y}_C}{Đ_{BC}}+\frac{\bar{y}_B}{Đ_{BC}}\right)\bar{N}_B \tag{5-244b}$$

Determine N_B from the first equation and N_A from the second.

$$\bar{N}_B = \frac{\bar{\rho}_m\dfrac{\Delta y_A}{\Delta z} + \left(\dfrac{\bar{y}_B}{Đ_{AB}}+\dfrac{\bar{y}_C}{Đ_{AC}}+\dfrac{\bar{y}_A}{Đ_{AC}}\right)\bar{N}_A}{\left(\dfrac{\bar{y}_A}{Đ_{AB}}-\dfrac{\bar{y}_A}{Đ_{AC}}\right)} \tag{5-245a}$$

$$\bar{N}_A = \frac{\bar{\rho}_m\dfrac{\Delta y_B}{\Delta z} + \left(\dfrac{\bar{y}_A}{Đ_{BA}}+\dfrac{\bar{y}_C}{Đ_{BC}}+\dfrac{\bar{y}_B}{Đ_{BC}}\right)\bar{N}_B}{\left(\dfrac{\bar{y}_B}{Đ_{BA}}-\dfrac{\bar{y}_B}{Đ_{BC}}\right)} \tag{5-245b}$$

Input these equations and the values for mole fractions at the boundaries, diffusivities, ρ_m from ideal gas law, and $\Delta z = \delta$ into a spreadsheet. Guess a value for $N_{A,guess}$, calculate $N_{A,calc}$ and $N_{B,calc}$, and use Goal Seek to make $N_{A,guess} - N_{A,calc} = 0$ by changing the value of $N_{A,guess}$. Then $N_C = -(N_A + N_B)$.

Results $N_{air} = -5.846 \times 10^{-5}$, $N_{H2} = 1.216 \times 10^{-4}$, and $N_{NH3} = -6.312 \times 10^{-5}$ kmol/(m²s). The transfer rates = $N \times$ (area of capillary tube). As expected, hydrogen diffuses in the positive direction and ammonia in the negative direction. The surprise is the substantial negative diffusion rate of air despite a Fickian driving force of zero. The air diffusion is caused by the friction of the ammonia.

Example 5-17 Maxwell-Stefan Diffusion Counter to Gradient Repeat Example 5-16, but bulb 2 at $z = \delta$ contains $y_{air} = 0.610$, $y_{H2} = 0.010$, and $y_{NH3} = 0.380$.

Solution The solution approach is identical to that of Example 5-16, and the same spreadsheet is used with different mole fractions in bulb 2.

Results $N_{air} = -5.561 \times 10^{-5}$, $N_{H2} = 1.1850 \times 10^{-4}$, and $N_{NH3} = -6.289 \times 10^{-5}$ kmol/(m²s). Since air and hydrogen have gradients in the same direction, extrapolation of binary Fickian diffusion would lead us to expect diffusion of these gases in the same direction. However, because of friction with ammonia, the diffusion of air is in the same direction as the ammonia.

The spreadsheet can be used for other ideal gas, ternary diffusion systems with equimolal counterdiffusion that are at either steady state or pseudo-steady state. Kmit and Shah [*Chem. Eng. Educ.* **30**(1): 14 (1996)] have a detailed discussion of when pseudo-steady state is valid.

Multicomponent Liquid Mixtures Most liquid mixtures are not ideal, and each $Đ_{ij}$ can be strongly composition-dependent in binary mixtures; moreover, the binary $Đ_{ij}$ are strongly affected in a multicomponent mixture (see Taylor and Krishna). Several theories have been developed for predicting multicomponent liquid-diffusion coefficients, but the necessity for extensive activity data, pure component and mixture volumes, mixture viscosity data, and tracer and binary diffusion coefficients has significantly limited the utility of the theories (see Poling et al.).

One particular case of multicomponent liquid diffusion that is somewhat tractable is the dilute diffusion of a solute in a homogeneous mixture (e.g., of A in $B + C$). Umesi and Danner [*Ind. Eng. Chem. Process Des. Dev.* **20:** 662 (1981)] compared the three equations given below for 49 ternary systems. All three equations were equivalent, giving average absolute deviations of 25 percent.

Perkins and Geankoplis [*Chem. Eng. Sci.* **24:** 1035 (1969)] suggested the following empirical equation as an extension of the Caldwell-Babb [4] equation, in order to take into account variations in viscosity in multicomponent mixtures.

$$D_{am}\mu_m^{0.8} = \sum_{\substack{j=1\\ j\neq A}}^{n} x_j D_{Aj}^{\circ}\mu_j^{0.8} \tag{5-246}$$

Cullinan [*Can. J. Chem. Eng.* **45:** 377 (1967)] extended Vignes' equation to multicomponent systems:

$$D_{am} = \prod_{\substack{j=1\\ j\neq A}}^{n}\left(D_{Aj}^{\circ}\right)^{x_j} \tag{5-247}$$

Leffler and Cullinan [13] extended their binary relation to an arbitrary multicomponent mixture:

$$D_{am}\mu_m = \prod_{\substack{j=1\\ j\neq A}}^{n}\left(D_{Aj}^{\circ}\mu_j\right)^{x_j} \tag{5-248}$$

where D_{Aj} is the dilute binary diffusion coefficient of A in j; D_{am} is the dilute diffusion of a through m; x_j is the mole fraction; μ_j is the viscosity of component j; and μ_m is the mixture viscosity.

Dilute multicomponent diffusion of gases in aqueous electrolyte solutions is of significant practical interest because many gas absorption processes use electrolyte solutions. Akita [*Ind. Eng. Chem. Fundam.* **10:** 89 (1981)] presents experimentally tested equations for this case.

Multicomponent diffusion of electrolytes is important in ion exchange. Graham and Dranoff [*Ind. Eng. Chem. Fundam.* **21:** 360, 365 (1982)] found that the M-S interaction coefficients reduce to limiting ion tracer diffusivities of each ion. Pinto and Graham [*AIChE J.* **32:** 291 (1986) and **33:** 436 (1987)] corrected for solvation effects in multicomponent diffusion in electrolyte solutions. They achieved excellent results for 1-1 electrolytes in water at 25°C up to concentrations of 4*M*.

DIFFUSION OF FLUIDS IN POROUS SOLIDS

Diffusion in porous solids is usually the most important factor controlling mass transfer in adsorption, ion exchange, drying, heterogeneous catalysis, leaching, and many other applications. Some of the applications of interest are outlined in Table 5-16. Applications of these equations are found in Secs. 16, 22, and 23.

Diffusion within the largest cavities of a porous medium is assumed to be similar to ordinary or bulk diffusion except that it is hindered by the pore walls [see Eq. (5-249)]. The tortuosity τ that expresses this hindrance was originally estimated from geometric arguments. Unfortunately, measured values are often an order of magnitude greater than those estimates. Thus, the effective diffusivity D_{eff} (and hence τ) is normally determined by comparing a diffusion model to experimental measurements. The normal range of tortuosities for silica gel, alumina, and other porous solids is $2 \le \tau \le 6$, but for activated carbon, $5 \le \tau \le 65$.

In small pores and at low pressures, the mean free path ℓ of the gas molecule (or atom) is significantly greater than the pore diameter d_{pore}. Its magnitude may be estimated from

$$\ell = \frac{3.2\mu}{P}\left(\frac{RT}{2\pi M}\right)^{1/2} \quad \text{m} \qquad (5\text{-}261)$$

As a result, collisions with the wall occur more frequently than with other molecules. This is referred to as the *Knudsen mode of diffusion* and is contrasted with ordinary or bulk diffusion, which occurs by intermolecular collisions. At intermediate pressures, both ordinary diffusion and Knudsen diffusion may be important [see Eqs. (5-252) and (5-253)].

For gases and vapors that adsorb on the porous solid, surface diffusion may be important, particularly at high surface coverage [see Eqs. (5-254) and (5-257)]. The mechanism of surface diffusion may be viewed as molecules hopping from one surface site to another. Thus, if adsorption is too strong, surface diffusion is impeded, while if adsorption is too weak, surface diffusion contributes insignificantly to the overall rate. Surface diffusion and bulk diffusion usually occur in parallel [see Eqs. (5-258) and (5-259)]. Although D_s is expected to be less than D_{eff}, the solute flux due to surface diffusion may be larger than that due to bulk diffusion if $\partial q_i/\partial z >> \partial C_i/\partial z$. This can occur when a component is strongly adsorbed and the surface coverage is high. For all that, surface diffusion is not well understood. The references in Table 5-15 should be consulted for further details. For multicomponent diffusion in porous media the M-S formulation should be employed for combinations of ordinary, Knudsen, and surface diffusion (see Krishna, *Gas Separ. Purific.* **7**(2): 91 (1993)).

INTERPHASE MASS TRANSFER

Transfer of material between phases is important in most separation processes in which two phases are involved. In this section, mass transfer between gas and liquid phases is discussed. The principles are easily applied to other phases. When one phase is pure, mass transfer in the pure phase

TABLE 5-16 Relations for Diffusion in Porous Solids

Mechanism	Equation		Applies to	References*
Bulk diffusion in pores	$D_{\text{eff}} = \dfrac{\varepsilon_p D}{\tau}$	(5-249)	Gases or liquids in large pores. $N_{Kn} = \ell/d_{\text{pore}} < 0.01$	[24]
Knudsen diffusion	$D_K = 48.5\, d_{\text{pore}}\left(\dfrac{T}{M}\right)^{1/2}$ in m²/s	(5-250)	Dilute (low pressure) gases in small pores. $N_{Kn} = \ell/d_{\text{pore}} > 10$	Geankoplis, [25, 26]
	$D_{K\text{eff}} = \dfrac{\varepsilon_p D_K}{\tau}$			
	$N_i = -D_K \dfrac{dC_i}{dz}$	(5-251)	" " " "	
Combined bulk and Knudsen diffusion	$D_{\text{eff}} = \left(\dfrac{1-\alpha x_A}{D_{\text{eff}}} + \dfrac{1}{D_{K\text{eff}}}\right)^{-1}$	(5-252)	" " " "	Geankoplis, [23, 26]
	$\alpha = 1 + \dfrac{N_B}{N_A}$		$N_A \neq N_B$	
	$D_{\text{eff}} = \left(\dfrac{1}{D_{\text{eff}}} + \dfrac{1}{D_{K\text{eff}}}\right)^{-1}$	(5-253)	$N_A = N_B$	
Surface diffusion	$J_{si} = -D_{S\text{eff}}\, \rho_p \left(\dfrac{dq_i}{dz}\right)$	(5-254)	Adsorbed gases or vapors	[23, 25, 26]
	$D_{S\text{eff}} = \dfrac{\varepsilon_p D_S}{\tau}$	(5-255)	" " " "	
	$D_{S\theta} = \dfrac{D_{S\theta=0}}{(1-\theta)}$	(5-256)	θ = fractional surface coverage ≤ 0.6	
	$D_S = D_S'(q)\exp\left(\dfrac{-E_S}{RT}\right)$	(5-257)	" " " "	
Parallel bulk and surface diffusion	$J = -\left[D_{\text{eff}}\left(\dfrac{dp_i}{dz}\right) + D_{S\text{eff}}\rho_p\left(\dfrac{dq_i}{dz}\right)\right]$	(5-258)	" " " "	[25]
	$J = -D_{\text{app}}\left(\dfrac{dp_i}{dz}\right)$	(5-259)	" " " "	
	$D_{\text{app}} = D_{\text{eff}} + D_{S\text{eff}}\, \rho_p \left(\dfrac{dq_i}{dp_i}\right)$	(5-260)	" " " "	

*Author names refer to the General References list at the beginning of the Mass Transfer subsection (pp. 5-41, 5-42). Bracketed numbers refer to the table reference lists (p. 5-42).

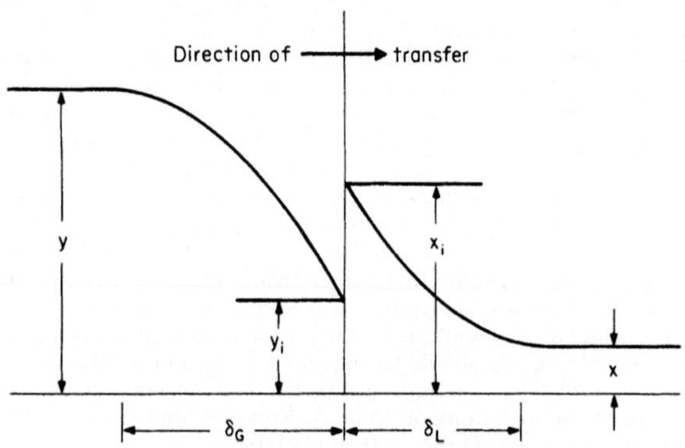

FIG. 5-24 Concentration gradients near a gas-liquid interface.

is not involved. For example, when a pure liquid is being evaporated into a gas, only the gas-phase mass transfer need be calculated. When phases are not pure, the gas phase consists of an insoluble carrier gas plus solute A, and the liquid phase consists of a nonvolatile solvent plus solute A. Thus, mass transfer in each phase is binary. When the resistance to mass transfer is much larger in one phase than in the other, mass transfer in the phase with low resistance may be neglected even though pure components are not involved. Understanding the nature and magnitudes of these resistances is one of the keys to performing reliable mass-transfer analyses.

Mass-Transfer Principles: Dilute Systems When material is transferred from one phase to another across an interface that separates the two, the resistance to mass transfer in each phase causes a concentration gradient in each, as shown in Fig. 5-24 for a gas-liquid interface. The concentrations of the diffusing material in the two phases immediately adjacent to the interface generally are unequal, even if expressed in the same units, but usually are assumed to reach equilibrium almost immediately when a gas and a liquid are brought into contact.

For systems with dilute solute concentrations in the gas and liquid phases, the rate of mass transfer is proportional to the difference between the solute's bulk concentration and its concentration at the gas-liquid interface. Thus

$$N_A = k'_G(p - p_i) = k'_L(c_i - c) \tag{5-262}$$

where N_A = mass-transfer rate, k'_G = gas-phase mass-transfer coefficient, k'_L = liquid-phase mass-transfer coefficient, p = solute partial pressure in bulk gas, p_i = solute partial pressure at interface, c = solute concentration in bulk liquid, and c_i = solute concentration in liquid at the interface. The mass-transfer coefficients k'_G and k'_L by definition are equal to the ratios of the molal mass flux N_A to the concentration driving forces $p - p_i$ and $c_i - c$, respectively.

An alternative expression for the rate of transfer in dilute systems is given by

$$N_A = k_G(y - y_i) = k_L(x_i - x) \tag{5-263}$$

where k_G = gas-phase mass-transfer coefficient, k_L = liquid-phase mass-transfer coefficient, y = mole-fraction solute in bulk-gas phase, y_i = mole-fraction solute in gas at interface, x = mole-fraction solute in bulk-liquid phase, and x_i = mole-fraction solute in liquid at interface. The mass-transfer coefficients defined by Eqs. (5-262) and (5-263) are related to each other as follows:

$$k_G = k'_G p_T \tag{5-264a}$$

$$k_L = k'_L \bar{\rho}_L \tag{5-264b}$$

where p_T = total system pressure employed *during the experimental determinations* of k'_G values and $\bar{\rho}_L$ = average molar density of the liquid phase. The coefficient k_G is relatively independent of the total system pressure and therefore is more convenient to use than k'_G, which is inversely proportional to the total system pressure.

The above equations may be used for finding the interfacial concentrations corresponding to any set of values of x and y provided the ratio of the individual coefficients is known. Thus

$$(y - y_i)/(x_i - x) = k_L/k_G = k'_L\bar{\rho}_L/k'_G p_T = L_M H_G/G_M H_L \tag{5-265}$$

where L_M = molar liquid velocity, G_M = molar gas velocity, H_L = height of 1 transfer unit based on liquid-phase resistance, Eq. (5-283), and H_G = height of 1 transfer unit based on gas-phase resistance, Eq. (5-281). The interfacial mole fractions y_i and x_i can be determined by solving Eq. (5-265) simultaneously with the equilibrium relation $y_i^* = F(x_i)$ to obtain y_i and x_i. The rate of transfer may then be calculated from Eq. (5-263). Equation (5-265) may be solved graphically if a plot is made of the equilibrium vapor and liquid compositions and a point representing the bulk concentrations x and y is located on this diagram. This construction is shown in Fig. 5-25, which represents a gas absorption situation.

If the equilibrium relation $y_i^* = F(x_i)$ is linear, not necessarily through the origin, the rate of transfer is proportional to the difference between the bulk concentration in one phase and the concentration (in that same phase), which would be in equilibrium with the bulk concentration in the second phase. One such difference is $y - y^*$, and another is $x^* - x$. In this case, there is no need to solve for the interfacial compositions, as may be seen from the following derivation.

The rate of mass transfer may be defined by the equation

$$N_A = K_G(y - y^*) = k_G(y - y_i) = k_L(x_i - x) = K_L(x^* - x) \tag{5-266}$$

where K_G = overall gas-phase mass-transfer coefficient, K_L = overall liquid-phase mass-transfer coefficient, y^* = vapor composition in equilibrium with x, and x^* = liquid composition in equilibrium with vapor of composition y. This equation can be rearranged to the formula

$$\frac{1}{K_G} = \frac{1}{k_G}\left(\frac{y - y^*}{y - y_i}\right) = \frac{1}{k_G} + \frac{1}{k_G}\left(\frac{y_i - y^*}{y - y_i}\right) = \frac{1}{k_G} + \frac{1}{k_L}\left(\frac{y_i - y^*}{x_i - x}\right) \tag{5-267}$$

in view of Eq. (5-265). Comparison of the last term in parentheses with the diagram of Fig. 5-25 shows that it is equal to the slope of the chord connecting the points (x, y^*) and (x_i, y_i). If the equilibrium curve is a straight line, then this term is the slope m. Thus

$$1/K_G = 1/k_G + m/k_L \tag{5-268}$$

When Henry's law is valid ($p_A = Hx_A$ or $p_A = H'C_A$), the slope m is

$$m = H/p_T = H'\bar{\rho}_L/p_T \tag{5-269}$$

where m is defined in terms of mole-fraction driving forces in Eq. (5-263).

If it is desired to calculate the rate of transfer from the overall concentration difference based on bulk-liquid compositions $x^* - x$, the appropriate overall coefficient K_L is related to the individual coefficients by the equation

$$1/K_L = 1/k_L + 1/(mk_G) \tag{5-270}$$

Conversion of these equations to a k'_G, k'_L basis can be accomplished readily by direct substitution of Eqs. (5-264a) and (5-264b).

Occasionally one will find k'_L or K'_L values reported in units (SI) of meters per second. The correct units for these values are kmol/[(s · m²)(kmol/m³)],

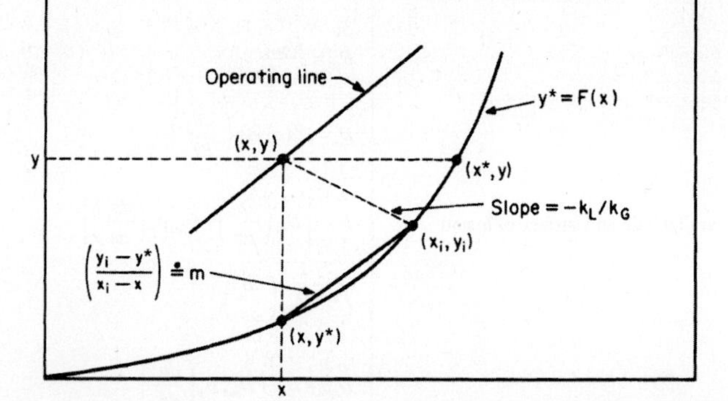

FIG. 5-25 Identification of concentrations at a point in a countercurrent absorption tower.

and Eq. (5-264*b*) is the correct equation for converting them to a mole-fraction basis.

When k'_G and K'_G values are reported in units (SI) of kmol/[(s · m^2)(kPa)], one must be careful, in converting them to a mole-fraction basis, to multiply by the total pressure actually employed in the original experiments and *not* by the total pressure of the system to be designed. This conversion is valid for systems in which Dalton's law of partial pressures ($p = yp_T$) is valid.

Comparison of Eqs. (5-268) and (5-270) shows that for systems in which the equilibrium line is straight, the overall mass-transfer coefficients are related to one another by the equation

$$K_L = mK_G \tag{5-271}$$

When the equilibrium curve is not straight, there is no strictly logical basis for the use of an overall transfer coefficient, since the value of m will be a function of position in the apparatus, as can be seen from Fig. 5-25. In such cases the rate of transfer should be calculated by solving for the interfacial compositions as described above. Experimentally observed rates of mass transfer often are expressed in terms of overall transfer coefficients even when the equilibrium lines are curved. This procedure is approximate, since the rates of transfer may not vary in direct proportion to the overall bulk concentration differences $y - y^*$ and $x^* - x$ at all concentration levels even though the rates may be proportional to the concentration difference in each phase taken separately, that is, $x_i - x$ and $y - y_i$.

In most types of separation equipment such as packed or spray towers, the interfacial area that is effective for mass transfer cannot be accurately determined. For this reason it is customary to report experimentally observed rates of transfer in terms of transfer coefficients based on a unit volume of the apparatus rather than on a unit of interfacial area. Such volumetric coefficients are designated as K_Ga, k_La, etc., where a represents the interfacial area per unit volume of the apparatus. Experimentally observed variations in the values of these volumetric coefficients with variations in flow rates, type of packing, etc., may be due as much to changes in the effective value of a as to changes in k. Calculation of the overall coefficients from the individual volumetric coefficients is done by means of the equations

$$1/(K_Ga) = 1/(k_Ga) + m/(k_La) \tag{5-272a}$$

$$1/(K_La) = 1/(k_La) + 1/(mk_Ga) \tag{5-272b}$$

Because of the wide variation in equilibrium, the variation in the values of m from one system to another can have an important effect on the overall coefficient and on the selection of the type of equipment to use. For example, if m is large, the liquid-phase part of the overall resistance might be extremely large where k_L might be relatively small. This kind of reasoning must be applied with caution, however, since species with different equilibrium characteristics are separated under different operating conditions. Thus, the effect of changes in m on the overall resistance to mass transfer may partly be counterbalanced by changes in the individual specific resistances as the flow rates are changed.

Mass-Transfer Principles: Concentrated Systems When solute concentrations in the gas and/or liquid phases are large, the equations derived above for dilute systems no longer are applicable. The correct equations to use for concentrated systems are as follows:

$$N_A = \hat{k}_G\,(y - y_i)/y_{BM} = \hat{k}_L\,(x_i - x)/x_{BM}$$

$$= \hat{K}_G(y - y^*)/y^*_{BM} = \hat{k}_L(x^* - x)/x^*_{BM} \tag{5-273}$$

where $N_B = 0$. In these and following equations, y_B represents the insoluble carrier gas in the gas phase and x_B represents the nonvolatile solvent in the liquid. Hence, $y_A = y$ and $x_A = x$ represent the solute in the gas and liquid phases, respectively. Thus, for example, it is understood that $1 - x = x_B$.

$$y_{BM} = \frac{(1-y)-(1-y_i)}{\ln[(1-y)/(1-y_i)]} \tag{5-274a}$$

$$y^*_{BM} = \frac{(1-y)-(1-y^*)}{\ln[(1-y)/(1-y^*)]} \tag{5-274b}$$

$$x_{BM} = \frac{(1-x)-(1-x_i)}{\ln[(1-x)/(1-x_i)]} \tag{5-275a}$$

$$x^*_{BM} = \frac{(1-x)-(1-x^*)}{\ln[(1-x)/(1-x^*)]} \tag{5-275b}$$

For concentrated systems the gas-phase $\hat{k}_G$ and liquid-phase $\hat{k}_L$ mass-transfer coefficients and the overall gas-phase $\hat{K}_G$ and liquid-phase $\hat{K}_L$ mass-transfer coefficients are defined later in Eqs. (5-278) to (5-280).

The factors y_{BM} and x_{BM} arise because in the diffusion of a solute through a stationary layer of fluid the resistance to diffusion varies in proportion to the concentration of the stationary fluid, approaching zero as the concentration of the fluid approaches zero. See Eq. (5-198).

The factors y^*_{BM} and x^*_{BM} cannot be justified on the basis of mass-transfer theory since they are based on overall resistances. These factors therefore are included in the equations by analogy with the corresponding film equations.

In dilute systems the logarithmic-mean insoluble-gas and nonvolatile-liquid concentrations approach unity, and Eq. (5-273) reduces to the dilute system formula, Eq. (5-266). For equimolar counterdiffusion (e.g., binary distillation), the log-mean factors are omitted. See Eq. (5-197).

Substitution of Eqs. (5-274) and (5-275) into Eq. (5-273) results in the following simplified formula:

$$N_A = \hat{k}_G \ln[(1-y_i)/(1-y)] = \hat{K}_G \ln[(1-y^*)/(1-y)] = \hat{k}_L \ln[(1-x)/(1-x_i)]$$

$$= \hat{K}_L \ln[(1-x)/(1-x^*)] \tag{5-276}$$

Note that the units of $\hat{k}_G$, $\hat{K}_G$, $\hat{k}_L$, and $\hat{K}_L$ are identical, i.e., kmol/(s · m^2) in SI units.

The interfacial gas and liquid compositions in concentrated systems can be determined from

$$(y - y_i)/(x_i - x) = \hat{k}_L y_{BM}/\hat{k}_G x_{BM}$$

$$= L_M H_G y_{BM}/G_M H_L x_{BM} = k_L/k_G \tag{5-277}$$

This equation is identical to Eq. (5-265) for dilute systems since $\hat{k}_G = k_G y_{BM}$ and $\hat{k}_L = k_L x_{BM}$, and y_{BM} and x_{BM} are both 1 in dilute systems. Note, however, that when $\hat{k}_G$ and $\hat{k}_L$ are given, the equation must be solved by trial and error, since x_{BM} contains x_i and y_{BM} contains y_i.

The overall gas-phase and liquid-phase mass-transfer coefficients for concentrated systems are computed from

$$\frac{1}{\hat{K}_G} = \frac{y_{BM}}{y^*_{BM}}\frac{1}{\hat{k}_G} + \frac{x_{BM}}{y^*_{BM}}\frac{1}{\hat{k}_L}\left(\frac{y_i - y^*}{x_i - x}\right) \tag{5-278}$$

$$\frac{1}{\hat{K}_L} = \frac{x_{BM}}{x^*_{BM}}\frac{1}{\hat{k}_L} + \frac{y_{BM}}{x^*_{BM}}\frac{1}{\hat{k}_G}\left(\frac{x^* - x_i}{y - y_i}\right) \tag{5-279}$$

When the equilibrium curve is a straight line, the terms in parentheses can be replaced by the slope m or $1/m$ as before. In this case, the overall mass-transfer coefficients for concentrated systems are related to one another by the equation

$$\hat{K}_L = m\hat{K}_G\left(x^*_{BM}/y^*_{BM}\right) \tag{5-280}$$

All these equations reduce to their dilute system equivalents as the inert concentrations approach unity in terms of mole fractions of inert concentrations in the fluids.

Height Equivalent to 1 Transfer Unit (HTU) In packed beds used for distillation, absorption, stripping, and extraction, it is convenient to represent mass transfer as the height of apparatus required to accomplish a separation of standard difficulty. The gas-phase HTU is the ratio of gas flow rate to gas-phase mass-transfer coefficient, and the liquid-phase HTU is the ratio of the liquid flow rate to liquid-phase coefficient. The following relations between the transfer coefficients and the values of HTU apply:

$$H_G = G_M/(k_Ga\,y_{BM}) = G_M/(\hat{k}_G\,a) \tag{5-281}$$

$$H_{OG} = G_M/(K_Ga\,y^*_{BM}) = G_M/(\hat{K}_G\,a) \tag{5-282}$$

$$H_L = L_M/(k_La\,x_{BM}) = L_M/(\hat{k}_L\,a) \tag{5-283}$$

$$H_{OL} = L_M/(K_La\,x^*_{BM}) = L_M/(\hat{K}_L\,a) \tag{5-284}$$

Frequently the HTU values are closer to constant than the mass-transfer coefficients.

The equations that express the addition of individual resistances in terms of HTUs, applicable to either dilute or concentrated systems, are

$$H_{OG} = \frac{y_{BM}}{y^*_{BM}} H_G + \frac{mG_M}{L_M} \frac{x_{BM}}{y^*_{BM}} H_L \tag{5-285}$$

$$H_{OL} = \frac{x_{BM}}{x^*_{BM}} H_L + \frac{L_M}{mG_M} \frac{y_{BM}}{x^*_{BM}} H_G \tag{5-286}$$

These equations are strictly valid only when the slope m of the equilibrium curve is constant, as noted previously.

Example 5-18 Conversion of Overall Mass Transfer Coefficient An overall mass-transfer coefficient was experimentally measured for SO_2 in water at 30°C and 1 atm, and it was found to be $K'_G = 0.40$ kmol/(h · m² · atm). Linear regression of the equilibrium data in Table 2-13 yielded $y = 43.727x - 0.0081$, $R^2 = 0.9971$. Based on 30 percent of the resistance to mass transfer residing in the liquid phase, find the value of the alternative, overall mass-transfer coefficient in the liquid phase K_L.

Solution $K'_G = 0.4$ kmol/(h · m² · atm) and $P = 1.0$ atm, $K_G = PK'_G = 0.4$ kmol/(h · m²). Since resistance $= 1/K_G$, $m/k_L = 0.3/K_G \rightarrow 1/k_L = 0.3/(K_G m)$. From the equilibrium data, $m = 43.727$ (slope of y versus x). Thus, $1/k_L = 0.3/[(0.40)(43.727)] = 0.01715$.

Similarly, $1/k_G = 0.7/K_G \rightarrow k_G = 1/1.75 = 0.5714$.

Then $1/K_L = 1/(mk_G) + 1/k_L = 1/(43.727)(0.5714) + 0.01715 = 0.0572$ and $K_L = 17.49$.

Note that $K_L = mK_G = 17.49$ kmol/(h · m²).

Number of Transfer Units (NTU) The total height of packing is HTU × NTU. The NTU required for a given separation is closely related to the number of theoretical stages or plates required to carry out the same separation in a stage or plate-type apparatus. For equimolal counterdiffusion, such as in a binary distillation, the number of overall gas-phase transfer units N_{OG} required for changing the composition of the vapor stream from y_1 to y_2 is

$$N_{OG} = \int_{y2}^{y1} \frac{dy}{y - y^*} \tag{5-287}$$

When diffusion is in one direction only, as in the absorption of a soluble component from an insoluble gas,

$$N_{OG} = \int_{y2}^{y1} \frac{y^*_{BM}\,dy}{(1-y)(y-y^*)} \tag{5-288}$$

For dilute systems Eq. (5-288) simplifies to Eq. (5-287).

If H_{OG} is constant, the total height of packing required is

$$h_T = H_{OG} N_{OG} \tag{5-289}$$

When it is known that H_{OG} varies appreciably within the tower, this term must be placed inside the integral in Eqs. (5-287) and (5-288) for accurate calculations of h_T. For example, the packed-tower design equation in terms of the overall gas-phase mass-transfer coefficient for absorption would be expressed as follows:

$$h_T = \int_{y2}^{y1} \left[\frac{G_M}{K_G a y^*_{BM}}\right] \frac{y^*_{BM}\,dy}{(1-y)(y-y^*)} \tag{5-290}$$

where the first term under the integral can be recognized as the HTU term. Convenient solutions of these equations for special cases are discussed later.

Film, Penetration, and Surface-Renewal Theories In certain simple situations, the mass-transfer coefficients can be calculated from first principles. The film, penetration, and surface-renewal theories are attempts to extend these theoretical calculations to more-complex situations. Although these theories are often not accurate, they provide useful physical pictures for variations in the mass-transfer coefficient.

For the special case of steady-state unidirectional diffusion of a component through an inert-gas film in an ideal-gas system, the rate of mass transfer is derived as

$$N_A = \frac{D_{AB} p_T}{RT\delta_G} \frac{y - y_i}{y_{BM}} = \frac{D_{AB} p_T}{RT\delta_G} \ln\frac{1-y_i}{1-y} \tag{5-291}$$

where δ_G = the "effective" thickness of a stagnant-gas layer which would offer a resistance to molecular diffusion equal to the experimentally observed resistance and R = the gas constant [Nernst, *Z. Phys. Chem.* **47:** 52 (1904); Whitman, *Chem. Mat. Eng.* **29:** 149 (1923), and Lewis and Whitman, *Ind. Eng. Chem.* **16:** 1215 (1924)].

According to this analysis, one can see that for gas absorption problems, which often exhibit unidirectional diffusion, the most appropriate driving-force expression is of the form $(y - y_i)/y_{BM}$, and the most appropriate mass-transfer coefficient is therefore $\hat{k}_G$. This concept recurs for all unidirectional diffusion problems. Comparing Eq. (5-273), $N_A = \hat{k}_G\,(y - y_i)/y_{BM}$, to Eq. (5-291), we obtain

$$\hat{k}_G = \frac{D_{AB} p_T}{RT\delta_G} = \frac{D_{AB} C}{\delta_G} \tag{5-292}$$

where C is the molar concentration of the stagnant gas film. Thus, with the film model the mass-transfer coefficient is inversely proportional to film thickness δ_G, which depends primarily on the hydrodynamics of the system and hence on the Reynolds (N_{Re}) and Schmidt (N_{Sc}) numbers. Thus, the gas-phase mass-transfer coefficient $\hat{k}_G$ depends principally upon the transport properties of the fluid (N_{Sc}) and the hydrodynamics of the particular system involved (N_{Re}). Correlations have been developed for different geometries in terms of the following dimensionless variables:

$$N_{Sh} = \hat{k}_G\, RTd/(p_T D_{AB}) = \hat{k}_G\, d/(CD_{AB}) = f(N_{Re}, N_{Sc}) \tag{5-293}$$

where the Sherwood number N_{Sh} and Reynolds number $N_{Re} = Gd/\mu_G$ are based on the characteristic length d appropriate to the geometry of the particular system; and $N_{Sc} = \mu_G/(\rho_G D_{AB})$ is the Schmidt number.

It is sometimes convenient to work in terms of the Stanton number $N_{St} = \hat{k}_G/G_M = k'_G p_{BM} y_{BM}/G_M$, instead of the Sherwood number. The Sherwood number can be written as

$$N_{Sh} = (\hat{k}_G/G_M) N_{Re} N_{Sc} = N_{St} N_{Re} N_{Sc} \tag{5-294}$$

Equations (5-293) and (5-294) can now be combined in the alternative functional form

$$N_{St} = \hat{k}_G/G_M = g(N_{Re}, N_{Sc}) \tag{5-295}$$

It is important to recognize that specific mass-transfer correlations can be derived only in conjunction with the investigator's particular assumptions concerning the numerical values of the effective interfacial area a of the packing.

The stagnant-film model assumes a steady state in which the local flux across each element of area is constant; i.e., there is no accumulation of the diffusing species within the film. For a liquid the film model predicts $\hat{k}_L = D_L C/\delta_L$. Higbie [*Trans. Am. Inst. Chem. Eng.* **31:** 365 (1935)] pointed out that industrial contactors often operate with repeated brief contacts between phases in which the contact times are too short for the steady state to be achieved. Higbie advanced the penetration theory that in a packed tower the liquid flows across each packing piece in laminar flow and is remixed at the points of discontinuity between the packing elements. Thus, a fresh liquid surface is formed at the top of each piece, and as it moves downward, it absorbs gas at a decreasing rate until it is mixed at the next discontinuity. If the velocity of the flowing stream is uniform over a very deep region of liquid (total thickness $\delta_T >> \sqrt{Dt}$), where the time-averaged liquid-phase mass-transfer coefficient $\hat{k}_L$ according to penetration theory is

$$\hat{k}_L = 2\,\bar{\rho}_L x_{BM} \sqrt{D_L/(\pi t)} \tag{5-296}$$

where D_L = liquid-phase diffusion coefficient and t = contact time. In practice, the contact time t is not known except in special cases in which the hydrodynamics are clearly defined. This is somewhat similar to the case of the stagnant-film theory in which the unknown quantity is the thickness of the stagnant layer δ.

Penetration theory predicts that $\hat{k}_L$ should vary by the square root of the molecular diffusivity, as compared with film theory, which predicts a first-power dependency on D. Various investigators have reported experimental powers of D ranging from 0.5 to 0.75, and the Chilton-Colburn analogy uses a ⅔ power.

Penetration theory often is used in analyzing absorption with chemical reaction because it makes no assumption about the depths of penetration of the various reacting species, and it gives a more accurate result when the diffusion coefficients of the reacting species are not equal. When the reaction process is very complex, however, penetration theory is more difficult to use than film theory, and the latter method normally is preferred.

Danckwerts [*Ind. Eng. Chem.* **42:** 1460 (1951)] proposed an extension of the penetration theory, called the *surface renewal theory*, which allows for the eddy motion in the liquid to bring masses of fresh liquid continually from the interior to the surface, where they are exposed to the gas for finite lengths of time before being replaced. Danckwerts assumed that every element of fluid has an equal chance of being replaced regardless of its age. The Danckwerts model gives

$$\hat{k}_L = \bar{\rho}_L x_{BM} \sqrt{D_L s} \tag{5-297}$$

where s = fractional rate of surface renewal. Both the penetration and the surface renewal theories predict a square-root dependency on D_L. Unfortunately, values of the surface renewal rate s generally are not available, which presents the same problems as do δ and t in the film and penetration models.

The predictions of correlations based on the film model often are nearly identical to predictions based on the penetration and surface renewal models. Thus, in view of its relative simplicity, the film model normally is preferred for purposes of discussion or calculation. Theoretical models have not proved adequate for making a priori predictions of mass-transfer rates in packed towers, and therefore empirical correlations such as those outlined later in Table 5-23 must be employed.

Effects of High Solute Concentrations on $\hat{k}_G$ and $\hat{k}_L$ The stagnant-film model indicates that $\hat{k}_G$ should be independent of y_{BM} and k_G should be inversely proportional to y_{BM}. The data of Vivian and Behrman [*Am. Inst. Chem. Eng. J.* **11:** 656 (1965)] for the absorption of ammonia from an inert gas strongly suggest that the film model's predicted trend is correct. This is another indication that the most appropriate rate coefficient to use in concentrated systems is $\hat{k}_G$ and the proper driving-force term is of the form $(y - y_i)/y_{BM}$.

The use of the rate coefficient $\hat{k}_L$ and the driving force $(x_i - x)/x_{BM}$ is also believed to be appropriate. For many practical absorption and stripping situations, the liquid-phase solute concentrations are low, thus making this assumption unimportant.

Effects of Total Pressure on $\hat{k}_G$ and $\hat{k}_L$ The influence of total system pressure on the rate of mass transfer from a gas to a liquid or to a solid has been shown to be the same as would be predicted from stagnant-film theory value of $\hat{k}_G = D_{AB}p_T/(RT\delta_G)$. Since the quantity $D_{AB}p_T$ is known to be relatively independent of the pressure, it follows that the rate coefficients $\hat{k}_G$, $k_G y_{BM}$, and $k'_G\, p_T y_{BM}$ $(= k'_G\, p_{BM})$ do not depend on the total pressure of the system, subject to the limitations discussed later.

Investigators of tower packings normally report $k'_G a$ values measured at very low inlet-gas concentrations, so that $y_{BM} = 1$, and at total pressures close to 100 kPa (~1 atm). Thus, the correct rate coefficient for use in packed-tower designs involving the use of the driving force $(y - y_i)/y_{BM}$ is obtained by multiplying the reported $k'_G a$ values by the value of p_T employed in the actual test unit (for example, 100 kPa) and *not* the total pressure of the system to be designed. In other words, one can determine $k'_G a$ in kmol/[(s · m³)(kPa)] for design pressure p_T (in kPa) from

$$k'_G a \text{ at design pressure } p_T = k'_G a \text{ at 1 atm} \times 101.3/p_T \tag{5-298}$$

One way to avoid a lot of confusion on this point is to convert the experimentally measured $k'_G a$ values to values of $\hat{k}_G a$ straightaway, before beginning the design calculations. A design based on the rate coefficient $\hat{k}_G a$ and the driving force $(y - y_i)/y_{BM}$ will be independent of the total system pressure with cautions for systems that have significant vapor-phase nonideality, that operate in the vicinity of the critical point, or that have total pressures higher than about 3040 to 4050 kPa (30 to 40 atm).

Experimental confirmations of the relative independence of $\hat{k}_G$ with respect to total pressure have been widely reported. Deviations do occur at extreme conditions. For example, Bretsznajder (*Prediction of Transport and Other Physical Properties of Fluids*, Pergamon Press, Oxford, 1971, p. 343) discusses the effects of pressure on the $D_{AB}p_T$ product and presents experimental data on the self-diffusion of CO_2 which show that the Dp product begins to decrease at a pressure of approximately 8100 kPa (80 atm). For reduced temperatures (T_r) higher than about 1.5, the deviations are relatively modest for pressures up to the critical pressure. However, deviations are large near the critical point. The effect of pressure on the gas-phase viscosity also is negligible for pressures below about 5060 kPa (50 atm).

For the liquid-phase mass-transfer coefficient $\hat{k}_L$, the effects of total system pressure can be ignored for all practical purposes. Thus, when $\hat{k}_G$ and $\hat{k}_L$ are used for the design of gas absorbers or strippers, the primary pressure effects to consider will be those that affect the equilibrium curves and the values of m. However, if pressure changes affect the hydrodynamics, $\hat{k}_G$, $\hat{k}_L$, and a can all change significantly.

Effects of Temperature on $\hat{k}_G$ and $\hat{k}_L$ The Stanton number relationship for gas-phase mass transfer in packed beds, Eq. (5-295), indicates that for a given system geometry the rate coefficient $\hat{k}_G$ depends on only the Reynolds number and the Schmidt number. Since the Schmidt number for a gas is approximately independent of temperature, the principal effect of temperature upon $\hat{k}_G$ arises from changes in the gas viscosity with changes in temperature. For normally encountered temperature ranges, these effects will be small owing to the fractional powers involved in Reynolds number terms (see Tables 5-16 to 5-23). Thus, for all practical purposes $\hat{k}_G$ is independent of temperature and pressure in the normal ranges of these variables.

For modest changes in temperature, the influence of temperature upon the interfacial area a may be neglected. For example, in experiments on the absorption of SO_2 in water, Whitney and Vivian [*Chem. Eng. Prog.* **45:** 323 (1949)] found no appreciable effect of temperature upon $\hat{k}_G a$ over the range from 10 to 50°C.

Whitney and Vivian found that the effect of temperature upon $\hat{k}_L a$ and Sherwood and Holloway [101] (see Table 5-23A) found that the effect of temperature upon H_L could be explained entirely by variations in the liquid-phase viscosity and diffusion coefficient with temperature. These effects can be very large and therefore must be carefully accounted for when using $\hat{k}_L$ or H_L data. For liquids, the mass-transfer coefficient $\hat{k}_L$ is correlated by Eqs. (5-297) and (5-299). Typically, the general form of the correlation for H_L is (see Table 5-23)

$$H_L = bN_{\text{Re}}^{a}\, N_{\text{Sc}}^{1/2} \tag{5-299}$$

where b is a proportionality constant and the exponent a may range from about 0.2 to 0.5 for different packings and systems. The liquid-phase diffusion coefficients may be corrected from a base temperature T_1 to another temperature T_2 by using the Einstein relation as recommended by Wilke [*Chem. Eng. Prog.* **45:** 218 (1949)]:

$$D_2 = D_1(T_2/T_1)(\mu_1/\mu_2) \tag{5-300}$$

The Einstein relation can be rearranged to relate Schmidt numbers at two temperatures:

$$N_{\text{Sc}2} = N_{\text{Sc}1}(T_1/T_2)(\rho_1/\rho_2)(\mu_2/\mu_1)^2 \tag{5-301}$$

Substitution of this relation into Eq. (5-299) shows that for a given geometry the effect of temperature on H_L can be estimated as

$$H_{L2} = H_{L1}(T_1/T_2)^{1/2}(\rho_1/\rho_2)^{1/2}(\mu_2/\mu_1)^{1-a} \tag{5-302}$$

In using these relations, note that for equal liquid flow rates

$$H_{L2}/H_{L1} = (\hat{k}_L a)_1/(\hat{k}_L a)_2 \tag{5-303}$$

Maxwell-Stefan Mass Transfer Equation (5-262) can be obtained by solving Fick's law for a thin stagnant film and then noting $\hat{k}_L = D_L C/\delta_L$ where δ is the unknown film thickness. Thus, the usual linear driving-force mass-transfer analysis is inherently a Fickian analysis that is excellent for binary mass transfer but can be problematic for multicomponent systems (see Taylor and Krishna). A M-S mass-transfer analysis can be derived by starting with Eq. (5-240), multiplying both sides of the equation by δ, and defining the M-S mass-transfer coefficient as $\not{k} = \mathcal{D}C/\delta$ (see Wesselingh and Krishna). For ternary mass transfer in the z direction the result is

$$\left(1+\frac{x_A}{\gamma_A}\frac{\partial \gamma_A}{\partial x_A}\right)\frac{dx_A}{d(z/\delta)} = -\frac{x_B N_{A,z} - x_A N_{B,z}}{\not{k}_{AB}} - \frac{x_C N_{A,z} - x_A N_{C,z}}{\not{k}_{AC}} \tag{5-304}$$

For ideal gases this result simplifies to

$$\frac{dx_A}{d(z/\delta)} = -\frac{y_B N_{A,z} - y_A N_{B,z}}{\not{k}_{AB}} - \frac{y_C N_{A,z} - y_A NC,z}{\not{k}_{AC}} \tag{5-305}$$

The resulting difference form of the mass-transfer equation for ideal ternary gas systems is

$$\Delta y_A = -\frac{\bar{y}_B \bar{N}_{A,z} - \bar{y}_A \bar{N}_{B,z}}{\not{k}_{AB}} - \frac{\bar{y}_C \bar{N}_{A,z} - \bar{y}_A \bar{N}_{C,z}}{\not{k}_{AC}} \tag{5-306}$$

For binary systems the M-S mass-transfer coefficient is related to the Fickian mass-transfer coefficient by

$$\not{k}_{AB} = \frac{k_{AB}\mathcal{D}_{AB}}{D_{AB}} = \frac{k_{AB}}{1+\dfrac{x_A}{\gamma_A}\dfrac{\partial \gamma_A}{\partial x_A}} \tag{5-307}$$

For ideal systems and at the infinite dilution limits $\mathcal{D}_{AB} = D_{AB}$ and $\not{k}_{AB} = k_{AB}$. Thus, for most gas systems and for dilute liquid systems, the standard mass-transfer correlations discussed next can be used to determine $\not{k}_{AB}$ which is then used in Eq. (5-305) or (5-306).

MASS-TRANSFER CORRELATIONS

Because of the tremendous importance of mass transfer in chemical engineering, a very large number of studies have determined mass-transfer coefficients. Tables 5-17 to 5-24 summarize a variety of different configurations to provide a sense of the range of correlations available. These

TABLE 5-17 Mass-Transfer Correlations for a Single Flat Plate or Disk—Transfer to or from Plate to Fluid

Situation	Correlation	Comments E = Empirical, S = Semiempirical, T = Theoretical	References*
A. Laminar, local, flat plate, forced flow	$N_{Sh,x} = \frac{k'x}{D} = 0.323\,(N_{Re,x})^{1/2}(N_{Sc})^{1/3}$ Coefficient 0.332 is a better fit.	[T] Low MT rates. Low mass-flux, constant property systems. $N_{Sh,x}$ is local k. Use with arithmetic difference in concentration. Coefficient 0.323 is Blasius' approximate solution. $N_{Re,x} = \frac{xu_\infty \rho}{\mu}$, x = length along plate	[57] p. 183 [66] p. 526 Sherwood, Pigford, & Wilke, p. 79
Laminar, average, flat plate, forced flow	$N_{Sh,avg} = \frac{k'_m L}{D} = 0.646(N_{Re,L})^{1/2}(N_{Sc})^{1/3}$ Coefficient 0.664 is a better fit. k'_m is mean mass-transfer coefficient for dilute systems.	$N_{Re,L} = \frac{Lu_\infty \rho}{\mu}$ 0.664 (Polhausen) is a better fit for $N_{Sc} > 0.6$, $N_{Re,x} < 3 \times 10^5$	Skelland, p. 110 [69] p. 480
j factors	$j_D = j_H = \frac{f}{2} = 0.664(N_{Re,L})^{-1/2}$	[S] Chilton-Colburn analogy. $N_{Sc} = 1.0$, f = drag coefficient. j_D is defined in terms of k'_m.	Skelland, p. 271
B. Laminar, local, flat plate, natural convection vertical plate	$N_{Sh,x} = \frac{k'x}{D} = 0.508\ N_{Sc}^{1/2}(0.952 + N_{Sc})^{-1/4} N_{Gr}^{1/4}$ $N_{Gr} = \frac{gx^3}{(\mu/\rho)^2}\left(\frac{\rho_\infty}{\rho_0} - 1\right)$	[T] Low MT rates. Dilute systems, $\Delta\rho/\rho << 1$. $N_{Gr}N_{Sc} < 10^8$. Use with arithmetic concentration difference. x = length from plate bottom.	Skelland, p. 120
C. Laminar, stationary disk	$N_{Sh} = \frac{k'd_{disk}}{D} = \frac{8}{\pi}$	[T] Stagnant fluid. Use arithmetic concentration difference.	Sherwood, Pigford, & Wilke, p. 240
Laminar, spinning disk	$N_{Sh} = \frac{k'd_{disk}}{D} = 0.879\ N_{Re}^{1/2} N_{Sc}^{1/3}$ $N_{Re} <\sim 10^4$	[T] Asymptotic solution for large N_{Sc}. $u = \omega d_{disk}/2$, ω = rotational speed, rad/s. Rotating disks are often used in electrochemical research.	[76] p. 60 Sherwood, Pigford, & Wilke, p. 240
D. Laminar, inclined, plate	$N_{Sh,avg} = 0.783\ N_{Re,film}^{1/9} N_{Sc}^{1/3}\left(\frac{x^3\rho^2 g \sin\alpha}{\mu^2}\right)^{2/9}$ $N_{Re,film} = \frac{4Q\rho}{\mu^2} < 2000$ $N_{Sh,avg} = \frac{k'_m x}{D}$ $\delta_{film} = \left(\frac{3\mu Q}{w\rho g \sin\alpha}\right)^{1/3}$ = film thickness	[T] Constant-property liquid film with low mass-transfer rates. Use arithmetic concentration difference. Newtonian fluid. Solute does not penetrate past region of linear velocity profile. Differences between theory and experiment. w = width of plate, δ_f = film thickness α = angle of inclination x = distance from start soluble surface.	Skelland, p. 130 Sherwood, Pigford, & Wilke, p. 209
E. Laminar and turbulent, flat plate, forced flow	$j_D = j_H = \frac{f}{2} = 0.037\,N_{Re,L}^{-0.2}$ $j_D = (k_G/G_M)N_{Sc}^{2/3}$ $j_H = (h'/C_p G)N_{Pr}^{2/3}$	[E] Chilton-Colburn analogies, $N_{Sc} = 1.0$, (gases), f = drag coefficient. $8000 < N_{Re} < 300{,}000$. Can apply analogy, $j_D = f/2$, to entire plate (including laminar portion) if average values are used.	[57] p. 193 Kirwan, p. 112 Sherwood, Pigford, & Wilke, p. 201 Skelland, p. 271 [59] [53]
F. Laminar and turbulent, flat plate, forced flow	$N_{Sh,avg} = 0.037\,N_{Sc}^{1/3}(N_{Re,L}^{0.8} - 15{,}500)$ to $N_{Re}, L = 320{,}000$ $N_{Sh,avg} = 0.037\,N_{Sc}^{1/3} \times \left(N_{Re,L}^{0.8} - N_{Re,Cr}^{0.8} + \frac{0.664}{0.037}N_{Re,Cr}^{1/2}\right)$ in range 3×10^5 to 3×10^6	[E] Use arithmetic concentration difference. $N_{Sh,avg} = \frac{k'_m L}{D}$, $N_{Sc} > 0.5$ Entrance effects are ignored. $N_{Re,Cr}$ is transition laminar to turbulent.	Kirwan, p. 112 Sherwood, Pigford, & Wilke, p. 201
G. Turbulent, local flat plate, natural convection, vertical plate Turbulent, average, flat plate, natural convection, vertical plate	$N_{Sh,x} = \frac{k'x}{D} = 0.0299\,N_{Gr}^{2/5} N_{Sc}^{7/15} \times (1 + 0.494\,N_{Sc}^{2/3})^{-2/5}$ $N_{Sh,avg} = 0.0249\,N_{Gr}^{2/5} N_{Sc}^{7/15} \times (1 + 0.494\,N_{Sc}^{2/3})^{-2/5}$ $N_{Gr} = \frac{gx^3}{(\mu/\rho)^2}\left(\frac{\rho_\infty}{\rho_0} - 1\right)$, $N_{Sh,avg} = \frac{k'_m L}{D}$	[S] Low solute concentration and low transfer rates. Use arithmetic concentration difference. $N_{Gr} > 10^{10}$ Assumes laminar boundary layer is small fraction of total.	Skelland, p. 225
H. Cross-corrugated plate (turbulence promoter for membrane systems)	$N_{Sh} = cN_{Re}^{a} N_{Sc}^{1/3}$	[E] Entrance turbulent channel For parallel flow and corrugations: $N_{SC} = 1483$, $a = 0.56$, $c = 0.268$ $N_{SC} = 4997$, $a = 0.50$, $c = 0.395$ Corrugations perpendicular to flow: $N_{SC} = 1483$, $a = 0.57$, $c = 0.368$ $N_{SC} = 4997$, $a = 0.52$, $c = 0.487$	[98]
I. Turbulent, spinning disk	$N_{Sh} = \frac{k'd_{disk}}{D} = 5.6\,N_{Re}^{1.1} N_{Sc}^{1/3}$ $6 \times 10^5 < N_{Re} < 2 \times 10^6$, $120 < N_{Sc} < 1200$	[E] Use arithmetic concentration difference. $u = \omega d_{disk}/2$ where ω = rotational speed, radians/s. $N_{Re} = \rho\omega d^2_{disk}/2\mu$	[42] Sherwood, Pigford, & Wilke, p. 241

TABLE 5-17 Mass-Transfer Correlations for a Single Flat Plate or Disk—Transfer to or from Plate to Fluid *(Continued)*

Situation	Correlation	Comments E = Empirical, S = Semiempirical, T = Theoretical	References*
J. Mass transfer to a flat-plate membrane in a stirred vessel	$N_{Sh} = \frac{k'd_{tank}}{D} = aN_{Re}^b N_{Sc}^c$ a depends on system. $a = 0.0443$ [40]; b is often 0.65 – 0.70 [67].	[E] Useful for laboratory dialysis, R.O., U.F., and microfiltration systems. Use arithmetic concentration difference. ω = stirrer speed, radians/s. $N_{Re} = \frac{\omega d_{tank}^2 \rho}{\mu}$ $b = 0.785$ [27]. c is often 0.33 but other values have been reported [67].	[27] [67] p. 965
K. Spiral wound type RO (seawater desalination)	$N_{Sh} = 0.210 N_{Re}^{2/3} N_{Sc}^{1/4}$ Or with slightly larger error, $N_{Sh} = 0.080 N_{Re}^{0.875} N_{Sc}^{1/4}$ $N_{Sh} = 0.065 N_{Re}^{0.875} N_{Sc}^{1/4}$	[E] Polyamide membrane. $p = 6.5$ MPa and TDS rejection = 99.8%. Recovery ratio 40%.	[108] [103], p. 113

*Author names refer to the General References list at the beginning of the Mass Transfer subsection (pp. 5-41, 5-42). Bracketed numbers refer to the table reference lists (p. 5-42).

TABLE 5-18 Mass-Transfer Correlations for Falling Films with a Free Surface in Wetted Wall Columns—Transfer between Gas and Liquid

Situation	Correlation	Comments E = Empirical, S = Semiempirical, T = Theoretical	References*
A. Laminar, vertical wetted wall column	$N_{Sh,avg} = \frac{k'_m x}{D} \approx 3.41 \frac{x}{\delta_{film}}$ (first term of infinite series) $\delta_{film} = \left(\frac{3\mu Q}{w\rho g}\right)^{1/3}$ = film thickness $N_{Re,film} = \frac{4Q\rho}{w\mu} < 20$	[T] Low rates MT. Use with log-mean concentration difference. Parabolic velocity distribution in films. w = film width (circumference in column) Derived for flat plates, used for tubes if $r_{tube}\left(\frac{\rho g}{2\sigma}\right)^{1/2} > 3.0$. σ = surface tension If $N_{Re,film} > 20$, surface waves and rates increase. An approximate solution $D_{apparent}$ can be used. Ripples are suppressed with a wetting agent good to $N_{Re} = 1200$.	Sherwood, Pigford, & Wilke, p. 78 Skelland, p. 137 Treybal, p. 50
B. Turbulent, vertical wetted wall column Better fit	$N_{Sh,avg} = \frac{k'_m d_t}{D} = 0.023 N_{Re}^{0.83} N_{Sc}^{0.44}$ A coefficient 0.0163 has also been reported using $N_{Re'}$, where $\upsilon = \upsilon$ of gas relative to liquid film. $N_{Sh,avg} = 0.0318 N_{Re}^{0.790} N_{Sc}^{0.5}$	[E] Use with log-mean concentration difference for correlations in B and D. N_{Re} is for gas. N_{Sc} for vapor in gas. $2000 < N_{Re} \le 35{,}000$, $0.6 \le N_{Sc} \le 2.5$. Use for gases, d_t = tube diameter. [S] Reevaluated data	[51], [57] p.181 Sherwood, Pigford, & Wilke, p. 211 Skelland, p. 265 Taylor & Krishna, p. 212 Treybal, p. 71 [44]
C. Turbulent, very short column	$N_{Sh} = 0.00283 N_{Re,g} N_{Sc,g}^{0.5} N_{Re,liq}^{0.08}$ $N_{Sh} = k_g(d_{tube} - 2\delta)/D$ $N_{Re,g} = \rho_g u_g (d_{tube} - 2\delta)/\mu_g$ $N_{Re,liq} = \rho_{liq} Q_{liq}/[\pi\mu(d_{tube} - 2\delta)]$	[E] Evaporation data $N_{Sh,g} = 11$ to 65, $N_{Re,g} = 2400$ to 9100 $N_{Re,liq} = 11$ to 480, $N_{Sc,g} = 0.62$ to 1.93 δ = film thickness	[43]
D. Turbulent, vertical wetted wall column with ripples	$N_{Sh,avg} = \frac{k'_m d_t}{D} = 0.00814 N_{Re}^{0.83} N_{Sc}^{0.44} \left(\frac{4Q\rho}{w\mu}\right)^{0.15}$ $30 \le \left(\frac{4Q\rho}{w\mu}\right) < 1200$ $N_{Sh,avg} = \frac{k'_m d_t}{D} = 0.023 N_{Re}^{0.8} N_{Sc}^{1/3}$	[E] For gas systems with rippling. Fits 5-17B for $\left(\frac{4Q\rho}{w\mu}\right) = 1000$ [E] "Rounded" approximation to include ripples. Includes solid-liquid mass-transfer data to find ⅓ coefficient on N_{Sc}. May use $N_{Re}^{0.83}$. Use for liquids. See also Table 5-19.	[64] Sherwood, Pigford, & Wilke, p. 213
E. Rectification in vertical wetted wall column with turbulent vapor flow, Johnstone and Pigford correlation	$N_{Sh,avg} = \frac{k'_G d_{col} p_{BM}}{D_v p} = 0.0328 (N'_{Re})^{0.77} N_{Sc}^{0.33}$ $3000 < N'_{Re} < 40{,}000, 0.5 < N_{Sc} < 3$ $N'_{Re} = \frac{d_{col} \upsilon_{rel} \rho_v}{\mu_v}$, υ_{rel} = gel velocity relative to liquid film $= \frac{3}{2} u_{avg}$ in film	[E] Use log-mean driving force at two ends of column. Based on four systems with gas-side resistance only. p_{BM} = log-mean partial pressure of nondiffusing species B in binary mixture. p = total pressure Modified form is used for structured packings (see Table 5-24H).	[63] Sherwood, Pigford, & Wilke, p. 214 [113]

*Author names refer to the General References list at the beginning of the Mass Transfer subsection (pp. 5-41, 5-42). Bracketed numbers refer to the table reference lists (p. 5-42).

TABLE 5-19 Mass-Transfer Correlations for Flow in Pipes and Ducts—Transfer Is from Wall to Fluid

Situation	Correlation	Comments E = Empirical, S = Semiempirical, T = Theoretical	References*
A. Tubes, laminar, fully developed parabolic velocity profile, developing concentration profile, constant wall concentration	$N_{Sh} = \frac{k'd_t}{D} = 3.66 + \frac{0.0668(d_t/x)N_{Re}N_{Sc}}{1+0.04[(d_t/x)N_{Re}N_{Sc}]^{2/3}}$	[T] Use log-mean concentration difference. For $\frac{x/d_t}{N_{Re}N_{Sc}} < 0.10, N_{Re} < 2100$. x = distance from tube entrance. Good agreement with experiment at values $10^4 > \frac{\pi}{4}\frac{d_t}{x} N_{Re}N_{Sc} > 10$	[57] p. 176 [66] p. 525 Skelland, p. 159
Fully developed concentration profile	$N_{Sh} = \frac{k'd_t}{D} = 3.66$	[T] $\frac{x/d_t}{N_{Re}N_{Sc}} > 0.1$	Skelland, p. 165
B. Tubes, approximate solution	$N_{Sh,x} = \frac{k'd_t}{D} = 1.077\left(\frac{d_t}{x}\right)^{1/3}(N_{Re}N_{Sc})^{1/3}$ $N_{Sh,avg} = \frac{k'd_t}{D} = 1.615\left(\frac{d_t}{L}\right)^{1/3}(N_{Re}N_{Sc})^{1/3}$	[T] For arithmetic concentration difference. $\frac{W}{\rho Dx} > 400$ Leveque's approximation: Concentration BL is thin. Assume velocity profile is linear. High mass velocity. Fits liquid data well.	Skelland, p. 166
C. Tubes, laminar, uniform plug velocity, developing concentration profile, constant wall concentration	$N_{Sh,avg} = \frac{1}{2}\frac{d_t}{L}N_{Re}N_{Sc}\left[\frac{1-4\sum_{j=1}^{\infty}a_j^{-2}\exp\left(\frac{-2a_j^2(x/r_t)}{N_{Re}N_{Sc}}\right)}{1+4\sum_{j=1}^{\infty}a_j^{-2}\exp\left(\frac{-2a_j^2(x/r_t)}{N_{Re}N_{Sc}}\right)}\right]$ Graetz solution for heat transfer written for MT.	[T] Use arithmetic concentration difference. Fits gas data well, for $\frac{W}{D\rho x} < 50$ (fit is fortuitous) $N_{Sh,avg} = (k'_m d_t)/D, a_1 = 2.405, a_2 = 5.520,$ $a_3 = 8.654, a_4 = 11.792, a_5 = 14.931.$ graphical solution are in reference.	[78] Skelland, p. 150
D. Laminar, fully developed parabolic velocity profile, constant mass flux at wall	$N_{Sh,x} = \left[\frac{11}{48} - \frac{1}{2}\sum_{j=1}^{\infty}\frac{\exp[-\lambda_j^2(x/r_t)/(N_{Re}N_{Sc})]}{C_j\lambda_j^4}\right]^{-1}$ j — λ_j^2 — C_j 1 — 25.68 — 7.630×10^{-3} 2 — 83.86 — 2.058×10^{-3} 3 — 174.2 — 0.901×10^{-3} 4 — 296.5 — 0.487×10^{-3} 5 — 450.9 — 0.297×10^{-3}	[T] For arithmetic concentration difference. $N_{Re} < 2100$ $N_{Sh,x} = \frac{k'd_t}{D}$ $N_{Re} = \frac{vd_t\rho}{\mu}$	[102] Skelland, p. 167
E. Laminar, alternate	$N_{Sh} = 4.36 + \frac{0.023(d_t/L)N_{Re}N_{Sc}}{1+0.0012(d_t/L)N_{Re}N_{Sc}}$	[T] $N_{Sh} = \frac{k'd_t}{D}$. Use log-mean concentration difference. $N_{Re} < 2100$	[57] p. 176
F. Laminar, fully developed concentration and velocity profile	$N_{Sh} = \frac{k'd_t}{D} = \frac{48}{11} = 4.3636$	[T] Use log-mean concentration difference. $N_{Re} < 2100$	Skelland, p. 167
G. Vertical tubes, laminar flow, forced and natural convection	$N_{Sh,avg} = 1.62 N_{Gz}^{1/3}\left[1 \pm 0.0742\frac{(N_{Gr}N_{Sc}d/L)^{3/4}}{N_{Gz}}\right]^{1/3}$	[T] Approximate solution. Use minus sign if forced and natural convection oppose each other. Good agreement with experiment. $N_{Gz} = \frac{N_{Re}N_{Sc}d}{L}, N_{Gr} = \frac{g\Delta\rho d^3}{\rho\nu^2}$	[94]
H. Hollow-fiber extraction inside fibers	$N_{Sh} = 0.5\, N_{Gz}, N_{Gz} < 6$ $N_{Sh} = 1.62 N_{Gz}^{0.5}, N_{Gz} \geq 6$	[E] Use arithmetic concentration difference.	[28]
I. Tubes, laminar, RO systems	$N_{Sh,avg} = \frac{k'_m d_t}{D} = 1.632\left(\frac{ud_t^2}{DL}\right)^{1/3}$	Use arithmetic concentration difference. Thin concentration polarization layer, not fully developed. $N_{Re} < 2000$, L = length tube.	[27]
J. Tubes and parallel plates, laminar RO	Graphical solutions for concentration polarization. Uniform velocity through walls.	[T]	[100]
K. Rotating annulus for reverse osmosis	For nonvortical flow: $N_{Sh} = 2.15\left[N_{Ta}\left(\frac{d}{r_i}\right)^{0.5}\right]^{0.18} N_{Sc}^{1/3}$ For vortical flow: $N_{Sh} = 1.05\left[N_{Ta}\left(\frac{d}{r_i}\right)^{0.5}\right] N_{Sc}^{1/3}$	[E,S] N_{Te} = Taylor number = $r_i\omega d/\nu$ r_i = inner cylinder radius ω = rotational speed, rad/s d = gap width between cylinders	[75]
L. Fully developed, parallel plates, laminar, parabolic velocity, developing concentration profile, constant wall concentration	$N_{Sh} = \frac{k'(2h)}{D} = 7.6$	[T] h = distance between plates. Use log-mean concentration difference $\frac{N_{Re}N_{Sc}}{x/(2h)} < 20$	Skelland, p. 177

TABLE 5-19 Mass-Transfer Correlations for Flow in Pipes and Ducts—Transfer Is from Wall to Fluid *(Continued)*

Situation	Correlation	Comments E = Empirical, S = Semiempirical, T = Theoretical	References*
M. same as 5-19L except constant mass flux at wall.	$N_{Sh} = \frac{k'(2h)}{D} = 8.23$	[T] Use log-mean concentration difference. $\frac{N_{Re}N_{Sc}}{x/(2h)} < 20$	Skelland, p. 177
N. Laminar flow, vertical parallel plates, forced and natural convection	$N_{Sh,avg} = 1.47 N_{Gz}^{1/3}\left[1 \pm 0.0989\frac{(N_{Gr}N_{Sc}h/L)^{3/4}}{N_{Gz}}\right]^{1/3}$	[T] Approximate solution. Use minus sign if forced and natural convection oppose each other. Good agreement with experiment. $N_{Gz} = \frac{N_{Re}N_{Sc}h}{L}$, $N_{Gr} = \frac{g\Delta\rho h^3}{\rho v^2}$	[94]
O. Parallel plates, laminar, RO systems	$N_{Sh,avg} = \frac{k'(2H_p)}{D} = 2.354\left(\frac{uH_p^2}{DL}\right)^{1/3}$	Thin concentration polarization layer. Short tubes, concentration profile not fully developed. Use arithmetic concentration difference.	[27]
P. Tubes, turbulent	$N_{Sh,avg} = \frac{k'_m d_t}{D} = 0.023 N_{Re}^{0.83} N_{Sc}^{1/3}$ $2100 < N_{Re} < 35{,}000$ $0.6 < N_{Sc} < 3000$ $N_{Sh,avg} = \frac{k'_m d_t}{D} = 0.023 N_{Re}^{0.83} N_{Sc}^{0.44}$ $2100 < N_{Re} < 35{,}000$ $0.6 < N_{Sc} < 2.5$	[E] Use with log-mean concentration difference at two ends of tube. Good fit for liquids. From wetted wall column and dissolution data—see Table 5-18B. [E] Evaporation of liquids. Use with log-mean concentration difference. Better fit for gases.	[57] p. 181 [78], Treybal, p. 72 [51], [57] p. 181 Kirwan, p. 112 Sherwood, Pigford, & Wilke, p. 211
Q. Tubes, turbulent	$N_{Sh} = \frac{k'd_t}{D} = 0.0096 N_{Re}^{0.913} N_{Sc}^{0.346}$	[E] $430 < N_{Sc} < 100{,}000$. Dissolution data. Use for high N_{Sc}.	McCabe, Smith, & Harriott, p. 668
R. Tubes, turbulent, smooth tubes, Reynolds analogy	$N_{Sh} = \frac{k'd_t}{D} = \left(\frac{f}{2}\right) N_{Re} N_{Sc}$ f = Fanning friction faction	[T] Use arithmetic concentration difference. N_{Sc} near 1.0 Turbulent core extends to wall. Of limited utility.	Geankoplis p. 474 [57] p. 171 Skelland, p. 239 Taylor & Krishna, p. 250
S. Tubes, turbulent, smooth tubes, Chilton-Colburn analogy	$j_D = j_H \le \frac{f}{2}$ If $\frac{f}{2} = 0.023 N_{Re}^{-0.2}$, $j_D = \frac{N_{Sh}}{N_{Re}N_{Sc}^{1/3}} = 0.023 N_{Re}^{-0.2}$ $N_{Sh} = \frac{k'd_t}{D}$, see Table 5-17G $j_D = j_H = F(N_{Re}$, geometry and B.C.)	[E] Use log-mean concentration difference. Relating j_D to $f/2$ approximate. N_{Pr} and N_{Sc} near 1.0. Low concentration. Results about 20% lower than experiment. $3 \times 10^4 \times N_{Re} \times 10^6$ [E] Good over wide ranges.	Bird, Stewart, & Lightfoot, pp. 400, 647 [38][40] Skelland, p. 264 Taylor & Krishna, p. 251 Geankoplis, p. 475 Bird, Stewart, & Lightfoot, p. 647 [38]
T. Tubes, turbulent, smooth tubes, constant surface concentration, Prandtl analogy	$N_{Sh} = \frac{k'd_t}{D} = \frac{(f/2)N_{Re}N_{Sc}}{1+5\sqrt{f/2}(N_{Sc}-1)}$ $\frac{f}{2} = 0.04 N_{Re}^{-0.25}$	[T] Use arithmetic concentration difference. Improvement over Reynolds analogy. Best for N_{Sc} near 1.0.	[57] p. 173 Skelland, p. 241
U. Tubes, turbulent, smooth tubes, Constant surface concentration, Von Karman analogy	$N_{Sh} = \frac{(f/2)N_{Re}N_{Sc}}{1+5\sqrt{f/2}\left\{(N_{Sc}-1)+\ln\left[1+\frac{5}{6}(N_{Sc}-1)\right]\right\}}$ $\frac{f}{2} = 0.04 N_{Re}^{-0.25}$	[T] Use arithmetic concentration difference. $N_{Sh} = k'd_t/D$. Improvement over Prandtl, $N_{Sc} < 25$.	[57] p. 173 Skelland, p. 243 Taylor & Krishna, p. 250 [111]
V. Turbulent flow, tubes	$N_{St} = \frac{N_{Sh}}{N_{Pe}} = \frac{N_{Sh}}{N_{Re}N_{Sc}} = 0.0149\, N_{Re}^{-0.12} N_{Sc}^{-2/3}$	[E] Smooth pipe data. Data fits within 4% except at $N_{Sc} > 20{,}000$, where experimental data is underpredicted. $N_{Sc} > 100$, $10^5 > N_{Re} > 2100$	[80]
W. Turbulent flow, noncircular ducts	$d_{eq} = \frac{4 \text{ cross-sectional area}}{\text{wetted perimeter}}$	Can be suspect for systems with sharp corners. Parallel plates: $d_{eq} = 4\frac{2hw}{2w+2h}$	Skelland, p. 289

*Author names refer to the General References list at the beginning of the Mass Transfer subsection (pp. 5-41, 5-42). Bracketed numbers refer to the table reference lists (p. 5-42).

TABLE 5-20 Mass-Transfer Correlations for Flow Past Submerged Objects

Situation	Correlation	Comments E = Empirical, S = Semiempirical, T = Theoretical	References*
A. Single sphere	$N_{Sh} = \dfrac{k'_G p_{BLM} RT d_s}{PD} = \dfrac{2r}{r - r_s}$ r/r_s: 2, 5, 10, 50, ∞ (asymptotic limit) N_{Sh}: 4.0, 2.5, 2.22, 2.04, 2.0	[T] Use with log-mean concentration difference. r = distance from sphere, r_s, d_s = radius and diameter of sphere. No convection.	Skelland, p. 18
B. Single sphere, creeping flow with forced convection	$N_{Sh} = \dfrac{k'd}{D} = [4.0 + 1.21(N_{Re}N_{Sc})^{2/3}]^{1/2}$ $N_{Sh} = \dfrac{k'd}{D} = a(N_{Re}N_{Sc})^{1/3}$ $a = 1.00 \pm 0.01$	[T] Use with log-mean concentration difference. Average over sphere. Numerical calculations. $(N_{Re}N_{Sc}) < 10{,}000$ $N_{Re} < 1.0$. Constant sphere diameter. Low mass-transfer rates. [T] Fit to above ignoring molecular diffusion. $1000 < (N_{Re}N_{Sc}) < 10{,}000$.	[33], Kirwan, p. 114 Sherwood, Pigford, & Wilke, p. 214 [76] p. 80 Sherwood, Pigford, & Wilke, p. 215
C. Single spheres, molecular diffusion, and forced convection, low flow rates	$N_{Sh} = 2.0 + AN_{Re}^{b}N_{Sc}^{c}$, $b = 1/2$, $c = 1/3$ $A = 0.5$ to 0.62 $A = 0.60$. $A = 0.95$. $A = 0.95$. $A = 0.544$. $A = 0.575$, $b = 0.5$, $c = 0.35$ $A = 0.552$, $b = 0.53$, $c = 1/3$	[E] Use with log-mean concentration difference. Average over sphere. Frössling Eq. ($A = 0.552$), $2 \le N_{Re} \le 800$, $0.6 \le N_{Sc} \le 2.7$. N_{Sh} lower than experimental at high N_{Re}. Ranz and Marshall $2 \le N_{Re} \le 200$, $0.6 \le N_{Sc} \le 2.5$. Modifications recommended [83] Liquids $2 \le N_{Re} \le 2000$. Graph Sherwood, Pigford, & Wilke, pp. 217–218. $100 \le N_{Re} \le 700$; $1{,}200 \le N_{Sc} \le 1525$. Use with arithmetic concentration difference. $N_{Sc} = 1$; $50 \le N_{Re} \le 350$. $N_{Sc} < 1$, $N_{Re} < 1$. $1.0 \le N_{Re} \le 48{,}000$, Gases: $0.6 \le N_{Sc} \le 2.7$	[57], p. 194; Kirwan, p. 114 Skelland, p. 276 Bird, Stewart, & Lightfoot, pp. 409, 647 [90] [83] Sherwood, Pigford, & Wilke, p. 217 Skelland, p. 276 [49], Geankoplis, p. 482 Sherwood, Pigford, & Wilke, p. 217 [93], Skelland, p. 276 [60], Skelland, p. 276 [53], Skelland, p. 276 Geankoplis, p. 482
D. Single cylinders, perpendicular flow	$N_{Sh} = \dfrac{k'd_s}{D} = AN_{Re}^{1/2}N_{Sc}^{1/3}$, $A = 0.82$ $A = 0.74$ $A = 0.582$ $j_D = 0.600(N_{Re})^{-0.487}$ $N_{Sh} = \dfrac{k'd_{cyl}}{D}$	[E] $100 < N_{Re} \le 3500$, $N_{Sc} = 1560$. $120 \le N_{Re} \le 6000$, $N_{Sc} = 2.44$. $300 \le N_{Re} \le 7600$, $N_{Sc} = 1200$. Use with arithmetic concentration difference. $50 \le N_{Re} \le 50{,}000$; gases, $0.6 \le N_{Sc} \le 2.6$; liquids; $1000 \le N_{Sc} \le 3000$. Data scatter ± 30%.	Skelland, p. 276 Skelland, p. 276 [104] Skelland, p. 276 Geankoplis, p. 486
E. Rotating cylinder in an infinite liquid, no forced flow	$j'_D = \dfrac{k'}{v} N_{Sc}^{0.644} = 0.0791 N_{Re}^{-0.30}$ Results presented graphically to $N_{Re} = 241{,}000$. $N_{Re} = \dfrac{v d_{cyl} \mu}{\rho}$ where $v = \dfrac{\omega d_{cyl}}{2}$ = peripheral velocity	[E] Used with arithmetic concentration difference. Useful geometry in electrochemical studies. $112 < N_{Re} \le 100{,}000$. $835 < N_{Sc} < 11490$ k' = mass-transfer coefficient, cm/s; ω = rotational speed, radian/s.	[46] Sherwood, Pigford, & Wilke, p. 238
F. Oblate spheroid, forced convection	$j_D = \dfrac{N_{Sh}}{N_{Re}N_{Sc}^{1/3}} = 0.74\, N_{Re}^{-0.5}$, $N_{Sh} = \dfrac{k'd_{ch}}{D}$ $N_{Re} = \dfrac{d_{ch} \upsilon \rho}{\mu}$, $d_{ch} = \dfrac{\text{total surface area}}{\text{perimeter normal to flow}}$ e.g., for cube with side length a, $d_{ch} = 1.27a$.	[E] Used with arithmetic concentration difference. $120 \le N_{Re} \le 6000$; standard deviation 2.1%. Eccentricities between 1:1 (spheres) and 3:1. Oblate spheroid is often approximated by drops.	Skelland, p. 284 [104]
G. Other objects, including prisms, cubes, hemispheres, spheres, and cylinders; forced convection Molecular diffusion limits	$j_D = 0.692 N_{Re,p}^{-0.486}$, $N_{Re,p} = \dfrac{\upsilon d_{ch} \rho}{\mu}$ Terms same as in 5-20E. $N_{Sh} = \dfrac{k'd_{ch}}{D} = A$	[E] Used with arithmetic concentration difference. Agrees with cylinder and oblate spheroid results, ±15%. Assumes molecular diffusion and natural convection are negligible. $500 \le N_{Re,p} \le 5000$. Turbulent. [T] Hard to reach limits in experiments. Spheres and cubes: $A = 2$, tetrahedrons: $A = 2\sqrt{6}$, octahedrons: $2\sqrt{2}$.	Kirwan, p. 115 Skelland, p. 285 [84] [85] Kirwan, p. 114
H. Shell side of microporous hollow fiber module for solvent extraction	$N_{Sh} = \beta[d_h(1-\varphi)/L] N_{Re}^{0.6} N_{Sc}^{0.33}$ $N_{Sh} = \dfrac{\bar{K} d_h}{D}$ $N_{Re} = \dfrac{d_h \upsilon \rho}{\mu}$, $\bar{K}$ = overall mass-transfer coefficient $\beta = 5.8$ for hydrophobic membrane. $\beta = 6.1$ for hydrophilic membrane.	[E] Use with log-mean concentration difference. d_h = hydraulic diameter = $\dfrac{4 \times \text{cross-sectional area of flow}}{\text{wetted perimeter}}$ φ = packing fraction of shell side. L = module length. Based on area of contact according to inside or outside diameter of tubes depending on location of interface between aqueous and organic phases. Can also be applied to gas-liquid systems with liquid on shell side.	[87] [103], p. 113

See Table 5-23 for flow in packed beds.

*Author names refer to the General References list at the beginning of the Mass Transfer subsection (pp. 5-41, 5-42). Bracketed numbers refer to the table reference lists (p. 5-42).

TABLE 5-21 Mass-Transfer Correlations for Drops, Bubbles, and Bubble Columns

Conditions	Correlations	Comments E = Empirical, S = Semiempirical, T = Theoretical	References*
A. Single liquid drop in immiscible liquid, drop formation, discontinuous (drop) phase coefficient	$\hat{k}_{d,f} = A\left(\frac{\rho_d}{M_d}\right)_{av}\left(\frac{D_d}{\pi t_f}\right)^{1/2}$ $A = 24/7$ (penetration theory), $A = 1.31$ (semiempirical value) $A = \left[\frac{24}{7}(0.8624)\right]$(extension by fresh surface elements)	[T,S] Use arithmetic mole fraction difference. Fits some, but not all, data. Low mass transfer rate. M_d = mean molecular weight of dispersed phase; t_f = formation time of drop $k_{L,d}$ = mean dispersed liquid phase MT. coefficient k mole/[s · m² (mole fraction)].	Skelland, p. 399
B. Single liquid drop in immiscible liquid, drop formation, continuous phase coefficient	$k_{L,c} = 0.386\times\left(\frac{\rho_c}{M_c}\right)_{av}\left(\frac{D_c}{t_f}\right)^{0.5}\left(\frac{\rho_c \sigma g_c}{\Delta\rho g t_f \mu_c}\right)^{0.407}\left(\frac{g t_f^2}{d_p}\right)^{0.148}$	[E] Average absolute deviation 11% for 20 data points for 3 systems.	Skelland, p. 402 [105] p. 434
C. Single liquid drop in immiscible liquid, free rise or fall, discontinuous phase coefficient, stagnant drops	$k_{L,d,m} = \frac{-d_p}{6t}\left(\frac{\rho_d}{M_d}\right)_{av}\ln\left\{\frac{6}{\pi^2}\sum_{j=1}^{\infty}\frac{1}{j^2}\exp\left[\left(\frac{-D_d j^2\pi^2 t}{(d_p/2)^2}\right)\right]\right\}$	[T] Use with log-mean mole fraction differences based on ends of column. t = rise time. No continuous phase resistance. Stagnant drops are likely if drop is very viscous, quite small, or is coated with surface active agent. $k_{L,d,m}$ = mean dispersed liquid MT coefficient.	Skelland, p. 404 [105] p. 435
D. Same as 5-21C	$\hat{k}_{L,d,m} = \frac{-d_p}{6t}\left(\frac{\rho_d}{M_d}\right)_{av}\ln\left[1-\frac{\pi D_d^{1/2}t^{1/2}}{d_p/2}\right]$	[S] See 5-21C. Approximation for fractional extractions less than 50%.	Skelland, p. 404 [105] p. 435
E. Same as 5-21C, continuous phase coefficient, stagnant drops, spherical	$N_{Sh} = \frac{k_{L,c,m}d_c}{D_c} = 0.74\left(\frac{\rho_c}{M_c}\right)_{av}N_{Re}^{1/2}(N_{Sc})^{1/3}$	[E] $N_{Re} = \upsilon_s d_p \rho_c/\mu_c, \upsilon_S$ = slip velocity between drop and continuous phase.	Skelland, p. 407 [104][105] p. 436
F. Single bubble or drop with surfactant. Stokes flow.	$N_{Sh} = 2.0 + \alpha N_{Pe}^{\beta}, N_{Sh} = 2rk/D$ $\alpha = \frac{5.49}{A+6.10} + \frac{A}{A+28.64}$ $\beta = \frac{0.35A+17.21}{A+34.14}$ $2r = 2$ to 50 μm, $A = 2.8E4$ to 7.0E5 $0.0026 < N_{Pe,s} < 340$, 2.1 $< N_{Ma} < 1.3E6$ $N_{Pe} = 1.0$ to 2.5×10^4, $N_{Re} = 2.2\times10^{-6}$ to 0.034	[T] A = surface retardation parameter $A = B\Gamma_o r/\mu D_s = N_{Ma}N_{Pe,s}$ $N_{Ma} = B\Gamma_o/\mu u$ = Marangoni no. Γ = surfactant surface concentration, $B = -\partial\sigma/\partial\Gamma$ = constant $N_{Pe,s}$ = surface Peclet number = ur/D_s D_s = surface diffusivity N_{Pe} = bulk Peclet number For $A >> 1$ acts like rigid sphere: $\beta \to 0.35, \alpha \to 1/2864 = 0.035$	[89]
G. See oblate spheroid	$N_{Sh} = \frac{k_{L,c,m}d_3}{D_c} = 0.74\left(\frac{\rho_c}{M_c}\right)_{av}(N_{Re,3})^{1/2}(N_{Sc,c})^{1/3}$ $N_{Re,3} = \frac{\upsilon_s d_3 \rho_c}{\mu_c}$	[E] Used with log-mean mole fraction. Differences based on ends of extraction column; 100 measured values ±2% deviation. Based on area oblate spheroid. υ_s = slip velocity, $d_3 = \frac{\text{total drop surface area}}{\text{perimeter normal to flow}}$	Skelland, pp. 285, 406, 407
H. Single liquid drop in immiscible liquid, free rise or fall, discontinuous phase coefficient, circulating drops	$\hat{k}_{L,d,circ} = -\frac{d_p}{6\theta}\left(\frac{\rho_d}{M_d}\right)_{av}\ln\left[1-\frac{R^{1/2}\pi D_d^{1/2}\theta^{1/2}}{d_p/2}\right]$	[E] Used with mole fractions for extraction less than 50%, $R \approx 2.25$.	Skelland, p. 405
I. Same as 5-21F	$N_{Sh} = \frac{\hat{k}_{L,d,circ}d_p}{D_d} = 31.4\left(\frac{\rho_d}{M_f}\right)_{av}\left(\frac{4D_d t}{d_p^2}\right)^{-0.34}N_{Sc,d}^{-0.125}\left(\frac{d_p \upsilon_s^2 \rho_c}{\sigma g_c}\right)^{-0.37}$	[E] Used with log-mean mole fraction difference. d_p = diameter of sphere with same volume as drop. $856 \le N_{Sc} \le 79{,}800$, $2.34 \le \sigma \le 4.8$ dynes/cm.	[105] p. 435 [106]
J. Liquid drop in immiscible liquid, free rise or fall, continuous phase coefficient, circulating single drops	$N_{Sh,c} = \frac{k'_{L,c}d_p}{D_d} = \left[2+0.463N_{Re,drop}^{0.484}N_{Sc,c}^{0.339}\left(\frac{d_p g^{1/3}}{D_c^{2/3}}\right)\right.$ $F = 0.281 + 1.615K + 3.73K^2 - 1.874K^3$ $K = N_{Re,drop}^{1/8}\left(\frac{\mu_c}{\mu_d}\right)^{1/4}\left(\frac{\mu_c \upsilon_s}{\sigma g_c}\right)^{1/6}$	[E] Used as an arithmetic concentration difference. $N_{Re,drop} = \frac{d_p \upsilon_s \rho_c}{\mu_c}$ Solid sphere form with correction factor F.	[61]
K. Same as 5-20J, circulating, single drop	$N_{Sh} = \frac{k_{L,c}d_p}{D_c} = 0.6\left(\frac{\rho_c}{M_c}\right)_{av}N_{Re,drop}^{1/2}N_{Sc,c}^{1/2}$	[E] Used as an arithmetic concentration difference. Low σ.	Skelland, p. 407
L. Same as 5-20J, circulating swarm of drops	$k_{L,c} = 0.725\left(\frac{\rho_c}{M_c}\right)_{av}N_{Re,drop}^{-0.43}N_{Sc,c}^{-0.58}\upsilon_s(1-\phi_d)$	[E] Used as an arithmetic concentration difference. Low σ, disperse-phase holdup of drop swarm. ϕ_d = volume fraction dispersed phase.	Skelland, p. 407 [105] p. 436
M. Liquid drops in immiscible liquid, free rise or fall, discontinuous phase coefficient, oscillating drops	$N_{Sh} = \frac{k_{L,d,osc}d_p}{D_d} = 0.32\left(\frac{\rho_d}{M_d}\right)_{av}\left(\frac{4D_d t}{d_p^2}\right)^{-0.14}N_{Re,drop}^{0.68}\left(\frac{\sigma^3 g_c^3 \rho_c^2}{g\mu_c^4\Delta\rho}\right)^{0.10}$	[E] Used with a log-mean mole fraction difference. Based on ends of extraction column. $N_{Re,drop} = \frac{d_p \upsilon_s \rho_c}{\mu_c}$, $411 \le N_{Re} \le 3114$ d_p = diameter of sphere with volume of drop. Average absolute deviation from data, 10.5%. Low interfacial tension (3.5–5.8 dyn), $\mu_c < 1.35$ centipoise.	Skelland, p. 406 [105] p. 435 [106]

TABLE 5-21 Mass-Transfer Correlations for Drops, Bubbles, and Bubble Columns *(Continued)*

Conditions	Correlations	Comments E = Empirical, S = Semiempirical, T = Theoretical	References*
N. Same as 5-21M	$k_{L,d,\mathrm{osc}} = \dfrac{0.00375 \upsilon_s}{1+\mu_d/\mu_c}$	[T] Use with log-mean concentration difference. Based on end of extraction column. No continuous phase resistance. $k_{L,d,\mathrm{osc}}$ in cm/s, v_s = drop velocity relative to continuous phase.	Sherwood, Pigford, & Wilke, p. 228 Skelland, p. 405
O. Single liquid drops in gas, gas side coefficient	$\dfrac{\hat{k}_g M_g d_p P}{D_{\mathrm{gas}} \rho_g} = 2 + A N_{\mathrm{Re},g}^{1/2} N_{\mathrm{Sc},g}^{1/3}$ $A = 0.552$ or 0.60 $N_{\mathrm{Re},g} = \dfrac{d_p \rho_g \upsilon_s}{\mu_g}$	[E] Used for spray drying (arithmetic partial pressure difference). v_s = slip velocity between drop and gas stream. Sometimes written with $M_g P/r_g = RT$.	[68] p. 388 [90]
P. Single water drop in air, liquid side coefficient	$k_L = 2\left(\dfrac{D_L}{\pi t}\right)^{1/2}$, short contact times $k_L = 10\dfrac{D_L}{d_p}$, long contact times	[T] Use arithmetic concentration difference. Penetration theory. t = contact time of drop. Gives plot for $k_G a$ also. Air-water system.	[68] p. 389
Q. Single bubbles of gas in liquid, continuous phase coefficient, very small bubbles	$N_{\mathrm{Sh}} = \dfrac{k_c' d_b}{D_c} = 1.0 (N_{\mathrm{Re}} N_{\mathrm{Sc}})^{1/3}$	[T] Solid-sphere Eq. $d_b < 0.1$ cm, k_c' is average over entire surface of bubble.	Sherwood, Pigford, & Wilke, p. 214
R. Same as 5-21Q, medium to large bubbles	$N_{\mathrm{Sh}} = \dfrac{k_c' d_b}{D_c} = 1.13 (N_{\mathrm{Re}} N_{\mathrm{Sc}})^{1/2}$	[T] Use arithmetic concentration difference. Droplet equation: $d_b > 0.5$ cm.	Sherwood, Pigford, & Wilke, p. 231
S. Same as 5-21R	$N_{\mathrm{Sh}} = \dfrac{k_c' d_b}{D_c} = 1.13 (N_{\mathrm{Re}} N_{\mathrm{Sc}})^{1/2} \left[\dfrac{d_b}{0.45 + 0.2 d_b}\right]$ $500 \le N_{\mathrm{Re}} \le 8000$	[S] Use arithmetic concentration difference. Modification of above (5-21R), $d_b > 0.5$ cm. No effect SAA for $d_p > 0.6$ cm.	[62], Sherwood, Pigford, & Wilke, p. 231
T. Taylor bubbles in single capillaries (square or circular)	$k_L a = 4.5 \left(\dfrac{D u_G}{L_{\mathrm{uc}}}\right)^{1/2} \dfrac{1}{d_c}$ Applicable $\left(\dfrac{u_G + u_L}{L_{\mathrm{slug}}}\right)^{0.5} > 3 (s^{-0.5})$	[E] Air-water L_{uc} = unit cell length, L_{slug} = slug length, d_c = capillary i.d. For most data $k_L a \pm 20\%$.	[110]
U. Gas-liquid mass transfer in monoliths	$k_L a \approx 0.1 \left(\dfrac{P}{V}\right)^{1/4}$ P/V = power/volume (kW/m^3), range = 100 to 10,000	[E] Each channel in monolith is a capillary. Results are in expected order of magnitude for capillaries based on 5-21T. k_L is larger than in stirred tanks.	[70]
V. Rising small bubbles of gas in liquid, continuous phase. Calderbank and Moo-Young correlation	$N_{\mathrm{Sh}} = \dfrac{k_c' d_b}{D_c} = 2 + 0.31 (N_{\mathrm{Gr}})^{1/3} N_{\mathrm{Sc}}^{1/3}, d_b < 0.25$ cm $N_{\mathrm{Ra}} = \dfrac{d_b^3 \lvert \rho_G - \rho_L \rvert g}{\mu_L D_L}$ = Raleigh number	[E] Use with arithmetic concentration difference. Valid for single bubbles or swarms. Independent of agitation as long as bubble size is constant. Recommended by [99]. Note that $N_{\mathrm{Ra}} = N_{\mathrm{Gr}} N_{\mathrm{Sc}}$.	[34] Geankoplis, p. 451 Kirwan, p. 119 Treybal, p. 156 [99]
W. Same as 5-21V, large bubbles	$N_{\mathrm{Sh}} = \dfrac{k_c' d_b}{D_c} = 0.42 (N_{\mathrm{Gr}})^{1/3} N_{\mathrm{Sc}}^{1/2}, d_b > 0.25$ cm $\dfrac{\text{Interfacial area}}{\text{volume}} = a = \dfrac{6 H_g}{d_b}$	[E] Use with arithmetic concentration difference. For large bubbles, k_c' is independent of bubble size and independent of agitation or liquid velocity. Resistance is entirely in liquid phase for most gas-liquid mass transfer. H_g = fractional gas holdup, volume gas/total volume.	[34] Geankoplis, p. 452 Kirwan, p. 119 [72] p. 249 [99]
X. Bubbles in bubble columns. Hughmark correlation	$N_{\mathrm{Sh}} = \dfrac{k_L d}{D} = 2 + b N_{\mathrm{Sc}}^{0.546} N_{\mathrm{Re}}^{0.779} \left(\dfrac{d g^{1/3}}{D^{2/3}}\right)^{0.116}$ $b = 0.061$ single gas bubbles $b = 0.0187$ swarms of bubbles $V_s = \dfrac{V_g}{\varphi_G} - \dfrac{V_L}{1 - \varphi_G}$	[E] d = bubble diameter Air–liquid. Recommended by [99] and Treybal. For swarms, calculate N_{Re} with slip velocity V_s. φ_G = gas holdup, V_G = superficial gas velocity Col. diameter = 0.025 to 1.1 m ρ_L' = 776 to 1696 kg/m^3, μ_L = 0.0009 to 0.152 Pa · s	[42] [61] Treybal, p. 144
Y. Bubbles in bubble column	$k_L = \dfrac{0.15 D}{d_{Vs}} \left(\dfrac{\nu}{D}\right)^{1/2} N_{\mathrm{Re}}^{3/4}$	[E] d_{Vs} = Sauter mean bubble diameter, $N_{\mathrm{Re}} = d_{Vs} u_G \rho_L / \mu_L$. Recommended by [36] based on experiments in industrial system.	[36] [97]

See Table 5-22 for agitated systems.
*Author names refer to the General References list at the beginning of the Mass Transfer subsection (pp. 5-41, 5-42). Bracketed numbers refer to the table reference lists (p. 5-42).

TABLE 5-22 Mass-Transfer Correlations for Particles, Drops, and Bubbles in Agitated Systems

Situation	Correlation	Comments E = Empirical, S = Semiempirical, T = Theoretical	References*
A. Solid particles suspended in agitated vessel containing vertical baffles, continuous phase coefficient	$\frac{k'_{LT} d_p}{D} = 2 + 0.6 N_{Re,T}^{1/2} N_{Sc}^{1/3}$ Replace v_{slip} with v_T = terminal velocity. Calculate Stokes' law terminal velocity $\upsilon_{Ts} = \frac{d_p^2 \lvert \rho_p - \rho_c \rvert g}{18\mu_c}$ and correction factor: $N_{Re,Ts}$: 1, 10, 100, 1000, 10,000, 100,000 υ_T/υ_{Ts}: 0.9, 0.65, 0.37, 0.17, 0.07, 0.023 Approximate: $k'_L = 2k'_{LT}$	[S] Use log-mean concentration difference. Modified Frossling equation: $N_{Re,Ts} = \frac{\upsilon_{Ts} d_p \rho_c}{\mu_c}$ (Reynolds number based on Stokes' law.) $N_{Re,T} = \frac{\upsilon_T d_p \rho_c}{\mu_c}$ (terminal velocity Reynolds number.) k'_L almost independent of d_p Harriott [56] suggests different correction procedures. Range k'_L/k'_{LT} is 1.5 to 8.0.	[56] Sherwood, Pigford, & Wilke, pp. 220–222 [83] [56]
B. Solid, neutrally buoyant particles, continuous phase coefficient	$N_{Sh} = \frac{k'_L d_p}{D} = 2 + 0.47 N_{Re,p}^{0.62} N_{Sc}^{0.36} \left(\frac{d_{imp}}{d_{tank}}\right)^{0.17}$ Graphical comparisons are in Ref. 88, p. 116. $N_{Sh} = 2 + 0.52 N_{Re,p}^{0.52} N_{Sc}^{1/3}$, $N_{Re,p} < 1.0$	[E] Use log-mean concentration difference. Density unimportant if particles are close to neutrally buoyant. Also used for drops. Geometric effect (d_{imp}/d_{tank}) is usually unimportant. Ref. [77] gives a variety of references on correlations. [E] E = energy dissipation rate per unit mass fluid $= \frac{P g_c}{V_{tank}\rho_c}$, P = power, $N_{Re,p} = \frac{E^{1/3} d_p^{4/3}}{\nu}$ [E] Small particles	Kirwan, p. 115 [77] p. 132 Treybal, p. 523 Kirwan, p. 116
C. Solid particles with significant density difference	$N_{Sh} = \frac{k'_L d_p}{D} = 2 + 0.44 \left(\frac{d_p \upsilon_{slip}}{\nu}\right)^{1/2} N_{Sc}^{0.38}$	[E] Use log-mean concentration difference. N_{Sh} standard deviation 11.1%. v_{slip} calculated by methods given in reference.	[77] [83]
D. Small solid particles, gas bubbles or liquid drops, d_p < 2.5 mm. Aerated mixing vessels	$N_{Sh} = \frac{k'_L d_p}{D} = 2 + 0.31 \left(\frac{d_p^3 \lvert \rho_p - \rho_c \rvert}{\mu_c D}\right)^{1/3}$	[E] Use log-mean concentration difference g = 9.80665 m/s². Second term RHS is free-fall or rise term. For large bubbles, see Table 5-21W.	[33] [50] p. 487 [72] p. 249
E. Highly agitated systems; solid particles, drops, and bubbles; continuous phase coefficient	$k'_L N_{Sc}^{2/3} = 0.13 \left[\frac{(P/V_{tank})\mu_c g_c}{\rho_c^2}\right]^{1/4}$	[E] Use arithmetic concentration difference. Use when gravitational forces overcome by agitation. Up to 60% deviation. Correlation prediction is low [77]. (P/V_{tank}) = power dissipated by agitator per unit volume liquid.	[34] Geankoplis, p. 489 [83]
F. Gas bubble swarms in sparged tank reactors	$k'_L a = \left(\frac{\nu}{g^2}\right)^{1/3} = C \left[\frac{P/V_L}{\rho(\nu g^4)^{1/3}}\right]^a \left[\frac{qG}{V_L}\left(\frac{\nu}{g^2}\right)^{1/3}\right]^b$ Rushton turbines: $C = 7.94 \times 10^{-4}$, $a = 0.62$, $b = 0.23$. Intermig® impellers: $C = 5.89 \times 10^{-4}$, $a = 0.62$, $b = 0.19$.	[E] Use arithmetic concentration difference. Done for biological system, O_2 transfer. $h_{tank}/D_{tank} = 2.1$; P = power, kW. V_L = liquid volume, m³. q_G = gassing rate, m³/s. $k'_L a$ = s⁻¹. Since a = m²/m³, ν = kinematic viscosity, m²/s. Low viscosity system. Better fit claimed with q_G/V_L than with u_G (see 5-22G and H).	[96]
G. Same as 5-22F, baffled tank with standard blade Rushton impeller	$k'_L a = 93.37 \left(\frac{P}{V_L}\right)^{0.76} u_G^{0.45}$	[E] Air-water. Same definitions as 5-22F. $0.005 < u_G < 0.025$, $3.83 < N < 8.33$, $400 < P/V_L < 7000$ $h = D_{tank}$ = 0.305 or 0.610 m. V_G = gas volume, m³, N = stirrer speed, rpm. Method assumes perfect liquid mixing.	[50] [73]
H. Same as 5-22G, bubbles	$\frac{k'_L a d_{imp}^2}{D} = 0.060 \left(\frac{d_{imp}^2 N\rho}{\mu_{eff}}\right)\left(\frac{d_{imp}^2 N^2}{g}\right)^{0.19}\left(\frac{\mu_{eff} u_G}{\sigma}\right)^{0.6}$	[E] Use arithmetic concentration difference. O_2 into aqueous glycerol solutions. O_2 into aqueous millet jelly solutions. Same definitions as 5-22G.	[73] [117]
I. Surface aerators for air-water contact	$\frac{k_L a}{N} = b N_p^{0.71} N_{Fr}^{0.48} N_{Re}^{0.82} \left(\frac{H}{d}\right)^{-0.54} \left(\frac{V}{d^3}\right)^{-1.08}$ $b = 7 \times 10^{-6}$, $N_p = P/(\rho N^3 d^5)$ $N_{Re} = N d^2 \rho_{liq}/\mu_{liq}$ $N_{Fr} = N^2 d/g$, P/V = 90 to 400 W/m³	[E] Three impellers: Pitched blade downflow turbine, pitched blade upflow turbine, standard disk turbine. Baffled cylindrical tanks 1.0- and 1.5-m id and 8.2 × 8.2-m square tank. Submergence optimized all cases. Good agreement with data. N = impeller speed, s⁻¹; d = impeller diameter, m; H = liquid height, m; V = liquid volume, m³; $k_L a$ = s⁻¹, g = acceleration gravity = 9.81 m/s²	[86]
J. Individual drops in LLE. Continuous phase coefficients	$Sh_{C,circ} \equiv (k_C d/D_{AC}) = (2/\pi^{0.5}) Pe_C^{0.5}$ $Pe_C = (d u_t/D_{AC})$ $Sh_{C,rigid} = 2.43 + \{0.773 Re_{drop}^{0.5} + 0.0103 Re_{drop}\} Sc_C^{0.33}$ $(Sh_C - Sh_{C,rigid})/(Sh_{C,circ} - Sh_{C,rigid}) = 1 - \exp[-0.00418 Pe_C^{0.42}]$	[E] Clean circulating drops, Steiner's results for $10 < Re_{drop} < 1200$, $190 < Sc_C < 241000$, and $1000 < Pe_C < 10^6$, [E] Rigid drops (common in dirty conditions), Steiner's results, $10^4 < Pe_C < 10^6$, Preferred for dirty conditions. [E] Preferred for clean conditions between circulating & rigid. Steiner's results	[107] [114]
K. Average drop size in LLE mixers	$\frac{d_p}{d_p^o} = 1 + 1.18\varphi_D \left(\frac{\sigma^2 g_c^2}{d_p^o \mu_c^2 g}\right)\left(\frac{\mu_c^4 g}{\Delta\rho\sigma^3 g_c^3}\right)^{0.62}\left(\frac{\Delta\rho}{\rho_c}\right)^{0.05}$	[E] Baffled mixers. LLE data. $\Delta\rho = \lvert\rho_C - \rho_D\rvert \cdot d_p^o$ is from, $\frac{d_p^{o3} \rho_C^2 g}{\mu_C^2} = 29 \left(\frac{V_{Liq\text{-}tank}^3 \rho_C^2 \mu_C g^4}{P^3 g_c^3}\right)^{0.32} \left(\frac{\sigma^3 \rho_C g_c^3}{\mu_C^4 g}\right)^{0.14}$	Treybal [114]

See also Table 5-21.

*Author names refer to the General References list at the beginning of the Mass Transfer subsection (pp. 5-41, 5-42). Bracketed numbers refer to the table reference lists (p. 5-42).

TABLE 5-23 Mass-Transfer Correlations for Fixed and Fluidized Beds

Transfer is to or from particles

Situation	Correlation	Comments E = Empirical, S = Semiempirical, T = Theoretical	References*
A. For gases, fixed and fluidized beds, Gupta and Thodos correlation	$j_H = j_D = \frac{2.06}{\varepsilon N_{Re}^{0.575}}, 90 \le N_{Re} \le A$ Equivalent: $N_{Sh} = \frac{k'd_s}{D} = \frac{2.06}{\varepsilon} N_{Re}^{0.425} N_{Sc}^{1/3}$ For other shapes: $\frac{\varepsilon j_D}{(\varepsilon j_D)_{sphere}} = 0.79$ (cylinder) or 0.71 (cube)	[E] For spheres. $N_{Re} = \frac{\upsilon_{super} d_p \rho}{\mu}$ $A = 2453$ (Skelland), $A = 4000$ [57]. For $N_{Re} > 1900$, $j_H = 1.05 j_D$. Heat transfer result is in absence of radiation. Graphical results are available for N_{Re} from 1900 to 10,300. $a = \frac{\text{surface area}}{\text{volume}} = 6(1-\varepsilon)/d_p$ For spheres, d_p = diameter. For nonspherical: $d_p = 0.567\sqrt{\text{Part. Surf. Area}}$	[54] [55] [57], p. 195 Skelland
B. For gases and liquids, fixed and fluidized beds	$j_D = \frac{0.4548}{\varepsilon N_{Re}^{0.4069}}, 10 \le N_{Re} \le 2000$ $j_D = \frac{N_{Sh}}{N_{Re} N_{Sc}^{1/3}}, N_{Sh} = \frac{k'd_s}{D}$	[E] Packed spheres, deep bed. Average deviation ±20%, $N_{Re} = d_p V_{super}\rho/\mu$. Can use for fluidized beds. $10 \le N_{Re} \le 4000$.	[46] Geankoplis, p. 484
C. For liquids, fixed bed, Wilson and Geankoplis correlation	$j_D = \frac{1.09}{\varepsilon N_{Re}^{2/3}}, 0.0016 < N_{Re} < 55$ $165 \le N_{Sc} \le 70{,}600, 0.35 < \varepsilon < 0.75$ Equivalent: $N_{Sh} = \frac{1.09}{\varepsilon} N_{Re}^{1/3} N_{Sc}^{1/3}$ $j_D = \frac{0.25}{\varepsilon N_{Re}^{0.31}}, 55 < N_{Re} < 1500, 165 \le N_{Sc} \le 10{,}690$ Equivalent: $N_{Sh} = \frac{0.25}{\varepsilon} N_{Re}^{0.69} N_{Sc}^{1/3}$	[E] Beds of spheres, $N_{Re} = \frac{d_p V_{super} \rho}{\mu}$ Deep beds. $N_{Sh} = \frac{k'd_s}{D}$	Geankoplis, p. 484 [57], p. 195 Skelland, p. 287 [115]
D. For liquids, fixed beds, Ohashi et al. correlation	$N_{Sh} = \frac{k'd_s}{D} = 2 + 0.51\left(\frac{E^{1/3} d_p^{4/3} \rho}{\mu}\right)^{0.60} N_{Sc}^{1/3}$ E = Energy dissipation rate per unit mass of fluid $= 50(1-\varepsilon)\varepsilon^2 C_{Do}\left(\frac{\upsilon_r^3}{d_p}\right), \text{m}^2/\text{s}^3$ $= \left[\frac{50(1-\varepsilon)C_D}{\varepsilon}\right]\left(\frac{\upsilon_{super}^3}{d_p}\right)$ General form: $N_{Sh} = 2 + K\left(\frac{E^{1/3} d_p^{4/3} \rho}{\mu}\right)^{\alpha} N_{Sc}^{\beta}$	[S] Applies to single particles, packed beds, two-phase tube flow, suspended bubble columns, and stirred tanks with different definitions of E. Correlates large amount of published data. Compares number of correlations, v_r = relative velocity, m/s. In packed bed, $v_r = v_{super}/\varepsilon$. C_{Do} = single particle drag coefficient at v_{super} calculated from $C_{Do} = AN_{Re_r}^{-m}$. N_{Re} — A — m 0 to 5.8 — 24 — 1.0 5.8 to 500 — 10 — 0.5 >500 — 0.44 — 0 Ranges for packed bed: $0.2 < \frac{E^{1/3} d_p^{4/3} \rho}{\mu} < 4600$ Compares different situations versus general correlation.	[81]
E. Electrolytic system. Pall rings. Transfer from fluid to rings.	Full liquid upflow, $N_{Re} d_e u/\nu = 80$ to 550: $N_{Sh} = k_L d_e/D = 4.1 N_{Re}^{0.39} N_{Sc}^{1/3}$ Irrigated liquid downflow (no gas flow): $N_{Sh} = 5.1 N_{Re}^{0.44} N_{Sc}^{1/3}$	[E] d_e = diameter of sphere with same surface area as Pall ring. Full liquid upflow agreed with literature values. Schmidt number dependence was assumed from literature values. In downflow, N_{Re} used superficial fluid velocity.	[52]
F. For liquids, fixed and fluidized beds	$\varepsilon j_D = \frac{1.1068}{N_{Re}^{0.72}}, 1.0 < N_{Re} \le 10$ $\varepsilon j_D = \frac{N_{Sh}}{N_{Re} N_{Sc}^{1/3}}, N_{Sh} = \frac{k'd_s}{D}$	[E] Spheres: $N_{Re} = \frac{d_p \upsilon_{super} \rho}{\mu}$	[45] Geankoplis, p. 484
G. For gases and liquids, fixed and fluidized beds, Dwivedi and Upadhyay correlation	$\varepsilon j_D = \frac{0.765}{N_{Re}^{0.82}} + \frac{0.365}{N_{Re}^{0.386}}$ Gases: $10 \le N_{Re} \le 15{,}000$. Liquids: $0.01 \le N_{Re} \le 15{,}000$. $N_{Re} = \frac{d_p \upsilon_{super} \rho}{\mu}, N_{Sh} = \frac{k'd_s}{D}$	[E] Deep beds of spheres, $j_D = \frac{N_{Sh}}{N_{Re} N_{Sc}^{1/3}}$ Best fit correlation at low conc. [39] Based on 20 gas studies and 17 liquid studies.	[45] [57] p. 196 [39]
H. For gases and liquids, fixed bed	$j_D = 1.17 N_{Re}^{-0.415}, 10 \le N_{Re} \le 2500$ $j_D = \frac{k'}{\upsilon_{av}} \frac{p_{BM}}{P} N_{Sc}^{2/3}$	[E] Spheres: Variation in packing that changes ε not allowed for. Extensive data referenced. $0.5 < N_{Sc} < 15{,}000$. Show comparisons with other results. $N_{Re} = \frac{d_p \upsilon_{super} \rho}{\mu}$	Sherwood, Pigford, & Wilke, p. 241

TABLE 5-23 Mass-Transfer Correlations for Fixed and Fluidized Beds *(Continued)*

Situation	Correlation	Comments E = Empirical, S = Semiempirical, T = Theoretical	References*
I. For liquids, fixed and fluidized beds, Rahman and Streat correlation	$N_{Sh} = \frac{0.86}{\varepsilon} N_{Re} N_{Sc}^{1/3}, 2 \le N_{Re} \le 25$	[E] Can be extrapolated to $N_{Re} = 2000$. $N_{Re} = d_p v_{super} \rho/\mu$. Done for neutralization of ion exchange resin.	[88]
J. Size exclusion chromatography of proteins	$N_{Sh} = \frac{k_L d}{D} = \frac{1.903}{\varepsilon} N_{Re}^{1/3} N_{Sc}^{1/3}$	[E] Slow mass transfer with large molecules. Aqueous solutions. Modest increase in N_{Sh} with increasing velocity.	[58]
K. For liquids and gases, Ranz and Marshall correlation	$N_{Sh} = \frac{k'd}{D} = 2.0 + 0.6 N_{Sc}^{1/3} N_{Re}^{1/2}$ $N_{Re} = \frac{d_p \upsilon_{super} \rho}{\mu}$	[E] Based on freely falling, evaporating spheres (see 5-20C). Has been applied to packed beds, prediction is low compared to experimental data. Limit of 2.0 at low N_{Re} is too high. Not corrected for axial dispersion.	[23] p. 214, [90] [112] [83]
L. For liquids and gases, Wakao and Funazkri correlation	$N_{Sh} = 2.0 + 1 N_{Sc}^{1.1/3} N_{Re}^{0.6}, 3 < N_{Re} < 10{,}000$ $N_{Sh} = \frac{k'_{film} d_p}{D}, N_{Re} = \frac{\rho_f \upsilon_{super} d_p}{\mu}$ $\frac{\varepsilon D_{axial}}{D} = 10 + 0.5 N_{Sc} N_{Re}$	[E] Correlate 20 gas studies and 16 liquid studies. Corrected for axial dispersion. Graphical comparison with data shown [23, p. 215], and [112]. D_{axial} is axial dispersion coefficient.	[23] p. 214 [112]
M. Semifluidized or expanded bed. Liquid-solid transfer.	$N_{Sh} = \frac{k_{film} d_p}{D} = 2 + 1.5(1-\varepsilon_L) N_{Re}^{1/3} N_{Sc}^{1/3}$ $N_{Re} = \rho_p d_p u/\mu\varepsilon_L$; $N_{Sc} = \mu/\rho D$	[E] ε_L = liquid-phase void fraction, ρ_p = particle density, ρ = fluid density, d_p = particle diameter. Fits expanded bed chromatography in viscous liquids.	[48] [116]
N. Mass-transfer structured packing and static mixers. Liquid with or without fluidized particles. Electrochemical	Fixed bed: $j' = 0.927 N_{Re'}^{0.572}, N'_{Re} < 219$ $j' = 0.443 N_{Re'}^{-0.435}, 219 < N'_{Re} < 1360$ Fluidized bed with particles: $j' = 6.02 N_{Re}^{-0.885}$, or $j' = 16.40 N_{Re'}^{-0.950}$ Natural convection: $N_{Sh} = 0.252(N_{Sc}N_{Gr})^{0.299}$ Bubble columns: Structured packing: $N_{St} = 0.105(N_{Re} N_{Fr} N_{Sc}^2)^{-0.268}$ Static mixer: $N_{St} = 0.157(N_{Re} N_{Fr} N_{Sc}^2)^{-0.298}$	[E] = Sulzer packings, $j' = \frac{k \cos\beta}{\upsilon} N_{Sc}^{2/3}$, β = corrugation incline angle. $N_{Re'} = v' d'_h \rho/\mu$, $v' = v_{super}/(\varepsilon \cos \beta)$, d'_h = channel side width. Particles enhance mass transfer in laminar flow for natural convection. Good fit with correlation of Ray et al., *Intl. J. Heat Mass Transfer* **41:** 1693 (1998). $N_{Gr} = g \Delta \rho Z^3 \rho/\mu^2$, Z = corrugated plate length. Bubble column results fit correlation of Neme et al., *Chem. Eng. Technol.* **20:** 297 (1997) for structured packing. N_{St} = Stanton number = kZ/D N_{Fr} = Froude number = v^2_{super}/gz	[35]
O. Liquid fluidized beds	$N_{Sh} = 0.250 N_{Re}^{0.023} N_{Ga}^{0.306} \left(\frac{\rho_s - \rho}{\rho}\right)^{0.282} N_{Sc}^{0.410}\ (\varepsilon < 0.85)$ $N_{Sh} = 0.304 N_{Re}^{-0.057} N_{Ga}^{0.332} \left(\frac{\rho_s - \rho}{\rho}\right)^{0.297} N_{Sc}^{0.404}\ (\varepsilon > 0.85)$ This can be simplified (with slight loss in accuracy at high ε) to $N_{Sh} = 0.245 N_{Ga}^{0.323} \left(\frac{\rho_s - \rho}{\rho}\right)^{0.300} N_{Sc}^{0.400}$	[E] Correlate amount of data from literature. Predicts very little dependence of N_{Sh} on velocity. Compare large number of published correlations. $N_{Sh} = \frac{k'_L d_p}{D}, N_{Re} = \frac{d_p \rho \upsilon_{super}}{\mu}, N_{Ga} = \frac{d_p^3 \rho^2 g}{\mu^2}$, $N_{Sc} = \frac{\mu}{\rho D}$ $1.6 < N_{Re} < 1320, 2470 < N_{Ga} < 4.42 \times 106$ $0.27 < \frac{\rho_s - \rho}{\rho} < 1.114, 305 < N_{Sc} < 1595$	[109]
P. Liquid film flowing over solid particles with air present, trickle bed reactors, fixed bed	$N_{Sh} = \frac{k_L}{aD} = 1.8 N_{Re}^{1/2} N_{Sc}^{1/3}, 0.013 < N_{Re} < 12.6$ two-phases, liquid trickle, no forced flow of gas. $N_{Sh} = 0.8 N_{Re}^{1/2} N_{Sc}^{1/3}$, one-phase, liquid only.	[E] $N_{Re} = \frac{L}{a\mu}$, irregular granules of benzoic acid, $0.29 \le d_p \le 1.45$ cm. L = superficial liquid flow rate, kg/m^{2}s. a = surface area/col. volume, m^2/m^3.	[95]
Q. Supercritical fluids in packed bed	$\frac{N_{Sh}}{(N_{Sc}N_{Gr})^{1/4}} = 0.5265 \left(\frac{N_{Re}^{1/2} N_{Sc}^{1/3}}{(N_{Sc}N_{Gr})^{1/4}}\right)^{1.6808} + 2.48 \left\lvert \left(\frac{N_{Re}^2 N_{Sc}^{1/3}}{N_{Gr}}\right)^{0.6439} - 0.8767 \right\rvert^{1.553}$	[E] Natural and forced convection. $0.3 < N_{Re} < 135$.	[74]
R. Co current gas-liquid flow in fixed beds	Downflow in trickle bed and up flow in bubble columns.	Literature review and meta-analysis. Analyzed both downflow and upflow. Recommendations for best mass and heat-transfer correlations (see reference).	[71]

NOTE: For $N_{Re} < 3$ convective contributions which are not included may become important. Use with logarithmic concentration difference (integrated form) or with arithmetic concentration difference (differential form).

*Author names refer to the General References list at the beginning of the Mass Transfer subsection (pp. 5-41, 5-42). Bracketed numbers refer to the table reference lists (p. 5-42).

TABLE 5-24 Mass-Transfer Correlations for Packed Two-Phase Contactors—Absorption, Distillation, Cooling Towers, and Extractors (Packing Is Inert)

<table>
<tr><th>Situation</th><th>Correlations</th><th>Comments E = Empirical,
S = Semiempirical, T = Theoretical</th><th>References*</th></tr>
<tr><td>A. Absorption, counter-current, liquid-phase coefficient H_L, Sherwood and Holloway correlation for random packings</td><td>

$$H_L = a_L \left(\frac{L}{\mu_L}\right)^n N_{Sc,L}^{0.5}, L = \text{lb/hr ft}^2$$

Ranges for 5-23-B (G and L)

<table>
<tr><th>Packing</th><th>a_G</th><th>b</th><th>c</th><th>G</th><th>L</th><th>a_L</th><th>n</th></tr>
<tr><td colspan="8">Raschig rings</td></tr>
<tr><td>⅜ inch</td><td>2.32</td><td>0.45</td><td>0.47</td><td>200–500</td><td>500–1500</td><td>0.00182</td><td>0.46</td></tr>
<tr><td>1</td><td>7.00</td><td>0.39</td><td>0.58</td><td>200–800</td><td>400–500</td><td>0.010</td><td>0.22</td></tr>
<tr><td>1</td><td>6.41</td><td>0.32</td><td>0.51</td><td>200–600</td><td>500–4500</td><td>—</td><td>—</td></tr>
<tr><td>2</td><td>3.82</td><td>0.41</td><td>0.45</td><td>200–800</td><td>500–4500</td><td>0.0125</td><td>0.22</td></tr>
<tr><td colspan="8">Berl saddles</td></tr>
<tr><td>½ inch</td><td>32.4</td><td>0.30</td><td>0.74</td><td>200–700</td><td>500–1500</td><td>0.0067</td><td>0.28</td></tr>
<tr><td>½</td><td>0.811</td><td>0.30</td><td>0.24</td><td>200–800</td><td>400–4500</td><td>—</td><td>—</td></tr>
<tr><td>1</td><td>1.97</td><td>0.36</td><td>0.40</td><td>200–800</td><td>400–4500</td><td>0.0059</td><td>0.28</td></tr>
<tr><td>1.5</td><td>5.05</td><td>0.32</td><td>0.45</td><td>200–1000</td><td>400–4500</td><td>0.0062</td><td>0.28</td></tr>
</table>

</td><td>[E] From experiments on desorption of sparingly soluble gases from water. Graphs: Sherwood, Pigford, & Wilke, p. 606. Equation is dimensional. Geankoplis states a typical value of n is 0.3 and has constants in kg, m, and s units for use in 5-24A and B with $\hat{k}_G$ in kg mole/sm² and $\hat{k}_L$ in kg mole/sm² (kg mol/m³). Constants for other packings are given by [79, p. 187] and Treybal, p. 239.

$$H_L = \frac{L_M}{\hat{k}_L a}$$

L_M = lbmol/hr ft², $\hat{k}_L$ = lbmol/hr ft², a = ft²/ft³, μ_L = lb/(h ft).
Range for 5-24A is $400 < L < 15{,}000$ lb/h ft²</td><td>[79] p. 187
Sherwood, Pigford, & Wilke, p. 606
[101], [114]
[113]</td></tr>
<tr><td>B. Absorption counter-current, gas-phase coefficient H_G, for random packing</td><td>

$$H_G = \frac{G_M}{\hat{k}_G a} = \frac{a_G (G)^b N_{Sc,v}^{0.5}}{(L)^c}$$

</td><td>[E] Based on ammonia-water-air data in Fellinger's 1941 MIT thesis. Curves: [79], p. 186 and Sherwood, Pigford, & Wilke, p. 607. Constants given in 5-24A. The equation is dimensional. G = lb/hr ft², G_M = lb mol/hr ft², $\hat{k}_G$ = lbmol/hr ft².</td><td>[79] p. 189
Sherwood, Pigford, & Wilke, p. 607
[114]</td></tr>
<tr><td>C. Absorption and and distillation, counter-current, gas and liquid individual coefficients and wetted surface area, Onda et al. correlation for random packings</td><td>

$$\frac{k'_G RT}{a_p D_G} = A\left(\frac{G}{a_p \mu_G}\right)^{0.7} N_{Sc,G}^{1/3} (a_p d'_p)^{-2.0}$$

$$k'_L \left(\frac{\rho_L}{\mu_L g}\right)^{1/3} = 0.0051\left(\frac{L}{a_w \mu_L}\right)^{2/3} N_{Sc,L}^{-1/2} (a_p d'_p)^{0.4}$$

k'_L = lbmol/hr ft² (lbmol/ft³) [kgmol/s m² (kgmol/m³)]

$$\frac{a_w}{a_p} = 1 - \exp\left\{-1.45\left(\frac{\sigma_c}{\sigma}\right)^{0.75}\left(\frac{L}{a_p \mu_L}\right)^{0.1} \times \left(\frac{L^2 a_p}{\rho_L^2 g}\right)^{-0.05}\left(\frac{L}{\rho_L \sigma a_p}\right)^{0.2}\right\}$$

</td><td>[E] Gas absorption and desorption from water and organics plus vaporization of pure liquids for Raschig rings, saddles, spheres, and rods. d'_p = nominal packing size, a_p = dry packing surface area/volume, a_w = wetted packing surface area/volume. Equations are dimensionally consistent, so any set of consistent units can be used.
σ = surface tension, dynes/cm.
A = 5.23 for packing ≥ 1/2 inch (0.012 m)
A = 2.0 for packing < 1/2 inch (0.012 m)
k'_G = lbmol/hr ft² atm [kg mol/s m² (N/m²)]
Critical surface tensions, σ_C = 61 (ceramic), 75 (steel), 33 (polyethylene), 40 (PVC), 56 (carbon) dynes/cm.

$$4 < \frac{L}{a_w \mu_L} < 400,\ 5 < \frac{G}{a_p \mu_G} < 1000$$

Most data ± 20% of correlation, some ± 50%. Graphical comparison with data in [82].</td><td>[31]
[68] p. 380
[82]
Taylor & Krishna, p. 355
[113]</td></tr>
<tr><td>D. Distillation and absorption, counter-current, random packings, modification of Onda correlation, Bravo and Fair correlation to determine interfacial area</td><td>Use Onda's correlations (5-24C) for k'_G and k'_L. Calculate:

$$H_G = \frac{G}{k'_G a_e P M_G}, H_L = \frac{L}{k'_L a_e \rho_L}, H_{OG} = H_G + \lambda H_L$$

$$\lambda = \frac{m}{L_M/G_M}$$

$$a_e = 0.489 a_p \left(\frac{\sigma^{0.5}}{Z^{0.4}}\right)(N_{Ca,L} N_{Re,G})^{0.392}$$

$$N_{Re,G} = \frac{6G}{a_p \mu_G}, N_{Ca,L} = \frac{L\mu_L}{\rho_L \sigma g_c} \text{(dimensionless)}$$

</td><td>[E] Used Bolles & Fair [30] database to determine new effective area a_e to use with Onda et al. [82] correlation. Same definitions as 5-24C. P = total pressure, atm; M_G = gas, molecular weight; m = local slope of equilibrium curve; L_M/G_M = slope operating line; Z = height of packing in feet.
Equation for a_e is dimensional. Fit to data for effective area quite good for distillation. Good for absorption at low values of ($N_{Ca,L} \times N_{Re,G}$), but correlation is too high at higher values of ($N_{Ca,L} \times N_{Re,G}$).</td><td>[31]</td></tr>
<tr><td>E. Absorption and distillation, countercurrent gas-liquid flow, random and structured packing. Determine H_L and H_G</td><td>

$$H_G = \left(\frac{0.226}{f_p}\right)\left(\frac{N_{Sc}}{0.660}\right)^b \left(\frac{G_x}{6.782}\right)^{-0.5}\left(\frac{G_y}{0.678}\right)^{0.35}$$

$$H_L = \left(\frac{0.357}{f_p}\right)\left(\frac{N_{Sc}}{372}\right)^{0.5}\left(\frac{G_x/\mu}{6.782/0.0008937}\right)^{0.3}$$

Relative transfer coefficients [69], f_p values are in table:

<table>
<tr><th>Size, in.</th><th>Ceramic Raschig rings</th><th>Ceramic Berl saddles</th><th>Metal Pall rings</th><th>Metal Intalox</th><th>Metal Hypac</th></tr>
<tr><td>0.5</td><td>1.52</td><td>1.58</td><td>—</td><td>—</td><td>—</td></tr>
<tr><td>1.0</td><td>1.20</td><td>1.36</td><td>1.61</td><td>1.78</td><td>1.51</td></tr>
<tr><td>1.5</td><td>1.00</td><td>—</td><td>1.34</td><td>—</td><td>—</td></tr>
<tr><td>2.0</td><td>0.85</td><td>—</td><td>1.14</td><td>1.27</td><td>1.07</td></tr>
</table>

Norton Intalox structured: 2T, f_p = 1.98; 3T, f_p = 1.94.</td><td>[S] H_G based on NH_3 absorption data (5-24B) for which $H_{G,base}$ = 0.226 m with $N_{Sc,base}$ = 0.660 at $G_{x,base}$ = 6.782 kg/(sm²) and $G_{y,base}$ = 0.678 kg/(sm²) with 1½ in. ceramic Raschig rings. The exponent b on N_{Sc} is reported as either 0.5 or as ⅔.

$$f_p = \frac{H_G \text{ for } NH_3 \text{ with } 1\tfrac{1}{2} \text{ Raschig rings}}{H_G \text{ for } NH_3 \text{ with desired packing}}$$

H_L based on O_2 desorption data (5-24A).
Base viscosity, μ_{base} = 0.0008937 kg/(ms).
H_L in m. $G_y < 0.949$ kg/(sm²), $0.678 < G_x < 6.782$ kg/(sm²).
Best use is for absorption and stripping. Limited use for organic distillation [113].</td><td>Geankoplis, pp. 686, 659
Sherwood, Pigford, & Wilke,
[113]</td></tr>
</table>

TABLE 5-24 Mass-Transfer Correlations for Packed Two-Phase Contactors—Absorption, Distillation, Cooling Towers, and Extractors (Packing Is Inert) *(Continued)*

<table>
<tr><th>Situation</th><th>Correlations</th><th>Comments E = Empirical,
S = Semiempirical, T = Theoretical</th><th>References*</th></tr>
<tr><td>F. Absorption, cocurrent downward flow, random packings, Reiss correlation</td><td>Air-oxygen-water results correlated by $k'_L a = 0.12 E_L^{0.5}$. Extended to other systems.
$k'_L a = 0.12 E_L^{0.5}\left(\frac{D_L}{2.4\times 10^5}\right)^{0.5}, E_L = \left(\frac{\Delta P}{\Delta L}\right)_{\text{2-phase}} \upsilon_L$
$\Delta p/\Delta L$ = 2-phase flow pressure loss, lbf/(ft$^2 \cdot$ ft)
$k'_G a = 2.0 + 0.91 E_G^{2/3}$ for NH_3, $E_g = (\Delta p/\Delta L)_{\text{2-phase}} \upsilon_g$</td><td>[E] Based on oxygen transfer from water to air 77°F. Liquid film resistance controls. (D_{water} @ 77°F = 2.4×10^{-5}). Equation is dimensional. Data was for thin-walled polyethylene Raschig rings. Correlation also fit data for spheres. Fit ±25%. See [91] for graph.
$k'_L a = s^{-1}$
D_L = cm/s, E_L = ft, lbf/s ft^3
v_L = superficial liquid velocity, ft/s; v_g = superficial gas velocity, ft/s
[E] Ammonia absorption into water from air at 70°F. Gas-film resistance controls. Thin-walled polyethylene Raschig rings and 1-inch Intalox saddles. Fit ±25%. See [91] for fit. Terms defined as above.</td><td>[91]
[95] p. 217

[91]</td></tr>
<tr><td>G. Absorption, stripping, distillation, counter-current, H_L, and H_G, random packings, Bolles and Fair correlation</td><td>For Raschig rings, Berl saddles, and spiral tile:
$H_L = \frac{\phi C_{\text{flood}}}{3.28} N_{\text{Sc},L}^{0.5}\left(\frac{Z}{3.05}\right)^{0.15}$
C_{flood} = 1.0 if below 40% flood—otherwise, use figure in [41] or [114].
$H_G = \frac{A\psi(d'_{\text{col}})^m Z^{0.33} N_{\text{Sc},G}^{0.5}}{\left[L\left(\frac{\mu L}{\mu_{\text{water}}}\right)^{0.16}\left(\frac{\rho_{\text{water}}}{\rho_L}\right)^{1.25}\left(\frac{\sigma_{\text{water}}}{\sigma_L}\right)^{0.8}\right]^n}$ Figures for ϕ and ψ in [42 and 43]. Ranges: $0.02 < \phi < 0.300$; $25 < \psi < 190$ m.</td><td>[E] Z = packed height, m of each section with its own liquid distribution. The original work is reported in English units. Cornell et al. (Ref. 41) review early literature. Improved fit of Cornell's ϕ values given by Bolles and Fair (Refs. [29], [30]) and [114].
A = 0.017 (rings) or 0.029 (saddles)
d'_{col} = column diameter in m (if diameter > 0.6 m, use d'_{col} = 0.6)
m = 1.24 (rings) or 1.11 (saddles)
n = 0.6 (rings) or 0.5 (saddles)
L = liquid rate, kg/(sm^2), μ_{water} = 1.0 Pa.s, ρ_{water} = 1000 kg/m^3, σ_{water} = 72.8 mN/m (72.8 dyn/cm).
H_G and H_L will vary from location to location. Design each section of packing separately.</td><td>[29, 30, 41]
[57] p. 428
[68] p. 381
Skelland, p. 353
[114] [113]</td></tr>
<tr><td>H. Distillation and absorption. Counter-current flow. Structured packings. Gauze-type with triangular flow channels, Bravo, Rocha, and Fair correlation</td><td>Equivalent channel:
$d_{eq} = Bh\left[\frac{1}{B+2S} + \frac{1}{2S}\right]$
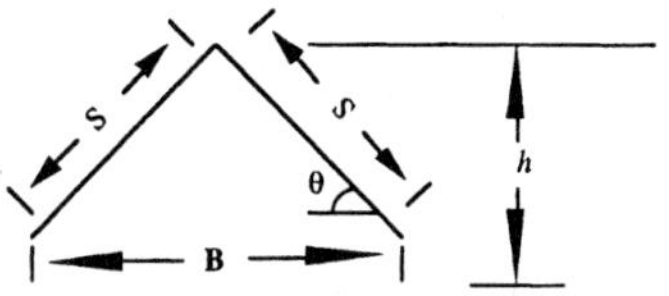

Use modified correlation for wetted wall column (See 5-18E)
$N_{\text{Sh},\upsilon} = \frac{k'_\upsilon d_{\text{eq}}}{D_\upsilon} = 0.0338 N_{\text{Re},\upsilon}^{0.8} N_{\text{Sc},\upsilon}^{0.333}$
$N_{\text{Re},\upsilon} = \frac{d_{\text{eq}}\rho_\upsilon(U_{\upsilon,\text{eff}} + U_{L,\text{eff}})}{\mu_\upsilon}$
Calculate k'_L from penetration model (use time for liquid to flow distance s).
$k'_L = 2(D_L U_{L,\text{eff}}/\pi S)^{1/2}$.</td><td>[T] Check of 132 data points showed average deviation 14.6% from theory. Johnstone and Pigford [Ref. 63] correlation (5-18E) has exponent on N_{Re} rounded to 0.8. Assume gauze packing is completely wet. Thus, $a_{\text{eff}} = a_p$ to calculate H_G and H_L. Same approach may be used generally applicable to sheet-metal packings, but they will not be completely wet and need to estimate transfer area. Fit to data shown in Ref. [32]. L = liquid flux, kg/s m^2, G = vapor flux, kg/s m^2.
$H_G = \frac{G}{k'_\upsilon a_p \rho_\upsilon}, H_L = \frac{L}{k'_L a_p \rho_L}$
effective velocities
$U_{\upsilon,\text{eff}} = \frac{U_{\upsilon,\text{super}}}{\varepsilon \sin\theta}, U_{L,\text{eff}} = \frac{3\Gamma}{2\rho_L}\left(\frac{\rho_L^2 g}{3\mu_L \Gamma}\right)^{0.333} \Gamma = \frac{L}{\text{Per}}$
$\text{Per} = \frac{\text{Perimeter}}{\text{Area}} = \frac{4S + 2B}{Bh}$</td><td>[32]
[47] pp. 310, 326
Taylor & Krishna, pp. 356, 362
[113]</td></tr>
<tr><td>I. Distillation and absorption, counter-current flow. Structured packing with corrugations. Rocha, Bravo, and Fair correlation.</td><td>$N_{\text{Sh},G} = \frac{k_g S}{D_g} = 0.054\, N_{\text{Re}}^{0.8} N_{\text{Sc}}^{0.33}$
$u_{\upsilon,\text{eff}} = \frac{u_{g,\text{super}}}{\varepsilon(1-h_L)\sin\theta}, u_{L,\text{eff}} = \frac{u_{\text{liq,super}}}{\varepsilon h_L \sin\theta},$
$k_L = 2\left(\frac{D_L C_E u_{L,\text{eff}}}{\pi S}\right)$
$H_{OG} = H_G + \lambda H_L = \frac{u_{g,\text{super}}}{k_g a_e} + \frac{\lambda u_{L,\text{super}}}{k_L a_e}$
Interfacial area:
$\frac{a_e}{a_p} = F_{\text{SE}} \frac{29.12(N_{we} N_{Fr})^{0.15} S^{0.359}}{N_{\text{Re},L}^{0.2} \varepsilon^{0.6} (1 - 0.93 \cos\gamma)(\sin\theta)^{0.3}}$
Packing factors:
<table><tr><th></th><th>a_p</th><th>ε</th><th>F_{SE}</th><th>θ</th></tr><tr><td>Flexi-pac 2</td><td>233</td><td>0.95</td><td>0.350</td><td>45°</td></tr><tr><td>Gempak 2A</td><td>233</td><td>0.95</td><td>0.344</td><td>45°</td></tr><tr><td>Intalox 2T</td><td>213</td><td>0.95</td><td>0.415</td><td>45°</td></tr><tr><td>Mellapak 350Y</td><td>350</td><td>0.93</td><td>0.350</td><td>45°</td></tr></table></td><td>[E, T] Modification of Bravo, Rocha, and Fair (5-24H). Same definitions as in (5-24H) unless defined differently here. Recommended by [113].
h_L = fractional hold-up of liquid
C_E = factor for slow surface renewal
$C_E \sim 0.9$
a_e = effective area/volume (1/m)
a_p = packing surface area/volume (1/m)
F_{SE} = surface enhancement factor
γ = contact angle; for sheet metal, $\cos\gamma = 0.9$ for $\sigma < 0.055$ N/m
$\cos\gamma = 5.211 \times 10^{-16.8356}$, $\sigma > 0.055$ N/m
$\lambda = \frac{m}{L/V}, m = \frac{dy}{dx}$ from equilibrium</td><td>[92], [113]</td></tr>
</table>

TABLE 5-24 Mass-Transfer Correlations for Packed Two-Phase Contactors—Absorption, Distillation, Cooling Towers, and Extractors (Packing Is Inert) *(Continued)*

Situation	Correlations	Comments E = Empirical, S = Semiempirical, T = Theoretical	References*
J. Rotating packed bed (Higee)	$\frac{k_L a d_p}{D a_p}\left(1-0.93\frac{V_o}{V_t}-1.13\frac{V_i}{V_t}\right)=0.65\,N_{Sc}^{0.5}$ $\times\left(\frac{L}{a_p\mu}\right)^{0.17}\left(\frac{d_p^3\rho^2 a_c}{\mu^2}\right)^{0.3}\left(\frac{L^2}{\rho a_p\sigma}\right)^{0.\text{[illegible]}}$ $500 \le N_{Sc} \le 1.2\text{ E}5$; $0.0023 \le L/(a_p\mu) \le 8.7$ $120 \le (d_p^3\rho_{a_c}^2)/\mu^2 \le 7.0\text{ E}7$; $3.7\text{ E}-6 \le \text{L}^2/(\rho a_p\sigma) \le 9.4\text{ E}-4$ $9.12 \le \frac{k_L a\, d_p}{D a_p} \le 2540$	[E] Studied oxygen desorption from water into N_2. Packing 0.22-mm-diameter stainless-steel mesh $\varepsilon = 0.954$, $a_p = 829$ (1/m), $h_{bed} = 2$ cm a = gas-liquid area/vol (1/m) L = liquid mass flux, kg/(m²s) a_c = centrifugal accel, m²/s V_i, V_o, V_t = volumes inside inner radius, between outer radius and housing, and total, respectively, m³. Coefficient (0.3) on centrifugal acceleration agrees with literature values (0.3–0.38).	[37]
K. High-voidage packings, cooling towers, splash-grid packings	$\frac{(Ka)_H V_{tower}}{L} = 0.07 + A'N'\left(\frac{L}{G_a}\right)^{-n'}$ A' and n' depend on deck type [65], $0.060 \le A' \le 0.135$, $0.46 \le n' \le 0.62$. General form fits the graphical comparisons Sherwood, Pigford, & Wilke, p. 286).	[E] General form. G_a = lb dry air/hr ft². L = lb/h ft², N' = number of deck levels. $(Ka)_H$ = overall enthalpy transfer coefficient = lb/(h)(ft³)$\left(\frac{\text{lb water}}{\text{lb dry air}}\right)$ V_{tower} = tower volume, ft³/ft². If normal packings are used, use absorption mass-transfer correlations.	[65] [79], p. 220 Sherwood, Pigford, & Wilke, p. 286

See also Sec. 14.

*Author names refer to the General References list at the beginning of the Mass Transfer subsection (pp. 5-41, 5-42). Bracketed numbers refer to the table reference lists (p. 5-42).

correlations include transfer to or from one fluid to either a second fluid or a solid. Many of the correlations are for k_L and k_G values obtained from dilute systems where $x_{BM} \approx 1.0$ and $y_{BM} \approx 1.0$. Each table is for a specific geometry or type of contactor, starting with flat plates (Table 5-17); then wetted wall columns (Table 5-18); flow in pipes and ducts (Table 5-19); submerged objects (Table 5-20); drops and bubbles (Table 5-21); agitated systems (Table 5-22); packed beds of particles for adsorption, ion exchange, and chemical reaction (Table 5-23); and finishing with packed bed two-phase contactors for distillation, absorption, and other unit operations (Table 5-24). For simple geometries, one may be able to determine a theoretical (T) form of the mass-transfer correlation. For very complex geometries, only an empirical (E) form can be found. In systems of intermediate complexity, semiempirical (S) correlations in which the form is determined from theory and the coefficients from experiment are often useful. Although the major limitations and constraints in use are usually included in the tables, obviously many details cannot be included in this summary form. Readers are strongly encouraged to check the references including the original paper before using the correlations in important situations. Note that even authoritative sources occasionally have typographical errors in the fairly complex correlation equations. The original papers will often include figures comparing the correlations with data.

Although extensive, these tables are not meant to be encyclopedic, and other sources such as Skelland, who extensively surveys older mass-transfer correlations, and Benitez, who surveys more recent correlations, should also be consulted. The extensive review of bubble column systems (see Table 5-21) by Shah et al. [*AIChE J.* **28:** 353 (1982)] includes estimation of bubble size, gas holdup, interfacial area $k_L a$, and the liquid dispersion coefficient. For correlations for particle-liquid mass transfer in stirred tanks (part of Table 5-22) see the review by Pangarkar et al. [*Ind. Eng. Chem. Res.* **41:** 4141 (2002)]. Mass-transfer correlations for membrane separators are reviewed by Sirkar [103]. For mass transfer in distillation, absorption, and extraction in packed beds (Table 5-24), see also the appropriate sections in this handbook and the review by Wang, Yuan, and Yu [*Ind. Eng. Chem. Res.* **44:** 8715 (2005)]. Mass transfer and interfacial area for absorption in packed beds for the specific problem of postcombustion carbon dioxide capture are reviewed by Mirzaei, Shamiri, and Aroua [*Rev. Chem. Engr.* **31:** 521 (2015)].

Since often several correlations are applicable, how does one choose the correlation to use? First, the engineer must determine which correlations are closest to the current situation. This involves recognizing the similarity of geometries, which is often challenging, and checking that the range of parameters in the correlation is appropriate. For example, the Bravo, Rocha, and Fair correlation for distillation with structured packings with triangular cross-sectional channels (Table 5-24H) uses the Johnstone and Pigford correlation for rectification in vertical wetted wall columns (Table 5-18E). Recognizing that this latter correlation pertains to a rather different application and geometry was a nontrivial step in the process of developing a correlation. If several correlations appear to be applicable, check to see if the correlations have been compared to one another and to the data. When a detailed comparison of correlations is not available, the following heuristics may be useful:

1. Mass-transfer coefficients are derived from models. They must be employed in a similar model. For example, if an arithmetic concentration difference was used to determine k, that k should only be used in a mass-transfer expression with an arithmetic concentration difference.
2. Semiempirical correlations are often preferred to purely empirical or purely theoretical correlations. Purely empirical correlations are dangerous to use for extrapolation. Purely theoretical correlations may predict trends accurately, but they can be several orders of magnitude off in the value of k.
3. Correlations with broader databases are often preferred.
4. The analogy between heat and mass transfer holds over wider ranges than the analogy between mass and momentum transfer. Good heat-transfer data (without radiation) can often be used to predict mass-transfer coefficients.
5. More recent data are often preferred to older data, since end effects are better understood, the new correlation often builds on earlier data and analysis, and better measurement techniques are often available.
6. With complicated geometries, the product of the interfacial area per volume and the mass-transfer coefficient is required. Correlations of ka_p or of HTU are more accurate than individual correlations of k and a_p since the measurements are simpler to determine the product ka_p or HTU.
7. Finally, if a mass-transfer coefficient looks too good to be true, it probably is incorrect.

Volumetric Mass-Transfer Coefficients $\hat{K}_G a$ and $\hat{K}_L a$ Experimental determinations of the individual mass-transfer coefficients k_G and k_L and of the effective interfacial area a involve the use of extremely difficult techniques, and therefore such data are not plentiful. More often, column experimental data are reported in terms of overall volumetric coefficients, which normally are defined as follows:

$$K'_G a = n_A/(h_T S_{pT} \Delta y^{\circ}_{1m}) \tag{5-308}$$

and

$$K'_L a = n_A/(h_T S\, \Delta x^{\circ}_{1m}) \tag{5-309}$$

where $K'_G a$ = overall volumetric gas-phase mass-transfer coefficient, $K'_L a$ = overall volumetric liquid-phase mass-transfer coefficient, n_A = overall rate of transfer of solute A, h_T = total packed depth in tower, S = tower cross-sectional area, p_T = total system pressure employed during the experiment, and Δx°_{1m} and Δy°_{1m} are defined as

$$\Delta y^{*}_{1m} = \frac{(y-y^*)_1-(y-y^*)_2}{\ln[(y-y^*)_1/(y-y^*)_2]} \tag{5-310}$$

and

$$\Delta x^{*}_{1m} = \frac{(x^*-x)_2-(x^*-x)_2}{\ln[(x^*-x)_2/(x^*-x)_1]} \tag{5-311}$$

where subscripts 1 and 2 refer to the bottom and top of the tower, respectively.

Experimental $K'_G a$ and $K'_L a$ data are available for most absorption and stripping operations of commercial interest (see Table 5-24 and Sec. 14). The solute concentrations employed in these experiments normally are very low, so that $K'_L a \doteq \hat{K}_L a$ and $K'_G a p_T \doteq \hat{K}_G a$, where p_T is the total pressure employed in the actual experimental-test system. Unlike the individual gas-film coefficient $\hat{k}_G a$, the overall coefficient $\hat{K}_G a$ will vary with the total system pressure except when the liquid-phase resistance is negligible (i.e., when either $m = 0$ or $\hat{K}_L a$ is very large, or both).

Extrapolation of $K'_G a$ data for absorption and stripping to conditions other than those for which the original measurements were made can be extremely risky, especially in systems involving chemical reactions in the liquid phase. One therefore would be wise to restrict the use of overall volumetric mass-transfer coefficient data to conditions not too far removed from those employed in the actual tests. The most reliable data for this purpose would be those obtained from an operating commercial unit of similar design.

Experimental values of H_{OG} and H_{OL} for a number of distillation systems of commercial interest are also readily available (e.g., see Table 5-24). Extrapolation of the data or the correlations to conditions that differ significantly from those used for the original experiments is risky. For example, pressure has a major effect on vapor density and thus can affect the hydrodynamics significantly. Changes in flow patterns affect both mass-transfer coefficients and interfacial area.

Analogies Analogies have been important in the study of mass transfer since Fick modeled his analysis of mass transfer on Fourier's analysis of heat transfer. If the underlying mechanisms for heat, mass, and momentum transfer are identical (e.g., transfer by eddies in turbulent flow), analogies are useful. If the underlying mechanisms are different (e.g., radiation in heat transfer), analogies do not apply. Reynolds developed an analogy (see Cussler for details) that is most commonly applied to turbulent flow in tubes (Table 5-19R)

$$\frac{k'}{v} = \frac{h}{\rho c_p v} = \frac{f}{2} \tag{5-312}$$

where h is the heat-transfer coefficient, c_p is the heat capacity, f is the Fanning friction factor, and v is a characteristic velocity. Since the Reynolds analogy is of limited utility, improved analogies for flow in tubes were developed by Prandtl (Table 5-19T) and Von Karman (Table 5-19U).

Chilton and Colburn [38] developed an empirical analogy that provided a better fit of experimental data. The general form of their analogy is

$$\frac{k'}{v}(N_{Sc})^{2/3} = \frac{h}{\rho c_p v}(N_{Pr})^{2/3} = \frac{f}{2} \quad \text{or} \quad j_D = j_H = f/2 \tag{5-313}$$

Specific applications are included in Tables 5-17A, 5-17E, and 5-19S.

The Chilton-Colburn analogy [38, 40, 68] is frequently used to develop estimates of the mass-transfer rates based on heat-transfer data. Extrapolation of experimental j_M or j_H data obtained with gases to predict liquid systems (and vice versa) should be approached with caution, however. When pressure-drop or friction-factor data are available, one may be able to place an upper bound on the rates of heat and mass transfer of $f/2$. In distillation columns there are more mass-transfer data than heat-transfer data, and the Chilton-Colburn analogy is used to estimate heat-transfer rates. The Chilton-Colburn analogy can be used for simultaneous heat and mass transfer as long as the concentration and temperature fields are independent [Venkatesan and Fogler, *AIChE J.* **50:** 1623 (2004)].

Effects of System Physical Properties on $\hat{k}_G$ and $\hat{k}_L$ When one is designing packed towers for nonreacting gas-absorption systems for which no experimental data are available, it is necessary to make corrections for differences in composition between the existing test data and the system in question. For example, ammonia-water test data (see Table 5-24B) can be used to estimate H_G, and the oxygen desorption data (see Table 5-24A) can be used to estimate H_L. The method for doing this is illustrated in Table 5-24E. There is some conflict on whether the value of the exponent for the Schmidt number is 0.5 or 2/3 [Yadav and Sharma, *Chem. Eng. Sci.* **34:** 1423 (1979)]. Despite this disagreement, this method is extremely useful, especially for absorption and stripping systems. If one is in doubt about the exponent, we recommend using 2/3, the value used in the Chilton-Colburn analogy.

Note that the influence of substituting solvents of widely differing viscosities upon the interfacial area a can be very large. One therefore should be cautious about extrapolating $\hat{k}_L a$ data to account for viscosity effects between different solvent systems.

Influence of Chemical Reactions on $\hat{k}_G$ and $\hat{k}_L$ When a chemical reaction occurs, the transfer rate may be influenced by the chemical reaction as well as by the purely physical processes of diffusion and convection within the two phases. Since this situation is common in gas absorption, gas absorption will be the focus of this discussion. One must consider the impacts of chemical equilibrium and reaction kinetics on the absorption rate in addition to accounting for the effects of gas solubility, diffusivity, and system hydrodynamics.

There is no sharp dividing line between pure physical absorption and absorption controlled by the rate of a chemical reaction. Most cases fall in an intermediate range in which the rate of absorption is limited both by the resistance to diffusion and by the finite velocity of the reaction. Even in these intermediate cases the equilibria between the various diffusing species involved in the reaction may affect the rate of absorption.

The gas-phase rate coefficient $\hat{k}_G$ is not affected by chemical reactions taking place in the liquid phase. If the liquid-phase chemical reaction is extremely fast and irreversible, the rate of absorption may be governed completely by the resistance to diffusion in the gas phase. In this case the absorption rate may be estimated by knowing only the gas-phase rate coefficient $\hat{k}_G$ or else the height of 1 gas-phase transfer unit $H_G = G_M/(\hat{k}_G a)$.

Note that the highest possible absorption rates will occur under conditions in which the liquid-phase resistance is negligible and the equilibrium back pressure of the gas over the solvent is zero. Such situations would exist, for instance, for NH_3 absorption into an acid solution, for SO_2 absorption into an alkali solution, for vaporization of water into air, and for H_2S absorption from a dilute-gas stream into a strong alkali solution, provided there is a large excess of reagent in solution to consume all the dissolved gas. This is known as the *gas-phase mass-transfer limited condition,* when both the liquid-phase resistance and the back pressure of the gas equal zero. Even when the reaction is sufficiently reversible to allow a small back pressure, the absorption may be gas-phase-controlled, and the values of $\hat{k}_G$ and H_G that would apply to a physical absorption process will govern the rate.

The liquid-phase rate coefficient $\hat{k}_L$ is strongly affected by fast chemical reactions and generally increases with increasing reaction rate. Indeed, the condition for zero liquid-phase resistance $(m/\hat{k}_L)$ implies that either the equilibrium back pressure is negligible or $\hat{k}_L$ is very large, or both. Frequently, even though reaction consumes the solute as it is dissolving, thereby enhancing both the mass-transfer coefficient and the driving force for absorption, the reaction rate is slow enough that the liquid-phase resistance must be taken into account. This may be due either to an insufficient supply of a second reagent or to an inherently slow chemical reaction.

In any event the value of $\hat{k}_L$ in the presence of a chemical reaction normally is larger than the value found when only physical absorption occurs, $\hat{k}_L^0$. This has led to the presentation of data on the effects of chemical reaction in terms of the *reaction factor* or *enhancement factor,* defined as

$$\phi = \hat{k}_L / \hat{k}_L^0 \geq 1 \tag{5-314}$$

where $\hat{k}_L$ = mass-transfer coefficient with reaction and $\hat{k}_L^0$ = mass-transfer coefficient for pure physical absorption. It is important to understand that when chemical reactions are involved, this definition of $\hat{k}_L$ is based on the driving force, defined as the difference between the concentration of *unreacted* solute gas at the interface and in the bulk of the liquid. A coefficient based on the total of both unreacted and reacted gas could have values *smaller* than the physical absorption mass-transfer coefficient $\hat{k}_L^0$.

When liquid-phase resistance is important, particular care should be taken in employing any given set of experimental data to ensure that the equilibrium data used conform with those employed by the original author in calculating values of $\hat{k}_L$ or H_L. Extrapolation to widely different concentration ranges or operating conditions should be made with caution, since the mass-transfer coefficient $\hat{k}_L$ may vary in an unexpected fashion, owing to changes in the apparent chemical reaction mechanism.

Generalized prediction methods for $\hat{k}_L$ and H_L do not apply when chemical reaction occurs in the liquid phase, and therefore one must use actual operating data for the particular system in question. A discussion of the various factors to consider in designing gas absorbers and strippers when chemical reactions are involved is presented by Astarita, Savage, and Bisio, *Gas Treating with Chemical Solvents,* Wiley, New York, 1983 and by Kohl and Nielsen, *Gas Purification,* 5th ed., Gulf Publishing, Houston, Tex., 1997.

Effective Interfacial Mass-Transfer Area a To determine the mass-transfer rate, one needs the interfacial area in addition to the mass-transfer coefficient. In a packed tower of constant cross-sectional area S the differential change in solute flow per unit time is given by

$$-d(G_M S y) = N_A a\, dV = N_A a S\, dh \tag{5-315}$$

where a = interfacial area effective for mass transfer per unit of packed volume and V = packed volume. Owing to incomplete wetting of the packing surfaces and to the formation of areas of stagnation in the liquid film, the effective area normally is significantly less than the total external area of the packing pieces.

For packed beds of particles, a can be estimated as shown in Table 5-23A. For packed beds in distillation, absorption, and so on in Table 5-24, the interfacial area per volume is usually included with the mass-transfer coefficient in the correlations for HTU. For agitated liquid-liquid systems, the interfacial area can be estimated from the dispersed phase holdup and mean drop size correlations. Godfrey, Obi, and Reeve [*Chem. Engr. Prog.* **85:** 61 (Dec. 1989)] summarize these correlations. For many systems,

$$\bar{d}_{\text{drop}}/d_{\text{imp}} = (\text{const})N_{\text{We}}^{-0.6} \quad \text{where} \quad N_{\text{We}} = \rho_c N^2 d_{\text{imp}}^3/\sigma \tag{5-316}$$

Piché, Grandjean, and Larachi [*Ind. Eng. Chem. Res.* **41:** 4911 (2002)] developed two correlations for reconciling the gas-liquid mass-transfer coefficient and interfacial area in randomly packed towers. The correlation for the interfacial area was a function of five dimensionless groups, and it yielded a relative error of 22.5 percent for 325 data points. That equation, when combined with a correlation for N_{Sh} as a function of four dimensionless groups, achieved a relative error of 24.4 percent, for 3455 data points for the product $k'_G a$.

The effective interfacial area depends on a number of factors, as discussed in a review by Charpentier [*Chem. Eng. J.* **11:** 161 (1976)]. Among these factors are (1) the shape and size of packing, (2) the packing material (for example, plastic generally gives smaller interfacial areas than either metal or ceramic), (3) the liquid mass velocity, and (4) for small-diameter towers, the column diameter.

Whereas the interfacial area generally increases with increasing liquid rate, it apparently is relatively independent of the superficial gas mass velocity below the flooding point. According to Charpentier's review, it appears valid to assume that the interfacial area is independent of the column height when specified in terms of unit packed volume (i.e., as a). Also, the existing data for chemically reacting gas-liquid systems (mostly aqueous electrolyte solutions) indicate that the interfacial area is independent of the chemical system. However, this situation may not hold true for systems involving large heats of reaction.

Rizzuti et al. [*Chem. Eng. Sci.* **36:** 973 (1981)] examined the influence of solvent viscosity upon the effective interfacial area in packed columns and concluded that for the systems studied the effective interfacial area a was proportional to the kinematic viscosity raised to the 0.7 power. Thus, the hydrodynamic behavior of a packed absorber is strongly affected by viscosity effects. Surface-tension effects also are important, as expressed in the work of Onda et al. [82] (see Table 5-24C).

Concluding Comment In developing correlations for the mass-transfer coefficients $\hat{k}_G$ and $\hat{k}_L$, various authors have assumed different but internally compatible correlations for the effective interfacial area a. It therefore would be inappropriate to mix the correlations of different authors unless it has been demonstrated that there is a valid area of overlap.

Section 6

Fluid and Particle Dynamics

James N. Tilton, Ph.D., P.E. *DuPont Fellow, Chemical and Bioprocess Engineering, E. I. du Pont de Nemours & Co.; Member, American Institute of Chemical Engineers; Registered Professional Engineer (Delaware)*

FLUID DYNAMICS

PARTICLE DYNAMICS

Nomenclature and Units*

In this listing, symbols used in this section are defined in a general way and appropriate SI units are given. Specific definitions, as denoted by subscripts, are stated at the place of application in the section. Some specialized symbols used in the section are defined only at the place of application. Some symbols have more than one definition; the appropriate one is identified at the place of application.

Symbol	Definition	SI units	U.S. Customary System units
a	Pressure wave velocity	m/s	ft/s
A	Area	m^2	ft^2
b	Wall thickness	m	in
b	Channel width	m	ft
c	Acoustic velocity	m/s	ft/s
c_f	Friction coefficient	Dimensionless	Dimensionless
C	Conductance	m^3/s	ft^3/s
Ca	Capillary number	Dimensionless	Dimensionless
C_0	Discharge coefficient	Dimensionless	Dimensionless
C_D	Drag coefficient	Dimensionless	Dimensionless
d	Diameter	m	ft
D	Diameter	m	ft
De	Dean number	Dimensionless	Dimensionless
E	Elastic modulus	Pa	lbf/in^2
Eo	Eotvos number	Dimensionless	Dimensionless
f	Fanning friction factor	Dimensionless	Dimensionless
f	Vortex shedding frequency	1/s	1/s
F	Force	N	lbf
F	Cumulative residence time distribution	Dimensionless	Dimensionless
Fr	Froude number	Dimensionless	Dimensionless
g	Acceleration of gravity	m/s^2	ft/s^2
G	Mass flux	$kg/(m^2 \cdot s)$	$lbm/(ft^2 \cdot s)$
h	Enthalpy per unit mass	J/kg	Btu/lbm
h	Liquid depth	m	ft
k	Ratio of specific heats	Dimensionless	Dimensionless
k	Kinetic energy of turbulence	J/kg	ft · lbf/lbm
K	Power law coefficient	$kg/(m \cdot s^{2-n})$	$lbm/(ft \cdot s^{2-n})$
l_v	Viscous losses per unit mass	J/kg	ft · lbf/lbm
L	Length	m	ft
$\dot{m}$	Mass flow rate	kg/s	lbm/s
M	Mass	kg	lbm
M	Mach number	Dimensionless	Dimensionless
M	Morton number	Dimensionless	Dimensionless
M_w	Molecular weight	kg/kgmol	lbm/lbmol
n	Power law exponent	Dimensionless	Dimensionless
N_b	Blend time number	Dimensionless	Dimensionless
N_P	Power number	Dimensionless	Dimensionless
N_Q	Pumping number	Dimensionless	Dimensionless
p	Pressure	Pa	lbf/in^2
q	Entrained flow rate	m^3/s	ft^3/s
Q	Volumetric flow rate	m^3/s	ft^3/s
Q	Throughput (vacuum flow)	$Pa \cdot m^3/s$	$(lbf/in^2) \cdot ft^3/s$
δQ	Heat input per unit mass	J/kg	Btu/lbm
r	Radial coordinate	m	ft
R	Radius	m	ft
R	Ideal gas universal constant	J/(kgmol · K)	Btu/(lbmol · °R)
R_i	Volume fraction of phase *i*	Dimensionless	Dimensionless
Re	Reynolds number	Dimensionless	Dimensionless
s	Density ratio	Dimensionless	Dimensionless
s	Entropy per unit mass	J/(kg · K)	Btu/(lbm · °R)
S	Slope	Dimensionless	Dimensionless
S	Pumping speed	m^3/s	ft^3/s
S	Surface area per unit volume	1/m	1/ft
Sr	Strouhal number	Dimensionless	Dimensionless
t	Time	s	s
t	Force per unit area	Pa	lbf/in^2
T	Absolute temperature	K	°R
u	Internal energy per unit mass	J/kg	Btu/lbm
u	Velocity	m/s	ft/s
U	Velocity	m/s	ft/s
υ	Velocity	m/s	ft/s
V	Velocity	m/s	ft/s
V	Volume	m^3	ft^3
We	Weber number	Dimensionless	Dimensionless
$\dot{W}_s$	Rate of shaft work	J/s	Btu/s
δW_s	Shaft work per unit mass	J/kg	Btu/lbm
x	Cartesian coordinate	m	ft
y	Cartesian coordinate	m	ft
z	Cartesian coordinate	m	ft
Z	Elevation	m	ft
	Greek Symbols		
α	Velocity profile factor	Dimensionless	Dimensionless
α	Included angle	rad	rad
β	Velocity profile factor	Dimensionless	Dimensionless
β	Bulk modulus of elasticity	Pa	lbf/in^2
$\dot{\gamma}$	Shear rate	1/s	1/s
Γ	Mass flow rate	kg/(m · s)	lbm/(ft · s)
Γ_{ij}	Components of rate of deformation tensor	1/s	1/s
Γ	Magnitude of rate of deformation tensor	1/s	1/s
δ	Boundary layer or film thickness	m	ft
δ_{ij}	Kronecker delta	Dimensionless	Dimensionless
ε	Pipe roughness	m	ft
ε	Void fraction	Dimensionless	Dimensionless
ε	Turbulent dissipation rate	J/(kg · s)	ft · lbf/(lbm · s)
θ	Residence time	s	S
θ	Angle	rad	rad
λ	Mean free path	m	ft
μ	Viscosity	Pa · s	lbm/(ft · s)
ν	Kinematic viscosity	m^2/s	ft^2/s
ρ	Density	kg/m^3	lbm/ft^3
σ	Surface tension	N/m	lbf/ft
σ	Cavitation number	Dimensionless	Dimensionless
σ_{ij}	Components of total stress tensor	Pa	lbf/in^2
τ	Shear stress	Pa	lbf/in^2
τ	Time period	s	s
τ_{ij}	Components of deviatoric stress tensor	Pa	lbf/in^2
Φ	Energy dissipation rate per unit volume	$J/(m^3 \cdot s)$	$ft \cdot lbf/(ft^3 \cdot s)$
ϕ	Angle of inclination	rad	rad
ω	Vorticity	1/s	1/s

*Note that with U.S. Customary System units, the conversion factor g_c may be required to make equations in this section dimensionally consistent; $g_c = 32.17\ (lbm \cdot ft)/(lbf \cdot s^2)$.

FLUID DYNAMICS

General References: Batchelor, G. K., *An Introduction to Fluid Dynamics*, Cambridge University, Cambridge, UK, 1967; Bird, R. B., W. E. Stewart, and E. N. Lightfoot, *Transport Phenomena*, 2d ed., Wiley, New York, 2002; Brodkey, R. S., *The Phenomena of Fluid Motions*, Addison-Wesley, Reading, Mass., 1967; Denn, M. M., *Process Fluid Mechanics*, Prentice-Hall, Englewood Cliffs, N.J., 1979; Govier, G. W., and K. Aziz, *The Flow of Complex Mixtures in Pipes*, Krieger, Huntington, N.Y., 1977; Landau, L. D., and E. M. Lifshitz, *Fluid Mechanics*, 2d ed., Pergamon, Oxford, 1987; Panton, R. L., *Incompressible Flow*, Wiley, New York, 1984; Schlichting, H., and K. Gersten, *Boundary Layer Theory*, 8th ed rev., Springer-Verlag, Berlin, 2003; Shames, I. H., *Mechanics of Fluids*, 3d ed., McGraw-Hill, New York, 1992; Streeter, V. L., *Handbook of Fluid Dynamics*, McGraw-Hill, New York, 1971; Streeter, V. L., and E. B. Wylie, *Fluid Mechanics*, 8th ed., McGraw-Hill, New York, 1985; Vennard, J. F., and R. L. Street, *Elementary Fluid Mechanics*, 5th ed., Wiley, New York, 1975; Whitaker, S., *Introduction to Fluid Mechanics*, Krieger, Malabar, Fla., 1981.

NATURE OF FLUIDS

Deformation and Stress A *fluid* is a substance that undergoes continuous deformation when subjected to a shear stress, as illustrated in Fig. 6-1. A fluid is bounded by two large parallel plates, of area A, separated by a small distance H. The bottom plate is held fixed. Application of a force F to the upper plate causes it to move at velocity V. The fluid continues to deform as long as the force is applied, unlike a solid, which would undergo only a finite deformation.

The force per unit area is the shear stress $\tau = F/A$. Within the fluid, a linear velocity profile $u = Vy/H$ is established; because of the *no-slip condition*, the fluid bounding the lower plate has zero velocity and the fluid bounding the upper plate moves at the plate velocity V. The velocity gradient $\dot{\gamma} = du/dy$ is the *shear rate* for this flow. Shear rates are usually reported in units of reciprocal seconds. The flow in Fig. 6-1 is a *simple shear flow*.

Viscosity The ratio of shear stress to shear rate is the viscosity μ.

$$\mu = \frac{\tau}{\dot{\gamma}} \tag{6-1}$$

The SI units of viscosity are kg/(m · s) or Pa · s (pascal-seconds). The cgs unit for viscosity is the poise (P); 1 Pa · s = 10 P = 1000 centipoise (cP) or 0.672 lbm/(ft · s). The terms *absolute viscosity* and *shear viscosity* are synonymous with the viscosity as used in Eq. (6-1). *Kinematic viscosity* $\nu \equiv \mu/\rho$ is the ratio of viscosity to density. The SI units of kinematic viscosity are m^2/s. The cgs unit stoke (St) = 1 cm^2/s.

Rheology In general, fluid flow patterns are more complex than the one shown in Fig. 6-1, as is the relationship between fluid deformation and stress. Rheology is the discipline of fluid mechanics which studies this relationship. One goal of rheology is to obtain *constitutive equations* by which stresses may be computed. For simplicity, fluids may be classified into rheological types in reference to the simple shear flow of Fig. 6-1. Complete definitions require extension to multidimensional flow. For more information, several good references are available, including R. B. Bird, R. C. Armstrong, and O. Hassager [*Dynamics of Polymeric Liquids*, vol. 1: *Fluid Mechanics*, Wiley, New York, 1977]; H. A. Barnes, J. F. Hutton, and K. Walters [*An Introduction to Rheology*, Elsevier, Amsterdam, 1989], C. W. Macosko [*Rheology Principles, Measurements and Applications*, Wiley-VCH, New York, 1994], and F. A. Morrison [*Understanding Rheology*, Oxford University, Oxford, 2001].

Fluids without elasticity do not undergo any reverse deformation when shear stress is removed, and they are called *purely viscous* fluids. The shear stress depends only on the rate of deformation, and not on the extent of deformation (strain). Those that exhibit both viscous and elastic properties are called *viscoelastic* fluids.

Purely viscous fluids are further classified into time-independent and time-dependent fluids. For time-independent fluids, the shear stress depends on only the instantaneous shear rate. The shear stress for time-dependent fluids depends on the history of the rate of deformation, as a result of structure or orientation buildup or breakdown during deformation.

A *rheogram* is a plot of shear stress versus shear rate for a fluid in simple shear flow. Rheograms for several types of time-independent fluids are shown in Fig. 6-2. The *newtonian* fluid rheogram is a straight line passing through the origin. The slope of the line is the viscosity. For a newtonian fluid, the viscosity is independent of shear rate and depends on temperature and perhaps pressure. By far, the newtonian fluid is the largest class of fluid of engineering importance. Gases and low-molecular-weight liquids are generally newtonian. Newton's law of viscosity is a rearrangement of Eq. (6-1) in which the viscosity is a constant:

$$\tau = \mu\dot{\gamma} = \mu\frac{du}{dy} \tag{6-2}$$

Fluids for which the viscosity varies with shear rate are called *nonnewtonian fluids*. For nonnewtonian fluids the viscosity, defined as the ratio of shear stress to shear rate, is often called the *apparent viscosity* to emphasize the distinction from newtonian behavior. Purely viscous, time-independent fluids, for which the apparent viscosity may be expressed as a function of shear rate, are called *generalized newtonian fluids*.

Nonnewtonian fluids include those for which a finite stress τ_y is required before continuous deformation occurs; these are called *yield-stress* materials. The *Bingham plastic* fluid is the simplest yield-stress material; its rheogram has a constant slope μ_∞, called the *infinite shear* viscosity.

$$\tau = \tau_y + \mu_\infty\dot{\gamma} \tag{6-3}$$

Highly concentrated suspensions of fine solid particles frequently exhibit Bingham plastic behavior.

Shear-thinning fluids are those for which the slope of the rheogram decreases with increasing shear rate. These fluids have also been called *pseudoplastic*, but this terminology is outdated. Many polymer melts and solutions, as well as some solids suspensions, are shear-thinning. Shear-thinning fluids without yield stresses are often fit to a power law model over a range of shear rates

$$\tau = K\dot{\gamma}^n \tag{6-4}$$

The apparent viscosity is

$$\mu = K\dot{\gamma}^{n-1} \tag{6-5}$$

The factor K is the consistency index or power law coefficient, and n is the power law exponent. The exponent n is dimensionless, while K has units of $kg/(m \cdot s^{2-n})$. For shear-thinning fluids, $n < 1$. The power law model typically provides a good fit to data over a range of one to two orders of magnitude in shear rate; behavior at very low and very high shear rates is often newtonian. Shear-thinning fluids with yield stresses are often fit to the Herschel-Bulkley model, which adds a yield stress to Eq. (6-4). Numerous other rheological model equations for shear-thinning fluids are in common use.

Dilatant, or shear-thickening, fluids show increasing viscosity with increasing shear rate. Over a limited range of shear rate, they may be fit to the power law model with $n > 1$. Dilatancy is rare, observed only in certain concentration ranges in some particle suspensions [G. W. Govier and K. Aziz, *The Flow of Complex Mixtures in Pipes*, Krieger, Huntington, N.Y., 1977, pp. 33–34]. Extensive discussions of dilatant suspensions, together with a listing of dilatant systems, are given by R. G. Green and R. G. Griskey [*Trans. Soc. Rheol.* **12**(1): 13–25 (1968)]; R. G. Griskey and R. G. Green [*AIChE J.* **17**: 725–728 (1971)]; and W. H. Bauer and E. A. Collins ["Thixotropy and Dilatancy," in F. R. Eirich, *Rheology*, vol. 4, Academic, New York, 1967].

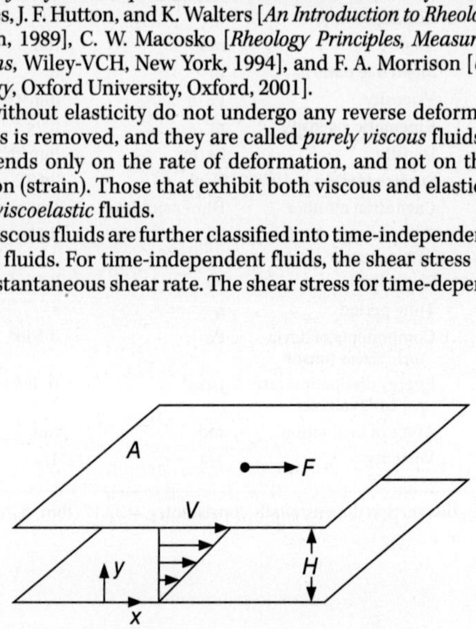

FIG. 6-1 Deformation of a fluid subjected to a shear stress.

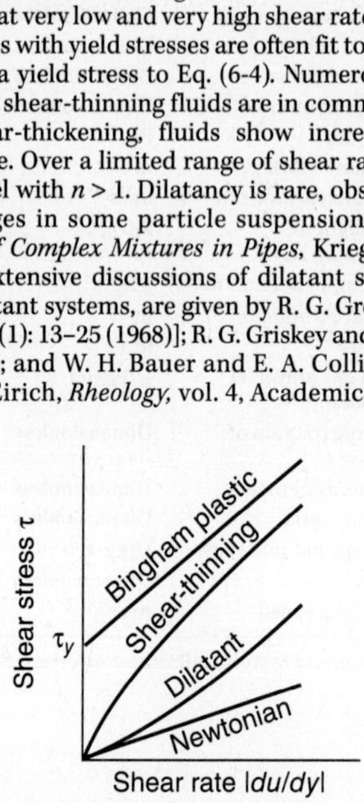

FIG. 6-2 Shear diagrams.

Time-dependent fluids are those for which structural rearrangements occur during deformation at a rate too slow to maintain equilibrium configurations. As a result, shear stress changes with duration of shear. *Thixotropic* fluids, such as mayonnaise, clay suspensions used as drilling muds, and some paints and inks, starting from rest, show decreasing shear stress with time at constant shear rate. A detailed description of thixotropic behavior and a list of thixotropic systems are found in W. H. Bauer and E. A. Collins ["Thixotropy and Dilatancy," in Eirich, *Rheology,* vol. 4, Academic, New York, 1967].

Rheopectic behavior is the opposite of thixotropy. Starting from rest, shear stress increases with time at a constant shear rate. Rheopectic behavior has been observed in bentonite sols, vanadium pentoxide sols, and gypsum suspensions in water [W. H. Bauer and E. A. Collins, "Thixotropy and Dilatancy," in Eirich, *Rheology,* vol. 4, Academic, New York, 1967] as well as in some polyester solutions [I. Steg and D. Katz, *J. Appl. Polym. Sci.* **9:** 3, 177 (1965)].

Viscoelastic fluids exhibit elastic recovery from deformation when stress is removed. Polymeric liquids comprise the largest group of fluids in this class. A property of viscoelastic fluids is the *relaxation time,* which is a measure of the time required for elastic effects to decay. Viscoelastic effects may be important with sudden changes in rates of deformation, as in flow startup and stop, rapidly oscillating flows, or flow through sudden expansions or contractions. In viscoelastic flows, normal stresses perpendicular to the direction of shear are different from those in the parallel direction. These give rise to such behaviors as the *Weissenberg effect,* in which fluid climbs up a rotating shaft, and *die swell,* where a stream of fluid issuing from a tube may expand to two or more times the tube diameter.

Analysis of viscoelastic flows is very difficult. Simple constitutive equations are unable to describe all the material behavior exhibited by viscoelastic fluids even in geometrically simple flows. More-complex constitutive equations may be more accurate, but become exceedingly difficult to apply, especially for complex geometries, even with advanced numerical methods. For good discussions of viscoelastic fluid behavior, including various types of constitutive equations, see R. B. Bird, R. C. Armstrong, and O. Hassager [*Dynamics of Polymeric Liquids,* vol. 1: *Fluid Mechanics,* Wiley, New York, 1977]; S. Middleman [*The Flow of High Polymers,* Interscience (Wiley), New York, 1968]; or G. Astarita and G. Marrucci [*Principles of Nonnewtonian Fluid Mechanics,* McGraw-Hill, New York, 1974].

Polymer processing depends heavily on the flow of nonnewtonian fluids. See the texts by S. Middleman [*Fundamentals of Polymer Processing,* McGraw-Hill, New York, 1977] and Z. Tadmor and C. Gogos [*Principles of Polymer Processing,* Wiley, New York, 1979].

There are a wide variety of instruments for measurement of newtonian viscosity as well as rheological properties of nonnewtonian fluids. They are described in C. W. Macosko [*Rheology Principles, Measurements and Applications,* Wiley-VCH, New York, 1994]; B. D. Coleman et al. [*Viscometric Flows of Nonnewtonian Fluids,* Springer-Verlag, Berlin, 1966]; and J. M. Dealy and K. F. Wissbrun [*Melt Rheology and Its Role in Plastics Processing,* Van Nostrand Reinhold, New York, 1990]. Measurement of rheological behavior requires well-characterized flows. Such *rheometric* flows are thoroughly discussed by G. Astarita and G. Marrucci [*Principles of Nonnewtonian Fluid Mechanics,* McGraw-Hill, New York, 1974].

KINEMATICS OF FLUID FLOW

Velocity *Kinematics* refers to the quantitative description of fluid motion or deformation. The rate of deformation depends on the distribution of velocity within the fluid. Fluid velocity **v** is a vector quantity, with three cartesian components υ_x, υ_y, and υ_z. The velocity vector is a function of spatial position and time. In a *steady* flow, the velocity is independent of time, while in *unsteady* flow **v** varies with time.

Streamlines, Pathlines, and Streaklines These are curves in a flow field which provide insight into the flow pattern. *Streamlines* are tangent at every point to the local instantaneous velocity vector. A *pathline* is the path followed by a material element of fluid. *Streaklines* are curves on which are found all the material particles that passed through a particular point in space at some earlier time. For example, a streakline is revealed by releasing smoke or dye at a point in a flow field. For steady flows, the streamlines, pathlines, and streaklines coincide. In two-dimensional incompressible flows, streamlines are contours of the *stream function.*

Many flows of practical importance, such as those in pipes and channels, are treated as *one-dimensional flows.* There is a single direction called the *flow direction;* velocity components perpendicular to this direction either are zero or are considered unimportant. In one type of one-dimensional flow, variations of quantities such as velocity, pressure, density, and temperature are considered only in the flow direction. More generally, one-dimensional flows have only one nonzero velocity component, which depends on only one coordinate direction, and this coordinate direction may or may not be the same as the flow direction.

Rate of Deformation Tensor For general three-dimensional flows, where all three velocity components may be important and may vary in all three coordinate directions, the concept of deformation previously introduced is generalized using the rate of deformation tensor $\Gamma = (1/2)[\nabla \mathbf{v} + (\nabla \mathbf{v})^T]$. In cartesian components,

$$\Gamma_{ij} = \frac{1}{2}\left(\frac{\partial v_i}{\partial x_j} + \frac{\partial v_j}{\partial x_i}\right) \tag{6-6}$$

where the subscripts *i* and *j* refer to the three coordinate directions. Some authors define the deformation rate tensor as twice that given by Eq. (6-6). For multidimensional flows of incompressible newtonian fluids, Eq. (6-2) may be generalized to

$$\tau = 2\mu\Gamma \tag{6-7}$$

with τ_{ij} the nine components of the viscous stress tensor. For generalized newtonian fluids in multidimensional flow, μ is a function of a scalar measure of the rate of deformation $\mu = \mu(\Gamma)$, where

$$\Gamma = \sqrt{\frac{1}{2}\Gamma:\Gamma} = \sqrt{\frac{1}{2}\sum_i\sum_j \Gamma_{ij}\Gamma_{ji}} \tag{6-8}$$

The components of the rate of deformation tensor and equations for the scalar Γ, in cartesian, cylindrical, and spherical coordinates, are found in Table 6-1. The table also provides the viscous stress components of generalized newtonian fluids and the differential balance equations for mass and momentum (see below) in the three coordinate systems.

Vorticity The relative motion between two points in a fluid can be decomposed into rigid body rotation and deformation. The rate of deformation tensor has been defined. Deformation includes uniform volumetric expansion (dilatation) and shear. Dilatation vanishes for incompressible flow. Rotation is described in cartesian coordinates by the vorticity tensor $\omega_{ij} = (1/2)[\partial\upsilon_i/\partial_{xj} - \partial\upsilon_j/\partial x_i]$. A related quantity is the vector of vorticity given by one-half the curl of the velocity. In two-dimensional flow in the *xy* plane, the vorticity ω is given by

$$\omega = \frac{1}{2}\left(\frac{\partial v_y}{\partial x} - \frac{\partial v_x}{\partial y}\right) \tag{6-9}$$

Here ω is the magnitude of the vorticity vector, which is directed along the *z* axis. An *irrotational* flow is one with zero vorticity. Irrotational flows have been widely studied because of their useful mathematical properties and applicability to flow regions where viscous effects may be neglected (*inviscid* flows).

Laminar and Turbulent Flow These terms refer to two distinct types of flow. In *laminar flow,* there are smooth streamlines and the fluid velocity components vary smoothly with position and time. The flow described in reference to Fig. 6-1 is laminar. In *turbulent flow,* streamlines are irregular, and the velocity fluctuates chaotically in time and space. For any given flow geometry, a dimensionless *Reynolds number* may be defined for a newtonian fluid as $Re = LU\rho/\mu$ where *L* and *U* are the characteristic length and velocity. Below a critical value of Re the flow is laminar, while above the critical value a transition to turbulent flow begins. The geometry-dependent critical Reynolds number is determined experimentally.

CONSERVATION EQUATIONS

Macroscopic and Microscopic Balances Three laws of physics are fundamental in fluid mechanics: *conservation of mass, conservation of momentum,* and *conservation of energy.* In addition, *conservation of moment of momentum* (angular momentum) and the *entropy inequality* (second law of thermodynamics) have occasional use. The momentum, moment of momentum, and energy conservation laws apply to inertial reference frames. Conservation principles may be applied to control volumes which may be of finite or differential size, resulting in either *algebraic* or *differential* conservation equations, respectively. These are often called *macroscopic* and *microscopic* balance equations.

Macroscopic Equations Figure 6-3 shows an arbitrary control volume of finite size V_a bounded by a surface of area A_a with an outwardly directed unit normal vector **n**. The volume is not necessarily fixed in space. Its boundary moves with velocity **w**. The fluid velocity is **v**.

Mass Balance Applied to the control volume, the principle of conservation of mass may be written as

$$\frac{d}{dt}\int_{V_a}\rho\, dV + \int_{A_a}\rho(\mathbf{v}-\mathbf{w})\cdot\mathbf{n}\, dA = 0 \tag{6-10}$$

This equation is also known as the *continuity* equation.

TABLE 6-1 Differential Equations of Motion, Newtonian Stress Constitutive Equation and Rate of Deformation Tensor, in Cartesian, Cylindrical, and Spherical Coordinates

Continuity Equation $\frac{D\rho}{Dt}=-\rho\nabla\cdot\mathbf{v}$

Cartesian

$$\frac{\partial\rho}{\partial t}+v_x\frac{\partial\rho}{\partial x}+v_y\frac{\partial\rho}{\partial y}+v_z\frac{\partial\rho}{\partial z}=-\rho\left(\frac{\partial v_x}{\partial x}+\frac{\partial v_y}{\partial y}+\frac{\partial v_z}{\partial z}\right)$$

Cylindrical

$$\frac{\partial\rho}{\partial t}+v_r\frac{\partial\rho}{\partial r}+\frac{v_\theta}{r}\frac{\partial\rho}{\partial\theta}+v_z\frac{\partial\rho}{\partial z}=-\rho\left[\frac{1}{r}\frac{\partial}{\partial r}(rv_r)+\frac{1}{r}\frac{\partial v_\theta}{\partial\theta}+\frac{\partial v_z}{\partial z}\right]$$

Spherical

$$\frac{\partial\rho}{\partial t}+v_r\frac{\partial\rho}{\partial r}+\frac{v_\theta}{r}\frac{\partial\rho}{\partial\theta}+\frac{v_\phi}{r\sin\theta}\frac{\partial\rho}{\partial\phi}=-\rho\left[\frac{1}{r^2}\frac{\partial}{\partial r}(r^2v_r)+\frac{1}{r\sin\theta}\frac{\partial}{\partial\theta}(v_\theta\sin\theta)+\frac{1}{r\sin\theta}\frac{\partial v_\phi}{\partial\phi}\right]$$

Cauchy Momentum Equation $\rho\frac{D\mathbf{v}}{Dt}=-\nabla p+\nabla\cdot\tau+\rho\mathbf{g}$

Cartesian

$$\rho\left(\frac{\partial v_x}{\partial t}+v_x\frac{\partial v_x}{\partial x}+v_y\frac{\partial v_x}{\partial y}+v_z\frac{\partial v_x}{\partial z}\right)=-\frac{\partial p}{\partial x}+\frac{\partial\tau_{xx}}{\partial x}+\frac{\partial\tau_{yx}}{\partial y}+\frac{\partial\tau_{zx}}{\partial z}+\rho g_x$$

$$\rho\left(\frac{\partial v_y}{\partial t}+v_x\frac{\partial v_y}{\partial x}+v_y\frac{\partial v_y}{\partial y}+v_z\frac{\partial v_y}{\partial z}\right)=-\frac{\partial p}{\partial y}+\frac{\partial\tau_{xy}}{\partial x}+\frac{\partial\tau_{yy}}{\partial y}+\frac{\partial\tau_{zy}}{\partial z}+\rho g_y$$

$$\rho\left(\frac{\partial v_z}{\partial t}+v_x\frac{\partial v_z}{\partial x}+v_y\frac{\partial v_z}{\partial y}+v_z\frac{\partial v_z}{\partial z}\right)=-\frac{\partial p}{\partial z}+\frac{\partial\tau_{xz}}{\partial x}+\frac{\partial\tau_{yz}}{\partial y}+\frac{\partial\tau_{zz}}{\partial z}+\rho g_z$$

Cylindrical

$$\rho\left(\frac{\partial v_r}{\partial t}+v_r\frac{\partial v_r}{\partial r}+\frac{v_\theta}{r}\frac{\partial v_r}{\partial\theta}-\frac{v_\theta^2}{r}+v_z\frac{\partial v_r}{\partial z}\right)=-\frac{\partial p}{\partial r}+\frac{1}{r}\frac{\partial}{\partial r}(r\tau_{rr})+\frac{1}{r}\frac{\partial\tau_{\theta r}}{\partial\theta}-\frac{\tau_{\theta\theta}}{r}+\frac{\partial\tau_{zr}}{\partial z}+\rho g_r$$

$$\rho\left(\frac{\partial v_\theta}{\partial t}+v_r\frac{\partial v_\theta}{\partial r}+\frac{v_\theta}{r}\frac{\partial v_\theta}{\partial\theta}+\frac{v_rv_\theta}{r}+v_z\frac{\partial v_\theta}{\partial z}\right)=-\frac{1}{r}\frac{\partial p}{\partial\theta}+\frac{1}{r^2}\frac{\partial}{\partial r}(r^2\tau_{r\theta})+\frac{1}{r}\frac{\partial\tau_{\theta\theta}}{\partial\theta}+\frac{\partial\tau_{z\theta}}{\partial z}+\rho g_\theta$$

$$\rho\left(\frac{\partial v_z}{\partial t}+v_r\frac{\partial v_z}{\partial r}+\frac{v_\theta}{r}\frac{\partial v_z}{\partial\theta}+v_z\frac{\partial v_z}{\partial z}\right)=-\frac{\partial p}{\partial z}+\frac{1}{r}\frac{\partial}{\partial r}(r\tau_{rz})+\frac{1}{r}\frac{\partial\tau_{\theta z}}{\partial\theta}+\frac{\partial\tau_{zz}}{\partial z}+\rho g_z$$

Spherical

$$\rho\left(\frac{\partial v_r}{\partial t}+v_r\frac{\partial v_r}{\partial r}+\frac{v_\theta}{r}\frac{\partial v_r}{\partial\theta}+\frac{v_\phi}{r\sin\theta}\frac{\partial v_r}{\partial\phi}-\frac{v_\theta^2+v_\phi^2}{r}\right)=-\frac{\partial p}{\partial r}+\frac{1}{r^2}\frac{\partial}{\partial r}(r^2\tau_{rr})$$
$$+\frac{1}{r\sin\theta}\frac{\partial}{\partial\theta}(\tau_{\theta r}\sin\theta)+\frac{1}{r\sin\theta}\frac{\partial\tau_{\phi r}}{\partial\phi}-\frac{\tau_{\theta\theta}+\tau_{\phi\phi}}{r}+\rho g_r$$

$$\rho\left(\frac{\partial v_\theta}{\partial t}+v_r\frac{\partial v_\theta}{\partial r}+\frac{v_\theta}{r}\frac{\partial v_\theta}{\partial\theta}+\frac{v_\phi}{r\sin\theta}\frac{\partial v_\theta}{\partial\phi}+\frac{v_rv_\theta}{r}-\frac{v_\phi^2\cot\theta}{r}\right)=-\frac{1}{r}\frac{\partial p}{\partial\theta}+\frac{1}{r^2}\frac{\partial}{\partial r}(r^2\tau_{r\theta})$$
$$+\frac{1}{r\sin\theta}\frac{\partial}{\partial\theta}(\tau_{\theta\theta}\sin\theta)+\frac{1}{r\sin\theta}\frac{\partial\tau_{\phi\theta}}{\partial\phi}+\frac{\tau_{r\theta}}{r}-\frac{\cot\theta}{r}\tau_{\phi\phi}+\rho g_\theta$$

$$\rho\left(\frac{\partial v_\phi}{\partial t}+v_r\frac{\partial v_\phi}{\partial r}+\frac{v_\theta}{r}\frac{\partial v_\phi}{\partial\theta}+\frac{v_\phi}{r\sin\theta}\frac{\partial v_\phi}{\partial\phi}+\frac{v_\phi v_r}{r}+\frac{v_\theta v_\phi}{r}\cot\theta\right)=-\frac{1}{r\sin\theta}\frac{\partial p}{\partial\phi}$$
$$+\frac{1}{r^2}\frac{\partial}{\partial r}(r^2\tau_{r\phi})+\frac{1}{r}\frac{\partial\tau_{\theta\phi}}{\partial\theta}+\frac{1}{r\sin\theta}\frac{\partial\tau_{\phi\phi}}{\partial\phi}+\frac{\tau_{r\phi}}{r}+\frac{2\cot\theta}{r}\tau_{\theta\phi}+\rho g_\phi$$

Stress Constitutive Equation for Newtonian and Generalized Newtonian Fluids

$$\tau=2\mu\Gamma-\frac{2}{3}\mu(\nabla\cdot\mathbf{v})\delta=\mu[\nabla\mathbf{v}+(\nabla\mathbf{v})^T]-\frac{2}{3}\mu(\nabla\cdot\mathbf{v})\delta$$

Cartesian

$$\tau_{xx}=\mu\left[2\frac{\partial v_x}{\partial x}-\frac{2}{3}\nabla\cdot\mathbf{v}\right]\quad\tau_{xy}=\tau_{yx}=\mu\left[\frac{\partial v_x}{\partial y}+\frac{\partial v_y}{\partial x}\right]\quad\tau_{xz}=\tau_{zx}=\mu\left[\frac{\partial v_x}{\partial z}+\frac{\partial v_z}{\partial x}\right]$$

$$\tau_{yy}=\mu\left[2\frac{\partial v_y}{\partial y}-\frac{2}{3}\nabla\cdot\mathbf{v}\right]\quad\tau_{zy}=\tau_{yz}=\mu\left[\frac{\partial v_z}{\partial y}+\frac{\partial v_y}{\partial z}\right]\quad\tau_{zz}=\mu\left[2\frac{\partial v_z}{\partial z}-\frac{2}{3}\nabla\cdot\mathbf{v}\right]$$

$$\nabla\cdot\mathbf{v}=\frac{\partial v_x}{\partial x}+\frac{\partial v_y}{\partial y}+\frac{\partial v_z}{\partial z}$$

Cylindrical

$$\tau_{rr}=\mu\left[2\frac{\partial v_r}{\partial r}-\frac{2}{3}\nabla\cdot\mathbf{v}\right]\quad\tau_{r\theta}=\tau_{\theta r}=\mu\left[r\frac{\partial}{\partial r}\left(\frac{v_\theta}{r}\right)+\frac{1}{r}\frac{\partial v_r}{\partial\theta}\right]\quad\tau_{rz}=\tau_{zr}=\mu\left[\frac{\partial v_r}{\partial z}+\frac{\partial v_z}{\partial r}\right]$$

$$\tau_{\theta\theta}=\mu\left[2\left(\frac{1}{r}\frac{\partial v_\theta}{\partial\theta}+\frac{v_r}{r}\right)-\frac{2}{3}\nabla\cdot\mathbf{v}\right]\quad\tau_{z\theta}=\tau_{\theta z}=\mu\left[\frac{1}{r}\frac{\partial v_z}{\partial\theta}+\frac{\partial v_\theta}{\partial z}\right]\quad\tau_{zz}=\mu\left[2\frac{\partial v_z}{\partial z}-\frac{2}{3}\nabla\cdot\mathbf{v}\right]$$

$$\nabla\cdot\mathbf{v}=\frac{1}{r}\frac{\partial}{\partial r}(rv_r)+\frac{1}{r}\frac{\partial v_\theta}{\partial\theta}+\frac{\partial v_z}{\partial z}$$

Spherical

$$\tau_{rr}=\mu\left[2\frac{\partial v_r}{\partial r}-\frac{2}{3}\nabla\cdot\mathbf{v}\right]\quad\tau_{r\theta}=\tau_{\theta r}=\mu\left[r\frac{\partial}{\partial r}\left(\frac{v_\theta}{r}\right)+\frac{1}{r}\frac{\partial v_r}{\partial\theta}\right]$$

$$\tau_{r\phi}=\tau_{\phi r}=\mu\left[\frac{1}{r\sin\theta}\frac{\partial v_r}{\partial\phi}+r\frac{\partial}{\partial r}\left(\frac{v_\phi}{r}\right)\right]$$

$$\tau_{\theta\theta}=\mu\left[2\left(\frac{1}{r}\frac{\partial v_\theta}{\partial\theta}+\frac{v_r}{r}\right)-\frac{2}{3}\nabla\cdot\mathbf{v}\right]\quad\tau_{\phi\theta}=\tau_{\theta\phi}=\mu\left[\frac{1}{r\sin\theta}\frac{\partial v_\theta}{\partial\phi}+\frac{\sin\theta}{r}\frac{\partial}{\partial\theta}\left(\frac{v_\phi}{\sin\theta}\right)\right]$$

$$\tau_{\phi\phi}=\mu\left[2\left(\frac{1}{r\sin\theta}\frac{\partial v_\phi}{\partial\phi}+\frac{v_r}{r}+\frac{v_\theta\cot\theta}{r}\right)-\frac{2}{3}\nabla\cdot\mathbf{v}\right]$$

$$\nabla\cdot\mathbf{v}=\frac{1}{r^2}\frac{\partial}{\partial r}(r^2v_r)+\frac{1}{r\sin\theta}\frac{\partial}{\partial\theta}(v_\theta\sin\theta)+\frac{1}{r\sin\theta}\frac{\partial v_\phi}{\partial\phi}$$

Navier-Stokes Equation $\rho\frac{D\mathbf{v}}{Dt}=-\nabla p+\mu\nabla^2\mathbf{v}+\rho\mathbf{g}$

Cartesian

$$\rho\left(\frac{\partial v_x}{\partial t}+v_x\frac{\partial v_x}{\partial x}+v_y\frac{\partial v_x}{\partial y}+v_z\frac{\partial v_x}{\partial z}\right)=-\frac{\partial p}{\partial x}+\mu\left(\frac{\partial^2v_x}{\partial x^2}+\frac{\partial^2v_x}{\partial y^2}+\frac{\partial^2v_x}{\partial z^2}\right)+\rho g_x$$

$$\rho\left(\frac{\partial v_y}{\partial t}+v_x\frac{\partial v_y}{\partial x}+v_y\frac{\partial v_y}{\partial y}+v_z\frac{\partial v_y}{\partial z}\right)=-\frac{\partial p}{\partial y}+\mu\left(\frac{\partial^2v_y}{\partial x^2}+\frac{\partial^2v_y}{\partial y^2}+\frac{\partial^2v_y}{\partial z^2}\right)+\rho g_y$$

$$\rho\left(\frac{\partial v_z}{\partial t}+v_x\frac{\partial v_z}{\partial x}+v_y\frac{\partial v_z}{\partial y}+v_z\frac{\partial v_z}{\partial z}\right)=-\frac{\partial p}{\partial z}+\mu\left(\frac{\partial^2v_z}{\partial x^2}+\frac{\partial^2v_z}{\partial y^2}+\frac{\partial^2v_z}{\partial z^2}\right)+\rho g_z$$

Cylindrical

$$\rho\left(\frac{\partial v_r}{\partial t}+v_r\frac{\partial v_r}{\partial r}+\frac{v_\theta}{r}\frac{\partial v_r}{\partial\theta}-\frac{v_\theta^2}{r}+v_z\frac{\partial v_r}{\partial z}\right)=-\frac{\partial p}{\partial r}$$
$$+\mu\left[\frac{\partial}{\partial r}\left(\frac{1}{r}\frac{\partial}{\partial r}(rv_r)\right)+\frac{1}{r^2}\frac{\partial^2v_r}{\partial\theta^2}-\frac{2}{r^2}\frac{\partial v_\theta}{\partial\theta}+\frac{\partial^2v_r}{\partial z^2}\right]+\rho g_r$$

$$\rho\left(\frac{\partial v_\theta}{\partial t}+v_r\frac{\partial v_\theta}{\partial r}+\frac{v_\theta}{r}\frac{\partial v_\theta}{\partial\theta}+\frac{v_rv_\theta}{r}+v_z\frac{\partial v_\theta}{\partial z}\right)=-\frac{1}{r}\frac{\partial p}{\partial\theta}$$
$$+\mu\left[\frac{\partial}{\partial r}\left(\frac{1}{r}\frac{\partial}{\partial r}(rv_\theta)\right)+\frac{1}{r^2}\frac{\partial^2v_\theta}{\partial\theta^2}+\frac{2}{r^2}\frac{\partial v_r}{\partial\theta}+\frac{\partial^2v_\theta}{\partial z^2}\right]+\rho g_\theta$$

$$\rho\left(\frac{\partial v_z}{\partial t}+v_r\frac{\partial v_z}{\partial r}+\frac{v_\theta}{r}\frac{\partial v_z}{\partial\theta}+v_z\frac{\partial v_z}{\partial z}\right)=-\frac{\partial p}{\partial z}$$
$$+\mu\left[\frac{1}{r}\frac{\partial}{\partial r}\left(r\frac{\partial v_z}{\partial r}\right)+\frac{1}{r^2}\frac{\partial^2v_z}{\partial\theta^2}+\frac{\partial^2v_z}{\partial z^2}\right]+\rho g_z$$

Spherical

$$\rho\left(\frac{\partial v_r}{\partial t}+v_r\frac{\partial v_r}{\partial r}+\frac{v_\theta}{r}\frac{\partial v_r}{\partial\theta}+\frac{v_\phi}{r\sin\theta}\frac{\partial v_r}{\partial\phi}-\frac{v_\theta^2+v_\phi^2}{r}\right)=-\frac{\partial p}{\partial r}$$
$$+\mu\left[\nabla^2v_r-\frac{2v_r}{r^2}-\frac{2}{r^2}\frac{\partial v_\theta}{\partial\theta}-\frac{2}{r^2}v_\theta\cot\theta-\frac{2}{r^2\sin\theta}\frac{\partial v_\phi}{\partial\phi}\right]+\rho g_r$$

$$\rho\left(\frac{\partial v_\theta}{\partial t}+v_r\frac{\partial v_\theta}{\partial r}+\frac{v_\theta}{r}\frac{\partial v_\theta}{\partial\theta}+\frac{v_\phi}{r\sin\theta}\frac{\partial v_\theta}{\partial\phi}+\frac{v_rv_\theta}{r}-\frac{v_\phi^2\cot\theta}{r}\right)=-\frac{1}{r}\frac{\partial p}{\partial\theta}$$
$$+\mu\left[\nabla^2v_\theta+\frac{2}{r^2}\frac{\partial v_r}{\partial\theta}-\frac{v_\theta}{r^2\sin^2\theta}-\frac{2\cos\theta}{r^2\sin^2\theta}\frac{\partial v_\phi}{\partial\phi}\right]+\rho g_\theta$$

$$\rho\left(\frac{\partial v_\phi}{\partial t}+v_r\frac{\partial v_\phi}{\partial r}+\frac{v_\theta}{r}\frac{\partial v_\phi}{\partial\theta}+\frac{v_\phi}{r\sin\theta}\frac{\partial v_\phi}{\partial\phi}+\frac{v_\phi v_r}{r}+\frac{v_\theta v_\phi}{r}\cot\theta\right)=-\frac{1}{r\sin\theta}\frac{\partial p}{\partial\phi}$$
$$+\mu\left[\nabla^2v_\phi-\frac{v_\phi}{r^2\sin^2\theta}+\frac{2}{r^2\sin\theta}\frac{\partial v_r}{\partial\phi}+\frac{2\cos\theta}{r^2\sin^2\theta}\frac{\partial v_\theta}{\partial\phi}\right]+\rho g_\phi$$

$$\nabla^2=\frac{1}{r^2}\frac{\partial}{\partial r}\left(r^2\frac{\partial}{\partial r}\right)+\frac{1}{r^2\sin\theta}\frac{\partial}{\partial\theta}\left(\sin\theta\frac{\partial}{\partial\theta}\right)+\frac{1}{r^2\sin^2\theta}\frac{\partial^2}{\partial\phi^2}$$

TABLE 6-1 Differential Equations of Motion, Newtonian Stress Constitutive Equation and Rate of Deformation Tensor, in Cartesian, Cylindrical, and Spherical Coordinates *(Continued)*

Rate of deformation tensor $\boldsymbol{\Gamma} = (1/2)[\nabla \mathbf{v} + (\Delta \mathbf{v})^T]$ and its magnitude $\Gamma = \sqrt{\frac{1}{2}\boldsymbol{\Gamma}:\boldsymbol{\Gamma}}$.
The rate of deformation is $\dot{\gamma} = 2\Gamma$.

Cartesian

$$\Gamma_{xx} = \frac{\partial v_x}{\partial x} \qquad \Gamma_{xy} = \Gamma_{yx} = \frac{1}{2}\left(\frac{\partial v_x}{\partial y} + \frac{\partial v_y}{\partial x}\right) \qquad \Gamma_{xz} = \Gamma_{zx} = \frac{1}{2}\left(\frac{\partial v_x}{\partial z} + \frac{\partial v_z}{\partial x}\right)$$

$$\Gamma_{yy} = \frac{\partial v_y}{\partial y} \qquad \Gamma_{zy} = \Gamma_{yz} = \frac{1}{2}\left(\frac{\partial v_z}{\partial y} + \frac{\partial v_y}{\partial z}\right) \qquad \Gamma_{zz} = \frac{\partial v_z}{\partial z}$$

$$(2\Gamma)^2 = 2\left[\left(\frac{\partial v_x}{\partial x}\right)^2 + \left(\frac{\partial v_y}{\partial y}\right)^2 + \left(\frac{\partial v_z}{\partial z}\right)^2\right] + \left[\frac{\partial v_y}{\partial x} + \frac{\partial v_x}{\partial y}\right]^2 + \left[\frac{\partial v_z}{\partial y} + \frac{\partial v_y}{\partial z}\right]^2 + \left[\frac{\partial v_x}{\partial z} + \frac{\partial v_z}{\partial x}\right]^2$$

Cylindrical

$$\Gamma_{rr} = \frac{\partial v_r}{\partial r} \qquad \Gamma_{r\theta} = \Gamma_{\theta r} = \frac{1}{2}\left[r\frac{\partial}{\partial r}\left(\frac{v_\theta}{r}\right) + \frac{1}{r}\frac{\partial v_r}{\partial \theta}\right] \qquad \Gamma_{rz} = \Gamma_{zr} = \frac{1}{2}\left[\frac{\partial v_r}{\partial z} + \frac{\partial v_z}{\partial r}\right]$$

$$\Gamma_{\theta\theta} = \frac{1}{r}\frac{\partial v_\theta}{\partial \theta} + \frac{v_r}{r} \qquad \Gamma_{z\theta} = \Gamma_{\theta z} = \frac{1}{2}\left[\frac{1}{r}\frac{\partial v_z}{\partial \theta} + \frac{\partial v_\theta}{\partial z}\right] \qquad \Gamma_{zz} = \frac{\partial v_z}{\partial z}$$

$$(2\Gamma)^2 = 2\left[\left(\frac{\partial v_r}{\partial r}\right)^2 + \left(\frac{1}{r}\frac{\partial v_\theta}{\partial \theta} + \frac{v_r}{r}\right)^2 + \left(\frac{\partial v_z}{\partial z}\right)^2\right] + \left[r\frac{\partial}{\partial r}\left(\frac{v_\theta}{r}\right) + \frac{1}{r}\frac{\partial v_r}{\partial \theta}\right]^2 + \left[\frac{1}{r}\frac{\partial v_z}{\partial \theta} + \frac{\partial v_\theta}{\partial z}\right]^2 + \left[\frac{\partial v_r}{\partial z} + \frac{\partial v_z}{\partial r}\right]^2$$

Spherical

$$\Gamma_{rr} = \frac{\partial v_r}{\partial r} \qquad \Gamma_{r\theta} = \Gamma_{\theta r} = \frac{1}{2}\left[r\frac{\partial}{\partial r}\left(\frac{v_\theta}{r}\right) + \frac{1}{r}\frac{\partial v_r}{\partial \theta}\right] \qquad \Gamma_{r\phi} = \Gamma_{\phi r} = \frac{1}{2}\left[\frac{1}{r\sin\theta}\frac{\partial v_r}{\partial \phi} + r\frac{\partial}{\partial r}\left(\frac{v_\phi}{r}\right)\right]$$

$$\Gamma_{\theta\theta} = \frac{1}{r}\frac{\partial v_\theta}{\partial \theta} + \frac{v_r}{r} \qquad \Gamma_{\phi\theta} = \Gamma_{\theta\phi} = \frac{1}{2}\left[\frac{1}{r\sin\theta}\frac{\partial v_\theta}{\partial \phi} + \frac{\sin\theta}{r}\frac{\partial}{\partial \theta}\left(\frac{v_\phi}{\sin\theta}\right)\right]$$

$$\Gamma_{\phi\phi} = \frac{1}{r\sin\theta}\frac{\partial v_\phi}{\partial \phi} + \frac{v_r}{r} + \frac{v_\theta \cot\theta}{r}$$

$$(2\Gamma)^2 = 2\left[\left(\frac{\partial v_r}{\partial r}\right)^2 + \left(\frac{1}{r}\frac{\partial v_\theta}{\partial \theta} + \frac{v_r}{r}\right)^2 + \left(\frac{1}{r\sin\theta}\frac{\partial v_\phi}{\partial \phi} + \frac{v_r}{r} + \frac{v_\theta \cot\theta}{r}\right)^2\right] + \left[r\frac{\partial}{\partial r}\left(\frac{v_\theta}{r}\right) + \frac{1}{r}\frac{\partial v_r}{\partial \theta}\right]^2$$
$$+ \left[\frac{\sin\theta}{r}\frac{\partial}{\partial \theta}\left(\frac{v_\phi}{\sin\theta}\right) + \frac{1}{r\sin\theta}\frac{\partial v_\theta}{\partial \phi}\right]^2 + \left[\frac{1}{r\sin\theta}\frac{\partial v_r}{\partial \phi} + r\frac{\partial}{\partial r}\left(\frac{v_\phi}{r}\right)\right]^2$$

The equations for $(2\Gamma)^2$ are taken from Deen, *Analysis of Transport Phenomena,* 2d ed., Oxford University Press, New York, 2012, p. 241.

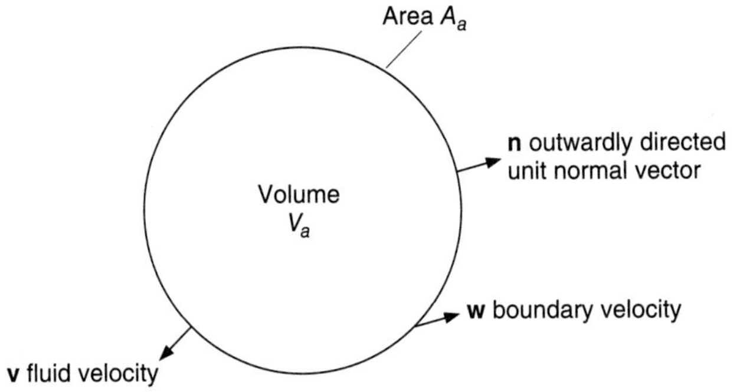

FIG. 6-3 Arbitrary control volume for application of conservation equations.

Simplified forms of Eq. (6-10) apply to special cases frequently found in practice. For a control volume fixed in space with one inlet of area A_1 through which an incompressible fluid enters the control volume at an average velocity V_1, and one outlet of area A_2 through which fluid leaves at an average velocity V_2, as shown in Fig. 6-4, the continuity equation becomes

$$V_1A_1 = V_2A_2 \tag{6-11}$$

The average velocity across a surface is given by

$$V = (1/A)\int_A v\,dA \tag{6-12}$$

where v is the local velocity component perpendicular to the surface. The volumetric flow rate Q is the product of average velocity and the cross-sectional area, $Q = VA$. The average *mass velocity* is $G = \rho V$. For steady flows through fixed control volumes with multiple inlets and/or outlets,

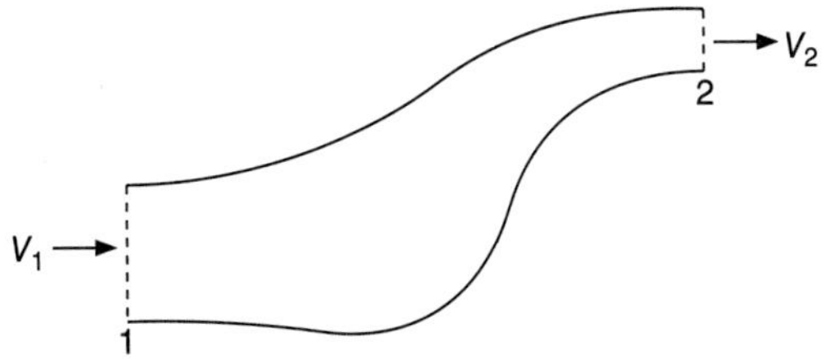

FIG. 6-4 Fixed control volume with one inlet and one outlet.

conservation of mass requires that the sum of inlet mass flow rates equals the sum of outlet mass flow rates. For incompressible flows through fixed control volumes, the sum of inlet flow rates (mass or volumetric) equals the sum of exit flow rates, whether the flow is steady or unsteady.

Momentum Balance Momentum is a vector quantity, and the momentum balance is a vector equation. It (and the energy balance, see below) are written in inertial reference frames. Where gravity is the only body force acting on the fluid, the linear momentum principle, applied to the arbitrary control volume of Fig. 6-3, results in the following expression.

$$\frac{d}{dt}\int_{V_a}\rho\mathbf{v}\,dV + \int_{A_a}\rho\mathbf{v}(\mathbf{v}-\mathbf{w})\cdot\mathbf{n}\,dA = \int_{V_a}\rho\mathbf{g}\,dV + \int_{A_a}\mathbf{n}\cdot\boldsymbol{\sigma}\,dA \tag{6-13}$$

Here $\mathbf{g}$ is the gravity vector and $\mathbf{n}\cdot\boldsymbol{\sigma}$ is the force per unit area exerted by the surroundings on the fluid in the control volume. The total stress tensor σ includes both pressure and viscous stress (see below). For the special case of steady flow at a mass flow rate $\dot{m}$ through a control volume fixed in space with one inlet and one outlet (Fig. 6-4), with the inlet and outlet velocity vectors perpendicular to planar inlet and outlet surfaces, giving average velocity vectors $\mathbf{V}_1$ and $\mathbf{V}_2$, the momentum equation becomes

$$\dot{m}(\beta_2\mathbf{V}_2 - \beta_1\mathbf{V}_1) = -p_1\mathbf{A}_1 - p_2\mathbf{A}_2 + \mathbf{F} + M\mathbf{g} \tag{6-14}$$

where M is the total mass of fluid in the control volume. The factor β arises from the averaging of the velocity across the area of the inlet or outlet surface. It is the ratio of the area average of the square of velocity magnitude to the square of the area average velocity magnitude. For a uniform velocity, $\beta = 1$. For turbulent flow, β is nearly unity, while for laminar pipe flow with a parabolic velocity profile, $\beta = 4/3$. Vectors $\mathbf{A}_1$ and $\mathbf{A}_2$ have magnitude equal to the areas of the inlet and outlet surfaces, respectively, and are outwardly directed normal to the surfaces. Vector $\mathbf{F}$ is the force exerted on the fluid by the nonflow boundaries of the control volume. Viscous contributions to the stress vector $\mathbf{n}\cdot\boldsymbol{\sigma}$ are neglected at the inlet and outlet surfaces, leaving only pressure forces there. Equation (6-14) may be generalized to multiple inlets and/or outlets. In such cases, a distinct flow rate $\dot{m}_i$ applies to each inlet or outlet i. To generalize the equation, $-p\mathbf{A}$ terms for each inlet and outlet, $-\dot{m}_i\beta_i\mathbf{V}_i$ terms for each inlet, and $\dot{m}_i\beta_i\mathbf{V}_i$ terms for each outlet are included.

Balance equations for angular momentum, or moment of momentum, may also be written. They are used less frequently than the linear momentum equations. See S. Whitaker [*Introduction to Fluid Mechanics,* Krieger, Huntington, N.Y., 1981]; J. O. Wilkes [*Fluid Mechanics for Chemical Engineers,* 2d ed., Prentice-Hall, Upper Saddle River, N.J., 2006]; and N. de Nevers [*Fluid Mechanics for Chemical Engineers,* 3d ed., McGraw-Hill, New York, 2005].

Total Energy Balance The total energy balance derives from the first law of thermodynamics. Applied to the arbitrary control volume of Fig. 6-3, it leads to an equation for the rate of change of the sum of internal, kinetic, and gravitational potential energy. Energy input to the control volume comes from work and heat flux at the boundary. The balance also includes energy input by relatively uncommon volumetric sources such as inductive heating, expressed as energy input per unit volume Q_V. In this equation, u is the internal energy per unit mass, υ is the magnitude of the velocity vector $\mathbf{v}$, Z is elevation, g is the gravitational acceleration, and $\mathbf{q}$ is the heat flux vector:

$$\frac{d}{dt}\int_{V_a}\rho\left(u+\frac{v^2}{2}+gZ\right)dV+\int_{A_a}\rho\left(u+\frac{v^2}{2}+gZ\right)(\mathbf{v}-\mathbf{w})\cdot\mathbf{n}\,dA$$
$$=\int_{A_a}(\mathbf{v}\cdot\mathbf{n}\cdot\boldsymbol{\sigma})\,dA-\int_{A_a}(\mathbf{q}\cdot\mathbf{n})\,dA+\int_{V_a}Q_V\,dV \tag{6-15}$$

The first integral on the right-hand side is the rate of work done on the fluid in the control volume by forces at the boundary. It includes both work done by moving solid boundaries and work done at flow entrances and exits. The work done by moving solid boundaries also includes that by surfaces such as pump impellers; this work is called *shaft work*; its rate is $\dot{W}_S$.

A useful simplification of the total energy equation applies to a particular set of assumptions. These are a control volume with fixed solid boundaries, except for those producing shaft work, steady-state conditions, and mass flow at a rate $\dot{m}$ through a single planar entrance and a single planar exit (Fig. 6-4), to which the velocity vectors are perpendicular and $Q_V = 0$. As with Eq. (6-14), viscous contributions to the stress vector $\mathbf{n}\cdot\boldsymbol{\sigma}$ are neglected at the entrance and exit surfaces.

$$h_1+\alpha_1\frac{V_1^2}{2}+gz_1=h_2+\alpha_2\frac{V_2^2}{2}+gz_2-\delta Q-\delta W_S \tag{6-16}$$

Here h is the enthalpy per unit mass, $h = u + p/\rho$. The shaft work per unit of mass flowing through the control volume is $\delta W_S = \dot{W}_S/\dot{m}$. Similarly, δQ is the heat input per unit of mass. The factor α is the ratio of the cross-sectional area average of the cube of the velocity to the cube of the average velocity. For a uniform velocity profile, $\alpha = 1$. In turbulent flow, α is usually assumed to equal unity; in turbulent pipe flow, it is typically about 1.07. For laminar flow in a circular pipe with a parabolic velocity profile, $\alpha = 2$.

Mechanical Energy Balance, Bernoulli Equation A balance equation for the sum of kinetic and potential energy may be obtained from the momentum balance by forming the scalar product with the velocity vector. The resulting equation, called the mechanical energy balance, contains a term accounting for the dissipation of mechanical energy into thermal energy by viscous forces. It is also derivable from the total energy equation in a way that reveals the relationship between the dissipation and entropy generation. The macroscopic mechanical energy balance for the arbitrary control volume of Fig. 6-3 may be written, with p = thermodynamic pressure, as

$$\frac{d}{dt}\int_{V_a}\rho\left(\frac{v^2}{2}+gZ\right)dV+\int_{A_a}\rho\left(\frac{v^2}{2}+gZ\right)(\mathbf{v}-\mathbf{w})\cdot\mathbf{n}\,dA$$
$$=\int_{V_a}p\nabla\cdot\mathbf{v}\,dV+\int_{A_a}(\mathbf{v}\cdot\mathbf{n}\cdot\boldsymbol{\sigma})\,dA-\int_{V_a}\Phi\,dV \tag{6-17}$$

The last term is the rate of viscous energy dissipation to internal energy, also called the rate of viscous losses. These losses are the origin of frictional pressure drop in fluid flow. S. Whitaker [*Introduction to Fluid Mechanics*, Krieger, Huntington N.Y., 1981]; R. B. Bird, W. E. Stewart, and E. N. Lightfoot [*Transport Phenomena*, 2d ed., Wiley, New York, 2002]; and W. M. Deen [*Analysis of Transport Phenomena*, 2d ed., Oxford University Press, Oxford, UK, 2012] provide expressions for the dissipation function Φ for newtonian fluids in terms of the local velocity gradients. However, when one is using macroscopic balance equations, the local velocity field within the control volume is usually unknown. For such cases additional information, which may come from empirical correlations, is needed.

For the same special conditions as for Eq. (6-16), except for the $Q_V = 0$ restriction, the mechanical energy equation reduces to

$$\alpha_1\frac{V_1^2}{2}+gZ_1+\delta W_S=\alpha_2\frac{V_2^2}{2}+gZ_2+\int_{p_1}^{p_2}\frac{dp}{\rho}+l_v \tag{6-18}$$

Here l_v is the energy dissipation per unit mass. This equation has been called the *engineering Bernoulli equation*. For an incompressible flow, Eq. (6-18) becomes

$$\frac{p_1}{\rho}+\alpha_1\frac{V_1^2}{2}+gZ_1+\delta W_S=\frac{p_2}{\rho}+\alpha_2\frac{V_2^2}{2}+gZ_2+l_v \tag{6-19}$$

The Bernoulli equation can be written for incompressible, inviscid flow along a streamline:

$$\frac{p_1}{\rho}+\frac{V_1^2}{2}+gZ_1=\frac{p_2}{\rho}+\frac{V_2^2}{2}+gZ_2 \tag{6-20}$$

Unlike the momentum equation, Eq. (6-14), the Bernoulli equation does not generalize to multiple inlets or outlets by mere addition of terms.

Microscopic Balance Equations Partial differential balance equations express the conservation principles at a point in space. The two most used equations, for mass and momentum, are presented here.

Mass Balance, Continuity Equation The continuity equation, expressing conservation of mass, may be written in vector notation as

$$\frac{\partial\rho}{\partial t}+\nabla\cdot(\rho\mathbf{v})=0 \tag{6-21}$$

In terms of the *substantial derivative*, $D/Dt = \partial/\partial t + \mathbf{v}\cdot\nabla$,

$$\frac{D\rho}{Dt}=-\rho\nabla\cdot\mathbf{v} \tag{6-22}$$

Also called the *material derivative*, D/Dt is the rate of change in a lagrangian reference frame, that is, following a material particle. For incompressible flow,

$$\nabla\cdot\mathbf{v}=0 \tag{6-23}$$

Equation (6-22) in cartesian, cylindrical, and spherical coordinates may be found in Table 6-1.

Stress Tensor The stress tensor is needed to completely describe the stress state for microscopic momentum balances in multidimensional flows. The components of the stress tensor σ_{ij} give the force in the j direction on a plane perpendicular to the i direction, using a sign convention defining a positive stress as one where the fluid with the greater i coordinate value exerts a force in the positive j direction on the fluid with the lesser i coordinate. Several references in fluid mechanics and continuum mechanics provide discussions, to various levels of detail, of stress in a fluid, e.g., R. Aris [*Vectors, Tensors and the Basic Equations of Fluid Mechanics*, Dover, New York, 1962]; W. M. Deen [*Analysis of Transport Phenomena*, 2d ed., Oxford University Press, Oxford, UK, 2012]; and J. C. Slattery [*Advanced Transport Phenomena*, Cambridge University Press, Cambridge, UK, 1999].

The stress has an isotropic contribution due to fluid pressure and a *deviatoric* contribution due to viscous deformation. The total stress is

$$\boldsymbol{\sigma}=-p\boldsymbol{\delta}+\boldsymbol{\tau} \tag{6-24}$$

where p is the pressure. The deviatoric stress for a newtonian fluid is

$$\boldsymbol{\tau}=2\mu\boldsymbol{\Gamma}+\left(\kappa-\frac{2}{3}\mu\right)(\nabla\cdot\mathbf{v})\boldsymbol{\delta} \tag{6-25}$$

The identity tensor components δ_{ij} are zero for $i \neq j$ and unity for $i = j$. There is uncertainty about the value of the *bulk viscosity* κ. Traditionally, Stokes' hypothesis, $\kappa = 0$, has been invoked. For incompressible flow, the value of bulk viscosity is immaterial as Eq. (6-25) reduces to Eq. (6-7).

Similar generalizations to multidimensional flow are necessary for non-newtonian constitutive equations. The components of the stress constitutive equation in cartesian, cylindrical, and spherical coordinates for newtonian and generalized newtonian fluids are shown in Table 6-1.

Cauchy Momentum and Navier-Stokes Equations The differential equation for conservation of momentum is called the Cauchy momentum equation.

$$\rho\frac{D\mathbf{v}}{Dt}=\rho\left[\frac{\partial\mathbf{v}}{\partial t}+\mathbf{v}\cdot\nabla\mathbf{v}\right]=-\nabla p+\nabla\cdot\boldsymbol{\tau}+\rho\mathbf{g} \tag{6-26}$$

For an incompressible newtonian fluid with constant viscosity, substitution of Eqs. (6-7) and (6-6) into Eq. (6-26) gives the *Navier-Stokes equation*

$$\rho\frac{D\mathbf{v}}{Dt}=\rho\left[\frac{\partial\mathbf{v}}{\partial t}+\mathbf{v}\cdot\nabla\mathbf{v}\right]=-\nabla p+\mu\nabla^2\mathbf{v}+\rho\mathbf{g} \tag{6-27}$$

The pressure and gravity terms in Eq. (6-27) may be combined by replacing the pressure p by the equivalent pressure $P = p + \rho gZ$, where Z is elevation. The left-hand side terms of the Navier-Stokes equation are the *inertial terms*, while the terms including viscosity μ are the *viscous terms*. Limiting cases under which the Navier-Stokes equations may be simplified include *creeping flows* in which the inertial terms are neglected, *inviscid flows* in which the viscous terms are neglected, and *boundary layer* and *lubrication* flows in which certain terms are neglected based on scaling arguments. Creeping flows are described by J. Happel and H. Brenner [*Low Reynolds Number Hydrodynamics*, Prentice-Hall, Englewood Cliffs, N.J., 1965] and L. G. Leal [*Advanced Transport Phenomena*, Cambridge University Press, Cambridge, UK, 2007]; inviscid flows by H. Lamb [*Hydrodynamics*, 6th ed., Dover, New York, 1945] and L. M. Milne-Thompson [*Theoretical Hydrodynamics*, 5th ed., Macmillan, New York, 1968]; boundary layer theory by H. Schlichting and K. Gersten [*Boundary Layer Theory*, 8th ed rev., Springer-Verlag, Berlin, 2003] and lubrication theory by G. K. Batchelor [*An Introduction to Fluid Dynamics*, Cambridge University, Cambridge, UK, 1967], M. M. Denn [*Process Fluid Mechanics*, Prentice-Hall, Englewood Cliffs, N.J., 1979], and

W. M. Deen [*Analysis of Transport Phenomena*, 2d ed., Oxford University Press, Oxford, UK, 2012].

Because the Navier-Stokes equations are first-order in pressure and second-order in velocity, their solution requires one pressure boundary condition and two velocity boundary conditions (for each velocity component) to completely specify the solution for a steady flow. The *no-slip* condition, which requires that the fluid velocity equal the velocity of any bounding solid surface, occurs in most problems. Specification of velocity is a type of boundary condition sometimes called a *Dirichlet condition.* Often boundary conditions involve stresses, and thus velocity gradients, rather than the velocities themselves. Specification of velocity derivatives is a *Neumann condition.* For example, at the boundary between a viscous liquid and a gas, it is often assumed that the liquid shear stresses are zero. In numerical solution of the Navier-Stokes equations, Dirichlet and Neumann, or *essential* and *natural,* boundary conditions may be satisfied by different means.

Fluid statics, discussed in Sec. 10 in reference to pressure measurement, is the branch of fluid mechanics in which the fluid velocity either is zero or is uniform and constant relative to an inertial reference frame. With velocity gradients equal to zero, the momentum equation reduces to a simple expression for the pressure field, $\nabla p = \rho \mathbf{g}$. Letting z be directed vertically upward, so that $g_z = -g$ where g is the gravitational acceleration (9.807 m/s^2 or 32.17 ft/s^2), the pressure field is given by

$$\frac{dp}{dz} = -\rho g \tag{6-28}$$

This equation applies to any incompressible or compressible static fluid. For an incompressible liquid, pressure varies linearly with depth.

The *force exerted on a submerged planar surface* of area A is given by $F = p_c A$ where p_c is the pressure at the geometric *centroid* of the surface. The *center of pressure,* the point of application of the net force, is always lower than the centroid. For details see, for example, I. H. Shames [*Mechanics of Fluids,* 3d ed., McGraw-Hill, New York, 1992] where may also be found discussion of forces on *curved surfaces, buoyancy,* and *stability of floating bodies.*

Examples Four examples follow, illustrating the application of the conservation equations.

Example 6-1 Force Exerted on a Reducing Bend An incompressible fluid flows through a reducing elbow (Fig. 6-5) situated in a horizontal plane. The inlet velocity V_1 is given, and pressures p_1 and p_2 are measured. By selecting the inlet and outlet surfaces 1 and 2 as shown, the continuity equation, Eq. (6-11), can be used to find the exit velocity $V_2 = V_1A_1/A_2$. The mass flow rate is obtained by $\dot{m} = \rho V_1 A_1$.

Assume that the velocity profile is nearly uniform so that β is approximately unity. The force exerted on the fluid by the bend has x and y components; these can be found from Eq. (6-14). The x component gives

$$F_x = \dot{m}(V_{2x} - V_{1x}) + p_1 A_{1x} + p_2 A_{2x}$$

while the y component gives

$$F_y = \dot{m}(V_{2y} - V_{1y}) + p_1 A_{1y} + p_2 A_{2y}$$

The velocity components are $V_{1x} = V_1$, $V_{1y} = 0$, $V_{2x} = V_2 \cos\theta$, and $V_{2y} = V_2 \sin\theta$. The area vector components are $A_{1x} = -A_1$, $A_{1y} = 0$, $A_{2x} = A_2 \cos\theta$, and $A_{2y} = A_2 \sin\theta$. Therefore, the force components may be calculated from

$$F_x = \dot{m}(V_2 \cos\theta - V_1) - p_1 A_1 + p_2 A_2 \cos\theta$$
$$F_y = \dot{m} V_2 \sin\theta + p_2 A_2 \sin\theta$$

The force acting on the fluid is **F**; the force exerted by the fluid on the bend is **−F**.

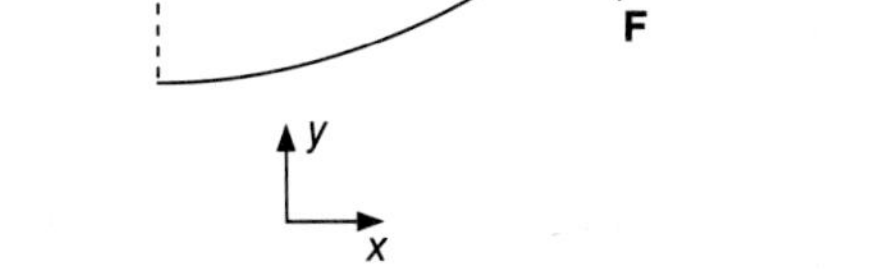

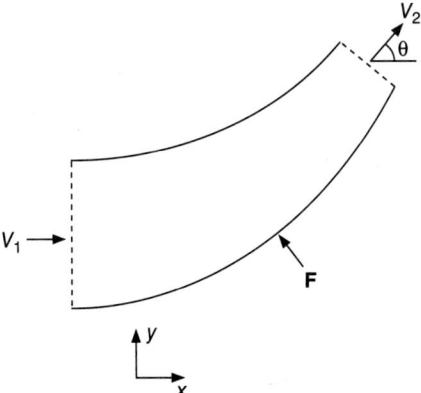

FIG. 6-5 Force at a reducing bend. **F** is the force exerted by the bend on the fluid. The force exerted by the fluid on the bend is **−F**.

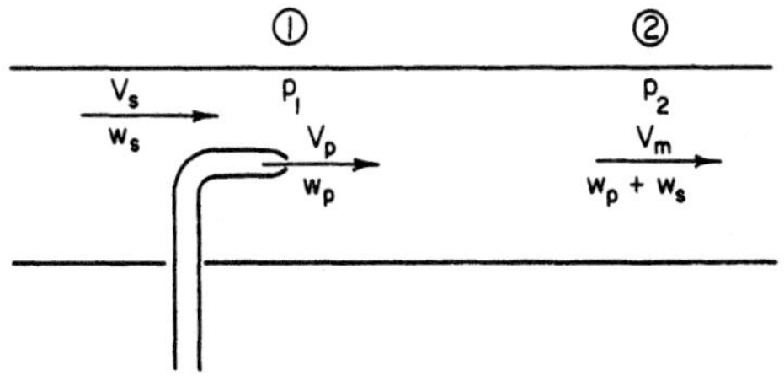

FIG. 6-6 Simplified ejector.

Example 6-2 Simplified Ejector Figure 6-6 shows a very simplified sketch of an ejector, a device that uses a high-velocity primary fluid to pump another (secondary) fluid. The continuity and momentum equations may be applied on the control volume with inlet and outlet surfaces 1 and 2, as indicated in the figure. The cross-sectional area is uniform, $A_1 = A_2 = A$. Let the mass flow rates and velocities of the primary and secondary fluids be $\dot{m}_p$, $\dot{m}_s$, V_p, and V_s. Assume for simplicity that the density is uniform. Conservation of mass gives $\dot{m}_2 = \dot{m}_p + \dot{m}_s$. The exit velocity is $V_2 = \dot{m}_2/(\rho A)$. The principal momentum exchange in the ejector occurs between the two fluids. Relative to this exchange, the force exerted by the walls of the device is small. Therefore, the force term F is neglected from the momentum equation. Written in the flow direction, assuming uniform velocity profiles, and using the extension of Eq. (6-14) for multiple inlets, it gives the pressure rise developed by the device:

$$(p_2 - p_1)A = (\dot{m}_p + \dot{m}_s)V_2 - \dot{m}_p V_p - \dot{m}_s V_s$$

Application of the momentum equation to ejectors of other types is discussed in C. E. Lapple (*Fluid and Particle Dynamics,* University of Delaware, Newark, 1951) and in Sec. 10 of this handbook.

Example 6-3 Venturi Flowmeter An incompressible fluid flows through the venturi flowmeter in Fig. 6-7. An equation is needed to relate the flow rate Q to the pressure drop measured by the manometer. This problem can be solved using the mechanical energy balance. In a well-made venturi, viscous losses are negligible, the pressure drop is the result of acceleration into the throat, and the flow rate predicted neglecting losses is quite accurate. The inlet area is A and the throat area is a. Gravity may be neglected even if the venturi is not horizontal, due to the small elevation change.

With control surfaces at 1 and 2 as shown in the figure, Eq. (6-19) in the absence of losses and shaft work gives

$$\frac{p_1}{\rho} + \frac{V_1^2}{2} = \frac{p_2}{\rho} + \frac{V_2^2}{2}$$

The continuity equation gives $V_2 = V_1A/a$, and $V_1 = Q/A$. The pressure drop measured by the manometer is $p_1 - p_2 = (\rho_m - \rho)g\,\Delta Z$. By substituting these relations into the energy balance and rearranging, the desired expression for the flow rate is found.

$$Q = \frac{1}{A}\sqrt{\frac{2(\rho_m - \rho)g\Delta Z}{\rho[(A/a)^2 - 1]}}$$

Example 6-4 Plane Poiseuille Flow Driven by a pressure gradient, an incompressible newtonian fluid with constant viscosity flows at a steady rate in the x direction between two very large stationary plates, as shown in Fig. 6-8. The flow is laminar. Cartesian coordinates are used for this rectangular geometry. The *fully developed* velocity profile is to be found. This is the velocity field in the region sufficiently far from the inlet and exit that the velocity is independent of x.

This problem requires use of the microscopic balance equations because the velocity is to be determined as a function of position. The boundary conditions result from the no-slip condition. All three velocity components must be zero at the plate surfaces $y = H/2$ and $y = -H/2$.

For fully developed flow, all velocity derivatives in the x direction vanish. Since the flow field is infinite in the z direction, all velocity derivatives in the z direction are zero. Therefore, velocity components are a function of y alone. It is also assumed that there is

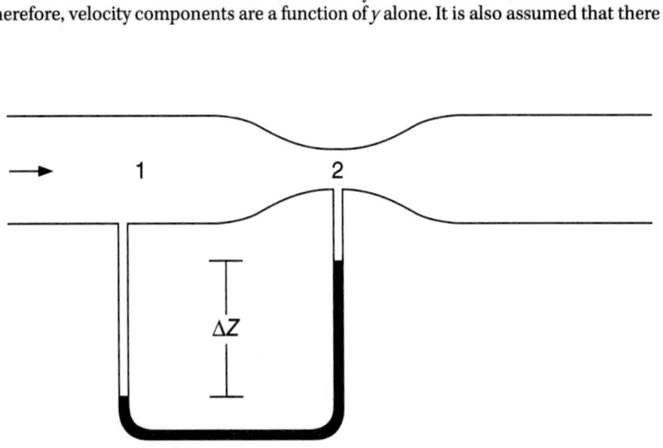

FIG. 6-7 Venturi flowmeter.

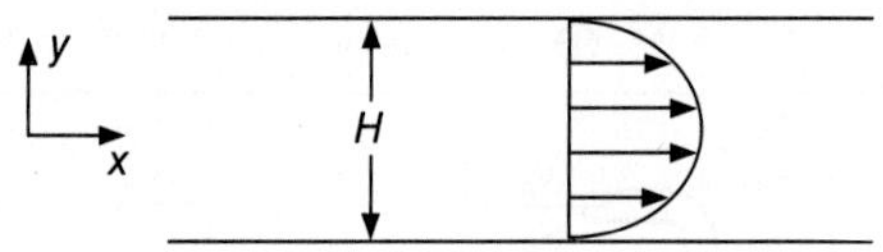

FIG. 6-8 Plane Poiseuille flow.

no flow in the z direction, so $\upsilon_z = 0$. The continuity equation from Table 6-1, with $\upsilon_z = 0$ and $\partial \upsilon_x/\partial_x = 0$, reduces to

$$\frac{dv_y}{dy} = 0$$

Since $\upsilon_y = 0$ at $y = \pm H/2$, the continuity equation integrates to $\upsilon_y = 0$. This is a direct result of the assumption of fully developed flow. The only nonzero velocity component is v_x, and it is a function only of y. The flow is one-dimensional.

The Navier-Stokes equations are greatly simplified with υ_y, υ_z, $\partial \upsilon_x/\partial x$, $\partial \upsilon_x/\partial z$, and $\partial \upsilon_x/\partial t$ all being zero. The three cartesian components from Table 6-1 are written in terms of the equivalent pressure P:

$$0 = -\frac{\partial P}{\partial x} + \mu\frac{\partial^2 v_x}{\partial y^2}$$

$$0 = -\frac{\partial P}{\partial y}$$

$$0 = -\frac{\partial P}{\partial z}$$

The latter two equations require that P be a function only of x, and therefore $\partial P/\partial x = dP/dx$. Inspection of the first equation then shows one term that is a function of only x and one that is a function of only y. This requires that both terms be constant. The pressure gradient $-dP/dx$ is constant. The x-component equation becomes

$$\frac{d^2 v_x}{dy^2} = \frac{1}{\mu}\frac{dP}{dx}$$

Two integrations give

$$v_x = \frac{1}{2\mu}\frac{dP}{dx}y^2 + C_1 y + C_2$$

The integration constants C_1 and C_2 are evaluated from the boundary conditions $\upsilon_x = 0$ at $y = \pm H/2$. The result is

$$v_x = \frac{H^2}{8\mu}\left(-\frac{dP}{dx}\right)\left[1 - \left(\frac{2y}{H}\right)^2\right]$$

This is a *parabolic* velocity distribution. The average velocity $V = (1/H)\int_{-H/2}^{H/2} v_x\,dy$ is

$$V = \frac{H^2}{12\mu}\left(-\frac{dP}{dx}\right)$$

INCOMPRESSIBLE FLOW IN PIPES AND CHANNELS

Mechanical Energy Balance The mechanical energy balance, Eq. (6-19), for *fully developed* incompressible flow in a straight circular pipe of constant diameter D reduces to

$$\frac{p_1}{\rho} + gZ_1 = \frac{p_2}{\rho} + gZ_2 + l_v \tag{6-29}$$

In terms of the equivalent pressure $P = p + \rho gZ$,

$$P_1 - P_2 = \rho l_v \tag{6-30}$$

The pressure drop due to frictional losses l_v is proportional to pipe length L for fully developed flow and may be denoted as the (positive) quantity $\Delta P \equiv P_1 - P_2$.

Friction Factor and Reynolds Number For a newtonian fluid in a smooth pipe, dimensional analysis relates the frictional pressure drop per unit length $\Delta P/L$ to the pipe diameter D, density ρ, viscosity μ, and average velocity V through two dimensionless groups, the *Fanning friction factor f* and the *Reynolds number* Re.

$$f \equiv \frac{D\Delta P}{2\rho V^2 L} \tag{6-31}$$

$$\mathrm{Re} \equiv \frac{DV\rho}{\mu} \tag{6-32}$$

For smooth pipe, the friction factor is a function of only the Reynolds number. In rough pipe, the relative roughness ε/D also affects the friction factor. Figure 6-9 plots f as a function of Re and ε/D. Values of ε for various

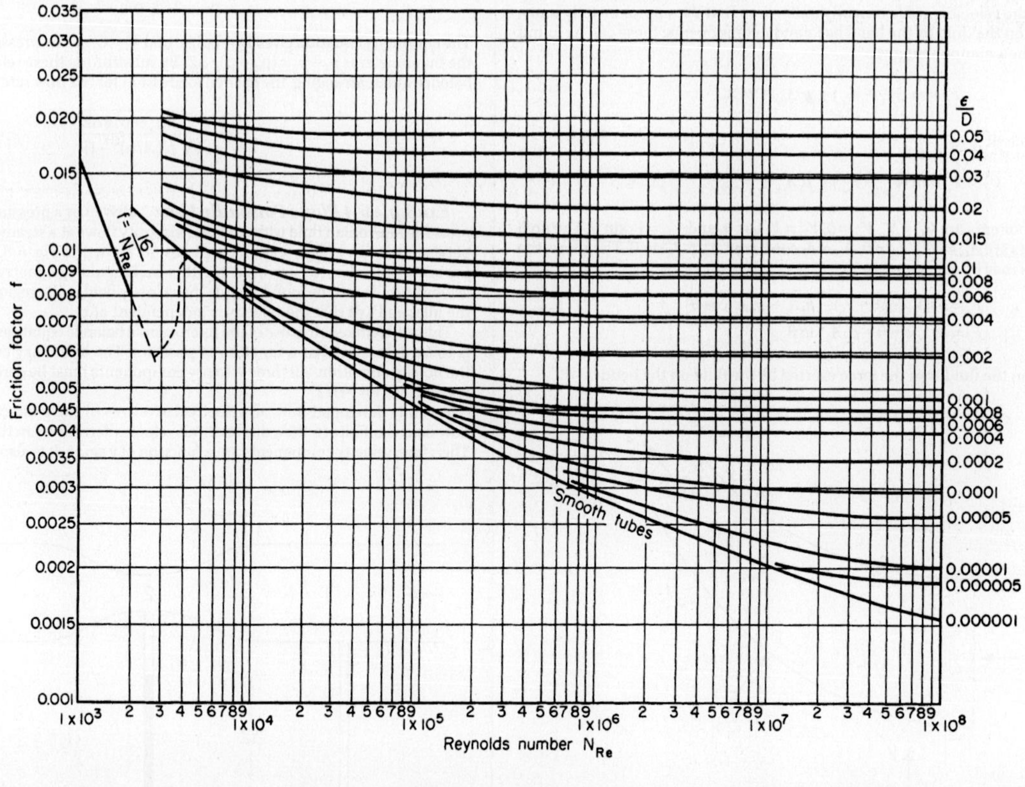

FIG. 6-9 Fanning friction factors. Reynolds number Re = $DV\rho/\mu$, where D = pipe diameter, V = velocity, ρ = fluid density, and μ = fluid viscosity. [Based on L. F. Moody, *Trans. ASME*, **66:** 671 (1944).]

TABLE 6-2 Values of Surface Roughness for Various Materials*

Material	Surface roughness ε, mm
Drawn tubing (brass, lead, glass, and the like)	0.00152
Commercial steel or wrought iron	0.0457
Asphalted cast iron	0.122
Galvanized iron	0.152
Cast iron	0.259
Wood stave	0.183–0.914
Concrete	0.305–3.05
Riveted steel	0.914–9.14

*From Moody, *Trans. Am. Soc. Mech. Eng.* **66:** 671–684 (1944); *Mech. Eng.* **69:** 1005–1006 (1947). Additional values of ε for various types or conditions of concrete wrought-iron, welded steel, riveted steel, and corrugated-metal pipes are given in Brater and King, *Handbook of Hydraulics,* 6th ed., McGraw-Hill, New York, 1976, pp. 6-12–6-13. To convert millimeters to feet, multiply by 3.281×10^{-3}.

materials are given in Table 6-2. The Fanning friction factor should not be confused with the Darcy friction factor used by L. F. Moody [*Trans. ASME,* **66:** 671 (1944)], which is four times greater. The Darcy-Weisbach equation is equivalent to Eq. (6-31). Using the momentum equation, the stress at the wall of the pipe may be expressed in terms of the friction factor:

$$\tau_w = f\frac{\rho V^2}{2} \tag{6-33}$$

Laminar and Turbulent Flow Below a *critical Reynolds number* of about 2100, the flow is laminar; over the range 2100 < Re < 5000 there is a transition to turbulent flow. Reliable correlations for the friction factor in transitional flow are not available. For laminar flow, the Hagen-Poiseuille equation

$$f = \frac{16}{\text{Re}} \quad \text{Re} \leq 2100 \tag{6-34}$$

may be derived from the Navier-Stokes equation and is in excellent agreement with experimental data. It may be rewritten in terms of volumetric flow rate $Q = V\pi D^2/4$ as

$$Q = \frac{\pi \Delta P D^4}{128 \mu L} \quad \text{Re} \leq 2100 \tag{6-35}$$

For turbulent flow in smooth tubes, the Blasius equation gives the friction factor accurately for a wide range of Reynolds number.

$$f = \frac{0.079}{\text{Re}^{0.25}} \quad 4000 < \text{Re} < 10^5 \tag{6-36}$$

The Colebrook formula [C. F. Colebrook, *J. Inst. Civ. Eng.* (*London*), **11:** 133–156 (1938–39)] gives a good approximation for the *f*-Re-(ε/*D*) data for rough pipes over the entire turbulent flow range:

$$\frac{1}{\sqrt{f}} = -4 \log\left[\frac{\varepsilon}{3.7D} + \frac{1.256}{\text{Re}\sqrt{f}}\right] \quad \text{Re} > 4000 \tag{6-37}$$

Equation (6-37) was used to construct the curves in the turbulent flow regime in Fig. 6-9. An equation by S. W. Churchill [*Chem. Eng.* **84**(24): 91–92 (Nov. 7, 1977)] closely approximating the Colebrook formula offers the advantage of being explicit in *f*:

$$\frac{1}{\sqrt{f}} = -4 \log\left[\frac{0.27\varepsilon}{D} + \left(\frac{7}{\text{Re}}\right)^{0.9}\right] \quad \text{Re} > 4000 \tag{6-38}$$

Churchill also provided a single equation that may be used for Reynolds numbers in laminar, transitional, and turbulent flow, closely fitting $f = 16/\text{Re}$ in the laminar regime and the Colebrook formula in the turbulent regime, and giving reasonable values in the transition regime, where the friction factor is uncertain.

$$f = 2\left[\left(\frac{8}{\text{Re}}\right)^{12} + \frac{1}{(A+B)^{3/2}}\right]^{1/12}$$

where

$$A = \left[2.457 \ln \frac{1}{(7/\text{Re})^{0.9} + 0.27\varepsilon/D}\right]^{16} \tag{6-39}$$

and

$$B = \left(\frac{37{,}530}{\text{Re}}\right)^{16}$$

In laminar flow, *f* is independent of ε/*D*. In turbulent flow, the friction factor for rough pipe follows the smooth tube curve for a range of Reynolds numbers (hydraulically smooth flow). For greater Reynolds numbers, *f* deviates from the smooth pipe curve, eventually becoming independent of Re. This region, often called *complete turbulence,* is frequently encountered in commercial pipe flows.

Two common pipe flow problems are calculation of pressure drop given the flow rate (or velocity) and calculation of flow rate (or velocity) given the pressure drop. When the flow rate is given, the Reynolds number may be calculated directly to determine the flow regime, so that the appropriate relations between *f* and Re can be selected. When the flow rate is specified and the flow is turbulent, Eq. (6-38), being explicit in *f,* may be preferable to Eq. (6-37), which is implicit in *f* and pressure drop.

When the pressure drop is given and the velocity and flow rate are to be determined, the Reynolds number cannot be computed directly, since the velocity is unknown. For such problems, it is useful to note that $\text{Re}\sqrt{f} = (D^{3/2}/\mu)\sqrt{\rho \Delta P/(2L)}$, appearing in the Colebrook equation (6-37), does not include velocity and so can be computed directly, so that *f* may be computed without iteration. Thus Eq. (6-37) is preferable to Eq. (6-38) or Eq. (6-39) when the pressure drop is given.

Velocity Profiles In laminar flow, the solution of the Navier-Stokes equation, corresponding to the Hagen-Poiseuille equation, gives the velocity v as a function of radial position r in a circular pipe of radius R in terms of the average velocity $V = Q/A$. The *parabolic* profile, with centerline velocity twice the average velocity, is shown in Fig. 6-10.

$$v = 2V\left(1 - \frac{r^2}{R^2}\right) \tag{6-40}$$

In turbulent flow, the velocity profile is more blunt, with a lower velocity gradient near the center and a steeper gradient near the wall. The region near the wall is described by a *universal* velocity profile, characterized by a *viscous sublayer,* a *buffer zone,* and a *turbulent core.*

Viscous sublayer

$$u_+ = y_+ \quad \text{for} \quad y_+ < 5 \tag{6-41}$$

Buffer zone

$$u_+ = 5.00 \ln y_+ - 3.05 \quad \text{for} \quad 5 < y_+ < 30 \tag{6-42}$$

Turbulent core

$$u_+ = 2.5 \ln y_+ + 5.5 \quad \text{for} \quad y_+ > 30 \tag{6-43}$$

Here, $u_+ = v/u_*$ is the dimensionless, time-averaged axial velocity, $u_* = \sqrt{\tau_w/\rho}$ is the *friction velocity,* and $\tau_w = f\rho V^2/2$ is the wall stress. The friction velocity is of the order of the root mean square velocity fluctuation perpendicular to the wall in the turbulent core. The dimensionless distance from the wall is $y_+ = yu_*\rho/\mu$. At sufficient Re, the universal velocity profile is valid in the wall region for any cross-sectional channel shape. For incompressible flow in constant-diameter circular pipes, $\tau_w = D\Delta P/4L$ where ΔP is the equivalent pressure drop in length L.

For rough pipes, the velocity profile in the turbulent core is given by

$$u_+ = 2.5 \ln y/\varepsilon + 8.5 \quad \text{for} \quad y_+ > 30 \tag{6-44}$$

when the dimensionless roughness $\varepsilon_+ = \varepsilon u_*\rho/\mu$ is greater than about 5 to 10; for smaller ε_+, the velocity profile in the turbulent core is unaffected by roughness.

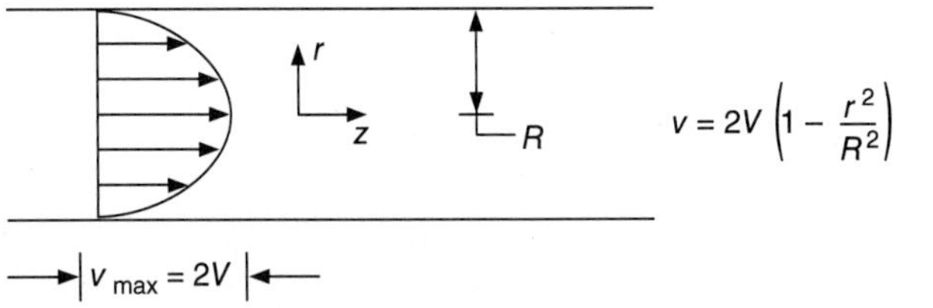

FIG. 6-10 Parabolic velocity profile for laminar flow in a pipe, with average velocity *V.*

For velocity profiles in the transition region, see Patel and Head [*J. Fluid Mech.* **38:** part 1, 181–201 (1969)] where profiles over the range 1500 < Re < 10,000 are reported.

Entrance and Exit Effects In the entrance region of a pipe, some distance is required for the flow to adjust from upstream conditions to the fully developed velocity profile. This distance depends on the Reynolds number and on the flow conditions upstream. For a uniform velocity profile at the pipe entrance, the computed length in laminar flow required for the centerline velocity to reach 99 percent of its fully developed value is [N. Dombrowski et al., *Can. J. Chem. Engr.* **71:** 472–476 (1993)]

$$L_{\text{ent}}/D = 0.370\exp(-0.148\,\text{Re}) + 0.0550\,\text{Re} + 0.260 \tag{6-45}$$

In turbulent flow, the entrance length is about

$$L_{\text{ent}}/D = 40 \tag{6-46}$$

The frictional losses in the entrance region are larger than those for the same length of fully developed flow. (See the subsection Frictional Losses in Pipeline Elements later.) At the pipe exit, the velocity profile also undergoes rearrangement, but the exit length is much shorter than the entrance length. At low Re, it is about one pipe radius. At Re > 100, the exit length is negligible.

Residence Time Distribution For the parabolic profile for laminar flow in a pipe, neglecting diffusion, the cumulative residence time distribution $F(\theta)$ is given by

$$\begin{aligned} F(\theta) &= 0 \quad \text{for} \quad \theta < \frac{\theta_{\text{avg}}}{2} \\ F(\theta) &= 1 - \frac{1}{4}\left(\frac{\theta_{\text{avg}}}{\theta}\right)^2 \quad \text{for} \quad \theta \geq \frac{\theta_{\text{avg}}}{2} \end{aligned} \tag{6-47}$$

where $F(\theta)$ is the fraction of material that resides in the pipe for less than time θ and θ_{avg} is the average residence time, $\theta = L/V$.

The residence time distribution in long transfer lines may be made narrower (more uniform) with the use of *flow inverters* or *static mixing elements.* These devices exchange fluid between the wall and central regions. Variations on the concept may be used to provide effective mixing of the fluid. See J. C. Godfrey ["Static Mixers," in N. Harnby et al., *Mixing in the Process Industries,* 2d ed., Butterworth Heinemann, Oxford, UK, 1992] and A. W. Etchells and C. F. Meyer ["Mixing in Pipelines," in E. L. Paul et al., *Handbook of Industrial Mixing,* Wiley Interscience, Hoboken, N.J., 2004].

The residence time distribution is narrower for helical coils than for straight pipes, because of the secondary flow which exchanges fluid between the wall and center regions. An equation for laminar flow in helical pipe coils by D. M. Ruthven [*Chem. Eng. Sci.* **26:** 1113–1121 (1971); **33:** 628–629 (1978)] gives

$$F(\theta) = 1 - \left(\frac{1}{4}\right)\left[\frac{\theta_{\text{avg}}}{\theta}\right]^{2.81} \quad \text{for} \quad 0.5 < \frac{\theta_{\text{avg}}}{\theta} < 1.63 \tag{6-48}$$

and agrees with the results of R. N. Trivedi and K. Vasudeva [*Chem. Eng. Sci.* **29:** 2291–2295 (1974)] for 0.6 < De < 6 and $0.0036 < D/D_c < 0.097$ where $\text{De} = \text{Re}\sqrt{D/D_c}$ is the *Dean number* and D_c is the diameter of curvature of the coil. Measurements by A. K. Saxena and K. D. P. Nigam [*Chem. Eng. Sci.* **34:** 425–426 (1979)] indicate that such a distribution will hold for De > 1.

In turbulent flow, axial mixing is usually described in terms of turbulent diffusion or dispersion coefficients, from which cumulative residence time distribution functions can be computed. J. T. Davies [*Turbulence Phenomena,* Academic, New York, 1972, p. 93] gives $D_L = 1.01\nu\text{Re}^{0.875}$ for the longitudinal dispersion coefficient. O. Levenspiel [*Chemical Reaction Engineering,* 2d ed., Wiley, New York, 1972, pp. 253–278] discusses the relations among various residence time distribution functions and the relation between dispersion coefficient and residence time distribution.

Noncircular Channels Calculation of frictional pressure drop in noncircular channels depends on whether the flow is laminar or turbulent and whether the channel is full or open. For *turbulent flow* in *ducts running full,* the *hydraulic diameter* D_H should be substituted for D in the friction factor and Reynolds number definitions, Eqs. (6-31) and (6-32). The hydraulic diameter is defined as four times the channel cross-sectional area divided by the wetted perimeter. For example, the hydraulic diameter for a circular pipe is $D_H = D$, for an annulus of inner diameter d and outer diameter D, $D_H = D - d$, for a rectangular duct of sides a, b, $D_H = ab/[2(a + b)]$. The **hydraulic radius** R_H is defined as *one-fourth* of the hydraulic diameter.

With the hydraulic diameter substituted for D in f and Re, Eqs. (6-36) through (6-38) are good approximations. Note that V appearing in f and Re is the actual average velocity $V = Q/A$; for noncircular pipes, it is *not* $Q/(\pi D_H^2/4)$. The pressure drop should be calculated from the friction factor for noncircular pipes. Equations relating Q to ΔP and D for circular pipes *may not be used* for noncircular pipes with D replaced by D_H because $V \neq Q/(\pi D_H^2/4)$.

Turbulent flow in noncircular channels is often accompanied by secondary flows perpendicular to the axial flow direction. These flows may cause the pressure drop to be slightly greater than that computed using the hydraulic diameter method. For data on pressure drop in annuli, see J. A. Brighton and J. B. Jones [*J. Basic Eng.* **86:** 835–842 (1964)]; T. H. Okiishi and G. K. Serovy [*J. Basic Eng.* **89:** 823–836 (1967)]; and C. J. Lawn and C. J. Elliot [*J. Mech. Eng. Sci.* **14:** 195–204 (1972)]. For rectangular ducts of large aspect ratio, R. B. Dean [*J. Fluids Eng.* **100:** 215–233 (1978)] found that the numerator of the exponent in the Blasius equation (6-37) should be increased to 0.0868. O. C. Jones [*J. Fluids Eng.* **98:** 173–181 (1976)] presents a method to improve the estimation of friction factors for rectangular ducts using a modification of the hydraulic diameter–based Reynolds number.

The hydraulic diameter method does not work well for *laminar flow* because the shape affects the flow resistance in a way that cannot be expressed as a function of only the ratio of cross-sectional area to wetted perimeter. For some shapes, the Navier-Stokes equations have been integrated to yield relations between flow rate and pressure drop. These relations may be expressed in terms of *equivalent diameters* D_E defined to make the relations reduce to the second form of the Hagen-Poiseuille equation, Eq. (6-35), that is, $D_E \equiv (128Q\mu L/\pi\,\Delta P)^{1/4}$. *Equivalent diameters are not the same as hydraulic diameters.* Equivalent diameters yield the correct relation between flow rate and pressure drop when substituted into Eq. (6-35), but not Eqs. (6-36) to (6-38) because $V \neq Q/(\pi D_E/4)$.

Ellipse, semiaxes a and b [H. Lamb, *Hydrodynamics,* 6th ed., Dover, New York, 1945, p. 587]:

$$D_E = \left(\frac{32a^3b^3}{a^2 + b^2}\right)^{1/4} \tag{6-49}$$

Rectangle, width a, height b [Owen, *Trans. Am. Soc. Civ. Eng.* **119:** 1157–1175 (1954)]:

$$D_E = \left(\frac{128ab^3}{\pi K}\right)^{1/4} \tag{6-50}$$

$a/b =$	1	1.5	2	3	4	5	10	∞
$K =$	28.45	20.43	17.49	15.19	14.24	13.73	12.81	12

Annulus, inner diameter D_1, outer diameter D_2 [H. Lamb, *Hydrodynamics,* 6th ed., Dover, New York, 1945, p. 587]:

$$D_E = \left\{(D_2^2 - D_1^2)\left[D_2^2 + D_1^2 - \frac{D_2^2 - D_1^2}{\ln(D_2/D_1)}\right]\right\}^{1/4} \tag{6-51}$$

For isosceles triangles and regular polygons, see E. M. Sparrow [*AIChE J.* **8:** 599–605 (1962)], L. W. Carlson and T. F. Irvine [*J. Heat Transfer* **83:** 441–444 (1961)], K. C. Cheng [*Proc. Third Int. Heat Transfer Conf.,* New York, **1:** 64–76 (1966)], and F. S. Shit [*Can. J. Chem. Eng.* **45:** 285–294 (1967)].

The critical Reynolds number for *transition from laminar to turbulent flow* in noncircular channels varies with channel shape. In rectangular ducts, $1900 < \text{Re}_c < 2800$ [R. W. Hanks and H.-C. Ruo, *Ind. Eng. Chem. Fundam.* **5:** 558–561 (1966)]. In triangular ducts, $1600 < \text{Re}_c < 1800$ [R. C. Cope and R. W. Hanks, *Ind. Eng. Chem. Fundam.* **11:** 106–117 (1972); P. Bandopadhayay and J. Hinwood, *J. Fluid Mech.* **59:** 775–783 (1973)].

Nonisothermal Flow For nonisothermal flow of *liquids,* the friction factor may be increased if the liquid is being cooled or may be decreased if the liquid is being heated, because of the effect of temperature on viscosity near the wall. In shell and tube heat-exchanger design, the recommended practice is to first estimate f using the bulk mean liquid temperature over the tube length. Then, in laminar flow, the result is divided by $(\mu_a/\mu_w)^{0.23}$ in the case of cooling or by $(\mu_a/\mu_w)^{0.38}$ in the case of heating. For turbulent flow, f is divided by $(\mu_a/\mu_w)^{0.11}$ in the case of cooling or by $(\mu_a/\mu_w)^{0.17}$ in case of heating. Here, μ_a is the viscosity at the average bulk temperature and μ_w is the viscosity at the average wall temperature [E. N. Seider and G. E. Tate, *Ind. Eng. Chem.* **28:** 1429–1435 (1936)]. In the case of rough commercial pipes, rather than heat-exchanger tubing, it is common for flow to be in the "complete" turbulence regime where f is independent of Re. In such cases, the friction factor should not be corrected for wall temperature. If the liquid density varies with temperature, the average bulk density should be used to calculate the pressure drop from the friction factor. In addition, a (usually small) correction may be applied for acceleration effects by adding the term $G^2[(1/\rho_2) - (1/\rho_1)]$ from the mechanical energy balance to the frictional pressure drop, where G is the mass velocity. This acceleration results from small compressibility effects associated with temperature-dependent density. E. B. Christiansen and G. E. Gordon [*AIChE J.* **15:** 504–507 (1969)] present equations and charts for frictional loss in laminar nonisothermal flow of newtonian and nonnewtonian liquids heated or cooled with constant wall temperature.

Frictional dissipation of mechanical energy can result in significant heating of fluids, particularly for very viscous liquids in small channels. Under adiabatic conditions, the bulk liquid temperature rise is given by $\Delta T = \Delta P / C_v \rho$ for incompressible flow through a channel of constant cross-sectional area. For flow of polymers, this amounts to about 4°C per 10 MPa (1500 psi) pressure drop, while for hydrocarbon liquids it is about 6°C per 10 MPa. The temperature rise in laminar flow is highly nonuniform, being concentrated near the pipe wall where most of the dissipation occurs. This may result in significant viscosity reduction near the wall, and greatly increased flow or reduced pressure drop, and a flattened velocity profile. Compensation should generally be made for the heat effect when ΔP exceeds 1.4 MPa (200 psi) for adiabatic walls or 3.5 MPa (500 psi) for isothermal walls [J. E. Gerrard et al., *Ind. Eng. Chem. Fundam.* **4:** 332–339 (1969)].

Open-Channel Flow For flow in *open channels,* the data are largely from experiments with water in turbulent flow, in channels of sufficient roughness that there is no Reynolds number effect. The hydraulic radius approach may be used to estimate a friction factor with which to compute friction losses. Under conditions of *uniform flow* where liquid depth and cross-sectional area do not vary significantly with position in the flow direction, there is a balance between gravitational forces and wall stress, or equivalently between frictional losses and potential energy change. The mechanical energy balance in terms of the friction factor and hydraulic diameter or hydraulic radius becomes

$$l_v = \frac{2fV^2L}{D_H} = \frac{fV^2L}{2R_H} = g(Z_1 - Z_2) \tag{6-52}$$

The hydraulic radius is the cross-sectional area divided by the wetted perimeter, where the wetted perimeter *does not include the free surface.* Letting $S = \sin\theta$ = channel slope (elevation loss per unit length of channel, θ = angle between channel and horizontal), Eq. (6-52) reduces to

$$V = \sqrt{\frac{2gSR_H}{f}} \tag{6-53}$$

The most often used friction correlation for open-channel flows is due to R. Manning [*Trans. Inst. Civ. Engrs. Ireland* **20:** 161 (1891)] and is equivalent to

$$f = \frac{29n^2}{R_H^{1/3}} \tag{6-54}$$

where n is the channel roughness, with dimensions of $(\text{length})^{1/6}$. Table 6-3 gives roughness values for several channel types.

TABLE 6-3 Average Values of *n* for Manning Formula, Eq. (6-54)

Surface	n, m$^{1/6}$	n, ft$^{1/6}$
Cast-iron pipe, fair condition	0.014	0.011
Riveted steel pipe	0.017	0.014
Vitrified sewer pipe	0.013	0.011
Concrete pipe	0.015	0.012
Wood-stave pipe	0.012	0.010
Planed-plank flume	0.012	0.010
Semicircular metal flumes, smooth	0.013	0.011
Semicircular metal flumes, corrugated	0.028	0.023
Canals and ditches		
Earth, straight and uniform	0.023	0.019
Winding sluggish canals	0.025	0.021
Dredged earth channels	0.028	0.023
Natural-stream channels		
Clean, straight bank, full stage	0.030	0.025
Winding, some pools and shoals	0.040	0.033
Same, but with stony sections	0.055	0.045
Sluggish reaches, very deep pools, rather weedy	0.070	0.057

SOURCE: Brater and King, *Handbook of Hydraulics,* 6th ed., McGraw-Hill, New York, 1976, p. 7-22. For detailed information, see Chow, *Open-Channel Hydraulics,* McGraw-Hill, New York, 1959, pp. 110–123.

For gradual changes in channel cross section and liquid depth, and for slopes less than 10°, the momentum equation for a rectangular channel of width b and liquid depth h may be written as a differential equation in the flow direction x.

$$\frac{dh}{dx}(1-\text{Fr}) - \text{Fr}\left(\frac{h}{b}\right)\frac{db}{dx} = S - \frac{fV^2(b+2h)}{2gbh} \tag{6-55}$$

For a given fixed flow rate $Q = Vbh$ and channel width profile $b(x)$, Eq. (6-55) may be integrated to determine the liquid depth profile $h(x)$. The dimensionless Froude number is $\text{Fr} = V^2/gh$. When Fr = 1, the flow is *critical;* when Fr < 1, the flow is *subcritical;* and when Fr > 1, the flow is *supercritical.* Surface disturbances move at a wave velocity $c = \sqrt{gh}$; they cannot propagate upstream in supercritical flows. The *specific energy* E_{sp} is nearly constant.

$$E_{sp} = h + \frac{V^2}{2g} \tag{6-56}$$

This equation is cubic in liquid depth. Below a minimum value of E_{sp} there are no real positive roots; above the minimum value there are two positive real roots. At this minimum value of E_{sp} the flow is critical; that is, $\text{Fr} = 1$, $V = \sqrt{gh}$, and $E_{sp} = (3/2)h$. Near critical flow conditions, wave motion and sudden depth changes called *hydraulic jumps* are likely. V. T. Chow [*Open Channel Hydraulics,* McGraw-Hill, New York, 1959] discusses the numerous surface profile shapes which may exist in nonuniform open-channel flows.

For flow over a *sharp-crested weir* of width b and height L, from a liquid depth H, the flow rate is given approximately by

$$Q = \frac{2}{3}C_d b\sqrt{2g}(H-L)^{3/2} \tag{6-57}$$

where $C_d \approx 0.6$ is a discharge coefficient. Flow through notched weirs is described under flowmeters in Sec. 10 of this handbook.

Nonnewtonian Flow For *isothermal laminar flow* of time-independent nonnewtonian liquids, integration of the Cauchy momentum equation yields the fully developed velocity profile and flow rate–pressure drop relations. For the *Bingham plastic* fluid described by Eq. (6-3), in a pipe of diameter D and a pressure drop per unit length of $\Delta P/L$, the flow rate is given by

$$Q = \frac{\pi D^3 \tau_w}{32\mu_\infty}\left[1 - \frac{4\tau_y}{3\tau_w} + \frac{\tau_y^4}{3\tau_w^4}\right] \tag{6-58}$$

where the wall stress is $\tau_w = D\Delta P/(4L)$. The velocity profile consists of a central nondeforming plug of radius $r_P = 2\tau_y/(\Delta P/L)$ and an annular deforming region. The velocity profile in the annular region is given by

$$v_z = \frac{1}{\mu_\infty}\left[\frac{\Delta P}{4L}(R^2 - r^2) - \tau_y(R - r)\right] \quad r_P \le r \le R \tag{6-59}$$

where r is the radial coordinate and R is the pipe radius. The velocity of the central, nondeforming plug is obtained by setting $r = r_P$ in Eq. (6-59). When Q is given and Eq. (6-58) is to be solved for τ_w and the pressure drop, multiple positive roots for the pressure drop may be found. The root corresponding to $\tau_w < \tau_y$ is physically unrealizable, as it corresponds to $r_P > R$ and the pressure drop is insufficient to overcome the yield stress.

For a *power law fluid,* Eq. (6-4), with constant properties K and n, the flow rate is given by

$$Q = \pi\left(\frac{\Delta P}{2KL}\right)^{1/n}\left(\frac{n}{1+3n}\right)R^{(1+3n)/n} \tag{6-60}$$

and the velocity profile by

$$v_z = \left(\frac{\Delta P}{2KL}\right)^{1/n}\left(\frac{n}{1+n}\right)[R^{(1+n)/n} - r^{(1+n)/n}] \tag{6-61}$$

Similar relations for other nonnewtonian fluids may be found in G. W. Govier and K. Aziz [*The Flow of Complex Mixtures in Pipes,* Krieger, Huntington, N.Y., 1977] and in R. B. Bird, R. C. Armstrong, and O. Hassager [*Dynamics of Polymeric Liquids,* vol. 1: *Fluid Mechanics,* Wiley, New York, 1977].

For steady laminar flow of any time-independent viscous fluid, at average velocity V in a pipe of diameter D, the Rabinowitsch-Mooney equation gives the shear rate at the pipe wall.

$$\dot{\gamma}_w = \frac{8V}{D}\left(\frac{1+3n'}{4n'}\right) \tag{6-62}$$

where n' is the slope of a plot of $D\Delta P/(4L)$ versus $8V/D$ on logarithmic coordinates,

$$n' = \frac{d \ln [D\Delta P/(4L)]}{d \ln (8V/D)} \tag{6-63}$$

By plotting capillary viscometry data in this way, they can be used directly for pressure drop design calculations, or to construct the rheogram for the fluid. For pressure drop calculation, the flow rate and diameter determine the velocity, from which $8V/D$ is calculated and $D\,\Delta P/(4L)$ read from the plot. For a newtonian fluid, $n' = 1$ and the shear rate at the wall is $\dot{\gamma} = 8V/D$. For a power law fluid, $n' = n$. To construct a rheogram, n' is obtained from the slope of the experimental plot at a given value of $8V/D$. The shear rate at the wall is given by Eq. (6-62), and the corresponding shear stress at the wall is $\tau_w = D\Delta P/(4L)$ read from the plot. By varying the value of $8V/D$, the shear stress versus shear rate plot can be constructed.

The generalized approach of A. B. Metzner and J. C. Reed [*AIChE J.* **1:** 434 (1955)] for time-independent nonnewtonian fluids uses a modified Reynolds number

$$\mathrm{Re}_{\mathrm{MR}} \equiv \frac{D^{n'} V^{2-n'} \rho}{K' 8^{n'-1}} \tag{6-64}$$

where K' satisfies

$$\frac{D\Delta P}{4L} = K' \left(\frac{8V}{D} \right)^{n'} \tag{6-65}$$

With this definition, $f = 16/\mathrm{Re}_{\mathrm{MR}}$ is automatically satisfied at the value of $8V/D$ where K' and n' are evaluated. For newtonian fluids, $K' = \mu$ and $n' = 1$; for power law fluids, $K' = K[(1 + 3n)/(4n)]^n$ and $n' = n$. For Bingham plastics, K' and n' are variable, given as a function of τ_w (A. B. Metzner, *Ind. Eng. Chem.* **49:** 1429–1432 [1957]).

$$K' = \tau_w^{1-n'} \left[\frac{\mu_\infty}{1 - 4\tau_y/3\tau_w + (\tau_y/\tau_w)^4/3} \right]^{n'} \tag{6-66}$$

$$n' = \frac{1 - 4\tau_y/(3\tau_w) + (\tau_y/\tau_w)^4/3}{1 - (\tau_y/\tau_w)^4} \tag{6-67}$$

For laminar flow of power law fluids in channels of noncircular cross section, see R. S. Schechter [*AIChE J.* **7:** 445–448 (1961)]; J. A. Wheeler and E. H. Wissler [*AIChE J.* **11:** 207–212 (1965)]; R. B. Bird, R. C. Armstrong, and O. Hassager [*Dynamics of Polymeric Liquids,* vol. 1: *Fluid Mechanics,* Wiley, New York, 1977]; and A. H. P. Skelland [*Nonnewtonian Flow and Heat Transfer,* Wiley, New York, 1967].

Steady, fully developed laminar flows of viscoelastic fluids in straight, constant-diameter pipes show no effects of viscoelasticity. The viscous component of the constitutive equation may be used to develop the flow rate–pressure drop relations, which apply downstream of the entrance region after viscoelastic effects have disappeared. A similar situation exists for time-dependent fluids in pipes of sufficient length.

The *transition to turbulent flow* begins at $\mathrm{Re}_{\mathrm{MR}}$ in the range of 2000 to 2500 [A. B. Metzner and J. C. Reed, *AIChE J.* **1:** 434 (1955)]. For Bingham plastic materials, K' and n' must be evaluated for the τ_w condition in question in order to determine $\mathrm{Re}_{\mathrm{MR}}$ and establish whether the flow is laminar. An alternative method for Bingham plastics is by R. W. Hanks [*AIChE J.* **9:** 306 (1963); **14:** 691 (1968)]; R. W. Hanks and D. R. Pratt [*Soc. Petrol. Engrs. J.* **7:** 342 (1967)]; and G. W. Govier and K. Aziz [*The Flow of Complex Mixtures in Pipes,* Krieger, Huntington, N.Y., 1977, pp. 213–215]. The transition from laminar to turbulent flow is influenced by *viscoelastic* properties [A. B. Metzner and M. G. Park, *J. Fluid Mech.* **20:** 291 (1964)] with the critical value of $\mathrm{Re}_{\mathrm{MR}}$ increased to beyond 10,000 for some materials. As a rough guide, the lower limit for the Fanning friction factor in laminar flow is ~0.01 for a wide range of rheological behavior.

For *turbulent flow of nonnewtonian fluids,* the design chart of D. W. Dodge and A. B. Metzner [*AIChE J.* **5:** 189 (1959)], Fig. 6-11, is widely used. K. C. Wilson and A. D. Thomas [*Can. J. Chem. Eng.* **63:** 539–546 (1985)] give friction factor equations for turbulent flow of power law fluids and Bingham plastic fluids.

Power law fluids:

$$\frac{1}{\sqrt{f}} = \frac{1}{\sqrt{f_N}} + 8.2 \frac{1-n}{1+n} + 1.77 \ln \left(\frac{1+n}{2} \right) \tag{6-68}$$

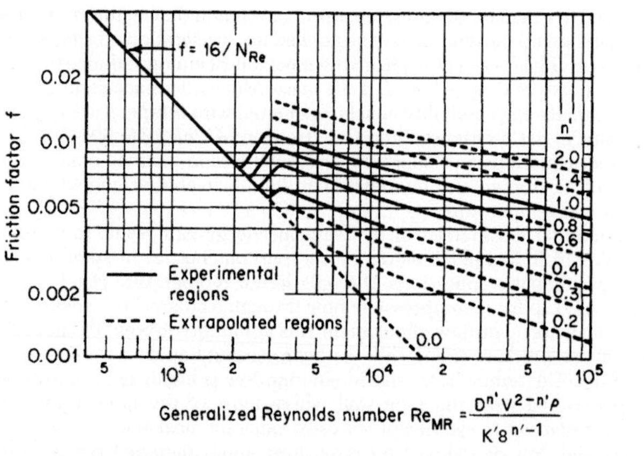

FIG. 6-11 Fanning friction factor for nonnewtonian flow. The abscissa is defined in Eq. (6-65). [From D. W. Dodge and A. B. Metzner, *Am. Inst. Chem. Eng. J.,* **5:** 189 (1959).]

where f_N is the friction factor for newtonian fluid evaluated at $\mathrm{Re} = DV\rho/\mu_{\mathrm{eff}}$ and where the effective viscosity is

$$\mu_{\mathrm{eff}} = K \left(\frac{3n+1}{4n} \right)^{n-1} \left(\frac{8V}{D} \right)^{n-1} \tag{6-69}$$

Bingham fluids:

$$\frac{1}{\sqrt{f}} = \frac{1}{\sqrt{f_N}} + 1.77 \ln \left(\frac{(1-\xi)^2}{1+\xi} \right) + \xi(10 + 0.884\xi) \tag{6-70}$$

where f_N is evaluated at $\mathrm{Re} = DV\rho/\mu_\infty$ and $\xi = \tau_y/\tau_w$. Iteration is required to use this equation since $\tau_w = f\rho V^2/2$.

Drag Reduction In turbulent flow, drag reduction can be achieved by adding soluble high-molecular-weight polymers even in extremely low concentration to newtonian liquids. The reduction in friction is generally believed to be associated with the extensional thickening viscoelastic nature of the solutions effective in the wall region. For a given polymer, there is a minimum molecular weight necessary to initiate drag reduction at a given flow rate, and a critical concentration above which drag reduction will not occur [O. K. Kim, R. C. Little, and R. Y. Ting, *J. Colloid Interface Sci.* **47:** 530–535 (1974)]. Drag reduction is reviewed by J. W. Hoyt [*J. Basic Eng.* **94:** 258–285 (1972)]; R. C. Little et al. [*Ind. Eng. Chem. Fundam.* **14:** 283–296 (1975)]; and P. S. Virk [*AIChE J.* **21:** 625–656 (1975)]. At maximum possible drag reduction in smooth pipes,

$$\frac{1}{\sqrt{f}} = -19 \log \left(\frac{50.73}{\mathrm{Re}\sqrt{f}} \right) \tag{6-71}$$

or approximately,

$$f = \frac{0.58}{\mathrm{Re}^{0.58}} \tag{6-72}$$

for 4000 < Re < 40,000. The actual drag reduction depends on the polymer system. For further details, see P. S. Virk [*AIChE J.* **21:** 625–656 (1975)]. More recently, K. D. Housiadas and A. N. Beris [*Int. J. Heat Fluid Flow* **42:** 49–67 (2013)] analyzed direct numerical simulation results to develop an expression for the drag reduction, defined as $1 - (f/f_s)$ where f and f_s are the Fanning friction factors for the polymer solution and the pure solvent, respectively, at the same Reynolds number. The expression depends on two parameters, a Weissenberg number and a limiting drag reduction (LDR). The *Weissenberg number* is the ratio of polymer relaxation time to a wall friction time scale. The LDR characterizes the extensional characteristics of the polymer solution rheology, and it depends on the polymer, polymer molecular weight, and polymer concentration. Note that the Reynolds number is defined based on the wall shear viscosity, which can be approximated by the laminar shear viscosity evaluated under the same shear rate conditions [K. D. Housiadas and A. N. Beris, *Int. J. Heat Fluid Flow* **42:** 49–67 (2013) and *Phys. Fluids* **16:** 1581–1586 (2004)].

Economic Pipe Diameter, Turbulent Flow The economic optimum pipe diameter may be computed so that the last increment of investment reduces the operating cost enough to produce the required minimum return on investment. For long cross-country pipelines, either alloy pipes of appreciable length and complexity or pipelines with control valves, detailed analyses of investment, and operating costs should be made. M. Peters and K. Timmerhaus [*Plant Design and Economics for Chemical Engineers,* 4th ed., McGraw-Hill, New York, 1991] provide a detailed method for determining the economic optimum size. For pipelines of the lengths usually encountered in chemical plants and petroleum refineries, simplified selection charts are often adequate. In many cases there is an economic optimum velocity that is nearly independent of diameter, which may be used to estimate the economic diameter from the flow rate. For low-viscosity liquids in Schedule 40 steel pipe, economic optimum velocity is typically in the range of 1.8 to 2.4 m/s (6 to 8 ft/s). For gases with density ranging from 0.2 to 20 kg/m^3 (0.012 to 1.2 lbm/ft^3), the economic optimum velocity is about 40 to 9 m/s (130 to 30 ft/s). Charts and rough guidelines for economic optimum size do not apply to multiphase flows.

Economic Pipe Diameter, Laminar Flow Pipelines for the transport of high-viscosity liquids are seldom designed purely on the basis of economics. More often the size is dictated by operability considerations such as available pressure drop, shear rate, or residence time distribution. M. Peters and K. Timmerhaus [*Plant Design and Economics for Chemical Engineers,* 4th ed., McGraw-Hill, New York, 1991] provide an economic pipe diameter chart for laminar flow. For nonnewtonian fluids, see A. H. P. Skelland [*Nonnewtonian Flow and Heat Transfer,* Wiley, New York, 1967].

Vacuum Flow When gas flows under high vacuum conditions or through very small openings, the continuum hypothesis is no longer appropriate if the channel dimension is not very large compared to the mean free path of the gas. When the mean free path is comparable to the channel dimension, flow is dominated by collisions of molecules with the wall, rather than by collisions between molecules. An approximate expression based on G. P. Brown et al. [*J. Appl. Phys.* **17:** 802–813 (1946)] for the mean free path is

$$\lambda = \left(\frac{2\mu}{p}\right)\sqrt{\frac{8RT}{\pi M_w}} \tag{6-73}$$

The Knudsen number Kn is the ratio of the mean free path to the channel dimension. For pipe flow, Kn = λ/D. *Molecular flow* is characterized by Kn > 1.0; *continuum viscous* (laminar or turbulent) flow is characterized by Kn < 0.01. *Transition* or *slip* flow applies over the range 0.01 < Kn < 1.0.

Vacuum flow is usually described with flow variables different from those used for normal pressures, often leading to confusion. *Pumping speed S* is the actual volumetric flow rate of gas through a flow cross section. *Throughput Q* is the product of pumping speed and absolute pressure. In SI, Q has units of Pa · m^3/s.

$$Q = Sp \tag{6-74}$$

The mass flow rate w is related to the throughput by using the ideal gas law.

$$w = \frac{M_w}{RT}Q \tag{6-75}$$

The relation between throughput and pressure drop $\Delta p = p_1 - p_2$ across a flow element is written in terms of the *conductance C. Resistance* is the reciprocal of conductance. Conductance has dimensions of volume per time.

$$Q = C\,\Delta p \tag{6-76}$$

The conductance of a series of flow elements is given by

$$\frac{1}{C} = \frac{1}{C_1} + \frac{1}{C_2} + \frac{1}{C_3} + \cdots \tag{6-77}$$

while for elements in parallel,

$$C = C_1 + C_2 + C_3 + \cdots \tag{6-78}$$

For a vacuum pump of speed S_p withdrawing from a vacuum vessel through a connecting line of conductance C, the pumping speed at the vessel is

$$S = \frac{S_p C}{S_p + C} \tag{6-79}$$

Molecular Flow Under molecular flow conditions, conductance is independent of pressure. It is proportional to $\sqrt{T/M_w}$, with the proportionality constant a function of geometry. For fully developed pipe flow,

$$C = \frac{\pi D^3}{8L}\sqrt{\frac{RT}{M_w}} \tag{6-80}$$

TABLE 6-4 Constants for Circular Annuli

D_2/D_1	K	D_2/D_1	K
0	1.00	0.707	1.254
0.259	1.072	0.866	1.430
0.500	1.154	0.966	1.675

For an orifice of area A,

$$C = 0.40A\sqrt{\frac{RT}{M_w}} \tag{6-81}$$

Conductance equations for several other geometries are given by J. L. Ryans and D. L. Roper [*Process Vacuum System Design and Operation,* chap. 2, McGraw-Hill, New York, 1986]. For a circular annulus of outer and inner diameters D_1 and D_2 and length L, the method of A. Guthrie and R. K. Wakerling [*Vacuum Equipment and Techniques,* McGraw-Hill, New York, 1949] may be written

$$C = 0.42K\frac{(D_1 - D_2)^2(D_1 + D_2)}{L}\sqrt{\frac{RT}{M_w}} \tag{6-82}$$

where K is a dimensionless constant with values given in Table 6-4.

For a short pipe of circular cross section, the conductance as calculated for an orifice from Eq. (6-81) is multiplied by a correction factor K which may be approximated as [E. H. Kennard, *Kinetic Theory of Gases,* McGraw-Hill, New York, 1938, pp. 306–308]

$$K = \frac{1}{1 + (L/D)} \quad \text{for} \quad 0 \le L/D \le 0.75 \tag{6-83}$$

$$K = \frac{1 + 0.8(L/D)}{1 + 1.90(L/D) + 0.6(L/D)^2} \quad \text{for} \quad L/D > 0.75 \tag{6-84}$$

For $L/D > 100$, the error in neglecting the end correction by using the fully developed pipe flow equation, Eq. (6-80), is less than 2 percent. For rectangular channels, see C. E. Normand [*Ind. Eng. Chem.* **40:** 783–787 (1948)].

H. S. Yu and E. M. Sparrow [*J. Basic Eng.* **70:** 405–410 (1970)] give a chart for slot seals with or without a sheet located in or passing through the seal, giving the mass flow rate as a function of the ratio of seal plate thickness to gap opening.

Slip Flow In the transition region between molecular flow and continuum viscous flow, the conductance for fully developed pipe flow is most easily obtained by the method of G. P. Brown et al. [*J. Appl. Phys.* **17:** 802–813 (1946)] which uses the parameter

$$X = \sqrt{\frac{8}{\pi}}\left(\frac{\lambda}{D}\right) = \left(\frac{2\mu}{p_m D}\right)\sqrt{\frac{RT}{M}} \tag{6-85}$$

where p_m is the arithmetic mean absolute pressure. A correction factor F, read from Fig. 6-12 as a function of X, is applied to the conductance for viscous flow.

$$C = F\frac{\pi D^4 p_m}{128\,\mu L} \tag{6-86}$$

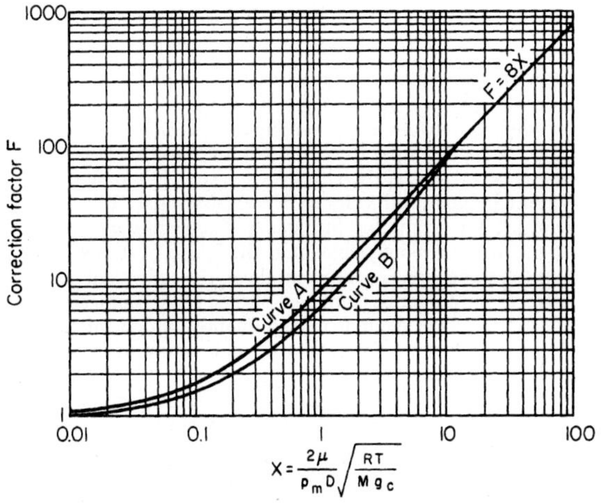

FIG. 6-12 Correction factor for Poiseuille's equation at low pressures. Curve *A:* experimental curve for glass capillaries and smooth metal tubes. [From G. P. Brown, et al., *J. Appl. Phys.*, **17**, 802 (1946).] Curve *B:* experimental curve for iron pipe. [From Riggle, *courtesy of E. I. du Pont de Nemours & Co.*]

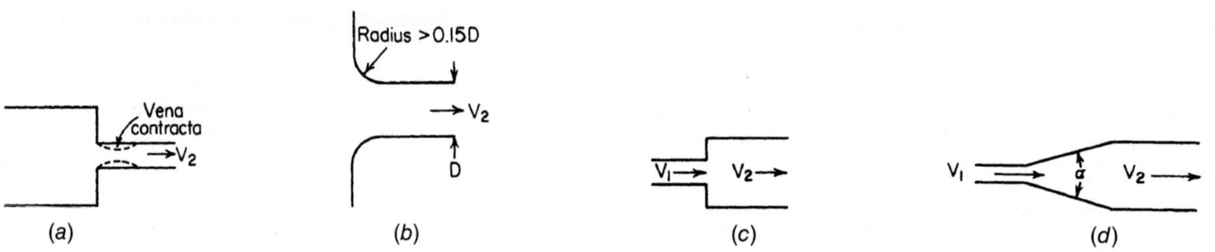

FIG. 6-13 Contractions and enlargements: (*a*) sudden contraction, (*b*) rounded contraction, (*c*) sudden enlargement, and (*d*) uniformly diverging duct.

For slip flow through *square channels,* see M. W. Milligan and H. J. Wilkerson [*J. Eng. Ind.* **95:** 370–372 (1973)]. For slip flow through *annuli,* see W. J. Maegley and A. S. Berman [*Phys. Fluids* **15:** 780–785 (1972)].

The *pump-down time* θ for evacuating a vessel in the absence of air in-leakage is given approximately by

$$\theta = \left(\frac{V_t}{S_0}\right) \ln\left(\frac{p_1 - p_0}{p_2 - p_0}\right) \tag{6-87}$$

where V_t = volume of vessel plus volume of piping between vessel and pump; S_0 = system speed as given by Eq. (6-79), assumed independent of pressure; p_1 = initial vessel pressure; p_2 = final vessel pressure; and p_0 = lowest pump intake pressure attainable with the pump in question. See S. Dushman and J. M. Lafferty [*Scientific Foundations of Vacuum Technique,* 2d ed., Wiley, New York, 1962].

The rate at which inert materials must be removed by a pumping system after the pump-down stage depends on the in-leakage of air at the various fittings, connections, etc. Air leakage is often correlated with system volume and pressure, but this approach is uncertain because the number and size of leaks may not correlate with system volume, and leakage is sensitive to maintenance quality. J. L. Ryans and D. L. Roper [*Process Vacuum System Design and Operation,* chap. 2, McGraw-Hill, New York, 1986] present a thorough discussion of air leakage.

FRICTIONAL LOSSES IN PIPELINE ELEMENTS

The viscous loss term in the mechanical energy balance for most cases is obtained experimentally. For many common fittings found in piping systems, such as expansions, contractions, elbows, and valves, data are available to estimate the losses. Substitution into the energy balance allows calculation of pressure drop. A common error is to assume that pressure drop and frictional losses are equivalent. Equation (6-19) shows that in addition to frictional losses, other factors such as shaft work and velocity or elevation change influence pressure drop.

Losses l_v for incompressible flow in sections of straight pipe of constant diameter may be calculated as previously described, using the Fanning friction factor:

$$l_v = \frac{\Delta P}{\rho} = \frac{2fV^2L}{D} \tag{6-88}$$

where ΔP = drop in equivalent pressure, $P = p + \rho gZ$, with p = pressure, ρ = fluid density, g = acceleration of gravity, and Z = elevation. Losses in the fittings of a piping network are frequently termed *minor losses* or *miscellaneous losses.* These descriptions are misleading because in process piping fitting losses may be greater than the losses in straight piping sections.

Equivalent Length and Velocity Head Methods Two methods are in common use for estimating fitting loss. The *equivalent length* method reports the losses in a piping element as the length of straight pipe which would have the same loss. For turbulent flows, the equivalent length is usually reported as a number of diameters of pipe of the same size as the fitting connection; L_e/D is given as a fixed quantity, independent of D. This approach tends to be most accurate for a single fitting size and loses accuracy with deviation from this size. For laminar flows, L_e/D correlations normally have size dependence through a Reynolds number term.

The other method is the *velocity head* method. The term $V^2/2g$ has dimensions of length and is commonly called a *velocity head* but this name is also used for $\rho V^2/2$. In the velocity head method, the losses are reported as a number of velocity heads K. Then the engineering Bernoulli equation for an incompressible fluid can be written as

$$p_1 - p_2 = \alpha_2\frac{\rho V_2^2}{2} - \alpha_1\frac{\rho V_1^2}{2} + \rho g(Z_2 - Z_1) + K\frac{\rho V^2}{2} \tag{6-89}$$

where V is the reference velocity upon which the velocity head loss coefficient K is based. For a section of straight pipe, $K = 4fL/D$.

Contraction and Entrance Losses For a *sudden contraction* at a sharp-edged entrance to a pipe or sudden reduction in cross-sectional area of a channel, as shown in Fig. 6-13*a,* the loss coefficient based on the downstream velocity V_2 is given for *turbulent flow* in Crane Co. [Tech. Paper 410 1980] approximately by

$$K = 0.5\left(1 - \frac{A_2}{A_1}\right) \tag{6-90}$$

Example 6-5 Entrance Loss Water, ρ = 1000 kg/m³, flows from a large vessel through a sharp-edged entrance into a pipe at a velocity in the pipe of 2 m/s. The flow is turbulent. Estimate the pressure drop from the vessel into the pipe.

With $A_2/A_1 \sim 0$, the viscous loss coefficient is $K = 0.5$ from Eq. (6-90). The mechanical energy balance, Eq. (6-19) with $V_1 = 0$ and $Z_2 - Z_1 = 0$ and assuming uniform flow ($\alpha_2 = 1$), becomes

$$p_1 - p_2 = \frac{\rho V_2^2}{2} + 0.5\frac{\rho V_2^2}{2} = 2000 + 1000 = 3000 \text{ Pa}$$

Note that the total pressure drop consists of 0.5 velocity head of frictional loss, and 1 velocity head of acceleration. The frictional contribution is a permanent loss of mechanical energy. The acceleration contribution is reversible; if the fluid were subsequently decelerated in a frictionless diffuser, a pressure rise would occur.

For a *trumpet-shaped* rounded entrance, with a radius of rounding greater than about 15 percent of the pipe diameter (Fig. 6-13*b*), the turbulent flow loss coefficient K is only about 0.1 [J. F. Vennard and R. L. Street, *Elementary Fluid Mechanics,* 5th ed., Wiley, New York, 1975, pp. 420–421]. Rounding of the inlet prevents formation of the *vena contracta,* reducing the resistance to flow.

For *laminar flow* the losses in sudden contraction may be estimated for area ratios $A_2/A_1 < 0.2$ by an equivalent additional pipe length L_e given by

$$L_e/D = 0.3 + 0.04 \text{ Re} \tag{6-91}$$

where D is the diameter of the smaller pipe and Re is the Reynolds number in the smaller pipe. For laminar flow in the entrance to rectangular ducts, see R. K. Shah [*J. Fluids Eng.* **100:** 177–179 (1978)]. For creeping flow, Re < 1, of power law fluids, the entrance loss is approximately $L_e/D = 0.3/n$ [D. V. Boger et al., *J. Nonnewtonian Fluid Mech.* **4:** 239–248 (1978)]. For viscoelastic fluid flow in circular channels with sudden contraction, a toroidal vortex forms upstream of the contraction plane. Such flows are reviewed by D. V. Boger [*Ann. Review Fluid Mech.* **19:** 157–182 (1987)].

For creeping flow through *conical converging channels,* the viscous pressure drop $\Delta p = \rho l_v$ may be computed by integration of the differential form of the Hagen-Poiseuille equation, Eq. (6-35), provided the angle of convergence is small. The result for a power law fluid is

$$\Delta p = 4K\left(\frac{3n+1}{4n}\right)^n\left(\frac{8V_2}{D_2}\right)^n\left\{\frac{1}{6n\tan(\alpha/2)}\left[1 - \left(\frac{D_2}{D_1}\right)^{3n}\right]\right\} \tag{6-92}$$

where D_1 = inlet diameter
D_2 = exit diameter
V_2 = velocity at the exit
α = total included angle

Equation (6-92) agrees with experimental data [Z. Kemblowski and T. Kiljanski, *Chem. Eng. J.* (*Lausanne*), **9:** 141–151 (1975)] for $\alpha < 11°$. For newtonian liquids, Eq. (6-95) simplifies to

$$\Delta p = \mu\left(\frac{32V_2}{D_2}\right)\left\{\frac{1}{6\tan(\alpha/2)}\left[1 - \left(\frac{D_2}{D_1}\right)^3\right]\right\} \tag{6-93}$$

For creeping flow through noncircular converging channels, the differential form of the Hagen-Poiseuille equation with equivalent diameter given by Eqs. (6-49) to (6-51) may be used, provided the convergence is gradual.

Expansion and Exit Losses For ducts of any cross section, the frictional loss for a *sudden enlargement* (Fig. 6-13*c*) with turbulent flow is given by the Borda-Carnot equation:

$$l_v = \frac{V_1^2 - V_2^2}{2} = \frac{V_1^2}{2}\left(1 - \frac{A_1}{A_2}\right)^2 \tag{6-94}$$

where V_1 = velocity in smaller duct
V_2 = velocity in larger duct
A_1 = cross-sectional area of smaller duct
A_2 = cross-sectional area of larger duct

Equation (6-94) is valid for incompressible flow. For compressible flows, see R. P. Benedict et al. [*J. Eng. Power* **98:** 327–334 (1976)]. For an infinite expansion, $A_1/A_2 = 0$, Eq. (6-97) shows that the *exit loss* from a pipe is 1 velocity head. This exit loss is due to the dissipation of the discharged jet; there is no pressure drop at the exit.

For creeping newtonian flow (Re < 1), the frictional loss due to a sudden enlargement should be obtained from the same equation for a sudden contraction, Eq. (6-91). Note, however, that D. V. Boger et al. [*J. Nonnewtonian Fluid Mech.* **4:** 239–248 (1978)] give an exit friction equivalent length of 0.12 diameter, increasing for power law fluids as the exponent decreases. For laminar flows at higher Reynolds numbers, the pressure drop is twice that given by Eq. (6-91). This results from the velocity profile factor α in the mechanical energy balance being 2.0 for the parabolic laminar velocity profile.

If the transition from a small to a large duct of any cross-sectional shape is accomplished by a *uniformly diverging duct* (see Fig. 6-13*d*) with a straight axis, the total frictional pressure drop can be computed by integrating the differential form of Eq. (6-88), $dl_v/dx = 2fV^2/D$ over the length of the expansion, provided the total angle α between the diverging walls is less than 7°. For angles between 7° and 45°, the loss coefficient may be estimated as 2.6 sin (α/2) times the loss coefficient for a sudden expansion; see W. B. Hooper [*Chem. Eng.* Nov. 7, 1988]. A. H. Gibson [*Hydraulics and Its Applications,* 5th ed., Constable, London, 1952, p. 93] recommends multiplying the sudden enlargement loss by 0.13 for 5° < α < 7.5° and by $0.0110\alpha^{1.22}$ for 7.5° < α < 35°. For angles greater than 35° to 45°, the losses are normally considered equal to those for a sudden expansion, although in some cases the losses may be greater. Expanding flow through standard pipe reducers should be treated as sudden expansions.

Trumpet-shaped enlargements for *turbulent flow* designed for constant decrease in velocity head per unit length were found by A. H. Gibson [*Hydraulics and Its Applications,* 5th ed., Constable, London, 1952, p. 95] to give 20 to 60 percent less frictional loss than straight taper pipes of the same length.

When *viscoelastic* liquids are extruded through a die at a low Reynolds number, the extrudate may expand to a diameter several times greater than the die diameter, whereas for a newtonian fluid the diameter expands only 10 percent. This phenomenon, called *die swell,* is most pronounced with short dies [W. W. Graessley et al., *Trans. Soc. Rheol.* **14:** 519–544 (1970)]. For velocity distribution measurements near the die exit, see D. D. Goulden and W. C. MacSporran [*J. Nonnewtonian Fluid Mech.* **1:** 183–198 (1976)] and B. A. Whipple and C. T. Hill [*AIChE J.* **24:** 664–671 (1978)]. At high flow rates, the extrudate becomes distorted, suffering *melt fracture,* a phenomenon reviewed by M. M. Denn [*Ann. Review Fluid Mech.* **22:** 13–34 (1990)]. A. V. Ramamurthy [*J. Rheol.* **30:** 337–357 (1986)] found a dependence of apparent stick-slip behavior in melt fracture to be dependent on the material of construction of the die.

Fittings and Valves For *turbulent flow,* the frictional loss for fittings and valves can be expressed by the equivalent length or velocity head methods. As fitting size is varied, *K* values are relatively more constant than L_e/D values, but since fittings generally do not achieve geometric similarity between sizes, *K* values tend to decrease with increasing fitting size. Table 6-5 gives *K* values for many types of fittings and valves.

Manufacturers of valves, especially control valves, express valve capacity in terms of a flow coefficient C_v, which gives the flow rate through the valve in gallons per minute of water at 60°F under a pressure drop of 1 lbf/in². It is related to *K* by

$$C_v = \frac{C_1 d^2}{\sqrt{K}} \tag{6-95}$$

where C_1 is a dimensional constant equal to 29.9 and *d* is the diameter of the valve connections in inches.

TABLE 6-5 Additional Frictional Loss for Turbulent Flow Through Fittings and Valves[a]

Type of fitting or valve	Additional friction loss, equivalent no. of velocity heads, *K*
45° ell, standard[b,c,d,e,f]	0.35
45° ell, long radius[c]	0.2
90° ell, standard[b,c,e,f,g,h]	0.75
Long radius[b,c,d,e]	0.45
Square or miter[h]	1.3
180° bend, close return[b,c,e]	1.5
Tee, standard, along run, branch blanked off[e]	0.4
Used as ell, entering run[g,i]	1.0
Used as ell, entering branch[c,g,i]	1.0
Branching flow[i,j,k]	1[l]
Coupling[c,e]	0.04
Union[e]	0.04
Gate valve,[b,e,m] open	0.17
¾ open	0.9
½ open	4.5
¼ open	24.0
Diaphragm valve, open	2.3
¾ open	2.6
½ open	4.3
¼ open	21.0
Globe valvee,[e,m]	
Bevel seat, open	6.0
½ open	9.5
Composition seat, open	6.0
½ open	8.5
Plug disk, open	9.0
¾ open	13.0
½ open	36.0
¼ open	112.0
Angle valve,[b,e] open	2.0
Y or blowoff valve,[b,m] open	3.0
Plug cock	
θ = 5°	0.05
θ = 10°	0.29
θ = 20°	1.56
θ = 40°	17.3
θ = 60°	206.0
Butterfly valve	
θ = 5°	0.24
θ = 10°	0.52
θ = 20°	1.54
θ = 40°	10.8
θ = 60°	118.0
Check valve,[b,e,m] swing	2.0
Disk	10.0
Ball	70.0
Foot valve[e]	15.0
Water meter,[h] disk	7.0
Piston	15.0
Rotary (star-shaped disk)	10.0
Turbine wheel	6.0

[a]Lapple, *Chem. Eng.* **56**(5): 96–104 (1949), general survey reference.
[b]"Flow of Fluids through Valves, Fittings, and Pipe," Tech. Pap. 410, Crane Co., 1969.
[c]Freeman, *Experiments upon the Flow of Water in Pipes and Pipe Fittings,* American Society of Mechanical Engineers, New York, 1941.
[d]Giesecke, *J. Am. Soc. Heat. Vent. Eng.* **32:** 461 (1926).
[e]*Pipe Friction Manual,* 3d ed., Hydraulic Institute, New York, 1961.
[f]Ito, *J. Basic Eng.* **82:** 131–143 (1960).
[g]Giesecke and Badgett, *Heat. Piping Air Cond.* **4**(6): 443–447 (1932).
[h]Schoder and Dawson, *Hydraulics,* 2d ed., McGraw-Hill, New York, 1934, p. 213.
[i]Hoopes, Isakoff, Clarke, and Drew, *Chem. Eng. Prog.* **44:** 691–696 (1948).
[j]Gilman, *Heat. Piping Air Cond.* **27**(4): 141–147 (1955).
[k]McNown, *Proc. Am. Soc. Civ. Eng.* **79,** Separate 258, 1–22 (1953); discussion, ibid., **80,** Separate 396, 19–45 (1954). For the effect of branch spacing on junction losses in dividing flow, see Hecker, Nystrom, and Qureshi, *Proc. Am. Soc. Civ. Eng., J. Hydraul. Div.* **103**(HY3): 265–279 (1977).
[l]This is pressure drop (including friction loss) between run and branch, based on velocity in the mainstream before branching. Actual value depends on the flow split, ranging from 0.5 to 1.3 if mainstream enters run and from 0.7 to 1.5 if mainstream enters branch.
[m]Lansford, *Loss of Head in Flow of Fluids through Various Types of 1½-in. Valves,* Univ. Eng. Exp. Sta. Bull. Ser. 340, 1943.

For *laminar* and *turbulent flow,* the "2K" method of W. B. Hooper [*Chem. Eng.* **88**(17), 96–100 (August 1981)] yields approximate fitting losses accounting for Re and fitting size of

$$K = \frac{K_1}{\text{Re}} + K_\infty\left(1 + \frac{1}{d}\right) \tag{6-96}$$

where d is the fitting size in inches and K_1 and K_∞ are shown in length and homogeneous equilibrium flow at 100.

Methods to calculate losses in *tee and wye junctions* for dividing and combining flow are given by D. S. Miller [*Internal Flow Systems,* 2d ed., chap. 13, British Hydrodynamics Research Association, Cranfield, UK, 1990], including effects of Reynolds number, angle between legs, area ratio, and radius. Junctions with more than three legs are also discussed. The sources of data for the loss coefficient charts are F. W. Blaisdell and P. W. Manson [*U.S. Dept. Agric. Res. Serv. Tech. Bull.* 1283 (August 1963)] for combining flow and A. Gardel [*Bull. Tech. Suisses Romande* **85**(9): 123–130 (1957); **85**(10): 143–148 (1957)] together with additional unpublished data for dividing flow.

D. S. Miller [*Internal Flow Systems,* 2d ed., chap. 13, British Hydrodynamics Research Association, Cranfield, UK, 1990] gives the most complete information on losses in *bends and curved pipes.* For turbulent flow in circular cross-section bends of constant area, as shown in Fig. 6-14*a,* a more accurate estimate of the loss coefficient K than that given in Tables 6-5 and 6-6 is

$$K = K^* C_{\text{Re}} C_o C_f \tag{6-97}$$

where K^*, given in Fig. 6-14*b*, is the loss coefficient for a smooth-walled bend at a Reynolds number of 10^6. The Reynolds number correction factor C_{Re} is given in Fig. 6-14*c*. For $0.7 < r/D < 1$ or for $K^* < 0.4$, use the C_{Re} value for $r/D = 1$. Otherwise, if $r/D < 1$, obtain C_{Re} from

$$C_{\text{Re}} = \frac{K^*}{K^* + 0.2(1 - C_{\text{Re},r/D=1})} \tag{6-98}$$

The correction C_o (Fig. 6-14*d*) accounts for the extra losses due to developing flow in the outlet tangent of the pipe, of length L_o. The total loss for the bend plus outlet pipe includes the bend loss K plus the straight pipe frictional loss in the outlet pipe $4fL_o/D$. Note that $C_o = 1$ for L_o/D greater than the termination of the curves on Fig. 6-14*d*, which indicate the distance at which fully developed flow in the outlet pipe is reached. Finally, the roughness correction is

$$C_f = \frac{f_{\text{rough}}}{f_{\text{smooth}}} \tag{6-99}$$

where f_{rough} is the friction factor for a pipe of diameter D with the roughness of the bend, at the bend inlet Reynolds number. Similarly, f_{smooth} is the friction factor for smooth pipe. For $\text{Re} > 10^6$ and $r/D \geq 1$, use the value of C_f for $\text{Re} = 10^6$.

Example 6-6 Losses with Fittings and Valves It is desired to calculate the liquid level in the vessel shown in Fig. 6-15 required to produce a discharge velocity of 2 m/s. The fluid is water at 20°C with $\rho = 1000$ kg/m³ and $\mu = 0.001$ Pa · s, and the butterfly valve is at $\theta = 10°$. The pipe is 2-in Schedule 40, with an inner diameter of 0.0525 m. The pipe roughness is 0.046 mm. Assuming the flow is turbulent and taking the velocity profile factor $\alpha = 1$, the engineering Bernoulli equation, Eq. (6-19), written between surfaces 1 and 2, where the pressures are both atmospheric and the fluid velocities are zero and $V = 2$ m/s, respectively, and there is no shaft work, simplifies to

$$gZ = \frac{V^2}{2} + l_v$$

Contributing to l_v are losses for the entrance to the pipe, the three sections of straight pipe, the butterfly valve, and the 90° bend. Note that no exit loss is used because the discharged jet is outside the control volume. Instead, the $V^2/2$ term accounts for the kinetic energy of the discharging stream. The Reynolds number in the pipe is

$$\text{Re} = \frac{DV\rho}{\mu} = \frac{0.0525 \times 2 \times 1000}{0.001} = 1.05 \times 10^5$$

From Fig. 6-9 or Eq. (6-37), at $\varepsilon/D = 0.046 \times 10^{-3}/0.0525 = 0.00088$, the friction factor is about 0.0054. The straight pipe losses are then

$$l_{v(\text{sp})} = \left(\frac{4fL}{D}\right)\frac{V^2}{2}$$
$$= \left(\frac{4 \times 0.0054 \times (1+1+1)}{0.0525}\right)\frac{V^2}{2}$$
$$= 1.23\frac{V^2}{2}$$

The losses from Table 6-5 in terms of velocity heads K are $K = 0.5$ for the sudden contraction and $K = 0.52$ for the butterfly valve. For the 90° standard radius ($r/D = 1$), the table gives $K = 0.75$. The value calculated using Table 6-6 is 0.38. The method of Eq. (6-97), using Fig. 6-14, gives

$$K = K^* C_{\text{Re}} C_o C_f$$
$$= 0.24 \times 1.24 \times 1.0 \times \left(\frac{0.0054}{0.0044}\right)$$
$$= 0.37$$

This value is close to the value from Table 6-6, and more accurate than the value in Table 6-5. The value $f_{\text{smooth}} = 0.0044$ is obtainable from Eq. (6-36) or Fig. 6-9.

The total losses are then

$$l_v = (1.23 + 0.5 + 0.52 + 0.37)\frac{V^2}{2} = 2.62\frac{V^2}{2}$$

and the liquid level Z is

$$Z = \frac{1}{g}\left(\frac{V^2}{2} + 2.62\frac{V^2}{2}\right) = 3.62\frac{V^2}{2g}$$
$$= \frac{3.62 \times 2^2}{2 \times 9.807} = 0.74 \text{ m}$$

Curved Pipes and Coils For flow through curved pipe or coil, a secondary circulation perpendicular to the main flow called the *Dean effect* occurs. This increases the friction relative to straight pipe flow and stabilizes laminar flow, delaying the transition Reynolds number to about

$$\text{Re}_{\text{crit}} = 2100\left(1 + 12\sqrt{\frac{D}{D_c}}\right) \tag{6-100}$$

where D_c is the coil diameter. Equation (6-100) is valid for $10 < D_c/D < 250$. The *Dean number* is defined as

$$\text{De} = \frac{\text{Re}}{(D_c/D)^{1/2}} \tag{6-101}$$

TABLE 6-6 2-K Method Friction Loss Parameters

		Fitting or valve		K_1	K_∞
Elbows	45°	Standard ($R/D = 1$) screwed, flanged, or welded		500	0.20
		Long radius ($R/D = 1.5$)		500	0.15
		Mitered	1 weld	500	0.25
			2 welds	500	0.15
	90°	Standard ($R/D = 1$) screwed		800	0.40
		Standard ($R/D = 1$) flanged or welded		800	0.25
		Long radius ($R/D = 1.5$)		800	0.20
		Mitered	1 weld	1000	1.15
			2 welds	800	0.35
			3 welds	800	0.30
			4 welds	800	0.27
			5 welds	800	0.25
	180°	Standard ($R/D = 1$), screwed		1000	0.60
		Standard ($R/D = 1$) flanged or welded		1000	0.35
		Long radius ($R/D = 1.5$)		1000	0.30
Tees	Used as elbows	Standard, screwed		500	0.70
		Long radius, screwed		800	0.40
		Standard, flanged or welded		800	0.80
		Stub-in-type branch		1000	1.00
	Run through	Screwed		200	0.10
		Flanged or welded		150	0.05
		Stub-in-type branch		100	0.00
Full open valves*	Gate, ball, or plug	Full line size, $\beta = 1.0$		300	0.10
		Reduced trim, $\beta = 0.9$		500	0.15
		Reduced trim, $\beta = 0.8$		1000	0.25
	Globe	Standard		1500	4.00
		Angle or Y-type		1000	2.00
	Diaphragm	Dam type		1000	2.00
	Butterfly			800	0.25
	Check	Lift		2000	10.0
		Swing		1500	1.50
		Tilting disk		1000	0.50

*For use as rough guide only. Consult manufacturer's data for greater accuracy.
source: Adapted from Hooper, *Chem. Eng.* Aug 24, 1981, pp. 96–100.

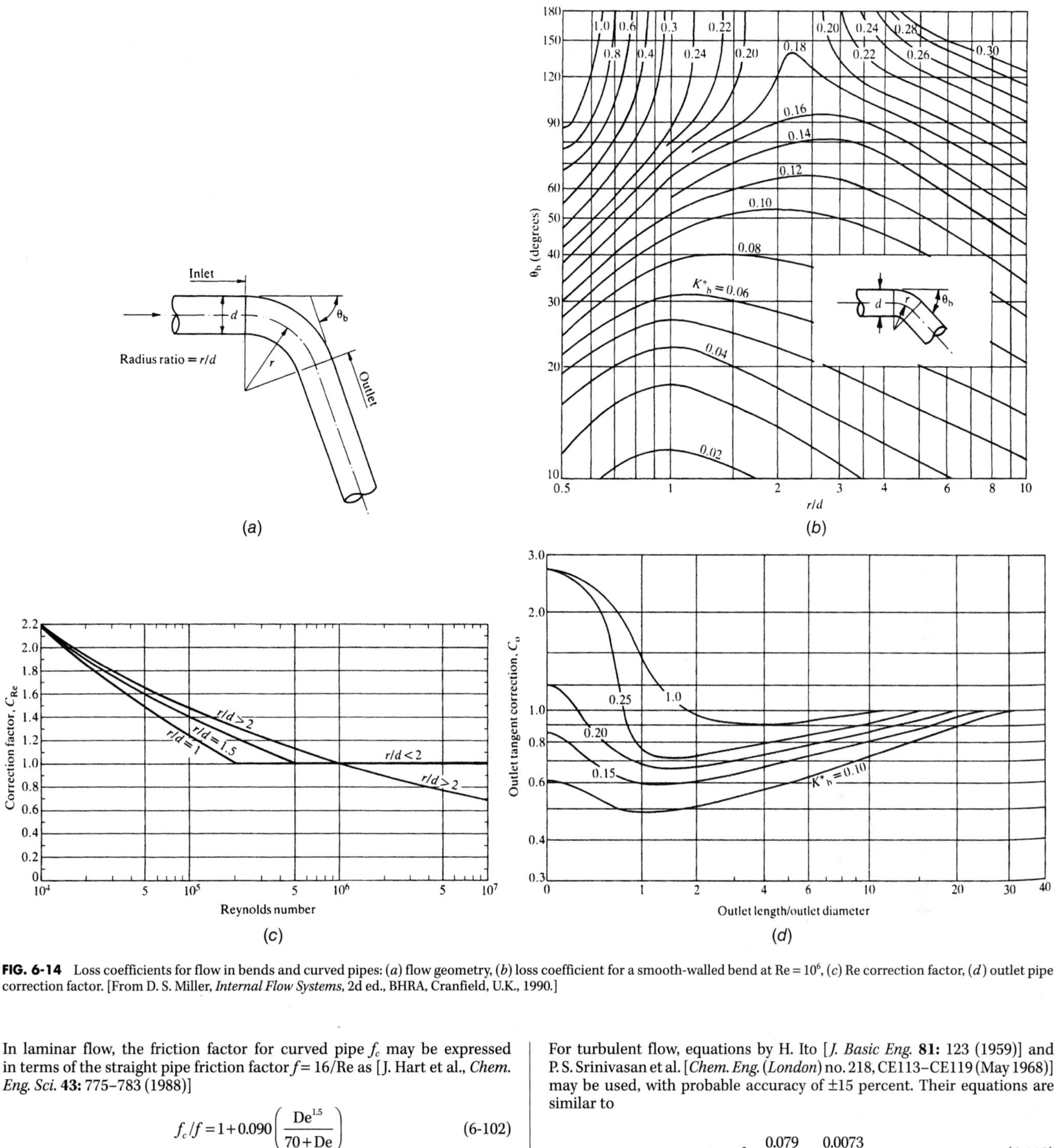

FIG. 6-14 Loss coefficients for flow in bends and curved pipes: (*a*) flow geometry, (*b*) loss coefficient for a smooth-walled bend at Re = 10^6, (*c*) Re correction factor, (*d*) outlet pipe correction factor. [From D. S. Miller, *Internal Flow Systems*, 2d ed., BHRA, Cranfield, U.K., 1990.]

In laminar flow, the friction factor for curved pipe f_c may be expressed in terms of the straight pipe friction factor $f = 16/\text{Re}$ as [J. Hart et al., *Chem. Eng. Sci.* **43:** 775–783 (1988)]

$$f_c/f = 1 + 0.090\left(\frac{\text{De}^{1.5}}{70 + \text{De}}\right) \tag{6-102}$$

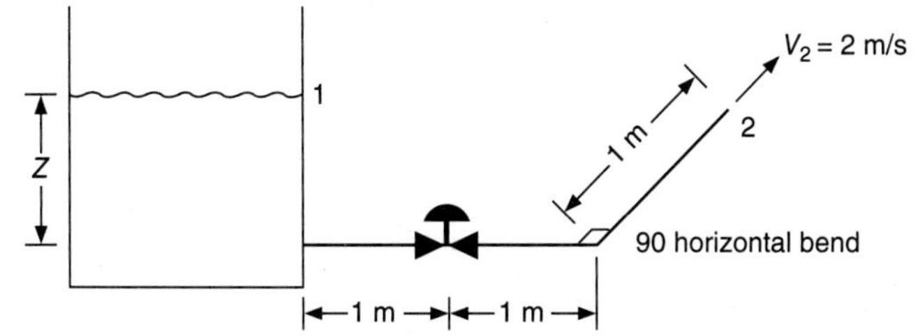

FIG. 6-15 Tank discharge example.

For turbulent flow, equations by H. Ito [*J. Basic Eng.* **81:** 123 (1959)] and P. S. Srinivasan et al. [*Chem. Eng.* (*London*) no. 218, CE113–CE119 (May 1968)] may be used, with probable accuracy of ±15 percent. Their equations are similar to

$$f_c = \frac{0.079}{\text{Re}^{0.25}} + \frac{0.0073}{\sqrt{(D_c/D)}} \tag{6-103}$$

The pressure drop for flow in *spirals* is discussed by P. S. Srinivasan et al. [*Chem. Eng.* (*London*) no. 218, CE113–CE119 (May 1968)] and S. Ali and C. V. Seshadri [*Ind. Eng. Chem. Process Des. Dev.* **10:** 328–332 (1971)]. For friction loss in laminar flow through *semicircular ducts,* see J. H. Masliyah and K. Nandakumar [*AIChE J.* **25:** 478–487 (1979)]; for curved channels of *square cross section,* see K. C. Cheng et al. [*J. Fluids Eng.* **98:** 41–48 (1976)].

For *nonnewtonian* (*power law*) *fluids* in coiled tubes, R. A. Mashelkar and G. V. Devarajan [*Trans. Inst. Chem. Eng.* (*London*), **54:** 108–114 (1976)] propose the correlation

$$f_c = (9.07 - 9.44n + 4.37n^2)(D/D_c)^{0.5}(\text{De}')^{-0.768 + 0.122n} \tag{6-104}$$

where De′ is a modified Dean number given by

$$\mathrm{De}' = \frac{1}{8}\left(\frac{6n+2}{n}\right)^n \mathrm{Re}_{MR}\sqrt{\frac{D}{D_c}} \tag{6-105}$$

and Re_{MR} is the Metzner-Reed Reynolds number, Eq. (6-64). This correlation was tested for the range De′ = 70 to 400, $D/D_c = 0.01$ to 0.135, and $n = 0.35$ to 1. See also D. R. Oliver and S. M. Asghar [*Trans. Inst. Chem. Eng.* (*London*), **53:** 181–186 (1975)].

Screens The pressure drop for incompressible flow across a screen of fractional free area α may be computed from

$$\Delta p = K\frac{\rho V^2}{2} \tag{6-106}$$

where ρ = fluid density
V = superficial velocity based on gross area of the screen
K = velocity head loss

$$K = \left(\frac{1}{C^2}\right)\left(\frac{1-\alpha^2}{\alpha^2}\right) \tag{6-107}$$

The discharge coefficient for the screen C with aperture D_s is given as a function of screen Reynolds number Re = $D_s(V/\alpha)\rho/\mu$ in Fig. 6-16 for *plain square-mesh screens,* α = 0.14 to 0.79. This curve fits most of the data within ±20 percent. In the laminar flow region, Re < 20, the discharge coefficient can be computed from

$$C = 0.1\sqrt{\mathrm{Re}} \tag{6-108}$$

Coefficients greater than 1.0 in Fig. 6-16 probably indicate partial pressure recovery downstream of the minimum aperture, due to rounding of the wires.

P. Grootenhuis [*Proc. Inst. Mech. Eng.* (*London*), **A168:** 837–846 (1954)] presents data indicating that for a series of screens, the total pressure drop equals the number of screens times the pressure drop for one screen, and is not affected by the spacing between screens or their orientation with respect to one another, and presents a correlation for frictional losses across plain square-mesh screens and sintered gauzes. Armour and Cannon [*AIChE J.* **14:** 415–420 (1968)] give a correlation based on a packed-bed model for plain, twill, and "dutch" weaves. For losses through monofilament fabrics see G. C. Pedersen [*Filtr. Sep.* **11:** 586–589 (1975)]. For screens *inclined at an angle* θ, use the normal velocity component V'

$$V' = V\cos\theta \tag{6-109}$$

[P. J. D. Carothers and W. D. Baines, *J. Fluids Eng.* **97:** 116–117 (1975)] in place of V in Eq. (6-109). This applies for Re > 500, C = 1.26, α ≤ 0.97, and 0 < θ < 45°, for square-mesh screens and diamond-mesh netting. Screens inclined at an angle to the flow direction also experience a tangential stress.

For *nonnewtonian* fluids in creeping flow, frictional loss across a square-woven or full-twill-woven screen can be estimated by considering the screen as a set of parallel tubes, each of diameter equal to the average minimum opening between adjacent wires, and length twice the diameter, without entrance effects [J. F. Carley and W. C. Smith, *Polym. Eng. Sci.* **18:** 408–415 (1978)]. For screen stacks, the losses of individual screens should be summed.

JET BEHAVIOR

A *free jet,* upon leaving an outlet, will entrain the surrounding fluid, expand, and decelerate. Total momentum is conserved as jet momentum is transferred to the entrained fluid. For practical purposes, a jet is considered free when its cross-sectional area is less than one-fifth of the total cross-sectional flow area of the region through which the jet is flowing [H. G. Elrod, *Heat. Piping Air Cond.* **26**(3): 149–155 (1954)], and the surrounding fluid is the same as the jet fluid. A *turbulent jet* in this discussion is considered to be a free jet issuing with Re > 2000. Additional discussion on the relation between Reynolds number and turbulence in jets is given by H. G. Elrod [*Heat. Piping Air Cond.* **26**(3): 149–155 (1954)]. G. N. Abramowitsch [*The Theory of Turbulent Jets,* MIT Press, Cambridge, Mass., 1963] and N. Rajaratnam [*Turbulent Jets,* Elsevier, Amsterdam, 1976] provide thorough discourses on turbulent jets. H. J. Hussein et al. [*J. Fluid Mech.* **258:** 31–75 (1994)] give extensive velocity data for a free jet, discussion of free jet experimentation, and comparison of data with momentum conservation equations.

A turbulent-free jet is normally considered to consist of four flow regions [G. L. Tuve, *Heat. Piping Air Cond.* **25**(1): 181–191 (1953); J. T. Davies, *Turbulence Phenomena,* Academic, New York, 1972, p. 93], as shown in Fig. 6-17:

1. Region of flow establishment, which is a short region of length about 6.4 nozzle diameters. The fluid in the conical core of the same length has a velocity about the same as the initial discharge velocity. The termination of this *potential core* occurs when the growing mixing (boundary) layer between the jet and the surroundings reaches the centerline of the jet.
2. A transition region that extends to about 8 nozzle diameters.
3. Region of established flow, which is the principal region of the jet. In this region, the velocity profile transverse to the jet is self-preserving when normalized by the decaying centerline velocity.
4. A terminal region where the residual centerline velocity reduces rapidly within a short distance. For air jets, the residual velocity will reduce to less than 0.3 m/s (1 ft/s), usually considered still air.

Several references quote 100 nozzle diameters for the length of the established flow region. However, this length is dependent on the initial velocity and Reynolds number.

Table 6-7 gives characteristics of *rounded-inlet circular jets* and *rounded-inlet infinitely wide slot jets* (aspect ratio > 15). The information in the table is for a homogeneous, incompressible air system under isothermal conditions. The table uses the following nomenclature:

B_0 = slot height
D_0 = circular nozzle opening
q = total jet flow at distance x
q_0 = initial jet flow rate
r = radius from circular jet centerline
y = transverse distance from slot jet centerline
V_c = centerline velocity
V_r = circular jet velocity at r
V_y = velocity at y

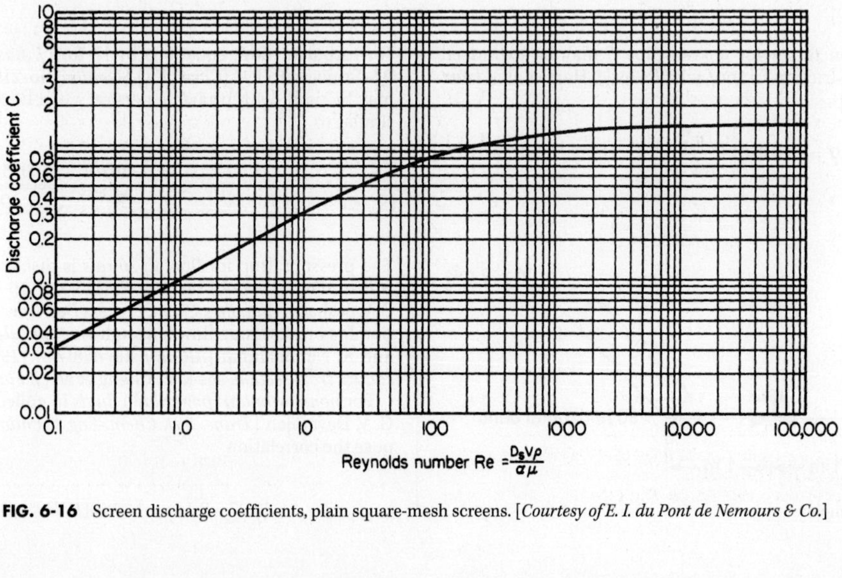

FIG. 6-16 Screen discharge coefficients, plain square-mesh screens. [*Courtesy of E. I. du Pont de Nemours & Co.*]

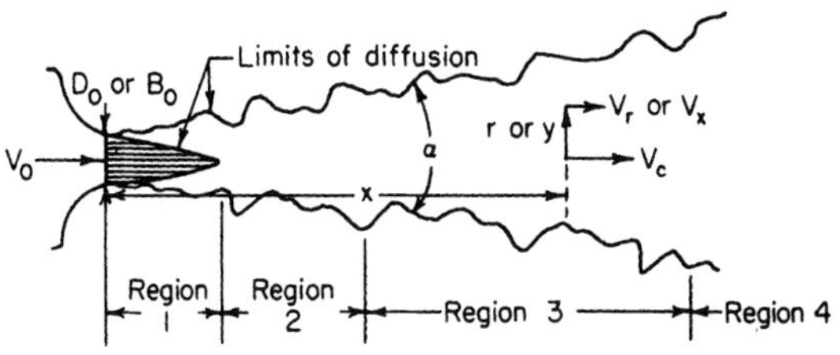

FIG. 6-17 Configuration of a turbulent free jet.

P. O. Witze [*Am. Inst. Aeronaut. Astronaut. J.* **12:** 417–418 (1974)] gives equations for the centerline velocity decay of different types of subsonic and supersonic circular free jets. Entrainment of surrounding fluid in the region of flow establishment is lower than in the region of established flow [see B. J. Hill, *J. Fluid Mech.* **51:** 773–779 (1972)]. Data of M. B. Donald and H. Singer [*Trans. Inst. Chem. Eng.* (*London*), **37:** 255–267 (1959)] indicate that jet angle and the coefficients given in Table 6-5 depend upon the fluids; for a water system, the jet angle for a circular jet is 14° and the entrainment ratio is about 70 percent of that for an air system. Most likely these variations are due to Reynolds number effects which are not taken into account in Table 6-7. J. H. Rushton [*AIChE J.* **26:** 1038–1041 (1980)] examined available published results for circular jets and found that the centerline velocity decay is given by

$$\frac{V_c}{V_0} = 1.41\,\mathrm{Re}^{0.135}\left(\frac{D_0}{x}\right) \tag{6-110}$$

TABLE 6-7 Turbulent Free-Jet Characteristics

Where both jet fluid and entrained fluid are air
Rounded-inlet circular jet
Longitudinal distribution of velocity along jet centerline*†
$\frac{V_c}{V_0} = K\frac{D_0}{x}$ for $7 < \frac{x}{D_0} < 100$
$K = 5$ for $V_0 = 2.5$ to 5.0 m/s
$K = 6.2$ for $V_0 = 10$ to 50 m/s
Radial distribution of longitudinal velocity†
$\log\left(\frac{V_c}{V_r}\right) = 40\left(\frac{r}{x}\right)^2$ for $7 < \frac{x}{D_0} < 100$
Jet angle*‡
$\alpha \simeq 20°$ for $\frac{x}{D_0} < 100$
Entrainment of surrounding fluid‡
$\frac{q}{q_0} = 0.32\frac{x}{D_0}$ for $7 < \frac{x}{D_0} < 100$
Rounded-inlet, infinitely wide slot jet
Longitudinal distribution of velocity along jet centerline‡
$\frac{V_c}{V_0} = 2.28\left(\frac{B_0}{x}\right)^{0.5}$ for $5 < \frac{x}{B_0} < 2000$ and $V_0 = 12$ to 55 m/s
Transverse distribution of longitudinal velocity‡
$\log\left(\frac{V_c}{V_x}\right) = 18.4\left(\frac{y}{x}\right)^2$ for $5 < \frac{x}{B_0} < 2000$
Jet angle‡
α is slightly larger than that for a circular jet
Entrainment of surrounding fluid‡
$\frac{q}{q_0} = 0.62\left(\frac{x}{B_0}\right)^{0.5}$ for $5 < \frac{x}{B_0} < 2000$

*Nottage, Slaby, and Gojsza, *Heat, Piping Air Cond.* **24**(1): 165–176 (1952).
†Tuve, *Heat, Piping Air Cond.* **25**(1): 181–191 (1953).
‡Albertson, Dai, Jensen, and Rouse, *Trans. Am. Soc. Civ. Eng.* **115:** 639–664 (1950), and Discussion, ibid., **115:** 665–697 (1950).

where $\mathrm{Re} = D_0V_0\rho/\mu$ is the initial jet Reynolds number. This result corresponds to a jet angle $\tan(\alpha/2)$ proportional to $\mathrm{Re}^{-0.135}$.

Characteristics of *rectangular jets* of various aspect ratios are given by H. G. Elrod [*Heat. Piping Air Cond.* **26**(3): 149–155 (1954)]. For *slot jets discharging into a moving fluid,* see A. S. Weinstein et al. [*J. Appl. Mech.* **23:** 437–443 (1967)]. *Coaxial jets* are discussed by W. Forstall and A. H. Shapiro [*J. Appl. Mech.* **17:** 399–408 (1950)], as are *double concentric jets* by N. A. Chigier and J. M. Beer [*J. Basic Eng.* **86:** 797–804 (1964)]. *Axisymmetric confined jets* are described by M. Barchilon and R. Curtet [*J. Basic Eng.* **86:** 777–787 (1964)]. *Restrained* turbulent jets of liquid discharging into air are described by J. T. Davies [*Turbulence Phenomena,* Academic, New York, 1972, p. 93]. These jets are inherently unstable and break up into drops after some distance. J. H. Lienhard and J. B. Day [*J. Basic Eng. Trans. ASME* pp. 515–522 (September 1970)] discuss the breakup of superheated liquid jets which flash upon discharge.

Density gradients affect the spread of a single-phase jet. A jet of lower density than the surroundings spreads more rapidly than a jet of the same density as the surroundings, and conversely, a denser jet spreads less rapidly. Additional details are given by W. R. Keagy and A. E. Weller [*Proc. Heat Transfer Fluid Mech. Inst.* ASME, pp. 89–98 (June 22–24, 1949)] and V. Cleeves and L. M. K. Boelter [*Chem. Eng. Prog.* **43:** 123–134 (1947)].

Few experimental data exist on *laminar jets* [see C. Gutfinger and R. Shinnar, *AIChE J.* **10:** 631–639 (1964)]. Theoretical analysis for velocity distributions and entrainment ratios in steady laminar flow is available in H. Schlichting and K. Gersten [*Boundary Layer Theory,* 8th ed rev., Springer-Verlag, Berlin, 2003] and in B. R. Morton [*Phys. Fluids* **10:** 2120–2127 (1967)]. The upper limit of the Reynolds number for stability of circular laminar jets may be in the range of 40 to 100. See, for example, P. J. Morris [*J. Fluid Mech.* **77:** 511–529 (1976)].

Theoretical analyses of jet flows for power law *nonnewtonian fluids* are given by J. Vlachopoulos and C. Stournaras [*AIChE J.* **21:** 385–388 (1975)], E. M. Mitwally [*J. Fluids Eng.* **100:** 363 (1978)], and K. Sridhar and G. W. Rankin [*J. Fluids Eng.* **100:** 500 (1978)].

FLOW THROUGH ORIFICES

Section 10 of this handbook describes the use of orifice meters for flow measurement. In addition, **orifices** are commonly found within pipelines as flow-restricting devices, in perforated-pipe distribution and return manifolds, and in perforated plates. Incompressible flow through an orifice in a pipeline, as shown in Fig. 6-18, is commonly described by the following equation for flow rate Q in terms of pressures P_1, P_2, and P_3; the orifice area A_o; the pipe cross-sectional area A; and the density ρ.

$$\begin{aligned} Q = v_o A_o &= C_o A_o \sqrt{\frac{2(P_1 - P_2)}{\rho[1-(A_o/A)^2]}} \\ &= C_o A_o \sqrt{\frac{2(P_1 - P_3)}{\rho(1-A_o/A)[1-(A_o/A)^2]}} \end{aligned} \tag{6-111}$$

The velocity based on the hole area is v_o. Pressure P_1 is the pressure upstream of the orifice, typically about 1 pipe diameter upstream, pressure P_2 is the pressure at the *vena contracta,* where the flow passes through a minimum area which is less than the orifice area, and pressure P_3 is the pressure downstream of the vena contracta after pressure recovery associated with deceleration of the fluid. The velocity of approach factor $1 - (A_o/A)^2$ accounts for the kinetic energy approaching the orifice, and the *orifice coefficient* or *discharge coefficient* C_o accounts for the vena contracta. The location of the vena contracta varies with A_o/A, but is about 0.7 pipe diameter for $A_o/A < 0.25$. The factor $1 - A_o/A$ accounts for pressure recovery. Pressure recovery is complete by about 4 to 8 pipe diameters downstream of the orifice. The permanent pressure drop is $P_1 - P_3$. When the orifice is at the end of pipe, discharging directly into a large chamber, there is negligible pressure recovery, the permanent pressure drop is $P_1 - P_2$, and the last equality in Eq. (6-111)

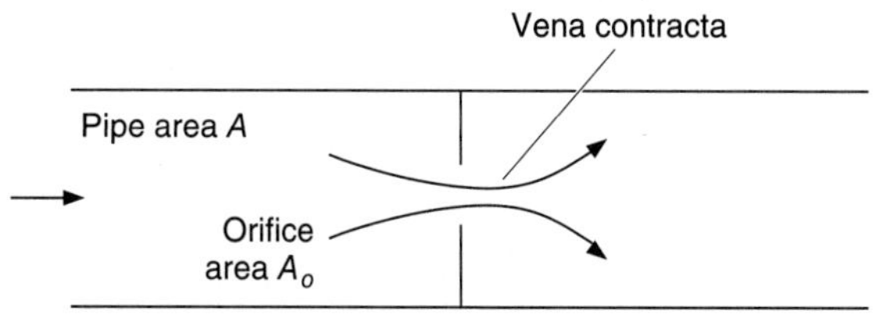

FIG. 6-18 Flow through an orifice.

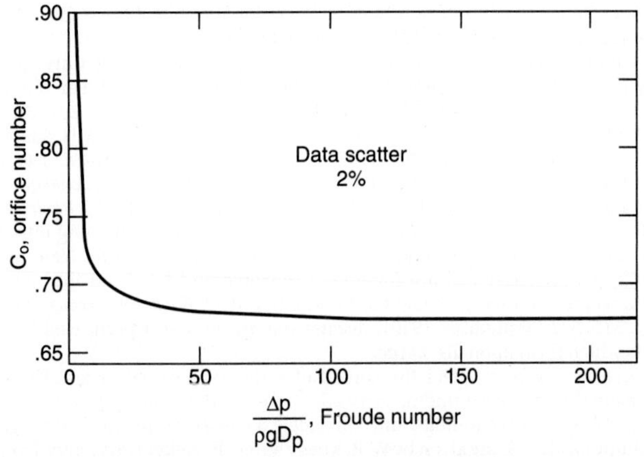

FIG. 6-19 Orifice coefficient vs. Froude number. [*Courtesy E. I. duPont de Nemours & Co.*]

does not apply. Instead, $P_2 = P_3$. Equation (6-111) may also be used for flow across a perforated plate with open area A_o and total area A. The location of the vena contracta and complete recovery would scale not with the vessel or pipe diameter in which the plate is installed, but with the hole diameter and pitch between holes.

The orifice coefficient has a value of about 0.62 at large Reynolds numbers ($Re = D_oV_o\rho/\mu > 20{,}000$), although values ranging from 0.60 to 0.70 are frequently used. At lower Reynolds numbers, the orifice coefficient varies with both Re and the area or diameter ratio. See Sec. 10 for more details.

When liquids discharge vertically downward from a pipe of diameter D_p, through orifices into gas, gravity increases the discharge coefficient. Figure 6-19 shows this effect, giving the discharge coefficient in terms of a modified Froude number, $Fr = \Delta p/(\rho g D_p)$.

The orifice coefficient deviates from its value for sharp-edged orifices when the orifice wall thickness exceeds about 75 percent of the orifice diameter. Some pressure recovery occurs within the orifice, and the orifice coefficient increases. Pressure drop across *segmental orifices* is roughly 10 percent greater than that for concentric circular orifices of the same open area.

COMPRESSIBLE FLOW

Flows are typically considered *compressible* when the density varies by more than 5 to 10 percent. In practice, compressible flows are normally limited to gases, supercritical fluids, and multiphase flows containing gases. Liquid flows are normally considered incompressible, except for certain calculations involved in *hydraulic transient* analysis (see following) where compressibility effects are important even for nearly incompressible liquids. Texts on compressible gas flow include those by A. H. Shapiro [*Dynamics and Thermodynamics of Compressible Fluid Flow,* vols. 1 and 2, Ronald Press, New York, 1953] and M. J. Zucrow and J. D. Hoffman [*Gas Dynamics,* vols. 1 and 2, Wiley, New York, 1976]. The most important chemical process applications of compressible flow are one-dimensional gas flows through nozzles or orifices and in pipelines. Multidimensional external flows are of interest mainly in aerodynamic applications.

Mach Number and Speed of Sound The *Mach number* $M = V/c$ is the ratio of fluid velocity V to the *speed of sound* or *acoustic velocity* c. The speed of sound is the propagation velocity of infinitesimal pressure disturbances and is derived from a momentum balance. The compression caused by the pressure wave is adiabatic and frictionless, and therefore isentropic.

$$c = \sqrt{\left(\frac{\partial p}{\partial \rho}\right)_s} \tag{6-112}$$

The partial derivative is taken at constant entropy s. For a perfect gas (ideal gas with constant heat capacity)

$$\left(\frac{\partial p}{\partial \rho}\right)_s = \frac{kRT}{M_w} \tag{6-113}$$

where k = ratio of specific heats, C_p/C_v
R = universal gas constant (8314 J/kgmol · K)
T = absolute temperature
M_w = molecular weight

Hence for a perfect gas,

$$c = \sqrt{\frac{kRT}{M_w}} \tag{6-114}$$

The Mach number is calculated by using the speed of sound evaluated at the local pressure and temperature. When $M = 1$, the flow is *critical* or *sonic,* and the velocity equals the local speed of sound. For *subsonic* flows, $M < 1$ while *supersonic* flows have $M > 1$. Compressibility effects are always important when the Mach number exceeds 0.1 to 0.2, but they may be important at lower Mach number; large density changes can occur in long pipelines at low velocity, for example.

Isothermal Gas Flow in Pipes and Channels Isothermal compressible flow is often approximated in long transport lines. Mach numbers are usually small, yet compressibility effects are important when the total pressure drop is a significant fraction of the absolute pressure. For an ideal gas, integration of the differential form of the momentum or mechanical energy balance equations, assuming a constant friction factor f over a length L of a channel of constant cross section and hydraulic diameter D_H, yields

$$p_1^2 - p_2^2 = G^2 \frac{RT}{M_w}\left[\frac{4fL}{D_H} + 2\ln\left(\frac{p_1}{p_2}\right)\right] \tag{6-115}$$

where the mass velocity $G = w/A = \rho V$ is the mass flow rate per unit cross-sectional area of the channel. The logarithmic term on the right-hand side accounts for the pressure change caused by acceleration of gas as its density decreases, while the first term is equivalent to the calculation of frictional losses using the density evaluated at the average pressure $(p_1 + p_2)/2$.

Solution of Eq. (6-115) for G and differentiation with respect to p_2 reveal a maximum mass flux $G_{max} = p_2\sqrt{M_w/(RT)}$ and a corresponding exit velocity $V_{2,max} = \sqrt{RT/M_w}$ and exit Mach number $M_2 = 1/\sqrt{k}$. This apparent *choking* condition is not physically meaningful because at such high velocities, and high rates of expansion, isothermal conditions are not maintained.

Adiabatic Frictionless Nozzle Flow In process plant pipelines, compressible flows are usually more nearly adiabatic than isothermal. Solutions for adiabatic flows through frictionless nozzles and in channels with constant cross section and constant friction factor are readily available.

Figure 6-20 illustrates adiabatic discharge of a perfect gas through a frictionless nozzle from a large chamber where velocity is effectively zero. The subscript 0 refers to the *stagnation* conditions in the chamber. More generally, stagnation conditions are the conditions that would be obtained by isentropically decelerating a gas flow to zero velocity. The minimum area section, or *throat,* of the nozzle is at the nozzle exit. The flow through the nozzle is isentropic because it is assumed frictionless (reversible) and adiabatic. In terms of the exit Mach number M_1, the following ratios between stagnation and exit conditions occur.

$$\frac{p_0}{p_1} = \left(1 + \frac{k-1}{2}M_1^2\right)^{k/(k-1)} \tag{6-116}$$

$$\frac{T_0}{T_1} = 1 + \frac{k-1}{2}M_1^2 \tag{6-117}$$

$$\frac{\rho_0}{\rho_1} = \left(1 + \frac{k-1}{2}M_1^2\right)^{1/(k-1)} \tag{6-118}$$

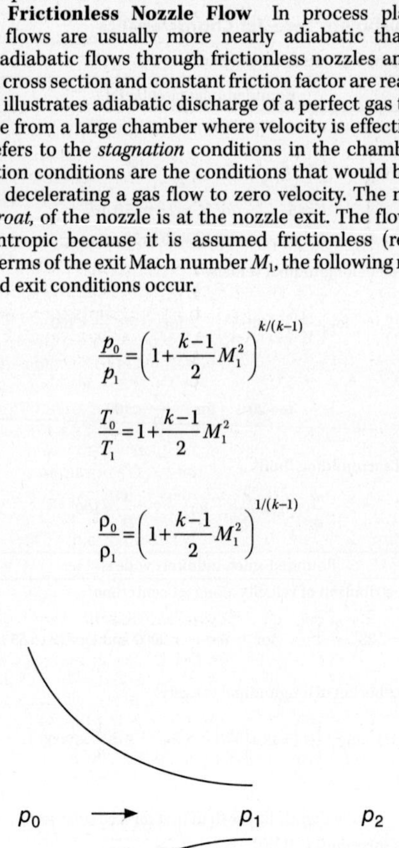

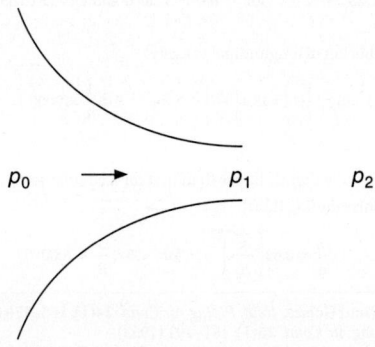

FIG. 6-20 Isentropic flow through a nozzle.

The mass velocity $G = w/A$ at the nozzle exit is given by

$$G = p_0\sqrt{\frac{kM_w}{RT_0}}\frac{M_1}{\left(1+\frac{k-1}{2}M_1^2\right)^{(k+1)/2(k-1)}} \qquad (6\text{-}119)$$

These equations are consistent with the isentropic relations for a perfect gas $p/p_0 = (\rho/\rho_0)^k$, $T/T_0 = (p/p_0)^{(k-1)/k}$. Equation (6-117) is valid for adiabatic flows with or without friction; it does not require isentropic flow. Equations (6-116), (6-118), and (6-119) do require isentropic flow.

The *exit Mach number M_1 may not exceed unity.* At $M_1 = 1$, the flow is said to be *choked, sonic,* or *critical.* When the flow is choked, the pressure at the exit is greater than the pressure of the surroundings into which the gas flow discharges. The pressure drops from the exit pressure to the pressure of the surroundings in the jet beyond the nozzle exit. Sonic flow conditions are denoted by *; sonic exit conditions are found by substituting $M_1 = M_1^* = 1$ into Eqs. (6-116) to (6-119).

$$\frac{p^*}{p_0} = \left(\frac{2}{k+1}\right)^{k/(k-1)} \qquad (6\text{-}120)$$

$$\frac{T^*}{T_0} = \frac{2}{k+1} \qquad (6\text{-}121)$$

$$\frac{\rho^*}{\rho_0} = \left(\frac{2}{k+1}\right)^{1/(k-1)} \qquad (6\text{-}122)$$

$$G^* = p_0\sqrt{\left(\frac{2}{k+1}\right)^{(k+1)/(k-1)}\left(\frac{kM_w}{RT_0}\right)} \qquad (6\text{-}123)$$

Under choked conditions, the exit velocity is $V = V^* = c^* = \sqrt{kRT^*/M_w}$, with sonic velocity evaluated at the exit temperature. For air ($k = 1.4$) the critical pressure ratio p^*/p_0 is 0.5283 and the critical temperature ratio $T^*/T_0 = 0.8333$. Thus, for air discharging from 300 K, the temperature drops by 50 K in accelerating to sonic velocity. This large temperature decrease results from the conversion of enthalpy into kinetic energy. With sufficient humidity, condensation could occur, invalidating the perfect gas analysis. As the discharged jet decelerates in the external stagnant gas, it recovers its initial enthalpy.

To determine the discharge conditions and rate through a nozzle from upstream pressure p_0 to external pressure p_2, as shown in Fig. 6-20, Eqs. (6-116) through (6-123) are used as follows. The critical pressure is first determined from Eq. (6-120). If $p_2 > p^*$, then the flow is subsonic (subcritical, unchoked). Then $p_1 = p_2$ and M_1 may be obtained from Eq. (6-116). Substitution of M_1 into Eqs. (6-117) through (6-120) then gives the exit temperature, density, and mass velocity. The exit velocity can be obtained from $V_1 = G/\rho_1$ or $V_1 = M_1c_1$ with c_1 evaluated at T_1.

On the other hand, if $p_2 \le p^*$, then the flow is choked, $M_1 = 1$, and $p_1 = p^*$. The temperature and density, respectively, are equal to T^* and ρ^* from Eqs. (6-121) and (6-122); the mass velocity is G^* obtained from Eq. (6-123). When the flow is choked, $G = G^*$ is independent of external downstream pressure. Reducing the downstream pressure will not increase the flow. The mass flow rate under choking conditions is directly proportional to the upstream pressure p_0.

Example 6-7 Flow Through Frictionless Nozzle Dry air at temperature $T_0 = 293$ K discharges through a frictionless nozzle to atmospheric pressure. Compute the discharge mass flux G, pressure, temperature, Mach number, and velocity at the exit. Consider two cases: (1) $p_0 = 7.0 \times 10^5$ Pa absolute and (2) $p_0 = 1.5 \times 10^5$ Pa absolute.

1. $p_0 = 7.0 \times 10^5$ Pa. For air with $k = 1.4$, the critical pressure ratio from Eq. (6-120) is $p^*/p_0 = 0.5283$ and $p^* = 0.5283 \times 7.0 \times 10^5 = 3.70 \times 10^5$ Pa. Since this is greater than the external atmospheric pressure $p_2 = 1.01 \times 10^5$ Pa, the flow is choked and the exit pressure is $p_1 = 3.70 \times 10^5$ Pa. The exit Mach number is 1.0, and the mass flux is equal to G^*, given by Eq. (6-123).

$$G^* = 7.0\times10^5 \times \sqrt{\left(\frac{2}{1.4+1}\right)^{(1.4+1)/(1.4-1)}\left(\frac{1.4\times29}{8314\times293}\right)} = 1650 \text{ kg/m}^2\cdot\text{s}$$

The exit temperature, because the flow is choked, is

$$T^* = \left(\frac{T^*}{T_0}\right)T_0 = \left(\frac{2}{1.4+1}\right)\times 293 = 244 \text{ K}$$

The exit velocity is $V = Mc = c^* = \sqrt{kRT^*/M_w} = 313$ m/s.

2. $p_0 = 1.5 \times 10^5$ Pa. In this case $p^* = 0.79 \times 10^5$ Pa, which is less than p_2. Hence, $p_1 = p_2 = 1.01 \times 10^5$ Pa. The flow is unchoked (subsonic). Equation (6-116) is solved for the Mach number.

$$\frac{1.5\times10^5}{1.01\times10^5} = \left(1+\frac{1.4-1}{2}M_1^2\right)^{1.4/(1.4-1)}$$
$$M_1 = 0.773$$

Substitution into Eq. (6-119) gives G.

$$G = 1.5\times10^5\times\sqrt{\frac{1.4\times29}{8{,}314\times293}}$$
$$\times\frac{0.773}{\left(1+\left(\frac{1.4-1}{2}\right)\times0.773^2\right)^{(1.4+1)/2(1.4-1)}} = 337 \text{ kg/m}^2\cdot\text{s}$$

The exit temperature is found from Eq. (6-117) to be 261.6 K. The exit velocity is

$$V = Mc = 0.773\times\sqrt{\frac{1.4\times8314\times261.6}{29}} = 250 \text{ m/s}$$

Adiabatic Flow with Friction in a Duct of Constant Cross Section Integration of the differential forms of the continuity, momentum, and total energy equations for a perfect gas, assuming a constant friction factor, leads to a set of simultaneous algebraic equations. These may be found in A. H. Shapiro [*Dynamics and Thermodynamics of Compressible Fluid Flow,* vols. 1 and 2, Ronald Press, New York, 1953] or M. J. Zucrow and J. D. Hoffman [*Gas Dynamics,* vols. 1 and 2, Wiley, New York, 1976]. C. E. Lapple's [*Trans. AIChE.* **39:** 395–432 (1943)] widely cited graphical presentation of the solution of these equations contained a subtle error, which was corrected by O. Levenspiel [*AIChE J.* **23:** 402–403 (1977)]. Levenspiel's graphical solutions are presented in Fig. 6-21. These charts refer to the physical situation illustrated in Fig. 6-22, where a perfect gas discharges from stagnation conditions in a large chamber through an isentropic nozzle followed by a duct of length L. The resistance parameter is $N = 4fL/D_H$, where f = Fanning friction factor and D_H = hydraulic diameter.

The exit Mach number M_2 may not exceed unity. $M_2 = 1$ corresponds to choked flow; sonic conditions may exist only at the pipe exit. The mass velocity G^* in the charts is the choked *mass flux for an isentropic nozzle* given by Eq. (6-123). For a pipe of finite length, the mass flux is less than G^* under choking conditions. The curves in Fig. 6-21 become vertical at the choking point, where flow becomes independent of downstream pressure.

The equations for nozzle flow, Eqs. (6-116) through (6-119), remain valid for the nozzle section even in the presence of the discharge pipe. The graphs in Fig. 6-21 are based on accurate calculations, but are difficult to interpolate precisely. While they are quite useful for qualitative insight and rough estimates, precise calculations are best done using the equations for one-dimensional adiabatic flow with friction, which are suitable for computer programming. Let subscripts 1 and 2 denote two points along a pipe of diameter D, point 2 being downstream of point 1. From a given point in the pipe, where the Mach number is M, the additional length of pipe required to accelerate the flow to sonic velocity ($M = 1$) is denoted L_{max} and may be computed from

$$\frac{4fL_{max}}{D_H} = \frac{1-M^2}{kM^2} + \frac{k+1}{2k}\ln\left(\frac{\frac{k+1}{2}M^2}{1+\frac{k-1}{2}M^2}\right) \qquad (6\text{-}124)$$

With L = length of pipe between points 1 and 2, the change in Mach number may be computed from

$$\frac{4fL}{D_H} = \left(\frac{4fL_{max}}{D}\right)_1 - \left(\frac{4fL_{max}}{D}\right)_2 \qquad (6\text{-}125)$$

and the pressures p_1 and p_2 are related to the pressure p^* where $M = 1$ by

$$\frac{p}{p^*} = \frac{1}{M}\sqrt{\frac{k+1}{2\left(1+\frac{k-1}{2}M^2\right)}} \qquad (6\text{-}126)$$

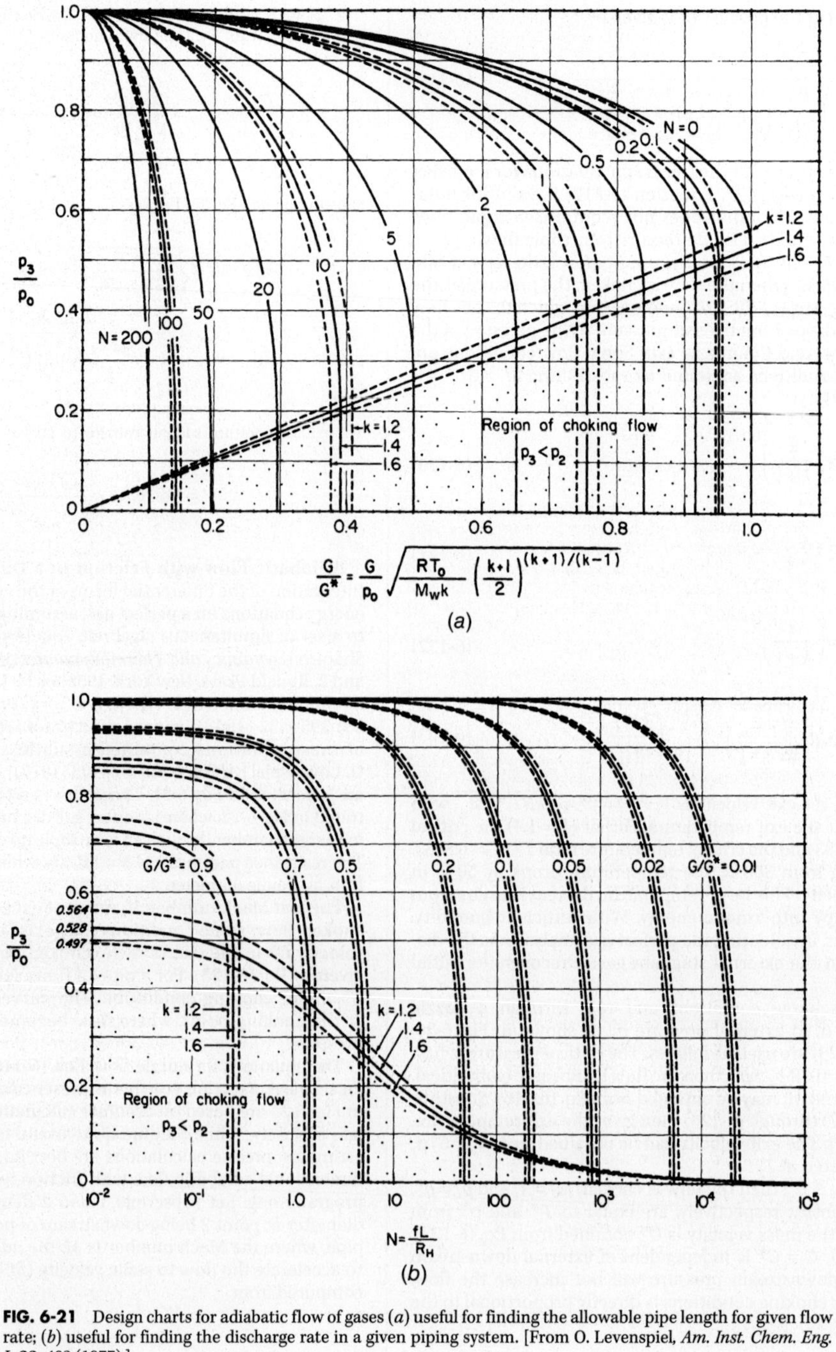

FIG. 6-21 Design charts for adiabatic flow of gases (*a*) useful for finding the allowable pipe length for given flow rate; (*b*) useful for finding the discharge rate in a given piping system. [From O. Levenspiel, *Am. Inst. Chem. Eng. J.,* **23:** 402 (1977).]

The additional frictional losses due to pipe fittings such as elbows may be added to the velocity head loss $N = 4fL/D_H$, using the same velocity head loss values as for incompressible flow. This works well for fittings that do not significantly reduce the channel cross-sectional area, but may cause large errors when the flow area is greatly reduced, for example, by restricting orifices. Compressible flow across restricting orifices is discussed in Sec. 10 of this handbook. An elbow near the exit of a pipeline may choke the flow even though the Mach number is less than unity due to the nonuniform velocity profile in the elbow. For an abrupt contraction rather than rounded nozzle inlet, an additional 0.5 velocity head should be added to *N*. This is a reasonable approximation for determining *G*, but note that it allocates the additional losses to the pipeline, even though they are actually incurred in the entrance. Do not include one velocity head exit loss in *N*. The kinetic energy at the exit is already accounted for in the integration of the balance equations.

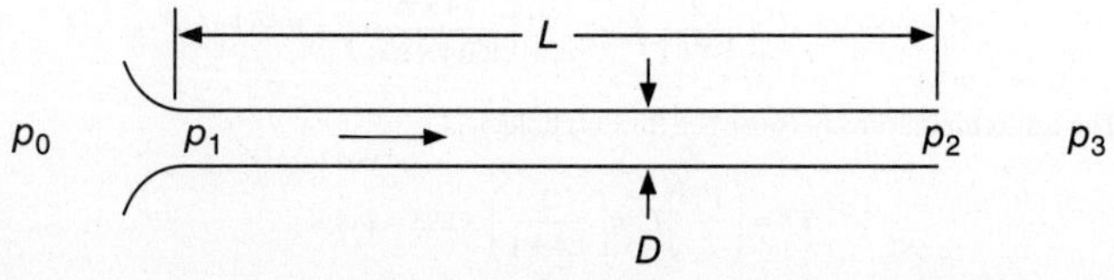

FIG. 6-22 Adiabatic compressible flow in a pipe with a well-rounded entrance.

Example 6-8 Compressible Flow with Friction Losses Calculate the discharge rate of air to the atmosphere from a reservoir at 10^6 Pa gauge and 20°C through 10 m of straight 2-in Schedule 40 steel pipe (inside diameter = 0.0525 m), and three 90° elbows. Assume 0.5 velocity head is lost for the elbows.

For commercial steel pipe, with a roughness of 0.046 mm, the friction factor for fully rough flow is about 0.00476, from Eq. (6-38). It remains to be verified that the Reynolds number is sufficiently large to assume fully rough flow. Assuming an abrupt entrance with 0.5 velocity head lost,

$$N = 4\times 0.00476\times\frac{10}{0.0525}+0.5+3\times 0.5 = 5.6$$

The pressure ratio p_3/p_0 is

$$\frac{1.01\times 10^5}{1\times 10^6+1.01\times 10^5} = 0.092$$

From Fig. 6-21*a* at $N = 5.6$, $p_3/p_0 = 0.092$, and $k = 1.4$ for air, the flow is seen to be choked. At the choke point with $N = 5.6$ the critical pressure ratio p_2/p_0 is about 0.25 and G/G^* is about 0.48. Equation (6-123) gives

$$G^* = 1.101\times 10^6\sqrt{\left(\frac{2}{1.4+1}\right)^{(1.4+1)/(1.4-1)}\left(\frac{1.4\times 29}{8{,}314\times 293.15}\right)} = 2600 \text{ kg/m}^2\cdot\text{s}$$

Multiplying by $G/G^* = 0.48$ yields $G = 1250$ kg/m$^2\cdot$s. The discharge rate is $w = GA = 1250\times\pi\times 0.0525^2/4 = 2.7$ kg/s. Numerical solution based on the isentropic nozzle equations and Eqs. (6-124) through (6-126) for the pipe gives $w = 2.71$ kg/s.

Before this solution is accepted, the Reynolds number should be checked. At the pipe exit, the temperature is given by Eq. (6-121) since the flow is choked. $T_2 = T^* = 244.6$ K. The viscosity of air at this temperature is about 1.6×10^{-5} Pa·s. Then

$$\text{Re} = \frac{DV\rho}{\mu} = \frac{DG}{\mu} = \frac{0.0525\times 1{,}250}{1.6\times 10^{-5}} = 4.1\times 10^6$$

At the beginning of the pipe, the temperature is greater, giving greater viscosity and a Reynolds number of 3.6×10^6. Over the entire pipe length, the Reynolds number is very large and the complete turbulence friction factor choice was indeed valid.

Once the mass flux G has been determined, Fig. 6-21*a* or 6-21*b* or Eqs. (6-124) through (6-126) can be used to determine the pressure at any point along the pipe.

Convergent/Divergent Nozzles (De Laval Nozzles) During frictionless adiabatic one-dimensional flow with changing cross-sectional area A, the following relations are obeyed:

$$\frac{dA}{A} = \frac{dp}{\rho V^2}(1-M^2) = \frac{1-M^2}{M^2}\frac{d\rho}{\rho} = -(1-M^2)\frac{dV}{V} \tag{6-127}$$

Equation (6-127) implies that in converging channels, subsonic flows are accelerated and the pressure and density decrease. In diverging channels, subsonic flows are decelerated as the pressure and density increase. In supersonic flows, the opposite is true. Diverging channels accelerate the flow, while converging channels decelerate the flow.

Figure 6-23 shows a converging/diverging nozzle. When p_2/p_0 is less than the critical pressure ratio (p^*/p_0), the flow will be subsonic in the converging portion of the nozzle, sonic at the throat, and supersonic in the diverging portion. At the throat, where the flow is critical and the velocity is sonic, the area is denoted A^*. The cross-sectional area and pressure vary with Mach number along the converging/diverging flow path according to the following equations for isentropic flow of a perfect gas:

$$\frac{A}{A^*} = \frac{1}{M}\left[\frac{2}{k+1}\left(1+\frac{k-1}{2}M^2\right)\right]^{(k+1)/2(k-1)} \tag{6-128}$$

$$\frac{p_0}{p} = \left(1+\frac{k-1}{2}M^2\right)^{k/(k-1)} \tag{6-129}$$

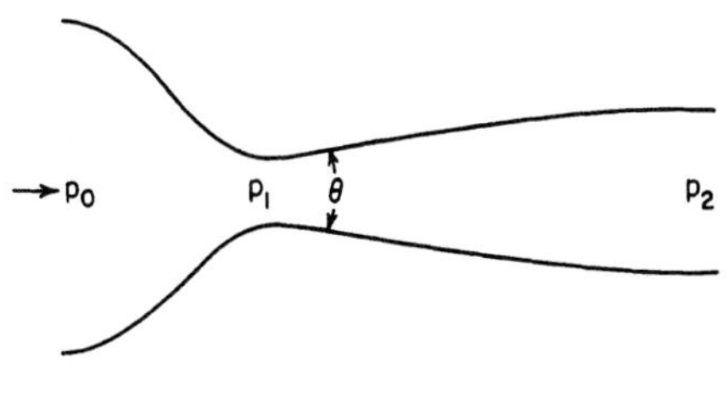

FIG. 6-23 Converging/diverging nozzle.

The temperature obeys the adiabatic flow equation for a perfect gas

$$\frac{T_0}{T} = 1+\frac{k-1}{2}M^2 \tag{6-130}$$

Equation (6-130) does not require frictionless (isentropic) flow. The sonic mass flux through the throat is given by Eq. (6-123). With A set equal to the nozzle exit area, the exit Mach number, pressure, and temperature may be calculated. Only if the exit pressure equals the ambient discharge pressure is the maximum possible expansion velocity reached in the nozzle. Expansion will be incomplete if the exit pressure exceeds the ambient discharge pressure. If the calculated exit pressure is less than the ambient discharge pressure, the nozzle is overexpanded and shocks within the expanding portion will result.

The shape of the converging section is a smooth trumpet shape similar to the simple converging nozzle. Special shapes of the diverging section are required to produce the maximum supersonic exit velocity. Shocks result if the divergence is too rapid, and excessive boundary layer friction occurs if the divergence is too shallow. See H. W. Liepmann and A. Roshko [*Elements of Gas Dynamics,* Wiley, New York, 1957, p. 284]. If the nozzle is to be used as a thrust device, the diverging section can be conical with a total included angle of 30° [G. P. Sutton and O. Biblarz, *Rocket Propulsion Elements,* 9th ed., Wiley, Hoboken, N.J., 2017]. To obtain large exit Mach numbers, slot-shaped rather than axisymmetric nozzles are used.

MULTIPHASE FLOW

Multiphase flows, even when restricted to simple pipeline geometry, are in general quite complex, with several features making them more complicated than single-phase flow. Flow pattern description is not merely an identification of laminar or turbulent flow. The relative quantities of the phases and the topology of the interfaces must be described. Because of phase density differences, vertical flow patterns are different from horizontal flow patterns, and horizontal flows are usually asymmetric. Even when phase equilibrium is achieved by good mixing in two-phase flow, the changing equilibrium state as pressure drops with distance, or as heat is added or lost, may require that interphase mass transfer, and changes in the relative amounts of the phases, be considered. C. T. Crowe [ed., *Multiphase Flow Handbook,* CRC Press, Boca Raton, Fla., 2006] presents multiphase flow fundamentals as well as information on flow in pipelines and process equipment.

G. B. Wallis [*One-dimensional Two-phase Flow,* McGraw-Hill, New York, 1969] and G. W. Govier and K. Aziz [*The Flow of Complex Mixtures in Pipes,* Krieger, Huntington, N.Y., 1977] present one-dimensional mass, momentum, mechanical energy, and total energy balance equations for two-phase flows. Such equations, for the most part, are used as a framework in which to interpret experimental data. Reliable prediction of multiphase flow behavior generally requires use of data or correlations. *Two-fluid modeling,* in which three-dimensional partial differential equations of motion are written for each phase, treating each as a continuum, occupying a volume fraction which is a continuous function of position, is a developing technique made possible by improved computational methods. For some relatively simple examples not requiring numerical computation, see J. R. A. Pearson [*Chem. Engr. Sci.* **49:** 727–732 (1994)]. Constitutive equations for two-fluid models are not yet sufficiently robust for accurate general-purpose two-phase flow computation, but may be quite good for particular classes of flows.

Liquids and Gases For cocurrent flow of liquids and gases in vertical (upflow), horizontal, and inclined pipes, a very large literature of experimental and theoretical work has been published, with less work on countercurrent and cocurrent vertical downflow. Much of the effort has been devoted to predicting flow patterns, pressure drop, and volume fractions of the phases, with emphasis on fully developed flow. In practice, many two-phase flows in process plants are not fully developed.

The most reliable methods for fully developed gas/liquid flows use *mechanistic models* to predict flow pattern, and they use different pressure drop and void fraction estimation procedures for each flow pattern. Such methods are too lengthy to include here and are well suited to incorporation into computer programs; commercial codes for gas/liquid pipeline flows are available. Some key references for mechanistic methods for flow pattern transitions and flow regime–specific pressure drop and void fraction methods include Y. Taitel and A. E. Dukler [*AIChE J.* **22:** 47–55 (1976)], D. Barnea et al. [*Int. J. Multiphase Flow* **6:** 217–225 (1980)], D. Barnea [*Int. J. Multiphase Flow* **12:** 733–744 (1986)], Y. Taitel et al. [*AIChE J.* **26:** 345–354 (1980)], G. B. Wallis [*One-dimensional Two-phase Flow,* McGraw-Hill, New York, 1969], and A. E. Dukler and M. G. Hubbard [*Ind. Eng. Chem. Fundam.* **14:** 337–347 (1975)]. See R. V. A. Oliemans and B. F. Pots in C. T. Crowe [ed., *Multiphase Flow Handbook,* CRC Press, Boca Raton, Fla., 2006] for a recent summary. For preliminary or approximate calculations, *flow pattern maps* and flow regime–independent empirical correlations are simpler and faster to use. Such methods for horizontal and vertical flows are provided in the following.

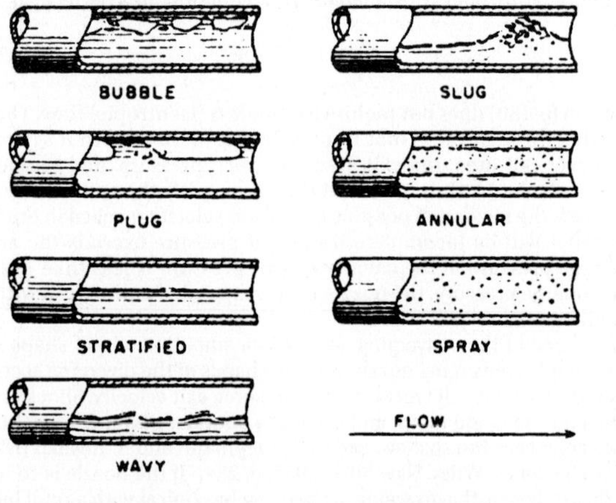

FIG. 6-24 Gas/liquid flow patterns in horizontal pipes. [From G. E. Alves, *Chem. Eng. Progr.*, **50**, 449–456 (1954).]

In *horizontal pipe,* flow patterns for fully developed flow have been reported in numerous studies. Transitions between flow patterns are gradual, and subjective owing to the visual interpretation of individual investigators. In some cases, statistical analysis of pressure fluctuations has been used to distinguish flow patterns. Figure 6-24 [G. E. Alves, *Chem. Eng. Progr.* **50:** 449–456 (1954)] shows seven flow patterns for horizontal gas/liquid flow. *Bubble flow* is prevalent at high ratios of liquid to gas flow rates. The gas is dispersed as bubbles which move at velocity similar to that of the liquid and tend to concentrate near the top of the pipe at lower liquid velocities. *Plug flow* describes a pattern in which alternate plugs of gas and liquid move along the upper part of the pipe. In *stratified flow,* the liquid flows along the bottom of the pipe, and the gas flows over a smooth liquid/gas interface. Similar to stratified flow, *wavy flow* occurs at greater gas velocities and has waves moving in the flow direction. When wave crests are sufficiently high to bridge the pipe, they form frothy slugs which move at much greater than the average liquid velocity. *Slug flow* can cause severe and sometimes dangerous vibrations in equipment because of impact of the high-velocity slugs against bends or other fittings. Slugs may also flood gas/liquid separation equipment.

In *annular flow,* liquid flows as a thin film along the pipe wall and gas flows in the core. Some liquid is entrained as droplets in the gas core. At very high gas velocities, nearly all the liquid is entrained as small droplets. This pattern is called *spray, dispersed,* or *mist flow.*

Approximate prediction of flow pattern may be quickly done using *flow pattern maps,* an example of which is shown in Fig. 6-25 [O. Baker, *Oil Gas J.* **53**(12): 185–190, 192–195 (1954)]. The Baker chart remains widely used; however, for critical calculations, the mechanistic model methods referenced previously or commercial software based on large proprietary databases, are generally preferred for their greater accuracy, especially for large

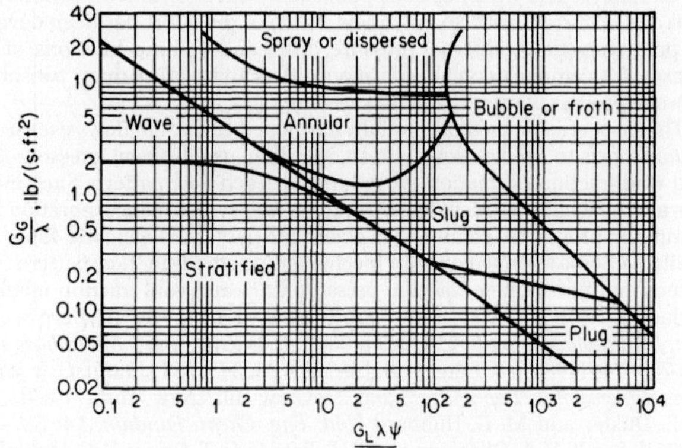

FIG. 6-25 Flow-pattern regions in cocurrent liquid/gas flow through horizontal pipes. To convert lbm/(ft²·s) to kg/(m²·s), multiply by 4.8824. [From O. Baker, *Oil Gas J.*, **53**[12], 185–190, 192, 195 (1954).]

pipe diameters and fluids with physical properties different from air/water at atmospheric pressure. In the chart,

$$\lambda = (\rho'_G \rho'_L)^{1/2} \tag{6-131}$$

$$\psi = \frac{1}{\sigma'}\left[\frac{\mu'_L}{(\rho'_L)^2}\right]^{1/3} \tag{6-132}$$

Here G_L and G_G are the liquid and gas mass velocities, respectively, μ'_L is the ratio of liquid viscosity to water viscosity, ρ'_G is the ratio of gas density to air density, ρ'_L is the ratio of liquid density to water density, and σ' is the ratio of liquid surface tension to water surface tension. The reference properties are at 20°C and atmospheric pressure, water density 1000 kg/m³ (62.3 lbm/ft³), air density 1.20 kg/m³ (0.075 lbm/ft³), water viscosity 0.001 Pa·s (1 cP), and surface tension 0.073 N/m (73 dyn/cm). The empirical parameters λ and ψ provide a crude accounting for physical properties. The Baker chart is dimensionally inconsistent since the dimensional quantity G_G/λ is plotted versus a dimensionless one, $G_L\lambda\psi/G_G$, and so must be used with G_G in lbm/(ft²·s) units on the ordinate. To convert to kg/(m²·s), multiply by 4.8824.

Approximate predictions of *pressure drop* for fully developed, incompressible horizontal gas/liquid flow may be made by using the method of R. W. Lockhart and R. C. Martinelli [*Chem. Eng. Prog.* **45:** 39–48 (1949)]. First, the pressure drops that would be expected for each of the two phases as if flowing alone in single-phase flow are calculated. The Lockhart-Martinelli parameter X is

$$X = \left[\frac{(\Delta p/L)_L}{(\Delta p/L)_G}\right]^{1/2} \tag{6-133}$$

The two-phase pressure drop may then be estimated from either of the single-phase pressure drops, using

$$\left(\frac{\Delta p}{L}\right)_{TP} = Y_L\left(\frac{\Delta p}{L}\right)_L \tag{6-134}$$

or

$$\left(\frac{\Delta p}{L}\right)_{TP} = Y_G\left(\frac{\Delta p}{L}\right)_G \tag{6-135}$$

where Y_L and Y_G are read from Fig. 6-26 as functions of X. The curve labels refer to the flow regime (laminar or turbulent) found for each of the phases flowing alone. In Fig. 6-26, the original curves for liquid viscous/gas turbulent and liquid turbulent/gas viscous have been collapsed onto a single curve. The Y_G curves are well fit by the following equation [J. O. Wilkes, *Fluid Mechanics for Chemical Engineers*, 2d ed., Prentice-Hall, Upper Saddle River, N.J., 2006].

$$Y_G = (1 + X^{2/n})^n \tag{6-136}$$

where the value of n is given in Table 6-8. Note that $Y_L = Y_G/X^2$.

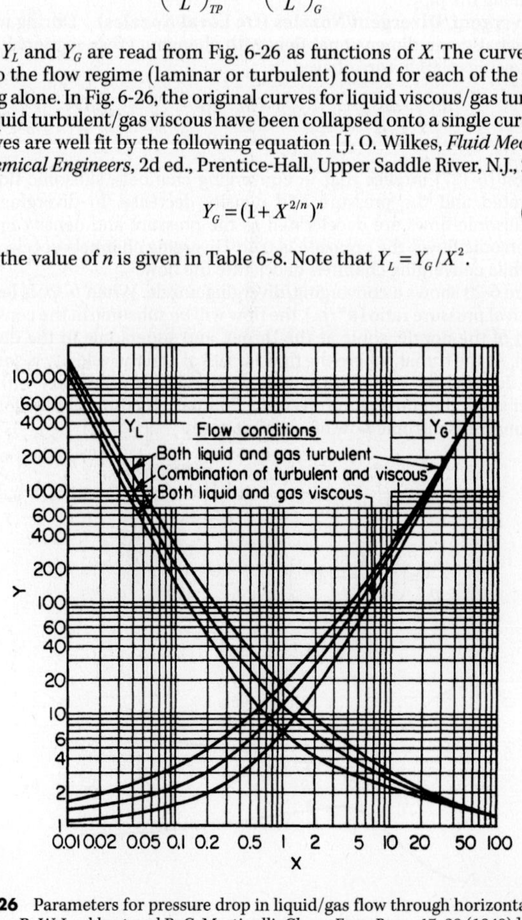

FIG. 6-26 Parameters for pressure drop in liquid/gas flow through horizontal pipes. [Based on R. W. Lockhart and R. C. Martinelli, *Chem. Engr. Prog.*, **45**, 39 (1949).]

TABLE 6-8 Parameter *n* for Lockhart-Martinelli Correlation

	Liquid	Gas	n
	Turbulent	Turbulent	4.12
	Viscous	Turbulent	3.61
	Turbulent	Viscous	3.56
$X < 1$	Viscous	Viscous	2.68
$X > 1$	Viscous	Viscous	3.27

R. W. Lockhart and R. C. Martinelli [*Chem. Eng. Prog.* **45:** 39–48 (1949)] correlated pressure drop data from pipes 25 mm in diameter or less within about ±50 percent. In general, the predictions are high for stratified, wavy, and slug flows and low for annular flow. The correlation can be applied to pipe diameters up to about 0.1 m (4 in) with about the same accuracy.

The *volume fraction,* sometimes called *holdup,* of each phase in two-phase flow is generally not equal to its volumetric flow rate fraction, because of velocity differences, or *slip,* between the phases. For each phase, denoted by subscript i, the relations among superficial velocity V_i, in situ velocity v_i, volume fraction R_i, phase volumetric flow rate Q_i, and pipe area A are

$$Q_i = V_i A = v_i R_i A \tag{6-137}$$

$$v_i = \frac{V_i}{R_i} \tag{6-138}$$

The *slip velocity* between gas and liquid is $v_s = v_G - v_L$. For two-phase gas/liquid flow, $R_L + R_G = 1$.

For fully developed incompressible horizontal gas/liquid flow, a quick estimate for R_L may be obtained from Fig. 6-27, as a function of the Lockhart-Martinelli parameter X. Indications are that liquid volume fractions may be overpredicted for liquids more viscous than water [G. E. Alves, *Chem. Eng. Progr.* **50:** 449–456 (1954)] and underpredicted for pipes larger than 25-mm diameter [O. Baker, *Oil Gas J.* **53**(12): 185–190, 192–195 (1954)]. J. O. Wilkes [*Fluid Mechanics for Chemical Engineers*, 2d ed., Prentice-Hall, Upper Saddle River, N.J., 2006] provides an estimate for R_L as a function of the Lockhart-Martinelli parameter.

$$R_L = 1 - \frac{1}{(1 + 0.0904X^{0.548})^{2.82}} \tag{6-139}$$

A method for predicting pressure drop and volume fraction for *nonnewtonian fluids* in annular flow has been proposed by F. G. Eisenberg and C. B. Weinberger [*AIChE J.* **25:** 240–245 (1979)]. S. K. Das et al. [*Can. J. Chem. Eng.* **70:** 431–437 (1992)] studied holdup in both horizontal and vertical gas/liquid flow with nonnewtonian liquids. S. I. Farooqi and J. F. Richardson [*Trans. Inst. Chem. Engrs.* **60:** 292–305, 323–333 (1982)] developed correlations for holdup and pressure drop for gas/nonnewtonian liquid horizontal flow. They used a modified Lockhart-Martinelli parameter for nonnewtonian liquid holdup. They found that two-phase pressure drop may actually be less than the single-phase liquid pressure drop with shear-thinning liquids in laminar flow.

Pressure drop data for a 1-in *feed tee* with the liquid entering the run and gas entering the branch are given by G. E. Alves [*Chem. Eng. Progr.* **50:** 449–456 (1954)]. Pressure drop and division of two-phase *annular flow in a tee* are discussed by A. E. Fouda and E. Rhodes [*Trans. Inst. Chem. Eng.* [*London*], **52:** 354–360 (1974)]. Flow through tees can result in unexpected flow splitting. Further reading on gas/liquid flow through tees may be found in R. F. Mudde et al. [*Int. J. Multiphase Flow* **19:** 563–573 (1993)], R. I. Issa and P. J. Oliveira [*Computers and Fluids* **23:** 347–372 (1994)], and B. J. Azzopardi and P. A. Smith [*Int. J. Multiphase Flow* **18:** 861–875 (1992)].

Results by J. M. Chenoweth and M. W. Martin [*Pet. Refiner* **34**(10): 151–155 (1955)] indicate that single-phase data for *fittings* and *valves* can be used in their correlation for two-phase pressure drop. L. T. Smith et al. [*J. Eng. Power* **99:** 343–347 (1977)] evaluated correlations for two-phase flow of steam/water and other gas/liquid mixtures through sharp-edged *orifices* meeting ASTM standards for flow measurement. The correlation of J. W. Murdock [*J. Basic Eng.* **84:** 419–433 (1962)] may be used for these orifices. See also D. B. Collins and M. Gacesa [*J. Basic Eng.* **93:** 11–21 (1971)] for measurements with steam and water beyond the limits of this correlation.

For pressure drop and holdup in *inclined pipe* with upward or downward flow, see Beggs and Brill [*J. Pet. Technol.* **25:** 607–617 (1973)] and R. V. A. Oliemans and B. F. M. Pots [C. T. Crowe, ed., *Multiphase Flow Handbook*, CRC Press, Boca Raton, Fla., 2006]; the mechanistic model methods referenced above may also be applied to inclined pipes. Up to 10° from horizontal, upward pipe inclination has little effect on holdup [G. A. Gregory, *Can. J. Chem. Eng.* **53:** 384–388 [1975)].

For fully developed incompressible *cocurrent upflow* of gases and liquids in *vertical pipes,* a variety of flow pattern terminologies have appeared in the literature; some of these have been summarized and compared by G. W. Govier et al. [*Can. J. Chem. Eng.* **35:** 58–70 (1957)]. One reasonable classification of patterns is illustrated in Fig. 6-28.

In *bubble flow,* gas is dispersed as bubbles throughout the liquid, with some tendency to concentrate toward the center of the pipe. In *slug flow,* the gas forms large *Taylor bubbles* of diameter nearly equal to the pipe diameter. A thin film of liquid surrounds the Taylor bubble. Between the Taylor bubbles are liquid slugs containing some bubbles. *Froth* or *churn flow* is characterized by strong intermittency and intense mixing, with neither phase easily described as continuous or dispersed. Churn flow may not be a fully developed flow pattern, but instead a large entry length for developing slug flow [L. Zao and A. E. Dukler, *Int. J. Multiphase Flow* **19:** 377–383 (1993); G. F. Hewitt and Jayanti, *Int. J. Multiphase Flow* **19:** 527–529 (1993)].

Ripple flow has an upward-moving wavy layer of liquid on the pipe wall; it may be thought of as a transition region to *annular, annular mist,* or *film flow,* in which gas flows in the core of the pipe while an annulus of liquid flows up the pipe wall. Some of the liquid is entrained as droplets in the gas core. *Mist flow* occurs when all the liquid is carried as fine drops in the gas phase; this pattern occurs at high gas velocities, typically 20 to 30 m/s (70 to 100 ft/s).

The correlation by Govier et al. [*Can. J. Chem. Eng.* **35:** 58–70 (1957)], Fig. 6-29, may be used for quick estimate of flow pattern. See R. V. A. Oliemans and B. F. M. Pots [C. T. Crowe, ed., *Multiphase Flow Handbook*, CRC Press, Boca Raton, Fla., 2006] for mechanistic predictions of flow pattern transitions.

Slip, or relative velocity between phases, occurs for both vertical and horizontal flow. No completely satisfactory, flow regime–independent correlation for volume fraction or holdup exists for vertical flow. Two frequently

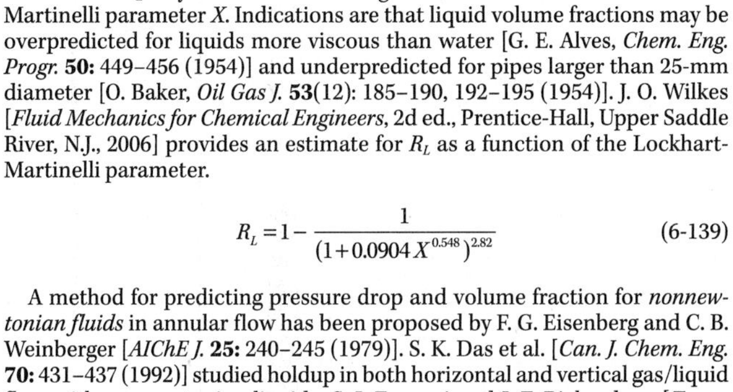

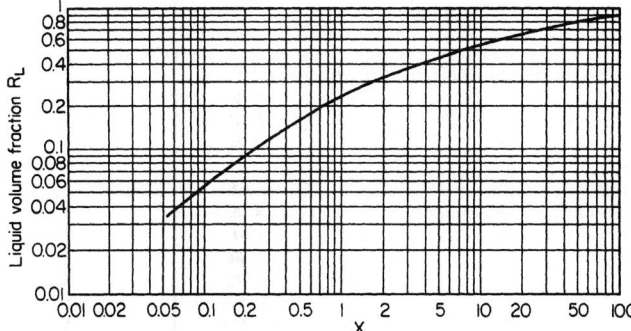

FIG. 6-27 Liquid volume fraction in liquid/gas flow through horizontal pipes. [From Lockhart and Martinelli, *Chem. Engr. Prog.*, **45,** 39 (1949).]

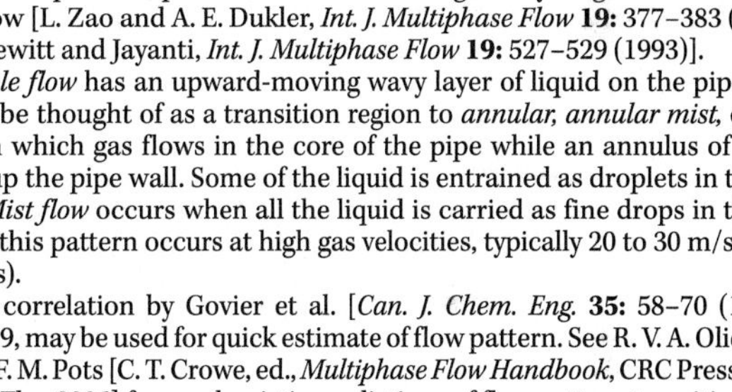

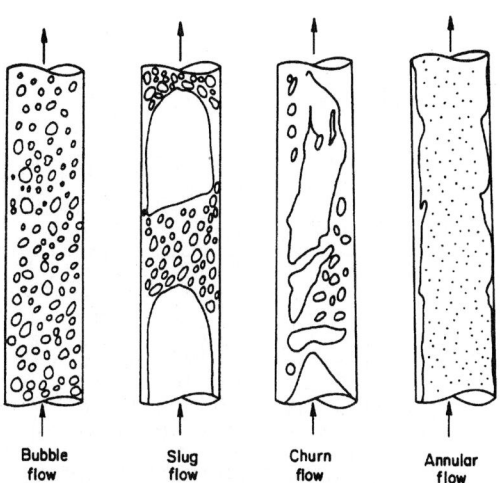

FIG. 6-28 Flow patterns in cocurrent upward vertical gas/liquid flow. [From Y. Taitel, D. Barnea, and A. E. Dukler, *AIChE J.*, **26,** 345–354 (1980). *Reproduced by permission of the American Institute of Chemical Engineers © 1980 AIChE. All rights reserved.*]

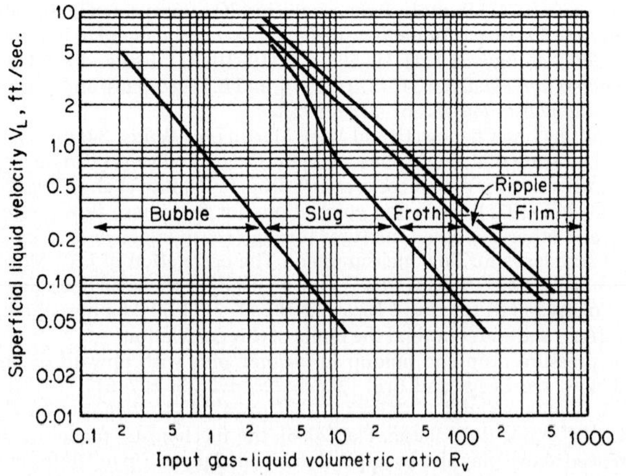

FIG. 6-29 Flow-pattern regions in cocurrent liquid/gas flow in upflow through vertical pipes. To convert ft/s to m/s, multiply by 0.3048. [From G. W. Govier, B. A. Radford, and J. S. C. Dunn, *Can. J. Chem. Eng.*, **35**, 58–70 (1957).]

used flow regime–independent methods are those by G. A. Hughmark and B. S. Pressburg [*AIChE J.* **7:** 677 (1961)] and G. A. Hughmark [*Chem. Eng. Prog.* **58**(4): 62 (April 1962)]. *Pressure drop* in *upflow* may be calculated by the procedure described in G. A. Hughmark [*Ind. Eng. Chem. Fundam.* **2:** 315–321 (1963)]. The mechanistic, flow regime–based methods are advisable for critical applications.

For *upflow* in *helically coiled tubes,* the flow pattern, pressure drop, and holdup can be predicted by the correlations of S. Banerjee et al. [*Can. J. Chem. Eng.* **47:** 445–453 (1969)] and K. Akagawa et al. [*Bull. JSME* **14:** 564–571 (1971)]. Correlations for flow patterns in *downflow* in vertical pipe are given by T. Oshinowo and M. E. Charles [*Can. J. Chem. Eng.* **52:** 25–35 (1974)] and D. Barnea et al. [*Chem. Eng. Sci.* **37:** 741–744 (1982)]. Use of *drift flux theory* for void fraction modeling in downflow is presented by N. N. Clark and R. L. C. Flemmer [*Chem. Eng. Sci.* **39:** 170–173 (1984)]. *Downward-inclined* two-phase flow data and modeling are given by D. Barnea et al. [*Chem. Eng. Sci.* **37:** 741–744 (1982)]. Data for *downflow* in *helically coiled tubes* are presented by C. Casper [*Chem. Ing. Tech.* **42:** 349–354 (1970)].

The entrance to a *drain* is flush with a horizontal surface, while the entrance to an *overflow* pipe is above the horizontal surface. When such pipes do not run full, considerable amounts of gas can be drawn down by the liquid. The amount of gas entrained is a function of pipe diameter, pipe length, and liquid flow rate as well as the drainpipe outlet boundary condition. Extensive data on air entrainment and liquid head above the entrance as a function of water flow rate for pipe diameters from 43.9 to 148.3 mm (1.7 to 5.8 in) and lengths from about 1.22 to 5.18 m (4.0 to 17.0 ft) are reported by A. A. Kalinske [*Univ. Iowa Stud. Eng. Bull.* **26:** 26–40 (1939–1940)]. For heads greater than the critical, the pipes will run full with no entrainment. The critical head h for flow of water in drains and overflow pipes is given in Fig. 6-30. Kalinske's results show little effect of the height of protrusion of overflow pipes when the protrusion height is greater than about one pipe diameter. For conservative design, N. G. McDuffie [*AIChE J.* **23:** 37–40 (1977)] recommends the following relation for minimum liquid height to prevent entrainment.

$$\mathrm{Fr} \le 1.6 \left(\frac{h}{D}\right)^2 \tag{6-140}$$

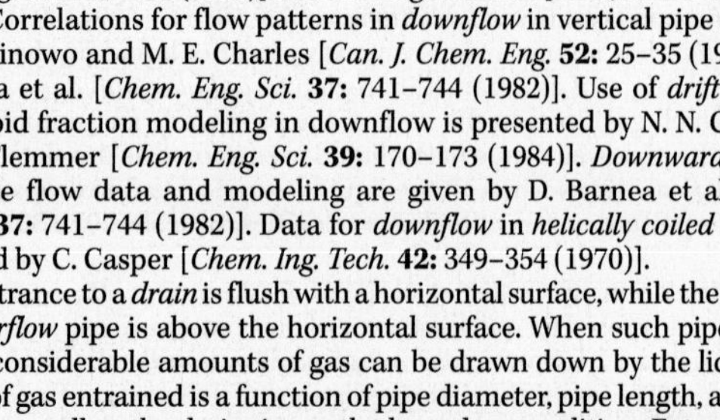

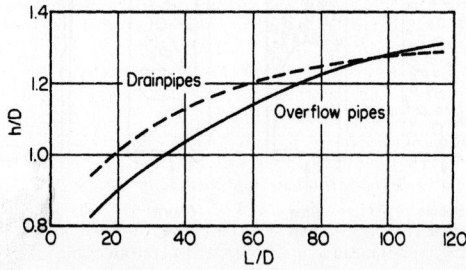

FIG. 6-30 Critical head for drain and overflow pipes. [From A. A. Kalinske, *Univ. Iowa Stud. Eng., Bull.* **26** (1939–1940).]

where the Froude number is defined by

$$\mathrm{Fr} \equiv \frac{V_L}{\sqrt{g(\rho_L - \rho_G)D/\rho_L}} \tag{6-141}$$

where g = acceleration due to gravity
V_L = liquid velocity in drainpipe
ρ_L = liquid density
ρ_G = gas density
D = pipe inside diameter
h = liquid height

For additional information, see L. L. Simpson [*Chem. Eng.* **75**(6): 192–214 (1968)] and J. N. Tilton and J. A. Garcia [*5th North American Conf. on Multiphase Technology*, Banff, pp. 119–133 (2006)]. A critical Froude number of 0.31 to ensure vented flow is widely cited. Experimental results [R. B.Thorpe, *3d Int. Conf. Multi-phase Flow,* The Hague, Netherlands, May 18–20, 1987, paper K2, and *4th Int. Conf. Multi-phase Flow,* Nice, France, June 19–21, 1989, paper K4] show hysteresis, with different critical Froude numbers for flooding and unflooding of drain pipes, and the influence of end effects. G. B Wallis et al. [*Trans. ASME J. Fluids Eng.* **99:** 405–413 (June 1977)] examine the conditions for horizontal discharge pipes to run full.

Flashing flow and *condensing flow* are two examples of multiphase flow with *phase change.* Flashing flow occurs when pressure drops below the bubble point pressure of a flowing liquid. A frequently used one-dimensional model for flashing flow through nozzles and pipes is the *homogeneous equilibrium model,* which assumes that both phases move at the same in situ velocity and maintain vapor/liquid equilibrium. At the *critical,* or *choking flow* condition, the velocity is sonic, evaluated from the derivative of pressure p with respect to mixture density at constant entropy.

$$c = \sqrt{\left(\frac{\partial p}{\partial \rho_m}\right)_s} \tag{6-142}$$

The critical velocity for flashing liquids is normally much less than that for gas flow. The critical mass flux is

$$G_{\text{crit}} = \rho_m \sqrt{\left(\frac{\partial p}{\partial \rho_m}\right)_s} \tag{6-143}$$

The mixture density is given in terms of the individual phase densities and the *quality* (mass flow fraction vapor) x by

$$\frac{1}{\rho_m} = \frac{x}{\rho_G} + \frac{1-x}{\rho_L} \tag{6-144}$$

Choked and unchoked flow situations arise in pipes and nozzles in the same fashion for homogeneous equilibrium flashing flow as for gas flow. For nozzle flow from stagnation pressure p_0 to exit pressure p_1, the mass flux is given by

$$G^2 = -2\rho_{m1}^2 \int_{p_0}^{p_1} \frac{dp}{\rho_m} \tag{6-145}$$

The integration is carried out over an isentropic flash path: flashes at constant entropy must be carried out to evaluate ρ_m as a function of p. Experience shows that isenthalpic flashes provide good approximations unless the liquid mass fraction is very small. Choking occurs when G obtained by Eq. (6-145) goes through a maximum at a value of p_1 greater than the external discharge pressure. The maximum flux will match the critical flux from Eq. (6-143). In such a case, the pressure at the nozzle exit equals the choking pressure.

For steady homogeneous flow in a pipe of diameter D, the differential form of the Bernoulli equation rearranges to

$$\frac{dp}{\rho_m} + g\,dz + \frac{G^2}{\rho_m} d\frac{1}{\rho_m} + 2f\frac{dx'}{D}\frac{G^2}{\rho_m^2} = 0 \tag{6-146}$$

where x' is distance along the pipe. Integration over a length L of pipe assuming constant friction factor f yields

$$G^2 = \frac{-\int_{p_1}^{p_2} \rho_m\,dp - g\int_{z_1}^{z_2} \rho_m^2\,dz}{\ln(\rho_{m1}/\rho_{m2}) + 2fL/D} \tag{6-147}$$

Frictional pipe flow is not isentropic. Strictly speaking, the flashes must be carried out at constant $h + V^2/2 + gZ$, where h is the enthalpy per unit mass of the two-phase flashing mixture. The effect of potential energy on the flashes is normally negligible, but the kinetic energy effects are not, for high-quality, high-velocity flows. The flash calculations are fully coupled with the integration of the Bernoulli equation; the velocity V must be known at every pressure p to evaluate ρ_m. Computational routines, employing the thermodynamic and material balance features of flow sheet simulators, are the most practical way to carry out such flashing flow calculations, particularly when multicomponent systems are involved. Equation (6-146) may be rearranged to give

$$\frac{dp}{dz} = -\frac{\dfrac{2fG^2}{\rho_m D} + \rho_m g \dfrac{dZ}{dx'}}{1 - \dfrac{G^2}{\rho^2}\dfrac{dp}{d\rho_m}} \tag{6-148}$$

The derivative $dp/d\rho_m$ is carried out at constant $h + V^2/2$. A singularity occurs at critical velocity where $G = G_{crit}$ and the denominator of Eq. (6-148) is zero. At the critical point, the derivative $dp/d\rho_m$ becomes equal to the partial derivative at constant entropy. Significant simplification arises when the mass fraction of liquid is large enough to neglect the effect of the $V^2/2$ term on the flash splits.

With flashes carried out along the appropriate thermodynamic paths, the formalism of Eqs. (6-142) through (6-148) applies to all homogeneous equilibrium compressible flows, including, for example, flashing flow, perfect gas flow, and nonideal gas flow.

Various *nonequilibrium* and *slip flow* models have been proposed as improvements on the homogeneous equilibrium flow model. See, for example, R. E. Henry and H. K. Fauske [*Trans. ASME J. Heat Transfer,* 179–187 (May 1971)]. Nonequilibrium and slip effects both increase computed mass flux for fixed pressure drop, compared to homogeneous equilibrium flow. For flow paths greater than about 100 mm (4 in), homogeneous equilibrium behavior appears to be the best assumption [H. G. Fischer et al., *Emergency Relief System Design Using DIERS Technology,* AIChE, New York, 1992]. For shorter flow paths, a reasonable estimate is to linearly interpolate (as a function of length) between *frozen flow* (constant quality, no flashing) at 0 length and homogeneous equilibrium flow at 100 mm (4 in).

In a series of papers by J. C. Leung and coworkers [*AIChE J.* **32:** 1743–1746 (1986); **33:** 524–527 (1987); **34:** 688–691 (1988); *J. Loss Prevention Proc. Ind.* **2**(2): 78–86 (April 1989); **31:** 27–32 (January 1990); *Trans. ASME J. Heat Transfer* **112:** 524–528, 528–530 (1990); **113:** 269–272 (1991)], approximate techniques have been developed for homogeneous equilibrium calculations based on pseudo–equation of state methods for flashing mixtures. Collectively known as the *omega method,* these developments are discussed in Sec. 23 of this handbook.

Less work has been done on *condensing flows.* Slip effects are more important for condensing than for flashing flows. M. Soliman et al. [*J. Heat Transfer* **90:** 267–276 (1968)] give a model for condensing vapor in *horizontal pipe.* They assume the condensate flows as an annular ring. The Lockhart-Martinelli correlation is used for the frictional pressure drop. To this pressure drop is added an acceleration term based on homogeneous flow, equivalent to $G^2 d(1/\rho_m)$ in Eq. (6-146). Pressure drop is computed by integration of the incremental pressure changes along the length of pipe.

For *condensing vapor* in *vertical downflow,* in which the liquid flows as a thin annular film, the frictional contribution to the pressure drop may be estimated based on the gas flow alone, using the friction factor plotted in Fig. 6-31, where Re_G is the Reynolds number for the gas flowing

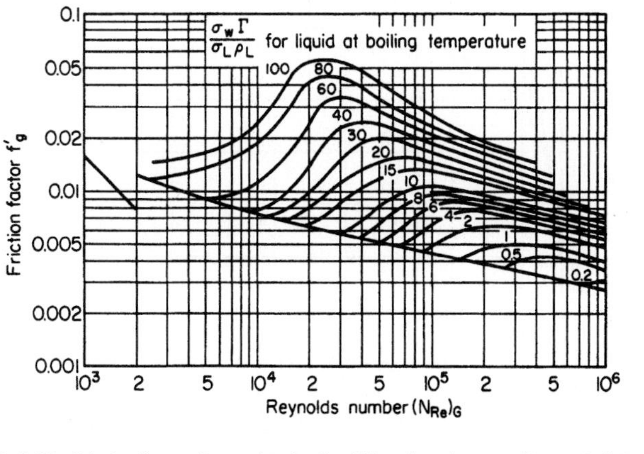

FIG. 6-31 Friction factors for condensing liquid/gas flow downward in vertical pipe. In this correlation Γ/ρ_L is in ft²/h. To convert ft²/h to m²/s, multiply by 0.00155. [From O. P. Bergelin et al., *Proc. Heat Transfer Fluid Mech. Inst.,* ASME, 1949, p. 19.]

alone [O. P. Bergelin et al., *Proc. Heat Transfer Fluid Mech. Inst.,* ASME, June 22–24, 1949, pp. 19–28].

$$-\frac{dp}{dz} = \frac{2f'_G \rho_G V_G^2}{D} \tag{6-149}$$

To this should be added the $G_G^2\, d(1/\rho_G)/dx$ term to account for velocity change effects.

Gases and Solids The flow of gases and solids (pneumatic transport) in *horizontal* pipe is usually classified as either *dilute phase* or *dense phase* flow. Unfortunately, there is no clear delineation between the two types of flow, and the *dense phase* description may take on more than one meaning, creating some confusion [T. M. Knowlton et al., *Chem. Eng. Progr.* **90**(4): 44–54 (April 1994)]. S. Dhodapkar et al. in C. T. Crowe [ed., *Multiphase Flow Handbook,* CRC Press, Boca Raton, Fla., 2006] describe pneumatic transport fundamentals and system design. For dilute phase flow, achieved at low solids-to-gas weight ratios (loadings) and high gas velocities, the solids may be fully suspended and fairly uniformly dispersed over the pipe cross section (homogeneous flow), particularly for low-density or small particle size solids. At lower gas velocities, the solids may bounce along the bottom of the pipe. With higher loadings and lower gas velocities, the particles may settle to the bottom of the pipe, forming dunes, with the particles moving from dune to dune. In dense phase conveying, solids tend to concentrate in the lower portion of the pipe at high gas velocity. As the gas velocity decreases, solids may first form dense moving strands, followed by slugs. Discrete plugs of solids may be created intentionally by timed injection of solids, or the plugs may form spontaneously. Eventually the pipe may become blocked. For more information on flow patterns, see J. M. Coulson and J. F. Richardson [*Chemical Engineering,* vol. 2, 2d ed., Pergamon, New York, 1968, p. 583]; C. Y. Wen and H. P. Simons [*AIChE J.* **5:** 263–267 (1959)]; and T. M. Knowlton et al. [*Chem. Eng. Progr.* **90**(4): 44–54 (April 1994)]. For the *minimum velocity* required to prevent formation of dunes or settled beds in *horizontal flow,* data are given by F. A. Zenz [*Ind. Eng. Chem. Fundam.* **3:** 65–75 (1964)], who presented a correlation for the minimum velocity required to keep particles from depositing on the bottom of the pipe. This rather tedious estimation procedure may also be found in G. W. Govier and K. Aziz [*The Flow of Complex Mixtures in Pipes,* Krieger, Huntington, N.Y., 1977], who provide additional references and discussion on transition velocities. In practice, the actual conveying velocities used in systems with loadings less than 10 are generally over 15 m/s (50 ft/s), while for high loadings (>20) they are generally less than 7.5 m/s (25 ft/s) and are roughly twice the actual solids velocity [C. Y. Wen and H. P. Simons [*AIChE J.* **5:** 263–267 (1959)].

Total *pressure drop* for horizontal flow includes acceleration effects at the entrance to the pipe and frictional effects beyond the entrance region. A great number of correlations for pressure gradient are available, none of which is applicable to all flow regimes. G. W. Govier and K. Aziz [*The Flow of Complex Mixtures in Pipes,* Krieger, Huntington, N.Y., 1977] review many of these and provide recommendations on when to use them. S. Dhodapkar et al. in C. T. Crowe [ed., *Multiphase Flow Handbook,* CRC Press, Boca Raton, Fla., 2006] recommend methods for computing pressure drop across the components of a pneumatic system. See also W. C. Yang [*AIChE J.* **24:** 548–552 (1978)].

For *upflow* of gases and solids in *vertical pipes,* the *minimum conveying velocity* for low loadings may be estimated as twice the terminal settling velocity of the largest particles. Equations for terminal settling velocity are found in the Particle Dynamics subsection following. *Choking* occurs as the velocity is dropped below the minimum conveying velocity and the solids are no longer transported, collapsing into solid plugs [T. M. Knowlton et al., *Chem. Eng. Progr.* **90**(4): 44–54 (April 1994)]. See T. N. Smith [*Chem. Eng. Sci.* **33:** 745–749 (1978)] for an equation to predict the onset of choking.

Total *pressure drop* for vertical upflow of gases and solids includes acceleration and frictional effects also found in horizontal flow, plus potential energy or hydrostatic effects. G. W. Govier and K. Aziz [*The Flow of Complex Mixtures in Pipes,* Krieger, Huntington, N.Y., 1977] review many of the pressure drop calculation methods and provide recommendations for their use. See also W. C. Yang [*AIChE J.* **24:** 548–552 (1978)].

Drag reduction has been reported for low loadings of small-diameter particles (< 60 μm), ascribed to damping of turbulence near the wall [S. J. Rossetti and R. P. Pfeffer, *AIChE J.* **18:** 31–39 (1972)].

For *dense phase* transport in *vertical pipes* of small diameter, see C. W. Sandy et al. [*Chem. Eng. Prog. Symp. Ser.* **66:** 105, 133–142 (1970)].

The *flow of bulk solids through restrictions and bins* is discussed in symposium articles [*J. Eng. Ind.* **91**(2) (1969)] and by Stepanoff [*Gravity Flow of Bulk Solids and Transportation of Solids in Suspension,* Wiley, New York, 1969]. Some problems encountered in discharge from bins include [T. M. Knowlton et al., *Chem. Eng. Progr.* **90**(4): 44–54 (April 1994)] flow stoppage due to *ratholing* or *arching, segregation* of fine and coarse particles, *flooding* upon collapse of ratholes, and poor *residence time distribution* when *funnel flow* occurs.

Solids and Liquids Slurry flow may be divided into two categories based on settling behavior (see A. W. Etchells in P. A. Shamlou [*Processing of Solid-Liquid Suspensions*, chap. 12, Butterworth-Heinemann, Oxford, UK, 1993]. *Nonsettling* slurries are made up of very fine, highly concentrated, or neutrally buoyant particles. These slurries are normally treated as pseudohomogeneous fluids, often quite viscous and usually nonnewtonian. Slurries of particles that tend to settle out rapidly are called *settling slurries* or *fast-settling slurries.* While in some cases positively buoyant solids are encountered, the present discussion will focus on solids which are denser than the liquid.

For *horizontal flow* of *fast-settling slurries,* the following rough description may be made [G. W. Govier and K. Aziz, *The Flow of Complex Mixtures in Pipes*, Krieger, Huntington, N.Y., 1977]. Ultrafine particles, 10 μm or smaller, are generally fully suspended, and the particle distributions are not influenced by gravity. Fine particles 10 to 100 μm are usually fully suspended, but gravity causes concentration gradients. Medium-size particles, 100 to 1000 μm, may be fully suspended at high velocity, but often form a moving deposit on the bottom of the pipe. Coarse particles, 1000 to 10,000 μm, are seldom fully suspended and are usually conveyed as a moving deposit. Ultracoarse particles larger than 10,000 μm are not suspended at normal velocities unless they are unusually light.

Figure 6-32, taken from G. W. Govier and K. Aziz [*The Flow of Complex Mixtures in Pipes*, Krieger, Huntington, N.Y., 1977] schematically indicates four flow pattern regions superimposed on a plot of pressure gradient versus mixture velocity $V_M = V_L + V_S = (Q_L + Q_S)/A$ where V_L and V_S are the superficial liquid and solid velocities, Q_L and Q_S are liquid and solid volumetric flow rates, and A is the pipe cross-sectional area. Also V_{M4} is the transition velocity above which a bed exists in the bottom of the pipe, part of which is stationary and part of which moves by *saltation,* with the upper particles tumbling and bouncing over one another, often with formation of dunes. With a broad particle-size distribution, the finer particles may be fully suspended. Near V_{M4}, the pressure gradient rapidly increases as V_M decreases. Above V_{M3}, the entire bed moves. Above V_{M2}, the solids are fully suspended; that is, there is no deposit, moving or stationary, on the bottom of the pipe. However, the concentration distribution of solids is asymmetric. This flow pattern is the most frequently used for fast-settling slurry transport. Typical mixture velocities are in the range of 1 to 4 m/s (3 to 13 ft/s). The minimum in the pressure gradient is found to be near V_{M2}. Above V_{M1}, the particles are symmetrically distributed, and the pressure gradient curve is nearly parallel to that for the liquid by itself.

The most important transition velocity, often regarded as the minimum transport or conveying velocity for settling slurries, is V_{M2}. The Durand equation [R. Durand, Minnesota Int. Hydraulics Conf., *Proc.,* 89, Int. Assoc. for Hydraulic Research (1953); R. Durand and E. Condolios, *Proc. Colloq. on the Hyd. Transport of Solids in Pipes, Nat. Coal Board* [*UK*], Paper IV, 39–35 (1952)] gives the minimum transport velocity as

$$V_{M2} = F_L[2gD(s-1)]^{0.5} \tag{6-150}$$

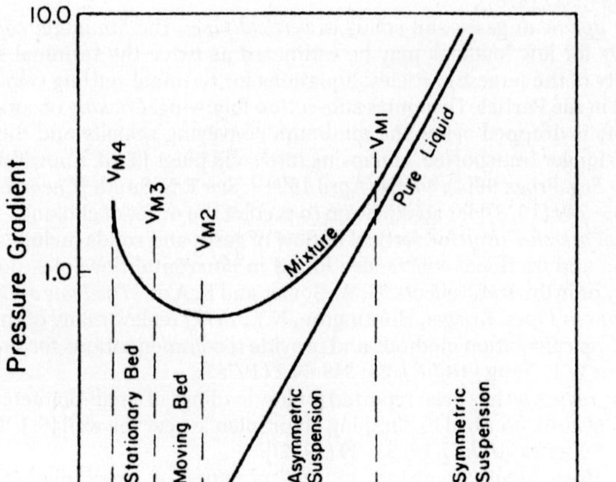

FIG. 6-32 Flow pattern regimes and pressure gradients in horizontal slurry flow. (From G. W. Govier and K. Aziz, *The Flow of Complex Mixtures in Pipes*, Van Nostrand Reinhold, New York, 1972.)

where g = acceleration of gravity
D = pipe diameter
$s = \rho_S/\rho_L$ = ratio of solid to liquid density
F_L = a factor influenced by particle size and concentration

Figure 6-33 gives Durand's empirical correlation for F_L as a function of particle diameter and the input, feed volume fraction solids, $C_S = Q_S/(Q_S + Q_L)$. Common practice for conservative design is to use $F_L = 1.5$. R. M. Turian et al. [*Powder Technol.* **51:** 35–47 (1987)] regressed an extensive data set using up to five parameters. The five-parameter result is shown below, but inclusion of all five parameters did not show significant improvement in RMS error.

$$F_L = 1.7951 C_S^{0.1087}(1-C_S)^{0.2501}\left\{\frac{D\rho_L[gD(s-1)]^{0.5}}{\mu}\right\}^{0.00179}\left(\frac{d}{D}\right)^{0.06623} \tag{6-151}$$

See A. P. Poloski et al. [*Can. J. Chem. Engr.* **88:** 182–189 (2010)] for data regressions focusing on small-particle (Archimedes number less than 80) transport.

No single correlation for *pressure drop* in horizontal solid/liquid flow has been found satisfactory for all particle sizes, densities, concentrations, and pipe sizes. However, with reference to Fig. 6-32, the following simplifications may be considered. The minimum pressure gradient occurs near V_{M2}, and for conservative purposes it is generally desirable to exceed V_{M2}. When V_{M2} is exceeded, a rough guide for pressure drop is 25 percent greater than that calculated assuming that the slurry behaves as a pseudohomogeneous fluid with the density of the mixture and the viscosity of the liquid. Above the transition velocity to symmetric suspension, V_{M1}, the pressure drop closely approaches the pseudohomogeneous pressure drop.

The following correlation by K. E. Spells [*Trans. Inst. Chem. Eng.* (*London*), **33:** 79–84 (1955)] may be used for V_{M1}.

$$V_{M1}^2 = 0.075\left(\frac{DV_{M1}\rho_M}{\mu}\right)^{0.775} gD_s(s-1) \tag{6-152}$$

where D = pipe diameter
D_S = particle diameter (such that 85 percent by weight of particles are smaller than D_S)
ρ_M = slurry mixture density
μ = liquid viscosity
$s = \rho_S/\rho_L$ = ratio of solid to liquid density

Between V_{M2} and V_{M1} the concentration of solids gradually becomes more uniform. This transition has been modeled by several authors as a concentration gradient where turbulent diffusion balances gravitational settling. See, for example, A. J. Karabelas [*AIChE J.* **23:** 426–434 (1977)].

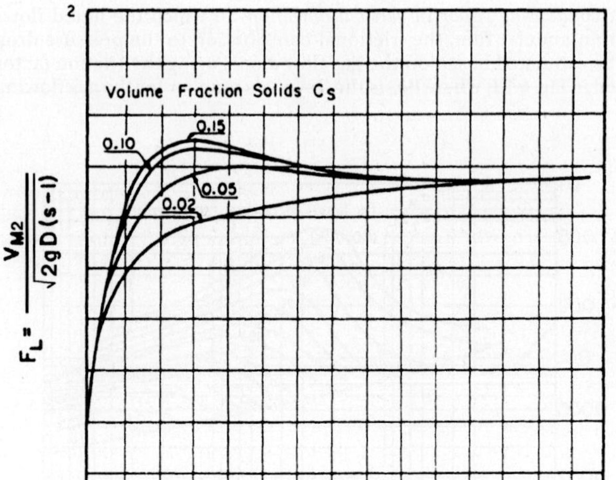

FIG. 6-33 Durand factor for minimum suspension velocity. [From G. W. Govier and K. Aziz, *The Flow of Complex Mixtures in Pipes*, Van Nostrand Reinhold, New York, 1972.]

Published correlations for pressure drop are frequently very complicated and tedious to use and may not offer significant accuracy advantages over the simple guide given here, and many of them are applicable only for velocities above V_{M2}. One that does include the effect of sliding beds is due to H. Gaessler [Doctoral Dissertation, Technische Hochshule, Karlsruhe, Germany, 1967] and reproduced by G. W. Govier and K. Aziz [*The Flow of Complex Mixtures in Pipes*, Krieger, Huntington, N.Y., 1977, pp. 668–669]. R. M. Turian and Yuan [*AIChE J.* **23:** 232–243 (1977); see also R. M. Turian and T. F. Yuan [*AIChE J.* **23:** 232–243 (1977)] segregated a large body of data into four flow regime groups and developed empirical correlations for predicting pressure drop in each flow regime.

The flow behavior of fiber suspensions is discussed by Bobkowicz and Gauvin [*Chem. Eng. Sci.* **22:** 229–241 (1967)], Bugliarello and Daily [*TAPPI,* **44:** 881–893 (1961)], and Daily and Bugliarello [*TAPPI* **44:** 497–512 (1961)].

In *vertical flow* of fast-settling slurries, the in situ concentration of solids with density greater than the liquid will exceed the feed concentration $C_S = Q_S/(Q_S + Q_L)$ for upflow and will be smaller than C_S for downflow. This results from slip between the phases. The *slip velocity,* the difference between the in situ average velocities of the two phases, is roughly equal to the terminal settling velocity of the solids in the liquid. Specification of the slip velocity for a pipe of a given diameter, along with the phase flow rates, allows calculation of in situ volume fractions, average velocities, and holdup ratios by simple material balances. Slip velocity may be affected by particle concentration and by turbulence conditions in the liquid. *Drift-flux theory,* a framework incorporating certain functional forms for empirical expressions for slip velocity, is described by G. B. Wallis [*One-Dimensional Two-Phase Flow,* McGraw-Hill, New York, 1969]. The correlation of J. F. Richardson and W. N. Zaki [*Trans. Instn. Chem. Engrs.* **32:** 35–53 (1954)] for sedimentation velocity may be expressed in drift-flux form

$$j_{LS} = v_\infty \varepsilon^n (1-\varepsilon) \tag{6-153}$$

where the drift velocity j_{LS} is the average in situ velocity of the liquid relative to the mixture velocity and v_∞ is the terminal settling velocity of a single particle. The slip velocity, $v_{LS} = v_L - v_S = (V_L/\varepsilon) - (V_S/[1-\varepsilon])$, the difference between the in situ velocities, is

$$v_{LS} = v_\infty \varepsilon^{n-1} \tag{6-154}$$

where V_L and V_S are the superficial velocities. The exponent n is a function of particle Reynolds number $\mathrm{Re} = dv_\infty\rho/\mu$ and the particle-to-pipe diameter ratio *d/D.*

$$\begin{aligned}
&\mathrm{Re} < 0.2 && n = 4.65 + 19.5\frac{d}{D} \\
&0.2 < \mathrm{Re} < 1 && n = \left(4.35 + 17.5\frac{d}{D}\right)\mathrm{Re}^{-0.03} \\
&1 < \mathrm{Re} < 200 && n = \left(4.45 + 18\frac{d}{D}\right)\mathrm{Re}^{-0.1} \\
&200 < \mathrm{Re} < 500 && n = 4.45\,\mathrm{Re}^{-0.1} \\
&\mathrm{Re} > 500 && n = 2.39
\end{aligned} \tag{6-155}$$

Minimum transport velocity for upflow for design purposes is usually taken as twice the particle settling velocity. *Pressure drop* in vertical pipe flow includes the effects of kinetic and potential energy (elevation) changes and friction. H. E. Rose and R. A. Duckworth [*The Engineer* **227**(5903): 392 (1969); **227**(5904): 430 (1969); **227**(5905): 478 (1969)]; see also G. W. Govier and K. Aziz [*The Flow of Complex Mixtures in Pipes,* Krieger, Huntington, N.Y., 1977, pp. 487–493] have developed a calculation procedure including all these effects, which may be applied not only to vertical solid/liquid flow, but also to gas/solid flow and to horizontal flow.

For fast-settling slurries, ensuring conveyance is usually the key design issue while pressure drop is somewhat less important. For *nonsettling slurries* conveyance is not an issue, because the particles do not separate from the liquid. Here, rheological behavior, which controls pressure drop, takes on critical importance. Further discussion of both fast-settling and nonsettling slurries may be found in C. A. Shook [in P. A. Shamlou, *Processing of Solid-Liquid Suspensions,* chap. 11, Butterworth-Heinemann, Oxford, UK, 1993].

FLUID DISTRIBUTION

Uniform fluid distribution is essential for efficient operation of many types of chemical-processing equipment. To obtain satisfactory distribution, proper consideration must be given to flow behavior in the distributor, flow conditions upstream and downstream of the distributor, and the distribution requirements of the equipment. This subsection provides guidelines for the design of various types of fluid distributors.

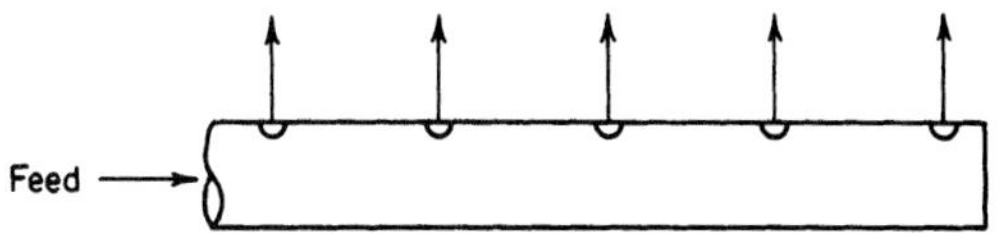

FIG. 6-34 Perforated-pipe distributor.

Perforated-Pipe Distributors The perforated pipe or sparger (Fig. 6-34) is a common type of distributor. As shown, the flow distribution is uniform; this is the case when the pressure drop across the holes is large compared to the pressure variation in the pipe due to velocity change and friction. Elevation changes may be also important when liquids are discharged into gases and vice versa. Sometimes hole size or spacing is varied to compensate for pressure variation along the pipe. In typical turbulent flow applications, in relatively short distributor pipes, inertial effects associated with velocity changes may exceed or even dominate frictional losses in determining the pressure distribution along the pipe. Application of the momentum or mechanical energy equations in such a case shows that the pressure inside the pipe increases with distance from the entrance. If the outlet holes are uniform in size and spacing, the discharge flow will be biased toward the closed end. Disturbances upstream of the distributor, such as pipe bends, may increase or decrease the flow to the holes at the beginning of the distributor. When frictional pressure drop dominates the inertial pressure recovery, the distribution is biased toward the feed end of the distributor.

For turbulent flow in a horizontal pipe distributor, with roughly uniform distribution, assuming a constant friction factor, the combined effect of friction and inertial (momentum) pressure recovery is given by

$$\Delta p = \left(\frac{4fL}{3D} - 2K\right)\frac{\rho V_i^2}{2} \qquad \text{(discharge manifolds)} \tag{6-156}$$

where Δp = net pressure drop over length of distributor
L = pipe length
D = pipe diameter
f = Fanning friction factor
V_i = distributor inlet velocity

The factor K would be 1 in the case of full momentum recovery, or 0.5 in the case of negligible viscous losses in the portion of flow that remains in the pipe after the flow divides at a takeoff point [M. M. Denn, *Process Fluid Mechanics,* Prentice-Hall, Englewood Cliffs, N.J., 1979]. Experimental data [B. G. Van der Hegge Zijnen, *Appl. Sci. Res.* **A3:** 144–162 (1951–1953)]; and B. J. Bailey [*J. Mech. Eng. Sci.* **17:** 338–347 (1975)], while scattered, show that K is probably close to 0.5 for discharge manifolds. For inertially dominated flows, Δp will be negative. For *return manifolds,* the recovery factor K is close to 1.0, and the pressure drop between the first hole and the exit is given by

$$\Delta p = \left(\frac{4fL}{3D} + 2K\right)\frac{\rho V_e^2}{2} \qquad \text{(return manifolds)} \tag{6-157}$$

where V_e is the pipe exit velocity.

One means to obtain a desired uniform distribution is to make the average pressure drop across the holes Δp_o large compared to the pressure variation over the length of pipe Δp. Then the relative variation in pressure drop across the holes will be small, and so will be the variation in flow. When the area of an individual hole is small compared to the cross-sectional area of the pipe, hole pressure drop may be expressed in terms of the discharge coefficient C_o and the velocity across the hole V_o as

$$\Delta p_o = \frac{1}{C_o^2}\frac{\rho V_o^2}{2} \tag{6-158}$$

Provided C_o is the same for all the holes, the *percent maldistribution,* defined as the percentage variation in flow between the first and last holes, may be estimated reasonably well for small maldistribution by [V. E. Senecal, *Ind. Eng. Chem.* **49:** 993–997 (1957)]

$$\%\ \text{maldistribution} = 100\left(1 - \sqrt{\frac{\Delta p_o - |\Delta p|}{\Delta p_o}}\right) \tag{6-159}$$

This equation shows that for 5 percent maldistribution, the pressure drop across the holes should be about 10 times the pressure variation over the length of the pipe. For discharge manifolds with $K = 0.5$ in Eq. (6-156), and with $4fL/3D << 1$, the pressure drop across the holes should be 10 times the inlet velocity head, $\rho V_i^2/2$ for 5 percent maldistribution. This leads to a simple design equation.

Discharge manifolds, $4fL/3D << 1$, 5% maldistribution:

$$\frac{V_o}{V_i} = \frac{A_p}{A_o} = \sqrt{10}C_o \quad (6\text{-}160)$$

Here A_p = pipe cross-sectional area and A_o is the *total* hole area of the distributor. Use of large ratios of hole velocity to pipe velocity promotes perpendicular discharge streams. In practice, there are many cases where the $4fL/3D$ term will be less than unity but not close to zero. In such cases, Eq. (6-160) will be conservative, while Eqs. (6-156), (6-158), and (6-159) will give more accurate design calculations. In cases where $4fL/(3D) > 2$, friction effects are large enough to render Eq. (6-160) nonconservative. When significant variations in f along the length of the distributor occur, calculations should be made by dividing the distributor into small enough sections that constant f may be assumed over each section.

For return manifolds with $K = 1.0$ and $4fL/(3D) << 1$, 5 percent maldistribution is achieved when the hole pressure drop is 20 times the pipe exit velocity head.

Return manifolds, $4fL/3D << 1$, 5% maldistribution:

$$\frac{V_o}{V_e} = \frac{A_p}{A_o} = \sqrt{20}C_o \quad (6\text{-}161)$$

When $4fL/3D$ is not negligible, Eq. (6-161) is not conservative and Eqs. (6-157), (6-158), and (6-159) should be used.

A common misconception is that good distribution is always provided by high pressure drop, so that increasing flow rate improves distribution by increasing the pressure drop. Conversely, it is mistakenly believed that turndown of flow through a perforated pipe will cause maldistribution. However, when both the pipe flow and the orifice flow are inertially dominated, changing the flow rate changes Δp and Δp_o in the same proportion, and the distribution uniformity is unchanged. Similarly, when pipe flow and orifice flow are both viscous-dominated, the flow rate has no effect on distribution.

Sometimes design for uniform velocity through uniformly sized and spaced orifices is impractical, because either the pressure drop required for an acceptable pipe diameter is too large, or the pipe diameter required for an acceptable pressure drop is too large. Some measures for such a situation include the following:

1. Taper the diameter of the distributor pipe so that the pipe velocity and velocity head remain constant along the pipe, thus substantially reducing inertial pressure variation.
2. Vary the hole size and/or the spacing between holes to compensate for the pressure variation along the pipe.
3. Feed or withdraw from both ends, reducing the pipe flow velocity head and required hole pressure drop by a factor of 4.

The orifice discharge coefficient C_o is usually taken to be about 0.62. However, C_o is dependent on the ratios of hole diameter to pipe diameter, pipe wall thickness to hole diameter, and pipe velocity to hole velocity. As long as all these are small, and the orifice Reynolds number is greater than about 100, the coefficient 0.62 is generally adequate.

Example 6-9 Pipe Distributor A 3-in Schedule 40 (inside diameter 7.793-cm) pipe is to be used as a distributor for a flow of 0.010 m³/s of water (ρ = 1000 kg/m³, μ = 0.001 Pa · s). The pipe is 0.7 m long and is to have 10 holes of uniform diameter and spacing along the length of the pipe. The distributor pipe is submerged. Calculate the required hole size to limit maldistribution to 5 percent, and estimate the pressure drop across the distributor.

The inlet velocity $V_i = Q/A_p = 4Q/(\pi D^2) = 2.10$ m/s, and the inlet Reynolds number is

$$\text{Re} = \frac{DV_i\rho}{\mu} = \frac{0.07793 \times 2.10 \times 1000}{0.001} = 1.64 \times 10^5$$

For commercial pipe with roughness ε = 0.046 mm, the friction factor is about 0.0043. Approaching the last hole, the flow rate, velocity, and Reynolds number are about one-tenth of their inlet values. At Re = 16,400 the friction factor f is about 0.0070. Using an average value of f = 0.0057 over the length of the pipe, $4fL/3D$ is 0.068 and may reasonably be neglected so that Eq. (6-158) may be used. With C_o = 0.62,

$$\frac{V_o}{V_i} = \frac{A_p}{A_o} = \sqrt{10}C_o = \sqrt{10} \times 0.62 = 1.96$$

With pipe cross-sectional area A_p = 0.00477 m², the total hole area is 0.00477/1.96 = 0.00243 m². The area and diameter of each hole are then 0.00243/10 = 0.000243 m² and 1.76 cm, respectively. With V_o/V_i = 1.96, the hole velocity is 1.96 × 2.10 = 4.12 m/s, and the pressure drop across the holes is obtained from Eq. (6-158).

$$\Delta p_o = \frac{1}{C_o^2}\frac{\rho V_o^2}{2} = \frac{1}{0.62^2} \times \frac{1000(4.12)^2}{2} = 22{,}100 \text{ Pa}$$

The hole pressure drop is 10 times the pressure variation in the pipe; the total pressure drop from the inlet of the distributor may be taken as approximately 22,100 Pa.

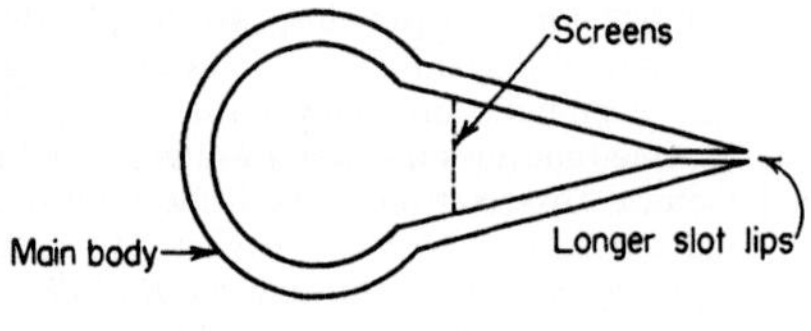

FIG. 6-35 Modified slot distributor.

Further detailed information on pipe distributors may be found in V. E. Senecal [*Ind. Eng. Chem.* **49:** 993–997 (1957)]. Much of the information on tapered manifold design has appeared in the pulp and paper literature [A. C. Spengos and R. B. Kaiser, *TAPPI* **46**(3): 195–200 (1963); G. D. Madeley, *Paper Technology* **9**(1): 35–39 (1968); J. Mardon et al., *TAPPI* **46**(3): 172–187 (1963); J. Mardon et al., *Pulp and Paper Magazine of Canada* **72**(11): 76–81 (November 1971); A. D. Trufitt, *TAPPI* **5**(11): 144–145 (1975)].

Slot Distributors These devices are generally used in sheeting dies for extrusion of films and coatings and in air knives for control of thickness of a material applied to a moving sheet. A simple slotted pipe for turbulent flow conditions may give severe maldistribution because of nonuniform discharge velocity, but also because this type of design does not readily give perpendicular discharge [A. Koestel and G. L. Tuve, *Heat. Piping Air Cond.* **20**(1): 153–157 (1948); V. E. Senecal, *Ind. Eng. Chem.* **49:** 993–997 (1957); A. Koestel and C. Y. Young, *Heat Piping Air Cond.* **23**(7): 111–115 (1951)]. For slots in tapered ducts where the duct cross-sectional area decreases linearly to zero at the far end, the discharge angle will be constant along the length of the duct (Koestel and Young). If the slot area is less than one-tenth of the pipe cross-sectional area, discharge will be perpendicular. As in the case of perforated-pipe distributors, pressure variation within the slot manifold and pressure drop across the slot must be carefully considered. In practice, the following methods may be used to keep the diameter of the pipe to a minimum consistent with good performance [V. E. Senecal, *Ind. Eng. Chem.* **49:** 993–997 (1957)].

1. Feed from both ends.
2. Modify the cross-sectional design (Fig. 6-35); the slot is thus farther away from the influence of feed stream velocity.
3. Increase pressure drop across the slot; this can be accomplished by lengthening the lips (Fig. 6-35).
4. Use screens (Fig. 6-35) to increase overall pressure drop across the slot.

Design of air knives is discussed by V. E. Senecal [*Ind. Eng. Chem.* **49:** 993–997 (1957)]. Design procedures for extrusion dies are presented by E. C. Bernhardt [*Processing of Thermoplastic Materials*, Rheinhold, New York, 1959, pp. 248–281].

Turning Vanes In applications such as ventilation, the discharge profile from slots can be improved by turning vanes. The tapered duct is the most amenable for turning vanes because the discharge angle remains constant. One way of installing the vanes is shown in Fig. 6-36. The vanes should have a depth twice the spacing [*Heating, Ventilating, Air Conditioning Guide*, vol. 38, American Society of Heating, Refrigerating and Air-Conditioning Engineers, 1960, pp. 282–283] and a curvature at the upstream end of the vanes of a circular arc which is tangent to the discharge angle θ of a slot without vanes and perpendicular at the downstream or discharge end of the vanes [A. Koestel and C. Y. Young, *Heat Piping Air Cond.* **23**(7): 111–115 (1951)]. The angle θ can be estimated from

$$\cot\theta = \frac{C_d A_s}{A_d} \quad (6\text{-}162)$$

where A_s = slot area
A_d = duct cross-sectional area at upstream end
C_d = discharge coefficient of slot

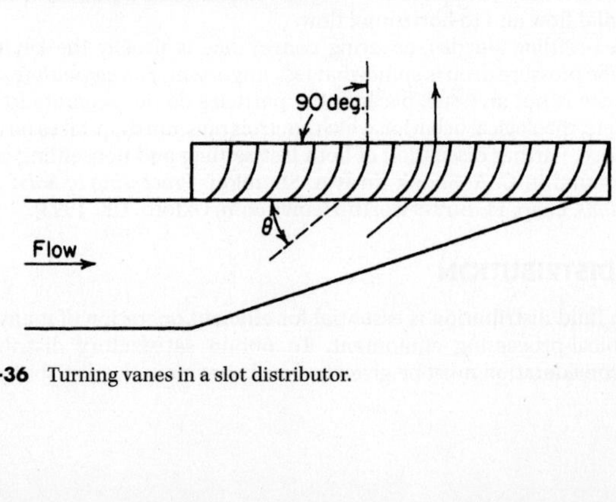

FIG. 6-36 Turning vanes in a slot distributor.

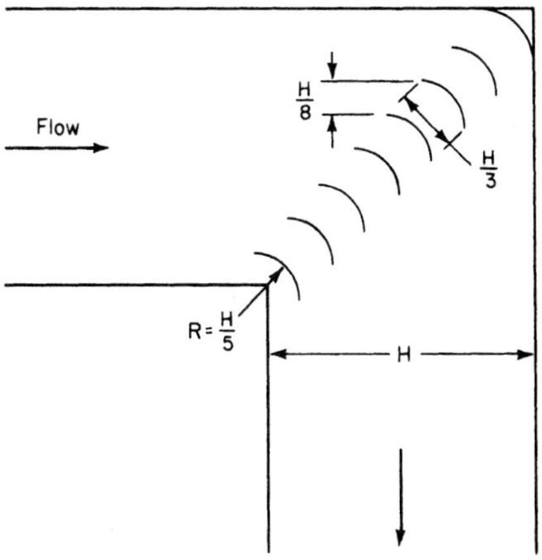

FIG. 6-37 Miter bend with vanes.

Vanes may be used to improve velocity distribution and reduce frictional loss in bends, when the ratio of bend turning radius to pipe diameter is less than 1.0. For a miter bend with low-velocity flows, simple circular arcs (Fig. 6-37) can be used, and with high-velocity flows, vanes of special airfoil shapes are required. For additional details and references, see E. Ower and R. C. Pankhurst [*The Measurement of Air Flow,* Pergamon, New York, 1977, p. 102]; R. C. Pankhurst and D. W. Holder [*Wind-Tunnel Technique,* Pitman, London, 1952, pp. 92–93]; H. Rouse [*Engineering Hydraulics,* Wiley, New York, 1950, pp. 399–401]; and R. Jorgensen [*Fan Engineering,* 7th ed., Buffalo Forge Co., Buffalo, N.Y., 1970, pp. 111, 117, 118].

Perforated Plates and Screens A nonuniform velocity profile in turbulent flow through channels or process equipment can be smoothed out to any desired degree by adding sufficient uniform resistance, such as perforated plates or screens across the flow channel, as shown in Fig. 6-38. R. L. Stoker [*Ind. Eng. Chem.* **38:** 622–624 (1946)] provides the following equation for the effect of a uniform resistance on velocity profile:

$$\frac{V_{2,\max}}{V} = \sqrt{\frac{(V_{1,\max}/V)^2 + \alpha_2 - \alpha_1 + \alpha_2 K}{1+K}} \qquad (6\text{-}163)$$

Here, V is the area average velocity, K is the number of velocity heads of pressure drop provided by the uniform resistance, $\Delta p = K\rho V^2/2$, and α is the velocity profile factor used in the mechanical energy balance, Eq. (6-18). It is the ratio of the area average of the cube of the velocity, to the cube of the area average velocity V. The shape of the exit velocity profile appears twice in Eq. (6-163), in $V_{2,\max}/V$ and α_2. Typically, K is on the order of 10, and the desired exit velocity profile is fairly uniform so that $\alpha_2 \sim 1.0$ may be appropriate. Downstream of the resistance, the velocity profile will gradually reestablish the fully developed profile characteristic of the Reynolds number and channel shape.

Screens and other flow restrictions may also be used to suppress stream swirl and turbulence [R. I. Loehrke and H. M. Nagib, *J. Fluids Eng.* **98:** 342–353 (1976)]. Contraction of the channel, as in a venturi, provides further reduction in turbulence level and flow nonuniformity.

Beds of Solids A suitable depth of solids can be used as a fluid distributor. A pressure drop of 10 velocity heads is typically used, here based on the superficial velocity through the bed. There are several substantial disadvantages to use of particle beds for flow distribution.

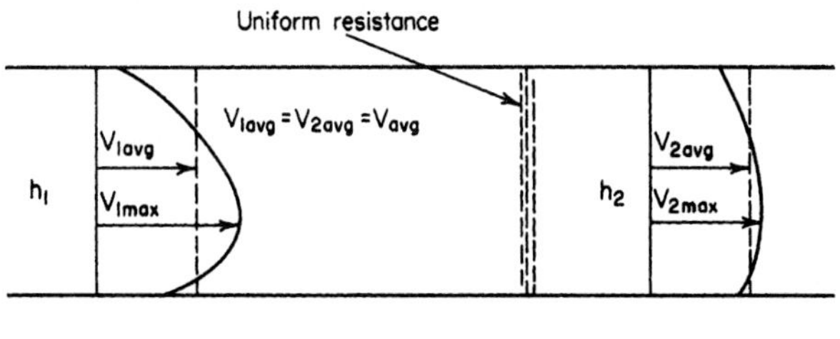

FIG. 6-38 Smoothing out a nonuniform profile in a channel.

Heterogeneity of the bed may actually worsen rather than improve distribution. Uniform flow may be found only downstream of the point in the bed where sufficient pressure drop has occurred to produce uniform flow. Therefore, inefficiency results when the bed also serves reaction or mass transfer functions, since portions of the bed are bypassed. In the case of trickle flow of liquid downward through column packings, inlet distribution is critical since the bed itself is relatively ineffective in distributing the liquid. Maldistribution of flow through packed beds also arises when the ratio of bed diameter to particle size is less than 10 to 30.

Other Flow-Straightening Devices Other devices designed to produce uniform velocity or reduce swirl, sometimes with reduced pressure drop, are available. These include both commercial devices of proprietary design and devices discussed in the literature. For pipeline flows, see the references under flow inverters and static mixing elements previously discussed in the Incompressible Flow in Pipes and Channels subsection. For large area changes, such as at the entrance to a vessel, it is sometimes necessary to diffuse the momentum of the inlet jet discharging from the feed pipe in order to produce a more uniform velocity profile within the vessel. Methods for this application exist, but are often proprietary.

FLUID MIXING

Mixing of fluids is a discipline of fluid mechanics. Fluid motion is used to accelerate the otherwise slow processes of diffusion and conduction to bring about uniformity of concentration and temperature, blend materials, facilitate chemical reactions, bring about intimate contact of multiple phases, and so on. A brief introduction and some references for further information are given here. E. L. Paul et al. [*Handbook of Industrial Mixing,* Wiley-Interscience, Hoboken, N.J., 2004], updated in S. M. Kresta et al. [*Advances in Industrial Mixing,* Wiley, Hoboken, N.J., 2016], is the most comprehensive and up-to-date reference. Textbooks include N. Harnby et al. [*Mixing in the Process Industries,* 2d ed., Butterworths, London, 1992]; J. Y. Oldshue [*Fluid Mixing Technology,* McGraw-Hill, New York, 1983]; G. B. Tatterson [*Fluid Mixing and Gas Dispersion in Agitated Tanks,* McGraw-Hill, New York, 1991]; V. W. Uhl and J. B. Gray [*Mixing,* vols. 1–3, Academic, New York, 1966, 1967, 1986]; and S. Nagata [*Mixing: Principles and Applications,* Wiley, New York, 1975]. A good, though dated, overview of stirred tank agitation is given in the series of articles from *Chemical Engineering* [pp. 110–114, Dec. 8, 1975; pp. 139–145, Jan. 5, 1976; pp. 93–100, Feb. 2, 1976; pp. 102–110, Apr. 26, 1976; pp. 144–150, May 24, 1976; pp. 141–148, July 19, 1976; pp. 89–94, Aug. 2, 1976; pp. 101–108, Aug. 30, 1976; pp. 109–112, Sept. 27, 1976; pp. 119–126, Oct. 25, 1976; pp. 127–133, Nov. 8, 1976].

Process mixing is commonly carried out in pipelines and vessels. The terms *radial mixing* and *axial mixing* are commonly used. Axial mixing refers to mixing of materials that pass a given point at different times and thus leads to *back-mixing.* For example, back-mixing or axial mixing occurs in stirred tanks where fluid elements entering the tank at different times are intermingled. Mixing of elements initially at different axial positions in a pipeline is axial mixing. Radial mixing occurs between fluid elements passing a given point at the same time, such as between fluids mixing in a pipeline tee.

Turbulent flow, by means of the chaotic eddy motion associated with velocity fluctuation, is conducive to rapid mixing and therefore is the preferred flow regime for mixing. *Laminar mixing* is carried out when high viscosity makes turbulent flow impractical.

Stirred Tank Agitation Turbine impeller agitators, of a variety of shapes, are used for stirred tanks, predominantly in turbulent flow. Figure 6-39 shows typical stirred tank configurations and time-averaged flow patterns for axial flow and radial flow impellers. In order to prevent formation of a *vortex,* four vertical baffles are normally installed. These cause top-to-bottom mixing and prevent mixing-ineffective swirling motion.

For a given impeller and tank geometry, the impeller Reynolds number determines the flow pattern in the tank:

$$\text{Re}_I = \frac{D^2 N\rho}{\mu} \qquad (6\text{-}164)$$

where D = impeller diameter, N = rotational speed, and ρ and μ are the liquid density and viscosity, respectively. Rotational speed N is typically reported in revolutions per minute, or revolutions per second in SI units. Radians per second are almost never used. Typically, $\text{Re}_I > 10^4$ is required for fully turbulent conditions throughout the tank. A wide transition region between laminar and turbulent flow occurs over the range $10 < \text{Re}_I < 10^4$.

The power P drawn by the impeller is made dimensionless in a group called the *power number*:

$$N_P = \frac{P}{\rho N^3 D^5} \qquad (6\text{-}165)$$

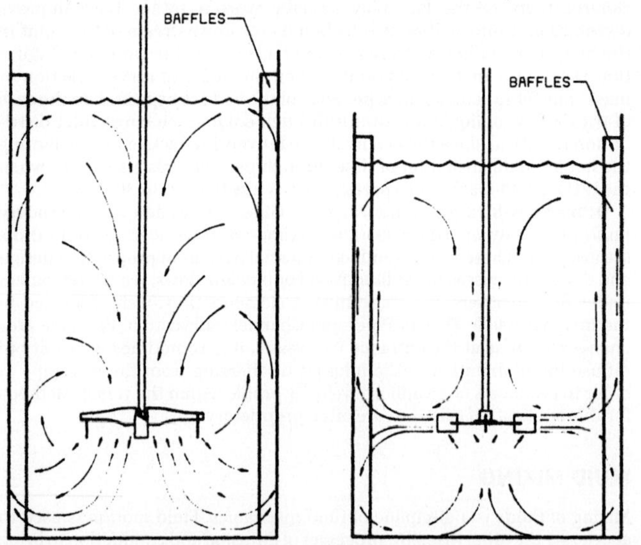

FIG. 6-39 Typical stirred tank configurations, showing time-averaged flow patterns for axial flow and radial flow impellers. [From J. Y. Oldshue, *Fluid Mixing Technology*, McGraw-Hill, New York, 1983.]

Figure 6-40 shows power number versus impeller Reynolds number for several impeller types. The similarity to the friction factor versus Reynolds number behavior for pipe flow is significant. In laminar flow, the power number is inversely proportional to the Reynolds number, reflecting the dominance of viscous forces over inertial forces. In turbulent flow, where inertial forces dominate, the power number is nearly constant.

The total volumetric flow rate Q discharged by an impeller is made dimensionless in a pumping number:

$$N_Q = \frac{Q}{ND^3} \tag{6-166}$$

Blend time t_b, the time required to achieve a specified standard deviation of concentration after injection of a tracer into a stirred tank, is made dimensionless by multiplying by the impeller rotational speed:

$$N_b = t_b N \tag{6-167}$$

Dimensionless pumping number and blend time are independent of Reynolds number under fully turbulent conditions. The magnitude of concentration fluctuations from the final well-mixed value in batch mixing decays exponentially with time.

The design of mixing equipment depends on the desired process result. There is often a tradeoff between operating cost, which depends mainly on power, and capital cost, which depends on agitator size and torque. For some applications bulk flow throughout the vessel is desired, while for others high local turbulence intensity is required. Multiphase systems introduce such design criteria as solids suspension and gas dispersion. In very viscous systems, helical ribbons, extruders, and other specialized equipment types are favored over turbine agitators.

Pipeline Mixing Mixing may be carried out with *mixing tees, inline* or *motionless mixing elements*, or in empty pipe. In the latter case, large pipe lengths may be required to obtain adequate mixing. Coaxially injected streams require lengths on the order of 100 pipe diameters.

Properly designed tee mixers, with due consideration given to mainstream and injected stream momentum, are capable of producing high degrees of uniformity in just a few diameters. L. J. Forney ["Jet Injection for Optimum Pipeline Mixing," in *Encyclopedia of Fluid Mechanics*, vol. 2., chap. 25, Gulf Publishing, Houston, Tex., 1986] provides a thorough discussion of tee mixing. Inline or motionless mixers are generally of proprietary commercial design, and they may be selected for viscous or turbulent, single-phase or multiphase mixing applications. They substantially reduce

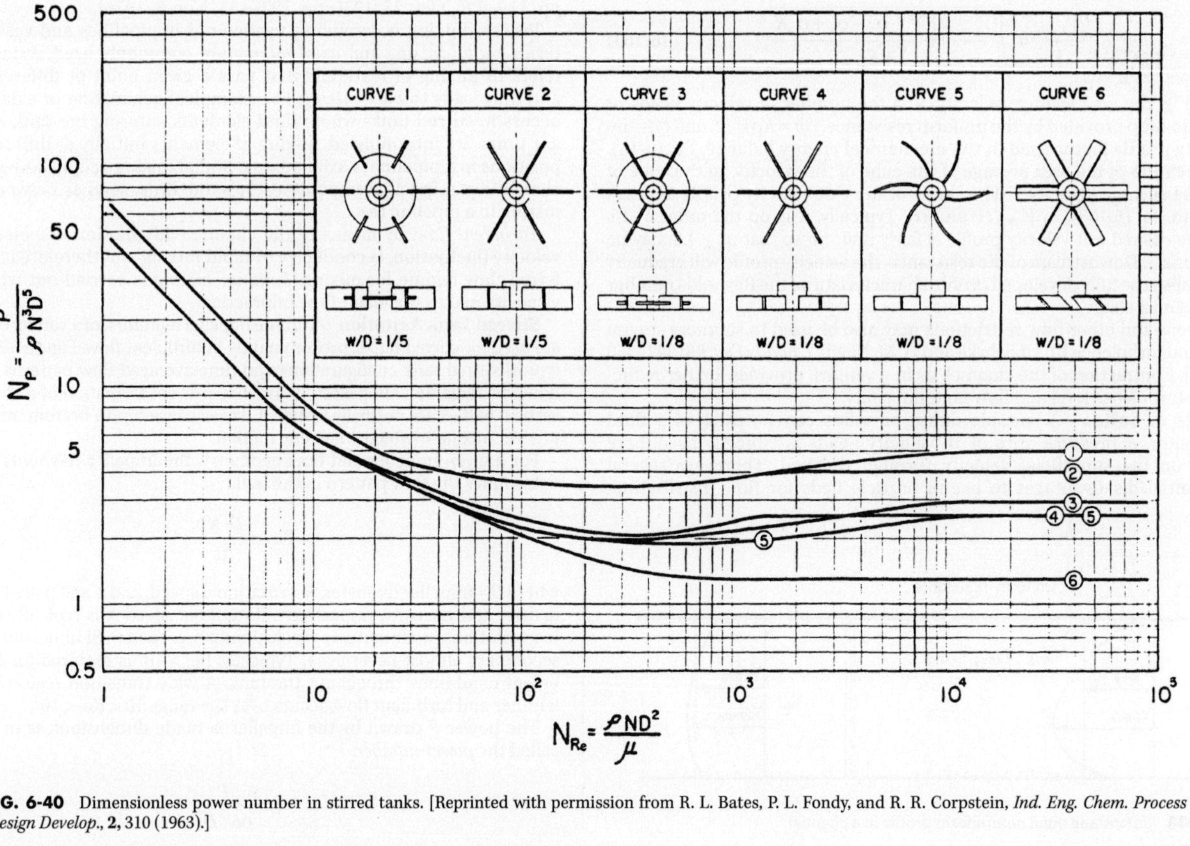

FIG. 6-40 Dimensionless power number in stirred tanks. [Reprinted with permission from R. L. Bates, P. L. Fondy, and R. R. Corpstein, *Ind. Eng. Chem. Process Design Develop.*, **2**, 310 (1963).]

required pipe length for mixing. See E. L. Paul et al. [*Handbook of Industrial Mixing,* Wiley-Interscience, Hoboken, N.J., 2004] for further information on static mixers.

TUBE BANKS

Pressure drop across tube banks may not be correlated by means of a single, simple friction factor—Reynolds number curve, owing to the variety of tube configurations and spacings encountered. The most common are staggered and in-line arrays, as shown in Fig. 6-41. Several investigators have allowed for configuration and spacing by incorporating spacing factors in their friction factor expressions or by using multiple friction factor plots. Heat exchanger design is the most important application for evaluating pressure drop for flow across tube banks. Commercial computer codes for heat-exchanger design are available which include features for estimating pressure drop across tube banks. These calculations are complicated by the fact that flow is not strictly normal to the tubes, even within each tube pass, as well as by the need to estimate turning losses in flow around baffles.

For flow normal to tube banks, the pressure drop is generally expressed in terms of the number of velocity heads lost, based on the maximum velocity, evaluated at the minimum free flow cross-sectional area.

$$\Delta p = K\frac{\rho V_{max}^2}{2} \tag{6-168}$$

The head loss factor becomes proportional to the number of tube rows when the number of rows is large. G. F. Hewitt et al. [*Process Heat Transfer,* CRC Press, Boca Raton, Fla., 1994] use

$$K = K_a + n_r K_r \tag{6-169}$$

where K_a, typically between 1 and 1.5, accounts for entrance and exit losses from the bundle and n_r and K_r are the number of rows, and loss factor per row, respectively. They provide graphs for determining K_r as a function of tube arrangement, pitch ratios, and Reynolds number. G. F. Hewitt [*Heat Exchanger Design Handbook, Part 2, Fluid Mechanics and Heat Transfer,* Begell House, New York, 2008] provides regression equations for K_r.

For *extended surfaces,* which include fins mounted perpendicular to the tubes or spiral-wound fins, pin fins, plate fins, and so on, friction data for the specific surface involved should be used. For details, see W. M. Kays and A. L. London [*Compact Heat Exchangers,* 2d ed., McGraw-Hill, New York, 1964]. If specific data are unavailable, the correlation by A. Y. Gunter and W. A. Shaw [*Trans. ASME* **67:** 643–660 (1945)] may be used as an approximation.

When a large temperature change occurs in a gas flowing across a tube bundle, gas properties should be evaluated at the mean temperature

$$T_m = T_t + K\Delta T_{lm} \tag{6-170}$$

where T_t = average tube-wall temperature
K = constant
ΔT_{lm} = log-mean temperature difference between gas and tubes

Values of K averaged from the recommendations of T. H. Chilton and R. P. Genereaux [*Trans. AIChE* **29:** 151–173 (1933)] and E. D. Grimison [*Trans. ASME* **59:** 583–594 (1937)] are as follows: for in-line tubes, 0.9 for cooling and −0.9 for heating; for staggered tubes, 0.75 for cooling and −0.8 for heating.

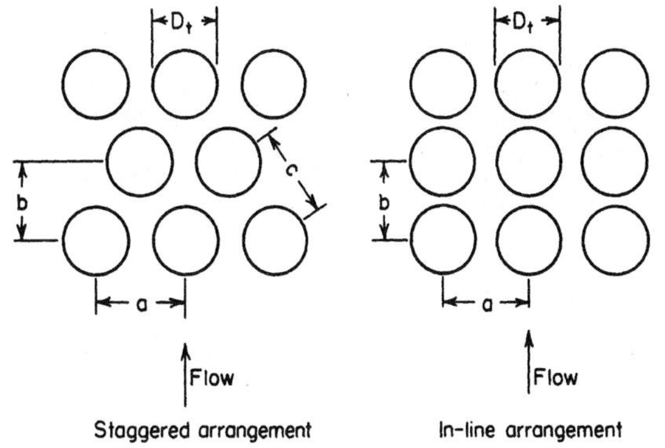

FIG. 6-41 Tube-bank configurations.

For nonisothermal flow of *liquids* across tube bundles, the friction factor is increased if the liquid is being cooled and is decreased if the liquid is being heated. The factors previously given for nonisothermal flow of liquids in pipes (Incompressible Flow in Pipes and Channels) should be used.

For two-phase gas/liquid horizontal crossflow through tube banks, the method of J. E. Diehl and C. H. Unruh [*Pet. Refiner* **37**(10): 124–128 (1958)] is available. For laminar flow of nonnewtonian fluids across tube banks, see D. Adams and K. J. Bell [*Chem. Eng. Prog.* **64;** *Symp. Ser.* **82:** 133–145 (1968)].

Flow-induced *tube vibration* occurs at critical fluid velocities through tube banks, and it is to be avoided because of the severe damage that can result. Methods to predict and correct vibration problems may be found in F. L. Eisinger [*Trans. ASME J. Pressure Vessel Tech.* **102:** 138–145 (May 1980)] and S. S. Chen [*J. Sound Vibration* **93:** 439–455 (1984)].

BEDS OF SOLIDS

Fixed Beds of Granular Solids Frictional pressure-drop prediction is complicated by the variety of granular materials and of their packing arrangement. For flow of a *single incompressible fluid* through an incompressible bed of granular solids, the pressure drop may be estimated by the correlation given in Fig. 6-42 [M. Leva, *Chem. Eng.* **56**(5): 115–117 (1949), or *Fluidization,* McGraw-Hill, New York, 1959]. The modified friction factor and Reynolds number are defined by

$$f_m = \frac{D_p \rho \phi_s \varepsilon^3 \Delta p}{2G^2 L(1-\varepsilon)^{3-n}} \tag{6-171}$$

$$\mathrm{Re}' \equiv \frac{D_p V \rho}{\mu} \tag{6-172}$$

where Δp = frictional pressure drop
L = depth of bed
D_p = average particle diameter, defined as diameter of a sphere of same volume as particle
ε = void fraction
n = exponent given in Fig. 6-42 as a function of Re′
ϕ_s = shape factor defined as the area of sphere of diameter D_p divided by the actual surface area of the particle
V = fluid superficial velocity based on the empty chamber cross section
ρ = fluid density
μ = fluid viscosity

In creeping flow (Re′ < 10),

$$f_m = \frac{100}{\mathrm{Re}'} \tag{6-173}$$

At high Reynolds numbers, the friction factor becomes nearly constant, approaching a value on the order of unity for most packed beds.

In terms of S, particle surface area per unit volume of bed is

$$D_p = \frac{6(1-\varepsilon)}{\phi_s S} \tag{6-174}$$

A simpler, widely used correlation is the Ergun equation [S. Ergun, *Chem. Eng. Progr.* **48:** 89–94 (1952)]

$$f_p = \frac{150}{\mathrm{Re}_p} + 1.75 \tag{6-175}$$

where the packed-bed friction factor and Reynolds number are defined, respectively, by

$$f_p = \frac{\Delta p D_p \varepsilon^3}{\rho V^2 L(1-\varepsilon)} \tag{6-176}$$

$$\mathrm{Re}_p = \frac{D_p V \rho}{(1-\varepsilon)\mu} \tag{6-177}$$

and the other variables are the same as those defined after Eq. (6-172) except that the equivalent particle diameter is given by

$$D_p = \frac{6}{a_V} \tag{6-178}$$

where a_V is the surface-to-volume ratio of the particle. Equation (6-178) reduces to the sphere diameter for spherical particles.

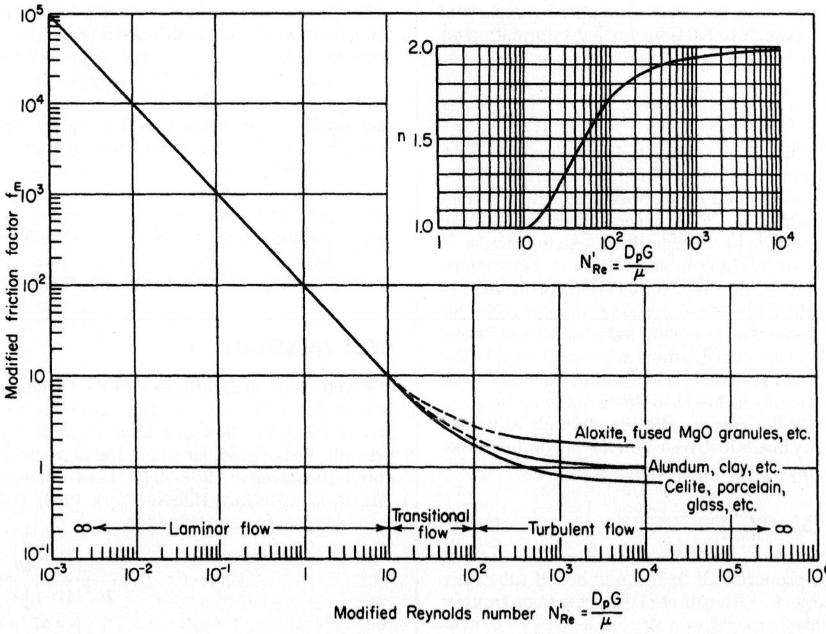

FIG. 6-42 Friction factor for beds of solids. [From M. Leva, *Fluidization*, McGraw-Hill, New York, 1959, p. 49.]

Porous Media Packed beds of granular solids are one type of the general class referred to as *porous media,* which include geological formations such as petroleum reservoirs and aquifers, manufactured materials such as sintered metals and porous catalysts, burning coal or char particles, and textile fabrics, to name a few. Pressure drop for incompressible flow across a porous medium has the same qualitative behavior as that for packed beds of solids. At low Reynolds numbers, viscous forces dominate and pressure drop is proportional to fluid viscosity and superficial velocity; and at high Reynolds numbers, pressure drop is proportional to fluid density and to the square of superficial velocity.

Creeping flow ($Re' < \approx 1$) through porous media is often described in terms of the *permeability k* and Darcy's law:

$$-\frac{\Delta p}{L} = \frac{\mu}{k} V \tag{6-179}$$

where V = superficial velocity. The SI units for permeability are square meters. Creeping flow conditions generally prevail in geological porous media. For *multidimensional flows* through *isotropic* porous media, the superficial velocity **V** and pressure gradient ∇p vectors replace the corresponding one-dimensional variables in Eq. (6-179).

$$\nabla p = -\frac{\mu}{k} \mathbf{V} \tag{6-180}$$

For isotropic homogeneous porous media (uniform permeability and porosity), the pressure for creeping incompressible single phase-flow may be shown to satisfy the Laplace equation:

$$\nabla^2 p = 0 \tag{6-181}$$

For *anisotropic* or *oriented* porous media, as are frequently found in geological media, permeability varies with direction and a ***permeability tensor*** **K** may be introduced [J. C. Slattery, *Momentum, Energy and Mass Transfer in Continua,* Krieger, Huntington, N.Y., 1981, p. 194]. Solutions for Darcy's law for several geometries of interest in petroleum reservoirs and aquifers, for both incompressible and compressible flows, are given in B. C. Craft and M. Hawkins [*Applied Petroleum Reservoir Engineering,* Prentice-Hall, Englewood Cliffs, N.J., 1959]. See also D. K. Todd [*Groundwater Hydrology,* 2d ed., Wiley, New York, 1980].

For granular solids of *mixed size* the average particle diameter may be calculated as

$$\frac{1}{D_p} = \sum_i \frac{x_i}{D_{p,i}} \tag{6-182}$$

where x_i = weight fraction of particles of size $D_{p,i}$.

For *isothermal compressible flow* of a gas with constant compressibility factor Z through a packed bed of granular solids, an equation similar to Eq. (6-115) for pipe flow may be derived:

$$p_1^2 - p_2^2 = \frac{2ZRG^2T}{M_w}\left[\ln\frac{v_2}{v_1} + \frac{2f_mL(1-\varepsilon)^{3-n}}{\phi_s^{3-n}\varepsilon^3 D_p}\right] \tag{6-183}$$

where p_1 = upstream absolute pressure
p_2 = downstream absolute pressure
R = gas constant
T = absolute temperature
M_w = molecular weight
v_1 = upstream specific volume of gas
v_2 = downstream specific volume of gas

For creeping flow of *power law* nonnewtonian fluids, the method of R. H. Christopher and S. Middleman [*Ind. Eng. Chem. Fundam.* **4:** 422–426 (1965)] may be used:

$$-\Delta p = \frac{150HLV^n(1-\varepsilon)^2}{D_p^2\phi_s^2\varepsilon^3} \tag{6-184}$$

$$H = \frac{K}{12}\left(9+\frac{3}{n}\right)^n \left[\frac{D_p^2\phi_s^2\varepsilon^4}{(1-\varepsilon)^2}\right]^{(1-n)/2} \tag{6-185}$$

where $V = G/\rho$ = superficial velocity, K and n = power law material constants, and all other variables are as defined after Eq. (6-172). This correlation is supported by data from R. H. Christopher and S. Middleman (1965), D. R. Gregory and R. G. Griskey [*AIChE J.* **13:** 122–125 (1967)], Y. H. Yu et al. [*Can. J. Chem. Eng.* **46:** 149–154 (1968)], N. Siskovic et al. [*AIChE J.* **17:** 281–285 (1971)], Z. Kemblowski and J. Mertl [*Chem. Eng. Sci.* **29:** 213–223 (1974)], and Z. Kemblowski and M. Dziuminski [*Rheol. Acta* **17:** 176–187 (1978)]. The measurements cover the range n = 0.50 to 1.60, and modified Reynolds number $Re' = 10^{-8}$ to 10, where

$$Re' = \frac{D_pV^{2-n}\rho}{H} \tag{6-186}$$

For the case $n = 1$ (newtonian fluid), Eqs. (6-184) and (6-185) give a pressure drop 25 percent less than that given by Eqs. (6-171) through (6-173).

For *viscoelastic fluids* see R. J. Marshall and A. B. Metzner [*Ind. Eng. Chem. Fundam.* **6:** 393–400 (1967)], D. R. Gregory and R. G. Griskey [*AIChE J.* **13:** 122–125 (1967)], and Z. Kemblowski and M. Dziubinski [*Rheol. Acta* **17:** 176–187 (1978)].

For gas flow through porous media with small pore diameters, the slip flow and molecular flow equations previously given (see the Vacuum Flow subsection) may be applied when the pore is of the same or smaller order as the mean free path.

Tower Packings For the flow of a *single fluid* through a bed of tower packing, pressure drop may be estimated using the preceding methods. See also Sec. 14 of this handbook. For *countercurrent gas/liquid flow* in commercial tower packings, both structured and unstructured, several sources of data and correlations for pressure drop and flooding are available. See, for example, R. F. Strigle [*Random Packings and Packed Towers, Design and Applications*, Gulf Publishing, Houston, Tex., 1989; *Chem. Eng. Prog.* **89**(8): 79–83 (August 1993)], G. A. Hughmark [*Ind. Eng. Chem. Fundam.* **25:** 405–409 (1986)], J. J. J. Chen [*Chem. Eng. Sci.* **40:** 2139–2140 (1985)], R. Billet and J. Mackowiak [*Chem. Eng. Technol.* **11:** 213–217 (1988)], H. Krehenwinkel and H. Knapp [*Chem. Eng. Technol.* **10:** 231–242 (1987)], A. Mersmann and A. Deixler [*Ger. Chem. Eng.* **9:** 265–276 (1986)]. Data and correlations for flooding and pressure drop for structured packings are given by J. R. Fair and J. R. Bravo [*Chem. Eng. Progr.* **86**(1): 19–29 (January 1990)].

Fluidized Beds When gas or liquid flows upward through a vertically unconstrained bed of particles, there is a minimum fluid velocity at which the particles will begin to move. Above this minimum velocity, the bed is said to be *fluidized.* Fluidized beds are widely used, in part because of their excellent mixing and heat- and mass-transfer characteristics. See Sec. 17 of this handbook for detailed information.

BOUNDARY LAYER FLOWS

In boundary layer flow, the flow far from the surface of an object is inviscid, and the effects of viscosity are manifest only in a thin region near the surface where steep velocity gradients occur to satisfy the no-slip condition at the solid surface. The thin layer where the velocity decreases from the inviscid, potential flow velocity to zero (relative velocity) at the solid surface is called the *boundary layer.* Boundary layer thickness is indefinite because the velocity asymptotically approaches the free-stream velocity at the outer edge. The boundary layer thickness is conventionally taken to be the distance at which the velocity equals 0.99 times the free-stream velocity. The boundary layer may be either laminar or turbulent. In the former case, the equations of motion may be simplified by scaling arguments. H. Schlichting and K. Gersten [*Boundary Layer Theory,* 8th ed. rev., Springer-Verlag, Berlin, 2003] is the most comprehensive source for information on boundary layer flows.

Flat Plate, Zero Angle of Incidence For flow over a wide, thin flat plate at zero angle of incidence with a uniform free-stream velocity, as shown in Fig. 6-43, the *critical Reynolds number* at which the boundary layer becomes turbulent is normally taken to be

$$\mathrm{Re}_x = \frac{xV\rho}{\mu} = 500{,}000 \qquad (6\text{-}187)$$

where V = free-stream velocity
ρ = fluid density
μ = fluid viscosity
x = distance from leading edge of plate

However, the transition Reynolds number depends on free-stream turbulence and may range from 3×10^5 to 3×10^6. The *laminar boundary layer* thickness δ is a function of distance from the leading edge:

$$\delta \approx 5.0x\mathrm{Re}_x^{-0.5} \qquad (6\text{-}188)$$

The total drag on the plate of length L and width b for a laminar boundary layer, including the drag on both surfaces, is

$$F_D = 1.328bL\rho V^2 \mathrm{Re}_L^{-0.5} \qquad (6\text{-}189)$$

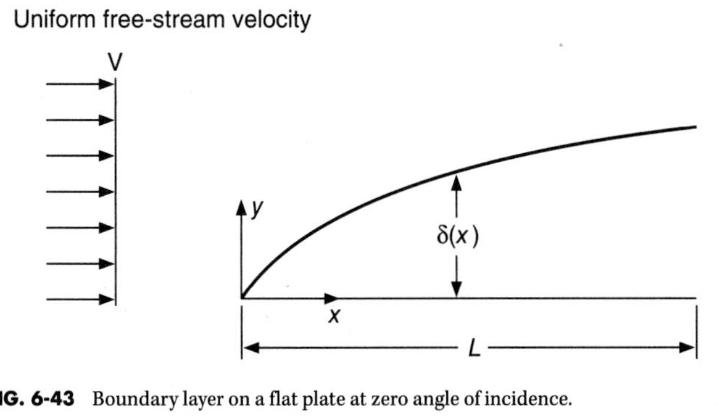

FIG. 6-43 Boundary layer on a flat plate at zero angle of incidence.

For *nonnewtonian power law fluids* [A. Acrivos et al., *AIChE J.* **6:** 312–317 (1960); Hsu, *AIChE J.* **15:** 367–370 (1969)],

$$F_D = CbL\rho V^2 \mathrm{Re}_L'^{\,-1/(1+n)} \qquad (6\text{-}190)$$

$n =$	0.2	0.3	0.4	0.5	0.6	0.7	0.8	0.9	1
$C =$	2.075	1.958	1.838	1.727	1.627	1.538	1.460	1.390	1.328

where $\mathrm{Re}_L' = \rho V^{2-n} L^n / K$ and K and n are the power law material constants (see Eq. (6-4)).

For a *turbulent boundary layer,* the thickness may be estimated as

$$\delta \approx 0.37x\mathrm{Re}_x^{-0.2} \qquad (6\text{-}191)$$

and the total drag force on both sides of the plate of length L is

$$F_D = \left[\frac{0.455}{(\log \mathrm{Re}_L)^{2.58}} - \frac{1{,}700}{\mathrm{Re}_L}\right]\rho bLV^2 \quad 5\times10^5 < \mathrm{Re}_L < 10^9 \qquad (6\text{-}192)$$

Here the second term accounts for the laminar leading edge of the boundary layer and assumes that the critical Reynolds number is 500,000.

Cylindrical Boundary Layer Laminar boundary layers on cylindrical surfaces, with flow parallel to the cylinder axis, are described by M. B. Glauert and M. J. Lighthill [*Proc. R. Soc. (London)*, **230A:** 188–203 (1955)], N. A. Jaffe and T. T. Okamura [*Z. Angew. Math. Phys.* **19:** 564–574 (1968)], and K. Stewartson [*Q. Appl. Math.* **13:** 113–122 (1955)]. For a turbulent boundary layer, the total drag may be estimated as

$$F_D = \bar{c}_f \pi r L \rho V^2 \qquad (6\text{-}193)$$

where r = cylinder radius, L = cylinder length, and the average friction coefficient is given by [F. M. White, *J. Basic Eng.* **94:** 200–206 (1972)]

$$\bar{c}_f = 0.0015 + \left[0.30 + 0.015\left(\frac{L}{r}\right)^{0.4}\right]\mathrm{Re}_L^{-1/3} \qquad (6\text{-}194)$$

for $\mathrm{Re}_L = 10^6$ to 10^9 and $L/r < 10^6$.

Continuous Flat Surface Boundary layers on continuous surfaces drawn through a stagnant fluid are shown in Fig. 6-44. Figure 6-44*a* shows the continuous flat surface [B. C. Sakiadis, *AIChE J.* **7:** 26–28, 221–225, 467–472 (1961)]. The critical Reynolds number for transition to turbulent flow may be greater than the 500,000 value for the finite flat-plate case discussed previously [F. K. Tsou et al., *J. Fluid Mech.* **26:** 145–161 (1966)]. For a laminar boundary layer, the thickness is given by

$$\delta = 6.37x\mathrm{Re}_x^{-0.5} \qquad (6\text{-}195)$$

and the total drag exerted on the two surfaces is

$$F_D = 1.776bL\rho V^2 \mathrm{Re}_L^{-0.5} \qquad (6\text{-}196)$$

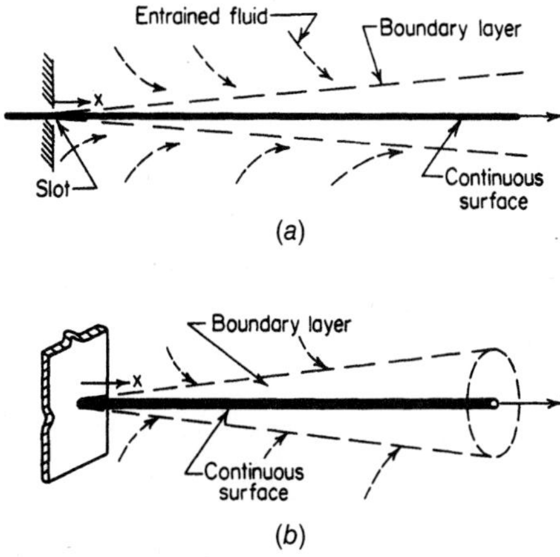

FIG. 6-44 Continuous surface: (*a*) continuous flat surface, (*b*) continuous cylindrical surface. [From B. C. Sakiadis, *Am. Inst. Chem. Eng. J.*, 7, 221, 467 (1961).]

The total flow rate of fluid entrained by the surface is

$$q = 3.232 Blv\, \mathrm{Re}_L^{-0.5} \qquad (6\text{-}197)$$

The theoretical velocity field was experimentally verified by F. K. Tsou et al. [*Int. J. Heat Mass Transfer* **10:** 219–235 (1967)] and A. Z. Szeri et al. [*J. Lubr. Technol.* **98:** 145–156 (1976)]. For *nonnewtonian power law fluids* see V. G. Fox et al. [*AIChE J.* **15:** 327–333 (1969)].

For a turbulent boundary layer, the thickness is given by

$$\delta = 1.01 x \mathrm{Re}_x^{-0.2} \qquad (6\text{-}198)$$

and the total drag on both sides by

$$F_D = 0.056 bL\rho V^2 \mathrm{Re}_L^{-0.2} \qquad (6\text{-}199)$$

and the total entrainment by

$$q = 0.252 Blv\, \mathrm{Re}_L^{-0.2} \qquad (6\text{-}200)$$

When the laminar boundary layer is a significant part of the total length of the object, the total drag should be corrected by subtracting a calculated turbulent drag for the length of the laminar section and then adding the laminar drag for the laminar section. F. K. Tsou et al. [*Int. J. Heat Mass Transfer* **10:** 219–235 (1967)] give an improved analysis of the turbulent boundary layer; their data indicate that Eq. (6-199) underestimates the drag by about 15 percent.

Continuous Cylindrical Surface The continuous surface shown in Fig. 6-44*b* is applicable, for example, for a wire drawn through a stagnant fluid [B. C. Sakiadis, *AIChE J.* **7:** 26–28, 221–225, 467–472 (1961); G. Vasudevan and S. Middleman, *AIChE J.* **16:** 614 (1970)]. The critical-length Reynolds number for transition is $\mathrm{Re}_x = 200{,}000$. The laminar boundary layer thickness, total drag, and entrainment flow rate may be obtained from Fig. 6-45. The normalized boundary layer thickness and integral friction coefficient are from G. Vasudevan and S. Middleman (1970), who used a similarity solution of the boundary layer equations. The drag force over a length x is given by

$$F_D = \bar{C}_f x \frac{\rho V^2}{2} 2\pi r_o \qquad (6\text{-}201)$$

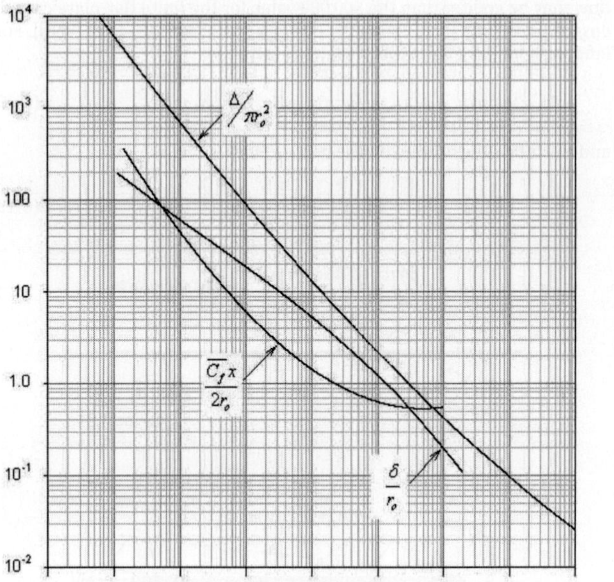

FIG. 6-45 Boundary layer parameters for continuous cylindrical surfaces. [$\Delta/\pi r_o^2$ is from B. C. Sakiadis, *Am. Inst. Chem. Engr. J.*, **7**, 467 (1961); $\bar{C}_f x/2r_o$ and δ/r_o are from G. Vasudevan and S. Middleman, *Am. Inst. Chem. Eng. J.*, **16**, 614 (1970).]

The entrainment flow rate is from B. C. Sakiadis [*AIChE J.* **7:** 26–28, 221–225, 467–472 (1961)], who used an integral momentum approximation rather than the exact similarity solution.

$$q = V\Delta \qquad (6\text{-}202)$$

Further laminar boundary layer analysis is given by L. J. Crane [*Z. Angew. Math. Phys.* **23:** 201–212 (1972)].

For a turbulent boundary layer, the total drag may be roughly estimated using Eqs. (6-193) and (6-194) for finite cylinders. Measured forces by Y. D. Kwon and D. C. Prevorsek [*J. Eng. Ind.* **101:** 73–79 (1979)] are greater than predicted this way.

The laminar boundary layer on *deforming continuous surfaces* with velocity varying with axial position is discussed by J. Vleggaar [*Chem. Eng. Sci.* **32:** 1517–1525 (1977)] and L. J. Crane [*Z. Angew. Math. Phys.* **26:** 619–622 (1975)].

VORTEX SHEDDING

When fluid flows past objects or through orifices or similar restrictions, vortices may periodically be shed downstream. Objects such as smokestacks, chemical processing columns, suspended pipelines, and electrical transmission lines can be subjected to damaging vibrations and forces due to the vortices, especially if the shedding frequency is close to a natural vibration frequency of the object. The shedding can also produce sound. See M. Z. von Krzywoblocki [*Appl. Mech. Rev.* **6:** 393–397 (1953)] and A. W. Marris [*J. Basic Eng.* **86:** 185–196 (1964)].

Development of a vortex street, or *von Kármán vortex street,* is shown in Fig. 6-46. Discussions of the vortex street may be found in R. L. Panton [*Incompressible Flow,* Wiley, New York, 1984]. The Reynolds number is

$$\mathrm{Re} = \frac{DV\rho}{\mu} \qquad (6\text{-}203)$$

where D = diameter of cylinder or effective width of object
V = free-stream velocity
ρ = fluid density
μ = fluid viscosity

For flow past a cylinder, the vortex street forms at Reynolds numbers above about 40. The vortices initially form in the wake, the point of formation moving closer to the cylinder as Re is increased. At a Reynolds number of 60 to 100, the vortices are formed from eddies attached to the cylinder surface. The vortices move at a velocity slightly less than V. The frequency of vortex shedding f is given in terms of the Strouhal number, which is approximately constant over a wide range of Reynolds numbers.

$$\mathrm{Sr} \equiv \frac{fD}{V} \qquad (6\text{-}204)$$

For $40 < \mathrm{Re} < 200$ the vortices are laminar and the Strouhal number has a nearly constant value of 0.2 for flow past a cylinder. Between Re = 200 and 400, Sr is no longer constant and the wake becomes irregular. Above about Re = 400 the vortices become turbulent, the wake is once again stable, and Sr remains constant at about 0.2 up to a Reynolds number of about 10^5. Above Re = 10^5 the vortex shedding is difficult to see in flow visualization experiments, but velocity measurements still show a strong spectral component at Sr = 0.2 (Panton, p. 392). The vortex street may disappear over the range $5 \times 10^5 < \mathrm{Re} < 3.5 \times 10^6$, but reestablishes above 3.5×10^6.

Vortex shedding exerts alternating lateral forces on a cylinder. Such forces may lead to severe vibration or mechanical failure of cylindrical elements such as heat-exchanger tubes, transmission lines, stacks, and columns when the vortex shedding frequency is close to resonant bending frequency. According to J. P. Den Hartog [*Proc. Nat. Acad. Sci.* **40:** 155–157 (1954)], the vortex shedding and cylinder vibration frequency will shift to the resonant frequency when the calculated shedding frequency is within 20 percent of the resonant frequency. The well-known Tacoma Narrows bridge collapse resulted from resonance between a torsional oscillation and

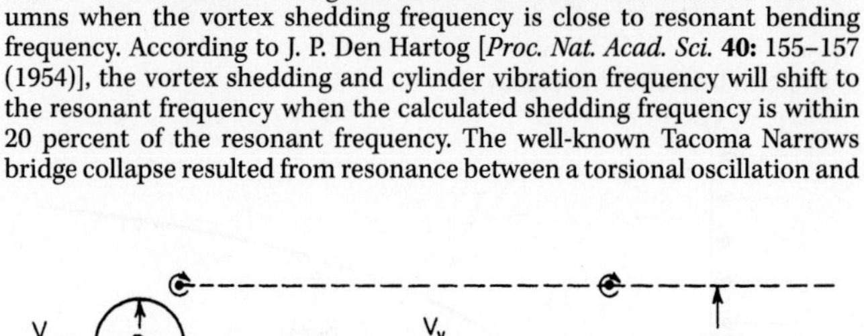

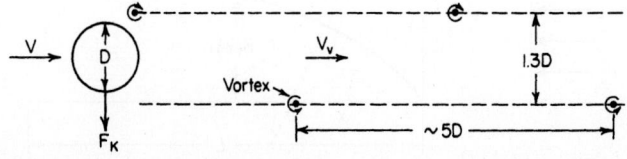

FIG. 6-46 Vortex street behind a cylinder.

vortex shedding (R. L. Panton, *Incompressible Flow,* Wiley, New York, 1984, p. 392). Spiral strakes are sometimes installed on tall stacks so that vortices at different axial positions are not shed simultaneously. The alternating lateral force F_K, sometimes called the *von Kármán force,* is given by [J. P. Den Hartog, *Mechanical Vibrations,* 4th ed., McGraw-Hill, New York, 1956, pp. 305–309]

$$F_K = C_K A \frac{\rho V^2}{2} \tag{6-205}$$

where C_K = von Kármán coefficient
A = projected area perpendicular to flow
ρ = fluid density
V = free-stream fluid velocity

For a cylinder, C_K = 1.7. For a vibrating cylinder, the effective projected area exceeds, but is always less than twice, the actual cylinder projected area [H. Rouse, *Engineering Hydraulics,* Wiley, New York, 1950].

The following references pertain to discussions of vortex shedding in specific structures: steel stacks [M. S. Osker and J. O. Smith, *Trans. ASME,* **78:** 1381–1391 (1956); J. O. Smith and J. H. McCarthy, *Mech. Eng.* **87:** 38–41 (1965)]; heat exchangers [F. L. Eisinger, *Trans. ASME J. Pressure Vessel Tech.* **102:** 138–145 (May 1980); S. S. Chen, *J. Sound Vibration* **93:** 439–455 (1984); N. R. Gainsboro, *Chem. Eng. Prog.* **64**(3): 85–88 (1968); "Flow-Induced Vibration in Heat Exchangers," *Symp. Proc.* ASME, New York, 1970]; suspended pipe lines [R. C. Baird, *Trans. ASME,* **77:** 797–804 (1955)]; and suspended cable [R. F. Steidel, *J. Appl. Mech.* **23:** 649–650 (1956)].

COATING FLOWS

In coating flows, liquid films are entrained on moving solid surfaces. For general discussions, see K. J. Ruschak [*Ann. Rev. Fluid Mech.* **17:** 65–89 (1985)], E. D. Cohen and E. B. Gutoff [*Modern Coating and Drying Technology,* VCH Publishers, New York, 1992], and S. Middleman [*Fundamentals of Polymer Processing,* McGraw-Hill, New York, 1977]. It is generally important to control the thickness and uniformity of the coatings.

In *dip coating,* or *free withdrawal coating,* a solid surface is withdrawn from a liquid pool, as shown in Fig. 6-47. It illustrates many of the features found in other coating flows as well. J. A. Tallmadge and C. Gutfinger [*Ind. Eng. Chem.* **59**(11): 19–34 (1967)] provide an early review of the theory of dip coating. The coating flow rate and film thickness are controlled by the withdrawal rate and the flow behavior in the meniscus region. For a withdrawal velocity V and an angle of inclination from the horizontal ϕ, the film thickness h may be estimated for low withdrawal velocities by

$$h\left(\frac{\rho g}{\sigma}\right)^{1/2} = \frac{0.944}{(1-\cos\varphi)^{1/2}}\mathrm{Ca}^{2/3} \tag{6-206}$$

where g = acceleration of gravity
Ca = $\mu V/\sigma$ = capillary number
μ = viscosity
σ = surface tension

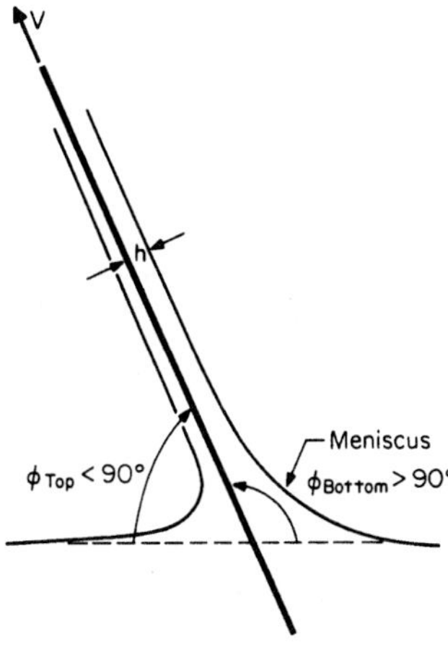

FIG. 6-47 Dip coating.

Equation (6-206) is asymptotically valid as Ca → 0 and agrees with experimental data up to capillary numbers in the range of 0.01 to 0.03. In practice, where high production rates require high withdrawal speeds, capillary numbers are usually too large for Eq. (6-206) to apply. Approximate analytical methods for larger capillary numbers have been obtained by numerous investigators, but none appears wholly satisfactory, and some are based on questionable assumptions [K. J. Ruschak, *Ann. Rev. Fluid Mech.* **17:** 65–89 (1985)]. With the availability of high-speed computers and the development of the field of computational fluid dynamics, numerical solutions accounting for two-dimensional flow aspects, along with gravitational, viscous, inertial, and surface tension forces, are now the most effective means to analyze coating flow problems.

Other common coating flows include premetered flows, such as *slide* and *curtain coating,* where the film thickness is an independent parameter that may be controlled within limits, and the curvature of the meniscus adjusts accordingly; the closely related *blade coating;* and *roll coating* and *extrusion coating.* See K. J. Ruschak [*Ann. Rev. Fluid Mech.* **17:** 65–89 (1985)], E. D. Cohen and E. B. Gutoff [*Modern Coating and Drying Technology,* VCH Publishers, New York, 1992], and S. Middleman [*Fundamentals of Polymer Processing,* McGraw-Hill, New York, 1977]. For dip coating of wires, see P. Tauguy et al. [*Int. J. Numerical Meth. Fluids* **4:** 441–475 (1984)].

Many coating flows are subject to instabilities that lead to defects. Three-dimensional flow instabilities lead to such problems as *ribbing.* Air entrainment is another common defect.

FALLING FILMS

Minimum Wetting Rate The minimum liquid rate required for complete wetting of a vertical surface is about 0.03 to 0.3 kg/m · s (0.02 to 0.2 lbm/ft · s) for water at room temperature. The minimum rate depends on the geometry and nature of the vertical surface, surface tension, and mass transfer between surrounding gas and the liquid. See A. B. Ponter et al. [*Int. J. Heat Mass Transfer* **10:** 349–359 (1967); *Trans. Inst. Chem. Eng. (London)* **45:** 345–352 (1967)], F. P. Stainthorpe and J. M. Allen [*Trans. Inst. Chem. Eng. (London)* **43:** 85–91 (1967)] and K. Watanabe et al. [*J. Chem. Eng. (Japan)* **8**(1): 75 (1975)].

Laminar Flow For films on *vertical flat surfaces,* as shown in Fig. 6-48, or vertical tubes with small film thickness compared to tube radius, laminar flow conditions prevail for Reynolds numbers less than about 2000, where the Reynolds number is given by

$$\mathrm{Re} = \frac{4\Gamma}{\mu} \tag{6-207}$$

where Γ = liquid mass flow rate per unit width of surface and μ = liquid viscosity. For a flat film surface, the film thickness δ is

$$\delta = \left(\frac{3\Gamma\mu}{\rho^2 g}\right)^{1/3} \tag{6-208}$$

and the average film velocity is

$$V = \frac{\Gamma}{\rho\delta} = \frac{g\rho\delta^2}{3\mu} \tag{6-209}$$

The downward velocity profile $u(x)$, where $x = 0$ at the liquid/gas interface and $x = \delta$ at the solid surface, is given by

$$u = 1.5V\left[1-\left(\frac{x}{\delta}\right)^2\right] \tag{6-210}$$

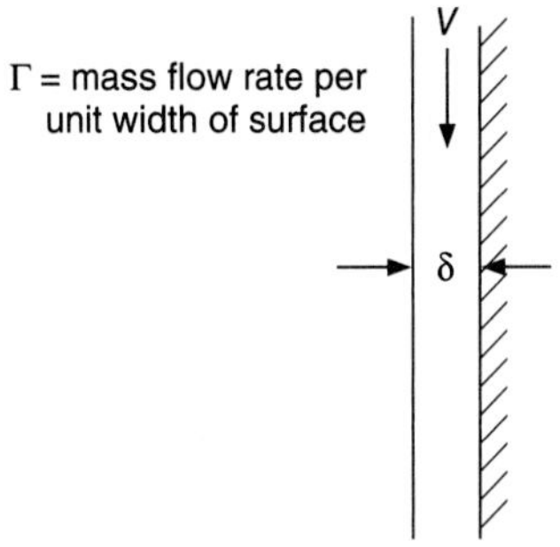

FIG. 6-48 Falling film.

These equations assume that there is no drag force at the gas/liquid interface, such as would be produced by gas flow. For a flat surface *inclined* at an angle θ with the horizontal, the preceding equations may be modified by replacing g by $g \sin \theta$. For films falling inside vertical tubes with film thickness up to and including the full pipe radius, see M. L. Jackson [*AIChE J.* **1:** 231–240 (1955)].

These equations have generally given good agreement with experimental results for low-viscosity liquids (< 0.005 Pa · s or < 0.5 cP) whereas Jackson found film thicknesses for higher-viscosity liquids (0.01 to 0.02 Pa · s, 10 to 20 cP) were significantly less than predicted by Eq. (6-208). At Reynolds numbers of 25 or greater, *surface waves* will be present on the liquid film. D. West and R. Cole [*Chem. Eng. Sci.* **22:** 1388–1389 (1967)] found that the surface velocity u is still within ±7 percent of that given by Eq. (6-210) even in wavy flow.

For laminar nonnewtonian film flow, see R. B. Bird et al. [*Dynamics of Polymeric Liquids,* vol. 1: *Fluid Mechanics,* Wiley, New York, 1977, pp. 215, 217], G. Astarita et al. [*Ind. Eng. Chem. Fundam.* **3:** 333–339 (1964)] and D. C. H. Cheng [*Ind. Eng. Chem. Fundam.* **13:** 394–395 (1974)].

Turbulent Flow In turbulent flow, Re > 2000, for vertical surfaces, the film thickness may be estimated to within ±25 percent using

$$\delta = 0.304 \left(\frac{\Gamma^{1.75} \mu^{0.25}}{\rho^2 g} \right)^{1/3} \tag{6-211}$$

Replace g by $g \sin \theta$ for a surface inclined at angle θ to the horizontal. The average film velocity is $V = \Gamma / \rho\delta$.

J. A. Tallmadge and C. Gutfinger [*Ind. Eng. Chem.* **59**(11): 19–34 (1967)] discuss prediction of drainage rates from liquid films on flat and cylindrical surfaces.

Effect of Surface Traction If drag is exerted on the surface of the film because of motion of the surrounding fluid, the film thickness will be reduced or increased, depending upon whether the drag acts with or against gravity. W. J. Thomas and S. Portalski [*Ind. Eng. Chem.* **50:** 1081–1088 (1958)], A. E. Dukler [*Chem. Eng. Prog.* **55**(10): 62–67 (1959)], and P. G. Kosky [*Int. J. Heat Mass Transfer,* **14:** 1220–1224 (1971)] have presented calculations of film thickness and film velocity. Film thickness data for falling water films with cocurrent and countercurrent airflow in pipes are given by L. Y. Zhivaikin [*Int. Chem. Eng.* **2:** 337–341 (1962)]. G. J. Zabaras et al. [*AIChE J.* **32:** 829–843 (1986)] and G. J. Zabaras and A. E. Dukler [*AIChE J.* **34:** 389–396 (1988)] present studies of film flow in vertical tubes with both cocurrent and countercurrent gas flow, including measurements of film thickness, wall shear stress, wave velocity, wave amplitude, pressure drop, and flooding point for countercurrent flow.

Flooding With countercurrent gas flow, a condition is reached with increasing gas rate for which flow reversal occurs and liquid is carried upward. The mechanism for this flooding condition most often has been attributed to waves either bridging the pipe or reversing direction to flow upward at flooding. However, the results of Zabaras and Dukler suggest that flooding may be controlled by flow conditions at the liquid inlet and that wave bridging or upward wave motion does not occur, at least for the 50.8-mm-diameter pipe used for their study. Flooding mechanisms are still incompletely understood. Under some circumstances, such as when the gas is allowed to develop its normal velocity profile in a "calming length" of pipe beneath the liquid draw-off, the gas superficial velocity at flooding will be increased, and increases with decreasing length of wetted pipe [G. F. Hewitt et al., *Proc. Two-Phase Flow Symp.,* University of Exeter Exeter, UK, paper 4H, AERE-4 4614 (1965)]. A bevel cut at the bottom of the pipe with an angle 30° from the vertical will increase the flooding velocity in small-diameter tubes at moderate liquid flow rates. If the gas approaches the tube from the side, the taper should be oriented with the point facing the gas entrance. Figures 6-49 and 6-50 give correlations for flooding in tubes with square and slant bottoms (courtesy Holmes, DuPont Co.) The superficial mass

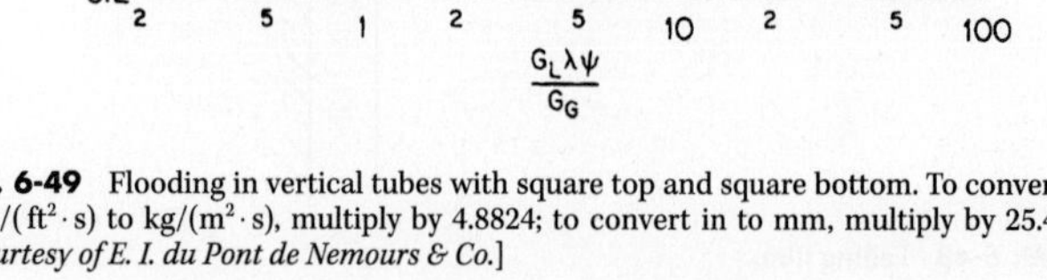

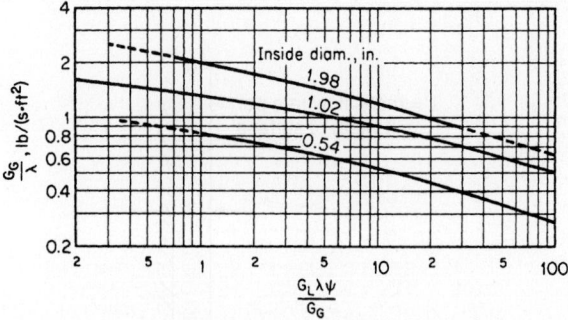

FIG. 6-49 Flooding in vertical tubes with square top and square bottom. To convert lbm/(ft² · s) to kg/(m² · s), multiply by 4.8824; to convert in to mm, multiply by 25.4. [*Courtesy of E. I. du Pont de Nemours & Co.*]

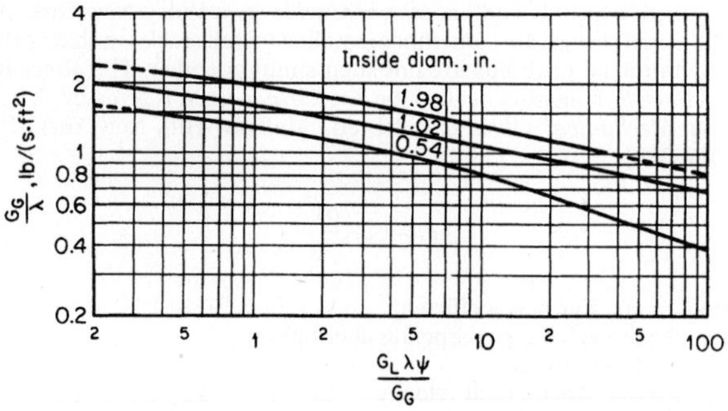

FIG. 6-50 Flooding in vertical tubes with square top and slant bottom. To convert lbm/(ft² · s) to kg/(m² · s), multiply by 4.8824; to convert in to mm, multiply by 25.4. [*Courtesy of E. I. du Pont de Nemours & Co.*]

velocities of gas and liquid G_G and G_L and the physical property parameters λ and ψ are the same as those defined for the Baker chart (Multiphase Flow subsection, Fig. 6-25). For tubes larger than 50 mm (2 in), flooding velocity appears to be relatively insensitive to diameter and the flooding curves for 1.98-in diameter may be used.

HYDRAULIC TRANSIENTS

Many transient flows of liquids may be analyzed by using the full time-dependent equations of motion for incompressible flow. However, some phenomena are controlled by the small compressibility of liquids. These phenomena are generally called *hydraulic transients.*

Water Hammer When liquid flowing in a pipe is suddenly decelerated to zero velocity by a fast-closing valve, a pressure wave propagates upstream to the pipe inlet, where it is reflected; a pounding of the line commonly known as *water hammer* is often produced. For an instantaneous flow stoppage of a truly incompressible fluid in an inelastic pipe, the pressure rise would be infinite. Finite compressibility of the fluid and elasticity of the pipe limit the pressure rise to a finite value. The Joukowski formula gives the maximum pressure rise as

$$\Delta p = \rho a \Delta V \tag{6-212}$$

where ρ = liquid density
ΔV = change in liquid velocity
a = pressure wave velocity

The wave velocity is given by

$$a = \frac{\sqrt{\beta/\rho}}{\sqrt{1 + (\beta/E)(D/b)}} \tag{6-213}$$

where β = liquid bulk modulus of elasticity
E = elastic modulus of pipe wall
D = pipe inside diameter
b = pipe wall thickness

The numerator gives the wave velocity for perfectly rigid pipe, and the denominator corrects for wall elasticity. This formula is for thin-walled pipes; for thick-walled pipes, the factor D/b is replaced by

$$2\frac{D_o^2 + D_i^2}{D_o^2 - D_i^2}$$

where D_o = pipe outside diameter
D_i = pipe inside diameter

Example 6-10 Response to Instantaneous Valve Closing Compute the wave speed and maximum pressure rise for instantaneous valve closing, with an initial velocity of 2.0 m/s, in a 4-in Schedule 40 steel pipe with elastic modulus 207 × 10⁹ Pa. Repeat for a plastic pipe of the same dimensions, with $E = 1.4 \times 10^9$ Pa. The liquid is water with $\beta = 2.2 \times 10^9$ Pa and $\rho = 1000$ kg/m³.

For the steel pipe, $D = 102.3$ mm, $b = 6.02$ mm, and the wave speed is

$$a = \frac{\sqrt{\beta/\rho}}{\sqrt{1 + (\beta/E)(D/b)}}$$

$$= \frac{\sqrt{2.2 \times 10^9 / 1000}}{\sqrt{1 + (2.2 \times 10^9 / 207 \times 10^9)(102.3/6.02)}}$$

$$= 1365 \text{ m/s}$$

The maximum pressure rise $\Delta p = \rho a\,\Delta V = 1000 \times 1365 \times 2.0 = 2.73 \times 10^6$ Pa
For the plastic pipe,

$$a = \frac{\sqrt{2.2\times10^9/1000}}{\sqrt{1+(2.2\times10^9/1.4\times10^9)(102.3/6.02)}}$$
$$= 282 \text{ m/s}$$
$$\Delta p = \rho a\,\Delta V = 1000\times282\times2.0 = 5.64\times10^5 \text{ Pa}$$

The maximum pressure surge is obtained when the valve closes in less time than the period τ required for the pressure wave to travel from the valve to the pipe inlet and back, a total distance of $2L$.

$$\tau = \frac{2L}{a} \tag{6-214}$$

The pressure surge will be reduced when the time of flow stoppage exceeds the pipe period τ, due to cancellation of direct and reflected waves. D. J. Wood and S. E. Jones [*Proc. Am. Soc. Civ. Eng., J. Hydraul. Div.* **99:** 167–178 (1973)] present charts for estimates of water-hammer pressure for different valve closure modes. V. L. Wylie and E. B. Streeter [*Hydraulic Transients,* McGraw-Hill, New York, 1978] describe several solution methods for hydraulic transients, including the method of characteristics, which is well suited to computer methods for accurate solutions. A rough approximation for the peak pressure for cases where the valve closure time t_c exceeds the pipe period τ is [R. L. Daugherty and J. B. Franzini, *Fluid Mechanics with Engineering Applications,* McGraw-Hill, New York, 1985]

$$\Delta p \approx \left(\frac{\tau}{t_c}\right)\rho a \Delta V \tag{6-215}$$

Successive reflections of the pressure wave between the pipe inlet and the closed valve result in alternating pressure increases and decreases, which are gradually attenuated by fluid friction and imperfect elasticity of the pipe. Periods of reduced pressure occur while the reflected pressure wave is traveling from inlet to valve. Degassing of the liquid may occur, as may vaporization if the pressure drops below the vapor pressure of the liquid. Gas and vapor bubbles decrease the wave velocity. Vaporization may lead to what is often called *liquid column separation;* subsequent collapse of the vapor pocket can result in pipe rupture.

In addition to water hammer induced by valve action, numerous other hydraulic transient flows are of interest, for example, those arising from starting or stopping of pumps; changes in power demand from turbines; reciprocating pumps; changing elevation of a reservoir; waves on a reservoir; turbine governor hunting; vibration of impellers or guide vanes in pumps, fans, or turbines; vibration of deformable parts such as valves; draft-tube instabilities due to vortexing; and unstable pump or fan characteristics [V. L. Wylie and E. B. Streeter, *Hydraulic Transients,* McGraw-Hill, New York, 1978]. Tube failure in heat exchangers is another cause of water hammer.

Pulsating Flow Reciprocating machinery (pumps and compressors) produces flow pulsations, which adversely affect flow meters and process control elements and can cause vibration and equipment failure, in addition to undesirable process results. Vibration and damage can result not only from the fundamental frequency of the pulse producer but also from higher harmonics. Multipiston double-acting units reduce vibrations. Pulsation dampeners are often added. Damping methods are described by M. W. Kellogg Co. [*Design of Piping Systems,* rev. 2d ed., Wiley, New York, 1965]. For liquid phase pulsation damping, gas-filled surge chambers, also known as accumulators, are commonly used; see Wylie and Streeter.

Commercial software programs are available for simulation of hydraulic transients. These may be used to analyze piping systems to reveal unsatisfactory behavior, and they allow the assessment of design changes such as increases in pipe wall thickness, changes in valve actuation, and addition of check valves, surge tanks, and pulsation dampeners.

Cavitation Loosely regarded as related to water hammer and hydraulic transients because it may cause similar vibration and equipment damage, *cavitation* is the collapse of vapor bubbles in flowing liquid. These bubbles may be formed where the local liquid pressure drops below the vapor pressure, or they may be injected into the liquid, as when steam is sparged into water. Local low-pressure zones may be produced by local velocity increases (in accordance with the Bernoulli equation) as in eddies or vortices, or near boundary contours; by rapid vibration of a boundary; by separation of liquid during water hammer; or by an overall reduction in static pressure, such as due to pressure drop in the suction line of a pump.

Collapse of vapor bubbles once they reach zones where the pressure exceeds the vapor pressure can cause objectionable noise and vibration and extensive erosion or pitting of the boundary materials. The critical cavitation number at inception of cavitation, denoted σ_i, is useful in correlating equipment performance data:

$$\sigma_i = \frac{p_i - p_v}{\rho V^2/2} \tag{6-216}$$

where p_i = static pressure in undisturbed flow
P_v = vapor pressure
ρ = liquid density
V = free-stream velocity of liquid

The value of the cavitation number for incipient cavitation for a specific piece of equipment is a characteristic of that equipment. Cavitation numbers for various head forms of cylinders, for disks, and for various hydrofoils are given by J. W. Holl and G. F. Wislicenus [*J. Basic Eng.* **83:** 385–398 (1961)] and for various surface irregularities by R. E. A. Arndt and A. T. Ippen [*J. Basic Eng.* **90:** 249–261 (1968)], J. W. Ball [*Proc. ASCE J. Constr. Div.* **89**(C02): 91–110 (1963)], and J. W. Holl [*J. Basic Eng.* **82:** 169–183 (1960)]. As a guide only, for blunt forms the cavitation number is generally in the range of 1 to 2.5, and for somewhat streamlined forms the cavitation number is in the range of 0.2 to 0.5. Critical cavitation numbers generally depend on a characteristic length dimension of the equipment in a way that has not been explained. This renders scale-up of cavitation data questionable.

Figure 6-51 [Y. Yan and R. B. Thorpe, *Int. J. Multiphase Flow* **16:** 1023–1045 (1990)] gives the critical cavitation number for flow through orifices. To use this cavitation number in Eq. (6-216), the pressure p_i is the orifice backpressure downstream of the vena contracta after full pressure recovery, and V is the average velocity through the orifice. Figure 6-51 includes data from J. P. Tullis and R. Govindarajan [*ASCE J. Hydraul. Div.* **HY13:** 417–430 (1973)] modified to use the same cavitation number definition; their data also include critical cavitation numbers for 30.50- and 59.70-cm pipes. Very roughly, compared with the 15.40-cm pipe, the cavitation number is about 20 percent greater for the 30.50-cm-diameter pipe and about 40 percent greater for the 59.70-cm-diameter pipe. Inception of cavitation appears to be related to release of dissolved gas and not merely vaporization of the liquid. K. Takahashi et al. [CAV 2001: 4th Symp. on Cavitation, June 20–23, 2001, Cal Tech, Pasadena] present data showing reduction in cavitation intensity by using multihole instead of single-hole orifice plates. For further discussion of cavitation, see Eisenberg and M. P. Tulin in V. L. Streeter [*Handbook of Fluid Dynamics,* McGraw-Hill, New York, 1961, Sec. 12].

TURBULENCE

Turbulent flow occurs when the Reynolds number exceeds a critical value above which laminar flow is unstable; the critical Reynolds number depends on the flow geometry. There is generally a transition regime between the critical Reynolds number and the Reynolds number at which the flow may be considered fully turbulent. In turbulent flow, variables such as velocity and pressure fluctuate chaotically; statistical methods are used to quantify turbulence.

Time Averaging For turbulent flows it is useful to define time-averaged and fluctuation values of flow variables. For example, the x-component velocity fluctuation v'_x is the difference between the instantaneous velocity v_x and the time-averaged velocity $\overline{v}_x$.

$$v'_x(x,y,z,t) = v_x(x,y,z,t) - \overline{v}_x(x,y,z) \tag{6-217}$$

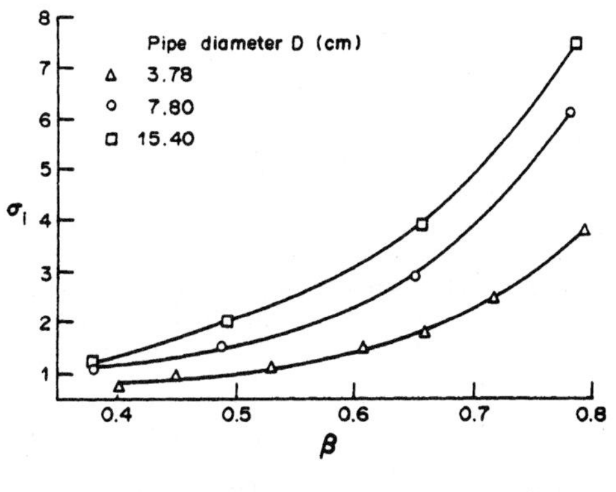

FIG. 6-51 Critical cavitation number versus diameter ratio β. [Reprinted from Y. Yan and R. B. Thorpe, "Flow regime transitions due to cavitation in the flow through an orifice," *Int. J. Multiphase Flow,* **16,** 1023–1045. Copyright © 1990, with kind permission from Elsevier Science, Ltd., The Boulevard, Langford Lane, Kidlington OX5 1GB, United Kingdom.]

The instantaneous and fluctuating velocity components are, in general, functions of spatial position and of time t. The time-averaged velocity $\overline{v_x}$ is independent of time for a *stationary* flow. Nonstationary processes may be considered where averages are defined over time scales long compared to the time scale of the turbulent fluctuations, but short compared to the characteristic time scale of the flow. The time average over a time interval $2T$ centered at time t of a turbulently fluctuating variable $\zeta(t)$ is defined as

$$\overline{\zeta(t)} = \frac{1}{2T}\int_{t-T}^{t+T} \zeta(\tau)\,d\tau \tag{6-218}$$

where τ = dummy integration variable. For stationary turbulence, $\bar{\zeta}$ does not vary with time.

The time average of a fluctuation is zero, $\overline{\zeta'} = \overline{\zeta - \bar{\zeta}} = 0$. Fluctuation magnitudes are quantified by root mean squares.

$$\tilde{v}'_x = \sqrt{\overline{(v'_x)^2}} \tag{6-219}$$

In *isotropic* turbulence, statistical measures of fluctuations are equal in all directions.

$$\tilde{v}'_x = \tilde{v}'_y = \tilde{v}'_z \tag{6-220}$$

In *homogeneous* turbulence, turbulence properties are independent of spatial position. The *kinetic energy of turbulence k* is given by

$$k = \frac{1}{2}\left(\tilde{v}'^2_x + \tilde{v}'^2_y + \tilde{v}'^2_z\right) \tag{6-221}$$

Velocity fluctuations ultimately dissipate their kinetic energy through viscous effects. Macroscopically, this energy dissipation requires pressure drop, or velocity decrease. The *energy dissipation rate* per unit mass is usually denoted ε. For steady flow in a pipe, the average energy dissipation rate per unit mass is given by

$$\varepsilon = \frac{2fV^3}{D} \tag{6-222}$$

where f = Fanning friction factor
D = pipe inside diameter
V = average velocity

The continuity equation and the Navier-Stokes equation for incompressible flow may be time-averaged; this process is also known as *Reynolds averaging*. The time-averaged continuity equation for an incompressible fluid is similar to the instantaneous continuity equation (6-23), but with the time averaged velocity $\bar{\mathbf{v}}$ replacing the instantaneous velocity **v**.

$$\nabla \cdot \bar{\mathbf{v}} = 0 \tag{6-223}$$

The Reynolds averaged Navier-Stokes (RANS) equation is similar to the instantaneous Navier-Stokes equation (6-27), except for the appearance of an additional term, the divergence of the *Reynolds stress*. This term arises from the averaging of the nonlinear $\mathbf{v}\cdot\nabla\mathbf{v}$ inertial term.

$$\rho\frac{D\bar{\mathbf{v}}}{Dt} = -\nabla\bar{p} + \mu\nabla^2\bar{\mathbf{v}} + \nabla\cdot\boldsymbol{\tau}^{(t)} + \rho\mathbf{g} \tag{6-224}$$

The Reynolds stress is

$$\boldsymbol{\tau}^{(t)} = -\rho\overline{\mathbf{v}'\mathbf{v}'} \tag{6-225}$$

Direct numerical simulation (DNS) under limited circumstances has been carried out to solve the unaveraged equations of motion and determine fluctuating velocity fields. Due to the extreme computational intensity, solutions to date have been limited to modest Reynolds numbers, generally in simple geometries, and sometimes with certain assumptions, such as periodicity in the flow direction. Since computational grids must be fine enough to resolve even the smallest eddies, the computational difficulty rapidly becomes prohibitive as the Reynolds number increases. Therefore, the solution of the equations of motion for turbulent flow is generally based on the time-averaged equations. This requires semiempirical models to express the Reynolds stresses in terms of time-averaged velocities. This is the *closure* problem of turbulence.

Closure Models A wide variety of closure models have been proposed. See S. B. Pope (*Turbulent Flows*, Cambridge University Press, Cambridge, UK, 2000) for descriptions of many of them. One class of closure model treats the Reynolds stress as analogous to the viscous stress, based on the proposition that turbulent eddy motion transports momentum in the same diffusive manner as the random motion of gas molecules. The Boussinesq approximation, simplified here for incompressible flow, introduces a scalar quantity called the *turbulent* or *eddy viscosity* μ_t, in analogy to Eqs. (6-7) and (6-24).

$$\boldsymbol{\tau}^{(t)} = -\rho\overline{\mathbf{v}'\mathbf{v}'} = -\frac{2}{3}k\boldsymbol{\delta} + \mu_t[\nabla\bar{\mathbf{v}} + (\nabla\bar{\mathbf{v}})^T] \tag{6-226}$$

To solve the equations of motion using the Boussinesq approximation, it is necessary to provide equations for the single scalar unknowns μ_t and k, rather than the nine unknown tensor components $\tau_{ij}^{(t)}$. Solutions to the time-averaged equations for turbulent flow using Eq. (6-226) are not equivalent to Navier-Stokes solutions for laminar flow because μ_t is not a constant. The turbulent viscosity model is physically realistic under some circumstances, but not in others, such as those with large elongation rates in the mean velocity field [S. B. Pope, *Turbulent Flows*, Cambridge University Press, Cambridge, UK, 2000].

Mixing length models estimate the turbulent viscosity as the product of the square of a mixing length ℓ_M and a time-averaged velocity gradient. The mixing length is geometry-dependent and must be specified to complete a model, rendering the model unsuitable for general predictive use. In boundary layer flows, the mixing length is assumed to be proportional to distance y from the wall, and $|\partial\overline{v_x}/\partial y|$ is chosen for the velocity gradient. The universal turbulent velocity profile near the pipe wall presented in the preceding subsection Incompressible Flow in Pipes and Channels may be developed using the Prandtl mixing length approximation for the eddy viscosity,

$$\mu_t = \rho l_P^2\left|\frac{d\overline{v_x}}{dy}\right| \tag{6-227}$$

where l_P is the Prandtl mixing length. The turbulent core of the universal velocity profile is obtained by assuming that the mixing length is proportional to the distance from the wall. The proportionality constant is one of two constants adjusted to fit experimental data.

The Prandtl mixing length concept is useful for shear flows parallel to walls, but is inadequate for multidimensional flows. A more advanced semiempirical model commonly used in numerical computations, and found in most commercial software for computational fluid dynamics (CFD; see the following subsection), is the k–ε model described by B. Launder and D. Spalding [*Lectures in Mathematical Models of Turbulence*, Academic, London, 1972]. In this model the eddy viscosity is assumed proportional to the ratio k^2/ε.

$$\mu_t = \rho C_\mu\frac{k^2}{\varepsilon} \tag{6-228}$$

where the value $C_\mu = 0.09$ is normally used. Semiempirical partial differential conservation equations for k and ε derived from the Navier-Stokes equations with simplifying closure assumptions are coupled with the equations of continuity and momentum. They may be written as

$$\frac{\partial}{\partial t}(\rho k) + \nabla\cdot(\rho\bar{\mathbf{v}}k) = \nabla\cdot\left(\frac{\mu_t}{\sigma_k}\nabla k\right) + \mu_t[\nabla\bar{\mathbf{v}} + (\nabla\bar{\mathbf{v}})^T]\,\nabla\bar{\mathbf{v}} - \rho\varepsilon \tag{6-229}$$

$$\frac{\partial}{\partial t}(\rho\varepsilon) + \nabla\cdot(\rho\bar{\mathbf{v}}\varepsilon) = \nabla\cdot\left(\frac{\mu_t}{\sigma_\varepsilon}\nabla\varepsilon\right) + C_{1\varepsilon}\frac{\varepsilon\mu_t}{k}[\nabla\bar{\mathbf{v}} + (\nabla\bar{\mathbf{v}})^T]\,\nabla\bar{\mathbf{v}} - C_{2\varepsilon}\frac{\rho\varepsilon^2}{k} \tag{6-230}$$

The values for the empirical constants $C_{1\varepsilon} = 1.44$, $C_{2\varepsilon} = 1.92$, $\sigma_k = 1.0$, and $\sigma_\varepsilon = 1.3$ are widely accepted [B. Launder and D. Spaulding, *The Numerical Computation of Turbulent Flows*, Imperial Coll. Sci. Tech. London, NTIS N74-12066 (1973)]. The k–ε model has proved reasonably accurate for many flows without highly curved streamlines or significant swirl. It usually underestimates flow separation and overestimates turbulence production by normal straining. The k–ε model is suitable for flows with high Reynolds numbers. See C. Virendra et al. [*AIAA J.* **23:** 1308–1319 (1984)] for a review of low Reynolds number k–ε models. Several other models based on turbulent viscosity exist. Some are two-equation model variants of the k–ε model, such as the k–ω model, where $\omega = \varepsilon/k$. A variant of the mixing length model is one using $k^{1/2}$ as a characteristic velocity so that $\mu_t \propto \rho k^{1/2}\ell_M$.

A second class of RANS closure models is based on solution of model transport equations for each of the independent components of $\boldsymbol{\tau}^{(t)}$ (due to symmetry, there are six independent components). One additional transport equation, usually for the dissipation rate ε, must also be solved. These models are computationally more intensive than turbulent viscosity-based models, but they are more accurate. Perhaps confusingly, these models are called Reynolds stress models, even though all RANS closure models are models for the Reynolds stress. A widely cited example is that of B. Launder et al. [*J. Fluid Mech.* **68:** 537–566 (1975)]. S. B. Pope [*Turbulent Flows*, Cambridge University Press, Cambridge, UK, 2000] may be consulted for a thorough discussion of Reynolds stress models.

A third class of models is *large eddy simulation* (LES). In LES, filtered versions of the equations of motion are derived, where the instantaneous velocity is decomposed into resolved (large-scale) and modeled (small-scale) motions. Lower-frequency eddies, with scales larger than the grid spacing, are resolved, while higher-frequency eddies, the subgrid fluctuations, are filtered out. Closure models are required for the subgrid-scale Reynolds stress. The Smagorinsky model, a one-equation mixing length model, is used in most commercial CFD (see Computational Fluid Dynamics below) codes that offer LES options and is also used in many academic and research CFD codes. See D. C. Wilcox [*Turbulence Modeling for CFD,* 2d ed., DCW Industries, La Cañada, Calif., 1998]. LES models are more accurate and more computationally intensive than Reynolds stress models. While they are not nearly as computationally expensive as direct numerical simulation, it is generally believed that highly accurate LES predictions of mean flow fields and large-scale motions can be obtained with practical computational grids.

A fourth class of models is based on solving transport equations, derived from the Navier-Stokes equation, for probability density functions. The PDF-based methods require solution of stochastic differential equations and generally are based on particle tracking methods. Closure models are not needed for the convective momentum transport that leads to the Reynolds stress in the RANS equation [S. B. Pope, *Turbulent Flows*, Cambridge University Press, Cambridge, UK, 2000].

Eddy Spectrum The energy that produces and sustains turbulence is extracted from velocity gradients in the mean flow, principally through vortex stretching. At Reynolds numbers well above the critical value there is a wide spectrum of eddy sizes, often described as a cascade of energy from the largest down to the smallest eddies. The largest eddies are of the order of the equipment size. The smallest are those for which viscous and inertial forces associated with the eddy velocity fluctuations are of the same order, so that turbulent fluctuations are rapidly damped out by viscous effects at smaller length scales. A distribution function may be plotted showing the distribution of kinetic energy with respect to eddy size. The peak of the distribution, that is, the eddy size containing the most kinetic energy, is much larger than the smallest eddies when the Reynolds number is large. The small eddies contain relatively little kinetic energy, but are responsible for most of the viscous dissipation. Large eddies, which extract energy from the mean flow velocity gradients, are generally anisotropic. At smaller length scales, the directionality of the mean flow exerts less influence, and *local isotropy* is approached. The range of eddy scales for which local isotropy holds is called the *equilibrium range.*

J. T. Davies [*Turbulence Phenomena,* Academic, New York, 1972] presents a good discussion of the spectrum of eddy lengths for well-developed isotropic turbulence. The smallest eddies, usually called *Kolmogorov eddies,* have a characteristic velocity fluctuation u_K given by

$$u_K = (\nu\varepsilon)^{1/4} \tag{6-231}$$

where ν = kinematic viscosity and ε = energy dissipation per unit mass [A. N. Kolmogorov, *Compt. Rend. Acad. Sci. URSS* **30:** 301; **32:** 16 (1941)]. The size of the Kolmogorov eddy scale is

$$l_K = (\nu^3/\varepsilon)^{1/4} \tag{6-232}$$

The Reynolds number for the Kolmogorov eddy, $\mathrm{Re}_K = l_K u_K/\nu$, is equal to unity. In the equilibrium range, which exists for well-developed turbulence and extends from the medium eddy sizes down to the smallest, the energy dissipation at the smaller length scales is supplied by turbulent energy drawn from the bulk flow and passed down the spectrum of eddy lengths according to the scaling rule

$$\varepsilon = \frac{u^3}{l} \tag{6-233}$$

For the energy-containing, eddy size

$$\varepsilon = \frac{u_e^3}{l_e} \tag{6-234}$$

For turbulent pipe flow, the friction velocity $\mu^* = \sqrt{\tau_w/\rho}$ used earlier in describing the universal turbulent velocity profile may be used as an estimate for u_e. Together with the Blasius equation for the friction factor from which ε may be obtained, Eq. (6-222), this provides an estimate for the energy-containing eddy size in turbulent pipe flow:

$$l_e = 0.05D\,\mathrm{Re}^{-1/8} \tag{6-235}$$

where D = pipe diameter and Re = pipe Reynolds number. Similarly, the Kolmogorov eddy size is

$$l_K = 4D\,\mathrm{Re}^{-0.78} \tag{6-236}$$

Most of the energy dissipation occurs on a length scale about 5 times the Kolmogorov eddy size.

The eddy spectrum is normally described using Fourier transform methods. The spectrum $E(\kappa)$ gives the turbulent kinetic energy contained in eddies of wave number between κ and $\kappa + d\kappa$, so that $K = \int_0^\infty E(\kappa)\,d\kappa$. The portion of the equilibrium range excluding the smallest eddies, those which are affected by dissipation, is the *inertial subrange.* The Kolmogorov law gives $E(\kappa) \propto \kappa^{-5/3}$ in the inertial subrange.

Texts for further reading on turbulent flow include S. B. Pope [*Turbulent Flows*, Cambridge University Press, Cambridge, UK, 2000], J. T. Davies [*Turbulence Phenomena,* Academic, New York, 1972], J. O. Hinze [*Turbulence,* McGraw-Hill, New York, 1975], H. Tennekes and J. L. Lumley [*A First Course in Turbulence,* MIT Press, Cambridge, Mass., 1972], and U. Frisch [*Turbulence,* Cambridge University Press, Cambridge, UK, 1995].

COMPUTATIONAL FLUID DYNAMICS

Computational fluid dynamics (CFD) emerged in the 1980s as a significant tool for fluid dynamics in both research and practice, enabled by rapid development in computer hardware and software. Commercial CFD software is widely available. Computational fluid dynamics normally refers to the numerical solution of the equations of continuity and momentum (e.g., Navier-Stokes equations) along with additional conservation equations for energy and material species in order to solve problems of nonisothermal flow, mixing, and chemical reaction.

Textbooks include C. A. J. Fletcher [*Computational Techniques for Fluid Dynamics,* vol. 1: *Fundamental and General Techniques,* and vol. 2: *Specific Techniques for Different Flow Categories,* Springer-Verlag, Berlin, 1988], C. Hirsch [*Numerical Computation of Internal and External Flows,* vol. 1: *Fundamentals of Numerical Discretization,* and vol. 2: *Computational Methods for Inviscid and Viscous Flows,* Wiley, New York, 1988], R. Peyret and T. D. Taylor [*Computational Methods for Fluid Flow,* Springer-Verlag, Berlin, 1990], C. Canuto et al. [*Spectral Methods in Fluid Dynamics,* Springer-Verlag, Berlin, 1988], R. H. Pletcher et al. [*Computational Fluid Mechanics and Heat Transfer,* 3d ed., CRC Press, Boca Raton, FL, 2013], and Patankar [*Numerical Heat Transfer and Fluid Flow,* Hemisphere, Washington, D.C., 1980].

A variety of numerical methods has been employed, but three basic steps are common.

1. *Subdivision or discretization of the flow domain into cells or elements.* Discretization produces a *mesh* and a set of *nodes* at which the flow variables are calculated. The equations of motion are solved approximately on a domain defined by the grid. The grid must be sufficiently refined to resolve flow features and to accurately fit the boundaries.

2. *Discretization of the governing equations.* In this step, the spatial partial derivatives are replaced by algebraic approximations written in terms of the nodal values of the dependent variables. Among the numerous spatial discretization methods, *finite difference, finite volume,* and *finite element* methods are the most common. The *finite difference* method estimates spatial derivatives in terms of the nodal values and spacing between nodes. The governing equations are then written in terms of the nodal unknowns at each interior node. *Finite volume* methods, related to finite difference methods, may be derived by a volume integration of the equations of motion, with application of the divergence theorem, reducing by one the order of the differential equations. Equivalently, macroscopic balance equations are written on each cell. *Finite element* methods are weighted residual techniques in which the unknown dependent variables are expressed in terms of *basis functions* interpolating among the nodal values. The basis functions are substituted into the equations of motion, resulting in error residuals which are multiplied by the weighting functions, integrated over the control volume, and set to zero to produce algebraic equations in terms of the nodal unknowns. Selection of the weighting functions defines the various finite element methods. For example, Galerkin's method uses the nodal interpolation basis functions as weighting functions. Each approach also has its own method for implementing *boundary conditions.* The end result after discretization of the equations and application of the boundary conditions for a steady flow is a set of algebraic equations for the nodal unknown variables. Discretization in time is also required for the time derivative terms in unsteady flow. The discretized equations represent an approximation of the exact equations, and their solution gives an approximation for the flow variables. The accuracy of the solution improves as the grid is *refined,* that is, as the number of nodal points is increased.

3. *Solution of the discretized equations.* Creeping flows with constant viscosity yield a linear matrix equation. Both direct and iterative solvers have been used. For most flows, the nonlinear inertial terms in the momentum equation are important, and the algebraic discretized equations are therefore nonlinear. Solution yields the nodal values of the unknowns. Various implicit and explicit methods for time integration have been employed.

A CFD method called the *lattice Boltzmann method* (LBM) models the fluid as a set of particles moving with discrete velocities on a discrete grid or lattice, rather than using discretization of the governing continuum partial differential equations. Reviews of the LBM include those by S. Chen and G. D. Doolen [*Ann. Rev. Fluid Mech.* **30:** 329 (1998)] and C. K. Aiden and J. R. Clausen [*Ann. Rev. Fluid Mech.* **42:** 439 (2010)]. Lattice Boltzmann approximations can be constructed that give the same macroscopic behavior as the Navier-Stokes equations. The method is currently used mainly in academic and research codes, rather than in general-purpose commercial CFD codes. There appear to be significant computational advantages to the lattice Boltzmann method, particularly with respect to parallel processing. For turbulent flows, direct numerical simulation, as well as turbulence models analogous to those described in the Turbulence section above, can be applied using LBM. This includes both unfiltered and filtered (subgrid scale) turbulence models, the latter described by S. Hou et al. [*Fields Institute Comm.* **6:** 151 (1996)]. Multiphase flow, heat transfer, species diffusion, and chemical reaction have been solved. LBM methods for multifluid flows and for flows with particulates are described by X. Shan and H. Chen [*Phys. Rev. E* **47:** 1815 (1993)] and Feng and Michaelides [*J. Comput. Phys.* **195:** 602–628 (2004)], respectively.

CFD solutions, especially for complex three-dimensional flows, generate very large quantities of solution data. Computer graphics have greatly improved the ability to examine CFD solutions and visualize flow.

CFD methods are used for incompressible and compressible, creeping, laminar and turbulent, newtonian and nonnewtonian, and isothermal and nonisothermal flows. Chemically reacting flows, particularly in the field of combustion, have been simulated. Solution accuracy must be considered from several perspectives. These include convergence of the algorithms for solving the nonlinear discretized equations and convergence with respect to refinement of the mesh so that the discretized equations better approximate the exact equations and, in some cases, so that the mesh more accurately fits the true geometry. The possibility that steady-state solutions are unstable should be considered. In addition to numerical sources of error, modeling errors are introduced in turbulent flow, where closure models are used to solve time-averaged equations of motion, as discussed previously. Most commercial CFD codes include the k-ε turbulence model, and often several other models such as the ones described previously under subsection Turbulence are included. Large eddy simulation (LES) methods for turbulent flow, described previously, are available in commercial CFD codes. Significant solution error is known to result in some problems from inadequacy of the turbulence model. Closure models for nonlinear chemical reaction source terms may also contribute to inaccuracy.

In its general sense, multiphase flow is not currently solvable by computational fluid dynamics. However, in certain cases reasonable solutions are possible. These include well-separated flows where the phases are confined to relatively well-defined regions separated by one or a few interfaces and flows in which a second phase appears as discrete particles of known size and shape whose motion may be approximately computed with drag coefficient formulations, or rigorously computed with refined meshes applying boundary conditions at the particle surface. *Two-fluid modeling,* in which the phases are treated as overlapping continua, with each phase occupying a volume fraction that is a continuous function of position (and time), is a useful approximation available in commercial software. See S. E. Elghobashi and T. W. Abou-Arab [*J. Physics Fluids* **26:** 931–938 (1983)] for a k-ε model for two-fluid systems.

DIMENSIONLESS GROUPS

For purposes of data correlation, model studies, and scale-up, it is useful to arrange variables into dimensionless groups. Table 6-9 lists many of the dimensionless groups commonly found in fluid mechanics problems, along with their physical interpretations and areas of application. More extensive tabulations may be found in J. P. Catchpole and G. Fulford [*Ind. Eng. Chem.* **58**(3): 46–60 (1966)] and G. Fulford and J. P. Catchpole [*Ind. Eng. Chem.* **60**(3): 71–78 (1968)].

TABLE 6-9 Dimensionless Groups and Their Significance

Name	Symbol	Formula	Physical interpretation	Comments
Archimedes number	Ar	$\dfrac{gL^3(\rho_p-\rho)\rho}{\mu^2}$	$\dfrac{\text{inertial forces} \times \text{buoyancy force}}{(\text{viscous forces})^2}$	Particle settling
Bingham number	Bm	$\dfrac{\tau_y L}{\mu_\infty V}$	$\dfrac{\text{yield stress}}{\text{viscous stress}}$	Flow of Bingham plastics = yield number Y
Bingham Reynolds number	Re_B	$\dfrac{LV\rho}{\mu_\infty}$	$\dfrac{\text{inertial forces}}{\text{viscous forces}}$	Flow of Bingham plastics
Blake number	B	$\dfrac{V\rho}{\mu(1-\epsilon)_s}$	$\dfrac{\text{inertial forces}}{\text{viscous forces}}$	Beds of solids
Bond number	Bo	$\dfrac{(\rho_L-\rho_G)L^2 g}{\sigma}$	$\dfrac{\text{gravitational force}}{\text{surface-tension force}}$	Atomization = Eotvos number Eo
Capillary number	Ca	$\dfrac{\mu V}{\sigma}$	$\dfrac{\text{viscous force}}{\text{surface-tension force}}$	Two-phase flows, free surface flows
Cauchy number	C	$\dfrac{\rho V^2}{\beta}$	$\dfrac{\text{inertial force}}{\text{compressibility force}}$	Compressible flow, hydraulic transients
Cavitation number	σ	$\dfrac{p-p_v}{\rho V^2/2}$	$\dfrac{\text{excess pressure above vapor pressure}}{\text{velocity head}}$	Cavitation
Dean number	De	$\dfrac{\text{Re}}{(Dc/D)^{1/2}}$	$\text{Reynolds number} \times \dfrac{\text{inertial force}}{\text{centrifugal force}}$	Flow in curved channels
Deborah number	De	$\lambda\omega$	$\dfrac{\text{fluid relaxation time}}{\text{flow characteristic time}}$	Viscoelastic flow
Drag coefficient	C_D	$\dfrac{F_D}{A\rho V^2/2}$	$\dfrac{\text{drag force}}{\text{projected area} \times \text{velocity head}}$	Flow around objects, particle settling
Elasticity number	El	$\dfrac{\lambda\mu}{\rho L^2}$	$\dfrac{\text{elastic force}}{\text{inertial force}}$	Viscoelastic flow
Euler number	Eu	$\dfrac{\Delta p}{\rho V^2}$	$\dfrac{\text{frictional pressure loss}}{2 \times \text{velocity head}}$	Fluid friction in conduits
Fanning friction factor	f	$\dfrac{D\Delta p}{2\rho V^2 L} = \dfrac{2\tau_w}{\rho V^2}$	$\dfrac{\text{wall shear stress}}{\text{velocity head}}$	Fluid friction in conduits Darcy

(*Continued*)

TABLE 6-9 Dimensionless Groups and Their Significance *(Continued)*

Name	Symbol	Formula	Physical interpretation	Comments
Froude number	Fr	$\frac{V^2}{gL}$	$\frac{\text{inertial force}}{\text{gravity force}}$	Often defined as $Fr = V/\sqrt{gL}$
Densometric Froude number	Fr′	$\frac{\rho V^2}{(\rho_d - \rho)gL}$	$\frac{\text{inertial force}}{\text{gravity force}}$	or $Fr' = \frac{V}{\sqrt{(\rho_d - \rho)gL/\rho}}$
Hedstrom number	He	$\frac{L^2\tau_y\rho}{\mu_\infty^2}$	Bingham Reynolds number × Bingham number	Flow of Bingham plastics
Hodgson number	H	$\frac{V'\omega\Delta p}{\bar{q}\,\bar{p}}$	$\frac{\text{time constant of system}}{\text{period of pulsation}}$	Pulsating gas flow
Mach number	M	$\frac{V}{c}$	$\frac{\text{fluid velocity}}{\text{sonic velocity}}$	Compressible flow
Newton number	Ne	$\frac{\Delta PD}{\rho V^2 L}$	2 × Fanning friction factor	
Ohnesorge number	Z	$\frac{\mu}{(\rho L\sigma)^{1/2}}$	$\frac{\text{viscous force}}{(\text{inertial force} \times \text{surface tension force})^{1/2}}$	$\text{Atomization} = \frac{\text{Wber number}}{\text{Reynolds number}}$
Peclet number	Pe	$\frac{LV}{D}$	$\frac{\text{convective transport}}{\text{diffusive transport}}$	Heat, mass transfer, mixing
Pipeline parameter	Pn	$\frac{aV_o}{2gH}$	$\frac{\text{maximum water} - \text{hammer pressure rise}}{2 \times \text{static pressure}}$	Water hammer
Power number	Po	$\frac{P}{\rho N^3 L^5}$	$\frac{\text{impeller drag force}}{\text{inertial force}}$	Agitation
Prandtl velocity ratio	υ+	$\frac{\upsilon}{(\tau_w/\rho)^{1/2}}$	velocity normalized by friction velocity	Turbulent flow near a wall, friction Velocity $= \sqrt{\tau_w/\rho}$
Reynolds number	Re	$\frac{LV\rho}{\mu}$	$\frac{\text{inertial force}}{\text{viscous force}}$	
Strouhal number	St	$\frac{f'L}{V}$	vortex shedding frequency × characteristic flow time scale	Vortex shedding, von Karman vortex streets
Weber number	We	$\frac{\rho V^2 L}{\sigma}$	$\frac{\text{inertial force}}{\text{surface tension force}}$	Bubble, drop formation

Nomenclature		SI units	Nomenclature		SI units
a	Wave speed	m/s	$\frac{P}{q}$	Power	Watts
A	Projected area	m	$\bar{q}$	Average volumetric flow rate	m³/s
c	Sonic velocity	m/s	s	Particle area/particle volume	1/m
D	Diameter of pipe	m	υ	Local fluid velocity	m/s
D_c	Diameter of curvature	m	V	Characteristic or average fluid velocity	m/s
D'	Diffusivity	m²/s	V'	System volume	m³
f'	Vortex shedding frequency	1/s	β	Bulk modulus	Pa
F_D	Drag force	N	ϵ	Void fraction	m³
g	Acceleration of gravity	m/s	λ	Fluid relaxation time	s
H	Static head	m	μ	Fluid viscosity	Pa·s
L	Characteristic length	m	μ_∞	Infinite shear viscosity (Bingham plastics)	Pa·s
N	Rotational speed	1/s	ρ	Fluid density	kg/m³
p	Pressure	Pa	ρ_G, ρ_L	Gas, liquid densities	kg/m³
$\frac{p_\upsilon}{p}$	Vapor pressure	Pa	ρ_d	Dispersed phase density	kg/m³
pw	Average static pressure	Pa	σ	Surface tension	N/m
Δp	Frictional pressure drop	Pa	ω	Characteristic frequency or reciprocal time scale of flow	1/s

PARTICLE DYNAMICS

General References: R. S. Brodkey, *The Phenomena of Fluid Motions*, Addison-Wesley, Reading, Mass., 1967; R. Clift, J. R. Grace, and M. E. Weber, *Bubbles, Drops and Particles*, Academic, New York, 1978; G. W. Govier and K. Aziz, *The Flow of Complex Mixtures in Pipes*, Van Nostrand Reinhold, New York, 1972, Krieger, Huntington, N.Y., 1977; C. E. Lapple et al., *Fluid and Particle Mechanics*, University of Delaware, Newark, 1951; V. G. Levich, *Physicochemical Hydrodynamics*, Prentice-Hall, Englewood Cliffs, N.J., 1962; C. Orr, *Particulate Technology*, Macmillan, New York, 1966; C. A. Shook and M. C. Roco, *Slurry Flow*, Butterworth-Heinemann, Boston, 1991; G. B. Wallis, *One-dimensional Two-phase Flow*, McGraw-Hill, New York, 1969.

DRAG COEFFICIENT

When relative motion exists between a particle and a surrounding fluid, the fluid will exert a drag force upon the particle. In steady flow, the drag force is

$$F_D = \frac{C_D A_p \rho u^2}{2} \tag{6-237}$$

where F_D = drag force
C_D = drag coefficient
A_P = projected particle area in direction of motion
ρ = density of surrounding fluid
u = relative velocity between particle and fluid

The drag force is exerted in a direction parallel to the relative velocity. Equation (6-237) defines the *drag coefficient.* For some solid bodies, such as aerofoils, a lift force component perpendicular to the velocity is also exerted. For free-falling particles, lift forces are generally unimportant. However, even spherical particles experience lift forces in shear flows near solid surfaces.

TERMINAL VELOCITY

A particle falling under the action of gravity will accelerate until the drag force balances gravitational force, after which it falls at its *terminal* or *free-settling velocity* u_t, given by

$$u_t = \sqrt{\frac{2gm_p(\rho_p - \rho)}{\rho \rho_p A_p C_D}} \tag{6-238}$$

where g = acceleration of gravity
m_p = particle mass
ρ_p = particle density

and the remaining symbols are as previously defined.

Settling particles may undergo fluctuating motions owing to vortex shedding, among other factors. Oscillation is enhanced with increasing separation between the mass and geometric centers of the particle. Variations in velocity are usually less than 10 percent. The drag force on a particle fixed in space with fluid moving is somewhat lower than the drag force on an oscillating freely settling particle in a stationary fluid at the same relative velocity.

Spherical Particles For spherical particles of diameter d_p, Eq. (6-238) becomes

$$u_t = \sqrt{\frac{4gd_p(\rho_p - \rho)}{3\rho C_D}} \tag{6-239}$$

The drag coefficient for rigid spherical particles is a function of particle Reynolds number $\text{Re}_p = d_p \rho u/\mu$ where μ = fluid viscosity, as shown in Fig. 6-52. At low Reynolds number, *Stokes' law* gives

$$C_D = \frac{24}{\text{Re}_p} \qquad \text{Re}_p < 0.1 \tag{6-240}$$

which may also be written

$$F_D = 3\pi\mu u d_p \qquad \text{Re}_p < 0.1 \tag{6-241}$$

and gives for the terminal settling velocity

$$u_t = \frac{gd_p^2(\rho_p - \rho)}{18\mu} \qquad \text{Re}_p < 0.1 \tag{6-242}$$

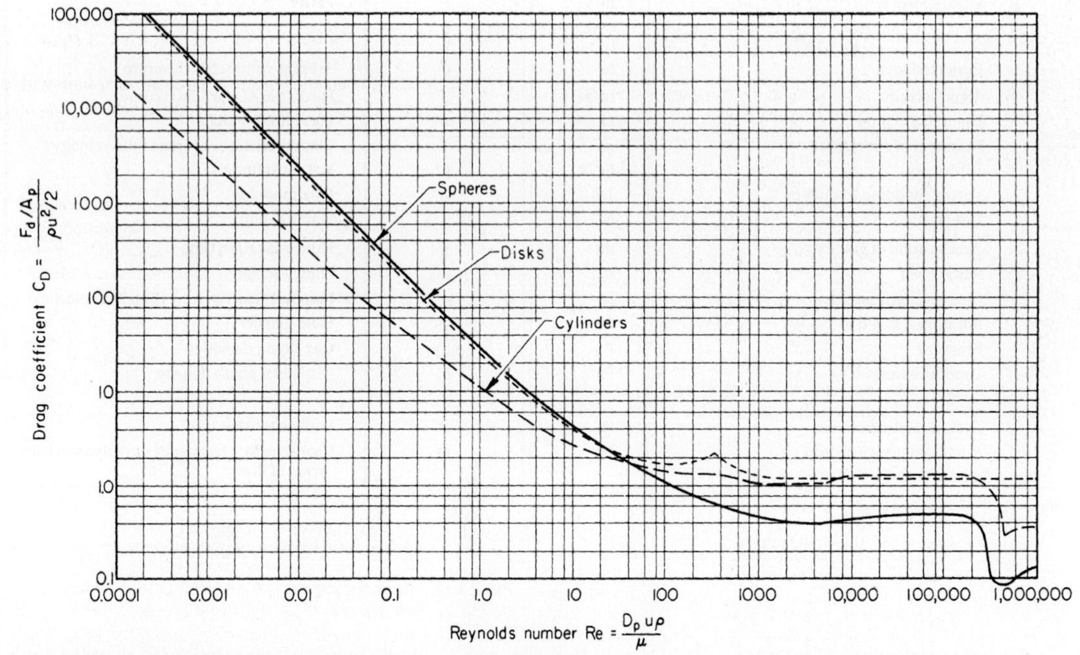

FIG. 6-52 Drag coefficients for spheres, disks, and cylinders: A_p = area of particle projected on a plane normal to direction of motion; C = overall drag coefficient, dimensionless; D_p = diameter of particle; F_d = drag or resistance to motion of body in fluid; Re = Reynolds number, dimensionless; u = relative velocity between particle and main body of fluid; μ = fluid viscosity; and ρ = fluid density. [From C. E. Lapple and C. B. Shepherd, *Ind. Eng. Chem.*, **32,** 605 (1940).]

In the *intermediate regime* ($0.1 < \text{Re}_p < 1000$), the drag coefficient may be estimated within 6 percent by

$$C_D = \left(\frac{24}{\text{Re}_p}\right)\left(1 + 0.14\text{Re}_p^{0.70}\right) \qquad 0.1 < \text{Re}_p < 1000 \tag{6-243}$$

In *Newton's law* regime, which covers the range $1000 < \text{Re}_p < 350{,}000$, $C_D = 0.445$, within 13 percent. In this region, Eq. (6-239) becomes

$$u_t = \sqrt{\frac{4gd_p(\rho_p - \rho)}{3\rho(0.445)}} \qquad 1000 < \text{Re}_p < 350{,}000 \tag{6-244}$$

Between about $\text{Re}_p = 350{,}000$ and 1×10^6, the drag coefficient drops dramatically in a *drag crisis* owing to the transition to turbulent flow in the boundary layer around the particle, which delays aft separation, resulting in a smaller wake and less drag. Beyond $\text{Re}_p = 1 \times 10^6$, the drag coefficient may be estimated from [R. Clift, J. R. Grace, and M. E. Weber, *Bubbles, Drops and Particles,* Academic, New York, 1978]

$$C_D = 0.19 - \frac{8 \times 10^4}{\text{Re}_p} \qquad \text{Re}_p > 1 \times 10^6 \tag{6-245}$$

Drag coefficients may be affected by turbulence in the free-stream flow; the drag crisis occurs at lower Reynolds numbers when the free stream is turbulent. L. B. Torobin and W. H. Guvin [*AIChE J.* **7:** 615–619 (1961)] found that the drag crisis Reynolds number decreases with increasing free-stream turbulence, reaching a value of 400 when the relative turbulence intensity, defined as $\sqrt{u'}/\overline{U}_R$, is 0.4. Here $\sqrt{u'}$ is the rms fluctuating velocity and $\overline{U}_R$ is the time-averaged relative velocity between the particle and the fluid.

For computing the terminal settling velocity, correlations for drag coefficient as a function of Archimedes number

$$\text{Ar} = \frac{gd_p^3(\rho_p - \rho)\rho}{\mu^2} \tag{6-246}$$

may be more convenient than C_D–Re_p correlations, because the latter are implicit in terminal velocity, and the settling regime is unknown. D. G. Karamanev [*Chem. Eng. Comm.* **147:** 75 (1996)] provides a correlation for drag coefficient for settling solid spheres in terms of Ar.

$$C_D = \frac{432}{\text{Ar}}(1 + 0.0470\text{Ar}^{2/3}) + \frac{0.517}{1 + 154\text{Ar}^{-1/3}} \tag{6-247}$$

This equation reduces to Stokes' law $C_D = 24/\text{Re}_p$ in the limit $\text{Ar} \to 0$ and is a fit to data up to about $\text{Ar} = 2 \times 10^{10}$, where it gives $C_D = 0.50$, slightly greater than Newton's law value above. For rising light spheres, which exhibit more energy dissipating lateral motion than do falling dense spheres, Karamanev found that Eq. (6-247) is followed up to $\text{Ar} = 13{,}000$ and that for $\text{Ar} > 13{,}000$, the drag coefficient is $C_D = 0.95$.

For particles settling in *nonnewtonian* fluids, correlations are given by Dallon and Christiansen [Preprint 24C, *Symposium on Selected Papers,* part III, 61st Ann. Mtg. AIChE, Los Angeles, Dec. 1–5, 1968] for spheres settling in shear-thinning liquids, and by S. Ito and T. Kajiuchi [*J. Chem. Eng. Japan,* **2**(1): 19–24 (1969)] and H. Pazwash and J. M. Robertson [*J. Hydraul. Res.* **13:** 35–55 (1975)] for spheres settling in Bingham plastics. A. N. Beris et al. [*J. Fluid Mech.* **158:** 219–244 (1985)] present a finite element calculation for creeping motion of a sphere through a Bingham plastic.

Nonspherical Rigid Particles The drag on a nonspherical particle depends upon its shape and orientation with respect to the direction of motion. The orientation in free fall as a function of Reynolds number is given in Table 6-10.

The drag coefficients for *disks* (flat side perpendicular to the direction of motion) and for *cylinders* (infinite length with axis perpendicular to the direction of motion) are given in Fig. 6-52 as a function of Reynolds number.

TABLE 6-10 Free-Fall Orientation of Particles

Reynolds number*	Orientation
0.1–5.5	All orientations are stable when there are three or more perpendicular axes of symmetry.
5.5–200	Stable in position of maximum drag.
200–500	Unpredictable. Disks and plates tend to wobble, while fuller bluff bodies tend to rotate.
500–200,000	Rotation about axis of least inertia, frequently coupled with spiral translation

*Based on diameter of a sphere having the same surface area as the particle.
SOURCE: From Becker, *Can. J. Chem. Eng.* **37:** 85–91 (1959).

The effect of length-to-diameter ratio for cylinders in Newton's law region is reported by J. G. Knudsen and D. L. Katz [*Fluid Mechanics and Heat Transfer,* McGraw-Hill, New York, 1958].

E. S. Pettyjohn and E. B. Christiansen [*Chem. Eng. Prog.* **44:** 157–172 (1948)] present correlations for the effect of particle shape on free-settling velocities of *isometric particles.* For $\text{Re} < 0.05$, the terminal or free-settling velocity is given by

$$u_t = K_1 \frac{gd_s^2(\rho_p - \rho)}{18\mu} \tag{6-248}$$

$$K_1 = 0.843 \log\left(\frac{\psi}{0.065}\right) \tag{6-249}$$

where ψ = sphericity, the surface area of a sphere having the same volume as the particle, divided by the actual surface area of the particle; d_s = equivalent diameter, equal to the diameter of the equivalent sphere having the same volume as the particle; and other variables are as previously defined.

In *Newton's law region,* the terminal velocity is given by

$$u_t = \sqrt{\frac{4d_s(\rho_p - \rho)g}{3K_3\rho}} \tag{6-250}$$

$$K_3 = 5.31 - 4.88\psi \tag{6-251}$$

Equations (6-248) to (6-251) are based on experiments on cube-octahedrons, octahedrons, cubes, and tetrahedrons for which the sphericity ψ ranges from 0.906 to 0.670, respectively. See also R. Clift, J. R. Grace, and M. E. Weber, *Bubbles, Drops and Particles,* Academic, New York, 1978. A graph of drag coefficient versus Reynolds number with ψ as a parameter may be found in Brown et al. [*Unit Operations,* Wiley, New York, 1950] and in G. W. Govier and K. Aziz [*The Flow of Complex Mixtures in Pipes,* Krieger, Huntington, N.Y., 1977].

For particles with $\psi < 0.67$, the correlations of Becker [*Can. J. Chem. Eng.* **37:** 85–91 (1959)] should be used. Reference to this paper is also recommended for *intermediate region* flow. Settling characteristics of nonspherical particles are discussed in chaps. 4 and 6 of R. Clift, J. R. Grace, and M. E. Weber, *Bubbles, Drops and Particles,* Academic, New York, 1978.

The terminal velocity of *axisymmetric particles in axial motion* can be computed from Bowen and Masliyah [*Can. J. Chem. Eng.* **51:** 8–15 (1973)] for low–Reynolds number motion:

$$u_t = \frac{V'}{K_2}\frac{gD_s^2(\rho_p - \rho)}{18\mu} \tag{6-252}$$

$$K_2 = 0.244 + 1.035\Sigma - 0.712\Sigma^2 + 0.441\Sigma^3 \tag{6-253}$$

where D_s = diameter of sphere with perimeter equal to maximum particle projected perimeter
V' = ratio of particle volume to volume of sphere with diameter D_s
Σ = ratio of surface area of particle to surface area of a sphere with diameter D_s

and other variables are as defined previously.

Hindered Settling When particle concentration increases, particle settling velocities decrease because of hydrodynamic interaction between particles and the upward motion of displaced liquid. The suspension viscosity increases. Hindered settling is normally encountered in sedimentation and transport of concentrated slurries. Below 0.1 percent volumetric particle concentration, there is less than 1 percent reduction in settling velocity. Several expressions have been given to estimate the effect of particle volume fraction on settling velocity. Maude and Whitmore [*Br. J. Appl. Phys.* **9:** 477–482 (1958)] give for uniformly sized spheres

$$u_t = u_{t0}(1 - c)^n \tag{6-254}$$

where u_t = terminal settling velocity
u_{t0} = terminal velocity of a single sphere (infinite dilution)
c = volume fraction solid in the suspension
n = function of Reynolds number $\text{Re}_p = d_p u_{t0}\rho/\mu$ as given in Fig. 6-53

In Stokes' law region ($\text{Re}_p < 0.3$) $n = 4.65$ and in Newton's law region ($\text{Re}_p > 1000$) $n = 2.33$. Equation (6-254) may be applied to particles of any size in a polydisperse system, provided the volume fraction corresponding to all the particles is used in computing terminal velocity [Richardson and Shabi, *Trans. Inst. Chem. Eng. (London)* **38:** 33–42 (1960)]. The concentration effect is greater for nonspherical and angular particles than for spherical particles [Steinour, *Ind. Eng. Chem.* **36:** 840–847 (1944)]. Theoretical developments for low–Reynolds number flow assemblages of spheres are given

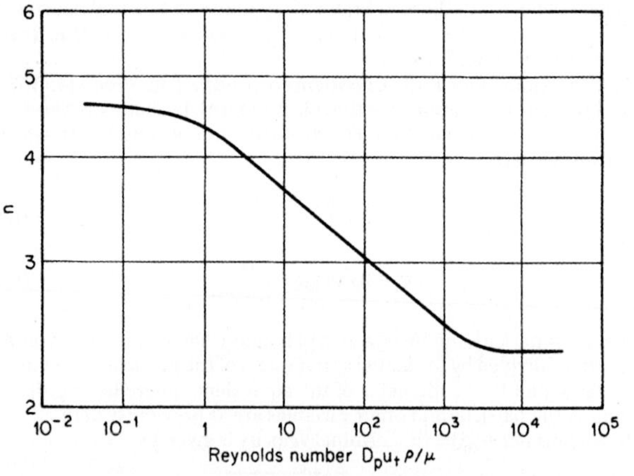

FIG. 6-53 Values of exponent n for use in Eq. (6-254). [From A. D. Maude and R. L. Whitmore, *Br. J. Appl. Phys.*, **9,** 481 (1958). *Courtesy of the Institute of Physics and the Physical Society.*]

by Happel and Brenner [*Low Reynolds Number Hydrodynamics,* Prentice-Hall, Englewood Cliffs, N.J., 1965] and Famularo and Happel [*AIChE J.* **11:** 981 (1965)] leading to an equation of the form

$$u_t = \frac{u_{t0}}{1 + \gamma c^{1/3}} \tag{6-255}$$

where γ is about 1.3. As particle concentration increases, resulting in interparticle contact, hindered settling velocities are difficult to predict. Thomas [*AIChE J.* **9:** 310 (1963)] provides an empirical expression reported to be valid over the range $0.08 < u_t/u_{t0} < 1$:

$$\ln\left(\frac{u_t}{u_{t0}}\right) = -5.9c \tag{6-256}$$

Time-Dependent Motion The time-dependent motion of particles is computed by application of Newton's second law, equating the rate of change of particle momentum to the net force acting on the particle. Rotation of particles may also be computed from the net torque. For large particles moving through low-density gases, it is usually sufficient to compute the force due to fluid drag from the relative velocity and the drag coefficient computed for steady flow conditions. For two- and three-dimensional problems, the velocity appearing in the particle Reynolds number and the drag coefficient is the amplitude of the relative velocity. The drag force, not the relative velocity, is resolved into vector components to compute the particle acceleration components. R. Clift, J. R. Grace, and M. E. Weber [*Bubbles, Drops and Particles,* Academic, New York, 1978] discuss the complexities that arise in the computation of transient drag forces on particles when the transient nature of the flow is important. Analytical solutions for the case of a single particle in creeping flow ($Re_p = 0$) are available. For example, the creeping motion of a spherical particle released from rest in a stagnant fluid is described by

$$\rho V \frac{dU}{dt} = g(\rho_p - \rho)V - 3\pi\mu d_p U - \frac{\rho}{2}V\frac{dU}{dt} - \left(\frac{3}{2}\right)d_p^2\sqrt{\pi\rho\mu}\int_0^t \frac{(dU/dt)_{t=s}\,ds}{\sqrt{t-s}} \tag{6-257}$$

Here, U = particle velocity, positive in the direction of gravity, and V = particle volume. The first term on the right-hand side is the net gravitational force on the particle, accounting for buoyancy. The second is the steady-state Stokes drag, Eq. (6-241). The third is the *added mass* or *virtual mass* term, which may be interpreted as the inertial effect of the fluid which is accelerated along with the particle. The volume of the added mass of fluid is one-half the particle volume. The last term, the *Basset force,* depends on the entire history of the transient motion, with past motions weighted inversely with the square root of elapsed time. R. Clift, J. R. Grace, and M. E. Weber [*Bubbles, Drops and Particles,* Academic, New York, 1978] provide integrated solutions. In *turbulent flows,* particle velocity will closely follow fluid eddy velocities when

$$\tau_0 >> \frac{d_p^2[(2\rho_p/\rho)+1]}{36\nu} \tag{6-258}$$

where τ_0 = oscillation period or eddy time scale, the right-hand side expression is the *particle relaxation time,* and ν = kinematic viscosity.

Gas Bubbles Drops and bubbles, unlike rigid solid particles, may undergo deformation and internal circulation. Figure 6-54 shows rise velocity data for air bubbles in stagnant water. In the figure, Eo = Eotvos number, $g(\rho_L - \rho_G)d_e/\sigma$, where ρ_L = liquid density, ρ_G = gas density, σ = surface tension, and the equivalent diameter d_e is the diameter of a sphere with volume equal to that of the bubble. Small bubbles (< 1-mm diameter) remain spherical and rise in straight lines. The presence of surface active materials

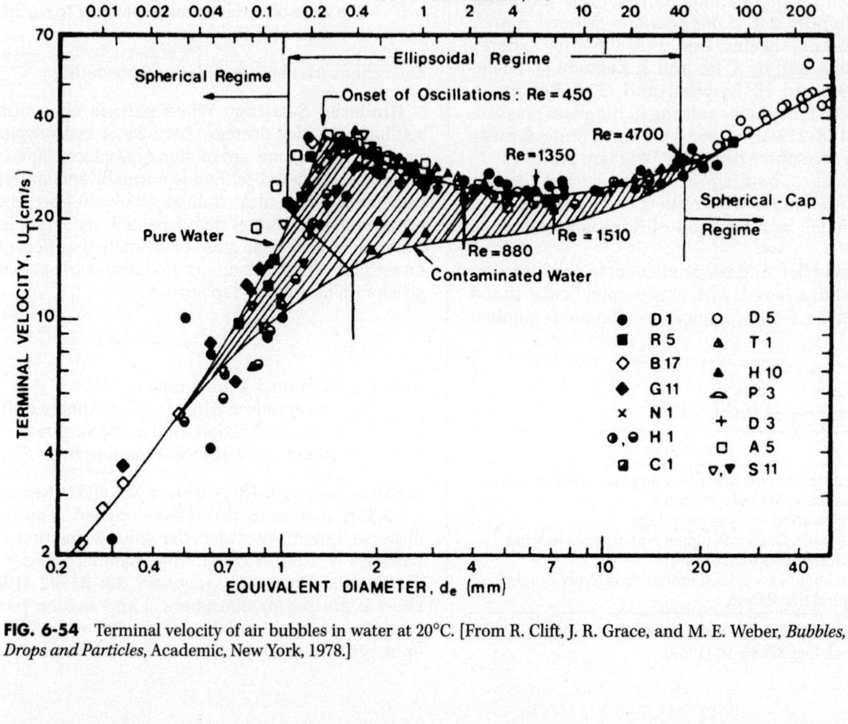

FIG. 6-54 Terminal velocity of air bubbles in water at 20°C. [From R. Clift, J. R. Grace, and M. E. Weber, *Bubbles, Drops and Particles,* Academic, New York, 1978.]

generally renders small bubbles rigid, and they rise roughly according to the drag coefficient and terminal velocity equations for spherical solid particles. Bubbles roughly in the range 2- to 8-mm diameter assume flattened, ellipsoidal shape, and rise in a zigzag or spiral pattern. This motion increases dissipation and drag, and the rise velocity may actually decrease with increasing bubble diameter in this region, characterized by rise velocities in the range of 20 to 30 cm/s. Large bubbles, > 8-mm diameter, are greatly deformed, assuming a mushroomlike, spherical cap shape. These bubbles are unstable and may break into smaller bubbles. Carefully purified water, free of surface active materials, allows bubbles to freely circulate even when they are quite small. Under creeping flow conditions $Re_b = d_b u_r \rho_L/\mu_L < 1$, where u_r = bubble rise velocity and μ_L = liquid viscosity, the bubble rise velocity may be computed analytically from the Hadamard-Rybczynski formula [V. G. Levich, *Physicochemical Hydrodynamics,* Prentice-Hall, Englewood Cliffs, N.J., 1962, p. 402]. When $\mu_G/\mu_L << 1$, which is normally the case, the rise velocity is 1.5 times the rigid sphere Stokes' law velocity. However, in practice, most liquids, including ordinary distilled water, contain sufficient surface active materials to render small bubbles rigid. Larger bubbles undergo deformation in both purified and ordinary liquids; however, the variation in rise velocity for large bubbles with degree of purity is quite evident in Fig. 6-54.

D. G. Karamanev [*Chem. Eng. Comm.* **147:** 75 (1996)] provides equations for bubble rise velocity based on the Archimedes number and on use of the bubble projected diameter d_h in the drag coefficient and the bubble equivalent diameter in Ar. The Archimedes number is as defined in Eq. (6-246) except that the density difference is liquid density minus gas density, and d_p is replaced by d_e.

$$u_t = \sqrt{\frac{8g}{6^{2/3}\pi^{1/3}C_D}}\,V^{1/6}\frac{d_e}{d_h} = \sqrt{\frac{4gd_e}{3C_D}}\,\frac{d_e}{d_h} \qquad (6\text{-}259)$$

$$C_D = \frac{432}{\text{Ar}}\left(1+0.0470\,\text{Ar}^{2/3}\right)+\frac{0.517}{1+154\,\text{Ar}^{-1/3}} \qquad \text{Ar} < 13{,}000 \qquad (6\text{-}260)$$

$$C_D = 0.95 \quad \text{Ar} > 13{,}000 \qquad (6\text{-}261)$$

$$\frac{d_e}{d_h} = \left(1+0.163\,Eo^{0.757}\right)^{-1/3} \qquad \text{Eo} < 40 \qquad (6\text{-}262)$$

$$\frac{d_e}{d_h} = 0.62 \qquad \text{Eo} > 40 \qquad (6\text{-}263)$$

Applied to air bubbles in water, these expressions give reasonable agreement with the contaminated water curve in Fig. 6-54.

Figure 6-55 gives the drag coefficient as a function of bubble or drop Reynolds number for air bubbles in water and water drops in air, compared with the standard drag curve for rigid spheres. Information on bubble motion in *nonnewtonian* liquids may be found in G. Astarita and G. Apuzzo [*AIChE J.* **1:** 815–820 (1965)]; P. H. Calderbank et al. [*Chem. Eng. Sci.* **25:** 235–256 (1970)]; and A. Acharya et al. [*Chem. Eng. Sci.* **32:** 863–872 (1977)].

Liquid Drops in Liquids Very small liquid drops in immiscible liquids behave as rigid spheres, and the terminal velocity can be approximated by use of the drag coefficient for solid spheres up to a Reynolds number of about 10 [Warshay et al., *Can. J. Chem. Eng.* **37:** 29–36 (1959)]. Between Reynolds numbers of 10 and 500, the terminal velocity exceeds that for rigid spheres owing to internal circulation. J. R. Grace et al. [*Trans. Inst. Chem. Eng.* **54:** 167–173 (1976)]; R. Clift, J. R. Grace, and M. E. Weber [*Bubbles, Drops and Particles,* Academic, New York, 1978, pp. 175–177] present a correlation for terminal velocity valid in the range

$$M < 10^{-3} \qquad \text{Eo} < 40 \qquad \text{Re} > 0.1 \qquad (6\text{-}264)$$

where M = Morton number = $g\mu^4\Delta\rho/\rho^2\sigma^3$
Eo = Eotvos number = $g\,\Delta\rho\, d^2/\sigma$
Re = Reynolds number = $du\rho/\mu$
$\Delta\rho$ = density difference between the phases
ρ = density of continuous liquid phase
d = drop diameter
μ = continuous liquid viscosity
σ = surface tension
u = relative velocity

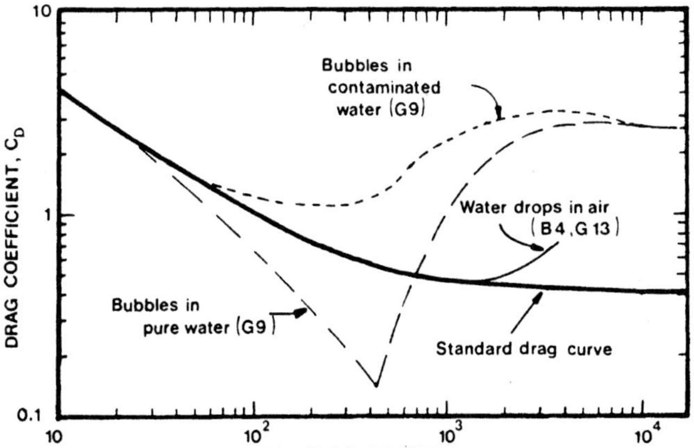

FIG. 6-55 Drag coefficient for water drops in air and air bubbles in water. Standard drag curve is for rigid spheres. [From R. Clift, J. R. Grace, and M. E. Weber, *Bubbles, Drops and Particles,* Academic, New York, 1978.]

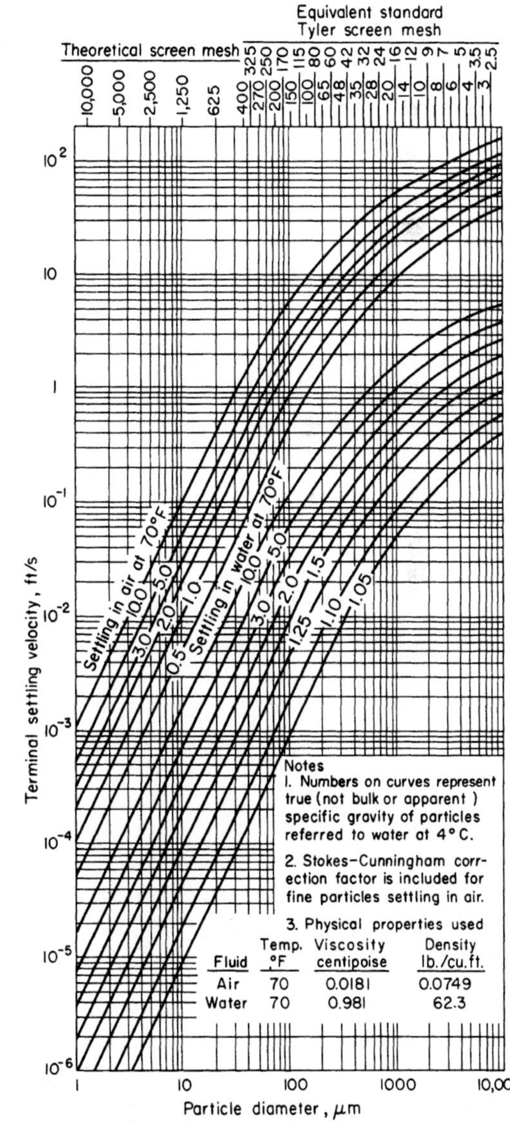

FIG. 6-56 Terminal velocities of spherical particles of different densities settling in air and water at 70°F under the action of gravity. To convert ft/s to m/s, multiply by 0.3048. [From C. E. Lapple et al., *Fluid and Particle Mechanics,* University of Delaware, Newark, 1951, p. 292.]

The correlation is represented by

$$J = 0.94H^{0.757} \qquad (2 < H \le 59.3) \tag{6-265}$$

$$J = 3.42H^{0.441} \qquad (H > 59.3) \tag{6-266}$$

where
$$H = \frac{4}{3}\text{Eo}M^{-0.149}\left(\frac{\mu}{\mu_w}\right)^{-0.14} \tag{6-267}$$

$$J = \text{Re}\, M^{0.149} + 0.857 \tag{6-268}$$

The terminal velocity may be evaluated explicitly from

$$u = \frac{\mu}{\rho d} M^{-0.149}(J - 0.857) \tag{6-269}$$

In Eq. (6-267), μ = viscosity of continuous liquid and μ_w = viscosity of water, taken as 0.0009 Pa·s (0.9 cP). This correlation neglects the effect of drop phase viscosity.

For drop velocities in nonnewtonian liquids, see V. Mhatre and R. C. Kinter [*Ind. Eng. Chem.* **51:** 865–867 (1959)]; G. Marrucci et al. [*AIChE J.* **16:** 538–541 (1970)]; and V. Mohan et al. [*Can. J. Chem. Eng.* **50:** 37–40 (1972)].

Liquid Drops in Gases Liquid drops falling in stagnant gases appear to remain spherical and follow the rigid sphere drag relationships up to a Reynolds number of about 100. Large drops will deform, with a resulting increase in drag, and in some cases will shatter. The largest water drop that will fall in air at its terminal velocity is about 8 mm (0.32 in) in diameter, with a corresponding velocity of about 9 m/s (30 ft/s). Drops shatter when the Weber number, defined as

$$\text{We} = \frac{\rho_G u^2 d}{\sigma} \tag{6-270}$$

exceeds a critical value. Here, ρ_G = gas density, u = drop velocity, d = drop diameter, and σ = surface tension. A value of $\text{We}_c = 13$ is often cited for the critical Weber number.

Terminal velocities for water drops in air have been correlated by E. X. Berry and M. R. Pranger [*J. Appl. Meteorol.* **13:** 108–113 (1974)] as

$$\text{Re} = \exp\,[-3.126 + 1.013 \ln N_D - 0.01912(\ln N_D)^2] \tag{6-271}$$

TABLE 6-11 Wall Correction Factor for Rigid Spheres in Stokes' Law Region

β*	k_w	β	k_w
0.0	1.000	0.4	0.279
0.05	0.885	0.5	0.170
0.1	0.792	0.6	0.0945
0.2	0.596	0.7	0.0468
0.3	0.422	0.8	0.0205

*β = particle diameter divided by vessel diameter.
SOURCE: From Haberman and Sayre, *David W. Taylor Model Basin Report* 1143, 1958.

for $2.4 < N_D < 10^7$ and $0.1 < \text{Re} < 3550$. The dimensionless group N_D (often called the *Best* number) [R. Clift, J. R. Grace, and M. E. Weber, *Bubbles, Drops and Particles,* Academic, New York, 1978] is given by

$$N_D = \frac{4\rho\Delta\rho g d^3}{3\mu^2} \tag{6-272}$$

and is proportional to the similar Archimedes and Galileo numbers.

Figure 6-56 gives calculated settling velocities for solid spherical particles settling in air or water using the standard drag coefficient curve for spherical particles. For fine particles settling in air, the *Stokes-Cunningham correction* has been applied to account for particle size comparable to the mean free path of the gas. The correction is less than 1 percent for particles larger than 16 μm settling in air. Smaller particles are also subject to *Brownian motion.* Motion of particles smaller than 0.1 μm is dominated by Brownian forces and gravitational effects are small.

Wall Effects When the diameter of a settling particle is significant compared to the diameter of the container, the settling velocity is reduced. For rigid spherical particles settling with Re < 1, the correction given in Table 6-11 may be used. The factor k_w is multiplied by the settling velocity obtained from Stokes' law to obtain the corrected settling rate. For values of diameter ratio β = particle diameter/vessel diameter less than 0.05, $k_w = 1/(1 + 2.1\beta)$ [F. A. Zenz and D. F. Othmer, *Fluidization and Fluid-Particle Systems,* Reinhold, New York, 1960, pp. 208–209]. In the range 100 < Re < 10,000, the computed terminal velocity for rigid spheres may be multiplied by k'_w to account for wall effects, where k'_w is given by [T. Z. Harmathy, *AIChE J.* **6:** 281 (1960)]

$$k'_w = \frac{1-\beta^2}{\sqrt{1+\beta^4}} \tag{6-273}$$

For gas bubbles in liquids, there is little wall effect for β < 0.1. For β > 0.1, see S. Uto and R. C. Kintner [*AIChE J.* **2:** 420–424 (1956)], C. C. Maneri and H. D. Mendelson [*Chem. Eng. Prog.* **64,** *Symp. Ser.* **82:** 72–80 (1968)], and R. Collins [*J. Fluid Mech.* **28:** part 1, 97–112 (1967)].

Section 7

Reaction Kinetics

Tiberiu M. Leib, Ph.D. *Principal Consultant, The Chemours Company (Retired); Fellow, American Institute of Chemical Engineers (Section Editor)*

Carmo J. Pereira, Ph.D., M.B.A. *DuPont Fellow, E. I. du Pont de Nemours and Company; Fellow, American Institute of Chemical Engineers (Section Editor)*

John R. Richards, Ph.D. *Research Fellow, E. I. du Pont de Nemours and Company (Retired); Fellow, American Institute of Chemical Engineers (Polymerization Reactions)*

The editors would like to thank Stanley M. Walas, Ph.D., Professor Emeritus (deceased), Department of Chemical and Petroleum Engineering, University of Kansas (Fellow, American Institute of Chemical Engineers), for editing this section in previous editions; Dennie T. Mah, M.S.Ch.E., Senior Consultant (retired), E. I. du Pont de Nemours and Company (Senior Member, American Institute of Chemical Engineers; Member, Industrial Electrolysis and Electrochemical Engineering; Member, The Electrochemical Society), for his contributions to the Electrochemical Reactions subsection in edition 8 that was reviewed and carried over to the current edition; and John Villadsen, Ph.D., Professor Emeritus, Department of Chemical Engineering, Technical University of Denmark, for his contributions to the Biocatalysis and Biochemical Reactions subsections in edition 8 that were reviewed and carried over to the current edition.

GAS-LIQUID REACTIONS

GAS-LIQUID-SOLID REACTIONS

DETERMINATION OF MECHANISM AND KINETICS

Nomenclature and Units

The component A is identified by the subscript a. Thus, the number of moles is n_a; the fractional conversion is X_a; the extent of reaction is ξ_a; the partial pressure is p_a; the rate of reaction is r_a; the molar flow rate is N_a; the volumetric flow rate is q; the reactor volume is V_r or simply V for batch reactors; the volumetric concentration is $C_a = n_a/V$ or $C_a = N_a/q$; the total pressure is P; and the temperature is T. Throughout this section, equations are presented without specification of units. Use of any consistent unit set is appropriate.

Following is a listing of typical nomenclature expressed in SI and U.S. Customary System units.

Symbol	Definition	SI units	U.S. Customary System units
$A, B, C, \ldots$	Names of substances		
a	Activity, gas-liquid interfacial area per unit volume of reactor		
BR	Batch reactor		
b	Estimate of kinetic parameters, vector		
C_a	Concentration of substance A	kg · mol/m³	lb · mol/ft³
CSTR	Continuous stirred tank reactor		
C_0	Initial or inlet concentration	kg · mol/m³	lb · mol/ft³
c_p	Heat capacity at constant pressure	kJ/(kg · mol · K)	Btu/(lb · mol · °F)
Δc_p	Heat capacity change in a reaction	kJ/(kg · mol · K)	Btu/(lb · mol · °F)
D	Diffusivity, dispersion coefficient	m²/s	ft²/s
D_e	Effective diffusivity	m²/s	ft²/s
D_K	Knudsen diffusivity	m²/s	ft²/s
E	Activation energy, enhancement factor for gas-liquid mass transfer with reaction, electrochemical cell potential		
F, F_1	Faraday constant, F statistic, monomer A mol fraction in polymer		
f_1, f_2	Efficiency of initiation in polymerization, monomer A and B mole fractions		
ΔG	Free energy change	kJ/kg · mol	Btu/lb · mol
Ha	Hatta number		
He	Henry constant for absorption of gas in liquid		
ΔH_r	Heat of reaction	kJ/kg · mol	Btu/lb · mol
I	Initiator species for polymerization, modified Bessel functions, electric current		
j	Electric current density	A/m²	
K_a	Adsorption constant		
K_e	Chemical equilibrium constant		
k	Specific rate constant of reaction, mass transfer coefficient		
L	Length of path in reactor	m	ft
LFSS	Lack of fit sum of squares		
M	Average molecular weight in polymers, dead polymer species, monomer species		
MWD	Molecular weight distribution in polymers		
m	Number of moles in electrochemical reaction	kg · mol	lb · mol
N	Molar flow rate, molar flux		
NCLD	Number chain length distribution		
NMWD	Number molecular weight distribution		
n	Number of stages in a network of CSTRs, reaction order, number of electrons in electrochemical reaction, number of experiments		
n_a	Number of moles of A	kg · mol	lb · mol
n_t	Total number of moles	kg · mol	lb · mol
P	Total pressure	Pa	psi
P	Live polymer species		
PD	Polydispersity in polymers		
PESS	Pure error sum of squares		
PFR	Plug flow reactor		
p	Number of kinetic parameters		
p_a	Partial pressure of substance A	Pa	psi
q	Volumetric flow rate	m³/s	ft³/s
Q	Electric charge	C	
Q	Live polymer species		
R	Radial position, radius, universal gas constant, initiator radical species		
Re	Reynolds number		

Symbol	Definition	SI units	U.S. Customary System units
RgSS	Regression sum of squares		
RSS	Residual sum of squares		
r_a, r_1, r_2	Rate of reaction per unit volume, reactivity ratios in polymerization		
S	Selectivity, stoichiometric matrix, objective function for parameter estimation		
SBR	Semibatch reactor		
Sc	Schmidt number		
Sh	Sherwood number		
ΔS	Entropy change	kJ/(kg · mol · K)	Btu/(lb · mol · °R)
s	Estimate of variance		
t	Time, t statistic		
u	Linear velocity	m/s	ft/s
V	Volume of reactor, variance-covariance matrix		
υ	Molar volume	m³/kg · mol	ft³/lb · mol
W	Condensation product species in polymerization		
WCLD	Weight chain length distribution		
WMWD	Weight molecular weight distribution		
w	Molecular weight of repeat unit in polymers		
X	Linear model matrix for parameter estimation, fractional conversion		
X_a	$1 - f_a = 1 - C_a/C_{a0}$ or $1 - n_a/n_{a0}$, fractional conversion of component A		
x	Axial position in a reactor, mole fraction in liquid		
Y	Yield, yield coefficient for biochemical reactions		
y	Mole fraction in gas, dependent variable		
z	x/L, normalized axial position		
	Greek letters		
α	Fraction of initial catalyst activity, probability of propagation for chain polymerization, conversion for step polymerization, confidence level		
β	r/R, normalized radial position, fraction of poisoned catalyst, kinetic parameter vector		
δ	Film thickness or boundary layer thickness, relative change in number of moles by reaction		
$\delta(t)$	Unit impulse input, Dirac function		
ε	Relative change in number of moles by reaction, residual error, catalyst particle porosity, current efficiency		
Φ	Weisz Prater parameter		
φ	Thiele modulus		
η	Effectiveness factor of porous catalyst, over potential in electrochemical reactions		
λ	Parameter for instantaneous gas-liquid reaction, moments in polymer chain length		
μ	Viscosity, biomass growth rate, average chain length or degree of polymerization in polymers		
ν	Stoichiometric coefficient, kinematic viscosity, fraction of surface covered by adsorbed species		
θ	Dimensionless time		
ρ	Density	kg/m³	lbm/ft³
σ	Variance		
τ	Residence time, tortuosity factor		
ξ	Extent of reaction		

Nomenclature and Units *(Continued)*

Symbol	Definition	SI units	U.S. Customary System units
	Subscripts		
act	Activation		
anode	At anode		
B	Bed		
cathode	At cathode		
cell	Electrochemical cell		
D	Diffusion, dispersion		
d	Deactivation		
e	Equilibrium		
f	Forward reaction, final, formation		
G	Gas		
i	Component i		
j	Reaction j, chain length in polymers, current		
L	Liquid		
m	Based on mass, mass transfer, chain length in polymers		
max	Maximum biomass growth rate, maximum extent of reaction		
n	Chain length in polymers, number average		
o	Oxidized		
obs	Observed		
p	Particle		
projected	Electrode projected area		
r	Reverse reaction, reduced		
S	Substrate		
s	Solid or catalyst, saturation, surface		
surf	Surface		
υ	Based on volume, vacant		
w	Weight average in polymers		
x	Biomass		
0	At initial or inlet conditions, as in C_{a0}, n_{a0}, V'_0, at reference temperature, in the bulk phase		
1/2	Half-life		
	Superscripts		
e	Equilibrium		
inst	Instantaneous		
o	At reference temperature		
T	Transposed matrix		

REFERENCES

General References: Amundson, *Mathematical Methods in Chemical Engineering—Matrices and Their Application,* Prentice-Hall International, New York, 1966; Aris, *Elementary Chemical Reactor Analysis,* Prentice-Hall, New York, 1969; Astarita, *Mass Transfer with Chemical Reaction,* Elsevier, New York, 1967; Bamford and Tipper (eds.), *Comprehensive Chemical Kinetics,* Elsevier, New York, 1969; Bird, Stewart, and Lightfoot, *Transport Phenomena,* 2d ed., Wiley, New York, 2002; Boudart, *Kinetics of Chemical Processes,* Prentice-Hall, New York, 1968; Boudart and Djega-Mariadassou, *Kinetics of Heterogeneous Catalytic Reactions,* Princeton University Press, Princeton, N.J., 1984; Brotz, *Fundamentals of Chemical Reaction Engineering,* Addison-Wesley, Boston, 1965; Butt, *Reaction Kinetics and Reactor Design,* Prentice-Hall, New York, 1980; Butt and Petersen, *Activation, Deactivation and Poisoning of Catalysts,* Academic Press, 1988; Capello and Bielski, *Kinetic Systems: Mathematical Description of Kinetics in Solution,* Wiley, 1972; Carberry, *Chemical and Catalytic Reaction Engineering,* McGraw-Hill, New York, 1976; Carberry and Varma (eds.), *Chemical Reaction and Reactor Engineering,* Marcel Dekker, New York, 1987; Chen, *Process Reactor Design,* Allyn & Bacon, Boston, 1983; Churchill, *The Interpretation and Use of Rate Data: The Rate Concept,* McGraw-Hill, New York, 1974; Cooper and Jeffreys, *Chemical Kinetics and Reactor Design,* Prentice-Hall, 1971; Cremer and Watkins (eds.), *Chemical Engineering Practice,* vol. 8: *Chemical Kinetics,* Butterworths, 1965; Davis and Davis, *Fundamentals of Chemical Reaction Engineering,* McGraw-Hill, New York, 2003; Delmon and Froment, *Catalyst Deactivation,* Elsevier, Amsterdam, Netherlands, 1980; Denbigh and Turner, *Chemical Reactor Theory: An Introduction,* Cambridge University Press, Cambridge, 1971; Denn, *Process Modeling,* Longman, New York, 1986; Fogler, *Elements of Chemical Reaction Engineering,* 4th ed., Prentice-Hall, 2006; Froment and Bischoff, *Chemical Reactor Analysis and Design,* Wiley, 1990; Froment and Hosten, "Catalytic Kinetics—Modeling," in *Catalysis—Science and Technology,* Springer Verlag, New York, 1981; Harriott, *Chemical Reactor Design,* Marcel Dekker, 2003; Hill, *An Introduction to Chemical Engineering Kinetics and Reactor Design,* 2d ed., Wiley, 1990; Holland and Anthony, *Fundamentals of Chemical Reaction Engineering,* Prentice-Hall, 1989; Kafarov, *Cybernetic Methods in Chemistry and Chemical Engineering,* Mir Publishers, Moscow, 1976; Laidler, *Chemical Kinetics,* Harper & Row, 1987; Lapidus and Amundson (eds.), *Chemical Reactor Theory—A Review,* Prentice-Hall, 1977; Levenspiel, *Chemical Reaction Engineering,* 3d ed., Wiley, 1999; Lewis (ed.), *Techniques of Chemistry,* vol. 4: *Investigation of Rates and Mechanisms of Reactions,* Wiley, 1974; Masel, *Chemical Kinetics and Catalysis,* Wiley, 2001; Naumann, *Chemical Reactor Design,* Wiley, 1987; Panchenkov and Lebedev, *Chemical Kinetics and Catalysis,* Mir Publishers, Moscow, 1976; Petersen, *Chemical Reaction Analysis,* Prentice-Hall, 1965; Rase, *Chemical Reactor Design for Process Plants: Principles and Case Studies,* Wiley, 1977; Rose, *Chemical Reactor Design in Practice,* Elsevier, Amsterdam, Netherlands, 1981; Satterfield, *Heterogeneous Catalysis in Practice,* McGraw-Hill, 1991; Schmidt, *The Engineering of Chemical Reactions,* Oxford University Press, 1998; Smith, *Chemical Engineering Kinetics,* McGraw-Hill, 1981; Steinfeld, Francisco, and Hasse, *Chemical Kinetics and Dynamics,* Prentice-Hall, 1989; Ulrich, *A Guide to Chemical Engineering Reactor Design and Kinetics,* Ulrich Research and Consulting, Durham, 1993; Van Santen and Neurock, *Molecular Heterogeneous Catalysis: A Conceptual and Computational Approach,* Wiley, 2006; Van Santen and Niemantsverdriet, *Chemical Kinetics and Catalysis, Fundamental and Applied Catalysis,* Plenum Press, New York, 1995; van't Riet and Tramper, *Basic Bioreactor Design,* Marcel Dekker, 1991; Walas, *Reaction Kinetics for Chemical Engineers,* McGraw-Hill, 1959, reprint, Butterworths, 1989; Walas, *Chemical Reaction Engineering Handbook of Solved Problems,* Gordon & Breach Publishers, Philadelphia, Pa., 1995; Westerterp, van Swaaij, and Beenackers, *Chemical Reactor Design and Operation,* Wiley, 1984.

References for Laboratory Reactors: Berty, Laboratory reactors for catalytic studies, in Leach (ed.), *Applied Industrial Catalysis,* vol. 1, Academic Press, 1983, pp. 41–57; Berty, *Experiments in Catalytic Reaction Engineering,* Elsevier, Amsterdam, 1999; Danckwerts, *Gas-Liquid Reactions,* McGraw-Hill, 1970; Hoffmann, "Industrial Process Kinetics and Parameter Estimation," in *ACS Advances in Chemistry* **109:** 519–534 (1972); Hoffman, "Kinetic Data Analysis and Parameter Estimation," in de Lasa (ed.), *Chemical Reactor Design and Technology,* Martinus Nijhoff, Boston, 1986, pp. 69–105; Horak and Pasek, *Design of Industrial Chemical Reactors from Laboratory Data,* Heiden, Philadelphia, Pa., 1978; Rase, *Chemical Reactor Design for Process Plants,* Wiley, 1977, pp. 195–259; Shah, *Gas-Liquid-Solid Reactor Design,* McGraw-Hill, 1979, pp. 149–179; Charpentier, "Mass Transfer Rates in Gas-Liquid Absorbers and Reactors," in Drew et al., eds., *Advances in Chemical Engineering,* vol. 11, Academic Press, 1981.

INTRODUCTION

The mechanism and corresponding kinetics provide the rate at which the chemical or biochemical species in the reactor system react at the prevailing conditions of temperature, pressure, composition, mixing, flow, heat, and mass transfer. Observable kinetics represent the true intrinsic chemical kinetics only when competing phenomena such as transport of mass and heat are not limiting the rates. The intrinsic chemical mechanism and kinetics are unique to the reaction system. Knowledge of the intrinsic kinetics therefore facilitates reactor selection, choice of optimal operating conditions, and reactor scale-up and design, when combined with understanding of the associated physical and transport phenomena for different reactor scales and types.

This section covers the following key aspects of reaction kinetics:

- Chemical mechanism of a reaction system and its relation to kinetics
- Intrinsic kinetic rates using equations that can be correlative, lumped, or based on detailed elementary kinetics
- Catalytic kinetics
- Effect of mass and heat transfer on kinetics in heterogeneous systems
- Intrinsic kinetic rates from experimental data and/or from theoretical calculations
- Kinetic parameter estimation

The use of reaction kinetics for analyzing and designing suitable reactors is discussed in Sec. 19.

BASIC CONCEPTS

MECHANISM

The mechanism describes the reaction steps and the relationship between the reaction rates of the chemical components. A single chemical reaction includes reactants *A, B,* ... and products *R, S,* ...

$$\nu_a A + \nu_b B + \cdots \Leftrightarrow \nu_r R + \nu_s S + \cdots \tag{7-1}$$

where ν_i are the *stoichiometric coefficients* of components *A, B,* ..., i.e., the relative number of molecules of *A, B,* ... that participate in the reaction. For instance, the HBr synthesis has the global stoichiometry $H_2 + Br_2 \Leftrightarrow 2HBr$.

The *stoichiometry* of the reaction defines the elemental balance (atoms of H and Br, for instance) and therefore relates the number of molecules of reactants and products participating in the reaction. The stoichiometric coefficients are not unique for a given reaction, but their ratios are unique. For instance, for the HBr synthesis above we could have written the stoichiometric equation $½H_2 + ½Br_2 \Leftrightarrow HBr$ as well.

Often several reactions occur simultaneously, resulting in a *network of reactions*. When the network is broken down into *elementary* or *single-event* steps (such as a single electron transfer), the network represents the *true mechanism* of the chemical transformations leading from initial reactants to final products through intermediates. The intermediates can be molecules, ions, free radicals, transition state complexes, and other moieties. A network of *global reactions*, with each reaction representing the combination of a number of elementary steps, does not represent the true mechanism of the chemical transformation but is still useful for global reaction rate calculations, albeit empirically. The stoichiometry can only be written in a unique manner for elementary reactions, since, as shown later, the reaction rate for elementary reactions is determined directly by the stoichiometry through the concept of the law of mass action.

REACTION RATE

The *specific rate* of consumption or production of any reaction species *i*, denoted r_i, is the rate of change of the number of molecules of species *i* with time per unit volume of reaction medium:

$$r_i = \frac{1}{V}\frac{dn_i}{dt} \tag{7-2}$$

The rate is negative when *i* represents a *reactant* (dn_i/dt is negative since n_i is decreasing with time) and positive when *i* represents a *product* (dn_i/dt is positive since n_i is increasing with time). The specific rate of a reaction, e.g., that in Eq. (7-1), is defined as

$$\begin{aligned} r &= -r_i/\nu_i \quad \text{for reactants} \\ r &= r_i/\nu_i \quad \text{for products} \end{aligned} \tag{7-3}$$

By this definition, the specific rate of reaction is uniquely defined, and its sign is always positive. Conversely, the rate of reaction of each component or species participating in the reaction is the specific reaction rate multiplied by the species' stoichiometric coefficient with the corrected sign (negative for reactants, positive for products).

CLASSIFICATION OF REACTIONS

Reactions can be classified in several ways. On the basis of *mechanism* they may be

1. *Irreversible*, i.e., the reverse reaction rate is negligible: $A + B \Rightarrow C + D$, e.g., CO oxidation: $CO + \frac{1}{2}O_2 \Rightarrow CO_2$
2. *Reversible*: $A + B \Rightarrow C + D$, e.g., the water-gas shift $CO + H_2O \Leftrightarrow CO_2 + H_2$
3. *Equilibrium*, a special case with zero net rate, i.e., with the forward and reverse reaction rates of a reversible reaction being equal. All reversible reactions, if left to go to completion, end in equilibrium.
4. *Networks of simultaneous reactions*, i.e., consecutive, parallel, complex (combination of consecutive and parallel reactions):

$$A + B \Rightarrow C + D \qquad C + E \Rightarrow F + G$$

for example two-step hydrogenation of acetylene to ethane

$$CH{\equiv}CH + H_2 \Rightarrow CH_2{=}CH_2 \qquad CH_2{=}CH_2 + H_2 \Rightarrow CH_3CH_3$$

A further classification is from the point of view of the number of reactant molecules participating in the reaction, or the *molecularity*:

1. *Unimolecular*: $A \Rightarrow B$, e.g., isomerization of ortho-xylene to para-xylene, *O*-xylene ⇒ *P*-xylene, or $A \Rightarrow B + C$, e.g., decomposition $CaCO_3 \Rightarrow CaO + CO_2$
2. *Bimolecular*: $A + B \Rightarrow C$ or $2A \Rightarrow B$ or $A + B \Rightarrow C + D$, e.g., $C_2H_4 + H_2 \Rightarrow C_2H_6$
3. *Trimolecular*: $A + B + C \Rightarrow D$ or $3A \Rightarrow B$

This last classification has fundamental meaning only when considering elementary reactions, i.e., reactions that constitute a single chemical transformation or a single event, such as a single electron transfer. For elementary reactions, molecularity is rarely higher than 2. Often elementary reactions are not truly unimolecular, since in order for the reaction to occur, energy is required and it can be obtained through collision with other molecules such as an inert solvent or gas. Thus the unimolecular reaction $A \Rightarrow B$ in reality could be represented as a bimolecular reaction $A + X \Rightarrow B + X$, i.e., *A* collides with *X* to produce *B* and *X*, and thus no net consumption of *X* occurs.

Reactions can be further classified according to the *phases* present. Examples for the more common cases are

1. *Homogeneous gas*, e.g., methane combustion
2. *Homogeneous liquid*, e.g., acid/base reactions to produce soluble salts
3. *Heterogeneous gas-solid*, e.g., HCN synthesis from NH_3, CH_4, and air on a solid catalyst
4. *Heterogeneous gas-liquid*, e.g., absorption of CO_2 in amine solutions
5. *Heterogeneous liquid-liquid*, e.g., reaction in immiscible organic and aqueous phases such as synthesis of adipic acid from cyclohexanone and nitric acid
6. *Heterogeneous liquid-solid*, e.g., reaction of limestone with sulfuric acid to make gypsum
7. *Heterogeneous solid-solid*, e.g., self-propagating high-temperature synthesis (SHS) of inorganic pure oxides
8. *Heterogeneous gas-liquid-solid*, e.g., catalytic Fischer-Tropsch synthesis of hydrocarbons on solid catalyst from CO and H_2
9. *Heterogeneous gas-liquid-liquid*, e.g., oxidations or hydrogenations with phase transfer catalysts

Reactions can also be classified with respect to the *mode of operation* in the reaction system (e.g. versus the mode of temperature control) as

1. *Isothermal constant volume* (batch)
2. *Isothermal constant pressure* (continuous)
3. *Adiabatic*
4. *Nonisothermal temperature-controlled (by cooling or heating), batch or continuous*

EFFECT OF CONCENTRATION ON RATE

The *concentration* of the reaction components determines the rate of reaction. For instance, for the irreversible reaction

$$pA + qB \Rightarrow rC + sD \tag{7-4}$$

the rate can be represented *empirically* as a power law function of the reactant concentrations such as

$$r = k\,C_a^m C_b^n \qquad C_i = \frac{n_i}{V} \tag{7-5}$$

The exponents *m* and *n* represent the *order of the reaction* with respect to components *A* and *B*, and the *sum* $m + n$ represents the *overall order of the reaction*. The order can be a positive, zero, or negative number indicating that the rate increases, is independent of, or decreases with an increase in a species concentration, respectively. The exponents can be whole (integral order) or fraction (fractional order). In Eq. (7-5) *k* is the *specific rate constant* of the reaction, and it is independent of concentrations for elementary reactions only. For global reactions consisting of several elementary steps, *k* may still be constant over a narrow range of compositions and operating conditions and therefore can be considered independent of concentration for limited practical purposes. A further complexity arises for nonideal chemical solutions where activities have to be used instead of concentrations. In this case the rate constant can be a function of composition even for elementary steps (see, for instance, Froment and Bischoff, *Chemical Reactor Analysis and Design*, 2d ed., Wiley, 1990).

When Eq. (7-4) represents a global reaction combining a number of elementary steps, then rate equation (7-5) represents an empirical correlation of the global or overall reaction rate. In this case, exponents *m* and *n* have no clear physical meaning other than indicating the overall effect of the various concentrations on rate, and they do not have any obvious relationship to the stoichiometric coefficients *p* and *q*. This is not so for elementary reactions, as shown in the next subsection. Also, as shown later, power law and rate expressions other than power law (e.g., hyperbolic) can be developed for specific reactions by starting with the mechanism of the elementary steps and making simplifying assumptions that are valid under certain conditions.

LAW OF MASS ACTION

As indicated above, the dependence of rate on concentration can be shown to be of the general form

$$r = kf(C_a, C_b, \ldots) \tag{7-6}$$

For elementary reactions, the *law of mass action* states that the rate is proportional to the concentrations of the reactants raised to the power of their respective molecularity. Thus for an elementary irreversible reaction such as Eq. (7-4) the rate equation is

$$r = k\,C_a^p\,C_b^q \tag{7-7}$$

Hence, the exponents p and q of Eq. (7-7) are the stoichiometric coefficients when the stoichiometric equation truly represents the mechanism of reaction, i.e., when the reactions are elementary. As discussed above, the exponents m and n in Eq. (7-5) identify the order of the reaction, while the stoichiometric coefficients p and q in Eq. (7-7) identify the molecularity—for elementary reactions these are the same.

EFFECT OF TEMPERATURE

The *Arrhenius equation* relates the specific rate constant to the absolute temperature

$$k = k_0 \exp\left(-\frac{E}{RT}\right) \tag{7-8}$$

where E is called the *activation energy* and k_0 is the *preexponential* factor. As seen from Eq. (7-8), the rate can be a very strongly increasing (exponential) function of temperature, depending on the magnitude of the activation energy E. This equation works well for elementary reactions, and it also works reasonably well for global reactions over a relatively narrow range of temperature in the absence of mass transfer limitations. The Arrhenius form represents an energy barrier on the reaction pathway between reactants and products that has to be overcome by the reactant molecules.

The Arrhenius equation can be derived from theoretical considerations using either of two competing theories: *collision theory* and *transition state theory*. A more accurate form of Eq. (7-8) includes an additional temperature factor

$$k = k_0 T^m \exp\left(-\frac{E}{RT}\right) \qquad 0 < m < 1 \tag{7-9}$$

but the T^m factor is often neglected because of the usually much stronger dependence on temperature of the exponential factor in Eq. (7-9), as m is usually small. When m is larger, as it can be for complex molecules, then the T^m term has to be taken into consideration. For more details, see Masel, *Chemical Kinetics and Catalysis*, Wiley, 2001; and Levenspiel, *Chemical Reaction Engineering*, 3d ed., Wiley, 1999.

HEAT OF REACTION

Chemical reactions are accompanied by evolution or absorption of energy. The enthalpy change (difference between the total enthalpy of formation of the products and that of the reactants) is called the *heat of reaction* ΔH_r:

$$\Delta H_r = (\nu_r H_{fr} + \nu_s H_{fs} + \cdots) - (\nu_a H_{fa} + \nu_b H_{fb} + \cdots) \tag{7-10}$$

where H_{fi} are the *enthalpies of formation* of components i. The reaction is *exothermic* if heat is produced by the reaction (negative heat of reaction) and *endothermic* if heat is consumed (positive heat of reaction). The heat of reaction depends upon the temperature range and the phases of the reactants and product. To estimate the dependence of the heat of reaction on temperature relative to a reference temperature T_0, the following expression can be used, provided there is no phase change:

$$\begin{gathered} \Delta H_r(T) = \Delta H_r(T_0) + \int_{T_0}^{T} \Delta c_p\, dT \\ \Delta c_p = (\nu_r c_{pr} + \nu_s c_{ps} + \cdots) - (\nu_a c_{pa} + \nu_b c_{pb} + \cdots) \end{gathered} \tag{7-11}$$

where the c_{pi} are the constant-pressure *heat capacities* of component i. The heat of reaction can be measured by using calorimetry, or it can be calculated by using a variety of thermodynamic methods out of the scope of this chapter (see Sec. 4 of this handbook, thermodynamic text books, and Bird, Stewart, and Lightfoot, *Transport Phenomena*, 2d ed., Wiley, New York, 2002). It is important to accurately capture the energy balance and its relation to the heat of reaction and heat capacities (see also Denn, *Process Modeling*, Longman, New York, 1986, for correct formulations). The coupling of the heat of reaction with the reaction rate often has a dominating effect on reactor selection and control, and on the laboratory reactor setup required to obtain accurate intrinsic kinetics and mechanisms. A more detailed discussion of this topic can be found in Sec. 19.

CHEMICAL EQUILIBRIUM

Often reactions or reaction steps in a network of reactions are at chemical equilibrium; i.e., the rate of the forward reaction equals the rate of the reverse reaction. For instance, for the reversible reaction

$$\nu_a A + \nu_b B \underset{k_r}{\overset{k_f}{\Leftrightarrow}} \nu_r R + \nu_s S \tag{7-12}$$

with mass action kinetics, the rate may be written as

$$r = r_f - r_r = k_f C_a^{\nu_a} C_b^{\nu_b} - k_r C_r^{\nu_r} C_s^{\nu_s} \tag{7-13}$$

At chemical equilibrium the forward and reverse reaction rates are equal:

$$r = r_f - r_r = 0 \tag{7-14}$$

The equilibrium constant K_e (based on volumetric concentrations) is defined as the ratio of the forward and reverse rate constants and is related to the composition at equilibrium as follows:

$$K_e = \frac{k_f}{k_r} = \frac{C_{r,e}^{\nu_r} C_{s,e}^{\nu_s}}{C_{a,e}^{\nu_a} C_{b,e}^{\nu_b}} \tag{7-15}$$

The equilibrium constant K_e can be calculated from the free energy change of the reaction. Using the van't Hoff relation, we obtain the dependence of K_e on temperature:

$$\frac{d(\ln K_e)}{dT} = \frac{\Delta H_r}{RT^2} \tag{7-16}$$

Integrating with respect to temperature, we obtain a form similar to the Arrhenius expression of the rate constant for a narrow range of temperature with ΔH_r assumed constant:

$$K_e = K_{e0} \exp\left(-\frac{\Delta H_r}{RT}\right) \tag{7-17}$$

A more general integral form of Eq. (7-16) is

$$\ln K_e(T) = \ln K_e(T_0) + \frac{1}{R}\int_{T_0}^{T} \frac{\Delta H_r(T_0) + \int_{T_0}^{T} \Delta c_p\, dT}{T^2}\, dT \tag{7-18}$$

When a reversible reaction is not at equilibrium, knowledge of K_e can be used to eliminate the rate constant of the reverse reaction by using Eq. (7-15) as follows:

$$r = k_f (C_a^{\nu_a} C_b^{\nu_b} - C_r^{\nu_r} C_s^{\nu_s} / K_e) \tag{7-19}$$

When several reversible reactions occur simultaneously, each reaction r_j is characterized by its equilibrium constant K_{ej}. When the K_{ej} are known, the composition at equilibrium can be calculated from a set of equations such as Eq. (7-15) for each reaction. At equilibrium, according to the principle of *microscopic reversibility* or *detailed balancing*, each reaction in the network is at equilibrium.

CONVERSION, EXTENT OF REACTION, SELECTIVITY, AND YIELD

Conversion of a reactant is the number of moles converted per initial or feed moles of a reactant. Thus for component A

$$X_a = 1 - \frac{n_a}{n_{a0}} \tag{7-20}$$

A *limiting reactant* is a reactant whose concentration at the start of the reaction is the least of all reactants relative to the required stoichiometric amount needed for complete conversion. For instance, for the single reaction in Eq. (7-12), A is the limiting reactant if the initial molar ratio of concentrations of A and B is less than the ratio of their stoichiometric coefficients:

$$\frac{n_{a0}}{n_{b0}} < \frac{\nu_a}{\nu_b} \tag{7-21}$$

Once the limiting reactant is depleted, the respective reaction stops even though other (nonlimiting) reactants may still be abundant.

For each reaction or each step in a network of reactions, a unique *extent of reaction* ξ that relates the composition of components that participate in the reaction to one another can be defined. For instance, for the single reaction in Eq. (7-1):

$$\xi = \frac{n_{a0} - n_a}{\nu_a} = \frac{n_{b0} - n_b}{\nu_b} = \cdots = -\frac{n_{r0} - n_r}{\nu_r} = -\frac{n_{s0} - n_s}{\nu_s} = \cdots \tag{7-22}$$

The extent of reaction is related to conversion as follows:

$$\xi = \frac{X_a n_{a0}}{\nu_a} = \frac{X_b n_{b0}}{\nu_b} = \cdots \tag{7-23}$$

When A is the limiting reactant as in Eq. (7-21), the maximum extent of reaction (with A fully converted) is

$$\xi_{\max} = \frac{n_{a0}}{\nu_a} \tag{7-24}$$

For multiple reactions with reactants participating in more than one reaction, it is more difficult to determine the limiting reactant, and often it is necessary to calculate the concentration as the reactions proceed to determine which reactant is consumed first. When the limiting reactant is depleted, all reactions that use this component as reactant stop, and the corresponding rates become zero.

Selectivity S of a product is the ratio of the rate of production of that product to the rate of production of all products combined. For a single reaction selectivity is trivial—if more than one product occurs, then the selectivity of each product is the ratio of the stoichiometric coefficient of that product to the sum of stoichiometric coefficients of all the products. Thus for the reaction in Eq. (7-1)

$$S_r = \frac{\nu_r}{\nu_r + \nu_s + \cdots} \tag{7-25}$$

The selectivity of product R for a network of reactions, with all the reactions making the various products included, is

$$S_r = \frac{r_r}{\sum\limits_{\text{all products } i} r_i} \tag{7-26}$$

For instance, for the network of reactions $A + B \overset{1}{\Rightarrow} C + D$ and $C + E \overset{2}{\Rightarrow} F + G$, the selectivity to product C is

$$S_C = \frac{r_c}{r_c + r_d + r_f + r_g} = \frac{r_1 - r_2}{(r_1 - r_2) + r_1 + r_2 + r_2} = \frac{r_1 - r_2}{2r_1 + r_2}$$

The *yield* Y of a product R with respect to a reactant A is the ratio of the rate of production of R to that of consumption of A:

$$Y_r = \frac{r_r}{-r_a} \tag{7-27}$$

For a single reaction the yield is trivial, and Eq. (7-27) simplifies to the ratio of the respective stoichiometric coefficients:

$$Y_r = \frac{\nu_r}{-\nu_a} \tag{7-28}$$

The yield quantifies the efficiency of the respective reactant utilization to make the desired products.

CONCENTRATION TYPES

Different concentration types are used for different reaction systems. For gas-phase reactions, *volumetric concentration* or *partial pressures* are equally useful, and these can be related by the thermodynamic equation of state. For instance, for ideal gases (approximation valid for gases at very low pressure)

$$PV = nRT \tag{7-29}$$

When it is applied to individual components in a constant-volume system,

$$p_i V = n_i RT \tag{7-30}$$

Using Eq. (7-5), we obtain the relationship between the volumetric concentrations and partial pressures:

$$C_i = \frac{p_i}{RT} \tag{7-31}$$

For an ideal gas, the total concentration is

$$C = \frac{P}{RT} \tag{7-32}$$

For higher pressure and nonideal gases, a *compressibility factor* z_i can be used:

$$p_i V = z_i n_i RT \quad \text{and} \quad C_i = \frac{p_i}{z_i RT} \tag{7-33}$$

Other relevant equations of state can also be used for both gases and liquids. This aspect is not in the scope of this section, and the reader is referred to Sec. 4 of this handbook.

Other concentration units include *mole fractions* for liquid x_i:

$$x_i = \frac{n_i}{\sum n_i} = \frac{C_i}{\sum C_i} = \frac{C_i}{C} \tag{7-34}$$

and for gas y_i

$$y_i = \frac{n_i}{\sum n_i} = \frac{C_i}{\sum C_i} = \frac{C_i}{C} = \frac{p_i}{\sum p_i} = \frac{p_i}{P} \tag{7-35}$$

The last two terms are only valid for an ideal gas.

STOICHIOMETRIC BALANCES

Single Reactions Equation (7-22) shows that for a single reaction, the number of moles and concentration of all other components can be calculated from the extent of reaction ξ or the conversion based on the limiting reactant, say A, X_a. In terms of number of moles n_i,

$$\begin{aligned}
n_a &= n_{a0} - \nu_a \xi = n_{a0}(1 - X_a) \\
n_b &= n_{b0} - \nu_b \xi = n_{b0}(1 - X_b) = n_{b0} - \frac{\nu_b}{\nu_a} n_{a0} X_a \\
&\cdots \\
n_r &= n_{r0} + \nu_r \xi = n_{r0} + \frac{\nu_r}{\nu_a} n_{a0} X_a \\
n_s &= n_{s0} + \nu_s \xi = n_{s0} + \frac{\nu_s}{\nu_a} n_{a0} X_a \\
&\cdots
\end{aligned} \tag{7-36}$$

Similarly, the number of moles of each component in terms of moles of A, n_a, is

$$\begin{aligned}
n_b &= n_{b0} - \frac{\nu_b}{\nu_a}(n_{a0} - n_a) \\
&\cdots \\
n_r &= n_{r0} + \frac{\nu_r}{\nu_a}(n_{a0} - n_a) \\
n_s &= n_{s0} + \frac{\nu_s}{\nu_a}(n_{a0} - n_a) \\
&\cdots
\end{aligned} \tag{7-37}$$

Change in number of moles by the reaction and change in temperature, pressure, and density affect the translation of stoichiometric balances from number of moles to volumetric concentrations. These relationships are different for gases and liquids. For instance, for *constant-density* systems (such as many liquid-phase isothermal reactions) or for *constant-temperature, constant-pressure* gas reaction with no change in number of moles, Eqs. (7-36) and (7-37) can be changed to volumetric concentration C_i by dividing each equation by the constant reaction volume V (e.g., in a batch reactor) and using Eq. (7-5). For example, for the single reaction in Eq. (7-4) with the rate in Eq. (7-5):

$$C_b = C_{b0} - \frac{q}{p}(C_{a0} - C_a)$$

$$C_r = C_{r0} + \frac{r}{p}(C_{a0} - C_a) \tag{7-38}$$

$$C_s = C_{s0} + \frac{s}{p}(C_{a0} - C_a)$$

$$r = kC_a^m\left[C_{b0} - \frac{q}{p}(C_{a0} - C_a)\right]^n = k[C_{a0}(1 - X_a)]^m\left(C_{b0} - \frac{q}{p}C_{a0}X_a\right)^n \tag{7-39}$$

It is best to represent all concentrations in terms of that of the limiting reactant.

Often there is a change in total number of moles due to reaction. Taking the general reaction in Eq. (7-1), in the gas phase the change in number of moles relative to moles of component A converted, δ_a, and the total number of moles can be calculated as follows:

$$\delta_a = \frac{\nu_q + \nu_s + \cdots - \nu_a - \nu_b - \cdots}{\nu_a}$$

$$n_0 = \sum n_{i0} \qquad n = \sum n_i \tag{7-40}$$

$$\frac{n}{n_0} = 1 + y_{a0}\,\delta_a X_a = 1 + \varepsilon_a X_a$$

$$\varepsilon_a = y_{a0}\delta_a$$

Here y_{a0} is the initial mol fraction of A. Using the ideal gas law, Eq. (7-29), the volume change depends on conversion as follows:

$$\frac{V}{V_0} = \frac{T}{T_0}\frac{P_0}{P}\frac{n}{n_0} = \frac{T}{T_0}\frac{P_0}{P}(1 + \varepsilon_a X_a) \tag{7-41}$$

Hence, for an isothermal constant-pressure ideal gas reaction system,

$$C_a = \frac{n_a}{V} = \frac{C_{a0}(1 - X_a)}{1 + \varepsilon_a X_a}$$

$$C_b = \frac{n_b}{V} = \frac{C_{b0} - \frac{\nu_b}{\nu_a}C_{a0}X_a}{1 + \varepsilon_a X_a}$$

$$\ldots \tag{7-42}$$

$$C_r = \frac{n_r}{V} = \frac{C_{r0} + \frac{\nu_r}{\nu_a}C_{a0}X_a}{1 + \varepsilon_a X_a}$$

$$C_s = \frac{n_s}{V} = \frac{C_{s0} + \frac{\nu_s}{\nu_a}C_{a0}X_a}{1 + \varepsilon_a X_a}$$

$$\ldots$$

Applying this to the reaction in Eq. (7-4) and rate in Eq. (7-5) gives

$$r = k\left[\frac{C_{a0}(1 - X_a)}{1 + \varepsilon_a X_a}\right]^m\left[\frac{C_{b0} - (q/p)C_{a0}X_a}{1 + \varepsilon_a X_a}\right]^n \tag{7-43}$$

Compare Eq. (7-43) to Eq. (7-39) where there is no change in number of moles.

Reaction Networks The analysis for single reactions can be extended to a network of reactions by defining an extent of reaction for each reaction, or by choosing a representative reactant concentration for each reaction step. For a complex network, the number of independent extents of reaction required to calculate the concentration of all components is equal to the number *of independent reactions*, which is less than or equal to the total number of reactions in the network. To calculate the number of independent reactions, and to form a set of independent reactions and corresponding independent set of concentrations or extents of reaction, we need to construct the *stoichiometric matrix* and determine its *rank*. The stoichiometric matrix is used to derive a relationship between the concentrations and the independent extents of reaction similar to that of a single reaction.

The stoichiometric matrix is the matrix of the stoichiometric coefficients of the reaction network with negative signs for reactants and positive signs for products. For instance, the hydrodechlorination of Freon 12 (CF_2Cl_2) can proceed with the following consecutive mechanism [Bonarowska et al., "Hydrodechlorination of CCl_2F_2 (CFC-12) over Silica-Supported Palladium-Gold Catalysts," *Appl. Catal. B: Environmental* **30:** 187–193 (2001)]:

$$CF_2Cl_2 + H_2 \Rightarrow CF_2ClH + HCl \qquad CF_2ClH + H_2 \Rightarrow CF_2H_2 + HCl$$

The stoichiometric matrix S for this network is

$$S = \begin{pmatrix} -1 & 1 & 0 & -1 & 1 \\ 0 & -1 & 1 & -1 & 1 \end{pmatrix}$$

The first row refers to the first reaction and the second row to the second reaction. The columns (species) are in the following order: 1-CF_2Cl_2, 2-CF_2ClH, 3-CF_2H_2, 4-H_2, and 5-HCl. The rank of a matrix is the largest square submatrix obtained by deleting rows and columns, whose determinant is not zero. The rank equals the number of independent reactions. This is also equivalent to stating that there are reactions in the network that are linear combinations of the independent reactions. The rank of S above is 2, since the *determinant* of the first 2×2 *submatrix* is not zero (there are other 2×2 submatrices that are not zero as well but it is sufficient to have at least one that is not zero):

$$S_1 = \begin{pmatrix} -1 & 1 \\ 0 & -1 \end{pmatrix} \qquad \det(S_1) = 1 \neq 0$$

Hence the two reactions are independent. Now if we add another step, which converts Freon 12 directly into the final hydrofluorocarbon CF_2H_2; $CF_2Cl_2 + 2H_2 \Rightarrow CF_2H_2 + 2HCl$, then the stoichiometric matrix becomes

$$S = \begin{pmatrix} -1 & 1 & 0 & -1 & 1 \\ 0 & -1 & 1 & -1 & 1 \\ -1 & 0 & 1 & -2 & 2 \end{pmatrix}$$

Since the last reaction is a linear combination of the first two (sum), it can be easily proved that the rank remains unchanged at 2. So to conclude, the concentrations of all components in this network can be expressed in terms of two, say H_2 and Freon 12, and the first two reactions form an independent reaction set. In case of more complicated networks, it may be difficult to determine the independent reactions by observation alone. In this case the *Gauss-Jordan decomposition* leads to a set of independent reactions (see, e.g., Amundson, *Mathematical Methods in Chemical Engineering—Matrices and Their Application,* Prentice-Hall International, New York, 1966).

For a network of reactions the general procedure is as follows:

1. Generate the reaction network by including all known reaction steps.
2. Generate the corresponding stoichiometric matrix.
3. Calculate the rank of the stoichiometric matrix which equals the number of independent reactions and independent component concentrations required to calculate all the remaining component concentrations.
4. For relatively simple networks, observation allows selection of reactions that are independent—for more complex systems use the Gauss-Jordan elimination to reduce the network to a set of independent (nonzero rows) reactions.
5. Select the independent concentration variables and independent reactions, and use these to calculate all other concentrations and reaction rates.

CATALYSIS

A *catalyst* is a material that increases the rate of both the forward and reverse reactions of a reaction step, with no net consumption or generation of catalyst by the reaction. A catalyst does not affect the reaction thermodynamics, i.e., the equilibrium composition or the heat of reaction. It does, however, affect the temperature sensitivity of the reaction rate by lowering the activation energy or the energy barrier on the reaction pathway from reactants to products. This allows the reaction to occur faster than the corresponding uncatalyzed reaction at a given temperature. Alternatively, catalytic reactions can proceed at lower temperatures than the corresponding noncatalytic reactions. For a network of reactions, the catalyst is often used to speed up desired reactions and/or to slow down undesired reactions for improved selectivity. On the basis of catalysis, reactions can be further classified into

1. Noncatalytic reactions, e.g., free-radical gas-phase reactions such as combustion of hydrocarbons.

2. Homogeneous catalytic reactions with the catalyst being dissolved in the same phase as the reactants and products in a homogeneous reaction medium. Here the catalyst is uniformly distributed throughout the system, e.g., the hydroformylation of olefins in the presence of dissolved Co or Rh carbonyls.

3. Heterogeneous catalytic reactions, with the catalyst, for instance, being a solid in contact with reactants and products in a gas-solid, gas-liquid-solid, or liquid-solid reaction system. Here the catalyst is not uniformly distributed, and the reaction occurring on the catalyst surface requires, for instance, adsorption of reactants and desorption of products from the solid surface, e.g., the catalytic cracking of gas oil to gasoline and lighter hydrocarbons.

Table 7-1 illustrates the enhancement of the reaction rates by the catalyst—this enhancement can be of many orders of magnitude.

TABLE 7-1 The Rate of Enhancement of Some Reactions in the Presence of a Catalyst

Reaction	Catalyst	Rate enhancement	Temperature, K
Ortho $H_2 \Rightarrow$ para H_2	Pt (solid)	10^{40}	300
$2NH_3 \Rightarrow N_2 + 3H_2$	Mo (solid)	10^{20}	600
$C_2H_4 + H_2 \Rightarrow C_2H_6$	Pt (solid)	10^{42}	300
$H_2 + Br_2 \Rightarrow 2HBr$	Pt (solid)	1×10^8	300
$2NO + 2H_2 \Rightarrow N_2 + 2H_2O$	Ru (solid)	3×10^{16}	500
$CH_3COH \Rightarrow CH_4 + CO$	I_2 (gas)	4×10^6	500
$CH_3CH_3 \Rightarrow C_2H_4 + H_2$	NO_2 (gas)	1×10^9	750
$(CH_3)_3COH \Rightarrow (CH_3)_2\,CH_2CH_2 + H_2O$	HBr (gas)	3×10^8	750

SOURCE: Masel, *Chemical Kinetics and Catalysis*, Wiley, 2001, Table 12.1.

IDEAL REACTORS

Reactions occur in reactors, and in addition to the intrinsic kinetics, observed reaction rates depend on the reactor type, scale, geometry, mode of operation, and operating conditions. Similarly, understanding of the reactor system used in the kinetic experiments is required to determine the reaction mechanism and intrinsic kinetics. In this section we address the effect of reactor type on observed rates. In Sec. 19 the effect of reactor type on performance (rates, selectivity, yield) is discussed in greater detail.

Material, energy, and momentum balances are essential to fully describe the performance of reactors, and often simplifying assumptions and phenomenological assumptions are needed especially for energy and momentum terms, as indicated in greater detail in Sec. 19 (see also Bird, Stewart, and Lightfoot, *Transport Phenomena*, 2d ed., Wiley, New York, 2002). Ideal reactors allow us to simplify the energy, momentum, and material balances, thus focusing the analysis on intrinsic kinetics. A useful classification of ideal reactor types is in terms of their concentration distributions versus reaction time and space. Three types of ideal reactors are considered in this section:

1. Ideal *batch reactors* (*BRs*) including *semibatch reactors* (*SBRs*)
2. Ideal *continuously stirred tank reactor* (*CSTR*), including single and multiple stages
3. *Plug flow reactor* (*PFR*) with and without *recycle*

The general form of a balance equation is

$$\text{input} + \text{sources} - \text{outputs} = \text{accumulation} \tag{7-44}$$

IDEAL BATCH REACTOR

Batch Reactor (BR) Ideal batch reactors (Fig. 7-1*a*) are tanks provided with agitation for uniform composition and temperature at all times. An ideal batch reactor can be operated under isothermal conditions (constant temperature), temperature-programmed mode (by controlling cooling rate according to a protocol), or adiabatic mode (with no heat crossing the reactor boundaries). In adiabatic mode the temperature is increasing, decreasing, or constant as the reaction proceeds for exothermic, endothermic, and thermally neutral reactions, respectively. In the ideal batch reactor, all the reactants are loaded into the reactor and well mixed by agitation before the conditions for reaction initiation (temperature and pressure) are reached; as the reaction proceeds, the concentration varies with time, but at any one time it is uniform throughout due to agitation.

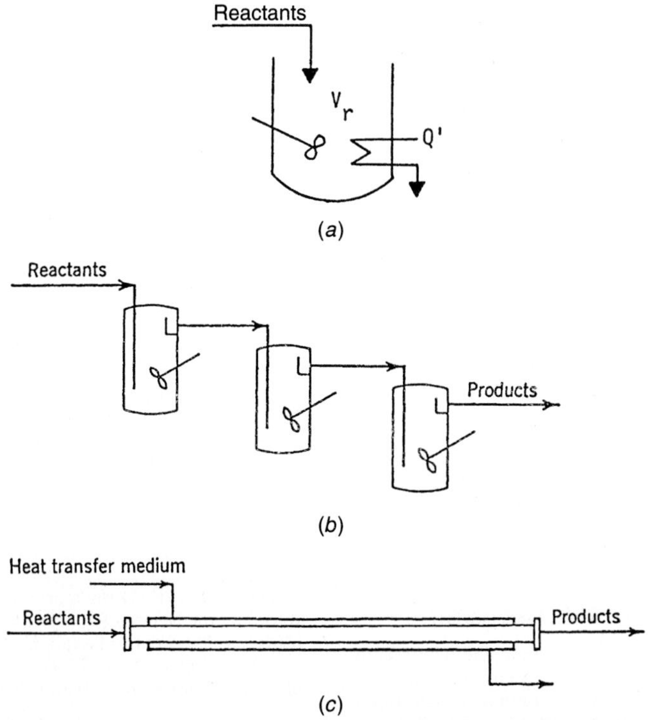

FIG. 7-1 Types of ideal reactors: (*a*) Batch or semibatch. (*b*) CSTR or series of CSTRs. (*c*) Plug flow.

Laboratory batch reactors can be single-phase (e.g., gas, liquid, etc.), multiphase (e.g., gas-liquid, gas-liquid-solid, etc.), and catalytic or noncatalytic. In this section we limit the discussion to operation at isothermal conditions. This eliminates the need to consider energy, and due to the uniform composition the component material balances are simple ordinary differential equations with time as the independent variable.

An ideal isothermal single-phase batch reactor in which a general reaction network takes place has the following general material balance equation:

$$\frac{dn_i}{dt} = V\sum_j \nu_{ij} r_j \qquad n_i = n_{i0} \qquad \text{at } t = 0 \tag{7-45}$$

The left-hand side is the accumulation term in moles per second of component i, and the right-hand side is the source term due to chemical reaction, also in moles per second, which includes all reactions j that consume or produce component i. The corresponding stoichiometric coefficients are represented in matrix form as ν_{ij} with a positive sign for products and a negative sign for reactants. This molar balance is valid for each component since we can multiply each side of the equation by the component molecular weight to obtain the true mass balance equation. In terms of conversion, Eq. (7-45) can be rewritten as

$$\frac{n_{i0}\,dX_i}{dt} = -V\sum \nu_{ij} r_j \qquad n_i = n_{i0} \qquad \text{at } t = 0 \tag{7-46}$$

and we can integrate this equation to get the *batch reaction time* or *batch residence time* τ_{BR} required to obtain a conversion X_i, starting with initial conversion X_{i0} and ending with final conversion X_{if}:

$$\tau_{BR} = -n_{i0}\int_{X_{i0}}^{X_{if}} \frac{dX_i}{V\sum \nu_{ij} r_j} \tag{7-47}$$

To integrate we need to represent all reaction rates r_j in terms of the conversion X_i. For a single reaction this is straightforward [see, e.g., Eq. (7-43)]. However, for a network of reactions, integration of a system of often nonlinear differential equations is required using implicit or semi-implicit methods. For references please see Sec. 3 of this handbook or any text on ordinary differential equations.

A special case of batch reactors is constant-volume or constant-density operation typical of liquid-phase reactions, with volume invariant with time:

$$\frac{dC_i}{dt} = \sum \nu_{ij} r_j \qquad C_i = C_{i0} \qquad \text{at } t = 0 \tag{7-48}$$

A typical concentration profile versus time for a reactant is shown in Fig. 7-2*a*. Integration of Eq. (7-48) gives the batch residence time

$$\tau_{BR} = \int_{C_{i0}}^{C_{if}} \frac{dC_i}{\sum \nu_{ij} r_j} \tag{7-49}$$

For instance, for a single reaction, Eq. (7-39) can be used to describe the reaction rate r_i in terms of one reactant concentration. For reaction networks, integration of a system of ordinary differential equations is required.

Semibatch Reactor (SBR) In semibatch operation, a gas of limited solubility or a liquid reactant may be fed in gradually as it is used up. An ideal isothermal single-phase semibatch reactor in which a general reaction network takes place has the following general material balance equation:

$$\frac{dn_i}{dt} = N_i + V\Sigma \nu_{ij} r_j \qquad n_i = n_{i0} \qquad \text{at } t = 0$$

$$N_i = N_i(t) \qquad \text{for} \qquad t_{s0i} \le t \le t_{sli} \tag{7-50}$$

The first term on the right-hand side of Eq. (7-50) is the molar feed rate of the components, which can be different for each component, hence the subscript i, and can vary with time. A typical concentration profile versus time for a reactant whose concentration is kept constant initially by controlling the feed rate is shown in Fig. 7-2*b*. Knowledge of the reaction kinetics allows these ordinary differential equations to be integrated to obtain the reactor composition versus time.

IDEAL CONTINUOUS STIRRED TANK REACTOR (CSTR)

In an ideal continuous stirred tank reactor, composition and temperature are uniform throughout just as in the ideal batch reactor. But this reactor also has a continuous feed of reactants and a continuous withdrawal of products and unconverted reactants, and the effluent composition and temperature are the same as those in the tank (Fig. 7-1*b*). A CSTR can be operated under transient conditions (due to variation in feed composition, temperature, cooling rate, etc., with time), or it can be operated under steady-state conditions. In this section we limit the discussion to isothermal conditions. This eliminates the need to consider energy balance equations, and due to the uniform composition, the component material balances are simple ordinary differential equations with time as the independent variable:

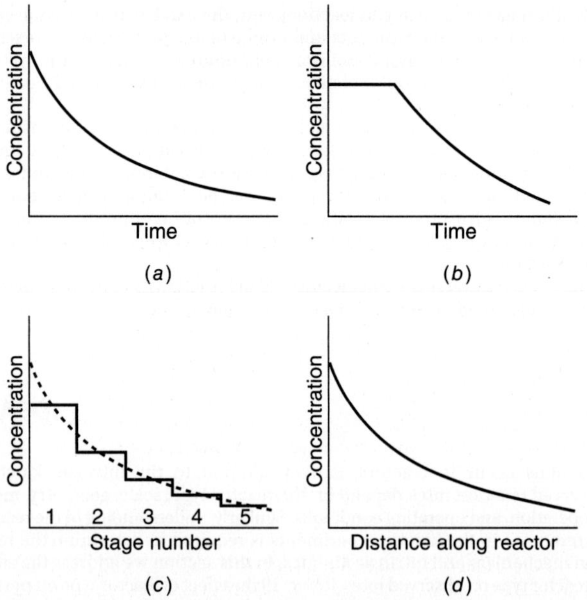

FIG. 7-2 Concentration profiles in ideal batch and continuous flow: (*a*) Batch time profile. (*b*) Semibatch time profile. (*c*) Five-stage CSTRs profile. (*d*) Plug flow distance profile.

$$\frac{dn_i}{dt} = N_{i0} - N_i + V\sum \nu_{ij} r_j \qquad n_i = n_{i0} \qquad \text{at } t = 0 \tag{7-51}$$

At steady state the differential equations simplify to algebraic equations, and the reactor volume is

$$V = -\frac{N_{i0} - N_i}{\sum \nu_{ij} r_j} \tag{7-52}$$

Equation (7-52) can be expressed in terms of volumetric concentration or in terms of conversions just as we did with the batch reactor. An *apparent residence time* based on feed conditions can be defined for a single-phase CSTR as follows:

$$\tau_{CSTR} = \frac{V}{q_0} \qquad N_{i0} = q_0 C_{i0} \qquad N_i = qC_i \tag{7-53}$$

In Eq. (7-53) the feed and effluent molar rates N_{i0} and N_i are expressed in terms of volumetric flow rates q_0 and q (inlet and outlet, respectively) and concentrations. Thus Eq. (7-52) can be rewritten as

$$\tau_{CSTR} = \frac{C_{i0} - (q/q_0)C_i}{\sum \nu_{ij} r_j} \tag{7-54}$$

Equation (7-54) allows calculation of the residence time required to achieve a given conversion or effluent composition. In the case of a network of reactions, knowing the reaction rates as a function of volumetric concentrations allows solution of the set of often nonlinear algebraic material balance equations using an implicit solver such as the multivariable Newton-Raphson method to determine the CSTR effluent concentration as a function of the residence time. As for batch reactors, for a single reaction all compositions can be expressed in terms of a component conversion or volumetric concentration, and Eq. (7-54) then becomes a single nonlinear algebraic equation solved by the Newton-Raphson method (for more details on this method please see Sec. 3 of this handbook or any relevant text).

A special case of Eq. (7-54) is a constant-density system (e.g., a liquid-phase reaction), with the true average residence time τ_{CSTR}

$$\tau_{CSTR} = \frac{C_{i0} - C_i}{\sum \nu_{ij} r_j} \qquad q = q_0 \tag{7-55}$$

When a number of such CSTRs are employed in series, the concentration profile is step-shaped if the abscissa is the total residence time or the stage number as indicated by a typical reactant concentration profile in Fig. 7-2*c*.

PLUG FLOW REACTOR (PFR)

In a *plug flow reactor* all portions of the feed stream move with the same radially uniform velocity along parallel streamlines and therefore have the same residence time; that is, there is no mixing in the axial direction but complete mixing radially (Fig. 7-1*c*). As the reaction proceeds, the concentration falls off with distance. A PFR can be operated under either transient conditions or steady-state conditions. In this section we limit the discussion to steady-state conditions. This eliminates the need to consider partial differential equations in time and space. We further limit the discussion to isothermal operation, which also eliminates the need for energy balance equations. Due to the radially uniform composition, the component material balances are simple ordinary differential equations with axial distance from inlet as the independent variable. An isothermal single-phase steady-state PFR in which a general reaction network takes place has the following general material balance equation:

$$\frac{dN_i}{dV} = \sum v_{ij} r_j \qquad N_i = N_{i0} \qquad \text{at } V = 0 \tag{7-56}$$

Note the similarity between the ideal batch and the plug flow reactors, Eqs. (7-45) and (7-56), respectively. In terms of conversion, Eq. (7-56) can be written as

$$N_{i0}\frac{dX_i}{dV} = -\sum v_{ij} r_j \tag{7-57}$$

Equation (7-57) can be integrated to calculate the reactor volume required to achieve a given conversion X_i:

$$V = -N_{i0}\int_{X_{i0}}^{X_{if}} \frac{dX_i}{\sum v_{ij} r_j} \tag{7-58}$$

An *apparent residence time* based on feed conditions can be defined for a single-phase PFR as follows:

$$\tau_{\text{PFR}} = \frac{V}{q_0} \tag{7-59}$$

Equation (7-58) becomes

$$\tau_{\text{PFR}} = -C_{i0}\int_{X_{i0}}^{X_{if}} \frac{dX_i}{\sum v_{ij} r_j} \tag{7-60}$$

Equation (7-60) is similar to that of the ideal batch reactor, Eq. (7-47), and the two reactor systems can be modeled in identical fashion.

For a constant-density system with no change in number of moles, the true residence time τ_{PFR} is

$$\tau_{\text{PFR}} = \int_{C_{i0}}^{C_{if}} \frac{dC_i}{\sum v_{ij} r_j} \tag{7-61}$$

This is identical to the corresponding ideal batch reactor, Eq. (7-49).

Ideal Recycle Reactor All reactors can sometimes be advantageously operated with recycling part of the product or intermediate streams. Heated or cooled recycle streams serve to moderate undesirable temperature gradients and they can be processed for changes in composition, such as separating products to remove equilibrium limitations, before being returned to the reactor. Say the recycle flow rate in a PFR is q_R and the fresh feed rate is q_0. With a fresh feed concentration of C_0 and a product concentration of C_2, the composite reactor feed concentration C_1 and the recycle ratio R are

$$C_1 = \frac{C_0 + RC_2}{1+R} \qquad R = \frac{q_R}{q_0} \tag{7-62}$$

The change in concentration across the reactor becomes

$$\Delta C = C_1 - C_2 = \frac{C_0 - C_2}{1+R} \tag{7-63}$$

Accordingly, the change in concentration (or in temperature) across the reactor can be made as small as desired by increasing the recycle ratio. Eventually, the reactor can become a well-mixed unit with essentially constant concentration and temperature, while substantial differences in composition will concurrently arise between the fresh feed inlet and the product withdrawal outlet, similar to a CSTR. Such an operation is useful for obtaining experimental data for analysis of rate equations. In the simplest case, where the product is recycled without change in composition, the reactor equation at constant density for a PFR with recycle is

$$\tau_{\text{PFR}} = (1+R)\int_{C_{i0}}^{C_{if}} \frac{dC_i}{\sum v_{ij} r_j} \tag{7-64}$$

Since $1 + R > 1$, recycling increases the residence time or reactor size required to achieve a given conversion.

EXAMPLES FOR SOME SIMPLE REACTIONS

Table 7-2 and Figs. 7-3 and 7-4 show the analytical solution of the integrals for two simple first-order reaction systems in an isothermal constant-volume batch reactor or plug flow reactor. Table 7-3 shows the analytical solution for the same reaction systems in an isothermal constant-density CSTR.

Section 19 discusses the advantages and disadvantages of CSTRs versus PFR and BR for various reaction systems.

TABLE 7-2 Consecutive and Parallel First-Order Reactions in an Isothermal Constant-Volume Ideal Batch or Plug Flow Reactor

The independent variable *t* is either the batch time or the plug flow residence time.

Reaction network	Material balances	Concentration profiles
$A \overset{1}{\Rightarrow} B \overset{2}{\Rightarrow} C$	$\frac{dC_a}{dt} = -k_1 C_a$	$C_a = C_{a0}e^{-k_1 t}$
	$\frac{dC_b}{dt} = k_1 C_a - k_2 C_b$	$C_b = C_{b0}e^{-k_2 t} + \frac{k_1 C_{a0}}{k_2 - k_1}(e^{-k_1 t} - e^{-k_2 t})$
	$\frac{dC_c}{dt} = k_2 C_b$	$C_c = C_{a0} + C_{b0} + C_{c0} - C_a - C_b$
$A \overset{1}{\Rightarrow} B$ $A \overset{2}{\Rightarrow} C$	$\frac{dC_a}{dt} = -(k_1 + k_2)C_a$	$C_a = C_{a0}e^{-(k_1+k_2)r}$
	$\frac{dC_b}{dt} = k_1 C_a$	$C_b = C_{b0} + \frac{k_1 C_{a0}}{k_2 + k_1}(1 - e^{(k_1+k_2)r})$
	$\frac{dC_c}{dt} = k_2 C_a$	$C_c = C_{a0} + C_{b0} + C_{c0} - C_a - C_b$

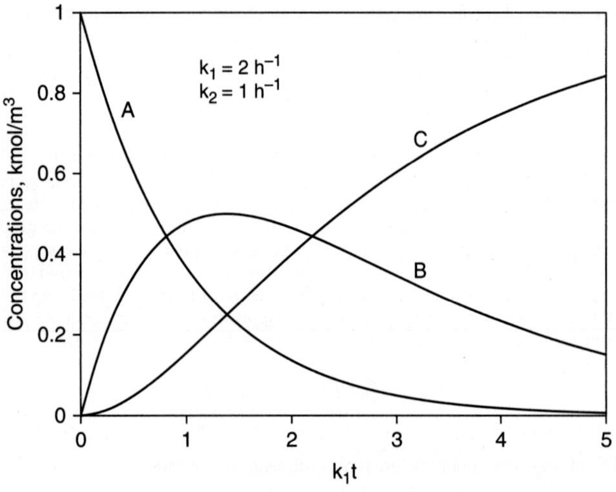

FIG. 7-3 Concentration profiles for the reactions $A \rightarrow B \rightarrow C$.

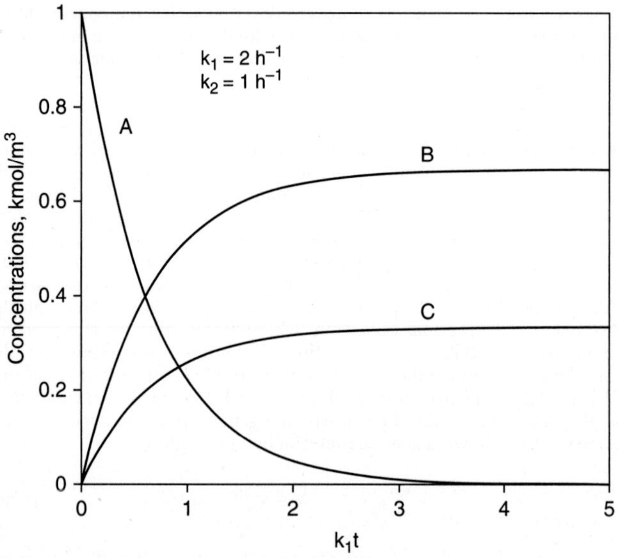

FIG. 7-4 Concentration profiles for the reactions $A \rightarrow B$ and $A \rightarrow C$.

TABLE 7-3 Consecutive and Parallel First-Order Reactions in an Isothermal Constant-Volume Ideal CSTR

Reaction network	Material balances	Concentration profiles
$A \xrightarrow{1} B \xrightarrow{2} C$	$C_{a0} - C_a - \tau k_1 C_a = 0$	$C_a = \dfrac{C_{a0}}{1+\tau k_1}$
	$C_{b0} - C_b + \tau(k_1 C_a - k_2 C_b)$	$C_a = \dfrac{C_{b0}}{1+\tau k_2} + \dfrac{\tau k_1 C_{a0}}{(1+\tau k_1)(1+\tau k_2)}$
	$C_{c0} - C_c + \tau k_2 C_b = 0$	$C_c = C_{a0} + C_{b0} + C_{c0} - C_a - C_b$
$A \xrightarrow{1} B$ $A \xrightarrow{2} C$	$C_{a0} - C_a - \tau(k_1 + k_2)C_a = 0$	$C_a = \dfrac{C_{a0}}{1+\tau(k_1+k_2)}$
	$C_{b0} - C_b + \tau k_1 C_a = 0$	$C_b = C_{b0}\dfrac{\tau k_1 C_{a0}}{1+\tau(k_1+k_2)}$
	$C_{c0} - C_c + \tau k_2 C_a = 0$	$C_c = C_{a0} + C_{b0} + C_{c0} - C_a - C_c$

KINETICS OF COMPLEX HOMOGENEOUS REACTIONS

Global or complex reactions are not usually well represented by mass action kinetics because the rate results from the combined effect of several simultaneous elementary reactions (each subject to mass action kinetics) that underlie the global reaction. The elementary steps include short-lived and unstable intermediate components such as free radicals, ions, molecules, transition complexes, etc.

The reason many global reactions between stable reactants and products have complex mechanisms is that these unstable intermediates have to be produced in order for the reaction to proceed at reasonable rates. Often simplifying assumptions lead to closed-form kinetic rate expressions even for very complex global reactions, but care must be taken when using these expressions since the simplifying assumptions are valid over limited ranges of compositions, temperature, and pressure. These assumptions can fail completely—in that case the full elementary reaction network has to be considered, and no closed-form kinetics can be derived to represent the complex system as a global reaction.

Typical simplifying assumptions include

- *Pseudo-steady-state* approximation for the intermediates; i.e., the concentration of these does not change during reaction
- Equilibrium for certain fast reversible reactions and completion of very fast irreversible steps
- *Rate-determining step*(*s*); i.e., the global reaction rate is determined by the rate(s) of the slowest step(s) in the reaction network composing the overall or global reaction

These simplifying assumptions allow elimination of some reaction steps, and representation of free radical and short-lived intermediates concentrations in terms of the concentration of the stable measurable components, resulting in complex non–mass action rate expressions.

Complex reactions can proceed through *chain* or *nonchain* reaction steps. In a *chain reaction*, the active unstable components are produced in an *initiation* step and are repeatedly regenerated through *propagation* steps, and only a small fraction of these are converted to stable components through a *termination* step. Free radicals are examples of such unstable components frequently encountered in chain reactions: free radicals are molecular fragments having one or more unpaired electrons, are usually short-lived (milliseconds), and are highly reactive. They are detectable spectroscopically, and some have been isolated. They occur as initiators and intermediates in such basic phenomena as oxidation, combustion, photolysis, and polymerization. Several examples of free radical mechanisms possessing nonintegral power law or hyperbolic rate equations are cited below. In a *nonchain reaction*, the unstable intermediate, such as an *activated complex* or *transition state complex*, reacts further to produce the products, and it is not regenerated through propagation but is continually made from reactants in stoichiometric quantities.

CHAIN REACTIONS

Phosgene Synthesis The global reaction $CO + Cl_2 \Rightarrow COCl_2$ proceeds through the following free radical mechanism:

$$Cl_2 \Leftrightarrow 2Cl\bullet$$
$$Cl\bullet + CO \Leftrightarrow COCl\bullet$$
$$COCl\bullet + Cl_2 \Rightarrow COCl_2 + Cl\bullet$$

Assuming the first two reactions are in equilibrium, expressions are found for the concentrations of the free radicals Cl• and COCl• in terms of the species CO, Cl_2, and $COCl_2$, and when these are substituted into the mass action rate expression of the third (rate-determining) reaction, the rate becomes

$$r_{COCl_2} = k(CO)(Cl_2)^{3/2} \tag{7-65}$$

Ozone Conversion to Oxygen in Presence of Chlorine The global reaction $2O_3 \overset{Cl_2}{\Rightarrow} 3O_2$ in the presence of Cl_2 proceeds through the following sequence:

$$Cl_2 + O_3 \Rightarrow ClO\bullet + ClO_2\bullet$$
$$ClO\bullet + O_3 \Rightarrow ClO_2\bullet + O_2$$
$$ClO_2\bullet + O_3 \Rightarrow ClO_3\bullet + O_2$$
$$ClO_3\bullet + O_3 \Rightarrow ClO_2\bullet + 2O_2$$
$$ClO_3\bullet + ClO_3\bullet \Rightarrow Cl_2 + 3O_2$$

The chain carriers ClO•, ClO_2•, and ClO_3• are assumed to attain pseudo-steady state. Then,

$$r_{O_3} = k(Cl_2)^{1/2}(O_3)^{3/2} \tag{7-66}$$

Hydrogen Bromide Synthesis The global reaction $H_2 + Br_2 \Rightarrow 2HBr$ proceeds through the following chain of reactions:

$$Br_2 \Leftrightarrow 2Br\bullet$$
$$Br\bullet + H_2 \Leftrightarrow HBr + H\bullet$$
$$H\bullet + Br_2 \Rightarrow HBr + Br\bullet$$

Assuming pseudo-steady state for the concentrations of the free radicals H• and Br•, the global rate equation becomes

$$r_{HBr} = \frac{k_1(H_2)(Br_2)^{1/2}}{k_2 + k_3(HBr)/(Br_2)} \tag{7-67}$$

Here the constants k_1, k_2, and k_3 are combinations of the kinetic parameters of the elementary steps.

Chain Polymerization For free radical polymerization, the following generic mechanism can be postulated:

$I\bullet$	$\overset{fk_d}{\rightarrow}$	$2R\bullet$		initiation
$R\bullet + M$	$\overset{k_i}{\rightarrow}$	$P_1\bullet$		
$P_n\bullet + M$	$\overset{k_p}{\rightarrow}$	$P_{n+1}\bullet$	$n,m = 1,2,\ldots$	propagation
$P_n\bullet + P_m\bullet$	$\overset{k_{tc}}{\rightarrow}$	M_{n+m}		termination by coupling
$P_n\bullet + P_m\bullet$	$\overset{k_{td}}{\rightarrow}$	$M_n + M_m$		termination by disproportionation

After writing the appropriate mass balances, and assuming the rates of formation of the free radicals $R\bullet$ and $P_n\bullet$ reach pseudo-steady state, the following polymerization rate is obtained:

$$r_p = -\frac{d(M)}{dt} = k_p(M)(P) = k_p\left(\frac{2fk_d}{k_{tc}+k_{td}}\right)(M)(I)^{1/2} \qquad P = \sum P_n \tag{7-68}$$

NONCHAIN REACTIONS

Nonchain reactions proceed through an active intermediate to the products. Many homogeneous nonchain reactions are also homogeneously catalyzed reactions (discussed below).

HOMOGENEOUS CATALYSIS

Homogeneous catalysts proceed through an activated or transition state complex between reactant(s) and catalysts that decomposes into products. Homogeneous catalysts are dissolved in the homogeneous reaction mixture and include acids/bases, metal salts, radical initiators, solvents, and enzymes.

Acid-Catalyzed Isomerization of Butene-1 Butene-1 isomerizes to butene-2 in the presence of an acid according to the global reaction

$$CH_3CH_2HC{=}CH_2 \overset{H^+}{\Rightarrow} CH_3HC{=}CHCH_3$$

Even though this appears to be a monomolecular reaction, it proceeds through the following mechanism:

$$CH_3CH_2HC{=}CH_2 + H^+ \overset{k_1}{\Leftrightarrow} \{CH_3CH_2HC{-}CH_2\}^+ \overset{k_2}{\Rightarrow} CH_3HC{=}CHCH_3 + H^+$$

(with H bridging the two carbons of the intermediate)

Assuming reaction 1 is in equilibrium, the reaction rate is

$$r = k_2K_1[H^+][CH_3CH_2HC{=}CH_2] \tag{7-69}$$

Enzyme Kinetics Enzymes are homogeneous catalysts, e.g., for cellular and enzymatic reactions. The enzyme E and the reactant S are assumed to form a complex ES that then dissociates into product P and releases the enzyme:

$$S + E \Leftrightarrow ES$$
$$ES \Rightarrow E + P$$

Assuming equilibrium for the first step results in the following rate, developed by Michaelis and Menten [*Biochem. Zeit.* **49:** 333 (1913)] and named *Michaelis-Menten kinetics,*

$$r_p = -r_s = \frac{k(S)}{K_m + (S)} \tag{7-70}$$

Here K_m is the inverse of the equilibrium constant for the first reaction.

AUTOCATALYSIS

In an autocatalytic reaction, a reactant reacts with a product to make more product. For the reaction to proceed, therefore, the product must be present initially in a batch or in the feed of a continuous reactor. Examples are cell

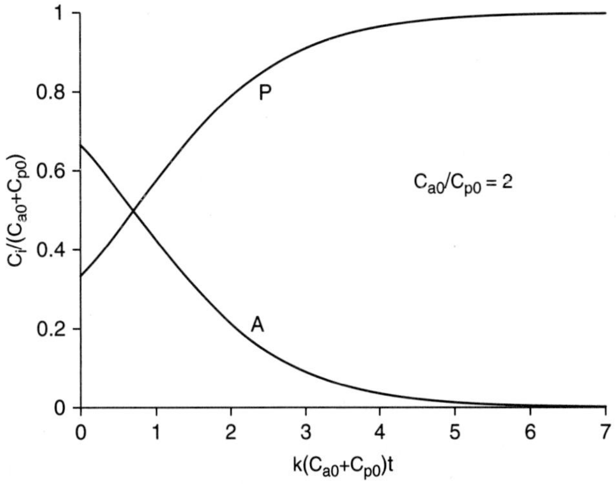

FIG. 7-5 Product concentration profile for the autocatalytic reaction $A + P \Rightarrow 2P$ with rate $r = kC_aC_p$.

growth in fermentation and combustion of fuels. For instance, the irreversible elementary reaction $A + P \Rightarrow 2P$ has the mass action kinetics

$$r = kC_aC_p \tag{7-71}$$

For an ideal batch reactor (see, e.g., Steinfeld, Francisco, and Hase, *Chemical Kinetics and Dynamics*, Prentice-Hall, 1989):

$$\frac{dC_a}{dt} = -kC_aC_p \qquad \frac{dC_p}{dt} = kC_aC_p \qquad C_a + C_p = C_{a0} + C_{p0} \tag{7-72}$$

Integration results in the following concentration profile:

$$C_p = \frac{C_{a0} + C_{p0}}{1 + (C_{a0}/C_{p0})e^{-k(Ca0+Cp0)t}} \qquad C_a = C_{a0} + C_{p0} - C_p \tag{7-73}$$

Figure 7-5 illustrates the dimensionless concentration profile for reactant A and product P, $C_i/(C_{a0} + C_{p0})$, for $C_{a0}/C_{p0} = 2$. If the initial concentration of P is lower than that of A (which is mostly the case for these autocatalytic reactions), then a maximum rate is indicated by the inflection point.

INTRINSIC KINETICS FOR FLUID-SOLID CATALYTIC REACTIONS

There are a large number of fluid-solid catalytic reactions, mostly gas-solid, including catalytic cracking, oxidation of polluting gases in automotive and power generation catalytic converters, partial oxidation synthesis reactions, HCN synthesis, chemical vapor deposition, etc. (see, e.g., Sec. 19 for more examples). Examples of solid catalysts include, among others, supported metals, transition metal oxides and sulfides, solid acids and bases, and immobilized homogeneous catalysts and enzymes. Solid catalysts can be a fine powder (suspended in a liquid or fluidized by a flowing gas), cylindrical, spherical, and more complex-shaped particles (in a packed bed), a thin layer of active components (on the walls of a monolith or a foam) and gauzes. The solid catalyst can be porous with active component distributed throughout the particle volume, or nonporous with active component present just on the catalyst external surface.

The analysis of Langmuir [*J. Am. Chem. Soc.* **40:** 1361 (1918)] and Hinshelwood (*Kinetics of Chemical Change*, Oxford, 1940) forms the basis for the simplified treatment of kinetics on heterogeneous catalysts. For a solid catalyzed reaction between gas-phase reactants A and B, the postulated mechanism may consist of the following steps in series:

1. The reactants from the gas adsorb to bond to active sites on the catalyst surface as molecules or dissociated atoms. The rate of adsorption is proportional to the partial pressure of reactants and to the fraction of uncovered surface sites ϑ_υ. More than one type of active site can be present. The adsorption isotherms such as the Langmuir isotherm relate the partial pressure of an adsorbed species to its surface coverage, and the form of this relationship is indicative of the type of adsorption process taking place (for more details, see Masel, *Chemical Kinetics and Catalysis*, Wiley, 2001).
2. The adsorbed species react on the surface to form adsorbed products. The rate of reaction between adsorbed species is proportional to their concentration on the surface.
3. The adsorbed products desorb into the gas. The rate of desorption of species A is proportional to the fraction of the surface covered by A, ϑ_a.

For instance, for the simple irreversible reaction $A + B \Rightarrow C + D$, the postulated mechanism is

$$\begin{aligned} A + \sigma &\Leftrightarrow A\sigma \\ B + \sigma &\Leftrightarrow B\sigma \\ A\sigma + B\sigma &\Rightarrow C\sigma + D\sigma \\ C\sigma &\Leftrightarrow C + \sigma \\ D\sigma &\Leftrightarrow D + \sigma \end{aligned} \tag{7-74}$$

$A\sigma$, $B\sigma$, $C\sigma$, and $D\sigma$ above are adsorbed species on the catalyst surface, and σ is an available active site. We will consider a variety of possible scenarios for this simple solid-catalyzed system. Note that the reaction rate for such systems is often expressed on a unit mass catalyst basis (r_m) instead of unit reaction volume basis (r_υ), and the latter is related to the former through the catalyst loading (mass catalyst/reaction volume) or bed density:

$$r_\upsilon = \frac{r_m M_{\text{cat}}}{V} = r_m \rho_B \tag{7-75}$$

ADSORPTION EQUILIBRIUM

Assuming equilibrium for all adsorption steps (e.g., the surface reaction is rate-limiting), the net rates of adsorption of reactants and product are all zero.

$$r_i = k_i p_i \vartheta_\upsilon - k_{-i}\vartheta_i = 0 \qquad i = a, b, c, d \tag{7-76}$$

A material balance on all sites (using the fraction of sites ϑ_υ and ϑ_i versus total number of sites) yields

$$\vartheta_\upsilon = 1 - \vartheta_a - \vartheta_b - \vartheta_c - \vartheta_d \tag{7-77}$$

and solving for the surface coverages gives

$$\begin{aligned} \vartheta_a &= \left(\frac{k_a}{k_{-a}}\right) p_a \vartheta_\upsilon = K_a p_a \vartheta_\upsilon \qquad \vartheta_b = \left(\frac{k_b}{k_{-b}}\right) p_b \vartheta_\upsilon = K_b p_b \vartheta_\upsilon \\ \vartheta_c &= \left(\frac{k_c}{k_{-c}}\right) p_c \vartheta_\upsilon = K_c p_c \vartheta_\upsilon \qquad \vartheta_d = \left(\frac{k_d}{k_{-d}}\right) p_d \vartheta_\upsilon = K_d p_d \vartheta_\upsilon \end{aligned} \tag{7-78}$$

The fraction of surface not covered is

$$\vartheta_\upsilon = \frac{1}{1 + K_a p_a + K_b p_b + K_c p_c + K_d p_d} \tag{7-79}$$

In the denominator, terms may be added for adsorbed inerts (say, $K_I p_I$) or other species that may be present. The rate of reaction or the rate-determining step is that between adsorbed reactant species:

$$r = k p_a p_b \vartheta_\upsilon^2 = \frac{k p_a p_b}{(1 + K_a p_a + K_b p_b + K_c p_c + K_d P_d)^2} \tag{7-80}$$

DISSOCIATION

A diatomic molecule A_2 may adsorb dissociatively as atoms

$$A_2 + 2\sigma \Rightarrow 2A\sigma \tag{7-81}$$

with the result

$$\vartheta_a = \sqrt{K_a p_a}\,\vartheta_\upsilon = \frac{\sqrt{K_a p_a}}{1 + \sqrt{K_a p_a} + K_b p_b + \cdots} \tag{7-82}$$

and the rate-determining step and its rate are

$$2A\sigma + B\sigma \Rightarrow \text{products} + 3\sigma \qquad r = k'\vartheta_a^2\vartheta_b = k p_a p_b \vartheta_\upsilon^3 \tag{7-83}$$

DIFFERENT SITES

When A and B adsorb on chemically different sites σ_1 and σ_2, the rate of the reaction

$$A + B \Rightarrow \text{nonadsorbed products} \tag{7-84}$$

with surface reaction controlling is

$$r = \frac{k p_a p_b}{(1 + K_a p_a)(1 + K_b p_b)} \tag{7-85}$$

CHANGE IN NUMBER OF MOLES

When the number of moles of product is larger than that of the reactants, extra sites are required:

$$A \Leftrightarrow M + N \tag{7-86}$$

$$A\sigma + \sigma \Rightarrow M\sigma + N\sigma \tag{7-87}$$

and the rate is

$$r = k'\left(\vartheta_a\vartheta_\upsilon - \frac{\vartheta_m\vartheta_n}{K}\right) = k\left(p_a - \frac{p_m p_n}{K}\right)\vartheta_\upsilon^2 = \frac{k(p_a - p_m p_n/K)}{(1+K_a p_a + K_m p_m + K_n p_n)^2} \quad (7\text{-}88)$$

REACTANT IN THE GAS PHASE

When A in the gas phase reacts directly with adsorbed B,

$$A + B\sigma \Rightarrow \text{products}$$

$$r = k' p_a \vartheta_b = k p_a p_b \vartheta_\upsilon = \frac{k p_a p_b}{1+\sum K_i p_i} \quad (7\text{-}89)$$

This mechanism is called *Ely-Rideal kinetics*.

CHEMICAL EQUILIBRIUM IN GAS PHASE

When A is not in adsorptive equilibrium but is in chemical equilibrium in the gas phase according to

$$A + B \Leftrightarrow M + N \quad p_a = \frac{p_m p_n}{K_e p_b} \quad (7\text{-}90)$$

this expression is substituted for p_a wherever it appears in the rate equation. If the rate-determining step is the surface reaction between adsorbed species, then

$$r = k p_a p_b \vartheta_\upsilon^2 = \frac{k p_m p_n / K_e}{1 + K_a(p_m p_n / K_e p_b) + K_b p_b + K_m p_m + K_n p_n} \quad (7\text{-}91)$$

Table 7-4 summarizes some examples of reactions where all substances are in adsorptive equilibrium and the surface reaction controls the rate. In Table 7-5, substance A is not in adsorptive equilibrium, and its rate of adsorption is controlling. Details of the derivations of these and some other equations are presented by Yang and Hougen [*Chem. Eng. Prog.* **46:** 146 (1950)], Walas (*Reaction Kinetics for Chemical Engineers,* McGraw-Hill, 1959; Butterworths, 1989, pp. 153–164), and Rase (*Chemical Reactor Design for Process Plants,* vol. 1, Wiley, 1977, pp. 178–191).

NO RATE-CONTROLLING STEP

All the relations developed above assume that only one step is controlling. In a reaction system, changing the operating conditions may shift the control from one step to another. At certain conditions, there may be no single step controlling. In that case all the reactions and their respective rates have to be considered, and the adsorbed species cannot be eliminated from the rate expressions to obtain a single closed-form kinetic rate.

LIQUID-SOLID CATALYTIC REACTIONS

A treatment analogous to that for fluid-solid catalysis can be derived for liquid-solid catalysis, with partial pressures replaced by liquid concentrations.

BIOCATALYSIS

Biochemical reactions such as *aerobic* and *anaerobic fermentations* occur in the presence of living organisms or *cells,* such as bacteria, algae, and yeast. These reactions can be considered as *biocatalyzed* by the organism. Thus in a typical bioreactor a *substrate* (such as glucose) is fed into the *fermenter* or *bioreactor* in the presence of an initial amount of cells. The desired product can be the cells themselves or one of the chemicals produced by the cell, called *metabolites.* In either case the cells multiply in the presence of the substrate, and the rate of production of cells is proportional to the concentration of the cells—hence this process is autocatalytic. In a batch reactor with an ample supply of substrate, this results in *exponential growth* of the culture. A typical cell or biomass growth rate function, called the *Monod kinetics,* is identical in form to the Michaelis-Menten enzyme kinetics in Eq. (7-70):

$$\mu = \frac{\mu_{\max} C_s}{K_s + C_s} \quad (7\text{-}92)$$

In Eq. (7-92) μ is the *specific growth rate* of the culture. It is measured in units of reciprocal time (h^{-1}) and it is related to the *volumetric growth rate* r_x of the culture: $r_x = C_x\mu$. This means that the true unit of μ is, e.g., (g biomass formed/h)/(g biomass present), where *g biomass* is the *dry weight* (DW) of the biomass, obtained after evaporation of the water content of the cell (which constitutes about 80 percent of the *wet biomass* weight). Similarly C_x has the unit, e.g., (g biomass)/(L medium volume). The variable C_s in Eq. (7-92) is the concentration of the *limiting substrate* in the medium (g/L). There are many substrates (including micronutrients) in the medium, but there is usually just one that determines the specific growth rate. This substrate is often a sugar (e.g., glucose), but it could also be O_2, a metal ion (Mg^{2+} etc.), or PO_4^{3-}, NH_4^+, ..., or perhaps a hormone. The limiting substrate may easily change during a batch fermentation, and then the rate expression will change.

The two parameters in Eq. (7-92) are the *maximum specific growth rate* $\mu_{\max}$ (h^{-1}) and the *saturation constant* K_s (g substrate/L). The value of K_s is obtained as the substrate concentration at which $\mu = \frac{1}{2}\mu_{\max}$. The growth rate versus limiting substrate concentration is shown in Fig. 7-6. The form of

TABLE 7-4 Surface-Reaction Controlling Table 7-3

Adsorptive equilibrium maintained for all participants.

Reaction	Special condition	Basic rate equation	Driving force	Adsorption term
1. $A \rightarrow M+N$	General case	$r = k\theta_a$	p_a	$1+K_a p_a+K_m p_m+K_n p_n$
$A \rightarrow M+N$	Sparsely covered surface	$r = k\theta_a$	p_a	1
$A \rightarrow M+N$	Fully covered surface	$r = k\theta_a$	1	1
2. $A \rightleftharpoons M$		$r = k_1\theta_a - k_{-1}\theta_m$	$p_a - \frac{p_m}{K}$	$1+K_a p_a+K_m p_m$
3. $A \rightleftharpoons M+N$	Adsorbed A reacts with vacant site	$r = k_1\theta_a\theta_v - k_{-1}\theta_m\theta_n$	$p_a - \frac{p_m p_n}{K}$	$(1+K_a p_a+K_m p_m+K_n p_n)^2$
4. $A_2 \rightleftharpoons M$	Dissociation of A_2 upon adsorption	$r = k_1\theta_a^2 - k_{-1}\theta_m\theta_v$	$p_a - \frac{p_m}{K}$	$(1+\sqrt{K_a p_a}+K_m p_m)^2$
5. $A+B \rightarrow M+N$	Adsorbed B reacts with A in gas but not with adsorbed A	$r = k\theta_a\theta_b$	$p_a p_b$	$(1+K_a p_a+K_b p_b+K_m p_m+K_n p_n)^2$
$A+B \rightarrow M+N$		$r = k p_a\theta_b$	$p_a p_b$	$1+K_a p_a+K_b p_b+K_m p_m+K_n p_n$
6. $A+B \rightleftharpoons M$		$r = k_1\theta_a\theta_b - k_{-1}\theta_m\theta_v$	$p_a p_b - \frac{p_m}{K}$	$(1+K_a p_a+K_b p_b+K_m p_m)^2$
7. $A+B \rightleftharpoons M+N$		$r = k_1\theta_a\theta_b - k_{-1}\theta_m\theta_n$	$p_a p_b - \frac{p_m p_n}{K}$	$(1+K_a p_a+K_b p_b+K_m p_m+K_n p_n)^2$
8. $A_2+B \rightleftharpoons M+N$	Dissociation of A_2 upon adsorption	$r = k_1\theta_a^2\theta_b - k_{-1}\theta_m\theta_n\theta_v$	$p_a p_b - \frac{p_m p_n}{K}$	$(1+\sqrt{K_a p_a}+K_b p_b+K_m p_m+K_n p_n)^3$

NOTE: The rate equation is:

$$r = \frac{k(\text{driving force})}{\text{adsorption term}}$$

When an inert substance I is adsorbed, the term $K_I p_I$ is to be added to the adsorption term.

SOURCE: From Walas, *Reaction Kinetics for Chemical Engineers,* McGraw-Hill, 1959; Butterworths, 1989, Table 7-3.

TABLE 7-5 Adsorption-Rate of Species A Controlling (Rapid Surface Reaction)

Reaction	Special condition	Basic rate equation	Driving force	Adsorption term
1. $A \rightarrow M+N$		$r = kp_a\theta_v$	p_a	$1+\frac{K_a p_m p_n}{K}+K_m p_m+K_n p_n$
2. $A \rightleftharpoons M$		$r=k\left(p_a\theta_v-\frac{\theta_a}{K_a}\right)$	$p_a-\frac{p_m}{K}$	$1+\frac{K_a p_m}{K}+K_m p_m$
3. $A \rightleftharpoons M+N$		$r=k\left(p_a\theta_v-\frac{\theta_a}{K_a}\right)$	$p_a-\frac{p_m p_n}{K}$	$1+\frac{K_a p_m p_n}{K}+K_m p_m+K_n p_n$
4. $A_2 \rightleftharpoons M$	Dissociation of A_2 upon adsorption	$r=k\left(p_a\theta_v^2-\frac{\theta_a^2}{K_a}\right)$	$p_a-\frac{p_m}{K}$	$\left(1+\sqrt{\frac{K_a p_m}{K}}+K_m p_m\right)^2$
5. $A+B \rightarrow M+N$	Unadsorbed A reacts with adsorbed B	$r = kp_a\theta_v$	p_a	$1+\frac{K_a p_m p_n}{Kp_b}+K_b p_b+K_m p_m+K_n p_n$
6. $A+B \rightleftharpoons M$		$r=k\left(p_a\theta_v-\frac{\theta_a}{K_a}\right)$	$p_a-\frac{p_m}{Kp_b}$	$1+\frac{K_a p_m}{Kp_b}+K_b p_b+K_m p_m$
7. $A+B \rightleftharpoons M+N$		$r=k\left(p_a\theta_v-\frac{\theta_a}{K_a}\right)$	$p_a-\frac{p_m p_n}{Kp_b}$	$1+\frac{K_a p_m p_n}{Kp_b}+K_b p_b+K_m p_m+K_n p_n$
8. $A_2+B \rightleftharpoons M+N$	Dissociation of A_2 upon adsorption	$r=k\left(p_a\theta_v^2-\frac{\theta_a^2}{K_a}\right)$	$p_a-\frac{p_m p_n}{Kp_b}$	$\left(1+\sqrt{\frac{K_a p_m p_n}{Kp_b}}+K_b p_b+K_m p_m+K_n p_n\right)^2$

NOTES: The rate equation is:

$$r=\frac{k(\text{driving force})}{\text{adsorption term}}$$

Adsorption rate of substance A is controlling in each case. When an inert substance I is adsorbed, the term $K_I p_I$ is to be added to the adsorption term.
SOURCE: From Walas, *Reaction Kinetics for Chemical Engineers*, McGraw-Hill, 1959; Butterworths, 1989, Table 7-4.

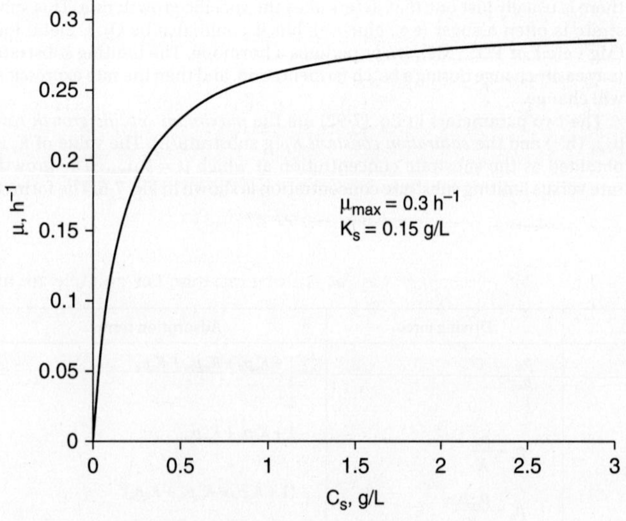

FIG. 7-6 The effect of substrate concentration on specific growth rate.

Eq. (7-92) is entirely empirical, but it incorporates two important features: (1) At high substrate concentration the whole cell machinery is involved in cell synthesis, and the specific growth rate reaches a maximum μ_{max}; (2) at low substrate concentration, formation of biomass is a first-order rate process (as in any other chemical reaction) and $\mu \rightarrow (\mu_{max}/K_s)C_s$. Note that for many commonly used microorganisms K_s is much smaller than the typical substrate concentration C_s. In batch cultivations K_s is several orders of magnitude smaller than C_s until the very end of the batch, and this is what gives the well-known *exponential growth* where $\mu = \mu_{max}$. Equation (7-93) shows the cell's material balance and its integral for a batch cultivation, and it applies after an initial *lag phase* when cell machinery is synthesized. Some typical glucose substrate values for K_s are 150 mg/L (*Saccharomyces cerevisiae*), 5 to 10 mg/L (lactic bacteria and *E. coli*), and less than 1 mg/L (filamentous fungi).

$$\frac{dC_x}{dt}=\mu C_x \qquad \mu=\mu_{max} \quad \text{for} \quad C_s \gg K_s \Rightarrow C_x = C_{xo}e^{-\mu_{max}t} \tag{7-93}$$

Equation (7-93) may have to be modified by subtraction of a *death-rate* term $\mu_d C_x$. Further, μ_d may well increase during the batch fermentation in which case the net growth rate of (viable) cells eventually becomes negative, and the concentration of (viable) cells will start to decrease.

FLUID-SOLID REACTIONS WITH MASS AND HEAT TRANSFER

GAS-SOLID CATALYTIC REACTIONS

The Langmuir-Hinshelwood mechanism of adsorption and reaction described above allowed us to relate the gas concentrations and partial pressures in the vicinity of the catalyst surface to the adsorbed species concentration at the active sites, which in turn determined the surface reaction rates. In practice, two additional mass transfer processes may need to be considered:

1. Diffusion to and from the bulk gas to the external catalyst surface, represented as an external mass transfer process across a film or boundary layer concentration gradient. For nonporous catalyst this is the only mass-transfer step.

2. Diffusion to and from the catalyst external surface through pores in a porous catalyst particle to active sites inside the catalyst particle where the adsorption and reaction occur, represented as intraparticle diffusion and modeled as a diffusion-reaction process.

External Mass Transfer In a reactor, the solid catalyst is deposited on the surface of narrow tubes (such as monolith), is packed as particles in a tube, or is suspended in slurry or in a fluidized bed as fine particles. For these systems, the bulk concentration of the gas phase approaches that on the catalyst surface if the mass transfer rate from bulk to surface is substantially faster than the reaction rates on the surface. This, however, is often not the case. The mechanism of mass transfer and reaction on the external catalyst surface includes the following consecutive steps:

1. Mass transfer of gas reactants from the bulk gas to the solid catalyst surface, also called *external mass transfer*

2. Adsorption, reaction on the surface, and desorption of products, e.g., Langmuir-Hinshelwood kinetics

3. Mass transfer of products from the catalyst surface to the bulk gas

At steady state all these rates are equal.

For example, for a first-order irreversible reaction $A \Rightarrow B$, the rate of mass transfer equals the rate of intrinsic reaction:

$$k_s a_s (C_a - C_{as}) = kC_{as} \tag{7-94}$$

Here a_s is the external particle surface area/volume of reactor. Eliminating the surface concentration C_{as} in terms of the observable bulk gas concentration C_a yields the overall specific rate of consumption of A:

$$r_{\text{obs}} = k_{\text{obs}} C_a = \frac{C_a}{1/(k_s a_s) + 1/k}$$

$$k_{\text{obs}} = \frac{1}{1/(k_s a_s) + 1/k} \tag{7-95}$$

Hence the observable overall rate constant k_{obs} is actually a combination of the mass transfer coefficient and the intrinsic rate coefficient; or in terms of resistances (in units of time) the overall resistance is the sum of the mass transfer and intrinsic kinetic resistances. For this first-order rate case, the overall behavior remains first-order in bulk gas concentration. The two limiting cases are mass transfer and kinetic (reaction) control, respectively:

$$k_{\text{obs}} = k_s a_s \quad \text{and} \quad r_{\text{obs}} = k_s a_s C_a \quad \text{for} \quad k_s a_s << k \quad \text{for mass transfer control} \tag{7-96}$$

$$k_{\text{obs}} = k \quad \text{and} \quad r_{\text{obs}} = kC_a \quad \text{for} \quad k_s a_s >> k \quad \text{for kinetic control}$$

The mass transfer coefficient depends on the geometry of the solid surface, on the hydrodynamic conditions in the vicinity of the catalyst (which are a function, e.g., of the reactor type, geometry, operating conditions, and flow regime), and it also depends on the diffusivity of the gas species. Correlations for the mass transfer coefficient are a large topic and outside the scope of this section. For more details see Bird, Stewart, and Lightfoot, *Transport Phenomena*, 2d ed., Wiley, New York, 2002, and relevant sections in this handbook. For non-first-order kinetics, a closed-form relationship (such as resistances in series description) cannot always be derived, but the steady-state assumption of the consecutive mass and reaction steps still applies.

Intraparticle Diffusion As indicated above, the larger the catalyst surface area per unit reactor volume a_s, the larger the overall reaction rate. For a fixed mass of catalyst, decreasing the particle size increases the total external surface area available for the reaction. Another way to increase the surface area is by providing a porous catalyst with lots of internal surface area. The internal structure of the catalyst determines how accessible these internal sites are to the gas-phase reactant and how easily the products can escape back to the bulk gas. The analysis is based on the pseudohomogeneous reaction-diffusion equation, with the gas reactant diffusing through the pores and reacting at active sites inside the catalyst particle. For a first-order irreversible reaction of species A in an infinite slab geometry, the diffusion-reaction equation describes the reactant concentration profile from the external surface to the center of the slab ($y = 0$) as follows:

$$D_{ea}\frac{d^2 C_{ay}}{dy^2} - kC_{ay} = 0 \qquad C_{ay}(L) = C_{as}$$

$$\frac{dC_{ay}}{dy}(0) = 0 \tag{7-97}$$

where D_{ea} is the effective diffusivity and L is the characteristic length, in this case half the thickness of the slab. The solution of the second-order ordinary differential equation provides $C_a(y)$ and can be used to calculate the reaction rate in the pellet under intraparticle diffusion conditions

$$r = \eta k C_{as} \tag{7-98}$$

The *effectiveness factor* η is defined as the ratio of the actual rate to that if the reaction occurred at the external surface concentration in the absence of intraparticle diffusion resistance C_{as}:

$$\eta = \frac{(1/L)\int_0^L r(C_{ay})dy}{r(C_{as})} = \frac{\text{observed rate with pore diffusion resistances}}{\text{rate at catalyst external surface condition}} \tag{7-99}$$

The effectiveness factor can be obtained as a function of a dimensionless parameter called the *Thiele modulus*, representing the ratio of the rate of reaction to the rate of intraparticle diffusion. For a first-order reaction, the Thiele modulus and the effectiveness factor derived by integration of Eq. (7-97) are shown below:

$$\eta = \frac{\tanh \varphi_{\text{slab}}}{\varphi_{\text{slab}}} \qquad \varphi_{\text{slab}} = L\sqrt{\frac{k}{D_{ea}}} \tag{7-100}$$

The effective diffusivity accounts for the decrease in gas diffusivity on account of the solid parts of the catalyst. The diffusion path is tortuous as gas molecules follow the pores. The simplest expression used to explain the effective gas diffusivity is

$$D_{ea} = \frac{\varepsilon_s}{\tau} D_a \tag{7-101}$$

where the porosity ε_s accounts for the fact that diffusion only occurs through the gas-filled part of the particle, and the tortuosity τ accounts for the effect of diffusion path length and contraction/expansion of pores along the diffusion path. The diffusion regime depends on the diffusing molecule, pore size, and operating conditions (concentration, temperature, pressure), and this can be visualized in Fig. 7-7. As indicated, the effective diffusion coefficient ranges over many orders of magnitude from very low values in the configuration regime (e.g., in zeolites) to high values approaching molecular diffusivity where the pores are large.

There is a large body of literature that deals with the proper definition of the diffusivity used in the intraparticle diffusion-reaction model, especially in multicomponent mixtures found in many practical reaction systems. The reader should consult references, e.g., Bird, Stewart, and Lightfoot, *Transport Phenomena,* 2d ed., Wiley, New York, 2002; Taylor and Krishna, *Multicomponent Mass Transfer,* Wiley, 1993; and Cussler, *Diffusion Mass Transfer in Fluid Systems,* Cambridge University Press, Cambridge, UK, 1997.

The larger the *characteristic length L,* the larger the Thiele modulus, the smaller the effectiveness factor, and the steeper the reactant concentration profile in the catalyst particle. A generalized characteristic length definition V_p/S_{px} (particle volume/external particle surface area) brings together the η-ϕ curves for a variety of particle shapes, as illustrated in Table 7-6 and Fig. 7-8 for slabs, cylinders, and spheres. Here I_0 and I_1 are the corresponding modified Bessel functions of the first kind.

Further generalization of the Thiele modulus and effectiveness factor for a general global reaction and various shapes is

$$\varphi = \frac{(V_p/S_{px})r_a(C_{as})}{\sqrt{2\int_{C_{ae}}^{C_{as}} D_{ea} r_a(C_{ay})dC_{ay}}} \tag{7-102}$$

In Eq. (7-102) component A is the limiting reactant. For example, for an nth-order irreversible reaction

$$\varphi = \frac{V_p}{S_{px}}\sqrt{\frac{(n+1)}{2}\frac{kC_{as}^{n-1}}{D_{ea}}} \tag{7-103}$$

This generalized Thiele modulus works well with the effectiveness factors for low and high values of the Thiele modulus, but it is not as accurate for intermediate values. However, these differences are not significant, given the uncertainties associated with measuring some of the other key parameters that go into the calculation of the Thiele modulus, e.g., the effective diffusivity and the intrinsic rate constant.

Effect of Intraparticle Diffusion on Observed Order and Activation Energy For an nth-order reaction in the limit of intraparticle diffusion control, i.e., large Thiele modulus, the effectiveness factor is

$$\eta = \frac{1}{\varphi} \tag{7-104}$$

the observed rate is

$$r_{\text{obs}} = \eta r = \frac{S_{px}}{V_p}\sqrt{\frac{2D_{ea}k}{n+1}}\, C_{as}^{(n+1)/2} \tag{7-105}$$

and the observed rate constant is

$$k_{\text{obs}} = \eta k = \frac{S_{px}}{V_p}\sqrt{\frac{2}{n+1} D_{ea0} e^{-(E_D/RT)} k_0 e^{-(E/RT)}} \tag{7-106}$$

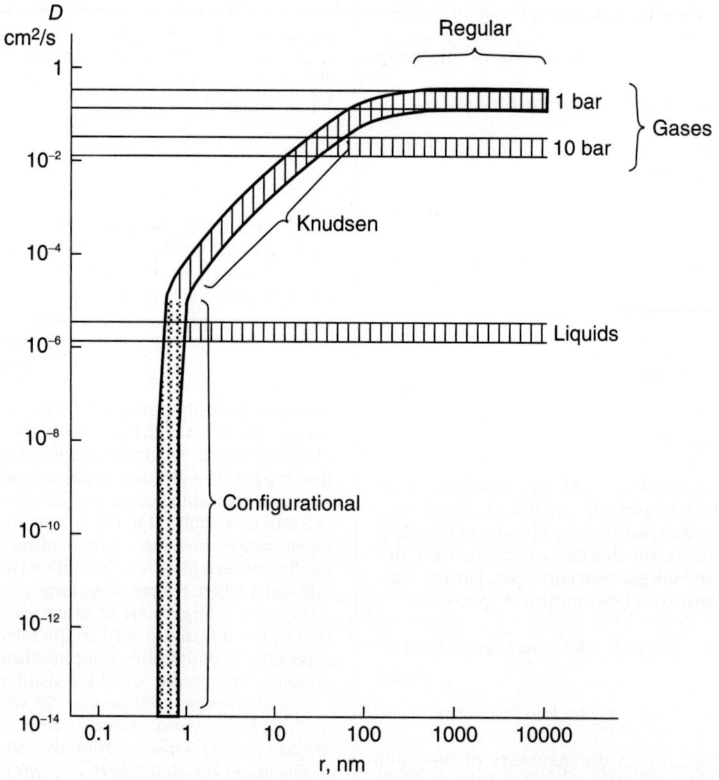

FIG. 7-7 Diffusion regimes in heterogeneous catalysts [From Weisz, *Trans. Fara. Soc.* **69:** 1696–1705 (1973); Froment and Bischoff, *Chemical Reactor Analysis and Design,* Wiley, 1990, Fig. 3.5.1-1.]

Hence, the observed order and activation energy differ from those of the intrinsic *n*th-order kinetics:

$$n_{\text{obs}} = \frac{n+1}{2} \qquad E_{\text{obs}} = \frac{E+E_D}{2} \approx \frac{E}{2} \tag{7-107}$$

Here E_D is the activation energy for diffusion; therefore, the intraparticle diffusion limitation lowers the apparent activation energy.

Note that the observed and intrinsic reaction order is the same under intraparticle diffusion control only for a first-order reaction.

Weisz and Prater ["Interpretation of Measurements in Experimental Catalysis," *Adv. Catal.* **6:** 144 (1954)] developed general estimates for the observed order and activation energy over the entire range of ϕ for an irreversible *n*th-order reaction:

$$n_{\text{obs}} = n + \frac{n-1}{2}\frac{d\ln\eta}{d\ln\varphi} \qquad E_{\text{obs}} = E + \frac{E-E_D}{2}\frac{d\ln\eta}{d\ln\varphi} \tag{7-108}$$

Weisz and Prater ["Interpretation of Measurements in Experimental Catalysis," *Adv. Catal.* **6:** 144 (1954)] also developed a general criterion for diffusion limitations, which can guide the lab analysis of rate data:

$$\text{If } \Phi = \left(\frac{3V_p}{S_{px}}\right)^2 \frac{r_{\text{obs}}}{D_{ea}C_{as}} \begin{cases} >> 1, & \text{then diffusion-limited} \\ << 1, & \text{then no intraparticle diffusion resistance} \end{cases} \tag{7-109}$$

TABLE 7-6 Effectiveness Factors versus Thiele Modulus for Different Shapes for a First-Order Reaction

Shape	V_p/S_{px}	Effectiveness factor η
Infinite slab	R	$\frac{\tanh\varphi}{\varphi}$
Infinite cylinder	$R/2$	$\frac{I_1(2\varphi)}{\varphi I_0(2\varphi)}$
Sphere	$R/3$	$\frac{1}{\varphi}\left(\frac{3}{\tanh 3\varphi} - \frac{1}{\varphi}\right)$

Effect of Intraparticle Diffusion for Reaction Networks For multiple reactions, intraparticle diffusion resistance can also affect the observed selectivity and yield. For example, for consecutive reactions, intraparticle diffusion resistance reduces the yield of the intermediate (often desired) product if both reactions have the same order. For parallel reactions, diffusion resistance reduces the selectivity for the higher-order reaction. For more details see, e.g., Carberry, *Chemical and Catalytic Reaction Engineering,* McGraw-Hill, 1976; and Levenspiel, *Chemical Reaction Engineering,* 3d ed., Wiley, 1999.

For more complex reactions, the effect of the intraparticle diffusion resistance on rate, selectivity, and yield depends on the particulars of the network. Also the use of the Thiele modulus–effectiveness factor relationships is not as easily applicable, and numerical solution of the diffusion-reaction equations may be required.

Intraparticle Diffusion and External Mass Transfer Resistance For typical industrial conditions, external mass transfer is important only if there is substantial intraparticle diffusion resistance. This subject has been discussed by Luss, "Diffusion-Reaction Interactions in Catalyst Pellets," in Carberry and Varma (eds.), *Chemical Reaction and Reactor Engineering,* Marcel Dekker, 1987. This, however, may not be the case for laboratory conditions, and care must be taken to include the proper data interpretation. For instance, for a spherical particle with both external and internal mass-transfer limitations and first-order reaction, an overall effectiveness factor η_t can be derived, indicating the series-of-resistances nature of external mass transfer followed by intraparticle diffusion reaction:

$$\frac{1}{\eta_t} = \frac{1}{\eta} + \frac{\varphi^2}{3Sh'} \tag{7-110}$$

$$Sh' = \frac{\varepsilon_s k_s R}{D_{ea}} \tag{7-111}$$

As indicated above, intraparticle diffusion lowers the apparent activation energy. The apparent activation energy is even further lowered under external mass transfer control. Figure 7-9 illustrates how the rate-controlling step changes with temperature, and as a result the dependence of the apparent first-order rate constant on temperature also changes, from a very strong dependence under kinetic control to being virtually independent of temperature under external mass transfer control.

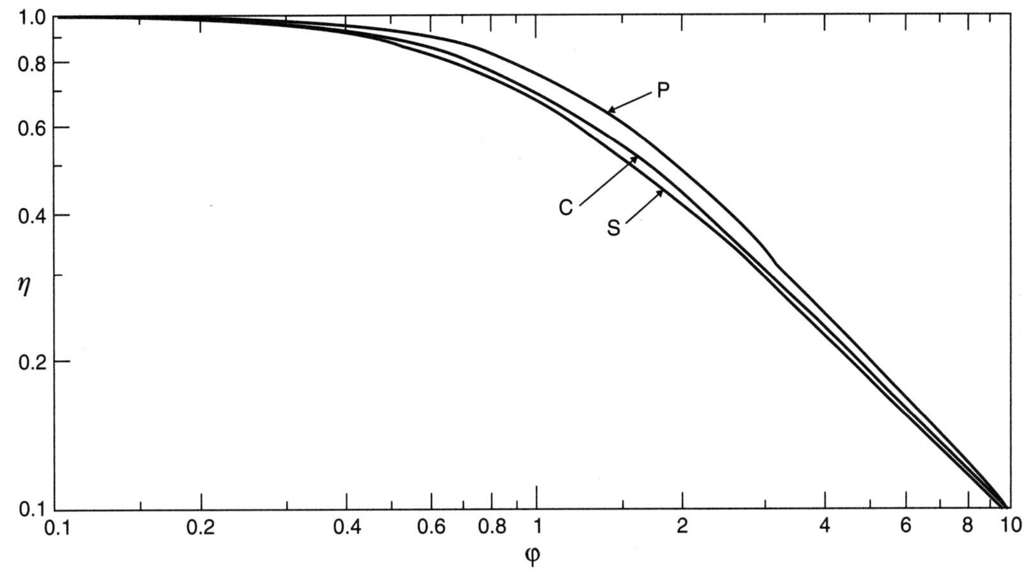

FIG. 7-8 Effectiveness factors versus Thiele modulus for a first order reaction in a slab (P), a cylinder (C), and a sphere (S). [Adapted from Fig. 1 in Aris and Rester, "The Effect of Shape on the Effectiveness Factor," *Chem. Eng. Sci.* **24:** 793 (1969).]

Note that in the limit of external mass transfer control, the activation energy $E_{obs} \rightarrow 0$, as can be shown when substituting Eq. (7-110) in Eq. (7-108). For more details on how to represent the combined effect of external and intraparticle diffusion on the effectiveness factor for more complex systems, see Luss, "Diffusion-Reaction Interactions in Catalyst Pellets," in Carberry and Varma (eds.), *Chemical Reaction and Reactor Engineering,* Marcel Dekker, 1987.

Heat Transfer Resistance A similar analysis regarding external and intraparticle heat transfer limitations leads to temperature gradients which add further complexity to the behavior of heterogeneous catalytic systems, including steady-state multiplicity. More details are given in Sec. 19.

Catalyst Deactivation The catalyst life may range from seconds to minutes to a few days to several years, as the active surface of a catalyst is degraded by chemical, thermal, or mechanical factors. *Chemical deactivation* occurs due to feed or product *poisoning* or *masking.* Poisoning may be due to compounds such as P, S, As, Na, and Bi and is generally considered irreversible. In some cases a reduced life is simply accepted, as in the case of slow accumulation of trace metals from feed to catalytic cracking; but in other cases the deactivation is too rapid. Sulfur and water are removed from feed to ammonia synthesis, sulfur from feed to platinum reforming, and arsenic from feed to SO_2 oxidation with platinum. Masking may be due to covering of the active sites by contaminants in either the feed or products. Examples of feed masking agents can include Si (from organic silicones) and rust. An example of product masking is coking. Catalysts may be regenerated depending on the cost-effectiveness of the regeneration process.

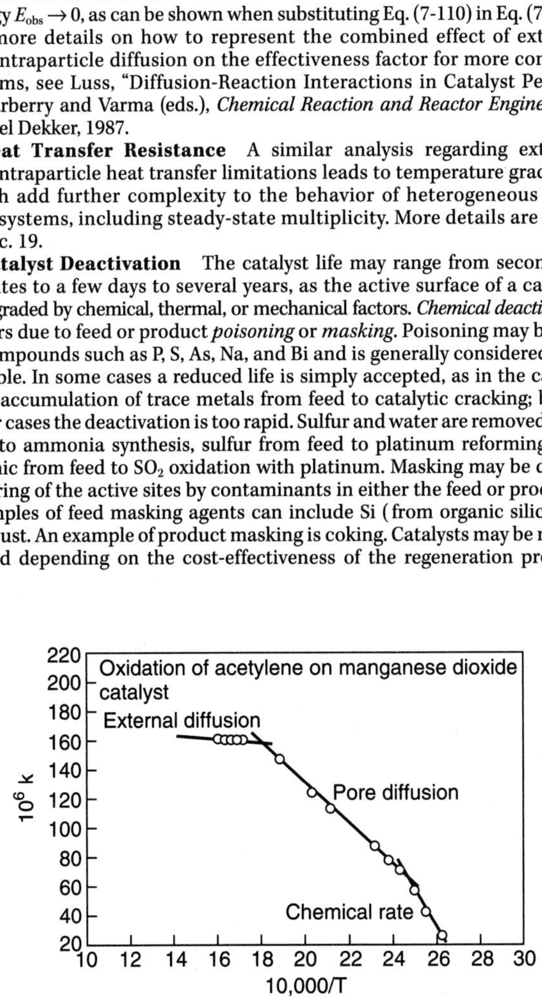

FIG. 7-9 Dependence of the rate-controlling step on temperature.

Catalyst regeneration sometimes is done in place; for instance, coke is burned off cracking catalyst or off nickel and nickel-molybdenum catalysts in a fluidized reactor/regenerator system. *Thermal deactivation* is primarily due to rearrangement of the active sites at high temperature due to *sintering.* Sintering results in agglomeration of active ingredients (lower dispersion). In most cases sintering is irreversible; however, Pt/Al_2O_3 catalysts have been regenerated in place by Cl_2 treatment. The catalyst also can be modified by additives, for instance, chromia to nickel to prevent sintering, and rhenium to platinum to reduce coking. *Mechanical deactivation* may be caused by attrition or erosion and subsequent loss of catalyst as fines. The attrition resistance of catalysts is related to the nature of the support and its porosity.

For additional references, see, e.g., Thomas, *Catalytic Processes and Proven Catalysts,* Academic, 1970; Butt and Petersen, *Activation, Deactivation and Poisoning of Catalysts,* Academic, 1988; and Delmon and Froment, *Catalyst Deactivation,* Elsevier, 1980.

The *activity* α at any time on stream may be simply defined as the ratio of the rate at time *t* to the rate with fresh catalyst

$$\alpha = \frac{r(t)}{r(t=0)} \tag{7-112}$$

The rate of destruction of active sites and pore structure can be expressed as a kinetic relation that can be generally first- or second-order. For instance, for a second-order dependence,

$$\frac{d\alpha}{dt} = -k_d \alpha^2 \tag{7-113}$$

and the corresponding integral is

$$\alpha = \frac{1}{1 + k_d t} \tag{7-114}$$

This type of deactivation mechanism often applies to catalyst sintering and coke deactivation. The deactivation rate constant is expected to have an Arrhenius dependence on temperature.

When the feedstock contains constant proportions of reactive impurities, the rate of decline may also depend on the concentration of the main reactant, e.g., for a power law rate

$$\frac{d\alpha}{dt} = -k_d \alpha^p C^q \tag{7-115}$$

This differential equation now must be solved simultaneously with a rate equation for the main reactant.

The deactivation rate constants are estimated by methods similar to those for finding constants of any rate equation, given suitable (α, t) data. There are different chemical deactivation mechanisms for catalyst pellets and two of the most common are described below. For more details see Butt and Petersen, *Activation, Deactivation and Poisoning of Catalysts,* Academic, 1988; and Froment and Bischoff, *Chemical Reactor Analysis and Design,* Wiley, 1990. In *uniform deactivation,* the poison is distributed uniformly throughout the pellet and degrades it gradually. In *pore mouth* (*shell progressive*) *poisoning,* the poison is so effective that it kills the active site as it enters the pore; hence complete deactivation begins at the mouth and moves gradually inward.

Uniform Deactivation When uniform deactivation occurs, the specific rate declines by a factor $1 - \beta$, where β is the fractional poisoning. Factor β is calculated from the poisoning rate, and it is often assumed to be proportional to the concentration of the poison in the bulk fluid. Then a power law rate equation becomes

$$r = k(1-\beta)\eta C_s^n \tag{7-116}$$

The effectiveness factor depends on β through the Thiele modulus

$$\varphi = \frac{V_p}{S_{px}}\sqrt{\frac{n+1}{2}\frac{k(1-\beta)C_s^{n-1}}{D_e}} \tag{7-117}$$

To find the effectiveness under poisoned conditions, this form of the Thiele modulus is substituted into the appropriate relation for the effectiveness factor. For example, for a first-order reaction in slab geometry, with $V_p/S_{px} = L$, the effectiveness factor is

$$\eta = \frac{\tanh\phi}{\phi} = \frac{\tanh\left(L\sqrt{k(1-\beta)/D_e}\right)}{L\sqrt{k(1-\beta)/D_e}} \tag{7-118}$$

Figure 7-10*a* shows the ratio of the effectiveness factor with fresh catalyst to that with uniform poisoning for the above case of first-order reaction in a slab. Hence for the same value of the Thiele modulus, the effectiveness factor is larger for the poisoned catalyst, or alternatively the poisoned catalyst Thiele modulus is smaller, resulting in a larger effectiveness factor than that of the fresh catalyst.

Pore Mouth (or Shell Progressive) Poisoning This mechanism occurs when the poisoning of a pore surface begins at the mouth of the pore and moves gradually inward. This is a moving boundary problem, and the pseudo-steady-state assumption is made that the boundary moves slowly compared with diffusion of poison and reactants to the active surface. Here β is the fraction of the pore that is deactivated. The poison diffuses through the dead zone and deposits at the interface between the dead and active zones. The reactants diffuse across the dead zone without reaction, followed by diffusion reaction in the active zone.

Figure 7-10*b* shows simulation results for the ratio of the effectiveness factor with no poisoning to that with pore mouth poisoning for the same first-order reaction in a slab shown above. Here as well the effectiveness factor of the poisoned catalyst is larger than that of the fresh catalyst.

GAS-SOLID NONCATALYTIC REACTIONS

Examples of gas-solid noncatalytic reactions include production of iron from iron ores, roasting of sulfide oxides, combustion of solid fuels, chlorination of Ti ores to make $TiCl_4$ in the production of TiO_2 pigments, incineration of waste, gasification of coal and biomass to produce syngas, and decomposition of solids to produce gases, e.g., solid propellants and explosives. The kinetic treatment of these reactions has to take into consideration external mass transfer and intraparticle diffusion just as in the case of gas-solid catalytic reactions. However, there are major differences, the primary one being consumption of the solid reactant, making the conditions inside the solid particle transient in nature, including change in unreacted particle size, particle density, porosity, etc. For more details see, e.g., Wen ["Noncatalytic Heterogeneous Solid-Fluid Reaction Models," *Ind. Eng. Chem.* **60**(9): 34–54 (1968)], Szekely [in Lapidus and Amundson (eds.), *Chemical Reactor Theory—A Review,* Prentice-Hall, 1977], Doraiswamy and Kulkarni [in Carberry and Varma (eds.), *Chemical Reaction and Reactor Engineering,* Marcel Dekker, 1987], and Levenspiel (*Chemical Reaction Engineering,* 3d ed., Wiley, 1999).

The basic steps are identical to those of catalytic gas-solid reactions. However, as indicated above, the process is transient (non-steady-state) due to change in particle size and properties as the reaction progresses.

Several models that describe gas-solid noncatalytic reactions are summarized in Table 7-7. The first two, the sharp interface and volume reaction models, are pseudohomogeneous, part of the class of shrinking core models, and can be treated by using the Thiele modulus and effectiveness factor concept. The last three are heterogeneous models. The following discussion is restricted to spherical particles.

Sharp Interface Model For a first-order reaction in gas and solid reactants,

$$A(g) + bB(s) \Rightarrow \text{products} \tag{7-119}$$

the rate of conversion of the solid B per unit area at the solid surface is

$$r_b = kC_{as}C_s \tag{7-120}$$

Assuming pseudo-steady state for the gas reactant A, and constant solid reactant B concentration C_{s0}, it can be shown that the observed rate per particle external surface area at any time t is

$$r_{a,\text{obs}} = \frac{C_{a0}}{\dfrac{1}{k_p} + \dfrac{R_p}{D_{ea}}\left(\dfrac{R_p}{R_s} - 1\right) + \dfrac{b}{kC_{s0}}\left(\dfrac{R_p}{R_s}\right)^2} \tag{7-121}$$

where R_p and R_s are the particle and solid core radii. Equation (7-121) represents three resistances in series for the gaseous reactant: external mass transfer R_{am}, diffusion in the reacted (ash) zone R_{ad}, and reaction at the unreacted solid-ash interface R_{ar} as indicated below (for details see Doraiswamy and Kulkarni [in Carberry and Varma (eds.), *Chemical Reaction and Reactor Engineering,* Marcel Dekker, 1987]):

$$\begin{gathered} R_{am} = 4\pi R_p^2 k_p (C_{a0} - C_{ap}) \\ \frac{1}{r^2}\frac{\partial}{\partial r}\left(D_{ea} r^2 \frac{\partial C_a}{\partial r}\right) = 0 \quad R_s \le r \le R_p \Rightarrow R_{ad} = \frac{4\pi R_p R_s D_{ea}}{R_p - R_s}(C_{ap} - C_{as}) \\ R_{ar} = 4\pi R_s^2 \frac{kC_{as}C_{s0}}{b} \quad C_{s0} = \frac{\rho_s}{MW_s} \end{gathered} \tag{7-122}$$

The moving boundary radius R_s is determined from a material balance that relates the unreacted solid volume to the observed reaction rate. Integration gives the time τ required to achieve a given conversion of the solid B, X_s:

$$\tau = \frac{R_p C_{s0}}{bC_{a0}}\left\{\frac{1}{3}\left(\frac{1}{k_p} - \frac{R_p}{D_{ea}}\right)X_s + \frac{R_p}{2D_{ea}}\left[1-(1-X_s)^{2/3}\right] + \frac{b}{C_{s0}k}\left[1-(1-X_s)^{1/3}\right]\right\}$$

$$-\frac{d}{dt}\left(\frac{4\pi R_s^3 C_{s0}}{3}\right) = 4\pi R_p^2 b r_{a,\text{obs}} \quad X_s = 1 - \left(\frac{R_s}{R_p}\right)^3 \tag{7-123}$$

Similar solutions can be obtained for other shapes [Doraiswamy and Kulkarni, in Carberry and Varma (eds.), *Chemical Reaction and Reactor Engineering,* Dekker, 1987]. Figure 7-11 shows typical concentration profiles for this case.

Volume Reaction Model A typical concentration profile for the volume reaction model is shown in Fig. 7-12.

A general transient model of diffusion reaction that uses the effective diffusivity concept described for gas-solid catalytic reactions can be derived here as well, e.g., for a spherical particle:

$$\frac{\partial(\varepsilon_s C_{as})}{\partial t} = \frac{1}{r^2}\frac{\partial}{\partial r}\left(D_{ea} r^2 \frac{\partial C_{as}}{\partial r}\right) - r_a \quad \frac{\partial C_s}{\partial t} = -r_s$$

$$C_{as} = C_{as0} \quad C_s = C_{s0} \quad \text{at} \quad t = 0 \tag{7-124}$$

$$\frac{\partial C_{as}}{\partial r} = 0 \quad \text{at} \quad r = 0 \quad D_{ea}\frac{\partial C_{as}}{\partial r}\Big|_{r=R_p} = k_p\left(C_a - C_{as}\big|_{r=R_p}\right)$$

$$\varepsilon_s = \varepsilon_{s0} + C_{s0}(\upsilon_s - \upsilon_p)\left(1 - \frac{C_s}{C_{s0}}\right) \quad \frac{D_{ea}}{D_{ea0}} = \left(\frac{\varepsilon_s}{\varepsilon_{s0}}\right)^\beta \quad \beta = 2\text{–}3 \tag{7-125}$$

Here the porosity and the effective diffusivity vary with conversion of solid; υ_s and υ_p are the reactant and product molar volumes. A Thiele modulus φ

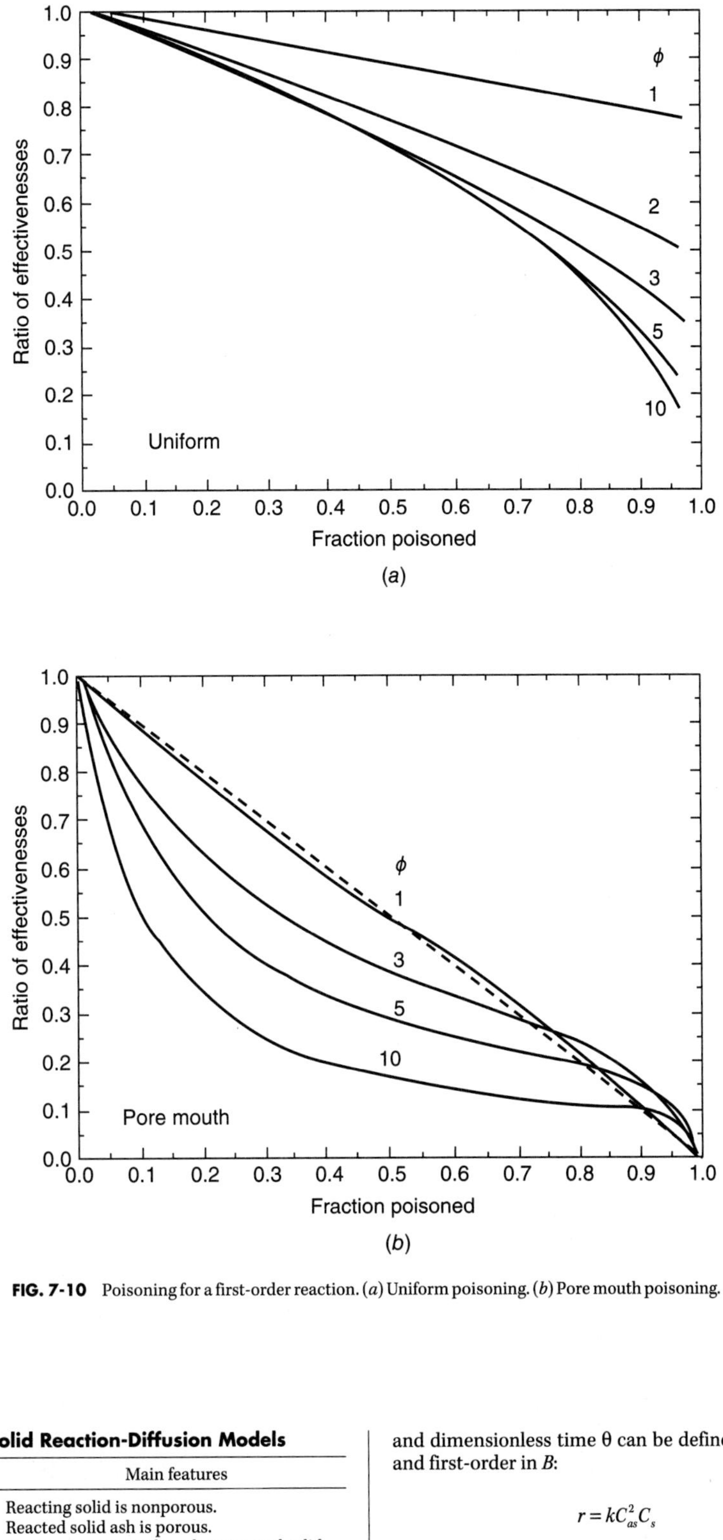

FIG. 7-10 Poisoning for a first-order reaction. (*a*) Uniform poisoning. (*b*) Pore mouth poisoning.

TABLE 7-7 Noncatalytic Gas-Solid Reaction-Diffusion Models

Model	Main features
Sharp interface model (SIM)	Reacting solid is nonporous. Reacted solid ash is porous. Reaction occurs at the ash-unreacted solid interface.
Volume reaction model	Reacting solid is also porous. Reaction occurs everywhere in the particle.
Grain model	Particle is divided into identical solid spherical grains. Each grain reacts according to the sharp interface model.
Crackling core model	Combination of SIM and grain model.
Nucleation model	Nucleation of metals in metal reduction reactions.

and dimensionless time θ can be defined, e.g., for a rate second-order in *A* and first-order in *B*:

$$r = kC_{as}^2 C_s \tag{7-126}$$

$$\varphi = R_p \sqrt{\frac{kC_{as}(R_p)C_{s0}}{D_{ea0}}} \qquad \theta = bkC_{as0}^2 t \tag{7-127}$$

For the given rate expression, Eqs. (7-124) to (7-127) can be numerically integrated, e.g., in Fig. 7-13 for reaction control and Fig. 7-14 for intraparticle diffusion control, both with negligible external mass transfer resistance; *x* is the fractional conversion.

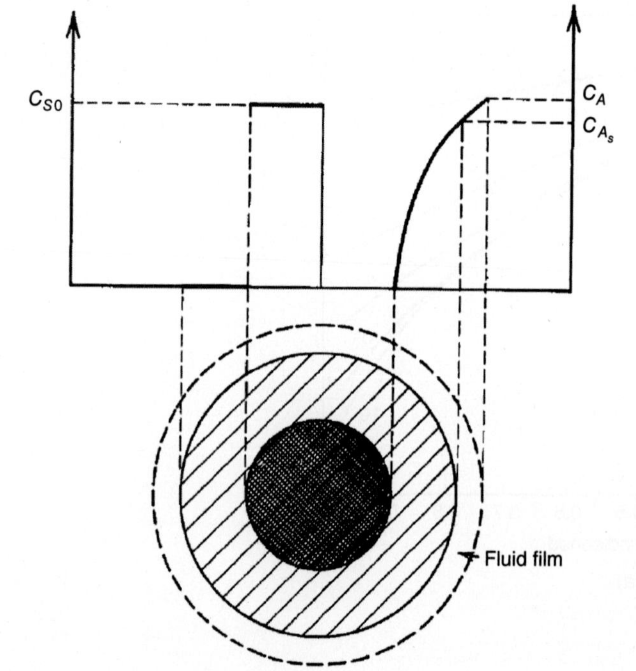

FIG. 7-11 Sharp interface model—concentration profiles [From Wen, "Noncatalytic Heterogeneous Solid-Fluid Reaction Models," *Ind. Eng. Chem.* **60**(9): 34–54 (1968), Fig. 1.]

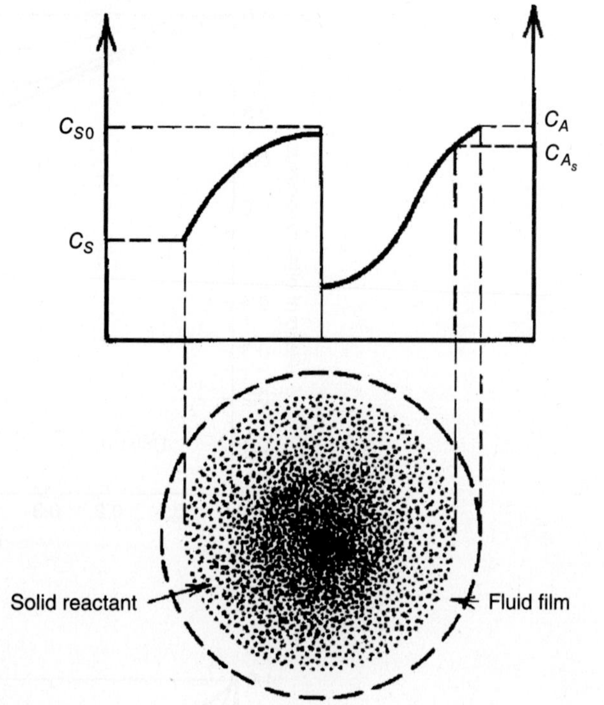

FIG. 7-12 Volume reaction model—concentration profiles [From Wen, "Noncatalytic Heterogeneous Solid-Fluid Reaction Models," *Ind. Eng. Chem.* **60**(9): 34–54 (1968), Fig. 3.]

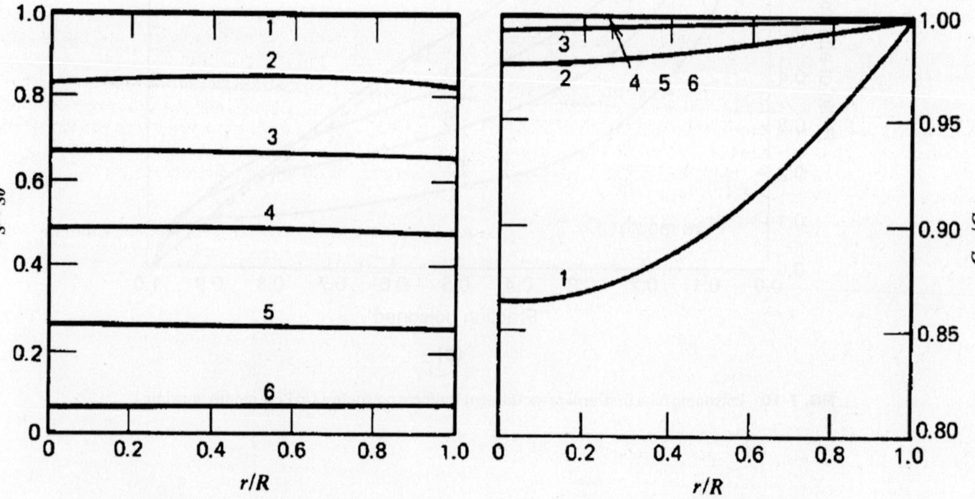

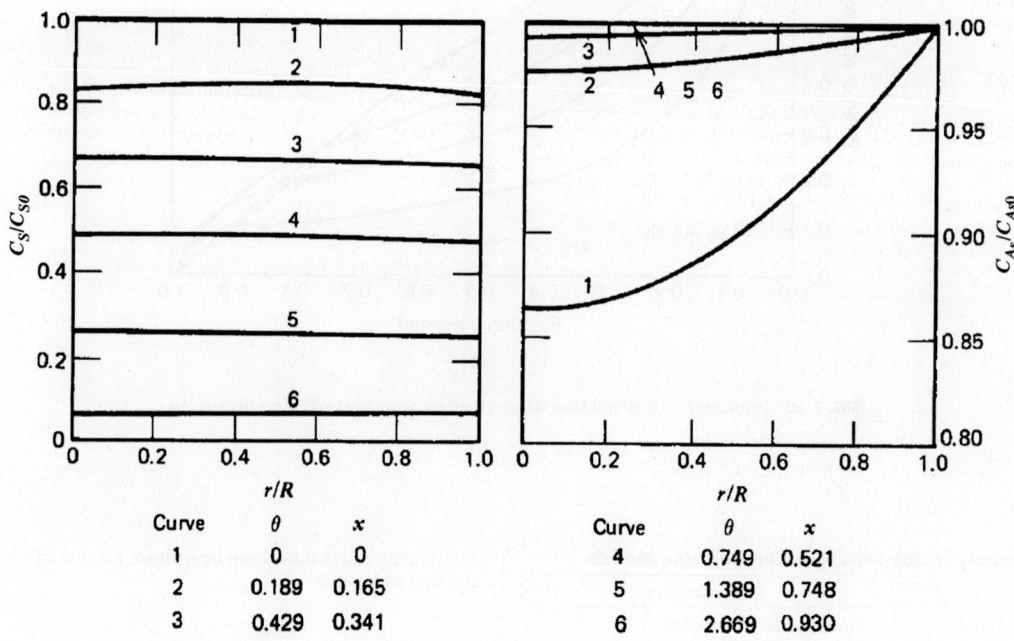

Curve	θ	x
1	0	0
2	0.189	0.165
3	0.429	0.341

Curve	θ	x
4	0.749	0.521
5	1.389	0.748
6	2.669	0.930

FIG. 7-13 Concentration profiles with reaction control $\phi = 1$, in the absence of gas particle mass transfer resistance. [From Wen, "Noncatalytic Heterogeneous Solid-Fluid Reaction Models," *Ind. Eng. Chem.* **60**(9): 34–54 (1968), Fig. 11.]

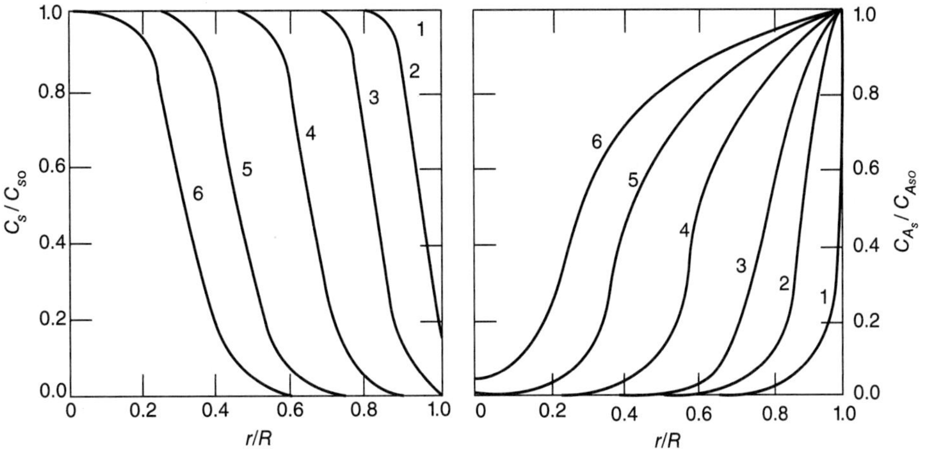

Curve	θ	x
1	0	0
2	1.84	0.151
3	4.88	0.417
4	10.16	0.708
5	16.40	0.892
6	20.56	0.957

FIG. 7-14 Concentration profiles with intraparticle diffusion control, $\phi = 70$, in absence of gas particle mass transfer resistance. [From Wen, "Noncatalytic Heterogeneous Solid-Fluid Reaction Models," *Ind. Eng. Chem.* **60**(9): 34–54 (1968), Fig. 12.]

GAS-LIQUID REACTIONS

Many industrial processes employ gas-liquid reactions that can be either noncatalytic or homogeneously catalyzed. These include, for instance, absorption of acid gases (SO_3, NO_2, CO_2), chlorinations (benzene, dodecane, toluene), oxidations (P-xylene to terephthalic acid, cyclohexane to cyclohexanone and cyclohexanol, acetaldehyde to acetic acid), hydrogenations (olefins, esters to fatty acids), and hydroformylation of olefins to alcohols, to name a few. See also Sec. 19 and Shah (*Gas-Liquid-Solid Reactor Design*, McGraw-Hill, 1979). These reactions include gas reactants dissolving in a liquid and reacting there with a liquid reactant. When determining the kinetics of such reactions from lab data, one needs to understand the mechanism and the controlling steps, just as in the case for heterogeneous gas-solid reactions. The simplest model is the two-film model. It involves the following consecutive steps for the gaseous reactant:

1. Mass transfer of the gas reactant from bulk gas to the gas-liquid interface across the gas film.
2. Mass transfer of the dissolved gas reactant to the bulk liquid across the liquid film—if the reaction is fast, the reaction will occur in the liquid film (in parallel with diffusion).
3. Reaction in the bulk liquid.

At the gas-liquid interface, the liquid and gas concentrations of the gaseous reactant are assumed to be at thermodynamic equilibrium.

For a volatile liquid reactant or a volatile product, these steps are essentially reversed. For a nonvolatile liquid reactant or product, only the reaction and diffusion in the liquid take place. Figure 7-15 describes the absorbing gas concentration profiles in a gas-liquid system (excluding the solid catalyst in this case).

For a general gas-liquid reaction:

$$A(g) + bB(l) \rightarrow \text{products} \tag{7-128}$$

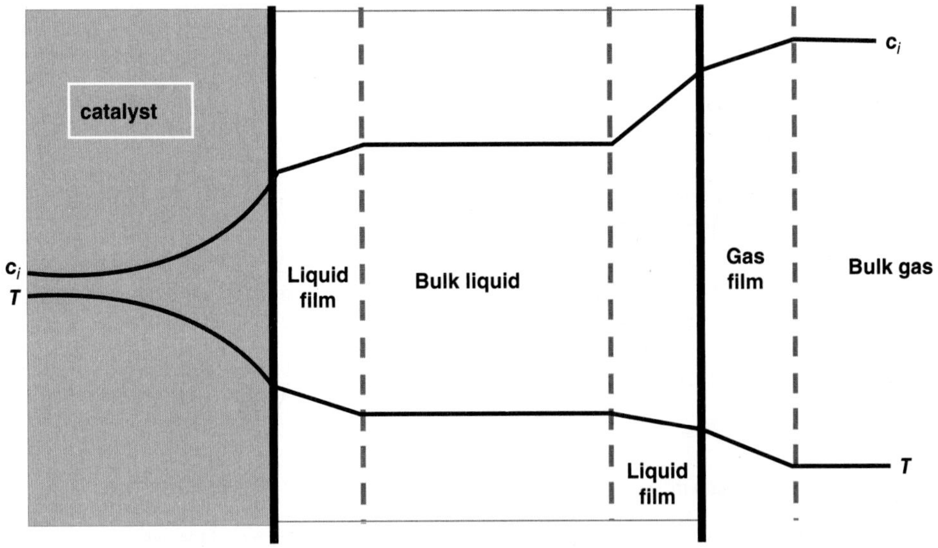

FIG. 7-15 Absorbing gas concentration and temperature profiles (exothermic reaction) in gas-liquid and gas-liquid-solid reactions.

the two-film pseudo-steady-state model is described by the following fluxes across the interface for the gaseous reactant A:

$$N_{aG}=\frac{D_{aG}}{\delta_G}(P_a-P_{ai})=k_G(P_a-P_{ai})=N_{aL}=\frac{D_{aL}}{\delta_L}(C_{Lai}-C_{La})=k_L(C_{Lai}-C_{La}) \tag{7-129}$$

Here the subscript L denotes liquid, G denotes gas, i denotes the gas-liquid interface (where the gas and liquid concentrations are in equilibrium). The thickness of the liquid and gas films is not a directly measurable quantity, and instead mass transfer coefficients are defined as indicated above. These depend on the diffusivity of the molecule, geometry, flow patterns and rates, and operating conditions; typical values of mass transfer coefficients can be viewed in Sec. 19. In addition to the two-film steady-state model, other more accurate, non-steady-state models have also been developed such as the surface renewal and penetration models (see, e.g., Astarita, *Mass Transfer with Chemical Reaction*, Elsevier, 1967). In many cases of industrial interest, mass-transfer resistance in the gas-film is negligible, especially considering that gas-phase diffusivities are 2 to 3 orders of magnitude larger for the same species than those in the liquid. Hence we drop the subscripts L and G from the concentrations since the concentrations considered are in the liquid phase only.

REACTION-DIFFUSION REGIMES

Depending on the relative rates of diffusion and reaction, the following diffusion-reaction regimes occur:

$t_D << t_r$ slow reaction regime with reaction control

$t_D >> t_r$ fast reaction regime with diffusion control

$t_D \simeq t_r$ both reaction and diffusion are important (7-130)

$$t_D=\frac{D_a}{k_L^2} \qquad t_r=\frac{C_{ai}-C_{ae}}{r(C_{ai})}$$

Here t_D and t_r are the diffusion and reaction times, respectively, C_{ai} is the concentration at the liquid side of the interface and C_{ae} is the reaction equilibrium concentration of the gaseous reactant A, and k_L is the mass transfer coefficient. For the *fast reaction regime*, diffusion and reaction occur simultaneously in the liquid film, while for the *slow reaction regime*, there is no reaction in the liquid film and the mass transfer can be considered to occur independently of reaction in a consecutive manner. For the slow reaction regime, the following subregimes can be defined:

$t_m << t_r$ slow reaction kinetic control

$t_m >> t_r$ slow reaction mass transfer control

$t_m \simeq t_r$ both reaction and mass transfer are important (7-131)

$$t_m=\frac{1}{k_L a}$$

Here t_m is the mass transfer time, and a is the *gas-liquid interfacial area per unit volume of reactor*. Only under *slow reaction kinetic control regime* can intrinsic kinetics be derived directly from lab data. Otherwise the intrinsic kinetics have to be extracted from the observed rate by using the mass transfer and diffusion-reaction equations, in a manner similar to those defined for catalytic gas-solid reactions. For instance, in the *slow reaction regime* for a first-order reaction,

$$r_{a,\text{obs}}=\frac{C_{ai}}{\text{He}_a/(k_G a)+1/(k_L a)+1/k} \tag{7-132}$$

$$k_{\text{obs}}=\frac{1}{\text{He}_a/(k_G a)+1(k_L a)+1/k} \tag{7-133}$$

Here He_a is the Henry constant for the gaseous solute A. For the fast reaction regime, instead of the effectiveness factor adjustment for the intrinsic reaction rate, it is customary to define an *enhancement factor* to describe the observed mass transfer enhancement by the reaction, defined as the ratio of mass transfer in the presence of reaction in the liquid to mass transfer in the absence of reaction:

$$E=k_L/k_{L0} \tag{7-134}$$

Solving the diffusion-reaction equation in the liquid, the enhancement factor can be related to the *Hatta number* Ha, which is similar to the Thiele modulus defined for heterogeneous gas-solid catalysts. Thus, the Hatta number and its relation to the controlling regime are

$$\text{Ha}=\sqrt{\frac{t_D}{t_R}}=\sqrt{\frac{\text{reaction rate in the film}}{\text{diffusion rate through the film}}}$$

Ha << 1 slow reaction regime (7-135)

Ha >> 1 fast reaction regime

For instance, for a first-order reaction in the gaseous reactant A (e.g., with large excess of liquid reactant B), using the film model, the following relates the enhancement factor to the Hatta number:

$$\text{Ha}=\delta_L\sqrt{\frac{k}{D_a}}=\sqrt{\frac{kD_a}{k_{L0}^2}} \qquad \text{for } C_b >> C_{ai} \tag{7-136}$$

$$E=\frac{\text{Ha}}{\tanh \text{Ha}}\left(1-\frac{C_a}{C_{ai}}\frac{1}{\cosh \text{Ha}}\right)>1 \tag{7-137}$$

When both A and B have comparable concentrations, then the enhancement factor is an increasing function of an additional parameter:

$$\lambda=\frac{D_b C_b}{bD_a C_{ai}} \tag{7-138}$$

In the limit of an irreversible instantaneous reaction, the reaction occurs at a plane in the liquid where the concentration of both reactants A and B is zero and the flux of A equals the flux of B. The criterion for an instantaneous reaction is

$$\text{Ha} >> \lambda \qquad E_\infty = 1+\lambda >> 1 \tag{7-139}$$

Figure 7-16 illustrates typical concentration profiles of A and B for the various diffusion-reaction regimes.

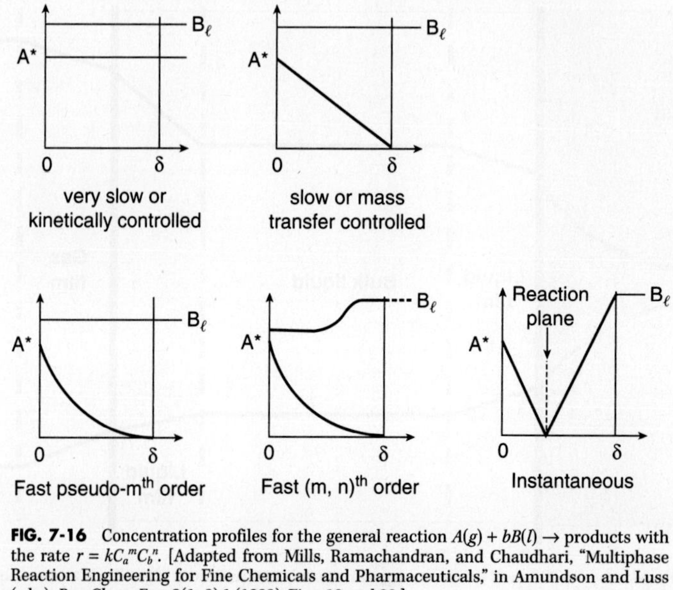

FIG. 7-16 Concentration profiles for the general reaction $A(g) + bB(l) \rightarrow$ products with the rate $r = kC_a^m C_b^n$. [Adapted from Mills, Ramachandran, and Chaudhari, "Multiphase Reaction Engineering for Fine Chemicals and Pharmaceuticals," in Amundson and Luss (eds.), *Rev. Chem. Eng.* **8**(1–2):1 (1992), Figs. 19 and 20.]

GAS-LIQUID-SOLID REACTIONS

GAS-LIQUID-SOLID CATALYTIC REACTIONS

Many solid catalyzed reactions take place with one of the reactants absorbing from the gas phase into the liquid and reacting with a liquid reactant on the surface or inside the pores of a solid catalyst (see Fig. 7-15). Examples include the Fischer-Tropsch synthesis of hydrocarbons from synthesis gas (CO and H_2) in the presence of Fe or Co-based heterogeneous catalysts, methanol synthesis from synthesis gas (H_2 + CO) in the presence of heterogeneous CuO/ZnO catalyst, and a large number of noble metal catalyzed hydrogenations among others. For a slow first-order reaction of a gaseous reactant, the concept of resistances in series can be expanded as follows, e.g., for a slurry reactor with fine catalyst powder:

$$r_{a,\text{obs}} = \frac{C_{ai}}{\dfrac{\text{He}_a}{k_G a} + \dfrac{1}{k_L a} + \dfrac{1}{k_s a_s} + \dfrac{1}{k}} \qquad k_{\text{obs}} = \frac{1}{\dfrac{\text{He}_a}{k_G a} + \dfrac{1}{k_L a} + \dfrac{1}{k_s a_s} + \dfrac{1}{k}} \tag{7-140}$$

Intraparticle diffusion resistance may become important when the particles are larger than the powders used in slurry reactors, such as for catalytic packed beds operating in trickle flow mode (downflow gas and liquid), in packed bubble column mode (upflow gas and liquid), or countercurrent mode (gas upflow and liquid downflow). For these the effectiveness factor concept for intraparticle diffusion resistance has to be considered in addition to the other resistances present. See more details in Sec. 19.

POLYMERIZATION REACTIONS

Polymers are high-molecular-weight compounds assembled by the linking of small molecules called *monomers*. Most polymerization reactions involve two or three phases, as indicated below. There are several excellent references dealing with *polymerization kinetics* and reactors, including Ray, *J. Macromol. Sci.-Revs. Macromol. Chem.* **C8:** 1 (1972); Ray in Lapidus and Amundson (eds.), *Chemical Reactor Theory—A Review*, Prentice-Hall, 1977; Tirrel et al. in Carberry and Varma (eds.), *Chemical Reaction and Reactor Engineering*, Marcel Dekker, 1987; Dotson et al., *Polymerization Process Modeling*, Wiley, 1995; and Meyer and Keurentjes (eds.), *Handbook of Polymer Reaction Engineering*, Wiley, 2005. An overview of polymerization modeling in industry can be found in Mueller et al., *Macromol. Reaction Eng.* **5:** 261 (2011). A general reference for polymers is Rodriguez et al., *Principles of Polymer Systems*, 6th ed., CRC Press, 2014.

Polymerization can be classified according to the main phase in which the reaction occurs as liquid (most polymerizations), vapor (e.g., Ziegler Natta polymerization of olefins), and solid phase (e.g., finishing of melt polymerization). Polymerization reactions can be further classified into

1. Bulk mass polymerization
   a. Polymer soluble in monomer
   b. Polymer insoluble in monomer
   c. Polymer swollen by monomer
2. Solution polymerization
   a. Polymer soluble in solvent
   b. Polymer insoluble in solvent
3. Suspension polymerization with initiator dissolved in monomer
4. Emulsion polymerization with initiator dissolved in dispersing medium

Polymerization can be catalytic or noncatalytic, and it can be homogeneously or heterogeneously catalyzed. Polymers that form from the liquid phase may remain dissolved in the remaining monomer or solvent, or they may precipitate. Sometimes beads are formed and remain in suspension; sometimes emulsions form. In some processes, solid polymers precipitate from a fluidized gas phase. Polymerization processes are also characterized by extremes in temperature, viscosity, and reaction times. For instance, many industrial polymers are mixtures with molecular weights of 10^4 to 10^7. In polymerization of styrene the viscosity increases by a factor of 10^6 as conversion increases from 0 to 60 percent. The adiabatic reaction temperature for complete polymerization of ethylene is 1800 K (3240°R). Initiators of the chain reactions have concentration as low as 10^{-8} g-mol/L, so they are highly sensitive to small concentrations of poisons and impurities.

Polymerization mechanism and kinetics require special treatment and special mathematical tools due to the very large number of similar reaction steps. Some polymerization types are briefly described next. In this subsection species names and concentrations are denoted by the same capital letters.

Bulk Polymerization The monomer and initiators are reacted with or without mixing (e.g., without mixing to make useful shapes directly, such as in poly(methyl methacrylate) acrylic glass sheet). Because of viscosity limitations, stirred bulk polymerization is not carried to completion. For instance, for addition polymerization, conversions as low as 30 to 60 percent are achieved, with the remaining monomer stripped out and recycled (e.g., in the case of polystyrene).

Suspension or Bead Polymerization Bulk reaction proceeds in droplets of 10 μm to 1000 μm diameter suspended in water or another medium and insulated from one another by some colloid. A typical suspending agent is polyvinyl alcohol dissolved in water. The polymerization can be done to high conversion. Temperature control is easy because of the moderating thermal effect of the water and its low viscosity. The suspensions sometimes are unstable, and agitation may be critical. Examples are polyvinyl acetate in methanol, copolymers of acrylates and methacrylates, and polyacrylonitrile in aqueous $ZnCl_2$ solution.

Emulsion Polymerization Emulsions have particles of 0.05 to 5.0 μm in diameter. The product is a stable latex, rather than a filterable suspension. Some latexes are usable directly, as in paints, or they may be coagulated by various means to produce very high-molecular-weight polymers. Examples are polyvinyl chloride and butadiene-styrene rubber.

Solution Polymerization These processes may retain the polymer in solution or precipitate it. Examples include polyethylene, the copolymerization of styrene and acrylonitrile in methanol, and the aqueous solution of acrylonitrile to precipitate polyacrylonitrile.

Polymer Characterization The physical properties of polymers depend largely on the *molecular weight distribution* (MWD), which can cover a wide range. Since it is impractical to fractionate the products and reformulate them into desirable ranges of molecular weights, immediate attainment of desired properties must be achieved through the correct choice of reactor type and operating conditions, notably of distributions of residence time and temperature. High viscosities influence those factors. For instance, high viscosities prevalent in bulk and melt polymerizations can be avoided by using solutions, suspensions, or emulsions. The interaction between the flow pattern in the reactor and the type of reaction affects the MWD. If the period during which the molecule is growing is short compared with the residence time in the reactor, the MWD in a batch reactor is broader than in a CSTR. This situation holds for many free radical and ionic polymerization processes where the reaction intermediates are very short lived. In cases where the growth period is the same as the residence time in the reactor, the MWD is narrower in batch than in CSTR. Polymerizations that have no termination step—for instance, polycondensations—are of this type. This topic is treated by Denbigh [*J. Applied Chem.* **1:** 227 (1951)] and Schork et al. [*Control of Polymerization Reactors*, Marcel Dekker, 1993].

Four types of MWD which are encountered in practice can be defined: (1) The *number chain length distribution* (NCLD), relating the chain length distribution to the number of molecules per unit volume; (2) the *weight chain length distribution* (WCLD) relating the chain length distribution to the weight of molecules per unit volume; (3) the *number molecular weight distribution* (NMWD) relating the chain length distribution to molecular weight; and (4) the *weight molecular weight distribution* (WMWD) relating the weight distribution to molecular weight. It is often not necessary to know the entire molecular weight distribution, and it is enough to know the first few averages. Two average molecular weights and corresponding average chain lengths are typically defined: the *number average molecular weight* M_n and the corresponding *number average chain length* μ_n; and the *weight average molecular weight* M_w and the corresponding *weight average chain length* μ_w. The ratio of the weight to the number average molecular weights or chain lengths is called *polydispersity* (PD) and describes the width of the molecular weight distribution.

$$M_n = \frac{\sum_{j=1}^{\infty}(jw)P_j}{\sum_{j=1}^{\infty}P_j} \qquad \mu_n = \frac{\sum_{j=1}^{\infty}jP_j}{\sum_{j=1}^{\infty}P_j} \qquad M_w = \frac{\sum_{j=1}^{\infty}(jw)^2 P_j}{\sum_{j=1}^{\infty}(jw)P_j} \qquad \mu_w = \frac{\sum_{j=1}^{\infty}j^2P_j}{\sum_{j=1}^{\infty}jP_j} \tag{7-141}$$

$$\text{PD} = \frac{M_w}{M_n} = \frac{\mu_w}{\mu_n}$$

The average chain lengths can be related to the kth moments λ_k of the distribution as follows:

$$\mu_n = \frac{\lambda_1}{\lambda_0} \quad \mu_w = \frac{\lambda_2}{\lambda_1} \quad \text{PD} = \frac{\lambda_0\lambda_2}{\lambda_1^2} \quad \text{where } \lambda_k = \sum_{j=1}^{\infty} j^k P_j \tag{7-142}$$

Here P_j is the concentration of the polymer with chain length j—the same symbol is also used for representing the polymer species P_j; w is the molecular weight of the repeating unit in the chain.

A factor in addition to the residence time distribution and temperature distribution that affects the molecular weight distribution is the type of chemical reaction (e.g., step or chain addition growth polymerization).

Two major polymerization mechanisms are considered: *chain growth* and *step growth*. In addition, polymerization can be *homopolymerization*—a single monomer is used—and *copolymerization* usually with two or more monomers with complementary functional groups.

Chain Growth Homopolymerization Mechanism and Kinetics Free radical and ionic polymerizations proceed through this type of mechanism, such as styrene polymerization. Here one monomer molecule is added to the chain in each step. The general reaction steps and corresponding rates can be written as follows:

$$
\begin{array}{lll}
I\cdot \xrightarrow{f k_d} 2R\cdot & & \text{initiation} \\
R\cdot + M \xrightarrow{k_i} P_1\cdot & & \\
P_n\cdot + M \xrightarrow{k_p} P_{n+1}\cdot & n,m=1,2,\ldots & \text{propagation} \\
P_n\cdot + M \xrightarrow{k_{fm}} M_n + P_1\cdot & & \text{chain transfer to monomer} \\
P_n\cdot + P_m\cdot \xrightarrow{k_{tc}} M_{n+m} & & \text{termination by coupling} \\
P_n\cdot + P_m\cdot \xrightarrow{k_{td}} M_n + M_m & & \text{termination by disproportionation}
\end{array}
\tag{7-143}
$$

Here *M*, *I*, and *R* are the monomer, initiator, and initiator radical concentrations, P_n is the *growing* or *live polymer* concentration of chain length *n*, *P* is the *total live polymer concentration*, $P=\sum_{n=1}^{\infty} P_n$, and M_n is the *dead* or *product polymer* of chain length *n*. Considering a well-stirred batch polymerization, the mass balances with initial conditions are

$$\frac{dI}{dt}=-k_d I \quad I(0)=I_0$$

$$\frac{dR}{dt}=2fk_d I-k_i MR \quad R(0)=0$$

$$\frac{dM}{dt}=-k_i MR-k_{fm}MP-k_p MP \quad M(0)=M_0$$

$$\frac{dP_1}{dt}=k_i MR-k_p MP_1+k_{fm}M(P-P_1)-(k_{tc}+k_{td})PP_1 \quad P_1(0)=0$$

$$\frac{dP_n}{dt}=k_p M(P_{n-1}-P_n)-k_{fm}MP_n-(k_{tc}+k_{td})PP_n \quad P_n(0)=0 \quad n>1$$

$$\frac{dM_n}{dt}=k_{fm}MP_n+k_{td}PP_n+\frac{k_{tc}}{2}\sum_{m=1}^{n-1}P_m P_{n-m} \quad M_n(0)=0 \quad n>1$$

Assuming reaction steps are independent of chain length and the pseudo-steady-state approximation for the radicals and the long chain hypothesis, that is, propagation terms are much larger than initiation or termination terms, using the *z transform* or *generating function* methods, leads to the following rates for monomer and initiator conversion and live polymer distribution and the instantaneous dead polymer degree of polymerization. The growing chains distribution is the *Flory-Schulz, geometric, or most probable distribution* [see, e.g., Ray, *J. Macromol. Sci.-Revs. Macromol. Chem.* **C8:** 1 (1972); Ray in Lapidus and Amundson (eds.), *Chemical Reactor Theory—A Review*, Prentice-Hall, 1977; Tirrel et al. in Carberry and Varma (eds.), *Chemical Reaction and Reactor Engineering*, Marcel Dekker, 1987; Schork et al., *Control of Polymerization Reactors*, Marcel Dekker, 1993; and Dotson et al., *Polymerization Process Modeling*, Wiley, 1995]:

$$P_n=P(1-\alpha)\alpha^{n-1} \qquad \alpha=\frac{k_p M}{(k_p+k_{fm})M+(k_{tc}+k_{td})P}$$

$$P=\left(\frac{2fk_d I}{k_{tc}+k_{td}}\right)^{1/2} \qquad I=I_0 e^{-k_d t} \tag{7-144}$$

$$r_p=-\frac{dM}{dt}=k_p\left(\frac{2fk_d}{k_{tc}+k_{td}}\right)^{1/2} M\, I^{1/2} \qquad \mu_n^{\text{inst}}=\frac{k_p M}{k_{fm}M+(k_{td}+\frac{1}{2}k_{tc})P}$$

Here r_p is the rate of polymerization, α is the probability of propagation, μ_n^{inst} is the number average instantaneous degree of polymerization, i.e., the number of monomer units in the dead polymer, and *f* is the initiator efficiency. Compare r_p in Eq. (7-144) with the simpler mechanism represented by Eq. (7-68). When chain transfer is the primary termination mechanism or when disproportionation is the primary termination method, then the *instantaneous polydispersity* $\text{PD}^{\text{inst}} = 2$. When coupling is the primary termination mechanism, then the instantaneous polydispersity $\text{PD}^{\text{inst}} = 1.5$. These are lower bounds on the *cumulative polydispersity* of the dead polymer MWD since polydispersities greater than these bounds occur due to variations in monomer, initiator, or temperature during the batch.

Mathematically, the infinite set of equations describing the rate of each chain length can be solved by using the *z transform* or the *generating function methods* (which are discrete methods), *continuous variable approximation method*, or the *method of moments* [see, e.g., Ray, *J. Macromol. Sci.-Revs. Macromol. Chem.* **C8:** 1 (1972); and Ray in Lapidus and Amundson (eds.), *Chemical Reactor Theory—A Review*, Prentice-Hall, 1977]. In general, the infinite set of equations must be solved numerically to obtain the full *cumulative molecular weight distribution* of dead chains due to changing initiator and monomer during the batch [see, e.g., Butté et al., *Macromol. Theory Simul.* **11**(1): 22, 2002]. The effect of compartmentalization on emulsion polymerization kinetics can be found in Wulkow and Richards, *Ind. Eng. Chem. Res.* **53**(18): 7275, 2014.

Typical ranges of the kinetic parameters for low conversion homopolymerization are given in Table 7-8. For more details, see Hutchinson in Meyer and Keurentjes (eds.), *Handbook of Polymer Reaction Engineering*, Wiley, 2005.

Step Growth Homopolymerization Mechanism and Kinetics Here any two growing chains can react with each other. The propagation mechanism is an infinite set of reactions

$$P_n+P_m \underset{k'_p}{\overset{k_p}{\rightleftarrows}} P_{n+m}+W \qquad n,m=1,2,\ldots \tag{7-145}$$

Here P_n is the *growing polymer* concentration of chain length *n*, and *W* is the condensation product concentration, e.g., water for polyesters. Some nylons are also produced through this mechanism. This is usually modeled under the simplifying assumption that the rate constants are independent of chain length which has proved fairly accurate. Assuming a well-stirred batch reactor, pure monomer with concentration that is charged initially, and that the condensation product is removed continuously, the reaction may be considered irreversible. The polymer chain balances and initial conditions are

$$\frac{dP_1}{dt}=-2k_p P_1 P, \quad P_1(0)=P_{10}$$

$$\frac{dP_n}{dt}=k_p\sum_{r=1}^{n-1}P_r P_{n-r}-2k_p P_n P, \quad P_n(0)=0, \quad n>1$$

By using the *z* transform or generating function methods, the polymer has a Flory-Schulz distribution:

$$P_n=P_{10}(1-\alpha)^2\alpha^{n-1},\ n\ge 1 \qquad k_{p,n,m}=k_p \qquad \alpha=\frac{P_{10}k_p t}{P_{10}k_p t+1}$$

$$\mu_n=\frac{1}{1-\alpha} \qquad \mu_w=\frac{1+\alpha}{1-\alpha} \qquad \text{PD}=\frac{\mu_w}{\mu_n}=1+\alpha \tag{7-146}$$

As the conditions approach complete conversion of functional groups (α approaches 1), the polydispersity approaches a value of PD = 2. For further details, see Gupta and Kumar, *Reaction Engineering of Step Growth Polymerization*, Springer, 1987; and Dotson et al., *Polymerization Process Modeling*, Wiley, 1995.

Chain Growth Copolymerization Copolymerization involves more than one monomer, usually two comonomers, as opposed to the single monomer involved in the chain growth and step homopolymerization schemes above. Examples are styrene and ethylene copolymers, some nylons, polyesters, and aramids. Here as well there are step growth and chain growth mechanisms, and these are much more complex [see, e.g., Ray, *J. Macromol.*

TABLE 7-8 Typical Ranges of Kinetic Parameters

Coefficient/concentration	Typical range
k_d, 1/s	10^{-6}–10^{-4}
f	0.4–0.9
k_p, L/(mol/s)	10^2–10^4
k_t, L/(mol/s)	10^6–10^8
k_{fm}/k_p	10^{-6}–10^{-4}
I, mol/L	10^{-4}–10^{-2}
M, mol/L	1–10

SOURCE: Hutchenson, "Typical Ranges of Kinetic Parameters," in *Handbook of Polymer Reaction Engineering*, Wiley, 2005, Table 4.1.

Sci.-Revs. Macromol. Chem. **C8:** 1 (1972); Ray in Lapidus and Amundson (eds.), *Chemical Reactor Theory—A Review*, Prentice-Hall, 1977; Schork, *Control of Polymerization Reactors*, Marcel Dekker, 1993].

Consider a free radical polymerization of two monomers A and B. The propagation reactions are

$$P_{n,m}\cdot + A \xrightarrow{k_{paa}} P_{n+1,m}\cdot$$
$$P_{n,m}\cdot + B \xrightarrow{k_{pab}} Q_{n,m+1}\cdot$$
$$Q_{n,m}\cdot + A \xrightarrow{k_{pba}} P_{n+1,m}\cdot$$
$$Q_{n,m}\cdot + B \xrightarrow{k_{pbb}} Q_{n,m+1}\cdot$$

Here $P_{n,m}$ and $Q_{n,m}$ are live chain concentrations with n monomer A units and m monomer B units having A and B units at the radical ends, respectively. Assuming the long chain hypothesis, that is, propagation terms are much larger than initiation or termination terms, the monomer and radical mass balances are

$$-\frac{dA}{dt} = k_{paa}PA + k_{pba}QA$$
$$-\frac{dB}{dt} = k_{pab}PB + k_{pbb}QB$$
$$\frac{dP}{dt} = k_{pba}QA - k_{pab}PB$$
$$\frac{dQ}{dt} = k_{pab}PB - k_{pba}QA$$

Here P and Q are total radical concentrations. Assuming pseudo-steady state radical concentrations results in the *copolymer equation* or *Mayo-Lewis equation*:

$$r_1 = \frac{k_{paa}}{k_{pab}} \qquad r_2 = \frac{k_{pbb}}{k_{pba}} \qquad F_1 = \frac{r_1(f_1)^2 + f_1 f_2}{r_1(f_1)^2 + 2f_1 f_2 + r_2(f_2)^2}$$

where r_1 and r_2 are the *reactivity ratios*, F_1 is the *instantaneous monomer mole fraction composition of A* in the polymer being formed, and f_1 and f_2 are the *instantaneous monomer mole fractions of A and B* in the reactor. The reactivity ratios can be used to understand the copolymer structure. If $r_1 = r_2 = 0$, then $F_1 = 1/2$, and the copolymer has an *alternating sequence distribution*. If $r_1 = r_2 = 1$, then $F_1 = f_1$ and the copolymer has a *random sequence distribution*.

BIOCHEMICAL REACTIONS

Mechanism and kinetics in biochemical systems describe the reactions that occur in living cells. Biochemical reactions involve two or three phases. For example, aerobic fermentation involves gas (air), liquid (water and dissolved nutrients), and solid (cells), as described in the Biocatalysis subsection above. Bioreactions convert feeds called *substrates* into more *cells* or *biomass* (cell growth), proteins, and *metabolic* products. Any of these can be the desired product in a commercial *fermentation*. For instance, methane is converted to biomass in a commercial process to supply fish meal to the fish farming industry. Ethanol, a metabolic product used in transportation fuels, is obtained by fermentation of corn-based or sugar cane–based sugars. There is a substantial effort to develop genetically modified biocatalysts that produce a desired metabolite at high rate and yield.

Bioreactions follow the same general laws that govern conventional chemical reactions, but the complexity of the mechanism is higher due to the close coupling of bioreactions and *enzymes* that are turned on (*expressed*) or turned off (*repressed*) by the cell depending on the conditions in the fermenter and in the cell. Thus the Monod rate expression in Eq. (7-92) can be used mainly to design bioreaction processes when the culture is in *balanced growth*, i.e., for steady-state cultivations or batch growth (in the exponential growth phase). After a sudden process upset (e.g., a sudden change in substrate concentration or pH), the *control network* of the cell that lies under the *mass flow network* is activated, and dramatic changes in the kinetics of product formation can occur. Table 7-9 summarizes key differences between biochemical and conventional chemical systems [see, e.g., Leib, Pereira, and Villadsen, "Bioreactors, A Chemical Engineering Perspective," *Chem. Eng. Sci.* **56:** 5485–5497 (2001)].

The network of bioreactions is called the *metabolic network*; the parallel and consecutive steps between key intermediates in the network are called *metabolic pathways*, and the determination of the mechanism and kinetics is called *metabolic flux analysis*. As for chemical systems, there are several levels of mechanistic and kinetic representation and analysis, listed in order of increasing complexity in Table 7-10.

TABLE 7-9 Biological versus Chemical Systems

- There is tighter control on conditions (e.g., pH, temperature, substrate and product concentrations, dissolved O_2 concentration, avoidance of contamination by foreign organisms).
- Pathways can be turned on/off by the microorganism through expression of certain enzymes depending on the substrate type and concentration and operating conditions, leading to a richness of behavior unparalleled in chemical systems.
- The global stoichiometry changes with operating conditions and feed composition; kinetics and stoichiometry obtained from steady-state (chemostat) data cannot be used reliably over a wide range of conditions, unless fundamental models are employed.
- Long-term adaptations (mutations) may occur in response to environment changes that can alter completely the product distribution.
- Only the substrates that maximize biomass growth are utilized even in the presence of multiple substrates.
- Cell energy balance requirements pose additional constraints on the stoichiometry that can make it very difficult to predict flux limitations.

Additional complexity can be included through cell population balances that account for the distribution of cell generations present in the fermenter through use of stochastic models. In this section we limit the discussion to simple black box and unstructured models. For more details on bioreaction systems, see, e.g., Villadsen, Nielsen, and Liden, *Bioreaction Engineering Principles*, 3d ed., Springer, 2011; Bailey and Ollis, *Biochemical Engineering Fundamentals*, 2d ed., McGraw-Hill, 1986; Blanch and Clark, *Biochemical Engineering*, Marcel Dekker, 1997; and Sec. 19.

Mechanism Stoichiometric balances are done on a C atom basis called *C-moles*, e.g., relative to the substrate (denoted by subscript s), and the corresponding stoichiometric coefficients Y_{si} (based on C-mole of the primary substrate) are called *yield coefficients*. For instance,

$$CH_2O + Y_{so}\,O_2 + Y_{sn}\,NH_3 + Y_{ss1}S_1 + \cdots$$
$$\Rightarrow Y_{sx}X + Y_{sc}\,CO_2 + Y_{sp1}\,P_1 + \cdots + Y_{sw}\,H_2O \qquad (7\text{-}147)$$

Here the reactants (substrates) are glucose (CH_2O), O_2, NH_3, and a sulfur-providing nutrient S_1, and the products are biomass X, CO_2, metabolic products P_i, and H_2O.

The products of bioreactions can be reduced or oxidized, and all feasible pathways have to be *redox neutral*. There are several cofactors that transfer redox power in a pathway or between pathways, each equivalent to the reducing power of a molecule of H_2, e.g., nicotinamide adenine dinucleotide (NADH), and these have to be included in the stoichiometric balances as H equivalents through redox balancing. For instance, for the reaction of glucose to glycerol ($CH_{8/3}O$), $\frac{1}{3}$NADH equivalent is consumed:

$$CH_2O + \frac{1}{3}NADH \Rightarrow CH_{8/3}O \qquad (7\text{-}148)$$

The stoichiometry in the biochemical literature often does not show H_2O produced by the reaction; however, for complete elemental balance, and for fermenter design, water has to be included, and this is easily done once an O_2 requirement has been determined based on a redox balance.

TABLE 7-10 Heirarchy of Kinetic Models in Biological Systems

- Stoichiometric black box models (similar to a single global chemical reaction) represent the biochemistry by a single global reaction with fixed stoichiometric or *yield coefficients* (limited to a narrow range of conditions). Black box models can be used over a wider range of conditions by establishing different sets of yield coefficient for different conditions. These are also needed to establish the quantitative amounts of various nutrients needed for the completion of the bioreaction.
- Unstructured models view the cell as a single component interacting with the fermentation medium, and each bioreaction is considered to be a global reaction, with a corresponding empirical rate expression.
- Structured models include information on individual reactions or groups of reactions occurring in the cell, and cell components such as DNA, RNA, and proteins are included in addition to the primary metabolites and substrates (see, e.g., the active cell model of Nielsen and Villadsen, *Bioreaction Engineering Principles*, 2d ed., Kluwer Academic/Plenum Press, 2003).
- Fundamental models include cell dimensions, transport of substrates and metabolites across the cell membrane, and the elementary cell bioreaction steps and their corresponding enzyme induction mechanism. In recent years further kinetic steps have been added to the above models which are based on the conversion of substrates to metabolites. Thus the kinetics of protein synthesis by *transcription* and *translation* from the *genome* add much further complexity to cell kinetics.

Likewise for simplicity, the other form of the cofactor [e.g., the oxidized form of the cofactor NADH in Eq. (7-148)] is usually left out. In addition to C balances, for aerobic systems cell respiration has to be accounted for as well through a stoichiometric equation:

$$\text{NADH} + 0.5\text{O}_2 \Rightarrow \text{H}_2\text{O} + (\text{P/O})\text{ATP} \tag{7-149}$$

where P/O is the stoichiometric yield factor of adenosine triphosphate (ATP) on NaDH, $Y_{\text{NaDH, ATP}}$. The associated free energy produced or consumed in each reaction is measured in ATP units. To obtain the stoichiometric coefficient of ATP for a given reaction in a pathway, one has to consult tables (or charts) of biochemical reactions. The ATP produced in one part of the metabolic network (for example, in catabolism) has to be consumed in another part of the network (for example, in anabolism, the cell building reactions). Thus for Eq. (7-148) the stoichiometric ATP requirement to convert one C-mole of glucose to one C-mole of glycerol is $\frac{1}{3}$. In calculations of the carbon flux distribution in different pathways this ATP requirement has to be added on the left-hand side of the equation. Again the other form of the cofactor ATP is usually left out to simplify the reaction equation.

Several metabolic pathways are repeated for many living cells, and these are split into two: *catabolic*, or energy-producing, and *anabolic*, or energy-consuming, the later producing building blocks such as amino acids and the resulting macromolecules such as proteins. Of course, the energy produced in catabolic steps has to be balanced by the energy consumed in anabolic steps. Catabolic pathways include the well-studied *glycolysis, TCA cycle, oxidative phosphorylation,* and *fermentative pathways*. For more details see Stephanopoulos, Aristidou, and Nielsen, *Metabolic Engineering: Principles and Methodologies*, Academic, 1998; and Villadsen, Nielsen, and Liden, *Bioreaction Engineering Principles*, 3d ed., Springer, 2011; Bailey and Ollis, *Biochemical Engineering Fundamentals*, 2d ed., McGraw-Hill, 1986.

Monod-Type Empirical Kinetics Many bioreactions show increased biomass growth rate with increasing substrate concentration at low substrate concentration for the limiting substrate, but no effect of substrate concentration at high concentrations. This behavior can be represented by the Monod equation, Eq. (7-92). Additional variations on the Monod equation are briefly illustrated below. For two essential substrates, the Monod equation can be modified as

$$\mu = \frac{\mu_{\max} C_{s1} C_{s2}}{(K_{s1} + C_{s1})(K_{s2} + C_{s2})} \tag{7-150}$$

This type of rate expression is often used in models for water treatment, and many environmental factors can be included (the effect of phosphate, ammonia, volatile fatty acids, etc.). The correlation between parameters in such complicated models is, however, severe, and very often a simple Monod model such as Eq. (7-92) with only one limiting substrate is sufficient.

When *substrate inhibition* occurs,

$$\mu = \frac{\mu_{\max} C_s}{K_s + C_s + K_i/C_s^2} \tag{7-151}$$

Typically O_2 is a substrate that in high concentrations leads to substrate inhibition, but a high concentration of the carbon source can also be inhibiting (e.g., in bioremediation of toxic waste a high concentration of the organic substrate can lead to severe inhibition or death of the microorganism).

When *product inhibition* is present,

$$\mu = \frac{\mu_{\max} C_s}{K_s + C_s}\left(1 - \frac{C_p}{C_{p\max}}\right) \tag{7-152}$$

Here the typical example is the inhibitor effect of ethanol on yeast growth. Considerable efforts are made by the biotech companies to develop yeast strains that are tolerant to high ethanol concentrations since this will give considerable savings in, e.g., production of biofuel ethanol by fermentation.

The various component reaction rates for a single reaction can be related to the growth rate by using the stoichiometric (yield) coefficients, e.g., from Eq. (7-147):

$$r_i = Y_{xi} \mu C_x = \frac{Y_{si}}{Y_{sx}} \mu C_x \tag{7-153}$$

Chemostat with Empirical Kinetics Using the CSTR equation, Eq. (7-54), for a constant-volume single reaction [e.g. Eq. (7-147)], the substrate, biomass, and product material balances are

$$\begin{aligned} &\frac{1}{Y_{sx}} \mu C_x + D(C_{s0} - C_s) = 0 \\ &\mu C_x - DC_p = 0 \rightarrow D = \mu \\ &\frac{Y_{sp}}{Y_{sx}} \mu C_x - DC_p = 0 \end{aligned} \tag{7-154}$$

Here C_{s0} is the feed substrate concentration, and D is the *dilution rate* which at steady state is equal to both the feed and effluent volumetric flow rates and to the specific growth rate. The effluent concentrations of substrate, biomass, and products can be calculated by using a suitable expression for the specific growth rate μ such as one of the relevant variants of the Monod kinetics described above.

ELECTROCHEMICAL REACTIONS

Electrochemical reactions involve coupling between chemical reactions and electric charge transfer. These reactions often include two or three phases, for instance, a gas (e.g., H_2 or O_2 evolved at the electrodes or fed as reactants), a liquid (the electrolyte solution), and solids (electrodes). Electrocatalysts may be employed to enhance the reaction for a particular desired product. Also an electrode material may be selected to retard an undesired reaction. Electrochemical reactions are heterogeneous reactions that occur at the surface of electrodes and involve the transfer of charge in the form of electrons as part of a chemical reaction. The electrochemical reaction can produce a chemical change by passing an electric current through the system (e.g., electrolysis), or reversely a chemical change can produce electric energy (e.g., using a battery or fuel cell to power an appliance). There are a variety of practical electrochemical reactions, some occurring naturally, such as corrosion, and others used in production of chemicals (e.g., the decomposition of HCl to produce Cl_2 and H_2, the production of caustic soda and chlorine from NaCl brine, the smelting of aluminum), electroplating, and energy generation or storage (e.g., fuel cells, batteries, flow cells, and photovoltaics). Electrochemical reactions are typically reversible and can be generally written as a *reduction-oxidation* (*redox*) couple:

$$O + ne^- \Leftrightarrow R$$

where O is an oxidized species and R is a reduced species. For instance, the corrosion process includes oxidation at the *anode*:

$$\text{Fe} \Leftrightarrow \text{Fe}^{2+} + 2e^-$$

and reduction at the *cathode*:

$$\text{O}_2 + 2\text{H}_2\text{O} + 4e^- \Leftrightarrow 4\text{OH}^-$$

The overall electrochemical reaction is the stoichiometric sum of the anode and cathode reactions:

$$2\text{Fe} + \text{O}_2 + 2\text{H}_2\text{O} \Leftrightarrow 2\text{Fe}^{2+} + 4\text{OH}^- \quad \text{(four-electron transfer process, } n = 4\text{)}$$

The anode and cathode reactions are closely coupled in that the electric charge is conserved; therefore, the overall production rate is a direct function of the electric charge passed per unit time, the electric current I.

Faraday's law relates the charge transferred by ions in the electrolyte and electrons in the external circuit to the moles of chemical species reacted (Newman and Thomas-Alvea, *Electrochemical Systems*, 3d ed., Wiley Interscience, 2004):

$$\begin{aligned} &Q = nmF \qquad F = 96{,}485 \text{ C/equiv} \\ &I = \frac{Q}{t} = \frac{\text{charge}}{\text{time}} \text{ A} \end{aligned} \tag{7-155}$$

where n is the number of equivalents per mole, m is the number of moles, F is the Faraday constant, Q is the charge, and t is time. The total current passed may represent several parallel electrochemical reactions; therefore, we designate a current efficiency for each chemical species. The *current efficiency* is the ratio of the theoretical electric charge (Coulombs) required for the amount of product obtained to the total amount of electrical charge passed through the electrochemical cell. The chemical species production rate (mass/time), w_i, is related to the total current passed I, the species *current efficiency* $\varepsilon_{\text{current},i}$, and the molecular weight of the chemical species MW_i:

$$\begin{aligned} &w_i = \frac{I \times \varepsilon_{\text{current},i} \times \text{MW}_i}{nF} = \frac{\text{mass}}{\text{time}} \\ &j = \frac{\text{current}}{\text{area}} = \frac{I}{A_{\text{projected}}} \end{aligned} \tag{7-156}$$

Since electrochemical reactions are heterogeneous at electrode surfaces, the current I is generally normalized by dividing it by the geometric or projected area of the electrode, resulting in the quantity known as the *current density* j, in units of kA/m^2.

The overall electrochemical cell *equilibrium potential* E^o_{cell}, as measured between the cathode and the anode, is related to the Gibbs free energy change for the overall electrochemical reaction:

$$\Delta G^o = \Delta H^o - T\Delta S^o = -nFE^o_{cell}$$

$$E^o_{cell} = -\frac{\Delta G^o}{nF} = E^o_{cathode} - E^o_{anode} \tag{7-157}$$

Each anode and cathode electrode reaction, or half-cell reaction, has an associated energy level or electric potential (volts) associated with it. Values of the standard equilibrium electrode reduction potentials E^o at unit activity and 25°C may be obtained from the literature (de Bethune and Swendeman, "Table of Electrode Potentials and Temperature Coefficients," *Encyclopedia of Electrochemistry*, Van Nostrand Reinhold, 1964; and de Bethune and Swendeman, *Standard Aqueous Electrode Potentials and Temperature Coefficients*, Hampel, 1964). The overall electrochemical cell *equilibrium potential* either can be obtained from ΔG values or is equal to the cathode half-cell potential minus the anode half-cell potential, as shown above.

The *Nernst equation* allows one to calculate the *equilibrium potential* E^{eq} when the activity of the reactants or products is not at unity:

$$\sum_i \nu_i M_i^{n_i} \rightarrow ne^-$$

$$E^{eq} = E^0 - \frac{RT}{nF}\ln(\prod a_i^{\nu_i}) \tag{7-158}$$

$$\left(\frac{\partial E}{\partial T}\right)_P = \frac{\Delta S}{nF}$$

where ν_i is the stoichiometric coefficient of chemical species i (positive for products; negative for reactants), M_i is the symbol for species i, n_i is the charge number of the species, a_i is the activity of the species, E^o is the *formal potential*, and $\prod$ represents the product of all respective activities raised to their stoichiometric powers as required by the reaction. Please note that if the value of the equilibrium potential is desired at another temperature, E^o must also be evaluated at the new temperature as indicated above.

Kinetic Control In 1905, Julius Tafel experimentally observed that when mass transport was not limiting, the *current density j* of electrochemical reactions exhibited the following behavior:

$$j = a'e^{\eta_{act}/b'} \quad \text{or} \quad \eta_{act} = a + b\log j$$

where the quantity η_{act} is known as the *activation overpotential* $(E - E^{eq})$ and is the difference between the actual electrode potential E and the reversible equilibrium potential of the electrochemical reaction E^{eq}. Thus the driving force for the electrochemical reaction is not the absolute potential; it is the activation overpotential η_{act}.

This relationship between the current density and activation overpotential has been further developed and resulted in the Butler-Volmer equation

$$r = \frac{j}{nF} = k_f C_o - k_r C_r \quad j = j_0\left(e^{-(\alpha nF/RT)\eta_{act}} - e^{[(1-\alpha)nF/RT]\eta_{act}}\right) \tag{7-159}$$

$$\eta_{act} = E - E^{eq}$$

Here the reaction rate r is defined per unit electrode area, moles per area per time, j_0 is the *equilibrium exchange current* when $E = E^{eq}$, η_{act} is the *activation overpotential*, α is the *transfer coefficient* for the cathode, and $1-\alpha$ for the anode. For large activation overpotentials, the *Tafel empirical equation* applies as indicated above:

$$\eta_{act} = a + b\log j \quad \text{for } \eta_{act} > 100 \text{ mV}, \; b = \text{Tafel slope} \tag{7-160}$$

For small activation overpotentials, linearization gives

$$j = j_0 \frac{nF}{RT}\eta_{act} \tag{7-161}$$

Mass Transfer Control In this case, the surface concentration at the electrodes differs significantly from the bulk electrolyte concentration. The Nernst equation applies to the surface concentrations (or activities in case of nonideal solutions):

$$E^{eq} = E^o - \frac{RT}{nF}\ln(\prod a_{i,surf}^{\nu_i}) \tag{7-162}$$

If mass transfer is limiting, then a *limiting current* is obtained for each chemical species i:

$$j_{i,lim} = \frac{nFD_iC_i}{\delta} = nFk_{L,i}C_i \tag{7-163}$$

where D_i is the diffusion coefficient, δ is the boundary layer thickness, and $k_{L,i}$ is the mass transfer coefficient of species i. The effect of mass transfer is included as follows:

$$j = j_0\left[\left(1 - \frac{j}{j_{c,lim}}\right)e^{-(\alpha nF/RT)\eta_{act}} - \left(1 - \frac{j}{j_{a,lim}}\right)e^{[(1-\alpha)nF/RT]\eta_{act}}\right]$$

$$\frac{C_{i,surf}}{C_i} = \left(1 - \frac{j}{j_{i,lim}}\right) \quad i = o, r \tag{7-164}$$

Ohmic Control The overall electrochemical reactor cell voltage may be dependent on the kinetic and mass transfer aspects of the electrochemical reactions; however, a third factor is the potential lost within the electrolyte as current is passing through this phase. The potential drops may become dominant and limit the electrochemical reactions, requiring an external potential to be applied to drive the reactions or significantly lower the delivered electric potential in power generation applications such as batteries and fuel cells.

Multiple Reactions With multiple reactions, the total current is the sum of the currents from the individual reactions with anodic currents positive and cathodic currents negative. This is called the *mixed potential principle*. For more details see Bard and Faulkner, *Electrochemical Methods: Fundamentals and Applications*, 2d ed., Wiley, 2001.

Additional references on electrochemical reaction kinetics and mechanism include Newman and Thomas-Alvea, *Electrochemical Systems*, 3d ed., Wiley Interscience, 2004; Bard and Faulkner, *Electrochemical Methods: Fundamentals and Applications*, 2d ed., Wiley, 2001; Bethune and Swendeman, "Table of Electrode Potentials and Temperature Coefficients," *Encyclopedia of Electrochemistry*, Van Nostrand Reinhold, New York, 1964, pp. 414–424; and Bethune and Swendeman, *Standard Aqueous Electrode Potentials and Temperature Coefficients*, C. A. Hampel Publisher, 1964. Discussion on electrochemical reactors is in Sec. 19.

DETERMINATION OF MECHANISM AND KINETICS

Laboratory data are the predominant source for reaction mechanism and kinetics in industrial practice. However, laboratory data intended for scoping and demonstration studies rather than for kinetic evaluation often have to be used, thus reducing the effectiveness and accuracy of the resulting kinetic model. The following are the steps required to obtain kinetics from laboratory data:

1. Develop initial guesses on mechanism, reaction time scale, and potential kinetic models from the literature, scoping experiments, similar chemistries, and computational chemistry calculations, when possible.
2. Select a suitable laboratory reactor type and scale, and analytical tools for kinetic measurements.
3. Develop a priori factorial experimental design or sequential experimental design.
4. When possible, provide ideal reactor conditions, e.g., good mechanical agitation in batch and CSTR, high-velocity flow in PFR.
5. Estimate the limiting diffusion-reaction regimes under the prevailing lab reactor conditions for heterogeneous reactions, and use the appropriate lab reactor model to interpret the data. When possible, operate the reactor under kinetic control.
6. Discriminate between competing mechanisms and kinetic rates by forcing maximum differentiation between competing hypotheses regarding mechanism and rates through the experimental design, and by obtaining the best fit of the kinetic data to the proposed kinetic forms.

LABORATORY REACTORS

Selection of the laboratory reactor type and size, and associated feed and product handling, control, and analytical schemes depends on the type of reaction, reaction time scales, and type of analytical methods required. The criteria for selection include equipment cost, ease of operation, ease of data analysis, accuracy, versatility, temperature uniformity and controllability, suitability for mixed phases, and to a lesser extent scale-up feasibility. Many configurations of laboratory reactors have been employed. Rase (*Chemical Reactor Design for Process Plants,* Wiley, 1977) and Shah (*Gas-Liquid-Solid Reactor Design,* McGraw-Hill, 1979) each have about 25 sketches, and Shah's bibliography has 145 items classified into 22 categories of reactor types. Jankowski et al. [*Chemische Technik* **30:** 441–446 (1978)] illustrate 25 different kinds of gradientless laboratory reactors for use with solid catalysts.

Laboratory reactors are of two main types:

1. Reactors used to obtain fundamental data on intrinsic chemical rates free of mass transfer resistances or other complications. Some of the gas-liquid lab reactors, for instance, employ known interfacial areas, thus avoiding the uncertainty regarding the area for gas to liquid mass transfer. When ideal behavior cannot be achieved, intrinsic kinetic estimates need to account for mass and heat transfer effects using the appropriate lab reactor model.

2. Reactors used to obtain scale-up data due to their similarity to the reactor intended for the pilot or commercial plant scale. How to scale down from the conceptual commercial or pilot scale to lab scale is a difficult problem in itself, and it is not possible to maintain all key features while scaling down.

The first type is often the preferred one—once the intrinsic kinetics are obtained at "ideal" lab conditions, scale-up is done by using models or correlations that describe large-scale reactor hydrodynamics coupled with the intrinsic kinetics. However, in some cases ideal conditions cannot be achieved, and the laboratory reactor has to be adequately modeled to account for mass and heat transfer and nonideal mixing effects to enable extraction of intrinsic kinetics. In addition, with homogeneous reactions, attention must be given to prevent wall-catalyzed reactions, which can result in observed kinetics that are fundamentally different from intrinsic homogeneous kinetics. This is a problem for scale-up, due to the high surface/volume ratio in small reactors versus the low surface/volume ratio in large-scale systems, resulting in widely different contributions of wall effects at different scales. Similar issues arise in bioreactors with the potential of undesirable wall growth of the biocatalyst cells masking the homogeneous growth kinetics. In catalytic reactions, certain reactor configurations may enhance undesirable homogeneous reactions, and the importance of these reactions may be different at larger scale, causing potential scale-up pitfalls.

The reaction rate is expressed in terms of chemical compositions of the reacting species, so ultimately the variation of composition with time or space must be found. The composition is determined in terms of a property that is measured by some instrument and calibrated. Among the measures that have been used are titration, pressure, refractive index, density, chromatography, spectrometry, polarimetry, conductimetry, absorbance, and magnetic resonance. Therefore, batch or semibatch data are converted to composition as a function of time (C, t), or to composition and temperature as functions of time (C, T, t), to prepare for kinetic analysis. In a CSTR and PFR at steady state, the rate and compositions in the effluent are observed as a function of residence time and operating conditions, e.g., temperature and pressure. In addition, for a PFR it is desirable to also obtain composition and temperature along the reactor.

TABLE 7-11 Laboratory Reactors

Reaction	Reactor
Homogeneous gas	Isothermal U-tube in temperature-controlled batch
Homogeneous liquid	Mechanically agitated batch or CSTR with jacketed cooling/heating
Catalytic gas-solid	Packed tube in furnace
	Isothermal U-tube in temperature-controlled bath
	Rotating basket with jacketed cooling/heating
	Internal recirculation (Berty) reactor with jacketed cooling/heating
Noncatalytic gas-solid	Packed tube in furnace
Liquid-solid	Packed tube in furnace
Gas-liquid	CSTR with jacketed cooling/heating
	Fixed interface CSTR
	Wetted wall
	Laminar jet
Gas-liquid-solid	Slurry CSTR with jacketed cooling/heating
	Packed bed with downflow, upflow, or countercurrent
Solid-solid	Packed tube in furnace

When a reaction has many reactive species (which may be the case even for apparently simple processes such as pyrolysis of ethane or synthesis of methanol), a factorial or sequential experimental design should be developed, and the data can be subjected to a *response surface analysis* (Box, Hunter, and Hunter, *Statistics for Experimenters,* 2d ed., Wiley Interscience, 2005; Davies, *Design and Analysis of Industrial Experiments,* Oliver & Boyd, 1954). This can result in a black box correlation or statistical model, such as a quadratic (limited to first- and second-order effects) expression for the variables x_1, x_2, and x_3:

$$r = k_1x_1 + k_2x_2 + k_3x_3 + k_{12}x_1x_2 + k_{13}x_1x_3 + k_{23}x_2x_3$$

Analysis of such statistical correlations may reveal the significant variables and interactions and may suggest potential mechanisms and kinetic models, say, of the Langmuir-Hinshelwood type, that could be analyzed in greater detail by a regression process. The variables x_i could be various parameters of heterogeneous processes as well as concentrations. An application of this method to isomerization of n-pentane is given by Kittrel and Erjavec [*Ind. Eng. Chem. Proc. Des. Dev.* **7:** 321 (1968)].

Table 7-11 summarizes laboratory reactor types that approach the three ideal concepts BR, CSTR, and PFR, classified according to reaction types.

For instance, Fig. 7-17 summarizes laboratory reactor types and typical hydrodynamic parameters for gas-liquid reactions.

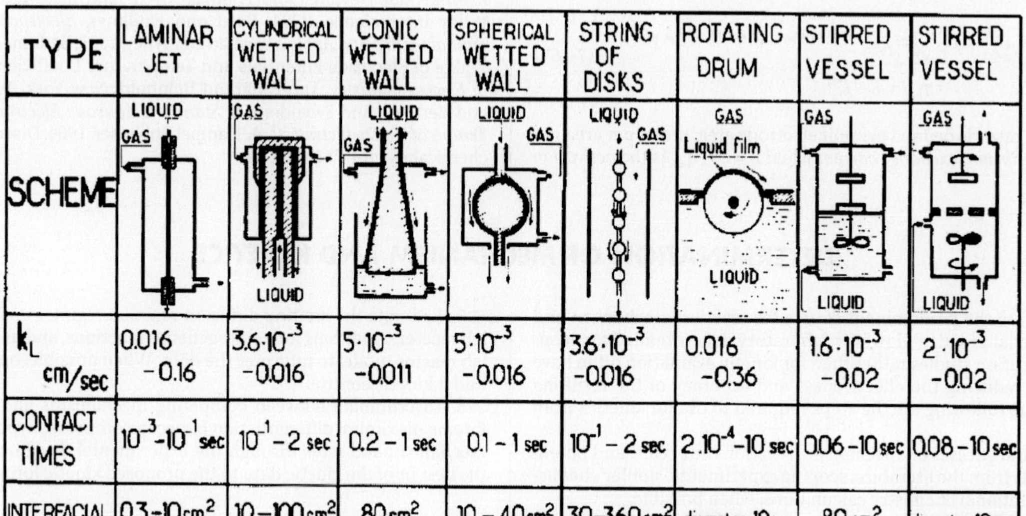
TYPE
LAMINAR JET
CYLINDRICAL WETTED WALL
CONIC WETTED WALL
SPHERICAL WETTED WALL
STRING OF DISKS
ROTATING DRUM
STIRRED VESSEL
STIRRED VESSEL
SCHEME
LIQUID
GAS
Liquid film
LIQUID
k_L cm/sec
0.016 - 0.16
3.6·10^-3 - 0.016
5·10^-3 - 0.011
5·10^-3 - 0.016
3.6·10^-3 - 0.016
0.016 - 0.36
1.6·10^-3 - 0.02
2·10^-3 - 0.02
CONTACT TIMES
10^-3-10^-1 sec
10^-1 - 2 sec
0.2 - 1 sec
0.1 - 1 sec
10^-1 - 2 sec
2.10^-4 - 10 sec
0.06 - 10 sec
0.08 - 10 sec
INTERFACIAL AREA
0.3-10 cm² high precision
10-100 cm² high precision
80 cm² high precision
10 - 40 cm² high precision
30-360 cm² moderate precision
diameter 10cm length 12cm high precision
80 cm² good precision
diameter 10cm length 15cm 2-30% open

FIG. 7-17 Principal types of laboratory reactors for gas-liquid reactions. [From Fig. 8 in Charpentier, "Mass Transfer Rates in Gas-Liquid Absorbers and Reactors," in Drew et al. (eds.), *Advances in Chemical Engineering,* vol. 11, Academic Press, 1981.]

Batch Reactors In the simplest kind of investigation, reactants can be loaded into a number of sealed tubes, kept in a constant temperature bath for various periods, shaken mechanically to maintain uniform composition, and analyzed. In terms of cost and versatility, the stirred batch reactor is the unit of choice for homogeneous reactions or heterogeneous slurry reactions including liquid-phase, gas-solid, gas-liquid, and gas-liquid-solid systems. For multiphase systems, some of the reactants can be semibatch or continuous. The batch reactor is especially suited to reactions with half-lives in excess of 10 min. Samples are taken at time intervals, and the reaction is stopped by cooling, by dilution, or by neutralizing a residual reactant such as an acid or base; analysis can then be made at a later time. Analytical methods that do not necessitate termination of reaction include nonintrusive measurements, e.g., (1) the amount of gas produced, (2) the gas pressure in a constant-volume vessel, (3) absorption of light, (4) electric or thermal conductivity, (5) polarography, (6) viscosity in polymerization, (7) pH and dissolved oxygen probes, and so on. The reactor may be operated under isothermal conditions, with the important effect of temperature determined from several isothermal runs, or the composition and temperature may be recorded simultaneously and the data regressed. On the laboratory scale, it is essential to ensure that a batch reactor is stirred to uniform composition, and for critical cases such as high viscosities this should be checked with tracer tests.

Flow Reactors CSTRs and other devices that require flow control are more expensive and difficult to operate. However, CSTRs and PFRs are the preferred laboratory reactors for steady-state operation. One of the benefits of CSTRs is their ability to operate isothermally and the fact that their mathematical representation is an algebraic equation thus making data analysis simpler. For laboratory research purposes, CSTRs are considered feasible for holding times of 1 to 4000 s, reactor volumes of 2 to 1000 cm^3 (0.122 to 61 in^3), and flow rates of 0.1 to 2.0 cm^3/s. Fast reactions and those in the gas phase are generally done in tubular flow reactors, just as they are often done on the commercial scale. Often it is not possible to measure compositions along a PFR, although temperatures can be measured using a thermowell with fixed or mobile thermocouple bundle. PFRs can be kept at nearly constant temperatures; small-diameter tubes immersed in a fluidized sand bed or molten salt can hold quite constant temperatures up to a maximum temperature of a few hundred degrees. PFRs may also be operated at near adiabatic conditions by providing dual radial temperature control to minimize the radial heat flux, with multiple axial zones. A recycle unit can be operated as a differential reactor with arbitrarily small conversion and temperature change. Test work in a tubular flow unit may be desirable if the intended commercial unit is of that type.

Multiphase Reactors Reactions between gas-liquid, liquid-liquid, and gas-liquid-solid phases are often tested in CSTRs. Other laboratory types are suggested by the commercial units depicted in appropriate sketches in Sec. 19 and in Fig. 7-17 [Charpentier, *Mass Transfer Rates in Gas-Liquid Absorbers and Reactors*, in Drew et al. (eds.), *Advances in Chemical Engineering*, vol. 11, Academic, 1981]. Liquids can be reacted with gases of low solubilities in stirred vessels, with the liquid charged first and the gas fed continuously at the rate of reaction or dissolution. Some of these reactors are designed to have known interfacial areas. Most equipment for gas absorption without reaction is adaptable to absorption with reaction. The many types of equipment for liquid-liquid extraction also are adaptable to reactions of immiscible liquid phases.

Solid Catalysts Processes with solid catalysts are affected by diffusion of heat and mass (1) within the pores of the pellet, (2) between the fluid and the particle, and (3) axially and radially within the packed bed. Criteria in terms of various dimensionless groups have been developed to tell when these effects are appreciable, and some of these were discussed above. For more details see Mears [*Ind. Eng. Chem. Proc. Des. Devel.* **10:** 541–547 (1971); *Ind. Eng. Chem. Fund.* **15:** 20–23 (1976)] and Satterfield (*Heterogeneous Catalysis in Practice*, McGraw-Hill, 1991, p. 491). For catalytic investigations, the rotating basket or fixed basket with internal recirculation is the standard device, usually more convenient and less expensive than equipment with external recirculation. In the fixed-basket type, an internal recirculation rate of 10 to 15 or so times the feed rate effectively eliminates external diffusional resistance and temperature gradients (see, e.g., Berty, *Experiments in Catalytic Reaction Engineering*, Elsevier, 1999). A unit holding 50 cm^3 (3.05 in^3) of catalyst can operate up to 800 K (1440°R) and 50 bar (725 psi). When deactivation occurs rapidly (in a few seconds during catalytic cracking, for instance), the fresh activity can be maintained with a transport reactor through which both reactants and fresh catalyst flow without slip and with short contact time. Since catalysts often are sensitive to traces of impurities, the time deactivation of the catalyst usually can be evaluated only with commercial feedstock. Physical properties of catalysts also may need to be checked periodically, including pellet size, specific surface, porosity, pore size and size distribution, effective diffusivity, and active metals content and dispersion. The effectiveness factor of a porous catalyst is found by measuring conversions with successively smaller pellets until no further change occurs. These topics are touched on by Satterfield (*Heterogeneous Catalysis in Industrial Practice*, McGraw-Hill, 1991).

TABLE 7-12 Calorimetric Methods

Adiabatic	Nonadiabatic
Accelerating rate calorimeter (ARC)	Reaction calorimeter (RC1) + IR
Vent sizing package (VSP) calorimeter	Differential scaning calorimeter (DSC)
	Thermal gravitometry (TG)
PHI-TEC	Isothermal calorimetry
Dewar	Differential thermal analysis (DTA)
Automatic pressure tracking adiabatic calorimeter (APTAC)	Differential microcalorimeters
	Advanced reaction system screening tool (ARSST)

To determine the deactivation kinetics, long-term deactivation studies at constant conditions and at different temperatures are required. In some cases, accelerated aging can be used to reduce the time required for the experimental work, by either increasing the feed flow rate (if the deactivation is a result of feed or product poisoning) or increasing the temperature above the standard reaction temperature. These approaches for accelerated aging require a good understanding of how the higher-temperature or rate-accelerated deactivation correlates with deactivation at commercial operating conditions.

Bioreactors There are several types of laboratory bioreactors:

1. Mechanically agitated batch/semibatch with pH control and nutrients or other species either fed at the start or added continuously based on a recipe or protocol.
2. CSTR to maintain a controlled *dilution rate* (feed rate). These require some means to separate the biocatalyst from the product and recycle to the reactor, such as centrifuge or microfiltration:
   a. *Chemostat* controls the flow or dilution rate to maintain a constant fermentation volume.
   b. *Turbidostat* controls the biomass or cell concentration.
   c. *pH-auxostat* controls pH in the effluent (same as pH in reactor).
   d. *Productostat* controls the effluent concentration of one of the metabolic products.

The preferred reactor for kinetics is the chemostat, but semibatch reactors are more often used owing to their simpler operation and wide use for process and biocatalyst development.

Calorimetry Another category of laboratory systems that can be used for kinetics includes *calorimeters*. These are primarily used to establish temperature effects and thermal runaway conditions, but can also be employed to determine reaction kinetics. Types of calorimeters are summarized in Table 7-12; for more details see Reid, "Differential Microcalorimeters," *J. Physics E: Scientific Instruments*, **9** (1976).

Additional methods of laboratory data acquisition are described in Masel, *Chemical Kinetics and Catalysis*, Wiley, 2001.

KINETIC PARAMETERS

The kinetic parameters are constants that appear in the intrinsic kinetic rate expressions and are required to describe the rate of a reaction or rates of a reaction network. For instance, for the simple global nth-order reaction with Arrhenius temperature dependence:

$$A \Rightarrow B \qquad r = kC_a^n \qquad k = k_0 e^{-E/RT} \tag{7-165}$$

The kinetic parameters are k_0, E, and n, and knowledge of these parameters and the prevailing concentration of A and temperature fully determines the reaction rate.

For a more complex expression such as the Langmuir-Hinshelwood rate for gas reaction on heterogeneous catalyst surface with equilibrium adsorption of reactants A and B on two different sites and nonadsorbing products, Eq. (7-85) can be rewritten as

$$r = \frac{k_0 e^{-E/RT} p_a p_b}{(1 + K_{a0} e^{-E_{aa}/RT})(1 + K_{b0} e^{-E_{ab}/RT})} \tag{7-166}$$

and the kinetic parameters are k_0, E, K_{a0}, E_{aa}, K_{b0}, and E_{ab}.

A number of factors limit the accuracy with which parameters needed for the design of commercial equipment can be determined. The kinetic parameters may be affected by inaccurate accounting for laboratory reactor heat and mass transport and hydrodynamics; correlations for these are typically determined under nonreacting conditions at ambient temperature and pressure and with nonreactive model fluids and may not be applicable or accurate at reaction conditions. Experimental uncertainty including errors in analysis, measurement, and control is also a contributing factor

TABLE 7-13 Comparison of Direct and Indirect Methods

Direct method	Indirect method
Advantages	Disadvantages
Get rate equation directly	Must infer rate equation
Easy to fit data to a rate law	Hard to analyze rate data
High confidence on final rate equation	Low confidence on final rate equation
Disadvantages	Advantages
Difficult experiment	Easier experiment
Need many runs	Can do a few runs and get important information
Not suitable for very fast or very slow reactions	Suitable for all reactions including very fast or very slow ones

SOURCE: Masel, *Chemical Kinetics and Catalysis,* Wiley, 2001, Table 3.2.

(see, e.g., Hoffman, "Kinetic Data Analysis and Parameter Estimation," in de Lasa (ed.), *Chemical Reactor Design and Technology,* Martinus Nijhoff, 1986).

DATA ANALYSIS METHODS

In this subsection we focus on the three main types of ideal reactors: BR, CSTR, and PFR. Laboratory data are usually in the form of concentrations or partial pressures versus batch time (batch reactors), concentrations or partial pressures versus distance from reactor inlet or residence time (PFR), or rates versus residence time (CSTR). Rates can also be calculated from batch and PFR data by differentiating the concentration versus time or distance data, respectively, usually by first smoothing the data to reduce noise in the calculated rates. It follows that a general classification of experimental methods is based on whether the data measure rates directly (differential or direct method) or indirectly (integral of indirect method). Table 7-13 shows the pros and cons of these methods.

Some simple reaction kinetics are amenable to analytical solutions and graphical linearized analysis to calculate the kinetic parameters from rate data. More complex systems require numerical solution of nonlinear systems of differential and algebraic equations coupled with nonlinear parameter estimation or regression methods.

Differential Data Analysis As indicated above, the rates can be obtained either directly from differential CSTR data or by differentiation of integral data. A common way of evaluating the kinetic parameters is by rearrangement of the rate equation, to make it linear in parameters (or some transformation of parameters) where possible. For example, in the case of the simple *n*th-order reaction in Eq. (7-165), taking the natural logarithm of both sides of the equation results in a linear relationship between the variables $\ln r$, $1/T$, and $\ln C_a$:

$$\ln r = \ln k_0 - \frac{E}{RT} + n \ln C_a \tag{7-167}$$

Multilinear regression can be used to find the constants k_0, E, and n. For constant-temperature (isothermal) data, Eq. (7-167) can be simplified by using the Arrhenius form as

$$\ln r = \ln k + n \ln C_a \tag{7-168}$$

and the kinetic parameters n and k can be determined as the intercept and slope of the best straight-line fit to the data, respectively, as shown in Fig. 7-18.

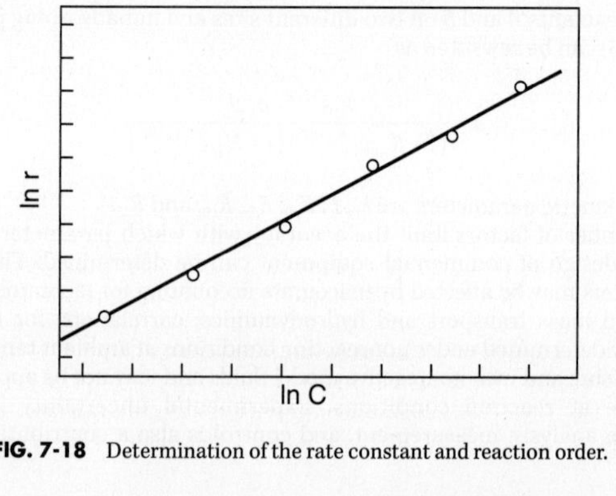

FIG. 7-18 Determination of the rate constant and reaction order.

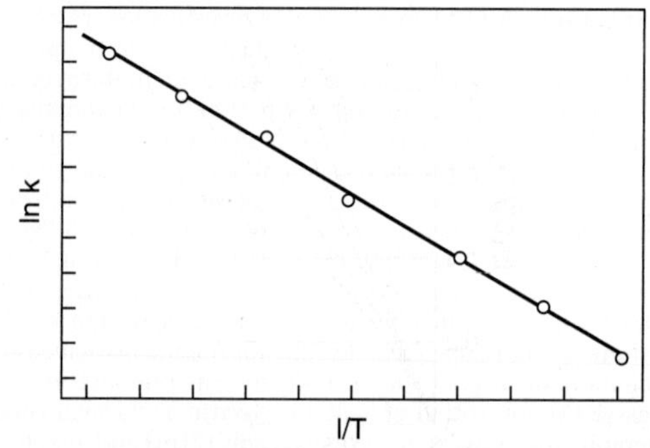

FIG. 7-19 Determination of the preexponential and activation energy.

The preexponential k_0 and activation energy E can be obtained from multiple isothermal data sets at different temperatures by using the linearized form of the Arrhenius equation

$$\ln k = \ln k_0 - \frac{E}{RT} \tag{7-169}$$

as shown in Fig. 7-19.

Integral Data Analysis Integral data such as from batch and PFR relate concentration to time or distance. Integration of the BR equation for an *n*th-order homogeneous constant-volume reaction yields

$$\ln \frac{C_{a0}}{C_a} = k\tau \quad \text{for } n=1$$
$$\left(\frac{C_{a0}}{C_a}\right)^{n-1} = 1 + k\tau(n-1)C_{a0}^{n-1} \quad \text{for } n \neq 1 \tag{7-170}$$

For the first-order case, the rate constant k can be obtained directly from the slope of the graph of the left-hand side of Eq. (7-170) versus batch time, as shown in Fig. 7-20.

For orders other than first, plotting the natural log of Eq. (7-170) can at least indicate if the order is larger or smaller than 1, as shown in Fig. 7-21.

The Half-Life Method The half-life is the batch time required to get 50 percent conversion. For an *n*th-order reaction,

$$\tau_{1/2} = \frac{\ln 2}{k} \quad \text{for } n=1$$
$$\tau_{1/2} = \frac{2^{n-1}-1}{(n-1)kC_{a0}^{n-1}} \quad \text{for } n \neq 1 \tag{7-171}$$

Thus for first-order reactions, the *half-life* is constant and independent of the initial reactant concentration and can be used directly to calculate the rate constant k. For non-first-order reactions, Eq. (7-171) can be linearized as follows:

$$\ln \tau_{1/2} = \ln \frac{2^{n-1}-1}{(n-1)k} - (n-1)\ln C_{a0} \quad \text{for } n \neq 1 \tag{7-172}$$

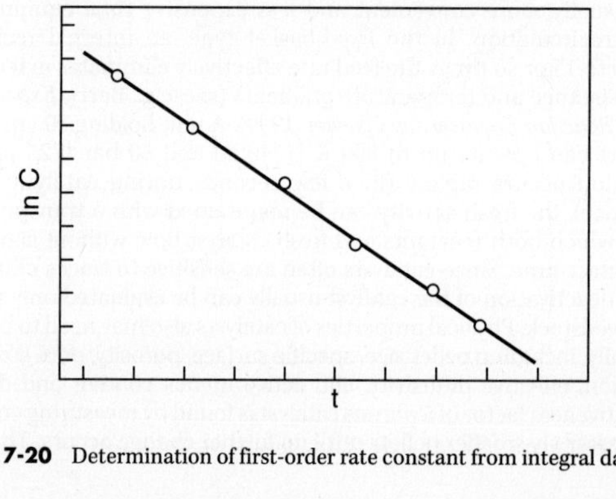

FIG. 7-20 Determination of first-order rate constant from integral data.

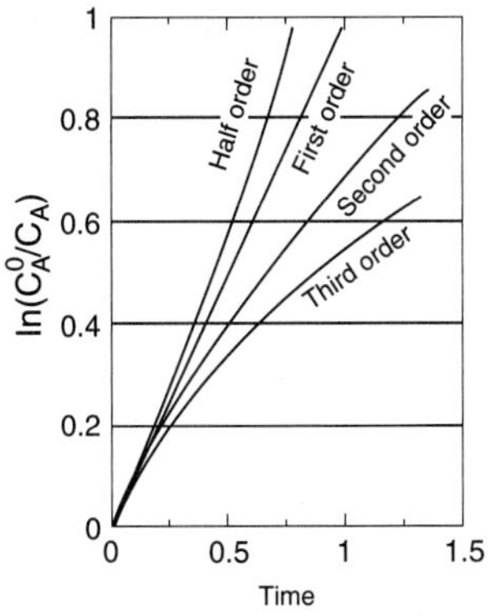

FIG. 7-21 Reaction behavior for *n*th-order reaction. (Masel, *Chemical Kinetics and Catalysis,* Wiley, 2001, Fig. 3.15.)

The reaction order *n* can be obtained from the slope and the rate constant *k* from the intercept of the plot of Eq. (7-172), shown in Fig. 7-22.

Complex Rate Equations The examples above are for special cases amenable to simple treatment. Complex rate equations and reaction networks with complex kinetics require individual treatment, which often includes both numerical solvers for the differential and algebraic equations describing the laboratory reactor used to obtain the data and linear or nonlinear parameter estimation.

PARAMETER ESTIMATION

The straightforward method to obtain kinetic parameters from data is the numerical fitting of the concentration data (e.g., from BR or PFR data) to integral equations, or the rate data (e.g., from a CSTR or from differentiation of BR or PFR data) to rate equations. This is done by parameter estimation methods described here. An excellent reference for experimental design and parameter estimation (illustrated for heterogeneous gas-solid reactions) is the review paper of Froment and Hosten, "Catalytic Kinetics—Modeling," in *Catalysis—Science and Technology,* Springer-Verlag, New York, 1981. Two previous papers devoted to this topic by Hofmann [in *Chemical Reaction Engineering, ACS Advances in Chemistry* **109:** 519–534 (1972); in de Lasa (ed.), *Chemical Reactor Design and Technology,* Martinus Nijhoff, 1985, pp. 69–105] are also very useful. As indicated above, the acquisition of kinetic data and parameter estimation can be a complex endeavor. It includes statistical design of experiments, laboratory equipment, computer-based data acquisition, complex analytical methods, and statistical evaluation of the data.

Regression is the procedure used to estimate the kinetic parameters by fitting kinetic model predictions to experimental data. When the parameters can be made to appear linear in the kinetic model (through transformations, grouping of parameters, and rearrangement), the regression is linear, and an accurate fit to data can be obtained, provided the form of the kinetic model represents well the reaction kinetics and the data are obtained over a broad range of temperature, pressure, and composition for statistically significant estimates. Often such linearization is not possible.

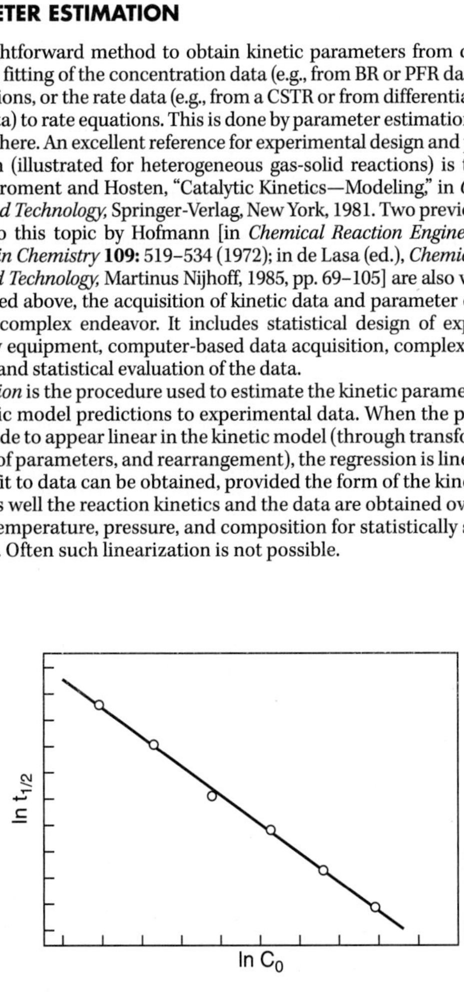

FIG. 7-22 Determination of reaction order and rate constant from half-life data.

Linear Models in Parameters, Single Reaction We adopt the terminology from Froment and Hosten, "Catalytic Kinetics—Modeling," in *Catalysis—Science and Technology,* Springer-Verlag, New York, 1981. For *n* observations (experiments) of the concentration vector *y* for a model linear in the parameter vector β of length $p < n$, the *residual error* ε is the difference between the measured data values and kinetic model-predicted values:

$$\varepsilon = y - X\beta = y - \hat{y} \tag{7-173}$$

The linear model is represented as a linear transformation of the parameter vector β through the model matrix *X*. Estimates *b* of the true parameters β are obtained by minimizing the *objective function* $S(\beta)$, *the sum of squares of the residual errors,* while varying the values of the parameters:

$$S(\beta) = \varepsilon^T\varepsilon = \sum_{i=1}^{n}(y_i - \hat{y}_i)^2 \xrightarrow{\beta} \text{Min} \tag{7-174}$$

This linear optimization problem, subject to constraints on the possible values of the parameters (e.g., requiring positive preexponentials, activation energies, etc.), can be solved to give the estimated parameters:

$$b = (X^TX)^{-1}X^Ty \tag{7-175}$$

When the error is *normally distributed* and has zero *mean* and *variance* σ^2, then the *variance-covariance matrix* $V(b)$ is defined as

$$V(b) = (X^TX)^{-1}\sigma^2 \tag{7-176}$$

An estimate for σ^2, denoted s^2, is

$$s^2 = \frac{\sum_{i=1}^{n}(y_i - \hat{y}_i)^2}{n-p} \tag{7-177}$$

When $V(b)$ is known from experimental observations, a *weighted objective function* should be used for optimization of the objective function:

$$S(\beta) = \varepsilon^T V^{-1}\varepsilon \xrightarrow{\beta} \text{Min} \tag{7-178}$$

and the estimates *b* are obtained as

$$b = (X^TV^{-1}X)^{-1}X^TV^{-1}y \tag{7-179}$$

The parameter fit is adequate if the *F test* is satisfied, that is, F_c, the calculated *F,* is larger than the tabulated statistical one at the *confidence level* of $1 - \alpha$:

$$F_c = \frac{\dfrac{\text{LFSS}}{n-p-n_e+1}}{\dfrac{\text{PESS}}{n_e-1}} \geq F(n-p-n_e+1, n_e-1; 1-\alpha) \tag{7-180}$$

$$\text{LFSS} = \sum_{i=1}^{n}(y_i - \hat{y}_i)^2 - \sum_{i=1}^{n_e}(y_i - \bar{y}_i)^2 \qquad \text{PESS} = \sum_{i=1}^{n_e}(y_i - \bar{y}_i)^2$$

Here $\bar{y}_i$ are the averaged values of the data for replicates, LFSS is the lack of fit sum of squares, PESS is the pure error sum of squares, and n_e is the number of replicate experiments. Equation (7-180) is valid if there are replicate experiments and PESS is known. Without replicates,

$$F_C = \frac{\dfrac{\text{RgSS}}{p}}{\dfrac{\text{RSS}}{n-p}} \geq F(p, n-p; 1-\alpha) \quad \text{RgSS} = \sum_{i=1}^{n}\hat{y}_i^2 \quad \text{RSS} = \sum_{i=1}^{n}(y_i - \hat{y}_i)^2 \tag{7-181}$$

Here RgSS is the regression sum of squares and RSS is the residual sum of squares. The *error bounds* on the parameter estimates are given by the *t statistics:*

$$b_i - t\left(n-p; 1-\frac{\alpha}{2}\right) \leq \beta_i \leq b_i + t\left(n-p; 1-\frac{\alpha}{2}\right) \tag{7-182}$$

An example of a linear model in parameters is Eq. (7-167), where the parameters are ln k_0, *E,* and *n,* and the linear regression can be used directly to estimate these.

Nonlinear Models in Parameters, Single Reaction In practice, the parameters appear often in nonlinear form in the rate expressions, requiring nonlinear regression. Nonlinear regression does not guarantee optimal parameter estimates even if the kinetic model adequately represents the true kinetics and the data range is adequate. Further, the statistical tests of model adequacy apply rigorously only to models linear in parameters, and can only be considered approximate for nonlinear models.

For a general nonlinear model $f(x_i, \beta)$, where x is the vector of the independent model variables and β is the vector of parameters,

$$\varepsilon = y - f(x,\beta) \tag{7-183}$$

An example of a model nonlinear in parameters is Eq. (7-166). Here it is not possible through any number of transformations to obtain a linear form in all the parameters k_0, E, K_{a0}, E_{aa}, K_{b0}, E_{ab}. Note that for some Langmuir-Hinshelwood rate expressions it is possible to linearize the model in parameters at isothermal conditions and obtain the kinetic constants for each temperature, followed by Arrhenius-type plots to obtain activation energies (see, e.g., Churchill, *The Interpretation and Use of Rate Data: The Rate Concept*, McGraw-Hill, 1974).

Minimization of the sum of squares of residuals does not result in a closed form for nonlinear parameter estimates as for the linear case; rather it requires an iterative numerical solution, and having a reasonable initial estimate for the parameter values and their feasible ranges is critical for success. Also, the minima in the residual sum of squares are local and not global. To increase the probability of approach to global minima that better represent the kinetics over a wide range of conditions, parameter estimation has to be repeated with a wide range of initial parameter guesses. The nonlinear regression procedure typically involves a *steepest descent* optimization search combined with *Newton's linearization* method when a minimum is approached, enhancing the convergence speed [e.g., the *Marquardt-Levenberg* or *Newton-Gauss* method; Marquardt, *J. Soc. Ind. Appl. Math.* **2:** 431 (1963)].

An integral part of the parameter estimation methodology is *mechanism discrimination*, i.e., selection of the best mechanism that would result in the best kinetic model. Nonlinear parameter estimation is an extensive topic and will not be further discussed here. For more details see Froment and Hosten, "Catalytic Kinetics—Modeling," in *Catalysis—Science and Technology*, Springer-Verlag, New York, 1981.

Network of Reactions The statistical parameter estimation for multiple reactions is more complex than for a single reaction. As indicated before, a single reaction can be represented by a single concentration [e.g., Eq. (7-39)]. With a network of reactions, there are a number of dependent variables equal to the number of stoichiometrically independent reactions, also called *responses*. In this case the objective function has to be modified. For details see Froment and Hosten, "Catalytic Kinetics—Modeling," in *Catalysis—Science and Technology*, Springer-Verlag, New York, 1981.

THEORETICAL METHODS

Prediction of Mechanism and Kinetics Reaction mechanisms for a variety of reaction systems can be predicted to some extent by following a set of heuristic rules derived from experience with a wide range of chemistries. For instance, Masel, *Chemical Kinetics and Catalysis*, Wiley, 2001, chap. 5, enumerates the rules for gas-phase chain and nonchain reactions including limits on activation energies for various elementary steps. Other reaction systems such as ionic reactions, and reactions on metal and acid surfaces, are also discussed by Masel, although these mechanisms are not as well understood. Nevertheless, the rules can lead to computer-generated mechanisms for complex systems such as homogeneous gas-phase combustion and partial oxidation of methane and higher hydrocarbons. Developments in computational chemistry methods allow, in addition to the derivation of most probable elementary mechanisms, prediction of thermodynamic and kinetic parameters for relatively small molecules in homogeneous gas-phase and liquid-phase reactions, and even for some heterogeneous catalytic systems. This is especially useful for complex kinetics where there is no easily discernible rate-determining step, and therefore no simple closed-form global reaction rate can be determined. In particular, estimating a large number of kinetic parameters from laboratory data requires a large number of experiments and use of intermediate reaction components that are not stable or not readily available. The nonlinear parameter estimation with many parameters is difficult, with no assurance that near global minima are actually obtained. For such complex systems, computational chemistry estimates are an attractive starting point, requiring experimental validation.

Computational chemistry includes a wide range of methods of varying accuracy and complexity, summarized in Table 7-14. Many of these methods have been implemented as software packages that require high-speed supercomputers or parallel computers to solve realistic reactions. For more details on computational chemistry, see, e.g., Cramer, *Essentials of Computational Chemistry: Theories and Models*, 2d ed., Wiley, 2004.

TABLE 7-14 Computational Chemistry Methods

Ab initio methods (no empirical parameters)
Electronic structure determination (time-independent Schrodinger equation)
Hartree-Fock (HF) with corrections
Quantum Monte Carlo (QMT)
Density functional theory (DFT)
Chemical dynamics determination (time-dependent Schrodinger equation)
Split operator technique
Multiconfigurational time-dependent Hartree-Fock method
Semiclassical method
Semiempirical methods (approximate parts of HF calculations such as two-electron integrals)
Huckel
Extended Huckel
Molecular mechanics (avoids quantum mechanical calculations)
Empirical methods (group contributions)
Polanyi linear approximation of activation energy

Lumping and Mechanism Reduction It is often useful to reduce complex reaction networks to a smaller reaction set that still maintains the key features of the detailed reaction network but with a much smaller number of representative species, reactions, and kinetic parameters. Simple examples were already given above for reducing simple networks into global reactions through assumptions such as pseudo-steady state, rate-limiting step, and equilibrium reactions.

In general, *mechanism reduction* can only be used over a limited range of conditions for which the simplified system simulates the original complete reaction network. This reduces the number of kinetic parameters that have to be either estimated from data or calculated by using computational chemistry. The simplified system also reduces the computation load for reactor scale-up, design, and optimization.

A type of mechanism reduction called *lumping* is typically performed on a reaction network that consists of a large number of similar reactions occurring between similar species, such as homologous series or molecules having similar functional groups. Such situations occur, for instance, in the oil refining industry, examples including catalytic reforming, catalytic cracking, hydrocracking, and hydrotreating. Lumping is done by grouping similar species, or molecules with similar functional groups, into pseudocomponents called *lumped species*. The behavior of the lumped system depends on the initial composition, the distribution of the rate constants in the detailed system, and the form of the rate equation. The two main issues in lumping are

1. Determination of the lump structure that simulates the detailed system over the required range of conditions
2. Determination of the kinetics of the lumped system from general knowledge about the type of kinetics and the overall range of parameters of the detailed system

Lumping has been applied extensively to first-order reaction networks [e.g., Wei and Kuo, "A Lumping Analysis in Monomolecular Reaction Systems," *I&EC Fundamentals* **8**(1): 114–123 (1969); Golikeri and Luss, "Aggregation of Many Coupled Consecutive First Order Reactions," *Chem. Eng. Sci.* **29:** 845–855 (1974)]. For instance, it has been shown that a lumped reaction network of first-order reactions can behave under certain conditions as a global second-order reaction. Where analytical solutions were not available, others, such as Golikeri and Luss, "Aggregation of Many Coupled Consecutive First Order Reactions," *Chem. Eng. Sci.* **29:** 845–855 (1974), developed bounds that bracketed the behavior of the lump for first-order reactions as a function of the initial composition and the rate constant distribution. Lumping has not been applied as successfully to nonlinear or non-first-order kinetics. More recent applications of lumping were published, including structure-oriented lumping that lumps similar structural groups, by Quann and Jaffe, "Building Useful Models of Complex Reaction Systems in Petroleum Refining," *Chem. Eng. Sci.* **51**(10): 1615–1635 (1996).

For other types of systems such as highly branched reaction networks for homogeneous gas-phase combustion and combined homogeneous and catalytic partial oxidation, mechanism reduction involves pruning branches and pathways of the reaction network that do not contribute significantly to the overall reaction. This pruning is done by using *sensitivity analysis*. See, e.g., Bui et al., "Hierarchical Reduced Models for Catalytic Combustion: H_2/Air Mixtures near Platinum Surfaces," *Combustion Sci. Technol.* **129**(1–6): 243–275 (1997).

Multiple Steady States, Oscillations, and Chaotic Behavior There are reaction systems whose steady-state behavior depends on the initial or starting conditions; i.e., for different starting conditions, different steady states can be reached at the same operating conditions. This behavior is called *steady-state multiplicity* and is often the result of the interaction of kinetic and transport phenomena. For some cases, the cause of the multiplicity is entirely reaction-related, as shown below. Associated with steady-state multiplicity is hysteresis (achieving different steady states depending on the direction in which a key parameter is varied), and higher-order instabilities such as self-sustained oscillations (repeated oscillations around a steady state) and chaotic behavior. The existence of multiple steady states may be relevant to analysis of laboratory data, since faster or slower rates may be observed at the same conditions depending on how the lab reactor is started up.

For example, CO oxidation on heterogeneous Rh catalyst exhibits hysteresis and multiple steady states, and one of the explained causes is the existence of two crystal structures for Rh, each with a different reactivity (Masel, *Chemical Kinetics and Catalysis*, Wiley, 2001, p. 38).

Another well-known example of chemistry-related instability includes the oscillatory behavior of the Bhelousov-Zhabotinsky reaction of malonic acid and bromate in the presence of homogeneous Ce catalyst having the overall reaction

$$\mathrm{HOOCCH_2COOH + HBrO \overset{Ce^{4+}}{\Rightarrow} products}$$

Ce can be in two oxidation states, Ce^{3+} and Ce^{4+}, and there are competing reaction pathways. Complex kinetic models are required to predict the oscillatory behavior, the most well known being that of Noyes [e.g., Showalter, Noyes, and Bar-Eli, *J. Chem. Phys.* **69**(6): 2514–2524 (1978)].

A large body of work has been done to develop criteria that determine the onset of chemistry- and transport-based instabilities. More details and transport-reaction coupling examples are discussed in Sec. 19.

SOFTWARE TOOLS

There are a number of useful software packages that enable efficient analysis of laboratory data for developing the mechanism and kinetics of reactions and for testing the kinetics by using simple reactor models. The reader is urged to search the Internet as some of these software packages change ownership or name. Worth mentioning are the Aspen Engineering Suite (Aspen), the MATLAB suite (Mathworks), the Chemkin software suite (Reaction Design), the NIST Chemical Kinetics database (NIST), the Thermal Safety Software suite (Cheminform St. Petersburg), and Gepasi for biochemical kinetics (freeware). The user is advised to experiment and validate any software package with known data and kinetics to ensure robustness and reliability.

Section 8

Process Control

Thomas F. Edgar, Ph.D. *Professor of Chemical Engineering, University of Texas—Austin (Section Editor, Fundamentals of Process Dynamics and Control, Process Measurements, Digital Technology for Process Control)*

Cecil L. Smith, Ph.D. *Principal, Cecil L. Smith Inc. (Batch Process Control, Telemetering and Transmission, Digital Technology for Process Control, Process Control and Plant Safety)*

B. Wayne Bequette, Ph.D. *Professor of Chemical and Biological Engineering, Rensselaer Polytechnic Institute (Unit Operations Control, Advanced Control Systems)*

Juergen Hahn, Ph.D. *Professor of Biomedical Engineering, Rensselaer Polytechnic Institute (Advanced Control Systems, Bioprocess Control)*

FUNDAMENTALS OF PROCESS DYNAMICS AND CONTROL

ADVANCED CONTROL SYSTEMS

UNIT OPERATIONS CONTROL

BATCH PROCESS CONTROL

BIOPROCESS CONTROL

TELEMETERING AND TRANSMISSION

DIGITAL TECHNOLOGY FOR PROCESS CONTROL

PROCESS MEASUREMENTS

CONTROLLERS, FINAL CONTROL ELEMENTS, AND REGULATORS

Nomenclature

Symbol	Definition
A	Area
A_a	Actuator area
A_c	Output amplitude limits
A_v	Amplitude of controlled variable
A_1	Cross-sectional area of tank
b	Controller output bias
B	Bottoms flow rate
B_i^*	Limit on control
c_A	Concentration of A
C	Cumulative sum
C_d	Discharge coefficient
C_i	Inlet concentration
C_i^*	Limit on control move
C_L	Specific heat of liquid
C_0	Integration constant
C_p	Process capability
C_r	Heat capacity of reactants
C_V	Valve flow coefficient
D	Distillate flow rate, disturbance
D_i^*	Limit on output
e	Error
E	Economy of evaporator
f	Function of time
F, f	Feed flow rate
F_L	Pressure recovery factor
g_c	Unit conversion constant
g_i	Algebraic inequality constraint
G	Transfer function
G_c	Controller transfer function
G_d	Disturbance transfer function
G_f	Feedforward controller transfer function
G_m	Sensor transfer function
G_p	Process transfer function
G_t	Transmitter transfer function
G_v	Valve transfer function
h_i	Algebraic equality constraints
h_1	Liquid head in tank
H	Latent heat of vaporization, control limit or threshold
i	Summation index
I_i	Impulse response coefficient
j	Time index
J	Objective function or performance index
k	Time index
k_f	Flow coefficient
k_r	Kinetic rate constant
K	Gain, slack parameter
K_c	Controller gain
K_d	Disturbance transfer function gain
K_m	Measurement gain
K_p	Process gain
K_u	Ultimate controller gain (stability)
L	Load variable
L_p	Sound pressure level
M	Manipulated variable
m_c	Number of constraints
M_v	Mass flow
M_r	Mass of reactants
M_w	Molecular weight
n	Number of data points, number of stages or effects
N	Number of inputs/outputs, model horizon
p	Proportional band (percent)
p_c	Vapor pressure
p_d	Actuator pressure
p_i	Pressure
p_u	Proportional band (ultimate)
q	Radiated energy flux
q_b	Energy flux to a black body
Q	Flow rate
r_c	Number of constraints
R	Equal-percentage valve characteristic
R_T	Resistance in temperature sensor
R_1	Valve resistance

Symbol	Definition
s	Laplace transform variable
s	Search direction
S_i	Step response coefficient
t	Time
T	Temperature, target
$T(s)$	Decoupler transfer function
T_b	Base temperature
T_f	Exhaust temperature
T_R	Reset time
U	Heat-transfer coefficient
u, U	Manipulated variable, controller output
V	Volume
V_s	Product value
w	Mass flow rate
w_i	Weighting factor
W	Steam flow rate
x	Mass fraction
$\bar{x}$	Sample mean
x_i	Optimization variable
x_T	Pressure drop ratio factor
X	Transform of deviation variable
y, Y	Process output, controlled variable, valve travel
Y_{sp}	Set point
z	Controller tuning law, expansion factor
z_i	Feed mole fraction (distillation)
Z	Compressibility factor
	Greek symbols
α	Digital filter coefficient
α_T	Temperature coefficient of resistance
β	Resistance thermometer parameter
γ	Ratio of specific heats
δ	Move suppression factor, shift in target value
Δ_q	Load step change
Δ_t	Time step
Δ_T	Temperature change
Δ_u	Control move
ε	Spectral emissivity, step size
ζ	Damping factor (second-order system)
θ	Time delay
λ	Relative gain array parameter, wavelength
Λ	Relative gain array
ξ	Deviation variable
ρ	Density
σ	Stefan-Boltzmann constant, standard deviation
Σ_t	Total response time
τ	Time constant
τ_d	Natural period of closed loop, disturbance time constant
τ_D	Derivative time (PID controller)
τ_F	Filter time constant
τ_I	Integral time (PID controller)
τ_P	Process time constant
τ_o	Period of oscillation
ϕ_{PI}	Phase lag
	Subscripts
A	Species A
b	Best
c	Controller
d	Disturbance
eff	Effective
f	Feedforward
i	Initial, inlet
L	Load, disturbance
m	Measurement or sensor
p	Process
s	Steady state
sp	Set-point value
t	Transmitter
u	Ultimate
v	Valve

FUNDAMENTALS OF PROCESS DYNAMICS AND CONTROL

GENERAL CONTROL SYSTEM

A process is shown in Fig. 8-1 with a manipulated input *U*, a load or disturbance input *D*, and a controlled output *Y*, which could be flow, pressure, liquid level, temperature, composition, or any other inventory, environmental, or quality variable that is to be held at a desired value identified as the set point Y_{sp}. The load may be a single variable or an aggregate of variables either acting independently or manipulated for other purposes, affecting the controlled variable much as the manipulated variable does. Changes in load may occur randomly as caused by changes in weather, diurnally with ambient temperature, manually when operators change production rate, stepwise when equipment is switched into or out of service, or cyclically as the result of oscillations in other control loops. Variations in load will drive the controlled variable away from the set point, requiring a corresponding change in the manipulated variable to bring it back. The manipulated variable must also change to move the controlled variable from one set point to another.

An open-loop system positions the manipulated variable either manually or on a programmed basis, without using any process measurements. This operation is acceptable for well-defined processes without disturbances. An automated transfer switch is provided to allow manual adjustment of the manipulated variable in case the process or the control system is not performing satisfactorily.

A closed-loop system uses the measurement of one or more process variables to move the manipulated variable to achieve control. Closed-loop systems may include feedforward, feedback, or both.

Feedback Control In a feedback control loop, the controlled variable is compared to the set point Y_{sp}, with the error *E* acted upon by the controller to move *U* in such a way as to minimize the error. This action is specifically negative feedback, in that an increase in error moves *U* so as to decrease the error. (Positive feedback would cause the error to expand rather than diminish and therefore does not regulate.) The action of the controller is selectable to allow use on process gains of both signs.

The controller has tuning parameters related to proportional, integral, derivative, lag, dead time, and sampling functions. A negative feedback loop will oscillate if the controller gain is too high; but if it is too low, control will be ineffective. The controller parameters must be properly related to the process parameters to ensure closed-loop stability while still providing effective control. This relationship is accomplished, first, by the proper selection of control modes to satisfy the requirements of the process and, second, by the appropriate tuning of those modes.

Feedforward Control A feedforward system uses measurements of disturbance variables to position the manipulated variable in such a way as to minimize any resulting deviation. The disturbance variables could be either measured loads or the set point, the former being more common. The feedforward gain must be set precisely to reduce the deviation of the controlled variable from the set point.

Feedforward control is usually combined with feedback control to eliminate any offset resulting from inaccurate measurements and calculations and unmeasured load components. The feedback controller can be used as a bias on the feedforward controller or in a multiplicative form.

Computer Control Computers have been used to replace analog PID controllers, either by setting set points of lower-level controllers in supervisory control or by driving valves directly in direct digital control. Single-station digital controllers perform PID control in one or two loops, including computing functions such as mathematical operations, characterization, lags, and dead time, with digital logic and alarms. Distributed control systems provide all these functions, with the digital processor shared among many control loops; separate processors may be used for displays, communications, file servers, and the like. A host computer may be added to perform high-level operations such as scheduling, optimization, and multivariable control. More details on computer control are provided later in this section.

PROCESS DYNAMICS AND MATHEMATICAL MODELS

GENERAL REFERENCES: Seborg, Edgar, Mellichamp, and Doyle, *Process Dynamics and Control*, Wiley, New York, 2016; Marlin, *Process Control*, McGraw-Hill, New York, 2000; Ogunnaike and Ray, *Process Dynamics Modeling and Control*, Oxford University Press, New York, 1994; Bequette, *Process Control: Modeling, Design and Simulation*, 2d ed., Prentice-Hall, Upper Saddle River, N.J., 2017.

Open-Loop versus Closed-Loop Dynamics It is common in industry to manipulate coolant in a jacketed reactor in order to control conditions in the reactor itself. A simplified schematic diagram of such a reactor control system is shown in Fig. 8-2. Assume that the reactor temperature is adjusted by a controller that increases the coolant flow in proportion to the difference between the desired reactor temperature and the temperature that is measured. The proportionality constant is K_c. If a small change in the temperature of the inlet stream occurs, then depending on the value of K_c, one might observe the reactor temperature responses shown in Fig. 8-3. The top plot shows the case for no control ($K_c = 0$), which is called the *open loop*, or the normal dynamic response of the process by itself. As K_c increases, several effects can be noted. First, the reactor temperature responds faster and faster. Second, for the initial increases in K_c, the maximum deviation in the reactor temperature becomes smaller. Both of these effects are desirable so that disturbances from normal operation have as small an effect as possible on the process under study. As the gain is increased further, eventually a point is reached where the reactor temperature oscillates indefinitely, which is undesirable. This point is called the *stability limit*, where $K_c = K_u$, the ultimate controller gain. Increasing K_c further causes the magnitude of the oscillations to increase, with the result that the control valve will cycle between full open and closed.

The responses shown in Fig. 8-3 are typical of the vast majority of regulatory loops encountered in the process industries. Figure 8-3 shows that there is an optimal choice for K_c, somewhere between 0 (no control) and K_u (stability limit). If one has a dynamic model of a process, then this model can be used to calculate controller settings. In Fig. 8-3, no time scale is given, but rather the figure shows relative responses. A well-designed controller might be able to speed up the response of a process by a factor of roughly 2 to 4. Exactly how fast the control system responds is determined by the dynamics of the process itself.

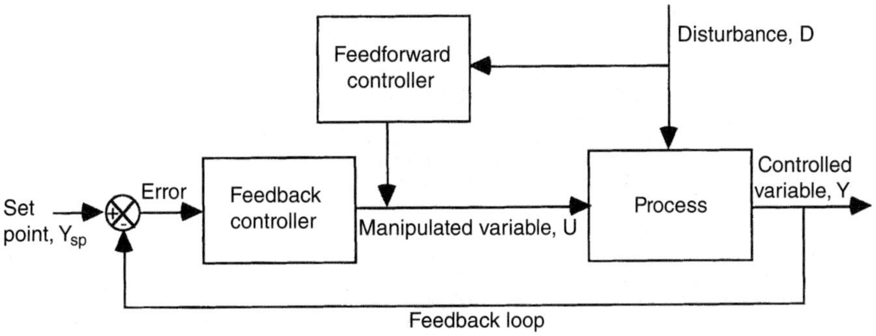

FIG. 8-1 Block diagram for feedforward and feedback control.

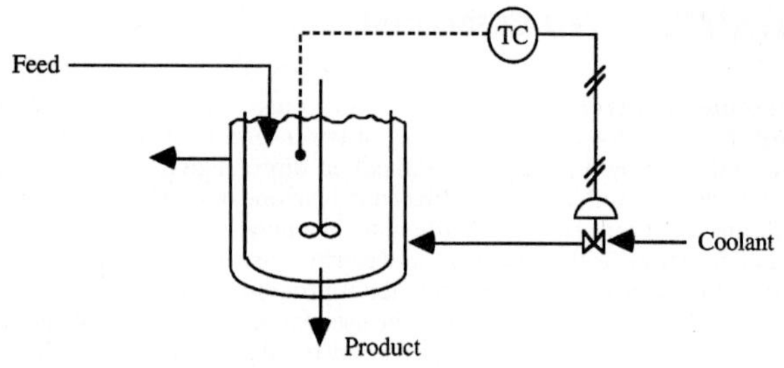

FIG. 8-2 Reactor control system.

Physical Models versus Empirical Models In developing a dynamic process model, two distinct approaches can be taken. The first involves models based on first principles, called *physical* or *first-principles* models, and the second involves empirical models. The conservation laws of mass, energy, and momentum form the basis for developing physical models. The resulting models typically involve sets of differential and algebraic equations that must be solved simultaneously. Empirical models, by contrast, involve postulating the form of a dynamic model, usually as a transfer function (input-output model), which is discussed below. This transfer function contains a number of parameters that need to be estimated from data. For the development of both physical and empirical models, the most expensive step normally involves verification of their accuracy in predicting plant behavior.

To illustrate the development of a physical model, a simplified treatment of the reactor, shown in Fig. 8-2, is used. It is assumed that the reactor is operating isothermally and that the inlet and exit volumetric flows and densities are the same. There are two components, A and B, in the reactor, and a single first-order reaction of $A \rightarrow B$ takes place. The inlet concentration of A, which we call c_i, varies with time. A dynamic mass balance for the concentration of A, denoted c_A, can be written as follows:

$$V\frac{dc_A}{dt} = Fc_i - Fc_A - kVc_A \tag{8-1}$$

In Eq. (8-1), the flow in of component A is Fc_i, the flow out is Fc_A, and the loss via reaction is k_rVc_A, where V = reactor volume and k_r = kinetic rate constant. In this example, c_i is the input, or forcing, variable and c_A is the output variable. If V, F, and k_r are constant, Eq. (8-1) can be rearranged by dividing by $F + k_rV$ so that it contains only two groups of parameters. The result is

$$\tau\frac{dc_A}{dt} = Kc_i - c_A \tag{8-2}$$

where $\tau = V/(F + k_rV)$ and $K = F/(F + k_rV)$. For this example, the resulting model is a first-order differential equation in which τ is called the time constant and K the process gain.

As an alternative to deriving Eq. (8-2) from a dynamic mass balance, one could simply postulate a first-order differential equation to be valid (empirical modeling). Then it would be necessary to estimate values for τ and K so that the postulated model described the reactor's dynamic response. The advantage of the physical model over the empirical model is that the physical model gives insight into how reactor parameters affect the values of τ and K, which in turn affects the dynamic response of the reactor.

Nonlinear versus Linear Models If V, F, and k_r are constant, then Eq. (8-1) is an example of a linear differential equation model. In a linear equation, the output and input variables and their derivatives appear to only the first power. If the rate of reaction were second-order, then the resulting dynamic mass balance would be

$$V\frac{dc_A}{dt} = Fc_i - Fc_A - k_rVc_A^2 \tag{8-3}$$

Since c_A appears in this equation to the second power, the equation is nonlinear.

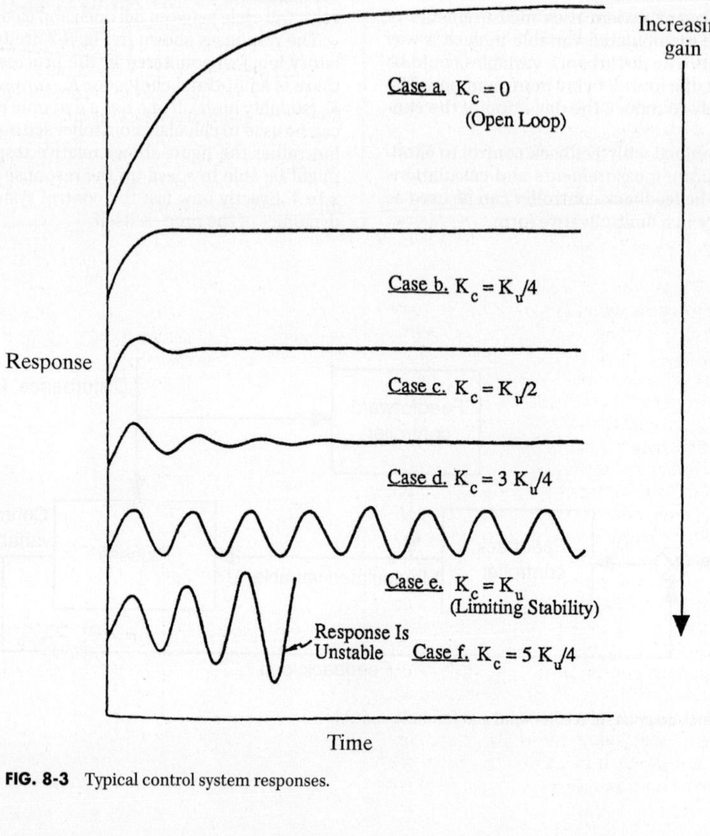

FIG. 8-3 Typical control system responses.

The difference between linear systems and nonlinear systems can be seen by considering the steady-state behavior of Eq. (8-1) compared to Eq. (8-3) (the left-hand side is zero; that is, $dc_A/dt = 0$). For a given change in c_i, Δc_i, the change in c_A calculated from Eq. (8-1), Δc_A, is always proportional to Δc_i, and the proportionality constant is K [see Eq. (8-2)]. The change in the output of a system divided by a change in the input to the system is called the *process gain.* Linear systems have constant process gains for all changes in the input. By contrast, Eq. (8-3) gives a Δc_A that varies with Δc_i, which is a function of the concentration levels in the reactor. Thus, depending on the reactor operating conditions, a change in c_i produces different changes in c_A. In this case, the process has a nonlinear gain. Systems with nonlinear gains are more difficult to control than linear systems that have constant gains.

Simulation of Dynamic Models Linear dynamic models are particularly useful for analyzing control system behavior. The insight gained through linear analysis is invaluable. However, accurate dynamic process models can involve large sets of nonlinear equations. Analytical solution of these models is not possible. Thus, in these cases, one must turn to simulation approaches to study process dynamics and the effect of process control. Equation (8-3) will be used to illustrate the simulation of nonlinear processes. If dc_A/dt on the left-hand side of Eq. (8-3) is replaced with its finite difference approximation, one gets

$$c_A(t+\Delta t) = \frac{c_A(t) + \Delta t \cdot \left[Fc_i(t) - Fc_A(t) - k_r V c_A^2(t) \right]}{V} \tag{8-4}$$

Starting with an initial value of c_A and given $c_i(t)$, Eq. (8-4) can be solved for $c_A(t + \Delta t)$. Once $c_A(t + \Delta t)$ is known, the solution process can be repeated to calculate $c_A(t + 2\Delta t)$, and so on. This approach is called the *Euler integration method*; while it is simple, it is not necessarily the best approach to numerically integrating nonlinear differential equations. As discussed in Sec. 3, more sophisticated approaches are available that allow much larger step sizes to be taken.

Increasingly sophisticated simulation packages (e.g., MATLAB, gPROMS) are being used to calculate the dynamic behavior of processes and to test control system behavior. These packages have good user interfaces, and they can handle stiff systems where some variables respond on a time scale that is significantly faster or slower than that of other variables. A simple Euler approach cannot effectively handle stiff systems, which frequently occur in chemical process models. See Sec. 3 of this handbook for more details.

Laplace Transforms When mathematical models are used to describe process dynamics in conjunction with control system analysis, the models generally involve linear differential equations. Laplace transforms are very effective for solving linear differential equations analytically. The key advantage of using Laplace transforms is that they convert differential equations to algebraic equations. The resulting algebraic equations are easier to solve than the original differential equations. When the Laplace transform is applied to a linear differential equation in time, the result is an algebraic equation in a new variable s, called the *Laplace variable.* To get the solution to the original differential equation, one needs to invert the Laplace transform. Details on these procedures are contained in process control textbooks, e.g., Seborg, Edgar, Mellichamp, and Doyle, *Process Dynamics and Control,* 4th ed., Wiley, New York, 2016. The use of Laplace transforms in process control analysis has decreased over the past 40 years, due to ever-improving simulation tools.

Transfer Functions and Block Diagrams A very convenient and compact method of representing the process dynamics of linear systems involves the use of transfer functions and block diagrams. A transfer function can be obtained by starting with a physical model, as discussed previously. If the physical model is nonlinear, first it needs to be linearized around an operating point. The resulting linearized model is then approximately valid in a region around this operating point. A transfer function is typically characterized by its process gain K and time constant s (τ_1, τ_2, etc.) to fit the process under study. In fitting the parameters, data can be generated by forcing the process. If step forcing is used, then the resulting response is called the *process reaction curve.* Block diagrams show how changes in an input variable affect an output variable. Block diagrams are a means of concisely representing the dynamics of a process under study. Since linearity is assumed in developing a block diagram, if more than one variable affects an output, the contributions from each can be added (see Seborg, Edgar, Mellichamp, and Doyle, *Process Dynamics and Control,* 4th ed., Wiley, New York, 2016).

Continuous versus Discrete Models The preceding discussion has focused on systems where variables change continuously with time. Most real processes have variables that are continuous, such as temperature, pressure, and flow. However, some processes involve discrete events, such as the starting or stopping of a pump. In addition, modern plants are controlled by digital computers, which are discrete. In controlling a process, a digital system samples variables at a fixed rate, and the resulting system is a sampled data system. From one sampling instant until the next, variables are assumed to remain fixed at their sampled values. Similarly, in controlling a process, a digital computer sends out signals to control elements, usually valves, at discrete instants of time. These signals remain fixed until the next sampling instant.

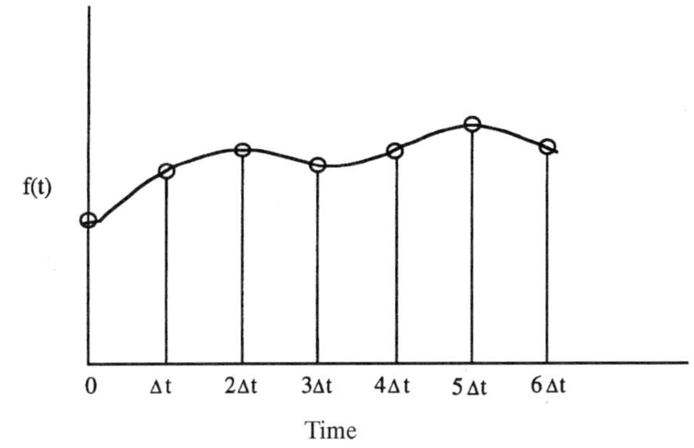

FIG. 8-4 Sampled data example.

Figure 8-4 illustrates the concept of sampling a continuous function. At integer values of the sampling rate Δt, the value of the variable to be sampled is measured and held until the next sampling instant. To deal with sampled data systems, the z transform has been developed, which is analogous to the Laplace transform for continuous systems. Utilization of z transforms has been discussed in various textbooks (e.g., Seborg, Edgar, Mellichamp, and Doyle, *Process Dynamics and Control,* Wiley, New York, 2016). Sampling a continuous variable results in a loss of information. However, in practical applications, sampling is fast enough that the loss is typically insignificant and the difference between continuous and discrete modeling is small in terms of its effect on control. Increasingly, model predictive controllers that make use of discrete dynamic models are being used in the process industries. The purpose of these controllers is to guide a process to optimum operating points. These model predictive control algorithms are typically run at much slower sampling rates than are used for basic control loops such as flow control or pressure control. The discrete dynamic models used are normally developed from data generated from plant testing, as discussed hereafter. For a detailed discussion of modeling sampled data systems, the interested reader is referred to textbooks on digital control (Åström and Wittenmark, *Computer Controlled Systems,* Prentice-Hall, Englewood Cliffs, N.J., 1997).

Process Characteristics in Transfer Functions Process characteristics can be expressed in the form of transfer functions. Consider a system involving flow out of a tank, shown in Fig. 8-5.

Proportional Element First, consider the outflow through the exit valve on the tank. If the flow through the line is turbulent, then Bernoulli's equation can be used to relate the flow rate through the valve to the pressure drop across the valve as

$$f_1 = k_f A_v \sqrt{2g_c(h_1 - h_0)} \tag{8-5}$$

where f_1 = flow rate, k_f = flow coefficient, A_v = cross-sectional area of the restriction, g_c = constant, h_1 = liquid head in tank (pressure at the base of the tank), and h_0 = atmospheric pressure. This relationship between flow and

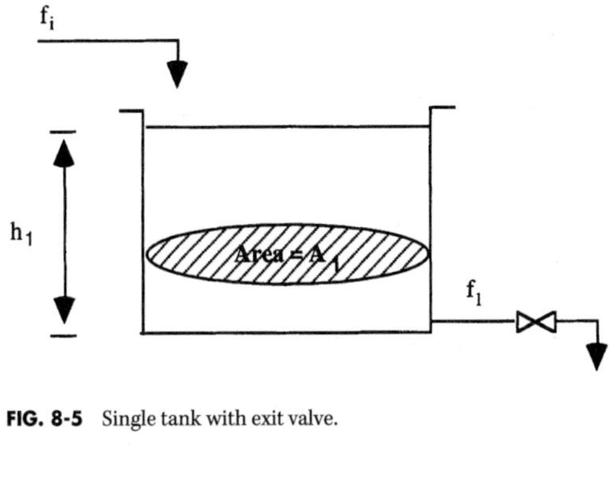

FIG. 8-5 Single tank with exit valve.

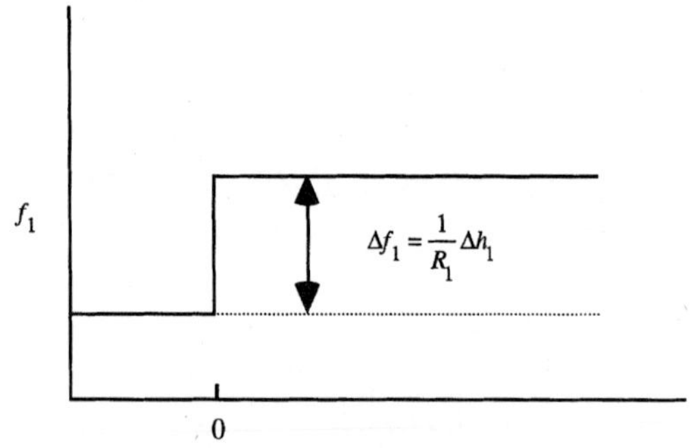

FIG. 8-6 Response of proportional element.

pressure drop across the valve is nonlinear, and it can be linearized around a particular operating point to give

$$f_1 - \bar{f}_1 = \left(\frac{1}{R_1}\right)(h_1 - \bar{h}_1) \qquad (8\text{-}6)$$

where $R_1 = \bar{f}_1/(g_c k_f^2 A_v^2)$ is called the *resistance* of the valve in analogy with an electrical resistance. This is an example of a pure gain system with no dynamics. In this case, the process gain is $K = 1/R_1$. Such a system has an instantaneous dynamic response, and for a step change in head, there is an immediate step change in flow, as shown in Fig. 8-6. The exact magnitude of the step in flow depends on the operating flow $\bar{f}_1$ as the definition of R_1 shows.

First-Order Lag (Time Constant Element) Next consider the system to be the tank itself. A dynamic mass balance on the tank gives

$$A_1 \frac{dh_1}{dt} = f_i - f_1 \qquad (8\text{-}7)$$

where A_1 is the cross-sectional area of the tank and f_i is the inlet flow. By substituting Eq. (8-6) into Eq. (8-7) one can develop the differential equation relating changes in h_1 to changes in f_i. The resulting input-output relationship is a first-order system, and it has a gain $K = R_1$ and a time constant $\tau_1 = R_1 A_l$. For a step change in f_i, h_1 follows a decaying exponential response from its initial value $\bar{h}_1$ to a final value of $\bar{h}_1 + R_1\,\Delta f_i$ (Fig. 8-7). At a time equal to τ_1, the transient in h_1 is 63 percent finished; and at $3\tau_1$, the response is 95 percent finished. These percentages are the same for all first-order processes. Thus, knowledge of the time constant of a first-order process gives insight into how fast the process responds to sudden input changes.

Capacity Element Now consider the case where the valve in Fig. 8-7 is replaced with a pump. In this case, it is reasonable to assume that the exit flow from the tank is independent of the level in the tank. For such a case, Eq. (8-7) still holds, except that f_l no longer depends on h_1. For changes in f_i, a differentiated equation relates changes in h_1 to changes in f_i, which is an example of a pure capacity process, also called an *integrating system.* The cross-sectional area of the tank is the chemical process equivalent of an electrical capacitor. If the inlet flow is step-forced while the outlet is held constant, then the level builds up linearly, as shown in Fig. 8-8. Eventually the liquid will overflow the tank.

Second-Order Element Consider the two-tank system shown in Fig. 8-9. For tank 1, the first-order dynamic model relates changes in f_1 to changes in f_i. Since f_i is the inlet flow to tank 2, a first-order dynamic model also relates changes in h_2 to changes in f_i, which has the same form as tank 1. The overall model is a second-order differential equation, which has a gain R_2 and two time constants A_1R_1 and A_2R_2. For two tanks with equal areas, a step change in f_i produces the S-shaped response in level in the second tank shown in Fig. 8-10.

General Second-Order Element Figure 8-3 illustrates the fact that closed-loop systems can exhibit oscillatory behavior. A general second-order transfer function that can exhibit oscillatory behavior is important for the study of automatic control systems. For a unit step input, the transient responses shown in Fig. 8-11 result. As can be seen, when $\zeta < 1$, the response oscillates; and when $\zeta < 1$, the response is S-shaped. Few open-loop chemical processes exhibit an oscillating response; most exhibit an S-shaped step response.

Distance-Velocity Lag (Dead-Time Element) The dead-time or time-delay element, commonly called a *distance-velocity lag,* is often encountered in process systems. For example, if a temperature-measuring element is located downstream from a heat exchanger, a time delay occurs before the

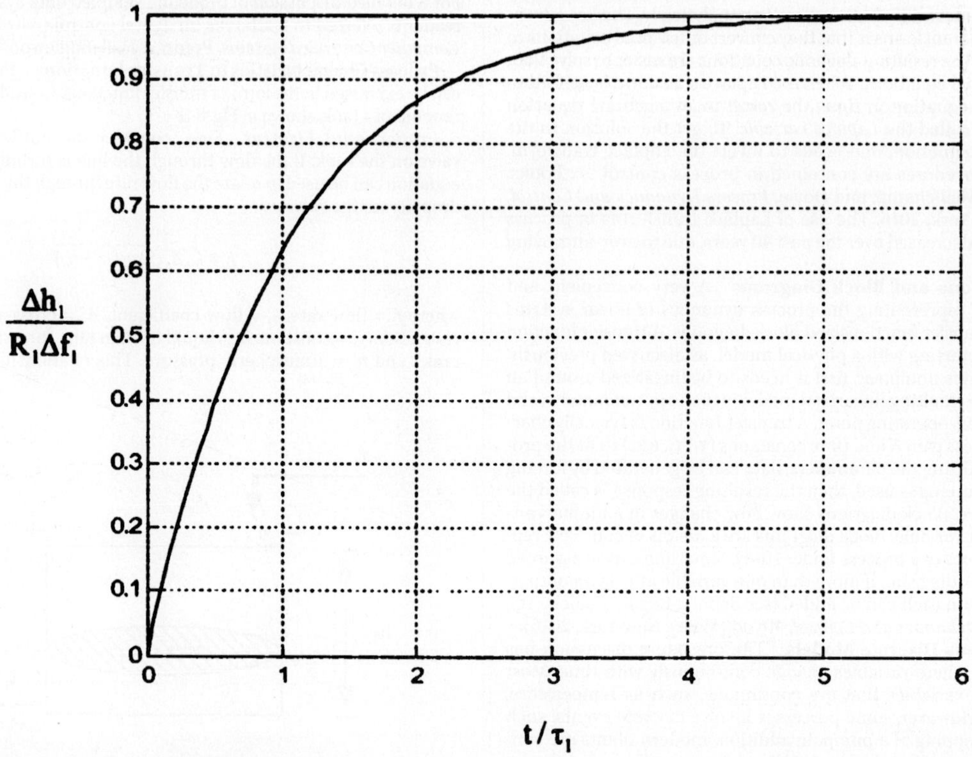

FIG. 8-7 Response of first-order system.

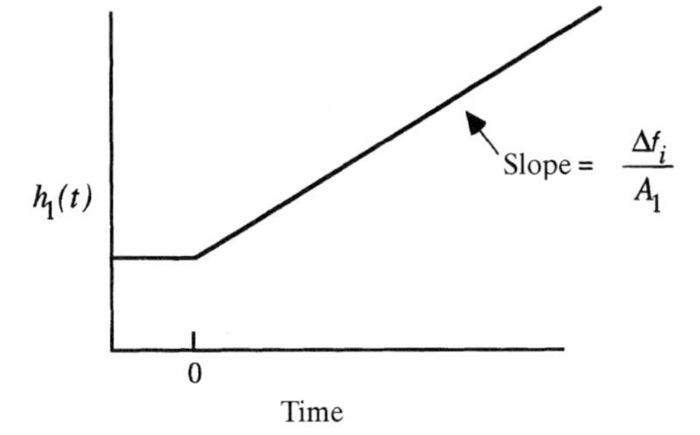

FIG. 8-8 Response of pure capacity system.

heated fluid leaving the exchanger arrives at the temperature measurement point. If some element of a system produces a dead time of θ time units, then an input to that unit $f(t)$ will be reproduced at the output as $f(t-\theta)$. The transient response of the element is shown in Fig. 8-12.

Higher-Order Lags If a process is described by a series of n first-order lags, the overall system response becomes proportionally slower with each lag added. The special case of a series of n first-order lags with equal time constants has a transfer function given by

$$G(s)=\frac{K}{(\tau s+1)^n} \tag{8-8}$$

The step response of this transfer function is shown in Fig. 8-13. Note that all curves reach about 60 percent of their final value at $t=n\tau$.

Higher-order systems can be approximated by a first- or second-order plus dead-time system for control system design.

Multi-input, Multioutput Systems The dynamic systems considered up to this point have been examples of single-input, single-output (SISO) systems. In chemical processes, one often encounters systems where one input can affect more than one output. For example, assume that one is studying a distillation tower in which both reflux and boil-up are manipulated for control purposes. If the output variables are the top and bottom product compositions, then each input affects both outputs. For this distillation example, the process is referred to as a 2 × 2 system to indicate the number of inputs and outputs. In general, *multi-input, multi-output* (MIMO) systems can have n inputs and m outputs with $n \neq m$, and they can be nonlinear. Such a system would be called an *$n \times m$ system.* A depiction of a 2 × 2 linear system is given in Fig. 8-14. Note that since linear systems are involved, the effects of the two inputs on each output are additive. In many process control systems, one input is selected to control one output in a MIMO system. For m outputs there would be m such selections. For this type of control strategy, one needs to consider which inputs and outputs to couple with feedback controllers, and this problem is referred to as *loop pairing.* Another important issue that arises involves interaction between control loops. When one loop makes a change in its manipulated variable, the change affects the other loops in the system. These changes are the direct result of the multivariable nature of the process. In some cases, the interaction can be so severe that overall control system performance is drastically reduced. Finally, some of the modern approaches to process control tackle the MIMO problem directly, and they simultaneously use all manipulated variables to control all output variables rather than pair one input to one output (see later section on multivariable control).

Fitting Dynamic Models to Experimental Data In developing empirical transfer functions, it is necessary to identify model parameters from experimental data. A number of approaches to process identification have been published. The simplest approach involves introducing a step test into the process and recording the response of the process, as illustrated in Fig. 8-15. The x's in the figure represent the recorded data. For purposes of illustration, the process under study will be assumed to be first-order with dead time and will have three parameters to characterize the response: gain K, time constant τ, and time delay θ.

An experimental response is shown in Fig. 8-15 for a set of model parameters K, τ, and θ fitted to the data. These parameters are calculated by using optimization to minimize the squared difference between the model predictions and the data, i.e., a least-squares approach. Let each measured data point be represented by y_j (measured response), t_j (time of measured response), $j = 1$ to n. Then the least-squares problem can be formulated as

$$\min_{\tau,\theta,K} \sum_{j=1}^{n} [y_j - \hat{y}(t_j)]^2 \tag{8-9}$$

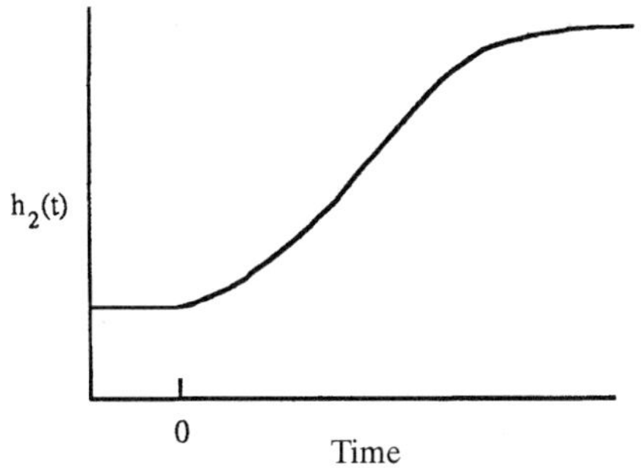

FIG. 8-10 Response of second-order system.

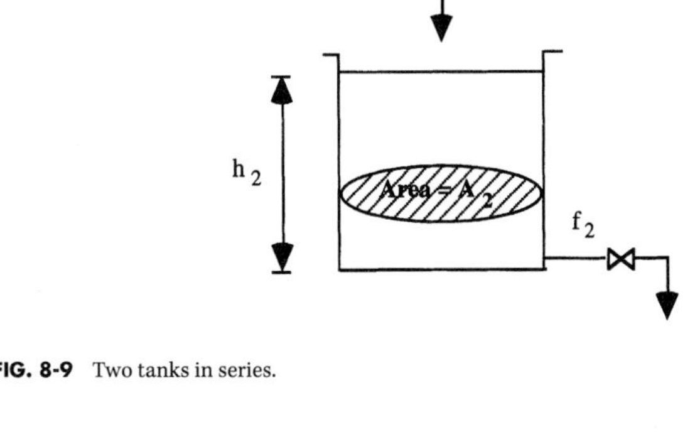

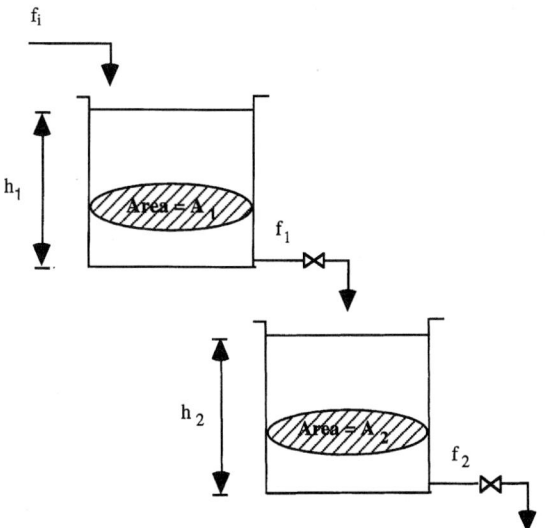

FIG. 8-9 Two tanks in series.

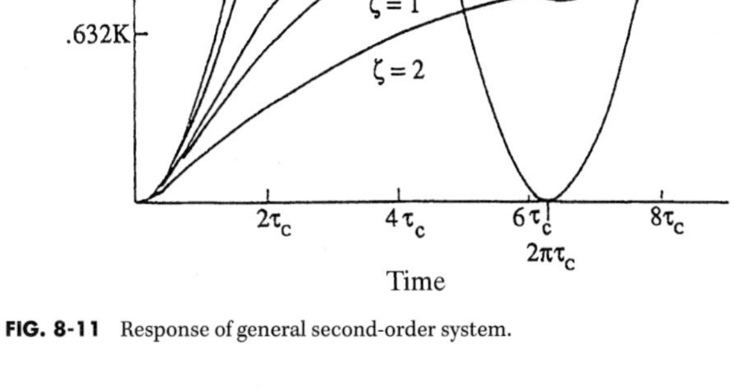

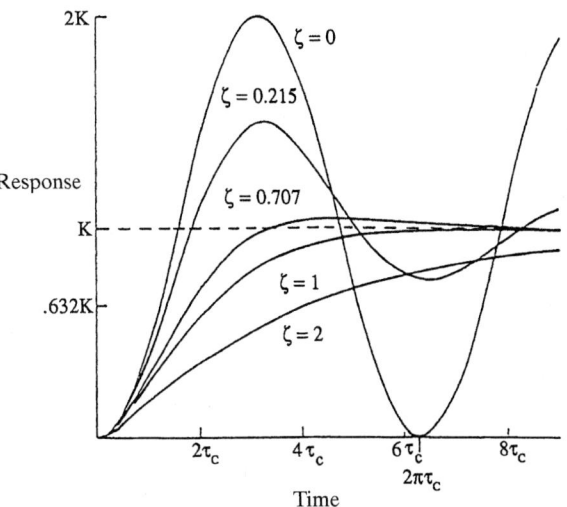

FIG. 8-11 Response of general second-order system.

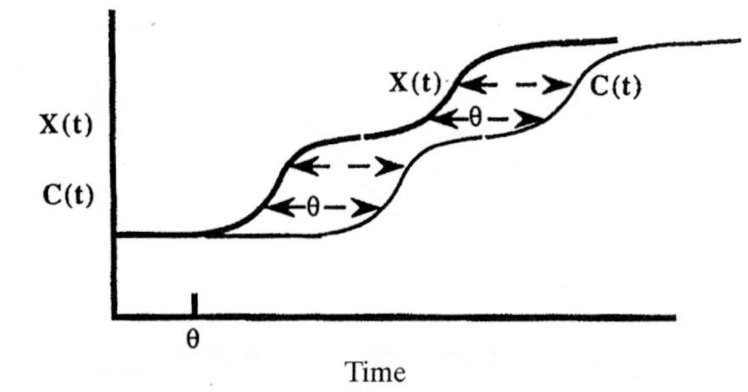

FIG. 8-12 Response of dead-time system.

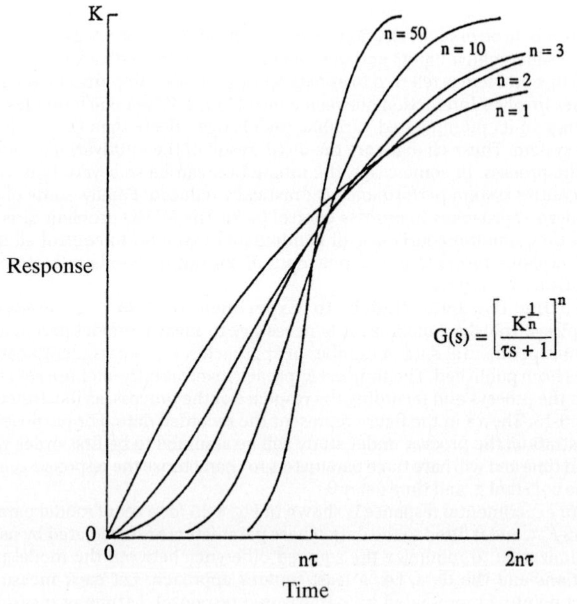

FIG. 8-13 Response of *n*th-order lags.

where $\hat{y}(t_j)$ is the predicted value of y at time t_j and n is the number of data points. This optimization problem can be solved to calculate the optimal values of K, τ, and θ. A number of software packages such as Excel Solver are available for minimizing Eq. (8-9).

One operational problem caused by step forcing is the fact that the process under study is moved away from its steady-state operating point. Plant managers may be reluctant to allow large steady-state changes, since normal production will be disturbed by the changes. As a result, alternative methods of forcing actual processes have been developed, and these included pulse testing and *pseudorandom binary signal* (PRBS) forcing. With pulse forcing, one introduces a step, and then after a period of time the input is returned to its original value. The result is that the process dynamics are excited, but after the forcing the process returns to its original steady state. PRBS forcing involves a series of pulses of fixed height and random duration, as shown in Fig. 8-16. The advantage of PRBS is that forcing can be concentrated on particular frequency ranges that are important for control system design.

Transfer function models assume linear, dynamic models are adequate, but chemical processes are known to exhibit nonlinear behavior. One could use the same type of optimization objective as given in Eq. (8-27) to determine parameters in nonlinear first-principles models, such as Eq. (8-3) presented earlier. Also, nonlinear input-output empirical models, such as neural network models, have recently been proposed for process applications. The key to the use of these nonlinear empirical models is to have high-quality process data, which allows the important nonlinearities to be identified.

FEEDBACK CONTROL SYSTEM CHARACTERISTICS

General References: Shinskey, *Process Control Systems*, 4th ed., McGraw-Hill, New York, 1996; Seborg, Edgar, Mellichamp, and Doyle, *Process Dynamics and Control*, 4th ed., Wiley, New York, 2016; Bequette, *Process Control: Modeling, Design and Simulation*, 2d ed., Prentice-Hall, Upper Saddle River, N.J., 2017. There are two objectives in applying feedback control: (1) regulate the controlled variable at the set point following changes in load and (2) respond to set-point changes, with the latter called *servo operation*. In fluid processes, almost all control loops must contend with variations in load, and therefore regulation is of primary importance. While most loops will operate continuously at fixed set points, frequent changes in set points can occur in flow loops and in batch production. The most common mechanism for achieving both objectives is feedback control, because it is the simplest and most universally applicable approach to the problem.

Closing the Loop The simplest representation of the closed feedback loop is shown in Fig. 8-17. Other versions of the block diagram can include a measurement or sensor block in the feedback path rather than including it in G_p (e.g., Seborg et al., 2016). The load is shown entering the process at the same point as the manipulated variable because that is the most common point of entry, and because, lacking better information, the elements in the path of the manipulated variable are the best estimates of those in the load path. The load rarely impacts directly on the controlled variable without passing through the dominant lag in the process. Where the load is unmeasured, its current value can be observed as the controller output required to keep the controlled variable Y at set point Y_{sp}.

If the loop is opened—either by placing the controller in manual operation or by setting its gains to zero—the load will have complete influence over the controlled variable, and the set point will have none. Only by closing the loop with controller gain as high as possible will the influence of

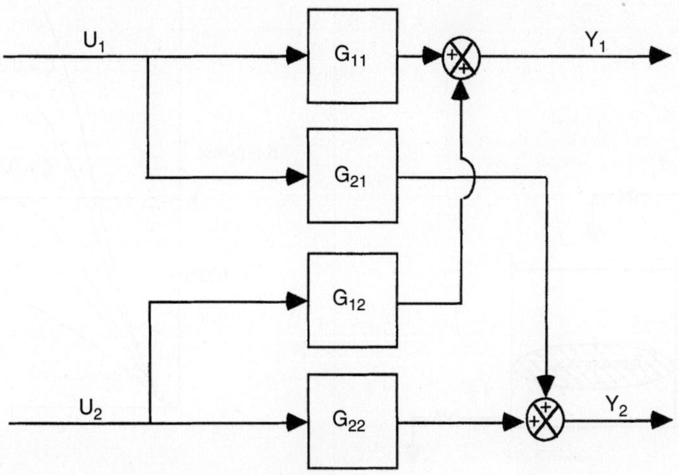

FIG. 8-14 Example of 2 × 2 transfer function.

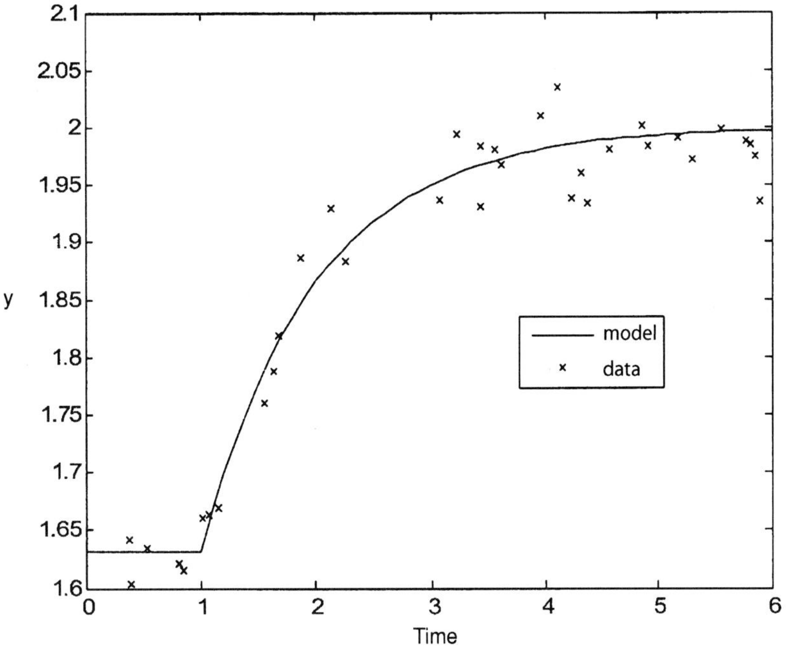

FIG. 8-15 Plot of experimental data and first-order model fit.

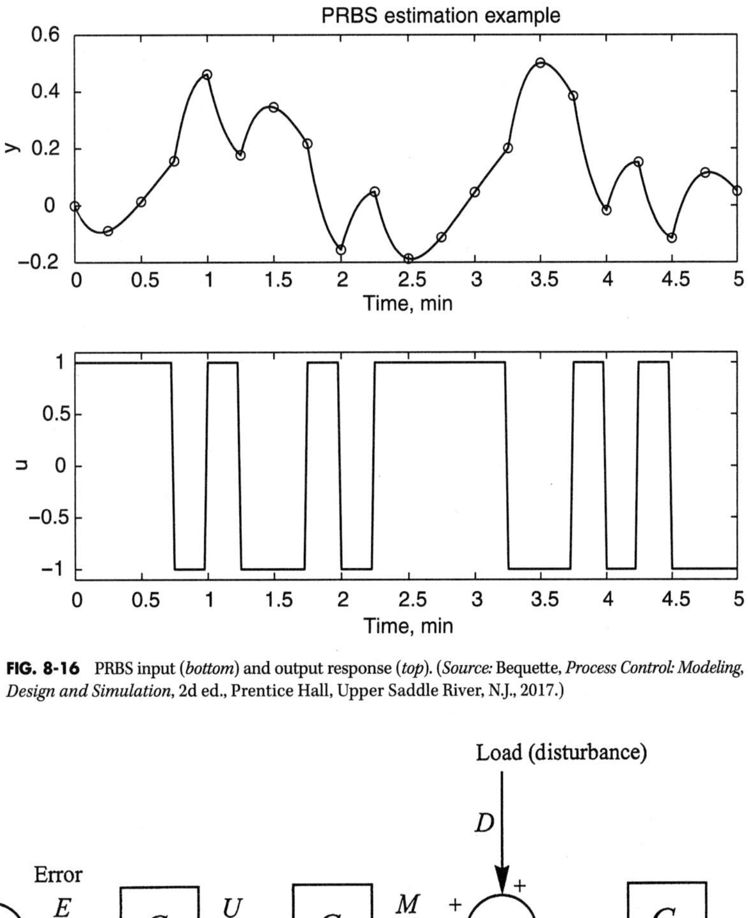

FIG. 8-16 PRBS input (*bottom*) and output response (*top*). (*Source:* Bequette, *Process Control: Modeling, Design and Simulation*, 2d ed., Prentice Hall, Upper Saddle River, N.J., 2017.)

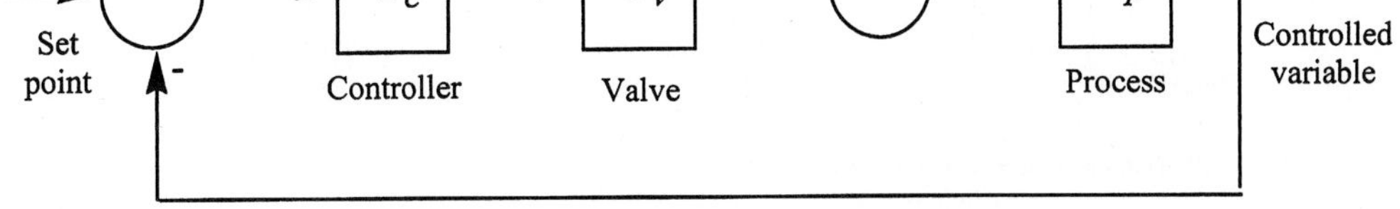

FIG. 8-17 Both load regulation and set-point response require high gains for the feedback controller.

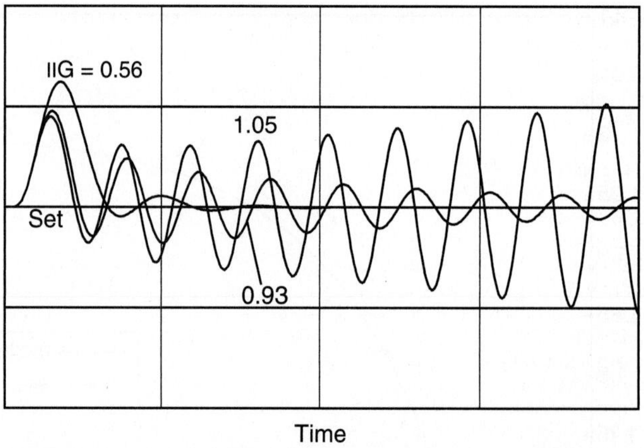

FIG. 8-18 Transition from well-damped load response to instability develops as loop gain increases.

the load be minimized and that of the set point be maximized. There is a practical limit to the controller gain, however, at the point where the controlled variable develops a uniform oscillation. This is defined as the *limit of stability*, and it is reached when the product of gains in the loop $\Pi G = G_c G_v G_p$ is equal to 1.0 at the period of the oscillation. If the gain of any element in the loop increases from this condition, oscillations will expand, creating a dangerous situation where safe limits of operation could be exceeded in a few cycles. Consequently, control loops should be left in a condition where the loop gain is less than 1.0 by a safe margin that allows for possible variations in process parameters. Figure 8-18 describes a load response under PID (proportional-integral-derivative) control where the loop is well damped at a loop gain of 0.56; loop gain is then increased to 0.93 and to 1.05, creating a lightly damped and then an expanding cycle, respectively.

In controller tuning, a choice must be made between performance and robustness. Performance is a measure of how well a given controller with certain parameter settings regulates a variable, relative to the best response that can be achieved for that particular process. Robustness is a measure of how small a change in a process parameter is required to bring the loop from its current state to the limit of stability ($\Pi G = 1.0$). The well-damped loop in Fig. 8-18 has a robustness of 79 percent, in that increasing the gain of any element in the loop by a factor of 1/0.56, or 1.79, would bring the loop to the limit of stability. Increasing controller performance by raising its gain can therefore be expected to decrease robustness. Both performance and robustness are functions of the dynamics of the process being controlled, the selection of the controller, and the tuning of the controller parameters.

On/Off Control An on/off controller is used for manipulated variables having only two states. They commonly control temperatures in homes, electric water heaters and refrigerators, and pressure and liquid level in pumped storage systems. On/off control is satisfactory where slow cycling is acceptable, because it always leads to cycling when the load lies between the two states of the manipulated variable. The cycle will be positioned symmetrically about the set point only if the load happens to be equidistant between the two states of the manipulated variable. The period of the symmetric cycle will be approximately 4θ, where θ is the dead time in the loop. If the load is not centered between the states of the manipulated variable, the period will tend to increase and the cycle will follow a sawtooth pattern.

Every on/off controller has some degree of dead band, also known as *lockup*, or differential gap. Its function is to prevent erratic switching between states, thereby extending the life of contacts and motors. Instead of changing states precisely when the controlled variable crosses the set point, the controller will change states at two different points for increasing and decreasing signals. The difference between these two switching points is the dead band (see Fig. 8-19); it increases the amplitude and period of the cycle, similar to the effects of dead time.

A three-state controller is used to drive either a pair of independent two-state actuators, such as heating and cooling valves, or a bidirectional motorized actuator. The controller is comprised of two on/off controllers, each with dead band, separated by a dead zone. While the controlled variable lies within the dead zone, neither output is energized. This controller can drive a motorized valve to the point where the manipulated variable matches the load, thereby avoiding cycling.

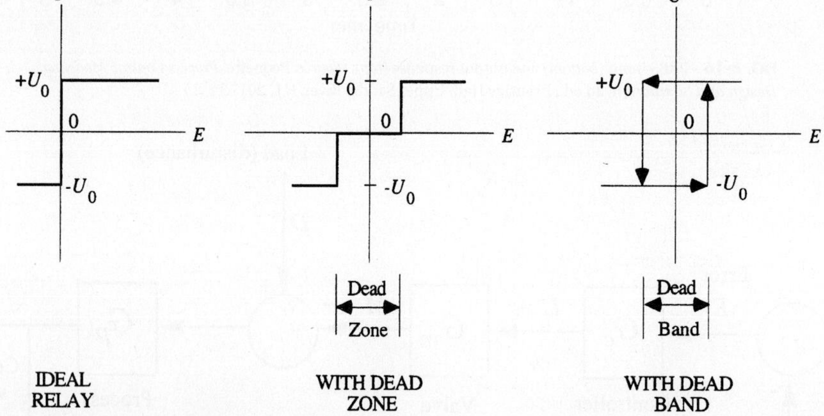

FIG. 8-19 On/off controller characteristics.

Proportional Control A proportional controller moves its output proportional to the deviation e between the controlled variable y and its set point y_{sp}:

$$u = K_c e + b = \frac{100}{p} e + b \qquad (8\text{-}10)$$

where $e = \pm(y - y_{sp})$, the sign selected to produce negative feedback. In some controllers, proportional gain K_c is introduced as a pure number; in others, it is set as $100/P$, where P is the proportional band in percent. The output bias b of the controller is also known as *manual reset*. The proportional controller is not a good regulator, because any change in output required to respond to a change in load results in a corresponding change in the controlled variable. To minimize the resulting offset, the bias should be set at the best estimate of the load, and the proportional band set as low as possible. Processes requiring a proportional band of more than a few percent may control with unacceptably large values of offset.

Proportional control is most often used to regulate liquid level, where variations in the controlled variable carry no economic penalty and where other control modes can easily destabilize the loop. It is actually recommended for controlling the level in a surge tank when manipulating the flow of feed to a critical downstream process. By setting the proportional band just under 100 percent, the level is allowed to vary over the full range of the tank capacity as inflow fluctuates, thereby minimizing the resulting rate of change of manipulated outflow. This technique is called *averaging level control.*

Proportional-plus-Integral (PI) Control Integral action eliminates the offset described above by moving the controller output at a rate proportional to the deviation from set point—the output will then not stop moving until the deviation is zero. Although available alone in an integral controller, it is most often combined with proportional action in a PI controller:

$$u = \frac{100}{P}\left(e + \frac{1}{\tau_I}\int e\,dt\right) + C_0 \qquad (8\text{-}11)$$

where τ_I is the integral time constant in minutes; in some controllers it is introduced as integral gain or reset rate $1/\tau_I$ in repeats per minute. The last term in the equation is the constant of integration C_0, the value of the controller output when integration begins. The PI controller is by far the most commonly used controller in the process industries.

Proportional-plus-Integral-plus-Derivative (PID) Control The derivative mode moves the controller output as a function of the rate of change of the controlled variable, which adds phase lead to the controller, increasing its speed of response. It is normally combined with proportional and integral modes. The noninteracting or ideal form of the PID controller appears functionally as

$$u = \frac{100}{p}\left(e + \frac{1}{\tau_I}\int e\,dt \pm \tau_D \frac{dy}{dt}\right) + C_0 \qquad (8\text{-}12)$$

where τ_D is the derivative time constant. Note that derivative action is applied to the controlled variable rather than to the deviation, as it should not be applied to the set point; the selection of the sign for the derivative term must be consistent with the action of the controller.

In some PID controllers, the integral and derivative terms are combined serially rather than in parallel, as done in the last equation. This results in interaction between these modes, such that the effective values of the controller parameters differ from their set values as follows:

$$\tau_{I,\text{eff}} = \tau_I + \tau_D$$
$$\tau_{D,\text{eff}} = \frac{1}{1/\tau_D + 1/\tau_I} \qquad (8\text{-}13)$$
$$K_c = \frac{100}{P}\left(1 + \frac{\tau_D}{\tau_I}\right)$$

The performance of the interacting controller is almost as high as that of the noninteracting controller on most processes, but the tuning rules differ because of the above relationships. Both controllers are in common use in digital systems.

There is always a gain limit placed upon the derivative vector—a value of 10 is typical. However, interaction decreases the derivative gain below this value by the factor $1 + \tau_D/\tau_I$, which is the reason for the decreased performance of the interacting PID controller. Sampling in a digital controller has a similar effect, limiting the derivative gain to the ratio of derivative time to the sample interval of the controller. Noise on the controlled variable is amplified by the derivative gain, preventing its use in controlling flow and liquid level. Derivative action is recommended for control of temperature and composition in multiple-capacity processes with little measurement noise.

Controller Comparison Figure 8-20 compares the step load response of a distributed lag without control, and with P, PI, and interacting PID control. A distributed lag is a process whose resistance and capacity are distributed throughout its length—a heat exchanger is characteristic of this class, its heat-transfer surface and heat capacity being uniformly distributed. Other examples include imperfectly stirred tanks and distillation columns—both trayed and packed. The signature of a distributed lag is its open-loop (uncontrolled) step response, featuring a relatively short dead time followed by a dominant lag called $\Sigma\tau$, which is the time required to reach 63.2 percent complete response.

The proportional controller is unable to return the controlled variable to the set point following the step load change, as a deviation is required to sustain its output at a value different from its fixed bias b. The amount of proportional offset produced as a fraction of the uncontrolled offset is $1/(1 + KK_c)$, where K is the steady-state process gain—in Fig. 8-20 that fraction is 0.13. Increasing K_c can reduce the offset, but with an accompanying loss in damping.

The PI and PID controller were tuned to produce a minimum *integrated absolute error* (IAE). Their response curves are similar in appearance to a gaussian distribution curve, but with a damped cycle in the trailing edge. The peak deviation of the PID response curve is only 0.12 times the uncontrolled offset, occurring at 0.36 $\Sigma\tau$; the peak deviation of the PI response curve is 0.21 times the uncontrolled offset, occurring at 0.48 $\Sigma\tau$. These values can be used to predict the load response of any distributed lag whose parameters K and $\Sigma\tau$ are known or can be estimated as described below.

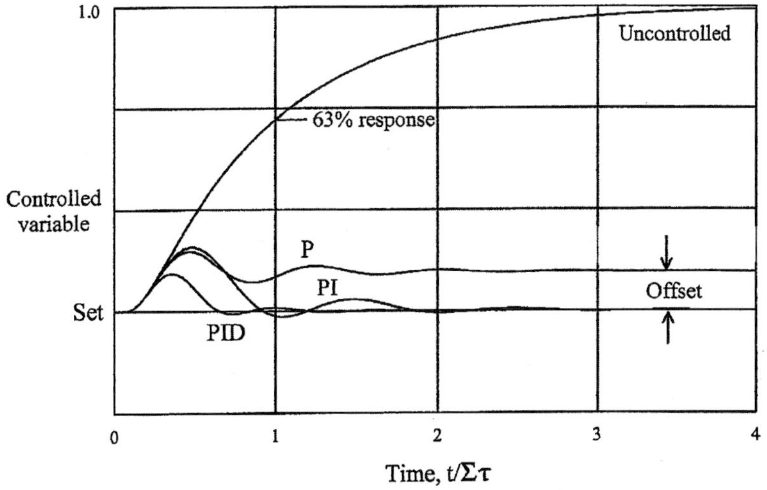

FIG. 8-20 Minimum-IAE tuning gives very satisfactory load response for a distributed lag.

CONTROLLER TUNING

The performance of a controller depends as much on its tuning as on its design. Tuning must be applied by the end user to fit the controller to the controlled process. There are many different approaches to controller tuning, based on the particular performance criteria selected, whether load or set-point changes are more important, whether the process is lag- or dead-time-dominant, and the availability of information about the process dynamics. The earliest definitive work in this field was done at Taylor Instrument Company by Ziegler and Nichols (*Trans. ASME*, p. 759, 1942), tuning PI and interacting PID controllers for optimum response to step load changes applied to lag-dominant processes. While these tuning rules are still in use, they do not apply to set-point changes, dead-time-dominant processes, or noninteracting PID controllers (Seborg, Edgar, Mellichamp, and Doyle, *Process Dynamics and Control*, Wiley, New York, 2016).

Controller Performance Criteria The most useful measures of controller performance in an industrial setting are the maximum deviation in the controlled variable resulting from a disturbance, and its integral. The disturbance could be to the set point or to the load, depending on the variable being controlled and its context in the process. The size of the deviation and its integral are proportional to the size of the disturbance (if the loop is linear at the operating point). While actual disturbances arising in a plant may appear to be random, the controller needs a reproducible test to determine how well it is tuned. The disturbance of choice for test purposes is the step, because it can be applied manually, and by containing all frequencies including zero it exercises all modes of the controller. (The step actually has the same frequency distribution as integrated white noise, a "random walk.") When tuned optimally for step disturbances, the controller will be optimally tuned for most other disturbances as well.

A step change in set point, however, may be a poor indicator of a loop's load response. For example, a liquid-level controller does not have to integrate to follow a set-point change, as its steady-state output is independent of the set point. Stepping a flow controller's set point is an effective test of its tuning, however, as its steady-state output is proportional to its set point. Other loops should be load-tested: simulate a load change from a steady state at zero deviation by transferring the controller to manual and stepping its output, and then immediately transferring back to automatic before a deviation develops.

Figure 8-21*a* and *b* shows variations in the response of a distributed lag to a step change in load for different combinations of proportional and integral settings of a PI controller. The maximum deviation is the most important criterion for variables that could exceed safe operating levels, such as steam pressure, drum level, and steam temperature in a boiler. The same rule can apply to product quality if violating specifications causes it to be rejected. However, if the product can be accumulated in a downstream storage tank, its average quality is more important, and this is a function of the deviation integrated over the residence time of the tank. Deviation in the other direction, where the product is better than specification, is safe but increases production costs in proportion to the integrated deviation because quality is given away.

For a PI or PID controller, the integrated deviation—better known as *integrated error* IE—is related to the controller settings

$$\text{IE} = \Delta u \frac{P\tau_I}{100} \tag{8-14}$$

where Δu is the difference in controller outputs between two steady states, as required by a change in load or set point. The proportional band P and

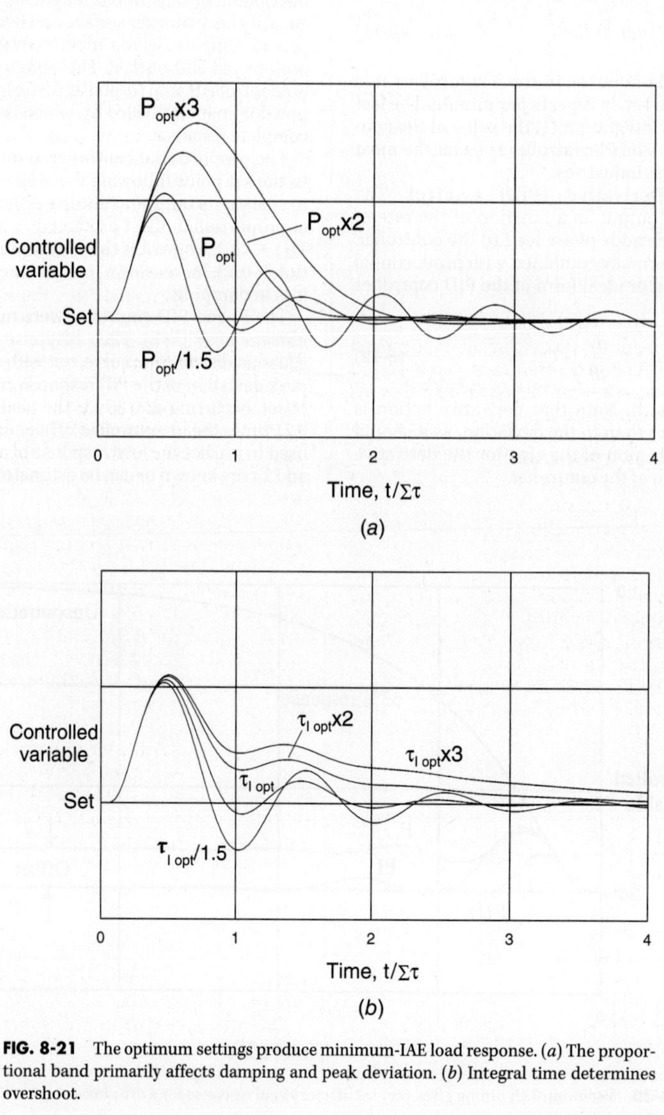

FIG. 8-21 The optimum settings produce minimum-IAE load response. (*a*) The proportional band primarily affects damping and peak deviation. (*b*) Integral time determines overshoot.

integral time τ_I are the indicated settings of the controller for PI and both interacting and noninteracting PID controllers. Although the derivative term does not appear in the relationship, its use typically allows a 50 percent reduction in integral time and therefore in IE. The integral time in the IE expression should be augmented by the sample interval if the controller is digital, the time constant of any filter used, and the value of any dead-time compensator.

It would appear, from the above, that minimizing IE is simply a matter of minimizing the P and τ_I settings of the controller. However, settings will be reached that produce excessive oscillations, such as shown in the lowest two response curves in Fig. 8-21*a* and *b*. It is preferable instead to find a combination of controller settings that minimizes *integrated absolute error* IAE, which for both load and set-point changes is a well-damped response with minimal overshoot. The curves designated P_{opt} and $\tau_{I,opt}$ in Fig. 8-21 are the same minimum-IAE response to a step change in load for a distributed-lag process under PI control. Because of the very small overshoot, the IAE will be only slightly larger than the IE. Loops that are tuned to minimize IAE tend to give responses that are close to minimum IE and with minimum peak deviation. The other curves in Fig. 8-21*a* and *b* describe the effects of individual adjustments to P and τ_I, respectively, around those optimum values and can serve as a guide to fine-tuning a PI controller.

The performance of a controller (and its tuning) must be based on what is achievable for a given process. The concept of best practical IE (IE_b) for a step change in load ΔL to a process consisting of dead time and one or two lags can be estimated (Shinskey, *Process Control Systems*, 4th ed., McGraw-Hill, New York, 1996) as

$$IE_b = \Delta L K_L \tau_L (1 - e^{-\theta/\tau_L}) \tag{8-15}$$

where K_L is the gain and τ_L the primary time constant in the load path, and θ the dead time in the manipulated path to the controlled variable. If the load or its gain is unknown, Δu and $K (= K_v K_p)$ may be substituted. If the process is non-self-regulating (i.e., an integrator), the relationship is

$$IE_b = \frac{\Delta L \theta^2}{\tau_1} \tag{8-16}$$

where τ_1 is the time constant of the process integrator. The peak deviation of the best practical response curve is

$$e_b = \frac{IE_b}{\theta + \tau_2} \tag{8-17}$$

where τ_2 is the time constant of a common secondary lag (e.g., in the measuring device).

The performance of any controller can be measured against this standard by comparing the IE it achieves in responding to a load change with the best practical IE. Maximum performance levels for PI controllers on lag-dominant processes lie in the 20 to 30 percent range, while for PID controllers they fall between 40 and 60 percent, varying with secondary lags.

Tuning Methods Based on Known Process Models The most accurate tuning rules for controllers have been based on simulation, where the process parameters can be specified and IAE and IE can be integrated during the simulation as an indication of performance. Controller settings are then iterated until a minimum IAE is reached for a given disturbance. Next these optimum settings are related to the parameters of the simulated process in tables, graphs, or equations, as a guide to tuning controllers for processes whose parameters are known (Seborg, Edgar, Mellichamp, and Doyle, *Process Dynamics and Control*, Wiley, New York, 2016). This is a multidimensional problem, however, in that the relationships change as a function of process type, controller type, and source of disturbance.

Table 8-1 summarizes rules cited by Shinskey for minimum-IAE load response for the most common controllers. The process gain and time constant τ_m are obtained from the product of G_v and G_p in Fig. 8-17. Derivative action is not effective for dead-time-dominant processes. Any secondary lag, sampling interval, or filter time constant should be added to dead time θ. A more recent set of tuning rules is called *internal model control* (IMC). Table 8-2 presents the PID controller tuning relations for the parallel form developed by Chien and Fruehauf [*Chem. Engr. Progress* **86**(10): 33 (1990)] for common types of process control models. For example, for model A, the control settings are $K_cK = \tau/\tau_c$ and $\tau_I = \tau_c$. The advantage of IMC is that it allows model uncertainty and trade-offs between performance and robustness to be considered with a closed-loop time constant τ_c. So specifying the value of one tuning parameter can numerically determine all three controller modes. For lag-dominant models with load changes, a modification due to Skogestad [*J. Process Control* **13:** 291 (2003)] is effective.

TABLE 8-1 Tuning Rules Using Known Process Parameters

Process	Controller	P	τ_I	τ_D
Dead-time-dominant	PI	$250K$	0.5θ	
Lag-dominant	PI	$106K\theta/\tau_m$	4.0θ	
	PID_n	$77K\theta/\tau_m$	1.8θ	0.45θ
	PID_i	$106K\theta/\tau_m$	1.5θ	0.55θ
Non-self-regulating	PI	$106\theta/\tau_1$	4.0θ	
	PID_n	$78\theta/\tau_1$	1.9θ	0.48θ
	PID_i	$108\theta/\tau_1$	1.6θ	0.58θ
Distributed lags	PI	$20K$	$0.50\ \Sigma\tau$	
	PID_n	$10K$	$0.30\ \Sigma\tau$	$0.09\ \Sigma\tau$
	PID_i	$15K$	$0.25\ \Sigma\tau$	$0.10\Sigma\tau$

NOTE: n = noninteracting, i = interacting controller modes.

TABLE 8-2 Tuning Rules Using Slope and Intercept

Controller	P	τ_I	τ_D
PI	$150\theta/\tau_1$	3.5θ	
PID_n	$75\theta/\tau_1$	2.1θ	0.63θ
PID_i	$113\theta/\tau_1$	1.8θ	0.70θ

NOTE: n = noninteracting, i = interacting controller modes.

The principal limitation to using these rules is that the true process parameters are often unknown. Steady-state gain K can be calculated from a process model, or determined from the steady-state results of a step test as $\Delta c/\Delta u$, as shown in Fig. 8-22. The test will not be viable, however, if the time constant of the process τ_m is longer than a few minutes, since five time constants must elapse to approach a steady state within 1 percent, and unexpected disturbances may intervene. Estimated dead time θ is the time from the step to the intercept of a straight line tangent to the steepest part of the response curve. The estimated time constant τ is the time from that point to 63 percent of the complete response. In the presence of a significant secondary lag, these results will not be completely accurate, however. The time for 63 percent response may be more accurately calculated as the residence time of the process: its volume divided by current volumetric flow rate.

Tuning Methods When Process Model Is Unknown Ziegler and Nichols developed two tuning methods for processes with unknown parameters. The open-loop method uses a step test without waiting for a steady state to be reached and is therefore applicable to very slow processes. Dead time is estimated from the intercept of the steepest tangent to the response curve in Fig. 8-22, whose slope is also used. If the process is non-self-regulating, the controlled variable will continue to follow this slope, changing by an amount equal to Δu in a time equal to its time constant τ_1. This time estimate τ_1 is used along with θ to tune controllers according to Table 8-2, applicable to lag-dominant processes. If the process is known to be a distributed lag, such as a heat exchanger, distillation column, or stirred tank, then better results will be obtained by converting the estimated values of θ and τ_1 to K and $\Sigma\tau$ and using Table 8-1. The conversion factors are $K = 7.5\theta/\tau_1$ and $\Sigma\tau = 7.0\theta$.

The Ziegler and Nichols closed-loop method requires forcing the loop to cycle uniformly under proportional control, by setting the integral time to maximum and derivative time to zero and reducing the proportional band until a constant-amplitude cycle results. The natural period τ_n of the cycle (the proportional controller contributes no phase shift to alter it) is used to set the optimum integral and derivative time constants. The optimum proportional band is set relative to the undamped proportional band P_u, which was found to produce the uniform oscillation. Table 8-3 lists the tuning rules for a lag-dominant process.

A uniform cycle can also be forced by using on/off control to cycle the manipulated variable between two limits. The period of the cycle will be close to τ_n if the cycle is symmetric; the peak-to-peak amplitude A_c of the controlled variable divided by the difference between the output limits A_u is a measure of process gain at that period and is therefore related to P_u for the proportional cycle:

$$P_u = 100 \frac{\pi}{4} \frac{A_c}{A_u} \tag{8-18}$$

The factor $\pi/4$ compensates for the square wave in the output. Tuning rules are given in Table 8-3.

Set-Point Response All the above tuning methods are intended to minimize IAE for step load changes. When applied to lag-dominant processes, the resulting controller settings produce excessive overshoot of set-point changes. This behavior has led to the practice of tuning to optimize

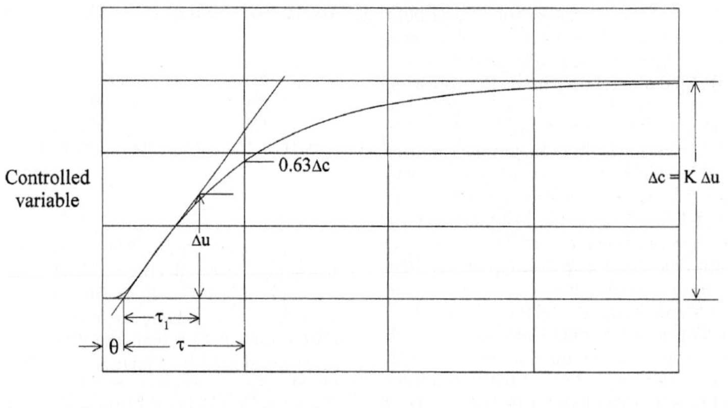

FIG. 8-22 If a steady state can be reached, gain K and time constant t can be estimated from a step response; if not, use t_1 instead.

TABLE 8-3 Tuning Rules Using Proportional Cycle

Controller	P	τ_I	τ_D
PI	$1.70P_u$	$0.81\tau_n$	
PID$_n$	$1.30P_u$	$0.48\tau_n$	$0.11\tau_n$
PID$_i$	$1.80P_u$	$0.38\tau_n$	$0.14\tau_n$

NOTE: n = noninteracting, i = interacting controller modes.

set-point response, which unfortunately degrades the load response of lag-dominant loops. An option has been available with some controllers to remove proportional action from set-point changes, which eliminates set-point overshoot but lengthens settling time. A preferred solution to this dilemma is available in many modern controllers which feature an independent gain adjustment for the set point, through which set-point response can be optimized after the controller has been tuned to optimize load response.

Figure 8-23 shows set-point and load responses of a distributed lag for both set-point and load tuning, including the effects of fractional set-point gain K_r. The set point was stepped at time 0, and the load stepped at time 2.4. With full set-point gain, the PI controller was tuned for minimum-IAE set-point response with $P = 29K$ and $\tau_I = \Sigma\tau$, compared to $P = 20K$ and $\tau_I = 0.50 \times \Sigma\tau$ for minimum-IAE load response. These settings increase its IE for load response by a factor of 2.9, and its peak deviation by 20 percent, over optimum load tuning. However, with optimum load tuning, that same set-point overshoot can be obtained with set-point gain $K_r = 0.54$. The effects of full set-point gain (1.0) and no set-point gain (0) are shown for comparison.

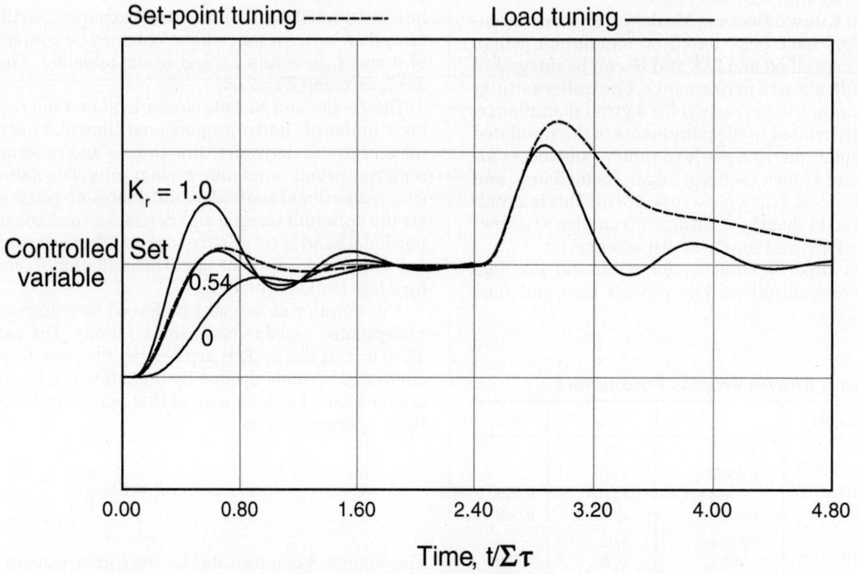

FIG. 8-23 Tuning proportional and integral settings to optimize set-point response degrades load response; using a separate set-point gain adjustment allows both responses to be optimized.

ADVANCED CONTROL SYSTEMS

BENEFITS OF ADVANCED CONTROL

The economics of most processes are determined by the steady-state operating conditions. Excursions from these steady-state conditions usually have a less important effect on the economics of the process, except when the excursions lead to off-specification products. To enhance the economic performance of a process, the steady-state operating conditions must be altered in a manner that leads to more efficient process operation.

The hierarchy shown in Fig. 8-24 indicates that process control activities consist of the following five levels:

Level 1: Measurement devices and actuators
Level 2: Safety, environmental/equipment protection
Level 3: Regulatory control
Level 4: Real-time optimization
Level 5: Planning and scheduling

Levels 4 and 5 clearly affect the process economics, as both levels are directed to optimizing the process in some manner. In contrast, levels 1, 2, and 3 would appear to have no effect on process economics. Their direct effect is indeed minimal, although indirectly they can have a major effect. Basically, these levels provide the foundation for all higher levels. A process cannot be optimized until it can be operated consistently at the prescribed targets. Thus, satisfactory regulatory control must be the first goal of any automation effort. In turn, the measurements and actuators provide the process interface for regulatory control.

For most processes, the optimum operating point is determined by a constraint. The constraint might be a product specification (a product stream can contain no more than 2 percent ethane); violation of this constraint causes off-specification product. The constraint might be an equipment limit (e.g., vessel pressure rating is 300 psig); violation of this constraint causes the equipment protection mechanism (pressure relief device) to activate. As the penalties are serious, violation of such constraints must be very infrequent.

If the regulatory control system were perfect, the target could be set exactly equal to the constraint (i.e., the target for the pressure controller could be set at the vessel relief pressure). However, no regulatory control system is perfect. Therefore, the value specified for the target must be on the safe side of the constraint, thus allowing the control system some operating margin. How much depends on the following:

1. *The performance of the control system (i.e., how effectively it responds to disturbances).* The faster the control system reacts to a disturbance, the closer the process can be operated to the constraint.

2. *The magnitude of the disturbances to which the control system must respond.* If the magnitude of the major disturbances can be reduced, the process can be operated closer to the constraint.

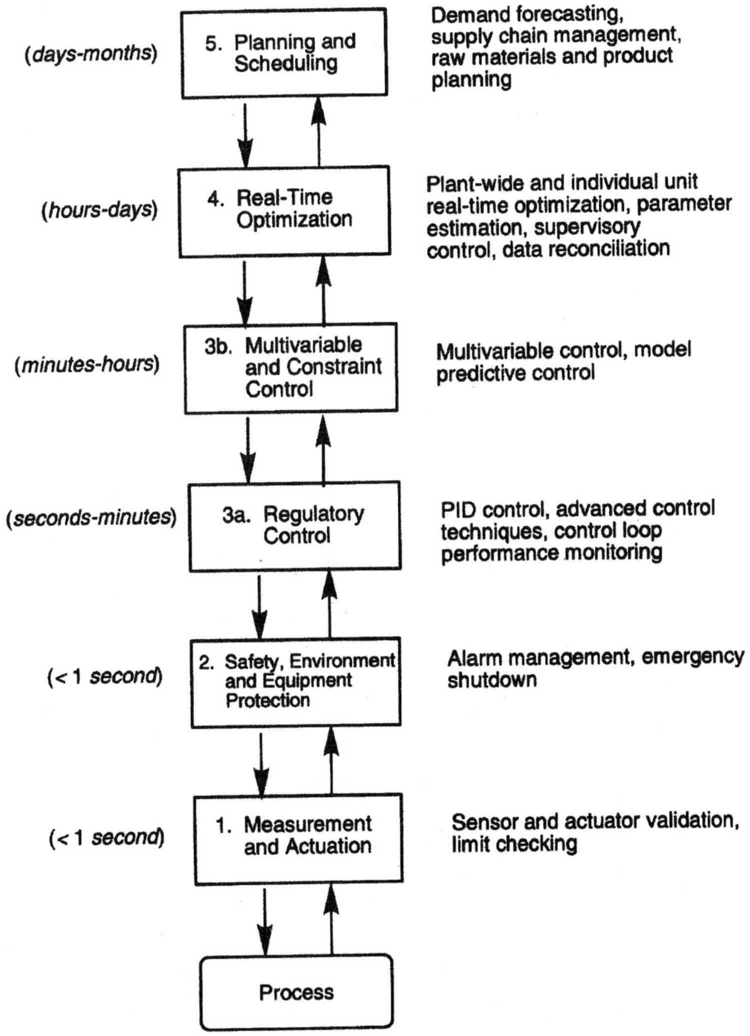

FIG. 8-24 The five levels of process control and optimization in manufacturing. Time scales are shown for each level. (*Source:* Seborg et al., *Process Dynamics and Control,* 3d ed., Wiley, New York, 2010.)

One measure of the performance of a control system is the variance of the controlled variable from the target. Both improving the control system and reducing the disturbances will lead to a lower variance in the controlled variable.

In a few applications, improving the control system leads to a reduction in off-specification product and thus improved process economics. However, in most situations, the process is operated sufficiently far from the constraint that very little, if any, off-specification product results from control system deficiencies. Management often places considerable emphasis on avoiding off-specification production, so consequently the target is actually set far more conservatively than it should be.

In most applications, simply improving the control system does not directly lead to improved process economics. Instead, the control system improvement must be accompanied by shifting the target closer to the constraint. There is always a cost of operating a process in a conservative manner. The cost may be a lower production rate, a lower process efficiency, a product giveaway, or other. When management places undue emphasis on avoiding off-specification production, the natural reaction is to operate very conservatively, thus incurring other costs.

The immediate objective of an advanced control effort is to reduce the variance in an important controlled variable. However, this effort must be coupled with a commitment to adjust the target for this controlled variable so that the process is operated closer to the constraint. In large-throughput (commodity) processes, very small shifts in operating targets can lead to large economic returns.

ADVANCED CONTROL TECHNIQUES

General References: Seborg, Edgar, Mellichamp, and Doyle, *Process Dynamics and Control*, 4th ed., Wiley, New York, 2016. Bequette, *Process Control: Modeling, Design and Simulation*, 2d ed., Prentice-Hall, Upper Saddle River, N.J., 2017. Shinskey, *Process Control Systems*, 4th ed., McGraw-Hill, New York, 1996. Ogunnaike and Ray, *Process Dynamics, Modeling, and Control*, Oxford University Press, New York, 1994.

While the single-loop PID controller is satisfactory in many process applications, it does not perform well for processes with slow dynamics, time delays, frequent disturbances, or multivariable interactions. We discuss several advanced control methods below that can be implemented via computer control, namely, feedforward control, cascade control, time-delay compensation, selective and override control, adaptive control, fuzzy logic control, and statistical process control.

Feedforward Control If the process exhibits slow dynamic response and disturbances are frequent, then the application of feedforward control may be advantageous. Feedforward (FF) control differs from feedback (FB) control in that the primary disturbance or load (D) is measured via a sensor and the manipulated variable (U) is adjusted so that deviations in the controlled variable from the set point are minimized or eliminated (see Fig. 8-25). By taking control action based on measured disturbances rather than controlled variable error, the controller can reject disturbances before they affect the controlled variable Y. To determine the appropriate settings for the manipulated variable, one must develop mathematical models that relate

1. The effect of the manipulated variable U on the controlled variable Y
2. The effect of the disturbance D on the controlled variable Y

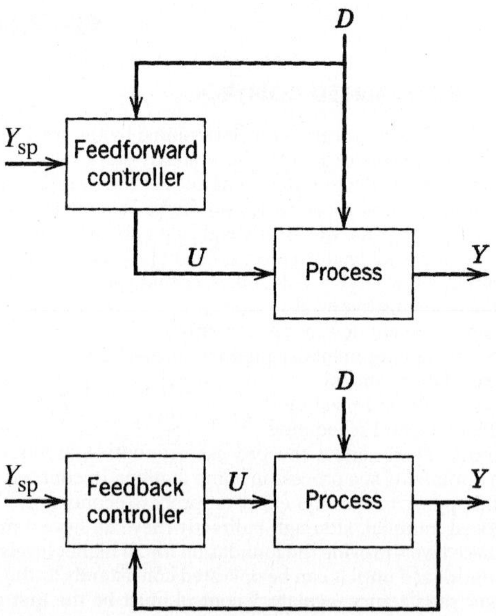

FIG. 8-25 Simplified block diagrams for feedforward and feedback control.

These models can be based on steady-state or dynamic analysis. The performance of the feedforward controller depends on the accuracy of both models. If the models are exact, then feedforward control offers the potential of perfect control (i.e., holding the controlled variable precisely at the set point at all times because of the ability to predict the appropriate control action). However, since most mathematical models are only approximate and since not all disturbances are measurable, it is standard practice to utilize feedforward control in conjunction with feedback control. Table 8-4 lists the relative advantages and disadvantages of feedforward and feedback control. By combining the two control methods, the strengths of both schemes can be utilized.

FF control therefore attempts to eliminate the effects of measurable disturbances, while FB control would correct for unmeasurable disturbances and modeling errors. This latter case is often referred to as *feedback trim*. These controllers have become widely accepted in the chemical process industries since the 1960s.

Design Based on Material and Energy Balances Consider a heat exchanger example (see Fig. 8-26) to illustrate the use of FF and FB control. The control objective is to maintain T_2, the exit liquid temperature, at the desired value (or set point) T_{2sp} despite variations in the inlet liquid flow rate F and inlet liquid temperature T_1. This is done by manipulating W, the steam flow rate. A feedback control scheme would entail measuring T_2, comparing T_2 to T_{2sp}, and then adjusting W. A feedforward control scheme requires measuring F and T_1 and adjusting W (knowing T_{2sp}), in order to control exit temperature T_2.

TABLE 8-4 Relative Advantages and Disadvantages of Feedforward and Feedback

Advantages	Disadvantages
Feedforward	
• Acts before the effect of a disturbance has been felt by the system • Is good for systems with large time constant or dead time • Does not introduce instability in the closed-loop response	• Requires direct measurement of all possible disturbances • Cannot cope with unmeasured disturbances • Is sensitive to process/model error
Feedback	
• Does not require identification and measurement of any disturbance for corrective action • Does not require an explicit process model • Is possible to design controller to be robust to process/model errors	• Control action not taken until the effect of the disturbance has been felt by the system • Is unsatisfactory for processes with large time constants and frequent disturbances • May cause instability in the closed-loop response

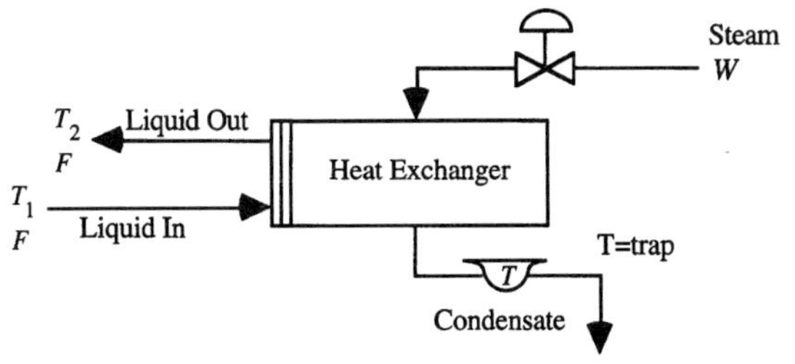

FIG. 8-26 A heat exchanger diagram.

Figure 8-27*a* and *b* shows the control system diagrams for FB and FF control. A feedforward control algorithm can be designed for the heat exchanger in the following manner. Using a steady-state energy balance and assuming no heat loss from the heat exchanger,

$$WH = FC(T_2 - T_1) \tag{8-19}$$

where H = latent heat of vaporization and C_L = specific heat of liquid

$$W = \frac{C_L}{H} F(T_2 - T_1) \tag{8-20}$$

or

$$W = K_1 F(T_2 - T_1) \tag{8-21}$$

with

$$K_1 = \frac{C_L}{H} \tag{8-22}$$

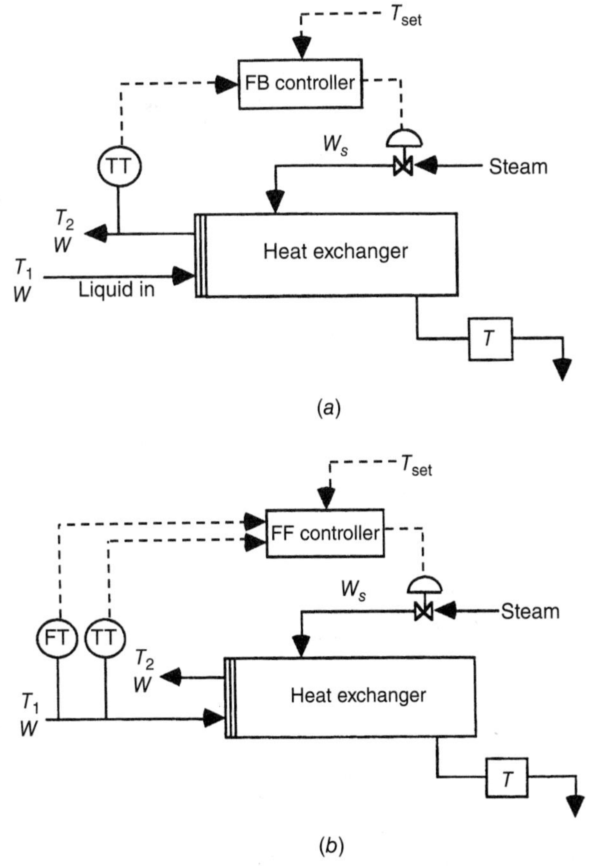

FIG. 8-27 (*a*) Feedback control of a heat exchanger. (*b*) Feedforward control of a heat exchanger.

Replace T_2 by T_{2sp}

$$W = K_1 F(T_{2sp} - T_1) \tag{8-23}$$

Equation (8-23) can be used in the FF calculation, assuming one knows the physical properties C_L and H. Of course, it is probable that the model will contain errors (e.g., unmeasured heat losses, incorrect C_L or H). Therefore, K_1 can be designated as an adjustable parameter that can be tuned. The use of a physical model for FF control is desirable because it provides a physical basis for the control law and gives an a priori estimate of what the tuning parameters should be. Note that such a model could be nonlinear [e.g., in Eq. (8-23), F and T_{2sp} are multiplied].

Block Diagram Analysis One shortcoming of this feedforward design procedure is that it is based on the steady-state characteristics of the process and, as such, neglects process dynamics (i.e., how fast the controlled variable responds to changes in the load and manipulated variables). Thus, it is often necessary to include "dynamic compensation" in the feedforward controller. The most direct method of designing the FF dynamic compensator is to use a block diagram of a general process, as shown in Fig. 8-28, where G_t represents the disturbance transmitter, G_f is the feedforward controller, G_d relates the disturbance to the controlled variable, G_v is the valve, G_p is the process, G_m is the output transmitter, and G_c is the feedback controller. All blocks correspond to transfer functions (input-output models).

For disturbance rejection ($D \neq 0$) we require that $Y = 0$, or zero error. Using each block as a "multiplier," there are two paths from D to Y: One involves the feedforward controller G_f and the other involves the disturbance model. Thus $0 = G_d D + G_t G_f G_v G_p D$; solving for the feedforward controller gives

$$G_f = \frac{-G_d}{G_t G_v G_p} \tag{8-24}$$

Suppose there are no dynamics in G_d and G_p and all models are constant gains ($G_c = K_c$ and $G_t = K_t$). Then G_f is equal to $-K_d/K_t K_v K_p = K_f$. If there are first-order dynamics in the process models for G_p (time constant τ_p) and G_d (time constant τ_d), then the feedforward controller is a lead-lag controller, which is a standard vendor controller configuration (Shinskey, *Process Control Systems: Application, Design, and Timing,* McGraw-Hill, New York, 1996). The lead-lag controller is comprised of a PD controller in series with a first-order filter.

The above FF controller can be implemented by using a digital computer. Figure 8-29*a* and *b* compares typical responses for PID FB control, steady-state FF control ($s = 0$), dynamic FF control, and combined FF/FB control. In practice, the engineer can tune K, τ_p, and τ_d in the field to improve the performance of the FF controller.

Other Considerations in Feedforward Control The tuning of feedforward and feedback control systems can be performed independently. In analyzing the block diagram in Fig. 8-28, note that G_f is chosen to cancel the effects of the disturbance D, as long as there are no model errors. For the feedback loop, therefore, the effects of D can also be ignored.

The feedback control stability limits will be unchanged for the FF + FB system. In general, the tuning of the FF/FB controller can be less conservative than for the case of FB alone, because smaller excursions from the set point will result.

For more information on feedforward/feedback control applications and design of such controllers, refer to the General References.

Cascade Control One of the disadvantages of using conventional feedback control for processes with large time lags or delays is that disturbances are not recognized until after the controlled variable deviates from its set point. In these processes, correction by feedback control is generally slow and results in long-term deviation from the set point. One way to improve the dynamic response to load changes is by using a secondary measurement point and a secondary controller; the secondary measurement point is located so that it recognizes the upset condition before the primary controlled variable is affected.

One such approach is called *cascade control,* which is routinely used in most modern computer control systems. Consider a chemical reactor, where reactor temperature is to be controlled by coolant flow to the jacket of the reactor. The reactor temperature can be influenced by changes in disturbance variables such as feed rate or feed temperature; a feedback controller could be employed to compensate for such disturbances by adjusting a valve on the coolant flow to the reactor jacket. However, suppose an increase occurs in the coolant temperature as a result of changes in the plant coolant system. This will cause a change in the reactor temperature measurement, although such a change will not occur quickly, and the corrective action taken by the controller will be delayed.

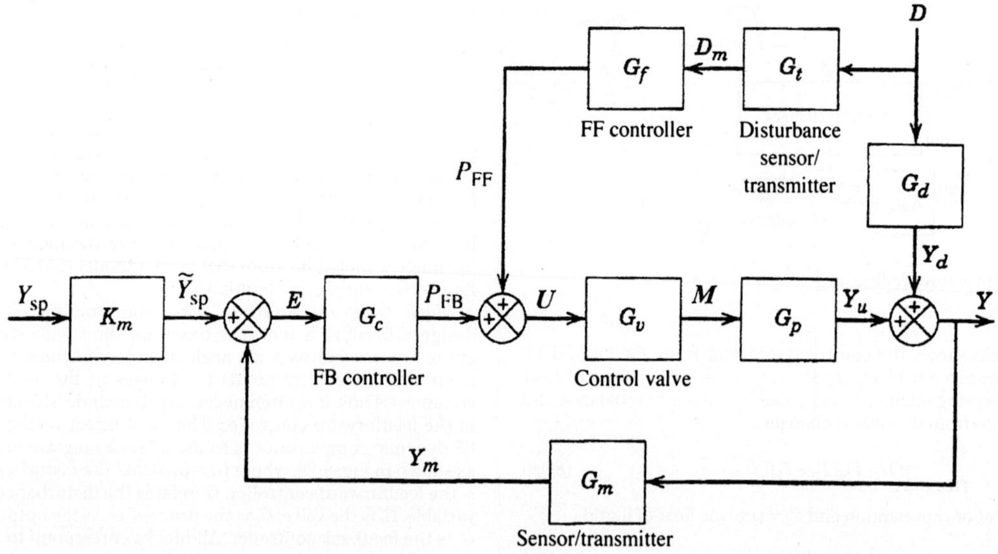

FIG. 8-28 A block diagram of a feedforward-feedback control system. (*Source:* Seborg et al., *Process Dynamics and Control*, 3d ed., Wiley, New York, 2010.)

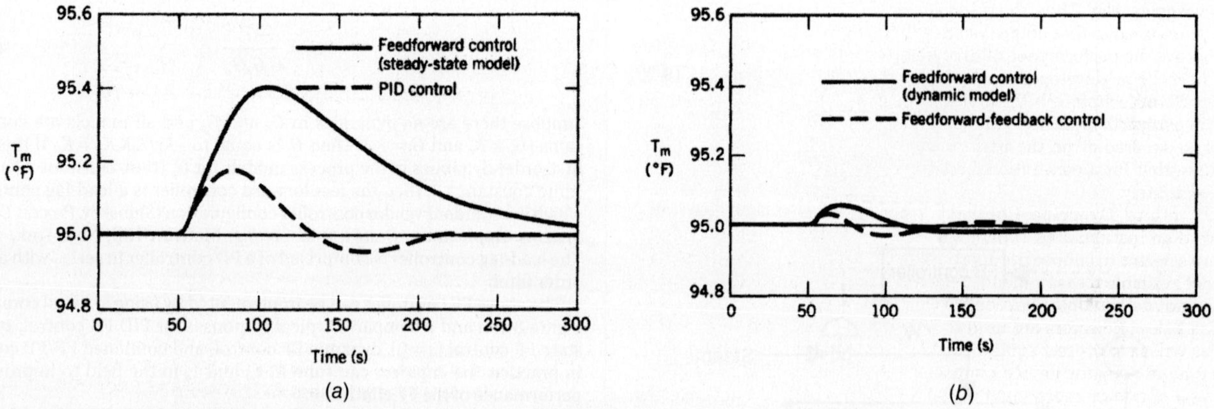

FIG. 8-29 (*a*) Comparison of FF (steady-state model) and PID FB control for disturbance change. (*b*) Comparison of FF (dynamic model) and combined FF/FB control.

Cascade control is one solution to this problem (see Fig. 8-30). Here the jacket temperature is measured, and an error signal is sent from this point to the coolant control valve; this reduces coolant flow, maintaining the heat-transfer rate to the reactor at a constant level and rejecting the disturbance. The cascade control configuration will also adjust the setting of the coolant control valve when an error occurs in the reactor temperature. The cascade control scheme shown in Fig. 8-30 contains two controllers. The primary controller is the reactor temperature controller. It measures the reactor temperature, compares it to the set point, and computes an output, which is the set point for the coolant flow rate controller. The secondary controller compares this set point to the coolant temperature measurement and adjusts the valve. The principal advantage of cascade control is that the secondary measurement (jacket temperature) is located closer to a potential disturbance in order to improve the closed-loop response.

Figure 8-31 shows the block diagram for a general cascade control system. In tuning of a cascade control system, the secondary controller (in the inner loop) is tuned first with the primary controller in manual. Often only a proportional controller is needed for the secondary loop, because offset in the secondary loop can be treated by using proportional-plus-integral action in the primary loop. When the primary controller is transferred to automatic, it can be tuned by using the techniques described earlier in this section. For more information on theoretical analysis of cascade control systems, see the General References for a discussion of applications of cascade control.

Time-Delay Compensation Time delays are a common occurrence in the process industries because of the presence of recycle loops, fluid-flow distance lags, and dead time in composition measurements resulting from

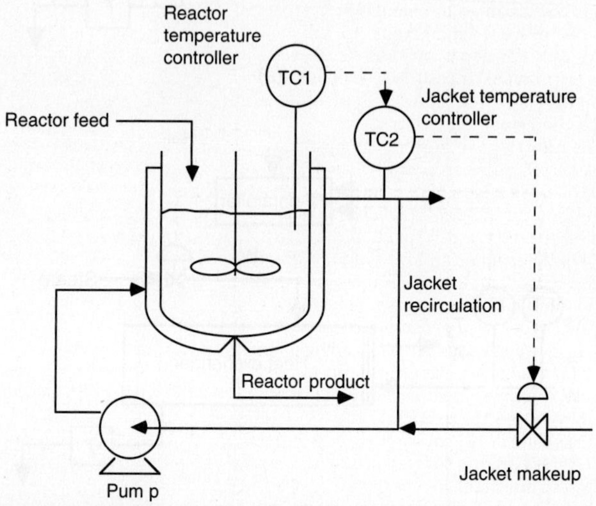

FIG. 8-30 Cascade control of an exothermic chemical reactor. (*Source:* Bequette, *Process Control: Modeling, Design and Simulation*, 2d ed., Prentice-Hall, Upper Saddle River, N.J., 2017.)

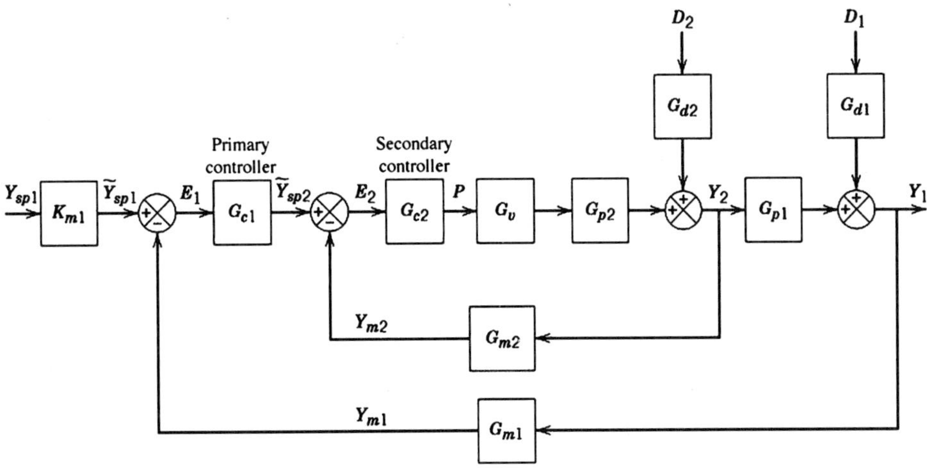

FIG. 8-31 Block diagram of the cascade control system.

use of chromatographic analysis. The presence of a time delay in a process severely limits the performance of a conventional PID control system, reducing the stability margin of the closed-loop control system. Consequently, the controller gain must be reduced below that which could be used for a process without delay. Thus, the response of the closed-loop system will be sluggish compared to that of the system with no time delay.

To improve the performance of time-delay systems, special control algorithms have been developed to provide time-delay compensation. The Smith predictor technique is the best-known algorithm; a related method is called the *analytical predictor*. Various investigators have found that, based on integral squared error, the performance of the Smith predictor can be better than that for a conventional controller, as long as the time delay is known accurately.

Selective and Override Control When there are more controlled variables than manipulated variables, a common solution to this problem is to use a selector to choose the appropriate process variable from among a number of available measurements. Selectors can be based on multiple measurement points, multiple final control elements, or multiple controllers, as discussed below. Selectors are used to improve the control system performance as well as to protect equipment from unsafe operating conditions.

One type of selector device chooses as its output signal the highest (or lowest) of two or more input signals. This approach is often referred to as *auctioneering*. On instrumentation diagrams, the symbol > denotes a high selector and < a low selector. For example, a high selector can be used to determine the hot-spot temperature in a fixed-bed chemical reactor. In this case, the output from the high selector is the input to the temperature controller. In an exothermic catalytic reaction, the process may run away due to disturbances or changes in the reactor. Immediate action should be taken to prevent a dangerous rise in temperature. Because a hot spot may potentially develop at one of several possible locations in the reactor, multiple (redundant) measurement points should be employed. This approach minimizes the time required to identify when a temperature has risen too high at some point in the bed.

The use of high or low limits for process variables is another type of selective control, called an *override*. The feature of antireset windup in feedback controllers is a type of override. Another example is a distillation column with lower and upper limits on the heat input to the column reboiler. The minimum level ensures that liquid will remain on the trays, while the upper limit is determined by the onset of flooding. Overrides are also used in forced-draft combustion control systems to prevent an imbalance between airflow and fuel flow, which could result in unsafe operating conditions.

Other types of selective systems employ multiple final control elements or multiple controllers. In some applications, several manipulated variables are used to control a single process variable (also called *split-range control*). Typical examples include the adjustment of both inflow and outflow from a chemical reactor to control reactor pressure or the use of both acid and base to control pH in wastewater treatment. In this approach, the selector chooses among several controller outputs which final control element should be adjusted.

Split-Range Control Split-range control is a common strategy for processes that must operate over a wide range of conditions. A batch reactor, for example, may need to be heated up from ambient temperature to a higher temperature, yet may need to be cooled once the exothermic reaction is initiated. A split-range controller allows both heating and cooling fluids to be admitted to the heat-transfer jacket, as shown in Fig. 8-32. This diagram illustrates the combination of cascade control (the output of the reactor temperature controller is the set point to the jacket temperature controller) and split-range control, where the jacket temperature controller output opens both the hot and cold glycol valves. The cold glycol valve is fully open at 0 percent controller output and fully closed at 50 percent controller output; similarly, the hot glycol valve is fully closed at 50 percent controller output (and below) and fully open at 100 percent controller output, as shown in Fig. 8-33. Notice that the cold glycol valve fails open while the hot glycol valve fails closed. Another common application of split-range control is pH control, where both control valves on both acid and base streams may be used to regulate pH.

Adaptive Control Process control problems inevitably require online tuning of the controller constants to achieve a satisfactory degree of control. If the process operating conditions or the environment changes significantly, the controller may have to be retuned. If these changes occur quite frequently, then adaptive control techniques should be considered. An *adaptive control system* is one in which the controller parameters are adjusted automatically to compensate for changing process conditions.

The subject of adaptive control is one of current interest. New algorithms continue to be developed, but these need to be field-tested before industrial acceptance can be expected. An adaptive controller is inherently nonlinear

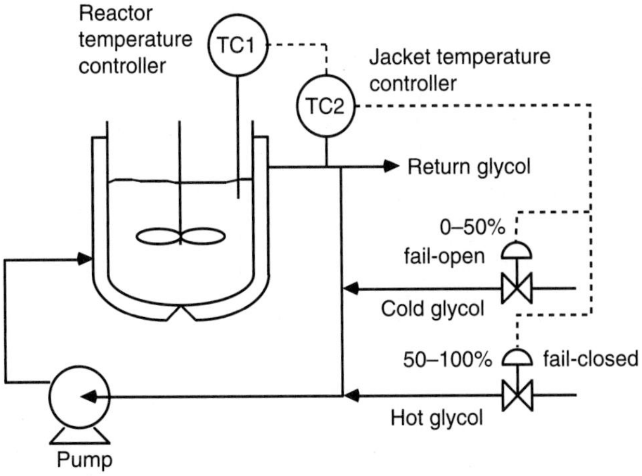

FIG. 8-32 Batch reactor temperature control. The jacket temperature controller has a split-range output, where the cold glycol valve is open during "cooling mode" and the hot glycol valve is open during "heating mode." (*Source:* Bequette, *Process Control: Modeling, Design and Simulation*, 2d ed., Prentice-Hall, Upper Saddle River, N.J., 2017.)

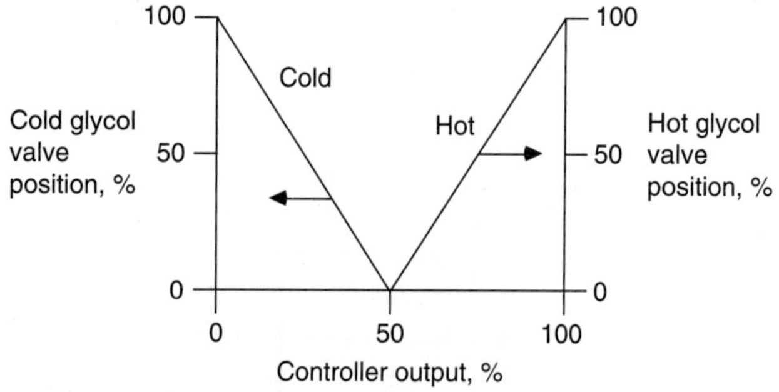

FIG. 8-33 Depiction of the split-range controller action. (*Source:* Bequette, *Process Control: Modeling, Design and Simulation,* 2d ed., Prentice-Hall, Upper Saddle River, N.J., 2017.)

and therefore more complicated than the conventional PID controller. One type of adaptive controller uses gain scheduling, where different controller settings are used for different operating conditions.

Fuzzy Logic Control The application of fuzzy logic to process control requires the concepts of fuzzy rules and fuzzy inference. A fuzzy rule, also known as a fuzzy IF-THEN statement, has the form

If x then y

if input1 = high

and input2 = low

then output = medium

Three functions are required to perform logical inferencing with fuzzy rules. The fuzzy AND is the product of a rule's input membership values, generating a weight for the rule's output. The fuzzy OR is a normalized sum of the weights assigned to each rule that contributes to a particular decision. The third function used is *defuzzification,* which generates a crisp final output. In one approach, the crisp output is the weighted average of the peak element values.

With a single feedback control architecture, information that is readily available to the algorithm includes the error signal, difference between the process variable and the set-point variable, change in error from previous cycles to the current cycle, changes to the set-point variable, change of the manipulated variable from cycle to cycle, and change in the process variable from past to present. In addition, multiple combinations of the system response data are available. As long as the irregularity lies in that dimension wherein fuzzy decisions are being based or associated, the result should be enhanced performance. This enhanced performance should be demonstrated in both the transient and steady-state response. If the system tends to have changing dynamic characteristics or exhibits nonlinearities, fuzzy logic control should offer a better alternative to using constant PID settings. Most fuzzy logic software begins building its information base during the autotune function. In fact, the majority of the information used in the early stages of system start-up comes from the autotune solutions.

In addition to single-loop process controllers, products that have benefited from the implementation of fuzzy logic are camcorders, elevators, antilock braking systems, and televisions with automatic color, brightness, and sound control. Sometimes fuzzy logic controllers are combined with pattern recognition software such as artificial neural networks (Jantzen, *Foundations of Fuzzy Control,* Wiley, New York, 2007; Blevins, Wojsznis, and Nixon, *Advanced Control Foundation: Tools Techniques and Applications,* International Society of Automation, Research Triangle Park, N.C., 2013).

EXPERT SYSTEMS

An *expert system* is a computer program that uses an expert's knowledge in a particular domain to solve a narrowly focused, complex problem. An offline system uses information entered manually and produces results in visual form to guide the user in solving the problem at hand. An online system uses information taken directly from process measurements to perform tasks automatically or instruct or alert operating personnel to the status of the plant.

Each expert system has a rule base created by the expert to respond as the expert would to sets of input information. Expert systems used for plant diagnostics and management usually have an open rule base, which can be changed and augmented as more experience accumulates and more tasks are automated. The "expert" in this case would be the person or persons having the deepest knowledge about the process, its problems, its symptoms, and remedies. Converting these inputs to meaningful outputs is the principal task in constructing a rule base. First-principles models (deep knowledge) produce the most accurate results, although heuristics are always required to establish limits. Often modeling tools such as artificial neural nets are used to develop relationships among the process variables.

A number of process control vendors offer comprehensive, object-oriented software environments for building and deploying expert systems. Advantages of such software include transforming complex real-time data to useful information through knowledge-based reasoning and analysis, monitoring for potential problems before they adversely impact operations, diagnosing root causes of time-critical problems to speed up resolution, and recommending or taking corrective actions to help ensure successful recovery.

MULTIVARIABLE AND MULTILOOP CONTROL

General References: Shinskey, *Process Control Systems,* 4th ed., McGraw-Hill, New York, 1996. Seborg, Edgar, Mellichamp, and Doyle, *Process Dynamics and Control,* 4th ed., Wiley, New York, 2016. McAvoy, *Interaction Analysis,* ISA, Research Triangle Park, N.C., 1983.

Process control books and journal articles tend to emphasize problems with a single controlled variable. In contrast, many processes require multivariable control with many process variables to be controlled. In fact, for virtually any important industrial process, at least two variables must be controlled: product quality and throughput. In this section, strategies for multivariable control are considered.

Three examples of simple multivariable control systems are shown in Fig. 8-34. The in-line blending system blends pure components A and B to produce a product stream with flow rate w and mass fraction of A denoted by x. Adjusting either inlet flow rate w_A or w_B affects both of the controlled variables w and x. For the pH neutralization process in Fig. 8-34*b*, liquid level h and exit stream pH are to be controlled by adjusting the acid and base flow rates w_A and w_B. Each of the manipulated variables affects both of the controlled variables. Thus, both the blending system and the pH neutralization process are said to exhibit strong process interactions. In contrast, the process interactions for the gas-liquid separator in Fig. 8-34*c* are not as strong because one manipulated variable, liquid flow rate L, has only a small and indirect effect on one of the controlled variables, pressure P.

Strong process interactions can cause serious problems if a conventional multiloop feedback control scheme (e.g., either PI or PID controllers) is employed. The process interactions can produce undesirable control loop interactions where the controllers fight one another. Also, it may be difficult to determine the best pairing of controlled and manipulated variables. For example, in the in-line blending process in Fig. 8-34*a*, should w be controlled with w_A and x with w_B, or vice versa?

Control Strategies for Multivariable Control Problems If a conventional multiloop control strategy performs poorly because of control loop interactions, a number of solutions are available:

1. Detune one or more of the control loops.
2. Choose different controlled or manipulated variables (or their pairings).
3. Use a multivariable control scheme (e.g., model predictive control).

Detuning a controller (e.g., using a smaller controller gain or a larger reset time) tends to reduce control loop interactions by sacrificing the performance for the detuned loops. This approach may be acceptable if some of the controlled variables are faster or less important than others.

The selection of controlled and manipulated variables is of crucial importance in designing a control system. In particular, a judicious choice may significantly reduce control loop interactions. For the blending process in Fig. 8-34*a*, a straightforward control strategy would be to control x by adjusting w_A, and w by adjusting w_B. But physical intuition suggests that it would be better to control x by adjusting the ratio $w_A/(w_A + w_B)$ and to control product flow rate w by the sum $w_A + w_B$. Thus, the new manipulated variables would be $U_1 = w_A/(w_A + w_B)$ and $U_2 = w_A + w_B$. In this control scheme, U_1 affects only x, and U_2 affects only w. Thus, the control loop interactions have been eliminated. Similarly, for the pH neutralization process in Fig. 8-34*b*, the control loop interactions would be greatly reduced if pH were controlled by $U_1 = w_A/(w_A + w_B)$ and liquid level h were controlled by $U_2 = w_A + w_B$.

Pairing of Controlled and Manipulated Variables A key decision in multiloop control system design is the pairing of manipulated and controlled variables. Suppose there are N controlled variables and N manipulated variables. Then $N!$ distinct single-loop control configurations exist. For example, if $N = 5$, then there are 120 different multiloop control schemes. In practice, many would be rejected based on physical insight or previous experience. But a smaller number (say, 5 to 15) may appear to be feasible, and further analysis would be warranted. Thus, it is very useful to have a simple method for choosing the most promising control configuration.

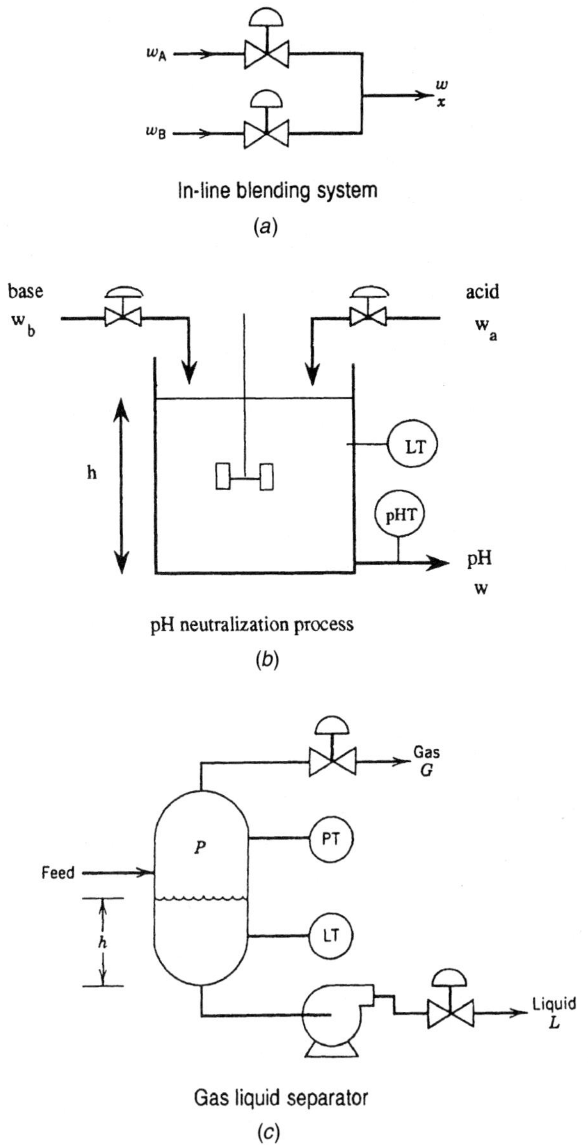

FIG. 8-34 Physical examples of multivariable control problems.

The most popular and widely used technique for determining the best controller pairing is the *relative gain array* (RGA) method [Bristol, "On a New Measure of Process Interaction," *IEEE Trans. Auto. Control* **AC-11:** 133 (1966)]. The RGA method provides two important items of information:

1. A measure of the degree of process interactions between the manipulated and controlled variables
2. A recommended controller pairing

An important advantage of the RGA method is that it requires minimal process information, namely, steady-state gains. Another advantage is that the results are independent of both the physical units used and the scaling of the process variables. The chief disadvantage of the RGA method is that it neglects process dynamics, which can be an important factor in the pairing decision. Thus, the RGA analysis should be supplemented with an evaluation of process dynamics via simulation. Although extensions of the RGA method that incorporate process dynamics have been reported, these extensions have not been widely applied.

RGA Method for 2 × 2 Control Problems To illustrate the use of the RGA method, consider a control problem with two inputs and two outputs. The more general case of $N \times N$ control problems is considered elsewhere (McAvoy, *Interaction Analysis*, ISA, Research Triangle Park, N.C., 1983). As a starting point, it is assumed that a linear, steady-state process model in Eqs. (8-25) and (8-26) is available, where U_1 and U_2 are steady-state values of the manipulated inputs; Y_1 and Y_2 are steady-state values of the controlled outputs; and the K values are steady-state gains. The Y and U variables are deviation variables from nominal steady-state values. This process model could be obtained in a variety of ways, such as by linearizing a theoretical model or by calculating steady-state gains from experimental data or a steady-state simulation.

$$Y_1 = K_{11}U_1 + K_{12}U_2 \qquad (8\text{-}25)$$

$$Y_2 = K_{21}U_1 + K_{22}U_2 \qquad (8\text{-}26)$$

By definition, the relative gain λ_{ij} between the ith manipulated variable and the jth controlled variable is defined as

$$\lambda_{ij} = \frac{\text{open-loop gain between } Y_i \text{ and } U_j}{\text{closed-loop gain between } Y_i \text{ and } U_j} \qquad (8\text{-}27)$$

where the open-loop gain is simply K_{ij} from Eqs. (8-25) and (8-26). The *closed-loop gain* is defined to be the steady-state gain between U_j and Y_i when the other control loop is closed and no offset occurs in the other controlled variable due to the presence of integral control action. The RGA for the 2 × 2 process is denoted by

$$\Lambda = \begin{bmatrix} \lambda_{11} & \lambda_{12} \\ \lambda_{21} & \lambda_{22} \end{bmatrix} \qquad (8\text{-}28)$$

The RGA has the important normalization property that the sum of the elements in each row and each column is exactly 1. Consequently, the RGA in Eq. (8-28) can be written as

$$\Lambda = \begin{bmatrix} \lambda & 1-\lambda \\ 1-\lambda & \lambda \end{bmatrix} \qquad (8\text{-}29)$$

where λ can be calculated from the following formula:

$$\lambda = \frac{1}{1 - K_{12}K_{21}/(K_{11}K_{22})} \qquad (8\text{-}30)$$

Ideally, the relative gains that correspond to the proposed controller pairing should have a value of 1 because Eq. (8-27) implies that the open- and closed-loop gains are then identical. If a relative gain equals 1, the steady-state operation of this loop will not be affected when the other control loop is changed from manual to automatic, or vice versa. Consequently, the recommendation for the best controller pairing is to pair the controlled and manipulated variables so that the corresponding relative gains are positive and close to 1.

RGA Example To illustrate use of the RGA method, consider the following steady-state version of a transfer function model for a pilot-scale, methanol-water distillation column [Wood and Berry, "Terminal Composition Control of a Binary Distillation Column," *Chem. Eng. Sci.* **28:** 1707 (1973)]: $K_{11} = 12.8$, $K_{12} = -18.9$, $K_{21} = 6.6$, and $K_{22} = -19.4$. It follows that $\lambda = 2$ and

$$\Lambda = \begin{pmatrix} 2 & -1 \\ -1 & 2 \end{pmatrix} \qquad (8\text{-}31)$$

Thus it is concluded that the column is fairly interacting and the recommended controller pairing is to pair Y_1 with U_1 and Y_2 with U_2.

MODEL PREDICTIVE CONTROL

General References: Qin and Badgwell, *Control Eng. Practice* **11:** 773 (2003). Camacho and Bordons, *Model Predictive Control*, 2d ed., Springer-Verlag, New York, 2004. Maciejowski, *Predictive Control with Constraints*, Prentice-Hall, Upper Saddle River, N.J., 2002. Bequette, *Process Control: Modeling, Design and Simulation*, 2d ed., Prentice-Hall, Upper Saddle River, N.J., 2017. Seborg, Edgar, Mellichamp, and Doyle, *Process Dynamics and Control*, 4th ed., Wiley, New York, 2016, chap. 20. Darby and Nikolaou, *Control Eng. Practice* **20:** 328–342 (2012).

The model-based control strategy that has been most widely applied in the process industries is *model predictive control* (MPC). It is a general method that is especially well suited for difficult multi-input, multi-output (MIMO) control problems where there are significant interactions between the manipulated inputs and the controlled outputs. Unlike other model-based control strategies, MPC can easily accommodate inequality constraints on input and output variables such as upper and lower limits and rate-of-change limits.

A key feature of MPC is that future process behavior is predicted by using a dynamic model and available measurements. The controller outputs are calculated to minimize the difference between the predicted process response and the desired response. At each sampling instant, the control calculations are repeated and the predictions updated based on current measurements. In typical industrial applications, the set point and target values for the MPC calculations are updated by using online optimization based on a steady-state model of the process.

The current widespread interest in MPC techniques was initiated by pioneering research performed by two industrial groups in the 1970s. Shell Oil (Houston, Tex.) reported its Dynamic Matrix Control (DMC) approach in 1979, while a similar technique, marketed as IDCOM, was published by a small French company ADERSA in 1978. Since then, there have been thousands of applications of these and related MPC techniques in oil refineries and petrochemical plants around the world. Thus, MPC has had a substantial impact and is currently the method of choice for difficult multivariable control problems in these industries. However, relatively few applications have been reported in other process industries, even though MPC is a very general approach that is not limited to a particular industry.

Advantages and Disadvantages of MPC Model predictive control offers a number of important advantages in comparison with conventional multiloop PID control:

1. It is a general control strategy for MIMO processes with inequality constraints on input and output variables.
2. It can easily accommodate difficult or unusual dynamic behavior such as large time delays and inverse responses.
3. Because the control calculations are based on optimizing control system performance, MPC can be readily integrated with online optimization strategies to optimize plant performance.
4. The control strategy can be easily updated online to compensate for changes in process conditions, constraints, or performance criteria.

But current versions of MPC have significant disadvantages in comparison with conventional multiloop control:

1. The MPC strategy is very different from conventional multiloop control strategies and thus initially unfamiliar to plant personnel.
2. The MPC calculations can be relatively complicated [e.g., solving a linear programming (LP) or quadratic programming (QP) problem at each sampling instant] and thus require a significant amount of computer resources and effort. These optimization strategies are described in the next section.
3. The development of a dynamic model from plant data is time-consuming, typically requiring days, or even weeks, of around-the-clock plant tests.
4. Because empirical models are generally used, they are valid only over the range of conditions considered during the plant tests.
5. MPC can perform poorly for some types of process disturbances, especially when the additive output disturbance assumption of DMC is used.

Because MPC has been widely used and has had considerable impact, there is a broad consensus that its advantages far outweigh its disadvantages.

Economic Incentives for Automation Projects Industrial applications of advanced process control strategies such as MPC are motivated by the need for improvements regarding safety, product quality, environmental standards, and economic operation of the process. One view of the economic incentives for advanced automation techniques is illustrated in Fig. 8-35.

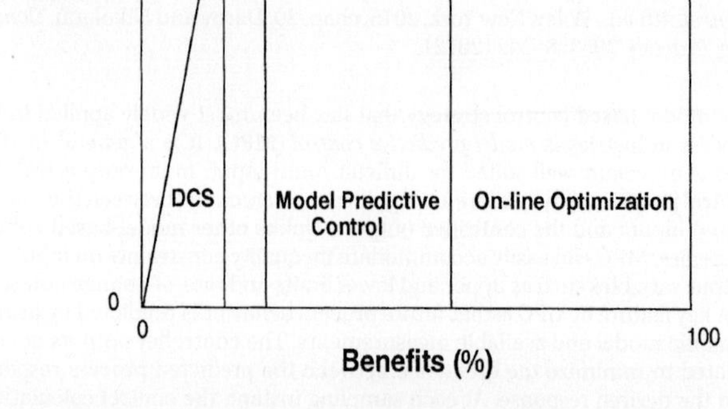

FIG. 8-35 Economic incentives for automation projects in the process industries.

Distributed control systems (DCSs) are widely used for data acquisition and conventional single-loop (PID) control. The addition of advanced regulatory control systems such as selective controls, gain scheduling, and time-delay compensation can provide benefits for a modest incremental cost. But experience has indicated that the major benefits can be obtained for relatively small incremental costs through a combination of MPC and online optimization. The results in Fig. 8-35 are shown qualitatively, rather than quantitatively, because the actual costs and benefits are application-dependent.

A key reason why MPC has become a major commercial and technical success is that there are numerous vendors who are licensed to market MPC products and install them on a turnkey basis. Consequently, even medium-size companies are able to take advantage of this new technology. Payout times of 3 to 12 months have been widely reported.

Basic Features of MPC Model predictive control strategies have a number of distinguishing features:

1. A dynamic model of the process is used to predict the future outputs over a prediction *horizon* consisting of the next P sampling periods.
2. A reference trajectory is used to represent the desired output response over the prediction horizon.
3. Inequality constraints on the input and output variables can be included as an option.
4. At each sampling instant, a control policy consisting of the next M control moves is calculated. The control calculations are based on minimizing a quadratic or linear performance index over the prediction horizon while satisfying the constraints.
5. The performance index is expressed in terms of future control moves and the predicted deviations from the reference trajectory.
6. A *receding horizon approach* is employed. At each sampling instant, only the first control move (of the M moves that were calculated) is actually implemented.
7. Then the predictions and control calculations are repeated at the next sampling instant.

These distinguishing features of MPC will now be described in greater detail.

Dynamic Model A key feature of MPC is that a dynamic model of the process is used to predict future values of the controlled outputs. There is considerable flexibility concerning the choice of the dynamic model. For example, a physical model based on first principles (e.g., mass and energy balances) or an empirical model developed from data could be employed. Also, the empirical model could be a linear model (e.g., transfer function, step response model, or state space model) or a nonlinear model (e.g., neural net model). However, most industrial applications of MPC have relied on linear empirical models, which may include simple nonlinear transformations of process variables.

The original formulations of MPC (i.e., DMC and IDCOM) were based on empirical linear models expressed in either step response or impulse response form. For simplicity, we consider only a single-input, single-output (SISO) model. However, the SISO model can be easily generalized to the MIMO models that are used in industrial applications. The step response model relating a single controlled variable y and a single manipulated variable u can be expressed as

$$\hat{y}(k+1) = y_0 + \sum_{i=1}^{N-1} S_i \Delta u(k-i+1) + S_N u(k-N+1) \qquad (8\text{-}32)$$

where $\hat{y}(k+1)$ is the predicted value of y at the $k+1$ sampling instant, $u(k)$ is the value of the manipulated input at time k, and the model parameters S_i are referred to as the *step response coefficients.* The initial value y_0 is assumed to be known. The change in the manipulated input from one sampling instant to the next is denoted by

$$\Delta u(k) = u(k) - u(k-1) \qquad (8\text{-}33)$$

The step response model is also referred to as a *discrete convolution model,* which is quite different from the first-order plus time delay model used for PID controller design.

In principle, the step response coefficients can be determined from the output response to a step change in the input. A typical response to a unit step change in input u is shown in Fig. 8-36. The step response coefficients S_i are simply the values of the output variable at the sampling instants, after the initial value y_0 has been subtracted. Theoretically, they can be determined from a single step response, but, in practice, a number of "bump tests" are required to compensate for unanticipated disturbances, process nonlinearities, and noisy measurements.

Horizons Step and impulse response models typically contain a large number of parameters because the model horizon N is usually quite large ($30 < N < 120$). In fact, these models are often referred to as *nonparametric models.*

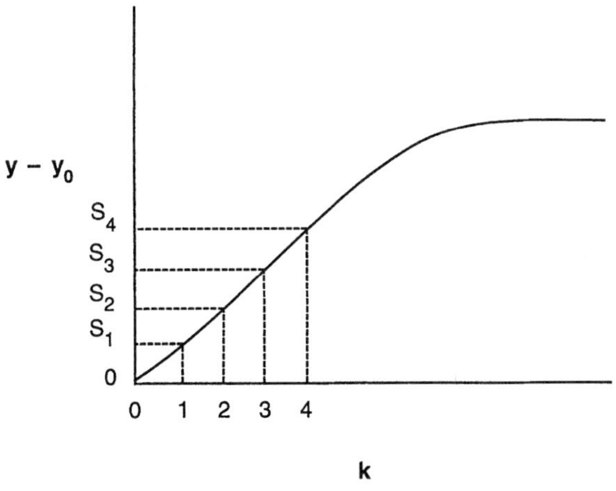

FIG. 8-36 Step response for a unit step change in the input.

The receding horizon feature of MPC is shown in the top portion of Fig. 8-37 with the current sampling instant denoted by k. Past and present input signals [$u(i)$ for $i \le k$] are used to predict the output at the next P sampling instants [$y(k+i)$ for $i = 1, 2, \ldots, P$]. The control calculations are performed to generate an M-step control policy [$u(k), u(k+1), \ldots, u(k+M-1)$], which optimizes a performance index. The first control action $u(k)$ is implemented. Then at the next sampling instant $k + 1$, illustrated in the bottom portion of Fig. 8-37, the new measurement is compared with the model predicted output, the model is updated to account for this difference, and the prediction and control calculations are repeated in order to determine $u(k + 1)$. Note that one of the key steps is compensating for the plant-model mismatch before the optimization problem is solved at step $k + 1$. In the original DMC formulation, a simple additive bias term was used, but this can lead to poor input disturbance rejection performance. See Muske and Badgwell ["Disturbance Modeling for Offset-Free Linear Model Predictive Control," *J. Proc. Control* **12**(5): 617–632 (2003)] for a more detailed discussion of how to improve disturbance rejection by using MPC. In Fig. 8-37 the reference trajectory (or target) is considered to be constant. Other possibilities include a gradual or step set-point change that can be generated by online optimization.

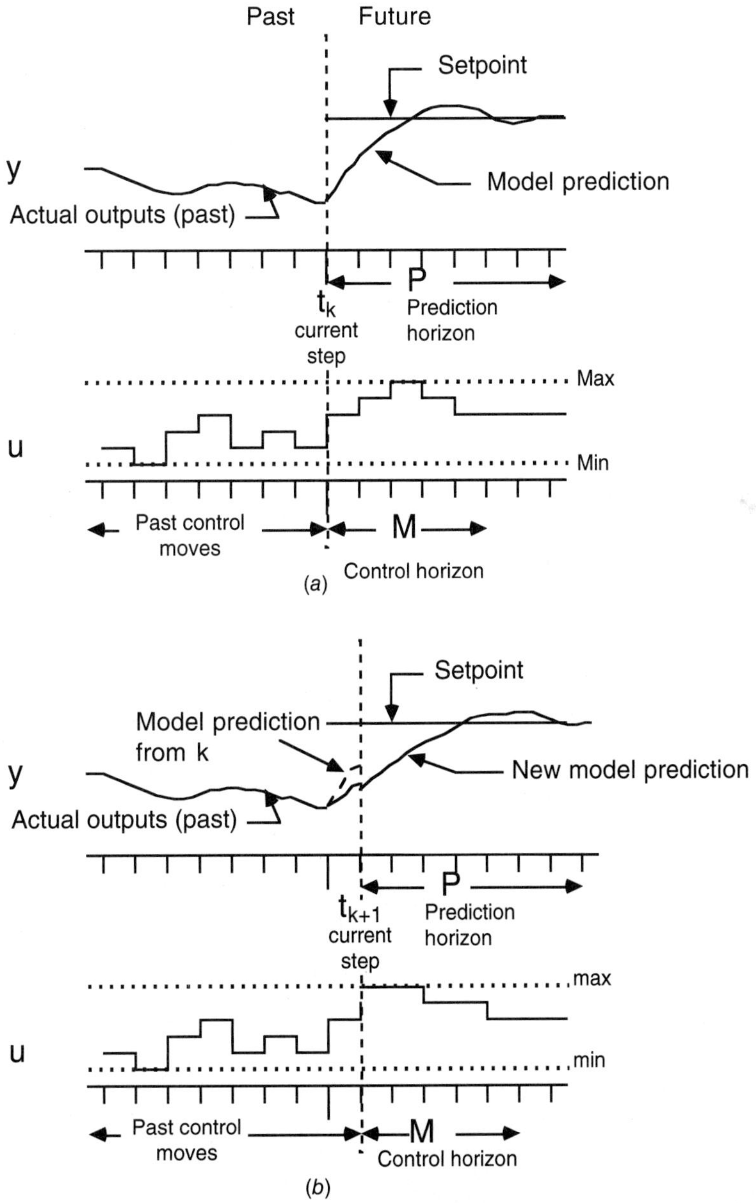

FIG. 8-37 Basic concept of model predictive control. (*a*) Top. Current and future control moves solved at time step k. (*b*) Bottom. After first control move is implemented at time step k, a new measurement is obtained and a new problem is solved at step $k + 1$.

Performance Index The performance index for MPC applications is usually a linear or quadratic function of the predicted errors and calculated future control moves. For example, the following quadratic performance index has been widely used:

$$\min_{\Delta \mathbf{u}(k)} J = \sum_{i=1}^{P} Q_i e^2(k+i) + \sum_{i=1}^{M} R_i \, \Delta u^2(k+i-1) \qquad (8\text{-}34)$$

where $e(k+i)$ is the predicted error from the set point and $\Delta\mathbf{u}(k)$ denotes the column vector of current and future control moves over the next M sampling instants:

$$\Delta\mathbf{u}(k) = [\Delta u(k), \Delta u(k+1), \ldots, \Delta u(k+M-1)]^T \qquad (8\text{-}35)$$

Equation (8-34) contains two types of design parameters that can also be used for tuning purposes. Weighting factor R_i penalizes large control moves, while weighting factor Q_i allows the predicted errors to be weighed differently at each time step, if desired.

Inequality Constraints Inequality constraints on the future inputs or their rates of change are widely used in the MPC calculations. For example, if both upper and lower limits are required, the constraints could be expressed as

$$u^-(k) \le u(k+j) \le u^+(k) \qquad \text{for } j = 0, 1, \ldots, M-1 \qquad (8\text{-}36)$$

$$\Delta u^-(k) \le \Delta u(k+j) \le \Delta u^+(k) \qquad \text{for } j = 0, 1, \ldots, M-1 \qquad (8\text{-}37)$$

where B_i and C_i are constants. Constraints on the predicted outputs $\hat{y}(k+j)$ are sometimes included as well:

$$y^-(k) \le \hat{y}(k+j) \le y^+(k) \qquad \text{for } j = 0, 1, \ldots, P \qquad (8\text{-}38)$$

It is possible that including output constraints will result in an infeasible solution, so any practical algorithm must relax the constraints until the solution becomes feasible. The minimization of the quadratic performance index in Eq. (8-34), subject to the constraints in Eqs. (8-36) to (8-38) and the step response model in Eq. (8-32), can be solved by efficient QP (quadratic programming) techniques. When the inequality constraints in Eqs. (8-36) to (8-38) are omitted, the optimization problem has an analytical solution (Camacho and Bordons, *Model Predictive Control*, 2d ed., Springer-Verlag, New York, 2004; Maciejowski, *Predictive Control with Constraints*, Prentice-Hall, Upper Saddle River, N.J., 2002). If the quadratic terms in Eq. (8-34) are replaced by linear terms, an LP (linear programming) problem results that can also be solved by using standard optimization methods. This MPC formulation for SISO control problems can easily be extended to MIMO problems.

Implementation Issues For a new MPC application, a cost/benefit analysis is usually performed prior to project approval. Then the steps involved in the implementation of MPC can be summarized as follows [Hokanson and Gerstle, "Dynamic Matrix Control Multivariable Controllers," in Luyben (ed.), *Practical Distillation Control*, Van Nostrand Reinhold, New York, 1992, p. 248; Qin and Badgwell, *Control Eng. Practice* **11:** 773 (2003)].

Step 1 Initial Controller Design The first step in MPC design is to select the controlled, manipulated, and measured disturbance variables. These choices determine the structure of the MPC system and should be based on process knowledge and control objectives. In typical applications the number of controlled variables ranges from 5 to 40, and the number of manipulated variables is typically between 5 and 20.

Step 2 Pretest Activity During the pretest activity, the plant instrumentation is checked to ensure that it is working properly and to decide whether additional sensors should be installed. The pretest data can be used to estimate the steady-state gain and approximate settling times for each input/output pair. This information is used to plan the full plant tests of step 3.

As part of the pretest, it is desirable to benchmark the performance of the existing control system for later comparison with MPC performance (step 8).

Step 3 Plant Tests The dynamic model for the MPC calculations is developed from data collected during special plant tests. The excitation for the plant tests usually consists of changing an input variable or a disturbance variable (if possible) from one value to another, using either a series of step changes with different durations or a pseudorandom binary sequence (PRBS). To ensure that sufficient data are obtained for model identification, each input variable is typically moved to a new value, 8 to 15 times during a plant test [Qin and Badgwell, *Control Eng. Practice* **11:** 773 (2003)]. Identification during closed-loop operation is becoming more common.

Step 4 Model Development The dynamic model is developed from the plant test data by selecting a model form (e.g., a step response model) and then estimating the model parameters. However, first it is important to eliminate periods of test data where plant upsets or other abnormal situations have occurred. Decisions to omit portions of the test data are based on visual inspection of the data, knowledge of the process, and experience. Parameter estimation is usually based on least-squares estimation.

Step 5 Control System Design and Simulation The preliminary control system design from step 1 is critically evaluated and modified, if necessary. Then the MPC design parameters are selected including the sampling periods, weighting factors, and control and prediction horizons. Next, the closed-loop system is simulated, and the MPC design parameters are adjusted, if necessary, to obtain satisfactory control system performance and robustness over the specified range of operating conditions.

Step 6 Operator Interface Design and Operator Training Operator training is important because MPC concepts such as predictive control, multivariable interactions, and constraint handling are very different from conventional regulatory control concepts. Thus, understanding why the MPC system responds as it does, especially for unusual operating conditions, can be a challenge for both operators and engineers.

Step 7 Installation and Commissioning After an MPC control system is installed, first it is evaluated in a "prediction mode." Model predictions are compared with measurements, but the process continues to be controlled by the existing control system. After the output predictions are judged to be satisfactory, the calculated MPC control moves are evaluated to determine if they are reasonable. Finally, the MPC software is evaluated during closed-loop operation with the calculated control moves implemented as set points to the DCS control loops. The MPC design parameters are tuned, if necessary. The commissioning period typically requires some troubleshooting and can take as long as, or even longer than, the plant tests of step 3.

Step 8 Measuring Results and Monitoring Performance The evaluation of MPC system performance is not easy, and widely accepted metrics and monitoring strategies are not available. However, useful diagnostic information is provided by basic statistics such as the means and standard deviations for both measured variables and calculated quantities, such as control errors and model residuals. Another useful statistic is the relative amount of time that an input is saturated or a constraint is violated, expressed as a percentage of the total time the MPC system is in service.

Integration of MPC and Online Optimization As indicated in Fig. 8-35, significant potential benefits can be realized by using a combination of MPC and online optimization. At present, most commercial MPC packages integrate the two methodologies in a hierarchical configuration such as the one shown in Fig. 8-38. The MPC calculations are performed quite often (e.g., every 1 to 10 min) and implemented as set points for PID control loops at the DCS level. The targets and constraints for the MPC calculations are generated by solving a steady-state optimization problem (LP or QP) based on a linear process model. These calculations may be performed as often as the MPC calculations. As an option, the targets and constraints for the LP or QP optimization can be generated from a nonlinear process model using a nonlinear optimization technique. These calculations tend to be performed less frequently (e.g., every 1 to 24 h) due to the complexity of the calculations and the process models.

The combination of MPC and frequent online optimization has been successfully applied in oil refineries and petrochemical plants around the world.

REAL-TIME PROCESS OPTIMIZATION

General References: Biegler, *Nonlinear Programming: Concepts, Algorithms, and Applications to Chemical Processes*, MPS-SIAM Series on Optimization, Philadelphia, Pa., 2010. Darby, Nikolaou, Jones, and Nicholson, "RTO: An Overview and Assessment of Current Practice," *J. Proc. Control* **21:** 874 (2011). Edgar, Himmelblau, and Lasdon, *Optimization of Chemical Processes*, 2d ed., McGraw-Hill, New York, 2001. Marlin and Hrymak, "Real-Time Optimization of Continuous Processes," *Chem. Proc. Cont. V, AIChE Symp. Ser.* **93**(316): 156 (1997). Narashimhan and Jordache, *Data Reconciliation and Gross Error Detection*, Gulf Publishing, Houston, Tex., 2000. Nocedal and Wright, *Numerical Optimization*, 2d ed., Springer, New York, 2006. Shobrys and White, "Planning, Scheduling, and Control Systems: Why They Cannot Work Together," *Comp. Chem. Engng.* **26:** 149 (2002). Timmons, Jackson, and White, "Distinguishing On-line Optimization Benefits from Those of Advanced Controls," *Hydrocarb. Proc.* **79**(6): 69 (2000).

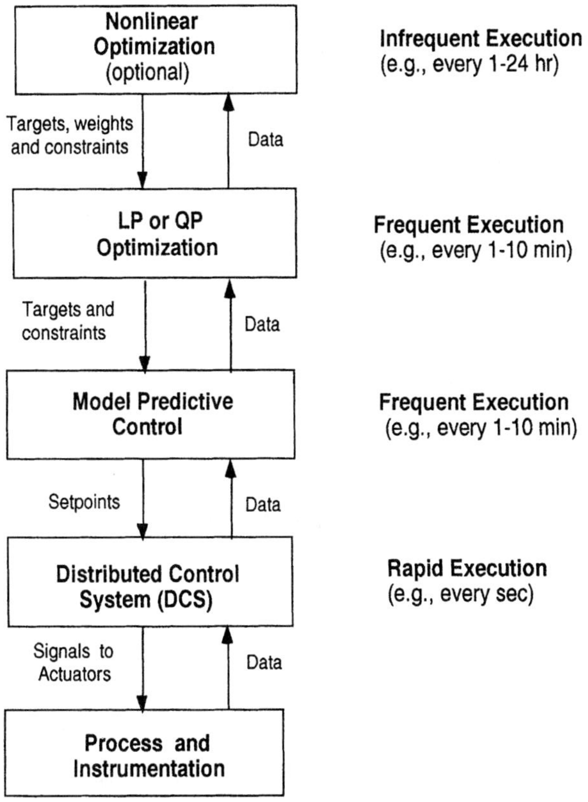

FIG. 8-38 Hierarchical control configuration for MPC and online optimization.

The chemical industry has undergone significant changes during the past 20 years due to the increased cost of energy and raw materials, more stringent environmental regulations, and intense worldwide competition. Modifications of both plant design procedures and plant operating conditions have been implemented to reduce costs and meet constraints. One of the most important engineering tools that can be employed in such activities is optimization. As plant computers have become more powerful, the size and complexity of problems that can be solved by optimization techniques have correspondingly expanded. A wide variety of problems in the operation and analysis of chemical plants (as well as many other industrial processes) can be solved by optimization. Real-time optimization means that the process operating conditions (set points) are evaluated on a regular basis and optimized, as shown earlier in level 4 in Fig. 8-22. Sometimes this is called *steady-state optimization* or *supervisory control.* This subsection examines the basic characteristics of optimization problems and their solution techniques and describes some representative benefits and applications in the chemical and petroleum industries.

Typical problems in chemical engineering process design or plant operation have many possible solutions. Optimization is concerned with selecting the best among the entire set of solutions by efficient quantitative methods. Computers and associated software make the computations involved in the selection manageable and cost-effective. Engineers work to improve the initial design of equipment and strive for enhancements in the operation of the equipment once it is installed in order to realize the greatest production, the greatest profit, the maximum cost, the least energy usage, and so on. In plant operations, benefits arise from improved plant performance, such as improved yields of valuable products (or reduced yields of contaminants), reduced energy consumption, higher processing rates, and longer times between shutdowns. Optimization can also lead to reduced maintenance costs, less equipment wear, and better staff utilization. It is helpful to systematically identify the objectives, constraints, and degrees of freedom in a process or a plant if such benefits as improved quality of designs, faster and more reliable troubleshooting, and faster decision making are to be achieved.

Optimization can take place at many levels in a company, ranging from a complex combination of plants and distribution facilities down through individual plants, combinations of units, individual pieces of equipment, subsystems in a piece of equipment, or even smaller entities. Problems that can be solved by optimization can be found at all these levels.

While process design and equipment specification are usually performed prior to the implementation of the process, optimization of operating conditions is carried out monthly, weekly, daily, hourly, or even every minute. Optimization of plant operations determines the set points for each unit at the temperatures, pressures, and flow rates that are the best in some sense. For example, the selection of the percentage of excess air in a process heater is quite critical and involves balancing the fuel/air ratio to ensure complete combustion and at the same time making maximum use of the heating potential of the fuel. Typical day-to-day optimization in a plant minimizes steam consumption or cooling water consumption, optimizes the reflux ratio in a distillation column, or allocates raw materials on an economic basis.

A *real-time optimization* (RTO) system determines set-point changes and implements them via the computer control system without intervention from unit operators. The RTO system completes all data transfer, optimization calculations, and set-point implementation before unit conditions change that may invalidate the computed optimum. In addition, the RTO system should perform all tasks without upsetting plant operations. Several steps are necessary for implementation of RTO, including determination of the plant steady state, data gathering and validation, updating of model parameters (if necessary) to match current operations, calculation of the new (optimized) set points, and the implementation of these set points.

To determine if a process unit is at steady state, a program monitors key plant measurements (e.g., compositions, product rates, feed rates, and so on) and determines if the plant is close enough to steady state to start the sequence. Only when all the key measurements are within the allowable tolerances is the plant considered steady and the optimization sequence started. Tolerances for each measurement can be tuned separately. Measured data are then collected by the optimization computer. The optimization system runs a program to screen the measurements for unreasonable data (gross error detection). This validity checking automatically modifies the model updating calculation to reflect any bad data or when equipment is taken out of service. Data validation and reconciliation (online or offline) is an extremely critical part of any optimization system.

The optimization system then may run a parameter-fitting case that updates model parameters to match current plant operation. The integrated process model calculates such items as exchanger heat-transfer coefficients, reactor performance parameters, furnace efficiencies, and heat and material balances for the entire plant. Parameter fitting allows for continual updating of the model to account for plant deviations and degradation of process equipment. After completion of the parameter fitting, the information regarding the current plant constraints, control status data, and economic values for feed products, utilities, and other operating costs is collected. The economic values are updated by the planning and scheduling department on a regular basis. The optimization system then calculates the optimized set points. The steady-state condition of the plant is rechecked after the optimization case is successfully completed. If the plant is still steady, then the values of the optimization targets are transferred to the process control system for implementation. After a line-out period, the process control computer resumes the steady-state detection calculations, restarting the cycle.

Essential Features of Optimization Problems The solution of optimization problems involves the use of various tools of mathematics, which is discussed in detail in Sec. 3. The formulation of an optimization problem requires the use of mathematical expressions. From a practical viewpoint, it is important to mesh properly the problem statement with the anticipated solution technique. Every optimization problem contains three essential categories:

1. An objective function to be optimized (revenue function, cost function, etc.)
2. Equality constraints (equations)
3. Inequality constraints (inequalities)

Categories 2 and 3 comprise the model of the process or equipment; category 1 is sometimes called the *economic model.*

No single method or algorithm of optimization exists that can be applied efficiently to all problems. The method chosen for any particular case will depend primarily on (1) the character of the objective function, (2) the nature of the constraints, and (3) the number of independent and dependent variables. Table 8-5 summarizes the six general steps for the analysis and solution of optimization problems (Edgar, Himmelblau, and Lasdon, *Optimization of Chemical Processes*, 2d ed., McGraw-Hill, New York, 2001). You do not have to follow the cited order exactly, but you should cover all the steps at some level of detail. Shortcuts in the procedure are allowable, and the easy steps can be performed first. Steps 1, 2, and 3 deal with the mathematical definition of the problem: identification of variables, specification of the objective function, and statement of the constraints. If the process to be optimized is very complex, it may be necessary to reformulate the problem so that it can be solved with reasonable effort. Later in

TABLE 8-5 Six Steps Used to Solve Optimization Problems

1. Analyze the process itself so that the process variables and specific characteristics of interest are defined (i.e., make a list of all the variables).
2. Determine the criterion for optimization, and specify the objective function in terms of the above variables together with coefficients. This step provides the performance model (sometimes called the economic model, when appropriate).
3. Develop via mathematical expressions a valid process or equipment model that relates the input/output variables of the process and associated coefficients. Include both equality and inequality constraints. Use well-known physical principles (mass balances, energy balances), empirical relations, implicit concepts, and external restrictions. Identify the independent and dependent variables (number of degrees of freedom).
4. If the problem formulation is too large in scope, (*a*) break it up into manageable parts and/or (*b*) simplify the objective function and model.
5. Apply a suitable optimization technique to the mathematical statement of the problem.
6. Check the answers and examine the sensitivity of the result to changes in the coefficients in the problem and the assumptions.

this section, we discuss the development of mathematical models for the process and the objective function (the economic model) in typical RTO applications.

Step 5 in Table 8-5 involves the computation of the optimum point. Quite a few techniques exist to obtain the optimal solution for a problem. We describe several classes of methods below. Over the past 15 years, substantial progress has been made in developing efficient and robust computational methods for optimization. Much is known about which methods are most successful. Virtually all numerical optimization methods involve iteration, and the effectiveness of a given technique can depend on a good first guess for the values of the variables at the optimal solution. After the optimum is computed, a sensitivity analysis for the objective function value should be performed to determine the effects of errors or uncertainty in the objective function, mathematical model, or other constraints.

Development of Process (Mathematical) Models Constraints in optimization problems arise from physical bounds on the variables, empirical relations, physical laws, and so on. The mathematical relations describing the process also comprise constraints. Two general categories of models exist:

1. Those based on physical theory
2. Those based on strictly empirical descriptions

Mathematical models based on physical and chemical laws (e.g., mass and energy balances, thermodynamics, chemical reaction kinetics) are frequently employed in optimization applications. These models are conceptually attractive because a general model for any system size can be developed before the system is constructed. On the other hand, an empirical model can be devised that simply correlates input/output data without any physiochemical analysis of the process. For these models, optimization is often used to fit a model to process data, using parameter estimation. One example is the yield matrix, where the percentage yield of each product in a unit operation is estimated for each feed component by using process data rather than employing a mechanistic set of chemical reactions.

Formulation of the Objective Function The formulation of objective functions is one of the crucial steps in the application of optimization to a practical problem. You must be able to translate the desired objective to mathematical terms. In the chemical process industries, the objective function often is expressed in units of currency per unit time (e.g., U.S. dollars per week, month, or year) because the normal industrial goal is to minimize costs or maximize profits subject to a variety of constraints.

A typical economic model involves the costs of raw materials, values of products, and costs of production as functions of operating conditions, projected sales figures, and the like. An objective function can be expressed in terms of these quantities; e.g., annual operating profit ($/yr) might be expressed as

$$J = \sum_s F_s V_s - \sum_r F_r C_r - \text{OC} \tag{8-39}$$

where J = profit/time

$\sum_s F_s V_s$ = sum of product flow rates times respective product values (income)

$\sum_r F_r C_r$ = sum of feed flows times respective unit costs

OC = operating costs/time

Unconstrained Optimization *Unconstrained optimization* refers to the case where no inequality constraints are present and all equality constraints can be eliminated by solving for selected dependent variables followed by substitution for them in the objective function. Very few realistic problems in process optimization are unconstrained. However, the availability of efficient unconstrained optimization techniques is important because these techniques must be applied in real time, and iterative calculations may require excessive computer time. Two classes of unconstrained techniques are single-variable optimization and multivariable optimization.

Single-Variable Optimization Many real-time optimization problems can be reduced to the variation of a single-variable optimization so as to maximize profit or some other overall process objective function. Some examples of single-variable optimization include optimizing the reflux ratio in a distillation column or the air/fuel ratio in a furnace. While most processes actually are multivariable processes with several operating degrees of freedom, often we choose to optimize only the most important variable in order to keep the strategy uncomplicated. One characteristic implicitly required in a single-variable optimization problem is that the objective function J be unimodal in variable x.

There are three classes of techniques that can be used efficiently for one-dimensional search: indirect, region elimination, and interpolation.

Indirect methods seek to solve the necessary condition $dJ/dx = 0$ by iteration, but these methods are not as popular as the second two classes. Region elimination methods include equal interval search, dichotomous search (or bisecting), Fibonacci search, and golden section. These methods do not use information on the shape of the function (other than its being unimodal) and thus tend to be rather conservative. The third class of techniques uses repeated polynomial fitting to predict the optimum. These interpolation methods tend to converge rapidly to the optimum without being very complicated. Two interpolation methods—quadratic and cubic interpolation—are used in many optimization packages.

Multivariable Optimization The numerical optimization of general nonlinear multivariable objective functions requires that efficient and robust techniques be employed. Efficiency is important since iteration is employed. For example, in multivariable "grid" search for a problem with four independent variables, an equally spaced grid for each variable is prescribed. For 10 values of each of the four variables, 10^4 total function evaluations would be required to find the best answer for the grid intersections; but this result may not be close enough to the true optimum and would require further search. A larger number of variables (say, 20) would require exponentially more computation, so grid search is a very inefficient method for most problems.

In multivariable optimization, the difficulty of dealing with multivariable functions is usually resolved by treating the problem as a series of one-dimensional searches. For a given starting point, a search direction s is specified, and the optimum is found by searching along that direction. The step size ε is the distance moved along s. Then a new search direction is determined, followed by another one-dimensional search. The algorithm used to specify the search direction depends on the optimization method selected.

There are two basic types of unconstrained optimization algorithms: (1) those requiring function derivatives and (2) those that do not. Here we give only an overview and refer the reader to Sec. 3 or the references for more details. The nonderivative methods are of interest in optimization applications because these methods can be readily adapted to the case in which experiments are carried out directly on the process. In such cases, an actual process measurement (such as yield) can be the objective function, and no mathematical model for the process is required. Methods that do not require derivatives are called *direct methods* and include sequential simplex (Nelder-Meade) and Powell's method. The sequential simplex method is quite satisfactory for optimization with two or three independent variables, is simple to understand, and is fairly easy to execute. Powell's method is more efficient than the simplex method and is based on the concept of conjugate search directions. This class of methods can be used in special cases but is not recommended for optimization involving more than 6 to 10 variables.

The second class of multivariable optimization techniques in principle requires the use of partial derivatives of the objective function, although finite difference formulas can be substituted for derivatives. Such techniques are called *indirect methods* and include the following classes:

1. Steepest descent (gradient) method
2. Conjugate gradient (Fletcher-Reeves) method
3. Newton's method
4. Quasi-Newton methods

The steepest descent method is quite old and utilizes the intuitive concept of moving in the direction in which the objective function changes the most. However, it is clearly not as efficient as the other three. Conjugate gradient utilizes only first-derivative information, as does steepest descent, but generates improved search directions. Newton's method requires second-derivative information but is very efficient, while quasi-Newton retains most of the benefits of Newton's method but utilizes only first-derivative information. All these techniques are also used with constrained optimization.

Constrained Optimization When constraints exist and cannot be eliminated in an optimization problem, more general methods must be employed than those described above, because the unconstrained optimum may correspond to unrealistic values of the operating variables. The general form of a nonlinear programming problem allows for a nonlinear objective function and nonlinear constraints, or

$$\begin{aligned} &\text{Minimize} && J(x_1, x_2, \ldots, x_n) && \\ &\text{Subject to} && h_i(x_1, x_2, \ldots, x_n) = 0 && i = 1, r_c \\ & && g_i(x_1, x_2, \ldots, x_n) > 0 && i = 1, m_c \end{aligned} \tag{8-40}$$

In this case, there are n process variables with r_c equality constraints and m_c inequality constraints. Such problems pose a serious challenge to performing optimization calculations in a reasonable amount of time. Typical constraints in chemical process optimization include operating conditions (temperatures, pressures, and flows have limits), storage capacities, and product purity specifications.

An important class of constrained optimization problems is one in which both the objective function and the constraints are linear. The solution of these problems is highly structured and can be obtained rapidly. The accepted procedure, linear programming (LP), has become quite popular in the past 20 years, solving a wide range of industrial problems. It is increasingly being used for online optimization. For processing plants, there are several different kinds of linear constraints that may arise, making the LP method of great utility.

1. Production limitation due to equipment throughput restrictions, storage limits, or market constraints.
2. Raw material (feedstock) limitation.
3. Safety restrictions on allowable operating temperatures and pressures.
4. Physical property specifications placed on the composition of the final product. For blends of various products, we usually assume that a composite property can be calculated through the mass-averaging of pure-component physical properties.
5. Material and energy balances of the steady-state model.

The optimum in linear programming lies at the constraint intersections. The simplex algorithm is a matrix-based numerical procedure for which many digital computer codes exist (Edgar, Himmelblau, and Lasdon, *Optimization of Chemical Processes*, 2d ed., McGraw-Hill, New York, 2001; Nash and Sofer, *Linear and Nonlinear Programming*, McGraw-Hill, New York, 1996). The algorithm can handle virtually any number of inequality constraints and any number of variables in the objective function, and it utilizes the observation that only the constraint boundaries need to be examined to find the optimum. In some instances, nonlinear optimization problems even with nonlinear constraints can be linearized so that the LP algorithm can be employed to solve them (called *successive linear programming*, or SLP). In the process industries, LP and SLP have been applied to a wide range of RTO problems, including refinery scheduling, olefins production, the optimal allocation of boiler fuel, and the optimization of a total plant.

Figure 8-39 gives an overview of which optimization algorithms are appropriate for certain types of RTO problems. No single NLP algorithm is best for every problem, so several solvers should be tested on a given application. See Biegler, *Nonlinear Programming: Concepts, Algorithms, and Applications to Chemical Processes*, MPS-SIAM Series on Optimization, Philadelphia, Pa., 2010.

Nonlinear Programming The most general case for optimization in Fig. 8-39 occurs when both the objective function and the constraints are nonlinear, a case referred to as nonlinear programming. While the ideas behind the search methods used for unconstrained multivariable problems are applicable, the presence of constraints complicates the solution procedure. All the methods discussed below have been utilized to solve nonlinear programming problems in the field of chemical engineering design and operations. Nonlinear programming is now used extensively in the area of real-time optimization.

A good overview of nonlinear programming is contained in Sec. 3 in this text. One popular NLP algorithm called the *generalized reduced gradient* (GRG) algorithm employs iterative linearization and is used in the Excel Solver. The CONOPT software package uses a reduced gradient algorithm that works well for large-scale problems and nonlinear constraints. Successive quadratic programming (SQP) solves a sequence of quadratic programs that approach the solution of the original NLP by linearizing the constraints and using a quadratic approximation to the objective function, which is used in MATLAB's constrained optimizer (fmincon). MINOS and NPSOL are SQP-based software packages originally developed in the 1980s. For large-scale NLPs, IPOPT is an interior point line search with barrier functions that is very effective (Biegler, *Nonlinear Programming: Concepts, Algorithms, and Applications to Chemical Processes*, SIAM, Philadelphia, Pa., 2010). Successive linear programming (SLP) is used less often for solving RTO problems and requires linear approximations of both the objective function and constraints.

Software libraries such as GAMS (General Algebraic Modeling System) or NAG (Numerical Algorithms Group) offer one or more NLP algorithms, but rarely are all algorithms available from a single source. Web sources that serve as comprehensive repositories of optimization packages include Optic Toolbox (http://www.i2c2.aut.ac.nz/Wiki/OPTI/), NEOS Server (http://www.neos-server.org/neos/) (Czyzyk et al., University of Wisconsin http://pages.cs.wisc.edu/~swright/PCx/, 2016), and COIN-OR (http://www.coin-or.org/) [Lougee-Heimer, The Common Optimization Interface for Operations Research, *IBM Journal of Research and Development* **47:** 57 (2003); Nocedal and Wright, *Numerical Optimization*, 2d ed., Springer, New York, 2006]. No single NLP algorithm is best for every problem, so several solvers should be tested on a given application.

Other optimization software packages are designed to solve mixed-integer problems (MIPs), with both discrete and continuous variables (MILP, MINLP) and also so-called *global optimization* problems that can converge to a nonconvex optimum. Impressive speed-ups in MIP solution have occurred during the past 20 years. See Biegler, *Nonlinear Programming: Concepts, Algorithms, and Applications to Chemical Processes*, MPS-SIAM Series on Optimization, Philadelphia, Pa., 2010 for more details.

Linear and nonlinear programming solvers have been interfaced to spreadsheet software, which has become a popular user interface for entering and manipulating numeric data. Spreadsheet software increasingly incorporates analytic tools that are accessible from the spreadsheet interface and permit access to external databases. For example, Microsoft Excel incorporates an optimization-based routine called Solver that operates on the values and formulas of a spreadsheet model. Current versions include LP and NLP solvers and mixed integer programming (MIP) capability for both linear and nonlinear problems. The user specifies a set of cell addresses to be independently adjusted (the decision variables), a set of formula cells whose values are to be constrained (the constraints), and a formula cell designated as the optimization objective.

Referring to Fig. 8-24, the highest level of process control, planning, and scheduling also employs optimization extensively, often with variables that are integer. Level 5 sets production goals to meet supply and logistics constraints and addresses time-varying capacity and workforce utilization decisions. Enterprise resource planning (ERP) and supply chain management (SCM) in level 5 refer to the links in a web of relationships involving retailing (sales), distribution, transportation, and manufacturing. Planning and scheduling usually operate over relatively long time scales and tend to be decoupled from the rest of the activities in lower levels. For example, all the refineries owned by an oil company are usually included in a comprehensive planning and scheduling model. This model can be optimized to obtain target levels and prices for interrefinery transfers, crude oil and product allocations to each refinery, production targets, inventory targets, optimal operating conditions, stream allocations, and blends for each refinery.

Some planning and scheduling problems are mixed integer optimization problems that involve both continuous and integer problems; whether to operate or use a piece of equipment is a binary (on/off) decision that arises in batch processing. Solution techniques for this type of problem include branch and bound methods and global search. This latter approach handles very complex problems with multiple optima by using algorithms such as tabusearch, scatter search, simulated annealing, and genetic evolutionary algorithms (see Edgar, Himmelblau, and Lasdon, *Optimization of Chemical Processes*, 2d ed., McGraw-Hill, New York, 2001).

STATISTICAL PROCESS CONTROL

In industrial plants, large numbers of process variables must be maintained within specified limits in order for the plant to operate properly. Excursions of key variables beyond these limits can have significant consequences for plant safety, the environment, product quality, and plant profitability. *Statistical process control* (SPC), also called *statistical quality control* (SQC), involves the application of statistical techniques to determine whether a process is operating normally or abnormally. Thus, SPC is a process monitoring technique that relies on *quality control charts* to monitor measured variables, especially product quality.

The basic SPC concepts and control chart methodology were introduced by Shewhart in the 1930s. The current widespread interest in SPC techniques began in the 1950s when they were successfully applied first in Japan and then elsewhere. Control chart methodologies are now widely used to monitor product quality and other variables that are measured infrequently or irregularly. The basic SPC methodology is described in introductory

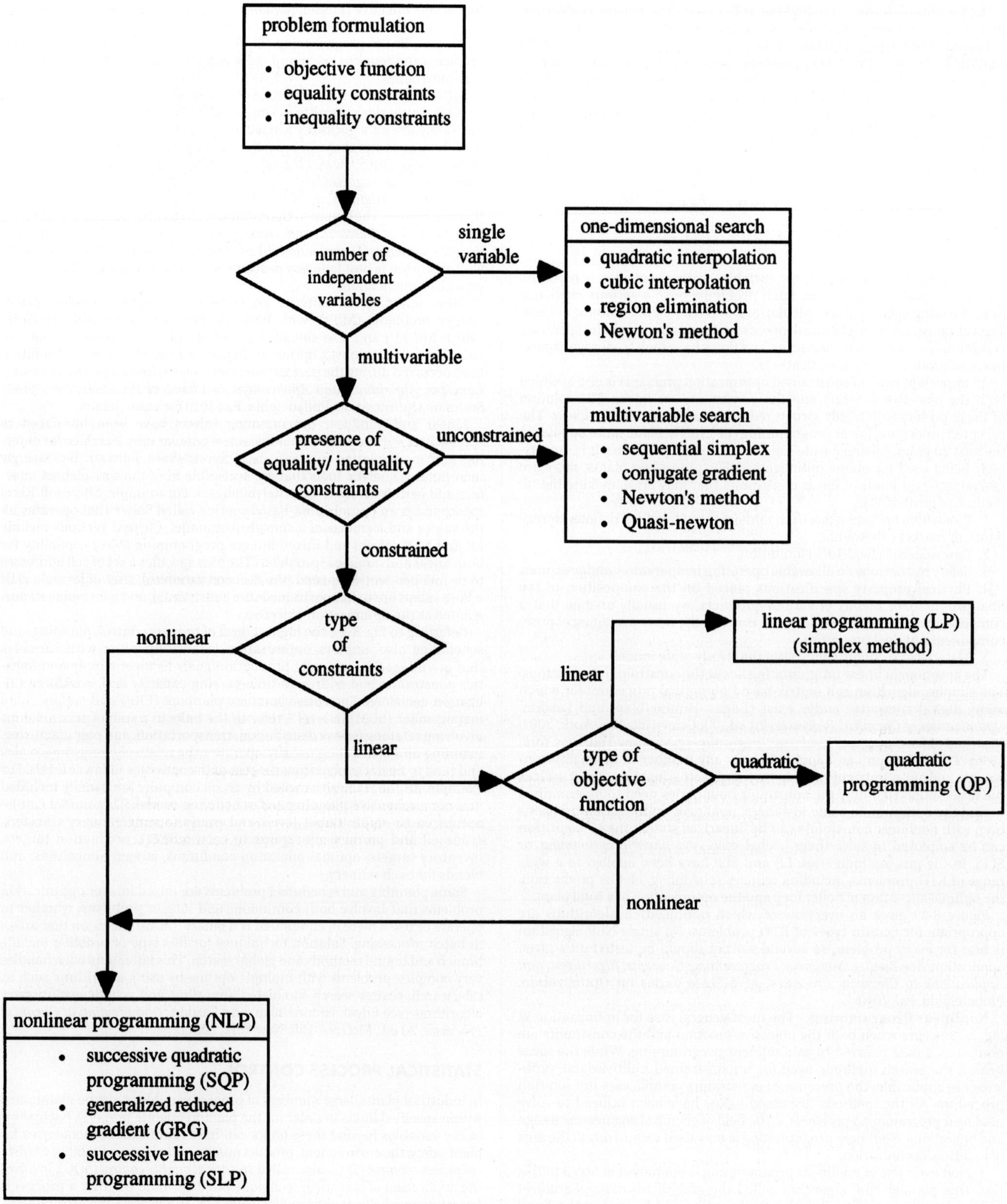

FIG. 8-39 Diagram for selection of optimization techniques with algebraic constraints and objective function.

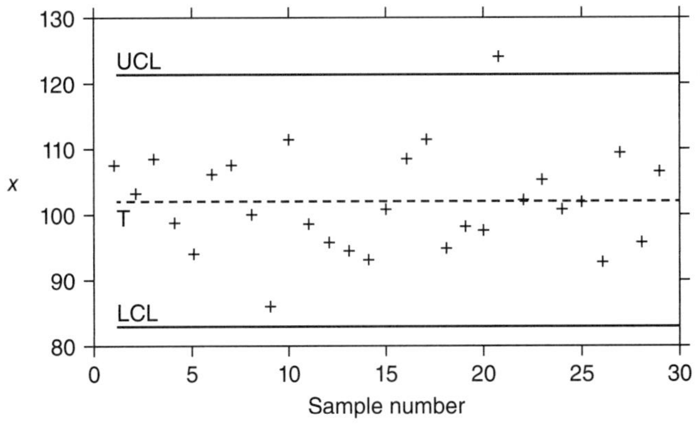

FIG. 8-40 The Shewhart chart. (*Source:* Seborg et al., *Process Dynamics and Control*, 3d ed., Wiley, New York, 2010.)

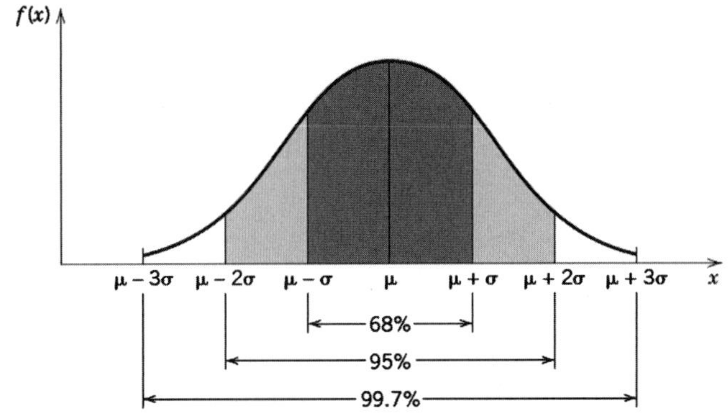

FIG. 8-41 Probabilities associated with the normal distribution. From Montgomery and Runger (2007). (*Source:* Seborg et al., *Process Dynamics and Control*, 3d ed., Wiley, New York, 2010.)

statistics textbooks (e.g., Montgomery and Runger, *Applied Statistics and Probability for Engineers*, 6th ed., Wiley, New York, 2013) and some process control textbooks (e.g., Seborg, Edgar, Mellichamp, and Doyle, *Process Dynamics and Control*, 4th ed., Wiley, New York, 2016).

An example of the most common control chart, the Shewhart chart, is shown in Fig. 8-40. It merely consists of measurements plotted versus sample number with control limits that indicate the range for normal process operation. The plotted data are either an individual measurement x or the sample mean $\bar{x}$ if more than one sample is measured at each sampling instant. The sample mean for k samples is calculated as

$$\bar{x} = \frac{1}{n}\sum_{j=1}^{n} x_j \tag{8-41}$$

The Shewhart chart in Fig. 8-40 has a *target* (T), an *upper control limit* (UCL), and a *lower control limit* (LCL). The target (or centerline) is the desired (or expected) value for $\bar{x}$ while the region between UCL and LCL defines the range of normal variability. If all the $\bar{x}$ data are within the control limits, the process operation is considered to be normal or "in a state of control." Data points outside the control limits are considered to be abnormal, indicating that the process operation is out of control. This situation occurs for the 21st sample. A single measurement slightly beyond a control limit is not necessarily a cause for concern. But frequent or large chart violations should be investigated to determine the root cause.

The major objective in SPC is to use process data and statistical techniques to determine whether the process operation is normal or abnormal. The SPC methodology is based on the fundamental assumption that normal process operation can be characterized by random variations around a mean value. The random variability is caused by the cumulative effects of a number of largely unavoidable phenomena such as electrical measurement noise, turbulence, and random fluctuations in feedstock or catalyst preparation. If this situation exists, the process is said to be *in a state of statistical control* (or *in control*), and the control chart measurements tend to be normally distributed about the mean value. By contrast, frequent control chart violations would indicate abnormal process behavior or an *out-of-control* situation. Then a search would be initiated to attempt to identify the *assignable cause* or the *special cause* of the abnormal behavior.

The control limits in Fig. 8-40 (UCL and LCL) are based on the assumption that the measurements follow a normal distribution. Figure 8-41 shows the probability distribution for a normally distributed random variable x with mean μ and standard deviation σ. There is a very high probability (99.7 percent) that any measurement is within 3 standard deviations of the mean. Consequently, the control limits for x are typically chosen to be $T \pm 3\hat{\sigma}$, where $\hat{\sigma}$ is an estimate of σ. This estimate is usually determined from a set of representative data for a period of time when the process operation is believed to be typical. For the common situation in which the plotted variable is the sample mean, its standard deviation is estimated.

Shewhart control charts enable average process performance to be monitored, as reflected by the sample mean. It is also advantageous to monitor process variability. Process variability within a sample of k measurements can be characterized by its range, standard deviation, or sample variance. Consequently, control charts are often used for one of these three statistics.

Western Electric Rules Shewhart control charts can detect abnormal process behavior by comparing individual measurements with control chart limits. In addition, the pattern of measurements can provide useful information. For example, if 10 consecutive measurements are all increasing, then it is very unlikely that the process is in a state of control. A wide variety of *pattern tests* (also called *zone rules*) can be developed based on the properties of the normal distribution. For example, the following excerpts from the *Western Electric Rules* (Western Electric Company, *Statistical Quality Control Handbook*, Delmar Printing Company, Charlotte, N.C., 1956; Montgomery and Runger, *Applied Statistics and Probability for Engineers*, 6th ed., Wiley, New York, 2013) indicate that the process is out of control if one or more of the following conditions occur:

1. One data point outside the 3σ control limits
2. Two out of three consecutive data points beyond a 2σ limit
3. Four out of five consecutive data points beyond a 1σ limit and on one side of the centerline
4. Eight consecutive points on one side of the centerline

Note that the first condition is the familiar Shewhart chart limits. Pattern tests can be used to augment Shewhart charts. This combination enables out-of-control behavior to be detected earlier, but the false-alarm rate is higher than that for Shewhart charts alone.

CUSUM Control Charts Although Shewhart charts with 3σ limits can quickly detect large process changes, they are ineffective for small, sustained process changes (e.g., changes smaller than 1.5σ). Alternative control charts have been developed to detect small changes such as the CUSUM control chart. They are often used in conjunction with Shewhart charts. The *cumulative sum* (CUSUM) is defined to be a running summation of the deviations of the plotted variable from its target. If the sample mean is plotted, the cumulative sum at sampling instant k, denoted $C(k)$, is

$$C(k) = \sum_{j=1}^{k} [\bar{x}(j) - T] \tag{8-42}$$

where T is the target for $\bar{x}$. During normal process operation, $C(k)$ fluctuates about zero. But if a process change causes a small shift in $\bar{x}$, then $C(k)$ will drift either upward or downward.

The CUSUM control chart was originally developed using a graphical approach based on *V masks*. However, for computer calculations, it is more convenient to use an equivalent algebraic version that consists of two recursive equations

$$C^+(k) = \max[0, \bar{x}(k) - (T + K) + C^+(k-1)] \tag{8-43}$$

$$C^-(k) = \max[0, (T - K) - \bar{x}(k) + C^-(k-1)] \tag{8-44}$$

where C^+ and C^- denote the sums for the high and low directions and K is a constant, the *slack parameter*. The CUSUM calculations are initialized by setting $C^+(0) = C - 0 = 0$. A deviation from the target that is larger than K increases either C^+ or C^-. A control limit violation occurs when either C^+ or C^- exceeds a specified control limit (or *threshold*) H. After a limit violation occurs, that sum is reset to zero or to a specified value.

The selection of the threshold H can be based on considerations of *average run length* (ARL), the average number of samples required to detect a disturbance of specified magnitude. For example, suppose that the objective is to be able to detect if the sample mean $\bar{x}$ has shifted from the target by a small amount δ. The slack parameter K is usually specified as $K = 0.5\delta$.

TABLE 8-6 Average Run Lengths for CUSUM Control Charts

Shift from target (in multiples of σ)	ARL for $H = 4\sigma$	ARL for $H = 5\sigma$
0	168.	465.
0.25	74.2	139.
0.50	26.6	38.0
0.75	13.3	17.0
1.00	8.38	10.4
2.00	3.34	4.01
3.00	2.19	2.57

Adapted from Ryan, *Statistical Methods for Quality Improvement*, 2d ed., Wiley, New York, 2000.

For the ideal situation (e.g., normally distributed, uncorrelated disturbances), ARL values have been tabulated for different values of δ, *K*, and *H*. Table 8-6 summarizes ARL values for two values of *H* and different values of δ. (The values of δ are usually expressed as multiples of the standard deviation σ.) The ARL values indicate the average number of samples before a change of δ is detected. Thus the ARL values for $\delta = 0$ indicate the average time between false alarms, i.e., the average time between successive CUSUM alarms when no shift in $\bar{x}$ has occurred. Ideally, we would like the ARL value to be very large for $\delta = 0$ and small for $\delta \neq 0$. Table 8-6 shows that as the magnitude of the shift δ increases, ARL decreases and thus the CUSUM control chart detects the change faster. Increasing the value of *H* from 4σ to 5σ increases all the ARL values and thus provides a more conservative approach.

The relative performance of the Shewhart and CUSUM control charts is compared in Fig. 8-42 for a set of simulated data for the tensile strength of a resin. It is assumed that the tensile strength *x* is normally distributed with a mean of $\mu = 70$ MPa and a standard deviation of $\sigma = 3$ MPa. A single measurement is available at each sampling instant. A constant ($\delta = 0.5\sigma = 1.5$) was added to $x(k)$ for $k \geq 10$ in order to evaluate each chart's ability to detect a small process shift. The CUSUM chart was designed using $K = 0.5\sigma$ and $H = 5\sigma$.

The Shewhart chart fails to detect the 0.5σ shift in *x* at $k = 10$. But the CUSUM chart quickly detects this change because a limit violation occurs at $k = 20$. The mean shift can also be detected by applying the Western Electric Rules in the previous section.

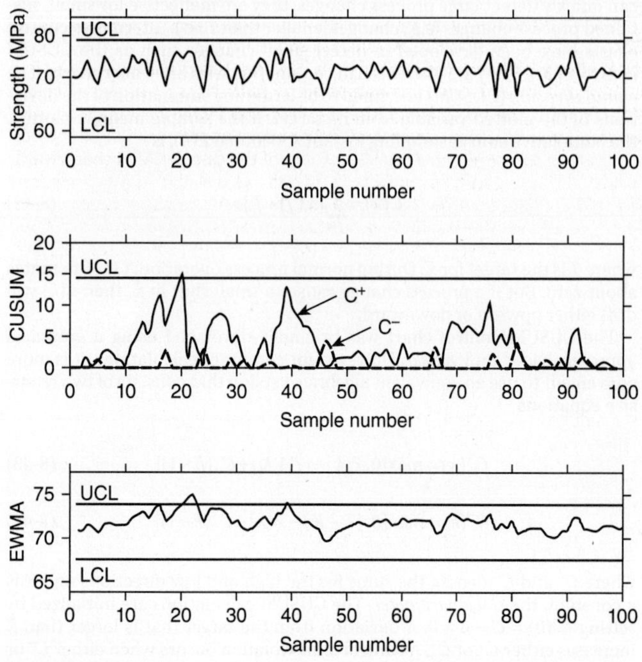

FIG. 8-42 Comparison of Shewhart (*top*), CUSUM (*middle*), and EWMA (*bottom*) control charts. (*Source:* Seborg et al., *Process Dynamics and Control*, 3d ed., Wiley, New York, 2010.)

Process Capability Indices Also known as *process capability ratios*, these provide a measure of whether an in-control process is meeting its product specifications. Suppose that a quality variable *x* must have a value between an *upper specification limit* (USL) and a *lower specification limit* (LSL) in order for product to satisfy customer requirements. The *capability index* C_p is defined as

$$C_p = \frac{\text{USL} - \text{LSL}}{6\sigma} \tag{8-45}$$

where σ is the standard deviation of *x*. Suppose that $C_p = 1$ and *x* is normally distributed. Based on the normal distribution, we would expect that 99.7 percent of the measurements satisfy the specification limits, or equivalently, we would expect that only 2700 out of 1 million measurements would lie outside the specification limits. If $C_p < 1$, the product specifications are satisfied; for $C_p > 1$, they are not. However, capability indices are applicable even when the data are not normally distributed.

A second capability index C_{pk} is based on average process performance $\bar{x}$ as well as process variability σ. It is defined as

$$C_{pk} = \frac{\text{Min}\,[\bar{x} - \text{LSL},\ \text{USL} - \bar{x}]}{3\sigma} \tag{8-46}$$

Although both C_p and C_{pk} are used, we consider C_{pk} to be superior to C_p for the following reason. If $\bar{x} = T$, the process is said to be "centered" and $C_{pk} = C_p$. But for $\bar{x} \neq T$, C_p does not change, even though the process performance is worse, while C_{pk} does decrease. For this reason, C_{pk} is preferred.

If the standard deviation σ is not known, it is replaced by an estimate $\hat{\sigma}$ in Eqs. (8-45) and (8-46). For situations where there is only a single specification limit, either USL or LSL, the definitions of C_p and C_{pk} can be modified accordingly.

In practical applications, a common objective is to have a capability index of 2.0 while a value greater than 1.5 is considered to be acceptable. If the C_{pk} value is too low, it can be improved by making a change that either reduces process variability or causes $\bar{x}$ to move closer to the target. These improvements can be achieved in a number of ways that include better process control, better process maintenance, reduced variability in raw materials, improved operator training, and process changes.

Six-Sigma Approach Product quality specifications continue to become more stringent as a result of market demands and intense worldwide competition. Meeting quality requirements is especially difficult for products that consist of a very large number of components and for manufacturing processes that consist of hundreds of individual steps. For example, the production of a microelectronic device typically requires 100 to 300 batch processing steps. Suppose that there are 200 steps and that each one must meet a quality specification for the final product to function properly. If each step is independent of the others and has a 99 percent success rate, the overall yield of satisfactory product is $(0.99)^{200} = 0.134$, or only 13.4 percent. This low yield is clearly unsatisfactory. Similarly, even when a processing step meets 3σ specifications (99.73 percent success rate), it will still result in an average of 2700 "defects" for every 1 million produced. Furthermore, the overall yield for this 200-step process is still only 58.2 percent.

The six-sigma approach was pioneered by the Motorola Corporation in the early 1980s as a strategy for achieving both six-sigma quality and continuous improvement. Since then, other large corporations have adopted companywide programs that apply the six-sigma approach to all their business operations, both manufacturing and nonmanufacturing. Thus, although the six-sigma approach is "data-driven" and based on statistical techniques, it has evolved into a broader management philosophy that has been implemented successfully by many large corporations. The six-sigma programs have also had a significant financial impact.

Multivariate Statistical Techniques For common SPC monitoring problems, two or more quality variables are important, and they can be highly correlated. For these situations, multivariable (or *multivariate)* SPC techniques can offer significant advantages over the single-variable methods discussed earlier. Multivariate monitoring based on the classical Hotelling T^2 statistic (Montgomery, *Introduction to Statistical Quality Control*, 7th ed., Wiley, New York, 2012) can be effective if the data are not highly correlated and the number of variables *p* is not large (for example, $p < 10$). Fortunately, alternative multivariate monitoring techniques such as *principal-component analysis* (PCA) and *partial least-squares* (PLS) methods have been developed that are very effective for monitoring problems with large numbers of variables and highly correlated data [Piovoso and Hoo (eds.), Special Issue of *IEEE Control Systems Magazine*, 22(5): 2002].

UNIT OPERATIONS CONTROL

PIPING AND INSTRUMENTATION DIAGRAMS

General References: Shinskey, *Process Control Systems*, 4th ed., McGraw-Hill, New York, 1996. Luyben, *Practical Distillation Control*, Van Nostrand Reinhold, New York, 1992. Luyben, Tyreus, and Luyben, *Plantwide Process Control*, McGraw-Hill, New York, 1999.

The piping and instrumentation (P&I) diagram provides a graphical representation of the control configuration of the process. P&I diagrams illustrate the measuring devices that provide inputs to the control strategy, the actuators that will implement the results of the control calculations, and the function blocks that provide the control logic. They may also include piping details such as line sizes and the location of hand valves and condensate traps.

The symbology for drawing P&I diagrams generally follows standards developed by the Instrumentation, Systems, and Automation Society (ISA). The chemicals, refining, and food industries generally follow this standard. The standards are updated from time to time, primarily because the continuing evolution in control system hardware and software provides additional capabilities for implementing control schemes. The ISA symbols are simple and represent a device or function as a circle containing its tag number and identifying the type of variable being controlled, e.g., pressure, and the function performed, e.g., control: PC-105. Examples of extensions of the ISA standard appear on the pages following.

Figure 8-43 presents a simplified P&I diagram for a temperature control loop that applies the ISA symbology. The measurement devices and most elements of the control logic are shown as circles:

1. TT102 is the temperature transmitter.
2. TC102 is the temperature controller.
3. TY102 is the current-to-pneumatic (I/P) transducer.

The symbol for the control valve in Fig. 8-43 is for a pneumatic modulating valve without a valve positioner.

Electronic (4- to 20-mA) signals are represented by dashed lines. In Fig. 8-43, these include the signal from the transmitter to the controller and the signal from the controller to the I/P transducer. Pneumatic signals are represented by solid lines with double crosshatching at regular intervals. The signal from the I/P transducer to the valve actuator is pneumatic.

The ISA symbology provides different symbols for different types of actuators. Furthermore, variations for the controller symbol distinguish control algorithms implemented in distributed control systems from those in panel-mounted single-loop controllers.

CONTROL OF HEAT EXCHANGERS

Steam-Heated Exchangers Steam, the most common heating medium, transfers its latent heat in condensing, causing heat flow to be proportional to steam flow. Thus a measurement of steam flow is essentially a measure of heat transfer. Consider raising a liquid from temperature T_1 to T_2 by condensing steam:

$$Q = WH = M_v C_L (T_2 - T_1) \tag{8-47}$$

where W and H are the mass flow of steam and its latent heat, M_v and C_L are the mass flow and specific heat of the liquid, and Q is the rate of heat transfer. The response of controlled temperature to steam flow is linear:

$$\frac{dT_2}{dW} = \frac{H}{M_v C_L} \tag{8-48}$$

However, the steady-state process gain described by this derivative varies inversely with liquid flow: adding a given increment of heat flow to a smaller flow of liquid produces a greater temperature rise.

Dynamically, the response of liquid temperature to a step in steam flow is that of a distributed lag, shown in Fig. 8-20 (uncontrolled); the time required to reach 63 percent complete response $\Sigma\tau$ is essentially the residence time of the fluid in the exchanger, which is its volume divided by its flow. The residence time then varies inversely with flow. Table 8-1 gives optimum settings for PI and PID controllers for distributed lags, the proportional band varying directly with steady-state gain, and integral and derivative settings directly with $\Sigma\tau$. Since both these parameters vary inversely with liquid flow, fixed settings for the temperature controller are optimal at only one flow rate.

The variability of the process parameters with flow causes variability in load response, as shown in Fig. 8-44. The PID controller was tuned for optimum (minimum-IAE) load response at 50 percent flow. Each curve represents the response of exit temperature to a 10 percent step in liquid flow, culminating at the stated flow. The 60 percent curve is overdamped, and the 40 percent curve is underdamped. The differences in gain are reflected in the amplitude of the deviation, and the differences in dynamics are reflected in the period of oscillation.

If steam flow is linear with controller output, as it is in Fig. 8-44, undamped oscillations will be produced when the flow decreases by one-third from the value at which the controller was optimally tuned—in this example at 33 percent flow. The stable operating range can be extended to one-half the original flow by using an equal-percentage (logarithmic) steam valve, whose gain varies directly with steam flow, thereby compensating for the variable process gain. Further extension requires increasing the integral setting and reducing the derivative setting from their optimum values. The best solution is to adapt all three PID settings to change inversely with measured flow, thereby keeping the controller optimally tuned for all flow rates.

Feedforward control can also be applied, as described previously under Advanced Control Techniques. The feedforward system solves Eq. (8-78) for the manipulated set point to the steam flow controller, first by subtracting inlet temperature T_1 from the output of the outlet temperature controller (in place of T_2) and then by multiplying the result by the dynamically compensated liquid flow measurement. If the inlet temperature is not subject to rapid or wide variation, it can be left out of the calculation. Feedforward is capable of a reduction in integrated error as much as 100-fold, but requires the use of a steam flow loop and lead-lag compensator to approach this effectiveness. Multiplication of the controller output by liquid flow varies the feedback loop gain directly proportional to flow, extending the

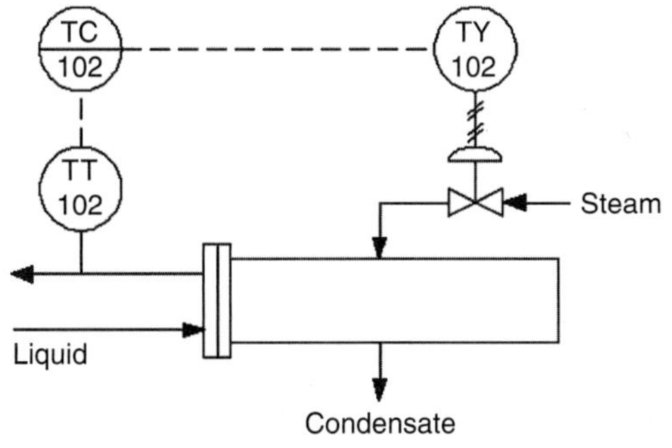

FIG. 8-43 Example of a simplified piping and instrumentation diagram.

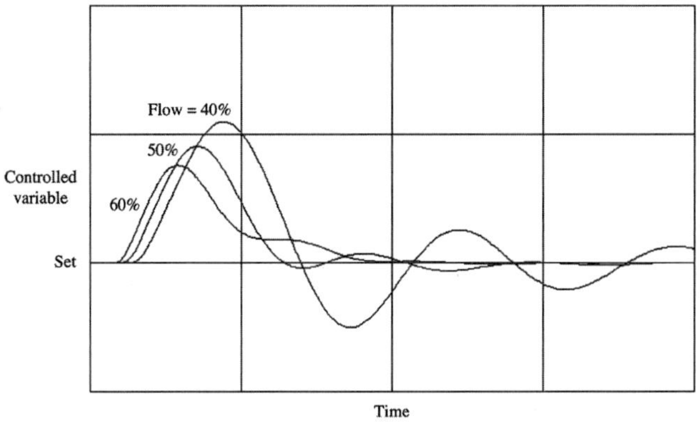

FIG. 8-44 The response of a heat exchanger varies with flow in both gain and dynamics; here the PID temperature controller was tuned for optimum response at 50 percent flow.

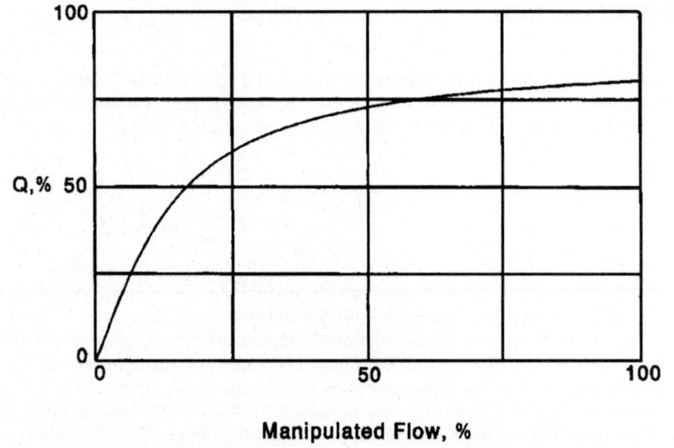

FIG. 8-45 Heat-transfer rate in sensible-heat exchange varies nonlinearly with flow of the manipulated fluid, requiring equal-percentage valve characterization.

stable operating range of the feedback loop much as the equal-percentage steam valve did without feedforward. (This system is eminently applicable to control of fired heaters in oil refineries, which commonly provide a circulating flow of hot oil to several distillation columns and are therefore subject to frequent disturbances.)

Steam flow is sometimes controlled by manipulating a valve in the condensate line rather than the steam line, because it is smaller and hence less costly. Heat transfer then is changed by raising or lowering the level of condensate flooding the heat-transfer surface, an operation that is slower than manipulating a steam valve. Protection also needs to be provided against an open condensate valve blowing steam into the condensate system.

Exchange of Sensible Heat When there is no change in phase, the rate of heat transfer is no longer linear with the flow of the manipulated stream, but is a function of the mean temperature difference ΔT_m:

$$Q = UA\,\Delta T_m = M_H C_H (T_{H1} - T_{H2}) = M_C C_C (T_{C2} - T_{C1}) \tag{8-48}$$

where U and A are the overall heat-transfer coefficient and area and subscripts H and C refer to the hot and cold fluids, respectively. An example would be a countercurrent cooler, where the hot-stream outlet temperature is controlled. Using the logarithmic mean temperature difference and solving for T_{H2} give

$$T_{H2} = T_{c1} + (T_{H1} - T_{c1}) \frac{1 - M_H C_H / M_C C_C}{\varepsilon - M_H C_H / M_C C_C} \tag{8-50}$$

where

$$\varepsilon = \exp\left[-UA\left(\frac{1}{M_c C_c} - \frac{1}{M_H C_H}\right)\right] \tag{8-51}$$

At a given flow of hot fluid, the heat-transfer rate is plotted as a function of coolant flow in Fig. 8-45, as a percentage of its maximum value (corresponding to $T_{C2} = T_{C1}$). The extreme nonlinearity of this relationship requires the use of an equal-percentage coolant valve for gain compensation. The variable dynamics of the distributed lag also apply, limiting the stable operating range in the same way as for the steam-heated exchanger.

Sensible-heat exchangers are also subject to variations in the temperature of the manipulated stream, an increasingly common problem where heat is being recovered at variable temperatures for reuse. Figure 8-46 shows a temperature controller (TC) setting a heat flow controller (QC) in cascade. A measurement of the manipulated flow is multiplied by its temperature difference across the heat exchanger to calculate the current heat-transfer rate, by using the right side of Eq. (8-47). Variations in supply temperature then appear as variations in calculated heat-transfer rate, which the QC can quickly correct by adjusting the manipulated flow. An equal-percentage valve is still required to linearize the secondary loop, but the primary loop of temperature-setting heat flow is linear. Feedforward can be added by multiplying the dynamically compensated flow measurement of the other fluid by the output of the temperature controller.

When a stream is manipulated whose flow is independently determined, such as the flow of a product or of a heat-transfer fluid from a fired heater, a three-way valve is used to divert the required flow to the heat exchanger. This does not alter the linearity of the process or its sensitivity to supply variations, and it even adds the possibility of independent flow variations. The three-way valve should have equal-percentage characteristics, and heat flow control may be even more beneficial.

DISTILLATION COLUMN CONTROL

Distillation columns have four or more closed loops—increasing with the number of product streams and their specifications—all of which interact with one another to some extent. Because of this interaction, there are many possible ways to pair manipulated and controlled variables through controllers and other mathematical functions, with widely differing degrees of effectiveness. Columns also differ from one another, so that no single rule of configuring control loops can be applied successfully to all. The following rules apply to the most common separations.

Controlling Quality of a Single Product If one of the products of a column is far more valuable than the other(s), its quality should be controlled to satisfy given specifications, and its recovery should be maximized by minimizing losses of its principal component in other streams. This is achieved by maximizing the reflux ratio consistent with flooding limits on trays, which means maximizing the flow of internal reflux or vapor, whichever is limiting. The same rule should be followed when heating and cooling have little value. A typical example is the separation of high-purity propylene from much lower-valued propane, usually achieved with the waste heat of quench water from the cracking reactors.

The most important factor affecting product quality is the material balance. In separating a feed stream F into distillate D and bottom B products, an overall mole flow balance must be maintained

$$F = D + B \tag{8-52}$$

as well as a balance on each component

$$Fz_i = Dy_i + Bx_i \tag{8-53}$$

where z, y, and x are mole fractions of component i in the respective streams. Combining these equations gives a relationship between the composition of the products and their relative portion of the feed:

$$\frac{D}{F} = 1 - \frac{B}{F} = \frac{z_i - x_i}{y_i - x_i} \tag{8-54}$$

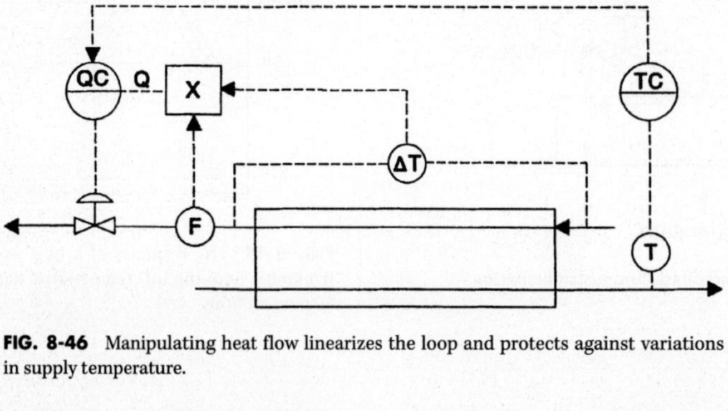

FIG. 8-46 Manipulating heat flow linearizes the loop and protects against variations in supply temperature.

From the above, it can be seen that control of either x_i or y_i requires both product flow rates to change with feed rate and feed composition.

Figure 8-47 shows a propylene-propane fractionator controlled at maximum boil-up by the differential pressure controller (DPC) across the trays. This loop is fast enough to reject upsets in the temperature of the quench water quite easily. Pressure is controlled by manipulating the heat-transfer surface in the condenser through flooding. If the condenser should become overloaded, pressure will rise above the set point, but this has no significant effect on the other control loops. Temperature measurements on this column are not helpful, as the difference between the component boiling points is too small. Propane content in the propylene distillate is measured by a chromatographic analyzer sampling the overhead vapor for fast response, and it is controlled by the analyzer controller (AC) manipulating the ratio of distillate to feed rates. The feedforward signal from the feed rate is dynamically compensated and nonlinearly characterized to account for variations in propylene recovery as the feed rate changes. Distillate flow can be measured and controlled more accurately than reflux flow by a factor equal to the reflux ratio, which in this column is typically between 10 and 20. Therefore, reflux flow is placed under accumulator level control (LC). Yet composition responds to the difference between internal vapor and reflux flow rates. To eliminate the lag inherent in the response of the accumulator level controller, reflux flow is driven by the subtractor in the direction opposite to distillate flow—this is essential to fast response of the composition loop. The gain of converting distillate flow changes to reflux flow changes can even be increased beyond −1, thereby changing the accumulator level loop from a lag into a dominant lead.

Controlling Quality of Two Products Where the two products have similar values, or where heating and cooling costs are comparable to product losses, the compositions of both products should be controlled. This introduces the possibility of strong interaction between the two composition loops, as they tend to have similar speeds of response. Interaction in most columns can be minimized by controlling distillate composition with reflux ratio and bottom composition with boil-up, or preferably boil-up/bottom flow ratio. These loops are insensitive to variations in feed rate, eliminating the need for feedforward control, and they also reject heat balance upsets quite effectively. Figure 8-48 shows a depropanizer controlled by reflux and boil-up ratios. The actual mechanism through which these ratios are manipulated is $D/(L + D)$ and $B/(V + B)$, where L is reflux flow and V is vapor boil-up, which decouples the temperature loops from the liquid-level loops. Column pressure here is controlled by flooding both the condenser and accumulator; however, there is no level controller on the accumulator, so this arrangement will not function with an overloaded condenser. Temperatures are used as indications of composition in this column because of the substantial differences in boiling points between propane and butanes. However, off-key components such as ethane do affect the accuracy of the relationship, so that an analyzer controller is used to set the top temperature controller (TC) in cascade.

If the products from a column are especially pure, even this configuration may produce excessive interaction between the composition loops. Then the composition of the less pure product should be controlled by manipulating its own flow; the composition of the remaining product should be controlled by manipulating the reflux ratio if it is the distillate, or the boil-up ratio if it is the bottom product.

Most sidestream columns have a small flow dedicated to removing an off-key impurity entering the feed, and that stream must be manipulated to control its content in the major product. For example, an ethylene fractionator separates its feed into a high-purity ethylene sidestream, an ethane-rich bottom product, and a small flow of methane overhead. This small flow must be withdrawn to control the methane content in the ethylene product. The key impurities may then be controlled in the same way as in a two-product column.

Most volatile mixtures have a relative volatility that varies inversely with column pressure. Therefore, their separation requires less energy at lower pressure, and savings in the range of 20 to 40 percent have been achieved. Column pressure can be minimized by floating on the condenser, i.e., by operating the condenser with minimal or no restrictions. In some columns, such as the propylene-propane splitter, pressure can be left uncontrolled. Where it cannot, the set point of the pressure controller can be indexed by an integral-only controller acting to slowly drive the pressure control valve toward a position just short of maximum cooling. In the case of a flooded condenser, the degree of reflux subcooling can be controlled in place of the condenser valve position. Where column temperatures are used to indicate product composition, their measurements must be pressure-compensated.

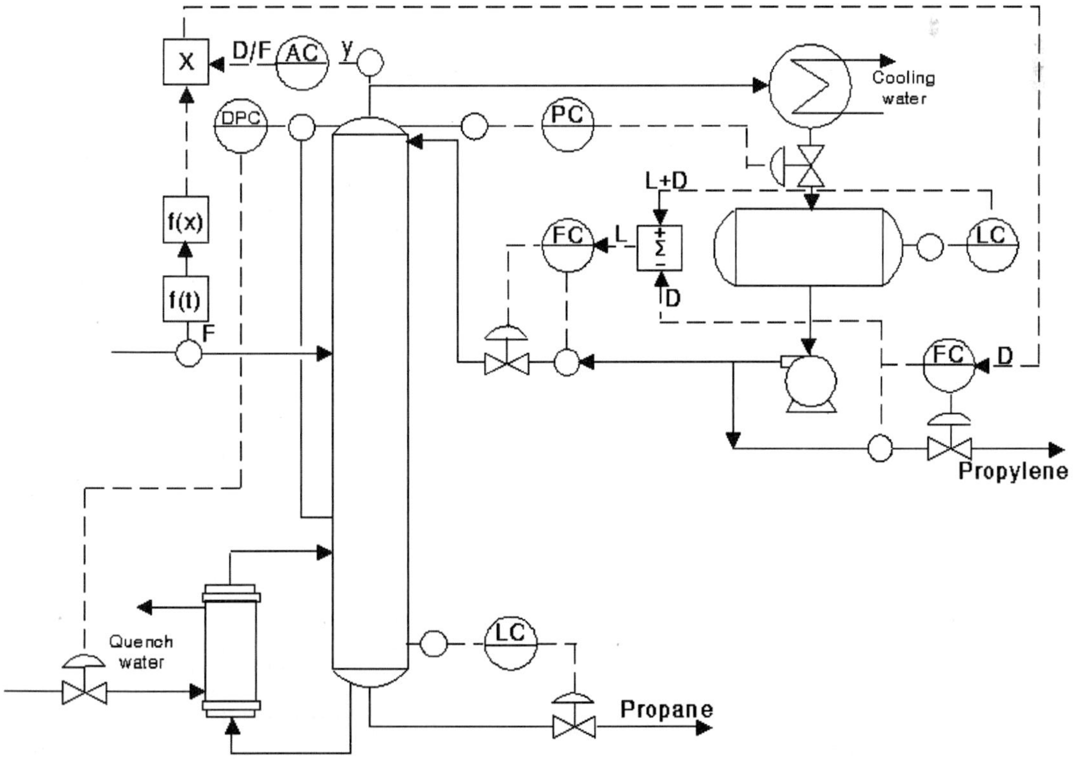

FIG. 8-47 The quality of high-purity propylene should be controlled by manipulating the material balance.

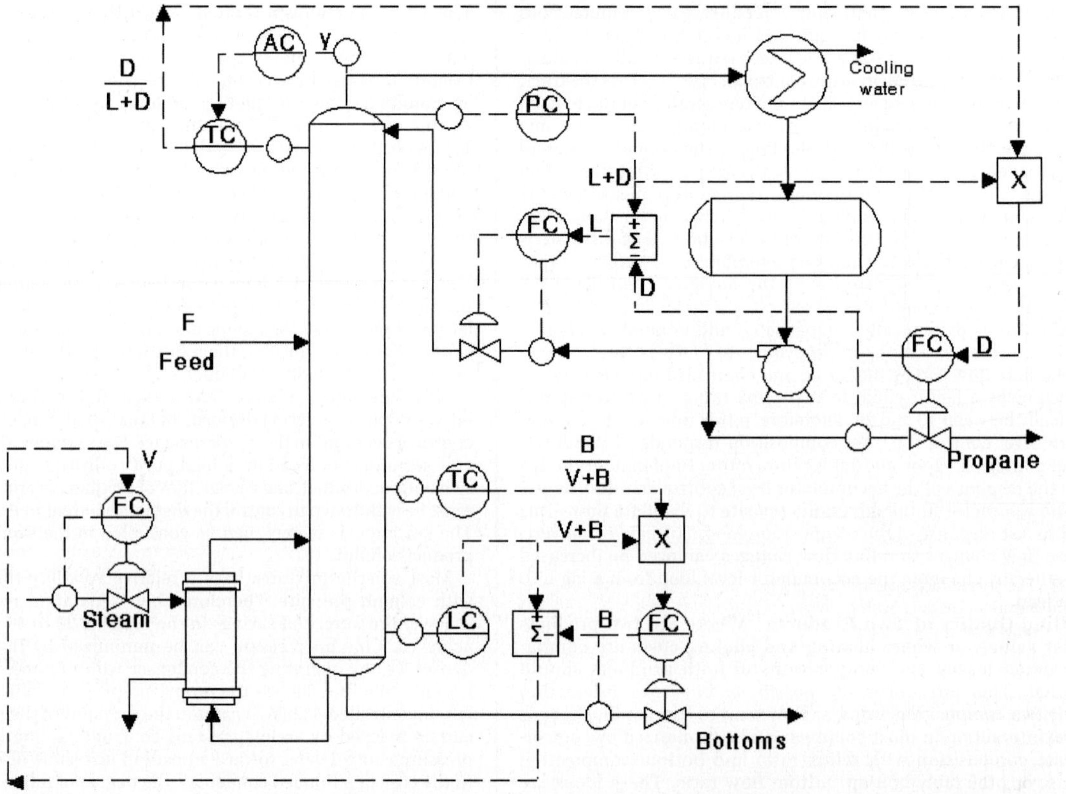

FIG. 8-48 Depropanizers require control of the quality of both products, here using the reflux ratio and boil-up ratio manipulation.

CHEMICAL REACTORS

Composition Control The first requirement for successful control of a chemical reactor is to establish the proper stoichiometry, i.e., to control the flow rates of the reactants in the proportions needed to satisfy the reaction chemistry. In a continuous reactor, this begins by setting ingredient flow rates in ratio to one another. However, because of variations in the purity of the feed streams and inaccuracy in flow metering, some indication of excess reactant such as pH or a composition measurement should be used to trim the ratios. Many reactions are incomplete, leaving one or more reactants unconverted. They are separated from the products of the reaction and recycled to the reactor, usually contaminated with inert components. While reactants can be recycled to complete conversion (extinction), inerts can accumulate to the point of impeding the reaction and must be purged from the system. Inerts include noncondensable gases that must be vented and nonvolatiles from which volatile products must be stripped.

If one of the reactants differs in phase from the others and the product(s), it may be manipulated to close the material balance on that phase. For example, a gas reacting with liquids to produce a liquid product may be added as it is consumed to control reactor pressure; a gaseous purge would be necessary. Similarly, a liquid reacting with a gas to produce a gaseous product could be added as it is consumed to control the liquid level in the reactor; a liquid purge would be required. Where a large excess of one reactant A is used to minimize side reactions, the unreacted excess is sent to a storage tank for recycling. Its flow from the recycle storage tank is set in the desired ratio to the flow of reactant B, with the flow of fresh A manipulated to control the recycle tank level if the feed is a liquid, or tank pressure if it is a gas. Some catalysts travel with the reactants and must be recycled in the same way.

With batch reactors, it may be possible to add all reactants in their proper quantities initially, if the reaction rate can be controlled by injection of initiator or adjustment of temperature. In semibatch operation, one key ingredient is flow-controlled into the batch at a rate that sets the production. This ingredient should not be manipulated for temperature control of an exothermic reactor, as the loop includes two dominant lags—concentration of the reactant and heat capacity of the reaction mass—and can easily go unstable. It also presents the unfavorable dynamic of inverse response—increasing feed rate may lower temperature by its sensible heat before the increased reaction rate raises temperature.

Temperature Control Reactor temperature should always be controlled by heat transfer. Endothermic reactions require heat and therefore are eminently self-regulating. Exothermic reactions produce heat, which tends to raise the reaction temperature, thereby increasing the reaction rate and producing greater heat. This positive feedback is countered by negative feedback in the cooling system, which removes more heat as the reactor temperature rises. Most continuous reactors have enough heat-transfer surface relative to reaction mass that negative feedback dominates and they are self-regulating. But most batch reactors do not, and they are therefore steady-state unstable. Unstable reactors can be controlled if their temperature controller gain can be set high enough, and if their cooling system has enough margin to accommodate the largest expected disturbance in heat load. Stirred-tank reactors are lag-dominant, and their dynamics allow a high controller gain, but plug flow reactors are dead-time-dominant, preventing their temperature controller from providing enough gain to overcome steady-state instability. Therefore, unstable plug flow reactors are also uncontrollable, their temperature tending to limit-cycle in a sawtooth wave. A stable reactor can become unstable as its heat-transfer surface fouls, or as the production rate is increased beyond a critical point (Shinskey, "Exothermic Reactors: The Stable, the Unstable, and the Uncontrollable," *Chem. Eng.*, pp. 54–59, March 2002).

Figure 8-49 shows the recommended system for controlling the temperature of an exothermic stirred-tank reactor, either continuous or batch. The circulating pump on the coolant loop is absolutely essential to effective temperature control in keeping dead time minimum and constant—without it, dead time varies inversely with cooling load, causing limit-cycling at low loads. Heating is usually required to raise the temperature to reaction conditions, although it is often locked out of a batch reactor once the initiator is introduced. The valves are operated in split range, the heating valve opening from 50 to 100 percent of controller output and the cooling valve opening from 50 to 0 percent. The cascade system linearizes the reactor temperature loop, speeds its response, and protects it from disturbances in the cooling system. The flow of heat removed per unit of coolant flow is directly proportional to the temperature rise of the coolant, which varies with both the temperature of the reactor and the rate of heat transfer from it. Using an

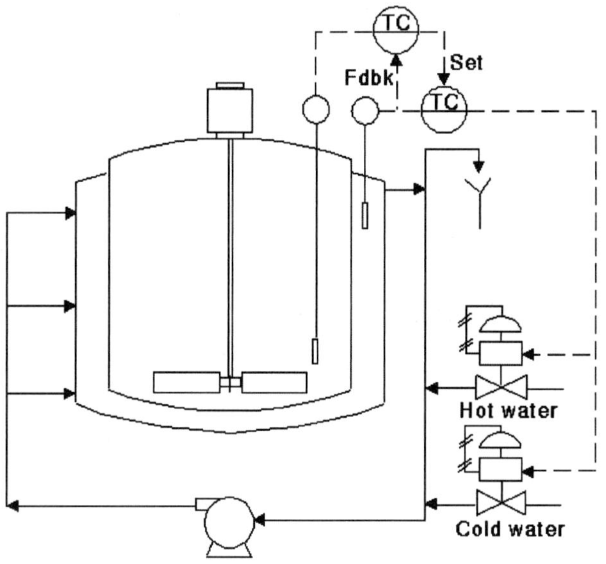

FIG. 8-49 The stirred-tank reactor temperature controller sets the coolant outlet temperature in cascade, with primary integral feedback taken from the secondary temperature measurement.

equal-percentage cooling valve helps compensate for this nonlinearity, but incompletely.

The flow of heat across the heat-transfer surface is linear with both temperatures, leaving the primary loop with a constant gain. Using the coolant exit temperature as the secondary controlled variable, as shown in Fig. 8-49, places the jacket dynamics in the secondary loop, thereby reducing the period of the primary loop. This is dynamically advantageous for a stirred-tank reactor because of the slow response of its large heat capacity. However, a plug flow reactor cooled by an external heat exchanger lacks this heat capacity, and requires the faster response of the coolant inlet temperature loop.

Performance and robustness are both improved by using the secondary temperature measurement as the feedback signal to the integral mode of the primary controller. (This feature may be available only with controllers that integrate by positive feedback.) This places the entire secondary loop in the integral path of the primary controller, effectively pacing its integral time to the rate at which the secondary temperature is able to respond. It also permits the primary controller to be left in the automatic mode at all times without integral windup.

The primary time constant of the reactor is

$$\tau_1 = \frac{M_r C_r}{UA} \tag{8-55}$$

where M_r and C_r are the mass and heat capacity of the reactants and U and A are the overall heat-transfer coefficient and area, respectively. The control system of Fig. 8-49 was tested on a pilot reactor where the heat-transfer area and mass could both be changed by a factor of 2, changing τ_1 by a factor of 4 as confirmed by observations of the rates of temperature rise. Yet neither controller required retuning as τ_1 varied. The primary controller should be PID and the secondary controller at least PI in this system (if the secondary controller has no integral mode, the primary will control with offset). Set-point overshoot in batch reactor control can be avoided by setting the derivative time of the primary controller higher than its integral time, but this is effective only with interacting PID controllers.

DRYING OPERATIONS

Controlling dryers is difficult, because online measurements of feed rate and composition and product composition are rarely available. Most dryers transfer moisture from wet feed into hot dry air in a single pass. The process is generally very self-regulating, in that moisture becomes progressively harder to remove from the product as it dries: this is known as *falling-rate drying*. Controlling the temperature of the air leaving a cocurrent dryer tends to regulate the moisture in the product, as long as the rate and moisture content of the feed and air are reasonably constant. However, at constant outlet air temperature, product moisture tends to rise with all three of these disturbance variables.

In the absence of moisture analyzers, regulation of product quality can be improved by raising the temperature of the exhaust air in proportion to the evaporative load. The evaporative load can be estimated by the loss in temperature of the air passing through the dryer in the steady state. Changes in load are first observed in upsets in exhaust temperature at a given inlet temperature; the controller then responds by returning the exhaust air to its original temperature by changing that of the inlet air.

Figure 8-50 illustrates the simplest application of this principle because of the linear relationship

$$T_0 = T_b + K\Delta T \tag{8-56}$$

where T_0 is the set point for exhaust temperature elevated above a base temperature T_b corresponds to zero-load operation and ΔT is the drop in air temperature from inlet to outlet. Coefficient K must be set to regulate product moisture over the expected range of evaporative load. If K is set too low, product moisture will increase with increasing load; if K is set too high, product moisture will decrease with increasing load. While K can be estimated from the model of a dryer, it does depend on the rate-of-drying curve for the product, its mean particle size, and whether the load variations are due primarily to changes in feed rate or feed moisture.

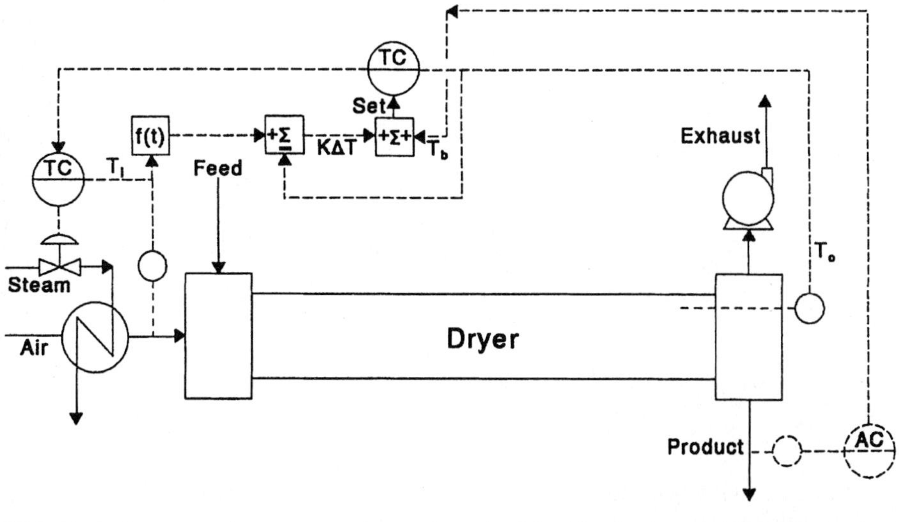

FIG. 8-50 Product moisture from a cocurrent dryer can be regulated through temperature control indexed to heat load.

It is important to have the most accurate measurement of exhaust temperature attainable. Note that Fig. 8-50 shows the sensor inserted into the dryer upstream of the rotating seal, because air infiltration there could cause the temperature in the exhaust duct to read low—even lower than the wet-bulb temperature, an impossibility without either substantial heat loss or outside-air infiltration.

The calculation of the exhaust temperature set point forms a positive feedback loop capable of destabilizing the dryer. For example, an increase in evaporative load causes the controller to raise the inlet temperature, which will in turn raise the calculated set point, calling for a further increase in inlet temperature. The gain in the set-point loop K is typically well below the gain of the exhaust temperature measurement responding to the same change in inlet temperature. Negative feedback then dominates in the steady state, but the response of the exhaust temperature measurement is delayed by the dryer. A compensating lag $f(t)$ is shown inserted in the set-point loop to prevent positive feedback from dominating in the short term, which could cause cycling. Lag time can be safely set equal to the integral time of the outlet-air temperature controller.

If product moisture is measured offline, analytical results can be used to adjust K and T_b manually. If an online analyzer is used, the analyzer controller would be most effective in adjusting the bias T_b, as is done in the figure.

While a rotary dryer is shown, commonly used for grains and minerals, this control system has been successfully applied to fluid-bed drying of plastic pellets, air-lift drying of wood fibers, and spray drying of milk solids. The air may be steam-heated as shown or heated by direct combustion of fuel, provided that a representative measurement of inlet air temperature can be made. If it cannot, then evaporative load can be inferred from a measurement of fuel flow, which then would replace ΔT in the set-point calculation.

If the feed flows countercurrent to the air, as is the case when drying granulated sugar, the exhaust temperature does not respond to variations in product moisture. For these dryers, the moisture in the product can better be regulated by controlling its temperature at the point of discharge. Conveyor-type dryers are usually divided into a number of zones, each separately heated with recirculation of air, which raises its wet-bulb temperature. Only the last two zones may require indexing of exhaust air temperature as a function of ΔT.

Batch drying, used on small lots such as pharmaceuticals, begins operation by blowing air at constant inlet temperature through saturated product in constant-rate drying, where ΔT is constant at its maximum value ΔT_c. When product moisture reaches the point where falling-rate drying begins, the exhaust temperature begins to rise. The desired product moisture will be reached at a corresponding exhaust temperature T_f, which is related to the temperature T_c observed during constant-rate drying, as well as to ΔT_c:

$$T_f = T_c + K\Delta T_c \tag{8-57}$$

The control system requires that the values of T_c and ΔT_c observed during the first minutes of operation be stored as the basis for the above calculation of end-point. When the exhaust temperature then reaches the calculated value of T_f, drying is terminated. Coefficient K can be estimated from models, but requires adjustment online to reach product specifications repeatedly. Products having different moisture specifications or particle size will require different settings of K, but the system does compensate for variations in feed moisture, batch size, air moisture, and inlet temperature. Some exhaust air may be recirculated to control the dew point of the inlet air, thereby conserving energy toward the end of the batch and when the ambient air is especially dry.

COMPRESSOR CONTROL

While it is usually more economical to condense a vapor stream and pump it than to vaporize a liquid stream and compress it, compressors are often used on recycle streams composed of light components (hydrogen, methane, etc.) or on streams with incondensable gases. Compressors that operate at a constant speed are often controlled using one of the strategies shown in Figure 8-51. Usually the "pinch" method will have lower operating costs than the "spillback" method. Variable speed control methods are shown in Fig. 8-52 for an electric motor drive and for a steam turbine drive. The operating costs of variable-speed compressors are generally lower than for fixed-speed systems, with perhaps a slightly higher initial capital cost.

When compressors are used on recycle streams, Douglas (*Conceptual Design of Chemical Processes*, McGraw-Hill, New York, 1988) recommends that they generally be operated "wide open," that is, at a constant maximum

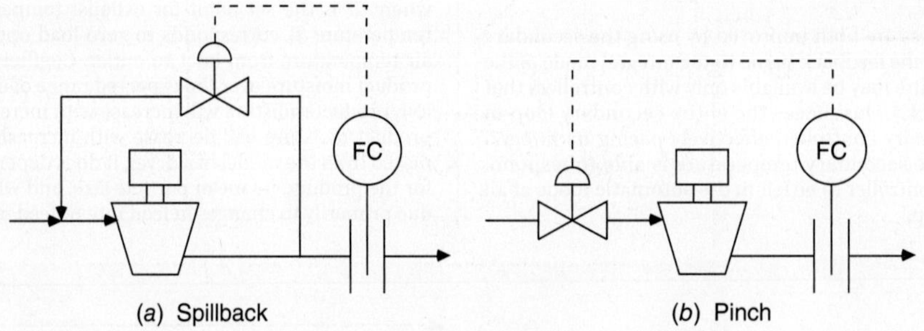

FIG. 8-51 Compressor control using a constant-speed compressor. (*Source:* Bequette, *Process Control: Modeling, Design and Simulation*, 2d ed., Prentice-Hall, Upper Saddle River, N.J., 2017.)

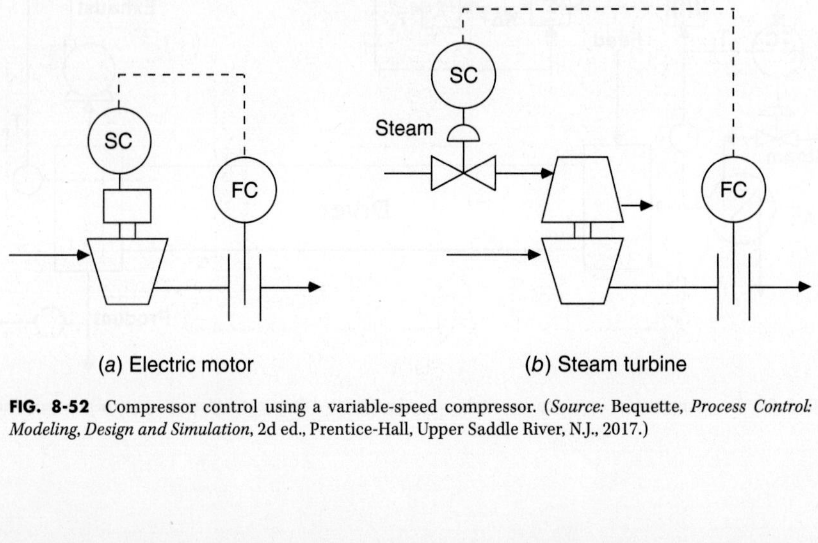

FIG. 8-52 Compressor control using a variable-speed compressor. (*Source:* Bequette, *Process Control: Modeling, Design and Simulation*, 2d ed., Prentice-Hall, Upper Saddle River, N.J., 2017.)

flow rate. This is so because the economic return from the increased overall yield usually outweighs the incremental operating cost of the compressor.

PLANTWIDE CONTROL

General References: Bequette, *Process Control: Modeling, Design and Simulation*, 2d ed., Prentice Hall, Upper Saddle River, N.J., 2017. Douglas, *Conceptual Design of Chemical Processes*, McGraw-Hill, New York, 1988. Seborg, Edgar, Mellichamp, and Doyle, III, *Process Dynamics and Control*, 4th ed., Wiley, New York, 2016. Luyben, Tyréus, and Luyben, *Plantwide Process Control*, McGraw-Hill, New York, 1999.

A process engineer must be concerned about the operation of an entire process plant or operating unit, and not just individual control loops or unit operations. The following questions must be addressed in development of a plantwide control strategy:

- What are the major operational objectives of the plant?
- What sensors should be paired with what manipulated inputs to form a control structure to achieve the major objectives?
- For any control valve or sensor failure, what steps will a process operator (and control system) need to take?

Much of this type of discussion occurs during the design or retrofit stage of a process. Here, process engineers develop and review process flow sheets and process and instrumentation diagrams. A detailed operational description of the proposed process and control strategy is developed, with consideration to all possible operating modes that can be selected by an operator, as well as to all equipment failure modes.

In this section we cannot do justice to the scope of work needed to develop a plantwide control strategy. We attempt to show some of the major issues in plantwide control by using an illustrative example.

HDA Process Example Toluene hydrodealkylation (HDA) is used for the production of benzene and is the basic case study process used in the textbook by Douglas (1988). A simplified process and instrumentation diagram for an HDA process is shown in Fig. 8-53. For clarity, a number of details are omitted, such as pumps on distillation reflux streams and furnace combustion air dampers.

The feed streams are high-purity (99.98 mol%) toluene and hydrogen (96 mol% hydrogen, 4 mol% methane). The objective is to produce a desired rate of benzene at a purity of 99.98 mol%. Because of coking considerations, a 5:1 hydrogen/aromatics ratio must be maintained at the reactor entrance. The reactor inlet pressure is to be maintained at just under 500 psig (the pressure of the hydrogen feed stream); the pressure drop between the feed stream mixing point and the flash drum is roughly 35 psi. The minimum reactor inlet temperature is 1150°F, while the maximum outlet temperature is 1300°F. The reactor exit stream must be immediately quenched to 1150°F to minimize secondary reactions.

The primary reaction is an irreversible reaction of toluene and hydrogen to produce benzene and methane, while the secondary reaction is the reversible reaction of benzene to form diphenyl and hydrogen:

$$C_7H_8 + H_2 \rightarrow C_6H_6 + CH_4$$

$$2C_6H_6 \leftrightarrow C_{12}H_{10} + H_2$$

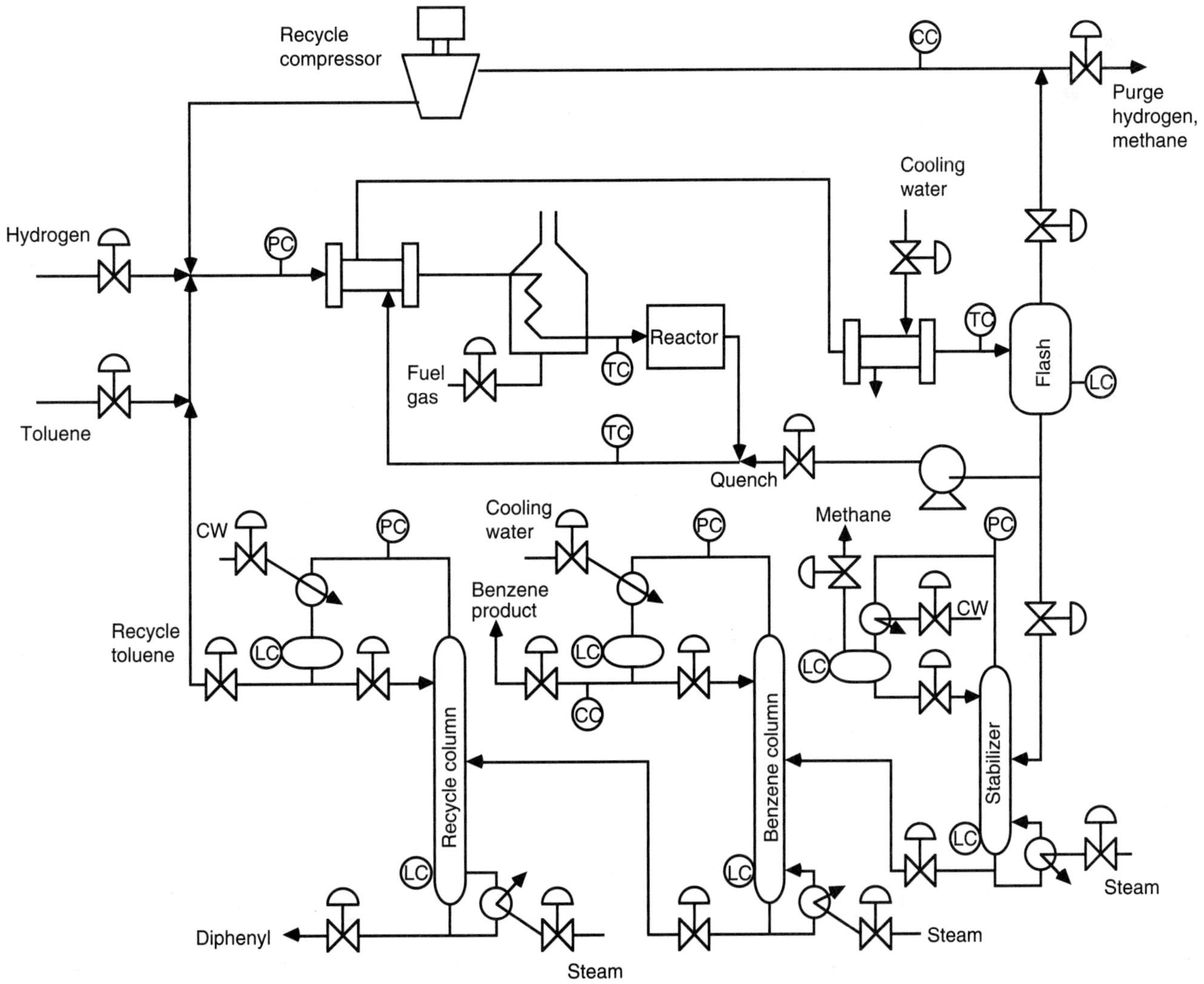

FIG. 8-53 HDA process flowsheet with actuators and measurements shown. (*Source:* Bequette, *Process Control: Modeling, Design and Simulation*, 2d ed., Prentice-Hall, Upper Saddle River, N.J., 2017.)

The selectivity decreases as the temperature increases; a purge stream is necessary to eliminate methane from the process system.

Use the following suggestions to guide control structure development and place control loops on Fig. 8-53.

- A possible phenomenon associated with processes that have recycle streams is known as the *snowball effect*, whereby a minor disturbance results in a drastic change in a stream flow rate. To avoid the snowball effect, Luyben et al. (1999) recommend placing a flow control loop on a stream in the recycle path. For this process, they suggest placing the recycle toluene stream on flow control.
- Douglas (1988) recommends operating recycle compressors "wide open" because the value of recovered components is usually higher than the additional compressor power cost.
- A tray temperature in each column can be used to "infer" a product composition. Dual composition control for the columns is not necessary, and this petrochemical plant has had good success by placing reflux streams under flow control.

There are 23 control degrees of freedom, since there are 23 control valves. It is not necessary that all these valves be used for feedback control; one or more valves may be held at a constant valve position.

A solution to the HDA problem was developed by Luyben et al. (1999) and shown in Fig. 8-54. The important control structure decisions to note are as follows:

- The total toluene flow (recycle + feed) is set to prevent the snowball effect (it is recommended that one loop in a recycle system be placed on flow control). Notice that this flow also sets the production rate of benzene.
- The makeup (feed) toluene flow is manipulated by the recycle column distillate receiver level controller. Intuition may suggest that there would be a significant lag between the feed flow, through the various unit operations, to the final distillation column. This is not the case, however, since the total toluene flow is regulated. Any change in toluene feed has an immediate effect on the recycle toluene flow and hence the distillate receiver level.
- The overhead valve from the flash drum is set wide open to allow maximum flow through the recycle compressor. The flash drum operating pressure is then dictated by pressure drop through the system, since the pressure is controlled at the reactor entrance.
- Since the diphenyl flow rate from the recycle column is low, the level is controlled by the reboiler heat duty (steam to the reboiler).
- The benzene purity is maintained using a cascade strategy, where the output of the composition controller is the set point to a tray temperature controller. This strategy yields a faster rejection of feed disturbances and provides satisfactory composition control when the composition sensor fails or needs recalibration.

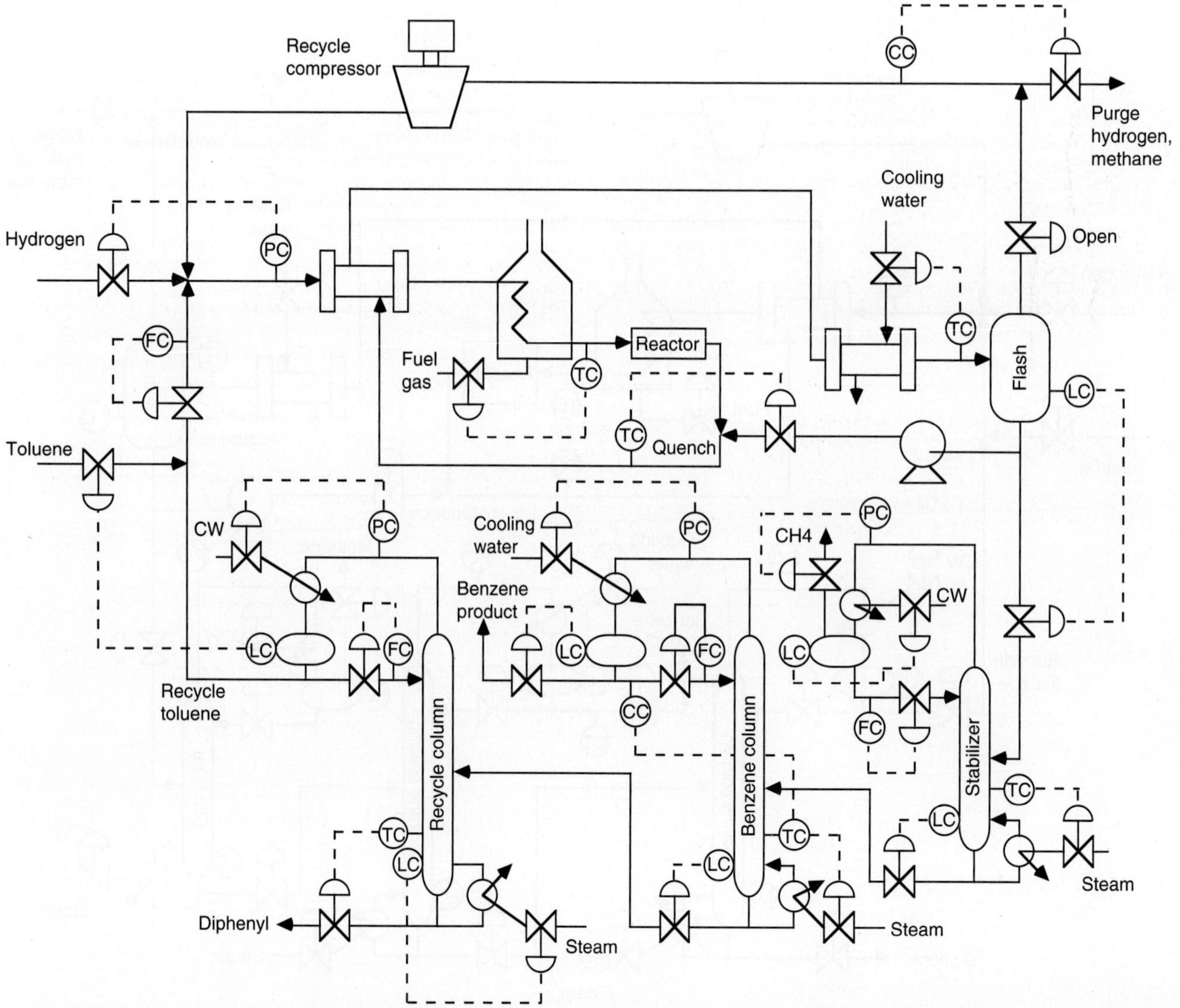

FIG. 8-54 HAD process flowsheet with control loops suggested by Luyben et al. (1999). (*Source:* Bequette, *Process Control: Modeling, Design and Simulation*, 2d ed., Prentice-Hall, Upper Saddle River, N.J., 2017.)

BATCH PROCESS CONTROL

General References: Rosenof and Ghosh, *Batch Process Automation*, Van Nostrand Reinhold, New York, 1987. Smith, *Control of Batch Processes*, Wiley, New York, 2014.

BATCH VERSUS CONTINUOUS PROCESSES

When one is categorizing process plants, the following two extremes can be identified:

1. *Commodity plants.* These plants are custom-designed to produce large amounts of a single product (or a primary product plus one or more secondary products). An example is a chlorine plant, where the primary product is chlorine and the secondary products are hydrogen and sodium hydroxide. Usually the margins (product value less manufacturing costs) for the products from commodity plants are small, so the plants must be designed and operated for best possible efficiencies. Although a few are batch, most commodity plants are continuous. Factors such as energy costs are life-and-death issues for such plants.

2. *Specialty plants.* These plants are capable of producing small amounts of a variety of products. Such plants are common in fine chemicals, pharmaceuticals, foods, and so on. In specialty plants, the margins are usually high, so factors such as energy costs are important but not life-and-death issues. As the production amounts are relatively small, it is not economically feasible to dedicate processing equipment to the manufacture of only one product. Instead, batch processing is utilized so that several products (perhaps hundreds) can be manufactured with the same process equipment. The key issue in such plants is to manufacture consistently each product in accordance with its specifications.

The above two categories represent the extremes in process configurations. The term *semibatch* designates plants in which some processing is continuous but other processing is batch. Even processes that are considered to be continuous can have a modest amount of batch processing. For example, the reformer unit within a refinery is thought of as a continuous process, but catalyst regeneration is possibly batch.

In a continuous process, the conditions within the process are largely the same from one day to the next. Variations in feed composition, plant utilities (e.g., cooling water temperature), catalyst activities, and other variables occur, but normally these changes either are about an average (e.g., feed compositions) or exhibit a gradual change over an extended period (e.g., catalyst activities). Summary data such as hourly averages, daily averages, and the like are meaningful in a continuous process.

In a batch process, the conditions within the process are continually changing. The technology for making a given product is contained in the product recipe that is specific to that product. Such recipes normally state the following:

1. *Raw material amounts.* This is the stuff needed to make the product.
2. *Processing instructions.* This is what must be done with the stuff to make the desired product.

This concept of a *recipe* is quite consistent with the recipes found in cookbooks. Sometimes the term *recipe* is used to designate only the raw material amounts and other parameters to be used in manufacturing a batch. Although appropriate for some batch processes, this concept is far too restrictive for others. For some products, the differences from one product to the next are largely physical as opposed to chemical. For such products, the processing instructions are especially important. The term *formula* is more appropriate for the raw material amounts and other parameters, with *recipe* designating the formula and the processing instructions. This concept of a recipe permits the following three different categories of batch processes to be identified:

1. *Cyclical batch.* Both the formula and the processing instructions are the same from batch to batch. Batch operations within processes that are primarily continuous often fall into this category. A batch catalyst regenerator within a reformer unit is cyclical batch.

2. *Multigrade.* The processing instructions are the same from batch to batch, but the formula can be changed to produce modest variations in the product. In a batch PVC plant, the different grades of PVC are manufactured by changing the formula. In a batch pulp digester, the processing of each batch or cook is the same, but at the start of each cook, the process operator is permitted to change the formula values for chemical-to-wood ratios, cook time, cook temperature, and so on.

3. *Flexible batch.* Both the formula and the processing instructions can change from batch to batch. Emulsion polymerization reactors are a good example of a flexible batch facility. The recipe for each product must detail both the raw materials required and how conditions within the reactor must be sequenced to make the desired product.

Of these, the flexible batch is by far the most difficult to automate and requires a far more sophisticated control system than either the cyclical batch or the multigrade batch facility.

Batches and Recipes Each batch of product is manufactured in accordance with a product recipe, which contains all information (formula and processing instructions) required to make a batch of the product (see Fig. 8-55). For each batch of product, there will be one and only one product recipe.

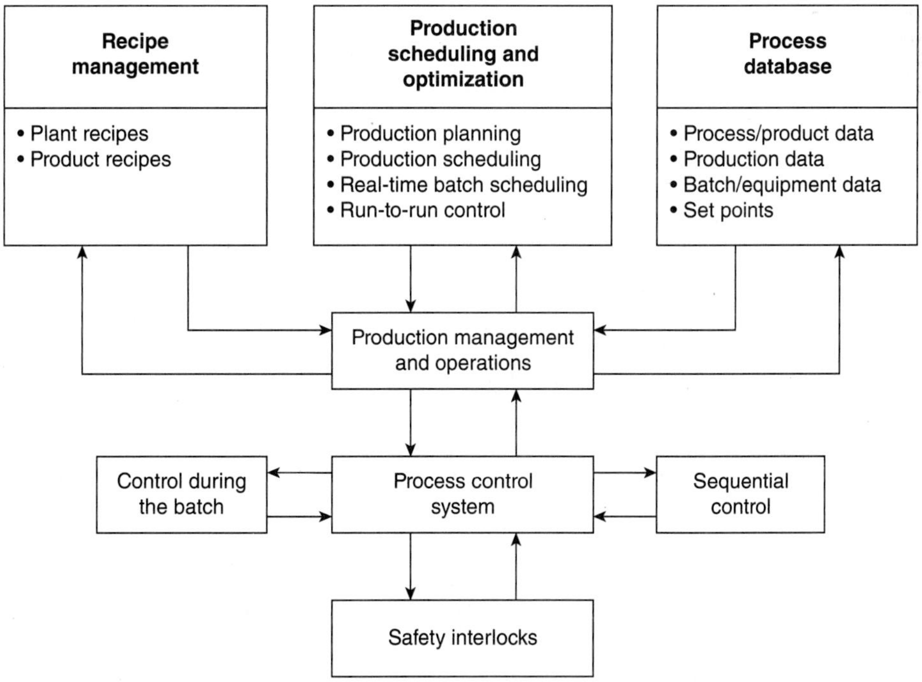

FIG. 8-55 Batch control system—a more detailed view. (*Source:* Seborg et al., *Process Dynamics and Control*, 3d ed., Wiley, New York, 2010.)

However, a given product recipe is normally used to make several batches of product. To uniquely identify a batch of product, each batch is assigned a unique identifier called the *batch ID*. Most companies adopt a convention for generating the batch ID, but this convention varies from one company to the next. In most batch facilities, more than one batch of product will be in some stage of production at any given time. The batches in progress may or may not be using the same recipe. The maximum number of batches that can be in progress at any given time is a function of the equipment configuration for the plant.

The existence of multiple batches in progress at a given time presents numerous opportunities for the process operator to make errors, such as charging a material to the wrong batch. Charging a material to the wrong batch is almost always detrimental to the batch to which the material is incorrectly charged. Unless this error is recognized quickly so that the proper change can be made, the error is also detrimental to the batch to which the charge was supposed to have been made.

Errors of this type arise from product issues, and are very common in batch facilities. Possibilities include the following:

1. Adding a chemical to a vessel at an inappropriate time. Steam is a common material that is usually innocuous, but if added to some reacting systems, leads to adverse consequences.
2. Adding a chemical other than the one specified by the product recipe. Charging materials from drums on a weigh scale is common in specialty batch plants. Is the drum on the scale the appropriate drum? Does the drum on the scale contain the chemical that it should contain? Plants institute procedures to guard against such errors, but are they foolproof and will they be followed?

Such errors usually lead to an off-specification batch, but the consequences could be more serious and could result in a hazardous condition.

A critical issue in batch reacting systems can be summarized as follows: Is the desired reaction proceeding at the desired rate? To obtain the desired reaction, the appropriate materials must be charged to the vessel. To obtain the desired rate, the reaction must initiate and proceed in the proper manner. However, reactions can stop abruptly. In emulsion reaction systems, the content within the reactor is an emulsion and one of the feeds is an emulation. Emulsions occasionally break, causing the reaction to stop. When this occurs, continuing to feed reactive materials leads to a hazardous situation.

Recipe management refers to the assumption of such duties by the control system. Each batch of product is tracked throughout its production, which may involve multiple processing operations on various pieces of processing equipment. Recipe management ensures that all actions specified in the product recipe are performed on each batch of product made in accordance with that recipe. As the batch proceeds from one piece of processing equipment to the next, recipe management is also responsible for ensuring that the proper type of process equipment is used and that this processing equipment is not currently in use by another batch.

By assuming such responsibilities, the control system greatly reduces the incidences where operator error results in off-specification batches. Such a reduction in error is essential to implement just-in-time production practices, where each batch of product is manufactured at the last possible moment. When a batch (or batches) is made today for shipment by overnight truck, there is insufficient time for producing another batch to make up for an off-specification batch.

Routing and Production Monitoring In some facilities, batches are individually scheduled. However, in most facilities, production is scheduled by product runs (also called *process orders*), where a run is the production of a stated quantity of a given product. From the stated quantity and the standard yield of each batch, the number of batches can be determined. A production run is normally a sequence of some number of batches of the same product.

In executing a production run, the following issues must be addressed (see Fig. 8-55):

1. *Processing equipment must be dedicated to making the run.* More than one run is normally in progress at a given time. The maximum number of runs simultaneously in progress depends on the equipment configuration of the plant. Routing involves determining which processing equipment will be used for each production run.
2. *Raw materials must be available.* When a production run is scheduled, the necessary raw materials must be allocated to the production run. As the individual batches proceed, the consumption of raw materials must be monitored for consistency with the allocation of raw materials to the production run.
3. *The production quantity for the run must be achieved by executing the appropriate number of batches.* The number of batches is determined from a standard yield for each batch. However, some batches may achieve yields higher than the standard yield, but other batches may achieve yields lower than the standard yield. The actual yields from each batch must be monitored, and significant deviations from the expected yields must be communicated to those responsible for scheduling production.

The last two activities are key components of production monitoring, although production monitoring may also involve other activities such as tracking equipment utilization.

Production Scheduling In this regard, it is important to distinguish between scheduling runs (sometimes called *long-term scheduling*) and assigning equipment to runs (sometimes called *routing* or *short-term scheduling*). As used here, *production scheduling* refers to scheduling runs and is usually a corporate-level as opposed to a plant-level function. Short-term scheduling or routing was previously discussed and is implemented at the plant level. The long-term scheduling is basically a material resources planning (MRP) activity involving the following:

1. *Forecasting.* Orders for long-delivery raw materials are issued at the corporate level based on the forecast for the demand for products. The current inventory of such raw materials is also maintained at the corporate level. This constitutes the resources from which products can be manufactured. Functions of this type are now incorporated into supply chain management.
2. *Orders for products.* Orders are normally received at the corporate level and then assigned to individual plants for production and shipment. Although the scheduling of some products is based on required product inventory levels, scheduling based on orders and shipping directly to the customer (usually referred to as just-in-time) avoids the costs associated with maintaining product inventories.
3. *Plant locations and capacities.* While producing a product at the nearest plant usually lowers transportation costs, plant capacity limitations sometimes dictate otherwise. Any company competing in the world economy needs the flexibility to accept orders on a worldwide basis and then assign them to individual plants to be filled. Such a function is logically implemented within the corporate-level information technology framework.

BATCH AUTOMATION FUNCTIONS

Automating a batch facility requires a spectrum of functions.

Interlocks Those provided for safety are properly called *safety interlocks* and are incorporated into the safety systems. Interlocks provided to avoid mistakes in processing the batch that do not lead to a process hazard are properly called *process interlocks*, and they are incorporated into the process controls. Terms such as *permissives* and *process actions* can be used so as to restrict the term *interlock* to a connection to safety (*interlock* is sometimes defined as a protective response initiated on the detection of a process hazard).

Discrete Device States Discrete devices such as two-position valves can be driven to either of two possible states. Such devices can be optionally outfitted with limit switches that indicate the state of the device. For two-position valves, the following combinations are possible:

1. No limit switches
2. One limit switch on the closed position
3. One limit switch on the open position
4. Two limit switches

In process control terminology, the discrete device driver is the software routine that generates the output to a discrete device such as a valve and also monitors the state feedback information to ascertain that the discrete device actually attains the desired state. Given the variety of discrete devices used in batch facilities, this logic must include a spectrum of capabilities. For example, valves do not instantly change states; instead each valve exhibits a travel time for the change from one state to another. To accommodate this characteristic of the field device, the processing logic within the discrete device driver must provide for a user-specified transition time for each field device. When equipped with limit switches, the potential states for a valve are as follows:

1. *Open.* The valve has been commanded to open, and the limit switch inputs are consistent with the open state.
2. *Closed.* The valve has been commanded to close, and the limit switch inputs are consistent with the closed state.
3. *Transition.* This is a temporary state that is possible only after the valve has been commanded to change state. The limit switch inputs are not consistent with the commanded state, but the transition time has not expired.
4. *Invalid.* The transition time has expired, and the limit switch inputs are not consistent with the commanded state for the valve.

The invalid state is an abnormal condition that is generally handled in a manner similar to process alarms. The transition state is not considered to be an abnormal state but may be implemented in either of the following ways:

1. *Drive and wait.* Further actions are delayed until the device attains its commanded state.
2. *Drive and proceed.* Further actions are initiated while the device is in the transition state.

The latter is generally necessary for devices with long travel times, such as flush-fitting reactor discharge valves that are motor-driven. Closing of such valves is normally done via drive and wait; however, drive and proceed is usually appropriate when opening the valve. Although two-state devices are most common, the need occasionally arises for devices with three or more states. For example, an agitator may be on high speed, on slow speed, or off.

Process States Batch processing usually involves imposing the proper sequence of states on the process. For example, a simple blending sequence might be as follows:

1. Transfer specified amount of material from tank A to tank R. The process state is "transfer from A."
2. Transfer specified amount of material from tank B to tank R. The process state is "transfer from B."
3. Agitate for specified time. The process state is "agitate without cooling."
4. Cool (with agitation) to specified target temperature. The process state is "agitate with cooling."

For each process state, the various discrete devices are expected to be in a specified device state. For process state "transfer from A," the device states might be as follows:

1. Tank A discharge valve: open
2. Tank R inlet valve: open
3. Tank A transfer pump: running
4. Tank R agitator: off
5. Tank R cooling valve: closed

For many batch processes, process state representations are a very convenient mechanism for representing the batch logic. A grid or table can be constructed, with the process states as rows and the discrete device states as columns (or vice versa). For each process state, the state of every discrete device is specified to be one of the following:

1. Device state 0, which may be valve closed, agitator off, and so on
2. Device state 1, which may be valve open, agitator on, and so on
3. No change or don't care

This representation is easily understandable by those knowledgeable about the process technology and is a convenient mechanism for conveying the process requirements to the control engineers responsible for implementing the batch logic.

Many batch software packages also recognize process states. A configuration tool is provided to define a process state. With such a mechanism, the batch logic does not need to drive individual devices but can simply command that the desired process state be achieved. The system software then drives the discrete devices to the device states required for the target process state. This normally includes the following:

1. Generating the necessary commands to drive each device to its proper state.
2. Monitoring the transition status of each device to determine when all devices have attained their proper states.
3. Continuing to monitor the state of each device to ensure that the devices remain in their proper states. Should any discrete device not remain in its target state, an appropriate response must be initiated.

Regulatory Control For most batch processes, the discrete logic requirements overshadow the continuous control requirements. Often the continuous control can be provided by simple loops for flow, pressure, level, and temperature. However, very sophisticated advanced control techniques are occasionally applied. As temperature control is especially critical in reactors, the simple feedback approach is replaced by model-based strategies that rival, if not exceed, the sophistication of advanced control loops in continuous plants.

In some installations, alternative approaches for regulatory control may be required. Where a variety of products are manufactured, the reactor may be equipped with alternative heat removal capabilities, including the following:

1. Jacket filled with cooling water. Jackets may be once-through or recirculating.
2. Heat exchanger in a pump-around loop.
3. Reflux condenser.

The heat removal capability to be used usually depends on the product being manufactured. Therefore, regulatory loops must be configured for each possible option, and sometimes for certain combinations of the possible options. These loops are enabled and disabled depending on the product being manufactured.

The interface between continuous controls and sequence logic (discussed shortly) is also important. For example, a feed might be metered into a reactor at a variable rate, depending on another feed or possibly on reactor temperature. However, the product recipe calls for a specified quantity of this feed. The flow must be totalized (i.e., integrated), and when the flow total attains a specified value, the feed must be terminated. The sequence logic must have access to operational parameters such as controller modes. That is, the sequence logic must be able to switch a controller to manual, automatic, or cascade. Furthermore, the sequence logic must be able to force the controller output to a specified value.

Sequence Logic Sequence logic must not be confused with discrete logic. Discrete logic is especially suitable for process interlocks; e.g., the reactor discharge valve must be closed for the feed valve to be opened. Sequence logic is used to force the process to attain the proper sequence of states. For example, a feed preparation might be to first charge A, then charge B, next mix, and finally cool. Although discrete logic can be used to implement sequence logic, other alternatives are often more attractive.

Sequence logic is often, but not necessarily, coupled with the concept of a process state. Basically, the sequence logic determines when the process should proceed from the current state to the next and sometimes what the next state should be.

Sequence logic must encompass both normal and abnormal process operations. Thus, sequence logic is often viewed as consisting of two distinct but related parts:

1. *Normal logic.* This sequence logic provides for the normal or expected progression from one process state to another.
2. *Failure logic.* This logic provides for responding to abnormal conditions, such as equipment failures.

Of these, the failure logic can easily be the most demanding. The simplest approach is to stop or hold on any abnormal condition and let the process operator sort things out. However, this is not always acceptable. Failures that lead to hazardous conditions require immediate action; waiting for the operator to decide what to do is not acceptable. The appropriate response to such situations is best determined in conjunction with the process hazards analysis.

No single approach has evolved as the preferred way to implement sequence logic. The approaches utilized include the following:

1. *Discrete logic.* Although sequence logic is different from discrete logic, sequence logic can be implemented using discrete logic capabilities. Simple sequences are commonly implemented as ladder diagrams in programmable logic controllers (PLCs). Sequence logic can also be implemented using the boolean logic functions provided by a distributed control system (DCS), although this approach is now infrequently pursued.
2. *Programming languages.* Traditional procedural languages do not provide the necessary constructs for implementing sequence logic. This necessitates one of the following:

   a. Special languages. The necessary extensions for sequence logic are provided by extending the syntax of the programming language. This is the most common approach within distributed control systems. The early implementations used BASIC as the starting point for the extensions; the later implementations used C as the starting point. A major problem with this approach is portability, especially from one manufacturer to the next but sometimes from one product version to the next within the same manufacturer's product line.

   b. Subroutine or function libraries. The facilities for sequence logic are provided via subroutines or functions that can be referenced from programs written in FORTRAN or C. This requires a general-purpose program development environment and excellent facilities to trap the inevitable errors in such programs.
3. *State machines.* This technology is commonly applied within the discrete manufacturing industries. However, its migration to process batch applications has been limited.
4. *Graphical implementations.* For sequence logic, the flowchart traditionally used to represent the logic of computer programs must be extended to provide parallel execution paths. Such extensions have been implemented in a graphical representation generally referred to as a *sequential function chart*, which is a derivative of an earlier technology known as *Grafcet*. As process engineers have demonstrated a strong dislike for ladder logic, most PLC manufacturers now provide sequential function charts either in addition to or as an alternative to ladder logic. Many DCS manufacturers also provide sequential function charts either in addition to or as an alternative to special sequence languages.

INDUSTRIAL APPLICATIONS

An industrial example requiring simple sequence logic is the effluent tank with two sump pumps illustrated in Fig. 8-56. There are two sump

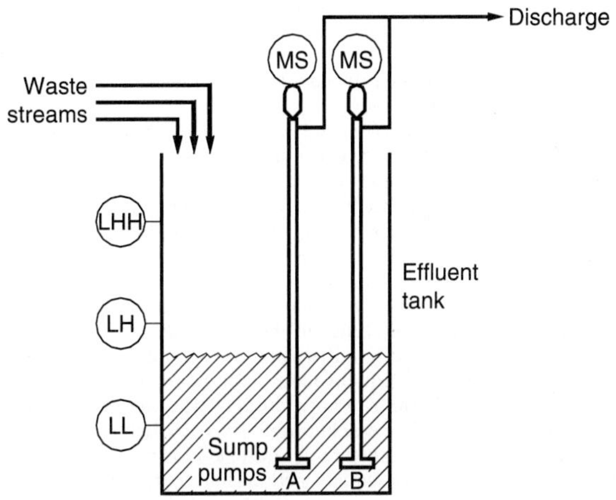

FIG. 8-56 Effluent tank process.

pumps, A and B. The tank is equipped with three level switches, one for low level (LL), one for high level (LH), and one for high-high level (LHH). All level switches actuate on rising level. The logic is to be as follows:

1. When level switch LH actuates, start one sump pump. This must alternate between the sump pumps. If pump A is started on this occasion, then pump B must be started on the next occasion.
2. When level switch LHH actuates, start the other sump pump.
3. When level switch LL deactuates, stop all sump pumps.

Once a sump pump is started, it is not stopped until level switch LL deactuates. With this logic, one, both, or no sump pump may be running when the level is between LL and LH. Either one or both sump pumps may be running when the level is between LH and LHH.

Figure 8-57*a* presents the ladder logic implementation of the sequence logic. Ladder diagrams were originally developed for representing hardwired logic, but are now widely used in PLCs. The vertical bar on the left provides the source of power; the vertical bar on the right is ground. If a coil is connected between the power source and ground, the coil will be energized. If a circuit consisting of a set of contacts is inserted between the power source and the coil, the coil will be energized only if power can flow through the circuit. This will depend on the configuration of the circuit and the states of the contacts within the circuit. Ladder diagrams are constructed as rungs, with each rung consisting of a circuit of contacts and an output coil.

Contacts are represented as vertical bars. A vertical bar represents a normally open contact; power flows through this contact only if the device with which the contact is associated is actuated (energized). Vertical bars separated by a slash represent a normally closed contact; power flows through this contact only if the device with which the contact is associated is not actuated. The level switches actuate on rising level. If the vessel level is below the location of the switch, the normally open contact is open and the normally closed contact is closed. If the level is above the location of the switch, the normally closed contact is closed and the normally open contact is open.

The first rung in Fig. 8-57*a* is for pump A. It will run if one (or more) of the following conditions is true:

1. Level is above LH and pump A is the lead pump. A coil (designated as LeadIsB) will be subsequently provided to designate the pump to be started next (called the *lead* pump). If this coil is energized, pump B is the lead pump. Hence, pump A is to be started at LH if this coil is not energized, hence the use of the normally closed contact on coil LeadIsB in the rung of ladder logic for pump A.
2. Level is above LHH.
3. Pump A is running and the level is above LL.

The second rung is an almost identical circuit for pump B. The difference is the use of the normally open contact on the coil lead is B.

When implemented as hardwired logic, ladder diagrams are truly parallel logic; i.e., all circuits are active at all instants of time. But when ladder diagrams are implemented in PLCs, the behavior is slightly different. The ladder logic is scanned very rapidly (on the order of 100 times per second), which gives the appearance of parallel logic. But within a scan of ladder logic, the rungs are executed sequentially. This permits constructs within ladder logic for PLCs that make no sense in hardwired circuits.

One such construct is for a "one-shot." Some PLCs provide this as a built-in function, but here it will be presented in terms of separate components. The one-shot is generated by the third rung of ladder logic in Fig. 8-57*a*. But first examine the fourth rung. The input LL drives the output coil LL1. This coil provides the state of level switch LL on the previous scan of ladder logic. This is used in the third rung to produce the one-shot. Output coil OneShot is energized if

1. LL is not actuated on this scan of ladder logic (note the use of the normally closed contact for LL)
2. LL was actuated on the previous scan of ladder logic (note the use of the normally open contact for LL1)

When LL deactuates, coil OneShot is energized for one scan of ladder logic. OneShot does not energize when LL actuates (a slight modification of the circuit would give a one-shot when LL actuates).

The one-shot is used in the fifth rung of ladder logic to toggle the lead pump. The output coil LeadIsB is energized provided that

1. LeadIsB is energized and OneShot is not energized. Once LeadIsB is energized, it remains energized until the next "firing" of the one-shot.
2. LeadIsB is not energized and OneShot is energized. This causes coil LeadIsB to change states each time the one-shot fires.

Ladder diagrams are ideally suited for representing discrete logic, such as required for interlocks. Sequence logic can be implemented via ladder logic, but usually with some tricks or gimmicks (the one-shot in Fig. 8-57*a* is such a gimmick). These are well known to those "skilled in the art" of PLC programming. But to others, they can be quite confusing.

Figure 8-57*b* provides a sequential function chart for the pumps. Sequential function charts consist of steps and transitions. A step consists of actions to be performed, as represented by statements. A transition consists of a logical expression. As long as the logical expression is false, the sequence logic remains at the transition. When the logical expression is true, the sequence logic proceeds to the step following the transition.

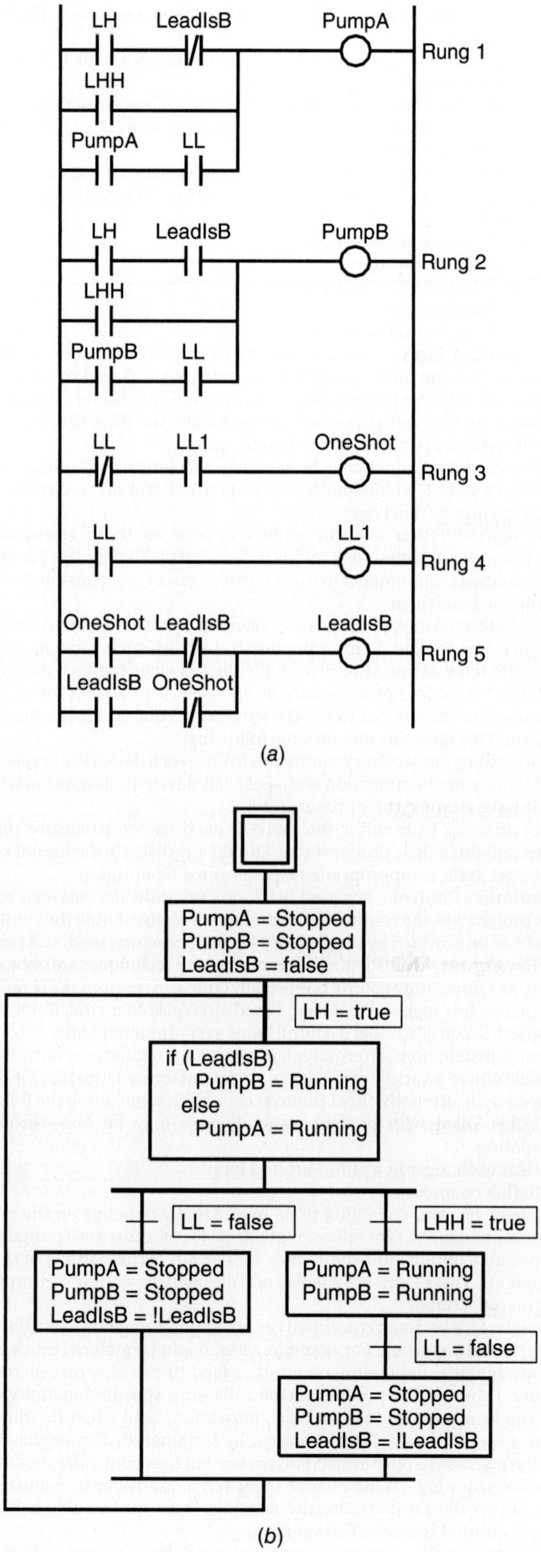

FIG. 8-57 (*a*) Ladder logic. (*b*) Sequence logic for effluent tank sump pumps.

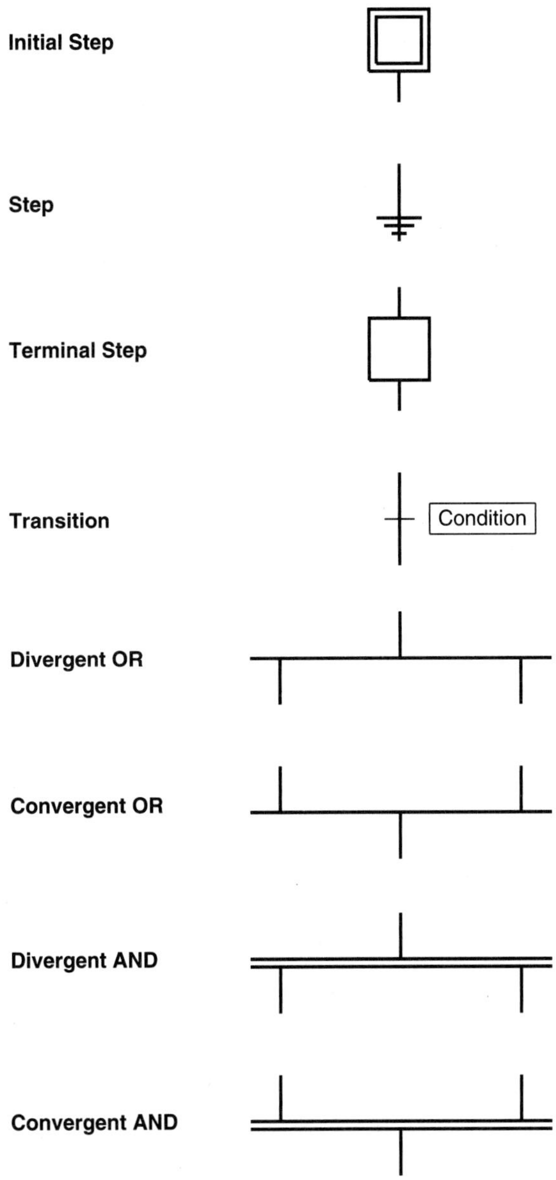

FIG. 8-58 Elements of sequential function charts.

The basic constructs of sequential function charts are presented in Fig. 8-58. The basic construct of a sequential function chart is the step-transition-step. But also note the constructs for OR and AND. At the divergent OR, the logic proceeds on only one of the possible paths, specifically, the one whose transition is the first to attain the true condition. At the divergent AND, the logic proceeds on all paths simultaneously, and all must complete to proceed beyond the convergent AND. This enables sequential function charts to provide parallel logic.

In the sequential function chart in Fig. 8-57*b* for the pumps, the logic is initiated with both pumps stopped and pump A as the lead pump. When LH actuates, the lead pump is started. A divergent OR is used to create two paths:

1. If LL deactuates, both pumps are stopped and the lead pump is swapped.
2. If LHH actuates, both pumps are started (one is already running). Both remain running until LL deactuates, at which time both are stopped. The logic then loops to the transition for LH actuating.

Although not illustrated here, programming languages (either custom sequence languages or traditional languages extended by libraries of real-time functions) are a viable alternative for implementing the logic for the pumps. Graphical constructs such as ladder logic and sequential function charts are appealing to those uncomfortable with traditional programming languages. But in reality, these are programming methodologies.

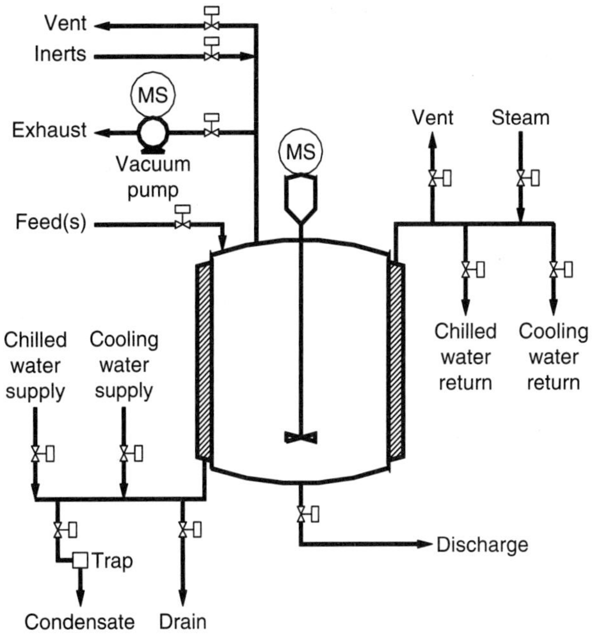

FIG. 8-59 Chemical reactor schematic.

BATCH REACTOR CONTROL

The reactors in flexible batch chemical plants usually present challenges. Many reactors have multiple mechanisms for heating and/or cooling. The reactor in Fig. 8-59 has three mechanisms:

1. Heat with steam.
2. Cool with cooling tower water.
3. Cool with chilled water.

Sometimes glycol is an option; occasionally liquid nitrogen is used to achieve even lower temperatures. Some jacket configurations require sequences for switching between the various modes for heating and cooling (the jacket has to be drained or depressurized before another medium can be admitted).

The reactor in Fig. 8-59 has three mechanisms for pressure control:

1. Vacuum
2. Atmospheric (using the vent and inert valves)
3. Pressure

Some reactors are equipped with multiple vacuum pumps, with different operating modes for moderate vacuum versus high vacuum. Sequence logic is usually required to start vacuum pumps and establish vacuum.

With three options for heating/cooling and three options for pressure, the reactor in Fig. 8-59 has nine combinations of operating modes. In practice, this is a low number. This number increases with features such as

1. Recirculation or pump-arounds containing a heater and/or cooler
2. Reflux condensers that can be operated at total reflux (providing only cooling) or such that some component is removed from the reacting system

These further increase the number of possible combinations. Some combinations may not make sense, may not be used in current production operations, or otherwise can be eliminated. However, the net number of combinations that must be supported tends to be large.

The order in which the systems are activated usually depends on the product being manufactured. Sometimes heating/cooling and pressure control are established simultaneously; sometimes heating/cooling is established first and then pressure control, and sometimes pressure control is established first and then heating/cooling. One has to be very careful when imposing restrictions. Suppose no current products establish pressure control first and then establish heating/cooling. But what about the next product to be introduced? After all, this is a flexible batch facility.

Such challenging applications for recipe management and sequence logic require a detailed analysis of the production equipment and the operations conducted within that production equipment. While applications such as the sump pumps in the effluent tank can be pursued without it, a structured approach is essential in flexible batch facilities.

BATCH PRODUCTION FACILITIES

Especially for flexible batch applications, the batch logic must be properly structured in order to be implemented and maintained in a reasonable manner. An underlying requirement is that the batch process equipment be properly structured. The following structure is appropriate for most batch production facilities.

Plant A plant is the collection of production facilities at a geographic site. The production facilities at a site normally share warehousing, utilities, and the like.

Equipment Suite An equipment suite is the collection of equipment available for producing a group of products. Normally, this group of products is similar in certain respects. For example, they might all be manufactured from the same major raw materials. Within the equipment suite, material transfer and metering capabilities are available for these raw materials. The equipment suite contains all the necessary types of processing equipment (reactors, separators, and so on) required to convert the raw materials to salable products. A plant may consist of only one suite of equipment, but large plants usually contain multiple equipment suites.

Process Unit or Batch Unit A process unit is a collection of processing equipment that can, at least at certain times, be operated in a manner completely independent of the remainder of the plant. A process unit normally provides a specific function in the production of a batch of product. For example, a process unit might be a reactor complete with all associated equipment (jacket, recirculation pump, reflux condenser, and so on). However, each feed preparation tank is usually a separate process unit. With this separation, preparation of the feed for the next batch can be started as soon as the feed tank is emptied for the current batch.

All but the very simplest equipment suites contain multiple process units. The minimum number of process units is one for each type of processing equipment required to make a batch of product. However, many equipment suites contain multiple process units of each type. In such equipment suites, multiple batches and multiple production runs can be in progress at a given time.

Item of Equipment An item of equipment is a hardware item that performs a specific purpose. Examples are pumps, heat exchangers, agitators, and the like. A process unit could consist of a single item of equipment, but most process units consist of several items of equipment that must be operated in harmony to achieve the function expected of the process unit.

Device A device is the smallest element of interest to batch logic. Examples of devices include measurement devices and final control elements.

STRUCTURED BATCH LOGIC

Flexible batch applications must be pursued by using a structured approach to batch logic. In such applications, the same processing equipment is used to make a variety of products. In most facilities, little or no proprietary technology is associated with the equipment itself; the proprietary technology is how this equipment is used to produce each of the products.

The primary objective of the structured approach is to separate cleanly the following two aspects of the batch logic:

Product Technology This encompasses the technology such as how to mix certain molecules to make other molecules. This technology ultimately determines the chemical and physical properties of the final product. The product recipe is the principal source for the product technology.

Process Technology The process equipment permits certain processing operations (e.g., heat to a specified temperature) to be undertaken. Each processing operation will involve certain actions (e.g., opening appropriate valves).

The need to keep these two aspects separated is best illustrated by a situation where the same product is to be made at different plants. While it is possible that the processing equipment at the two plants is identical, this is rarely the case. Suppose one plant uses steam for heating its vessels, but the other uses a hot oil system as the source of heat. When a product recipe requires that material be heated to a specified temperature, each plant can accomplish this objective, but each will go about it in quite different ways. The ideal case for a product recipe is as follows:

1. It contains all the product technology required to make a product.
2. It contains no equipment-dependent information, i.e., no process technology.

In the previous example, such a recipe would simply state that the product must be heated to a specified temperature. Whether heating is undertaken with steam or hot oil is irrelevant to the product technology. By restricting the product recipe to a given product technology, the same product recipe can be used to make products at different sites. At a given site, the specific approach to be used to heat a vessel is important. The traditional approach is for an engineer at each site to expand the product recipe into a document that explains in detail how the product is to be made at the specific site. This document goes by various names, although *standard operating procedure* or *SOP* is a common one. Depending on the level of detail to which it is written, the SOP could specify exactly which valves must be opened to heat the contents of a vessel. Thus, the SOP is site-dependent and contains both product technology and process technology.

In structuring the logic for a flexible batch application, the following organization permits product technology to be cleanly separated from process technology:

- A recipe consists of a formula and one or more processing operations. Ideally, only product technology is contained in a recipe.
- A processing operation consists of one or more phases. Ideally, only product technology is contained in a processing operation.
- A phase consists of one or more actions. Ideally, only process technology is contained in a phase.

In this structure, the recipe and processing operations would be the same at each site that manufactures the product. However, the logic that comprises each phase would be specific to a given site. In the earlier heating example, each site would require a phase to heat the contents of the vessel. However, the logic within the phase at one site would accomplish the heating by opening the appropriate steam valves, while the logic at the other site would accomplish the heating by opening the appropriate hot oil valves.

Usually the critical part of structuring batch logic is the definition of the phases. There are two ways to approach this:

1. Examine the recipes for the current products for commonality, and structure the phases to reflect this commonality.
2. Examine the processing equipment to determine what processing capabilities are possible, and write phases to accomplish each possible processing capability.

There is the additional philosophical issue of whether to have a large number of simple phases with few options each, or a small number of complex phases with numerous options. The issues are analogous to structuring a complex computer program into subprograms. Each possible alternative has advantages and disadvantages.

As the phase contains no product technology, the implementation of a phase must be undertaken by those familiar with the process equipment. Furthermore, they should undertake this on the basis that the result will be used to make a variety of products, not just those that are initially contemplated. The development of the phase logic must also encompass all equipment-related safety issues. The phase should accomplish a clearly defined objective, so the implementers should be able to thoroughly consider all relevant issues in accomplishing this objective. The phase logic is defined in detail, implemented in the control system, and then thoroughly tested. Except when the processing equipment is modified, future modifications to the phase should be infrequent. The result should be a very dependable module that can serve as a building block for batch logic.

Even for flexible batch applications, a comprehensive menu of phases should permit most new products to be implemented by using currently existing phases. By reusing existing phases, numerous advantages accrue:

1. The engineering effort to introduce a new recipe at a site is reduced.
2. The product is more likely to be on-spec the first time, thus avoiding the need to dispose of off-spec product.
3. The new product can be supplied to customers sooner, hopefully before competitors can supply the product.

There is also a distinct advantage in maintenance. When a problem with a phase is discovered and the phase logic is corrected, the correction is effectively implemented in all recipes that use the phase. If a change is implemented in the processing equipment, the affected phases must be modified accordingly and then thoroughly tested. These modifications are also effectively implemented in all recipes that use these phases.

BIOPROCESS CONTROL

General References: Alford, "Bioprocess Control: Advances and Challenges," *Computers and Chemical Engineering* **30**(10-12): 1464–1475 (2006). Buckbee and Alford, *Automation Applications in Bio-Pharmaceuticals*, ISA, Research Triangle Park, N.C., 2008. Dochain (ed.), *Automatic Control of Bioprocesses*, Wiley, New York, 2008. Doyle, Srinivasan, and Bonvin, "Run-to-Run Control Strategy for Diabetes Management," *Proceedings IEEE Engineering in Medicine and Biology Conference*, Istanbul, Turkey, 2001. Edgar, Himmelblau, and Lasdon, *Optimization of Chemical Processes*, 2d ed.,

McGraw-Hill, New York, 2001. Monod, *Recherches sur la Croissance des Cultures Bacteriennes*, Hermann, Paris, 1942. Rawlings and Mayne, *Model Predictive Control Theory and Design*, Nob Hill, Madison, Wisc., 2009. Riggs and Karim, *Chemical and Bio-Process Control*, 3d ed., Ferret Publishing, Austin, Tex., 2007.

CHALLENGES OF BIOPROCESS CONTROL

Bioprocesses form a subcategory of processes where the reaction and/or separation is taking place via biological microorganisms such as bacteria, yeasts, or fungi. What distinguishes bioprocesses from other processes from a control point of view is that most of the processes are batch processes and that it is challenging to obtain measurements for many of the process variables of interest. As such, batch process control and model predictive control are often used to control the trajectory of the controlled variables during a batch, and run-to-run control plays a role to counter long-term drifts from one batch to the next. Also, inferential measurements, via observers, Kalman filters, or correlations, play a key role for bioprocess control as important process variables need to be predicted rather than directly measured.

BIOPROCESSES

Bioprocesses usually require one or more reaction steps in addition to separation processes, similar to what is done for traditional chemical processes. The reaction steps often require microorganisms, and in this case the process is referred to as *fermentation*. Fermentation can be performed as a continuous or a batch process, affecting the dynamics and control of the operation. Separations can include processes found in traditional chemical processes that can handle solids; these unit operations are often found in bioprocesses but can also include operations such as crystallization.

Fermentation Fermentation is a unit operation involving biochemical reactions in the presence of microorganisms. The microorganisms convert one or more of the reactants to desired products. Commonly used microorganisms are yeasts, bacteria, or fungi. Controlling operating conditions during fermentation is a key challenge as the microorganisms have a small range of operating conditions where they perform optimally. Furthermore, deviations from the desired operating conditions can lead to die-off of the microorganisms, which stops the fermentation process. Typical controlled variables for fermenters are temperature, pH, nutrient concentration, and dissolved oxygen. Manipulated variables often include sterile airflow, nutrient feed, pressure, cooling water flow, and agitation rate. Operating modes for bioreactors are discontinuous (or batch) mode, semicontinuous (or fed-batch), continuous (or chemostat), or sequencing batch reactors, which is a combination of the aforementioned. As a number of fermentation processes operate in some form of a batch mode, batch control is an important consideration. Batch control can take different forms, such as optimization of the set-point trajectories during a batch or batch-to-batch control from one batch to the next.

One aspect related to modeling and control that sets fermenters apart from regular chemical reactors is that the reaction rates describing the process are in the form of Monod kinetics [Gernaey, Jeppsson, Vanrolleghem, and Copp (eds.), *Benchmarking of Control Strategies for Wastewater Treatment Plants*, IWA Publishing, London, 2014] or modifications thereof, due to the presence of microorganisms.

Separations Final product concentration requirements are often greater than 99 percent purity. However, products produced via fermentation are characterized by relatively low yields, often in the 1 to 15 percent range. As such, fermentation is almost always followed by some form of separation process. These separation processes must take into account product size, density, vapor pressure, or solubility. Common separation processes for bioproducts are distillation, solvent extraction, stripping, reverse osmosis, ultrafiltration, centrifugation, flotation, or evaporation. The choice of separation technology used is dependent upon the feedstock and product characteristics.

CHARACTERISTICS OF BIOPROCESS CONTROL

One key characteristic of bioprocesses is that they are mostly batch processes, which require control of time-varying trajectories of the controlled variables during a batch as well as adjustments from one batch to the next via run-to-run control. Furthermore, bioprocesses are often characterized by a lack of measurement capabilities for controlled variables, which requires the use of inferential measurements.

Batch Control As discussed in an earlier section on batch process control, the product is made in discrete batches by sequentially performing a number of processing steps in a defined order on the raw materials and intermediate products. For example, the first step in fermentation is usually an inoculation step where a small number of culture cells are transferred to the fermenter, which is followed by an exponential growth phase until either a component becomes limiting (oxygen, carbon source) or inhibiting (substrate, product, by-products, etc.). Furthermore, many bioprocesses also include a step in which the culture is switched from a "growth" mode to a "product synthesis" mode, often triggered by a programmed shift in temperature or by a chemical addition. Last, batch fermentation involves draining the vessel, product separation, drying, and packaging. Large production runs are achieved by repeating the process. The term *recipe* has a range of definitions in batch processing, but in general a recipe is a procedure with the set of data, operations, and control steps to manufacture a particular grade of product. A formula is the list of recipe parameters, which includes the raw materials, processing parameters, and product outputs. A recipe procedure has operations for both normal and abnormal conditions. Each operation contains resource requests for certain ingredients (and their amounts). The operations in the recipe can adjust set points and turn equipment on and off. The complete production run for a specific recipe is called a *campaign* (multiple batches). A production run consists of a specified number of batches using the same raw materials and making the same product to satisfy customer demand. The accumulated batches are called a *lot* [Smith, *Control of Batch Processes*, Wiley, New York, 2014; Rosenof and Ghosh, *Batch Process Automation: Theory and Practice*, Van Nostrand Reinhold, New York, 1987].

In multigrade batch processing, the instructions remain the same from batch to batch, but the formula can be changed to yield modest variations in the product. In flexible batch processing, both the formula (recipe parameters) and the processing instructions can change from batch to batch. The recipe for each product must specify both the raw materials required and how conditions within the reactor are to be sequenced in order to make the desired product.

Many batch plants, especially those used to manufacture pharmaceuticals, are certified by the International Standards Organization (ISO). ISO 9000 (and related standards 9001–9004) states that every manufactured product should have an established, documented procedure, and the manufacturer should be able to document that the procedure was followed. Companies must pass periodic audits to maintain ISO 9000 status. Both ISO 9000 and the U.S. Food and Drug Administration (FDA) require that only a certified recipe be used. Thus, if the operation of a batch becomes "abnormal," then performing any unusual corrective action to bring it back within the normal limits is not an option. Also, if a slight change in the recipe apparently produces superior batches, the improvement cannot be implemented unless the entire recipe is recertified. The FDA typically requires product and raw material tracking, so that product abnormalities can be traced back to their sources.

Batch Process Control Hierarchy Functional control activities for batch process control are summarized below in four categories: batch sequencing and logic control, control during the batch, run-to-run control, and batch production management [Bonvin, "Optimal Operation of Batch Reactors—A Personal View," *Journal of Process Control* **8:** 355 (1998); Pekny and Reklaitis, "Towards the Convergence of Theory and Practice: A Technology Guide for Scheduling/Planning Methodology," Foundations of Computer-Aided Process Operations (FOCAPO) Conference Proceedings, *AIChE Symposium Series* **94:** 91 (1998); Seborg, Edgar, Mellichamp, and Doyle, *Process Dynamics and Control*, 4th ed., Wiley, New York, 2016].

In batch sequencing and logic control, sequencing of control steps that follow the recipe involves, for example, mixing of ingredients, heating, waiting for a reaction to complete, cooling, or discharging the resulting product. Transfer of materials to and from batch tanks or reactors includes metering of materials as they are charged (as specified by each recipe) as well as transfer of materials at the completion of the process operation. In addition to discrete logic for the control steps, logic is needed for safety interlocks to protect personnel, equipment, and the environment from unsafe conditions. Process interlocks ensure that process operations can only occur in the correct time sequence.

Feedback control of flow rate, temperature, pressure, composition, and level, including advanced control strategies, falls under control during the batch, which is also called "within-the-batch" control [Bonvin, "Optimal Operation of Batch Reactors—A Personal View," *Journal of Process Control* **8:** 355 (1998); Pekny and Reklaitis, "Towards the Convergence of Theory and Practice: A Technology Guide for Scheduling/Planning Methodology," Foundations of Computer-Aided Process Operations (FOCAPO) Conference Proceedings, *AIChE Symposium Series* **94:** 91 (1998)]. In sophisticated applications, this requires defining an operating trajectory for the batch (i.e., temperature or flow rate as a function of time). In simpler cases, it involves tracking of set points of the controlled variables, which includes ramping the controlled variables up and down and/or holding them for a prescribed time. Detection of when the batch operations should be terminated (endpoint) may be performed by inferential measurements of product quality, if direct measurement is not feasible.

Run-to-run control (also called *batch-to-batch*) is a supervisory function based on offline product quality measurements at the end of a run. Operating conditions and profiles for the batch are adjusted between runs to improve the product quality by using tools such as optimization. Batch production management entails advising the plant operator of process status and how to interact with the recipes and the sequential, regulatory, and discrete controls. Complete information (recipes) is maintained for manufacturing each product grade, including the names and amounts of ingredients, process variable set points, ramp rates, processing times, and sampling procedures. Other database information includes batches produced on a shift, daily, or weekly basis as well as material and energy balances. Scheduling of process units is based on availability of raw materials and equipment and customer demand.

The ability to handle recipe changes after a recipe has started is a challenging aspect of batch control systems. Often it is desirable to change the grade of the batch to meet product demand, or to change the resources used by the batch after the batch has started. Because not every batch of product is always good, there needs to be special-purpose control recipes to fix, rework, blend, or dispose of bad batches, if that is allowable. It is important to be able to respond to unusual situations by creating special-purpose recipes and still meet the demand. This procedure is referred to as *reactive scheduling*. When ample storage capacity is available, the normal practice has been to build up large inventories of raw materials and ignore the inventory carrying cost. However, improved scheduling can be employed to minimize inventory costs, which implies that supply chain management techniques may be necessary to implement the schedule [Pekny and Reklaitis, "Towards the Convergence of Theory and Practice: A Technology Guide for Scheduling/Planning Methodology," Foundations of Computer-Aided Process Operations (FOCAPO) Conference Proceedings, *AIChE Symposium Series* **94:** 91 (1998)]. More details on batch control can be found in the earlier subsection Batch Process Control.

Run-to-Run Control Run-to-run control, also referred to as batch-to-batch control, is a technique that seeks to address slow-moving drifts from one batch to the next. As batch processes are very common for biological production processes, run-to-run control plays an important role for process automation of these systems [Teixeira, Clemente, Cunha, Carrondo, and Oliveira, "Bioprocess Iterative Batch-to-Batch Optimization Based on Hybrid Parametric/Nonparametric Models," *Biotechnology Progress* **22:** 247–258 (2006)].

Although there are many different types of run-to-run controllers, most controllers utilize a similar structure which includes a process model, an observer, and a control law. The process model is used to relate the measurable inputs and states to the desired product qualities. Next, an observer is used to estimate the model parameters while filtering process noise and unmodeled dynamics. Finally, the control law specifies how the input recipe should be modified to keep the process on target. This is usually a plant inversion or a deadbeat control law [Bonvin, Srinivasan, and Ruppen, *Dynamic Optimization in the Batch Chemical Industry*, Proceedings CPC-VI, Tucson, Ariz., 2001; Edgar, Butler, Campbell, Pfeiffer, Bode, Hwang, Balakrishnan, and Hahn, "Automatic Control in Microelectronics Manufacturing: Practices, Challenges, and Possibilities," *Automatica* **36**(11): 1567–1603 (2000); Alford, "Bioprocess Control: Advances and Challenges," *Computers and Chemical Engineering* **30**(10-12): 1464–1475 (2006)].

Two commonly used run-to-run algorithms are the *exponentially weighted moving average* (EWMA) controller and a variation of the *optimizing adaptive quality controller* (OAQC) [Del Castillo and Yey, "An Adaptive Run-to-Run Optimizing Controller for Linear and Nonlinear Semiconductor Processes," *IEEE Transactions on Semiconductor Manufacturing* **11**(2): 285–295 (1996)]. In addition to these two run-to-run controllers, other run-to-run control algorithms include the predictive corrector controller, knowledge-based interactive controller, model predictive run-to-run controller, time-based exponentially weighted moving average controller, and the generic cell controller. For further details on the implementation of these algorithms, see Åström and Wittenmark, *Adaptive Control*, 2d ed., Addison Wesley, Reading, Mass., 1994.

The *single-input single-output* (SISO) EWMA run-to-run controller assumes a linear model with constant gain and a bias that only drifts slowly over time. As the bias can change with time, it is estimated by weighting the most recent observations more heavily than older observations.

In contrast to the EWMA run-to-run controller, the optimizing adaptive quality controller estimates both the gain and the bias recursively [Åström and Wittenmark, *Adaptive Control*, 2d ed., Addison Wesley, Reading, Mass., 1994; Del Castillo and Yey, "An Adaptive Run-to-Run Optimizing Controller for Linear and Nonlinear Semiconductor Processes," *IEEE Transactions on Semiconductor Manufacturing* **11**(2): 285–2950 (1996)].

Inferential Sensors via Estimation A small number of online sensors are commonly used in bioprocesses such as temperature, fermenter pressure, gas flow rates, agitation rate, dissolved oxygen, and pH. Some fermentation rates include a few additional probes, e.g., dissolved carbon dioxide, oxidation reduction, and optical density.

The lack of direct measurements of certain controlled variables is one of the challenges of bioprocess control problems [Dochain (ed.), *Automatic Control of Bioprocesses*, Wiley, New York, 2008]. For example, a number of fermentation processes rely on the concentration of active yeast in the fermenter; however, available online measurement techniques cannot distinguish between active yeast and inactive/dead yeast. As such, the concentration of active yeast has to be inferred from other available measurements. Inferring the quantity of one variable by measuring other variables is performed via an inferential sensor, sometimes also called a *soft sensor*. Examples of soft sensors are Kalman filters, moving horizon estimators, Luenberger observers, or, in simpler cases, algebraic relationships. The one aspect that all these inferential sensors have in common is that they are based upon the solution of an inverse problem that computes the variables of interest from available measurements via a model connecting the inferred and measured variables. The quality of this model is very important because an inferential sensor based upon a poor model will result in poor predictions and will degrade control quality.

Optimization of Set-Point Trajectory during a Batch Batch control requires that specific trajectories of the controlled variables be followed in order to maximize product yield, minimize cost and use of raw materials, and/or meet environmental requirements. Often the set-point trajectories are determined based upon heuristics and prior knowledge of a process. If a model of the process is available, then the trajectories can be computed in order to meet specific requirements. For example, the trajectory of the temperature set point can be computed for a fermentation process in order to maximize the product concentration at the end of the batch. The implementation of this trajectory can then be performed via batch control, model predictive control, or regular PID control.

Model Predictive Control *Model predictive control* (MPC) is a model-based control technique often used for bioprocess control. It is the most popular technique for handling multivariable control problems with multiple inputs and multiple outputs (MIMO) and can also accommodate inequality constraints on the inputs or outputs such as upper and lower limits. All these problems are addressed by MPC by solving an optimization problem, and therefore no complicated override control strategy is required [Seborg, Edgar, Mellichamp, and Doyle, *Process Dynamics and Control*, 4th ed., Wiley, New York, 2016].

Usually, MPC implementations change set points in order to move the plant to a desired steady state; the actual control changes are implemented by PID controllers in response to the set points. MPC is one of the most popular advanced control strategies and, among its other uses, can be used to determine optimal set-point trajectories that should be followed over the course of a batch process.

Complex interactions between different manipulated and controlled variables are common in bioprocesses, which makes MPC a popular control strategy for these types of processes. Furthermore, software packages that implement MPC solutions are now available from a variety of vendors which make their adoption straightforward with adequate engineering personnel. More details on MPC can be found in the earlier subsection Advanced Control Systems.

BIOPROCESS CONTROL APPLICATIONS

Biofuels Fuel-grade ethanol is produced in one of two ways, using either the wet mill or dry mill process. Wet milling involves separating the grain kernel into different components: germ, fiber, protein, and starch. The starch then undergoes the processes of liquifaction, saccharification, and fermentation separately. The focus here is on dry-mill–based plants, where the entire grain kernel is ground into flour, and the flour is transported into the cook tank allowing for complete water penetration. Then, the starch inside the flour is hydrolyzed into dextrin as it passes through the liquifaction tank, as shown in Fig. 8-60. Dextrin, which is a mixture of different

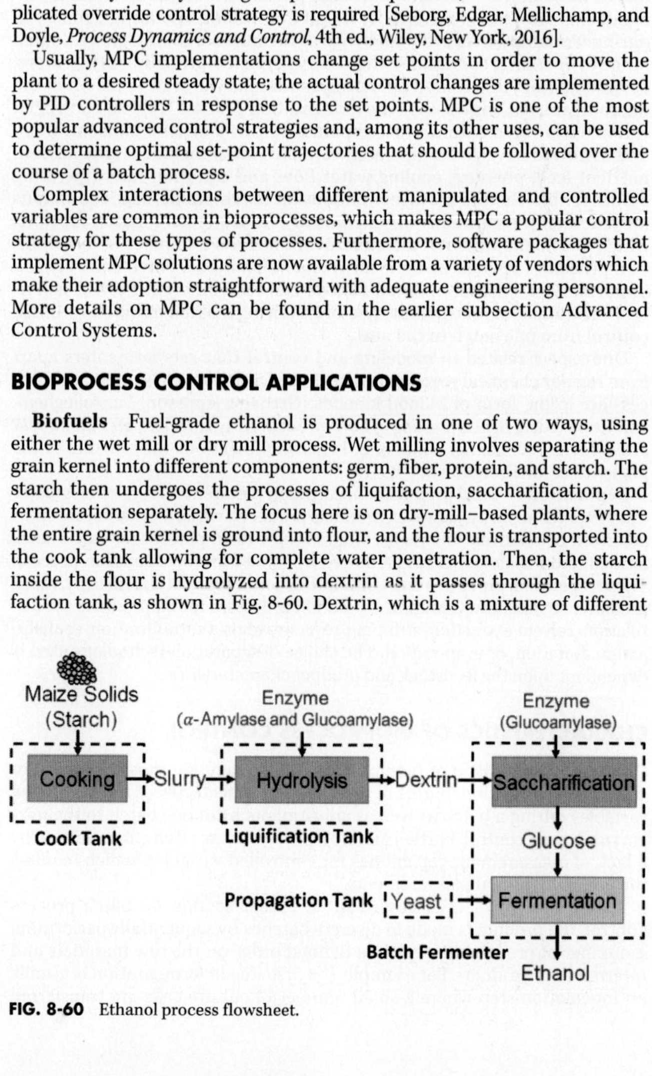

FIG. 8-60 Ethanol process flowsheet.

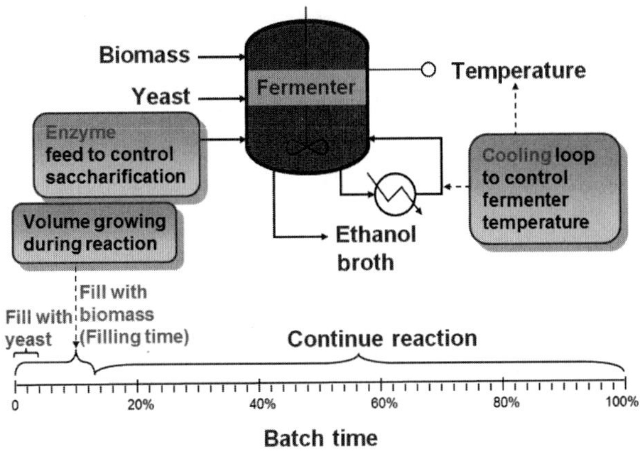

FIG. 8-61 Ethanol fermentation batch control strategy.

long-chain sugars (maltose, D_3, D_4 to D_7), then enters the batch fermenter and undergoes a *simultaneous saccharification and fermentation* (SSF) process in response to the addition of enzymes and of yeast from the propagation tank [Schweiger, Sayyar-Rodsari, Bartee, and Axelrud, "Plant-wide Optimization of an Ethanol Plant Using Parametric Hybrid Models," *49th IEEE Conference on Decision and Control,* Atlanta, Georgia, pp. 3896–3901, 2010].

A general overview of the SSF process and a timeline describing batch operation are shown in Fig. 8-61 [Dai, Word, and Hahn, "Modeling and Dynamic Optimization of Fuel-grade Ethanol Fermentation Using Fed-batch Process," *Control Eng. Practice* **22:** 231–241, 2014]. Measurements of the SSF process include ethanol and dextrose concentrations at a few regular time intervals and temperatures at a much higher frequency, usually every few minutes. The two key manipulated variables are the cooling water flow rate and the enzyme addition rate. The cooling system is operating during the entire fermentation, while the enzyme is added only during the filling phase, which ranges from the first 10 to 20 percent of the batch time.

The cooling water flow rate is usually controlled by a regular feedback controller, e.g., a PI controller, based upon the temperature measurement. However, note that it is often not the goal of this control scheme to keep the temperature in the fermenter constant over the course of a batch; instead the temperature should follow a preset temperature trajectory where the trajectory is either predetermined to maximize ethanol yield or based upon past experience with the process. In contrast to the temperature control, the enzyme addition rate is often manually controlled, and enzyme is only added during the filling phase. It is not uncommon for the enzyme addition rate to be constant during either part or the entire filling phase. The exact trajectory to follow for the enzyme addition rate is usually determined by experience with a particular plant.

Pharmaceuticals Production of pharmaceuticals is one area in which bioprocesses play a key role. There are two major stages involved in pharmaceutical manufacturing: the production of active ingredients and the conversion of active ingredients to drugs suitable for administration. The main manufacturing steps in pharmaceutical processing are (1) preparation of process intermediates, (2) introduction of functional groups, (3) esterification and coupling, (4) separation, and (5) final product purification. Additional production processes can include granulation, drying, tablet pressing/coating, filling, or packaging; however, none of these additional steps involve bioprocesses [Buckbee and Alford, *Automation Applications in Bio-Pharmaceuticals,* ISA, Research Triangle Park, N.C., 2008].

The production of monoclonal antibodies is now the largest product of the pharmaceutical industries with annual sales exceeding $45 billion. Approximately 40 percent of all pharmaceuticals in R&D are now biopharmaceuticals, and the trend is that this number will increase further in the future [Rader and Langer, "Upstream Single-Use Bioprocessing Systems Future Market Trends and Growth Assessment," *BioProcess International Magazine* **10**(2): 1218, 2012].

Wastewater Treatment Wastewater treatment plants are needed to meet stringent environmental regulations controlling water quality. Bioprocesses play an important role in wastewater treatment. The key challenges for wastewater treatment facilities are large perturbations in the wastewater flow rate, load, and composition. Most treatment facilities are operated continuously.

A number of models for wastewater treatment plants exist that have been extensively used for developing control strategies. The *activated sludge models* (ASMs) have been developed since the 1980s and applied to full-scale wastewater treatment plants for optimization. These models are quite generic, but can be tailored for particular treatment facilities once parameters related to sludge production and nutrients in the effluent have been estimated [Mogens, Gujer, Mino, and van Loosdrecht, *Activated Sludge Models ASM1, ASM2, ASM2d and ASM3,* IWA Publishing, London, 2000]. ASM1 was the first developed model and formed the foundation for a variety of extensions. The most popular models are ASM2, ASM2d, and ASM3P and they focus on improving prediction of nitrogen and phosphorus removal over ASM1 [Szilveszter, Ráduly, Ábrahám, Lányi, and Niculae, "Mathematical Models for Domestic Wastewater Treatment Process," *Environmental Engineering and Management Journal* **9**(5): 629–636, 2010].

While a number of control strategies have been proposed in the literature for improved and more efficient operation of wastewater treatment plants [Gernaey, Jeppsson, Vanrolleghem, and Copp (Eds.), *Benchmarking of Control Strategies for Wastewater Treatment Plants,* IWA Publishing, London, 2014], the evaluation and comparison of these strategies is nontrivial. Reasons for this are that the influent can vary significantly, but also that the biological/biochemical phenomena are not well characterized, and that time constants of the process range from a few minutes to several days. Nevertheless, a significant amount of work has been done on controlling wastewater treatment plants; see Gernaey, Jeppsson, Vanrolleghem, and Copp (eds.), *Benchmarking of Control Strategies for Wastewater Treatment Plants,* IWA Publishing, London, 2014 for more details.

Food Processing Almost all production facilities for food processing require bioprocesses that need to be appropriately operated and controlled. One area where bioprocesses play a critical role is fermentation for the production of beer and other alcoholic beverages. One indicator of the importance of this industry is that brewers in the United States produced 196 million barrels of beer in 2012 [*Brewers Almanac 2013,* The Beer Institute, March 7, 2014], which resulted in sales of over $100 billion. The brewing process consists of several steps which include malting, milling, mashing, lautering, boiling, fermenting, conditioning, filtering, and packaging. The three most commonly used fermentation methods are warm, cool, and spontaneous fermentation, where the fermentation may take place in an open or closed fermenting vessel.

Production of snack foods, which had a worldwide market of $66 billion in 2003 [Nikolaou, "Control of Snack Food Manufacturing Systems: Potato Chips and Micro-Chips are More Similar than Commonly Believed," *IEEE Control Systems Magazine,* Special issue on Process Control, **26**(4): 40–54, 2006], is another area that involves bioprocesses. A relatively small number of processes are used, which include extrusion, frying, baking, and drying. Food processing generally involves serial operations of these unit operations. One of the challenges for control of food processing is a limited number of intermediate variables that assess product quality, which prevents subsequent steps from counteracting disturbances that have occurred at earlier unit operations as the effect of these disturbances on the current state of the (unfinished) product are not known. Similarly, raw materials can vary widely from one batch to the next, which poses a control challenge.

TELEMETERING AND TRANSMISSION

Digital technology is used for data collection, feedback control, and all other information processing requirements in production facilities. Such systems must acquire data from a variety of measurement devices, and control systems must drive final control elements. Until the turn of this century, digital systems relied on the same technology as their analog predecessors, primarily analog signal transmission in the form of current loops. Digital transmission permits values to be transmitted in engineering units, and it is gradually displacing analog signal transmission.

FIELDBUS

Fieldbus is the generic name of a family of industrial communications networks intended for data acquisition and control applications in industrial facilities, both process and manufacturing. IEC 61158, "Digital Data Communications for Measurement and Control," defines five different communications technologies, each considered to be a fieldbus. Most expect a standard to define one and only one way of doing something, but the computer industry does things differently. Consequently, one person's "fieldbus" might not be the same as another person's.

The delays in developing the standard led to a slow acceptance of fieldbus technology by the user community. Most plants have preferred suppliers for various technologies, including measurement devices, controllers, valves, variable-frequency drives (VFDs), etc. Users were understandably reluctant to proceed with digital communications until they could be confident that their preferred suppliers could provide as a standard offering an interface to the user's preferred fieldbus technology.

To some extent, the multiple fieldbus technologies reflect the difference in the mixture of I/O points. In the manufacturing industries, 90 percent or more of the I/O points are discrete; *numerical controllers* and *programmable logic controllers* (PLCs) are preferred. In the continuous process industries, 50 percent or more of the I/O points are analog; *distributed control systems* (DCSs) are preferred. In batch facilities, discrete I/O exceeds analog I/O, and both DCSs and PLCs are used as controllers.

In process facilities in the United States, Foundation Fieldbus technology is preferred; in Europe, Profibus technology is preferred. Most manufacturers of smart transmitters, smart valves, VFDs, etc. can supply interfaces for either (and usually a current loop interface as well). Most plants select one technology for their fieldbus, although it is possible to have both Foundation Fieldbus and Profibus interfaces on a given control system.

Eventually fieldbus technology will displace all other forms of data transmission within process facilities. A fieldbus of some type is generally preferred for new construction. Smart transmitters and smart valves gained widespread acceptance long before the fieldbus, so most were installed using current loops. The installed base of electronic current loops is large, but current loop interfaces will gradually disappear.

The major advantages of fieldbus technologies lie in installation and commissioning. With fewer wires and connections, fieldbus installations start with fewer errors that can be located and corrected more rapidly. Once a fieldbus installation is up and running, modifications and additions are much easier, saving both time and money.

Current loop transmission requires dedicated circuits to connect each measurement device and final control element to the controller. Fieldbus permits a single circuit, often called a *segment* or *trunk*, to be used to communicate with a mixture of measurement devices and final control elements, each of which connects to the cable as a "drop" or "node." A fieldbus segment is a shielded twisted-pair wire that serves two purposes: two-way digital signal transmission and DC power for devices with low power requirements. These two-wire fieldbus devices are analogous to two-wire current loop transmitters (explained subsequently). Devices with high power requirements are supported as four-wire fieldbus devices. Wireless communications technology is available for fieldbus, but power must be separately provided.

The fieldbus communications protocols provide for 32 drops or nodes. However, power considerations usually limit the number of devices on a segment to about one-half of this. So that the electronic equipment cannot be a source of ignition, the electrical classifications of most process applications require intrinsically safe circuits (the alternative of explosion-proof or purged enclosures is prohibitively expensive). The approach from current loops of using barriers to limit the amount of energy in the field circuit can also be applied to fieldbus. However, this permits only four devices to be connected to a segment, which increases the number of segments and the costs. Alternate approaches permit a larger number of devices to be connected to a segment, but with some increase in wiring complexity.

Each fieldbus segment must be connected to the data acquisition and control system through an interface card or coupler. These are very different between Foundation Fieldbus and Profibus. For Foundation Fieldbus the power supply is separate from the interface card; for Profibus the power supply is part of the coupler. Communication on Profibus is a master-slave arrangement; the devices (the slaves) can respond only to communications from the data acquisition and control system (the master). Communications on Foundation Fieldbus permit devices to communicate with other devices on the network as well as with the data acquisition and control system.

The data acquisition and control system must be told what devices are on each segment of the fieldbus (not exactly "plug and play"). For each device on a segment, a *device description* (DD) file specifies the input and output variables, the available functions, and other parameters. Originally, Foundation Fieldbus and Profibus used identical text-based files that can be downloaded from a web site. Supporting additional features necessitated the incorporation of extensions into the files, but these are specific to Foundation Fieldbus (EDDL for Electronic Device Description Language) and Profibus (FDT for field device tool). Suppliers of data acquisition and control systems have somewhat insulated the user by supporting both, but the user is responsible for verifying that each field device is compatible with the requirements imposed by the fieldbus in use.

INDUSTRIAL INTERNET OF THINGS AND CLOUD COMPUTING

The impact of the Internet on retail, banking, and other consumer-oriented industries has been dramatic. Will it similarly impact the process industries? The Industrial Internet of Things (IIoT) envisions connecting the various islands of automation within industrial facilities to each other and to the cloud. Time will tell if this offers real opportunities in the process industries or is largely hype.

Fieldbus technology provides communications between a data acquisition and control system and the various field devices, primarily measurement devices and final control elements. Industrial Ethernet is commonly used in process facilities to communicate from one automation system to another (called *machine-to-machine* or *M2M* communications). Although targeted to distinct layers in the automation hierarchy, the fieldbus and industrial Ethernet technologies overlap and could eventually merge.

An often cited benefit is the application of "big data" analytics to the large quantity of data that can be acquired. The IIoT coupled with the cloud promises to lower the barriers to exploring the potential of such analytics. The analytical programs are available for installation within a company's computing systems. For something that is promising but not proved, a major obstacle is implementing the required software by IT people who already have much to do. The alternative is to prove the worth of the endeavor by using software available through the cloud. Especially at the exploratory stage of an endeavor, purchasing processing and storage from the cloud lowers initial costs and does not incur long-term commitments.

The initial application of the IIoT will be activities that impact operational efficiency, such as dynamic scheduling within a batch facility. Such activities can be pursued with minimal impact on the current structures and practices within the industry. However, the IIoT makes it possible to rethink the industry's approach to many activities. Is it possible to control a process plant using software and computing from the cloud? At the time of this writing, such capability is not available, but with the current processing and communications speeds, all technical issues, including nonstop computing, could be resolved. But providing process control from the cloud is a major departure from the current practice of pushing all processing to the lowest possible level in the hierarchy. Discussions regarding the pros and cons will be lengthy, heated, and possibly more emotional than technical. Process control must stay in the mainstream of computing technology, so the discussions will commence.

ANALOG SIGNAL TRANSMISSION

For analog signal transmission the transmitter converts the value in engineering units to an analog signal that can be transmitted over some distance. The signal ranges are as follows:

Pneumatic: 3 to 15 psi.

Electronic: 4 to 20 mA. The result is a circuit, known as a *current loop*, that is less susceptible than voltage to electrical interference from the power equipment present in industrial facilities.

Pneumatic transmission has largely disappeared. Electronic transmission will eventually suffer the same fate, but at the time of this writing the installed base is too large to be ignored.

To transmit a value, the source or transmitter converts the process value (temperature, pressure, valve opening, etc.) to a scaled value that can be transmitted as a signal. The lower range of the signal is 4 mA and corresponds to the lower-range value of the process value in engineering units. The upper range of the signal is 20 mA and corresponds to the upper-range value of the process value in engineering units. The relationship between the process value and the signal value is linear. For modern electronic equipment, the departure from linearity is rarely an issue.

Converting the input signal to an engineering value is a three-step process:

1. *Convert current to voltage.* Current is used for transmission, but all internal signal processing is based on voltages. Inserting a 250-Ω *range resistor* into the circuit converts the 4- to 20-mA signal to a 1- to 5-V input. A nominally more complex circuit can convert the signal to a 0 to 5-V input.

2. *Digitize the voltage input.* The analog-to-digital (A/D) converter translates the input voltage to a short (16-bit) integer known as the *raw value.* A/D converters may be unipolar (voltage input must be positive) or bipolar (voltage input may be positive or negative). Of greater importance in process applications is the resolution, which is the number of data bits in the converted value. A 12-bit bipolar A/D converter provides 11 data bits plus a sign bit. A 12-bit unipolar A/D converter provides 12 data bits. The resolution provided by 11 data bits is 1 part in 2^{11} (2048), which is approximately 0.05 percent. Although adequate for older analog or dumb measurement devices, higher resolutions are appropriate for the microprocessor-based smart transmitters. Especially in control applications that use the derivative

mode of control, the resolution should be consistent with the repeatability of the measurement device, which is generally better than its accuracy.

3. *Convert raw value to engineering units.* Digital systems do this in software. The conversion software requires two parameters: the lower-range value (corresponds to 4 mA) and the upper-range value (corresponds to 20 mA). The relationship between the raw value and the engineering value is assumed to be linear, but with one exception. For inputs from head-type flow meters, the lower- and upper-range values are in flow units, but the 4- to 20-mA signal may be the sensed value for the differential pressure. To convert the raw value to flow units, a square root must be incorporated into the computations, and process control systems continue to provide this option. However, most smart differential pressure transmitters can be configured to "transmit in flow units," which means the square root is incorporated into the calculations within the transmitter and the 4- to 20-mA signal is the measured value of the flow.

Some, but not all, "5-V A/D converters" have an input range of 0 to 5.12 V. For an A/D converter with 11 data bits, 5.12 V converts to a raw value of 2047. A 5-V input converts to 2000, giving a resolution of 1 part in 2000 (exactly 0.05 percent) over the 0- to 5-V range. This practice will be assumed in subsequent discussions, but is not universal.

Converting a 4- to 20-mA signal to a 1- to 5-V input has consequences for the resolution. For an A/D converter with 11 data bits, 1 V converts to a digital value of 400. Thus, the range for the digital value is 400 to 2000, making the effective input resolution 1 part in 1600 (0.0625 percent). Sometimes this is expressed as a resolution of 10.6 bits, where $2^{10.6} = 1600$. Using slightly more complex input circuitry to convert the 4- to 20-mA signal to a 0- to 5-V input gives a resolution of 1 part in 2000.

For most measurement devices *two-wire* transmitters are possible. The measurement device is powered over the same circuit used for signal transmission. For measurement devices requiring higher power levels, *four-wire* transmitters (power is separate from the current loop) must be installed.

The electrical classification for most areas of a process entail intrinsically safe circuits (the alternative of explosion-proof or purged enclosures is costly). For two-wire transmitters, the usual practice is to install barriers to limit the amount of power in the field part of the circuit.

Especially when the measured value is the input to a controller, the consequences of a failure known as a "broken wire" must be assessed. For current loops, the result is an open circuit, causing the measured variable to fail to its lower-range value. The consequences may be undesirable. For a fired heater that is heating material to a target temperature, failure of the temperature measurement to its lower-range value drives the output of the combustion control logic to the maximum possible firing rate.

In such applications, the analog transmission signal could be inverted, with the upper range corresponding to 4 mA and the lower range corresponding to 20 mA. On an open circuit, the measured variable fails to its upper range. For the fired heater, failure of the measured variable to its upper range drives the output of the combustion control logic to the minimum firing rate.

An even better approach is for the receiver to distinguish between 4 mA and 0 mA. The former means that the measured value is equal to its lower-range value. The latter is an open circuit, meaning that the measured value is unknown. On an open circuit, the conversion software can be configured to fail to lower-range value, fail to upper-range value, or hold the last value (freeze). If the measured value is the input to a controller, the controller behavior on an invalid value of the measured variable could be to fail to lower output limit, fail to upper output limit, or hold the last value.

MICROPROCESSOR-BASED SMART TRANSMITTERS

In all aspects of process control the preference is to express everything in engineering units. Although slower than most desired, alternatives to analog signal transmission have become common. Several examples will be presented, beginning with smart transmitters.

Smart transmitters offer several benefits over dumb transmitters:

1. They check on the internal electronics, such as verifying that the voltage levels of internal power supplies are within specifications.

2. They check on environmental conditions within the instruments, such as verifying that the case temperature is within specifications.

3. They perform compensation of the measured value for conditions within the instrument, such as compensating the output of a pressure transmitter for the temperature within the transmitter. Smart transmitters are much less affected by temperature and pressure variations than are conventional transmitters.

4. They perform compensation of the measured value for other process conditions, such as compensating the output of a capacitance level transmitter for variations in process temperature.

5. They linearize the output of the transmitter. Functions such as square root extraction of the differential pressure for a head-type flow meter can be done within the instrument instead of within the control system.

6. They configure the transmitter from a remote location, such as changing the span of the transmitter output.

7. They do automatic recalibration of the transmitter. Although this is highly desired by users, the capabilities, if any, in this respect depend on the type of measurement.

Improved performance is the main incentive to install smart transmitters. With regard to signal transmission, smart transmitters can be purchased with the following:

Current loop output. This option permits a smart transmitter to be a direct replacement for a conventional transmitter. The HART (Highway Addressable Remote Transducer) communication protocol can be used over the current loop circuit for remote configuration via a central system or a handheld configuration tool. The value of the measured variable can be retrieved by HART, but the current loop input provides the value for process control, data acquisition, etc.

Network interface. Most manufacturers provide a fieldbus interface in the form of Foundation Fieldbus and/or Profibus. Some also support Modbus/TCP, an Ethernet version of the older Modbus serial communications protocol.

The network interface is clearly the preferred option, but can be pursued only where a fieldbus is installed.

THERMOCOUPLES AND RESISTANCE TEMPERATURE DETECTORS (RTDs)

Input systems that provide temperatures in engineering units have been available for some time. From a marketing perspective, compatibility with systems designed for current loop inputs has been absolutely necessary. However, this will gradually disappear.

Thermocouples present two issues for input signal processing:

Low-level signals. Thermocouple outputs rarely exceed 30 mV. In addition, the output could be zero or even negative. Processing such low-level signals requires a bipolar A/D converter front-ended by an amplifier to boost the signal to voltage levels. Such equipment is available, but with increased complexity and cost.

Reference junction compensation. The output from a thermocouple depends on the temperatures at the process junction and the reference junction. Either the reference junction temperature must be sensed, or the reference junction must be maintained at a known temperature. Terminal strips with a temperature sensor embedded for the reference junction temperature are manufactured specifically for this purpose.

Although the input equipment for thermocouples is specialized, thermocouples are easily multiplexed. They are ideally suited for applications where a large number of temperatures must be read on a relatively slow scan rate. Consider monitoring the temperature in a cold storage warehouse to ensure it never exceeds a specified value. Uniform temperatures cannot be assumed, so thermocouples must be installed in several locations. Scan rates such as once per minute are acceptable. For such applications, multiplexers that can accept 100 or more points can be purchased. Although the base cost is large, the cost to add one additional thermocouple input is nominal, making the cost per thermocouple very attractive.

RTDs present different issues:

Bridge network. Sensing the resistance of the RTD requires a bridge network.

Three-wire installations. For plant applications, three-wire RTD installations are the norm. Both complicate multiplexing RTD inputs. Basically, two options are available:

Temperature transmitter. The temperature can be transmitted as a 4- to 20-mA current signal or via fieldbus.

RTD input card. Such cards typically accept between 4 and 16 RTDs. Usually the cost per point is less than a dedicated transmitter for each RTD. All inputs to the card must be from the same type of RTD, and all must be converted to the same temperature units (degrees Celsius or degrees Fahrenheit), but rarely are these restrictions a problem in an application.

For thermocouples, similar input cards are available. For both thermocouples and RTDs, advancements in electronics permit the conversion of the input to temperature units, including linearization, reference junction compensation (thermocouples), etc., to be performed on the input card. The following approaches permit such cards to be readily integrated into systems designed for current loop inputs:

Emulate an input from a current loop. The input card converts the computed value for the temperature to a short integer raw value, using the lower- and upper-range values for the measurement range for the thermocouple (0 to 750°C for type J) or RTD (−200 to 800°C).

Express temperature as an integer value to 0.1°C or 0.1°F. For example, a temperature of 123.4°C gives a raw value of 1234. The value of the temperature is provided in engineering units with a resolution of 0.1°. A resolution of 0.1°C is acceptable in most applications, but not all. In aqueous-based processes such as fermentations, a typical measurement range is 0 to 100°C.

Expressing the temperature to 0.1°C gives a resolution of 1 part in 1000 for this measurement range. Using a current loop and an A/D converter with 12 data bits gives a resolution of 1 part in 4000, or 0.025°C. The latter is more consistent with the repeatability of the RTDs normally installed in such processes.

For fieldbus installations, the counterpart to a thermocouple or RTD input card is a multipoint temperature module. The network interface frees the modules from the restrictions (such as expressing the temperature as a short integer value) resulting from being compatible with current loop inputs.

MULTIVALUE MEASUREMENT DEVICES

Some measurement devices produce more than one value. Here are two examples:

Chromatographs. Since it is a sampling instead of a continuous analyzer, the results pertain to a sample withdrawn from the process. Chromatographs occasionally produce a single value such as the ratio of two key components in the sample. But the more common result is a composition analysis for the sample, with the number of values depending on the number of components of interest. The values are associated in that they pertain to a specific sample. In addition, capturing information such as the analysis time (normally the time the sample was injected) is also a requirement.

Coriolis meters. If only the mass flow is of interest, a current loop interface is adequate, and most commercial products provide such an option. But in addition to the flow, the Coriolis meter also senses the temperature and density. Coriolis meters can also sense viscosity, provided this option is purchased. Unlike the chromatograph, these are continuous variables. However, providing current loops for all variables is inconvenient, and most manufacturers do not provide such an option.

For chromatographs, the traditional interface was a serial interface, either RS-232 for point-to-point or RS-485 for multidropped. The major issue is the protocol (the character or byte sequence for transmitting the data). Some were proprietary protocols, requiring custom software in the data acquisition or control system. Some use the MODBUS protocol, developed in 1979 for reading and writing the registers in a PLC. In the context of MODBUS, a register is a 16-bit (short integer) storage location. Transmitting values in engineering units as floating-point numbers requires two consecutive registers but eliminates the need to somehow express values as short integers.

Except in legacy systems, serial interfaces have been replaced almost entirely by network interfaces. Most suppliers provide one or more of the following network interfaces:

1. MODBUS/TCP is essentially an Ethernet version of the MODBUS serial protocol.
2. Foundation Fieldbus.
3. Profibus.

FILTERING AND SMOOTHING

A signal received from a process transmitter generally contains the following components distinguished by their frequency [frequencies are measured in hertz (Hz), with 60-cycle ac being a 60-Hz frequency):

1. *Low-frequency process disturbances.* The control system is expected to react to these disturbances.
2. *High-frequency process disturbances.* The frequency of these disturbances is beyond the capability of the control system to effectively react.
3. *Measurement noise.*
4. *Stray electrical pickup, primarily 50- or 60-cycle ac.*

The objective of filtering and smoothing is to remove the last three components, leaving only the low-frequency process disturbances. Normally this has to be accomplished by using the proper combination of analog and digital filters. Sampling a continuous signal results in a phenomenon often referred to as *aliasing* or *foldover*. When a signal is sampled at a frequency ω_s, all frequencies higher than $\omega_s/2$ cannot be represented at their original frequency. Instead, they are present in the sampled signal with their original amplitude but at a lower-frequency harmonic. Because of the aliasing or foldover issues, a combination of analog and digital filtering is usually required. The sampler (i.e., the A/D converter) must be preceded by an analog filter that rejects those high-frequency components such as stray electrical pickup that would result in foldover when sampled. In commercial products, analog filters are normally incorporated into the input processing hardware by the manufacturer. The software then permits the user to specify digital filtering to remove any undesirable low-frequency components.

On the analog side, the filter is often the conventional resistor-capacitor or *RC* filter. However, other possibilities exist. For example, one type of A/D converter is called an *integrating A/D* because the converter basically integrates the input signal over a fixed interval of time. By making the interval $\frac{1}{60}$ s, this approach provides excellent rejection of any 60-Hz electrical noise.

On the digital side, the input processing software generally provides for smoothing via the exponentially weighted moving average, which is the digital counterpart to the *RC* network analog filter. The smoothing equation is

$$y_i = \alpha x_i + (1-\alpha) y_{i-1} \tag{8-58}$$

where x_i = current value of input
y_i = current output from filter
y_{i-1} = previous output from filter
α = filter coefficient ($0 \le \alpha \le 1$)

The degree of smoothing is determined by the filter coefficient α, with $\alpha = 1$ being no smoothing and $\alpha = 0$ being infinite smoothing (no effect of new measurements). The filter coefficient α is related to the filter time constant τ_F and the sampling interval Δt by

$$\alpha = 1 - \exp\left(\frac{-\Delta t}{\tau_F}\right) \tag{8-59}$$

or by the approximation

$$\alpha = \frac{\Delta t}{\Delta t + \tau_F} \tag{8-60}$$

Another approach to smoothing is to use the arithmetic moving average, which is represented by

$$y_i = \frac{\sum_{j=1}^{n} x_{i+1-j}}{n} \tag{8-61}$$

The term *moving* is used because the filter software maintains a storage array with the previous n values of the input. When a new value is received, the oldest value in the storage array is replaced with the new value, and the arithmetic average is recomputed. This permits the filtered value to be updated each time a new input value is received.

In process applications, determining τ_F (or α) for the exponential filter and n for the moving average filter is often done merely by observing the behavior of the filtered value. If the filtered value is "bouncing," the degree of smoothing (that is, τ_F or n) is increased. This can easily lead to an excessive degree of filtering, which will limit the performance of any control system that uses the filtered value. The degree of filtering is best determined from the frequency spectrum of the measured input, but such information is rarely available for process measurements.

DIGITAL TECHNOLOGY FOR PROCESS CONTROL

General References: Auslander and Ridgely, *Design and Implementation of Real-Time Software for the Control of Mechanical Systems*, Prentice-Hall, Upper Saddle River, N.J., 2002. Hughes, *Programmable Controllers*, ISA, Research Triangle Park, N.C., 2005. Johnson, *Process Control Instrumentation Technology*, 8th ed., Prentice-Hall, Upper Saddle River, N.J., 2005. Kopetz, *Real-Time Systems: Design Principles for Distributed Embedded Applications*, 2d ed., Springer, Berlin, 2011. Lipták, *Instrument Engineers Handbook*, 4th ed., CRC, Boca Raton, Fla., 2011. Petruzella, *Programmable Logic Controllers*, 4th ed., McGraw-Hill, New York, 2010. Thompson and Shaw, *Industrial Data Communications*, 5th ed., ISA, Research Triangle Park, N.C., 2016.

Since the 1970s, process controls have evolved from pneumatic analog technology to electronic analog technology to microprocessor-based controls. Electronic and pneumatic controllers have now virtually disappeared from process control systems, which are dominated by programmable electronic systems based on microprocessor technology.

HIERARCHY OF INFORMATION SYSTEMS

Coupling digital controls with networking technology permits information to be passed from level to level within a corporation at high rates of speed. This technology is capable of presenting the measured variable from a flow transmitter installed in a plant in a remote location anywhere in the world to the company headquarters in less than 1 s.

A hierarchical representation of the information flow within a company leads to a better understanding of how information is passed from one layer to the next. Such representations can be developed in varying degrees of detail, and most companies have developed one that describes their specific practices. The following hierarchy consists of five levels, as shown in Fig. 8-24.

Measurement Devices and Final Control Elements This lowest layer couples the control and information systems to the process. The measurement devices provide information on the current conditions within the process. The final control elements permit control decisions to be imposed on the process. Although traditionally analog, smart transmitters and smart valves based on microprocessor technology are now beginning to dominate this layer.

Safety and Environmental/Equipment Protection The level 2 functions play a critical role by ensuring that the process is operating safely and satisfies environmental regulations. Process safety relies on the principle of multiple protection layers that involve groupings of equipment and human actions. One layer includes process control functions, such as alarm management during abnormal situations, and safety instrumented systems for emergency shutdowns. The safety equipment (including sensors and block valves) operates independently of the regular instrumentation used for regulatory control in level 3. Sensor validation techniques can be employed to confirm that the sensors are functioning properly.

Regulatory Controls The objective of this layer is to operate the process at or near the targets supplied by a higher layer in the hierarchy. To achieve consistent process operations, a high degree of automatic control is required from the regulatory layer. The direct result is a reduction in variance in the key process variables. More uniform product quality is an obvious benefit. However, consistent process operation is a prerequisite for optimizing the process operations. To ensure success for the upper-level functions, the first objective of any automation effort must be to achieve a high degree of regulatory control.

Real-Time Optimization Determining the most appropriate targets for the regulatory layer is the responsibility of the RTO layer. Given the current production and quality targets for a unit, RTO determines how the process can be best operated to meet them. Usually this optimization has a limited scope, being confined to a single production unit or possibly even a single unit operation within a production unit. RTO translates changes in factors such as current process efficiencies, current energy costs, cooling medium temperatures, and so on to changes in process operating targets so as to optimize process operations.

Production Controls The nature of the production control logic differs greatly between continuous and batch plants. A good example of production control in a continuous process is refinery optimization. From the assay of the incoming crude oil, the values of the various possible refined products, the contractual commitments to deliver certain products, the performance measures of the various units within a refinery, and the like, it is possible to determine the mix of products that optimizes the economic return from processing this crude. The solution of this problem involves many relationships and constraints and is solved with techniques such as linear programming.

In a batch plant, production control often takes the form of routing or short-term scheduling. For a multiproduct batch plant, determining the long-term schedule is basically a *manufacturing resource planning* (MRP) problem, where the specific products to be manufactured and the amounts to be manufactured are determined from the outstanding orders, the raw materials available for production, the production capacities of the process equipment, and other factors. The goal of the MRP effort is the long-term schedule, which is a list of the products to be manufactured over a specified time (often one week). For each product on the list, a target amount is also specified. To manufacture this amount usually involves several batches. The term *production run* often refers to the sequence of batches required to make the target amount of product, so in effect the long-term schedule is a list of production runs.

Most multiproduct batch plants have more than one piece of equipment of each type. Routing refers to determining the specific pieces of equipment that will be used to manufacture each run on the long-term production schedule. For example, the plant might have five reactors, eight neutralization tanks, three grinders, and four packing machines. For a given run, a rather large number of routes are possible. Furthermore, rarely is only one run in progress at a given time. The objective of routing is to determine the specific pieces of production equipment to be used for each run on the long-term production schedule. Given the dynamic nature of the production process (equipment failures, insertion/deletion of runs into the long-term schedule, etc.), the solution of the routing problem continues to be quite challenging.

Corporate Information Systems Terms such as *management information systems* (MIS), *enterprise resource planning* (ERP), *supply chain management* (SCM), and *information technology* (IT) are frequently used to designate the upper levels of computer systems within a corporation. From a control perspective, the functions performed at this level are normally long-term and/or strategic. For example, in a processing plant, long-term contracts are required with the providers of the feedstocks. A forecast must be developed for the demand for possible products from the plant. This demand must be translated to needed raw materials, and then contracts executed with the suppliers to deliver these materials on a relatively uniform schedule.

DIGITAL HARDWARE IN PROCESS CONTROL

Digital control technology was first applied to process control in 1959, using a single central computer (and analog backup for reliability). In the mid-1970s, a microcomputer-based process control architecture referred to as a *distributed control system* (DCS) was introduced and rapidly became a commercial success. A DCS consists of some number of microprocessor-based nodes that are interconnected by a digital communications network, often called a *data highway*. Today the DCS is still dominant, but there are other options for carrying out computer control, such as single-loop controllers, programmable logic controllers, and personal computer controllers. A brief review of each type of controller device is given in the following section, Controllers, Final Control Elements, and Regulators, along with more details on controller hardware options.

Distributed Control System Figure 8-62 depicts a representative distributed control system. The DCS consists of many commonly used components, including multiplexers (MUXs), single-loop and multiple-loop controllers, PLCs, and smart devices. A system includes some of or all the following components:

1. *Control network.* The control network is the communication link between the individual components of a network. Coaxial cable and, more recently, fiber-optic cable have often been used, with Ethernet protocols becoming more common (100 mbit/s or higher). A redundant pair of cables (dual redundant highway) is normally supplied to reduce the possibility of link failure.

2. *Workstations.* Workstations are the most powerful computers in the system, capable of performing functions not normally available in other units. A workstation acts both as an arbitrator unit to route internodal communications and as the database server. An operator interface is supported, and various peripheral devices are coordinated through the workstations. Computationally intensive tasks, such as real-time optimization or model predictive control, can be implemented in a workstation. Operators supervise and control processes from these workstations. Operator stations may be connected directly to printers for alarm logging, printing reports, or process graphics.

3. *Remote control units (RCUs).* These components are used to implement basic control functions such as PID control. Some RCUs may be configured to acquire or supply set points to single-loop controllers. Radio telemetry (wireless) may be installed to communicate with MUX units located at great distances.

4. *Application stations.* These separate computers run application software such as databases, spreadsheets, financial software, and simulation software via an OPC interface. OPC is an acronym for object linking and embedding for process control, a software architecture based on standard interfaces. These stations can be used for e-mail and as web servers, for remote diagnosis configuration, and even for operation of devices that have an IP (Internet Protocol) address. Applications stations can communicate with the main database contained in online mass storage systems. Typically hard disk drives are used to store active data, including online and historical databases and nonmemory resident programs. Memory resident programs are also stored to allow loading at system start-up.

5. *Fieldbuses and smart devices.* An increasing number of field-mounted devices are available that support digital communication of the process I/O in addition to, or in place of, the traditional 4- to 20-mA current signal. These devices have greater functionality, resulting in reduced setup time, improved control, combined functionality of separate devices, and control valve diagnostic capabilities. Digital communication also allows the control system to become completely distributed where, e.g., a PID control algorithm could reside in a valve positioner or in a sensor/transmitter.

DISTRIBUTED DATABASE AND THE DATABASE MANAGER

A database is a centralized location for data storage. The use of databases enhances system performance by maintaining complex relations between data elements while reducing data redundancy. A database may be built based on the relational model, the entity relationship model, or some other model. The database manager is a system utility program or programs

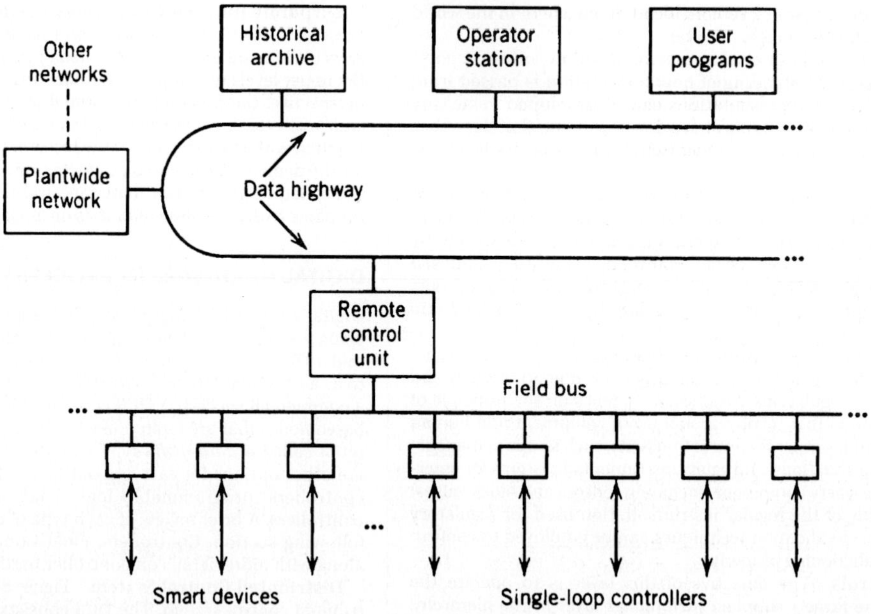

FIG. 8-62 A DCS using a broadband (high-bandwidth) data highway and fieldbus connected to a single remote control unit that operates smart devices and single-loop controllers.

acting as the gatekeeper to the databases. All functions retrieving or modifying data must submit a request to the manager. Information required to access the database includes the tag name of the database entity, often referred to as a *point*, the attributes to be accessed, and the values, if they are to be modified. The database manager maintains the integrity of the databases by executing a request only when not processing other conflicting requests.

To allow flexibility, the database manager must also perform point addition or deletion. However, the ability to create a point type or to add or delete attributes of a point type is not normally required because, unlike other data processing systems, a process control system normally involves a fixed number of point types and related attributes. For example, analog and binary input and output types are required for process I/O points. Related attributes for these point types include tag names, values, and hardware addresses. Different system manufacturers may define different point types using different data structures.

Data Historian An historical database is built similar to an online database. Unlike their online counterparts, the information stored in a historical database is not normally accessed directly by other subsystems for process control and monitoring. Periodic reports and long-term trends are generated based on the archived data. The reports are often used for planning and system performance evaluations such as statistical process (quality) control. The trends may be used to detect process drifts or to compare process variations at different times.

The historical data are sampled at user-specified intervals. A typical process plant contains a large number of data points, but it is not feasible to store data for all points at all times. The user determines if a data point should be included in the list of archived points. Most systems provide archive-point menu displays. The sampling periods are normally some multiple of their base scan frequencies. However, some systems allow historical data sampling of arbitrary intervals. This is necessary when intermediate virtual data points that do not have the scan frequency attribute are involved. The archive point lists are continuously scanned by the historical database software. Online databases are polled for data, and the times of data retrieval are recorded with the data obtained. To conserve storage space, different data compression techniques are employed by various manufacturers, although the greatly reduced costs of data storage make this less critical.

PROCESS CONTROL LANGUAGES

Originally, software for process control utilized high-level programming languages such as FORTRAN and BASIC. Some companies have incorporated libraries of software routines for these languages, but others have developed specialty languages characterized by natural language statements. The most widely adopted user-friendly approach is the fill-in-the-forms or table-driven *process control languages* (PCLs). Typical PCLs include function block diagrams, ladder logic, and programmable logic. The core of these languages is a number of basic function blocks or software modules, such as analog in, digital in, analog out, digital out, PID, summer, and splitter. Using a module is analogous to calling a subroutine in conventional FORTRAN or C programs.

In general, each module contains one or more inputs and an output. The programming involves connecting outputs of function blocks to inputs of other blocks via the graphical-user interface. Some modules may require additional parameters to direct module execution. Users are required to fill in templates to indicate the sources of input values, the destinations of output values, and the parameters for forms/tables prepared for the modules. The source and destination blanks may specify process I/O channels and tag names when appropriate. To connect modules, some systems require filling in the tag names of modules originating or receiving data.

Many DCSs allow users to write custom code (similar to BASIC) and attach it to data points, so that the code is executed each time the point is scanned. The use of custom code allows many tasks to be performed that cannot be carried out by standard blocks.

All process control languages contain PID control blocks of different forms. Other categories of function blocks include

1. *Logical operators.* AND, OR, and exclusive OR (XOR) functions.
2. *Calculations.* Algebraic operations such as addition, multiplication, square root extraction, or special function evaluation.
3. *Selectors.* Min and max functions, transferring data in a selected input to the output or the input to a selected output.
4. *Comparators.* Comparison of two analog values and transmission of a binary signal to indicate whether one analog value exceeds the other.
5. *Timers.* Delayed activation of the output for a programmed duration after activation by the input signal.
6. *Process dynamics.* Emulation of a first-order process lag (or lead) and time delay.

PROCESS MEASUREMENTS

General References: Agrawal, *Fiber-Optic Communication Systems*, 4th ed., Wiley, Hoboken, N.J., 2010. Baker, *Flow Measurement Handbook*, Cambridge University Press, New York, 2000. Borden and Friedmann (eds.), *Control Valves*, ISA, Research Triangle Park, N.C., 1998. Dakin and Culshaw (eds.), *Optical Fiber Sensors: Applications, Analysis, and Future Trends*, vol. 4, Artech House, Norwood, Mass., 1997. Johnson, *Process Control Instrumentation*

Technology, 8th ed., Prentice-Hall, Upper Saddle River, N.J., 2007. Lipták (ed.), *Instrument Engineers' Handbook*, 4th ed., *Process Measurement and Analysis*, vol. 1; *Process Control*, vol. 2., CRC Press, Boca Raton, Fla., 2006. Nichols, *On-Line Process Analyzers*, Wiley, New York, 2010. Scott, *Industrial Process Sensors*, CRC Press, Boca Raton, Fla., 2007. Seborg, Edgar, Mellichamp, and Doyle, *Process Dynamics and Control*, 4th ed., Wiley, New York, 2016. Spitzer, *Flow Measurement*, 2d ed., ISA, Research Triangle Park, N.C., 2001.

GENERAL CONSIDERATIONS

Process measurements encompass the application of the principles of metrology to the process in question. The objective is to obtain values for the current conditions within the process and to make this information available in a form usable by the control system, process operators, or management information systems. The term *measured variable* or *process variable* designates the process condition that is being determined.

Process measurements fall into two categories:

1. *Continuous measurements.* An example of a continuous measurement is a level measurement device that determines the liquid level in a tank (e.g., in meters).

2. *Discrete measurements.* An example of a discrete measurement is a level switch that indicates the presence or absence of liquid at the location at which the level switch is installed.

In continuous processes, most process control applications rely on continuous measurements. In batch processes, many of the process control applications utilize discrete as well as continuous measurements. In both types of processes, the safety interlocks and process interlocks rely largely on discrete measurements.

Continuous Measurements In most applications, continuous measurements provide more information than discrete measurements. Basically, discrete measurements involve a yes/no decision, whereas continuous measurements may entail considerable signal processing.

The components of a typical continuous measurement device are as follows:

1. *Sensor.* This component produces a signal that is related in a known manner to the process variable of interest. The sensors in use today are primarily of the electrical analog variety, and the signal is in the form of a voltage, a resistance, a capacitance, or some other directly measurable electrical quantity. Prior to the mid-1970s, instruments tended to use sensors whose signal was mechanical and thus compatible with pneumatic technology. Since that time, the fraction of sensors that are electronic and digital has grown considerably, often eliminating the need for analog-to-digital conversion.

2. *Signal processing.* The signal from a sensor is usually related in a nonlinear fashion to the process variable of interest. For the output of the measurement device to be linear with respect to the process variable of interest, linearization is required. Furthermore, the signal from the sensor might be affected by variables other than the process variable. In this case, additional variables must be sensed, and the signal from the sensor compensated to account for the other variables. For example, reference junction compensation is required for thermocouples (except when used for differential temperature measurements).

3. *Transmitter.* The measurement device output must be a signal that can be transmitted over some distance. Where electronic analog transmission is used, the low range on the transmitter output is 4 mA, and the upper range is 20 mA. Microprocessor-based transmitters (often referred to as *smart transmitters*) are usually capable of transmitting the measured variable digitally in engineering units.

Accuracy and Repeatability Definitions of terminology pertaining to process measurements can be obtained from standards available from the Instrumentation, Systems, and Automation Society (ISA) and from the Scientific Apparatus Makers Association [now Measurement, Control, and Automation Association (MCAA)], both of which are updated periodically. An appreciation of accuracy and repeatability is especially important. Some applications depend on the accuracy of the instrument, but other applications depend on repeatability. Excellent accuracy implies excellent repeatability; however, an instrument can have poor accuracy but excellent repeatability. In some applications, this is acceptable, as discussed below.

Range and Span A continuous measurement device is expected to provide credible values of the measured value between a lower range and an upper range. The difference between the upper range and the lower range is the span of the measurement device. The maximum value for the upper range and the minimum value for the lower range depend on the principles on which the measurement device is based and on the design chosen by the manufacturer of the measurement device. If the measured variable is greater than the upper range or less than the lower range, the measured variable is said to be *out of range* or the measurement device is said to be *overranged.*

Accuracy Accuracy refers to the difference between the measured value and the true value of the measured variable. Unfortunately, the true value is never known, so in practice accuracy refers to the difference between the measured value and an accepted standard value for the measured variable.

Accuracy can be expressed in four ways:

1. As an absolute difference in the units of the measured variable
2. As a percentage of the current reading
3. As a percentage of the span of the measured variable
4. As a percentage of the upper range of the span

For process measurements, accuracy as a percentage of span is the most common.

Manufacturers of measurement devices always state the accuracy of the instrument. However, these statements always provide specific or reference conditions at which the measurement device will perform with the stated accuracy, with temperature and pressure most often appearing in the reference conditions. When the measurement device is applied at other conditions, the accuracy is affected. Manufacturers usually also provide some statements on how accuracy is affected when the conditions of use deviate from the referenced conditions in the statement of accuracy. Although appropriate calibration procedures can minimize some of these effects, rarely can they be totally eliminated. It is easily possible for such effects to cause a measurement device with a stated accuracy of 0.25 percent of span at reference conditions to ultimately provide measured values with accuracies of 1 percent or less. Microprocessor-based measurement devices usually provide better accuracy than do the traditional electronic measurement devices.

In practice, most attention is given to accuracy when the measured variable is the basis for billing, such as in custody transfer applications. However, whenever a measurement device provides data to any type of optimization strategy, accuracy is very important.

Repeatability Repeatability refers to the difference between the measurements when the process conditions are the same. This can also be viewed from the opposite perspective. If the measured values are the same, repeatability refers to the difference between the process conditions.

For regulatory control, repeatability is of major interest. The basic objective of regulatory control is to maintain uniform process operation. Suppose that on two different occasions, it is desired that the temperature in a vessel be 800°C. The regulatory control system takes appropriate actions to bring the measured variable to 800°C. The difference between the process conditions at these two times is determined by the repeatability of the measurement device.

In the use of temperature measurement for control of the separation in a distillation column, repeatability is crucial but accuracy is not. Composition control for the overhead product is often based on a measurement of the temperature on one of the trays in the rectifying section. A target would be provided for this temperature. However, at periodic intervals, a sample of the overhead product is analyzed in the laboratory and the information is provided to the process operator. Should this analysis be outside acceptable limits, the operator would adjust the set point for the temperature. This procedure effectively compensates for an inaccurate temperature measurement; however, the success of this approach requires good repeatability from the temperature measurement.

Dynamics of Process Measurements Especially where the measurement device is incorporated into a closed-loop control configuration, dynamics are important. The dynamic characteristics depend on the nature of the measurement device, and on the nature of components associated with the measurement device (e.g., thermowells and sample conditioning equipment). The term *measurement system* designates the measurement device and its associated components.

The following dynamics are commonly exhibited by measurement systems:

- *Time constants.* Where there is a capacity and a throughput, the measurement device response will exhibit a time constant. For example, any temperature measurement device has a thermal capacity (mass times heat capacity) and a heat flow term (heat-transfer coefficient and area). Both the temperature measurement device and its associated thermowell will exhibit behavior reflecting its time constants.
- *Dead time.* Probably the best example of a measurement device that exhibits pure dead time (time delay) is the chromatograph, because the analysis is not available for some time after a sample is injected. Additional dead time results from the transportation lag within the sample system. Even continuous analyzer installations can exhibit dead time from the sample system.
- *Underdamped.* Measurement devices with mechanical components often have a natural harmonic and can exhibit underdamped behavior. The displacer type of level measurement device is capable of such behavior.

While the manufacturers of measurement devices can supply some information on the dynamic characteristics of their devices, interpretation is often difficult. Measurement device dynamics are quoted on varying bases, such as rise time, time to 63 percent response, settling time, etc. Even where the time to 63 percent response is quoted, it might not be safe to assume that the measurement device exhibits first-order behavior.

Where the manufacturer of the measurement device does not supply the associated equipment (thermowells, sample conditioning equipment, etc.), the user must incorporate the characteristics of these components to obtain the dynamics of the measurement system. An additional complication is that most dynamic data are stated for configurations involving reference materials such as water and air. The nature of the process material will affect the dynamic characteristics. For example, a thermowell will exhibit different characteristics when immersed in a viscous organic emulsion versus when immersed in water. It is often difficult to extrapolate the available data to process conditions of interest.

Similarly, it is often impossible, or at least very difficult, to experimentally determine the characteristics of a measurement system under the conditions where it is used. It is certainly possible to fill an emulsion polymerization reactor with water and determine the dynamic characteristics of the temperature measurement system. However, it is not possible to determine these characteristics when the reactor is filled with the emulsion under polymerization conditions.

The primary impact of unfavorable measurement dynamics is on the performance of closed-loop control systems. This explains why most control engineers are very concerned with minimizing measurement dynamics, even though the factors considered in dynamics are often subjective.

Selection Criteria The selection of a measurement device entails a number of considerations given below, some of which are almost entirely subjective.

1. *Measurement span.* The measurement span required for the measured variable must lie entirely within the instrument's envelope of performance.
2. *Performance.* Depending on the application, accuracy, repeatability, or perhaps some other measure of performance is appropriate. Where closed-loop control is contemplated, speed of response must be included.
3. *Reliability.* Data available from the manufacturers can be expressed in various ways and at various reference conditions. Often, previous experience with the measurement device within the purchaser's organization is weighted most heavily.
4. *Materials of construction.* The instrument must withstand the process conditions to which it is exposed. This encompasses considerations such as operating temperatures, operating pressures, corrosion, and abrasion. For some applications, seals or purges may be necessary.
5. *Prior use.* For the first installation of a specific measurement device at a site, training of maintenance personnel and purchases of spare parts might be necessary.
6. *Potential for releasing process materials to the environment.* Fugitive emissions are receiving ever-increasing attention. Exposure considerations, both immediate and long-term, for maintenance personnel are especially important when the process fluid is either corrosive or toxic.
7. *Electrical classification.* Article 500 of the National Electric Code provides for the classification of the hazardous nature of the process area in which the measurement device will be installed. If the measurement device is not inherently compatible with this classification, suitable enclosures must be purchased and included in the installation costs.
8. *Physical access.* Subsequent to installation, maintenance personnel must have physical access to the measurement device for maintenance and calibration. If additional structural facilities are required, they must be included in the installation costs.
9. *Invasive or noninvasive.* The insertion of a probe can result in fouling problems and a need for maintenance. Probe location must be selected carefully for good accuracy and minimal fouling.
10. *Cost.* There are two aspects of the cost:
    a. Initial purchase and installation (capital cost).
    b. Recurring costs (operational expense). This encompasses instrument maintenance, instrument calibration, consumables (e.g., titrating solutions must be purchased for automatic titrators), and any other costs entailed in keeping the measurement device in service.

Calibration Calibration entails the adjustment of a measurement device so that the value from the measurement device agrees with the value from a standard. The International Standards Organization (ISO) has developed a number of standards specifically directed to calibration of measurement devices. Furthermore, compliance with the ISO 9000 standards requires that the working standard used to calibrate a measurement device be traceable to an internationally recognized standard such as those maintained by the National Institute of Standards and Technology (NIST).

Within most companies, the responsibility for calibrating measurement devices is delegated to a specific department. Often, this department may also be responsible for maintaining the measurement device. The specific calibration procedures depend on the type of measurement device. The frequency of calibration is normally predetermined, but earlier action may be dictated if the values from the measurement device become suspect.

Calibration of some measurement devices involves comparing the measured value with the value from the working standard. Pressure and differential pressure transmitters are calibrated in this manner. Calibration of analyzers normally involves using the measurement device to analyze a specially prepared sample whose composition is known. These and similar approaches can be applied to most measurement devices.

Flow is an important measurement whose calibration presents some challenges. When a flow measurement device is used in applications such as custody transfer, provision is made to pass a known flow through the meter. However, such a provision is costly and is not available for most in-process flow meters. Without such a provision, a true calibration of the flow element itself is not possible. For orifice meters, calibration of the flow meter normally involves calibration of the differential pressure transmitter, and the orifice plate is usually only inspected for deformation, abrasion, etc. Similarly, calibration of a magnetic flow meter normally involves calibration of the voltage measurement circuitry, which is analogous to calibration of the differential pressure transmitter for an orifice meter.

In the next subsection we cover the major types of measurement devices used in the process industries, principally the "big five" measurements: temperature, flow rate, pressure, level, and composition, along with online physical property measurement techniques. Table 8-7 summarizes the different options under each of the principal measurements.

TEMPERATURE MEASUREMENTS

Measurement of the hotness or coldness of a body or fluid is commonplace in the process industries. Temperature-measuring devices utilize systems with properties that vary with temperature in a simple, reproducible manner and thus can be calibrated against known references (sometimes called *secondary thermometers*). The three dominant measurement devices used in automatic control are thermocouples, resistance thermometers, and pyrometers, and they are applicable over different temperature regimes.

Thermocouples Temperature measurements using thermocouples are based on the discovery by Seebeck in 1821 that an electric current flows in a continuous circuit of two different metallic wires if the two junctions are at different temperatures. Suppose A and B are the two metals, and T_1 and T_2 are the temperatures of the junctions. Let T_1 and T_2 be the reference junction (cold junction) and the measuring junction, respectively. If the thermoelectric current i flows from A to B, metal A is customarily referred to as

TABLE 8-7 Online Measurement Options for Process Control

Temperature	Flow	Pressure	Level	Composition
Thermocouple	Orifice	Liquid column	Float-activated	Gas-liquid chromatography (GLC)
Resistance temperature detector (RTD)	Venturi	Elastic element	Chain gauge	Mass spectrometry (MS)
Filled-system thermometer	Rotameter	Bourdon tube	Lever	Magnetic resonance analysis (MRA)
Bimetal thermometer	Turbine	Bellow	Magnetically coupled	Infrared (IR) spectroscopy
Pyrometer	Vortex-shedding	Diaphragm	Head devices	Raman spectroscopy
Total radiation	Ultrasonic	Strain gauges	Bubble tube	Ultraviolet (uv) spectroscopy
Photoelectric	Magnetic	Piezoresistive transducers	Electrical (conductivity)	Thermal conductivity
Ratio	Thermal mass	Piezoelectric transducers	Sonic	Refractive index (RI)
Laser	Coriolis	Optical fiber	Laser	Capacitance probe
Surface acoustic wave	Target		Radiation	Surface acoustic wave
Semiconductor			Radar	Electrophoresis
				Electrochemical
				Paramagnetic
				Chemi/bioluminescence
				Tunable diode laser absorption

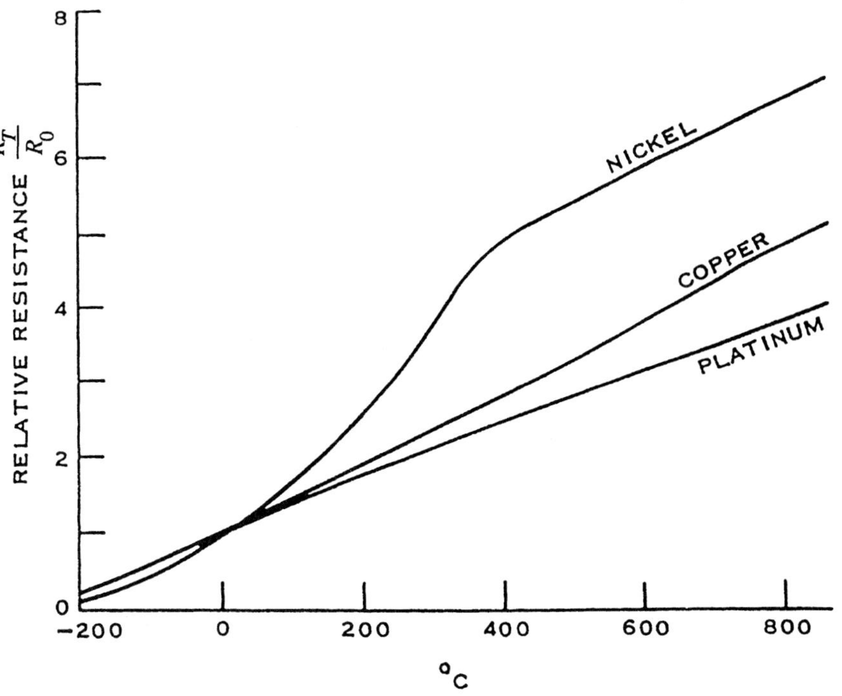

FIG. 8-63 Typical resistance thermometer curves for platinum, copper, and nickel wire, where R_T = resistance at temperature T and R_0 = resistance at 0°C.

thermoelectrically positive to metal B. Metal pairs used for thermocouples include platinum-rhodium (the most popular and accurate), chromel-alumel, copper-constantan, and iron-constantan. The thermal emf is a measure of the difference in temperature between T_2 and T_1. In control systems the reference junction is usually located at the emf-measuring device. The reference junction may be held at constant temperature such as in an ice bath or a thermostated oven, or it may be at ambient temperature but electrically compensated (cold junction compensated circuit) so that it appears to be held at a constant temperature.

Resistance Thermometers The resistance thermometer depends upon the inherent characteristics of materials to change in electrical resistance when they undergo a change in temperature. Industrial resistance thermometers are usually constructed of platinum, copper, or nickel, and more recently semiconducting materials such as thermistors are being used. Basically, a resistance thermometer is an instrument for measuring electrical resistance that is calibrated in units of temperature instead of in units of resistance (typically ohms). Several common forms of bridge circuits are employed in industrial resistance thermometry, the most common being the Wheatstone bridge. A *resistance thermometer detector* (RTD) consists of a resistance conductor (metal) that generally shows an increase in resistance with temperature. The following equation typically represents the variation of resistance with temperature (°C or K):

$$R_T = R_0(1 + a_1T + a_2T^2 + \cdots + a_nT^n) \tag{8-62}$$
$$R_0 = \text{resistance at } 0°\text{C}$$

The temperature coefficient of resistance α_T is expressed as

$$\alpha_T = \frac{1}{R_T}\frac{dR_T}{dT} \tag{8-63}$$

For most metals α_T is positive. For many pure metals, the coefficient is essentially constant and stable over large portions of their useful range. Typical resistance versus temperature curves for platinum, copper, and nickel are given in Fig. 8-63, with platinum usually the metal of choice. Platinum has a useful range of −200 to 800°C, while nickel and copper are more limited. Detailed resistance versus temperature tables are available from the National Institute of Standards and Technology (NIST) and suppliers of resistance thermometers. Table 8-8 gives recommended temperature measurement ranges for thermocouples and RTDs. Resistance thermometers are receiving increased usage because they are about 10 times more accurate than thermocouples. Note that Fig. 8-63 shows a temperature region where the change in resistance is nearly linear, a desirable characteristic.

Thermistors Thermistors are nonlinear temperature-dependent resistors, and normally only the materials with negative temperature coefficient of resistance (NTC type) are used. The resistance is related to temperature as

$$R_T = R_{T_r} \exp\left[\beta\left(\frac{1}{T} - \frac{1}{T_r}\right)\right] \tag{8-64}$$

where T_r is a reference temperature, which is generally 298 K. Thus

$$\alpha_T = \frac{1}{R_T}\frac{dR_T}{dT} \tag{8-65}$$

The value of β is on the order of 4000, so at room temperature (298 K), α_T = −0.045 for thermistor and 0.0035 for 100-Ω platinum RTD. Compared with RTDs, NTC-type thermistors are advantageous in that the detector dimension can be made small, the resistance value is higher (less affected by the resistances of the connecting leads), and it has higher temperature sensitivity and low thermal inertia of the sensor. Disadvantages of thermistors to RTDs include nonlinear characteristics and low measuring temperature range.

TABLE 8-8 Recommended Temperature Measurement Ranges for RTDs and Thermocouples

Resistance Thermometer Detectors (RTDs)	
100-V Pt	−200–+800°C
120-V Ni	-80–+320°C
Thermocouples	
Type B	0–+1700°C
Type E	0–+900°C
Type J	0–+750°C
Type K	0–+1260°C
Type N	0–+1300°C
Type R	870–+1450°C
Type S	960–+1450°C
Type T	-170–+350°C

Filled-System Thermometers The filled-system thermometer is designed to provide an indication of temperature some distance removed from the point of measurement. The measuring element (bulb) contains a gas or liquid that changes in volume, pressure, or vapor pressure with temperature. This change is communicated through a capillary tube to a Bourdon tube or other pressure- or volume-sensitive device. The Bourdon tube responds so as to provide a motion related to the bulb temperature. Those systems that respond to volume changes are completely filled with a liquid. Systems that respond to pressure changes either are filled with a gas or are partially filled with a volatile liquid. Changes in gas or vapor pressure with changes in bulb temperatures are carried through the capillary to the Bourdon. The latter bulbs are sometimes constructed so that the capillary is filled with a non-volatile liquid.

Fluid-filled bulbs deliver enough power to drive controller mechanisms and even directly actuate control valves. These devices are characterized by large thermal capacity, which sometimes leads to slow response, particularly when they are enclosed in a thermal well for process measurements. Filled-system thermometers are used extensively in industrial processes for a number of reasons. The simplicity of these devices allows rugged construction, minimizing the possibility of failure with a low level of maintenance, and inexpensive overall design of control equipment. In case of system failure, the entire unit must be replaced or repaired.

As they are normally used in the process industries, the sensitivity and percentage of span accuracy of these thermometers are generally the equal of those of other temperature-measuring instruments. Sensitivity and absolute accuracy are not the equal of those of short-span electrical instruments used in connection with resistance-thermometer bulbs. Also the maximum temperature is somewhat limited.

Bimetal Thermometers *Thermostatic bimetal* can be defined as a composite material made up of strips of two or more metals fastened together. This composite, because of the different expansion rates of its components, tends to change curvature when subjected to a change in temperature. With one end of a straight strip fixed, the other end deflects in proportion to the temperature change, the square of the length, and inversely as the thickness, throughout the linear portion of the deflection characteristic curve. If a bimetallic strip is wound into a helix or a spiral and one end is fixed, the other end will rotate when heat is applied. For a thermometer with uniform scale divisions, a bimetal must be designed to have linear deflection over the desired temperature range. Bimetal thermometers are used at temperatures ranging from 580°C down to −180°C and lower. However, at the low temperatures the rate of deflection drops off quite rapidly. Bimetal thermometers do not have long-time stability at temperatures above 430°C.

Pyrometers Planck's distribution law gives the radiated energy flux $q_b(\lambda,T)\,d\lambda$ in the wavelength range λ to $\lambda + d\lambda$ from a black surface:

$$q_b(\lambda,T) = \frac{C_1}{\lambda^5}\frac{1}{e^{C_2/\lambda T}-1} \tag{8-66}$$

where $C_1 = 3.7418 \times 10^{10}\ \mu\text{W}\cdot\mu\text{m}^4\cdot\text{cm}^{-2}$ and $C_2 = 14{,}388\ \mu\text{m}\cdot\text{K}$.

If the target object is a black body and if the pyrometer has a detector that measures the specific wavelength signal from the object, then the temperature of the object can be accurately estimated from Eq. (8-66). While it is possible to construct a physical body that closely approximates black body behavior, most real-world objects are not black bodies. The deviation from a black body can be described by the spectral emissivity

$$\varepsilon_T = \frac{q(T)}{q_b(T)} \tag{8-67}$$

where $q(\lambda,T)$ is the radiated energy flux from a real body in the wavelength range λ to $\lambda + d\lambda$ and $0 < \varepsilon_{\lambda,T} < 1$. Integrating Eq. (8-66) over all wavelengths gives the Stefan-Boltzmann equation

$$q_b(T) = \int_0^{\infty} q_b(\lambda,T)\,d\lambda = \sigma T^4 \tag{8-68}$$

where σ is the Stefan-Boltzmann constant. Similar to Eq. (8-63), the emissivity ε_T for the total radiation is

$$\varepsilon_T = \frac{q(T)}{q_b(T)} \tag{8-69}$$

where $q(T)$ is the radiated energy flux from a real body with emissivity ε_T.

Total Radiation Pyrometers In total radiation pyrometers, the thermal radiation is detected over a large range of wavelengths from the object at high temperature. The detector is normally a thermopile, which is built by connecting several thermocouples in series to increase the temperature measurement range. The pyrometer is calibrated for black bodies, so the indicated temperature T_p should be converted for non–black body temperature.

Photoelectric Pyrometers Photoelectric pyrometers belong to the class of band radiation pyrometers. The thermal inertia of thermal radiation detectors does not permit the measurement of rapidly changing temperatures. For example, the smallest time constant of a thermal detector is about 1 ms while the smallest time constant of a photoelectric detector can be about 1 or 2 s. Photoelectric pyrometers may use photoconductors, photodiodes, photovoltaic cells, or vacuum photocells. Photoconductors are built from glass plates with thin film coatings of 1-μm thickness, using PbS, CdS, PbSe, or PbTe. When the incident radiation has the same wavelength as the materials are able to absorb, the captured incident photons free photoelectrons, which form an electric current. Photodiodes in germanium or silicon are operated with a reverse-bias voltage applied. Under the influence of the incident radiation, their conductivity as well as their reverse saturation current is proportional to the intensity of the radiation within the spectral response band from 0.4 to 1.7 μm for Ge and from 0.6 to 1.1 μm for Si. Because of the above characteristics, the operating range of a photoelectric pyrometer can be either spectral or in a specific band. Photoelectric pyrometers can be applied for a specific choice of the wavelength.

Disappearing Filament Pyrometers Disappearing filament pyrometers can be classified as spectral pyrometers. The brightness of a lamp filament is changed by adjusting the lamp current until the filament disappears against the background of the target, at which point the temperature is measured. Because the detector is the human eye, it is difficult to calibrate for online measurements.

Ratio Pyrometers The ratio pyrometer is also called the *two-color pyrometer*. Two different wavelengths are utilized for detecting the radiated signal. If one uses Wien's law applicable for small values of λT, the detected signals from spectral radiant energy flux emitted at wavelengths λ_1 and λ_2 with emissivities ε_{λ_1} and ε_{λ_2} can be calculated, along with their ratio.

$$S_{\lambda_1} = KC_1\varepsilon_{\lambda_1}\lambda_1^{-5}\exp\left(\frac{-C_2}{\lambda_1 T}\right) \tag{8-70}$$

$$S_{\lambda_2} = KC_1\varepsilon_{\lambda_2}\lambda_2^{-5}\exp\left(\frac{-C_2}{\lambda_2 T}\right) \tag{8-71}$$

The ratio of signals S_{λ_1} and S_{λ_2} is

$$\frac{S_{\lambda_1}}{S_{\lambda_2}} = \frac{\varepsilon_{\lambda_1}}{\varepsilon_{\lambda_2}}\left(\frac{\lambda_2}{\lambda_1}\right)^5 \exp\left[\frac{C_2}{T}\left(\frac{1}{\lambda_2}-\frac{1}{\lambda_1}\right)\right] \tag{8-72}$$

Nonblack or nongray bodies are characterized by the wavelength dependence of their spectral emissivity. Let T_c be defined as the temperature of the body corresponding to the temperature of a black body. If the ratio of its radiant intensities at wavelengths λ_1 and λ_2 equals the ratio of the radiant intensities of the nonblack body, whose temperature is to be measured at the same wavelength, then Wien's law gives

$$\frac{\varepsilon_{\lambda_1}\exp(-C_2/\lambda_1 T)}{\varepsilon_{\lambda_2}\exp(-C_2/\lambda_2 T)} = \frac{\exp(-C_2/\lambda_1 T_c)}{\exp(-C_2/\lambda_2 T_c)} \tag{8-73}$$

where T is the true temperature of the body. Rearranging Eq. (8-73) gives

$$T = \left[\frac{\ln(\varepsilon_{\lambda_1}/\varepsilon_{\lambda_2})}{C_2(1/\lambda_1 - 1/\lambda_2)} + \frac{1}{T_c}\right]^{-1} \tag{8-74}$$

For black or gray bodies, Eq. (8-74) reduces to

$$\frac{S_{\lambda_1}}{S_{\lambda_2}} = \left(\frac{\lambda_2}{\lambda_1}\right)^5 \exp\left[\frac{C_2}{T}\left(\frac{1}{\lambda_2}-\frac{1}{\lambda_1}\right)\right] \tag{8-75}$$

Thus by measuring the ratio of S_{λ_1} and S_{λ_2}, temperature T can be estimated.

Accuracy of Pyrometers Most of the temperature estimation methods for pyrometers assume that the object either is a gray body or has known emissivity values. The emissivity of the nonblack body depends on the internal state or the surface geometry of the objects. Also the medium through which the thermal radiation passes is not always transparent. These inherent uncertainties of the emissivity values make the accurate estimation of the temperature of the target objects difficult. Proper selection of the pyrometer and accurate emissivity values can provide a high level of accuracy. The fact that the pyrometer can be protected from the high-temperature environment makes it attractive.

PRESSURE MEASUREMENTS

Pressure, defined as force per unit area, is usually expressed in terms of familiar units of weight-force and area or the height of a column of liquid which produces a like pressure at its base. Process pressure-measuring devices may be divided into three groups: (1) those based on the measurement of the height of a liquid column, (2) those based on the measurement of the distortion of an elastic pressure chamber, and (3) electrical sensing devices.

Liquid-Column Methods Liquid-column pressure-measuring devices are those in which the pressure being measured is balanced against the pressure exerted by a column of liquid. If the density of the liquid is known, the height of the liquid column is a measure of the pressure. Most forms of liquid-column pressure-measuring devices are commonly called *manometers*. When the height of the liquid is observed visually, the liquid columns are contained in glass or other transparent tubes. The height of the liquid column may be measured in length units or calibrated in pressure units. Depending on the pressure range, water and mercury are the liquids most frequently used. Because the density of the liquid used varies with temperature, the temperature must be taken into account for accurate pressure measurements.

Elastic-Element Methods Elastic element pressure-measuring devices are those in which the measured pressure deforms some elastic material (usually metallic) within its elastic limit, the magnitude of the deformation being approximately proportional to the applied pressure. These devices may be loosely classified into three types: Bourdon tube, bellows, and diaphragm.

Bourdon Tube Probably the most frequently used process pressure-indicating device is the C-spring Bourdon tube pressure gauge. Gauges of this general type are available in a wide variety of pressure ranges and materials of construction. Materials are selected on the basis of pressure range, resistance to corrosion by the process materials, and effect of temperature on calibration. Gauges calibrated with pressure, vacuum, compound (combination pressure and vacuum), and suppressed-zero ranges are available.

Bellows The bellows element is an axially elastic cylinder with deep folds or convolutions. The bellows may be used unopposed, or it may be restrained by an opposing spring. The pressure to be measured may be applied either to the inside or to the space outside the bellows, with the other side exposed to atmospheric pressure. For measurement of absolute pressure either the inside or the space outside of the bellows can be evacuated and sealed. Differential pressures may be measured by applying the pressures to opposite sides of a single bellows or to two opposing bellows.

Diaphragm Diaphragm elements may be classified into two principal types: those that utilize the elastic characteristics of the diaphragm and those that are opposed by a spring or other separate elastic element. The first type usually consists of one or more capsules, each composed of two diaphragms bonded together by soldering, brazing, or welding. The diaphragms are flat or corrugated circular metallic disks. Metals commonly used in diaphragm elements include brass, phosphor bronze, beryllium copper, and stainless steel. Ranges are available from fractions of an inch of water to over 800 in (200 kPa) gauge. The second type of diaphragm is used for containing the pressure and exerting a force on the opposing elastic element. The diaphragm is a flexible or slack diaphragm of rubber, leather, impregnated fabric, or plastic. Movement of the diaphragm is opposed by a spring which determines the deflection for a given pressure. This type of diaphragm is used for the measurement of extremely low pressure, vacuum, or differential pressure.

Electrical Methods Electrical methods for pressure measurement include strain gauges, piezoresistive transducers, and piezoelectric transducers.

Strain Gauges When a wire or other electrical conductor is stretched elastically, its length is increased and its diameter is decreased. Both of these dimensional changes result in an increase in the electrical resistance of the conductor. Devices utilizing resistance-wire grids for measuring small distortions in elastically stressed materials are commonly called *strain gauges*. Pressure-measuring elements utilizing strain gauges are available in a wide variety of forms. They usually consist of one of the elastic elements described earlier to which one or more strain gauges have been attached to measure the deformation. There are two basic strain gauge forms: bonded and unbonded. Bonded strain gauges are bonded directly to the surface of the elastic element whose strain is to be measured. The unbonded strain gauge transducer consists of a fixed frame and an armature that moves with respect to the frame in response to the measured pressure. The strain gauge wire filaments are stretched between the armature and frame. The strain gauges are usually connected electrically in a Wheatstone bridge configuration.

Strain gauge pressure transducers are manufactured in many forms for measuring gauge, absolute, and differential pressures and vacuum. Full-scale ranges from 25.4 mm of water to 10,134 MPa are available. Strain gauges bonded directly to a diaphragm pressure-sensitive element usually have an extremely fast response time and are suitable for high-frequency dynamic pressure measurements.

Piezoresistive Transducers A variation of the conventional strain gauge pressure transducer uses bonded single-crystal semiconductor wafers, usually silicon, whose resistance varies with strain or distortion. Transducer construction and electrical configurations are similar to those using conventional strain gauges. A permanent magnetic field is applied perpendicular to the resonating sensor. An alternating current causes the resonator to vibrate, and the resonant frequency is a function of the pressure (tension) of the resonator. The principal advantages of piezoresistive transducers are a much higher bridge voltage output and smaller size. Full-scale output voltages of 50 to 100 mV/V of excitation are typical. Some newer devices provide digital rather than analog output.

Piezoelectric Transducers Certain crystals produce a potential difference between their surfaces when stressed in appropriate directions. Piezoelectric pressure transducers generate a potential difference proportional to a pressure-generated stress. Because of the extremely high electrical impedance of piezoelectric crystals at low frequency, these transducers are usually not suitable for measurement of static process pressures.

FLOW MEASUREMENTS

Flow, defined as volume per unit of time at specified temperature and pressure conditions, is generally measured by positive displacement or rate meters. The term *positive displacement meter* applies to a device in which the flow is divided into isolated measured volumes when the number of fillings of these volumes is counted in some manner. The term *rate meter* applies to all types of flow meters through which the material passes without being divided into isolated quantities. Movement of the material is usually sensed by a primary measuring element that activates a secondary device. The flow rate is then inferred from the response of the secondary device by means of known physical laws or from empirical relationships.

The principal classes of flow-measuring instruments used in the process industries are variable-head, variable-area, positive-displacement, and turbine instruments; mass flow meters; vortex-shedding and ultrasonic flow meters; magnetic flow meters; and more recently Coriolis mass flow meters. Head meters are covered in detail in Sec. 5.

Orifice Meter The most widely used flow meter involves placing a fixed-area flow restriction (an orifice) in the pipe carrying the fluid. This flow restriction causes a pressure drop which can be related to flow rate. The sharp-edge orifice is popular because of its simplicity, low cost, and the large amount of research data on its behavior. For the orifice meter, the flow rate Q_a for a liquid is given by

$$Q_a = \frac{C_d A_2}{\sqrt{1-(A_2/A_1)^2}} \cdot \sqrt{\frac{2(p_1 - p_2)}{\rho}} \tag{8-76}$$

where $p_1 - p_2$ is the pressure drop, ρ is the density, A_1 is the pipe cross-sectional area, A_2 is the orifice cross-sectional area, and C_d is the discharge coefficient. The discharge coefficient C_d varies with the Reynolds number at the orifice and can be calibrated with a single fluid, such as water (typically $C_d \approx 0.6$). If the orifice and pressure taps are constructed according to certain standard dimensions, quite accurate (about 0.4 to 0.8 percent error) values of C_d may be obtained. Also note that the standard calibration data assume no significant flow disturbances such as elbows and valves for a certain minimum distance upstream of the orifice. The presence of such disturbances close to the orifice can cause errors of as much as 15 percent. Accuracy in measurements limits the meter to a flow rate range of 3:1. The orifice has a relatively large permanent pressure loss that must be made up by the pumping machinery.

Venturi Meter The venturi tube operates on exactly the same principle as the orifice [see Eq. (8-76)]. Discharge coefficients of venturis are larger than those for orifices and vary from about 0.94 to 0.99. A venturi gives a definite improvement in power losses over an orifice and is often indicated for measuring very large flow rates, where power losses can become economically significant. The initial higher cost of a venturi over an orifice may thus be offset by reduced operating costs.

Rotameter A rotameter consists of a vertical tube with a tapered bore in which a float changes position with the flow rate through the tube. For a given flow rate, the float remains stationary because the vertical forces of differential pressure, gravity, viscosity, and buoyancy are balanced. The float position is the output of the meter and can be made essentially linear with flow rate by making the tube area vary linearly with the vertical distance.

Turbine Meter If a turbine wheel is placed in a pipe containing a flowing fluid, its rotary speed depends on the flow rate of the fluid. A turbine can be designed whose speed varies linearly with flow rate. The speed can be measured accurately by counting the rate at which turbine blades pass a given point, using magnetic pickup to produce voltage pulses. By feeding these pulses to an electronic pulse rate meter one can measure the flow rate by summing the pulses during a timed interval. Turbine meters are available with full-scale flow rates ranging from about 0.1 to 30,000 gal/min for liquids and 0.1 to 15,000 ft^3/min for air. Nonlinearity can be less than 0.05 percent in the larger sizes. Pressure drop across the meter varies with the square of the flow rate and is about 3 to 10 psi at full flow. Turbine meters can follow flow transients quite accurately since their fluid/mechanical time constant is on the order of 2 to 10 ms.

Vortex-Shedding Flow Meters These flow meters take advantage of vortex shedding, which occurs when a fluid flows past a nonstreamlined object (a blunt body). The flow cannot follow the shape of the object and separates from it, forming turbulent vortices or eddies at the object's side surfaces. As the vortices move downstream, they grow in size and are eventually shed or detached from the object. Shedding takes place alternately at either side of the object, and the rate of vortex formation and shedding is directly proportional to the volumetric flow rate. The vortices are counted and used to develop a signal linearly proportional to the flow rate. The digital signals can easily be totaled over an interval of time to yield the flow rate. Accuracy can be maintained regardless of density, viscosity, temperature, or pressure when the Reynolds number is greater than 10,000. There is usually a low flow cutoff point below which the meter output is clamped at zero. This flow meter is recommended for use with relatively clean, low-viscosity liquids, gases, and vapors, and rangeability of 10:1 to 20:1 is typical. A sufficient length of straight-run pipe is necessary to prevent distortion in the fluid velocity profile.

Ultrasonic Flow Meters An ultrasonic flow meter is based upon the variable time delays of received sound waves which arise when a flowing liquid's rate of flow is varied. Two fundamental measurement techniques, depending upon liquid cleanliness, are generally used. In the first technique two opposing transducers are inserted in a pipe so that one transducer is downstream from the other. These transducers are then used to measure the difference between the velocity at which the sound travels with the direction of flow and the velocity at which it travels against the direction of flow. The differential velocity is measured either by (1) direct time delays using sound wave burst or (2) frequency shifts derived from beat-together, continuous signals. The frequency measurement technique is usually preferred because of its simplicity and independence of the liquid static velocity. A relatively clean liquid is required to preserve the uniqueness of the measurement path.

In the second technique, the flowing liquid must contain scatters in the form of particles or bubbles which will reflect the sound waves. These scatters should be traveling at the velocity of the liquid. A Doppler method is applied by transmitting sound waves along the flow path and measuring the frequency shift in the returned signal from the scatters in the process fluid. This frequency shift is proportional to liquid velocity.

Magnetic Flow Meters The principle behind these flow meters is Faraday's law of electromagnetic inductance. The magnitude of the voltage induced in a conductive medium moving at right angles through a magnetic field is directly proportional to the product of the magnetic flux density, the velocity of the medium, and the path length between the probes. A minimum value of fluid conductivity is required to make this approach viable. The pressure of multiple phases or undissolved solids can affect the accuracy of the measurement if the velocities of the phases are different from that for straight-run pipe. Magmeters are very accurate over wide flow ranges and are especially accurate at low flow rates. Typical applications include metering viscous fluids, slurries, or highly corrosive chemicals. Because magmeters should be filled with fluid, the preferred installation is in vertical lines with flow going upward. However, magmeters can be used in tight piping schemes where it is impractical to have long pipe runs, typically requiring lengths equivalent to five or more pipe diameters.

Coriolis Mass Flow Meters Coriolis mass flow meters utilize a vibrating tube in which Coriolis acceleration of a fluid in a flow loop can be created and measured. They can be used with virtually any liquid and are extremely insensitive to operating conditions, with pressure ranges over 100:1. These meters are more expensive than volumetric meters and range in size from $\frac{1}{16}$ to 6 in. Due to the circuitous path of flow through the meter, Coriolis flow meters exhibit higher than average pressure drops. The meter should be installed so that it will remain full of fluid, with the best installation in a vertical pipe with flow going upward. There is no Reynolds number limitation with this meter, and it is quite insensitive to velocity profile distortions and swirl, hence there is no requirement for straight piping upstream. Coriolis flow meters are popular for custody measurement in the pipeline because they are very accurate but are more expensive than other flow meters.

Thermal Mass Flow Meters The trend in the chemical process industries is toward increased usage of mass flow meters that are independent of changes in pressure, temperature, viscosity, and density. Thermal mass meters are widely used in semiconductor manufacturing and in bioprocessing for control of low flow rates (called *mass flow controllers*, or MFCs). MFCs measure the heat loss from a heated element, which varies with flow rate, with an accuracy of ±1 percent. Capacitance probes measure the dielectric constant of the fluid and are useful for flow measurements of slurries and other two-phase flows.

LEVEL MEASUREMENTS

The measurement of *level* can be defined as the determination of the location of the interface between two fluids, separable by gravity, with respect to a fixed reference plane. The most common level measurement is that of the interface between a liquid and a gas. Other level measurements frequently encountered are the interface between two liquids, between a granular or fluidized solid and a gas, and between a liquid and its vapor.

A commonly used basis for classification of level devices is as follows: float-actuated, displacer, and head devices, and a miscellaneous group which depends mainly on fluid characteristics.

Float-Actuated Devices Float-actuated devices are characterized by a buoyant member which floats at the interface between two fluids. Because a significant force is usually required to move the indicating mechanism, float-actuated devices are generally limited to liquid-gas interfaces. By properly weighting the float, they can be used to measure liquid-liquid interfaces. Float-actuated devices may be classified on the basis of the method used to couple the float motion to the indicating system, as discussed below.

Chain or Tape Float Gauge In these types of gauges, the float is connected to the indicating mechanism by means of a flexible chain or tape. These gauges are commonly used in large atmospheric storage tanks. The gauge-board type is provided with a counterweight to keep the tape or chain taut. The tape is stored in the gauge head on a spring-loaded reel. The float is usually a pancake-shaped hollow metal float, with guide wires from top to bottom of the tank to constrain it.

Lever and Shaft Mechanisms In pressurized vessels, float-actuated lever and shaft mechanisms are frequently used for level measurement. This type of mechanism consists of a hollow metal float and lever attached to a rotary shaft which transmits the float motion to the outside of the vessel through a rotary seal.

Magnetically Coupled Devices A variety of float-actuated level devices have been developed which transmit the float motion by means of magnetic coupling. Typical of this class of devices are magnetically operated level switches and magnetic-bond float gauges. A typical magnetic-bond float gauge consists of a hollow magnet-carrying float that rides along a vertical nonmagnetic guide tube. The follower magnet is connected and drives an indicating dial similar to that on a conventional tape float gauge. The float and guide tube are in contact with the measured fluid and come in a variety of materials for resistance to corrosion and to withstand high pressures or vacuum. Weighted floats for liquid-liquid interfaces are available.

Head Devices A variety of devices utilize hydrostatic head as a measure of level. As in the case of displacer devices, accurate level measurement by hydrostatic head requires an accurate knowledge of the densities of both heavier-phase and lighter-phase fluids. The majority of this class of systems utilize standard pressure and differential pressure measuring devices.

Bubble Tube Systems The commonly used bubble tube system sharply reduces restrictions on the location of the measuring element. To eliminate or reduce variations in pressure drop due to the gas flow rate, a constant differential regulator is commonly employed to maintain a constant gas flow rate. Because the flow of gas through the bubble tube prevents entry of the process liquid into the measuring system, this technique is particularly useful with corrosive or viscous liquids, liquids subject to freezing, and liquids containing entrained solids.

Electrical Methods Two electrical characteristics of fluids—conductivity and dielectric constant—are frequently used to distinguish between two phases for level measurement purposes. An application of electrical conductivity is the fixed-point level detection of a conductive liquid such as high and low water levels. A voltage is applied between two electrodes inserted into the vessel at different levels. When both electrodes are immersed in the liquid, current flows. Capacitance-type level measurements are based on the fact that the electrical capacitance between two electrodes varies with the dielectric constant of the material between them. A typical continuous level measurement system consists of a rod electrode positioned vertically in a vessel, the other electrode usually being the metallic vessel wall. The electrical capacitance between the electrodes is a measure of the height of the interface along the rod electrode. The rod is usually conductively insulated from process fluids by a coating of plastic. The dielectric constants of most liquids and solids are markedly higher than those of gases and vapors (by a factor of 2 to 5). The dielectric constant of water and other polar liquids is 10 to 20 times that of hydrocarbons and other nonpolar liquids.

Thermal Methods Level-measuring systems may be based on the difference in thermal characteristics between the fluids, such as temperature or thermal conductivity. A fixed-point level sensor based on the difference in thermal conductivity between two fluids consists of an electrically heated thermistor inserted into the vessel. The temperature of the thermistor and consequently its electrical resistance increase as the thermal conductivity of the fluid in which it is immersed decreases. Because the thermal conductivity of liquids is markedly higher than that of vapors, such a device can be used as a point level detector for the liquid-vapor interface.

Sonic Methods A fixed-point level detector based on sonic propagation characteristics is available for detection of a liquid-vapor interface. This device uses a piezoelectric transmitter and receiver, separated by a short gap. When the gap is filled with liquid, ultrasonic energy is transmitted across the gap, and the receiver actuates a relay. With a vapor filling the gap, the transmission of ultrasonic energy is insufficient to actuate the receiver.

Laser Level Transmitters These are designed for bulk solids, slurries, and opaque liquids. A laser near the vessel top fires a short pulse of light down to the surface of the process liquid, where it reflects back to a detector at the vessel top. A timing circuit measures the elapsed time and calculates the fluid depth. Lasers are attractive because lasers have no false echoes and can be directed through tight spaces.

Radar Level Transmitters Radar systems operate by beaming microwaves downward, from either a horn or parabolic dish located on top of the vessel. The signal reflects off the fluid surface back to the source after it detects a change in dielectric constant from the vapor to the fluid. The round-trip time is proportional to the distance to the fluid level. Guided-wave radar systems provide a rigid probe or flexible cable to guide the microwave down the height of the tank and back. Guided-wave radar is much more efficient than open-air radar because the guide provides a more focused energy path.

PHYSICAL PROPERTY MEASUREMENTS

Physical property measurements are sometimes equivalent to composition analyzers, because the composition can frequently be inferred from the measurement of a selected physical property.

Density and Specific Gravity For binary or pseudo-binary mixtures of liquids or gases or a solution of a solid or gas in a solvent, the density is a function of the composition at a given temperature and pressure. Specific gravity is the ratio of the density of a noncompressible substance to the density of water at the same physical conditions. For nonideal solutions, empirical calibration will give the relationship between density and composition. Several types of measuring devices are described below.

Liquid Column Density may be determined by measuring the gauge pressure at the base of a fixed-height liquid column open to the atmosphere. If the process system is closed, then a differential pressure measurement is made between the bottom of the fixed-height liquid column and the vapor over the column. If vapor space is not always present, the differential pressure measurement is made between the bottom and top of a fixed-height column with the top measurement being made at a point below the liquid surface.

Displacement There are a variety of density measurement devices based on displacement techniques. A hydrometer is a constant-weight, variable-immersion device. The degree of immersion, when the weight of the hydrometer equals the weight of the displaced liquid, is a measure of the density. The hydrometer is adaptable to manual or automatic use. Another modification includes a magnetic float suspended below a solenoid, the varying magnetic field maintaining the float at a constant distance from the solenoid. Change in position of the float, resulting from a density change, excites an electrical system which increases or decreases the current through the solenoid.

Direct Mass Measurement One type of densitometer measures the natural vibration frequency and relates the amplitude to changes in density. The density sensor is a V-shaped tube that is held stationary at its node points and allowed to vibrate at its natural frequency. At the curved end of the V is an electrochemical device that periodically strikes the tube. At the other end of the V, the fluid is continuously passed through the tube. Between strikes, the tube vibrates at its natural frequency. The frequency changes directly in proportion to changes in density. A pickup device at the curved end of the V measures the frequency and electronically determines the fluid density. This technique is useful because it is not affected by the optical properties of the fluid. However, particulate matter in the process fluid can affect the accuracy.

Radiation-Density Gauges Gamma radiation may be used to measure the density of material inside a pipe or process vessel. The equipment is basically the same as for level measurement, except that here the pipe or vessel must be filled over the effective, irradiated sample volume. The source is mounted on one side of the pipe or vessel and the detector on the other side with appropriate safety radiation shielding surrounding the installation. Cesium 137 is used as the radiation source for path lengths under 610 mm (24 in) and cobalt 60 above 610 mm. The detector is usually an ionization gauge. The absorption of the gamma radiation is a function of density. Since the absorption path includes the pipe or vessel walls, an empirical calibration is used. Appropriate corrections must be made for the source intensity decay with time.

Viscosity Continuous viscometers generally measure either the resistance to flow or the drag or torque produced by movement of an element (moving surface) through the fluid. Each installation is normally applied over a narrow range of viscosities. Empirical calibration over this range allows use on both newtonian and nonnewtonian fluids. One such device uses a piston inside a cylinder. The hydrodynamic pressure of the process fluid raises the piston to a preset height. Then the inlet valve closes, the piston is allowed to free-fall, and the time of travel (typically a few seconds) is a measure of viscosity. Other geometries include the rotation of a spindle inside a sample chamber and a vibrating probe immersed in the fluid. Because viscosity depends on temperature, the viscosity measurement must be thermostated with a heater or cooler.

Refractive Index When light travels from one medium (e.g., air or glass) into another (e.g., a liquid), it undergoes a change of velocity and, if the angle of incidence is not 90°, a change of direction. For a given interface, angle, temperature, and wavelength of light, the amount of deviation or refraction will depend on the composition of the liquid. If the sample is transparent, the normal method is to measure the refraction of light transmitted through the glass-sample interface. If the sample is opaque, the reflectance near the critical angle at a glass-sample interface is measured. In an online refractometer, the process fluid is separated from the optics by a prism material. A beam of light is focused on a point in the fluid which creates a conic section of light at the prism, striking the fluid at different angles (greater than or less than the critical angle). The critical angle depends on the species concentrations; as the critical angle changes, the proportions of reflected and refracted light change. A photodetector produces a voltage signal proportional to the light refracted, when compared to a reference signal. Refractometers can be used with opaque fluids and in streams that contain particulates.

Dielectric Constant The dielectric constant of material represents its ability to reduce the electric force between two charges separated in space. This property is useful in process control for polymers, ceramic materials, and semiconductors. Dielectric constants are measured with respect to vacuum (1.0); typical values range from 2 (benzene) to 33 (methanol) to 80 (water). The value for water is higher than that for most plastics. A measuring cell is made of glass or some other insulating material and is usually doughnut-shaped, with the cylinders coated with metal, which constitute the plates of the capacitor.

Thermal Conductivity All gases and vapor have the ability to conduct heat from a heat source. At a given temperature and physical environment, radiation and convection heat losses will be stabilized, and the temperature of the heat source will mainly depend on the thermal conductivity and thus the composition of the surrounding gases. Thermal conductivity analyzers normally consist of a sample cell and a reference cell, each containing a combined heat source and detector. These cells are normally contained in a metal block with two small cavities in which the detectors are mounted. The sample flows through the sample cell cavity past the detector. The reference cell is an identical cavity with a detector through which a known gas flows. The combined heat source and detectors are normally either wire filaments or thermistors heated by a constant current. Because their resistance is a function of temperature, the sample detector resistance will vary with sample composition while the reference detector resistance will remain constant. The output from the detector bridge will be a function of sample composition.

CHEMICAL COMPOSITION ANALYZERS

Chemical composition is generally the most challenging online measurement. Before the era of online analyzers, messengers were required to deliver samples to the laboratory for analysis and to return the results to the control room. The long time delay involved prevented process adjustment from being made, affecting product quality. The development of online analyzers has automated this approach and reduced the analysis time. However, manual sampling is still frequently employed, especially in the specialty chemical industry where few instruments are commercially available. It is not unusual for a chemical composition analysis system to cost over $100,000, so it is important to assess the payback of such an investment versus the cost of manual sampling. Potential quality improvements can be an important consideration.

A number of composition analyzers used for process monitoring and control require chemical conversion of one or more sample components preceding quantitative measurement. These reactions include formation of suspended solids for turbidimetric measurement, formation of colored materials for colorimetric detection, selective oxidation or reduction for electrochemical measurement, and formation of electrolytes for measurement by electrical conductance. Some nonvolatile materials may be separated and measured by gas chromatography after conversion to volatile derivatives.

Chromatographic Analyzers These analyzers are widely used for the separation and measurement of volatile compounds and of compounds that can be quantitatively converted to volatile derivatives. The compounds to be measured are separated by placing a portion of the sample in a chromatographic column and carrying the compounds through the column with a gas stream, called *gas chromatography*, or GC. As a result of the different affinities of the sample components for the column packing, the compounds emerge successively as binary mixtures with the carrier gas. A detector at the column outlet measures a specific physical property that can be related to the concentrations of the compounds in the carrier gas. Both the concentration peak height and the peak height-time integral, i.e., peak area, can be related to the concentration of the compound in the original sample. The two detectors most commonly used for process chromatographs are the thermal conductivity detector and the hydrogen flame ionization detector. Thermal conductivity detectors, discussed earlier, require calibration for the thermal response of each compound. Hydrogen flame ionization detectors are more complicated than thermal conductivity detectors but are capable of 100 to 10,000 times greater sensitivity for hydrocarbons and organic compounds. For ultrasensitive detection of trace impurities, carrier gases must be specially purified.

Typically, all components can be analyzed in a 5- to 10-min time period (although miniaturized GCs are faster). *High-performance liquid chromatography* (HPLC) can be used to measure dissolved solute levels, including proteins.

Infrared Analyzers Many gaseous and liquid compounds absorb infrared radiation to some degree. The degree of absorption at specific wavelengths depends on the molecular structure and concentration. There are two common detector types for nondispersive infrared analyzers. These analyzers normally have two beams of radiation, an analyzing and a reference beam. One type of detector consists of two gas-filled cells separated by a diaphragm. As the amount of infrared energy absorbed by the detector gas in one cell changes, the cell pressure changes. This causes movement in the diaphragm, which in turn causes a change in capacitance between the diaphragm and a reference electrode. This change in electrical capacitance is measured as the output. The second type of detector consists of two thermopiles or two bolometers, one in each of the two radiation beams. The infrared radiation absorbed by the detector is measured by a differential thermocouple output or a resistance thermometer (bolometer) bridge circuit.

There are two common detector types for nondispersive analyzers. These analyzers normally have two beams of radiation, an analyzing beam and a reference beam. One type of detector consists of two gas-filled cells separated by a diaphragm. As the amount of infrared energy absorbed by the detector gas in one cell changes, the cell pressure changes. This causes movement in the diaphragm, which in turn causes a change in capacitance between the diaphragm and a reference electrode. This change in electrical capacitance is measured as the output. The second type of detector consists of two thermopiles or two bolometers, one in each of the two radiation beams. The infrared radiation absorbed by the detector is measured by a differential thermocouple output or a resistance thermometer (bolometer) bridge circuit. With gas-filled detectors, a chopped light system is normally used in which one side of the detector sees the source through the analyzing beam and the other side sees through the reference beam, alternating at a frequency of a few hertz.

Ultraviolet and Visible-Radiation Analyzers Many gas and liquid compounds absorb radiation in the near-ultraviolet or visible region. For example, organic compounds containing aromatic and carbonyl structural groups are good absorbers in the ultraviolet region. Also many inorganic salts and gases absorb in the ultraviolet or visible region. In contrast, straight-chain and saturated hydrocarbons, inert gases, air, and water vapor are essentially transparent. Process analyzers are designed to measure the absorbance in a particular wavelength band. The desired band is normally isolated by means of optical filters. When the absorbance is in the visible region, the term *colorimetry* is used. A phototube is the normal detector. Appropriate optical filters are used to limit the energy reaching the detector to the desired level and the desired wavelength region. Because absorption by the sample is logarithmic, if a sufficiently narrow wavelength region is used, an exponential amplifier is sometimes used to compensate and produce a linear output.

Paramagnetism A few gases including O_2, NO, and NO_2 exhibit paramagnetic properties as a result of unpaired electrons. In a nonuniform magnetic field, paramagnetic gases, because of their magnetic susceptibility, tend to move toward the strongest part of the field, thus displacing diamagnetic gases. Paramagnetic susceptibility of these gases decreases with temperature. These effects permit measurement of the concentration of the strongest paramagnetic gas, oxygen. An oxygen analyzer uses a dumbbell suspended in the magnetic field which is repelled or attracted toward the magnetic field depending on the magnetic susceptibility of the gas.

Other Analyzers *Mass spectroscopy* (MS) determines the partial pressures of gases in a mixture of directing ionized gases into a detector under a vacuum (10^{-6} torr), and the gas-phase composition is then monitored more or less continuously based on the molecular weight of the species (Nichols, 2010). Sometimes GC is combined with MS to obtain a higher level of discrimination of the components present. Fiber-optic sensors are attractive options (although higher-cost) for acquiring measurements in harsh environments such as high temperature or pressure. The transducing technique used by these sensors is optical and does not involve electric signals, so they are immune to electromagnetic interference. Raman spectroscopy uses fiber optics and involves pulsed light scattering by molecules. It has a wide variety of applications in process control [Workman, Koch, and Veltkamp, *Anal. Chem.* **75:** 2859 (2003)].

Significant advances have occurred during the past decade to miniaturize the size of the measurement system in order to make online analysis economically feasible and to reduce time delays that often are present in analyzers. Recently, chemical sensors have been placed on microchips, even those requiring multiple physical, chemical, and biochemical steps (such as electrophoresis) in the analysis. This device has been called *lab-on-a-chip.* The measurements of chemical composition can be direct or indirect, the latter case referring to applications in which some property of the process stream is measured (such as refractive index) and then related to composition of a particular component.

ELECTROANALYTICAL INSTRUMENTS

Conductometric Analysis Solutions of electrolytes in ionizing solvents (e.g., water) conduct current when an electrical potential is applied across electrodes immersed in the solution. Conductance is a function of ion concentration, ionic charge, and ion mobility. Conductance measurements are ideally suited for measurement of the concentration of a single strong electrolyte in dilute solutions. At higher concentrations conductance becomes a complex, nonlinear function of concentration requiring suitable calibration for quantitative measurements.

Measurement of pH The primary detecting element in pH measurement is the glass electrode. A potential is developed at the pH-sensitive glass membrane as a result of differences in hydrogen ion activity in the sample and a standard solution contained within the electrode. This potential measured relative to the potential of the reference electrode gives a voltage which is expressed as pH. Instrumentation for pH measurement is among the most widely used process measurement devices. Rugged electrode systems and highly reliable electronic circuits have been developed for this use.

After installation, the majority of pH measurement problems are sensor-related, mostly on the reference side, including junction plugging, poisoning, and depletion of electrolyte. For the glass (measuring electrode), common difficulties are broken or cracked glass, coating, and etching or abrasion. Symptoms such as drift, sluggish response, unstable readings, and inability to calibrate are indications of measurement problems. Online diagnostics such as impedance measurements, wiring checks, and electrode temperature are now available in most instruments. Other characteristics that can be measured offline include efficiency or slope and asymmetry potential (offset), which indicate whether the unit should be cleaned or changed [McMillan and Cameron, *Advanced Measurement and Control,* 3d ed., ISA, Research Triangle Park, N.C., 2005].

Specific-Ion Electrodes In addition to the pH glass electrode specific for hydrogen ions, a number of electrodes that are selective for the measurement of other ions have been developed. This selectivity is obtained through the composition of the electrode membrane (glass, polymer, or liquid-liquid) and the composition of the electrode. These electrodes are subject to interference from other ions, and the response is a function of the total ionic strength of the solution. However, electrodes have been designed to be highly selective for specific ions, and when properly used, these provide valuable process measurements.

MOISTURE MEASUREMENT

Moisture measurements are important in the process industries because moisture can foul products, poison reactions, damage equipment, or cause explosions. Moisture measurements include both absolute moisture methods and relative-humidity methods. The absolute methods provide a primary output that can be directly calibrated in terms of dew point temperature, molar concentration, or weight concentration. Loss of weight on heating is the most familiar of these methods. The relative-humidity methods provide a primary output that can be more directly calibrated in terms of percentage of saturation of moisture.

Dew Point Method For many applications the dew point is the desired moisture measurement. When concentration is desired, the relation between water content and dew point is well known and available. The dew point method requires an inert surface whose temperature can be adjusted and measured, a sample gas stream flowing past the surface, a manipulated variable for adjusting the surface temperature to the dew point, and a means of detecting the onset of condensation.

Although the presence of condensate can be detected electrically, the original and most often used method is the optical detection of change in light reflection from an inert metallic-surface mirror. Some instruments measure the attenuation of reflected light at the onset of condensation. Others measure the increase of light dispersed and scattered by the condensate instead of, or in addition to, the reflected-light measurement. Surface cooling is obtained with an expendable refrigerant liquid, conventional mechanical refrigeration, or thermoelectric cooling. Surface-temperature measurement is usually made with a thermocouple or a thermistor.

Piezoelectric Method A piezoelectric crystal in a suitable oscillator circuit will oscillate at a frequency dependent on its mass. If the crystal has a stable hygroscopic film on its surface, the equivalent mass of the crystal varies with the mass of water sorbed in the film. Thus the frequency of oscillation depends on the water in the film. The analyzer contains two such crystals in matched oscillator circuits. Typically, valves alternately direct the sample to one crystal and a dry gas to the other on a 30-s cycle. The oscillator frequencies of the two circuits are compared electronically, and the output is the difference between the two frequencies. This output is then representative of the moisture content of the sample. The output frequency is usually converted to a variable dc voltage for meter readout and recording. Multiple ranges are provided for measurement from about 1 ppm to near saturation. The dry reference gas is preferably the same as the sample except for the moisture content of the sample. Other reference gases which are adsorbed in a manner similar to the dried sample gas may be used. The dry gas is usually supplied by an automatic dryer. The method requires a vapor sample to the detector. Mist striking the detector destroys the accuracy of measurement until it vaporizes or is washed off the crystals. Water droplets or mist may destroy the hygroscopic film, thus requiring crystal replacement. Vaporization or gas-liquid strippers may sometimes be used for the analysis of moisture in liquids.

Capacitance Method Several analyzers utilize the high dielectric constant of water for its detection in solutions. The alternating electric current through a capacitor containing all or part of the sample between the capacitor plates is measured. Selectivity and sensitivity are enhanced by increasing the concentration of moisture in the cell by filling the capacitor sample cell with a moisture-specific sorbent as part of the dielectric. This both increases the moisture content and reduces the amount of other interfering sample components. Granulated alumina is the most frequently used sorbent. These detectors may be cleaned and recharged easily and with satisfactory reproducibility if the sorbent itself is uniform.

Oxide Sensors Aluminum oxide can be used as a sensor for moisture analysis. A conductivity call has one electrode node of aluminum, which is anodized to form a thin film of aluminum oxide, followed by coating with a thin layer of gold (the opposite electrode). Moisture is selectively adsorbed through the gold layer and into the hygroscopic aluminum oxide layer, which in turn determines the electrical conductivity between gold and aluminum oxide. This value can be related to ppm water in the sample. This sensor can operate between near vacuum to several hundred atmospheres, and it is independent of flow rate (including static conditions). Temperature, however, must be carefully monitored. A similar device is based on phosphorous pentoxide. Moisture content influences the electric current between two inert metal electrodes, which are fabricated as a helix on the inner wall of a tubular nonconductive sample cell. For a constant dc voltage applied to the electrodes, current flow is proportional to moisture. The moisture is absorbed into the hygroscopic phosphorous pentoxide, where the current electrolyzes the water molecules into hydrogen and oxygen. This sensor will handle moisture up to 1000 ppm and 6-atm pressure. As with the aluminum oxide ion, temperature control is very important.

Photometric Moisture Analysis This analyzer requires a light source, a filter wheel rotated by a synchronous motor, a sample cell, a detector to measure the light transmitted, and associated electronics. Water has two absorption bands in the near-infrared region at 1400 and 1900 nm. This analyzer can measure moisture in liquid or gaseous samples at levels from 5 ppm up to 100 percent, depending on other chemical species in the sample. Response time is less than 1 s, and samples can be run up to 300°C and 400 psig.

OTHER TRANSDUCERS

Other types of transducers used in process measurements include mechanical drivers such as gear trains and electrical drivers such as a differential transformer or a Hall effect (semiconductor-based) sensor.

Gear Train Rotary motion and angular position are easily transduced by various types of gear arrangements. A gear train in conjunction with a mechanical counter is a direct and effective way to obtain a digital readout of shaft rotations. The numbers on the counter can mean anything desired, depending on the gear ratio and the actuating device used to turn the shaft. A pointer attached to a gear train can be used to indicate a number of revolutions or a small fraction of a revolution for any specified pointer rotation.

Differential Transformer These devices produce an ac electrical output from linear movement of an armature. They are very versatile in that they can be designed for a full range of output with any range of armature travel up to several inches. The transformers have one or two primaries and two secondaries connected to oppose each other. With an ac voltage applied to the primary, the output voltage depends on the position of the armature and the coupling. Such devices produce accuracies of 0.5 to 1.0 percent of full scale and are used to transmit forces, pressures, differential pressures, or weights up to 1500 m. They can also be designed to transmit rotary motion.

Hall Effect Sensors Some semiconductor materials exhibit a phenomenon in the presence of a magnetic field which is adaptable to sensing devices. When a current is passed through one pair of wires attached to a semiconductor, such as germanium, another pair of wires properly attached and oriented with respect to the semiconductor will develop a voltage proportional to the magnetic field present and the current in the other pair of wires. Holding the exciting current constant and moving a permanent magnet near the semiconductor produce a voltage output proportional to the movement of the magnet. The magnet may be attached to a process variable measurement device, which moves the magnet as the variable changes. Hall effect devices provide high speed of response, excellent temperature stability, and no physical contact.

SAMPLING SYSTEMS FOR PROCESS ANALYZERS

The sampling system consists of all the equipment required to present a process analyzer with a clean representative sample of a process stream and to dispose of that sample. When the analyzer is part of an automatic control loop, the reliability of the sampling system is as important as the reliability of the analyzer or the control equipment. Sampling systems have several functions. The sample must be withdrawn from the process, transported, conditioned, introduced to the analyzer, and disposed. Probably the most common problem in sample system design is the lack of realistic information concerning the properties of the process material at the sampling point. Another common problem is the lack of information regarding the conditioning required so that the analyzer may utilize the sample without malfunction for long periods. Some samples require enough conditioning and treating that the sampling systems become equivalent to miniature online processing plants. These systems possess many of the same fabrication, reliability, and operating problems as small-scale pilot plants except that the sampling system must generally operate reliably for much longer periods.

Selecting the Sampling Point The selection of the sampling point is based primarily on supplying the analyzer with a sample whose composition or physical properties are pertinent to the control function to be performed. Other considerations include selecting locations that provide representative homogeneous samples with minimum transport delay, locations which collect a minimum of contaminating material, and locations that are accessible for test and maintenance procedures.

Sample Withdrawal from Process A number of considerations are involved in the design of sample withdrawal devices which will provide representative samples. For example, in a horizontal pipe carrying process fluid, a sample point on the bottom of the pipe will collect a maximum count of rust, scale, or other solid materials being carried along by the process fluid. In a gas stream, such a location will also collect a maximum amount of liquid contaminants. A sample point on the top side of a pipe will, for liquid streams, collect a maximum amount of vapor contaminants being carried along. Bends in the piping that produce swirls or cause centrifugal concentration of the denser phase may cause maximum contamination to be at unexpected locations. Two-phase process materials are difficult to sample for a total-composition representative sample.

A typical method for obtaining a sample of process fluid well away from vessel or pipe walls is an eduction tube inserted through a packing gland. This sampling method withdraws a liquid sample and vaporizes it for transport to the analyzer location. The transport lag time from the end of the probe to the vaporizer is minimized by using tubing having a small internal volume compared with pipe and valve volumes.

This sample probe may be removed for maintenance and reinstalled without shutting down the process. The eduction tube is made of material that will not corrode so that it will slide through the packing gland even after long periods of service. There may be a small amount of process fluid leakage until the tubing is withdrawn sufficiently to close the gate valve. A swaged ferrule on the end of the tube prevents accidental ejection of the eduction tube prior to removal of the packing gland. The section of pipe surrounding the eduction tube and extending into the process vessel provides mechanical protection for the eduction tube.

Sample Transport Transport time—the time elapsed between sample withdrawal from the process and its introduction into the analyzer—should be minimized, particularly if the analyzer is an automatic analyzer-controller. Any sample transport time in the analyzer-controller loop must be treated as equivalent to process dead time in determining conventional feedback controller settings or in evaluating controller performance. Reduction in transport time usually means transporting the sample in the vapor state.

Design considerations for sample lines are as follows:

1. The structural strength or protection must be compatible with the area through which the sample line runs.
2. Line size and length must be small enough to meet transport time requirements without excessive pressure drop or excessive bypass of sample at the analyzer input.
3. Line size and internal surface quality must be adequate to prevent clogging by the contaminants in the sample.
4. The prevention of a change of state of the sample may require installation, refrigeration, or heating of the sample line.
5. Sample line material must be such as to minimize corrosion due to the sample or the environment.

Sample Conditioning Sample conditioning usually involves the removal of contaminants or some deleterious component from the sample mixture and/or the adjustment of temperature, pressure, and flow rate of the sample to values acceptable to the analyzer. Some of the more common contaminants that must be removed are rust, scale, corrosion products, deposits due to chemical reactions, and tar. In sampling some process streams, the material to be removed may include the primary process product such as a polymer or the main constituent of the stream such as oil. In other cases, the material to be removed is present in trace quantities, e.g., water in an online chromatograph sample that can damage the chromatographic column packing. When contaminants or other materials that will hinder analysis represent a large percentage of the stream composition, their removal may significantly alter the integrity of the sample. In some cases, removal must be done as part of the analysis function so that removed material can be accounted for. In other cases, proper calibration of the analyzer output will suffice.

CONTROLLERS, FINAL CONTROL ELEMENTS, AND REGULATORS

General References: ANSI/ISA-75.25.01, *Test Procedure for Control Valve Response Measurement from Step Inputs*, ISA, Research Triangle Park, N.C., 2000. Blevins and Nixon, *Control Loop Foundation-Batch and Continuous Processes*, ISA, Research Triangle Park, N.C., 2011. Borden and Friedman (eds.), *Control Valves*, ISA, Research Triangle Park, N.C., 1998. Johnson, *Process Control Instrumentation Technology*, 8th ed., Prentice Hall, Upper Saddle River, N.J., 2005. Kinsler and Frey, *Fundamentals of Acoustics*, 4th ed., Wiley, New York, 1999. Liptak, *Instrument Engineering Handbook*, CRC Press, Boca Raton, Fla., 2005. McMillan and Considine, *Process/Industrial Instruments and Controls Handbook*, 5th ed., McGraw-Hill, New York, 1999. Michaelides and Crowe, *Multiphase Flow Handbook*, 2d ed., CRC Press, Boca Raton, Fla., 2016. Norton and Karczub, *Fundamentals of Noise and Vibration Analysis for Engineers*, 2d ed., Cambridge University Press, London, 2003. Skousen, *Valve Handbook*, 3d ed., McGraw-Hill, New York, 2011. *National Electrical Code Handbook*, 13th ed., National Fire Protection Association, Inc., Quincy, Mass., 2014.

External control of the process is achieved by devices that are specially designed, selected, and configured for the intended process control application. The text that follows covers three very common function classifications of process control devices: controllers, final control elements, and regulators.

The process controller is the "master" of the process control system. It accepts a set point and other inputs and generates an output or outputs that it computes from a rule or set of rules that is part of its internal configuration. The controller output serves as an input to another controller or, more often, as an input to a final control element. The final control element typically is a device that affects the flow in the piping system of the process. The final control element serves as an interface between the process controller and the process. Control valves and adjustable-speed pumps are the principal types discussed.

Regulators, though not controllers or final control elements, perform the combined function of these two devices (controller and final control element) along with the measurement function commonly associated with the process variable transmitter. The uniqueness, control performance, and widespread use of the regulator make it deserving of a functional grouping of its own.

PNEUMATIC, ELECTRONIC, AND DIGITAL CONTROLLERS

Pneumatic Controllers The pneumatic controller is an automatic controller that uses variable air pressure as input and output signals. An air supply is also required to "power" the mechanical components of the controller and provide an air source for the controller output signal. Pneumatic controllers were first available in the early 1940s but are now rarely used for large-scale industrial control projects. Pneumatic controllers are still used where cost, ruggedness, or the installation requires an all-pneumatic solution.

Pneumatic process transmitters are used to produce a pressure signal that is proportional to the calibrated range of the measuring element. Of the transmitter range 0 to 100 percent is typically represented by a 0.2- to 1.0-bar (3- to 15-psig) pneumatic signal. This signal is sent through tubing to the pneumatic controller process variable feedback connection. The process variable feedback can also be sensed directly in cases where the sensing element has been incorporated into the controller design. Controllers with integral sensing elements are available that sense pressure, differential pressure, temperature, and level.

The pneumatic controller is designed so that 0 to 100 percent output is also represented by 0.2 to 1.0 bar (3 to 15 psig). The output signal is sent through tubing to the control valve or other final control element. Most pneumatic controllers provide a manual control mode where the output pressure is manually set by operating personnel. The controller design also provides a mechanism to adjust the set point.

Early controller designs required "balancing" of the controller output prior to switching to or from automatic and manual modes. This procedure minimized inadvertent disturbance to the process caused by potentially large differences between the automatic and manual output levels. Later designs featured "bumpless" or "procedureless" automatic-to-manual transfer.

Although the pneumatic controller is often used in single-loop control applications, cascade strategies can be implemented where the controller design supports input of external or remote set-point signals. A balancing procedure is typically required to align the remote set point with the local set point before the controller is switched into cascade mode.

Almost all pneumatic controllers include indicators for process variable, set point, and output. Many controller designs also feature integral chart recorders. There are versions of the pneumatic controller that support combinations of proportional, integral, and derivative actions.

The pneumatic controller can be installed into panel boards that are adjacent to the process being controlled or in a centrally located control room. Field-mountable controllers can be installed directly onto the control valve, a nearby pipe stand, or wall in close proximity to the control valve and/or measurement transmitter.

If operated on clean, dry plant air, pneumatic controllers offer good performance and are extremely reliable. In many cases, however, plant air is neither clean nor dry. A poor-quality air supply will cause unreliable performance of pneumatic controllers, pneumatic field measurement devices, and final control elements. The main shortcoming of the pneumatic controller is its lack of flexibility when compared to modern electronic controller designs. Increased range of adjustability, choice of alternative control algorithms, the communication link to the control system, and other features and services provided by the electronic controller make it a superior choice in most of today's applications. Controller performance is also affected by the time delay induced by pneumatic tubing runs. For example, a 100-m run of 6.35-mm ($\frac{1}{4}$-in) tubing will typically cause 5 s of apparent process dead time, which will limit the control performance of fast processes such as flows and pressures.

Pneumatic controllers continue to be used in areas where it would be hazardous to use electronic equipment, such as locations with flammable or explosive atmospheres or other locations where compressed air is available but where access to electrical services is limited or restricted.

Electronic (Digital) Controllers Almost all the electronic process controllers used today are microprocessor-based, digital devices. In the transition from pneumatic to electronic controllers, a number of analog controller designs were available. Due to the inflexible nature of the analog designs, these controllers have been almost completely replaced by digital designs. The microprocessor-based controllers contain, or have access to, input/output (I/O) interface electronics that allow various types of signals to enter and leave the controller's processor.

The resolution of the analog I/O channels of the controller varies by manufacturer and age of the design. The 12- to 14-bit conversion resolution of the analog input channels is quite common. Conversion resolution of the analog output channels is typically 10- to 12-bit. Newer designs support up to 16-bit input and output resolution. Although 10-bit output resolution had been considered satisfactory for many years, it has recently been identified as a limitation of control performance. This limitation has emerged as the performance of control valve actuators has improved and the use of other high-resolution field devices, such as variable-speed pump drives, has become more prevalent. These improvements have been driven by the need to deliver higher operating efficiencies and improved product specifications through enhanced process control performance.

Sample rates for the majority of digital controllers are adjustable and range from 1 sample every 5 s to 10 samples per second. Some controller designs have fixed sample rates that fall within the same range. Hardwired low-pass filters are usually installed on the analog inputs to the controller to help protect the sampler from aliasing errors.

The real advantage of digital controllers is the substantial flexibility offered by a number of different configuration schemes. The simplest form of configuration is seen in a controller design that features a number of user-selectable control strategies. These strategies are customized by setting "tunable" parameters within the strategy. Another common configuration scheme uses a library of function blocks that can be selected and combined to form the desired control strategy. Each function block has adjustable parameters. Additional configuration schemes include text-based scripting languages, higher-level languages such as BASIC or C, and ladder logic.

Some digital controller designs allow the execution rates of control strategy elements to be set independently of one another and independently of the I/O subsystem sample rate. Data passed from control element to subsystems that operate at slower sample or execution rates present additional opportunities for timing and aliasing errors.

Distributed Control Systems Some knowledge of the *distributed control system* (DCS) is useful in understanding electronic controllers. A DCS is a process control system with sufficient performance to support large-scale, real-time process applications. The DCS has (1) an operations workstation with input devices, such as a keyboard, mouse, track ball, or other similar device, and a display device, such as a CRT or LCD panel; (2) a controller subsystem that supports various types of controllers and controller functions; (3) an I/O subsystem for converting analog process input signals to digital data and digital data back to analog output signals; (4) a higher-level computing platform for performing process supervision, historical data trending and archiving functions, information processing, and analysis; and (5) communication networks to tie the DCS subsystems, plant areas, and other plant systems together. The component controllers used in the controller subsystem portion of the DCS can be of various types and include multiloop controllers, programmable logic controllers, personal computer controllers, single-loop controllers, and fieldbus controllers. The type of electronic controller utilized depends on the size and functional characteristics of the process application being controlled. Personal computers are increasingly being used as DCS operation workstations or interface stations in place of custom-built machines. This is due to the low cost and high performance of the PC. See the earlier subsection Digital Technology for Process Control.

Multiloop Controllers The multiloop controller is a DCS network device that uses a single 32-bit microprocessor to provide control functions to many process loops. The controller operates independently of the other devices on the DCS network and can support from 20 to 500 loops. Data acquisition capability for 1000 analog and discrete I/O channels or more can also be provided by this controller. The I/O is typically processed through a subsystem that is connected to the controller through a dedicated bus or network interface. The multiloop controller contains a variety of function blocks (for example, PID, totalizer, lead/lag compensator, ratio control, alarm, sequencer, and boolean) that can be "soft-wired" together to form complex control strategies. The multiloop controller also supports additional configuration schemes including text-based scripting languages, higher-level languages such as BASIC or C, and, to a limited extent, ladder logic. The multiloop controller, as part of a DCS, communicates with other controllers and human/machine interface (HMI) devices also on the DCS network.

Programmable Logic Controllers The *programmable logic controller* (PLC) originated as a solid-state, and far more flexible, replacement for the hardwired relay control panel and was first used in the automotive industry for discrete manufacturing control. Today, PLCs are primarily used to implement boolean logic functions, timers, and counters. Some PLCs offer a limited number of math functions and PID control. PLCs are often used with on/off input and output devices such as limit or proximity switches, solenoid-actuated process control valves, and motor switch gear. PLCs vary greatly in size with the smallest supporting less than 128 I/O channels and the largest supporting more than 1023 I/O channels. Very small PLCs combine processor, I/O, and communications functions into a single, self-contained unit. For larger PLC systems, hardware modules such as the power supply, processor module, I/O modules, communication module, and backplane are specified based on the application. These systems support multiple I/O backplanes that can be chained together to increase the I/O count available to the processor. Discrete I/O modules are available that support high-current motor loads and general-purpose voltage and current loads. Other modules support analog I/O and special-purpose I/O for servomotors, stepping motors, high-speed pulse counting, resolvers, decoders, displays, and keyboards. PLC I/O modules often come with indicators to determine the status of key I/O channels. When used as an alternative to a DCS, the PLC is programmed with a handheld or computer-based loader. The PLC is typically programmed with ladder logic or a high-level computer language such as BASIC, FORTRAN, or C. Programmable logic controllers use 16- or 32-bit microprocessors and offer some form of point-to-point serial communications such as RS-232C, RS-485, or networked communication such as Ethernet with proprietary or open protocols. PLCs typically execute the boolean or ladder logic configuration at high rates; 10-ms execution intervals are common. This does not necessarily imply that the analog I/O or PID control functions are executed at the same rate. Many PLCs execute the analog program at a much slower rate. Manufacturers' specifications must be consulted.

Personal Computer Controller Because of its high performance at low cost and its unexcelled ease of use, application of the personal computer (PC) as a platform for process controllers is growing. When configured to perform scan, control, alarm, and data acquisition (SCADA) functions and combined with a spreadsheet or database management application, the PC controller can be a low-cost, basic alternative to the DCS or PLC. Using the PC for control requires installation of a board into the expansion slot in the computer, or the PC can be connected to an external I/O module by using a standard communication port on the PC. The communication is typically achieved through a serial interface (RS-232, RS-422, or IEEE-488), universal serial bus (USB), or Ethernet. The controller card/module supports 16- or 32-bit microprocessors. Standardization and high volume in the PC market have produced a large selection of hardware and software tools for PC controllers.

The PC can also be interfaced to a DCS to perform advanced control or optimization functions that are not available within the standard DCS function library.

Single-Loop Controller The *single-loop controller* (SLC) is a process controller that produces a single output. SLCs can be pneumatic, analog electronic, or microprocessor-based. Pneumatic SLCs are discussed in the pneumatic controller section, and analog electronic SLC is not discussed because it has been virtually replaced by the microprocessor-based design. The microprocessor-based SLC uses an 8- or 16-bit microprocessor with a small number of digital and analog process input channels with control logic for the I/O incorporated within the controller. Analog inputs and outputs are available in the standard ranges (1 to 5 V dc and 4 to 20 mA dc). Direct process inputs for temperature sensors (thermistor RTD and thermocouple types) are available. Binary outputs are also available. The face of the SLC has some form of visible display and pushbuttons that are used to view or adjust control values and configuration. SLCs are available for mounting in panel openings as small as 48 × 48 mm (1.9 × 1.9 in).

The processor-based SLC allows the user to select from a set of predefined control strategies or to create a custom control strategy by using a set of control function blocks. Control function blocks include PID, on/off, lead/lag, adder/subtractor, multiply/divide, filter functions, signal selector, peak detector, and analog track. SLCs feature auto/manual transfer switching, multi-set-point self-diagnostics, gain scheduling, and perhaps also time sequencing. Most processor-based SLCs have self-tuning or auto-tuning PID control algorithms. Sample times for the microprocessor-based SLCs vary from 0.1 to 0.5 s. Low-pass analog electronic filters are usually installed on the process inputs to minimize aliasing errors caused by high-frequency content in the process signal. Input filter time constants are typically in the range of 0.1 to 1 s. Microprocessor-based SLCs may be made part of a DCS by using the communication port (RS-488 is common) on the controller or may be operated in a stand-alone mode independently of the DCS.

Fieldbus Controller Fieldbus technology is a network-based communications system that interconnects measurement and control equipment such as sensors, actuators, and controllers. Advanced fieldbus systems, intended for process control applications, such as Foundation Fieldbus, enable digital interoperability among these devices and have a built-in capability to distribute the control application across the network. Several manufacturers have made available Foundation Fieldbus devices that support process controller functionality. These controllers, known as fieldbus controllers, typically reside in the final control element or measurement transmitter, but can be designed into any fieldbus device. A suitable communications interface connects the fieldbus segment to the distributed control system. When the control strategy is configured, all or part of the strategy may be loaded into the fieldbus devices. The remaining part of the control strategy would reside in the DCS itself. The distribution of the control function depends on the processing capacity of the fieldbus devices, the control strategy, and where it makes sense to perform these functions. Linearization of a control valve could be performed in the digital valve positioner (controller), for example. Temperature and pressure compensation of a flow measurement could be performed in the flow transmitter processor. The capability of fieldbus devices varies greatly. Some devices will allow instances of control system function blocks to be loaded and executed, while other devices allow the use of only preconfigured function blocks. Fieldbus controllers are typically configured as single-loop PID controllers, but cascade or other complex control strategies can be configured depending on the capability of the fieldbus device. Fieldbus devices that have native support for process control functions do not necessarily implement the PID algorithm in the same way. It is important to understand these differences so that the controller tuning will deliver the desired closed-loop characteristics. The functionality of fieldbus devices is projected to increase as the controller market develops.

Controller Reliability and Application Trends Critical process control applications demand a high level of reliability from the electronic controller. Some methods that improve the reliability of electronic controllers include (1) focusing on robust circuit design using quality components; (2) using redundant circuits, modules, or subsystems where necessary; (3) using small backup systems when needed; (4) reducing repair time and using more powerful diagnostics; and (5) distributing functionality to more independent modules to limit the impact of a failed module. Currently, the trend in process control is away from centralized process control and toward an increased number of small distributed control or PLC systems. This trend will put emphasis on the evolution of the fieldbus controller and continued growth of the PC-based controller. Also, as hardware and software improve, the functionality of the controller will increase, and the supporting hardware will be physically smaller. Hence, the traditional lines between the DCS and the PLC will become less distinct as systems become capable of supporting either function set.

Controller Performance and Process Dynamics The design of a control loop must take the control objectives into account. What do you want this loop to do? And under what operating conditions? There may be control applications that require a single control objective and others that have several control objectives. Control objectives may include such requirements as minimum variance control at steady state, maximum speed of recovery from a major disturbance, maximum speed of set-point response where overshoot and ringing are acceptable, critically damped set-point response with no overshoot, robustness issues, and special start-up or shutdown requirements. The control objectives will define not only the tuning requirements of the controller, but also, to a large extent, the allowable dynamic parameters of the field instruments and process design. Process dynamics alone can prevent control objectives from being realized. Tuning of the controller cannot compensate for an incompatible process or unrealistic control objectives. For most controllers, the difference between the set-point and process feedback signal—the error—is the input to the PID algorithm. The calculated PID output is sent back to the final control element. Every component between the controller output and the process feedback is considered by the controller as the "process" and will directly affect the dynamics and ultimately the performance of the system. This includes not only the dynamics of the physical process, but also the dynamics of the field instruments, signal conditioning equipment, and controller signal processing elements such as filters, scaling, and linearization routines. The choice of final control element can significantly affect the dynamics of the system. If the process dynamics are relatively slow, with time constants of a few minutes or longer, most control valves are fast enough that their contribution to the overall process time response will be negligible. In cases where the process time constants are only a few seconds, the control valve dynamics may become the dominant lag in the overall response. Excessive filtering in the field-sensing devices may also mask the true process dynamics and potentially limit control performance. Often, the design of a control loop and the tuning of the controller are a compromise between a number of different control objectives. When a compromise is unacceptable, gain scheduling or other adaptive tuning routine may be necessary to match the controller response to the most appropriate control objective.

When one is tuning a controller, the form of the PID algorithm must be known. The three common forms of the PID algorithm are parallel or noninteracting, classical or interacting, and the ISA Standard form. In *most* cases, a controller with any of these PID forms can be tuned to produce the desired closed-loop response. The actual tuning parameters will be different. The units of the tuning parameters also affect their value. The controller gain parameter is typically represented as a pure gain (K_c), acting on the error, or as *proportional band* (PB). In cases where the proportional band parameter is used, the equivalent controller gain is equal to 100 divided by the proportional band and represents the percent span that the error must traverse to produce a 100 percent change in controller output. The proportional band is always applied to the controller error in terms of percent of span and percent of output. Controllers that use a gain tuning parameter commonly scale the error into percent span and use a percent output basis. In some controllers, the error is scaled by using a separate parameter into percent span prior to the PID algorithm. The gain parameter can also be applied to the error in engineering units. Even though most controller outputs are scaled as a percent, in cascade strategies the controller output may need to be scaled to the same span at the slave loop set point. In this case, the controller gain may in fact be required to calculate the controller output in terms of the slave loop engineering units.

The execution rate of a digital controller should be sufficiently fast, compared to the process dynamics, to produce a response that closely approximates that of an analog controller with the same tuning. A general rule of thumb is that the execution interval should be at least 3 times faster than the dominant lag of the process or about 10 times faster than the closed-loop time constant. The controller can be used when the sample rates are slower than this recommendation, but the controller output will consist of a series of coarse steps as compared to a smooth response. This may create stability problems. Some integral and derivative algorithms may not be accurate when the time-based tuning parameters approach the controller execution interval. The analog inputs of the controller are typically protected from aliasing errors through the use of one- or two-pole analog filters. Faster sample rates allow a smaller antialiasing filter and improved input response characteristics. Some controllers or I/O subsystems oversample the analog inputs with respect to the controller execution interval and then process the input data through a digital filter. This technique can produce a more representative measurement with less quantization noise.

Differences in the PID algorithm, controller parameters, units, and other fundamental control functions highlight the importance of understanding the structure of the controller and the requirement of sufficiently detailed documentation. This is especially important for the controller but is also important for the field instruments, final control elements, and devices that have the potential to affect the signal characteristics.

CONTROL VALVES

A control valve consists of a valve, an actuator, and usually one or more valve control devices. The valves discussed in this section are applicable to throttling control (i.e., where flow through the valve is regulated to any desired amount between maximum and minimum limits). Other valves such as check, isolation, and relief valves are addressed in the next subsection. As defined, control valves are automatic control devices that modify the fluid flow rate as specified by the controller.

Valve Types Types of valves are categorized according to their design style. These styles can be grouped into type of stem motion—linear or rotary. The valve stem is the rod, shaft, or spindle that connects the actuator with the closure member (i.e., a movable part of the valve that is positioned in the flow path to modify the rate of flow). Movement of either type of stem is known as *travel*. The major categories are described briefly below.

Globe and Angle The most common linear stem-motion control valve is the *globe valve*. The name comes from the globular cavities around the port. In general, a port is any fluid passageway, but often the reference is to the passage that is blocked off by the closure member when the valve is closed. In globe valves, the closure member is called a *plug*. A popular construction is a cage-guided plug, as illustrated in Fig. 8-64. In many such designs, openings in the cage provide the flow control orifices. The valve seat is the zone of contact between the moving closure member and the stationary valve body, which shuts off the flow when the valve is closed. Often the seat in the body is on a replaceable part known as a *seat ring*. This stationary seat can also be designed as an integral part of the cage. Plugs may also be port-guided by wings or a skirt that fits snugly into the seat-ring bore.

One distinct advantage of cage guiding is the use of balanced plugs in single-port designs. In the balanced design (Fig. 8-64), note that both the top and bottom of the plug are subjected to the same downstream pressure when the valve is closed. Leakage via the plug-to-cage clearance is prevented by a plug seal.

The plug, cage, seat ring, and associated seals are known as the *trim*. A key feature of globe valves is that they allow maintenance of the trim via a removable bonnet without removing the valve body from the line. Bonnets are typically bolted on but may be threaded in smaller sizes.

Angle valves are an alternate form of the globe valve. They often share the same trim options and have the top-entry bonnet style. Angle valves can eliminate the need for an elbow but are especially useful when direct impingement of the process fluid on the body wall is to be avoided. Sometimes it is not practical to package a long trim within a globe body, so an angle body is used. Some angle bodies are self-draining, which is an important feature for dangerous fluids.

Butterfly The classic design of butterfly valves is shown in Fig. 8-65. Its chief advantage is high capacity in a small package and a very low initial cost. Much of the size and cost advantage is due to the wafer body design, which is clamped between two pipeline flanges. In the simplest design,

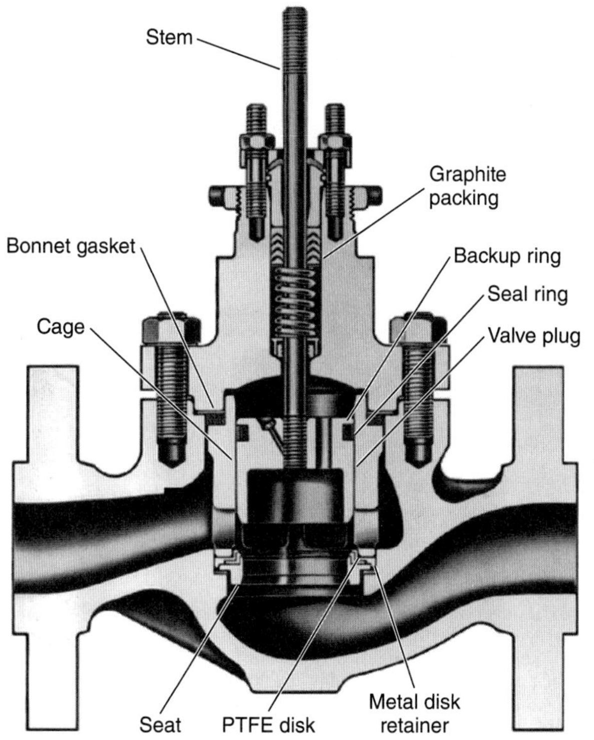

FIG. 8-64 Cage-guided balanced plug globe valve with polymer seat and plug seal. (*Courtesy Fisher Controls International LLC.*)

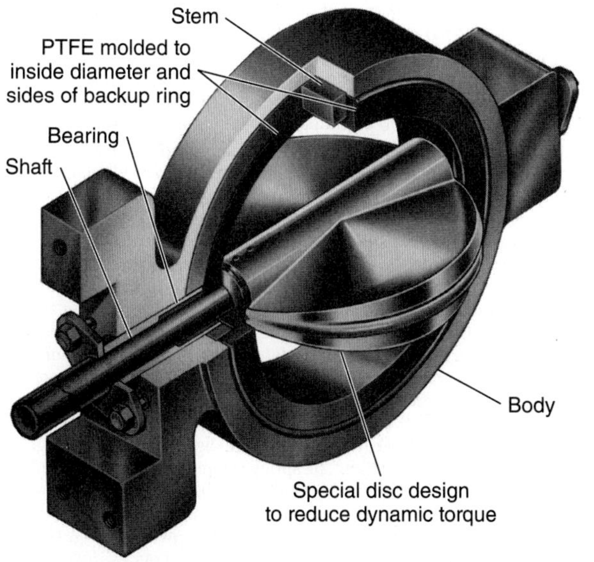

FIG. 8-65 Partial cutaway of wafer-style lined butterfly valve. (*Courtesy Fisher Controls International LLC.*)

there is no seal as such, merely a small clearance gap between the disc OD and the body ID. Often a true seal is provided by a resilient material in the body that is engaged via an interference fit with the disc. In a lined butterfly valve, this material covers the entire body ID and extends around the body ends to eliminate the need for pipeline joint gaskets. In a fully lined valve, the disc is also coated to minimize corrosion or erosion.

A high-performance butterfly valve has a disc that is offset from the shaft centerline. This eccentricity causes the seating surface to move away from the seal once the disc is out of the closed position, reducing friction and seal wear. It is also known as an *eccentric disc valve*; the advantage of the butterfly valve is improved shutoff while maintaining high ultimate capacity at a reasonable cost. This cost advantage relative to other design styles is particularly true in sizes above 6-in nominal pipe size (NPS). Improved shutoff is due to advances in seal technologies, including polymer, flexing metal, combination metal with polymer inserts, and so on, many utilizing pressure assist.

Ball Ball valves get their name from the shape of the closure member. One version uses a full spherical member with a cylindrical bore through it. The ball is rotated one-quarter turn from the full-closed to the full-open position. If the bore is the same diameter as the mating-pipe fitting ID, the valve is referred to as *full-bore.* If the hole is undersized, the ball valve is considered to be a venturi style. A segmented ball is a portion of a hollow sphere that is large enough to block the port when closed. Segmented balls often have a V-shaped contour along one edge, which provides a desirable flow characteristic (see Fig. 8-66). Both full ball and segmented ball valves are known for their low resistance to flow when full open. Shutoff leakage is minimized through the use of flexing or spring-loaded elastomeric or metal seals. Bodies are usually in two or three pieces or have a removable retainer to facilitate installing seals. End connections are usually flanged or threaded in small sizes, although segmented ball valves are offered in wafer style also.

Plug There are two substantially different rotary valve design categories referred to as *plug valves.* The first consists of a cylindrical or slightly conical plug with a port through it. The plug rotates to vary the flow much as a ball valve does. The body is top-entry but is geometrically simpler than a globe valve and thus can be lined with fluorocarbon polymer to protect against corrosion. These plug valves have excellent shutoff but are generally not for modulating service due to high friction. A variation of the basic design (similar to the eccentric butterfly disc) only makes sealing contact in the closed position and is used for control.

The other rotary plug design is portrayed in Fig. 8-67. The seating surface is substantially offset from the shaft, producing a ball-valve-like motion with the additional cam action of the plug into the seat when closing. In reverse flow, high-velocity fluid motion is directed inward, impinging on itself and only contacting the plug and seat ring.

Multiport This term refers to any valve or manifold of valves with more than one inlet or outlet. For throttling control, the three-way body is used for blending (two inlets, one outlet) or as a diverter (one inlet, two outlets). A three-way valve is most commonly a special globelike body with special trim that allows flow both over and under the plug. Two rotary valves and a pipe tee can also be used. Special three-, four-, and five-way ball valve designs are used for switching applications.

Special Application Valves

Digital Valves True digital valves consist of discrete solenoid-operated flow ports that are sized according to binary weighing. The valve can be designed with sharp-edged orifices or with streamlined nozzles that can be used for flow metering. Precise control of the throttling control orifice is the strength of the digital valve. Digital valves are mechanically complicated and expensive, and they have considerably reduced maximum flow capacities compared to the globe and rotary valve styles.

Cryogenic Service Valves designed to minimize heat absorption for throttling liquids and gases below 80 K are called *cryogenic service valves.* These valves are designed with small valve bodies to minimize heat absorption and long bonnets between the valve and actuator to allow for extra layers of insulation around the valve. For extreme cases, vacuum jacketing can be constructed around the entire valve to minimize heat influx.

High Pressure Valves used for pressures nominally above 760 bar (11,000 psi, pressures above ANSI Class 4500) are often custom-designed for specific applications. Normally, these valves are of the plug type and use specially hardened plug and seat assemblies. Internal surfaces are polished, and internal corners and intersecting bores are smoothed to reduce high localized stresses in the valve body. Steam loops in the valve body are available to raise the body temperature to increase the ductility and impact strength of the body material.

High-Viscous Process Used most extensively by the polymer industry, the valve for high-viscous fluids is designed with smooth finished internal passages to prevent stagnation and polymer degradation. These valves are available with integral body passages through which a heat-transfer fluid is pumped to keep the valve and process fluid heated.

Pinch The industrial equivalent of controlling flow by pinching a soda straw is the pinch valve. Valves of this type use fabric-reinforced elastomer sleeves that completely isolate the process fluid from the metal parts in the valve. The valve is actuated by applying air pressure directly to the outside of the sleeve, causing it to contract or pinch. Another method is to pinch the sleeve with a linear actuator with a specially attached foot. Pinch valves are used extensively for corrosive material service and erosive slurry service. This type of valve is used in applications with pressure drops up to 10 bar (145 psi).

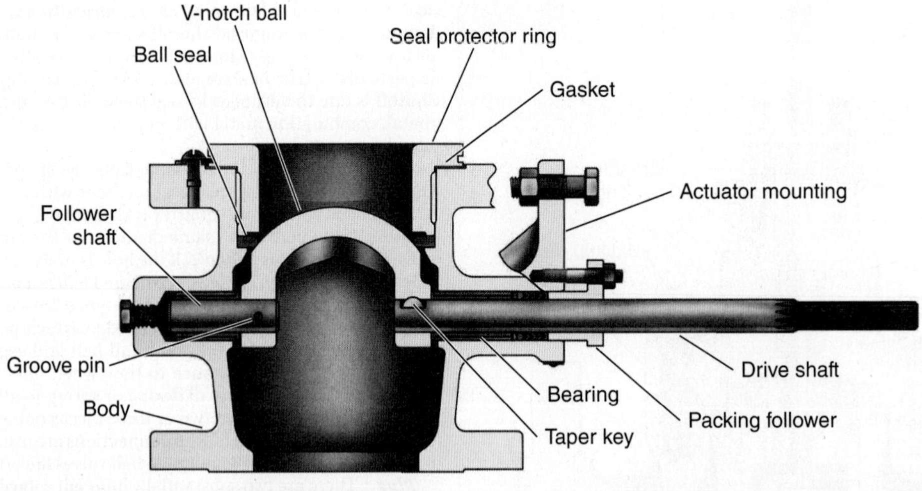

FIG. 8-66 Segmented ball valve. Partial view of actuator mounting shown 90° out of position. (*Courtesy Fisher Controls International LLC.*)

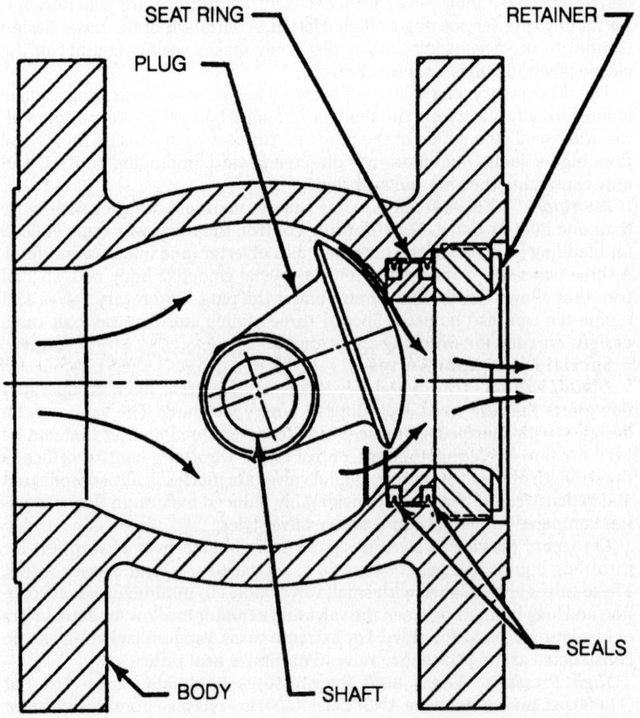

FIG. 8-67 Eccentric plug valve shown in erosion-resistant reverse-flow direction. Shaded components can be made of hard metal or ceramic materials. (*Courtesy Fisher Controls International LLC.*)

Fire-Rated Valves that handle flammable fluids may have additional safety-related requirements for minimal external leakage, minimal internal (downstream) leakage, and operability during and after a fire. Being fire-rated does not mean being totally impervious to fire, but a sample valve must meet particular specifications such as those of the American Petroleum Institute (API) 607, Factory Mutual Research Corp. (FM) 7440, or the British Standard 5146 under a simulated fire test. Due to very high flame temperature, metal seating (either primary or as a backup to a burned-out elastomer) is mandatory.

Solids Metering The control valves described earlier are primarily used for the control of fluid (liquid or gas) flow. Sometimes these valves, particularly the ball, butterfly, or sliding gate valves, are used to throttle dry or slurry solids. More often, special throttling mechanisms such as venturi ejectors, conveyers, knife-type gate valves, or rotating vane valves are used. The particular solids-metering valve hardware depends on the volume, density, particle shape, and coarseness of the solids to be handled.

Actuators An actuator is a device that applies the force (torque) necessary to cause a valve's closure member to move. Actuators must overcome pressure and flow forces as well as friction from packing, bearings or guide surfaces, and seals; and must provide the seating force. In rotary valves, maximum friction occurs in the closed position, and the moment necessary to overcome it is referred to as *breakout torque.* The rotary valve shaft torque generated by steady-state flow and pressure forces is called *dynamic torque.* It may tend to open or close the valve depending on valve design and travel. Dynamic torque per unit pressure differential is largest in butterfly valves at roughly 70° open. In linear stem-motion valves, the flow forces should not exceed the available actuator force, but this is usually accounted for by default when the seating force is provided.

Actuators often provide a fail-safe function. In the event of an interruption in the power source, the actuator will place the valve in a predetermined safe position, usually either full-open or full-closed. Safety systems are often designed to trigger local fail-safe action at specific valves to cause a needed action to occur, which may not be a complete process or plant shutdown.

Actuators are classified according to their power source. The nature of these sources leads naturally to design features that make their performance characteristics distinct.

Pneumatic Despite the availability of more sophisticated alternatives, the pneumatically driven actuator is still by far the most popular type. Historically the most common has been the spring and diaphragm design (Fig. 8-68). The compressed air input signal fills a chamber sealed by an elastomeric diaphragm. The pressure force on the diaphragm plate causes a spring to be compressed and the actuator stem to move. This spring provides the fail-safe function and contributes to the dynamic stiffness of the actuator. If the accompanying valve is "push down to close," the actuator depicted in Fig. 8-68 will be described as "air to close" or synonymously as fail-open. A slightly different design yields "air to open" or fail-closed action. The spring is typically precompressed to provide a significant available force in the failed position (e.g., to provide seating load). The spring also provides a proportional relationship between the force generated by air pressure and the stem position. The pressure range over which a spring and diaphragm actuator strokes in the absence of valve forces is known as the *bench set.* The chief advantages of spring and diaphragm actuators are their high reliability, low cost, adequate dynamic response, and fail-safe action—all of which are inherent in their simple design.

Motion Conversion Actuator power units with translational output can be adapted to rotary valves that generally need 90° or less rotation. A lever is attached to the rotating shaft, and a link with pivoting means on the end connects to the linear output of the power unit, an arrangement similar to an internal combustion engine crankshaft, connecting rod, and piston. When the actuator piston, or more commonly the diaphragm plate, is designed to tilt, one pivot can be eliminated. Scotch yoke and rack-and-pinion arrangements are also commonly used, especially with piston power units. Friction and the changing mechanical advantage of these motion conversion mechanisms mean the available torque may vary greatly with travel. One notable exception is vane-style rotary actuators whose offset "piston" pivots, giving direct rotary output.

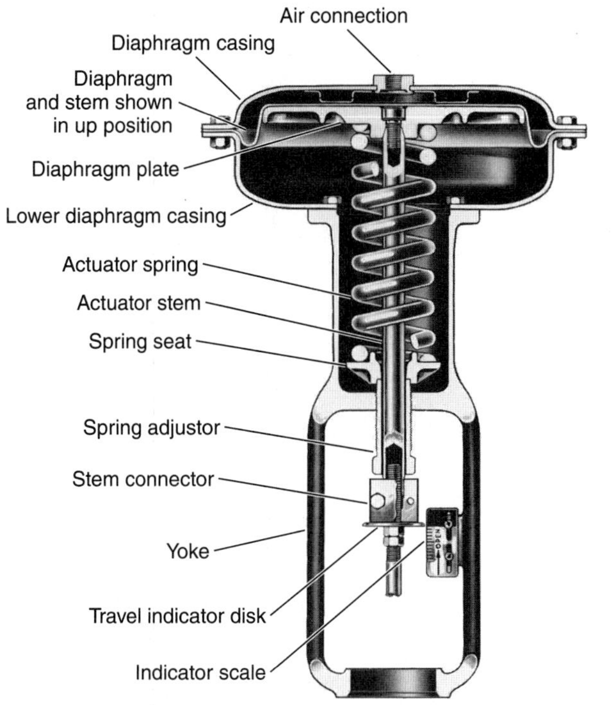

FIG. 8-68 Spring and diaphragm actuator with an "up" fail-safe mode. Spring adjuster allows slight alteration of bench set. (*Courtesy Fisher Controls International LLC.*)

Hydraulic The design of typical hydraulic actuators is similar to that of double-acting piston pneumatic types. One key advantage is the high pressure [typically 35 to 70 bar (500 to 1000 psi)], which leads to high thrust in a smaller package. The incompressible nature of the hydraulic oil means these actuators have very high dynamic stiffness. The incompressibility and small chamber size connote fast stroking speed and good frequency response. The disadvantages include high initial cost, especially when considering the hydraulic supply. Maintenance is much more difficult than with pneumatics, especially on the hydraulic positioner.

Electrohydraulic actuators have similar performance characteristics and cost/maintenance ramifications. The main difference is that they contain their own electric-powered hydraulic pump. The pump may run continuously or be switched on when a change in position is required. Their main application is remote sites without an air supply when a fail-safe spring return is needed.

Electric The most common electric actuators use a typical motor—three-phase ac induction, capacitor-start split-phase induction, or dc. Normally the motor output passes through a large gear reduction and, if linear motion output is required, a ball screw or thread. These devices can provide large thrust, especially given their size. Lost motion in the gearing system does create backlash, but if not operating across a thrust reversal, this type of actuator has very high stiffness. Usually the gearing system is self-locking, which means that forces on the closure member cannot move it by spinning a nonenergized motor. This behavior is called a *lock-in-last-position fail-safe mode.* Some gear systems (e.g., low-reduction spur gears) can be back-driven. A solenoid-activated mechanical brake or locking current to motor field coils is added to provide lock-in-last-position fail-safe mode. A battery backup system for a dc motor can guard against power failures. Otherwise, an electric actuator is not acceptable if fail-open/closed action is mandatory. Using electric power requires environmental enclosures and explosion protection, especially in hydrocarbon processing facilities; see the full discussion in the subsection Valve Control Devices.

Unless sophisticated speed control power electronics is used, position modulation is achieved via bang-zero-bang control. Mechanical inertia causes overshoot, which is (1) minimized by braking and/or (2) hidden by adding dead band to the position control. Without these provisions, high starting currents would cause motors to overheat from constant "hunting" within the position loop. Travel is limited with power interruption switches or with force (torque) electromechanical cutouts when the closed position is against a mechanical stop (e.g., a globe valve). Electric actuators are often used for on/off service. Stepper motors can be used instead, and they, as their name implies, move in fixed incremental steps. Through gear reduction, the typical number of increments for 90° rotation ranges from 5000 to 10,000; hence positioning resolution at the actuator is excellent.

An electromagnetic solenoid can be used to directly actuate the plug on very small linear stem-motion valves. A solenoid is usually designed as a two-position device, so this valve control is on/off. Special solenoids with position feedback can provide proportional action for modulating control. Force requirements of medium-sized valves can be met with piloted plug designs, which use process pressure to assist the solenoid force. Piloted plugs are also used to minimize the size of common pneumatic actuators, especially when there is need for high seating load.

Manual A manually positioned valve is by definition not an automatic control valve, but it may be involved with process control. For rotary valves, the manual operator can be as simple as a lever, but a wheel driving a gear reduction is necessary in larger valves. Linear motion is normally created with a wheel turning a screw-type device. A manual override is usually available as an option for the powered actuators listed above. For spring-opposed designs, an adjustable travel stop will work as a one-way manual input. In more complex designs, the handwheel can provide loop control override via an engagement means. Some gear reduction systems of electric actuators allow the manual positioning to be independent of the automatic positioning without declutching. In practice, most control valves have a bypass line with a manual valve that can be adjusted when the control valve fails or is taken out of service, as shown in Fig. 8-69.

OTHER PROCESS VALVES

In addition to the throttling control valve, other types of process valves can be used to manipulate a process.

Valves for On/Off Applications Valves are often required for service that is primarily nonthrottling. Valves in this category, depending on the service requirements, may be of the same design as the types used for throttling control or, as in the case of gate valves, different in design. Valves in this category usually have tight shutoff when they are closed and low pressure drops when they are wide open. The on/off valve can be operated manually, such as by handwheel or lever; or automatically, with pneumatic or electric actuators.

Batch Batch process operation is an application requiring on/off valve service. Here the valve is opened and closed to provide reactant, catalyst, or product to and from the batch reactor. Like the throttling control valve, the valve used in this service must be designed to open and close thousands of times. For this reason, valves used in this application are often the same valves used in continuous throttling applications. Ball valves are especially useful in batch operations. The ball valve has a straight-through flow passage that reduces pressure drop in the wide-open state and provides tight shutoff capability when closed. In addition, the segmented ball valve provides for shearing action between the ball and the ball seat that promotes closure in slurry service.

Isolation A means for pressure-isolating control valves, pumps, and other piping hardware for installation and maintenance is another common application for an on/off valve. In this application, the valve is required to have tight shutoff so that leakage is stopped when the piping system is under repair. As the need to cycle the valve in this application is far less than that of a throttling control valve, the wear characteristics of the valve are less important. Also, because many are required in a plant, the isolation valve needs to be reliable, simple in design, and simple in operation.

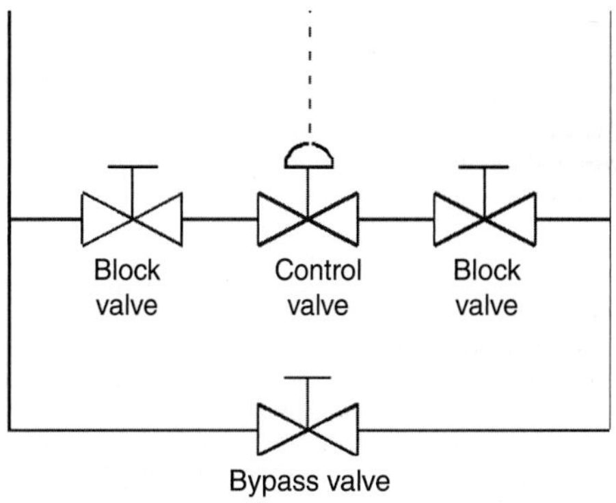

FIG. 8-69 Control valve bypass.

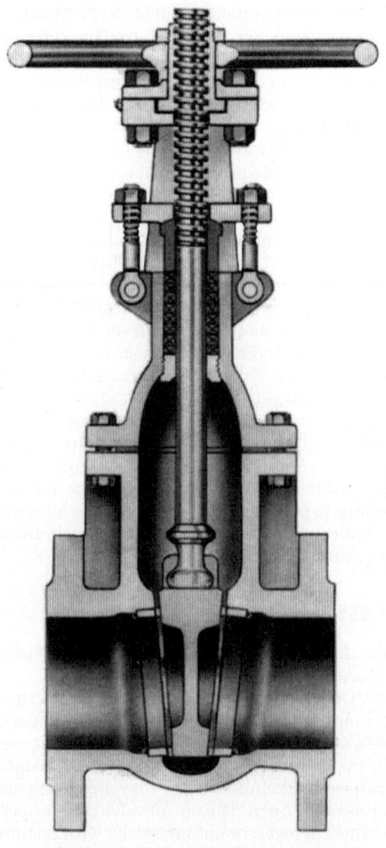

FIG. 8-70 Gate valve. (*Courtesy Crane Valves.*)

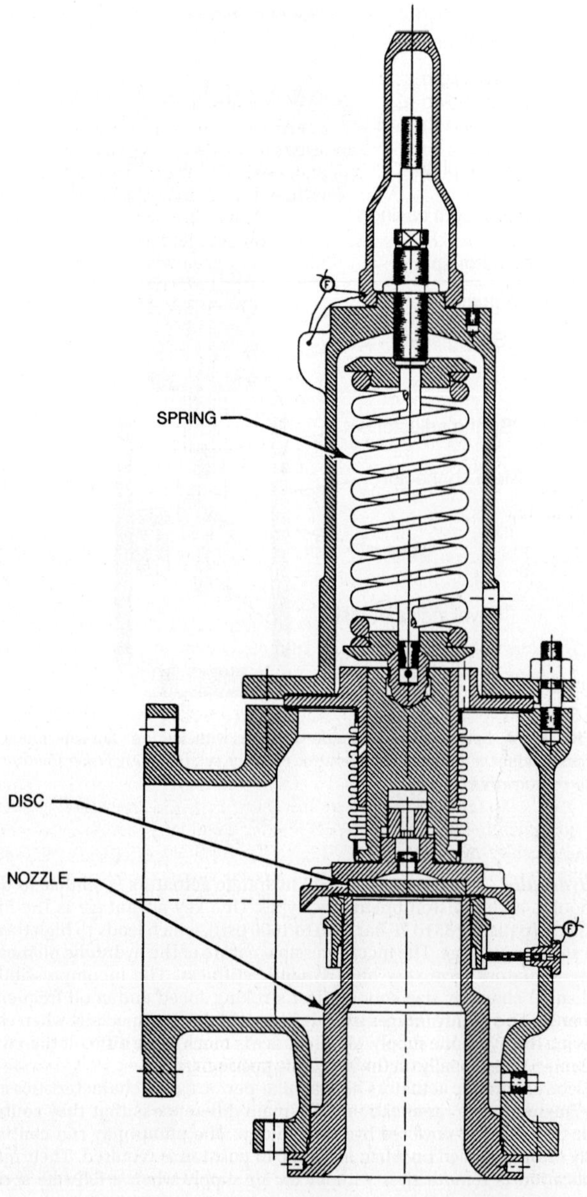

FIG. 8-71 Relief valve. (*Courtesy Teledyne Fluid Systems, Farris Engineering.*)

The gate valve, shown in Fig. 8-70, is the most widely used valve in this application. The gate valve is composed of a gatelike disc that moves perpendicular to the flow stream. The disc is moved up and down by a threaded screw that is rotated to effect disc movement. Because the disc is large and at right angles to the process pressure, a large seat loading for tight shutoff is possible. Wear produced by high seat loading during the movement of the disc prohibits the use of the gate valve for throttling applications.

Pressure Relief Valves The pressure relief valve is an automatic pressure-relieving device designed to open when normal conditions are exceeded and to close again when normal conditions are restored. Within this class there are relief valves, pilot-operated pressure relief valves, and safety valves. Relief valves (see Fig. 8-71) have spring-loaded discs that close a main orifice against a pressure source. As pressure rises, the disc begins to rise off the orifice and a small amount of fluid passes through the valve. Continued rise in pressure above the opening pressure causes the disc to open the orifice in a proportional fashion. The main orifice reduces and closes when the pressure returns to the set pressure. Additional sensitivity to overpressure conditions can be improved by adding an auxiliary pressure relief valve (pilot) to the basic pressure relief valve. This combination is known as a *pilot-operated pressure relief valve.*

The safety valve is a pressure relief valve that is designed to open fully, or pop, with only a small amount of pressure over the rated limit. Where conventional safety valves are sensitive to downstream pressure and may have unsatisfactory operating characteristics in variable backpressure applications, pressure-balanced safety relief valve designs are available to minimize the effect of downstream pressure on performance.

Application and sizing of pressure relief valves, pilot-operated pressure relief valves, and safety valves for use on pressure vessels are found in the ASME Boiler and Pressure Vessel Code, Section VIII, Division 1, "Rules for Construction of Pressure Vessels," Paragraphs UG-125 through UG-137.

Check Valves The purpose of a check valve is to allow relatively unimpeded flow in the desired direction but to prevent flow in the reverse direction. Two common designs are swing-type and lift-type check valves, the names of which denote the motion of the closure member. In the forward direction, flow forces overcome the weight of the member or a spring to open the flow passage. With reverse pressure conditions, flow forces drive the closure member into the valve seat, thus providing shutoff.

VALVE DESIGN CONSIDERATIONS

Functional requirements and the properties of the controlled fluid determine which valve and actuator types are best for a specific application. If demands are modest and no unique valve features are required, the valve design style selection may be determined solely by cost. If so, general-purpose globe or angle valves provide exceptional value, especially in sizes less than 3-in NPS and hence are very popular. Beyond type selection, there are many other valve specifications that must be determined properly to ultimately yield improved process control.

Materials and Pressure Ratings Valves must be constructed from materials that are sufficiently immune to corrosive or erosive action by the process fluid. Common body materials are cast iron, steel, stainless steel, high-nickel alloys, and copper alloys such as bronze. Trim materials need better corrosion and erosion resistance due to the higher fluid velocity in the throttling region. High hardness is desirable in erosive and cavitating applications. Heat-treated and precipitation-hardened stainless steels are common. High hardness is also good for guiding, bearing, and seating surfaces; cobalt-chromium alloys are utilized in cast or wrought form and frequently as welded overlays called *hard facing.* In less stringent situations, chrome plating, heat-treated nickel coatings, and nitriding are used. Tungsten carbide and ceramic trim are warranted in extremely erosive services. See Sec. 25, Materials of Construction, for specific material properties.

Since the valve body is a pressurized vessel, it is usually designed to comply with a standardized system of pressure ratings. Two common systems are described in the standards ASME B16.34 and EN 12516. Internal pressure limits under these standards are divided into broad classes, with specific limits being a function of material and temperature. Manufacturers also assign their own pressure ratings based on internal design rules. A common insignia is *250 WOG*, which means a pressure rating of 250 psig (~17 bar) in water, oil, or gas at ambient temperature. The subsection Storage and Process Vessels in Sec. 10 provides introductory information on compliance of pressure vessel design to industry codes (e.g., ASME Boiler and Pressure Vessel Code, Section VIII; ASME B31.3 Chemical Plant and Petroleum Refinery Piping).

Valve bodies are also standardized to mate with common piping connections: flanged, butt-welded end, socket-welded end, and screwed end. Dimensional information for some of these joints and class pressure-temperature ratings are included in Sec. 10, Process Plant Piping. Control valves have their own standardized face-to-face dimensions that are governed by ANSI/ISA Standards S75.08 and S75.22. Butterfly valves are also governed by API 609 and Manufacturers Standardization Society (MSS) SP-67 and SP-68.

Sizing Throttling control valves must be selected to pass the required flow rate, given expected pressure conditions. Sizing is not merely matching the end connection size with surrounding piping; it is a key step in ensuring that the process can be properly controlled. Sizing methods range from simple models based on elementary fluid mechanics to very complex models when unusual thermodynamics or nonideal behaviors occur. Basic sizing practices have been standardized (for example, ANSI-75.01.01) and are implemented as PC-based programs by manufacturers. The following is a discussion of very basic sizing equations and the associated physics.

Regardless of the particular process variable being controlled (e.g., temperature, level, pH), the output of a control valve is the flow rate. The throttling valve performs its function of manipulating the flow rate by virtue of being an adjustable resistance to flow. Flow rate and pressure conditions are normally known when a process is designed, and the valve resistance range must be matched accordingly. In the tradition of orifice and nozzle discharge coefficients, this resistance is embodied in the valve flow coefficient C_V. By applying the principles of conservation of mass and energy, the mass flow rate w kg/h is given for a liquid by

$$w = 27.3 C_V \sqrt{\rho(p_1 - p_2)} \tag{8-77}$$

where p_1 and p_2 are upstream and downstream static pressures, in bar, respectively. The density of the fluid ρ is expressed in kilograms per cubic meter. This equation is valid for nonvaporizing, turbulent flow conditions for a valve with no attached fittings.

While Eq. (8-77) gives the relationship between pressure and flow from a macroscopic point of view, it does not explain what is going on inside the valve. Valves create a resistance to flow by restricting the cross-sectional area of the flow passage and by forcing the fluid to change direction as it passes through the body and trim. The conservation of mass principle dictates that, for steady flow, the product of density, average velocity, and cross-sectional area remains a constant. The average velocity of the fluid stream at the minimum restriction in the valve is therefore much higher than that at the inlet. Note that due to the abrupt nature of the flow contraction that forms the minimum passage, the main fluid stream may separate from the passage walls and form a jet that has an even smaller cross section, the so-called vena contracta. The ratio of minimum stream area to the corresponding passage area is called the *contraction coefficient.* As the fluid expands from the minimum cross-sectional area to the full passage area in the downstream piping, large amounts of turbulence are generated. Direction changes can also induce significant amounts of turbulence.

Figure 8-72 is an illustration of how the mean pressure changes as fluid moves through a valve. Some of the potential energy that was stored in the fluid by pressurizing it (e.g., the work done by a pump) is first converted to the kinetic energy of the fast-moving fluid at the vena contracta. Some of that kinetic energy turns into the kinetic energy of turbulence. As the turbulent eddies break down into smaller and smaller structures, viscous effects ultimately convert all the turbulent energy to heat. Therefore, a valve converts fluid energy from one form to another.

For many valve constructions, it is reasonable to approximate the fluid transition from the valve inlet to the minimum cross section of the flow stream as an isentropic or lossless process. Using this approximation, the minimum pressure p_{VC} can be estimated from the Bernoulli relationship. See Sec. 6, Fluid and Particle Dynamics, for more background information. Downstream of the vena contracta, the flow is definitely not lossless due to

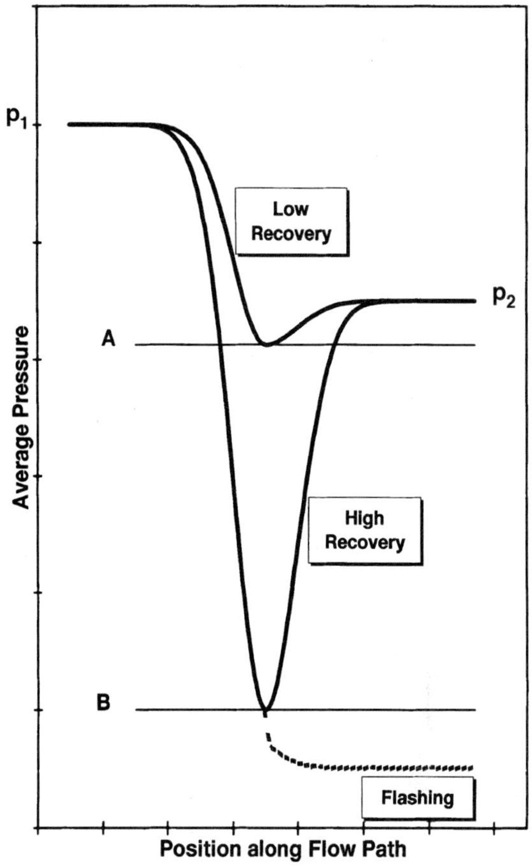

FIG. 8-72 Generic depictions of average pressure at subsequent cross sections throughout a control valve. The F_L values selected for illustration are 0.9 and 0.63 for low and high recovery, respectively. Internal pressure in the high-recovery valve is shown as a dashed line for flashing conditions ($p_2 < p_v$) with $p_v = B$.

all the turbulence that is generated. As the flow passage area increases and the fluid slows down, some of the kinetic energy of the fluid is converted back to pressure energy as pressure recovers. The energy that is permanently lost via turbulence accounts for the permanent pressure or head loss of the valve. The relative amount of pressure that is recouped determines whether the valve is considered to be high- or low-recovery. The flow passage geometry at and downstream of the vena contracta primarily determines the amount of recovery. The amount of recovery is quantified by the liquid pressure recovery factor F_L

$$F_L = \sqrt{\frac{p_1 - p_2}{p_1 - p_{vc}}} \tag{8-78}$$

Under some operating conditions, sufficient pressure differential may exist across the valve to cause the vena contracta pressure to drop to the saturation pressure (also known as the *vapor pressure*) of the liquid. If this occurs, a portion of the liquid will vaporize, forming a two-phase, compressible mixture within the valve. If sufficient vapor forms, the flow may *choke.* When a flow is choked, any increase in pressure differential across the valve no longer produces an increase in flow through the valve.

The vena contracta condition at choked flow for pure liquids has been shown to be

$$p_{vc} = F_F p_v \tag{8-79}$$

where

$$F_F = 0.96 - 0.28\sqrt{\frac{p_v}{p_c}} \tag{8-80}$$

and p_{vc} is the absolute vena contracta pressure under choked conditions, F_F is the liquid critical pressure ratio factor, p_v is the absolute vapor pressure of the liquid at inlet temperature, and p_c is the absolute thermodynamic critical pressure of the fluid.

Equations (8-78) and (8-79) can be used together to determine the pressure differential across the valve at the point where the flow chokes.

$$\Delta p_{\text{choked}} = F_L^2 (p_1 - F_F p_v) \tag{8-81}$$

The pressure recovery factor is a constant for any given valve at a given opening. The value of this factor can be established by flow test and is published by the valve manufacturer. If the actual pressure differential across the valve is greater than the choked pressure differential of Eq. (8-81), then Δp_{choked} should be used in Eq. (8-77) to determine the correct valve size. A more complete presentation of sizing relationships is given in ANSI 75.01.01, including provisions for pipe reducers and Reynolds number effects.

Equations (8-77) to (8-81) are restricted to incompressible fluids. For gases and vapors, the fluid density is dependent on pressure. For convenience, compressible fluids are often assumed to follow the ideal gas law model. Deviations from ideal behavior are corrected for, to first order, with nonunity values of compressibility factor Z (see Sec. 2, Physical and Chemical Data, for definitions and data for common fluids). For compressible fluids

$$w = 94.8 C_v p_1 Y \sqrt{\frac{x M_w}{T_1 Z}} \tag{8-82}$$

where p_1 is in bar absolute, T_1 is inlet temperature in K, M_w is the molecular weight, and x is the dimensionless pressure drop ratio $(p_1 - p_2)/p_1$. The expansion factor Y accounts for changes in the fluid density as the fluid passes through the valve. It is dependent on pressure drop and valve geometry. Experimental data have shown that for small departures in the ratio of specific heat from that of air (1.4), a simple linear relationship can be used to represent the expansion factor:

$$Y = 1 - \frac{1.4x}{3 x_T \gamma} \quad \text{for} \quad x \le \frac{x_T \gamma}{1.4} \tag{8-83}$$

where γ is the ratio of specific heats and x_T is an experimentally determined factor for a specific valve and is the largest value of x that contributes to flow (i.e., values of x greater than x_T do not contribute to flow).

The terminal value of x, x_T, results from a phenomenon known as *choking*. Given a nozzle geometry with fixed inlet conditions, the mass flow rate will increase as p_2 is decreased up to a maximum amount at the critical pressure drop. The velocity at the vena contracta has reached sonic, and a standing shockwave has formed. This shock causes a step change in pressure as flow passes through it, and further reduction in p_2 does not increase mass flow. Thus x_T relates to the critical pressure drop ratio and also accounts for valve geometry effects. The value of x_T varies with flow path geometry and is supplied by the valve manufacturer. In the choked case,

$$x > \frac{x_T y}{1.4} \quad \text{and} \quad Y = 0.67 \tag{8-84}$$

Noise Control Sound is a fluctuation of air pressure that can be detected by the human ear. Sound travels through any fluid (e.g., the air) as a compression/expansion wave. This wave travels radially outward in all directions from the sound source. The pressure wave induces an oscillating motion in the transmitting medium that is superimposed on any other net motion it may have. These waves are reflected, refracted, scattered, and absorbed as they encounter solid objects. Sound is transmitted through solids in a complex array of types of elastic waves. Sound is characterized by its amplitude, frequency, phase, and direction of propagation.

Sound strength is therefore location-dependent and is often quantified as a sound pressure level L_p in decibels based on the root mean square (rms) sound pressure value p_s, where

$$L_p = 10 \log \left(\frac{p_s}{p_{\text{reference}}} \right)^2 \tag{8-85}$$

For airborne sound, the reference pressure is 2×10^{-5} Pa (29×10^{-1} psi), which is nominally the human threshold of hearing at 1000 Hz. The corresponding sound pressure level is 0 dB. A voice in conversation is about 50 dB, and a jackhammer operator is subject to 100 dB. Extreme levels such as a jet engine at takeoff might produce 140 dB at a distance of 3 m, which is a pressure amplitude of 200 Pa (29×10^{-3} psi). These examples demonstrate both the sensitivity and the wide dynamic range of the human ear.

Traveling sound waves carry energy. Sound intensity I is a measure of the power passing through a unit area in a specified direction and is related to p_s. Measuring sound intensity in a process plant gives clues to the location of the source. As one moves away from the source, the fact that the energy is spread over a larger area requires that the sound pressure level decrease. For example, doubling one's distance from a point source reduces L_p by 6 dB. Viscous action from the induced fluid motion absorbs additional acoustic energy. However, in free air, this viscous damping is negligible over short distances (on the order of 1 m).

Noise is a group of sounds with many nonharmonic frequency components of varying amplitudes and random phase. The turbulence generated by a throttling valve creates noise. As a valve converts potential energy to heat, some of the energy becomes acoustic energy as an intermediate step. Valves handling large amounts of compressible fluid through a large pressure change create the most noise because more total power is being transformed. Liquid flows are noisy only under special circumstances, as will be seen in the next subsection. Due to the random nature of turbulence and the broad distribution of length and velocity scales of turbulent eddies, valve-generated sound is usually random, broad-spectrum noise. The total sound pressure level from two such statistically uncorrelated sources is (in decibels)

$$L_p = 10 \log \frac{(p_{s1})^2 + (p_{s2})^2}{(p_{\text{reference}})^2} \tag{8-86}$$

For example, two sources of equal strength combine to create an L_p that is 3 dB higher.

While noise is annoying to listen to, the real reasons for being concerned about noise relate to its impact on people and equipment. Hearing loss can occur due to long-term exposure to moderately high—or even short exposure to very high—noise levels. The U.S. Occupational Safety and Health Act (OSHA) has specific guidelines for permissible levels and exposure times. The human ear has a frequency-dependent sensitivity to sound. When the effect on humans is the criterion, L_p measurements are weighted to account for the ear's response. This so-called A-weighted scale is defined in ANSI S1.4 and is commonly reported as L_{pA}.

There are two approaches to fluid-generated noise control—source or path treatment. Path treatment means absorbing or blocking the transmission of noise after it has been created. The pipe itself is a barrier. The sound pressure level inside a standard schedule pipe is roughly 40 to 60 dB higher than on the outside. Thicker-walled pipe reduces levels somewhat more, and adding acoustical insulation on the outside of the pipe reduces ambient levels up to 10 dB per inch of thickness. Since noise propagates relatively unimpeded inside the pipe, barrier approaches require the entire downstream piping system to be treated in order to be totally effective. In-line silencers place absorbent material inside the flow stream, thus reducing the level of the internally propagating noise. Noise reductions up to 25 dB can be achieved economically with silencers.

The other approach to valve noise problems is the use of quiet trim. Two basic strategies are used to reduce the initial production of noise—dividing the flow stream into multiple paths and using several flow resistances in series. Sound pressure level L_p is proportional to mass flow and is dependent on vena contracta velocity. If each path is an independent source, it is easy to show from Eq. (8-86) that P_s^2 is inversely proportional to the number of passages; additionally, smaller passage size shifts the predominant spectral content to higher frequencies, where structural resonance may be less of a problem. Series resistances or multiple stages can reduce maximum velocity and/or produce backpressure to keep jets issuing from multiple passages from acting independently. While some of the basic principles are understood, predicting noise for a particle flow passage requires some empirical data as a basis. Valve manufacturers have developed noise prediction methods for the valves they build. ANSI/ISA-75.17 is a public-domain methodology for standard (non–low-noise) valve types, although treatment of some multistage, multipath types is underway. Low-noise hardware consists of special cages in linear stem valves, perforated domes or plates and multichannel inserts in rotary valves, and separate devices that use multiple fixed restrictions.

Cavitation and Flashing From the discussion of pressure recovery it was seen that the pressure at the vena contracta can be much lower than the downstream pressure. If the pressure on a liquid falls below its vapor pressure p_v, the liquid will vaporize. Due to the effect of surface tension, this vapor phase will first appear as bubbles. These bubbles are carried downstream with the flow, where they collapse if the pressure recovers to a value above p_v. This pressure-driven process of vapor bubble formation and collapse is known as *cavitation*.

Cavitation has three negative side effects in valves—noise and vibration, material removal, and reduced flow. The bubble collapse process is a violent asymmetric implosion that forms a high-speed microjet and induces pressure waves in the fluid. This hydrodynamic noise and the mechanical vibration that it can produce are far stronger than other noise generation sources in liquid flows. If implosions occur adjacent to a solid component, minute pieces of material can be removed, which, over time, will leave a rough, cinderlike surface.

The presence of vapor in the vena contracta region puts an upper limit on the amount of liquid that will pass through a valve. A mixture of vapor

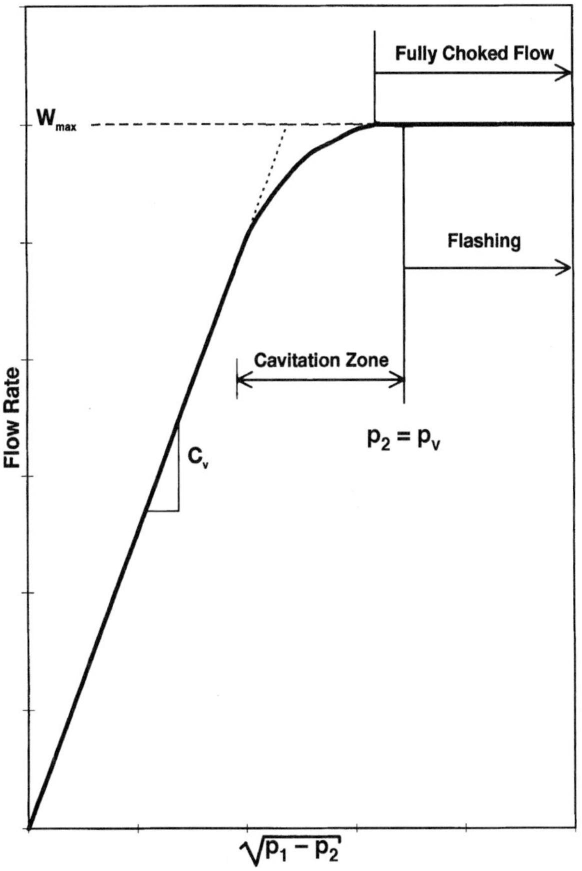

FIG. 8-73 Liquid flow rate versus pressure drop (assuming constant p_1 and p_v).

and liquid has a lower density than that of the liquid alone. While Eq. (8-77) is not applicable to two-phase flows because pressure changes are redistributed due to varying density and the two phases do not necessarily have the same average velocity, it does suggest that lower density reduces the total mass flow rate. Figure 8-73 illustrates a typical flow rate/pressure drop relationship. As with compressible gas flow at a given p_1, flow increases as p_2 is decreased until the flow chokes (i.e., no additional fluid will pass). The transition between incompressible and choked flow is gradual because, within the convoluted flow passages of valves, the pressure is actually an uneven distribution at each cross section and consequently vapor formation zones increase gradually. In fact, isolated zones of bubble formation or incipient cavitation often occur at pressure drops well below that at which a reduction in flow is noticeable. The similarity between liquid and gas choking is not serendipitous; it is surmised that the two-phase fluid is traveling at the mixture's sonic velocity in the throat when choked. Complex fluids with components having varying vapor pressures and/or entrained noncondensable gases (e.g., crude oil) will exhibit soft vaporization/implosion transitions.

There are several methods to reduce cavitation or at least its negative side effects. Material damage is slowed by using harder materials and by directing the cavitating stream away from passage walls (e.g., with an angle body flowing down). Sometimes the system can be designed to place the valve in a higher p_2 location or add downstream resistance, which creates backpressure. A low recovery valve has a higher minimum pressure for a given p_2 and so is a means to eliminate the cavitation itself, not just its side effects. In Fig. 8-72, if $p_v < B$, neither valve will cavitate substantially. For $p_v > B$ but $p_v < A$, the high recovery valve will cavitate substantially, but the low recovery valve will not. Special anticavitation trims are available for globe and angle valves and more recently for some rotary valves. These trims use multiple contraction/expansion stages or other distributed resistances to boost F_L to values sometimes near unity.

If p_2 is below p_v, the two-phase mixture will continue to vaporize in the body outlet and/or downstream pipe until all liquid phase is gone, a condition known as *flashing*. The resulting huge increase in specific volume leads to high velocities, and any remaining liquid droplets acquire much of the higher vapor-phase velocity. Impingement of these droplets can produce material damage, but it differs from cavitation damage because it exhibits a smooth surface. Hard materials and directing the two-phase jets away from solid surfaces are means to avoid this damage.

Seals, Bearings, and Packing Systems In addition to their control function, valves often need to provide shutoff. FCI 70-2-1998 and IEC 60534-4 recognize six standard classifications and define their as-shipped qualification tests. Class I is an amount agreed to by user and supplier with no test needed. Classes II, III, and IV are based on an air test with maximum leakage of 0.5 percent, 0.1 percent, and 0.01 percent of rated capacity, respectively. Class V restricts leakage to 5×10^{-6} mL of water per second per millimeter of port diameter per bar differential. Class VI allows 0.15 to 6.75 mL/min of air to escape depending on port size; this class implies the need for interference-fit elastomeric seals. With the exception of class V, all classes are based on standardized pressure conditions that may not represent actual conditions. Therefore, it is difficult to estimate leakage in service. Leakage normally increases over time as seals and seating surfaces become nicked or worn. Leak passages across the seat-contact line, known as *wire drawing*, may form and become worse over time—even in hard metal seats under sufficiently high pressure differentials.

Polymers used for seat and plug seals and internal static seals include PTFE (polytetrafluoroethylene) and other fluorocarbons, polyethylene, nylon, polyether-ether-ketone, and acetal. Fluorocarbons are often carbon- or glass-filled to improve mechanical properties and heat resistance. Temperature and chemical compatibility with the process fluid are the key selection criteria. Polymer-lined bearings and guides are used to decrease friction, which lessens dead band and reduces actuator force requirements. See Sec. 25, Materials of Construction, for properties.

Packing forms the pressure-tight seal, where the stem protrudes through the pressure boundary. Packing is typically made from PTFE or, for high temperature, a bonded graphite. If the process fluid is toxic, more sophisticated systems such as dual packing, live-loaded, or a flexible metal bellows may be warranted. Packing friction can significantly degrade control performance. Pipe, bonnet, and internal-trim joint gaskets are typically a flat sheet composite. Gaskets intended to absorb dimensional mismatch are typically made from filled spiral-wound flat stainless-steel wire with PTFE or graphite filler. The use of asbestos in packing and gaskets has been largely eliminated.

Flow Characteristics The relationship between valve flow and valve travel is called the *valve flow characteristic*. The purpose of flow characterization is to make loop dynamics independent of load, so that a single controller tuning remains optimal for all loads. Valve gain is one factor affecting loop dynamics. In general, gain is the ratio of change in output to change in input. The input of a valve is travel y, and the output is flow w. Since pressure conditions at the valve can depend on flow (hence travel), valve gain is

$$\frac{dw}{dy} = \frac{\partial w}{\partial C_V}\frac{dC_V}{dy} + \frac{\partial w}{\partial p_1}\frac{dp_1}{dy} + \frac{\partial w}{\partial p_2}\frac{dp_2}{dy} \tag{8-87}$$

An inherent valve flow characteristic is defined as the relationship between flow rate and travel, under constant-pressure conditions. Since the rightmost two terms in Eq. (8-87) are zero in this case, the inherent characteristic is necessarily also the relationship between flow coefficient and travel.

Figure 8-74 shows three common inherent characteristics. A linear characteristic has a constant slope, meaning the inherent valve gain is a constant. The most popular characteristic is equal-percentage ("=%" in the Fig. 8-74), which gets its name from the fact that equal changes in travel produce equal-percentage changes in the existing flow coefficient. In other words, the slope of the curve is proportional to C_V, or equivalently that inherent valve gain is proportional to flow. The equal-percentage characteristic can be expressed mathematically by

$$C_V(y) = (\text{rated } C_V)\exp\left[\left(\frac{y}{\text{rated } y} - 1\right)\ln R\right] \tag{8-88}$$

This expression represents a set of curves parameterized by R. Note that $C_V(y = 0)$ equals (rated C_V)/R rather than zero; real equal-percentage characteristics deviate from theory at some small travel to meet shutoff requirements. An equal-percentage characteristic provides perfect compensation for a process where the gain is inversely proportional to flow (e.g., liquid pressure). *Quick opening* does not have a standardized mathematical definition. Its shape arises naturally from high-capacity plug designs used in on/off service globe valves. Frequently, pressure conditions at the valve will change with flow rate. This so-called *process influence* [the rightmost two terms on the right-hand side of Eq. (8-87)] combines with inherent gain to express the installed valve gain. The flow versus travel relationship for a specific set of conditions is called the *installed flow characteristic*. Typically, valve Δp decreases with load, since pressure losses in the piping system

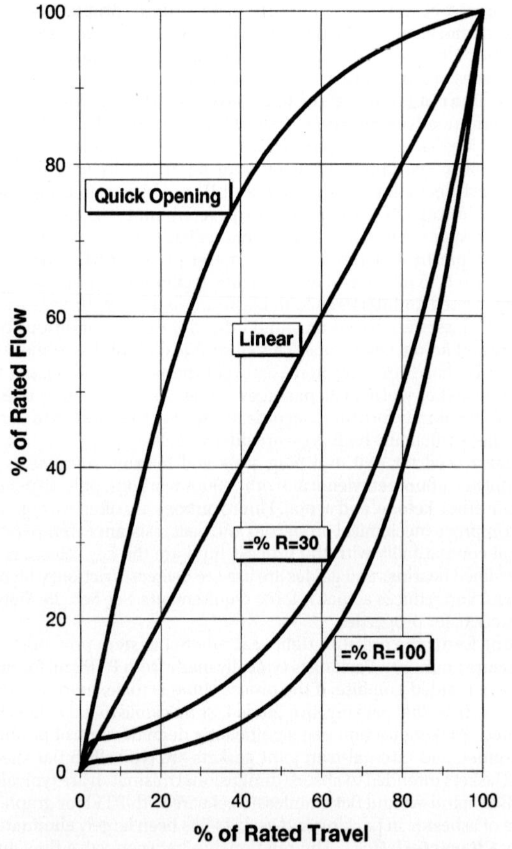

FIG. 8-74 Typical inherent flow characteristics.

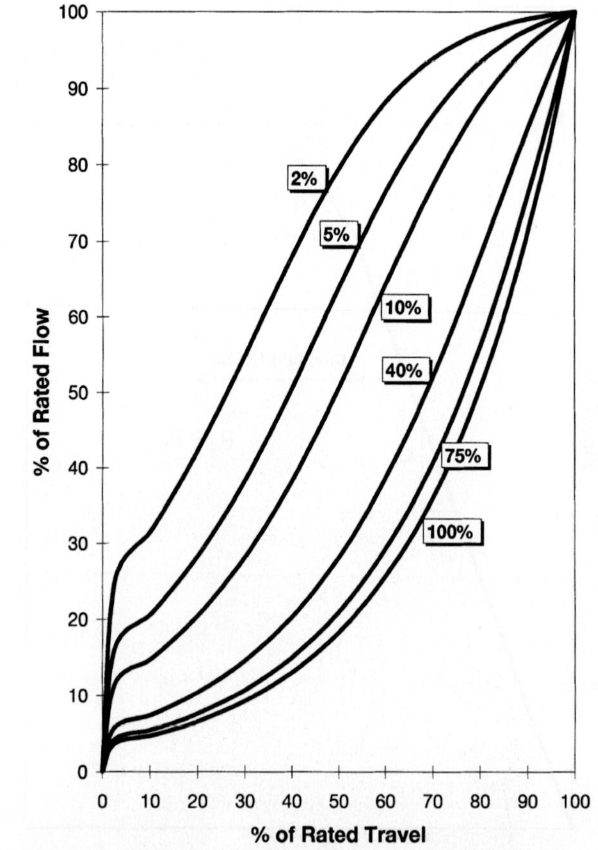

FIG. 8-75 Installed flow characteristic as a function of percent of total system head allocated to the control valve (assuming constant-head pump, no elevation head loss, and an R equal to 30 equal-percentage inherent characteristic).

increase with flow. Figure 8-75 illustrates how allocation of total system head to the valve influences the installed flow characteristics. For a linear or quick-opening characteristic, this transition toward a concave down shape would be more extreme. This effect of typical process pressure variation, which causes equal-percentage characteristics to have fairly constant installed gain, is one reason the equal-percentage characteristic is the most popular.

Due to clearance flow, flow force gradients, seal friction, and the like, flow cannot be throttled to an arbitrarily small value. Installed rangeability is the ratio of maximum to minimum controllable flow. The actuator and positioner, as well as the valve, influence the installed rangeability. *Inherent rangeability* is defined as the ratio of the largest to the smallest C_V within which the characteristic meets specified criteria (see ISA 75.11). The R value in the equal-percentage definition is a theoretical rangeability only. While high installed rangeability is desirable, it is also important not to oversize a valve; otherwise, turndown (ratio of maximum normal to minimum controllable flow) will be limited.

Sliding stem valves are characterized by altering the contour of the plug when the port and plug determine the minimum (controlling) flow area. Passage area versus travel is also easily manipulated in characterized cage designs. Inherent rangeability varies widely, but typical values are 30 for contoured plugs and 20 to 50 for characterized cages. While these types of valves can be characterized, the degree to which manufacturers conform to the mathematical ideal is revealed by plotting measured C_V versus travel. Note that ideal equal-percentage will plot as a straight line on a semilog graph. Custom characteristics that compensate for a specific process are possible.

Rotary stem-valve designs are normally offered only in their naturally occurring characteristic, since it is difficult to appreciably alter this. If additional characterization is required, the positioner or controller may be characterized. However, these approaches are less direct, since it is possible for device nonlinearity and dynamics to distort the compensation.

VALVE CONTROL DEVICES

Devices mounted on the control valve that interface various forms of input signals, monitor and transmit valve position, or modify valve response are valve control devices. In some applications, several auxiliary devices are used together on the same control valve. For example, mounted on the control valve, one may find a current-to-pressure transducer, a valve positioner, a volume booster relay, a solenoid valve, a trip valve, a limit switch, a process controller, and/or a stem position transmitter. Figure 8-76 shows a valve positioner mounted on the yoke leg of a spring and diaphragm actuator.

As most throttling control valves are still operated by pneumatic actuators, the control valve device descriptions that follow relate primarily to devices that are used with pneumatic actuators. The functions of hydraulic and electrical counterparts are very similar. Specific details on a particular valve control device are available from the vendor of the device.

Valve Positioners The valve positioner, when combined with an appropriate actuator, forms a complete closed-loop valve position control system. This system makes the valve stem conform to the input signal coming from the process controller in spite of force loads that the actuator may encounter while moving the control valve. Usually, the valve positioner is contained in its own enclosure and is mounted on the control valve.

The key parts of the positioner/actuator system, shown in Fig. 8-77*a*, are (1) an input conversion network, (2) a stem position feedback network, (3) a summing junction, (4) an amplifier network, and (5) an actuator.

The input conversion network shown is the interface between the input signal and the summer. This block converts the input current or pressure (from an I/P transducer or a pneumatic process controller) to a voltage, electric current, force, torque, displacement, or other particular variable that can be directly used by the summer. The input conversion usually contains a means to adjust the slope and offset of the block to provide for a means of spanning and zeroing the positioner during calibration. In addition, means for changing the sense (known as "action") of the input/output characteristic are often addressed in this block. Also exponential, logarithmic, or other predetermined characterization can be put in this block to provide a characteristic that is useful in offsetting or reinforcing a nonlinear valve or process characteristic.

The stem position feedback network converts stem travel to a useful form for the summer. This block includes the feedback linkage which varies with

FIG. 8-76 Valve and actuator with valve positioner attached. (*Courtesy Fisher Controls International LLC.*)

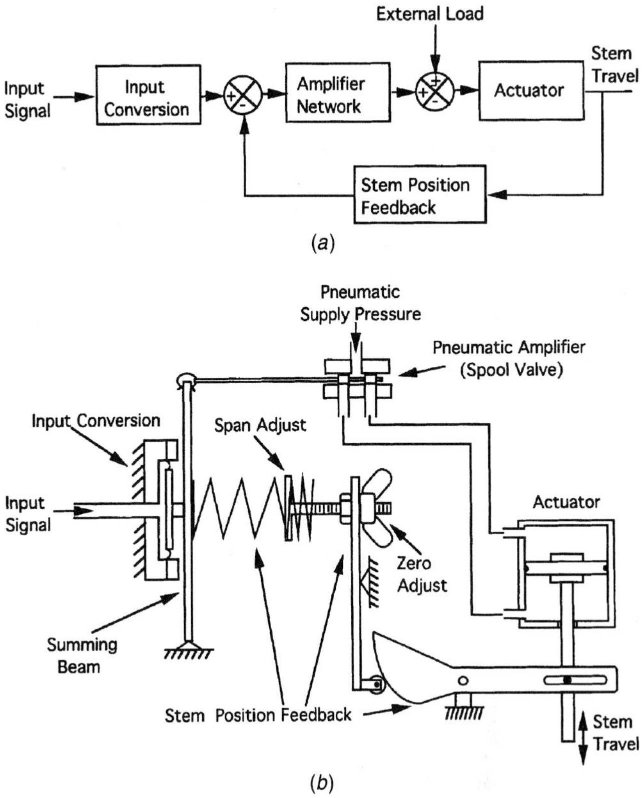

FIG. 8-77 Positioner/actuators. (*a*) Generic block diagram. (*b*) Example of a pneumatic positioner/actuator.

actuator type. Depending on positioner design, the stem position feedback network can provide span and zero and characterization functions similar to that described for the input conversion block.

The amplifier network provides signal conversion and suitable static and dynamic compensation for good positioner performance. Control from this block usually reduces to a form of proportional or proportional plus derivative control. The output from this block in the case of a pneumatic positioner is a single connection to the spring and diaphragm actuator or two connections for push/pull operation of a springless piston actuator. The action of the amplifier network and the action of the stem position feedback can be reversed together to provide for reversed positioner action.

By design, the gain of the amplifier network shown in Fig. 8-77*a* is made very large. Large gain in the amplifier network means that only a small proportional deviation will be required to position the actuator through its active range of travel. This means that the signals into the summer track very closely and that the gain of the input conversion block and the stem position feedback block determine the closed-loop relationship between the input signal and the stem travel.

Large amplifier gain also means that only a small amount of additional stem travel deviation will result when large external force loads are applied to the actuator stem. For example, if the positioner's amplifier network has a gain of 50 (and assuming that high packing box friction loads require 25 percent of the actuator's range of thrust to move the actuator), then only 25 percent/50 (or 0.5 percent deviation) between input signal and output travel will result due to valve friction.

Figure 8-77*b* is an example of a pneumatic positioner/actuator. The input signal is a pneumatic pressure that (1) moves the summing beam, which (2) operates the spool valve amplifier, which (3) provides flow to and from the piston actuator, which (4) causes the actuator to move and continue moving until (5) the feedback force returns the beam to its original position and stops valve travel at a new position. Typical positioner operation is thereby achieved.

Static performance measurements related to positioner/actuator operation include the conformity, measured accuracy, hysteresis, dead band, repeatability, and locked stem pressure gain. Definitions and standardized test procedures for determining these measurements can be found in ISA-S75.13, "Method of Evaluating the Performance of Positioners with Analog Input Signals and Pneumatic Output."

Dynamics of Positioner-Based Control Valve Assemblies Control valve assemblies are complete, functional units that include the valve body, actuator, positioner, if so equipped, associated linkages, and any auxiliary equipment such as current to pneumatic signal transducers and air supply pressure regulators. Although performance information such as frequency response, sensitivity, and repeatability data may be available for a number of these components individually, it is the performance of the entire assembly that will ultimately determine how well the demand signal from the controller output is transferred through the control valve to the process. The valve body, actuator, and positioner combination is typically responsible for the majority of the control valve assembly's dynamic behavior. On larger actuators, the air supply pressure regulator capacity or other airflow restrictions may limit the control valve assembly's speed of response.

The control valve assembly response can usually be characterized quite well by using a first-order plus dead-time response model. The control valve assembly will also exhibit backlash, stiction, and other nonlinear behavior. During normal operation of a control loop, the controller usually makes small output changes from one second to the next. Typically this change is less than 1 percent. With very small controller output changes, e.g., less than 0.1 percent, the control valve assembly may not move at all. As the magnitude of the controller output change increases, eventually the control valve will move. At the threshold of movement, the positional accuracy and repeatability of the control valve are usually quite poor. The speed of response may be quite slow and may occur after a number of seconds of dead time. This poor performance is due to the large backlash and stiction effects relative to the requested movement and the small output change of the positioner. With a further increase in the magnitude of the controller output steps, the behavior of the control valve typically becomes more repeatable and "linear." Dead time usually drops to only a fraction of a second, and the first-order time constant becomes faster. For much larger steps in the controller output, e.g., over 10 percent, the positioner and air supply equipment may be unable to deliver the necessary air volume to maintain the first-order response. In this case, the control valve will exhibit very little dead time, but will be rate-limited and will ramp toward the requested position. It is within the linear region of motion that the potential for the best control performance exists.

When one is specifying a control valve for process control applications, in addition to material, style, and size information, the dynamic response characteristics and maximum allowable dead band (sum of backlash, stiction, and hysteresis effects) *must* be stated. The requirement for the control valve assembly's speed of response is ultimately determined by the dynamic characteristics of the process and the control objectives. Typically, the equivalent first-order time constant specified for the control valve assembly should be at least 5 times faster than the desired controller closed-loop time constant. If this requirement is not met, the tuning of the control loop must be slowed down to accommodate the slow control valve response; otherwise, control robustness and stability may be compromised. The dead band of the control valve assembly is typically the determining factor for control resolution and frequently causes control instability in the form of a "limit" cycle. The controller output will typically oscillate across a range that is one to two times the magnitude of the control valve dead band. This is very dependent on the nature of the control valve nonlinearities, the process dynamics, and the controller tuning. The magnitude of the process limit cycle is determined by the size of the control valve dead band multiplied by the installed gain of the control valve. For this reason, a high-performance control valve assembly, e.g., with only 0.5 percent dead band, may cause an unacceptably large process limit cycle if the valve is oversized and has a high installed gain. For typical process control applications, the installed gain of the control valve should be in the range of 0.5 to 2 percent of the process variable span per percent of controller output. The total dead band of the control valve assembly should be less than 1 percent. For applications that require more precise control, the dead band and possibly the installed gain of the control valve must be reduced. Specialized actuators are available that are accurate down to 0.1 percent or less. At this level of performance, however, the design of the valve body, bearings, linkages, and seals starts to become a significant source of dead band.

Positioner/Actuator Stiffness Minimizing the effect of dynamic loads on valve stem travel is an important characteristic of the positioner/actuator. Stem position must be maintained in spite of changing reaction forces caused by valve throttling. These forces can be random (buffeting force) or can result from a negative-slope force/stem travel characteristic (negative gradient); either could result in valve stem instability and loss of control. To reduce and eliminate the effect of these forces, the effective stiffness of the positioner/actuator must be made sufficiently high to maintain stationary control of the valve stem.

The air spring effect is added to the spring stiffness and results from adiabatic expansion and compression of air in the actuator casing. Numerically, the small perturbation value for air spring stiffness in newtons per meter is given by

$$\text{Air spring rate} = \frac{\gamma p_a A_a^2}{V} \tag{8-89}$$

where γ is the ratio of specific heats (1.4 for air), p_a is the actuator pressure in pascals absolute, A_a is the actuator pressure area in square meters, and V is the internal actuator volume in cubic meters.

Positioner Application Positioners are widely used on pneumatic valve actuators. Often they provide improved process loop control because they reduce valve-related nonlinearity. Dynamically, positioners maintain their ability to improve control valve performance for sinusoidal input frequencies up to about one-half of the positioner bandwidth. At input frequencies greater than this, the attenuation in the positioner amplifier network increases, and valve nonlinearity begins to affect final control element performance more significantly. Because of this, the most successful use of the positioner occurs when the positioner response bandwidth is greater than twice that of the most dominant time lag in the process loop.

Some typical examples in which the dynamics of the positioner are sufficiently fast to improve process control are the following:

1. *In a distributed control system (DCS) process loop with an electronic transmitter.* The DCS controller and the electronic transmitter have time constants that are dominant over the positioner response. Positioner operation is therefore beneficial in reducing valve-related nonlinearity.

2. *In a process loop with a pneumatic controller and a large process time constant.* Here the process time constant is dominant, and the positioner will improve the linearity of the final control element. Some common processes with large time constants that benefit from positioner application are liquid level, temperature, large-volume gas pressure, and mixing.

3. *Additional situations in which valve positioners are used:*

a. On springless actuators where the actuator is not usable for throttling control without position feedback.

b. When split ranging is required to control two or more valves sequentially. In the case of two valves, the smaller control valve is calibrated to open in the lower half of the input signal range, and a larger valve is calibrated to open in the upper half of the input signal range. Calibrating the input command signal range in this way is known as *split-range operation* and increases the practical range of throttling process flows over that of a single valve.

c. In open-loop control applications best static accuracy is needed. On occasion, positioner use can degrade process control. Such is the case when the process controller, process, and process transmitter have time constants that are similar to or smaller than that of the positioner/actuator. This situation is characterized by low process controller proportional gain (gain < 0.5), and hunting or limit cycling of the process variable is observed. Improvements here can be made by doing one of the following:

1. Install a dominant first-order, low-pass filter in the loop ahead of the positioner and retune the process loop. This should allow increased proportional gain in the process loop and reduce hunting. Possible means for adding the filter include adding it to the firmware of the DCS controller, by adding an external *RC* network on the output of the process controller or by enabling the filter function in the input of the positioner, if it is available. Also, some transducers, when connected directly to the actuator, form a dominant first-order lag that can be used to stabilize the process loop.

2. Select a positioner with a faster response characteristic.

Processor-Based Positioners When designed around an electronic microcontroller, the valve positioner [now commonly referred to as a *digital valve controller* (DVC)] takes on additional functionality that provides convenience and performance enhancements over the traditional design. The most common form of processor-based positioner, shown in Fig. 8-77, is a digitally communicating stem position controller that operates by using the fundamental blocks shown in Fig. 8-77*a*. A local display is part of the positioner and provides tag information, command input and travel, servo tuning parameters, and diagnostic information. Often auxiliary sensors are integrated into the device to provide increased levels of functionality and performance. Sensed variables can include actuator pressure, relay input pressure, relay valve position, board temperature, or a discrete input. A 4- to 20-mA valve travel readback circuit is also common. The travel sensor is based on a potentiometer or can be a noncontacting type such as a variable capacitance sensor, Hall effect sensor, or GMR device. Some positioners require a separate connection to an ac or dc supply voltage, but the majority of the designs are "loop-powered," which means that they receive power either through the current input (for positioners that require a 4- to 20-mA analog input signal) or through the digital communications link when the control signal is a digital signal.

Processor-based positioners support automatic travel calibration and automatic response tuning for quick commissioning of the final control element. Features of this type of valve positioner include compensators for improved static and dynamic travel response; diagnostics for evaluating positioner, actuator, and valve health; and the capability to be polled from remote locations through a PC-based application or through a handheld communicator attached to the field wiring. Capability to support custom firmware for special valve applications, such as emergency safety shutdown, is also a characteristic of the processor-based design.

Digital Field Communications To provide increased data transmission capability between valve-mounted devices and the host control system, manufacturers are providing digital network means in their devices. The field networks, commonly known as fieldbuses, compete fiercely in the marketplace and have varying degrees of flexibility and specific application strengths. A prospective fieldbus customer is advised to study the available bus technologies and to make a selection based on needs, and not be seduced by the technology itself.

Generally, a fieldbus protocol must be nonproprietary ("open") so that different vendors of valve devices can design their bus interface to operate properly on the selected fieldbus network. Users demand that the devices be "interoperable" so that the device will work with other devices on the same segment or can be substituted with a device from an alternate manufacturer. International standardization of some of the protocols is currently underway (for example, IEC 61158), whereas others are sponsored by user groups or foundations that provide democratic upgrades to the standard as well as network compliance testing.

The physical wiring typically used is the plant standard twisted-pair wiring for 4- to 20-mA instrumentation. Because of the networking capability of the bus, more than one device can be supported on a single pair of wires, and thus wiring requirements are reduced. Compared to a host level bus such as Ethernet, fieldbuses exhibit slower communication rates, have longer transmission distance capability (1 to 2 km), use standard two-wire installation, are capable of multidrop busing, can support bus-powered devices, do not have redundant modes of bus operation, and are available for intrinsically safe installations. Devices on the fieldbus network may be either powered by the bus itself or powered separately.

The simplest digital networks available today support discrete sensors and on/off actuators, including limit switches and motor starters. Networks of this type have fast cycle times and are often used as an alternative to PLC discrete I/O. More sophisticated field networks are designed to support process automation, more complex process transmitters, and throttling valve actuators. These process-level networks are fundamentally continuous and analoglike in operation, and data computation is floating-point. They support communication of materials of construction, calibration and commissioning, device and loop level diagnostics (including information displays outlining corrective action), and unique manufacturer-specific functionality. Some process networks are able to automatically detect, identify, and assign an address to a new device added to the network, thus reducing labor, eliminating addressing errors, and indicating proper network function immediately after the connection is made. Final control elements operated by the process-level network include I/P transducers, motorized valves, digital valve controllers, and transmitters.

A particular field network protocol known as HART (highway addressable remote transducer) is the most widely used field network protocol. It is estimated that as of 2004 there are more than 14 million HART-enabled devices installed globally and that 70 percent of all processor-based process measurement and control instruments installed each year use HART communications. HART's popularity is based on its similarity to the traditional 4- to 20-mA field signaling and thus represents a safe, controlled transition to digital field communications without the risk often associated with an abrupt change to a totally digital fieldbus. With this protocol, the digital communications occur over the same two wires that provide the 4- to 20-mA process control signal without disrupting the process signal. The protocol uses the frequency-shift keying (FSK) technique (see Fig. 8-78) where two individual frequencies, one representing the mark and the other representing the space, are superimposed on the 4- to 20-mA current signal. As the average value of the signals used is zero, there is no dc offset value added to the 4- to 20-mA signal. The HART protocol is principally a master/slave protocol which means that a field device (slave) speaks only when requested by a master device. In this mode of operation, the slave can update the master at a rate of twice per second. An optional communication mode, *burst mode*, allows a HART slave device to continuously broadcast updates without stimulus requests from the master device. Update rates of 3 to 4 updates per second are typical in the burst mode of operation.

HART-enabled devices are provided by the valve device manufacturer at little or no additional cost. The HART network is compatible with existing 4- to 20-mA applications using current plant personnel and practices, provides for a gradual transition from analog to fully digital protocols, and is provided by the valve device manufacturer at little or no additional cost. Contact the HART Communication Foundation for additional information.

Wireless digital communication to and from the final control element is not yet commercially available but is presently being investigated by more than one device manufacturer. The positive attribute of a wireless field network is the reduced cost of a wireless installation compared to a wired installation. Hurdles for wireless transmissions include security from non-network sources, transmission reliability in the plant environment, limited bus speed, and the conservative nature of the process industry relative to change. Initial installations of wireless networks are supporting secondary variables and diagnostics; in the future primary control of processes with large time constants and finally general application to process control are expected. Both point-to-point and mesh architectures are commercialized at the device level. Mesh architectures rely on the other transmitting devices in the area to receive and then pass on any data transmission, thus rerouting communications around sources of interference. Two unlicensed spread spectrum radio bands are the main focus for current wireless development: 900 MHz and 2.4 GHz. The 900-MHz band is unique to North America and has better propagation and penetrating properties than the 2.4-GHz band. The 2.4-GHz band is a worldwide band and has wider channels, allowing much higher data rates. The spread spectrum technique uses multiple frequencies within the radio band to transmit data. Spread spectrum is further divided into the direct sequence technique, where the device changes frequency many times per data bit, and the frequency-hopping technique, where the device transmits a data packet on one frequency and then changes to a different frequency. Because of the rapid growth expected in this decade, the prospective wireless customer is encouraged to review up-to-date literature to determine the state of field wireless commercialization as it applies to her or his specific application.

Diagnostic Capability The rapid proliferation of communicating, processor-based digital valve controllers over the last decade has led to a corresponding rise in diagnostic capability at the control valve. Diagnosing control valve health is critical to plant operation as maintenance costs can be reduced by identifying the valves that are candidates for repair. Less time is spent during plant shutdown repairing valves that do not need repair, which ultimately results in increased online operating time. Valve diagnostics can detect and flag a failed valve more quickly than by any other means, and they can be configured to cause the valve to move to its fail-safe position on detection of specified fault conditions. The diagnostic-enabled positioner, when used with its host-based software application, can pinpoint exact components in a given final control element that have failed and can recommend precise maintenance procedures to follow to remedy the fault condition.

The state variables that provide valve position control are used to diagnose the health of the final control element. In addition, some digital valve controller designs integrate additional sensors into their construction to provide increased diagnostic capability. For example, pressure sensors are provided to detect supply pressure, actuator pressure (upper and lower cylinder pressures in the case of a springless piston actuator), and internal pilot pressure. Also, the position of the pneumatic relay valve is available in some designs to provide quiescent flow data used for leak detection in the actuator.

Valve diagnostics are divided into two types: online and offline. Offline diagnostics are those diagnostics that occur when the control valve is bypassed or otherwise isolated from the process. The offline diagnostic routine manipulates the travel command to the valve and records the corresponding valve travel, actuator pressure, and servo drive value. These parameters are plotted in various combinations to provide hysteresis plus dead-band information, actuator operating pressure as a function of travel, valve friction, servo drive performance, valve seating load, positioner calibration endpoints, and dynamic response traces. Small- and large-amplitude step inputs as well as large slow ramps (exceeding 100 percent of the input range) are common offline test waveforms generated by the diagnostic as command inputs for

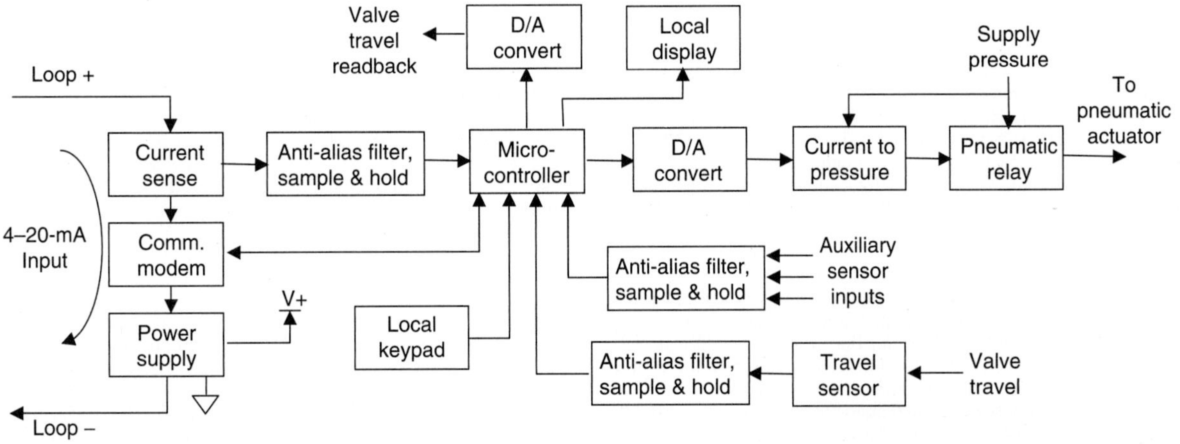

FIG. 8-78 Generic loop-powered digital valve controller.

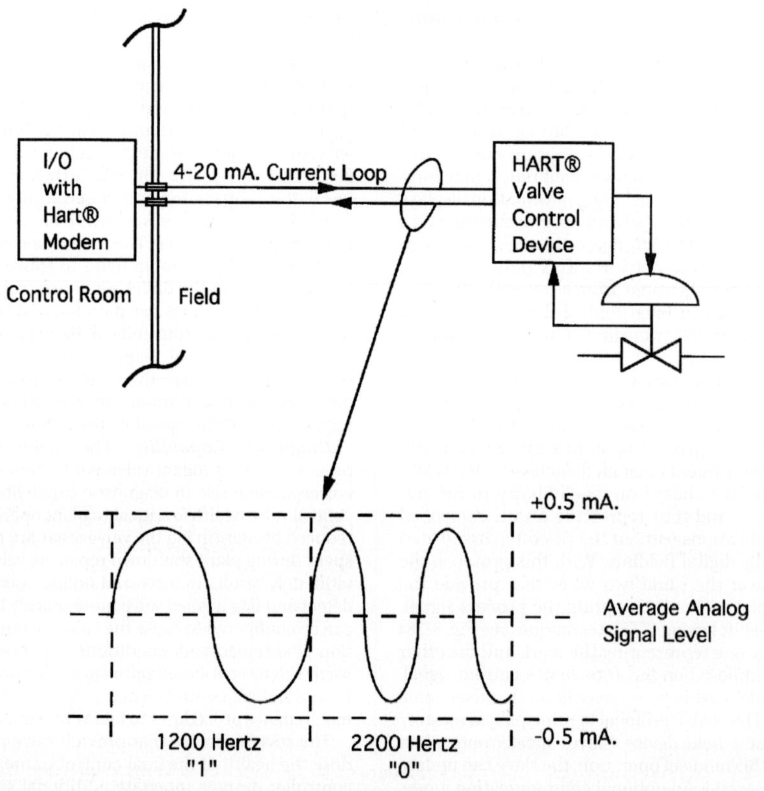

FIG. 8-79 Hybrid point-to-point communications between the control room and the control valve device.

offline diagnostic tests. Figure 8-79 is an example of one offline diagnostic test performed on a small globe valve actuated by a spring and diaphragm actuator. During this test the command input, travel, actuator pressure, and servo drive level are recorded and plotted as they result from a command input that is slowly ramped by the diagnostic routine (Fig. 8-80*a*). This diagnostic is extremely useful in detecting problems with the valve/actuator system and can flag potential problems with the final control element before catastrophic failure occurs. For example, Fig. 8-80*b* indicates the overall tracking capability of the control valve, and Fig. 8-80*c* indicates the pressure operating range of the actuator and the amount of frictional force resulting from the combined effects of valve packing and valve plug contact. Figure 8-80*d* displays the level of servo drive required to stroke the valve from one end of travel to the other. The composite operative health of the control valve is determined through comparison of the empirical levels presented in Fig. 8-80 with the manufacturers' recommendations. Recommended maintenance actions result from this comparison.

Online diagnostics are diagnostics that monitor and evaluate conditions at the control valve during normal throttling periods (i.e., during valve-in-service periods). Online diagnostics monitor mean levels and disturbances generated in the normal operation of the valve and typically do not force or generate disturbances on the valve's operation. For example, an online diagnostic can calculate travel deviation relative to the input command and flag a condition where the valve travel has deviated beyond a preset band. Such an event, if it exists for more than a short time, indicates that the valve has lost its ability to track the input command within specified limits. Additional diagnostics could suggest that the feedback linkage has ceased functioning, or that the valve has stuck, or that some other specific malfunction is the cause of excess travel deviation. The manufacturer of the positioner diagnostic incorporates default limits into the host software application that are used to determine the relative importance of a specific deviation. To quickly indicate the severity of a problem detected by a diagnostic routine, a red, yellow, or green, or "advise, maintenance now, or failed," indication is presented on the user-interface screen for the valve problem diagnosed. Help notes and recommended remedial action are available by pointing and clicking on the diagnostic icon presented on the user's display.

Event-triggered recording is an online diagnostic technique supported in digital valve controllers (DVCs). Functionally a triggering event, such as a valve coming off a travel stop or a travel deviation alert, starts a time-series recording of selected variables. A collection of variables such as the input command, stem travel, actuator pressure, and drive command is stored for several minutes before and after the triggered event. These variables are then plotted as time series for immediate inspection or are stored in memory for later review. Event-triggering diagnostics are particularly useful in diagnosing valves that are closed or full-open for extended periods. In this case the event-triggered diagnostic focuses on diagnostic rich data at the time the valve is actually in operation and minimizes the recording of flat-line data with little diagnostic content. Other online diagnostics detected by DVC manufacturers include excess valve friction, supply pressure failure, relay operation failure, broken actuator spring, current to pressure module failure, actuator diaphragm leaking, and shifted travel calibration.

Safety shutdown valves, which are normally wide open and operate infrequently, are expected to respond to a safety trip command reliably and without fault. To achieve the level of reliability required in this application, the safety valve must be periodically tested to ensure positive operation under safety trip conditions. To test the operation of the shutdown system without disturbing the process, the traditional method is to physically lock the valve stem in the wide-open position and then to electrically operate the pneumatic shutdown solenoid valve. Observing that the pneumatic solenoid valve has properly vented the actuator pressure to zero, the actuator is seen as capable of applying sufficient spring force to close the valve, and a positive safety valve test is indicated. The pneumatic solenoid valve is then returned to its normal electrical state, the actuator pressure returns to full supply pressure, and the valve stem lock mechanism is removed. This procedure, though necessary to enhance process safety, is time-consuming and takes the valve out of service during the locked stem test. Digital valve controllers are able to validate the operation of a safety shutdown valve by using an online diagnostic referred to as a *partial stroke test*. The partial stroke test can be substituted for the traditional test method described above and does not require the valve to be locked in the wide-open position to perform the test. In a fashion similar to that shown in Fig. 8-80*a* (the partial stroke diagnostic), the system physically ramps the command input to the positioner from the wide-open position to a new position, pauses at the new position for a few seconds, and then ramps the command input back to the wide-open position. During this time, the valve travel measurement is monitored and compared to the input command. If the travel measurement deviates for the input by more than a fixed amount for the configured period of time, the valve is considered to have failed the test and a failed-test message is communicated to the host system. Also during this test, the actuator

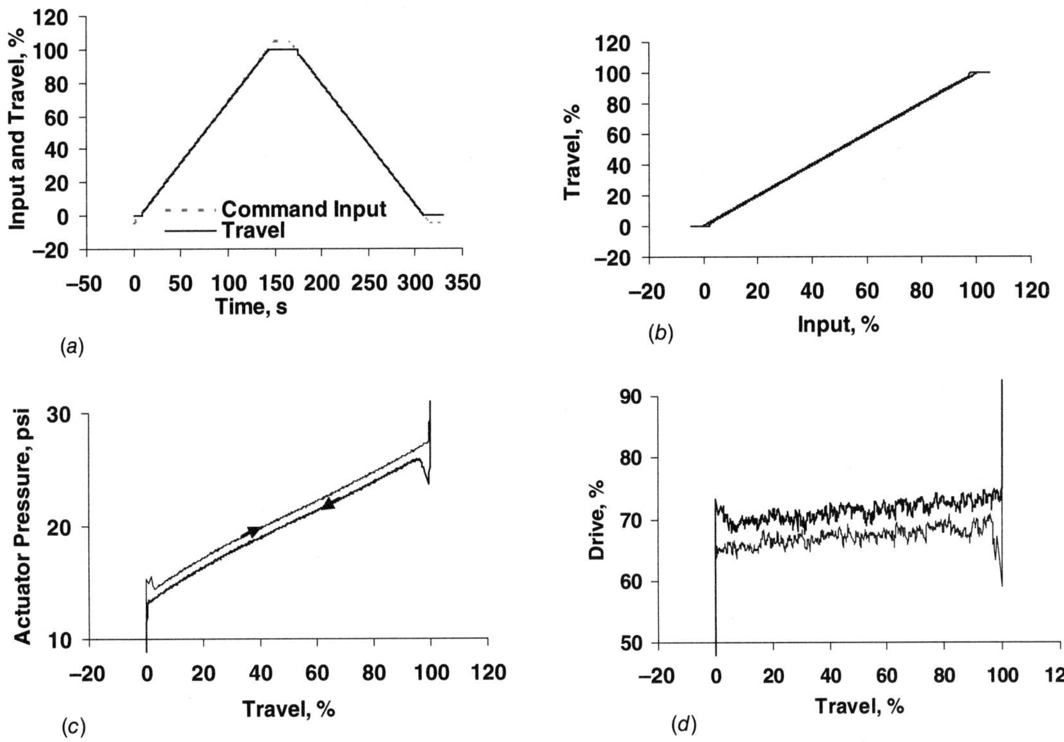

FIG. 8-80 Offline valve diagnostic scan showing results of a diagnostic ramp. (*a*) The command input and resulting travel. (*b*) The dynamic scan. (*c*) The valve signature. (*d*) The servo drive versus travel plot. The hysteresis shown in the valve signature results from sliding friction due to valve packing and valve plug contact.

pressure required to move the valve is detected via a dedicated pressure sensor. If the thrust (pressure) required to move the valve during the partial stroke test exceeds the predefined thrust limit for this test, then the control valve is determined to have a serious sticking problem, the test is immediately aborted, and the valve is flagged as needing maintenance. The partial stroke test can be automated to perform on a periodic basis, for instance, once a week; or it can be initialized by operator request at any time. The amount of valve travel that occurs during the partial stroke test is typically limited to a minimum valve position of 70 percent open or greater. This limit is imposed to prevent the partial stroking of the safety valve from significantly affecting the process flow through the valve. Comparison of partial stroke curves from past tests can indicate the gradual degradation of valve components. Use of "overlay" graphics, identification of unhealthy shifts in servo drive, increases in valve friction, and changes in dynamic response provide information leading to a diagnosis of needed maintenance.

In addition to device-level diagnostics, networked final control elements, process controllers, and transmitters can provide "loop" level diagnostics that can detect loops that are operating below expectations. Process variability, time in a limit (saturated) condition, and time in the wrong control mode are metrics used to detect problems in process loop operation.

Transducers The current-to-pressure transducer (I/P transducer) is a conversion interface that accepts a standard 4- to 20-mA input current from the process controller and converts it to a pneumatic output in a standard pneumatic pressure range [normally 0.2 to 1.0 bar (3 to 15 psig) or, less frequently, 0.4 to 2.0 bar (6 to 30 psig)]. The output pressure generated by the transducer is connected directly to the pressure connection on a spring-opposed diaphragm actuator or to the input of a pneumatic valve positioner.

Figure 8-81*a* is the schematic of a basic I/P transducer. The transducer shown is characterized by (1) an input conversion that generates an angular displacement of the beam proportional to the input current, (2) a pneumatic amplifier stage that converts the resulting angular displacement to pneumatic pressure, and (3) a pressure area that serves as a means to return the beam to very near its original position when the new output pressure is achieved. The result is a device that generates a pressure output that tracks the input current signal. The transducer shown in Fig. 8-81*a* is used to provide pressure to small load volumes (normally 4.0 in^3 or less), such as a positioner or booster input. With only one stage of pneumatic amplification, the flow capacity of this transducer is limited and not sufficient to provide responsive load pressure directly to a pneumatic actuator.

The flow capacity of the transducer can be increased by adding a booster relay such as the one shown in Fig. 8-81*b*. The flow capacity of the booster relay is nominally 50 to 100 times that of the nozzle amplifier shown in Fig. 8-81*a* and makes the combined transducer/booster suitably responsive to operate pneumatic actuators. This type of transducer is stable for all sizes of load volume and produces measured accuracy (see ANSI/ISA-51.1, "Process Instrumentation Terminology," for the definition of measured accuracy) of 0.5 to 1.0 percent of span.

Better measured accuracy results from the transducer design shown in Fig. 8-81*c*. In this design, pressure feedback is taken at the output of the booster relay stage and fed back to the main summer. This allows the transducer to correct for errors generated in the pneumatic booster as well as errors in the I/P conversion stage. Also, particularly with the new analog electric and digital versions of this design, PID control is used in the transducer control network to give extremely good static accuracy, fast dynamic response, and reasonable stability into a wide range of load volumes (small instrument bellows to large actuators). Also environmental factors such as temperature change, vibration, and supply pressure fluctuation affect this type of transducer the least. Even a perfectly accurate I/P transducer cannot compensate for stem position errors generated by friction, backlash, and varying force loads coming from the actuator and valve. To do this compensation, a different control valve device—the valve positioner—is required.

Booster Relays The booster relay is a single-stage power amplifier having a fixed gain relationship between the input and output pressures. The device is packaged as a complete stand-alone unit with pipe thread connections for input, output, and supply pressure. The booster amplifier shown in Fig. 8-81*b* shows the basic construction of the booster relay. Enhanced versions are available that provide specific features such as (1) variable gain to split the output range of a pneumatic controller to operate more than one valve or to provide additional actuator force; (2) low hysteresis for relaying measurement and control signals; (3) high flow capacity for increased actuator stroking speed; and (4) arithmetic, logic, or other compensation functions for control system design.

A particular type of booster relay, called a *dead-band booster*, is designed to be used exclusively between the output of a valve positioner and the input to a pneumatic actuator. It provides extra flow capacity to stroke the actuator faster than with the positioner alone. The dead-band booster is designed intentionally with a large dead band (approximately 5 percent of the input span), elastomer seats for tight shutoff, and an adjustable bypass

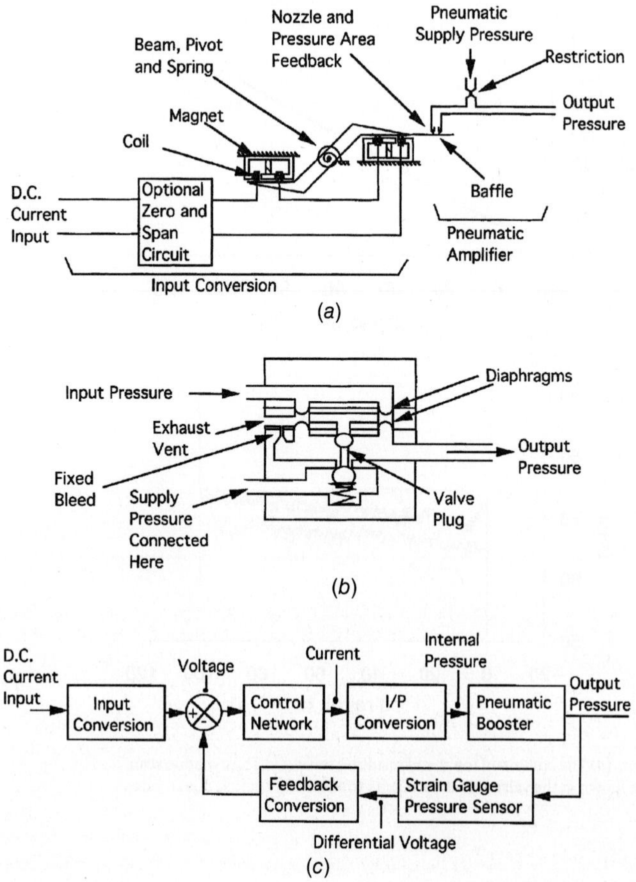

FIG. 8-81 Current-to-pressure transducer component parts. (*a*) Direct-current-pressure conversion. (*b*) Pneumatic booster amplifier (relay). (*c*) Block diagram of a modern I/P transducer.

valve connected between the input and output of the booster. The bypass valve is tuned to provide the best compromise between increased actuator stroking speed and positioner/actuator stability.

With the exception of the dead-band booster, the application of booster relays has diminished somewhat by the increased use of current-to-pressure transducers, electropneumatic positioners, and electronic control systems. Transducers and valve positioners serve much the same functionality as the booster relay in addition to interfacing with the electronic process controller.

Solenoid Valves The electric solenoid valve has two output states. When sufficient electric current is supplied to the coil, an internal armature moves against a spring to an extreme position. This motion causes an attached pneumatic or hydraulic valve to operate. When current is removed, the spring returns the armature and the attached solenoid valve to the deenergized position. An intermediate pilot stage is sometimes used when additional force is required to operate the main solenoid valve. Generally, solenoid valves are used to pressurize or vent the actuator casing for on/off control valve application and safety shutdown applications.

Trip Valves The trip valve is part of a system used where a specific valve action (i.e., fail up, fail down, or lock in last position) is required when the pneumatic supply pressure to the control valve falls below a preset level. Trip systems are used primarily on springless piston actuators requiring fail-open or fail-closed action. An air storage or "volume" tank and a check valve are used with the trip valve to provide power to stroke the valve when supply pressure is lost. Trip valves are designed with hysteresis around the trip point to avoid instability when the trip pressure and the reset pressure settings are too close to the same value.

Limit Switches and Stem Position Transmitters Travel limit switches, position switches, and valve position transmitters are devices that detect the component's relative position, when mounted on the valve, actuator, damper, louver, or other throttling element. The switches are used to operate alarms, signal lights, relays, solenoid valves, or discrete inputs into the control system. The valve position transmitter generates a 4- to 20-mA output that is proportional to the position of the valve.

FIRE AND EXPLOSION PROTECTION

Electrical equipment and wiring methods can be sources of ignition in environments with combustible concentrations of gas, liquid, dust, fibers, or flyings. Most of the time it is possible to locate the electronic equipment away from these hazardous areas. However, where electric or electronic valve-mounted instruments must be used in areas where there is a hazard of fire or explosion, the equipment and installation must meet requirements for safety. Articles 500 through 504 of the National Electrical Code cover the definitions and requirements for electrical and electronic equipment used in class I (flammable gases or vapors), divisions 1 and 2; class II (combustible dust), divisions 1 and 2; and class III (ignitable fibers or flyings), divisions 1 and 2. Division 1 locations are locations with hazardous concentrations of gases, vapors, or combustible dust under normal operating conditions; hazardous concentration of gases, vapors, or combustible dust that occur frequently due to repair, maintenance, or leakage; or hazardous due to the presence of easily ignitable fibers or materials producing combustible flyings during handling, manufacturing, or use. Division 2 locations are locations that normally do not have ignitable concentrations of gases, vapors, or combustible dust. Division 2 locations might become hazardous through failure of ventilating equipment; adjacent proximity to a class I, division 1 location where ignitable concentrations of gases or vapors might occasionally exist; through dust accumulations on or in the vicinity of the electrical equipment sufficient to interfere with the safe dissipation of heat or by abnormal operation or failure of electrical equipment; or when easily ignitable fibers are stored or handled other than in the process of manufacture. An alternate method used for class I hazardous locations is the European "zone" method described in IEC 60079-10, "Electrical Apparatus for Explosive Gas Atmospheres." The zone designation for class I locations has been adapted by the NEC as an alternate method and is defined in Article 505 of the NEC.

Acceptable protection techniques for electrical and electronic valve accessories used in specific class and division locations include explosion-proof enclosures; intrinsically safe circuits; nonincendive circuits, equipment, and components; dust-ignition-proof enclosures; dust-tight enclosures; purged and pressurized enclosures; oil immersion for current-interrupting contacts; and hermetically sealed equipment. Details of these techniques can be found in the *National Electrical Code Handbook*, available from the National Fire Protection Association.

Certified testing and approval for control valve devices used in hazardous locations is normally procured by the manufacturer of the device. The manufacturer typically goes to a third-party laboratory for testing and certification. Applicable approval standards are available from CSA, CENELEC, FM, SAA, and UL.

Environmental Enclosures Enclosures for valve accessories are sometimes required to provide protection from specific environmental conditions. The National Electrical Manufacturers Association (NEMA) provides descriptions and test methods for equipment used in specific environmental conditions in NEMA 250. IEC 60529, "Degrees of Protection Provided by Enclosures (IP Code)," describes the European system for classifying the degrees of protection provided by the enclosures of electrical equipment. Rain, windblown dust, hose-directed water, and external ice formation are examples of environmental conditions covered by these enclosure standards.

Of growing importance is the electronic control valve device's level of immunity to, and emission of, electromagnetic interference in the chemical valve environment. *Electromagnetic compatibility* (EMC) for control valve devices is presently mandatory in the European Community and is specified in International Electrotechnical Commission (IEC) 61326, "Electrical Equipment for Measurement Control and Laboratory Use—EMC Requirements." Test methods for EMC testing are found in the series IEC 61000-4, "EMC Compatibility (EMC), Testing and Measurement Techniques." Somewhat more stringent EMC guidelines are found in the German document NAMUR NE21, "Electromagnetic Compatibility of Industrial Process and Laboratory Control Equipment."

ADJUSTABLE-SPEED PUMPS

An alternative to throttling a process with a process control valve and a fixed-speed pump is by adjusting the speed of the process pump and not using a throttling control valve at all. Pump speed can be varied by using variable-speed prime movers such as turbines, motors with magnetic or hydraulic couplings, and electric motors. Each of these methods of modulating pump speed has its own strengths and weaknesses, but all offer energy savings and dynamic performance advantages over throttling with a control valve.

The centrifugal pump directly driven by a variable-speed electric motor is the most commonly used hardware combination for adjustable-speed pumping.

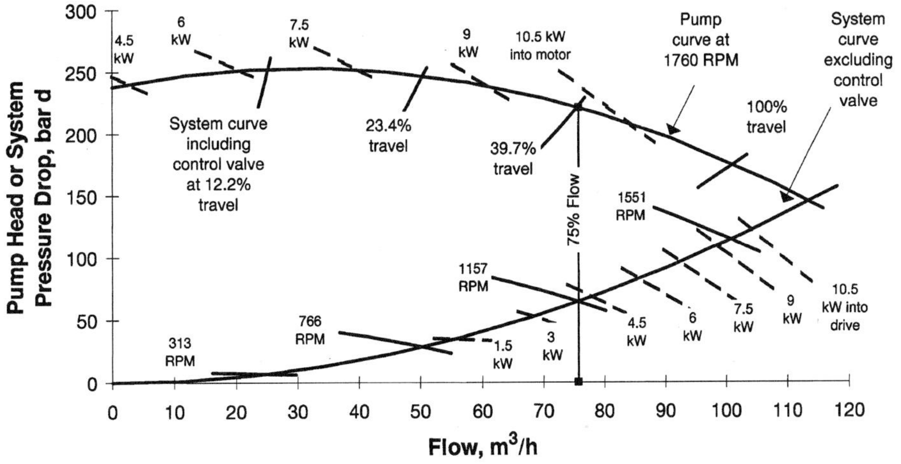

FIG. 8-82 Pressure, flow, and power for throttling a process using a control valve and a constant-speed pump compared to throttling with an adjustable-speed pump.

The motor is operated by an electronic motor speed controller whose function is to generate the voltage or current waveform required by the motor to make the speed of the motor track the input command signal from the process controller.

The most popular form of motor speed control for adjustable-speed pumping is the voltage-controlled *pulse-width-modulated* (PWM) frequency synthesizer and ac squirrel-cage induction motor combination. The flexibility of application of the PWM motor drive and its 90+ percent electrical efficiency along with the proven ruggedness of the traditional ac induction motor make this combination popular.

From an energy consumption standpoint, the power required to maintain steady process flow with an adjustable-speed pump system (three-phase PWM drive and a squirrel-cage induction motor driving a centrifugal pump on water) is less than that required with a conventional control valve and a fixed-speed pump. Figure 8-82 shows this to be the case for a system where 100 percent of the pressure loss is due to flow velocity losses. At 75 percent flow, the figure shows that using the constant-speed pump/control valve results in a 10.1-kW rate, while throttling with the adjustable-speed pump and not using a control valve results in a 4.1-kW rate. This trend of reduced energy consumption is true for the entire range of flows, although amounts vary.

From a dynamic response standpoint, the electronic adjustable-speed pump has a dynamic characteristic that is more suitable in process control applications than those characteristics of control valves. The small amplitude response of an adjustable-speed pump does not contain the dead band or the dead time commonly found in the small amplitude response of the control valve. Nonlinearities associated with friction in the valve and discontinuities in the pneumatic portion of the control valve instrumentation are not present with electronic variable-speed drive technology. As a result, process control with the adjustable-speed pump does not exhibit limit cycles, problems related to low controller gain, and generally degraded process loop performance caused by control valve nonlinearities.

Unlike the control valve, the centrifugal pump has poor or nonexistent shutoff capability. A flow check valve or an automated on/off valve may be required to achieve shutoff requirements. This requirement may be met by automating an existing isolation valve in retrofit applications.

REGULATORS

A regulator is a compact device that maintains the process variable at a specific value in spite of disturbances in load flow. It combines the functions of the measurement sensor, controller, and final control element in one self-contained device. Regulators are available to control pressure, differential pressure, temperature, flow, liquid level, and other basic process variables. They are used to control the differential across a filter press, heat exchanger, or orifice plate. Regulators are used for monitoring pressure variables for redundancy, flow check, and liquid surge relief.

Regulators may be used in gas blanketing systems to maintain a protective environment above any liquid stored in a tank or vessel as the liquid is pumped out. When the temperature of the vessel is suddenly cooled, the regulator maintains the tank pressure and protects the walls of the tank from possible collapse. Regulators are known for their fast dynamic response. The absence of time delay that often comes with more sophisticated control systems makes the regulator useful in applications requiring fast corrective action.

Regulators are designed to operate on the process pressures in the pipeline without any other sources of energy. Upstream and downstream pressures are used to supply and exhaust the regulator. Exhausting is connected back to the downstream piping so that no contamination or leakage to the external environment occurs. This makes regulators useful in remote locations where power is not available or where external venting is not allowed.

The regulator is limited to operating on processes with clean, nonslurry process fluids. The small orifice and valve assemblies contained in the regulator can plug and malfunction if the process fluid that operates the regulator is not sufficiently clean.

Regulators are normally not suited to systems that require constant set-point adjustment. Although regulators are available with capability to respond to remote set-point adjustment, this feature adds complexity to the regulator and may be better addressed by a control valve–based system. In the simplest of regulators, tuning of the regulator for best control is accomplished by changing a spring, an orifice, or a nozzle.

Self-Operated Regulators Self-operated regulators are the simplest form of regulator. This regulator (see Fig. 8-83) is composed of a main throttling valve, a diaphragm or piston to sense pressure, and a spring. The self-contained regulator is completely operated by the process fluid, and no outside control lines or pilot stage is used. In general, self-operated regulators are simple in construction, are easy to operate and maintain, and are usually stable devices. Except for some of the pitot-tube types, self-operated regulators have very good dynamic response characteristics. This is so because any change in the controlled variable registers directly and immediately upon the main diaphragm to produce a quick response to the disturbance.

The disadvantage of the self-operated regulator is that it is not generally capable of maintaining a set point as load flow is increased. Because of the proportional nature of the spring and diaphragm-throttling effect, offset from set point occurs in the controlled variable as flow increases. Reduced set-point offset with increasing load flow can be achieved by adding a pitot tube to the self-operated regulator. The tube is positioned somewhere near the vena contracta of the main regulator valve. As flow through the valve increases, the measured feedback pressure from the pitot tube drops below the control pressure. This causes the main valve to open or boost more than it would if the static value of control pressure were acting on the diaphragm. The resultant effect keeps the control pressure closer to the set point and thus prevents a large drop in process pressure during high-load-flow conditions.

Pilot-Operated Regulators Another category of regulators uses a pilot stage to provide the load pressure on the main diaphragm. This pilot is a regulator itself that has the ability to multiply a small change in downstream pressure into a large change in pressure applied to the regulator diaphragm. Due to this high-gain feature, pilot-operated regulators can achieve a dramatic improvement in steady-state accuracy over that achieved with a self-operated regulator.

The main limitation of the pilot-operated regulator is stability. When the gain in the pilot amplifier is raised too much, the loop can become unstable and oscillate or hunt.

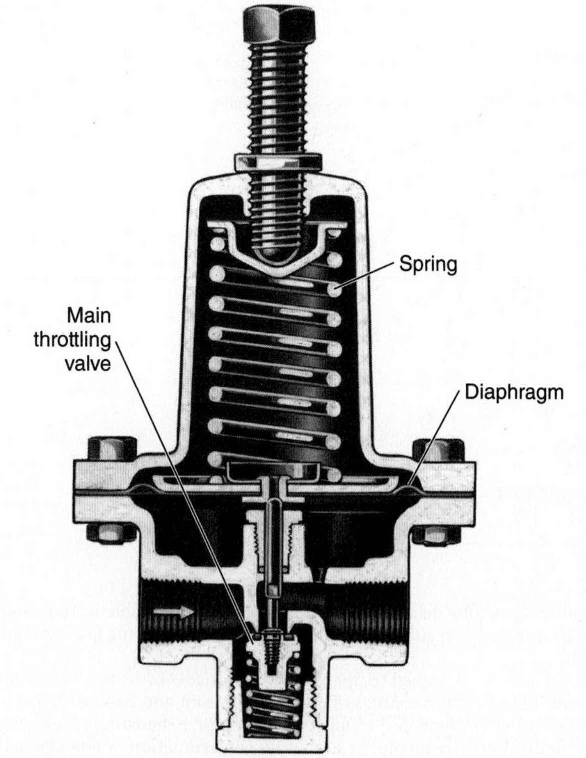

FIG. 8-83 Self-operated regulators.

PROCESS CONTROL AND PLANT SAFETY

General Reference: *Guidelines for Safe Automation of Chemical Processes*, AIChE Center for Chemical Process Safety, New York, 1993.

Accidents in chemical plants make headline news, especially when there is loss of life or the general public is affected in even the slightest way. This increases the public's concern and may lead to government action. The terms *hazard* and *risk* are defined as follows:

- *Hazard.* A potential source of harm to people, property, or the environment.
- *Risk.* Possibility of injury, loss, or an environmental accident created by a hazard.

Safety is the freedom from hazards and thus the absence of any associated risks. Unfortunately, absolute safety cannot be realized, and so the objective is to reduce risks *as low as reasonably practicable* (ALARP).

The design and implementation of safety systems must be undertaken with a view to two issues:

- *Regulatory.* The safety system must be consistent with all applicable codes and standards as well as "generally accepted good engineering practices."
- *Technical.* Just meeting all applicable regulations and "following the crowd" do not relieve a company of its responsibilities. The safety system must work.

The regulatory environment will continue to change. As of this writing, the key regulatory instrument is OSHA 29 CFR 1910.119, "Process Safety Management of Highly Hazardous Chemicals," which pertains to process safety management within plants in which chemicals deemed to be highly hazardous are present.

In addition to government regulation, industry groups and professional societies are producing documents ranging from standards to guidelines. Two applicable standards are IEC 61508, "Functional Safety of Electrical/Electronic/Programmable Electronic Safety-related Systems," and IEC 61511, "Functional Safety—Safety Instrumented Systems for the Process Industry Sector" (the U.S. counterpart is ANSI/ISA S84.00.01-2004, "Application of Safety Instrumented Systems for the Process Industries"). *Guidelines for Safe Automation of Chemical Processes* (1993) from the American Institute of Chemical Engineers' Center for Chemical Process Safety provides comprehensive coverage of the various aspects of safety; and although short on specifics, it is very useful to operating companies developing their own specific safety practices (i.e., it does not tell you what to do, but it helps you decide what is proper for your plant).

The ultimate responsibility for safety rests with the operating company; OSHA 1910.119 is clear on this. Each company is expected to develop (and enforce) its own practices in the design, installation, testing, and maintenance of safety systems. Fortunately, some companies make these documents public. Monsanto's *Safety System Design Practices* was published in its entirety in the proceedings of the International Symposium and Workshop on Safe Chemical Process Automation, Houston, Texas, September 27–29, 1994 (available from the American Institute of Chemical Engineers' Center for Chemical Process Safety).

ROLE OF AUTOMATION IN PLANT SAFETY

As microprocessor-based controls displaced hardwired electronic and pneumatic controls, the impact on plant safety has definitely been positive. When automated procedures replace manual procedures for routine operations, the probability of human errors leading to hazardous situations is lowered, especially in batch facilities. The enhanced capability for presenting information to the process operators in a timely manner and in the most meaningful form increases the operator's awareness of current conditions in the process. Process operators are expected to exercise due diligence in the supervision of the process, and timely recognition of an abnormal situation reduces the likelihood that the situation will progress to the hazardous state. Figure 8-84 depicts the layers of safety protection in a typical chemical plant. Although microprocessor-based process controls enhance plant safety, their primary objective is efficient process operation. Manual operations are automated to reduce variability, to minimize the time required, to increase productivity, and so on. Remaining competitive in the world market demands that the plant be operated in the best manner possible, and microprocessor-based process controls provide numerous functions that make this possible. Safety is never compromised in the effort to increase competitiveness; although not a primary objective, enhanced safety is a by-product of the process control function. By attempting to maintain process conditions at or near their design values, the process controls also attempt to prevent abnormal conditions from developing within the process.

Although process controls can be viewed as a protective layer, this is really a by-product and not the primary function. Where the objective of a function is specifically to reduce risk, the implementation is normally not within the process controls. Instead, logic is separately provided for the specific purpose of reducing risk.

INTEGRITY OF PROCESS CONTROL SYSTEMS

Ensuring the integrity of process controls involves hardware issues, software issues, and human issues. Of these, the hardware issues are usually the easiest to assess and the software issues the most difficult.

The hardware issues are addressed by providing various degrees of redundancy, by providing multiple sources of power and/or an uninterruptible power supply, and the like. The manufacturers of process controls provide a variety of configuration options. Where the process is inherently safe and infrequent shutdowns can be tolerated, nonredundant configurations are acceptable. For more-demanding situations, an appropriate requirement might be that no single component failure be able to render the process

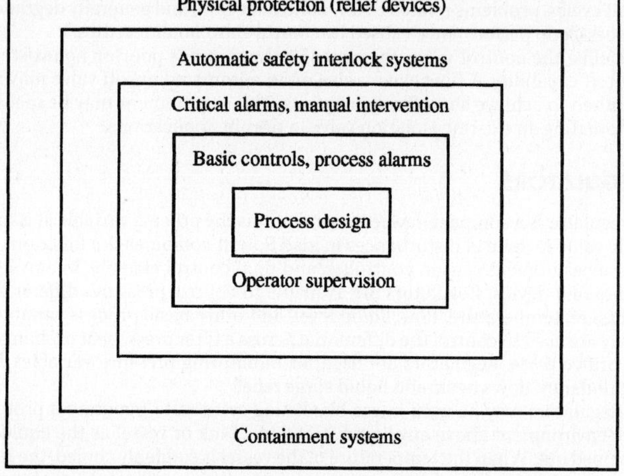

FIG. 8-84 Layers of safety protection in chemical plants.

control system inoperable. For the very critical situations, triple-redundant controls with voting logic might be appropriate. The difficulty lies in assessing what is required for a given process.

Another difficulty lies in assessing the potential for human errors. If redundancy is accompanied with increased complexity, the resulting increased potential for human errors must be taken into consideration. Redundant systems require maintenance procedures that can correct problems in one part of the system while the remainder of the system is in full operation. When maintenance is conducted in such situations, the consequences of human errors can be rather unpleasant.

The use of programmable systems for process control presents some possibilities for failures that do not exist in hardwired electromechanical implementations. Probably of greatest concern are latent defects or "bugs" in the software, either the software provided by the supplier or the software developed by the user. The source of this problem is very simple. There is no methodology available that can be applied to obtain absolute assurance that a given set of software is completely free of defects. Increased confidence in a set of software is achieved via extensive testing, but no amount of testing results in absolute assurance that there are no defects. This is especially true of real-time systems, where the software can easily be exposed to a sequence of events that was not anticipated. Just because the software performs correctly for each event individually does not mean that it will perform correctly when two (or more) events occur at nearly the same time. This is further complicated by the fact that the defect may not be in the programming; it may be in how the software was designed to respond to the events.

The testing of any collection of software is made more difficult as the complexity of the software increases. Software for process control has become progressively complex, mainly because the requirements have become progressively demanding. To remain competitive in the world market, processes must be operated at higher production rates, within narrower operating ranges, closer to equipment limits, and so on. Demanding applications require sophisticated control strategies, which translate to more-complex software. Even with the best efforts of both supplier and user, complex software systems are unlikely to be completely free of defects.

CONSIDERATIONS IN IMPLEMENTATION OF SAFETY INTERLOCK SYSTEMS

Where hazards can arise within a process that put plant workers and/or the general public at risk, some form of protective system is required. *Process hazards analysis* (PSA) addresses the various issues, ranging from assessment of the process hazards to ensuring the integrity of the protective equipment installed to cope with the hazards. Logic must be provided for the specific purpose of taking the process to a state where the hazardous condition cannot exist.

The *safety instrumented system* (SIS) provides the automatic actions that are deemed appropriate from the process hazards analysis. Prior to 1970, the logic was implemented in hardwired electrical circuits that were separate from and very different from the process control logic. The introduction of *programmable logic controllers* (PLCs) in the 1970s created the option to implement the logic in a *programmable electronic system* (PES). At approximately the same time, the increased attention to plant safety expanded the logic required within the safety instrumented system. More and more companies began to implement the logic pertaining to safety in PLCs, but with one caveat: the safety logic must be implemented in a manner that was entirely separate from the process controls (see Fig. 8-84). Separation also extends to the measurement devices and final control elements.

To some extent, this reflected the organizational structure. The "safety group" responsible for the safety logic relied on electrical personnel skilled in implementing discrete logic in the form of *relay ladder logic* (RLL), and thus were comfortable with PLCs. The "process control group" implemented the automatic controls as reflected in the piping and instrumentation (P&I) diagram using proportional-integral-derivative (PID) controllers and preferred *distributed control systems* (DCSs). The complete separation of the safety instrumented system in the form of a PLC from the process control system in the form of a DCS allows each group to proceed in whatever manner it deems most appropriate; that is, each uses its preferred "box."

Modifications to the process controls are more frequent than modifications to the safety instrumented system. Proponents of physically separating the safety instrumented system from the process controls cite the following benefits:

1. The possibility of a change to the process controls leading to an unintentional change to the safety instrumented system is eliminated.
2. The possibility of a human error in the maintenance of the process controls having consequences for the safety instrumented system is eliminated.
3. Management of change is simplified.
4. Administrative procedures for software version control are more manageable.

When PLCs are used in lieu of hardwired circuits for the safety instrumented system, the safety group must develop and strictly enforce administrative procedures to address the following issues:

1. Version controls for the PLC program must be implemented and rigidly enforced. Revisions to the program must be reviewed in detail and thoroughly tested before implementation in the PLC. The various versions must be clearly identified so that there can be no doubt as to what logic is provided by each version of the program.
2. The version of the program that is currently being executed by the PLC must be known with absolute certainty. It must be impossible for a revised version of the program undergoing testing to be downloaded to the PLC.

Constant vigilance is required to prevent lapses in such administrative procedures.

The complete separation of safety from process control usually works well in continuous processes, of which the refining industry is most influential. Continuous processes produce the same product (or possibly different grades of the product), generally meaning that the safety functions do not change. Batch processes present greater challenges to separating the safety logic from the process control logic. Conditions change throughout the batch, often presenting different requirements for the safety functions. Especially in flexible-batch facilities, the nature of the product for one batch could be very different from that for another batch, which often impacts the safety functions. In simple cases, the safety system can make the necessary adjustments based on the states of certain final control elements, but an interface between the process control system and the safety instrumented system is becoming increasingly necessary.

Physical separation is widely practiced within the process industries, but is not mandated by the standards. This opens the possibility for "integrated safety" in which the safety and control functions are implemented within the same "box." Although it is commonly described as "integrated," the current focus is to logically separate the safety functions from the control functions in much the same manner as when physically separated. This raises a number of issues: technical issues (non–safety-related activities must not compromise the safety functions), managerial issues (which group owns the common "box"), supplier issues (mainly commercial), etc. Strong feelings have been expressed on both sides of the physical versus logical separation issue. However, parameters such as *probability of failure on demand* (defined shortly) can be computed to quantify the risk reduction provided by both options. These promise to introduce quantitative data into the discussion, but computing such parameters requires many numbers, some of which are of debatable precision.

SAFETY INSTRUMENTED FUNCTION

In the latest standards, *safety instrumented function* (SIF) is the term for what is commonly called a *safety interlock*, which can be defined as a protective response initiated on the detection of a process hazard. The *safety instrumented system* (SIS) consists of the measurement devices, logic solvers, and final control elements that recognize the hazard and initiate an appropriate response. Most interlocks consist of one or more logic conditions that detect out-of-limit process conditions and respond by driving the final control elements to the safe states.

The potential that the logic within the interlock could contain a defect or bug is a strong incentive to keep it simple. Within process plants, most interlocks are implemented with discrete logic, which usually means either hardwired electromechanical devices or PLCs.

The discrete logic within process plants can be broadly classified as follows:

1. *Safety interlocks.* These are designed to protect the public, plant personnel, and possibly plant equipment from process hazards. These are implemented within the safety instrumented system.
2. *Process interlocks* (those who associate *interlock* with safety prefer terms such as *process actions*). These are designed to prevent process conditions that would unduly stress equipment (perhaps leading to minor damage), lead to off-specification product, and so on. Basically, the process interlocks address situations whose consequences do not meet the requirements to be process hazards, but instead lead to a monetary loss, possibly even a short plant shutdown. Situations resulting in minor equipment damage that can be quickly repaired do not generally require a safety interlock; however, a process interlock might be appropriate.

Implementation of process interlocks within process control systems is perfectly acceptable. Furthermore, it is also permissible (and probably advisable) for responsible operations personnel to be authorized to bypass or ignore a process interlock. Bypassing or ignoring safety interlocks by operations personnel is simply not permitted. When this is necessary for actions such as verifying that the interlock continues to be functional, such situations must be infrequent and incorporated into the design of the interlock.

The process hazards analysis is conducted by an experienced, multidisciplinary team that examines the process design, plant equipment, operating procedures, and so on, using techniques such as *hazard and operability* (HAZOP), *failure mode and effect analysis* (FMEA), and others. The process hazards analysis recommends appropriate measures to reduce the risk, including (but not limited to) the safety interlocks to be implemented in the safety instrumented system.

The process hazards analysis must identify the safety interlocks required for a process and to provide the following for each:

1. The hazard that is to be addressed by the safety interlock
2. The category for the safety interlock (the category reflects the severity of the consequences of the hazard)
3. The logic for the safety interlock, including inputs from measurement devices and outputs to final control elements

The specific categories for safety interlocks used within a company are completely at the discretion of the company, but most use categories that distinguish among the following:

1. Hazards that pose a risk to the public (consequences extend to off-site)
2. Hazards that could lead to injury of company personnel (consequences are confined to the plant site)
3. Hazards that could result in major equipment damage and consequently lengthy plant downtime

Such categories reflect the severity of the consequences should the interlock fail to perform as intended. The requirement for redundancy (complete redundancy, redundant sensors only, etc.) could be associated with each category, but the standards contemplate a different approach.

The standards define the *probability of failure on demand* (PFD) as the probability that a hardware or software failure causes the safety instrumented system to not respond as required. The standards define four safety integrity levels:

Level 1: Probability of failure on demand between 0.1 and 0.01 (availability of 90 to 99 percent)

Level 2: Probability of failure on demand between 0.01 and 0.001 (availability of 99 to 99.9 percent)

Level 3: Probability of failure on demand between 0.001 and 0.0001 (availability of 99.9 to 99.99 percent)

Level 4: Probability of failure on demand less than 0.0001 (availability of 99.99 percent or better)

For each process hazard identified in the process hazards analysis, a safety instrumented function (or safety interlock) must be defined and assigned to a safety integrity level. The safety integrity level determines issues such as the need for redundancy in the safety instrumented system (complete redundancy, redundant measurements only, etc.).

Instead of associating the requirements for redundancy with each category, a safety integrity level is associated with each category. The safety interlocks assigned to a category must be implemented in a hardware and software configuration whose safety integrity level meets or exceeds the safety integrity level for that category.

Diversity is recognized as a useful approach to reduce the number of defects. The team that conducts the process hazards analysis does not implement the safety interlocks but provides the specifications for the safety interlocks to another organization for implementation. This organization reviews the specifications for each safety interlock, seeking clarifications as necessary from the process hazards analysis team and bringing any perceived deficiencies to the attention of the process hazards analysis team.

Diversity can be used to further advantage in redundant configurations. Where redundant measurement devices are required, different technology can be used for each. Where redundant logic is required, one can be programmed and one hardwired. Reliability of the interlock systems has two aspects:

1. It must react, should the hazard arise.
2. It must not react when there is no hazard.

Emergency shutdowns often pose risks in themselves, and therefore they should be undertaken only when truly appropriate. The need to avoid extraneous shutdowns is not motivated by a desire simply to avoid disruption in production operations.

TESTING

As part of the detailed design of each safety interlock, written test procedures must be developed for the following purposes:

1. Ensure that the initial implementation complies with the requirements defined by the process hazards analysis team.
2. Ensure that the interlock (hardware, software, and I/O) continues to function as designed. The design must also determine the time interval over which this must be done. Often these tests must be done with the plant in full operation.

The former is the responsibility of the implementation team and is required for the initial implementation and following any modification to the interlock. The latter is the responsibility of plant maintenance, with plant management responsible for seeing that it is done at the specified interval of time.

Execution of each test must be documented, showing when it was done, by whom, and the results. Failures must be analyzed for possible changes in the design or implementation of the interlock. These tests must encompass the complete interlock system, from the measurement devices through the final control elements. Merely simulating inputs and checking the outputs is not sufficient. The tests must duplicate the process conditions and operating environments as closely as possible. The measurement devices and final control elements are exposed to process and ambient conditions and thus are usually the most likely to fail. Valves that remain in the same position for extended periods may stick in that position and not operate when needed. The easiest component to test is the logic; however, this is the least likely to fail.

ALARMS

The purpose of an alarm is to alert the process operator to a process condition that requires immediate attention. An alarm is said to occur whenever the abnormal condition is detected and the alert is issued. An alarm is said to return to normal when the abnormal condition no longer exists. Analog alarms can be defined on measured variables, calculated variables, controller outputs, and the like. For analog alarms, the following possibilities exist:

1. *High/low alarms.* A high alarm is generated when the value is greater than or equal to the value specified for the high-alarm limit. A low alarm is generated when the value is less than or equal to the value specified for the low-alarm limit.
2. *Deviation alarms.* An alarm limit and a target are specified. A high-deviation alarm is generated when the value is greater than or equal to the target plus the deviation alarm limit. A low-deviation alarm is generated when the value is less than or equal to the target minus the deviation alarm limit.
3. *Trend or rate-of-change alarms.* A limit is specified for the maximum rate of change, usually specified as a change in the measured value per minute. A high-trend alarm is generated when the rate of change of the variable is greater than or equal to the value specified for the trend alarm limit. A low-trend alarm is generated when the rate of change of the variable is less than or equal to the negative of the value specified for the trend alarm limit.

Most systems permit multiple alarms of a given type to be configured for a given value. For example, configuring three high alarms provides a high alarm, a high-high alarm, and a high-high-high alarm.

One operational problem with analog alarms is that noise in the variable can cause multiple alarms whenever its value approaches a limit. This can be avoided by defining a dead band on the alarm. For example, a high alarm would be processed as follows:

1. *Occurrence.* The high alarm is generated when the value is greater than or equal to the value specified for the high-alarm limit.
2. *Return to normal.* The high-alarm return to normal is generated when the value is less than or equal to the high-alarm limit less the dead band.

As the degree of noise varies from one input to the next, the dead band must be individually configurable for each alarm.

Discrete alarms can be defined on discrete inputs, limit switch inputs from on/off actuators, and so on. For discrete alarms, the following possibilities exist:

1. *Status alarms.* An expected or normal state is specified for the discrete value. A status alarm is generated when the discrete value is other than its expected or normal state.
2. *Change-of-state alarm.* A change-of-state alarm is generated on any change of the discrete value.

The expected sequence of events on an alarm is basically as follows:

1. The alarm occurs. This usually activates an audible annunciator.
2. The alarm occurrence is acknowledged by the process operator. When all alarms have been acknowledged, the audible annunciator is silenced.
3. Corrective action is initiated by the process operator.
4. The alarm condition returns to normal.

However, additional requirements are imposed at some plants. Sometimes the process operator must acknowledge the alarm's return to normal. Some plants require that the alarm occurrence be reissued if the alarm remains in the occurred state longer than a specified time. Consequently, some "personalization" of the alarming facilities is done.

When alarms were largely hardware-based (i.e., the panel alarm systems), the purchase and installation of the alarm hardware imposed a certain discipline on the configuration of alarms. With digital systems, the suppliers have made it extremely easy to configure alarms. In fact, it is sometimes easier to configure alarms on a measured value than not to configure the alarms. Furthermore, the engineer assigned the responsibility for defining alarms is often paranoid that an abnormal process condition will go undetected because an alarm has not been configured. When alarms are defined on every measured and calculated variable, the result is an excessive number of alarms, most of which are duplicative and unnecessary.

The accident at the Three Mile Island nuclear plant clearly demonstrated that an alarm system can be counterproductive. An excessive number of alarms can distract the operator's attention from the real problem that needs to be addressed. Alarms that merely tell the operator something that is already known do the same. In fact, a very good definition of a nuisance alarm is one that informs the operator of a situation of which the operator is already aware. The problem with applying this definition lies in determining what the operator already knows.

Unless some discipline is imposed, engineering personnel, especially where contractors are involved, will define far more alarms than plant operations require. This situation may be addressed by simply setting the alarm limits to values such that the alarms never occur. However, changes in alarms and alarm limits are changes from the perspective of management of change. It is prudent to impose the necessary discipline to avoid an excessive number of alarms. Potential guidelines are as follows:

1. For each alarm, a specific action is expected from the process operator. Operator actions such as call maintenance are inappropriate with modern systems. If maintenance needs to know, modern systems can inform maintenance directly.

2. Alarms should be restricted to abnormal situations for which the process operator is responsible. A high alarm on the temperature in one of the control system cabinets should not be issued to the process operator. Correcting this situation is the responsibility of maintenance, not the process operator.

3. Process operators are expected to be exercising normal surveillance of the process. Therefore, alarms are not appropriate for situations known to the operator either through previous alarms or through normal process surveillance. The "sleeping operator" problem can be addressed by far more effective means than the alarm system.

4. When the process is operating normally, no alarms should be triggered. Within the electric utility industry, this design objective is known as *darkboard.* Application of darkboard is especially important in batch plants, where much of the process equipment is operated intermittently.

Another serious distraction to a process operator is the multiple-alarm event, where a single event within the process results in multiple alarms. When the operator must individually acknowledge each alarm, considerable time can be lost in silencing the obnoxious annunciator before the real problem is addressed. Air-handling systems are especially vulnerable to this, where any fluctuation in pressure (e.g., resulting from a blower trip) can cause a number of pressure alarms to occur.

Point alarms (high alarms, low alarms, status alarms, etc.) are especially vulnerable to the multiple-alarm event. This can be addressed in one of two ways:

1. *Ganging alarms.* Instead of individually issuing the point alarms, all alarms associated with a certain aspect of the process are configured to give a single trouble alarm. The responsibility rests entirely with the operator to determine the nature of the problem.

2. *Intelligent alarms.* Logic is incorporated into the alarm system to determine the nature of the problem and then issue a single alarm to the process operator.

While the intelligent alarm approach is clearly preferable, substantial process analysis is required to support intelligent alarming. Meeting the following two objectives is quite challenging:

1. The alarm logic must consistently detect abnormal conditions within the process.

2. The alarm logic must not issue an alert to an abnormal condition when in fact none exists.

Often the latter case is more challenging than the former. Logically, the intelligent alarm effort must be linked to the process hazards analysis. Developing an effective intelligent alarming system requires substantial commitments of effort, involving process engineers, control systems engineers, and production personnel. Methodologies such as expert systems can facilitate the implementation of an intelligent alarming system, but they must still be based on a sound analysis of the potential process hazards.

Complaints regarding an excessive number of alarm occurrences, alarm acknowledgment fatigue on the part of operators, etc., arose with early digital controls and continue to the current day. The practice may suggest otherwise, but the methodology of designing, implementing, and maintaining an effective alarm system is well known. These practices are commonly followed in facilities with a heavy influence by regulatory agencies, the result being a very effective alarm system. In other facilities, achieving an effective alarm system requires production management to make adequate resources available for reviewing and analyzing the proposed alarm configurations and modifications thereof. The alarm system competes with other endeavors for resources of time and money, and in the absence of pressures from a regulatory agency, resources will be allocated to what is perceived to be most financially rewarding.

HUMAN FACTORS

General Reference: *Investigation Report: Refinery Explosion and Fire, Texas City, Texas,* U.S. Chemical Safety and Hazard Investigation Board. Report no. 2005-04-I-TX. March 2007. Available at: http://www.csb.gov/assets/1/19/csbfinalreportbp.pdf

Human factors play an extremely important role in process safety. For example, the explosion that occurred at the BP Texas City refinery in 2005 was due to miscommunication, failure to meet blowdown drum safety standards, miscalibrated and nonfunctioning sensors and alarms, a history of violating proper start-up operating procedures, and poor siting of temporary office trailers. In this disaster, an isomerization unit was being restarted after a process shutdown. The unit had a number of failing and miscalibrated sensors and alarms, and proper start-up protocols were not followed. Also, there was miscommunication between operators at shift change, and there was inadequate supervision available. A separation column became overfilled, because of lack of knowledge of actual liquid level in the tower. A heavy, hot liquid was then sent to a local flash drum—and the resulting heavy vapor was released to the atmosphere. This vapor ignited when a truck engine backfired, causing an explosion that destroyed several trailers next to the unit, killing 15 workers. An investigation revealed that there was a history of plant safety violations and that one of the problems was due to the reward structure for plant managers, who would typically be in the position for only 1 or 2 years before being promoted to other positions in the company. Also, there was a history of violating the unit start-up procedure during many previous start-ups over a 5-year period. Start-ups and shut-downs are the most dangerous times, with an incident rate 10 times that of normal steady-state operation. See the CSB report on the Texas City explosion for more details.

Section 9

Process Economics

James R. Couper, D.Sc. *Professor Emeritus, The Ralph E. Martin Department of Chemical Engineering, University of Arkansas—Fayetteville (Section Editor)*

Darryl W. Hertz, B.S. *Senior Manager, Value Improvement Group, KBR, Houston, Texas (Front-End Loading, Value-Improving Practices)*

(Francis) Lee Smith, Ph.D., M. Eng. *Principal, Wilcrest Consulting Associates, LLC, Katy, Texas; Partner and General Manager, Albutran USA, LLC, Katy, Texas (Front-End Loading, Value-Improving Practices)*

The considerable contribution of Dr. Richard Ulrich, Professor Emeritus, The Ralph E. Martin Department of Chemical Engineering, University of Arkansas, Fayetteville, in editing a major part of this section is deeply appreciated.

PROFITABILITY

OTHER ECONOMIC TOPICS

CAPITAL PROJECT EXECUTION AND ANALYSIS

GLOSSARY

Nomenclature and Units

Symbol	Definition	Units
A_1	Annual conversion expense at production rate 1	$
A_{TC}	Annual capital outlay	$
B	Constant	Dimensionless
C	Cost of equipment	$
C_B	Base cost of carbon steel exchanger	$
CE	Chemical engineering cost index	Dimensionless
$(C_{FC})_{BL}$	Battery-limits fixed capital investment	$
$(C_{EQ})_{DEL}$	Delivered equipment cost	$
C_{HE}	Purchased equipment cost of heat exchanger	$
C_L	Cost of labor	$
COEs	Cash operating expenses	$
C_P	Equipment cost in base year	$
cP	Viscosity in centipoise	cP
D	Depreciation	$
DCFROR	Discounted cash flow rate of return	%
EBIT	Earnings before interest and taxes	$
e	Naperian logarithm base	2.718
F	Future value, future worth, future amount	$
F	Heat exchanger efficiency factor	Dimensionless
F_B	Heat exchanger design type	Dimensionless
FCI	Fixed capital investment	$
FE	Fixed expenses	$
FEL	Front-end loading	
FIFO	First in, first out (inventory)	
FOB	Free on board	
F_M	Material of construction cost factor	Dimensionless
F_P	Design pressure cost factor	Dimensionless
f_1, f_2, f_3	Inflation factors for years 1, 2, and 3	Dimensionless
f'	Declining-balance factor	Dimensionless
I	Investment	$
IRS	Internal Revenue Service	
i	Nominal interest	%
i_{eff}	Effective interest	%
K	Factor for cost index	Dimensionless
LIFO	Last in, first out (inventory)	
M	Annual raw material expense	$
MACRS	Modified Accelerated Cost Recovery System	
m	Number of interest periods per year	Varies
m, n, p, q	Constants or exponent	Dimensionless
N	Annual labor requirement	Operators per shift per year
n	Number of years, depreciation	Years
P	Principal, present value, present worth	$
PC	Personal computer	
POP	Payout period (no interest)	Years
$POP + I$	Payout period plus interest	Years
Q	Energy transferred	Btu/h
R_1, R_2	Annual production rates	lb/yr
S	Salvage value or equipment capacity	Various
SL	Straight-line depreciation	
Sp gr	Specific gravity	Dimensionless
TE	Total expenses	$
T_c	Combined incremental tax rate	%
T_f	Incremental federal income tax rate	%
T_s	Incremental state income tax rate	%
U	Annual utility expenses	$
UAC	Uniform annual cost	$
U_D	Overall heat-transfer coefficient	$Btu/(h \cdot ft^2 \cdot °F)$
V_e	Asset value at end of year	$
V_i	Asset value at beginning of year	$
VE	Variable expenses	$
VIP	Value-improving practice	
X	Plant capacity	tons/day
Y	Operating labor	operator-hour/ton per processing step

General References: Allen, D. H., *Economic Evaluation of Projects*, 3d ed., Institution of Chemical Engineers, Rugby, England, 1991. Baasel, W. D., *Chemical Engineering Plant Design*, 2d ed., Van Nostrand Reinhold, New York, 1989. Brown, T. R., *Hydrocarbon Processing*, October 2000, pp. 93–100. Brown, T. R., *Engineering Economics and Engineering Design for Process Engineers*, CRC Press, Boca Raton, Fla., 2007. Brown, T. R., and S. Sonali, "Project Optimization Through Engineering," *Chem. Eng.*, pp. 51–58 (July 2014). Canada, J. R., and J. A. White, *Capital Investment Decision: Analysis for Management and Engineering*, 2d ed., Prentice-Hall, Englewood Cliffs, N.J., 1980. Popper, H. *Modern Cost Engineering*, McGraw-Hill, New York, 1979. Couper, J. R., and W. H. Rader, *Applied Finance and Economic Analysis for Scientists and Engineers*, Van Nostrand Reinhold, New York, 1986. Couper, J. R., and O. T. Beasley, *The Chemical Process Industries: Function and Economics*, Marcel Dekker, New York, 2001. Couper, J. R., *Process Engineering Economics*, Marcel Dekker, New York, 2003. Garrett, D. E., *Chemical Engineering Economics*, Van Nostrand Reinhold, New York, 1989. Grant, E. L., and W. G. Ireson, *Engineering Economy*, 2d ed., Wiley, New York, 1950. Grant, E. L., W. G. Ireson, and R. S. Leavenworth, *Engineering Economy*, 8th ed., Wiley, New York, 1990. Hackney, J. W., and K. K. Humphreys (eds.), *Control and Management of Capital Projects*, 2d ed., McGraw-Hill, New York, 1992. Hill, D. A., and L. E. Rockley, *Secrets of Successful Financial Management*, Heinemann, London, 1990. Holland, F. A., F. A. Watson, and J. E. Wilkerson, *Introduction to Process Economics*, 2d ed., Wiley, London, 1983. K. K. Humphreys, F. C. Jelen, and J. H. Black (eds.), *Cost and Optimization Engineering*, 3d ed., McGraw-Hill, New York, 1991. Institution of Chemical Engineers, *A Guide to Capital Cost Estimation*, 3d ed., Rugby, England, 1990. Kharbanda, O. P., and E. A. Stallworthy, *Capital Cost Estimating in the Process Industries*, 2d ed., Butterworth-Heinemann, London, 1988. Merrill Lynch, *How to Read an Annual Report*, New York, 1997. Nickerson, C. B., *Accounting Handbook for Non Accountants*, 2d ed., CBI Publishing, Boston, 1979. Ostwald, P. F., *Engineering Cost Estimating*, 3d ed., Prentice-Hall, Englewood Cliffs, N.J., 1991. Park, W. R., and D. E. Jackson, *Cost Engineering Analysis*, 2d ed., Wiley, New York, 1984. Peters, M. S., and K. D. Timmerhaus, *Plant Design and Economics for Chemical Engineers*, 6th ed., McGraw-Hill, New York, 2003. Popper, H. (ed.), *Modern Cost Estimating Techniques*, McGraw-Hill, New York, 1970. Rose, L. M., *Engineering Investment Decisions: Planning under Uncertainty*, Elsevier, Amsterdam, 1976. Thorne, H. C., and J. B. Weaver (eds.), *Investment Appraisal for Chemical Engineers*, American Institute of Chemical Engineers, New York, 1991. Ulrich, G., and P. T. Vasudevan, *Chemical Engineering Process Design and Economics*, CRC Press, Boca Raton, Fla., 2004. Valle-Riestra, J. F., *Project Evaluation in the Chemical Process Industries*, McGraw-Hill, New York, 1983. Wells, G. L., *Process Engineering with Economic Objectives*, Wiley, New York, 1973. Woods, D. R., *Process Design and Engineering*, Prentice-Hall, Englewood Cliffs, N.J., 1993.

GENERAL COMMENTS

One of the most confusing aspects of process engineering economics is the nomenclature used by various authors and companies. In this part of Sec. 9, generic, descriptive terms have been used. Further, an attempt has been made to bring together most of the methods currently in use for project evaluation and to present them in such a way as to make them amenable to modern computational techniques. Most of the calculations can be performed on handheld calculators equipped with scientific function keys. For calculations requiring greater sophistication than that of handheld calculators, algorithms may be solved by using such programs as MATHCAD, TKSOLVER, etc. Spreadsheets are also used whenever the solution to a problem lends itself to this technique.

The nomenclature in process economics has been developed by accountants, engineers, and others such that there is no one correct set of nomenclature. Often it seems confusing, but one must question what is meant by a certain term since companies have adopted their own language. A Glossary of terms is included at the end of this section to assist the reader in understanding the nomenclature. Further, abbreviations of terms such as DCFRR (discounted cash flow rate of return) are used to reduce the wordiness. The number of letters and numbers used to define a variable has been limited to five. The parentheses are removed whenever the letter group is used to define a variable for a computer. Also, a general symbol is defined for a type variable and is modified by mnemonic subscript, e.g., an annual cash quantity, annual capital outlay A_{TC}, \$/year. Wherever a term like this is introduced, it is defined in the text.

It is impossible to allow for all possible variations of equation requirements, but it is hoped that the nomenclature presented will prove adequate for most purposes and will be capable of logical extension to other more specialized requirements.

ACCOUNTING AND FINANCIAL CONSIDERATIONS

PRINCIPLES OF ACCOUNTING

Accounting has been defined as the art of recording business transactions in a systematic manner. It is the language of business and is used to communicate financial information. Conventions that govern accounting are fairly simple, but their application is complex. In this section, the basic principles are illustrated by a simple example and applied to analyzing a company report. The fair allocation of costs requires considerable technical knowledge of operations, so a close liaison between process engineers and accountants in a company is desirable.

In simplest terms, assets that are the economic resources of a company are balanced against equities that are claims against the firm. In equation form,

$$\text{Assets} = \text{Equities}$$

or

$$\text{Assets} = \text{Liabilities} + \text{Owners' Equity}$$

This dual aspect has led to the double-entry bookkeeping system in use today. Any transaction that takes place causes changes in the accounting equation. An increase in assets must be accompanied by one of the following:

- An increase in liabilities
- An increase in stockholders' equity
- An increase in assets

A change in one part of the equation due to an economic transaction must be accompanied by an equal change in another place—therefore, the term *double-entry bookkeeping*. On a page of an account, the left-hand side is designated the debit side, and the right-hand side is the credit side. This convention holds regardless of the type of account. Therefore, for every economic transaction, there is an entry on the debit side balanced by the same entry on the credit side.

All transactions in their original form (receipts and invoices) are recorded chronologically in a journal. The date of the transaction together with an account title and a brief description of the transaction is entered. Table 9-1 is an example of a typical journal page for a company. Journal entries are transferred to a ledger in a process called *posting*. Separate ledger accounts, such as a revenue account, expense account, liability account, or asset account, may be set up for each major transaction. Table 9-2 shows an example of a typical ledger page. The number of ledger accounts depends on the information that management needs to make decisions. Periodically, perhaps on a monthly basis but certainly on a yearly basis, the ledger sheets are closed and balanced. The ledger sheets are then intermediate documents between journal records and balance sheets, income statements, and retained earnings statements, and they provide information for management and various government reports. For example, a consolidated income statement can be prepared for the ledger, revenue, and expense accounts. In like manner, the asset and liability accounts provide information for balance sheets.

FINANCIAL STATEMENTS

A basic knowledge of accounting and financial statements is necessary for a chemical professional to be able to analyze a firm's operation and to communicate with accountants, financial personnel, and managers. Financial reports of a company are important sources of information used by management, owners, creditors, investment bankers, and financial analysts. All publicly held companies are required to submit annual reports to the Securities and Exchange Commission. As with any field a certain basic nomenclature is used to be able to understand the financial operation of a company.

TABLE 9-1 Typical Journal Page

Date	Explanation	LP	Debit	Credit
200X				
Mar 1	Cash	1	$95,000	
	J. Jones, Capital	2		$95,000
Mar 4	Property	4	5,000	
	Cash	1		3,000
	Mortgage	3		2,000
Mar 11	Remodeling Bldg.	5	7,800	
	Cash	1		7,800
Mar 13	Equipment	6	62,300	
	Cash	1		10,000
	Note Payable	3		52,300
Apr 4	To J. Jones	2	2,500	
	Cash	1		2,500

SOURCE: J. R. Couper, *Process Engineering Economics,* Dekker, New York, 2003. By permission of Taylor & Francis Books, Inc., Boca Raton, Fla.

It should be emphasized that companies may also have their own internal nomenclature, but some terms are universally accepted. In this section, the common terminology is used.

A financial report contains two important documents—the balance sheet and the income statement. Two other documents that appear in the financial report are the accumulated retained earnings and the changes in working capital. All these documents are discussed in the following sections, using a fictitious company.

Balance Sheet The balance sheet represents an accounting view of the financial status of a company on a particular date. Table 9-3 is an example of a balance sheet for a company. The date frequently used by corporations is December 31 (calendar year), although some companies are now using June 30 or September 30 as the closing date, depending on when the company closes it books. It is as if the company's operation were frozen in time on that date. The term *consolidated* means that all the balance sheet and income statement data include information from the parent as well as subsidiary operations. The balance sheet consists of two parts: assets are the items that the company owns, and liabilities and stockholders' equity are what the company owes to creditors and stockholders. Although the balance sheet has two sides, it is not part of the double-entry accounting system. The balance sheet is not an account but a statement of claims against company assets on the date of the reporting period. The claims are the creditors and the stockholders. Therefore, the total assets must equal the total liabilities plus the stockholders' equity.

Assets are classified as current, fixed, or intangibles. *Current assets* include cash, cash equivalents, marketable securities, accounts receivable, inventories, and prepaid expenses. Cash and cash equivalents are those items that can be easily converted to cash. Marketable securities are securities that a company holds that also may be converted to cash. Accounts receivable are the amounts due to a company from customers from material that has been delivered but has not been collected as yet. Customers are given 30, 60, or 90 days in which to pay; however, some customers fail to pay bills on time or may not be able to pay at all. An allowance is made for doubtful accounts. The amount is deducted from the accounts receivables. Inventories include the cost of raw materials, goods in process, and product on hand. Prepaid expenses include insurance premiums paid, charges for leased equipment, and charges for advertising that are paid prior to the receipt of the benefit from these items. The sum of all the above items is the total current assets. The term *current* refers to the fact that these assets are easily converted within a year, or more likely in a shorter time, say, 90 days.

Fixed assets are items that have a relatively long life such as land, buildings, and manufacturing equipment. The sum of these items is the total property, plant, and equipment. From this total, accumulated depreciation is subtracted and the result is net property and equipment. Last, an item referred to as intangibles includes a variety of items such as patents, licenses, intellectual capital, and goodwill. Intangibles are difficult to evaluate since they have no physical existence; e.g., goodwill is the value of the company's name and reputation. The sum of the total current assets, net property, and intangibles is the total assets.

Liabilities are the obligations that the company owes to creditors and stockholders. Current liabilities are obligations that come due within a year and include accounts payable (money owed to creditors for goods and services), notes payable (money owed to banks, corporations, or other lenders), accrued expenses (salaries and wages to employees, interest on borrowed funds, fees due to professionals, etc.), income taxes payable, current part of long-term debt, and other current liabilities due within the year.

Long-term liabilities are the amounts due after 1 year from date of the financial report. They include deferred income taxes that a company is permitted to postpone due to accelerated depreciation to encourage investment (but they must be paid at some time in the future) and bonds and notes that do not have to be paid within the year but at some later date. The sum of the current and long-term liabilities is the total liabilities.

Stockholders' equity is the interest that all stockholders have in a company and is a liability with respect to the company. This category includes preferred and common stock as well as additional paid-in capital (the amount that stockholders paid above the par value of the stock) and retained earnings. These are earnings from accumulated profit that a company earns and are used for reinvestment in the company. The sum of these items is the stockholders' equity.

On a balance sheet, the sum of the total liabilities and the stockholders' equity must equal the total assets, hence the term *balance sheet.* Comparing balance sheets for successive years, one can follow changes in various items that will indicate how well the company manages its assets and meets its obligations.

Income Statement An income statement shows the revenue and the corresponding expenses for the year and serves as a guide for how the company may do in the future. Often income statements may show how the company performed for the last two or three years. Table 9-4 is an example of a consolidated income statement.

Net sales are the primary source of revenue from goods and services. This figure includes the amount reported after returned goods, discounts, and allowances for price reductions are taken into account. Cost of sales represents all the expenses to convert raw materials to finished products. The major components of these expenses are direct material, direct labor, and overhead. If the cost of sales is subtracted from net sales, the result is the gross margin. One of the most important items on the income statement is depreciation and amortization. Depreciation is an allowance the

TABLE 9-2 Typical Ledger Page

			Cash: Account 01			
200X Mar 1 Capital	J-1	$95,000	Mar 1 Mar 11 Mar 13 Apr 4	Property Remodeling Equipment J. Jones	J-1 J-1 J-1 J-1	$3,000 7,800 10,000 2,500
			Capital: Account 02			
Apr 4 Cash to J. Jones	J-1	$2,500	Mar 1	Capital	J-1	$95,000
			Accounts Payable: Account 03			
			Mar 4 Mar 13	Mortgage Note Payable	J-1 J-1	$2,000 52,300
			Property and Building: Account 04			
Mar 4 Mar 11	J-1 J-1	$5,000 7,800				
			Equipment: Account 05			
Mar 13	J-1	$62,300				

SOURCE: J. R. Couper, *Process Engineering Economics,* Dekker, New York, 2003. By permission of Taylor & Francis Books, Inc., Boca Raton, Fla.

TABLE 9-3 Consolidated Balance Sheet* (December 31)

Assets	2005	2004
Current assets		
Cash	$63,000	$51,000
Marketable securities	41,000	39,000
Accounts receivable†	135,000	126,000
Inventories	149,000	153,000
Prepaid expenses	3,200	2,500
Total current assets	$391,200	$371,500
Fixed assets		
Land	35,000	35,000
Buildings	101,000	97,500
Machinery	278,000	221,000
Office equipment	24,000	19,000
Total fixed assets	$438,000	$372,500
Less accumulated depreciation	128,000	102,000
Net fixed assets	$310,000	$270,500
Intangibles	4,500	4,500
Total assets	$705,700	$646,500
Liabilities	2005	2004
Current liabilities		
Accounts payable	$92,300	$81,300
Notes payable	67,500	59,500
Accrued expenses payable	23,200	26,300
Federal income taxes payable	18,500	17,500
Total current liabilities	$201,500	$184,600
Long-term liabilities		
Debenture bonds, 10.3% due in 2015	110,000	110,000
Debenture bonds, 11.5% due in 2007	125,000	125,000
Deferred income taxes	11,600	10,000
Total liabilities	$448,100	$429,600
Stockholder's equity		
Preferred stock, 5% cumulative		
$5 par value—200,000 shares	$10,000	$10,000
Common stock, $1 par value		
2000 28,000,000 shares	32,000	28,000
2000X 32,000,000 shares		
Capital surplus	8,000	6,000
Accumulated retained earnings	207,600	172,900
Total stockholder's equity	**$257,600**	**$216,900**
Total liabilities and stockholder's equity	**$705,700**	**$646,500**

*All amounts in thousands of dollars.
†Includes an allowance for doubtful accounts.
SOURCE: J. R. Couper, *Process Engineering Economics,* Dekker, New York, 2003. By permission of Taylor & Francis Books, Inc., Boca Raton, Fla.

federal government permits for the wear and tear as well as the obsolescence of plant and equipment and is treated as an expense. Amortization is the decline in value of intangible assets such as patents, franchises, and goodwill. Selling, general, and administrative expenses include the marketing salaries, advertising expenses, travel, executive salaries, as well as office and payroll expenses. When depreciation, amortization, and the sales and administrative expenses are subtracted from the gross margin, the result is the operating income. Dividends and interest income received by the company are then added. Next interest expense earned by the stockholders and income taxes are subtracted, yielding the term income before extraordinary loss. It is the expenses a company may incur for unusual and infrequent occasions. When all the above items are added or subtracted from the operating income, net income (or loss) is obtained. This latter term is the *bottom line* often referred to in various reports.

TABLE 9-4 Consolidated Income Statement (December 31)

	2005	2004
Net sales (revenue)	$932,000	$850,000
Cost of sales and operating expenses		
Cost of goods sold	692,000	610,000
Depreciation and amortization	40,000	36,000
Sales, general, and administrative expenses	113,500	110,000
Operating profit	$86,500	$94,000
Other income (expenses)		
Dividends and interest income	10,000	7,000
Interest expense	(22,000)	(22,000)
Income before provision for income taxes	$74,500	$79,000
Provision for federal income taxes	24,500	26,000
Net profit for year	$50,000	$53,000

SOURCE: J. R. Couper, *Process Engineering Economics,* Dekker, New York, 2003. By permission of Taylor & Francis Books, Inc., Boca Raton, Fla.

TABLE 9-5 Accumulated Retained Earnings Statement* (December 31)

	2005	2004
Balance as of January 1	$172,900	$141,850
Net profit for year	50,000	53,000
Total for year	$222,900	$194,850
Less dividends paid on:		
Preferred stock	700	700
Common stock	14,600	21,250
Balance December 31	$207,600	$172,900

*All amounts in thousands of dollars.
SOURCE: J. R. Couper, *Process Engineering Economics,* Dekker, New York, 2003. By permission of Taylor & Francis Books, Inc., Boca Raton, Fla.

Accumulated Retained Earnings This is an important part of the financial report because it shows how much money has been retained for growth and how much has been paid as dividends to stockholders. When the accumulated retained earnings increase, the company has greater value. The calculation of this value of the retained earnings begins with the previous year's balance. To that figure add the net profit after taxes for the year. Dividends paid to stockholders are then deducted, and the result is the accumulated retained earnings for the year. See Table 9-5.

Concluding Remarks One of the most important sections of an annual report is the *notes.* These contain any liabilities that a company may have due to impending litigation that could result in charges or expenses not included in the annual report.

OTHER FINANCIAL TERMS

Profit margin is the ratio of net income to total sales, expressed as a percentage or sometimes quoted as the ratio of profit before interest and taxes to sales, expressed as a percentage. *Operating margin* is obtained by subtracting operating expenses from gross profit expressed as a percentage of sales. *Net worth* is the difference between total assets and total liabilities plus stockholders' equity. *Working capital* is the difference between total current assets and current liabilities.

FINANCIAL RATIOS

There are many financial ratios of interest to financial analysts. A brief discussion of some of these ratios follows; however, a more complete discussion may be found in Couper (2003).

Liquidity ratios are a measure of a company's ability to pay its short-term debts. *Current ratio* is obtained by dividing the current assets by the current liabilities. Depending on the economic climate, this ratio is 1.5 to 2.0 for the chemical process industries, but some companies operate closer to 1.0. The *quick ratio* is another measure of liquidity and is cash plus marketable securities divided by the current liabilities and is slightly greater than 1.0.

Leverage ratios are an indication of the company's overall debt burden. The *debt/total assets ratio* is determined by dividing the total debt by total assets expressed as a percentage. The industry average is 35 percent. *Debt/equity ratio* is another such ratio. The higher these ratios, the greater the financial risk since if an economic downturn did occur, it might be difficult for a company to meet the creditors' demands. The *times interest earned* is a measure of the extent to which profit could decline before a company is unable to pay interest charges. The ratio is calculated by dividing the *earnings before interest and taxes* (EBIT) by interest charges. The *fixed-charge coverage* is obtained by dividing the income available for meeting fixed charges by the fixed charges.

Activity ratios are a measure of how effectively a firm manages its assets. There are two inventory/turnover ratios in common use today. The *inventory/sales ratio* is found by dividing the inventory by the sales. Another method is to divide the cost of sales by inventory. The average collection period measures the number of days that customers' invoices remain unpaid. Fixed assets and total assets turnover indicate how well the fixed and total assets of the firm are being used.

Profitability ratios are used to determine how well income is being managed. The *gross profit margin* is found by dividing the gross profits by the net sales, expressed as a percentage. The *net operating margin* is equal to the earnings before interest and taxes divided by net sales. Another measure, the *profit margin on sales,* is calculated by dividing the net profit after taxes by net sales. The *return on total assets* ratio is the net profit after taxes divided by the total assets expressed as a percentage. The *return on equity ratio* is the net income after taxes and interest divided by stockholders' equity.

Table 9-6 shows the financial ratios for Tables 9-3 and 9-4. Table 9-7 is a summary of selected financial ratios and industry averages.

TABLE 9-6 Financial Ratios for Tables 9-3 and 9-4

Liquidity
Current ratio = $391,200/$201,500 = 1.94
Cash ratio = $391,200 – 149,000/$201,500 = 1.20
Leverage
Debt/assets ratio = [($448,100 – 201,500)/$705,700] × 100 = 35%
Times interest earned = $74,500 – 22,000/$22,000 = 4.39
Fixed-charge coverage = $86,500/$22,000 = 3.93
Activity
Inventory turnover = $932,000/$149,000 = 6.25
Average collection period = $135,000/($932,000/365) = 52.8 days
Fixed-assets turnover = $932,000/$438,000 = 2.13
Total-assets turnover = $932,000/$705,700 = 1.32
Profitability
Gross profit margin = [($932,000 – 692,000)/$932,000] × 100 = 25.8%
Net operating margin = $74,500/$932,000 × 100 = 7.99%
Profit margin on sales = $50,000/$932,000 × 100 = 5.36%
Return on net worth (return on equity) = [$50,000/($705,700 – 448,100)] × 100 = 19.4%
Return on total assets = ($50,000/$705,700) × 100 = 7.09%

RELATIONSHIP BETWEEN BALANCE SHEETS AND INCOME STATEMENTS

There is a relationship between these two documents because information obtained from each is used to calculate the returns on assets and equity. Figure 9-1 is an operating profitability tree for a fictitious company and contains the fixed and variable expenses as reported on internal company reports, such as the manufacturing expense sheet. Figure 9-2 is a financial family tree for the same company depicting the relationship between values in the income statement and in the balance sheet.

FINANCING ASSETS BY DEBT AND/OR EQUITY

The various options for obtaining funds to finance new projects are not a simple matter. Significant factors such as the state of the economy, inflation, a company's present indebtedness, and the cost of capital will affect the decision. Should a company incur more long-term debt, or should it seek new venture capital from equity sources? A simple yes or no answer will not suffice because the financial decision is complex. One consideration is the company's position with respect to leverage. If a company has a large proportion of its debt in bonds and preferred stock, the common stock is highly leveraged. Should the earnings decline, say, by 10 percent, the dividends available to common stockholders might be wiped out. The company also might not be able to cover the interest on its bonds without dipping into the accumulated earnings. A high debt/equity ratio illustrates the fundamental weakness of companies with a large amount of debt. When low-interest financing is available, such as for large government projects, the return-on-equity evaluations are used. Such leveraging is tantamount to transferring money from one pocket to another; or, to put it another way, a company may find itself borrowing from itself. In the chemical process industries, debt/equity ratios of 0.3 to 0.5 are common for industries that are capital-intensive (Couper and Beasley, 2001). Much has been written on the strategies of financing a corporate venture. The correct strategy has to be evaluated from the standpoint of what is best for the company. It must maintain a debt/equity ratio similar to those of successful companies in the same line of business.

COST OF CAPITAL

The cost of capital is what it costs a company to borrow money from all sources, such as loans, bonds, and preferred and common stock. It is an important consideration in determining a company's minimum acceptable rate of return on an investment. A company must make more than the cost of capital to pay its debts and make a profit. From profits, a company pays dividends to the stockholders. If a company ignores the cost of capital to increase dividends to the stockholders, then management is not meeting its obligations to pay off outstanding debts.

A sample calculation of the after-tax weighted cost of capital is found in Table 9-8. Each debt item is divided by the total debt, and that result is multiplied by the after-tax yield to maturity that equals the after-tax weighted average cost of that debt item contributing to the cost of capital. The information to estimate the cost of capital may be obtained from the annual report, the 10K, or the 10Q reports.

WORKING CAPITAL

The accounting definition of *working capital* is total current assets minus total current liabilities. This information can be found from the balance sheet. Current assets consist chiefly of cash, marketable securities, accounts receivable, and inventories; current liabilities include accounts payable, short-term debts, and the part of the long-term debt currently due. The accounting definition is in terms of the entire company.

For economic evaluation purposes, another definition of working capital is used. It is the funds, in addition to the fixed capital, that a company must contribute to a project. It must be adequate to get the plant in operation and to meet subsequent obligations when they come due. Working capital is not a one-time investment that is known at the project inception, but varies with the sales level and other factors. The relationship of working capital to other project elements may be viewed in the cash flow model (see Fig. 9-9). Estimation of an adequate amount of working capital is found in the section Capital Investment.

INVENTORY EVALUATION AND COST CONTROL

Under ordinary circumstances, inventories are priced (valued) at some form of cost. The problem in valuating inventory lies in "determining what costs are to be identified with inventories in a given situation" (Nickerson, 1979).

Valuation of materials can be made by using the

- Cost of a specific lot
- Average cost
- Standard cost

Under cost of a specific lot, those lots to be valuated must be identified by referring to related invoices. Many companies use the average cost for valuating inventories. The average used should be weighted by the quantities purchased rather than by an average purchase price. The average cost method tends to spread the effects of short-run price changes and has a tendency to level out profits in those industries that use raw materials whose prices are volatile. For many manufacturing companies, inventory valuation is an important consideration varying in degree of importance. Inventories that are large are subject to significant fluctuations from time to time in size and mix and in prices, costs, and values.

TABLE 9-7 Selected Financial Ratios

Item	Equation for calculation	Industry average
Liquidity		
Current ratio	Current assets/current liabilities	1.5–2.0
Cash ratio	Current assets – inventory/current liabilities	1.0–1.5
Leverage		
Debt to total assets	Total debt/total assets	30–40%
Times interest earned	Profit before taxes plus interest charges/interest charges	7.0–8.0
Fixed-charge coverage	Income available for meeting fixed charges/fixed charges	6.0
Activity		
Inventory turnover	Sales or revenue/inventory	7.0
Average collection period	Receivables/sales per day	40–60 days
Fixed assets turnover	Sales/fixed assets	2–4
Total assets turnover	Sales/total assets	1–2
Profitability		
Gross profit margin	Net sales – cost of goods sold/sales	25–40%
Net operating margin	Net operating profit before taxes/sales	10–15%
Profit margin on sales	Net profit after taxes/sales	5–8%
Return on net worth (return on equity)	Net profit after taxes/net worth	15%
Return on total assets	Net profit after taxes/total assets	7–10%

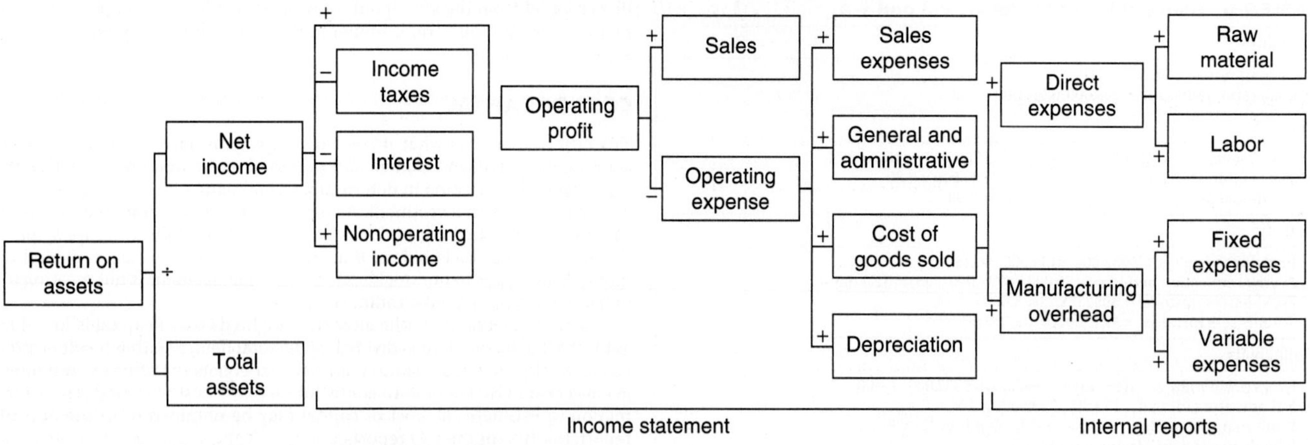

FIG. 9-1 Operating profitability tree. (*Source: Adapted from Couper, 2003.*)

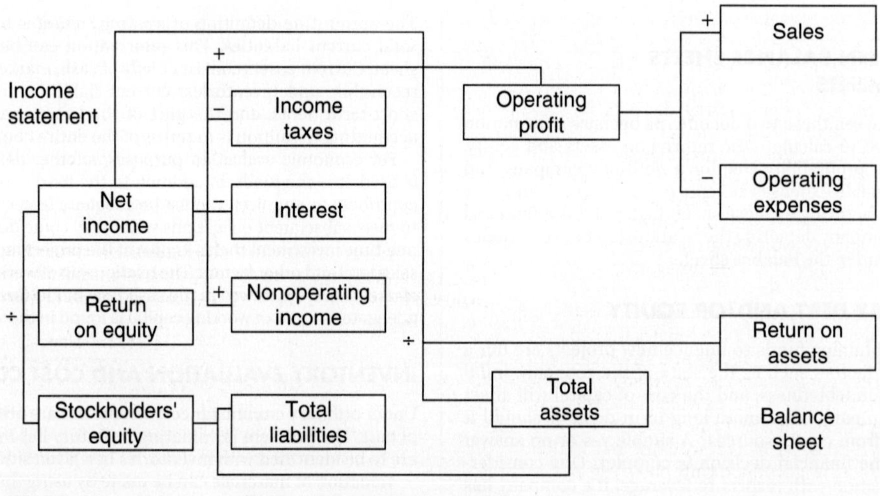

FIG. 9-2 Financial family tree. (*Source: Adapted from Couper, 2003.*)

Materials are valuated in accordance with their acquisition. Some companies use the *first-in, first-out* (FIFO) basis. Materials are used in order of their acquisition to minimize losses from deterioration. Another method is *last-in, first-out* (LIFO) in which materials coming in are the first to leave storage for use. The method used depends on a number of factors. Accounting texts discuss the pros and cons of each method, often giving numerical examples. Some items to consider are income tax considerations and cash flow that motivate management to adopt conservative valuation policies. Tax savings may accrue using one method compared to the other, but they may not be permanent. Whatever method is selected, consistency is important so that comparability of reported figures may be maintained from one time period to another. It is management's responsibility to make the decision regarding the method used. In some countries, government regulations control the method to be used. There are several computer software programs that permit the user to organize, store, search, and manage inventory from a desktop computer.

TABLE 9-8 Cost of Capital Illustration

Balance sheet 12/31/XX	Debt, $M	After-tax yield to maturity, %	After-tax weighted average cost, %
Long-term debt			
Revolving account	5.0	4.5	0.02
$4\frac{3}{8}$% debentures	12.0	4.0	0.05
$6\frac{1}{2}$% debentures	3.4	4.7	0.02
$6\frac{3}{4}$% debentures	9.4	4.2	0.04
$7\frac{1}{2}$% debentures	74.5	4.2	0.30
$9\frac{3}{8}$% loan	125.0	4.4	0.53
Other	23.2	4.4	0.10
Total long-term debt	252.5		1.06
Deferred taxes	67.7	0.0	0
Reserves	16.1	0.0	0
Preferred stock	50.0	8.6	0.42
Shareholders' equity	653.9	15.6	9.80
Total debt	1,040.2		11.28

Each debt item in $M divided by the total debt times the after-tax yield to maturity equals the after-tax weighted average cost contributing to the cost of capital.

SOURCE: Private communication.

BUDGETS AND COST CONTROL

A budget is an objective expressed in monetary terms for planning and controlling the resources of a company. Budgeted numbers are objectives, not achievements. A comparison of actual expenses with budgeted (cost standards) figures is used for control at the company, plant, departmental, or project level. A continuing record of performance should be maintained to provide the data for preparing future budgets (Nickerson, 1979). Often when a company compares actual results with cost standards or budgeted figures, a need for improving operations will surface. For example, if repairs to equipment continuously exceed the budgeted amount, perhaps it is time to consider replacement of that older equipment with a newer, more efficient model. Budgets are usually developed for a 1-year period; however, budgets for various time frames are frequently prepared. For example, in planning future operations, an intermediate time period of, say, 5 years may be appropriate, or for long-range planning the time period selected may be 10 years.

A cost control system is used

- To provide early warning of uneconomical or excessive costs in operations
- To provide relevant feedback to the personnel responsible for developing budgets
- To develop cost standards
- To promote a sense of cost consciousness
- To summarize progress

Budgetary models based upon mathematical equations are available to determine the effect of changes in variables. There are numerous sources extant in the literature for these models.

CAPITAL COST ESTIMATION

TOTAL CAPITAL INVESTMENT

The *total capital investment* includes funds required to purchase land, design and purchase equipment, structures, and buildings as well as to bring the facility into operation (Couper, 2003). The following is a list of items constituting the total capital investment:

Land
Fixed capital investment
Offsite capital
Allocated capital
Working capital
Start-up expenses
Other capital items (interest on borrowed funds prior to start-up; catalysts and chemicals; patents, licenses, and royalties; etc.)

Land Land is often acquired by a company some time prior to the building of a manufacturing facility. When a project is committed to be built on this land, the value of the land becomes part of that facility's capital investment.

Fixed Capital Investment When a firm considers the manufacture of a product, a capital cost estimate is prepared. These estimates are required for a variety of reasons such as feasibility studies, the selection of alternative processes or equipment, etc., to provide information for planning capital appropriations, or to enable a contractor to bid on a project. Included in the fixed capital investment is the cost of purchasing, delivery, and installation of manufacturing equipment, piping, automatic controls, buildings, structures, insulation, painting, site preparation, environmental control equipment, and engineering and construction costs. The fixed capital investment is significant in developing the economics of a process since this figure is used in estimating operating expenses and calculating depreciation, cash flow, and project profitability. The estimating method used should be the best, most accurate means consistent with the time and money available to prepare the estimate.

Classification of Estimates There are two broad classes of estimates: grass roots and battery limits. Grass-roots estimates include the entire facility, starting with site preparation, buildings and structures, processing equipment, utilities, services, storage facilities, railroad yards, docks, and plant roads. A battery-limits estimate is one in which an imaginary boundary is drawn around the proposed facility to be estimated. It is assumed that all materials, utilities, and services are available in the quality and quantity required to manufacture a product. Only costs within the boundary are estimated.

Quality of Estimates Capital cost estimation is more art than science. An estimator must use considerable judgment in preparing the estimate, and as the estimator gains experience, the accuracy of the estimate improves. There are several types of fixed capital cost estimates:

- *Order-of-magnitude (ratio estimate).* Rule-of-thumb methods based on cost data from similar-type plants are used. The probable accuracy is −30 percent to +50 percent.
- *Study estimate (factored estimate).* This type requires knowledge of preliminary material and energy balances as well as major equipment items. It has a probable accuracy of −25 to +30 percent.
- *Preliminary estimate (budget authorization estimate).* More details about the process and equipment, e.g., design of major plant items, are required. The accuracy is probably −20 to +25 percent.
- *Definitive estimate (project control estimate).* The data needed for this type of estimate are more detailed than those for a preliminary estimate and include the preparation of specifications and drawings. The probable accuracy is −10 to +15 percent.
- *Detailed estimate (firm estimate).* Complete specifications, drawings, and site surveys for the plant construction are required, and the estimate has an accuracy of −5 to +10 percent.

Detailed information requirements for each type of estimate may be found in Fig. 9-3.

In periods of high inflation, the results of various estimates and accuracy may overlap. At such times, four categories may be more suitable, namely, study, preliminary, definitive, and detailed categories. At present, some companies employing the *front-end loading* (FEL) process for project definition and execution use three categories:

Project stage	Accuracy
Conceptual	±40%
Feasibility	±25%
Definition	±10%

For more information on the FEL process, see Capital Project Execution and Analysis near the end of Sec. 9.

Scope The *scope* is a document that defines a project. It contains words, drawings, and costs. A scope should answer the following questions clearly:

What product is being manufactured?
How much is being produced?
What is the quality of the product?
Where is the product to be produced?
What is the quality of the estimate?
What is the basis for the estimate?
What are the knowns and unknowns with respect to the project?

Before an estimate can be prepared, it is essential to prepare a scope. It may be as simple as a single page, such as for an order-of-magnitude estimate, or several large manuals, for a detailed estimate. As the project moves forward from inception to a detailed estimate, the scope must be revised and updated to provide the latest information. Changes during the progress of a project are inevitable, but a well-defined scope prepared in advance can help minimize costly changes. If a scope is properly defined, the following results:

An understanding between those who prepared the scope (engineering) and those who accept it (management)

A document that indicates clearly what is provided in terms of technology, quality, schedule, and cost

A basis in enough detail to be used in controlling the project and its costs to permit proper evaluation of any proposed changes

A device to permit subsequent evaluation of the performance compared to the intended performance

A document to control the detailed estimate for the final design and construction

Equipment Cost Data The foundation of a fixed capital investment estimate is the equipment cost data. From this information, through the application of factors or percentages based upon the estimator's experience, the fixed capital investment is developed.

Cost data are reported as purchased, delivered, or installed cost. Purchased cost is the price of the equipment FOB at the manufacturer's plant, whereas delivered cost is the purchased price plus the delivery charge to the purchaser's plant FOB. Installed cost means the equipment has been purchased, delivered, uncrated, and placed on a foundation in the purchaser's operating department but does not include costs for piping, electrical, instrumentation, insulation, etc. Perhaps a better name might be *set-in-place cost.*

It is essential to have up-to-date, reliable cost data since the engineer producing the estimate starts with this information and develops the fixed capital cost estimate. The estimator must know the source of the data, the basis for the data, the date, potential errors, and the range over which the data apply. There are many sources of graphical equipment cost data in the literature, but some are old and the latest published data were in the early 1990s. There have been no significant cost data published recently. To obtain current cost data, one should solicit bids from vendors; however, it is essential to impress on the vendor that the information is to be used for preliminary estimates. A disadvantage of using vendor sources is that there is a chance of compromising proprietary information.

Cost-capacity plots of equipment indicate a straight-line relationship on a log-log plot. Figure 9-4 is an example of such a plot. A convenient method of presenting these data is in equation format:

$$C_2 = C_1\left(\frac{S_2}{S_1}\right)^n \qquad (9\text{-}1)$$

where C_1 = cost of equipment of capacity S_1
C_2 = cost of equipment of capacity S_2
n = exponent that may vary between 0.4 and 1.2 depending on type of equipment

Equation (9-1) is known as the *six-tenths rule* since the average value of n for all equipment is about 0.6. D. S. Remer and L. H. Chai (*Chemical Engineering Progress*, August 1990, pp. 77–82) published an extensive list of six-tenths data. Figure 9-5 shows how the exponent may vary from 0.4 to 0.9 for a given equipment item. Data accuracy is the highest in the narrow, middle range of capacity, but at either end of the plot, the error is great. These errors occur when one correlates cost data with one independent variable when more than one variable is necessary to represent the data; or when pressure, temperature, materials of construction, or design features vary considerably.

ESTIMATING INFORMATION GUIDE

Information Either Required or Available

	Estimate types →	Order of magnitude (ratio)	Study (factored)	Preliminary (budget authorization)	Definitive (project control)	Detailed (firm)
Site	Location		•	•	•	•
	General description		•	•	•	•
	Site survey			•	•	•
	Geotechnical report			•	•	•
	Site plot plan and contours				•	•
	Well-developed site facilities					•
Process flow	Rough sketches		•			
	Preliminary			•		
	Engineered				•	•
Equipment	Rough sizes and construction		•	•		
	Engineered specifications				•	•
	Vessel data sheets				•	•
	General arrangement			•	•	
	Final arrangement					•
Buildings and structures	Rough sizes and construction		•	•		
	Foundation sketches			•	•	
	Architectural and construction			•	•	•
	Preliminary structural design			•	•	
	General arrangements and elevations				•	•
	Detailed drawings					•
Utilities and services	Rough quantities		•			
	Preliminary heat balance			•		
	Preliminary flow sheets			•		
	Engineered heat balance				•	•
	Engineered flow sheets				•	•
	Detailed drawings					•
Piping and insulation	Preliminary flow sheets		•	•		
	Engineered flow sheets				•	•
	Piping layouts and schedules					•
	Insulation rough specifications			•		
	Insulation applications				•	•
	Insulation details					•
Instrumentation	Preliminary list			•		
	Engineered list				•	•
	Detail drawings					•
Electrical	Rough motor list and sizes		•	•		
	Engineered list and sizes				•	•
	Substation number and size			•	•	•
	Preliminary specifications			•	•	
	Distribution specifications					•
	Preliminary interlocks and controls				•	
	Engineered single-line diagrams				•	•
	Detailed drawings					•
Work-hours	Engineering and drafting		•	•	•	•
	Construction supervision					•
	Craft labor				•	•
Project scope	Product, capacity, location, utilities, and services; Building requirements, process, storage, and handling	•	•	•	•	•

FIG. 9-3 Information guide for preparing estimates. (*Source*: Perry's Chemical Engineers' Handbook, *5th ed., McGraw-Hill, New York, 1973.*)

A convenient way to display cost-capacity data is by algorithms. They are readily adaptable for computerized cost estimation programs. Algorithm modifiers in equation format may be used to account for temperature, pressure, material of construction, equipment type, etc. Equation (9-2) is an example of obtaining the cost of a shell-and-tube heat exchanger by using such modifiers.

$$C_{HE} = KC_BF_DF_MF_P \qquad (9\text{-}2)$$

where C_{HE} = purchased equipment cost
K = factor for cost index based upon a base year
C_B = base cost of a carbon-steel floating-head exchanger, 150-psig design pressure
F_D = design-type cost factor if different from that in C_B
F_M = material-of-construction cost factor
F_P = design pressure cost factor

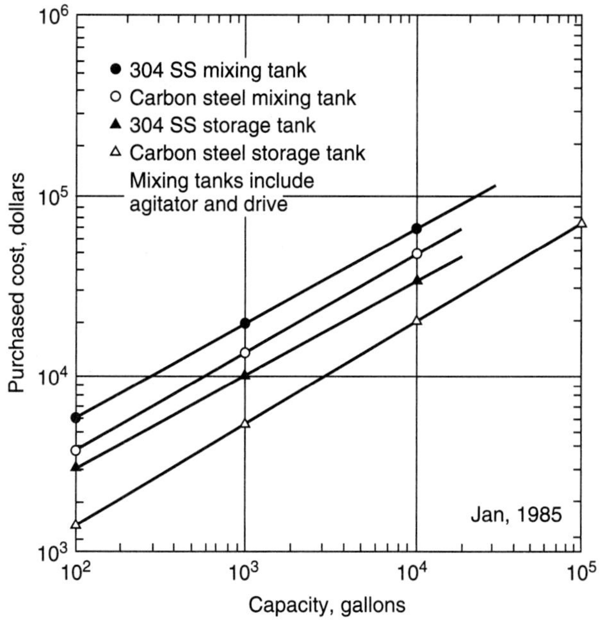

FIG. 9-4 Cost-capacity plot.

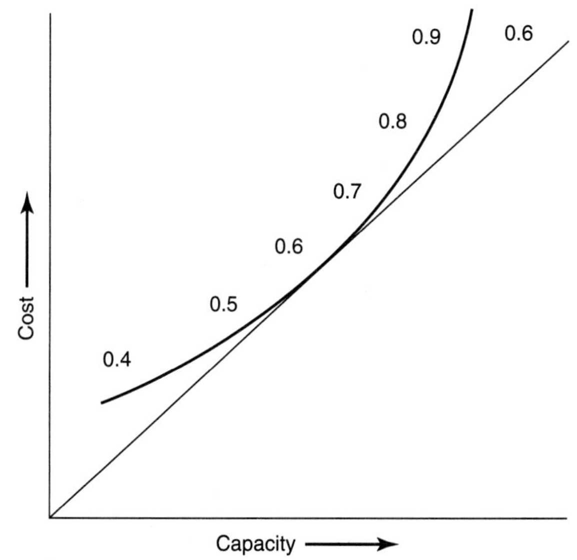

FIG. 9-5 Variation of *n* on cost-capacity plot.

Each cost factor is obtained from equations or tables from Couper (2003, App. C) and have been updated to third-quarter 2002.

Cost Indices Cost data are given as of a specific date and can be converted to more recent costs through the use of cost indices. In general, the indices are based upon constant dollars in a base year and actual dollars in a specific year. In this way, with the proper application of the index, the effect of inflation (or deflation) and price increases by multiplying the historical cost by the ratio of the present cost index divided by the index applicable in the historical year. Labor, material, construction costs, energy prices, and product prices all change at different rates. Most cost indices represent national averages, and local averages may vary considerably. Table 9-9 is a list of selected values of three cost indices of significance in the chemical process industries. The chemical engineering (CECI) index, which is most commonly used by the CPI, is found in each issue of *Chemical Engineering* magazine. The *Oil and Gas Journal* reports the *Nelson-Farrar Refinery* indices in the first issue of each quarter. The base years selected for each index are generally periods of low inflation so that the index is stable. The derivation of base values is referred to in the respective publications.

TABLE 9-9 Selected Chemical Engineering Indices

Year base	CE index (CECI) 1957–1959 = 100	Year base	CE index (CECI)
2000	394.1	2008	575.4
2001	394.3	2009	521.9
2002	395.6	2010	550.8
2003	402.0	2011	585.7
2004	444.2	2012	584.6
2005	468.2	2013	567.3
2006	496.7	2014	576.1
2007	525.4	2015 (est.)	550

SOURCE: *Chemical Engineering*, Nov. 2015, p. 76.

A cost index is used to project a cost from a base year to another selected year. The following equation is used:

$$\text{Cost at } \Theta_2 = \text{Cost at } \Theta_1 \left(\frac{\text{index at } \Theta_2}{\text{index at } \Theta_1} \right) \tag{9-3}$$

Example 9-1 Use of Cost Index A centrifuge cost \$95,000 in 1999. What is the cost of the same centrifuge in the third quarter of 2004? Use the CE index.

Solution:

CE index in 1999 = 390.6

CE index in 3d quarter 2004 = 457.4

Cost in 2004 = cost in 1999 × (CE index in 3d quarter 2004/CE index in 1999)

$$= \$95{,}000\left(\frac{457.4}{390.6}\right) = \$111{,}200$$

Inflation When costs are to be projected into the future due to inflation, it is a highly speculative exercise, but it is necessary for estimating investment costs, operating expenses, etc. Inflation is the increase in price of goods without a corresponding increase in productivity. A method for estimating an inflated cost is

$$C_i = (1 + f_1)(1 + f_2)(1 + f_3)\, C_P \tag{9-4}$$

where C_i = inflated cost
f_1 = inflation rate the first year
f_2 = inflation rate the second year
f_3 = inflation rate the third year
C_P = cost in a base year

The assumed inflation factors *f* are obtained from federal economic reports, financial sources such as banks and investment houses, and news media. These factors must be reviewed periodically to update estimates.

Example 9-2 Inflation A dryer today costs \$475,000. The projected inflation rates for the next 3 years are 3, 4.2, and 4.7 percent. Calculate the projected cost in 3 years.

Solution:

$$C_i = (1 + f_1)(1 + f_2)(1 + f_3)C_P$$

$$= (1.030)(1.042)(1.047)(\$475{,}000) = \$533{,}800$$

Equipment Sizing Before equipment costs can be obtained, it is necessary to determine equipment size from material and energy balances. For preliminary estimates, rules of thumb may be used; but for definitive and detailed estimates, detailed equipment calculations must be made.

Example 9-3 Equipment Sizing and Costing Oil at 490,000 lb/h is to be heated from 100 to 170°F with 145,000 lb/h of kerosene initially at 390°F from another section of a plant. The oil enters at 20 psig and the kerosene at 25 psig. The physical properties are

Oil: 0.85 sp gr, 3.5 cP at 135°F, 0.49 sp ht

Kerosene: 0.82 sp gr, 0.45 cP, 0.61 sp ht

Estimate the cost of an all-carbon-steel exchanger in late 2004. Assume an exchanger consisting of a TEMA E-shell with an even number of tube passes.

Solution:
Energy required to heat oil stream (490,000)(0.49)(170 − 100) = 16,807,000 Btu/h

$$\text{Exit kerosene temperature } T = 390 - \left(\frac{490{,}000}{145{,}000}\right)\left(\frac{0.49}{0.61}\right)(170-100)$$
$$= 200°\text{F}$$
$$\text{LMTD} = \frac{220-100}{\ln(220/100)} = 152°\text{F}$$

Calculate the exchanger efficiency factor F_T.

$$S = \frac{170-100}{390-100} = 0.241$$
$$R = \frac{390-200}{170-100} = 2.71$$

Using these values of R and S in Sec. 11, Fig. 11.4(a), $F_T = 0.88$. Since the factor must be greater than 0.75, the exchanger is satisfactory. Therefore, $\Delta T = (F)$ (LMTD) = (0.88)(152) = 134°F.

Assume $U_D = 50$ Btu/(h · ft² · °F).

$$Q = U_D A\,\Delta T = 16{,}800{,}000 = (50)(A)(134)$$
$$A = 2510 \text{ ft}^2$$

Use the cost algorithm cited above.

$$C_B = \exp[8.821 - 0.30863 \ln A + 0.0681(\ln A)^2]$$
$$= \exp[8.821 - 0.30863(7.83) + 0.0681(61.3)] = \$39{,}300 \text{ base cost}$$

$F_D = 1.0$
$F_M = 1.0$ for materials of this type
$F_P = 1.00$ since this exchange is operating below 4 bar

$$K = 1.218\,(\text{CE index 4th qtr 2004/CE index 1st qtr 2003}) = 1.218\left(\frac{463}{406}\right) = 1.389$$

Therefore, $C_{HE} = KC_BF_DF_MF_P = (1.389)(39{,}300)(1.0)(1.0)(1.0) = \$54{,}600$.

Quality of an Estimate Capital cost is more art than science. An estimator must use a great deal of judgment in the preparation of an estimate. Estimates may be classified base k on the quality and the amount of information available. The American Association of Cost Engineers proposed the following:

Estimate type	Accuracy range
Order of magnitude	−30 to +50%
Budget	−15 to +30%
Definitive	−5 to +15%

Many companies have a fourth type between budget and definitive called an "authorization" estimate with a range of −10 to +20 percent. Other companies have a fifth category or detailed estimate.

Estimate type	Accuracy range
Order of magnitude	−30 to +50%
Study	−25 to +30%
Preliminary	−20 to +25%
Definitive	−10 to +15%
Detailed	−5 to +10%

Estimation of Fixed Capital Investment

Order-of-Magnitude Methods The *ratio method* will give the fixed capital investment per gross annual sales; however, most of these data are from the 1960s, and no recent data have been published. The ratio above is called the *capital ratio*, often used by financial analysts. The reciprocal of the capital ratio is the *turnover ratio* that for various businesses ranges from 4 to 0.3. The chemical industry has an average of about 0.4 to 0.5. The ratio method of obtaining fixed capital investment is rapid but suitable only for order-of-magnitude estimates.

The *exponential method* may be used to obtain a rapid capital cost for a plant based upon existing company data or from published sources such as those of D. S. Remer and L. H. Chai, *Chemical Engineering*, April 1990, pp. 138–175. In the method known as the seven-tenths rule, the cost-capacity data for process plants may be correlated by a logarithmic plot similar to the six-tenths plot for equipment. Remer and Chai compiled exponents for a variety of processes and found that the exponents ranged from 0.6 to 0.8. When the data are used to obtain a capital cost for a different-size plant, the estimated capital must be for the same process.

The equation is

$$\text{Cost of plant B} = \text{Cost of plant A}\left(\frac{\text{capacity of plant B}}{\text{capacity of plant A}}\right)^{0.7} \qquad (9\text{-}5)$$

Cost indices may be used to correct costs for time changes.

Example 9-4 Seven-Tenths Rule A company is considering the manufacture of 150,000 tons annually of ethylene oxide by the direct oxidation of ethylene. According to Remer and Chai (1990), the cost-capacity exponent for such a plant is 0.67. A subsidiary of the company built a 100,000-ton annual capacity plant for $70 million fixed capital investment in 1996. Using the seven-tenths rule, estimate the cost of the proposed new facility in the third quarter 2004.

Solution:

$$\text{Cost}_{150} = \text{Cost}_{100}\left(\frac{\text{Cap}_{150}}{\text{Cap}_{100}}\right)^{0.67}\left(\frac{\text{CE index 3Q 2004}}{\text{CE index 1996}}\right)$$
$$= (\$70{,}000{,}000)\left(\frac{150{,}000}{100{,}000}\right)^{0.67}\left(\frac{457.4}{381.7}\right) = \$110{,}000{,}000$$

Study Method The single-factor method begins with collecting the delivered cost of various items of equipment and applying one factor to obtain the battery-limits (BL) fixed capital (FC) investment or total capital investment as follows:

$$(C_{FC})_{BL} = f\Sigma(C_{EQ})_{DEL} \qquad (9\text{-}6)$$

where $(C_{FC})_{BL}$ = battery-limits fixed capital investment or total capital investment
$(C_{EQ})_{DEL}$ = purchased equipment costs

The single factors include piping, automatic controls, insulation, painting, electrical work, engineering costs, etc. (Couper, 2003). Table 9-10 shows the Lang factors for various types of processing plants. The boundaries between the classifications are not clear-cut, and considerable judgment is required in the selection of the appropriate factors.

Preliminary Estimate Methods A refinement of the Lang factor method is the Hand method. The Hand factors are found in Table 9-11. Equipment is grouped in categories, such as heat exchangers and pumps, and then a factor is applied to each group to obtain the installed cost; finally the groups are summed to give the battery-limits installed cost. Wroth compiled a more detailed list of installation factors; a selection of these can be found in Table 9-12. The Lang and Hand methods start with purchased equipment costs whereas the Wroth method begins with delivered equipment costs, so delivery charges must be included in the Lang and Hand methods. At best the Lang and Hand methods will yield study quality estimates, and the Wroth method might yield a preliminary quality estimate.

TABLE 9-10 Lang Factors

	Lang factors	
Type of plant	Fixed capital investment	Total capital investment
Solid processing plant	4.0	4.7
Solid-fluid processing	4.3	5.0
Fluid processing	5.0	6.0

SOURCE: Adapted from M. S. Peters, K. D. Timmerhaus, and R. West, *Plant Design and Economics for Chemical Engineers*, 5th ed., McGraw-Hill, New York, 2003.

TABLE 9-11 Hand Factors

Equipment type	Factor
Fractionating columns	4.0
Pressure vessels	4.0
Heat exchangers	3.5
Fired heaters	2.0
Pumps	4.0
Compressors	2.5
Instruments	4.0
Miscellaneous	2.5

SOURCE: Adapted from W. E. Hand, *Petroleum Refiner*, September 1958, pp. 331–334.

TABLE 9-12 Selected Wroth Factors

Equipment	Factor
Blender	2.0
Blowers and fans	2.5
Centrifuge	2.0
Compressors	
Centrifugal (motor-driven)	2.0
Centrifugal (steam-driven, including turbine)	2.0
Reciprocating (steam and gas)	2.3
Reciprocating (motor-driven less motor)	2.3
Ejectors, vacuum	2.5
Furnaces (packaged units)	2.0
Heat exchangers	4.8
Instruments	4.1
Motors, electric	3.5
Pumps	
Centrifugal (motor-driven less motor)	7.0
Centrifugal (steam-driven including turbine)	6.5
Positive-displacement (less motor)	5.0
Reactors (factor as appropriate, equivalent-type equipment)	—
Refrigeration (packaged units)	2.5
Tanks	
Process	4.1
Storage	3.5
Fabricated and field-erected 50,000+ gal	2.0
Towers (columns)	4.0

Abstracted from W. F. Wroth, *Chemical Engineering*, October 17, 1960, p. 204.

Example 9-5 Fixed Capital Investment Using the Lang and Hand Methods The following is a list of the purchased equipment costs for a proposed processing unit:

Heat exchangers	$620,000
Distillation towers and internals	975,000
Receivers	320,000
Accumulator drum	125,000
Pumps and motors	220,000
Automatic controls	275,000
Miscellaneous equipment	150,000

Assume delivery charges are 5 percent of the purchased price. Estimate the fixed capital investment 2 years into the future, using the Lang, Hand, and Wroth methods. The inflation rates are 3.5 percent for the first year and 4.0 percent for the second.

Solution:

Equipment	Purchased equipment cost	Delivered equipment cost
Heat exchangers	$620,000	$651,000
Distillation towers, internals	975,000	1,024,000
Receivers	320,000	336,000
Accumulator drum	125,000	131,000
Pumps and motors	220,000	231,000
Automatic controls	275,000	289,000
Miscellaneous equipment	150,000	158,000
Total	$2,685,000	$2,820,000

Lang method: The Lang factor for a fluid processing unit starting with purchased equipment costs is 5.0. Therefore, fixed capital investment is $2,820,000 × 5.0 × 1.035 × 1.040 = $15,177,000.

Hand method: The Hand method begins with purchased equipment costs, and factors are applied from Table 9-11.

Hand method:

Equipment	Purchased equipment cost	Hand factor	Purchased equipment installed cost
Heat exchangers	$620,000	3.5	$2,170,000
Distillation towers, internals	975,000	4.0	3,900,000
Receivers	320,000	2.5*	800,000
Accumulator drum	125,000	2.5*	313,000
Pumps and motors	220,000	4.0	880,000
Automatic controls	275,000	4.0	1,100,000
Miscellaneous	150,000	2.5	375,000
Total	$2,685,000		$9,538,000

The asterisk on the receivers and accumulators indicates that if these vessels are pressure vessels, a factor of 4.0 should be used instead of 2.5. The total purchased equipment installed is $9,538,000 for non-pressure vessels and the delivered cost is $10,015,000. Therefore, the fixed capital investment installed would be $10,015,000 × 1.035 × 1.040 = $10,780,000. Using pressure vessels increases the total purchased equipment cost $667,000; therefore, the fixed capital investment for this case including inflation would be $10,780,000 × 1.05 × 1.035 × 1.04 = $11,534,000.

Therefore, the summary of the fixed capital investment by the various methods is

Lang	$15,177,000
Hand	11,534,000

Experience has shown that the fixed capital investment by the Lang method is generally higher than that of the other methods. Whatever figure is reported to management, it is advisable to state the potential accuracy of these methods.

Garrett (1989) developed a similar method based upon a variety of equipment modules, starting with purchase equipment costs obtained from plots and applying factors for materials of construction and plant location. The method provides all supporting and connecting equipment to make the equipment installation operational.

T. R. Brown developed guidelines for the preparation of order-of-magnitude and study capital cost estimates based upon the Lang and Hand methods. Brown modified the Lang and Hand methods for cost of services, environmental equipment, materials of construction, instrumentation, and location factors. He found that the modified Hand and Garrett module factor methods gave results within 3.5 percent.

It is necessary to refer to pages 50–54 of Brown's book (op. cit.) in order to implement Brown's method. The Brown method is adequate for FEL/VIP studies, which are explained later in this section.

Other multiple-factor methods that have been published in the past are those by C. E. Chilton, *Cost Estimation in the Process Industries*, McGraw-Hill, New York, 1960; M. S. Peters, K. D. Timmerhaus, and R. E. West, *Plant Design and Economics for Chemical Engineers*, 5th ed., McGraw-Hill, New York, 2003; C. A. Miller, *Chem. Eng.*, Sept. 13, 1965, pp. 226–236; and F. A. Holland, F. A. Watson, and V. K. Wilkinson, *Chem. Eng.*, Apr. 1, 1974, pp. 71–76. These methods produced preliminary quality estimates. Most companies have developed their own in-house multiple-factor methods for preliminary cost estimation.

Step-counting methods are based upon a number of processing steps or "functional units." The concept was first introduced by H. E. Wessel, *Chem. Eng.*, 1952, p. 209. Subsequently, R. D. Hill, *Petrol. Refin.*, **35**(8): 106–110, August 1956; F. C. Zevnik and R. L. Buchanan, *Chem. Eng. Progress*, **59**(2): 70–77, February 1963; and J. H. Taylor, *Eng. Process Econ.*, **2:** 259–267, 1977, further developed the step-counting method.

A step or functional unit is a significant process step including all process equipment and ancillary equipment necessary for operating the unit. A functional unit may be a unit operation, unit process, or separation in which mass and energy are transferred. The sum of all functional units is the total fixed capital investment. Pumping and heat exchangers are considered as part of a functional unit. In-process storage is generally ignored except for raw materials, intermediates, or products. Difficulties are encountered in applying the method due to defining a step. This takes practice and experience. If equipment has been omitted from a step, the resulting estimate is seriously affected. These methods are reported to yield estimates of study quality or at best preliminary quality.

Definitive Estimate Methods Modular methods are an extension of the multiple-factor methods and have been proposed by several authors. One of the most comprehensive methods and one of the earliest was that of K. M. Guthrie, *Chem. Eng.*, **76:** 114–142, Mar. 24, 1969. It began with equipment FOB equipment costs, and through the use of factors listed in Table 9-13, the module material cost was obtained. Labor for erection and setting equipment was added to the material cost as well as indirect costs for freight, insurance, engineering, and field expenses to give a total module cost. Such items as contingencies, contractors' fees, auxiliaries, site development land, and industrial buildings were added if applicable. Since any plant consists of equipment modules, these are summed to give the total fixed capital investment. Unfortunately, the factors and data are old, but the concept is useful. See Table 9-14. T. R. Brown, *Hydrocarbon Processing*, October 2000, pp. 93–100, made modifications to the Garrett method.

Another method, called the *discipline method*, mentioned by L. R. Dysert, *Cost Eng.* **45**(6), June 6, 2003, is similar to the models of Guthrie and Garrett. It uses equipment factors to generate separate costs for each of the "disciplines" associated with the installation of equipment, such as installation labor, concrete, structural steel, and piping, to obtain direct field costs for each type of equipment, e.g., heat exchangers, towers, and reactors.

Modular methods, depending on the amount of detail provided, will yield preliminary quality estimates.

Detailed Estimate Method For estimates in the detailed category, a code of account needs to be used to prevent oversight of certain significant items in the capital cost. See Table 9-15. Each item in the code is estimated and provides the capital cost estimate; then this estimate serves for cost control during the construction phase of a project.

TABLE 9-13 Guthrie Method Factors

Details	Exchangers			Vessels				
	Furnaces	Shell and tube	Air-cooled	Vertical	Horizontal	Pump and driver	Compressor and driver	Tanks
FOB equipment	1.00	1.00	1.00	1.00	1.00	1.00	1.00	1.00
Piping	0.18	0.46	0.18	0.61	0.42	0.30	0.21	
Concrete	0.10	0.05	0.02	0.10	0.06	0.04	0.12	
Steel		0.03		0.08				
Instruments	0.04	0.10	0.05	0.12	0.06	0.03	0.08	
Electrical	0.02	0.02	0.12	0.05	0.05	0.31	0.16	
Insulation		0.05		0.08	0.05	0.03	0.03	
Paint			0.01	0.01	0.01	0.01	0.01	
Total materials = M	1.34	1.71	1.38	2.05	1.65	1.72	1.61	1.20
Erection and setting (L)	0.30	0.63	0.38	0.95	0.59	0.70	0.58	0.13
χ, excluding site preparation and auxiliaries ($M + L$)	1.64	2.34	1.76	3.00	2.24	2.42	2.19	1.33
Freight, insurance, taxes, engineering, home office, construction		0.08		0.08	0.08	0.08	0.08	0.08
Overhead or field expense	0.60	0.95	0.70	1.12	0.92	0.97	0.97	
Total module factor	2.24	3.37	2.46	4.20	3.24	3.47	3.24	1.41

SOURCE: From K. M. Guthrie, *Chem. Eng.*, **76**, 114–142 (Mar. 24, 1969). Based on FOB equipment cost = 100 (carbon steel).

Comments on Significant Cost Items

Piping This cost includes the cost of the pipe, installation labor, valves, fittings, supports, and miscellaneous items necessary to complete installation of all pipes in the process. The accuracy of the estimates can be seriously in error by the improper application of estimating techniques to this component. Many pipe estimating methods are extant in the literature.

Two general methods have been used to estimate piping costs when detailed flow sheets are not available. One method is to use a percentage of the FOB equipment costs or a percentage of the fixed capital investment. Typical figures are 80 to 100 percent of the FOB equipment costs or 20 to 30 percent of the fixed capital investment. This method is used for preliminary estimates. Another group of methods such as the Dickson N method

TABLE 9-14 Selected Garrett Module Factors

Equipment type (carbon steel unless otherwise noted)	Module factor
Agitators: dual-bladed turbines/single-blade propellers	2.0
Agitated tanks	2.5
Air conditioning	1.46
Blender, ribbon	2.0
Blowers, centrifugal	2.5
Centrifuges: solid-bowl, screen-bowl, pusher, stainless steel	2.0
Columns: distillation, absorption, etc.	
Horizontal	3.05
Vertical	4.16
Compressors: low-, medium-, high-pressure	2.6 avg.
Coolers, quenchers	2.7
Crystallizers	2.6 avg.
Drives/motors	
Electric, for fans, compressors, pumps	1.5
Electric for other units	2.0
Gasoline	2.0
Turbine: gas and steam	3.5
Dryers	
Fluid bed, spray	2.7
Rotary	2.3
Dust collectors	
Bag filters	2.2
Cyclones, multiclones	3.0
Evaporators, single-effect stainless steel	
Falling film	2.3
Forced circulation	2.9
Fans	2.2
Filters	
Belt, rotary drum and leaf, tilting pan	2.4
Others	2.8
Furnaces	2.1
Heat exchangers	
Air-cooled	2.2
Double-pipe	1.8
Shell-and-tube	3.2
Mills	
Hammer	2.8
Ball, rod	2.3 avg.
Pumps	
Centrifugal	5.0
Reciprocating	3.3
Turbine	1.8
Reactors, jacketed, no agitator	
304 SS	1.8
Glass-lined	2.1
Mild steel	2.3
Vacuum equipment	2.2

SOURCE: Adapted from Garrett (1989).

TABLE 9-15 Code of Accounts

Category number	Direct capital cost account titles
010	Equipment items
020	Instrument items
030	Setting and testing equipment
040	Setting and testing instruments
050	Piling
060	Excavation
070	Foundations
080	Supports, platforms, and structures
090	Other building items
100	Fire protection and sprinklers
110	Piping
120	Ductwork
130	Electrical and wiring
140	Site preparation
150	Sewers, drains, and plumbing
160	Underground piping
170	Yards, roads, and fencing
180	Railroads
190	Insulation
200	Painting
210	Walls, masonry, roofs, and roofing
220	Spares
230	Lump-sum contracts
	Distributives
500	Site burden
510	Direct labor burden
530	Construction equipment, tools, and supplies
550	Rental and servicing construction equipment and tools
580	Premium wages and overtime—contractor
670	Temporary facilities
740	Cancellation charges
750	Abandoned design
760	Self-insured losses
790	Unvouchered liabilities
800	In-house engineering
810	Outside engineering
870	Undeveloped design allowances
880	Distributives transferred to expense
890	Contingencies—capital items
	Expense
900	Dismantling
910	Sales and use taxes
920	Repairs expense
930	Relocation and modification expense
940	Start-up relocation and modification expense
990	Contingencies

SOURCE: Private communication.

(R. A. Dickson, *Chem. Eng.,* **57:** 123–135, Nov. 1947), estimating by weight, estimating by cost per joint, etc., requires a detailed piping takeoff from either PID or piping drawings with piping specifications, material costs, labor expenses, etc. These methods are used for definitive or detailed estimates where accuracy of 10 to 15 percent is required. The takeoff methods must be employed with great care and accuracy by an experienced engineer. A detailed breakdown by plant type for process piping costs is presented in Peters et al. (2003) and in *Perry's Chemical Engineers' Handbook*, 6th ed., McGraw-Hill, New York, 1984.

Electrical This item consists of transformers, wiring, switching gear, as well as instrumentation and control wiring. The installed costs of the electrical items may be estimated as 20 to 40 percent of the delivered equipment costs or 5 to 10 percent of the fixed capital investment for preliminary estimates. As with piping estimation, the process design must be well along toward completion before detailed electrical takeoffs can be made.

Buildings and Structures The cost of the erection of buildings and structures in a chemical process plant as well as the plumbing, heating and ventilation, and miscellaneous building service items may be estimated as 20 to 50 percent of delivered equipment costs or as 10 to 20 percent of the fixed capital investment for a preliminary estimate.

Yards, Railroad Sidings, Roads, etc. This investment includes roads, railroad spurs, docks, and fences. A reasonable figure for preliminary estimates is 15 to 20 percent of the FOB equipment cost or 3 to 7 percent of the fixed capital investment for a preliminary estimate.

Service Facilities For a process plant, utility services such as steam, water, electric power, fuel, compressed air, shop facilities, and a cafeteria require capital expenditures. The cost of these facilities lumped together may be 10 to 20 percent of the fixed capital investment for a preliminary estimate. (*Note:* Buildings, yards, and service facilities must be well defined to obtain a definitive or detailed estimate.)

Environmental Control and Waste Disposal These items are treated as a separate expenditure and are difficult to estimate due to the variety and complexity of the process requirements. Pollution control equipment is generally included as part of the process design. Couper (2003) and Peters and Timmerhaus (2003) mention that at present there are no general guidelines for estimating these expenditures.

Computerized Cost Estimation With the advent of powerful personal computers (PCs) and software packages, capital cost estimates advanced from large mainframe computers to the PCs. The reasons for using computer cost estimation and economic evaluation packages are time saved on repetitive calculations and reduction in mathematical errors. Numerous computer simulation software packages have been developed over the past two decades. Examples of such software are those produced by ASPEN, ICARUS, CHEMCAD, SUPERPRO, PRO II, HYSYS, etc.; but most do not contain cost estimation software packages. ICARUS developed a PC cost estimation and economic evaluation package called Questimate. This system built a cost estimate from design and equipment cost modules, bulk items, site construction, piping and ductwork, buildings, electrical equipment, instruments, etc., developing worker-hours for engineering and fieldwork costs. This process is similar to quantity takeoff methods to which unit costs are applied. A code of accounts is also provided.

ASPEN acquired ICARUS in 2000 and developed Process Evaluator based on Questimate that is used for conceptual design, known as *front-end loading* (FEL). More information on FEL and *value-improving process* (VIP) is found later in Sec. 9. Basic and detailed estimates are coupled with a business decision framework in ASPENTECH ICARUS 2000.

Many companies have developed their own factored estimates using computer spreadsheets based upon their in-house experience and cost database information that they have developed from company project history. For detailed estimates, the job is outsourced to design-construction companies that have the staff to perform those estimates.

Whatever package is used, it is recommended that computer-generated costs be spot-checked for reasonable results using a handheld calculator, since errors do occur. Some commercial software companies will develop cost estimation databases in cooperation with a company for site-specific costs.

Contingency This is a provision for unforeseen events that experience has demonstrated are likely to occur. Contingencies are of two types: process and project contingency. In the former, there are uncertainties in

Equipment and performance
Integration of old and new process steps
Scaling up to a large-scale plant size
Accurate definition of certain process parameters, such as severity of process conditions, number of recycles, process blocks and equipment, multiphase streams, and unusual separations

No matter how much time and effort are spent preparing estimates, there is a chance of errors occurring due to

Engineering errors and omissions
Cost and labor rate changes
Construction problems
Estimating inaccuracies
Miscellaneous "unforeseens"
Weather-related problems
Strikes by fabricators, transportation, and construction personnel

For preliminary estimates, a 15 to 20 percent project contingency should be applied if the process information is firm. As the quality of the estimate moves to definitive and detailed, the contingency value may be lowered to 10 to 15 percent and 5 to 10 percent, respectively. Experience has shown that the smaller the dollar value of the project, the higher the contingency should be.

Offsite Capital These facilities include all structures, equipment, and services that do not enter into the manufacture of a product but are important to the functioning of the plant. Such capital items might be steam-generating and electrical-generating and distribution facilities, well-water cooling tower, and pumping stations for water distribution. Service capital might be auxiliary buildings, such as warehouses, service roads, railroad spurs, material storage, fire protection equipment, and security systems. For estimating purposes, the following percentages of the fixed capital investment might be used:

Small modification of offsites, 1 to 5 percent
Restructuring of offsites, 5 to 15 percent
Major expansion of offsites, 15 to 45 percent
Grass-roots plants, 45 to 150 percent

Allocated Capital This is capital that is shared due to its proportionate share use in a new facility. Such items include intermediate chemicals, utilities, services and sales, administration, research, and engineering overhead.

Working Capital Working capital is the funds necessary to conduct day-to-day company business. These are funds required to purchase raw materials, supplies, etc. It is continuously liquidated and rejuvenated from the sale of products or services. If an adequate amount of working capital is available, management has the necessary flexibility to cover expenses in case of strikes, delays, fires, etc. Several methods are available for estimating an adequate amount of working capital. They may be broadly classified into percentage and inventory methods. The percentage methods are satisfactory for study and preliminary capital estimates. The percentage methods are of two types: percentage based on capital investment and percentage based upon sales. In the former method, 15 to 25 percent of the total capital investment may be sufficient for preliminary estimates. In the case of certain specialty chemicals where the raw materials are expensive, it is perhaps better to use the percentage of sales method. Such chemicals as flavors, fragrances, perfumes, etc., are in this category. Experience has shown that 15 to 45 percent of sales has been used with 30 to 35 percent being a reasonable average value.

Start-Up Expenses *Start-up expenses* are defined as the total costs directly related to bringing a new manufacturing facility onstream. Start-up time is the time span between the end of construction and the beginning of normal operation. Normal operation is operation at a certain percentage of design capacity or a specified number of days of continuous operation or the ability to make product of a specified purity. Start-up costs are part of the total capital investment and include labor, materials, and overhead for design modifications or changes due to errors on the part of engineering, contractors, costs of tests, final alterations, and adjustments. These items cannot be included as contingency because it is known that such work will be necessary before the project is completed. Experience has shown that start-up costs are a percentage of the battery-limits fixed capital investment of the order on average of 3 percent.

Depending on the tax laws in effect, not all start-up costs can be expensed and a portion must be capitalized. Start-up costs can reduce the after-tax earnings during the early years of a project because of a delay in the start-up of production causing a loss of earnings. Construction changes are items of capital cost, and production start-up costs are expensed as an operating expense.

Other Capital Items Paid-up royalties and licenses are considered part of the capital investment since these are replacements for capital to perform process research and development. The initial catalyst and chemical charge, especially for noble metal catalysts and/or in electrolytic processes, is a large amount. These materials are considered to have a life of 1 year. If funds must be borrowed for a new facility, then the interest on borrowed funds during the construction period is capitalized; otherwise, the interest is part of the operating expense.

MANUFACTURING/OPERATING EXPENSES

The estimation of manufacturing expenses has received less attention in the open literature than the estimation of capital requirements. Operating expenses are estimated from proprietary company files. In this section, methods for estimating the elements that constitute operating expenses are presented. Operating expenses consist of the expense of manufacturing a product, packaging and shipping, as well as general overhead expense. These are described later in this section. Figure 9-6 shows an example of a typical manufacturing expense sheet.

RAW MATERIAL EXPENSE

Estimates of the amount of raw material consumed can be obtained from the process material balance. Normally, the raw material expense is the largest expense item in the manufacture of a product. Since yields in a chemical reaction determine the quantity of raw materials consumed, assumed yields may be used to obtain approximate exploratory estimates if possible ranges are given. The prices of the raw materials are published in various trade journals that list material according to form, grade, method of delivery, unit of measure, and cost per unit. The *Chemical Marketing Reporter* is a typical source of these prices. The prices are generally higher than quotations from suppliers, and these latter should be used whenever possible. It may be possible for a company to negotiate the price of a raw material based upon large-quantity use on a long-term basis. With the amount of material used from the material balance and the price of the raw material, the following information can be obtained: annual material consumption, annual material expense, as well as the consumption and expense per unit of product.

Occasionally, by-products may be produced, and if there is a market for these materials, a credit can be given. By-products are treated in the same manner as raw materials and are entered into the manufacturing expense sheet as a credit. If by-products are intermediates for which no market exists, they may be credited to downstream or subsequent operations at a value equivalent to their value as a replacement, or no credit may be obtained.

DIRECT EXPENSES

These are the expenses that are directly associated with the manufacture of a product, e.g., utilities, labor, and maintenance.

Utilities The utility requirements are obtained from the material and energy balances. Utilities include steam, electricity, cooling water, fuel, compressed air, and refrigeration. The current utility prices can be obtained from company plant accounting or from the plant utility supervisor. This person might be able to provide information concerning rate prices for the near future. As requirements increase, the unit cost declines. If large incremental amounts are required, e.g., electricity, it may be necessary to tie the company's utility line to a local utility as a floating source.

With the current energy demands increasing, the unit costs of all utilities are increasing. Any prices quoted need to be reviewed periodically to determine their effect on plant operations. A company utility supervisor is a good source of future price trends. Unfortunately, there are no shortcuts for estimating and projecting utility prices. Utilities are the third largest expense item in the manufacture of a product, behind raw materials and labor.

Operating Labor The most reliable method for estimating labor requirements is to prepare a table of shift, weekend, and vacation coverage. For round-the-clock operation of a continuous process, one operator per shift requires 4.2 operators, if it is assumed that 21 shifts cover the operation and each operator works five 8-h shifts per week. For batch or semicontinuous operation, it is advisable to prepare a labor table, listing the number of tasks and the number of operators required per task, paying particular attention to primary processing steps such as filtration and distillation that may have several items of equipment per step.

Labor rates may be obtained from the union contract or from a company labor relations supervisor. This person will know the current labor rates and any potential labor rate increases in the near future. One should not forget shift differential and overtime charges. Once the number of operators per shift has been established, the annual labor expense and unit expense may be estimated. Wessel (*Chem. Eng.*, **59:** 209–210, July 1952) developed a method for estimating labor requirements for various types of chemical processes in the United States. The equation is applicable for a production rate of 2 to 2000 tons/day (2000 lb/ton).

$$\log Y = -0.783 \log X + 1.252 + B \tag{9-7}$$

where Y = operating labor, operator h/ton per processing step
X = plant capacity, tons/day
B = constant depending upon type of process
+ 0.132 (for batch operations that have minimum labor requirements)
+ 0 (for operations with average labor requirements)
− 0.167 (for a well-instrumented continuous process)

A processing step is one in which a unit operation occurs; e.g., a filtration step might consist of a feed (precoat) tank, pump, filter, and receiver so a processing step may have several items of equipment. By using a flow sheet, the number of processing steps may be counted. The Wessel equation does not take into account changes in labor productivity, but this information can be obtained from each issue of *Chemical Engineering*. Labor productivity varies widely in various sections of this country but even more widely in foreign countries.

Ulrich and Vasudevan (2004) developed a table for estimating labor requirements from flow sheets and drawings of the process. Consideration is given to the type and arrangement of equipment, multiplicity of units, and amount of process control equipment. This method is easier to use than the Wessel method and has been updated in a new edition of the original text.

Supervision The approximate expense for supervision of operations depends on process complexity, but 15 to 30 percent of the operating labor expense is reasonable.

Payroll Charges This item includes workers' compensation, social security premiums, unemployment taxes, paid vacations, holidays, and some part of health and dental insurance premiums. The figure has steadily declined from 1980 and now is 30 to 40 percent of operating labor plus supervision expenses.

Maintenance The maintenance expense consists of two components, namely, materials and labor, approximately 60 and 40 percent, respectively. Company records are the best information sources; however, a value of 6 to 10 percent of the fixed capital investment is a reasonable figure. Processes with a large amount of rotating equipment or that operate at extremes of temperature and/or pressure have higher maintenance requirements.

Miscellaneous Direct Expenses These items include operating supplies, clothing and laundry, laboratory expenses, royalties, environmental control expenses, etc.

Item	Basis	Percent
Operating supplies	Operating labor	5–7
Clothing and laundry	Operating labor	10–15
Laboratory expenses	Operating labor	10–20
Royalties and patents	Sales	1–5

Environmental Control Expense Wastes from manufacturing operations must be disposed of in an environmentally acceptable manner. This direct expense is borne by each manufacturing department. Some companies have their own disposal facilities, or they may contract with a firm that handles the disposal operation. Regardless of how the wastes are handled, there is an expense. Published data are found in the open literature, some of which have been published by Couper (2003).

INDIRECT EXPENSES

These indirect expenses consist of two major items; depreciation and plant indirect expenses.

Depreciation The Internal Revenue Service allows a deduction for the "exhaustion, wear and tear and normal obsolescence of equipment used in the trade or business." (This topic is treated more fully later in this section.) Briefly, for manufacturing expense estimates, straight-line depreciation is used, and accelerated methods are employed for cash flow analysis and profitability calculations.

Plant Indirect Expenses These expenses cover a wide range of items such as property taxes, personal and property liability insurance premiums, fire protection, plant safety and security, maintenance of plant roads, yards and docks, plant personnel staff, and cafeteria expenses (if one is available). A quick estimate of these expenses based upon company records is of the order of 2 to 4 percent of the fixed capital investment. Hackney presented a method for estimating these expenses based upon a capital investment factor, and a labor factor, but the result is high.

TOTAL MANUFACTURING EXPENSE

The total manufacturing expense for a product is the sum of the raw materials and direct and indirect expenses.

TOTAL OPERATING EXPENSES

PRODUCT:	PLASTICIZER X			
TOTAL SALES ($/YR):	7200000			
RATED CAPACITY (MM LBS/YR):	12			
LOCATION:				
FIXED CAPITAL INVESTMENT :	800000			
LAND	25000			
WORKING CAPITAL	120000			
OPERATING HOURS (HRS/YR):				
DATE:				
BY:				
RAW MATERIALS:				
MATERIAL	UNIT	ANNUAL QUANTITY	$/UNIT	$/YEAR
A AND B	LB	12000000	.23	2760000
				0
				0
				0
GROSS MATERIAL EXPENSE				2760000
BY-PRODUCTS:	UNIT	ANNUAL QUANTITY	$/UNIT	$/YEAR
				0
				0
BY-PRODUCT CREDIT				0
NET MATERIAL EXPENSE				2760000
DIRECT EXPENSES:				
	UNIT	ANNUAL QUANTITY	$/UNIT	$/YEAR
UTILITIES:				
steam, low pressure				0
steam, medium pressure				0
steam, hgh pressure	LB	60000000	.003	180000
GROSS STEAM EXPENSE				180000
STEAM CREDIT				0
NET STEAM EXPENSE				180000
electricity	KWH	3000000	.035	105000
cooling water	GALLONS	72000000	.000045	3240
fuel gas				0
other:				0
city water	GALLONS	360000000	.0002	72000
TOTAL UTILITIES COST				540240
LABOR:				
men per shift	4			
annual labor rate per shift	25000			25000
TOTAL LABOR COSTS				100000
SUPERVISION:				
% total of labor expense				0
SUPERVISION EXPENSE =				18000
PAYROLL CHARGES, FRINGE BENEFITS:				
% total of labor expense	40			
PAYROLL EXPENSE				47200
MAINTENANCE				
% of fixed capital investment	8			
MAINTENANCE EXPENSE				64000
SUPPLIES:				
% of operating labor				0
SUPPLIES EXPENSE				1800
LABORATORY:				
laboratory hours per year	900			
cost per hour	30			
TOTAL LABORATORY EXPENSE				27000

FIG. 9-6 Total operating expense sheet.

ROYALTIES			
WASTE DISPOSAL:			0
tons per year			
waste charge per ton			0
WASTE DISPOSAL EXPENSE			0
OTHER:			
laundry	6000		6000
TOTAL DIRECT EXPENSE			804240
TOTAL DIRECT + NET MATERIAL COSTS			3564240
INDIRECT EXPENSES:			
DEPRECIATION			
% of fixed capital investment	100		
life of project (yrs)	7		
DEPRECIATION			114000
PLANT INDIRECT EXPENSES			
% of fixed capital investment	5		
PLANT INDIRECT EXPENSES			40000
TOTAL INDIRECT EXPENSES			154000
TOTAL MANUFACTURING EXPENSE:			3718240
PACKAGING, SHIPPING EXPENSE			
rated capacity per	12000000		
dollars per unit	.005		
PACKAGING AND SHIPPING EXPENSE			60000
TOTAL PRODUCTION EXPENSE			3778240
GENERAL OVERHEAD EXPENSES			
percent of annual sales		5	
GENERAL OVERHEAD EXPENSES			360000
TOTAL OPERATING EXPENSE			4138240

FIG. 9-6 *(Continued)*

PACKAGING AND SHIPPING EXPENSES

The packaging expense depends on how the product is sold. The package may vary from small containers to fiber packs to lever packs, or the product may be shipped via tank truck, tank car, or pipeline. Each product must be considered and the expense of the container included on a case-by-case basis. The shipping expense includes the in-plant movement to warehousing facilities. Product delivery expenses are difficult to estimate because products are shipped in various amounts to numerous destinations. Often these expenses come under the heading of freight allowed in the sale of a product.

TOTAL PRODUCT EXPENSE

The sum of the total manufacturing expense and the packaging and in-plant shipping expense is the total product expense.

GENERAL OVERHEAD EXPENSE

This expense is often separated from the manufacturing expenses. It includes the expense of maintaining sales offices throughout the country, staff engineering departments, and research and development facilities and administrative offices. All manufacturing departments are expected to share in these expenses so an appropriate charge is made for each product varying between 6 and 15 percent of the product's annual revenue. The wide range in percentage will vary depending on the amount of customer service required due to the nature of the product.

TOTAL OPERATING EXPENSE

The sum of the total product expense and the general overhead expense is the total operating expense. This item ultimately becomes part of the operating expense on the income statement.

RAPID MANUFACTURING EXPENSE ESTIMATION

Holland et al. (1983) developed an expression for estimating annual manufacturing expenses for production rates other than the base case based upon fixed capital investment, labor requirements, and utility expense.

$$A_1 = mC_{\text{fci}} + nC_L N_1 + pU_1 \tag{9-8}$$

where C_{fci} = cost of fixed capital investment, \$
C_L = cost of labor, \$ per operator per shift
N_1 = annual labor requirements, operators/shift/year at rate 1
U_1 = annual utility expenses at production rate 1
A_1 = annual conversion expense at rate 1
m, n, p = constants obtained from company records in consistent units

Equation (9-8) can be modified to include raw materials by adding a term qM_1, where q = a constant and M_1 = annual raw material expense at rate 1. See also Table 9-16.

SCALE-UP OF MANUFACTURING EXPENSES

If it is desired to estimate the annual manufacturing expense at some rate other than a base case, the following modification may be made:

$$A_2 = mC_{\text{fci}}\left(\frac{R_2}{R_1}\right)^{0.7} + nC_L N_1\left(\frac{R_2}{R_1}\right)^{0.25} + pU_1\left(\frac{R_2}{R_1}\right) + qM_1\left(\frac{R_2}{R_1}\right) \tag{9-9}$$

where A_2 = annual manufacturing expense at production rate 2
R_1 = production rate 1
R_2 = production rate 2

Equation (9-9) may also be used to calculate data for a plot of manufacturing expense as a function of annual production rate, as shown in Fig. 9-7. Plots of these data show that the manufacturing expense per unit of production decreases with increasing plant size. The first term in Eq. (9-9) reflects the increase in the capital investment by using the 0.7 power for variations in production rates. Labor varies as the 0.25 power for continuous operations based upon experience. Utilities and raw materials are essentially in direct proportion to the amount of product manufactured, so the exponent of these terms is 1.

TABLE 9-16 Typical Labor Requirements for Various Equipment

Equipment	Laborers per unit per shift
Blowers and compressor	0.1–0.2
Centrifuge	0.25–0.50
Crystallizer, mechanical	0.16
Dryers	
Rotary	0.5
Spray	1.0
Tray	0.5
Evaporator	0.25
Filters	
Vacuum	0.125–0.25
Plate and frame	1.0
Rotary and belt	0.1
Heat exchangers	0.1
Process vessels, towers (including auxiliary pumps and exchangers)	0.2–0.5
Reactors	
Batch	1.0
Continuous	0.5

Adapted from G. D. Ulrich, *A Guide to Chemical Engineering Process Design and Economics*, Wiley, New York, 1984.

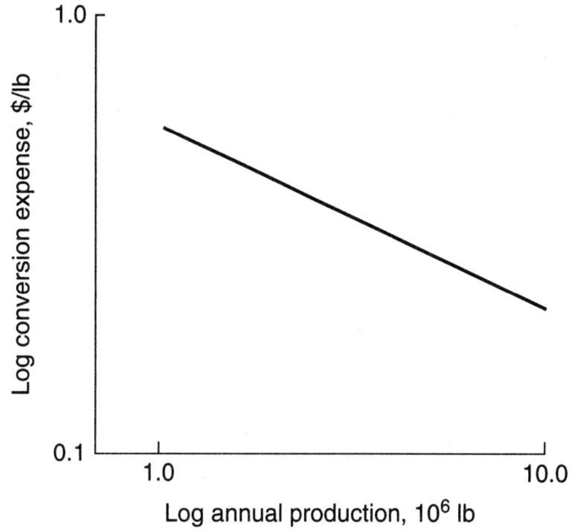

FIG. 9-7 Annual conversion expense as a function of production rate.

FACTORS THAT AFFECT PROFITABILITY

DEPRECIATION

According to the Internal Revenue Service (IRS), *depreciation* is defined as an allowance for the decrease in value of a property over a period of time due to wear and tear, deterioration, and normal obsolescence. The intent is to recover the cost of an asset over a period of time. It begins when a property is placed in a business or trade for the production of income and ends when the asset is retired from service or when the cost of the asset is fully recovered, whichever comes first. Depreciation and taxes are irrevocably tied together. It is essential to be aware of the latest tax law changes because the rules governing depreciation will probably change. Over the past 70 to 80 years, there have been many changes in the tax laws of which depreciation is a major component. Couper (2003) discussed the history and development of depreciation accounting. Accelerated depreciation was introduced in the early 1950s to stimulate investment and the economy. It allowed greater depreciation rates in the early years of a project when markets were not well established, manufacturing facilities were coming onstream, and expenses were high due to bringing the facility up to design capacity.

The current methods for determining annual depreciation charges are the straight-line depreciation and the Modified Accelerated Cost Recovery System (MACRS). In the straight-line method, the cost of an asset is distributed over its expected useful life such that the annual charge is

$$D = \frac{I+S}{n} \text{ \$ per year} \tag{9-10}$$

where D = annual depreciation charge
I = investment
n = number of years
S = salvage value

The MACRS went into effect in January 1987 (Couper, 2003) with six asset recovery periods: 3, 5, 7, 10, 15, and 20 years. It is based upon the declining-balance method. The equation for the declining-balance method is

$$V_e = V_i(1-f) \tag{9-11}$$

where V_i = value of asset at beginning of year
V_e = value of asset at end of year
f = declining-balance factor

For 150 percent declining balance f = 1.5, and for 200 percent f = 2.0. These factors are applied to the previous year's remaining balance. It is evident that the declining-balance method will not recover the asset that the IRS permits. Therefore, a combination of the declining-balance and straight-line methods forms the basis for the MACRS method.

Class lives for selected industries are found in Couper (2003), but most chemical processing equipment falls in the 5-year category and petroleum processing equipment in the 7-year category. For those assets with class lives less than 10 years, a 200 percent declining-balance method with a switch to straight-line method in the later years is used. The IRS adopted a half-year convention for both depreciation methods. Under this convention, a property placed in service is considered to be only one-half year irrespective of when during the year the property was placed in service. Table 9-17 is a listing of the class lives, and Table 9-18 contains factors with the half-year convention for both the MACRS and straight-line methods.

Depreciation is entered as an indirect expense on the manufacturing expense sheet based upon the straight-line method. However, when one is determining the after-tax cash flow, straight-line depreciation is removed from the manufacturing expense and the MACRS depreciation is entered. This is illustrated under the section on cash flow.

There are certain terms that apply to depreciation:

- *Depreciation reserve* is the accumulated depreciation at a specific time.
- *Book value* is the original investment minus the accumulated depreciation.
- *Service life* is the time period during which an asset is in service and is economically feasible.
- *Salvage value* is the net amount of money obtained from the sale of a used property over and above any charges involved in the removal and sale of the property.
- *Scrap value* implies that the asset has no further useful life and is sold for the amount of scrap material in it.
- *Economic life* is the most likely period of successful operation before a need arises for subsequent investment in additional equipment as the result of product or process obsolescence or equipment due to wear and tear.

DEPLETION

Depletion is concerned with the diminution of natural resources. Generally depletion does not enter into process economic studies. Rules for determining the amount of depletion are found in the IRS Publication 535.

AMORTIZATION

Amortization is the ratable deduction for the cost of an intangible property over its useful life, perhaps a 15-year life, via straight-line calculations. An example of an intangible property is a franchise, patent, trademark, etc. Two IRS publications, Form 4562 and Publication 535 (1999), established the regulations regarding amortization.

TABLE 9-17 Depreciation Class Lives and MACRS Recovery Periods

Asset class	Description of asset	Class life, yr	MACRS recovery period, yr
00.12	Information systems	6	5
00.4	Industrial steam and electric generation and/or distribution systems	22	15
13.3	Petroleum refining	16	10
20.3	Manufacture of vegetable oils and vegetable oil products	18	10
20.5	Manufacture of food and beverages	4	3
22.4	Manufacture of textile yams	8	5
22.5	Manufacture of nonwoven fabrics	10	7
26.1	Manufacture of pulp and paper	13	7
28.0	Manufacture of chemicals and allied products	9.5	5
30.1	Manufacture of rubber products	14	7
30.2	Manufacture of finished plastic products	11	7
32.1	Manufacture of glass products	14	7
32.2	Manufacture of cement	20	15
32.3	Manufacture of other stone and clay products	15	7
33.2	Manufacture of primary nonferrous metals	14	7
32.4	Manufacture of primary steel mill products	15	7
49.223	Substitute natural gas-coal gasification	18	10
49.25	Liquefied natural gas plant	22	15

SOURCE: "How to Depreciate Property," Publication 946, Internal Revenue Service, U.S. Department of Treasury, Washington, 1999.

TAXES

Most major corporations pay the federal tax rate of 21 percent on their annual gross earnings. In addition, some states have a stepwise corporate income tax rate. State income tax is deductible as an expense item before the calculation of the federal tax. If T_s is the incremental tax rate and T_f is the incremental federal tax, both expressed as decimals, then the combined incremental rate T_c is

$$T_c = T_s + (1 - T_s)T_f \qquad (9\text{-}12)$$

If the federal rate is 21 percent and the state rate is 7 percent, then the combined rate is

$$T_c = 0.07 + (1 - 0.07)(0.21) = 0.265$$

Therefore, the combined tax rate is 26.5 percent.

TIME VALUE OF MONEY

In business, money is either borrowed or loaned. If money is loaned, there is the risk that it may not be repaid. From the lender's standpoint, the funds could have been invested somewhere else and made a profit; therefore, the interest charged for the loan is compensation for the forgone profit. The borrower may look upon this interest as the cost of renting money. The amount of interest charged depends on the scarcity of money, the size of the loan, the length of the loan period, the risk that the lender feels that the loan may not be repaid, and the prevailing economic conditions. Engineers involved in the presentation and/or the evaluation of an investment of funds in a venture, therefore, need to understand the time value of money and how it is applied in the evaluation of projects.

The amount of the loan is called the *principal P*. The longer the time for which the money is loaned, the greater the total amount of interest paid. The future amount of the money F is greater than the principal or *present worth P*. The relationship between F and P depends upon the type of interest used. Table 9-19 is a summary of the nomenclature used in time value of money calculations.

Simple Interest The relationship between F and P is $F = P(1 + \text{interest})$. The interest is charged on the original loan and not on the unpaid balance (Couper and Rader, 1986). The interest is paid at the end of each time interval. Although the simple-interest concept still exists, it is seldom used in business.

Discrete Compound Interest In financial transactions, loans or deposits are made using compound interest. The interest is not withdrawn but is added to the principal for that time period. In the next time period, the interest is calculated upon the principal plus the interest from

TABLE 9-19 Interest Nomenclature

Symbol	Definition
F	Future sum Future value Future worth Future amount
P	Principal Present worth Present value Present amount
A	End of period payment in a uniform series

TABLE 9-18 Depreciation Rates for Straight-Line and MACRS Methods

Straight-line, half-year convention

Year	3	5	7	10	15	20
1	16.67%	10.00%	7.14%	5.0%	3.33%	2.5%
2	33.33	20.00	14.29	10.0	6.67	5.0
3	33.33	20.00	14.29	10.0	6.67	5.0
4	16.67	20.00	14.28	10.0	6.67	5.0
5		20.00	14.29	10.0	6.67	5.0
6		10.00	14.28	10.0	6.67	5.0
7			14.29	10.0	6.67	5.0
8			7.14	10.0	6.66	5.0
9				10.0	6.67	5.0
10				10.0	6.66	5.0
11				5.0	6.67	5.0
12					6.66	5.0
13					6.67	5.0
14					6.66	5.0
15					6.67	5.0
16					3.33	5.0
17						5.0
18						5.0
19						5.0
20						5.0
21						2.5

MACRS*, half-year convention

Year	3	5	7	10	15	20
1	33.33%	20.00%	14.29%	10.00%	5.00%	3.750%
2	44.45	32.00	24.49	18.00	9.50	7.219
3	14.81	19.20	17.49	14.40	8.55	6.677
4	7.41	11.52	12.49	11.52	7.70	6.177
5		11.52	8.93	9.22	6.93	5.713
6		5.76	8.92	7.37	6.23	5.285
7			8.93	6.55	5.90	4.888
8			4.46	6.55	5.90	4.522
9				6.56	5.91	4.462
10				6.55	5.90	4.461
11				3.28	5.91	4.462
12					5.90	4.461
13					5.91	4.462
14					5.90	4.461
15					5.91	4.462
16					2.95	4.461
17						4.462
18						4.461
19						4.462
20						4.461
21						2.231

*General depreciation system. Declining-balance switching to straight-line. Recovery periods 3, 5, 7, 10, 15, and 20 years.
SOURCE: "How to Depreciate Property," Publication 946, Internal Revenue Service, U.S. Department of Treasury, Washington, 1999.

the preceding time period. This process illustrates compound interest. In equation format,

$$\text{Year 1: } P + Pi = P(1+i) = F_1 \quad (9\text{-}13a)$$

$$\text{Year 2: } P + Pi(1+i) = P(1+i)^2 = F_2 \quad (9\text{-}13b)$$

$$\text{Year } n\text{: } P(1+i)^n = F \quad (9\text{-}13c)$$

An interest rate quoted on an annual basis is called *nominal interest.* However, interest may be payable on a semiannual, quarterly, monthly, or daily basis. To determine the amount compounded, the following equation applies:

$$F = P\left(1 + \frac{i}{m}\right)^{mn} \quad (9\text{-}14)$$

where m = number of interest periods per year
n = number of years
i = nominal interest

Interest calculated for a given time period is known as *discrete compound interest,* with *discrete* referring to a discrete time period. Table 9-20 contains 5 and 6 percent discrete interest factors.

Examples of the use of discrete factors for various applications are found in Table 9-21, assuming that the present time is when the first funds are expended.

Continuous Compound Interest In some companies, namely, petroleum, petrochemical, and chemical companies, money transactions occur hourly or daily, or essentially continuously. The receipts from sales and services are invested immediately upon receipt. The interest on this cash flow is continuously compounded. To use continuous compounding when evaluating projects or investments, one assumes that cash flows continuously.

In continuous compounding, the year is divided into an infinite number of periods. Mathematically, the limit of the interest term is

$$\lim_{n\to\infty}\left(1 + \frac{r}{m}\right)^{mn} = e^{rn} \quad (9\text{-}15)$$

where n = number of years
m = number of interest periods per year
r = nominal interest rate
e = base for naperian logarithms

The numerical difference between discrete compound interest and continuous compound interest is small, but when large sums of money are involved, the difference may be significant. Table 9-22 is an abbreviated continuous interest table, assuming that time zero is when start-up occurs. A summary of the equations for discrete compound and continuous compound interest is found in Table 9-23.

Compounding and Discounting When money is moved forward in time from the present to a future time, the process is called *compounding.* The effect of compounding is that the total amount of money increases with time due to interest. *Discounting* is the reverse process, i.e., a sum of money moved backward in time. Figure 9-8 is a sketch of this process. The time periods are years, and the interest is normally on an annual basis using end-of-year money flows. The longer the time before money is received, the less it is worth at present.

Effective Interest Rates When an interest rate is quoted, it is nominal interest that is stated. These quotes are on an annual basis, however, when compounding occurs that is not the actual or effective interest. According to government regulations, an effective rate APY must be stated also. The effective interest is calculated by

$$i_{\text{eff}} = \left(1 + \frac{i}{m}\right)^{(m)(1)} - 1 \quad (9\text{-}16)$$

The time period for calculating the effective interest rate is 1 year.

Example 9-6 Effective Interest Rate A person is quoted an 8.33 percent nominal interest rate on a 4-year loan compounded monthly. Determine the effective interest rate.

Solution:

$$i_{\text{eff}} = \left(\frac{1+i}{m}\right)^{(m)(1)} = \left(\frac{1+0.833}{12}\right)^{(12)(1)} - 1 = (1.00694)^{12} - 1$$
$$= 1.0865 - 1 = 0.865$$

The effective interest rate is 8.65 percent.

CASH FLOW

Cash flow is the amount of funds available to a company to meet current operating expenses. Cash flow may be expressed on a before- or after-tax basis. *After-tax cash flow* is defined as the net profit (income) after taxes plus depreciation. It is an integral part of the *net present worth* (NPW) and discounted cash flow profitability calculations.

The cash flow diagram, also referred to as a cash flow model (Fig. 9-9), shows the relationship between revenue, cash operating expenses, depreciation, and profit. This diagram is similar in many respects to a process flow diagram, but it is in dollars. Revenue is generated from the sale of a product manufactured in "operations." Working capital is replenished from sales and may be considered to be in dynamic equilibrium with operations. Leaving the operations box is a stream, "cash operating expenses." It includes all the cash expenses incurred in the operation but does not include the noncash item depreciation. Since depreciation is an allowance, it is reported on the operating expense sheet, in accordance with the tax laws, as an operating expense item. (See the section Operating Expense Estimation.) Depreciation is an internal expense, and this allowance is retained within the company. If the cash operating expenses are subtracted from the revenue, the result is the operating income. If depreciation is subtracted from the operating income, the net profit before taxes results. Federal income taxes are then deducted from the net profit before taxes, giving the net profit after taxes. When depreciation and net profit after taxes are summed, the result is the after-tax cash flow. The terminology in Fig. 9-9 is consistent with that found in most company income statements in company annual reports.

An equation can be developed for cash flow as follows:

$$\text{CF} = D + (R - C^* - D)(1 - t) \quad (9\text{-}17)$$

where CF = after-tax cash flow
D = depreciation
R = revenue
C^* = cash operating expenses
t = tax rate

Example 9-7 is a sample calculation of the after-tax cash flow and the tabulated results.

Example 9-7 After-Tax Cash Flow The revenue from the manufacture of a product in the first year of operation is \$9.0 million, and the cash operating expenses are \$4.5 million. Depreciation on the invested capital is \$1.7 million. If the federal income tax rate is 35 percent, calculate the after-tax cash flow.

Solution: The resulting after-tax cash flow is \$3.52 million. See Fig. 9-10.

Cumulative Cash Position Table To organize cash flow calculations, it is suggested that a cumulative cash position table be prepared by using an electronic spreadsheet. For this discussion, time zero is assumed to be at project start-up. Expenditures for land and equipment occurred prior to time zero and represent negative cash flows. At time zero, working capital is charged to the project as a negative cash flow. Start-up expenses are charged in the first year, and positive cash flow from the sale of product as net income after taxes plus depreciation begins, reducing the negative cash position. This process continues until the project is terminated. At that time, adjustments are made to recover land and working capital. An example of a cumulative cash position table is Table 9-24.

When equipment is added for plant expansions to an existing facility, it may be more convenient to use time zero when the first expenditures occur. The selection of either time base is satisfactory for economic analysis as long as consistency is maintained.

TABLE 9-20 Discrete Compound Interest Factors*

	Single payment		Uniform annual series				Single payment		Uniform annual series				
	Compound-amount factor	Present-worth factor	Sinking-fund factor	Capital-recovery factor	Compound-amount factor	Present-worth factor	Compound-amount factor	Present-worth factor	Sinking-fund factor	Capital-recovery factor	Compound-amount factor	Present-worth factor	
n	Given P, to find F $(1+i)^n$	Given F, to find P $\frac{1}{(1+i)^n}$	Given F, to find A $\frac{i}{(1+i)^n-1}$	Given P, to find A $\frac{i(1+i)^n}{(1+i)^n-1}$	Given A, to find F $\frac{(1+i)^n-1}{i}$	Given A, to find P $\frac{(1+i)^n-1}{i(1+i)^n}$	Given P, to find F $(1+i)^n$	Given F, to find P $\frac{1}{(1+i)^n}$	Given F, to find A $\frac{i}{(1+i)^n-1}$	Given P, to find A $\frac{i(1+i)^n}{(1+i)^n-1}$	Given A, to find F $\frac{(1+i)^n-1}{i}$	Given A, to find P $\frac{(1+i)^n-1}{i(1+i)^n}$	n
	5% Compound Interest Factors						6% Compound Interest Factors						
1	1.050	0.9524	1.00000	1.05000	1.000	0.952	1.060	0.9434	1.00000	1.06000	1.000	0.943	1
2	1.103	.9070	0.48780	0.53780	2.050	1.859	1.124	.8900	0.48544	0.54544	2.060	1.833	2
3	1.158	.8638	.31721	.36721	3.153	2.723	1.191	.8396	.31411	.37411	3.184	2.673	3
4	1.216	.8227	.23201	.28201	4.310	3.546	1.262	.7921	.22859	.28859	4.375	3.465	4
5	1.276	.7835	.18097	.23097	5.526	4.329	1.338	.7473	.17740	.23740	5.637	4.212	5
6	1.340	.7462	.14702	.19702	6.802	5.076	1.419	.7050	.14336	.20336	6.975	4.917	6
7	1.407	.7107	.12282	.17282	8.142	5.786	1.504	.6651	.11914	.17914	8.394	5.582	7
8	1.477	.6768	.10472	.15472	9.549	6.463	1.594	.6274	.10104	.16104	9.897	6.210	8
9	1.551	.6446	.09069	.14069	11.027	7.108	1.689	.5919	.08702	.14702	11.491	6.802	9
10	1.629	.6139	.07940	.12950	12.578	7.722	1.791	.5584	.07587	.13587	13.181	7.360	10
11	1.710	.5847	.07039	.12039	14.207	8.306	1.898	.5268	.06679	.12679	14.972	7.887	11
12	1.796	.5568	.06283	.11283	15.917	8.863	2.012	.4970	.05928	.11928	16.870	8.384	12
13	1.886	.5303	.05646	.10646	17.713	9.394	2.133	.4688	.05296	.11296	18.882	8.853	13
14	1.980	.5051	.05102	.10102	19.599	9.899	2.261	.4423	.04758	.10758	21.015	9.295	14
15	2.079	.4810	.04634	.09634	21.579	10.380	2.397	.4173	.04296	.10296	23.276	9.712	15
16	2.183	.4581	.04227	.09227	23.657	10.838	2.540	.3936	.03895	.09895	25.673	10.106	16
17	2.292	.4363	.03870	.08870	25.840	11.274	2.693	.3714	.03544	.09544	28.213	10.477	17
18	2.407	.4155	.03555	.08555	28.132	11.690	2.854	.3503	.03236	.09236	30.906	10.828	18
19	2.527	.3957	.03275	.08275	30.539	12.085	3.026	.3305	.02962	.08962	33.760	11.158	19
20	2.653	.3769	.03024	.08024	33.066	12.462	3.207	.3118	.02718	.08718	36.786	11.470	20
21	2.786	.3589	.02800	.07800	35.719	12.821	3.400	.2942	.02500	.08500	39.993	11.764	21
22	2.925	.3418	.02597	.07597	38.505	13.163	3.604	.2775	.02305	.08305	43.392	12.042	22
23	3.072	.3256	.02414	.07414	41.430	13.489	3.820	.2618	.02128	.08128	46.996	12.303	23
24	3.225	.3101	.02247	.07247	44.502	13.799	4.049	.2470	.01968	.07968	50.816	12.550	24
25	3.386	.2953	.02095	.07095	47.727	14.094	4.292	.2330	.01823	.07823	54.865	12.783	25
26	3.556	.2812	.01956	.06956	51.113	14.375	4.549	.2198	.01690	.07690	59.156	13.003	26
27	3.733	.2678	.01829	.06829	54.669	14.643	4.822	.2074	.01570	.07570	63.706	13.211	27
28	3.920	.2551	.01712	.06712	58.403	14.898	5.112	.1956	.01459	.07459	68.528	13.406	28
29	4.116	.2429	.01605	.06605	62.323	15.141	5.418	.1846	.01358	.07358	73.640	13.591	29
30	4.322	.2314	.01505	.06505	66.439	15.372	5.743	.1741	.01265	.07265	79.058	13.765	30
31	4.538	.2204	.01413	.06413	70.761	15.593	6.088	.1643	.01179	.07179	84.802	13.929	31
32	4.765	.2099	.01328	.06328	75.299	15.803	6.453	.1550	.01100	.07100	90.890	14.084	32
33	5.003	.1999	.01249	.06249	80.064	16.003	6.841	.1462	.01027	.07027	97.343	14.230	33
34	5.253	.1904	.01176	.06176	85.067	16.193	7.251	.1379	.00960	.06960	104.184	14.368	34
35	5.516	.1813	.01107	.06107	90.320	16.374	7.686	.1301	.00897	.06897	111.435	14.498	35
40	7.040	.1420	.00828	.05828	120.800	17.159	10.286	.0972	.00646	.06646	154.762	15.046	40
45	8.985	.1113	.00626	.05626	159.700	17.774	13.765	.0727	.00470	.06470	212.744	15.456	45
50	11.467	.0872	.00478	.05478	209.348	18.256	18.420	.0543	.00344	.06344	290.336	15.762	50
55	14.636	.0683	.00367	.05367	272.713	18.633	24.650	.0406	.00254	.06254	394.172	15.991	55
60	18.679	.0535	.00283	.05283	353.584	18.929	32.988	.0303	.00188	.06188	533.128	16.161	60
65	23.840	.0419	.00219	.05219	456.798	19.161	44.145	.0227	.00139	.06139	719.083	16.289	65
70	30.426	.0329	.00170	.05170	588.529	19.343	59.076	.0169	.00103	.06103	967.932	16.385	70
75	38.833	.0258	.00132	.05132	756.654	19.485	79.057	.0126	.00077	.06077	1,300.949	16.456	75
80	49.561	.0202	.00103	.05103	971.229	19.596	105.796	.0095	.00057	.06057	1,746.600	16.509	80
85	63.254	.0158	.00080	.05080	1,245.087	19.684	141.579	.0071	.00043	.06043	2,342.982	16.549	85
90	80.730	.0124	.00063	.05063	1,594.607	19.752	189.465	.0053	.00032	.06032	3,141.075	16.579	90
95	103.035	.0097	.00049		2,040.694	19.806	253.546	.0039	.00024	.06024	4,209.104	16.601	95
100	131.501	.0076	.00038	.05038	2,610.025	19.848	339.302	.0029	.00018	.06018	5,638.368	16.618	100

*Factors presented for two interest rates only. By using the appropriate formulas, values for other interest rates may be calculated.

TABLE 9-21 Examples of the Use of Compound Interest Table

Given: \$2500 is invested now at 5 percent.
Required: Accumulated value in 10 years (i.e., the amount of a given principal).

Solution:

$$F = P(1+i)^n = \$2500 \times 1.05^{10}$$
$$\text{Compound-amount factor} = (1+i)^n = 1.05^{10} = 1.629$$
$$F = \$2500 \times 1.629 = \$4062.50$$

Given: \$19,500 will be required in 5 years to replace equipment now in use.
Required: With interest available at 3 percent, what sum must be deposited in the bank at present to provide the required capital (i.e., the principal which will amount to a given sum)?

Solution:

$$P = F\frac{1}{(1+i)^n} = \$19{,}500\frac{1}{1.03^5}$$
$$\text{Present-worth factor} = 1/(1+i)^n = 1/1.03^5 = 0.8626$$
$$P = \$19{,}500 \times 0.8626 = \$16{,}821$$

Given: \$50,000 will be required in 10 years to purchase equipment.
Required: With interest available at 4 percent, what sum must be deposited each year to provide the required capital (i.e., the annuity which will amount to a given fund)?

Solution:

$$A = F\frac{i}{(1+i)^n - 1} = \$50{,}000\frac{0.04}{1.04^{10} - 1}$$
$$\text{Sinking-fund factor} = \frac{i}{(1+i)^n - 1} = \frac{0.04}{1.04^{10} - 1} = 0.08329$$
$$A = \$50{,}000 \times 0.08329 = \$4{,}164$$

Given: \$20,000 is invested at 10 percent interest.
Required: Annual sum that can be withdrawn over a 20-year period (i.e., the annuity provided by a given capital).

Solution:

$$A = P\frac{i(1+i)^n}{(1+i)^n - 1} = \$20{,}000\frac{0.10 \times 1.10^{20}}{1.10^{20} - 1}$$
$$\text{Capital-recovery factor} = \frac{i(1+i)^n}{(1+i)^n - 1} = \frac{0.10 \times 1.10^{20}}{1.10^{20} - 1} = 0.11746$$
$$A = \$20{,}000 \times 0.11746 = \$2349.20$$

Given: \$500 is invested each year at 8 percent interest.
Required: Accumulated value in 15 years (i.e., amount of an annuity).

Solution:

$$F = A\frac{(1+i)^n - 1}{i} = \$500\frac{1.08^{15} - 1}{0.08}$$
$$\text{Compound-amount factor} = \frac{(1-i)^n - 1}{i} = \frac{1.08^{15} - 1}{0.08} = 27.152$$
$$F = \$500 \times 27.152 = \$13{,}576$$

Given: \$8000 is required annually for 25 years.
Required: Sum that must be deposited now at 6 percent interest.

Solution:

$$P = A\frac{(1+i)^n - 1}{i(1+i)^n} = \$8000\frac{1.06^{25} - 1}{0.06 \times 1.06^{25}}$$
$$\text{Present-worth factor} = \frac{(1+i)^n - 1}{i(1+i)^n} = \frac{1.06^{25} - 1}{0.06 \times 1.06^{25}} = 12.783$$
$$P = \$8000 \times 12.783 = \$102{,}264$$

Example 9-8 Cumulative Cash Position Table (Time Zero at Start-Up) A specialty chemical company is considering the manufacture of an additive for use in the plastics industry. The following is a list of production, sales, and cash operating expenses.

Year	Production, Mlb	Sales, \$1000	Cash operating expenses, \$1000
1	40	20,000	10,320
2	42	21,000	10,800
3	45	23,400	11,520
4	48	24,960	12,240
5	50	27,500	13,470
6	50	28,000	13,970
7	47	23,500	13,175
8	45	21,600	12,645
9	40	18,800	11,320
10	35	15,750	9,995

Land for the project is available at \$300,000. The fixed capital investment was estimated to be \$12,000,000. A working capital of \$1,800,000 is needed initially for the venture. Start-up expenses based upon past experience are estimated to be \$750,000. The project qualifies under IRS guidelines as a 5-year class life investment. The company uses MACRS depreciation with the half-year convention. At the conclusion of the project, the land and working capital are returned to management. Develop a cash flow analysis for this project, using a cumulative cash position table (Table 9-24).

Cumulative Cash Position Plot A pictorial representation of the cumulative cash flows as a function of time is the cumulative cash position plot. All expenditures for capital as well as revenue from sales are plotted as a function of time. Figure 9-11 is such an idealized plot showing time zero at start-up in part a and time zero when the first funds are expended in part b. It should be understood that the plots have been idealized for illustration purposes. Expenditures are usually stepwise, and accumulated cash flow from sales is seldom a straight line but more likely a curve with respect to time.

TABLE 9-22 Condensed Continuous Interest Table*

Factors for determining zero-time values for cash flows which occur at other than zero time.

Compounding of Cash Flows Which Occur:	1%	5%	10%	15%	20%	25%	30%	35%	40%	50%	60%	70%	80%	90%	100%
A. In an Instant															
$\frac{1}{2}$ year before	1.005	1.025	1.051	1.078	1.105	1.133	1.162	1.191	1.221	1.284	1.350	1.419	1.492	1.568	1.649
1 " "	1.010	1.051	1.105	1.162	1.221	1.284	1.350	1.419	1.492	1.649	1.822	2.014	2.226	2.460	2.718
$1\frac{1}{2}$ " "	1.015	1.078	1.162	1.252	1.350	1.455	1.568	1.690	1.822	2.117	2.460	2.858	3.320	3.857	4.482
2 " "	1.020	1.105	1.221	1.350	1.492	1.649	1.822	2.014	2.226	2.718	3.320	4.055	4.953	6.050	7.389
3 " "	1.030	1.162	1.350	1.568	1.822	2.117	2.460	2.858	3.320	4.482	6.050	8.166	11.023	14.880	20.086
B. Uniformly until Zero Time															
From $\frac{1}{2}$ year before to 0 time	1.002	1.013	1.025	1.038	1.052	1.065	1.079	1.093	1.107	1.136	1.166	1.197	1.230	1.263	1.297
" 1 " " " " "	1.005	1.025	1.052	1.079	1.107	1.136	1.166	1.197	1.230	1.297	1.370	1.448	1.532	1.622	1.718
" $1\frac{1}{2}$ " " " " "	1.008	1.038	1.079	1.121	1.166	1.213	1.263	1.315	1.370	1.489	1.622	1.769	1.933	2.117	2.321
" 2 " " " " "	1.010	1.052	1.107	1.166	1.230	1.297	1.370	1.448	1.532	1.718	1.933	2.182	2.471	2.805	3.194
" 3 " " " " "	1.015	1.079	1.166	1.263	1.370	1.489	1.622	1.769	1.933	2.321	2.805	3.412	4.176	5.141	6.362
Discounting of Cash Flows Which Occur:															
C. In an Instant															
1 year later	.990	.951	.905	.861	.819	.779	.741	.705	.670	.606	.549	.497	.449	.407	.360
2 " "	.980	.905	.819	.741	.670	.606	.549	.497	.449	.368	.301	.247	.202	.165	.135
3 " "	.970	.861	.741	.638	.549	.472	.407	.350	.301	.223	.165	.122	.091	.067	.050
4 " "	.961	.819	.670	.549	.449	.368	.301	.247	.202	.135	.091	.061	.041	.027	.018
5 " "	.951	.779	.606	.472	.368	.286	.223	.174	.135	.082	.050	.030	.018	.011	.007
10 years later	.905	.606	.368	.223	.135	.082	.050	.030	.018	.007	.002	.001	—	—	—
15 " "	.861	.472	.223	.105	.050	.024	.011	.005	.002	.001	—	—	—	—	—
20 " "	.819	.368	.135	.050	.018	.007	.002	.001	—	—	—	—	—	—	—
25 " "	.779	.286	.082	.024	.007	.002	.001	—	—	—	—	—	—	—	—
D. Uniformly over Individual Years															
1st year	.995	.975	.952	.929	.906	.885	.864	.844	.824	.787	.752	.719	.688	.659	.632
2nd "	.985	.928	.861	.799	.742	.689	.640	.595	.552	.477	.413	.357	.309	.268	.232
3rd "	.975	.883	.779	.688	.608	.537	.474	.419	.370	.290	.226	.177	.139	.109	.086
4th "	.966	.840	.705	.592	.497	.418	.351	.295	.248	.176	.124	.088	.062	.044	.032
5th "	.956	.799	.638	.510	.407	.326	.260	.208	.166	.106	.068	.044	.028	.018	.012
6th year	.946	.760	.577	.439	.333	.254	.193	.147	.112	.065	.037	.022	.013	.007	.004
7th "	.937	.723	.522	.378	.273	.197	.143	.103	.075	.039	.020	.011	.006	.003	.002
8th "	.928	.687	.473	.325	.224	.154	.108	.073	.050	.024	.011	.005	.002	.001	.001
9th "	.918	.654	.428	.280	.183	.120	.078	.051	.034	.014	.006	.003	.001	—	—
10th "	.909	.622	.387	.241	.150	.093	.058	.036	.022	.009	.003	.001	—	—	—
E. Uniformly over 5-Year Periods															
1st 5 years	.975	.885	.787	.704	.632	.571	.518	.472	.432	.367	.317	.277	.245	.220	.199
6th through 10th year	.928	.689	.477	.332	.232	.164	.116	.082	.058	.030	.016	.008	.004	.002	.001
11th through 15th year	.883	.537	.290	.157	.086	.047	.026	.014	.008	.002	.001	—	—	—	—
16th through 20th year	.840	.418	.176	.074	.032	.013	.006	.002	.001	—	—	—	—	—	—
21st through 25th year	.799	.326	.106	.035	.012	.004	.001	—	—	—	—	—	—	—	—
F. Declining to Nothing at Constant Rate															
1st 5 years	.983	.922	.852	.791	.736	.687	.643	.603	.568	.506	.456	.413	.377	.347	.320
" 10 "	.968	.852	.736	.643	.568	.506	.456	.413	.377	.320	.278	.245	.219	.198	.180
" 15 "	.952	.791	.643	.536	.456	.394	.347	.309	.278	.231	.198	.172	.153	.137	.124
" 20 "	.936	.736	.568	.456	.377	.320	.278	.245	.219	.180	.153	.133	.117	.105	.095
" 25 "	.922	.687	.506	.394	.320	.269	.231	.203	.180	.147	.124	.108	.095	.085	.077

*From tables compiled by J. C. Gregory, The Atlantic Refining Co.

TABLE 9-23 Summary of Discrete and Compound Interest Equations

Factor	Find	Given	Discrete compounding		Continuous compounding	
Single payment						
Compound amount	F	P	$F=P(1+i)^n$	$P(F/P\ i,n)$	$F=P(e^{rn})$	$P(F/P\ r,n)^{\infty}$
Present worth	P	F	$P=F\left[\frac{1}{(1+i)^n}\right]$	$F(P/F\ i,n)$	$P=F(e^{-rn})$	$F(P/F\ r,n)^{\infty}$
Uniform series						
Compound amount	F	A	$F=A\left[\frac{(1+i)^n-1}{i}\right]$	$A(F/A\ i,n)$	$F=A\left(\frac{e^{rn}-1}{e^r-1}\right)$	$F(F/A\ r,n)^{\infty}$
Sinking fund	A	F	$A=F\left[\frac{i}{(1+i)^n-1}\right]$	$F(A/F\ i,n)$	$A=F\left(\frac{e^r-1}{e^{rn}-1}\right)$	$F(A/F\ r,n)^{\infty}$
Present worth	P	A	$P=A\left[\frac{(1+i)^n-1}{i(1+i)^n}\right]$	$A(F/A\ i,n)$	$P=A\left[\frac{e^{rn}-1}{e^{rn}(e^r-1)}\right]$	$A(P/A\ r,n)^{\infty}$
Capital recovery	A	P	$A=P\left[\frac{i(1+i)^n}{(1+i)^n-1}\right]$	$P(A/P\ i,n)$	$A=P\left[\frac{e^{rn}(e^r-1)}{e^{rn}-1}\right]$	$A(P/A\ r,n)^{\infty}$

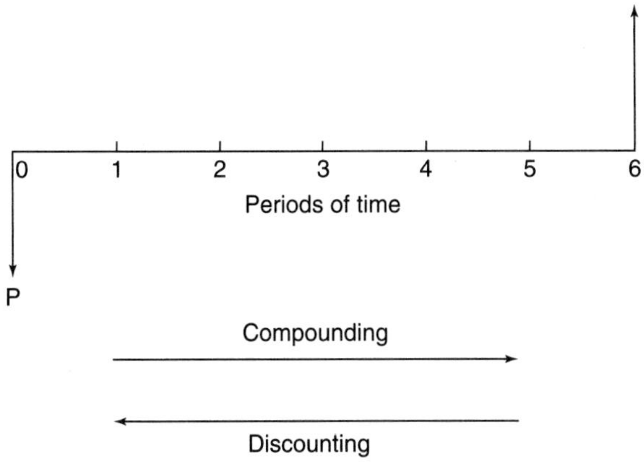

FIG. 9-8 Compounding-discounting diagram.

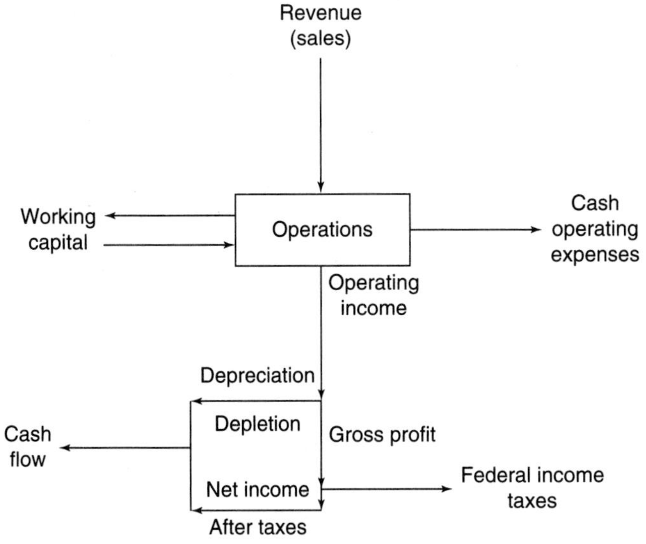

FIG. 9-9 Cash flow model.

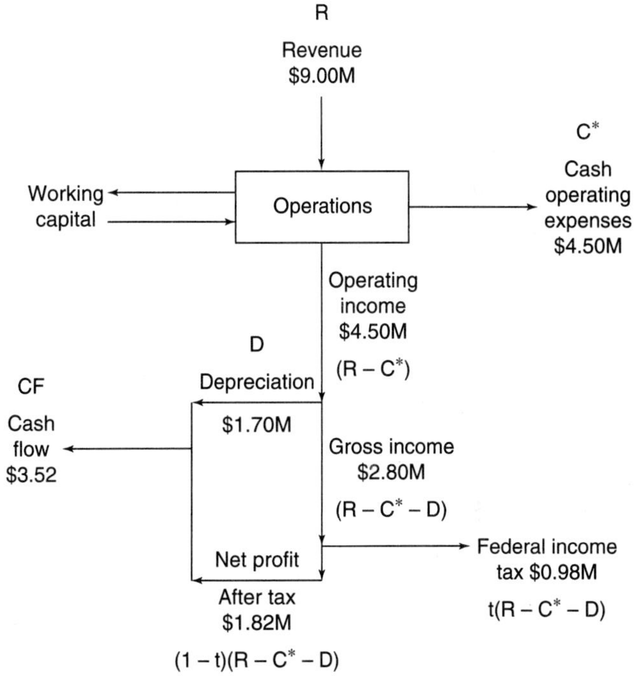

FIG. 9-10 Cash flow model for Example 9-7. M = million.

Time Zero at Start-Up Prior to time zero, expenditures are made for land, fixed capital investment, and working capital. It is assumed that land had been purchased by the company at some time in the past, and a parcel is allocated for the project under consideration. Land is allocated instantaneously to the project sometime prior to the purchase of equipment and construction of the plant. The fixed capital investment is purchased and installed over a period of time prior to start-up. For the purpose of this presentation, it is assumed that it occurs uniformly over a period of time. Both land and fixed capital investment are compounded to time zero by using the appropriate compound interest factors. At time zero, working capital is charged to the project. Start-up expenses are entered in the first year of operation after start-up. After time zero, start-up occurs and then manufacturing begins and income is generated, so cash flow begins to accumulate if the process is sound. At the end of the project life, land and working capital are recovered instantaneously.

TABLE 9-24 Cash Flow Analysis for Example 9-8

Year	−2	−2 to 0	0	1	2	3	4	5	6	7	8	9	10	End 10
Production Mlb/yr				40	42	45	48	50	50	47	45	40	35	
All money is $1000														
Land	−300													
Fixed capital investment $1000		−12,000												
Working capital			−1,800											
Start-up expenses			−750											
Total capital investment			14,250											
Sales				20,000	21,000	23,400	24,960	27,500	28,000	23,500	21,600	18,800	15,750	
Cash operating expenses				10,320	10,800	11,520	12,240	13,470	13,970	13,175	12,645	11,320	9,995	
Operating income				9,680	10,200	11,880	12,720	14,030	14,030	10,325	8,955	7,480	5,755	
Depreciation				2,400	3,840	2,300	1,390	1,380	690	0	0	0	0	
Net income before taxes				7,280	6,360	9,580	11,330	12,650	13,340	10,325	8,955	7,480	5,755	
Income tax				3,640	2,226	3,353	3,966	4,428	4,669	3,614	3,134	2,618	2,014	
Net income after taxes				3,640	4,134	6,227	7,364	8,222	8,671	6,711	5,821	4,862	3,741	
Depreciation				2,400	3,840	2,300	1,390	1,380	690	0	0	0	0	
After-tax cash flow	−300	−12,000	−2,550	6,040	7,974	8,527	8,754	9,602	9,361	6,711	5,821	4,882	3,741	
Cumulative cash flow	−300	−12,300	−14,850	−8,810	−836	7,691	16,445	26,047	35,408	42,119	47,940	52,802	56,543	
Capital recovery														2,100
End of project value														58,643

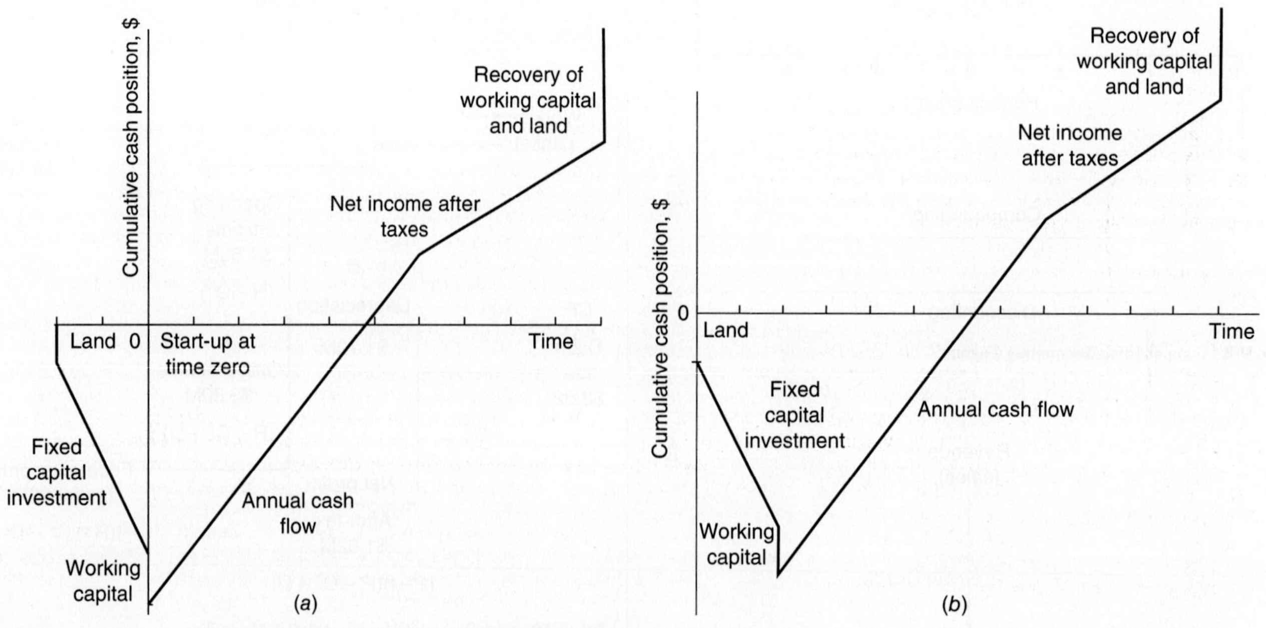

FIG. 9-11 Typical cumulative cash position plot. (*a*) Time zero is start-up. (*b*) Time zero occurs when first funds are spent.

PROFITABILITY

In the free enterprise system, companies are in business to make a profit. Management has the responsibility of investing in ventures that are financially attractive by increasing earnings, providing attractive rates of return, and increasing the value added of the company. Every viable business has limitations on the capital available for investment; therefore, it will invest in the most economically attractive ventures. The objectives and goals of a company are developed by management. Corporate objectives may include one or several of the following: maximize return on investment, maximize return on stockholders' equity, maximize aggregate earnings, maximize common stock prices, increase market share, increase economic value, increase earnings per share of stock, and increase market value added. These objectives are the ones most frequently listed by executives.

To determine the worthiness of a venture, quantitative and qualitative measures of profitability are considered.

QUANTITATIVE MEASURES OF PROFITABILITY

When a company invests in a venture, the investment must earn more than the cost of capital for it to be worthwhile. A profitability estimate is an attempt to quantify the desirability of taking a risk in a venture.

The *minimum acceptable rate of return* (MARR) for a venture depends on a number of factors such as interest rate, cost of capital, availability of capital, degree of risk, economic project life, and other competing projects. Management will adjust the MARR depending on any of the above factors to screen out the more attractive ventures. When a company invests in a venture, the investment must earn more than the cost of capital and should be able to pay dividends.

Although there have been many quantitative measures suggested through the years, some did not take into account the time value of money. In today's economy, the following measures are the ones most companies use:

Payout period (POP) plus interest
Net present worth (NPW)
Discounted cash flow rate of return (DCFROR)

Payout Period Plus Interest Payout period (POP) is the time that will be required to recover the depreciable fixed capital investment from the accrued after-tax cash flow of a project with no interest considerations. In equation format

$$\text{Payout period} = \frac{\text{depreciable fixed capital investment}}{\text{after-tax cash flow}} \tag{9-18}$$

This model does not take into account the time value of money, and no consideration is given to cash flows that occur in a project's later years after the depreciable investment has been recovered. A variation on this method includes interest, called payout period plus interest (POP + I); and the net effect is to increase the payout period. This variation accounts for the time value of money.

$$\text{Payout period plus interest (POP} + I) = \left(\frac{\text{depreciable fixed capital investment}}{\text{after-tax cash flow}}\right)_i \tag{9-19}$$

Neither of these methods makes provision for including land and working capital, and no consideration is given to cash flows that occur in a project's later years after the depreciable fixed investment has been recovered for projects that earn most of their profit in the early years.

Net Present Worth In the net present worth method, an arbitrary time frame is selected as the basis of calculation. This method is the measure many companies use, as it reflects properly the time value of money and its effect on profitability. In equation form

$$\text{Net present worth (NPW)} = \text{present worth of all cash inflows} - \text{present worth of all investment items} \tag{9-20}$$

When the NPW is calculated according to Eq. (9-20), if the result is positive, the venture will earn more than the interest (discount) rate used; conversely, if the NPW is negative, the venture earns less than that rate.

Discounted Cash Flow In the discounted cash flow method, all the yearly after-tax cash flows are discounted or compounded to time zero depending upon the choice of time zero. The following equation is used to solve for the interest rate i, which is the discounted cash flow rate of return (DCFROR).

$$\text{DCFROR} = \sum_{0}^{n} (\text{after-tax cash flows}) = 0 \tag{9-21}$$

Equation (9-21) may be solved graphically or analytically by an iterative trial-and-error procedure for the value of i, which is the discounted cash flow rate of return. It has also been known as the *profitability index*. For a project to be profitable, the interest rate must exceed the cost of capital.

The effect of interest on the cash position of a project is shown in Fig. 9-12. As interest increases, the time to recover the capital expenditures is increased.

In the chemical business, operating net profit and cash flow are received on a nearly continuous basis. Therefore, there is justification for using the condensed continuous interest tables, such as Table 9-22, in discounted cash flow calculations.

Example 9-9 Profitability Calculations Example 9-8 data are used to demonstrate these calculations. Calculate the following:

a. Payout period (POP)
b. Payout period with interest (POP + I)
c. NPW at a 30 percent interest rate
d. DCF rate of return

Solution:

a. From Table 9-25, the second column is the cash flow by years with no interest. The payout period occurs where the cumulative cash flow is equal to the fixed capital investment, $12,000,000 or 1.7 years.

b. In Table 9-25, the payout period at 30 percent interest occurs at 2.4 years.

c. The results of the present worth calculations for 20, 30, and 40 percent interest rates are tabulated. At 30 percent interest, the net present worth is $4,782,000, and since it is a positive figure, this means the project will earn more than 30 percent interest.

d. Discounted cash flow rate of return is determined by interpolating in Table 9-25. At 30 percent interest the net present worth is positive, and at 40 percent interest it is negative. By definition, the DCFROR occurs when the summation of the net present worth equals zero. This occurs at an interest of 33.9 percent.

QUALITATIVE MEASURES

In addition to quantitative measures, there are certain qualitative measures or intangible factors that may affect the ultimate investment decision. Those most frequently mentioned by management are employee morale, employee safety, environmental constraints, legal constraints, product liability, corporate image, and management goals. Attempts have been made to quantify these intangibles by using an ordinal or a ranking system, but most have had little or no success. Couper (2003) discussed in greater detail the effect of qualitative measures on the decision-making process.

SENSITIVITY ANALYSIS

Whenever an economic study is prepared, the marketing, capital investment, and operating expense data used are estimates, and therefore a degree of uncertainty exists. Questions arise such as, What if the capital investment is 15 percent greater than the value reported? A sensitivity analysis is used to determine the effect of percentage changes in pertinent variables on the profitability of the project. Such an analysis indicates which variables are most susceptible to change and need further study.

Break-Even Analysis Break-even analysis is a simple form of sensitivity analysis and is a useful concept that can be of value to managers. *Break-even* refers to the point in an operation where income just equals expenses. Figure 9-13 is a pictorial example of the results of a

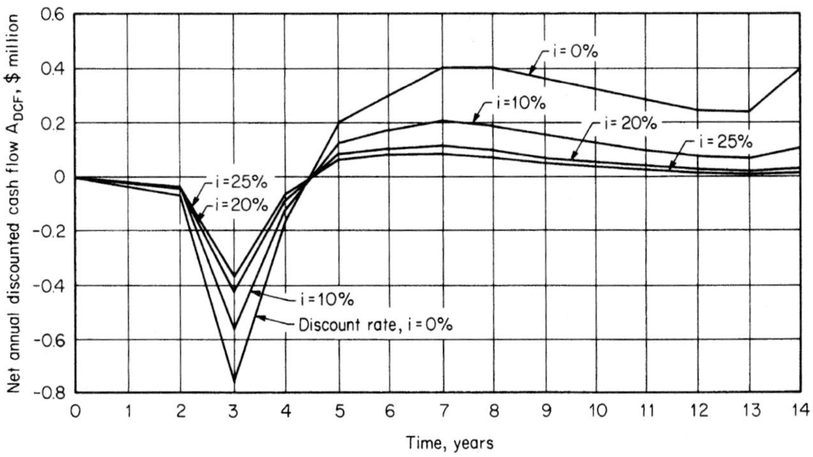

FIG. 9-12 Effect of interest rate on cash flow (time zero occurs when first funds are expanded).

TABLE 9-25 Profitability Analysis for Example 9-9

Time	Cash flow	20% interest factors	Present worth 20% interest	30% interest factors	Present worth 30% interest	40% interest factors	Present worth 40% interest
−2	−300	1.492	−448	1.822	−547	2.226	−668
−2 to 0	−12,000	1.230	−14,760	1.370	−16,440	1.532	−18,384
0	−2,250	1.000	−2,250	1.000	−2,250	1.000	−2,250
1	6,040	0.906	5,472	0.864	5,219	0.824	4,977
2	7,974	0.742	5,917	0.640	5,103	0.552	4,402
3	8,527	0.608	5,184	0.474	4,042	0.370	3,155
4	8,754	0.497	4,351	0.351	3,073	0.248	2,171
5	9,602	0.407	3,908	0.260	2,497	0.166	1,594
6	9,361	0.333	3,117	0.193	1,807	0.112	1,048
7	6,711	0.273	1,832	0.143	960	0.075	503
8	5,821	0.224	1,304	0.106	617	0.050	291
9	4,862	0.183	890	0.078	379	0.034	165
10	3,471	0.150	561	0.058	217	0.022	82
End 10	2,100	0.135	284	0.050	105	0.018	284
Net present worth			15,362		4,782		−2,360
Discounted cash flow rate of return		33.90%					

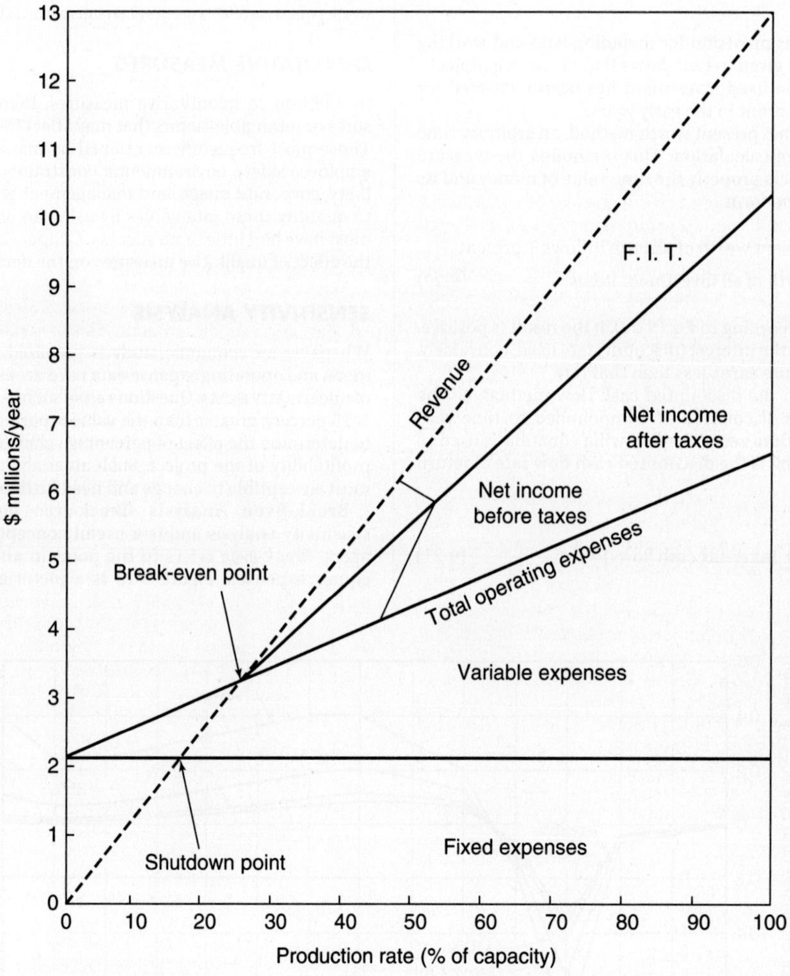

FIG. 9-13 Break-even plot.

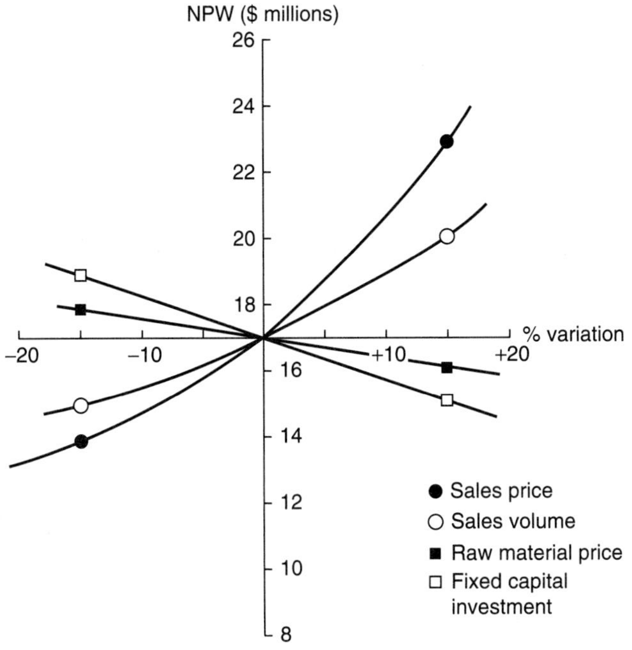

FIG. 9-14 Strauss plot.

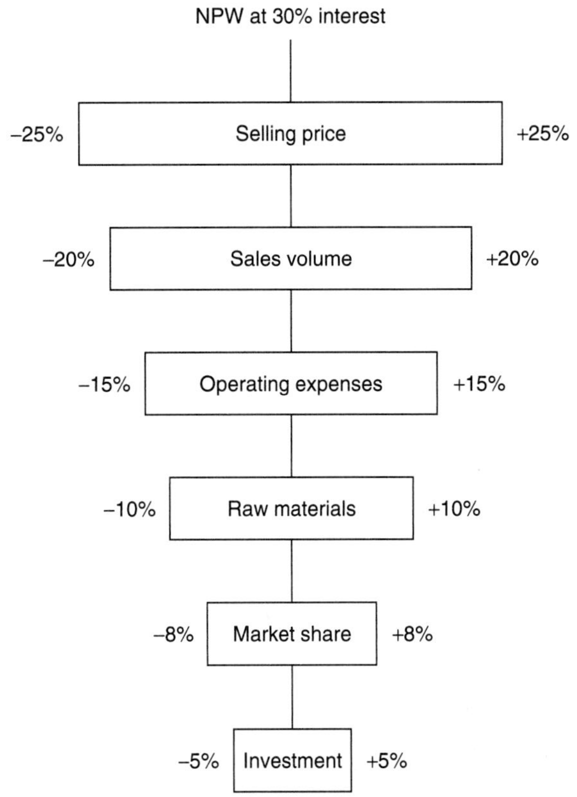

FIG. 9-15 Typical tornado plot. (*Source: Adapted from Couper, 2003.*)

break-even analysis, showing that the break-even point is at 26 percent of production capacity. Management wants to do better than just break even; therefore, such plots can be used as a profit planning tool, for product pricing, production operating level, incremental equipment costs, etc. Another significant point is the shutdown point where revenue just equals the fixed expenses. Therefore, if a proposed operation can't make fixed expenses, it should be shut down.

Strauss Plot R. Strauss (*Chem. Eng.,* pp. 112–116, Mar. 25, 1968) developed a sensitivity plot, in Fig. 9-14, in which the ordinate is a measure of profitability and the abscissa is the change in a variable greater than (or less than) the value used in the base case. Where the abscissa crosses the ordinate is the result of the base case of NPW, return, annual worth, etc. The slope of a line on this "spider" plot is the degree of change in profitability resulting from a change in a variable, selling price, sales volume, investment, etc. The length of the line represents the sensitivity of the variable and its degree of uncertainty. Positive-slope lines are income-related, and negative-slope lines are expense-related. A spreadsheet is useful in developing data for this "what if" plot since numerous scenarios must be prepared to develop the plot.

Tornado Plot Another graphical sensitivity analysis is the "tornado" plot. Its name is derived from the shape of the resulting envelope. As in other methods, a base case is solved first, usually expressing the profitability as the net present worth. In Fig. 9-15, the NPW is a vertical line, and variations in each selected variable above and below the base case are solved and plotted. In this figure, the variables of selling price, sales volume, operating expenses, raw material expenses, share of the market, and investment are plotted. It is apparent that the selling price and sales volume are the critical factors affecting the profitability. A commercial computer program known as @RISK® developed by the Palisade Corporation, Newfield, N.Y., may be used to prepare a tornado plot.

Relative Sensitivity Plot Another type of analysis developed by J. C. Agarwal and I. V. Klumpar (*Chem. Eng.*, pp. 66–72, Sept. 29, 1975) is the relative sensitivity plot. The variables studied are related to those in the base case, and the resulting plot is the relative profitability.

Although sensitivity analyses are easy to prepare and they yield useful information for management, there is a serious disadvantage. Only one variable at a time can be studied. Frequently, there are synergistic effects among variables; e.g., in marketing, variables such as sales volume, selling price, and market share may have a synergistic effect, and that effect cannot be taken into account. Other interrelated variables such as fixed capital investment, maintenance, and other investment-based items also cannot be represented properly. These disadvantages lead to another management tool—uncertainty analysis.

UNCERTAINTY ANALYSIS

This analysis allows the user to account for variable interaction that is another level of sophistication. Two terms need clarification—uncertainty and risk. *Uncertainty* is exactly what the word means—not certain. *Risk,* however, implies that the probability of achieving a specific outcome is known within certain confidence limits.

Since sensitivity analysis has the shortcoming of being able to inspect only one variable at a time, the next step is to use probability risk analysis, generally referred to as the Monte Carlo technique. R. C. Ross (*Chem. Eng.,* pp. 149–155, Sept. 20, 1971), P. Macalusa (BYTE, pp. 179–192, March 1984), and D. B. Hertz (*Harvard Bus. Rev.,* pp. 96–108, Jan.-Feb. 1968) have written classic articles on the use of the Monte Carlo technique in uncertainty analysis. These articles incorporate subjective probabilities and assumptions of the distribution of errors into the analysis. Each variable is represented by a probability distribution model. Figure 9-16 is a pictorial representation of the steps in the Monte Carlo simulation. The first step is to gather enough data to develop a reasonable probability model. Not all variables follow the normal distribution curve, but perhaps sales volume and sales-related variables do. Studies have shown that capital investment estimates are best represented by a beta distribution. Next the task is to select random values from the various models by using a random number generator and from these data calculate a profitability measure such as NPW or rate of return. The procedure is repeated a number of times to generate a plot of the probability of achieving a given profitability versus profitability. Figure 9-17 is a typical plot. Once the analysis has been performed, the next task is to interpret the results. Management must understand what the results mean and the reliability of the results. Experience can be gained only by performing uncertainty analyses, not just one or two attempts, to develop confidence in the process. The stakes may be high enough to spend time and learn the method. Software companies such as @RISK or SAS permit the user to develop probability models and perform the Monte Carlo analysis. The results may be plotted as the probability of achieving at least a given return or of achieving less than the desired profitability.

FEASIBILITY ANALYSIS

A feasibility analysis is prepared for the purpose of determining that a proposed investment meets the minimum requirements established by management.

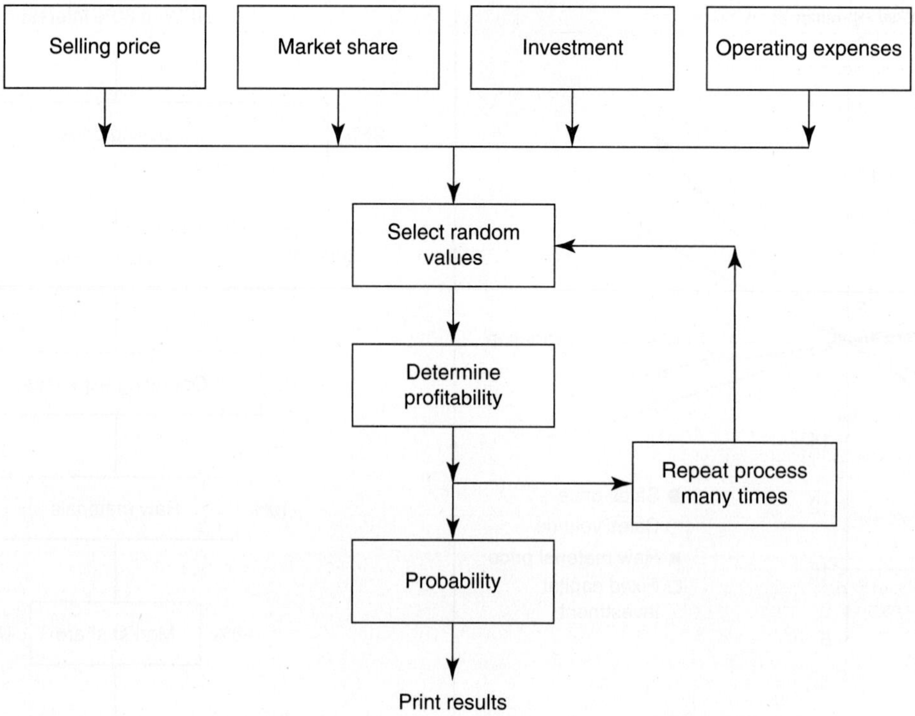

FIG. 9-16 Schematic diagram of Monte Carlo simulation.

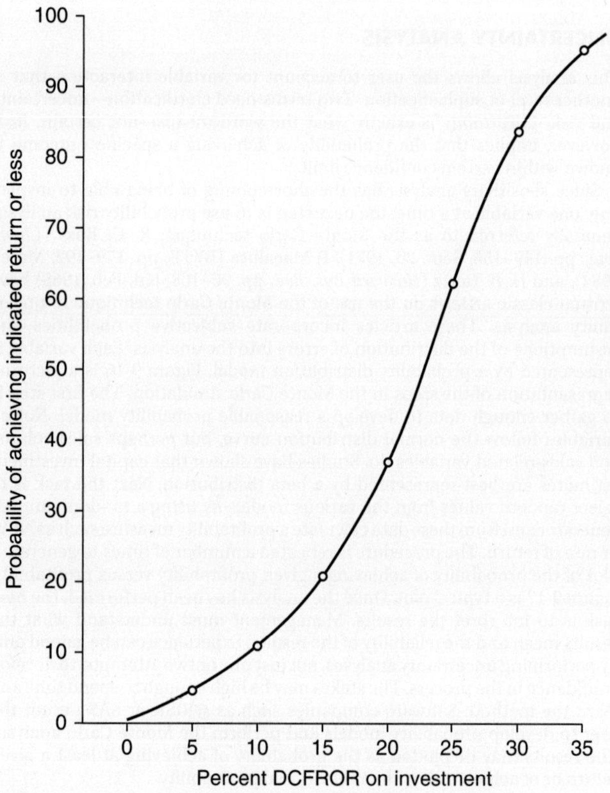

FIG. 9-17 Probability curve for Monte Carlo simulation.

TABLE 9-26 Checklist of Required Information for a Feasibility Analysis

Fixed capital investment
Working capital requirements
Total capital investment
Total manufacturing expense
Packaging and in-plant expense
Total product expense
General overhead expense
Total operating expense
Marketing data
Cash flow analysis
Project profitability
Sensitivity analysis
Uncertainty analysis

It should be in sufficient detail to provide management with the facts required to make an investment decision. All the basic information has been discussed in considerable detail in the earlier parts of Sec. 9.

The minimum information required should include, but not be limited to, that in Table 9-26. Forms and spreadsheets are the most succinct method to present the information. The forms should state clearly the fund amounts and the date that each estimate was performed. The forms may be developed so that data for other scenarios may be reported by extending the tables to the right of the page. It is suggested that blank lines be included for any additional information. Finally the engineer preparing the feasibility analysis should make recommendations based upon management's guidelines.

The development of the information required for Table 9-26 was discussed previously in Sec. 9 with the exception of marketing information. An important document for a feasibility analysis is the marketing data so that the latest income projections can be included for management's consideration. As a minimum, the tabulation of sales volume, sales prices, and market share both domestically and globally should be included. Table 9-27 shows a sample of such marketing information.

Other templates may be prepared for total capital investment, working capital, total product expense, general overhead expense, and cash flow. Table 9-28 may be used to organize cash flow data by showing investment, operating expenses, cash flow, and cumulative cash flow.

TABLE 9-27 Marketing Data Template

Project title:
Basis: Sales and marketing data are not inflated (20__ dollars)

		20__		20__		20__
	Amount	%Total	Amount	%Total	Amount	%Total
Total market						
Units						
Average realistic price, $/unit						
Value, $						
Estimated product sales (with AR)						
Units						
Average realistic price, $/unit						
Value, $						
Current product sales (without AR)						
Units						
Average realistic price, $/unit						
Value, $						
Incremental product sales						
Units						
Average realistic price, $/unit						
Value, $						
Current product sales displaced by improved product sales						
Units						
Value, $						
Total improved product sales						
Units						
Value, $						

NOTE: Table extends to the right to accommodate the number of project years.
AR = appropriation request.

TABLE 9-28 Cash Flow Analysis Template

	Cash flow summary		
	200X	200Y	200Z, etc.*
Investment			
Land			
Fixed capital investment			
Offsite capital			
Allocated capital			
Working capital			
Start-up expenses			
Interest			
Catalysts and chemicals			
Licenses, patents, etc.			
Total capital investment			
Income statement			
Income			
Expenses			
Cash operating expenses			
Depreciation			
Total operating expenses			
Operating income			
Net income before taxes			
Federal income taxes			
Net income after taxes			
Cash flow			
Capital recovery			
Cumulative cash flow			

*Table may be extended to the right to accommodate the number of years of the project.

OTHER ECONOMIC TOPICS

COMPARISON OF ALTERNATIVE INVESTMENTS

Engineers are often confronted with making choices between alternative equipment, designs, procedures, plans, or methods. The courses of action require different amounts of capital and different operating expenses. Some basic concepts must be considered before attempting to use mathematical methods for a solution. It is necessary to clearly define the alternatives and their merits. Flow of money takes the form of expenditures or income. Savings from operations are considered as income or as a reduction in operating expenses. Income taxes and inflation as well as a reasonable return on the investment must be included. Money spent is negative and money earned or saved is positive.

Expenditures are of two kinds; instantaneous like land, working capital and capital recovery or uniformly continuous for plant investment, operating expenses, etc. A methodology involving after-tax cash flow is developed to reduce all the above to a manageable format.

In an earlier part of this section, after-tax cash flow was defined as

$$\mathrm{CF} = D + (R - C^* - D)(1 - t) \tag{9-17}$$

where CF = after-tax cash flow
D = depreciation
R = revenue
C^* = cash operating expenses
t = tax rate

Several methods are available for determining the choice among alternatives:

Net present worth
Rate of return
Capitalized cost
Cash flow
Uniform annual cost

Humphreys in Jelen and Black, *Cost and Optimization Engineering* (1991), has shown that each of these methods would result in the same decision, but the numerical results will differ.

Net Present Worth Method The NPW method allows the conversion of all money flows to be discounted to the present time. Appropriate interest factors are applied depending on how and when the cash flow enters a venture. They may be instantaneous, as in the purchase of capital equipment, or uniform, as in operating expenses. The alternative with the more positive NPW is the one to be preferred. In some instances, the alternatives may have different lives so the cost analysis must be for the least common multiple number of years. For example, if alternative A has a 2-year life and alternative B has a 3-year life, then 6 years is the least common multiple. The rate of return, capitalized cost, cash flow, and uniform annual cost methods avoid this complication.

Rate of return and capitalized cost methods are discussed at length in Humphreys (1991).

Cash Flow Method Cash flows for each case are determined, and the case that generates the greater cash flow is the preferred one.

Uniform Annual Cost (UAC) Method In the uniform annual cost method, the cost is determined over the entire estimated project life. The least common multiple does not have to be calculated, as in the NPW method. This is the advantage of the UAC method; however, the result obtained by this method is more meaningful than the results obtained by other methods.

The UAC method begins with a calculation for each alternative. If discrete interest is used, the annual cost C is found by multiplying the present worth P by the appropriate discrete interest factor, found in Table 9-20, for the number of years n and the interest rate i. If continuous interest is preferred, the UAC equation is

$$\text{UAC} = \frac{\text{NPW}}{(\text{years of life})\,(\text{continuous interest factor})} \tag{9-22}$$

The continuous interest factor may be found from continuous interest equations or from the continuous interest table, Table 9-29. In this table time zero is the present, and all cash flows are discounted back to the present. Note that there are three sections to this table, depending on the cash flow: uniform, instantaneous, or declining uniformly to zero. One enters the table with the argument $R \times T$, where R is the interest rate expressed as a whole number and T is the time in years to obtain a factor. This factor is then used to calculate the present worth of the cash flow item. All cash flows are summed algebraically, giving the net present worth, which is substituted in Eq. (9-22). This procedure is followed for both alternatives, and the alternative that yields the more positive UAC (or the least negative) value is the preferred alternative. In Eq. (9-22) the "factor" is always the uniform factor that annualizes all the various cash flows.

This method of comparing alternatives is demonstrated in Example 9-10.

Example 9-10 Choice among Alternatives Two filters are considered for installation in a process to remove solids from a liquid discharge stream to meet environmental requirements. The equipment is to be depreciated over a 7-year period by the straight-line method. The income tax rate is 35 percent, and 15 percent continuous interest is to be used. Assume that the service life is 7 years and there is no capital recovery. Data for the two systems are as follows:

System	B	C
Fixed investment	$18,000	$30,000
Annual operating expenses	14,200	4,800

Which alternative is preferred?

Solution:

System B:

Year	Item	Cash flow, $	Factor	PW, $
0	Investment	−18,000	1.0	−18,000
0–7	Contribution to cash flow from depreciation	(0.35)(18,000)	0.6191	+3,900
0–7	Contribution to cash flow from operating expense	(1 − 0.35)(7)(14,200)	0.6191	−40,000
			NPW B	−54,100

$$\text{UAC}_B = \frac{\text{NPW}}{(\text{years of life})(\text{uniform factor})} = \frac{-\$54{,}100}{(7)(0.6191)} = -\$12{,}484$$

System C:

Year	Item	Cash flow, $	Factor	PW, $
0	Investment	−30,000	1.0	−30,000
0–7	Contribution to cash flow from depreciation	(0.35)(30,000)	0.6191	+6,500
0–10	Contribution to cash flow from operating expense	(1 − 0.35)(10)(4,800)	0.6191	−19,316
			NPW C	−42,816

$$\text{UAC}_C = \frac{\text{NPW}}{(\text{years of life})(\text{uniform factor})} = \frac{-\$42{,}816}{(10)(0.6191)} = -\$6{,}916$$

Alternative C is preferred because it has the more positive UAC.

REPLACEMENT ANALYSIS

During the lifetime of a physical asset, continuation of its use may make it a candidate for replacement. In this type of analysis, a replacement is intended to supplant a similar item performing the same service without plant or equipment expansion. In a chemical plant, *replacement* usually refers to a small part of the processing equipment such as a heat exchanger, filter, or compressor. If the replacement is required due to "physical" deterioration, there is no question of whether to replace the item, but the entire plant may be shut down if it is not replaced. The problem then becomes whether the equipment should be replaced like for like or whether an alternative should be chosen that may be different in cost and/or efficiency. If the replacement is due to technical obsolescence, the timing of the replacement may be important, especially if a plant expansion may be imminent in the near future. Whatever the situation, the replaced item should not present a bottleneck to the processing. The engineer should understand replacement theory to determine if alternative equipment is adequate for the job but with different costs and timing.

Certain terminology has been developed to identify the equipment under consideration. The item in place is called the *defender*, and the candidate for replacement is called the *challenger*. This terminology and methodology was reported by E. L. Grant and W. G. Ireson in *Engineering Economy*, Wiley, New York, 1950. To apply this method, there are certain rules. The value of the defender asset is a sunk cost and is irrelevant except insofar as it affects cash flow from depreciation for the rest of its life and a tax credit for the book loss if it is replaced sooner than its depreciation life. A capital cost for the defender is the net capital recovery forgone and the tax credit from the book loss of the defender asset that was not realized. The UAC method will be used and will be computed for each case, using the time period most favorable to each. For the defender it is 1 year, and for the challenger it is the full economic life. The UAC for the challenger is handled in the same manner as in the comparison of alternatives. The method is demonstrated in Example 9-11.

Example 9-11 Replacement Analysis A 3-year-old reciprocating compressor is being considered for replacement. Its original cost was $150,000, and it was being depreciated over a 7-year period by the straight-line method. If it is replaced now, the net proceeds from its sale are $50,000, and it is believed that 1 year from now they will be $35,000. A new centrifugal compressor can be installed for $160,000, which would save the company $2000 per year in operating expenses for the 10-year life. At the end of the 10th year, its net proceeds are estimated to be zero. The 7-year depreciation applies also to the centrifugal compressor. A 35 percent tax rate may be assumed. The company requires a 15 percent after-tax return on an investment of this type. Should the present compressor be replaced now?

Solution: The UAC method will be used as a basis for comparison. It is assumed that all money flows are continuous, and continuous interest will be used.

Defender case: The basis for this unit will be 1 year. If it is not replaced now, the rules listed above indicate that there is an equivalent of a capital cost for two benefits forgone (given up). They are

1. Net proceeds now at 3 years of $50,000
2. Tax credit for the loss not realized

Thus net loss forgone = book value at the end of 3 years minus net capital recovery, or

$$\text{NLF}_3 = \text{BV}_3 - \text{NCR}_3$$

where NLF_3 = net loss forgone at end of 3 years
BV_3 = book value at end of 3 years
NCR_3 = net capital recovery at end of 3 years

$$\text{NLF}_3 = \$150{,}000\left(1 - \frac{1}{7}\right) - \$50{,}000 = \$35{,}714$$

TABLE 9-29 Factors for Continuous Discounting

$R \times T$	0	1	2	3	4	5	6	7	8	9
					Uniform					
0	1.0000	.9950	.9901	.9851	.9803	.9754	.9706	.9658	.9610	.9563
10	.9516	.9470	.9423	.9377	.9332	.9286	.9241	.9196	.9152	.9107
20	.9063	.9020	.8976	.8933	.8891	.8848	.8806	.8764	.8722	.8681
30	.8639	.8598	.8558	.8517	.8477	.8437	.8398	.8359	.8319	.8281
40	.8242	.8204	.8166	.8128	.8090	.8053	.8016	.7979	.7942	.7906
50	.7869	.7833	.7798	.7762	.7727	.7692	.7657	.7622	.7588	.7554
60	.7520	.7486	.7453	.7419	.7386	.7353	.7320	.7288	.7256	.7224
70	.7192	.7160	.7128	.7097	.7066	.7035	.7004	.6974	.6944	.6913
80	.6883	.6854	.6824	.6795	.6765	.6736	.6707	.6679	.6650	.6622
90	.6594	.6566	.6538	.6510	.6483	.6455	.6428	.6401	.6374	.6348
100	.6321	.6295	.6269	.6243	.6217	.6191	.6166	.6140	.6115	.6090
110	.6065	.6040	.6015	.5991	.5967	.5942	.5918	.5894	.5871	.5847
120	.5823	.5800	.5777	.5754	.5731	.5708	.5685	.5663	.5640	.5618
130	.5596	.5574	.5552	.5530	.5509	.5487	.5466	.5444	.5423	.5402
140	.5381	.5361	.5340	.5320	.5299	.5279	.5259	.5239	.5219	.5199
150	.5179	.5160	.5140	.5121	.5101	.5082	.5063	.5044	.5025	.5007
160	.4988	.4970	.4951	.4933	.4915	.4897	.4879	.4861	.4843	.4825
170	.4808	.4790	.4773	.4756	.4738	.4721	.4704	.4687	.4671	.4654
180	.4637	.4621	.4604	.4588	.4572	.4555	.4539	.4523	.4507	.4492
190	.4476	.4460	.4445	.4429	.4414	.4399	.4383	.4368	.4353	.4338
200	.4323	.4309	.4294	.4279	.4265	.4250	.4236	.4221	.4207	.4193
210	.4179	.4165	.4151	.4137	.4123	.4109	.4096	.4082	.4069	.4055
220	.4042	.4029	.4015	.4002	.3989	.3976	.3963	.3950	.3937	.3925
230	.3912	.3899	.3887	.3874	.3862	.3849	.3837	.3825	.3813	.3801
240	.3789	.3777	.3765	.3753	.3741	.3729	.3718	.3706	.3695	.3683
250	.3672	.3660	.3649	.3638	.3627	.3615	.3604	.3593	.3582	.3571
260	.3560	.3550	.3539	.3528	.3518	.3507	.3496	.3486	.3476	.3465
270	.3455	.3445	.3434	.3424	.3414	.3404	.3394	.3384	.3374	.3364
280	.3354	.3344	.3335	.3325	.3315	.3306	.3296	.3287	.3277	.3268
290	.3259	.3249	.3240	.3231	.3222	.3212	.3203	.3194	.3185	.3176
300	.3167	.3158	.3150	.3141	.3132	.3123	.3115	.3106	.3098	.3089
310	.3080	.3072	.3064	.3055	.3047	.3039	.3030	.3022	.3014	.3006
320	.2998	.2990	.2982	.2974	.2966	.2958	.2950	.2942	.2934	.2926
330	.2919	.2911	.2903	.2896	.2888	.2880	.2873	.2865	.2858	.2850
340	.2843	.2836	.2828	.2821	.2814	.2807	.2799	.2792	.2785	.2778
350	.2771	.2764	.2757	.2750	.2743	.2736	.2729	.2722	.2715	.2709
360	.2702	.2695	.2688	.2682	.2675	.2669	.2662	.2655	.2649	.2642
370	.2636	.2629	.2623	.2617	.2610	.2604	.2598	.2591	.2585	.2579
380	.2573	.2567	.2560	.2554	.2548	.2542	.2536	.2530	.2524	.2518
390	.2512	.2506	.2500	.2495	.2489	.2483	.2477	.2471	.2466	.2460
400	.2454	.2449	.2443	.2437	.2432	.2426	.2421	.2415	.2410	.2404
					Instantaneous					
0	1.0000	.9900	.9802	.9704	.9608	.9512	.9418	.9324	.9231	.9139
10	.9048	.8958	.8869	.8781	.8694	.8607	.8521	.8437	.8353	.8270
20	.8187	.8106	.8025	.7945	.7866	.7788	.7711	.7634	.7558	.7483
30	.7408	.7334	.7261	.7189	.7118	.7047	.6977	.6907	.6839	.6771
40	.6703	.6637	.6570	.6505	.6440	.6376	.6313	.6250	.6188	.6126
50	.6065	.6005	.5945	.5886	.5827	.5769	.5712	.5655	.5599	.5543
60	.5488	.5434	.5379	.5326	.5273	.5220	.5169	.5117	.5066	.5016
70	.4966	.4916	.4868	.4819	.4771	.4724	.4677	.4630	.4584	.4538
80	.4493	.4449	.4404	.4360	.4317	.4274	.4232	.4190	.4148	.4107
90	.4066	.4025	.3985	.3946	.3906	.3867	.3829	.3791	.3753	.3716
100	.3679	.3642	.3606	.3570	.3535	.3499	.3465	.3430	.3396	3362
110	.3329	.3296	.3263	.3230	.3198	.3166	.3135	.3104	.3073	.3042
120	.3012	.2982	.2952	.2923	.2894	.2865	.2837	.2808	.2780	.2753
130	.2725	.2698	.2671	.2645	.2618	.2592	.2567	.2541	.2516	.2491
140	.2466	.2441	.2417	.2393	.2369	.2346	.2322	.2299	.2276	.2254
150	.2231	.2209	.2187	.2165	.2144	.2122	.2101	.2080	.2060	.2039
160	.2019	.1999	.1979	.1959	.1940	.1920	.1901	.1882	.1864	.1845
170	.1827	.1809	.1791	.1773	.1755	.1738	.1720	.1703	.1686	.1670
180	.1653	.1637	.1620	.1604	.1588	.1572	.1557	.1541	.1526	.1511
190	.1496	.1481	.1466	.1451	.1437	.1423	.1409	.1395	.1381	.1367
200	.1353	.1340	.1327	.1313	.1300	.1287	.1275	.1262	.1249	.1237
210	.1225	.1212	.1200	.1188	.1177	.1165	.1153	.1142	.1130	.1119
220	.1108	.1097	.1086	.1075	.1065	.1054	.1044	.1033	.1023	.1013
230	.1003	.0993	.0983	.0973	.0963	.0954	.0944	.0935	.0926	.0916
240	.0907	.0898	.0889	.0880	.0872	.0863	.0854	.0846	.0837	.0829
250	.0821	.0813	.0805	.0797	.0789	.0781	.0773	.0765	.0758	.0750
260	.0743	.0735	.0728	.0721	.0714	.0707	.0699	.0693	.0686	.0679
270	.0672	.0665	.0659	.0652	.0646	.0639	.0633	.0627	.0620	.0614
280	.0608	.0602	.0596	.0590	.0584	.0578	.0573	.0567	.0561	.0556
290	.0550	.0545	.0539	.0534	.0529	.0523	.0518	.0513	.0508	.0503
300	.0498	.0493	.0488	.0483	.0478	.0474	.0469	.0464	.0460	.0455
310	.0450	.0446	.0442	.0437	.0433	.0429	.0424	.0420	.0416	.0412

TABLE 9-29 Factors for Continuous Discounting *(Continued)*

$R \times T$	0	1	2	3	4	5	6	7	8	9
					Instantaneous					
320	.0408	.0404	.0400	.0396	.0392	.0388	.0384	.0380	.0376	.0373
330	.0369	.0365	.0362	.0358	.0354	.0351	.0347	.0344	.0340	.0337
340	.0334	.0330	.0327	.0324	.0321	.0317	.0314	.0311	.0308	.0305
350	.0302	.0299	.0296	.0293	.0290	.0287	.0284	.0282	.0279	.0276
360	.0273	.0271	.0268	.0265	.0263	.0260	.0257	.0255	.0252	.0250
370	.0247	.0245	.0242	.0240	.0238	.0235	.0233	.0231	.0228	.0226
380	.0224	.0221	.0219	.0217	.0215	.0213	.0211	.0209	.0207	.0204
390	.0202	.0200	.0198	.0196	.0194	.0193	.0191	.0189	.0187	.0185
400	.0183	.0181	.0180	.0178	.0176	.0174	.0172	.0171	.0169	.0167
					Declining Uniformly to 0					
0	1.0000	.9968	.9934	.9902	.9867	.9836	.9803	.9771	.9739	.9707
10	.9675	.9643	.9612	.9580	.9549	.9518	.9487	.9457	.9426	.9396
20	.9365	.9335	.9305	.9275	.9246	.9216	.9187	.9158	.9129	.9100
30	.9071	.9042	.9013	.8985	.8957	.8929	.8901	.8873	.8845	.8817
40	.8790	.8763	.8735	.8708	.8681	.8655	.8628	.8601	.8575	.8549
50	.8522	.8496	.8470	.8445	.8419	.8393	.8368	.8343	.8317	.8292
60	.8267	.8242	.8218	.8193	.8169	.8144	.8120	.8096	.8072	.8048
70	.8024	.8000	.7977	.7953	.7930	.7906	.7883	.7860	.7837	.7814
80	.7792	.7769	.7746	.7724	.7702	.7679	.7657	.7635	.7613	.7591
90	.7570	.7548	.7526	.7505	.7484	.7462	.7441	.7420	.7399	.7378
100	.7358	.7337	.7316	.7296	.7275	.7255	.7235	.7215	.7195	.7175
110	.7155	.7135	.7115	.7096	.7076	.7057	.7038	.7018	.6999	.6980
120	.6961	.6942	.6923	.6905	.6886	.6867	.6849	.6830	.6812	.6794
130	.6776	.6757	.6739	.6721	.6704	.6686	.6668	.6650	.6633	.6615
140	.6598	.6581	.6563	.6546	.6529	.6512	.6495	.6478	.6461	.6445
150	.6428	.6411	.6395	.6378	.6362	.6345	.6329	.6313	.6297	.6281
160	.6265	.6249	.6233	.6217	.6202	.6186	.6170	.6155	.6139	.6124
170	.6109	.6093	.6078	.6063	.6048	.6033	.6018	.6003	.5988	.5973
180	.5959	.5944	.5929	.5915	.5900	.5886	.5872	.5857	.5843	.5829
190	.5815	.5801	.5787	.5773	.5759	.5745	.5731	.5718	.5704	.5690
200	.5677	.5663	.5650	.5636	.5623	.5610	.5596	.5583	.5570	.5557
210	.5544	.5531	.5518	.5505	.5492	.5480	.5467	.5454	.5442	.5429
220	.5417	.5404	.5392	.5379	.5367	.5355	.5342	.5330	.5318	.5306
230	.5294	.5282	.5270	.5258	.5246	.5234	.5223	.5211	.5199	.5188
240	.5176	.5165	.5153	.5142	.5130	.5119	.5108	.5096	.5085	.5074
250	.5063	.5052	.5041	.5029	.5018	.5008	.4997	.4986	.4975	.4964
260	.4953	.4943	.4932	.4922	.4911	.4900	.4890	.4879	.4869	.4859
270	.4848	.4838	.4828	.4818	.4807	.4797	.4787	.4777	.4767	.4757
280	.4747	.4737	.4727	.4717	.4707	.4698	.4688	.4678	.4669	.4659
290	.4649	.4640	.4630	.4621	.4611	.4602	.4592	.4583	.4574	.4564
300	.4555	.4546	.4537	.4527	.4518	.4509	.4500	.4491	.4482	.4473
310	.4464	.4455	.4446	.4438	.4429	.4420	.4411	.4402	.4394	.4385
320	.4376	.4368	.4359	.4351	.4342	.4334	.4325	.4317	.4308	.4300
330	.4292	.4283	.4275	.4267	.4259	.4251	.4242	.4234	.4226	.4218
340	.4210	.4202	.4194	.4186	.4178	.4170	.4162	.4154	.4147	.4139
350	.4131	.4123	.4115	.4108	.4100	.4092	.4085	.4077	.4070	.4062
360	.4055	.4047	.4040	.4032	.4025	.4017	.4010	.4003	.3995	.3988
370	.3981	.3973	.3966	.3959	.3952	.3945	.3937	.3930	.3923	.3916
380	.3909	.3902	.3895	.3888	.3881	.3874	.3867	.3860	.3854	.3847
390	.3840	.3833	.3826	.3820	.3813	.3806	.3799	.3793	.3786	.3779
400	.3773	.3766	.3760	.3753	.3747	.3740	.3734	.3727	.3721	.3714

SOURCE: Adapted and abridged from Couper, 2003.

$$\text{Depreciation for 4th year} = \$150{,}000\left(\frac{1}{7}\right) = \$21{,}429$$

$$\text{NLF}_4 = \text{BV}_4 - \text{NCR}_4 = \$150{,}000\left(1-\frac{4}{7}\right) - \$35{,}000 = \$29{,}286$$

Year	Item	Cash flow, $	Factor	PW, $
At 0[a]	Tax credit for net loss forgone	(0.35)(−35,714)	1.0	−12,500
At 0[a]	Net cash recovery forgone	−50,000	1.0	−50,000
0–1	Contribution to CF from depreciation	(0.35)(−21,429)	0.9286[b]	6,965
0–1	Contribution to CF from operating expense	(1 − 0.35)(−15,000)	0.9286[b]	−9,054
End 1	Tax credit for net loss	(0.35)(29,286)	0.8607[c]	8,822
End 1	Net cash	35,000	0.8607[c]	30,125
			NPW	−25,642

[a]For the defender case, 0 year is the end of the third year.
[b]From Table 9-29, uniform section, for the argument $R \times T = 15 \times 1 = 15$.
[c]From Table 9-29, instantaneous section, for the argument $R \times T = 15 \times 1 = 15$.

$$\text{UAC} = \frac{\text{NPW}}{(\text{years of life})(\text{uniform factor})} = \frac{-\$25{,}642}{(1)(0.9286)} = -\$27{,}614$$

Challenger case:

Year	Item	Cash flow, $	Factor	PW, $
0	First cost	−160,000	1.0	−160,000
0–7	Contribution to CF from depreciation	(0.35)(160,000)	0.6191[d]	56,000
0–10	Contribution	(1 − 0.35)(10)(−13,000)	0.5179[e]	−43,763
			NPW	−147,763

[d]From Table 9-29, uniform section, $R \times T = 15 \times 7 = 105$.
[e]From Table 9-29, uniform section, $R \times T = 15 \times 10 = 150$.

$$\text{UAC} = \frac{\text{NPW}}{(\text{years of life})(\text{uniform factor})} = \frac{-\$147{,}763}{(10)(0.5179)} = -\$28{,}531$$

The UAC for the defender case is less negative (more positive) than that for the challenger case; therefore, the defender should not be replaced now. But there will be a time in the near future when the defender should be replaced, as maintenance and deterioration will increase.

OPPORTUNITY COST

Opportunity cost refers to the cost or value that is forgone or given up because a proposed investment is undertaken, often used as a base case.

Perhaps the term should be lost opportunity. For example, the profit from production in obsolete facilities is an opportunity cost of replacing them with more efficient ones. In cost analysis on investments, an incremental approach is often used, and if it is applied properly, the correct cost analysis will result.

INTERACTIVE SYSTEMS

If the TE does not pass through a minimum or maximum, but continues to decline or to increase with the number of equipment items or equipment size, the next step is to look at the flow sheet for equipment upstream or downstream from the selected item. It may be necessary to group two or more items and treat them as one in the analysis. Such a system is said to be *interactive*, since more than one item affects the optimum results. An example of such an interactive system is the removal of nitrogen from helium in a natural gas stream. Carbon adsorption is a method for removing nitrogen, but compressors are also required since this is a high-pressure process. If one attempts to find the optimum operating pressure, optimizing on compressor pressure will not result in an optimum condition; and conversely, optimizing on the size of the carbon bed will not yield an optimum. This is an example of an interactive system. Therefore, to find the optimum pressure, both the size of the carbon bed and the compressor pressure must be considered together.

CAPITAL PROJECT EXECUTION AND ANALYSIS

Front-end loading (FEL) and *value-improving practices* (VIPs) are closely integrated project management practices that are performed primarily during the early stages of a project's life cycle. They have been proved statistically by Independent Project Analysis, Inc. (IPA) to be the most effective means for improving project profitability. FEL and VIPs have very different characteristics, which are presented in the following sections. Properly performed together, they both contribute to maximizing project performance by ensuring that matters which influence project profitability are considered in the most productive manner and at the optimal time.

FRONT-END LOADING

General References:

1. Porter, James B., E.I. DuPont de Nemours and Company, "DuPont's Role in Capital Projects," *Proceedings of Government/Industry Forum: The Owner's Role in Project Management and Pre-Project Planning*, Washington, D.C.: The National Academies Press, 2002.
2. KBR Compilation of clients' project terminology, 1995–2016.
3. Adapted from Paulson, Boyd C., "Designing to Reduce Construction Costs," *Journal of the Construction Division*, **102**(4): 587–592 (1976).
4. KBR FEL Experience, 2000 through 2016.
5. KBR internal compilation of client parameters, IPA parameters, and KBR FEL experience.
6. Independent Project Analysis, Inc., *Best Practical Requirements for IPA's FEL Index*, 2016.
7. Merrow, E.W., Independent Project Analysis, Inc., 25th IPMA World Congress, Redefining Project Management for Major Projects, October 2011, and IPA website; http://www.ipaglobal.com/webinar-the-7-deadly-sins-in-industrial-megaprojects, 2016.
8. Independent Project Analysis, Inc., "The Changing Role of Design Contractors: Their Effective Use in Project Definition," November 1993, data updated 2016.
9. Independent Project Analysis, Inc. website; http://www.ipaglobal.com/webinar-the-7-deadly-sins-in-industrial-megaprojects, 2016.
10. Merrow, E. W., Independent Project Analysis, Inc., ECC Conference, Using PES™ Database of Projects Authorized after 1992, September 1998, data updated 2016.
11. Construction Industry Institute, University of Texas at Austin, CII Special Publication 268-3, *Adding Value Through Front-End Planning*, 2012.
12. Merrow, E. W., Independent Project Analysis, Inc., 32d Annual Engineering & Construction Contracting Conference, September 2000, data updated 2016.
13. IPA and KBR FEL Experience, 2000 through 2016.

Introduction *Front-end loading* (FEL) is a specialized and adaptable work process that translates business opportunities into a technical reality by developing a sufficiently defined project scope of work and execution plan that satisfies the intended business objectives. The product of the FEL process is also a design-basis package of customized information used to support the production of detailed engineering design documents. Completion of the FEL design-basis package is required for the project *final investment decision* (FID) or project authorization. Project authorization is that point in the project life cycle where the owner organization commits the majority of the project's capital investment and contracts. FID is also the point where overall project risks have been identified and sufficiently mitigated to support project funding approval.

The term *front-end loading* was first coined by the DuPont Company in 1987 and is commonly used throughout the chemical, refining, and oil and gas industries (Porter, James B., E.I. DuPont de Nemours and Company). FEL starts when a project idea is first conceived by a group with a company such as research and development, facilities planning, or project planning. After the initial concept is organized into a project strategy, organized and collaborative interaction and development are required among the various project stakeholders to create and assemble a project design-basis package suitable for subsequent authorization.

Typically at the end of each FEL phase, there are decision gates often called *stage gates*. These decision gates are formally established by the operating company authorizing or funding the project. These formal gates allow for continuity across the enterprise for authorization of additional funding for the next phase of engineering and project definition. FID often follows company corroboration that the project has achieved or exceeded the minimum company requirements for level of definition, risk exposure, capital cost, investment rate of return, and execution planning. Figure 9-18 illustrates the typical decision gates or stage gates for capital projects.

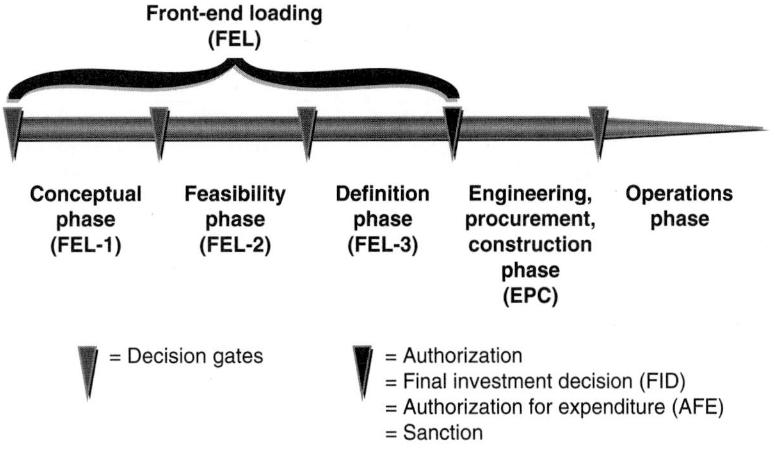

FIG. 9-18 FEL phases and decision gates.

Emphasis for Each FEL Phase There is a different emphasis for each FEL phase that builds on the previous phase deliverables and findings to produce a product or package that is sufficiently detailed to support the project FID.

Conceptual Phase (FEL-1) In FEL-1, the emphasis is to determine the basic economic viability of the conceptual project before committing to more-definitive engineering and study expense. The FEL-1 phase also emphasizes confirmation that the preferred conceptual process flow scheme and supporting economics fulfill the business case that caused its initiation. This phase also involves determining the suitable site or short list of candidate sites for the project that have the best combination of location attributes to best support the planned construction and operational environment that will fulfill the project business drivers. For each candidate site, a conceptual plot plan is generated that confirms the area requirements for processing units and systems, area to support construction operations, and operations and maintenance activities once the facility is operational. Once this work is defined adequately, generation of a representative cost estimate and milestone schedule for the entire project is possible.

Feasibility Phase (FEL-2) In FEL-2, the emphasis is to determine the best technical and economic flow scheme, associated technology, and support systems required to provide the necessary annual production rate at the sales quality required. In this phase, the final site is selected and preliminary geotechnical surveys are completed. A preliminary plot plan is generated that confirms the site area and infrastructure are adequate. This plot plan must be sufficiently detailed to support the FEL-2 cost estimate.

Definition Phase (FEL-3) In FEL-3, the emphasis is on achieving a "best practical" level of project definition that includes a good-quality project estimate, schedule, and a detailed EPC phase project execution plan that includes buy-in from key owner operations and maintenance staff. FEL-3 also involves process optimization to determine the best flow scheme and support systems combination. This optimum includes consideration of the plot plan and equipment arrangements for the entire facility. Process optimization cannot be done in isolation and depends on significant interaction with owner operations, maintenance, and construction experts to produce the best results. This level of project definition and planning for the post-FID project is normally required in order to present to management a candidate project that has the right combination of overall risk and projected economic performance, and thereby secure the project FID.

FEL Terminology Various FEL terms and definitions are used by operating and engineering companies today, which are often points of confusion. This situation results from differing terms for each FEL phase, differing levels of project definition ascribed to each phase, and differing project-level and discipline-level deliverables expected at the end of each FEL phase. Table 9-30 provides some insight to the differing FEL terminology in use today. Since these terms change periodically, due diligence is important to confirm definitions of FEL terms as they relate to the planned project work. Depending on what FEL phase is involved and how the owner organization has defined it, information expected prior to starting the work can be significantly more or less defined than originally expected. Therefore, clarity must be sought and confirmed as to the amount of work required versus originally planned to produce the intended deliverables and their level of thoroughness or detail.

Project Changes Project changes are a reality in every project phase. The influence of changes on capital projects is a function of when those changes are identified and incorporated into the project scope of work. The earlier the change is considered and incorporated into the project

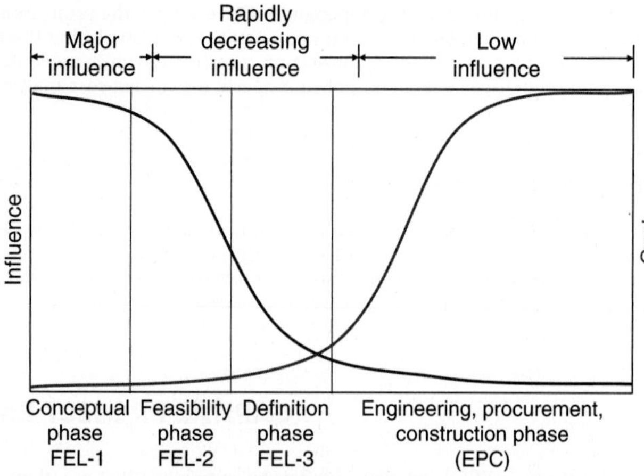

FIG. 9-19 Project life cycle cost-influence curve.

scope, the greater its potential influence on the project's profitability and the greater the ease of incorporating the change. Conversely, the later a change is identified, the more negative the impact on the project. Later changes such as those in the EPC phase are far more expensive to implement, are more disruptive, and are considered very undesirable. This disruption often delays completion of engineering deliverables which have strong statistical linkage to downstream procurement and construction work.

Figure 9-19 shows how quickly this influence curve changes as the typical project progresses. This is why proactively seeking changes during FEL is far more advantageous to project profitability than allowing those needed changes to be "discovered" during later project phases. This also means that potentially beneficial changes (value improvements) must be sought during FEL. If not, their likelihood of being cost effective to implement during the EPC phase may be very low. This is also why seeking owner operations, maintenance, and construction experience during FEL offers significant profitability advantages over practices which bring such experience onto the project team following FEL.

The FEL project team should proactively seek value improvement alternatives that challenge the project premises, scope, and design until such time that implementing those alternatives loses their economic and technical advantage. By doing so, such value improvements will not develop into costly corrections, which surface later, during the EPC phase.

Goals and Objectives of FEL The FEL work process enables a nearly constant consideration of changes as the work progresses. FEL phases also consider the long-term or life cycle cost implications of every aspect of the design and project execution. Predictability of equipment and process system life cycle costs must always be balanced with operations and maintenance preferences. Additional important goals and objectives of FEL projects are listed below.

- Develop a well-defined and acceptably profitable project.
- Define the primary technical and financial drivers for capital project investment.

TABLE 9-30 Project Life Cycle Terminology

FEL-1	FEL-2	FEL-3	EPC	Operations
Conceptual	Feasibility	Definition	EPC	Operations
Early pre-FEED*	Late pre-FEED*	FEED*	Execute	Operation
Appraise	Select	Define	Project execution	Operate
Identify	Selection	Develop	Execution	Start-up and operation
Business planning	Facility planning	Project planning	Implementation	Operate/evaluate
Assess opportunities	Evaluate alternatives	Development	EPC/start-up	
Assessment	Development	Validation		
Concept development		Scope finalization		
Feasibility		AFD†		
Appraisal/conceptual				
Assess				

*Front-end engineering design
†Authority For Development
SOURCE: KBR compilation of clients' project terminology, 1995–2016.

- Challenge baseline premises and purposely seek out and evaluate alternatives and opportunities.
- Minimize changes during the EPC, turnover, and start-up phases.
- Reduce project schedule and capital cost.
- Reduce business and project execution risk.
- Balance project technical, financial, and operational profitability drivers.

Comparison of FEL Projects and EPC Projects FEL projects are very different from EPC projects. Engineers and project managers having significant experience with projects only in the EPC phase often are unfamiliar with the significant differences and challenges of the FEL phase. One of the most important, but most subtle, aspects of FEL is the demand for more highly experienced staff and more sophisticated analysis tools when compared to EPC projects. This is so because of the need in FEL to create, analyze, and implement improvements to what many might already consider a "good" design.

In spite of its relatively short duration, FEL proactively seeks to implement the best possible design and execution plan. This changing environment requires people of many widely differing disciplines and functions to work together and communicate effectively. An integrated project team always seems to perform best during FEL, if it has well-established, informal, and personal interfaces between project groups and organizations. Additional important attributes of FEL phases are listed below.

- Focuses on owner's business objectives
- Emphasizes proactive elimination of low or zero value scope
- Demands significant access to and interaction with downstream experience such as operations, maintenance, construction, commissioning, and start-up
- Demands more experienced staff when compared to EPC projects
- Has far shorter time scale than EPC projects
- Focuses on overall project profitability rather than only cost, schedule, and work hours
- Demands higher level and frequency of communication than EPC projects
- Has far more interfaces than EPC projects
- Has owner-contractor management interfaces that are informal, close, and effective
- Has projects co-led by the owner and contractor which are more successful
- Requires greater development of personal relationships that result in respect and trust than EPC projects
- Demands continuous realignment of owner's desires and requirements with contractor needs

Table 9-31 lists further differences between FEL and EPC projects. Understanding these many differences is very important in that awareness of them and the driving forces behind them will prepare the project team member for the challenging and rewarding environment of FEL projects.

Parameters of FEL Phases Other important aspects of each phase of FEL are cost estimate accuracy, cumulative engineering hours spent, and the contingency assigned to the cost estimate. Table 9-32 lists the typical parameters encountered industrywide. For the capital cost estimate, each operating company may request a slightly different accuracy, which is often project-specific. What is important is the level of engineering required to support such estimating accuracy. This determination is the responsibility of both the owner and the engineering contractor. Agreement on this is critical prior to initiating project work.

FEL engineering hours spent are typically proportional to the project scope and equipment count. These engineering hours also vary widely between small and large projects apart from equipment count and project scope. Further, they tend to be proportional to the project complexity, extent of interfaces with other parties, whether new or emerging technology is being applied, or whether higher throughput capacities are being applied than previously commercially demonstrated. Projects such as these may require additional engineering to achieve the desired estimate accuracy, level of engineering definition, and project contingency desired.

Best FEL Project Performance Characteristics Overall project performance can be enhanced by ensuring that the following characteristics are emphasized during the FEL phases.

- Methodical business and project execution planning
- Effective integration of owner and engineering contractor workforces
- Projects with an integrated management team (owner and engineering contractor) have the lowest number of design changes at any project stage
- Engineering contractor should be brought onto the project in early FEL phases
- Clear roles for project team members which relate to the expertise of both owner and contractor staff
- Effective personal communication between owner and contractor organizations and their project team representatives, ensuring extensive site and manufacturing input
- Schedule and cost goals set by an integrated business and technical project team composed of owner and contractor representatives

Best FEL performance targets should be actively pursued by the project management team because the level of FEL definition is a leading indicator of post-FID project performance. When benchmarking projects, consultants such as IPA measure numerous aspects of the project to allow statistical comparison to previously completed projects. IPA's FEL Index is one of the most important such indices. The target for this index is the "Best Practical" rating whose characteristics or indicators are described briefly below.

IPA defines their Best Practical FEL Index rating by ensuring that the project has identified all potential project risks and has captured those risks within the project's authorization cost estimate. Based on industry historical performance, the project is, therefore, well positioned to achieve a better level of capital effectiveness with lower costs than industry norms. Their FEL Index is measured by the level of definition of the three equally weighted categories listed below.

1. Site-specific factors
2. EPC phase project execution planning
3. Level of engineering completeness

Figure 9-20 provides more details of what is measured in each FEL Index category.

Figure 9-21 illustrates the benefit of improving FEL performance or thoroughness on overall project costs and shows that these economic benefits increase as the project size increases.

Figure 9-22 illustrates the benefit of good FEL performance on cost growth and schedule slip. Here the negative impacts become evident if the FEL effort is compromised. IPA statistics indicate that significant project financial and schedule benefits can be realized by implementing a thorough FEL effort, prior to the EPC phase.

Figure 9-23 shows how the level of FEL definition can drive the EPC phase critical path schedule predictability.

Figure 9-24 presents the benefits of having an integrated project team during FEL on the overall project performance. IPA project data indicate that a well-integrated FEL team can produce significantly better project performance in terms of lower capital investment and more predictable execution schedule, when compared to projects where FEL teams were not properly integrated. An integrated FEL project team produces fewer late changes in the EPC phase. The performance of the individual team member is best measured by the performance of the entire integrated FEL project team. FEL is a team sport, not an individual performance. This illustrates the benefits for each engineering team member working closely with other team members, to produce the most profitable overall project results.

Investment in FEL for Best Project Performance The cost and schedule required to optimally complete the FEL phase are always under pressure and must be justified. This is especially true for "fast-track" projects where the time pressures can be significant. The Construction Industry Institute (CII) has shown that projects with good scope definition (i.e., well-defined FEL deliverables) prior to FID outperformed projects with poorly defined FEL scope.

This is also supported by Table 9-33 where a CII study of 62 industrial projects showed that a higher level of FEL effort resulted in a reduction

TABLE 9-31 FEL Projects versus EPC Projects

Parameter	FEL	EPC
Project state	Undefined	Defined
Changes	Actively seeks changes	Actively resists changes
Impact of change	Low	High
Opportunity for change	High	Low
Contract type	Typically reimbursable	Typically lump sum
Value improvement potential	High	Low
Client participation	Encouraged	Discouraged
Philosophy	Information driven	Deliverable driven

SOURCE: KBR FEL Experience, 2000 through 2016.

TABLE 9-32 Parameters of FEL Phases

Parameter	FEL-1 (Conceptual)	FEL-2 (Feasibility)	FEL-3 (Definition)
Cost estimate accuracy	+/−40%	+/−20–40%	+/−10–20%
Cumulative engineering hours spent	1–5%	5–15%	15–30%
Contingency	15–20%	10–15%	8–12%

SOURCE: KBR internal compilation of client parameters, IPA parameters, and KBR FEL experience.

Site-Specific Factors	*Plot Plans and Unit Configurations*	• All equipment has been placed. • All large-bore piping and one-line drawings for smaller bore piping have been provided. • Plot plans have been approved by the owner. • All major pipe routings have been identified and inspected. • Plot plans include major equipment, piping, and associated utilities.
	Soils and Hydrology	• Comprehensive understanding of the site is evident including knowledge of overhead and underground interferences and contamination issues. • Full geotechnical investigation completed for load-bearing capacity and soil contamination.
	Environmental Regulatory Requirements	• Grassroots or Greenfield projects – all known requirements addressed – all permit applications submitted to agencies. • Revamp projects – all environmental and building permits received or have been incontrovertibly established that permits are not needed to execute the project.
	Health and Safety Requirements	• Completed detailed, but not necessarily final, Hazard and Operability (HAZOP) reviews using advanced P&IDs and detailed equipment layouts. • HAZOP recommendations incorporated into design and EPC Phase cost estimate. • HAZOPs conducted with full participation of plant operations and maintenance personnel.
EPC Phase Project Execution Planning		• Full project team members assigned and all understand their roles and responsibilities. • Detailed EPC Phase Project Execution Plan with project charter, communication plans, and turnaround strategy. • Detailed EPC contracting strategy. • Detailed and integrated resource-loaded critical-path project schedule including effects of equipment delivery dates, interferences, successor and predecessor activities, and turnover and commissioning sequences for start-up.
Level of Engineering Completeness		• Project engineering 15% to 30% complete. • Owner/operator provided extensive input into design at time of estimate preparation. • For revamp projects, process tie-ins field-verified and approved by owner operations staff. • Completed electrical single lines and process control strategy.

FIG. 9-20 IPA's FEL index-best practical parameters.

in total EPC phase project design and construction cost of as much as 20 percent versus the FID authorized cost estimate. Further, the study indicated a reduction also in total project design and construction schedule by as much as 39 percent versus the FID authorized schedule.

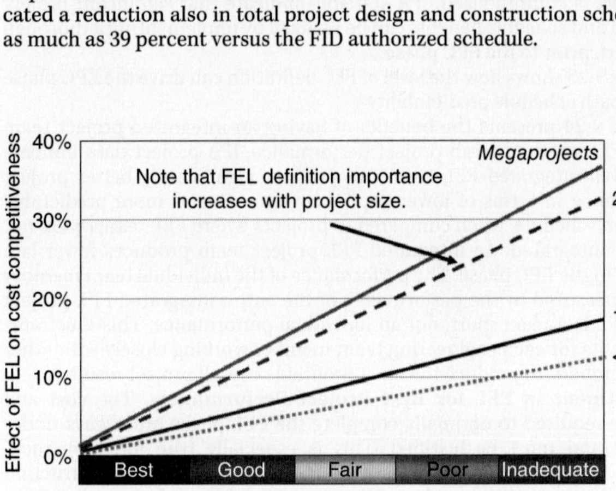

FIG. 9-21 Importance of FEL definition on project cost.

The level of definition of a project during the FEL phases has a direct influence on the ultimate project's outcome in terms of the number and impacts of changes in the EPC phase. This level of FEL performance translates to fewer major changes during detail engineering, construction, commissioning, and start-up. These conclusions are depicted by Fig. 9-25 where IPA has defined a major late change to mean changes made after the start of detail engineering and involving impacts greater than either 0.5 percent of the total project capital investment or 1 month in the critical path project schedule.

This graph illustrates why better project performance is produced through proactively seeking profitability-improving changes as early as possible. One of the reasons for this observation is that operation, maintenance, and construction expertise must be incorporated into the project team in every FEL phase.

This means that the design engineer should be working closely with these "real-world" experts as they design processes and their support systems. This also means that in order to improve overall project performance, achieving the best practical or highest level of definition during FEL is critical. Finally, this high level of definition results in a reduced number of changes during the EPC phase. These observations should be the critical goals of all project teams.

Typical FEL Deliverables and Level of Definition FEL phases are most accurately defined by the product or deliverables for each FEL phase. This is a very good means to determine in which phase(s) the project is actually conducting work. FEL phase deliverables are always customized to

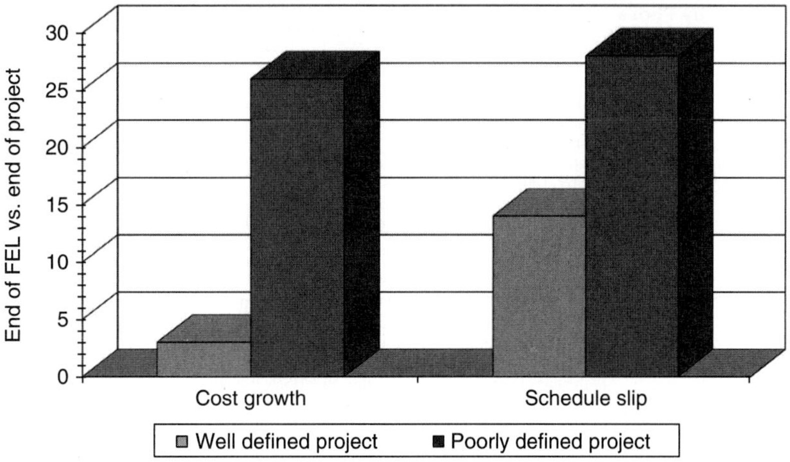

FIG. 9-22 FEL quality vs. project cost and schedule performance.

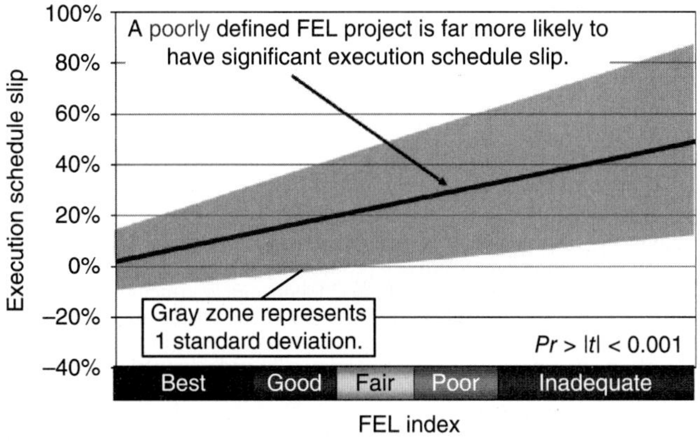

FIG. 9-23 FEL drives schedule predictability.

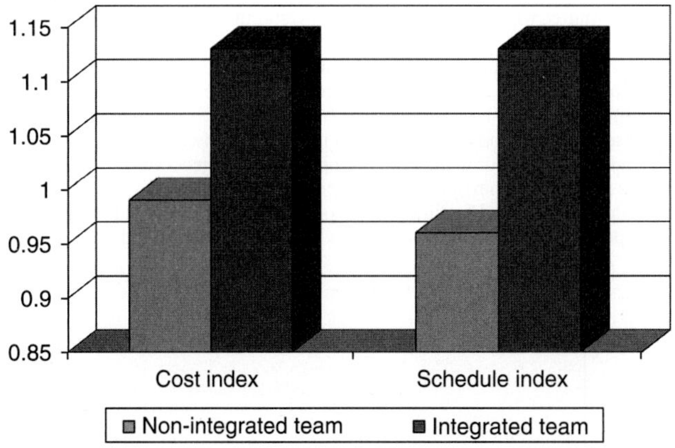

FIG. 9-24 Team integration vs. project cost and schedule performance.

suit the particular project's scope and business drivers. However, there are certain FEL deliverables and their levels of definition that actually define the FEL phase themselves. These deliverables are listed in Table 9-34.

Since there is no such thing as a "standard" FEL, every FEL project team member should be very aware of which deliverables or end products are required by those who have commissioned their work. Therefore, all project team members must understand the detailed expectations for each deliverable for each FEL phase. They should also confirm at the outset what information will be available to them prior to starting their work, the expected deliverable format, content, level of detail, and due dates. Further, the splits of work or division of work (who will do which aspect of the work) must be well understood by all as they are likely to be interdependent.

Today, it is common to see multiple operating companies form a joint venture to authorize major projects. It is also common for multiple engineering contractors to form joint ventures to execute the FEL and/or EPC phases of those major projects. These situations require an even more heightened understanding of information flow, project interfaces, deliverables work requiring multiple disciplines (i.e., discipline interdependence), divisions of work, who will be reviewing the work products, and what their expectations are.

TABLE 9-33 Project Performance vs. Level of FEL Effort

Level of FEL effort	Cost	Schedule
High	−4%	−13%
Medium	−2%	+8%
Low	+16%	+26%
Difference	20%	39%

SOURCE: Construction Industry Institute, University of Texas at Austin, CII Special Publication 268-3, *Adding Value Through Front-End Planning*, 2012.

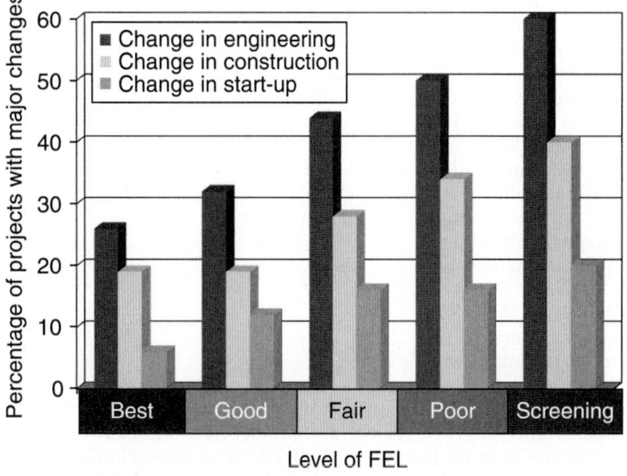

FIG. 9-25 Good FEL drives late charges down.

TABLE 9-34 Typical FEL Deliverables by Phase

FEL-1 (Conceptual phase)	FEL-2 (Feasibility phase)	FEL-3 (Definition phase)
Strategic business assessment	Updated strategic business assessment	Updated strategic business assessment
Preliminary project milestone schedule (level 1)	Level 2 EPC phase project schedule	Level 3 EPC phase project schedule
Key technology selected and risk identified	Cost estimate (+/−25%)	Completed environment permit submittal
Market assessment for feed, products, and capacity	Overall project execution strategy	EPC phase project execution plan
Potential sites identified and under evaluation	Contracting and purchasing strategies	EPC phase cash flow plan
Cost estimate (+/−40%)	Permitting and regulatory compliance plan	EPC phase training, commissioning, and startup plans
Block flow diagrams completed	Degree of economic off-site pre-assembly (modularization) determined	EPC phase contracting plans
Process cases identified	Soil survey and report	EPC phase materials management plan
Critical long lead equipment identified	Project alternatives analysis	EPC phase safety process and quality management plan
Construction strategy	Process flow diagrams (PFDs) for selected option(s)	EPC phase cost estimate (+/−10%)
FEL-1 phase VIP reports	Preliminary utility flow diagrams and balances	Finalized utility flow diagrams and balances
	Preliminary equipment list and equipment load sheets	P&IDs—includes owner review and preliminary hazards analysis changes
	Process hazards analysis (PFD level)	Preliminary process hazards analysis (P&ID level)
	Materials of construction identified and MSDs issued	Plot plans and critical equipment layouts—includes owner, operations, and maintenance review changes
	FEL-2 phase VIP reports	Equipment list and datasheets—includes owner review changes
		Single-line electrical diagrams
		Control system summary and control room layout
		Materials of construction confirmed
		Material selection diagrams (MSDs) completed
		FEL-3 phase VIP reports

SOURCE: IPA and KBR FEL Experience, 2000–2016.

VALUE-IMPROVING PRACTICES

GENERAL REFERENCES:

1. Independent Project Analysis, Inc. (IPA), http://www.ipaglobal.com.
2. Construction Industry Institute (CII), University of Texas at Austin, http://www.construction-institute.org.
3. Independent Project Analysis, Inc., *Best Practical Requirements for IPA's FEL Index*, data updated 2016.
4. KBR Value-Improving Practices Program, 1995 through 2016.
5. PDRI: Project Definition Rating Index—Industrial Projects, Construction Industry Institute (CII), University of Texas at Austin, http://www.construction-institute.org.
6. Society of American Value Engineers International (SAVE), http://www.value-eng.org.
7. McCuish, J. D., and J. J. Kaufman, *Value Management & Value Improving Practices*, Pinnacle Results LLC, 2016, http://pinnacleresults.com.

Introduction Value-improving practices (VIPs) are formal structured work processes applied to capital projects to improve profitability or value above that which can be attained through the application of good engineering and project management practices. VIPs provide an objective forum for formal analyses of project characteristics and features and are performed by small multidisciplinary teams of subject matter experts and conducted at optimum times during the engineering design and development of capital projects.

VIPs that have been statistically verified by Independent Project Analysis, Inc. (IPA) benchmarking of capital projects are listed below. Each has a different purpose and focus, but all produce project profitability improvements that the project team cannot achieve on their own.

- Classes of Facility Quality
- Technology Selection
- Process Simplification
- Constructability
- Customization of Standards and Specifications
- Energy Optimization
- Predictive Maintenance
- Waste Minimization
- Process Reliability Simulation
- Value Engineering
- Design to Capacity
- 3D-CAD

Application of VIPs to capital projects has been statistically proved to significantly improve project profitability according to IPA and the Construction Industry Institute (CII). IPA data presented in Fig. 9-26 have been gathered from thousands of capital projects since 1987 and illustrate that up to 10 percent reduction in project CAPEX can be realized in high-performing projects by conducting VIPs while also conducting rigorous FEL work processes—7.5 percent from VIPs and 2.5 percent from FEL. The better-performing projects are often referred to as Best Practical or Best in Class projects and represent the upper 20 percent of projects benchmarked by IPA.

Some organizations' VIP programs have documented far higher CAPEX reduction in addition to project profitability improvements from OPEX reduction, critical path schedule improvement, improved facility throughput, and improved facility availability. These improved results are often due to continual adaptation and improvement of the VIP work processes to maintain their relevance and ability to technically and economically improve projects above that accomplished through good project management practices (KBR VIP Program).

Characteristics of VIPs VIPs are often described by their characteristics as listed below.

- Out-of-the-ordinary practices are used to improve cost, schedule, and/or reliability of capital projects.
- Statistical links exist between the use of the practice and better project performance which are demonstrated, systematic, repeatable, and proven correlations.
- Formal and documented practices involve repeatable work processes.
- All involve stated explicit support from the owner's corporate executive team.
- These VIPs must be performed by a trained experienced VIP facilitator—someone who is not a full-time member of the project team.
- VIPs are used primarily during FEL project phases.
- All involve a formal facilitated workshop to confirm the value gained by the project and to formally approve vip team recommendations.

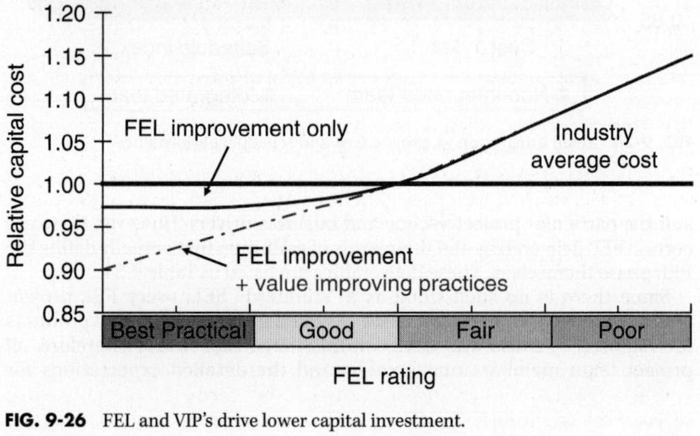

FIG. 9-26 FEL and VIP's drive lower capital investment.

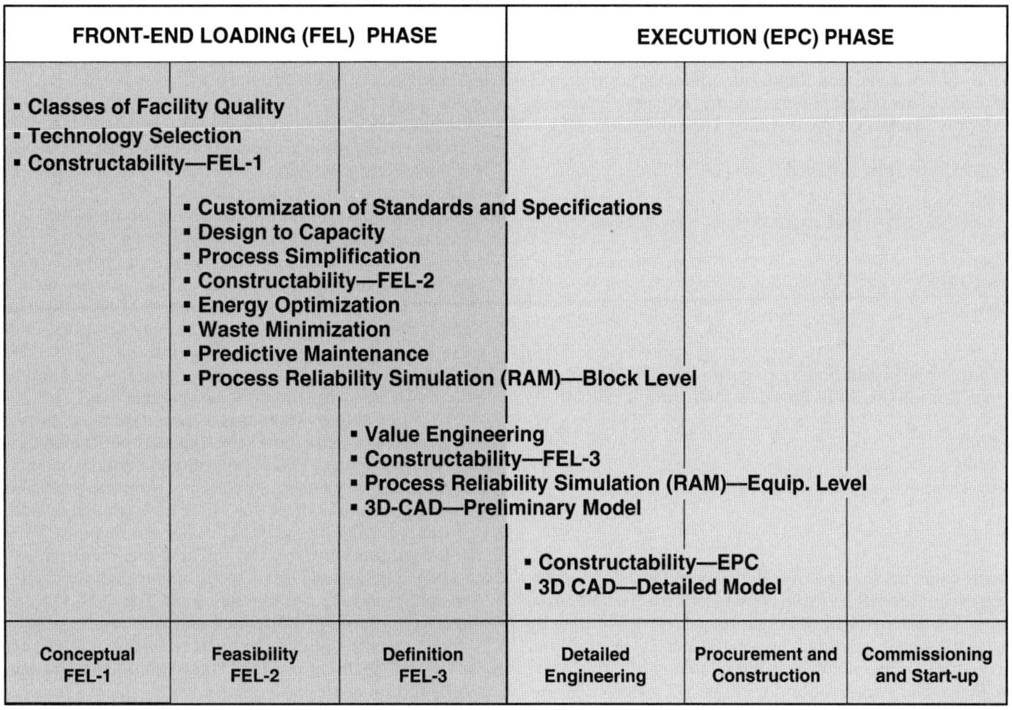

FIG. 9-27 Implementation timing for value-improving practices.

VIPs are also further clarified by what they are *not*, as listed below.

- Just "good engineering"
- Simple brainstorming or strategy sessions
- Business as usual
- A special look at some aspect of the project
- Process flow diagram (PFD) reviews
- Piping & instrumentation diagram (P&ID) reviews
- Safety reviews
- Audits
- Project readiness reviews

VIP Selection and Implementation Execution of VIPs must be deliberately and carefully planned in the initial phase of the project just after the project kick-off. VIPs should be selected from those offered from both the engineering contractor and the owner organizations. It is important to ensure that those selecting the VIPs are fully aware of the project scope and proven benefits for each VIP. They should also be aware of the resources, preparation, and time required to implement each VIP. During the VIP selection meeting, the project management team and the selected experienced VIP facilitator should

- Confirm which VIPs should be applied to the project and when
- Incorporate the planned VIPs into the project scope of work and schedule
- Determine the required workshop resources and best combination of engineering, operations, maintenance, construction, and other expertise for each selected VIP workshop team

The implementation of VIPs to any project must have the explicit commitment of the owner organization and be evident in the integrated project management team. This is so because the VIPs inherently seek project changes that are deemed profitability improvements beyond that evident at the start of the VIP effort. In other words, VIPs drive change that improves project performance, but must be supported by the owner organization to provide the additional resources and time to incorporate the value improvement into the project scope and execution plan. Experience has shown that economic benefit from VIPs is directly proportional to the owner's active and sustained support for implementing the VIPs and especially incorporating the VIP recommendations into the project design, scope, and execution plan.

Experience has also shown that each VIP should be conducted at an optimum time in the project life cycle for best results. Therefore, it is important to incorporate the optimal VIP timing into the project schedule. Figure 9-27 presents the optimal times in the project life cycle for conducting VIPs.

When capital projects are benchmarked by third-party organizations such as IPA or through CII's Project Definition Rating Index (PDRI), the implementation of applicable VIPs to the project is an important part of that benchmarking process.

Sources of Expertise VIP workshops should be planned and led by a trained and experienced facilitator who has significant experience in conducting such workshops. Technical expertise for VIP workshops should be a combination of senior project team members and subject matter experts from the owner's organization, the engineering contractor's organization, licensed technology providers, and any key fabrication or installation subcontractors to be used. It is important for the VIP team makeup to include subject matter experts not involved in the project to provide a "cold eyes" or unbiased perspective to the VIP team. Figure 9-28 illustrates the best balance of expertise for VIP workshops.

VIP Descriptions The VIP descriptions below provide a basic understanding of the important elements of each practice. Customization of VIPs beyond the basics is an important means to keep these practices able to

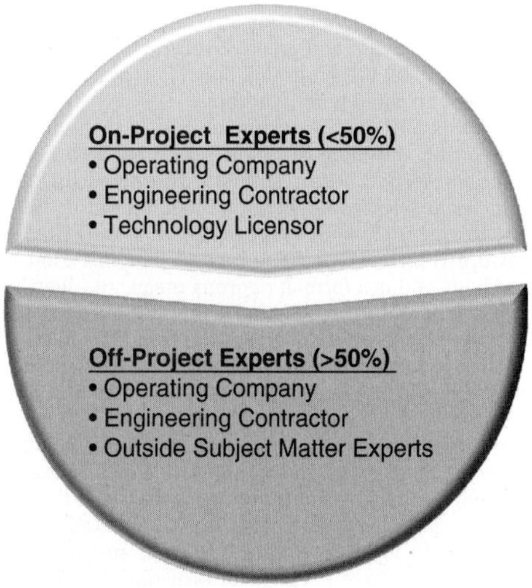

FIG. 9-28 Ideal team makeup.

consistently produce results above those of good engineering and management practices.

Classes of Facility Quality VIP The Classes of Facility Quality VIP determines with the owner organization the design philosophies that will produce the desired levels of overall plant performance and associated risk consistent with the highest long-term profitability. The minimum categories considered include:

- Capital Expense (CAPEX) or Total Installed Costs (TIC)
- Planned Facility Life
- Expandability
- Level of Automation
- Equipment Selection
- Operating Expense (OPEX)
- Environmental Controls
- Capacity
- Technology

Some organizations have applied additional categories listed below to provide a greater level of definition (KBR VIP Program, 2007 through 2016).

- Energy Intensity
- Raw Material Flexibility
- Product Mix Flexibility
- Product Availability
- Product Quality
- Project Schedule
- Site Integration

This VIP individually confirms the best overall design philosophy for the project team, for each of the parameters listed above. Here, the designer first learns by category how risk-averse the owner organization is with respect to how the facility is to be designed, operated, and maintained. For example, if the plant is to have only commercially proven technology that has been in operation within the owner organization for at least 2 years, then the engineer will need to confirm prior to starting the process design what technologies have been so used and in what services.

Alternatively, if the plant must be equipped with environmental controls that must be demonstrated Best Available Control Technology (BACT) on all waste streams, then this performance level and the associated unit operations and equipment must be well understood before the process design is begun. The results of this VIP are used by the project management team to update their Project Execution Plan for each FEL phase. The Classes of Facility Quality VIP provides the best results when conducted just following project kick-off in the FEL-1 phase.

Technology Selection VIP The Technology Selection VIP is the application of evaluation criteria aligned with the project's business objectives to identify manufacturing and processing technology that may be superior to that currently used. The goal is to ensure that the technology suite finally selected is the most competitive available. This requires a systematic search—both inside and outside the operating company's organization—to identify emerging technology alternatives.

This formal facilitated process is also meant to ensure due diligence for all parties involved and that all emerging and near-commercial alternative technologies for accomplishing a particular processing function are objectively considered. This VIP should not be confused with the routine engineering practice of evaluating and selecting the best processing and equipment options for a given project application. Its focus should be on the new and emerging technology for the intended function. This VIP is most commonly applied at the unit operation level, although it has also been successfully applied at the major equipment level and to help select competing processing schemes (KBR VIP Program). This VIP is particularly effective for combating the "not invented here" syndrome.

The goals of this VIP are to document which technology evaluation criteria are applicable and then to conduct a formal technology screening and evaluation assessment. The result is a prioritized listing of technology options for each selected application or function for the project. The preferred time to execute this VIP is the midpoint in the FEL-1 phase.

Process Simplification VIP The Process Simplification VIP uses the Value Methodology and is a formal, rigorous means to identify opportunities to eliminate or combine process and utility system steps or equipment, ultimately reducing investment and operating costs. The focus is the reduction of installed costs and critical path schedule while balancing these value improvements with expected facility operability, flexibility, and overall life cycle costs.

The Process Simplification VIP is far more than just evaluating and simplifying processing steps. This very productive VIP ensures that low- or zero-value system and equipment functions included in the project scope are challenged by subject matter experts and eliminated, if possible. This VIP tries to systematically differentiate "wants" from "needs" and remove the wants. It can be especially effective for providing a neutral professional environment for identifying and challenging sacred cows and then removing them. Removal of these low- or zero-value functions yields significant profitability improvements to the overall project. Process Simplification results in

- Reduced capital expense (CAPEX)
- Improved critical-path schedule
- Reduced process inventory
- Increased yields
- Reduced operating and maintenance expense (OPEX)
- Increased productivity
- Incremental capacity gains
- Reduced utility and support systems requirements
- Reduced waste generation

Process Simplification is executed in a formal workshop with a trained experienced facilitator. This VIP should always include key participants from each of the project owner's organizations, the engineering contractor organization, key third-party technology licensors, and equipment or systems vendors, where possible. One or more "cold eyes" reviewers or subject matter experts, who have extensive experience, should also be included to provide an objective and unbiased perspective.

This VIP also provides a means for integrating overall plantwide systems. The Process Simplification VIP is typically performed during the FEL-2 phase after the preliminary PFDs and heat and material balances become available. However, for very large and complex projects, considerable value has been gained by also performing this VIP at the midpoint or later in the FEL-1 phase.

Constructability VIP The Constructability VIP is the systematic implementation of the latest engineering, procurement, and construction concepts and lessons learned consistent with the facility's intended operations and maintenance requirements. The goal is to enhance construction safety, scope, predictability, cost, schedule, and quality.

Since constructability has seen widespread implementation in industry for over 30 years, in order for constructability to remain consistent with the definition of a VIP (i.e., above what project teams can do on their own), something formal must be added to the standard work process. One large engineering and construction company has enhanced the standard constructability work process to include a formal facilitated workshop in each project phase. This formal VIP workshop seeks profitability improvements above those already identified by the project team in the course of their normal work (KBR VIP Program, 2000 through 2016). These formal VIP workshops and the standard work process are mutually additive, flexible, and compatible.

The traditional constructability work process includes the following characteristics:

- An ongoing structured work process that starts in the FEL-1 phase and continues through facility start-up
- Systematic implementation of the latest engineering, procurement, and construction lessons learned
- Optimized use of operations, maintenance, engineering, procurement, key vendors, and construction knowledge and experience
- Enhanced achievement of project objectives
- Construction experts working with engineering and procurement resulting in profitability improvements in construction safety, cost, schedule, and quality

The enhanced Constructability VIP workshops add the following to the traditional work process:

- Conducts formal facilitated workshops in every engineering phase of the project focused on the pertinent aspects of that phase
- Identifies additional design and execution options that improve project profitability above those already being considered by the traditional constructability work process
- Involves a detailed review of planning, design, procurement, fabrication, and installation functions to achieve the best overall project safety performance, lowest CAPEX, and the shortest reasonable schedule
- Includes considerations for operability and maintainability
- Uses on-project and off-project subject matter experts

Conducting a formal Constructability VIP workshop in the FEL-1 phase should focus on confirming the best overall project construction strategies. These strategies should cover the overall plot plan, equipment arrangements, potential congestion areas, construction and turnaround laydown areas, site access for equipment and modules, economic driving forces for modularization, heavy lifts, procurement limitations, fabrication and transport limitations, area labor limitations, and coordination with existing structures and facilities. The conceptual plot plan and satellite views of the site should be used.

Conducting a formal Constructability VIP workshop in the FEL-2 phase should focus on more specific topics of layout optimization using a preliminary plot plan, equipment layouts, and satellite views of the site. Considerations should include optimum site layout in terms of construction laydown areas; optimum equipment arrangement to reduce piping and steel for structures and pipe racks; specific sizes and weights for modules; which components will be included in each module; module sequencing; crane locations for heavy lifts; equipment requiring early purchase to support the project critical path schedule; further analysis of procurement limitations; fabrication limitations; area labor availability; and pre-commissioning, commissioning, and start-up considerations. A preliminary 3D model should also be used, if available.

Conducting a formal Constructability VIP workshop in the FEL-3 phase focuses on even more detail for what was discussed above. If available, the preliminary 3D model will have more details than in FEL-2, especially if the project is a revamp or expansion. For these projects, incorporating tie-in details from laser scanning surveys into the 3D model should be strongly considered. This will then allow a more realistic determination of the scope of work that should be done before, during, and after a major planned maintenance turnaround.

In the detail engineering stage of the EPC phase, considerable detailed information will be available regarding engineering design and procurement, construction, and commissioning plans and schedules. In addition, the detailed 3D model and plot plans are recommended to be used as visual depictions of equipment and package systems footprints based on final vendor/supplier data as well as near-final pipe routings, steel detail, cable details, etc. The best timing for the Constructability VIP workshops is usually about three weeks ahead of the formal 3D model reviews, to allow workshop recommendations to be incorporated into the model before the formal reviews. Constructability VIP workshops should be formal and facilitated and should draw on personnel from operations, maintenance, and construction in addition to project and owner organization representation.

Customization of Standards and Specifications VIP The Customization of Standards and Specifications VIP is a direct and systematic method to improve project value by selecting aspects of applicable codes, standards, and specifications most appropriate for the project. The goal is to make needed changes to meet project performance requirements by ensuring that the codes, standards, and specifications selected do not exceed the minimum required for the project consistent with owner operation goals. Figure 9-29 shows the typical hierarchy of project guidelines and specifications and the focus areas of this VIP.

This VIP is beyond typical good engineering practices and should not be confused with ongoing systematic improvements to corporate standards and specifications, nor with required identification of applicable equipment and materials procurement specifications to be used for the project. This formal VIP takes a combination of project owner and engineering contractor corporate specifications and aggressively seeks profitability improvements consistent with the project's goals and limitations. This VIP promotes the procurement of off-the-shelf equipment over equipment customized for the project. This VIP is best performed early in the FEL-2 phase and should include project team members involved from the project owner and engineering contractor as well as appropriate suppliers of major packaged subsystems, modularized equipment, etc.

Energy Optimization VIP The Energy Optimization VIP is the systematic evaluation and economic optimization of energy use within a process or multiple subunits within a larger process or facility. This optimization starts by using the "Pinch" technology branch of process energy integration (energy pinch) to identify better process energy exchange options.

Energy pinch (usually just called *pinch*) is a methodology for the conceptual design of process heating, utility, and power systems. Pinch identifies the maximum theoretical energy use within a process, while minimizing the use of plant utilities. Such minimization is achieved by reusing energy via heat exchange between process streams. Once the minimum theoretical energy requirements and applicable process options have been determined, a formal facilitated workshop follows to determine which process options are operationally and economically supported by the owner's stakeholders.

This methodology is most profitably applied to processes where energy costs are a large fraction of the OPEX and to optimize high-value complex mass flow circuits, such as refinery crude unit pump-around circuits, refinery hydrogen networks (hydrogen pinch), and wastewater minimization (water pinch). This VIP is commonly applied to both Greenfield (new or grassroots plants) and Brownfield projects (revamps and expansions). This VIP should be implemented in the FEL-2 phase when preliminary PFDs and heat and material balances are available.

Predictive Maintenance VIP The Predictive Maintenance VIP is the proactive use of sensors and associated controls to monitor machinery mechanical "health," using both current-state and historical trends. This allows more effective planning of shutdowns and maintenance, thereby detecting equipment abnormalities and diagnosing potential problems before they cause permanent equipment damage. Examples include

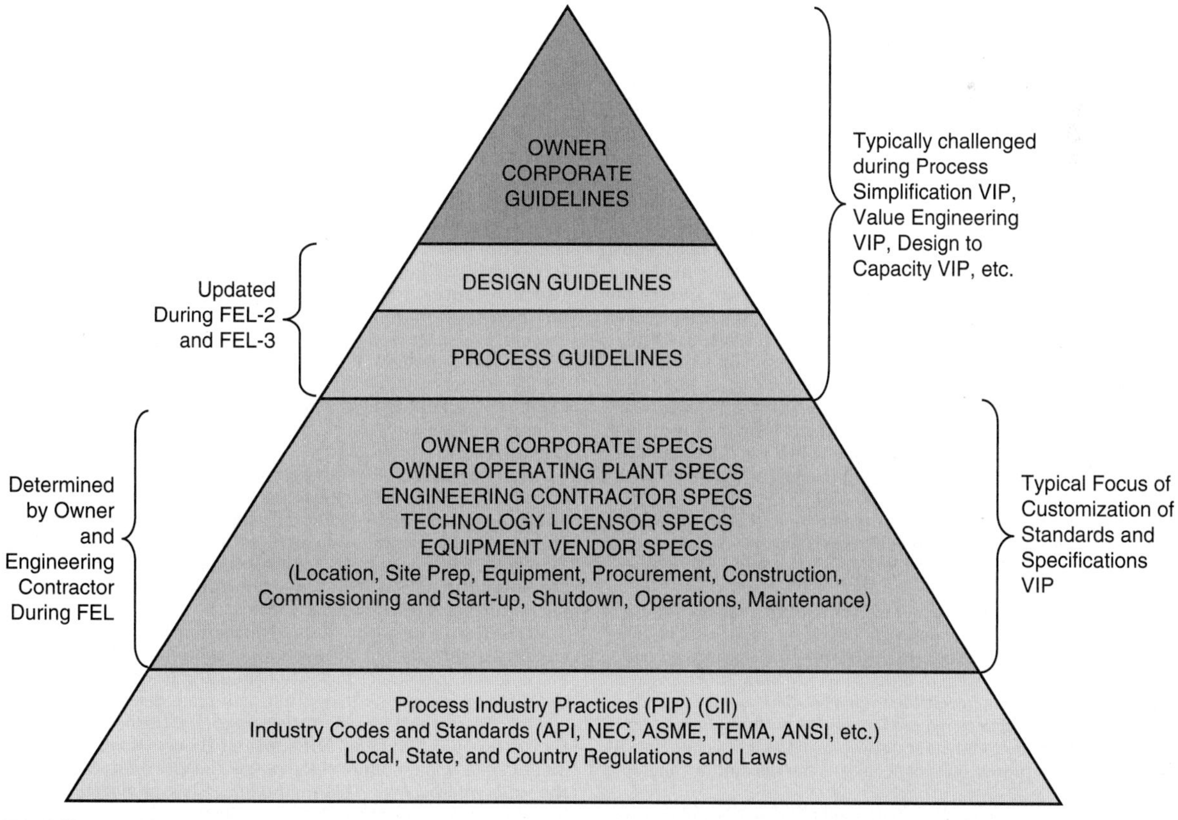

FIG. 9-29 Typical hierarchy of project guidelines and specifications.

real-time corrosion monitoring and equipment vibration monitoring. This additional instrumentation is generally economically justified in the case of critical equipment items and key operations.

Predictive maintenance reduces maintenance costs, improves the confidence of extending time between turnarounds, improves reliability, and provides a more predictable maintenance schedule for key process equipment. It also minimizes the amount of remaining equipment life that is lost through using only preventive maintenance practices. Preventive maintenance is an older practice which is limited to periodic inspections and repairs to avoid unplanned equipment breakdowns.

For the Predictive Maintenance VIP to be effective, maintenance personnel from the project owner's organization must be involved in determining key predictive maintenance requirements. Suppliers of critical equipment items (e.g., compressors, turbines, large pumps) are also important participants in this process. Today, this VIP is considered by some operating companies and engineering contractors to be standard practice. Whether standard practice or not, consideration of the provisions required for portable and permanent predictive maintenance technology should be initiated in the FEL-2 phase and concluded prior to purchase of the equipment systems involved.

Waste Minimization VIP The Waste Minimization VIP involves a formal analysis of each process stream to identify ways to eliminate or reduce the waste streams generated by a chemical processing facility. For those waste streams not eliminated, reduced, or converted into saleable by-products, this VIP provides a formal method to identify ways to manage the resulting waste streams. Waste Minimization incorporates environmental requirements into the facility design and combines life cycle environmental benefits and positive economic returns through energy reduction, reduced end-of-pipe treatment requirements, and improved yield of products from feedstocks or raw materials. The waste reduction hierarchy used is as follows:

- Eliminate or minimize the generation of waste at the source (i.e., source reduction).
- Recycle by use, reuse, or reclamation those potential waste streams or materials that cannot be eliminated or minimized (i.e., recycle/reuse).
- Treat generated wastes to reduce volume, toxicity, or mobility prior to storage or disposal (i.e., end-of-pipe treatment).

This VIP is executed in a formal workshop with an experienced facilitator and with owner organization and engineering contractor representatives involved. A cold-eyes reviewer with extensive experience should also be included to add a nonbiased perspective. Today, this VIP is considered by many operating companies and engineering contractors to be standard practice. However, there still appears to be benefit in confirming the best handling, transportation, and final disposition of wastes generated. The best results are achieved when implemented in the FEL-2 phase.

Process Reliability Simulation VIP The Process Reliability Simulation VIP is the use of *reliability, availability, and maintainability* (RAM) computer simulation modeling of the process and the mechanical reliability of the facility. A principal goal is to optimize the engineering design in terms of life cycle cost, thereby maximizing the project's potential profitability. The objective is to determine the optimum relationships between maximum production rates and design and operational factors. Process reliability simulation is also applied for safety and environmental purposes, since it considers the consequences of specific equipment failures and failure modes.

This VIP is typically led by an engineer experienced in plant operations and use of the RAM simulation modeling software. The VIP should also directly involve the owner organization since they would most often supply the historical operating and maintenance information required for the development of the RAM model. This process provides the project team with a more effective means of assessing, early in the design, the economic impact of changes in process design, the technical and economic justification for installed spare equipment, identification of bottlenecks in the system, simulation of key operating scenarios, and training and maintenance requirements of a facility.

The Process Reliability Simulation VIP should be initiated in the FEL-2 phase to produce a block-level RAM model. Based on the results of that model, a more detailed equipment-level RAM model should be developed starting in the FEL-3 phase.

Value Engineering VIP The Value Engineering VIP applies the value methodology, which is a systematic and structured team approach to analyze the functions and essential characteristics of a program, project, process, technology, work process, or system in order to satisfy those functions and essential characteristics at the lowest life cycle cost. The value methodology helps achieve balance between required functions, performance, quality, safety, and scope with the cost and other resources necessary to accomplish those requirements. The proper balance results in the maximum value for the project.

This VIP tries to systematically differentiate wants from needs and remove the wants. It also tests for non-income-producing investments including redundancy, overdesign, manufacturing add-ons, upgraded materials of construction, and customized designs versus vendor standards. The Value Engineering VIP also ensures that low- or zero-value functions or equipment included in the project scope are challenged to be the highest value possible for the project. Removal of these low- or zero-value functions from the project scope, if possible, has been proven to yield significant economic improvements to the overall project. These can encompass the following:

- Misalignment of unit or system capacity or operations capability with respect to the overall facility
- Overly conservative assumptions of the basic design data
- Overly conservative interpretation of how the facilities will be used during peak, seasonal, or upset conditions
- Traditional design, layout, and operations approaches
- Preinvestment included in the project scope that may not be value added
- Overdesign of equipment or systems to provide uneconomic added flexibility

The Value Engineering VIP is executed in a formal workshop with an experienced technical workshop facilitator/leader. Both the owner organization and the engineering contractor are always involved. Third-party licensors and equipment/systems vendors should be included where applicable. A cold-eyes reviewer with extensive experience is also included to add a nonbiased perspective.

This VIP leverages the growing accumulation of more detailed project knowledge to test the value of earlier more generalized scope assumptions. It also tests the presumed added value of different stakeholder requirements, which have influenced the evolution of the project scope.

This highly adaptable VIP results in reduced CAPEX, improved critical-path schedule, reduced process inventory, increased yields, reduced OPEX, increased productivity, incremental capacity gains, reduced utility and support systems requirements, and reduced waste generation. The Value Engineering VIP is easily the most productive of the VIPs in terms of CAPEX reduction and the most adaptive. Results are best achieved when conducted early in the FEL-3 phase.

Design to Capacity VIP The Design to Capacity VIP systematically evaluates the maximum capacity of processing systems, major equipment, ancillary piping, valving, instrumentation, and associated engineering calculations and guidelines. This VIP reduces project capital investment by confirming minimum required system and equipment capacities and flexibility necessary to meet project business objectives. Design to Capacity drills down to each process system and subsystem and scrutinizes the design of each equipment item. The goal is to improve life cycle costs or project profitability by eliminating preinvestment and overdesign not supported by project economics. This VIP produces the best results when conducted in the FEL-2 phase as a facilitated workshop with both owner and engineering contractor representation.

3D-CAD VIP The 3D-CAD VIP is the creation of a detailed three-dimensional (3D) computer-aided design (CAD) model depicting the proposed process and associated equipment along with the optimized plant layout and specific equipment arrangements and orientations.

The principal benefit is to produce an electronic model of the intended facility to enable project teams and owner operations and maintenance staff to review and agree on the planned plant layout and equipment arrangements. The goals of this VIP are to reduce engineering and construction rework, improve operability and maintainability, and confirm the incorporation into the design of advantageous human factors (a.k.a. ergonomics) focused on ease of operation and maintenance.

Application of this VIP during the FEL phases involves creation of conceptual or preliminary level models to gain early stakeholder acceptance of plant layouts, multilevel structures such as modules, tie-in point details for revamp or expansion projects, and improved inputs to project cost estimate generation.

Application in the EPC phase involves more detailed 3D models based on specific equipment supplier information where engineering and construction rework can be reduced by supporting interference checks by the engineering disciplines involved prior to issuing fabrication or construction drawings. Development of 3D models throughout the project life cycle has become standard practice for most operating companies and major engineering and construction contractors.

VIPs That Apply the Value Methodology Many of the VIPs are conducted only once in a project at a "sweet spot" where maximum benefit is found. For some, they are conducted in each project phase using the same overall methodology, but with more detailed information and a slightly different mix of subject matter experts. The best such example is the Constructability VIP. Other VIPs such as Process Simplification and Value Engineering apply a more rigorous and adaptive methodology known as the *value methodology*. This methodology has produced excellent results in industry for more than 65 years [Society of American Value Engineers International (SAVE)].

Some organizations have applied the value methodology to additional VIPs such as Design to Capacity, Technology Selection, Waste Minimization, and Constructability to produce far better results (McCuish and Kaufman, 2016). Other organizations have applied VIPs in combination to better solve highly complex situations evident in large or complex projects where the owner is a joint venture or the engineering contractor is a joint venture (KBR VIP Program). For those VIPs where the value methodology is to be applied, a typical sequential approach is described below.

Value Methodology Job Plan The value methodology job plan focuses on the analysis of the "function" of each unit, unit operation, system, and equipment of the process—how it is expected to perform and to what expected degree of efficiency. Project execution functions are also identified and analyzed to validate them and identify value improvements. This analysis covers all aspects of the project including execution strategies, permitting issues, engineering and design issues, unresolved issues, construction execution issues, and project risks. Before the formal VIP workshop begins, the goals, objectives, and scheduled time for the workshop must be agreed to by the integrated project management team. The workshop facilitator/leader must ensure that all the information required for the workshop is available and the workshop team members are briefed on the objectives, methodology, and expectations.

The Formal Workshop The formal workshop should always be structured to make maximum use of the multidisciplinary team's time and effort. Such workshops typically require no less than 2 days and as many as 5 days depending on the size and complexity of the project. The required workshop length should be determined by the VIP facilitator. The workshop includes the following six phases of a typical "job plan" that are supported by the Society of American Value Engineers International.

Information Phase Important project background information is reviewed to confirm understanding of the basis for design, project constraints, owner cost and reliability targets, and the sensitivity of expected plant capital and operating costs. Discussions of these issues' validity and basis are completed during the morning of the first workshop day. Experience consistently shows that the information phase overlaps significantly with the function analysis phase and the creative phase. That is, workshop attendees tend to generate ideas in the form of questions as they are being informed of the project basis and when updating the *function analysis system technique* (FAST) diagram. Such spontaneous idea generation should be encouraged.

Function Analysis Phase The team analyzes the project scope to understand and clarify the required processing and execution functions. To enable this, a draft project FAST diagram is prepared prior to the workshop. This function analysis diagramming illustrates the logical or functional relationships and dependencies between different process systems and project activities and their associated costs. In the workshop, the team reviews, discusses, and modifies the draft "overall project" FAST diagram to identify the highest-value functions that offer the greatest potential to improve project value or profitability. These highest-value functions present targets of opportunity where brainstorming should be focused.

Creative or Speculation Phase Once the pertinent information and issues have been reviewed and the important functions of each process and project step identified, the team is encouraged to speculate or brainstorm for alternative means to accomplish the targeted functions. A creative environment is strictly maintained to encourage the unconstrained flow of new ideas and to encourage the team to strive for fresh, innovative ideas. All ideas generated are listed—no ideas are deleted. Most importantly, no analysis or judgment of ideas is allowed in this phase. Ideas are then generated as the team reviews process and utility flow diagrams and plot plans with simultaneous focus on the project functions previously identified. In this way, the "targets of opportunity" or functions within the project FAST diagram allow the team to focus all their effort on the targets most likely to reveal realistic value improvement ideas.

Evaluation or Analytical Phase The team reviews the ideas against relevant project criteria such as potential impact on long-term economics, impact on plant operations and maintenance costs, effect on capital cost, applicability to the project scope of work, technical and execution risks associated with implementation, impact on project schedule, and cost and resources required to implement the improvement. Each workshop has specific criteria against which proposed alternatives are judged. The ideas are weighted, sorted, grouped, linked, and ranked so that the best of the technically viable ideas are efficiently identified for further study.

Development or Proposal Phase The VIP team selects and expands the ideas ranked highest (i.e., best potential value to be gained) to obtain additional technical and economic insights and information to support those ideas. Each selected idea is assigned to a workshop team member best equipped to render an objective analysis of the idea when compared to the current project scope and premises. This analysis involves written proposals where potential benefits, costs, and risks are estimated in the workshop. Experience indicates that having the VIP team perform this stage within the formal workshop produces the best results. The written proposals are then presented to the other workshop team members by those who prepared them and are discussed sufficiently to reach consensus on whether the ideas retain sufficient technical and economic merit to become workshop value recommendations.

Presentation Phase The workshop team presents orally and with supporting available materials the agreed value recommendations to the project stakeholders. These stakeholders are often the integrated project management team. Ideally, higher-level owner stakeholders arrive near the end of the workshop to hear the workshop team's recommendations. The owner stakeholders then approve only those recommendations that pass muster and authorize the project team to begin the implementation effort. For team recommendations that offer insufficient value improvement, the owner stakeholder rationales for not approving those recommendations are noted. In some cases, the owner has given workshop team members the authority to approve or reject workshop team recommendations. Often, this approval is conditional on early validation by owner subject matter experts. Approval is meant to provide support to the project team for additional resources and schedule consideration to incorporate fully the value improvements into the project scope of work.

Reporting and Follow-Up After completing the formal VIP workshop, the workshop facilitator/leader completes and issues the written VIP workshop report for the project record. During this time, the project management team assigns each approved recommendation to a member of the project team and estimates the engineering time and resources required to incorporate the improvement into the project scope of work. The assigned team member evaluates further each workshop recommendation to confirm both technical and economic viability. Important considerations are the potential impact on the project schedule and the timely development and completion of key project deliverables. If the recommendation remains attractive following this evaluation, it is presented to the owner management team for final approval. If approved, the project scope is revised accordingly via the project Management of Change protocol. The management team then communicates the results of the VIP effort within the integrated project team. This follow-up action plan creates a very positive and cost-conscious attitude within the project team that leads to further improvements in project value.

GLOSSARY

Accounts payable The value of purchased goods and services that are being used but have not been paid.

Accounts receivable Credit extended to customers, usually on a 30-day basis. Cash is set aside to take care of the probability that some customers may not pay their bills.

Accrual basis The accounting method that recognizes revenues and disbursement of funds by receipt of bills or orders and not by cash flow, distinguished from cash basis.

Administrative expense An overhead expense due to the general direction of a company beyond the plant level. It includes administrative and office salaries, rent, auditing, accounting, legal, central purchasing and engineering, etc., expenses.

Allocation of expenses A procedure whereby overhead expenses and other indirect charges are assigned back to processing units or to products on what is expected to be an equitable basis. All allocations are somewhat arbitrary.

Amortization Often used interchangeably with *depreciation*, but there is a slight difference depending on whether the life of an asset is known. If the period of time is known to be usually more than a year, this annual expense is amortization; however, if the life is estimated, then it is depreciation.

Annual net sales Pounds of product sold times the net selling price. Net means that any allowances have been subtracted from the gross selling price.

Annual report Management's report to the stockholders and other interested parties at the end of a year of operation showing the status of the company, its activities, funds, income, profits, expenses, and other information.

Appurtenances The auxiliaries to either process or nonprocess equipment: piping, electrical, insulation, instrumentation, etc.

Assets The list of money on hand, marketable securities, monies due, investments, plants, properties, intellectual property, inventory, etc., at cost or market value, whichever is smaller. The assets are what a company (or person) owns.

Balance sheet This is an accounting, historical tabulation of assets, liabilities, and stockholders' equity for a company. The assets must equal the liabilities plus the stockholders' equity.

Battery limit A geographic boundary defining the coverage of a specific project. Usually it takes in the manufacturing area of a proposed plant, including all process equipment but excluding provision for storage, site preparation, utilities, administrative buildings, or auxiliary facilities.

Bonds When one purchases a bond, the company (or person) acquires an interest in debt and becomes a creditor of the company. The purchaser receives the right to receive regular interest payments and the subsequent repayment of the principal.

Book value Current investment value on the company books as the original installed cost less depreciation accruals.

Book value of common stock Net worth of a firm divided by the number of shares of common stock issued at the time of a report.

Break-even chart An economic production chart depicting total revenue and total expenses as functions of operation of a processing facility.

Break-even point The percentage of capacity at which income equals all fixed and variable expenses at that level of operation.

By-product A product made as a consequence of the production of a main product. The by-product may have a market value or a value as a raw material.

Capacity The estimated maximum level of production on a sustained basis.

Capital ratio Ratio of capital investment to sales dollars; the reciprocal of capital turnover.

Capital recovery The process by which original investment in a project is recovered over its life.

Capital turnover The ratio of sales dollars to capital investment; the reciprocal of capital ratio.

Cash Money that is on hand to pay for operating expenses, e.g., wages, salaries, raw materials, supplies, etc., to maintain a liquid financial position.

Cash basis The accounting basis whereby revenue and expense are recorded when cash is received and paid, distinguished from accrual basis.

Cash flow Net income after taxes plus depreciation (and depletion) flowing into the company treasury.

Code of accounts A system in which items of expense or fixed capital such as equipment and material are identified with numerical figures to facilitate accounting and cost control.

Common stock Money paid into a corporation for the purchase of shares of common stock that becomes the permanent capital of the firm. Common stockholders have the right to transfer ownership and may sell the stock to individuals or firms. Common stockholders have the right to vote at annual meetings on company business or may do so by proxy.

Compound interest The interest charges under the condition that interest is charged on previous interest plus principal.

Contingencies An allowance for unforeseeable elements of cost in fixed investment estimates that previous experience has shown to exist.

Continuous compounding A mathematical procedure for evaluating compound interest based upon continuous interest function rather than discrete interest periods.

Conversion expense The expense of converting raw materials to finished product.

Corporation In 1819, defined by Chief Justice Marshall of the Supreme Court as "an artificial being, invisible, intangible and existing only in contemplation of law." It exists by the grace of a state, and the laws of a state govern the procedure for its formation.

Cost of capital The cost of borrowing money from all sources, namely, loans, bonds, and preferred and common stock. It is expressed as an interest rate.

Cost center For accounting purposes, a grouping of equipment and facilities comprising a product manufacturing system.

Cost of sales The sum of the fixed and variable (direct and indirect) expenses for manufacturing a product and delivering it to a customer.

Decision or decision making A program of action undertaken as a result of (1) an established policy or (2) an analysis of variables that can be altered to influence a final result.

Depletion A provision in the tax regulations that allows a business to charge as current expense a noncash expense representing the portion of limited natural resources consumed in the conduct of business.

Depreciation A reasonable allowance by the Internal Revenue Service for the exhaustion, wear and tear, and normal obsolescence of equipment used in a trade or business. The property must have a useful life of more than 1 year. Depreciation is a noncash expense deductible from income for tax purposes.

Design to cost A management technique to achieve system designs that meet cost parameters. Cost as a design parameter is considered on a continuous basis as part of a system's development and production processes.

Direct expense An expense directly associated with the production of a product such as utilities, labor, and maintenance.

Direct labor expense The expense of labor involved in the manufacture of a product or in the production of a service.

Direct material expense The expense associated with materials consumed in the manufacture of a product or the production of a service.

Distribution expense Expense including advertising, preparation of samples, travel, entertainment, freight, warehousing, etc., to distribute a sample or product.

Dollar volume Dollar's worth of a product manufactured per unit of time.

Earnings The difference between income and operating expenses.

Economic life The period of commercial use of a product or facility. It may be limited by obsolescence, physical life of equipment, or changing economic conditions.

Economic value added The period dollar profit above the cost of capital. It is a means to measure an organization's value and a way to determine how management's decisions contribute to the value of a company.

Effective interest The true value of interest computed by equations for the compound interest rate for a period of 1 year.

Equity The owner's actual capital held by a company for its operations.

Escalation A provision in actual or estimated cost for an increase in equipment cost, material, labor, expenses, etc., over those specified in an original estimate or contract due to inflation.

External funds Capital obtained by selling stocks or bonds or by borrowing.

FEL (front-end loading) The process by which a company develops a detailed definition of the scope of a capital project that meets corporate business objectives.

FIFO (first in, first out) The valuation of raw material and supplies inventory, meaning first into the company or process is the first used or out.

Financial expense The charges for use of borrowed funds.

Fixed assets The real or material facilities that represent part of the capital in an economic venture.

Fixed capital Item including the equipment and buildings.

Fixed expense An expense that is independent of the rate of output, e.g., depreciation and plant indirect expenses.

Fringe benefits Employee welfare benefits; expenses of employment over and above compensation for actual time worked, such as holidays, vacations, sick leave, and insurance.

Full-cost accounting Method of pricing goods and services to reflect their true costs, including production, use, recycling, and disposal.

Future worth The expected value of capital in the future according to some predetermined method of computation.

Goods manufactured, cost of Total expense (direct and indirect expenses) including overhead charges.

Goods-in-process inventory The holdup of product in a partially finished state.

Goods sold, cost of The total of all expenses before income taxes that is deducted from income (revenue).

Grass-roots plant A complete plant erected on new site including land, site preparation, battery-limits facilities, and auxiliary facilities.

Gross domestic product An indicator of a country's economic activity. It is the sum of all goods and services produced by a nation within its borders.

Gross margin (profit) Total revenue minus cost of goods manufactured.

Gross national product An economic indicator of a country's economic activity. It is the sum of all the goods and services produced by a nation both within and outside its borders.

Income Profit before income taxes or gross income from sales before deduction of expenses.

Income statement The statement of earnings of a firm as approximated by accounting practices, usually covering a 1-year period.

Income tax The tax imposed on corporate profits by the federal and/or state governments.

Indirect expenses Part of the manufacturing expense of a product not directly related to the amount of product manufactured, e.g., depreciation, local taxes, and insurance.

Internal funds Capital available from depreciation and accumulated retained earnings.

Inventory The quantity of raw materials and/or supplies held in a process or in storage.

Last in, first out (LIFO) The valuation of raw materials and supplies, meaning the last material into a process or storage is the first used or out.

Leverage The influence of debt on the earning rate of a company.

Liabilities An accounting term for capital owed by a company.

Life cycle cost Cost of development, acquisition, support, and disposal of a system over its full life.

Manufacturing expense The sum of the raw material, labor, utilities, maintenance, depreciation, local taxes, etc., expenses. It is the sum of the direct and indirect (fixed and variable) manufacturing expenses.

Marginal cost The incremental cost of making one additional unit without additional investment in facilities.

Market capitalization The product of the number of shares of common stock outstanding and the share price.

Market value added A certain future economic value added for a company. It is the present value of the future economic value (EVA) generated by a company. It is a measure of how much value a firm has created.

Minimum acceptable rate of return (MARR) The level of return on investment, at or above the cost of capital, chosen as acceptable for discounting or cutoff purposes.

Net sales price Gross sales price minus freight adjustments.

Net worth The sum of the stockholders' investment plus surplus, or total assets minus total liabilities.

Nominal interest The number applied loosely to describe the annual interest rate.

Obsolescence The occurrence of decreasing value of physical assets due to technological changes rather than physical deterioration.

Operating expense The sum of the manufacturing expenses for a product and the general, administrative, and selling expenses.

Operating margin The gross margin minus the general, administrative, and selling expenses.

Opportunity cost The estimate of values that are forgone by undertaking one alternative instead of another one.

Payout time (payback period) The time to recover the fixed capital investment from profit plus depreciation. It is usually after taxes but not always.

Preferred stock Stock having claims that it commands over common stock, with the preference related to dividends. The holders of such stock receive dividends before any distribution is made to common stockholders. Preferred stockholders usually do not have voting rights as common stockholders do.

Present worth The value at some datum time (present time) of expenditures, costs, profits, etc., according to a predetermined method of computation. It is the current value of cash flow obtained by discounting.

Production rate The amount of product manufactured in a given time period.

Profitability A term generally applied in a broad sense to the economic feasibility of a proposed venture or an ongoing operation. It is generally considered to be related to return on investment.

Rate of return on investment The efficiency ratio relating profit or cash flow to investment.

Replacement A new facility that takes the place of an older facility with no increase in capacity.

Revenue The net sales received from the sale of a product or a service to a customer.

Sales, administration, research, and engineering expenses (SARE) Overhead expenses incurred as a result of maintaining sales offices and administrative offices and the expense of maintaining research and engineering departments. This item is usually expressed as a percentage of annual net sales.

Sales volume The amount of sales expressed in pounds, gallons, tons, cubic feet, etc., per unit of time.

Salvage value The value that can be realized from equipment or other facilities when taken out of service and sold.

Selling expense Salaries and commissions paid to sales personnel.

Simple interest The interest charges in any time period that is charged on only the principal.

Sinking fund An accounting procedure computed according to a specified procedure to provide capital to replace an asset.

Surplus The excess of earnings over expenses that is not distributed to stockholders.

Tax credit The amount available to a firm as part of its annual return because of deductible expenses for tax purposes. Examples have been research and development expenses, energy tax credit, etc.

Taxes In a manufacturing cost statement, usually property taxes. In an income statement, usually federal and state income taxes.

Time value of money The expected interest rate that capital should or would earn. Money has value with respect to time.

Total operating investment The fixed capital investment, backup capital, auxiliary capital, utilities and services capital, and working capital.

Utilities and services capital Electrical substations, plant sewers, water distribution facilities, and occasionally the steam plant.

Value added The difference between the raw material expense and the selling price of that product.

Value-improving practices (VIPs) Formal structured practices applied to capital projects to improve profitability ("or value") above that which is attained through the application of proven good engineering and project management practices.

Variable expense Any expense that varies directly with production output.

Working capital In the accounting sense, the current assets minus the current liabilities. It consists of the total amount of money invested in raw materials, supplies, goods in process, product inventories, accounts receivable, and cash minus those liabilities due within 1 year.

Section 10

Transport and Storage of Fluids

Meherwan P. Boyce, Ph.D., P.E. *(Deceased) Chairman and Principal Consultant, The Boyce Consultancy Group, LLC; Fellow, American Society of Mechanical Engineers (U.S.); Fellow, National Academy Forensic Engineers (U.S.); Fellow, Institution of Mechanical Engineers (U.K.); Fellow, Institution of Diesel and Gas Turbine Engineers (U.K.); Registered Professional Engineer (Texas), Chartered Engineer (U.K.); Sigma Xi, Tau Beta Pi, Tau Beta Pi, Phi Kappa Phi. Section Editor, Perry's Chemical Engineering Handbook, Measurement of Flow, Pumps and Compressors, Piping, Storage and Process Vessels, 7th ed.; and Joint Section Editor, 8th and 9th eds.*

Victor H. Edwards, Ph.D., P.E. *Principal, VHE Technical Analysis; Fellow and Life Member, American Institute of Chemical Engineers; Member, American Association for the Advancement of Science, American Chemical Society, National Society of Professional Engineers; Life Member, New York Academy of Sciences; Registered Professional Engineer (Texas), Phi Lambda Upsilon, Sigma Tau; Joint Section Editor, 8th and 9th eds.*

Roy A. Grichuk, P.E. *Piping Director, Fluor, BSME, P.E.; Member, American Society of Mechanical Engineers, B31 Main Committee, B31MTC Committee, and B31.3 Committee; Registered Professional Engineer (Texas) (Piping)*

Hugh D. Kaiser, P.E., B.S., M.B.A. *Principal Engineer, WSP USA; Fellow, American Institute of Chemical Engineers; Registered Professional Engineer (Indiana, Nebraska, Oklahoma, and Texas) (Storage and Process Vessels)*

Ronnie Montgomery, *Technical Manager, Process Control Systems, IHI Engineering and Construction International Corporation; Member, Process Industries Practices, Process Controls Function Team; Member, International Society of Automation (Flow Measurement)*

MEASUREMENT OF FLOW

PUMPS AND COMPRESSORS

COMPRESSORS

PROCESS PLANT PIPING

STORAGE AND PROCESS VESSELS

Nomenclature and Units

In this list, symbols used in the section are defined in a general way, and appropriate SI and U.S. Customary System (USCS) units are given. Specific definitions, as denoted by subscripts, are stated at the place of application in the section. Some specialized symbols used in the section are defined only at the place of application.

Symbol	Definition	SI units	U.S. customary units
A	Area	m^2	ft^2
A	Factor for determining minimum value of R_1		
A_∞	Free-stream speed of sound		
a	Area	m^2	ft^2
a	Duct or channel width	m	ft
a	Coefficient, general		
B	Height	m	ft
b	Duct or channel height	m	ft
b	Coefficient, general		
C	Coefficient, general		
C	Conductance	m^3/s	ft^3/s
C	Sum of mechanical allowances (thread or groove depth) plus corrosion or erosion allowances	mm	in
C	Cold-spring factor		
C	Constant		
C_a	Capillary number	Dimensionless	Dimensionless
C_1	Estimated self-spring or relaxation factor		
c_p	Constant-pressure specific heat	J/(kg · K)	Btu/(lb · °R)
c_v	Constant-volume specific heat	J/(kg · K)	Btu/(lb · °R)
D	Diameter	m	ft
D, D_0	Outside diameter of pipe	mm	in
d	Diameter	m	ft
E	Modulus of elasticity	N/m^2	lbf/ft^2
E	Quality factor		
E_a	As-installed Young's modulus	MPa	kip/in^2 (ksi)
E_c	Casting quality factor		
E_j	Joint quality factor		
E_m	Minimum value of Young's modulus	MPa	kip/in^2 (ksi)
F	Force	N	lbf
F	Friction loss	(N · m)/kg	(ft · lbf)/lb
F	Correction factor	Dimensionless	Dimensionless
f	Frequency	Hz	1/s
f	Friction factor	Dimensionless	Dimensionless
f	Stress-range reduction factor		
G	Mass velocity	kg/(s · m^2)	lb/(s · ft^2)
g	Local acceleration due to gravity	m/s^2	ft/s^2
g_c	Dimensional constant	1.0 (kg · m)/(N · s^2)	32.2 (lb · ft)/(lbf · s^2)
H	Depth of liquid	m	ft
H, h	Head of fluid, height	m	ft
H_{ad}	Adiabatic head	N · m/kg	lbf · ft/lbm
h	Flexibility characteristic		
h	Height of truncated cone; depth of head	m	in
i	Specific enthalpy	J/kg	Btu/lb
i	Stress-intensification factor		
i_i	In-plane stress-intensification factor		
i_o	Out-plane stress intensification factor		
I	Electric current	A	A
J	Mechanical equivalent of heat	1.0 (N · m)/J	778 (ft · lbf)/Btu
K	Index, constant or flow parameter		
K	Fluid bulk modulus of elasticity	N/m^2	lbf/ft^2
K_1	Constant in empirical flexibility equation		
k	Ratio of specific heats	Dimensionless	Dimensionless
k	Flexibility factor		
k	Adiabatic exponent c_p/c_v		
L	Length	m	ft
L	Developed length of piping between anchors	m	ft
L	Dish radius	m	in
M	Molecular weight	kg/mol	lb/mol
M_i, m_i	In-plane bending moment	N · mm	in · lbf
M_o	Out-plane bending moment	N · mm	in · lbf
M_t	Torsional moment	N · mm	in · lbf
M_∞	Free stream Mach number		
m	Mass	kg	lb
m	Thickness	m	ft
N	Number of data points or items	Dimensionless	Dimensionless
N	Frictional resistance	Dimensionless	Dimensionless
N	Equivalent full temperature cycles		
N_S	Strouhal number	Dimensionless	Dimensionless
N_{De}	Dean number	Dimensionless	Dimensionless
N_{Fr}	Froude number	Dimensionless	Dimensionless
N_{Re}	Reynolds number	Dimensionless	Dimensionless
N_{We}	Weber number	Dimensionless	Dimensionless
NPSH	Net positive suction head	m	ft
n	Polytropic exponent		
n	Pulsation frequency	Hz	1/s
n	Constant, general		
n	Number of items	Dimensionless	Dimensionless
P	Design gauge pressure	kPa	lbf/in^2
P_{ad}	Adiabatic power	kW	hp
p	Pressure	Pa	lbf/ft^2
p	Power	kW	hp
Q	Heat	J	Btu
Q	Volume	m^3	ft^3
Q	Volume rate of flow (liquids)	m^3/h	gal/min
Q	Volume rate of flow (gases)	m^3/h	ft^3/min (cfm)
q	Volume flow rate	m^3/s	ft^3/s
R	Gas constant	8314 J/(K · mol)	1545 (ft · lbf)/(mol · °R)
R	Radius	m	ft
R	Electrical resistance	Ω	Ω
R	Head reading	m	ft
R	Range of reaction forces or moments in flexibility analysis	N or N · mm	lbf or in · lbf
R	Cylinder radius	m	ft
R	Universal gas constant	J/(kg · K)	(ft · lbf)/(lbm · °R)
R_a	Estimated instantaneous reaction force or moment at installation temperature	N or N · mm	lbf or in · lbf
R_m	Estimated instantaneous maximum reaction force or moment at maximum or minimum metal temperature	N or N · mm	lbf or in · lbf
R_1	Effective radius of miter bend	mm	in

Nomenclature and Units *(Continued)*

Symbol	Definition	SI units	U.S. customary units
r	Radius	m	ft
r	Pressure ratio	Dimensionless	Dimensionless
r_c	Critical pressure ratio		
r_k	Knuckle radius	m	in
r_2	Mean radius of pipe using nominal wall thickness $\bar{T}$	mm	in
S	Specific surface area	m^2/m^3	ft^2/ft^3
S	Fluid head loss	Dimensionless	Dimensionless
S	Specific energy loss	m/s^2	lbf/lb
S	Speed	m^3/s	ft^3/s
S	Basic allowable stress for metals, excluding factor E, or bolt design stress	MPa	kip/in^2 (ksi)
S_A	Allowable stress range for displacement stress	MPa	kip/in^2 (ksi)
S_E	Computed displacement-stress range	MPa	kip/in^2 (ksi)
S_L	Sum of longitudinal stresses	MPa	kip/in^2 (ksi)
S_T	Allowable stress at test temperature	MPa	kip/in^2 (ksi)
S_b	Resultant bending stress	MPa	kip/in^2 (ksi)
S_c	Basic allowable stress at minimum metal temperature expected	MPa	kip/in^2 (ksi)
S_h	Basic allowable stress at maximum metal temperature expected	MPa	kip/in^2 (ksi)
S_t	Torsional stress	MPa	kip/in^2 (ksi)
s	Specific gravity		
s	Specific entropy	J/(kg·K)	Btu/(lb·°R)
T	Temperature	K (°C)	°R (°F)
T_s	Effective branch-wall thickness	mm	in
$\bar{T}$	Nominal wall thickness of pipe	mm	in
$\bar{T}_b$	Nominal branch-pipe wall thickness	mm	in
$\bar{T}_h$	Nominal header-pipe wall thickness	mm	in
t	Head or shell radius	mm	in
t	Pressure design thickness	mm	in
t	Time	s	s
t_m	Minimum required thickness, including mechanical, corrosion, and erosion allowances	mm	in
t_r	Pad or saddle thickness	mm	in
U	Straight-line distance between anchors	m	ft
u	Specific internal energy	J/kg	Btu/lb
u	Velocity	m/s	ft/s
V	Velocity	m/s	ft/s
V	Volume	m^3	ft^3

Symbol	Definition	SI units	U.S. customary units
υ	Specific volume	m^3/kg	ft^3/lb
W	Work	N·m	lbf·ft
W	Weight	kg	lb
w	Weight flow rate	kg/s	lb/s
x	Weight fraction	Dimensionless	Dimensionless
x	Distance or length	m	ft
x	Value of expression $[(p_2/p_1)^{(k-1/k)} - 1]$		
Y	Expansion factor	Dimensionless	Dimensionless
y	Distance or length	m	ft
y	Resultant of total displacement strains	mm	in
Z	Section modulus of pipe	mm^3	in^3
Z	Vertical distance	m	ft
Z_e	Effective section modulus for branch	mm^3	in^3
Z	Gas-compressibility factor	Dimensionless	Dimensionless
z	Vertical distance	m	ft
	Greek symbols		
α	Viscous-resistance coefficient	$1/m^2$	$1/ft^2$
α	Angle	°	°
σ	Half-included angle	°	°
α, β, θ	Angles	°	°
β	Inertial-resistance coefficient	1/m	1/ft
β	Ratio of diameters	Dimensionless	Dimensionless
Γ	Liquid loading	kg/(s·m)	lb/(s·ft)
Γ	Pulsation intensity	Dimensionless	Dimensionless
δ	Thickness	m	ft
ε	Wall roughness	m	ft
ε	Voidage—fractional free volume	Dimensionless	Dimensionless
η	Viscosity, nonnewtonian fluids	Pa·s	lb/(ft·s)
η_{ad}	Adiabatic efficiency		
η_p	Polytropic efficiency		
θ	Angle	°	°
λ	Molecular mean free-path length	m	ft
μ	Viscosity	Pa·s	lb/(ft·s)
ν	Kinematic viscosity	m^2/s	ft^2/s
ρ	Density	kg/m^3	lb/ft^3
σ	Surface tension	N/m	lbf/ft
σ_c	Cavitation number	Dimensionless	Dimensionless
τ	Shear stress	N/m^2	lbf/ft^2
ϕ	Shape factor	Dimensionless	Dimensionless
ϕ	Angle	°	°
ϕ	Flow coefficient		
ψ	Pressure coefficient		
ψ	Sphericity	Dimensionless	Dimensionless

MEASUREMENT OF FLOW

General References: ASME, *Performance Test Code on Compressors and Exhausters,* PTC 10-1997, American Society of Mechanical Engineers (ASME), New York, 1997. Norman A. Anderson, *Instrumentation for Process Measurement and Control,* 3d ed., CRC Press, Boca Raton, Fla., 1997. Roger C. Baker, *Flow Measurement Handbook: Industrial Designs, Operating Principles, Performance, and Applications,* Cambridge University Press, Cambridge, UK, 2000. Roger C. Baker, *An Introductory Guide to Flow Measurement,* ASME, New York, 2003. Howard S. Bean, ed., *Fluid Meters—Their Theory and Application—Report of the ASME Research Committee on Fluid Meters,* 6th ed., ASME, New York, 1971. Douglas M. Considine, Editor-in-Chief, *Process/Industrial Instruments and Controls Handbook,* 4th ed., McGraw-Hill, New York, 1993. Bela G. Liptak, Editor-in-Chief, *Process Measurement and Analysis,* 4th ed., CRC Press, Boca Raton, Fla., 2003. Richard W. Miller, *Flow Measurement Engineering Handbook,* 3d ed., McGraw-Hill, New York, 1996. Ower and Pankhurst, *The Measurement of Air Flow,* Pergamon, Oxford, UK, 1966. Brian Price et al., *Engineering Data Book,* 12th ed., Gas Processors Suppliers Association, Tulsa, Okla., 2004. David W. Spitzer, *Flow Measurement,* 2d ed., International Society of Automation, Research Triangle Park, N.C., 2001. David W. Spitzer, *Industrial Flow Measurement,* 3d ed., International Society of Automation, Research Triangle Park, N.C., 2005.

INTRODUCTION

The flow rate of fluids is a critical variable in most chemical engineering applications, ranging from flows in the process industries to environmental flows and to flows within the human body. *Flow* is defined as mass flow or volume flow per unit of time at specified temperature and pressure conditions for a given fluid. This subsection deals with the techniques of measuring the pressure, temperature, velocities, and flow rates of flowing fluids. For more detailed discussion of these variables, consult Sec. 8, which introduces methods of measuring flow rate, temperature, and pressure. This subsection builds on the coverage in Sec. 8 with emphasis on measurement of the flow of fluids.

PROPERTIES AND BEHAVIOR OF FLUIDS

Transportation and the storage of fluids (gases and liquids) involve the understanding of the properties and behavior of fluids. The study of fluid dynamics is the study of fluids and their motion in a force field.

Flows can be classified into two major categories: (1) incompressible flow and (2) compressible flow. Most liquids fall into the incompressible flow category, while most gases are compressible. A *perfect fluid* can be defined as a fluid that is nonviscous and nonconducting. Fluid flow, compressible or incompressible, can be classified by the ratio of the inertial forces to the viscous forces. This ratio is represented by *Reynolds number* N_{Re}. At a low Reynolds number, the flow is considered to be laminar, and at high Reynolds numbers, the flow is considered to be turbulent. The limiting types of flow are the inertialess flow, sometimes called *Stokes' flow,* and the inviscid flow that occurs at an infinitely large Reynolds number. The Reynolds number (dimensionless) for flow in a pipe is given as

$$N_{Re} = \frac{\rho VD}{\mu} \tag{10-1}$$

where ρ is the density of the fluid, V the velocity, D the diameter, and μ the viscosity of the fluid. In fluid motion where the friction forces interact with the inertia forces, it is important to consider the ratio of the viscosity μ to the density ρ. This ratio is known as the *kinematic viscosity* ν. Tables 10-1 and 10-2 give the kinematic viscosity for several fluids. A flow is considered to be *adiabatic* when there is no transfer of heat between the fluid and its surroundings. An *isentropic* flow is one in which the entropy of each fluid element remains constant.

To fully understand the mechanics of flow, the following definitions explain the behavior of various types of fluids in both their static and flowing states.

A perfect fluid is a nonviscous, nonconducting fluid. An example of this type of fluid would be a fluid that has a very small viscosity and conductivity and is at a high Reynolds number. An ideal gas is one that obeys the equation of state:

$$\frac{P}{\rho} = RT \tag{10-2}$$

where P = pressure, ρ = density, R is the gas constant per unit mass, and T = absolute temperature.

A flowing fluid is acted upon by many forces that result in changes in pressure, temperature, stress, and strain. A fluid is said to be *isotropic* when the relations between the components of stress and those of the rate of strain are the same in all directions. The fluid is said to be *newtonian* when this relationship is linear. These pressures and temperatures must be fully understood so that the entire flow picture can be described.

The *static pressure* in a fluid has the same value in all directions and can be considered as a scalar point function. It is the pressure of a flowing fluid. It is normal to the surface on which it acts and at any given point has the same magnitude irrespective of the orientation of the surface. The static pressure arises because of the random motion in the fluid of the molecules that make up the fluid. In a diffuser or nozzle, there is an increase or decrease in the static pressure due to the change in velocity of the moving fluid.

Total pressure is the pressure that would occur if the fluid were brought to rest in a reversible adiabatic process. Many texts and engineers use the words *total* and *stagnation* to describe the flow characteristics interchangeably. To be accurate, the stagnation pressure is the pressure that would occur if the fluid were brought to rest isentropically.

Total pressure will change in a fluid only if shaft work or the work of extraneous forces is introduced. Therefore, total pressure would increase in the impeller of a compressor or pump; it would remain constant in the diffuser. Similarly, total pressure would decrease in the turbine impeller but would remain constant in the nozzles.

Static temperature is the temperature of the flowing fluid. Like static pressure, it arises because of the random motion of the fluid molecules. Static temperature is, in most practical installations, impossible to measure since it can be measured only by a thermometer or thermocouple at rest relative to the flowing fluid that is moving with the fluid. Static temperature will increase in a diffuser and decrease in a nozzle.

Total temperature is the temperature that would occur if the fluid were brought to rest in a reversible adiabatic manner. Just like its counterpart *total pressure, total* and *stagnation temperatures* are used interchangeably by many test engineers.

TABLE 10-1 Density, Viscosity, and Kinematic Viscosity of Water and Air in Terms of Temperature

Temperature		Water			Air at a pressure of 760 mm Hg (14.696 lbf/in^2)		
(°C)	(°F)	Density ρ (lbf·s^2/ft^4)	Viscosity μ × 10^6 (lbf·s/ft^2)	Kinematic viscosity ν × 10^6 (ft^2/s)	Density ρ (lbf·s^2/ft^4)	Viscosity μ × 10^6 (lbf·s/ft^2)	Kinematic viscosity ν × 10^6 (ft^2/s)
−20	−4	—	—	—	0.00270	0.326	122
−10	14	—	—	—	0.00261	0.338	130
0	32	1.939	37.5	19.4	0.00251	0.350	140
10	50	1.939	27.2	14.0	0.00242	0.362	150
20	68	1.935	21.1	10.9	0.00234	0.375	160
40	104	1.924	13.68	7.11	0.00217	0.399	183
60	140	1.907	9.89	5.19	0.00205	0.424	207
80	176	1.886	7.45	3.96	0.00192	0.449	234
100	212	1.861	5.92	3.19	0.00183	0.477	264

Conversion factors: 1 kp·s^2/m^4 = 0.01903 lbf·s^2/ft^4 (= slug/ft^3)
1 lbf·s^2/ft^4 = 32.1719 lb/ft^3 (lb = lb mass; lbf = lb force)
1 kp·s^2/m^4 = 9.80665 kg/m^3 (kg = kg mass; kp = kg force)
1 kg/m^3 = 16.02 lb/ft^3

TABLE 10-2 Kinematic Viscosity

Liquid	Temperature °C	°F	$\nu \times 10^6$ (ft²/s)
Glycerine	20	68	7319
Mercury	0	32	1.35
Mercury	100	212	0.980
Lubricating oil	20	68	4306
Lubricating oil	40	104	1076
Lubricating oil	60	140	323

Dynamic temperature and *dynamic pressure* are the difference between the total and static conditions.

$$P_d = P_T - P_s \tag{10-3}$$

$$T_d = T_T - T_s \tag{10-4}$$

where subscript d refers to dynamic, T to total, and s to static.

Another helpful formula is

$$P_K = \frac{1}{2}\rho V^2 \tag{10-5}$$

For incompressible fluids, $P_K = P_d$.

TOTAL TEMPERATURE

For most points requiring temperature monitoring, either thermocouples or *resistive thermal detectors* (RTDs) can be used. Each type of temperature transducer has its own advantages and disadvantages, and both should be considered when temperature is to be measured. Since there is considerable confusion in this area, a short discussion of the two types of transducers is necessary.

Thermocouples The various types of thermocouples provide transducers suitable for measuring temperatures from −330 to 5000°F (−201 to 2760°C). Thermocouples function by producing a voltage proportional to the temperature differences between two junctions of dissimilar metals. By measuring this voltage, the temperature difference can be determined. It is assumed that the temperature is known at one of the junctions; therefore, the temperature at the other junction can be determined. Since the thermocouples produce a voltage, no external power supply is required to the test junction; however, for accurate measurement, a reference junction is required. For a temperature monitoring system, reference junctions must be placed at each thermocouple, or similar thermocouple wire must be installed from the thermocouple to the monitor where there is a reference junction. Properly designed thermocouple systems can be accurate to approximately ±2°F (±1°C).

Resistive Thermal Detectors RTDs determine temperature by measuring the change in resistance of an element due to temperature. Platinum is generally utilized in RTDs because it remains mechanically and electrically stable, resists contaminations, and can be highly refined. The useful range of platinum RTDs is −454 to 1832°F (−270 to 1000°C). Since the temperature is determined by the resistance in the element, any type of electrical conductor can be utilized to connect the RTD to the indicator; however, an electric current must be provided to the RTD. A properly designed temperature monitoring system utilizing RTDs can be accurate to ±0.02°F (±0.01°C).

STATIC TEMPERATURE

Since this temperature requires the thermometer or thermocouple to be at rest relative to the flowing fluid, it is impractical to measure. However, it can be calculated from the measurement of total temperature and total and static pressure.

$$T_S = \frac{T_o}{\left(\dfrac{P_o}{P_S}\right)^{(k-1)/k}} \tag{10-6}$$

DRY- AND WET-BULB TEMPERATURES

The moisture content or humidity of air has an important effect on the properties of the gaseous mixture. Steam in air at any relative humidity less than 100 percent must exist in a superheated condition. The saturation temperature corresponding to the actual partial pressure of the steam in air is called the *dew point*. This term arose from the fact that when air at less than 100 percent relative humidity is cooled to the temperature at which it becomes saturated, the air has reached the minimum temperature to which it can be cooled without precipitation of the moisture (dew). *Dew point* can also be defined as that temperature at which the weight of steam associated with a certain weight of dry air is adequate to saturate that weight of air.

The *dry-bulb temperature* of air is the temperature indicated by an ordinary thermometer. In contrast to dry-bulb, or air, temperature, the term *wet-bulb temperature of the air*, or simply *wet-bulb temperature*, is employed. When a thermometer, with its bulb covered by a wick wetted with water, is moved through air unsaturated with water vapor, the water evaporates in proportion to the capacity of the air to absorb the evaporated moisture, and the temperature indicated by the thermometer drops below the dry-bulb, or air, temperature. The equilibrium temperature finally reached by the thermometer is known as the *wet-bulb temperature*. The purpose in measuring both the dry-bulb and wet-bulb temperatures of the air is to find the exact humidity characteristics of the air from the readings obtained, either by calculation or by use of a psychrometric chart. Instruments for measuring wet-bulb and dry-bulb temperatures are known as *psychrometers*, which are defined in Sec. 12. For other methods of measuring the moisture content of gases, see Sec. 8.

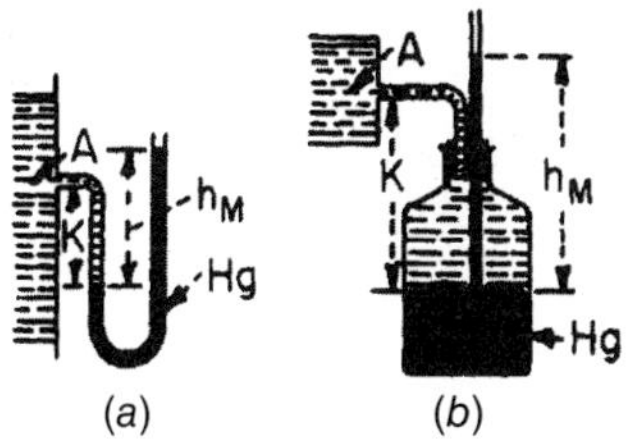

FIG. 10-1 Open manometers.

PRESSURE MEASUREMENTS

Pressure is defined as the force per unit area. Pressure devices measure with respect to the ambient atmospheric pressure: The *absolute pressure* P_a is the pressure of the fluid (gauge pressure) plus the atmospheric pressure.

Process pressure-measuring devices may be divided into three groups:

1. Those based on the height of a liquid column (manometers)
2. Those based on the measurement of the distortion of an elastic pressure chamber (mechanical pressure gauges such as Bourdon-tube gauges and diaphragm gauges)
3. Electric sensing devices (strain gauges, piezoresistive transducers, and piezoelectric transducers)

This subsection contains an expanded discussion of manometric methods. See Sec. 8 for other methods.

Liquid-Column Manometers The *height*, or *head*, $p_n = \rho hg/g_c$ to which a fluid rises in an open vertical tube attached to an apparatus containing a liquid is a direct measure of the pressure at the point of attachment and is frequently used to show the level of liquids in tanks and vessels. This same principle can be applied with U tube gauges (Fig. 10-1*a*) and equivalent devices (such as that shown in Fig. 10-1*b*) to measure pressure in terms of the head of a fluid other than the one under test. Most of these gauges may be used either as *open* or as *differential manometers*. The manometric fluid that constitutes the measured liquid column of these gauges may be any liquid immiscible with the fluid under pressure. For high vacuums or for high pressures and large pressure differences, the gauge liquid is a high-density liquid, generally mercury; for low pressures and small pressure differences, a low-density liquid (e.g., alcohol, water, or carbon tetrachloride) is used.

The *open U tube* (Fig. 10-1*a*) and the *open gauge* (Fig. 10-1*b*) each show a reading h_M, which represents m (ft) of manometric fluid. If the interface of the manometric fluid and the fluid of which the pressure is wanted is K m (ft) below the point of attachment A, ρ_A is the density of the latter fluid at A, and ρ_M is that of the manometric fluid, then gauge pressure p_A (lb$_f$/ft²) at A is

$$p_A = (h_M\rho_M - K\rho_A)(g/g_c) \tag{10-7}^*$$

where g = local acceleration due to gravity and g_c = dimensional constant. The head h_A at A as meters (feet) of the fluid at that point is

$$h_A = h_M(\rho_M/\rho_A) - K \tag{10-8}^*$$

When a gas pressure is measured, unless it is very high, ρ_A is so much smaller than ρ_M that the terms involving K in these formulas are negligible.

*The line leading from the pressure tap to the gauge is assumed to be filled with fluid of the same density as that in the apparatus at the location of the pressure tap. If this is not the case, ρ_A is the density of the fluid actually filling the gauge line, and the value given for h_A must be multiplied by ρ_A/ρ, where ρ is the density of the fluid whose head is being measured.

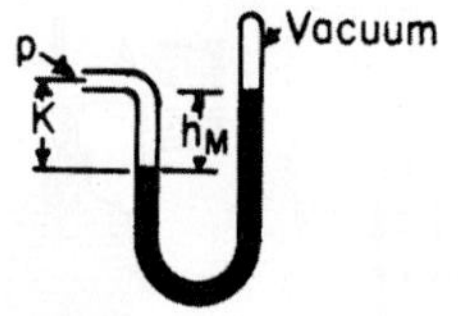

FIG. 10-2 Closed U tube.

Closed U Tubes Closed U tubes (Fig. 10-2) using mercury as the manometric fluid serve to measure directly the absolute pressure p of a fluid, provided that the space between the closed end and the mercury is substantially a perfect vacuum.

The *mercury barometer* (Fig. 10-3) indicates directly the absolute pressure of the atmosphere in terms of the height of the mercury column. Normal (standard) barometric pressure is 101.325 kPa by definition. Equivalents of this pressure in other units are 760 mm mercury (at 0°C), 29.921 in.Hg (at 0°C), 14.696 lbf/in^2, and 1 atm. For cases in which barometer readings, when expressed by the height of a mercury column, must be corrected to standard temperature (usually 0°C), appropriate temperature correction factors are given in ASME, *Performance Test Code,* 1997, pp. 23–26, and Weast, *Handbook of Chemistry and Physics,* 65th ed., Chemical Rubber, Cleveland, Ohio, 1985, pp. E36–E37.

Tube Size for Manometers To avoid capillary error, the tube diameter should be sufficiently large and the manometric fluids of such densities that the effect of capillarity is negligible in comparison with the gauge reading. The effect of capillarity is practically negligible for tubes with inside diameters of 12.7 mm (0.5 in) or larger (see ASME, *Performance Test Code,* 1997, p. 15). Small diameters are generally permissible for U tubes because the capillary displacement in one leg tends to cancel that in the other leg.

The capillary rise in a small vertical open tube of circular cross section dipping into a pool of liquid is given by

$$h = \frac{4\sigma g_c \cos\theta}{gD(\rho_1 - \rho_2)} \tag{10-9}$$

Here σ = surface tension, D = inside diameter, ρ_1 and ρ_2 are the densities of the liquid and gas (or light liquid), respectively, g = local acceleration due to gravity, g_c = dimensional constant, and θ = the contact angle subtended by the heavier fluid. For most organic liquids and water, the contact angle θ is zero against glass, provided the glass is wet with a film of the liquid; for mercury against glass, θ =140° (*International Critical Tables,* vol. 4, McGraw-Hill, New York, 1928, pp. 434–435). For further discussion of capillarity, see Schwartz, *Ind. Eng. Chem.* **61**(1): 10–21 (1969).

Multiplying Gauges To attain the requisite precision in measurement of small pressure differences by liquid-column manometers, means must often be devised to magnify the readings. The inclined U tube (Fig. 10-4) and the draft gauge may give 10-fold multiplication. The two-fluid U tube can magnify small pressure measurements by as much as 30-fold (*Perry's Chemical Engineers' Handbook,* 8th ed., McGraw-Hill, New York, 2008, p. 10-9). In general, the greater the multiplication, the more elaborate must be the precautions in the use of the gauge if the gain in precision is not to be illusory.

1. *Change of manometric fluid.* In open manometers, choose a fluid of lower density. In differential manometers, choose a fluid such that the difference between its density and that of the fluid being measured is as small as possible.

2. *Inclined U tube* (Fig. 10-4). If the reading R m (ft) is taken as shown and R_0 m (ft) is the zero reading, by making the substitution $h_M = (R - R_0) \sin\theta$, the formulas of preceding paragraphs give $p_A - p_B$ when the corresponding upright U tube is replaced by an inclined one. For precise work, the gauge should be calibrated because of possible variations in tube diameter and slope.

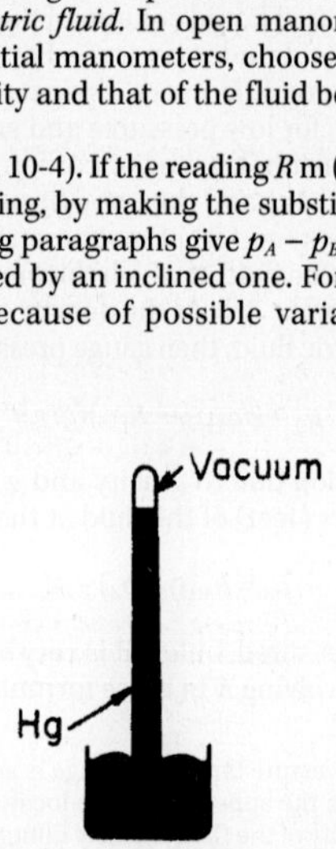

FIG. 10-3 Mercury barometer.

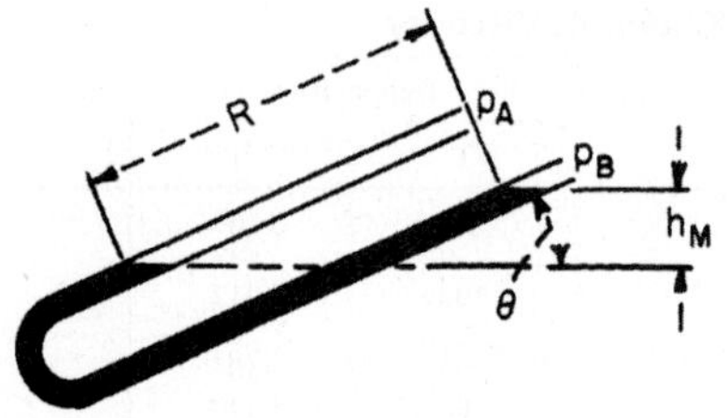

FIG. 10-4 Inclined U tube.

Several *micromanometers,* based on the liquid-column principle and possessing extreme precision and sensitivity, have been developed for measuring minute gas pressure differences and for calibrating low-range gauges. Some of these micromanometers are available commercially. These micromanometers are free from errors due to capillarity and, aside from checking the micrometer scale, require no calibration.

Mechanical Pressure Gauges The *Bourdon-tube gauge* indicates pressure by the amount of flection under internal pressure of an oval tube bent in an arc of a circle and closed at one end. These gauges are commercially available for all pressures below atmospheric and for pressures up to 700 MPa (about 100,000 lbf/in^2) above atmospheric. Details on Bourdon-type gauges are given by Harland [*Mach. Des.* **40**(22): 69–74 (Sept. 19, 1968)].

A *diaphragm gauge* depends for its indication on the deflection of a diaphragm, usually metallic, when subjected to a difference of pressure between the two faces. These gauges are available for the same general purposes as Bourdon gauges but are not usually employed for high pressures. The aneroid barometer is a type of diaphragm gauge.

Small *pressure transducers with flush-mounted diaphragms* are commercially available for the measurement of either steady or fluctuating pressures up to 100 MPa (about 15,000 lbf/in^2). The metallic diaphragms are as small as 4.8 mm ($\frac{3}{16}$ in) in diameter. The transducer is mounted on the apparatus containing the fluid whose pressure is to be measured so that the diaphragm is flush with the inner surface of the apparatus. Deflection of the diaphragm is measured by unbonded strain gauges and recorded electrically.

With nonnewtonian fluids the pressure measured at the wall with non-flush-mounted pressure gauges may be in error (see the subsection Static Pressure).

Bourdon and diaphragm gauges that show both pressure and vacuum indications on the same dial are called *compound gauges.*

Conditions of Use Bourdon tubes should not be exposed to temperatures over about 65°C (about 150°F) unless they are specifically designed for such operation. When the pressure of a hotter fluid is to be measured, some type of liquid seal should be used to keep the hot fluid from the tube. In using either a Bourdon or a diaphragm gauge to measure gas pressure, if the gauge is below the pressure tap of the apparatus so that liquid can collect in the lead, then the gauge reading will be too high by an amount equal to the hydrostatic head of the accumulated liquid.

For measuring pressures of corrosive fluids, slurries, and similar process fluids which may foul Bourdon tubes, a *chemical gauge,* consisting of a Bourdon gauge equipped with an appropriate flexible diaphragm to seal off the process fluid, may be used. The combined volume of the tube and the connection between the diaphragm and the tube is filled with an inert liquid. These gauges are available commercially.

Further details on pressure-measuring devices can be found in Sec. 8.

Calibration of Gauges Simple *liquid-column manometers* do not require calibration if they are so constructed as to minimize errors due to capillarity (see the subsection Liquid-Column Manometers). If the scales used to measure the readings have been checked against a standard, the accuracy of the gauges depends solely on the precision of determining the position of the liquid surfaces. Hence liquid-column manometers are primary standards used to calibrate other gauges.

For *high pressures* and, with commercial mechanical gauges, even for quite moderate pressures, a deadweight gauge (see ASME, *Performance Test Code,* 1997, pp. 36–41) is commonly used as the primary standard because it is safer and more convenient than use of manometers. When manometers are used as high-pressure standards, an extremely high mercury column may be avoided by connecting a number of the usual U tubes in series. Multiplying gauges are standardized by comparing them with a micromanometer. The procedure in the calibration of a gauge consists merely of connecting it, in parallel with a standard gauge, to a reservoir wherein constant pressure may be maintained. Readings of the unknown gauge are then made for various reservoir pressures as determined by the standard.

Calibration of *high-vacuum gauges* is described by Sellenger [*Vacuum* **18**(12): 645–650 (1968)].

STATIC PRESSURE

Local Static Pressure In a moving fluid, the local static pressure is equal to the pressure on a surface which moves with the fluid or to the normal pressure (for newtonian fluids) on a stationary surface that parallels the flow. The pressure on such a surface is measured by making a small hole perpendicular to the surface and connecting the opening to a pressure-sensing element (Fig. 10-5*a*). The hole is known as a *piezometer opening* or *pressure tap*.

Measurement of local static pressure is frequently difficult or impractical. If the channel is so small that introduction of any solid object disturbs the flow pattern and increases the velocity, there will be a reduction and redistribution of the static pressure. If the flow is in straight parallel lines, aside from the fluctuations of normal turbulence, the flat disk (Fig. 10-5*b*) and the bent tube (Fig. 10-5*c*) give satisfactory results when properly aligned with the stream. Slight misalignments can cause serious errors. The diameter of the disk should be 20 times its thickness and 40 times the static opening; the face must be flat and smooth, with the knife edges made by bevelling the underside. The piezometer tube, such as that in Fig. 10-5*c*, should have openings with size and spacing as specified for a pitot-static tube (Fig. 10-9).

Readings given by open straight tubes (Fig. 10-5*d* and *e*) are too low due to flow separation. Readings of closed tubes oriented perpendicular to the axis of the stream and provided with side openings (Fig. 10-5*e*) may be low by as much as two velocity heads.

Average Static Pressure In most cases, the object of a static pressure measurement is to obtain a suitable average value for substitution in Bernoulli's theorem or in an equivalent flow formula. This can be done simply only when the flow is in straight lines parallel to the confining walls, such as in straight ducts at sufficient distance downstream from bends (2 diameters) or other disturbances. For such streams, the sum of the static head and gravitational potential head is the same at all points in a cross section taken perpendicular to the axis of flow. Thus the exact location of a piezometer opening about the periphery of such a cross section is immaterial, provided its elevation is known. However, in stating the static pressure, the custom is to give the value at the elevation corresponding to the centerline of the stream.

With flow in curved passages or with swirling flow, determination of a true average static pressure is, in general, impractical. In metering, straightening vanes are often placed upstream of the pressure tap to eliminate swirl. Figure 10-6 shows various flow equalizers and straighteners.

Specifications for Piezometer Taps The size of a static opening should be small compared with the diameter of the pipe and yet large compared with the scale of surface irregularities. For reliable results, it is essential that (1) the surface in which the hole is made be substantially smooth and parallel to the flow for some distance on either side of the opening and (2) the opening be flush with the surface and possess no "burr" or other irregularity around its edge. Rounding of the edge is often employed to ensure absence of a burr. Pressure readings will be high if the tap is inclined upstream, is rounded excessively on the upstream side, has a burr on the downstream side, or has an excessive countersink or recess. Pressure readings will be low if the tap is inclined downstream, is rounded excessively on the downstream side, has a burr on the upstream side, or protrudes into the flow stream. Errors resulting from these faults can be large.

Recommendations for *pressure-tap dimensions* are summarized in Table 10-3. Data from several references were used in arriving at these composite values. The length of a pressure-tap opening prior to any enlargement in the tap channel should be at least 2 tap diameters, preferably 3 or more.

A *piezometer ring* is a toroidal manifold into which are connected several sidewall static taps located around the perimeter of a common cross section. Its intent is to give an average pressure if differences in pressure exist around the perimeter other than those due to static head. However, there is generally no assurance that a true average is provided thereby. The principal advantage of the ring is that use of several holes in place of a single hole reduces the possibility of completely plugging the static openings.

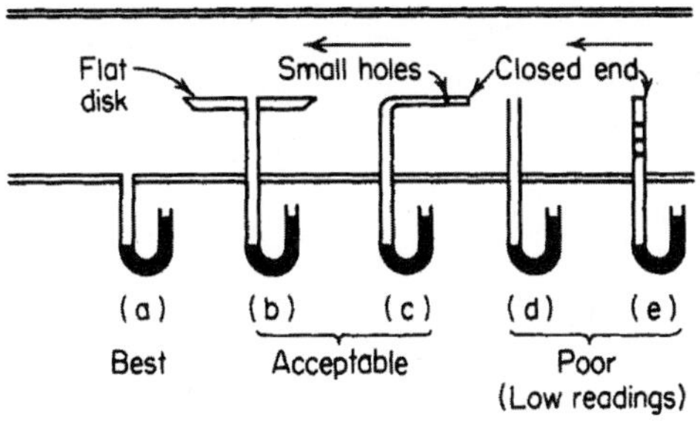

FIG. 10-5 Measurement of static pressure.

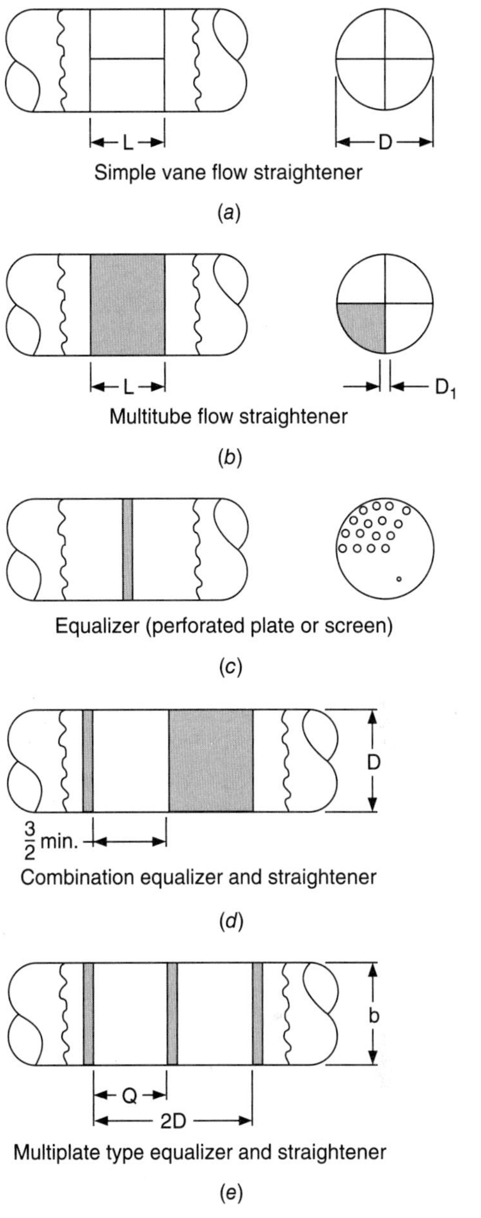

FIG. 10-6 Flow equalizers and straighteners. [*Power Test Code 10, Compressors and Exhausters,* ASME, 1997]

For information on prediction of static-hole error, see Shaw, *J. Fluid Mech.* **7:** 550–564 (1960); Livesey, Jackson, and Southern, *Aircr. Eng* **34:** 43–47 (February 1962).

For nonnewtonian fluids, pressure readings with taps may also be low because of fluid-elasticity effects. This error can be largely eliminated by using flush-mounted diaphragms.

For information on the pressure-hole error for nonnewtonian fluids, see Han and Kim, *Trans. Soc. Rheol.* **17:** 151–174 (1973); Novotny and Eckert, *Trans. Soc. Rheol.* **17:** 227–241 (1973); and Higashitani and Lodge, *Trans. Soc. Rheol.* **19:** 307–336 (1975).

TABLE 10-3 Pressure-Tap Holes

Nominal inside pipe diameter, in	Maximum diameter of pressure tap, mm (in)	Radius of hole-edge rounding, mm (in)
1	3.18 (1/8)	<0.40 (1/64)
2	6.35 (1/4)	0.40 (1/64)
3	9.53 (3/8)	0.40–0.79 (1/64 – 1/32)
4	12.7 (1/2)	0.79 (1/32)
8	12.7 (1/2)	0.79–1.59 (1/32–1/16)
16	19.1 (3/4)	0.79–1.59 (1/32–1/16)

VELOCITY MEASUREMENTS

Measurement of flow can be based on the measurement of velocity in ducts or pipes by using devices such as pitot tubes and hot wire anemometers. The local velocity is measured at various sections of a conduit and then averaged for the area under consideration.

$$\frac{\dot{m}}{\rho} = A \times V = \dot{Q} \tag{10-10}$$

where $\dot{m}$ = mass flow rate, lbm/s, kg/s
ρ = density, lbm/ft^3, kg/m^3
A = area, ft^2, m^2
V = velocity, ft/s, m/s
$\dot{Q}$ = volumetric flow rate, ft^3/s, m^3/s

Equation (10-10) shows that the fluid density directly affects the relationship between mass flow rate and both the velocity and volumetric flow rates. Liquid temperature affects liquid density and hence volumetric flow rate at a constant mass flow rate. Liquid density is relatively insensitive to pressure. Both temperature and pressure affect gas density and thus volumetric flow rate.

Variables Affecting Measurement Flow measurement methods may sense local fluid velocity, volumetric flow rate, total or cumulative volumetric flow (the integral of volumetric flow rate with respect to elapsed time), mass flow rate, and total mass flow.

Velocity Profile Effects Many variables can influence the accuracy of specific flow measurement methods. For example, the velocity profile in a closed conduit affects many types of flow-measuring devices. The velocity of a fluid varies from zero at the wall and at other stationary solid objects in the flow channel to a maximum at a distance from the wall. In the entry region of a conduit, the velocity field may approach plug flow and a constant velocity across the conduit, dropping to zero only at the wall. As a newtonian fluid progresses down a pipe, a velocity profile develops that is parabolic for laminar flow [Eq. (6-41)] and that approaches plug flow for highly turbulent flow. Once a steady flow profile has developed, the flow is said to be *fully developed*; the length of conduit necessary to achieve fully developed flow is called the *entrance region*. For long cylindrical, horizontal pipe ($L < 40D$, where D is the inside diameter of the pipe and L is the upstream length of pipe), the velocity profile becomes fully developed. Velocity profiles in flowing fluids are discussed in greater detail in Sec. 6 (p. 6-XX).

For steady-state, isothermal, single-phase, uniform, fully developed newtonian flow in straight pipes, the velocity is greatest at the center of the channel and symmetric about the axis of the pipe. Of those flowmeters that are dependent on the velocity profile, they are usually calibrated for this type of flow. Thus any disturbances in flow conditions can affect flowmeter readings.

Upstream and downstream disturbances in the flow field are caused by valves, elbows, and other types of fittings. Two upstream elbows in two perpendicular planes will impart swirl in the fluid downstream. Swirl, similar to atypical velocity profiles, can lead to erroneous flow measurements. Although the effect is not as great as in upstream flow disturbances, downstream flow disturbances can also lead to erroneous flow measurements.

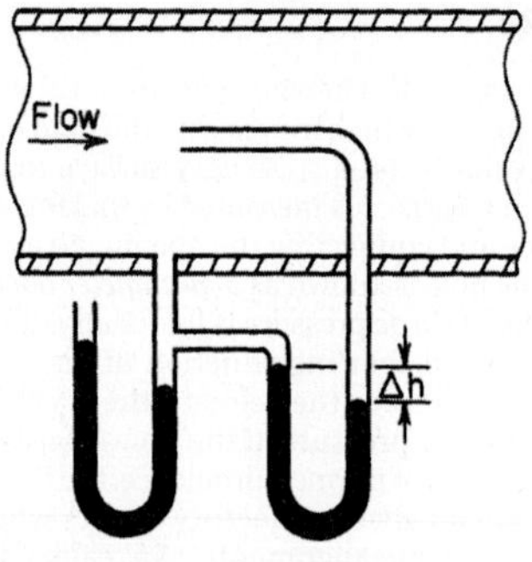

FIG. 10-7 Pitot tube with sidewall static tap.

Other Flow Disturbances Other examples of deviations from fully developed, single-phase newtonian flow include nonnewtonian flow, pulsating flow, cavitation, multiphase flow, boundary layer flows, and nonisothermal flows. See Sec. 6.

Pitot Tubes The combination of pitot tubes in conjunction with sidewall static taps measures local or point velocities by measuring the difference between the total pressure and the static pressure. The pitot tube shown in Fig. 10-7 consists of an impact tube whose opening faces directly into the stream to measure impact pressure, plus one or more sidewall taps to measure local static pressure.

Dynamic pressure may be measured by use of a pitot tube that is a simple impact tube. These tubes measure the pressure at a point where the velocity of the fluid is brought to zero. Pitot tubes must be parallel to the flow. The pitot tube is sensitive to yaw or angle attack. In general, angles of attack over 10° should be avoided. In cases where the flow direction is unknown, it is recommended to use a Kiel probe. Figure 10-8 shows a Kiel probe. This probe will read accurately to an angle of about 22° with the flow.

The combined pitot-static tube shown in Fig. 10-9 consists of a jacketed impact tube with one or more rows of holes, 0.51 to 1.02 mm (0.02 to 0.04 in) in diameter, in the jacket to measure the static pressure. Velocity V_0 m/s (ft/s) at the point where the tip is located is given by

$$V_0 = C\sqrt{2g_c \Delta h} = C\sqrt{2g_c(p_T - p_S)/\rho_0} \tag{10-11}$$

where C = coefficient, dimensionless; g_c = dimensional constant; Δh = dynamic pressure ($\Delta h_s g/g_c$), expressed in (N · m)/kg [(ft · lbf)/lb or ft of fluid flowing]; Δh_s = differential height of static liquid column corresponding

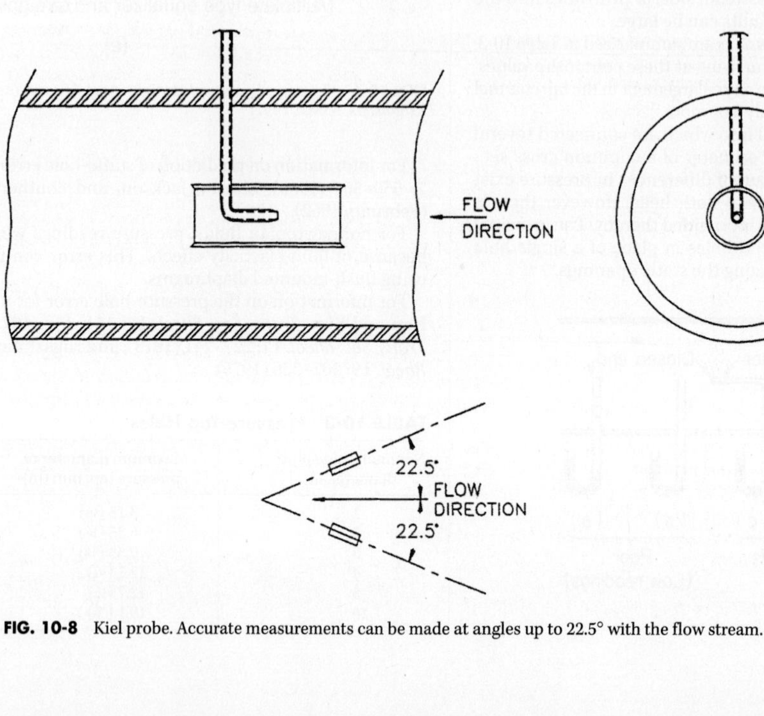

FIG. 10-8 Kiel probe. Accurate measurements can be made at angles up to 22.5° with the flow stream.

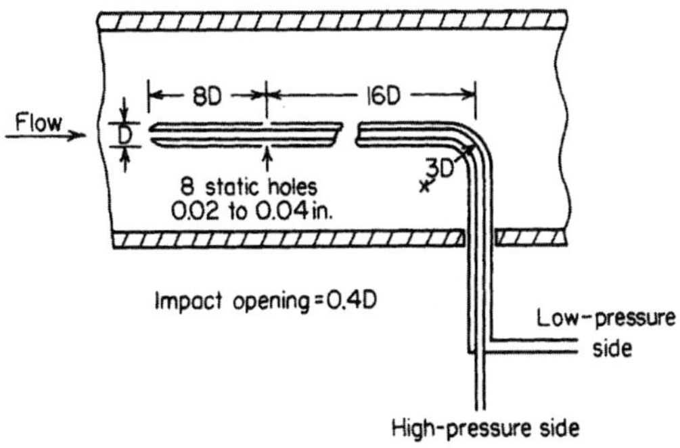

FIG. 10-9 Pitot-static tube.

to Δh; g = local acceleration due to gravity; g_c = dimensional constant; p_i = impact pressure; p_0 = local static pressure; and ρ_0 = fluid density measured at pressure p_0 and the local temperature. With gases at velocities above 60 m/s (about 200 ft/s), compressibility becomes important, and the following equation should be used:

$$V_0 = C\sqrt{\frac{2g_c k}{k-1}\left(\frac{p_0}{\rho_0}\right)\left[\left(\frac{p_i}{p_0}\right)^{(k-1)/k} - 1\right]} \tag{10-12}$$

where k is the ratio of specific heat at constant pressure to that at constant volume. (See ASME, *Report of the ASME Research Committee on Fluid Meters*, p. 105.) Coefficient C is usually close to 1.00 (±0.01) for simple pitot tubes (Fig. 10-7) and generally ranges between 0.98 and 1.00 for pitot-static tubes (Fig. 10-9).

There are certain limitations on the range of usefulness of pitot tubes. With gases, the differential is very small at low velocities; e.g., at 4.6 m/s (15.1 ft/s) the differential is only about 1.30 mm (0.051 in) of water (20°C) for air at 1 atm (20°C), which represents a lower limit for 1 percent error even when one uses a micromanometer with a precision of 0.0254 mm (0.001 in) of water. The equation does not apply for Mach numbers greater than 0.7 because of the interference of shock waves. For supersonic flow, local Mach numbers can be calculated from a knowledge of the dynamic and true static pressures. The *free stream Mach number* M_∞ is defined as the ratio of the speed of the stream V_∞ to the speed of sound in the free stream:

$$A_\infty = \sqrt{\left(\frac{\partial P}{\partial \rho}\right)_{s=c}} \tag{10-13}$$

$$M_\infty = \frac{V_\infty}{\sqrt{\left(\frac{\partial P}{\partial \rho}\right)_{s=c}}} \tag{10-14}$$

where s is the entropy. For isentropic flow, this relationship and pressure can be written as

$$M_\infty = \frac{V_\infty}{\sqrt{kRT_s}} \tag{10-15}$$

The relationships between total and static temperature and pressure are given by the following:

$$\frac{T_T}{T_S} = 1 + \frac{k-1}{2}M^2 \tag{10-16}$$

$$\frac{P_T}{P_S} = \left(1 + \frac{k-1}{2}M^2\right)^{(k-1)/k} \tag{10-17}$$

With *liquids* at low velocities, the effect of the Reynolds number upon the coefficient is important. The coefficients are appreciably less than unity for Reynolds numbers less than 500 for pitot tubes and for Reynolds numbers less than 2300 for pitot-static tubes [see Folsom, *Trans. Am. Soc. Mech. Eng.* **78:** 1447–1460 (1956)]. Reynolds numbers here are based on the probe outside diameter. Operation at low Reynolds numbers requires prior calibration of the probe.

The pitot-static tube is also more sensitive to *yaw* or *angle of attack* than is the simple pitot tube because of the sensitivity of the static taps to orientation. The error involved is strongly dependent on the exact probe dimensions. In general, angles greater than 10° should be avoided if the velocity error is to be 1 percent or less.

Disturbances upstream of the probe can cause large errors, in part because of the turbulence generated and its effect on the static-pressure measurement. A calming section of at least 50 pipe diameters is desirable. If this is not possible, the use of straightening vanes or a honeycomb is advisable.

The effect of *pulsating flow* on pitot-tube accuracy is treated by E. Ower in his article "On the response of a vane anemometer to an air-stream of pulsating speed," *The London and Edinburgh, and Dublin Philosophical Magazine and Journal of Science* Series 7, **23**(157) (1937), now available online https://doi.org/10.1080/14786443708561870. For sinusoidal velocity fluctuations, the ratio of indicated velocity to actual mean velocity is given by the factor $\sqrt{1+\lambda^2/2}$, where λ is the velocity excursion as a fraction of the mean velocity ±50 percent, and pulsations greater than ±20 percent should be damped to avoid errors greater than 1 percent. The error increases as the frequency of flow oscillations approaches the natural frequency of the pitot tube and the density of the measuring fluid approaches the density of the process fluid [see Horlock and Daneshyar, *J. Mech. Eng. Sci.* **15:** 144–152 (1973)].

Pressures substantially lower than true impact pressures are obtained with pitot tubes in turbulent flow of dilute polymer solutions [see Halliwell and Lewkowicz, *Phys. Fluids* **18:** 1617–1625 (1975)].

Special Tubes A variety of special forms of the pitot tube have evolved. Richard Gilman Folsom [*Trans. Am. Soc. Mech. Eng.* **78:** 1447–1460 (1956)] gives a description of many of these special types of pitot tubes together with a comprehensive bibliography. Included are the impact tube for *boundary-layer* measurements and *shielded total-pressure tubes*. The latter are insensitive to angle of attack up to 40°.

Chue [*Prog. Aerosp. Sci.* **16:** 147–223 (1975)] reviews the use of the pitot tube and allied pressure probes for impact pressure, static pressure, dynamic pressure, flow direction and local velocity, skin friction, and flow measurements.

A reversed pitot tube, also known as a *pitometer*, has one pressure opening facing upstream and the other facing downstream. Coefficient C for this type is on the order of 0.85. This gives about a 40 percent increase in pressure differential compared with standard pitot tubes and is an advantage at low velocities. There are commercially available very compact types of pitometers which require relatively small openings for their insertion into a duct.

The *pitot-venturi* flow element is capable of developing a pressure differential 5 to 10 times that of a standard pitot tube. This is accomplished by employing a pair of concentric venturi elements in place of the pitot probe. The low-pressure tap is connected to the throat of the inner venturi, which in turn discharges into the throat of the outer venturi. For a discussion of performance and application of this flow element, see Stoll, *Trans. Am. Soc. Mech. Eng.* **73:** 963–969 (1951).

Traversing for Mean Velocity The mean velocity in a duct can be obtained by dividing the cross section into a number of equal areas, finding the local velocity at a representative point in each, and averaging the results. In the case of *rectangular passages*, the cross section is usually divided into small squares or rectangles, and the velocity is found at the center of each. In circular pipes, the cross section is divided into several equal annular areas, as shown in Fig. 10-10. Readings of velocity are made at the intersections of a diameter and the set of circles which bisect the annuli and the central circle.

For an N-point traverse on a circular cross section, make readings on each side of the cross section at

$$100 \times \sqrt{(2n-1)/N} \text{ percent} \qquad (n = 1, 2, 3 \text{ to } N/2)$$

of the pipe radius from the center. Traversing several diameters spaced at equal angles about the pipe is required if the velocity distribution is unsymmetrical. With a normal velocity distribution in a circular pipe, a 10-point traverse theoretically gives a mean velocity 0.3 percent high; a 20-point traverse, 0.1 percent high.

For normal velocity distribution in straight circular pipes at locations preceded by runs of at least 50 diameters without pipe fittings or other obstructions, the graph in Fig. 10-10 shows the ratio of mean velocity V to velocity at the center u_{max} plotted versus the Reynolds number, where D = inside pipe diameter, ρ = fluid density, and μ = fluid viscosity, all in consistent units. Mean velocity is readily determined from this graph and a pitot reading at the center of the pipe if the quantity $Du_{max}\rho/\mu$ is less than 2000 or greater than 5000. The method is unreliable at intermediate values of the Reynolds number.

Methods for determining the mean flow rate from probe measurements under nonideal conditions are described by Mandersloot, Hicks, and Langejan [*Chem. Eng.* (*London*), no. 232, CE370–CE380 (1969)].

The *hot-wire anemometer* consists essentially of an electrically heated fine wire (generally platinum) exposed to the gas stream whose velocity

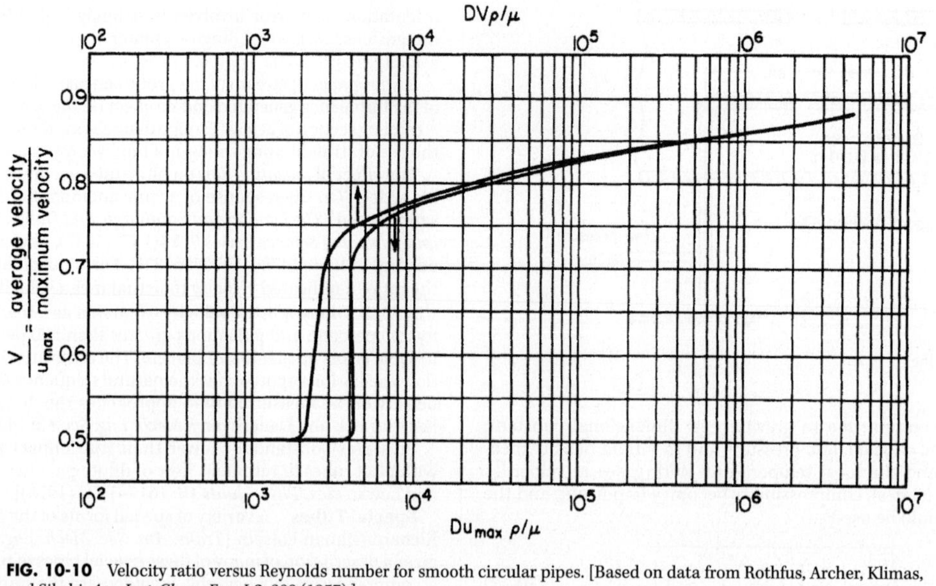

FIG. 10-10 Velocity ratio versus Reynolds number for smooth circular pipes. [Based on data from Rothfus, Archer, Klimas, and Sikchi, *Am. Inst. Chem. Eng. J.* 3: 208 (1957).]

is being measured. An increase in fluid velocity, other things being equal, increases the rate of heat flow from the wire to the gas, thereby tending to cool the wire and alter its electrical resistance. In a constant-current anemometer, the gas velocity is determined by measuring the resulting wire resistance; in the constant-resistance type, the gas velocity is determined from the current required to keep the wire temperature, and thus the resistance, constant. The difference in the two types is primarily in the electric circuits and instruments employed.

The hot-wire anemometer, with suitable calibration, can accurately measure velocities from about 0.15 m/s (0.5 ft/s) to supersonic velocities and can detect velocity fluctuations with frequencies up to 200,000 Hz. Fairly rugged, inexpensive units can be built for the measurement of mean velocities in the range of 0.15 to 30 m/s (about 0.5 to 100 ft/s). More elaborate, compensated units are commercially available for use in unsteady flow and turbulence measurements. In calibrating a hot-wire anemometer, it is preferable to use the same gas, temperature, and pressure as will be encountered in the intended application. In this case the quantity $I^2R_w/\Delta t$ can be plotted against $\sqrt{V}$, where I = hot-wire current, R_w = hot-wire resistance, Δt = difference between the wire temperature and the gas bulk temperature, and V = mean local velocity. A procedure is given by Wasan and Baid [*Am. Inst. Chem. Eng. J.* **17:** 729–731 (1971)] for use when it is impractical to calibrate with the same gas composition or conditions of temperature and pressure. Andrews, Bradley, and Hundy [*Int. J. Heat Mass Transfer* **15:** 1765–1786 (1972)] give a calibration correlation for measurement of small gas velocities. The hot-wire anemometer is treated in considerable detail in the following articles: The I.S.V.R. Constant Temperature **Hot Wire Anemometer**, ISVR Tech. Rep. No. 66. University of Southampton, United Kingdom. Davis, J. G. (1977). *IEEE Trans. Biomed. Eng.* **24:** 484–486. **Dean**, J. W., Brennan, J. A., Mann, D. B., and Kneebone, C. H. (1971). Natl. Bur. Stand. (U.S.), Tech. Note 606. **Dean**, J. W., Brennan ...; in Ladenburg et al., art. F-2; Grant and Kronauer, *Symposium on Measurement in Unsteady Flow*, ASME, New York, 1962, pp. 44–53; ASME, *Report of Research Committee on Fluid Meters*, op. cit., pp. 105–107; and Compte-Bellot, *Ann. Rev. Fluid Mech.* **8:** 209–231 (1976).

The hot-wire anemometer can be modified for liquid measurements, although difficulties are encountered because of bubbles and dirt adhering to the wire. See the articles for hydraulic flows by Stevens, Borden, and Strausser, David Taylor Model Basin Rep. 953, December 1956; Middlebrook and Piret, in the journal of *Industrial and Engineering Chemistry Research* **42:** 1511–1513 (1950); and Piret et al., *Ind. Eng. Chem.* **39:** 1098–1103 (1947).

The *hot-film anemometer* has been developed for applications in which use of the hot-wire anemometer presents problems. It consists of a platinum-film sensing element deposited on a glass substrate. Various geometries can be used. The most common involves a wedge with a 30° included angle at the end of a tapered rod. The wedge is commonly 1 mm (0.039 in) long and 0.2 mm (0.0079 in) wide on each face. Compared with the hot wire, it is less susceptible to fouling by bubbles or dirt when used in liquids, has greater mechanical strength when used with gases at high velocities and high temperatures, and can give a higher signal-to-noise ratio. For additional information see Ling and Hubbard, *J. Aeronaut. Sci.* **23:** 890–891 (1956); and Ling, *J. Basic Eng.* **82:** 629–634 (1960).

The *heated-thermocouple anemometer* measures gas velocity from the cooling effect of the gas stream flowing across the hot junctions of a thermopile supplied with constant electric power input. Alternate junctions are maintained at ambient temperature, thus compensating for the effect of ambient temperature. For details see Bunker, *Proc. Instrum. Soc. Am.* **9:** paper 54-43-2 (1954).

A glass-coated bead *thermistor anemometer* can be used for the measurement of low fluid velocities, down to 0.001 m/s (0.003 ft/s) in air and 0.0002 m/s (0.0007 ft/s) in water [see Murphy and Sparks, *Ind. Eng. Chem. Fundam.* **7:** 642–645 (1968)].

The *laser-Doppler anemometer* measures local fluid velocity from the change in frequency of radiation, between a stationary source and a receiver, due to scattering by particles along the wave path. A laser is commonly used as the source of incident illumination. The measurements are essentially independent of local temperature and pressure. This technique can be used in many different flow systems with transparent fluids containing particles whose velocity is actually measured. For a brief review of the laser-Doppler technique see Goldstein, *Appl. Mech. Rev.* **27:** 753–760 (1974). For additional details see Durst, Melling, and Whitelaw, *Principles and Practice of Laser-Doppler Anemometry*, Academic, New York, 1976.

FLOWMETERS

In the process industries, flow measurement devices are the largest market in the process instrumentation field. Two web sites for process equipment and instrumentation, www.globalspec.com and www.thomasnet.com, both list more than 800 companies that offer flow measurement products. There are more than 100 types of flowmeters commercially available. The aforementioned web sites not only facilitate selection and specification of commercial flowmeters, but also provide electronic access to manufacturers' technical literature.

Devices that measure flow can be categorized in two areas as follows:

1. All types of measuring devices in which the material passes without being divided into isolated quantities. Movement of the material is usually sensed by a primary measuring element which activates a secondary device. The flow rate is then inferred from the response of the secondary device by means of known physical laws or from empirical relationships.

2. A positive-displacement meter, which applies to a device in which the flow is divided into isolated measured volumes. The number of fillings of these known volumes is measured with respect to time.

The most common application of flow measurement in process plants is flow in pipes, ducts, and tubing. Table 10-4 lists widely used flowmeters for these closed conduits as well as the two major classes of open-channel flowmeters. Table 10-4 also lists many other types of flowmeters that are discussed later in this subsection.

TABLE 10-4 Comparison of Flowmeter Technologies

Flowmeter technology	Accuracy* (+/−)	Turndown	Fluids†	Pipe sizes,‡ in	Maximum pressure,‡ psig	Temperature range,‡ °F	Pipe run	Relative pressure loss
Differential Pressure Meters								
Pitot		8:1	L, G	1 to 96	2.76E-4 to 200	−200 to 750		L
Averaging pitot	1% R	8:1	L, G, S	0.25 to 72	8800	−20 to 2370		L
Orifice							Long	
Square-edged	0.5 to 1.5% R	4:1	L, G, S	0.5 to 40	8800	−4 to 2300		M
Eccentric	2% R	4:1	L, G, S	0.5 to 40	8800	−4 to 2300		M
Segmental	2% R	4:1	L, G, S, SL	0.5 to 40	8800	−4 to 2300		M
Orifice and multi-variable flow transmitter	0.5 to 1% R	10:1	L, G, S	> 0.5	4000	1000		M
Venturi	0.5 to 1.5% R	10:1	L, G, S, SL	1 to 120	8800	−4 to 2300		L
Flow nozzle	0.5 to 2% R	8:1	L, G, S	2 to 80	>1000	<1000		M
V cone	0.5% F	10:1	L, G, S, SL	0.25 to 120	6000	To 400		M
Wedge	0.5 to 5% R	10:1	L, G, S, SL	0.5 to 24	>600	‡		M
Velocity Meters								
Correlation	0.5% R	10:1	L, G, SL	1 to 60	Piping limits	To 600	Long	L
Electromagnetic	0.2 to 2% R	10:1	L	0.15 to 60	5000	−40 to 350	Short	L
Propeller	2% R	15:1	L	2 to 12	230	0 to 300		M
Turbine	0.15 to 1% R	10:1	L, G	0.5 to 30	6000	−450 to 600	Short	M
Ultrasonic Doppler	1 to 30% R	50:1	L, G, SL	0.5 to 200	6000	−40 to 250	Long	L
Ultrasonic transit time	0.5 to 5% R	Down to zero flow	L, G	1 to 540	6000	−40 to 650	Long	L
Vortex	0.5 to 2% R	20:1	L, G, S	0.5 to 16	1500	−330 to 800	Short	M
Mass Meters								
Coriolis	0.1 to 0.3% R	10:1 to 80:1	L, G	0.06 to 12	5700	−400 to 800	None	L, M
Thermal (for gases)	1% F	50:1	G	0.125 to 8	4500	32 to 572	Short	L
Thermal (for liquids)	0.5% F	50:1	L	0.06 to 0.25	4500	40 to 165	Short	L
Volumetric	0.15 to 2% R	10:1	L	0.25 to 16	2000	−40 to 600	None	M to H
Variable area	1 to 5% F	10:1	L, G	0.125 to 6	6000	<1000	None	M
Open-Channel Flowmeters								
Weirs	2 to 5% R	25:1	L	Wide range	NA	NA		L
Flumes	3 to 10% R	40:1	L	Wide range	NA	NA		L

*F = full scale, R = rate.
†L = liquid, G = gas, S = steam, SL = slurry.
‡Dependent on the material selection and application. Readers should consult manufacturers for current capabilities.
Adapted from J. Pomroy, *Chemical Engineering,* pp. 94–102, May 1996; J. W. Dolenc, *Chemical Engineering Progress,* pp. 22–32, Jan. 1996; R. C. Baker, *Introductory Guide to Flow Measurement,* American Society of Mechanical Engineers, New York, 2003; R. W. Miller, *Flow Measurement Engineering Handbook,* 3d ed., McGraw-Hill, New York, 1996; D. W. Spitzer, *Industrial Flow Measurement,* 3d ed., The Instrumentation, Systems, and Automation Society, Research Triangle Park, N.C., 2005; and manufacturers' literature at www.globalspec.com.

This subsection summarizes selection and installation of flowmeters, including the measurement of pressure and velocities of fluids when the flow measurement technique requires it.

INDUSTRY GUIDELINES AND STANDARDS

Because flow measurement is important, many engineering societies and trade organizations have developed flow-related guidelines, standards, and other publications (Table 10-5). The reader should consult the appropriate standards when specifying, installing, and calibrating flow measurement systems.

There are also numerous articles in scholarly journals, trade magazines, and manufacturers' literature related to flow measurement.

Different types of flowmeters vary markedly in their degrees of sensitivity to flow disturbances. In the most extreme cases, obtaining highly accurate flow measurements with certain types of flowmeters may require 60*D* upstream straight pipe and 20*D* downstream. Valves can be particularly problematic because their effects on a flowmeter vary with valve position. Numerous types of flow straighteners or conditioners, as shown in Fig. 10-6, can significantly reduce the required run of straight pipe upstream of a given flowmeter.

TABLE 10-5 Guidelines, Standards, and Other Publications Related to Flow Measurement

Technical society	Number of guidelines and standards*
American Gas Association (AGA)	2
American Petroleum Institute (API)	11
American Society of Heating, Refrigeration, and Air Conditioning Engineers (ASHRAE)	5
American Society of Mechanical Engineers (ASME)	18
ASTM International (ASTM)	17
British Standards Institution (BSI)	100
Deutsches Institut fur Normung E. V. (DIN)	48
International Electrotechnical Commission (IEC)	6
Instrumentation, Systems, and Automation Society (ISA)	3
International Organization for Standardization (ISO)	212
SAE International (SAE)	6

*Number of documents identified by searching for *flow measurement* on http://global.ihs.com, the web site of a clearinghouse of industry guidelines, codes, and standards.

CLASSIFICATION OF FLOWMETERS

Table 10-4 lists the major classes of flowmeters, along with common examples of each. Brief descriptions are provided in this subsection, followed by more details in subsequent subsections.

Differential Pressure Meters Differential pressure meters or head meters measure the change in pressure across a special flow element. The differential pressure increases with increasing flow rate. The pitot tubes described previously work on this principle. Other examples include orifices [see also Eqs. (6-111) and (8-102) and Fig. 10-11], nozzles (Fig. 10-16), targets, venturis (see also Sec. 8 and Fig. 10-14), and elbow meters. Averaging pitot tubes produce a pressure differential that is based on multiple measuring points across the flow path.

Differential pressure meters are widely used. Temperature, pressure, and density affect gas density and readings of differential pressure meters. For that reason, many commercial flowmeters that are based on measurement of differential pressure often have integral temperature and absolute pressure measurements in addition to differential pressure. They also frequently have automatic temperature and pressure compensation.

Velocity Meters Velocity meters measure fluid velocity. Examples include electromagnetic, propeller, turbine, ultrasonic Doppler, ultrasonic transit time, and vortex meters. Section 8 describes the principles of operation of electromagnetic, turbine, ultrasonic, and vortex flowmeters.

Mass Meters Mass flowmeters measure the rate of mass flow through a conduit. Examples include Coriolis flowmeters and thermal mass flowmeters. Coriolis flowmeters can measure fluid density simultaneously with mass flow rate. This permits calculation of the volumetric flow rate as well. Section 8 includes brief descriptions of Coriolis and thermal mass flowmeters.

Volumetric Meters Volumetric meters (also called *positive-displacement flowmeters*) are devices that mechanically divide a fluid stream into discrete, known volumes and count the number of volumes that pass through the device. See Spitzer (2005, op. cit.).

Variable-Area Meters Variable-area meters, which are also called *rotameters*, offer popular and inexpensive flow measurement devices. These meters employ a float inside a tube that has an internal cross-sectional area which increases with distance upward in the flow path through the tube. As the flow rate increases, the float rises in the tube to provide a larger area for the flowing fluid to pass.

Open-Channel Flow Measurement Open-channel flow measurements are usually based on measurement of liquid level in a flow channel constructed of a specified geometry. The two most common flow channels used are weirs and flumes. See Spitzer, op. cit., 2005.

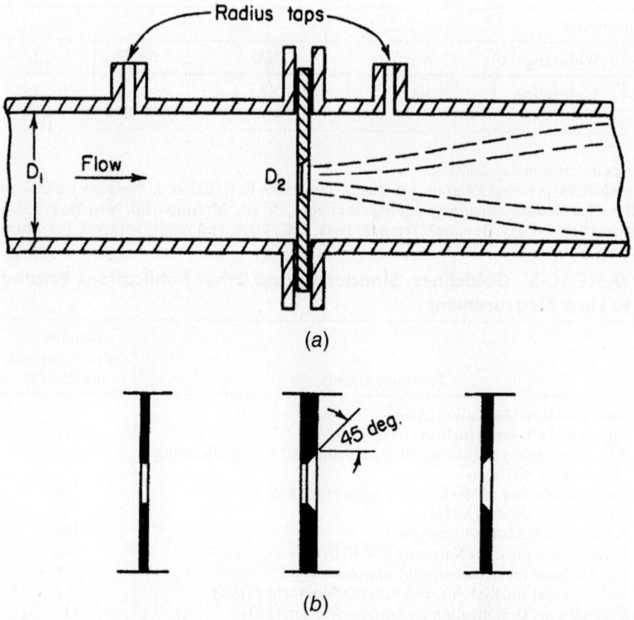

FIG. 10-11 Square-edged or sharp-edged orifices. The plate at the orifice opening must not be thicker than one-thirtieth of the pipe diameter, one-eighth of the orifice diameter, or one-fourth of the distance from the pipe wall to the edge of the opening. (*a*) Pipeline orifice. (*b*) Types of plates.

DIFFERENTIAL PRESSURE FLOWMETERS

General Principles If a constriction is placed in a closed channel carrying a stream of fluid, there will be an increase in velocity, and hence an increase in kinetic energy, at the point of constriction. From an energy balance, as given by Bernoulli's theorem [see Sec. 6, subsection Energy Balance, Eq. (6-16)], there must be a corresponding reduction in pressure. The rate of discharge from the constriction can be calculated by knowing this pressure reduction, the area available for flow at the constriction, the density of the fluid, and the coefficient of discharge *C*. The coefficient of discharge is defined as the ratio of actual flow to the theoretical flow and makes allowance for stream contraction and frictional effects. The metering characteristics of commonly used differential pressure meters are reviewed and grouped by Halmi [*J. Fluids Eng.* **95:** 127–141 (1973)].

The term *static head* generally denotes the pressure in a fluid due to the head of fluid above the point in question. Its magnitude is given by the application of Newton's law (force = mass × acceleration). In the case of *liquids* (constant density), the static head P_h (lbf/ft²) is given by

$$P_h = h\rho g/g_c \tag{10-18}$$

where h = head of liquid above the point, m (ft); ρ = liquid density; g = local acceleration due to gravity; and g_c = dimensional constant.

The head developed in a compressor or pump is the energy force per unit mass. In the measuring systems it is often misnamed as feet while the units are really ft · lb_f/lbm or kilojoules (kJ).

For a compressor or turbine, it is represented by the following relationship:

$$E = U_1V_{\theta_1} - U_2V_{\theta_2} \tag{10-19}$$

where U is the blade speed and V_θ is the tangential velocity component of absolute velocity. This equation is known as the *Euler equation*.

Orifice Meters A *square-edged* or *sharp-edged* orifice, as shown in Fig. 10-11, is a clean-cut square-edged hole with straight walls perpendicular to the flat upstream face of a thin plate placed crosswise to the channel. The stream issuing from such an orifice attains its minimum cross section (vena contracta) at a distance downstream of the orifice which varies with the ratio β of orifice to pipe diameter (see Fig. 10-12).

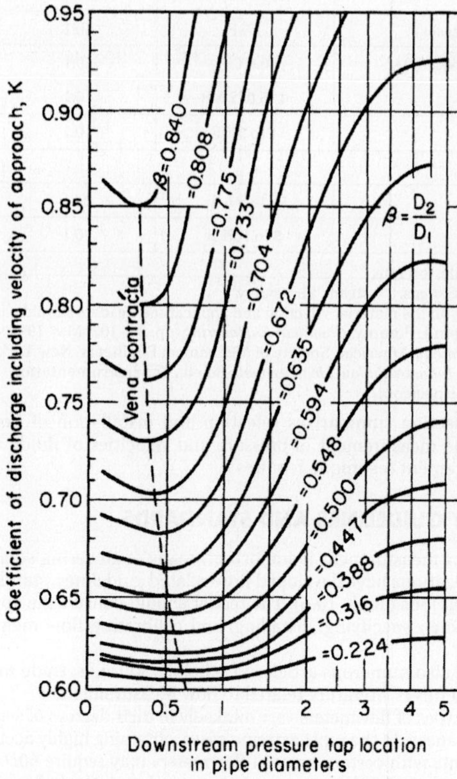

FIG. 10-12 Coefficient of discharge for square-edged circular orifices for $N_{Re} > 30{,}000$ with the upstream tap located between 1 and 2 pipe diameters from the orifice plate. [Spitzglass, *Trans. Am. Soc. Mech. Eng.* **44:** 919 (1922).]

For a centered circular orifice in a pipe, the pressure differential is customarily measured between one of the following pressure-tap pairs. Except in the case of flange taps, all measurements of distance from the orifice are made from the upstream face of the plate.

1. *Corner taps.* Static holes are drilled, one in the upstream and one in the downstream flange, with the openings as close as possible to the orifice plate.
2. *Radius taps.* Static holes are located 1 pipe diameter upstream and 0.5 pipe diameter downstream from the plate.
3. *Pipe taps.* Static holes are located 2½ pipe diameters upstream and 8 pipe diameters downstream from the plate.
4. *Flange taps.* Static holes are located 25.4 mm (1 in) upstream and 25.4 mm (1 in) downstream from the plate.
5. *Vena contracta taps.* The upstream static hole is 0.5 to 2 pipe diameters from the plate. The downstream tap is located at the position of minimum pressure (see Fig. 10-12).

Radius taps are best from a practical standpoint; the downstream pressure tap is located at about the mean position of the vena contracta, and the upstream tap is sufficiently far upstream to be unaffected by distortion of the flow in the immediate vicinity of the orifice (in practice, the upstream tap can be as much as 2 pipe diameters from the plate without affecting the results). Vena contracta taps give the largest differential head for a given rate of flow but are inconvenient if the orifice size is changed from time to time. Corner taps offer the sometimes great advantage that the pressure taps can be built into the plate carrying the orifice. Thus the entire apparatus can be quickly inserted in a pipeline at any convenient flanged joint without having to drill holes in the pipe.

Flange taps are similarly convenient since by merely replacing standard flanges with special orifice flanges, suitable pressure taps are made available. Pipe taps give the lowest differential pressure, the value obtained being close to the permanent pressure loss.

The practical working equation for flow rate ($\dot{m}$) of discharge, adopted by the ASME Research Committee on Fluid Meters for use with either gases or liquids, is

$$\dot{m} = q_1\rho_1 = CYA_2\sqrt{\frac{2g_c(p_1-p_2)\rho_1}{1-\beta^4}} = KYA_2\sqrt{2g_c(p_1-p_2)\rho_1} \qquad (10\text{-}20)$$

where A_2 = cross-sectional area of throat; C = coefficient of discharge, dimensionless; g_c = dimensional constant; $K = C/\sqrt{1-\beta^4}$, dimensionless; p_1, p_2 = pressure at upstream and downstream static pressure taps, respectively; q_1 = volumetric rate of discharge measured at upstream pressure and temperature; $\dot{m}$ = weight rate of discharge; Y = expansion factor, dimensionless; β = ratio of throat diameter to pipe diameter, dimensionless; and ρ_1 = density at upstream pressure and temperature.

For the case of subsonic flow of a gas ($r_c < r < 1.0$), the expansion factor Y for orifices is approximated by

$$Y = 1 - [(1-r)/k]\,(0.41 + 0.35\beta^4) \qquad (10\text{-}21)$$

where r = ratio of downstream to upstream static pressure (p_2/p_1), k = ratio of specific heats (c_p/c_v), and β = diameter ratio. (See also Fig. 10-15.) Values of Y for supercritical flow of a gas ($r < r_c$) through orifices are given by Benedict [*J. Basic Eng.* **93:** 121–137 (1971)]. For the case of *liquids*, expansion factor Y is unity, and Eq. (10-25) should be used, since it allows for any difference in elevation between the upstream and downstream taps.

Coefficient of discharge C for a given orifice type is a function of the Reynolds number N_{Re} (based on orifice diameter and velocity) and diameter ratio β. At Reynolds numbers greater than about 30,000, the coefficients are substantially constant. For square-edged or sharp-edged concentric circular orifices, the value will fall between 0.595 and 0.620 for vena contracta or radius taps for β up to 0.8 and for flange taps for β up to 0.5. Figure 10-12 gives the coefficient of discharge K, including the velocity-of-approach factor $1/\sqrt{1-\beta^4}$, as a function of β and the location of the downstream tap. Precise values of K are given in ASME, *PTC*, 1997, pp. 20–39, for flange taps, radius taps, vena contracta taps, and corner taps. Precise values of C are given in ASME, *Report of the Research Committee on Fluid Meters,* 1971, pp. 202–207, for the first three types of taps.

The discharge coefficient of sharp-edged orifices was shown by Benedict, Wyler, and Brandt [*J. Eng. Power* **97:** 576–582 (1975)] to increase with edge roundness. Typical as-purchased orifice plates may exhibit deviations on the order of 1 to 2 percent from ASME values of the discharge coefficient.

In the transition region (N_{Re} between 50 and 30,000), the coefficients are generally higher than the above values. Although calibration is generally advisable in this region, the curves given in Fig. 10-13 for corner and vena

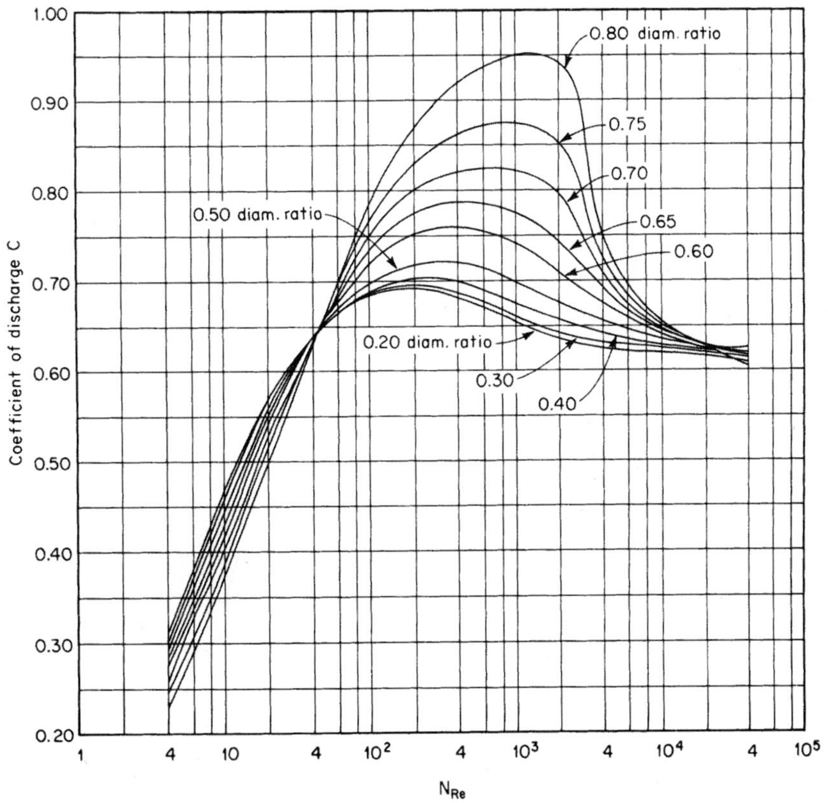

FIG. 10-13 Coefficient of discharge for square-edged circular orifices with corner taps. [Tuve and Sprenkle, *Instruments* **6:** 201 (1933).]

contracta taps can be used as a guide. In the laminar-flow region ($N_{Re} < 50$), the coefficient C is proportional to $\sqrt{N_{Re}}$. For $1 < N_{Re} < 100$, Johansen [*Proc. R. Soc. (London)*, **A121:** 231–245 (1930)] presents discharge coefficient data for sharp-edged orifices with corner taps. For $N_{Re} < 1$, Miller and Nemecek [ASME Paper 58-A-106 (1958)] present correlations giving coefficients for sharp-edged orifices and short-pipe orifices (L/D from 2 to 10). For short-pipe orifices (L/D from 1 to 4), Dickerson and Rice [*J. Basic Eng.* **91:** 546–548 (1969)] give coefficients for the intermediate range ($27 < N_{Re} < 7000$). See also the subsection Contraction and Entrance Losses.

Permanent pressure loss across a concentric circular orifice with radius or vena contracta taps can be approximated for turbulent flow by

$$(p_1 - p_4)/(p_1 - p_2) = 1 - \beta^2 \tag{10-22}$$

where p_1, p_2 = upstream and downstream pressure-tap readings, respectively, p_4 = fully recovered pressure (4 to 8 pipe diameters downstream of the orifice), and β = diameter ratio. See ASME, *PTC,* 1997, fig. 5.

See Benedict, *J. Fluids Eng.* **99:** 245–248 (1977), for a general equation for pressure loss for orifices installed in pipes or with plenum inlets. Orifices show higher loss than nozzles or ventura's. Permanent pressure loss for laminar flow depends on the Reynolds number in addition to β. See Alvi, Sridharan, and Rao, *J. Fluids Eng.* **100**(3): 99–307 (1978).

For the case of *critical flow* through a square- or sharp-edged concentric circular orifice (where $r \le r_c$, as discussed earlier in this subsection), use Eqs. (10-29), (10-30), and (10-31) as given for critical-flow nozzles. However, unlike with nozzles, the flow through a sharp-edged orifice continues to increase as the downstream pressure drops below that corresponding to the critical pressure ratio r_c. This is due to an increase in the cross section of the vena contracta as the downstream pressure is reduced, giving a corresponding increase in the coefficient of discharge. At $r = r_c$, C is about 0.75, while at $r = 0$, C has increased to about 0.84. See Benedict, *J. Basic Eng.* **93:** 99–120 (1971).

Measurements by Harris and Magnall [*Trans. Inst. Chem. Eng. (London)*, **50:** 61–68 (1972)] with a venturi ($\beta = 0.62$) and orifices with radius taps ($\beta = 0.60$ to 0.75) indicate that the discharge coefficient for nonnewtonian fluids, in the range of N_{Re} (generalized Reynolds number) from 3500 to 100,000, is approximately the same as for newtonian fluids at the same Reynolds number.

Quadrant-edge orifices have holes with rounded edges on the upstream side of the plate. The quadrant-edge radius is equal to the thickness of the plate at the orifice location. The advantages claimed for this type versus the square- or sharp-edged orifice are constant-discharge coefficients extending to lower Reynolds numbers and less possibility of significant changes in coefficient because of erosion or other damage to the inlet shape.

Values of discharge coefficient C and Reynolds number limits for constant C are presented in Table 10-6, based on Ramamoorthy and Seetharamiah [*J. Basic Eng.* **88:** 9–13 (1966)] and Bogema and Monkmeyer (*J. Basic Eng.* **82:** 729–734 (1960)]. At Reynolds numbers above those listed for the upper limits, the coefficients rise abruptly. As Reynolds numbers decrease below those listed for the lower limits, the coefficients pass through a hump and then drop off. According to Bogema, Spring, and Ramamoorthy [*J. Basic Eng.* **84:** 415–418 (1962)], the hump can be eliminated by placing a fine-mesh screen about 3 pipe diameters upstream of the orifice. This reduces the lower N_{Re} limit to about 500.

Permanent pressure loss across quadrant-edge orifices for turbulent flow is somewhat lower than that given by Eq. (10-22). See Alvi, Sridharan, and Rao, loc. cit., for values of the discharge coefficient and permanent pressure loss in laminar flow.

Slotted orifices offer significant advantages over a standard square-edged orifice with an identical open area for homogeneous gases or liquids [Morrison and Hall, *Hydrocarbon Processing* **79:** 12, 65–72 (2000)]. The slotted orifice flowmeter only requires compact header configurations with very short upstream pipe lengths and maintains accuracy in the range of 0.25 percent with no flow conditioner. Permanent head loss is less than or equal to that of a standard orifice that has the same β ratio. Discharge coefficients for the slotted orifice are much less sensitive to swirl or to axial velocity profiles. A slotted orifice plate can be a "drop-in" replacement for a standard orifice plate.

Segmental and *eccentric orifices* are frequently used for gas metering when there is a possibility that entrained liquids or solids would otherwise accumulate in front of a concentric circular orifice. This can be avoided if the opening is placed on the lower side of the pipe. For liquid flow with entrained gas, the opening is placed on the upper side. The pressure taps should be located on the opposite side of the pipe from the opening.

Coefficient C for a square-edged eccentric circular orifice (with opening tangent to pipe wall) varies from about 0.61 to 0.63 for β values from 0.3 to 0.5, respectively, and pipe Reynolds numbers > 10,000 for either vena contracta or flange taps (where β = diameter ratio). For square-edged segmental orifices, coefficient C falls generally between 0.63 and 0.64 for $0.3 \le \beta \le 0.5$ and pipe Reynolds numbers > 10,000, for vena contracta or flange taps, where β = diameter ratio for an equivalent circular orifice = $\sqrt{\alpha}$ (α = ratio of orifice to pipe cross-sectional areas). Values of expansion factor Y are slightly higher than for concentric circular orifices, and the location of the vena contracta is moved farther downstream as compared with concentric circular orifices. For further details, see ASME, *Report of the Research Committee on Fluid Meters*, 1971, pp. 210–213.

For permanent pressure loss with segmental and eccentric orifices with laminar pipe flow see Lakshmana Rao and Sridharan, *Proc. Am. Soc. Civ. Eng., J. Hydraul. Div.* **98**(HY 11): 2015–2034 (1972).

Annular orifices can also be used to advantage for gas metering when there is a possibility of entrained liquids or solids and for liquid metering with entrained gas present in small concentrations. Coefficient K was found by Bell and Bergelin [*Trans. Am. Soc. Mech. Eng.* **79:** 593–601 (1957)] to range from about 0.63 to 0.67 for annulus Reynolds numbers in the range of 100 to 20,000, respectively, for values of $2L/(D - d)$ less than 1 where L = thickness of orifice at outer edge, D = inside pipe diameter, and d = diameter of orifice disk. The *annulus Reynolds number* is defined as

$$N_{Re} = (D - d)\,(G/\mu) \tag{10-23}$$

where G = mass velocity ρV through orifice opening and μ = fluid viscosity. The above coefficients were determined for β values (= d/D) in the range of 0.95 to 0.996 and with pressure taps located 19 mm (¾ in) upstream of the disk and 230 mm (9 in) downstream in a 5.25-in-diameter pipe.

Venturi Meters The standard Herschel-type venturi meter consists of a short length of straight tubing connected at either end to the pipeline by conical sections (see Fig. 10-14). Recommended proportions (ASME, *PTC,* 1997, p. 17) are entrance cone angle $\alpha_1 = 21 \pm 2°$, exit cone angle $\alpha_2 = 5$ to 15°, throat length = 1 throat diameter, and upstream tap located 0.25 to 0.5 pipe diameter upstream of the entrance cone. The straight and conical sections should be joined by smooth curved surfaces for best results. *Rate of discharge* of either gases or liquids through a venturi meter is given by Eq. (10-20).

For the flow of *gases,* expansion factor Y, which allows for the change in gas density as it expands adiabatically from p_1 to p_2, is given by

$$Y = \sqrt{r^{2/k}\left(\frac{k}{k-1}\right)\left(\frac{1-r^{(k-1)/k}}{1-r}\right)\left(\frac{1-\beta^4}{1-\beta^4 r^{2/k}}\right)} \tag{10-24}$$

for venturi meters and flow nozzles, where $r = p_2/p_1$ and k = specific heat ratio c_p/c_v. Values of Y computed from Eq. (10-24) are given in Fig. 10-15 as a function of r, k, and β.

TABLE 10-6 Discharge Coefficients for Quadrant-Edge Orifices

			Limiting N_{Re}^* for constant coefficient[†]	
β	$C^‡$	$K^‡$	Lower	Upper
0.225	0.770	0.771	5000	60,000
0.400	0.780	0.790	5000	150,000
0.500	0.824	0.851	4000	200,000
0.600	0.856	0.918	3000	120,000
0.630	0.885	0.964	3000	105,000

*Based on pipe diameter and velocity.
†For a precision of about ±0.5 percent.
‡Can be used with corner taps, flange taps, or radius taps.

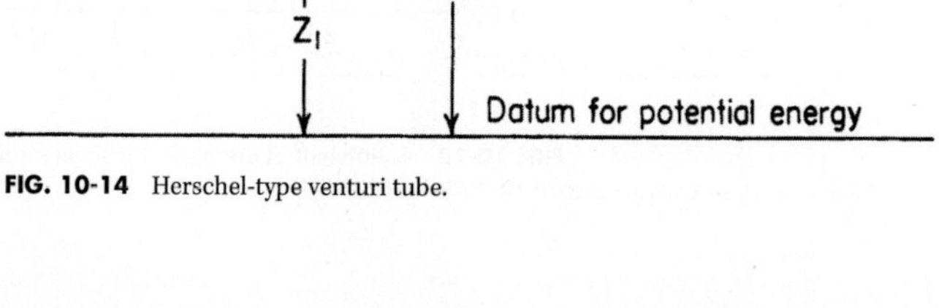

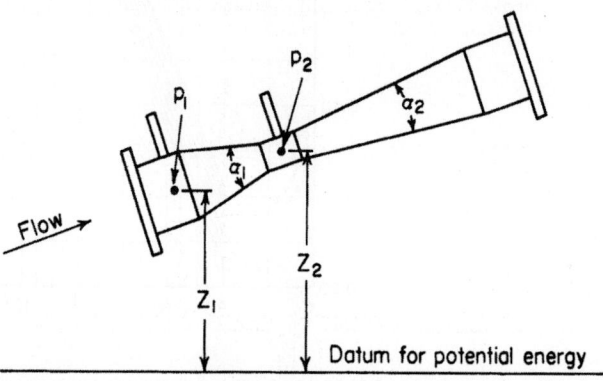

FIG. 10-14 Herschel-type venturi tube.

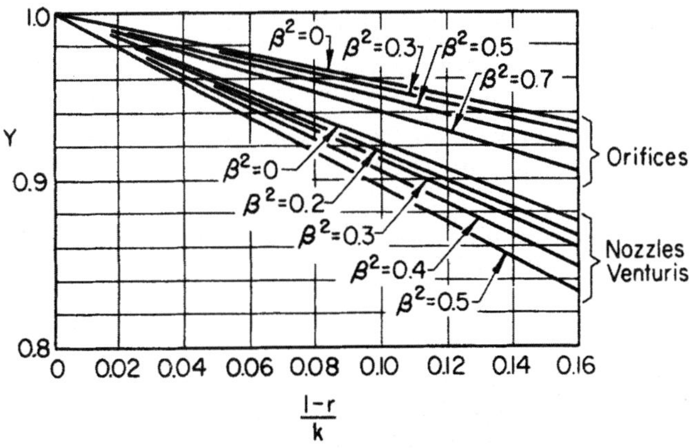

FIG. 10-15 Values of expansion factor Y for orifices, nozzles, and venturis.

For the flow of *liquids,* expansion factor Y is unity. The change in potential energy in the case of an inclined or vertical venturi meter must be allowed for. Equation (10-20) is accordingly modified to give

$$\dot{m} = q_1\rho = CA_2\sqrt{\frac{[2g_c(p_1 - p_2) + 2g\rho(Z_1 - Z_2)]\rho}{1-\beta^4}} \tag{10-25}$$

where g = local acceleration due to gravity and Z_1, Z_2 = vertical heights above an arbitrary datum plane corresponding to the centerline pressure-reading locations for p_1 and p_2, respectively.

The value of the *discharge coefficient C* for a *Herschel-type venturi meter* depends on the Reynolds number and to a minor extent on the size of the venturi, increasing with diameter. A plot of C versus pipe Reynolds number is given in ASME, *PTC,* 1997, p. 19. A value of 0.984 can be used for pipe Reynolds numbers larger than 200,000.

Permanent pressure loss for a Herschel-type venturi tube depends on the diameter ratio β and discharge cone angle α_2. It ranges from 10 to 15 percent of the pressure differential $p_1 - p_2$ for small angles (5° to 7°) and from 10 to 30 percent for large angles (15°), with the larger losses occurring at low values of β (see ASME, *PTC,* 1997, p. 12). See Benedict, *J. Fluids Eng.* **99:** 245–248 (1977), for a general equation for pressure loss for venturis installed in pipes or with plenum inlets.

For flow measurement of *steam and water mixtures* with a Herschel-type venturi in 2½-in- and 3-in-diameter pipes, see Collins and Gacesa, *J. Basic Eng.* **93:** 11–21 (1971).

A variety of *short-tube* venturi meters are available commercially. They require less space for installation and are generally (although not always) characterized by a greater pressure loss than the corresponding Herschel-type venturi meter. Discharge coefficients vary widely for different types, and individual calibration is recommended if the manufacturer's calibration is not available. Results of tests on the Dall flow tube are given by Miner [*Trans. Am. Soc. Mech. Eng.* **78:** 475–479 (1956)] and Dowdell [*Instrum. Control Syst.* **33:** 1006–1009 (1960)]; and on the Gentile flow tube (also called the Beth flow tube or Foster flow tube) by Hooper [*Trans. Am. Soc. Mech. Eng.* **72:** 1099–1110 (1950)].

The use of a *multiventuri system* (in which an inner venturi discharges into the throat of an outer venturi) to increase both the differential pressure for a given flow rate and the signal-to-loss ratio is described by Klomp and Sovran [*J. Basic Eng.* **94:** 39–45 (1972)].

Flow Nozzles A simple form of flow nozzle is shown in Fig. 10-16. It consists essentially of a short cylinder with a flared approach section.

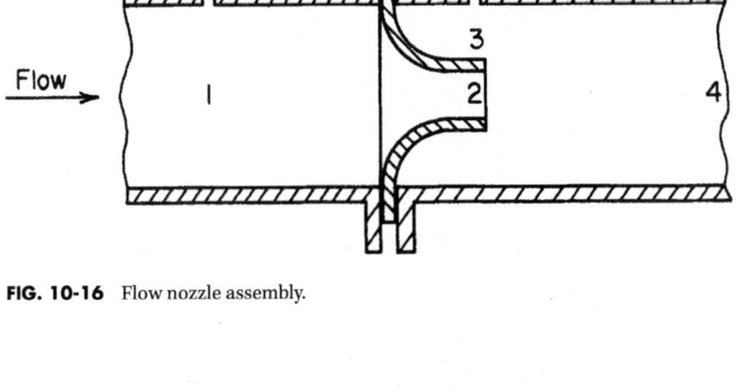

FIG. 10-16 Flow nozzle assembly.

The approach cross section is preferably elliptical but may be conical. Recommended contours for long-radius flow nozzles are given in ASME, *PTC,* 1997, p. 13. In general, the length of the straight portion of the throat is about one-half the throat diameter, the upstream pressure tap is located about 1 pipe diameter from the nozzle inlet face, and the downstream pressure tap is about ½ pipe diameter from the inlet face. For subsonic flow, the pressures at points 2 and 3 will be practically identical. If a conical inlet is preferred, the inlet and throat geometry specified for a Herschel-type venturi meter can be used, omitting the expansion section.

The *rate of discharge* through a flow nozzle for subcritical flow can be determined by the equations given for venturi meters, Eq. (10-20) for gases and Eq. (10-25) for liquids. The expansion factor Y for nozzles is the same as that for venturi meters [Eq. (10-24), Fig. 10-15]. The value of the discharge coefficient C depends primarily on the pipe Reynolds number and to a lesser extent on the diameter ratio β. Curves of recommended coefficients for long-radius flow nozzles with pressure taps located 1 pipe diameter upstream and ½ pipe diameter downstream of the inlet face of the nozzle are given in ASME, *PTC,* 1997, p. 15. In general, coefficients range from 0.95 at a pipe Reynolds number of 10,000 to 0.99 at 1,000,000.

The performance characteristics of pipe wall-tap nozzles (Fig. 10-16) and throat-tap nozzles are reviewed by Wyler and Benedict [*J. Eng. Power* **97:** 569–575 (1975)].

Permanent pressure loss across a subsonic flow nozzle is approximated by

$$p_1 - p_4 = \frac{1-\beta^2}{1+\beta^2}(p_1 - p_2) \tag{10-26}$$

where p_1, p_2, p_4 = static pressures measured at the locations shown in Fig. 10-16 and β = ratio of nozzle throat diameter to pipe diameter, dimensionless. Equation (10-26) is based on a momentum balance assuming constant fluid density (see Lapple et al., *Fluid and Particle Mechanics,* University of Delaware, Newark, 1951, p. 13).

See Benedict, R. P., "On the Determination and Combination of Loss. Coefficients for Compressible Fluid Flows," *J. Eng. Power, ASME Trans.,* vol. 88, for a general equation for pressure loss for nozzles installed in pipes or with plenum inlets. Nozzles show higher loss than venturis. Permanent pressure loss for laminar flow depends on the Reynolds number in addition to β. For details, see Alvi, Sridharan, and Lakshamana Rao, *J. Fluids Eng.* **100:** 299–307 (1978).

Critical Flow Nozzle For a given set of upstream conditions, the rate of discharge of a gas from a nozzle will increase for a decrease in the absolute pressure ratio p_2/p_1 until the linear velocity in the throat reaches that of sound in the gas at that location. The value of p_2/p_1 for which the acoustic velocity is just attained is called the *critical pressure ratio* r_c. The actual pressure in the throat will not fall below $p_1 r_c$ even if a much lower pressure exists downstream.

The critical pressure ratio r_c can be obtained from the following theoretical equation, which assumes a perfect gas and a frictionless nozzle:

$$r_c^{(1-k)/k} + \left(\frac{k-1}{2}\right)\beta^4 r_c^{2/k} = \frac{k+1}{2} \tag{10-27}$$

This reduces, for $\beta \le 0.2$, to

$$r_c = \left(\frac{2}{k+1}\right)^{k/(k-1)} \tag{10-28}$$

where k = ratio of specific heats c_p/c_v and β = diameter ratio. A table of values of r_c as a function of k and β is given in ASME, *Report of the Research Committee on Fluid Meters,* New York, 1971, p. 68. For small values of β, $r_c = 0.487$ for $k = 1.667$, $r_c = 0.528$ for $k = 1.40$, $r_c = 0.546$ for $k = 1.30$, and $r_c = 0.574$ for $k = 1.15$.

Under *critical flow conditions,* only the upstream conditions p_1, υ_1, and T_1 need be known to determine the flow rate, which, for $\beta \le 0.2$, is given

$$\dot{m}_{\max} = CA_2\sqrt{g_c k\left(\frac{p_1}{\upsilon_1}\right)\left(\frac{2}{k+1}\right)^{(k+1)/(k-1)}} \tag{10-29}$$

For a *perfect gas,* this corresponds to

$$\dot{m}_{\max} = CA_2 p_1\sqrt{g_c k\left(\frac{M}{RT_1}\right)\left(\frac{2}{k+1}\right)^{(k+1)/(k-1)}} \tag{10-30}$$

For air, Eq. (10-30) reduces to

$$\dot{m}_{\max} = C_1 CA_2 p_1/\sqrt{T_1} \tag{10-31}$$

where A_2 = cross-sectional area of throat; C = coefficient of discharge, dimensionless; g_c = dimensional constant; k = ratio of specific heats c_p/c_v; M = molecular weight; p_1 = pressure on upstream side of nozzle; R = gas constant;

T_1 = absolute temperature on upstream side of nozzle; v_1 = specific volume on upstream side of nozzle; C_1 = dimensional constant, 0.0405 SI unit (0.533 U.S. Customary System unit); and $\dot{m}_{max}$ = maximum-weight flow rate.

Discharge coefficients for critical flow nozzles are, in general, the same as those for subsonic nozzles. See Grace and Lapple, *Trans. Am. Soc. Mech. Eng.* **73:** 639–647 (1951); and Szaniszlo, *J. Eng. Power* **97:** 521–526 (1975). Arnberg, Britton, and Seidl [*J. Fluids Eng.* **96:** 111–123 (1974)] present discharge coefficient correlations for circular-arc venturi meters at critical flow. For the calculation of the flow of natural gas through nozzles under critical-flow conditions, see Johnson, *J. Basic Eng.* **92:** 580–589 (1970).

Elbow Meters A pipe elbow can be used as a flowmeter for liquids if the differential centrifugal head generated between the inner and outer radii of the bend is measured by means of pressure taps located midway around the bend. Equation (10-25) can be used, except that the pressure difference term $p_1 - p_2$ is now taken to be the differential centrifugal pressure and β is taken as zero, if one assumes no change in cross section between the pipe and the bend. The discharge coefficient should preferably be determined by calibration, but as a guide it can be estimated within ±6 percent for circular pipe for Reynolds numbers greater than 10^5 from $C = 0.98\sqrt{R_c/2D}$, where R_c = radius of curvature of the centerline and D = inside pipe diameter in consistent units. See Murdock, Foltz, and Gregory, *J. Basic Eng.* **86:** 498–506 (1964); or ASME, *Report of the Research Committee on Fluid Meters*, 1971, pp. 75–77.

Accuracy Square-edged orifices and venturi tubes have been so extensively studied and standardized that reproducibility within 1 to 2 percent can be expected between standard meters when new and clean. This is therefore the order of reliability to be had, if one assumes (1) accurate measurement of meter differential, (2) selection of the coefficient of discharge from recommended published literature, (3) accurate knowledge of fluid density, (4) accurate measurement of critical meter dimensions, (5) smooth upstream face of orifice, and (6) proper location of the meter with respect to other flow-disturbing elements in the system. Care must also be taken to avoid even slight corrosion or fouling during use.

Presence of *swirling flow* or an *abnormal velocity distribution* upstream of the metering element can cause serious metering error unless calibration in place is employed or sufficient straight pipe is inserted between the meter and the source of disturbance. Table 10-7 gives the minimum lengths of straight pipe required to avoid appreciable error due to the presence of certain fittings and valves either upstream or downstream of an orifice or nozzle. These values were extracted from plots presented by Sprenkle [*Trans. Am. Soc. Mech. Eng.* **67:** 345–360 (1945)]. Table 10-7 also shows the reduction in spacing made possible by the use of straightening vanes between the fittings and the meter. Entirely adequate straightening vanes can be provided by fitting a bundle of thin-wall tubes within the pipe. The center-to-center distance between tubes should not exceed one-fourth of the pipe diameter, and the bundle length should be at least 8 times this distance.

TABLE 10-7 Locations of Orifices and Nozzles Relative to Pipe Fittings

Distances in pipe diameters D_1

Type of fitting upstream	$\frac{D_2}{D_1}$	Distance, upstream fitting to orifice: Without straightening vanes	Distance, upstream fitting to orifice: With straightening vanes	Distance, vanes to orifice	Distance, nearest downstream fitting from orifice
Single 90° ell, tee, or cross used as ell	0.2	6			2
	0.4	6			
	0.6	9	9		
	0.8	20	12	8	4
2 short-radius 90° ells in form of S	0.2	7			2
	0.4	8	8		
	0.6	13	10	6	
	0.8	25	15	11	4
2 long- or short-radius 90° ells in perpendicular planes	0.2	15	9	5	2
	0.4	18	10	6	
	0.6	25	11	7	
	0.8	40	13	9	4
Contraction or enlargement	0.2	8	Vanes have no advantage		2
	0.4	9			
	0.6	10			
	0.8	15			4
Globe valve or stop check	0.2	9	9	5	2
	0.4	10	10	6	
	0.6	13	10	6	
	0.8	21	13	9	4
Gate valve, wide open, or plug cocks	0.2	6	Same as globe valve		2
	0.4	6			
	0.6	8			
	0.8	14			4

The distances specified in Table 10-7 will be conservative if applied to venturi meters. For specific information on requirements for venturi meters, see a discussion by Pardoe appended to Sprenkle [*Trans. Am. Soc. Mech. Eng.* **67:** 345–360 (1945)]. Extensive data on the effect of installation on the coefficients of venturi meters are given elsewhere by Pardoe [*Trans. Am. Soc. Mech. Eng.* **65:** 337–349 (1943)].

As a general rule, a concentric orifice plate is the most economical solution to flow measurement for most applications if the calculated permanent pressure loss is acceptable and the upstream and downstream piping configuration provides for a stabilized flow pattern. Unique designs such as the conditioning orifice plate offered by Emerson Rosemount Pak or a flow tube with an integral flow stabilization design such as the FlowPak flow tube assembly offered by Fluidic Techniques provide optional considerations for areas where straight lengths of upstream and downstream piping are problematic.

1. Conditioning orifice plate (see Fig. 10-17)

This copyrighted design requires only 2 pipe diameters upstream and downstream from a flow disturbance and is suitable for most liquids, gases, and steam. Refer to the Emerson Rosemount web site for more details.

2. The high head recovery FlowPak (see Fig. 10-18) is a patented flow tube design with a uniquely designed flow-conditioning translineal flow plate which eliminates the requirement of straight pipe between the upstream disturbances while providing the known properties of the flow tube characteristics including constant discharge coefficient, low uncertainty, and low permanent pressure loss. Refer to the Fluidics Techniques web site for more details.

In the presence of *flow pulsations,* the indications of head meters such as orifices, nozzles, and venturis will often be undependable for several reasons. First, the measured pressure differential will tend to be high, since the pressure differential is proportional to the square of the flow rate for a head meter, and the square root of the mean differential pressure is always greater than the mean of the square roots of the differential pressures. Second, there is a phase shift as the wave passes through the metering restriction, which can affect the differential. Third, pulsations can be set up in the manometer leads themselves. Frequency of the pulsation also plays a part. At low frequencies, the meter reading can generally faithfully follow the flow pulsations, but at high frequencies it cannot. This is due to inertia of the fluid in the manometer leads or of the manometric fluid, whereupon the meter would give a reading intermediate between the maximum and minimum flows but having no readily predictable relation to the mean flow. Pressure transducers with flush-mounted diaphragms can be used together

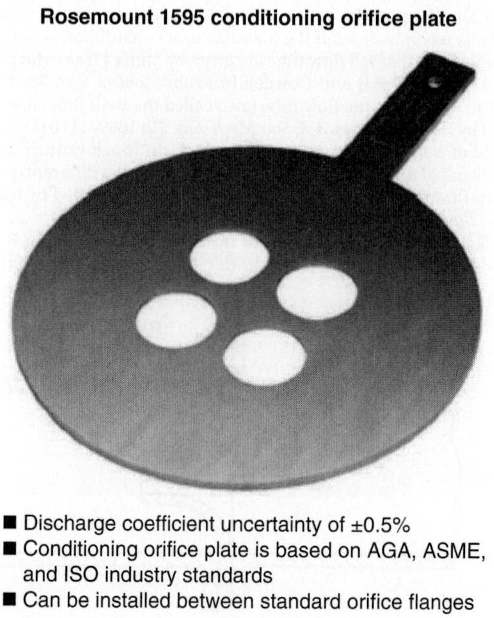
Rosemount 1595 conditioning orifice plate
■ Discharge coefficient uncertainty of ±0.5%
■ Conditioning orifice plate is based on AGA, ASME, and ISO industry standards
■ Can be installed between standard orifice flanges

FIG. 10-17 Rosemont 1595 Conditioning Orifice Plate. (*Courtesy of Rosemont.*)

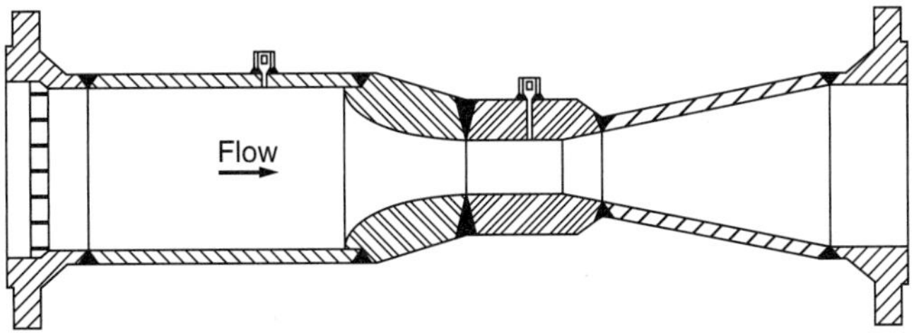

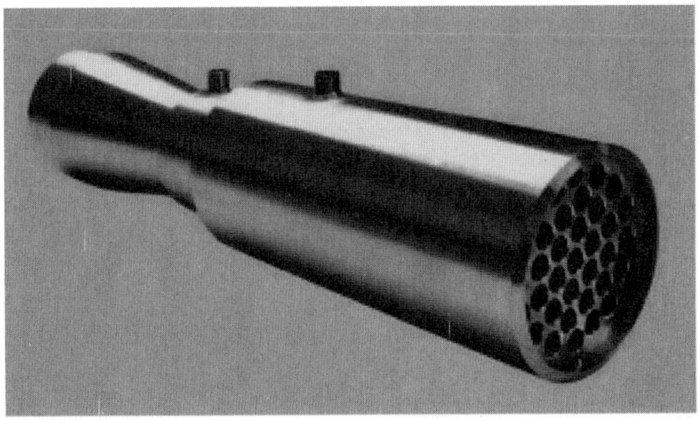

FIG. 10-18 High Head Recovery FloPak. (*Courtesy of Fluidic Technology.*)

with high-speed recording equipment to provide accurate records of the pressure profiles at the upstream and downstream pressure taps, which can then be analyzed and translated into a mean flow rate.

The rather general practice of producing a steady differential reading by placing restrictions in the manometer leads can result in a reading which, under a fixed set of conditions, may be useful in control of an operation but which has no readily predictable relation to the actual average flow. If calibration is employed to compensate for the presence of pulsations, then complete reproduction of operating conditions, including source of pulsations and waveform, is necessary to ensure reasonable accuracy.

According to Head [*Trans. Am. Soc. Mech. Eng.* **78:** 1471–1479 (1956)], a pulsation intensity limit of $\Gamma = 0.1$ is recommended as a practical pulsation threshold below which the performance of all types of flowmeters will differ negligibly from steady-flow performance (an error of less than 1 percent in flow due to pulsation). The peak-to-trough flow variation Γ is expressed as a fraction of the average flow rate. According to the *Report of the ASME Research Committee on Fluid Meters* (pp. 34–35), the fractional metering error E for *liquid flow* through a head meter is given by

$$(1 + E)^2 = 1 + \Gamma^2/8 \tag{10-32}$$

When the pulsation amplitude is such as to result in a greater-than-permissible metering error, consideration should be given to installation of a pulsation damper between the source of pulsations and the flowmeter. References to methods of pulsation damper design are given in the subsection Unsteady-State Behavior.

Pulsations are most likely to be encountered in discharge lines from reciprocating pumps or compressors and in lines supplying steam to reciprocating machinery. For *gas flow,* a combination involving a surge chamber and a constriction in the line can be used to damp out the pulsations to an acceptable level. The surge chamber is generally located as close to the pulsation source as possible, with the constriction between the surge chamber and the metering element. This arrangement can be used for either a suction or a discharge line. For such an arrangement, the metering error has been found to be a function of the *Hodgson number* N_H, which is defined as

$$N_H = Qn\,\Delta p_s / qp_s \tag{10-33}$$

where Q = volume of surge chamber and pipe between metering element and pulsation source; n = pulsation frequency; Δp_s = permanent pressure drop between metering element and surge chamber; q = average volume flow rate, based on gas density in the surge chamber; and p_s = pressure in surge chamber.

TABLE 10-8*a* Minimum Hodgson Numbers

Simplex double-acting compressor

s	N_H	s	N_H
0.167	1.31	0.667	0.60
0.333	1.00	0.833	0.43
0.50	0.80	1.00	0.34

TABLE 10-8*b* Minimum Hodgson Numbers

Duplex double-acting compressor		Triplex double-acting compressor	
s	N_H	s	N_H
0.167	1.00	0.167	0.85
0.333	0.70	0.333	0.30
0.50	0.30	0.50	0.15
0.667	0.10	0.667	0.06
0.833	0.05	0.833	0.00
1.00	0.00	1.00	0.00

Herning and Schmid [*Z. Ver. Dtsch. Ing.* **82:** 1107–1114 (1938)] presented charts for a simplex double-acting compressor for the prediction of metering error as a function of the Hodgson number and *s,* the ratio of piston discharge time to total time per stroke. Table 10-8*a* gives the minimum Hodgson numbers required to reduce the metering error to 1 percent as given by the charts (for specific heat ratios between 1.28 and 1.37). Schmid [*Z. Ver. Dtsch. Ing.* **84:** 596–598 (1940)] presented similar charts for a duplex double-acting compressor and a triplex double-acting compressor for a specific heat ratio of 1.37. Table 10-8*b* gives the minimum Hodgson numbers corresponding to a 1 percent metering error for these cases. The value of $Q\,\Delta p_s$ can be calculated from the appropriate Hodgson number, and appropriate values of Q and Δp_s selected so as to satisfy this minimum requirement.

VELOCITY METERS

Anemometers An anemometer may be any instrument for the measurement of gas velocity, e.g., a pitot tube, but usually the term refers to one of the following types.

The *vane anemometer* is a delicate revolution counter with jeweled bearings, actuated by a small windmill, usually 75 to 100 mm (about 3 to 4 in) in diameter, constructed of flat or slightly curved radially disposed vanes. Gas velocity is determined by using a stopwatch to find the time interval required to pass a given number of meters (feet) of gas as indicated by the counter. The velocity so obtained is inversely proportional to gas density. If the original calibration was carried out in a gas of density ρ_0 and the density of the gas stream being metered is ρ_1, the true gas velocity can be found as follows: From the calibration curve for the instrument, find $V_{t,0}$ corresponding to the quantity $V_m\sqrt{\rho_1/\rho_0}$, where V_m = measured velocity. Then the actual velocity $V_{t,1}$ is equal to $V_{t,0}\sqrt{\rho_0/\rho_1}$. In general, when working with air, the effects of atmospheric density changes can be neglected for all velocities above 1.5 m/s (about 5 ft/s). In all cases, care must be taken to hold the anemometer well away from one's body or from any object not normally present in the stream.

Vane anemometers can be used for gas velocity measurements in the range of 0.3 to 45 m/s (about 1 to 150 ft/s), although a given instrument generally has about a 20-fold velocity range. Bearing friction has to be minimized in instruments designed for accuracy at the low end of the range, while ample rotor and vane rigidity must be provided for measurements at the higher velocities. Vane anemometers are sensitive to shock and cannot be used in corrosive atmospheres. Therefore, accuracy is questionable unless a recent calibration has been done and the history of the instrument subsequent to calibration is known. For additional information, see Ower et al., chap. 8.

Turbine Flowmeters They consist of a straight flow tube containing a turbine which is free to rotate on a shaft supported by one or more bearings and located on the centerline of the tube. Means are provided for magnetic detection of the rotational speed, which is proportional to the volumetric flow rate. Its use is generally restricted to clean, noncorrosive fluids. Additional information on construction, operation, range, and accuracy can be obtained from Baker, *Flow Measurement Handbook,* 2000, pp. 215–252; Miller, *Flow Measurement Engineering Handbook,* 1996; and Spitzer, 2005, pp. 303–317.

The *current meter* is generally used for measuring velocities in open channels such as rivers and irrigation channels. There are two types, the cup meter and the propeller meter. The former is more widely used. It consists of six conical cups mounted on a vertical axis pivoted at the ends and free to rotate between the rigid arms of a U-shaped clevis to which a vaned tailpiece is attached. The wheel rotates because of the difference in drag for the two sides of the cup, and a signal proportional to the revolutions of the wheel is generated. The velocity is determined from the count over a period of time. The current meter is generally useful in the range of 0.15 to 4.5 m/s (about 0.5 to 15 ft/s) with an accuracy of ±2 percent. For additional information see Creager and Justin, *Hydroelectric Handbook,* 2d ed., Wiley, New York, 1950, pp. 42–46.

Other important classes of velocity meters include electromagnetic flowmeters and ultrasonic flowmeters. Both are described in Sec. 8.

MASS FLOWMETERS

General Principles There are two main types of mass flowmeters: (1) the so-called true mass flowmeter, which responds directly to mass flow rate, and (2) the inferential mass flowmeter, which commonly measures volume flow rate and fluid density separately. A variety of types of true mass flowmeters have been developed, including the following: (*a*) the Magnus-effect mass flowmeter, (*b*) the axial-flow, transverse-momentum mass flowmeter, (*c*) the radial-flow, transverse-momentum mass flowmeter, (*d*) the gyroscopic transverse-momentum mass flowmeter, and (*e*) the thermal mass flowmeter. Type *b* is the basis for several commercial mass flowmeters, one version of which is briefly described here.

Axial-Flow Transverse-Momentum Mass Flowmeter This type is also referred to as an *angular-momentum* mass flowmeter. One embodiment of its principle involves the use of axial flow through a driven impeller and a turbine in series. The impeller imparts angular momentum to the fluid, which in turn causes a torque to be imparted to the turbine, which is restrained from rotating by a spring. The torque, which can be measured, is proportional to the rotational speed of the impeller and the mass flow rate.

Inferential Mass Flowmeter There are several types in this category, including the following:

1. *Head meters with density compensation.* Head meters such as orifices, venturis, or nozzles can be used with one of a variety of densitometers [e.g., based on (*a*) buoyant force on a float, (*b*) hydraulic coupling, (*c*) voltage output from a piezoelectric crystal, or (*d*) radiation absorption]. The signal from the head meter, which is proportional to ρV^2 (where ρ = fluid density and V = fluid velocity), is multiplied by ρ given by the densitometer. The square root of the product is proportional to the mass flow rate.
2. *Head meters with velocity compensation.* The signal from the head meter, which is proportional to ρV^2, is divided by the signal from a velocity meter to give a signal proportional to the mass flow rate.
3. *Velocity meters with density compensation.* The signal from the velocity meter (e.g., turbine meter, electromagnetic meter, or sonic velocity meter) is multiplied by the signal from a densitometer to give a signal proportional to the mass flow rate.

Coriolis Mass Flowmeter This type, described in Sec. 8, offers simultaneous direct measurement of both mass flow rate and fluid density. The Coriolis flowmeter is insensitive to upstream and downstream flow disturbances, but its performance is adversely affected by the presence of even a few percent of a gas when measuring a liquid flow.

VARIABLE-AREA METERS

General Principles The underlying principle of an ideal area meter is the same as that of a head meter of the orifice type (see subsection Orifice Meters). The stream to be measured is throttled by a constriction, but instead of observing the variation with flow of the differential head across an orifice of fixed size, the constriction of an area meter is so arranged that its size is varied to accommodate the flow while the differential head is held constant.

A simple example of an area meter is a gate valve of the rising-stem type provided with static-pressure taps before and after the gate and a means for measuring the stem position. In most common types of area meters, the variation of the opening is automatically brought about by the motion of a weighted piston or float supported by the fluid. Two different cylinder- and piston-type area meters are described in the *Report of ASME Research Committee on Fluid Meters,* pp. 82–83.

Rotameters The rotameter, an example of which is shown in Fig. 10-19, has become one of the most popular flowmeters in the chemical-process industries. It consists essentially of a plummet, or "float," which is free to move up or down in a vertical, slightly tapered tube having its small end down. The fluid enters the lower end of the tube and causes the float to rise until the annular area between the float and the wall of the tube is such that the pressure drop across this constriction is just sufficient to support the float. Typically, the tapered tube is of glass and carries etched upon it a nearly linear scale on which the position of the float may be visually noted as an indication of the flow.

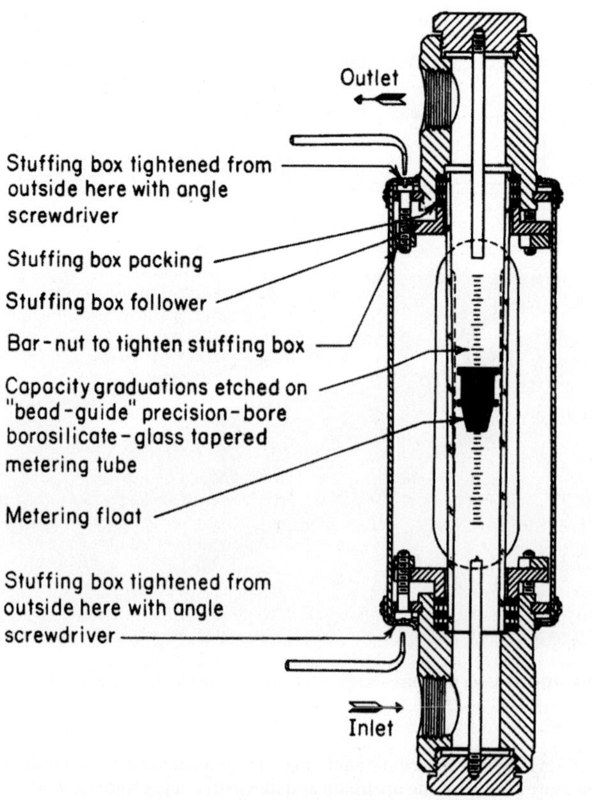

FIG. 10-19 Rotameter.

Interchangeable precision-bore glass tubes and metal metering tubes are available. Rotameters have proved satisfactory both for gases and for liquids at high and low pressures. A single instrument can readily cover a 10-fold range of flow, and by providing floats of different densities a 200-fold range is practicable. Rotameters are available with pneumatic, electric, and electronic transmitters for actuating remote recorders, integrators, and automatic flow controllers (see Considine, pp. 4-35 to 4-36, and Sec. 8 of this text).

Rotameters require no straight runs of pipe before or after the point of installation. Pressure losses are substantially constant over the whole flow range. In experimental work, for greatest precision, a rotameter should be calibrated with the fluid to be metered. However, most modern rotameters are precision-made so that their performance closely corresponds to a master calibration plot for the type in question. Such a plot is supplied with the meter upon purchase.

According to Head [*Trans. Am. Soc. Mech. Eng.* **76:** 851–862 (1954)], the flow rate through a rotameter can be obtained from

$$\dot{m} = q\rho = KD_f\sqrt{\frac{W_f(\rho_f-\rho)\rho}{\rho_f}} \tag{10-34}$$

and

$$K = \phi\left[\frac{D_t}{D_f}, \frac{\mu}{\sqrt{\frac{W_f(\rho_f-\rho)\rho}{\rho_f}}}\right] \tag{10-35}$$

where $\dot{m}$ = weight flow rate; q = volume flow rate; ρ = fluid density; K = flow parameter, $m^{1/2}/s$ ($ft^{1/2}/s$); D_f = float diameter at constriction; W_f = float weight; ρ_f = float density; D_t = tube diameter at point of constriction; and μ = fluid viscosity. The appropriate value of K is obtained from a composite correlation of K versus the parameters shown in Eq. (10-35) corresponding to the float shape being used. The relation of D_t to the rotameter reading is also required for the tube taper and size being used.

The ratio of flow rates for two different fluids A and B at the same rotameter reading is given by

$$\frac{\dot{m}_A}{\dot{m}_B} = \frac{K_A}{K_B}\sqrt{\frac{(\rho_f-\rho_A)\rho_A}{(\rho_f-\rho_B)\rho_B}} \tag{10-36}$$

A measure of self-compensation, with respect to weight rate of flow, for fluid density changes can be introduced through the use of a float with a density twice that of the fluid being metered, in which case an increase of 10 percent in ρ will produce a decrease of only 0.5 percent in w for the same reading. The extent of immunity to changes in fluid viscosity depends on the shape of the float.

According to Baird and Cheema [*Can. J. Chem. Eng.* **47:** 226–232 (1969)], the presence of square-wave pulsations can cause a rotameter to overread by as much as 100 percent. The higher the pulsation frequency, the less the float oscillation, although the error can still be appreciable even when the frequency is high enough that the float is virtually stationary. Use of a damping chamber between the pulsation source and the rotameter will reduce the error.

Additional information on rotameter theory is presented by Fischer [*Chem. Eng.* **59**(6): 180–184 (1952)], Coleman [*Trans. Inst. Chem. Eng.* **34:** 339–350 (1956)], and McCabe, Smith, and Harriott (*Unit Operations of Chemical Engineering,* 4th ed., McGraw-Hill, New York, 1985, pp. 202–205).

TWO-PHASE SYSTEMS

It is generally preferable to meter each of the individual components of a two-phase mixture separately prior to mixing, since it is difficult to meter such mixtures accurately. Problems arise because of fluctuations in composition with time and variations in composition over the cross section of the channel. Information on metering of such mixtures can be obtained from the following sources.

Gas-Solid Mixtures Carlson, Frazier, and Engdahl [*Trans. Am. Soc. Mech. Eng.* **70:** 65–79 (1948)] describe the use of a *flow nozzle* and a *square-edged orifice* in series for the measurement of both the gas rate and the solids rate in the flow of a finely divided solid-in-gas mixture. The nozzle differential is sensitive to the flow of both phases, whereas the orifice differential is not influenced by the solids flow.

Farbar [*Trans. Am. Soc. Mech. Eng.* **75:** 943–951 (1953)] describes how a *venturi meter* can be used to measure solids flow rate in a gas-solids mixture when the gas rate is held constant. Separate calibration curves (solids flow versus differential) are required for each gas rate of interest.

Cheng, Tung, and Soo [*J. Eng. Power* **92:** 135–149 (1970)] describe the use of an *electrostatic probe* for measurement of solids flow in a gas-solids mixture.

Goldberg and Boothroyd [*Br. Chem. Eng.* **14:** 1705–1708 (1969)] describe several types of solids-in-gas flowmeters and give an extensive bibliography.

Gas-Liquid Mixtures An empirical equation was developed by Murdock [*J. Basic Eng.* **84:** 419–433 (1962)] for the measurement of gas-liquid mixtures using *sharp-edged orifice* plates with radius, flange, or pipe taps.

An equation for use with *venturi meters* was given by Chisholm [*Br. Chem. Eng.* **12:** 454–457 (1967)]. A procedure for determining steam quality via pressure drop measurement with upflow through either venturi meters or sharp-edged orifice plates was given by Collins and Gacesa [*J. Basic Eng.* **93:** 11–21 (1971)].

Liquid-Solid Mixtures Liptak [*Chem. Eng.* **74**(4): 151–158 (1967)] discusses a variety of techniques that can be used for the measurement of solids-in-liquid suspensions or slurries. These include metering pumps, weigh tanks, magnetic flowmeter, ultrasonic flowmeter, gyroscope flowmeter, etc.

Shirato, Gotoh, Osasa, and Usami [*J. Chem. Eng. Japan* **1:** 164–167 (January 1968)] present a method for determining the mass flow rate of suspended solids in a liquid stream wherein the liquid velocity is measured by an electromagnetic flowmeter and the flow of solids is calculated from the pressure drops across each of two vertical sections of pipe of different diameter through which the suspension flows in series.

FLOWMETER SELECTION

Web sites for process equipment and instrumentation, such as www.globalspec.com and www.thomasnet.com, are valuable tools when one is selecting a flowmeter. These search engines can scan the flowmeters manufactured by more than 800 companies for specific products that meet the user's specifications. Table 10-4 was based in part on information from these web sites. Note that the accuracies claimed are achieved only under ideal conditions when the flowmeters are clean, properly installed, and calibrated for the application.

TABLE 10-9 Flowmeter Classes

Class I: Flowmeters with wetted moving parts	Class II: Flowmeters with no wetted moving parts
Positive displacement Turbine Variable-area	Differential pressure Vortex Target Thermal
Class III: Obstructionless flowmeters	**Class IV: Flowmeters with sensors mounted external to the pipe**
Coriolis mass Electromagnetic Ultrasonic	Clamp-on ultrasonic Correlation

Adapted from Spitzer, op. cit., 2005.

The purpose of this subsection is to summarize the preferred applications as well as the advantages and disadvantages of some of the common flowmeter technologies.

Table 10-9 divides flowmeters into four classes. Flowmeters in class I depend on wetted moving parts that can wear, plug, or break. The potential for catastrophic failure is a disadvantage. However, in clean fluids, class I flowmeters have often proved reliable and stable when properly installed, calibrated, and maintained.

Class II flowmeters have no wetted moving parts to break and are thus not subject to catastrophic failure. However, the flow surfaces such as orifice plates may wear, eventually biasing flow measurements. Other disadvantages of some flowmeters in this class include high pressure drop and susceptibility to plugging. Very dirty and abrasive fluids should be avoided.

Because class III flowmeters have neither moving parts nor obstructions to flow, they are suitable for dirty and abrasive fluids provided that appropriate materials of construction are available.

Class IV flowmeters have sensors mounted external to the pipe, and would thus seem to be ideal, but problems of accuracy and sensitivity have been encountered in early devices. These comparatively new technologies are under development, and these problems may be overcome in the future.

Section 8 outlines the following criteria for selection of measurement devices: measurement span, performance, reliability, materials of construction, prior use and potential for releasing process materials to the environment, electrical classification, physical access, invasive or non-invasive, and life-cycle cost.

Spitzer, *Industrial Flow Measurement,* 2005, cites four intended end uses of the flowmeter: rate indication, control, totalization, and alarm. Thus high accuracy may be important for rate indication, while control may just need good repeatability. Volumetric flow or mass flow indication is another choice.

Baker, *Flow Measurement Handbook,* 2003, identifies the type of fluid (liquid or gas, slurry, multiphase), special fluid constraints (clean or dirty, hygienic, corrosive, abrasive, high flammability, low lubricity, fluids causing scaling). He lists the following flowmeter constraints: accuracy or measurement uncertainty, diameter range, temperature range, pressure range, viscosity range, flow range, pressure loss caused by the flowmeter, sensitivity to installation, sensitivity to pipework supports, sensitivity to pulsation, whether the flowmeter has a clear bore, availability of a clamp-on version, response time, and ambient conditions. Finally, Baker identifies these environmental considerations: ambient temperature, humidity, exposure to weather, level of electromagnetic radiation, vibration, tamperproof for domestic use, and classification of area requiring explosion proof, intrinsic safety, etc.

Note that the accuracies cited in Table 10-4 can be achieved by those flowmeters only under ideal conditions of application, installation, and calibration. This subsection has given only an introduction to issues to consider in the choice of a flowmeter for a given application. See Baker, 2003; Miller, 1996; and Spitzer, 2005, for further guidance. To further refine choices, obtain application-specific data from flowmeter vendors.

PUMPS AND COMPRESSORS

General References: Meherwan P. Boyce, P.E., *Centrifugal Compressors: A Basic Guide*, Pennwell Books, Tulsa, Okla., 2002; Royce N. Brown, *Compressors: Selection and Sizing*, 3d ed., Gulf Professional Publishing, Houston, Tex., 2005; James Corley, "The Vibration Analysis of Pumps: A Tutorial," *Fourth International Pump Symposium*, Texas A & M University, Houston, Tex., May 1987; John W. Dufor and William E. Nelson, *Centrifugal Pump Sourcebook*, McGraw-Hill, New York, 1992; *Engineering Data Book*, 12th ed., vol. I, Secs. 12 and 13, Gas Processors Suppliers Association, Tulsa, Okla., 2004; Paul N. Garay, P.E., *Pump Application Desk Book*, Fairmont Press, Lilburn, Ga., 1993; *Process Pumps*, IIT Fluid Technology Corporation, 1992; Igor J. Karassik et al., *Pump Handbook*, 3d ed., McGraw-Hill, New York, 2001; Val S. Lobanoff and Robert R. Ross, *Centrifugal Pumps: Design and Application*, 2d ed., Gulf Professional Publishing, Houston, Tex., 1992; A. J. Stephanoff, *Centrifugal and Axial Flow Pumps: Theory, Design, and Application*, 2d ed., Krieger Publishing, Melbourne, Fla., 1992.

INTRODUCTION

The following subsections deal with pumps and compressors. A pump or compressor is a physical contrivance that is used to deliver fluids from one location to another through conduits. The term *pump* is used when the fluid is a liquid, while the term *compressor* is used when the fluid is a gas. The basic requirements to define the application are suction and delivery pressures, pressure loss in transmission, and flow rate. Special requirements may exist in food, pharmaceutical, nuclear, and other industries that impose material selection requirements of the pump.

The primary means of transfer of energy to the fluid that causes flow are gravity, displacement, centrifugal force, electromagnetic force, transfer of momentum, mechanical impulse, and a combination of these energy transfer mechanisms. Displacement and centrifugal force are the most common energy transfer mechanisms in use.

Pumps and compressors are designed per technical specifications and standards developed over years of operating and maintenance experience. Table 10-10 lists some of these standards for pumps and compressors and for related equipment such as lubrication systems and gearboxes which, if not properly specified, could lead to many operational and maintenance problems with the pumps and compressors. These standards specify design, construction, maintenance, and testing details such as terminology, material selection, shop inspection and tests, drawings, clearances, construction procedures, and so on.

Three major types of pumps are discussed here: (1) positive-displacement, (2) dynamic (kinetic), and (3) lift. Piston pumps are positive-displacement pumps. The most common centrifugal pumps are of dynamic type; ancient bucket-type pumps are lift pumps. Canned pumps are also becoming popular in the petrochemical industry because of the drive to minimize fugitive emissions. Figure 10-20 shows pump classification.

TERMINOLOGY

Displacement Discharge of a fluid from a vessel by partially or completely displacing its internal volume with a second fluid or by mechanical means is the principle upon which a great many fluid-transport devices operate. Included in this group are reciprocating-piston and diaphragm machines, rotary-vane and gear types, fluid piston compressors, acid eggs, and air lifts.

The large variety of displacement-type fluid-transport devices makes it difficult to list characteristics common to each. However, for most types it is correct to state that (1) they are adaptable to high-pressure operation, (2) the flow rate through the pump is variable (auxiliary damping systems may be employed to reduce the magnitude of pressure pulsation and flow variation), (3) mechanical considerations limit maximum throughputs, and (4) the devices are capable of efficient performance at extremely low-volume throughput rates.

Centrifugal Force Centrifugal force is applied by means of the centrifugal pump to a liquid. Though the physical appearance of the many types of centrifugal pumps and compressors varies greatly, the basic function of each is the same, i.e., to produce kinetic energy by the action of centrifugal force and then to convert this energy to pressure by efficiently reducing the velocity of the flowing fluid.

In general, centrifugal fluid-transport devices have these characteristics: (1) discharge is relatively free of pulsation; (2) mechanical design lends itself to high throughputs, and capacity limitations are rarely a problem; (3) the devices are capable of efficient performance over a wide range of pressures and capacities even at constant-speed operation; (4) discharge pressure is a function of fluid density; and (5) these are relatively small high-speed devices and less costly.

A device that combines the use of centrifugal force with mechanical impulse to produce an increase in pressure is the axial-flow compressor or pump. In this device the fluid travels roughly parallel to the shaft through a series of alternately rotating and stationary radial blades having airfoil cross sections. The fluid is accelerated in the axial direction by mechanical impulses from the rotating blades; concurrently, a positive-pressure gradient in the radial direction is established in each stage by centrifugal force. The net pressure rise per stage results from both effects.

Electromagnetic Force When the fluid is an electrical conductor, as is the case with molten metals, it is possible to impress an electromagnetic field around the fluid conduit in such a way that a driving force that will cause flow to be created. Such pumps have been developed for the handling of heat-transfer liquids, especially for nuclear reactors.

Transfer of Momentum Deceleration of one fluid (motivating fluid) in order to transfer its momentum to a second fluid (pumped fluid) is a principle commonly used in the handling of corrosive materials, in pumping from inaccessible depths, or for evacuation. Jets and eductors are in this category.

Absence of moving parts and simplicity of construction have frequently justified the use of jets and eductors. However, they are relatively inefficient devices. When air or steam is the motivating fluid, operating costs may be several times the cost of alternative types of fluid-transport equipment. In addition, environmental considerations in today's chemical plants often inhibit their use.

Mechanical Impulse The principle of mechanical impulse when applied to fluids is usually combined with one of the other means of imparting motion. As mentioned earlier, this is the case in axial-flow compressors and pumps. The turbine or regenerative-type pump is another device that functions partially by mechanical impulse.

Measurement of Performance The amount of useful work that any fluid-transport device performs is the product of (1) the mass rate of fluid flow through it and (2) the total pressure differential measured immediately before and after the device, usually expressed in the height of column of fluid equivalent under adiabatic conditions. The first of these quantities is normally referred to as *capacity,* and the second is known as *head.*

Capacity This quantity is expressed in the following units. In SI units, capacity is expressed in cubic meters per hour (m^3/h) for both liquids and gases. In U.S. Customary System units it is expressed in U.S. gallons per minute (gal/min) for liquids and in cubic feet per minute (ft^3/min) for gases. Since all these are volume units, the density or specific gravity must be used for conversion to mass rate of flow. When gases are being handled, capacity must be related to a pressure and a temperature, usually the conditions prevailing at the machine inlet. It is important to note that all heads and other terms in the following equations are expressed in height of column of liquid.

TABLE 10-10 Standards Governing Pumps and Compressors

ASME Standards, American Society of Mechanical Engineers, New York
- B73.1-2001, *Specification for Horizontal End Suction Centrifugal Pumps for Chemical Process*
- B73.2-2003, *Specification for Vertical In-Line Centrifugal Pumps for Chemical Process*
- PTC 10, 1997 *Test Code on Compressors and Exhausters*
- PTC 11, 1984 *Fans*
- B19.3-1991, *Safety Standard for Compressors for Process Industries*

API Standards, American Petroleum Institute, Washington
- API Standard 610, *Centrifugal Pumps for Petroleum, Petrochemical, and Natural Gas Industries,* Adoption of ISO 13709, October 2004
- API Standard 613, *Special Purpose Gear Units for Petroleum, Chemical and Gas Industry Services,* February 2003
- API Standard 614, *Lubrication, Shaft-Sealing, and Control-Oil Systems and Auxiliaries for Petroleum, Chemical and Gas Industry Services,* April 1999
- API Standard 616, *Gas Turbines for the Petroleum, Chemical, and Gas Industry Services,* August 1998
- API Standard 617, *Axial and Centrifugal Compressors and Expanders— Compressors for Petroleum, Chemical, and Gas Industry Services,* June 2003
- API Standard 618, *Reciprocating Compressors for Petroleum, Chemical, and Gas Industry Services,* June 1995
- API Standard 619, *Rotary-Type Positive Displacement Compressors for Petroleum, Petrochemical, and Natural Gas Industries,* December 2004
- API Standard 670, *Machinery Protection Systems,* November 2003
- API Standard 671, *Special Purpose Couplings for Petroleum, Chemical, and Gas Industry Services,* October 1998
- API Standard 672, *Packaged, Integrally Geared, Centrifugal Air Compressors for Petroleum, Chemical, and Gas Industry Services,* March 2004
- API Standard 673, *Centrifugal Fans for Petroleum, Chemical, and Gas Industry Services,* October 2002
- API Standard 674, *Positive Displacement Pumps—Reciprocating,* June 1995
- API Standard 675, *Positive Displacement Pumps—Controlled Volume,* March 2000
- API Standard 677, *General Purpose Gear Units for Petroleum, Chemical, and Gas Industry Services,* April 2006
- API Standard 680, *Packaged Reciprocating Plant and Instrument Air Compressors for General Refinery Services,* October 1987
- API Standard 681, *Liquid Ring Vacuum Pumps and Compressors for Petroleum, Chemical, and Gas Industry Services,* June 2002
- API Standard 682, *Pumps—Shaft Sealing Systems for Centrifugal and Rotary Pumps,* September 2004
- API Standard 685, *Sealless Centrifugal Pumps for Petroleum, Heavy Duty Chemical, and Gas Industry Services,* October 2000

Hydraulic Institute, Parsippany, N.J. (www.pumps.org)
- ANSI/HI Pump Standards, 2005 (covers centrifugal, vertical, rotary, and reciprocating pumps)

National Fire Protection Association, Quincy, Mass. (www.nfpa.org) Standards for pumps used in fire protection systems

PUMPS

Total Dynamic Head The total dynamic head H of a pump is the total discharge head h_d minus the total suction head h_s.

Total Suction Head This is the reading h_{gs} of a gauge at the suction flange of a pump (corrected to the pump centerline), plus the barometer reading and the velocity head h_{vs} at the point of gauge attachment:

$$h_s = h_{gs} + \text{atm} + h_{vs} \tag{10-37}$$

If the gauge pressure at the suction flange is less than atmospheric, requiring use of a vacuum gauge, this reading is used for h_{gs} in Eq. (10-38) with a negative sign.

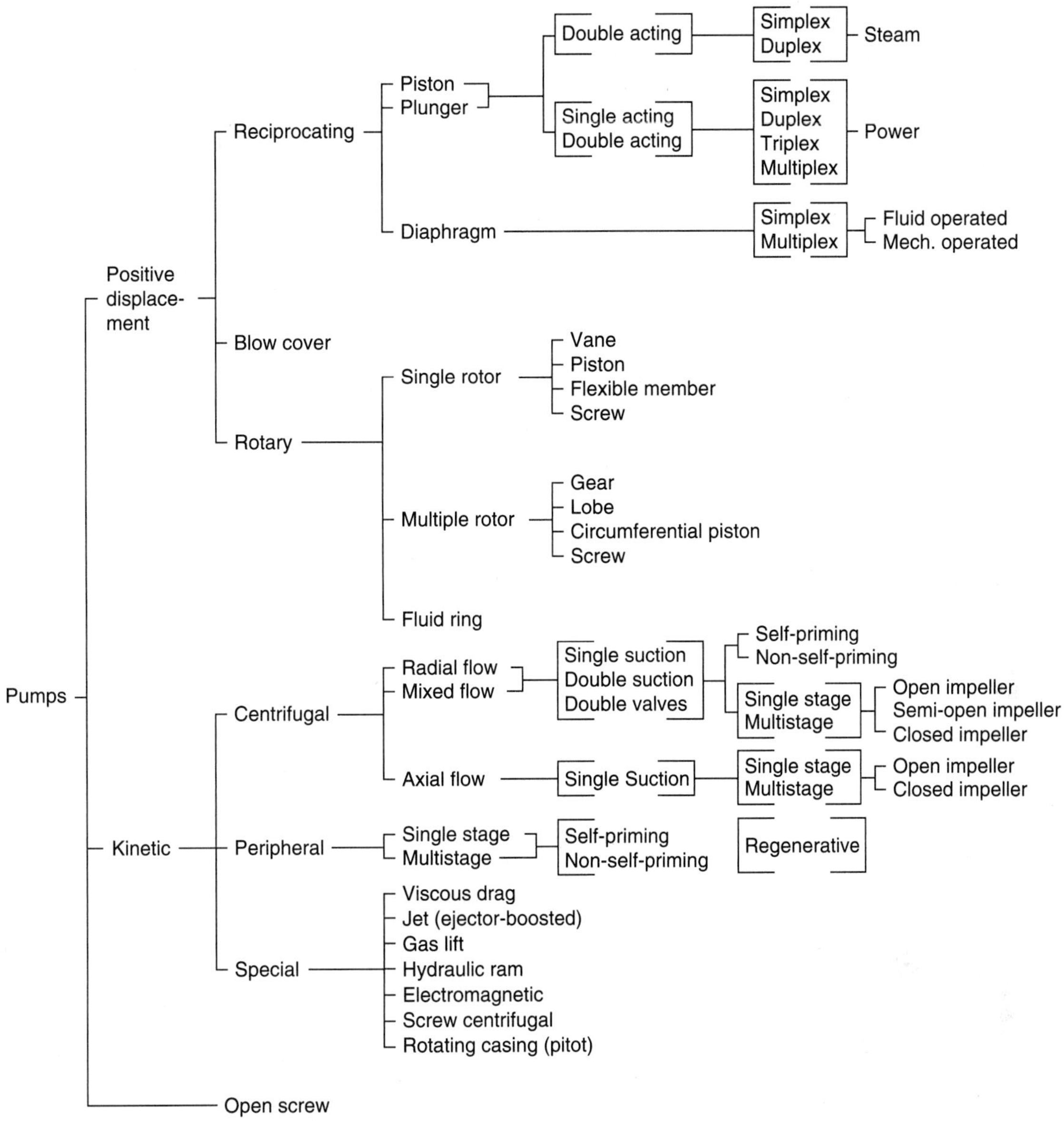

FIG. 10-20 Classification of pumps. (*Courtesy of Hydraulic Institute.*)

Before installation it is possible to estimate the total suction head as follows:

$$h_s = h_{ss} - h_{fs} \tag{10-38}$$

where h_{ss} = static suction head and h_{fs} = suction friction head.

Static Suction Head The static suction head h_{ss} is the vertical distance measured from the free surface of the liquid source to the pump centerline plus the absolute pressure at the liquid surface.

Total Discharge Head The total discharge head h_d is the reading h_{gd} of a gauge at the discharge flange of a pump (corrected to the pump centerline*), plus the barometer reading and the velocity head h_{vd} at the point of gauge attachment:

$$h_d = h_{gd} + \text{atm} + h_{vd} \tag{10-39}$$

Again, if the discharge gauge pressure is below atmospheric, the vacuum-gauge reading is used for h_{gd} in Eq. (10-39) with a negative sign.

*On vertical pumps, the correction should be made to the eye of the suction impeller.

Before installation it is possible to estimate the total discharge head from the static discharge head h_{sd} and the discharge friction head h_{fd} as follows:

$$h_d = h_{sd} + h_{fd} \tag{10-40}$$

Static Discharge Head The static discharge head h_{sd} is the vertical distance measured from the free surface of the liquid in the receiver to the pump centerline,* plus the absolute pressure at the liquid surface. *Total static head* h_{ts} is the difference between discharge and suction static heads.

Velocity Since most liquids are practically incompressible, the relation between the quantity flowing past a given point at a given time and the volume flow rate is expressed as follows:

$$Q = AV_{avg} \tag{10-41}$$

This relationship in SI units is as follows:

$$V_{avg}(\text{for circular conduits}) = 3.54Q/d^2 \tag{10-42}$$

where V_{avg} = average velocity of flow, m/s; Q = quantity of flow, m^3/h; and d = inside diameter of conduit, cm.

This same relationship in U.S. Customary System (USCS) units is

$$V_{avg}(\text{for circular conduits}) = 0.409Q/d^2 \tag{10-43}$$

where V_{avg} = average velocity of flow, ft/s; Q = volume flow rate, gal/min; and d = inside diameter of conduit, in.

Velocity Head This is the force generated by the pump and is given in $\text{ft}\cdot\text{lb}_f/\text{lb}_m$, the vertical height that the pump can maintain.

$$H = V^2/2g_c = \text{ft}\cdot\text{lb}_f/\text{lb}_m \tag{10-44}$$

where g_c = the gravitational constant = 32.2 $\text{ft}\cdot\text{lb}_m/\text{lb}_f\cdot\text{s}^2$

In the SI system the head is in meters.

Viscosity (See Sec. 6 for further information.) In flowing liquids, the existence of internal friction or the internal resistance to relative motion of the fluid particles must be considered. This resistance is called *viscosity.* Frictional losses in pipes increase with higher viscosity. Viscosity decreases with the rising temperature of the fluid. The increase in viscosity of fluids will increase the pump power required for the same head and capacity and will reduce the efficiency of the pump.

Friction Head This is the pressure required to overcome the resistance to flow in pipe and fittings. It is dealt with in detail in Sec. 6.

Work Performed in Pumping To cause liquid to flow, work must be expended. A pump may raise the liquid to a higher elevation, force it into a vessel at higher pressure, provide the head to overcome pipe friction, or perform any combination of these. Regardless of the service required of a pump, all energy imparted to the liquid in performing this service must be accounted for; consistent units for all quantities must be employed in arriving at the work or power performed.

When arriving at the performance of a pump, it is customary to calculate its *power output,* which is the product of (1) the total dynamic head and (2) the mass of liquid pumped in a given time. In SI units power is expressed in kilowatts; horsepower is the conventional unit used in the United States.

In SI units,

$$\text{kW} = HQ\rho/3.670\times10^5 \tag{10-45}$$

where kW is the pump power output, kW; H = total dynamic head, $\text{N}\cdot\text{m/kg}$ (column of liquid); Q = capacity, m^3/h; and ρ = liquid density, kg/m^3.

When the total dynamic head H is expressed in pascals, then

$$\text{kW} = HQ\rho/3.599\times10^6 \tag{10-46}$$

In USCS units,

$$\text{hp} = HQs/3.960\times10^3 \tag{10-47}$$

where hp is the pump power output, hp; H = total dynamic head, $\text{lbf}\cdot\text{ft/lbm}$ (column of liquid); Q = capacity, U.S. gal/min; and s = liquid specific gravity.

When the total dynamic head hp is expressed in pounds force per square inch, then

$$\text{hp} = HQ/1.714\times10^3 \tag{10-48}$$

The *power input* to a pump is greater than the *power output* because of internal losses resulting from friction, leakage, etc. The efficiency of a pump is therefore defined as

$$\text{Pump efficiency} = (\text{power output})/(\text{power input}) \tag{10-49}$$

PUMP SELECTION

When one is selecting pumps for any service, it is necessary to know the liquid to be handled, total dynamic head, suction and discharge heads, and, in most cases, temperature, viscosity, vapor pressure, and specific gravity. In the chemical industry, the task of pump selection is frequently further complicated by the presence of solids in the liquid and liquid corrosion characteristics requiring special materials of construction. Solids may accelerate erosion and corrosion, have a tendency to agglomerate, or require delicate handling to prevent undesirable degradation.

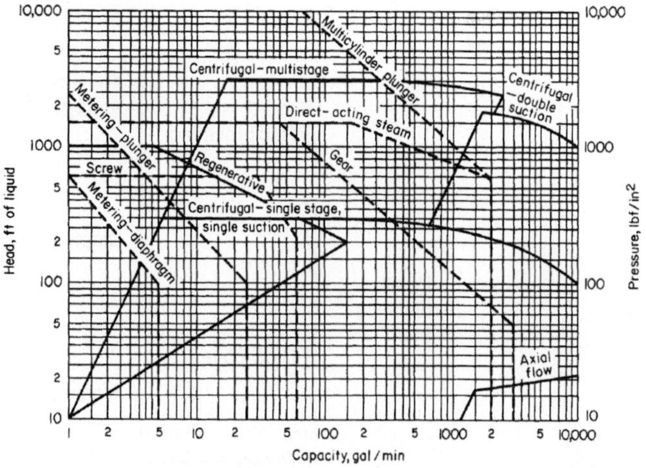

FIG. 10-21 Pump coverage chart based on normal ranges of operation of commercially available types. Solid lines: use left ordinate, head scale. Broken lines: use right ordinate, pressure scale. To convert gallons per minute to cubic meters per hour, multiply by 0.2271; to convert feet to meters, multiply by 0.3048; and to convert pounds-force per square inch to kilopascals, multiply by 6.895.

Range of Operation Because of the wide variety of pump types and the number of factors that determine the selection of any one type for a specific installation, the designer must first eliminate all but those types of reasonable possibility. Since range of operation is always an important consideration, Fig. 10-21 should be of assistance. The boundaries shown for each pump type are at best approximate. In most cases, following Fig. 10-21 will select the pump that is best suited for a given application. Reciprocating pumps and rotary pumps such as gear and roots rotor-type pumps are examples of positive-displacement pumps. Displacement pumps provide high heads at low capacities which are beyond the capability of centrifugal pumps. Displacement pumps achieve high pressure with low velocities and are thus suited for high-viscosity service and slurry.

The centrifugal pump operates over a very wide range of flows and pressures. The axial pump is best suited for low heads but high flows. Both the centrifugal and axial-flow pumps impart energy to the fluid by the rotational speed of the impeller and the velocity it imparts to the fluid.

NET POSITIVE SUCTION HEAD

Net positive suction head available $(\text{NPSH})_A$ is the difference between the total absolute suction pressure at the pump suction nozzle when the pump is running and the vapor pressure at the flowing liquid temperature. All pumps require the system to provide adequate $(\text{NPSH})_A$. In a positive-displacement pump the $(\text{NPSH})_A$ should be large enough to open the suction valve, to overcome the friction losses within the pump liquid end, and to overcome the liquid acceleration head.

Suction Limitations of a Pump Whenever the pressure in a liquid drops below the vapor pressure corresponding to its temperature, the liquid will vaporize. When this happens within an operating pump, the vapor bubbles will be carried along to a point of higher pressure, where they suddenly collapse. This phenomenon is known as *cavitation.* Cavitation in a pump should be avoided, as it is accompanied by metal removal, vibration, reduced flow, loss in efficiency, and noise. When the absolute suction pressure is low, cavitation may occur in the pump inlet and damage may result in the pump suction and on the impeller vanes near the inlet edges. To avoid this phenomenon, it is necessary to maintain a *required net positive suction head* $(\text{NPSH})_R$, which is the equivalent total head of liquid at the pump centerline less the vapor pressure p. Each pump manufacturer publishes curves relating $(\text{NPSH})_R$ to capacity and speed for each pump.

When a pump installation is being designed, the available net positive suction head $(\text{NPSH})_A$ must be equal to or greater than the $(\text{NPSH})_R$ for the desired capacity. The $(\text{NPSH})_A$ can be calculated as follows:

$$(\text{NPSH})_A = h_{ss} - h_{fs} - p \tag{10-50}$$

If $(\text{NPSH})_A$ is to be checked on an existing installation, it can be determined as follows:

$$(\text{NPSH})_A = \text{atm} + h_{gs} - p + h_{vs} \tag{10-51}$$

Practically, the NPSH required for operation without cavitation and vibration in the pump is somewhat greater than the theoretical. The actual $(NPSH)_R$ depends on the characteristics of the liquid, total head, pump speed, capacity, and impeller design. Any suction condition which reduces $(NPSH)_A$ below that required to prevent cavitation at the desired capacity will produce an unsatisfactory installation and can lead to mechanical difficulty.

The following two equations usually provide an adequate design margin between $(NPSH)_A$ and $(NPSH)_R$:

$$(NPSH)_A = (NPSH)_R + 5 \text{ ft} \tag{10-52}$$

$$(NPSH)_A = 1.35(NPSH)_R \tag{10-53}$$

Use the larger value of $(NPSH)_A$ calculated with Eqs. (10-52) and (10-53).

NPSH Requirements for Other Liquids NPSH values depend on the fluid being pumped. Since water is considered a standard fluid for pumping, various correction methods have been developed to evaluate NPSH when pumping other fluids. The most recent of these corrective methods has been developed by the Hydraulic Institute and is shown in Fig. 10-22.

The chart shown in Fig. 10-22 is for pure liquids. Extrapolation of data beyond the ranges indicated in the graph may not produce accurate results. Figure 10-22 shows the variation of vapor pressure and NPSH reductions for various hydrocarbons and hot water as a function of temperature. Certain rules apply while using this chart. When using the chart for hot water, if the NPSH reduction is greater than one-half of the NPSH required for cold water, deduct one-half of cold water NPSH to obtain the corrected NPSH required. However, if the value read on the chart is less than one-half of cold water NPSH, deduct this chart value from the cold water NPSH to obtain the corrected NPSH.

Example 10-1 NPSH Calculation Suppose a selected pump requires a minimum NPSH of 16 ft (4.9 m) when pumping cold water. What will be the NPSH limitation to pump propane at 55°F (12.8°C) with a vapor pressure of 120 psi (8.274 bar)? Using the chart in Fig. 10-22, NPSH reduction for propane gives 9.5 ft (2.9 m). This is greater than one-half of the cold water NPSH of 16 ft (4.9 m). The corrected NPSH is therefore 8 ft (2.4 m) or one-half of the cold water NPSH.

PUMP SPECIFICATIONS

Pump specifications depend on numerous factors but mostly on application. Typically, the following factors should be considered while preparing a specification:

1. Application, scope, and type
2. Service conditions

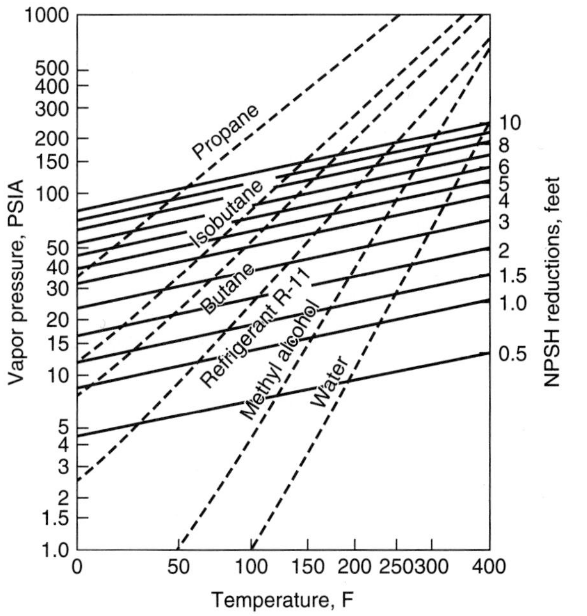

FIG. 10-22 NPSH reductions for pumps handling hydrocarbon liquids and high-temperature water. This chart has been constructed from test data obtained using the liquids shown. (*Hydraulic Institute Standards.*)

3. Operating conditions
4. Construction application-specific details and special considerations
   a. Casing and connections
   b. Impeller details
   c. Shaft
   d. Stuffing box details—lubrications, sealing, etc.
   e. Bearing frame and bearings
   f. Baseplate and couplings
   g. Materials
   h. Special operating conditions and miscellaneous items

Table 10-11 is based on the API and ASME codes and illustrates a typical specification for centrifugal pumps.

POSITIVE-DISPLACEMENT PUMPS

Positive-displacement pumps and those that approach positive displacement will ideally produce whatever head is impressed upon them by the system restrictions to flow. The maximum head attainable is determined by the power available in the drive (slippage neglected) and the strength of the pump parts. A pressure relief valve on the discharge side should be set to open at a safe pressure for the casing and the internal components of the pump such as piston rods, cylinders, crankshafts, and other components which would be pressurized. In the case of a rotary pump, the total dynamic head developed is uniquely determined for any given flow by the speed at which it rotates.

In general, overall efficiencies of positive-displacement pumps are higher than those of centrifugal equipment because internal losses are minimized. However, the flexibility of each piece of equipment in handling a wide range of capacities is somewhat limited.

Positive-displacement pumps may be of either the *reciprocating* or the *rotary* type. In all positive-displacement pumps, a cavity or cavities are alternately filled and emptied of the pumped fluid by the action of the pump.

Reciprocating Pumps There are three classes of reciprocating pumps: *piston pumps, plunger pumps,* and *diaphragm pumps.* Basically, the action of the liquid-transferring parts of these pumps is the same, with a cylindrical piston, or plunger, or bucket or a round diaphragm being caused to pass or flex back and forth in a chamber. The device is equipped with valves for the inlet and discharge of the liquid being pumped, and the operation of these valves is related in a definite manner to the motions of the piston. In all modern-design reciprocating pumps, the suction and discharge valves are operated by pressure difference. That is, when the pump is on its suction stroke and the pump cavity is increasing in volume, the pressure is lowered within the pump cavity, permitting the higher suction pressure to open the suction valve and allowing liquid to flow into the pump. At the same time, the higher discharge-line pressure holds the discharge valve closed. Likewise, on the discharge stroke, as the pump cavity is decreasing in volume, the higher pressure developed in the pump cavity holds the suction valve closed and opens the discharge valve to expel liquid from the pump into the discharge line.

The *overall efficiency* of these pumps varies from about 50 percent for the small pumps to about 90 percent or more for the larger sizes.

Reciprocating pumps may be of *single-cylinder* or *multicylinder* design. Multicylinder pumps have all cylinders in parallel for increased capacity. Piston-type pumps may be single-acting or double-acting; i.e., pumping may be accomplished from one end or both ends of the piston. Plunger pumps are always single-acting. The tabulation in Table 10-12 provides data on the flow variation of reciprocating pumps of various designs.

Piston Pumps There are two ordinary types of piston pumps: simplex double-acting pumps and duplex double-acting pumps.

Simplex Double-Acting Pumps These pumps may be direct-acting (i.e., direct-connected to a steam cylinder) or power-driven (through a crank and flywheel from the crosshead of a steam engine).

Duplex Double-Acting Pumps These pumps differ primarily from those of the simplex type in having two cylinders whose operation is coordinated. They may be direct-acting, steam-driven, or power-driven with crank and flywheel.

Plunger Pumps These differ from piston pumps in that they have one or more constant-diameter plungers reciprocating through packing glands and displacing liquid from cylinders in which there is considerable radial clearance. They are always single-acting, in the sense that only one end of the plunger is used in pumping the liquid. Plunger pumps are available with one, two, three, four, five, or even more cylinders. Simplex and duplex units are often built in a horizontal design. Those with three or more cylinders are usually of vertical design.

Diaphragm Pumps These pumps perform similarly to piston and plunger pumps, but the reciprocating driving member is a flexible diaphragm fabricated of metal, rubber, or plastic. The chief advantage of this arrangement is the elimination of all packing and seals exposed to the liquid

TABLE 10-11 Typical Pump Specification

Specification	Description
1.0	Scope: This specification covers horizontal, end suction, vertically split, single-stage centrifugal pumps with top centerline discharge and "back pullout" feature.
2.0	Service Conditions: Pump shall be designed to operate satisfactorily with a reasonable service life when operated either intermittently or continuously in typical process applications.
3.0	Operating Conditions: Capacity ____ U.S. gallons per minute ____ Head (__ ft total head) (__ psig). Speed __ r/min Suction Pressure (__ ft head) (positive) (lift) (__ psig) Liquid to be handled ________ Specific gravity _____ Viscosity (______) Temperature of liquid at inlet ____ °F Solids content ____% ____ Max. size
4.0	Pump Construction: 4.1 *Casing.* Casing shall be vertically split with self-venting top centerline discharge, with an integral foot located directly under the casing for added support. All casings shall be of the "back pullout" design with suction and discharge nozzles cast integrally. Casings shall be provided with bosses in suction and discharge nozzles, and in bottom of casing for gauge taps and drain tap. (Threaded taps with plugs shall be provided for these features.) 4.2 *Casing Connections.* Connections shall be ANSI flat-faced flanges. [Cast iron (125) (250) psig rated] [Duron metal, steel, alloy steel (150) (300) psig rated] 4.3 *Casing Joint Gasket.* A confined-type nonasbestos gasket suitable for corrosive service shall be provided at the casing joint. 4.4 *Impeller.* Fully-open impeller with front edge having contoured vanes curving into the suction for minimum NPSH requirements and maximum efficiency shall be provided. A hex head shall be cast in the eye of the impeller to facilitate removal, and eliminate need for special impeller removing tool. All impellers shall have radial "pump-out" vanes on the back side to reduce stuffing box pressure and aid in eliminating collection of solids at stuffing box throat. Impellers shall be balanced within A.N.S.I. guidelines to ISO tolerances. 4.4.1 *Impeller Clearance Adjustment.* All pumps shall have provisions for adjustment of axial clearance between the leading edge of the impeller and casing. This adjustment shall be made by a precision microdial adjustment at the outboard bearing housing, which moves the impeller forward toward the suction wall of the casing. 4.5 *Shafts.* Shafts shall be suitable for hook-type sleeve. Shaft material shall be (SAE 1045 steel on Duron and 316 stainless steel pumps) or (AISI 316 stainless steel on CD-4MCu pumps and #20 stainless steel pumps). Shaft deflection shall not exceed .005 at the vertical center-line of the impeller. 4.6 *Shaft Sleeve.* Renewable hook-type shaft sleeve that extends through the stuffing box and gland shall be provided. Shaft sleeve shall be (316 stainless steel), (#20 stainless steel) or (XH-800 Ni-chrome-boron coated 316 stainless steel with coated surface hardness of approximately 800 Brinell). 4.7 *Stuffing Box.* Stuffing box shall be suitable for packing, single (inside or outside) or double-inside mechanical seal without modifications. Stuffing box shall be accurately centered by machined rabbit fits on case and frame adapter. 4.7.1 *Packed Stuffing Box.* The standard packed stuffing box shall consist of five rings of graphited nonasbestos packing; a stainless steel packing base ring in the bottom of the box to prevent extrusion of the packing past the throat; a Teflon seal cage, and a two-piece 316 stain-less steel packing gland to ensure even pressure on the packing. Ample space shall be provided for repacking the stuffing box. 4.7.1.1 *Lubrication-Packed Stuffing Box.* A tapped hole shall be provided in the stuffing box directly over the seal cage for lubrication and cooling of the packing. Lubrication liquid shall be supplied (from an external source) (through a by-pass line from the pump discharge nozzle). 4.7.2 *Stuffing Box with Mechanical Seal.* Mechanical seal shall be of the (single inside) (single outside) (double inside) (cartridge) type and (balanced) (unbalanced). Stuffing box is to be (standard) (oversize) (oversize tapered). Suitable space shall be provided in the standard and oversized stuffing box for supplying a (throttle bushing) (dilution control bushing) with single seals. Throttle bushings and dilution control bushings shall be made of (glass-filled Teflon) (a suitable metal material). 4.7.2.1 *Lubrication—Stuffing Box with Mechanical Seals.* Suitable tapped connections shall be provided to effectively lubricate, cool, flush, quench, etc., as required by the application or recommendations of the mechanical seal manufacturer.
4.8	Bearing Frame and Bearings: 4.8.1 *Bearing Frame.* Frames shall be equipped with axial radiating fins extending the length of the frame to aid in heat dissipation. Frame shall be provided with ductile iron outboard bearing housing. Both ends of the frame shall be provided with lip-type oil seals and labyrinth-type deflectors of metallic reinforced synthetic rubber to prevent the entrance of contaminants. 4.8.2 *Bearings.* Pump bearings shall be heavy-duty, antifriction ball-type on both ends. The single row inboard bearing, nearest the impeller, shall be free to float within the frame and shall carry only radial load. The double row outboard bearing (F4-G1 and F4-I1) or duplex angular contact bearing (F4-H1), coupling end, shall be locked in place to carry radial and axial thrust loads. Bearings shall be designed for a minimum life of 20,000 hours in any normal pump operating range.
4.9	*Bearing Lubrication.* Ball bearings shall be oil-mist—lubricated by means of a slinger. The oil slinger shall be mounted on the shaft between the bearings to provide equal lubrication to both bearings. Bulls-eye oil-sight glasses shall be provided on both sides of the frame to provide a positive means of checking the proper oil level from either side of the pump. A tapped and plugged hole shall also be provided in both sides of the frame to mount bottle-type constant-level oilers where desired. A tapped and plugged hole shall be provided on both sides for optional straight-through oil cooling device.
5.0	Baseplate and Coupling: 5.1 *Baseplate.* Baseplates shall be rigid and suitable for mounting pump and motor. Baseplates shall be of channel steel construction. 5.2 *Coupling.* Coupling shall be flexible-spacer type. Coupling shall have at least three-and-one-half-inch spacer length for ease of rotating element removal. Both coupling hubs shall be provided with flats 180° apart to facilitate removal of impeller. Coupling shall not require lubrication.*
6.0	Mechanical Modifications Required for High Temperature: 6.1 *Modifications Required, Temperature Range 250–350°F.* Pumps for operation in this range shall be provided with a water-jacketed stuffing box. 6.2 *Modifications Required, Temperature Range 351–550°F (Maximum).* Pumps for operation in this range shall be provided with a water-jacketed stuffing box and a water-cooled bearing frame.
7.0	Materials: Pump materials shall be selected to suit the particular service requirements. 7.1 *Cast Iron—316 SS Fitted.* 15″ only; pump shall have cast iron casing and stuffing box cover. 316 SS metal impeller; shaft shall be 1045 steel with 316 SS sleeve. 7.2 *All Duron Metal.* All pump materials shall be Duron metal. Shaft shall be 1045 steel, with 316 SS sleeve. 316 SS metal impeller optional. 7.3 *All AISI 316 Stainless Steel.* All pump materials shall be AISI 316 stainless steel. Shaft should be 1045 steel, with 316 SS sleeve. 7.4 *All #20 Stainless Steel.* All pump materials shall be #20 SS stainless steel. Shaft shall be 316 SS, with #20 SS sleeve. 7.5 *All CD-4MCu.* All pump materials shall be CD-4MCu. Shaft shall be 316 SS, with #20 SS sleeve.
8.0	Miscellaneous: 8.1 *Nameplates.* All nameplates and other data plates shall be stainless steel, suitably secured to the pump. 8.2 *Hardware.* All machine bolts, stud nuts, and cap screws shall be of the hex-head type. 8.3 *Rotation.* Pump shall have clockwise rotation viewed from its driven end. 8.4 *Parts Numbering.* Parts shall be completely identified with a numerical system (no alphabetical letters) to facilitate parts inventory control and stocking. Each part shall be properly identified by a separate number, and those parts that are identical shall have the same number to effect minimum spare parts inventory.

*Omit if not applicable.

TABLE 10-12 Flow Variation of Reciprocating Pumps

Number of cylinders	Single- or double-acting	Flow variation per stroke from mean, percent
Single	Single	+220 to −100
Single	Double	+60 to −100
Duplex	Single	+24.1 to −100
Duplex	Double	+6.1 to −21.5
Triplex	Single and double	+1.8 to −16.9
Quintuplex	Single	+1.8 to −5.2

being pumped. This is an important asset for equipment required to handle hazardous or toxic liquids.

Low-capacity diaphragm pumps are designed for metering service and employ a plunger working in oil to actuate a metallic or plastic diaphragm. Built for pressures in excess of 6.895 MPa (1000 lbf/in^2) with flow rates up to about 1.135 m^3/h (5 gal/min) per cylinder, such pumps possess all the characteristics of plunger-type metering pumps with the added advantage that the pumping head can be mounted in a remote (even a submerged) location entirely separate from the drive.

Figure 10-23 shows a high-capacity 22.7 m^3/h (100 gal/min) pump with actuation provided by a mechanical linkage.

Rotary Pumps In rotary pumps the liquid is displaced by rotation of one or more members within a stationary housing. Because internal clearances, although minute, are a necessity in all but a few special types, capacity decreases somewhat with increasing pump differential pressure. Therefore, these pumps are not truly positive-displacement pumps. However, for many other reasons they are considered as such.

The selection of materials of construction for rotary pumps is critical. The materials must be corrosion-resistant, compatible when one part is running against another, and capable of some abrasion resistance.

Gear Pumps When two or more impellers are used in a rotary-pump casing, the impellers will take the form of toothed-gear wheels as in Fig. 10-24, of helical gears, or of lobed cams. In each case, these impellers rotate with extremely small clearance between them and between the surfaces of the impellers and the casing. In Fig. 10-24, the two toothed impellers rotate as indicated by the arrows; the suction connection is at the bottom. The pumped liquid flows into the spaces between the impeller teeth as these cavities pass the suction opening. The liquid is then carried around the casing to the discharge opening, where it is forced out of the impeller teeth mesh. The arrows indicate this flow of liquid.

Rotary pumps are available in two general classes: interior-bearing and exterior-bearing. The *interior-bearing type* is used for handling liquids of a lubricating nature, and the *exterior-bearing type* is used with nonlubricating liquids. The interior-bearing pump is lubricated by the liquid being pumped, and the exterior-bearing type is oil-lubricated.

The use of spur gears in gear pumps will produce in the discharge pulsations having a frequency equivalent to the number of teeth on both gears multiplied by the speed of rotation. The amplitude of these disturbances is a function of tooth design. The pulsations can be reduced markedly by the use of rotors with helical teeth. This in turn introduces end thrust, which can be eliminated by the use of double-helical or herringbone teeth.

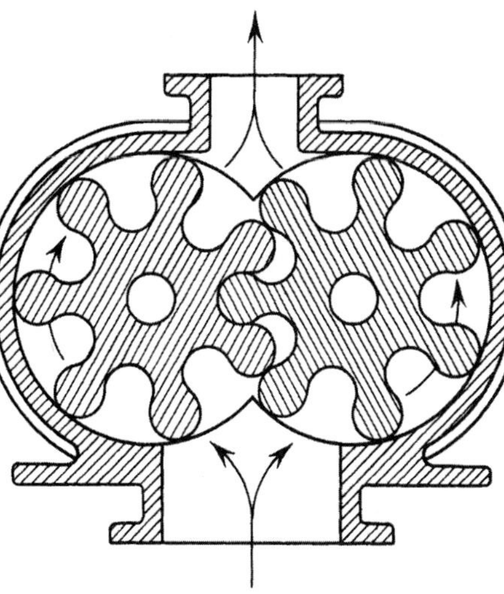

FIG. 10-24 Positive-displacement gear-type rotary pump.

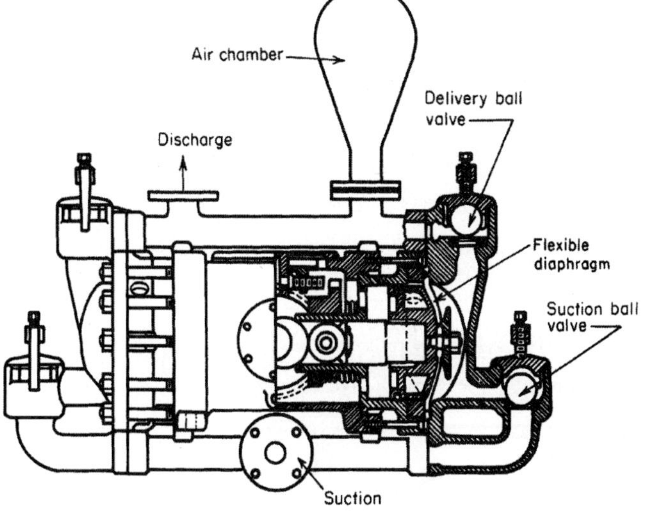

FIG. 10-23 Mechanically actuated diaphragm pump.

Screw Pumps A modification of the helical gear pump is the screw pump. A screw pump delivers and increases the pressure of slightly lubricating liquids. Both gear and screw pumps are positive-displacement pumps. Figure 10-25 illustrates a two-rotor version in which the liquid is fed to either the center or the ends, depending on the direction of rotation, and progresses axially in the cavities formed by the meshing threads or teeth. In three-rotor versions, the center rotor is the driving member while the other two are driven. Figure 10-26 shows still another arrangement, in which a metal rotor of unique design rotates without clearance in an elastomeric stationary sleeve.

Screw pumps, because of multiple dams that reduce slip, are well adapted for producing higher pressure rises, for example, 6.895 MPa (1000 lbf/in^2), especially when handling viscous liquids such as heavy oils. The all-metal pumps are generally subject to the same limitations on handling abrasive solids as conventional gear pumps. In addition, the wide bearing spans usually demand that the liquid have considerable lubricity to prevent metal-to-metal contact.

Among the liquids handled by rotary pumps are mineral oils, vegetable oils, animal oils, greases, glucose, viscose, molasses, paints, varnish, shellac, lacquers, alcohols, catsup, brine, mayonnaise, sizing, soap, tanning liquors, vinegar, and ink. Some screw-type units are specially designed for the gentle handling of large solids suspended in the liquid.

CENTRIFUGAL PUMPS

The centrifugal pump is the type most widely used in the chemical industry for transferring liquids of all types—raw materials, materials in manufacture, and finished products—as well as for general services of water supply, boiler feed, condenser circulation, condensate return, etc. These pumps are available through a vast range of sizes, in capacities from 0.5 m^3/h to 2×10^4 m^3/h (2 gal/min to 10^5 gal/min), and for discharge heads (pressures) from a few meters to approximately 48 MPa (7000 lbf/in^2).

The primary advantages of a centrifugal pump are simplicity, low first cost, uniform (nonpulsating) flow, small floor space, low maintenance expense, quiet operation, and adaptability for use with a motor or a turbine drive.

A centrifugal pump, in its simplest form, consists of an impeller rotating within a casing. The *impeller* consists of a number of blades, either open or shrouded, mounted on a shaft that projects outside the casing. Its axis of rotation may be either horizontal or vertical, to suit the work to be done. *Closed-type,* or *shrouded,* impellers are generally the most efficient. *Open* or *semi-open* impellers are used for viscous liquids or for liquids containing solid materials and on many small pumps for general service. Impellers may be of the *single-suction* or *double-suction* type—single if the liquid enters from one side, double if it enters from both sides.

Casings There are three general types of casings, but each consists of a chamber in which the impeller rotates, provided with inlet and exit for the liquid being pumped. The simplest form is the *circular casing,* consisting of an annular chamber around the impeller; no attempt is made to overcome the losses that will arise from eddies and shock when the liquid leaving the impeller at relatively high velocities enters this chamber. Such casings are seldom used.

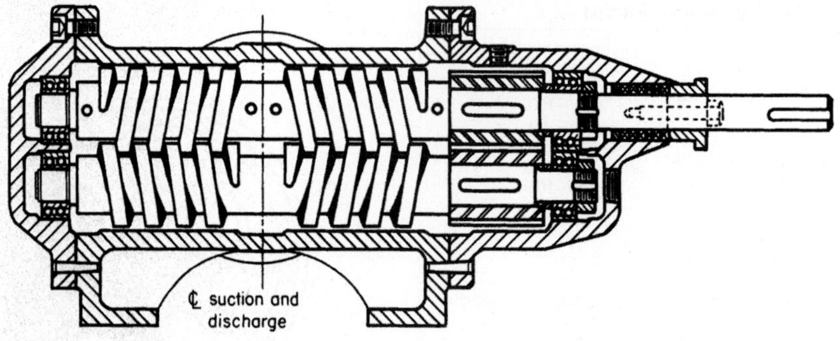

FIG. 10-25 Two-rotor screw pump. (*Courtesy of Warren Quimby Pump Co.*)

Volute casings take the form of a spiral increasing uniformly in cross-sectional area as the outlet is approached. The volute efficiently converts the velocity energy imparted to the liquid by the impeller into pressure energy.

A third type of casing is used in *diffuser-type* or turbine pumps. In this type, *guide vanes* or *diffusers* are interposed between the impeller discharge and the casing chamber. Losses are kept to a minimum in a well-designed pump of this type, and improved efficiency is obtained over a wider range of capacities. This construction is often used in multistage high-head pumps.

Action of a Centrifugal Pump Briefly, the action of a centrifugal pump may be shown by Fig. 10-27. Power from an outside source is applied to shaft *A*, rotating the impeller *B* within the stationary casing *C*. The blades of the impeller in revolving produce a reduction in pressure at the entrance or eye of the impeller. This causes liquid to flow into the impeller from the suction pipe *D*. This liquid is forced outward along the blades at increasing tangential velocity. The velocity head it has acquired when it leaves the blade tips is changed to pressure head as the liquid passes into the volute chamber and then out the discharge *E*.

Centrifugal Pump Characteristics Figure 10-28 shows a typical characteristic curve of a centrifugal pump. It is important to note that at any fixed speed the pump will operate along this curve and at no other points. For instance, on the curve shown, at 45.5 m^3/h (200 gal/min) the pump will generate 26.5-m (87-ft) head. If the head is increased to 30.48 m (100 ft), then 27.25 m^3/h (120 gal/min) will be delivered. It is not possible to reduce the capacity to 27.25 m^3/h (120 gal/min) at 26.5-m (87-ft) head unless the discharge is throttled so that 30.48 m (100 ft) is actually generated within the pump. On pumps with variable-speed drivers such as steam turbines, it is possible to change the characteristic curve, as shown by Fig. 10-29.

As shown in Eq. (10-44), the head depends on the velocity of the fluid, which in turn depends on the capability of the impeller to transfer energy to the fluid. This is a function of the fluid viscosity and the impeller design. It is important to remember that the head produced will be the same for any liquid of the same viscosity. The pressure rise, however, will vary in proportion to the specific gravity.

For quick pump selection, manufacturers often give the most essential performance details for a whole range of pump sizes. Figure 10-30 shows typical performance data for a range of process pumps based on suction and discharge pipes and impeller diameters. The performance data consist of the pump flow rate and the head. Once a pump meets a required specification, then more-detailed performance data for the particular pump can be easily found based on the curve reference number. Figure 10-31 shows a more detailed pump performance curve that includes, in addition to pump head and flow, the brake horsepower required, NPSH required, number of vanes, and pump efficiency for a range of impeller diameters.

If detailed manufacturer-specified performance curves are not available for a different size of the pump or operating condition, then a best estimate of the off-design performance of pumps can be obtained through similarity relationship or the affinity laws:

1. Capacity Q is proportional to impeller rotational speed N.
2. Head h varies as square of the impeller rotational speed.
3. Brake horsepower (BHP) varies as the cube of the impeller rotational speed.

These equations can be expressed mathematically and appear in Table 10-13.

System Curves In addition to the pump design, the operational performance of a pump depends on factors such as the downstream load characteristics, pipe friction, and valve performance. Typically, head and flow follow the following relationship:

$$\frac{(Q_2)^2}{(Q_1)^2} = \frac{h_2}{h_1} \tag{10-54}$$

where subscript 1 refers to the design condition and subscript 2 to the actual conditions. The above equation indicates that head will change as the square of the water flow rate.

Figure 10-32 shows the schematic of a pump, moving a fluid from tank A to tank B, both of which are at the same level. The only force that the pump has to overcome in this case is the pipe friction, variation of which with fluid flow rate is also shown in the figure. On the other hand, for the use shown in Fig. 10-33, the pump in addition to pipe friction should overcome head due to the difference in elevation between tanks A and B. In this case, elevation head is constant, whereas the head required to overcome friction depends on the flow rate. Figure 10-34 shows the pump performance requirement of a valve opening and closing.

Pump Selection One of the parameters that is extremely useful in selecting a pump for a particular application is specific speed N_s. Specific speed of a pump can be evaluated based on its design speed, flow, and head:

$$N_S = \frac{NQ^{1/2}}{H^{3/4}} \tag{10-55}$$

where N = rpm, Q is flow rate in gpm, and H is head in ft · lbf/lbm.

Specific speed is a parameter that defines the speed at which impellers of geometrically similar design have to be run to discharge 1 gal/min against a 1-ft head. In general, pumps with a low specific speed have a low capacity; and high specific speed, high capacity. Specific speeds of different types of pumps are shown in Table 10-14 for comparison.

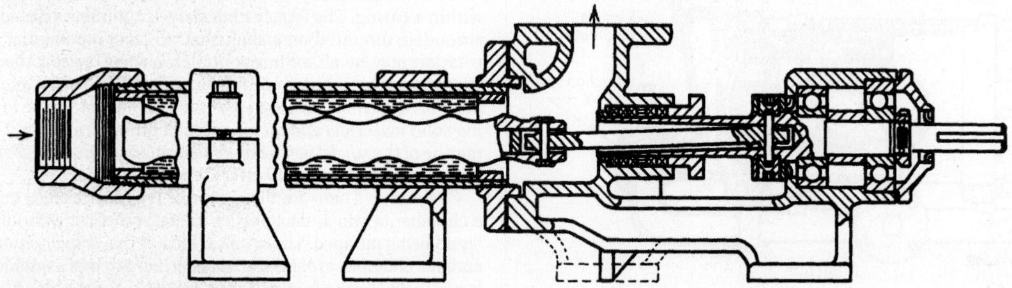

FIG. 10-26 Single-rotor screw pump with an elastomeric lining. (*Courtesy of Moyno Pump Division, Robbins & Myers, Inc.*)

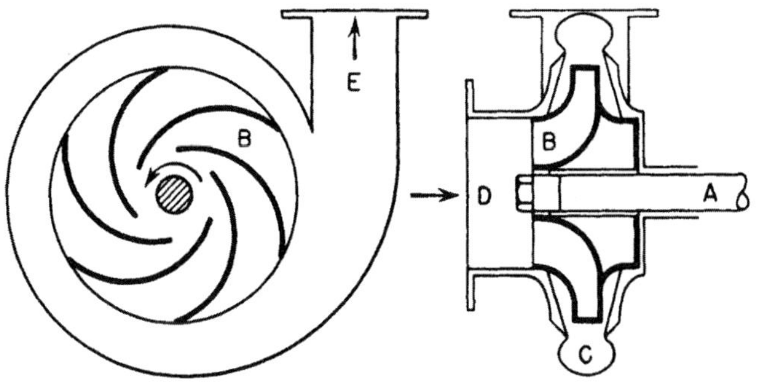

FIG. 10-27 A simple centrifugal pump.

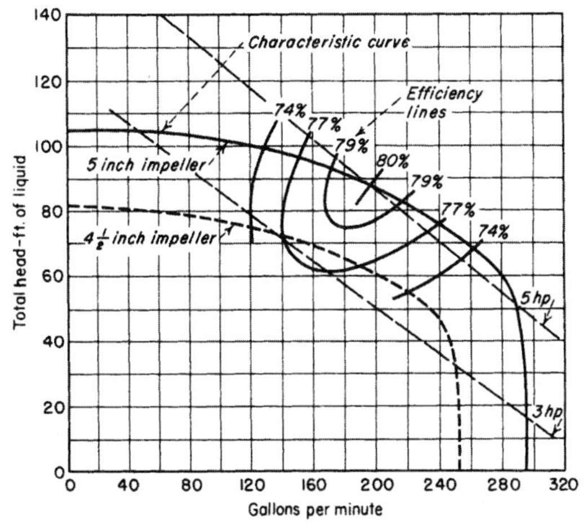

FIG. 10-28 Characteristic curve of a centrifugal pump operating at a constant speed of 3450 r/min. To convert gallons per minute to cubic meters per hour, multiply by 0.2271; to convert feet to meters, multiply by 0.3048; to convert horsepower to kilowatts, multiply by 0.746; and to convert inches to centimeters, multiply by 2.54.

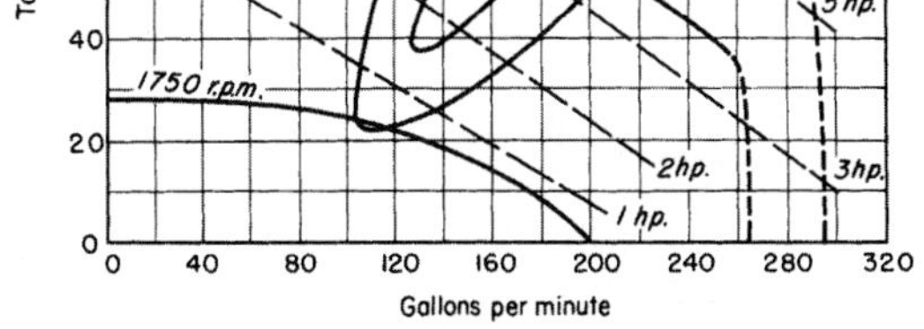

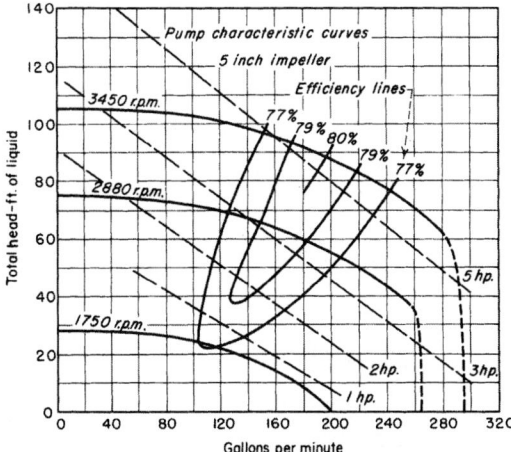

FIG. 10-29 Characteristic curve of a centrifugal pump at various speeds. To convert gallons per minute to cubic meters per hour, multiply by 0.2271; to convert feet to meters, multiply by 0.3048; to convert horsepower to kilowatts, multiply by 0.746; and to convert inches to centimeters, multiply by 2.54.

Another parameter that helps in evaluating the pump suction limitations, such as cavitation, is the suction-specific speed, S:

$$S = \frac{NQ^{1/2}}{(\text{NPSH})^{3/4}} \tag{10-56}$$

Typically, for single-suction pumps, suction-specific speed above 11,000 is considered excellent. Below 7000 is poor and 7000 to 9000 is of an average design. Similarly, for double-suction pumps, suction-specific speed above 14,000 is considered excellent, below 7000 is poor, and 9000 to 11,000 is average.

Figure 10-35 shows the schematic of specific-speed variation for different types of pumps. The figure clearly indicates that as the specific speed increases, the ratio of the impeller outer diameter D_1 to inlet or eye diameter D_2 decreases, tending to become unity for pumps of axial-flow type.

Typically, axial-flow pumps are of high-flow and low-head type and have a high specific speed. On the other hand, purely radial pumps are of high head and low flow rate capability and have a low specific speed. Obviously, a pump with a moderate flow and head has an average specific speed.

A typical pump selection chart such as shown in Fig. 10-36 calculates the specific speed for given flow, head, and speed requirements. Based on the calculated specific speed, the optimal pump design is indicated.

Process Pumps This term is usually applied to single-stage pedestal-mounted units with single-suction overhung impellers and with a single packing box. These pumps are ruggedly designed for ease in dismantling and accessibility, with mechanical seals or packing arrangements, and are built specially to handle corrosive or otherwise difficult-to-handle liquids.

Specifically, but not exclusively for the chemical industry, most pump manufacturers now build to national standards *horizontal and vertical process pumps*. ASME Standards B73.1-2001 and B73.2-2003 apply to the horizontal (Fig. 10-37) and vertical in-line (Fig. 10-38) pumps, respectively.

The horizontal pumps are available for capacities up to 900 m^3/h (4000 gal/min); the vertical in-line pumps, for capacities up to 320 m^3/h (1400 gal/min). Both horizontal and vertical in-line pumps are available for heads up to 120 m (400 ft). The intent of each ANSI specification is that pumps from all vendors for a given nominal capacity and total dynamic head at a given rotative speed shall be dimensionally interchangeable with respect to mounting, size, and location of suction and discharge nozzles, input shaft, baseplate, and foundation bolts.

The vertical in-line pumps, although relatively new additions, are finding considerable use in chemical and petrochemical plants in the United States. An inspection of the two designs will make clear the relative advantages and disadvantages of each.

Chemical pumps are available in a variety of materials. Metal pumps are the most widely used. Although they may be obtained in iron, bronze, and iron with bronze fittings, an increasing number of pumps of ductile-iron, steel, and nickel alloys are being used. Pumps are also available in glass, glass-lined iron, carbon, rubber, rubber-lined metal, ceramics, and a variety of plastics, such units usually being employed for special purposes.

Sealing the Centrifugal Chemical Pump Engineers who specify an appropriate mechanical sealing system on their pumps can significantly improve the energy efficiency of a manufacturing plant as it is estimated that around 10 percent of electric power is used for pumping equipment. Regulatory bodies and engineers are focused on improving the energy efficiency of pumps and pumping systems. Choosing the right mechanical seal is one of the most effective ways of doing this. The purpose of a mechanical seal is to seal the process fluid—whether it is toxic or expensive, the objective is to keep it within the system and pipework to avoid its seeping out and resulting in a cost for lost process fluid and cleanup. Seals not only prevent process fluid contamination and leakage to the external atmosphere but are an important part of conserving energy within the system.

Mechanical seals on pumps are probably the most delicate components, and using seal flush plans to change the environment that the seals operate in and flourish enables them to provide reliable operation. Flush plans are formalized by the American Petroleum Institute in its Standard API-682, where they are detailed in standardized formats.

Current practice demands that packing boxes be designed to accommodate both packing and mechanical seals. With either type of seal, one consideration is of paramount importance in chemical service: the liquid present at the sealing surfaces must be free of solids. Consequently, it is necessary to provide a secondary compatible liquid to flush the seal or packing whenever the process liquid is not absolutely clean.

The use of *packing seals* requires the continuous escape of liquid past the seal to minimize and to carry away the frictional heat developed. If the effluent is toxic or corrosive, quench glands or catch pans are usually employed. Although packing can be adjusted with the pump operating, leaking mechanical seals require shutting down the pump to correct the leak. Properly applied and maintained *mechanical seals* usually show no visible

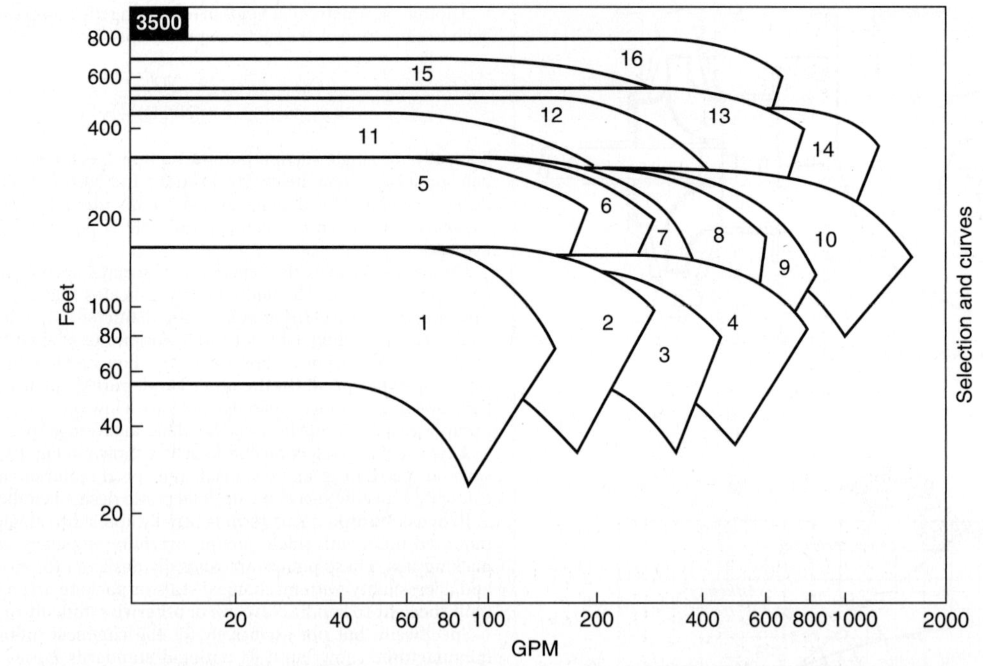

Range No.	Pump		Curve	Range No.	Pump		Curve
1	1.5 × 6 E	731 Plus	A-8475	9	4 × 3 × 8.5	731 Plus	A-8969
2	3 × 1.5 × 6	731 Plus	A-6982	10	6 × 4 × 8.5	731 Plus	A-8547
3	3 × 2 × 6	731 Plus	A-8159	11	2 × 1 × 10 E	731 Plus	A-8496
4	4 × 3 × 6	731 Plus	A-8551	12	3 × 1.5 × 11 E	731 Plus	A-8543
5	1.5 × 1 × 8	731 Plus	A-8153	13	3 × 2 × 11	731 Plus	A-8456
6	3 × 1.5 × 8	731 Plus	A-8155	14	4 × 3 × 11	731 Plus	A-7342
7	3 × 1.5 × 8.5 E	731 Plus	A-8529	15	3 × 1.5 × 13 E	731 Plus	A-8492
8	3 × 2 × 8.5 E	731 Plus	A-8506	16	3 × 2 × 13	731 Plus	A-7338

FIG. 10-30 Performance curves for a range of open impeller pumps.

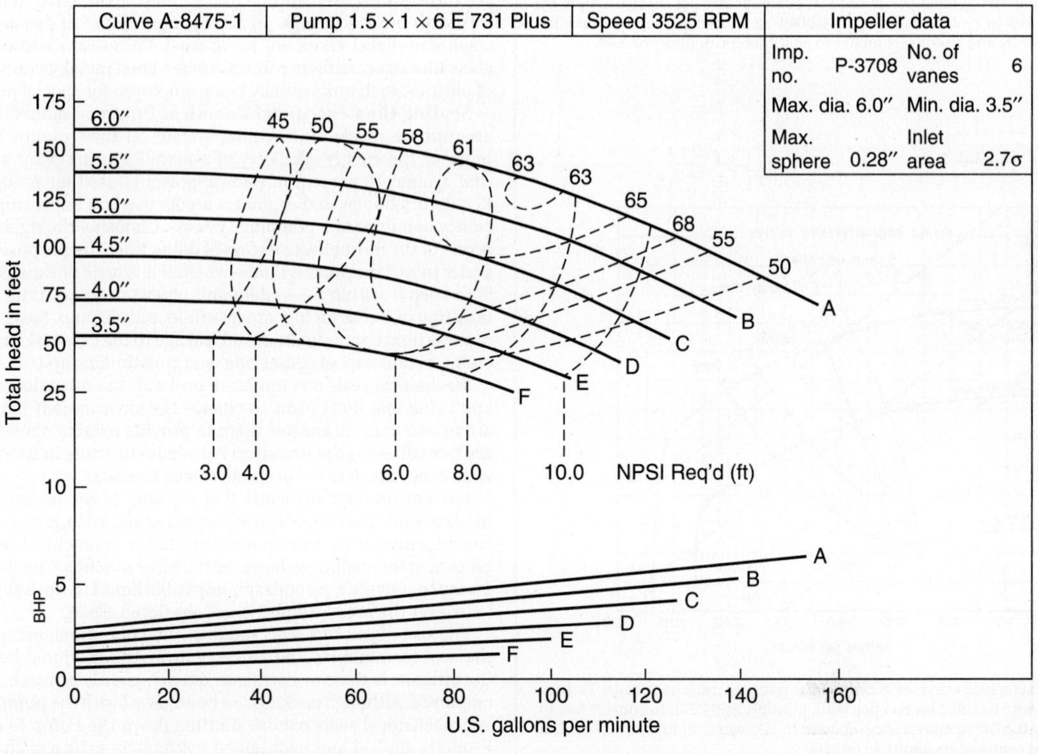

FIG. 10-31 Typical pump performance curve. The curve is shown for water at 85°F. If the specific gravity of the fluid is other than unity, BHP must be corrected.

TABLE 10-13 The Affinity Laws

	Constant impeller diameter	Constant impeller speed
Capacity	$\frac{Q_1}{Q_2}=\frac{N_1}{N_2}$	$\frac{Q_1}{Q_2}=\frac{D_1}{D_2}$
Head	$\frac{H_1}{H_2}=\frac{(N_1)^2}{(N_2)^2}$	$\frac{h_1}{h_2}=\frac{(D_1)^2}{(D_2)^2}$
Brake horsepower	$\frac{\text{BHP}_1}{\text{BHP}_2}=\frac{(N_1)^3}{(N_2)^3}$	$\frac{\text{BHP}_1}{\text{BHP}_2}=\frac{(P_1)^3}{(P_2)^3}$

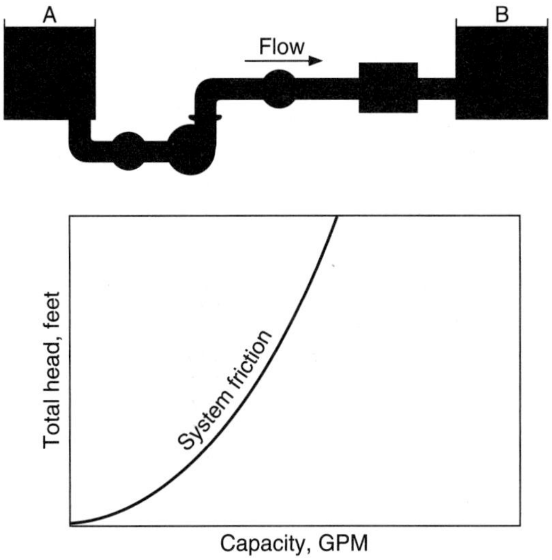

FIG. 10-32 Variation of total head versus flow rate to overcome friction.

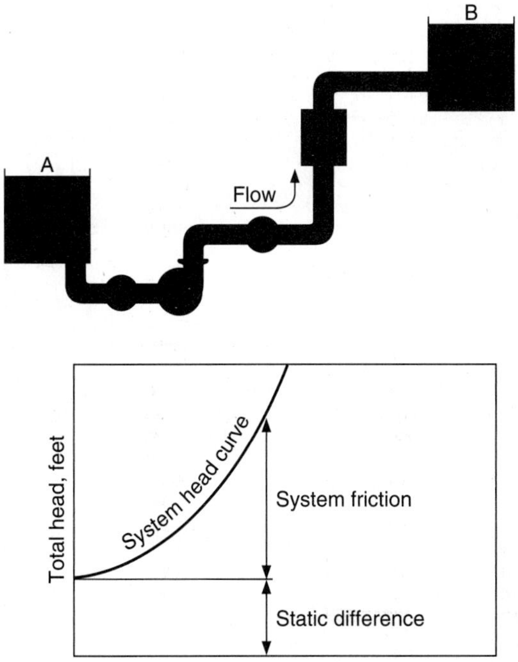

FIG. 10-33 Variation of total head as a function of flow rate to overcome both friction and static head.

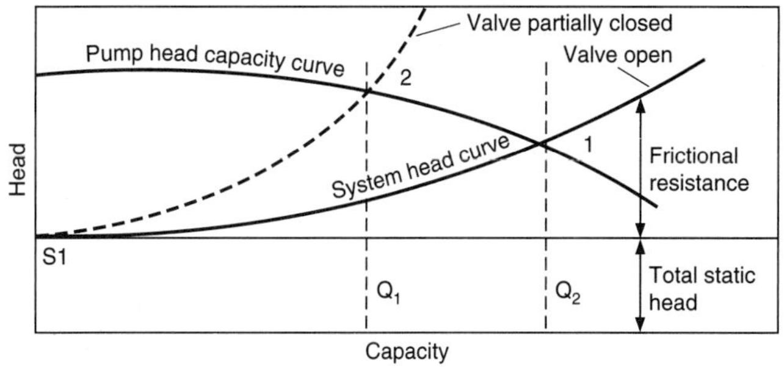

FIG. 10-34 Typical steady-state response of a pump system with a valve fully and partially open.

leakage. In general, owing to the more effective performance of mechanical seals, they have gained almost universal acceptance.

Double-Suction, Single-Stage Pumps These pumps are used for general water supply and circulating service and for chemical service when liquids that are noncorrosive to iron or bronze are being handled. They are available for capacities from about 5.7 m^3/h (25 gal/min) up to as high as 1.136×10^4 m^3/h (50,000 gal/min) and heads up to 304 m (1000 ft). Such units are available in iron, bronze, and iron with bronze fittings. Other materials increase the cost; when they are required, a standard chemical pump is usually more economical.

Close-Coupled Pumps Pumps equipped with a built-in electric motor or sometimes steam-turbine-driven (i.e., with pump impeller and driver on the same shaft) are known as *close-coupled pumps* (Fig. 10-39). Such units are extremely compact and are suitable for a variety of services for which standard iron and bronze materials are satisfactory. They are available in capacities up to about 450 m^3/h (2000 gal/min) for heads up to about 73 m (240 ft). Two-stage units in the smaller sizes are available for heads to around 150 m (500 ft).

Canned-Motor Pumps These pumps (Fig. 10-40) command considerable attention in the chemical industry. They are close-coupled units in which the cavity housing the motor rotor and the pump casing are interconnected. As a result, the motor bearings run in the process liquid, and all seals are eliminated. Because the process liquid is the bearing lubricant, abrasive solids cannot be tolerated. Standard single-stage canned-motor pumps are available for flows up to 160 m^3/h (700 gal/min) and heads up to 76 m (250 ft). Two-stage units are available for heads up to 183 m (600 ft). Canned-motor pumps are being widely used for handling organic solvents, organic heat-transfer liquids, and light oils as well as many clean toxic or hazardous liquids or for installations in which leakage is an economic problem.

Vertical Pumps In the chemical industry, the term *vertical process pump* (Fig. 10-41) generally applies to a pump with a vertical shaft having a length from drive end to impeller of approximately 1 m (3.1 ft) minimum to 20 m (66 ft) or more. Vertical pumps are used as either *wet-pit pumps* (immersed) or *dry-pit pumps* (externally mounted) in conjunction with stationary or mobile tanks containing difficult-to-handle liquids. They have the following advantages: the liquid level is above the impeller, and the pump is thus self-priming; and the shaft seal is above the liquid level and is not wetted by the pumped liquid, which simplifies the sealing task. When no bottom connections are permitted on the tank (a safety consideration for highly corrosive or toxic liquid), the vertical wet-pit pump may be the only logical choice.

These pumps have the following disadvantages: intermediate or line bearings are generally required when the shaft length exceeds about 3 m (10 ft) in order to avoid shaft resonance problems; these bearings must be lubricated whenever the shaft is rotating. Since all wetted parts must be corrosion-resistant, low-cost materials may not be suitable for the shaft, column, etc. Maintenance is costlier since the pumps are larger and more difficult to handle.

TABLE 10-14 Specific Speeds of Different Types of Pumps

Pump type	Specific speed range
Below 2000	Process pumps and feed pumps
2000–5000	Turbine pumps
4000–10,000	Mixed-flow pumps
9000–15,000	Axial-flow pumps

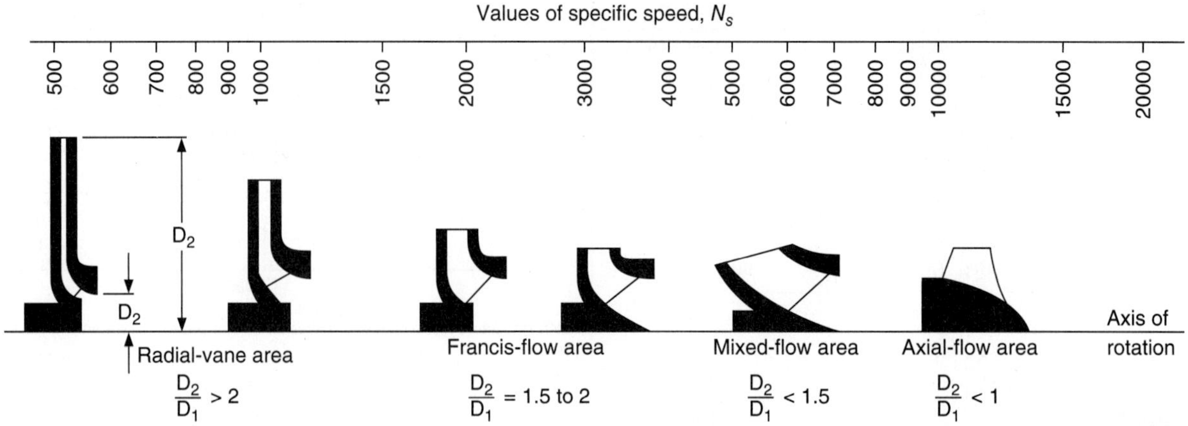

FIG. 10-35 Specific speed variations of different types of pump.

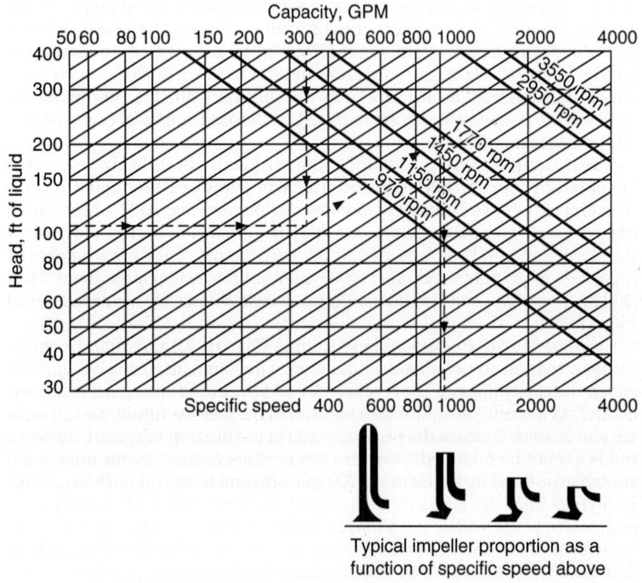

FIG. 10-36 Relationships between specific speed, rotative speed, and impeller proportions. (Worthington Pump Inc., *Pump World*, vol. 4, no. 2, 1978.)

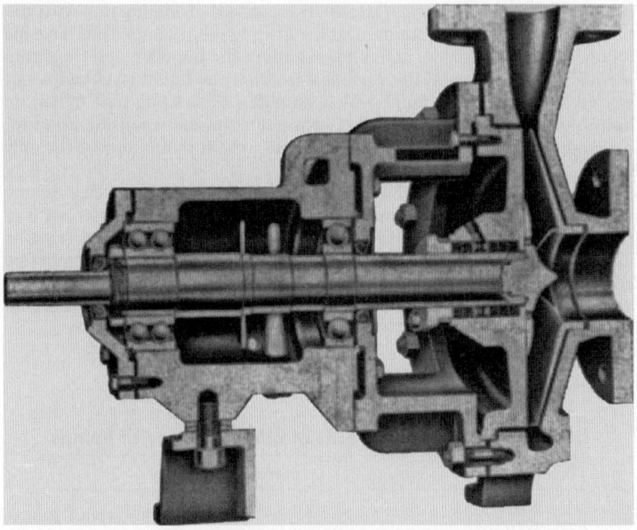

FIG. 10-37 Horizontal process pump conforming to American Society of Mechanical Engineers Standard B73.1-2001.

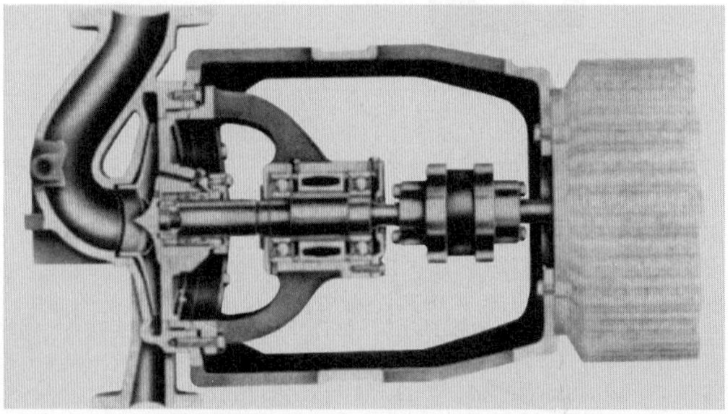

FIG. 10-38 Vertical in-line process pump conforming to ASME standard B73.2-2003. The pump shown is driven by a motor through flexible coupling. Not shown but also conforming to ASME Standard B73.2-2003 are vertical in-line pumps with rigid couplings and with no coupling (impeller-mounted on an extended motor shaft).

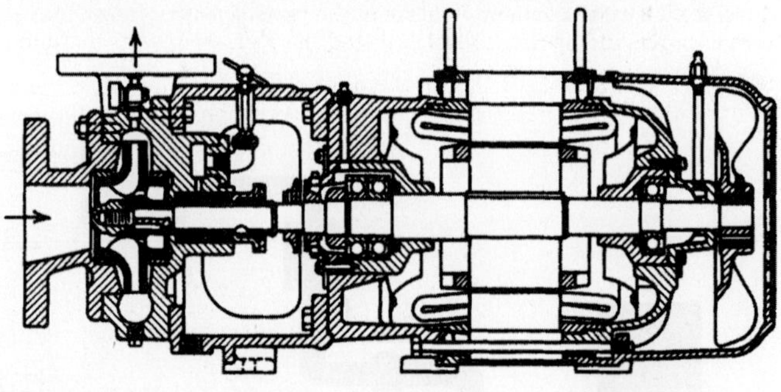

FIG. 10-39 Close-coupled pump.

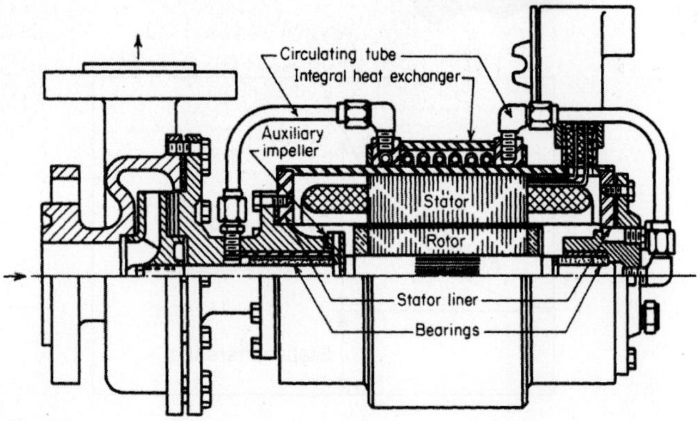

FIG. 10-40 Canned-motor pump. (*Courtesy of Chempump Division, Crane Co.*)

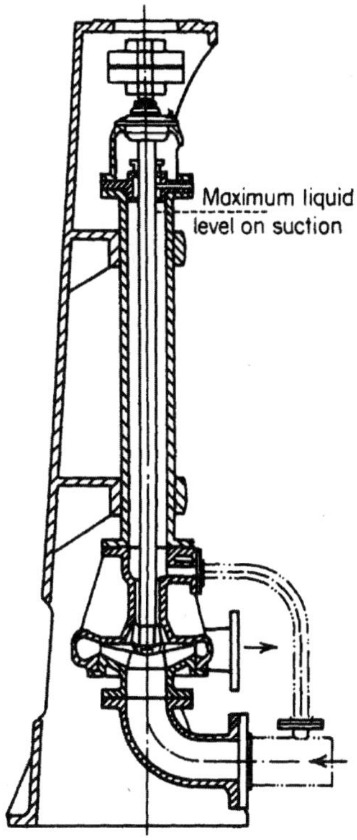

FIG. 10-41 Vertical process pump for dry-pit mounting. (*Courtesy of Lawrence Pumps, Inc.*)

For abrasive service, vertical cantilever designs requiring no line or foot bearings are available. Generally, these pumps are limited to about a 1-m (3.1-ft) maximum shaft length. Vertical pumps are also used to pump waters to reservoirs. One such application in the Los Angeles water basin has 14 four-stage pumps, each pump requiring 80,000 hp to drive it.

Sump Pumps These are small single-stage vertical pumps used to drain shallow pits or sumps. They are of the same general construction as vertical process pumps but are not designed for severe operating conditions.

Multistage Centrifugal Pumps These pumps are used for services requiring heads (pressures) higher than can be generated by a single impeller. All impellers are in series, the liquid passing from one impeller to the next and finally to the pump discharge. The total head then is the summation of the heads of the individual impellers. Deep-well pumps, high-pressure water supply pumps, boiler-feed pumps, fire pumps, and charge pumps for refinery processes are examples of multistage pumps required for various services.

Multistage pumps may be of the *volute type* (Fig. 10-42), with single- or double-suction impellers (Fig. 10-43), or of the *diffuser type* (Fig. 10-44). They may have horizontally split casings or, for extremely high pressures, 20 to 40 MPa (3000 to 6000 lbf/in^2), vertically split barrel-type exterior casings with inner casings containing diffusers, interstage passages, etc.

PROPELLER AND TURBINE PUMPS

Axial-Flow (Propeller) Pumps These pumps (Fig. 10-45) are essentially very high-capacity, low-head units. Normally they are designed for flows in excess of 450 m^3/h (2000 gal/min) against heads of 15 m (50 ft) or less. They are used to great advantage in closed-loop circulation systems in which the pump casing becomes merely an elbow in the line. A common installation is for calandria circulation. A characteristic curve of an axial-flow pump is given in Fig. 10-46.

Turbine Pumps The term *turbine pump* is applied to units with mixed-flow (part axial and part centrifugal) impellers. Such units are available in capacities from 20 m^3/h (100 gal/min) upward for heads up to about 30 m (100 ft) per stage. Turbine pumps are usually vertical.

A common form of turbine pump is the vertical pump, which has the pump element mounted at the bottom of a column that serves as the discharge pipe (see Fig. 10-47). Such units are immersed in the liquid to be pumped and are commonly used for wells, condenser circulating water, large-volume drainage, etc. Another form of the pump has a shell surrounding the pumping element which is connected to the intake pipe. In this form, the pump is used on condensate service in power plants and for process work in oil refineries.

Regenerative Pumps Also referred to as turbine pumps because of the shape of the impeller, regenerative pumps employ a combination of mechanical impulse and centrifugal force to produce heads of several hundred meters (feet) at low volumes, usually less than 20 m^3/h (100 gal/min). The impeller, which rotates at high speed with small clearances, has many short radial passages milled on each side at the periphery. Similar channels are milled in the mating surfaces of the casing. Upon entering, the liquid is directed into the impeller passages and proceeds in a spiral pattern around the periphery, passing alternately from the impeller to the casing and receiving successive impulses as it does so. Figure 10-48 illustrates a typical performance characteristic curve.

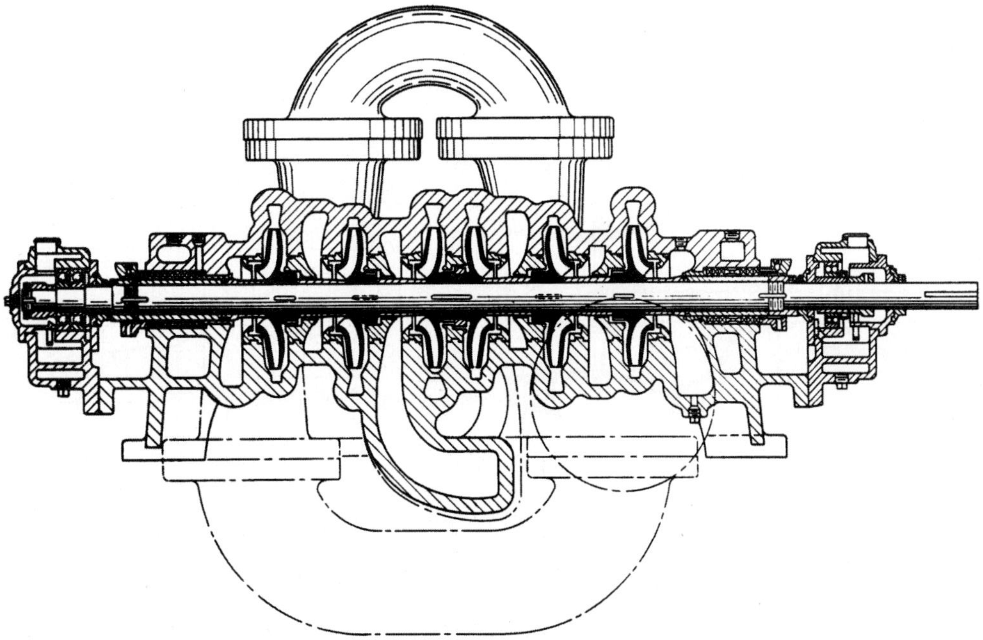

FIG. 10-42 Six-stage volute-type pump.

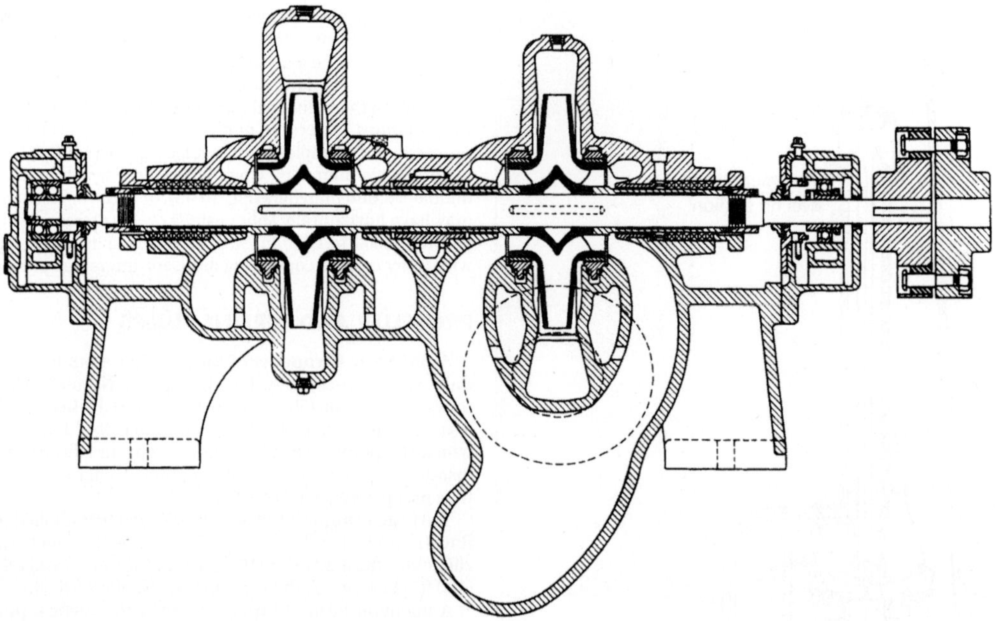

FIG. 10-43 Two-stage pump having double-suction impellers.

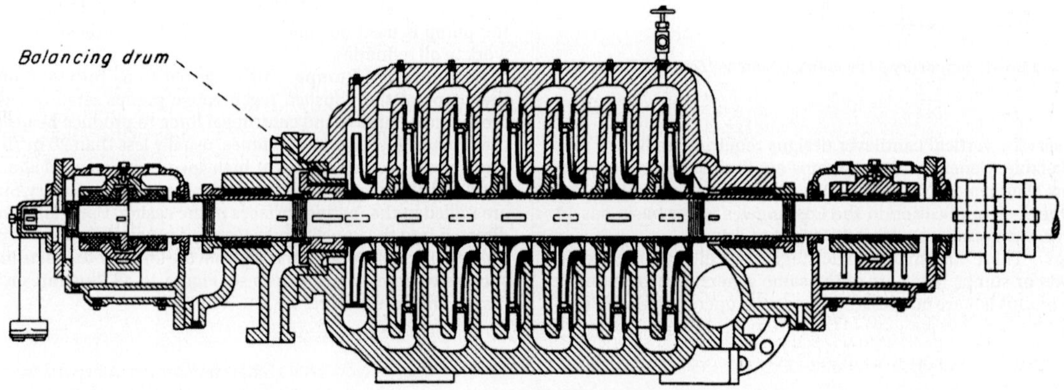

FIG. 10-44 Seven-stage diffuser-type pump.

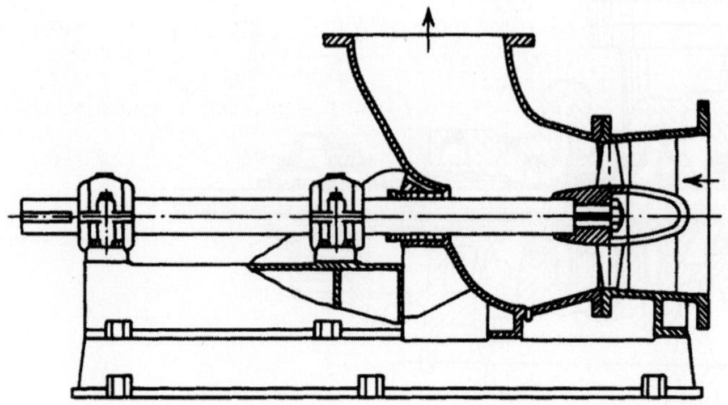

FIG. 10-45 Axial-flow elbow-type propeller pump. (*Courtesy of Lawrence Pumps, Inc.*)

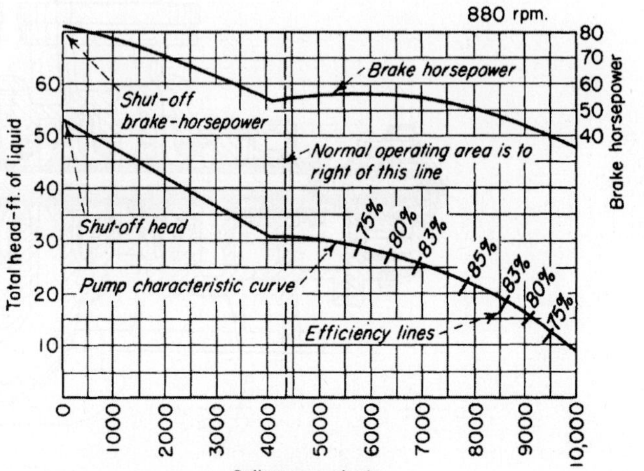

FIG. 10-46 Characteristic curve of an axial-flow pump. To convert gallons per minute to cubic meters per hour, multiply by 0.2271; to convert feet to meters, multiply by 0.3048; and to convert horsepower to kilowatts, multiply by 0.746.

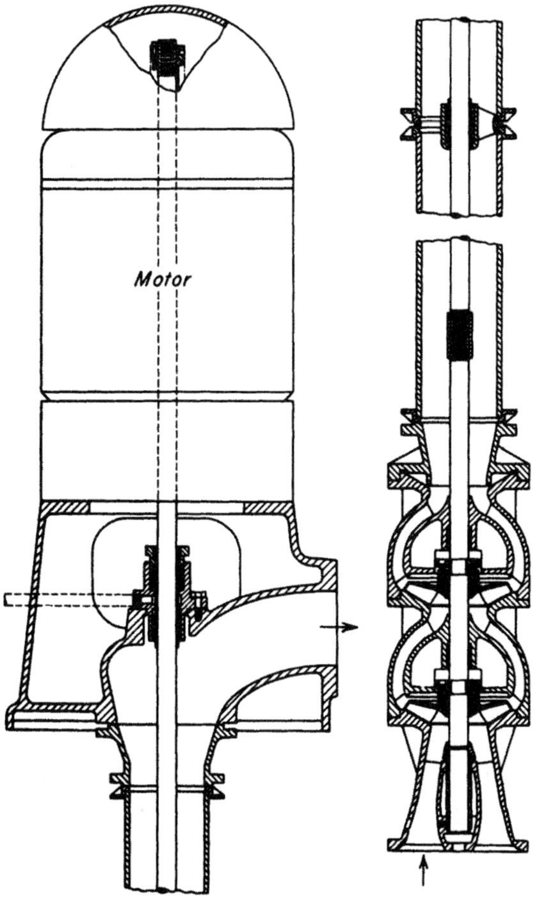

FIG. 10-47 Vertical multistage turbine, or mixed-flow, pump.

These pumps are particularly useful when low volumes of low-viscosity liquids must be handled at higher pressures than are normally available with centrifugal pumps. Close clearances limit their use to clean liquids. For very high heads, multistage units are available.

JET PUMPS

Jet pumps are a class of liquid-handling devices that makes use of the momentum of one fluid to move another.

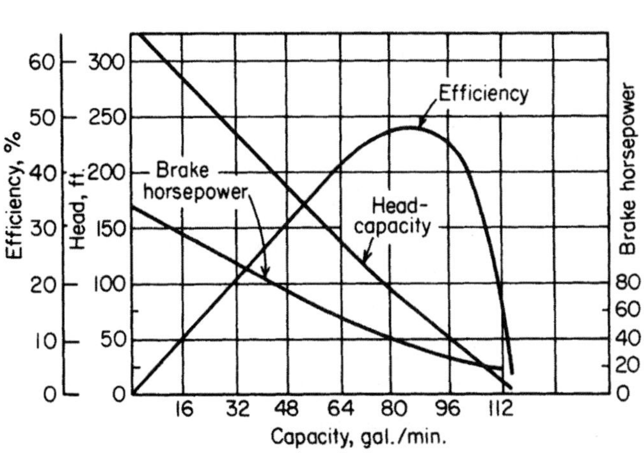

FIG. 10-48 Characteristic curves of a regenerative pump. To convert gallons per minute to cubic meters per hour, multiply by 0.2271; to convert feet to meters, multiply by 0.3048; and to convert horsepower to kilowatts, multiply by 0.746.

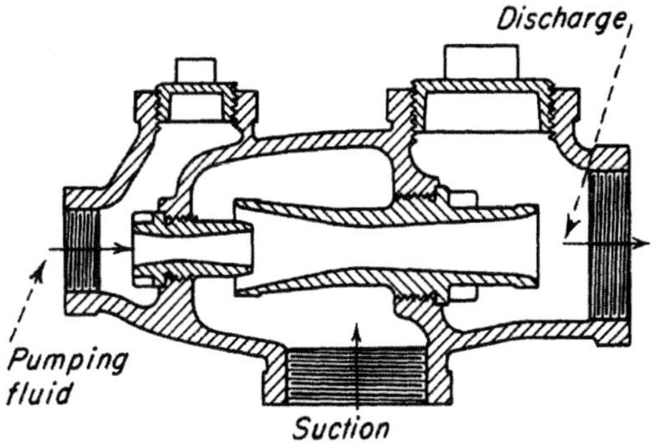

FIG. 10-49 Simple ejector using a liquid-motivating fluid.

Ejectors and *injectors* are the two types of jet pumps of interest to chemical engineers. The ejector, also called the *siphon, exhauster,* or *eductor,* is designed for use in operations in which the head pumped against is low and is less than the head of the fluid used for pumping. The injector is a special type of jet pump, operated by steam and used for boiler feed and similar services, in which the fluid being pumped is discharged into a space under the same pressure as that of the steam being used to operate the injector.

Figure 10-49 shows a simple design for a jet pump of the ejector type. The pumping fluid enters through the nozzle at the left and passes through the venturi nozzle at the center and out of the discharge opening at the right. As it passes into the venturi nozzle, it develops a suction that causes some of the fluid in the suction chamber to be entrained with the stream and delivered through this discharge.

The efficiency of an ejector or jet pump is low, being only a few percent. The head developed by the ejector is also low except in special types. The device has the disadvantage of diluting the fluid pumped by mixing it with the pumping fluid. In steam injectors for boiler feed and similar services in which the heat of the steam is recovered, efficiency is close to 100 percent.

The simple ejector or siphon is widely used, in spite of its low efficiency, for transferring liquids from one tank to another, for lifting acids, alkalies, or solid-containing liquids of an abrasive nature, and for emptying sumps.

PUMP DIAGNOSTICS

Pump problems vary over a large range depending on the type of pumps and the use of the pumps. They can be classified in the following manner by pump type and service:

1. *Positive-displacement pumps—reciprocating pump* problems can be classified into the following categories:
   a. Compressor valve problems: plate valves, feather valves, concentric disk valves, relief valves
   b. Piston and rod assembly: piston rings, cylinder chatter, cylinder cooling, piston-rod packing
   c. Lubrication system
2. *Positive-displacement pumps—gear-type and roots-type* problems can be classified into the following categories:
   a. Rotor dynamic problems: vibration problems, gear problems or roots rotor problems, bearing and seal problems
   b. Lubrication systems
3. *Continuous flow pumps such as centrifugal pump* problems can be classified into the following categories:
   a. Cavitation
   b. Capacity flow
   c. Motor overload
   d. Impeller
   e. Bearings and seals
   f. Lubrication systems

Table 10-15 classifies different types of centrifugal pump–related problems, their possible causes, and corrective actions that can be taken to solve some of the more common issues. These problems in the table are classified into three major categories: cavitation, flow capacity, and motor overload for these types of pumps. The use of vibration monitoring to diagnose pump and compressor problems is discussed at the end of the subsection on compressor problems.

TABLE 10-15 Pump Problems

Possible causes	Corrective action
Cavitating-Type Problems	
Plugged suction screen.	Check for indications of the presence of screen. Remove and clean screen.
Piping gaskets with undersized IDs installed, a very common problem in small pumps.	Install proper-sized gaskets.
Column tray parts or ceramic packing lodged in the impeller eye.	Remove suction piping and debris.
Deteriorated impeller eye due to corrosion.	Replace impeller and overhaul pump.
Flow rate is high enough above design that NPSH for flow rate has increased above NPSH.	Reduce flow rate to that of design.
Lined pipe collapsed at gasket area or ID due to buildup of corrosion products between liner and carbon-steel pipe.	Replace deteriorated piping.
Poor suction piping layout, too many elbows in too many planes, a tee branch almost directly feeding the suction of the other pump, or not enough straight run before the suction flange of the pump.	Redesign piping layout, using fewer elbows and laterals for tees, and have five or more straight pipe diameters before suction flange.
Vertical pumps experience a vortex formation due to loss of submergence required by the pump. Observe the suction surface while the pump is in operation, if possible.	Review causes of vortexing. Consider installation of a vortex breaker such as a bell mouth umbrella or changes to sump design.
Spare pump begins to cavitate when attempt is made to switch it with the running pump. The spare is "backed off" by the running pump because its shutoff head is less than the head produced by the running pump. This is a frequent problem when one pump is turbine-driven and one is motor-driven.	Throttle discharge of running pump until spare can get in system. Slow down running pump if it is a turbine or variable-speed motor.
Suction piping configuration causes adverse fluid rotation when approaching impeller.	Install sufficient straight run of suction piping, or install vanes in piping to break up pre-rotation.
Velocity of the liquid is too high as it approaches the impeller eye.	Install larger suction piping or reduce flow through pump.
Pump is operating at a low-flow-producing suction recirculation in the impeller eye. This results in a sound like cavitation.	Install bypass piping back to suction vessel to increase flow through pump. Remember bypass flow may have to be as high as 50 percent of design flow.
Capacity-Type Problems	
Check the discharge block valve opening first. It may be partially closed and thus the problem.	Open block valve completely.
Wear-ring clearances are excessive (closed impeller design).	Overhaul pump. Renew wear rings if clearance is about twice design value for energy and performance reasons.
Impeller-to-case or head clearances are excessive (open impeller design).	Reposition impeller to obtain correct clearance.
Air leaks into the system if the pump suction is below atmospheric pressure.	Take actions as needed to eliminate air leaks.
Increase in piping friction to the discharge vessel due to the following:	Take the following actions:
1. Gate has fallen off the discharge valve stem.	1. Repair or replace gate valve.
2. Spring is broken in the spring-type check valve.	2. Repair valve by replacing spring.
3. Check valve flapper pin is worn, and flapper will not swing open.	3. Overhaul check valve; restore proper clearance to pin and flapper bore.
4. Lined pipe collapsing.	4. Replace damaged pipe.
5. Control valve stroke improperly set, causing too much pressure drop.	5. Adjust control valve stroke as necessary.
Suction and/or discharge vessel levels are not correct, a problem mostly seen in lower-speed pumps.	Calibrate level controllers as necessary.
Motor running backward or impeller of double suction design is mounted backward. Discharge pressure developed in both cases is about one-half design value.	Check for proper rotation and mounting of impeller. Reverse motor leads if necessary.
Entrained gas from the process lowering NPSH available.	Reduce entrained gas in liquid by process changes as needed.
Polymer or scale buildup in discharge nozzle areas.	Shut down pump and remove scale or deposits.
Mechanical seal in suction system under vacuum is leaking air into system, causing pump curve to drop.	Change percentage balance of seal faces or increase spring tension.
The pump may have formed a vortex at high flow rates or low liquid level. Does the vessel have a vortex breaker? Does the incoming flow cause the surface to swirl or be agitated?	Reduce flow to design rates. Raise liquid level in suction vessel. Install vortex breaker in suction vessel.
Variable-speed motor running too slowly.	
Bypassing is occurring between volute channels in a double-volute pump casing due to a casting defect or extreme erosion.	Adjust motor speed as needed.
The positions of impellers are not centered with diffuser vanes. Several impellers will cause vibration and lower head output.	Overhaul pump; repair eroded area.
When the suction system is under vacuum, the spare pump has difficulty getting into system.	Overhaul pump; reposition individual impellers as needed. Reposition whole rotor by changing thrust collar locator spacer.
Certain pump designs use an internal bypass orifice port to alter head-flow curve. High liquid velocities often erode the orifice, causing the pump to go farther out on the pump curve. The system head curve increase corrects the flow back up the curve.	Install a positive-pressure steam (from running pump) to fill the suction line from the block valve through the check valve.
Replacement impeller is not correct casting pattern; therefore NPSH required is different.	Overhaul pump, restore orifice to correct size.
Volute and cutwater area of casing is severely eroded.	Overhaul pump; replace casing or repair by welding. Stress-relieve after welding as needed.
Overload Problems	
Polymer buildup between wear surfaces (rings or vanes).	Remove buildup to restore clearances.
Excessive wear ring (closed impeller) or cover-case clearance (open impeller).	Replace wear rings or adjust axial clearance of open impeller. In severe cases, cover or case must be replaced.
Pump circulating excessive liquid back to suction through a breakdown bushing or a diffuser gasket area.	Overhaul pump, replacing parts as needed.
Minimum-flow loop left open at normal rates, or bypass around control valve is open.	Close minimum-flow loop or control valve bypass valve.
Discharge piping leaking under liquid level in sump-type design.	Inspect piping for leakage. Replace as needed.
Electrical switch gear problems cause one phase to have low amperage.	Check out switch gear and repair as necessary.
Specific gravity is higher than design specification.	Change process to adjust specific gravity to design value, or throttle pump to reduce horsepower requirements. This will not correct problem with some vertical turbine pumps that have a flat horsepower-required curve.
Pump motor not sized for end of curve operation.	Replace motor with one of larger size, or reduce flow rate.
Open impeller has slight rub on casing. Most often occurs in operations from 250 to 400°F due to piping strain and differential growth in the pump.	Increase clearance of impeller to casing.
A replacement impeller was not trimmed to the correct diameter.	Remove impeller from pump and turn to correct diameter.

COMPRESSORS

A compressor is a device that pressurizes a working fluid. One of the basic purposes of using a compressor is to compress the fluid and to deliver it at a pressure higher than its original pressure. Compression is required for a variety of purposes, some of which are listed below:

1. To provide air for combustion
2. To transport process fluid through pipelines
3. To provide compressed air for driving pneumatic tools
4. To circulate process fluid within a process

Different types of compressors are shown in Fig. 10-50. Positive-displacement compressors are used for intermittent flow in which successive volumes of fluid are confined in a closed space to increase their pressures. Rotary compressors provide continuous flow. In rotary compressors, rapidly rotating parts (impellers) accelerate fluid to a high speed; this velocity is then converted to additional pressure by gradual deceleration in the diffuser or volute which surrounds the impeller. Positive-displacement compressors can be further classified as either reciprocating or rotary type, as shown in Fig. 10-50. The reciprocating compressor has a piston having a reciprocating motion within a cylinder. The rotary positive-displacement compressors have rotating elements whose positive action results in compression and displacement. The rotary positive-displacement compressors can be further subdivided into sliding vane, liquid piston, straight lobe, and helical lobe compressors. The continuous flow compressors (Fig. 10-50) can be classified as either dynamic compressors or ejectors. Ejectors entrain the in-flowing fluid by using a high-velocity gas or steam jet and then convert the velocity of the mixture to pressure in a diffuser. The dynamic compressors have rotating elements, which accelerate the in-flowing fluid and convert the velocity head to pressure head, partially in the rotating elements and partially in the stationary diffusers or blade. The dynamic compressors can be further subdivided into centrifugal, axial-flow, and mixed-flow compressors. The main flow of gas in the centrifugal compressor is radial. The flow of gas in an axial compressor is axial, and the mixed-flow compressor combines some characteristics of both centrifugal and axial compressors.

It is not always obvious what type of compressor is needed for an application. Of the many types of compressors used in the process industries, some of the more significant are the centrifugal, axial, rotary, and reciprocating compressors. They fall into three categories, as shown in Fig. 10-51.

For very high flows and low pressure ratios, an axial-flow compressor would be best. Axial-flow compressors usually have a higher efficiency, as seen in Fig. 10-52, but a smaller operating region than does a centrifugal machine. Centrifugal compressors operate most efficiently at medium flow rates and high pressure ratios. Rotary and reciprocating compressors (positive-displacement machines) are best used for low flow rates and high pressure ratios. The positive-displacement compressors and, in particular, reciprocating compressors were the most widely used in the process and pipeline industries up to and through the 1960s.

In turbomachinery the centrifugal flow and axial-flow compressors are the ones used for compressing gases. Positive-displacement compressors such as reciprocating, gear type, or lobe type are widely used in the industry for many other applications such as slurry pumping.

The performance characteristics of a single stage of the three main types of compressors are given in Table 10-16. The pressure ratios of the axial and centrifugal compressors have been classified into three groups: industrial, aerospace, and research.

The industrial pressure ratio is low because the operating range needs to be large. The *operating range* is defined as the range between the surge point and the choke point. The surge point is the point at which the flow is reversed in the compressor. The choke point is the point at which the flow has reached Mach = 1.0, the point where no more flow can get through the unit, a "stone wall." When surge occurs, the flow is reversed, and so are all the forces acting on the compressor, especially the thrust forces. Surge can lead to total destruction of the compressor. Thus surge is a region that must be avoided. Choke conditions cause a large drop in efficiency, but do not lead to destruction of the unit. Note that with the increase in pressure ratio and the number of stages, the operating range is narrowed in axial-flow and centrifugal compressors.

Compressor Selection To select the most satisfactory compression equipment, engineers must consider a wide variety of types, each of which offers peculiar advantages for particular applications. Among the major factors to be considered are the flow rate, head or pressure, temperature limitations, method of sealing, method of lubrication, power consumption, serviceability, and cost.

To be able to decide which compressor best fits the job, the engineer must analyze the flow characteristics of the units. The following dimensionless numbers describe the flow characteristics.

The *Reynolds number* is the ratio of the inertia forces to the viscous forces

$$N_{\text{Re}} = \frac{\rho VD}{\mu} \tag{10-57}$$

where ρ is the density of the gas, V is the velocity of the gas, D is the diameter of the impeller, and μ is the viscosity of the gas.

The *specific speed* compares the adiabatic head and flow rate in geometrically similar machines at various speeds.

$$N_s = \frac{N\sqrt{\dot{Q}}}{H_{\text{ad}}^{3/4}} \tag{10-58}$$

where N is the speed of rotation of the compressor, $\dot{Q}$ is the volume flow rate, and H is the adiabatic head.

The specific diameter compares head and flow rates in geometrically similar machines at various diameters

$$D_S = \frac{DH^{1/4}}{\sqrt{\dot{Q}}} \tag{10-59}$$

The flow coefficient is the capacity of the flow rate of the machine

$$\phi = \frac{\dot{Q}_1}{ND^3} \tag{10-60}$$

The pressure coefficient is the pressure or the pressure rise of the machine

$$\Psi = \frac{H}{N^2D^2} \tag{10-61}$$

In selecting the machines of choice, the use of specific speed and diameter best describes the flow. Figure 10-53 shows the characteristics of the three types of compressors. Other considerations in chemical plant service such as problems with gases which may be corrosive or have abrasive solids in suspension must be dealt with. Gases at elevated temperatures may create a potential explosion hazard, while air at the same temperatures may be handled quite normally; minute amounts of lubricating oil or water may contaminate the process gas and so may not be permissible, and for continuous-process use, a high degree of equipment reliability is required, since frequent shutdowns for inspection or maintenance cannot be tolerated.

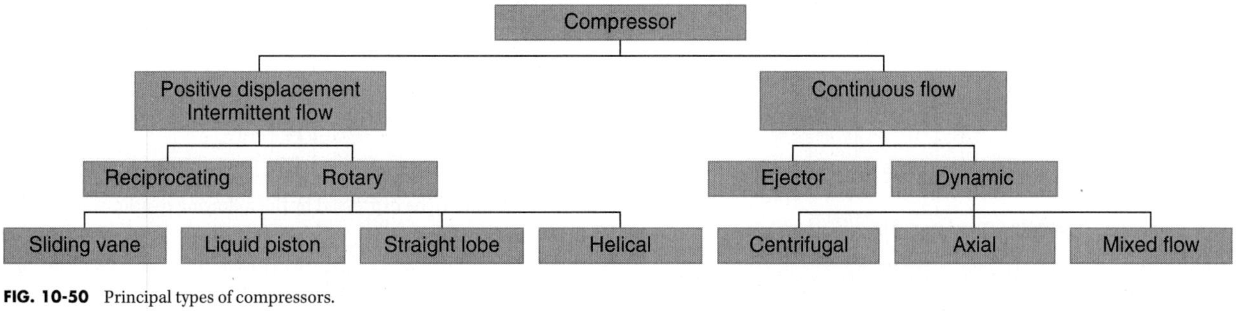

FIG. 10-50 Principal types of compressors.

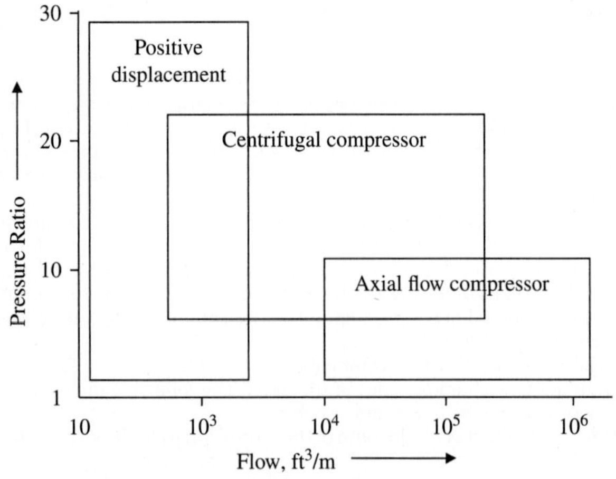

FIG. 10-51 Performance characteristics of different types of compressors.

COMPRESSION OF GASES

Theory of Compression In any continuous compression process, the relation of absolute pressure p to volume V is expressed by

$$pV^n = C = \text{constant} \tag{10-62}$$

The plot of pressure versus volume for each value of exponent n is known as the *polytropic* curve. Since the work W performed in proceeding from p_1 to p_2 along any polytropic curve (Fig. 10-54) is

$$W = \int_1^2 p\,dV \tag{10-63}$$

it follows that the amount of work required is dependent on the polytropic curve involved and increases with increasing values of n. The path requiring the least amount of input work is $n = 1$, which is equivalent to *isothermal* compression. For *adiabatic* compression (i.e., no heat is being added or taken away during the process), $n = k$ = ratio of specific heat at constant pressure to that at constant volume.

Since most compressors operate along a polytropic path approaching the adiabatic, compressor calculations are generally based on the adiabatic curve.

Some formulas based on the adiabatic equation and useful in compressor work are as follows:

Pressure, volume, and *temperature* relations for perfect gases:

$$P_2/P_1 = (V_1/V_2)^k \tag{10-64}$$

$$T_2/T_1 = (V_1/V_2)^{k-1} \tag{10-65}$$

$$P_2/P_1 = (T_2/T_1)^{k/(k-1)} \tag{10-66}$$

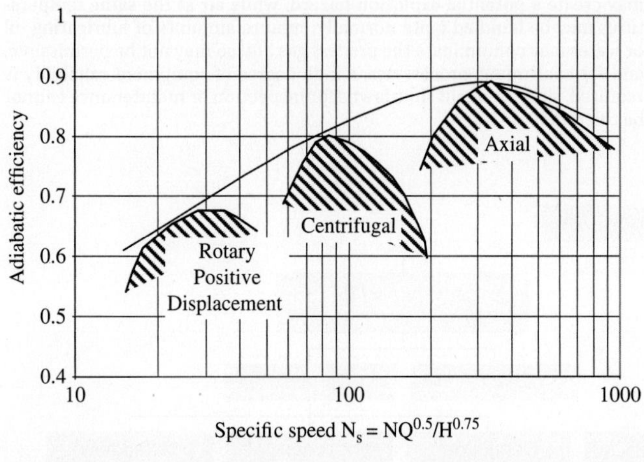

FIG. 10-52 Variation of adiabatic efficiency with specific speed for the three types of compressors.

Adiabatic Calculations Adiabatic head is expressed as follows: In SI units,

$$H_{ad} = \frac{k \times ZRT_1}{k-1}\left[\left(\frac{P_2}{P_1}\right)^{(k-1)/k} - 1\right] \tag{10-67}$$

where H_{ad} = adiabatic head, N·m/kg; R = gas constant = 53.35 ft-lbf/(°R lbm) (BTU system), 287.074 = J/(kg·°K) (metric system); T_1 = inlet gas temperature, °R, °K; P_1 = absolute inlet pressure, lb_f/ft^2, kPa; and P_2 = absolute discharge pressure, lb_f/ft^2, kPa.

In USCS units,

$$H_{ad} = \frac{k}{k-1} ZRT_1\left[\left(\frac{P_2}{P_1}\right)^{(k-1)/k} - 1\right] \tag{10-68}$$

where H_{ad} = adiabatic head, ft·lbf/lbm; R = gas constant, $(\text{ft}\cdot\text{lb}_f)/(\text{lbm}\cdot°\text{R})$ = 1545/molecular weight; T_1 = inlet gas temperature, °R; P_1 = absolute inlet pressure, lb_f/in^2; and P_2 = absolute discharge pressure, lb_f/in^2.

The *work* expended on the gas during compression is equal to the product of the adiabatic head and the mass flow of gas handled. Therefore, the adiabatic power is as follows:

In SI units,

$$kW_{ad} = \frac{\dot{m}H_{ad}}{10^3} = \frac{k \times ZWRT_1}{k-1}\left[\left(\frac{P_2}{P_1}\right)^{(k-1)/k} - 1\right] \tag{10-69}$$

$$kW_{ad} = 2.78 \times 10^{-4}\frac{k}{k-1} Q_1P_1\left[\left(\frac{P_2}{P_1}\right)^{(k-1)/k} - 1\right] \tag{10-70}$$

where kW_{ad} = power, kW; $\dot{m}$ = mass flow, kg/s; and $\dot{Q}_1$ = volume rate of gas flow, m^3/h, at compressor inlet conditions.

In USCS units,

$$hp_{ad} = \frac{\dot{m}H_{ad}}{550} = \frac{k}{k-1}\frac{ZWRT_1}{550}\left[\left(\frac{P_2}{P_1}\right)^{(k-1)/k} - 1\right] \tag{10-71}$$

or

$$hp_{ad} = \left[\frac{k}{k-1}\right]\frac{\dot{Q}_1P_1}{3600}\left[\left(\frac{P_2}{P_1}\right)^{(k-1)/k} - 1\right] \tag{10-72}$$

where hp_{ad} = power, hp; $\dot{m}$ = mass flow, lb/s; and $\dot{Q}_1$ = volume rate of gas flow, ft^3/min.

Adiabatic discharge temperature is

$$T_2 = T_1(P_2/P_1)^{(k-1)/k} \tag{10-73}$$

The work in a compressor under ideal conditions as previously shown occurs at constant entropy. The actual process is a polytropic process as shown in Fig. 10-54 and given by the equation of state PV^n = constant.

Adiabatic efficiency is given by the following relationship:

$$\eta_{ad} = \frac{\text{ideal work}}{\text{actual work}} \tag{10-74}$$

In terms of the change in total temperatures, the relationship can be written as

$$\eta_{ad} = \frac{T_2 - T_1}{T_{2a} - T_1} \tag{10-75}$$

where T_{2a} is the total actual discharge temperature of the gas. The adiabatic efficiency can be represented in terms of the total pressure change:

$$\eta_{ad} = \frac{\left(\frac{p_2}{p_1}\right)^{(k-1)/k} - 1}{\left(\frac{p_2}{p_1}\right)^{(n-1)/n} - 1} \tag{10-76}$$

Polytropic head can be expressed by the following relationship:

$$H_{poly} = \frac{n}{n-1} ZRT_1\left[\left(\frac{p_2}{p_1}\right)^{(n-1)/n} - 1\right] \tag{10-77}$$

TABLE 10-16 Performance Characteristics of Compressors

Types of compressors	Pressure ratio per stage: Industrial	Aerospace	Research	Efficiency, %	Operating range surge – choke, %
Positive displacement	Up to 30	—	—	75–82	—
Centrifugal	1.2–1.9	2.0–7.0	13	75–87 25	Large
Axial	1.05–1.3	1.1–1.45	2.1	80–91	Narrow 3–10

Likewise, for polytropic efficiency, which is often considered as the small stage efficiency, or the hydraulic efficiency

$$\eta_{pc} = \frac{(k-1)/k}{(n-1)/n} \tag{10-78}$$

Polytropic efficiency is the limited value of the isentropic efficiency as the pressure ratio approaches 1.0, and the value of the polytropic efficiency is higher than the corresponding adiabatic efficiency.

A characteristic of polytropic efficiency is that the polytropic efficiency of a multistage unit is equal to the stage efficiency if each stage has the same efficiency.

If the compression cycle approaches the isothermal condition, pV = constant, as is the case when several stages with intercoolers are used, then a simple approximation of the power is obtained from the following formula:

In SI units,

$$\text{kW} = 2.78 \times 10^{-4}\, \dot{Q}_1 p_1 \ln p_2/p_1 \tag{10-79}$$

In USCS units,

$$\text{hp} = 4.4 \times 10^{-3}\, \dot{Q}_1 p_1 \ln p_2/p_1 \tag{10-80}$$

Reciprocating Compressors Reciprocating compressors are used mainly when high-pressure head is required at a low flow. Reciprocating compressors are furnished in either single-stage or multistage types. The number of stages is determined by the required compressor ratio p_2/p_1. The compression ratio per stage is generally limited to 4, although low-capacity units are furnished with compression ratios of 8 and even higher. Generally, the maximum compression ratio is determined by the maximum allowable discharge-gas temperature.

Single-acting air-cooled and water-cooled air compressors are available in sizes up to about 75 kW (100 hp). Such units are available in one, two, three, or four stages for pressure as high as 24 MPa (3500 lbf/in^2). These machines are seldom used for gas compression because of the difficulty of preventing gas leakage and contamination of the lubricating oil.

The compressors most commonly used for compressing gases have a crosshead to which the connecting rod and piston rod are connected. This provides a straight-line motion for the piston rod and permits simple packing to be used. Figure 10-55 illustrates a simple single-stage machine of this type having a double-acting piston. Either single-acting (Fig. 10-56) or double-acting pistons (Fig. 10-57) may be used, depending on the size of the machine and the number of stages. In some machines double-acting pistons are used in the first stages and single-acting in the later stages.

On multistage machines, intercoolers are provided between stages. These heat exchangers remove the heat of compression from the gas and reduce its temperature to approximately the temperature existing at the compressor intake. Such cooling reduces the volume of gas going to the high-pressure cylinders, reduces the power required for compression, and keeps the temperature within safe operating limits.

Figure 10-58 illustrates a two-stage compressor end such as might be used on the compressor illustrated in Fig. 10-55.

Compressors with horizontal cylinders such as illustrated in Figs. 10-55 to 10-57 are most commonly used because of their accessibility. However, machines are also built with vertical cylinders and other arrangements such as right-angle (one horizontal and one vertical cylinder) and V-angle.

Compressors up to around 75 kW (100 hp) usually have a single center-throw crank, as illustrated in Fig. 10-55. In larger sizes compressors are commonly of duplex construction with cranks on each end of the shaft (see Fig. 10-59). Some large synchronous motor-driven units are of four-corner construction; i.e., they are of double-duplex construction with two connecting rods from each of the two crank throws (see Fig. 10-60). Steam-driven compressors have one or more steam cylinders connected directly by piston rod or tie rods to the gas-cylinder piston or crosshead.

Valve Losses Above piston speeds of 2.5 m/s (500 ft/min), suction and discharge valve losses begin to exert significant effects on the actual internal compression ratio of most compressors, depending on the valve port area available. The obvious results are high temperature rise and higher power requirements than might be expected. These effects become more pronounced with higher-molecular-weight gases. Valve problems can be a very major contributor to downtime experienced by these machines.

Control Devices In many installations the use of gas is intermittent, and some means of controlling the output of the compressor is therefore necessary. In other cases, constant output is required despite variations in discharge pressure, and the control device must operate to maintain a constant compressor speed. Compressor capacity, speed, or pressure may be varied in accordance with requirements. The nature of the control device will depend on the function to be regulated. Regulation of pressure, volume, temperature, or some other factor determines the type of regulation required and the type of the compressor driver.

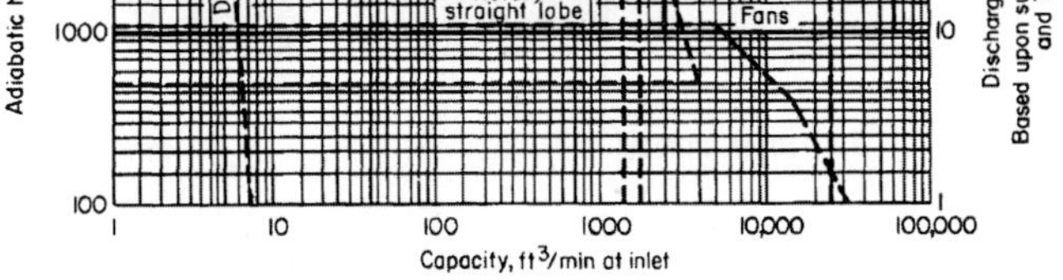

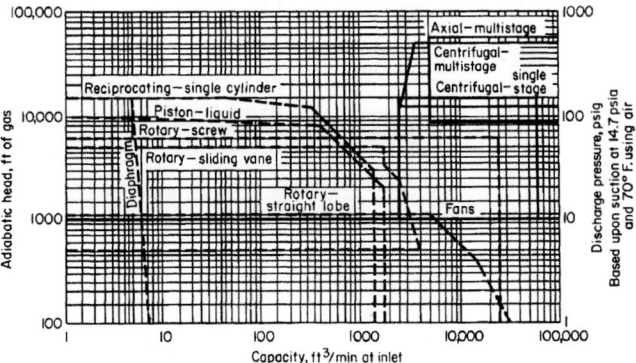

FIG. 10-53 Compressor coverage chart based on the normal range of operation of commercially available types shown. Solid lines: use left ordinate, head. Broken lines: use right ordinate, pressure. To convert cubic feet per minute to cubic meters per hour, multiply by 1.699; to convert feet to meters, multiply by 0.3048; and to convert pounds-force per square inch to kilopascals, multiply by 6.895; (°F − 32)5/9 = °C.

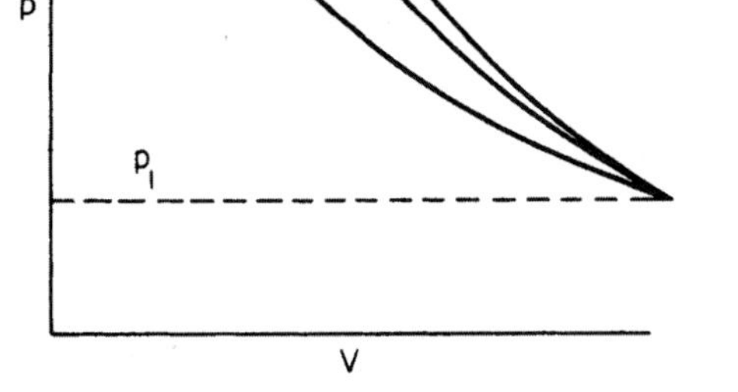

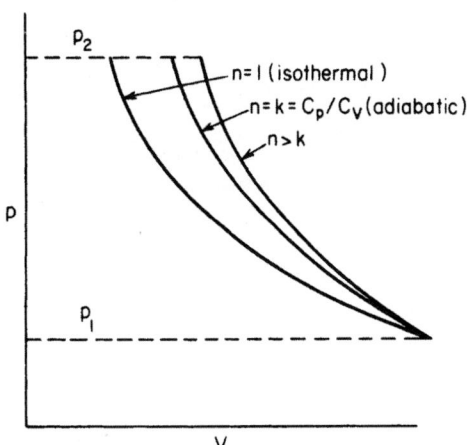

FIG. 10-54 Polytropic compression curves.

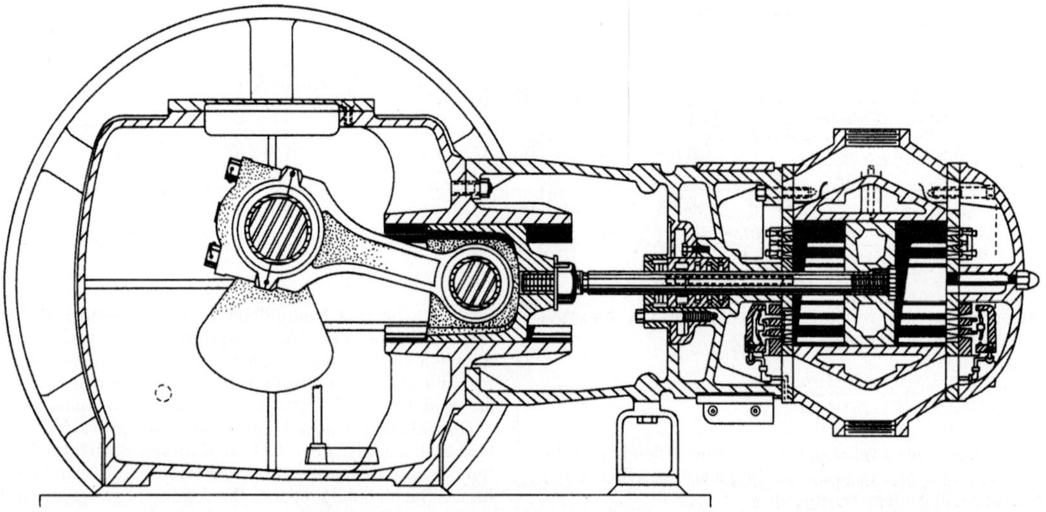

FIG. 10-55 Typical single-stage, double-acting water-cooled compressor.

The most common control requirement is regulation of capacity. Many capacity controls, or *unloading devices,* as they are usually termed, are actuated by the pressure on the discharge side of the compressor. A falling pressure indicates that gas is being used faster than it is being compressed and that more gas is required. A rising pressure indicates that more gas is being compressed than is being used and that less gas is required.

An obvious method of controlling the capacity of a compressor is to vary the speed. This method is applicable to units driven by variable-speed drivers such as steam pistons, steam turbines, gas engines, diesel engines, etc. In these cases, the regulator actuates the steam-admission or fuel-admission valve on the compressor driver and thus controls the speed.

Motor-driven compressors usually operate at constant speed, and other methods of controlling the capacity are necessary. On reciprocating compressors discharging into receivers, up to about 75 kW (100 hp), two types of control are usually available. These are automatic start-and-stop control and constant-speed control.

Automatic start-and-stop control, as its name implies, stops or starts the compressor by means of a pressure-actuated switch as the gas demand varies. It should be used only when the demand for gas will be intermittent.

Constant-speed control should be used when the gas demand is fairly constant. With this type of control, the compressor runs continuously but compresses only when gas is needed. Three methods of unloading the compressor with this type of control are in common use: (1) *closed suction unloaders,* (2) *open inlet-valve unloaders,* and (3) *clearance unloaders.* The closed suction unloader consists of a pressure-actuated valve which shuts off the compressor intake. Open inlet-valve unloaders (see Fig. 10-61) operate to hold the compressor inlet valves open and thereby prevent compression. Clearance unloaders (see Fig. 10-62) consist of pockets or small reservoirs that are opened when unloading is desired. The gas is compressed into them on the compression stroke and reexpands into the cylinder on the return stroke, thus preventing the compression of additional gas.

It is sometimes desirable to have a compressor equipped with both constant-speed and automatic start-and-stop control. When this is done, a switch allows immediate selection of either type.

Motor-driven reciprocating compressors above about 75 kW (100 hp) in size are usually equipped with a step control. This is in reality a variation of constant-speed control in which unloading is accomplished in a series of steps, varying from full load down to no load. *Three-step control* (full load, one-half load, and no load) is usually accomplished with inlet-valve unloaders. *Five-step control* (full load, three-fourths load, one-half load, one-fourth load, and no load) is accomplished by means of clearance pockets (see Fig. 10-63). On some machines, inlet valve and clearance-control unloading are used in combination.

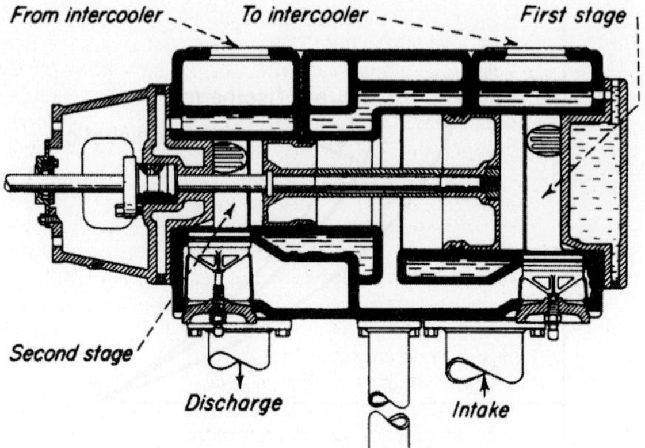

FIG. 10-56 Two-stage, single-acting opposed piston in a single step-type cylinder.

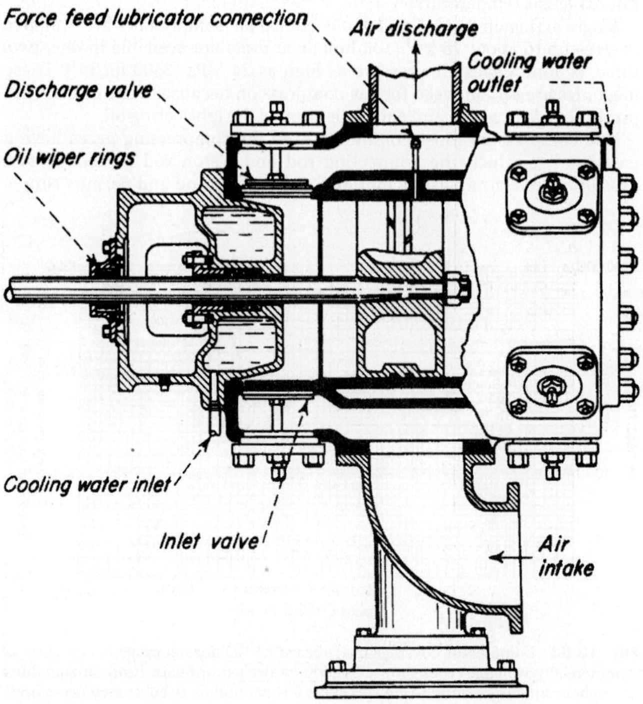

FIG. 10-57 Typical double-acting compressor piston and cylinder.

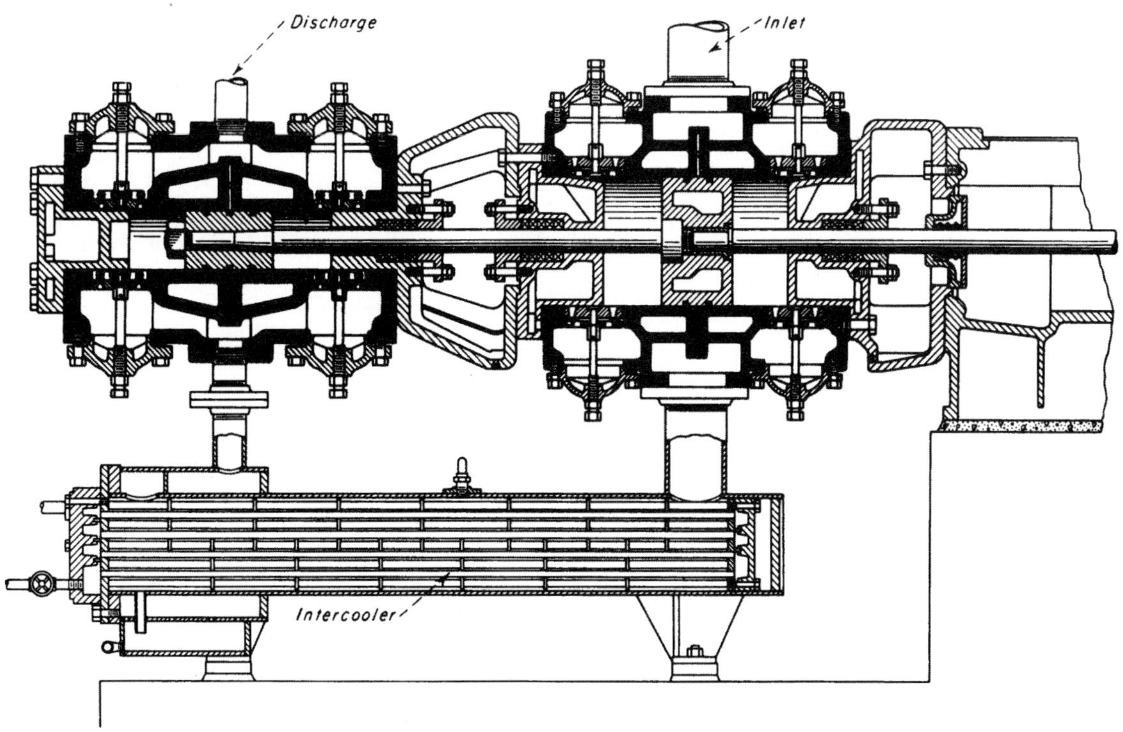

FIG. 10-58 Two-stage double-acting compressor cylinders with intercooler.

Although such control devices are usually automatically operated, manual operation is satisfactory for some services. When manual operation is provided, it often consists of a valve or valves to open and close clearance pockets. In some cases, a movable cylinder head is provided for variable clearance in the cylinder (see Fig. 10-64).

When no capacity control or unloading device is provided, it is necessary to provide bypasses between the inlet and discharge so that the compressor can be started against no load (see Fig. 10-65).

Nonlubricated Cylinders Most compressors use oil to lubricate the cylinder. In some processes, however, the slightest oil contamination is objectionable. For such cases a number of manufacturers furnish a "nonlubricated" cylinder (see Fig. 10-66). The piston on these cylinders is equipped with piston rings of graphitic carbon or Teflon* as well as pads or rings of the same material to maintain proper clearance between the piston and the cylinder. Plastic packing of a type that requires no lubricant is used on the stuffing box. Although oil-wiper rings are used on the piston rod where it leaves the compressor frame, minute quantities of oil might conceivably enter the cylinder on the rod. If even such small amounts of oil are objectionable, an extended cylinder connecting piece can be furnished. This simply lengthens the piston rod enough that no portion of the rod can alternately enter the frame and the cylinder.

In many cases, a small amount of gas leaking through the packing is objectionable. Special connecting pieces are furnished between the cylinder and the frame, which may be either single-compartment or double-compartment. These may be furnished gastight and vented back to the suction or filled with a sealing gas or fluid and held under a slight pressure.

High-Pressure Compressors There is a definite trend in the chemical industry toward the use of high-pressure compressors with discharge pressures of 34.5 to 172 MPa (5000 to 25,000 lbf/in^2) and with capacities from 8.5×10^3 to 42.5×10^3 m^3/h (5000 to 25,000 ft^3/min). These require special design, and a complete knowledge of the characteristics of the gas is necessary. In most cases, these types of applications use the barrel-type centrifugal compressor.

The gas usually deviates considerably from the perfect-gas laws, and in many cases temperature or other limitations necessitate a thorough engineering study of the problem. These compressors usually have five, six, seven, or eight stages, and the cylinders must be properly proportioned to meet the various limitations involved and to balance the load among the various stages. In many cases, scrubbing or other processing is carried on between stages. High-pressure cylinders are steel forgings with single-acting plungers (see Fig. 10-67). The compressors are usually designed so that the pressure load against the plunger is opposed by one or more single-acting pistons of the lower-pressure stages. Piston-rod packing is usually of the segmental-ring metallic type. Accurate fitting and correct lubrication are very important. High-pressure compressor valves are designed for the conditions involved. Extremely high-grade engineering and skill are necessary.

Piston-Rod Packing Proper piston-rod packing is important. Many types are available, and the most suitable is determined by the gas handled and the operating conditions for a particular unit.

There are many types and compositions of soft packing, semimetallic packing, and metallic packing. In many cases, metallic packing is to be recommended. A typical low-pressure packing arrangement is shown in Fig. 10-68. A high-pressure packing arrangement is shown in Fig. 10-69.

When wet, volatile, or hazardous gases are handled or when the service is intermittent, an auxiliary packing gland and soft packing are usually employed (see Fig. 10-70).

*®Du Pont tetrafluoroethylene fluorocarbon resin.

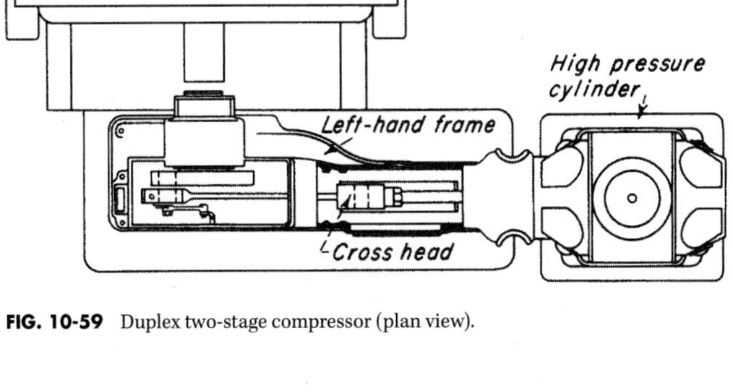

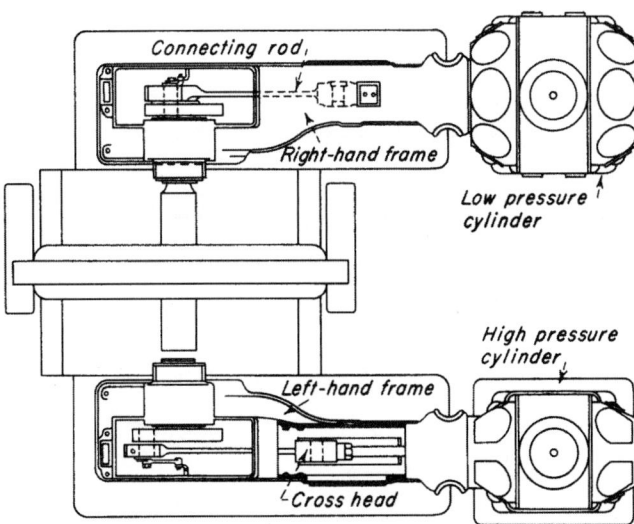

FIG. 10-59 Duplex two-stage compressor (plan view).

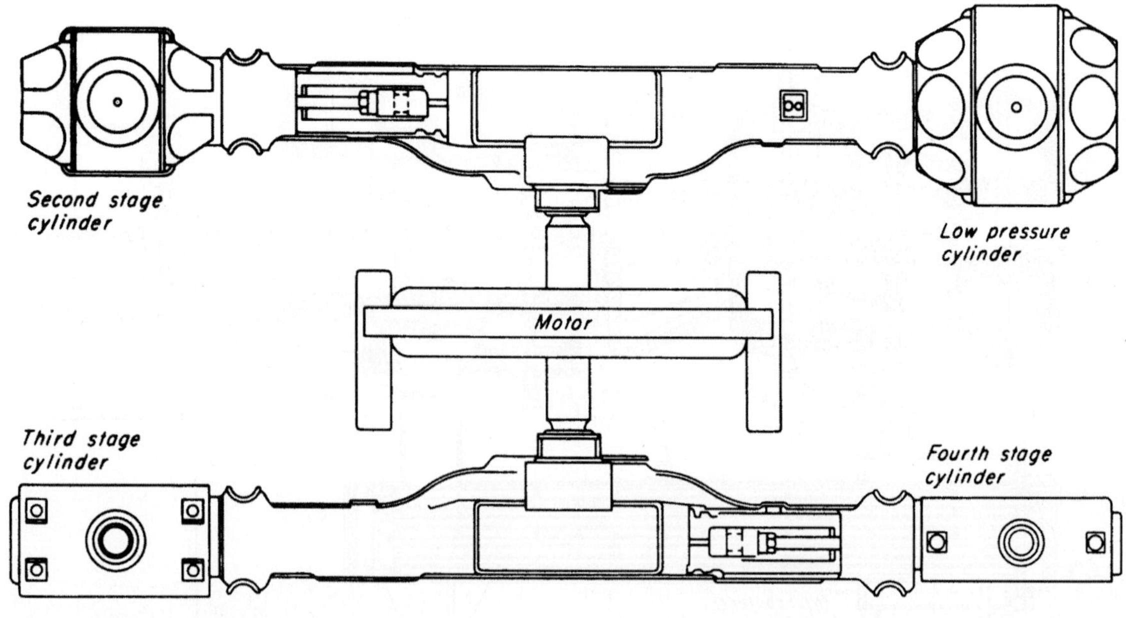

FIG. 10-60 Four-corner four-stage compressor (plan view).

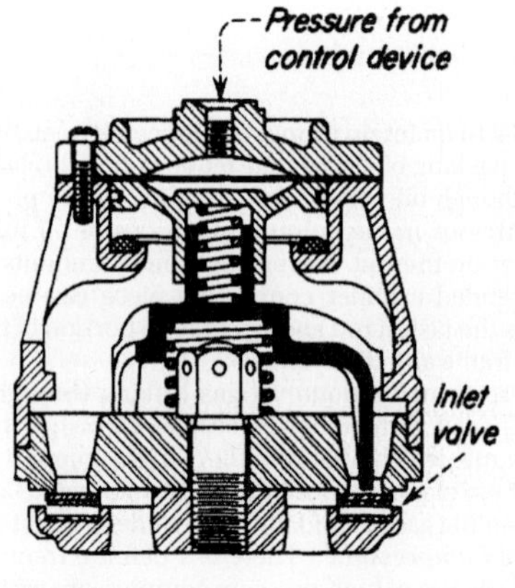

FIG. 10-61 Inlet-valve unloader.

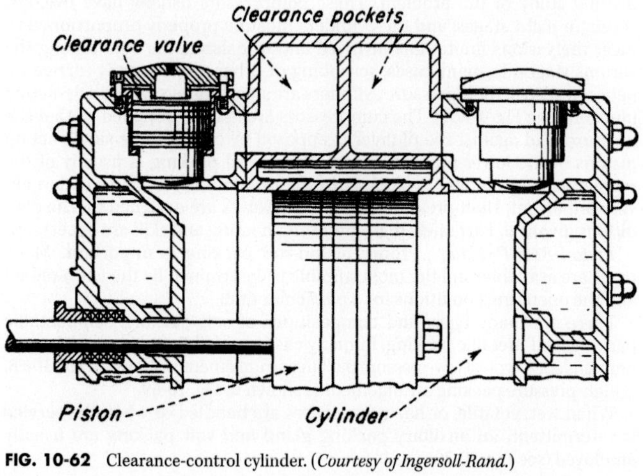

FIG. 10-62 Clearance-control cylinder. (*Courtesy of Ingersoll-Rand.*)

Metallic Diaphragm Compressors These (Fig. 10-71) are available for small quantities [up to about 17 m^3/h (10 ft^3/min)] for compression ratios as high as 10:1 per stage. Temperature rise is not a serious problem, as the large wall area relative to the gas volume permits sufficient heat transfer to approach isothermal compression. These compressors possess the advantage of having no seals for the process gas. The diaphragm is actuated hydraulically by a plunger pump.

FANS AND BLOWERS

Fans are used for low pressures where generally the delivery pressure is less than 3.447 kPa (0.5 lb/in^2), and blowers are used for higher pressures.

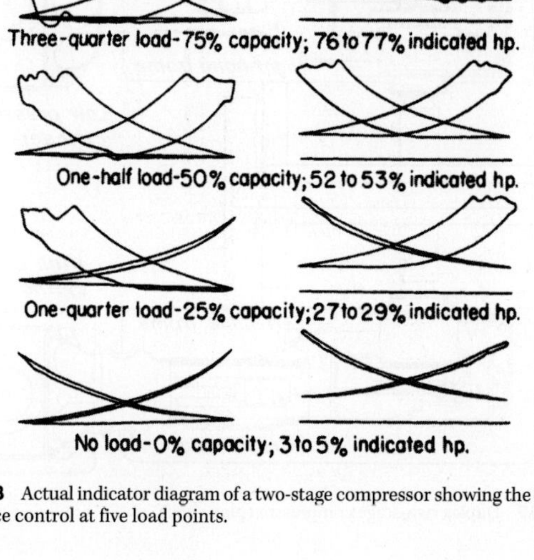

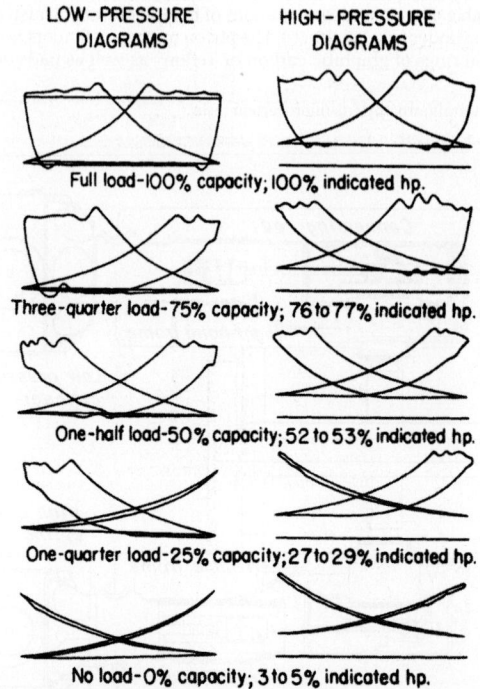

FIG. 10-63 Actual indicator diagram of a two-stage compressor showing the operation of clearance control at five load points.

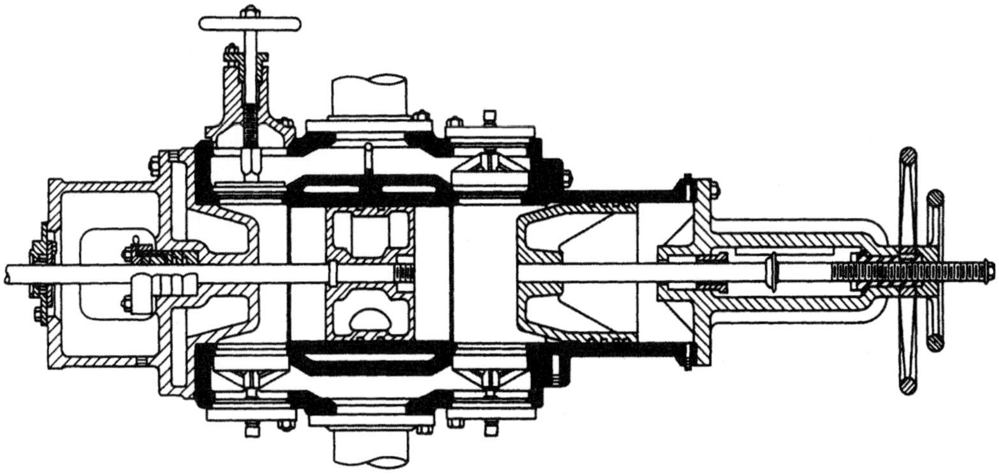

FIG. 10-64 Sectional view of a cylinder equipped with a hand-operated valve lifter on one end and a variable-volume clearance pocket at other end.

However, they are usually below delivery pressures of 10.32 kPa (1.5 lbf/in^2). These units can be either centrifugal or the axial-flow type.

Fans and blowers are used for many types of ventilating work such as air-conditioning systems. In large buildings, blowers are often used owing to the high delivery pressures needed to overcome the pressure drop in the ventilation system. Most of these blowers are of the centrifugal type. Blowers are also used to supply draft air to boilers and furnaces. Fans are used to move large volumes of air or gas through ducts, supplying air for drying, conveying material suspended in the gas stream, removing fumes, condensing towers and other high-flow, low-pressure applications.

Axial-Flow Fans These are designed to handle very high flow rates and low pressure. The disc-type fans are similar to a household fan. They are usually for general circulation or exhaust work without ducts.

Starting compressor	Stopping compressor
Start with A and D open	Close - - - - C
Close - - - - - - D	Close - - - B
Close - - - - - - - A	Open - - - - A and D
Open - - - - - - - B	
Slowly open - - - C	

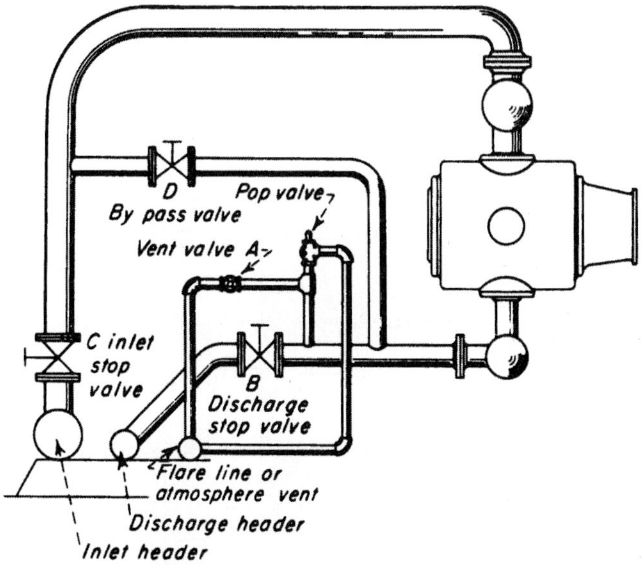

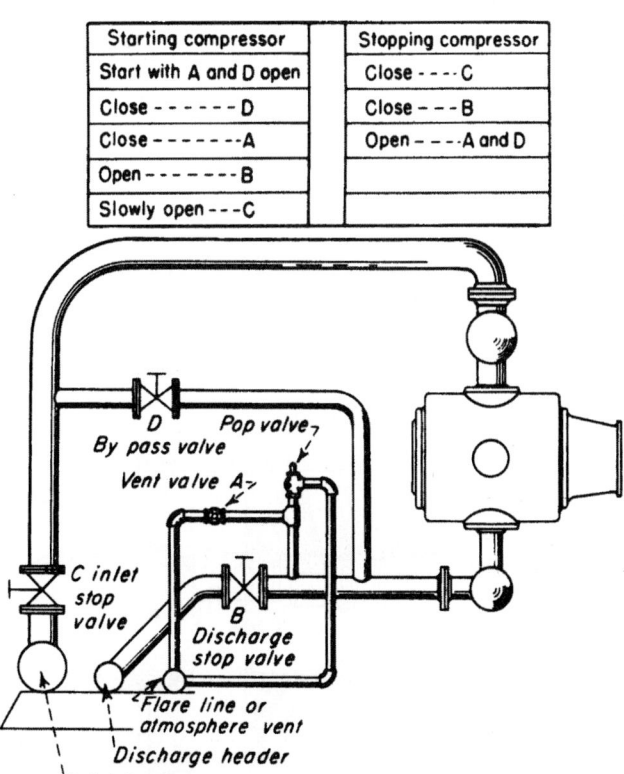

FIG. 10-65 Bypass arrangement for a single-stage compressor. On multistage machines, each stage is bypassed in a similar manner. Such an arrangement is necessary for no-load starting.

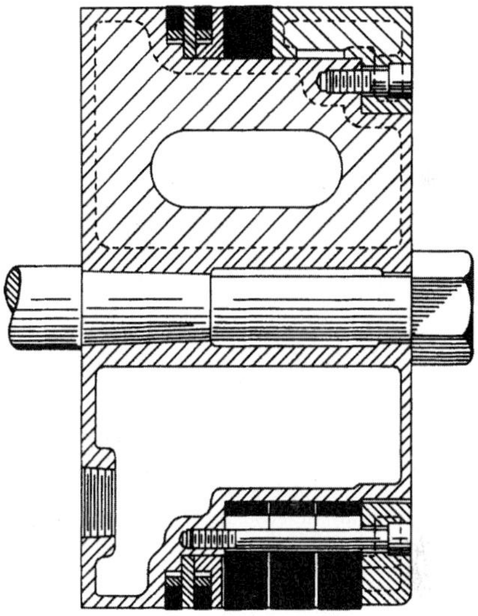

FIG. 10-66 Piston equipped with carbon piston and wearing rings for a nonlubricated cylinder.

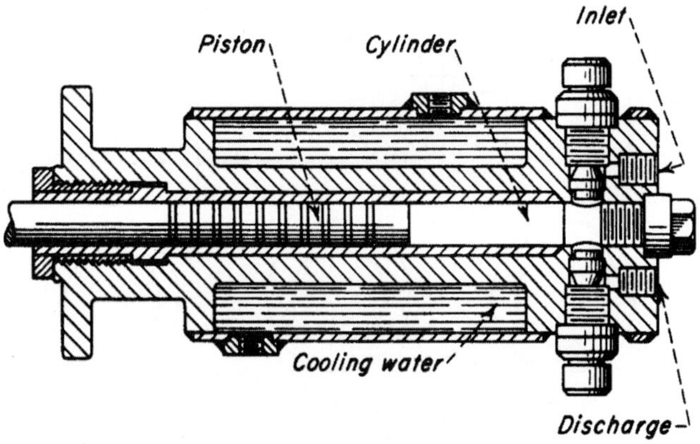

FIG. 10-67 Forged-steel single-acting high-pressure cylinder.

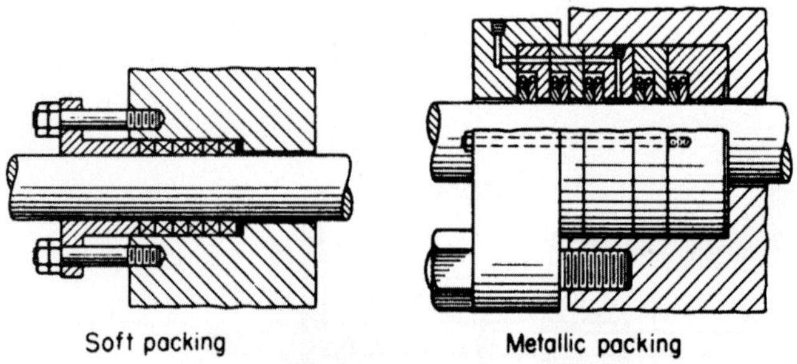

FIG. 10-68 Typical packing arrangements for low-pressure cylinders.

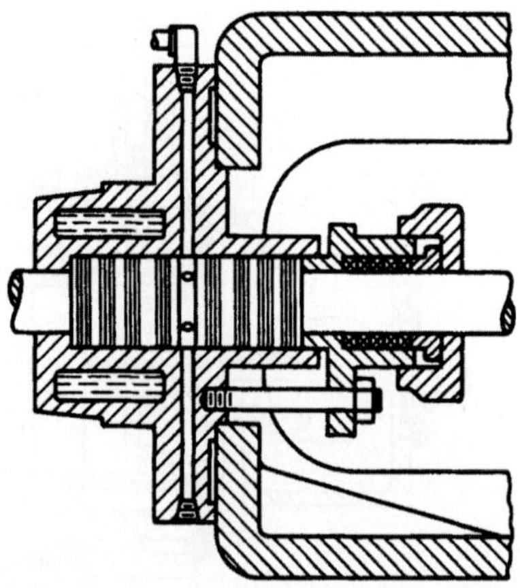

FIG. 10-70 Soft packing in an auxiliary stuffing box for handling gases.

The so-called propeller-type fans with blades that are aerodynamically designed (as seen in Fig. 10-72) can consist of two or more stages. The air in these fans enters in an axial direction and leaves in an axial direction. The fans usually have inlet guide vanes followed by a rotating blade, followed by a stationary (stator) blade.

Centrifugal Blowers These blowers have air or gases entering in the axial direction and being discharged 90° from the entrance. These blowers have three types of blades: radial or straight blades, forward-curved blades, and backward-curved blades (Figs. 10-73 to 10-75).

Radial blade blowers as seen in Fig. 10-73 are usually used in large-diameter or high-temperature applications. The blades being radial in direction have very low stresses compared to the backward- or forward-curved blades. The rotors have anywhere from 4 to 12 blades and usually operate at low speeds. These fans are used in exhaust work especially for gases at high temperature and with suspensions in the flow stream.

Forward-Curved Blade Blowers These blowers discharge the gas at a very high velocity. The pressure supplied by this blower is lower than that produced in the other two blade characteristics. The number of blades in such a rotor can be large—up to 50 blades—and the speed is high—usually 3600 to 1800 rpm in 60-cycle countries and 3000 to 1500 rpm in 50-cycle countries.

Backward-Curved Blade Blowers These blowers are used when a higher discharge pressure is needed. It is used over a wide range of applications. Both the forward and backward curved blades do have much higher stresses than the radial blade blower.

The centrifugal blower produces energy in the airstream by the centrifugal force and imparts a velocity to the gas by the blades. Forward-curved blades impart the greatest velocity to the gas. The scroll-shaped volute diffuses the air and creates an increase in the static pressure by reducing the gas velocity. The change in total pressure occurs in the impeller—this is usually a small change. The static pressure is increased in both the impeller and

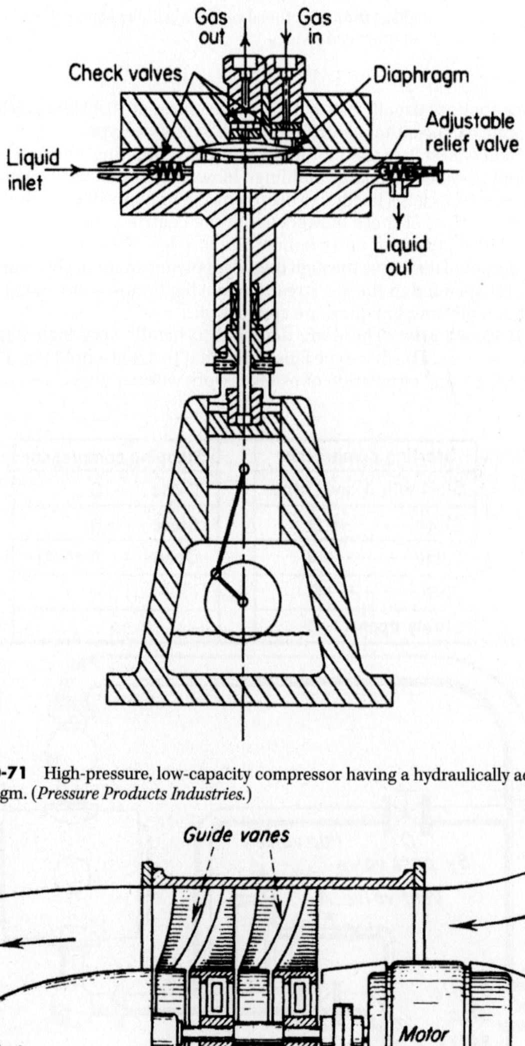

FIG. 10-71 High-pressure, low-capacity compressor having a hydraulically actuated diaphragm. (*Pressure Products Industries.*)

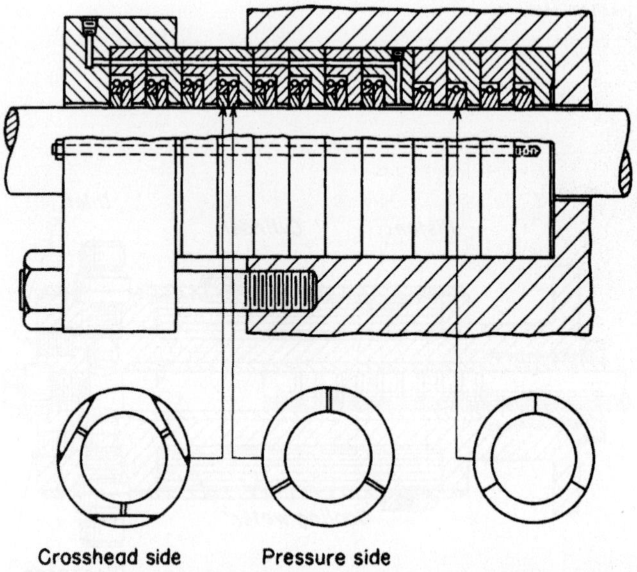

FIG. 10-69 Typical packing arrangement, using metallic packing, for high-pressure cylinders.

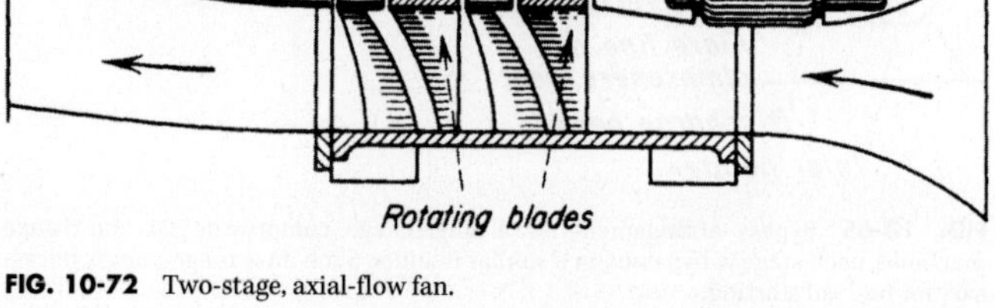

FIG. 10-72 Two-stage, axial-flow fan.

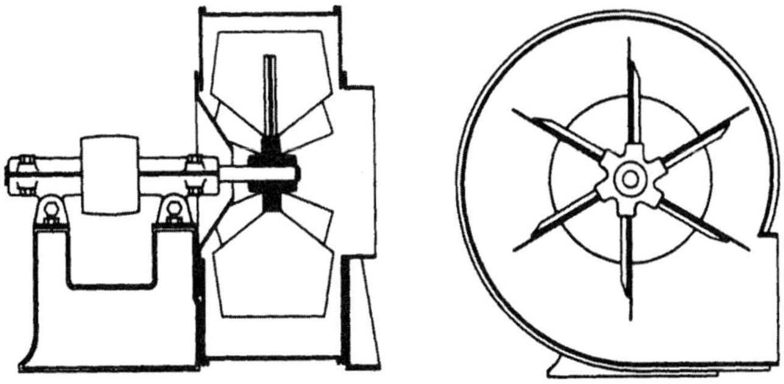

FIG. 10-73 Straight-blade, or steel-plate, fan.

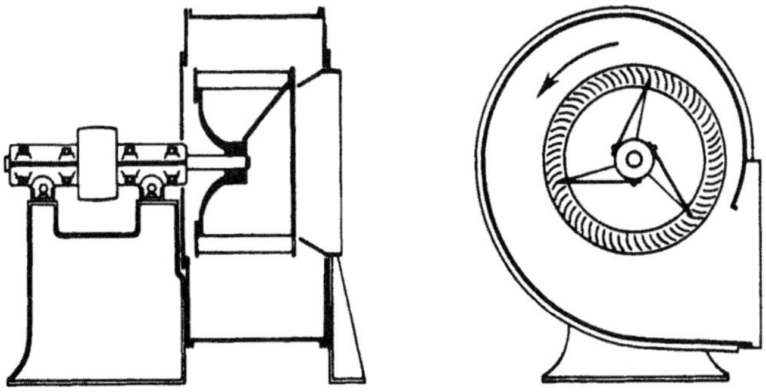

FIG. 10-74 Forward-curved blade, or "scirocco"-type, fan.

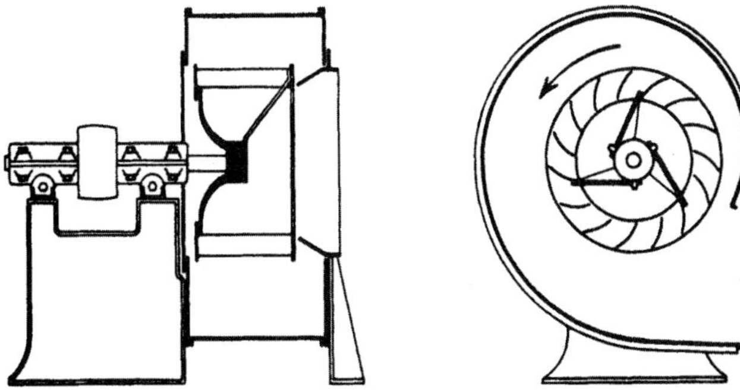

FIG. 10-75 Backward-curved-blade fan.

the diffuser section. Operating efficiencies of the fan range from 40 to 80 percent. The discharge total pressure is the sum of the static pressure and the velocity head.

The power needed to drive the fan can be computed as follows:

$$\text{Power (kW)} = 2.72 \times 10^{-5} \dot{Q}P \tag{10-81}$$

where $\dot{Q}$ is the fan volume (m^3/h) and P is the total discharge pressure in centimeters of water column.

In USCS units,

$$\text{Power (hp)} = 1.57 \times 10^{-4} \dot{Q}P \tag{10-82}$$

where hp is the fan power output, hp; $\dot{Q}$ is the fan volume, ft^3/min; and p is the fan operating pressure, inches of water column.

$$\text{Efficiency} = \frac{\text{air power output}}{\text{shaft power input}} \tag{10-83}$$

Fan Performance The performance of a centrifugal fan varies with changes in conditions such as temperature, speed, and density of the gas being handled. It is important to keep this in mind in using the catalog data of various fan manufacturers, since such data are usually based on stated

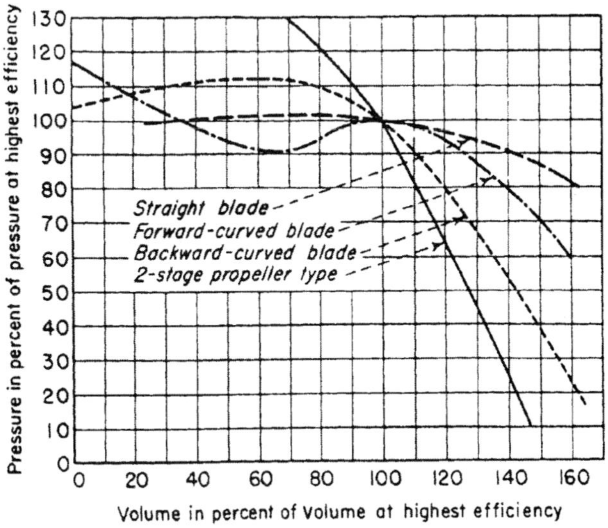

FIG. 10-76 Approximate characteristic curves of various types of fans.

standard conditions. Corrections must be made for variations from these standards. The usual variations are as follows:

When speed varies, (1) capacity varies directly as the speed ratio, (2) pressure varies as the square of the speed ratio, and (3) horsepower varies as the cube of the speed ratio.

When the temperature of air or gas varies, horsepower and pressure vary inversely as the absolute temperature, with speed and capacity being constant. See Fig. 10-76.

When the density of air or gas varies, horsepower and pressure vary directly as the density, with speed and capacity being constant.

CONTINUOUS-FLOW COMPRESSORS

Continuous-flow compressors are machines where the flow is continuous, unlike positive-displacement machines where the flow is fluctuating. Continuous-flow compressors are also classified as turbomachines. These types of machines are widely used in the chemical and petroleum industry for many services. They are also used extensively in many other industries such as the iron and steel industry, pipeline boosters, and on offshore platforms for reinjection compressors. Continuous-flow machines are usually much smaller and produce much less vibration than their counterpart, positive-displacement units.

Centrifugal Compressors The flow in a centrifugal compressor enters the impeller in an axial direction and exits in a radial direction.

In a typical centrifugal compressor, the fluid is forced through the impeller by rapidly rotating impeller blades. The velocity of the fluid is converted to pressure, partly in the impeller and partly in the stationary diffusers. Most of the velocity leaving the impeller is converted to pressure energy in the diffuser, as shown in Fig. 10-77. It is normal practice to design the compressor so that one-half the pressure rise takes place in the impeller and the other

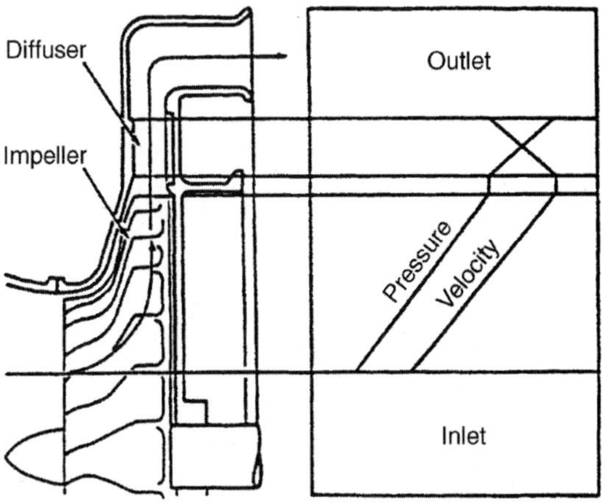

FIG. 10-77 Pressure and velocity through a centrifugal compressor.

half in the diffuser. The diffuser consists of a vaneless space, a vane that is tangential to the impeller, or a combination of both. These vane passages diverge to convert the velocity head to pressure energy.

Centrifugal compressors in general are used for higher pressure ratios and lower flow rates compared to lower-stage pressure ratios and higher flow rates in axial compressors. The pressure ratio in a single-stage centrifugal compressor varies according to the industry and application. In the petrochemical industry, the single-stage pressure ratio is about 1.2:1. Centrifugal compressors used in the aerospace industry, usually as a compressor of a gas turbine, have pressure ratios from 3:1 to as high as 9:1 per stage.

In the petrochemical industry, the centrifugal compressors consist mainly of casings with multiple stages. In many instances, multiple casings are also used, and to reduce the power required to drive these multiple casings, there are intercoolers between them. Each casing can have up to nine stages. In some cases, intercoolers are also used between single stages of compressor to reduce the power required for compression. These compressors are usually driven by gas turbines, steam turbines, and electric motors. Speed-increasing gears may be used in conjunction with these drivers to obtain the high speeds at which many of these units operate. Rotative speeds as high as 50,000 rpm are not uncommon. Most of the petrochemical units run between 9000 and 15,000 rpm.

The compressor's operating range is between two major regions, as seen in Fig. 10-78, which is a performance map of a centrifugal compressor. These two regions are *surge,* which is the lower flow limit of stable operation, and *choke* or *stonewall,* which is the maximum flow through the compressor at a given operating speed. The centrifugal compressor's operating range between surge and choke is reduced as the pressure ratio per stage is increased or a number of stages are added.

A compressor is said to be in *surge* when the main flow through the compressor reverses its direction. Surge is often symptomized by excessive vibration and a large audible sound. This flow reversal is accompanied with a very violent change in energy, which causes a reversal of the thrust force. The surge process is cyclic, and if it is allowed to cycle for some time, irreparable damage can occur to the compressor. In a centrifugal compressor, surge is usually initiated at the exit of the impeller or at the diffuser entrance for impellers producing a pressure ratio of less than 3:1. For higher pressure ratios, the initiation of surge can occur in the inducer.

A centrifugal compressor impeller can have three types of blades at the exit of the impeller: forward-curved, backward-curved, and radial blades. Forward-curved blades are not often used in a centrifugal compressor's impeller because of the very high-velocity discharge at the compressor that would require conversion of the high velocity to a pressure head in the diffuser, which would be accompanied by high losses. Radial blades are used in impellers of high pressure ratio since the stress levels are minimal. Backward-curved blades give the highest efficiency and the largest operating margin of any of the various types of blades in an impeller. Most centrifugal compressors in the petrochemical industry use backward-curved impellers because of the higher efficiency and larger operating range.

Process compressors have impellers with very low pressure ratio impellers and thus large surge-to-choke margins. The common method of classifying process-type centrifugal compressors is based on the number of impellers and the casing design. Sectionalized casing types have impellers that are usually mounted on the extended motor shaft, and similar sections are bolted together to obtain the desired number of stages. Casing material is either steel or cast iron. These machines require minimum supervision and maintenance and are quite economical in their operating range. The sectionalized casing design is used extensively in supplying air for combustion in ovens and furnaces.

The horizontally split type has casings split horizontally at the midsection and the top. The bottom halves are bolted and doweled together. This design type is preferred for large multistage units. The internal parts such as shaft, impellers, bearings, and seals are readily accessible for inspection and repairs by removing the top half. The casing material is cast iron or cast steel.

Barrel casings are used for high pressures in which the horizontally split joint is inadequate. This type of compressor consists of a barrel into which a compressor bundle of multiple stages is inserted. The bundle is itself a horizontally split casing compressor.

Compressor Configuration To properly design a centrifugal compressor, one must know the operating conditions—the type of gas, its pressure, temperature, and molecular weight. One must also know the corrosive properties of the gas so that proper metallurgical selection can be made. Gas fluctuations due to process instabilities must be pinpointed so that the compressor can operate without surging.

Centrifugal compressors for industrial applications have relatively low pressure ratios per stage. This condition is necessary so that the compressors can have a wide operating range while stress levels are kept at a minimum. Because of the low pressure ratios for each stage, a single machine may have a number of stages in one "barrel" to achieve the desired overall pressure ratio. Figure 10-79 shows some of the many configurations. These are some factors to consider when selecting a configuration to meet plant needs:

1. Intercooling between stages can considerably reduce the power consumed.
2. Back-to-back impellers allow for a balanced rotor thrust and minimize overloading of the thrust bearings.
3. Cold inlet or hot discharge at the middle of the case reduces oil-seal and lubrication problems.

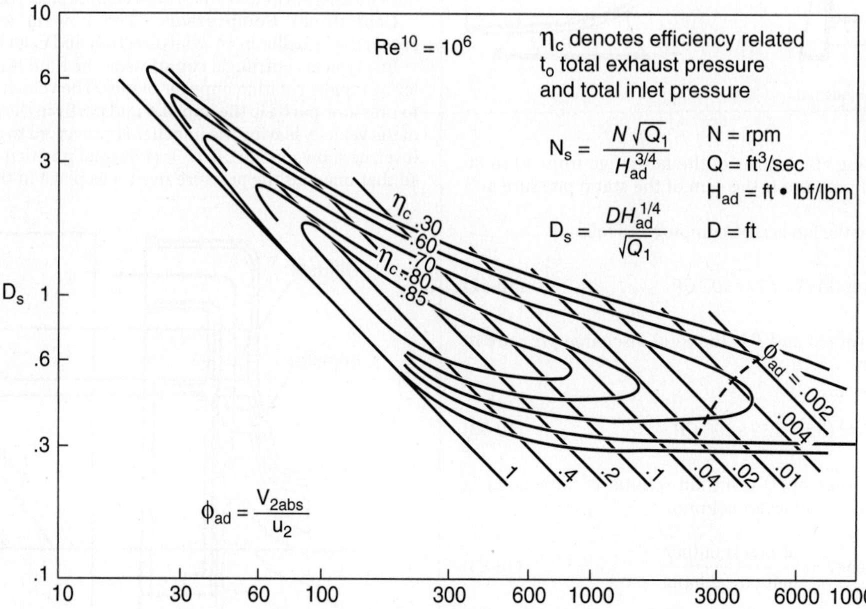

FIG. 10-78 Centrifugal compressor map. (O. E. Balje, "A Study of Reynolds Number Effects in Turbomachinery," *Journal of Engineering for Power, ASME Trans.,* vol. 86, series A, p. 227.)

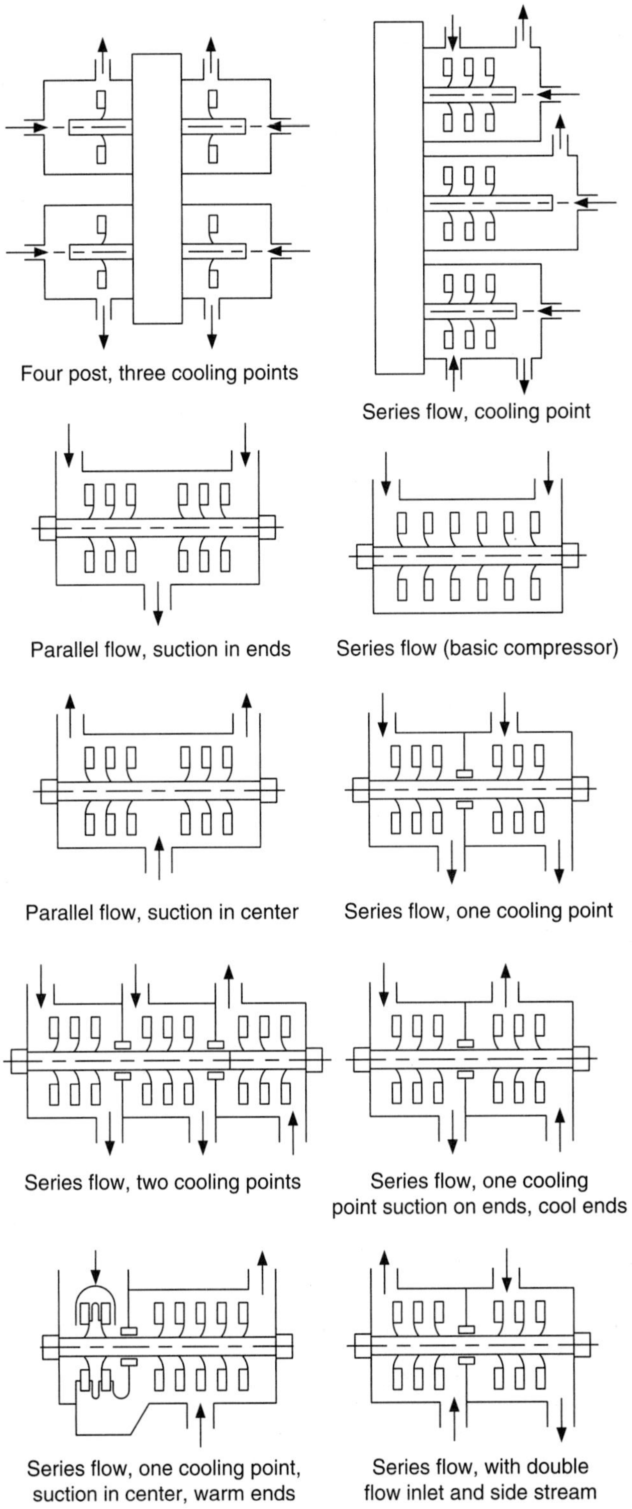

FIG. 10-79 Various configurations of centrifugal compressors.

4. Single inlet or single discharge reduces external piping problems.
5. Balance planes that are easily accessible in the field can appreciably reduce field-balancing times.
6. Balance piston with no external leakage will greatly reduce wear on the thrust bearings.
7. Hot and cold sections of the case that are adjacent to each other will reduce thermal gradients and thus reduce case distortion.
8. Horizontally split casings are easier to open for inspection than vertically split ones, reducing maintenance time.
9. Overhung rotors present an easier alignment problem because shaft-end alignment is necessary only at the coupling between the compressor and driver.
10. Smaller, high-pressure compressors that do the same job will reduce foundation problems but will have greatly reduced operational range.

Impeller Fabrication Centrifugal compressor impellers are either shrouded or unshrouded. Open, shrouded impellers that are mainly used in single-stage applications are made by investment-casting techniques or by three-dimensional milling. Such impellers are used, in most cases, for the high-pressure-ratio stages. The shrouded impeller is commonly used in the process compressor because of its low-pressure-ratio stages. The low tip stresses in this application make it a feasible design. Figure 10-80 shows several fabrication techniques. The most common type of construction is seen in (a) and (b) where the blades are fillet-welded to the hub and shroud. In (b), the welds are full penetration. The disadvantage in this type of construction is the obstruction of the aerodynamic passage. In (c), the blades are partially machined with the covers and then butt-welded down the middle. For backward lean-angled blades, this technique has not been very successful, and there has been difficulty in achieving a smooth contour around the leading edge.

Figure 10-80(d) illustrates a slot-welding technique and is used where blade-passage height is too small (or the backward lean angle too high) to permit conventional fillet welding. In (e), an electron-beam technique is shown. Its major disadvantage is that electron-beam welds should preferably be stressed in tension but, for the configuration of (e), they are in shear. The configurations of (g) through (j) use rivets. Where the rivet heads protrude into the passage, aerodynamic performance is reduced. Riveted impellers were used in the 1960s—they are very rarely used now. Elongation of these rivets occurs at certain critical surge conditions and can lead to major failures.

Materials for fabricating these impellers are usually low-alloy steels, such as AISI 4140 or AISI 4340. AISI 4140 is satisfactory for most applications; AISI 4340 is used for large impellers requiring higher strengths. For corrosive gases, AISI 410 stainless steel (about 12 percent chromium) is used. Monel K-500 is employed in halogen gas atmospheres and oxygen compressors because of its resistance to sparking. Titanium impellers have been applied to chlorine service. Aluminum-alloy impellers have been used in great numbers, especially at lower temperatures (below 300°F). With new developments in aluminum alloys, this range is increasing. Aluminum and titanium are sometimes selected because of their low density. This low density can cause a shift in the critical speed of the rotor, which may be advantageous.

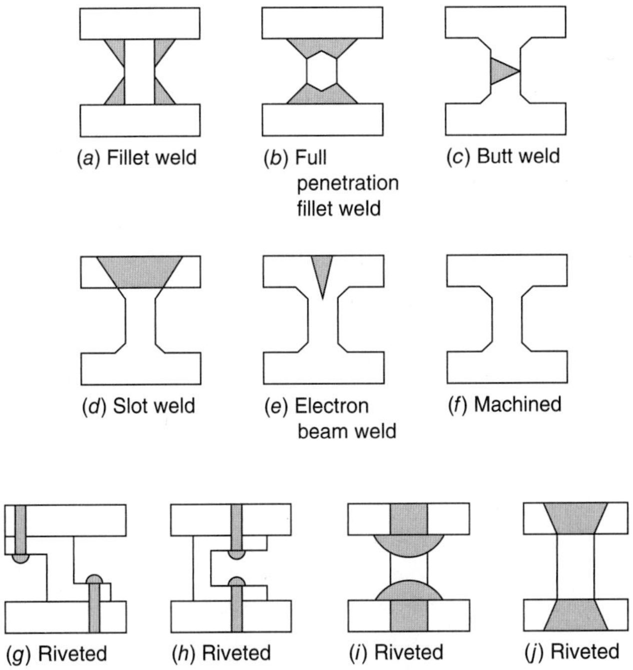

FIG. 10-80 Several fabrication techniques for centrifugal impellers.

Axial-Flow Compressors Axial-flow compressors are used mainly as compressors for gas turbines. They are also used in the steel industry as blast furnace blowers and in the chemical industry for large nitric acid plants. They are mainly used for applications where the head required is low and the flow large.

Figure 10-81 shows a typical axial-flow compressor. The rotating element consists of a single drum to which are attached several rows of decreasing-height blades having airfoil cross sections. Between each rotating blade row is a stationary blade row. All blade angles and areas are designed precisely for a given performance and high efficiency. The use of multiple stages permits overall pressure increases up to 30:1. The efficiency in an axial-flow compressor is higher than that in the centrifugal compressor.

The pressure ratio per casing can be comparable with those of centrifugal equipment, although flow rates are considerably higher for a given casing diameter because of the greater area of the flow path. The pressure ratio per stage is less than that in a centrifugal compressor. The pressure ratio per stage in industrial compressors is between 1:05 and 1:15, and for aero turbines, 1.1 and 1.2.

The axial-flow compressors used in gas turbines vary depending on the type of turbine. The industrial-type gas turbine has an axial-flow compressor of rugged construction. These units have blades that have low aspect ratio (R = blade height/blade chord) with minimum streamline curvation, and the shafts are supported on sleeve-type bearings. The industrial gas turbine compressor has also a lower pressure ratio per stage (stage = rotor + stationary blade), giving a low blade loading. This also gives a larger operating range than its counterpart, the aero axial gas turbine compressor, but considerably less than the centrifugal compressor.

The axial-flow compressors in aero gas turbines are heavily loaded. The aspect ratio of the blades, especially the first few stages, can be as high as 4.0, and the effect of streamline curvature is substantial. The streamline configuration is a function of the annular passage area, the camber and thickness distribution of the blade, and the flow angles at the inlet and outlet of the blades. The shafts on these units are supported on antifriction bearings (roller or ball bearings).

The operation of the axial-flow compressor is a function of the rotational speed of the blades and the turning of the flow in the rotor. The stationary blades (stator) are used to diffuse the flow and convert the velocity increased in the rotor to a pressure increase. One rotor and one stator make up a stage in a compressor. One additional row of fixed blades (inlet guide vanes) is frequently used at the compressor inlet to ensure that air enters the first-stage rotors at the desired angle. In addition to the stators, another diffuser at the exit of the compressor further diffuses the gas and, in the case of gas turbines, controls its velocity entering the combustor. The axial-flow compressor has a much smaller operating range "surge to choke" than its counterpart in the centrifugal compressor. Because of the steep characteristics of the head/flow capacity curve, the surge point is usually within 10 percent of the design point.

The axial-flow compressor has three distinct stall phenomena. Rotating stall and individual blade stall are aerodynamic phenomena. Stall flutter is an aeroelastic phenomenon. Rotating stall (propagating stall) consists of large stall zones covering several blade passages and propagates in the direction of the rotor and at some fraction of rotor speed. The number of stall zones and the propagating rates vary considerably. Rotating stall is the most prevalent type of stall phenomenon. Individual blade stall occurs when all the blades around the compressor annulus stall simultaneously without the occurrence of the stall propagation mechanism. The phenomena of stall flutter are caused by self-excitation of the blade and are aeroelastic. It must be distinguished from classic flutter, since classic flutter is a coupled torsional-flexural vibration that occurs when the freestream velocity over an airfoil section reaches a certain critical velocity. Stall flutter, however, is a phenomenon that occurs due to the stalling of the flow around a blade. Blade stall causes Karman vortices in the airfoil wake. Whenever the frequency of the vortices coincides with the natural frequency of airfoil, flutter will occur. Stall flutter is a major cause of compressor blade failure.

Positive-Displacement Compressors Positive-displacement compressors are essentially constant-volume machines with variable discharge pressures. These machines can be divided into two types:

1. Rotary compressors
2. Reciprocating compressors

Many users consider rotary compressors, such as the "Rootes"-type blower, as turbomachines because their behavior in terms of the rotor dynamics is very close to centrifugal and axial-flow machinery. Unlike

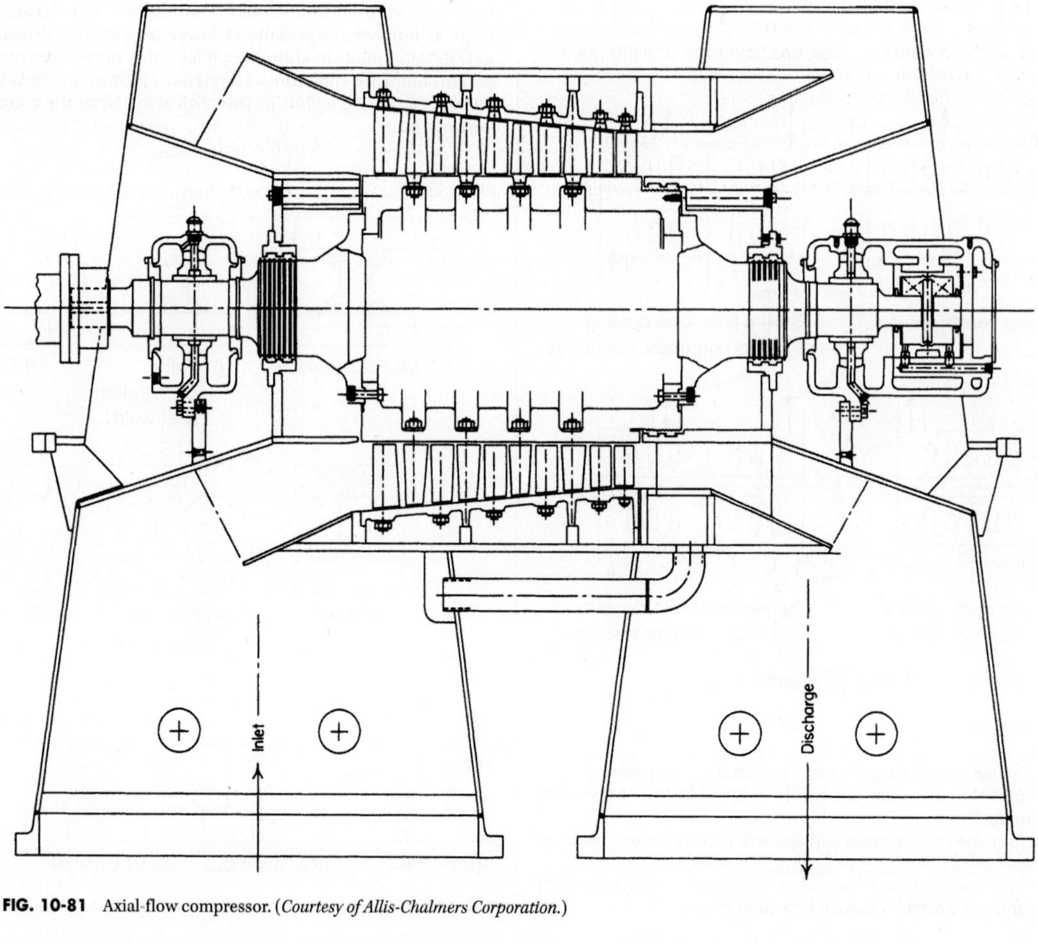

FIG. 10-81 Axial-flow compressor. (*Courtesy of Allis-Chalmers Corporation.*)

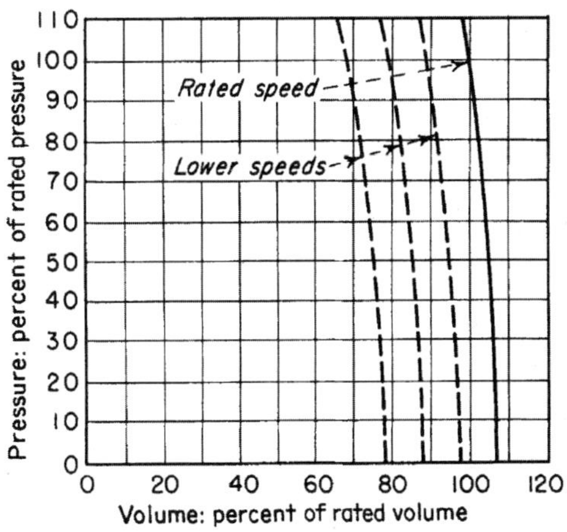

FIG. 10-82 Approximate performance curves for a rotary positive-displacement compressor. The safety valve in discharge line or bypass must be set to operate at a safe value determined by construction.

the reciprocating machines, the rotary machines do not have a very high vibration problem but, like the reciprocating machines, they are positive-displacement machines.

Rotary Compressors Rotary compressors are machines of the positive-displacement type. Such units are essentially constant-volume machines with variable discharge pressure. The volume can be varied only by changing the speed or by bypassing or wasting some of the capacity of the machine. The discharge pressure will vary with the resistance on the discharge side of the system. A characteristic curve typical of the form produced by these rotary units is shown in Fig. 10-82. Rotary compressors are generally classified as of the straight-lobe type, screw type, sliding-vane type, and liquid-piston type.

Straight-Lobe Type This type is illustrated in Fig. 10-83. Such units are available for pressure differentials up to about 83 kPa (12 lbf/in^2) and capacities up to 2.549×10^4 m^3/h (15,000 ft^3/min). Sometimes multiple units are operated in series to produce higher pressures; individual-stage pressure differentials are limited by the shaft deflection, which must necessarily be kept small to maintain rotor and casing clearance.

Screw Type This type of rotary compressor, as shown in Fig. 10-84, is capable of handling capacities up to about 4.248×10^4 m^3/h (25,000 ft^3/min) at pressure ratios of 4:1 and higher. Relatively small-diameter rotors allow rotative speeds of several thousand rpm. Unlike the straight-lobe rotary machine, it has male and female rotors whose rotation causes the axial progression of successive sealed cavities. These machines are staged with intercoolers when such an arrangement is advisable. Their high-speed operation usually necessitates the use of suction- and discharge-noise suppressors. The bearings used are sleeve-type bearings. Due to the side pressures experienced, tilting pad bearings are highly recommended.

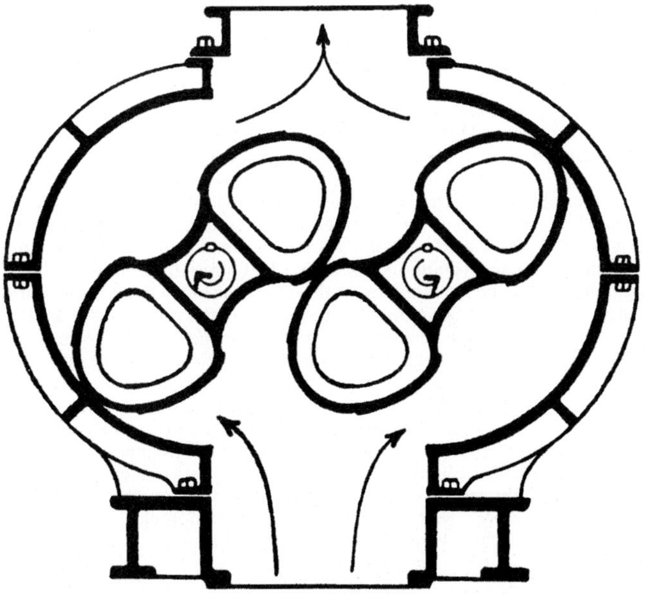

FIG. 10-83 Two-impeller type of rotary positive-displacement blower.

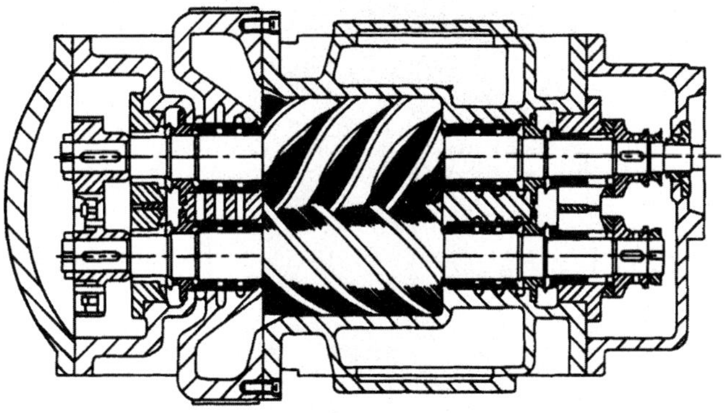

FIG. 10-84 Screw-type rotary compressor.

Sliding-Vane Type This type is illustrated in Fig. 10-85. These units are offered for operating pressures up to 0.86 MPa (125 lbf/in^2) and in capacities up to 3.4×10^3 m^3/h (2000 ft^3/min). Generally, pressure ratios per stage are limited to 4:1. Lubrication of the vanes is required, and the air or gas stream therefore contains lubricating oil.

Liquid-Piston Type This type is illustrated in Fig. 10-86. These compressors are offered as single-stage units for pressure differentials up to about 0.52 MPa (75 lbf/in^2) in the smaller sizes and capacities up to 6.8×10^3 m^3/h (4000 ft^3/min) when used with a lower pressure differential. Staging is employed for higher pressure differentials. These units have found wide application as vacuum pumps on wet-vacuum service. Inlet and discharge ports are located in the impeller hub. As the vaned impeller rotates, centrifugal force drives the sealing liquid against the walls of the elliptical housing, causing the air to be successively drawn into the vane cavities and expelled against discharge pressure. The sealing liquid must be externally cooled unless it is used in a once-through system. A separator is usually employed in the discharge line to minimize carryover of entrained liquid. Compressor capacity can be considerably reduced if the gas is highly soluble in the sealing liquid.

The liquid-piston type of compressor has been of particular advantage when hazardous gases are being handled. Because of the gas-liquid contact and because of the much greater liquid specific heat, the gas temperature rise is very small.

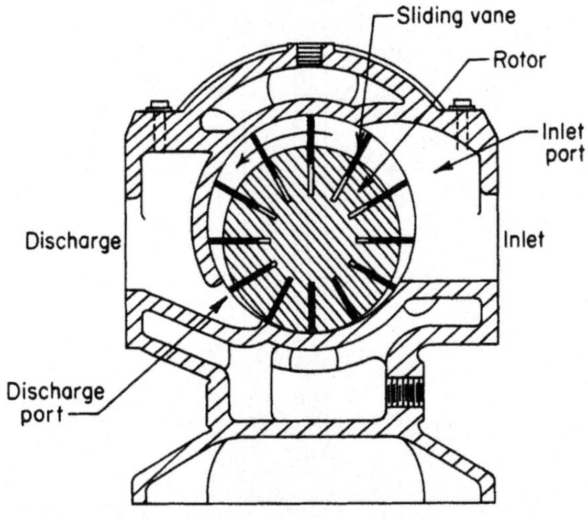

FIG. 10-85 Sliding-vane type of rotary compressor.

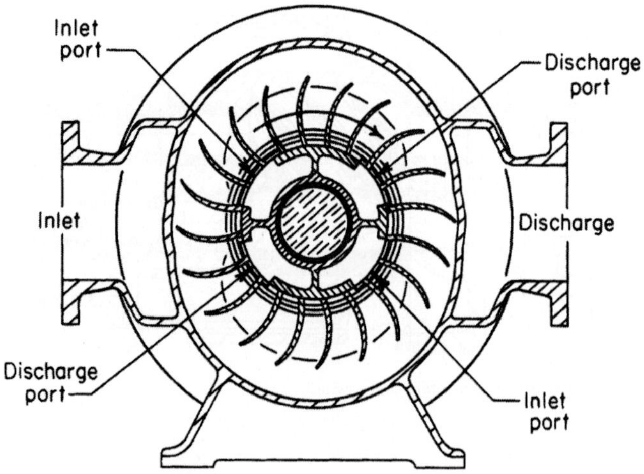

FIG. 10-86 Liquid-piston type of rotary compressor.

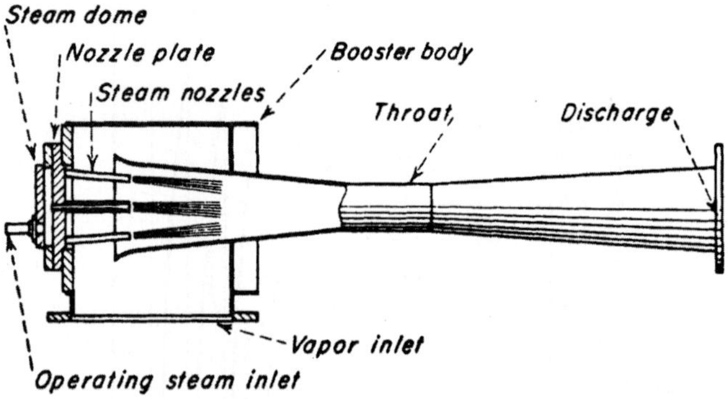

FIG. 10-88 Booster ejector with multiple steam nozzles.

EJECTORS

An ejector is a simplified type of vacuum pump or compressor that has no pistons, valves, rotors, or other moving parts. Figure 10-87 illustrates a steam-jet ejector. It consists essentially of a nozzle which discharges a high-velocity jet across a suction chamber that is connected to the equipment to be evacuated. The gas is entrained by the steam and carried into a venturi-shaped diffuser which converts the velocity energy to pressure energy. Figure 10-88 shows a large ejector, sometimes called a *booster ejector*, with multiple nozzles. Nozzles are devices in subsonic flow that have a decreasing area and accelerate the flow. They convert pressure energy to velocity energy.

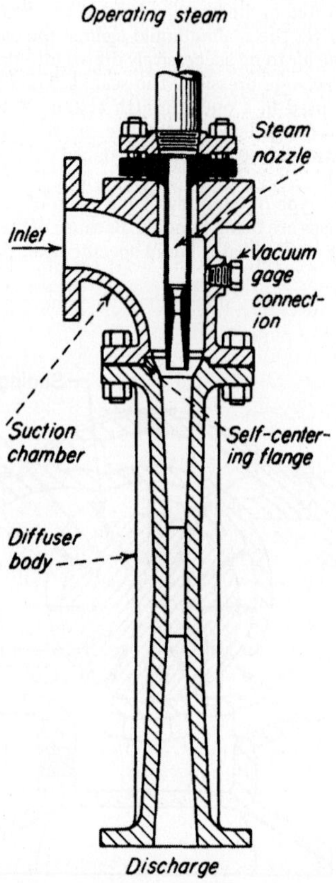

FIG. 10-87 Typical steam-jet ejector.

A minimum area is reached when velocity reaches sonic flow. In supersonic flow, the nozzle is an increasing area device. A diffuser in subsonic flow has an increasing area and converts velocity energy to pressure energy. A diffuser in supersonic flow has a decreasing area.

Two or more ejectors may be connected in series or stages. Also, a number of ejectors may be connected in parallel to handle larger quantities of gas or vapor.

Liquid- or air-cooled condensers are usually used between stages. Liquid-cooled condensers may be of either the direct-contact (barometric) or the surface type. By condensing vapor the load on the following stage is reduced, thus minimizing its size and reducing consumption of motive gas. Likewise, a precondenser installed ahead of an ejector reduces its size and consumption if the suction gas contains vapors that are condensable at the temperature condition available. An *aftercondenser* is frequently used to condense vapors from the final stage, although this does not affect ejector performance.

Ejector Performance The performance of any ejector is a function of the area of the motive-gas nozzle and venturi throat, pressure of the motive gas, suction and discharge pressures, and ratios of specific heats, molecular weights, and temperatures. Figure 10-89, based on the assumption of *constant-area mixing*, is useful in evaluating single-stage ejector performance for compression ratios up to 10 and area ratios up to 100 (see Fig. 10-90 for notation).

For example,* assume that it is desired to evacuate air at 2.94 lb_f/in^2 with a steam ejector discharging to 14.7 lb_f/in^2 with available steam pressure of 100 lb_f/in^2. Entering the chart at $p_{03}/p_{0b} = 5.0$, at $p_{0b}/p_{0a} = 2.94/100 = 0.0294$ the optimum area ratio is 12. Proceeding horizontally to the left, w_b/w_a is approximately 0.15 lb of air per 1 lb of steam. This value must be corrected for the temperature and molecular weight differences of the two fluids by Eq. (10-84).

$$w/w_a = w_b/w_a\sqrt{T_{0a}M_b/T_{0b}M_a} \tag{10-84}$$

In addition, there are empirical correction factors which should be applied. Laboratory tests show that for ejectors with constant-area mixing the actual entrainment and compression ratios will be approximately 90 percent of the calculated values and even less at very small values of p_{0b}/p_{0a}. This compensates for ignoring wall friction in the mixing section and irreversibilities in the nozzle and diffuser. In theory, each point on a given design curve of Fig. 10-89 is associated with an optimum ejector for prevailing operating conditions. Adjacent points on the same curve represent theoretically different ejectors for the new conditions, the difference being that for each ratio of p_{0b}/p_{0a} there is an optimum area for the exit of the motive-gas nozzle. In practice, however, a segment of a given curve for constant A_2/A_t represents the performance of a single ejector satisfactorily for estimating purposes, provided that the suction pressure lies within 20 to 130 percent of the design suction pressure and the motive pressure within 80 to 120 percent of the design motive pressure. Thus the curves can be used to select an optimum ejector for the design point and to estimate its performance at off-design conditions within the limits noted. Final ejector selection should, of course, be made with the assistance of a manufacturer of such equipment.

Uses of Ejectors For the operating range of steam-jet ejectors in vacuum applications, see the subsection Vacuum Systems below.

*All data are given in USCS units since the charts are in these units. Conversion factors to SI units are given on the charts.

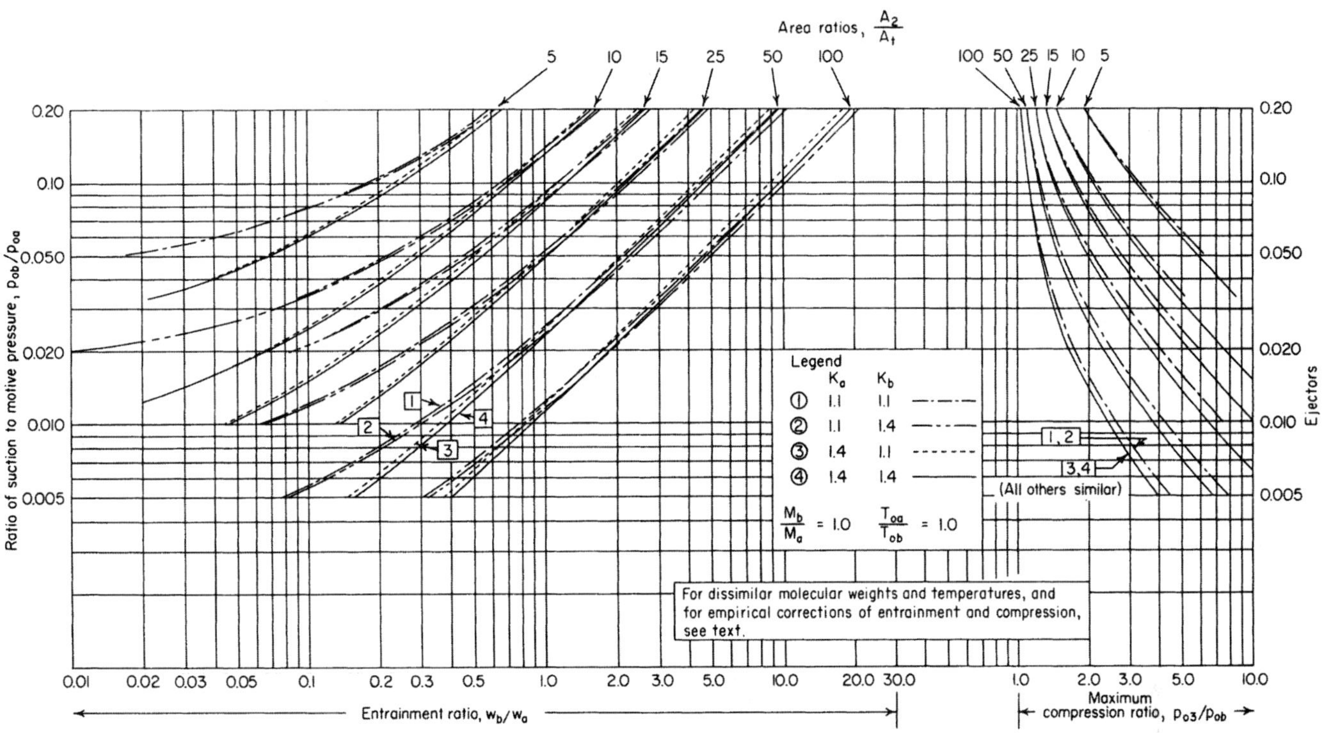

FIG. 10-89 Design curves for optimum single-stage ejectors. [DeFrate and Hoerl, *Chem. Eng. Prog.* **55,** Symp. Ser. 21, 46 (1959). Reprinted with permission, American Institute of Chemical Engineers. All rights reserved.]

The choice of the most suitable type of ejector for a given application depends upon the following factors:

1. *Steam pressure.* Ejector selection should be based on the minimum pressure in the supply line selected to serve the unit.
2. *Water temperature.* Selection is based on the maximum water temperature.
3. *Suction pressure and temperature.* Overall process requirements should be considered. Selection is usually governed by the minimum suction pressure required (the highest vacuum).
4. *Capacity required.* Again, overall process requirements should be considered, but selection is usually governed by the capacity required at the minimum process pressure.

Ejectors are easy to operate and require little maintenance. Installation costs are low. Since they have no moving parts, they have long life, sustained efficiency, and low maintenance cost. Ejectors are suitable for handling practically any type of gas or vapor. They are also suitable for handling wet or dry mixtures or gases containing sticky or solid matter such as chaff or dust.

Ejectors are available in many materials of construction to suit process requirements. If the gases or vapors are not corrosive, the diffuser is usually constructed of cast iron and the steam nozzle of stainless steel. For more corrosive gases and vapors, many combinations of materials such as bronze, various stainless-steel alloys, and other corrosion-resistant metals, carbon, and glass can be used.

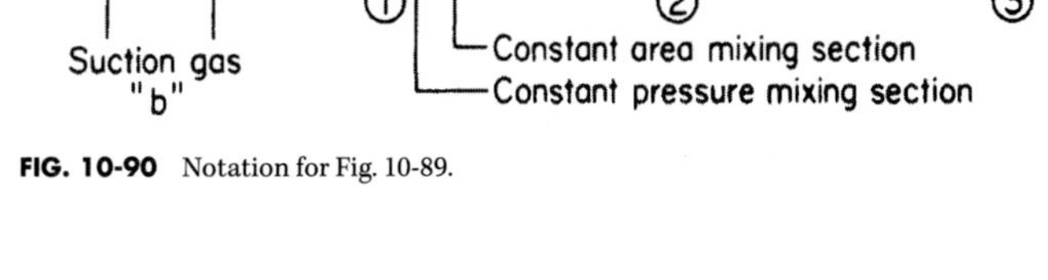

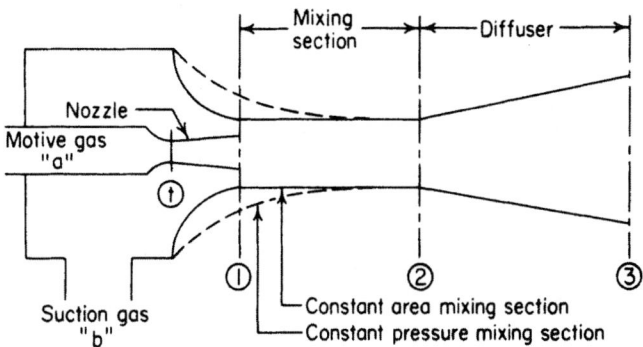

FIG. 10-90 Notation for Fig. 10-89.

VACUUM SYSTEMS

Figure 10-91 illustrates the level of vacuum normally required to perform many of the common manufacturing processes. The attainment of various levels is related to available equipment in Fig. 10-92.

Vacuum Equipment The equipment shown in Fig. 10-92 has been discussed elsewhere in this section with the exception of the *diffusion pump.* Figure 10-93 depicts a typical design. A liquid of low absolute vapor pressure is boiled in the reservoir. The vapor is ejected at high velocity in a downward direction through multiple jets and is condensed on the walls, which are cooled by the surrounding coils. Molecules of the gas being pumped enter the vapor stream and are driven downward by collisions with the vapor molecules. The gas molecules are removed through the discharge line by a backing pump such as a rotary oil-sealed unit.

Diffusion pumps operate at very low pressures. The ultimate vacuum attainable depends somewhat on the vapor pressure of the pump liquid at the temperature of the condensing surfaces. By providing a cold trap between the diffusion pump and the region being evacuated, pressures as low as 10^{-7} mmHg absolute are achieved in this manner. Liquids used for diffusion pumps are mercury and oils of low vapor pressure. Silicone oils have excellent characteristics for this service.

SEALING OF ROTATING SHAFTS

Seals are very important and often critical components in large rotating machinery especially on high-pressure and high-speed equipment. The principal sealing systems used between the rotor and stationary elements fall into two main categories: (1) noncontacting seals and (2) face seals. These seals are an integral part of the rotating system; they affect the dynamic operating characteristics of the machine. The stiffness and damping factors will be changed by the seal geometry and pressures. In operation the rotating shafts have both radial and axial movement. Therefore, any seal must be flexible and compact to ensure maximum sealing with minimum effect on rotor dynamics.

Noncontacting Seals Noncontacting seals are used extensively in gas service in high-speed rotating equipment. These seals have good mechanical reliability and minimum impact on the rotor dynamics of the system. They are not positive sealing. There are two types of noncontacting seals: (1) labyrinth seals and (2) ring seals.

Labyrinth Seals The labyrinth is one of the simplest of the many sealing devices. It consists of a series of circumferential strips of metal extending from the shaft or from the bore of the shaft housing to form a cascade

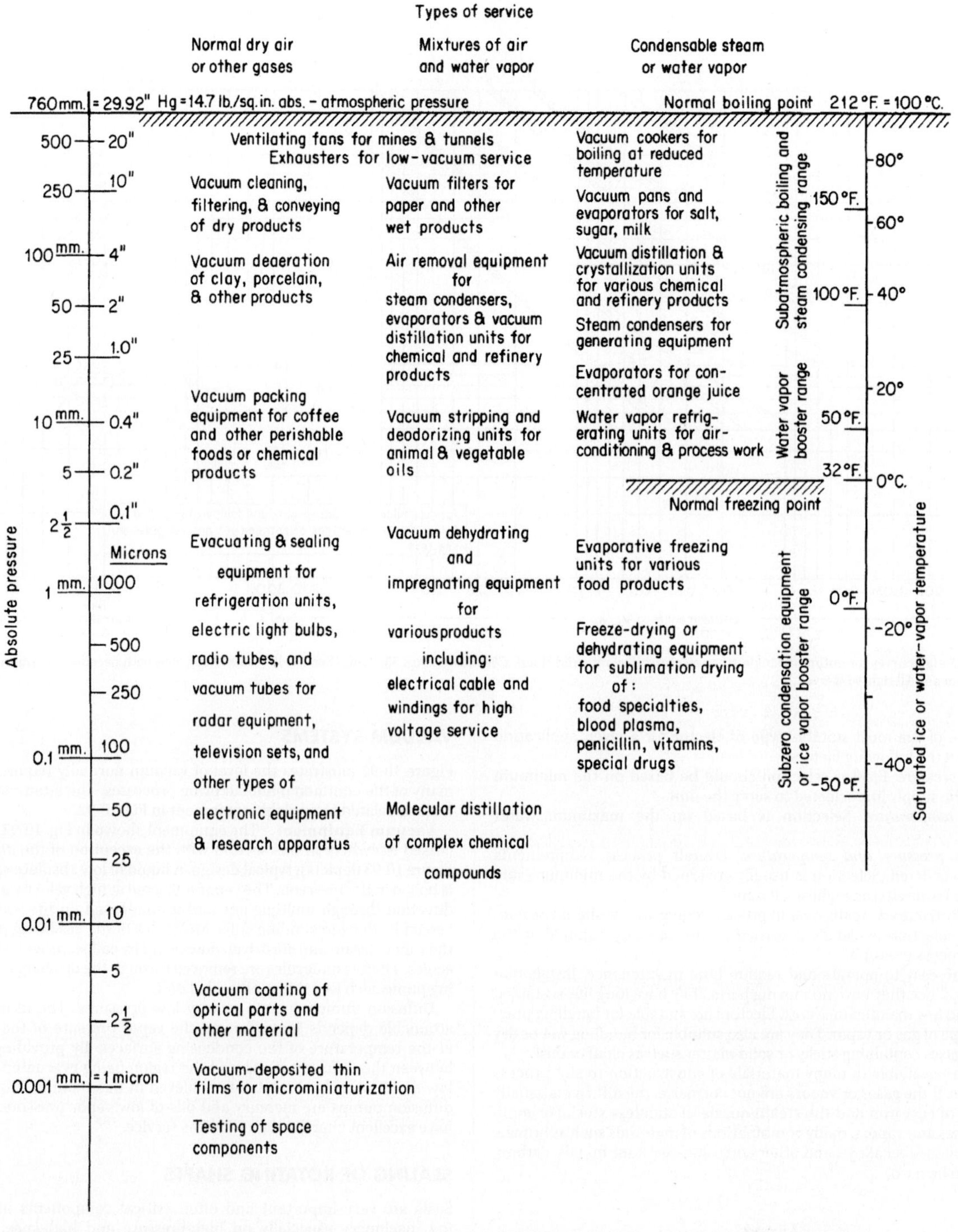

FIG. 10-91 Vacuum levels normally required to perform common manufacturing processes. (*Courtesy of* Compressed Air *magazine.*)

of annular orifices. Labyrinth seal leakage is greater than that of clearance bushings, contact seals, or film riding seals.

The major advantages of labyrinth seals are their simplicity, reliability, tolerance to dirt, system adaptability, very low shaft power consumption, material selection flexibility, minimal effect on rotor dynamics, back-diffusion reduction, integration of pressure, lack of pressure limitations, and tolerance to gross thermal variations. The major disadvantages are the high leakage; loss of machine efficiency; increased buffering costs; tolerance to ingestion of particulates with resulting damage to other critical items such as bearings; the possibility of the cavity clogging due to low gas velocities or back diffusion; and the inability to provide a simple seal system that meets OSHA or EPA standards. Because of some of the foregoing disadvantages, many machines are being converted to other types of seals.

Labyrinth seals are simple to manufacture and can be made from conventional materials. Early designs of labyrinth seals used knife-edge seals and relatively large chambers or pockets between the knives. These relatively long knives are easily subject to damage. The modern, more functional, and more reliable labyrinth seals consist of sturdy, closely spaced lands. Some labyrinth seals are shown in Fig. 10-94. Figure 10-94*a* is the simplest form of the seal. Figure 10-94*b* shows a grooved seal; it is more difficult to manufacture but produces a tighter seal. Figure 10-94*c* and Fig. 10-94*d* are rotating labyrinth-type seals. Figure 10-94*e* shows a simple labyrinth seal with a buffered gas for which pressure must be maintained above the process gas pressure and the outlet pressure (which can be greater than or less than the atmospheric pressure). The buffered gas produces a fluid barrier to the process gas. The eductor sucks gas from the vent near the atmospheric end.

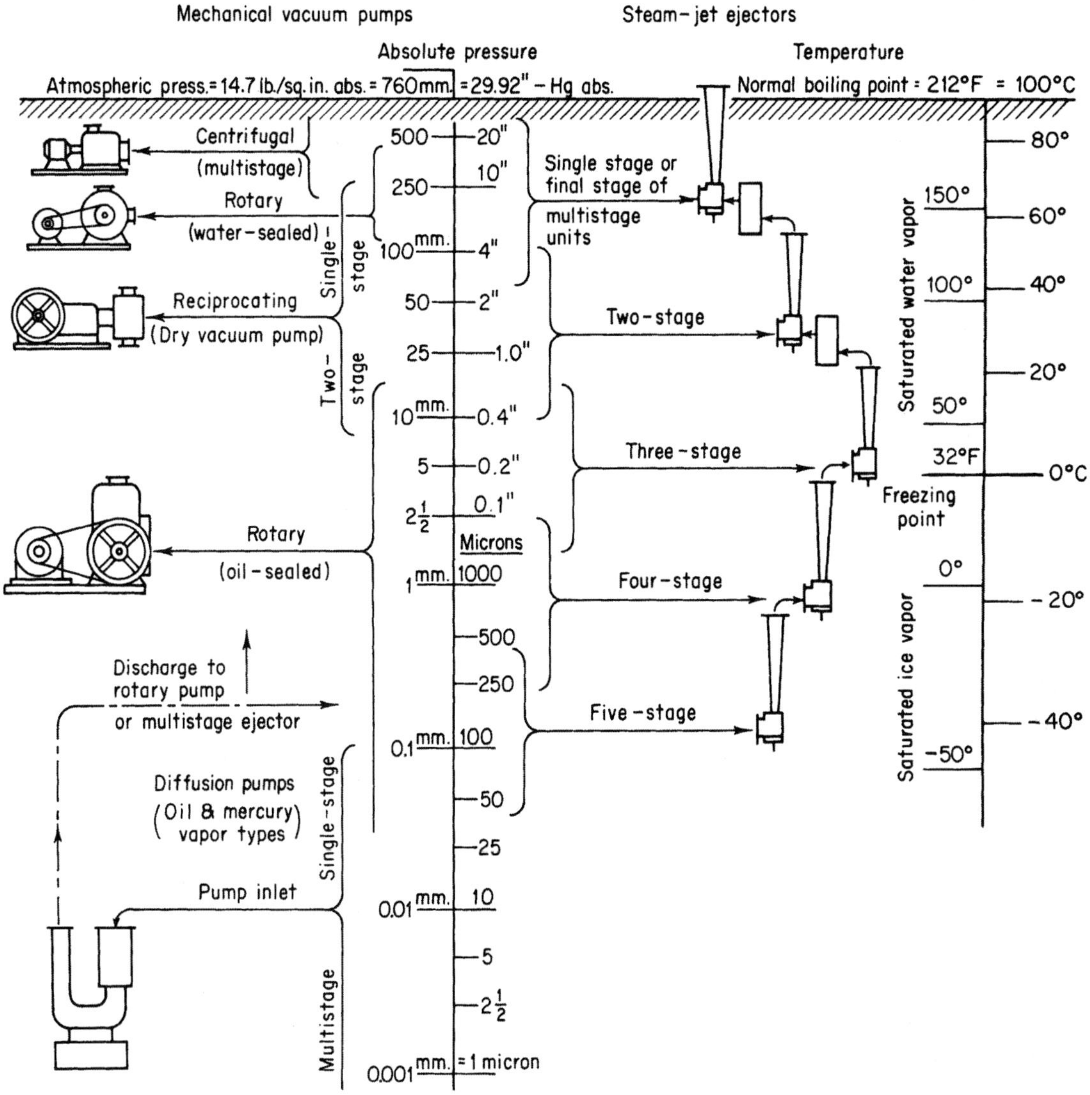

FIG. 10-92 Vacuum levels attainable with various types of equipment. (*Courtesy of* Compressed Air *magazine.*)

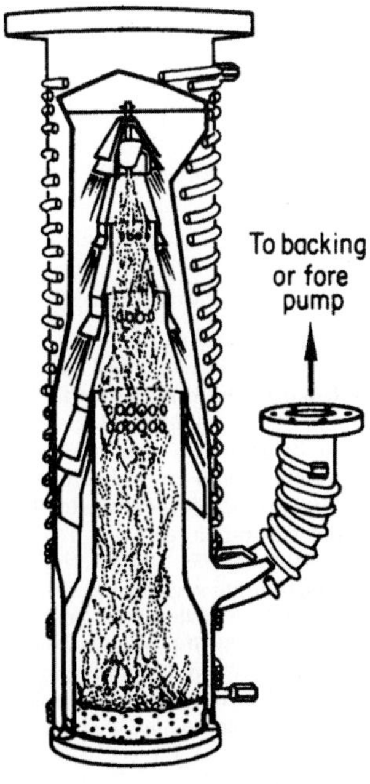

FIG. 10-93 Typical diffusion pump. (*Courtesy of* Compressed Air *magazine.*)

Figure 10-94*f* shows a buffered, stepped labyrinth. The step labyrinth gives a tighter seal. The matching stationary seal is usually manufactured from soft materials such as babbitt or bronze, while the stationary or rotating labyrinth lands are made from steel. This composition enables the seal to be assembled with minimal clearance. The lands can therefore cut into the softer materials to provide the necessary running clearances for adjusting to the dynamic excursions of the rotor. To maintain maximum sealing efficiency, it is essential that the labyrinth lands maintain sharp edges in the direction of the flow.

Leakage past these labyrinths is approximately inversely proportional to the square root of the number of labyrinth lands. This translates to the following relationship if leakage is to be cut in half in a four-labyrinth seal: The number of labyrinths would have to be increased to 16. The Elgi leakage formula can be modified and written as

$$m_\ell = AK\left[\frac{(g/V_o)(P_o - P_n)}{n + \ln(P_n/P_o)}\right]^{1/2} \tag{10-85}$$

where m_ℓ = leakage
A = leakage area of single throttling
K = labyrinth constant ($K = 0.9$ for straight labyrinths, $K = 0.75$ for staggered labyrinths)
P_o = absolute pressure before the labyrinth
P_n = absolute pressure after the last labyrinth
V_o = specific volume before the labyrinth
n = number of lands

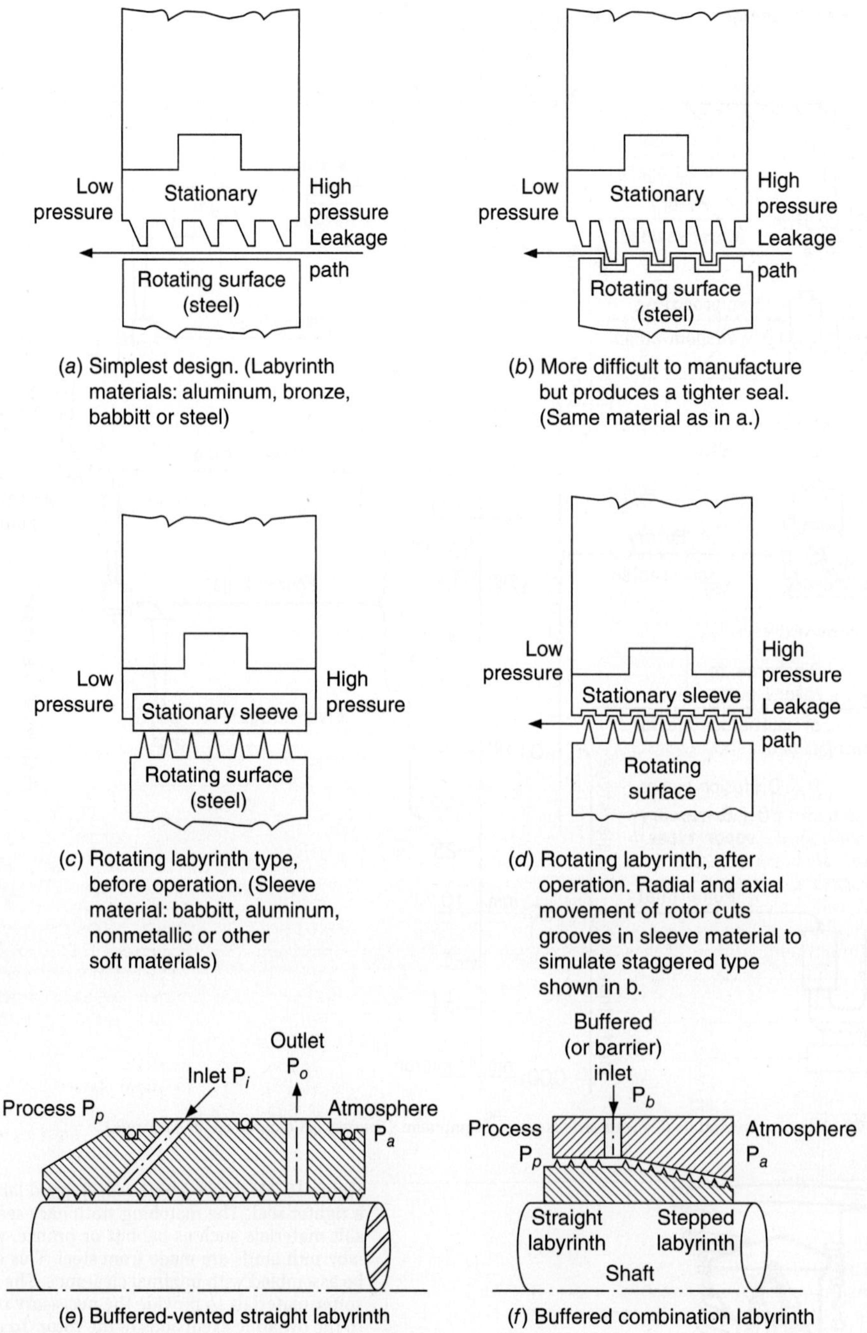

FIG. 10-94 Various configurations of labyrinth seals.

The leakage of a labyrinth seal can be kept to a minimum by providing (1) minimum clearance between the seal lands and the seal sleeve, (2) sharp edges on the lands to reduce the flow discharge coefficient, and (3) grooves or steps in the flow path for reducing dynamic head carryover from stage to stage.

The labyrinth sleeve can be flexibly mounted to permit radial motion for self-aligning effects. In practice, a radial clearance under 0.008 is difficult to achieve.

Ring Seals The restrictive ring seal is essentially a series of sleeves in which the bores form a small clearance around the shaft. Thus, the leakage is limited by the flow resistance in the restricted area and controlled by the laminar or turbulent friction. There are two types of ring seals: (1) fixed seal rings and (2) floating seal rings. The floating rings permit a much smaller leakage; they can be either the segmented type, as shown in Fig. 10-95*a*, or the rigid type, as shown in Fig. 10-95*b*.

Fixed Seal Rings The fixed seal ring consists of a long sleeve affixed to a housing in which the shaft rotates with small clearances. Long assemblies must be used to keep leakage within a reasonable limit. Long seal assemblies aggravate alignment and rubbing problems, thus requiring shafts to operate below their capacity. The fixed bushing seal operates with appreciable eccentricity and, combined with large clearances, produces large leakages, thus making this kind of seal impractical where leakage is undesirable.

Floating Seal Rings Clearance seals that are free to move in a radial direction are known as *floating seals*. The floating characteristics permit them to move freely, thus avoiding severe rubs. Due to differential thermal expansion between the shaft and bushing, the bushings should be made of material with a higher coefficient of thermal expansion. This is achieved by shrinking the carbon into a metallic retaining ring with a coefficient of expansion that equals or exceeds that of the shaft material. It is advisable in high-shearing applications to lock the bushings against rotation.

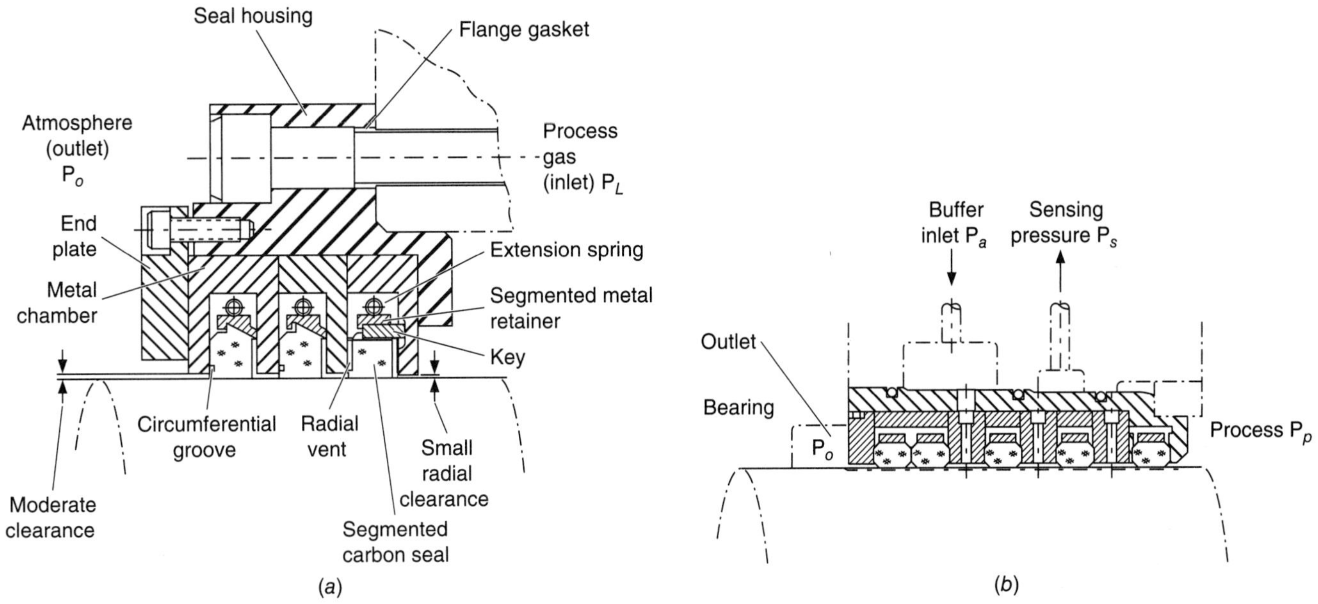

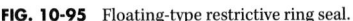

FIG. 10-95 Floating-type restrictive ring seal.

Buildup of dirt and other foreign material lodged between the seal ring and seat will create an excessive spin and damage on the floating seal ring unit. It is therefore improper to use soft material such as babbitt and silver as seal rings.

Packing Seal A common type of rotating shaft seal consists of packing composed of fibers which are first woven, twisted, or braided into strands and then formed into coils, spirals, or rings. To ensure initial lubrication and to facilitate installation, the basic materials are often impregnated. Common materials are braided and twisted rubber and duck, flax, jute, and metallic braids. The so-called plastic packings can be made up with varying amounts of fiber combined with a binder and lubricant for high-speed applications. The maximum temperatures that base materials of packings withstand and still give good service are as follows:

	°C	°F
Flax	38	100
Cotton	93	200
Duck and rubber	149	300
Rubber	177	350
Metallic (lead-based)	218	425
Metallic (aluminum-based)	552	1025
Metallic (copper-based)	829	1525

Packing may not provide a completely leak-free seal. With shaft surface speeds less than approximately 2.5 m/s (500 ft/min), the packing may be adjusted to seal completely. However, for higher speeds some leakage is required for lubrication, friction reduction, and cooling.

Application of Packing Coils and spirals are cut to form closed or nearly closed rings in the stuffing box. Clearance between ends should be sufficient to allow for fitting and possible expansion due to increased temperature or liquid absorption of the packing while in operation.

The correct form of the ring joint depends on materials and service requirements. Braided and flexible metallic packings usually have butt or square joints (Fig. 10-96*a*). With other packing material, service experience indicates that rings cut with bevel or skive joints (Fig. 10-96*b*) are more satisfactory. A slight advantage of the bevel joint over the butt joint is that the bevel permits a certain amount of sliding action, thus absorbing a portion of ring expansion.

In the manufacture of packings, the proper grade and type of *lubricant* are usually impregnated for each service for which the packing is recommended. However, it may be desirable to replenish the lubricant during the normal life of the packing. Lack of lubrication causes packing to become hard and lose its resiliency, thus increasing friction, shortening packing life, and increasing operating costs.

An effective auxiliary device frequently used with packing and rotary shafts is the *seal cage* (or *lantern ring*), shown in Fig. 10-97. The seal cage provides an annulus around the shaft for the introduction of a lubricant, oil, grease, etc. The seal cage is also used to introduce liquid for cooling, to prevent the entrance of atmospheric air, or to prevent the infiltration of abrasives from the process liquid.

The chief advantage of packing over other types of seals is the ease with which it can be adjusted or replaced. Most equipment is designed so that disassembly of major components is not required to remove or add packing rings. The major disadvantages of a packing-type seal are (1) its short life, (2) the requirement for frequent adjustment, and (3) the need for some leakage to provide lubrication and cooling.

Mechanical Face Seals This type of seal forms a running seal between flat precision-finished surfaces. It is an excellent seal against leakages. The sealing surfaces are planes perpendicular to the rotating shaft, and the forces that hold the contact faces are parallel to the shaft axis. For a seal to function properly, there are four sealing points:

1. Stuffing box face
2. Leakage down the shaft
3. Mating ring in the gland plate
4. Dynamic faces

Mechanical Seal Selection Many factors govern the selection of seals. These factors apply to any type of seal:

1. Product
2. Seal environment
3. Seal arrangement
4. Equipment
5. Secondary packing
6. Seal face combinations
7. Seal gland plate
8. Main seal body

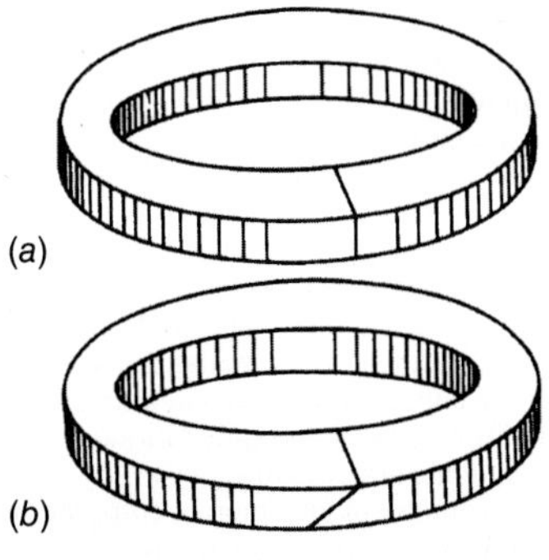

FIG. 10-96 Butt (*a*) and skive (*b*) joints for compression packing rings.

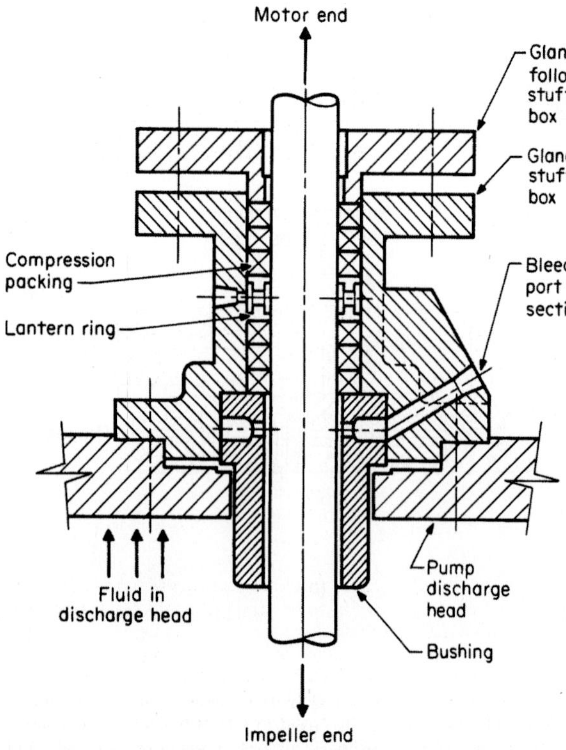

FIG. 10-97 Seal cage or lantern ring. (*Courtesy of Crane Packing Co.*)

Product Physical and chemical properties of the liquid or gas being sealed place constraints on the type of material, design, and arrangement of the seal.

Pressure. Pressure affects the choice of material and whether balanced or unbalanced seal design can be used. Most unbalanced seals are good up to 100 psig stuffing box pressure. Over 100 psig, balanced seals should be used.

Temperature. The temperature of the liquid being pumped is important because it affects the seal face material selection as well as the wear life of the seal face.

Lubricity. In any mechanical seal design, there is rubbing motion between the dynamic seal faces. This rubbing motion is often lubricated by the fluid being pumped. Most seal manufacturers limit the speed of their seals to 90 ft/s (30 m/s). This is primarily due to centrifugal forces acting on the seal, which tend to restrict the seal's axial flexibility.

Abrasion. If there are entrained solids in the liquid, it is desirable to have a flushed single inside type with a face combination of very hard material.

Corrosion. This affects the type of seal body: what spring material, what face material, and what type of elastomer or gasket material. The corrosion rate will affect the decision of whether to use a single- or multiple-spring design because the spring can usually tolerate a greater amount of corrosion without weakening appreciably.

Seal Environment The design of the seal environment is based on the product and the four general parameters that regulate it:

1. Pressure control
2. Temperature control
3. Fluid replacement
4. Atmospheric air elimination

Seal Arrangement There are four types of seal arrangements:

1. Double seals are standard with toxic and lethal products, but maintenance problems and seal design contribute to poor reliability. The double face-to-face seal may be a better solution.

2. Do not use a double seal in dirty service—the inside seal will hang up.

3. API standards for balanced and unbalanced seals are good guidelines; too low a pressure for a balanced seal may encourage face liftoff.

4. Arrangement of the seal will determine its success more than the vendor. Over 100 arrangements are available.

Equipment The geometry of the pump or compressor is very important in seal effectiveness. Different pumps with the same shaft diameter and total differential head can present different sealing problems.

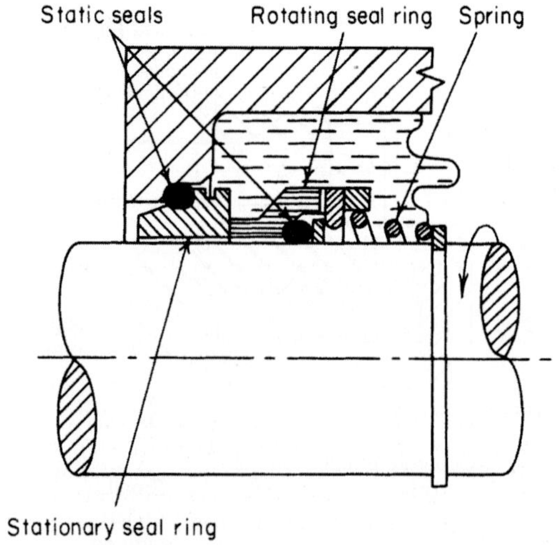

FIG. 10-98 Mechanical-seal components.

Secondary Packing Much greater emphasis should be placed on secondary packing, especially if Teflon is used. A wide variation in performance is seen between various seal vendors, and depending on the seal arrangement there can be differences in mating ring packing.

Seal Face Combinations The dynamics of seal faces are better understood today. Seal face combinations have come a long way in the past 8 to 10 years. Stellite is being phased out of the petroleum and petrochemical applications. Better grades of ceramic are available, the cost of tungsten has come down, and relapping of tungsten is available near most industrial areas. Silicon carbide is being used in abrasive service.

Seal Gland Plate The seal gland plate is caught in between the pump vendor and the seal vendor. Special glands should be furnished by seal vendors, especially if they require heating, quenching, and drain with a floating-throat bushing. Gland designs are complex and may have to be revisited, especially if seals are changed.

Main Seal Body The term *seal body* makes reference to all rotating parts on a pusher seal, excluding the shaft packing and seal ring. In many cases it is the chief reason to avoid a particular design for a particular service.

Basically, most mechanical seals have the following components, as seen in Fig. 10-98.

1. Rotating seal ring
2. Stationary seal ring
3. Spring devices to provide pressure
4. Static seals

A loading device such as a spring is needed to ensure that the sealing surfaces are kept closed in the event of loss of hydraulic pressure. The amount of the load on the sealing area is determined by the degree of "seal balance." Figure 10-99 shows what seal balance means. A completely balanced seal

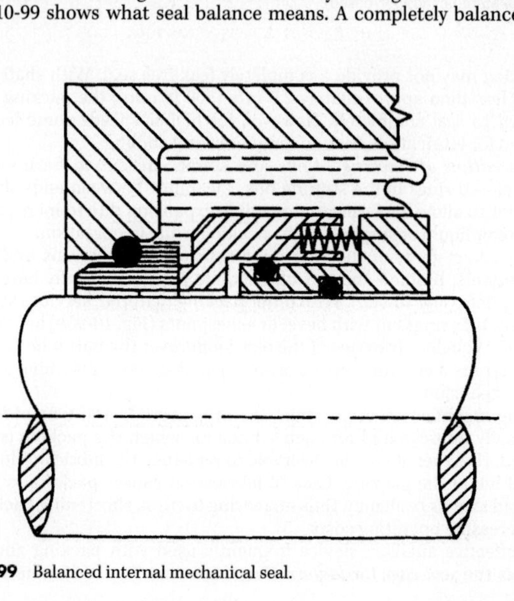

FIG. 10-99 Balanced internal mechanical seal.

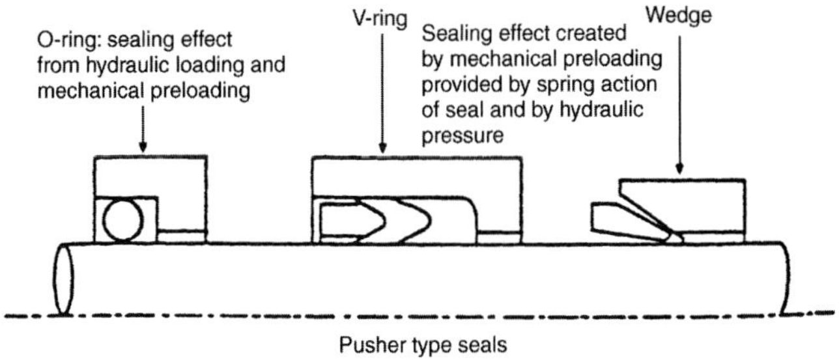

FIG. 10-100 Various types of shaft-sealing elements.

exists when the only force exerted on the sealing surfaces is the spring force; i.e., hydraulic pressure does not act on the sealing surface. The type of spring depends on the space available, loading characteristics, and seal environment. Based on these considerations, either a single spring or multiple springs can be used. In small axial space, Belleville springs, finger washers, or curved washers can be used.

Shaft-sealing elements can be split up into two groups. The first type may be called *pusher-type* seals and includes the O-ring, V-ring, U-cup, and wedge configurations. Figure 10-100 shows some typical pusher-type seals. The second type is the bellow-type seals, which differ from the pusher-type seals in that they form a static seal between themselves and the shaft.

Internal and External Seals Mechanical seals are classified broadly as internal or external. *Internal seals* (Fig. 10-101) are installed with all seal components exposed to the fluid sealed. The advantages of this arrangement are (1) the ability to seal against high pressure, since the hydrostatic force is normally in the same direction as the spring force; (2) protection of seal parts from external mechanical damage; and (3) reduction in the shaft length required.

For high-pressure installations, it is possible to balance partially or fully the hydrostatic force on the rotating member of an internal seal by using a stepped shaft or shaft sleeve (Fig. 10-99). This method of relieving face pressure is an effective way of decreasing power consumption and extending seal life.

When abrasive solids are present and it is not permissible to introduce appreciable quantities of a secondary flushing fluid into the process, double internal seals are sometimes used (Fig. 10-102). Both sealing faces are protected by the flushing fluid injected between them even though the inward flow is negligible.

External seals (Fig. 10-103) are installed with all seal components protected from the process fluid. The advantages of this arrangement are that (1) fewer critical materials of construction are required, (2) the installation and setting are somewhat simpler because of the exposed position of the parts, and (3) stuffing-box size is not a limiting factor. Hydraulic balancing is accomplished by proper proportioning of the seal face and secondary seal diameters.

Throttle Bushings These bushings (Fig. 10-104) are commonly used with single internal or external seals when solids are present in the fluid and the inflow of a flushing fluid is not objectionable. These close-clearance bushings are intended to serve as flow restrictions through which the maintenance of a small inward flow of flushing fluid prevents the entrance of a process fluid into the stuffing box.

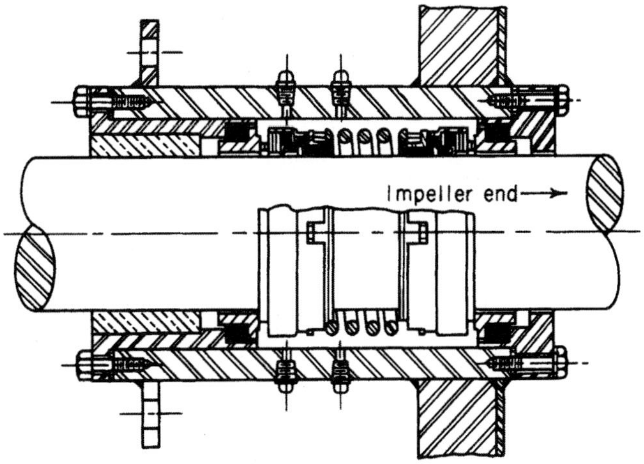

FIG. 10-102 Internal bellows-type double mechanical seal.

A typical complex seal utilizes both the noncontact and mechanical aspects of sealing. Figure 10-105 shows such a seal with its two major elements. This type of seal will normally have buffering via a labyrinth seal and a positive shutdown device. For shutdown, the carbon ring is tightly sandwiched between the rotating seal ring and the stationary sleeve, with gas pressure to prevent gas from leaking out when no oil pressure is available.

In operation the seal oil pressure is about 30 to 50 psi over the process gas pressure. The high-pressure oil enters the top and completely fills the seal cavity. A small percentage is forced across the carbon ring seal faces. The rotative speed of the carbon ring can be anywhere between zero and full rotational speed. Oil crossing the seal faces contacts the process gas and therefore is "contaminated oil." The contaminated oil leaves through the contaminated oil drain to a degassifier for purification. The majority of the oil flows through the uncontaminated seal oil drain.

Materials Springs and other metallic components are available in a wide variety of alloys and are usually selected on the basis of temperature and corrosion conditions. The use of a particular mechanical seal is frequently restricted by the temperature limitations of the organic materials used in the static seals. Most elastomers are limited to about 121°C (250°F). Teflon will withstand temperatures of 260°C (500°F) but softens appreciably above 204°C (400°F). Glass-filled Teflon is dimensionally stable up to 232 to 260°C (450 to 500°F).

One of the most common elements used for seal faces is carbon. Although compatible with most process media, carbon is affected by strong oxidizing agents, including fuming nitric acid, hydrogen chloride, and high-temperature air [above 316°C (600°F)]. Normal mating-face materials for carbon are tungsten or chromium carbide, hard steel, stainless steel, or one of the cast irons.

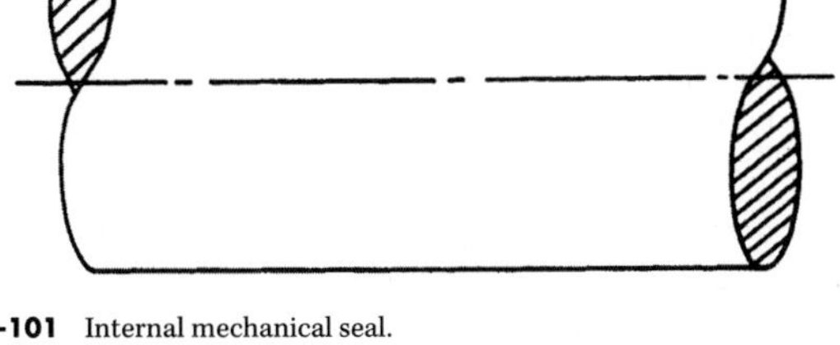

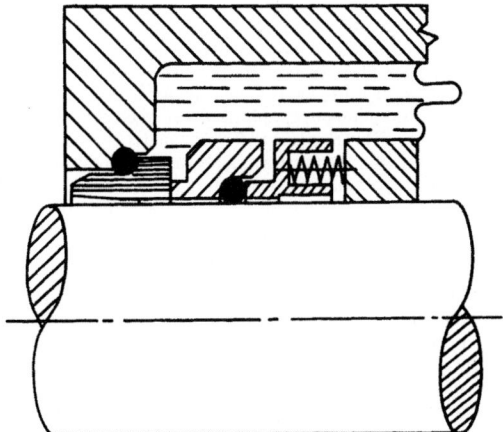

FIG. 10-101 Internal mechanical seal.

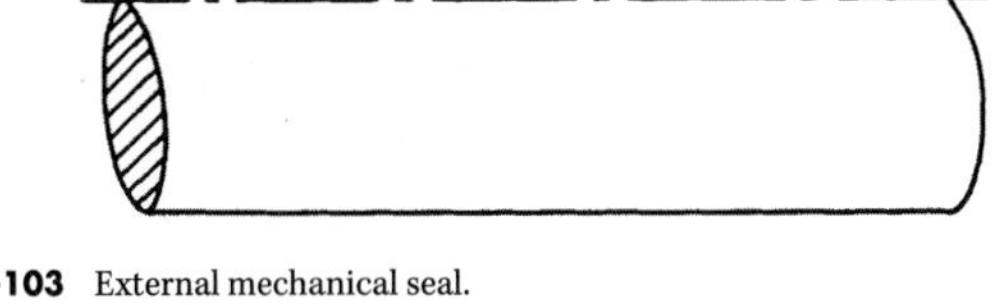

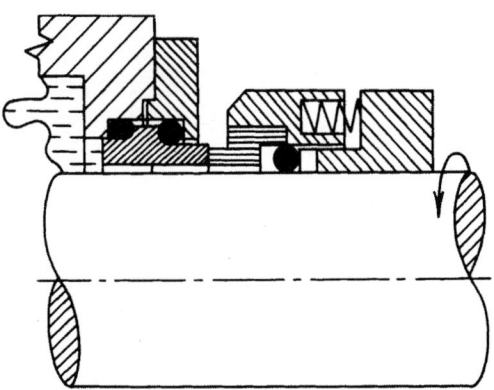

FIG. 10-103 External mechanical seal.

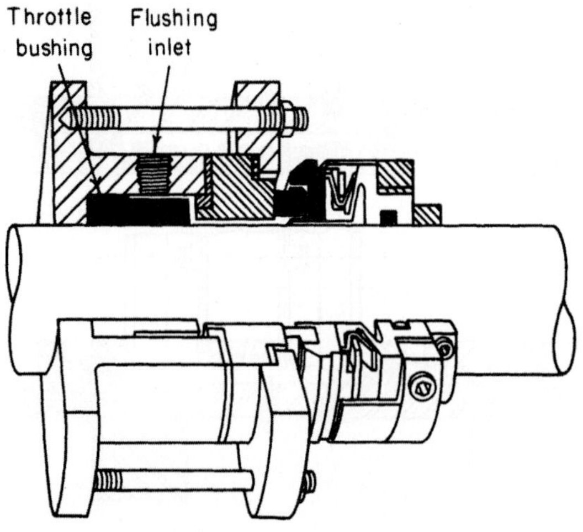

FIG. 10-104 External mechanical seal and throttle bushing.

Other sealing-face combinations that have been satisfactory in corrosive service are carbide against carbide, ceramic against ceramic, ceramic against carbon, and carbon against glass. The ceramics have also been mated with the various hard-facing alloys. When one is selecting seal materials, the possibility of galvanic corrosion must also be considered.

BEARINGS

Many factors enter into the selection of the proper design for bearings. These are some of the factors:

1. Shaft speed range
2. Maximum shaft misalignment that can be tolerated
3. Critical speed analysis and the influence of bearing stiffness on this analysis
4. Loading of the compressor impellers

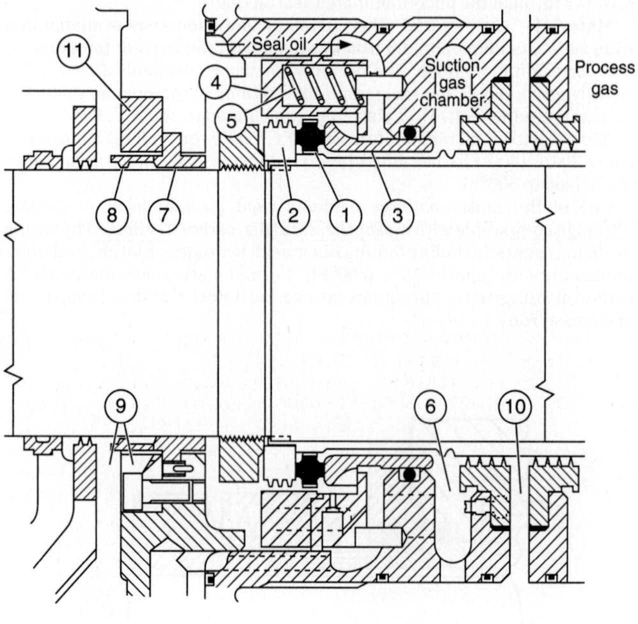

1. Rotating carbon ring
2. Rotating seal ring
3. Stationary sleeve
4. Spring retainer
5. Spring
6. Gas and contaminated oil drain
7. Floating babbitt-faced steel ring
8. Seal wiper ring
9. Seal oil drain line
10. Buffer gas injection port
11. Bypass orifice

FIG. 10-105 Mechanical contact shaft seal.

5. Oil temperatures and viscosity
6. Foundation stiffness
7. Axial movement that can be tolerated
8. Type of lubrication system and its contamination
9. Maximum vibration levels that can be tolerated

Types of Bearings Figure 10-106 shows a number of different types of journal bearings. A description of a few of the pertinent types of journal bearings is given here.

1. *Plain journal.* The bearing is bored with equal amounts of clearance (on the order of 0.0015 to 0.002 in per inch of journal diameter) between the journal and bearing.
2. *Circumferential grooved bearing.* Normally the oil groove is one-half the bearing length. This configuration provides better cooling but reduces load capacity by dividing the bearing into two parts.
3. *Cylindrical bore bearings.* This is another common bearing type used in turbines. It has a split construction with two axial oil-feed grooves at the split.
4. *Pressure or pressure dam.* Used in many places where bearing stability is required, this bearing is a plain journal bearing with a pressure pocket cut in the unloaded half. This pocket is approximately $\frac{1}{32}$ in deep with a width 50 percent of the bearing length. This groove or channel covers an arc of 135° and terminates abruptly in a sharp edge at the dam edge. The direction or rotation is such that the oil is pumped down the channel toward the sharp edge. Pressure dam bearings are for one direction of rotation. They can be used in conjunction with cylindrical bore bearings as shown in Fig. 10-106.
5. *Lemon bore or elliptical.* This bearing is bored with shims split line, which are removed before installation. The resulting shape approximates an ellipse with the major axis clearance approximately twice the minor axis clearance. Elliptical bearings are for both directions of rotation.
6. *Three-lobe bearing.* The three-lobe bearing is not commonly used in turbomachines. It has a moderate load-carrying capacity and can be operated in both directions.
7. *Offset halves.* In principle, this bearing acts very similar to a pressure dam bearing. Its load-carrying capacity is good. It is restricted to one direction of rotation.
8. *Tilt-pad bearings.* This bearing is the most common bearing type in today's machines. It consists of several bearing pads posed around the circumference of the shaft. Each pad is able to tilt to assume the most effective working position. This bearing also offers the greatest increase in fatigue life because of the following advantages:
   - Thermal conductive backing material dissipates heat developed in the oil film.
   - A thin babbitt layer can be centrifugally cast with a uniform thickness of about 0.005 in. Thick babbitts greatly reduce bearing life. Babbitt thickness in the neighborhood of 0.01 in reduces the bearing life by more than one-half.
   - Oil film thickness is critical in bearing stiffness calculations. In a tilt-pad bearing, one can change this thickness in a number of ways: change the number of pads; direct the load on or in between the pads; change the axial length of the pad.

The previous list contains some of the most common types of journal bearings. They are listed in order of growing stability. All the bearings designed for increased stability are obtained at higher manufacturing costs and reduced efficiency. The antiwhirl bearings all impose a parasitic load on the journal, which causes higher-power losses to the bearings and in turn requires higher oil flow to cool the bearing.

Thrust Bearings The most important function of a thrust bearing is to resist the unbalanced force in a machine's working fluid and to maintain the rotor in its position (within prescribed limits). A complete analysis of the thrust load must be conducted. As mentioned earlier, compressors with back-to-back rotors reduce this load greatly on thrust bearings. Figure 10-107 shows a number of thrust bearing types. Plain, grooved thrust washers are rarely used with any continuous load, and their use tends to be confined to cases where the thrust load is of very short duration or possibly occurs at standstill or low speed only. Occasionally, this type of bearing is used for light loads (less than 50 lb/in^2), and in these circumstances the operation is probably hydrodynamic due to small distortions present in the nominally flat bearing surface.

When significant continuous loads have to be taken on a thrust washer, it is necessary to machine into the bearing surface a profile to generate a fluid film. This profile can be either a tapered wedge or occasionally a small step.

The tapered-land thrust bearing, when properly designed, can take and support a load equal to that of a tilt-pad thrust bearing. With perfect alignment, it can match the load of even a self-equalizing tilt-pad thrust bearing that pivots on the back of the pad along a radial line. For variable-speed operation, tilt-pad thrust bearings, as shown in Fig. 10-108, are advantageous when compared to conventional taper-land bearings. The pads are free to pivot to form a proper angle for lubrication over a wide speed range. The self-leveling feature equalizes individual pad loadings and reduces the sensitivity to shaft misalignments that may occur during service. The major

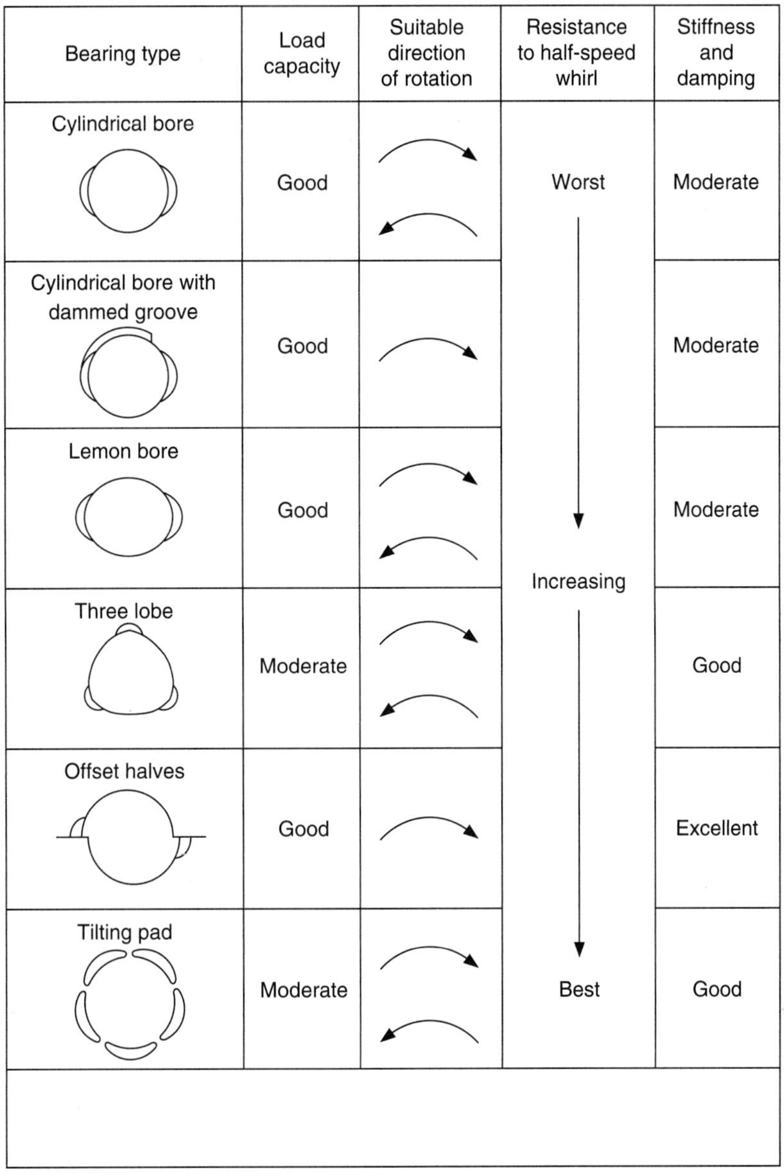
Bearing type
Load capacity
Suitable direction of rotation
Resistance to half-speed whirl
Stiffness and damping
Cylindrical bore
Good
Worst
Moderate
Cylindrical bore with dammed groove
Good
Moderate
Lemon bore
Good
Moderate
Increasing
Three lobe
Moderate
Good
Offset halves
Good
Excellent
Tilting pad
Moderate
Best
Good

FIG. 10-106 Comparison of general bearing types.

drawback of this bearing type is that standard designs require more axial space than a nonequalizing thrust bearing.

The thrust-carrying capacity can be greatly improved by maintaining pad flatness and removing heat from the loaded zone. By the use of high-thermal-conductivity backing materials with proper thickness and proper support, the maximum continuous thrust limit can be increased to 1000 psi or more. This new limit can be used to increase the factor of safety and improve the surge capacity of a given size bearing or reduce the thrust bearing size and consequently the losses generated for a given load.

Since the higher-thermal-conductivity material (copper or bronze) is a much better bearing material than the conventional steel backing, it is possible to reduce the babbitt thickness to 0.010 to 0.030 in. Embedded thermocouples and RTDs will signal distress in the bearing if properly positioned. Temperature-monitoring systems have been found to be more accurate than axial-position indicators, which tend to have linearity problems at high temperatures.

In a change from steel backing to copper backing, a different set of temperature limiting criteria should be used. Figure 10-109 shows a typical set of curves for the two backing materials. This chart also shows that drain oil temperature is a poor indicator of bearing operating conditions because there is very little change in drain oil temperature from low load to failure load.

Thrust Bearing Power Loss The power consumed by various thrust bearing types is an important consideration in any system. Power losses must be accurately predicted so that turbine efficiency can be computed and the oil supply system properly designed.

Figure 10-110 shows a typical power consumption in thrust bearings as a function of unit speed. The total power loss is usually about 0.8 to 10 percent of the total rate power of the unit. New vector lube bearings reduce the horsepower loss by as much as 30 percent. In large vertical pumps, thrust bearings take not only the load caused by the fluid but also the load caused by the weight of the entire assembly (shaft and impellers). In some large pumps these could be about 60 ft (20 m) high and weigh 16 US tons. The thrust bearing for such a pump is over 5 ft. (1.7 m) in diameter with each thrust pad weighing more than 110 lb. (50 kg). In such cases, the entire pump assembly is first floated before the unit is started.

CENTRIFUGAL COMPRESSOR PROBLEMS

Compressors in process gas applications suffer from many problems. The following are some of the major categories in which these problems fall (see Meherwan P. Boyce, *Centrifugal Compressors: A Basic Guide*, PennWell, Nashua, N.H., 2003):

1. Compressor fouling
2. Compressor failures
3. Impeller problems

Bearing type		Load capacity	Suitable direction of rotation	Tolerance of changing load/speed	Tolerance of misalignment	Space requirement
Plain washer		Poor	⇄	Good	Moderate	Compact
Taper land	Bidirectional	Moderate	⇄	Poor	Poor	Compact
	Unidirectional	Good	→	Poor	Poor	Compact
Tilting pad	Bidirectional	Good	⇄	Good	Good	Greater
	Unidirectional	Good	→	Good	Good	Greater

FIG. 10-107 Comparison of thrust bearing types.

4. Rotor thrust problems
5. Seal and bearing problems
6. Bearing maintenance

Compressor Fouling Centrifugal compressors, especially in the process gas applications, suffer greatly from fouling. Fouling is the deposit and nonuniform accumulation of debris in the gas on internal compressor surfaces. Fouling is due to the carryover of liquids and other debris from the suction knockout drums. Debris can roughen compressor surfaces. Polymerization can occur also due to changes in process conditions. In wet gas compressors, ethylene plant cracked gas compressors, and polyethylene recycle compressors, the temperature of the gas must be kept below the threshold temperature that would initiate polymerization. The buildup will usually occur on the hub and the shroud with a larger buildup on the shroud at the elbow of the impeller on closed-face impellers. There is also a buildup on the blades, with the buildup usually more on the pressure side than on the suction side. Often the buildup is the heaviest on the pressure side at the blade exit where there is also flow separation.

Techniques to Prevent Fouling in Process Gas Compressors

1. *Condition monitoring of compressor aerodynamic and mechanical parameters.* Vibration monitoring could also alert the operator to fouling problems.

2. *Process control.* Accurate control of process conditions can prevent fouling in applications where polymers can form. Control of temperature is usually the most important. The following are examples of applications that can be affected by excessive process temperature:

a. Ethylene cracked gas
b. Linear low-density polyethylene
c. High-density propylene
d. Fluid catalytic cracker off-gas (wet gas)
e. Thermal cracker off-gas (wet gas)
f. Coker gas

The temperature below which fouling can be prevented varies with each process, compressor, and application. Monitoring of process conditions is

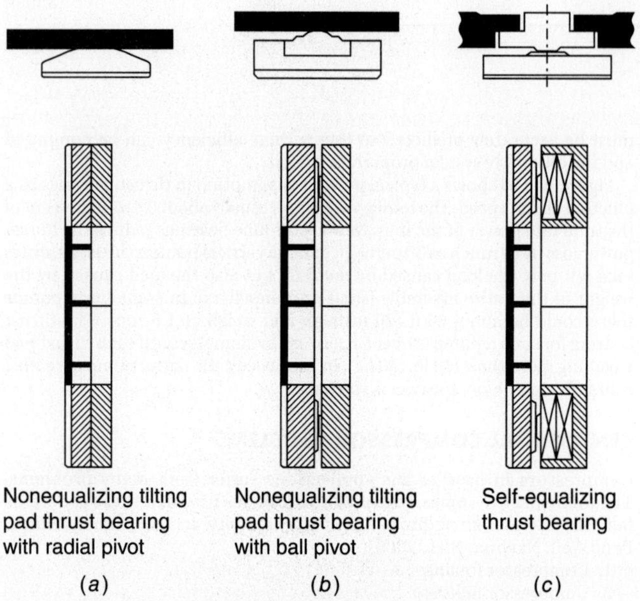

FIG. 10-108 Various types of thrust bearings.

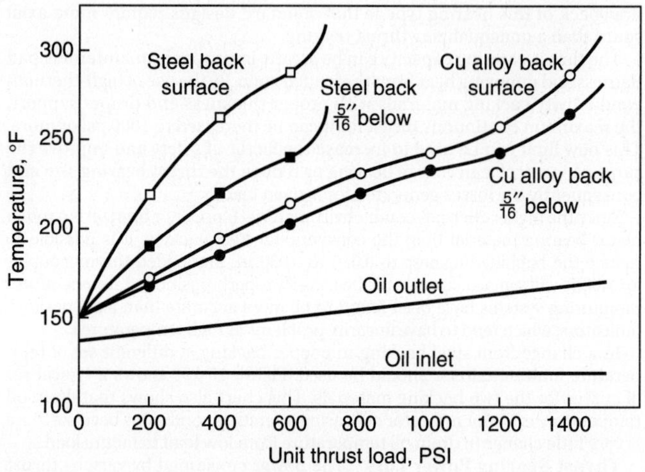

FIG. 10-109 Thrust bearing temperature characteristics.

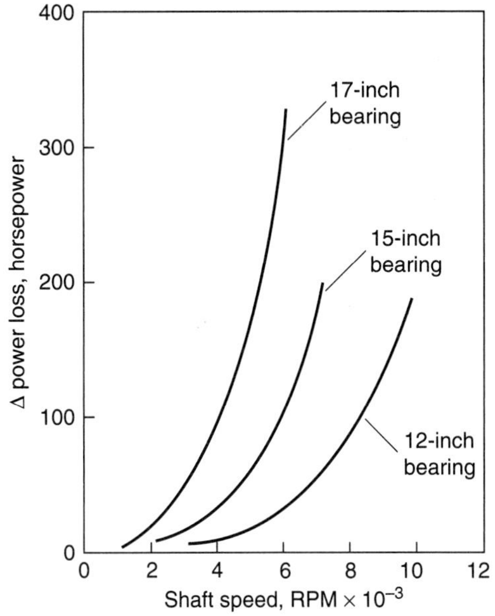

FIG. 10-110 Difference in total-power-loss data test minus catalog frictional losses versus shaft speed for 6 × 6 pad double-element thrust bearings.

necessary to establish a threshold temperature in each case. In some cases, fouling cannot be prevented with an existing compressor. It may be necessary to modify the aerodynamic design and/or add cooling.

3. *On-line solvent injection.* On-line solvent injection is very successful in various processes. The objective of this measure is to *continuously* inject a small amount of solvent to reduce the friction coefficient of the blade and impeller surface and thus prevent fouling of the surface. The injection should be done from the start; otherwise, the foulant could be dislodged and moved downstream, creating a major problem. The downstream areas are much smaller so foulant lodging there could create a blockage.

Air Compressors Inlet Filter In air compressors filter selection is an important factor in preventing fouling. Most high-efficiency air filters have a triple-stage filtration system. Also these filters often have rain shades to prevent water from entering the filters. Site conditions play a very important part in the selection of the filters.

Compressor Blade Coating Coatings protect blades against oxidation, corrosion, and cracking problems. Coatings guard the base metal of the compressor from attack. Other benefits include reduced thermal fatigue from cyclic operation, increased surface smoothness, reduced erosion, and reduced heat flux loading when one is considering thermal barriers. Coatings increase resistance to spalling from light impacts. Coatings also extend compressor life, better endure operational conditions, and serve as a sacrificial layer by allowing the coating to be restripped and recoated on the same base metal.

Compressor Failures In the process industry there are three types of compressors that have very different maintenance problems:

1. Barrel-type compressor
2. Horizontal split-casing centrifugal compressor with closed-face impellers
3. Air integral gear-type compressor with open-face impellers

Barrel-Type Compressors Barrel-type compressors are being utilized in the process industry to an increased extent because the barrel design confines gases more effectively than horizontally split cases. This becomes a critical consideration in two areas: high-pressure and low-molecular-weight gas compression. API-617, *Centrifugal Compressors for General Refinery Services,* requires a barrel design based on the molecular percentage of hydrogen contained in the process gas and the discharge pressure. API-617 defines high-pressure compressors as units in which the partial pressure of the hydrogen exceeds 200 psig, and it specifies that these units must be vertically (radially) split. Hydrogen partial pressure is given by the following relationship (in absolute pressures):

$$P_{H_2} = P(\%H_2/100) \tag{10-86}$$

Maximum casing working pressure for axially split compressors (psig) is

$$P_{casing} = \frac{200}{\%H_2/100} \tag{10-87}$$

The casings should include a minimum of a ⅛-in corrosion allowance. Casing strength and rigidity should limit change of alignment to 0.002 in if it is caused by the worst combination of pressure, torque, or allowable piping stress.

The barrel design is essentially a compressor placed inside a pressure vessel. For higher pressures some manufacturers have merely "beefed up" lower-pressure barrel designs, while others have perfected unique designs such as the "shear ring" head design. All these designs make extensive use of elastomer O-rings as sealing devices. There are several inherent maintenance problems with barrel-type compressors:

Handling: Barrel-type machines must be removed from their foundations for total maintenance. Because barrel machines weigh up to 30 tons, the handling problems can be formidable.

Inner casing alignment: Since this type of compressor consists of a bundle contained within the pressure walls of the barrel, alignment and positive positioning are often very poor, and the bundle is free to move to a certain extent. Bundle length is critical. Interstage leakage may occur if the bundle length is not correct. Assembly errors can be particularly detrimental in the case of a stacked diaphragm design, and care must be exercised to maintain proper impeller-diaphragm positioning. Since the bundle is subjected to discharge pressure on one end and suction pressure on the other, a force builds up that is transmitted from diaphragm to diaphragm, causing high loading on the inlet wall.

Internal leakage: The discharge and suction compartments of the inner bundle on a straight-through flow design are normally separated by a single O-ring. Compressors with side nozzles can have several bundles of O-rings. Excessive bundle-to-barrel clearance may cause leakage past the O-rings. O-rings are frequently pinched and cut across the suction nozzle opening in the barrel, a condition that is hard to prevent and doubly hard to detect if it occurs. Pressure differentials in excess of 400 to 500 psi, even using good design practice, can cause extrusion and failure of the O-rings. In many cases backup rings to the O-rings have been added to prevent failures. Grooves with O-ring ribbons have been added to the horizontal joints of the bundles of almost all the machines to prevent interstage leakage.

Bearing bracket alignment: In contrast to horizontally split compressors where the bearing brackets are normally an integral part of the lower casing half, in barrel machines bearing brackets are bolted to the barrel heads. Both the bearing brackets *and* the head are removed during the disassembly operation, thus requiring *all* internal alignment to be reestablished each time maintenance work is performed.

Material problems: To limit the physical size of the case or pressure vessel, to limit the rotor bearing span, and to maximize the number of stages within the heavy barrel, the gas path of a barrel is "squeezed" to a greater extent than in a horizontally split machine. This means the diaphragms and inlet guide vanes are intricate shapes with very small openings. Plain gray cast iron is normally used for these shapes because of casting ease and for other economic reasons. The gray iron is not strong enough in many instances to withstand the pressure differentials imposed on it, resulting in failures. Inlet guide vanes have been especially troublesome. On several occasions inlet guide vanes have been fabricated from wrought stainless and carbon steel materials. Replacement diaphragms and inlet guide vanes cast of nodular iron have also been used to alleviate some of these material problems.

Impeller Problems The high-speed rotation of the impeller of a centrifugal compressor imparts the vital aerodynamic velocity to the flow within the gas path. The buffeting effects of the gas flow can cause fatigue failures in the conventional fabricated shrouded impeller due to vibration-induced alternating stresses. These may be of the following types:

1. Resultant vibration in a principal mode
2. Forced-undamped vibration, associated with aerodynamic buffeting or high acoustic energy levels

The vibratory mode most frequently encountered is of the plate type and involves either the shroud or the disc. Fatigue failure generally originates at the impeller outside diameter, adjacent to a vane often due to the vibratory motion of the shroud or disc. The fatigue crack propagates inward along the nodal line, and finally a section of the shroud or disc tears out.

To eliminate failures of the covered impellers when operating at high density levels, the impellers are frequently scalloped between vanes at the outside diameter. The consequent reduction in disc friction also causes a small increase in impeller efficiency. However, there may be a slight reduction in overall efficiency due to higher losses in the diffuser. The advantages of scalloped impellers from a mechanical point of view are large. Several rotors have been salvaged by scalloping the wheels after a partial failure has occurred.

Rotor Thrust Problems Thrust loads in compressors due to aerodynamic forces are affected by impeller geometry, by pressure rise through the compressor, and by internal leakage due to labyrinth clearances. The impeller thrust is calculated by using correction factors to account for internal leakage, and a balance piston size is selected to compensate for the impeller thrust load. These common assumptions are made in the calculation:

1. The radial pressure distribution along the outside of the disc cover is essentially balanced.
2. Only the "eye" area is effective in producing thrust.
3. The pressure differential applied to the eye area is equal to the difference between the static pressure at the impeller tip, corrected for the pumping action of the disc, and the total pressure at inlet.

These "common assumptions" are grossly erroneous and can be disastrous when applied to high-pressure, barrel-type compressors where a large part of the impeller-generated thrust is compensated by a balance piston. The actual thrust is about 50 percent more than the calculations indicate. The error is less when the thrust is compensated by opposed impellers, because the mistaken assumptions offset one another.

The magnitude of the thrust is considerably affected by leakage at the impeller labyrinth seals. Increased leakage here produces increased thrust independent of balancing piston labyrinth seal clearance or leakage.

The thrust errors are further compounded in the design of the balancing piston, labyrinths, and bleed line. API-617, *Centrifugal Compressors*, specifies that a separate pressure tap connection must be provided to indicate the pressure in the balance chamber. It also specifies that (1) the balance line should be sized to handle balance piston labyrinth gas leakage at twice the initial clearance without exceeding the load ratings of the thrust bearing and (2) thrust bearings for compressors should be selected at no more than 50 percent of the bearing manufacturer's rating. The leaks and the consequential pressure change across the balance piston destabilize the entire rotor system.

Journal Bearing Failures With high-speed machines, simple bearing failures are rare unless they are caused by faulty alignment, distortion, wrong clearance, or dirt. More common are failures caused by vibrations and rotor whirls. Some of these originate in the bearings; others can be amplified or attenuated by the bearings, the bearing cases, and the bearing support structure.

During inspection, all journal bearings should be closely inspected. If the machine has not suffered from excessive vibration or lubrication problems, the bearings can be reinstalled and utilized.

Four places should be checked for wear during inspection periods:

1. Babbitted shoe surface
2. Pivoting shoe surface and seat in retaining ring
3. Seal ring bore or end plates
4. The shoe thickness at the pivot point or across ball *and* socket (all shoes should be within 0.0005 percent of the same thickness)

Thrust Bearing Failures The tilting-pad-type thrust bearings are used in most major pieces of rotating equipment under the general term *Kingsbury type.* A thrust bearing failure is one of the worst things that can happen to a machine, since it often wrecks the machine. To evaluate the reliability of a thrust bearing arrangement, one must first consider how a failure is initiated and evaluate the merits of the various designs.

Failures caused by bearing overload during normal operation (design error) are rare today, but still far more thrust failures occur than one would expect, considering all the precautions taken by the bearing designer. The causes in the following list are roughly in sequence of importance:

1. *Fluid slugging.* Passing a slug of fluid through a turbine or compressor can increase the thrust to many times its normal level.
2. *Buildup of solids in rotor and/or stator passages ("plugging" of turbine blades).* This problem should be noticed from performance or pressure distribution in the machine (first-stage pressure) long before the failure occurs.
3. *Off-design operation.* This arises especially from backpressure (vacuum), inlet pressure, extraction pressure, or moisture. Many failures are caused by overload, off-design speed, and flow fluctuations.
4. *Compressor surging.* This problem occurs especially in double-flow machines.
5. *Gear coupling thrust.* This is a frequent cause of failure, especially of upstream thrust bearings. The thrust is caused by friction in the loaded teeth that opposes thermal expansion. Therefore, thrust can get very high, since it has no relation to the normal thrust caused by pressure distribution inside the machine. The coupling thrust may act either way, adding to or subtracting from normal thrust. Much depends on tooth geometry and coupling quality.
6. *Dirt in oil.* This is a common cause of failures, especially when combined with other factors. The oil film at the end of the oil wedge is only a small fraction of 0.001 in thick. If dirt goes through, it can cause the film to rupture, and the bearing may burn out. Therefore, very fine filtering of the oil is required.
7. *Momentary loss of oil pressure.* This type of failure is usually encountered while switching filters or coolers, or in some instances when the dc pump does not come on-line when the main pumps fail.

The thrust bearings must be closely maintained. This type of bearing consists of pivoted segments or pads (usually six) against which the thrust collar revolves, forming a wedge-shaped oil film. This film plus minute misalignment of the thrust collar and the bearing pads causes movement and wear of the various bearing parts. The erroneous thrust calculations discussed earlier cause the bearing to be loaded more heavily than desired. This accelerates the wear problem. There are seven wear points in the bearing. All these points must be checked for wear:

1. The soft babbitted shoe face
2. The hardened steel shoe insert face
3. The face of the hardened steel upper leveling plate
4. The outer edge of the upper leveling plate
5. The upper edge of lower leveling plate
6. The pivot point of the lower leveling plate
7. The inner face of the base ring

To protect thrust bearings, accurate and reliable instrumentation is now available to monitor thrust bearings well enough to ensure safe continuous operation and to prevent catastrophic failure in the event of an upset to the system.

Temperature sensors, such as resistance temperature detectors (RTDs), thermocouples, and thermistors, can be installed directly in the thrust bearing to measure metal temperature.

Axial proximity probes are another means of monitoring rotor position and the integrity of the thrust bearing. This method detects thrust collar runout and rotor movement. In most cases this ideal positioning of the probes is not possible. Many times the probes are indexed to the rotor or other convenient locations and thus do not truly show the movement of the rotor with respect to the thrust bearing.

A critical installation should have the metal temperature sensors in the thrust pad. Axial proximity probes may be used as a backup system. If metal temperatures are high and the rate of change of those temperatures begins to alter rapidly, thrust bearing failure should be anticipated.

Compressor Seal Problems The extent of the leakage past the seals where the shaft comes through the casing frequently limits the running time of the compressor; yet the seals and the seal systems are not given adequate treatment in the maintenance manuals or in the operating instructions furnished by the compressor manufacturer.

Shaft seals are divided into the following categories by API Standard 617:

- Labyrinth
- Restrictive carbon rings
- Mechanical (contact) type
- Liquid-film or floating bushing type
- Liquid-film type with pump bushings

The first two seal categories are usually operated dry, and the last three categories require seal oil consoles either separately or as part of the lube system. Each of these seal designs has its own characteristics and maintenance difficulties.

Oil flows and critical clearances are not spelled out well in either the operating instructions or the maintenance manuals. Because of this, several maintenance technique improvements are needed:

1. *Radial clearances.* Radial clearance between the bushing and the shaft as well as the length of the bushing must be selected to obtain minimum leakage without exceeding fluid temperature limitations.
2. *Quality control.* The flatness, parallelism, and surface finish of the mating sleeve faces must be carefully controlled to obtain maximum seal effectiveness.
3. *Axial clearances.* Axial clearance between the bushing or sleeve and the housing is critical. There should be 12 to 15 mils of clearance per bushing between the bushing or sleeve and the housing; where the sleeves are mounted back to back, there will be 25 to 30 mils of clearance total for the seal.
4. *Seal design.* In higher-pressure seals, more than one outboard (i.e., high differential) sleeve may be used. Generally, it is desirable to use a single sleeve because the inboard sleeve operates with up to 80 percent of the total pressure drop across it. The outer sleeve with the lower differential causes lubrication and cooling problems that can shorten the life of one or both sleeves.
5. *Guidelines.* These should be explicit in indicating oil flow rates and the interaction of various components.
6. *Rules of thumb.* There are a few rules of thumb that help in understanding seal operation and maintenance.
   a. The oil flow rate will vary (1) *directly* with the differential pressure and the wetted perimeter of the sleeve; (2) with the *cube* of the radial clearance; (3) with the *square* of the eccentricity of the sleeve and shaft; (4) *inversely* with oil viscosity, temperature, and length of the sleeve.
   b. Shear work done on the sealing fluid during its passage through the sleeve raises its temperature to a much higher level than may be expected.

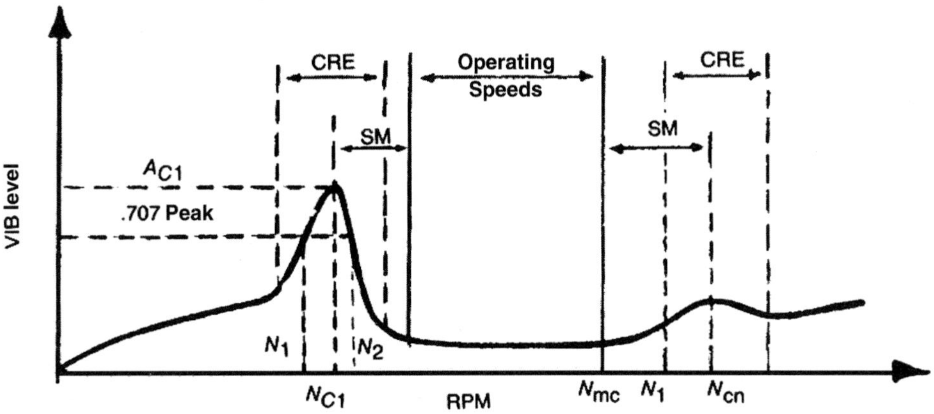

N_{C1} = Rotor 1st critical, center frequency, cycles per minute
N_{cn} = Critical speed, nth
N_+ = Trip speed
N_{mc} = Maximum continuous speed, 105 percent
N_1 = Initial (lesser) speed at .707 × peak amplitude (critical)
N_2 = Final (greater) speed at .707 × peak amplitude (critical)
$N_2 - N_1$ = Peak width at the "half-power" point

AF = Amplification factor $= \dfrac{N_{C1}}{N_2 - N_1}$
SM = Separation margin
CRE = Critical response envelope
N_{C1} = Amplitude @ N_{C1}
A_{cn} = Amplitude @ N_{cn}

FIG. 10-111 Rotor response plot. This plot is Figure 7 in API Standard 617, *Centrifugal Compressors for General Refinery Services*, 4th ed., 1979. (*Courtesy American Petroleum Institute.*)

ROTOR DYNAMICS

The rotating elements consist of the impeller and the shaft. The shaft should be made of one-piece, heat-treated forged steel, with the shaft ends tapered for coupling fits. Interstage sleeves should be renewable and made of material which is corrosion-resistant in the specified service. The rotor shaft sensing area observed by the noncontact probes should be concentric with the bearing journals and free of any scratches, marks, or other surface discontinuity. The surface finish should be 16 to 32 μin root mean square, and the area should be demagnetized and treated. Electromechanical runout should not exceed 25 percent of the maximum allowed peak-to-peak vibration amplitude or 0.25 mil, whichever is greater. Although not mentioned in the standard, chrome plating of the shaft in the sensing area is unacceptable. Maximum vibration should not exceed 2.0 mils as given by

$$\text{Vib}_{\max} = \sqrt{\frac{12{,}000}{\text{rpm}}} + 0.25\sqrt{\frac{12{,}000}{\text{rpm}}} \qquad (10\text{-}88)$$

(Vibration) (runout)

At the trip speed of the driver (105 percent for a gas turbine), the vibration should not exceed this level by more than 0.5 mil.

The impellers can be an open-faced (stationary shroud) or closed-face (rotating shroud) design. As long as the tip velocities are below 1000 ft/s, closed-face impellers can be used. The standards allow the impellers to be welded, riveted, milled, or cast. Riveted impellers are unacceptable, especially if the impeller loading is high. Impellers are to be assembled on the shaft with a shrink fit with or without key. Shrink fits should be carefully done because excessive shrink fits can cause a problem known as *hysteresis whirl*. In compressors where the impellers require their thrust to be balanced, a balance drum is acceptable and preferred.

The high-speed pumps or compressors must operate in a region away from any critical speed. The *amplification factor* AF used to indicate the severity of the critical speed is given by the relationship

$$\text{AF} = \frac{\text{critical speed}}{\text{differential between RMS (0.707) of the final and initial speed}} \qquad (10\text{-}89)$$

$$\text{AF} = \frac{N_{C1}}{N_2 - N_1} \qquad (10\text{-}90)$$

where $N_2 - N_1$ is the rpm corresponding to the 0.707 (root mean square) of the peak critical amplitude between the final and the initial critical speed.

The amplification factor should be below 8 and preferably above 5. A rotor response plot is shown in Fig. 10-111. The operational speed for units operating below their first critical speed should be at least 20 percent below the critical speed. For units operating above their first critical speed, the operational speed must be at least 15 percent above the critical speed and/or 20 percent below any critical speed. The preferred bearings for the various types of installation are tilting-shoe radial bearings and the self-equalizing tilting-pad thrust bearings. Radial and thrust bearings should be equipped with embedded temperature sensors to detect pad surface temperatures.

VIBRATION MONITORING

One of the major factors that causes pump failure is vibration, which usually causes seal damage and oil leakage. Vibration in pumps is caused by numerous factors such as cavitation, impeller unbalance, loose pump impeller on the impeller shaft, loose bearings, seals, and pipe pulsations.

As the mechanical integrity of the pump system changes, the amplitudes of vibration levels change. In some cases, to identify the source of vibration, pump speed may have to be varied, as these problems are frequency- or resonance-dependent. Pump impeller imbalance and cavitation are related to this category.

It is advisable in most of these cases to use accelerometers. Displacement probes will not give the high-frequency signals and velocity probes because their mechanical design is very directional and prone to deterioration. Figure 10-112 shows the signal from the various types of probes.

Typically, large-amplitude vibration occurs when the frequency of vibration coincides with that of the natural frequency of the pump system. This results in a catastrophic operating condition that should be avoided. If the natural frequency is close to the upper end of the operating speed range, then the pump system should be stiffened to reduce vibration. On the other hand, if the natural frequency is close to the lower end of the operating range, the unit should be made more flexible. During startup, the pump system may go through its system natural frequency, and vibration can occur. Continuous operation at this operating point should be avoided.

ASME recommends periodic monitoring of all pumps. The pump vibration level should fall within the prescribed limits. The reference vibration level is measured during acceptance testing. This level is specified by the manufacturer.

During periodic maintenance, the vibration level should not exceed the alert level (see Table 10-17). If the measured level exceeds the alert level, then preventive maintenance should be performed, by diagnosing the cause of vibration and reducing the vibration level prior to continued operation.

Typical problems and their vibration frequency ranges are shown in Fig. 10-113.

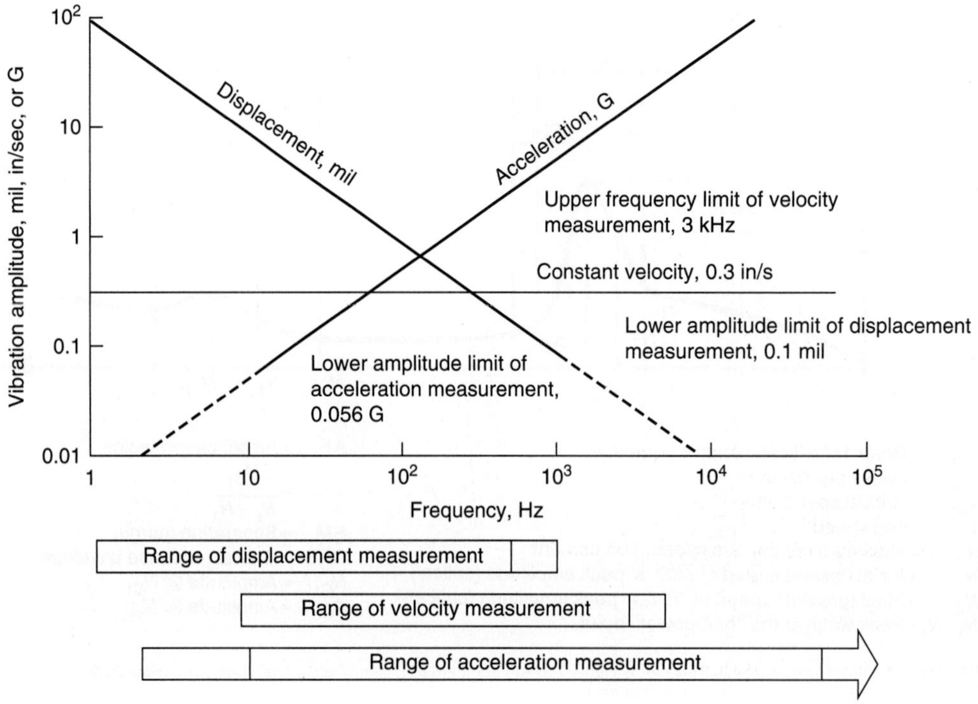

FIG. 10-112 Limitations on machinery vibrations analysis systems and transducers.

TABLE 10-17 Alert Levels

Reference value, mils	Alert, mils, μm	Action required, mils, μm
$V_r < 0.5$	1.0	1.5
$0.5 < V_r < 2.0$	$2V_r$	$3V_r$
$2.0 < V_r < 5.0$	$2 + V_r$	$4 + V_r$
$5.0 < V_r$	$1.4V_r$	$1.8V_r$

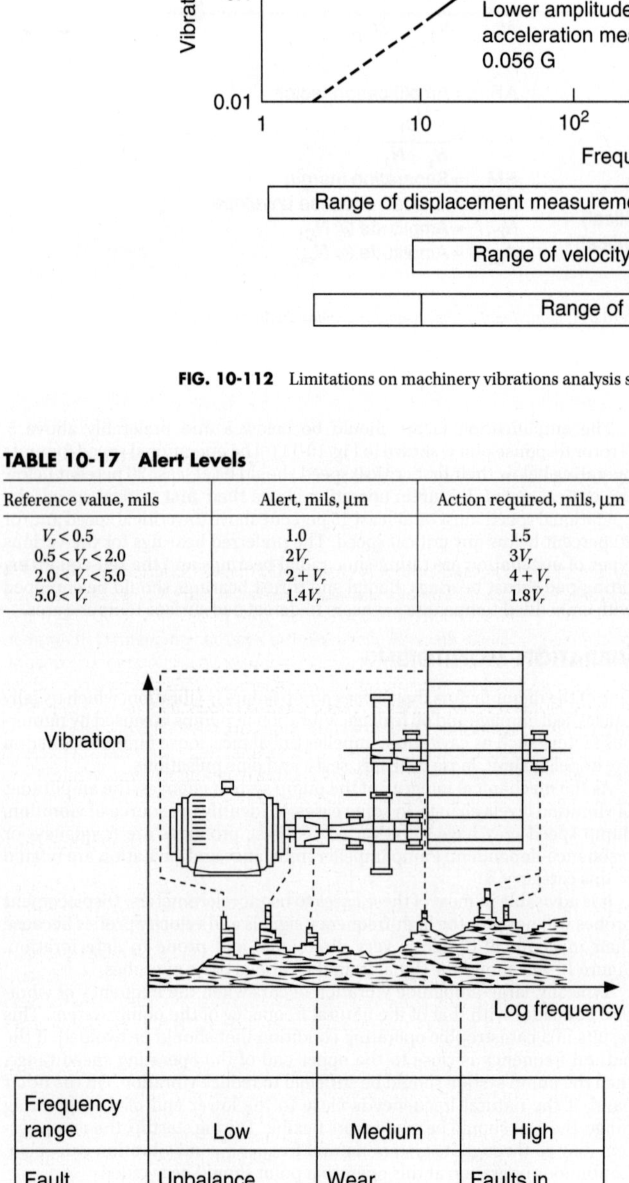

Vibration
Log frequency
Frequency range
Low
Medium
High
Fault to be detected
Unbalance
Misalignment
Bent shaft
Oil whirl
Eccentricity
Wear
Faults in gears
Faults in rolling element bearings

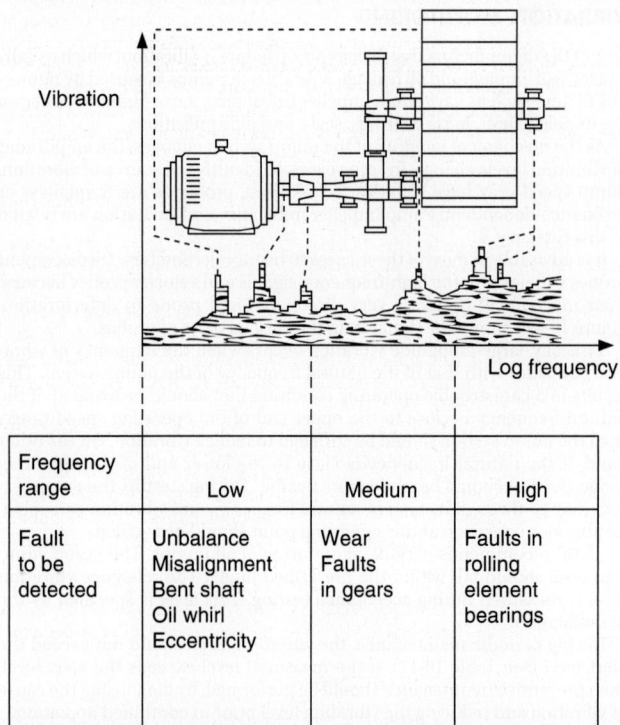

FIG. 10-113 Frequency range of typical machinery faults.

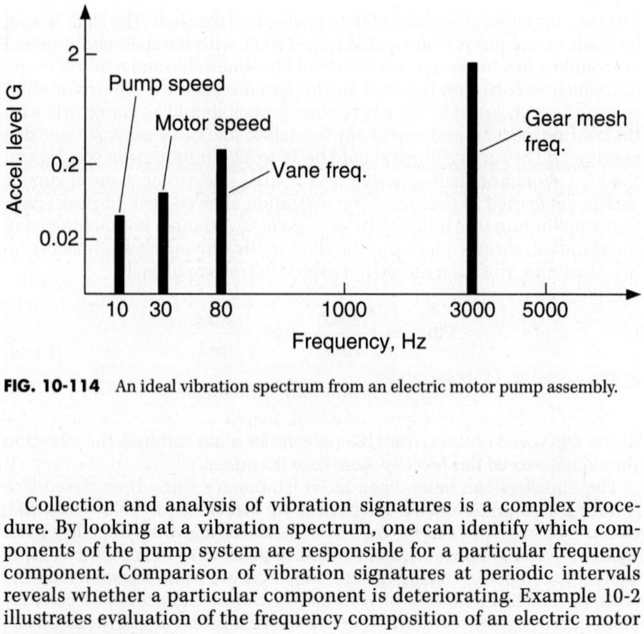

FIG. 10-114 An ideal vibration spectrum from an electric motor pump assembly.

Collection and analysis of vibration signatures is a complex procedure. By looking at a vibration spectrum, one can identify which components of the pump system are responsible for a particular frequency component. Comparison of vibration signatures at periodic intervals reveals whether a particular component is deteriorating. Example 10-2 illustrates evaluation of the frequency composition of an electric motor gear pump system.

Example 10-2 Vibration Consider an electric motor rotating at 1800 rpm driving an eight-vane centrifugal pump rotating at 600 rpm. For this 3:1 speed reduction, assume a gearbox having two gears of 100 and 300 teeth. Since 60 Hz is 1 rpm,

Motor frequency = 1800/60 = 30 Hz
Pump frequency = 600/60 = 10 Hz
Gear mesh frequency = 300 teeth × 600 rpm = 3000 Hz
Vane frequency = 8 × 600 rpm = 80 Hz

Ideal vibration spectra for this motor-gear pump assembly would appear as shown in Fig. 10-114.

Figure 10-115 shows actual pump vibration spectra. In the figure, several amplitude peaks occur at several frequencies.

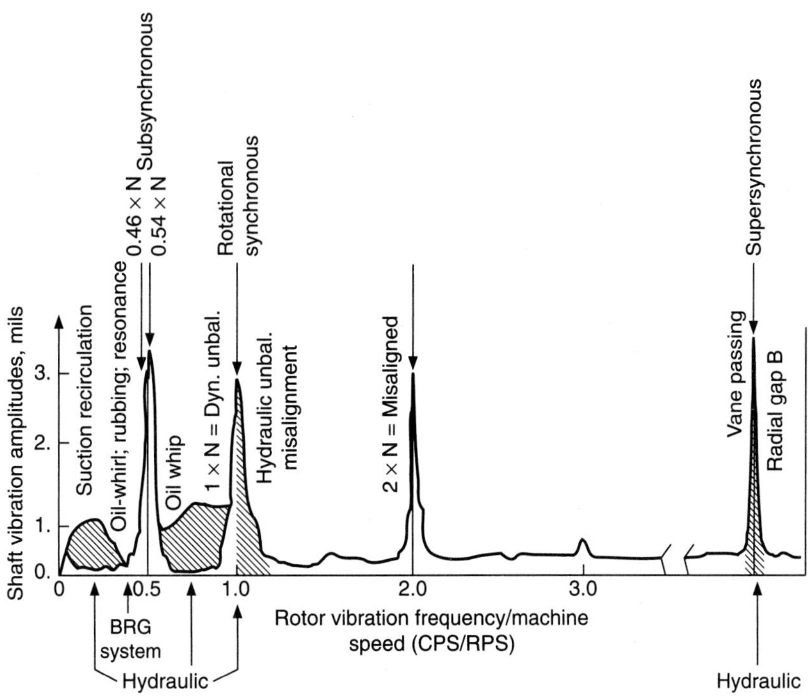

FIG. 10-115 An actual pump vibration spectrum.

PROCESS PLANT PIPING

INTRODUCTION

This section provides general comments that are pertinent to the design of process plant piping. It is intended to provide a convenient summary of commonly used information from various sources. It is not intended to serve as a comprehensive source of requirements or as a substitute for referenced codes, standards, and specifications. It is intended that qualified designers obtain copies of all applicable codes, standards, and specifications and thoroughly review all pertinent requirements of these documents prior to execution of work.

CODES AND STANDARDS

Units: Pipe and Tubing Sizes and Ratings In this subsection, pipe and tubing sizes are generally quoted in units of inches. To convert inches to millimeters, multiply by 25.4. Ratings are given in pounds. To convert pounds to kilograms, multiply by 0.454.

Pressure-Piping Codes The code for pressure piping (ASME B31) consists of a number of sections which collectively constitute the code. Table 10-18 shows the status of the B31 code as of July 2016. The sections are published as separate documents for simplicity and convenience. The sections differ extensively.

The *Process Piping* code (ASME B31.3) is a subsection of the ASME code for *Pressure Piping* B31. It was derived from a merging of the code groups for chemical plant (B31.6) and petroleum refinery (B31.3) piping into a single committee. Some of the significant requirements of ASME B31.3, *Process Piping* (2014 edition), are summarized in the following presentation.

Where the word *code* is used in this subsection of the text without other identification, it refers to the B31.3 section of ASME B31. The code has been extensively quoted in this subsection with the permission of the publisher. The code is published by, and copies are available from, the American Society of Mechanical Engineers (ASME), Three Park Avenue, New York, NY 10016–5990.

National Standards The American Society of Mechanical Engineers and the American Petroleum Institute (API) have established dimensional standards for the most widely used piping components. Lists of these standards as well as specifications for pipe and fitting materials and testing methods of the American Society for Testing and Materials (ASTM), American Welding Society (AWS) specifications, and standards of the Manufacturers Standardization Society of the Valve and Fittings Industry (MSS) can be found in the ASME B31 code sections. Many of these standards contain pressure-temperature ratings which will be of assistance to engineers in their design function. The use of published standards does not eliminate the need for engineering judgment. For example, although the code calculation formulas recognize the need to provide an allowance for corrosion, the standard rating tables for valves, flanges, fittings, etc., do not incorporate a corresponding allowance. Judgments regarding the suitability of these components are left to the designer.

The introduction to the code sets forth engineering requirements deemed necessary for the safe design and construction of piping systems. While safety is the basic consideration of the code, this factor alone will not necessarily govern final specifications for any pressure piping system.

Designers are cautioned that the code is not a design handbook and does not do away with the need for competent engineering judgment.

Government Regulations: OSHA Sections of the ASME B31 code have been adopted with certain reservations or revisions by some state and local authorities as local codes.

The specific requirements for piping systems in certain services have been promulgated as Occupational Safety and Health Administration (OSHA) regulations. These rules and regulations will presumably be revised and supplemented from time to time and may include specific requirements not addressed by the ASME B31 sections.

International Regulations ASME piping codes have been widely used throughout the world for the design of facilities falling within their defined scopes. Although the use of ASME codes is widely acceptable in areas outside the United States, it is essential to identify additional local or national codes or standards that may apply. Such documents may require qualified third-party review and approval of project specifications, facility design, fabrication, material documentation, inspection, and testing. For example, within the European Community, such requirements are imposed by the Pressure Equipment Directive 97/23/EC (also known as the PED). These requirements must be recognized early in the project to avoid costly error.

CODE CONTENTS AND SCOPE

The code prescribes minimum requirements for materials, design, fabrication, assembly, support, erection, examination, inspection, and testing of piping systems subject to pressure or vacuum. The scope of

TABLE 10-18 Status of ASME B31 Code for Pressure Piping

ASME B31.3 section number, latest issue, and title	General scope and application
B31.1-2014, *Power Piping*	Addresses piping typically found in electric power generating stations, industrial and institutional plants, geothermal heating systems, and central and district heating and cooling systems.
B31.2, *Fuel Gas Piping*	Withdrawn as a National Standard and replaced by ANSI/NFPA Z223.1 includes piping between the outlet of the consumer's meter and the outlet of the first pressure-containing valve upstream of the gas utilization device.
B31.3-2014, *Process Piping*	Addresses piping typically found in petroleum refineries; chemical, pharmaceutical, textile, paper, semiconductor, and cryogenic plants; and related processing plants and terminals.
B31.4-2016, *Pipeline Transportation Systems for Liquids and Slurries and Other Liquids*	Addresses piping transporting products which are predominately liquid between plants and terminals. Included are terminals, tank farms, and pumping, regulating, and metering stations.
B31.5-2013, *Refrigeration Piping and Heat Transfer Components*	Addresses refrigeration piping in packaged units and in commercial and public buildings.
B31.7, *Nuclear Power Piping*	Withdrawn; see *ASME Boiler and Pressure Vessel Code*, Sec. 3.
B31.8-2014, *Gas Transmission and Distribution Piping Systems*	Addresses piping transporting products which are predominately gas between sources and terminals. Included are gas-gathering pipelines and compressor, regulating, and metering stations.
B31.8S-2014, *Managing System Integrity of Gas Pipelines*	Addresses development and implementation of integrity management systems for gas pipelines.
B31.9-2014, *Building Services Piping*	Addresses piping outside the scope of ASME B31.1 that is typically found in multiunit residences and in industrial, institutional, commercial, and public buildings.
B31.11-2002, *Slurry Transportation Piping Systems*	Addresses piping transporting aqueous slurries between plants and terminals and within terminals, pumping, and regulating stations.
B31.12-2014, *Hydrogen Piping and Pipelines*	Addresses piping design, construction, operation, and maintenance requirements for piping, pipelines, and distribution systems in hydrogen service.
B31E-2008, *Standard for the Seismic Design and Retrofit of Above-Ground Piping Systems*	Addresses a more explicit and structured guidance for seismic design of new piping systems, as well as retrofit of existing systems.
B31G-2012, *Manual for Determining the Remaining Strength of Corroded Pipelines: A Supplement to ASME B31 Code for Pressure Piping*	Addresses methods of determining the remaining strength of corroded pipelines that are within the scope of the *ASME B31 Code for Pressure Piping.*
B31J-2008, *Standard Test Method for Determining Stress Intensification Factors (i-Factors) for Metallic Piping Components*	Addresses a uniform approach to the development of SIFs for standard, nonstandard, and proprietary piping components and joints of all types.
B31Q-2014, *Pipeline Personnel Qualification*	Addresses general and specific requirements for the qualification of pipeline personnel.
B31T-2012, *Standard Toughness Requirements for Piping*	Addresses requirements for evaluating the suitability of materials used in piping systems for piping that may be subject to brittle failure due to low-temperature service conditions.

Addenda are issued at intervals between publication of complete editions. Information on the latest issues can be obtained from the American Society of Mechanical Engineers, Three Park Avenue, New York, NY 10016-5990; or on the World Wide Web at ASME.org.

the piping covered by B31.3 is illustrated in Fig. 10-116. It applies to all fluids including fluidized solids and to all services except as noted in the figure.

The code also excludes piping systems designed for internal gauge pressures at or above zero but less than 0.105 MPa (15 lbf/in^2) provided the fluid handled is nonflammable, nontoxic, and not damaging to human tissues, and its design temperature is from −29°C (−20°F) through 186°C (366°F). Refer to the code for definitions of nonflammable and nontoxic.

Some of the more significant requirements of ASME B31.3 (2014 edition) have been summarized and incorporated in this section of the text. For a more comprehensive treatment of code requirements, engineers are referred to the B31.3 code and the standards referenced therein.

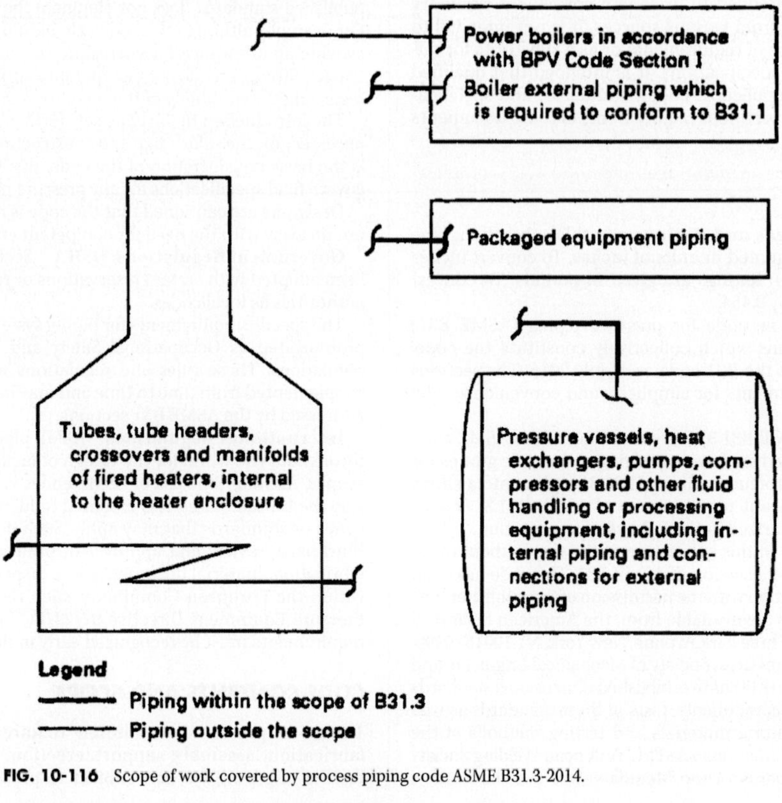

FIG. 10-116 Scope of work covered by process piping code ASME B31.3-2014.

SELECTION OF PIPE SYSTEM MATERIALS

The selection of material to resist deterioration in service is outside the scope of the B31.3 code (see Sec. 25). Experience has, however, resulted in the following material considerations extracted from the code with the permission of the publisher, the American Society of Mechanical Engineers, New York.

General Considerations* Following are some general considerations which should be evaluated when selecting and applying materials in piping:

1. The possibility of exposure of the piping to fire and the melting point, degradation temperature, loss of strength at elevated temperature, and combustibility of the piping material under such exposure
2. The susceptibility to brittle failure or failure from thermal shock of the piping material when exposed to fire or to firefighting measures, and possible hazards from fragmentation of the material in the event of failure
3. The ability of thermal insulation to protect piping against failure under fire exposure (e.g., its stability, fire resistance, and ability to remain in place during a fire)
4. The susceptibility of the piping material to crevice corrosion under backing rings, in threaded joints, in socket-welded joints, and in other stagnant, confined areas
5. The possibility of adverse electrolytic effects if the metal is subject to contact with a dissimilar metal
6. The compatibility of lubricants or sealants used on threads with the fluid service
7. The compatibility of packing, seals, and O-rings with the fluid service
8. The compatibility of materials, such as cements, solvents, solders, and brazing materials, with the fluid service
9. The chilling effect of sudden loss of pressure on highly volatile fluids as a factor in determining the lowest expected service temperature
10. The possibility of pipe support failure resulting from exposure to low temperatures (which may embrittle the supports) or high temperatures (which may weaken them)
11. The compatibility of materials, including sealants, gaskets, lubricants, and insulation, used in strong oxidizer fluid service (e.g., oxygen or fluorine)
12. The possibility of adverse effects from microbiologically influenced corrosion (MIC) or its remediation

Specific Material Considerations—Metals* Following are some specific considerations which should be evaluated when applying certain metals in piping.

1. *Irons—cast, malleable, and high silicon (14.5 percent).* Their lack of ductility and their sensitivity to thermal and mechanical shock.
2. *Carbon steel, and low and intermediate alloy steels.*
   a. The possibility of embrittlement when handling alkaline or strong caustic fluids.
   b. The possible conversion of carbides to graphite during long time exposure to temperatures above 427°C (800°F) of carbon steels, plain nickel steel, carbon-manganese steel, manganese-vanadium steel, and carbon-silicon steel.
   c. The possible conversion of carbides to graphite during long time exposure to temperatures above 468°C (875°F) of carbon-molybdenum steel, manganese-molybdenum-vanadium steel, and chromium-vanadium steel.
   d. The advantages of silicon-killed carbon steel (0.1 percent silicon minimum) for temperatures above 482°C (900°F).
   e. The possibility of damage due to hydrogen exposure at elevated temperature (see API RP941); possible hydrogen damage (blistering) at lower temperatures under exposure to aqueous acid solutions.†
   f. The possibility of stress corrosion cracking when exposed to cyanides, acids, acid salts, or wet hydrogen sulfide; a maximum hardness limit is usually specified (see NACE MR0175, or MR0103, and RP0472).†
   g. The possibility of sulfidation in the presence of hydrogen sulfide at elevated temperatures.
3. *High-alloy (stainless) steels.*
   a. The possibility of stress corrosion cracking of austenitic stainless steels exposed to media such as chlorides and other halides either internally or externally; the latter can result from improper selection or application of thermal insulation, or from use of marking inks, paints, labels, tapes, adhesives, and other accessory materials containing chlorides or other halides.
   b. The susceptibility to intergranular corrosion of austenitic stainless steels sensitized by exposure to temperatures between 427 and 871°C (800 and 1600°F); as an example, stress corrosion cracking of sensitized metal at room temperature by polythionic acid (reaction of oxidizable sulfur compound, water, and air); stabilized or low-carbon grades may provide improved resistance (see NACE RP0170).†
   c. The susceptibility to intercrystalline attack of austenitic stainless steels on contact with liquid metals (including aluminum, antimony, bismuth, cadmium, gallium, lead, magnesium, tin, and zinc) or their compounds.
   d. The brittleness of ferritic stainless steels at room temperature after service at temperature above 371°C (700°F).
4. *Nickel and nickel-base alloys.*
   a. The susceptibility to grain boundary attack of nickel and nickel-base alloys not containing [OL] chromium when exposed to small quantities of sulfur at temperatures above 316°C (600°F).
   b. The susceptibility to grain boundary attack of nickel-base alloys containing chromium at temperatures above 593°C (1100°F) under reducing conditions and above 760°C (1400°F) under oxidizing conditions.
   c. The possibility of stress corrosion cracking of nickel-copper Alloy 400 in hydrofluoric acid vapor in the presence of air, if the alloy is highly stressed (including residual stresses from forming or welding).
5. *Aluminum and aluminum alloys.*
   a. The compatibility with aluminum of thread compounds used in aluminum threaded joints to prevent seizing and galling.
   b. The possibility of corrosion from concrete, mortar, lime, plaster, or other alkaline materials used in buildings or structures.
   c. The susceptibility of Alloys 5083, 5086, 5154, and 5456 to exfoliation or intergranular attack; and the upper temperature limit of 66°C (150°F) shown in Appendix A to avoid such deterioration.
6. *Copper and copper alloys.*
   a. The possibility of dezincification of brass alloys.
   b. The susceptibility to stress corrosion cracking of copper-based alloys exposed to fluids such as ammonia or ammonium compounds.
   c. The possibility of unstable acetylide formation when exposed to acetylene.
7. *Titanium and titanium alloys.* The possibility of deterioration of titanium and its alloys above 316°C (600°F).
8. *Zirconium and zirconium alloys.* The possibility of deterioration of zirconium and zirconium alloys above 316°C (600°F).
9. *Tantalum.* Above 299°C (570°F), the possibility of reactivity of tantalum with all gases except the inert gases. Below 299°C, the possibility of embrittlement of tantalum by nascent (monatomic) hydrogen (but not molecular hydrogen). Nascent hydrogen is produced by galvanic action or as a product of corrosion by certain chemicals.
10. *Metals with enhanced properties.* The possible loss of strength, in a material whose properties have been enhanced by heat treatment, during long, continued exposure to temperatures above the tempering temperature.
11. The desirability of specifying some degree of production impact testing, in addition to the weld procedure qualification tests, when using materials with limited low-temperature service experience below the minimum temperature stated in ASME B31.3 Table A-1.

Specific Material Considerations—Nonmetals Following are some considerations to be evaluated when applying nonmetals in piping. Refer to Tables 10-19, 10-20, and 10-21 for typical temperature limits.

1. *Static charges.* Because of the possibility of producing hazardous electrostatic charges in nonmetallic piping and metallic piping lined with nonmetals, consideration should be given to grounding the metallic components of such systems conveying nonconductive fluids.
2. *Thermoplastics.* If thermoplastic piping is used aboveground for compressed air or other compressed gases, special precautions should be observed. In determining the needed safeguarding for such services, the energetics and the specific failure mechanism need to be evaluated. Encasement of the plastic piping in shatter-resistant material may be considered.
3. *Borosilicate glass.* Take into account its lack of ductility and its sensitivity to thermal and mechanical shock.

METALLIC PIPING SYSTEM COMPONENTS

Metallic pipe systems comprise the majority of applications. Metallic pipe, tubing, and pipe fittings are divided into two main categories: seamless and welded. Both have advantages and disadvantages in terms of economy

*Extracted from ASME B31.3-2014, Section F323, with permission of the publisher, the American Society of Mechanical Engineers, New York.
†Titles of referenced documents are:
API RP941, *Steels for Hydrogen Service at Elevated Temperatures and Pressures in Petroleum Refineries and Petrochemical Plants*
NACE MR0175, *Sulfide Stress-Cracking Resistant Metallic Materials for Oil Field Equipment*
NACE MR0103, *Materials Resistant to Sulfide Stress Cracking in Corrosive Petroleum Refining Environments*
NACE RP0472, *Methods and Controls to Prevent In-Service Cracking of Carbon Steel (P-1) Welds in Corrosive Petroleum Refining Environments*
NACE RP0170, *Protection of Austenitic Stainless Steel in Refineries Against Stress Corrosion Cracking by Use of Neutralizing Solutions During Shutdown*

TABLE 10-19 Hydrostatic Design Stresses (HDS) and Recommended Temperature Limits for Thermoplastic Pipe

				Recommended Temperature Limits, °F*†		Hydrostatic Design Stress, ksi, at			
ASTM Spec. No.	Pipe Designation	Material Designation	Cell Class	Minimum	Maximum	73°F‡	100°F	180°F	200°F
...	PR	ABS	43232	−40	176	...	...	...	...
D2846	SDR11	CPVC4120	23447	...	180	2.0	...	0.5	...
F441	Sch. 40	CPVC4120	23447	73	200	2.0	...	0.5	...
F441	Sch. 80	CPVC4120	23447	73	200	2.0	...	0.5	...
F442	SDR-PR	CPVC4120	23447	73	200	2.0	1.64	0.5	...
D3309	SDR11	PB2110	...	73	200	1.0	...	0.5	...
D2239	SIDR-PR	PE1404	...	73	...	0.40	...	...	...
D2239	SIDR-PR	PE2305	...	73	...	0.50	...	...	...
D2239	SIDR-PR	PE2306	...	73	...	0.63	...	...	...
D2239	SIDR-PR	PE2406	...	73	...	0.63	...	...	...
D2239	SIDR-PR	PE3306	...	73	...	0.63	...	...	...
D2239	SIDR-PR	PE3406	...	73	...	0.63	...	...	...
D2239	SIDR-PR	PE3408	...	73	...	0.80	...	...	...
D2447	Sch. 40 and 80	PE1404	...	73	...	0.40	...	...	...
D2447	Sch. 40 and 80	PE2305	...	73	...	0.50	...	...	...
D2447	Sch. 40 and 80	PE2306	...	73	...	0.63	...	...	...
D2447	Sch. 40 and 80	PE2406	...	73	...	0.63	...	...	...
D2447	Sch. 40 and 80	PE3306	...	73	...	0.63	...	...	...
D2447	Sch. 40 and 80	PE3406	...	73	...	0.63	...	...	...
D2737	SDR7.3, SDR9, SDR11	PE2305	...	73	...	0.50	...	...	...
D2737	SDR7.3, SDR9, SDR11	PE2306	...	73	...	0.63	...	...	...
D2737	SDR7.3, SDR9, SDR11	PE2406	...	73	...	0.63	...	...	...
D2737	SDR7.3, SDR9, SDR11	PE3306	...	73	...	0.63	...	...	...
D2737	SDR7.3, SDR9, SDR11	PE3406	...	73	...	0.63	...	...	...
D2737	SDR7.3, SDR9, SDR11	PE3408	...	73	...	0.80	...	...	...
D3035	DR-PR	PE1404	...	73	...	0.40	...	...	...
D3035	DR-PR	PE2606	...	73	...	0.63	...	...	...
D3035	DR-PR	PE2708	...	73	...	0.80	...	...	...
D3035	DR-PR	PE3608	...	73	...	0.80	...	...	...
D3035	DR-PR	PE3708	...	73	...	0.80	...	...	...
D3035	DR-PR	PE3710	...	73	...	1.00	...	...	...
D3035	DR-PR	PE4608	...	73	...	0.80	...	...	...
D3035	DR-PR	PE4708	...	73	...	0.80	...	...	...
D3035	DR-PR	PE4710	...	73	...	1.00	...	...	...
F714	SDR-PR	PE1404	...	73	...	0.40	...	...	...
F714	SDR-PR	PE2606	...	73	...	0.63	...	...	...
F714	SDR-PR	PE2708	...	73	...	0.80	...	...	...
F714	SDR-PR	PE3608	...	73	...	0.80	...	...	...
F714	SDR-PR	PE3708	...	73	...	0.80	...	...	...
F714	SDR-PR	PE3710	...	73	...	1.00	...	...	...
F714	SDR-PR	PE4608	...	73	...	0.80	...	...	...
F714	SDR-PR	PE4708	...	73	...	0.80	...	...	...
F714	SDR-PR	PE4710	...	73	...	1.00	...	...	...
F2788/F2788M	SDR/DR-PR	PEX0006	...	−58	230	0.63	...	0.40	0.31
F2788/F2788M	SDR/DR-PR	PEX0008	...	−58	230	0.80	...	0.40	0.31
F2389	SDR6, SDR7.3, SDR11	PP	...	0	210	0.63	0.50	0.20	...
D1785	Sch. 40, 80, 120	PVC1120	12454	73	...	2.00	...	...	...
D1785	Sch. 40, 80, 120	PVC1220	12454	73	...	2.00	...	...	...
D1785	Sch. 40, 80, 120	PVC2120	14333	73	...	2.00	...	...	...
D1785	Sch. 40, 80, 120	PVC2116	14333	73	...	1.60	...	...	...
D1785	Sch. 40, 80, 120	PVC2112	14333	73	...	1.25	...	...	...
D1785	Sch. 40, 80, 120	PVC2110	14333	73	...	1.00	...	...	...
D2241	PR (SDR series)	PVC1120	12454	73	...	2.00	...	...	...
D2241	PR (SDR series)	PVC1220	12454	73	...	2.00	...	...	...
D2241	PR (SDR series)	PVC2120	14333	73	...	2.00	...	...	...
D2241	PR (SDR series)	PVC2116	14333	73	...	1.60	...	...	...
D2241	PR (SDR series)	PVC2112	14333	73	...	1.25	...	...	...
D2241	PR (SDR series)	PVC2110	14333	73	...	1.00	...	...	...

NOTES:
*These recommended limits are for low pressure applications with water and other fluids that do not significantly affect the properties of the thermoplastic. The upper temperature limits are reduced at higher pressures, depending on the combination of fluid and expected service life. Lower temperature limits are affected more by the environment, safeguarding, and installation conditions than by strength.
†These recommended limits apply only to materials listed. Manufacturers should be consulted for temperature limits on specific types and kinds of materials not listed.
‡Use these hydrostatic design stress (HDS) values at all lower temperatures.

and function. Specifications governing the production of these products dictate the permissible mechanical and dimensional variations, and code design calculations account for these variations.

Seamless Pipe and Tubing Seamless pipe and tubing may be formed by various methods. A common technique involves piercing solid round forgings, followed by rolling and drawing. Other techniques include forging and boring, extrusion, and static and centrifugal casting. Piercing frequently produces pipe with a less uniform wall thickness and concentricity of bore than is the case with products produced by other methods. Since seamless products have no weld joints, there is no reduction of strength due to weld joint efficiency.

Welded Pipe and Tubing These products are typically made by forming strips or plate into cylinders and seam-welding by various methods. Manufacturing by welding permits the production of larger-diameter pipe than is possible with seamless manufacturing methods, as well as larger diameter/wall thickness ratios. While strip and plate thickness may be more closely controlled than is possible for some seamless products, the specifications governing production are not always more stringent for welded products.

Weld quality has the potential of making the weld weaker than the base material. Depending on the welding method and the degree of nondestructive examination required by the product specification or dictated by the designer, the code assigns a joint efficiency ranging from 60 to 100 percent of the strength of the base material. Although some welding methods have the potential of producing short sections of partially fused joints that may develop into small leaks in corrosive conditions, proper matching of the

TABLE 10-20 Recommended Temperature Limits for Thermoplastics Used as Linings*

Materials	Minimum		Maximum	
	°C	°F	°C	°F
PFA	−198	−325	260	500
PTFE	−198	−325	260	500
FEP	−198	−325	204	400
ECTFE	−198	−325	171	340
ETFE	−198	−325	149	300
PVDF	−18	0	135	275
PP	−18	0	107	225
PVDC	−18	0	79	175

*Reprinted from ASME B31.3-2014, Table A323.4.3, with permission of the publisher, the American Society of Mechanical Engineers, New York. All rights reserved.

NOTE: These temperature limits are based on material tests and do not necessarily reflect evidence of successful use as piping component linings in specific fluid services at these temperatures. The designer should consult the manufacturer for specific applications, particularly as temperature limits are approached.

Abbreviations for plastics: ABS, acrylonitrile-butadiene-styrene; CPVC, chlorinated poly vinyl chloride; ECTFE, ethylene-chlorotrifluoroethylene; ETFE, ethylene-tetrafluoroethylene; PB, polybutylene; PE, polyethylene; PFA, perfluoroalkoxy copolymer; FEP, perfluoro (ethylene-propylene) copolymer; PTFE, polytetrafluoroethylene; PP, polypropylene; PVC, poly vinyl chloride; PVDC, poly vinylidene chloride; PVDF, poly vinylidene fluoride.

weld method and the type and extent of examination will result in highly reliable joints that are suitable for use in critical services. Welds must be considered when developing specifications for bending, flaring, or expanding welded pipe or tubing.

Tubing Tubing sizes typically reflect the actual outside diameter of the product. Pipe is manufactured to nominal diameters, which are not the same as the actual outside diameters for sizes 12 in and smaller. Facilities within the scope of the ASME B31 codes nearly exclusively use pipe, rather than tubing, for applications external to equipment. Tubing is commonly classified as suitable for either mechanical or pressure applications. Tubing is available in size and wall thickness combinations not normally produced as pipe. Tubing wall thickness (gauge) is specified as either average wall or minimum wall. Minimum wall is more costly than average wall, and because of closer tolerances on thickness and diameter, tubing of either gauge system is generally more costly than pipe. Tubing having outside diameters of 2⅜, 2⅞, 3½, and 4½ in are commonly available; however, these sizes are generally considered to be nonstandard for typical piping applications.

Table 10-22 gives some of the more common standard size and wall-thickness combinations together with capacity and weight.

Methods of Joining Pipe Piping joints must be reliably leak-tight and provide adequate mechanical strength to resist external loads due to thermal expansion, weight, wind, seismic activity, and other factors. Joints for pipe buried in soil may be subjected to unique external loads resulting from thermal expansion and contraction, settlement, and other factors. Joint designs that permit rotation about an axis perpendicular to the longitudinal axis of the pipe may be advantageous in certain situations.

Disassembly frequency and ease should be considered when selecting joining methods. Ideally the method for joining piping system components provides minimum installed cost, maintains its integrity throughout the lifetime of the facility, provides restraint against axial thrust due to internal pressure, provides strength against external loads equal to that of the pipe, permits unrestricted flow with minimum pressure drop, and is free from crevices that may be detrimental to the product or contribute to corrosion or erosion problems.

Joint design and selection generally involve compromising between the ideal and practical. A number of manufacturers produce patented or "proprietary" joints that embody many ideal characteristics. Some are excellent products and are well suited to special applications. Valves and fittings are often available with proprietary joints that have gained wide acceptance; however, consideration should be given to the possible impact on product delivery time and cost.

Welded Joints The most widely used joint in piping systems is the *butt-weld joint* (Fig. 10-117). In all ductile pipe metals which can be welded, pipe, elbows, tees, laterals, reducers, caps, valves, flanges, and V-clamp joints are available in all sizes and wall thicknesses with ends prepared for butt welding. Joint strength equal to the original pipe (except for work-hardened pipes which are annealed by the welding), unimpaired flow pattern, and generally unimpaired corrosion resistance more than compensate for the necessary careful alignment, skilled labor, and equipment required.

Plain-end pipe used for socket-weld joints (Fig. 10-118) is available in all sizes, but fittings and valves with socket-weld ends are limited to sizes 3 in and smaller, for which the extra cost of the socket is outweighed by much easier alignment and less skill needed in welding.

Socket-welded joints are not as resistant to externally applied bending moments as are butt-welded joints, are not easily examined by volumetric nondestructive examination methods such as radiography and ultrasonic, and should not be used where crevice corrosion has been determined to be of concern. However, they are widely used in sizes 2 in and smaller and are quite satisfactory for most applications when used within the limits established by code restrictions and good engineering judgment. Components with socket-welded ends are generally specified as requiring compliance with ASME B16.11, *Forged-Fittings, Socket-Welding and Threaded.*

Branch Connections Branch connections may be made with manufactured tees, fabricated reinforced and nonreinforced branch connections (Fig. 10-119), or manufactured integrally reinforced branch connections. Butt-welded fittings offer the best opportunity for nondestructive examination; however, branch connections are commonly specified for branches smaller than the header, and often best satisfy the design and economic requirements. Design of fabricated branch connections is addressed in the subsection Pressure Design of Metallic Components: Wall Thickness. Integrally reinforced fittings are generally specified as requiring compliance with Manufacturer's Standardization Society specification MSS SP-97, *Integrally Reinforced Forged Branch Outlet Fittings—Socket Welding, Threaded, and Butt Welding Ends.*

Threaded Joints Pipe with *taper-pipe-thread* ends (Fig. 10-120), per ASME B1.20.1, is available 12 in and smaller, subject to minimum-wall limitations. Fittings and valves with taper-pipe-thread ends are available in most pipe metals.

Principal use of threaded joints is in sizes 2 in and smaller, in metals for which the most economically produced walls are thick enough to withstand pressure and corrosion after reduction in thickness due to threading. For threaded joints over 2 in, assembly difficulty and cost of tools increase rapidly. Careful alignment, required at the start of assembly and during rotation of the components, as well as variation in length produced by diametral tolerances in the threads, severely limits preassembly of the components. Threading is not a precise machining operation, and filler materials known as "pipe dope" are necessary to block the spiral leakage path.

Threads notch the pipe and cause loss of strength and fatigue resistance. Enlargement and contraction of the flow passage at threaded joints creates turbulence; sometimes corrosion and erosion are aggravated at the point where the pipe has already been thinned by threading. The tendency of pipe wrenches to crush pipe and fittings limits the torque available for tightening threaded joints. For low-pressure systems, a slight rotation in the joint may be used to impart flexibility to the system, but this same rotation, unwanted, may cause leaks to develop in higher-pressure systems. In some metals, galling occurs when threaded joints are disassembled.

Straight Pipe Threads These are confined to lightweight couplings in sizes 2 in and smaller (Fig. 10-121). Manufacturers of threaded pipe ship it with such couplings installed on one end of each pipe. The joint obtained is inferior to that obtained with taper threads. The code limits the joint shown in Fig. 10-121 to 1.0 MPa (150 lbf/in^2) gauge maximum, 186°C (366°F) maximum, and to nonflammable, nontoxic fluids.

When both components of a threaded joint are of weldable metal, the joint may be *seal-welded* as shown in Fig. 10-122. Seal welds may be used only to prevent leakage of threaded joints. They are not considered as contributing any strength to the joint. This type of joint is limited to new construction and is not suitable as a repair procedure, since pipe dope in the threads would interfere with welding. Careful consideration should be given to the

TABLE 10-21 Recommended Temperature Limits for Reinforced Thermosetting Resin Pipe*

Materials		Recommended temperature limits			
		Minimum		Maximum	
Resin	Reinforcing	°C	°F	°C	°F
Epoxy	Glass fiber	−29	−20	149	300
Phenolic	Glass fiber	−29	−20	149	300
Furan	Carbon	−29	−20	93	200
Furan	Glass fiber	−29	−20	93	200
Polyester	Glass fiber	−29	−20	93	200
Vinyl ester	Glass fiber	−29	−20	93	200

*Reprinted from ASME B31.3-2014, Table A323.4.2C, with permission of the publisher, the American Society of Mechanical Engineers, New York. All rights reserved.

NOTE: These temperature limits apply only to materials listed and do not reflect evidence of successful use in specific fluid services at these temperatures. The designer should consult the manufacturer for specific applications, particularly as the temperature limits are approached.

TABLE 10-22 Properties of Steel Pipe

					Cross-sectional area		Circumference, ft, or surface, ft²/ft of length		Capacity at 1 ft/s velocity		
Nominal pipe size, in	Outside diameter, in	Schedule no.	Wall thickness, in	Inside diameter, in	Metal, in²	Flow, ft²	Outside	Inside	U.S. gal/min	lb/h water	Weight of plain-end pipe, lb/ft
⅛	0.405	10S	0.049	0.307	0.055	0.00051	0.106	0.0804	0.231	115.5	0.19
		40ST, 40S	.068	.269	.072	.00040	.106	.0705	.179	89.5	.24
		80XS, 80S	.095	.215	.093	.00025	.106	.0563	.113	56.5	.31
¼	0.540	10S	.065	.410	.097	.00092	.141	.107	.412	206.5	.33
		40ST, 40S	.088	.364	.125	.00072	.141	.095	.323	161.5	.42
		80XS, 80S	.119	.302	.157	.00050	.141	.079	.224	112.0	.54
⅜	0.675	10S	.065	.545	.125	.00162	.177	.143	.727	363.5	.42
		40ST, 40S	.091	.493	.167	.00133	.177	.129	.596	298.0	.57
		80XS, 80S	.126	.423	.217	.00098	.177	.111	.440	220.0	.74
½	0.840	5S	.065	.710	.158	.00275	.220	.186	1.234	617.0	.54
		10S	.083	.674	.197	.00248	.220	.176	1.112	556.0	.67
		40ST, 40S	.109	.622	.250	.00211	.220	.163	0.945	472.0	.85
		80XS, 80S	.147	.546	.320	.00163	.220	.143	0.730	365.0	1.09
		160	.188	.464	.385	.00117	.220	.122	0.527	263.5	1.31
		XX	.294	.252	.504	.00035	.220	.066	0.155	77.5	1.71
¾	1.050	5S	.065	.920	.201	.00461	.275	.241	2.072	1036.0	0.69
		10S	.083	.884	.252	.00426	.275	.231	1.903	951.5	0.86
		40ST, 40S	.113	.824	.333	.00371	.275	.216	1.665	832.5	1.13
		80XS, 80S	.154	.742	.433	.00300	.275	.194	1.345	672.5	1.47
		160	.219	.612	.572	.00204	.275	.160	0.917	458.5	1.94
		XX	.308	.434	.718	.00103	.275	.114	0.461	230.5	2.44
1	1.315	5S	.065	1.185	.255	.00768	.344	.310	3.449	1725	0.87
		10S	.109	1.097	.413	.00656	.344	.287	2.946	1473	1.40
		40ST, 40S	.133	1.049	.494	.00600	.344	.275	2.690	1345	1.68
		80XS, 80S	.179	0.957	.639	.00499	.344	.250	2.240	1120	2.17
		160	.250	0.815	.836	.00362	.344	.213	1.625	812.5	2.84
		XX	.358	0.599	1.076	.00196	.344	.157	0.878	439.0	3.66
1¼	1.660	5S	.065	1.530	0.326	.01277	.435	.401	5.73	2865	1.11
		10S	.109	1.442	0.531	.01134	.435	.378	5.09	2545	1.81
		40ST, 40S	.140	1.380	0.668	.01040	.435	.361	4.57	2285	2.27
		80XS, 80S	.191	1.278	0.881	.00891	.435	.335	3.99	1995	3.00
		160	.250	1.160	1.107	.00734	.435	.304	3.29	1645	3.76
		XX	.382	0.896	1.534	.00438	.435	.235	1.97	985	5.21
1½	1.900	5S	.065	1.770	0.375	.01709	.497	.463	7.67	3835	1.28
		10S	.109	1.682	0.614	.01543	.497	.440	6.94	3465	2.09
		40ST, 40S	.145	1.610	0.800	.01414	.497	.421	6.34	3170	2.72
		80XS, 80S	.200	1.500	1.069	.01225	.497	.393	5.49	2745	3.63
		160	.281	1.338	1.429	.00976	.497	.350	4.38	2190	4.86
		XX	.400	1.100	1.885	.00660	.497	.288	2.96	1480	6.41
2	2.375	5S	.065	2.245	0.472	.02749	.622	.588	12.34	6170	1.61
		10S	.109	2.157	0.776	.02538	.622	.565	11.39	5695	2.64
		40ST, 40S	.154	2.067	1.075	.02330	.622	.541	10.45	5225	3.65
		80ST, 80S	.218	1.939	1.477	.02050	.622	.508	9.20	4600	5.02
		160	.344	1.687	2.195	.01552	.622	.436	6.97	3485	7.46
		XX	.436	1.503	2.656	.01232	.622	.393	5.53	2765	9.03
2½	2.875	5S	.083	2.709	0.728	.04003	.753	.709	17.97	8985	2.48
		10S	.120	2.635	1.039	.03787	.753	.690	17.00	8500	3.53
		40ST, 40S	.203	2.469	1.704	.03322	.753	.647	14.92	7460	5.79
		80XS, 80S	.276	2.323	2.254	.02942	.753	.608	13.20	6600	7.66
		160	.375	2.125	2.945	.02463	.753	.556	11.07	5535	10.01
		XX	.552	1.771	4.028	.01711	.753	.464	7.68	3840	13.69
3	3.500	5S	.083	3.334	0.891	.06063	.916	.873	27.21	13,605	3.03
		10S	.120	3.260	1.274	.05796	.916	.853	26.02	13,010	4.33
		40ST, 40S	.216	3.068	2.228	.05130	.916	.803	23.00	11,500	7.58
		80XS, 80S	.300	2.900	3.016	.04587	.916	.759	20.55	10,275	10.25
		160	.438	2.624	4.213	.03755	.916	.687	16.86	8430	14.32
		XX	.600	2.300	5.466	.02885	.916	.602	12.95	6475	18.58
3½	4.0	5S	.083	3.834	1.021	.08017	1.047	1.004	35.98	17,990	3.48
		10S	.120	3.760	1.463	.07711	1.047	0.984	34.61	17,305	4.97
		40ST, 40S	.226	3.548	2.680	.06870	1.047	0.929	30.80	15,400	9.11
		80XS, 80S	.318	3.364	3.678	.06170	1.047	0.881	27.70	13,850	12.50
4	4.5	5S	.083	4.334	1.152	.10245	1.178	1.135	46.0	23,000	3.92
		10S	.120	4.260	1.651	.09898	1.178	1.115	44.4	22,200	5.61
		40ST, 40S	.237	4.026	3.17	.08840	1.178	1.054	39.6	19,800	10.79
		80XS, 80S	.337	3.826	4.41	.07986	1.178	1.002	35.8	17,900	14.98
		120	0.438	3.624	5.58	0.07170	1.178	0.949	32.2	16,100	19.00
		160	.531	3.438	6.62	.06647	1.178	0.900	28.9	14,450	22.51
		XX	.674	3.152	8.10	.05419	1.178	0.825	24.3	12,150	27.54

TABLE 10-22 Properties of Steel Pipe *(Continued)*

					Cross-sectional area		Circumference, ft, or surface, ft^2/ft of length		Capacity at 1 ft/s velocity		
Nominal pipe size, in	Outside diameter, in	Schedule no.	Wall thickness, in	Inside diameter, in	Metal, in^2	Flow, ft^2	Outside	Inside	U.S. gal/min	lb/h water	Weight of plain-end pipe, lb/ft
5	5.563	5S	.109	5.345	1.87	.1558	1.456	1.399	69.9	34,950	6.36
		10S	.134	5.295	2.29	.1529	1.456	1.386	68.6	34,300	7.77
		40ST, 40S	.258	5.047	4.30	.1390	1.456	1.321	62.3	31,150	14.62
		80XS, 80S	.375	4.813	6.11	.1263	1.456	1.260	57.7	28,850	20.78
		120	.500	4.563	7.95	.1136	1.456	1.195	51.0	25,500	27.04
		160	.625	4.313	9.70	.1015	1.456	1.129	45.5	22,750	32.96
		XX	.750	4.063	11.34	.0900	1.456	1.064	40.4	20,200	38.55
6	6.625	5S	.109	6.407	2.23	.2239	1.734	1.677	100.5	50,250	7.60
		10S	.134	6.357	2.73	.2204	1.734	1.664	98.9	49,450	9.29
		40ST, 40S	.280	6.065	5.58	.2006	1.734	1.588	90.0	45,000	18.97
		80XS, 80S	.432	5.761	8.40	.1810	1.734	1.508	81.1	40,550	28.57
		120	.562	5.501	10.70	.1650	1.734	1.440	73.9	36,950	36.39
		160	.719	5.187	13.34	.1467	1.734	1.358	65.9	32,950	45.34
		XX	.864	4.897	15.64	.1308	1.734	1.282	58.7	29,350	53.16
8	8.625	5S	.109	8.407	2.915	.3855	2.258	2.201	173.0	86,500	9.93
		10S	.148	8.329	3.941	.3784	2.258	2.180	169.8	84,900	13.40
		20	.250	8.125	6.578	.3601	2.258	2.127	161.5	80,750	22.36
		30	.277	8.071	7.265	.3553	2.258	2.113	159.4	79,700	24.70
		40ST, 40S	.322	7.981	8.399	.3474	2.258	2.089	155.7	77,850	28.55
		60	.406	7.813	10.48	.3329	2.258	2.045	149.4	74,700	35.64
		80XS, 80S	.500	7.625	12.76	.3171	2.258	1.996	142.3	71,150	43.39
		100	.594	7.437	14.99	.3017	2.258	1.947	135.4	67,700	50.95
		120	.719	7.187	17.86	.2817	2.258	1.882	126.4	63,200	60.71
		140	.812	7.001	19.93	.2673	2.258	1.833	120.0	60,000	67.76
		XX	.875	6.875	21.30	.2578	2.258	1.800	115.7	57,850	72.42
		160	.906	6.813	21.97	.2532	2.258	1.784	113.5	56,750	74.69
10	10.75	5S	.134	10.482	4.47	.5993	2.814	2.744	269.0	134,500	15.19
		10S	.165	10.420	5.49	.5922	2.814	2.728	265.8	132,900	18.65
		20	.250	10.250	8.25	.5731	2.814	2.685	257.0	128,500	28.04
		30	.307	10.136	10.07	.5603	2.814	2.655	252.0	126,000	34.24
		40ST, 40S	.365	10.020	11.91	.5475	2.814	2.620	246.0	123,000	40.48
		80S, 60XS	.500	9.750	16.10	.5185	2.814	2.550	233.0	116,500	54.74
		80	.594	9.562	18.95	.4987	2.814	2.503	223.4	111,700	64.43
		100	.719	9.312	22.66	.4729	2.814	2.438	212.3	106,150	77.03
		120	.844	9.062	26.27	.4479	2.814	2.372	201.0	100,500	89.29
		140, XX	1.000	8.750	30.63	.4176	2.814	2.291	188.0	94,000	104.13
		160	1.125	8.500	34.02	.3941	2.814	2.225	177.0	88,500	115.64
12	12.75	5S	0.156	12.438	6.17	.8438	3.338	3.26	378.7	189,350	20.98
		10S	0.180	12.390	7.11	.8373	3.338	3.24	375.8	187,900	24.17
		20	0.250	12.250	9.82	.8185	3.338	3.21	367.0	183,500	33.38
		30	0.330	12.090	12.88	.7972	3.338	3.17	358.0	179,000	43.77
		ST, 40S	0.375	12.000	14.58	.7854	3.338	3.14	352.5	176,250	49.56
		40	0.406	11.938	15.74	.7773	3.338	3.13	349.0	174,500	53.52
		XS, 80S	0.500	11.750	19.24	.7530	3.338	3.08	338.0	169,000	65.42
		60	0.562	11.626	21.52	.7372	3.338	3.04	331.0	165,500	73.15
		80	0.688	11.374	26.07	.7056	3.338	2.98	316.7	158,350	88.63
		100	0.844	11.062	31.57	.6674	3.338	2.90	299.6	149,800	107.32
		120, XX	1.000	10.750	36.91	.6303	3.338	2.81	283.0	141,500	125.49
		140	1.125	10.500	41.09	.6013	3.338	2.75	270.0	135,000	139.67
		160	1.312	10.126	47.14	.5592	3.338	2.65	251.0	125,500	160.27
14	14	5S	0.156	13.688	6.78	1.0219	3.665	3.58	459	229,500	23.07
		10S	0.188	13.624	8.16	1.0125	3.665	3.57	454	227,000	27.73
		10	0.250	13.500	10.80	0.9940	3.665	3.53	446	223,000	36.71
		20	0.312	13.376	13.42	0.9750	3.665	3.50	438	219,000	45.61
		30, ST	0.375	13.250	16.05	0.9575	3.665	3.47	430	215,000	54.57
		40	0.438	13.124	18.66	0.9397	3.665	3.44	422	211,000	63.44
		XS	0.500	13.000	21.21	0.9218	3.665	3.40	414	207,000	72.09
		60	0.594	12.812	25.02	0.8957	3.665	3.35	402	201,000	85.05
		80	0.750	12.500	31.22	0.8522	3.665	3.27	382	191,000	106.13
		100	0.938	12.124	38.49	0.8017	3.665	3.17	360	180,000	130.85
		120	1.094	11.812	44.36	0.7610	3.665	3.09	342	171,000	150.79
		140	1.250	11.500	50.07	0.7213	3.665	3.01	324	162,000	170.21
		160	1.406	11.188	55.63	0.6827	3.665	2.93	306	153,000	189.11
16	16	5S	0.165	15.670	8.21	1.3393	4.189	4.10	601	300,500	27.90
		10S	0.188	15.624	9.34	1.3314	4.189	4.09	598	299,000	31.75
		10	0.250	15.500	12.37	1.3104	4.189	4.06	587	293,500	42.05
		20	0.312	15.376	15.38	1.2985	4.189	4.03	578	289,000	52.27
		30, ST	0.375	15.250	18.41	1.2680	4.189	3.99	568	284,000	62.58
		40, XS	0.500	15.000	24.35	1.2272	4.189	3.93	550	275,000	82.77
		60	0.656	14.688	31.62	1.1766	4.189	3.85	528	264,000	107.50
		80	0.844	14.312	40.19	1.1171	4.189	3.75	501	250,500	136.61
		100	1.031	13.938	48.48	1.0596	4.189	3.65	474	237,000	164.82
		120	1.219	13.562	56.61	1.0032	4.189	3.55	450	225,000	192.43
		140	1.438	13.124	65.79	0.9394	4.189	3.44	422	211,000	223.64
		160	1.594	12.812	72.14	0.8953	4.189	3.35	402	201,000	245.25

TABLE 10-22 Properties of Steel Pipe *(Continued)*

Nominal pipe size, in	Outside diameter, in	Schedule no.	Wall thickness, in	Inside diameter, in	Cross-sectional area: Metal, in^2	Cross-sectional area: Flow, ft^2	Circumference, ft, or surface, ft^2/ft of length: Outside	Circumference, ft, or surface, ft^2/ft of length: Inside	Capacity at 1 ft/s velocity: U.S. gal/min	Capacity at 1 ft/s velocity: lb/h water	Weight of plain-end pipe, lb/ft
18	18	5S	0.165	17.670	9.25	1.7029	4.712	4.63	764	382,000	31.43
		10S	0.188	17.624	10.52	1.6941	4.712	4.61	760	379,400	35.76
		10	0.250	17.500	13.94	1.6703	4.712	4.58	750	375,000	47.39
		20	0.312	17.376	17.34	1.6468	4.712	4.55	739	369,500	58.94
		ST	0.375	17.250	20.76	1.6230	4.712	4.52	728	364,000	70.59
		30	0.438	17.124	24.16	1.5993	4.712	4.48	718	359,000	82.15
		XS	0.500	17.000	27.49	1.5763	4.712	4.45	707	353,500	93.45
		40	0.562	16.876	30.79	1.5533	4.712	4.42	697	348,500	104.67
		60	0.750	16.500	40.64	1.4849	4.712	4.32	666	333,000	138.17
		80	0.938	16.124	50.28	1.4180	4.712	4.22	636	318,000	170.92
		100	1.156	15.688	61.17	1.3423	4.712	4.11	602	301,000	207.96
		120	1.375	15.250	71.82	1.2684	4.712	3.99	569	284,500	244.14
		140	1.562	14.876	80.66	1.2070	4.712	3.89	540	270,000	274.22
		160	1.781	14.438	90.75	1.1370	4.712	3.78	510	255,000	308.50
20	20	5S	0.188	19.624	11.70	2.1004	5.236	5.14	943	471,500	39.78
		10S	0.218	19.564	13.55	2.0878	5.236	5.12	937	467,500	46.06
		10	0.250	19.500	15.51	2.0740	5.236	5.11	930	465,000	52.73
		20, ST	0.375	19.250	23.12	2.0211	5.236	5.04	902	451,000	78.60
		30, XS	0.500	19.000	30.63	1.9689	5.236	4.97	883	441,500	104.13
		40	0.594	18.812	36.21	1.9302	5.236	4.92	866	433,000	123.11
		60	0.812	18.376	48.95	1.8417	5.236	4.81	826	413,000	166.40
		80	1.031	17.938	61.44	1.7550	5.236	4.70	787	393,500	208.87
		100	1.281	17.438	75.33	1.6585	5.236	4.57	744	372,000	256.10
		120	1.500	17.000	87.18	1.5763	5.236	4.45	707	353,500	296.37
		140	1.750	16.500	100.3	1.4849	5.236	4.32	665	332,500	341.09
		160	1.969	16.062	111.5	1.4071	5.236	4.21	632	316,000	397.17
24	24	5S	0.218	23.564	16.29	3.0285	6.283	6.17	1359	679,500	55.37
		10, 10S	0.250	23.500	18.65	3.012	6.283	6.15	1350	675,000	63.41
		20, ST	0.375	23.250	27.83	2.948	6.283	6.09	1325	662,500	94.62
		XS	0.500	23.000	36.90	2.885	6.283	6.02	1295	642,500	125.49
		30	0.562	22.876	41.39	2.854	6.283	5.99	1281	640,500	140.68
		40	0.688	22.624	50.39	2.792	6.283	5.92	1253	626,500	171.29
		60	0.969	22.062	70.11	2.655	6.283	5.78	1192	596,000	238.35
		80	1.219	21.562	87.24	2.536	6.283	5.64	1138	569,000	296.58
		100	1.531	20.938	108.1	2.391	6.283	5.48	1073	536,500	367.39
		120	1.812	20.376	126.3	2.264	6.283	5.33	1016	508,000	429.39
		140	2.062	19.876	142.1	2.155	6.283	5.20	965	482,500	483.12
		160	2.344	19.312	159.5	2.034	6.283	5.06	913	456,500	542.13
30	30	5S	0.250	29.500	23.37	4.746	7.854	7.72	2130	1,065,000	79.43
		10, 10S	0.312	29.376	29.10	4.707	7.854	7.69	2110	1,055,000	98.93
		ST	0.375	29.250	34.90	4.666	7.854	7.66	2094	1,048,000	118.65
		20, XS	0.500	29.000	46.34	4.587	7.854	7.59	2055	1,027,500	157.53
		30	0.625	28.750	57.68	4.508	7.854	7.53	2020	1,010,000	196.08

5S, 10S, 40S, and 80S are extracted from *Stainless Steel Pipe*, ASME B36.19M-2004, with permission of the publisher, the American Society of Mechanical Engineers, New York. All rights reserved. ST = standard wall, XS = extra strong wall, XX = double extra strong wall, and Schedules 10 through 160 are extracted *from Welded and Seamless Wrought Steel Pipe*, ASME B36.10M-2004, with permission of the same publisher. Refer to these standards for a more comprehensive listing of material sizes and wall thicknesses. Decimal thicknesses for respective pipe sizes represent their nominal or average wall dimensions. Mill tolerances as high as ±12½ percent are permitted.

Plain-end pipe is produced by a square cut. Pipe is also shipped from the mills threaded, with a threaded coupling on one end, or with the ends beveled for welding, or grooved or sized for patented couplings.

To convert inches to millimeters, multiply by 25.4; to convert square inches to square millimeters, multiply by 645; to convert feet to meters, multiply by 0.3048; to convert square feet to square meters, multiply by 0.0929; to convert pounds per foot to kilograms per meter, multiply by 1.49; to convert gallons to cubic meters, multiply by 3.7854×10^{-3}; and to convert pounds to kilograms, multiply by 0.4536.

suitability of threaded joints when joining metals having significantly different coefficients of expansion. Thermal expansion and temperature cycling may eventually result in leakage.

To assist in assembly and disassembly of both threaded and welded systems, *union joints* (Fig. 10-123) are used. They comprise metal-to-metal or gasketed seats drawn together by a shouldered straight thread nut and are available both in couplings for joining two lengths of pipe and on the ends of some fittings. On threaded piping systems in which disassembly is not contemplated, union joints installed at intervals permit future further tightening of threaded joints. Tightening of heavy unions yields tight joints even if the pipe is slightly misaligned at the start of tightening.

Flanged Joints For sizes larger than 2 in when disassembly is contemplated, the flanged joint (Fig. 10-124) is the most widely used. Figures 10-125 and 10-126 illustrate the wide variety of types and facings available. Though

FIG. 10-117 Butt weld.

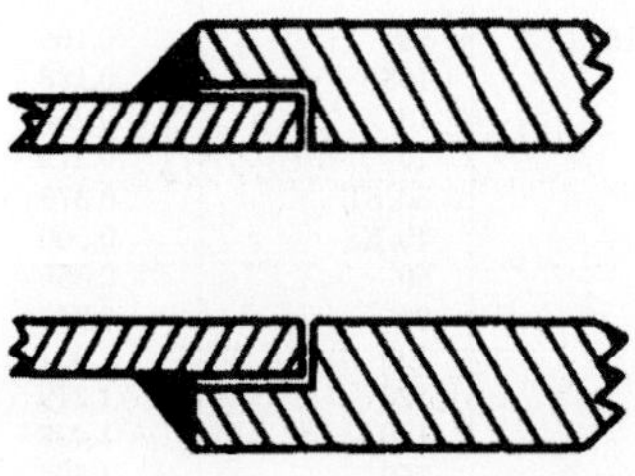

FIG. 10-118 Socket weld.

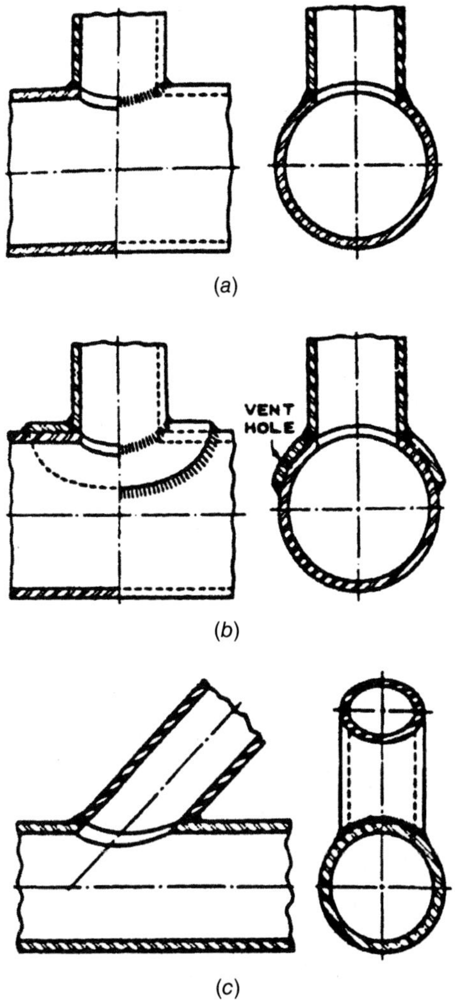

FIG. 10-119 Branch welds. (*a*) Without added reinforcement. (*b*) With added reinforcement. (*c*) Angular branch.

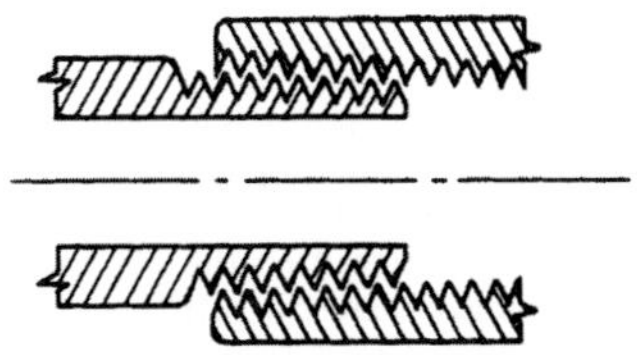

FIG. 10-120 Taper pipe thread.

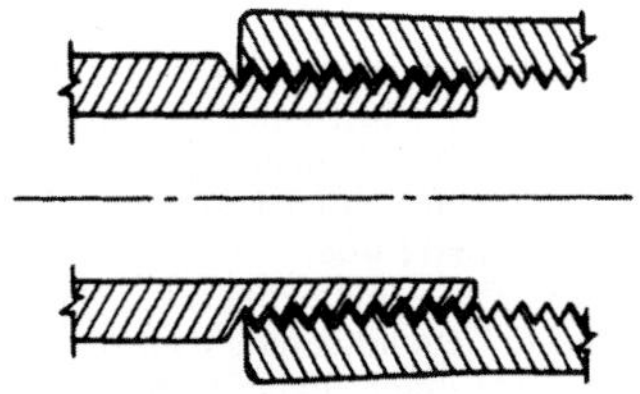

FIG. 10-121 Taper pipe to straight coupling thread.

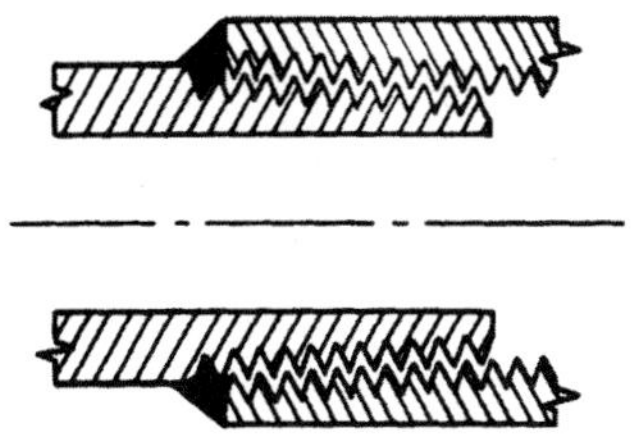

FIG. 10-122 Taper pipe thread seal-welded.

flanged joints consume a large volume of metal, precise machining is required only on the facing. Flanged joints do not impose severe diametral tolerances on the pipe. Alignment tolerances required for flanged joints are reasonably achieved with quality construction practices, and in comparison, with taper threaded joints, required wrench sizes are smaller and sealing is more easily and reliably obtained.

Manufacturers offer *flanged-end pipe* in only a few metals. Otherwise, flanges are attached to pipe by various types of joints (Fig. 10-125). The lap joint involves a modification of the pipe which may be formed from the pipe itself or by welding a ring or a lap-joint stub end to it. *Flanged-end fittings* and valves are available in all sizes of most pipe metals.

Of the flange types shown in Fig. 10-125, welding-neck flanges offer the highest mechanical strength and are the type most suitable for extreme temperatures and cyclic loading. Regardless of the type selected, designers must be aware that the flange's capability to resist external bending moments and maintain its seal does not necessarily match the bending moment capability of the attached pipe.

When selecting the flange type, the designer should review the usage restrictions contained in each section of the ASME B31 code. Each of the other types shown provides significant fabrication and economic advantages and is suitable for many of the routine applications. Lap-joint flanges permit adjustment of the bolt-hole orientation and can greatly simplify construction when configurations are complex and bolt-hole orientations are difficult to ensure.

Dimensions for alloy and carbon-steel pipe flanges are given in Tables 10-23 through 10-27. The dimensions were extracted from *Pipe Flanges and Flanged Fittings*, ASME B16.5–2013, with permission of the publisher, the American Society of Mechanical Engineers, New York. Class 400 and 2500 are not included due to the limited use of these ratings. If needed, the user can refer to ASME B16.5 for dimensions. Dimensions for cast-iron flanges are provided in *Cast Iron Pipe Flanges and Flanged Fittings*, ASME B16.1. Bolt patterns and bolt sizes for Class 125 cast-iron flanges match the ASME B16.5 Class 150 flange dimensions, and bolt patterns for Class 250 cast-iron flanges match the ASME B16.5 Class 300 flange dimensions.

When mating with cast-iron flanged fittings or valves, steel pipe flanges are often preferred to cast-iron flanges because they permit welded rather than screwed assembly to the pipe and because cast-iron pipe flanges, not being reinforced by the pipe, are not so resistant to abuse as flanges cast integrally on cast-iron fittings.

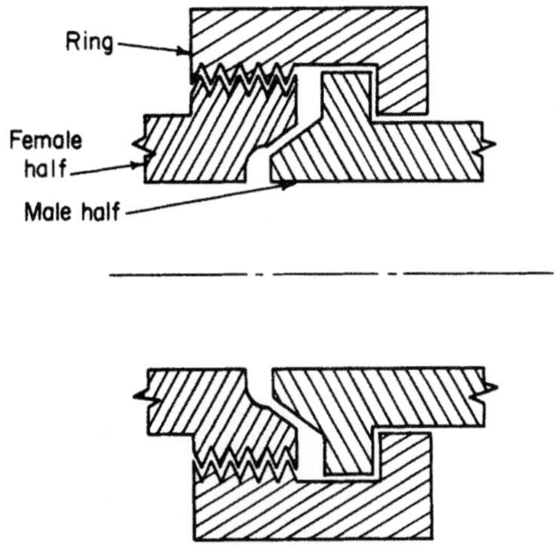

FIG. 10-123 Union.

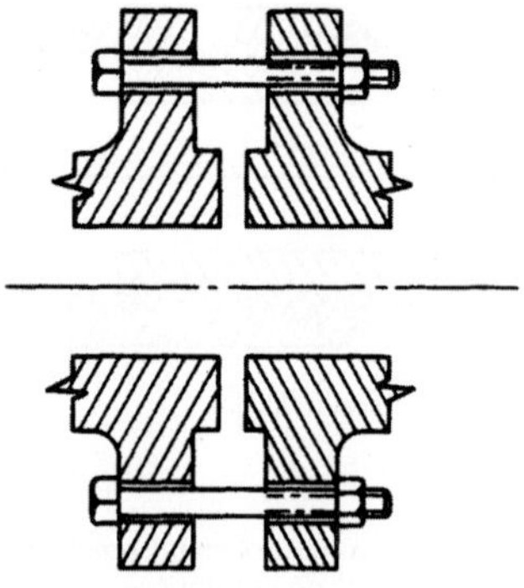

FIG. 10-124 Flanged joint.

Facing of flanges for alloy and carbon-steel pipe and fittings is shown in Fig. 10-126; Class 125 cast-iron pipe and fitting flanges have flat faces, which with full-face gaskets minimize bending stresses; Class 250 cast-iron pipe and fitting flanges have 1.5-mm (1/16-in) raised faces (wider than on steel flanges) for the same purpose. Carbon-steel and ductile- (nodular-) iron lap-joint flanges are widely used as backup flanges with stub ends in piping systems of austenitic stainless steel and other expensive materials to reduce costs (see Fig. 10-125). The code prohibits the use of ductile-iron flanges at temperatures above 343°C (650°F). When the type of facing affects the length through the hub dimension of flanges, correct dimensions for commonly used facings can be determined from the dimensional data in Tables 10-23 through 10-27.

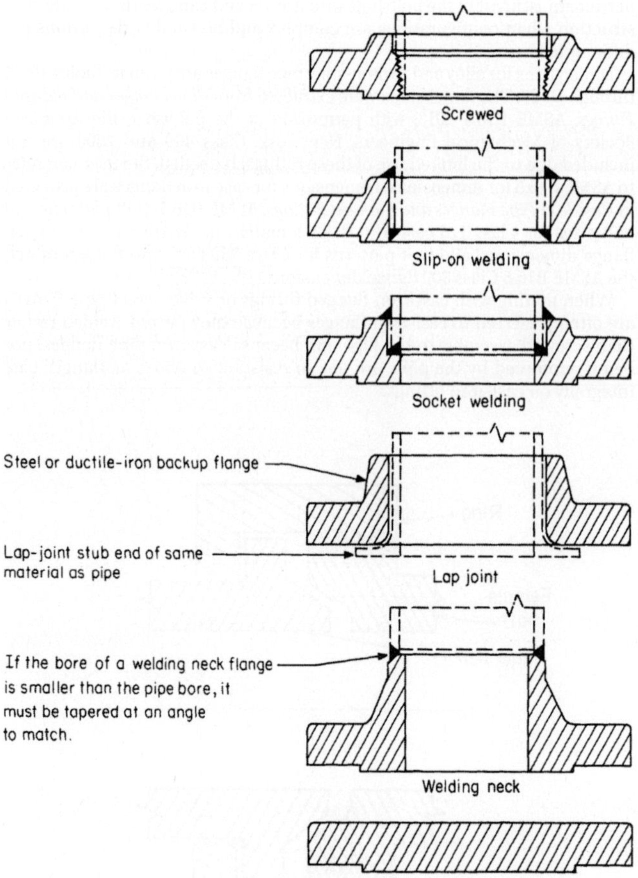

FIG. 10-125 Types of carbon and alloy steel flanges.

Gaskets Gaskets must resist corrosion by the fluids handled. The more expensive male-and-female or tongue-and-groove facings may be required to seat hard gaskets adequately. With these facings the gasket generally cannot blow out. Flanged joints, by placing the gasket material under heavy compression and permitting only edge attack by the fluid handled, can use gasket materials which in other joints might not satisfactorily resist the fluid handled.

Standards to which flanges are manufactured (e.g., ASME B16.1, ASME B16.5, ASME B16.47) typically specify a standard surface finish for the gasket seal area. Flange rating and type, flange size, flange facing type, gasket style, commodity, design conditions, and bolting must all be considered to ensure proper seating of the gasket and reliable performance. Unless the user is familiar with gasket design and the particular application being considered, it is highly recommended that the gasket manufacturer be consulted regarding gasket selection. Upon request, gasket manufacturers typically provide assistance in determining the proper material selection and the proper gasket style to ensure an economical choice and a reliable system.

When appropriate for the commodity, elastomer sheet gaskets without fillers are generally the least expensive gasket type. They are typically limited to Class 150 and temperatures below 120°C (250°F). Composition sheet gaskets are somewhat more expensive than elastomer sheet gaskets. They are generally composed of an elastomer binder with fiber filler. Their use is generally limited to Class 150 and Class 300, and depending on the filler selected the upper temperature limit typically ranges between 205°C (400°F) and 370°C (700°F). Nonelastomer sheet gaskets, such as graphite sheet gaskets, are generally somewhat more expensive than composition sheet gaskets. Their use is typically limited to Class 150 and Class 300, and the upper temperature limit may be 535°C (1000°F) or higher. Spiral-wound gaskets with graphite, PTFE, or other filler are generally appropriate for applications more demanding than those handled by sheet gaskets. They are generally more expensive than sheet gaskets and are commonly used in Class 150 through Class 1500 services (and higher) class ratings. Because of their breadth of capabilities and the advantages of standardizing, they are often used when less expensive gaskets would suffice. The solid metal outer ring on spiral-wound gaskets serves to center the gasket and provide blowout resistance. With the proper filler, spiral-wound gaskets and some sheet gaskets provide good sealing under fire conditions.

Ring Joint Flanges Ring joint (RTJ) flanges provide sealing capability for pressure-temperature combinations higher than those for which spiral-wound gaskets are typically used. Depending on the service, use of RTJ flanges is often considered in Class 900 and higher applications. RTJ flange facings and gaskets are more expensive than the spiral-wound counterparts. The ring material must be softer than the flange seating surface and corrosion-resistant to the service. They provide good resistance to leakage under fire conditions. RTJ flanges must be separated in the axial direction to permit insertion and removal of the gasket.

Bolting Bolt strength requirements are addressed to some extent by the code and by code-referenced flange standards. Bolts are categorized by the code as high strength, intermediate strength, and low strength. Bolting materials having allowable stresses meeting or exceeding those of ASTM A193 Grade B7 are categorized as high strength. Bolting materials having specified minimum yield strengths of 207 MPa (30 ksi) or less are categorized as low strength. ASTM A307 Grade B is a commonly used specification for low-strength bolting.

The suitability of the strength of any bolting throughout the required temperature range should be verified by the designer. Verification of the suitability of intermediate-strength bolting for the intended joint design is required prior to its use. The code restricts the use of low-strength bolting to nonmetallic gaskets and flanges rated ASME B16.5 Class 300 and lower having bolt temperatures at −29 to 204°C (−20 to 400°F) inclusive. Low-strength bolting is not permitted for use under severe cyclic conditions as defined by the code.

Except when bolting brittle flange materials such as gray cast iron, the code permits the use of high-strength bolting for any style of flanged joint and gasket type of combination. Per the code, if either mating flange is specified in accordance with ASME B16.1, ASME B16.24, MSS SP-42, or MSS SP-51, then the bolting material shall be no stronger than low-strength unless both mating flanges have a flat face and a full-face gasket is used. Exception to this requirement is permitted if sequence and torque limits for bolt-up are specified, with consideration given to sustained and occasional loads, and displacement strains. When both flanges are flat face and the gasket is full face extending to the outside diameter of the flange, intermediate-strength and high-strength bolts may be used.

Miscellaneous Mechanical Joints

Packed-Gland Joints These joints (Fig. 10-127) require no special end preparation of pipe but do require careful control of the diameter of the pipe. Thus the supplier of the pipe should be notified when packed-gland joints are to be used. Cast- and ductile-iron pipe, fittings, and valves are

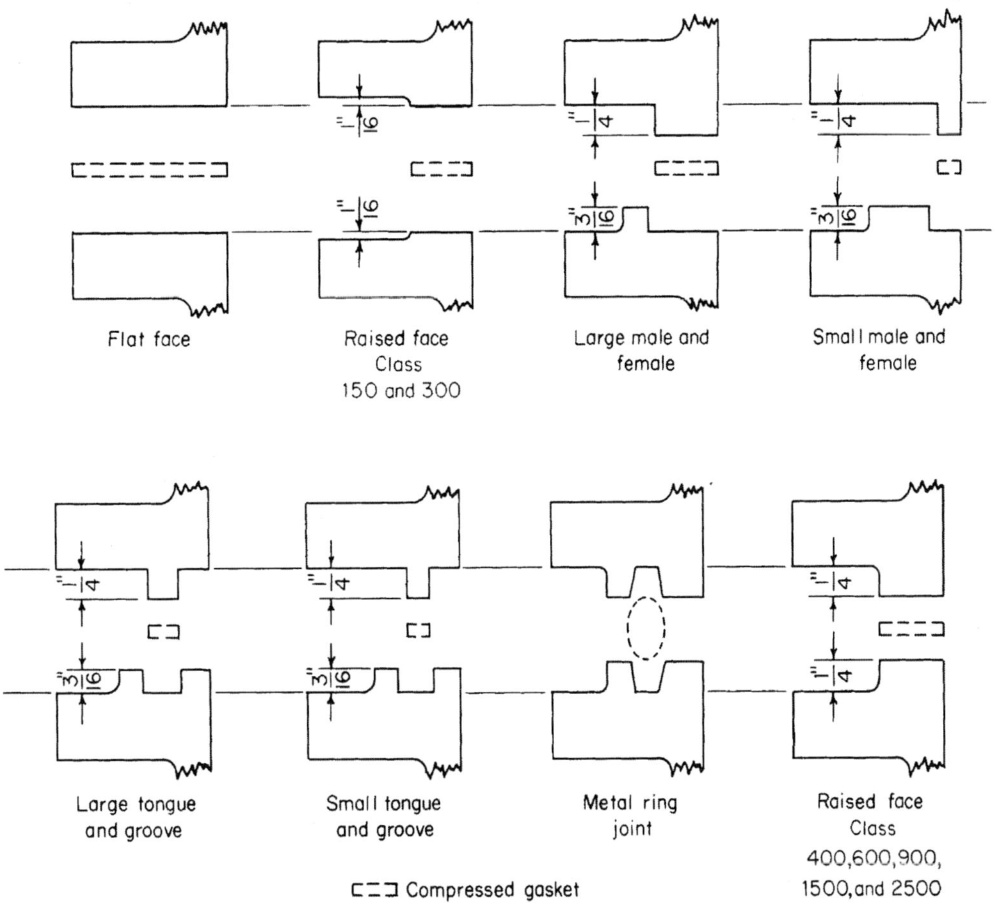

FIG. 10-126 Flange facings, illustrated on welding-neck flanges. (On small male-and-female facings the outside diameter of the male face is less than the outside diameter of the pipe, so this facing does not apply to screwed or slip-on flanges. A similar joint can be made with screwed flanges and threaded pipe by projecting the pipe through one flange and recessing it in the other. However, pipe thicker than Schedule 40 is required to avoid crushing gaskets.) To convert inches to millimeters, multiply by 25.4.

available with the bell cast on one or more ends. Glands, bolts, and gaskets are shipped with the pipe. Couplings equipped with packed glands at each end, known as *Dresser couplings*, are available in several metals. The joints can be assembled with small wrenches and unskilled labor, in limited space, and if necessary, under water.

Packed-gland joints are designed to take the same hoop stress as the pipe. They do not resist bending moments or axial forces tending to separate the joints, but yield to them to an extent indicated by the manufacturer's allowable-angular-deflection and end movement specifications. Further angular or end movement produces leakage, but end movement can be limited by harnessing or bridling with a combination of rods and welded clips or clamps, or by anchoring to existing or new structures. The crevice between the bell and the spigot may promote corrosion. The joints are widely used in underground lines. They are not affected by limited earth settlement, and friction of the earth is often adequate to prevent end separation. When disassembly by moving pipe axially is not practical, packed-joint couplings which can be slid entirely onto one of the two lengths joined are available.

Poured Joints Figure 10-128 illustrates a poured joint design. With regard to performance and ease of installation, most other joint designs are preferable to poured joints, and their use can generally be avoided.

TABLE 10-23 Dimensions of ASME B16.5 Class 150 Flanges*

All dimensions in inches

Nominal pipe size	Outside diameter of flange	Thickness of flange, minimum	Diameter of bolt circle	Diameter of bolts	No. of bolts	Length through hub: Threaded slip-on socket welding	Length through hub: Lap joint	Length through hub: Welding neck	Length through hub: ANSI B16.1, screwed (125-lb)
1	4.25	0.56	3.12	½	4	0.69	0.69	2.19	0.69
1½	5.00	0.69	3.88	½	4	0.88	0.88	2.44	0.88
2	6.00	0.75	4.75	⅝	4	1.00	1.00	2.50	1.00
3	7.50	0.94	6.00	⅝	4	1.19	1.19	2.75	1.19
4	9.00	0.94	7.50	⅝	8	1.31	1.31	3.00	1.31
6	11.00	1.00	9.50	¾	8	1.56	1.56	3.50	1.56
8	13.50	1.12	11.75	¾	8	1.75	1.75	4.00	1.75
10	16.00	1.19	14.25	⅞	12	1.94	1.94	4.00	1.94
12	19.00	1.25	17.00	⅞	12	2.19	2.19	4.50	2.19
16	23.50	1.44	21.25	1	16	2.50	3.44	5.00	2.50
20	27.50	1.69	25.00	1⅛	20	2.88	4.06	5.69	2.88
24	32.00	1.88	29.50	1¼	20	3.25	4.38	6.00	3.25

*Dimensions from ASME B16.5-2013, unless otherwise noted. Reprinted from ASME B16.5-2013 by permission of the American Society of Mechanical Engineers, New York. All rights reserved. To convert inches to millimeters, multiply by 25.4.

TABLE 10-24 Dimensions of ASME B16.5 Class 300 Flanges*

All dimensions in inches

Nominal pipe size	Outside diameter of flange	Thickness of flange, minimum	Diameter of bolt circle	Diameter of bolts	No. of bolts	Length through hub: Threaded slip-on socket welding	Length through hub: Lap joint	Length through hub: Welding neck	Length through hub: ANSI B16.1, screwed (Class 250)
1	4.88	0.69	3.50	⅝	4	1.06	1.06	2.44	0.88
1½	6.12	0.81	4.50	¾	4	1.19	1.19	2.69	1.12
2	6.50	0.88	5.00	⅝	8	1.31	1.31	2.75	1.25
3	8.25	1.12	6.62	¾	8	1.69	1.69	3.12	1.56
4	10.00	1.25	7.88	¾	8	1.88	1.88	3.38	1.75
6	12.50	1.44	10.62	¾	12	2.06	2.06	3.88	1.94
8	15.00	1.62	13.00	⅞	12	2.44	2.44	4.38	2.19
10	17.50	1.88	15.25	1	16	2.62	3.75	4.62	2.38
12	20.50	2.00	17.75	1⅛	16	2.88	4.00	5.12	2.56
16	25.50	2.25	22.50	1¼	20	3.25	4.75	5.75	2.88
20	30.50	2.50	27.00	1¼	24	3.75	5.50	6.38	
24	36.00	2.75	32.00	1½	24	4.19	6.00	6.62	

*Dimensions from ASME B16.5-2013, unless otherwise noted. Reprinted from ASME B16.5-2013 by permission of the American Society of Mechanical Engineers, New York. To convert inches to millimeters, multiply by 25.4.

TABLE 10-25 Dimensions of ASME B16.5 Class 600 Flanges*

All dimensions in inches

Nominal pipe size	Outside diameter of flange	Thickness of flange, minimum	Diameter of bolt circle	Diameter of bolts	No. of bolts	Length through hub: Threaded slip-on socket welding	Length through hub: Lap joint	Length through hub: Welding neck
1	4.88	0.69	3.50	⅝	4	1.06	1.06	2.44
1½	6.12	0.88	4.50	¾	4	1.25	1.25	2.75
2	6.50	1.00	5.00	⅝	8	1.44	1.44	2.88
3	8.25	1.25	6.62	¾	8	1.81	1.81	3.25
4	10.75	1.50	8.50	⅞	8	2.12	2.12	4.00
6	14.00	1.88	11.50	1	12	2.62	2.62	4.62
8	16.50	2.19	13.75	1⅛	12	3.00	3.00	5.25
10	20.00	2.50	17.00	1¼	16	3.38	4.38	6.00
12	22.00	2.62	19.25	1¼	20	3.62	4.62	6.12
16	27.00	3.00	23.75	1¼	20	4.19	5.50	7.00
20	32.00	3.50	28.50	1⅝	24	5.00	6.50	7.50
24	37.00	4.00	33.00	1⅞	24	5.50	7.25	8.00

*Dimensions from ASME B16.5-2013. Reprinted from ASME B16.5-2013 by permission of the American Society of Mechanical Engineers, New York. To convert inches to millimeters, multiply by 25.4.

TABLE 10-26 Dimensions of ASME B16.5 Class 900 Flanges*

All dimensions in inches

Nominal pipe size	Outside diameter of flange	Thickness of flange, minimum	Diameter of bolt circle	Diameter of bolts	No. of bolts	Length through hub: Threaded slip-on socket welding	Length through hub: Lap joint	Length through hub: Welding neck
1	Use Class 1500 dimensions in these sizes.							
1½								
2								
3	9.50	1.50	7.50	⅞	8	2.12	2.12	4.00
4	11.50	1.75	9.25	1⅛	8	2.75	2.75	4.50
6	15.00	2.19	12.50	1⅛	12	3.38	3.38	5.50
8	18.50	2.50	15.50	1⅜	12	4.00	4.50	6.38
10	21.50	2.75	18.50	1⅜	16	4.25	5.00	7.25
12	24.00	3.12	21.00	1⅜	20	4.62	5.62	7.88
16	27.75	3.50	24.25	1⅝	20	5.25	6.50	8.50
20	33.75	4.25	29.50	2	20	6.25	8.25	9.75
24	41.00	5.50	35.50	1½	20	8.00	10.50	11.50

*Dimensions from ASME B16.5-2013. Reprinted from ASME B16.5-2013 by permission of the American Society of Mechanical Engineers, New York. To convert inches to millimeters, multiply by 25.4.

TABLE 10-27 Dimensions of ASME B16.5 Class 1500 Flanges*

All dimensions in inches

Nominal pipe size	Outside diameter of flange	Thickness of flange, minimum	Diameter of bolt circle	Diameter of bolts	No. of bolts	Length through hub: Threaded slip-on socket welding	Length through hub: Lap joint	Length through hub: Welding neck
1	5.88	1.12	4.00	⅞	4	1.62	1.62	2.88
1½	7.00	1.25	4.88	1	4	1.75	1.75	3.25
2	8.50	1.50	6.50	⅞	8	2.25	2.25	4.00
3	10.50	1.88	8.00	1¼	8	2.88	2.88	4.62
4	12.25	2.12	9.50	1¼	8	3.56	3.56	4.88
6	15.50	3.25	12.50	1⅜	12	4.69	4.69	6.75
8	19.00	3.62	15.50	1⅝	12	5.62	5.62	8.38
10	23.00	4.25	19.00	1⅞	12	6.25	7.00	10.00
12	26.50	4.88	22.50	2	16	7.12	8.62	11.12
16	32.50	5.75	27.75	2½	16		10.25	12.25
20	38.75	7.00	32.75	3	16		11.50	14.00
24	46.00	8.00	39.00	3½	16		13.00	16.00

*Dimensions from ASME B16.5-2013. Reprinted from ASME B16.5-2013 by permission of the American Society of Mechanical Engineers, New York. To convert inches to millimeters, multiply by 25.4.

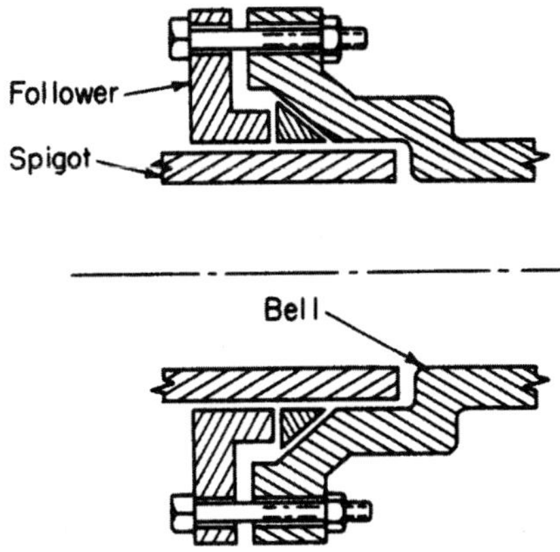

FIG. 10-127 Packed-gland joint.

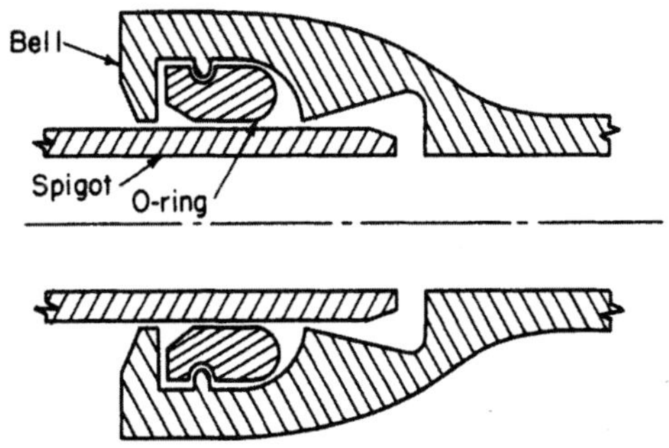

FIG. 10-129 Push-on joint.

Push-on Joints These joints (Fig. 10-129) require diametral control of the end of the pipe. They are used for brittle and nonmetallic materials. Pipe, fittings, and valves are furnished with the bells on one or more ends.

Push-on joints do not resist bending moments or axial forces tending to separate the joints but yield to them to an extent limited by the manufacturer's allowable-angular-deflection and end-movement specifications. End movement can be limited by harnessing or bridling with a combination of rods and clamps, or by anchoring to existing or new structures. Some manufacturers offer O-rings with metallic embedments that grip the pipe and prevent axial separation under internal pressure loading. The joints are widely used on underground lines. They are not affected by limited earth settlement, and friction of the earth is often adequate to prevent end separation. A lubricant is used on the O-ring during assembly. After this disappears, the O-ring bonds somewhat to the spigot and disassembly is very difficult. Disassembly for maintenance is accomplished by cutting the pipe and reassembly by use of a coupling with a packed-gland joint on each end.

Expanded Joints These joints (Fig. 10-130) are confined to the smaller pipe sizes and ductile metals. Various proprietary designs are available in which either the pipe is expanded into the coupling or the coupling is crimped down onto the pipe. In some designs, the seal between the pipe and coupling is metal to metal, while in others elastomer O-rings are employed. Joints of these types typically are quickly and easily made with specialized equipment, and they may be particularly attractive in maintenance applications since no welding is involved. The designer should clearly understand the limitations of the joint design and should verify the success of its long-term service in similar applications.

Grooved Joints These joints (Fig. 10-131) are divided into two classes: cut grooves and rolled grooves. Rolled grooves are preferred because, compared with cut grooves, they are easier to form and reduce the metal wall less. However, they slightly reduce the flow area. They are limited to thin walls of ductile material, while cut grooves, because of their reduction of the pipe wall, are limited to thicker walls. In the larger pipe sizes, some commonly used wall thicknesses are too thick for rolled grooves but too thin for cut grooves. The thinning of the walls impairs resistance to corrosion and erosion but not to internal pressure, because the thinned area is reinforced by the coupling.

Control of outside diameter is important. Permissible minus tolerance is limited, since it impairs the grip of the couplings. Plus tolerance makes it necessary to cut the cut grooves more deeply, increasing the thinning of the wall. Plus tolerance is not a problem with rolled grooves, since they are confined to walls thin enough that the couplings can compress the pipe. Pipe is available from vendors already grooved and also with heavier-wall grooved ends welded on.

Grooved joints resist axial forces tending to separate the joints. Angular deflection, up to the limit specified by the manufacturer, may be used to absorb thermal expansion and to permit the piping to be laid on uneven ground. Grooved joints provide quick and easy assembly and disassembly when compared with flanges, but may require greater support than welded joints.

Gaskets are self-sealing against both internal and external pressure and are available in a wide variety of elastomers. However, successful performance of an elastomer as a flange gasket does not necessarily mean equally satisfactory performance in a grooved joint, since exposure to the fluid in the latter is much greater and hardening has a greater unfavorable effect. It is advisable to select coupling material that is suitably corrosion-resistant with respect to the service; but with proper gasket style it may be permissible to use a coupling material that might otherwise be unacceptable with respect to fluid contamination.

V-Clamp Joints These joints (Fig. 10-132) are attached to the pipe by butt-weld or expanded joints. Theoretically, there is only one relative position of the parts in which the conical surfaces of the clamp are completely in contact with the conical surfaces of the stub ends. In actual practice, there is considerable flexing of the stub ends and the clamp; also complete contact is not required. This permits use of elastomeric gaskets as well as metal gaskets. Fittings are also available with integral conical shouldered ends.

Conical ends vary from machined forgings to roll-formed tubing, and clamps vary from machined forgings to bands to which several roll-formed channels are attached at their centers by spot welding. A hinge may be inserted in the band as a substitute for one of the draw bolts. Latches may also be substituted for draw bolts.

Compared with flanges, V-clamp joints use less metal, require less labor for assembly, and are less likely to leak under wide-range rapid temperature cycling. However, they are more susceptible to failure or damage from overtightening. They are widely used for high-alloy piping subject to periodic cleaning or relocation. Manufactured as forgings, they are used in carbon steel with metal gaskets for very high pressures. They resist both axial strain and bending moments. Each size of each type of joint is customarily rated by the vendor for both internal pressure and bending moment.

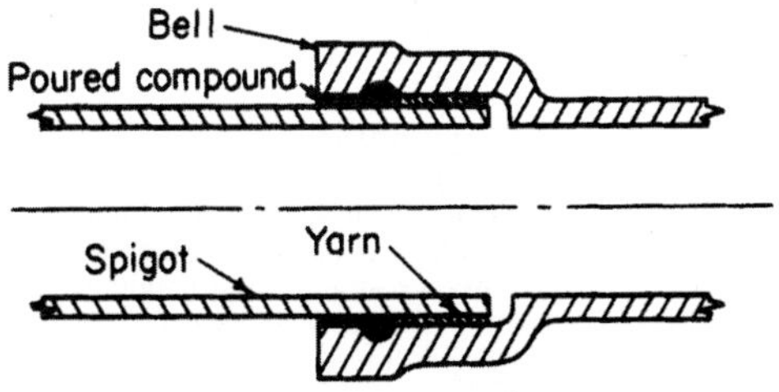

FIG. 10-128 Poured joint.

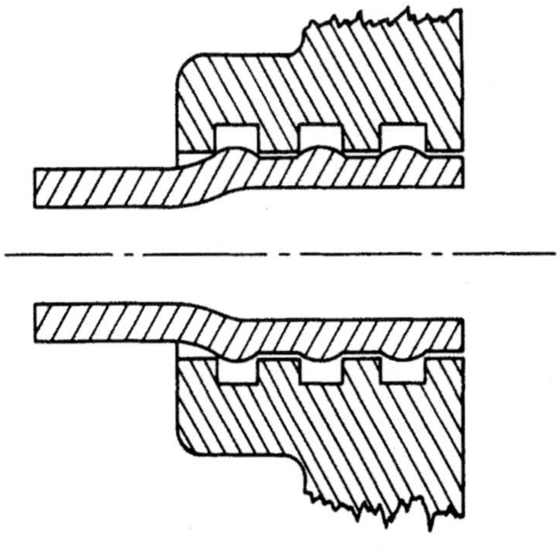

FIG. 10-130 Expanded joint.

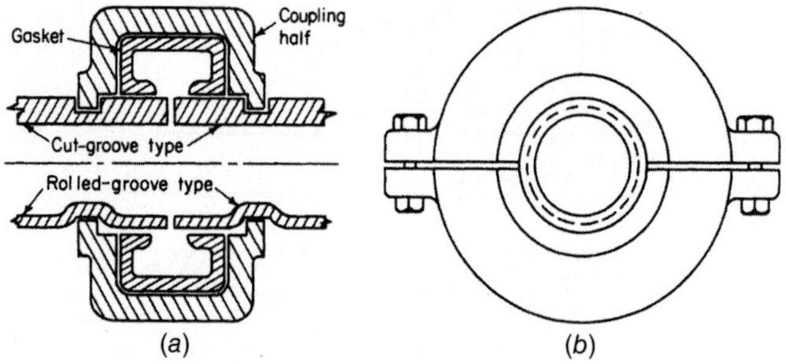

FIG. 10-131 Grooved joint. (*a*) Section. (*b*) End view.

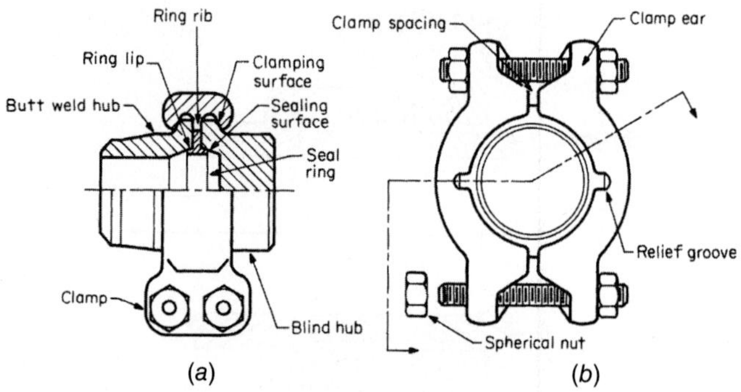

FIG. 10-133 Seal-ring joint. (*Courtesy of Gray Tool Co.*)

Seal Ring Joints These joints (Fig. 10-133) consist of hubs that are attached to pipe ends by welding. Joints of this type are proprietary, and their pressure/temperature ratings and external loading capabilities are established by the manufacturers. Variations of this design are offered by various manufacturers. Many of these designs have been widely used in critical high-pressure/temperature applications. They are particularly cost-effective in high-pressure alloy material applications.

Pressure-Seal Joints These joints (Fig. 10-134) are used for pressures of ASME Class 600 and higher. They use less metal than flanged joints but require much greater machining of surfaces. There are several designs, in all of which the increasing fluid pressure increases the force holding the sealing surfaces against each other. These joints are widely used as bonnet joints in carbon and alloy steel valves.

Tubing Joints *Flared-fitting joints* (see Fig. 10-135) are used for ductile tubing when the ratio of wall thickness to the diameter is small enough to permit flaring without cracking the inside surface. The tubing must have a smooth interior surface. The three-piece type avoids torsional strain on the tubing and minimizes vibration fatigue on the flared portion of the tubing. More labor is required for assembly, but the fitting is more resistant to temperature cycling than other tubing fittings and is less likely to be damaged by overtightening. Its efficiency is not impaired by repeated assembly and disassembly. Size is limited because of the large number of machined surfaces. The nut and, in the three-piece type, the sleeve need not be of the same material as the tubing. For these fittings, less control of tubing diameter is required.

Compression-Fitting Joints These joints (Fig. 10-136) are used for ductile tubing with thin walls. The outside of the tubing must be clean and smooth. Assembly consists only of inserting the tubing and tightening the nut. These are the least costly tubing fittings but are not resistant to vibration or temperature cycling.

Bite-Type-Fitting Joints These joints (Fig. 10-137) are used when the tubing has too high a ratio of wall thickness to diameter for flaring, when the tubing lacks sufficient ductility for flaring, and for low assembly-labor cost. The outside of the tubing must be clean and smooth. Assembly consists in merely inserting the tubing and tightening the nut. The sleeve must be considerably harder than the tubing yet still ductile enough to be diametrally compressed and must be as resistant to corrosion by the fluid handled as the tubing. The fittings are resistant to vibration but not to wide-range rapid temperature cycling. Compared with flared fittings, they are less suited for repeated assembly and disassembly, require closer diametral control of the tubing, and are more susceptible to damage from overtightening. They are widely used for oil-filled hydraulic systems at all pressures.

O-ring Seal Joints These joints (Fig. 10-138) are also used for applications requiring heavy-wall tubing. The outside of the tubing must be clean and smooth. The joint may be assembled repeatedly, and as long as the tubing is not damaged, leaks can usually be corrected by replacing the O-ring and the antiextrusion washer. This joint is used extensively in oil-filled hydraulic systems.

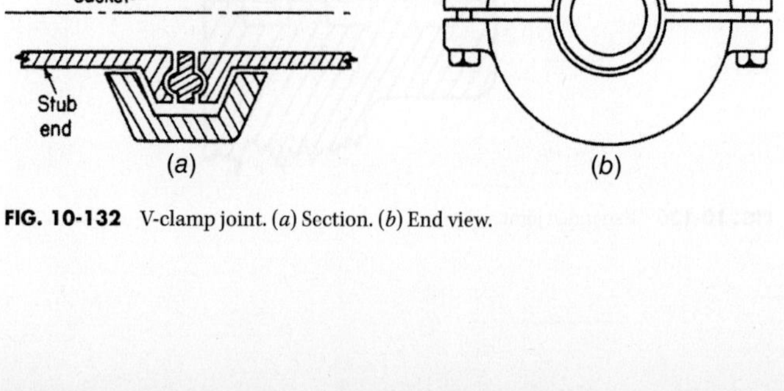

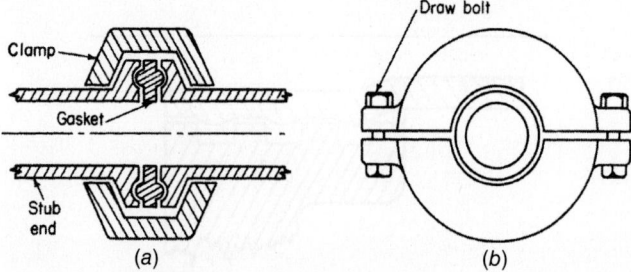

FIG. 10-132 V-clamp joint. (*a*) Section. (*b*) End view.

Soldered Joints These joints (Fig. 10-139) require precise control of the diameter of the pipe or tubing and of the cup or socket in the fitting in order to cause the solder to draw into the clearance between the cup and the tubing by capillary action (Fig. 10-139). Extrusion provides this diametral control, and the joints are most widely used in copper. A 50 percent lead, 50 percent tin solder is used for temperatures up to 93°C (200°F). Careful cleaning of the outside of the tubing and inside of the cup is required.

Heat for soldering is usually obtained from torches. The high conductivity of copper makes it necessary to use large flames for the larger sizes, and for this reason the location in which the joint will be made must be carefully considered. Soldered joints are most widely used in sizes 2 in and smaller for which heat requirements are less burdensome. Soldered joints should not be used in areas where plant fires are likely because exposure to fires results in rapid and complete failure of the joints. Properly made, the joints are completely impervious. The code permits the use of soldered joints only for Category D fluid service and then only if the system is not subject to severe cyclic conditions.

Silver Brazed Joints These are similar to soldered joints except that a temperature of about 600°C (1100°F) is required. A 15 percent silver, 80 percent copper, 5 percent phosphorus solder are used for copper and copper alloys, while 45 percent silver, 15 percent copper, 16 percent zinc, 24 percent cadmium solders are used for copper, copper alloys, carbon steel, and alloy steel. Silver-brazed joints are used for temperatures up to 200°C (400°F). Cast-bronze fittings and valves with preinserted rings of 15 percent silver, 80 percent copper, 5 percent phosphorus brazing alloy are available.

Silver-brazed joints are used when temperature or the combination of temperature and pressure is beyond the range of soldered joints. They are also more reliable in the event of plant fires and are more resistant to vibration. If they are used for fluids that are flammable, toxic, or damaging to human tissue, appropriate safeguarding is required by the code. There are OSHA regulations governing the use of silver brazing alloys containing cadmium and other toxic materials.

Pipe Fittings and Bends Directional changes in piping systems are typically made with bends or welded fittings. Bends are made as either hot bends or cold bends. Cold bending is done at temperatures below the material transformation temperature. Depending on the material and the amount of strain involved, annealing or stress relief may be required after bending. The bend radius that may be achieved for pipe of a given size, material, and thickness depends on the bending machine capabilities and bending procedures used. When contemplating bending, the bending limitations should be reviewed with the pipe fabricators being considered for the project. Because bends are not generally made to radii as small as those of standard butt-weld or socket-weld fittings, the use of bends must be considered during piping layout. Wall thinning resulting from bending must also be considered when specifying the wall thickness of material to be bent. A detailed bending specification that addresses all aspects of bending, including requirements for bending procedure specifications, availability of bending procedure qualification records and bending operator qualification records, the range of bends covered by a single bending procedure qualification, in-process nondestructive examination requirements (including minimum wall thickness verification), dimensional tolerance requirements, etc., should be part of the bending agreement. Some bending operations and subsequent heat treatment can result in tenacious oxide formation on certain materials (such as 9Cr-1Mo-V). Removal of this oxide by conventional means such as abrasive blasting may be very difficult. Methods of avoiding this formation or of removing it should be discussed prior to bending when

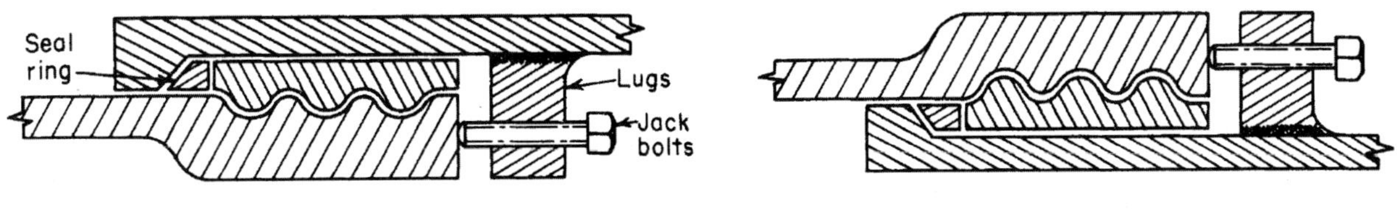

FIG. 10-134 Pressure-seal joint.

the application requires a high level of cleanliness, such as is the case with steam supply lines to turbines.

Elbow Fittings These fittings may be cast, forged, or hot- or cold-formed from short pieces of pipe or made by welding together pieces of miter-cut pipe. The thinning of pipe during the forming of elbows is compensated for by starting with heavier walls.

Flow in bends and elbow fittings is more turbulent than in straight pipe, thus increasing corrosion and erosion. This can be countered by selecting a component with greater radius of curvature, thicker wall, or smoother interior contour, but this is seldom economical in miter elbows.

Compared with elbow fittings, bends with a centerline radius of three or five nominal pipe diameters save the cost of joints and reduce pressure drop. It is sometimes difficult to nest bends of unequal pipe size when they lie in the same plane.

Flanged fittings are used when pipe is likely to be dismantled for frequent cleaning or extensive revision, or for lined piping systems. They are also used in areas where welding is not permitted. Cast fittings are usually flanged. Table 10-28 gives dimensions for flanged fittings.

Dimensions of carbon and alloy steel *butt-welding fittings* are shown in Table 10-29. Butt-welding fittings are available in the wall thicknesses shown in Table 10-22. Larger sizes and other wall thicknesses are also available. Schedule 5 and Schedule 10 stainless-steel butt-welding fittings are available with extensions for expanding into stainless-steel hubs mechanically locked in carbon-steel ASME B16.5 dimension flanges. The use of expanded joints (Fig. 10-130) is restricted by the code.

Depending on the size, forged fittings are available with socket-weld (Fig. 10-118) or screwed ends in sizes 4 in and smaller; however, 2 in is the upper size limit normally used. ASME B16.11 gives minimum dimensions for Class 3000, 6000, and 9000 socket-weld fittings, and for Class 2000, 3000, and 6000 threaded fittings. The use of socket-weld and threaded fittings is restricted by the code.

Steel forged fittings with screwed ends may be installed without pipe dope in the threads and seal-welded (Fig. 10-122) to secure bubble-tight joints.

ASME B16.3–2011 gives pressure ratings and dimensions for Class 150 and Class 300 *malleable-iron threaded fittings.* Primary usage is 2 in and smaller; however, Class 150 fittings are available in 6 in and smaller, and Class 300 fittings are available in 3 in and smaller. Malleable-iron fittings are generally less expensive than forged carbon-steel fittings, but cannot be seal-welded. Threaded ends are typically female; however, male threads or a combination of male and female is available in some fittings. Among other restrictions, the code does not permit the use of malleable iron in severe cyclic conditions, in situations subject to thermal or mechanical shock, or in any fluid service below −29°C (−20°F) or above 343°C (650°F). It also does not permit its use in flammable fluid service at temperatures above 149°C (300°F) or at gauge pressures above 2.76 MPa (400 lbf/in^2). ASME B16.3 ratings for Class 150 fittings are 2.07 MPa (300 lbf/in^2) at 66°C (150°F) and below and 1.03 MPa (150 lbf/in^2) at 186°C (366°F). ASME B16.3 ratings for Class 300 fittings are size-dependent, but at least 6.89 MPa (1000 lbf/in^2) at 66°C (150°F) and below and 2.07 MPa (300 lbf/in^2) at 288°C (550°F).

ASME B16.4–2011 gives pressure ratings and dimensions for Class 125 and Class 250 *gray-iron (cast-iron) threaded fittings.* Threaded fittings in both classes are available in sizes 12 in and smaller; however, consideration should be given to other types of end connections prior to using threaded fittings in sizes larger than 2 in. Threaded ends are typically female. Cast-iron fittings are less expensive than forged carbon-steel fittings, but cannot be seal-welded. The code places significant restrictions on the use of cast iron, and its use is typically limited to low-pressure, noncritical, nonflammable services. Its brittle nature should be considered before using it for compressed gas services. The minimum permissible design temperature is 29°C (−20°F). ASME B16.4 ratings for Class 125 fittings are 1.21 MPa (175 lbf/in^2) at 66°C (150°F) and below and 0.86 MPa (125 lbf/in^2) at 178°C (353°F). ASME B16.4 ratings for Class 250 fittings are 2.76 MPa (400 lbf/in^2) at 66°C (150°F) and below and 1.72 MPa (250 lbf/in^2) at 208°C (406°F).

Tees Tees may be cast, forged, or hot- or cold-formed from plate or pipe. Tees are typically stocked with both header (run) ends of the same size. In general, run ends of different sizes are not typically stocked or specified; however, occasionally run ends of different sizes are specified in threaded or socket-welded sizes. Branch connections may be full size or reducing sizes. Branch reductions two sizes smaller than the header are routinely stocked, and it is not typically difficult to purchase reducing tees with branches as small as those listed in ASME B16.9 (i.e., approximately one-half the header size). Economics, stress intensification factors, and nondestructive examination requirements typically dictate the branch connection type.

Reducers Reducers may be cast, forged, or hot- or cold-formed from pipe or plate. End connections may be concentric or eccentric, that is, tangent to the same plane at one point on their circumference. For pipe supported by hangers, concentric reducers permit maintenance of the same hanger length; for pipe laid on structural steel, eccentric reducers permit maintaining the same elevation of top of steel. Eccentric reducers with the common tangent plane on the bottom side permit complete drainage of branched horizontal piping systems. With the common tangent plane on the top side, they permit liquid flow in horizontal lines to sweep the line free of gas or vapor.

Reducing Elbow Fittings These permit change of direction and concentric size reduction in the same fitting.

Valves Valve bodies may be cast, forged, machined from bar stock, or fabricated from welded plate. Steel valves are available with screwed or socket-weld ends in the smaller sizes. Bronze and brass screwed-end valves are widely used for low-pressure service in steel systems. Table 10-30 gives contact-surface-of-face to contact-surface-of-face dimensions for flanged ferrous valves and end-to-end dimensions for butt-welding ferrous valves. Drilling of end flanges is shown in Tables 10-23 to 10-27. Bolt holes are located so that the stem is equidistant from the centerline of two bolt holes. Even if removal for maintenance is not anticipated, flanged valves

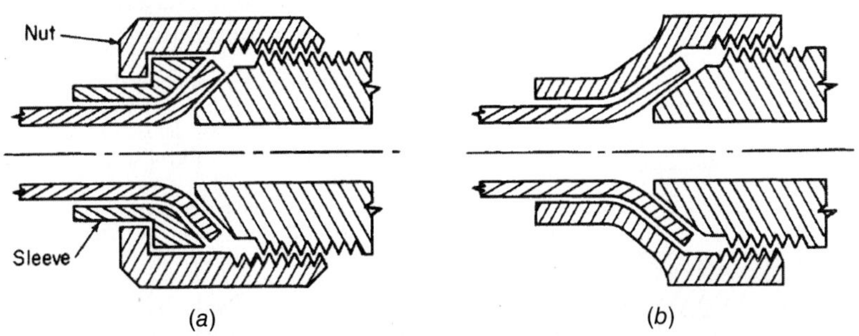

FIG. 10-135 Flared-fitting joint. (*a*) Three-piece. (*b*) Two-piece.

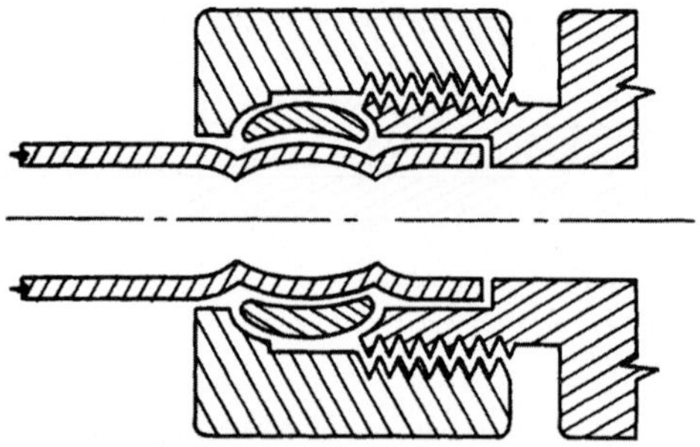

FIG. 10-136 Compression-fitting joint.

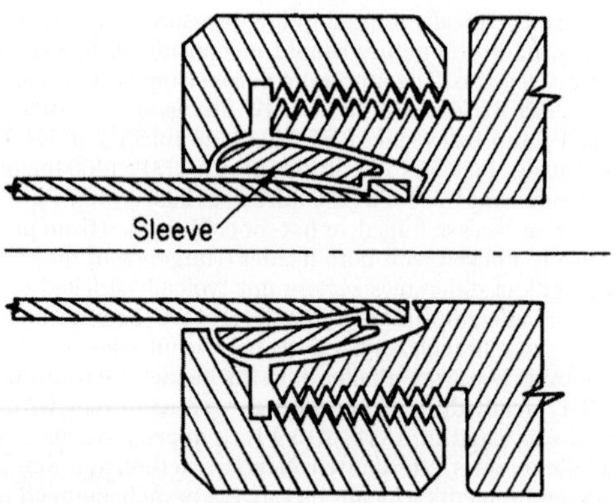

FIG. 10-137 Bite-type-fitting joint.

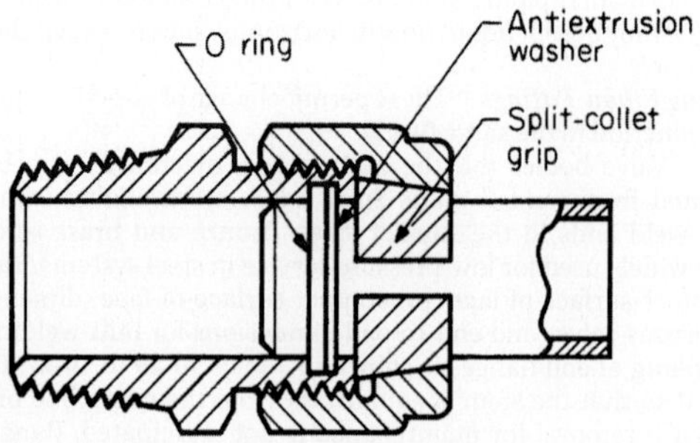

FIG. 10-138 O-ring seal joint. (*Courtesy of the Lenz Co.*)

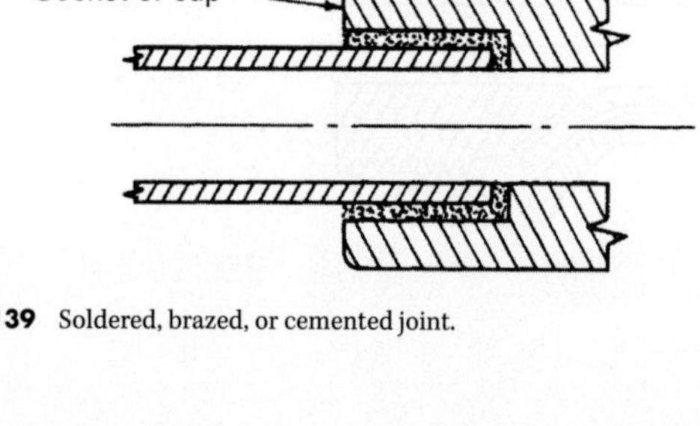

FIG. 10-139 Soldered, brazed, or cemented joint.

are frequently used instead of butt-welding-end valves because they permit insertion of blanks for isolating sections of a loop piping system.

Ferrous valves are also available in nodular (ductile) iron, which has tensile strength and yield point approximately equal to those of cast carbon steel at temperatures of 343°C (650°F) and below and only slightly less elongation.

Valves serve not only to regulate the flow of fluids but also to isolate piping or equipment for maintenance without interrupting other connected units. Valve designers attempt to minimize body distortion due to pressure, changes in temperature, and applied loads. The sealing mechanisms of certain valve designs are inherently more tolerant of these factors than are others. The selection of valve type and materials of construction should provide a valve that functions reliably and that is acceptably tight across the sealing surfaces for the lowest lifetime cost. Valve manufacturers are a valuable source of information when evaluating the suitability of specific designs. The principal types are named, described, compared, and illustrated with line diagrams in subsequent subsections. In the line diagrams, the operating stem is shown in solid black, direction of flow by arrows on a thin solid line, and motion of valve parts by arrows on a dotted line. Moving parts are drawn with solid lines in the nearly closed position and with dotted lines in the fully open position. Packing is represented by an X in a square.

Gate Valves These valves are designed in two types. The wedge-shaped-gate, *inclined-seat* type (Fig. 10-140) is most commonly used. The wedge gate may be solid or flexible (partly cut into halves by a plane at right angles to the pipe) or split (completely cleft by such a plane). Flexible and split wedges minimize galling of the sealing surfaces by distorting more easily to match angularly misaligned seats. In the double-disk *parallel-seat* type, an inclined-plane device mounted between the disks converts stem force to axial force, pressing the disks against the seats after the disks have been positioned for closing. This gate assembly distorts automatically to match both angular misalignment of the seats and longitudinal shrinkage of the valve body on cooling.

During opening and closing, some parallel-seat designs are more subject to vibration resulting from fluid flow than are wedge gates. Specific applications should be discussed with the manufacturer. In some applications it may be advisable to use a small bypass around the in-line valve to help lower opening and closing forces and to relieve binding between the gate and the seat due to high differential pressure or temperature. Double-disk parallel-seat valves should be installed with the stem essentially vertical unless otherwise recommended by the manufacturer. All wedge gate valves are equipped with tongue-and-groove guides to keep the gate sealing surfaces from clattering on the seats and marring them during opening and closing. Depending on the velocity and density of the fluid stream being sheared, these guiding surfaces may be specified as cast, machined, or hard-surfaced and ground.

Gate valves may have nonrising stems, inside-screw rising stems, or outside-screw rising stems, listed in order of decreasing exposure of the stem threads to the fluid handled. Rising-stem valves require greater space, but the position of the stem visually indicates the position of the gate. Indication is clearest on the outside-screw rising-stem valves, and on these the stem threads and thrust collars may be lubricated, reducing operating effort. The stem connection to the gate assembly prevents the stem from rotating.

Gate valves are used to minimize pressure drop in the open position and to stop the flow of fluid rather than to regulate it. The problem, when the valve is closed, of pressure buildup in the bonnet from cold liquids expanding or chemical action between fluid and bonnet should be solved by a relief valve or by notching the upstream seat ring.

Globe Valves (Fig. 10-141) These are designed as either inside-screw rising stem or outside-screw rising stem. In most designs the disk is free to rotate on the stem; this prevents galling between the disk and the seat. Various designs are used to maintain alignment between the disk and the seat, and to keep the fluid flow from vibrating or rotating the disk. Disks are typically guided either by the valve stem or against the valve body. Body guiding reduces side thrust loads on the stem. The suitability of each design can be determined by reviewing specific applications with valve manufacturers.

Disk shapes are commonly flat or conical. Conical designs provide either line or area contact between the seat and disk, and are generally more suitable than flat disks for high pressures and temperatures. Needle-type disks provide better flow control and are commonly available in valves 1 in and smaller.

For certain valve designs, sizes, and applications, globe valves may be installed with the stem in the horizontal position; however, unless approved by the manufacturer, the stem orientation should be vertical. Globe valves are not symmetric with respect to flow. Generally, globe valves are installed with pressure under the seat when the valve is in the closed position, and with the flow direction coming from under the seat. Opposite-flow direction provides pressure-assisted seating, lower seating torque requirements,

TABLE 10-28 Dimensions of Flanged Fittings*

All dimensions in inches

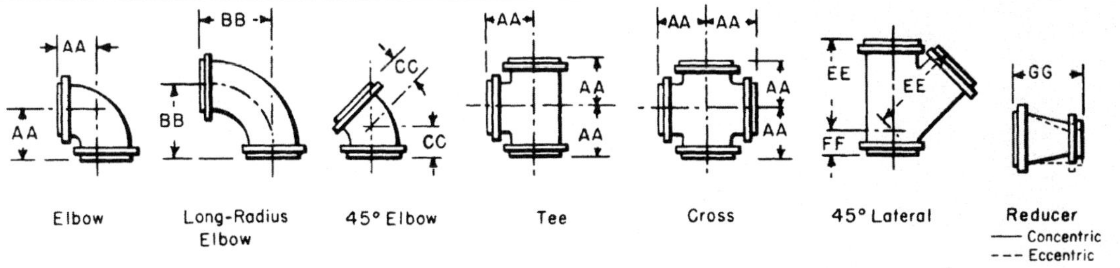

Nominal pipe size	ASME B16.5, Class 150 ASME B16.1, Class 125						ASME B16.5, Class 300 ASME B16.1, Class 250						ASME B16.5, Class 600				
	AA	BB	CC	EE	FF	GG	AA	BB	CC	EE	FF	GG	AA	CC	EE	FF	GG
½													3.25	2.00	5.75	1.75	5.00
¾													3.75	2.50	6.75	2.00	5.00
1	3.50	5.00	1.75	5.75	1.75	4.50	4.00	5.00	2.25	6.50	2.00	4.50	4.25	2.50	7.25	2.25	5.00
1¼	3.75	5.50	2.00	6.25	1.75	4.50	4.25	5.50	2.50	7.25	2.25	4.50	4.50	2.75	8.00	2.50	5.00
1½	4.00	6.00	2.25	7.00	2.00	4.50	4.50	6.00	2.75	8.50	2.50	4.50	4.75	3.00	9.00	2.75	5.00
2	4.50	6.50	2.50	8.00	2.50	5.00	5.00	6.50	3.00	9.00	2.50	5.00	5.75	4.25	10.25	3.50	6.00
2½	5.00	7.00	3.00	9.50	2.50	5.50	5.50	7.00	3.50	10.50	2.50	5.50	6.50	4.50	11.50	3.50	6.75
3	5.50	7.75	3.00	10.00	3.00	6.00	6.00	7.75	3.50	11.00	3.00	6.00	7.00	5.00	12.75	4.00	7.25
3½	6.00	8.50	3.50	11.50	3.00	6.50	6.50	8.50	4.00	12.50	3.00	6.50	7.50	5.50	14.00	4.50	7.75
4	6.50	9.00	4.00	12.00	3.00	7.00	7.00	9.00	4.50	13.50	3.00	7.00	8.50	6.00	16.50	4.50	8.75
5	7.50	10.25	4.50	13.50	3.50	8.00	8.00	10.25	5.00	15.00	3.50	8.00	10.00	7.00	19.50	6.00	10.25
6	8.00	11.50	5.00	14.50	3.50	9.00	8.50	11.50	5.50	17.50	4.00	9.00	11.00	7.50	21.00	6.50	11.25
8	9.00	14.00	5.50	17.50	4.50	11.00	10.00	14.00	6.00	20.50	5.00	11.00	13.00	8.50	24.50	7.00	13.25
10	11.00	16.50	6.50	20.50	5.00	12.00	11.50	16.50	7.00	24.00	5.50	12.00	15.50	9.50	29.50	8.00	15.75
12	12.00	19.00	7.50	24.50	5.50	14.00	13.00	19.00	8.00	27.50	6.00	14.00	16.50	10.00	31.50	8.50	16.75
14	14.00	21.50	7.50	27.00	6.00	16.00	15.00	21.50	8.50	31.00	6.50	16.00	17.50	10.75	34.25	9.00	17.75
16	15.00	24.00	8.00	30.00	6.50	18.00	16.50	24.00	9.50	34.50	7.50	18.00	19.50	11.75	38.50	10.00	19.75
18	16.50	26.50	8.50	32.00	7.00	19.00	18.00	26.50	10.00	37.50	8.00	19.00	21.50	12.25	42.00	10.50	21.75
20	18.00	29.00	9.50	35.00	8.00	20.00	19.50	29.00	10.50	40.50	8.50	20.00	23.50	13.00	45.50	11.00	23.75
24	22.00	34.00	11.00	40.50	9.00	24.00	22.50	34.00	12.00	47.50	10.00	24.00	27.50	14.75	53.00	13.00	27.75

Nominal pipe size	ASME B16.5, Class 900					ASME B16.5, Class 1500				
	AA	CC	EE	FF	GG	AA	CC	EE	FF	GG
½	Use Class 1500 dimensions in these sizes					4.25	3.00			
¾						4.50	3.25			
1						5.00	3.50	9.00	2.50	5.00
1¼						5.50	4.00	10.00	3.00	5.75
1½						6.00	4.25	11.00	3.50	6.25
2						7.25	4.75	13.25	4.00	7.25
1½						8.25	5.25	15.25	4.50	8.25
3	7.50	5.50	14.50	4.50	7.75	9.25	5.75	17.25	5.00	9.25
4	9.00	6.50	17.50	5.50	9.25	10.75	7.25	19.25	6.00	10.75
5	11.00	7.50	21.00	6.50	11.25	13.25	8.75	23.25	7.50	13.75
6	12.00	8.00	22.50	6.50	12.25	13.88	9.38	24.88	8.12	14.50
8	14.50	9.00	27.50	7.50	14.75	16.38	10.88	29.88	9.12	17.00
10	16.50	10.00	31.50	8.50	16.75	19.50	12.00	36.00	10.25	20.25
12	19.00	11.00	34.50	9.00	17.75	22.25	13.25	40.75	12.00	23.00
14	20.25	11.50	36.50	9.50	19.00	24.75	14.25	44.00	12.50	25.75
16	22.25	12.50	40.75	10.50	21.00	27.25	16.25	48.25	14.75	28.25
18	24.00	13.25	45.50	12.00	24.50	30.25	17.75	53.25	16.50	31.50
20	26.00	14.50	50.25	13.00	26.50	32.75	18.75	57.75	17.75	34.00
24	30.50	18.00	60.00	15.50	30.50	38.25	20.75	67.25	20.50	39.75

*Outline drawings show a ¼-in (6.5-mm) raised face machined onto the flange for ASME Class 400 and higher. ASME B16.1 Class 250 and ASME B16.5 Classes 150 and 300 have a 1⁄16-in (1.5-mm) raised face. ASME B16.1 Class 125 has no raised face. See Tables 10-23 through 10-29 for flange drillings. Dimensions for Class 400 and Class 600 fittings are identical for sizes ½ through 3½ in. Dimensions for Class 900 and Class 1500 fittings are identical for sizes ½ through 2½ in. To convert inches to millimeters, multiply by 25.4. The dimensions were extracted from *Cast Iron Pipe Flanges and Flanged Fittings*, ASME B16.1-1998, and *Pipe Flanges and Flanged Fittings*, ASME B16.5-2003, with permission of the publisher, the American Society of Mechanical Engineers, New York. All rights reserved.

TABLE 10-29 Butt-Welding Fittings*

All dimensions in inches

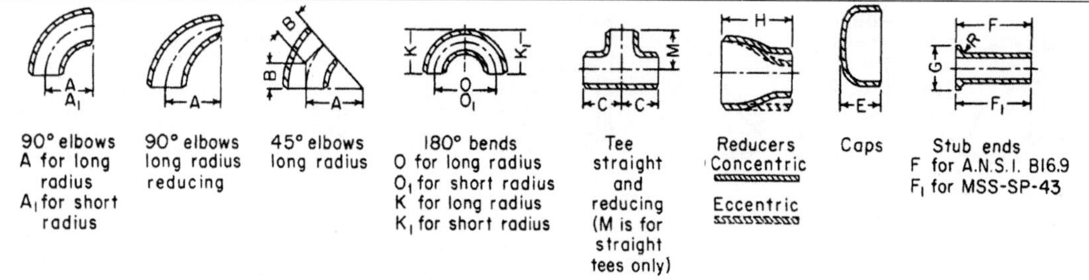

Pipe size	A	K	A1	K1	B	O	O1	M, C	H	E†	G	F	F1	R‡
½	1.50	1.88			0.62	3.00		1.00		1.00	1.38	3.00	2.00	0.12
¾	1.50	2.00			0.75	3.00		1.12	1.50	1.00	1.69	3.00	2.00	0.12
1	1.50	2.19	1.00	1.62	0.88	3.00	2.00	1.50	2.00	1.50	2.00	4.00	2.00	0.12
1¼	1.88	2.75	1.25	2.06	1.00	3.75	2.50	1.88	2.00	1.50	2.50	4.00	2.00	0.19
1½	2.25	3.25	1.50	2.44	1.12	4.50	3.00	2.25	2.50	1.50	2.88	4.00	2.00	0.25
2	3.00	4.19	2.00	3.19	1.38	6.00	4.00	2.50	3.00	1.50	3.62	6.00	2.50	0.31
2½	3.75	5.19	2.50	3.94	1.75	7.50	5.00	3.00	3.50	1.50	4.12	6.00	2.50	0.31
3	4.50	6.25	3.00	4.75	2.00	9.00	6.00	3.38	3.50	2.00	5.00	6.00	2.50	0.38
3½	5.25	7.25	3.50	5.50	2.25	10.50	7.00	3.75	4.00	2.50	5.50	6.00	3.00	0.38
4	6.00	8.25	4.00	6.25	2.50	12.00	8.00	4.12	4.00	2.50	6.19	6.00	3.00	0.44
5	7.50	10.31	5.00	7.75	3.12	15.00	10.00	4.88	5.00	3.00	7.31	8.00	3.00	0.44
6	9.00	12.31	6.00	9.31	3.75	18.00	12.00	5.62	5.50	3.50	8.50	8.00	3.50	0.50
8	12.00	16.31	8.00	12.31	5.00	24.00	16.00	7.00	6.00	4.00	10.62	8.00	4.00	0.50
10	15.00	20.38	10.00	15.38	6.25	30.00	20.00	8.50	7.00	5.00	12.75	10.00	5.00	0.50
12	18.00	24.38	12.00	18.38	7.50	36.00	24.00	10.00	8.00	6.00	15.00	10.00	6.00	0.50
14	21.00	28.00	14.00	21.00	8.75	42.00	28.00	11.00	13.00	6.50	16.25	12.00	6.00	0.50
16	24.00	32.00	16.00	24.00	10.00	48.00	32.00	12.00	14.00	7.00	18.50	12.00	6.00	0.50
18	27.00	36.00	18.00	27.00	11.25	54.00	36.00	13.50	15.00	8.00	21.00	12.00	6.00	0.50
20	30.00	40.00	20.00	30.00	12.50	60.00	40.00	15.00	20.00	9.00	23.00	12.00	6.00	0.50
24	36.00	48.00	24.00	36.00	15.00	72.00	48.00	17.00	20.00	10.50	27.25	12.00	6.00	0.50

*Extracted from *Factory-Made Wrought Buttwelding Fittings,* ASME B16.9-2012, with permission of the publisher, the American Society of Mechanical Engineers, New York. All rights reserved. O and K dimensions of 2.25 and 1.69 in, respectively, may be furnished for NPS ¾ at the manufacturer's option.

†For wall thicknesses greater than extra heavy, E is greater than shown here for sizes 2 in and larger.

‡For MSS SP-43 type B stub ends, which are designed to be backed up by slip-on flanges, $R = \frac{1}{32}$ in for 4 in and smaller and $\frac{1}{16}$ in for 6 through 12 in. To convert inches to millimeters multiply by 25.4.

TABLE 10-30 Dimensions of Valves*

All dimensions in inches

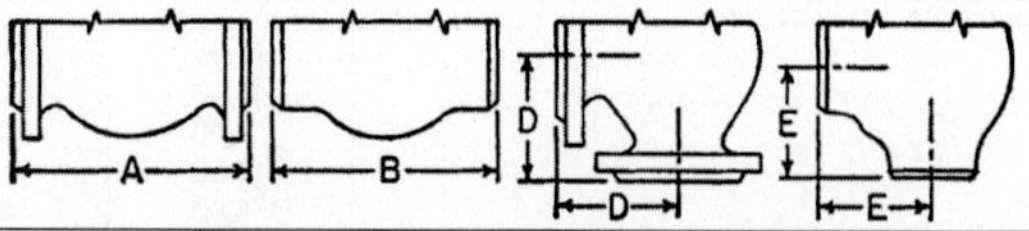

	Class 125 cast iron					Class 150 steel, MSS-SP-42 through 12-in size					Class 250 cast iron		
	Flanged end					Flanged end	Welding end	Flanged end and welding end			Flanged end		
	Gate					Gate	Gate						
Nominal valve size	Solid wedge *A*	Double disk A	Globe and lift check A	Angle and lift check D	Swing check A	Solid wedge and double disk *A*	Solid wedge and double disk *B*	Globe and lift check A and B	Angle and lift check D and E	Swing check A and B	Gate solid wedge and double disk A	Globe, lift check, and swing check A	Angle and lift check D
¼						4	4	4	2	4			
⅜						4	4	4	2	4			
½						4¼	4¼	4¼	2¼	4¼			
¾						4⅝	4⅝	4⅝	2½	4⅝			
1						5	5	5	2¾	5			
1¼						5½	5½	5½	3	5½			
1½						6½	6½	6½	3¼	6½			
2	7	7				7	8½	8	4	8	8½		
2½	7½	7½				7½	9½	8½	4¼	8½	9½		
3	8	8			8	8	11⅛	9½	4¾	9½	11f		
3½	8½	8½			8½	†					12		
4	9	9	8	4	9½	9	12	11½	5¾	11½	15	10½	
5	10	10	8½	4¼	11½	10	15	14	7	13	15⅞	11½	
6	10½	10½	9½	4¾	13	10½	15⅞	16	8	14	16½	12½	5¼
8	11½	11½	11½	5¾	14	11½	16½	19½	9¾	19½	18	14	5¾
10	13	13	13	6½	19½	13	18	24½	12¼	24½	19¾	15¾	6¼
12	14	14	14	7	24½	14	19¾	27½	13¾	27½	22½	17½	7
													7⅞
14	15	†	19½	9¾	27½	15	22½	31	15½	31	24	21	8¾
16	16	†	24½	12¼	31	16	24	36	18	†	26	24½	10½
18	17	†	27½	13¾	†	17	26			†	28	28	12¼
20	18	†	31	15½	†	18	28			†		†	14
24	20	†	36	18	†	20	32			†	31	†	

TABLE 10-30 Dimensions of Valves *(Continued)*

Nominal valve size	Class 300 steel				Class 600 steel						
	Flanged end and welding end				Flanged end and welding end						
	Gate				Gate					Angle and lift check	
	Solid wedge and double disk A and B	Globe and lift check A and B	Angle and lift check D and E	Swing check A and B	Solid wedge A and B	Double disk A and B	Short pattern‡ B	Regular globe, regular lift check, swing check A and B	Short pattern‡ globe, short pattern lift check B	Regular D and E	Short pattern E
½	5½	6	3		6½			6½		3¼	
¾	6	7	3½		7½			7½		3¾	
1	6½§	8	4	8½	8½	8½	5¼	8½	5¼	4¼	
1d	7§	8½	4¼	9	9	9	5e	9	5¾	4½	
1½	7½	9	4½	9½	9½	9½	6	9½	6	4¾	
2	8½	10½	5¼	10½	11½	11½	7	11½	7	5¾	4¼
2½	9½	11½	5¾	11½	13	13	8½	13	8½	6½	5
3	11⅛	12½	6¼	12½	14	14	10	14	10	7	6
4	12	14	7	14	17	17	12	17	12	8½	7
5	15	15¾	7⅞	15¾	20	20	15	20	15	10	8½
6	15⅞	17½	8¾	17½	22	22	18	22	18	11	10
8	16½	22	11	21	26	26	23	26	23	13	
10	18	24½	12¼	24½	31	31	28	31	28	15½	
12	19¾	28	14	28	33	33	32	33	32	16½	
14	30			†	35	35	35	†			
16	33			†	39	39	39	†			
18	36			†	43	43	43	†			
20	39			†	47	47	47	†			
22	43			†	51	51		†			
24	45			†	55	55	55	†			

Nominal valve size	Class 900 steel							Class 1500 steel				
	Flanged end and welding end							Flanged end and welding end				
	Gate					Angle and lift check		Gate				
	Solid wedge A and B	Double disk A and B	Short pattern‡ B	Regular globe, regular lift check, swing check A and B	Short pattern‡ globe, short pattern lift check B	Regular D and E	Short pattern E	Solid wedge A and B	Double disk A and B	Short pattern‡ B	Globe, lift check, swing check A and B	Angle and lift check D and E
¾				9		4½					9	4½
1	10		5½	10		5		10		5½	10	5
1¼	11		6½	11		5½		11		6½	11	5½
1½	12		7	12		6		12		7	12	6
2	14½	14½	8½	14½		7¼		14½	14½	8½	14½	7¼
2½	16½	16½	10	16½		8¼		16½	16½	10	16½	8¼
3	15	15	12	15	12	7½		18½	18½	12	18½	9¼
4	18	18	14	18	14	9	6	21½	21½	16	21½	10¾
5	22	22	17	22	17	11	7	26½	26½	19	26½	13¼
6	24	24	20	24	20	12	8½	27¾	27¾	22	27¾	13⅞
8	29	29	26	29	26	14½	10	32¾	32¾	28	32¾	16⅜
10	33	33	31	33	31	16½	13	39	39	34	39	19½
12	38	38	36	38	36	19	15½	44½	44½	39	44½	22¼
14	40½	40½	39	40½	39	20¼	18	49½	49½	42	49½	24¾
16	44½	44½	43				19½	54½	54½	47		
18	48	48	†					60½	60½	53		
20	52	52	†					65½	65½	58		
24	61	61	†					76½	76½			

NOTE: Outline drawings for flanged valves show ¼-in raised face machined onto flange, as for Class 600 cast-steel valves; Class 150 and 300 cast-steel valves and Class 250 cast-iron valves have 1/16-in raised faces; Class 125 cast-iron has no raised faces.

*Extracted from *Face-to-Face and End-to-End Dimensions of Valves,* ASME B16.10, with permission of the publisher, the American Society of Mechanical Engineers, New York. All rights reserved. To convert inches to millimeters, multiply by 25.4.

†Not shown in ANSI B16.10 but commercially available.

‡These dimensions apply to pressure-seal or flangeless bonnet valves only.

§Solid wedge only.

and higher opening torques, and may result in blockage in dirty services. Consult the manufacturer before installing globe valves in the opposite-flow direction.

Pressure drop through globe valves is much greater than that for gate valves. In Y-type globe valves, the stem and seat are at about 45° to the pipe instead of 90°. This reduces pressure drop but presents design challenges with regard to disk alignment.

Globe valves in horizontal lines prevent complete drainage.

Angle Valves These valves are similar to globe valves; the same bonnet, stem, and disk are used for both (Fig. 10-142). They combine an elbow fitting and a globe valve into one component with a substantial saving in pressure drop.

Diaphragm Valves These valves are limited to pressures of approximately 50 lbf/in^2 (Fig. 10-143). The fabric-reinforced diaphragms may be made from natural rubber, from a synthetic rubber, or from natural or synthetic rubbers faced with Teflon* fluorocarbon resin. The simple shape of the body makes lining it economical. Elastomers have shorter lives as diaphragms than as linings because of flexing but still provide satisfactory service. Plastic bodies, which have low moduli of elasticity compared with metals, are practical in diaphragm valves since alignment and distortion are minor problems.

These valves are excellent for fluids containing suspended solids and can be installed in any position. Models are available in which the dam is very low, reducing pressure drop to a negligible quantity and permitting complete drainage in horizontal lines. However, drainage can be obtained with any model simply by installing it with the stem horizontal. The only maintenance required is replacement of the diaphragm, which can be done very quickly without removing the valve from the line.

*Du Pont PTFE fluorocarbon resin.

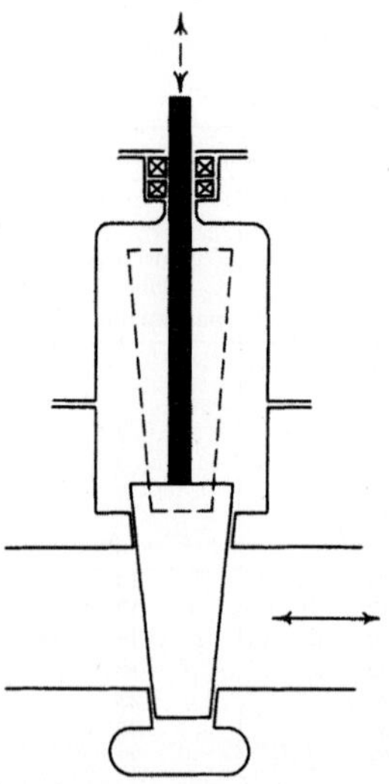

FIG. 10-140 Gate valve.

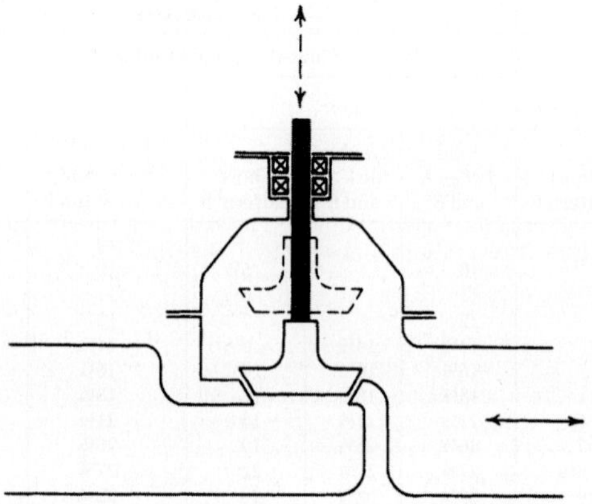

FIG. 10-141 Globe valve.

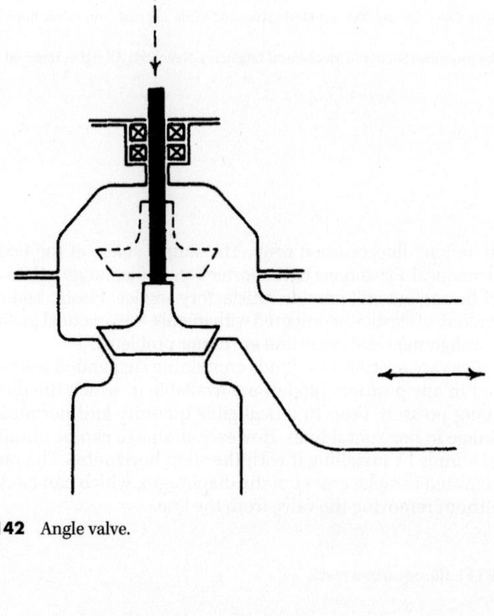

FIG. 10-142 Angle valve.

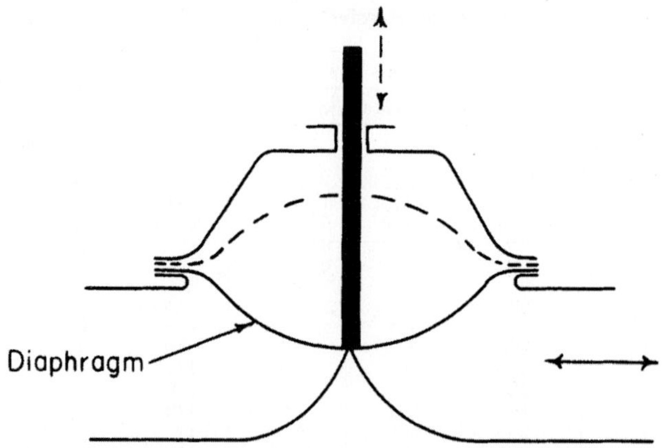

FIG. 10-143 Diaphragm valve.

Plug Valves These valves (Fig. 10-144) are typically limited to temperatures below 260°C (500°F) since differential expansion between the plug and the body results in seizure. The size and shape of the port divide these valves into different types. In order of increasing cost, they are short venturi, reduced rectangular port; long venturi, reduced rectangular port; full rectangular port; and full round port.

In lever-sealed plug valves, tapered plugs are used. The plugs are raised by turning one lever, rotated by another lever, and reseated by the first lever. *Lubricated* plug valves may use straight or tapered plugs. The tapered plugs may be raised slightly, to reduce turning effort, by injection of the lubricant, which also acts as a seal. Plastic is used in nonlubricated plug valves as a body liner, a plug coating, or port seals in the body or on the plug.

In plug valves other than lever-sealed plug valves, the contact area between plug and body is large, and gearing is usually used in sizes 6 in and larger to minimize operating effort. There are several lever-sealed plug valves incorporating mechanisms which convert the rotary motion of a handwheel into sequenced motion of the two levers.

For lubricated plug valves, the lubricant must have limited viscosity change over the range of operating temperature, must have low solubility in the fluid handled, and must be applied regularly. There must be no chemical reaction between the lubricant and the fluid which would harden or soften the lubricant or contaminate the fluid. For these reasons, lubricated plug valves are most often used when there are a large number handling the same or closely related fluids at approximately the same temperature.

Lever-sealed plug valves are used for throttling service. Because of the large contact area between plug and body, if a plug valve is operable, there is little likelihood of leakage when closed, and the handle position is a clearly visible indication of the valve position.

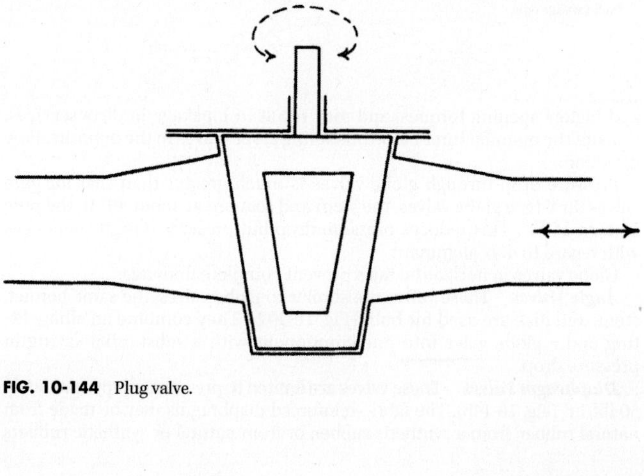

FIG. 10-144 Plug valve.

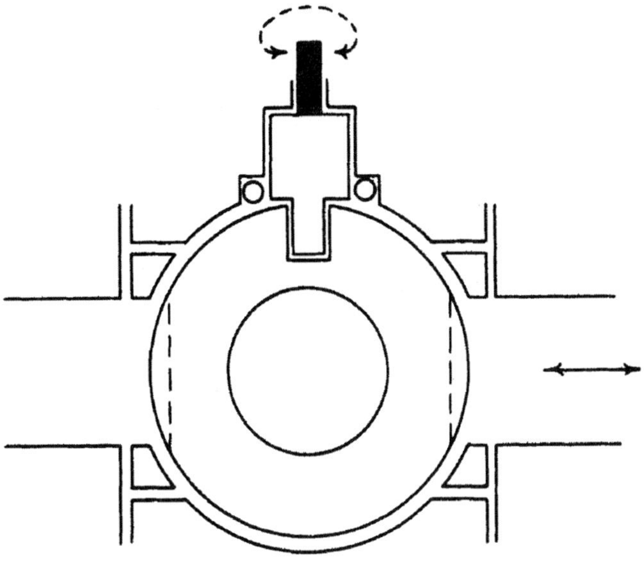

FIG. 10-145 Ball valve; floating ball.

Ball Valves Ball valves are of two primary designs: floating ball (Fig. 10-145) and trunnion-mounted ball (Fig. 10-146). In floating ball designs, the ball is supported by the downstream seat. In trunnion ball designs, the ball is supported by the trunnion, and the seat loads are less than those in floating ball valves. Because of operating torque and shutoff pressure ratings, trunnion ball valves are available in larger sizes and higher pressure ratings than floating ball valves.

Both floating and trunnion-mounted designs are available with other design variations that include metal seated valves, soft seated valves, top-entry valves, end-entry valves, and split body valves. Valves in all these design variations are available as either full port or reduced port. *Port* refers to the round-bore fluid flow area through the ball. Full port valves have a bore that is approximately equal to the inside diameter of the mating pipe. Reduced port valves have a bore that is approximately equal to the inside diameter of pipe one size smaller than full bore.

A variety of soft seat materials are available, including PTFE and nylon. Since the shutoff pressure capability of ball valves is limited by the load capabilities of the seat material, the upper temperature limit of soft seated valves is limited by the seat material selection. The shutoff pressure rating of soft seated valves typically declines rapidly with increasing temperature, and the shutoff rating is often less than the body pressure rating. Metal seated valves do not share this characteristic.

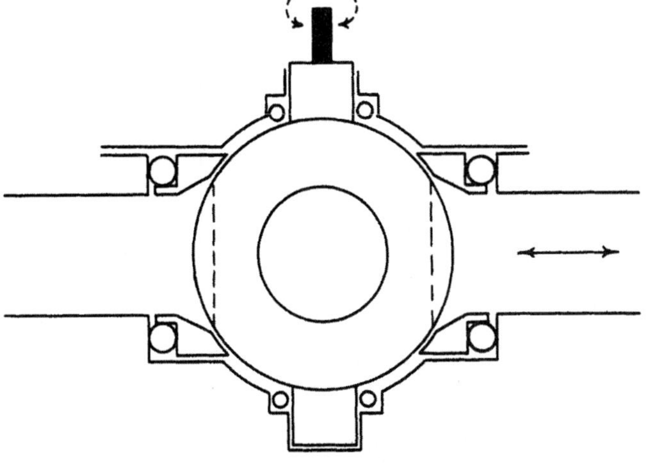

FIG. 10-146 Ball valve; trunnion-mounted ball.

For equal port size, ball valves share the low pressure drop characteristics of gate valves. Also as is the case with gate valves, consideration must be given to venting the valve when the expansion of fluid trapped within the body cavity could overpressurize the valve body. Some seat designs are inherently self-venting to either the upstream or downstream side of the valve. In floating ball valves, venting may result in a unidirectional valve that seats against flow in only one direction.

With split body designs, internals must be removed by separating the valve body in the axial direction of the mating pipe. Top-entry design permits removal of the internals through the top of the valve. When valves are butt-welded, top entry may be specified to permit repair without removing the valve from the piping. Top-entry valves are significantly more expensive than split body and end-entry valves, and full port valves are more expensive than reduced port in any body design. Metal seated valves are significantly more expensive than soft seated valves and are typically used only when other types of valves are unsuitable for the application.

Butterfly Valves These valves (Fig. 10-147) occupy less space, are much lighter than other types of block valves, and are available in body styles that include wafer, lugged (drilled-through or tapped), and flanged. They are available in ASME Class 900 and lower pressure ratings. The maximum size available varies with pressure rating. Valves in Class 150 are available in sizes exceeding 60-in diameter. Within limits, they may be used for throttling. Their relatively high pressure drop must be considered during design.

Like ball valves, butterfly valves are fully opened in one-quarter turn and are therefore well suited to automation. *Butterfly Valves: Double Flange, Lug- and Wafer-Type*, API 609, is one of the standards commonly used to specify butterfly valves. API 609 defines two major categories of butterfly valves: Category A and Category B. Category A valves are typically soft seated valves with shell pressure ratings that may be less than the flange rating of the valve. They are typically used for utility services and are commonly referred to as *utility* valves. Category B may be soft seated or metal seated, and must have shell pressure ratings equal to the full pressure rating of the valve flange, and seat ratings that essentially meet the shell rating within the temperature capability of the seat material. Within Category B, valves may be further divided into concentric shaft, double-offset shaft, and triple-offset shaft designs. *Offset* refers to the position of the shaft with respect to the seat area. With minor exception, double- and triple-offset valve designs are metal-to-metal seated. They are distinguished from other designs by their exceptional seat tightness (often "zero" leakage) that is maintained throughout the life of the valve. Their tightness exceeds the seat tightness capability and reliability of wedge-type gate valves. Double-offset valves minimize rubbing between the disk and the seat, and triple-offset valves virtually eliminate rubbing. Although double- and triple-offset valves are more expensive than other butterfly valve designs, because of their weight they are often more economical than gate valves for some combinations of pressure class, size, and materials.

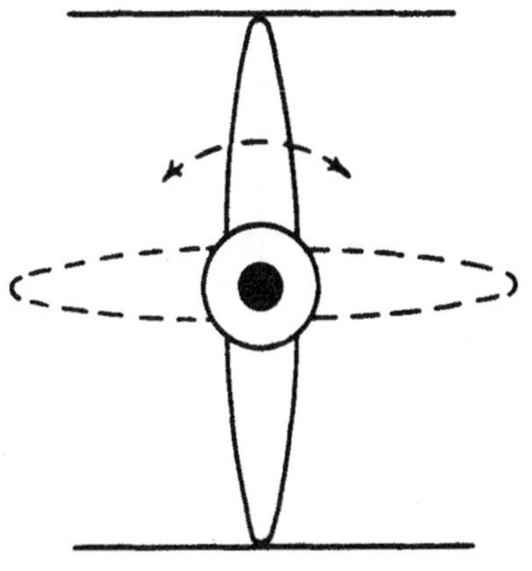

FIG. 10-147 Butterfly valve.

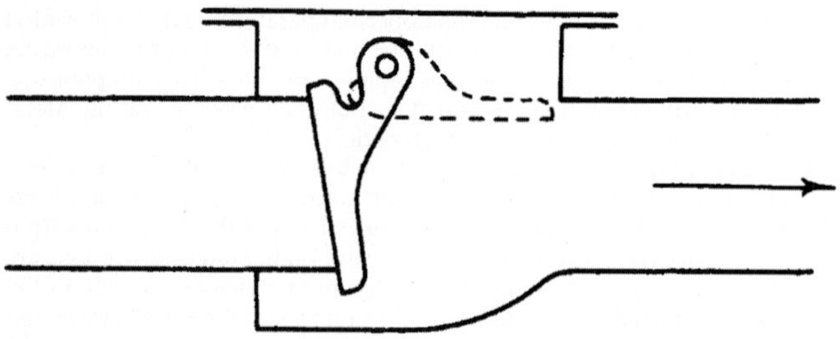

FIG. 10-148 Swing check valve.

Check Valves These valves are used to prevent reversal of flow. They must be located where flow turbulence or instability does not result in chatter (high-frequency opening and closing of the valve) and in systems designed to prevent sudden high-velocity flow reversal which results in slamming upon closure. Many valve manufacturers can provide application advice.

Swing Check Valves These valves (Fig. 10-148) are normally designed for use in horizontal lines or in vertical lines with normally upward flow. Since their seating force is primarily due to pipeline pressure, they may not seal as tightly at low pressures as at higher pressures. When suitable, nonmetallic seats may be used to minimize this problem.

Lift Check Valves These valves (Figs. 10-149 through 10-151) are made in three styles. Vertical lift check valves are for installation in vertical lines with flow normally upward. Globe (or piston) valves with a 90° bonnet (Fig. 10-149) are for installation in horizontal lines, although inclined bonnet versions (approximately 45°) with spring assist may be used in vertical lines with normally upward flow. Globe and angle check valves often incorporate mechanisms to control the opening or closing rate of the piston, or to promote full opening under low-flow conditions. In some designs, spring-assisted closure is available, but this increases pressure drop. Lift check valves should not be used when the fluid contains suspended solids. Ball check valves having designs similar to those in Figs. 10-150 and 10-151 are available in sizes 2 in and smaller. They promote even wear of the seat area and are more suitable for viscous services, or services with limited solids.

Tilting-Disk Check Valves These valves (Fig. 10-152) may be installed in a horizontal line or in lines in which the flow is vertically upward. The pivot point is located so that the distribution of pressure in the fluid handled speeds the closing but arrests slamming. Compared with swing check valves of the same size, pressure drop is less at low velocities but greater at high velocities.

Closure at the instant of flow reversal is most nearly attained with tilting-disk, dual-plate, and specialty axial-flow check valves. However, quick closure is not the solution to all noise, shock, and water hammer problems. External dashpots are available when a controlled rate of closure is desired. Nonmetallic seats are also available.

Dual-Plate Check Valves These valves (Fig. 10-152) occupy less space, are much lighter than other types of check valves, and are available in body

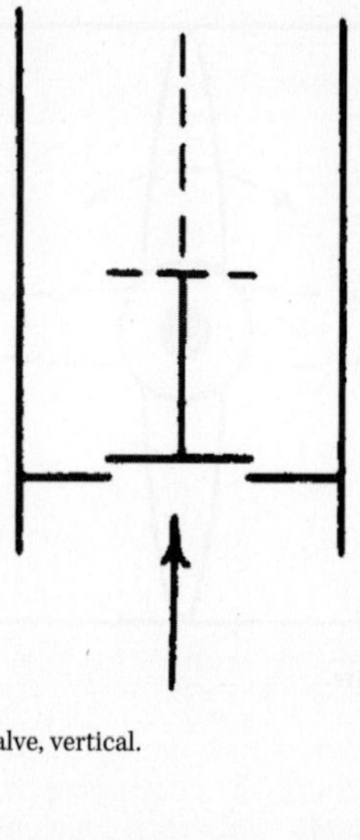

FIG. 10-149 Lift check valve, vertical.

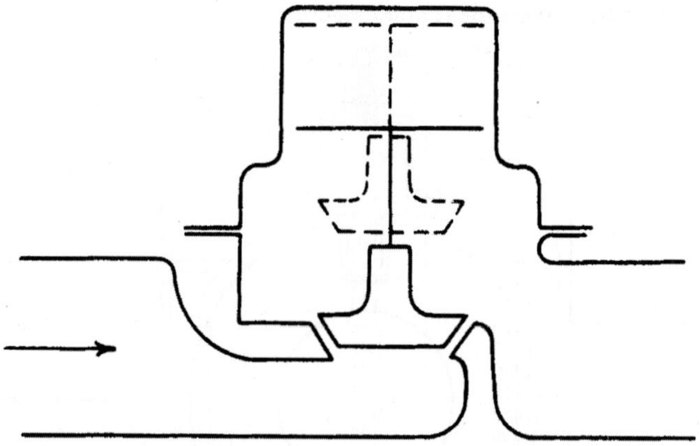

FIG. 10-150 Lift check valve, globe.

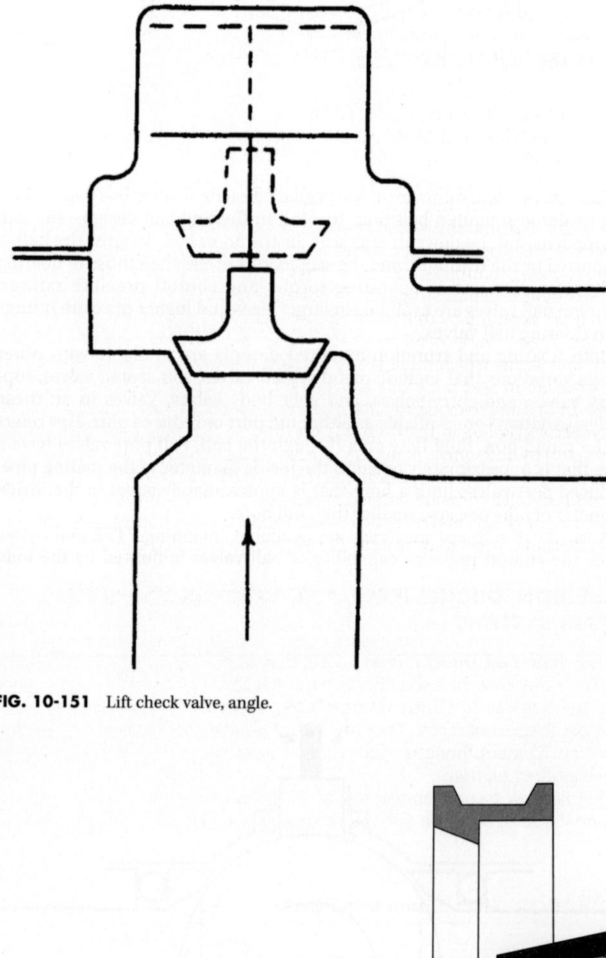

FIG. 10-151 Lift check valve, angle.

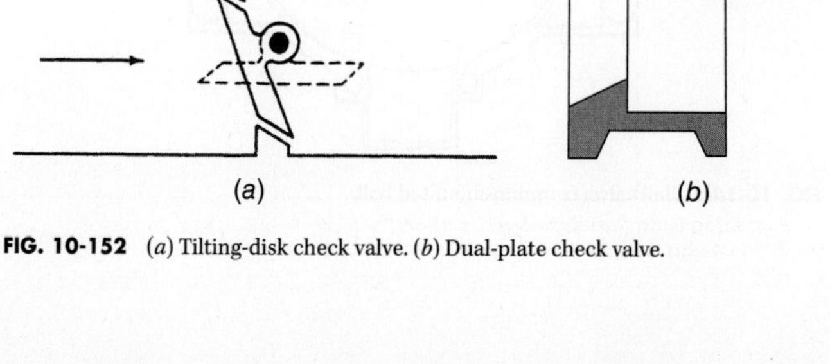

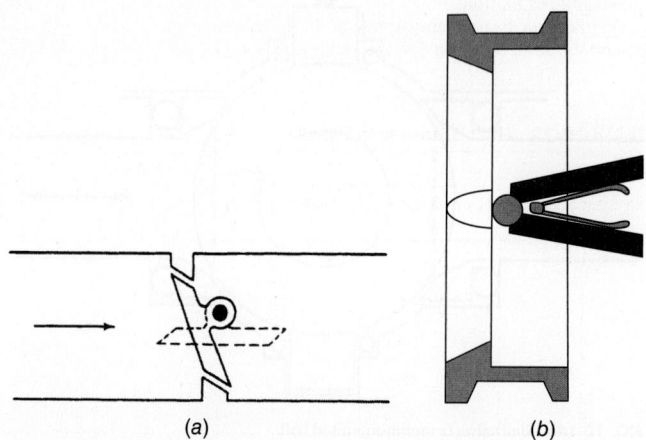

FIG. 10-152 (*a*) Tilting-disk check valve. (*b*) Dual-plate check valve.

TABLE 10-31 Dimensions of Ductile-Iron Pipe*

Standard thickness for internal pressure†

		Rated water working pressure, lbf/in²‡									
		150		200		250		300		350	
Pipe size, in	Outside diameter, in	Thickness, in	Thickness class	Thickness, in	Thickness class	Thickness, in	Thickness class	Thickness, in	Thickness class	Thickness, in	Thickness class
3	3.96	0.25	51	0.25	51	0.25	51	0.25	51	0.25	51
4	4.80	0.26	51	0.26	51	0.26	51	0.26	51	0.26	51
6	6.90	0.25	50	0.25	50	0.25	50	0.25	50	0.25	50
8	9.05	0.27	50	0.27	50	0.27	50	0.27	50	0.27	50
10	11.10	0.29	50	0.29	50	0.29	50	0.29	50	0.29	50
12	13.20	0.31	50	0.31	50	0.31	50	0.31	50	0.31	50
14	15.30	0.33	50	0.33	50	0.33	50	0.33	50	0.33	50
16	17.40	0.34	50	0.34	50	0.34	50	0.34	50	0.34	50
18	19.50	0.35	50	0.35	50	0.35	50	0.35	50	0.35	50
20	21.60	0.36	50	0.36	50	0.36	50	0.36	50	0.39	51
24	25.80	0.38	50	0.38	50	0.38	50	0.41	51	0.44	52
30	32.00	0.39	50	0.39	50	0.43	51	0.47	52	0.51	53
36	38.30	0.43	50	0.43	50	0.48	51	0.53	52	0.58	53
42	44.50	0.47	50	0.47	50	0.53	51	0.59	52	0.65	53
48	50.80	0.51	50	0.51	50	0.58	51	0.65	52	0.72	53
54	57.56	0.57	50	0.57	50	0.65	51	0.73	52	0.81	53

*Extracted from *Ductile Iron Pipe, Centrifugally Cast,* AWWA C151/A21.51-2002, with permission of the publisher, the American Water Works Association, Denver, Colo.

†To convert from inches to millimeters, multiply by 25.4; to convert pounds-force per square inch to megapascals, multiply by 0.006895.

‡These pipe walls are adequate for the rated working pressure plus a surge allowance of 100 lbf/in². For the effect of laying conditions and depth of bury, see ANSI A21.51.

styles that include wafer, lugged (drilled-through or tapped), and flanged. They are available in all ASME pressure classes. The maximum size available varies with pressure rating. Valves in Class 150 are available in sizes 60 in or larger. They are available with either metallic or nonmetallic seats. Pressure drop is greater than that in a fully open swing check valve. Plate closure is spring-assisted, and the rate of closure can be controlled with proper spring selection. High-performance valves with fast closure rates are available to address water hammer problems. They typically weigh as little as 15 to 30 percent as much as swing check valves. Because of their weight they are often more economical than other types of check valves.

Valve Trim Various alloys are available for valve parts such as seats, disks, and stems which must retain smooth finish for successful operation. The problem in seat materials is fivefold: (1) resistance to corrosion by the fluid handled and to oxidation at high temperatures, (2) resistance to erosion by suspended solids in the fluid, (3) prevention of galling (seizure at point of contact) by differences in material or hardness or both, (4) maintenance of high strength at high temperature, and (5) avoidance of distortion.

Standard valve trims are defined by standards such as API 600 and API 602. Elastomer or plastic inserts may be specified to achieve bubble-tight shutoff. Valve manufacturers may be consulted for recommended trims.

CAST IRON, DUCTILE IRON, AND HIGH-SILICON IRON PIPING SYSTEMS

Cast Iron and Ductile Iron Cast iron and ductile iron provide more metal for less cost than steel in piping systems and are widely used in low-pressure services in which internal and external corrosion may cause a considerable loss of metal. They are widely used for underground water distribution. Cement lining is available at a nominal cost for handling water causing tuberculation.

Ductile iron has an elongation of 10 percent or more compared with essentially nil elongation for cast iron and for all practical purposes has supplanted cast iron as a cast piping material. It is usually centrifugally cast. This manufacturing method improves tensile strength and reduces porosity. Ductile-iron pipe is manufactured to AWWA C151/A21.51-2002 and is available in nominal sizes from 3 through 64 in. Wall thicknesses are specified by seven standard thickness classes. Table 10-31 gives the outside diameter and standard thickness for various rated water working pressures for centrifugally cast ductile-iron pipe. The required wall thickness for underground installations increases with internal pressure, depth of laying, and weight of vehicles operating over the pipe. It is reduced by the degree to which the soil surrounding the pipe provides uniform support along the pipe and around the lower 180°. Tables are provided in AWWA C151/A21.51 for determining wall-thickness-class recommendations for various installation conditions. The poured joint (Fig. 10-128) has been almost entirely superseded by the mechanical joint (Fig. 10-127) and the push-on joint (Fig. 10-129), which are better suited to wet trenches, bad weather, and unskilled labor. Such joints also minimize strain on the pipe from ground settlement. Lengths vary between 5 and 6 m (between 18 and 20 ft), depending on the supplier. Stock fittings are designed for 1.72 MPa (250 lbf/in²) cast iron or 2.41 MPa (350 lbf/in²) ductile iron in sizes through 12 in and for 1.0 and 1.72 MPa (150 and 250 lbf/in²) cast iron or 2.41 MPa (350 lbf/in²) ductile iron in sizes 14 in and larger. Stock fittings include 22½° and 11¼° bends. Ductile-iron pipe is also supplied with flanges that match the dimensions of Class 125 flanges shown in ASME B16.1 (see Table 10-23). These flanges are assembled to the pipe barrel by threaded joints.

High-Silicon Iron Pipe and fittings are cast products of material typically conforming to ASTM A518. Nominal silicon content is 14.5 percent, and nominal carbon content is approximately 0.85 percent. This material is corrosion-resistant to most chemicals, highly abrasion-resistant, and suitable for applications to 260°C (500°F). Applications are primarily gravity drain. Pipe and fittings are available under the trade name Duriron.

Pipe and fitting sizes typically available are shown in Table 10-32. Pipe and fitting dimensions commonly conform to ASTM A861. Bell and spigot

TABLE 10-32 High-Silicon Iron Pipe*

	No hub (MJ) ends				Bell-and-spigot ends			
Size, inside diam., in	Outside diam., in	Wall thickness, in	Standard length, ft	Est. weight per piece, lb	Outside diam., in	Wall thickness, in	Standard length, ft	Est. weight per piece, lb
1½	2 3/16	5/16	7	46			15/16	
2	2 11/16	5/16	7	56	2 11/16		7	62
3	3 49/64	5/16	7	79	3 25/32	5/16	7	90
4	4 49/64	5/16	7	100	4 25/32	5/16	7	114
6					6 11/16	13/32	7	169
8					9	13/32	7	234
10					11¾	⅝	7	340
12					13¼	⅝	5	470
15					16¼	⅞	5	800

*Extracted from *Standard Specification for High-Silicon Iron Pipe and Fittings,* ASTM A861-2004, with permission of the American Society for Testing and Materials, West Conshohocken, Pa.

connections are sealed with chemical-resistant rope packing and molten lead. Mechanical joint connections are made with TFE gaskets and stainless steel clamps.

The coefficient of linear expansion of this alloy in the temperature range of 21 to 100°C (70 to 212°F) is 12.2×10^{-6}/°C (6.8×10^{-6}/°F), which is slightly above that of cast iron (National Bureau of Standards). Since this material has practically no elasticity, the need for expansion joints should be considered. Connections for flanged pipe, fittings, valves, and pumps are made to ASME B16.1, Class 125.

The use of high-silicon iron in flammable-fluid service or in Category M fluid service is prohibited by the code.

NONFERROUS METAL PIPING SYSTEMS

Aluminum Seamless aluminum pipe and tube are produced by extrusion in essentially pure aluminum and in several alloys; 6-, 9-, and 12-m (20-, 30-, and 40-ft) lengths are available. Alloying and mill treatment improve physical properties, but welding reduces them. Essentially pure aluminum has an ultimate tensile strength of 58.6 MPa (8500 lbf/in^2) subject to a slight increase by mill treatment which is lost during welding. Alloy 6061, which contains 0.25 percent copper, 0.6 percent silicon, 1 percent magnesium, and 0.25 percent chromium, has an ultimate tensile strength of 124 MPa (18,000 lbf/in^2) in the annealed condition, 262 MPa (38,000 lbf/in^2), mill-treated as 6061-T6, and 165 MPa (24,000 lbf/in^2) at welded joints. Extensive use is made of alloy 1060, which is 99.6 percent pure aluminum, for hydrogen peroxide; of alloy 3003, which contains 1.2 percent manganese, for high-purity chemicals; and of alloys 6063 and 6061 for many other services. Alloy 6063 is the same as 6061 minus the chromium and has slightly lower mechanical properties.

Aluminum is not embrittled by low temperatures and is not subject to external corrosion when exposed to normal atmospheres. At 200°C (400°F) its strength is less than one-half that at room temperature. It is attacked by alkalies, by traces of copper, nickel, mercury, and other heavy-metal ions, and by prolonged contact with wet insulation. It suffers from galvanic corrosion when coupled to copper, nickel, or lead-base alloys but not when coupled to galvanized iron.

Aluminum pipe schedules conform to those in Table 10-22. Consult suppliers for available sizes and schedules.

Threaded aluminum fittings are seldom recommended for process piping. Wrought fittings with welding ends (see Table 10-28 for dimensions) and with grooved joint ends are available. Wrought 6061-T6 flanges with dimensions per Table 10-23 are also available. Consult suppliers on the availability of cast flanges and fittings. Castings manufactured in accordance with ASTM B26 are available in several grades. Refer to Table 10-28 for dimensions. The low strength and modulus of elasticity of aluminum must be considered when using flanged connections.

Aluminum-body diaphragm and ball valves are used extensively.

Copper and Copper Alloys Seamless pipe and tubing manufactured from copper, bronze, brass, and copper-nickel alloys are produced by extrusion. The availability of pipe or tubing depends on the metallurgy and size. Copper tubing is widely used for plumbing, steam tracing, compressed air, instrument air, and inert gas applications. Copper tubing specifications are generally segregated into three types: water and general service, refrigeration service (characterized by cleanliness requirements), and drain/waste/vent (DWV) service. Tubing is available in the annealed or hard-drawn condition. Hard-drawn products are available only as straight lengths. Annealed products are often available in coils or as straight lengths. Suppliers should be consulted regarding wall thickness availability and standard lengths; however, many products are available in 3.0-m (10-ft) and 6.1-m (20-ft) straight lengths. Coil lengths are generally 12.2 m (40 ft) to 30 m (100 ft).

ASTM copper tubing specifications for water and general-purpose applications include B75 and B88 (Table 10-33). Tubing, fittings, and solders specified for potable water services must always be approved by the appropriate authority, such as the National Sanitation Foundation (NSF). ASTM copper tubing specifications for refrigeration services include B68 and B280 (Table 10-34). ASTM specifications for DWV services include B306 (1¼ through 8-in OD). Pipe conforming to the diameters shown in Table 10-35 is available as ASTM B42 and B302. ASTM B302 has diameters matching ASTM B42, but is available with thinner walls (Table 10-36). Red brass pipe is available as ASTM B43.

Joints are typically soldered, silver-brazed, or mechanical. When using flare, compression, or other mechanical joint fittings, consideration must be given to the fitting manufacturer's recommendations regarding hardness, minimum and maximum wall thickness, finish requirements, and diameter tolerances. Flanges and flanged fittings are seldom used since soldered and brazed joints can be easily disassembled. Brass and bronze valves are available with female ends for soldering.

Seventy percent copper, 30 percent nickel and 90 percent copper, 10 percent nickel ASTM B466 are available as seamless pipe (ASTM B466) or welded pipe (ASTM B467) and welding fittings for handling brackish water in Schedule 10 and regular copper pipe thicknesses.

Nickel and Nickel Alloys A wide range of ferrous and nonferrous nickel and nickel-bearing alloys are available. They are usually selected because of their improved resistance to chemical attack or their superior resistance to the effects of high temperature. In general terms their cost

TABLE 10-33 Dimensions, Weights, and Tolerances in Diameter and Wall Thickness for Nominal or Standard Copper Water Tube Sizes (ASTM B88-2014)*

All tolerances are plus and minus except as otherwise noted.

Nominal standard size, in	Outside diameter, in	Average outside diameter[A] tolerance, in: Annealed	Average outside diameter[A] tolerance, in: Drawn	Wall thickness and tolerances, in: Type K Wall thickness	Type K Tolerance[B]	Type L Wall thickness	Type L Tolerance[B]	Type M Wall thickness	Type M Tolerance[B]	Theoretical weight, lb/ft: Type K	Type L	Type M
¼	0.375	0.002	0.001	0.035	0.0035	0.030	0.003	*C*	*C*	0.145	0.126	*C*
3/3	0.500	0.0025	0.001	0.049	0.005	0.035	0.004	0.025	0.002	0.269	0.198	0.145
½	0.625	0.0025	0.001	0.049	0.005	0.040	0.004	0.028	0.003	0.344	0.285	0.204
⅝	0.750	0.0025	0.001	0.049	0.005	0.042	0.004	*C*	*C*	0.418	0.362	*C*
¾	0.875	0.003	0.001	0.065	0.006	0.045	0.004	0.032	0.003	0.641	0.455	0.328
1	1.125	0.0035	0.0015	0.065	0.006	0.050	0.005	0.035	0.004	0.839	0.655	0.465
1¼	1.375	0.004	0.0015	0.065	0.006	0.055	0.006	0.042	0.004	1.04	0.884	0.682
1½	1.625	0.0045	0.002	0.072	0.007	0.060	0.006	0.049	0.005	1.36	1.14	0.940
2	2.125	0.005	0.002	0.083	0.008	0.070	0.007	0.058	0.006	2.06	1.75	1.46
2½	2.625	0.005	0.002	0.095	0.010	0.080	0.008	0.065	0.006	2.93	2.48	2.03
3	3.125	0.005	0.002	0.109	0.011	0.090	0.009	0.072	0.007	4.00	3.33	2.68
3½	3.625	0.005	0.002	0.120	0.012	0.100	0.010	0.083	0.008	5.12	4.29	3.58
4	4.125	0.005	0.002	0.134	0.013	0.110	0.011	0.095	0.010	6.51	5.38	4.66
5	5.125	0.005	0.002	0.160	0.016	0.125	0.012	0.109	0.011	9.67	7.61	6.66
6	6.125	0.005	0.002	0.192	0.019	0.140	0.014	0.122	0.012	13.9	10.2	8.92
8	8.125	0.006	+0.002 −0.004	0.271	0.027	0.200	0.020	0.170	0.017	25.9	19.3	16.5
10	10.125	0.008	+0.002 −0.006	0.338	0.034	0.250	0.025	0.212	0.021	40.3	30.1	25.6
12	12.125	0.008	+0.002 −0.006	0.405	0.040	0.280	0.028	0.254	0.025	57.8	40.4	36.7

*Extracted from ASTM B88-2014 with permission of the publisher, the American Society for Testing and Materials, West Conshohocken, Pa.

[A] The average outside diameter of a tube is the average of the maximum and minimum outside diameter, as determined at any one cross section of the tube.

[B] Maximum deviation at any one point.

[C] Indicates that the material is not generally available or that no tolerance has been established.

TABLE 10-34a Standard Dimensions and Weights, and Tolerances in Diameter and Wall Thickness for Coil Lengths (ASTM B280-2016*) Copper for Refrigeration Tubing

Standard size, in	Outside diameter, in (mm)	Wall thickness, in (mm)	Weight, lb/ft (kg/m)	Tolerances	
				Average[A] outside diameter, plus and minus, in (mm)	Wall[B] thickness, plus and minus, in (mm)
⅛	0.125 (3.18)	0.030 (0.762)	0.0347 (0.0516)	0.002 (0.051)	0.003 (0.08)
3/16	0.187 (4.75)	0.030 (0.762)	0.0575 (0.0856)	0.002 (0.051)	0.003 (0.08)
¼	0.250 (6.35)	0.030 (0.762)	0.0804 (0.120)	0.002 (0.051)	0.0025 (0.08)
5/16	0.312 (7.92)	0.032 (0.813)	0.109 (0.162)	0.002 (0.051)	0.003 (0.08)
⅜	0.375 (9.52)	0.032 (0.813)	0.134 (0.199)	0.002 (0.051)	0.003 (0.08)
½	0.500 (12.7)	0.032 (0.813)	0.182 (0.271)	0.002 (0.051)	0.003 (0.08)
⅝	0.625 (15.9)	0.035 (0.889)	0.251 (0.373)	0.002 (0.051)	0.004 (0.011)
¾	0.750 (19.1)	0.042 (0.889)	0.305 (0.454)	0.0025 (0.064)	0.004 (0.011)
¾	0.750 (19.1)	0.042 (1.07)	0.362 (0.539)	0.0025 (0.064)	0.004 (0.011)
⅞	0.875 (22.3)	0.045 (1.14)	0.455 (0.677)	0.003 (0.076)	0.004 (0.11)
1⅛	1.125 (28.6)	0.050 (1.27)	0.665 (0.975)	0.0035 (0.089)	0.005 (0.13)
1⅜	1.375 (34.9)	0.055 (1.40)	0.884 (1.32)	0.004 (0.10)	0.006 (0.15)
1⅝	1.625 (41.3)	0.060 (1.52)	1.14 (1.70)	0.0045 (0.11)	0.006 (0.15)

*Extracted from ASTM B280-2016 with permission of the publisher, the American Society for Testing and Materials, West Conshohocken, Pa.
[A]The average outside diameter of a tube is the average of the maximum and minimum outside diameters as determined at any one cross section of the tube.
[B]The tolerances listed represent the maximum deviation at any point.

TABLE 10-34b Standard Dimensions and Weights, and Tolerances in Diameter and Wall Thickness for Straight Lengths (ASTM B280-2016*)

Note 1—Applicable to drawn temper tube only.

Standard size, in	Outside diameter, in (mm)	Wall thickness, in (mm)	Weight, lb/ft (kg/m)	Tolerances	
				Average[A] outside diameter, plus and minus, in (mm)	Wall[B] thickness, plus and minus, in (mm)
¼	0.250 (6.35)	0.025 (0.635)	0.068 (0.102)	0.001(0.025)	0.0025 (0.06)
⅜	0.375 (9.52)	0.030 (0.762)	0.126 (0.187)	0.001 (0.025)	0.0035 (0.08)
⅜	0.500 (12.7)	0.035 (0.889)	0.198 (0.295)	0.001 (0.025)	0.004 (0.09)
3/2	0.625 (15.9)	0.040 (1.02)	0.285 (0.424)	0.001 (0.025)	0.004 (0.10)
¾	0.750 (19.1)	0.042 (1.07)	0.362 (0.539)	0.001 (0.025)	0.004 (0.11)
⅞	0.875 (22.3)	0.045 (1.14)	0.455 (0.677)	0.001 (0.025)	0.004 (0.11)
1⅛	1.125 (28.6)	0.050 (1.27)	0.655 (0.975)	0.0015 (0.038)	0.004 (0.13)
1⅜	1.375 (34.9)	0.055 (1.40)	0.884 (1.32)	0.0015 (0.038)	0.006 (0.11)
1⅝	1.625 (41.3)	0.060 (1.52)	1.14 (1.70)	0.002 (0.051)	0.006 (0.11)
2⅛	2.125 (54.0)	0.070 (1.78)	1.75 (2.60)	0.002 (0.051)	0.007 (0.15)
2⅝	2.625 (66.7)	0.080 (2.03)	2.48 (3.69)	0.002 (0.051)	0.008 (0.15)
3⅛	3.125 (79.4)	0.090 (2.29)	3.33 (4.96)	0.002 (0.051)	0.009 (0.23)
3⅝	3.625 (92.1)	0.100 (2.54)	4.29 (6.38)	0.002 (0.051)	0.010 (0.25)
4⅛	4.125 (105)	0.110 (2.79)	5.38 (8.01)	0.002 (0.051)	0.011 (0.28)

*Extracted from ASTM B280-2016, with permission of the publisher, the American Society for Testing and Materials, West Conshohocken, Pa.
[A]The average outside diameter of a tube is the average of the maximum and minimum outside diameters as determined at any one cross section of the tube.
[B]The tolerances listed represent the maximum deviation at any point.

and corrosion resistance are somewhat a function of their nickel content. The 300 Series stainless steels are the most generally used. Some other frequently used alloys are listed in Table 10-37 together with their nominal compositions. For metallurgical and corrosion resistance data, see Sec. 25.

Titanium Seamless pipe is available as ASTM B861 and welded pipe as ASTM B862. Both standards offer numerous grades of unalloyed and alloyed materials. While the alloys often have higher tensile strengths, corrosion resistance may be sacrificed. Forged or wrought fittings and forged or cast valves are available. For many applications, elastomer-lined valves having carbon-steel or ductile-iron bodies and titanium trim offer an economical alternative to valves with titanium bodies. Titanium pipe is available with wall thicknesses conforming to many of those listed in Table 10-22, including Schedule 10S and Standard Weight. Properly selected and specified, titanium can be a good choice for seawater systems such as offshore fire water systems. Seamless and welded tubing is manufactured to ASTM B338; however, availability may be limited.

Flexible Metal Hose Flexible hoses provide flexible connections for conveying gases or liquids, wherever rigid pipes are impractical. There are two basic types of flexible hose: corrugated hoses and interlocked hoses. These flexible hoses can absorb vibrations and noise. They can also provide some flexibility for misaligned rigid piping or equipment during construction. Corrugated or interlocked thin brass, bronze, Monel, aluminum, and steel tubes are covered with flexible braided-wire jackets to form flexible metal hose. Both tube and braid are brazed or welded to pipe-thread, union, or flanged ends. Failures are often the result of corrosion of the braided-wire jacket or of a poor jacket-to-fitting weld. Maximum recommended temperature for bronze hose is approximately 230°C (450°F). Metal thickness is much less than for straight tube for the same pressure-temperature conditions; so accurate data on corrosion and erosion are required to make proper selection.

NONMETALLIC PIPE AND METALLIC PIPING SYSTEMS WITH NONMETALLIC LININGS

Cement-Lined Carbon-Steel Pipe Cement-lined carbon-steel pipe is made by lining steel pipe with special cement. The cement lining prevents pickup of iron by the fluid handled, corrosion of the metal by brackish or saline water, and growth of tuberculation. Various grades of cement are available, and the proper grade should be selected to match the application.

Cement-lined pipe in sizes smaller than 1½ in is not generally recommended. Cement-lined carbon-steel pipe can be supplied with butt-welded or flanged ends. Butt-welded construction involves the use of special joint grouts at the weld joint and controlled welding procedures. See Table 10-38.

American Water Works Association Standard C205 addresses shop-applied cement mortar lining of pipe sizes 4 in and larger. Fittings are available as cement mortar–lined butt-weld or flanged carbon steel, flanged cast iron, or flanged ductile iron. AWWA C602 addresses in-place (i.e., in situ or field) application of cement mortar lining for pipe sizes 4 in and larger.

Concrete Pipe Concrete piping and nonmetallic piping such as PVC, RTR, and HDPE are commonly used for buried gravity drain and pressurized applications. Common applications for both include construction culverts and forced water mains, sewage, industrial waste, and stormwater systems. Some of the factors to be considered when deciding whether to use concrete or nonmetallic piping include local code requirements, pipe size, soil and commodity corrosivity, commodity temperature and pressure, resistance to tuberculin growth, traffic and burial loads, soil conditions and bedding requirements, groundwater level and buoyancy issues, suitability of available joining methods, ability of joints to resist internal pressure thrust without the use of thrust blocks, availability of pressure-rated and non-pressure-rated fittings, shipping weight, load capacity of available construction equipment, requirements for special equipment such as fusion bonding

TABLE 10-35 Copper and Red-Brass Pipe (ASTM B42-2015a and B43-2015)*: Standard Dimensions, Weights, and Tolerances

Standard pipe size, in	Nominal outside diameter, in (mm)	Average outside diameter tolerances, in (mm), all minus†	Nominal wall thickness, in (mm)	Tolerance, in (mm)‡	Theoretical weight, lb/ft (kg/m)	Theoretical weight, lb/ft (kg/m)
			Regular pipe		Red brass	Copper
⅛	0.405 (10.3)	0.004 (0.10)	0.062 (1.57)	0.004 (0.10)	0.253 (0.376)	0.259 (0.385)
¼	0.540 (13.7)	0.004 (0.10)	0.082 (2.08)	0.005 (0.13)	0.447 (0.665)	0.457 (0.680)
⅜	0.675 (17.1)	0.005 (0.13)	0.090 (2.29)	0.005 (0.13)	0.627 (0.933)	0.641 (0.954)
½	0.840 (21.3)	0.005 (0.13)	0.107 (2.72)	0.006 (0.15)	0.934 (1.39)	0.955 (1.42)
¼	1.050 (26.7)	0.006 (0.15)	0.114 (2.90)	0.006 (0.15)	1.27 (1.89)	1.30 (1.93)
1	1.315 (33.4)	0.006 (0.15)	0.126 (3.20)	0.007 (0.18)	1.78 (2.65)	1.82 (2.71)
1⅕	1.660 (42.2)	0.006 (0.15)	0.146 (3.71)	0.008 (0.20)	2.63 (3.91)	2.69 (4.00)
1½	1.900 (48.3)	0.006 (0.15)	0.150 (3.81)	0.008 (0.20)	3.13 (4.66)	3.20 (4.76)
2	2.375 (60.3)	0.008 (0.20)	0.156 (3.96)	0.009 (0.23)	4.12 (6.13)	4.22 (6.28)
2½	2.875 (73.0)	0.008 (0.20)	0.187 (4.75)	0.010 (0.25)	5.99 (8.91)	6.12 (9.11)
3	3.500 (88.9)	0.010 (0.25)	0.219 (5.56)	0.012 (0.30)	8.56 (12.7)	8.76 (13.0)
3½	4.000 (102)	0.010 (0.25)	0.250 (6.35)	0.013 (0.33)	11.2 (16.7)	11.4 (17.0)
4	4.500 (114)	0.012 (0.30)	0.250 (6.35)	0.014 (0.36)	12.7 (18.9)	12.9 (19.2)
5	5.562 (141)	0.014 (0.36)	0.250 (6.35)	0.014 (0.36)	15.8 (23.5)	16.2 (24.1)
6	6.625 (168)	0.016 (0.41)	0.250 (6.35)	0.014 (0.36)	19.0 (28.3)	19.4 (28.9)
8	8.625 (219)	0.020 (0.51)	0.312 (7.92)	0.022 (0.56)	30.9 (46.0)	31.6 (47.0)
10	10.750 (273)	0.022 (0.56)	0.365 (9.27)	0.030 (0.76)	45.2 (67.3)	46.2 (68.7)
12	12.750 (324)	0.024 (0.61)	0.375 (9.52)	0.030 (0.76)	55.3 (82.3)	56.5 (84.1)
			Extra strong pipe		Red brass	Copper
⅛	0.405 (10.3)	0.004 (0.10)	0.100 (2.54)	0.006 (0.15)	0.363 (0.540)	0.371 (0.552)
¼	0.540 (13.7)	0.004 (0.10)	0.123 (3.12)	0.007 (0.18)	0.611 (0.909)	0.625 (0.930)
⅜	0.675 (17.1)	0.005 (0.13)	0.127 (3.23)	0.007 (0.18)	0.829 (1.23)	0.847 (1.26)
½	0.840 (21.3)	0.005 (0.13)	0.149 (3.78)	0.008 (0.20)	1.23 (1.83)	1.25 (1.86)
¾	1.050 (26.7)	0.006 (0.15)	0.157 (3.99)	0.009 (0.23)	1.67 (2.48)	1.71 (2.54)
1	1.315 (33.4)	0.006 (0.15)	0.182 (4.62)	0.010 (0.25)	2.46 (3.66)	2.51 (3.73)
1¼	1.660(42.2)	0.006 (0.15)	0.194 (4.93)	0.010 (0.25)	3.39 (5.04)	3.46 (5.15)
1½	1.900 (48.3)	0.006 (0.15)	0.203 (5.16)	0.011 (0.28)	4.10 (6.10)	4.19 (6.23)
2	2.375 (60.3)	0.008 (0.20)	0.221 (5.61)	0.012 (0.30)	5.67 (8.44)	5.80 (8.63)
2½	2.875 (73.0)	0.008 (0.20)	0.280 (7.11)	0.015 (0.38)	8.66 (12.9)	8.85 (13.2)
3	3.500 (88.9)	0.010 (0.25)	0.304 (7.72)	0.016 (0.41)	11.6 (17.3)	11.8 (17.6)
3½	4.000 (102)	0.010 (0.25)	0.321 (8.15)	0.017 (0.43)	14.1 (21.0)	14.4 (21.4)
4	4.500 (114)	0.012 (0.30)	0.341 (8.66)	0.018 (0.46)	16.9 (25.1)	17.3 (25.7)
5	5.562 (141)	0.014 (0.36)	0.375 (9.52)	0.019 (0.48)	23.2 (34.5)	23.7 (35.3)
6	6.625 (168)	0.016 (0.41)	0.437 (11.1)	0.027 (0.69)	32.2 (47.9)	32.9 (49.0)
8	8.625 (219)	0.020 (0.51)	0.500 (12.7)	0.035 (0.89)	48.4 (72.0)	49.5 (73.7)
10	10.750 (273)	0.022 (0.56)	0.500 (12.7)	0.040 (1.0)	61.1 (90.9)	62.4 (92.9)

*Copyright American Society for Testing and Materials, West Conshohocken, PA; reprinted/adapted with permission. All tolerances are plus and minus except as otherwise indicated.
†The average outside diameter of a tube is the average of the maximum and minimum outside diameters as determined at any one cross section of the tube.
‡Maximum deviation at any one point.

TABLE 10-36 Hard-Drawn Copper Threadless Pipe (ASTM B302)*

	Nominal dimensions, in (mm)					Tolerances, in (mm)	
Standard pipe size, in	Outside diameter	Inside diameter	Wall thickness	Cross-sectional area of bore, in² (cm²)	Nominal weight, lb/ft (kg/m)	Average outside diameter, all minus†	Wall thickness, plus and minus
¼	0.540 (13.7)	0.410 (10.4)	0.065 (1.65)	0.132 (0.852)	0.376 (0.559)	0.004 (0.10)	0.0035 (0.089)
⅜	0.675 (17.1)	0.545 (13.8)	0.065 (1.65)	0.233 (1.50)	0.483 (0.719)	0.004 (0.10)	0.004 (0.10)
½	0.840 (21.3)	0.710 (18.0)	0.065 (1.65)	0.396 (2.55)	0.613 (0.912)	0.005 (0.13)	0.004 (0.10)
¾	1.050 (26.7)	0.920 (23.4)	0.065 (1.65)	0.665 (4.29)	0.780 (1.16)	0.005 (0.13)	0.004 (0.10)
1	1.315 (33.4)	1.185 (30.1)	0.065 (1.65)	1.10 (7.10)	0.989 (1.47)	0.005 (0.13)	0.004 (0.10)
1¼	1.660 (42.2)	1.530 (38.9)	0.065 (1.65)	1.84 (11.9)	1.26 (1.87)	0.006 (0.15)	0.004 (0.10)
1½	1.900 (48.3)	1.770 (45.0)	0.065 (1.65)	2.46 (15.9)	1.45 (2.16)	0.006 (0.15)	0.004 (0.10)
2	2.375 (60.3)	2.245 (57.0)	0.065 (1.65)	3.96 (25.5)	1.83 (272)	0.007 (0.18)	0.006 (0.15)
2½	2.875 (73.0)	2.745 (69.7)	0.065 (1.65)	5.92 (38.2)	2.22 (3.30)	0.007 (0.18)	0.006 (0.15)
3	3.500 (88.9)	3.334 (84.7)	0.083 (2.11)	8.73 (56.3)	3.45 (5.13)	0.008 (0.20)	0.007 (0.18)
3½	4.000 (102)	3.810 (96.8)	0.095 (2.41)	11.4 (73.5)	4.52 (6.73)	0.008 (0.20)	0.007 (0.18)
4	4.500 (114)	4.286 (109)	0.107 (2.72)	14.4 (92.9)	5.72 (8.51)	0.010 (0.25)	0.009 (0.23)
5	5.562 (141)	5.298 (135)	0.132 (3.40)	22.0 (142)	8.73 (13.0)	0.012 (0.30)	0.010 (0.25)
6	6.625 (168)	6.309 (160)	0.158 (4.01)	31.3 (202)	12.4 (18.5)	0.014 (0.36)	0.010 (0.25)
8	8.625 (219)	8.215 (209)	0.205 (5.21)	53.0 (342)	21.0 (31.2)	0.018 (0.46)	0.014 (0.36)
10	10.750 (273)	10.238 (260)	0.256 (6.50)	82.3 (531)	32.7 (48.7)	0.018 (0.46)	0.016 (0.41)
12	12.750 (324)	12.124 (308)	0.313 (7.95)	115 (742)	47.4 (70.5)	0.018 (0.46)	0.020 (0.51)

*Copyright American Society for Testing and Materials, West Conshohocken, PA; reprinted/adapted with permission.
†The average outside diameter of a tube is the average of the maximum and minimum outside diameters, as determined at any one cross section of the tube.

TABLE 10-37 Common Nickel and Nickel-Bearing Alloys

Common trade name or registered trademark	Code designation	Alloy no.	ASTM specification (pipe)	Nominal composition, %										
				Ni	Cr	Mo	Fe	C[a]	Si[a]	Mn	Cu	Cb	Co	W
Type 304 stainless steel		S30400	A312	8	18		BAL	0.08	1.00	2.0				
Type 304L stainless steel		S30403	A312	8	18		BAL	0.03	1.00	2.0				
Type 316 stainless steel		S31600	A312	12	16	2	BAL	0.08	1.00	2.0				
Type 316L stainless steel		S31603	A312	12	16	2	BAL	0.03	1.00	2.0				
Carpenter 20cb[b]	Ni-Cr-Fe-Mo-Cu-Cb stabilized	N08020	B464	33	20	2.5	38.5	0.06		2.0	3	1		
Incoloy 800[c]	Ni-Fe-Cr	N08800	B407	32.5	21		46	0.05	0.5	0.8	0.4			
Incoloy 825[c]	Ni-Fe-Cr-Mo-Cu	N08825	B423	42	21.5	3	30	0.03	0.2	0.5	2.2			
Hastelloy C-276[d]	Ni-Mo-Cr low carbon	N10276	B575[e]	54	15	16	5	0.02	0.08	1			2.5	4
Hastelloy B-2[d]	Ni-Mo	N10001	B333[e]	64	1	28	2	0.02	0.1	1				
Inconel 625[c]	Ni-Cr-Mo-Cb	N06625	B444	61	21.5	9	2.5	0.05	0.2	0.2		4		
Inconel 600[c]	Ni-Cr-Fe	N06600	B167	76	15.5		8	0.08	0.2	0.5	0.2			
Monel 400[c]	Ni-Cu	N04400	B165	66			1.2	0.20	0.2	1	31.5			
Nickel 200[c]	Ni	N02200	B161	99+			0.2	0.08						
Hastelloy G[d]	Ni-Cr-Fe-Mo-Cu	N06007	B622	42	22.2	6.5	19.5	0.05	1	1.5	2	2.2[f]	2.5[a]	1[a]

[a]Maximum.
[b]Registered trademark, Carpenter Technology Corp.
[c]Registered trademark, Huntington Alloys, Inc.
[d]Registered trademark, Cabot Corp.
[e]Plate.
[f]Cb + Ta.

machines, contractor's experience and labor skill level requirements, and final installed cost.

Non-reinforced-concrete culvert pipe for gravity drain applications is manufactured to ASTM C14 in strength Classes 1, 2, and 3. It is available with internal diameters 4 through 36 in. Reinforced concrete culvert for gravity drain applications is manufactured to ASTM C76 with internal diameters of 12 through 144 in. Metric sizes are manufactured to ASTM C14M and C76M. Joints are typically bell and spigot (or a similar variation) with rubber gaskets.

Concrete pressure pipe is typically custom-designed to three different specifications. Each design provides a cement mortar lining or concrete interior. It is advisable to consult manufacturers regarding the most appropriate specification for a given application and the availability of fittings. The names of some manufacturers can be obtained through the American Concrete Pressure Pipe Association. American Water Works Association standard AWWA C300 addresses steel cylinder reinforced-concrete pressure pipe in inside-diameter sizes 30 through 144 in. AWWA C301 addresses prestressed reinforced pipe with a steel cylinder wrapped with steel wire. Inside-diameter sizes are 16 through 144 in. AWWA C302 addresses circumferentially reinforced pipe without a steel cylinder or prestress. Inside-diameter sizes are 12 through 144 in, with continuous pressure ratings to 0.38 MPa (55 lbf/in^2) and total pressure (including surge) to 0.45 MPa (65 lbf/in^2). AWWA C303 addresses reinforced pipe with a steel cylinder helically wrapped with steel bar. Inside-diameter sizes are 10 through 60 in, with pressure ratings to 2.7 MPa (400 lbf/in^2) working pressure. Joints are typically bell and spigot (or a similar variation) with rubber gaskets. In addition to the gasket, grouting is used on the exterior and interior of the joint to seal otherwise exposed steel.

Glass Pipe and Fittings These are made from heat- and chemical-resistant borosilicate glass in accordance with ASTM E-438 Type 1, Class A. This glass is resistant to chemical attack by almost all products, which makes its resistance much more comprehensive than that of other well-known materials. It is highly resistant to water, saline solutions, organic substances, halogens such as chlorine and bromine, and many acids. Only a few chemicals can cause noticeable corrosion of the glass surface, such as hydrofluoric acid, concentrated phosphoric acid, and strong caustic solutions at elevated temperatures. Some important physical properties are as follows:

Mean linear expansion coefficient between 20 and 300°C: $[(3.3 \pm 0.1) \times 10^{-6}]/\text{K}$
Mean thermal conductivity between 20 and 200°C: 1.2 W/(m · K)
Mean specific heat capacity between 20 and 100°C: 0.8 kJ/(kg · K)
Mean specific heat capacity between 101 and 200°C: 0.9 kJ/(kg · K)
Density at 20°C: 2.23 kg/dm^3

Flanged glass pipe with conical ends (Fig. 10-153) should be used in applications requiring a pressure rating or that are expected to see thermal cycling. Flanged ends are normally required to mate with other glass components (vessels, coil heat exchangers, etc). The flanges are specially designed plastic backing flanges that are cushioned from the glass by either molded plastic or fiber inserts. The liquid seal is provided by means of a gasket that is gripped between the grooved pipe ends.

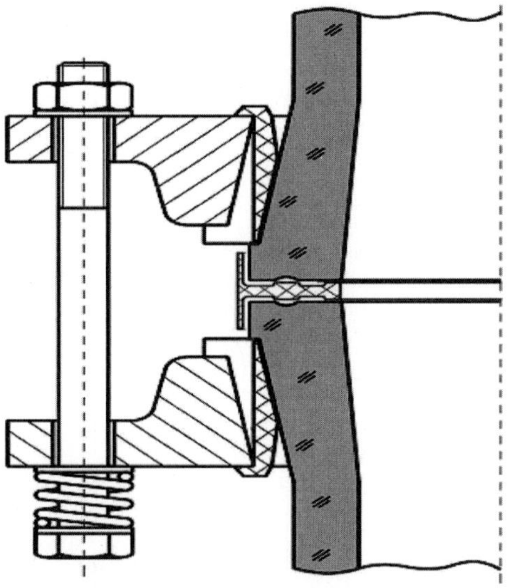

FIG. 10-153 Flanged joint with conical ends. (*Adapted with permission of De Dietrich Process Systems, Mountainside, N.J.*)

TABLE 10-38 Cement-Lined Carbon-Steel Pipe*

Standard pipe size, in	Inside diam. after lining, in	Typical thickness of lining, in	Weight, per ft, lb	Standard pipe size, in	Inside diam. after lining, in	Typical thickness of lining, in	Weight per ft, lb
				3	2.70	0.13	8.3
				4	3.60	0.16	12.0
				6	5.40	0.25	24.0
1½	1.40	0.09	3.0	8	7.40	0.25	32.0
2	1.80	0.10	4.1	10	9.40	0.30	43.0
2½	2.20	0.10	6.6	12	11.40	0.30	55.0

*To convert inches to millimeters, multiply by 25.4; to convert pounds per foot to kilograms per meter, multiply by 1.49.

TABLE 10-39 Dimensions for Glass Pipe and Flanged Joints (see Fig. 10-154)*

Nominal pipe size, mm	D1	D2	D3	D4	Type
15	16.8	28.6	23	15.5–17.5	A
25	26.5	42.2	34	25–27	A
40	38.5	57.4	48	36.5–39.75	A
50	50.5	70	60.5	48–52	A
80	76	99.2	88	72–78	A
100	104.5	132.6	120.5	97.6–110	A
150	154	185	172	150–156	A
200	203	235	220	197–205	B
300	300	340	321	299–303	B

*Adapted with permission of De Dietrich Process Systems, Mountainside, N.J.

Glass pipe is made in the sizes shown in Table 10-39. Depending on the nominal diameter, lengths range from 0.075 to 3 m. Design pressure ranges from −0.10 MPa (−14.5 lbf/in^2) vacuum to 0.40 MPa (58 lbf/in^2) for nominal diameters of 15 through 50 mm, 3 MPa (43 lbf/in^2) for nominal diameter of 80 mm, 0.20 MPa (29 lbf/in^2) for nominal diameters of 100 and 150 mm, and 0.10 MPa (14.5 lbf/in^2) for nominal diameters of 200 and 300 mm. Maximum permissible thermal shock, as a general guide, is 120 K. Maximum operating temperature is 200°C (248°F). A complete line of fittings is available, and special parts can be made to order. Thermal expansion stresses should be completely relieved by tied PTFE corrugated expansion joints and offsets. Temperature rating may be limited by joint design and materials.

Beaded pipe is used for process waste lines and vent lines. Some applications have also been made in low-pressure and vacuum lines. Beaded end pipe is available in nominal diameters of 1½ through 6 in. For operating conditions, manufacturers should be consulted. In this system the ends of the pipe and fittings are formed into a bead, as shown in Fig. 10-155. The coupling consists of a stainless steel outer shell, a rubber collar, and a TFE liner gasket. When the single coupling nut is tightened, the thick rubber sleeve forces the TFE gasket against the glass to make the seal.

Glass-Lined Steel Pipe and Fittings This pipe is fully resistant to all acids except hydrofluoric and concentrated phosphoric acids at temperatures up to 120°C (248°F). It is also resistant to alkaline solutions at moderate temperatures. Glass-lined steel pipe can be used at temperatures up to 220°C (428°F) under some exposure conditions provided there are no excessive temperature changes. The operating pressure rating of commonly available systems is 1 MPa (145 lbf/in^2). The glass lining is approximately 1.6 mm (1⁄16 in) thick. It is made by lining Schedule 40 steel pipe. Fittings are available in glass-lined cast steel. Standard nominal pipe sizes available are 1½ through 8 in. Larger-diameter pipe up to 48 in is available on a custom-order basis. A range of lengths are generally available. See Table 10-40 for dimensional data. Steel split flanges drilled to ANSI B16.5 Class 150 dimensions along with PTFE envelope gaskets are used for the assembly of the system.

Fused Silica or Fused Quartz Containing 99.8 percent silicon dioxide, fused silica or fused quartz can be obtained as opaque or transparent pipe and tubing. The melting point is 1710°C (3100°F). Tensile strength is approximately 48 MPa (7000 lbf/in^2); specific gravity is about 2.2. The pipe and tubing can be used continuously at temperatures up to 1000°C (1830°F) and intermittently up to 1500°C (2730°F). The material's chief assets are noncontamination of most chemicals in high-temperature service, thermal-shock resistance, and high-temperature electrical insulating characteristics.

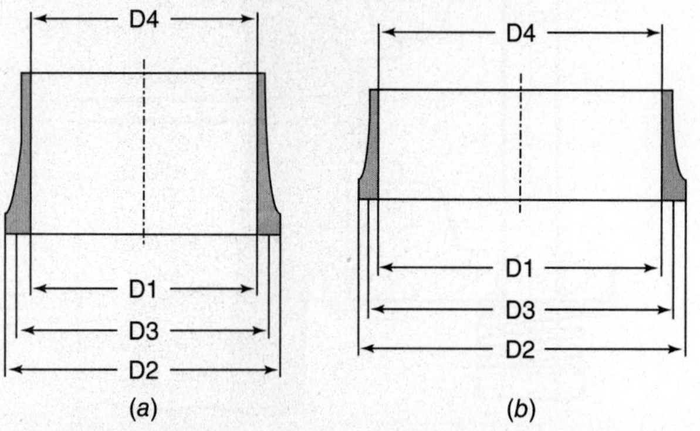

FIG. 10-154 Flanged pipe ends. (*Adapted with permission of De Dietrich Process Systems, Mountainside, N.J.*)

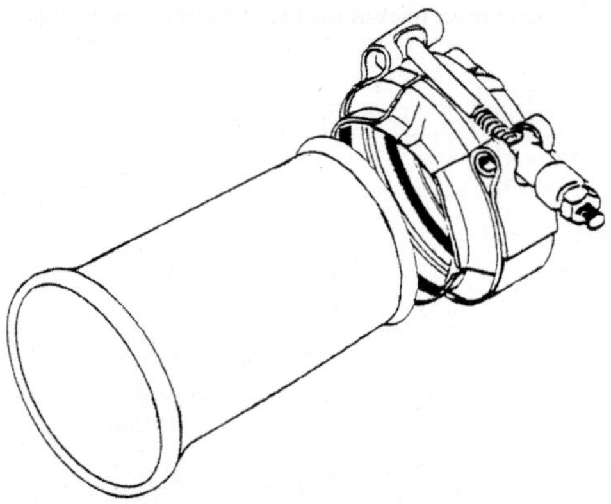

FIG. 10-155 Joint with beaded pipe ends. (*Adapted with permission of De Dietrich Process Systems, Mountainside, N.J.*)

Transparent tubing is available in inside diameters from 1 to 125 mm in a range of wall thicknesses. Satin-surface tubing is available in inside diameters from 1⁄18 to 2 in, and sand-surface pipe and tubing are available in ½- to 24-in inside diameters and lengths up to 6 m (20 ft). Sand-surface pipe and tubing are obtainable in wall thicknesses varying from ⅛ to 1 in. Pipe and tubing sections in both opaque and transparent fused silica or fused quartz can be readily machine-ground to special tolerances for pressure joints or other purposes. Also, fused-silica piping and tubing can be reprocessed to meet special design requirements. Manufacturers should be consulted for specific details.

Plastic-Lined Steel Pipe Use of a variety of polymeric materials as liners for steel pipe rather than as piping systems solves problems which the relatively low tensile strength of the polymer at elevated temperature and high thermal expansion, compared with steel, would produce. The steel outer shell permits much wider spacing of supports, reliable flanged joints, and higher pressure and temperature in the piping. The size range is 1 through 12 in. The systems are flanged with 125-lb cast-iron, 150-lb ductile-iron, and 150- and 300-lb steel flanges. The linings are factory-installed in both pipe and fittings. Lengths are available up to 6 m (20 ft). Lined ball, diaphragm, and check valves and plug cocks are available.

One method of manufacture consists of inserting the liner into an oversize, approximately Schedule 40 steel tube and swaging the assembly to produce iron-pipe-size outside diameter, firmly engaging the liner which projects from both ends of the pipe. Flanges are then screwed onto the pipe, and the projecting liner is hot-flared over the flange faces nearly to the bolt holes. In another method, the liner is pushed into steel pipe having cold-flared laps backed up by flanges at the ends and then is hot-flared over the faces of the laps. Pipe lengths made by either method may be shortened in the field and reflared with special procedures and tools. Square and tapered spacers are furnished to adjust for small discrepancies in assembly.

Liner types available are suitable for a wide variety of chemical services, including acids, alkalies, and various solvents. All liners are permeable to some degree, and manufacturers use various methods to vent gas out of the interspace between the liner and casing. All plastics are subject to *environmental stress cracking* (ESC). ESC can occur even when the liner is chemically resistant to the service. Lined pipe manufacturers should always be consulted regarding liner selections and service applications. Also consult manufacturers regarding vacuum service limits.

TABLE 10-40 Glass-Lined Steel Pipe*

		Range of standard lengths, in	
Size, in	Outside diameter, in	Minimum	Maximum
1½	1.900	6	60
2	2.375	6	84
3	3.500	6	120
4	4.500	6	120
6	6.625	6	120
8	8.625	6	120

*Adapted with permission of De Dietrich Process Systems, Mountainside, N.J. Other manufacturers may offer different standard lengths.

Polyvinylidene Chloride Liners Polyvinylidene chloride liners have excellent resistance to hydrochloric acid. Maximum temperature is 80°C (175°F). Polyvinylidene chloride is also known as Saran, a product of Dow Chemical Co.

Polypropylene Liners Polypropylene liners are used in sulfuric acid service. At 10 to 30 percent concentration the upper temperature limit is 93°C (200°F). Polypropylene is also suitable for higher concentrations at lower temperatures.

Kynar Liners Kynar (Pennwalt Chemicals Corp.) polyvinylidene fluoride liners are used for many chemicals, including bromine and 50 percent hydrochloric acid.

PTFE and PFA Lined Steel Pipe These are available in sizes from 1 through 12 in and in lengths through 6 m (20 ft). Experience has determined that practical upper temperature limits are approximately 204°C (400°F) for PTFE (polytetrafluoroethylene) and PFA (perfluoroalkoxy) and 149°C (300°F) for FEP (fluoroethylene polymer); Class 150 and 300 ductile-iron or steel flanged lined fittings and valves are used. The nonadhesive properties of the liner make it ideal for handling sticky or viscous substances. The thickness of the lining varies from 1.5 to 3.8 mm (60 to 150 mil), depending on the pipe size. Only flanged joints are used.

Rubber-Lined Steel Pipe This pipe is made in lengths up to 6 m (20 ft) with seamless, straight seam-welded and some types of spiral-welded pipe using various types of natural and synthetic adhering rubber. The type of rubber is selected to provide the most suitable lining for the specific service. In general, soft rubber is used for abrasion resistance, semihard for general service, and hard for the more severe service conditions. Multiple-ply lining and combinations of hard and soft rubber are available. The thickness of lining ranges from 3.2 to 6.4 mm (⅛ to ¼ in) depending on the service, the type of rubber, and the method of lining. Cast-steel, ductile-iron, and cast-iron flanged fittings are available rubber-lined. The fittings are usually purchased by the vendor since absence of porosity on the inner surface is essential. Pipe is flanged before rubber lining, and welding elbows and tees may be incorporated at one end of the length of pipe, subject to the conditions that the size of the pipe and the location of the fittings be such that the operator doing the lining can place a hand on any point on the interior surface of the fitting. Welds must be ground smooth on the inside, and a radius is required at the inner edge of the flange face.

The rubber lining is extended out over the face of flanges. With hard-rubber lining, a gasket is required. With soft-rubber lining, a release coating or a polyethylene sheet is required in place of a gasket to avoid bonding of the lining of one flange to the lining on the other and to permit disassembly of the flanged joint. Also, for pressures over 0.86 MPa (125 lbf/in^2), the tendency of soft-rubber linings to extrude out between the flanges may be prevented by terminating the lining inside the bolt holes and filling the balance of the space between the flange faces with a spacer of the proper thickness. Hard-rubber-lined gate, diaphragm, and swing check valves are available. In gate valves, the stem, wedge assembly, and seat rings—and in the check valves, the hinge pin, flapper arm, disk, and seat ring—must be made of metal resistant to the solution handled.

Plastic Pipe In contrast to other piping materials, plastic pipe is free from internal and external corrosion, is easily cut and joined, and does not cause galvanic corrosion when coupled to other materials. Allowable stresses and upper temperature limits are low. Normal operation is in the creep range. Fluids for which a plastic is not suited penetrate and soften it rather than dissolve surface layers. Coefficients of thermal expansion are high.

Plastic pipe or tubing may be used for a wide variety of services. As with all nonmetallic materials, code restrictions limit the applications in which their use is permitted. In general, their use in flammable or toxic service is limited. Plastic tubing of various types may be used for instrument air-signal connections; however, as is the case with all nonmetallic applications, the need for fire resistance must be considered. When it is used in specialized applications such as potable water or underground fire water, care should be taken to ensure that the specified products are certified by appropriate agencies such as the National Sanitation Foundation and Factory Mutual.

Support spacing must be much closer than for carbon steel. As temperature increases, the allowable stress for many plastic pipes decreases very rapidly, and heat from sunlight or adjacent hot uninsulated equipment has a marked effect. Many plastics deteriorate with exposure to ultraviolet light if not provided with a UV-resistant coating or other surface barrier. Successful economical underground use of plastic pipe does not necessarily indicate similar economies outdoors aboveground.

Methods of joining include threaded joints with IPS dimensions, solvent-welded joints, heat-fused joints, and insert fittings. Schedules 40 and 80 (see Table 10-22) have been used as a source for standardized dimensions at joints. Some plastics are available in several grades with allowable stresses varying by a factor of 2 to 1. For the same plastic, ½-in Schedule 40 pipe of the strongest grade may have 4 times the allowable internal pressure of the weakest grade of a 2-in Schedule 40 pipe. For this reason, the plastic-pipe industry is shifting to standard dimension ratios (approximately the same ratio of diameter to wall thickness over a wide range of pipe sizes).

ASTM and the Plastics Pipe Institute have established identifications for plastic pipe in which the first group of letters identifies the plastic, the two following numbers identify the grade of that plastic, and the last two numbers represent the design stress in the nearest lower 0.7 MPa (100 lbf/in^2) unit at 23°C (73.4°F).

Polyethylene The Plastics Pipe Institute (www.plasticpipe.org) is an excellent source of information regarding specification, design, fabrication, and testing of polyethylene piping. Polyethylene (PE) pipe and tubing are available in sizes 48 in and smaller. They have excellent resistance at room temperature to salts, sodium and ammonium hydroxides, and sulfuric, nitric, and hydrochloric acids. *High-density polyethylene* (HDPE) is often used for underground fire water. Pipe and tubing are produced by extrusion from resins whose density varies with the manufacturing process. Physical properties and therefore wall thickness depend on the particular resin used. In some products, about 3 percent carbon black is added to provide resistance to ultraviolet light. Use of higher-density resin reduces splitting and pinholing in service and increases the strength of the material and the maximum service temperature.

ASTM D2104 covers PE pipe in sizes ½ through 6 in with IPS Schedule 40 outside and inside diameters for insert-fitting joints. ASTM D2239 covers six standard dimension ratios of pipe diameter to wall thickness in sizes ½ through 6 in. ASTM D2447 covers sizes ½ through 12 in with IPS Schedule 40 and 80 outside and inside diameters for use with heat-fusion socket-type and butt-type fittings. ASTM D3035 covers 10 standard dimension ratios of pipe sizes from ½ through 24 in with IPS outside diameters. All these specifications cover various PE materials. Hydrostatic design stresses within the recommended temperature limits are given in App. B, Table B-1, of the code. The hydrostatic design stress is the maximum tensile hoop stress due to internal hydrostatic water pressure that can be applied continuously with a high degree of certainty that failure of the pipe will not occur. Both manufacturers and the Plastic Pipe Institute publish literature describing design calculations required to determine the required wall thickness.

Polyethylene water piping is not damaged by freezing. Pipe and tubing 2 in and smaller are shipped in coils several hundred feet in length.

Clamped-insert joints (Fig. 10-156) are used for flexible plastic pipe up through the 2-in size. Friction between the pipe and the spud is developed both by forcing the spud into the pipe and by tightening the clamp. For the larger sizes, which have thicker walls, these methods cannot develop adequate friction. Insert joints also have high pressure drop. Stainless steel bands are available. Inserts are available in nylon, polypropylene, and a variety of metals.

Joints of all sizes may be made with heat fusion techniques. Fused joints may be made with either electrofusion or conventional heat fusion. Electrofusion joints are made with fittings that have embedded heating wires. Conventional fusion joints are made with special machines that trim the pipe ends, apply heat, and then force them together to form a bond. Consult the manufacturer regarding the sizes for which electrofusion fittings are available.

A significant use for PE and PP pipe is the technique of rehabilitating deteriorated pipe lines by lining them with plastic pipe. Lining an existing pipe with plastic pipe has a large cost advantage over replacing the line, particularly if replacement of the old line would require excavation.

Polyvinyl Chloride Polyvinyl chloride (PVC) pipe and tubing are available with socket fittings for solvent-cemented joints in sizes 24 in and smaller. PVC with gasketed bell and spigot joints is available in sizes 4 through 48 in. Chlorinated polyvinyl chloride (CPVC) pipe and tubing are available with socket fittings for solvent-cemented joints in sizes 4 in and smaller. PVC and CPVC are suitable for a variety of chemical services and

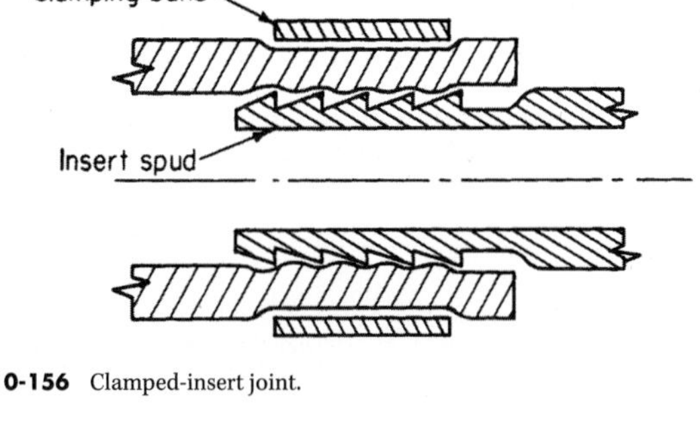

FIG. 10-156 Clamped-insert joint.

are commonly used for potable water. Consult manufacturers or the Plastics Pipe Institute for chemical resistance data. Hydrostatic design stresses within the temperature limits are given in App. B, Table B-1, of the code.

ASTM D1785 covers sizes from ⅛ through 12 in of PVC pipe in IPS Schedules 40, 80, and 120, except that Schedule 120 starts at ½ in and is not IPS for sizes from ½ through 3 in. ASTM D2241 covers sizes ⅛ through 36 in but with IPS outside diameter and seven standard dimension ratios: 13.5, 17, 21, 26, 32.5, 41, and 64.

ASTM D2513 specifies requirements for thermoplastic materials for buried fuel gas and relining applications. Materials addressed include PE, PVC, and *crosslinked polyethylene* (PEX). Tubing sizes covered are ¼ through 1¼ in. IPS sizes covered are ½ through 12 in. Size availability depends upon the material.

Solvent-cemented joints (Fig. 10-139) are standard, but screwed joints are sometimes used with Schedule 80 pipe. Cemented joints must not be disturbed for 5 min and achieve full strength in 1 day. A wide variety of valve types are available in PVC and CPVC.

Polypropylene Polypropylene (PP) pipe and fittings have excellent resistance to most common organic and mineral acids and their salts, strong and weak alkalies, and many organic chemicals. They are available in sizes ½ through 6 in, in Schedules 40 and 80, but are not covered as such by ASTM specifications.

Reinforced-Thermosetting-Resin (RTR) Pipe Glass-reinforced epoxy resin has good resistance to nonoxidizing acids, alkalies, saltwater, and corrosive gases. The glass reinforcement is many times stronger at room temperature than plastics are, does not lose strength with increasing temperature, and reinforces the resin effectively up to 149°C (300°F). (See Table 10-21 for temperature limits.) The glass reinforcement is located near the outside wall, protected from the contents by a thick wall of resin and protected from the atmosphere by a thin wall of resin. Stock sizes are 2 through 12 in.

Pipe is supplied in 6- and 12-m (20- and 40-ft) lengths. It is more economical for long, straight runs than for systems containing numerous fittings. When the pipe is sawed to nonfactory lengths, it must be sawed very carefully to avoid cracking the interior plastic zone. A two-component cement may be used to bond lengths into socket couplings or flanges or cemented-joint fittings. Curing of the cement is temperature-sensitive; it sets to full strength in 45 min at 93°C (200°F), in 12 h at 38°C (100°F), and in 24 h at 10°C (50°F). Extensive use is made of shop-fabricated flanged preassemblies. Only flanged joints are used to connect to metallic piping systems. Compared with that of other plastics, the ratio of fitting cost to pipe cost is high. Cemented-joint fittings and flanged fittings are available. Internally lined flanged metallic valves are used.

RTR is more flexible than metallic pipe and consequently requires closer support spacing. While the recommended spacing varies among manufacturers and with the type of product, Table 10-41 gives typical hanger-spacing ranges. The pipe fabricator should be consulted for recommended hanger spacing on the specific pipe-wall construction being used.

Epoxy resin has a higher strength at elevated temperatures than polyester resins but is not as resistant to attack by some fluids. Some glass-reinforced epoxy-resin pipe is made with a polyester-resin liner. The coefficient of thermal expansion of glass-reinforced resin pipe is higher than that for carbon steel but much less than that for plastics.

Glass-reinforced polyester is the most widely used reinforced-resin system. A wide choice of polyester resins are available. The bisphenol resins resist strong acids as well as alkaline solutions. The size range is 2 through at least 36 in; the temperature range is shown in Table 10-21. Diameters are not standardized. Adhesive-cemented socket joints and hand-layup reinforced butt joints are used. For the latter, reinforcement consists of layers of glass cloth saturated with adhesive cement.

DESIGN OF PIPING-SYSTEMS

Safeguarding *Safeguarding* may be defined as the provision of protective measures as required to ensure the safe operation of a proposed piping system. General considerations to be evaluated should include (1) the hazardous properties of the fluid, (2) the quantity of fluid that could be released by a piping failure, (3) the effect of a failure (such as possible loss of cooling water) on overall plant safety, (4) evaluation of effects on a reaction with the environment (i.e., possibility of a nearby source of ignition), (5) the probable extent of exposure of operating or maintenance personnel, and (6) the relative inherent safety of the piping by virtue of materials of construction, methods of joining, and history of service reliability.

Evaluation of safeguarding requirements might include engineered protection against possible failures such as thermal insulation, armor, guards, barricades, and damping for protection against severe vibration, water hammer, or cyclic operating conditions. Simple means to protect people and property such as shields for valve bonnets, flanged joints, and sight glasses should not be overlooked. The necessity for means to shut off or control flow in the event of a piping failure such as block valves or excess-flow valves should be examined.

Classification of Fluid Services The code applies to piping systems as illustrated in Fig. 10-116, but two categories of fluid services are segregated for special consideration as follows:

Category D *Category D fluid service* is defined as a fluid service to which all the following apply: (1) the fluid handled is nonflammable and nontoxic; (2) the design gage pressure does not exceed 150 psi (1.0 MPa); and (3) the design temperature is between −20°F (−29°C) and 366°F (186°C).

Category M *Category M fluid service* is defined as a fluid service in which a single exposure to a very small quantity of a toxic fluid, caused by leakage, can produce serious irreversible harm to persons on breathing or bodily contact, even when prompt restorative measures are taken.

The code assigns to the owner the responsibility for identifying those fluid services which are in Categories D and M. The design and fabrication requirements for Class M toxic-service piping are beyond the scope of this text. See ASME B31.3-2014, chap. VIII.

Design Conditions Definitions of the temperatures, pressures, and various forces applicable to the design of piping systems are as follows:

Design Pressure The design pressure of a piping system shall not be less than the pressure at the most severe condition of coincident internal or external pressure and temperature resulting in the greatest required component thickness or rating.

Design Temperature The design temperature is the material temperature representing the most severe condition of coincident pressure and temperature. For uninsulated metallic pipe with fluid below 65°C (150°F), the metal temperature is taken as the fluid temperature.

With fluid at or above 65°C (150°F) and without external insulation, the metal temperature is taken as a percentage of the fluid temperature unless a lower temperature is determined by test or calculation. For pipe, threaded and welding-end valves, fittings, and other components with a wall thickness comparable with that of the pipe, the percentage is 95 percent; for flanges and flanged valves and fittings, 90 percent; for lap-joint flanges, 85 percent; and for bolting, 80 percent.

With external insulation, the metal temperature is taken as the fluid temperature unless service data, tests, or calculations justify lower values. For internally insulated pipe, the design metal temperature shall be calculated or obtained from tests.

Ambient Influences If cooling results in a vacuum, the design must provide for external pressure or a vacuum breaker installed; also provision must be made for thermal expansion of contents trapped between or in closed valves. Nonmetallic or nonmetallic-lined pipe may require protection when the ambient temperature exceeds the design temperature.

Occasional variations of pressure or temperature, or both, above operating levels are characteristic of certain services. If the following criteria are met, such variations need not be considered in determining pressure-temperature design conditions. Otherwise, the most severe conditions of coincident pressure and temperature during the variation shall be used to determine design conditions. (Application of pressures exceeding pressure-temperature ratings of valves may, under certain conditions, cause loss of seat tightness or difficulty of operation. Such an application is the owner's responsibility.)

All the following criteria must be met:

1. The piping system shall have no pressure-containing components of cast iron or other nonductile metal.
2. Nominal pressure stresses shall not exceed the yield strength at temperature (see S_y data in ASME Boiler Pressure Vessel Code, Sec. II, Part D).
3. Combined longitudinal stresses S_L shall not exceed the limits established in the code (see pressure design of piping components for S_L limitations).
4. The total number of pressure-temperature variations above the design condition shall not exceed 1000 during the life of the piping system.
5. Occasional variations above design conditions shall remain within one of the following limits for pressure design:
   - When the variation lasts no more than 10 h at any one time and no more than 100 h/yr, it is permissible to exceed the pressure rating or the allowable stress for pressure design at the temperature of the increased condition by not more than 33 percent.

TABLE 10-41 Typical Hanger-Spacing Ranges Recommended for Reinforced-Thermosetting-Resin Pipe

Nominal pipe size, in	2	3	4	6	8	10	12
Hanger-spacing range, ft	5–8	6–9	6–10	8–11	9–13	10–14	11–15

NOTE: Consult pipe manufacturer for recommended hanger spacing for the specific RTR pipe being used. Tabulated values are based on a specific gravity of 1.25 for the contents of the pipe. To convert feet to meters, multiply by 0.3048.

- When the variation lasts no more than 50 h at any one time and not more than 500 h/yr, it is permissible to exceed the pressure rating or the allowable stress for pressure design at the temperature of the increased condition by not more than 20 percent.

Dynamic Effects Design must provide for impact (hydraulic shock, etc.), wind (exposed piping), earthquake, discharge reactions, and vibrations (of piping arrangement and support).

Weight Effects Weight considerations include (1) live loads (contents, ice, and snow), (2) dead loads (pipe, valves, insulation, etc.), and (3) test loads (test fluid).

Thermal Expansion and Contraction Effects Thermal expansion and thermal contraction loads occur when a piping system is prevented from free thermal expansion or contraction as a result of anchors and restraints or undergoes large, rapid temperature changes or unequal temperature distribution because of an injection of cold liquid striking the wall of a pipe carrying hot gas.

Effects of Support, Anchor, and Terminal Movements The effects of movements of piping supports, anchors, and connected equipment shall be taken into account in the design of piping. These movements may result from the flexibility and/or thermal expansion of equipment, supports, or anchors; and from settlement, tidal movements, or wind sway.

Reduced Ductility The harmful effects of reduced ductility shall be taken into account in the design of piping. The effects may, for example, result from welding, heat treatment, forming, bending, or low operating temperatures, including the chilling effect of sudden loss of pressure on highly volatile fluids. Low ambient temperatures expected during operation shall be considered.

Cyclic Effects Fatigue due to pressure cycling, thermal cycling, and other cyclic loading shall be considered in the design.

Air Condensation Effects At operating temperatures below −191°C (−312°F) in ambient air, condensation and oxygen enrichment occur. These shall be considered in selecting materials, including insulation, and adequate shielding and/or disposal shall be provided.

Design Criteria: Metallic Pipe The code uses three different approaches to design:

1. It provides for the use of dimensionally standardized components at their published pressure-temperature ratings.
2. It provides design formulas and maximum stresses.
3. It prohibits the use of materials, components, or assembly methods in certain conditions.

Components Having Specific Ratings These are listed in ASME, API, and industry standards. These ratings are acceptable for design pressures and temperatures unless limited in the code. A list of component standards is given in Appendix E of the ASME B31.3 code. Table 10-42 lists pressure-temperature ratings for flanges, flanged fittings, and flanged valves, and it has been extracted from ASME B16.5 with permission of the publisher, the American Society of Mechanical Engineers, New York. Only a few of the more common materials of construction of piping are reproduced here. See ASME B16.5 for other materials. Flanged joints, flanged valves in the open position, and flanged fittings may be subjected to system hydrostatic tests at a pressure not to exceed

TABLE 10-42a Pressure–Temperature Ratings for Group 1.1 Materials (Carbon Steel)*

Nominal designation	Forgings	Castings	Plates
C–Si	A 105 (1)	A 216 Gr. WCB (1)	A 515 Gr. 70 (1)
C–Mn–Si	A 350 Gr. LF2 (1)		A 516 Gr. 70 (1), (2)
C–Mn–Si–V	A 350 Gr. LF6 Cl. 1 (4)		A 537 Cl. 1 (3)
3½ NI	A 350 Gr. LF 3		

Class temp., °F	150	300	600	900	1500
	Working pressures by classes, psig				
−20 to 100	285	740	1480	2220	3705
200	260	680	1360	2035	3395
300	230	655	1310	1965	3270
400	200	635	1265	1900	3170
500	170	605	1205	1810	3015
600	140	570	1135	1705	2840
650	125	550	1100	1650	2745
700	110	530	1060	1590	2655
750	95	505	1015	1520	2535
800	80	410	825	1235	2055
850	65	320	640	955	1595
900	50	230	460	690	1150
950	35	135	275	410	685
1000	20	85	170	255	430

NOTES:
(1) Upon prolonged exposure to temperatures above 800°F, the carbide phase of steel may be converted to graphite. It is permissible, but not recommended for prolonged use above 800°F.
(2) Not to be used over 850°F.
(3) Not to be used over 700°F.
(4) Not to be used over 500°F.

TABLE 10-42b Pressure–Temperature Ratings for Group 1.5 Materials (Carbon, ½Mo Steel)

Nominal designation	Forgings	Castings	Plates
C–½Mo	A 182 Gr. F1 (1)		A 204 Gr. A (1) A 204 Gr. B (1)

Class temp., °F	150	300	600	900	1500
	Working pressures by classes, psig				
−20 to 100	265	695	1395	2090	3480
200	260	695	1395	2090	3480
300	230	685	1375	2060	3435
400	200	660	1325	1985	3310
500	170	640	1285	1925	3210
600	140	605	1210	1815	3025
650	125	590	1175	1765	2940
700	110	570	1135	1705	2840
750	95	530	1065	1595	2660
800	80	510	1015	1525	2540
850	65	485	975	1460	2435
900	50	450	900	1350	2245
950	35	280	560	845	1405
1000	20	165	330	495	825

NOTE:
(1) Upon prolonged exposure to temperatures above 875°F, the carbide phase of carbon-molybdenum steel may be converted to graphite. It is permissible, but not recommended for prolonged use above 875°F.

TABLE 10-42c Pressure–Temperature Ratings for Group 2.1 Materials (Type 304 Stainless Steel)

Nominal designation	Forgings	Castings	Plates
18Cr–8Ni	A 182 Gr. F304 (1) A 182 Gr. F304H	A 351 Gr. CF3 (2) A 351 Gr. CF8 (1)	A 240 Gr. 304 (1) A 240 Gr. 304H

Class temp., °F	150	300	600	900	1500
	Working pressures by classes, psig				
−20 to 100	275	720	1440	2160	3600
200	230	600	1200	1800	3000
300	205	540	1075	1615	2690
400	190	495	995	1490	2485
500	170	465	930	1395	2330
600	140	440	885	1325	2210
650	125	430	865	1295	2160
700	110	420	845	1265	2110
750	95	415	825	1240	2065
800	80	405	810	1215	2030
850	65	395	790	1190	1980
900	50	390	780	1165	1945
950	35	380	765	1145	1910
1000	20	355	710	1065	1770
1050	...	325	650	975	1630
1100	...	255	515	770	1285
1150	...	205	410	615	1030
1200	...	165	330	495	825
1250	...	135	265	400	670
1300	...	115	225	340	565
1350	...	95	185	280	465
1400	...	75	150	225	380
1450	...	60	115	175	290
1500	...	40	85	125	205

NOTES:
(1) At temperatures over 1000°F, use only when the carbon content is 0.04 percent or higher.
(2) Not to be used over 800°F.

TABLE 10-42*d* Pressure–Temperature Ratings for Group 2.2 Materials (Type 316 Stainless Steel)

Nominal designation	Forgings	Castings	Plates
16Cr–12Ni–2Mo	A 182 Gr. F316 (1) A 182 Gr. F316H	A 351 Gr. CF3M (2) A 351 Gr. CF8M (1)	A 240 Gr. 316 (1) A 240 Gr. 316H
18Cr–13Ni–3Mo	A 182 Gr. F317 (1)		A 240 Gr. 317 (1)
19Cr–10Ni–3Mo		A 351 Gr. CG8M (3)	

	Working pressures by classes, psig				
Class temp., °F	150	300	600	900	1500
−20 to 100	275	720	1440	2160	3600
200	235	620	1240	1860	3095
300	215	560	1120	1680	2795
400	195	515	1025	1540	2570
500	170	480	955	1435	2390
600	140	450	900	1355	2255
650	125	440	885	1325	2210
700	110	435	870	1305	2170
750	95	425	855	1280	2135
800	80	420	845	1265	2110
850	65	420	835	1255	2090
900	50	415	830	1245	2075
950	35	385	775	1160	1930
1000	20	365	725	1090	1820
1050	...	360	720	1080	1800
1100	...	305	610	915	1525
1150	...	235	475	710	1185
1200	...	185	370	555	925
1250	...	145	295	440	735
1300	...	115	235	350	585
1350	...	95	190	290	480
1400	...	75	150	225	380
1450	...	60	115	175	290
1500	...	40	85	125	205

NOTES:
(1) At temperatures over 1000°F, use only when the carbon content is 0.04 percent or higher.
(2) Not to be used over 850°F.
(3) Not to be used over 1000°F.

the hydrostatic-shell test pressure. Flanged valves in the closed position may be subjected to a system hydrostatic test at a pressure not to exceed 110 percent of the 100°F rating of the valve unless otherwise limited by the manufacturer.

Pressure-temperature ratings for soldered tubing joints are given in ASME B16.22-2013.

Components without Specific Ratings Components such as pipe and butt-welding fittings are generally furnished in nominal thicknesses. Fittings are rated for the same allowable pressures as pipe of the same nominal thickness and, along with pipe, are rated by the rules for pressure design and other provisions of the code.

Limits of Calculated Stresses Due to Sustained Loads and Displacement Strains

1. *Internal pressure stresses.* Stresses due to internal pressure shall be considered safe when the wall thickness of the piping component, including any reinforcement, meets the requirements of the pressure design of components defined by the ASME B31.3 code.

2. *External pressure stresses.* Stresses due to external pressure shall be considered safe when both the wall thickness of the piping component and its means of stiffening meet the requirements of the pressure design of components defined by the ASME B31.3 code.

3. *Longitudinal stresses S_L.* The sum of longitudinal stresses S_L in any component in a piping system, due to sustained loads such as pressure and weight, shall not exceed the product S_hW, where S_h is the basic allowable stress at maximum metal temperature expected during the displacement cycle under analysis and W is the weld joint strength reduction factor.

4. *Allowable displacement stress range S_A.* The computed displacement stress range S_E in a piping system shall not exceed the allowable displacement stress range S_A.

$$S_A = f(1.25S_c + 0.25S_h) \tag{10-91}$$

When S_h is greater than S_L, the difference between them may be added to the term $0.25S_h$ in Eq. (10-91). In that case, the allowable stress range is calculated by

$$S_A = f[1.25(S_c + S_h) - S_L] \tag{10-92}$$

where S_A = allowable displacement stress range
f = stress range factor (see Fig. 10-157)
S_c = basic allowable stress at minimum metal temperature expected during displacement cycle under analysis
S_h = basic allowable stress at maximum metal temperature expected during displacement cycle under analysis
S_L = longitudinal stresses, including pressure and weight

5. *Weld joint strength reduction factor W.* It is very important to include the weld joint strength reduction factor W in the design consideration. Especially, at elevated temperatures, the long-term strength of weld joints may be lower than the long-term strength of the base material. The weld joint strength reduction factor only applies at specific weld locations.

Pressure Design of Metallic Components External-pressure stress evaluation of piping is the same as for pressure vessels. But an important difference exists when one is establishing design pressure and wall thickness for internal pressure as a result of the ASME Boiler and Pressure Vessel Code's requirement that the relief-valve setting be not higher than the design pressure. For vessels this means that the design is for a pressure 10 percent more or less above the intended maximum operating pressure to avoid popping or leakage from the valve during normal operation. However, on piping the design pressure and temperature are taken as the maximum intended operating pressure and coincident temperature combination which results in the maximum thickness. The temporary increased operating conditions listed under "Design Criteria" cover temporary operation at pressures that cause relief valves to leak or open fully. Allowable stresses for nearly 1000 materials are contained in the code.

For *straight metal pipe under internal pressure* the formula for minimum required wall thickness t_m is applicable for:

1. $t < D/6$. The internal pressure design thickness for straight pipe shall be not less than that calculated in accordance with the equation below. The more conservative Barlow and Lamé equations may also be used. Equation (10-93) includes a factor Y varying with material and temperature to account for the redistribution of circumferential stress which occurs under steady-state creep at high temperature and permits slightly lesser thickness at this range.

$$t_m = \frac{PD_o}{2(SE + PY)} + C \tag{10-93}$$

where (in consistent units)
P = design pressure
D_o = outside diameter of pipe
C = sum of allowances for corrosion, erosion, and any thread or groove depth. For threaded components the depth is h of ASME B1.20.1, and for grooved components the depth is the depth removed (0.02 in when no tolerance is specified).
SE = allowable stress
S = basic allowable stress for materials, excluding casting, joint, or structural-grade quality factors

TABLE 10-42*e* Pressure–Temperature Ratings for Group 2.3 Materials (Type 304L and 316L Stainless Steel)

Nominal designation	Forgings	Castings	Plates
16Cr–12Ni–2Mo	A 182 Gr. F316L		A 240 Gr. 316L
18Cr–8Ni	A 182 Gr. F304L (1)		A 240 Gr. 304L (1)

	Working pressures by classes, psig				
Class temp., °F	150	300	600	900	1500
−20 to 100	230	600	1200	1800	3000
200	195	510	1020	1535	2555
300	175	455	910	1370	2280
400	160	420	840	1260	2100
500	150	395	785	1180	1970
600	140	370	745	1115	1860
650	125	365	730	1095	1825
700	110	360	720	1080	1800
750	95	355	705	1060	1765
800	80	345	690	1035	1730
850	65	340	675	1015	1690

NOTE:
(1) Not to be used over 800°F.

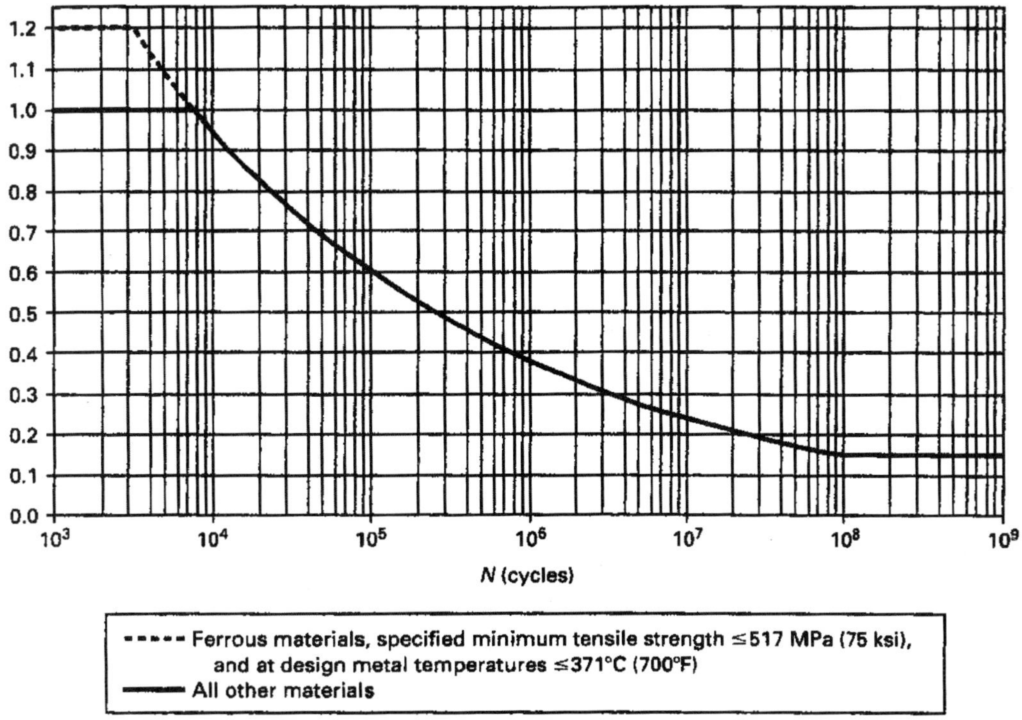

FIG. 10-157 Stress range reduction factor *f*. (*Reproduced from ASME B31.3-2004 with permission of the publisher, the American Society of Mechanical Engineers, New York. All rights reserved.*)

E = quality factor, which is one or the product of more than one of the following quality factors: casting quality factor E_c, joint quality factor E_j (see Fig. 10-158)

W = weld joint strength reduction factor in accordance with ASME B31.3.

Y = coefficient having value in Table 10-43 valid for $t < D/6$ and for materials shown

t_m = minimum required thickness, in, including mechanical, corrosion, and erosion allowances

2. $t \geq D/6$ or $P/SE > 0.385$. Calculation of pressure design thickness for straight pipe requires special consideration of factors such as theory of failure, effects of fatigue, and thermal stress.

For flanges of nonstandard dimensions or for sizes beyond the scope of the approved standards, design shall be in accordance with the requirements of the ASME Boiler and Pressure Vessel Code, Sec. VIII, except that requirements for fabrication, assembly, inspection testing, and the pressure and temperature limits for materials of the Piping Code are to prevail. Countermoment flanges of flat face or otherwise providing a reaction outside the bolt circle are permitted if designed or tested in accordance with code requirements under pressure-containing components "not covered by standards and for which design formulas or procedures are not given."

Test Conditions The shell pressure test for flanged fittings shall be at a pressure no less than 1.5 times the 38°C (100°F) pressure rating rounded off to the next higher 1-bar (25 psi) increment.

In accordance with listed standards, *blind flanges* may be used at their pressure-temperature ratings. The minimum thickness of nonstandard blind flanges shall be the same as for a bolted flat cover, in accordance with the rules of the ASME Boiler and Pressure Vessel Code, Sec. VIII.

Operational blanks shall be of the same thickness as blind flanges or may be calculated by the following formula (use consistent units):

$$t = d\sqrt{3P/16SE} + C \tag{10-94}$$

where d = inside diameter of gasket for raised- or flat (plain)-face flanges, or gasket pitch diameter for retained gasketed flanges

P = internal design pressure or external design pressure

S = applicable allowable stress

E = quality factor

C = sum of the mechanical allowances

Valves must comply with the applicable standards listed in Table 326.1 and App. E of the code and with the allowable pressure-temperature limits established thereby but not beyond the code-established service or materials limitations. Special valves must meet the same requirements as for countermoment flanges.

The code contains no specific rules for the design of *fittings* other than as branch openings. Ratings established by recognized standards are acceptable, however. ASME Standard B16.5 for steel-flanged fittings incorporates a 1.5 shape factor and thus requires the entire fitting to be 50 percent heavier than a simple cylinder in order to provide reinforcement for openings and/or general shape. ASME B16.9 for butt-welded fittings, on the other hand, requires only that the fittings be able to withstand the calculated bursting strength of the straight pipe with which they are to be used.

The thickness of *pipe bends* shall be determined as for straight pipe, provided the bending operation does not result in a difference between maximum and minimum diameters greater than 8 and 3 percent of the nominal outside diameter of the pipe for internal and external pressure, respectively.

The maximum allowable internal pressure for multiple miter bends shall be the lesser value calculated from Eqs. (10-95) and (10-96). These equations are not applicable when θ exceeds 22.5°.

$$P_m = \frac{SEW(T-C)}{r^2} = \frac{T-C}{(T-C)+0.643\tan\theta\sqrt{r_2(T-C)}} \tag{10-95}$$

$$P_m = \frac{SEW(T-C)}{r^2} = \frac{R_1 - r_2}{R_1 - 0.5r_2} \tag{10-96}$$

where the nomenclature is the same as for straight pipe except as follows (see Fig. 10-159):

S = stress value for material

C = sum of mechanical allowances

r_2 = mean radius of pipe

R_1 = effective radius of miter bend, defined as the shortest distance from the pipe centerline to the intersection of the planes of adjacent miter joints

θ = angle of miter cut

α = angle of change in direction at miter joint = 2θ

T = pipe wall thickness

W = weld joint strength reduction factor

E = quality factor

No.	Type of joint		Type of seam	Examination	Factor, E_j
1	Furnace butt weld, continuous weld		Straight	As required by listed specification	0.60 [Note (1)]
2	Electric resistance weld	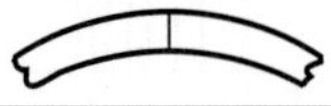	Straight or spiral	As required by listed specification	0.85 [Note (1)]
3	Electric fusion weld				
	(a) Single butt weld		Straight or spiral	As required by listed specification or this Code	0.80
	(with or without filler metal)	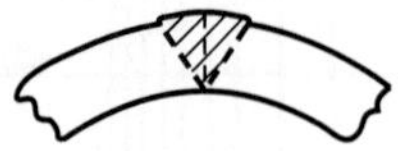		Additionally spot radiographed per para. 341.5.1 ASME B31.3	0.90
				Additionally 100% radiographed per para. 344.5.1 and Table 341.3.2 ASME B31.3	1.00
	(b) Double butt weld		Straight or spiral [except as provided in 4 below]	As required by listed specification or this Code	0.85
	(with or without filler metal)	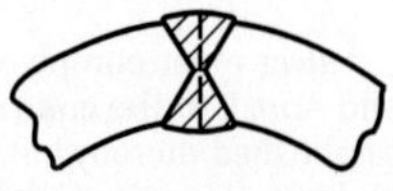		Additionally Spot radiographed per para. 341.5.1 ASME B31.3	0.90
				Additionally 100% radiographed per para. 344.5.1 and Table 341.3.2 ASME B31.3	1.00
4	Per specific specification				
	API 5L	Submerged arc weld (SAW) Gas metal arc weld (GMAW) Combined GMAW, SAM 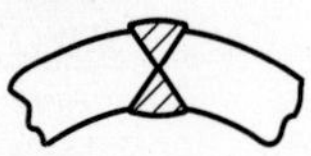	Straight (with one or two seams) or spiral	As required by specification	0.95
				Additionally 100% radiographed in accordance with para. 344.5.1 and Table 341.3.2 ASME B31.3	1.00

FIG. 10-158 Longitudinal weld joint quality factor E_j. [NOTE (1): It is not permitted to increase the joint quality factor by additional examination for joint 1 or 2.] *(Reproduced from ASME B31.3 with permission of the publisher, the American Society of Mechanical Engineers, New York. All rights reserved.)*

TABLE 10-43 Values of Coefficient *Y* for *t* < *D*/6

Material	Temperature, °C (°F) 482 (900) and below	510 (950)	538 (1000)	566 (1050)	593 (1100)	621 (1150)	649 (1200)	677 (1250) and above
Ferritic steels	0.4	0.5	0.7	0.7	0.7	0.7	0.7	0.7
Austenitic steels	0.4	0.4	0.4	0.4	0.5	0.7	0.7	0.7
Nickel alloys UNS Nos. N06617, N08800, N08810, and N08825	0.4	0.4	0.4	0.4	0.4	0.4	0.5	0.7
Gray iron	0.0	…	…	…	…	…	…	…
Other ductile metals	0.4	0.4	0.4	0.4	0.4	0.4	0.4	0.4

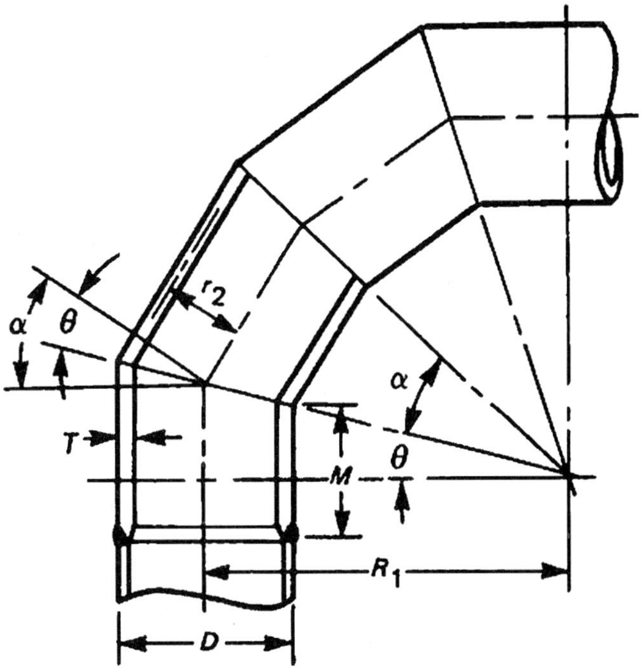

FIG. 10-159 Nomenclature for miter bends. (*Extracted from the Process Piping Code, ASME B31.3-2014, with permission of the publisher, the American Society of Mechanical Engineers, New York. All rights reserved.*)

For compliance with the code, the value of R_1 shall not be less than that given by Eq. (10-97):

$$R_1 = A/\tan\theta + D/2 \qquad (10\text{-}97)$$

where A has the following empirical values:

t, in	A
$t \le 0.5$	1.0
$0.5 < t < 0.88$	$2(T - C)$
$t \ge 0.88$	$[2(T - C)/3] + 1.17$

Piping branch connections involve the same considerations as pressure-vessel nozzles. However, outlet size in proportion to piping header size is unavoidably much greater for piping. The current Piping Code rules for calculation of branch-connection reinforcement are similar to those of the ASME Boiler and Pressure Vessel Code, Sec. VIII, Division I-2014 for a branch with axis at right angles to the header axis. If the branch connection makes an angle β with the header axis from 45 to 90°, the Piping Code requires that the area to be replaced be increased by dividing it by sin β. In such cases the half width of the reinforcing zone measured along the header axis is similarly increased, except that it may not exceed the outside diameter of the header. Some details of commonly used reinforced branch connections are given in Fig. 10-160.

The rules provide that a branch connection has adequate strength for pressure if a fitting (tee, lateral, or cross) is in accordance with an approved standard and is used within the pressure-temperature limitations or if the connection is made by welding a coupling or half coupling (wall thickness not less than the branch anywhere in reinforcement zone or less than extra heavy or 3000 lb) to the run and provided the ratio of branch to run diameters is not greater than one-fourth and that the branch is not greater than 2 in nominal diameter.

Dimensions of extra-heavy couplings are given in the *Steel Products Manual* published by the American Iron and Steel Institute. In ASME B16.11-2014, 2000-lb couplings were superseded by 3000-lb couplings.

ASME B31.3 states that the reinforcement area for resistance to external pressure is to be at least one-half of that required to resist internal pressure.

The code provides no guidance for analysis but requires that external and internal *attachments* be designed to avoid flattening of the pipe, excessive localized bending stresses, or harmful thermal gradients, with further emphasis on minimizing stress concentrations in cyclic service.

The code provides design requirements for *closures* which are flat, ellipsoidal, spherically dished, hemispherical, conical (without transition knuckles), conical convex to pressure, toriconical concave to pressure, and toriconical convex to pressure.

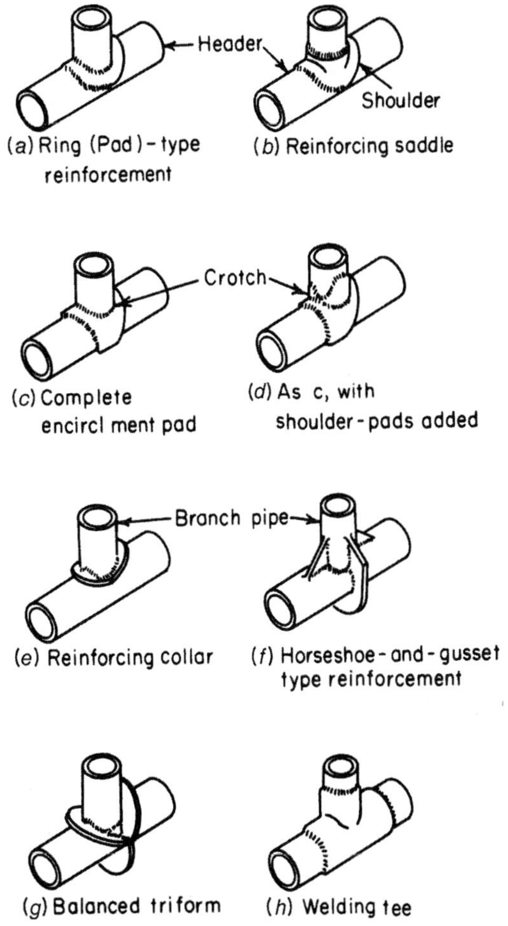

FIG. 10-160 Types of reinforcement for branch connections. (From Kellogg, *Design of Piping Systems*, Wiley, New York, 1965.)

Openings in closures over 50 percent in diameter are designed as flanges in flat closures and as reducers in other closures. Openings of not over one-half of the diameter are to be reinforced as branch connections.

Thermal Expansion and Flexibility: Metallic Piping ASME B31.3 requires that piping systems have sufficient flexibility to prevent thermal expansion or contraction or the movement of piping supports or terminals from causing (1) failure of piping supports from overstress or fatigue; (2) leakage at joints; or (3) detrimental stresses or distortions in piping or in connected equipment (pumps, turbines, or valves, for example), resulting from excessive thrusts or movements in the piping.

To ensure that a system meets these requirements, the computed displacement–stress range S_E shall not exceed the allowable stress range S_A [Eqs. (10-91) and (10-92)], the reaction forces R_m [Eq. (10-104)] shall not be detrimental to supports or connected equipment, and movement of the piping shall be within any prescribed limits.

Displacement Strains Strains result from piping being displaced from its unrestrained position:

1. *Thermal displacements.* A piping system will undergo dimensional changes with any change in temperature. If it is constrained from free movement by terminals, guides, and anchors, then it will be displaced from its unrestrained position.

2. *Reaction displacements.* If the restraints are not considered rigid and there is a predictable movement of the restraint under load, this may be treated as a compensating displacement.

3. *Externally imposed displacements.* Externally caused movement of restraints will impose displacements on the piping in addition to those related to thermal effects. Such movements may result from causes such as wind sway or temperature changes in connected equipment.

Total Displacement Strains Thermal displacements, reaction displacements, and externally imposed displacements all have equivalent effects on the piping system and must be considered together in determining the total displacement strains in a piping system.

Expansion strains may be taken up in three ways: by bending, by torsion, or by axial compression. In the first two cases, maximum stress occurs at the extreme fibers of the cross section at the critical location. In the third case, the entire cross-sectional area over the entire length for practical purposes is equally stressed.

Bending or torsional flexibility may be provided by bends, loops, or offsets; by corrugated pipe or expansion joints of the bellows type; or by other devices permitting rotational movement. These devices must be anchored or otherwise suitably connected to resist end forces from fluid pressure, frictional resistance to pipe movement, and other causes.

Axial flexibility may be provided by expansion joints of the slip-joint or bellows types, suitably anchored and guided to resist end forces from fluid pressure, frictional resistance to movement, and other causes.

Displacement Stresses Stresses may be considered proportional to the total displacement strain only if the strains are well distributed and not excessive at any point. The methods outlined here and in the code are applicable only to such a system. Poor distribution of strains (unbalanced systems) may result from the following:

1. Highly stressed small-size pipe runs in series with large and relatively stiff pipe runs
2. Local reduction in size or wall thickness or local use of a material having reduced yield strength (for example, girth welds of substantially lower strength than the base metal)
3. A line configuration in a system of uniform size in which expansion or contraction must be absorbed largely in a short offset from the major portion of the run

If unbalanced layouts cannot be avoided, appropriate analytical methods must be applied to ensure adequate flexibility. If the designer determines that a piping system does not have adequate inherent flexibility, additional flexibility may be provided by adding bends, loops, offsets, swivel joints, corrugated pipe, expansion joints of the bellows or slip-joint type, or other devices. Suitable anchoring must be provided.

As contrasted with stress from sustained loads such as internal pressure or weight, displacement stresses may be permitted to cause limited overstrain in various portions of a piping system. When the system is operated initially at its greatest displacement condition, any yielding reduces stress. When the system is returned to its original condition, there occurs a redistribution of stresses which is referred to as *self-springing*. It is similar to cold springing in its effects.

Stresses resulting from thermal strain tend to diminish with time. However, the algebraic difference in displacement condition and in either the original (as-installed) condition or any anticipated condition with a greater opposite effect than the extreme displacement condition remains substantially constant during any one cycle of operation. This difference is defined as the *displacement-stress range*, and it is a determining factor in the design of piping for flexibility. See Eqs. (10-91) and (10-92) for the allowable stress range S_A and Eq. (10-99) for the computed stress range S_E.

Cold Spring The intentional deformation of piping during assembly to produce a desired initial displacement and stress is called *cold spring*. For pipe operating at a temperature higher than that at which it was installed, cold spring is accomplished by fabricating it slightly shorter than design length. Cold spring is beneficial in that it serves to balance the magnitude of stress under initial and extreme displacement conditions. When cold spring is properly applied, there is less likelihood of overstrain during initial operation; hence, it is recommended especially for piping materials of limited ductility. There is also less deviation from as-installed dimensions during initial operation, so that hangers will not be displaced as far from their original settings.

Inasmuch as the service life of a system is affected more by the range of stress variation than by the magnitude of stress at a given time, no credit for cold spring is permitted in stress-range calculations. However, in calculating the thrusts and moments when actual reactions as well as their range of variations are significant, credit is given for cold spring.

Values of thermal expansion coefficients to be used in determining total displacement strains for computing the stress range are determined from Table 10-44 as the algebraic difference between the value at the design maximum temperature and that at the design minimum temperature for the thermal cycle under analysis.

Values for Reactions Values of thermal displacements to be used in determining total displacement strains for the computation of reactions on supports and connected equipment shall be determined as the algebraic difference between the value at design maximum (or minimum) temperature for the thermal cycle under analysis and the value at the temperature expected during installation.

The as-installed and maximum or minimum moduli of elasticity E_a and E_m, respectively, shall be taken as the values shown in Table 10-45.

Poisson's ratio may be taken as 0.3 at all temperatures for all metals.

The allowable stress range for displacement stresses S_A and permissible additive stresses shall be as specified in Eqs. (10-91) and (10-92) for systems primarily stressed in bending and/or torsion. For pipe or piping components containing longitudinal welds, the basic allowable stress S may be used to determine S_A.

Nominal thicknesses and outside diameters of pipe and fittings shall be used in flexibility calculations.

In the absence of more directly applicable data, the flexibility factor k and stress intensification factor i shown in Table 10-46 may be used in flexibility calculations in Eq. (10-100). For piping components or attachments (such as valves, strainers, anchor rings, and bands) not covered in the table, suitable stress intensification factors may be assumed by comparison of their significant geometry with that of the components shown.

Requirements for Analysis No formal analysis of adequate flexibility is required in systems which (1) are duplicates of successfully operating installations or replacements without significant change of systems with a satisfactory service record; (2) can readily be judged adequate by comparison with previously analyzed systems; or (3) are of uniform size, have no more than two points of fixation, have no intermediate restraints, and fall within the limitations of empirical Eq. (10-98)*

$$\frac{Dy}{(L-U)^2} \le K_1 \qquad (10\text{-}98)$$

where D = outside diameter of pipe, in (mm)
y = resultant of total displacement strains, in (mm), to be absorbed by the piping system
L = developed length of piping between anchors, ft (m)
U = anchor distance, straight line between anchors, ft (m)
K_1 = 208,000 S_A/E_{ar}, (mm/m)^2
= 30 S_A/E_{ar} (in/ft)^2
= 208.0 for SI units listed in parentheses

1. All systems not meeting these criteria shall be analyzed by simplified, approximate, or comprehensive methods of analysis appropriate for the specific case.
2. Approximate or simplified methods may be applied only if they are used in the range of configurations for which their adequacy has been demonstrated.
3. Acceptable comprehensive methods of analysis include analytical and chart methods that provide an evaluation of the forces, moments, and stresses caused by displacement strains.
4. Comprehensive analysis shall take into account stress-intensification factors for any component other than straight pipe. Credit may be taken for the extra flexibility of such a component.

In calculating the flexibility of a piping system between anchor points, the system shall be treated as a whole. The significance of all parts of the line and of all restraints introduced for the purpose of reducing moments and forces on equipment or small branch lines and also the restraint introduced by support friction shall be recognized. Consider all displacements over the temperature range defined by the operating and shutdown conditions.

Flexibility Stresses Bending and torsional stresses shall be computed using the as-installed modulus of elasticity E_a and then combined in accordance with Eq. (10-99) to determine the computed displacement stress range S_E, which shall not exceed the allowable stress range S_A [Eqs. (10-91) and (10-92)]:

$$S_E = \sqrt{S_b^2 + 4S_t^2} \qquad (10\text{-}99)$$

where S_a = axial stress range due to displacement strains
= i_aF_a/A_p
A_p = cross-sectional area of pipe; nominal thickness and outside diameters of pipe
F_a = axial force range between any two conditions being evaluated
i_a = axial stress intensification factor
i_t = torsional stress intensification factor
S_b = resultant bending stress, lbf/in^2 (MPa)
S_t = $M_t/2Z$ = torsional stress, lbf/in^2 (MPa)
M_t = torsional moment, in · lbf (N · mm)
Z = section modulus of pipe, in^3 (mm^3)

**Warning:* No general proof can be offered that this equation will yield accurate or consistently conservative results. It is not applicable to systems used under severe cyclic conditions. It should be used with caution in configurations such as unequal leg U bends ($L/U > 2.5$) or near-straight sawtooth runs, or for large thin-wall pipe ($i \ge 5$), or when extraneous displacements (not in the direction connecting anchor points) constitute a large part of the total displacement. There is no assurance that terminal reactions will be acceptably low even if a piping system falls within the limitations of Eq. (10-98).

TABLE 10-44 Thermal Coefficients, USCS Units, for Metals

Mean coefficient of linear thermal expansion between 70°F and indicated temperature, μin/(in · °F)

	Material															
Temp., °F	Carbon steel carbon–moly–low–chrome (through 3Cr–Mo)	5Cr–Mo through 9Cr–Mo	Austenitic stainless steels 18Cr–8Ni	12Cr 17Cr 27Cr	25Cr–20Ni	UNS N04400 Monel 67Ni–30Cu	3½Ni	Copper and copper alloys	Aluminum	Gray cast iron	Bronze	Brass	70Cu–30Ni	UNS N08XXX series Ni–Fe–Cr	UNS N06XXX series Ni–Cr–Fe	Ductile iron
−450	. . .	. . .	. . .	. . .	. . .	. . .	. . .	6.30	. . .	. . .	. . .	. . .	. . .	. . .	. . .	. . .
−425	. . .	. . .	. . .	. . .	. . .	. . .	. . .	6.61	. . .	. . .	. . .	. . .	. . .	. . .	. . .	. . .
−400	. . .	. . .	. . .	. . .	. . .	. . .	. . .	6.93	. . .	. . .	. . .	. . .	. . .	. . .	. . .	. . .
−375	. . .	. . .	. . .	. . .	. . .	. . .	. . .	7.24	. . .	. . .	. . .	. . .	. . .	. . .	. . .	. . .
−350	. . .	. . .	. . .	. . .	. . .	. . .	. . .	7.51	. . .	. . .	. . .	. . .	. . .	. . .	. . .	. . .
−325	5.00	4.70	8.15	4.30	. . .	5.55	4.76	7.74	9.90	. . .	8.40	8.20	6.65	. . .	. . .	. . .
−300	5.07	4.77	8.21	4.36	. . .	5.72	4.90	7.94	10.04	. . .	8.45	8.24	6.76	. . .	. . .	. . .
−275	5.14	4.84	8.28	4.41	. . .	5.89	5.01	8.11	10.18	. . .	8.50	8.29	6.86	. . .	. . .	. . .
−250	5.21	4.91	8.34	4.47	. . .	6.06	5.15	8.26	10.33	. . .	8.55	8.33	6.97	. . .	. . .	. . .
−225	5.28	4.98	8.41	4.53	. . .	6.23	5.30	8.40	10.47	. . .	8.60	8.37	7.08	. . .	. . .	. . .
−200	5.35	5.05	8.47	4.59	. . .	6.40	5.45	8.51	10.61	. . .	8.65	8.41	7.19	. . .	. . .	4.65
−175	5.42	5.12	8.54	4.64	. . .	6.57	5.52	8.62	10.76	. . .	8.70	8.46	7.29	. . .	. . .	4.76
−150	5.50	5.20	8.60	4.70	. . .	6.75	5.59	8.72	10.90	. . .	8.75	8.50	7.40	. . .	. . .	4.87
−125	5.57	5.26	8.66	4.78	. . .	6.85	5.67	8.81	11.08	. . .	8.85	8.61	7.50	. . .	. . .	4.98
−100	5.65	5.32	8.75	4.85	. . .	6.95	5.78	8.89	11.25	. . .	8.95	8.73	7.60	. . .	. . .	5.10
−75	5.72	5.38	8.83	4.93	. . .	7.05	5.83	8.97	11.43	. . .	9.05	8.84	7.70	. . .	. . .	5.20
−50	5.80	5.45	8.90	5.00	. . .	7.15	5.88	9.04	11.60	. . .	9.15	8.95	7.80	. . .	. . .	5.30
−25	5.85	5.51	8.94	5.05	. . .	7.22	5.94	9.11	11.73	. . .	9.23	9.03	7.87	. . .	. . .	5.40
0	5.90	5.56	8.98	5.10	. . .	7.28	6.00	9.17	11.86	. . .	9.32	9.11	7.94	. . .	. . .	5.50
25	5.96	5.62	9.03	5.14	. . .	7.35	6.08	9.23	11.99	. . .	9.40	9.18	8.02	. . .	. . .	5.58
50	6.01	5.67	9.07	5.19	. . .	7.41	6.16	9.28	12.12	. . .	9.49	9.26	8.09	. . .	. . .	5.66
70	6.07	5.73	9.11	5.24	. . .	7.48	6.25	9.32	12.25	. . .	9.57	9.34	8.16	. . .	7.13	5.74
100	6.13	5.79	9.16	5.29	. . .	7.55	6.33	9.39	12.39	. . .	9.66	9.42	8.24	. . .	7.20	5.82
125	6.19	5.85	9.20	5.34	. . .	7.62	6.36	9.43	12.53	. . .	9.75	9.51	8.31	. . .	7.25	5.87
150	6.25	5.92	9.25	5.40	. . .	7.70	6.39	9.48	12.67	. . .	9.85	9.59	8.39	. . .	7.30	5.92
175	6.31	5.98	9.29	5.45	. . .	7.77	6.42	9.52	12.81	. . .	9.93	9.68	8.46	. . .	7.35	5.97
200	6.38	6.04	9.34	5.50	8.79	7.84	6.45	9.56	12.95	5.75	10.03	9.76	8.54	7.90	7.40	6.02
225	6.43	6.08	9.37	5.54	8.81	7.89	6.50	9.60	13.03	5.80	10.05	9.82	8.58	8.01	7.44	6.08
250	6.49	6.12	9.41	5.58	8.83	7.93	6.55	9.64	13.12	5.84	10.08	9.88	8.63	8.12	7.48	6.14
275	6.54	6.15	9.44	5.62	8.85	7.98	6.60	9.68	13.20	5.89	10.10	9.94	8.67	8.24	7.52	6.20
300	6.60	6.19	9.47	5.66	8.87	8.02	6.65	9.71	13.28	5.93	10.12	10.00	8.71	8.35	7.56	6.25
325	6.65	6.23	9.50	5.70	8.89	8.07	6.69	9.74	13.36	5.97	10.15	10.06	8.76	8.46	7.60	6.31
350	6.71	6.27	9.53	5.74	8.90	8.11	6.73	9.78	13.44	6.02	10.18	10.11	8.81	8.57	7.63	6.37
375	6.76	6.30	9.56	5.77	8.91	8.16	6.77	9.81	13.52	6.06	10.20	10.17	8.85	8.69	7.67	6.43
400	6.82	6.34	9.59	5.81	8.92	8.20	6.80	9.84	13.60	6.10	10.23	10.23	8.90	8.80	7.70	6.48
425	6.87	6.38	9.62	5.85	8.92	8.25	6.83	9.86	13.68	6.15	10.25	10.29	. . .	8.82	7.72	6.57
450	6.92	6.42	9.65	5.89	8.92	8.30	6.86	9.89	13.75	6.19	10.28	10.35	. . .	8.85	7.75	6.66
475	6.97	6.46	9.67	5.92	8.92	8.35	6.89	9.92	13.83	6.24	10.30	10.41	. . .	8.87	7.77	6.75
500	7.02	6.50	9.70	5.96	8.93	8.40	6.93	9.94	13.90	6.28	10.32	10.47	. . .	8.90	7.80	6.85
525	7.07	6.54	9.73	6.00	8.93	8.45	6.97	9.97	13.98	6.33	10.35	10.53	. . .	8.92	7.82	6.88
550	7.12	6.58	9.76	6.05	8.93	8.49	7.01	9.99	14.05	6.38	10.38	10.58	. . .	8.95	7.85	6.92
575	7.17	6.62	9.79	6.09	8.93	8.54	7.04	10.1	14.13	6.42	10.41	10.64	. . .	8.97	7.88	6.95
600	7.23	6.66	9.82	6.13	8.94	8.58	7.08	10.04	14.20	6.47	10.44	10.69	. . .	9.00	7.90	6.98
625	7.28	6.70	9.85	6.17	8.94	8.63	7.12	. . .	. . .	6.52	10.46	10.75	. . .	9.02	7.92	7.02
650	7.33	6.73	9.87	6.20	8.95	8.68	7.16	. . .	. . .	6.56	10.48	10.81	. . .	9.05	7.95	7.04
675	7.38	6.77	9.90	6.23	8.95	8.73	7.19	. . .	. . .	6.61	10.50	10.86	. . .	9.07	7.98	7.08
700	7.44	6.80	9.92	6.26	8.96	8.78	7.22	. . .	. . .	6.65	10.52	10.92	. . .	9.10	8.00	7.11
725	7.49	6.84	9.95	6.29	8.96	8.83	7.25	. . .	. . .	6.70	10.55	10.98	. . .	9.12	8.02	7.14
750	7.54	6.88	9.99	6.33	8.96	8.87	7.29	. . .	. . .	6.74	10.57	11.04	. . .	9.15	8.05	7.18
775	7.59	6.92	10.02	6.36	8.96	8.92	7.31	. . .	. . .	6.79	10.60	11.10	. . .	9.17	8.08	7.22
800	7.65	6.96	10.05	6.39	8.97	8.96	7.34	. . .	. . .	6.83	10.62	11.16	. . .	9.20	8.10	7.25
825	7.70	7.00	10.08	6.42	8.97	9.01	7.37	. . .	. . .	6.87	10.65	11.22	. . .	9.22	. . .	7.27

TABLE 10-44 Thermal Coefficients, USCS Units, for Metals *(Continued)*

Mean coefficient of linear thermal expansion between 70°F and indicated temperature, μin/(in·°F)

	Material														
Temp., °F	Carbon steel carbon–moly–low–chrome (through 3Cr–Mo)	5Cr–Mo through 9Cr–Mo	Austenitic stainless steels 18Cr–8Ni	12Cr 17Cr 27Cr	25Cr–20Ni	UNS N04400 Monel 67Ni–30Cu	3½Ni	Copper and copper alloys	Aluminum	Gray cast iron	Bronze	Brass	70Cu–30Ni	UNS N08XXX series Ni–Fe–Cr	UNS N06XXX series Ni–Cr–Fe	Ductile iron
850	7.75	7.03	10.11	6.46	8.98	9.06	7.40	...	...	6.92	10.67	11.28	...	9.25	...	7.31
875	7.79	7.07	10.13	6.49	8.99	9.11	7.43	...	...	6.96	10.70	11.34	...	9.27	...	7.34
900	7.84	7.10	10.16	6.52	9.00	9.16	7.45	...	...	7.00	10.72	11.40	...	9.30	...	7.37
925	7.87	7.13	10.19	6.55	9.05	9.21	7.47	...	...	7.05	10.74	11.46	...	9.32	...	7.41
950	7.91	7.16	10.23	6.58	9.10	9.25	7.49	...	...	7.10	10.76	11.52	...	9.35	...	7.44
975	7.94	7.19	10.26	6.60	9.15	9.30	7.52	...	...	7.14	10.78	11.57	...	9.37	...	7.47
1000	7.97	7.22	10.29	6.63	9.18	9.34	7.55	...	...	7.19	10.80	11.63	...	9.40	...	7.50
1025	8.01	7.25	10.32	6.65	9.20	9.39	...	...	...	...	10.83	11.69	...	9.42	...	...
1050	8.05	7.27	10.34	6.68	9.22	9.43	...	...	...	...	10.85	11.74	...	9.45	...	...
1075	8.08	7.30	10.37	6.70	9.24	9.48	...	...	...	...	10.88	11.80	...	9.47	...	...
1100	8.12	7.32	10.39	6.72	9.25	9.52	...	...	...	...	10.90	11.85	...	9.50	...	...
1125	8.14	7.34	10.41	6.74	9.29	9.57	...	...	...	...	10.93	11.91	...	9.52	...	...
1150	8.16	7.37	10.44	6.75	9.33	9.61	...	...	...	...	10.95	11.97	...	9.55	...	...
1175	8.17	7.39	10.46	6.77	9.36	9.66	...	...	...	...	10.98	12.03	...	9.57	...	...
1200	8.19	7.41	10.48	6.78	9.39	9.70	...	...	...	...	11.00	12.09	...	9.60	...	...
1225	8.21	7.43	10.50	6.80	9.43	9.75	...	...	...	...	...	...	...	9.64	...	...
1250	8.24	7.45	10.51	6.82	9.47	9.79	...	...	...	...	...	...	...	9.68	...	...
1275	8.26	7.47	10.53	6.83	9.50	9.84	...	...	...	...	...	...	...	9.71	...	...
1300	8.28	7.49	10.54	6.85	9.53	9.88	...	...	...	...	...	...	...	9.75	...	...
1325	8.30	7.51	10.56	6.86	9.53	9.92	...	...	...	...	...	...	...	9.79	...	...
1350	8.32	7.52	10.57	6.88	9.54	9.96	...	...	...	...	...	...	...	9.83	...	...
1375	8.34	7.54	10.59	6.89	9.55	10.00	...	...	...	...	...	...	...	9.86	...	...
1400	8.36	7.55	10.60	6.90	9.56	10.04	...	...	...	...	...	...	...	9.90	...	...
1425	...	...	10.64	...	...	...	...	...	...	...	...	...	...	9.94	...	...
1450	...	...	10.68	...	...	...	...	...	...	...	...	...	...	9.98	...	...
1475	...	...	10.72	...	...	...	...	...	...	...	...	...	...	10.01	...	...
1500	...	...	10.77	...	...	...	...	...	...	...	...	...	...	10.05	...	...

GENERAL NOTE: For Code references to this table, see para. 319.3.1, ASME B31.3. These data are for use in the absence of more applicable data. It is the designer's responsibility to verify that materials are suitable for the intended service at the temperatures shown.

TABLE 10-45 Modulus of Elasticity, USCS Units, for Metals

Material	E = modulus of elasticity, Msi (millions of psi), at temperature, °F																				
	−425	−400	−350	−325	−200	−100	70	200	300	400	500	600	700	800	900	1000	1100	1200	1300	1400	1500
Ferrous Metals																					
Gray cast iron	...	...	...	...	...	...	13.4	13.2	12.9	12.6	12.2	11.7	11.0	10.2	...	...	...	...	...	...	...
Carbon steels, $C \leq 0.3\%$	31.9	...	...	31.4	30.8	30.2	29.5	28.8	28.3	27.7	27.3	26.7	25.5	24.2	22.4	20.4	18.0	...	...	...	...
Carbon steels, $C \leq 0.3\%$	31.7	...	...	31.2	30.6	30.0	29.3	28.6	28.1	27.5	27.1	26.5	25.3	24.0	22.2	20.2	17.9	15.4	...	...	...
Carbon-moly steels	31.7	...	...	31.1	30.5	29.9	29.2	28.5	28.0	27.4	27.0	26.4	25.3	23.9	22.2	20.1	17.8	15.3	...	...	...
Nickel steels, Ni 2%–9%	30.1	...	...	29.6	29.1	28.5	27.8	27.1	26.7	26.1	25.7	25.2	24.6	23.0	...	...	...	...	...	...	...
Cr-Mo steels, Cr ½%–2%	32.1	...	...	31.6	31.0	30.4	29.7	29.0	28.5	27.9	27.5	26.9	26.3	25.5	24.8	23.9	23.0	21.8	20.5	18.9	...
Cr-Mo steels, Cr 2¼%–3%	33.1	...	...	32.6	32.0	31.4	30.6	29.8	29.4	28.8	28.3	27.7	27.1	26.3	25.6	24.6	23.7	22.5	21.1	19.4	...
Cr-Mo steels, Cr 5%–9%	33.4	...	...	32.9	32.3	31.7	30.9	30.1	29.7	29.0	28.6	28.0	27.3	26.1	24.7	22.7	20.4	18.2	15.5	12.7	...
Chromium steels, Cr 12%, 17%, 27%	31.8	...	...	31.2	30.7	30.1	29.2	28.5	27.9	27.3	26.7	26.1	25.6	24.7	22.2	21.5	19.1	16.6	...	...	...
Austenitic steels (TP304, 310, 316, 321, 347)	30.8	...	...	30.3	29.7	29.0	28.3	27.6	27.0	26.5	25.8	25.3	24.8	24.1	23.5	22.8	22.1	21.2	20.2	19.2	18.1
Copper and Copper Alloys (UNS Nos.)																					
Comp. and leaded Sn-bronze (C83600, C92200)	...	...	...	14.8	14.6	14.4	14.0	13.7	13.4	13.2	12.9	12.5	12.0	...	...	...	...	...	...	...	...
Naval brass, Si- and Al-bronze (C46400, C65500, C95200, C95400)	...	...	...	15.9	15.6	15.4	15.0	14.6	14.4	14.1	13.8	13.4	12.8	...	...	...	...	...	...	...	...
Copper (C11000)	...	...	...	16.9	16.6	16.5	16.0	15.6	15.4	15.0	14.7	14.2	13.7	...	...	...	...	...	...	...	...
Copper, red brass, Al-bronze (C10200, C12000, C12200, C12500, C14200, C23000, C61400)	...	...	...	18.0	17.7	17.5	17.0	16.6	16.3	16.0	15.6	15.1	14.5	...	...	...	...	...	...	...	...
90Cu 10Ni (C70600)	...	...	...	19.0	18.7	18.5	18.0	17.6	17.3	16.9	16.6	16.0	15.4	...	...	...	...	...	...	...	...
Leaded Ni-bronze	...	...	...	20.1	19.8	19.6	19.0	18.5	18.2	17.9	17.5	16.9	16.2	...	...	...	...	...	...	...	...
80Cu-20Ni (C71000)	...	...	...	21.2	20.8	20.6	20.0	19.5	19.2	18.8	18.4	17.8	17.1	...	...	...	...	...	...	...	...
70Cu-30Ni (C71500)	...	...	...	23.3	22.9	22.7	22.0	21.5	21.1	20.7	20.2	19.6	18.8	...	...	...	...	...	...	...	...
Nickel and Nickel Alloys (UNS Nos.)																					
Monel 400 N04400	28.3	...	...	27.8	27.3	26.8	26.0	25.4	25.0	24.7	24.3	24.1	23.7	23.1	22.6	22.1	21.7	21.2	...	...	...
Alloys N06007, N08320	30.3	...	...	29.5	29.2	28.6	27.8	27.1	26.7	26.4	26.0	25.7	25.3	24.7	24.2	23.6	23.2	22.7	...	...	...
Alloys N08800, N08810, N06002	31.1	...	...	30.5	29.9	29.4	28.5	27.8	27.4	27.1	26.6	26.4	25.9	25.4	24.8	24.2	23.8	23.2	...	...	...
Alloys N06455, N10276	32.5	...	...	31.6	31.3	30.6	29.8	29.1	28.6	28.3	27.9	27.6	27.1	26.5	25.9	25.3	24.9	24.3	...	...	...
Alloys N02200, N02201, N06625	32.7	...	...	32.1	31.5	30.9	30.0	29.3	28.8	28.5	28.1	27.8	27.3	26.7	26.1	25.5	25.1	24.5	...	...	...
Alloy N06600	33.8	...	...	33.2	32.6	31.9	31.0	30.2	29.9	29.5	29.0	28.7	28.2	27.6	27.0	26.4	25.9	25.3	...	...	...
Alloy N10001	33.9	...	...	33.3	32.7	32.0	31.1	30.3	29.9	29.5	29.1	28.8	28.3	27.7	27.1	26.4	26.0	25.3	...	...	...
Alloy N10665	34.2	...	...	33.3	33.0	32.3	31.4	30.6	30.1	29.8	29.4	29.0	28.6	27.9	27.3	26.7	26.2	25.6	...	...	...
Unalloyed Titanium																					
Grades 1, 2, 3, and 7	...	...	...	...	...	...	15.5	15.0	14.6	14.0	13.3	12.6	11.9	11.2	...	...	...	...	...	...	...
Aluminum and Aluminum Alloys (UNS Nos.)																					
Grades 443, 1060, 1100, 3003, 3004, 6061, 6063 (A24430, A91060, A91100, A93003, A93004, A96061, A96063)	11.4	...	...	11.1	10.8	10.5	10.0	9.6	9.2	8.7											
Grades 5052, 5154, 5454, 5652 (A95052, A95154, A95454, A95652)	11.6	...	...	11.3	11.0	10.7	10.2	9.7	9.4	8.9											
Grades 356, 5083, 5086, 5456 (A03560, A95083, A95086, A95456)	11.7	...	...	11.4	11.1	10.8	10.3	9.8	9.5	9.0											

TABLE 10-46 Flexibility Factor, *k*, and Stress Intensification Factor, *i*

Description	Flexibility factor, k	Stress Intensification Factor [Notes (1), (2)]		Flexibility characteristic, h	Sketch
		Out-of-plane, i_o	In-plane, i_i		
Welding elbow or pipe bend [Notes (1), (3)–(6)]	$\frac{1.65}{h}$	$\frac{0.75}{h^{2/3}}$	$\frac{0.9}{h^{2/3}}$	$\frac{\bar{T} R_1}{r_2^2}$	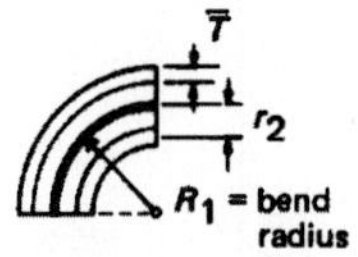
Closely spaced miter bend $s < r_2 (1 + \tan\theta)$ [Notes (1), (3), (4), (6)]	$\frac{1.52}{h^{5/6}}$	$\frac{0.9}{h^{2/3}}$	$\frac{0.9}{h^{2/3}}$	$\frac{\cot\theta}{2}\left(\frac{s\bar{T}}{r_2^2}\right)$	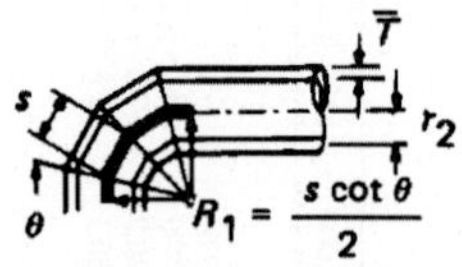
Single miter bend or widely spaced miter bend $s \geq r_2 (1 + \tan\theta)$ [Notes (1), (3), (6)]	$\frac{1.52}{h^{5/6}}$	$\frac{0.9}{h^{2/3}}$	$\frac{0.9}{h^{2/3}}$	$\frac{1+\cot\theta}{2}\left(\frac{\bar{T}}{r_2}\right)$	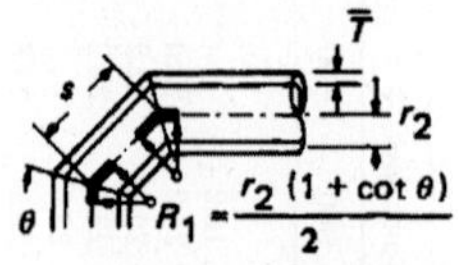
Welding tee in accordance with ASME B16.9 [Notes (1), (3), (5), (7), (8)]	1	$\frac{0.9}{h^{2/3}}$	$\frac{3}{4}i_o + \frac{1}{4}$	$3.1\frac{\bar{T}}{r_2}$	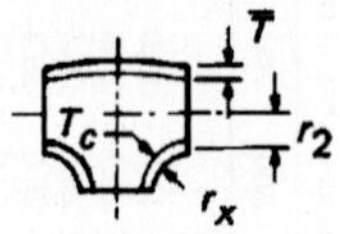
Reinforced fabricated tee with pad or saddle [Notes (1), (3), (8), (9), (10)]	1	$\frac{0.9}{h^{2/3}}$	$\frac{3}{4}i_o + \frac{1}{4}$	$\frac{\left(\bar{T} + \frac{1}{2}\bar{T}_r\right)^{2.5}}{\bar{T}^{1.5} r_2}$	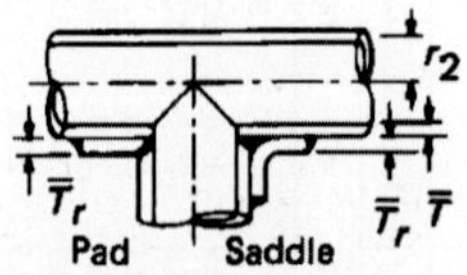
Unreinforced fabricated tee [Notes (1), (3), (8), (10)]	1	$\frac{0.9}{h^{2/3}}$	$\frac{3}{4}i_o + \frac{1}{4}$	$\frac{\bar{T}}{r_2}$	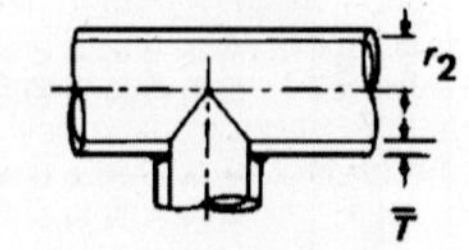
Extruded welding tee with $r_x \geq 0.05 D_b$ $T_c < 1.5\bar{T}$ [Notes (1), (3), (8)]	1	$\frac{0.9}{h^{2/3}}$	$\frac{3}{4}i_o + \frac{1}{4}$	$\left(1 + \frac{r_x}{r_2}\right)\frac{\bar{T}}{r_2}$	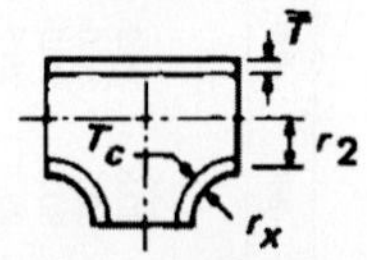
Welded-in contour insert [Notes (1), (3), (7), (8)]	1	$\frac{0.9}{h^{2/3}}$	$\frac{3}{4}i_o + \frac{1}{4}$	$3.1\frac{\bar{T}}{r_2}$	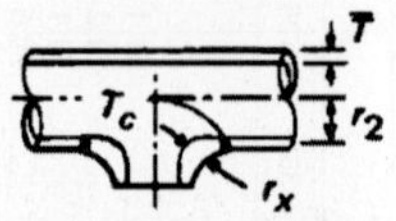
Branch welded-on fitting (integrally reinforced) [Notes (1), (3), (10), (11)]	1	$\frac{0.9}{h^{2/3}}$	$\frac{0.9}{h^{2/3}}$	$3.3\frac{\bar{T}}{r_2}$	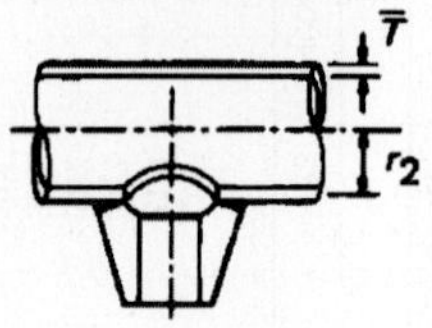

TABLE 10-46 Flexibility Factor, *k*, and Stress Intensification Factor, *i* *(Continued)*

Description	Flexibility factor, *k*	Stress intensification factor, *i*
Butt welded joint, reducer, or weld neck flange	1	1.0
Double-welded slip-on flange	1	1.2
Fillet or socket weld	1	1.3 [Note (12)]
Lap joint flange (with ASME B16.9 lap joint stub)	1	1.6
Threaded pipe joint or threaded flange	1	2.3
Corrugated straight pipe, or corrugated or creased bend [Note (13)]	5	2.5

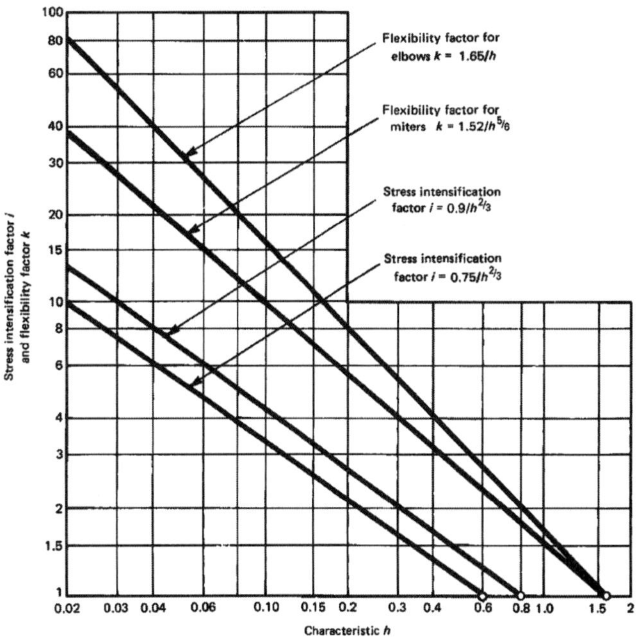

Chart A

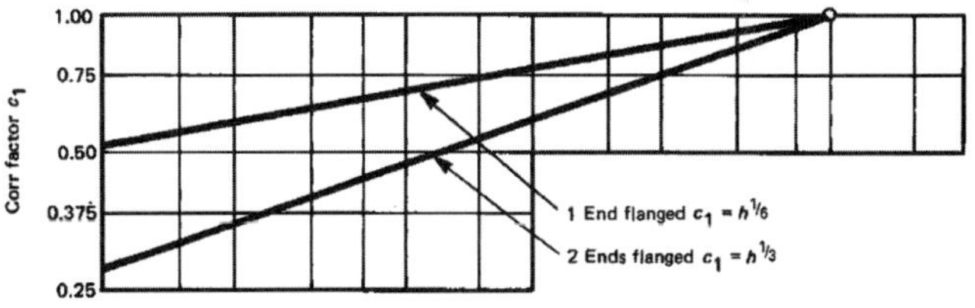

Chart B

GENERAL NOTE: Stress intensification and flexibility factor data in Table 10-46 are for use in the absence of more directly applicable data (see para. 319.3.6). Their validity has been demonstrated for $D/\bar{T} \le 100$.

NOTES:

(1) The flexibility factor, k, in the Table applies to bending in any plane; also see para. 319.3.6. The flexibility factors, k, and stress intensification factors, i, shall apply over the effective arc length (shown by heavy centerlines in the illustrations) for curved and miter bends, and to the intersection point for tees.

(2) A single intensification factor equal to $0.9/h^{2/3}$ may be used for both i_i and i_o if desired.

(3) The values of k and i can be read directly from Chart A by entering with the characteristic h computed from the formulas given above. Nomenclature is as follows:

- D_b = outside diameter of branch
- R_1 = bend radius of welding elbow or pipe bend
- r_x = see definition in para. 304.3.4(c)
- r_2 = mean radius of matching pipe
- s = miter spacing at centerline
- $\bar{T}$ = for elbows and miter bends, the nominal wall thickness of the fitting
  = for tees, the nominal wall thickness of the matching pipe
- T_c = crotch thickness of branch connections measured at the center of the crotch where shown in the illustrations
- $\bar{T}_r$ = pad or saddle thickness
- θ = one-half angle between adjacent miter axes

(4) Where flanges are attached to one or both ends, the values of k and i in the Table shall be corrected by the factors C_1, which can be read directly from Chart B, entering with the computed h.

(5) The designer is cautioned that cast buttwelded fittings may have considerably heavier walls than that of the pipe with which they are used. Large errors may be introduced unless the effect of these greater thicknesses is considered.

(6) In large diameter thin-wall elbows and bends, pressure can significantly affect the magnitudes of k and i. To correct values from the Table, divide k by

$$1+6\left(\frac{P_j}{E_j}\right)\left(\frac{r_2}{\bar{T}}\right)^{7/3}\left(\frac{R_1}{r_2}\right)^{1/3}$$

divide i by

$$1+3.25\left(\frac{P_j}{E_j}\right)\left(\frac{r_2}{\bar{T}}\right)^{5/2}\left(\frac{R_1}{r_2}\right)^{2/3}$$

For consistency, use kPa and mm for SI metric, and psi and in. for U.S. customary notation.

(7) If $r_x \ge \frac{1}{8} D_b$ and $T_c \ge 1.5\bar{T}$, a flexibility characteristic of $4.4\bar{T}/r_2$ may be used.

(8) Stress intensification factors for branch connections are based on tests with at least two diameters of straight run pipe on each side of the branch centerline. More closely loaded branches may require special consideration.

(9) When $\bar{T}_r$ is $> 1\frac{1}{2}\bar{T}$, use $h = 4\bar{T}/r_2$.

(10) The out-of-plane stress intensification factor (SIF) for a reducing branch connection with branch-to-run diameter ratio of $0.5 < d/D < 1.0$ may be nonconservative. A smooth concave weld contour has been shown to reduce the SIF. Selection of the appropriate SIF is the designer's responsibility.

(11) The designer must be satisfied that this fabrication has a pressure rating equivalent to straight pipe.

(12) For welds to socket welded fittings, the stress intensification factor is based on the assumption that the pipe and fitting are matched in accordance with ASME B16.11 and a fillet weld is made between the pipe and fitting. For welds to socket welded flanges, the stress intensification factor is based on the weld geometry and has been shown to envelope the results of the pipe to socket welded fitting tests. Blending the toe of the fillet weld smoothly into the pipe wall has been shown to improve the fatigue performance of the weld.

(13) Factors shown apply to bending. Flexibility factor for torsion equals 0.9.

The resultant bending stresses S_b to be used in Eq. (10-99) for elbows and miter bends shall be calculated in accordance with Eq. (10-100), with moments as shown in Fig. 10-161:

$$S_b = \frac{\sqrt{(i_i M_i)^2 + (i_o M_o)^2}}{Z_e} \qquad (10\text{-}100)$$

where S_b = resultant bending stress, lbf/in² (MPa)
i_i = in-plane stress intensification factor from Table 10-46
i_o = out-plane stress intensification factor from Table 10-46
M_i = in-plane bending moment, in · lbf (N · mm)
M_o = out-of-plane bending moment, in · lbf (N · mm)
Z = section modulus of pipe, in³ (mm³)

The resultant bending stresses S_b to be used in Eq. (10-99) for branch connections shall be calculated in accordance with Eqs. (10-101) and (10-102), with moments as shown in Fig. 10-162.

For header (legs 1 and 2):

$$S_b = \frac{\sqrt{(i_i M_i)^2 + (i_o M_o)^2}}{Z} \qquad (10\text{-}101)$$

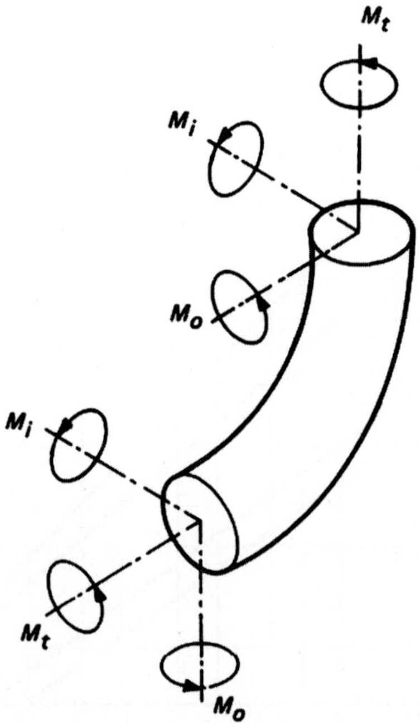

FIG. 10-161 Moments in bends. (*Extracted from the Process Piping Code, B31.3-2014, with permission of the publisher, the American Society of Mechanical Engineers, New York. All rights reserved.*)

For branch (leg 3):

$$S_b = \frac{\sqrt{(i_i M_i)^2 + (i_o M_o)^2}}{Z_e} \qquad (10\text{-}102)$$

where S_b = resultant bending stress, lbf/in² (MPa)
Z_e = effective section modulus for branch, in³ (mm³)

$$Z_e = \pi r_2^2 T_s \qquad (10\text{-}103)$$

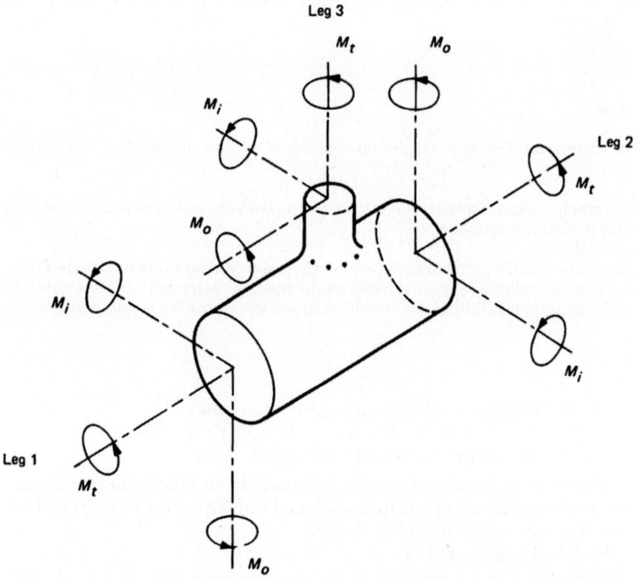

FIG. 10-162 Moments in branch connections. (*Extracted from the Process Piping Code, B31.3-2004, with permission of the publisher, the American Society of Mechanical Engineers, New York. All rights reserved.*)

r_2 = mean branch cross-sectional radius, in (mm)
T_s = effective branch wall thickness, in (mm) [lesser of $\overline{T}_h$ and $(i_o)(\overline{T}_b)$]
$\overline{T}_h$ = thickness of pipe matching run of tee or header exclusive of reinforcing elements, in (mm)
$\overline{T}_b$ = thickness of pipe matching branch, in (mm)
i_o = out-of-plane stress–intensification factor (Table 10-46)
i_i = in-plane stress–intensification factor (Table 10-46)

Allowable stress range S_A and permissible additive stresses shall be computed in accordance with Eqs. (10-91) and (10-92).

Required Weld Quality Assurance Any weld at which S_E exceeds $0.8S_A$ for any portion of a piping system, and the equivalent number of cycles N exceeds 7000, shall be fully examined in accordance with the requirements for severe cyclic service (presented later in this section).

Reactions: Metallic Piping Reaction forces and moments to be used in the design of restraints and supports and in evaluating the effects of piping displacements on connected equipment shall be based on the reaction range R for the extreme displacement conditions, considering the range previously defined for reactions and using E_a. The designer shall consider instantaneous maximum values of forces and moments in the original and extreme displacement conditions as well as the reaction range in making these evaluations.

Maximum Reactions for Simple Systems For two-anchor systems without intermediate restraints, the maximum instantaneous values of reaction forces and moments may be estimated from Eqs. (10-104) and (10-105).

1. *For extreme displacement conditions* R_m. The temperature for this computation is the design maximum or design minimum temperature as previously defined for reactions, whichever produces the larger reaction:

$$R_m = R\left(1 - \frac{2C}{3}\right)\frac{E_m}{E_a} \qquad (10\text{-}104)$$

where C = cold-spring factor varying from 0 for no cold spring to 1.0 for 100 percent cold spring. (The factor ⅔ is based on experience, which shows that specified cold spring cannot be fully ensured even with elaborate precautions.)
E_a = modulus of elasticity at installation temperature, lbf/in² (MPa)
E_m = modulus of elasticity at design maximum or design minimum temperature, lbf/in² (MPa)
R = range of reaction forces or moments (derived from flexibility analysis) corresponding to the full displacement-stress range and based on reaction force, E_a, lbf or on moment in · lbf (N or N · mm)
R_m = estimated instantaneous maximum reaction force or moment at design maximum or design minimum temperature, reaction force lbf or moment in · lbf (N or N · mm)

2. *For original condition* R_a. The temperature for this computation is the expected temperature at which the piping is to be assembled.

$$R_a = CR \quad \text{or} \quad C_1 R \quad \text{whichever is greater} \qquad (10\text{-}105)$$

where nomenclature is as for Eq. (10-104) and

$C_1 = 1 - S_h E_a / S_E E_m$
= estimated self-spring or relaxation factor (use 0 if value of C_1 is negative)
R_a = estimated instantaneous reaction force or moment at installation temperature, lbf or in · lbf (N or N · mm)
S_E = computed displacement-stress range, lbf/in² (MPa); see Eq. (10-99)
S_h = see Eq. (10-91)

Maximum Reactions for Complex Systems For multianchor systems and for two-anchor systems with intermediate restraints, Eqs. (10-104) and (10-105) are not applicable. Each case must be studied to estimate the location, nature, and extent of local overstrain and its effect on stress distribution and reactions.

Acceptable comprehensive methods of analysis are analytical, model-test, and chart methods, which evaluate for the entire piping system under consideration the forces, moments, and stresses caused by bending and torsion from a simultaneous consideration of terminal and intermediate restraints to thermal expansion and include all external movements transmitted under thermal change to the piping by its terminal and

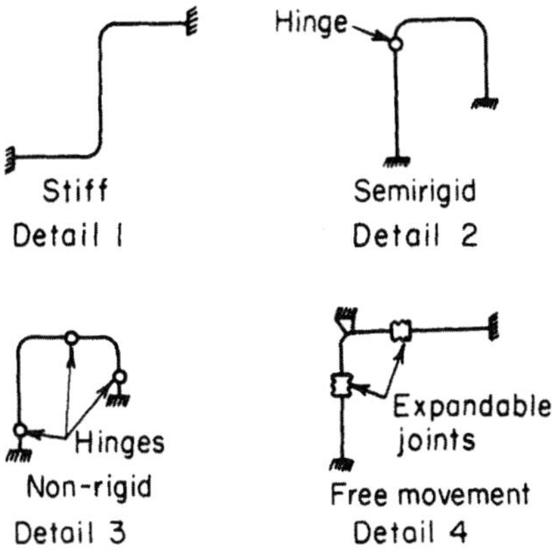

FIG. 10-163 Flexibility classification for piping systems. (From Kellogg, *Design of Piping Systems,* Wiley, New York, 1965.)

intermediate attachments. Correction factors, as provided by the details of these rules, must be applied for the stress intensification of curved pipe and branch connections and may be applied for the increased flexibility of such component parts.

Expansion Joints All the foregoing applies to "stiff piping systems," i.e., systems without expansion joints (see detail 1 of Fig. 10-163). When space limitations, process requirements, or other considerations result in configurations of insufficient flexibility, the capacity for deflection within allowable stress range limits may be increased successively by the use of one or more hinged bellows expansion joints, i.e., semirigid (detail 2) and nonrigid (detail 3) systems, and expansion effects essentially eliminated by a free-movement joint (detail 4) system. Expansion joints for semirigid and nonrigid systems are restrained against longitudinal and lateral movement by the hinges with the expansion element under bending movement only and are known as *rotation* or *hinged* joints (see Fig. 10-164). Semirigid systems are limited to one plane; nonrigid systems require a minimum of three joints for two-dimensional and five joints for three-dimensional expansion movement.

Joints similar to that shown in Fig. 10-164, except with two pairs of hinge pins equally spaced around a gimbal ring, achieve similar results with fewer joints.

Expansion joints for free-movement systems can be designed for axial or offset movement alone, or for combined axial and offset movements (see Fig. 10-165). For offset movement alone, the end load due to pressure and weight can be transferred across the joint by tie rods or structural members (see Fig. 10-166). For axial or combined movements, anchors must be provided to absorb the unbalanced pressure load and to force bellows to deflect.

Commercial bellows elements are usually light-gauge (of the order of 0.05 to 0.10 in thick) and are available in stainless and other alloy steels, copper, and other nonferrous materials. Multiply bellows, bellows with external reinforcing rings, and toroidal contour bellows are available for higher pressures. Since bellows elements are ordinarily rated for strain ranges that involve repetitive yielding, predictable performance is ensured only by adequate fabrication controls and knowledge of the potential fatigue performance of each design. The attendant cold work can affect

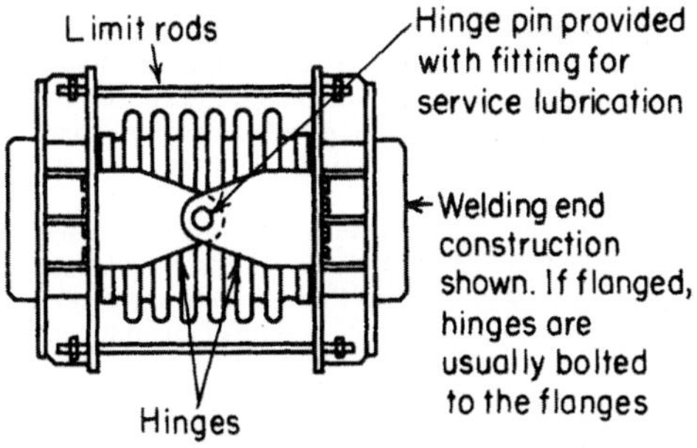

FIG. 10-164 Hinged expansion joint. (From Kellogg, *Design of Piping Systems,* Wiley, New York, 1965.)

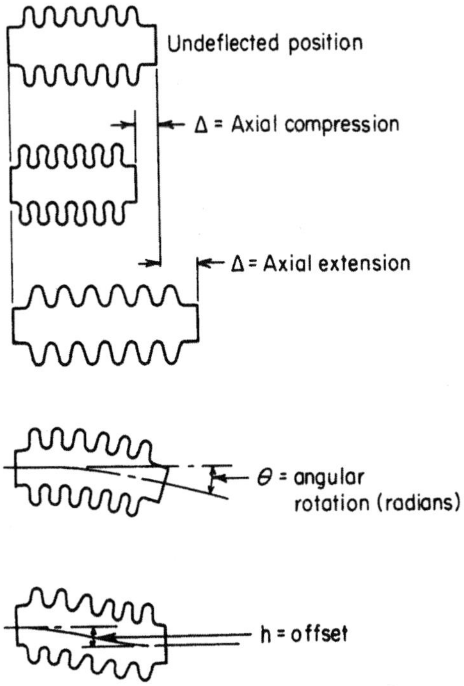

FIG. 10-165 Action of expansion bellows under various movements. (From Kellogg, *Design of Piping Systems,* Wiley, New York, 1965.)

corrosion resistance and promote susceptibility to corrosion fatigue or stress corrosion; joints in a horizontal position cannot be drained and have frequently undergone pitting or cracking due to the presence of condensate during operation or off-stream. For low-pressure essentially nonhazardous service, nonmetallic bellows of fabric-reinforced rubber or special materials are sometimes used. For corrosive service PTFE bellows may be used.

Because of the inherently greater susceptibility of expansion bellows to failure from unexpected corrosion, failure of guides to control joint movements, etc., it is advisable to examine critically their design choice in comparison with a stiff system.

Slip-type expansion joints (Fig. 10-167) substitute packing (ring or plastic) for bellows. Their performance is sensitive to adequate design with respect to guiding to prevent binding and the adequacy of stuffing boxes and attendant packing, sealant, and lubrication. Anchors must be provided for the

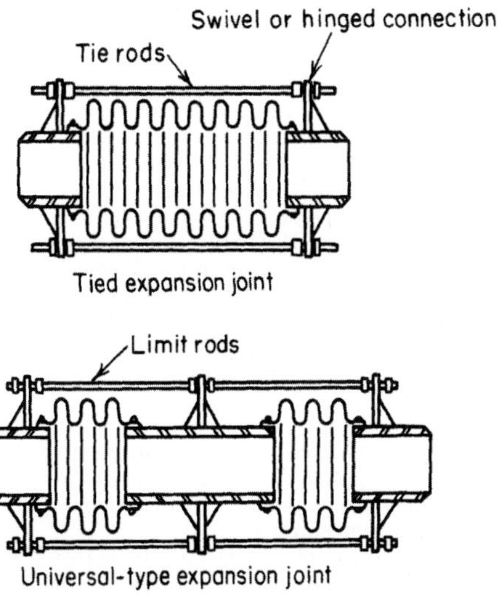

FIG. 10-166 Constrained-bellows expansion joints. (From Kellogg, *Design of Piping Systems,* Wiley, New York, 1965.)

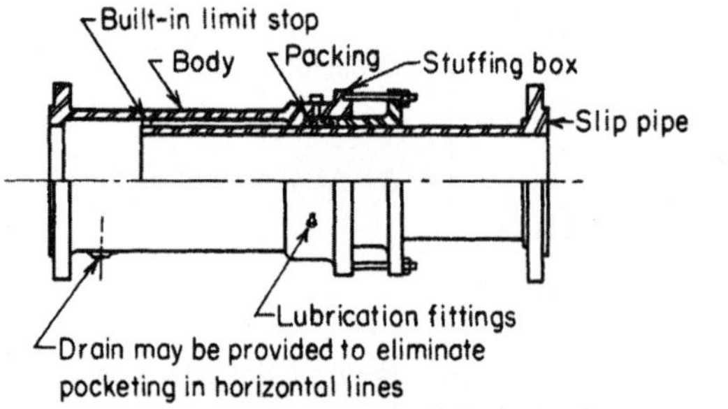

FIG. 10-167 Slip-type expansion joint. (From Kellogg, *Design of Piping Systems*, Wiley, New York, 1965.)

unbalanced pressure force and for the friction forces to move the joint. The latter can be much higher than the elastic force required to deflect a bellows joint. Rotary packed joints, ball joints, and other special joints can absorb end load.

Corrugated pipe and corrugated and creased bends are also used to decrease stiffness.

Pipe Supports Loads transmitted by piping to attached equipment and supporting elements include weight, temperature- and pressure-induced effects, vibration, wind, earthquake, shock, and thermal expansion and contraction. The design of supports and restraints is based on concurrently acting loads (if it is assumed that wind and earthquake do not act simultaneously).

Resilient and constant-effort-type supports shall be designed for maximum loading conditions including test unless temporary supports are provided.

Though not specified in the code, supports for discharge piping from relief valves must be adequate to withstand the jet reaction produced by their discharge.

The code states further that pipe-supporting elements shall (1) avoid excessive interference with thermal expansion and contraction of pipe which is otherwise adequately flexible; (2) be such that they do not contribute to leakage at joints or excessive sag in piping, requiring drainage; (3) be designed to prevent overstress, resonance, or disengagement due to variation of load with temperature; also, so that combined longitudinal stresses in the piping shall not exceed the code allowable limits; (4) be such that a complete release of the piping load will be prevented in the event of spring failure or misalignment, weight transfer, or added load due to test during erection; (5) be of steel or wrought iron; (6) be of alloy steel or protected from temperature when the temperature limit for carbon steel may be exceeded; (7) not be cast iron except for roller bases, rollers, anchor bases, etc., under mainly compression loading; (8) not be malleable or nodular iron except for pipe clamps, beam clamps, hanger flanges, clips, bases, and swivel rings; (9) not be wood except for supports mainly in compression when the pipe temperature is at or below ambient; and (10) have threads for screw adjustment which shall conform to ASME B1.1.

A supporting element used as an anchor shall be designed to maintain an essentially fixed position.

To protect terminal equipment or other (weaker) portions of the system, restraints (such as anchors and guides) shall be provided where necessary to control movement or to direct expansion into those portions of the system that are adequate to absorb them. The design, arrangement, and location of restraints shall ensure that expansion-joint movements occur in the directions for which the joint is designed. In addition to the other thermal forces and moments, the effects of friction in other supports of the system shall be considered in the design of such anchors and guides.

Anchors for Expansion Joints Anchors (such as those of the corrugated, omega, disk, or slip type) shall be designed to withstand the algebraic sum of the forces at the maximum pressure and temperature at which the joint is to be used. These forces are as follows:

1. Pressure thrust, which is the product of the effective thrust area and the maximum pressure to which the joint will be subjected during normal operation. (For slip joints the effective thrust area shall be computed by using the outside diameter of the pipe. For corrugated, omega, or disk-type joints, the effective thrust area shall be that area recommended by the joint manufacturer. If this information is unobtainable, the effective area shall be computed by using the maximum inside diameter of the expansion-joint bellows.)
2. The force required to compress or extend the joint in an amount equal to the calculated expansion movement.
3. The force required to overcome the static friction of the pipe in expanding or contracting on its supports, from installed to operating position. The length of pipe considered should be that located between the anchor and the expansion joint.

Support Fixtures Hanger rods may be pipe straps, chains, bars, or threaded rods that permit free movement for thermal expansion or contraction. Sliding supports shall be designed for friction and bearing loads. Brackets shall be designed to withstand movements due to friction in addition to other loads. Spring-type supports shall be designed for weight load at the point of attachment, to prevent misalignment, buckling, or eccentric loading of springs, and provided with stops to prevent spring overtravel. Compensating-type spring hangers are recommended for high-temperature and critical-service piping to make the supporting force uniform with appreciable movement. Counterweight supports shall have stops to limit travel. Hydraulic supports shall be provided with safety devices and stops to support load in the event of loss of pressure. Vibration dampers or sway braces may be used to limit vibration amplitude.

The code requires that the safe load for threaded hanger rods be based on the root area of the threads. This, however, assumes concentric loading. When hanger rods move to a non-vertical position so that the load is transferred from the rod to the supporting structure via the edge of one flat of the nut on the rod, it is necessary to consider the root area to be reduced by one-third. If a clamp is connected to a vertical line to support its weight, then it is recommended that shear lugs be welded to the pipe, or that the clamp be located below a fitting or flange, to prevent slippage. Consideration shall be given to the localized stresses induced in the piping by the integral attachment. Typical pipe supports are shown in Fig. 10-168.

Much piping is supported from structures installed for other purposes. It is common practice to use beam formulas for tubular sections to determine stress, maximum deflection, and maximum slope of piping in *spans between supports.* When piping is supported from structures installed for that sole purpose and those structures rest on driven piles, detailed calculations are usually made to determine maximum permissible spans. Limits imposed on maximum slope to make the contents of the line drain to the lower end require calculations made on the weight per foot of the empty line. To avoid interference with other components, maximum deflection should be limited to 25.4 mm (1 in).

Pipe hangers are essentially frictionless but require taller pipe-support structures which cost more than structures on which pipe is laid. Devices that reduce friction between laid pipe subject to thermal movement and its supports are used to accomplish the following:

1. Reduce loads on anchors or on equipment acting as anchors.
2. Reduce the tendency of pipe acting as a column loaded by friction at supports to buckle sideways off supports.
3. Reduce nonvertical loads imposed by piping on its supports so as to minimize the cost of support foundations.
4. Reduce longitudinal stress in pipe.

Linear bearing surfaces made of fluorinated hydrocarbons or of graphite and also rollers are used for this purpose.

Design Criteria: Nonmetallic Pipe In using a nonmetallic material, designers must satisfy themselves as to the adequacy of the material and its manufacture, considering such factors as strength at design temperature, impact- and thermal-shock properties, toxicity, methods of making connections, and possible deterioration in service. Rating information, based usually on ASTM standards or specifications, is generally available from the manufacturers of these materials. Particular attention should be given to provisions for the thermal expansion of nonmetallic piping materials, which may be as much as 5 to 10 times that of steel (Table 10-47). Special consideration should be given to the strength of small pipe connections to piping and equipment and to the need for extra flexibility at the junction of metallic and nonmetallic systems.

Table 10-48 gives values for the modulus of elasticity for nonmetals; however, no specific stress-limiting criteria or methods of stress analysis are presented. Stress-strain behavior of most nonmetals differs considerably from that of metals and is less well defined for mathematic analysis. The piping system should be designed and laid out so that flexural stresses resulting from displacement due to expansion, contraction, and other movement are minimized. This concept requires special attention to supports, terminals, and other restraints.

Displacement Strains The concepts of strain imposed by restraint of thermal expansion or contraction and by external movement described for metallic piping apply in principle to nonmetals. Nevertheless, the assumption that stresses throughout the piping system can be predicted from these strains because of fully elastic behavior of the piping materials is not generally valid for nonmetals.

In thermoplastics and some thermosetting resins, displacement strains are not likely to produce immediate failure of the piping, but may result in detrimental distortion. Especially in thermoplastics, progressive deformation may occur upon repeated thermal cycling or on prolonged exposure to elevated temperature.

In brittle nonmetallics (such as porcelain, glass, impregnated graphite) and some thermosetting resins, the materials show rigid behavior and develop high displacement stresses up to the point of sudden breakage due to overstrain.

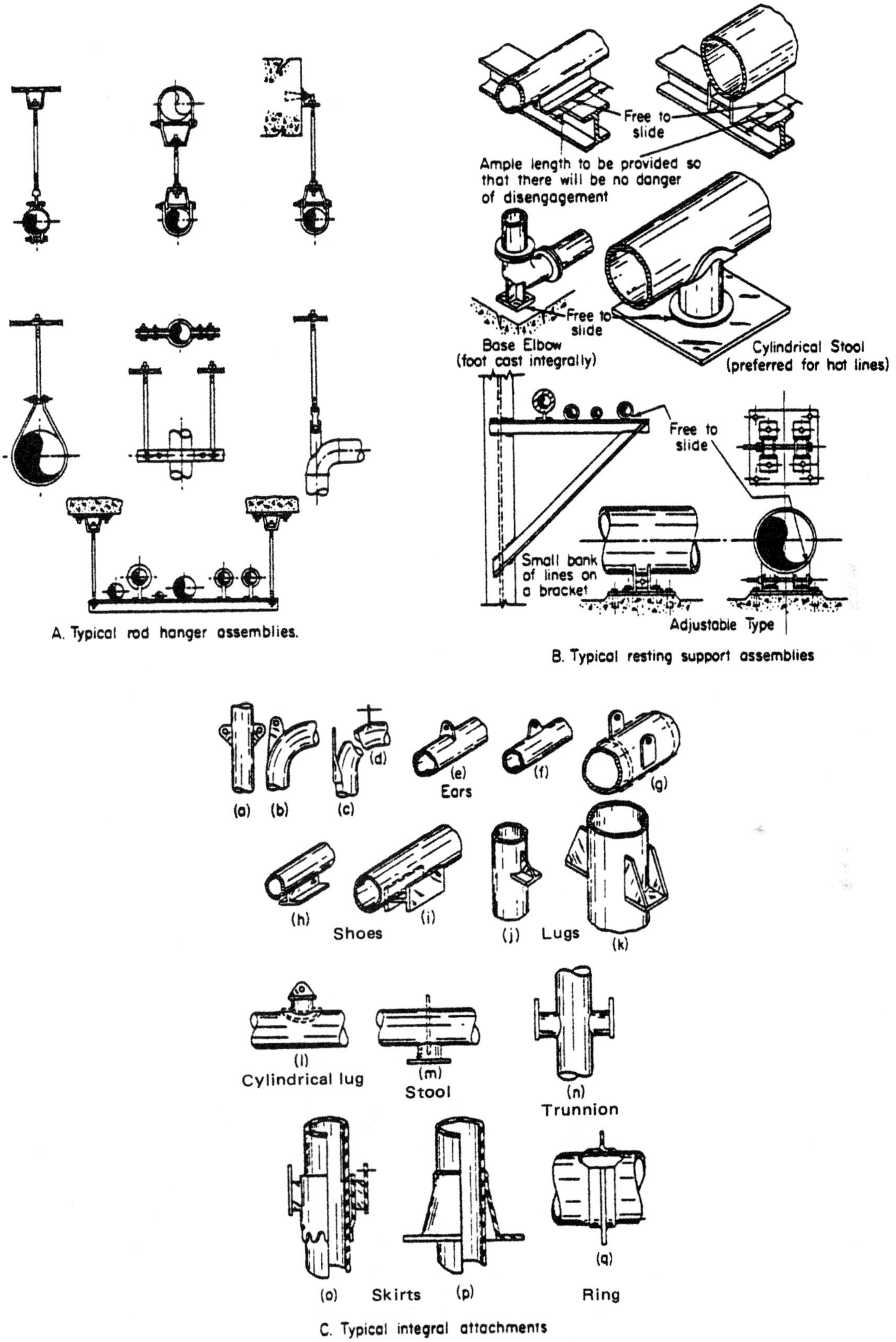

FIG. 10-168 Typical pipe supports and attachments. (From Kellogg, *Design of Piping Systems,* Wiley, New York, 1965.)

TABLE 10-47 Thermal Expansion Coefficients, Nonmetals

Material description	Mean coefficients (divide table values by 10^6)			
	in/in, °F	Range, °F	mm/mm, °C	Range, °C
Thermoplastics				
Acetal AP2012	2	. . .	3.6	. . .
Acrylonitrile-butadiene-styrene				
ABS 1208	60	. . .	108	. . .
ABS 1210	55	45–55	99	7–13
ABS 1316	40	. . .	72	. . .
ABS 2112	40	. . .	72	. . .
Cellulose acetate butyrate				
CAB MH08	80	. . .	144	. . .
CAB S004	95	. . .	171	. . .
Chlorinated poly(vinyl chloride)				
CPVC 4120	35	. . .	63	. . .
Polybutylene PB 2110	72	. . .	130	. . .
Polyether, chlorinated	45	. . .	81	. . .
Polyethylene				
PE 1404	100	46–100	180	8–38
PE 2305	90	46–100	162	8–38
PE 2306	80	46–100	144	8–38
PE 3306	70	46–100	126	8–38
PE 3406	60	46–100	108	8–38
Polyphenylene POP 2125	30	. . .	54	. . .
Polypropylene				
PP1110	48	33–67	86	1–19
PP1208	43	. . .	77	. . .
PP2105	40	. . .	72	. . .
Poly(vinyl chloride)				
PVC 1120	30	23–37	54	−5 to +3
PVC 1220	35	34–40	63	1–4
PVC 2110	50	. . .	90	. . .
PVC 2112	45	. . .	81	. . .
PVC 2116	40	37–45	72	3–7
PVC 2120	30	. . .	54	. . .
Poly(vinylidene fluoride)	79	. . .	142	. . .
Poly(vinylidene chloride)	100	. . .	180	. . .
Polytetrafluoroethylene	55	73–140	99	23–60
Poly(fluorinated ethylene propylene)	46–58	73–140	83–104	23–60
Poly(perfluoroalkoxy alkane)	67	70–212	121	21–100
Poly(perfluoroalkoxy alkane)	94	212–300	169	100–149
Poly(perfluoroalkoxy alkane)	111	300–408	200	149–209
Reinforced Thermosetting Resins and Reinforced Plastic Mortars				
Glass-epoxy, centrifugally cast	9–13	. . .	16–23.5	. . .
Glass-polyester, centrifugally cast	9–15	. . .	16–27	. . .
Glass-polyester, filament-wound	9–11	. . .	16–20	. . .
Glass-polyester, hand lay-up	12–15	. . .	21.5–27	. . .
Glass-epoxy, filament-wound	9–13	. . .	16–23.5	. . .
Other Nonmetallic Materials				
Borosilicate glass	1.8	. . .	3.25	. . .

GENERAL NOTES:
For Code references to this table, see para. A319.3.1, ASME B31.3-2014. Reprinted from ASME B31.3-2014 by permission of the American Society of Mechanical Engineers, New York. All rights reserved. These data are for use in the absence of more applicable data. It is the designer's responsibility to verify that materials are suitable for the intended service at the temperatures shown.
Individual compounds may vary from the values shown. Consult manufacturer for specific values for products.

Elastic Behavior The assumption that displacement strains will produce proportional stress over a sufficiently wide range to justify an elastic-stress analysis often is not valid for nonmetals. In brittle nonmetallic piping, strains initially will produce relatively large elastic stresses. The total displacement strain must be kept small, however, since overstrain results in failure rather than plastic deformation. In plastic and resin nonmetallic piping, strains generally will produce stresses of the overstrained (plastic) type even at relatively low values of total displacement strain.

FABRICATION, ASSEMBLY, AND ERECTION

Welding, Brazing, or Soldering Code requirements dealing with fabrication are more detailed for welding than for other methods of joining, since welding is the predominant method of construction and the method used for the most demanding applications. The code requirements for welding processes and operators are essentially the same as covered in the subsection on pressure vessels (i.e., qualification to Sec. IX of the ASME Boiler and Pressure Vessel Code) except that welding processes are not restricted, the material grouping (P number) must be in accordance with ASME B31.3 App. A-1 (A-1M Metric), and welding positions are related to pipe position. The code also permits one fabricator to accept welders or welding operators qualified by another employer without requalification when welding pipe by the same or equivalent procedure. The code may require that the welding procedure qualification include low-temperature toughness testing (see Table 10-49).

Filler metal is required to conform with the requirements of Sec. IX. Backing rings (of ferrous material), when used, shall be of weldable quality

TABLE 10-48 Modulus of Elasticity, Nonmetals

Material description	*E*, ksi (73.4°F)	*E*, MPa (23°C)
Thermoplastics [Note (1)]		
Acetal	410	2,830
ABS, Type 1210	250	1,725
ABS, Type 1316	340	2,345
CAB	120	825
PVC, Type 1120	420	2,895
PVC, Type 1220	410	2,825
PVC, Type 2110	340	2,345
PVC, Type 2116	380	2,620
Chlorinated PVC	420	2,895
Chlorinated polyether	160	1,105
PE2606	100	690
PE2706	100	690
PE3608	125	860
PE3708	125	860
PE3710	125	860
PE4708	130	895
PE4710	130	895
Polypropylene	120	825
Poly(vinylidene chloride)	100	690
Poly(vinylidene fluoride)	194	1,340
Poly(tetrafluoroethylene)	57	395
Poly(fluorinated ethylene propylene)	67	460
Poly(perfluoroalkoxy alkane)	100	690
Thermosetting Resins, Axially Reinforced		
Epoxy-glass, centrifugally cast	1,200–1,900	8,275–13,100
Epoxy-glass, filament-wound	1,100–2,000	7,585–13,790
Polyester-glass, centrifugally cast	1,200–1,900	8,275–13,100
Polyester-glass, hand lay-up	800–1,000	5,515–6,895
Other		
Borosilicate glass	9,800	67,570

GENERAL NOTE: For Code references to this table, see para. A319.3.2, ASME B31.3-2014. Reprinted from ASME B31.3-2014 by permission of the American Society of Mechanical Engineers, New York. All rights reserved. These data are for use in the absence of more applicable data. It is the designer's responsibility to verify that materials are suitable for the intended service at the temperatures shown.

NOTE:

(1) The modulus of elasticity data shown for thermoplastics are based on short-term tests. The manufacturer should be consulted to obtain values for use under long-term loading.

with sulfur limited to 0.05 percent. Backing rings of nonferrous and nonmetallic materials may be used provided they are proved satisfactory by procedure qualification tests and provided their use has been approved by the designer.

The code requires internal alignment within the dimensional limits specified in the welding procedure and the engineering design without specific dimensional limitations. Internal trimming is permitted for correcting internal misalignment provided such trimming does not result in a finished wall thickness before welding of less than required minimum wall thickness t_m. When necessary, weld metal may be deposited on the inside or outside of the component to provide alignment or sufficient material for trimming.

Table 10-51 is a summary of the code acceptance criteria (limits on imperfections) for welds. The defects referred to are illustrated in Fig. 10-169.

Brazing procedures, brazers, and brazing operators must be qualified in accordance with the requirements of Part QB, Sec. IX, ASME Code. Qualification is not required for Category D fluid service not exceeding 93°C (200°F), unless specified in the engineering design. The clearance between surfaces to be joined by brazing or soldering shall be no larger than is necessary to allow complete capillary distribution of the filler metal.

The only requirement for solderers is that they follow the procedure in ASTM B823–02, *Standard Practice for Making Capillary Joints by Soldering of Copper and Copper Alloy Tube and Fittings.*

Bending and Forming Pipe may be bent to any radius for which the bend-arc surface will be free of cracks and substantially free of buckles. The use of qualified bends that are creased or corrugated is permitted. Bending may be done by any hot or cold method that produces a product meeting code and service requirements, and that does not have an adverse effect on the essential characteristics of the material. Hot bending and hot forming must be done within a temperature range consistent with material characteristics, end use, and postoperation heat treatment. Postbend heat treatment may be required for bends in some materials; its necessity is dependent on the type of bending operation and the severity of the bend. Postbend heat treatment requirements are defined in the code.

Piping components may be formed by any suitable hot or cold pressing, rolling, forging, hammering, spinning, drawing, or other method. Thickness after forming shall not be less than required by design. Special rules cover the forming and pressure design verification of flared laps.

The development of fabrication facilities for bending pipe to the radius of commercial butt-welding long-radius elbows and forming flared metallic (Van Stone) laps on pipe is an important technique in reducing welded-piping costs. These techniques save both the cost of the ell or stub end and that of the welding operation required to attach the fitting to the pipe.

Preheating and Heat Treatment Preheating and postoperation heat treatment are used to avert or relieve the detrimental effects of the high temperature and severe thermal gradients inherent in the welding of metals. In addition, heat treatment may be needed to relieve residual stresses created during the bending or forming of metals. The code provisions shown in Tables 10-52 and 10-53 represent basic practices that are acceptable for most applications of welding, bending, and forming, but they are not necessarily suitable for all service conditions. The specification of more or less stringent preheating and heat-treating requirements is a function of those responsible for the engineering design.

Refer to the code for rules establishing the thickness to be used in determining PWHT (post weld heat treatment) requirements for configurations other than butt welds (e.g., fabricated branch connections and socket welds).

Joining Nonmetallic Pipe All joints should be made in accordance with procedures complying with the manufacturer's recommendations and code requirements. General welding and heat fusion procedures are described in ASTM D-2657. ASTM D2855 describes general solvent-cementing procedures.

Depending on size, thermoplastic piping can be joined with mechanical joints, solvent-cemented joints, hot-gas welding, or heat fusion procedures. Mechanical joints are frequently bell-and-spigot joints which employ an elastomer O-ring gasket. Joints of this type are generally not self-restrained against internal pressure thrust.

Thermosetting resin pipe can be joined with mechanical joints or adhesive-bonded joints. Mechanical joints are generally a variation of gasketed bell-and-spigot joints and may be either nonrestrained or self-restrained. Adhesive-bonded joints are typically bell-and-spigot or butt-and-strap. Butt-and-strap joints join piping components with multiple layers of resin-saturated glass reinforcement.

Assembly and Erection Flanged-joint faces shall be aligned to the design plane to within ¹⁄₁₆ in/ft (0.5 percent) maximum measured across any diameter, and flange bolt holes shall be aligned to within 3.2-mm (⅛-in) maximum offset. Flanged joints involving flanges with widely differing mechanical properties shall be assembled with extra care, and tightening to a predetermined torque is recommended.

The use of flat washers under bolt heads and nuts is a code requirement when assembling nonmetallic flanges. It is preferred that the bolts extend completely through their nuts; however, a lack of complete thread engagement not exceeding one thread is permitted by the code. The assembly of cast-iron bell-and-spigot piping is covered in AWWA Standard C600.

Screwed joints that are intended to be seal-welded shall be made up dry without any thread compound.

When one is installing conductive nonmetallic piping and metallic pipe with nonmetallic linings, consideration should be given to the need to provide electrical continuity throughout the system and to grounding requirements. This is particularly critical in areas with potentially explosive atmospheres.

EXAMINATION, INSPECTION, AND TESTING

This subsection provides a general synopsis of code requirements. It should not be viewed as comprehensive.

Examination and Inspection The code differentiates between examination and inspection. *Examination* applies to quality-control functions performed by personnel of the piping manufacturer, fabricator, or erector. *Inspection* applies to functions performed for the owner by the authorized inspector.

The authorized inspector shall be designated by the owner and shall be the owner, an employee of the owner, an employee of an engineering or scientific organization, or an employee of a recognized insurance or inspection company acting as the owner's agent. The inspector shall not represent or be an employee of the piping erector, the manufacturer, or the fabricator unless the owner is also the erector, the manufacturer, or the fabricator.

TABLE 10-49 Requirements for Low-Temperature Toughness Tests for Metals*

These toughness requirements are in addition to tests required by the material specification.

	Type of material	Column A Design minimum temperature at or above min. temp. in Table A-1 of ASME B31.3-2014 or Table 10-50		Column B Design minimum temperature below min. temp. in Table A-1 of ASME B31.3-2014 or Table 10-50
Listed Materials	1 Gray cast iron	A-1 No additional requirements		B-1 No additional requirements
	2 Malleable and ductile cast iron: carbon steel per Note (1)	A-2 No additional requirements		B-2 Materials designated in Box 2 shall not be used.
		(a) Base metal	(b) Weld metal and heat affected zone (HAZ) [Note (2)]	
	3 Other carbon steels, low and intermediate alloy steels, high alloy ferritic steels, duplex stainless steels	A-3 (a) No additional requirements	A-3 (b) Weld metal deposits shall be impact tested per para. 323.3.[†] If design min. temp. < −29°C (−20°F), except as provided in Notes (3) and (5), and except as follows: for materials listed for Curves C and D of Table 10-50, where corresponding welding-consumables are qualified by impact testing at the design minimum temperature or lower in accordance with the applicable AWS specification, additional testing is not required.	B-3 Except as provided in Notes (3) and (5), heat treat base metal per applicable ASTM specification listed in para. 323.3.2[‡]; then impact test base metal, weld deposits, and HAZ per para. 323.3[†] [see Note (2)]. When materials are used at design min. temp. below the assigned curve as permitted by Notes (2) and (3) of Table 10-50, weld deposits and HAZ shall be impact tested [see Note (2)].
	4 Austenitic stainless steels	A-4 (a) If: (1) carbon content by analysis >0.1%; or (2) material is not in solution heat treated condition; then, impact test per para. 323.3[†] for design min. temp. < −29°C (−20°F) except as provided in Notes (3) and (6)	A-4 (b) Weld metal deposits shall be impact tested per para 323.3.[†] If design min. temp. < −29°C (−20°F) except as provided in para. 323.2.2[b] and in Notes (3) and (6)	B-4 Base metal and weld metal deposits shall be impact tested per para. 323.3[‡]. See Notes (2),(3), and (6).
	5 Austenitic ductile iron, ASTM A 571	A-5 (a) No additional requirements	A-5 (b) Welding is not permitted	B-5 Base metal shall be impact tested per para 323.3.[†] Do not use < −196°C (−320°F). Welding is not permitted.
	6 Aluminum, copper, nickel, and their alloys; unalloyed titanium	A-6 (a) No additional requirements	A-6 (b) No additional requirements unless filler metal composition is outside the range for base metal composition; then test per column B-6	B-6 Designer shall be assured by suitable tests [see Note (4)] that base metal, weld deposits, and HAZ are suitable at the design min. temp.
Unlisted Materials	7 An unlisted material shall conform to a published specification. Where composition, heat treatment, and product form are comparable to those of a listed material, requirements for the corresponding listed material shall be met. Other unlisted materials shall be qualified as required in the applicable section of column B.			

NOTES:

(1) Carbon steels conforming to the following are subject to the limitations in Box B-2; plates per ASTM A 36, A 283, and A 570; pipe per ASTM A 134 when made from these plates; structural shapes in accordance with ASTM A992; and pipe per ASTM A 53 Type F and API 5L Gr. A25 butt weld.

(2) Impact tests that meet the requirements of Table 323.3.1,[†] which are performed as part of the weld procedure qualification, will satisfy all requirements of para. 323.2.2,[‡] and need not be repeated for production welds.

(3) Impact testing is not required if the design minimum temperature is below −29°C (−20°F) but at or above −104°C (−155°F) and the Stress Ratio defined in Fig. 323.2.2B[‡] does not exceed 0.3.

(4) Tests may include tensile elongation, sharp-notch tensile strength (to be compared with unnotched tensile strength), and/or other tests, conducted at or below design minimum temperature. See also para. 323.3.4.[‡]

(5) Impact tests are not required when the maximum obtainable Charpy specimen has a width along the notch of less than 2.5 mm (0.098 in). Under these conditions, the design minimum temperature shall not be less than the lower of −48°C (−55°F) or the minimum temperature for the material in Table A-1.

(6) Impact tests are not required when the maximum obtainable Charpy specimen has a width along the notch of less than 2.5 mm (0.098 in).

*Table 10-49 and notes have been extracted (with minor modifications) from *Process Piping,* ASME B31.3-2014, with permission of the publisher, the American Society of Mechanical Engineers, New York. All rights reserved.

[†]Refer to the referenced code paragraph for impact testing methods and acceptance.

[‡]Refer to the referenced code paragraph for comments regarding circumstances under which impact testing can be excluded.

The authorized inspector shall have a minimum of 10 years' experience in the design, fabrication, or inspection of industrial pressure piping. Each 20 percent of satisfactory work toward an engineering degree accredited by the Engineers' Council for Professional Development shall be considered equivalent to 1 year's experience, up to 5 years total.

It is the owner's responsibility, exercised through the authorized inspector, to verify that all required examinations and testing have been completed and to inspect the piping to the extent necessary to be satisfied that it conforms to all applicable requirements of the code and the engineering design. This verification may include certifications and records pertaining to materials, components, heat treatment, examination and testing, and qualifications of operators and procedures. The authorized inspector may delegate the performance of inspection to a qualified person.

Inspection does not relieve the manufacturer, the fabricator, or the erector of responsibility for providing materials, components, and skill in accordance with requirements of the code and the engineering design, performing all required examinations, and preparing records of examinations and tests for the inspector's use.

Examination Methods The code establishes the types of examinations for evaluating various types of imperfections (see Table 10-54).

Personnel performing examinations other than visual shall be qualified in accordance with applicable portions of SNT TC-1A, *Recommended Practice for Nondestructive Testing Personnel Qualification and Certification.* Procedures shall be qualified as required in Para. T-150, Art. 1, Sec. V of the ASME Code. Limitations on imperfections shall be in accordance with the engineering design but shall at least meet the requirements of the code

TABLE 10-50 Tabular Values for Minimum Temperatures Without Impact Testing for Carbon Steel Materials*

Nominal thickness [Note (6)]		Design minimum temperature							
		Curve A [Note (2)]		Curve B [Note (3)]		Curve C [Note (3)]		Curve D	
mm	in	°C	°F	°C	°F	°C	°F	°C	°F
6.4	0.25	−9.4	15	−28.9	−20	−48.3	−55	−48.3	−55
7.9	0.3125	−9.4	15	−28.9	−20	−48.3	−55	−48.3	−55
9.5	0.375	−9.4	15	−28.9	−20	−48.3	−55	−48.3	−55
10.0	0.394	−9.4	15	−28.9	−20	−48.3	−55	−48.3	−55
11.1	0.4375	−6.7	20	−28.9	−20	−41.7	−43	−48.3	−55
12.7	0.5	−1.1	30	−28.9	−20	−37.8	−36	−48.3	−55
14.3	0.5625	2.8	37	−21.7	−7	−35.0	−31	−45.6	−50
15.9	0.625	6.1	43	−16.7	2	−32.2	−26	−43.9	−47
17.5	0.6875	8.9	48	−12.8	9	−29.4	−21	−41.7	−43
19.1	0.75	11.7	53	−9.4	15	−27.2	−17	−40.0	−40
20.6	0.8125	14.4	58	−6.7	20	−25.0	−13	−38.3	−37
22.2	0.875	16.7	62	−3.9	25	−23.3	−10	−36.7	−34
23.8	0.9375	18.3	65	−1.7	29	−21.7	−7	−35.6	−32
25.4	1.0	20.0	68	0.6	33	−19.4	−3	−34.4	−30
27.0	1.0625	22.2	72	2.2	36	−18.3	−1	−33.3	−28
28.6	1.125	23.9	75	3.9	39	−16.7	2	−32.2	−26
30.2	1.1875	25.0	77	5.6	42	−15.6	4	−30.6	−23
31.8	1.25	26.7	80	6.7	44	−14.4	6	−29.4	−21
33.3	1.3125	27.8	82	7.8	46	−13.3	8	−28.3	−19
34.9	1.375	28.9	84	8.9	48	−12.2	10	−27.8	−18
36.5	1.4375	30.0	86	9.4	49	−11.1	12	−26.7	−16
38.1	1.5	31.1	88	10.6	51	−10.0	14	−25.6	−14
39.7	1.5625	32.2	90	11.7	53	−8.9	16	−25.0	−13
41.3	1.625	33.3	92	12.8	55	−8.3	17	−23.9	−11
42.9	1.6875	33.9	93	13.9	57	−7.2	19	−23.3	−10
44.5	1.75	34.4	94	14.4	58	−6.7	20	−22.2	−8
46.0	1.8125	35.6	96	15.0	59	−5.6	22	−21.7	−7
47.6	1.875	36.1	97	16.1	61	−5.0	23	−21.1	−6
49.2	1.9375	36.7	98	16.7	62	−4.4	24	−20.6	−5
50.8	2.0	37.2	99	17.2	63	−3.3	26	−20.0	−4
51.6	2.0325	37.8	100	17.8	64	−2.8	27	−19.4	−3
54.0	2.125	38.3	101	18.3	65	−2.2	28	−18.9	−2
55.6	2.1875	38.9	102	18.9	66	−1.7	29	−18.3	−1
57.2	2.25	38.9	102	19.4	67	−1.1	30	−17.8	0
58.7	2.3125	39.4	103	20.0	68	−0.6	31	−17.2	1
60.3	2.375	40.0	104	20.6	69	0.0	32	−16.7	2
61.9	2.4375	40.6	105	21.1	70	0.6	33	−16.1	3
63.5	2.5	40.6	105	21.7	71	1.1	34	−15.6	4
65.1	2.5625	41.1	106	21.7	71	1.7	35	−15.0	5
66.7	2.625	41.7	107	22.8	73	2.2	36	−14.4	6
68.3	2.6875	41.7	107	22.8	73	2.8	37	−13.9	7
69.9	2.75	42.2	108	23.3	74	3.3	38	−13.3	8
71.4	2.8125	42.2	108	23.9	75	3.9	39	−13.3	8
73.0	2.875	42.8	109	24.4	76	4.4	40	−12.8	9
74.6	2.9375	42.8	109	25.0	77	4.4	40	−12.2	10
76.2	3.0	43.3	110	25.0	77	5.0	41	−11.7	11

(see Tables 10-51 and 10-54) for the specific type of examination. Repairs shall be made as applicable.

Visual Examination This consists of observation of the portion of components, joints, and other piping elements that are or can be exposed to view before, during, or after manufacture, fabrication, assembly, erection, inspection, or testing.

The examination includes verification of code and engineering design requirements for materials and components, dimensions, joint preparation, alignment, welding, bonding, brazing, bolting, threading and other joining methods, supports, assembly, and erection.

Visual examination shall be performed in accordance with Art. 9, Sec. V of the ASME Code.

Magnetic-Particle Examination This examination shall be performed in accordance with Art. 7, Sec. V of the ASME Code.

Liquid-Penetrant Examination This examination shall be performed in accordance with Art. 6, Sec. V of the ASME Code.

Radiographic Examination The following definitions apply to radiography required by the code or by the engineering design:

1. *Random radiography* applies only to girth and miter groove welds. It is radiographic examination of the complete circumference of a specified percentage of the girth butt welds in a designated lot of piping.

2. Unless otherwise specified in engineering design, *100 percent radiography* applies only to girth welds, miter groove welds, and fabricated branch connections that utilize butt-type welds to join the header and the branch. The design engineer may, however, elect to designate other types of welds as requiring 100 percent radiography. By definition, 100 percent radiography requires radiographic examination of the complete length of all such welds in a designated lot of piping.

3. *Spot radiography* is the practice of making a single-exposure radiograph at a point within a specified extent of welding. Required coverage for a single-spot radiograph is as follows:

- For longitudinal welds, at least 150 mm (6 in) of weld length.
- For girth, miter, and branch welds in piping 2½ in NPS and smaller, a single elliptical exposure which encompasses the entire weld circumference, and in piping larger than 2½ in NPS, at least 25 percent of the inside circumference or 150 mm (6 in), whichever is less.

Radiography of components other than castings and of welds shall be in accordance with Art. 2, Sec. V of the ASME Code. Limitations on imperfections in components other than castings and welds shall be as stated in Table 10-51 for the degree of radiography involved.

Ultrasonic Examination Ultrasonic examination of welds shall be in accordance with Art. 4, Sec. V of the ASME Code, except that the modifications stated in Para. 336.6.1 of the code shall be substituted for T434.2.1 and T434.3. Refer to the code for additional requirements.

Type and Extent of Required Examination The intent of examinations is to provide the examiner and the inspector with reasonable

TABLE 10-50 Tabular Values for Minimum Temperatures Without Impact Testing for Carbon Steel Materials* *(Continued)*

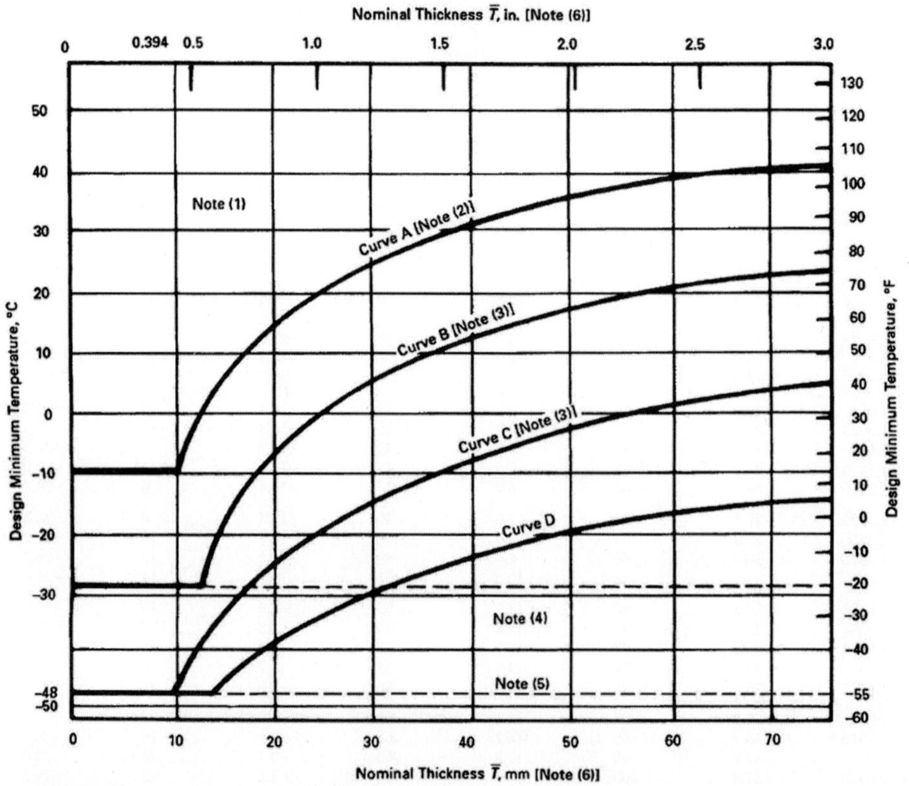

NOTES:

(1) Any carbon steel material may be used to a minimum temperature of −29°C (−20°F) for Category D Fluid Service.

(2) X Grades of API SL, and ASTM A 381 materials, may be used in accordance with Curve B if normalized or quenched and tempered.

(3) The following materials may be used in accordance with Curve D if normalized:

(*a*) ASTM A 516 Plate, all grades

(*b*) ASTM A 671 Pipe, Grades CE55, CE60, and all grades made with A 516 plate

(*c*) ASTM A 672 Pipe, Grades E55, E60, and all grades made with A 516 plate

(4) A welding procedure for the manufacture of pipe or components shall include impact testing of welds and HAZ for any design minimum temperature below −29°C (−20°F), except as provided in Table 10-49, A-3(b).

(5) Impact testing in accordance with para. 323.3† is required for any design minimum temperature below −48°C (−55°F), except as permitted by Note (3) in Table 10-49.‡

(6) For blind flanges and blanks, *T* shall be ¼ of the flange thickness.

*Table 10-50 and notes have been extracted (with minor modifications) from *Process Piping*, ASME B31.3-2014, with permission of the publisher, the American Society of Mechanical Engineers, New York. All rights reserved.

†Refer to the referenced code paragraph for impact testing methods and acceptance.

‡Refer to the referenced code paragraph for comments regarding circumstances under which impact testing can be excluded.

assurance that the requirements of the code and the engineering design have been met. For P-number 3, 4, and 5 materials, examination shall be performed after any heat treatment has been completed.

Examination Normally Required* Piping in normal fluid service shall be examined to the extent specified herein or to any greater extent specified in the engineering design. Acceptance criteria are as stated in the code for Normal Fluid Service unless more stringent requirements are specified.

1. *Visual examination.* At least the following shall be examined in accordance with code requirements:

a. Sufficient materials and components, selected at random, to satisfy the examiner that they conform to specifications and are free from defects.

b. At least 5 percent of fabrication. For welds, each welder's and welding operator's work shall be represented.

c. One hundred percent of fabrication for longitudinal welds, except those in components made in accordance with a listed specification. Longitudinal welds required to have a joint efficiency of 0.9 must be spot-radiographed to the extent of 300 mm (12 in) in each 30 m (100 ft) of weld for each welder or welding operator. Acceptance criteria shall comply with code radiography acceptance criteria for Normal Fluid Service.

*Extracted (with minor editing) from *Process Piping*, ASME B31.3-2014, paragraph 341, with permission of the publisher, the American Society of Mechanical Engineers, New York.

d. Random examination of the assembly of threaded, bolted, and other joints to satisfy the examiner that they conform to the applicable code requirements for erection and assembly. When pneumatic testing is to be performed, all threaded, bolted, and other mechanical joints shall be examined.

e. Random examination during erection of piping, including checking of alignment, supports, and cold spring.

f. Examination of erected piping for evidence of defects that would require repair or replacement, and for other evident deviations from the intent of the design.

2. *Other examination*

a. Not less than 5 percent of circumferential butt and miter groove welds shall be examined fully by random radiography or random ultrasonic examination in accordance with code requirements established for these methods. The welds to be examined shall be selected to ensure that the work product of each welder or welding operator doing the production welding is included. They shall also be selected to maximize coverage of intersections with longitudinal joints. When a circumferential weld with intersecting longitudinal weld(s) is examined, at least the adjacent 38 mm (1½ in) of each intersecting weld shall be examined. In-process examination in accordance with code requirements may be substituted for all or part of the radiographic or ultrasonic examination on a weld-for-weld basis if specified in the engineering design or specifically authorized by the Inspector.

b. Not less than 5 percent of all brazed joints shall be examined by in-process examination in accordance with the code definition of in-process

TABLE 10-51 Acceptance Criteria for Welds—Visual and Radiographic Examination

Criteria (A to M) for Types of Welds and for Service Conditions [Note (1)]											Examination Methods	
Normal and Category M Fluid Service			Severe Cyclic Conditions			Category D Fluid Service						
Type of Weld			Type of Weld			Type of Weld						
Girth, Miter Groove & Branch Connection [Note (2)]	Longitudinal Groove [Note (3)]	Fillet [Note (4)]	Girth, Miter Groove & Branch Connection [Note (2)]	Longitudinal Groove [Note (3)]	Fillet [Note (4)]	Girth and Miter Groove	Longitudinal Groove [Note (3)]	Fillet [Note (4)]	Branch Connection [Note (2)]	Weld Imperfection	Visual	Radiography
A	A	A	A	A	A	A	A	A	A	Crack	✓	✓
A	A	A	A	A	A	C	A	N/A	A	Lack of fusion	✓	✓
B	A	N/A	A	A	N/A	C	A	N/A	A	Incomplete penetration	✓	✓
E	E	N/A	D	D	N/A	N/A	N/A	N/A	N/A	Rounded Indications	. . .	✓
G	G	N/A	F	F	N/A	N/A	N/A	N/A	N/A	Elongated indications	. . .	✓
H	A	H	A	A	A	I	A	H	H	Undercutting	✓	✓
A	A	A	A	A	A	A	A	A	A	Surface porosity or exposed slag inclusion [Note (5)]	✓	. . .
N/A	N/A	N/A	J	J	J	N/A	N/A	N/A	N/A	Surface finish	✓	. . .
K	K	N/A	K	K	N/A	K	K	N/A	K	Concave surface, concave root, or burn-through	✓	✓
L	L	L	L	L	L	M	M	M	M	Weld reinforcement or internal protrusion	✓	. . .

examination, the joints to be examined being selected to ensure that the work of each brazer making the production joints is included.

3. *Certifications and records.* The examiner shall be assured, by examination of certifications, records, and other evidence, that the materials and components are of the specified grades and that they have received required heat treatment, examination, and testing. The examiner shall provide the Inspector with a certification that all the quality control requirements of the code and of the engineering design have been carried out.

Examination—Category D Fluid Service* Piping and piping elements for Category D fluid service as designated in the engineering design shall be visually examined in accordance with code requirements for visual examination to the extent necessary to satisfy the examiner that components, materials, and workmanship conform to the requirements of this code and the engineering design. Acceptance criteria shall be in accordance with code requirements and criteria in Table 10-51, for Category D fluid service, unless otherwise specified.

Examination—Severe Cyclic Conditions* Piping to be used under severe cyclic conditions shall be examined to the extent specified herein or to any greater extent specified in the engineering design. Acceptance criteria shall be in accordance with code requirements and criteria in Table 10-51, for severe cyclic conditions, unless otherwise specified.

1. *Visual examination.* The requirements for Normal Fluid Service apply with the following exceptions.

   a. All fabrication shall be examined.

   b. All threaded, bolted, and other joints shall be examined.

   c. All piping erection shall be examined to verify dimensions and alignment. Supports, guides, and points of cold spring shall be checked to ensure that movement of the piping under all conditions of startup, operation, and shutdown will be accommodated without undue binding or unanticipated constraint.

2. *Other examination.* All circumferential butt and miter groove welds and all fabricated branch connection welds comparable to those recognized by the code (see Fig. 10-119) shall be examined by 100 percent radiography or 100 percent ultrasonic (if specified in engineering design) in accordance with code requirements. Socket welds and branch connection welds which are not radiographed shall be examined by magnetic-particle or liquid-penetrant methods in accordance with code requirements.

3. In-process examination in accordance with the code definition, supplemented by appropriate nondestructive examination, may be substituted for the examination required in 2 above on a weld-for-weld basis if specified in the engineering design or specifically authorized by the Inspector.

4. *Certification and records.* The requirements established by the code for Normal Fluid Service apply.

Impact Testing In specifying materials, it is critical that the low-temperature limits of materials and impact testing requirements of the applicable code edition be clearly understood. In the recent past, code criteria governing low-temperature limits and requirements for impact testing have undergone extensive revision. The code contains extensive criteria detailing when impact testing is required and describing how it is to be performed. Because of the potentially changing requirements and the complexity of the code requirements, this text does not attempt to provide a comprehensive treatment of this subject or a comprehensive presentation of the requirements of the current code edition. Some of the general guidelines are provided here; however, the designer should consult the code to clearly understand additional requirements and special circumstances under which impact testing may be omitted. These exclusions permitted by the code may be significant in selecting materials or establishing material requirements.

In general, materials conforming to ASTM specifications listed by the code may be used at temperatures down to the lowest temperature listed for that material in ASME B31.3 Table A-1, and A-1M. When welding or other operations are performed on these materials, additional low-temperature toughness tests (impact testing) may be required. Refer to Table 10-51 for a general summary of these requirements.

Pressure Testing Prior to initial operation, installed piping shall be pressure-tested to ensure tightness except as permitted for Category D fluid service described later. The pressure test shall be maintained for a sufficient time to determine the presence of any leaks but not less than 10 min.

If repairs or additions are made following the pressure tests, the affected piping shall be retested except that, in the case of minor repairs or additions, the owner may waive retest requirements when precautionary measures are taken to ensure sound construction.

*Extracted (with minor editing) from *Process Piping,* ASME B31.3-2014, paragraph 341, with permission of the publisher, the American Society of Mechanical Engineers, New York.

TABLE 10-51 Acceptance Criteria for Welds—Visual and Radiographic Examination *(Continued)*

Criterion Value Notes for Table 10-51

Criterion		
Symbol	Measure	Acceptable Value Limits [Note (6)]
A	Extent of imperfection	Zero (no evident imperfection)
B	Cumulative length of incomplete penetration	≤38 mm (1.5 in.) in any 150 mm (6 in.) weld length or 25% of total weld length, whichever is less
C	Cumulative length of lack of fusion and incomplete penetration	≤38 mm (1.5 in.) in any 150 mm (6 in.) weld length or 25% of total weld length, whichever is less
D	Size and distribution of rounded indications	See BPV Code, Section VIII, Division 1, Appendix 4 [Note (10)]
E	Size and distribution of rounded indications	For $\bar{T}_w$ ≤ 6 mm (¼ in.), limit is same as D [Note (10)] For $\bar{T}_w$ > 6 mm (¼ in.), limit is 1.5 × D [Note (10)]
F	Elongated indications	
	Individual length	≤ $\bar{T}_w/3$
	Individual width	≤2.5 mm (3/32 in.) and ≤ $\bar{T}_w/3$
	Cumulative length	≤ $\bar{T}_w$ in any 12 $\bar{T}_w$ weld length [Note (10)]
G	Elongated indications	
	Individual length	≤2 $\bar{T}_w$
	Individual width	≤3 mm (⅛ in.) and ≤ $\bar{T}_w/2$
	Cumulative length	≤4 $\bar{T}_w$ in any 150 mm (6 in.) weld length [Note (10)]
H	Depth of undercut	≤1 mm (1/32 in.) and ≤ $\bar{T}_w/4$
	Cumulative length of internal and external undercut	≤38 mm (1.5 in.) in any 150 mm (6 in.) weld length or 25% of total weld length, whichever is less
I	Depth of undercut	≤1.5 mm (1/16 in.) and [≤ $\bar{T}_w/4$ or 1 mm (1/32 in.)]
	Cumulative length of internal and external undercut	≤38 mm (1.5 in.) in any 150 mm (6 in.) weld length or 25% of total weld length, whichever is less
J	Surface roughness	≤12.5 µm (500 µin.) R_a in accordance with ASME B46.1
K	Depth of surface concavity, root concavity, or burn-through	Total joint thickness, incl. weld reinf., ≥ $\bar{T}_w$ [Notes (7) and (11)]
L	Height of reinforcement or internal protrusion [Note (8)] in any plane through the weld shall be within limits of the applicable height value in the tabulation at right, except as provided in Note (9). Weld metal shall merge smoothly into the component surfaces.	For $\bar{T}_w$, mm (in.) — Height, mm (in.) ≤6 (¼) — ≤1.5 (1/16) >6 (¼), ≤13 (½) — ≤3 (⅛) >13 (½), ≤25 (1) — ≤4 (5/32) >25 (1) — ≤5 (3/16)
M	Height of reinforcement or internal protrusion [Note (8)] as described in L. Note (9) does not apply.	Limit is twice the value applicable for L above

GENERAL NOTES:

(a) Weld imperfections are evaluated by one or more of the types of examination methods given, as specified in paras. 341.4.1, 341.4.2, 341.4.3, and M341.4, or by the engineering design.

(b) "N/A" indicates the Code does not establish acceptance criteria or does not require evaluation of this kind of imperfection for this type of weld.

(c) Check (✓) indicates examination method generally used for evaluating this kind of weld imperfection.

(d) Ellipsis (. . .) indicates examination method not generally used for evaluating this kind of weld imperfection.

NOTES:

(1) Criteria given are for required examination. More stringent criteria may be specified in the engineering design. Other methods of examination may be specified by engineering design to supplement the examination required by the code. The extent of supplementary examination and any acceptance criteria differing from those specified by the code shall be specified by engineering design. Any examination method recognized by the code may be used to resolve doubtful indications. Acceptance criteria shall be those established by the code for the required examination.

(2) Longitudinal groove weld includes straight and spiral seam. Criteria are not intended to apply to welds made in accordance with component standards recognized by the code (ref. ASME.B31.3 Table A1 and Table 326.1); however, alternative leak test requirements dictate that all component welds be examined (see code for specific requirements).

(3) Fillet weld includes socket and seal welds, and attachment welds for slip-on flanges, branch reinforcement, and supports.

(4) Branch connection weld includes pressure containing welds in branches and fabricated laps.

(5) These imperfections are evaluated only for welds ≤5 mm (3/16 in) in nominal thickness.

(6) Where two limiting values are separated by *and* the lesser of the values determines acceptance. Where two sets of values are separated by *or* the larger value is acceptable. $\bar{T}_w$ is the nominal wall thickness of the thinner of two components joined by a butt weld.

(7) Tightly butted unfused root faces are unacceptable.

(8) For groove welds, height is the lesser of the measurements made from the surfaces of the adjacent components; both reinforcement and internal protrusion are permitted in a weld. For fillet welds, height is measured from the theoretical throat defined by the code; internal protrusion does not apply.

(9) For welds in aluminum alloy only, internal protrusion shall not exceed the following values:
(a) for thickness ≤2 mm (5/64 in): 1.5 mm (1/16 in)
(b) for thickness >2 mm and ≤6 mm (¼ in): 2.5 mm (3/32 in)
For external reinforcement and for greater thicknesses, see the tabulation for Symbol L.

*Table 10-51 and notes have been extracted (with minor modifications) from Section 341 of *Process Piping*, ASME B31.3–2004, with permission of the publisher, the American Society of Mechanical Engineers, New York. All rights reserved.

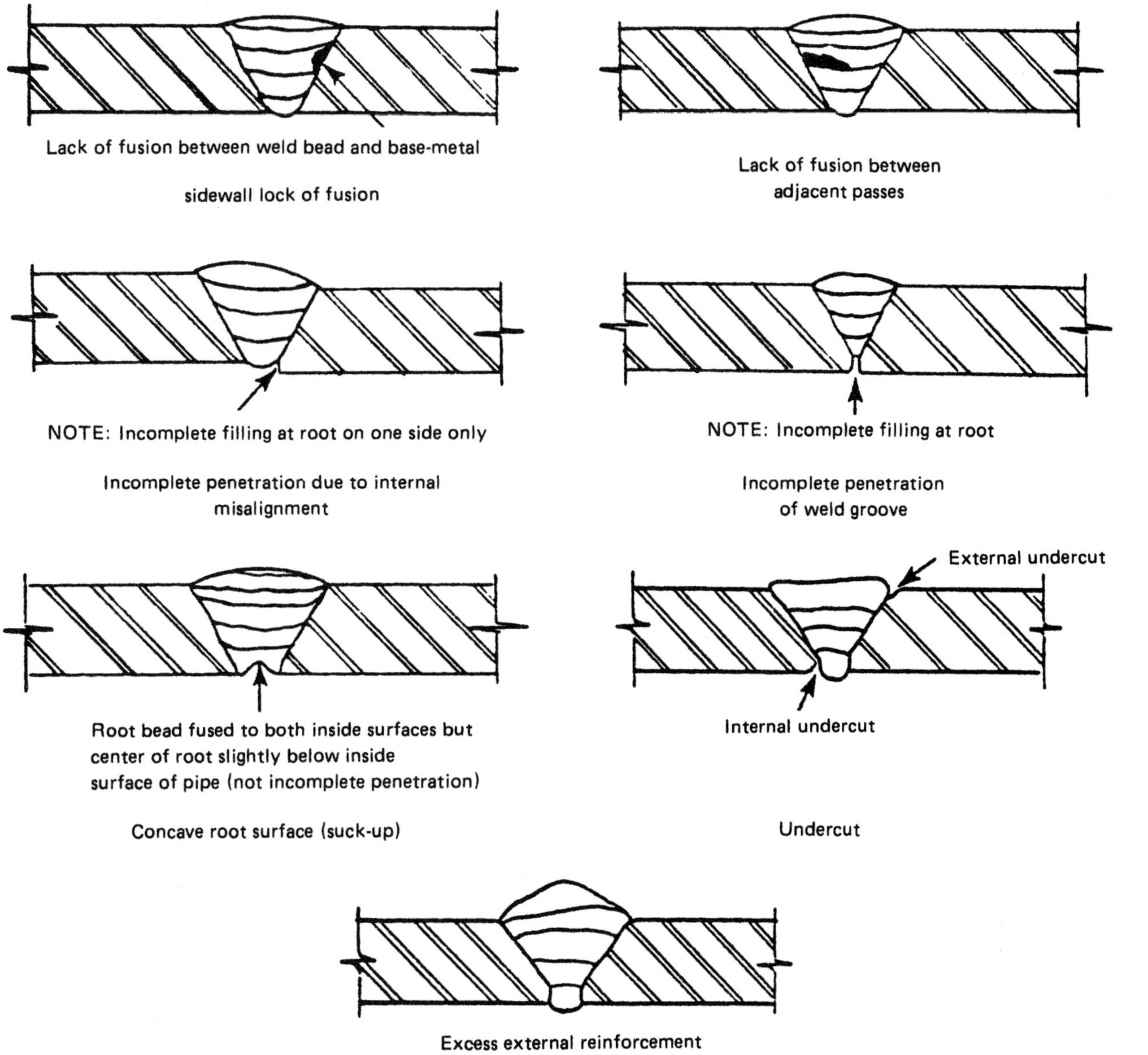

FIG. 10-169 Typical weld imperfections. (*Extracted from Process Piping, ASME B31.3-2014, with permission of the publisher, the American Society of Mechanical Engineers, New York. All rights reserved.*)

When pressure tests are conducted at metal temperatures near the ductile-to-brittle transition temperature of the material, the possibility of brittle fracture shall be considered.

The test shall be hydrostatic, using water, with the following exceptions. If there is a possibility of damage due to freezing or if the operating fluid or piping material would be adversely affected by water, any other suitable nontoxic liquid may be used. If a flammable liquid is used, its flash point shall not be less than 50°C (120°F), and consideration shall be given to the test environment.

The hydrostatic-test pressure at any point in the system shall be as follows:

1. Not less than 1½ times the design pressure.
2. For a design temperature above the test temperature, the minimum test pressure shall be as calculated by the following formula:

$$P_T = 1.5PS_T/S \qquad (10\text{-}106)$$

where P_T = test hydrostatic gauge pressure, MPa (lbf/in^2)
P = internal design pressure, MPa (lbf/in^2)
S_T = allowable stress at test temperature, MPa (lbf/in^2)
S = allowable stress at design temperature, MPa (lbf/in^2)

If the test pressure as so defined would produce a stress in excess of the yield strength at test temperature, the test pressure may be reduced to the maximum pressure that will not exceed the yield strength at test temperature. If the test liquid in the system is subject to thermal expansion, precautions shall be taken to avoid excessive pressure.

A preliminary air test at not more than 0.17 MPa (25 lbf/in^2) gauge pressure may be made prior to hydrostatic test in order to locate major leaks.

If hydrostatic testing is not considered practicable by the owner, a pneumatic test in accordance with the following procedure may be substituted, using air or another nonflammable gas.

If the piping is tested pneumatically, the test pressure shall be 110 percent of the design pressure. When pneumatically testing nonmetallic materials, ensure that the materials are suitable for compressed gas. Pneumatic testing involves a hazard due to the possible release of energy stored in compressed gas. Therefore, particular care must be taken to minimize the chance of brittle failure of metals and thermoplastics. The test temperature is important in this regard and must be considered when material is chosen in the original design. Any pneumatic test shall include a preliminary check at not more than 0.17 MPa (25 lbf/in^2) gauge pressure. The pressure shall be increased gradually in steps providing sufficient time to allow the piping to equalize strains during

TABLE 10-52 Preheat Temperatures

Base metal P-No. [Note (1)]	Base Metal Group	Greater material thickness mm	Greater material thickness in.	Additional limits [Note (2)]	Required minimum temperature °C	Required minimum temperature °F
1	Carbon steel	≤25	≤1	%C > 0.30 [Note (3)]	10	50
		>25	>1	%C ≤ 0.30 [Note (3)]	10	50
		>25	>1	%C > 0.30 [Note (3)]	95	200
3	Alloy steel, Cr ≤ ½%	≤13	≤½	SMTS ≤ 450 MPa (65 ksi)	10	50
		>13	>½	SMTS ≤ 450 MPa (65 ksi)	95	200
		All	All	SMTS > 450 MPa (65 ksi)	95	200
4	Alloy steel, ½% < Cr ≤ 2%	All	All	None	120	250
5A	Alloy steel	All	All	SMTS ≤ 414 MPa (60 ksi)	150	300
		All	All	SMTS > 414 MPa (60 ksi)	200	400
5B	Alloy steel	All	All	SMTS ≤ 414 MPa (60 ksi)	150	300
		All	All	SMTS > 414 MPa (60 ksi)	200	400
		≤13	≤½	%Cr > 6.0 [Note (3)]	200	400
6	Martensitic stainless steel	All	All	None	200 [Note (4)]	400 [Note (4)]
9A	Nickel alloy steel	All	All	None	120	250
9B	Nickel alloy steel	All	All	None	150	300
10I	27Cr steel	All	All	None	150 [Note (5)]	300 [Note (5)]
15E	9Cr–1Mo–V CSEF steel	All	All	None	200	400
. . .	All other materials	. . .	. . .	None	10	50

NOTES:
(1) P-Nos. and Group Nos. from BPV Code, Section IX, QW/QB-422. Reprinted from ASME Boiler and Pressure Vessel Code, Section IX by permission of the American Society of Mechanical Engineers, New York. All rights reserved.
(2) SMTS = Specified Minimum Tensile Strength.
(3) Composition may be based on ladle or product analysis or in accordance with specification limits.
(4) Maximum interpass temperature 315°C (600°F).
(5) Maintain interpass temperature between 150°C and 230°C (300°F and 450°F).
Extracted from Process Piping, ASME B31.3-2014 with permission of the publisher, the American Society of Mechanical Engineers, New York. All rights reserved.

TABLE 10-53 Postweld Heat Treatment

P-No. and Group No. (BPV Code Section IX, QW/QB-420)	Holding Temperature Range, °C (°F) [Note (1)]	Minimum Holding Time at Temperature for Control Thickness [Note (2)]: Up to 50 mm (2 in.)	Minimum Holding Time at Temperature for Control Thickness [Note (2)]: Over 50 mm (2 in.)
P-No. 1, Group Nos. 1–3	595 to 650 (1100 to 1200)	1 h/25 mm (1 hr/in.); 15 min min.	2 hr plus 15 min for each additional 25 mm (in.) over 50 mm (2 in.)
P-No. 3, Group Nos. 1 and 2	595 to 650 (1100 to 1200)		
P-No. 4, Group Nos. 1 and 2	650 to 705 (1200 to 1300)		
P-No. 5A, Group No. 1	675 to 760 (1250 to 1400)		
P-No. 5B, Group No. 1	675 to 760 (1250 to 1400)		
P-No. 6, Group Nos. 1–3	760 to 800 (1400 to 1475)		
P-No. 7, Group Nos. 1 and 2 [Note (3)]	730 to 775 (1350 to 1425)		
P-No. 8, Group Nos. 1–4	PWHT not required unless required by WPS		
P-No. 9A, Group No. 1	595 to 650 (1100 to 1200)		
P-No. 9B, Group No. 1	595 t o 650 (1100 to 1200)		
P-No. 10H, Group No. 1	PWHT not required unless required by WPS. If done, see Note (4).		
P-No. 10I, Group No. 1 [Note (3)]	730 to 815 (1350 to 1500)		
P-No. 11A	550 to 585 (1025 to 1085) [Note (5)]		
P-No. 15E, Group No. 1	705 to 775 (1300 to 1425) [Notes (6) and (7)]	1 h/25 mm (1 hr/in.); 30 min min.	1 h/25 mm (1 hr/in.) up to 125 mm (5 in.) plus 15 min for each additional 25 mm (in.) over 125 mm (5 in.)
P-No. 62	540 to 595 (1000 to 1100)	. . .	See Note (8)
All other materials	PWHT as required by WPS	In accordance with WPS	In accordance with WPS

GENERAL NOTE: The exemptions for mandatory PWHT are defined in ASME B31.1-2016, Table 331.1.3. Reprinted from ASME B31.1-2016 by permission of the American Society of Mechanical Engineers, New York. All rights reserved.
NOTES:
(1) The holding temperature range is further defined in para. 331.1.6(c), ASME B31.1-2016, and in Table 331.1.2.
(2) The control thickness is defined in para. 331.1.3.
(3) Cooling rate shall not be greater than 55°C (100°F) per hour in the range above 650°C (1200°F), after which the cooling rate shall be sufficiently rapid to prevent embrittlement.
(4) If PWHT is performed after welding, it shall be within the following temperature ranges for the specific alloy, followed by rapid cooling:
Alloys S31803 and S32205 — 1020°C to 1100°C (1870°F to 2010°F)
Alloy S32550 — 1040°C to 1120°C (1900°F to 2050°F)
Alloy S32750 — 1025°C to 1125°C (1880°F to 2060°F)
All others — 980°C to 1040°C (1800°F to 1900°F)
(5) Cooling rate shall be >165°C (300°F)/h to 315°C (600°F)/h.
(6) The minimum PWHT holding temperature may be 675°C (1250°F) for nominal material thicknesses [see para. 331.1.3(c)] ≤13 mm (½ in.).
(7) The Ni + Mn content of the filler metal shall not exceed 1.2% unless specified by the designer, in which case the maximum temperature to be reached during PWHT shall be the A_1 (lower transformation or lower critical temperature) of the filler metal, as determined by analysis and calculation or by test, but not exceeding 800°C (1470°F). If the 800°C (1470°F) limit was not exceeded but the A_1 of the filler metal was exceeded or if the composition of the filler metal is unknown, the weld must be removed and replaced. It shall then be rewelded with compliant filler metal and subjected to a compliant PWHT. If the 800°C (1470°F) limit was exceeded, the weld and the entire area affected by the PWHT will be removed and, if reused, shall be renormalized and tempered prior to reinstallation.
(8) Heat treat within 14 days after welding. Hold time shall be increased by 1.2 h for each 25 mm (1 in.) over 25 mm (1 in.) thickness.
Cool to 425°C (800°F) at a rate ≤280°C (500°F).

TABLE 10-54 Types of Examination for Evaluating Imperfections*

Type of imperfection	Type of examination			
	Visual	Liquid-penetrant or magnetic-particle	Ultrasonic or radiographic: Random	Ultrasonic or radiographic: 100%
Crack	X	X	X	X
Incomplete penetration	X		X	X
Lack of fusion	X		X	X
Weld undercutting	X			
Weld reinforcement	X			
Internal porosity			X	X
External porosity	X			
Internal slag inclusions			X	X
External slag inclusions	X			
Concave root surface	X		X	X

*Extracted from the *Chemical Plant and Petroleum Refinery Piping Code*, ANSI B31.3-1980, with permission of the publisher, the American Society of Mechanical Engineers, New York. All rights reserved. For code acceptance criteria (limits on imperfections) for welds see Table 10-51.

test and to check for leaks. Once test pressure has been achieved, the pressure shall be reduced to design pressure prior to examining for leakage.

At the owner's option, a piping system used only for Category D fluid service as defined in the subsection Classification of Fluid Service may be tested at the normal operating conditions of the system during or prior to initial operation by examining for leaks at every joint not previously tested. A preliminary check shall be made at not more than 0.17 MPa (25 lbf/in^2) gauge pressure when the contained fluid is a gas or a vapor. The pressure shall be increased gradually in steps providing sufficient time to allow the piping to equalize strains during testing and to check for leaks.

Tests alternative to those required by these provisions may be applied under certain conditions described in the code.

Piping required to have a sensitive leak test shall be tested by the gas and bubble formation testing method specified in Art. 10, App. I, Sec. V of the ASME Boiler and Pressure Vessel Code or by another method demonstrated to have equal or greater sensitivity. The sensitivity of the test shall be at least $(100\ Pa \cdot mL)/s\ [(10^{-3}\ atm \cdot mL)/s]$ under test conditions.

Records shall be kept of each piping installation during the testing.

COST COMPARISON OF PIPING SYSTEMS

Piping may represent as much as 25 percent of the cost of a chemical-process plant. The installed cost of piping systems varies widely with the materials of construction and the complexity of the system. A study of piping costs shows that the most economical choice of material for a simple straight piping run may not be the most economical for a complex installation made up of many short runs involving numerous fittings and valves. The economics also depend heavily on the pipe size and fabrication techniques employed. Fabrication methods such as bending to standard long-radius-elbow dimensions and machine-flaring lap joints have a large effect on the cost of fabricating pipe from ductile materials suited to these techniques. Cost reductions of as high as 35 percent are quoted by some custom fabricators utilizing advanced techniques; however, the basis for pricing comparisons should be carefully reviewed.

Figure 10-170 is based on data extracted from a comparison of the installed cost of piping systems of various materials published by Dow Chemical Co. The chart shows the relative cost ratios for systems of various materials based on two installations, one consisting of 152 m (500 ft) of 2-in pipe in a complex piping arrangement and the other of 305 m (1000 ft) of 2-in pipe in a straight-run piping arrangement. Figure 10-170 is based on field-fabrication construction techniques using welding stubs, the method commonly used by contractors. A considerably different ranking would result from using other construction methods such as machine-formed lap joints and bends in place of welding elbows. Piping cost experience shows that it is difficult to generalize and reflect accurate piping cost comparisons. Although the prices for many of the metallurgies shown in Fig. 10-170 are very volatile even over short periods, Fig. 10-170 may still be used as a reasonable initial estimate of the relative cost of metallic materials. The cost of nonmetallic materials and lined metallic materials versus solid alloy materials should be carefully reviewed prior to material selection. For an accurate comparison the cost for each type of material must be estimated individually on the basis of the actual fabrication and installation methods that will be used, pipe sizes, and the conditions anticipated for the proposed installation.

FORCES OF PIPING ON PROCESS MACHINERY AND PIPING VIBRATION

The reliability of process rotating machinery is affected by the quality of the process piping installation. Excessive external forces and moments upset casing alignment and can reduce clearance between motor and casing. Further, the bearings, seals, and coupling can be adversely affected, resulting in repeated failures that may be correctly diagnosed as misalignment, and may have excessive piping forces as the root causes. Most turbine and compressor manufacturers have prescribed specification or will follow NEMA standards for allowable nozzle loading. For most of the pumps, API or ANSI pump standards will be followed when evaluating the pump nozzle loads. Pipe support restraints need to be placed at the proper locations to protect the machinery nozzles during operation.

Prior to any machinery alignment procedure, it is imperative to check for machine pipe strain. This is accomplished by the placement of dial indicators on the shaft and then loosening the hold-down bolts. Movements of greater than 1 mil are considered indication of a pipe strain condition.

This is an important practical problem area, as piping vibration can cause considerable downtime or even pipe failure.

Pipe vibration is caused by

1. Internal flow (pulsation)
2. Plant machinery (such as compressors, pumps)

Pulsation can be problematic and difficult to predict. Pulsations are also dependent on acoustic resonance characteristics. For reciprocating equipment, such as reciprocating compressors and pumps, in some cases, an analog (digital) study needs to be performed to identify the deficiency in the piping and pipe support systems as well as to evaluate the performance of the machine during operation. The study will also provide recommendations on how to improve the machine and piping system's performance.

When a pulsation frequency coincides with a mechanical or acoustic resonance, severe vibration can result. A common cause for pulsation is the presence of flow control valves or pressure regulators. These often operate with high pressure drops (i.e., high flow velocities), which can result in the generation of severe pulsation. Flashing and cavitation can also contribute.

Modern-day piping design codes can model the vibration situation, and problems can thus be resolved in the design phases.

HEAT TRACING OF PIPING SYSTEMS

Heat tracing is used to maintain pipes and the material that pipes contain at temperatures above the ambient temperature. Two common uses of heat tracing are to prevent water pipes from freezing and maintain fuel oil pipes at high enough temperatures that the viscosity of the fuel oil will allow easy pumping. Heat tracing is also used to prevent the condensation of a liquid from a gas and to prevent the solidification of a liquid commodity.

A heat-tracing system is often more expensive on an installed cost basis than the piping system it is protecting, and it will also have significant operating costs. A study of heat-tracing costs by a major chemical company showed installed costs of \$31/ft to \$142/ft and yearly operating costs of \$1.40/ft to \$16.66/ft. In addition to being a major cost, the heat-tracing system is an important component of the reliability of a piping system. A failure in the heat-tracing system will often render the piping system inoperable. For example, with a water freeze protection system, the piping system may be destroyed by the expansion of water as it freezes if the heat-tracing system fails.

The vast majority of heat-traced pipes are insulated to minimize heat loss to the environment. A heat input of 2 to 10 W/ft is generally required to prevent an insulated pipe from freezing. With high wind speeds, an uninsulated pipe could require well over 100 W/ft to prevent freezing. Such a high heat input would be very expensive.

Heat tracing for insulated pipes is generally only required for the period when the material in the pipe is not flowing. The heat loss of an insulated pipe is very small compared to the heat capacity of a flowing fluid. Unless the pipe is extremely long (several thousands of feet), the temperature drop of a flowing fluid will not be significant.

There are three major methods of avoiding heat tracing:

1. Change the ambient temperature around the pipe to a temperature that will avoid low-temperature problems. Burying water pipes below the frost line or running them through a heated building are the two most common examples of this method.

2. Empty a pipe after it is used. Arranging the piping such that it drains itself when not in use can be an effective method of avoiding the need for heat tracing. Some infrequently used lines can be pigged or blown out with compressed air. This technique is not recommended for commonly used lines due to the high labor requirement.

3. Arrange a process such that some lines have continuous flow; this can eliminate the need for tracing these lines. This technique is generally not recommended because a failure that causes a flow stoppage can lead to blocked or broken pipes.

Some combination of these techniques may be used to minimize the quantity of traced pipes. However, the majority of pipes containing fluids that must be kept above the minimum ambient temperature are generally going to require heat tracing.

Types of Heat-Tracing Systems Industrial heat-tracing systems are generally fluid systems or electrical systems. In fluid systems, a pipe or tube called the *tracer* is attached to the pipe being traced, and a warm fluid is

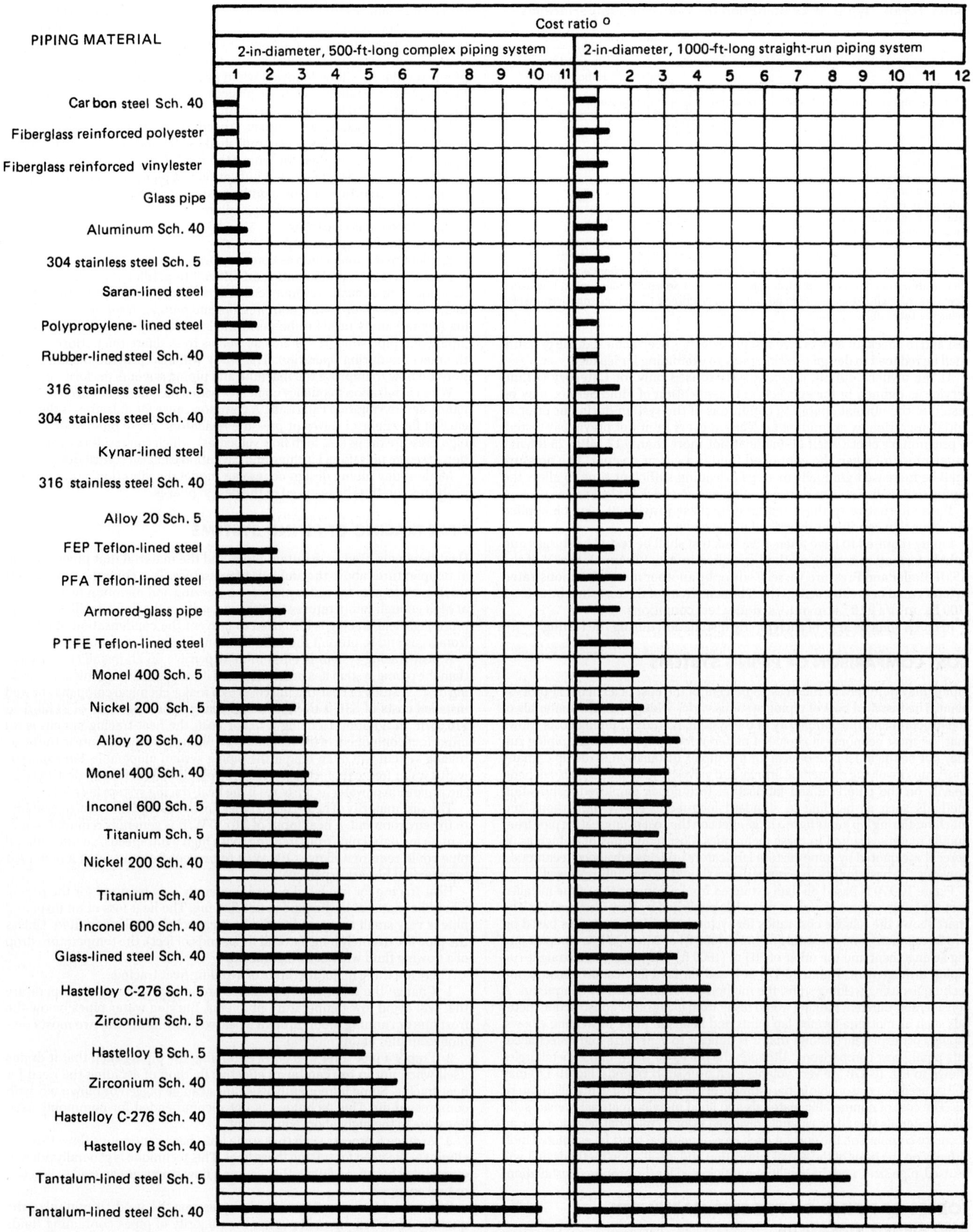

FIG. 10-170 Cost rankings and cost ratios for various piping materials. This figure is based on field-fabrication construction techniques using welding stubs, as this is the method most often employed by contractors. A considerably different ranking would result from using other construction methods, such as machined-formed lap joints, for the alloy pipe. °Cost ratio = (cost of listed item)/(cost of Schedule 40 carbon-steel piping system, field-fabricated by using welding stubs). (*Extracted with permission from* Installed Cost of Corrosion Resistant Piping, *copyright 1977, Dow Chemical Co.*)

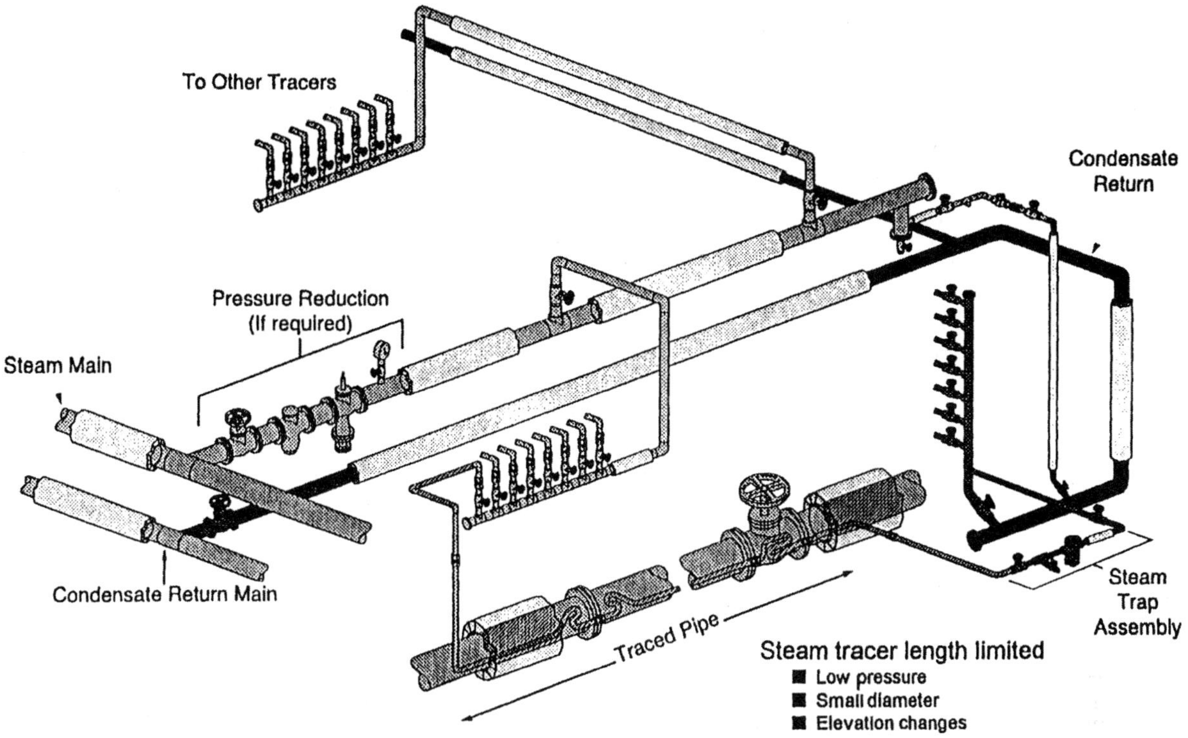

FIG. 10-171 Steam tracing system.

put through it. The tracer is placed under the insulation. Steam is by far the most common fluid used in the tracer, although ethylene glycol and more exotic heat-transfer fluids are used. In electrical systems, an electrical heating cable is placed against the pipe under the insulation.

Fluid Tracing Systems Steam tracing is the most common type of industrial pipe tracing. In 1960, over 95 percent of industrial tracing systems were steam-traced. By 1995, improvements in electric heating technology increased the electric share to 30 to 40 percent. Fluid systems other than steam are rather uncommon and account for less than 5 percent of tracing systems.

Half-inch copper tubing is commonly used for steam tracing. Three-eighths-inch tubing is also used, but the effective circuit length is then decreased from 150 ft to about 60 ft. In some corrosive environments, stainless steel tubing is used.

In addition to the tracer, a steam tracing system (Fig. 10-171) consists of steam supply lines to transport steam from the existing steam lines to the traced pipe, a steam trap to remove the condensate and hold back the steam, and in most cases a condensate return system to return the condensate to the existing condensate return system. In the past, a significant percentage of condensate from steam tracing was simply dumped to drains, but increased energy costs and environmental rules have caused almost all condensate from new steam tracing systems to be returned. This has significantly increased the initial cost of steam tracing systems.

Applications requiring accurate temperature control are generally limited to electric tracing. For example, chocolate lines cannot be exposed to steam temperatures, or the product will degrade; and if caustic soda is heated above 65°C (150°F), it becomes extremely corrosive to carbon-steel pipes.

For some applications, either steam or electricity is simply not available, and this makes the decision. It is rarely economic to install a steam boiler just for tracing. Steam tracing is generally considered only when a boiler already exists or is going to be installed for some other primary purpose. Additional electric capacity can be provided in most situations for reasonable costs. It is considerably more expensive to supply steam from a long distance than it is to provide electricity. Unless steam is available close to the pipes being traced, the automatic choice is usually electric tracing.

For most applications, particularly in processing plants, either steam tracing or electric tracing could be used, and the correct choice is dependent on the installed costs and the operating costs of the competing systems.

Economics of Steam Tracing versus Electric Tracing The question of the economics of various tracing systems has been examined thoroughly. All these papers have concluded that electric tracing is generally less expensive to install and significantly less expensive to operate. Electric tracing has significant cost advantages in terms of installation because less labor is required than for steam tracing. However, it is clear that there are some special cases where steam tracing is more economical.

The two key variables in the decision to use steam tracing or electric tracing are the temperature at which the pipe must be maintained and the distance to the supply of steam and a source of electric power.

Table 10-55 shows the installed costs and operating costs for 400 ft of 4-in pipe, maintained at four different temperatures, with supply lengths of 100 ft for both electricity and steam and $25/h labor.

These are the major advantages of a steam tracing system:

1. *High heat output.* Due to its high temperature, a steam tracing system provides a large amount of heat to the pipe. There is a very high heat-transfer rate between the metallic tracer and a metallic pipe. Even with damage to the insulation system, there is very little chance of a low-temperature failure with a steam tracing system.

2. *High reliability.* Many things can go wrong with a steam tracing system, but very few of the potential problems lead to a heat-tracing failure. Steam traps fail, but they usually fail in the open position, allowing for a continuous flow of steam to the tracer. Other problems such as steam leaks that can cause wet insulation are generally prevented from becoming heat-tracing failures by the extremely high heat output of a steam tracer. Also, a tracing tube is capable of withstanding a large amount of mechanical abuse without failure.

3. *Safety.* While steam burns are fairly common, there are generally fewer safety concerns than with electric tracing.

4. *Common usage.* Steam tracing has been around for many years, and many operators are familiar with the system. Because of this familiarity, failures due to operator error are not very common.

These are the weaknesses of a steam tracing system:

1. *High installed costs.* The incremental piping required for the steam supply system and the condensate return system must be installed and insulated, and in the case of the supply system, additional steam traps are often required.

TABLE 10-55 Steam versus Electric Tracing*

Temperature maintained	TIC		Ratio S/E	TOC		Ratio S/E
	Steam	Electric		Steam	Electric	
50°F	22,265	7,733	2.88	1,671	334	5.00
150°F	22,265	13,113	1.70	4,356	1,892	2.30
250°F	22,807	17,624	1.29	5,348	2,114	2.53
400°F	26,924	14,056	1.92	6,724	3,942	1.71

*Specifications: 400 ft of 44-in pipe, $25/h labor, $0.07/kWh, $4.00/1000 lb steam, 100-ft supply lines. TIC = total installed cost; TOC = total operating costs.

The tracer itself is not expensive, but the labor required for installation is relatively high. Studies have shown that steam tracing systems typically cost from 50 to 150 percent more than a comparable electric tracing system.

2. *Energy inefficiency.* A steam tracing system's total energy use is often more than 20 times the actual energy requirement to keep the pipe at the desired temperature. The steam tracer itself puts out significantly greater energy than required. The steam traps use energy even when they are properly operating and waste large amounts of energy when they fail in the open position, which is the common failure mode. Steam leaks waste large amounts of energy, and both the steam supply system and the condensate return system use significant amounts of energy.

3. *Poor temperature control.* A steam tracing system offers very little temperature control capability. The steam is at a constant temperature (50 psig steam is 300°F) usually well above that desired for the pipe. The pipe will reach an equilibrium temperature somewhere between the steam temperature and the ambient temperature. However, the section of pipe against the steam tracer will effectively be at the steam temperature. This is a serious problem for temperature-sensitive fluids such as food products. It also represents a problem with fluids such as bases and acids, which are not damaged by high temperatures but often become extremely corrosive to piping systems at higher temperatures.

4. *High maintenance costs.* Leaks must be repaired and steam traps must be checked and replaced if they have failed. Numerous studies have shown that, due to the energy lost through leaks and failed steam traps, an extensive maintenance program is an excellent investment. Steam maintenance costs are so high that for low-temperature maintenance applications, total steam operating costs are sometimes greater than electric operating costs, even if no value is placed on the steam.

Electric Tracing An electric tracing system (see Fig. 10-172) consists of an electric heater placed against the pipe under the thermal insulation, the supply of electricity to the tracer, and any control or monitoring system that may be used (optional). The supply of electricity to the tracer usually consists of an electrical panel and electrical conduit or cable trays. Depending on the size of the tracing system and the capacity of the existing electrical system, an additional transformer may be required.

Advantages of Electric Tracing

1. *Lower installed and operating costs.* Most studies have shown that electric tracing is less expensive to install and less expensive to operate. This is true for most applications. However, for some applications, the installed costs of steam tracing are equal to or less than those of electric tracing.

2. *Reliability.* In the past, electric heat tracing had a well-deserved reputation for poor reliability. However, since the introduction of self-regulating heaters in 1971, the reliability of electric heat tracing has improved dramatically. Self-regulating heaters cannot destroy themselves with their own heat output. This eliminates the most common failure mode of polymer-insulated constant-wattage heaters. Also, the technology used to manufacture mineral-insulated cables, high-temperature electric heat tracing, has improved significantly, and this has improved their reliability.

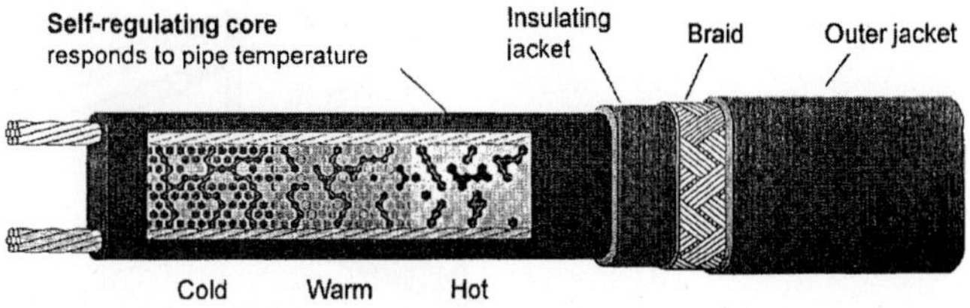

FIG. 10-173 Self-regulating heating cable.

3. *Temperature control.* Even without a thermostat or any control system, an electric tracing system usually provides better temperature control than a steam tracing system. With thermostatic or electronic control, very accurate temperature control can be achieved.

4. *Safety.* The use of self-regulating heaters and ground leakage circuit breakers has answered the safety concerns of most engineers considering electric tracing. Self-regulating heaters eliminate the problems from high-temperature failures, and ground leakage circuit breakers minimize the danger of an electrical fault to ground, causing injury or death.

5. *Monitoring capability.* One question often asked about any heat-tracing system is, "How do I know it is working?" Electric tracing now has available almost any level of monitoring desired. The temperature at any point can be monitored with both high and low alarm capability. This capability has allowed many users to switch to electric tracing with a high degree of confidence.

6. *Energy efficiency.* Electric heat tracing can accurately provide the energy required for each application without the large additional energy use of a steam system. Unlike steam tracing systems, other parts of the system do not use significant amounts of energy.

Disadvantages of Electric Tracing

1. *Poor reputation.* In the past, electric tracing has been less than reliable. Due to past failures, some operating personnel are unwilling to take a chance on any electric tracing.

2. *Design requirements.* A slightly higher level of design expertise is required for electric tracing than for steam tracing.

3. *Lower power output.* Since electric tracing does not provide a large multiple of the required energy, it is less forgiving to problems such as damaged insulation or below design ambient temperatures. Most designers include a 10 to 20 percent safety factor in the heat loss calculation to cover these potential problems. Also, a somewhat higher than required design temperature is often specified to provide an additional safety margin. For example, many water systems are designed to maintain 50°F to prevent freezing.

Types of Electric Tracing *Self-regulating electric tracing* (see Fig. 10-173) is by far the most popular type of electric tracing. The heating element in

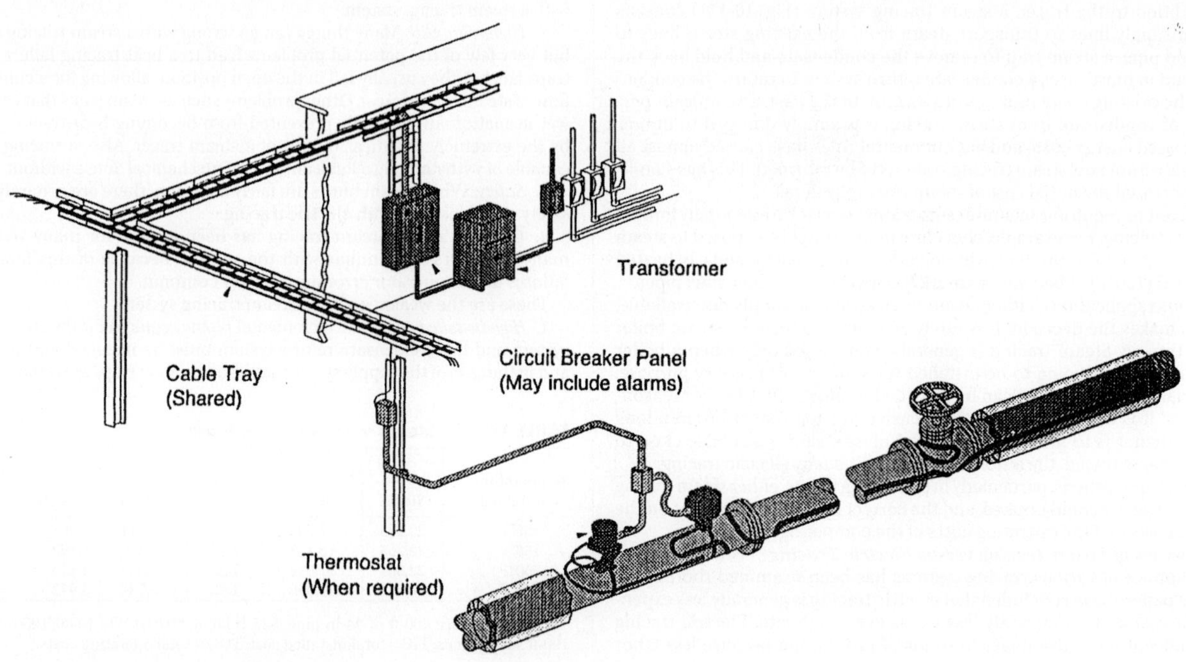

FIG. 10-172 Electrical beat tracing system.

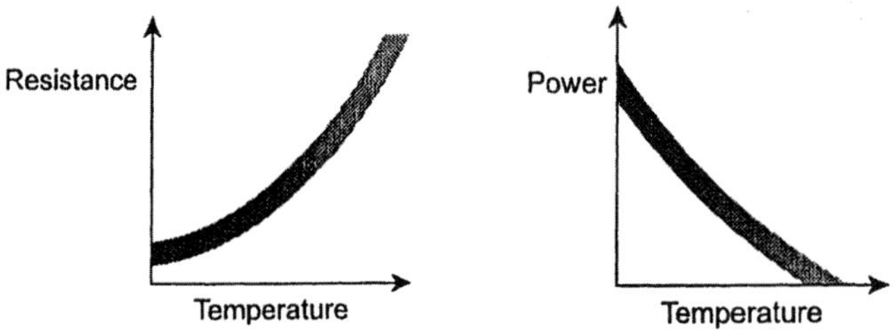

FIG. 10-174 Self-regulation.

a self-regulating heater is a conductive polymer between the bus wires. This conductive polymer increases its resistance as its temperature increases. The increase in resistance with temperature causes the heater to lower its heat output at any point where its temperature increases (Fig. 10-174). This self-regulating effect eliminates the most common failure mode of constant-wattage electric heaters, which is destruction of the heater by its own heat output.

Because self-regulating heaters are parallel heaters, they may be cut to length at any point without changing their power output per unit of length. This makes them much easier to deal with in the field. They may be terminated, teed, or spliced in the field with hazardous-area-approved components.

MI Cables Mineral insulated cables (Fig. 10-175) are the electric heat tracers of choice for high-temperature applications. High-temperature applications are generally considered to maintain temperatures above 250°F or exposure temperatures above 420°F where self-regulating heaters cannot be used. MI cable consists of one or two heating wires, magnesium oxide insulation (whence it gets its name) and an outer metal sheath. Today the metal sheath is generally Inconel. This eliminates both the corrosion problems with copper sheaths and the stress cracking problems with stainless steel.

MI cables can maintain temperatures up to 1200°F and withstand exposure up to 1500°F. The major disadvantage of MI cable is that it must be factory-fabricated to length. It is very difficult to terminate or splice the heater in the field. This means pipe measurements are necessary before the heaters are ordered. Also, any damage to an MI cable generally requires a complete new heater. It's not as easy to splice in a good section as with self-regulating heaters.

Polymer-Insulated Constant-Wattage Electric Heaters These are slightly cheaper than self-regulating heaters, but they are generally being replaced with self-regulating heaters due to inferior reliability. These heaters tend to destroy themselves with their own heat output when they are overlapped at valves or flanges. Since overlapping self-regulating heaters is the standard installation technique, it is difficult to prevent this technique from being used on the similar-looking constant-wattage heaters.

TABLE 10-56 Effect of Supply Lengths

Ratio of steam TIC to electric TIC maintained at 150°F

	Electric supply length		
Steam supply length, ft	40 ft	100 ft	300 ft
40	1.1	1.0	0.7
100	1.9	1.7	1.1
300	4.9	4.2	2.9

SECT (Skin-Effect Current Tracing) This is a special type of electric tracing employing a tracing pipe, usually welded to the pipe being traced, that is used for extremely long lines. With SECT circuits, up to 10 mi can be powered from one power point. All SECT systems are specially designed by heat-tracing vendors.

Impedance Tracing This uses the pipe being traced to carry the current and generate the heat. Less than 1 percent of electric heat-tracing systems use this method. Low voltages and special electrical isolation techniques are used. Impedance heating is useful when extremely high heat densities are required, such as when a pipe containing aluminum metal must be melted from room temperature on a regular basis. Most impedance systems are specially designed by heat-tracing vendors.

Choosing the Best Tracing System Some applications require either steam tracing or electric tracing regardless of the relative economics. For example, a large line that is regularly allowed to cool and needs to be quickly heated would require steam tracing because of its much higher heat output capability. In most heat-up applications, steam tracing is used with heat-transfer cement, and the heat output is increased by a factor of up to 10. This is much more heat than would be practical to provide with electric tracing. For example, a ½-in copper tube containing 50 psig steam with heat-transfer cement would provide over 1100 Btu/(h·ft) to a pipe at 50°F. This is over 300 W/ft or more than 15 times the output of a high-powered electric tracer.

Table 10-55 shows that electric tracing has a large advantage in terms of cost at low temperatures and smaller but still significant advantages at higher temperatures. Steam tracing does relatively better at higher temperatures because steam tracing supplies significantly more power than is necessary to maintain a pipe at low temperatures. Table 10-55 indicates that there is very little difference between the steam tracing system at 50°F and the system at 250°F. However, the electric system more than doubles in cost between these two temperatures because more heaters, higher-powered heaters, and higher-temperature heaters are required.

The effect of supply lengths on a 150°F system can be seen from Table 10-56. Steam supply pipe is much more expensive to run than electrical conduit. With each system having relatively short supply lines (40 ft each) the electric system has only a small cost advantage (10 percent, or a ratio of 1.1). This ratio is 2.1 at 50°F and 0.8 at 250°F. However, as the supply lengths increase, electric tracing has a large cost advantage.

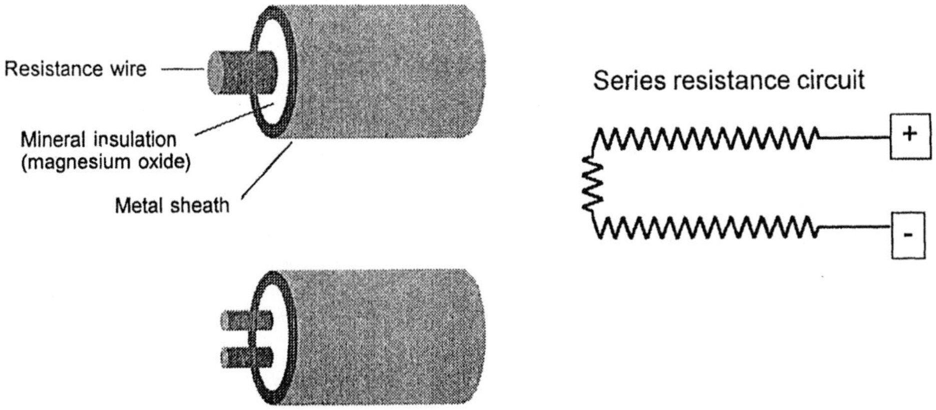

FIG. 10-175 Mineral insulated cable (MI cable).

STORAGE AND PROCESS VESSELS

STORAGE OF LIQUIDS

Atmospheric Tanks The term *atmospheric tank* as used here applies to any tank that is designed to operate at pressures from atmospheric through 3.45 kPag (0.5 psig). It may be either open to the atmosphere or enclosed. Atmospheric tanks may be either shop-fabricated or field-erected. The most common application is storage of fuel for transportation and power generation. See Pressure Tanks and Pressure Vessels later in this subsection.

Shop-Fabricated Storage Tanks A shop-fabricated storage tank is a storage tank constructed at a tank manufacturer's plant and transported to a facility for installation. In general, tanks 190 m^3 (50,000 gal) and under can be shop-fabricated and shipped in one piece to an installation site. Shop-fabricated storage tanks may be either *underground storage tanks* (USTs) or *aboveground storage tanks* (ASTs).

USTs versus ASTs For decades, USTs were the standard means of storing petroleum and chemicals in quantities of 190 m^3 (50,000 gal) or less. However, during the 1990s, many industrial and commercial facilities shifted to ASTs for hazardous product storage. Reasons include the ability to visually monitor the storage tank as well as to avoid perceived risks and myriad regulations surrounding underground storage tanks. Nonetheless, AST installations are also subject to certain regulations and codes, particularly fire codes.

The choice of USTs or ASTs is driven by numerous factors. Local authorities having jurisdiction may allow only USTs. Limited real estate may also preclude the use of ASTs. In addition, ASTs are subject to minimum distance separations from one another and from buildings, property lines, and public ways. ASTs are visually monitorable, yet may be aesthetically undesirable. Other considerations are adequate protection from spills, vandalism, and vehicular damage. Additionally, central design elements regarding product transfer and system functionality must be taken into account.

USTs are subject to myriad EPA regulations and fire codes. Further, a commonly cited drawback is the potential for unseen leaks and subsequent environmental damage and cleanup. However, advances in technology have addressed these concerns. Corrosion protection and leak detection are now standard in all UST systems. Sophisticated tank and pipe secondary containment systems have been developed to meet the EPA's secondary containment mandate for underground storage of nonpetroleum chemicals. In-tank monitoring devices for tracking inventory, tank integrity testing equipment, statistical inventory reconciliation analysis, leak-free dry-disconnect pipe and hose joints for loading/unloading, and in-tank fill shutoff valves are just a few of the many pieces of equipment which have surfaced as marketplace solutions.

Properly designed and installed in accordance with industry standards, regulations, and codes, both UST and AST systems are reliable and safe. Because of space limitations and the prevalence of ASTs at plant sites, only ASTs will be discussed further.

Aboveground Storage Tanks Aboveground storage tanks are classified as either field-erected or shop-fabricated. The latter are typically 190 m^3 (50,000 gal) or less and may be shipped over the highway, while larger tanks are more economically erected in the field. Whereas field-erected tanks likely constitute the majority of total AST storage *capacity*, shop-fabricated ASTs constitute the majority of the total *number* of ASTs in existence today. Most of these shop-fabricated ASTs store hazardous liquids at atmospheric pressure and have 45-m^3 (12,000-gal) capacity or smaller.

Shop-fabricated ASTs can be designed and fabricated as pressure vessels, but are more typically vented to atmosphere. They are oriented for either horizontal or vertical installation and are made in either cylindrical or rectangular form. Tanks are often secondarily contained and may also include insulation for fire safety or temperature control. Compartmented ASTs are also available. Over 90 percent of the atmospheric tank applications store some sort of hydrocarbon. Within that, a majority are used to store motor fuels.

Fire Codes Many chemicals are hazardous and may be subject to fire codes. All hydrocarbon tanks are classified as hazardous. Notably, with the increase in the use of ASTs at private fleet fueling facilities in the 1990s, fire codes were rapidly modified to address safety concerns. In the United States, two principal organizations publish fire codes for underground storage tanks, with each state adopting all or part of the respective codes.

The *National Fire Protection Association* (NFPA) has developed several principal codes pertaining to the storage of flammable and combustible liquids:

NFPA 30, *Flammable and Combustible Liquids Code* (2015)

NFPA 30A, *Code for Motor Fuel Dispensing Facilities and Repair Garages* (2015)

NFPA 31, *Standard for the Installation of Oil Burning Equipment* (2016)

The *International Code Council* (ICC) was formed by the consolidation of three formerly separate fire code organizations: International Conference of Building Officials (ICBO), which had published the Uniform Fire Code under its fire service arm, the International Fire Code Institute (IFCI); Building Officials and Code Administrators (BOCA), which had published the National Fire Prevention Code; and Southern Building Congress Code International (SBCCI), which had published the Standard Fire Prevention Code. When the three groups merged in 2000, in part to develop a common fire code, the individual codes became obsolete; however, they are noted above since references to them may periodically surface. The consolidated code is **IFC-2015,** *International Fire Code.*

The *Canadian Commission on Building and Fire Codes* (CCBFC) developed a recommended model code to permit adoption by various regional authorities. The National Research Council of Canada publishes the model code document *National Fire Code of Canada* (2010).

Standards Third-party standards for AST fabrication have evolved over the past two decades, as have recommended practice guidelines for the installation and operations of AST systems. Standards developed by *Underwriters Laboratories* (UL) have been the most predominant of guidelines—in fact, ASTs are often categorized according to the UL standard that they meet, such as "a UL 142 tank."

Underwriters Laboratories Inc. standards for steel ASTs storing hazardous liquids include the following:

UL 142, *Steel Aboveground Tanks for Flammable and Combustible Liquids,* 9th ed., covers aboveground, steel atmospheric tanks for storage of noncorrosive, stable, and hazardous liquids that have a specific gravity not exceeding that of water. UL 142 applies to single-wall and double-wall horizontal carbon-steel and stainless steel tanks up to 190 m^3 (50,000 gal), with a maximum diameter of 3.66 m (12 ft) and a maximum length-to-diameter ratio of 8 to 1. A formula from *Roark's Formulas for Stress and Strain* has been incorporated within UL 58 to calculate minimum steel wall thicknesses. UL 142 also applies to vertical tanks up to 10.7 m (35 ft) in height. UL 142 has been the primary AST standard since its development in 1922.

Tanks covered by these standards can be fabricated in cylindrical or rectangular configurations. The standard covers secondary contained tanks, either of dual-wall construction or a tank in a steel or concrete dike or bund. It also provides listings for special AST construction, such as those used under generators for backup power. These tanks are fabricated, inspected, and tested for leakage before shipment from the factory as completely assembled vessels.

UL 142 provides details for steel type, wall thickness, compartments and bulkheads, manways, and other fittings and appurtenances. Issues relating to leakage, venting, and the ability of the tank to withstand the development of internal pressures encountered during operation and production leak testing are also addressed.

These requirements do not apply to large field-erected storage tanks covered by the *Standard for Welded Steel Tanks for Oil Storage,* **API 650,** or the *Specification for Field-Welded Tanks for Storage of Production Liquids,* **API 12D**; or the *Specification for Shop-Welded Tanks for Storage of Production Liquids,* **API 12F.**

UL 2085, *Protected Aboveground Tanks for Flammable and Combustible Liquids,* covers shop-fabricated, aboveground atmospheric protected tanks intended for storage of stable, hazardous liquids that have a specific gravity not greater than that of water and that are compatible with the material and construction of the tank.

The tank construction is intended to limit the heat transferred to the primary tank when the AST is exposed to a 2-h hydrocarbon pool fire of 1093°C (2000°F). The tank must be insulated to withstand the test without leakage and with an average maximum temperature rise on the primary tank not exceeding 127°C (260°F). Temperatures on the inside surface of the primary tank cannot exceed 204°C (400°F).

UL 2085 also provides criteria for resistance against vehicle impact, ballistic impact, and fire hose impact. These tanks are also provided with integral secondary containment intended to prevent any leakage from the primary tank.

UL 2080, *Fire Resistant Tanks for Flammable and Combustible Liquids,* is similar to UL 2085 tanks, except with an average maximum temperature rise on the primary tank limited to 427°C (800°F) during a 2-h pool fire. Temperatures on the inside surface of the primary tank cannot exceed 538°C (1000°F).

UL 2244, *Standard for Aboveground Flammable Liquid Tank Systems,* covers factory-fabricated, pre-engineered aboveground atmospheric tank systems intended for dispensing hazardous liquids, such as gasoline or diesel fuel, into motor vehicles, generators, or aircraft.

UL 80, *Steel Tanks for Oil-Burner Fuel,* covers the design and construction of welded, atmospheric steel tanks with a maximum capacity of 0.23 to 2.5 m^3 (60 to 660 gal) intended for unenclosed installation inside of buildings or for outside aboveground applications as permitted by the *Standard for Installation of Oil-Burning Equipment,* **NFPA 31,** primarily for the storage and supply of fuel oil for oil burners.

UL 2245, *Below-Grade Vaults for Flammable Liquid Storage Tanks,* covers below-grade vaults intended for the storage of hazardous liquids in an aboveground atmospheric tank. Below-grade vaults, constructed of a minimum of 150 mm (6 in) of reinforced concrete or other equivalent noncombustible material, are designed to contain one aboveground tank, which can be a compartment tank. Adjacent vaults may share a common wall. The lid of the vault may be at grade or below. Vaults provide a safe means to install hazardous tanks so that the system is accessible to the operator without unduly exposing the public.

Southwest Research Institute (SwRI) standards for steel ASTs storing hazardous liquids include the following:

SwRI 93-01, *Testing Requirements for Protected Aboveground Flammable Liquid/Fuel Storage Tanks,* includes tests to evaluate the performance of ASTs under fire, hose stream, ballistics, heavy vehicular impact, and different environments. This standard requires pool-fire resistance similar to that of UL 2085.

SwRI 97-04, *Standard for Fire Resistant Tanks,* includes tests to evaluate the performance of ASTs under fire and hose stream. This standard is similar to UL 2080 in that the construction is exposed to a 2-h hydrocarbon pool fire of 1093°C (2000°F). However, SwRI 97-04 is concerned only with the integrity of the tank after the 2-h test and is not concerned with the temperature inside the tank from heat transfer. As a result, UL 142 tanks have been tested to the SwRI standard and passed. Secondary containment with insulation is not necessarily an integral component of the system.

Underwriters Laboratories of Canada (ULC) publishes a number of standards for aboveground tanks and accessories. All the following pertain to the aboveground storage of hazardous liquids such as gasoline, fuel oil, or similar products with a relative density not greater than 1.0:

ULC S601, *Shop Fabricated Steel Aboveground Horizontal Tanks for Flammable and Combustible Liquids,* covers single- and double-wall cylindrical horizontal steel atmospheric tanks. These requirements do not cover tanks of capacities greater than 200 m^3 (52,800 gal).

ULC S630, *Shop Fabricated Steel Aboveground Vertical Tanks for Flammable and Combustible Liquids,* covers single- and double-wall cylindrical vertical steel atmospheric tanks.

ULC S643, *Shop Fabricated Steel Utility Tanks for Flammable and Combustible Liquids,* covers single- and double-wall cylindrical horizontal steel atmospheric tanks.

ULC S653-94, *Aboveground Steel Contained Tank Assemblies for Flammable and Combustible Liquids,* covers steel contained tank assemblies.

ULC S655, *Aboveground Protected Tank Assemblies for Flammable and Combustible Liquids,* covers shop-fabricated primary tanks that provided with secondary containment and protective encasement and are intended for stationary installation and use in accordance with

1. *National Fire Code of Canada,* Part 4
2. CAN/CSA-B139, *Installation Code for Oil Burning Equipment*
3. *The Environmental Code of Practice for Aboveground Storage Tank Systems Containing Petroleum Products*
4. The requirements of the authority having jurisdiction

ULC/ORD C142.20, *Secondary Containment for Aboveground Flammable and Combustible Liquid Storage Tanks,* covers secondary containments for aboveground primary tanks.

ULC S602, *Aboveground Steel Tanks for the Storage of Combustible Liquids Intended to Be Used as Heating and/or Generator Fuels,* covers the design and construction of tanks of the atmospheric-type, intended for installation inside or outside buildings. This standard covers single-wall tanks and tanks with secondary containment, having a maximum capacity of 2.5 m^3 (660 gal).

The *Petroleum Equipment Institute* (PEI) has developed a recommended practice for AST system installation: **PEI-RP200, RP 200,** *Recommended Practices for Installation of Aboveground Storage Systems for Motor Vehicle Fueling.*

The *American Petroleum Institute* (API) has developed a series of standards and specifications involving ASTs:

API 12F, *Shop Welded Tanks for Storage of Production Liquids*

RP 12R1, *Setting, Maintenance, Inspection, Operation, and Repair of Tanks in Production Service*

RP 575, *Inspection of Atmospheric and Low Pressure Storage Tanks*

RP 579, *Fitness-For-Service*

API 650, *Welded Tanks for Oil Storage,* now applies to welded aluminum alloy storage tanks as well as welded steel tanks.

API 652, *Lining of Aboveground Storage Tank Bottoms*

API 653, *Tank Inspection, Repair, Alteration, and Reconstruction*

API 2350, *Overfill Protection for Storage Tanks in Petroleum Facilities* (overfill is the primary cause of AST releases)

The *American Water Works Association* (AWWA) has many standards dealing with water handling and storage. A list of its publications is given in the *AWWA Handbook* (annually). AWWA D100, *Standard for Steel Tanks—Standpipes, Reservoirs, and Elevated Tanks for Water Storage,* contains rules for design and fabrication. Although AWWA tanks are intended for water, they could be used for the storage of other liquids.

The *Steel Tank Institute* (STI) publishes construction standards and recommended installation practices pertaining to ASTs fabricated to STI technologies. STI's recommended installation practices are notable for their applicability to similar respective technologies:

SP001-06, *Standard for Inspection of In-Service Shop Fabricated Aboveground Tanks for Storage of Combustible and Flammable Liquids*

R912-00, *Installation Instructions for Shop Fabricated Aboveground Storage Tanks for Flammable, Combustible Liquids*

F921, *Standard for Aboveground Tanks with Integral Secondary Containment*
Standard for Fire Resistant Tanks (Flameshield)
Standard for Fireguard Thermally Insulated Aboveground Storage Tanks

F911, *Standard for Diked Aboveground Storage Tanks*

The *National Association of Corrosion Engineers* (NACE International) has developed the following to protect the soil side of bottoms of on-grade carbon-steel storage tanks: **NACE RP0193-2001,** *Standard Recommended Practice—External Cathodic Protection of On-Grade Metallic Storage Tank Bottoms.*

Environmental Regulations A key *U.S. Environmental Protection Agency* (U.S. EPA) requirement for certain aboveground storage facilities is the development and submittal of Spill Prevention Control and Countermeasure (SPCC) Plans within 40 CFR 112, the Oil Pollution Prevention regulation, which in turn is part of the *Clean Water Act* (CWA). SPCC Plans and Facility Response Plans pertain to facilities which may discharge oil into groundwater or storm runoff, which in turn may flow into navigable waters. Enacted in 1973, these requirements were principally used by owners of large, field-fabricated aboveground tanks predominant at that time, although the regulation applied to all bulk containers larger than 2.5 m^3 (660 gal) and included the requirement for a Professional Engineer to certify the spill plan.

In July 2002, the U.S. EPA issued a final rule amending 40 CFR 112 which included differentiation of shop-fabricated from field-fabricated ASTs. The rule also includes new subparts outlining the requirements for various classes of oil, revises the applicability of the regulation, amends the requirements for completing SPCC Plans, and makes other modifications.

The revised rule also states that all bulk storage container installations must provide a secondary means of containment for the entire capacity of the largest single container, with sufficient freeboard to contain precipitation, and that such containment must be sufficiently impervious to contain discharged oil. The U.S. EPA encourages the use of industry standards to comply with the rules. Many owners of shop-fabricated tanks have opted for double-wall tanks built to STI or UL standards as a means to comply with this requirement.

State and Local Jurisdictions Due to the manner in which aboveground storage tank legislation was promulgated in 1972 for protection of surface waters from oil pollution, state environmental agencies did not receive similar jurisdiction as they did within the underground storage tank rules. Nonetheless, many state environmental agencies, state fire marshals, or Weights and Measures departments—including Minnesota, Florida, Wisconsin, Virginia, Oklahoma, Missouri, Maryland, Delaware, and Michigan—are presently regulating aboveground storage tanks through other means. Other regulations exist for hazardous chemicals and should be consulted for specific requirements.

Aboveground Storage Tank Types and Options Most hydrocarbon storage applications use carbon steel as the most economical and available material that provides suitable strength and compatibility for the specific storage application. For vertical tanks installed on grade, corrosion protection can be given to exterior tank bottoms in contact with soil. The interior of the tank can incorporate special coatings and linings (e.g., polymer, glass, or other metals). Some chemical storage applications require the storage tank be made from a stainless steel or nickel alloy. *Fiberglass-reinforced plastic* (FRP), polyethylene, or polypropylene may be used for nonflammable storage in smaller sizes. Suppliers can be contacted to verify the appropriate material to be used.

As stated earlier, shop-fabricated ASTs are often categorized according to the standards to which the tanks are fabricated, e.g., a UL 142 or UL 2085 tank. That said, however, there are defined categories such as diked tanks, protected tanks, fire-resistant tanks, and insulated tanks. It is critical that the tank be specified for the given application, code requirements, and/or owner/operator preferences—and that the tank contractor and/or manufacturer be made aware of this.

Cylindrical or rectangular tanks storing flammable and combustible liquids (UL 142 ASTs) will normally comply with UL 142. The Seventh Edition published in 1993 was particularly notable, as it incorporated secondary containment designs (diking or steel secondary containment tanks) and rectangular tank designs. The latest edition is the Ninth Edition (2006).

Rectangular tanks became a desirable option for small tanks, typically less than 7.6 m^3 (2000 gal), as operators liked the accessibility of the flat top to perform operations and maintenance, without the need for special ladders or catwalks.

Post-tensioned concrete is frequently used for tanks to about 57,000 m^3 (15×10^6 gal), usually containing water. Their design is treated in detail by Creasy (*Pre-stressed Concrete Cylindrical Tanks,* Wiley, New York, 1961). For the most economical design of large open tanks at ground levels, he recommends limiting vertical height to 6 m (20 ft). Seepage can be a problem if unlined concrete is used with some liquids (e.g., gasoline).

Elevated tanks can supply a large flow when required, but pump capacities need be only for average flow. Thus, they may save on pump and piping investment. They also provide flow after pump failure, an important consideration for fire systems.

Open tanks may be used to store materials that will not be harmed by water, weather, or atmospheric pollution. Otherwise, a roof, either fixed or floating, is required. *Fixed roofs* are usually either domed or coned. Large tanks have coned roofs with intermediate supports. Since negligible pressure is involved, snow and wind are the principal design loads. Local building codes often give required values.

Fixed-roof atmospheric tanks require *vents* to prevent pressure changes which would otherwise result from temperature changes and withdrawal or addition of liquid. **API Standard 2000,** *Venting Atmospheric and Low Pressure Storage Tanks,* gives practical rules for conservative vent design. The principles of this standard can be applied to fluids other than petroleum products. Excessive losses of volatile liquids, particularly those with flash points below 38°C (100°F), may result from the use of open vents on fixed-roof tanks. Sometimes vents are connected to a manifold and lead to a vent tank, or the vapor may be removed by a *vapor recovery unit* (VRU).

An effective way of preventing vent loss is to use one of the many types of *variable-volume tanks.* These are built under **API Standard 650.** They may have floating roofs of the double-deck or the single-deck type. There are lifter-roof types in which the roof either has a skirt moving up and down in an annular liquid seal or is connected to the tank shell by a flexible membrane. A fabric expansion chamber housed in a compartment on top of the tank roof also permits variation in volume.

Floating roofs must have a seal between the roof and the tank shell. If not protected by a fixed roof, they must have drains for the removal of water, and the tank shell must have a "wind girder" to avoid distortion. An industry has developed to retrofit existing tanks with floating roofs. Many details on the various types of tank roofs are given in manufacturers' literature. Figure 10-176 shows roof types. These roofs cause less condensation buildup and are highly recommended.

Fire-Rated or Insulated ASTs: Protected and Fire-Resistant These ASTs have received much attention within the fire regulatory community, particularly for motor fuel storage and dispensing applications and generator base tanks. National model codes have been revised to allow this type of storage aboveground.

An insulated tank can be a protected tank, built to third-party standards UL 2085 and/or SwRI 93-01, or a fire-resistant tank built to UL 2080 or SwRI 97-04. Protected tanks were developed in line with NFPA requirements and terminology, while fire-resistant ASTs were developed in line with Uniform Fire Code (now International Fire Code) requirements and terminology. Both protected tanks and fire-resistant tanks must pass a 1093°C (2000°F), 2-h fire test.

The insulation properties of many fire-rated ASTs marketed today are typically provided by concrete; i.e., the primary steel tank is surrounded by concrete. Due to the weight of concrete, this design is normally limited to small tanks. Another popular AST technology meeting all applicable code requirements for insulated tanks and fabricated to UL 2085 is a tank that utilizes a lightweight monolithic thermal insulation in between two walls of steel to minimize heat transfer from the outer tank to the inner tank and to make tank handling easier.

A *secondary containment AST* to prevent contamination of our environment has become a necessity for all hazardous liquid storage, regardless of its chemical nature, in order to minimize liability and protect neighboring property. A number of different regulations exist, but the regulations with the greatest impact are fire codes and the U.S. EPA SPCC rules for oil storage.

In 1991, the Spill Prevention Control and Countermeasure (SPCC) rule proposed a revision to require secondary containment that was impermeable for at least 72 h following a release. The 2003 promulgated EPA SPCC rule no longer mandates a 72-h containment requirement, instead opting to require means to contain releases until they can be detected and removed. Nonetheless, the need for impermeable containment continues to position steel as a material of choice for shop-fabricated tanks. However, release prevention barriers made from plastic or concrete can also meet U.S. EPA requirements when periodically inspected for integrity.

Diked ASTs Fire codes dictate that hazardous liquid tanks have spill control in the form of dike, remote impounding, or small secondary containment tanks. The dike must contain the content of the largest tank to prevent hazardous liquids from endangering the public and property. Traditional bulk storage systems include multiple tanks within a concrete or earthen dike wall.

From a shop-fabricated tank manufacturer's perspective, a diked AST generally refers to a steel tank within a factory-fabricated steel box, or dike. An example of a diked AST is the STI F911 standard, providing an open-topped steel rectangular dike and floor as support and secondary containment of a UL 142 steel tank. The dike will contain 110 percent of the tank capacity; as rainwater may already have collected in the dike, the additional 10 percent acts as freeboard should a catastrophic failure dump a full tank's contents into a dike. Many fabricators offer steel dikes with shields to prevent rainwater from collecting.

A *double-wall AST* of steel fulfills the same function as a diked AST with a rain shield. Double-wall designs consist of a steel wrap over a horizontal or vertical steel storage tank. The steel wrap provides an intimate, secondary containment over the primary tank. One such design is the Steel Tank Institute's F921 standard, based upon UL 142–listed construction for the primary tank, outer containment, associated tank supports, or skids.

Venting of ASTs is critical, since they are exposed to greater ambient temperature fluctuations than are USTs. Properly designed and sized venting, both normal (atmospheric) and emergency, is required. Normal vents permit the flow of air or inert gas into and out of the tank. The vent line must be large enough to accommodate the maximum filling or withdrawal rates without exceeding the allowable stress for the tank.

Fire codes' recommended installation procedures also detail specifics on pressure/vacuum venting devices and vent line flame arresters. For example, codes mandate different ventilation requirements for Class I-A liquids versus Class I-B or I-C liquids. Tank vent piping is generally not connected to a manifold unless required for special purposes, such as pollution control or vapor recovery. As always, local codes must be followed.

Emergency venting prevents an explosion during a fire or another emergency. All third-party laboratory standards except UL 80 include emergency relief provisions, since these tanks are designed for atmospheric pressure conditions.

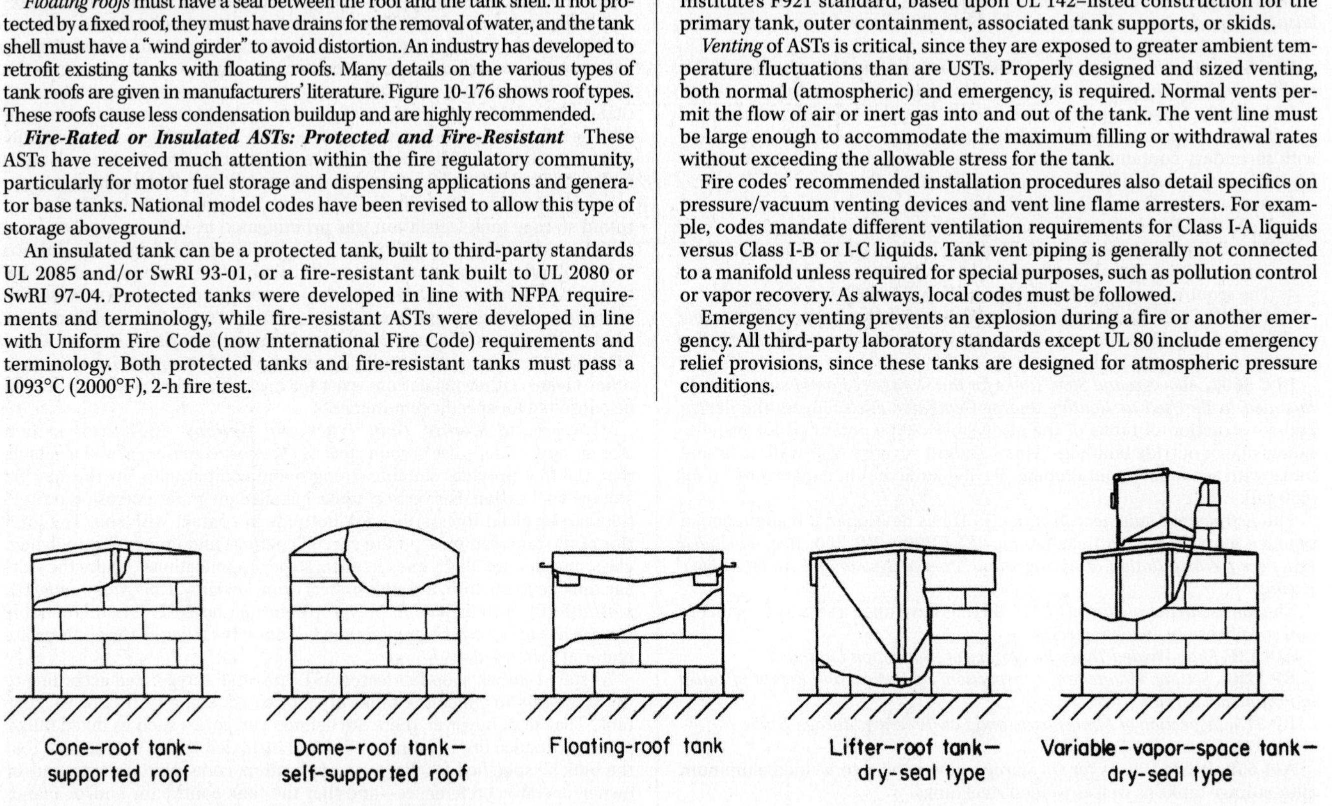

FIG. 10-176 Some types of atmospheric storage tanks.

Separation distances are also important. Aboveground storage tanks must be separated from buildings, property lines, fuel dispensers, and delivery trucks in accordance with the level of safety the tank design provides, depending on whether they are constructed of traditional steel or are vault/fire-resistant.

For most chemical storage tanks, codes such as NFPA 30 and IFC give specific separation distances. For motor vehicle fueling applications, the codes are more stringent on separation requirements due to a greater exposure of the public to the hazards. Hence codes such as NFPA 30A establish variable separation distances depending on whether the facility is private or public.

Separation distance requirements may dictate whether a tank buyer purchases a traditional steel UL 142 tank, a fire-resistant tank, or a tank in a vault. For example, NFPA 30, NFPA 30A, and the IFC codes allow UL 2085 tanks to be installed closer to buildings and property lines, thereby reducing the real estate necessary to meet fire codes.

Under NFPA 30A, dispensers may be installed directly over vaults or upon fire-resistant tanks at fleet-type installations, whereas a 7.6- to 15.2-m (25- to 50-ft) separation distance is required at retail-type service station installations. The IFC only allows gasoline and diesel to be dispensed from ASTs, designed with a 2-h fire rating. Non-2-h fire-rated UL 142 tanks dispensing diesel can be installed if permitted by local codes.

Maintenance and Operations Water in any storage system can cause myriad problems from product quality to corrosion caused by trace contaminants and microbial action. Subsequently, any operations and maintenance program must include a proactive program of monitoring for and removal of water. Other operations and maintenance procedures include periodic integrity testing and corrosion control for vertical tank bottoms. Additional guidance is available from organizations such as API, Petroleum Equipment Institute (PEI), ASTM International, *National Oilheat Research Alliance* (NORA), and STI. Also see the STI document *Keeping Water Out of Your Storage System* (http://www.steeltank.com/library/pubs/waterinfueltanks.pdf).

Integrity testing and visual inspection requirements are discussed in the SPCC requirements, Subpart B, Para. 112.8c(6). Chemical tanks storing toluene and benzene are subject to the rule in addition to traditional fuels. But a good inspection program is recommended regardless of applicable regulations. Both visual inspection and another testing technique are required. Comparison records must be kept, and frequent inspections must be made of the outside of the tank and system components for signs of deterioration, discharge, or accumulation of oil inside diked areas.

For inspection of large field-erected tanks, API 653, *Tank Inspection, Repair, Alteration, and Reconstruction*, is referenced by the U.S. EPA. A certified inspector must inspect tanks. U.S. EPA references the Steel Tank Institute Standard SP001-06, *Standard for Inspection of In-Service Shop Fabricated Aboveground Tanks for Storage of Combustible and Flammable Liquids*, as an industry standard that may assist an owner or operator with the integrity testing and inspection of shop-fabricated tanks. The STI SP001-06 standard includes inspection techniques for all types of shop-fabricated tanks—horizontal cylindrical, vertical, and rectangular. SP001-06 also addresses tanks that rest directly on the ground or on release prevention barriers, tanks that are elevated on supports, and tanks that are either single- or double-wall using a risk-based approach.

Pressurized Tanks Vertical cylindrical tanks constructed with domed or coned roofs, which operate at pressures above several hundred pascals (a few pounds per square foot) but which are still relatively close to atmospheric pressure, can be built according to API Standard 650. The pressure force acting against the roof is transmitted to the shell, which may have sufficient weight to resist it. If not, the uplift will act on the tank bottom. The strength of the bottom, however, is limited, and if it is not sufficient, an anchor ring or a heavy foundation must be used. In the larger sizes, uplift forces limit this style of tank to very low pressures.

As the size or the pressure goes up, curvature on all surfaces becomes necessary. Tanks in this category, up to and including a pressure of 103.4 kPag (15 psig), can be built according to API Standard 620. Shapes used are spheres, ellipsoids, toroidal structures, and circular cylinders with torispherical, ellipsoidal, or hemispherical heads. The ASME Boiler and Pressure Vessel Code, Sec. VIII-1 (2015), although not required below 15 psig (103.4 kPag), is also useful for designing such tanks.

Tanks that could be subjected to vacuum should be provided with vacuum-breaking valves or be designed for vacuum (external pressure). The BPVC contains design procedures.

Calculation of Tank Volume A tank may be a single geometric element, such as a cylinder, a sphere, or an ellipsoid. It may also have a compound form, such as a cylinder with hemispherical ends or a combination of a toroid and a sphere. To determine the volume, each geometric element usually must be calculated separately. Calculations for a full tank are usually simple, but calculations for partially filled tanks may be complicated.

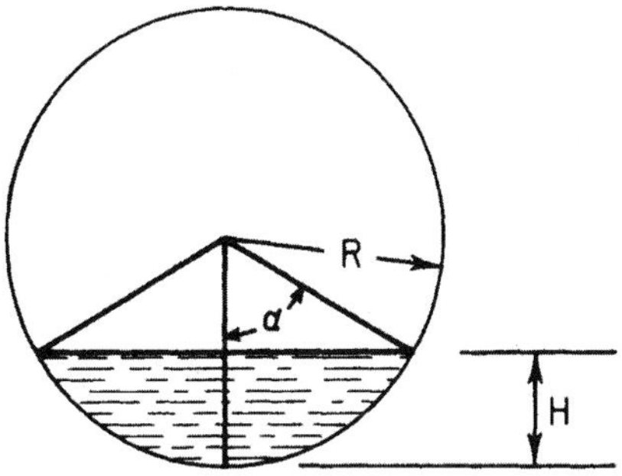

FIG. 10-177 Calculation of partially filled horizontal tanks. H = depth of liquid; R = radius; D = diameter; L = length; α = half of the included angle; and $\cos\alpha = 1 - H/R = 1 - 2H/D$.

To calculate the volume of a *partially filled horizontal cylinder,* refer to Fig. 10-177. Calculate the angle α in degrees. Any units of length can be used, but they must be the same for H, R, and L. The liquid volume

$$V = LR^2\left(\frac{\alpha}{57.30} - \sin\alpha\ \cos\alpha\right) \qquad (10\text{-}107)$$

This formula may be used for any depth of liquid between zero and the full tank, provided the algebraic signs are observed. If H is greater than R, then $\sin\alpha\cos\alpha$ will be negative and thus will add numerically to $\alpha/57.30$. Table 10-57 gives liquid volume, for a partially filled horizontal cylinder, as a fraction of the total volume, for the dimensionless ratio H/D or $H/2R$.

The *volumes of heads* must be calculated separately and added to the volume of the cylindrical portion of the tank. The four types of heads most frequently used are the standard dished head,* torispherical or ASME head, ellipsoidal head, and hemispherical head. Dimensions and volumes for all four types are given in *Lukens Spun Heads,* Lukens Inc., Coatesville, Pennsylvania. Approximate volumes can also be calculated by the formulas in Table 10-58. Consistent units must be used in these formulas.

A partially filled horizontal tank requires the determination of the partial volume of the heads. The Lukens catalog gives approximate volumes for partially filled (axis horizontal) standard ASME and ellipsoidal heads. A formula for *partially filled heads* (excluding conical), by Doolittle [*Ind. Eng. Chem.* **21:** 322–323 (1928)], is

$$V = 0.215H^2(3R - H) \qquad (10\text{-}108)$$

where in consistent units V = volume, R = radius, and H = depth of liquid. Doolittle made some simplifying assumptions that affect the volume given

* The standard dished head does not comply with the ASME BPVC.

TABLE 10-57 Volume of Partially Filled Horizontal Cylinders

H/D	Fraction of volume	H/D	Fraction of volume	H/D	Fraction of volume	H/D	Fraction of volume
0.01	0.00169	0.26	0.20660	0.51	0.51273	0.76	0.81545
.02	.00477	.27	.21784	.52	.52546	.77	.82625
.03	.00874	.28	.22921	.53	.53818	.78	.83688
.04	.01342	.29	.24070	.54	.55088	.79	.84734
.05	.01869	.30	.25231	.55	.56356	.80	.85762
.06	.02450	.31	.26348	.56	.57621	.81	.86771
.07	.03077	.32	.27587	.57	.58884	.82	.87760
.08	.03748	.33	.28779	.58	.60142	.83	.88727
.09	.04458	.34	.29981	.59	.61397	.84	.89673
.10	.05204	.35	.31192	.60	.62647	.85	.90594
.11	.05985	.36	.32410	.61	.63892	.86	.91491
.12	.06797	.37	.33636	.62	.65131	.87	.92361
.13	.07639	.38	.34869	.63	.66364	.88	.93203
.14	.08509	.39	.36108	.64	.67590	.89	.94015
.15	.09406	.40	.37353	.65	.68808	.90	.94796
.16	.10327	.41	.38603	.66	.70019	.91	.95542
.17	.11273	.42	.39858	.67	.71221	.92	.96252
.18	.12240	.43	.41116	.68	.72413	.93	.96923
.19	.13229	.44	.42379	.69	.73652	.94	.97550
.20	.14238	.45	.43644	.70	.74769	.95	.98131
.21	.15266	.46	.44912	.71	.75930	.96	.98658
.22	.16312	.47	.46182	.72	.77079	.97	.99126
.23	.17375	.48	.47454	.73	.78216	.98	.99523
.24	.18455	.49	.48727	.74	.79340	.99	.99831
.25	.19550	.50	.50000	.75	.80450	1.00	1.00000

TABLE 10-58 Volumes of Heads*

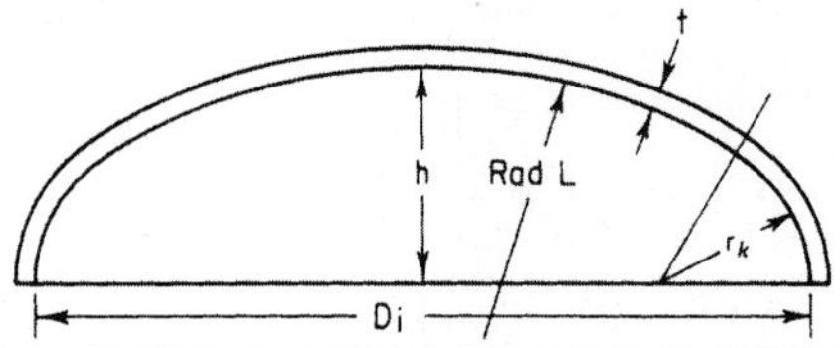

Type of head	Knuckle radius r_k	h	L	Volume	% Error	Remarks
Standard dished	Approx. $3t$		Approx. D_i	Approx. $0.050D_i^3 + 1.65tD_i^2$	±10	h varies with t
Torispherical or ASME	$0.06L$		D_i	$0.0809D_i^3$	±0.1	r_k must be the larger of $0.06L$ and $3t$
Torispherical or ASME	$3t$		D_i	Approx. $0.513hD_i^2$	±8	
Ellipsoidal				$\pi D_i^2 h/6$	0	
Ellipsoidal		$D_i/4$		$\pi D_i^3/24$	0	Standard proportions
Hemispherical		$D_i/2$	$D_i/2$	$\pi D_i^3/12$	0	
Conical				$\pi h(D_i^2 + D_i d + d^2)/12$	0	Truncated cone h = height d = diameter at small end

*Use consistent units.

by the equation, but the equation is satisfactory for determining the volume as a fraction of the entire head. This fraction, calculated by Doolittle's formula, is given in Table 10-59 as a function of H/D_i (H is the depth of liquid, and D_i is the inside diameter). Table 10-59 can be used for standard dished, torispherical, ellipsoidal, and hemispherical heads with an error of less than 2 percent of the volume of the entire head. The error is zero when $H/D_i = 0$, 0.5, and 1.0. Table 10-59 cannot be used for conical heads.

When a tank volume cannot be calculated or when greater precision is required, *calibration* may be necessary. This is done by draining (or filling) the tank and measuring the volume of liquid. The measurement may be made by weighing, by a calibrated fluid meter or by repeatedly filling small measuring tanks which have been calibrated by weight.

Container Materials and Safety Storage tanks are made of almost any structural material. Steel and reinforced concrete are most widely used. Plastics and glass-reinforced plastics are used for tanks up to about 230 m^3 (60,000 gal). Resistance to corrosion, light weight, and lower cost are their advantages. Plastic and glass coatings are also applied to steel tanks. Aluminum and other nonferrous metals are used when their special properties are required. When expensive metals such as tantalum are required, they may be applied as tank linings or as clad metals.

Some grades of steel listed by API and AWWA Standards are of lower quality than is customarily used for pressure vessels. The stresses allowed by these standards are also higher than those allowed by the ASME Pressure Vessel Code. Small tanks containing nontoxic substances are not particularly hazardous and can tolerate a reduced factor of safety. Tanks containing highly toxic substances and very large tanks containing any substance can be hazardous. The designer must consider the magnitude of the hazard. The possibility of brittle behavior of ferrous metal should be taken into account in specifying materials (see subsection Safety in Design).

Volume 1 of National Fire Codes (NFPA, Quincy, Massachusetts) contains recommendations (NFPA 30) for venting, drainage, and dike construction of tanks for *flammable liquids.*

TABLE 10-59 Volume of Partially Filled Heads on Horizontal Tanks*

H/D_i	Fraction of volume	H/D_i	Fraction of volume	H/D_i	Fraction of volume	H/D_i	Fraction of volume
0.02	0.0012	0.28	0.1913	0.52	0.530	0.78	0.8761
.04	.0047	.30	.216	.54	.560	.80	.8960
.06	.0104	.32	.242	.56	.590	.82	.9145
.08	.0182	.34	.268	.58	.619	.84	.9314
.10	.0280	.36	.295	.60	.648	.86	.9467
.12	.0397	.38	.323	.62	.677	.88	.9603
.14	.0533	.40	.352	.64	.705	.90	.9720
.16	.0686	.42	.381	.66	.732	.92	.9818
.18	.0855	.44	.410	.68	.758	.94	.9896
.20	.1040	.46	.440	.70	.784	.96	.9953
.22	.1239	.48	.470	.72	.8087	.98	.9988
.24	.1451	.50	.500	.74	.8324	1.00	1.0000
.26	.1676			.76	.8549		

*Based on Eq. (10-108).

Container Insulation Tanks containing materials above atmospheric temperature may require insulation to reduce the loss of heat. Almost any of the commonly used insulating materials can be employed. Calcium silicate, glass fiber, mineral wool, cellular glass, and plastic foams are among those used. Tanks exposed to weather must have jackets or protective coatings, usually asphalt, to keep water out of the insulation.

Tanks with contents at lower than atmospheric temperature may require insulation to minimize heat absorption. The insulation must have a vapor barrier on the outside to prevent condensation of moisture from reducing its effectiveness. The insulation techniques presently used for refrigerated systems can be applied (see subsection Low-Temperature and Cryogenic Storage).

Tank Supports Large vertical atmospheric steel tanks may be built on a base of about 150 cm (6 in) of sand, gravel, or crushed stone if the subsoil has adequate bearing capacity. It can be level or slightly coned, depending on the shape of the tank bottom. The porous base provides drainage in case of leaks. A few feet beyond the tank perimeter the surface should drop about 1 m (3 ft) to ensure proper drainage of the subsoil. API Standard 650, App. B, and API Standard 620, App. C, give recommendations for tank foundations.

The bearing pressure of the tank and contents must not exceed the *bearing capacity* of the soil. Local building codes usually specify allowable soil loading. These are some approximate bearing capacities:

	kPa	psf
Soft clay (can be crumbled between fingers)	100	2,000
Dry fine sand	200	4,000
Dry fine sand with clay	300	6,000
Coarse sand	300	6,000
Dry hard clay (requires a pick to dig it)	350	7,000
Gravel	400	8,000
Rock	1000–4000	20,000–80,000

For high, heavy tanks, a foundation ring may be needed. Prestressed concrete tanks are sufficiently heavy to require foundation rings. Foundations must extend below the frost line. Some tanks that are not flat-bottomed may also be supported by soil if it is suitably graded and drained. When soil does not have adequate bearing strength, it may be excavated and backfilled with a suitable soil, or piles capped with a concrete mat may be required.

Spheres, spheroids, and toroids use steel or concrete saddles or are supported by columns. Some may rest directly on soil. Horizontal cylindrical tanks should have two rather than multiple saddles to avoid indeterminate load distribution. Small horizontal tanks are sometimes supported by legs. Most tanks must be designed to resist the reactions of the saddles or legs, and they may require reinforcing. Neglect of this can cause collapse. Tanks without stiffeners usually need to make contact with the saddles on at least 2.1 rad (120°) of their circumference. An elevated steel tank may have either a circle of steel columns or a large central steel standpipe. Concrete tanks usually have concrete columns. Tanks are often supported by buildings.

Pond and Underground Storage Low-cost liquid materials, if they will not be damaged by rain or atmospheric pollution, may be stored in *ponds.* A pond may be excavated or formed by damming a ravine. To prevent

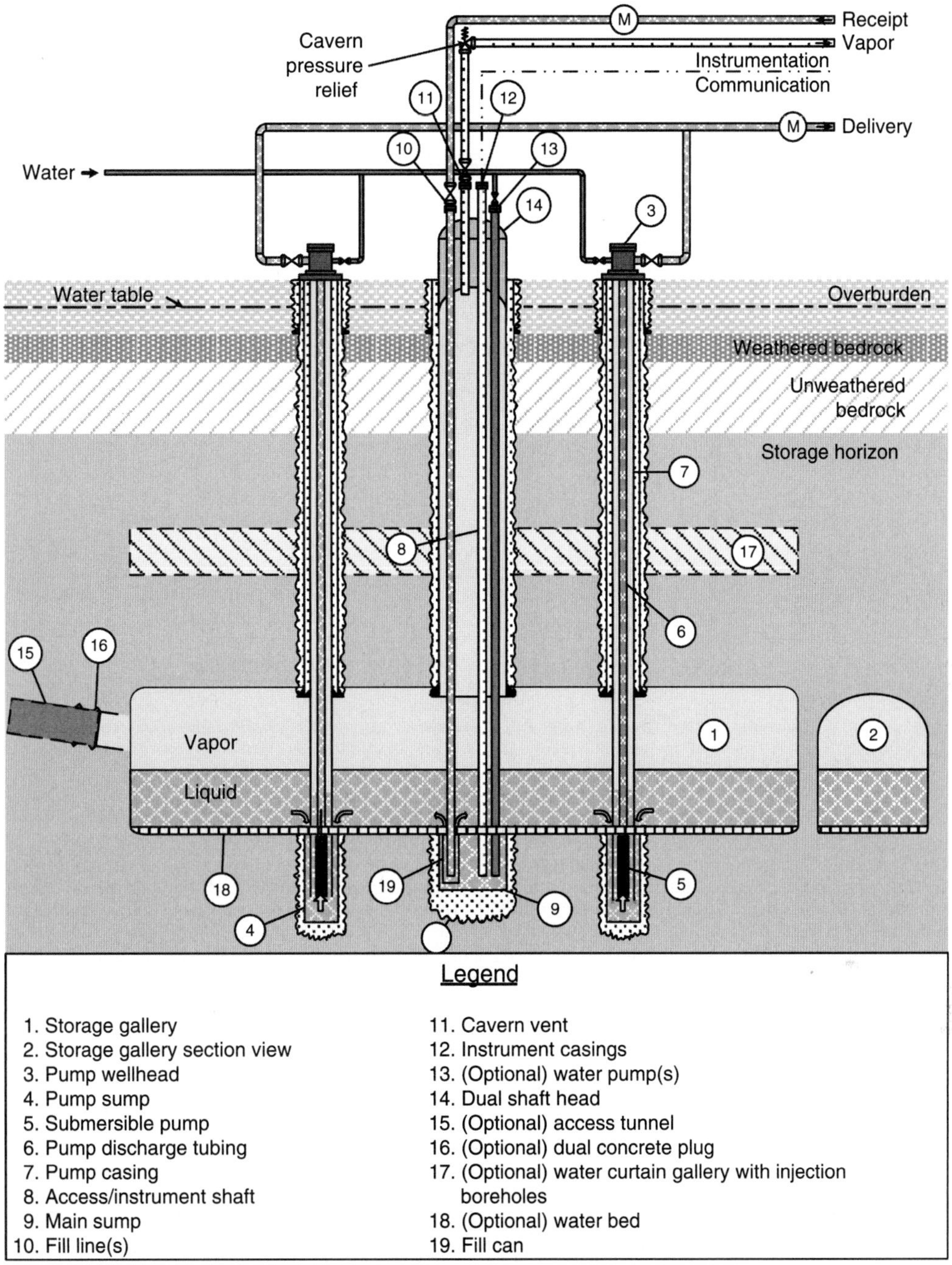

FIG. 10-178 Mined cavern.

loss by seepage, the soil which will be submerged may require treatment to make it sufficiently impervious. This can also be accomplished by lining the pond with concrete, polymeric membrane, or another barrier. Detection and mitigation of seepage is especially necessary if the pond contains material that could contaminate present or future water supplies.

Underground Cavern Storage Large volumes of liquids and gases are often stored below ground in artificial caverns as an economical alternative to aboveground tanks and other modes of storage. The stored fluids must tolerate water, brine, and other contaminants that are usually present to some degree in the cavern. The liquids that are most commonly stored are *natural gas liquids* (NGLs), LPG, crude oil, and refined petroleum products. Gases commonly stored are natural gas and hydrogen. If fluids are suitable for cavern storage, this method may be less expensive, safer, and more secure than other storage modes.

There are two types of caverns used for storing liquids. *Hard rock* (mined) caverns are constructed by mining rock formations such as shale, granite, limestone, and many other types of rock. *Solution-mined* caverns are constructed by dissolution processes, i.e., solution mining or leaching a mineral deposit, most often salt (sodium chloride). The salt deposit may take the form of a massive salt dome or thinner layers of bedded salt that are stratified between layers of rock. Hard rock and solution-mined caverns have been constructed in the United States and many other parts of the world.

Mined Caverns Caverns mined in hard rock are generally situated 100 to 150 m (325 to 500 ft) below ground level. These caverns are constructed by excavating rock with conventional drill-and-blast mining methods. The excavated cavern consists of a grouping of interconnecting tunnels or storage "galleries." Mined caverns have been constructed for volumes ranging from as little as 3200 to 800,000 m^3 [20,000 to 5 million API barrels* (bbl)]. Figure 10-178 illustrates a typical mined cavern for liquid storage.

*One API barrel = 42 U.S. gal = 5.6146 ft^3 = 0.159 m^3.

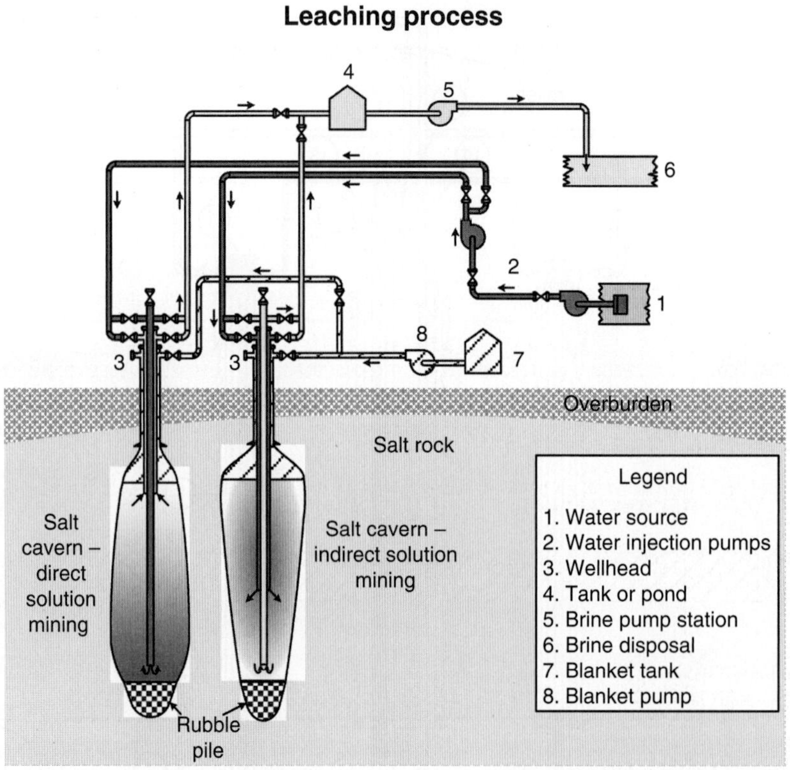

FIG. 10-179 Cavern leaching process.

Hard rock caverns are designed so that the internal storage pressure at all times is less than the hydrostatic head of the water contained in the surrounding rock matrix. Thus, the depth of a cavern determines its maximum allowable operating pressure. Groundwater that continuously seeps into hard rock caverns in permeable formations is periodically pumped out of the cavern. The maximum operating pressure of the cavern is established after a thorough geological and hydrogeological evaluation is made of the rock formation and the completed cavern is pressure-tested.

Salt Caverns Salt caverns are constructed in both *domal salt,* more commonly referred to as "salt domes," and *bedded salt,* which consists of a body of salt sandwiched between layers of rock. The greatest total volume of underground liquid storage in the United States is stored in salt dome caverns. A salt dome is a large body, mostly consisting of sodium chloride, which over geologic time moved upward thousands of feet from extensive halite deposits deep below the earth's crust. There are numerous salt domes in the United States and other parts of the world [see Harben, P. W., and R. L. Bates, "Industrial Minerals Geology and World Deposits," *Metal Bulletin Plc,* UK, pp. 229–234 (1990)]. An individual salt dome may exceed 1 mi in diameter and contain many storage caverns. The depth to the top of a salt dome may range from a few hundred to several thousand feet, although depths to about 460 m (1500 ft) are commercially viable for cavern development. The extent of many salt domes allows for caverns of many different sizes and depths to be developed. The extensive nature of salt domes has allowed the development of caverns as large as 5.7×10^6 m^3 (36 million bbl) (U.S. DOE Bryan Mound Strategic Petroleum Reserve) and larger; however, cavern volumes of 159,000 to 1.59×10^6 m^3 (1 to 10 million bbl) are more common for liquid storage.

The benefits of salt are its high compressive strength of 13.8 to 27.6 MPa (2000 to 4000 psi), its impermeability to hydrocarbon liquids and gases, and its non–chemically reactive (inert) nature. Due to the impervious nature of salt, the maximum allowed storage pressure gradient in this type of cavern is greater than that of a mined cavern. A typical storage pressure gradient for liquids is about 18 kPa/m of depth (0.80 psi/ft) to the bottom of the well casing. Actual maximum and minimum allowable operating pressure gradients are determined from geologic evaluations and rock mechanics studies. Typical depths to the top of a salt cavern may range from 500 to 4000 ft (about 150 to 1200 m). Therefore, the maximum storage pressure (2760 to 32,060 kPag, or 400 to 3200 psig) usually exceeds the vapor pressure of all commonly stored hydrocarbon liquids. Higher-vapor-pressure products such as ethylene or ethane cannot be stored in relatively shallow caverns.

Salt caverns are developed by solution mining, a process (leaching) in which water is injected to dissolve the salt. Approximately 7 to 10 units of freshwater are required to leach 1 unit of cavern volume. Figure 10-179 illustrates the leaching process for two caverns. Modern salt dome caverns are shaped as relatively tall, slender cylinders. The leaching process produces nearly saturated brine from the cavern. Brine may be disposed into nearby disposal wells or offshore disposal fields, or it may be supplied to nearby plants as a feedstock for manufacturing of caustic (NaOH) and chlorine (Cl_2). The final portion of the produced brine is retained and stored in artificial surface ponds or tanks to be used to displace the stored liquid from the cavern.

Salt caverns are usually developed using a single well, although some employ two or more wells. The well consists of a series of concentric casings that protect the water table and layers of rock and sediments (overburden) that lie above the salt dome. The innermost well casing is referred to as the last cemented or well "production" casing and is cemented in place, sealing the cavern and protecting the surrounding geology. Once the last cemented casing is in place, a bore hole is drilled from the bottom of the well, through the salt to the design cavern depth. For single-well leaching, two concentric tubing strings are then suspended in the well. A liquid, such as diesel or propane, or a gas, such as nitrogen, is then injected through the outer annular space and into the top of the cavern to act as a "blanket" to prevent undesired leaching of the top of the cavern. Water is then injected into one of the suspended tubing strings, and brine is withdrawn from the other. During the leaching process, the water is injected initially for 30 to 60 days into the innermost tubing and into the inner annulus for the remaining time. The tubing strings are periodically raised upward to control the cavern shape. A typical salt dome cavern may require 9 to 30 months of leaching time, whereas smaller, bedded salt caverns may be developed in a shorter time frame.

Brine-Compensated Storage As the stored product is pumped into the cavern, brine is displaced into an aboveground brine storage reservoir. To withdraw the product from the cavern, brine is pumped back into the cavern, displacing the stored liquid. This method of product transfer is termed *brine-compensated,* and caverns that operate in this fashion remain liquid-filled at all times. Figure 10-180 illustrates brine-compensated storage operations.

Uncompensated Storage Hard rock caverns and a few bedded salt caverns do not use brine for product displacement. This type of storage operation is referred to as *pump out* or *uncompensated* storage operations. When the cavern is partially empty of liquid, the void space is filled with the vapor that is in equilibrium with the stored liquid. When liquid is introduced into the

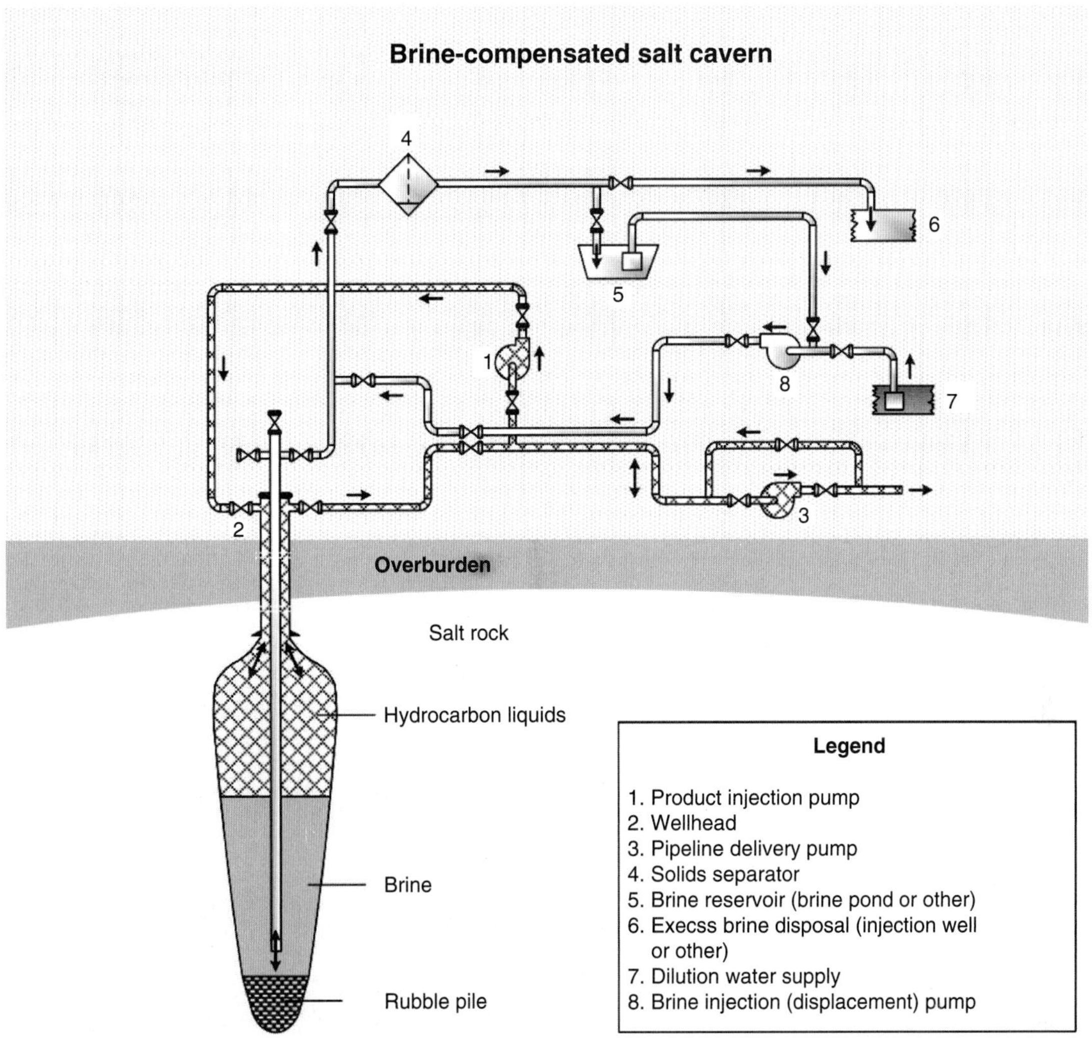

FIG. 10-180 Brine-compensated storage.

cavern, it compresses and condenses this saturated vapor phase. In some cases, vapor may be vented to the surface where it may be refrigerated, liquefied, and recycled to the cavern.

Submersible pumps or vertical line shaft pumps are used for withdrawing the stored liquid. Vertical line shaft pumps are suited for depths of no more than several hundred feet. Figure 10-178 illustrates an example of uncompensated storage operations.

Underground chambers are also constructed in frozen earth (see subsection Low-Temperature and Cryogenic Storage). Underground tunnel or tank storage is often the most practical way of storing hazardous or radioactive materials, such as proposed at Yucca Mountain, Nevada. A cover of 30 m (100 ft) of rock or dense earth can exert a pressure of about 690 kPa (100 lbf/in^2).

The storage of natural gas in depleted aquifers and petroleum reservoirs is another mode of underground storage. This type of storage requires that a number of wells be drilled into the underground storage zone at different locations and depths determined from geologic analysis.

STORAGE OF GASES

Gas Holders Gas is sometimes stored in expandable gas holders of either the liquid-seal or dry-seal type. The liquid-seal holder is a familiar sight. It has a cylindrical container, closed at the top, and varies its volume by moving it up and down in an annular water-filled seal tank. The seal tank may be staged in several lifts (as many as five). Seal tanks have been built in sizes up to 280,000 m^3 (9.9 × 10^6 ft^3). The dry-seal holder has a rigid top attached to the sidewalls by a flexible fabric diaphragm which permits it to move up and down. It does not involve the weight and foundation costs of the liquid-seal holder. Additional information on gas holders can be found in *Gas Engineers Handbook*, Industrial Press, New York, 1966.

Solution of Gases in Liquids Certain gases will dissolve readily in liquids. In some cases in which the quantities are not large, this may be a practical storage procedure. Examples of gases that can be handled in this way are ammonia in water, acetylene in acetone, and hydrogen chloride in water. Whether this method is used depends mainly on whether the end use requires the anhydrous or the liquid state. Pressure may be either atmospheric or elevated. The solution of acetylene in acetone is also a safety feature because of the instability of acetylene.

Storage in Pressure Vessels, Bottles, and Pipelines The distinction between pressure vessels, bottles, and pipes is arbitrary. They can all be used for storing gases under pressure. A storage pressure vessel is usually a permanent installation. Storing a gas under pressure not only reduces its volume but also in many cases liquefies it at ambient temperature. Some gases in this category are carbon dioxide, several petroleum gases, chlorine, ammonia, sulfur dioxide, and some types of Freon or Suva. Pressure tanks are frequently installed underground.

Liquefied petroleum gas (LPG) is the subject of API Standard 2510, *The Design and Construction of Liquefied Petroleum Gas Installations at Marine and Pipeline Terminals, Natural Gas Processing Plants, Refineries, and Tank Farms*. This standard in turn refers to:

1. National Fire Protection Association (NFPA) Standard 58, *Standard for the Storage and Handling of Liquefied Petroleum Gases*
2. NFPA Standard 59, *Standard for the Storage and Handling of Liquefied Petroleum Gases at Utility Gas Plants*
3. NFPA Standard 59A, *Standard for the Production, Storage, and Handling of Liquefied Natural Gas (LNG)*

The API Standard gives considerable information on the construction and safety features of such installations. It also recommends minimum distances from property lines. The user may wish to obtain added safety by increasing these distances.

The term *bottle* is usually applied to a pressure vessel that is small enough to be conveniently portable. Bottles range from about 57 L (2 ft^3) down to CO_2 capsules of about 16.4 mL (1 in^3). Bottles are convenient for small quantities of many gases, including air, hydrogen, nitrogen, oxygen, argon, acetylene, Freon, and petroleum gas. Some are one-time-use disposable containers.

Pipelines A pipeline is not ordinarily a storage device. Sections of pipe have been connected in series and in parallel, buried underground, and used for storage. This avoids the necessity of providing foundations, and the earth protects the pipe from extremes of temperature. The economics of such an installation would be doubtful if it were designed to the same stresses as a pressure vessel. Storage is also obtained by increasing the pressure in operating pipelines and thus having a similar impact as a tank.

Low-Temperature and Cryogenic Storage This type is used for gases that liquefy under pressure at atmospheric temperature. In cryogenic storage the gas is at, or near to, atmospheric pressure and remains liquid because of low temperature. A system may also operate with a combination of pressure and reduced temperature. The term *cryogenic* usually refers to temperatures below −101°C (−150°F). Some gases, however, liquefy between −101°C and ambient temperatures. The principle is the same, but cryogenic temperatures create different problems with insulation and construction materials.

The liquefied gas must be maintained at or below its boiling point. Refrigeration can be used, but the usual practice is to cool by evaporation. The quantity of liquid evaporated is minimized by insulation. The vapor may be vented to the atmosphere (this may be prohibited due to emissions limitations), it may be compressed and reliquefied, or it may be consumed as fuel.

At very low temperatures with liquid air and similar substances, the tank may have double walls with the interspace evacuated. The well-known Dewar flask is an example. Large tanks and even pipelines are now built this way. An alternative is to use double walls without vacuum but with an insulating material in the interspace. Perlite and plastic foams are two insulating materials employed in this way. Sometimes both insulation and vacuum are used.

Materials Materials for liquefied-gas containers must be suitable for the temperatures, and they must not become embrittled. Some carbon steels can be used down to −59°C (−75°F), and low-alloy steels to −101°C (−150°F) and sometimes −129°C (−200°F). Below these temperatures austenitic stainless steel (AISI 300 series) and aluminum are the principal materials. (See discussion of brittle fracture on p. 10-139.)

Low temperatures involve problems of *differential thermal expansion.* With the outer wall at ambient temperature and the inner wall at the liquid boiling point, relative movement must be accommodated. Proprietary systems accomplish this. The Gaz Transport of France reduces dimensional change by using a thin inner liner of Invar. Another patented French system accommodates this change by means of the flexibility of thin metal which is creased. The creases run in two directions, and the form of the crossings of the creases is a feature of the system.

Low-temperature tanks may be installed in-ground to take advantage of the insulating value of the earth. Frozen-earth storage is also used. The frozen earth forms the tank. Some installations using this technique have been unsuccessful because of excessive heat absorption.

Cavern Storage Gases are also stored below ground in salt caverns. The most common type of gas stored in caverns is natural gas, although hydrogen and air have also been stored. Hydrogen storage requires special consideration in selecting metallurgy for the wellhead and the wellbore casings. Air is stored for the purpose of providing compressed air energy for peak shaving power plants. Two such plants are in operation, one in the United States (McIntosh, Alabama), the other in Huntorf, Germany. A discussion of the Alabama plant is presented in *History of First U.S. Compressed Air Energy Storage (CAES) Plant,* vol. 1, *Early CAES Development,* Electric Power Research Institute (EPRI), Palo Alto, Calif., 1992.

Since salt caverns contain brine and other contaminants, the type of gas to be stored should not be sensitive to the presence of contaminants. If the gas is determined suitable for cavern storage, then cavern storage may not offer only economic benefits and enhanced safety and security; salt caverns also offer relatively high rates of deliverability compared to reservoir and aquifer storage fields. Solution-mined gas storage caverns in salt formations operate as *uncompensated* storage—no fluid is injected into the well to displace the compressed gas.

Surface gas handling facilities for storage caverns typically include custody transfer measurement for receipt and delivery, gas compressors, and gas dehydration equipment. When compressors are required for cavern injection and/or withdrawal, banks of positive-displacement-type compressors are commonly used, since this compressor type is well suited for handling the highly variable compression ratios and flow rates associated with cavern injection and withdrawal operations. Cavern withdrawal operations typically involve single- or dual-stage pressure reduction stations and full or partial gas dehydration. Large pressure throttling requirements often require heating the gas upon withdrawal and immediately before throttling, and injection of methanol or other liquid desiccant may be necessary to help control hydrate formation.

An in-depth discussion of natural gas storage in underground caverns may be found in *Gas Engineering and Operating Practices, Supply,* Book S-1, Part 1, *Underground Storage of Natural Gas,* and Part 2, Chap. 2, "Leached Caverns," American Gas Association, Arlington, Va., 1990.

Additional References API Recommended Practice 1114, *Design of Solution-Mined Underground Storage Facilities,* January 2013. API 1115, *Operation of Solution-Mined Underground Storage Facilities,* Washington, September 1994. Stanley J. LeFond, *Handbook of World Salt Resources, Monographs in Geoscience,* Department of Geology, Columbia University, New York, 1969. *SME Mining Engineering Handbook,* 2d ed., vol. 2, The Society for Mining, Metallurgy, and Exploration, Littleton, Colorado, 1992.

COST OF STORAGE FACILITIES

Contractors' bids offer the most reliable information on cost. Order-of-magnitude costs, however, may be required for preliminary studies. One way of estimating them is to obtain cost information from similar facilities and scale it to the proposed installation. Costs of steel storage tanks and vessels have been found to vary approximately as the 0.6 to 0.7 power of their weight [see Happel, *Chemical Process Economics,* Wiley, 1958, p. 267; also Williams, *Chem. Eng.* **54**(12): 124 (1947)]. All estimates based on the costs of existing equipment must be corrected for changes in the price index from the date when the equipment was built. Considerable uncertainty is involved in adjusting data more than a few years old.

Based on a survey in 1994 for storage tanks, the prices for field-erected tanks are for multiple-tank installations erected by the contractor on foundations provided by the owner. Some cost information on tanks is given in various references cited in Sec. 9. Cost data vary considerably from one reference to another. (See Figs. 10-181 to 10-183.)

Prestressed (post-tensioned) concrete tanks cost about 20 percent more than steel tanks of the same capacity. Once installed, however, concrete tanks require very little maintenance. A true comparison with steel would therefore require evaluating the maintenance cost of both types.

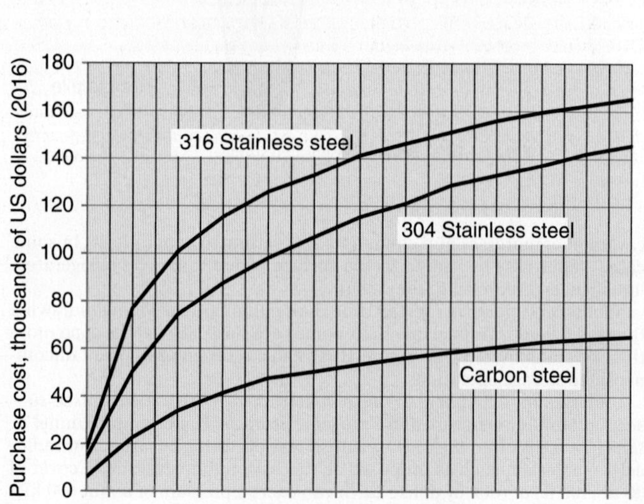

FIG. 10-181 Cost of shop-fabricated tanks in mid-1980 with ¼-in walls. Multiplying factors on carbon-steel costs for other materials are: carbon steel, 1.0; rubber-lined carbon steel, 1.5; aluminum, 1.6; glass-lined carbon steel, 4.5; and fiber-reinforced plastic, 0.75 to 1.5. Multiplying factors on type 316 stainless-steel costs for other materials are: 316 stainless steel, 1.0; Monel, 2.0; Inconel, 2.0; nickel, 2.0; titanium, 3.2; and Hastelloy C, 3.8. Multiplying factors for wall thicknesses different from ¼ in are:

Thickness, in	Carbon Steel	304 Stainless Steel	316 Stainless Steel
½	1.4	1.8	1.8
¾	2.1	2.5	2.6
1	2.7	3.3	3.5

To convert gallons to cubic meters, multiply by 3.785×10^{-3}.

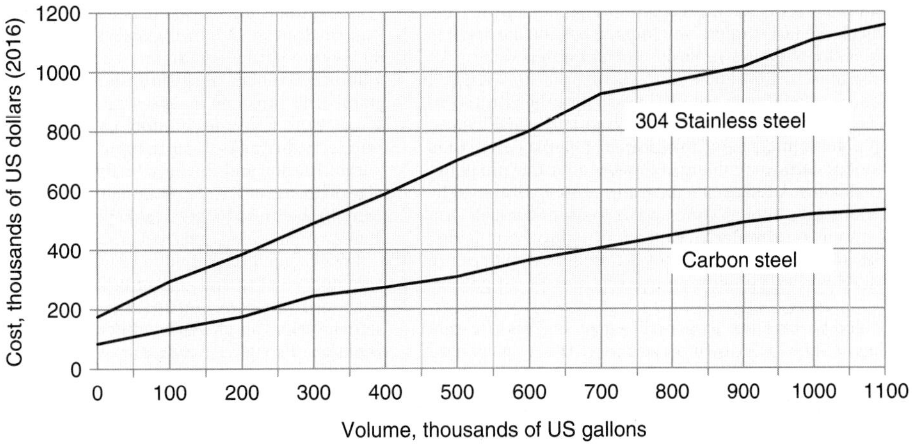

FIG. 10-182 Cost (±30 percent) of field-erected, domed, flat-bottom API 650 tanks, October 2016, includes concrete foundation and typical nozzles, ladders, and platforms. 1 gal = 0.003785 m^3.

BULK TRANSPORT OF FLUIDS

Transportation is often an important part of product cost. Bulk transportation may provide significant savings. When there is a choice between two or more forms of transportation, the competition may result in rate reduction. Transportation is subject to considerable regulation, which will be discussed in some detail under specific headings.

Pipelines For quantities of fluid that an economic investigation indicates are sufficiently large and continuous to justify the investment, pipelines are one of the lowest-cost means of transportation. They have been built up to 1.22 m (48 in) or more in diameter and about 3200 km (2000 mi) in length for oil, gas, and other products. Water is usually not transported more than 160 to 320 km (100 to 200 mi), but the conduits may be much greater than 1.22 m (48 in) in diameter. Open canals are also used for water transportation.

Petroleum pipelines before 1969 were built to ASA (now ASME) Standard B31.4 for liquids and Standard B31.8 for gas. These standards were seldom mandatory because few states adopted them. The U.S. Department of Transportation (DOT), which now has responsibility for pipeline regulation, issued Title 49, Part 192—Transportation of Natural Gas and Other Gas by Pipeline: Minimum Safety Standards, and Part 195—Transportation of Liquids by Pipeline. These contain considerable material from B31.4 and B31.8. They allow generally higher stresses than the ASME Boiler and Pressure Vessel Code would allow for steels of comparable strength. The enforcement of their regulations is presently left to the states and is therefore somewhat uncertain.

Pipeline pumping stations usually range from 16 to 160 km (10 to 100 mi) apart, with maximum pressures up to 6900 kPa (1000 lbf/in^2) and velocities up to 3 m/s (10 ft/s) for liquid. Gas pipelines have higher velocities and may have greater spacing of stations.

Tanks Tank cars (single and multiple tanks), tank trucks, portable tanks, drums, barrels, carboys, and cans are used to transport fluids. Interstate transportation is regulated by the DOT. There are other regulating agencies—state, local, and private. Railroads make rules determining what they will accept, some states require compliance with DOT specifications on intrastate movements, and tunnel authorities as well as fire chiefs apply restrictions. Water shipments involve regulations of the U.S. Coast Guard. The American Bureau of Shipping sets rules for design and construction which are recognized by insurance underwriters.

The most pertinent *DOT regulations* (*Code of Federal Regulations,* Title 18, Parts 171–179 and 397) were published by R. M. Graziano (then agent and attorney for carriers and freight forwarders) in his tariff titled *Hazardous Materials Regulations of the Department of Transportation* (1978). New tariffs identified by number are issued at intervals, and interim revisions are sent out. Agents change at intervals.

Graziano's tariff lists many regulated (dangerous) commodities (Part 172, DOT regulations) for transportation. This includes those that are poisonous, flammable, oxidizing, corrosive, explosive, radioactive, and compressed gases. Part 178 covers specifications for all types of containers from carboys to large portable tanks and tank trucks. Part 179 deals with tank-car construction.

Thickness, in	Carbon steel	304 Stainless steel	316 Stainless steel
½	1.4	1.8	1.8
¾	2.1	2.5	2.6
1	2.7	3.3	3.5

To convert gallons to cubic meters, multiply by 3.786 × 10^{-3}.

The Association of American Railroads (AAR) publication *Specifications for Tank Cars* covers many requirements beyond the DOT regulations.

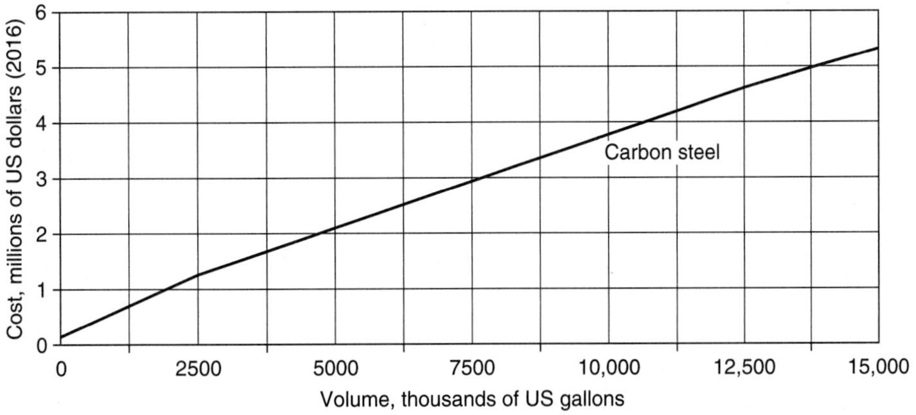

FIG. 10-183 Cost (±30 percent) of field-erected, floating roof tanks, October 2016, includes concrete foundation and typical nozzles, ladders, and platforms. 1 gal = 0.003785 m^3.

Some additional details are given later. Because of frequent changes, it is always necessary to check the latest rules. The *shipper,* not the carrier, has the ultimate responsibility for shipping in the correct container.

Tank Cars These range in size from about 7.6 to 182 m^3 (2000 to 48,000 gal), and a car may be single-unit or multiunit. The DOT now limits them to 130 m^3 (34,500 gal) and 120,000 kg (263,000 lb) gross weight. Large cars usually result in lower investment per cubic meter and take lower shipping rates. Cars may be insulated to reduce heating or cooling of the contents. Certain liquefied gases may be carried in insulated cars; temperatures are maintained by evaporation (see subsection Low-Temperature and Cryogenic Storage). Cars may be heated by steam coils or by electricity. Some products are loaded hot, solidify in transport, and are melted for removal. Some low-temperature cargoes must be unloaded within a given time (usually 30 days) to prevent pressure buildup.

Tank cars are classified as pressure or general-purpose. Pressure cars have relief-valve settings of 517 kPa (75 lbf/in^2) and above. Those designated as general-purpose cars are, nevertheless, pressure vessels and may have relief valves or rupture disks. The DOT specification code number indicates the type of car. For instance, 105A500W indicates a pressure car with a test pressure of 3447 kPa (500 lbf/in^2) and a relief-valve setting of 2585 kPa (375 lbf/in^2). In most cases, loading and unloading valves, safety valves, and vent valves must be in a dome or an enclosure.

Companies shipping dangerous materials sometimes build tank cars with metal thicker than required by the specifications in order to reduce the possibility of leakage during a wreck or fire. The punching of couplers or rail ends into heads of tanks is a hazard.

Older tank cars have a center sill or beam running the entire length of the car. Most modern cars have no continuous sill, only short stub sills at each end. Cars with full sills have tanks anchored longitudinally at the center of the sill. The anchor is designed to be weaker than either the tank shell or the doubler plate between anchor and shell. Cars with stub sills have similar safeguards. Anchors and other parts are designed to meet AAR requirements.

The impact forces on car couplers put high stresses on sills, anchors, and doublers. This may start fatigue cracks in the shell, particularly at the corners of welded doubler plates. With brittle steel in cold weather, such cracks sometimes cause complete rupture of the tank. Large end radii on the doublers and tougher steels will reduce this hazard. Inspection of older cars can reveal cracks prior to failure.

A difference between tank cars and most pressure vessels is that tank cars are designed in terms of the theoretical ultimate or bursting strength of the tank. The test pressure is usually 40 percent of the bursting pressure (sometimes less). The safety valves are set at 75 percent of the test pressure. Thus, the maximum operating pressure is usually 30 percent of the bursting pressure. This gives a nominal factor of safety of 3.3, compared with 3.5 for Division 1 of the ASME Boiler and Pressure Vessel Code.

The DOT rules require that pressure cars have relief valves designed to limit pressure to 82.5 percent (with certain exceptions) of test pressure (110 percent of maximum operating pressure) when exposed to fire. Appendix A of AAR Specifications deals with the flow capacity of relief devices. The formulas apply to cars in the upright position with the device discharging vapor. They may not protect the car adequately when it is overturned and the device is discharging liquid.

Appendix B of AAR Specifications deals with the certification of facilities. Fabrication, repairing, testing, and specialty work on tank cars must be done in certified facilities. The AAR certifies shops to build cars of certain materials, to do test work on cars, or to make certain repairs and alterations.

Tank Trucks These trucks may have single, compartmented, or multiple tanks. Many of their requirements are similar to those for tank cars, except that thinner shells are permitted in most cases. Trucks for nonhazardous materials are subject to few regulations other than the normal highway laws governing all motor vehicles. But trucks carrying hazardous materials must comply with DOT regulations, Parts 173, 177, 178, and 397. Maximum weight, axle loading, and length are governed by state highway regulations. Many states have limits in the vicinity of 31,750 kg (70,000 lb) total weight, 14,500 kg (32,000 lb) for tandem axles, and 18.3 m (60 ft) or less overall length. Some allow tandem trailers.

Truck cargo tanks (for dangerous materials) are built under Part 173 and Subpart J of Part 178, DOT regulations. This includes Specifications MC-306, MC-307, MC-312, and MC-331. MC-331 is required for compressed gas. Subpart J requires tanks for pressures above 345 kPa (50 lbf/in^2) in one case and 103 kPa (15 lbf/in^2) in another to be built according to the ASME Boiler and Pressure Vessel Code. A particular issue of the code is specified.

Because of the demands of highway service, the DOT specifications have a number of requirements in addition to the ASME Code. These include design for impact forces and rollover protection for fittings.

Portable tanks, drums, or bottles are shipped by rail, ship, air, or truck. Portable tanks containing hazardous materials must conform to DOT regulations, Parts 173 and 178, Subpart H.

Some tanks are designed to be shipped by trailer and transferred to railcars or ships (see following discussion).

Marine Transportation Seagoing *tankers* are for high tonnage. The traditional tanker uses the ship structure as a tank. It is subdivided into a number of tanks by means of transverse bulkheads and a centerline bulkhead. More than one product can be carried. An elaborate piping system connects the tanks to a pumping plant which can discharge or transfer the cargo. Harbor and docking facilities appear to be the only limit to tanker size. The largest crude oil tanker size to date is about 560,000 deadweight tons. In the United States, tankers are built to specifications of the American Bureau of Shipping and the U.S. Coast Guard.

Low-temperature liquefied gases are shipped in special ships with insulation between the hull and an inner tank. The largest LNG carrier's capacity is about 145,000 m^3. Poisonous materials are shipped in separate tanks built into the ship. This prevents tank leakage from contaminating harbors. Separate tanks are also used to transport pressurized gases.

Barges are used on inland waterways. Popular sizes are up to 16 m (52½ ft) wide by 76 m (250 ft) long, with 2.6-m (8½-ft) to 4.3-m (14-ft) draft. Cargo requirements and waterway limitations determine the design. Use of barges of uniform size facilitates rafting them together.

Portable tanks may be stowed in the holds of conventional cargo ships or special container ships, or they may be fastened on deck.

Container ships have guides in the hold and on deck which hold boxlike containers or tanks. The tank is latched to a trailer chassis and hauled to shipside. A movable gantry, sometimes permanently installed on the ship, hoists the tank from the trailer and lowers it into the guides on the ship. This system achieves large savings in labor, but its application is sometimes limited by lack of agreement between ship operators and unions.

Portable tanks for regulated commodities in marine transportation must be designed and built under Coast Guard regulations (see discussion under Pressure Vessels).

Materials of Construction for Bulk Transport Because of the more severe service, construction materials for transportation usually are more restricted than for storage. Most large pipelines are constructed of steel conforming to API Specification 5L or 5LX. Most tanks (cars, etc.) are built of pressure-vessel steels or AAR specification steels, with a few made of aluminum or stainless steel. Carbon-steel tanks may be lined with rubber, plastic, nickel, glass, or other materials. In many cases this is practical and cheaper than using a stainless-steel tank. Other materials for tank construction may be proposed and used if approved by the appropriate authorities (AAR and DOT).

PRESSURE VESSELS

This discussion of pressure vessels is intended as an overview of the codes most frequently used for the design and construction of pressure vessels. Chemical engineers who design or specify pressure vessels should determine the federal and local laws relevant to the problem and then refer to the most recent issue of the pertinent code or standard before proceeding. Laws, codes, and standards are frequently changed.

A *pressure vessel* is a closed container of limited length (in contrast to the indefinite length of piping). Its smallest dimension is considerably larger than the connecting piping, and it is subject to pressures above 7 or 14 kPa (1 or 2 lbf/in^2). It is distinguished from a boiler, which in most cases is used to generate steam for use external to itself.

Code Administration The American Society of Mechanical Engineers has written the ASME Boiler and Pressure Vessel Code (BPVC), which contains rules for the design, fabrication, and inspection of boilers and pressure vessels. The ASME Code is an American National Standard. Most states in the United States and all Canadian provinces have passed legislation which makes the ASME Code or certain parts of it their legal requirement. Only a few jurisdictions have adopted the code for all vessels. The others apply it to certain types of vessels or to boilers. States employ inspectors (usually under a chief boiler inspector) to enforce code provisions. The authorities also depend a great deal on insurance company inspectors to see that boilers and pressure vessels are maintained in a safe condition.

The ASME Code is written by a large committee and many subcommittees, composed of engineers appointed by the ASME. The Code Committee meets regularly to review the code and consider requests for its revision, interpretation, or extension. *Interpretation* and *extension* are accomplished through "code cases." The decisions are published in *Mechanical Engineering.* Code cases are also mailed to those who subscribe to the service. A typical code case might be the approval of the use of a metal which is not presently on the list of approved code materials. Inquiries relative to code cases should be addressed to the secretary of the ASME Boiler and Pressure Vessel Committee, American Society of Mechanical Engineers, New York.

A new edition of the code is issued every 3 years. Between editions, alterations are handled by issuing semiannual addenda, which may be purchased by subscription. The ASME considers any issue of the code to be adequate and safe, but some government authorities specify certain issues of the code as their legal requirement.

Inspection Authority The National Board of Boiler and Pressure Vessel Inspectors is composed of the chief inspectors of states and municipalities in the United States and Canadian provinces who have made any part of the Boiler and Pressure Vessel Code a legal requirement. This board promotes uniform enforcement of boiler and pressure-vessel rules. One of the board's important activities is to provide examinations for, and commissioning of, inspectors. Inspectors so qualified and employed by an insurance company, state, municipality, or Canadian province may inspect a pressure vessel and permit it to be stamped ASME—NB (National Board). An inspector employed by a vessel user may authorize the use of only the ASME stamp. The ASME Code Committee authorizes fabricators to use the various ASME stamps. The stamps, however, may be applied to a vessel only with the approval of the inspector.

The ASME Boiler and Pressure Vessel Code consists of eleven sections as follows:

I. Power Boilers
II. Materials
  a. Ferrous
  b. Nonferrous
  c. Welding rods, electrodes, and filler metals
  d. Properties
III. Rules for Construction of Nuclear Power Plant Components
IV. Heating Boilers
V. Nondestructive Examination
VI. Rules for Care and Operation of Heating Boilers
VII. Guidelines for the Care of Power Boilers
VIII. Pressure Vessels
IX. Welding and Brazing Qualifications
X. Fiber-Reinforced Plastic Pressure Vessels
XI. Rules for In-service Inspection of Nuclear Power Plant Components

Pressure vessels (as distinguished from boilers) are involved with Secs. II, III, V, VIII, IX, X, and XI. Section VIII, Division 1, is the Pressure Vessel Code as it existed in the past (and will continue). Division 1 was brought out as a means of permitting higher design stresses while ensuring at least as great a degree of safety as in Division 1. These two divisions plus Secs. III and X will be discussed briefly here. They refer to Secs. II and IX.

ASME Code Section VIII, Division 1 Most pressure vessels used in the process industry in the United States are designed and constructed in accordance with Sec. VIII, Division 1 (see Fig. 10-184). This division is divided into three subsections followed by appendices.

Introduction The Introduction contains the scope of the division and defines the responsibilities of the user, the manufacturer, and the inspector. The scope defines pressure vessels as containers for the containment of pressure. It specifically excludes vessels having an internal pressure not exceeding 103 kPa (15 lbf/in^2) and further states that the rules are applicable for pressures not exceeding 20,670 kPa (3000 lbf/in^2). For higher pressures it is usually necessary to deviate from the rules in this division.

The scope covers many other less basic exclusions, and inasmuch as the scope is occasionally revised, except for the most obvious cases, it is prudent to review the current issue before specifying or designing pressure vessels to this division. Any vessel that meets all the requirements of this division may be stamped with the code *U* symbol even though exempted from such stamping.

Subsection A This subsection contains the general requirements applicable to all materials and methods of construction. Design temperature and pressure are defined here, and the loadings to be considered in design are specified. For stress failure and yielding, this section of the code uses the maximum-stress theory of failure as its criterion.

This subsection refers to the tables elsewhere in the division in which the maximum allowable tensile stress values are tabulated. The basis for the establishment of these allowable stresses is defined in detail in App. P; however, as the safety factors used were very important in establishing the various rules of this division, note that the safety factors for internal-pressure loads are 3.5 on ultimate strength and 1.6 or 1.5 on yield strength, depending on the material. For external-pressure loads on cylindrical shells, the safety factors are 3 for both elastic buckling and plastic collapse. For other shapes subject to external pressure and for longitudinal shell compression, the safety factors are 3.5 for both elastic buckling and plastic collapse. Longitudinal compressive stress in cylindrical elements is limited in this subsection by the lower of either stress failure or buckling failure.

Internal-pressure design rules and formulas are given for cylindrical and spherical shells and for ellipsoidal, torispherical (often called ASME heads), hemispherical, and conical heads. The formulas given assume membrane-stress failure, although the rules for heads include consideration for buckling failure in the transition area from cylinder to head (knuckle area).

Longitudinal joints in cylinders are more highly stressed than circumferential joints, and the code takes this fact into account. When forming heads, there is usually some thinning from the original plate thickness in the knuckle area, and it is prudent to specify the minimum allowable thickness at this point.

Unstayed flat heads and covers can be designed by very specific rules and formulas given in this subsection. The stresses caused by pressure on these members are bending stresses, and the formulas include an allowance for additional edge moments induced when the head, cover, or blind flange is attached by bolts. Rules are provided for quick-opening closures because of the risk of incomplete attachment or opening while the vessel is pressurized. Rules for braced and stayed surfaces are also provided.

External-pressure failure of shells can result from overstress at one extreme or from elastic instability at the other or at some intermediate loading. The code provides the solution for most shells by using a number of charts. One chart is used for cylinders where the shell diameter-to-thickness ratio and the length-to-diameter ratio are the variables. The rest of the charts depict curves relating the geometry of cylinders and spheres to allowable stress by curves which are determined from the modulus of elasticity, tangent modulus, and yield strength at temperatures for various materials or classes of materials. The text of this subsection explains how the allowable stress is determined from the charts for cylinders, spheres, and hemispherical, ellipsoidal, torispherical, and conical heads.

Frequently cost savings for cylindrical shells can result from reducing the effective length-to-diameter ratio and thereby reducing shell thickness. This can be accomplished by adding circumferential stiffeners to the shell. Rules are included for designing and locating the stiffeners.

Openings are always required in pressure-vessel shells and heads. Stress intensification is created by the existence of a hole in an otherwise symmetric section. The code compensates for this by an area-replacement method. It takes a cross section through the opening, and it measures the area of the metal of the required shell that is removed and replaces it in the cross section by additional material (shell wall, nozzle wall, reinforcing plate, or weld) within certain distances of the opening centerline. These rules and formulas for calculation are included in Subsection A.

When a cylindrical shell is drilled for the insertion of multiple tubes, the shell is significantly weakened and the code provides rules for tube-hole patterns and the reduction in strength that must be accommodated.

Fabrication tolerances are covered in this subsection. The tolerances permitted for shells for external pressure are much closer than those for internal pressure because the stability of the structure is dependent on the symmetry. Other paragraphs cover repair of defects during fabrication, material identification, heat treatment, and impact testing.

Inspection and testing requirements are covered in detail. Most vessels are required to be hydrostatically tested (generally with water) at 1.3 times the maximum allowable working pressure. Some enameled (glass-lined) vessels are permitted to be hydrostatically tested at lower pressures. Pneumatic tests are permitted and are carried to at least 1.25 times the maximum allowable working pressure, and there is provision for proof testing when the strength of the vessel or any of its parts cannot be computed with satisfactory assurance of accuracy. Pneumatic or proof tests are rarely conducted because release of the stored energy of compression of the test gas can cause an explosion upon failure of the vessel under test.

Pressure-relief device requirements are defined in Subsection A. Set point and maximum pressure during relief are defined according to the service, the cause of overpressure, and the number of relief devices. Safety, safety relief, relief valves, rupture disk, rupture pin, and rules on tolerances for the relieving point are given.

Testing, certification, and installation rules for relieving devices are extensive. Every chemical engineer responsible for the design or operation of process units should become very familiar with these rules. The pressure-relief device paragraphs are the only parts of Sec. VIII, Division 1, that are concerned with the installation and ongoing operation of the facility; all other rules apply only to the design and manufacture of the vessel.

Subsection B This subsection contains rules pertaining to the methods of fabrication of pressure vessels. Part UW is applicable to welded vessels. Service restrictions are defined. Lethal service is for *lethal substances,* defined as poisonous gases or liquids of such a nature that a very small amount of the gas or the vapor of the liquid mixed or unmixed with air is dangerous to life when inhaled. It is stated that it is the user's responsibility to advise the designer or manufacturer if the service is lethal. All vessels in lethal service shall have all butt-welded joints fully radiographed, and when practical, joints shall be butt-welded. All vessels fabricated of carbon steel or low-alloy steel shall be postweld-heat-treated.

Low-temperature service is defined as being below −29°C (−20°F), and impact testing of many materials is required. The code is restrictive in the type of welding permitted.

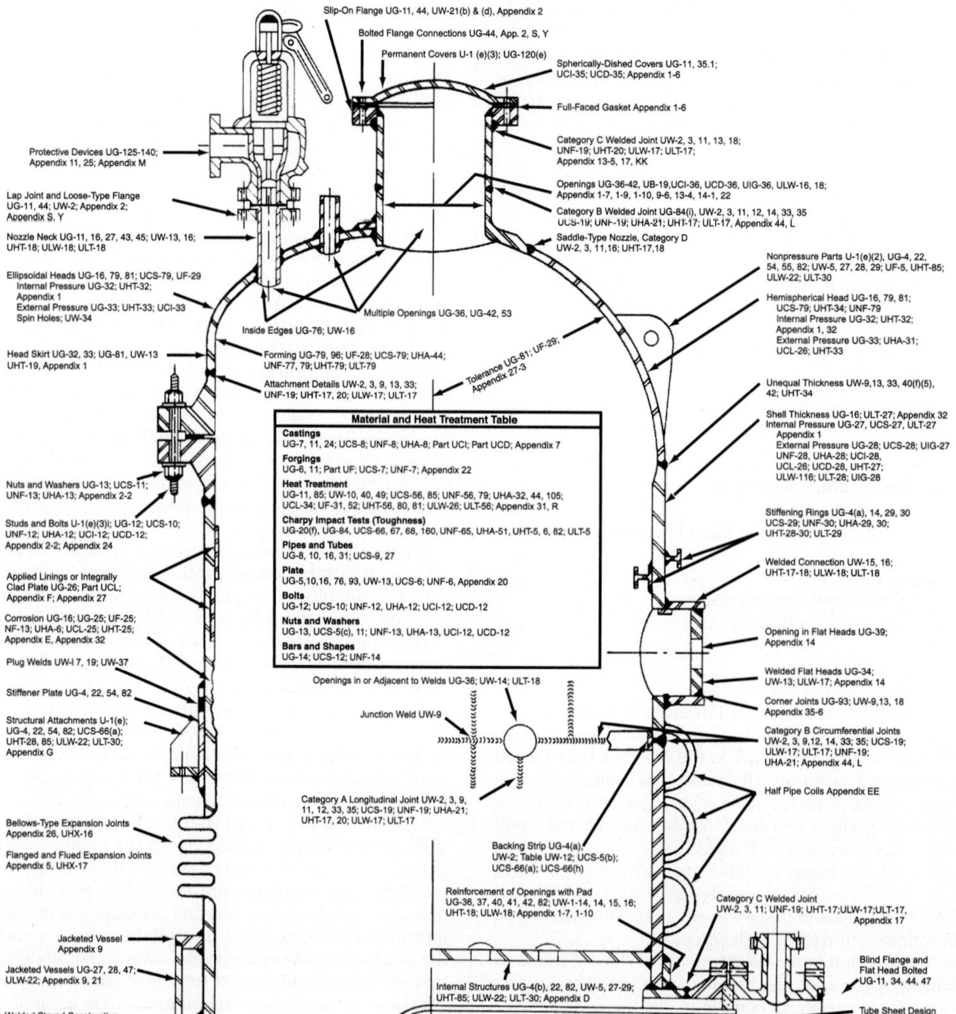

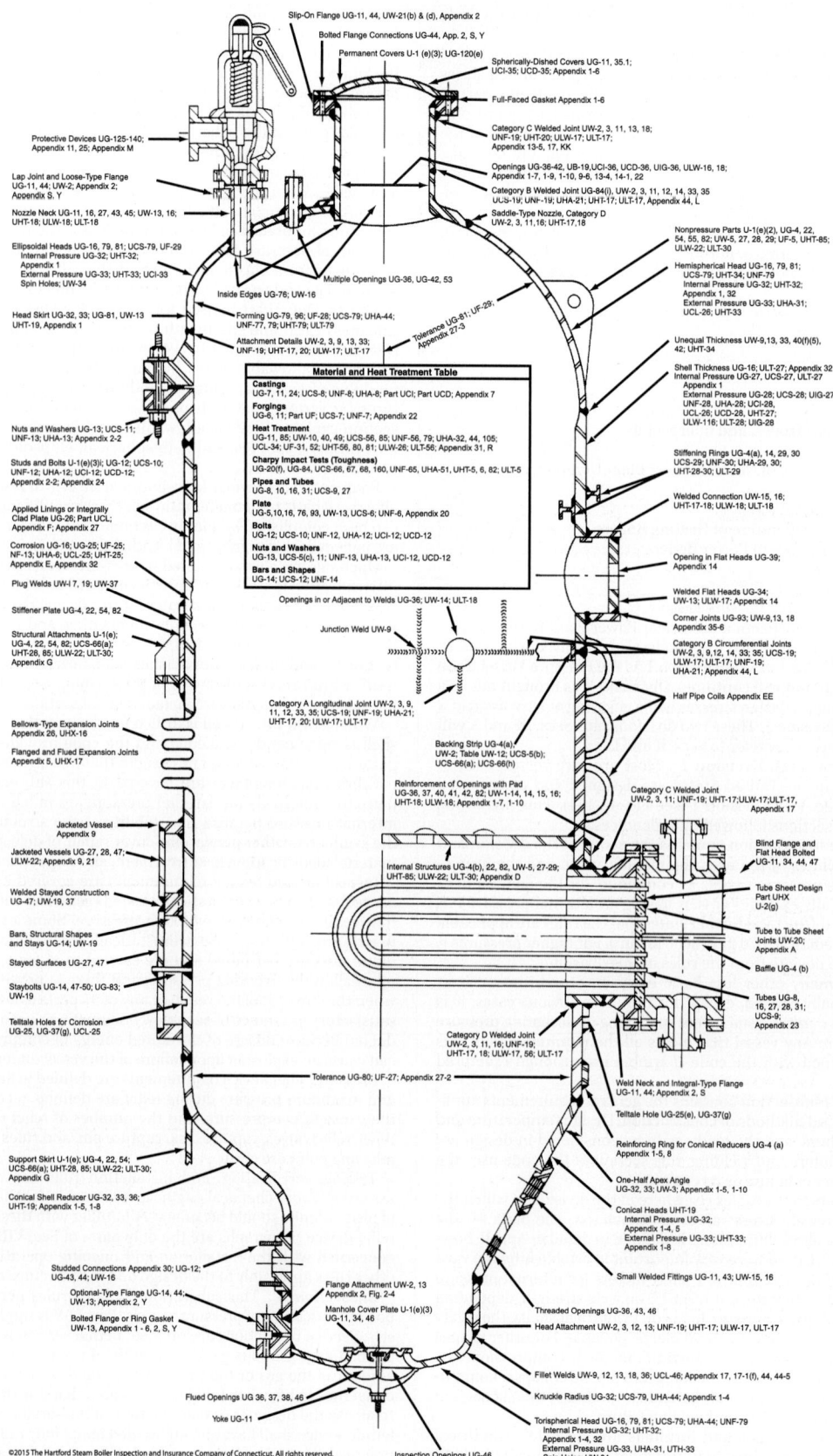

FIG. 10-184 Quick reference guide to ASME Boiler and Pressure Vessel Code Section VIII, Division 1 (2014 edition). (*Reprinted with permission of publisher, HSB Global Standards, Hartford, Conn.*)

ORGANIZATION

Introduction	Scope, General, Referenced Standards, Units
Subsection A	Part UG-General requirements for all construction and all materials
Subsection B	Requirements for methods of fabrication
Part UW	Welding
Part UF	Forging
Part UB	Brazing
Subsection C	Requirements for classes of materials
Part UCS	Carbon and low alloy steels
Part UNF	Nonferrous materials
Part UHA	High alloy steels
Part UCI	Cast iron
Part UCL	Clad plate and corrosion resistant liners
Part UCD	Cast ductile iron
Part UHT	Ferritic steels with tensile properties enhanced by heat treatment
Part ULW	Layered Construction
Part ULT	Low Temperature Materials
Part UHX	Rules for shell and tube heat exhangers
Part UIG	Requirements for pressure vessels constructed of Impregnated Graphite

Mandatory Appendices 1-44

Nonmandatory Appendices A, C, D-H, DD, EE, FF, GG, HH, JJ, KK, LL, MM, NN, K-M, P, R-T, W, Y

Endnotes

GENERAL NOTES

Governing Code Editions and Cases for Pressure Vessels and Parts .... Appendix 43
Code Jurisdiction for Piping U-1
Design Pressure .... UG-21; UG-98; App. 3; UCI-3; UCD-3
Design Temperature .... UG-20; UCL-24; UCD-3; App. C
Dimpled or Embossed Assemblies .... App. 17; UW-19
Inspector's Responsibility U-2; UG-90
Loadings .... UG-22; App. G, H
Low Temperature Service .... UG-20(f); UW-2; UCS-66-68; .... UNF-65; UHA-51; Part ULT
Manufacturer's Responsibility U-2; UG-90
User's Responsibility .... Appendix NN
Material, General UG-4, 10, 11, 15
- a. Bolts and Studs UG-12
- b. Castings UG-7
- c. Forgings UG-6
- d. Nuts and Washers UG-13
- e. Pipes and Tubes UG-8
- f. Plates UG-5
- g. Rod and Bar UG-14
- h. Standard Parts UG-11, 44
- i. Welding UG-9

Material Identification, Marking, and Certification UG-77, 93, 94
Material Tolerances UG-16
Nameplates, Stamping and Reports ....UG-115-120; .... UHT-115; ULW-115; ULT-115; App. W & 18
Nondestructive Examination
- a. Liquid Penetrant .... App. 8
- b. Magnetic Particle .... App. 6
- c. Radiography UW-51, 52
- d. Ultrasonic .... UW-53, App. 12

Porosity Charts ....App. 4
Pressure Tests .... UG-99, 100, 101; UW-50 .... UCI-99; UCD-99; ULT-99; App. 44
Quality Control System .... U-2; App. 10
Quick Actuating Closures .... U-1; UG-35; App. FF
Service Restrictions UW-2; UB-3; UCI-2; UCL-3; UCD-2
Stress, Maximum Allowable ..UG-23; App. P; UCS-23; UNF-23; UHA-23; UCI-23; UCL-23; UCD-23; UHT-23; ULT-23
Units .... U-4; App. 33, GG
User's Responsibility....................... UG-125; U-2; App. KK, NN
Welding Preheat .... App. R
Welding Qualifications .... UW-26 thru 29; UNF-95; UHA-52; UHT-82; ULT-82

PARA	REQUIREMENTS/REMARKS	PT	MT	UT	RT
UG-24	General requirements for castings	X	X	X	X
UG-93	General requirements for the inspection of all materials	X	X		
UW-11	Radiographic and ultrasonic examination required for pressure vessels and vessel parts (also see UW-51, UW-52, UW-53) [*UW-11 (e)(2)]	X*	X*	X	X
UW-13	Surface examination for offset head-to-shell butt joint	X	X		
UW-42	Surface weld metal buildup	X	X		X
UW-50	Welds on pneumatically tested vessels	X	X		
UF-31	Vessels fabricated from SA-372 forging material to be liquid quenched and tampered	X	X		
UF-32	Finished welds after postweld heat treatment	X	X		
UF-37	Repair welds in forgings	X	X		X
UF-55	Vessels constructed of SA-372 Class VIII material			X	
UCS-56	Surface examination following weld repairs	X	X	X	X
UCS-57	Examination in addition to UW-11 for butt welded joints on carbon and low alloy steel pressure vessels and vessel parts			X	X
UCS-68	Exemption from impact testing when RT is performed	X	X		
UNF-56	Surface examination following weld repair in nonferrous materials	X	X		
UNF-57	Examination in addition to UW-11 for pressure vessels and vessel parts constructed of nonferrous materials			X	X
UNF-58	All groove and fillet wels in vessels constructed of certain nonferrious materials	X			
UHA-33	Exceptions for radiographic examinations of high alloy steel vessels			X	X

Qualifications of Personnel Performing Section VIII, Division 1 Nondestructive Examinations			
Examination	**Reference**	**Examination**	**Reference**
PT	Appendix 8 (UG-103)	UT	UW-53
MT	Appendix 6 (UG-103)	RT	UW-51

PARA	REQUIREMENTS/REMARKS	PT	MT	UT	RT
UHA-34	Austenitic chromium-nickel alloy steel but and fillet welds	X			
UCI-78	Repair of defects in cast iron pressure vessels and vessel parts	X	X	X	X
UCL-35	Vessels or parts of vessels constructed of clad plate and those having applied corrosion resistant linings				X
UCL-36	Chromium stainless steel cladding or lining				X
UCD-78	Repair of defects in cast ductile iron pressure vessels and vessel parts				X
UHT-57	Pressure vessels or vessel parts constructed of ferritic steels having tensile properties enhanced by heat treatment	X	X		X
UHT-83	Metal removal accomplished by methods involving melting on pressure vessels and vessel parts constructed of ferritic steels haivng tensile properies enhanced by heat treatment	X	X		
UHT-85	Removal of temporary welds on pressure vessels and vessel parts constructed of ferritic steels having tensile properties enhanced by heat treatment	X	X		
ULW-51	Inner shells and inner heads of layered pressure vessels				X
ULW-52	Welded joints in the layers of layered pressure vessels		X	X	X
ULW-53	Step welded girth joints in the layers of layered pressure vessels		X	X	
ULW-54	Butt welded joints in layered pressure vessels				X
ULW-55	Flat head and tube sheet welded joints in layered pressure vessels		X	X	X
ULW-56	Nozzle and communicating chamber welded joints in layered pressure vessels	X	X		
ULW-57	Random spot examinations and repairs of welds in layered pressure vessels		X	X	
ULT-57	Welds in pressure vessels and vessel parts constructed of materials having increased design stress values due to low temperature applications	X	X	X	

FIG. 10-184 *(Continued)*

Unfired steam boilers with design pressures exceeding 345 kPa (50 lbf/in^2) have restrictive rules on welded-joint design, and all butt joints require full radiography.

Pressure vessels subject to direct firing have special requirements relative to welded-joint design and postweld heat treatment.

This subsection includes rules governing welded-joint designs and the degree of radiography, with efficiencies for welded joints specified as functions of the quality of joint. These efficiencies are used in the formulas in Subsection A for determining vessel thicknesses.

Details are provided for head-to-shell welds, tube sheet-to-shell welds, and nozzle-to-shell welds. Acceptable forms of welded stay-bolts and plug and slot welds for staying plates are given here.

Rules for the welded fabrication of pressure vessels cover welding processes, manufacturer's record keeping on welding procedures, welder qualification, cleaning, fit-up alignment tolerances, and repair of weld defects. Procedures for postweld heat treatment are detailed. Checking the procedures and welders and radiographic and ultrasonic examination of welded joints are covered.

Requirements for vessels fabricated by forging in Part UF include unique design requirements with particular concern for stress risers, fabrication, heat treatment, repair of defects, and inspection. Vessels fabricated by brazing are covered in Part UB. Brazed vessels cannot be used in lethal service, for unfired steam boilers, or for direct firing. Permitted brazing processes and testing of brazed joints for strength are covered. Fabrication and inspection rules are also included.

Subsection C This subsection contains requirements pertaining to classes of materials. Carbon steels and low-alloy steels are governed by Part UCS, nonferrous materials by Part UNF, high-alloy steels by Part UHA, and steels with tensile properties enhanced by heat treatment by Part UHT. Each of these parts includes tables of maximum allowable stress values for all code materials for a range of metal temperatures. These stress values include appropriate safety factors. Rules governing the application, fabrication, and heat treatment of the vessels are included in each part.

Part UHT also contains more stringent details for nozzle welding that are required for some of these high-strength materials. Part UCI has rules for cast-iron construction, Part UCL has rules for welded vessels of clad plate as lined vessels, and Part UCD has rules for ductile-iron pressure vessels.

A relatively recent addition to the code is Part ULW, which contains requirements for vessels fabricated by layered construction. This type of construction is most frequently used for high pressures, usually in excess of 13,800 kPa (2000 lbf/in^2).

There are several methods of layering in common use: (1) thick layers shrunk together; (2) thin layers, each wrapped over the other and the longitudinal seam welded by using the prior layer as backup; and (3) thin layers spirally wrapped. The code rules are written for either thick or thin layers. Rules and details are provided for all the usual welded joints and nozzle reinforcement. Supports for layered vessels require special consideration, in that only the outer layer could contribute to the support. For lethal service, only the inner shell and inner heads need comply with the requirements in Subsection B. Inasmuch as radiography would not be practical for inspection of many of the welds, extensive use is made of magnetic-particle and ultrasonic inspection. When radiography is required, the code warns the inspector that indications sufficient for rejection in single-wall vessels may be acceptable. Vent holes are specified through each layer down to the inner shell to prevent buildup of pressure between layers in the event of leakage at the inner shell.

Mandatory Appendices These include a section on supplementary design formulas for shells not covered in Subsection A. Formulas are given for thick shells, heads, and dished covers. Another appendix gives very specific rules, formulas, and charts for the design of bolted-flange connections. The nature of these rules is such that they are readily computer-programmable, and most flanges now are computer-designed. One appendix includes only the charts used for calculating shells for external pressure discussed previously. Jacketed vessels are covered in a separate appendix in which very specific rules are given, particularly for the attachment of the jacket to the inner shell. Other appendices cover inspection and quality control.

Nonmandatory Appendices These cover a number of subjects, primarily suggested good practices and other aids in understanding the code and in designing with the code. Several current nonmandatory appendixes will probably become mandatory.

Figure 10-184 illustrates a pressure vessel with the applicable code paragraphs noted for the various elements. Additional important paragraphs are referenced at the bottom of the figure.

ASME BPVC Section VIII, Division 2 Paragraph AG-100e of Division 2 states:

> In relation to the rules of Division 1 of Section VIII, these rules of Division 2 are more restrictive in the choice of materials which may be used but permit higher design stress intensity values to be employed in the range of temperatures over which the design stress intensity value is controlled by the ultimate strength or the yield strength; more precise design procedures are required and some common design details are prohibited; permissible fabrication procedures are specifically delineated and more complete testing and inspection are required.

Most Division 2 vessels fabricated to date have been large or intended for high pressure and, therefore, expensive when the material and labor savings resulting from smaller safety factors have been greater than the additional engineering, administrative, and inspection costs.

The organization of Division 2 differs from that of Division 1.

Part AG This part gives the scope of the division, establishes its jurisdiction, and sets forth the responsibilities of the user and the manufacturer. Of particular importance is the fact that no upper limitation in pressure is specified and that a user's design specification is required. The user or the user's agent shall provide requirements for intended operating conditions in such detail as to constitute an adequate basis for selecting materials and designing, fabricating, and inspecting the vessel. The user's design specification shall include the method of supporting the vessel and any requirement for a fatigue analysis. If a fatigue analysis is required, the user must provide information in sufficient detail that an analysis for cyclic operation can be made.

Part AM This part lists permitted individual construction materials, applicable specifications, special requirements, design stress intensity values, and other property information. Of particular importance are the ultrasonic test and toughness requirements. Among the properties for which data are included are thermal conductivity and diffusivity, coefficient of thermal expansion, modulus of elasticity, and yield strength. The design stress intensity values include a safety factor of 3 on ultimate strength at temperature or 1.5 on yield strength at temperature.

Part AD This part contains requirements for the design of vessels. The rules of Division 2 are based on the maximum-shear theory of failure for stress failure and yielding. Higher stresses are permitted when wind or earthquake loads are considered. Any rules for determining the need for fatigue analysis are given here.

Rules for the design of shells of revolution under internal pressure differ from the Division 1 rules, particularly the rules for formed heads when plastic deformation in the knuckle area is the failure criterion. Shells of revolution for external pressure are determined on the same criterion, including safety factors, as in Division 1. Reinforcement for openings uses the same area-replacement method as Division 1; however, in many cases the reinforcement metal must be closer to the opening centerline.

The rest of the rules in Part AD for flat heads, bolted and studded connections, quick-actuating closures, and layered vessels essentially duplicate Division 1. The rules for support skirts are more definitive in Division 2.

Part AF This part contains requirements governing the fabrication of vessels and vessel parts.

Part AR This part contains rules for pressure-relieving devices.

Part AI This part contains requirements controlling inspection of the vessel.

Part AT This part contains testing requirements and procedures.

Part AS This part covers requirements for stamping and certifying the vessel and vessel parts.

Appendices Appendix 1 defines the basis used for defining stress intensity values. Appendix 2 contains external-pressure charts, and App. 3 has the rules for bolted-flange connections; these two are exact duplicates of the equivalent appendices in Division 1.

Appendix 4 gives definitions and rules for stress analysis for shells, flat and formed heads, and tube sheets, layered vessels, and nozzles including discontinuity stresses. Of particular importance are Table 4-120.1, Classification of Stresses for Some Typical Cases, and Fig. 4-130.1, Stress Categories and Limits of Stress Intensity. These are very useful in that they clarify a number of paragraphs and simplify stress analysis.

Appendix 5 contains rules and data for stress analysis for cyclic operation. Except in short-cycle batch processes, pressure vessels are usually subject to few cycles in their projected lifetime, and the endurance-limit data used in the machinery industries are not applicable. Curves are given for a broad spectrum of materials, covering a range from 10 to 1 million cycles with allowable stress values as high as 650,000 lbf/in^2. This low-cycle fatigue has been developed from strain-fatigue work in which stress values are obtained by multiplying the strains by the modulus of elasticity. Stresses of this magnitude cannot occur, but strains do. The curves given have a factor of safety of 2 on stress or 20 on cycles.

Appendix 6 contains the requirements of experimental stress analysis, Appendix 8 has acceptance standards for radiographic examination, Appendix 9 covers nondestructive examination, Appendix 10 gives rules for capacity conversions for safety valves, and App. 18 details quality control system requirements.

The remaining appendices are nonmandatory but useful to engineers working with the code.

General Considerations Most pressure vessels for the chemical-process industry will continue to be designed and built to the rules of BPVC, Section VIII, Division 1. While the rules of Section VIII, Division 2, will frequently provide thinner elements, the cost of the engineering analysis, stress analysis and higher-quality construction, material control, and inspection required by these rules frequently exceeds the savings from the use of thinner walls.

Additional ASME Code Considerations

ASME BPVC Section III: Nuclear Power Plant Components This section of the code includes vessels, storage tanks, and concrete containment vessels as well as other nonvessel items.

ASME BPVC Section X: Fiberglass–Reinforced-Plastic Pressure Vessels This section is limited to four types of vessels: bag-molded and centrifugally cast, each limited to 1000 kPa (150 lbf/in^2); filament-wound with cut filaments limited to 10,000 kPa (1500 lbf/in^2); and filament-wound with uncut filaments limited to 21,000 kPa (3000 lbf/in^2). Operating temperatures are limited to the range from +66°C (150°F) to −54°C (−65°F). Low modulus of elasticity and other property differences between metal and plastic required that many of the procedures in Section X be different from those in the sections governing metal vessels. The requirement that at least one vessel of a particular design and fabrication shall be tested to destruction has prevented this section from being widely used. The results from the combined fatigue and burst test must give the design pressure a safety factor of 6 to the burst pressure.

Safety in Design Designing a pressure vessel in accordance with the code, under most circumstances, will provide adequate safety. In the code's own words, however, the rules "cover minimum construction requirements for the design, fabrication, inspection, and certification of pressure vessels." The significant word is *minimum.* The *ultimate responsibility* for safety rests with the user and the designer. They must decide whether anything beyond code requirements is necessary. The code cannot foresee and provide for all the unusual conditions to which a pressure vessel might be exposed. If it tried to do so, the majority of pressure vessels would be unnecessarily restricted. Some of the conditions that a vessel might encounter are unusually low temperatures, unusual thermal stresses, stress ratcheting caused by thermal cycling, vibration of tall vessels excited by von Karman vortices caused by wind, very high pressures, runaway chemical reactions, repeated local overheating, explosions, exposure to fire, exposure to materials that rapidly attack the metal, containment of extremely toxic materials, and very large sizes of vessels. Large vessels, although they may contain nonhazardous materials, by their very size, could create a serious hazard if they burst. The failure of the Boston molasses tank in 1919 killed 12 people. For pressure vessels which are outside code jurisdiction, there are sometimes special hazards in very high-strength materials and plastics. There may be many others which the designers should recognize if they encounter them.

Metal fatigue, when it is present, is a serious hazard. BPVC Section VIII, Division 1, mentions rapidly fluctuating pressures. Division 2 and Section III do require a fatigue analysis. In extreme cases, vessel contents may affect the fatigue strength (endurance limit) of the material. This is *corrosion fatigue.* Although most ASME Code materials are not particularly sensitive to corrosion fatigue, even they may suffer an endurance limit loss of 50 percent in some environments. High-strength heat-treated steels, on the other hand, are very sensitive to corrosion fatigue. It is not unusual to find some of these which lose 75 percent of their endurance in corrosive environments. In fact, in corrosion fatigue many steels do not have an endurance limit. The curve of stress versus cycles to failure (*S/N* curve) continues to slope downward regardless of the number of cycles.

Brittle fracture is probably the most insidious type of pressure-vessel failure. Without brittle fracture, a pressure vessel could be pressurized approximately to its ultimate strength before failure. With brittle behavior some vessels have failed well below their design pressures (which are about 25 percent of the theoretical bursting pressures). To reduce the possibility of brittle behavior, Division 2 and Section III require impact tests.

The subject of brittle fracture has been understood only since about 1950, and knowledge of some of its aspects is still inadequate. A notched or cracked plate of pressure-vessel steel, stressed at 66°C (150°F), would elongate and absorb considerable energy before breaking. It would have a ductile or plastic fracture. As the temperature is lowered, a point is reached at which the plate would fail in a brittle manner with a flat fracture surface and almost no elongation. The transition from ductile to brittle fracture actually takes place over a temperature range, but a point in this range is selected as the *transition temperature.* One of the ways of determining this temperature is the Charpy impact test (see ASTM Specification E-23). After the transition temperature has been determined by laboratory impact tests, it must be correlated with service experience on full-size plates. The literature on brittle fracture contains information on the relation of impact tests to service experience on some carbon steels.

A more precise but more elaborate method of dealing with the ductile-brittle transition is the *fracture-analysis diagram.* This uses a transition known as the *nil-ductility temperature* (NDT), which is determined by the drop-weight test (ASTM Standard E208) or the drop-weight tear test (ASTM Standard E436). The application of this diagram is explained in two papers by Pellini and Puzak [*Trans. Am. Soc. Mech. Eng.* 429 (October 1964); *Welding Res. Counc. Bull.* 88, 1963].

BPVC Section VIII, Division 1, is lax with respect to brittle fracture. It allows the use of many steels down to −29°C (−20°F) without a check on toughness. Occasional brittle failures show that some vessels are operating below the nil-ductility temperature, i.e., the lower limit of ductility. Division 2 has solved this problem by requiring impact tests in certain cases. Tougher grades of steel, such as the SA516 steels (in preference to SA515 steel), are available for a small price premium. Stress relief, steel made to fine-grain practice, and normalizing all reduce the hazard of brittle fracture.

Nondestructive testing of both the plate and the finished vessel is important to safety. In the analysis of fracture hazards, it is important to know the size of the flaws that may be present in the completed vessel. The four most widely used methods of examination are radiographic, magnetic-particle, liquid-penetrant, and ultrasonic.

Radiographic examination is either by *x-rays* or by *gamma radiation.* The former has greater penetrating power, but the latter is more portable. Few x-ray machines can penetrate beyond 300-mm (12-in) thickness.

Ultrasonic techniques use vibrations with a frequency between 0.5 and 20 MHz transmitted to the metal by a transducer. The instrument sends out a series of pulses. These show on a cathode-ray screen as they are sent out and again when they return after being reflected from the opposite side of the member. If there is a crack or an inclusion along the way, it will reflect part of the beam. The initial pulse and its reflection from the back of the member are separated on the screen by a distance which represents the thickness. The reflection from a flaw will fall between these signals and indicate its magnitude and position. Ultrasonic examination can be used for almost any thickness of material from a fraction of an inch to several feet. Its use is dependent on the shape of the body because irregular surfaces may give confusing reflections. Ultrasonic transducers can transmit pulses normal to the surface or at an angle. Transducers transmitting pulses that are oblique to the surface can solve a number of special inspection problems.

Magnetic-particle examination is used only on magnetic materials. Magnetic flux is passed through the part in a path parallel to the surface. Fine magnetic particles, when dusted over the surface, will concentrate near the edges of a crack. The sensitivity of magnetic-particle examination is proportional to the sine of the angle between the direction of the magnetic flux and the direction of the crack. To be sure of picking up all cracks, it is necessary to probe the area in two directions.

Liquid-penetrant examination involves wetting the surface with a fluid that penetrates open cracks. After the excess liquid has been wiped off, the surface is coated with a material which will reveal any liquid that has penetrated the cracks. In some systems a colored dye will seep out of cracks and stain whitewash. Another system uses a penetrant that becomes fluorescent under ultraviolet light.

Each of these four popular methods has its advantages. Frequently, best results are obtained by using more than one method. Magnetic particles or liquid penetrants are effective on surface cracks. Radiography and ultrasonic examination are necessary for subsurface flaws. *No known method of nondestructive testing can guarantee the absence of flaws.* There are other less widely used methods of examination. Among these are eddy current, electrical resistance, acoustics, and thermal testing. *Nondestructive Testing Handbook* [Robert C. McMaster (ed.), Ronald, New York, 1959] gives information on many testing techniques.

The *eddy-current technique* involves an alternating-current coil along and close to the surface being examined. The electrical impedance of the coil is affected by flaws in the structure or changes in composition. Commercially, the principal use of eddy-current testing is for the examination of tubing. It could, however, be used for testing other things.

The *electrical resistance method* involves passing an electric current through the structure and exploring the surface with voltage probes. Flaws, cracks, or inclusions will cause a disturbance in the voltage gradient on the surface. Railroads have used this method for many years to locate transverse cracks in rails.

The *hydrostatic test* is, in one sense, a method of examination of a vessel. It can reveal gross flaws, inadequate design, and flange leaks. Many believe that a hydrostatic test guarantees the safety of a vessel. This is not necessarily so. A vessel that has passed a hydrostatic test is probably safer than one that has not been tested. It can, however, still fail in service, even on the next application of pressure. Care in material selection, examination, and fabrication does more to guarantee vessel integrity than the hydrostatic test.

The ASME Codes recommend that hydrostatic tests be run at a temperature that is usually above the nil-ductility temperature of the material.

This is, in effect, a pressure-temperature treatment of the vessel. When tested in the relatively ductile condition above the nil-ductility temperature, the material will yield at the tips of cracks and flaws and at points of high residual weld stress. This procedure will actually reduce the residual stresses and cause a redistribution at crack tips. The vessel will then be in a safer condition for subsequent operation. This procedure is sometimes referred to as *notch nullification.*

It is possible to design a hydrostatic test in such a way that it probably will be a proof test of the vessel. This usually requires, among other things, that the test be run at a temperature as low as and preferably lower than the minimum operating temperature of the vessel. Proof tests of this type are run on vessels built of ultrahigh-strength steel to operate at cryogenic temperatures.

Other Regulations and Standards Pressure vessels may come under many types of regulation, depending on where they are and what they contain. Although many states have adopted the ASME Boiler and Pressure Vessel Code, either in total or in part, any state or municipality may enact its own requirements. The federal government regulates some pressure vessels through the Department of Transportation, which includes the Coast Guard. If pressure vessels are shipped into foreign countries, they may face additional regulations.

Pressure vessels carried aboard United States–registered ships must conform to rules of the *U.S. Coast Guard.* Subchapter F of Title 46, *Code of Federal Regulations,* covers marine engineering. Of this, Parts 50 through 61 and 98 include pressure vessels. Many of the rules are similar to those in the ASME Code, but there are differences.

The *American Bureau of Shipping* (ABS) has rules that insurance underwriters require for the design and construction of pressure vessels which are a permanent part of a ship. Pressure cargo tanks may be permanently attached and come under these rules. Such tanks supported at several points are independent of the ship's structure and are distinguished from integral cargo tanks such as those in a tanker. ABS has pressure-vessel rules in two of its publications. Most of them are in *Rules for Building and Classing Steel Vessels.*

Standards of Tubular Exchanger Manufacturers Association (TEMA) These standards give recommendations for the construction of tubular heat exchangers. Although TEMA is not a regulatory body and there is no legal requirement for the use of its standards, they are widely accepted as a good basis for design. By specifying TEMA standards, one can obtain adequate equipment without having to write detailed specifications for each piece. TEMA gives formulas for the thickness of tube sheets. Such formulas are not in ASME Codes. (See further discussion of TEMA in Sec. 11.)

Vessels with Unusual Construction High pressures create design problems. ASME BPVC Section VIII, Division 1, applies to vessels rated for pressures up to 20,670 kPa (3000 lbf/in^2). Division 2 is unlimited. At high pressures, special designs not necessarily in accordance with the code are sometimes used. At such pressures, a vessel designed for ordinary low-carbon-steel plate, particularly in large diameters, would become too thick for practical fabrication by ordinary methods. The alternatives are to make the vessel of high-strength plate, use a solid forging, or use multilayer construction.

High-strength steels with tensile strengths over 1380 MPa (200,000 lbf/in^2) are limited largely to applications for which weight is very important. Welding procedures are carefully controlled, and preheat is used. These materials are brittle at almost any temperature, and vessels must be designed to prevent brittle fracture. Flat spots and variations in curvature are avoided. Openings and changes in shape require appropriate design. The maximum permissible size of flaws is determined by fracture mechanics, and the method of examination must ensure as much as possible that larger flaws are not present. All methods of nondestructive testing may be used. Such vessels require the most sophisticated techniques in design, fabrication, and operation.

Solid forgings are frequently used in construction for pressure vessels above 20,670 kPa (3000 lbf/in^2) and even lower. Almost any shell thickness can be obtained, but most of them range between 50 and 300 mm (2 and 12 in). The ASME Code lists forging materials with tensile strengths from 414 to 930 MPa (from 60,000 to 135,000 lbf/in^2). Brittle fracture is a possibility, and the hazard increases with thickness. Furthermore, some forging alloys have nil-ductility temperatures as high as 121°C (250°F). A forged vessel should have an NDT at least 17°C (30°F) below the design temperature. In operation, it should be slowly and uniformly heated at least to the NDT before it is subjected to pressure. During construction, nondestructive testing should be used to detect dangerous cracks or flaws. Section VIII of the ASME Code, particularly Division 2, gives design and testing techniques.

As the size of a forged vessel increases, the sizes of ingot and handling equipment become larger. The cost may increase faster than the weight. The problems of getting sound material and avoiding brittle fracture also become more difficult. Some of these problems are avoided by use of *multilayer construction.* In this type of vessel, the heads and flanges are made of forgings, and the cylindrical portion is built up by a series of layers of thin material. The thickness of these layers may be between 3 and 50 mm (⅛ and 2 in), depending on the type of construction. There is an inner lining which may be different from the outer layers.

Although there are multilayer vessels as small as 380-mm (15-in) inside diameter and 2400 mm (8 ft) long, their principal advantage applies to the larger sizes. When properly made, a multilayer vessel is probably safer than a vessel with a solid wall. The layers of thin material are tougher and less susceptible to brittle fracture, have lower probability of defects, and have the statistical advantage of a number of small elements instead of a single large one. The heads, flanges, and welds, of course, have the same hazards as other thick members. Proper attention is necessary to avoid cracks in these members.

There are several assembly techniques. One technique frequently used is to form successive layers in half cylinders and butt-weld them over the previous layers. In doing this, the welds are staggered so that they do not fall together. This type of construction usually uses plates from 6 to 12 mm (¼ to ½ in) thick. Another method is to weld each layer separately to form a cylinder and then shrink it over the previous layers. Layers up to about 50-mm (2-in) thickness are assembled in this way. A third method of fabrication is to wind the layers as a continuous sheet. This technique is used in Japan. The Wickel construction, fabricated in Germany, uses helical winding of interlocking metal strip. Each method has its advantages and disadvantages, and the choice will depend on circumstances.

Because of the possibility of voids between layers, it is preferable not to use multilayer vessels in applications where they will be subjected to fatigue. Inward thermal gradients (inside temperature lower than outside temperature) are also undesirable.

Articles on these vessels have been written by Fratcher [*Pet. Refiner* **34**(11): 137 (1954)] and by Strelzoff, Pan, and Miller [*Chem. Eng.* **75**(21): 143–150 (1968)].

Vessels for high-temperature service may be beyond the temperature limits of the stress tables in the ASME Codes. BPVC Section VIII, Division 1, makes provision for construction of pressure vessels up to 650°C (1200°F) for carbon and low-alloy steel and up to 815°C (1500°F) for stainless steels (300 series). If a vessel is required for temperatures above these values and above 103 kPa (15 lbf/in^2), it would be necessary, in a code state, to get permission from the state authorities to build it as a special project. Above 815°C (1500°F), even the 300 series stainless steels are weak, and creep rates increase rapidly. If the metal that resists the pressure operates at these temperatures, the vessel pressure and size will be limited. The vessel must also be expendable because its life will be short. Long exposure to high temperature may cause the metal to deteriorate and become brittle. Sometimes, however, economics favor this type of operation.

One way to circumvent the problem of low metal strength is to use a metal inner liner surrounded by insulating material, which in turn is confined by a pressure vessel. The liner, in some cases, may have perforations that will allow pressure to pass through the insulation and act on the outer shell, which is kept cool to obtain normal strength. The liner has no pressure differential acting on it and, therefore, does not need much strength. Ceramic linings are also useful for high-temperature work.

Lined vessels are used for many applications. Any type of lining can be used in an ASME Code vessel, provided it is compatible with the metal of the vessel and the contents. Glass, rubber, plastics, rare metals, and ceramics are a few types. The lining may be installed separately, or if a metal is used, it may be in the form of clad plate. The cladding on plate can sometimes be considered as a stress-carrying part of the vessel.

A *ceramic lining* when used with high temperature acts as an insulator so that the steel outer shell is at a moderate temperature while the temperature at the inside of the lining may be very high. Ceramic linings may be of unstressed brick, or prestressed brick, or cast in place. Cast ceramic linings or unstressed brick may develop cracks and is used when the contents of the vessel will not damage the outer shell. They are usually designed so that the high temperature at the inside will expand them sufficiently to make them tight in the outer (and cooler) shell. This, however, is not usually sufficient to prevent some penetration by the product.

Prestressed-brick linings can be used to protect the outer shell. In this case, the bricks are installed with a special thermosetting-resin mortar. After lining, the vessel is subjected to internal pressure and heat. This expands the steel vessel shell, and the mortar expands to take up the space. The pressure and temperature must be at least as high as the maximum that will be encountered in service. After the mortar has set, reduction of pressure and temperature will allow the vessel to contract, putting the brick in compression. The upper temperature limit for this construction is about 190°C (375°F). The installation of such linings is highly specialized work done by a few companies. Great care is usually exercised in operation to protect the vessel from exposure to asymmetrical temperature gradients. Side nozzles and other unsymmetrical designs are avoided insofar as possible.

Concrete pressure vessels may be used in applications that require large sizes. Such vessels, if made of steel, would be too large and heavy to ship. Through the use of post-tensioned (prestressed) concrete, the vessel is fabricated on site. In this construction, the reinforcing steel is placed in tubes or plastic covers, which are cast into the concrete. Tension is applied to the steel after the concrete has acquired most of its strength.

Concrete nuclear reactor vessels, of the order of magnitude of 15-m (50-ft) inside diameter and length, have inner linings of steel which confine the pressure. After fabrication of the liner, the tubes for the cables or wires are put in place and the concrete is poured. High-strength reinforcing steel is used. Because there are thousands of reinforcing tendons in the concrete vessel, there is a statistical factor of safety. The failure of 1 or even 10 tendons would have little effect on the overall structure.

Plastic pressure vessels have the *advantages of chemical resistance* and light weight. Above 103 kPa (15 lbf/in^2), with certain exceptions, they must be designed according to ASME BPVC Section X (see Storage of Gases) and are confined to the three types of approved code construction. Below 103 kPa (15 lbf/in^2), any construction may be used. Even in this pressure range, however, the code should be used for guidance. Solid plastics, because of low strength and creep, can be used only for the lowest pressures and sizes. A stress of a few hundred pounds-force per square inch is the maximum for most plastics. To obtain higher strength, the filled plastics or filament-wound vessels, specified by the code, must be used. Solid-plastic parts, however, are often employed inside a steel shell, particularly for heat exchangers.

Graphite and ceramic vessels are used fully armored; that is, they are enclosed within metal pressure vessels. These materials are also used for boxlike vessels with backing plates on the sides. The plates are drawn together by tie bolts, thus putting the material in compression so that it can withstand low pressure.

ASME Code Developments ASME BPVC Section VIII (2015) has been reorganized into three classes. Class 1 is for low-pressure vessels employing spot radiography. Class 2 is for vessels requiring full radiography. Class 3 is for vessels experiencing fatigue. Material stress levels similar to those of competing vessel codes from Europe and Asia are included.

Vessel Codes Other than ASME Different design and construction rules are used in other countries. Chemical engineers concerned with pressure vessels outside the United States must become familiar with local pressure-vessel laws and regulations. *Boilers and Pressure Vessels,* an international survey of design and approval requirements published by the British Standards Institution, Maylands Avenue, Hemel Hempstead, Hertfordshire, England, in 1975, gives pertinent information for 76 political jurisdictions.

The British Code (British Standards) and the German Code (*A. D. Merkblätter*) in addition to the ASME Code are most commonly permitted, although Netherlands, Sweden, and France also have codes. The major difference between the codes lies in factors of safety and in whether ultimate strength is considered. BPVC, Section VIII, Division 1, vessels are generally heavier than vessels built to the other codes; however, the differences in allowable stress for a given material are less in the higher-temperature (creep) range.

Engineers and metallurgists have developed alloys to comply economically with individual codes. In Germany, where design stress is determined from yield strength and creep-rupture strength and no allowance is made for ultimate strength, steels that have a very high yield strength/ultimate strength ratio are used.

Other differences between codes include different bases for the design of reinforcement for openings and the design of flanges and heads. Some codes include rules for the design of heat-exchanger tube sheets, while others (ASME Code) do not. The Dutch Code (*Grondslagen*) includes very specific rules for calculation of wind loads, while the ASME Code leaves this entirely to the designer.

There are also significant differences in construction and inspection rules. Unless engineers make a detailed study of the individual codes and keep current, they will be well advised to make use of responsible experts for any of the codes.

Vessel Design and Construction The ASME Code lists a number of loads that must be considered in designing a pressure vessel. Among them are impact, weight of the vessel under operating and test conditions, superimposed loads from other equipment and piping, wind and earthquake loads, temperature-gradient stresses, and localized loadings from internal and external supports. In general, the code gives no values for these loads or methods for determining them, and no formulas are given for determining the stresses from these loads. Engineers must be knowledgeable in mechanics and strength of materials to solve these problems.

Some of the problems are treated by Brownell and Young, *Process Equipment Design,* Wiley, New York, 1959. ASME papers treat others, and a number of books published by the ASME are collections of papers on pressure-vessel design: *Pressure Vessels and Piping Design: Collected Papers, 1927–1959; Pressure Vessels and Piping Design and Analysis,* four volumes; and *International Conference: Pressure Vessel Technology,* published annually.

Throughout the year the Welding Research Council publishes bulletins which are final reports from projects sponsored by the council, important papers presented before engineering societies, and other reports of current interest which are not published in *Welding Research.* A large number of the published bulletins are pertinent for vessel designers.

Care of Pressure Vessels Protection against *excessive pressure* is largely taken care of by code requirements for relief devices. Exposure to fire is also covered by the code. The code, however, does not provide for the possibility of local overheating and weakening of a vessel in a fire. Insulation reduces the required relieving capacity and also reduces the possibility of local overheating.

A pressure-reducing valve in a line leading to a pressure vessel is not adequate protection against overpressure. Its failure will subject the vessel to full line pressure.

Vessels that have an operating cycle which involves the solidification and remelting of solids can develop excessive pressures. A solid plug of material may seal off one end of the vessel. If heat is applied at that end to cause melting, the expansion of the liquid can build up a high pressure and possibly result in yielding or rupture. Solidification in connecting piping can create similar problems.

Some vessels may be exposed to a runaway chemical reaction or even an explosion. This requires relief valves, rupture disks, or, in extreme cases, a frangible roof design or barricade (the vessel is expendable). A vessel with a large rupture disk needs anchors designed for the jet thrust when the disk blows.

Vacuum must be considered. It is nearly always possible that the contents of a vessel might contract or condense sufficiently to subject it to an internal vacuum. If the vessel cannot withstand the vacuum, it must have vacuum-breaking valves.

Improper operation of a process may result in the vessel's *exceeding design temperature.* Proper control is the only solution to this problem. Maintenance procedures can also cause excessive temperatures. Sometimes the contents of a vessel may be burned out with torches. If the flame impinges on the vessel shell, overheating and damage may occur.

Excessively low temperature may involve the hazard of brittle fracture. A vessel that is out of use in cold weather could be at a subzero temperature and well below its nil-ductility temperature. In startup, the vessel should be warmed slowly and uniformly until it is above the NDT. A safe value is 38°C (100°F) for plate if the NDT is unknown. The vessel should not be pressurized until this temperature is exceeded. Even after the NDT has been passed, excessively rapid heating or cooling can cause high thermal stresses.

Corrosion is probably the greatest threat to vessel life. Partially filled vessels frequently have severe pitting at the liquid-vapor interface. Vessels usually do not have a corrosion allowance on the outside. Lack of protection against the weather or against the drip of corrosive chemicals can reduce vessel life. Insulation may contain damaging substances. Chlorides in insulating materials can cause cracking of stainless steels. Water used for hydrotesting should be free of chlorides.

Pressure vessels should be *inspected periodically.* No rule can be given for the frequency of these inspections. Frequency depends on operating conditions. If the early inspections of a vessel indicate a low corrosion rate, intervals between inspections may be lengthened. Some vessels are inspected at 5-year intervals; others, as frequently as once a year. Measurement of corrosion is an important inspection item. One of the most convenient ways of measuring thickness (and corrosion) is to use an ultrasonic gauge. The location of the corrosion and whether it is uniform or localized in deep pits should be observed and reported. Cracks, any type of distortion, and leaks should be observed. Cracks are particularly dangerous because they can lead to sudden failure. Insulation is usually left in place during inspection of insulated vessels. If, however, severe external corrosion is suspected, the insulation should be removed. All forms of nondestructive testing are useful for examinations.

There are many ways in which a pressure vessel can suffer *mechanical damage.* The shells can be dented or even punctured; they can be dropped or have hoisting cables improperly attached; bolts can be broken; flanges are bent by excessive bolt tightening; gasket contact faces can be scratched and dented; rotating paddles can drag against the shell and cause wear; and a flange can be bolted up with a gasket half in the groove and half out. Most of these forms of damage can be prevented by taking care and using common sense. If damage is repaired by straightening, as with a dented shell, it may be necessary to stress-relieve the repaired area. Some steels are susceptible to embrittlement by aging after severe straining. A safer procedure is to cut out the damaged area and replace it.

The National Board Inspection Code, published by the National Board of Boiler and Pressure Vessel Inspectors, Columbus, Ohio, is helpful. Any repair, however, is acceptable if it is made in accordance with the rules of the Pressure Vessel Code.

Care in *reassembling* the vessel is particularly important. Gaskets should be properly located, particularly if they are in grooves. Bolts should be

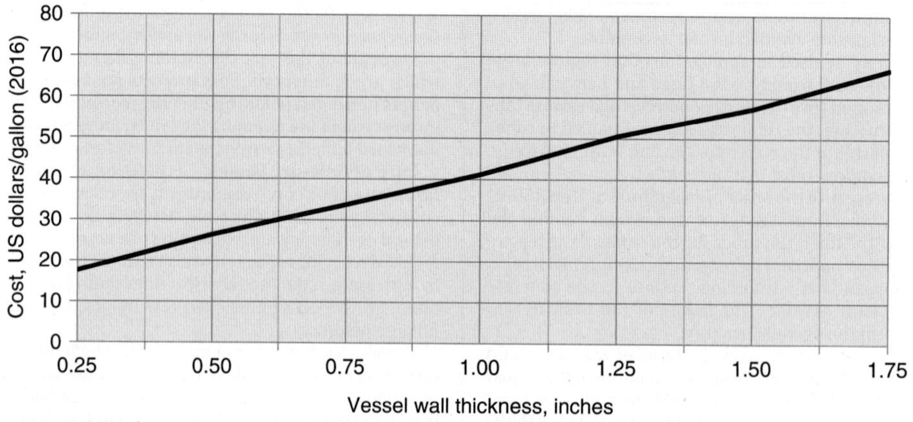

FIG. 10-185 Carbon-steel pressure-vessel cost as a function of wall thickness. 1 gal = 0.003875 m^3; 1 in = 0.0254 m. (*Courtesy of E. S. Fox, Ltd.*)

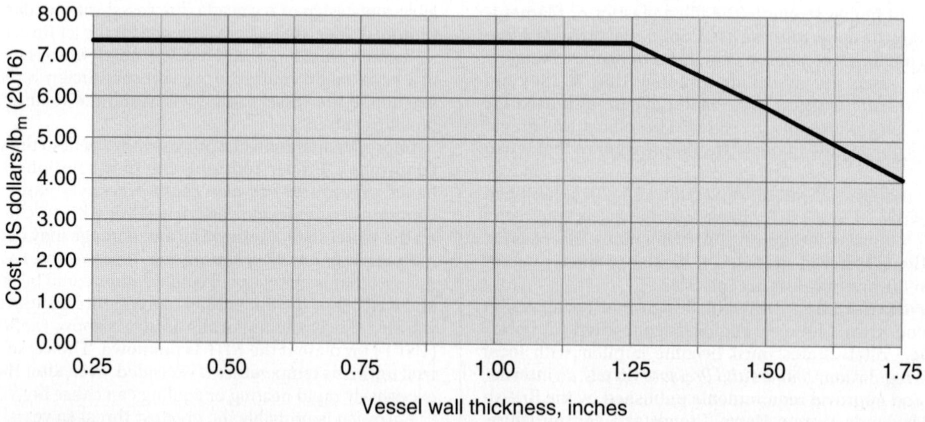

FIG. 10-186 Carbon-steel pressure-vessel cost as a function of wall thickness. 1 gal = 0.003875 m^3; 1 in = 0.0254 m; 1 lb = 0.4536 kg. (*Courtesy of E. S. Fox, Ltd.*)

tightened in proper sequence. In some critical cases and with large bolts, it is necessary to control bolt tightening by torque wrenches, micrometers, patented bolt-tightening devices, or heating bolts. After assembly, vessels are sometimes given a hydrostatic test.

Pressure-Vessel Cost and Weight Figure 10-185 can be used for estimating carbon-steel vessel cost when a weight estimate is not available and Fig. 10-186 with a weight estimate. Weight and cost include skirts and other supports. The cost is based on several 2016 pressure vessels. Costs are for vessels not of unusual design. Complicated vessels could cost considerably more. Guthrie [*Chem. Eng.* **76**(6): 114–142 (1969)] also gives pressure-vessel cost data.

When vessels have complicated construction (large, heavy bolted connections, support skirts, etc.), it is preferable to estimate their weight and apply a unit cost in dollars per pound.

Pressure-vessel weights are obtained by calculating the cylindrical shell and heads separately and then adding the weights of nozzles and attachments. Steel has a density of 7817 kg/m^3 (488 lb/ft^3). Metal in heads can be approximated by calculating the area of the blank (disk) used for forming the head. The required diameter of the blank can be calculated by multiplying the head outside diameter by the approximate factors given in Table 10-60. These factors make no allowance for the straight flange which is a cylindrical extension that is formed on the head. The blank diameter obtained from these factors must be increased by twice the length of straight flange, which is usually 1½ to 2 in but can be up to several inches in length. Manufacturers' catalogs give weights of heads.

Forming a head thins it in certain areas. To obtain the required minimum thickness of a head, it is necessary to use a plate that is initially thicker. Table 10-61 gives allowances for additional thickness.

Nozzles and flanges may add considerably to the weight of a vessel. Their weights can be obtained from manufacturers' catalogs (Taylor Forge Division of Gulf & Western Industries, Inc., Tube Turns Inc., Ladish Co., Lenape Forge, and others). Other parts such as skirts, legs, support brackets, and other details must be calculated.

TABLE 10-60 Factors for Estimating Diameters of Blanks for Formed Heads

	Ratio d/t	Blank diameter factor
ASME head	Over 50	1.09
	30–50	1.11
	20–30	1.15
Ellipsoidal head	Over 20	1.24
	10–20	1.30
Hemispherical head	Over 30	1.60
	18–30	1.65
	10–18	1.70

d = head diameter
t = nominal minimum head thickness

TABLE 10-61 Extra Thickness Allowances for Formed Heads*

	Extra thickness, in		
	ASME and ellipsoidal		
Minimum head thickness, in	Head OD up to 150 in incl.	Head OD over 150 in	Hemispherical
Up to 0.99	1/16	1/8	3/16
1 to 1.99	1/8	1/8	3/8
2 to 2.99	1/4	1/4	5/8

*Lukens, Inc.

Section 11

Heat-Transfer Equipment

Richard L. Shilling, P.E., B.E.M.E. *Senior Engineering Consultant, Heat Transfer Research, Inc.; American Society of Mechanical Engineers (Section Editor, Cryogenic Heat Exchangers, Shell-and-Tube Heat Exchangers, Hairpin/Double-Pipe Heat Exchangers, Air-Cooled Heat Exchangers, Heating and Cooling of Tanks, Fouling and Scaling, Heat Exchangers for Solids, Thermal Insulation, Thermal Design of Evaporators, Evaporators)*

Patrick M. Bernhagen, P.E., B.S. *Director of Sales—Fired Heater, Amec Foster Wheeler North America Corp.; API Subcommittee on Heat Transfer Equipment API 530, 536, 560, and 561 (Compact and Nontubular Heat Exchangers)*

William E. Murphy, Ph.D., P.E. *Professor of Mechanical Engineering, University of Kentucky; American Society of Heating, Refrigerating, and Air-Conditioning Engineers; American Society of Mechanical Engineers; International Institute of Refrigeration (Air Conditioning)*

Predrag S. Hrnjak, Ph.D. *Will Stoecker Res. Professor of Mechanical Science and Engineering, University of Illinois at Urbana-Champaign; Principal Investigator—U of I Air Conditioning and Refrigeration Center; Assistant Professor, University of Belgrade; International Institute of Chemical Engineers; American Society of Heat, Refrigerating, and Air Conditioning Engineers (Refrigeration)*

David Johnson, P.E., M.Ch.E. *Retired (Thermal Design of Heat Exchangers, Condensers, Reboilers)*

THERMAL DESIGN OF HEAT-TRANSFER EQUIPMENT

The prior and substantial contributions of Frank L. Rubin (Section Editor, Sixth Edition) and Dr. Kenneth J. Bell (Thermal Design of Heat Exchangers, Condensers, Reboilers), Dr. Thomas M. Flynn (Cryogenic Processes), and F. C. Standiford (Thermal Design of Evaporators, Evaporators), who were authors for the Seventh Edition, are gratefully acknowledged.

TEMA-STYLE SHELL-AND-TUBE HEAT EXCHANGERS

HAIRPIN/DOUBLE-PIPE HEAT EXCHANGERS

AIR-COOLED HEAT EXCHANGERS

COMPACT AND NONTUBULAR HEAT EXCHANGERS

HEAT EXCHANGERS FOR SOLIDS

THERMAL INSULATION

AIR CONDITIONING

REFRIGERATION

EVAPORATORS

THERMAL DESIGN OF HEAT-TRANSFER EQUIPMENT

INTRODUCTION TO THERMAL DESIGN

Designers commonly use computer software to design heat exchangers. The best sources of such software are Heat Transfer Research, Inc. (HTRI), and Heat Transfer and Fluid Flow Services (HTFFS), a division of ASPENTECH. These companies develop proprietary correlations based on their research and provide software that utilizes these correlations. However, it is important that engineers understand the fundamental principles underlying the framework of the software. Therefore, design methods for several important classes of process heat-transfer equipment are presented in later subsections of this section. Mechanical descriptions and specifications of equipment are given in this subsection and should be read in conjunction with the use of this subsequent design material. However, it is impossible to present here a comprehensive treatment of heat exchanger selection, design, and application. The best general references in this field are Hewitt, Shires, and Bott, *Process Heat Transfer*, CRC Press, Boca Raton, FL, 1994; and Schlünder (ed.), *Heat Exchanger Design Handbook*, Begell House, New York, 2002.

Approach to Heat Exchanger Design The proper use of basic heat-transfer knowledge in the design of practical heat-transfer equipment is an art. Designers must be constantly aware of the differences between the idealized conditions for and under which the basic knowledge was obtained and the real conditions of the mechanical expression of their design and its environment. The result must satisfy process and operational requirements (such as availability, flexibility, and maintainability) and do so economically. An important part of any design process is to consider and offset the consequences of error in the basic knowledge, in its subsequent incorporation into a design method, in the translation of design into equipment, or in the operation of the equipment and the process. Heat exchanger design is not a highly accurate art under the best of conditions.

The design of a process heat exchanger usually proceeds through the following steps:

1. Process conditions (stream compositions, flow rates, temperatures, pressures) must be specified.
2. Required physical properties over the temperature and pressure ranges of interest must be obtained.
3. The type of heat exchanger to be employed is chosen.
4. A preliminary estimate of the size of the exchanger is made, using a heat-transfer coefficient appropriate to the fluids, process, and equipment.
5. A first design is chosen, complete in all details necessary to carry out the design calculations.
6. The design chosen in step 5 is evaluated, or *rated*, as to its ability to meet the process specifications with respect to both heat transfer and pressure drop.
7. On the basis of the result of step 6, a new configuration is chosen if necessary and step 6 is repeated. If the first design was inadequate to meet the required heat load, it is usually necessary to increase the size of the exchanger while still remaining within specified or feasible limits of pressure drop, tube length, shell diameter, etc. This will sometimes mean going to multiple-exchanger configurations. If the first design more than meets heat load requirements or does not use all the allowable pressure drop, a less expensive exchanger can usually be designed to fulfill process requirements.
8. The final design should meet process requirements (within reasonable expectations of error) at lowest cost. The lowest cost should include operation and maintenance costs and credit for ability to meet long-term process changes, as well as installed (capital) cost.

Exchangers should not be selected entirely on a lowest-first-cost basis, which frequently results in future penalties.

Overall Heat-Transfer Coefficient The basic design equation for a heat exchanger is

$$dA = dQ/U\Delta T \tag{11-1}$$

where dA is the element of surface area required to transfer an amount of heat dQ at a point in the exchanger where the overall heat-transfer coefficient is U and where the overall bulk temperature difference between the two streams is ΔT. The overall heat-transfer coefficient is related to the individual film heat-transfer coefficients and fouling and wall resistances by Eq. (11-2). Basing U_o on the outside surface area A_o results in

$$U_o = \frac{1}{1/h_o + R_{do} + xA_o/k_w A_{wm} + (1/h_t + R_{dt})A_o/A_t} \tag{11-2}$$

Equation (11-1) can be formally integrated to give the outside area required to transfer the total heat load Q_T:

$$A_o = \int_0^{Q_T} \frac{dQ}{U_o \Delta T} \tag{11-3}$$

To integrate Eq. (11-3), U_o and ΔT must be known as functions of Q. For some problems, U_o varies strongly and nonlinearly throughout the exchanger. In these cases, it is necessary to evaluate U_o and ΔT at several intermediate values and numerically or graphically integrate. For many practical cases, it is possible to calculate a constant mean overall coefficient U_{om} from Eq. (11-2) and define a corresponding mean value of ΔT_m, such that

$$A_o = Q_T/U_{om}\,\Delta T_m \tag{11-4}$$

Care must be taken that U_o does not vary too strongly, that the proper equations and conditions are chosen for calculating the individual coefficients, and that the mean temperature difference is the correct one for the specified exchanger configuration.

Mean Temperature Difference The temperature difference between the two fluids in the heat exchanger will, in general, vary from point to point. The *mean temperature difference* (ΔT_m or MTD) can be calculated from the terminal temperatures of the two streams if the following assumptions are valid:

1. All elements of a given fluid stream have the same thermal history in passing through the exchanger.*
2. The exchanger operates at steady state.
3. The specific heat is constant for each stream (or if either stream undergoes an isothermal phase transition).
4. The overall heat-transfer coefficient is constant.
5. Heat losses are negligible.

Countercurrent or Cocurrent Flow If the flow of the streams is either *completely* countercurrent or completely cocurrent or if one or both streams are isothermal (condensing or vaporizing a pure component with negligible pressure change), then the correct MTD is the *logarithmic-mean temperature difference* (LMTD), defined as

$$\text{LMTD} = \Delta T_{\text{lm}} = \frac{(t_1' - t_2'') - (t_2' - t_1'')}{\ln\left(\dfrac{t_1' - t_2''}{t_2' - t_1''}\right)} \tag{11-5a}$$

for *countercurrent flow* (Fig. 11-1*a*) and

$$\text{LMTD} = \Delta T_{\text{lm}} = \frac{(t_1' - t_1'') - (t_2' - t_2'')}{\ln\left(\dfrac{t_1' - t_1''}{t_2' - t_2''}\right)} \tag{11-5b}$$

for *cocurrent flow* (Fig. 11-1*b*).

If U is not constant but a linear function of ΔT, then the correct value of $U_{om}\Delta T_m$ to use in Eq. (11-4) is [Colburn, *Ind. Eng. Chem.* **25:** 873 (1933)]

$$U_{om}\,\Delta T_m = \frac{U_o''(t_1' - t_2'') - U_o'(t_2' - t_1'')}{\ln\left(\dfrac{U_o''(t_1' - t_2'')}{U_o'(t_2' - t_1'')}\right)} \tag{11-6a}$$

*This assumption is vital but is usually omitted or less satisfactorily stated as "each stream is well mixed at each point." In a heat exchanger with substantial bypassing of the heat-transfer surface, e.g., a typical baffled shell-and-tube exchanger, this condition is not satisfied. However, the error is offset to some degree if the same MTD formulation used in reducing experimental heat-transfer data to obtain the basic correlation is used in applying the correlation to design a heat exchanger. The compensation is not in general exact, and insight and judgment are required in the use of the MTD formulations. Particularly, in the design of an exchanger with a very close temperature approach, bypassing may result in an exchanger that is inefficient and even thermodynamically incapable of meeting specified outlet temperatures.

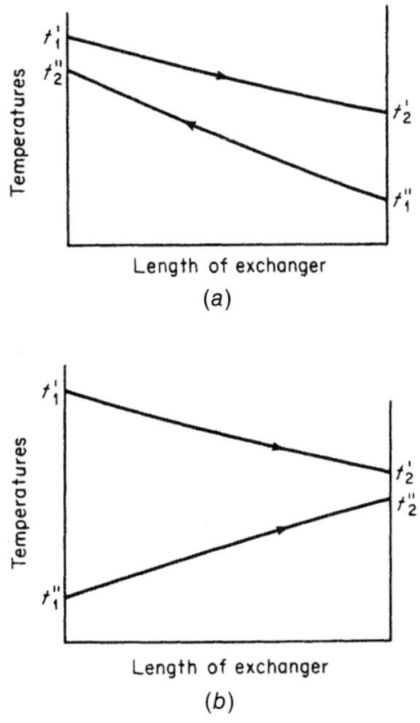

FIG. 11-1 Temperature profiles in heat exchangers. (*a*) Countercurrent. (*b*) Cocurrent.

for *countercurrent flow*, where U_o'' is the overall coefficient evaluated when the stream temperatures are t_1' and t_2'' and U_o' is evaluated at t_2' and t_1''. The corresponding equation for *cocurrent flow* is

$$U_{om}\Delta T_m = \frac{U_o''(t_1' - t_1'') - U_o'(t_2' - t_2'')}{\ln\left(\dfrac{U_o''(t_1' - t_1'')}{U_o'(t_2' - t_2'')}\right)} \tag{11-6b}$$

where U_o' is evaluated at t_2' and t_2'' and U_o'' is evaluated at t_1' and t_1''. To use these equations, it is necessary to calculate two values of U_o.

The use of Eq. (11-6) will frequently give satisfactory results even if U_o is not strictly linear with temperature difference.

Reversed, Mixed, or Cross-Flow If the flow pattern in the exchanger is not completely countercurrent or cocurrent, it is necessary to apply a *correction factor* F_T by which the LMTD is multiplied to obtain the appropriate MTD. These corrections have been mathematically derived for flow patterns of interest, still by making assumptions 1 to 5 [see Bowman, Mueller, and Nagle, *Trans. Am. Soc. Mech. Eng.* **62:** 283 (1940) or Hewitt, Shires, and Bott, *Process Heat Transfer*, CRC Press, Boca Raton, FL, 1994]. For a common flow pattern, the 1-2 exchanger (Fig. 11-2), the correction factor F_T is given in Fig. 11-4*a*, which is also valid for finding F_T for a 1-2 exchanger in which the shell-side flow direction is reversed from that shown in Fig. 11-2. Figure 11-4*a* is also applicable with negligible error to exchangers with one shell pass and any number of tube passes. Values of F_T less than 0.8 (0.75 at the very lowest) are generally unacceptable because the exchanger configuration chosen is inefficient; the chart is difficult to read accurately; and even a small violation of the first assumption underlying the MTD will invalidate the mathematical derivation and lead to a thermodynamically inoperable exchanger.

Correction factor charts are also available for exchangers with more than one shell pass provided by a longitudinal shell-side baffle. However, these exchangers are seldom used in practice because of mechanical complications in their construction. Also thermal and physical leakages across the longitudinal baffle further reduce the mean temperature difference and are not properly incorporated into the correction factor charts. Such charts are useful, however, when it is necessary to construct a multiple-shell exchanger train such as that shown in Fig. 11-3 and are included here for two, three, four, and six *separate, identical* shells and two or more tube passes per shell in Fig. 11-4*b, c, d,* and *e*. If only one tube pass per shell is required, the piping can and should be arranged to provide pure countercurrent flow, in which case the LMTD is used with no correction.

Cross-flow exchangers of various kinds are also important and require correction to be applied to the LMTD calculated by assuming countercurrent flow. Several cases are given in Fig. 11-4*f, g, h, i,* and *j*.

Many other MTD correction factor charts have been prepared for various configurations. The F_T charts are often employed to make approximate corrections for configurations even in cases for which they are not completely valid.

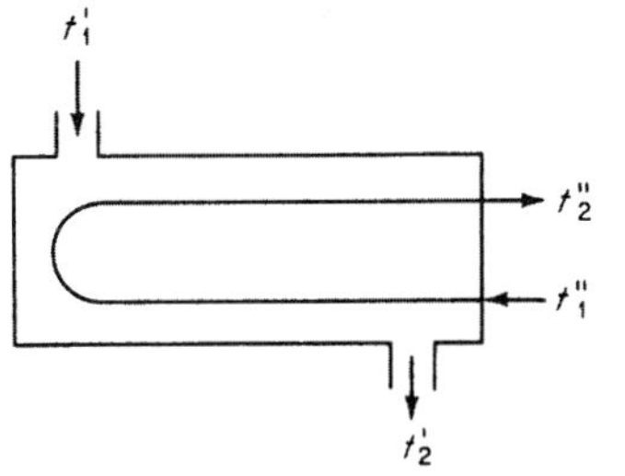

FIG. 11-2 Diagram of a 1-2 exchanger (one well-baffled shell pass and two tube passes with an equal number of tubes in each pass).

THERMAL DESIGN FOR SINGLE-PHASE HEAT TRANSFER

Double-Pipe Heat Exchangers The design of double-pipe heat exchangers is straightforward. It is generally conservative to neglect natural convection and entrance effects in turbulent flow. In laminar flow, natural convection effects can increase the theoretical Graetz prediction by a factor of 3 or 4 for fully developed flows. Pressure drop is calculated by using the correlations given in Sec. 6.

If the inner tube is longitudinally finned on the outside surface, the equivalent diameter is used as the characteristic length in both the Reynolds number and the heat-transfer correlations. The fin efficiency must also be known to calculate an *effective* outside area to use in Eq. (11-2).

Fittings contribute strongly to the pressure drop on the annulus side. General methods for predicting this are not reliable, and manufacturer's data should be used when available.

Double-pipe exchangers are often piped in complex series-parallel arrangements on both sides. The MTD to be used has been derived for some of these arrangements and is reported in Kern (*Process Heat Transfer*, McGraw-Hill, New York, 1950). More complex cases may require trial-and-error balancing of the heat loads and rate equations for subsections or even for individual exchangers in the bank.

Baffled Shell-and-Tube Exchangers The method given here is based on the research summarized in Final Report, Cooperative Research Program on Shell and Tube Heat Exchangers, *Univ. Del. Eng. Exp. Sta. Bull.* 5 (June 1963). The method assumes that the shell-side heat-transfer and pressure-drop characteristics are equal to those of the ideal tube bank corresponding to the cross-flow sections of the exchanger, modified for the distortion of flow pattern introduced by the baffles and the presence of leakage and bypass flow through the various clearances required by mechanical construction.

It is assumed that process conditions and physical properties are known and the following are known or specified: tube outside diameter D_o, tube geometric arrangement (unit cell), shell inside diameter D_s, shell outer tube limit D_{otl}, baffle cut l_c, baffle spacing l_s, and number of sealing strips N_{ss}. The effective tube length between tube sheets L may be either specified or calculated after the heat-transfer coefficient has been determined. If additional specific information (e.g., tube-baffle clearance) is available, the exact values (instead of estimates) of certain parameters may be used in the calculation with some improvement in accuracy. To complete the rating, it is necessary to know also the tube material and wall thickness or inside diameter.

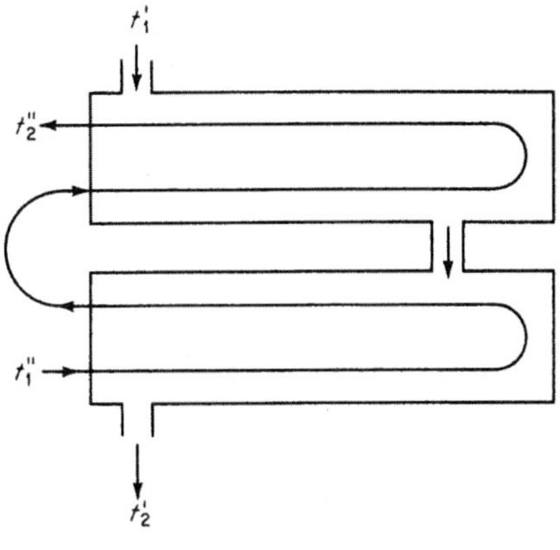

FIG. 11-3 Diagram of a 2-4 exchanger (two separate identical well-baffled shells and four or more tube passes).

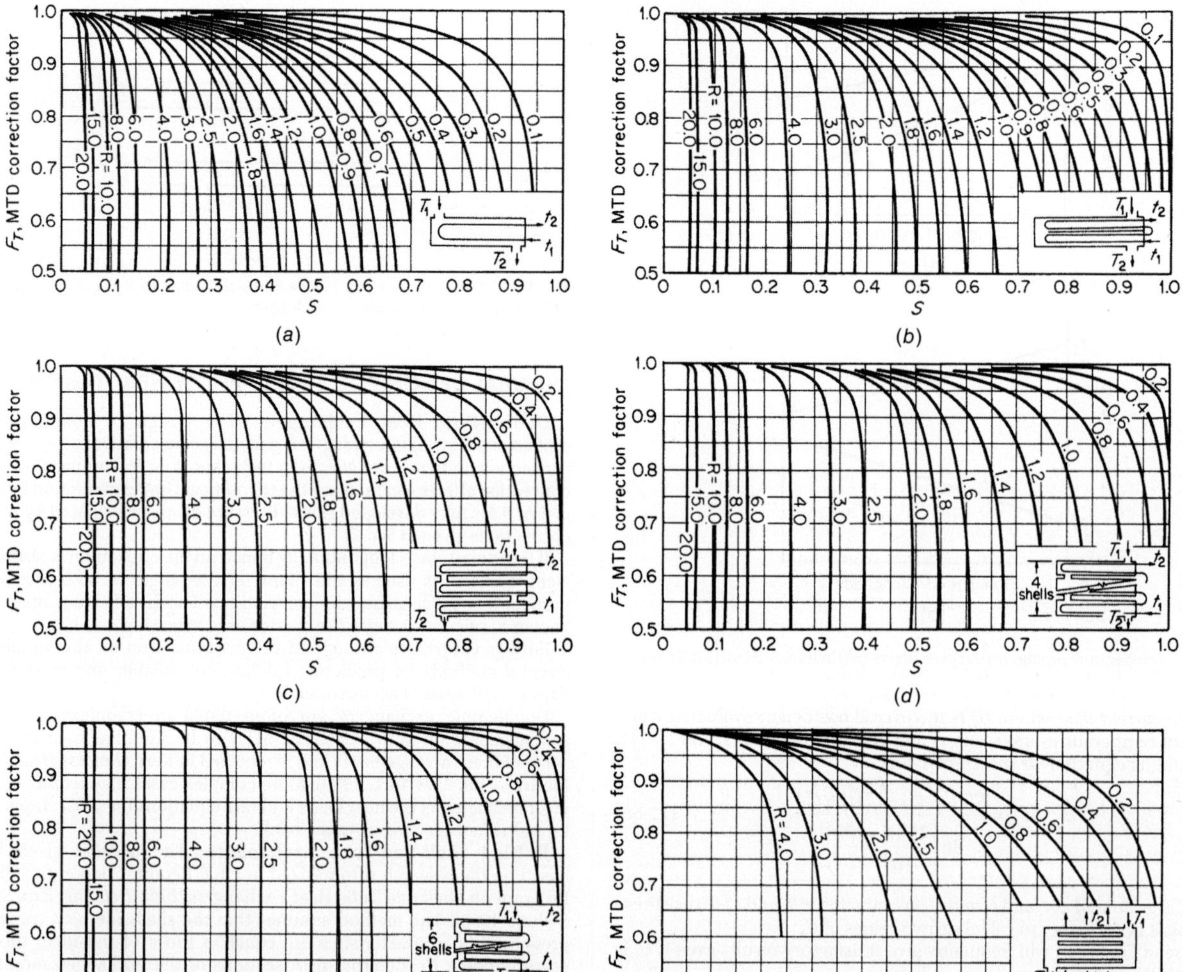

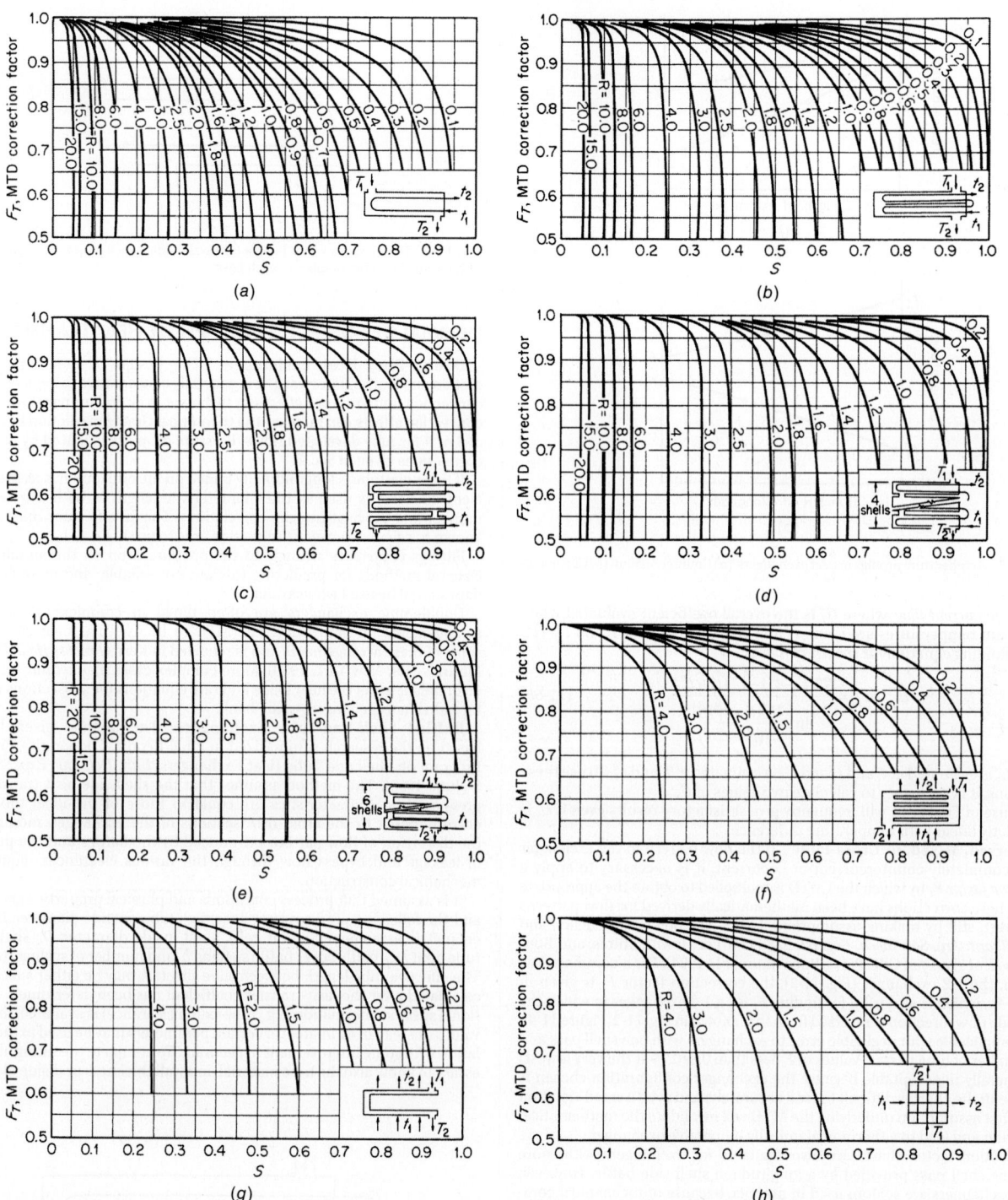

FIG. 11-4 LMTD correction factors for heat exchangers. In all charts, $R = (T_1 - T_2)/(t_2 - t_1)$ and $S = (t_2 - t_1)/(T_1 - t_1)$. (*a*) One shell pass, two or more tube passes. (*b*) Two shell passes, four or more tube passes. (*c*) Three shell passes, six or more tube passes. (*d*) Four shell passes, eight or more tube passes. (*e*) Six shell passes, twelve or more tube passes. (*f*) Cross-flow, one shell pass, one or more parallel rows of tubes. (*g*) Cross-flow, two passes, two rows of tubes; for more than two passes, use $F_T = 1.0$. (*h*) Cross-flow, one shell pass, one tube pass, both fluids unmixed. (*i*) Cross-flow (drip type), two horizontal passes with U-bend connections (trombone type). (*j*) Cross-flow (drip type), helical coils with two turns.

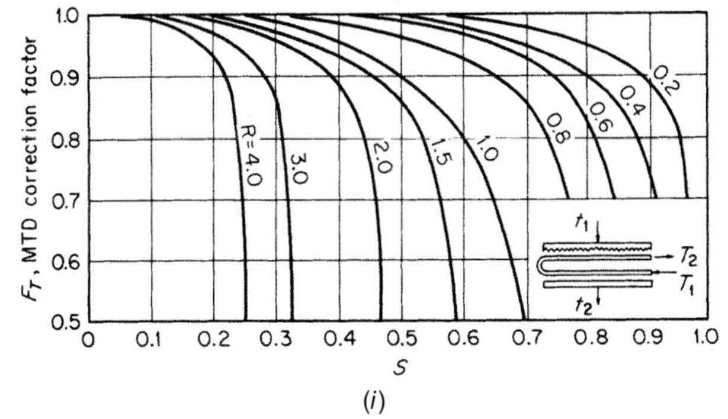

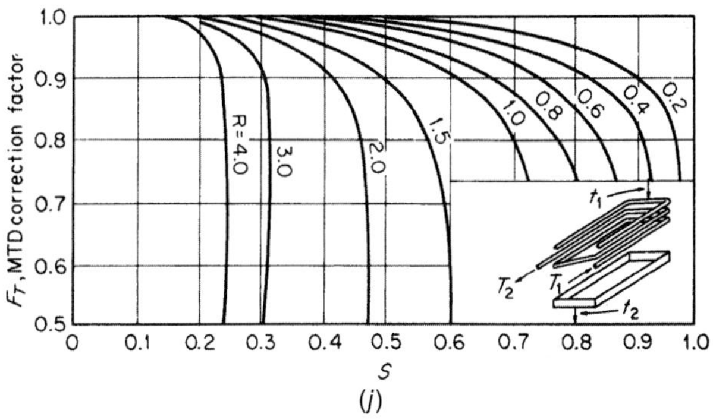

FIG. 11-4 *(Continued)*

This rating method, though apparently generally the best in the open literature, is not extremely accurate. An exhaustive study by Palen and Taborek [*Chem. Eng. Prog. Symp. Ser.* 92, **65:** 53 (1969)] showed that this method predicted shell-side coefficients from about 50 percent low to 100 percent high, while the pressure drop range was from about 50 percent low to 200 percent high. The mean error for heat transfer was about 15 percent low (safe) for all Reynolds numbers, while the mean error for pressure drop was from about 5 percent low (unsafe) at Reynolds numbers above 1000 to about 100 percent high at Reynolds numbers below 10.

Calculation of Shell-Side Geometric Parameters

1. *Total number of tubes in exchanger* N_t. If this is not known by direct count, estimate by using Eq. (11-74) or (11-75).

2. *Tube pitch parallel to flow* p_p *and normal to flow* p_n. These quantities are needed only for estimating other parameters. If a detailed drawing of the exchanger is available, it is better to obtain these other parameters by direct count or calculation. The pitches are described by Fig. 11-5 and read therefrom for common tube layouts.

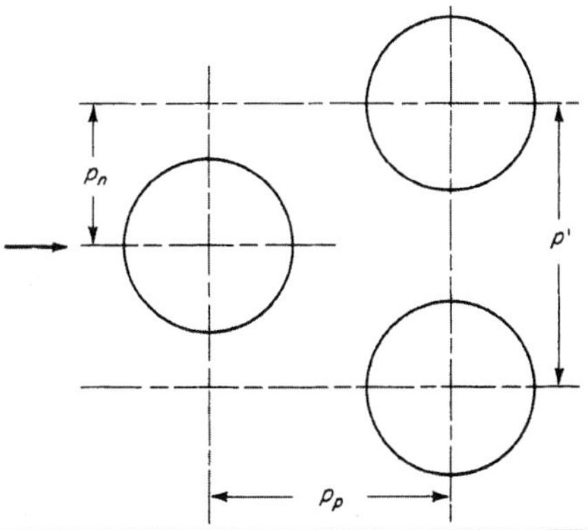

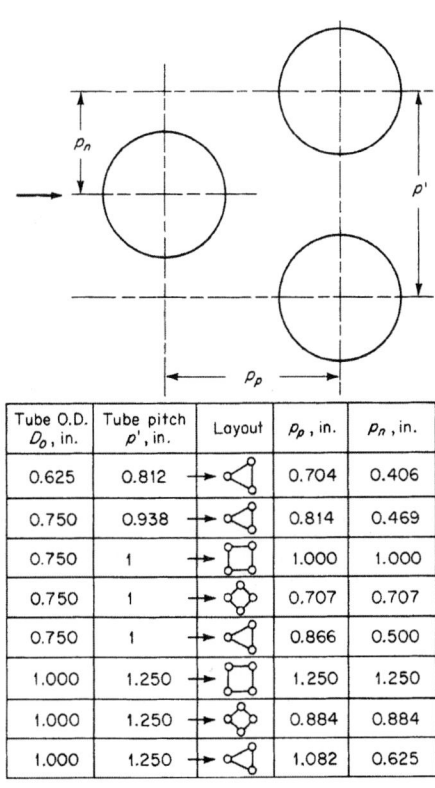

Tube O.D. D_o, in.	Tube pitch p', in.	Layout	p_p, in.	p_n, in.
0.625	0.812		0.704	0.406
0.750	0.938		0.814	0.469
0.750	1		1.000	1.000
0.750	1		0.707	0.707
0.750	1		0.866	0.500
1.000	1.250		1.250	1.250
1.000	1.250		0.884	0.884
1.000	1.250		1.082	0.625

FIG. 11-5 Values of tube pitch for common tube layouts. To convert inches to meters, multiply by 0.0254. Note that D_o, p', p_p, and p_n have units of inches.

3. *Number of tube rows crossed in one cross-flow section* N_c. Count from the exchanger drawing or estimate from

$$N_c = \frac{D_s[1-2(l_c/D_s)]}{p_p} \tag{11-7}$$

4. *Fraction of total tubes in cross-flow* F_c

$$F_c = \frac{1}{\pi}\left[\pi + 2\frac{D_s - 2l_c}{D_{otl}}\sin\left(\cos^{-1}\frac{D_s-2l_c}{D_{otl}}\right) - 2\cos^{-1}\frac{D_s-2l_c}{D_{otl}}\right] \tag{11-8}$$

where F_c is plotted in Fig. 11-6. This figure is strictly applicable only to split-ring, floating-head construction but may be used for other situations with minor error.

5. *Number of effective cross-flow rows in each window* N_{cw}

$$N_{cw} = \frac{0.8l_c}{p_p} \tag{11-9}$$

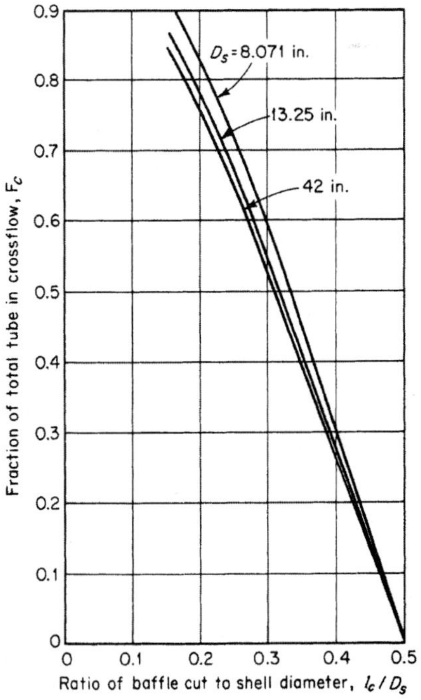

FIG. 11-6 Estimation of fraction of tubes in cross-flow F_c [Eq. (11-8)]. To convert inches to meters, multiply by 0.0254. Note that l_c and D_s have units of inches.

6. *Cross-flow area at or near centerline for one cross-flow section S_m*

a. For rotated and in-line square layouts:

$$S_m = l_s \left[D_s - D_{otl} + \frac{D_{otl} - D_o}{p_n}(p' - D_o) \right] \quad m^2 (ft^2) \qquad (11\text{-}10a)$$

b. For triangular layouts:

$$S_m = l_s \left[D_s - D_{otl} + \frac{D_{otl} - D_o}{p'}(p' - D_o) \right] \quad m^2 (ft^2) \qquad (11\text{-}10b)$$

7. *Fraction of cross-flow area available for bypass flow F_{bp}*

$$F_{bp} = \frac{(D_s - D_{otl}) l_s}{S_m} \qquad (11\text{-}11)$$

8. *Tube-to-baffle leakage area for one baffle S_{tb}.* Estimate from

$$S_{tb} = b D_o N_T (1 + F_c) \quad m^2 (ft^2) \qquad (11\text{-}12)$$

where $b = (6.223)(10^{-4})$ (SI) or $(1.701)(10^{-4})$ (USCS). These values are based on the Tubular Exchanger Manufacturers Association (TEMA) Class R construction which specifies ⅓₂-in diametral clearance between tube and baffle. Values should be modified if extra tight or loose construction is specified or if clogging by dirt is anticipated.

9. *Shell-to-baffle leakage area for one baffle S_{sb}.* If diametral shell-baffle clearance δ_{sb} is known, then S_{sb} can be calculated from

$$S_{sb} = \frac{D_s \delta_{sb}}{2} \left[\pi - \cos^{-1}\left(1 - \frac{2l_c}{D_s}\right) \right] \quad m^2 (ft^2) \qquad (11\text{-}13)$$

where the value of the term $\cos^{-1}(1 - 2l_c/D_s)$ is in radians and is between 0 and $\pi/2$. Shell-to-baffle leakage area S_{sb} is plotted in Fig. 11-7, based on TEMA Class R standards. Since pipe shells are generally limited to diameters below 24 in, the larger sizes are shown by using the rolled-shell specification. Allowance should be made for especially tight or loose construction.

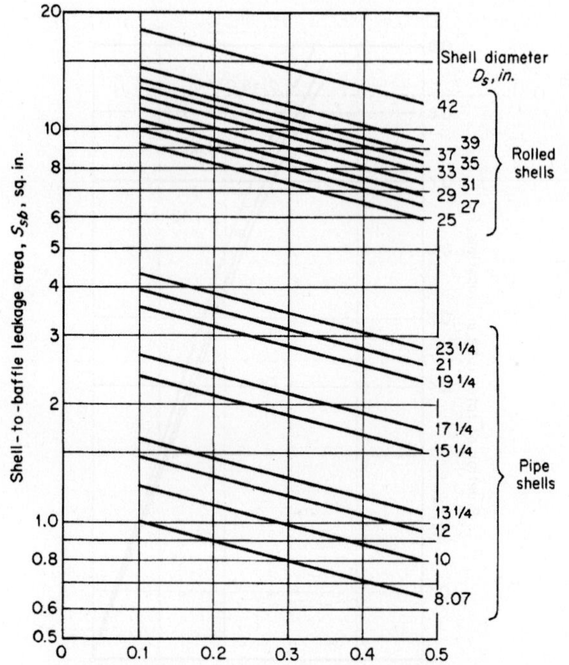

FIG. 11-7 Estimation of shell-to-baffle leakage area [Eq. (11-13)]. To convert inches to meters, multiply by 0.0254; to convert square inches to square meters, multiply by $(6.45)(10^{-4})$. Note that l_c and D_s have units of inches.

10. *Area for flow through window S_w.* This area is obtained as the difference between the gross window area S_{wg} and the window area occupied by tubes S_{wt}:

$$= S_{wg} - S_{wt} \qquad (11\text{-}14)$$

$$S_{wg} = \frac{D_s^2}{4} \left[\cos^{-1}\left(1 - 2\frac{l_c}{D_s}\right) - \left(1 - 2\frac{l_c}{D_s}\right) \sqrt{1 - \left(1 - 2\frac{l_c}{D_s}\right)^2} \right] \quad m^2 (ft^2) \qquad (11\text{-}15)$$

S_{wg} is plotted in Fig. 11-8; S_{wt} can be calculated from

$$S_{wt} = (N_T/8)(1 - F_c)\pi D_o^2 \quad m^2 (ft^2) \qquad (11\text{-}16)$$

11. *Equivalent diameter of window D_w [required only if laminar flow, defined as $(N_{Re})_s \le 100$, exists]*

$$D_w = \frac{4S_w}{(\pi/2) N_T (1 - F_c) D_o + D_s \theta_b} \quad m (ft) \qquad (11\text{-}17)$$

where θ_b is the baffle-cut angle given by

$$\theta_b = 2\cos^{-1}\left(1 - \frac{2l_c}{D_s}\right) \quad rad \qquad (11\text{-}18)$$

12. *Number of baffles N_b*

$$N_b = \frac{L - 2le}{l_s} + 1 \qquad (11\text{-}19)$$

where *le* is the entrance/exit baffle spacing, often different from the central baffle spacing. The effective tube length L must be known to calculate N_b, which is needed to calculate the shell-side pressure drop. In designing an exchanger, the shell-side coefficient may be calculated and the required exchanger length for heat transfer obtained before N_b is calculated.

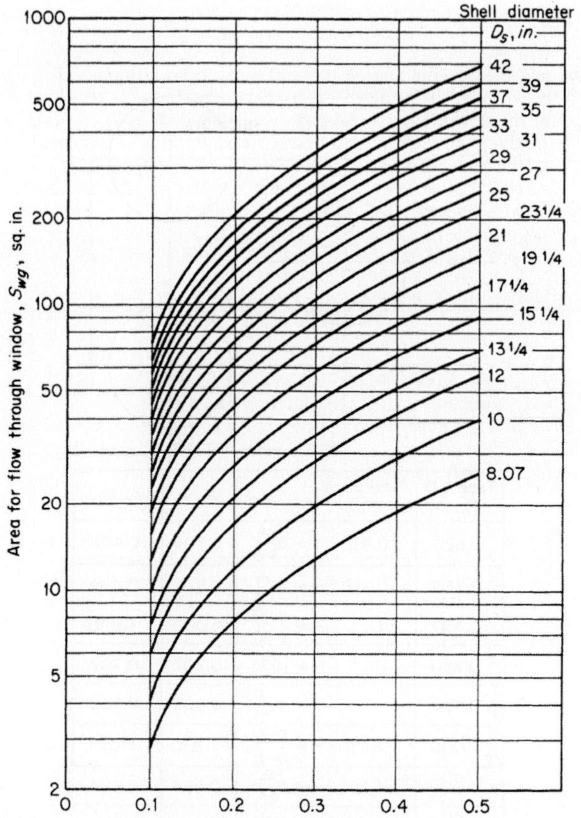

FIG. 11-8 Estimation of window cross-flow area [Eq. (11-15)]. To convert inches to meters, multiply by 0.0254. Note that l_c and D_s have units of inches.

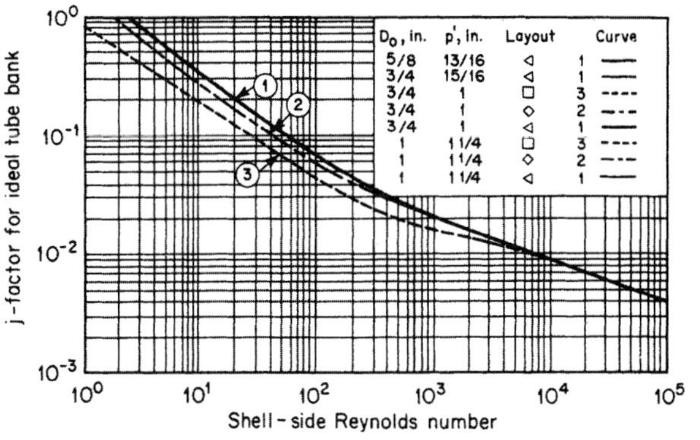

FIG. 11-9 Correlation of *j* factor for ideal tube bank. To convert inches to meters, multiply by 0.0254. Note that p' and D_o have units of inches.

Shell-Side Heat-Transfer Coefficient Calculation

1. Calculate the *shell-side Reynolds number* $(N_{Re})_s$

$$(N_{Re})_s = D_o W / \mu_b S_m \tag{11-20}$$

where W = mass flow rate and μ_b = viscosity at bulk temperature. The arithmetic mean bulk shell-side fluid temperature is usually adequate to evaluate all bulk properties of the shell-side fluid. For large temperature ranges or for viscosity that is very sensitive to temperature change, special care must be taken, such as using Eq. (11-6).

2. Find j_k from the ideal-tube bank curve for a given tube layout at the calculated value of $(N_{Re})_s$, using Fig. 11-9, which is adapted from ideal-tube bank data obtained at Delaware by Bergelin et al. [*Trans. Am. Soc. Mech. Eng.* **74:** 953 (1952)] and the Grimison correlation [*Trans. Am. Soc. Mech. Eng.* **59:** 583 (1937)].

3. Calculate the shell-side *heat-transfer coefficient for an ideal tube bank* h_k.

$$h_k = j_k c \frac{W}{S_m}\left(\frac{k}{c\mu}\right)^{2/3}\left(\frac{\mu_b}{\mu_w}\right)^{0.14} \tag{11-21}$$

where c is the specific heat, k is the thermal conductivity, and μ_w is the viscosity evaluated at the mean surface temperature.

4. Find the correction factor for baffle configuration effects J_c from Fig. 11-10.
5. Find the correction factor for baffle leakage effects J_l from Fig. 11-11.
6. Find the correction factor for bundle-bypassing effects J_b from Fig. 11-12.
7. Find the correction factor for adverse temperature-gradient buildup at low Reynolds number J_r:
   a. If $(N_{Re})_s < 100$, find J_r^* from Fig. 11-13, given N_b and $N_c + N_{cw}$.
   b. If $(N_{Re})_s \leq 20$, $J_r = J_r^*$.
   c. If $20 < (N_{Re})_s < 100$, find J_r from Fig. 11-14, given J_r^* and $(N_{Re})_s$.
8. Calculate the *shell-side heat-transfer coefficient for the exchanger* h_s from

$$h_s = h_k J_c J_l J_b J_r \tag{11-22}$$

Shell-Side Pressure Drop Calculation

1. Find f_k from the ideal-tube bank friction factor curve for the given tube layout at the calculated value of $(N_{Re})_s$, using Fig. 11-15*a* for triangular and rotated square arrays and Fig. 11-15*b* for in-line square arrays. These curves are adapted from Bergelin et al. and Grimison.

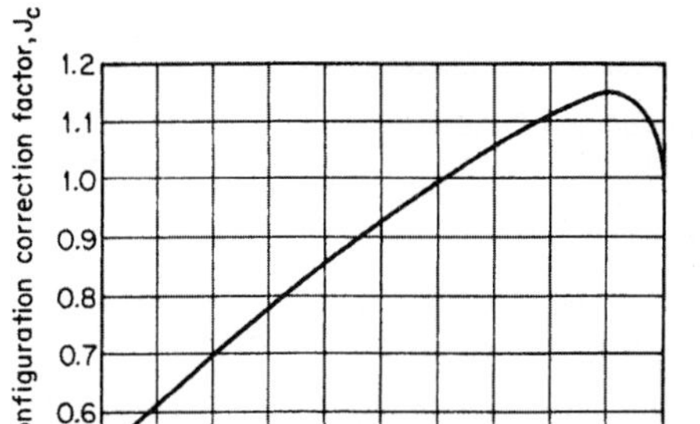

FIG. 11-10 Correction factor for baffle configuration effects.

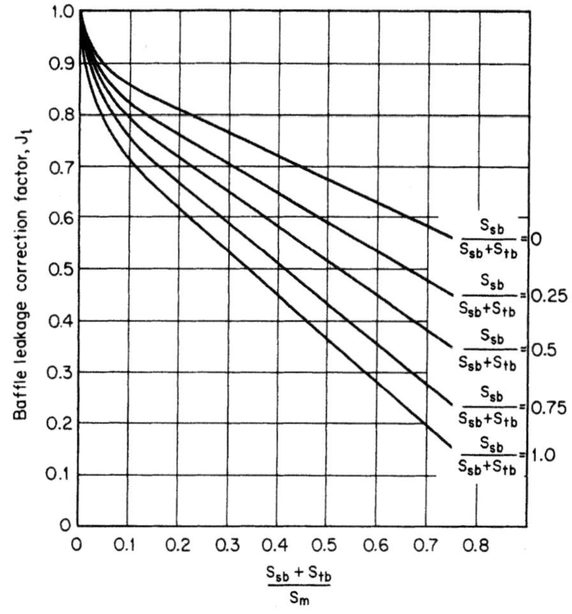

FIG. 11-11 Correction factor for baffle leakage effects.

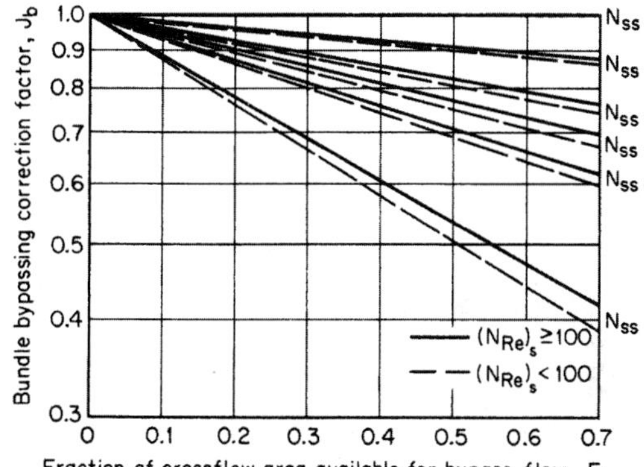

FIG. 11-12 Correction factor for bypass flow.

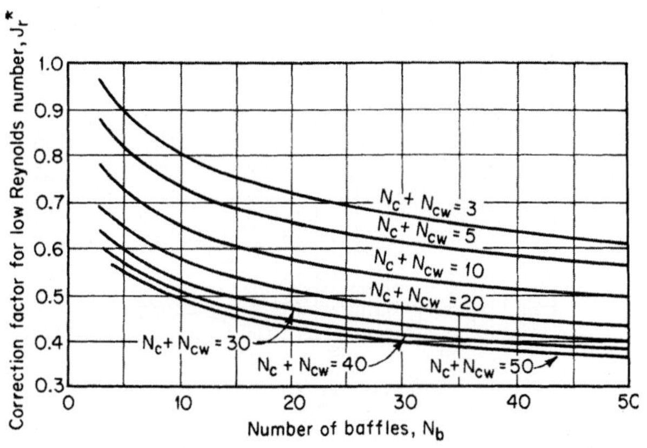

FIG. 11-13 Basic correction factor for adverse temperature gradient at low Reynolds numbers.

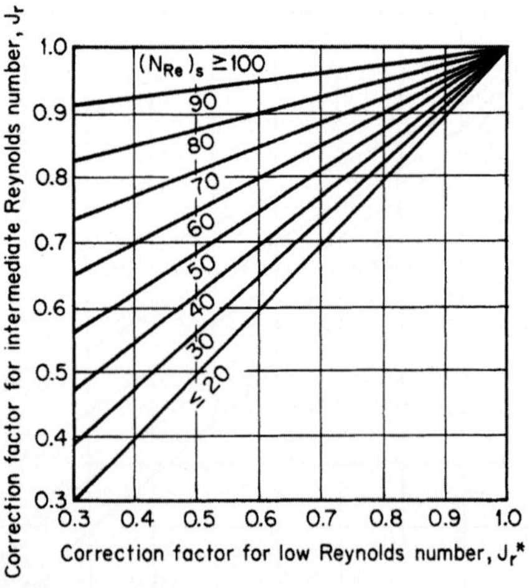

FIG. 11-14 Correction factor for adverse temperature gradient at intermediate Reynolds numbers.

2. Calculate the *pressure drop for an ideal cross-flow section.*

$$\Delta P_{bk} = b\frac{f_k W^2 N_c}{\rho S_m^2}\left(\frac{\mu_w}{\mu_b}\right)^{0.14} \tag{11-23}$$

where $b = (2.0)(10^{-3})$ (SI) or $(9.9)(10^{-5})$ (USCS).

3. Calculate the *pressure drop for an ideal window section.* If $(N_{Re})_s \geq 100$,

$$\Delta P_{wk} = b\frac{W^2(2+0.6N_w)}{S_m S_w \rho} \tag{11-24a}$$

where $b = (5)(10^{-4})$ (SI) or $(2.49)(10^{-5})$ (USCS).

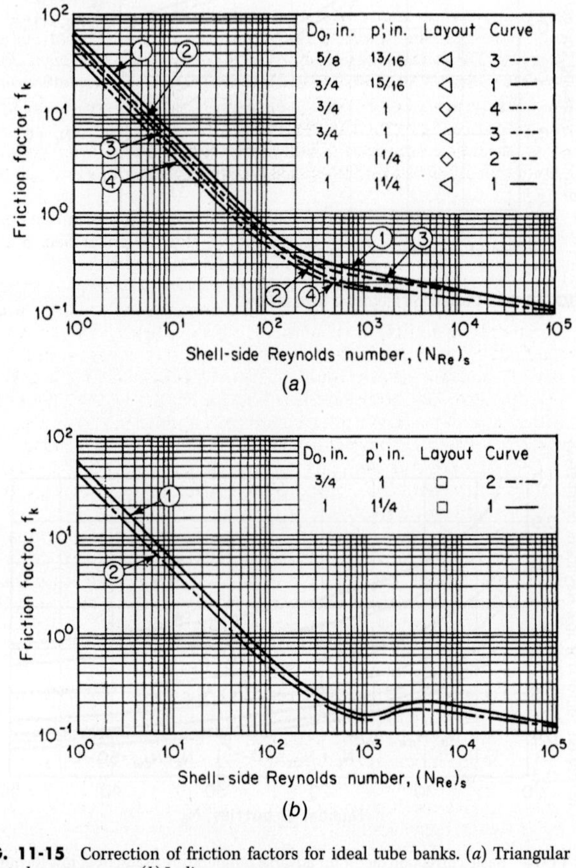

FIG. 11-15 Correction of friction factors for ideal tube banks. (*a*) Triangular and rotated square arrays. (*b*) In-line square arrays.

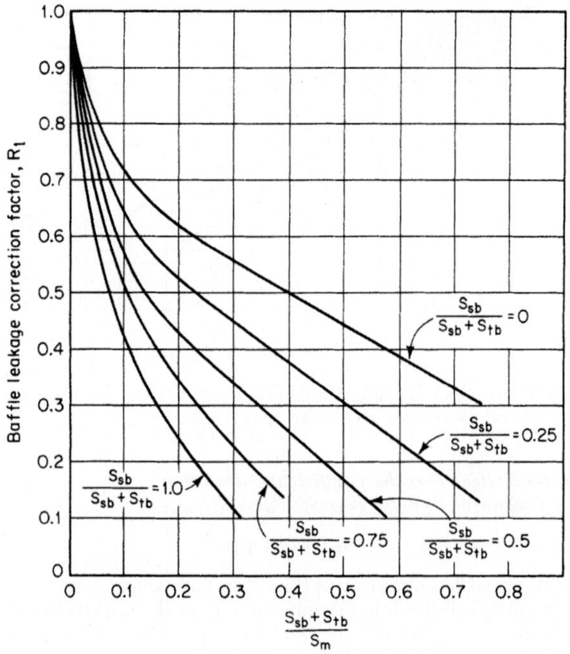

FIG. 11-16 Correction factor for baffle leakage effect on pressure drop.

If $(N_{Re})_s < 100$,

$$\Delta P_{wk} = b_1 \frac{\mu_b W}{S_m S_w \rho}\left(\frac{N_{cw}}{p' - D_o} + \frac{l_s}{D_w^2}\right) + b_2 \frac{W^2}{S_m S_w \rho} \tag{11-24b}$$

where $b_1 = (1.681)(10^{-5})$ (SI) or $(1.08)(10^{-4})$ (USCS), and $b_2 = (9.99)(10^{-4})$ (SI) or $(4.97)(10^{-5})$ (USCS).

4. Find the correction factor for the effect of baffle leakage on pressure drop R_l from Fig. 11-16. Curves shown are not to be extrapolated beyond the points shown.

5. Find the correction factor for bundle bypass R_b from Fig. 11-17.

6. Calculate the *pressure drop across the shell side* (excluding nozzles). Units for pressure drop are lbf/ft².

$$\Delta P_s = [(N_b - 1)(\Delta P_{bk})R_b + N_b\,\Delta P_{wk}]R_l + 2\Delta P_{bk}R_b\left(1 + \frac{N_{cw}}{N_c}\right) \tag{11-25}$$

The values of h_s and ΔP_s calculated by this procedure are for clean exchangers and are intended to be as accurate as possible, not conservative. A fouled exchanger will generally give lower heat-transfer rates, as reflected by the dirt resistances incorporated into Eq. (11-2), and higher pressure drops. Some estimate of *fouling effects* on pressure drop may be made by using the methods just given and assuming that the fouling deposit blocks the leakage and possibly the bypass areas. The fouling may also decrease the clearance between tubes and significantly increase the pressure drop in cross-flow.

THERMAL DESIGN OF CONDENSERS

Single-Component Condensers

Mean Temperature Difference In condensing a single component at its saturation temperature, the entire resistance to heat transfer on the condensing side is generally assumed to be in the layer of condensate. A mean condensing coefficient is calculated from the appropriate correlation and combined with the other resistances in Eq. (11-2). The overall coefficient is then used with the LMTD (no F_T correction is necessary for isothermal condensation) to give the required area, even though the condensing coefficient and hence U are not constant throughout the condenser.

If the vapor is *superheated* at the inlet, the vapor may first be desuperheated by sensible heat transfer from the vapor. This occurs if the surface temperature is above the saturation temperature, and a single-phase heat-transfer correlation is used. If the surface is below the saturation temperature, condensation will occur directly from the superheated vapor,

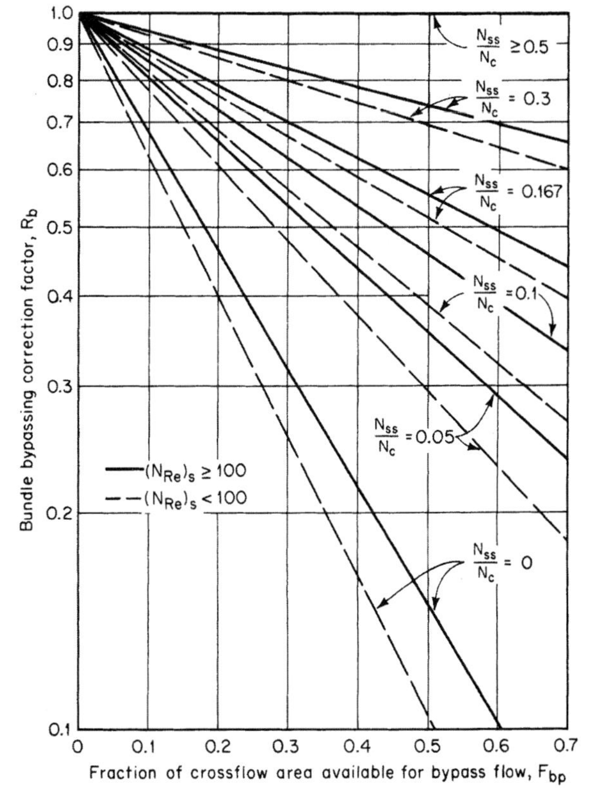

FIG. 11-17 Correction factor on pressure drop for bypass flow.

and the effective coefficient is determined from the appropriate condensation correlation, using the saturation temperature in the LMTD. To determine whether condensation will occur directly from the superheated vapor, calculate the surface temperature by assuming single-phase heat transfer.

$$T_{\text{surface}} = T_{\text{vapor}} - \frac{U}{h}(T_{\text{vapor}} - T_{\text{coolant}}) \tag{11-26}$$

where h is the sensible heat-transfer coefficient for the vapor, U is calculated by using h, and both are on the same area basis. If $T_{\text{surface}} > T_{\text{saturation}}$, no condensation occurs at that point and the heat flux is actually higher than if $T_{\text{surface}} \leq T_{\text{saturation}}$ and condensation did occur. It is generally conservative to design a pure-component desuperheater-condenser as if the entire heat load were transferred by condensation, using the saturation temperature in the LMTD.

The design of an integral *condensate subcooling section* is more difficult, especially if close temperature approach is required. The condensate layer on the surface is subcooled on average by one-third to one-half of the temperature drop across the film, and this is often sufficient if the condensate is not reheated by raining through the vapor. If the condensing-subcooling process is carried out inside tubes or in the shell of a vertical condenser, the single-phase subcooling section can be treated separately, giving an area that is added onto that needed for condensation. If the subcooling is achieved on the shell side of a horizontal condenser by flooding some of the bottom tubes with a weir or level controller, then the rate and heat-balance equations must be solved for each section to obtain the area required.

Pressure drop on the condensing side reduces the final condensing temperature and the MTD and should always be checked. In designs requiring close approach between inlet coolant and exit condensate (subcooled or not), underestimation of pressure drop on the condensing side can lead to an exchanger that cannot meet specified terminal temperatures. Since pressure drop calculations in two-phase flows such as condensation are relatively inaccurate, designers must consider carefully the consequences of a larger than calculated pressure drop.

Horizontal In-Shell Condensers The mean *condensing coefficient* for the outside of a bank of horizontal tubes is calculated from Eq. (5-87) for a single tube, corrected for the number of tubes in a vertical row. For undisturbed laminar flow over all the tubes, Eq. (5-88) is, for realistic condenser sizes, overly conservative because of rippling, splashing, and turbulent flow (*Process Heat Transfer*, McGraw-Hill, New York, 1950). Kern proposed an exponent of $-1/6$ on the basis of experience, while Freon-11 data of Short and Brown (*General Discussion on Heat Transfer*, Institute of Mechanical Engineers, London, 1951) indicate independence of the number of tube rows. It seems reasonable to use no correction for inviscid liquids and Kern's correction for viscous condensates. For a cylindrical tube bundle, where N varies, it is customary to take N equal to two-thirds of the maximum or centerline value.

Baffles in a horizontal in-shell condenser are oriented with the cuts vertical to facilitate drainage and eliminate the possibility of flooding in the upward cross-flow sections. *Pressure drop* on the vapor side can be estimated by the data and method of Diehl and Unruh [*Pet. Refiner* **36**(10): 147 (1957); **37**(10): 124 (1958)].

High vapor velocities across the tubes enhance the condensing coefficient. There is no correlation in the open literature to permit designers to take advantage of this. Since the vapor flow rate varies along the length, an incremental calculation procedure would be required in any case. In general, the pressure drops required to gain significant benefit are above those allowed in most process applications.

Vertical In-Shell Condensers Condensers are often designed so that condensation occurs on the outside of vertical tubes. Equation (5-79) is valid as long as the condensate film is laminar. When it becomes turbulent, Eq. (5-86) or Colburn's equation [*Trans. Am. Inst. Chem. Eng.* **30:** 187 (1933–1934)] may be used.

Some judgment is required in the use of these correlations because of the construction features of the condenser. The tubes must be supported by baffles, usually with maximum cut (45 percent of the shell diameter) and maximum spacing to minimize pressure drop. The flow of the condensate is interrupted by the baffles, which may draw off or redistribute the liquid and which will also cause some splashing of free-falling drops onto the tubes.

For *subcooling*, a liquid inventory may be maintained in the bottom end of the shell by means of a weir or a liquid-level-controller. The subcooling heat-transfer coefficient is given by the correlations for natural convection on a vertical surface [Eq. (5-42)], with the pool assumed to be well mixed (isothermal) at the subcooled condensate exit temperature. Pressure drop may be estimated by the shell-side procedure.

Horizontal In-Tube Condensers Condensation of a vapor inside horizontal tubes occurs in kettle and horizontal thermosiphon reboilers and in air-cooled condensers. In-tube condensation also offers certain advantages for condensation of multicomponent mixtures, discussed in the subsection Multicomponent Condensers. The various in-tube correlations are closely connected to the *two-phase flow pattern* in the tube [*Chem. Eng. Prog. Symp. Ser.* **66**(102): 150 (1970)]. At low flow rates, when gravity dominates the flow pattern, Eq. (5-87) may be used. At high flow rates, the flow and heat transfer are governed by vapor shear on the condensate film, and Eq. (5-86) is valid. A simple and generally conservative procedure is to calculate the coefficient for a given case by both correlations and use the *larger* one.

Pressure drop during condensation inside horizontal tubes can be computed by using the correlations for two-phase flow given in Sec. 6 and neglecting the pressure recovery due to deceleration of the flow.

Vertical In-Tube Condensation Vertical tube condensers are generally designed so that vapor and liquid flow cocurrently downward; if pressure drop is not a limiting consideration, this configuration can result in higher heat-transfer coefficients than for shell-side condensation and has particular advantages for multicomponent condensation. If gravity controls, the mean heat-transfer coefficient for condensation is given by Eq. (5-79). If vapor shear controls, Eq. (5-86) is applicable. It is generally conservative to calculate the coefficients by both methods and choose the *higher* value. The pressure drop can be calculated by using the Lockhart-Martinelli method [*Chem. Eng. Prog.* **45:** 39 (1945)] for friction loss, neglecting momentum and hydrostatic effects.

Vertical in-tube condensers are often designed for *reflux* **or** *knock-back application* in reactors or distillation columns. In this case, vapor flow is upward, countercurrent to the liquid flow on the tube wall; the vapor shear acts to thicken and retard the drainage of the condensate film, reducing the coefficient. Neither the fluid dynamics nor the heat transfer is well understood in this case, but Soliman, Schuster, and Berenson [*J. Heat Transfer* **90:** 267–276 (1968)] discuss the problem and suggest a computational method. The Diehl-Koppany correlation [*Chem. Eng. Prog. Symp. Ser.* 92, **65** (1969)] may be used to estimate the maximum allowable vapor velocity at the tube inlet. If the vapor velocity is great enough, the liquid film will be carried upward; this design has been employed in a few cases in which only part of the stream is to be condensed. This velocity cannot be accurately computed, and a very conservative (high) outlet velocity must be used if unstable flow and flooding are to be avoided; 3 times the vapor velocity given by the Diehl-Koppany correlation for incipient flooding has been suggested as the design value for completely stable operation.

Multicomponent Condensers

Thermodynamic and Mass-Transfer Considerations *Multicomponent vapor mixture* includes several different cases: all the components may be liquids at the lowest temperature reached in the condensing side, or there may be components that dissolve substantially in the condensate even though their boiling points are below the exit temperature, or one or more components may be both noncondensible and nearly insoluble.

Multicomponent condensation always involves sensible-heat changes in the vapor and liquid along with the latent-heat load. Compositions of both phases in general change through the condenser, and *concentration gradients* exist in both phases. Temperature and concentration profiles and transport rates at a point in the condenser usually cannot be calculated, but the binary cases have been treated: condensation of one component in the presence of a completely insoluble gas [Colburn and Hougen, *Ind. Eng. Chem.* **26:** 1178–1182 (1934); and Colburn and Edison, *Ind. Eng. Chem.* **33:** 457–458 (1941)] and condensation of a binary vapor [Colburn and Drew, *Trans. Am. Inst. Chem. Eng.* **33:** 196–215 (1937)]. It is necessary to know or calculate diffusion coefficients for the system, and a reasonable approximate method to avoid this difficulty and the reiterative calculations is desirable. To integrate the point conditions over the total condensation requires the temperature, composition enthalpy, and flow-rate profiles as functions of the heat removed. These are calculated from component thermodynamic data if the vapor and liquid are assumed to be in equilibrium at the local vapor temperature. This assumption is not exactly true, since the condensate and the liquid-vapor interface (where equilibrium does exist) are intermediate in temperature between the coolant and the vapor.

In calculating the condensing curve, it is generally assumed that the vapor and liquid flow collinearly and in intimate contact so that composition equilibrium is maintained between the total streams at all points. If, however, the condensate drops out of the vapor (as can happen in horizontal shell-side condensation) and flows to the exit without further interaction, the remaining vapor becomes excessively enriched in light components with a decrease in condensing temperature and in the temperature difference between vapor and coolant. The result may be not only a small reduction in the amount of heat transferred in the condenser but also an inability to condense totally the light ends even at reduced throughput or with the addition of more surface. To prevent the liquid from segregating, in-tube condensation is preferred in critical cases.

Thermal Design If the controlling resistance for heat and mass transfer in the vapor is sensible-heat removal from the cooling vapor, the following design equation is obtained:

$$A = \int_0^{Q_T} \frac{1 + U'Z_H/h_{sv}}{U'(T_v - T_c)}\,dQ \qquad (11\text{-}27)$$

U' is the overall heat-transfer coefficient between the vapor-liquid interface and the coolant, including condensate film, dirt and wall resistances, and coolant. The condensate film coefficient is calculated from the appropriate equation or correlation for pure vapor condensation for the geometry and flow regime involved, using mean liquid properties. The ratio of the sensible heat removed from the vapor-gas stream to the total heat transferred is Z_H; this quantity is obtained from thermodynamic calculations and may vary substantially from one end of the condenser to the other, especially when removing vapor from a noncondensible gas. The sensible-heat-transfer coefficient for the vapor-gas stream h_{sv} is calculated by using the appropriate correlation or design method for the geometry involved, neglecting the presence of the liquid. As the vapor condenses, this coefficient decreases and must be calculated at several points in the process. And T_v and T_c are temperatures of the vapor and the coolant, respectively. This procedure is similar in principle to that of Ward [*Petro/Chem. Eng.* **32**(11): 42–48 (1960)]. It may be nonconservative for condensing steam and other high-latent-heat substances, in which case it may be necessary to increase the calculated area by 25 to 50 percent.

Pressure drop on the condensing side may be estimated by judicious application of the methods suggested for pure-component condensation, taking into account the generally nonlinear decrease of vapor-gas flow rate with heat removal.

THERMAL DESIGN OF REBOILERS

For a *single-component reboiler design*, attention is focused upon the mechanism of heat and momentum transfer at the hot surface. In *multicomponent systems*, the light components are preferentially vaporized at the surface, and the process becomes limited by their rate of diffusion. The net effect is to decrease the effective temperature difference between the hot surface and the bulk of the boiling liquid. If one attempts to vaporize too high a fraction of the feed liquid to the reboiler, then the temperature difference between surface and liquid is reduced to the point that nucleation and vapor generation on the surface are suppressed and heat transfer to the liquid proceeds at the lower rate associated with single-phase natural convection. The only safe procedure in design for wide boiling-range mixtures is to vaporize such a limited fraction of the feed that the boiling point of the remaining liquid mixture is still at least 5.5°C (10°F) below the surface temperature. Positive flow of the unvaporized liquid through and out of the reboiler should be provided.

Kettle Reboilers It has been generally assumed that kettle reboilers operate in the pool boiling mode, but with a lower peak heat flux because of vapor binding and blanketing of the upper tubes in the bundle. There is some evidence that vapor generation in the bundle causes a high circulation rate through the bundle. The result is that, at the lower heat fluxes, the kettle reboiler actually gives higher heat-transfer coefficients than a single tube. Present understanding of the recirculation phenomenon is insufficient to take advantage of this in design. Available nucleate pool boiling correlations are only very approximate, failing to account for differences in the nucleation characteristics of different surfaces. Equation (5-97b) may be used for single components or narrow boiling-range mixtures at low fluxes. For hydrocarbons not listed, approximation may be made assuming *n*-Pentane. Experimental heat-transfer coefficients for pool boiling of a given liquid on a given surface should be used if available. The bundle *peak heat flux* is a function of tube bundle geometry, especially of tube-packing density. But in the absence of better information, Eq. (5-99) may be used for this purpose.

A general method for analyzing kettle reboiler performance is given by Fair and Klip, *Chem. Eng. Prog.* **79**(3): 86 (1983). It is effectively limited to computer application.

Kettle reboilers are generally assumed to require negligible pressure drop. It is important to provide good longitudinal liquid flow paths within the shell so that the liquid is uniformly distributed along the entire length of the tubes and excessive local vaporization and vapor binding are avoided.

This method may also be used for the thermal design of *horizontal thermosiphon reboilers.* The recirculation rate and pressure profile of the thermosiphon loop can be calculated by the methods of Fair [*Pet. Refiner* **39**(2): 105–123 (1960)].

Vertical Thermosiphon Reboilers Vertical thermosiphon reboilers operate by natural circulation of the liquid from the still through the downcomer to the reboiler and of the two-phase mixture from the reboiler through the return piping. The flow is induced by the hydrostatic pressure imbalance between the liquid in the downcomer and the two-phase mixture in the reboiler tubes. Thermosiphons do not require any pump for recirculation and are generally regarded as less likely to foul in service because of the relatively high two-phase velocities obtained in the tubes. Heavy components are not likely to accumulate in the thermosiphon, but they are more difficult to design satisfactorily than kettle reboilers, especially in vacuum operation. Several shortcut methods have been suggested for thermosiphon design, but they must generally be used with caution. The method due to Fair (1960), based upon two-phase flow correlations, is the most complete in the open literature but requires a computer for practical use. Fair also suggests a shortcut method that is satisfactory for preliminary design and can be reasonably done by hand.

Forced-Recirculation Reboilers In forced-recirculation reboilers, a pump is used to ensure circulation of the liquid past the heat-transfer surface. Forced-recirculation reboilers may be designed so that boiling occurs inside vertical tubes, inside horizontal tubes, or on the shell side. For forced boiling inside vertical tubes, Fair's method may be employed, by making only the minor modification that the recirculation rate is fixed and does not need to be balanced against the pressure available in the downcomer. Excess pressure required to circulate the two-phase fluid through the tubes and back into the column is supplied by the pump, which must develop a positive pressure increase in the liquid.

Fair's method may also be modified to design forced-recirculation reboilers with horizontal tubes. In this case the hydrostatic head pressure effect through the tubes is zero but must be considered in the two-phase return lines to the column.

The same procedure may be applied in principle to design of forced-recirculation reboilers with shell-side vapor generation. Little is known about two-phase flow on the shell side, but a reasonable estimate of the friction pressure drop can be made from the data of Diehl and Unruh [*Pet. Refiner* **36**(10): 147 (1957); **37**(10): 124 (1958)]. No void-fraction data are available to permit accurate estimation of the hydrostatic or acceleration terms. These may be roughly estimated by assuming homogeneous flow.

THERMAL DESIGN OF EVAPORATORS

Heat duties of evaporator heating surfaces are usually determined by conventional heat and material balance calculations. Heating surface areas are normally, but not always, taken as those in contact with the material being evaporated. It is the heat transfer ΔT that presents the greatest difficulty in deriving or applying heat-transfer coefficients. The total ΔT between heat source and heat sink is never all available for heat transfer. Since energy usually is carried to and from an evaporator body or effect by condensible vapors, loss in pressure represents a loss in ΔT. Such losses include pressure drop through entrainment separators, friction in vapor piping, and acceleration losses into and out of the piping. The latter loss has often been overlooked, even though it can be many times greater than the friction loss. Similarly, friction and acceleration losses past the heating surface, such as in a falling film evaporator, cause a loss of ΔT that may or may not have been included in the heat transfer ΔT when reporting experimental results. Boiling-point rise, the difference between the boiling point of the solution and the condensing point of the solvent at the same pressure, is another loss. Experimental data are almost always corrected for boiling-point rise, but plant data are suspect when based on temperature measurements because vapor at the point of measurement may still contain some superheat, which represents but a very small fraction of the heat given up when the vapor condenses but may represent a substantial fraction of the actual net ΔT available for heat transfer. A loss of ΔT that must be considered in forced-circulation evaporators is that due to temperature rise through the heater, a consequence of the heat being absorbed there as sensible heat. A further loss may occur when the heater effluent flashes as it enters the vapor-liquid separator. Some of the liquid may not reach the surface and flash to equilibrium with the vapor pressure in the separator, instead of recirculating to the heater, raising the average temperature at which heat is absorbed and further reducing the net ΔT. Whether these ΔT losses are allowed for in the heat-transfer coefficients reported depends on the method of measurement. Simply basing the liquid temperature on the measured vapor head pressure may ignore both—or only the latter if temperature rise through the heater is estimated separately from known heat input and circulation rate. In general, when one is calculating the overall heat-transfer coefficients from individual-film coefficients, all these losses must be allowed for, while when using reported overall coefficients, care must be exercised to determine which losses may already have been included in the heat transfer ΔT.

Forced-Circulation Evaporators In evaporators of this type in which hydrostatic head prevents boiling at the heating surface, *heat-transfer coefficients* can be predicted for forced-convection sensible heating using Eqs. (5-54) and (5-57). The liquid film coefficient is improved if boiling is not completely suppressed. When only the film next to the wall is above the boiling point, Boarts, Badger, and Meisenberg [*Ind. Eng. Chem.* **29:** 912 (1937)] found that results could be correlated by Eq. (5-50) by using a constant of 0.0278 instead of 0.023. In such cases, the course of the liquid temperature can still be calculated from known circulation rate and heat input.

When the bulk of the liquid is boiling in part of the tube length, the film coefficient is even higher. However, the liquid temperature starts dropping as soon as full boiling develops, and it is difficult to estimate the course of the temperature curve. It is certainly safe to estimate heat transfer on the basis that no bulk boiling occurs. Fragen and Badger [*Ind. Eng. Chem.* **28:** 534 (1936)] obtained an *empirical correlation* of overall heat-transfer coefficients in this type of evaporator, based on the ΔT at the heater inlet:

In USCS units

$$U = 2020 D^{0.57} (V_s)^{3.6/L} / \mu^{0.25} \, \Delta T^{0.1} \qquad (11\text{-}28)$$

where D = mean tube diameter, V_s = inlet velocity, L = tube length, and μ = liquid viscosity. This equation is based primarily on experiments with copper tubes of 0.022 m (0.866 in) outside diameter, 0.00165 m (0.065 in or 16 gauge) wall thickness, and 2.44 m (8 ft) long, but it includes some work with 0.0127 m (½ in) tubes 2.44 m (8 ft) long and 0.0254 m (1 in) tubes 3.66 m (12 ft) long.

Long-Tube Vertical Evaporators In the rising-film version of this type of evaporator, there is usually a nonboiling zone in the bottom section and a boiling zone in the top section. The length of the nonboiling zone depends on heat-transfer characteristics in the two zones and on pressure drop during two-phase flow in the boiling zone. The work of Martinelli and coworkers [Lockhart and Martinelli, *Chem. Eng. Prog.* **45:** 39–48 (January 1949); and Martinelli and Nelson, *Trans. Am. Soc. Mech. Eng.* **70:** 695–702 (August 1948)] permits a prediction of pressure drop, and a number of correlations are available for estimating film coefficients of heat transfer in the two zones. In estimating pressure drop, integrated curves similar to those presented by Martinelli and Nelson are the easiest to use. The curves for pure water are shown in Figs. 11-18 and 11-19, based on the assumption that the flow of both vapor and liquid would be turbulent if each were flowing alone in the tube. Similar curves can be prepared if one or both flows are laminar or if the properties of the liquid differ appreciably from the properties of pure water. The *acceleration pressure drop* ΔP_a is calculated from the equation

$$\Delta P_a = b r_2 G^2 / 32.2 \qquad (11\text{-}29)$$

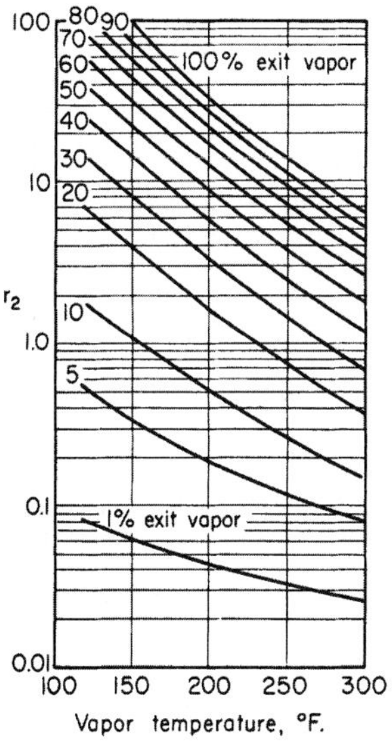

FIG. 11-18 Acceleration losses in boiling flow. °C = (°F − 32)/1.8.

where $b = (2.6)(10^7)$ (SI) and $b = 1.0$ (USCS) and using r_2 from Fig. 11-18. The frictional pressure drop is derived from Fig. 11-19, which shows the ratio of two-phase pressure drop to the entering liquid flowing alone.

Pressure drop due to hydrostatic head can be calculated from liquid holdup R_l. For nonfoaming dilute aqueous solutions, R_l can be estimated from $R_l = 1/[1 + 2.5(V/L)(\rho_l/\rho_v)^{1/2}]$. Liquid holdup, which represents the ratio of liquid-only velocity to actual liquid velocity, also appears to be the principal determinant of the convective coefficient in the boiling zone (Dengler, Sc.D. thesis, MIT, 1952). In other words, the convective coefficient is that calculated from Eq. (5-57) by using the liquid-only velocity divided by R_l in the Reynolds number. Nucleate boiling augments convective heat transfer, primarily when ΔT values are high and the convective coefficient is low [Chen, *Ind. Eng. Chem. Process Des. Dev.* **5:** 322 (1966)].

Film coefficients for the *boiling of liquids other than water* have been investigated. Coulson and McNelly [*Trans. Inst. Chem. Eng.* **34:** 247 (1956)] derived the following relation, which also correlated the data of Badger and coworkers [*Chem. Metall. Eng.* **46:** 640 (1939); *Chem. Eng.* **61**(2): 183 (1954); and *Trans. Am. Inst. Chem. Eng.* **33:** 392 (1937); **35:** 17 (1939); **36:** 759 (1940)] on water:

$$N_{Nu} = (1.3 + bD)(N_{Pr})_l^{0.9} (N_{Re})_l^{0.23} (N_{Re})_g^{0.34} \left(\frac{\rho_l}{\rho_g}\right)^{0.25} \left(\frac{\mu_g}{\mu_l}\right) \qquad (11\text{-}30)$$

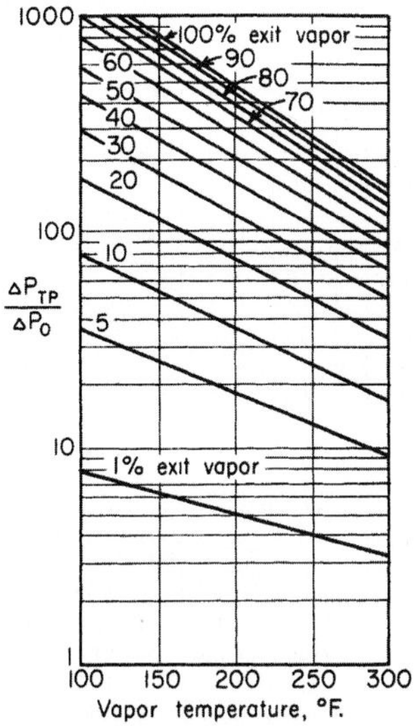

FIG. 11-19 Friction pressure drop in boiling flow. °C = (°F − 32)/1.8.

where b = 128 (SI) or 39 (USCS), N_{Nu} = Nusselt number based on liquid thermal conductivity, D = tube diameter, and the remaining terms are dimensionless groupings of the liquid Prandtl number, liquid Reynolds number, vapor Reynolds number, and ratios of densities and viscosities. The Reynolds numbers are calculated on the basis of each fluid flowing by itself in the tube.

Additional corrections must be applied when the fraction of vapor is so high that the remaining liquid does not wet the tube wall or when the velocity of the mixture at the tube exits approaches sonic velocity. McAdams, Woods, and Bryan (*Trans. Am. Soc. Mech. Eng.*, 1940), Dengler and Addoms (Dengler, Sc.D. thesis, MIT, 1952), and Stroebe, Baker, and Badger [*Ind. Eng. Chem.* **31:** 200 (1939)] encountered dry-wall conditions and reduced coefficients when the weight fraction of vapor exceeded about 80 percent. Schweppe and Foust [*Chem. Eng. Prog.* **49:** Symp. Ser. 5, 77 (1953)] and Harvey and Foust [*Chem. Eng. Prog.* **49:** Symp. Ser. 5, 91 (1953)] found that "sonic choking" occurred at surprisingly low flow rates.

The simplified method of calculation outlined includes no allowance for the *effect of surface tension.* Stroebe, Baker, and Badger found that by adding a small amount of surface-active agent the boiling-film coefficient varied inversely as the square of the surface tension. Coulson and Mehta [*Trans. Inst. Chem. Eng.* **31:** 208 (1953)] found the exponent to be −1.4. The higher coefficients at low surface tension are offset to some extent by a higher pressure drop, probably because the more intimate mixture existing at low surface tension causes the liquid fraction to be accelerated to a velocity closer to that of the vapor. The pressure drop due to acceleration ΔP_a derived from Fig. 11-18 allows for some slippage. In the limiting case, such as might be approached at low surface tension, the acceleration pressure drop in which "fog" flow is assumed (no slippage) can be determined from the equation

$$\Delta P'_a = \frac{y(V_g - V_l)G^2}{g_c} \tag{11-31}$$

where y = fraction vapor by weight
V_g, V_l = specific volume gas, liquid
G = mass velocity

While the foregoing methods are valuable for detailed evaporator design or for evaluating the effect of changes in conditions on performance, they are cumbersome to use when making preliminary designs or cost estimates. Figure 11-20 gives the general range of *overall long-tube vertical (LTV) evaporator heat-transfer coefficients* usually encountered in commercial practice. The higher coefficients are encountered when evaporating dilute solutions and the lower range when evaporating viscous liquids. The dashed curve represents the approximate lower limit, for liquids with viscosities of about 0.1 Pa·s (100 cP). The LTV evaporator does not work well at low temperature differences, as indicated by the results shown in Fig. 11-21 for seawater in 0.051-m (2-in), 0.0028-m (12-gauge) brass tubes 7.32 m (24 ft) long (W. L. Badger Associates, Inc., U.S. Department of the Interior, Office of Saline Water, Rep. 26, December 1959, OTS Publ. PB 161290). The feed was at its boiling point at the vapor head pressure, and feed rates varied from 0.025 to 0.050 kg/(s·tube) [200 to 400 lb/(h·tube)] at the higher temperature to 0.038 to 0.125 kg/(s·tube) [300 to 1000 lb/(h·tube)] at the lowest temperature.

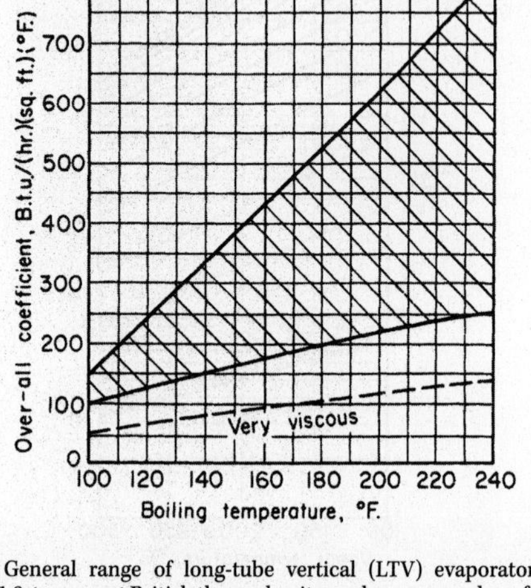

FIG. 11-20 General range of long-tube vertical (LTV) evaporator coefficients. °C = (°F − 32)/1.8; to convert British thermal units per hour squared per foot per degree Fahrenheit to joules per square meter per second per kelvin, multiply by 5.6783.

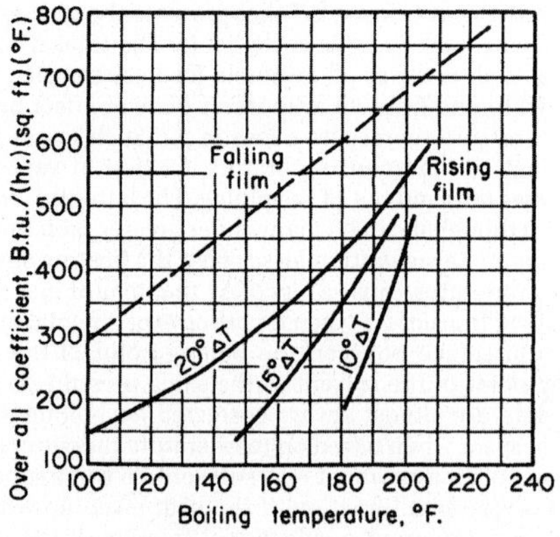

FIG. 11-21 Heat-transfer coefficients in LTV seawater evaporators. °C = (°F − 32)/1.8; to convert British thermal units per hour per square foot per degree Fahrenheit to joules per square meter per second per kelvin, multiply by 5.6783.

Falling-film evaporators find their widest use at low temperature differences—also at low temperatures. Under most operating conditions encountered, heat transfer is almost all by pure convection, with a negligible contribution from nucleate boiling. Film coefficients on the condensing side may be estimated from Dukler's correlation [*Chem. Eng. Prog.* **55:** 62 (1950)]. The same Dukler correlation presents curves covering falling-film heat transfer to nonboiling liquids that are equally applicable to the falling-film evaporator [Sinek and Young, *Chem. Eng. Prog.* **58:** 12, 74 (1962)]. Kunz and Yerazunis [*J. Heat Transfer* **8:** 413 (1969)] have since extended the range of physical properties covered, as shown in Fig. 11-22. The boiling point in the tubes of such an evaporator is higher than that in the vapor head because of both frictional pressure drop and the head needed to accelerate the vapor to the tube-exit velocity. These factors, which can easily be predicted, make the overall apparent coefficients somewhat lower than those for nonboiling conditions. Figure 11-21 shows overall apparent heat-transfer coefficients determined in a falling-film seawater evaporator using the same tubes and flow rates as for the rising-film tests (W. L. Badger Associates, Inc., U.S. Department of the Interior, Office of Saline Water, Rep. 26, December 1959, OTS Publ. PB 161290).

Short-Tube Vertical Evaporators Coefficients can be estimated by the same detailed method described for recirculating LTV evaporators. Performance is primarily a function of temperature level, temperature difference, and viscosity. While liquid level can also have an important influence, this is usually encountered only at levels lower than considered safe in commercial operation. *Overall heat-transfer coefficients* are shown in Fig. 11-23 for a basket-type evaporator (one with an annular downtake) when boiling water with 0.051-m (2-in) outside-diameter 0.0028-m wall (12-gauge), 1.22-m- (4-ft-) long steel tubes [Badger and Shepard, *Chem. Metall. Eng.* **23:** 281 (1920)]. Liquid level was maintained at the top tube sheet. Foust, Baker, and Badger [*Ind. Eng. Chem.* **31:** 206 (1939)] measured recirculating velocities and heat-transfer coefficients in the same evaporator except with 0.064-m (2.5-in) 0.0034-m-wall (10-gauge), 1.22-m- (4-ft-) long tubes and temperature differences from 7°C to 26°C (12°F to 46°F). In the normal range of liquid levels, their results can be expressed as

$$U_c = \frac{b(\Delta T_c)^{0.22} N_{pr}^{0.4}}{(V_g - V_l)^{0.37}} \tag{11-32}$$

where b = 153 (SI) or 375 (USCS) and the subscript c refers to true liquid temperature, which under these conditions was about 0.56°C (1°F) above the vapor head temperature. This work was done with water.

No detailed tests have been reported for the performance of propeller calandrias. Not enough is known regarding the performance of the propellers themselves under the cavitating conditions usually encountered to permit prediction of circulation rates. In many cases, it appears that the propeller does no good in accelerating heat transfer over the transfer for natural circulation (Fig. 11-23).

Miscellaneous Evaporator Types *Horizontal-tube evaporators* operating with partially or fully submerged heating surfaces behave in much the same way as short-tube verticals, and heat-transfer coefficients are of the same order of magnitude. Some test results for water were published by Badger [*Trans. Am. Inst. Chem. Eng.* **13:** 139 (1921)]. When operating unsubmerged, their heat-transfer performance is roughly comparable to that of the falling-film vertical tube evaporator. Condensing coefficients inside the

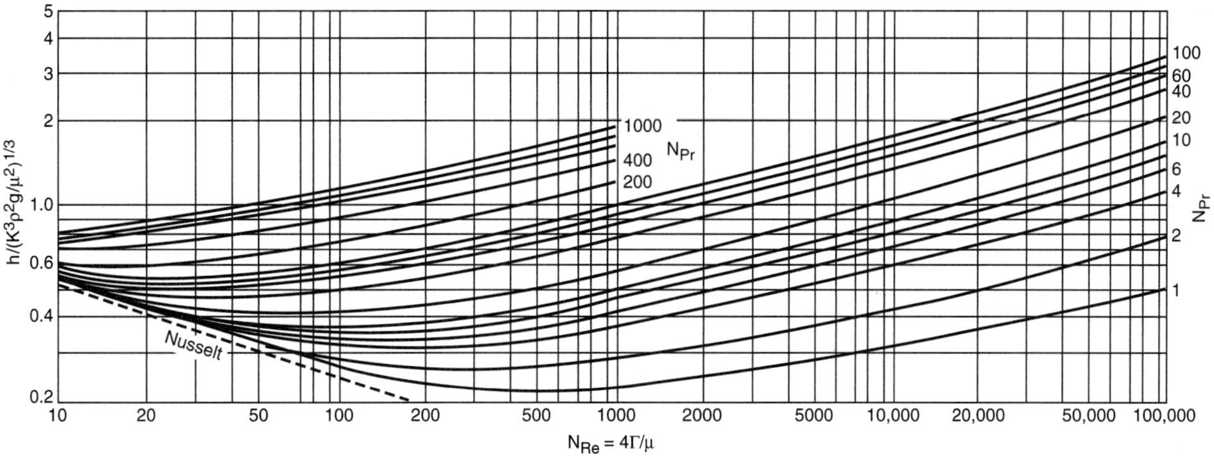

FIG. 11-22 Kunz and Yerazunis correlation for falling-film heat transfer.

tubes can be derived from Nusselt's theory which, based on a constant-heat flux rather than a constant film ΔT, gives

$$\frac{h}{(k^3\rho^2 g/\mu^2)^{1/3}} = 1.59(4\Gamma/\mu)^{-1/3} \tag{11-33a}$$

For the boiling side, a correlation based on seawater tests gives

$$\frac{h}{(k^3\rho^2 g/\mu^2)^{1/3}} = 0.0147(4\Gamma/\mu)^{1/3}(D)^{-1/3} \tag{11-33b}$$

where Γ is based on feed rate per unit length of the top tube in each vertical row of tubes and D is in meters.

Heat-transfer coefficients in clean coiled-tube evaporators for seawater are shown in Fig. 11-24 [Hillier, *Proc. Inst. Mech. Eng. (London)*, **1B**(7): 295 (1953)]. The tubes were of copper.

Heat-transfer coefficients in *agitated-film evaporators* depend primarily on liquid viscosity. This type is usually justifiable only for very viscous materials. Figure 11-25 shows general ranges of overall coefficients [Hauschild, *Chem. Ing. Tech.* **25:** 573 (1953); Lindsey, *Chem. Eng.* **60**(4): 227 (1953); and Leniger and Veldstra, *Chem. Ing. Tech.* **31:** 493 (1959)]. When used with nonviscous fluids, a wiped-film evaporator having fluted external surfaces can exhibit very high coefficients (Lustenader et al., *Trans. Am. Soc. Mech. Eng.* Paper 59-SA-30, 1959), although at a probably unwarranted first cost.

Heat Transfer from Various Metal Surfaces In an early work, Pridgeon and Badger [*Ind. Eng. Chem.* **16:** 474 (1924)] published test results on copper and iron tubes in a horizontal-tube evaporator that indicated an extreme *effect of surface cleanliness* on heat-transfer coefficients. However, the high degree of cleanliness needed for high coefficients was difficult to achieve, and the tube layout and liquid level were changed during the course of the tests so as to make direct comparison of results difficult. Other workers have found little or no effect of conditions of surface or tube material on boiling-film coefficients in the range of commercial operating conditions [Averin, *Izv. Akad. Nauk SSSR Otd. Tekh. Nauk*, no. 3, p. 116, 1954; and Coulson and McNelly, *Trans. Inst. Chem. Eng.* **34:** 247 (1956)].

Work in connection with desalination of seawater has shown that *specially modified surfaces* can have a profound effect on heat-transfer coefficients in evaporators. Figure 11-26 (Alexander and Hoffman, Oak Ridge National Laboratory TM-2203) compares overall coefficients for some of these surfaces when boiling freshwater in 0.051-m (2-in) tubes 2.44 m (8 ft) long at atmospheric pressure in both upflow and downflow. The area basis used was the nominal outside area. Tube 20 was a smooth 0.0016-m (0.062-in) wall aluminum brass tube that had accumulated about 6 years of fouling in seawater service and exhibited a fouling resistance of about $(2.6)(10^{-5})$ $(m^2 \cdot s \cdot K)/J$ [0.00015 $(ft^2 \cdot h \cdot °F)/Btu$]. Tube 23 was a clean aluminum tube with 20 spiral corrugations of 0.0032-m (⅛-in) radius on a 0.254-m (10-in) pitch indented into the tube. Tube 48 was a clean copper tube that had 50 longitudinal flutes pressed into the wall (General Electric double-flute profile, Diedrich, U.S. Patent 3,244,601, Apr. 5, 1966). Tubes 47 and 39 had a specially patterned porous sintered-metal deposit on the boiling side to promote nucleate boiling (Minton, U.S. Patent 3,384,154, May 21, 1968). Both of these tubes also had steam-side coatings to promote dropwise condensation—parylene for tube 47 and gold plating for tube 39.

Of these special surfaces, only the *double-fluted tube* has seen extended services. Most of the gain in heat-transfer coefficient is due to the condensing side; the flutes tend to collect the condensate and leave the lands bare [Carnavos, *Proc. First Int. Symp. Water Desalination* **2:** 205 (1965)].

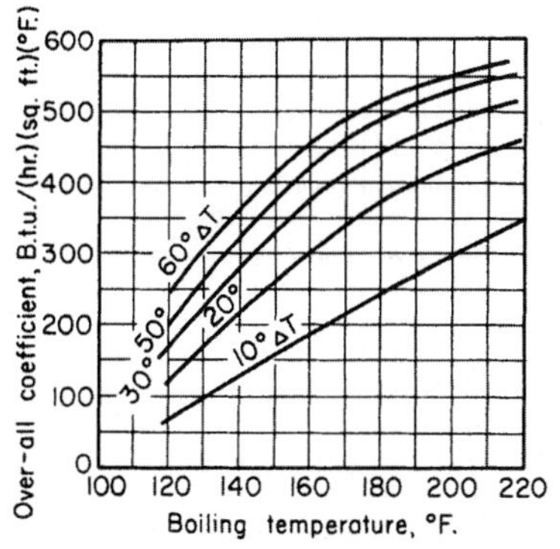

FIG. 11-23 Heat-transfer coefficients for water in short-tube evaporators. °C = (°F − 32)/1.8; to convert British thermal units per hour per square foot per degree Fahrenheit to joules per square meter per second per kelvin, multiply by 5.6783.

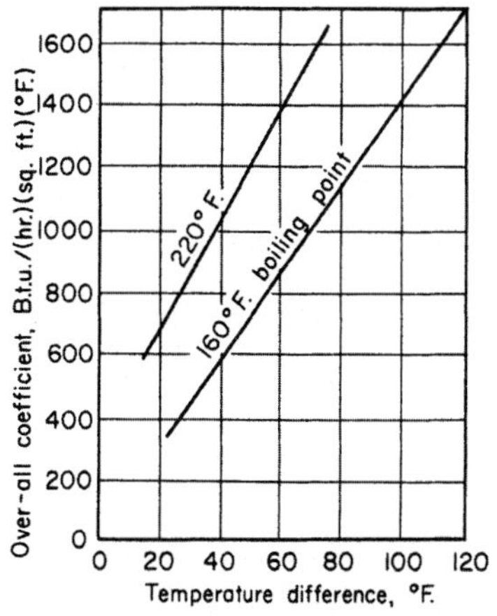

FIG. 11-24 Heat-transfer coefficients for seawater in coil-tube evaporators. °C = (°F − 32)/1.8; to convert British thermal units per hour per square foot per degree Fahrenheit to joules per square meter per second per kelvin, multiply by 5.6783.

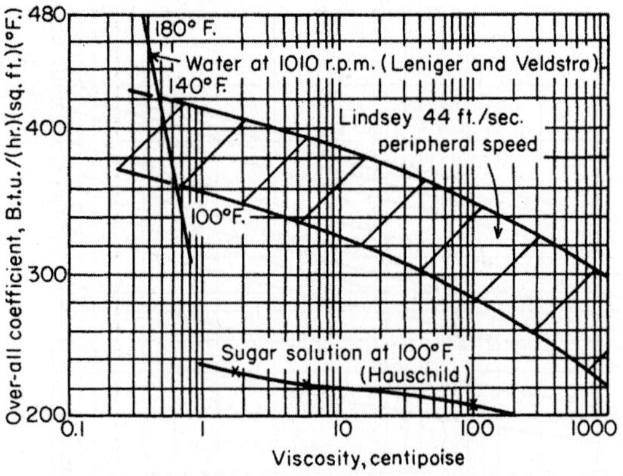

FIG. 11-25 Overall heat-transfer coefficients in agitated-film evaporators. °C = (°F − 32)/1.8; to convert British thermal units per hour per square foot per degree Fahrenheit to joules per square meter per second per kelvin, multiply by 5.6783; to convert centipoises to pascal-seconds, multiply by 10^{-3}.

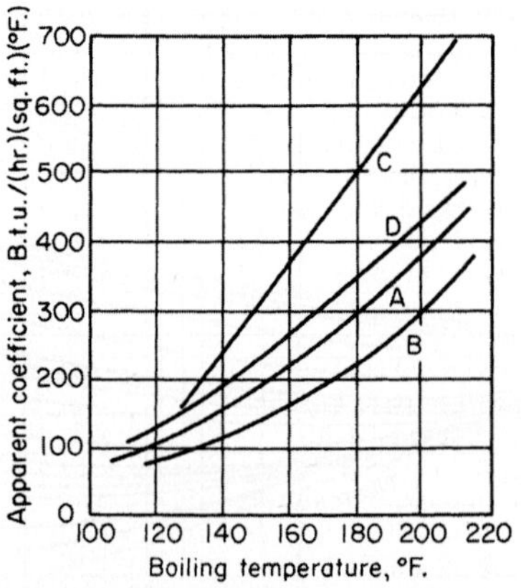

FIG. 11-27 Kerr's tests with full-sized sugar evaporators. °C = (°F − 32)/1.8; to convert British thermal units per hour per square foot per degree Fahrenheit to joules per square meter per second per kelvin, multiply by 5.6783.

The condensing-film coefficient (based on the actual outside area, which is 28 percent greater than the nominal area) may be approximated from the equation

$$h = b\left(\frac{k^3\rho^2 g}{\mu^2}\right)^{1/3}\left(\frac{\mu\lambda}{L}\right)^{1/3}\left(\frac{q}{A}\right)^{-0.833} \tag{11-34a}$$

where $b = 2100$ (SI) or 1180 (USCS). The boiling-side coefficient (based on actual inside area) for salt water in downflow may be approximated from the equation

$$h = 0.035(k^3\rho^2 g/\mu^2)1/3(4\Gamma/\mu)^{1/3} \tag{11-34b}$$

The boiling-film coefficient is about 30 percent lower for pure water than it is for salt water or seawater. There is as yet no accepted explanation for the superior performance in salt water. This phenomenon is also seen in evaporation from smooth tubes.

Effect of Fluid Properties on Heat Transfer Most of the heat-transfer data reported in the preceding paragraphs were obtained with water or with dilute solutions having properties close to those of water. Heat transfer with other materials will depend on the type of evaporator used. For forced-circulation evaporators, methods have been presented to calculate the effect of changes in fluid properties. For natural-circulation evaporators, *viscosity* is the most important variable as far as aqueous solutions are concerned. Badger (*Heat Transfer and Evaporation*, Chemical Catalog, New York, 1926, pp. 133–134) found that, as a rough rule, overall heat-transfer coefficients varied in inverse proportion to viscosity if the boiling film was the main resistance to heat transfer. When handling molasses solutions in a forced-circulation evaporator in which boiling was allowed to occur in the tubes, Coates and Badger [*Trans. Am. Inst. Chem. Eng.* **32:** 49 (1936)] found that from 0.005 to 0.03 Pa·s (5 to 30 cP) the overall heat-transfer coefficient could be represented by $U = b/\mu_f^{1.24}$, where $b = 2.55$ (SI) or 7043 (USCS). Fragen and Badger [*Ind. Eng. Chem.* **28:** 534 (1936)] correlated overall coefficients on sugar and sulfite liquor in the same evaporator for viscosities to 0.242 Pa·s (242 cP) and found a relationship that included the viscosity raised only to the 0.25 power.

Little work has been published on the effect of viscosity on heat transfer in the long-tube vertical evaporator. Cessna, Leintz, and Badger [*Trans. Am. Inst. Chem. Eng.* **36:** 759 (1940)] found that the overall coefficient in the nonboiling zone varied inversely as the 0.7 power of viscosity (with sugar solutions). Coulson and Mehta [*Trans. Inst. Chem. Eng.* **31:** 208 (1953)] found the exponent to be −0.44, and Stroebe, Baker, and Badger arrived at an exponent of −0.3 for the effect of viscosity on the film coefficient in the boiling zone.

Kerr (Louisiana *Agr. Exp. Sta. Bull.* 149) obtained plant data shown in Fig. 11-27 on various types of full-sized evaporators for cane sugar. These are invariably forward-feed evaporators concentrating to about 50° Brix, corresponding to a viscosity on the order of 0.005 Pa·s (5 cP) in the last effect. In Fig. 11-27 curve *A* is for short-tube verticals with central downtake, *B* is for standard horizontal-tube evaporators, *C* is for Lillie evaporators (which were horizontal-tube machines with no liquor level but having recirculating liquor showered over the tubes), and *D* is for long-tube vertical evaporators. These curves show apparent coefficients, but sugar solutions have boiling-point rises low enough to not affect the results noticeably. Kerr also obtained the data shown in Fig. 11-28 on a laboratory short-tube vertical evaporator with 0.44- by 0.61-m (1¾- by 24-in) tubes. This work was done with sugar juices boiling at 57°C (135°F) and an 11°C (20°F) temperature difference.

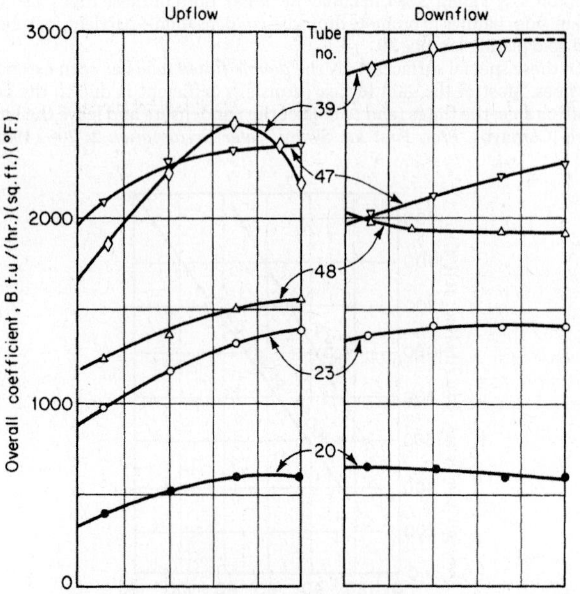

FIG. 11-26 Heat-transfer coefficients for enhanced surfaces. °C = (°F − 32)/1.8; to convert British thermal units per hour per square foot per degree Fahrenheit to joules per square meter per second per kelvin, multiply by 5.6783. (*By permission from Oak Ridge National Laboratory TM-2203.*)

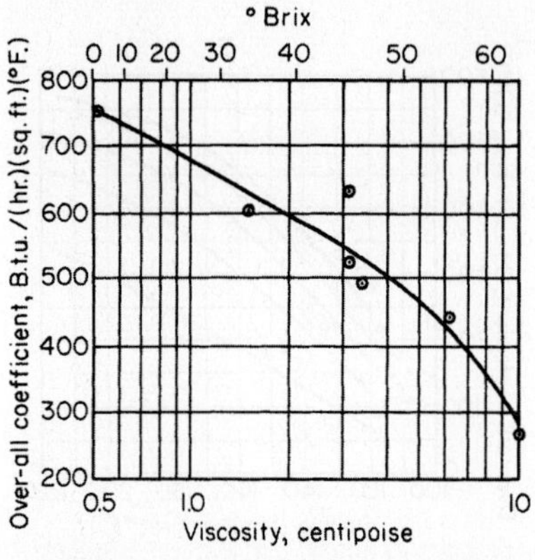

FIG. 11-28 Effect of viscosity on heat transfer in short-tube vertical evaporator. To convert centipoises to pascal-seconds, multiply by 10^{-3}; to convert British thermal units per hour per square foot per degree Fahrenheit to joules per square meter per second per kelvin, multiply by 5.6783.

Effect of Noncondensibles on Heat Transfer Most of the heat transfer in evaporators occurs not from pure steam but from vapor evolved in a preceding effect. This vapor usually contains inert gases—from air leakage if the preceding effect was under vacuum, from air entrained or dissolved in the feed, or from gases liberated by decomposition reactions. To prevent these inerts from seriously impeding heat transfer, the gases must be channeled past the heating surface and vented from the system while the gas concentration is still quite low. The influence of inert gases on heat transfer is due partially to the effect on ΔT of lowering the partial pressure and hence condensing temperature of the steam. The primary effect, however, results from the formation at the heating surface of an insulating blanket of gas through which the steam must diffuse before it can condense. The latter effect can be treated as an added resistance or fouling factor equal to 6.5×10^{-5} times the local mole percent inert gas (in $J^{-1} \cdot s \cdot m^2 \cdot K$) [Standiford, *Chem. Eng. Prog.* **75**: 59–62 (July 1979)]. The effect on ΔT is readily calculated from Dalton's law. Inert-gas concentrations may vary by a factor of 100 or more between vapor inlet and vent outlet, so these relationships should be integrated through the tube bundle.

CRYOGENIC HEAT EXCHANGERS

Most cryogenic fluids behave similar to room-temperature fluids as far as thermal/hydraulic design is concerned. For tubular exchangers, the Colburn equation and the ordinary Dittus-Boelter correlation, with slight modification, work well for single-phase cryogenic fluids as long as the critical point for the fluid is not closely approached. For critical-point calculations, a number of publications may be found in the literature to direct the designer along the best path forward for the fluid under consideration. For example, the use of an unmodified Dittus-Boelter correlation appears to give satisfactory results under swirl flow conditions for flows near the critical point.

Beyond the normal good design practice of the experienced thermal designer, a primary factor influencing the type of heat exchanger used in cryogenic applications is the ability of the exchanger geometry to handle large temperature changes and potential cyclic operation. Two popular geometries for these applications are coiled tube exchangers and bayonet exchangers.

The coiled tube heat exchanger is well suited for designs where heat transfer between more than two fluid streams is needed. It is also capable of handling high pressures while accommodating the thermal expansion/contraction issues of low-temperature operation.

The bayonet heat exchanger is also highly suited for extremes of high pressures and differential expansion. For large duties, it is simple to design a double tube sheet shell and tube heat exchanger with each tube consisting of the scabbard/bayonet tube combination of the bayonet exchanger. Segmental baffles or extended surface may be used on the outside of the scabbard tube bundle for the shell-side fluid. For cryogenic vaporization, a wire wrapped on the outside of each bayonet tube with the cold fluid entering in the bayonet tubes and exiting while vaporizing along the annulus between the bayonet tube and the scabbard pipe produces as efficient a vaporization performance as is possible in a tubular exchanger.

A third compact-type exchanger sometimes used for single-phase applications is the plate-fin heat exchanger. Again, however, the issue is to take care that the construction can handle the high differential expansion/contraction forces that the process may inflict upon the exchanger.

In most cryogenic processes, there is at least one service where heat is transferred between a cryogen and a fluid which can easily freeze on the tube wall. If cyclic operation is not desired (where the service is shut down at regular intervals or flow is transferred to a second exchanger while thawing occurs in the first), the best approach is to determine the thickness of frozen fluid needed to produce a skin temperature (at the solid-to-liquid interface) equal to the freezing point temperature of the fluid where further freezing will cease. The heat exchanger is then designed to accommodate this amount of solid buildup without flow blockage or unacceptable deterioration of heat transfer. For this reason, in these applications, the cryogenic fluid is almost always located inside the tubes while the heating stream with freezing potential is placed on the outside where a larger tube pitch can be used to create the space needed for the amount of solid buildup that will occur at thermodynamic equilibrium. When properly designed, these heat exchangers will operate free of problems at steady-state conditions for as long as necessary.

BATCH OPERATIONS: HEATING AND COOLING OF VESSELS

Nomenclature (Use consistent units.) A = heat-transfer surface; C, c = specific heats of hot and cold fluids, respectively; L_0 = flow rate of liquid added to tank; M = mass of fluid in tank; T, t = temperature of hot and cold fluids, respectively; T_1, t_1 = temperatures at beginning of heating or cooling period or at inlet; T_2, t_2 = temperature at end of period or at outlet; T_0, t_0 = temperature of liquid added to tank; U = coefficient of heat transfer; and W, w = flow rate through external exchanger of hot and cold fluids, respectively.

Applications One typical application in heat transfer with batch operations is the heating of a reactor mix, maintaining temperature during a reaction period, and then cooling the products after the reaction is complete. This subsection is concerned with the heating and cooling of such systems in either unknown or specified periods.

The technique for deriving expressions relating time for heating or cooling agitated batches to coil or jacket area, heat-transfer coefficients, and the heat capacity of the vessel contents was developed by Bowman, Mueller, and Nagle [*Trans. Am. Soc. Mech. Eng.* **62:** 283–294 (1940)] and extended by Fisher [*Ind. Eng. Chem.* **36:** 939–942 (1944)] and Chaddock and Sanders [*Trans. Am. Inst. Chem. Eng.* **40:** 203–210 (1944)] to external heat exchangers. Kern (*Process Heat Transfer*, McGraw-Hill, New York, 1950, chap. 18) collected and published the results of these investigators.

The assumptions made were that (1) U is constant for the process and over the entire surface, (2) liquid flow rates are constant, (3) specific heats are constant for the process, (4) the heating or cooling medium has a constant inlet temperature, (5) agitation produces a uniform batch fluid temperature, (6) no partial phase changes occur, and (7) heat losses are negligible. The developed equations are as follows. If any of the assumptions do not apply to a system being designed, new equations should be developed or appropriate corrections made. Heat exchangers are counterflow except for the 1-2 exchangers, which are one-shell-pass, two-tube-pass, parallel-flow counterflow.

Coil-in-Tank or Jacketed Vessel: Isothermal Heating Medium

$$\ln(T_1 - t_1)/(T_1 - t_2) = UA\theta/Mc \tag{11-35}$$

Cooling-in-Tank or Jacketed Vessel: Isothermal Cooling Medium

$$\ln (T_1 - t_1)/(T_2 - t_1) = UA\theta/MC \tag{11-35a}$$

Coil-in-Tank or Jacketed Vessel: Nonisothermal Heating Medium

$$\ln\frac{T_1 - t_1}{T_1 - t_2} = \frac{WC}{Mc}\left(\frac{K_1 - 1}{K_1}\right)\theta \tag{11-35b}$$

where $K_1 = e^{UA/WC}$.

Coil-in-Tank: Nonisothermal Cooling Medium

$$\ln\frac{T_1 - t_1}{T_2 - t_1} = \frac{wc}{MC}\left(\frac{K_2 - 1}{K_2}\right)\theta \tag{11-35c}$$

where $K_2 = e^{UA/wc}$.

External Heat Exchanger: Isothermal Heating Medium

$$\ln\frac{T_1 - t_1}{T_1 - t_2} = \frac{wc}{MC}\left(\frac{K_2 - 1}{K_2}\right)\theta \tag{11-35d}$$

External Exchanger: Isothermal Cooling Medium

$$\ln\frac{T_1 - t_1}{T_2 - t_1} = \frac{WC}{MC}\left(\frac{K_1 - 1}{K_1}\right)\theta \tag{11-35e}$$

External Exchanger: Nonisothermal Heating Medium

$$\ln\frac{T_1 - t_1}{T_2 - t_1} = \left(\frac{K_3 - 1}{M}\right)\left(\frac{wWC}{K_3 wc - WC}\right)\theta \tag{11-35f}$$

where $K_3 = e^{UA(1/WC - 1/wc)}$.

External Exchanger: Nonisothermal Cooling Medium

$$\ln\frac{T_1 - t_1}{T_2 - t_1} = \left(\frac{K_4 - 1}{M}\right)\left(\frac{Wwc}{K_4 wc - WC}\right)\theta \tag{11-35g}$$

where $K_4 = e^{UA(1/WC - 1/wc)}$.

External Exchanger with Liquid Continuously Added to Tank: Isothermal Heating Medium

$$\ln\left[\frac{t_1 - t_0 - \dfrac{w}{L_0}\left(\dfrac{K_2 - 1}{K_2}\right)(T_1 - t_1)}{t_2 - t_0 - \dfrac{w}{L_0}\left(\dfrac{K_2 - 1}{K_2}\right)(T_1 - t_2)}\right] = \left[\frac{w}{L_0}\left(\frac{K_2 - 1}{K_2}\right) + 1\right]\ln\frac{M + L_0\theta}{M} \tag{11-35h}$$

If the addition of liquid to the tank causes an average endothermic or exothermic heat of solution $\pm q_s$ J/kg (Btu/lb) of makeup, it may be included by adding $\pm q_s/c_0$ to both the numerator and the denominator of the left side. The subscript 0 refers to the makeup.

External Exchanger with Liquid Continuously Added to Tank: Isothermal Cooling Medium

$$\ln\left[\frac{T_0-T_1-\dfrac{W}{L_0}\left(\dfrac{K_1-1}{K_1}\right)(T_1-t_1)}{T_0-T_2-\dfrac{W}{L_0}\left(\dfrac{K_1-1}{K_1}\right)(T_2-t_1)}\right]=\left[1-\frac{W}{L_0}\left(\frac{K_1-1}{K_1}\right)\right]\ln\frac{M+L_0\theta}{M} \tag{11-35i}$$

The heat-of-solution effects can be included by adding $\pm q_s/C_0$ to both the numerator and the denominator of the left side.

External Exchanger with Liquid Continuously Added to Tank: Nonisothermal Heating Medium

$$\ln\left[\frac{t_0-t_1+\dfrac{wWC(K_5-1)(T_1-t_1)}{L_0(K_5WC-wc)}}{t_0-t_2+\dfrac{wWC(K_5-1)(T_1-t_2)}{L_0(K_5WC-wc)}}\right]=\left[\frac{wWC(K_5-1)}{L_0(K_5WC-wc)}+1\right]\ln\frac{M+L_0\theta}{M} \tag{11-35j}$$

where $K_5 = e^{(UA/wc)(1-wc/WC)}$.

The heat-of-solution effects can be included by adding $\pm q_s/c_0$ to both the numerator and the denominator of the left side.

External Exchanger with Liquid Continuously Added to Tank: Nonisothermal Cooling Medium

$$\ln\left[\frac{T_0-T_1+\dfrac{Wwc(K_6-1)(T_1-t_1)}{L_0(K_6wc-WC)}}{T_0-T_2+\dfrac{Wwc(K_6-1)(T_2-t_1)}{L_0(K_6wc-WC)}}\right]=\left[\frac{Wwc(K_6-1)}{L_0(K_6wc-WC)}+1\right]\ln\frac{M+L_0\theta}{M} \tag{11-35k}$$

where $K_6 = e^{(UA/WC)(1-WC/wc)}$.

The heat-of-solution effects can be included by adding $\pm q_s/C_0$ to both the numerator and the denominator of the left side.

Heating and Cooling Agitated Batches: 1-2 Parallel Flow-Counterflow

$$\frac{UA}{wc}=\frac{1}{\sqrt{R^2+1}}\ln\left[\frac{2-S(R+1-\sqrt{R^2+1})}{2-S(R+1-\sqrt{R^2+1})}\right] \tag{11-35l}$$

$$R=\frac{T_1-T_2}{t'-t}=\frac{wc}{WC} \quad\text{and}\quad S=\frac{t'-t}{T_1-t}$$

$$\frac{2-S(R+1-\sqrt{R^2+1})}{2-S(R+1+\sqrt{R^2+1})}=e^{(UA/wc)\sqrt{R^2+1}}=K_7$$

$$S=\frac{2(K_7-1)}{K_7(R+1+R^2+1)-(R+1-R^2+1)} \tag{11-35m}$$

External 1-2 Exchanger: Heating

$$\ln[(T_1-t_1)/(T_1-t_2)]=(Sw/M)\theta \tag{11-35n}$$

External 1-2 Exchanger: Cooling

$$\ln[(T_1-t_1)/(T_2-t_1)]=S(wc/MC)\theta \tag{11-35o}$$

The cases of multipass exchangers with liquid continuously added to the tank are covered by Kern, as cited earlier. An alternative method for all multipass-exchanger gases, including those presented as well as cases with two or more shells in series, is as follows:

1. Determine UA for using the applicable equations for counterflow heat exchangers.
2. Use the initial batch temperature T_1 or t_1.
3. Calculate the outlet temperature from the exchanger of each fluid. (This will require trial-and-error methods.)
4. Note the F_T correction factor for the corrected mean temperature difference. (See Fig. 11-4.)
5. Repeat steps 2, 3, and 4 by using the final batch temperature T_2 and t_2.
6. Use the average of the two values for F, then increase the required multipass UA as follows:

$$UA(\text{multipass}) = UA(\text{counterflow})/F_T$$

In general, values of F_T below 0.8 are uneconomical and should be avoided. The value of F_T can be raised by increasing the flow rate of either or both of the flow streams. Increasing flow rates to give values well above 0.8 is a matter of economic justification.

If F_T varies widely from one end of the range to the other, F_T should be determined for one or more intermediate points. The average should then be determined for each step which has been established and the average of these taken for use in step 6.

Effect of External Heat Loss or Gain If heat loss or gain through the vessel walls cannot be neglected, equations that include this heat transfer can be developed by using energy balances similar to those used for the derivations of equations given previously. Basically, these equations must be modified by adding a heat-loss or heat-gain term.

A simpler procedure, which is probably acceptable for most practical cases, is to adjust the ratio of UA or θ either up or down in accordance with the required modification in total heat load over time θ.

Another procedure, which is more accurate for the external-heat exchanger cases, is to use an equivalent value for MC (for a vessel being heated) derived from the following energy balance:

$$Q=(Mc)_e(t_2-t_1)=Mc(t_2-t_1)+U'A'(MTD')\theta \tag{11-35p}$$

where Q is the total heat transferred over time θ, $U'A'$ is the heat-transfer coefficient for heat loss times the area for heat loss, and MTD' is the mean temperature difference for the heat loss.

A similar energy balance would apply to a vessel being cooled.

Internal Coil or Jacket Plus External Heat Exchanger This case can be most simply handled by treating it as two separate problems: M is divided into two separate masses M_1 and $M - M_1$, and the appropriate equations given earlier are applied to each part of the system. Time θ, of course, must be the same for both parts.

Equivalent-Area Concept The preceding equations for batch operations, particularly Eq. (11-35), can be applied for the calculation of heat loss from tanks which are allowed to cool over an extended time. However, different surfaces of a tank, such as the top (which would not be in contact with the tank contents) and the bottom, may have coefficients of heat transfer different from those of the vertical-tank walls. The simplest way to resolve this difficulty is to use an equivalent area A_e in the appropriate equations, where

$$A_e=A_bU_b/U_s+A_tU_t/U_s+A_s \tag{11-35q}$$

and the subscripts b, s, and t refer to the bottom, sides, and top, respectively. Usually U is taken as U_s. Table 11-1 lists typical values for U_s and expressions for A_e for various tank configurations.

Nonagitated Batches Cases in which vessel contents are vertically stratified, rather than uniform in temperature, have been treated by Kern (*Process Heat Transfer*, McGraw-Hill, New York, 1950, chap. 18). These are of little practical importance except for tall, slender vessels heated or cooled with external exchangers. The result is that a smaller exchanger is required than for an equivalent agitated batch system that is uniform.

Storage Tanks The equations for batch operations with agitation may be applied to storage tanks even though the tanks are not agitated. This approach gives conservative results. The important cases (non-steady-state) are as follows:

1. *Tanks cool; contents remain liquid.* This case is relatively simple and can be easily handled by the equations given earlier.
2. *Tanks cool, contents partially freeze, and solids drop to bottom or rise to top.* This case requires a two-step calculation. The first step is handled as in case 1. The second step is calculated by assuming an isothermal system at the freezing point. It is possible, given time and a sufficiently low ambient temperature, for tank contents to freeze solid.
3. *Tanks cool and partially freeze; solids form a layer of self-insulation.* This complex case, which has been known to occur with heavy hydrocarbons and mixtures of hydrocarbons, has been discussed by Stuhlbarg [*Pet. Refiner* **38:** 143 (Apr. 1, 1959)]. The contents in the center of such tanks have been known to remain warm and liquid even after several years of cooling.

It is very important that a melt-out riser be installed whenever tank contents are expected to freeze on prolonged shutdown. The purpose is to provide a molten chimney through the crust for relief of thermal expansion or cavitation if fluids are to be pumped out or recirculated through an external exchanger. An external heat tracer, properly located, will serve the same purpose but may require greater remelt time before pumping can be started.

THERMAL DESIGN OF TANK COILS

The thermal design of tank coils involves the determination of the area of heat-transfer surface required to maintain the contents of the tank at a constant temperature or to raise or lower the temperature of the contents by a specified magnitude over a fixed time.

TABLE 11-1 Typical Values for Use with Eqs. (11-36) to (11-44)*

Application	Fluid	U_s	A_s
Tanks on legs, outdoors, not insulated	Oil	3.7	$0.22\,A_t + A_b + A_s$
	Water at 150°F	5.1	$0.16\,A_t + A_b + A_s$
Tanks on legs, outdoors, insulated 1 in.	Oil	0.45	$0.7\,A_t + A_b + A_s$
	Water	0.43	$0.67\,A_t + A_b + A_s$
Tanks on legs, indoors, not insulated	Oil	1.5	$0.53\,A_t + A_b + A_s$
	Water	1.8	$0.35\,A_t + A_b + A_s$
Tanks on legs, indoors, insulated 1 in.	Oil	0.36	$0.8\,A_t + A_b + A_s$
	Water	0.37	$0.73\,A_t + A_b + A_s$
Flat-bottom tanks,† outdoors, not insulated	Oil	3.7	$0.22\,A_t + A_s + 0.43\,D_t$
	Water	5.1	$0.16\,A_t + A_s + 0.31\,D_t$
Flat-bottom tanks,† outdoors, insulated 1 in.	Oil	0.36	$0.7\,A_t + A_s + 3.9\,D_t$
	Water	0.37	$0.16\,A_t + A_s + 3.7\,D_t$
Flat-bottom tanks, indoors, not insulated	Oil	1.5	$0.53\,A_t + A_s + 1.1\,D_t$
	Water	1.8	$0.35\,A_t + A_s + 0.9\,D_t$
Flat-bottom tanks, indoors, insulated 1 in.	Oil	0.36	$0.8\,A_t + A_s + 4.4\,D_t$
	Water	0.37	$0.73\,A_t + A_s + 4.5\,D_t$

*Based on typical coefficients.
†The ratio $(t - t_g)(t - t')$ assumed at 0.85 for outdoor tanks. °C = (°F − 32)/1.8; to convert British thermal units per hour-square foot-degrees Fahrenheit to joules per square meter-second-kelvins, multiply by 5.6783.

Nomenclature A = area; A_b = area of tank bottom; A_c = area of coil; A_e = equivalent area; A_s = area of sides; A_t = area of top; A_1 = equivalent area receiving heat from external coils; A_2 = equivalent area not covered with external coils; c = heat capacity of liquid phase of tank contents; D_t = diameter of tank; F = design (safety) factor; h = film coefficient; h_a = coefficient of ambient air; h_c = coefficient of coil; h_h = coefficient of heating medium; h_i = coefficient of liquid phase of tank contents or tube-side coefficient referred to outside of coil; h_z = coefficient of insulation; k = thermal conductivity; k_g = thermal conductivity of ground below tank; M = mass of tank contents when full; t = temperature; t_a = temperature of ambient air; t_d = temperature of dead-air space; t_f = temperature of contents at end of heating; t_g = temperature of ground below tank; t_h = temperature of heating medium; t_0 = temperature of contents at beginning of heating; U = overall coefficient; U_b = coefficient at tank bottom; U_c = coefficient of coil; U_d = coefficient of dead air to the tank contents; U_i = coefficient through insulation; U_s = coefficient at sides; U_t = coefficient at top; and U_2 = coefficient at area A_2.

Typical coil coefficients are listed in Table 11-2. More exact values can be calculated by using the methods for natural convection or forced convection given elsewhere in this section.

Maintenance of Temperature Tanks are often maintained at temperature with internal coils if the following equations are assumed to be applicable:

$$q = U_s A_e (T - t') \tag{11-36}$$

and

$$A_c = q / U_c(\text{MTD}) \tag{11-36a}$$

These make no allowance for unexpected shutdowns. One method of allowing for shutdown is to add a safety factor to Eq. (11-36*a*).

In the case of a tank maintained at temperature with internal coils, the coils are usually designed to cover only a portion of the tank. The temperature t_d of the dead-air space between the coils and the tank is obtained from

$$U_d A_1 (t_d - t) = U_2 A_2 (t - t') \tag{11-37}$$

The heat load is

$$q = U_d A_1 (t_d - t) + A_1 U_i (t_d - t') \tag{11-38}$$

The coil area is

$$A_c = \frac{qF}{U_c (t_h - t_d)_m} \tag{11-39}$$

where F is a safety factor.

Heating

Heating with Internal Coil from Initial Temperature for Specified Time

$$Q = Mc(t_f - t_o) \tag{11-40}$$

$$A_c = \left[\frac{Q}{\theta_h} + U_s A_e \left(\frac{t_f + t_o}{2} - t'\right)\right]\left[\frac{1}{U_c[t_h - (t_f + t_o)/2]}\right](F) \tag{11-41}$$

TABLE 11-2 Overall Heat-Transfer Coefficients for Coils Immersed in Liquids

U Expressed as Btu/(h · ft² · °F)

Substance inside coil	Substance outside coil	Coil material	Agitation	U
Steam	Water	Lead	Agitated	70
Steam	Sugar and molasses solutions	Copper	None	50–240
Steam	Boiling aqueous solution			600
Cold water	Dilute organic dye intermediate	Lead	Turboagitator at 95 r.p.m.	300
Cold water	Warm water	Wrought iron	Air bubbled into water surrounding coil	150–300
Cold water	Hot water	Lead	0.40 r.p.m. paddle stirrer	90–360
Brine	Amino acids		30 r.p.m.	100
Cold water	25% oleum at 60°C.	Wrought iron	Agitated	20
Water	Aqueous solution	Lead	500 r.p.m. sleeve propeller	250
Water	8% NaOH		22 r.p.m.	155
Steam	Fatty acid	Copper (pancake)	None	96–100
Milk	Water		Agitation	300
Cold water	Hot water	Copper	None	105–180
60°F water	50% aqueous sugar solution	Lead	Mild	50–60
Steam and hydrogen at 1500 lb./sq. in.	60°F. water	Steel		100–165
Steam 110–146 lb./sq. in. gage	Vegetable oil	Steel	None	23–29
Steam	Vegetable oil	Steel	Various	39–72
Cold water	Vegetable oil	Steel	Various	29–72

NOTES: Chilton, Drew, and Jebens [*Ind. Eng. Chem.* **36:** 510 (1944)] give film coefficients for heating and cooling agitated fluids using a coil in a jacketed vessel.

Because of the many factors affecting heat transfer, such as viscosity, temperature difference, and coil size, the values in this table should be used primarily for preliminary design estimates and checking calculated coefficients.

°C = (°F − 32)/1.8; to convert British thermal units per hour-square foot-degrees Fahrenheit to joules per square meter-second-kelvins, multiply by 5.6783.

where θ_h is the length of heating period. This equation may also be used when the tank contents have cooled from t_f to t_o and must be reheated to t_f. If the contents cool during a time θ_c, the temperature at the end of this cooling period is obtained from

$$\ln\left(\frac{t_f - t'}{t_o - t'}\right) = \frac{U_s A_e \theta_c}{Mc} \tag{11-42}$$

Heating with External Coil from Initial Temperature Specified Time The temperature of the dead-air space is obtained from

$$U_d A_1[t_d - 0.5(t_f - t_o)] = U_2 A_2[0.5(t_f - t_o) - t'] + Q/\theta_h \tag{11-43}$$

The heat load is

$$q = U_i A_1(t_d - t') + U_2 A_2[0.5(t_f - t_o) - t'] + Q/\theta_h \tag{11-44}$$

The coil area is obtained from Eq. (11-39).

The safety factor used in the calculations is a matter of judgment based on confidence in the design. A value of 1.10 is normally not considered excessive. Typical design parameters are shown in Tables 11-1 and 11-2.

HEATING AND COOLING OF TANKS

Tank Coils *Pipe tank coils* are made in a wide variety of configurations, depending upon the application and shape of the vessel. *Helical* and *spiral* coils are most commonly shop-fabricated, while the *hairpin* pattern is generally field-fabricated. The helical coils are used principally in process tanks and pressure vessels when large areas for rapid heating or cooling are required. In general, heating coils are placed low in the tank, and cooling coils are placed high or distributed uniformly throughout the vertical height.

Stocks that tend to solidify on cooling require uniform coverage of the bottom or agitation. A maximum spacing of 0.6 m (2 ft) between turns of 50.8-mm (2-in) and larger pipe and a close approach to the tank wall are recommended. For smaller pipe or for low-temperature heating media, closer spacing should be used. In the case of the common hairpin coils in vertical cylindrical tanks, this means adding an encircling ring within 152 mm (6 in) of the tank wall (see Fig. 11-29*a* for this and other typical coil layouts). The coils should be set directly on the bottom or raised not more than 50.8 to 152 mm (2 to 6 in), depending upon the difficulty of remelting the solids, in order to permit free movement of product within the vessel. The coil inlet should be above the liquid level (or an internal melt-out riser installed) to provide a molten path for liquid expansion or venting of vapors.

Coils may be sloped to facilitate drainage. When it is impossible to do so and remain close enough to the bottom to get proper remelting, the coils should be blown out after use in cold weather to avoid damage by freezing.

Most coils are firmly clamped (but not welded) to supports. *Supports* should allow expansion but be rigid enough to prevent uncontrolled motion (see Fig. 11-29*b*). Nuts and bolts should be securely fastened. Reinforcement of the inlet and outlet connections through the tank wall is recommended, since bending stresses due to thermal expansion are usually high at such points.

In general, 50.8- and 63.4-mm (2- and 2½-in) coils are the most economical for shop fabrication and 38.1- and 50.8-mm (1½- and 2-in) for field fabrication. The tube-side heat-transfer coefficient, high-pressure, or layout problems may lead to the use of smaller-size pipe.

The wall thickness selected varies with the service and material. Carbon-steel coils are often made from schedule 80 or heavier pipe to allow for corrosion. When stainless-steel or other high-alloy coils are not subject to corrosion or excessive pressure, they may be of schedule 5 or 10 pipe to keep costs at a minimum, although high-quality welding is required for these thin walls to ensure trouble-free service.

Methods for calculating heat loss from tanks and the sizing of tank coils have been published by Stuhlbarg [*Pet. Refiner* **38:** 143 (April 1959)].

Fin-tube coils are used for fluids that have poor heat-transfer characteristics to provide greater surface for the same configuration at reduced cost or when temperature-driven fouling is to be minimized. Fin tubing is not generally used when bottom coverage is important. *Fin-tube tank heaters* are compact, prefabricated bundles which can be brought into tanks through manholes. These are normally installed vertically with longitudinal fins to produce good convection currents. To keep the heaters low in the tank, they can be installed horizontally with helical fins or with perforated longitudinal fins to prevent entrapment. Fin tubing is often used for heat-sensitive material because of the lower surface temperature for the same heating medium, resulting in a lesser tendency to foul.

Plate or *panel* coils made from two metal sheets with one or both embossed to form passages for a heating or cooling medium can be used in lieu of pipe coils. Panel coils are relatively lightweight, easy to install, and easily removed for cleaning. They are available in a range of standard sizes and in both flat and curved patterns. Process tanks have been built by using panel coils for the sides or bottom. A serpentine construction is generally utilized when liquid flows through the unit. Header-type construction is used with steam or other condensing media.

Standard *glass coils* with 0.18 to 11.1 m² (2 to 120 ft²) of heat-transfer surface are available. Also available are plate-type units made of *impervious graphite.*

Teflon Immersion Coils Immersion coils made of Teflon fluorocarbon resin are available with 2.5-mm- (0.10-in-) ID tubes to increase overall heat-transfer efficiency. The flexible bundles are available with 100, 160, 280, 500, and 650 tubes with standard lengths varying in 0.6-m (2-ft) increments between 1.2 and 4.8 m (4 and 16 ft). These coils are most commonly used in metal-finishing baths and are adaptable to service in reaction vessels, crystallizers, and tanks where corrosive fluids are used.

Bayonet Heaters A bayonet-tube element consists of an outer tube and an inner tube. These elements are inserted into tanks and process vessels for heating and cooling purposes. Often the outer tube is of expensive alloy or nonmetallic (e.g., glass, impervious graphite), while the inner tube is of carbon steel. In glass construction, elements with 50.8- or 76.2-mm (2- or 3-in) glass pipe [with lengths to 2.7 m (9 ft)] are in contact with the external fluid, with an inner tube of metal.

External Coils and Tracers Tanks, vessels, and pipelines can be equipped for heating or cooling purposes with external coils. These are generally 9.8 to 19 mm (⅜ to ¾ in) so as to provide good distribution over the surface and are often of soft copper or aluminum, which can be bent by hand to the contour of the tank or line. When it is necessary to avoid "hot spots," the tracer is so mounted that it does not touch the tank.

External coils spaced away from the tank wall exhibit a coefficient of around 5.7 W/(m²·°C) [1 Btu/(h·ft² of coil surface·°F)]. Direct contact with the tank wall produces higher coefficients, but these are difficult to predict since they are strongly dependent upon the degree of contact. The use of *heat-transfer cements* does improve performance. These puttylike materials of high thermal conductivity are troweled or caulked into the space between the coil and the tank or pipe surface.

Costs of the cements (in 1960) varied from 37 to 63 cents per pound, with requirements running from about 0.27 lb/ft of ⅜-in outside-diameter tubing to 1.48 lb/ft of 1-in pipe. Panel coils require ½ to 1 lb/ft². A rule of thumb for preliminary estimating is that the per-foot installed cost of tracer with cement is about double that of the tracer alone.

Jacketed Vessels Jacketing is often used for vessels needing frequent cleaning and for glass-lined vessels that are difficult to equip with internal coils. The jacket eliminates the need for the coil yet gives a better overall coefficient than external coils do. However, only a limited heat-transfer area is available. The conventional jacket is of simple construction and is frequently used. It is most effective with a condensing vapor. A liquid heat-transfer fluid does not maintain uniform flow characteristics in such a jacket. Nozzles, which set up a swirling motion in the jacket, are effective in improving

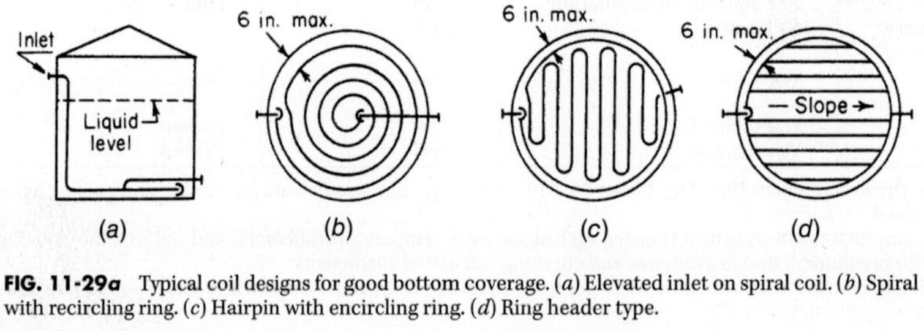

FIG. 11-29*a* Typical coil designs for good bottom coverage. (*a*) Elevated inlet on spiral coil. (*b*) Spiral with recircling ring. (*c*) Hairpin with encircling ring. (*d*) Ring header type.

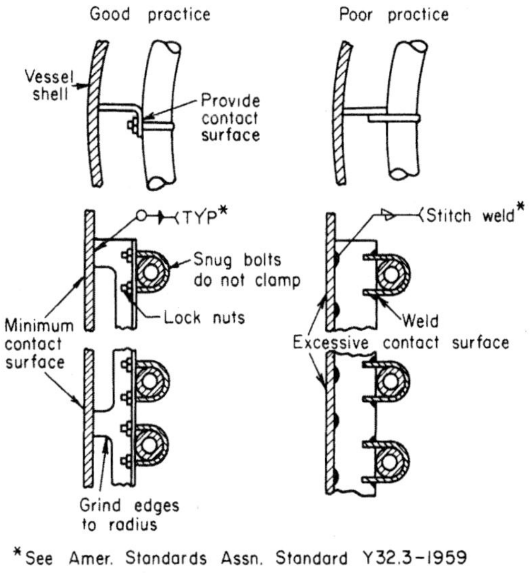

FIG. 11-29b Right and wrong ways to support coils. [*Chem. Eng.*, 172 (May 16, 1960).]

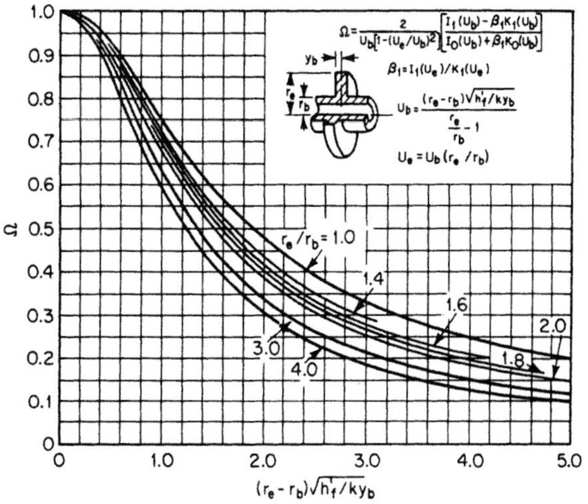

FIG. 11-30b Efficiencies for annular fins of constant thickness.

heat transfer. Wall thicknesses are often high unless reinforcement rings are installed.

Spiral baffles, which are sometimes installed for liquid services to improve heat transfer and prevent channeling, can be designed to serve as reinforcements. A spiral-wound channel welded to the vessel wall is an alternative to the spiral baffle which is more predictable in performance, since cross-baffle leakage is eliminated, and is reportedly lower in cost [Feichtinger, *Chem. Eng.* **67:** 197 (Sept. 5, 1960)].

The half-pipe jacket is used when high jacket pressures are required. The flow pattern of a liquid heat-transfer fluid can be controlled and designed for effective heat transfer. The dimple jacket offers structural advantages and is the most economical for high jacket pressures. The low volumetric capacity produces a fast response to temperature changes.

EXTENDED OR FINNED SURFACES

Finned-Surface Application Extended or finned surfaces are often used when one film coefficient is substantially lower than the other, the goal being to make $h_oA_{oe} \approx h_iA_i$. A few typical fin configurations are shown in Fig. 11-30*a*. Longitudinal fins are used in double-pipe exchangers. Transverse fins are used in cross-flow and shell-and-tube configurations.

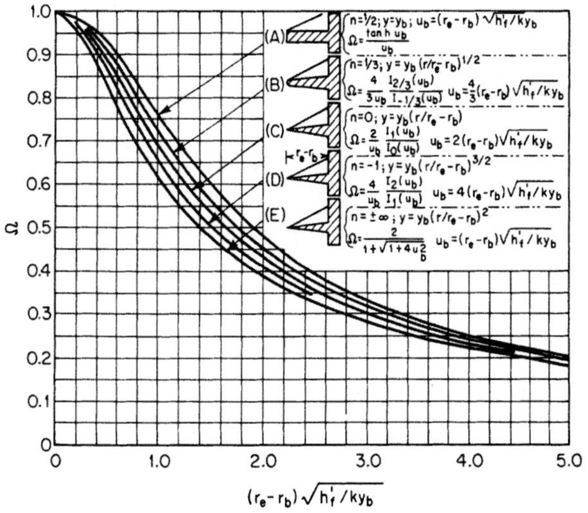

FIG. 11-30a Efficiencies for several longitudinal fin configurations.

High transverse fins are used mainly with low-pressure gases; low fins are used for boiling and condensation of nonaqueous streams as well as for sensible-heat transfer. Finned surfaces have proved to be a successful means of controlling temperature-driven fouling such as coking and scaling. Fin spacing should be great enough to avoid entrapment of particulate matter in the fluid stream (5-mm minimum spacing).

The area added by the fin is not as efficient for heat transfer as bare tube surface owing to resistance to conduction through the fin. The effective heat-transfer area is

$$A_{oe} = A_{uf} + A_f\Omega \qquad (11\text{-}45)$$

The fin efficiency is found from mathematically derived relations, in which the film heat-transfer coefficient is assumed to be constant over the entire fin and temperature gradients across the thickness of the fin have been neglected (see Kraus, *Extended Surfaces*, Spartan Books, Baltimore, Md., 1963). The efficiency curves for some common fin configurations are given in Fig. 11-30*a* and *b*.

High Fins To calculate heat-transfer coefficients for cross-flow to a transversely finned surface, it is best to use a correlation based on experimental data for that surface. Such data are not often available, and a more general correlation must be used, making allowance for the possible error. Probably the best general correlation for bundles of finned tubes is given by Schmidt [*Kaltetechnik* **15:** 98–102, 370–378 (1963)]:

$$hD_r/k = K(D_r\rho V'_{max}/\mu)^{0.625}R_f^{-0.375}N_{Pr}^{1/3} \qquad (11\text{-}46)$$

where $K = 0.45$ for staggered tube arrays and 0.30 for in-line tube arrays; D_r is the root or base diameter of the tube; V'_{max} is the maximum velocity through the tube bank, i.e., the velocity through the minimum flow area between adjacent tubes; and R_f is the ratio of the total outside surface area of the tube (including fins) to the surface of a tube having the same root diameter but without fins.

Pressure drop is particularly sensitive to geometric parameters, and available correlations should be extrapolated to geometries different from those on which the correlation is based only with great caution and conservatism. The best correlation is that of Robinson and Briggs [*Chem. Eng. Prog.* **62:** Symp. Ser. 64, 177–184 (1966)].

Low Fins Low-finned tubing is generally used in shell-and-tube configurations. For sensible-heat transfer, only minor modifications are needed to permit the shell-side method given earlier to be used for both heat transfer and pressure [see Briggs, Katz, and Young, *Chem. Eng. Prog.* **59**(11): 49–59 (1963)]. For condensing on low-finned tubes in horizontal bundles, the Nusselt correlation is generally satisfactory for low-surface-tension [$\sigma < (3)(10^{-6})$ N/m (30 dyn/cm)] condensates; fins of finned surfaces should not be closely spaced for high-surface-tension condensates (notably water), which do not drain easily.

The modified Palen-Small method can be employed for reboiler design using finned tubes, but the maximum flux is calculated from A_o, the total outside heat-transfer area including fins. The resulting value of q_{max} refers to A_o.

FOULING AND SCALING

Fouling refers to any change in the solid boundary separating two heat-transfer fluids, whether by dirt accumulation or other means, which results in a decrease in the rate of heat transfer occurring across that boundary. Fouling may be classified by mechanism into six basic categories:

1. *Corrosion fouling.* The heat-transfer surface reacts chemically with elements of the fluid stream, producing a less conductive corrosion layer on all or part of the surface.
2. *Biofouling.* Organisms present in the fluid stream are attracted to the warm heat-transfer surface where they attach, grow, and reproduce. The two subgroups are microbiofoulants such as slime and algae and macrobiofoulants such as snails and barnacles.
3. *Particulate fouling.* Particles held in suspension in the flow stream will deposit out on the heat-transfer surface in areas of sufficiently lower velocity.
4. *Chemical reaction fouling* (e.g., coking). Chemical reaction of the fluid takes place on the heat-transfer surface, producing an adhering solid product of reaction.
5. *Precipitation fouling* (e.g., scaling). A fluid containing some dissolved material becomes supersaturated with respect to this material at the temperatures seen at the heat-transfer surface. This results in a crystallization of the material which "plates out" on the warmer surface.
6. *Freezing fouling.* Overcooling of a fluid below the fluid's freezing point at the heat-transfer surface causes solidification and coating of the heat-transfer surface.

Control of Fouling Once the combination of mechanisms contributing to a particular fouling problem is recognized, methods to substantially reduce the fouling rate may be implemented. For the case of *corrosion fouling,* the common solution is to choose a less corrosive material of construction, balancing material cost with equipment life. In cases of *biofouling,* the use of copper alloys and/or chemical treatment of the fluid stream to control organism growth and reproduction is the most common solution.

In the case of *particulate fouling,* one of the more common types, ensuring a sufficient flow velocity and minimizing areas of lower velocities and stagnant flows to help keep particles in suspension are the most common means of dealing with the problem. For water, the recommended tube-side minimum velocity is about 0.9 to 1.0 m/s. This may not always be possible for moderate to high-viscosity fluids where the resulting pressure drop can be prohibitive.

Special care should be taken in the application of any velocity requirement to the shell side of segmental-baffled bundles due to the many different flow streams and velocities present during operation, the unavoidable existence of high-fouling areas of flow stagnation, and the danger of flow-induced tube vibration. In general, shell-side particulate fouling will be greatest for segmentally baffled bundles in the regions of low velocity, and the TEMA fouling factors (which are based upon the use of this bundle type) should be used. However, since the 1940s, there have been a host of successful, low-fouling exchangers developed, some tubular and others not, which have in common the elimination of the cross-flow plate baffle and provide practically no regions of flow stagnation at the heat-transfer surface. Some examples are the plate and frame exchanger, the spiral-plate exchanger, and the twisted tube exchanger, all of which have dispensed with baffles altogether and use the heat-transfer surface itself for bundle support. The general rule for these designs is to provide between 25 and 30 percent excess surface to compensate for potential fouling, although this can vary in special applications.

For the remaining classifications—*polymerization, precipitation,* and *freezing*—fouling is the direct result of temperature extremes at the heat-transfer surface and is reduced by reducing the temperature difference between the heat-transfer surface and the bulk-fluid stream. Conventional wisdom says to increase velocity, thus increasing the local heat-transfer coefficient to bring the heat-transfer surface temperature closer to the bulk-fluid temperature. However, due to a practical limit on the amount of heat-transfer coefficient increase available by increasing velocity, this approach, although better than nothing, is often not satisfactory by itself.

A more effective means of reducing the temperature difference is by using, in concert with adequate velocities, some form of extended surface. As discussed by Shilling (*Proceedings of the 10th International Heat Transfer Conference*, Brighton, U.K., **4:** 423, 1994), this will tend to reduce the temperature extremes between the fluid and heat-transfer surface and not only reduce the rate of fouling but also make the heat exchanger generally less sensitive to the effects of any fouling that does occur. In cases where unfinned tubing in a triangular tube layout would not be acceptable because fouling buildup and eventual mechanical cleaning are inevitable, an extended surface should be used only when the exchanger construction allows access for cleaning.

Fouling Transients and Operating Periods Three common behaviors are noted in the development of a fouling film over time. One is *asymptotic fouling* in which the speed of fouling resistance increase decreases over time as it approaches some asymptotic value beyond which no further fouling can occur. This is commonly found in temperature-driven fouling. A second behavior is *linear fouling* in which the increase in fouling resistance follows a straight line over the time of operation. This could be experienced in a case of severe particulate fouling where the accumulation of dirt during the time of operation did not appreciably increase velocities to mitigate the problem. The third behavior, *falling rate fouling,* is neither linear nor asymptotic but instead lies somewhere between these two extremes. The rate of fouling decreases with time but does not appear to approach an asymptotic maximum during the time of operation. This is the most common type of fouling in the process industry and is usually the result of a combination of different fouling mechanisms occurring together.

The optimum operating period between cleanings depends upon the rate and type of fouling, the heat exchanger used (i.e., baffle type, use of extended surface, and velocity and pressure drop design constraints), and the ease with which the heat exchanger may be removed from service for cleaning. As noted above, care must be taken in the use of fouling factors for exchanger design, especially if the exchanger configuration has been selected specifically to minimize fouling accumulation. An oversurfaced heat exchanger which will not foul enough to operate properly can be almost as much of a problem as an undersized exchanger. This is especially true in steam-heated exchangers where the ratio of design MTD to minimum achievable MTD is less than U_{clean} divided by U_{fouled}.

Removal of Fouling Deposits Chemical removal of fouling can be achieved in some cases by weak acid, special solvents, and so on. Other deposits adhere weakly and can be washed off by periodic operation at very high velocities or by flushing with a high-velocity steam or water jet or using a sand-water slurry. These methods may be applied to both the shell side and tube side without pulling the bundle. Many fouling deposits, however, must be removed by positive mechanical action such as rodding, turbining, or scraping the surface. These techniques may be applied inside of tubes without pulling the bundle but can be applied on the shell side only after bundle removal. Even then there is limited access because of the tube pitch, and rotated square or large triangular layouts are recommended. In many cases, it has been found that designs developed to minimize fouling often develop a fouling layer which is more easily removed.

Fouling Resistances There are no published methods for predicting fouling resistances a priori. The accumulated experience of exchanger designers and users was assembled more than 40 years ago based primarily upon segmental-baffled exchanger bundles and may be found in the *Standards of Tubular Exchanger Manufacturers Association* (TEMA). In the absence of other information, the fouling resistances contained therein may be used.

TYPICAL HEAT-TRANSFER COEFFICIENTS

Typical overall heat-transfer coefficients are given in Tables 11-3 through 11-8. Values from these tables may be used for preliminary estimating purposes. They should not be used in place of the design methods described elsewhere in this section, although they may serve as a useful check on the results obtained by those design methods.

THERMAL DESIGN FOR SOLIDS PROCESSING

Solids in divided form, such as powders, pellets, and lumps, are heated and/or cooled in chemical processing for a variety of objectives such as solidification or fusing (Sec. 11), drying and water removal (Sec. 20), solvent recovery (Secs. 13 and 20), sublimation (Sec. 17), chemical reactions (Sec. 20), and oxidation. For process and mechanical-design considerations, see the referenced sections.

Thermal design concerns itself with sizing the equipment to effect the heat transfer necessary to carry on the process. The design equation is the familiar one basic to all modes of heat transfer, namely,

$$A = Q/U\Delta t \tag{11-47}$$

where A = effective heat-transfer surface, Q = quantity of heat required to be transferred, Δt = temperature difference of the process, and U = overall heat-transfer coefficient. It is helpful to define the modes of heat transfer and the corresponding overall coefficient, U_{co}, as U_{co} = overall heat-transfer coefficient for (indirect through-a-wall) *conduction,* U_{cv} = overall heat-transfer coefficient for the little-used *convection* mechanism, U_{ct} = heat-transfer coefficient for the *contactive* mechanism in which the gaseous-phase heat carrier passes directly through the solids bed, and U_{ra} = heat-transfer coefficient for *radiation.*

There are two general methods for determining numerical values for U_{co}, U_{cv}, U_{ct}, and U_{ra}. One is by analysis of actual operating data. Values so obtained are used on geometrically similar systems of a size not too different from the equipment from which the data were obtained. The second

TABLE 11-3 Typical Overall Heat-Transfer Coefficients in Tubular Heat Exchangers

U = Btu/(°F · ft^2 · h)

Shell side	Tube side	Design U	Includes total dirt
Liquid-liquid media			
Aroclor 1248	Jet fuels	100–150	0.0015
Cutback asphalt	Water	10–20	.01
Demineralized water	Water	300–500	.001
Ethanol amine (MEA or DEA) 10–25% solutions	Water or DEA, or MEA solutions	140–200	.003
Fuel oil	Water	15–25	.007
Fuel oil	Oil	10–15	.008
Gasoline	Water	60–100	.003
Heavy oils	Heavy oils	10–40	.004
Heavy oils	Water	15–50	.005
Hydrogen-rich reformer stream	Hydrogen-rich reformer stream	90–120	.002
Kerosene or gas oil	Water	25–50	.005
Kerosene or gas oil	Oil	20–35	.005
Kerosene or jet fuels	Trichlorethylene	40–50	.0015
Jacket water	Water	230–300	.002
Lube oil (low viscosity)	Water	25–50	.002
Lube oil (high viscosity)	Water	40–80	.003
Lube oil	Oil	11–20	.006
Naphtha	Water	50–70	.005
Naphtha	Oil	25–35	.005
Organic solvents	Water	50–150	.003
Organic solvents	Brine	35–90	.003
Organic solvents	Organic solvents	20–60	.002
Tall oil derivatives, vegetable oil, etc.	Water	20–50	.004
Water	Caustic soda solutions (10–30%)	100–250	.003
Water	Water	200–250	.003
Wax distillate	Water	15–25	.005
Wax distillate	Oil	13–23	.005
Condensing vapor-liquid media			
Alcohol vapor	Water	100–200	.002
Asphalt (450°F.)	Dowtherm vapor	40–60	.006
Dowtherm vapor	Tall oil and derivatives	60–80	.004
Dowtherm vapor	Dowtherm liquid	80–120	.0015
Gas-plant tar	Steam	40–50	.0055
High-boiling hydrocarbons V	Water	20–50	.003
Low-boiling hydrocarbons A	Water	80–200	.003
Hydrocarbon vapors (partial condenser)	Oil	25–40	.004
Organic solvents A	Water	100–200	.003
Organic solvents high NC, A	Water or brine	20–60	.003
Organic solvents low NC, V	Water or brine	50–120	.003
Kerosene	Water	30–65	.004
Kerosene	Oil	20–30	.005
Naphtha	Water	50–75	.005
Naphtha	Oil	20–30	.005
Stabilizer reflux vapors	Water	80–120	.003
Steam	Feed water	400–1000	.0005
Steam	No. 6 fuel oil	15–25	.0055
Steam	No. 2 fuel oil	60–90	.0025
Sulfur dioxide	Water	150–200	.003
Tall-oil derivatives, vegetable oils (vapor)	Water	20–50	.004
Water	Aromatic vapor-stream azeotrope	40–80	.005
Gas-liquid media			
Air, N_2, etc. (compressed)	Water or brine	40–80	.005
Air, N_2, etc., A	Water or brine	10–50	.005
Water or brine	Air, N_2 (compressed)	20–40	.005
Water or brine	Air, N_2, etc., A	5–20	.005
Water	Hydrogen containing natural-gas mixtures	80–125	.003
Vaporizers			
Anhydrous ammonia	Steam condensing	150–300	.0015
Chlorine	Steam condensing	150–300	.0015
Chlorine	Light heat-transfer oil	40–60	.0015
Propane, butane, etc.	Steam condensing	200–300	.0015
Water	Steam condensing	250–400	.0015

NC = noncondensable gas present.
V = vacuum.
A = atmospheric pressure.
Dirt (or fouling factor) units are (h · ft^2 · °F)/Btu.
To convert British thermal units per hour-square foot-degrees Fahrenheit to joules per square meter-second-kelvins, multiply by 5.6783; to convert hours per square foot-degree Fahrenheit-British thermal units to square meters per second-kelvin-joules, multiply by 0.1761.

TABLE 11-4 Typical Overall Heat-Transfer Coefficients in Refinery Service

Btu/(°F · ft^2 · h)

Fluid	API gravity	Fouling factor (one stream)	Reboiler, steam-heated	Condenser, water-cooled*	Exchangers, liquid to liquid (tube-side fluid designation appears below)			Reboiler (heating liquid designated below)			Condenser (cooling liquid designated below)			
					C	G	H	C	G†	K	D	F	G	J
A Propane		0.001	160	95	85	85	80	110	95	35				
B Butane		.001	155	90	80	75	75	105	90	35	80	55	40	30
C 400°F. end-point gasoline	50	.001	120	80	70	65	60	65	50	30				
D Virgin light naphtha	70	.001	140	85	70	55	55	75	60	35	75			
E Virgin heavy naphtha	45	.001	95	75	65	55	50	55	45	30	70	50	35	30
F Kerosene	40	.001	85	60	60	55	50		45	25		50	35	30
G Light gas oil	30	.002	70	50	60	50	50		40	25	70	45	30	30
H Heavy gas oil	22	.003	60	45	55	50	45	50	40	20	70	40	30	20
J Reduced crude	17	.005			55	45	40							
K Heavy fuel oil (tar)	10	.005			50	40	35							

Fouling factor, water side 0.0002; heating or cooling streams are shown at top of columns as C, D, F, G, etc.; to convert British thermal units per hour-square foot-degrees Fahrenheit to joules per square meter-second-kelvins, multiply by 5.6783; to convert hours per square foot-degree Fahrenheit-British thermal units to square meters per second-kelvin-joules, multiply by 0.1761.

*Cooler, water-cooled, rates are about 5 percent lower.

†With heavy gas oil (H) as heating medium, rates are about 5 percent lower.

TABLE 11-5 Overall Coefficients for Air-Cooled Exchangers on Bare-Tube Basis

Btu/(°F · ft² · h)

Condensing	Coefficient	Liquid cooling	Coefficient
Ammonia	110	Engine-jacket water	125
Freon-12	70	Fuel oil	25
Gasoline	80	Light gas oil	65
Light hydrocarbons	90	Light hydrocarbons	85
Light naphtha	75	Light naphtha	70
Heavy naphtha	65	Reformer liquid streams	70
Reformer reactor effluent	70	Residuum	15
Low-pressure steam	135	Tar	7
Overhead vapors	65		

Gas cooling	Operating pressure, lb./sq. in. gage	Pressure drop, lb./sq. in.	Coefficient
Air or flue gas	50	0.1 to 0.5	10
	100	2	20
	100	5	30
Hydrocarbon gas	35	1	35
	125	3	55
	1000	5	80
Ammonia reactor stream			85

Bare-tube external surface is 0.262 ft²/ft.

Fin-tube surface/bare-tube surface ratio is 16.9.

To convert British thermal units per hour-square foot-degrees Fahrenheit to joules per square meter-second-kelvins, multiply by 5.6783; to convert pounds-force per square inch to kilopascals, multiply by 6.895.

method is predictive and is based on the material properties and certain operating parameters. Relative values of the coefficients for the various modes of heat transfer at temperatures up to 980°C (1800°F) are as follows (Holt, Paper 11, Fourth National Heat Transfer Conference, Buffalo, 1960):

Convective	1
Radiant	2
Conductive	20
Contactive	200

Because heat-transfer equipment for solids is generally an adaptation of a primarily material-handling device, the area of heat transfer is often small in relation to the overall size of the equipment. Also peculiar to solids heat transfer is that the Δt varies for the different heat-transfer mechanisms. With a knowledge of these mechanisms, the Δt term generally is readily estimated from temperature limitations imposed by the burden characteristics and/or the construction.

Conductive Heat Transfer Heat-transfer equipment in which heat is transferred by conduction is constructed so that the solids load (burden) is separated from the heating medium by a wall.

For a high proportion of applications, Δt is the log-mean temperature difference. Values of U_{co} are reported in Secs. 11, 15, 17, and 19.

A *predictive* equation for U_{co} is

$$U_{co} = \left(\frac{h}{h - 2c\alpha/d_m}\right)\left(\frac{2c\alpha}{d_m}\right) \qquad (11\text{-}48)$$

where h = wall film coefficient, c = volumetric heat capacity, d_m = depth of the burden, and α = thermal diffusivity. Relevant thermal properties of various materials are given in Table 11-9. For details of terminology, equation development, numerical values of terms in typical equipment and use, see Holt [*Chem. Eng.* **69:** 107 (Jan. 8, 1962)].

Equation (11-48) is applicable to burdens in the solid, liquid, or gaseous phase, either static or in laminar motion; it is applicable to solidification equipment and to divided-solids equipment such as metal belts, moving trays, stationary vertical tubes, and stationary-shell fluidizers.

Fixed- (or packed-) bed operation occurs when the fluid velocity is low or the particle size is large so that fluidization does not occur. For such operation, Jakob (*Heat Transfer*, vol. 2, Wiley, New York, 1957) gives

$$hD_t/k = b_1 b D_t^{0.17}(D_p G/\mu)^{0.83}(c\mu/k) \qquad (11\text{-}49a)$$

where b_1 = 1.22 (SI) or 1.0 (USCS), $h = U_{co}$ = overall coefficient between the inner container surface and the fluid stream,

$$b = 0.2366 + 0.0092\left(\frac{D_p}{D_t}\right) - 4.0672\left(\frac{D_p}{D_t}\right)^2 + 18.229\left(\frac{D_p}{D_t}\right)^3 - 11.837\left(\frac{D_p}{D_t}\right)^4 \qquad (11\text{-}49b)$$

where D_p = particle diameter, D_t = vessel diameter (note that D_p/D_t has units of feet per foot in the equation), G = superficial mass velocity, k = fluid

TABLE 11-6 Panel Coils Immersed in Liquid: Overall Average Heat-Transfer Coefficients*

U expressed in Btu/(h · ft² · °F)

Hot side	Cold side	Clean-surface coefficients: Natural convection	Clean-surface coefficients: Forced convection	Design coefficients, considering usual fouling in this service: Natural convection	Design coefficients, considering usual fouling in this service: Forced convection
Heating applications:					
Steam	Watery solution	250–500	300–550	100–200	150–275
Steam	Light oils	50–70	110–140	40–45	60–110
Steam	Medium lube oil	40–60	100–130	35–40	50–100
Steam	Bunker C or No. 6 fuel oil	20–40	70–90	15–30	60–80
Steam	Tar or asphalt	15–35	50–70	15–25	40–60
Steam	Molten sulfur	35–45	45–55	20–35	35–45
Steam	Molten paraffin	35–45	45–55	25–35	40–50
Steam	Air or gases	2–4	5–10	1–3	4–8
Steam	Molasses or corn sirup	20–40	70–90	15–30	60–80
High temperature hot water	Watery solutions	115–140	200–250	70–100	110–160
High temperature heat-transfer oil	Tar or asphalt	12–30	45–65	10–20	30–50
Dowtherm or Aroclor	Tar or asphalt	15–30	50–60	12–20	30–50
Cooling applications:					
Water	Watery solution	110–135	195–245	65–95	105–155
Water	Quench oil	10–15	25–45	7–10	15–25
Water	Medium lube oil	8–12	20–30	5–8	10–20
Water	Molasses or corn sirup	7–10	18–26	4–7	8–15
Water	Air or gases	2–4	5–10	1–3	4–8
Freon or ammonia	Watery solution	35–45	60–90	20–35	40–60
Calcium or sodium brine	Watery solution	100–120	175–200	50–75	80–125

*Tranter Manufacturing, Inc.

NOTE: To convert British thermal units per hour-square foot-degrees Fahrenheit to joules per square meter-second-kelvins, multiply by 5.6783.

TABLE 11-7 Jacketed Vessels: Overall Coefficients

Jacket fluid	Fluid in vessel	Wall material	Overall U* Btu/(h · ft² · °F)	Overall U* J/(m² · s · K)
Steam	Water	Stainless steel	150–300	850–1700
Steam	Aqueous solution	Stainless steel	80–200	450–1140
Steam	Organics	Stainless steel	50–150	285–850
Steam	Light oil	Stainless steel	60–160	340–910
Steam	Heavy oil	Stainless steel	10–50	57–285
Brine	Water	Stainless steel	40–180	230–1625
Brine	Aqueous solution	Stainless steel	35–150	200–850
Brine	Organics	Stainless steel	30–120	170–680
Brine	Light oil	Stainless steel	35–130	200–740
Brine	Heavy oil	Stainless steel	10–30	57–170
Heat-transfer oil	Water	Stainless steel	50–200	285–1140
Heat-transfer oil	Aqueous solution	Stainless steel	40–170	230–965
Heat-transfer oil	Organics	Stainless steel	30–120	170–680
Heat-transfer oil	Light oil	Stainless steel	35–130	200–740
Heat-transfer oil	Heavy oil	Stainless steel	10–40	57–230
Steam	Water	Glass-lined CS	70–100	400–570
Steam	Aqueous solution	Glass-lined CS	50–85	285–480
Steam	Organics	Glass-lined CS	30–70	170–400
Steam	Light oil	Glass-lined CS	40–75	230–425
Steam	Heavy oil	Glass-lined CS	10–40	57–230
Brine	Water	Glass-lined CS	30–80	170–450
Brine	Aqueous solution	Glass-lined CS	25–70	140–400
Brine	Organics	Glass-lined CS	20–60	115–340
Brine	Light oil	Glass-lined CS	25–65	140–370
Brine	Heavy oil	Glass-lined CS	10–30	57–170
Heat-transfer oil	Water	Glass-lined CS	30–80	170–450
Heat-transfer oil	Aqueous solution	Glass-lined CS	25–70	140–400
Heat-transfer oil	Organics	Glass-lined CS	25–65	140–370
Heat-transfer oil	Light oil	Glass-lined CS	20–70	115–400
Heat-transfer oil	Heavy oil	Glass-lined CS	10–35	57–200

*Values listed are for moderate nonproximity agitation. CS = carbon steel.

thermal conductivity, μ = fluid viscosity, and c = fluid specific heat. Other correlations are those of Leva [*Ind. Eng. Chem.* **42:** 2498 (1950)]:

$$h = 0.813 \frac{k}{D_t} e^{-6D_p/D_t} \left(\frac{D_pG}{\mu}\right)^{0.90} \quad \text{for} \quad \frac{D_p}{D_t} < 0.35 \tag{11-50a}$$

$$h = 0.125 \frac{k}{D_t} \left(\frac{D_pG}{\mu}\right)^{0.75} \quad \text{for} \quad 0.35 < \frac{D_p}{D_t} < 0.60 \tag{11-50b}$$

and Calderbank and Pogerski [*Trans. Inst. Chem. Eng.* (London), **35:** 195 (1957)]:

$$hD_p/k = 3.6(D_pG/\mu\epsilon_\upsilon)^{0.365} \tag{11-51}$$

where ϵ_υ = fraction voids in the bed.

A technique for calculating radial temperature gradients in a packed bed is given by Smith (*Chemical Engineering Kinetics*, McGraw-Hill, New York, 1956).

Fluidization occurs when the fluid flow rate is great enough that the pressure drop across the bed equals the weight of the bed. As stated previously, the solids film thickness adjacent to the wall d_m is difficult to measure and/or predict. Wen and Fau [*Chem. Eng.* **64**(7): 254 (1957)] give for *external walls*

$$h = bk(c_s\rho_s)^{0.4}(G\eta/\mu N_f)^{0.36} \tag{11-51a}$$

where b = 0.29 (SI) or 11.6 (USCS), c_s = heat capacity of solid, ρ_s = particle density, η = fluidization efficiency (Fig. 11-31) and N_f = bed expansion ratio (Fig. 11-32). *For internal walls*, Wen and Fau give

$$h_i = bhG^{-0.37} \tag{11-51b}$$

TABLE 11-8 External Coils; Typical Overall Coefficients*

U expressed in Btu/(h · ft² · °F)

Type of coil	Coil spacing, in*	Fluid in coil	Fluid in vessel	Temp. range, °F	U† without cement	U with heat-transfer cement
⅜ in o.d. copper tubing attached with bands at 24-in spacing	2	5 to 50 lb./sq. in. gage steam	Water under light agitation	158–210	1–5	42–46
	3⅛			158–210	1–5	50–53
	6¼			158–210	1–5	60–64
	12½ or greater			158–210	1–5	69–72
⅜ in o.d. copper tubing attached with bands at 24-in spacing	2	50 lb./sq. in. gage steam	No. 6 fuel oil under light agitation	158–258	1–5	20–30
	3⅛			158–258	1–5	25–38
	6¼			158–240	1–5	30–40
	12½ or greater			158–238	1–5	35–46
Panel coils		50 lb./sq. in. gage steam	Boiling water	212	29	48–54
		Water	Water	158–212	8–30	19–48
		Water	No. 6 fuel oil	228–278	6–15	24–56
			Water	130–150	7	15
			No. 6 fuel oil	130–150	4	9–19

Data courtesy of Thermon Manufacturing Co.

*External surface of tubing or side of panel coil facing tank.

†For tubing, the coefficients are more dependent upon tightness of the coil against the tank than upon either fluid. The low end of the range is recommended.

NOTE: To convert British thermal units per hour-square foot-degrees Fahrenheit to joules per square meter-second-kelvins, multiply by 5.6783; to convert inches to meters, multiply by 0.0254; and to convert pounds-force per square inch to kilopascals, multiply by 6.895.

TABLE 11-9 Thermal Properties of Various Materials as Affecting Conductive Heat Transfer

Material	Thermal conductivity, B.t.u./(hr.)(sq. ft.)(°F./ft.)	Volume specific heat, B.t.u./(cu. ft.)(°F.)	Thermal diffusivity, sq. ft./hr.
Air	0.0183	0.016	1.143
Water	0.3766	62.5	0.0755
Double steel plate, sand divider	0.207	19.1	0.0108
Sand	0.207	19.1	0.0108
Powdered iron	0.0533	12.1	0.0044
Magnetite iron ore	0.212	63	0.0033
Aerocat catalysts	0.163	20	0.0062
Table salt	0.168	12.6	0.0133
Bone char	0.0877	16.9	0.0051
Pitch coke	0.333	16.2	0.0198
Phenolformaldehyde resin granules	0.0416	10.5	0.0042
Phenolformaldehyde resin powder	0.070	10	0.0070
Powdered coal	0.070	15	0.0047

To convert British thermal units per hour-square foot-degrees Fahrenheit to joules per meter-second-kelvins, multiply by 1.7307; to convert British thermal units per cubic foot-degrees Fahrenheit to joules per cubic meter-kelvins, multiply by $(6.707)(10^4)$; and to convert square feet per hour to square meters per second, multiply by $(2.581)(10^{-5})$.

where $b = 0.78$ (SI) or 9 (USCS), h_i is the coefficient for internal walls, and h is calculated from Eq. (11-51*a*). The minimum fluidizing velocity G_{mf} is defined by

$$G_{mf} = \frac{b\rho_g^{1.1}(\rho_s - \rho_g)^{0.9} D_p^2}{\mu} \qquad (11\text{-}51c)$$

where $b = (1.23)(10^{-2})$ (SI) or $(5.23)(10^5)$ (USCS).

Wender and Cooper [*Am. Inst. Chem. Eng. J.* **4:** 15 (1958)] developed an empirical correlation for *internal walls:*

$$\frac{hD_p/k}{1-\epsilon_v}\left(\frac{k}{c\rho}\right)^{0.43} = bC_R\left(\frac{D_pG}{\mu}\right)^{0.23}\left(\frac{c_s}{c_g}\right)^{0.80}\left(\frac{\rho_s}{\rho_g}\right)^{0.66} \qquad (11\text{-}52a)$$

where $b = (3.51)(10^{-4})$ (SI) or 0.033 (USCS) and C_R = correction for displacement of the immersed tube from the axis of the vessel (see the reference). For external walls:

$$\frac{hD_p}{k_g(1-\epsilon_v)(c_s\rho_s/c_g\rho_g)} = f(1 + 7.5e^{-x}) \qquad (11\text{-}52b)$$

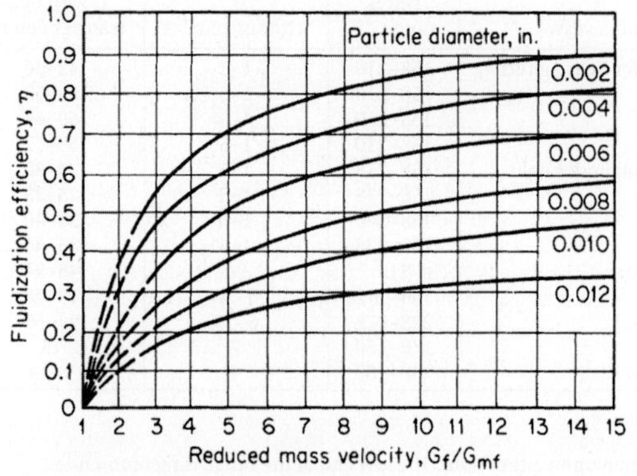

FIG. 11-31 Fluidization efficiency.

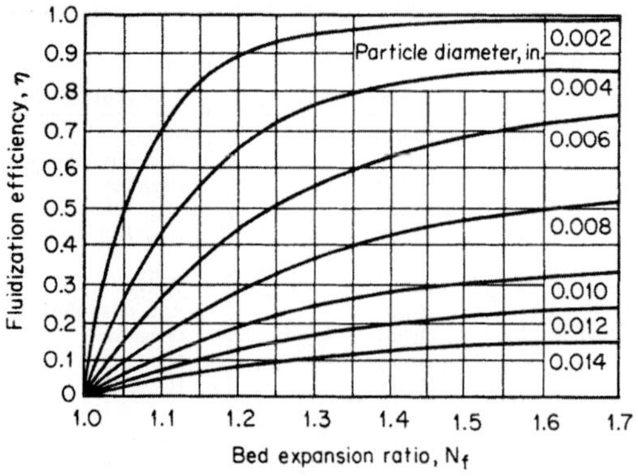

FIG. 11-32 Bed expansion ratio.

where $x = 0.44L_Hc_s/D_tc_g$ and f is given by Fig. 11-33. An important feature of this equation is the inclusion of the ratio of bed depth to vessel diameter L_H/D_t.

For *dilute fluidized beds* on the shell side of an unbaffled tubular bundle, Genetti and Knudsen [*Inst. Chem. Eng. (London)* Symp. Ser. 3:172 (1968)] obtained the relation:

$$\frac{hD_p}{k} = \frac{5\phi(1-\epsilon_v)}{\left[1 + \frac{580}{N_{Re}}\left(\frac{k_s}{D_p^{1.5}C_s\rho_s g^{0.5}}\right)\left(\frac{\rho_s}{\rho_g}\right)^{1.1}\left(\frac{G_{mf}}{G}\right)^{7/3}\right]^2} \qquad (11\text{-}53a)$$

where ϕ = particle surface area per area of sphere of same diameter. When particle transport occurred through the bundle, the heat-transfer coefficients could be predicted by

$$j_H = 0.14(N_{Re}/\phi)^{-0.68} \qquad (11\text{-}53b)$$

In Eqs. (11-53*a*) and (11-53*b*), N_{Re} is based on particle diameter and superficial fluid velocity.

Zenz and Othmer (*Fluidization and Fluid Particle Systems*, Reinhold, original from University of Michigan, 1960) give an excellent summary of fluidized bed-to-wall heat-transfer investigations.

Solidification involves heavy heat loads transferred essentially at a steady temperature difference. It also involves the varying values of liquid- and solid-phase thickness and thermal diffusivity. When these are substantial and/or in the case of a liquid flowing over a changing solid layer interface, Siegel and Savino (ASME Paper 67-WA/Ht-34, November 1967) offer equations and charts for prediction of the layer-growth time. For solidification (or melting) of a slab or a semi-infinite bar, initially at its transition temperature, the position of the interface is given by the one-dimensional Newmann's method given in Carslaw and Jaeger (*Conduction of Heat in Solids*, Clarendon Press, Oxford, 1959).

Later work by Hashem and Sliepcevich [*Chem. Eng. Prog.* **63:** 79, 35, 42 (1967)] offers more accurate second-order finite-difference equations.

The heat-transfer rate is found to be substantially higher under conditions of *agitation.* The heat transfer is usually said to occur by combined conductive and convective modes. A discussion and an explanation are given by Holt [*Chem. Eng.* **69**(1): 110 (1962)]. Prediction of U_{co} by Eq. (11-48) can be accomplished by replacing α by α_e, the effective thermal diffusivity of the bed. To date so little work has been performed in evaluating the effect of mixing parameters that few predictions can be made. However, for agitated liquid-phase devices, Eq. (18-19) is applicable. Holt [*Chem. Eng.* **69**(1): 110 (1962)] shows that this equation can be converted for solids heat transfer to yield

$$U_{co} = a'c_sD_t^{-0.3}N^{0.7}(\cos\omega)^{0.2} \qquad (11\text{-}54)$$

where D_t = agitator or vessel diameter; N = turning speed, r/min; ω = effective angle of repose of the burden; and a' = proportionality constant. This is applicable for such devices as agitated pans, agitated kettles, spiral conveyors, and rotating shells.

The solids passage time through *rotary devices* is given by Saemann [*Chem. Eng. Prog.* **47:** 508, (1951)]:

$$\theta = 0.318L\sin\omega/S_rND_t \qquad (11\text{-}55a)$$

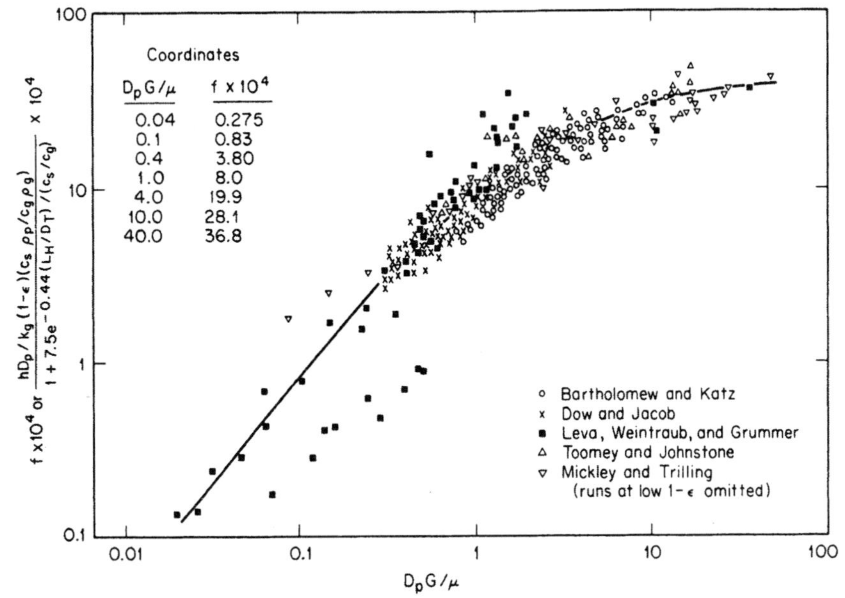

FIG. 11-33 The f factor for Eq. (11-52*b*).

and by Marshall and Friedman [*Chem. Eng. Prog.* **45:** 482–493, 573–588 (1949)]:

$$\theta = (0.23L/S_r N^{0.9} D_t) \pm (0.6BLG/F_a) \qquad (11\text{-}55b)$$

where the second term of Eq. (11-55*b*) is positive for counterflow of air, negative for concurrent flow, and zero for indirect rotary shells. From these equations a predictive equation is developed for rotary-shell devices, which is analogous to Eq. (11-54):

$$U_{co} = \frac{b' c_s D_t N^{0.9} Y}{(\Delta t) L \sin \omega} \qquad (11\text{-}56)$$

where θ = solids-bed passage time through the shell, min; S_r = shell slope; L = shell length; Y = percent fill; and b' = proportionality constant.

Vibratory devices which constantly agitate the solids bed maintain a relatively constant value for U_{co} such that

$$U_{co} = a' c_s \alpha_e \qquad (11\text{-}57)$$

with U_{co} having a nominal value of 114 J/(m²·s·K) [20 Btu/(h·ft²·°F)].

Contactive (Direct) Heat Transfer Contactive heat-transfer equipment is so constructed that the particulate burden in solid phase is directly exposed to and permeated by the heating or cooling medium (Sec. 20). The carrier may either heat or cool the solids. A large amount of the industrial heat processing of solids is effected by this mechanism. Physically, these can be classified into packed beds and various degrees of agitated beds from dilute to dense fluidized beds.

The temperature difference for heat transfer is the log-mean temperature difference when the particles are large and/or the beds packed, or the difference between the inlet fluid temperature t_3 and average exhausting fluid temperature t_4, expressed as $\Delta_3 t_4$, for small particles. The use of the log mean for packed beds has been confirmed by Thodos and Wilkins (Second American Institute of Chemical Engineers-IIQPR Meeting, Paper 30D, Tampa, Fla., May 1968). When fluid and solid flow directions are axially concurrent and particle size is small, as in a vertical-shell fluid bed, the temperature of the exiting solids t_2 (which is also that of exiting gas t_4) is used as $\Delta_3 t_2$, as shown by Levenspiel, Olson, and Walton [*Ind. Eng. Chem.* **44:** 1478 (1952)], Marshall [*Chem. Eng. Prog.* **50:** Monogr. Ser. 2, 77 (1954)], Leva (*Fluidization*, McGraw-Hill, New York, 1959), and Holt (Fourth Int. Heat Transfer Conf. Paper 11, American Institute of Chemical Engineers-American Society of Mechanical Engineers, Buffalo, N.Y., 1960). This temperature difference is also applicable for well-fluidized beds of small particles in crossflow as in various vibratory carriers.

The *packed-bed* to *fluid heat-transfer coefficient* has been investigated by Baumeister and Bennett [*Am. Inst. Chem. Eng. J.* **4:** 69 (1958)], who proposed the equation

$$j_H = (h/cG)(c\mu/k)^{2/3} = aN_{Re}^m \qquad (11\text{-}58)$$

where N_{Re} is based on particle diameter and superficial fluid velocity. Values of a and m are as follows:

D_t/D_p (dimensionless)	a	m
10.7	1.58	−0.40
16.0	0.95	−0.30
25.7	0.92	−0.28
>30	0.90	−0.28

Glaser and Thodos [*Am. Inst. Chem. Eng. J.* **4:** 63 (1958)] give a correlation involving individual particle shape and bed porosity. Kunii and Suzuki [*Int. J. Heat Mass Transfer* **10:** 845 (1967)] discuss heat and mass transfer in packed beds of fine particles.

Particle-to-fluid heat-transfer coefficients in gas *fluidized beds* are predicted by the relation (F. A. Zenz and D. F. Othmer, *Fluidization and Fluid Particle Systems*, Reinhold, original from University of Michigan, 1960)

$$\frac{hD_p}{k} = 0.017(D_p G_{mf}/\mu)^{1.21} \qquad (11\text{-}59a)$$

where G_{mf} is the superficial mass velocity at incipient fluidization.

A more general equation is given by Frantz [*Chem. Eng.* **69**(20): 89 (1962)]:

$$\frac{hD_p}{k} = 0.015(D_p G/\mu)^{1.6}(c\mu/k)^{0.67} \qquad (11\text{-}59b)$$

where h is based on true gas temperature.

Bed-to-wall coefficients in dilute-phase transport generally can be predicted by an equation of the form of Eq. (5-50). For example, Bonilla et al. (American Institute of Chemical Engineers Heat Transfer Symp., Atlantic City, N.J., December 1951) found for 1- to 2-µm chalk particles in water up to 8 percent by volume that the coefficient on Eq. (5-50) is 0.029 where k, ρ, and c were arithmetic weighted averages and the viscosity was taken equal to the coefficient of rigidity. Farber and Morley [*Ind. Eng. Chem.* **49:** 1143 (1957)] found the coefficient on Eq. (5-50) to be 0.025 for the upward flow of air transporting silica-alumina catalyst particles at rates less than 2 kg solids/kg air (2 lb solids/lb air). Physical properties used were those of the transporting gas. See Zenz and Othmer (*Fluidization and Fluid Particle Systems*, Reinhold, original from University of Michigan, 1960) for additional details covering wider porosity ranges.

The thermal performance of *cylindrical rotating shell* units is based upon a volumetric heat-transfer coefficient

$$U_{ct} = \frac{Q}{V_r(\Delta t)} \qquad (11\text{-}60a)$$

where V_r = volume. This term indirectly includes an area factor so that thermal performance is governed by a cross-sectional area rather than by a heated area. Use of the heated area is possible, however:

$$U_{ct} = \frac{Q}{(\Delta_3 t_2)A} \quad \text{or} \quad U_{ct} = \frac{Q}{(\Delta_3 t_4)A} \tag{11-60b}$$

For *heat transfer directly to solids,* predictive equations give directly the volume V or the heat-transfer area A, as determined by heat balance and air flow rate. For devices with gas flow normal to a fluidized-solids bed,

$$A = \frac{Q}{\Delta t_p (c\rho_g)(F_g)} \tag{11-61}$$

where $\Delta t_p = \Delta_3 t_4$ as explained above, $c\rho$ = volumetric specific heat, and F_g = gas flow rate. For air, $c\rho$ at normal temperature and pressure is about 1100 J/(m^3·K) [0.0167 Btu/(ft^3·°F)]; so

$$A = \frac{bQ}{(\Delta_3 t_4)F_g} \tag{11-62}$$

where b = 0.0009 (SI) or 60 (USCS). Another such equation—for stationary vertical-shell and some horizontal rotary-shell and pneumatic-transport devices in which the gas flow is parallel with and directionally concurrent with the fluidized bed—is the same as Eq. (11-62) with $\Delta_3 t_4$ replaced by $\Delta_3 t_2$. If the operation involves drying or chemical reaction, the heat load Q is much greater than for sensible-heat transfer only. Also the gas flow rate to provide moisture carry-off and stoichiometric requirements must be considered and simultaneously provided. A good treatise on the latter is given by Pinkey and Plint (*Miner. Process.* June 1968, p. 17).

Evaporative cooling is a special patented technique that often can be advantageously employed in cooling solids by contactive heat transfer. The drying operation is terminated *before* the desired final moisture content is reached, and the solids temperature is at a moderate value. The cooling operation involves contacting the burden (preferably fluidized) with air at normal temperature and pressure. The air adiabatically absorbs and carries off a large part of the moisture and, in doing so, picks up heat from the warm (or hot) solids particles to supply the latent heat demand of evaporation. For entering solids at temperatures of 180°C (350°F) and less with normal heat capacity values of 0.85 to 1.0 kJ/(kg·K) [0.2 to 0.25 Btu/(lb·°F)], the effect can be calculated by the following procedure:

1. Use 285 m^3 (1000 ft^3) of air flow at normal temperature and pressure at 40 percent relative humidity to carry off 0.45 kg (1 lb) of water [latent heat 2326 kJ/kg (1000 Btu/lb)] and to lower temperature by 22°C to 28°C (40°F to 50°F).
2. Use the lowered solids temperature as t_3 and calculate the remainder of the heat to be removed in the regular manner by Eq. (11-62). The required air quantity for (2) must be equal to or greater than that for (1).

When the solids heat capacity is higher (as is the case for most organic materials), the temperature reduction is inversely proportional to the heat capacity.

A nominal result of this technique is that the required air flow rate and equipment size are about two-thirds of that when evaporative cooling is not used. See Sec. 20 for equipment available.

Convective Heat Transfer Equipment using the true convective mechanism when the heated particles are mixed with (and remain with) the cold particles is used so infrequently that performance and sizing equations are not available. Such a device is the pebble heater as described by Norton (*Chem. Metall. Eng.* July 1946). For operation data, see Sec. 9.

Convective heat transfer is often used as an adjunct to other modes, particularly to the conductive mode. It is often more convenient to consider the agitative effect a performance improvement influence on the thermal diffusivity factor α, modifying it to α_e, the effective value.

A pseudo-convective heat-transfer operation is one in which the heating gas (generally air) is passed over a bed of solids. Its use is almost exclusively limited to drying operations (see Sec. 12, tray and shelf dryers). The operation, sometimes termed *direct,* is more akin to the conductive mechanism. For this operation, Tsao and Wheelock [*Chem. Eng.* **74**(13): 201 (1967)] predict the heat-transfer coefficient when radiative and conductive effects are absent by

$$h = bG^{0.8} \tag{11-63}$$

where b = 14.31 (SI) or 0.0128 (USCS), h = convective heat transfer, and G = gas flow rate.

The *drying rate* is given by

$$K_{cv} = \frac{h(T_d - T_w)}{\lambda} \tag{11-64}$$

where K_{cv} = drying rate, for constant-rate period, kg/(m^2·s) [lb/(h·ft^2)]; T_d and T_w = respective dry-bulb and wet-bulb temperatures of the air; and λ = latent heat of evaporation at temperature T_w. Note here that the temperature-difference determination of the operation is a simple linear one and of a steady-state nature. Also note that the operation is a function of the air flow rate. Further, the solids are granular with a fairly uniform size, have reasonable capillary voids, are of a firm texture, and have the particle surface wetted.

The coefficient h is also used to predict (in the constant-rate period) the total overall air-to-solids heat-transfer coefficient U_{cv} by

$$1/U_{cv} = 1/h + x/k \tag{11-65}$$

where k = solids thermal conductivity and x is evaluated from

$$x = \frac{z(X_c - X_o)}{X_c - X_e} \tag{11-65a}$$

where z = bed (or slab) thickness and is the total thickness when drying and/or heat transfer is from one side only but is one-half of the thickness when drying and/or heat transfer is simultaneously from both sides; X_o, X_c, and X_e are, respectively, the initial (or feed-stock), critical, and equilibrium (with the drying air) moisture contents of the solids, all in kg H_2O/kg dry solids (lb H_2O/lb dry solids). This coefficient is used to predict the *instantaneous* drying rate

$$-\frac{W}{A}\frac{dX}{d\theta} = \frac{U_{cv}(T_d - T_w)}{\lambda} \tag{11-66}$$

By rearrangement, this can be made into a design equation as follows:

$$A = -\frac{W\lambda(dX/d\theta)}{U_{cv}(T_d - T_w)} \tag{11-67}$$

where W = weight of dry solids in the equipment, λ = latent heat of evaporation, and θ = drying time. The reader should refer to the full reference article by Tsao and Wheelock for other solids conditions qualifying the use of these equations.

Radiative Heat Transfer Heat-transfer equipment using the radiative mechanism for divided solids is constructed as a "table" which is stationary, as with trays, or moving, as with a belt, and/or agitated, as with a vibrated pan, to distribute and expose the burden in a plane parallel to (but not in contact with) the plane of the radiant-heat sources. Presence of air is not necessary (see Sec. 12 for vacuum-shelf dryers and Sec. 22 for resublimation). In fact, if air in the intervening space has a high humidity or CO_2 content, it acts as an energy absorber, thereby depressing the performance.

For the radiative mechanism, the temperature difference is evaluated as

$$D_t = T_e^4 - T_r^4 \tag{11-68}$$

where T_e = absolute temperature of the radiant-heat source, K (°R), and T_r = absolute temperature of the bed of divided solids, K (°R).

Numerical values for U_{ra} for use in the general design equation may be calculated from experimental data by

$$U_{ra} = \frac{Q}{A(T_e^4 - T_r^4)} \tag{11-69}$$

The literature to date offers practically no such values. However, enough proprietary work has been performed to present a reliable evaluation for the comparison of mechanisms (see Introduction: Modes of Heat Transfer).

For the radiative mechanism of heat transfer to solids, the rate equation for parallel-surface operations is

$$q_{ra} = b(T_e^4 - T_r^4)i_f \tag{11-70}$$

where b = (5.67)(10^{-8}) (SI) or (0.172)(10^{-8}) (USCS), q_{ra} = radiative heat flux, and i_f = an interchange factor which is evaluated from

$$1/i_f = 1/e_s + 1/e_r - 1 \tag{11-70a}$$

where e_s = coefficient of emissivity of the source and e_r = "emissivity" (or "absorptivity") of the receiver, which is the divided-solids bed. For the emissivity values, particularly of the heat source e_s, an important consideration is the wavelength at which the radiant source emits as well as the flux density of the emission. Data for these values are available from Polentz [*Chem. Eng.* **65**(7): 137; (8): 151 (1958)] and Adlam (*Radiant Heating,* Industrial Press, New York, p. 40). Both give radiated flux density versus wavelength at varying temperatures. Often the seemingly cooler but longer wavelength source is the better selection.

Emitting sources are (1) pipes, tubes, and platters carrying steam, 2100 kPa (300 lbf/in^2); (2) electric conducting glass plates, 150°C to 315°C (300°F to 600°F) range; (3) lightbulb type (tungsten-filament resistance heater); (4) modules of refractory brick for gas burning at high temperatures and high fluxes; and (5) modules of quartz tubes, also operable at high temperatures and fluxes. For some emissivity values see Table 11-10.

For *predictive work*, where U_{ra} is desired for sizing, this can be obtained by dividing the flux rate q_{ra} by Δt:

$$U_{ra} = q_{ra}/(T_e^4 - T_r^4) = i_f b \quad (11\text{-}71)$$

where $b = (5.67)(10^{-8})$ (SI) or $(0.172)(10^{-8})$ (USCS). Hence

$$A = \frac{Q}{U_{ra}(T_e^4 - T_r^4)} \quad (11\text{-}72)$$

where A = bed area of solids in the equipment.

TABLE 11-10 Normal Total Emissivity of Various Surfaces

A. Metals and Their Oxides

Surface	t, °F.*	Emissivity*
Aluminum		
Highly polished plate, 98.3% pure	440–1070	0.039–0.057
Polished plate	73	0.040
Rough plate	78	0.055
Oxidized at 1110°F	390–1110	0.11–0.19
Aluminum-surfaced roofing	100	0.216
Calorized surfaces, heated at 1110°F.		
Copper	390–1110	0.18–0.19
Steel	390–1110	0.52–0.57
Brass		
Highly polished:		
73.2% Cu, 26.7% Zn	476–674	0.028–0.031
62.4% Cu, 36.8% Zn, 0.4% Pb, 0.3% Al	494–710	0.033–0.037
82.9% Cu, 17.0% Zn	530	0.030
Hard rolled, polished:		
But direction of polishing visible	70	0.038
But somewhat attacked	73	0.043
But traces of stearin from polish left on	75	0.053
Polished	100–600	0.096
Rolled plate, natural surface	72	0.06
Rubbed with coarse emery	72	0.20
Dull plate	120–660	0.22
Oxidized by heating at 1110°F	390–1110	0.61–0.59
Chromium; see Nickel Alloys for Ni-Cr steels	100–1000	0.08–0.26
Copper		
Carefully polished electrolytic copper	176	0.018
Commercial, emeried, polished, but pits remaining	66	0.030
Commercial, scraped shiny but not mirror-like	72	0.072
Polished	242	0.023
Plate, heated long time, covered with thick oxide layer	77	0.78
Plate heated at 1110°F	390–1110	0.57
Cuprous oxide	1470–2010	0.66–0.54
Molten copper	1970–2330	0.16–0.13
Gold		
Pure, highly polished	440–1160	0.018–0.035
Iron and steel		
Metallic surfaces (or very thin oxide layer):		
Electrolytic iron, highly polished	350–440	0.052–0.064
Polished iron	800–1880	0.144–0.377
Iron freshly emeried	68	0.242
Cast iron, polished	392	0.21
Wrought iron, highly polished	100–480	0.28
Cast iron, newly turned	72	0.435
Polished steel casting	1420–1900	0.52–0.56
Ground sheet steel	1720–2010	0.55–0.61
Smooth sheet iron	1650–1900	0.55–0.60
Cast iron, turned on lathe	1620–1810	0.60–0.70
Oxidized surfaces:		
Iron plate, pickled, then rusted red	68	0.612
Completely rusted	67	0.685
Rolled sheet steel	70	0.657
Oxidized iron	212	0.736
Cast iron, oxidized at 1100°F	390–1110	0.64–0.78
Steel, oxidized at 1100°F	390–1110	0.79
Smooth oxidized electrolytic iron	260–980	0.78–0.82
Iron oxide	930–2190	0.85–0.89
Rough ingot iron	1700–2040	0.87–0.95

Surface	t, °F.*	Emissivity*
Sheet steel, strong rough oxide layer	75	0.80
Dense shiny oxide layer	75	0.82
Cast plate:		
Smooth	73	0.80
Rough	73	0.82
Cast iron, rough, strongly oxidized	100–480	0.95
Wrought iron, dull oxidized	70–680	0.94
Steel plate, rough	100–700	0.94–0.97
High temperature alloy steels (see Nickel Alloys)		
Molten metal		
Cast iron	2370–2550	0.29
Mild steel	2910–3270	0.28
Lead		
Pure (99.96%), unoxidized	260–440	0.057–0.075
Gray oxidized	75	0.281
Oxidized at 390°F	390	0.63
Mercury	32–212	0.09–0.12
Molybdenum filament	1340–4700	0.096–0.292
Monel metal, oxidized at 1110°F	390–1110	0.41–0.46
Nickel		
Electroplated on polished iron, then polished	74	0.045
Technically pure (98.9% Ni, + Mn), polished	440–710	0.07–0.087
Electropolated on pickled iron, not polished	68	0.11
Wire	368–1844	0.096–0.186
Plate, oxidized by heating at 1110°F	390–1110	0.37–0.48
Nickel oxide	1200–2290	0.59–0.86
Nickel alloys		
Chromnickel	125–1894	0.64–0.76
Nickelin (18–32 Ni; 55–68 Cu; 20 Zn), gray oxidized	70	0.262
KA-2S alloy steel (8% Ni; 18% Cr), light silvery, rough, brown, after heating	420–914	0.44–0.36
After 42 hr. heating at 980°F	420–980	0.62–0.73
NCT-3 alloy (20% Ni; 25% Cr), brown, splotched, oxidized from service	420–980	0.90–0.97
NCT-6 alloy (60% Ni; 12% Cr), smooth, black, firm adhesive oxide coat from service	520–1045	0.89–0.82
Platinum		
Pure, polished plate	440–1160	0.054–0.104
Strip	1700–2960	0.12–0.17
Filament	80–2240	0.036–0.192
Wire	440–2510	0.073–0.182
Silver		
Polished, pure	440–1160	0.0198–0.0324
Polished	100–700	0.0221–0.0312
Steel, see Iron		
Tantalum filament	2420–5430	0.194–0.31
Tin—bright tinned iron sheet	76	0.043 and 0.064
Tungsten		
Filament, aged	80–6000	0.032–0.35
Filament	6000	0.39
Zinc		
Commercial, 99.1% pure, polished	440–620	0.045–0.053
Oxidized by heating at 750°F.	750	0.11
Galvanized sheet iron, fairly bright	82	0.228
Galvanized sheet iron, gray oxidized	75	0.276

B. Refractories, Building Materials, Paints, and Miscellaneous

Surface	t, °F.*	Emissivity*
Asbestos		
Board	74	0.96
Paper	100–700	0.93–0.945
Brick		
Red, rough, but no gross irregularities	70	0.93
Silica, unglazed, rough	1832	0.80
Silica, glazed, rough	2012	0.85
Grog brick, glazed	2012	0.75
See Refractory Materials below.		

Surface	t, °F.*	Emissivity*
Carbon		
T-carbon (Gebr. Siemens) 0.9% ash (this started with emissivity at 260°F. of 0.72, but on heating changed to values given)	260–1160	0.81–0.79
Carbon filament	1900–2560	0.526
Candle soot	206–520	0.952
Lampblack-waterglass coating	209–362	0.959–0.947

TABLE 11-10 Normal Total Emissivity of Various Surfaces *(Continued)*

B. Refractories, Building Materials, Paints, and Miscellaneous

Surface	t, °F.*	Emissivity*	Surface	t, °F.*	Emissivity*
Same	260–440	0.957–0.952	Oil paints, sixteen different, all colors	212	0.92–0.96
Thin layer on iron plate	69	0.927	Aluminum paints and lacquers		
Thick coat	68	0.967	10% Al, 22% lacquer body, on rough or		
Lampblack, 0.003 in. or thicker	100–700	0.945	smooth surface	212	0.52
Enamel, white fused, on iron	66	0.897	26% Al, 27% lacquer body, on rough or		
Glass, smooth	72	0.937	smooth surface	212	0.3
Gypsum, 0.02 in. thick on smooth or			Other Al paints, varying age and Al		
blackened plate	70	0.903	content	212	0.27–0.67
Marble, light gray, polished	72	0.931	Al lacquer, varnish binder, on rough plate	70	0.39
Oak, planed	70	0.895	Al paint, after heating to 620°F	300–600	0.35
Oil layers on polished nickel (lube oil)	68		Paper, thin		
Polished surface, alone		0.045	Pasted on tinned iron plate	66	0.924
+0.001-in. oil		0.27	On rough iron plate	66	0.929
+0.002-in. oil		0.46	On black lacquered plate	66	0.944
+0.005-in. oil		0.72	Plaster, rough lime	50–190	0.91
Infinitely thick oil layer		0.82	Porcelain, glazed	72	0.924
Oil layers on aluminum foil (linseed oil)			Quartz, rough, fused	70	0.932
Al foil	212	0.087†	Refractory materials, 40 different	1110–1830	
+1 coat oil	212	0.561	poor radiators		0.65 – 0.75
+2 coats oil	212	0.574			0.70
Paints, lacquers, varnishes			good radiators		0.80 – 0.85
Snowhite enamel varnish or rough iron					0.85 – 0.90
plate	73	0.906	Roofing paper	69	0.91
Black shiny lacquer, sprayed on iron	76	0.875	Rubber		
Black shiny shellac on tinned iron sheet	70	0.821	Hard, glossy plate	74	0.945
Black matte shellac	170–295	0.91	Soft, gray, rough (reclaimed)	76	0.859
Black lacquer	100–200	0.80–0.95	Serpentine, polished	74	0.900
Flat black lacquer	100–200	0.96–0.98	Water	32–212	0.95–0.963
White lacquer	100–200	0.80–0.95			

*When two temperatures and two emissivities are given, they correspond, first to first and second to second, and linear interpolation is permissible. °C = (°F − 32)/1.8.
†Although this value is probably high, it is given for comparison with the data by the same investigator to show the effect of oil layers. See Aluminum, Part A of this table.

These are important considerations in the application of the foregoing equations:

1. Since the temperature of the emitter is generally known (preselected or readily determined in an actual operation), the absorptivity value e_r is the unknown. This absorptivity is partly a measure of the ability of radiant heat to penetrate the body of a solid particle (or a moisture film) instantly, as compared with diffusional heat transfer by conduction. Such instant penetration greatly reduces processing time and case-hardening effects. Moisture release and other mass transfer, however, still progress by diffusional means.

2. In one of the major applications of radiative devices (drying), the surface-held moisture is a good heat absorber in the 2- to 7-μm wavelength range. Therefore, the absorptivity, color, and nature of the solids are of little importance.

3. For drying, it is important to provide a small amount of venting air to carry away the water vapor. This is needed for two reasons. First, water vapor is a good absorber of 2- to 7-μm energy. Second, water vapor accumulation depresses further vapor release by the solids. If the air over the solids is kept fairly dry by venting, very little heat is carried off, because dry air does not absorb radiant heat.

4. For some of the devices, when the overall conversion efficiency has been determined, the application is primarily a matter of computing the required heat load. It should be kept in mind, however, that there are two conversion efficiencies that must be differentiated. One measure of efficiency is that with which the source converts input energy to output radiated energy. The other is the overall efficiency that measures the proportion of input energy that is actually absorbed by the solids. This latter is, of course, the one that really matters.

Other applications of radiant-heat processing of solids are the toasting, puffing, and baking of foods and the low-temperature roasting and preheating of plastic powder or pellets. Since the determination of heat loads for these operations is not well established, bench and pilot tests are generally necessary. Such processes require a fast input of heat and higher heat fluxes than can generally be provided by indirect equipment. Because of this, infrared-equipment size and space requirements are often much lower.

Although direct contactive heat transfer can provide high temperatures and heat concentrations and at the same time be small in size, its use may not always be preferable because of undesired side effects such as drying, contamination, case hardening, shrinkage, off color, and dusting.

When radiating and receiving surfaces are not in parallel, as in rotary-kiln devices, and the solids burden bed may be only intermittently exposed and/or agitated, the calculation and procedures become very complex, with photometric methods of optics requiring consideration. The following equation for heat transfer, which allows for convective effects, is commonly used by designers of *high-temperature furnaces:*

$$q_{ra} = Q/A = b\sigma\,[(T_g/100)^4 - (T_s/100)^4] \quad (11\text{-}73)$$

where $b = 5.67$ (SI) or 0.172 (USCS); Q = total furnace heat transfer; σ = an emissivity factor with recommended values of 0.74 for gas, 0.75 for oil, and 0.81 for coal; A = effective area for absorbing heat (here the solids burden exposed area); T_g = exiting combustion gas absolute temperature; and T_s = absorbing surface temperature.

In rotary devices, reradiation from the exposed shell surface to the solids bed is a major design consideration. A treatise on furnaces, including radiative heat-transfer effects, is given by Ellwood and Danatos [*Chem. Eng.* **73**(8): 174 (1966)]. For discussion of radiation heat-transfer computational methods, heat fluxes obtainable, and emissivity values, see Schornshort and Viskanta (ASME Paper 68-H 7-32), Sherman (ASME Paper 56-A-111), and the following subsection.

SCRAPED-SURFACE EXCHANGERS

Scraped-surface exchangers have a rotating element with spring-loaded scraper blades to scrape the inside surface (Fig. 11-34). Generally a double-pipe construction is used; the scraping mechanism is in the inner pipe,

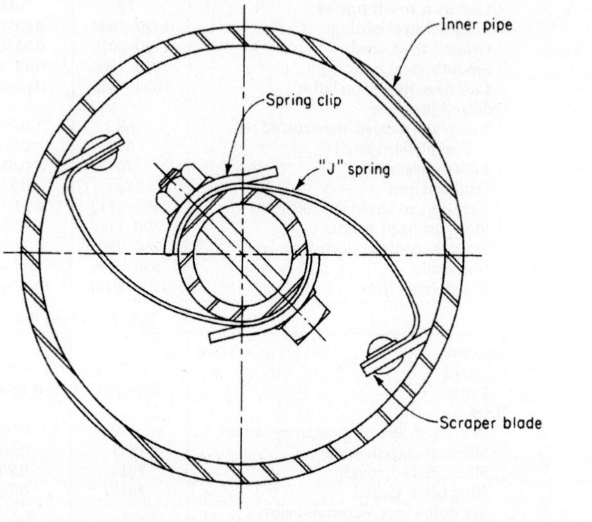

FIG. 11-34 Scraper blade of scraped-surface exchanger. (*Henry Vogt Machine Co., Inc.*)

where the process fluid flows; and the cooling or heating medium is in the outer pipe. The most common size has 6-in inside and 8-in outside pipes. Also available are 3- by 4-in, 8- by 10-in, and 12- by 14-in sizes (in × 25.4 = mm). These double-pipe units are commonly connected in series and arranged in double stands.

For *chilling* and *crystallizing* with an evaporating refrigerant, a 27-in shell with seven 6-in pipes is available (Henry Vogt Machine Co.). In direct contact with the scraped surface is the process fluid which may deposit crystals upon chilling or be extremely fouling or of very high viscosity. Motors, chain drives, appropriate guards, and so on are required for the rotating element. For chilling service with a refrigerant in the outer shell, an accumulator drum is mounted on top of the unit.

Scraped-surface exchangers are particularly suitable for heat transfer with crystallization, heat transfer with severe fouling of surfaces, heat transfer with solvent extraction, and heat transfer of high-viscosity fluids. They are extensively used in paraffin-wax plants and in petrochemical plants for crystallization.

TEMA-STYLE SHELL-AND-TUBE HEAT EXCHANGERS

TYPES AND DEFINITIONS

TEMA-style shell-and-tube-type exchangers constitute the bulk of the unfired heat-transfer equipment in chemical-process plants, although increasing emphasis has been developing in other designs. These exchangers are illustrated in Fig. 11-35, and their features are summarized in Table 11-11.

TEMA Numbering and Type Designation Recommended practice for the designation of TEMA-style shell-and-tube heat exchangers by numbers and letters has been established by the Tubular Exchanger Manufacturers Association. This information from the sixth edition of the TEMA Standards is reproduced in the following paragraphs.

It is recommended that heat exchanger size and type be designated by numbers and letters.

1. *Size.* Sizes of shells (and tube bundles) shall be designated by numbers describing shell (and tube-bundle) diameters and tube lengths as follows:

2. *Diameter.* The nominal diameter shall be the inside diameter of the shell in inches, rounded off to the nearest integer. For kettle reboilers the nominal diameter shall be the port diameter followed by the shell diameter, each rounded off to the nearest integer.

3. *Length.* The nominal length shall be the tube length in inches. Tube length for straight tubes shall be taken as the actual overall length. For U tubes the length shall be taken as the straight length from end of tube to bend tangent.

4. *Type.* Type designation shall be by letters describing stationary head, shell (omitted for bundles only), and rear head, in that order, as indicated in Fig. 11-1.

Typical Examples

1. Split-ring floating heat exchanger with removable channel and cover, single-pass shell, 591-mm (23¼-in) inside diameter with tubes 4.9 m (16 ft) long. SIZE 23–192 TYPE AES.

2. U-tube exchanger with bonnet-type stationary head, split-flow shell, 483-mm (19-in) inside diameter with tubes 21-m (7-ft) straight length. SIZE 19–84 TYPE GBU.

3. Pull-through floating-heat-kettle type of reboiler having stationary head integral with tube sheet, 584-mm (23-in) port diameter and 940-mm (37-in) inside shell diameter with tubes 4.9-m (16-ft) long. SIZE 23/37–192 TYPE CKT.

4. Fixed-tube sheet exchanger with removable channel and cover, bonnet-type rear head, two-pass shell, 841-mm (33⅛-in) diameter with tubes 2.4 m (8 ft) long. SIZE 33–96 TYPE AFM.

5. Fixed-tube sheet exchanger having stationary and rear heads integral with tube sheets, single-pass shell, 432-mm (17-in) inside diameter with tubes 4.9 m (16 ft) long. SIZE 17–192 TYPE CEN.

Functional Definitions Heat-transfer equipment can be designated by type (e.g., fixed tube sheet, outside packed head, etc.) or by function (chiller, condenser, cooler, etc.). Almost any type of unit can be used to perform any of or all the listed functions. Many of these terms have been defined by Donahue [*Pet. Process.* 103 (March 1956)].

Equipment	Function
Chiller	Cools a fluid to a temperature below that obtainable if only water were used as a coolant. It uses a refrigerant such as ammonia or Freon.
Condenser	Condenses a vapor or mixture of vapors, either alone or in the presence of a noncondensible gas.
Partial condenser	Condenses vapors at a point high enough to provide a temperature difference sufficient to preheat a cold stream of process fluid. This saves heat and eliminates the need for providing a separate preheater (using flame or steam).
Final condenser	Condenses the vapors to a final storage temperature of approximately 37.8°C (100°F). It uses water cooling, which means that the transferred heat is lost to the process.
Cooler	Cools liquids or gases by means of water.
Exchanger	Performs a double function: (1) heats a cold fluid by (2) using a hot fluid which it cools. None of the transferred heat is lost.
Heater	Imparts sensible heat to a liquid or a gas by means of condensing steam or Dowtherm.
Reboiler	Connected to the bottom of a fractionating tower, it provides the reboil heat necessary for distillation. The heating medium may be either steam or a hot-process fluid.
Thermosiphon reboiler	Natural circulation of the boiling medium is obtained by maintaining sufficient liquid head to provide for circulation.
Forced-circulation reboiler	A pump is used to force liquid through the reboiler.
Steam generator	Generates steam for use elsewhere in the plant by using the available high-level heat in tar or a heavy oil.
Superheater	Heats a vapor above the saturation temperature.
Vaporizer	A heater that vaporizes part of the liquid.
Waste-heat boiler	Produces steam; similar to steam generator, except that the heating medium is a hot gas or liquid produced in a chemical reaction.

GENERAL DESIGN CONSIDERATIONS

Selection of Flow Path In selecting the flow path for two fluids through an exchanger, several general approaches are used. The tube-side fluid is more corrosive or dirtier or at a higher pressure. The shell-side fluid is a liquid of high viscosity or a gas.

When alloy construction for one of the two fluids is required, a carbon-steel shell combined with alloy tube-side parts is less expensive than alloy in contact with the shell-side fluid combined with carbon-steel headers.

Cleaning of the inside of tubes is more readily done than cleaning of exterior surfaces.

For gauge pressures in excess of 2068 kPa (300 lbf/in^2) for one of the fluids, the less expensive construction has the high-pressure fluid in the tubes.

For a given pressure drop, higher heat-transfer coefficients are obtained on the shell side than on the tube side.

Heat exchanger shutdowns are most often caused by fouling, corrosion, and erosion.

Construction Codes "Rules for Construction of Pressure Vessels, Division 1," which is part of Section VIII of the ASME Boiler and Pressure Vessel Code (American Society of Mechanical Engineers), serves as a construction code by providing minimum standards. New editions of the code are usually issued every 3 years. Interim revisions are made semiannually in the form of addenda. Compliance with ASME Code requirements is mandatory in much of the United States and Canada. Originally these rules were not prepared for heat exchangers. However, the welded joint between tube sheet and shell of the fixed-tube-sheet heat exchanger is now included. A nonmandatory appendix on from the tube to tube-sheet joints is also included. Additional rules for heat exchangers are being developed.

Standards of Tubular Exchanger Manufacturers Association, 6th ed., 1978 (commonly referred to as the TEMA Standards), serve to supplement and define the ASME Code for all shell-and-tube type of heat exchanger applications (other than double-pipe construction). TEMA Class R design is "for the generally severe requirements of petroleum and related processing applications. Equipment fabricated in accordance with these standards is designed for safety and durability under the rigorous service and maintenance conditions in such applications." TEMA Class C design is "for the generally moderate requirements of commercial and general process applications," while TEMA Class B is "for chemical process service."

The mechanical design requirements are identical for all three classes of construction. The differences between the TEMA classes are minor and were listed by Rubin [*Hydrocarbon Process* **59:** 92 (June 1980)].

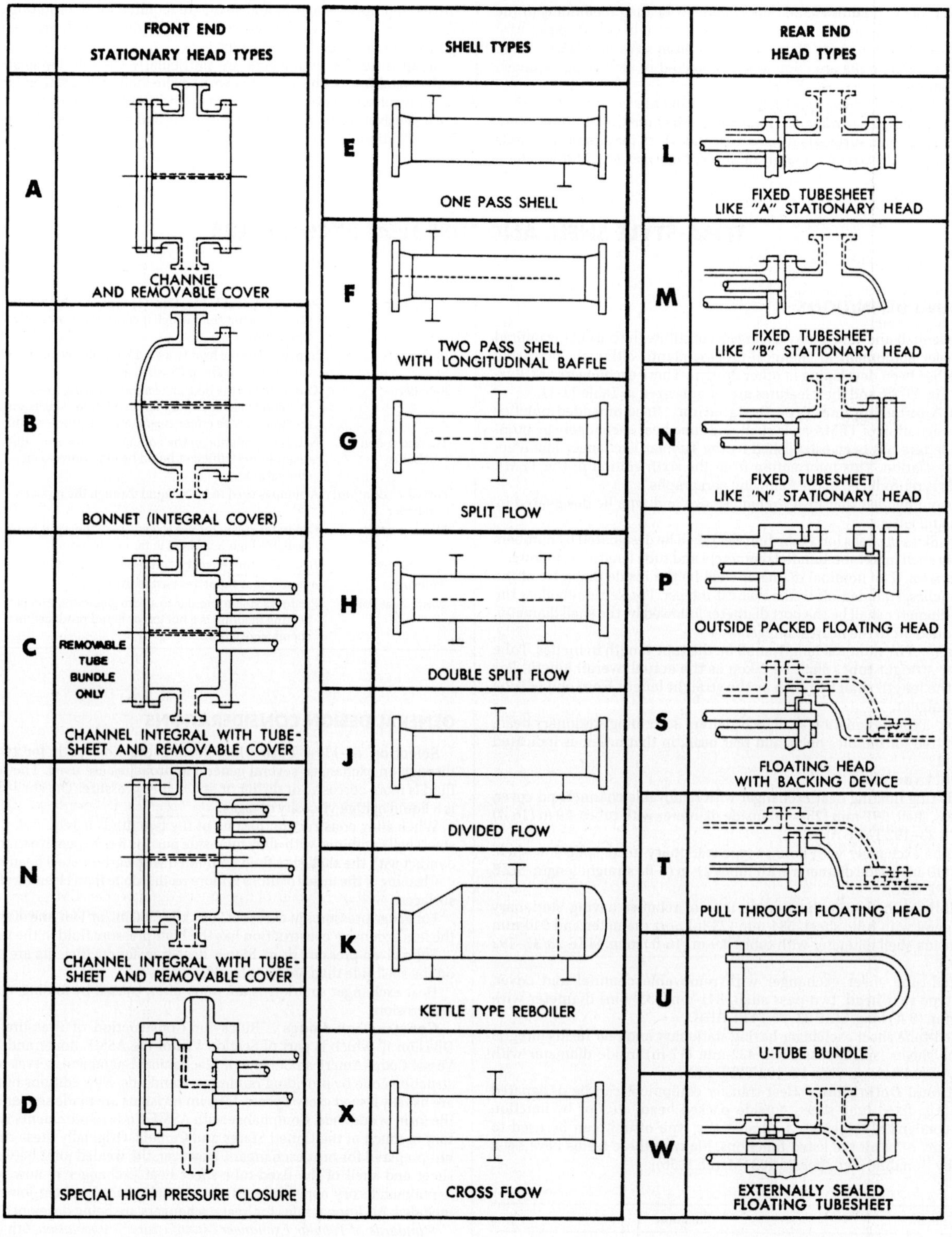

FIG. 11-35 TEMA-type designations for shell-and-tube heat exchangers. (*Standards of Tubular Exchanger Manufacturers Association, 6th ed., 1978.*)

TABLE 11-11 Features of TEMA Shell-and-Tube-Type Exchangers*

Type of design	Fixed tube sheet	U-tube	Packed lantern-ring floating head	Internal floating head (split backing ring)	Outside-packed floating head	Pull-through floating head
T.E.M.A. rear-head type	L or M or N	U	W	S	P	T
Relative cost increases from A (least expensive) through E (most expensive)	B	A	C	E	D	E
Provision for differential expansion	Expansion joint in shell	Individual tubes free to expand	Floating head	Floating head	Floating head	Floating head
Removable bundle	No	Yes	Yes	Yes	Yes	Yes
Replacement bundle possible	No	Yes	Yes	Yes	Yes	Yes
Individual tubes replaceable	Yes	Only those in outside row†	Yes	Yes	Yes	Yes
Tube cleaning by chemicals inside and outside	Yes	Yes	Yes	Yes	Yes	Yes
Interior tube cleaning mechanically	Yes	Special tools required	Yes	Yes	Yes	Yes
Exterior tube cleaning mechanically:						
Triangular pitch	No	No‡	No‡	No‡	No‡	No‡
Square pitch	No	Yes	Yes	Yes	Yes	Yes
Hydraulic-jet cleaning:						
Tube interior	Yes	Special tools required	Yes	Yes	Yes	Yes
Tube exterior	No	Yes	Yes	Yes	Yes	Yes
Double tube sheet feasible	Yes	Yes	No	No	Yes	No
Number of tube passes	No practical limitations	Any even number possible	Limited to one or two passes	No practical limitations§	No practical limitations	No practical limitations§
Internal gaskets eliminated	Yes	Yes	Yes	No	Yes	No

NOTE: Relative costs A and B are not significantly different and interchange for long lengths of tubing.
*Modified from page a-8 of the Patterson-Kelley Co. Manual No. 700A, Heat Exchangers.
†U-tube bundles have been built with tube supports which permit the U-bends to be spread apart and tubes inside of the bundle replaced.
‡Normal triangular pitch does not permit mechanical cleaning. With a wide triangular pitch, which is equal to 2 (tube diameter plus cleaning lane)/$\sqrt{3}$, mechanical cleaning is possible on removable bundles. This wide spacing is infrequently used.
§For odd number of tube side passes, floating head requires packed joint or expansion joint.

Among the topics of the TEMA Standards are nomenclature, fabrication tolerances, inspection, guarantees, tubes, shells, baffles and support plates, floating heads, gaskets, tube sheets, channels, nozzles, end flanges and bolting, material specifications, and fouling resistances.

Shell and Tube Heat Exchangers for General Refinery Services, API Standard 660, 4th ed., 1982, is published by the American Petroleum Institute to supplement both the TEMA Standards and the ASME Code. Many companies in the chemical and petroleum processing fields have their own standards to supplement these various requirements. *The Interrelationships between Codes, Standards, and Customer Specifications for Process Heat Transfer Equipment* is a symposium volume which was edited by F. L. Rubin and published by ASME in December 1979. (See discussion of pressure vessel codes in Sec. 6.)

Design pressures and temperatures for exchangers usually are specified with a margin of safety beyond the conditions expected in service. Design pressure is generally about 172 kPa (25 lbf/in^2) greater than the maximum expected during operation or at pump shutoff. Design temperature is commonly 14°C (25°F) greater than the maximum temperature in service.

Tube Bundle Vibration Damage from tube vibration has become an increasing problem as plate baffled heat exchangers are designed for higher flow rates and pressure drops. The most effective method of dealing with this problem is the avoidance of cross-flow by use of tube support baffles which promote only longitudinal flow. However, even then, strict attention must be paid to the bundle area under the shell inlet nozzle where flow is introduced through the side of the shell. TEMA has devoted an entire section in its standards to this topic. In general, the mechanisms of tube vibration are as follows:

Vortex Shedding The vortex-shedding frequency of the fluid in cross-flow over the tubes may coincide with a natural frequency of the tubes and excite large resonant vibration amplitudes.

Fluid-Elastic Coupling Fluid flowing over tubes causes them to vibrate with a whirling motion. The mechanism of fluid-elastic coupling occurs when a "critical" velocity is exceeded and the vibration then becomes self-excited and grows in amplitude. This mechanism frequently occurs in process heat exchangers which suffer vibration damage.

Pressure Fluctuation Turbulent pressure fluctuations which develop in the wake of a cylinder or are carried to the cylinder from upstream may provide a potential mechanism for tube vibration. The tubes respond to the portion of the energy spectrum that is close to their natural frequency.

Acoustic Coupling When the shell-side fluid is a low-density gas, acoustic resonance or coupling develops when the standing waves in the shell are in phase with vortex shedding from the tubes. The standing waves are perpendicular to the axis of the tubes and to the direction of cross-flow. Damage to the tubes is rare. However, the noise can be extremely painful.

Testing Upon completion of shop fabrication and also during maintenance operations, it is desirable hydrostatically to test the shell side of tubular exchangers so that visual examination of tube ends can be made. Leaking tubes can be readily located and serviced. When leaks are determined without access to the tube ends, it is necessary to reroll or reweld all the tube–tube-sheet joints with possible damage to the satisfactory joints.

Testing for leaks in heat exchangers was discussed by Rubin [*Chem. Eng.* **68:** 160–166 (July 24, 1961)].

Performance testing of heat exchangers is described in the American Institute of Chemical Engineers' *Standard Testing Procedure for Heat Exchangers*, Sec. 1, "Sensible Heat Transfer in Shell-and-Tube-Type Equipment."

PRINCIPAL TYPES OF CONSTRUCTION

Figure 11-36 shows details of the construction of the TEMA types of shell-and-tube heat exchangers. These and other types are discussed in the following paragraphs.

Fixed-Tube-Sheet Heat Exchangers Fixed-tube-sheet exchangers (Fig. 11-36*b*) are used more often than any other type, and the frequency of use has been increasing in recent years. The tubesheets are welded to the shell. Usually these extend beyond the shell and serve as flanges to which the tube-side headers are bolted. This construction requires that the shell-and-tube-sheet materials be weldable to each other.

When such welding is not possible, a "blind"-gasket type of construction is utilized. The blind gasket is not accessible for maintenance or replacement once the unit has been constructed. This construction is used for steam surface condensers, which operate under vacuum.

The tube-side header (or channel) may be welded to the tubesheet, as shown in Fig. 11-35 for type C and N heads. This type of construction is less costly than types B and M or A and L and still offers the advantage that tubes may be examined and replaced without disturbing the tube-side piping connections.

There is no limitation on the number of tube-side passes. Shell-side passes can number one or more, although shells with more than two shell-side passes are rarely used.

Tubes can completely fill the heat exchanger shell. Clearance between the outermost tubes and the shell is only the minimum necessary for fabrication. Between the inside of the shell and the baffles some clearance must be provided so that baffles can slide into the shell. Fabrication tolerances then require some additional clearance between the outside of the baffles and the outermost tubes. The edge distance between the *outer tube limit* (OTL) and the baffle diameter must be sufficient to prevent vibration of the tubes from breaking through the baffle holes. The outermost tube must be contained within the OTL. Clearances between the inside shell

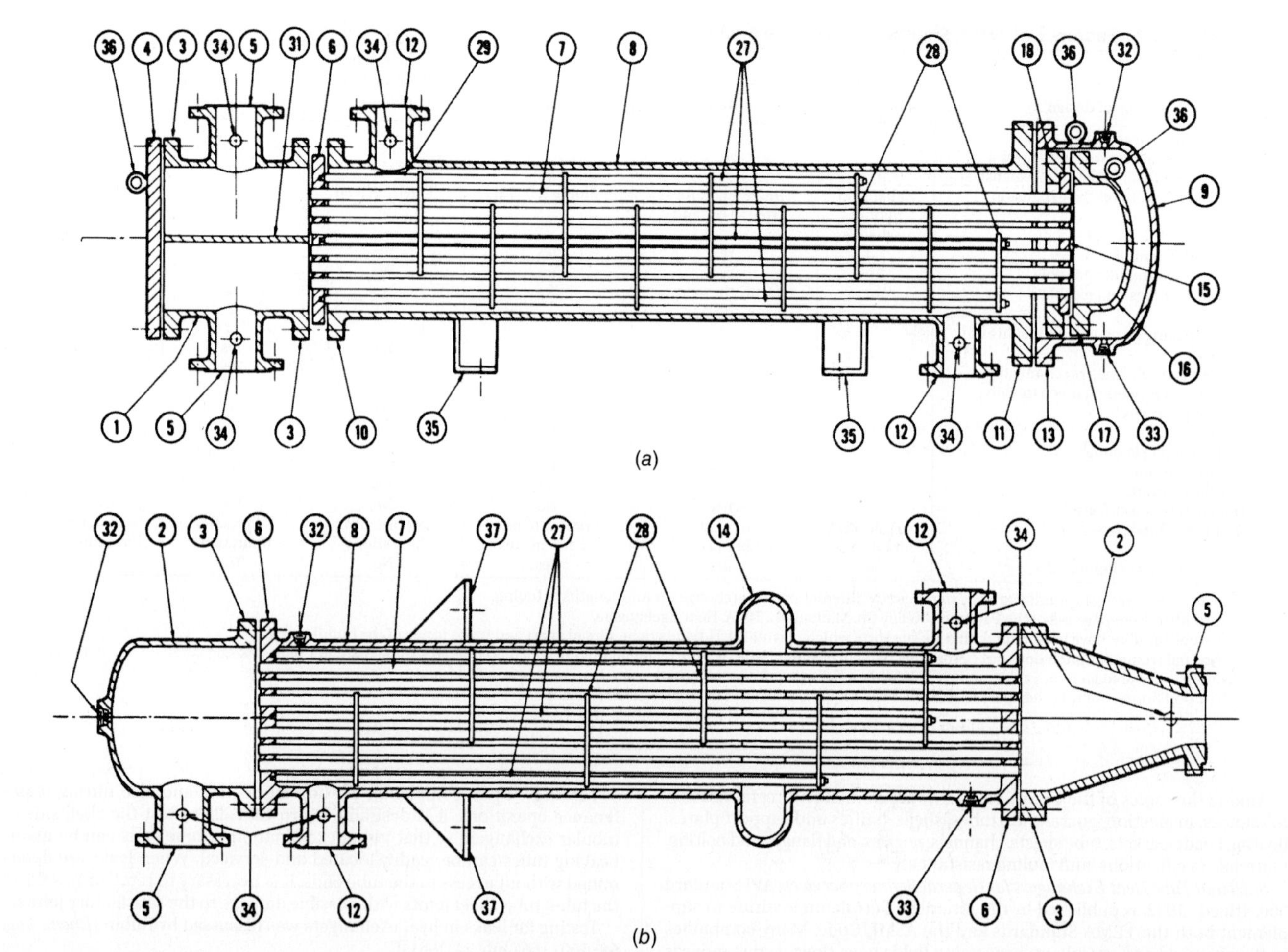

FIG. 11-36 Heat exchanger component nomenclature. (*a*) Internal-floating-head exchanger (with floating-head backing device). Type AES. (*b*) Fixed-tube-sheet exchanger. Type BEM. (*Standards of the Tubular Exchanger Manufacturers Association, 6th ed., 1978.*)

diameter and OTL are 13 mm (½ in) for 635 mm (25 in) inside-diameter (ID) shells and up, 11 mm (7/16 in) for 254 through 610 mm (10 through 24 in) pipe shells, and slightly less for smaller-diameter pipe shells.

Tubes can be replaced. Tube-side headers, channel covers, gaskets, etc., are accessible for maintenance and replacement. Neither the shell-side baffle structure nor the blind gasket is accessible. During tube removal, a tube may break within the shell. When this occurs, it is most difficult to remove or to replace the tube. The usual procedure is to plug the appropriate holes in the tubesheets.

Differential expansion between the shell and the tubes can develop because of differences in length caused by thermal expansion. Various types of expansion joints are used to eliminate excessive stresses caused by expansion. The need for an expansion joint is a function of both the amount of differential expansion and the cycling conditions to be expected during operation. A number of types of expansion joints are available (Fig. 11-37).

1. *Flat plates.* Two concentric flat plates with a bar at the outer edges. The flat plates can flex to make some allowance for differential expansion. This design is generally used for vacuum service and gauge pressures below 103 kPa (15 lbf/in^2). All are subject to severe stress during differential expansion.

2. *Flanged-only heads.* The flat plates are flanged (or curved). The diameter of these heads is generally 203 mm (8 in) or greater than the shell diameter. The welded joint at the shell is subject to the stress referred to before, but the joint connecting the heads is subjected to less stress during expansion because of the curved shape.

3. *Flared shell or pipe segments.* The shell may be flared to connect with a pipe section, or a pipe may be halved and quartered to produce a ring.

4. *Formed heads.* A pair of dished-only or elliptical or flanged and dished heads can be used. These are welded together or connected by a ring. This type of joint is similar to the flanged-only-head type but apparently is subject to less stress.

5. *Flanged and flued heads.* A pair of flanged-only heads is provided with concentric reverse flue holes. These heads are relatively expensive because of the cost of the flue operation. The curved shape of the heads reduces the amount of stress at the welds to the shell and also connecting the heads.

6. *Toroidal.* The toroidal joint has a mathematically predictable smooth stress pattern of low magnitude, with maximum stresses at sidewalls of the corrugation and minimum stresses at top and bottom.

The foregoing designs were discussed as ring expansion joints by Kopp and Sayre, "Expansion Joints for Heat Exchangers" (ASME Misc. Pap., **6:** 211). All are statically indeterminate but are subjected to analysis by introducing various simplifying assumptions. Some joints in current industrial use are of lighter wall construction than is indicated by the method of this paper.

7. *Bellows.* Thin-wall bellows joints are produced by various manufacturers. These are designed for differential expansion and are tested for axial and transverse movement as well as for cyclical life. Bellows may be of stainless steel, nickel alloys, or copper. (Aluminum, Monel, phosphor bronze, and titanium bellows have been manufactured.) Welding nipples of the same composition as the heat exchanger shell are generally furnished. The bellows may be hydraulically formed from a single piece of metal or may consist of welded pieces. External insulation covers of carbon steel are often provided to protect the light-gauge bellows from damage. The cover also prevents insulation from interfering with movement of the bellows (see item 8).

8. *Toroidal bellows.* For high-pressure service, the bellows type of joint has been modified so that movement is taken up by thin-wall small-diameter bellows of a toroidal shape. The thickness of parts under high pressure is reduced considerably (see item 6).

Improper handling during manufacture, transit, installation, or maintenance of the heat exchanger equipped with the thin-wall-bellows type or toroidal type of expansion joint can damage the joint. In larger units these light-wall joints are particularly susceptible to damage, and some designers prefer the use of the heavier walls of formed heads.

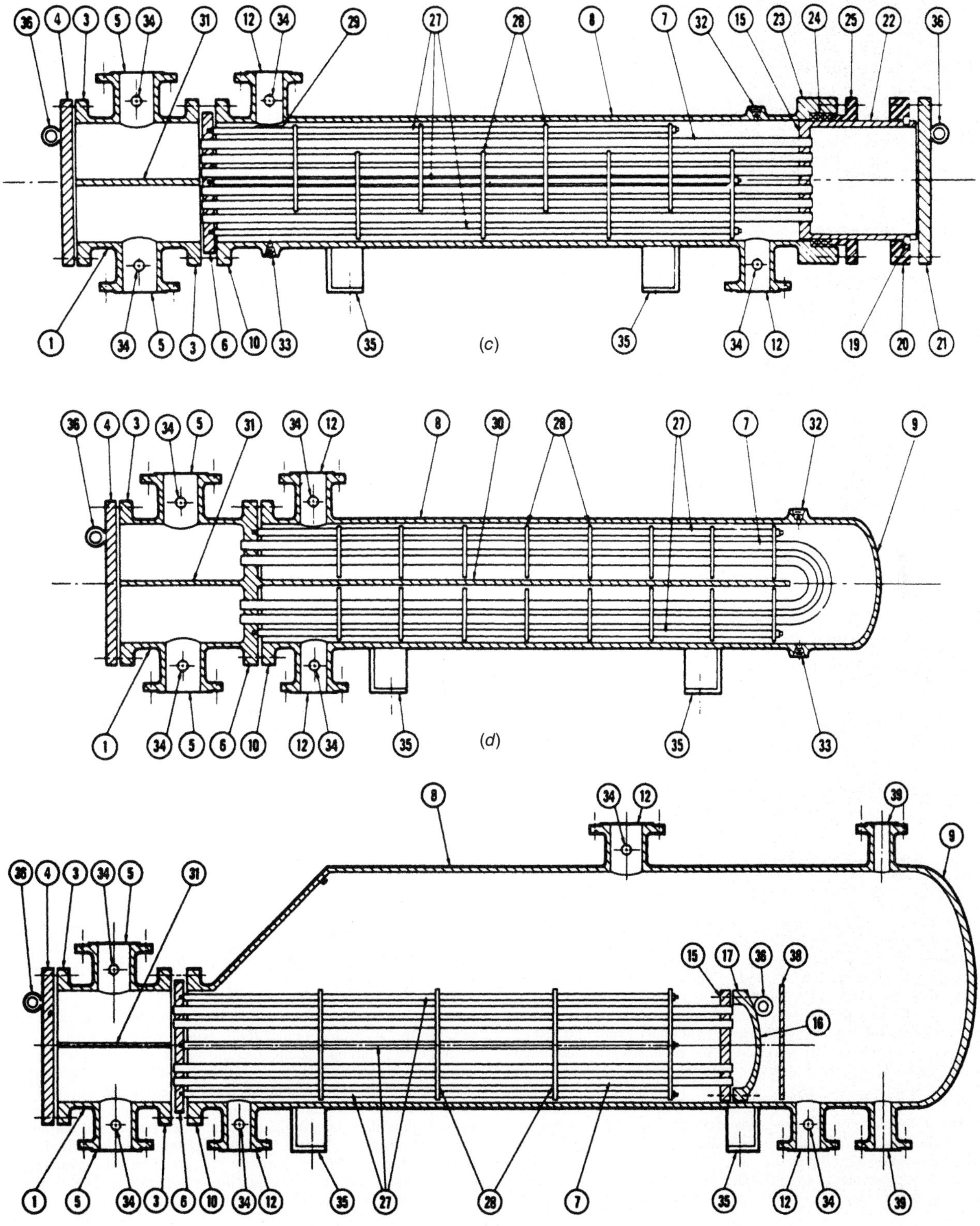

FIG. 11-36 (*Continued*) Heat exchanger component nomenclature. (*c*) Outside packed floating-head exchanger. Type AEP. (*d*) U-tube heat exchanger. Type CFU. (*e*) Kettle-type floating-head reboiler. Type AKT. (*Standards of the Tubular Exchanger Manufacturers Association, 6th ed., 1978.*)

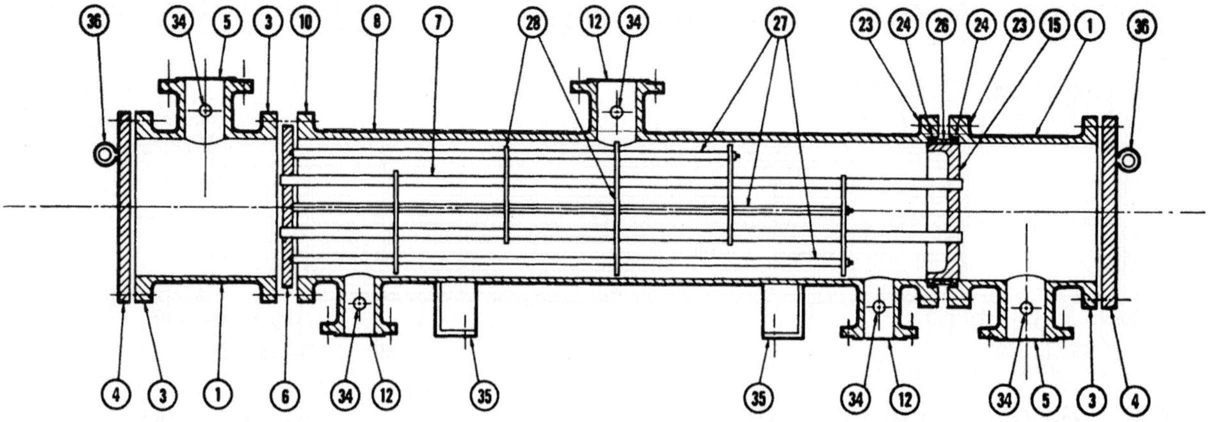

(*f*)

1. Stationary Head—Channel
2. Stationary Head—Bonnet
3. Stationary Head Flange—Channel or Bonnet
4. Channel Cover
5. Stationary Head Nozzle
6. Stationary Tubesheet
7. Tubes
8. Shell
9. Shell Cover
10. Shell Flange—Stationary Head End
11. Shell Flange—Rear Head End
12. Shell Nozzle
13. Shell Cover Flange
14. Expansion Joint
15. Floating Tubesheet
16. Floating Head Cover
17. Floating Head Flange
18. Floating Head Backing Device
19. Split Shear Ring
20. Slip-on Backing Flange
21. Floating Head Cover—External
22. Floating Tubesheet Skirt
23. Packing Box Flange
24. Packing
25. Packing Gland
26. Lantern Ring
27. Tie Rods and Spacers
28. Transverse Baffles or Support Plates
29. Impingement Plate
30. Longitudinal Baffle
31. Pass Partition
32. Vent Connection
33. Drain Connection
34. Instrument Connection
35. Support Saddle
36. Lifting Lug
37. Support Bracket
38. Weir
39. Liquid Level Connection

FIG. 11-36 (*Continued*) Heat exchanger component nomenclature. (*f*) Exchanger with packed floating tube sheet and lantern ring. Type AJW. (*Standards of the Tubular Exchanger Manufacturers Association, 6th ed., 1978.*)

Chemical-plant exchangers requiring expansion joints most commonly have used the flanged-and flued-head type. There is a trend toward more common use of the light-wall bellows type.

U-Tube Heat Exchanger (Fig. 11-36*d*) The tube bundle consists of a stationary tubesheet, U tubes (or hairpin tubes), baffles or support plates, and appropriate tie rods and spacers. The tube bundle can be removed from the heat exchanger shell. A tube-side header (stationary head) and a shell with integral shell cover, which is welded to the shell, are provided. Each tube is free to expand or contract without any limitation being placed upon it by the other tubes.

The U-tube bundle has the advantage of providing minimum clearance between the outer tube limit and the inside of the shell for any of the removable tube bundle constructions. Clearances are of the same magnitude as for fixed-tube-sheet heat exchangers.

The number of tube holes in a given shell is less than that for a fixed-tube-sheet exchanger because of limitations on bending tubes of a very short radius.

The U-tube design offers the advantage of reducing the number of joints. In high-pressure construction this feature becomes of considerable importance in reducing both initial and maintenance costs. The use of U-tube construction has increased significantly with the development of hydraulic tube cleaners, which can remove fouling residues from both the straight and the U-bend portions of the tubes.

Mechanical cleaning of the inside of the tubes was described by John [*Chem. Eng.* **66:** 187–192 (Dec. 14, 1959)]. Rods and conventional mechanical tube cleaners cannot pass from one end of the U tube to the other. Power-driven tube cleaners, which can clean both the straight legs of the tubes and the bends, are available.

Hydraulic jetting with water forced through spray nozzles at high pressure for cleaning tube interiors and exteriors of removal bundles is reported by Canaday ("Hydraulic Jetting Tools for Cleaning Heat Exchangers," ASME Pap. 58-A-217, unpublished).

The tank suction heater, as illustrated in Fig. 11-38, contains a U-tube bundle. This design is often used with outdoor storage tanks for heavy fuel oils, tar, molasses, and similar fluids whose viscosity must be lowered to permit easy pumping. Usually the tube-side heating medium is steam. One end of the heater shell is open, and the liquid being heated passes across the outside of the tubes. Pumping costs can be reduced without heating the entire contents of the tank. Bare-tube and integral low-fin tubes are provided with baffles. Longitudinal fin-tube heaters are not baffled. Fins are most often used to minimize the fouling potential in these fluids.

Kettle-type reboilers, evaporators, etc., are often U-tube exchangers with enlarged shell sections for vapor-liquid separation. The U-tube bundle replaces the floating-heat bundle of Fig. 11-36*e*.

The U-tube exchanger with copper tubes, cast-iron header, and other parts of carbon steel is used for water and steam services in office buildings, schools, hospitals, hotels, etc. Nonferrous tubesheets and admiralty or 90-10 copper-nickel tubes are the most frequently used substitute materials. These standard exchangers are available from a number of manufacturers at costs far below those of custom-built process industry equipment.

Packed Lantern-Ring Exchanger (Fig. 11-36*f*) This construction is the least costly of the straight-tube removable bundle types. The shell- and tube-side fluids are each contained by separate rings of packing separated by a lantern ring and are installed at the floating tubesheet. The lantern ring is provided with weep holes. Any leakage passing the packing goes through

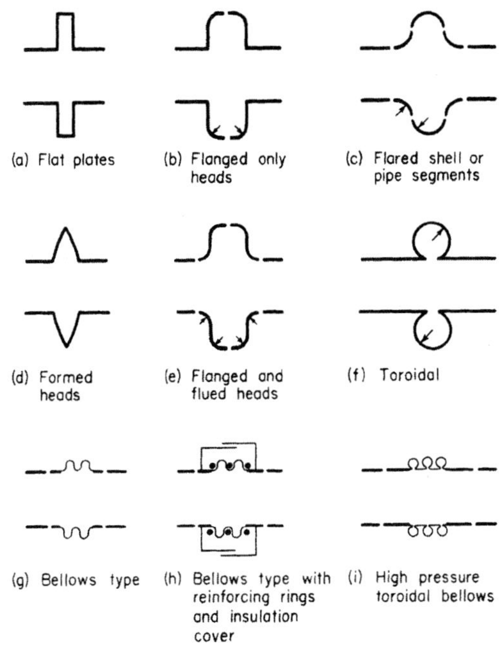

FIG. 11-37 Expansion joints.

the weep holes and then drops to the ground. Leakage at the packing will not result in mixing within the exchanger of the two fluids.

The width of the floating tubesheet must be great enough to allow for the packings, the lantern ring, and differential expansion. Sometimes a small skirt is attached to a thin tubesheet to provide the required bearing surface for packings and the lantern ring.

The clearance between the outer tube limit and the inside of the shell is slightly larger than that for fixed-tube-sheet and U-tube exchangers. The use of a floating-tube-sheet skirt increases this clearance. Without the skirt the clearance must make allowance for tube-hole distortion during tube rolling near the outside edge of the tubesheet or for tube-end welding at the floating tubesheet.

The packed lantern ring construction is generally limited to design temperatures below 191°C (375°F) and to the mild services of water, steam, air, lubricating oil, etc. Design gauge pressure does not exceed 2068 kPa (300 lbf/in^2) for pipe shell exchangers and is limited to 1034 kPa (150 lbf/in^2) for 610- to 1067-mm- (24- to 42-in-) diameter shells.

Outside Packed Floating-Head Exchanger (Fig. 11-36*c*) The shell-side fluid is contained by rings of packing, which are compressed within a stuffing box by a packing follower ring. This construction was frequently used in the chemical industry, but in recent years usage has decreased. The removable-bundle construction accommodates differential expansion between shell and tubes and is used for shell-side service up to 4137 kPa gauge pressure (600 lbf/in^2) at 316°C (600°F). There are no limitations upon the number of tube-side passes or upon the tube-side design pressure and

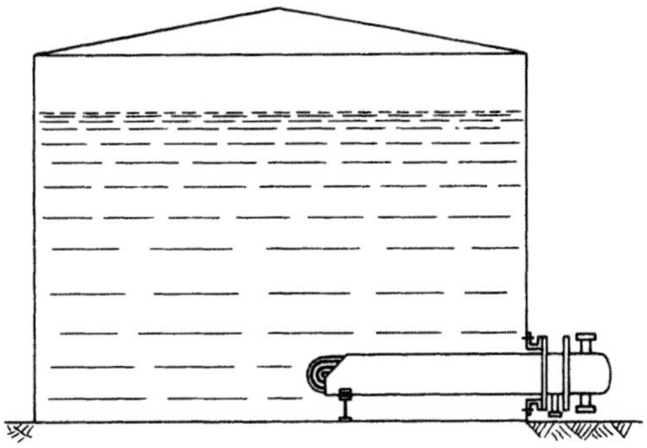

FIG. 11-38 Tank suction heater.

temperature. The outside packed floating-head exchanger was the most commonly used type of removable-bundle construction in chemical-plant service.

The floating-tube-sheet skirt, where in contact with the rings of packing, has a fine machine finish. A split shear ring is inserted into a groove in the floating-tube-sheet skirt. A slip-on backing flange, which in service is held in place by the shear ring, bolts to the external floating-head cover.

The floating-head cover is usually a circular disk. With an odd number of tube-side passes, an axial nozzle can be installed in such a floating-head cover. If a side nozzle is required, the circular disk is replaced by either a dished head or a channel barrel (similar to Fig. 11-36*f*) bolted between floating-head cover and floating-tube-sheet skirt.

The outer tube limit approaches the inside of the skirt but is farther removed from the inside of the shell than for any of the previously discussed constructions. Clearances between shell diameter and bundle OTL are 22 mm (⅞ in) for small-diameter pipe shells, 44 mm (1¾ in) for large-diameter pipe shells, and 58 mm (2$^1/_{16}$ in) for moderate-diameter plate shells.

Internal Floating-Head Exchanger (Fig. 11-36*a*) The internal floating-head design is used extensively in petroleum refinery service, but in recent years there has been a decline in usage.

The tube bundle is removable, and the floating tubesheet moves (or floats) to accommodate differential expansion between shell and tubes. The outer tube limit approaches the inside diameter of the gasket at the floating tubesheet. Clearances (between shell and OTL) are 29 mm (1⅛ in) for pipe shells and 37 mm (1$^7/_{16}$ in) for moderate-diameter plate shells.

A split backing ring and bolting usually hold the floating-head cover at the floating tubesheet. These are located beyond the end of the shell and within the larger-diameter shell cover. The shell cover, split backing ring, and floating-head cover must be removed before the tube bundle can pass through the exchanger shell.

With an even number of tube-side passes, the floating-head cover serves as return cover for the tube-side fluid. With an odd number of passes, a nozzle pipe must extend from the floating-head cover through the shell cover. Provision for both differential expansion and tube bundle removal must be made.

Pull-Through Floating-Head Exchanger (Fig. 11-36*e*) Construction is similar to that of the internal floating-head split-backing-ring exchanger except that the floating-head cover bolts directly to the floating tubesheet. The tube bundle can be withdrawn from the shell without removing either shell cover or floating-head cover. This feature reduces maintenance time during inspection and repair.

The large clearance between the tubes and the shell must provide for both the gasket and the bolting at the floating-head cover. This clearance is about 2 to 2½ times that required by the split-ring design. Sealing strips or dummy tubes are often installed to reduce bypassing of the tube bundle.

Falling-Film Exchangers Falling-film shell-and-tube heat exchangers have been developed for a wide variety of services and are described by Sack [*Chem. Eng. Prog.* **63:** 55 (July 1967)]. The fluid enters at the top of the vertical tubes. Distributors or slotted tubes put the liquid in film flow in the inside surface of the tubes, and the film adheres to the tube surface while falling to the bottom of the tubes. The film can be cooled, heated, evaporated, or frozen by means of the proper heat-transfer medium outside the tubes. Tube distributors have been developed for a wide range of applications. Fixed tubesheets, with or without expansion joints, and outside packed head designs are used.

Principal advantages are the high rate of heat transfer, no internal pressure drop, short time of contact (very important for heat-sensitive materials), easy accessibility to tubes for cleaning, and, in some cases, prevention of leakage from one side to another.

These falling-film exchangers are used in various services as described in the following paragraphs.

Liquid Coolers and Condensers Dirty water can be used as the cooling medium. The top of the cooler is open to the atmosphere for access to tubes. These can be cleaned without shutting down the cooler by removing the distributors one at a time and scrubbing the tubes.

Evaporators These are used extensively for the concentration of ammonium nitrate, urea, and other chemicals sensitive to heat when minimum contact time is desirable. Air is sometimes introduced in the tubes to lower the partial pressure of liquids whose boiling points are high. These evaporators are built for pressure or vacuum and with top or bottom vapor removal.

Absorbers These have a two-phase flow system. The absorbing medium is put in film flow during its fall downward on the tubes as it is cooled by a cooling medium outside the tubes. The film absorbs the gas that is introduced into the tubes. This operation can be cocurrent or countercurrent.

Freezers By cooling the falling film to its freezing point, these exchangers convert a variety of chemicals to the solid phase. The most common application is the production of sized ice and paradichlorobenzene. Selective freezing is used for isolating isomers. By melting the solid material and refreezing in several stages, a higher degree of purity of product can be obtained.

TUBE-SIDE CONSTRUCTION

Tube-Side Header The tube-side header (or stationary head) contains one or more flow nozzles.

The *bonnet* (Fig. 11-35B) bolts to the shell. It is necessary to remove the bonnet to examine the tube ends. The fixed-tube-sheet exchanger of Fig. 11-36*b* has bonnets at both ends of the shell.

The *channel* (Fig. 11-35A) has a removable channel cover. The tube ends can be examined by removing this cover without disturbing the piping connections to the channel nozzles. The channel can bolt to the shell as shown in Fig. 11-36*a* and *c*. Type C and type N channels of Fig. 11-35 are welded to the tubesheet. This design is comparable in cost with the bonnet but has the advantages of permitting access to the tubes without disturbing the piping connections and of eliminating a gasketed joint.

Special High-Pressure Closures (Fig. 11-35D) The channel barrel and the tubesheet are generally forged. The removable channel cover is seated in place by hydrostatic pressure, while a shear ring subjected to shearing stress absorbs the end force. For pressures above 6205 kPa (900 lbf/in^2) these designs are generally more economical than bolted constructions, which require larger flanges and bolting as pressure increases in order to contain the end force with bolts in tension. Relatively light-gauge internal pass partitions are provided to direct the flow of tube-side fluids but are designed only for the differential pressure across the tube bundle.

Tube-Side Passes Most exchangers have an even number of tube-side passes. The fixed-tube-sheet exchanger (which has no shell cover) usually has a return cover without any flow nozzles, as shown in Fig. 11-35M; types L and N are also used. All removable-bundle designs (except for the U tube) have a floating-head cover directing the flow of tube-side fluid at the floating tubesheet.

Tubes Standard heat exchanger tubing has ¼-, ⅜-, ½-, ⅝-, ¾-, 1-, 1¼-, and 1½-in outside diameter (in × 25.4 = mm). Wall thickness is measured in *Birmingham wire gauge* (BWG) units. A comprehensive list of tubing characteristics and sizes is given in Table 11-12. The most commonly used tubes in chemical plants and petroleum refineries are 19- and 25-mm (¾- and 1-in) outside diameter (OD). Standard tube lengths are 8, 10, 12, 16, and 20 ft, with 20 ft now the most common (ft × 0.3048 = m).

Manufacturing tolerances for steel, stainless-steel, and nickel-alloy tubes are such that the tubing is produced to either average or minimum wall thickness. Seamless carbon-steel tube of minimum wall thickness may vary from 0 to 20 percent above the nominal wall thickness. Average-wall seamless tubing has an allowable variation of ±10 percent. Welded carbon-steel tube is produced to closer tolerances (0 to +18 percent on minimum wall; ±9 percent on average wall). Tubing of aluminum, copper, and their alloys can be drawn easily and usually is made to minimum wall specifications.

Common practice is to specify the *exchanger surface* in terms of total external square feet of tubing. The effective outside heat-transfer surface is based on the length of tubes measured between the inner faces of tubesheets. In most heat exchangers, there is little difference between the total and the effective surface. Significant differences are usually found in high-pressure and double-tube-sheet designs.

Integrally finned tube, which is available in a variety of alloys and sizes, is being used in shell-and-tube heat exchangers. The fins are radially extruded from thick-walled tube to a height of 1.6 mm (1/16 in) spaced at 1.33 mm (19 fins per inch) or to a height of 3.2 mm (⅛ in) spaced at 2.3 mm (11 fins per inch). The external surface is approximately 2½ times the outside surface of a bare tube with the same outside diameter. Also available are 0.93-mm-high (0.037-in-high) fins spaced 0.91 mm (28 fins per inch) with an external surface about 3.5 times the surface of the bare tube. Bare ends of nominal tube diameter are provided, while the fin height is slightly less than this diameter. The tube can be inserted into a conventional tube bundle and rolled or welded to the tubesheet by the same means used for bare tubes. An integrally finned tube rolled into a tubesheet with double serrations and flared at the inlet is shown in Fig. 11-39. Internally finned tubes have been manufactured but have limited application.

Longitudinal fins are commonly used in double-pipe exchangers upon the outside of the inner tube. U-tube and conventional removable tube bundles are also made from such tubing. The ratio of external to internal surface generally is about 10:1 or 15:1.

Transverse fins upon tubes are used in low-pressure gas services. The primary application is in air-cooled heat exchangers (as discussed under that heading), but shell-and-tube exchangers with these tubes are in service.

Rolled Tube Joints Expanded tube—tube-sheet joints are standard. Properly rolled joints have uniform tightness to minimize tube fractures, stress corrosion, tube-sheet ligament pushover and enlargement, and dishing of the tubesheet. Tubes are expanded into the tubesheet for a length of two tube diameters, or 50 mm (2 in), or tube-sheet thickness minus 3 mm (⅛ in). Generally tubes are rolled for the last of these alternatives. The expanded portion should never extend beyond the shell-side face of the tubesheet, since removing such a tube is extremely difficult. Methods and tools for tube removal and tube rolling were discussed by John [*Chem. Eng.* **66:** 77–80 (Dec. 28, 1959)], and rolling techniques by Bach [*Pet. Refiner* **39:** 8, 104 (1960)].

Tube ends may be projecting, flush, flared, or beaded (listed in order of usage). The flare or bell-mouth tube end is usually restricted to water service in condensers and serves to reduce erosion near the tube inlet.

For moderate general process requirements at gauge pressures less than 2058 kPa (300 lbf/in^2) and less than 177°C (350°F), tube-sheet holes without grooves are standard. For all other services with expanded tubes, at least two grooves in each tube hole are common. The number of grooves is sometimes changed to one or three in proportion to tube-sheet thickness.

Expanding the tube into the *grooved tube holes* provides a stronger joint but results in greater difficulties during tube removal.

Welded Tube Joints When suitable materials of construction are used, the tube ends may be welded to the tubesheets. Welded joints may be seal-welded "for additional tightness beyond that of tube rolling" or may be strength-welded. Strength-welded joints have been found satisfactory in very severe services. Welded joints may or may not be rolled before or after welding.

The variables in tube-end welding were discussed in two unpublished papers (Emhardt, "Heat Exchanger Tube-to-Tubesheet Joints," ASME Pap. 69-WA/HT-47; and Reynolds, "Tube Welding for Conventional and Nuclear Power Plant Heat Exchangers," ASME Pap. 69-WA/HT-24), which were presented at the November 1969 meeting of the American Society of Mechanical Engineers.

Tube-end rolling before welding may leave lubricant from the tube expander in the tube hole. Fouling during normal operation followed by maintenance operations will leave various impurities in and near the tube ends. Satisfactory welds are rarely possible under such conditions, since tube-end welding requires extreme cleanliness in the area to be welded.

Tube *expansion after welding* has been found useful for low and moderate pressures. In high-pressure service tube rolling has not been able to prevent leakage after weld failure.

Double-Tube-Sheet Joints This design prevents the passage of either fluid into the other because of leakage at the tube–tube-sheet joints, which are generally the weakest points in heat exchangers. Any leakage at these joints admits the fluid to the gap between the tubesheets. Mechanical design, fabrication, and maintenance of double-tube-sheet designs require special consideration.

SHELL-SIDE CONSTRUCTION

Shell Sizes Heat exchanger shells are generally made from standard-wall steel pipe in sizes up to 305-mm (12-in) diameter; from 9.5-mm (⅜-in) wall pipe in sizes from 356 to 610 mm (14 to 24 in); and from steel plate rolled at discrete intervals in larger sizes. Clearances between the outer tube limit and the shell are discussed elsewhere in connection with the different types of construction.

The following formulas may be used to estimate tube counts for various bundle sizes and tube passes. The estimated values include the removal of tubes to provide an entrance area for shell nozzle sizes of one-fifth the shell diameter. Due to the large effect from other parameters such as design pressure/corrosion allowance, baffle cuts, seal strips, and so on, these are to be used as estimates only. Exact tube counts are part of the design package of most reputable exchanger design software and are normally used for the final design.

Triangular tube layouts with pitch equal to 1.25 times the tube outside diameter:

$C = 0.75(D/d) - 36$, where D = bundle OD and d = tube OD.

Range of accuracy: $-24 \le C \le 24$.

$$\text{1 Tube Pass: } N_t = 1298. + 74.86C + 1.283C^2 - 0.0078C^3 - 0.0006C^4 \quad (11\text{-}74a)$$

$$\text{2 Tube Pass: } N_t = 1266. + 73.58C + 1.234C^2 - 0.0071C^3 - 0.0005C^4 \quad (11\text{-}74b)$$

$$\text{4 Tube Pass: } N_t = 1196. + 70.79C + 1.180C^2 - 0.0059C^3 - 0.0004C^4 \quad (11\text{-}74c)$$

$$\text{6 Tube Pass: } N_t = 1166. + 70.72C + 1.269C^2 - 0.0074C^3 - 0.0006C^4 \quad (11\text{-}74d)$$

Square tube layouts with pitch equal to 1.25 times the tube outside diameter:

$C = (D/d) - 36$, where D = bundle OD and d = tube OD.

Range of accuracy: $-24 \le C \le 24$.

$$\text{1 Tube Pass: } N_t = 593.6 + 33.52C + 0.3782C^2 - 0.0012C^3 + 0.0001C^4 \quad (11\text{-}75a)$$

$$\text{2 Tube Pass: } N_t = 578.8 + 33.36C + 0.3847C^2 - 0.0013C^3 + 0.0001C^4 \quad (11\text{-}75b)$$

$$\text{4 Tube Pass: } N_t = 562.0 + 33.04C + 0.3661C^2 - 0.0016C^3 + 0.0002C^4 \quad (11\text{-}75c)$$

$$\text{6 Tube Pass: } N_t = 550.4 + 32.49C + 0.3873C^2 - 0.0013C^3 + 0.0001C^4 \quad (11\text{-}75d)$$

TABLE 11-12 Characterstics of Tubing (From *Standards of the Tubular Exchanger Manufacturers Association*, 8th Ed., 1999; 25 North Broadway, Tarrytown, N.Y.)

Tube O.D., in.	B.W.G. gage	Thickness, in.	Internal area, in.^{2}	Sq. ft. external surface per foot length	Sq. ft. internal surface per foot length	Weight per ft. length steel, lb*	Tube I.D., in.	Moment of inertia, in.^{4}	Section modulus, in.^{3}	Radius of gyration, in.	Constant $C^\dagger$	O.D. I.D.	Transverse metal area, in.^{3}
1/4	22	0.028	0.0296	0.0654	0.0508	0.066	0.194	0.00012	0.00098	0.0791	46	1.289	0.0195
	24	0.022	0.0333	0.0654	0.0539	0.054	0.206	0.00010	0.00083	0.0810	52	1.214	0.0158
	26	0.018	0.0360	0.0654	0.0560	0.045	0.214	0.00009	0.00071	0.0823	56	1.168	0.0131
	27	0.016	0.0373	0.0654	0.0571	0.040	0.218	0.00008	0.00065	0.0829	58	1.147	0.0118
3/8	18	0.049	0.0603	0.0982	0.0725	0.171	0.277	0.00068	0.0036	0.1166	94	1.354	0.0502
	20	0.035	0.0731	0.0982	0.0798	0.127	0.305	0.00055	0.0029	0.1208	114	1.230	0.0374
	22	0.028	0.0799	0.0982	0.0835	0.104	0.319	0.00046	0.0025	0.1231	125	1.176	0.0305
	24	0.022	0.0860	0.0982	0.0867	0.083	0.331	0.00038	0.0020	0.1250	134	1.133	0.0244
1/2	16	0.065	0.1075	0.1309	0.0969	0.302	0.370	0.0021	0.0086	0.1555	168	1.351	0.0888
	18	0.049.	0.1269	0.1309	0.1052	0.236	0.402	0.0018	0.0071	0.1604	198	1.244	0.0694
	20	0.035	0.1452	0.1309	0.1126	0.174	0.430	0.0014	0.0056	0.1649	227	1.163	0.0511
	22	0.028	0.1548	0.1309	0.1162	0.141	0.444	0.0012	0.0046	0.1672	241	1.126	0.0415
5/8	12	0.109	0.1301	0.1636	0.1066	0.601	0.407	0.0061	0.0197	0.1865	203	1.536	0.177
	13	0.095	0.1486	0.1636	0.1139	0.538	0.435	0.0057	0.0183	0.1904	232	1.437	0.158
	14	0.083	0.1655	0.1636	0.1202	0.481	0.459	0.0053	0.0170	0.1939	258	1.362	0.141
	15	0.072	0.1817	0.1636	0.1259	0.426	0.481	0.0049	0.0156	0.1972	283	1.299	0.125
	16	0.065	0.1924	0.1636	0.1296	0.389	0.495	0.0045	0.0145	0.1993	300	1.263	0.114
	17	0.058	0.2035	0.1636	0.1333	0.352	0.509	0.0042	0.0134	0.2015	317	1.228	0.103
	18	0.049	0.2181	0.1636	0.1380	0.302	0.527	0.0037	0.0119	0.2044	340	1.186	0.089
	19	0.042	0.2299	0.1636	0.1416	0.262	0.541	0.0033	0.0105	0.2067	359	1.155	0.077
	20	0.035	0.2419	0.1636	0.1453	0.221	0.555	0.0028	0.0091	0.2090	377	1.126	0.065
3/4	10	0.134	0.1825	0.1963	0.1262	0.833	0.482	0.0129	0.0344	0.2229	285	1.556	0.259
	11	0.120	0.2043	0.1963	0.1335	0.808	0.510	0.0122	0.0326	0.2267	319	1.471	0.238
	12	0.109	0.2223	0.1963	0.1393	0.747	0.532	0.0116	0.0309	0.2299	347	1.410	0.219
	13	0.095	0.2463	0.1963	0.1466	0.665	0.560	0.0107	0.0285	0.2340	384	1.339	0.195
	14	0.083	0.2679	0.1963	0.1529	0.592	0.584	0.0098	0.0262	0.2376	418	1.284	0.174
	15	0.072	0.2884	0.1963	0.1587	0.522	0.606	0.0089	0.0238	0.2411	450	1.238	0.153
	16	0.065	0.3019	0.1963	0.1623	0.476	0.620	0.0083	0.0221	0.2433	471	1.210	0.140
	17	0.058	0.3157	0.1963	0.1660	0.429	0.634	0.0076	0.0203	0.2455	492	1.183	0.126
	18	0.049	0.3339	0.1963	0.1707	0.367	0.652	0.0067	0.0178	0.2484	521	1.150	0.108
	20	0.035	0.3632	0.1963	0.1780	0.268	0.680	0.0050	0.0134	0.2531	567	1.103	0.079
7/8	10	0.134	0.2894	0.2291	0.1589	1.062	0.607	0.0221	0.0505	0.2662	451	1.442	0.312
	11	0.120	0.3167	0.2291	0.1662	0.969	0.635	0.0208	0.0475	0.2703	494	1.378	0.285
	12	0.109	0.3390	0.2291	0.1720	0.893	0.657	0.0196	0.0449	0.2736	529	1.332	0.262
	13	0.095	0.3685	0.2291	0.1793	0.792	0.685	0.0180	0.0411	0.2778	575	1.277	0.233
	14	0.083	0.3948	0.2291	0.1856	0.703	0.709	0.0164	0.0374	0.2815	616	1.234	0.207
	15	0.072	0.4197	0.2291	0.1914	0.618	0.731	0.0148	0.0337	0.2850	655	1.197	0.182
	16	0.065	0.4359	0.2291	0.1950	0.563	0.745	0.0137	0.0312	0.2873	680	1.174	0.165
	17	0.058	0.4525	0.2291	0.1987	0.507	0.759	0.0125	0.0285	0.2896	706	1.153	0.149
	18	0.049	0.4742	0.2291	0.2034	0.433	0.777	0.0109	0.0249	0.2925	740	1.126	0.127
	20	0.035	0.5090	0.2291	0.2107	0.314	0.805	0.0082	0.0187	0.2972	794	1.087	0.092
1	8	0.165	0.3526	0.2618	0.1754	1.473	0.670	0.0392	0.0784	0.3009	550	1.493	0.433
	10	0.134	0.4208	0.2618	0.1916	1.241	0.732	0.0350	0.0700	0.3098	656	1.366	0.365
	11	0.120	0.4536	0.2618	0.1990	1.129	0.760	0.0327	0.0654	0.3140	708	1.316	0.332
	12	0.109	0.4803	0.2618	0.2047	1.038	0.782	0.0307	0.0615	0.3174	749	1.279	0.305
	13	0.095	0.5153	0.2618	0.2121	0.919	0.810	0.0280	0.0559	0.3217	804	1.235	0.270
	14	0.083	0.5463	0.2618	0.2183	0.814	0.834	0.0253	0.0507	0.3255	852	1.199	0.239
	15	0.072	0.5755	0.2618	0.2241	0.714	0.856	0.0227	0.0455	0.3291	898	1.168	0.210
	16	0.065	0.5945	0.2618	0.2278	0.650	0.870	0.0210	0.0419	0.3314	927	1.149	0.191
	18	0.049	0.6390	0.2618	0.2361	0.498	0.902	0.0166	0.0332	0.3367	997	1.109	0.146
	20	0.035	0.6793	0.2618	0.2435	0.361	0.930	0.0124	0.0247	0.3414	1060	1.075	0.106
1¼	7	0.180	0.6221	0.3272	0.2330	2.059	0.890	0.0890	0.1425	0.3836	970	1.404	0.605
	8	0.165	0.6648	0.3272	0.2409	1.914	0.920	0.0847	0.1355	0.3880	1037	1.359	0.562
	10	0.134	0.7574	0.3272	0.2571	1.599	0.982	0.0742	0.1187	0.3974	1182	1.273	0.470
	11	0.120	0.8012	0.3272	0.2644	1.450	1.010	0.0688	0.1100	0.4018	1250	1.238	0.426
	12	0.109	0.8365	0.3272	0.2702	1.330	1.032	0.0642	0.1027	0.4052	1305	1.211	0.391
	13	0.095	0.8825	0.3272	0.2775	1.173	1.060	0.0579	0.0926	0.4097	1377	1.179	0.345
	14	0.083	0.9229	0.3272	0.2838	1.036	1.084	0.0521	0.0833	0.4136	1440	1.153	0.304
	16	0.065	0.9852	0.3272	0.2932	0.824	1.120	0.0426	0.0682	0.4196	1537	1.116	0.242
	18	0.049	1.0423	0.3272	0.3016	0.629	1.152	0.0334	0.0534	0.4250	1626	1.085	0.185
	20	0.035	1.0936	0.3272	0.3089	0.455	1.180	0.0247	0.0395	0.4297	1706	1.059	0.134
1½	10	0.134	1.1921	0.3927	0.3225	1.957	1.232	0.1354	0.1806	0.4853	1860	1.218	0.575
	12	0.109	1.2908	0.3927	0.3356	1.621	1.282	0.1159	0.1545	0.4933	2014	1.170	0.476
	14	0.083	1.3977	0.3927	0.3492	1.257	1.334	0.0931	0.1241	0.5018	2180	1.124	0.369
	16	0.065	1.4741	0.3927	0.3587	0.997	1.370	0.0756	0.1008	0.5079	2300	1.095	0.293
2	11	0.120	2.4328	0.5236	0.4608	2.412	1.760	0.3144	0.3144	0.6660	3795	1.136	0.709
	12	0.109	2.4941	0.5236	0.4665	2.204	1.782	0.2904	0.2904	0.6697	3891	1.122	0.648
	13	0.095	2.5730	0.5236	0.4739	1.935	1.810	0.2586	0.2588	0.6744	4014	1.105	0.569
	14	0.083	2.6417	0.5236	0.4801	1.701	1.834	0.2300	0.2300	0.6784	4121	1.091	0.500

*Weights are based on low-carbon steel with a density of 0.2836 lb/cu. in. For other metals multiply by the following factors: aluminum, 0.35; titanium, 0.58; A.I.S.I. 400 Series S/steels, 0.99; A.I.S.I. 300 Series S/steels, 1.02; aluminum bronze, 1.04; aluminum brass, 1.06; nickel-chrome-iron, 1.07; Admiralty, 1.09; nickel, 1.13; nickel-copper, 1.12; copper and cupro-nickels, 1.14.

†Liquid velocity $= \dfrac{\text{lb per tube hour}}{C \times \text{sp gr of liquid}}$ ft/s (sp gr of water at 60°F = 1.0)

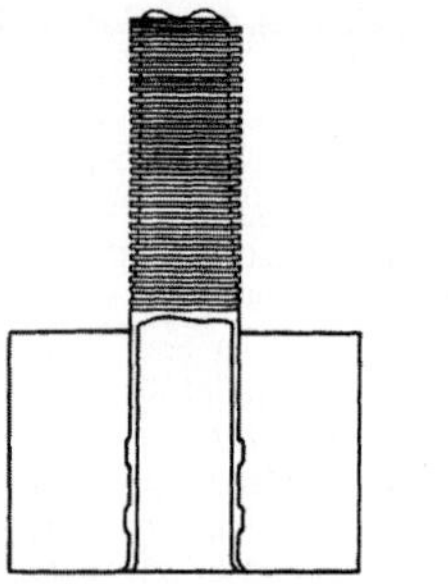

FIG. 11-39 Integrally finned tube rolled into tube sheet with double serrations and flared inlet. (*Woverine Division, UOP, Inc.*)

Shell-Side Arrangements The *one-pass shell* (Fig. 11-35E) is the most commonly used arrangement. Condensers from single-component vapors often have the nozzles moved to the center of the shell for vacuum and steam services.

A solid longitudinal baffle is provided to form a two-pass shell (Fig. 11-35F). It may be insulated to improve thermal efficiency. (See further discussion on baffles.) A two-pass shell can improve thermal effectiveness at a cost lower than for two shells in series.

For *split flow* (Fig. 11-35G), the longitudinal baffle may be solid or perforated. The latter feature is used with condensing vapors.

A *double-split-flow* design is shown in Fig. 11-35H. The longitudinal baffles may be solid or perforated.

The divided flow design (Fig. 11-35J) mechanically is like the one-pass shell except for the addition of a nozzle. Divided flow is used to meet low-pressure-drop requirements.

The *kettle reboiler* is shown in Fig. 11-35K. When nucleate boiling is to be done on the shell side, this common design provides adequate dome space for separation of vapor and liquid above the tube bundle and surge capacity beyond the weir near the shell cover.

BAFFLES AND TUBE BUNDLES

The *tube bundle* is the most important part of a tubular heat exchanger. The tubes generally constitute the most expensive component of the exchanger and are the one most likely to corrode. Tubesheets, baffles, or support plates, tie rods, and usually spacers complete the bundle.

Minimum *baffle spacing* is generally one-fifth of the shell diameter and not less than 50.8 mm (2 in). Maximum baffle spacing is limited by the requirement to provide adequate support for the tubes. The maximum unsupported tube span in inches equals $74d^{0.75}$ (where d is the tube OD in inches). The unsupported tube span is reduced by about 12 percent for aluminum, copper, and their alloys.

Baffles are provided for heat-transfer purposes. When shell-side baffles are not required for heat-transfer purposes, as may be the case in condensers or reboilers, tube supports are installed.

Segmental Baffles Segmental or cross-flow baffles are standard. Single, double, and triple segmental baffles are used. Baffle cuts are illustrated in Fig. 11-40. The double segmental baffle reduces cross-flow velocity

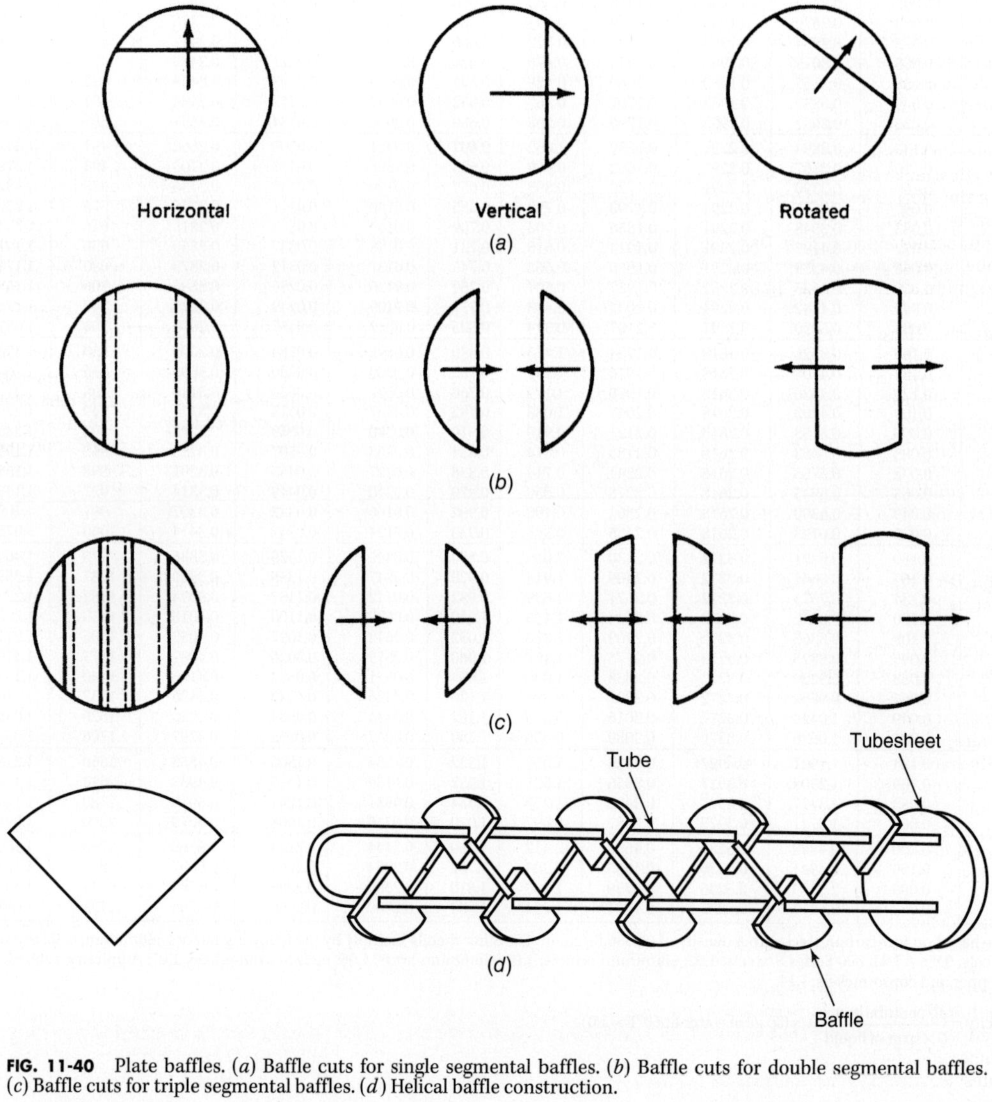

FIG. 11-40 Plate baffles. (*a*) Baffle cuts for single segmental baffles. (*b*) Baffle cuts for double segmental baffles. (*c*) Baffle cuts for triple segmental baffles. (*d*) Helical baffle construction.

for a given baffle spacing. The triple segmental baffle reduces both cross-flow and long-flow velocities and has been identified as the "window-cut" baffle.

Baffle cuts are expressed as the ratio of segment opening height to shell inside diameter. Cross-flow baffles with horizontal cut are shown in Fig. 11-36*a, c,* and *f.* This arrangement is not satisfactory for horizontal condensers, since the condensate can be trapped between baffles, or for dirty fluids in which the dirt might settle out. Vertical-cut baffles are used for side-to-side flow in horizontal exchangers with condensing fluids or dirty fluids. Baffles are notched to ensure complete drainage when the units are taken out of service. (These notches permit some bypassing of the tube bundle during normal operation.)

Tubes are most commonly arranged on an equilateral triangular pitch. Tubes are arranged on a square pitch primarily for mechanical cleaning purposes in removable-bundle exchangers.

Maximum baffle cut is limited to about 45 percent for single segmental baffles so that every pair of baffles will support each tube. Tube bundles are generally provided with baffles cut so that at least one row of tubes passes through all the baffles or support plates. These tubes hold the entire bundle together. In pipe-shell exchangers with a horizontal baffle cut and a horizontal pass rib for directing tube-side flow in the channel, the maximum baffle cut, which permits a minimum of one row of tubes to pass through all baffles, is approximately 33 percent in small shells and 40 percent in larger pipe shells.

Maximum shell-side heat-transfer rates in forced convection are apparently obtained by cross-flow of the fluid at right angles to the tubes. To maximize this type of flow, some heat exchangers are built with segmental-cut baffles and with "no tubes in the window" (or the baffle cutout). Maximum baffle spacing may thus equal maximum unsupported-tube span, while conventional baffle spacing is limited to one-half of this span.

The maximum baffle spacing for no tubes in the window of single segmental baffles is unlimited when intermediate supports are provided. These are cut on both sides of the baffle and therefore do not affect the flow of the shell-side fluid. Each support engages all the tubes; the supports are spaced to provide adequate support for the tubes.

Rod Baffles Rod or bar baffles have either rods or bars extending through the lanes between rows of tubes. A baffle set can consist of a baffle with rods in all the vertical lanes and another baffle with rods in all the horizontal lanes between the tubes. The shell-side flow is uniform and parallel to the tubes. Stagnant areas do not exist.

One device uses four baffles in a baffle set. Only half of either the vertical or the horizontal tube lanes in a baffle have rods. The new design apparently provides a maximum shell-side heat-transfer coefficient for a given pressure drop.

Tie Rods and Spacers Tie rods are used to hold the baffles in place with spacers, which are pieces of tubing or pipe placed on the rods to locate the baffles. Occasionally baffles are welded to the tie rods, and spacers are eliminated. Properly located tie rods and spacers serve both to hold the bundle together and to reduce bypassing of the tubes.

In very large fixed-tube-sheet units, in which concentricity of shells decreases, baffles are occasionally welded to the shell to eliminate bypassing between the baffle and the shell.

Metal baffles are standard. Occasionally plastic baffles are used either to reduce corrosion or in vibratory service, in which metal baffles may cut the tubes.

Impingement Baffle The tube bundle is customarily protected against impingement by the incoming fluid at the shell inlet nozzle when the shell-side fluid is at a high velocity, is condensing, or is a two-phase fluid. Minimum entrance area about the nozzle is generally equal to the inlet nozzle area. Exit nozzles also require adequate area between the tubes and the nozzles. A full bundle without any provision for shell inlet nozzle area can increase the velocity of the inlet fluid by as much as 300 percent with a consequent loss in pressure.

Impingement baffles are generally made of rectangular plate, although circular plates are more desirable. Rods and other devices are sometimes used to protect the tubes from impingement. To maintain a maximum tube count, the impingement plate is often placed in a conical nozzle opening or in a dome cap above the shell.

Impingement baffles or flow distribution devices are recommended for axial tube-side nozzles when entrance velocity is high.

Vapor Distribution Relatively large shell inlet nozzles, which may be used in condensers under low pressure or vacuum, require provision for uniform vapor distribution.

Tube-Bundle Bypassing Shell-side heat-transfer rates are maximized when bypassing of the tube bundle is at a minimum. The most significant bypass stream is generally between the outer tube limit and the inside of the shell. The clearance between tubes and shell is at a minimum for fixed-tube-sheet construction and is greatest for straight-tube removable bundles.

Arrangements to reduce tube-bundle bypassing include these:

Dummy tubes. These tubes do not pass through the tubesheets and can be located close to the inside of the shell.

Tie rods with spacers. These hold the baffles in place but can be located to prevent bypassing.

Sealing strips. These longitudinal strips either extend from baffle to baffle or may be inserted in slots cut into the baffles.

Dummy tubes or tie rods with spacers may be located within the pass partition lanes (and between the baffle cuts) to ensure maximum bundle penetration by the shell-side fluid.

When tubes are omitted from the tube layout to provide the entrance area about an impingement plate, the need for sealing strips or other devices to cause proper bundle penetration by the shell-side fluid is increased.

Helical Baffles An increasingly popular variant to the segmental baffle is the helical baffle. These are quadrant-shaped plate baffles installed at an angle to the axial bundle centerline to produce a pseudo-spiraling flow down the length of the tube bundle (Fig. 11-40*d*). This baffle has the advantage of producing shell-side heat-transfer coefficients similar to those of the segmental baffle with much less shell-side pressure loss for the same size of shell. In the case of equal pressure drops, the helical baffle exchanger will be smaller than the segmental baffle exchanger; or, for identical shell sizes, the helical baffle exchanger will permit a much higher throughput of flow for the same process inlet/outlet temperatures.

A great amount of proprietary research has been conducted by a few companies into the workings of helical baffled heat exchangers. The only known open literature method for estimating helical baffle performance has been "Comparison of Correction Factors for Shell-and-Tube Heat Exchangers with Segmental or Helical Baffles" by Stehlik, Nemcansky, Kral, and Swanson [*Heat Transfer Engineering* **15**(1): 55–65 (1994)].

Unique design variables for helical baffles include the baffle angle, adjacent baffle contact diameter (which sets the baffle spacing and is usually about one-half of the shell ID), and the number of baffle starts (i.e., number of intermediate baffle starts). Of course, consideration is also given to the tube layout, tube pitch, use of seal strips, and all the other configuration characteristics common to any plate baffled bundle.

A helical baffle bundle built in this way produces two distinct flow regions. The area outside of the adjacent baffle contact diameter tends to produce a stable helical cross-flow. However, inside the diameter where adjacent baffles touch is a second region where vortical flow is induced but in which the intensity of the rotational component tends to decrease as one approaches the center of the bundle. For a fixed flow rate and helix angle, this tendency may be minimized by proper selection of the baffle contact diameter. With the correct selection, stream temperatures may be made to be close to uniform across the bundle cross section through the shell. However, below a critical velocity (for the baffle configuration and fluid state), the tendency for nonuniformity of temperatures increases as velocity decreases until ever-increasing portions of the central core surface area pinch out with respect to temperature and become ineffective for further heat transfer.

The design approach involves varying the baffle spacing for the primary purpose of balancing the flows in the two regions and maximizing the effectiveness of the total surface area. In many cases, a shallower helix angle is chosen in conjunction with the baffle spacing in order to minimize the central core component while still achieving a reduced overall bundle pressure drop.

Longitudinal Flow Baffles In fixed-tube-sheet construction with multipass shells, the baffle is usually welded to the shell as positive assurance against bypassing results. Removable tube bundles have a sealing device between the shell and the longitudinal baffle. Flexible light-gauge sealing strips and various packing devices have been used. Removable U-tube bundles with four tube-side passes and two shell-side passes can be installed in shells with the longitudinal baffle welded in place.

In split-flow shells, the longitudinal baffle may be installed without a positive seal at the edges if design conditions are not seriously affected by a limited amount of bypassing.

Fouling in petroleum refinery service has necessitated rough treatment of tube bundles during cleaning operations. Many refineries avoid the use of longitudinal baffles, since the sealing devices are subject to damage during cleaning and maintenance operations.

CORROSION IN HEAT EXCHANGERS

Some of the special considerations in regard to heat exchanger corrosion are discussed in this subsection. An extended presentation in Sec. 23 covers corrosion and its various forms as well as materials of construction.

Materials of Construction The most common material of construction for heat exchangers is carbon steel. Stainless-steel construction throughout is sometimes used in chemical-plant service and on rare occasions in petroleum refining. Many exchangers are constructed from dissimilar metals. Such combinations are functioning satisfactorily in certain

TABLE 11-13 Dissimilar Materials in Heat-Exchanger Construction

Part	Relative use	1	2	3	4	5	6
	Relative cost	A	B	C	D	C	E
Tubes		●	●	●	●	●	●
Tube sheets			●	●	●	●	●
Tube-side headers				●	●		
Baffles					●	●	●
Shell							●

Carbon steel replaced by an alloy when ● appears.
Relative use: from 1 (most popular) through 6 (least popular) combinations.
Relative cost: from A (least expensive) to E (most expensive).

services. Extreme care in their selection is required since electrolytic attack can develop.

Carbon-steel and alloy combinations appear in Table 11-13. "Alloys" in chemical- and petrochemical-plant service in approximate order of use are stainless-steel series 300, nickel, Monel, copper alloy, aluminum, Inconel, stainless-steel series 400, and other alloys. In petroleum refinery service, the frequency order shifts, with copper alloy (for water-cooled units) in first place and low-alloy steel in second place. In some segments of the petroleum industry, copper alloy, stainless series 400, low-alloy steel, and aluminum are becoming the most commonly used alloys.

Copper-alloy tubing, particularly inhibited admiralty, is generally used with cooling water. Copper-alloy tubesheets and baffles are generally of naval brass.

Aluminum alloy (and in particular alclad aluminum) tubing is sometimes used in water service. The alclad alloy has a sacrificial aluminum-alloy layer metallurgically bonded to a core alloy.

Tube-side headers for water service are made in a wide variety of materials: carbon steel, copper alloy, cast iron, and lead-lined or plastic-lined or specially painted carbon steel.

Bimetallic Tubes When corrosive requirements or temperature conditions do not permit the use of a single alloy for the tubes, bimetallic (or duplex) tubes may be used. These can be made from almost any possible combination of metals. Tube sizes and gauges can be varied. For thin gauges the wall thickness is generally divided equally between the two components. In heavier gauges the more expensive component may comprise from one-fifth to one-third of the total thickness.

The component materials comply with applicable ASTM specifications, but after manufacture the outer component may increase in hardness beyond specification limits, and special care is required during the tube-rolling operation. When the harder material is on the outside, precautions must be taken to expand the tube properly. When the inner material is considerably softer, rolling may not be practical unless ferrules of the soft material are used.

To eliminate galvanic action, the outer tube material may be stripped from the tube ends and replaced with ferrules of the inner tube material. When the end of a tube with a ferrule is expanded or welded to a tubesheet, the tube-side fluid can contact only the inner tube material, while the outer material is exposed to the shell-side fluid.

Bimetallic tubes are available from a small number of tube mills and are manufactured only on special order and in large quantities.

Clad Tubesheets Usually tubesheets and other exchanger parts are made of a solid metal. Clad or bimetallic tubesheets are used to reduce costs or because no single metal is satisfactory for the corrosive conditions. The alloy material (e.g., stainless steel, Monel) is generally bonded or clad to a carbon-steel backing material. In fixed-tube-sheet construction, a copper alloy–clad tubesheet can be welded to a steel shell, while most copper-alloy tubesheets cannot be welded to steel in a manner acceptable to ASME Code authorities.

Clad tubesheets in service with carbon-steel backer material include stainless-steel types 304, 304L, 316, 316L, and 317, Monel, Inconel, nickel, naval rolled brass, copper, admiralty, silicon bronze, and titanium. Naval rolled brass and Monel clad on stainless steel are also in service.

Ferrous-alloy-clad tubesheets are generally prepared by a weld overlay process in which the alloy material is deposited by welding upon the face of the tubesheet. Precautions are required to produce a weld deposit free of defects, since these may permit the process fluid to attack the base metal below the alloy. Copper-alloy-clad tubesheets are prepared by brazing the alloy to the carbon-steel backing material.

Clad materials can be prepared by bonding techniques, which involve rolling, heat treatment, explosive bonding, etc. When properly manufactured, the two metals do not separate because of thermal expansion differences encountered in service. Applied tube-sheet facings prepared by tack welding at the outer edges of alloy and base metal or by bolting together the two metals are in limited use.

Nonmetallic Construction Shell-and-tube exchangers are available with glass tubes 14 mm (0.551 in) in diameter and 1 mm (0.039 in) thick with tube lengths from 2.015 m (79.3 in) to 4.015 m (158 in). Steel shell exchangers have a maximum design pressure of 517 kPa (75 lbf/in^2). Glass shell exchangers have a maximum design gauge pressure of 103 kPa (15 lbf/in^2). Shell diameters are 229 mm (9 in), 305 mm (12 in), and 457 mm (18 in). Heat-transfer surface ranges from 3.16 to 51 m^2 (34 to 550 ft^2). Each tube is free to expand, since a Teflon sealer sheet is used at the tube–tube sheet joint.

Impervious graphite heat exchanger equipment is made in a variety of forms, including outside packed-head shell-and-tube exchangers. They are fabricated with impervious graphite tubes and tube-side headers and metallic shells. Single units containing up to 1300 m^2 (14,000 ft^2) of heat-transfer surface are available.

Teflon heat exchangers of special construction are described later in this section.

Fabrication Expanding the tube into the tubesheet reduces the tube wall thickness and work-hardens the metal. The induced stresses can lead to *stress corrosion.* Differential expansion between tubes and shell in fixed-tube-sheet exchangers can develop stresses, which lead to stress corrosion.

When austenitic stainless-steel tubes are used for corrosion resistance, a close fit between the tube and the tube hole is recommended to minimize work hardening and the resulting loss of corrosion resistance.

To facilitate removal and replacement of tubes, it is customary to roller-expand the tubes to within 3 mm (⅛ in) of the shell-side face of the tubesheet. A 3-mm- (⅛-in-) long gap is thus created between the tube and the tube hole at this tube-sheet face. In some services this gap has been found to be a focal point for corrosion.

It is standard practice to provide a chamfer at the inside edges of tube holes in tubesheets to prevent cutting of the tubes and to remove burrs produced by drilling or reaming the tubesheet. In the lower tubesheet of vertical units, this chamfer serves as a pocket to collect material, dirt, etc., and acts as a corrosion center.

Adequate venting of exchangers is required both for proper operation and to reduce corrosion. Improper venting of the water side of exchangers can cause alternate wetting and drying and accompanying chloride concentration, which is particularly destructive to the series 300 stainless steels.

Certain corrosive conditions require that special consideration be given to complete drainage when the unit is taken out of service. Particular consideration is required for the upper surfaces of tubesheets in vertical heat exchangers, for sagging tubes, and for shell-side baffles in horizontal units.

SHELL-AND-TUBE EXCHANGER COSTS

Basic costs of shell-and-tube heat exchangers made in the United States of carbon-steel construction in 1958 are shown in Fig. 11-41.

Cost data for shell-and-tube exchangers from 15 sources were correlated and found to be consistent when scaled by the Marshall and Swift index [Woods et al., *Can. J. Chem. Eng.* **54:** 469–489 (December 1976)].

Costs of shell-and-tube heat exchangers can be estimated from Fig. 11-41 and Tables 11-14 and 11-15. These 1960 costs should be updated by use of the Marshall and Swift Index, which appears in each issue of *Chemical Engineering.* Note that during periods of high and low demand for heat exchangers the prices in the marketplace may vary significantly from those determined by this method.

Small heat exchangers and exchangers bought in small quantities are likely to be more costly than indicated.

Standard heat exchangers (which are sometimes off-the-shelf items) are available in sizes ranging from 1.9 to 37 m^2 (20 to 400 ft^2) at costs lower than for custom-built units. Steel costs are approximately one-half, admiralty tube-side costs are two-thirds, and stainless costs are three-fourths of those for equivalent custom-built exchangers.

Kettle-type reboiler costs are 15 to 25 percent greater than for equivalent internal floating-head or U-tube exchangers. The higher extra cost is applicable with relatively large kettle-to-port-diameter ratios and with increased internals (e.g., vapor-liquid separators, foam breakers, sight glasses).

To estimate exchanger costs for varying construction details and alloys, first determine the base cost of a similar heat exchanger of basic construction (carbon steel, Class R, 150 lbf/in^2) from Fig. 11-41. From Table 11-14, select appropriate extras for higher pressure rating and for alloy construction of tubesheets and baffles, shell and shell cover, and channel and floating-head cover. Compute these extras in accordance with the notes at the bottom of the table. For tubes other than welded carbon steel, compute the extra cost by multiplying the exchanger surface by the appropriate cost per square foot from Table 11-15.

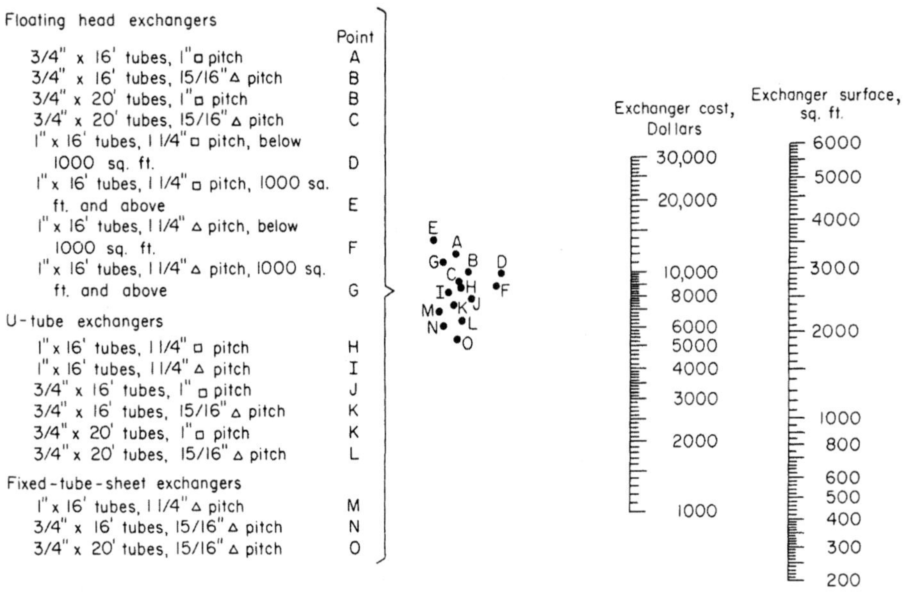

FIG. 11-41 Costs of basic exchangers—all steel, TEMA Class R, 150 lbf/in², 1958. To convert pounds-force per square inch to kilopascals, multiply by 6.895; to convert square feet to square meters, multiply by 0.0929; to convert inches to millimeters, multiply by 25.4; and to convert feet to meters, multiply by 0.3048.

TABLE 11-14 Extras for Pressure and Alloy Construction and Surface and Weights*

Percent of steel base price, 1500-lbf/in² working pressure

	Shell diameters, in												
	12	14	16	18	20	22	24	27	30	33	36	39	42
Pressure†													
300 lbf/in²	7	7	8	8	9	9	10	11	11	12	13	14	15
450 lbf/in²	18	19	20	21	22	23	24	27	29	31	32	33	35
600 lbf/in²	28	29	31	33	35	37	39	40	41	32	44	45	50
Alloy													
All-steel heat exchanger	100	100	100	100	100	100	100	100	100	100	100	100	100
Tube sheets and baffles													
Naval rolled brass	14	17	19	21	22	22	22	22	22	23	24	24	25
Monel	24	31	35	37	39	39	40	40	41	41	41	41	42
1¼ Cr, ½ Mo	6	7	7	7	8	8	8	8	9	10	10	10	11
4–6 Cr, ½ Mo	19	22	24	25	26	26	26	25	25	25	26	26	26
11–13 Cr (stainless 410)	21	24	26	27	27	27	27	27	27	27	27	27	28
Stainless 304	22	27	29	30	31	31	31	31	30	30	30	31	31
Shell and shell cover													
Monel	45	48	51	52	53	52	52	51	49	47	45	44	44
1¼ Cr, ½ Mo	20	22	24	25	25	25	24	22	20	19	18	17	17
4–6 Cr, ½ Mo	28	31	33	35	35	35	34	32	30	28	27	26	26
11–13 Cr (stainless 410)	29	33	35	36	36	36	35	34	32	30	29	27	27
Stainless 304	32	34	36	37	38	37	37	35	33	31	30	29	28
Channel and floating-head cover													
Monel	40	42	42	43	42	41	40	37	34	32	31	40	30
1¼ Cr, ½ Mo	23	24	24	25	24	24	23	22	21	21	21	20	20
4–6 Cr, ½ Mo	36	37	38	38	37	36	34	31	29	27	26	25	24
11–13 Cr (stainless 410)	37	38	39	39	38	37	35	32	30	28	27	26	25
Stainless 304	37	39	39	39	38	37	36	33	31	29	28	26	26
Surface													
Surface, ft², internal floating head, ¾-in OD by 1-in square pitch, 16 ft 0 in, tube‡	251	302	438	565	726	890	1040	1470	1820	2270	2740	3220	3700
1-in OD by 1¼-in square pitch, 16-ft 0-in tube§	218	252	352	470	620	755	876	1260	1560	1860	2360	2770	3200
Weight, lb, internal floating head, 1-in OD, 14 BWG tube	2750	3150	4200	5300	6600	7800	9400	11,500	14,300	17,600	20,500	24,000	29,000

*Modified from E. N. Sieder and G. H. Elliot, *Pet. Refiner* **39**(5): 223 (1960).

†Total extra is 0.7 × pressure extra on shell side plus 0.3 × pressure extra on tube side.

‡Fixed-tube-sheet construction with ¾-in OD tube on 15/16-in triangular pitch provides 36 percent more surface.

§Fixed-tube-sheet construction with 1-in OD tube on 1¼-in triangular pitch provides 18 percent more surface.

For an all-steel heat exchanger with mixed design pressures the total extra for pressure is 0.7 × pressure extra on shell side plus 0.3 × pressure extra tube side.

For an exchanger with alloy parts and a design pressure of 150 lbf/in², the alloy extras are added. For shell and shell cover the combined alloy-pressure extra is the alloy extra times the shell-side pressure extra/100. For channel and floating-head cover the combined alloy-pressure extra is the alloy extra times the tube-side pressure extra/100. For tube sheets and baffles the combined alloy-pressure extra is the alloy extra times the higher-pressure extra times 0.9/100. (The 0.9 factor is included since baffle thickness does not increase because of pressure.)

NOTE: To convert pounds-force per square inch to kilopascals, multiply by 6.895; to convert square feet to square meters, multiply by 0.0929; and to convert inches to millimeters, multiply by 25.4.

TABLE 11-15 Base Quantity Extra Cost for Tube Gauge and Alloy
Dollars per square foot

	¾-in OD tubes			1-in OD tubes		
	16 BWG	14 BWG	12 BWG	16 BWG	14 BWG	12 BWG
Carbon steel	0	0.02	0.06	0	0.01	0.07
Admiralty	0.78	1.20	1.81	0.94	1.39	2.03
(T-11) 1¼ Cr, ½ Mo	1.01	1.04	1.11	0.79	0.82	0.95
(T-5) 4–6 Cr	1.61	1.65	1.74	1.28	1.32	1.48
Stainless 410 welded	2.62	3.16	4.12	2.40	2.89	3.96
Stainless 410 seamless	3.10	3.58	4.63	2.84	3.31	4.47
Stainless 304 welded	2.50	3.05	3.99	2.32	2.83	3.88
Stainless 304 seamless	3.86	4.43	5.69	3.53	4.08	5.46
Stainless 316 welded	3.40	4.17	5.41	3.25	3.99	5.36
Stainless 316 seamless	7.02	7.95	10.01	6.37	7.27	9.53
90-10 cupronickel	1.33	1.89	2.67	1.50	2.09	2.90
Monel	4.25	5.22	6.68	4.01	4.97	6.47
Low fin						
Carbon steel	0.22	0.23		0.18	0.19	
Admiralty	0.58	0.75		0.70	0.87	
90-10 cupronickel	0.72	0.96		0.86	1.06	

NOTE: To convert inches to millimeters, multiply by 25.4.

When points for 20-ft-long tubes do not appear in Fig. 11-41, use 0.95 times the cost of the equivalent 16-ft-long exchanger. Length variation of steel heat exchangers affects costs by approximately $1 per square foot. Shell diameters for a given surface are approximately equal for U-tube and floating-head construction.

Low-fin tubes (¹⁄₁₆-in-high fins) provide 2.5 times the surface per lineal foot. The surface required should be divided by 2.5; then use Fig. 11-41 to determine the basic cost of the heat exchanger. Actual surface times extra costs (from Table 11-15) should then be added to determine cost of the fin-tube exchanger.

HAIRPIN/DOUBLE-PIPE HEAT EXCHANGERS

PRINCIPLES OF CONSTRUCTION

Hairpin heat exchangers (often also referred to as "double pipes") are characterized by a construction form that imparts a U-shaped appearance to the heat exchanger. In its classical sense, the term *double pipe* refers to a heat exchanger consisting of a pipe within a pipe, usually of a straight-leg construction with no bends. However, due to the need for removable bundle construction and the ability to handle differential thermal expansion while avoiding the use of expansion joints (often the weak point of the exchanger), the current U-shaped configuration has become the standard in the industry (Fig. 11-42). A further departure from the classical definition comes when more than one pipe or tube is used to make a tube bundle, complete with tubesheets and tube supports similar to the TEMA-style exchanger.

Hairpin heat exchangers consist of two shell assemblies housing a common set of tubes and interconnected by a return-bend cover referred to as the *bonnet.* The shell is supported by means of bracket assemblies designed to cradle both shells simultaneously. These brackets are configured to permit the modular assembly of many hairpin sections into an exchanger bank for inexpensive future expansion capability and for providing the very long thermal lengths demanded by special process applications.

The bracket construction permits support of the exchanger without fixing the supports to the shell. This provides for thermal movement of the shells within the brackets and prevents the transfer of thermal stresses into the process piping. In special cases the brackets may be welded to the shell. However, this is usually avoided due to the resulting loss of flexibility in field installation and equipment reuse at other sites and an increase in piping stresses.

The hairpin heat exchanger, unlike the removable-bundle TEMA styles, is designed for bundle insertion and removal from the return end rather than from the tubesheet end. This is accomplished by means of removable split rings which slide into grooves machined around the outside of each tubesheet and lock the tubesheets to the external closure flanges. This provides a distinct advantage in maintenance since bundle removal takes place at the exchanger end farthest from the plant process piping without disturbing any gasketed joints of this piping.

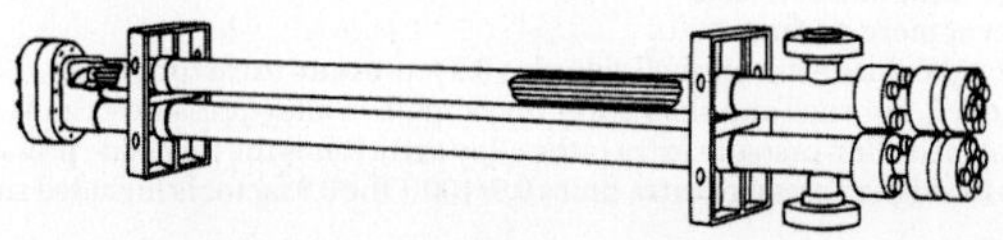

FIG. 11-42 Double-pipe exchanger section with longitudinal fins. (*Brown Fin-tube Co.*)

FINNED DOUBLE PIPES

The design of the classical single-tube double-pipe heat exchanger is an exercise in pure longitudinal flow with the shell-side and tube-side coefficients differing primarily due to variations in flow areas. Adding longitudinal fins gives the more common double-pipe configuration (Table 11-16). Increasing the number of tubes yields the *multitube* hairpin.

MULTITUBE HAIRPINS

For years, the slightly higher mechanical design complexity of the hairpin heat exchanger relegated it to only the smallest process requirements with shell sizes not exceeding 100 mm. In the early 1970s the maximum available sizes were increased to between 300 and 400 mm depending upon the manufacturer. At present, due to recent advances in design technology, hairpin exchangers are routinely produced in shell sizes between 51 mm (2 in) and 762 mm (30 in) for a wide range of pressures and temperatures and have been made in larger sizes as well. Table 11-17 gives common hairpin tube counts and areas for 19-mm- (¾-in-) OD tubes arranged on a 24-mm (15/16-in) triangular tube layout.

The hairpin width and the centerline distance of the two legs (shells) of the hairpin heat exchanger are limited by the outside diameter of the closure flanges at the tubesheets. This diameter, in turn, is a function of the design pressures. As a general rule, for low to moderate design pressures (less than 15 bar), the center-to-center distance is approximately 1.5 to

TABLE 11-16 Double-Pipe Hairpin Section Data

Shell pipe OD		Inner pipe OD		Fin height		Fin count	Surface-area-per-unit length	
mm	in	mm	in	mm	in	(max)	sq m/m	sq ft/ft
60.33	2.375	25.4	1.000	12.7	0.50	24	0.692	2.27
88.9	3.500	48.26	1.900	12.7	0.50	36	1.07	3.51
114.3	4.500	48.26	1.900	25.4	1.00	36	1.98	6.51
114.3	4.500	60.33	2.375	19.05	0.75	40	1.72	5.63
114.3	4.500	73.03	2.875	12.70	0.50	48	1.45	4.76
141.3	5.563	88.9	3.500	17.46	0.6875	56	2.24	7.34
168.3	6.625	114.3	4.500	17.46	0.6875	72	2.88	9.44

TABLE 11-17 Multitube Hairpin Section Data

	Shell OD		Shell thickness		Tube count	Surface area for 6.1-m (20-ft) nominal length	
Size	mm	in	mm	in	19 mm	m^2	ft^2
03-MT	88.9	3.500	5.49	0.216	4	2.98	32.1
04-MT	114.3	4.500	6.02	0.237	9	6.73	72.4
05-MT	141.3	5.563	6.55	0.258	14	10.5	113.2
06-MT	168.3	6.625	7.11	0.280	22	16.7	179.6
08-MT	219.1	8.625	8.18	0.322	42	32.0	344.3
10-MT	273.1	10.75	9.27	0.365	70	54.0	581.3
12-MT	323.9	12.75	9.53	0.375	109	84.7	912.1
14-MT	355.6	14.00	9.53	0.375	136	107	1159
16-MT	406.4	16.00	9.53	0.375	187	148	1594
18-MT	457.2	18.00	9.53	0.375	230	182	1960
20-MT	508.0	20.00	9.53	0.375	295	235	2529
22-MT	558.8	22.00	9.53	0.375	367	294	3162
24-MT	609.6	24.00	9.53	0.375	450	362	3902
26-MT	660.4	26.00	9.53	0.375	526	427	4591
28-MT	711.2	28.00	9.53	0.375	622	508	5463
30-MT	762.0	30.00	11.11	0.4375	733	602	6475

1.8 times the shell outside diameter, with this ratio decreasing slightly for the larger sizes.

One interesting consequence of this fact is the inability to construct a hairpin tube bundle having the smallest radius bends common to a conventional U-tube, TEMA shell, and tube bundle. In fact, in the larger hairpin sizes the tubes might be better described as curved rather than bent. The smallest U-bend diameters are greater than the outside diameter of shells less than 300 mm in size. The U-bend diameters are greater than 300 mm in larger shells. As a general rule, mechanical tube cleaning around the radius of a U-bend may be accomplished with a flexible shaft-cleaning tool for bend diameters greater than 10 times the tube's inside diameter. This permits the tool to pass around the curve of the tube bend without binding.

In all these configurations, maintaining longitudinal flow on both the shell side and the tube side allows the decision for placement of a fluid stream on either one side or the other to be based upon design efficiency (mass flow rates, fluid properties, pressure drops, and velocities), and not because there is any greater tendency to foul on one side than the other. Experience has shown that, in cases where fouling is influenced by flow velocity, overall fouling in tube bundles is less in properly designed longitudinal flow bundles where areas of low velocity can be avoided without flow-induced tube vibration.

This same freedom of stream choice is not as readily applied when a segmental baffle is used. In those designs, the baffle's creation of low velocities and stagnant flow areas on the outside of the bundle can result in increased shell-side fouling at various locations of the bundle. The basis for choosing the stream side in those cases will be similar to the common shell-and-tube heat exchanger. At times a specific selection of stream side must be made regardless of the tube support mechanism in expectation of an unresolvable fouling problem. However, this is often the exception rather than the rule.

DESIGN APPLICATIONS

One benefit of the hairpin exchanger is its ability to handle high tube-side pressures at a lower cost than other removable-bundle exchangers. This is due in part to the lack of pass partitions at the tubesheets which complicate the gasketing design process. Present mechanical design technology has allowed the building of dependable, removable-bundle, hairpin multitubes at tube-side pressures of 825 bar (12,000 psi).

The best-known use of the hairpin exchanger is its operation in true countercurrent flow, which yields the most efficient design for processes that have a close temperature approach or temperature cross. However, maintaining countercurrent flow in a tubular heat exchanger usually implies one tube pass for each shell pass. As recently as 30 years ago, the lack of inexpensive, multiple-tube pass capability often diluted the advantages gained from countercurrent flow.

The early attempts to solve this problem led to investigations into the area of heat-transfer augmentation. This familiarity with augmentation techniques inevitably led to improvements in the efficiency and capacity of the small heat exchangers. The result has been the application of the hairpin heat exchanger to the solution of unique process problems, such as dependable, once-through, convective boilers offering high-exit qualities, especially in cases of process temperature crosses.

AIR-COOLED HEAT EXCHANGERS

INTRODUCTION TO AIR-COOLED HEAT EXCHANGERS

Atmospheric air has been used for many years to cool and condense fluids in areas of water scarcity. During the 1960s the use of air-cooled heat exchangers grew rapidly in the United States and elsewhere. In Europe, where seasonal variations in ambient temperatures are relatively small, air-cooled exchangers are used for the greater part of process cooling. In some new plants all cooling is done with air. Increased use of air-cooled heat exchangers has resulted from lack of available water, significant increases in water costs, and concern for water pollution.

Air-cooled heat exchangers include a tube bundle, which generally has spiral-wound fins upon the tubes, and a fan, which moves air across the tubes and is provided with a driver. Electric motors are the most commonly used drivers; typical drive arrangements require a V belt or a direct right-angle gear. A plenum and structural supports are basic components. Louvers are often used.

A bay generally has two tube bundles installed in parallel. These may be in the same or different services. Each bay is usually served by two (or more) fans and is furnished with a structure, a plenum, and other attendant equipment.

The location of air-cooled heat exchangers must take into consideration the large space requirements and the possible recirculation of heated air because of the effect of prevailing winds upon buildings, fired heaters, towers, various items of equipment, and other air-cooled exchangers. Inlet air temperature at the exchanger can be significantly higher than the ambient air temperature at a nearby weather station. See *Air-Cooled Heat Exchangers for General Refinery Services,* API Standard 661, 2d ed., January 1978, for information on refinery process air-cooled heat exchangers.

Forced and Induced Draft The forced-draft unit, illustrated in Fig. 11-43, pushes air across the finned-tube surface. The fans are located below the tube bundles. The induced-draft design has the fan above the bundle, and the air is pulled across the finned-tube surface. In theory, a primary advantage of the forced-draft unit is that less power is required. This is true when the air temperature rise exceeds 30°C (54°F).

Air-cooled heat exchangers are generally arranged in banks with several exchangers installed side by side. The height of the bundle aboveground

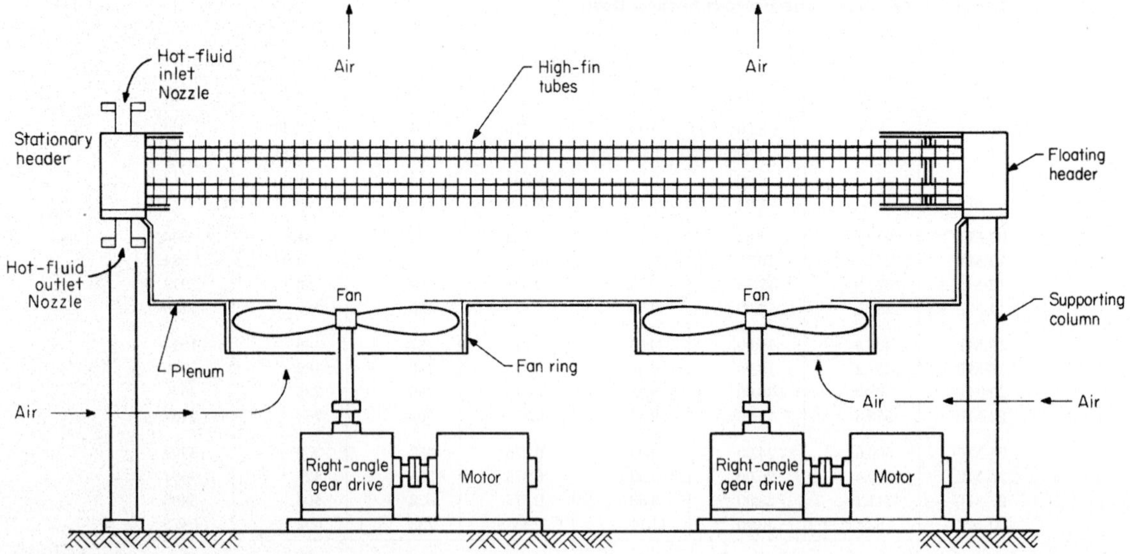

FIG. 11-43 Forced-draft air-cooled heat exchanger. [*Chem. Eng.* 114 (Mar. 27, 1978).]

must be one-half of the tube length to produce an inlet velocity equal to the face velocity. This requirement applies both to ground-mounted exchangers and to those pipe-rack-installed exchangers which have a fire deck above the pipe rack.

The forced-draft design offers better accessibility to the fan for on-stream maintenance and fan blade adjustment. The design also provides a fan and V-belt assembly, which are not exposed to the hot-air stream that exits from the unit. Structural costs are lower, and mechanical life is longer.

Induced-draft design provides more even distribution of air across the bundle, since air velocity approaching the bundle is relatively low. This design is better suited for exchangers designed for a close approach of product outlet temperature to ambient-air temperature.

Induced-draft units are less likely to recirculate the hot exhaust air, since the exit air velocity is several times that of the forced-draft unit. Induced-draft design more readily permits the installation of the air-cooled equipment above other mechanical equipment such as pipe racks or shell-and-tube exchangers.

In a service in which sudden temperature change would cause upset and loss of product, the induced-draft unit gives greater protection in that only a fraction of the surface (as compared with the forced-draft unit) is exposed to rainfall, sleet, or snow.

Tube Bundle The principal parts of the tube bundle are the finned tubes and the header. Most commonly used is the plug header, which is a welded box illustrated in Fig. 11-44. The finned tubes are described in a subsequent paragraph. The components of a tube bundle are identified in the figure.

The second most commonly used header is a cover-plate header. The cover plate is bolted to the top, bottom, and end plates of the header. Removing the cover plate provides direct access to the tubes without the necessity of removing individual threaded plugs.

Other types of headers include the bonnet-type header, which is constructed similarly to the bonnet of shell-and-tube heat exchangers; manifold-type headers, which are made from pipe and have tubes welded into the manifold; and billet-type headers, made from a solid piece of material with machined channels for distributing the fluid. Serpentine-type tube bundles are sometimes used for very viscous fluids. A single continuous flow path through pipe is provided.

Tube bundles are designed to be rigid and self-contained and are mounted so that they expand independently of the supporting structure.

The face area of the tube bundle is its length times width. The net free area for air flow through the bundle is about 50 percent of the face area of the bundle.

The standard air *face velocity* (FV) is the velocity of standard air passing through the tube bundle and generally ranges from 1.5 to 3.6 m/s (300 to 700 ft/min).

Tubing The 25.4-mm (1-in) OD tube is most commonly used. Fin heights vary from 12.7 to 15.9 mm (0.5 to 0.625 in), fin spacing from 3.6 to 2.3 mm (7 to 11 per linear inch), and tube triangular pitch from 50.8 to

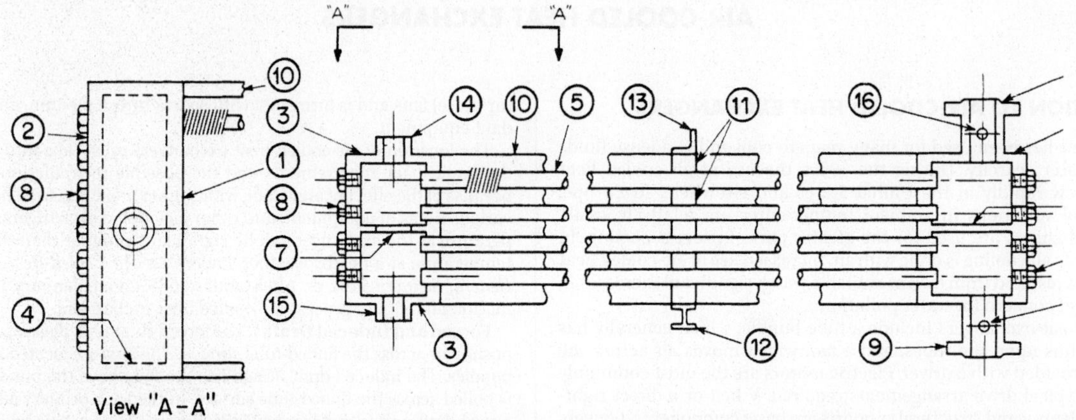

FIG. 11-44 Typical construction of a tube bundle with plug headers: (1) tube sheet; (2) plug sheet; (3) top and bottom plates; (4) end plate; (5) tube; (6) pass partition; (7) stiffener; (8) plug; (9) nozzle; (10) side frame; (11) tube spacer; (12) tube-support cross member; (13) tube keeper; (14) vent; (15) drain; (16) instrument connection. (*API Standard 661.*)

63.5 mm (2.0 to 2.5 in). The ratio of extended surface to bare-tube outside surface varies from about 7 to 20. The 38-mm (1½-in) tube has been used for flue gas and viscous oil service. Tube size, fin heights, and fin spacing can be further varied.

Tube lengths vary and may be as great as 18.3 m (60 ft). When tube length exceeds 12.2 m (40 ft), three fans are generally installed in each bay. Frequently used tube lengths vary from 6.1 to 12.2 m (20 to 40 ft).

Finned-Tube Construction The following are descriptions of commonly used finned-tube constructions (Fig. 11-45).

1. *Embedded.* The rectangular-cross-section aluminum fin is wrapped under tension and mechanically embedded in a groove 0.25 ± 0.05 mm (0.010 ± 0.002 in) deep, spirally cut into the outside surface of a tube.
2. *Integral (or extruded).* An aluminum outer tube from which fins have been formed by extrusion is mechanically bonded to an inner tube or liner.
3. *Overlapped footed.* L-shaped aluminum fin is wrapped under tension over the outside surface of a tube, with the tube fully covered by the overlapped feet under and between the fins.
4. *Footed.* L-shaped aluminum fin is wrapped under tension over the outside surface of a tube with the tube fully covered by the feet between the fins.
5. *Bonded.* Tube fins are bonded to the outside surface by hot-dip galvanizing, brazing, or welding.

Typical metal design temperatures for these finned-tube constructions are 399°C (750°F) embedded, 288°C (550°F) integral, 232°C (450°F) overlapped footed, and 177°C (350°F) footed.

Tube ends are left bare to permit insertion of the tubes into appropriate holes in the headers or tubesheets. Tube ends are usually roller-expanded into these tube holes.

Fans Axial-flow fans are large-volume, low-pressure devices. Fan diameters are selected to give velocity pressures of approximately 2.5 mm (0.1 in) of water. Total fan efficiency (fan, driver, and transmission device) is about 75 percent, and fan drives usually have a minimum of 95 percent mechanical efficiency.

Usually fans are provided with four or six blades. Larger fans may have more blades. Fan diameter is generally slightly less than the width of the bay.

At the fan-tip speeds required for economic performance, a large amount of noise is produced. The predominant source of noise is vortex shedding at the trailing edge of the fan blade. Noise control of air-cooled exchangers is required by the Occupational Safety and Health Act (OSHA). API Standard 661 (*Air-Cooled Heat Exchangers for General Refinery Services,* 2d ed., January 1978) has the purchaser specifying sound-pressure-level (SPL) values per fan at a location designated by the purchaser and also specifying sound-power-level (PWL) values per fan. These are designated at the following octave-band-center frequencies: 63, 125, 250, 1000, 2000, 4000, 8000, and also the dBa value (the dBa is a weighted single-value sound pressure level).

Reducing the fan-tip speed results in a straight-line reduction in air flow while the noise level decreases. The API Standard limits fan-tip speed to 61 m/s (12,000 ft/min) for typical constructions. Fan design changes that reduce noise include increasing the number of fan blades, increasing the width of the fan blades, and reducing the clearance between fan tip and fan ring.

Both the quantity of air and the developed static pressure of fans in air-cooled heat exchangers are lower than indicated by fan manufacturers' test data, which are applicable to testing-facility tolerances and not to heat exchanger constructions.

The axial-flow fan is inherently a device for moving a consistent volume of air when blade setting and speed of rotation are constant. Variation in the amount of air flow can be obtained by adjusting the blade angle of the fan and the speed of rotation. The blade angle can be (1) permanently fixed, (2) hand-adjustable, or (3) automatically adjusted. Air delivery and power are a direct function of blade pitch angle.

Fan mounting should provide a minimum of one-half to three-fourths diameter between fan and ground on a forced-draft heat exchanger and one-half diameter between tubes and fan on an induced-draft cooler.

Fan blades can be made of aluminum, molded plastic, laminated plastic, carbon steel, stainless steel, and Monel.

Fan Drivers Electric motors or steam turbines are most commonly used. These connect with gears or V belts. (Gas engines connected through gears and hydraulic motors either direct-connected or connected through gears are in use.) Fans may be driven by a prime mover such as a compressor with a V-belt takeoff from the flywheel to a jack shaft and then through a gear or V belt to the fan. Direct motor drive is generally limited to small-diameter fans.

V-belt drive assemblies are generally used with fans 3 m (10 ft) and less in diameter and motors of 22.4 kW (30 hp) and less.

Right-angle gear drive is preferred for fans over 3 m (10 ft) in diameter, for electric motors over 22.4 kW (30 hp), and with steam-turbine drives.

Fan Ring and Plenum Chambers The air must be distributed from the circular fan to the rectangular face of the tube bundle. The air velocity at the fan is between 3.8 and 10.2 m/s (750 and 2000 ft/in). The plenum-chamber depth (from fan to tube bundle) is dependent upon the fan dispersion angle (Fig. 11-46), which should have a maximum value of 45°.

The fan ring is made to commercial tolerances for the relatively large-diameter fan. These tolerances are greater than those upon closely machined fan rings used for small-diameter laboratory-performance testing. Fan performance is directly affected by this increased clearance between the blade tip and the ring, and adequate provision in design must be made for the reduction in air flow. API Standard 661 requires that fan-tip clearance be a maximum of 0.5 percent of the fan diameter for diameters between 1.9 and 3.8 m (6.25 and 12.5 ft). Maximum clearance is 9.5 mm (⅜ in) for smaller fans and 19 mm (¾ in) for larger fans.

The depth of the fan ring is critical. Worsham (ASME Pap. 59-PET-27, Petroleum Mechanical Engineering Conference, Houston, 1959) reports an increase in flow varying from 5 to 15 percent with the same power consumption when the depth of a fan ring was doubled. The percentage increase was proportional to the volume of air and static pressure against which the fan was operating.

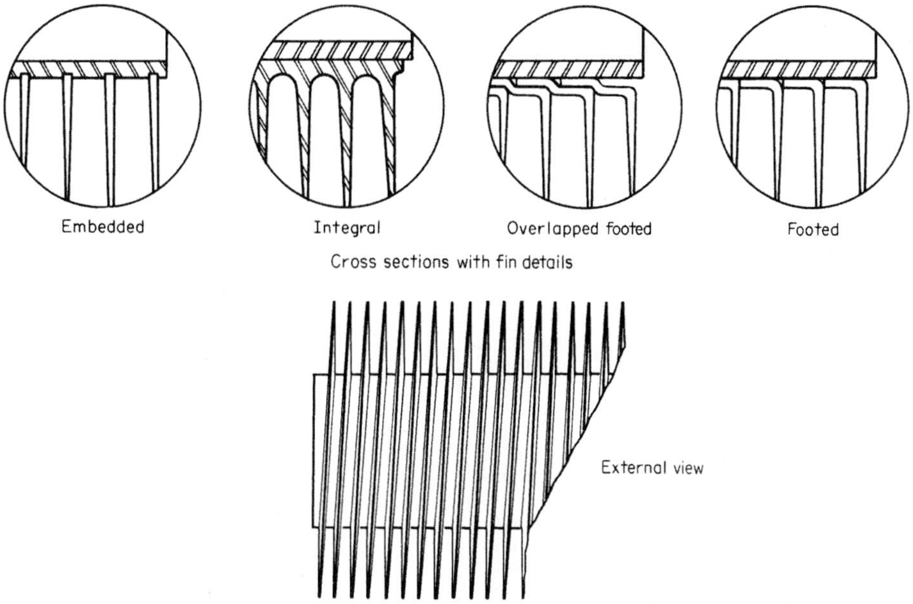

FIG. 11-45 Finned-tube construction.

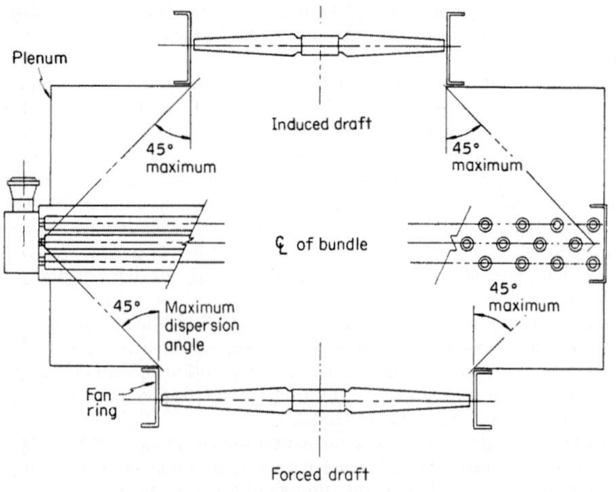

FIG. 11-46 Fan dispersion angle. (*API Standard 661.*)

When a selection is made, the stall-out condition, which develops when the fan cannot produce any more air regardless of power input, should be considered.

Air Flow Control Process operating requirements and weather conditions are considered in determining the method of controlling air flow. The most common methods include simple on-off control, on-off step control (in the case of multiple-driver units), two-speed-motor control, variable-speed drivers, controllable fan pitch, manually or automatically adjustable louvers, and air recirculation.

Winterization is the provision of design features, procedures, or systems for air-cooled heat exchangers to avoid process-fluid operating problems resulting from low-temperature inlet air. These include fluid freezing, pour point, wax formation, hydrate formation, laminar flow, and condensation at the dew point (which may initiate corrosion). The freezing points for some commonly encountered fluids in refinery service are benzene, 5.6°C (42°F); *p*-xylene, 15.5°C (55.9°F); cyclohexane, 6.6°C (43.8°F); phenol, 40.9°C (105.6°F); monoethanolamine, 10.3°C (50.5°F); and diethanolamine, 25.1°C (77.2°F). Water solutions of these organic compounds are likely to freeze in air-cooled exchangers during winter service. Paraffinic and olefinic gases (C_1 through C_4) saturated with water vapor form hydrates when cooled. These hydrates are solid crystals which can collect and plug exchanger tubes.

Air flow control in some services can prevent these problems. Cocurrent flow of air and process fluid during winter may be adequate to prevent problems. (Normal design has countercurrent flow of air and process fluid.) In some services when the hottest process fluid is in the bottom tubes, which are exposed to the lowest-temperature air, winterization problems may be eliminated.

Following are references which deal with problems in low-temperature environments: Brown and Benkley, "Heat Exchangers in Cold Service—A Contractor's View," *Chem. Eng. Prog.* **70:** 59–62 (July 1974); Franklin and Munn, "Problems with Heat Exchangers in Low Temperature Environments," *Chem. Eng. Prog.* **70:** 63–67 (July 1974); Newell, "Air-Cooled Heat Exchangers in Low Temperature Environments: A Critique," *Chem. Eng. Prog.* **70:** 86–91 (October 1974); Rubin, "Winterizing Air Cooled Heat Exchangers," *Hydrocarbon Process* **59:** 147–149 (October 1980); Shipes, "Air-Cooled Heat Exchangers in Cold Climates," *Chem. Eng. Prog.* **70:** 53–58 (July 1974).

Air Recirculation Recirculation of air which has been heated as it crosses the tube bundle provides the best means of preventing operating problems due to low-temperature inlet air. Internal recirculation is the movement of air within a bay so that the heated air which has crossed the bundle is directed by a fan with reverse flow across another part of the bundle. Wind skirts and louvers are generally provided to minimize the entry of low-temperature air from the surroundings. Contained internal recirculation uses louvers within the bay to control the flow of warm air in the bay, as illustrated in Fig. 11-47. Note that low-temperature inlet air has access to the tube bundle.

External recirculation is the movement of the heated air within the bay to an external duct, where this air mixes with inlet air, and the mixture serves as the cooling fluid within the bay. Inlet air does not have direct access to the tube bundle; an adequate mixing chamber is essential. Recirculation over the end of the exchanger is illustrated in Fig. 11-48. Over-the-side recirculation also is used. External recirculation systems maintain the desired low temperature of the air crossing the tube bundle.

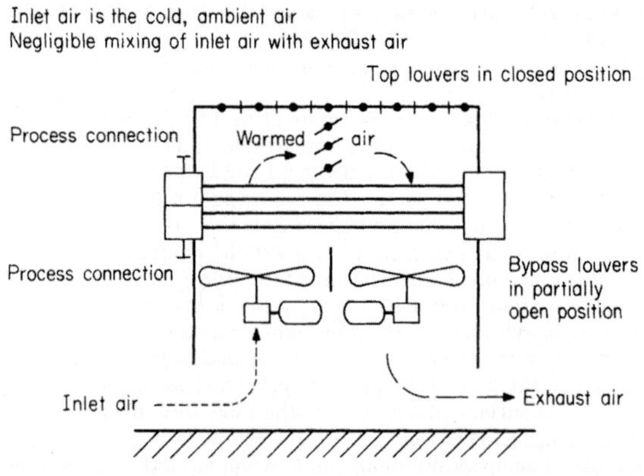

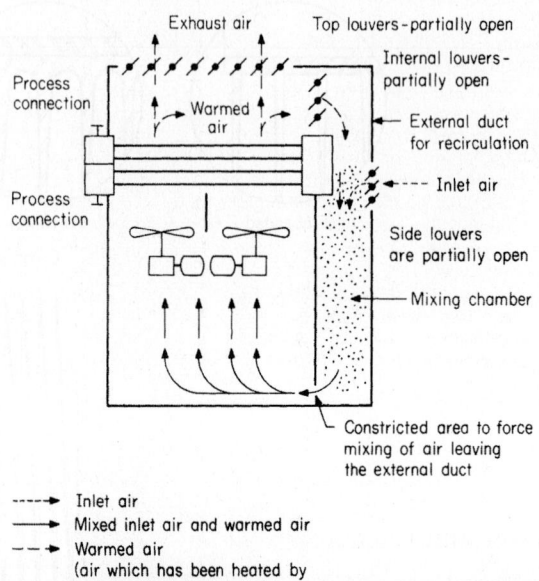

FIG. 11-47 Contained internal recirculation (with internal louvers). [*Hydrocarbon Process* 59: 148–149 (October 1980).]

Trim Coolers Conventional air-cooled heat exchangers can cool the process fluid to within 8.3°C (15°F) of the design dry-bulb temperature. When a lower process outlet temperature is required, a trim cooler is installed in series with the air-cooled heat exchanger. The water-cooled trim cooler can be designed for a 5.6°C to 11.1°C (10°F to 20°F) approach to the wet-bulb temperature, which in the United States is about 8.3°C (15°F) less than the dry-bulb temperature. In arid areas the difference between dry- and wet-bulb temperatures is much greater.

Humidification Chambers The air-cooled heat exchanger is provided with humidification chambers in which the air is cooled to a close approach to the wet-bulb temperature before entering the finned-tube bundle of the heat exchanger.

Evaporative Cooling The process fluid can be cooled by using evaporative cooling with the sink temperature approaching the wet-bulb temperature.

Steam Condensers Air-cooled steam condensers have been fabricated with a single tube-side pass and several rows of tubes. The bottom row has a higher temperature difference than the top row, since the air has been heated as it crosses the rows of tubes. The bottom row condenses all the entering steam before the steam has traversed the length of the tube. The top row, with a lower-temperature driving force, does not condense all the entering steam. At the exit header, uncondensed steam flows from the top row into the bottom row. Since noncondensible gases are always present in steam, these accumulate within the bottom row because steam is entering from both ends of the tube. Performance suffers.

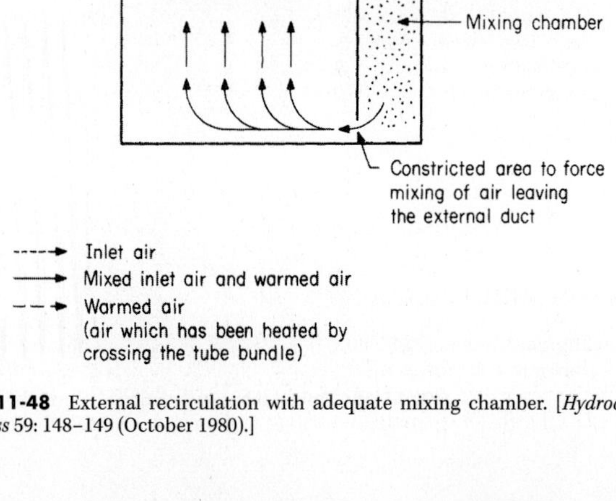

FIG. 11-48 External recirculation with adequate mixing chamber. [*Hydrocarbon Process* 59: 148–149 (October 1980).]

TABLE 11-18 Air-Cooled Heat-Exchanger Costs (1970)

Surface (bare tube), sq. ft.	500	1000	2000	3000	5000
Cost for 12-row-deep bundle, dollars/square foot	9.0	7.6	6.8	5.7	5.3
Factor for bundle depth:					
6 rows	1.07	1.07	1.07	1.12	1.12
4 rows	1.2	1.2	1.2	1.3	1.3
3 rows	1.25	1.25	1.25	1.5	1.5

Base: Bare-tube external surface 1 in. o.d. by 12 B.W.G. by 24 ft. 0 in. steel tube with 8 aluminum fins per inch ⅝-in. high. Steel headers. 150 lb./sq. in. design pressure. V-belt drive and explosion-proof motor. Bare-tube surface 0.262 sq. ft./ft. Fin-tube surface/bare-tube surface ratio is 16.9.

Factors:	
20 ft. tube length	1.05
30 ft. tube length	0.95
18 B.W.G. admiralty tube	1.04
16 B.W.G. admiralty tube	1.12

NOTE: To convert feet to meters, multiply by 0.3048; to convert square feet to square meters, multiply by 0.0929; and to convert inches to millimeters, multiply by 25.4.

Various solutions have been used. These include orifices to regulate the flow into each tube, a "blow-through steam" technique with a vent condenser, complete separation of each row of tubes, and inclined tubes.

Air-Cooled Overhead Condensers *Air-cooled overhead condensers* (AOCs) have been designed and installed above distillation columns as integral parts of distillation systems. The condensers generally have inclined tubes, with air flow over the finned surfaces induced by a fan. Prevailing wind affects both structural design and performance.

AOCs provide the additional advantages of reducing ground-space requirements and piping and pumping requirements and of providing smoother column operation.

The downflow condenser is used mainly for nonisothermal condensation. Vapors enter through a header at the top and flow downward. The reflux condenser is used for isothermal and small-temperature-change conditions. Vapors enter at the bottom of the tubes.

AOC usage first developed in Europe but became more prevalent in the United States during the 1960s. A state-of-the-art article was published by Dehne [*Chem. Eng. Prog.* **64:** 51 (July 1969)].

Air-Cooled Heat Exchanger Costs The cost data in Table 11-18 are unchanged from those published in the 1963 edition of this text. In 1969 Guthrie [*Chem. Eng.* **75:** 114 (Mar. 24, 1969)] presented cost data for field-erected air-cooled exchangers. These costs are only 25 percent greater than those of Table 11-18 and include the costs of steel stairways, indirect subcontractor charges, and field erection charges. Since minimal field costs would be this high (i.e., 25 percent of purchase price), the basic costs appear to be unchanged. (Guthrie indicated a cost band of ±25 percent.) Preliminary design and the cost estimation of air-cooled heat exchangers have been discussed by J. E. Lerner ["Simplified Air Cooler Estimating," *Hydrocarbon Process* **52:** 93–100 (February 1972)].

Design Considerations

1. *Design dry-bulb temperature.* The typically selected value is the temperature which is equaled or exceeded 2½ percent of the time during the warmest consecutive 4 months. Since air temperatures at industrial sites are frequently higher than those used for these weather data reports, it is good practice to add 1°C to 3°C (2°F to 6°F) to the tabulated value.

2. *Air recirculation.* Prevailing winds and the locations and elevations of buildings, equipment, fired heaters, etc., require consideration. All air-cooled heat exchangers in a bank are of one type, i.e., all forced-draft or all induced-draft. Banks of air-cooled exchangers must be placed far enough apart to minimize air recirculation.

3. *Wintertime operations.* In addition to the previously discussed problems of winterization, provision must be made for heavy rain, strong winds, freezing of moisture upon the fins, etc.

4. *Noise.* Two identical fans have a noise level 3 dBa higher than one fan, while eight identical fans have a noise level 9 dBa higher than a single fan. Noise level at the plant site is affected by the exchanger position, reflective surfaces near the fan, hardness of these surfaces, and noise from adjacent equipment. The extensive use of air-cooled heat exchangers contributes significantly to plant noise level.

5. *Ground area and space requirements.* Comparisons of the overall space requirements for plants using air cooling versus water cooling are not consistent. Some air-cooled units are installed above other equipment—pipe racks, shell-and-tube exchangers, etc. Some plants avoid such installations because of safety considerations, as discussed later.

6. *Safety.* Leaks in air-cooled units are transmitted directly to the atmosphere and can cause fire hazards or toxic-fume hazards. However, the large air flow through an air-cooled exchanger greatly reduces any concentration of toxic fluids. Segal [*Pet. Refiner* **38:** 106 (April 1959)] reports that air-fin coolers "are not located over pumps, compressors, electrical switchgear, control houses and, in general, the amount of equipment such as drums and shell-and-tube exchangers located beneath them are minimized."

Pipe-rack-mounted air-cooled heat exchangers with flammable fluids generally have concrete fire decks which isolate the exchangers from the piping.

7. *Atmospheric corrosion.* Air-cooled heat exchangers should not be located where corrosive vapors and fumes from vent stacks will pass through them.

8. *Air-side fouling.* Air-side fouling is generally negligible.

9. *Process-side cleaning.* Either chemical or mechanical cleaning on the inside of the tubes can readily be accomplished.

10. *Process-side design pressure.* The high-pressure process fluid is always in the tubes. Tube-side headers are relatively small compared with water-cooled units when the high pressure is generally on the shell side. High-pressure design of rectangular headers is complicated. The plug-type header is normally used for design gauge pressures to 13,790 kPa (2000 lbf/in^2) and has been used to 62,000 kPa (9000 lbf/in^2). The use of threaded plugs at these pressures creates problems. Removable cover plate headers are generally limited to gauge pressures of 2068 kPa (300 lbf/in^2). The expensive billet-type header is used for high-pressure service.

11. *Bond resistance.* Vibration and thermal cycling affect the bond resistance of the various types of tubes in different manners and thus affect the amount of heat transfer through the fin tube.

12. *Approach temperature.* The approach temperature, which is the difference between the process-fluid outlet temperature and the design dry-bulb air temperature, has a practical minimum of 8°C to 14°C (15°F to 25°F). When a lower process-fluid outlet temperature is required, an air humidification chamber can be provided to reduce the inlet air temperature toward the wet-bulb temperature. A 5.6°C (10°F) approach is feasible. Since typical summer wet-bulb design temperatures in the United States are 8.3°C (15°F) lower than dry-bulb temperatures, the outlet process-fluid temperature can be 3°C (5°F) below the dry-bulb temperature.

13. *Mean-temperature-difference (MTD) correction factor.* When the outlet temperatures of both fluids are identical, the MTD correction factor for a 1:2 shell-and-tube exchanger (one pass shell side, two or more passes tube side) is approximately 0.8. For a single-pass, air-cooled heat exchanger the factor is 0.91. A two-pass exchanger has a factor of 0.96, while a three-pass exchanger has a factor of 0.99 when passes are arranged for counterflow.

14. *Maintenance cost.* Maintenance for air-cooled equipment as compared with shell-and-tube coolers (complete with cooling-tower costs) indicates that air-cooling maintenance costs are approximately 0.3 to 0.5 times those for water-cooled equipment.

15. *Operating costs.* Power requirements for air-cooled heat exchangers can be lower than at the summer design condition provided that an adequate means of air flow control is used. The annual power requirement for an exchanger is a function of the means of air flow control, the exchanger service, the air-temperature rise, and the approach temperature.

When the mean annual temperature is 16.7°C (30°F) lower than the design dry-bulb temperature and when both fans in a bay have automatically controllable pitch of fan blades, the annual power required has been found to be 22, 36, and 54 percent, respectively, of that needed at the design condition for three process services [Frank L. Rubin, "Power Requirements Are Lower for Air-Cooled Heat Exchangers with AV Fans," *Oil Gas J.*, pp. 165–167 (Oct. 11, 1982)]. Alternatively, when fans have two-speed motors, these deliver one-half of the design flow of air at half speed and use only one-eighth of the power of the full-speed condition.

COMPACT AND NONTUBULAR HEAT EXCHANGERS

COMPACT HEAT EXCHANGERS

With equipment costs rising and limited available plot space, compact heat exchangers are gaining a larger portion of the heat exchanger market. Numerous types use special enhancement techniques to achieve the required heat transfer in smaller plot areas and, in many cases, require lower initial investment. As with all items that afford a benefit, a series of restrictions limit the effectiveness or application of these special heat exchanger products. In most products discussed, some of these considerations are presented, but a thorough review with reputable suppliers of these products is the only positive way to select a compact heat exchanger. The following guidelines will assist in prequalifying one of these.

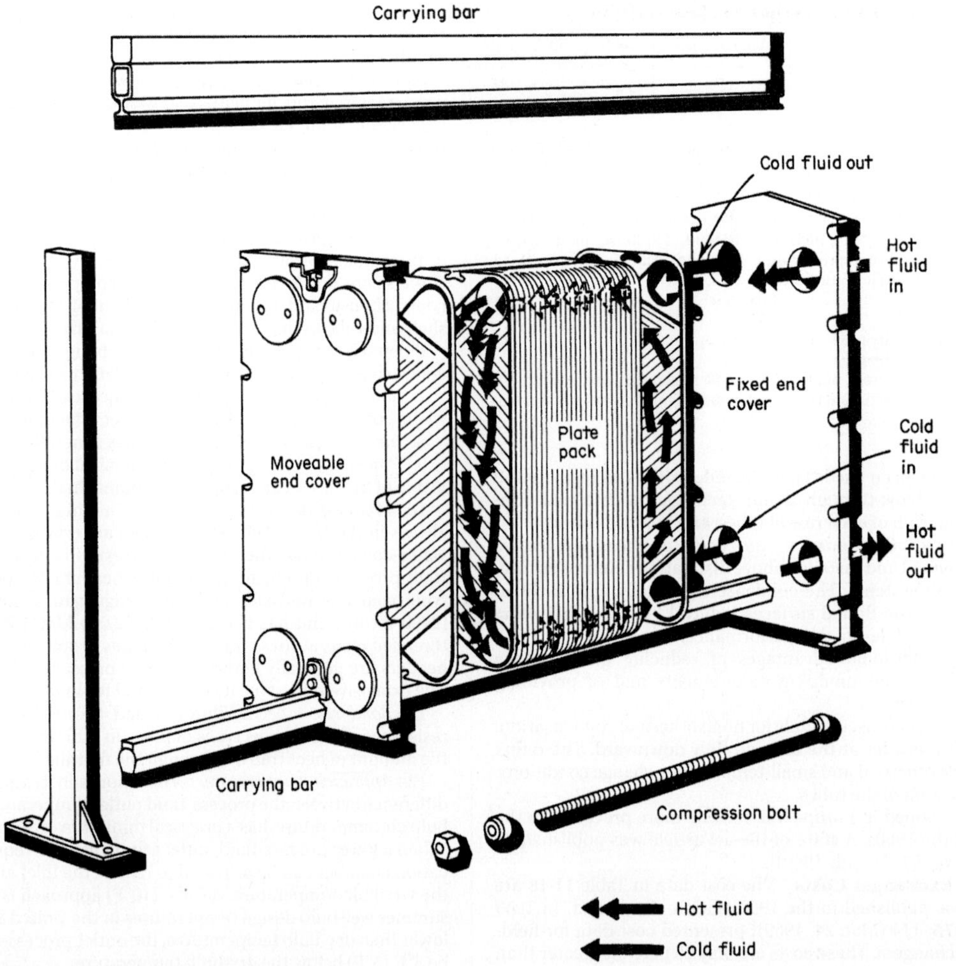

FIG. 11-49 Plate-and-frame heat exchanger. Hot fluid flows down between alternate plates, and cold fluid flows up between alternate plates. (*Thermal Division, Alfa-Laval, Inc.*)

PLATE-AND-FRAME EXCHANGERS

There are two major types: gasketed and welded-plate heat exchangers. Each is discussed individually.

GASKETED-PLATE EXCHANGERS

Description This type is the fastest growing of the compact exchangers and the most recognized (see Fig. 11-49). A series of corrugated alloy material channel plates, bounded by elastomeric gaskets, are hung off and guided by longitudinal carrying bars, then compressed by large-diameter tightening bolts between two pressure-retaining frame plates (cover plates). The frame and channel plates have portholes which allow the process fluids to enter alternating flow passages (the space between two adjacent-channel plates). Gaskets around the periphery of the channel plate prevent leakage to the atmosphere and prevent process fluids from coming in contact with the frame plates. No interfluid leakage is possible in the port area because of a dual-gasket seal.

The frame plates are typically epoxy-painted carbon-steel material and can be designed per most pressure vessel codes. Design limitations are found in Table 11-19. The channel plates are always an alloy material with 304SS as a minimum (see Table 11-19 for other materials).

Channel plates are typically 0.4 to 0.8 mm thick and have corrugation depths of 2 to 10 mm. Special wide-gap (WG PHE) plates are available, in limited sizes, for slurry applications with depths of approximately 16 mm. The channel plates are compressed to achieve metal-to-metal contact for pressure-retaining integrity. These narrow gaps and the high number of contact points that change the fluid flow direction combine to create a very high turbulence between the plates. This means high individual heat-transfer coefficients (up to 14,200 W/m^2·°C), but very high pressure drops per length as well. To compensate, the channel plate lengths are usually short, most under 2 m and few over 3 m in length. In general, the same pressure drops as conventional exchangers are used without loss of the enhanced heat transfer.

Expansion of the initial unit is easily performed in the field without special considerations. The original frame length typically has an additional capacity of 15 to 20 percent more channel plates (i.e., surface area). In fact, if a known future capacity is available during fabrication stages, a longer carrying bar could be installed, and increasing the surface area would be easily handled later. When the expansion is needed, simply loosen the carrying bolts, pull back the frame plate, add the additional channel plates, and tighten the frame plate. API 662 Part I now covers the design of these types of heat exchangers.

Applications Most plate heat exchanger (PHE) applications historically have been liquid-liquid services, but there has been significant development in the use of PHE for vacuum condensing as well as evaporation and column reboiling. Industrial users typically have chevron-style channel plates while some food applications are washboard style.

Fine particulate slurries in concentrations up to 70 percent by weight are possible with standard channel spacings. Wide-gap units are used with larger particle sizes. Typical particle size should not exceed 75 percent of the *single* plate (not total channel) gap.

Close temperature approaches and tight temperature control possible with PHEs and the ability to sanitize the entire heat-transfer surface easily were a major benefit in the food industry.

Multiple services in a single frame are possible.

Gasket selection is one of the most critical and limiting factors in PHE use. Table 11-20 gives some guidelines for fluid compatibility. Even trace

TABLE 11-19 Compact Exchanger Applications Guide

Design conditions	G. PHE	W. PHE	WG. PHE	BHE	DBL	MLT	STE	CP	SHE	THE
Design temperature, °C	180	150	150	185	+500	+500	+500	450	+400	+500
Minimum metal temp., °C	−30	−30	−30	−160	−160	−160	−160	−160	−160	−160
Design pressure, MPa	3	2.5	0.7	3.1	+20	+20	+20	4.2	10	+20
Inspect for leakage	Yes	Partial	Yes	No	Yes	Yes	Yes	Partial	Yes	Yes
Mechanical cleaning	Yes	Yes/no	Yes	No	Yes	Yes	Yes/no	Yes	Yes	Yes
Chemical cleaning	Yes	Yes	Yes	Yes	Yes	Yes	Yes	Yes	Yes	Yes
Expansion capability	Yes	Yes	Yes	No	No	No	No	No	No	No
Repair	Yes	Yes/no	Yes	No	Yes	Yes	Partial	Yes	Partial	Yes
Temperature cross	Yes	Yes	Yes	Yes	Yes	Yes	Yes	Yes	Yes	No*
Surface area/unit, m^2	2800	900	250	50	10	150	60	850	450	High
Holdup volume	Low	Low	Low	Low	Med	Med	Low	Low	Med	High
Materials	**G. PHE**	**W. PHE**	**WG. PHE**	**BHE**	**DBL**	**MLT**	**STE**	**CP**	**SHE**	**THE**
Mild steel	No	No	No	No	Yes	Yes	Yes	No	Yes	Yes
Stainless	Yes	Yes	Yes	Yes	Yes	Yes	Yes	Yes	Yes	Yes
Titanium	Yes	Yes	Yes	No	Yes	Yes	Yes	Yes	Yes	Yes
Hastalloy	Yes	Yes	No	No	Yes	Yes	Yes	Yes	Yes	Yes
Nickel	Yes	Yes	No	No	Yes	Yes	Yes	Yes	Yes	Yes
Alloy 20	Yes	Yes	No	No	Yes	Yes	Yes	Yes	Yes	Yes
Incoloy 825	Yes	Yes	No	No	Yes	Yes	Yes	Yes	Yes	Yes
Monel	Yes	Yes	No	No	Yes	Yes	Yes	Yes	Yes	Yes
Impervious graphite	Yes	No	No	No	No	No	No	No	No	Yes
Service	**G. PHE**	**W. PHE**	**WG. PHE**	**BHE**	**DBL**	**MLT**	**STE**	**CP**	**SHE**	**THE**
Clean fluids	A	A	A	A	A	A	A	A	A	A
Gasket incompatibility	D	A/D	D	A	A	A	A	A	A	A
Medium viscosity	A/B	A/B	A/B	B	A	A	A/B	A/B	A	A
High viscosity	A/B	A/B	A/B	D	A	A	A/B	A/B	A	A
Slurries and pulp (fine)	B/D	D	A/B	C	A	A/B	C	B	A	A/D
Slurries and pulp (coarse)	D	D	B	D	A	B/C	D	B	A	A/D
Refrigerants	D	A	D	A	A	A	B/C	A	A	A
Thermal fluids	D	A/B	D	A/B	A	A	C	A	A	A
Vent condensers	D	D	D	D	A/D	A	A	B/C	A	A
Process condenser	B	C	D	D	A/D	A	A	A/B	B	A
Vacuum reboil/condenser	A	D	B	D	A/D	B	A	B/C	B	A/C
Evaporator	A	C	C	A	B	B	A	B/C	C	A
Tight temp. control	A	A	A	A	A	A	A	B	A	C
High scaling	B	B	A	D	A	A/B	B/C	B	B	A/D

*Multipass.
Adapted from Alfa-Laval literature.
A—Very good C—Fair
B—Good D—Poor

fluid components need to be considered. The higher the operating temperature and pressure, the shorter the anticipated gasket life. *Always* consult the supplier on gasket selection and obtain an estimated or guaranteed lifetime.

TABLE 11-20 Elastomer Selection Guide

	Uses	Avoid
Nitrile (NBR)	Oil resistant Fat resistant Food stuffs Mineral oil Water	Oxidants Acids Aromatics Alkalies Alcohols
Resin cured butyl (IIR)	Acids Lyes Strong alkalies Strong phosphoric acid Dilute mineral acids Ketones Amines Water	Fats and fatty acids Petroleum oils Chlorinated hydrocarbons Liquids with dissolved chlorine Mineral oil Oxygen rich demin. water Strong oxidants
Ethylene-propylene (EPDM)	Oxidizing agents Dilute acids Amines Water (Mostly any IIR fluid)	Oils Hot & conc. acids Very strong oxidants Fats & fatty acids Chlorinated hydrocarbons
Viton (FKM, FPM)	Water Petroleum oils Many inorganic acids (Most all NBR fluids)	Amines Ketones Esters Organic acids Liquid ammonia

The major applications are, but not limited to, the following:

Temperature cross applications	(lean/rich solvent)
Close approaches	(freshwater/seawater)
Viscous fluids	(emulsions)
Sterilized surface required	(food, pharmaceutical)
Polished surface required	(latex, pharmaceutical)
Future expansion required	
Space restrictions	
Barrier coolant services	(closed-loop coolers)
Slurry applications	(TiO_2, Kaolin, precipitated calcium carbonate, and beet sugar raw juice)

Design Plate exchangers are becoming so commonplace that there is now an API 662 document available for the specification of these products. In addition, commercial computer programs are available from HTRI among others. Standard channel-plate designs, unique to each manufacturer, are developed with limited modifications of each plate's corrugation depths and included angles. Manufacturers combine their different style plates to custom-fit each service. Due to the possible combinations, it is impossible to present a way to exactly size PHEs. However, it is possible to estimate areas for new units and to predict performance of existing units with different conditions (chevron-type channel plates are presented).

The fixed length and limited corrugation included angles on channel plates makes the NTU (number of heat transfer units) method of sizing practical. (Water-like fluids are assumed for the following examples.)

$$\text{NTU} = \frac{\Delta t \text{ of either side}}{\text{LMTD}} \tag{11-76}$$

Most plates have NTU values of 0.5 to 4.0, with 2.0 to 3.0 as the most common (multipass shell-and-tube exchangers are typically less than 0.75). The more closely the fluid profile matches that of the channel plate, the smaller the required surface area. Attempting to increase the service NTU beyond the plate's NTU capability causes oversurfacing (inefficiency).

True sizing from scratch is impractical since a pressure balance on a channel-to-channel basis, from channel closest to inlet to farthest, must be achieved and when mixed plate angles are used; this is quite a challenge. Computer sizing is not just a benefit, it is a necessity for supplier's selection. Averaging methods are recommended to perform any sizing calculations.

From the APV heat-transfer handbook *Design and Application of Paraflow-Plate Heat Exchangers* and J. Marriott's article "Where and How to Use Plate Heat Exchangers," *Chemical Engineering,* April 5, 1971, there are the equations for plate heat transfer.

$$\text{Nu} = \frac{h\text{De}}{k} = 0.28 \times (\text{Re})^{0.65} \times (\text{Pr})^{0.4} \tag{11-77}$$

where De = 2 × depth of single-plate corrugation. Also

$$G = \frac{W}{\text{Np} \times w \times \text{De}} \tag{11-78}$$

The width of the plate w is measured from inside to inside of the channel gasket. If it is not available, use the tear-sheet drawing width and subtract 2 times the bolt diameter and subtract another 50 mm. For depth of corrugation ask the supplier, or take the compressed plate pack dimension, divide by the number of plates, and subtract the plate thickness from the result. The *number of passages* Np is the number of plates minus 1, then divided by 2.

Typical overall coefficients to start a rough sizing are listed below. Use these in conjunction with the NTU calculated for the process. The closer the NTU matches the plate (say, between 2.0 and 3.0), the higher the range of listed coefficients that can be used. The narrower (smaller) the depth of corrugation, the higher the coefficient (and pressure drop), but also the lower the ability to carry through any particulate.

Water/water	5700–7400 W/(m²·°C)
Steam/water	5700–7400 W/(m²·°C)
Glycol/glycol	2300–4000 W/(m²·°C)
Amine/amine	3400–5000 W/(m²·°C)
Crude/emulsion	400–1700 W/(m²·°C)

Pressure drops typically can match conventional tubular exchangers. Again from the APV handbook, an average correlation is as follows:

$$\Delta P = \frac{2fG^2L}{g\rho\text{De}} \tag{11-79}$$

where $f = 2.5(G\,\text{De}/\mu)^{-0.3}$
g = gravitational constant

Fouling factors are typically one-tenth of TEMA values or an oversurfacing of 10 to 20 percent is used (J. Kerner, "Sizing Plate Exchangers," *Chemical Engineering,* November 1993).

LMTD is calculated as a 1 pass-1 pass shell and tube with no F correction factor required in most cases.

Overall coefficients are determined as for shell-and-tube exchangers; that is, sum all the resistances, then invert. The resistances include the hot-side coefficient, the cold-side coefficient, the fouling factor (usually only a total value, not individual values per fluid side), and the wall resistance.

WELDED- AND BRAZED-PLATE EXCHANGERS

The title of this group of plate exchangers has been used for a great variety of designs for various applications from normal gasketed-plate exchanger services to air-preheater services on fired heaters or boilers. The intent here is to discuss more traditional heat exchanger designs, not the heat recovery designs on fired equipment flue-gas streams. Many similarities exist between these products, but the manufacturing techniques are quite different due to the normal operating conditions these units experience.

To overcome the gasket limitations, PHE manufacturers have developed welded-plate exchangers. There are numerous approaches to this solution: weld plate pairs together with the other fluid side conventionally gasketed; weld up both sides but use a horizonal stacking-of-plates method of assembly; entirely braze the plates together with copper or nickel brazing; diffusion bond and then pressure-form plates and bond-etched passage plates. The act of welding the plates together removed one of the largest limitations of the plate-and-frame: the many gaskets and their compatibility limitations to process fluids.

Most methods of welded-plate manufacturing do not allow for inspection of the heat-transfer surface or mechanical cleaning of that surface, and they have limited ability to repair or plug off damaged channels. Consider these limitations when the fluid is heavily fouling, has solids, or in general, the need for repair or the potential of plugging for severe services.

One of the previous types has an additional issue of the brazing material to consider for fluid compatibility. The brazing compound entirely coats both fluids' heat-transfer surfaces.

The second type, a Compabloc (CP) from Alfa-Laval AB, has the advantage of removable cover plates, similar to air-cooled exchanger headers, to observe both fluids' surface areas. The fluids flow at 90° angles to each other on a horizontal plane. LMTD correction factors approach 1.0 for Compabloc just as for the other welded and gasketed PHEs. Hydroblast cleaning of Compabloc surfaces is also possible. The Compabloc has higher operating conditions than PHEs or W-PHE.

The performances and estimating methods of welded PHEs match those of gasketed PHEs in most cases, but normally the Compabloc, with larger depth of corrugations, can be lower in overall coefficient. Some extensions of the design operating conditions are possible with welded PHEs; most notable is that cryogenic applications are possible. Pressure vessel code acceptance is available on most units.

Applications of welded plate exchangers, especially the Compabloc type, are increasingly being accepted in the chemical industry as a reliable process heat exchanger. Typical applications include, but not limited to, these:

1. Crude oil preheat trains
2. Oil and gas facilities, crude and gas sweetening applications
3. Process condensers and reboilers on chemical industry distillation columns
4. Refined product coolers and primary petrochemical manufacturing heat exchangers
5. Pharmaceutical condensers
6. Other areas where space and weight are primary concerns

COMBINATION WELDED-PLATE EXCHANGERS

Plate exchangers are well known for their high efficiency but suffer from limitations on operating pressure. Several companies have rectified this limitation by placing the welded plate exchanger inside a pressure vessel to withstand the pressure, such as Alfa Laval's Packinox design. One popular application is the feed effluent exchange in a catalytic reforming plant for oil refineries. Large volumes of gases with some liquids require cross-exchange to feed a reactor system. Close temperature approaches and lower pressure drops are required as well as a very clean service. These combined units provide an economic alternative to shell-and-tube exchangers.

SPIRAL-PLATE HEAT EXCHANGER (SHE)

Description The *spiral-plate heat exchanger* (SHE) may be one exchanger selected primarily for its virtues and not for its initial cost. SHEs offer high reliability and on-line performance in many severely fouling services such as slurries.

The SHE is formed by rolling two strips of plate, with welded-on spacer studs, upon each other into clock-spring shape. This forms two passages. Passages are sealed off on one end of the SHE by welding a bar to the plates; hot and cold fluid passages are sealed off on opposite ends of the SHE. A single rectangular flow passage is now formed for each fluid, producing very high shear rates compared to tubular designs. Removable covers are provided on each end to access and clean the entire heat-transfer surface. Pure countercurrent flow is achieved, and the LMTD correction factor is essentially = 1.0.

Since there are no dead spaces in a SHE, the helical flow pattern combines to entrain any solids and create high turbulence, creating a self-cleaning flow passage.

There are no thermal expansion problems in spirals. Since the center of the unit is not fixed, it can torque to relieve stress.

The SHE can be expensive when only one fluid requires a high-alloy material. Since the heat-transfer plate contacts both fluids, it is required to be fabricated out of the higher alloy. SHEs can be fabricated out of any material that can be cold-worked and welded.

The channel spacings can be different on each side to match the flow rates and pressure drops of the process design. The spacer studs are also adjusted in their pitch to match the fluid characteristics.

As the operating pressure or design conditions require, the plate thickness increases from a minimum of 2 mm to a maximum (as required by pressure) up to 10 mm if the shell is integrally rolled. For high pressures, the SHE is inserted in a standard pressure vessel with cover flanges. This means relatively thick material separates the two fluids compared to the tubing of conventional exchangers. Pressure vessel code conformance is a common request. API 664 was recently issued covering this type of heat exchanger and its use.

Applications The most common applications that fit SHE are slurries. The single rectangular channel provides an ideal geometry to sweep the surface clear of blockage and causes none of the distribution problems associated with other exchanger types. A localized restriction causes an increase in local velocity which aids in keeping the unit free-flowing, eliminating plugging problems that plague other heat exchanger types. Only fibers that are long and stringy cause SHE to have a blockage it cannot clear itself.

As an additional antifoulant measure, SHEs have been coated with a phenolic lining. This provides some degree of corrosion protection as well, but this is not guaranteed due to pinholes in the lining process.

There are three types of SHE to fit different applications:

Type I is the spiral-spiral flow pattern. It is used for all heating and cooling services and can accommodate temperature crosses such as lean/rich services in one unit. The removable covers on each end enable access to one side at a time to perform maintenance on that fluid side. Never remove a cover with one side under pressure as the unit will telescope out like a collapsible cup.

Type II units are the condenser and reboiler designs. One side is spiral flow, and the other side is in cross-flow. These SHEs provide very stable designs for vacuum condensing and reboiling services. A SHE can be fitted with special mounting connections for reflux-type vent-condenser applications. The vertically mounted SHE directly attaches onto the column or tank.

Type III units are a combination of the type I and type II where part is in spiral flow and part is in cross-flow. This SHE can condense and subcool in a single unit.

The unique channel arrangement has been used to provide on-line cleaning, by switching fluid sides to clean the fouling (caused by the fluid that previously flowed there) off the surface. Phosphoric acid coolers use pond water for cooling and both sides foul; water, as you expect, and phosphoric acid deposit crystals. By reversing the flow sides, the water dissolves the acid crystals and the acid clears up the organic fouling. SHEs are also used as oleum coolers, sludge coolers/heaters, slop oil heaters, and in other services where multiple-flow-passage designs have not performed well.

Design A thorough article by P. E. Minton of Union Carbide called "Designing Spiral-Plate Heat Exchangers" appeared in *Chemical Engineering* on May 4, 1970. It covers the design in detail. Also an article in *Chemical Engineering Progress* titled "Applications of Spiral Plate Heat Exchangers" by A. Hargis, A. Beckman, and J. Loicano appeared in July 1967 and provides formulas for heat-transfer and pressure-drop calculations.

Spacings are from 6.35 to 31.75 mm (in 6.35-mm increments) with 9.5 mm the most common. Stud densities are 60 × 60 to 110 × 110 mm, with the former the most common. The width (measured to the spiral flow passage) is from 150 to 2500 mm (in 150-mm increments). By varying the spacing and the width, separately for each fluid, velocities can be maintained at optimum rates to reduce fouling tendencies or utilize the allowable pressure drop most effectively. Diameters can reach 2500 mm. The total surface areas exceed 465 m². Materials that work harder are not suitable for spirals since hot-forming is not possible and heat treatment after forming is impractical.

$$\mathrm{Nu} = \frac{H\,\mathrm{De}}{k} = 0.0315\,(\mathrm{Re})^{0.8}(\mathrm{Pr})^{0.25}(\mu/\mu_w)^{0.17} \qquad (11\text{-}80)$$

where De = 2 × spacing and flow area = width × spacing

$$\Delta P = \frac{LV^2\rho}{1.705\mathrm{E}03} \times 1.45 \quad (1.45 \text{ for } 60\text{-}\times 60\text{-mm studs}) \qquad (11\text{-}81)$$

The LMTD and overall coefficient are calculated as in the PHE section above.

BRAZED-PLATE-FIN HEAT EXCHANGERS

Brazed-aluminum-plate-fin heat exchangers (or *core exchangers* or *cold boxes*, as they are sometimes called) were first manufactured for the aircraft industry during World War II. In 1950, the first tonnage air-separation plant with these compact, lightweight, reversing heat exchangers began producing oxygen for a steel mill. Aluminum-plate-fin exchangers are used in the process and gas-separation industries, particularly for services below −45°C.

Core exchangers are made up of a stack of rectangular sheets of aluminum separated by a wavy, usually perforated, aluminum fin. Two ends are sealed off to form a passage (see Fig. 11-50). The layers have the wavy fins and sealed ends alternating at 90° to each. Aluminum half-pipe-type headers are attached to the open ends to route the fluids into the alternating passages. Fluids usually flow at this same 90° angle to one another. Variations in the fin height, number of passages, and length and width of the prime sheet allow for the core exchanger to match the needs of the intended service.

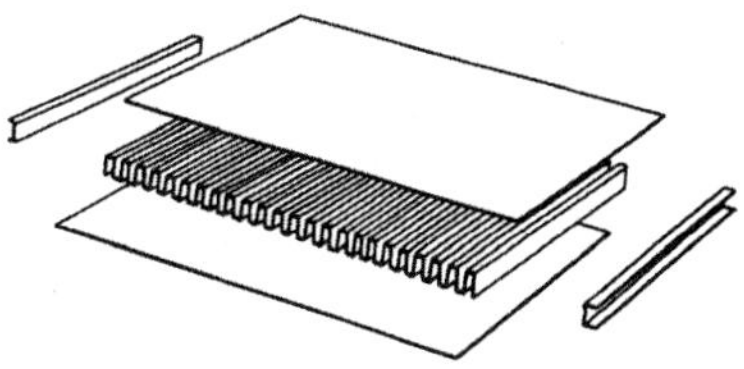

FIG. 11-50 Exploded view of a typical plate-fin arrangement. (*Trane Co.*)

Design conditions range in pressures from full vacuum to 96.5 bar g and in temperatures from −269°C to 200°C. This is accomplished while meeting the quality standards of most pressure vessel codes. API 662 Part 2 has been developed for this type of heat exchanger.

Design and Application Brazed plate heat exchangers have two design standards available. One is ALPEMA, the Aluminum Plate-Fin Heat Exchanger Manufacturers' Association, and the other is the API 662 document for plate heat exchangers.

Applications are varied for this highly efficient, compact exchanger. Mainly it is seen in the cryogenic fluid services of air-separation plants, in refrigeration trains as in ethylene plants, and in natural-gas processing plants. Fluids can be all vapor, liquid, condensing, or vaporizing. Multifluid exchangers and multiservice cores, that is, one exchanger with up to 10 different fluids, are common for this type of product. Cold boxes are a group of cores assembled into a single structure or module, prepiped for minimum field connections. (Data were obtained from ALTEC International, now Chart Industries. For detailed information refer to *GPSA Engineering Handbook*, Sec. 9.)

PLATE-FIN TUBULAR EXCHANGER (PFE)

Description These shell-and-tube exchangers are designed to use a group of tightly spaced plate fins to increase the shell-side heat-transfer performance as fins do on double-pipe exchangers. In this design, a series of very thin plates (fins), usually of copper or aluminum, are punched to the same pattern as the tube layout, spaced very close together, and mechanically bonded to the tube. Fin spacing is 315 to 785 FPM (fins per meter) with 550 FPM most common. The fin thicknesses are 0.24 mm for aluminum and 0.19 mm for copper. Surface-area ratios over bare prime-tube units can be 20:1 to 30:1. The cost of the additional plate-fin material, without a reduction in shell diameter in many cases, and increased fabrication has to be offset by the total reduction of plot space and prime tube surface area. The more costly the prime tube or plot space, the better the payout for this design. A rectangular tube layout is normally used with *no tubes in the window* (NTIW). The window area (where no tubes are) of the plate fins is cut out. This causes a larger shell diameter for a given tube count compared to conventional tubular units. A dome area on the top and bottom of the inside of the shell has been created for the fluid to flow along the tube length. To exit the unit, the fluid must flow across the plate-finned tube bundle with extremely low pressure loss. The units from the outside and from the tube side appear as in any conventional shell-and-tube exchanger.

Applications Two principal applications are rotating equipment oil coolers and compressor intercoolers and after-coolers. Although seemingly different applications, both rely on the shell-side finning to enhance the heat transfer of low heat-transfer characteristic fluids, viscous oils, and gases. By the nature of the fluids and their applications, both are clean servicing. The tightly spaced fins would be a maintenance problem otherwise.

Design The economics usually work out in the favor of gas coolers when the centrifugal machine's flow rate reaches about 5000 scfm. The pressure loss can be kept to 7.0 kPa in most cases. When the ratio of $A_t h_t$ to $A_s h_s$ is 20:1, this is another point to consider in these plate-fin designs. Vibration is practically impossible with this design, and uses in reciprocating compressors are possible because of this.

Marine and hydraulic-oil coolers use these characteristics to enhance the coefficient of otherwise poorly performing fluids. The higher metallurgies in marine applications such as 90/10 Cu-Ni allow the higher cost of plate-fin design to be offset by the reduced amount of alloy material being used. On small hydraulic coolers, these fins usually allow one to two size smaller coolers for the package and save skid space and initial cost.

Always check on metallurgy compatibility and cleanliness of the shell-side fluid! (Data are provided by Bos-Hatten and ITT-Standard.)

PRINTED-CIRCUIT HEAT EXCHANGERS

These are a variation of the welded or brazed plate heat exchangers that uses a chemical etching process to form the flow channels and diffusion bonding technique to secure the plates together. These units have the high heat-transfer characteristics and extended operating conditions that welded or brazed units have, but the diffusion process makes the bond the same strength as that of the prime plate material. The chemical etching, similar to that used in printed circuitry, allows greater flexibility in flow channel patterns than any other heat exchanger. This type of heat exchanger is perhaps the most compact design of all due to the infinite variations in passage size, layout, and direction.

Headers are welded on the core block to direct the fluids into the appropriate passages. The all-metal design allows very high operating conditions for both temperature and pressure. The diffusion bonding provides a near-homogeneous material for fluids that are corrosive or require high purity.

These exchangers can handle gases, liquids, and two-phase applications. They have the greatest potential in cryogenic, refrigeration, gas processing, and corrosive chemical applications. Other applications are possible with the exception of fluids containing solids: the narrow passages, as in most plate exchangers, are conducive to plugging.

SPIRAL-TUBE EXCHANGER (STE)

Description These exchangers are typically a series of stacked helical-coil tubes connected to manifolds, then inserted into a casing or shell. They have many advantages similar to those of spiral-plate designs, such as avoiding differential expansion problems, acceleration effects of the helical flow increasing the heat-transfer coefficient, and compactness of plot area. They are typically selected because of their economical design.

The most common form has both sides in helical flow patterns, pure countercurrent flow is followed, and the LMTD correction factor approaches 1.0. Temperature crosses are possible in single units. As with the spiral-plate unit, different configurations are possible for special applications.

Tube material includes any that can be formed into a coil, but usually copper, copper alloys, and stainless steel are most common. The casing or shell material can be cast iron, cast steel, cast bronze, fabricated steel, stainless steel, and other high-alloy materials. Units are available with pressure vessel code conformance.

The data provided here have been supplied by Graham Mfg. for their units called Heliflow.

Applications The common Heliflow applications are tank-vent condensers, sample coolers, pump-seal coolers, and steam-jet vacuum condensers. Instant water heaters, glycol/water services, and cryogenic vaporizers use the spiral tube's ability to reduce thermally induced stresses caused in these applications.

Many other applications are well suited for spiral tube units, but many believe only small surface areas are possible with these units. Graham Mfg. states that units are available to 60 m^2. Their ability to polish the surfaces, double-wall the coil, use finned coil, and insert static mixers, among other configurations in design, make them quite flexible. Tube-side design pressures can be up to 69,000 kPa. A cross-flow design on the external surface of the coil is particularly useful in steam-jet ejector condensing service. These Heliflow units can be made very cost-effective, especially in small units. The main difference, compared to spiral plate, is that the tube side cannot be cleaned except chemically and that multiple flow passages make tube-side slurry applications (or fouling) impractical.

Design The fluid flow is similar to that of the spiral-plate exchangers, but through parallel tube passages. Graham Mfg. has a liquid-liquid sizing pamphlet available from its local distributor. An article by M. A. Noble, J. S. Kamlani, and J. J. McKetta ("Heat Transfer in Spiral Coils," *Petroleum Engineer*, April 1952, p. 723) discusses sizing techniques.

The tube-side fluid must be clean or at least chemically cleanable. With a large number of tubes in the coil, cleaning of inside surfaces is not totally reliable. Fluids that attack stressed materials such as chlorides should be reviewed as to proper coil-material selection. Fluids that contain solids can be a problem due to erosion of relatively thin coil materials, unlike for the thick plates in spiral-plate units and multiple, parallel fluid passages compared to a single passage in spiral-plate units.

GRAPHITE HEAT EXCHANGERS

Impervious graphite exchangers now come in a variety of geometries to suit the particular requirements of the service. They include cubic block form, drilled cylinder block, shell-and-tube, and plate-and-frame.

Description Graphite is one of three crystalline forms of carbon. The other two are diamond and charcoal. Graphite has a hexagonal crystal structure, diamond is cubic, and charcoal is amorphous. Graphite is inert to most chemicals and resists corrosion attack. It is, however, porous and it must be impregnated with a resin sealer to be used. Two main resins used are phenolic and PTFE with furan (one currently being phased out of production). Selection of resins includes chemical compatibility, operating temperatures, and type of unit to be used. For proper selection, consult with a graphite supplier.

Shell-and-tube units in graphite were started by Karbate in 1939. The European market started using block design in the 1940s. Both technologies utilize the high thermal conductivity of the graphite material to compensate for the poor mechanical strength. The thicker materials needed to sustain pressure do not adversely impede the heat transfer. Maximum design pressures range from 0.35 to 1.0 kPa depending on the type and size of exchanger. Design temperature is dependent on the fluid and resin selection, the maximum is 230°C.

In all situations, the graphite heat-transfer surface is contained within a metal structure or a shell (graphite-lined on process side) to maintain the design pressure. For shell-and-tube units, the design is a packed floating tubesheet at both ends within a shell and channel. For stacked block design, the standardize blocks are glued together with special adhesives and compressed within a framework that includes manifold connections for each fluid. The cylindrical block unit is a combination of the above two with blocks glued together and surrounded by a pressure-retaining shell. Pressure vessel code conformance of the units is possible due to the metallic components of these designs. Since welding of graphite is not possible, the selection and application of the adhesives used are critical to the proper operation of these units. Tube–tubesheet joints are glued since rolling of tubes into tubesheet is not possible. The packed channels and gasketed manifold connections are two areas of additional concern when one is selecting sealants for these units.

Applications and Design The major applications for these units are in the acid-related industries. Sulfuric, phosphoric, and hydrochloric acids require either very costly metals or impervious graphite. Usually graphite is the more cost-effective material used. Applications are increasing in the herbicide and pharmaceutical industries as new products with chlorine and fluorine compounds expand. Services are coolers, condensers, and evaporators, basically all services requiring this material. Types of units are shell-and-tube, block-type (circular and rectangular), and plate-and-frame type of exchangers. The designs of the shell-and-tube units are the same as any others, but the design characteristics of tubes, spacing, and thickness are unique to the graphite design. The block and plate and frame also can be evaluated by using techniques previously addressed; but again the unique characteristics of the graphite materials require input from a reputable supplier. Most designs will need the supplier to provide the most cost-effective design for the immediate and future operation of the exchangers. Also consider the entire system design as some condensers and/or evaporators can be integral with their associated column.

CASCADE COOLERS

Cascade coolers are a series of standard pipes, usually manifolded in parallel and connected in series by vertically or horizontally oriented U bends. Process fluid flows inside the pipe entering at the bottom, and water trickles from the top downward over the external pipe surface. The water is collected from a trough under the pipe sections, cooled, and recirculated over the pipe sections. The pipe material can be any of the metallics and also glass, impervious graphite, and ceramics. The tube-side coefficient and pressure drop are as in any circular duct. The water coefficient (with Re number less than 2100) is calculated from the following equation by W. H. McAdams, T. B. Drew, and G. S. Bays, Jr., from *ASME Trans.* **62:** 627–631 (1940).

$$h = 218 \times (G'/D_o)^{1/3} \quad (\text{W/m}^2\cdot{}^\circ\text{C}) \tag{11-82}$$

where $G' = m/(2L)$

m = water rate, kg/h

L = length of each pipe section, m

D_o = outside diameter of pipe, m

LMTD corrections are per Fig. 11-4*i* or *j* depending on U-bend orientation.

BAYONET-TUBE EXCHANGERS

This type of exchanger gets its name from its design, which is similar to a bayonet sword and its associated scabbard or sheath. The bayonet tube is a smaller-diameter tube inserted into a larger-diameter tube that has been capped at one end. The fluid flow typically enters the inner tube, exiting, hitting the cap of the larger tube, and returning to the opposite direction in the annular area. The design eliminates any thermal expansion problems. It also creates a unique nonfreeze-type tube side for steam heating of cryogenic fluids; the inner tube steam keeps the annulus condensate from freezing against the cold shell-side fluid. This design can be expensive on a surface-area basis due to the need for a double-channel design, and only the outer tube surface is used to transfer heat. LMTD calculations for nonisothermal fluid are quite extensive, and those applications are far too few to attempt to define it. The heat transfer is like the annular calculation of a double-pipe unit. The shell side is a conventional baffled shell-and-tube design. A rigorous treatment of the design of bayonet exchangers, "Understanding Bayonet Heat Exchangers" by Richard L. Shilling, is available through Heat Transfer Research, Inc.

ATMOSPHERIC SECTIONS

These consist of a rectangular bundle of tubes in similar fashion to air cooler bundles, placed just under the cooled water distribution section of a cooling tower. It, in essence, combines the exchanger and cooling tower into a single piece of equipment. This design is practical only for single-service cooler/condenser applications, and expansion capabilities are not provided. The process fluid flows inside the tubes, and the cooling tower provides cool water that flows over the outside of the tube bundle. Water quality is critical for these applications to prevent fouling or corrosive attack on the outside of the tube surfaces and to prevent blockage of the spray nozzles. The initial and operating costs are lower than those for a separate cooling tower and exchanger. Principal applications now are in the HVAC, refrigeration, and industrial systems. Sometimes these are called *wet surface air coolers.*

$$h = 1729\,[(\text{m}^2/\text{h})/\text{face area m}^2]^{1/3} \qquad (11\text{-}83)$$

NONMETALLIC HEAT EXCHANGERS

Another growing field is that of nonmetallic heat exchanger designs which typically are of the shell-and-tube or coiled-tubing type. The graphite units were previously discussed, but numerous other materials are available. The materials include Teflon, PVDF, glass, ceramic, and others as the need arises.

When using these types of products, one should consider the following topics and discuss the application openly with experienced suppliers.

1. The tube-to-tubesheet joint, how is it made? Many use O-rings to add another material to the selection process. Preference should be given to a fusing technique of similar material.
2. What size tube or flow passage is available? Small tubes plug unless filtration is installed. The size of filtering is needed from the supplier.
3. These materials are very sensitive to temperature and pressure. Thermal or pressure shocks must be avoided.
4. Thermal conductivity of these materials is very low and affects the overall coefficient. When several materials are compatible, explore all of them, as final cost is not always the same as raw material costs.

HEAT EXCHANGERS FOR SOLIDS

This section describes equipment for heat transfer to or from solids by the indirect mode. Such equipment is constructed so that the solids load (burden) is separated from the heat-carrier medium by a wall; the two phases are never in direct contact. Heat transfer is by conduction based on diffusion laws. Equipment in which the phases are in direct contact is covered in other sections of this text, principally in Sec. 20.

Some of the devices covered here handle the solids burden in a static or laminar-flow bed. Other devices can be considered as continuously agitated kettles in their heat-transfer aspect. For the latter, unit-area performance rates are higher.

Computational and graphical methods for predicting performance are given for both major heat-transfer aspects in Sec. 10. In solids heat processing with indirect equipment, the engineer should remember that the heat-transfer capability of the wall is many times that of the solids burden. Hence the solids properties and bed geometry govern the rate of heat transfer. This is more fully explained earlier in this section. Only limited resultant (not predictive) and "experience" data are given here.

EQUIPMENT FOR SOLIDIFICATION

A frequent operation in the chemical field is the removal of heat from a material in a molten state to effect its conversion to the solid state. When the operation is carried on batchwise, it is termed *casting,* but when done continuously, it is termed *flaking.* Because of rapid heat transfer and temperature variations, jacketed types are limited to an initial melt temperature of 232°C (450°F). Higher temperatures [to 316°C (600°F)] require extreme care in jacket design and cooling-liquid flow pattern. Best performance and greatest capacity are obtained by (1) holding precooling to the minimum and (2) optimizing the cake thickness. The latter cannot always be done from the heat-transfer standpoint, as size specifications for the end product may dictate thickness.

Table Type This is a simple flat metal sheet with slightly upturned edges and jacketed on the underside for coolant flow. For many years this was the mainstay of food processors. Table types are still widely used when production is done in small batches, when considerable batch-to-batch variation occurs, for pilot investigation, and when the cost of continuous devices is unjustifiable. Slab thicknesses are usually in the range of 13 to 25 mm (½ to 1 in). These units are homemade, with no standards available. Initial cost is low, but operating labor is high.

Agitated-Pan Type A natural evolution from the table type is a circular flat surface with jacketing on the underside for coolant flow and the added feature of a stirring means to sweep over the heat-transfer surface. This device is the agitated-pan type (Fig. 11-51). It is a batch-operation device. Because of its age and versatility, it still serves a variety of heat-transfer operations for the chemical-process industries. While the most prevalent designation is agitated-pan dryer (in this mode, the burden is heated rather than cooled), considerable use is made of it for solidification applications. In this field, it is particularly suitable for processing burdens that change phase (1) slowly, by "thickening," (2) over a wide temperature range, (3) to an amorphous solid form, or (4) to a soft semigummy form (versus the usual hard crystalline structure).

The stirring produces the end product in the desired divided-solids form. Hence, it is frequently termed a *granulator* or a *crystallizer.* A variety of factory-made sizes in various materials of construction are available. Initial cost is modest, while operating cost is rather high (as is true of all batch devices), but the ability to process "gummy" burdens and/or simultaneously effect two unit operations often yields an economical application.

Vibratory Type This construction (Fig. 11-52) takes advantage of the burden's special needs and the characteristic of vibratory actuation. A flammable burden requires the use of an inert atmosphere over it and a suitable nonhazardous fluid in the jacket. The vibratory action permits construction of rigid self-cleaning chambers with simple flexible connections. When solidification has been completed and vibrators started, the intense vibratory motion of the whole deck structure (as a rigid unit) breaks free the friable cake [up to 76 mm (3 in) thick], shatters it into lumps, and conveys it up over the dam to discharge. Heat-transfer performance is good, with overall coefficient U of about 68 W/(m^2·°C) [12 Btu/(h·ft^2·°F)] and values of heat flux q on the order of 11,670 W/m^2 [3700 Btu/(h·ft^2)]. Application of timing-cycle controls and a surge hopper for the discharge solids facilitates automatic operation of the caster and continuous operation of subsequent equipment.

Belt Types The patented metal-belt type (Fig. 11-53*a*), termed the "water-bed" conveyor, features a thin wall, a well-agitated fluid side for a thin water film (there are no rigid welded jackets to fail), a stainless-steel or Swedish-iron conveyor belt "floated" on the water with the aid of guides, no removal knife, and cleanability. It is mostly used for cake thicknesses of 3.2 to 15.9 mm (⅛ to ⅝ in) at speeds up to 15 m/min (50 ft/min), with

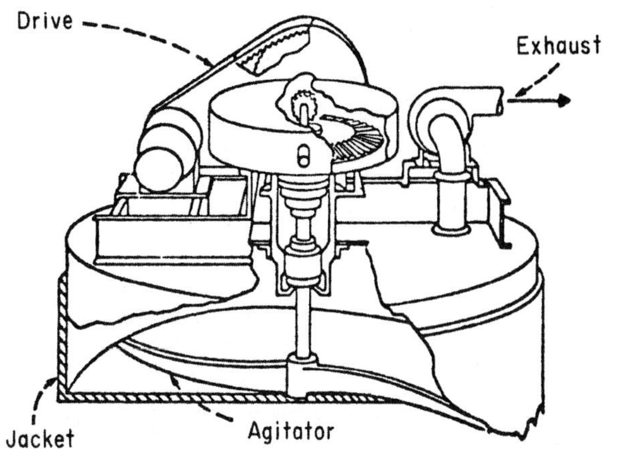

FIG. 11-51 Heat-transfer equipment for solidification (with agitation); agitated-pan type.

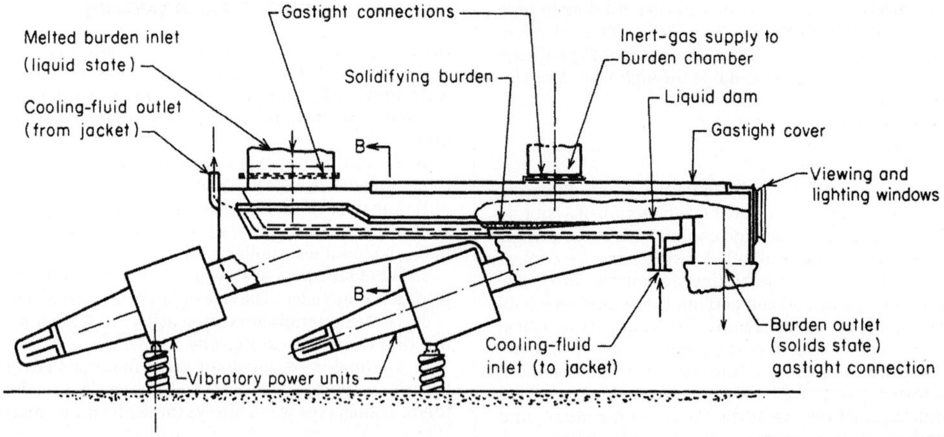

FIG. 11-52 Heat-transfer equipment for batch solidification; vibrating-conveyor type. (*Courtesy of Jeffrey Mfg. Co.*)

45.7-m (150-ft) pulley centers common. For 25- to 32-mm (1- to 1¼-in) cake, another belt on top to give two-sided cooling is frequently used. Applications are in food operations for cooling to harden candies, cheeses, gelatins, margarines, gums, etc.; and in chemical operations for solidification of sulfur, greases, resins, soaps, waxes, chloride salts, and some insecticides. Heat transfer is good, with sulfur solidification showing values of $q = 5800$ W/m^2 [1850 Btu/(h·ft^2)] and $U = 96$ W/(m^2·°C) [17 Btu/(h·ft^2·°F)] for a 7.9-mm ($^5/_{16}$-in) cake.

The submerged metal belt (Fig. 11-53*b*) is a special version of the metal belt to meet the peculiar handling properties of pitch in its solidification process. Although adhesive to a dry metal wall, pitch will not stick to the submerged wetted belt or rubber edge strips. Submergence helps to offset the very poor thermal conductivity through two-sided heat transfer.

A fairly recent application of the water-cooled metal belt to solidification duty is shown in Fig. 11-54. The operation is termed *pastillizing* from the form of the solidified end product, termed *pastilles.* The novel feature is a one-step operation from the molten liquid to a fairly uniformly sized and shaped product without intermediate operations on the solid phase.

Another development features a nonmetallic belt [*Plast. Des. Process* **13** (July 1968)]. When rapid heat transfer is the objective, a glass-fiber, Teflon-coated construction in a thickness as little as 0.08 mm (0.003 in) is selected for use. No performance data are available, but presumably the thin belt permits rapid heat transfer while taking advantage of the nonsticking property of Teflon. Another development [*Food Process. Mark.* **69** (March 1969)] is extending the capability of belt solidification by providing use of subzero temperatures.

Rotating-Drum Type This type (Fig. 11-55*a* and *b*) is not an adaptation of a material-handling device (though volumetric material throughput is a first consideration) but is designed specifically for heat-transfer service. It is well engineered, established, and widely used. The twin-drum type (Fig. 11-55*b*) is best suited to thin [0.4- to 6-mm (1/64- to ¼-in)] cake production. For temperatures to 149°C (300°F) the coolant water is piped in and siphoned out. Spray application of coolant water to the inside is employed for high-temperature work, permitting feed temperatures to at least 538°C (1000°F), or double those values for jacketed equipment. Vaporizing refrigerants are readily applicable for very low temperature work.

The burden must have a definite solidification temperature to ensure proper pickup from the feed pan. This limitation can be overcome by side feeding through an auxiliary rotating spreader roll. Application limits are further extended by special feed devices for burdens having oxidation-sensitive and/or supercooling characteristics. The standard double-drum model turns downward, with adjustable roll spacing to control sheet thickness. The newer twin-drum model (Fig. 11-55*b*) turns upward and, though subject to variable cake thickness, handles viscous and indefinite solidification temperature-point burden materials well.

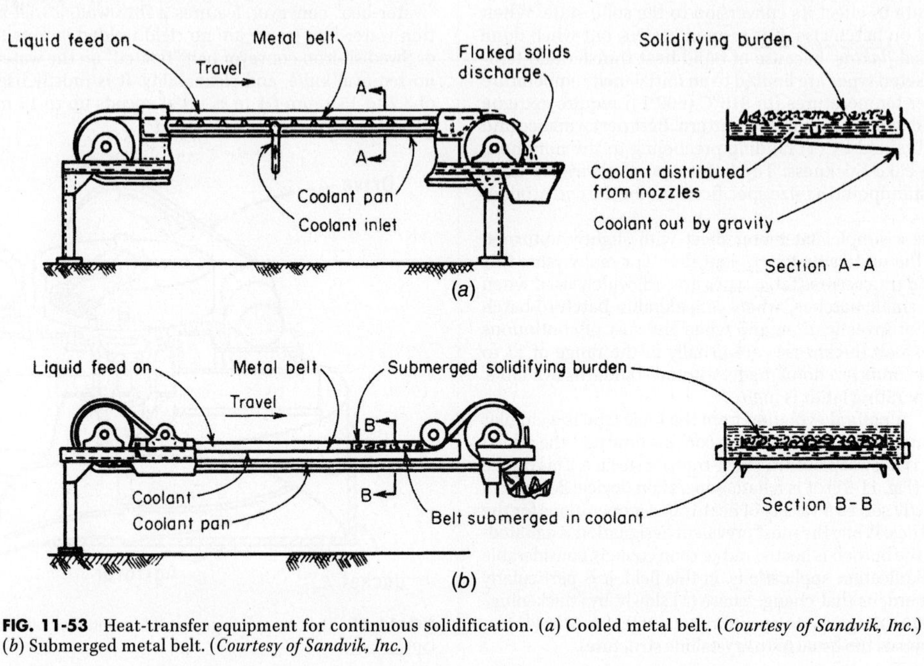

FIG. 11-53 Heat-transfer equipment for continuous solidification. (*a*) Cooled metal belt. (*Courtesy of Sandvik, Inc.*) (*b*) Submerged metal belt. (*Courtesy of Sandvik, Inc.*)

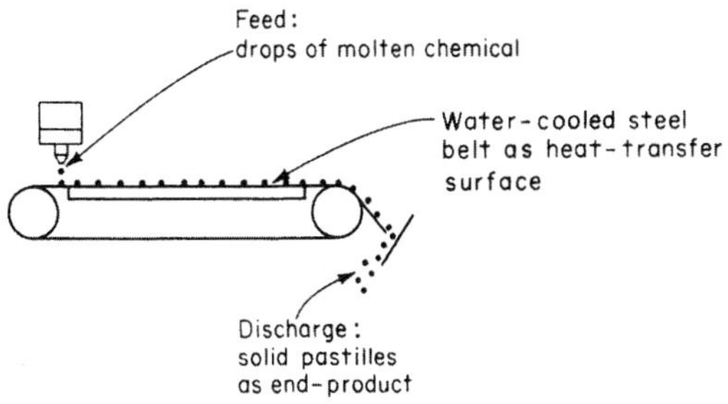

FIG. 11-54 Heat-transfer equipment for solidification; belt type for the operation of pastillization. (*Courtesy of Sandvik, Inc.*)

Drums have been successfully applied to a wide range of chemical products, both inorganic and organic, pharmaceutical compounds, waxes, soaps, insecticides, food products to a limited extent (including lard cooling), and even flake-ice production. A novel application is that of using a water-cooled roll to pick up from a molten-lead bath and turn out a 1.2-m- (4-ft-) wide continuous sheet, weighing 4.9 kg/m² (1 lb/ft²), which is ideal for a sound barrier. This technique is more economical than other sheeting methods [*Mech. Eng.* **631** (March 1968)].

Heat-transfer performance of drums, in terms of reported heat flux, is: for an 80°C (176°F) melting-point wax, 7880 W/m² [2500 Btu/(h·ft²)]; for a 130°C (266°F) melting-point organic chemical, 20,000 W/m² [6500 Btu/(h·ft²)]; and for high-melting-point [318°C (604°F)] caustic soda (water-sprayed in drum), 95,000 to 125,000 W/m² [30,000 to 40,000 Btu/(h·ft²)], with overall coefficients of 340 to 450 W/(m²·°C) [60 to 80 Btu/(h·ft²·°F)]. An innovation that is claimed often to increase these performance values by as much as 300 percent is the addition of hoods to apply impinging streams of heated air to the solidifying and drying solids surface as the drums carry it upward [*Chem. Eng.* **74:** 152 (June 19, 1967)]. Similar rotating-drum indirect heat-transfer equipment is also extensively used for drying duty on liquids and thick slurries of solids (see Sec. 20).

Rotating-Shelf Type The patented Roto-shelf type (Fig. 11-55*c*) features (1) a large heat-transfer surface provided over a small floor space and in a small building volume, (2) easy floor cleaning, (3) nonhazardous machinery, (4) stainless-steel surfaces, (5) good control range, and (6) substantial capacity by providing as needed 1 to 10 shelves operated in parallel. It is best suited for thick-cake production and burden materials having an indefinite solidification temperature. Solidification of liquid sulfur into 13- to 19-mm- (½- to ¾-in-) thick lumps is a successful application. Heat transfer, by liquid-coolant circulation through jackets, limits feed temperatures to 204°C (400°F). Heat-transfer rate, controlled by the thick cake rather than by equipment construction, should be equivalent to the belt type. Thermal performance is aided by applying water sprayed directly to the burden top to obtain two-sided cooling.

EQUIPMENT FOR FUSION OF SOLIDS

The thermal duty here is the opposite of solidification operations. The indirect heat-transfer equipment suitable for one operation is not suitable for the other because of the material-handling aspects rather than the thermal aspects. Whether the temperature of transformation is a definite or ranging one is of little importance in the selection of equipment for fusion. The burden is much agitated, but the beds are deep. Only fair overall coefficient values may be expected, although heat flux values are good.

Horizontal-Tank Type This type (Fig. 11-56*a*) is used to transfer heat for melting or cooking dry powdered solids, rendering lard from meat-scrap solids, and drying divided solids. Heat-transfer coefficients are 17 to 85 W/(m²·°C) [3 to 15 Btu/(h·ft²·°F)] for drying and 28 to 140 W/(m²·°C) [5 to 25 Btu/(h·ft²·°F)] for vacuum and/or solvent recovery.

Vertical Agitated-Kettle Type Shown in Fig. 11-56*b*, this type is used to cook, melt to the liquid state, and provide or remove reaction heat for solids that vary greatly in "body" during the process so that material handling is a real problem. The virtues are simplicity and 100 percent cleanability. These often outweigh the poor heat-transfer aspect. These devices are available from the small jacketed type illustrated to huge cast-iron direct-underfired bowls for calcining gypsum. Temperature limits vary with construction; the simpler jackets allow temperatures to 371°C (700°F) (as with Dowtherm), which is not true of all jacketed equipment.

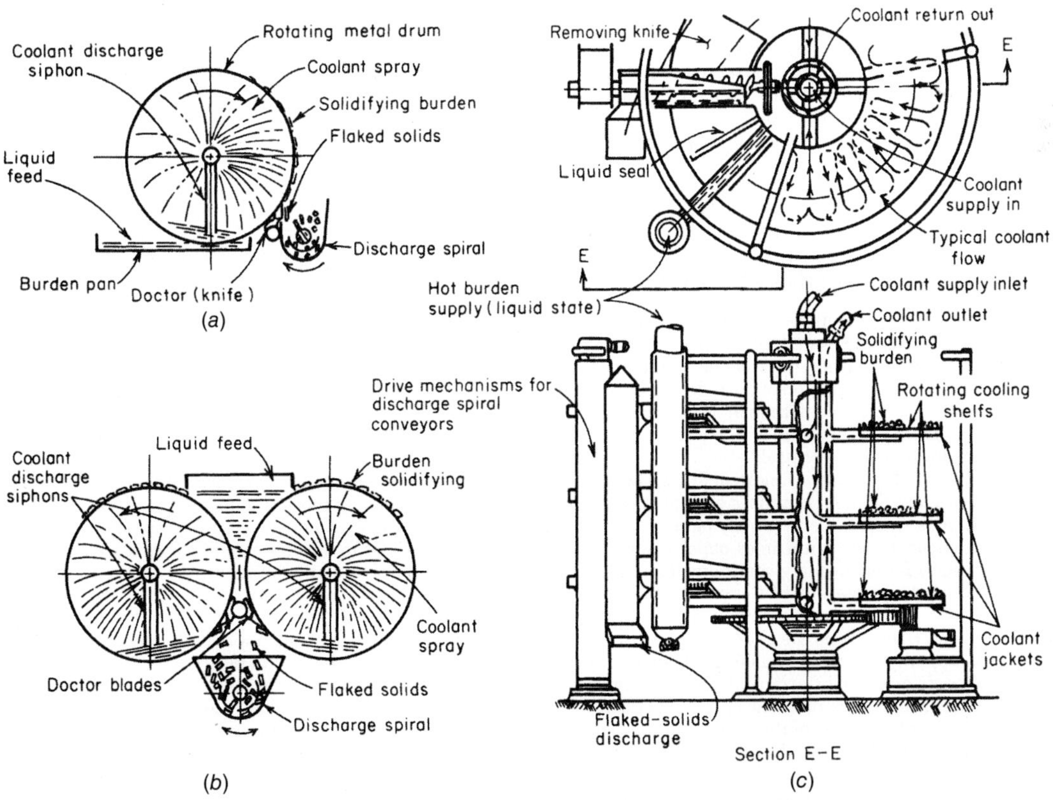

FIG. 11-55 Heat-transfer equipment for continuous solidification. (*a*) Single drum. (*b*) Twin drum. (*c*) Roto shelf. (*Courtesy of Buflovak Division, Blaw-Knox Food & Chemical Equipment, Inc.*)

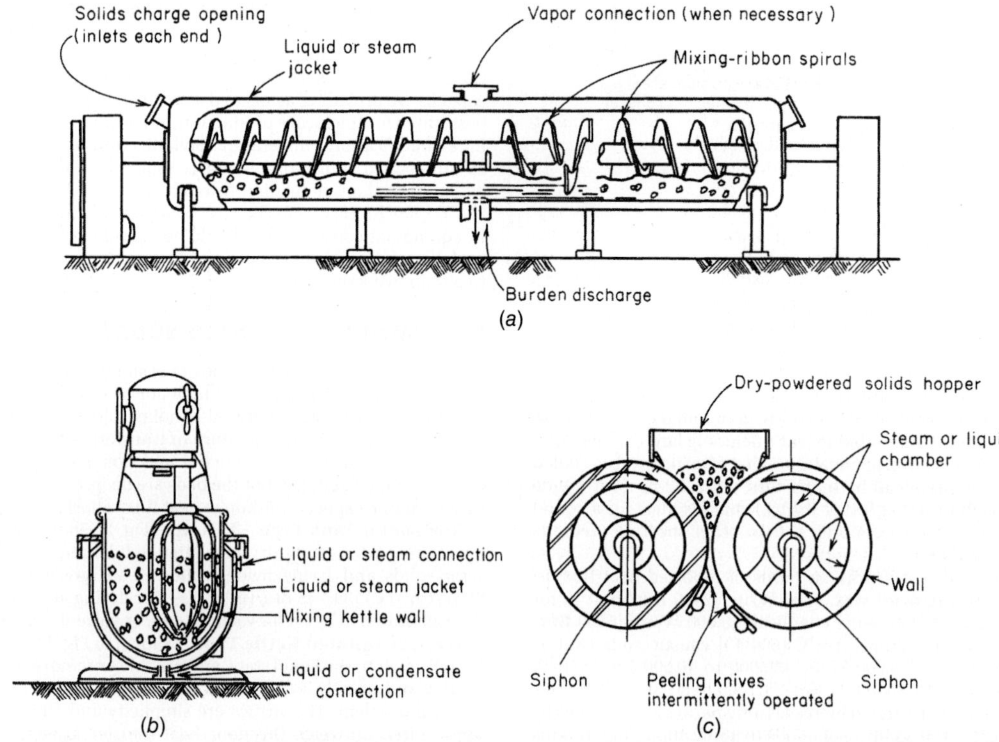

FIG. 11-56 Heat-transfer equipment for fusion of solids. (*a*) Horizontal-tank type. (*Courtesy of Struthers Wells Corp.*) (*b*) Agitated kettle. (*Courtesy of Read-Standard Division, Capital Products Co.*) (*c*) Double-drum mill. (*Courtesy of Farrel-Birmingham Co.*)

Mill Type Figure 11-56*c* shows one model of roll construction used. Note the ruggedness, as it is a *power device* as well as one for *indirect* heat transfer, employed to knead and heat a mixture of dry powdered-solid ingredients with the objective of reacting and reforming via fusion to a consolidated product. In this compounding operation, frictional heat generated by the kneading may require heat-flow reversal (by cooling). Heat-flow control and temperature-level considerations often predominate over heat-transfer performance. Power and mixing considerations, rather than heat transfer, govern. The two-roll mill shown is employed in compounding raw plastic, rubber, and rubberlike elastomer stocks. Multiple-roll mills less knives (termed *calenders*) are used for continuous sheet or film production in widths up to 2.3 m (7.7 ft). Similar equipment is employed in the chemical compounding of inks, dyes, paint pigments, and the like.

HEAT-TRANSFER EQUIPMENT FOR SHEETED SOLIDS

Cylinder Heat-Transfer Units Sometimes called "can" dryers or drying rolls, these devices are differentiated from drum dryers in that they are used for solids in flexible continuous-sheet form, whereas drum dryers are used for liquid or paste forms. The construction of the individual cylinders, or drums, is similar in most respects to that of drum dryers. Special designs are used to obtain uniform distribution of steam within large drums when uniform heating across the drum surface is critical.

A cylinder dryer may consist of one large cylindrical drum, such as the so-called Yankee dryer, but more often it comprises a number of drums arranged so that a continuous sheet of material may pass over them in series. Typical of this arrangement are Fourdrinier paper machine dryers, cellophane dryers, slashers for textile piece goods and fibers, etc. The multiple cylinders are arranged in various ways. Generally, they are staggered in two horizontal rows. In any one row, the cylinders are placed close together. The sheet material contacts the undersurface of the lower rolls and passes over the upper rolls, contacting 60 to 70 percent of the cylinder surface. The cylinders may also be arranged in a single horizontal row, in more than two horizontal rows, or in one or more vertical rows. When it is desired to contact only one side of the sheet with the cylinder surface, unheated guide rolls are used to conduct the sheeting from one cylinder to the next. For sheet materials that shrink on processing, it is frequently necessary to drive the cylinders at progressively slower speeds through the dryer. This requires elaborate individual electric drives on each cylinder.

Cylinder dryers usually operate at atmospheric pressure. However, the Minton paper dryer is designed for operation under vacuum. The drying cylinders are usually heated by steam, but occasionally single cylinders may be gas-heated, as in the case of the Pease blueprinting machine. Upon contacting the cylinder surface, wet sheet material is first heated to an equilibrium temperature somewhere between the wet-bulb temperature of the surrounding air and the boiling point of the liquid under the prevailing total pressure. The heat-transfer resistance of the vapor layer between the sheet and the cylinder surface may be significant.

These cylinder units are applicable to almost any form of sheet material that is not injuriously affected by contact with steam-heated metal surfaces. They are used chiefly when the sheet possesses certain properties such as a tendency to shrink or lacks the mechanical strength necessary for most types of continuous-sheeting air dryers. Applications are to dry films of various sorts, paper pulp in sheet form, paper sheets, paperboard, textile piece goods and fibers, etc. In some cases, imparting a special finish to the surface of the sheet may be an objective.

The *heat-transfer performance capacity* of cylinder dryers is not easy to estimate without a knowledge of the sheet temperature, which, in turn, is difficult to predict. According to published data, steam temperature is the largest single factor affecting capacity. Overall evaporation rates based on the total surface area of the dryers cover a range of 3.4 to 23 kg water/(h · m^2) [0.7 to 4.8 lb water/(h · ft^2)].

The value of the *coefficient of heat transfer* from steam to sheet is determined by the conditions prevailing on the inside and on the surface of the dryers. Low coefficients may be caused by (1) poor removal of air or other noncondensibles from the steam in the cylinders, (2) poor removal of condensate, (3) accumulation of oil or rust on the interior of the drums, and (4) accumulation of a fiber lint on the outer surface of the drums. In a test reported by Lewis et al. [*Pulp Pap. Mag. Can.* **22** (February 1927)] on a sulfite-paper dryer, in which the actual sheet temperatures were measured, a value of 187 W/(m^2 · °C) [33 Btu/(h · ft^2 · °F)] was obtained for the coefficient of heat flow between the steam and the paper sheet.

Operating-cost data for these units are meager. Power costs may be estimated by assuming 1 hp per cylinder for diameters of 1.2 to 1.8 m (4 to 6 ft). Data on labor and maintenance costs are also lacking.

The size of commercial cylinder dryers covers a wide range. The individual rolls may vary in diameter from 0.6 to 1.8 m (2 to 6 ft) and up to 8.5 m (28 ft) in width. In some cases, the width of rolls decreases throughout the dryer in order to conform to the shrinkage of the sheet. A single-cylinder dryer, such as the Yankee dryer, generally has a diameter between 2.7 and 4.6 m (9 and 15 ft).

HEAT-TRANSFER EQUIPMENT FOR DIVIDED SOLIDS

Most equipment for this service is some adaptation of a *material-handling* device whether or not the transport ability is desired. The old vertical tube and the vertical shell (fluidizer) are exceptions. Material-handling problems, plant transport needs, power, and maintenance are prime considerations in equipment selection and frequently overshadow heat-transfer and capital-cost considerations. Material handling is generally the most important aspect. Material-handling characteristics of the divided solids may vary during heat processing. The body changes are usually important in drying, occasionally significant for heating, and only on occasion important for cooling. The ability to minimize the effects of changes is a major consideration in equipment selection. Dehydration operations are better performed on contactive apparatus (see Sec. 12) that provides air to carry off released water vapor before a semiliquid form develops.

Some types of equipment are convertible from heat removal to heat supply by simply changing the temperature level of the fluid or air. Other types require an auxiliary change. Still others require constructional changes. Temperature limits for the equipment generally vary with the thermal operation. The kind of thermal operation has a major effect on heat-transfer values. For drying, overall coefficients are substantially higher in the presence of substantial moisture for the constant-rate period than in finishing. However, a stiff "body" occurrence due to moisture can prevent a normal "mixing" with an adverse effect on the coefficient.

Fluidized-Bed Type Known as the *cylindrical fluidizer*, this operates with a bed of *fluidized solids* (Fig. 11-57). It is an indirect heat-transfer version of the contactive type in Sec. 17. An application disadvantage is the need for batch operation unless some short circuiting can be tolerated. Solids-cooling applications are few, as they can be more effectively accomplished by the fluidizing gas via the contactive mechanism that is referred to in Sec. 11. Heating applications are many and varied. These are subject to one shortcoming, which is the dissipation of the heat input by carry-off in the fluidizing gas. Heat-transfer performance for the indirect mode to solids has been outstanding, with overall coefficients in the range of 570 to 850 W/(m^2·°C) [100 to 150 Btu/(h·ft^2·°F)]. This device with its thin film does for solids what the falling-film and other thin-film techniques do for fluids, as shown by Holt (Pap. 11, 4th National Heat-Transfer Conference, August 1960). In a design innovation with high heat-transfer capability, heat is supplied indirectly to the fluidized solids through the walls of in-bed, horizontally placed, finned tubes [Petrie, Freeby, and Buckham, *Chem. Eng. Prog.* **64**(7): 45 (1968)].

Moving-Bed Type This concept uses a single-pass tube bundle in a vertical shell with the divided solids flowing by gravity in the tubes. It is little used for solids. A major difficulty in divided-solids applications is the problem of charging and discharging with uniformity. A second is poor heat-transfer rates. Because of these limitations, this tube bundle type is not the workhorse for solids that it is for liquid- and gas-phase heat exchange.

However, there are applications in which the nature of a specific chemical reactor system requires indirect heating or cooling of a moving bed of divided solids. One of these is the segregation process which through a gaseous reaction frees chemically combined copper in an ore to a free copper form which permits easy, efficient subsequent recovery [Pinkey and Plint, *Miner. Process.* pp. 17–30 (June 1968)]. The apparatus construction and principle of operation are shown in Fig. 11-58. The functioning is abetted by a novel heat-exchange provision of a fluidized sand bed in the jacket. This provides a much higher unit heat-input rate (coefficient value) than would the usual low-density hot-combustion-gas flow.

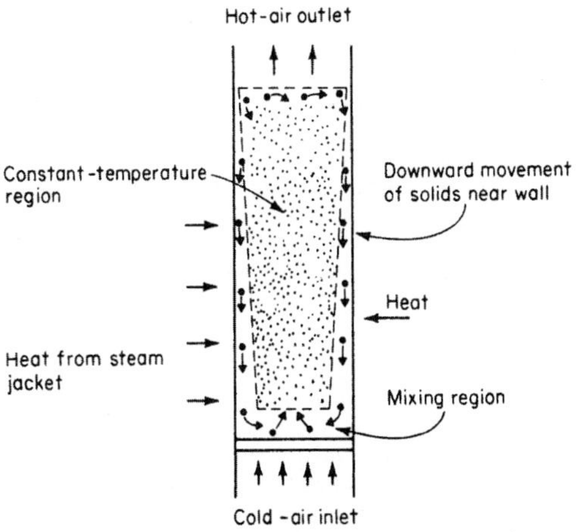

FIG. 11-57 Heat-transfer equipment for divided solids; stationary vertical-shell type. The indirect fluidizer.

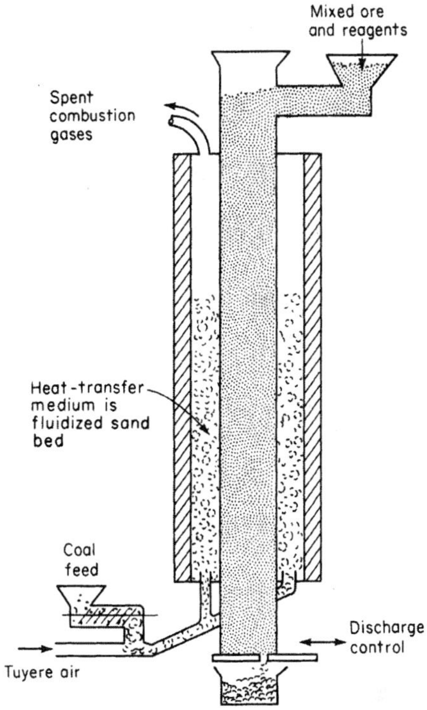

FIG. 11-58 Stationary vertical-tube type of indirect heat-transfer equipment with divided solids inside tubes, laminar solids flow, and steady-state heat conditions.

Agitated-Pan Type This device (Fig. 11-52) is not an adaptation of a material-handling device but was developed many years ago primarily for heat-transfer purposes. As such, it has found wide application. In spite of its batch operation with high attendant labor costs, it is still used for processing divided solids when no phase change is occurring. Simplicity and easy cleanout make the unit a wise selection for handling small, experimental, and even some production runs when quite a variety of burden materials are heat-processed. Both heating and cooling are feasible with it, but greatest use has been for drying [see Sec. 12 and Uhl and Root, *Chem. Eng. Prog.* **63**(7): 8 (1967)]. Because it can be readily covered (as shown in the illustration) and a vacuum drawn or special atmosphere provided, this device features versatility to widen its use. For drying granular solids, the heat-transfer rate ranges from 28 to 227 W/(m^2·°C) [5 to 40 Btu/(h·ft^2·°F)]. For atmospheric applications, thermal efficiency ranges from 65 to 75 percent. For vacuum applications, it is about 70 to 80 percent. These devices are available from several sources, fabricated of various metals used in chemical processes.

Kneading Devices These are closely related to the agitated pan but differ as being primarily mixing devices with heat transfer a secondary consideration. Heat transfer is provided by jacketed construction of the main body and is effected by a coolant, hot water, or steam. These devices are applicable for the compounding of divided solids by mechanical rather than chemical action. Application is largely in the pharmaceutical and food processing industries. For a more complete description, illustrations, performance, and power requirements, refer to Sec. 19.

Shelf Devices Equipment having heated and/or cooled shelves is available but is little used for divided-solids heat processing. Most extensive use of stationary shelves is for freezing of packaged solids for food industries and for freeze drying by sublimation (see Sec. 22).

Rotating-Shell Devices These (see Fig. 11-59) are installed horizontally, whereas stationary-shell installations are vertical. Material-handling aspects are of greater importance than thermal performance. Thermal results are customarily given in terms of overall coefficient on the basis of the total area provided, which varies greatly with the design. The effective use, chiefly percent fill factor, varies widely, affecting the reliability of

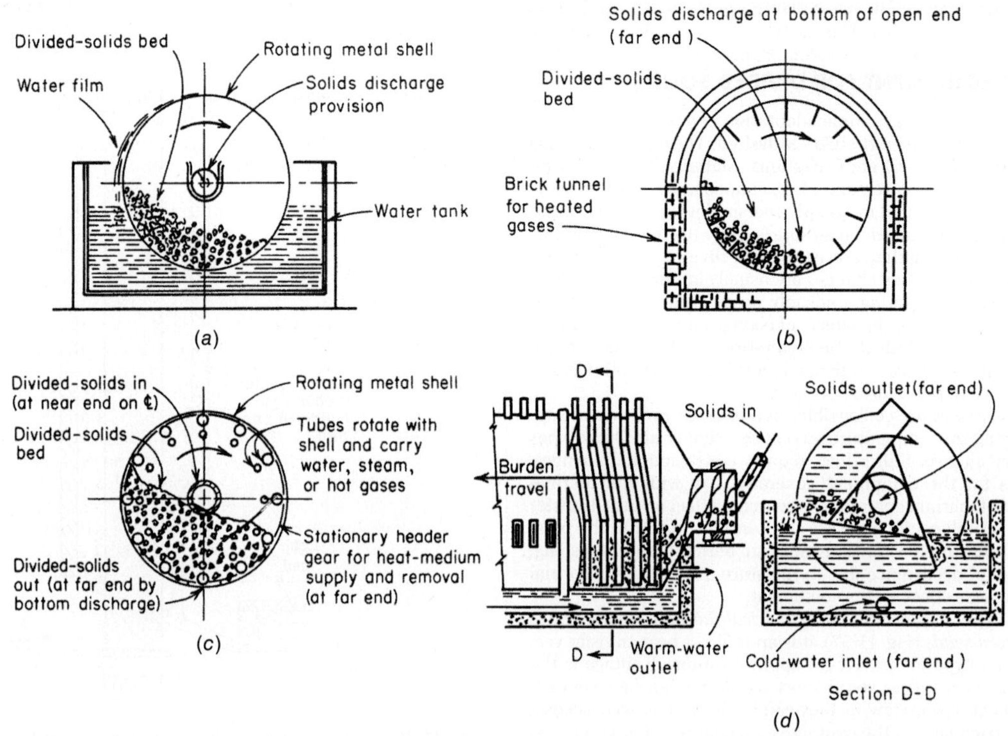

FIG. 11-59 Rotating shells as indirect heat-transfer equipment. (*a*) Plain. (*Courtesy of BSP Corp.*) (*b*) Flighted. (*Courtesy of BSP Corp.*) (*c*) Tubed. (*d*) Deep-finned type. (*Courtesy of Link-Belt Co.*)

stated coefficient values. For performance calculations see Sec. 10 on heat-processing theory for solids. These devices are variously used for cooling, heating, and drying and are the workhorses for heat-processing divided solids in the large-capacity range. Different modifications are used for each of the three operations.

The *plain* type (Fig. 11-59*a*) features simplicity and yet versatility through various end-construction modifications enabling wide and varied applications. Thermal performance is strongly affected by the "body" characteristics of the burden because of its dependency for material handling on frictional contact. Hence, performance ranges from well-agitated beds with good thin-film heat-transfer rates to poorly agitated beds with poor thick-film heat-transfer rates. Temperature limits in application are (1) low-range cooling with shell dipped in water, 400°C (750°F) and less; (2) intermediate cooling with forced circulation of tank water, to 760°C (1400°F); (3) primary cooling, above 760°C (1400°F), water copiously sprayed and loading kept light; (4) low-range heating, below steam temperature, hot-water dip; and (5) high-range heating by tempered combustion gases or ribbon radiant-gas burners.

The *flighted* type (Fig. 11-59*b*) is a first-step modification of the plain type. The simple flight addition improves heat-transfer performance. This type is most effective on semifluid burdens which slide readily. Flighted models are restricted from applications in which soft-cake sticking occurs, breakage must be minimized, and abrasion is severe. A special flighting is one having the cross section compartmented into four lesser areas with ducts between. Hot gases are drawn through the ducts en route from the outer oven to the stack to provide about 75 percent more heating surface, improving efficiency and capacity with a modest cost increase. Another similar unit has the flights made in a triangular-duct cross section with hot gases drawn through.

The *tubed-shell* type (Fig. 11-59*c*) is basically the same device more commonly known as a *steam-tube rotary dryer* (see Sec. 20). The rotation, combined with slight inclination from the horizontal, moves the shell-side solids through it continuously. This type features good mixing with the objective of increased heat-transfer performance. Tube-side fluid may be water, steam, or combustion gas. Bottom discharge slots in the shell are used so that heat-transfer-medium supply and removal can be made through the ends; these restrict wide-range loading and make the tubed type inapplicable for floody materials. These units are seldom applicable for sticky, soft-caking, scaling, or heat-sensitive burdens. They are not recommended for abrasive materials. This type has high thermal efficiency because heat loss is minimized. *Heat-transfer coefficient* values are as follows: water, 34 W/(m^2·°C) [6 Btu/(h·ft^2·°F)]; steam, same, with heat flux reliably constant at 3800 W/m^2 [1200 Btu/(h·ft^2)]; and gas, 17 W/(m^2·°C) [3 Btu/(h·ft^2·°F)], with a high temperature difference. Although from the preceding discussion the device may seem rather limited, it is nevertheless widely used for drying, with condensing steam predominating as the heat-carrying fluid. But with water or refrigerants flowing in the tubes, it is also effective for cooling operations. The units are custom-built by several manufacturers in a wide range of sizes and materials. A few fabricators that specialize in this type of equipment have accumulated a vast store of data for determining application sizing.

The patented *deep-finned* type in Fig. 11-59*d* is named the *Rotofin cooler.* It features loading with a small layer thickness, excellent mixing to give a good effective diffusivity value, and a thin fluid-side film. Unlike other rotating-shell types, it is installed horizontally, and the burden is moved positively by the fins acting as an Archimedes spiral. Rotational speed and spiral pitch determine travel time. For cooling, this type is applicable to both secondary and intermediate cooling duties. Applications include solids in small lumps [9 mm (¾ in)] and granular size [6 mm and less (¼ to 0 in)] with no larger pieces to plug the fins, solids that have a free-flowing body characteristic with no sticking or caking tendencies, and drying of solids that have a low moisture and powder content unless special modifications are made for substantial vapor and dust handling. Thermal performance is very good, with overall coefficients to 110 W/(m^2·°C) [20 Btu/(h·ft^2·°F)], with one-half of these coefficients nominal for cooling based on the total area provided (nearly double those reported for other indirect rotaries).

Conveyor-Belt Devices The metal-belt type (Fig. 11-55) is the only device in this classification of material-handling equipment that has had serious effort expended on it to adapt it to indirect heat-transfer service with divided solids. It features a lightweight construction of a large area with a thin metal wall. Indirect-cooling applications have been made with poor thermal performance, as could be expected with a static layer. Auxiliary plowlike mixing devices, which are considered an absolute necessity to secure any worthwhile results for this service, restrict applications.

Spiral-Conveyor Devices Figure 11-60 illustrates the major adaptations of this widely used class of material-handling equipment to indirect heat-transfer purposes. These conveyors can be considered for heat-transfer purposes as continuously agitated kettles. The adaptation of Fig. 11-60*d* offers a batch-operated version for evaporation duty. For this service, all are package-priced and package-shipped items requiring few, if any, auxiliaries.

The *jacketed solid-flight* type (Fig. 11-60*a*) is the standard low-cost (parts-basis-priced) material-handling device, with a simple jacket added and employed for secondary-range heat transfer of an incidental nature. Heat-transfer coefficients are as low as 11 to 34 W/(m²·°C) [2 to 6 Btu/(h·ft²·°F)] on sensible heat transfer and 11 to 68 W/(m²·°C) [2 to 12 Btu/(h·ft²·°F)] on drying because of substantial static solids-side film.

The *small-spiral–large-shaft* type (Fig. 11-60*b*) is inserted in a solids-product line as pipe banks are in a fluid line, solely as a heat-transfer device. It features a thin burden ring carried at a high rotative speed and subjected to two-sided conductance to yield an estimated heat-transfer coefficient of 285 W/(m²·°C) [50 Btu/(h·ft²·°F)], thereby ranking thermally next to the shell-fluidizer type. This device for powdered solids is comparable with the Votator of the fluid field.

Figure 11-60*c* shows a fairly new spiral device with a medium-heavy annular solids bed and having the combination of a jacketed, stationary outer shell with moving paddles that carry the heat-transfer fluid. A unique feature of this device to increase volumetric throughput, by providing an overall greater temperature drop, is that the heat medium is supplied to and withdrawn from the rotor paddles by a parallel piping arrangement in the rotor shaft. This is a unique flow arrangement compared with the usual series flow. In addition, the rotor carries burden-agitating spikes which give it the trade name of Porcupine Heat-Processor (*Chem. Equip. News,* April 1966; and Uhl and Root, *AIChE Prepr.* **21,** 11th National Heat-Transfer Conference, August 1967).

The *large-spiral hollow-flight* type (Fig. 11-60*d*) is an adaptation, with external bearings, full fill, and salient construction points as shown, that is highly versatile in application. Heat-transfer coefficients are 34 to 57 W/(m²·°C) [6 to 10 Btu/(h·ft²·°F)] for poor, 45 to 85 W/(m²·°C) [8 to 15 Btu/(h·ft²·°F)] for fair, and 57 to 114 W/(m²·°C) [10 to 20 Btu/(h·ft²·°F)] for wet conductors. A popular version of this employs two such spirals in one material-handling chamber for a pugmill agitation of the deep solids bed. The spirals are seldom heated. The shaft and shell are heated.

Another deep-bed spiral-activated solids-transport device is shown by Fig. 11-60*e*. The flights carry a heat-transfer medium as well as the jacket. A unique feature of this device, which is purported to increase heat-transfer capability in a given equipment space and cost, is the dense-phase fluidization of the deep bed that promotes agitation and moisture removal on drying operations.

Double-Cone Blending Devices The original purpose of these devices was mixing (see Sec. 19). Adaptations have been made; so many models now are primarily for indirect heat-transfer processing. A jacket on the shell carries the heat-transfer medium. The mixing action, which breaks up agglomerates (but also causes some degradation), provides very effective burden exposure to the heat-transfer surface. On drying operations, the vapor release (which in a static bed is a slow diffusional process) takes place relatively quickly. To provide vapor removal from the burden chamber, a hollow shaft is used. Many of these devices carry the hollow-shaft feature a step further by adding a rotating seal and drawing a vacuum. This increases thermal performance notably and makes the device a natural for solvent-recovery operations.

These devices are replacing the older tank and spiral-conveyor devices. Better provisions for speed and ease of fill and discharge (without powered

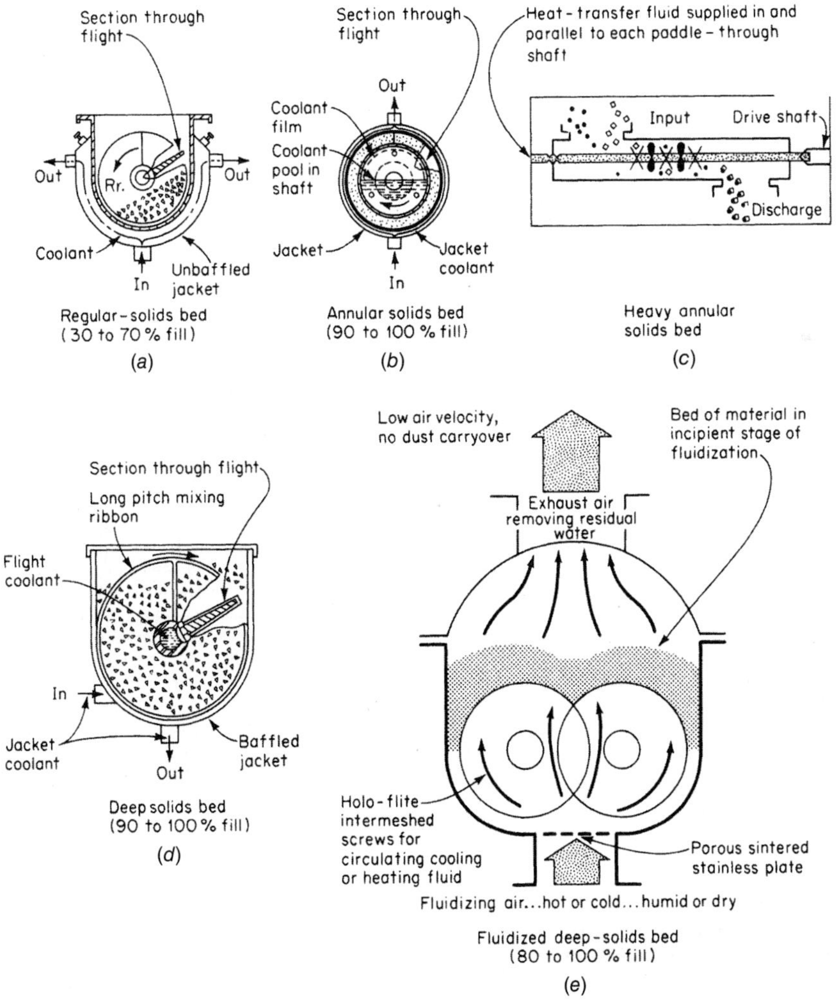

FIG. 11-60 Spiral-conveyor adaptations as heat-transfer equipment. (*a*) Standard jacketed solid flight. (*Courtesy of Jeffrey Mfg. Co.*) (*b*) Small spiral, large shaft. (*Courtesy of Fuller Co.*) (*c*) "Porcupine" medium shaft. (*Courtesy of Bethlehem Corp.*) (*d*) Large spiral, hollow flight. (*Courtesy of Rietz Mfg. Co.*) (*e*) Fluidized-bed large spiral, helical flight. (*Courtesy of Western Precipitation Division, Joy Mfg. Co.*)

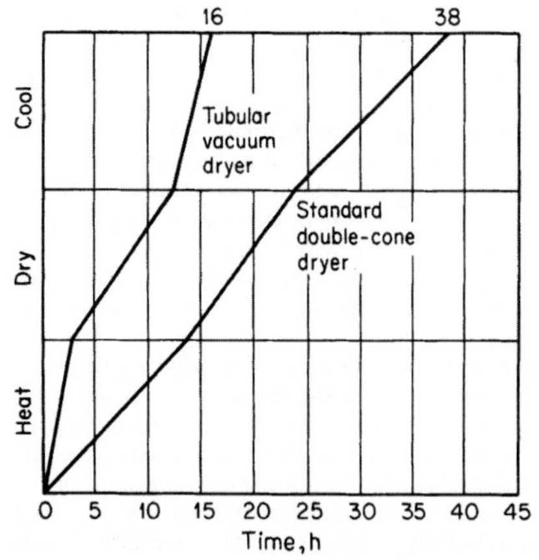

FIG. 11-61 Performance of tubed blender heat-transfer device.

rotation) minimize downtime to make this batch-operated device attractive. Heat-transfer coefficients ranging from 28 to 200 W/(m^2·°C) [5 to 35 Btu/(h·ft^2·°F)] are obtained. However, if caking on the heat-transfer walls is serious, then values may drop to 5.5 or 11 W/(m^2·°C) [1 or 2 Btu/(h·ft^2·°F)], constituting a misapplication. The double cone is available in a fairly wide range of sizes and construction materials. The users are the fine-chemical, pharmaceutical, and biological-preparation industries.

A novel variation is a cylindrical model equipped with a tube bundle to resemble a shell-and-tube heat exchanger with a bloated shell [*Chem. Process.* **20** (Nov. 15, 1968)]. Conical ends provide for redistribution of burden between passes. The improved heat-transfer performance is shown by Fig. 11-61.

Vibratory-Conveyor Devices Figure 11-62 shows the various adaptations of vibratory material-handling equipment for indirect heat-transfer service on divided solids. The basic vibratory-equipment data are given in Sec. 21. These indirect heat-transfer adaptations feature simplicity, nonhazardous construction, nondegradation, nondusting, no wear, ready conveying-rate variation [1.5 to 4.5 m/min (5 to 15 ft/min)], and good heat-transfer coefficient—115 W/(m^2·°C) [20 Btu/(h·ft^2·°F)] for sand. They usually require feed rate and distribution auxiliaries. They are suited for heating and cooling of divided solids in powdered, granular, or moist forms but no sticky, liquefying, or floody ones. Terminal-temperature differences less than 11°C (20°F) on cooling and 17°C (30°F) on heating or drying operations are seldom practical. These devices are for medium and light capacities.

The *heavy-duty jacketed* type (Fig. 11-62*a*) is a special custom-built adaptation of a heavy-duty vibratory conveyor shown in Fig. 11-60. Its application is to continuously cool the crushed material [from about 177°C (350°F)] produced by the vibratory-type "caster" of Fig. 11-53. It does not have the liquid dam and is made in longer lengths that employ L, switchback, and S arrangements on one floor. The capacity rate is 27,200 to 31,700 kg/h (30 to 35 ton/h) with heat-transfer coefficients in the order of 142 to 170 W/(m^2·°C) [25 to 30 Btu/(h·ft^2·°F)]. For heating or drying applications, it employs steam to 414 kPa (60 lbf/in^2).

The *jacketed* or *coolant-spraying* type (Fig. 11-62*b*) is designed to ensure a very thin, highly agitated liquid-side film and the same initial coolant temperature over the entire length. It is frequently employed for transporting substantial quantities of hot solids, with cooling as an incidental consideration. For heating or drying applications, hot water or steam at a gauge pressure of 7 kPa (1 lbf/in^2) may be employed. This type is widely used because of its versatility, simplicity, cleanability, and good thermal performance.

The *light-duty jacketed* type (Fig. 11-62*c*) is designed for use of air as a heat carrier. The flow through the jacket is highly turbulent and is usually counterflow. On long installations, the air flow is parallel to every two sections for greater heat-carrying capacity and a fairly uniform surface temperature. The outstanding feature is that a wide range of temperature control is obtained by merely changing the heat-carrier temperature level from as low as atmospheric moisture condensation will allow to 204°C (400°F). On heating operations, a very good thermal efficiency can be obtained by insulating the machine and recycling the air. While the heat-transfer rating is good, the heat-removal capacity is limited. Cooler units are often used in series with like units operated as dryers or when clean water is unavailable. Drying applications are for heat-sensitive [49°C to 132°C (120°F to 270°F)] products; when temperatures higher than steam at a gauge pressure of 7 kPa (1 lbf/in^2) can provide are wanted but heavy-duty equipment is too costly; when the

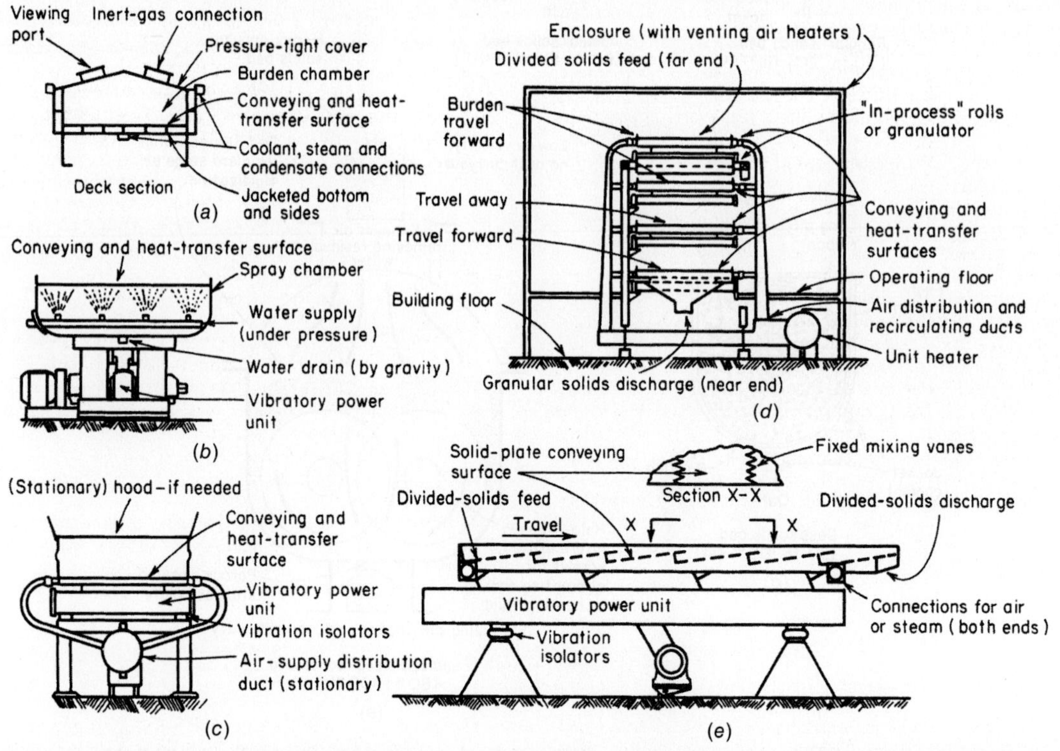

FIG. 11-62 Vibratory-conveyor adaptations as indirect heat-transfer equipment. (*a*) Heavy-duty jacketed for liquid coolant or high-pressure steam. (*b*) Jacketed for coolant spraying. (*c*) Light-duty jacketed construction. (*d*) Jacketed for air or steam in tiered arrangement. (*e*) Jacketed for air or steam with Mix-R-Step surface. (*Courtesy of Jeffrey Mfg. Co.*)

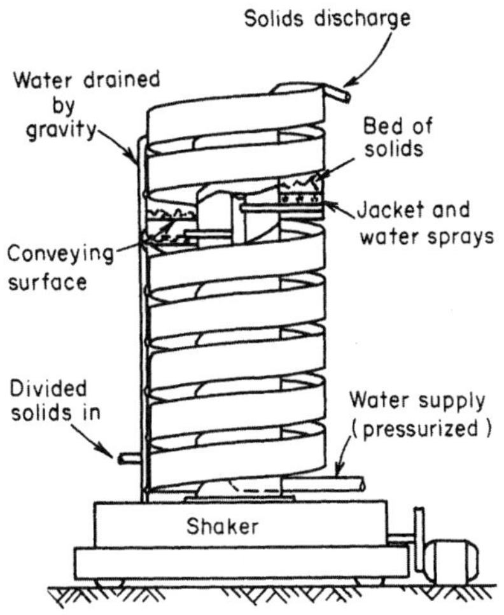

FIG. 11-63 Elevator type as heat-transfer equipment. (*Courtesy of Carrier Conveyor Corp.*)

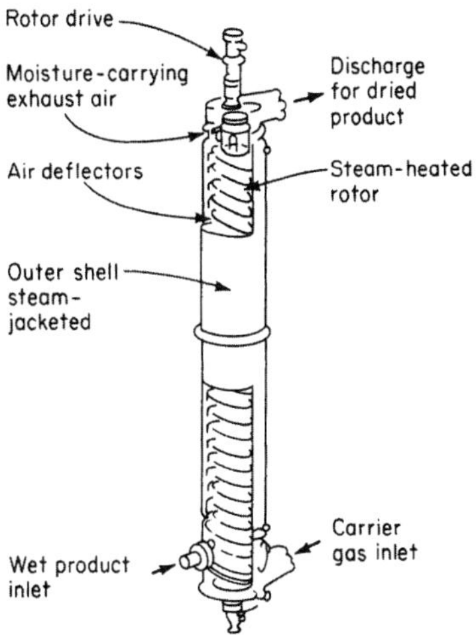

FIG. 11-64 A pneumatic transport adaptation for heat-transfer duty. (*Courtesy of Werner & Pfleiderer Corp.*)

jacket corrosion hazard of steam is unwanted; when headroom space is at a premium; and for highly abrasive burden materials such as fritted or crushed glasses and porcelains.

The *tiered arrangement* (Fig. 11-62*d*) employs the units of Fig. 11-62 with either air or steam at a gauge pressure of 7 kPa (1 lbf/in^2) as a heat medium. These are custom-designed and built to provide a large amount of heat-transfer surface in a small space with the minimum of transport and to provide a complete processing system. These receive a damp material, resize while in process by granulators or rolls, finish dry, cool, and deliver to packaging or tableting. The applications are primarily in the fine chemical, food, and pharmaceutical manufacturing fields.

The *Mix-R-Step* type in Fig. 11-62*e* is an adaptation of a vibratory conveyor. It features better heat-transfer rates, practically doubling the coefficient values of the standard flat surface and trebling heat-flux values, as the layer depth can be increased from the normal 13 to 25 and 32 mm (½ to 1 and 1¼ in). It may be provided on decks jacketed for air, steam, or water spray. It is also often applicable when an infrared heat source is mounted overhead to supplement the indirect or as the sole heat source.

Elevator Devices The *vibratory elevating-spiral* type (Fig. 11-63) adapts divided-solids-elevating material-handling equipment to heat-transfer service. It features a large heat-transfer area over a small floor space and employs a reciprocating shaker motion to effect transport. Applications, layer depth, and capacities are restricted, as burdens must be of such "body" character as to convey uphill by the microhopping transport principle. The type lacks self-emptying ability. Complete washdown and cleaning is a feature not inherent in any other elevating device. A typical application is the cooling of a low-density plastic powder at the rate of 544 kg/h (1200 lb/h).

Another elevator adaptation is that for a *spiral-type elevating* device developed for ground cement and thus limited to fine powdery burdens. The spiral operates inside a cylindrical shell, which is externally cooled by a falling film of water. The spiral not only elevates the material in a thin layer against the wall but keeps it agitated to achieve high heat-transfer rates. Specific operating data are not available [*Chem. Eng. Prog.* **68**(7): 113 (1968)]. The falling-water film, besides being ideal thermally, by virtue of no jacket pressure very greatly reduces the hazard that the cooling water may contact the water-sensitive burden in process. Surfaces wet by water are accessible for cleaning. A fair range of sizes are available, with material-handling capacities to 60 ton/h.

Pneumatic Conveying Devices See Sec. 21 for descriptions, ratings, and design factors of these devices. Use is primarily for transport purposes, and heat transfer is a very secondary consideration. Applications have largely been for plastics in powder and pellet forms.

By modifications, needed cooling operations have been simultaneously effected with transport to stock storage [*Plast. Des. Process* **28** (December 1968)].

Heat-transfer aspects and performance were studied and reported on by Depew and Farbar (ASME Pap. 62-HT-14, September 1962). Heat-transfer coefficient characteristics are similar to those shown in Sec. 11 for the indirectly heated fluid bed. Another frequent application on plastics is a small, rather incidental but necessary amount of drying required for plastic pellets and powders on receipt when shipped in bulk to the users. Pneumatic conveyors modified for heat transfer can handle this readily.

A pneumatic transport device designed primarily for heat-sensitive products is shown in Fig. 11-64. This was introduced into the United States after 5 years' use in Europe [*Chem. Eng.* **76:** 54 (June 16, 1969)].

Both the shell and the rotor carry steam as a heating medium to effect indirect transfer as the burden briefly contacts those surfaces rather than from the transport air, as is normally the case. The rotor turns slowly (1 to 10 r/min) to control, by deflectors, product distribution and prevent caking on walls. The carrier gas can be inert, as nitrogen, and also recycled through appropriate auxiliaries for solvent recovery. Application is limited to burdens that (1) are fine and uniformly grained for the pneumatic transport, (2) dry very fast, and (3) have very little, if any, sticking or decomposition characteristics. Feeds can carry 5 to 100 percent moisture (dry basis) and discharge at 0.1 to 2 percent. Wall temperatures range from 100 to 170°C (212 to 340°F) for steam and lower for a hot-water heat source. Pressure drops are on the order of 500 to 1500 mmH_2O (20 to 60 inH_2O). Steam consumption approaches that of a contractive-mechanism dryer down to a low value of 2.9 kg steam/kg water (2.9 lb steam/lb water). Available burden capacities are 91 to 5900 kg/h (200 to 13,000 lb/h).

Vacuum-Shelf Types These are very old devices, being a version of the table type. Early-day use was for drying (see Sec. 12). Heat transfer is slow even when supplemented by vacuum, which is 90 percent or more of present-day use. The newer vacuum blender and cone devices are taking over many applications. The slow heat-transfer rate is quite satisfactory in a major application, freeze drying, which is a sublimation operation (see Sec. 22 for description) in which the water must be retained in the solid state during its removal. Then slow diffusional processes govern. Another extensive application is in freezing packaged foods for preservation purposes.

Available sizes range from shelf areas of 0.4 to 67 m^2 (4 to 726 ft^2). These are available in several manufacturers' standards, either as system components or with auxiliary gear as packaged systems.

THERMAL INSULATION

Materials or combinations of materials which have air- or gas-filled pockets or void spaces that retard the transfer of heat with reasonable effectiveness are thermal insulators. Such materials may be particulate and/or fibrous, with or without binders, or may be assembled, such as multiple heat-reflecting surfaces that incorporate air- or gas-filled void spaces.

The ability of a material to retard the flow of heat is expressed by its thermal *conductivity* (for unit thickness) or *conductance* (for a specific thickness). Low values for thermal conductivity or conductance (or high thermal resistivity or resistance value) are characteristics of thermal insulation.

TABLE 11-21 Thicknesses of Piping Insulation

			Insulation, nominal thickness														
in mm	Outer diameter		1 25		1½ 38		2 51		2½ 64		3 76		3½ 89		4 102		
			Approximate wall thickness														
Nominal iron-pipe size, in	in	mm	in	mm	in	mm	in	mm	in	mm	in	mm	in	mm	in	mm	
½	0.84	21	1.01	26	1.57	40	2.07	53	2.88	73	3.38	86	3.88	99	4.38	111	
¾	1.05	27	0.90	23	1.46	37	1.96	50	2.78	71	3.28	83	3.78	96	4.28	109	
1	1.32	33	1.08	27	1.58	40	2.12	54	2.64	67	3.14	80	3.64	92	4.14	105	
1¼	1.66	42	0.91	23	1.66	42	1.94	49	2.47	63	2.97	75	3.47	88	3.97	101	
1½	1.90	48	1.04	26	1.54	39	2.35	60	2.85	72	3.35	85	3.85	98	4.42	112	
2	2.38	60	1.04	26	1.58	40	2.10	53	2.60	66	3.10	79	3.60	91	4.17	106	
2½	2.88	73	1.04	26	1.86	47	2.36	60	2.86	73	3.36	85	3.92	100	4.42	112	
3	3.50	89	1.02	26	1.54	39	2.04	52	2.54	65	3.04	77	3.61	92	4.11	104	
3½	4.00	102	1.30	33	1.80	46	2.30	58	2.80	71	3.36	85	3.86	98	4.36	111	
4	4.50	114	1.04	26	1.54	39	2.04	52	2.54	65	3.11	79	3.61	92	4.11	104	
4½	5.00	127	1.30	33	1.80	46	2.30	58	2.86	73	3.36	85	3.86	98	4.48	114	
5	5.56	141	0.99	25	1.49	38	1.99	51	2.56	65	3.06	78	3.56	90	4.18	106	
6	6.62	168	0.96	24	1.46	37	2.02	51	2.52	64	3.02	77	3.65	93	4.15	105	
7	7.62	194			1.52	39	2.02	51	2.52	64	3.15	80	3.65	93	4.15	105	
8	8.62	219			1.52	39	2.02	51	2.65	67	3.15	80	3.65	93	4.15	105	
9	9.62	244			1.52	39	2.15	55	2.65	67	3.15	80	3.65	93	4.15	105	
10	10.75	273			1.58	40	2.08	53	2.58	66	3.08	78	3.58	91	4.08	104	
11	11.75	298			1.58	40	2.08	53	2.58	66	3.08	78	3.58	91	4.08	104	
12	12.75	324			1.58	40	2.08	53	2.58	66	3.08	78	3.58	91	4.08	104	
14	14.00	356			1.46	37	1.96	50	2.46	62	2.96	75	3.46	88	3.96	101	
Over 14, up to and including 36					1.46	37	1.96	50	2.46	62	2.96	75	3.46	88	3.96	101	

Heat is transferred by radiation, conduction, and convection. Radiation is the primary mode and can occur even in a vacuum. The amount of heat transferred for a given area is relative to the temperature differential and emissivity from the radiating to the absorbing surface. Conduction is due to molecular motion and occurs within gases, liquids, and solids. The tighter the molecular structure, the higher the rate of transfer. As an example, steel conducts heat at a rate approximately 600 times that of typical thermal insulation materials. Convection is due to mass motion and occurs only in fluids. The prime purpose of a thermal insulation system is to minimize the amount of heat transferred.

INSULATION MATERIALS

Materials Thermal insulations are produced from many materials or combinations of materials in various forms, sizes, shapes, and thicknesses. The most commonly available materials fall within the following categories:

Fibrous or cellular—mineral: Alumina, asbestos, glass, perlite, rock, silica, slag, or vermiculite

Fibrous or cellular—organic: Cane, cotton, wood, and wood bark (cork)

Cellular organic plastics. Elastomer, polystyrene, polyisocyanate, polyisocyanurate, and polyvinyl acetate

Cements: Insulating and/or finishing

Heat-reflecting metals (reflective): Aluminum, nickel, stainless steel

Available forms. Blanket (felt and batt), block, cements, loose fill, foil and sheet, formed or foamed in place, flexible, rigid, and semirigid.

The actual thicknesses of piping insulation differ from the nominal values. Dimensional data of ASTM Standard C585 appear in Table 11-21.

Thermal Conductivity (*K* Factor) Depending on the type of insulation, the thermal conductivity (*K* factor) can vary with age, manufacturer, moisture content, and temperature. Typical published values are shown in Fig. 11-65. Mean temperature is equal to the arithmetic average of the temperatures on both sides of the insulating material.

Actual system heat loss (or gain) will normally exceed calculated values because of projections, axial and longitudinal seams, expansion-contraction openings, moisture, workers' skill, and physical abuse.

Finishes Thermal insulations require an external covering (finish) to provide protection against entry of water or process fluids, mechanical damage, and ultraviolet degradation of foamed materials. In some cases the finish can reduce the flame-spread rating and/or provide fire protection.

The finish may be a coating (paint, asphaltic, resinous, or polymeric), a membrane (coated felt or paper, metal foil, or laminate of plastic, paper, foil, or coatings), or sheet material (fabric, metal, or plastic).

Finishes for systems operating below 2°C (35°F) must be sealed and retard vapor transmission. Those from 2°C (35°F) through 27°C (80°F) should retard vapor transmission (to prevent surface condensation), and those above 27°C (80°F) should prevent water entry and allow moisture to escape.

Metal finishes are more durable, require less maintenance, reduce heat loss, and, if uncoated, increase the surface temperature on hot systems.

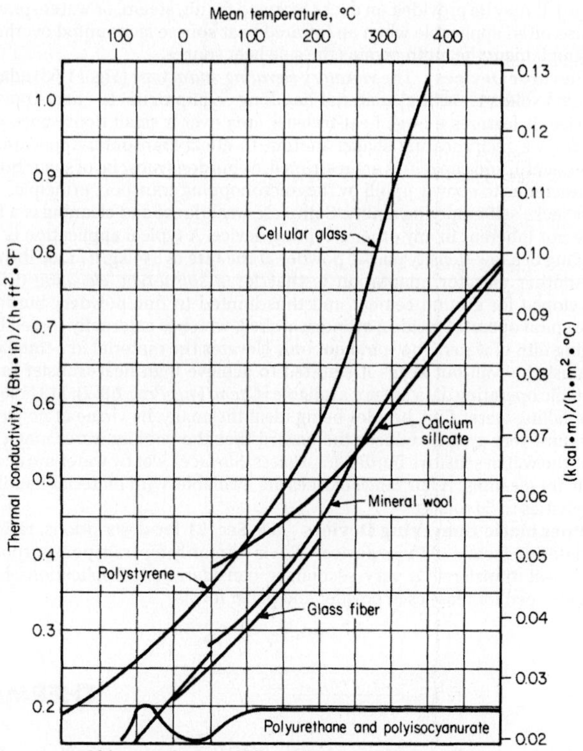

FIG. 11-65 Thermal conductivity of insulating materials.

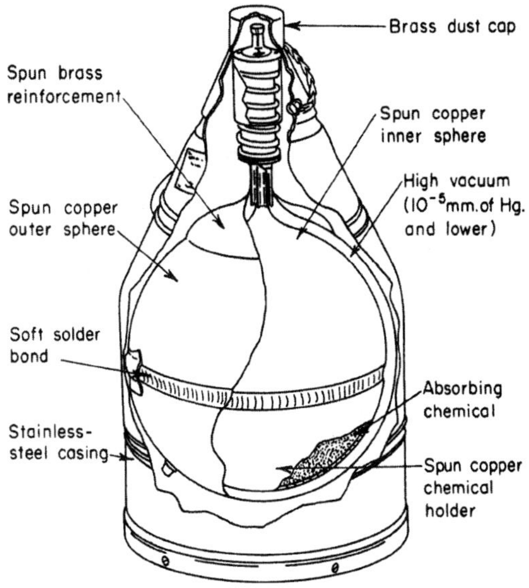

FIG. 11-66 Dewar flask.

SYSTEM SELECTION

A combination of insulation and finish produces the thermal insulation system. Selection of these components depends on the purpose for which the system is to be used. No single system performs satisfactorily from the cryogenic through the elevated-temperature range. Systems operating below freezing have a low vapor pressure, and atmospheric moisture is pushed into the insulation system, while the reverse is true for hot systems. Some general guidelines for system selection follow.

Cryogenic [−273 to −101°C (−459 to −150°F)] High Vacuum This technique is based on the Dewar flask, which is a double-walled vessel with reflective surfaces on the evacuated side to reduce radiation losses. Figure 11-66 shows a typical laboratory-size Dewar. Figure 11-67 shows a semiportable type. Radiation losses can be further reduced by filling the cavity with powders such as perlite or silica prior to pulling the vacuum.

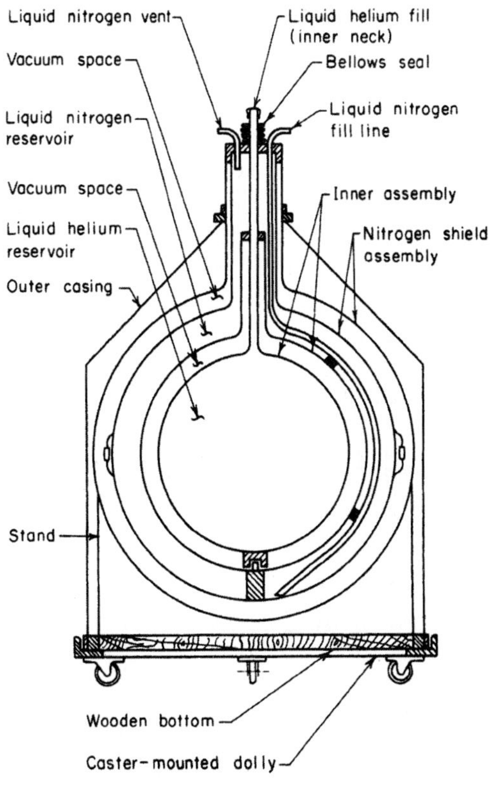

FIG. 11-67 Hydrogen bottle.

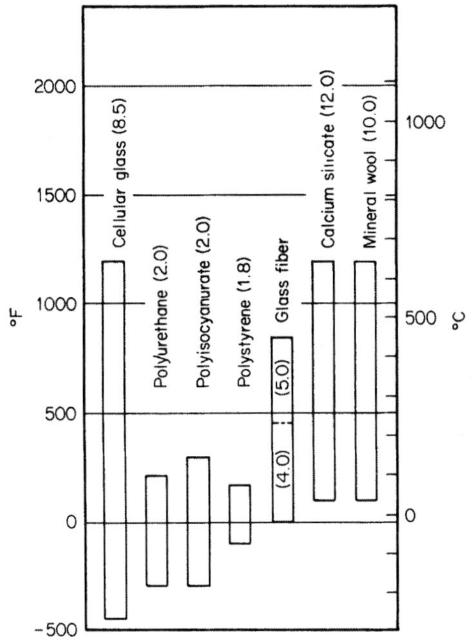

FIG. 11-68 Insulating materials and applicable temperature ranges.

Multilayer Multilayer systems consist of series of radiation-reflective shields of low emittance separated by fillers or spacers of very low conductance and exposed to a high vacuum.

Foamed or Cellular Cellular plastics such as polyurethane and polystyrene do not hold up or perform well in the cryogenic temperature range because of permeation of the cell structure by water vapor, which in turn increases the heat-transfer rate. Cellular glass holds up better and is less permeable.

Low Temperature [−101 to −1°C (−150 to +30°F)] Cellular glass, glass fiber, polyurethane foam, and polystyrene foam are frequently used for this service range. A vapor-retarder finish with a perm rating less than 0.02 is required. In addition, it is good practice to coat all contact surfaces of the insulation with a vapor-retardant mastic to prevent moisture migration when the finish is damaged or is not properly maintained. Closed-cell insulation should not be relied on as the vapor retarder. Hairline cracks can develop, cells can break down, glass-fiber binders are absorbent, and moisture can enter at joints between all materials.

Moderate and High Temperature [over 2°C (36°F)] Cellular or fibrous materials are normally used. See Fig. 11-68 for nominal temperature range. Nonwicking insulation is desirable for systems operating below 100°C (212°F).

Other Considerations *Autoignition* can occur if combustible fluids are absorbed by wicking-type insulations. *Chloride stress corrosion* of austenitic stainless steel can occur when chlorides are concentrated on metal surfaces at or above approximately 60°C (140°F). The chlorides can come from sources other than the insulation. Some calcium silicates are formulated to exceed the requirements of the MIL-I-24244A specification. *Fire resistance* of insulations varies widely. Calcium silicate, cellular glass, glass fiber, and mineral wool are fire-resistant but do not perform equally under actual fire conditions. A steel jacket provides protection, but aluminum does not.

Traced pipe performs better with a nonwicking insulation which has low thermal conductivity. Underground systems are very difficult to keep dry permanently. Methods of insulation include factory-preinsulated pouring types and conventionally applied types. Corrosion can occur under wet insulation. A protective coating, applied directly to the metal surface, may be required.

ECONOMIC THICKNESS OF INSULATION

Optimal economic insulation thickness may be determined by various methods. Two of these are the minimum total cost method and the incremental cost method (or marginal cost method). The minimum total cost method involves actual calculations of lost energy and insulation costs for each insulation thickness. The thickness producing the lowest total cost is the optimal economic solution. The optimum thickness is determined to be the point where the last dollar invested in insulation results in exactly $1 in energy cost savings ("ETI—Economic Thickness for Industrial Insulation," Conservation Pap. 46, Federal Energy Administration, August 1976). The incremental cost method provides a simplified and direct solution for the least-cost thickness.

The total cost method does *not* in general provide a satisfactory means for making most insulation investment decisions, since an economic return on investment is required by investors and the method does not properly consider this factor. Return on investment is considered by Rubin ("Piping Insulation—Economics and Profits," in *Practical Considerations in Piping Analysis,* ASME Symposium, vol. 69, 1982, pp. 27–46). The incremental method used in this reference requires that each incremental ½ in of insulation provide the predetermined return on investment. The minimum thickness of installed insulation is used as a base for calculations. The incremental installed capital cost for each additional ½ in of insulation is determined. The energy saved for each increment is then found. The value of this energy varies directly with the temperature level [e.g., steam at 538°C (1000°F) has a greater value than condensate at 100°C (212°F)]. The final increment selected for use is required either to provide a satisfactory return on investment or to have a suitable payback period.

Recommended Thickness of Insulation Indoor insulation thickness appears in Table 11-22, and outdoor thickness appears in Table 11-23. These selections were based upon calcium silicate insulation with a suitable aluminum jacket. However, the variation in thickness for fiberglass, cellular glass, and rock wool is minimal. Fiberglass is available for maximum temperatures of 260, 343, and 454°C (500, 650, and 850°F). Rock wool, cellular glass, and calcium silicate are used up to 649°C (1200°F).

The tables were based upon the cost of energy at the end of the first year, a 10 percent inflation rate on energy costs, a 15 percent interest cost, and a present-worth pretax profit of 40 percent per annum on the last increment of insulation thickness. Dual-layer insulation was used for 3½-in and greater

TABLE 11-22 Indoor Insulation Thickness, 80°F Still Ambient Air*

		Minimum pipe temperature, °F							
		Energy cost, $/million Btu							
Pipe size, in	Insulation thickness, in	1	2	3	4	5	6	7	8
¾	1½	950	600	550	400	350	300	250	250
	2				1100	1000	900	800	750
	2½				1750	1050	950	850	800
	3								1200
1	1½	1200	800	600	500	450	400	350	300
	2			1200	1000	900	800	700	700
	2½					1200	1050	1000	900
	3						1100	1150	950
1½	1½	1100	750	550	450	400	400	350	300
	2			1000	850	700	650	600	500
	2½				1050	900	800	750	650
	3						1150	1100	1000
2	1½	1050	700	500	450	400	350	300	300
	2			1050	850	750	700	600	600
	2½			1100	950	1000	750	700	650
	3				1200	1050	950	850	800
3	1½	950	650	500	400	350	300	300	250
	2		1100	900	700	600	550	500	450
	2½			1050	850	750	650	500	500
	3				1050	950	800	750	700
4	1½	950	600	500	400	350	300	300	250
	2		1100	850	700	600	550	500	450
	2½			1200	1000	850	750	700	650
	3				1050	900	800	750	700
	3½							1150	1050
6	1½	600	350	300	250	250	200	200	200
	2		1100	850	700	600	550	500	500
	2½			900	800	650	600	550	550
	3			1150	1000	850	750	700	600
	3½						1100	1000	900
	4								1200
8	2		1000	800	650	550	500	450	400
	2½		1050	850	700	600	550	500	450
	3				1050	900	800	750	700
	3½					1200	1100	1000	900
	4							1150	1100
10	2		1100	850	700	650	550	500	450
	2½		1200	900	750	700	600	550	500
	3			1050	900	750	700	600	550
	3½					1200	1050	950	900
	4								1200
12	2	1150	750	600	500	400	400	350	300
	2½		1000	800	650	550	500	450	400
	3			1200	1000	900	800	700	650
	3½					1200	1100	1000	900
	4						1150	1050	950
	4½						1200	1100	1000
14	2	1050	650	550	450	400	350	300	300
	2½		1000	800	650	550	500	450	400
	3			1100	950	800	700	650	600
	3½					1150	1000	950	850
	4					1200	1050	1000	900
	4½						1200	1100	1000

TABLE 11-22 Indoor Insulation Thickness, 80°F Still Ambient Air* *(Continued)*

		Minimum pipe temperature, °F							
		Energy cost, $/million Btu							
Pipe size, in	Insulation thickness, in	1	2	3	4	5	6	7	8
16	2	950	650	500	400	350	300	300	300
	2½		1000	800	700	600	550	500	450
	3		1200	950	800	700	600	550	500
	3½					1150	1050	950	850
	4					1200	1100	1000	900
	4½						1150	1050	950
18	2	1000	650	500	400	350	350	300	300
	2½		950	750	600	550	500	450	400
	3		1150	900	750	650	550	500	500
	3½					1200	1100	1000	900
	4						1150	1050	950
	4½						1200	1100	1000
20	2	1050	700	550	450	400	350	350	300
	2½		1000	800	600	550	500	450	400
	3		1150	900	750	650	550	500	500
	3½						1100	1000	950
	4						1150	1050	1000
	4½							1200	1100
24	2	950	600	500	400	350	300	300	250
	2½		1150	900	750	650	550	500	450
	3			1050	900	750	700	600	550
	3½					1100	1000	900	800
	4					1150	1050	950	850
	4½						1150	1050	950

*Aluminum-jacketed calcium silicate insulation with an emissivity factor of 0.05. To convert inches to millimeters, multiply by 25.4, to convert dollars per 1 million British thermal units to dollars per 1 million kilojoules, multiply by 0.948, °C = 5/9 (°F − 32).

TABLE 11-23 Outdoor Insulation Thickness, 7.5-mi/h Wind, 60°F Air*

		Minimum pipe temperature, °F							
		Energy cost, $/million Btu							
Pipe size, in	Thickness, in	1	2	3	4	5	6	7	8
¾	1	450	300	250	250	200	200	150	150
	1½	800	500	400	300	250	250	200	200
	2			1150	950	850	750	700	650
	2½			1100	1000	900	800	750	700
1	1	400	300	250	200	200	150	150	150
	1½	1000	650	500	400	350	300	300	250
	2			1100	900	800	700	600	600
	2½				1200	1050	950	850	800
	3					1100	1000	900	850
1½	1	350	250	200	200	150	150	150	150
	1½	900	600	450	350	300	300	250	250
	2		100	850	700	600	550	500	450
	2½			1150	950	800	750	700	600
	3					1200	1050	1000	900
2	1	350	250	200	150	150	150	150	150
	1½	900	550	450	400	300	300	250	250
	2		1150	900	750	650	600	550	500
	2½			1000	850	750	650	600	550
	3				1050	950	850	750	700
3	1	300	200	150	150	150	150	150	150
	1½	750	500	400	300	250	250	250	200
	2		950	750	600	500	450	400	350
	2½		1150	950	750	650	600	500	500
	3			1150	1000	850	750	650	600
	3½								1150
4	1	250	200	150	150	150	150	150	150
	1½	750	500	350	300	250	250	200	200
	2		950	750	600	500	450	400	350
	2½			1050	900	700	650	600	550
	3			1100	950	750	700	650	600
	3½						1200	1100	1000

TABLE 11-23 Outdoor Insulation Thickness, 7.5-mi/h Wind, 60°F Air* *(Continued)*

		Minimum pipe temperature, °F							
		Energy cost, $/million Btu							
Pipe size, in	Thickness, in	1	2	3	4	5	6	7	8
6	1	250	150	150	150	150	150	150	150
	1½	450	300	200	200	150	150	150	150
	2		900	700	600	500	450	400	350
	2½		1050	800	650	600	500	450	400
	3			1050	900	750	700	600	550
	3½					1150	1050	950	850
	4							1200	1150
	4½								1200
8	1	250	200	150	150	150	150	150	150
	2		850	650	550	450	400	350	350
	2½		900	700	600	500	450	400	400
	3			1100	950	800	750	700	600
	3½					1150	1000	950	850
	4							1050	1000
10	2	200	150	150	150	150	150	150	150
	2½		1000	800	650	550	500	450	400
	3		1200	950	800	700	600	550	500
	3½					1100	1000	900	800
	4							1150	1050
	4½							1200	1100
12	1½	250	150	150	150	150	150	150	150
	2	950	600	500	400	350	300	250	250
	2½		900	700	550	500	400	400	350
	3			1100	900	800	700	650	550
	3½					1100	1000	900	850
	4					1150	1050	950	900
	4½					1200	1100	1000	950
14	1½	250	150	150	150	150	150	150	150
	2	850	550	400	350	300	250	250	250
	2½		850	650	550	500	400	400	400
	3			1000	850	700	650	550	500
	3½				1200	1000	950	850	800
	4					1050	1000	900	850
	4½						1100	1000	950
16	1½	250	150	150	150	150	150	150	150
	2	800	500	350	300	300	250	250	200
	2½		900	700	550	500	450	400	350
	3		1000	850	700	600	500	450	400
	3½				1200	1000	950	850	800
	4					1100	1000	900	850
	4½					1150	1000	950	900
18	1½	250	150	150	150	150	150	150	150
	2	850	550	400	350	300	250	250	200
	2½		800	650	500	450	400	350	350
	3		1000	800	650	550	500	450	400
	3½					1100	1000	900	850
20	1½	150	150	150	150	150	150	150	150
	2	900	550	450	350	300	300	250	250
	2½		850	650	550	450	400	350	350
	3		1000	800	650	550	500	450	400
	3½					1150	1050	950	900
	4					1200	1100	1000	950
	4½						1200	1100	1050
24	1½	150	150	150	150	150	150	150	150
	2	800	500	400	300	250	250	200	200
	2½		950	750	650	550	500	450	400
	3		1150	950	750	650	600	550	500
	3½				1150	1000	900	800	750
	4				1200	1050	950	850	800
	4½						1050	950	850
	5								

*Aluminum-jacketed calcium silicate insulation with an emissivity factor of 0.05. To convert inches to millimeters, multiply by 25.4; to convert miles per hour to kilometers per hour, multiply by 1.609; and to convert dollars per 1 million British thermal units to dollars per 1 million kilojoules, multiply by 0.948; °C = 5/9 (°F − 32).

thicknesses. The tables and a full explanation of their derivation appear in a paper by F. L. Rubin ("Piping Insulation—Economics and Profits," in *Practical Considerations in Piping Analysis*, ASME Symposium, vol. 69, 1982, pp. 27–46). Alternatively, the selected thicknesses have a payback period on the last nominal ½-in increment of 1.44 years as presented in a later paper by Rubin ["Can You Justify More Piping Insulation?" *Hydrocarbon Process* 152–155 (July 1982)].

Example 11-1 For 24-in pipe at 371°C (700°F) with an energy cost of $4/million Btu, select 2-in thickness for indoor and 2½-in thickness for outdoor locations. [A 2½-in thickness would be chosen at 399°C (750°F) indoors and 3½-in outdoors.]

Example 11-2 For 16-in pipe at 343°C (650°F) with energy valued at $5/million Btu, select 2½-in insulation indoors [use 3-in thickness at 371°C (700°F)]. Outdoors choose 3-in insulation [use 3½-in dual-layer insulation at 538°C (1000°F)].

Example 11-3 For 12-in pipe at 593°C (1100°F) with an energy cost of $6/million Btu, select 3½-in thickness for an indoor installation and 4½-in thickness for an outdoor installation.

INSTALLATION PRACTICE

Pipe Depending on the diameter, pipe is insulated with cylindrical half, third, or quarter sections or with flat segmental insulation. Fittings and valves are insulated with preformed insulation covers or with individual pieces cut from sectional straight pipe insulation.

Method of Securing Insulation with factory-applied jacketing may be secured with adhesive on the overlap, staples, tape, or wire, depending on the type of jacket and the outside diameter. Insulation which has a separate jacket is wired or banded in place before the jacket (finish) is applied.

Double Layer Pipe expansion is a significant factor at temperatures above 600°F (316°C). Above this temperature, insulation should be applied in a double layer with all joints staggered to prevent excessive heat loss and high surface temperature at joints opened by pipe expansion. This procedure also minimizes thermal stresses in the insulation.

Finish Covering for cylindrical surfaces ranges from asphalt-saturated or saturated and coated organic and asbestos paper, through laminates of such papers and plastic films or aluminum foil, to medium-gauge aluminum, galvanized steel, or stainless steel. Fittings and irregular surfaces may be covered with fabric-reinforced mastics or preformed metal or plastic covers. Finish selection depends on function and location. Vapor-barrier finishes may be in sheet form or a mastic, which may or may not require reinforcing, depending on the method of application; and additional protection may be required to prevent mechanical abuse and/or provide fire resistance. Criteria for selecting other finishes should include protection of insulation against water entry, mechanical abuse, or chemical attack. Appearance, life-cycle cost, and fire resistance may also be determining factors. Finish may be secured with tape, adhesive, bands, or screws. Fasteners which will penetrate vapor-retarder finishes should not be used.

Tanks, Vessels, and Equipment Flat, curved, and irregular surfaces such as tanks, vessels, boilers, and breechings are normally insulated with flat blocks, beveled lags, curved segments, blankets, or spray-applied insulation. Since no general procedure can apply to all materials and conditions, it is important that manufacturers' specifications and instructions be followed for specific insulation applications.

Method of Securing On small-diameter cylindrical vessels, the insulation may be secured by banding around the circumference. On larger cylindrical vessels, banding may be supplemented with angle-iron ledges to support the insulation and prevent slipping. On large flat and cylindrical surfaces, banding or wiring may be supplemented with various types of welded studs or pins. Breather springs may be required with bands to accommodate expansion and contraction.

Finish The materials are the same as for pipe and should satisfy the same criteria. Breather springs may be required with bands.

Additional References: *ASHRAE Handbook and Product Directory: Fundamentals,* American Society of Heating, Refrigerating and Air Conditioning Engineers, Atlanta, Ga., 1981. Turner and Malloy, *Handbook of Thermal Insulation Design Economics for Pipes and Equipment,* Krieger, New York, 1980. Turner and Malloy, *Thermal Insulation Handbook,* McGraw-Hill, New York, 1981.

AIR CONDITIONING

INTRODUCTION

Air conditioning is the process of treating air to simultaneously control its temperature, humidity, cleanliness, and distribution to meet the requirements of the conditioned spaces. Detailed discussions of various air cleaning and air distribution systems can be found in the *HVAC Applications* volume of the ASHRAE Handbook (American Society of Heating, Refrigerating, and Air-Conditioning Engineers Inc., 1791 Tullie Circle NE, Atlanta, Ga.). Air conditioning applications may include human comfort as well as the maintenance of proper conditions for manufacturing, processing, or preserving a wide variety of material and equipment. Industrial environments may use localized air conditioning to maintain safe working conditions for the health and efficiency of workers, even when overall space conditions cannot be made entirely comfortable for economical or other practical reasons.

COMFORT AIR CONDITIONING

Human comfort is influenced primarily by air temperature and humidity, local air velocity, radiant heat exchange, clothing insulation value, and metabolic rate. Chapter 48 of the 2012 ASHRAE Handbook *HVAC Applications* has an extensive discussion of noise control, another important consideration in air conditioning system design. Chapter 9 of the 2013 ASHRAE Handbook *HVAC Fundamentals* relates ambient air temperature and moisture content to human comfort, accounting for a wide variety of clothing and activity levels. It also provides results from extensive research efforts that address the impact of air velocity, radiative exchange, vertical temperature variations, age, gender, and a variety of other factors. The standard that addresses human comfort design criteria is ASHRAE Standard 55, *Thermal Environmental Conditions for Human Occupancy*. Because of the differences typically found in a given conditioned space regarding occupant manner of dress, age, gender, activity levels, and personal preferences, an 80 percent occupant comfort satisfaction rate is about the maximum that can realistically be obtained.

INDUSTRIAL AIR CONDITIONING

Industrial buildings should be designed according to their intended use. For instance, the manufacture or processing of hygroscopic materials (e.g., paper, textiles, and foods) will require tight control of humidity. The production of many electronic components often requires clean rooms with stringent limitations on particulate matter in the air. The processing of many fresh foods requires low temperatures, while the ambient in a facility for manufacturing refractories or forged metal products might be acceptable at much higher temperatures. Chapter 14 of the 2011 ASHRAE Handbook *HVAC Applications* provides extensive tables of suggested temperature and humidity conditions for many industrial air conditioning applications as well as special space conditioning considerations that may be needed for a wide variety of industrial processes.

VENTILATION

Odors or pollutants arising from occupants, cooking, or building material outgassing in residential or commercial buildings must be controlled to maintain a pleasant and safe living or working environment. Acceptable air quality can be maintained by localized exhaust of pollutants at their source, dilution with outdoor air that is free of such pollutants, or a combination of the two processes. Recommended outdoor air requirements for different types of nonresidential buildings are given in ASHRAE Standard 62.1, *Ventilation for Acceptable Indoor Air Quality*. Ventilation air requirements will vary because of the amounts of pollutant produced by different occupant activities and structural materials, but the ventilation rates are typically between 15 and 25 cfm of outdoor air per person for non-manufacturing commercial environments. Outdoor ventilation air requires much more energy to condition than the recirculated air from the nearly constant temperature conditioned space. Occupancy sensors (which typically detect CO_2 levels) are often used to reduce space conditioning costs by regulating the amount of outdoor air when space occupancy may be highly variable, such as in schools, theaters, or office complexes occupied less than 50 h per week. Local or state building codes may restrict how ventilation systems must be designed for fire or smoke control.

Industrial air conditioning systems must often address harmful gases, vapors, dusts, or fumes that are released into the work environment. These contaminants are best controlled by exhaust systems located near the source before they can enter the working environment. Dilution ventilation may be acceptable where nontoxic contaminants come from widely dispersed points. Combinations of local exhaust and dilution ventilation may provide the least expensive installation. Dilution alone may not be appropriate for cases involving toxic materials or large volumes of contaminants, or where the employees must work near the contaminant source. Chapter 32 of the ASHRAE Handbook *HVAC Applications* provides extensive information on the design and efficacy of a variety of exhaust systems. Safety codes from OSHA or local government bodies may have requirements that must take priority.

AIR CONDITIONING EQUIPMENT

Basically, an air conditioning system consists of a fan unit that forces air through a series of devices which act upon the air to clean it, increase or decrease its temperature, and increase or decrease its water vapor content. Air conditioning equipment can generally be classified into two broad types: central (also called field-erected) and unitary.

CENTRAL COOLING AND HEATING SYSTEMS

At least a dozen types of central air conditioning and air distribution systems are commonly used in commercial and industrial applications. They usually have large cooling and heating equipment located in a central location from which many different spaces or zones are served. Chilled-water coils or direct-expansion refrigerant coils are most commonly used for cooling the airstream. Spray washers using chilled water are sometimes used where continuous humidity control and air cleaning are especially important. Steam or hot-water coils usually provide the heating effect where the steam or hot water is generated in a boiler. Humidification may be provided by steam injection into the airstream, target-type water nozzles, pan humidifiers, air washers, or sprayed coils. Air cleaning is most commonly provided by cleanable or throwaway filters. Electronic air cleaners may be used when a low air pressure drop is important. Air handling units are available in capacities up to 50,000 ft^3/min with the cooling/heating coils, filters, and humidity control systems in a prefabricated package. These units can be located on a rooftop of a low-rise structure and connected to the chiller and boiler for fast field installation with minimal design of components.

The principal types of refrigeration equipment used in central systems are reciprocating (up to 300 tons); helical rotary (up to 750 tons); absorption (up to 2000 tons); and centrifugal (up to 10,000 tons). The mechanical drives are most commonly electric motors, but larger systems may use turbines or engines depending on the system size and the availability or cost of various fuels. The heat rejected from the condensers usually calls for wet cooling towers for larger systems or air-cooled condensers for smaller units. Modular condensing units with the compressor(s), direct expansion condensers with fans, and chilled-water heat exchangers are available in capacities up to several hundred tons.

UNITARY REFRIGERANT-BASED AIR CONDITIONING SYSTEMS

Unitary systems range from window-mounted air conditioners and heat pumps to residential and small commercial systems to commercial self-contained systems. The various types of unitary systems are described in detail in the *HVAC Systems and Equipment* volume of the ASHRAE Handbook. A detailed analysis of the proposed installation is necessary to select the type of air conditioning equipment that is best for an application. Each type of system has its own particular advantages and disadvantages. Important factors to be considered in the selection of air conditioning equipment are the required precision of temperature and humidity control; investment, owning, and operating costs; and space requirements. Building characteristics are also important, such as whether it is new or existing, multiple-story, size, available space for ducts, etc. For example, rooftop air conditioners or low-profile water source units may offer advantages for existing buildings where extensive air ducts would be difficult to install. A central system would usually be employed for large industrial processes where precise temperature and humidity control may be required.

LOAD CALCULATION

The first step in the design of most air conditioning systems is to determine the peak load at suitably severe design operating conditions. Since both outdoor and indoor temperatures influence the size of the equipment, the designer must exercise good judgment in selecting appropriate conditions for sizing the system. The efficiency of most systems is highest at full-load conditions, so oversizing cooling equipment will increase the first cost as well as operating costs. The most severe historical local outdoor conditions should almost never be used for sizing a system, or it will operate at a small fraction of its full capacity almost all the time. ASHRAE has compiled extensive weather data for over 6000 locations around the world based on 30 years of hourly recordings at each site. Data are presented in statistical format, such as dry-bulb temperatures exceeded 2.0, 1.0, or 0.4 percent of the time; the mean coincident wet-bulb temperature at those dry-bulb conditions; wind speeds exceeded 5, 2.5, and 1.0 percent of the time, and similarly compiled dew point temperatures. These data provide a good understanding of the number of hours per year that the capacity of the system may be exceeded. The statistical data permit loads to be computed for peak sensible or latent conditions, as well as for sizing cooling towers or other humidity-sensitive equipment. Due to the size of this weather data set, an abbreviated data set is included in the ASHRAE Handbook *Fundamentals,* with the complete data set available on the CD that comes with the Handbook. In addition, the weather data have been standardized in ANSI/ASHRAE Standard 169. These data allow for different design conditions to be used for critical applications (such as a hospital) or for much less critical applications (such as an exercise gym).

After appropriate design temperature and humidity conditions are selected, the next step is to calculate the space cooling load. The sensible heat load consists of (1) transmission through the building exterior envelope; (2) solar and sky radiation through windows and skylights; (3) heat gains from infiltration of outside air; (4) heat gains from people, lights, appliances, and power equipment (including the A/C fan motors); and (5) heat from materials brought in at higher than room temperature. The latent heat load accounts for moisture (1) given off from people, appliances, and products and (2) from infiltration of outside air. The total space load is the sum of the sensible and latent loads. The total cooling equipment load consists of the total space load plus the sensible and latent loads from the outside ventilation air.

The procedure for load calculation in nonresidential buildings should account for thermal storage in the mass of the structure, occupancy patterns, and other uses of energy that affect the load. The load can be strongly dependent on the use of the building. For example, lighting, computers, and copy equipment might be major load components for an office building that will require cooling even in winter. Load calculations are now performed almost exclusively with computer software. Basic loads for simple buildings can be determined using spreadsheet software, while large buildings with variable occupancy, large internal loads, and extensive window areas usually require much more elaborate computer models. Most computer models incorporate hourly weather data so load variations and changes in equipment performance with outdoor conditions can both be properly accounted for. Chapter 19 of the ASHRAE Handbook *Fundamentals* presents a summary of the various types of computerized load models that are currently available.

REFERENCES

2011 ASHRAE Handbook. *Heating, Ventilating, and Air-Conditioning Applications.* American Society of Heating, Refrigerating and Air-Conditioning Engineers, Inc., Atlanta, Ga.

2012 ASHRAE Handbook. *Heating, Ventilating, and Air-Conditioning Systems and Equipment.* American Society of Heating, Refrigerating and Air-Conditioning Engineers, Inc., Atlanta, Ga.

2013 ASHRAE Handbook. *Fundamentals.* American Society of Heating, Refrigerating and Air-Conditioning Engineers, Inc., Atlanta, Ga.

ANSI/ASHRAE Standard 55-2013. *Thermal Environmental Conditions for Human Occupancy.* American Society of Heating, Refrigerating and Air-Conditioning Engineers, Inc., Atlanta, Ga.

ANSI/ASHRAE/IESNA Standard 62.1-2013. *Ventilation for Acceptable Indoor Air Quality.* American Society of Heating, Refrigerating and Air-Conditioning Engineers, Inc., Atlanta, Ga.

ANSI/ASHRAE Standard 169-2013. *Weather Data for Building Design Standards.* American Society of Heating, Refrigerating and Air-Conditioning Engineers, Inc., Atlanta, Ga.

REFRIGERATION

INTRODUCTION

Refrigeration is a process in which heat is transferred from a lower- to a higher-temperature level by doing work on a system. In some systems heat transfer is used to provide the energy to drive the refrigeration cycle. All refrigeration systems are heat pumps ("pump energy from a lower to a higher potential"). The term *heat pump* is mostly used to describe refrigeration system applications where heat rejected to the condenser is of primary interest.

There are many means to obtain the refrigerating effect, but here three are discussed: mechanical vapor refrigeration cycles, absorption, and steam-jet cycles due to their significance for industry.

Basic Principles Since refrigeration is the practical application of the thermodynamics, comprehending the basic principles of thermodynamics is crucial for a full understanding of refrigeration. Section 4 includes a through approach to the theory of thermodynamics. Since our goal is to understand refrigeration processes, cycles are of crucial interest.

The Carnot refrigeration cycle is reversible and consists of adiabatic (isentropic due to reversible character) compression (1-2), isothermal rejection of heat (2-3), adiabatic expansion (3-4), and isothermal addition of heat (4-1). The temperature-entropy diagram is shown in Fig. 11-69. The Carnot cycle is an unattainable ideal which serves as a standard of comparison, and it provides a convenient guide to the temperatures that should be maintained to achieve maximum effectiveness.

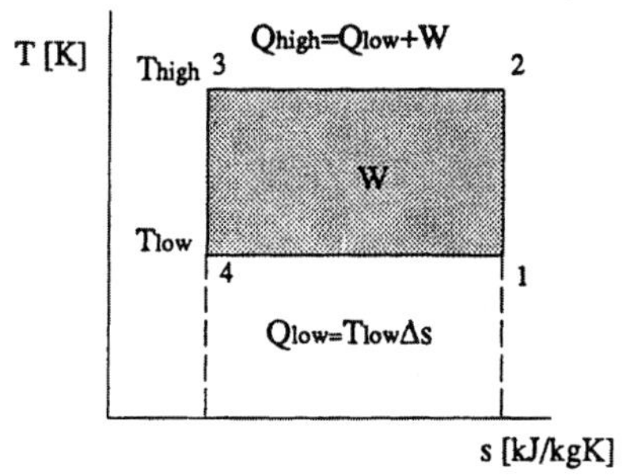

FIG. 11-69 Temperature–entropy diagram of the Carnot cycle.

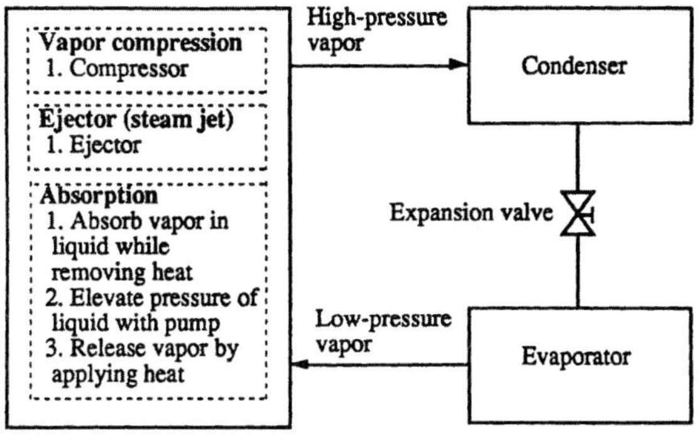

FIG. 11-70 Methods of transforming low-pressure vapor into high-pressure vapor in refrigeration systems (*Stoecker, Refrigeration, and Air-Conditioning.*)

The measure of the system performance is the *coefficient of performance* (COP). For refrigeration applications COP_R is the ratio of heat removed from the low-temperature level (Q_{low}) to the energy input (W):

$$COP_R = \frac{Q_{low}}{W} \tag{11-84}$$

For the heat pump (HP) operation, heat rejected at the high temperature Q_{high} is the objective, thus

$$COP_{HP} = \frac{Q_{high}}{W} = \frac{Q+W}{W} = COP_R + 1 \tag{11-85}$$

For a Carnot cycle (where $\Delta Q = T\,\Delta s$), the COP for the refrigeration application becomes (note that T is absolute temperature [K])

$$COP_R = \frac{T_{low}}{T_{high} - T_{low}} \tag{11-86}$$

and for heat pump application

$$COP_{HP} = \frac{T_{high}}{T_{high} - T_{low}} \tag{11-87}$$

The COP in real refrigeration cycles is always less than that for the ideal (Carnot) cycle, and there is constant effort to achieve this ideal value.

Basic Refrigeration Methods Three basic methods of refrigeration (mentioned above) use similar processes for obtaining the refrigeration effect: evaporation in the evaporator, condensation in the condenser where heat is rejected to the environment, and expansion in a flow restrictor. The main difference lies in the way compression is being done (Fig. 11-70): using mechanical work (in compressor), thermal energy (for absorption and desorption), or pressure difference (in ejector).

In Fig. 11-71 basic refrigeration systems are displayed in greater detail. A more elaborate approach is presented in the text.

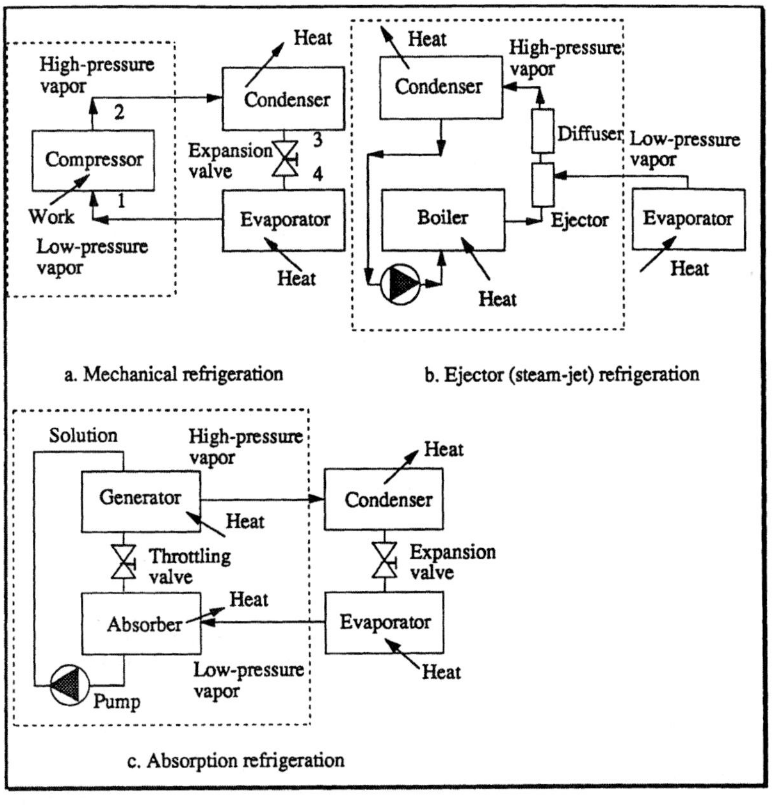

FIG. 11-71 Basic refrigeration systems.

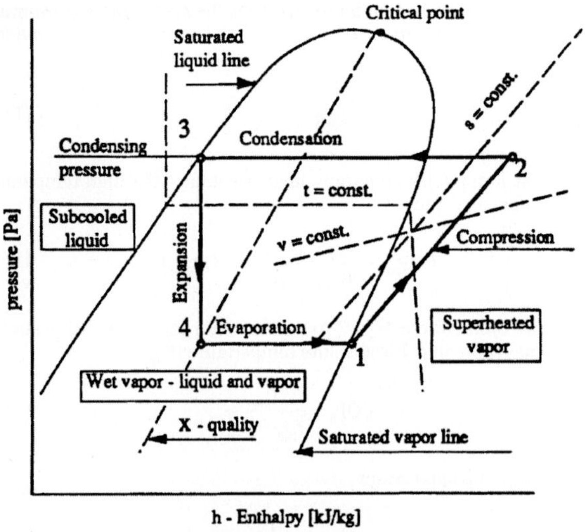

FIG. 11-72 The pressure-enthalpy diagram for the vapor compression cycle.

MECHANICAL REFRIGERATION (VAPOR COMPRESSION SYSTEMS)

Vapor Compression Cycles The most widely used refrigeration principle is vapor compression. Isothermal processes are realized through isobaric evaporation and condensation in the tubes. The standard vapor compression refrigeration cycle (counterclockwise Rankine cycle) is marked in Fig. 11-71*a* by 1, 2, 3, 4.

Work that could be obtained in a turbine is small, and a turbine is substituted for an expansion valve. For reasons of proper compressor function, wet compression is substituted for compression of dry vapor.

Although the *T–s* diagram is very useful for thermodynamic analysis, the pressure enthalpy diagram is used much more in refrigeration practice because both evaporation and condensation are isobaric processes so that the heat exchanged is equal to the enthalpy difference $\Delta Q = \Delta h$. For the ideal, isentropic compression, the work could be also presented as enthalpy difference $\Delta W = \Delta h$. The vapor compression cycle (Rankine) is presented in Fig. 11-72 in *p-h* coordinates.

Figure 11-73 presents the actual versus standard vapor-compression cycle. In reality, flow through the condenser and evaporator must be accompanied by a pressure drop. There is always some subcooling in the condenser and superheating of the vapor entering the compressor-suction line, both due to continuing process in the heat exchangers and the influence of the

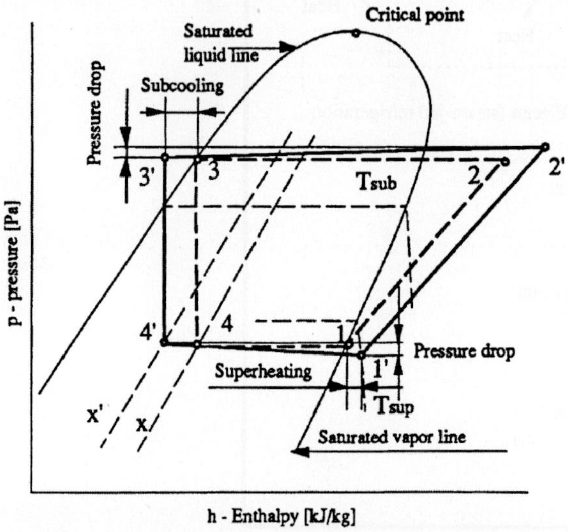

FIG. 11-73 Actual vapor compression cycle compared with standard cycle.

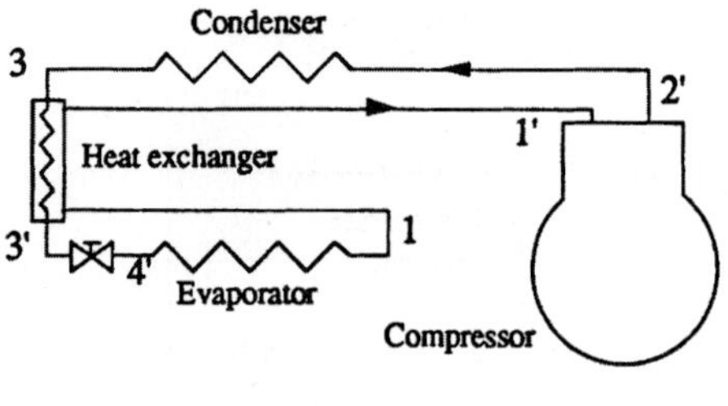

FIG. 11-74 Refrigeration system with a heat exchanger to subcool the liquid from the condenser.

environment. Subcooling and superheating are usually desirable to ensure only liquid enters the expansion device. Superheating is recommended as a precaution against droplets of liquid being carried over into the compressor.

There are many ways to increase cycle efficiency (COP). Some of them are better suited to one, but not for the other refrigerant. Sometimes, for the same refrigerant, the impact on COP could be different for various temperatures. One typical example is the use of a liquid-to-suction heat exchanger (Fig. 11-74).

The suction vapor coming from the evaporator could be used to subcool the liquid from the condenser. Graphic interpretation in the *T–s* diagram for such a process is shown in Fig. 11-75. The result of the use of suction line heat exchanger is to increase the refrigeration effect ΔQ and to increase the work by ΔW. The change in COP is then

$$\Delta COP = COP' - COP = \frac{Q + \Delta Q}{P + \Delta P - Q/P} \tag{11-88}$$

When dry, or superheated, vapor is used to subcool the liquid, the COP in R12 systems will increase, and the COP in NH_3 systems will decrease. For R22 systems it could have both effects, depending on the operating regime. Generally, this measure is advantageous (COP is improved) for fluids with high specific heat of liquid (less inclined saturated-liquid line on the *p-h* diagram), small heat of evaporation h_{fg} when vapor-specific heat is low (isobars in superheated regions are steep) and when the difference between evaporation and condensation temperatures is high. Measures to increase COP should be studied for every refrigerant. Sometimes the purpose of the suction-line heat exchanger is not only to improve the COP, but also to ensure that only the vapor reaches the compressor, particularly in the case of a malfunctioning expansion valve.

The system shown in Fig. 11-74 is direct expansion where dry or slightly superheated vapor leaves the evaporator. Such systems are predominantly used in small applications because of their simplicity and light weight. For the systems where efficiency is crucial (large industrial systems), recirculating systems (Fig. 11-76) are more appropriate.

Ammonia refrigeration plants are almost exclusively built as recirculating systems. The main advantage of recirculating versus direct expansion systems is better utilization of evaporator surface area. The diagram reflecting the influence of quality on the local heat-transfer coefficients is shown

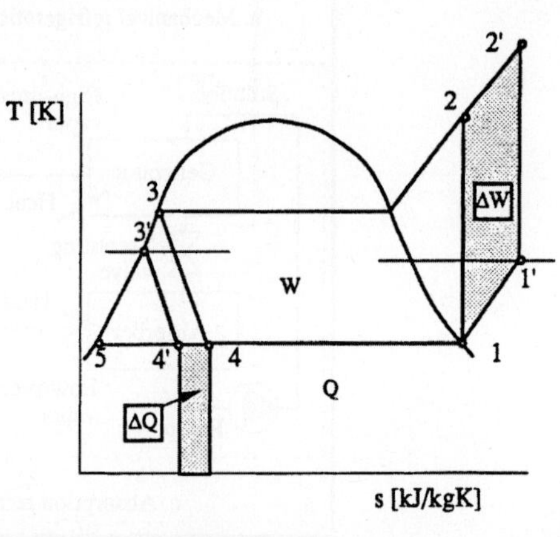

FIG. 11-75 Refrigeration system with a heat exchanger to subcool the liquid from the condenser.

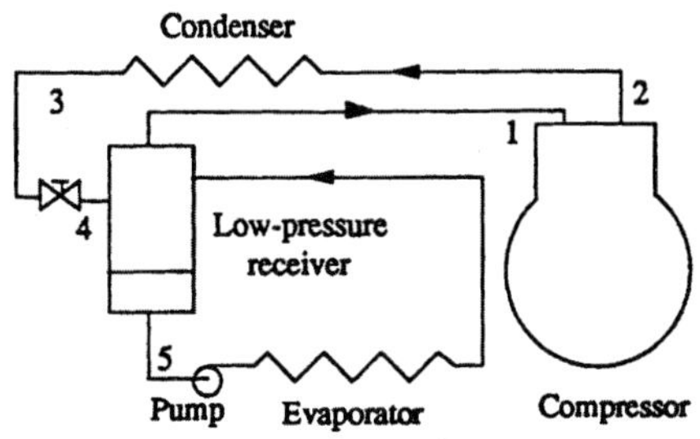

FIG. 11-76 Recirculation system.

in Fig. 11-89. It is clear that heat-transfer characteristics will be better if the outlet quality is lower than 1. Circulation could be achieved either by pumping (mechanical or gas) or by using gravity (thermosyphon effect: density of pure liquid at the evaporator entrance is higher than density of the vapor-liquid mixture leaving the evaporator). The circulation ratio (ratio of actual mass flow rate to the evaporated mass flow rate) is higher than 1 and up to 5. Higher values are not recommended due to a small increase in heat-transfer rate for a significant increase in pumping costs.

Multistage Systems When the evaporation and condensing pressure (or temperature) difference is large, it is prudent to separate compression into two stages. The use of multistage systems opens up the opportunity to use flash-gas removal and intercooling as measures to improve system performance. One typical two-stage system with two evaporating temperatures and both flash-gas removal and intercooling is shown in Fig. 11-77. The purpose of the flash-tank intercooler is to (1) separate vapor created in the expansion process, (2) cool superheated vapor from compressor discharge, and (3) eventually separate existing droplets at the exit of the medium-temperature evaporator. The first measure will decrease the size of the low-stage compressor because it will not wastefully compress the portion of flow which cannot perform the refrigeration, and the second measure will decrease the size of the high-stage compressor due to lowering the specific volume of the vapor from the low-stage compressor discharge, positively affecting operating temperatures of the high-stage compressor due to the cooling effect.

If the refrigerating requirement at a low-evaporating temperature is Q_l and at the medium level is Q_m, then the mass flow rates (m_1 and m_m, respectively) needed are

$$m_1 = \frac{Q_l}{h_1 - h_8} = \frac{Q_l}{h_1 - h_7} \tag{11-89}$$

$$m_m = \frac{Q_m}{h_3 - h_6} \tag{11-90}$$

The mass flow rate at the flash-tank inlet m_i consists of three components ($m_i = m_1 + m_{sup} + m_{flash}$):

m_1 = liquid at p_m feeding low-temperature evaporator

m_{sup} = liquid at p_m to evaporate in flash tank to cool superheated discharge

m_{flash} = flashed refrigerant, used to cool remaining liquid

The vapor component is

$$m_{flash} = x_m \times m_i \tag{11-91}$$

and the liquid component is

$$(1 - x_m) \times m_i = m_1 + m_{sup} \tag{11-92}$$

The liquid part of flow to cool superheated compressor discharge is determined by

$$m_{sup} = \frac{Q_l}{h_1 - h_8} \times \frac{h_2 - h_3}{h_3 - h_7} = m_1 \times \frac{h_2 - h_3}{h_{fgm}} \tag{11-93}$$

Since the quality x_m is

$$x_m = \frac{h_6 - h_7}{h_3 - h_7} \tag{11-94}$$

The mass flow rate through the condenser and high-stage compressor m_h is finally

$$m_h = m_m + m_i \tag{11-94a}$$

The optimum intermediate pressure for the two-stage refrigeration cycle is determined as the geometric mean between evaporation pressure p_l and condensing pressure p_h (Fig. 11-78):

$$p_m = \sqrt{\frac{p_h}{p_l}} \tag{11-95}$$

based on equal pressure ratios for low- and high-stage compressors. The optimum interstage pressure is slightly higher than the geometric mean of the suction and the discharge pressures, but, due to the very flat optimum of power versus interstage pressure relation geometric mean, it is widely accepted for determining the intermediate pressure. The required pressure of the intermediate-level evaporator may dictate interstage pressure other than determined as optimal.

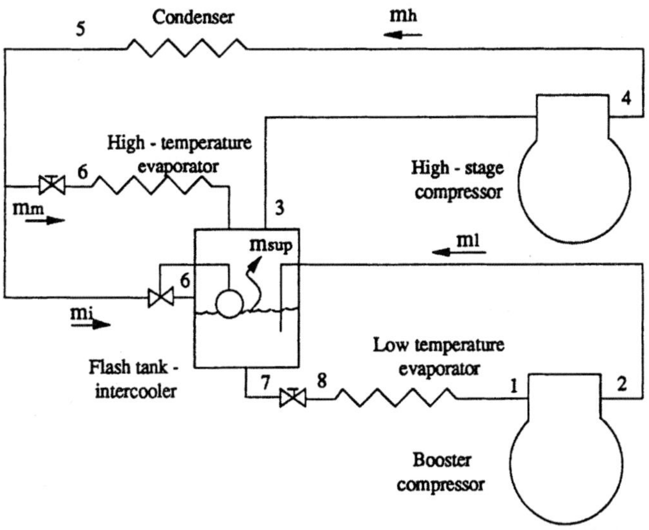

FIG. 11-77 Typical two-stage system with two evaporating temperatures, flash-gas removal, and intercooling.

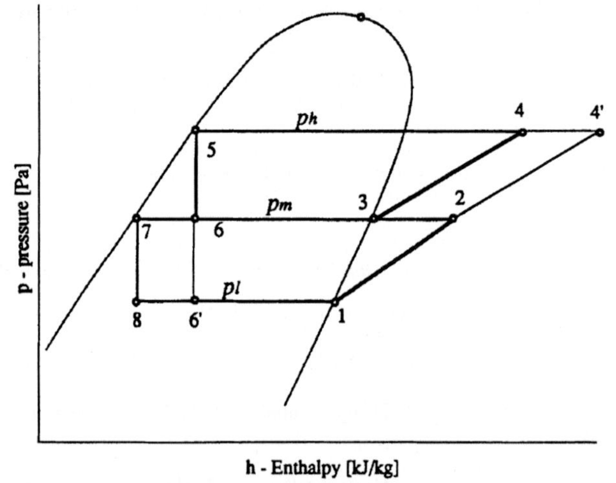

FIG. 11-78 Pressure–enthalpy diagram for typical two-stage system with two evaporating temperatures, flash-gas removal, and intercooling.

Two-stage systems should be seriously considered when the evaporating temperature is below −20°C. Such designs will save on power and reduce compressor discharge temperatures, but will increase the initial cost.

Cascade System This is a reasonable choice in cases where the evaporating temperature is very low (below −60°C). When condensing pressures are to be in the rational limits, the same refrigerant has a high specific volume at very low temperatures, requiring a large compressor. The evaporating pressure may be below atmospheric, which could cause moisture and air infiltration into the system if there is a leak. In other words, when the temperature difference between the medium that must be cooled and the environment is too high to be served with one refrigerant, it is wise to use different refrigerants in the high and low stages. Figure 11-79 shows a cascade system schematic diagram. There are basically two independent systems linked via a heat exchanger: the evaporator of the high-stage system and the condenser of the low-stage system.

EQUIPMENT

Compressors These could be classified by one criterion (the way the increase in pressure is obtained) as positive-displacement and dynamic types, as shown in Fig. 11-80 (see Sec. 10 for drawings and mechanical description of the various types of compressors). *Positive-displacement compressors* (PDCs) are the machines that increase the pressure of the vapor by reducing the volume of the chamber. Typical PDCs are reciprocating (in a variety of types) or rotary as screw (with one and two rotors), vane, scroll, and so on. Centrifugal compressors or turbocompressors are machines where the pressure is raised, converting some of the kinetic energy obtained by a rotating mechanical element which continuously adds angular momentum to a steadily flowing fluid, similar to a fan or pump.

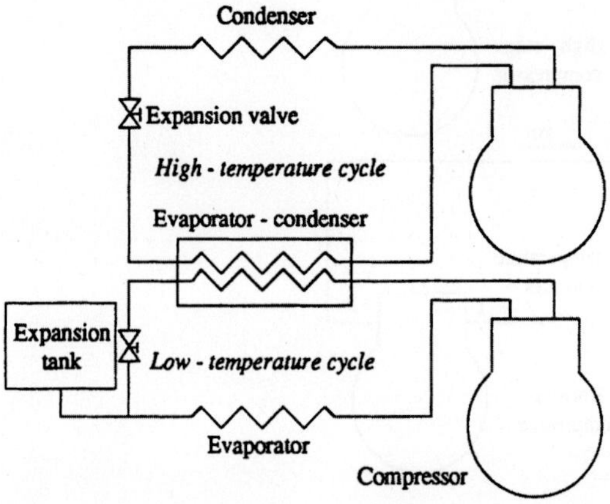

FIG. 11-79 Cascade system.

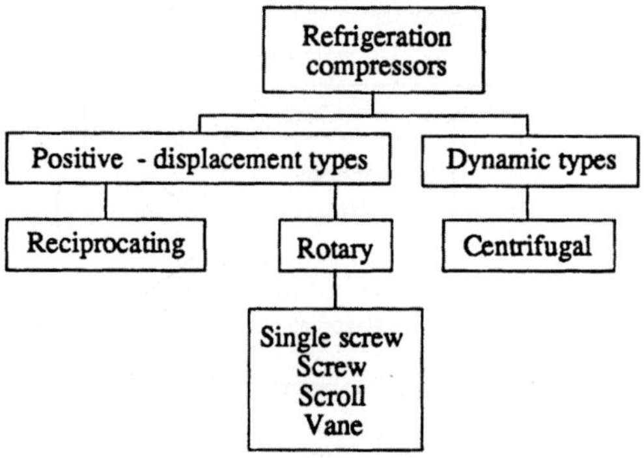

FIG. 11-80 Types of refrigeration compressors.

Generally, reciprocating compressors dominate in the range up to 300-kW refrigeration capacity. Centrifugal compressors are more accepted for the range over 500 kW, while screw compressors are in between with a tendency to go toward smaller capacities. The vane and the scroll compressors are finding their places primarily in very low-capacity range (domestic refrigerators and air conditioners), although vane compressors could be found in industrial compressors. Frequently, screw compressors operate as boosters, for the base load, while reciprocating compressors accommodate the variation of capacity in the high stage. The major reason for such design is the advantageous operation of screw compressors near full load and in design conditions, while reciprocating compressors seem to have better efficiencies at part-load operation than screw compressors.

Using other criteria, compressors are classified as *open, semihermetic* (*accessible*), or *hermetic.* Open type is characterized by shaft extension out of the compressor, where it is coupled to the driving motor. When the electric motor is in the same housing with the compressor mechanism, it could be either hermetic or accessible (semihermetic). Hermetic compressors have welded enclosures, not designed to be repaired, and are generally manufactured for smaller capacities (seldom over 30 kW), while semihermetic or an accessible type is located in the housing which is tightened by screws. Semihermetic compressors have all the advantages of hermetic (no sealing of moving parts, e.g., no refrigerant leakage at the seal shaft, no external motor mounting, no coupling alignment) and could be serviced, but it is more expensive.

Compared to other applications, refrigeration capacities in the chemical industry are usually high. That leads to wide use of centrifugal, screw, or high-capacity rotary compressors. Most centrifugal and screw compressors use economizers to minimize power and suction volume requirements. Generally, there is far greater use of open-drive type of compressors in the chemical plants than in air conditioning, commercial, or food refrigeration. Very frequently, compressor lube oil systems are provided with auxiliary oil pumps, filters, coolers, and other equipment to permit maintenance and repair without shutdown.

Positive-Displacement Compressors *Reciprocating compressors* are built in different sizes (up to about 1-MW refrigeration capacity per unit). Modern compressors are high-speed, mostly direct-coupled, single-acting, from 1 to mostly 8, and occasionally up to 16 cylinders.

Two characteristics of compressors for refrigeration are the most important: refrigerating capacity and power. Typical characteristics are as presented in Fig. 11-81.

Refrigerating capacity Q_e is the product of the mass flow rate of refrigerant m and the refrigerating effect R which is (for isobaric evaporation) $R = h_{\text{evaporator outlet}} - h_{\text{evaporator inlet}}$. Power P required for the compression, necessary for the motor selection, is the product of mass flow rate m and work of compression W. The latter is, for the isentropic compression, $W = h_{\text{discharge}} - h_{\text{suction}}$. Both of these characteristics could be calculated for the ideal (without losses) and for the actual compressor. Ideally, the mass flow rate is equal to the product of the compressor displacement V_i per unit time and the gas density ρ: $m = V_i \times \rho$. The compressor displacement rate is the volume swept through by the pistons (product of the cylinder number n and volume of cylinder V = stroke × $d^2\pi/4$) per second. In reality, the actual compressor delivers less refrigerant.

The ratio of the actual flow rate (entering compressor) to the displacement rate is the volumetric efficiency η_{va}. The volumetric efficiency is less than unity due to reexpansion of the compressed vapor in clearance volume, pressure drop (through suction and discharge valves, strainers, manifolds, etc.), internal gas leakage (through the clearance between piston rings and cylinder walls, etc.), valve inefficiencies, and expansion of the vapor in the suction cycle caused by the heat exchanged (hot cylinder walls, oil, motor, etc.).

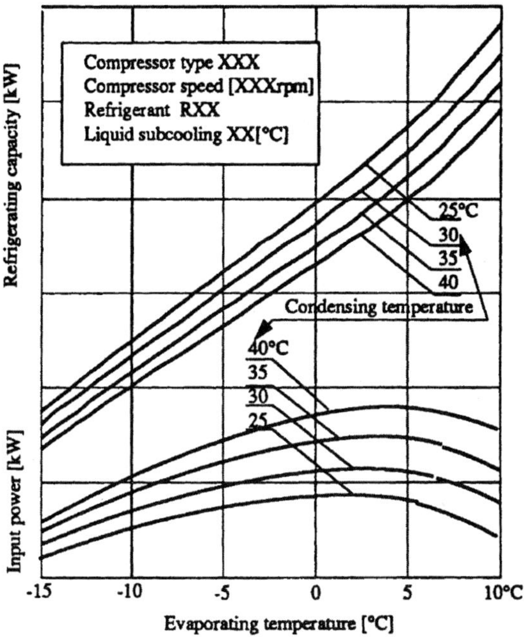

FIG. 11-81 Typical capacity and power-input curves for reciprocating compressor.

Similar to volumetric efficiency, isentropic (adiabatic) efficiency η_a is the ratio of the work required for isentropic compression of the gas to work input to the compressor shaft. The adiabatic efficiency is less than 1 mainly due to pressure drop through the valve ports and other restricted passages and the heating of the gas during compression.

Figure 11-82 presents the compression on a pressure-volume diagram for an ideal compressor with clearance volume (thin lines) and actual compressor (thick lines). Compression in an ideal compressor without clearance is extended using dashed lines to the points I_d (end of discharge), line $I_d - I_s$ (suction), and I_s (beginning of suction). The area surrounded by the lines of compression, discharge, reexpansion, and intake presents the work needed for compression. The actual compressor only appears to demand less work for compression due to smaller area in the *p-V* diagram. The mass flow rate for an ideal compressor is higher, which cannot be seen in the diagram. In reality, an actual compressor will have adiabatic compression and reexpansion and higher discharge and lower suction pressures due to pressure drops in valves and lines. The slight increase in the pressure at the beginning of the discharge and suction is due to forces needed to initially open valves.

When the suction pressure is lowered, the influence of the clearance will increase, causing in the extreme cases the entire volume to be used for reexpansion, which drives the volumetric efficiency to zero.

There are various options for capacity control of reciprocating refrigeration compressors:

1. Open the suction valves by some external force (oil from the lubricating system, discharge gas, electromagnets, ...).
2. Gas bypass: return the discharge gas to suction (within the compressor or outside the compressor).
3. Control the suction pressure by throttling in the suction line.
4. Control the discharge pressure.
5. Add reexpansion volume.
6. Change the stroke.
7. Change the compressor speed.

The first method is used most frequently. The next preference is for the last method, mostly used in small compressors due to problems with speed control of electric motors. Other means of capacity control are very seldom utilized due to thermodynamic inefficiencies and design difficulties. Energy losses in a compressor, when capacity regulation is provided by lifting the suction valves, are due to friction of gas flowing into and out of the unloaded cylinder. This is shown in Fig. 11-83 where the comparison is made for ideal partial-load operation, reciprocating, and screw compressors.

Rotary compressors are also PDC types, but where refrigerant flow rotates during compression. Unlike the reciprocating type, rotary compressors have a built-in volume ratio which is defined as the volume in the cavity when the suction port is closed ($V_s = m \times \upsilon_s$) over the volume in the cavity when the discharge port is uncovered ($V_d = m \times \upsilon_d$). The built-in volume ratio determines for a given refrigerant and conditions the pressure ratio, which is

$$\frac{p_d}{p_s} = \left(\frac{\upsilon_s}{\upsilon_d}\right)^n \tag{11-96}$$

where n represents the politropic exponent of compression.

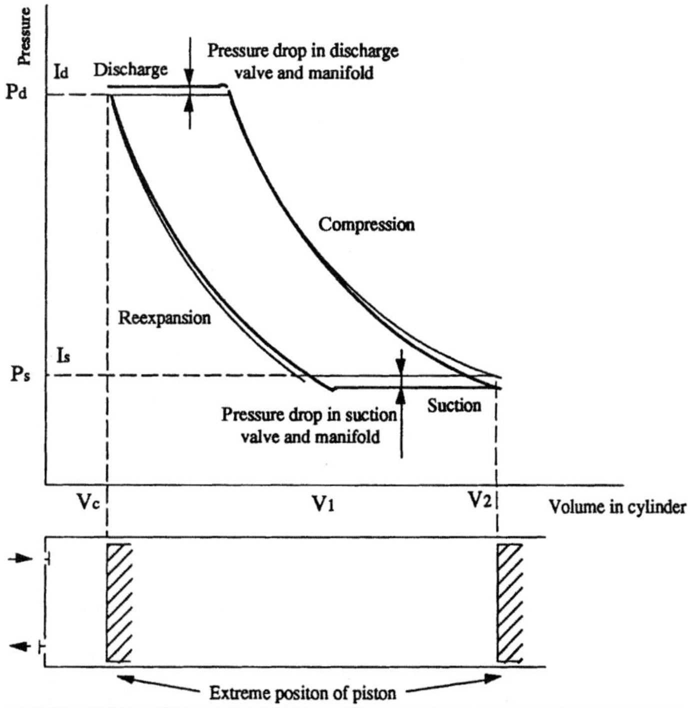

FIG. 11-82 Pressure–volume diagram of an ideal (thin line) and actual (thick line) reciprocating compressor.

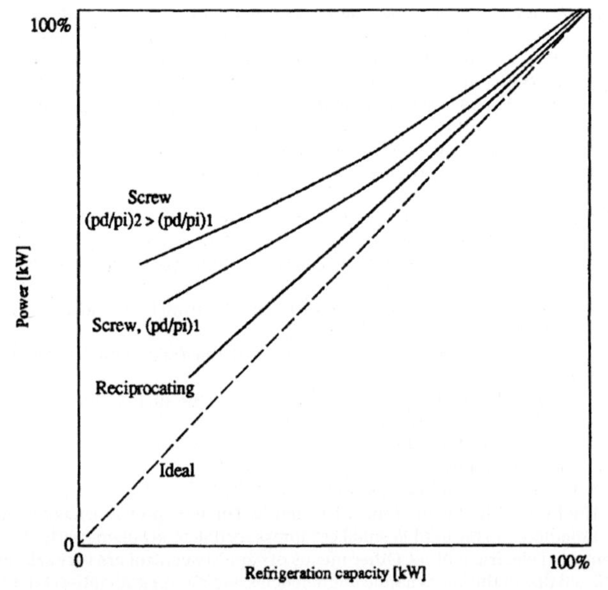

FIG. 11-83 Typical power-refrigeration capacity data for different types of compressors during partial unloaded operation.

In other words, in a reciprocating compressor the discharge valve opens when the pressure in the cylinder is slightly higher than the pressure in the high-pressure side of the system, while in rotary compressors the discharge pressure will be established only by inlet conditions (p_s, V_s) and the built-in volume ratio regardless of the system discharge pressure. Very seldom are the discharge and system (condensing) pressures equal, causing the situation shown in Fig. 11-84. When condensing pressure p is lower than discharge p_d, shown as case (*a*), "overcompression" will cause energy losses presented by the horn on the diagram. If the condensing pressure is higher, in the moment when the discharge port uncovers there will be flow of refrigerant backward into the compressor, causing losses shown in Fig. 11-84*b*; and the last stage will be only discharge without compression. The case when the compressor discharge pressure is equal to the condensing pressure is shown in Fig. 11-84*c*.

Double helical rotary (twin) screw compressors consist of two mating, helically grooved rotors (male and female) with asymmetric profile, in a housing formed by two overlapped cylinders, with inlet and outlet ports. Developed relatively recently (in the 1930s), the first twin-screw compressors were used for air, and later (in the 1950s) became popular for refrigeration. Screw compressors have some advantages over reciprocating compressors (fewer moving parts and greater compactness) but also some drawbacks (lower efficiency at off-design conditions, as discussed above, higher manufacturing cost due to complicated screw geometry, large separators and coolers for oil which is important as a sealant). Figure 11-85 shows the oil circuit of a screw compressor. Oil cooling could be provided by water, glycol, or refrigerant either in the heat exchanger utilizing thermosyphon effect or using the direct expansion concept.

To overcome some inherent disadvantages, screw compressors have been initially used predominantly as booster (low-stage) compressors, and following development in capacity control and decreasing prices, they are now widely used for high-stage applications. There are several methods for capacity regulation of screw compressors. One is variable-speed drive, but a more economical first-cost concept is a slide valve that is used in some form by practically all screw compressors.

The slide is located in the compressor casting below the rotors, allowing internal gas recirculation without compression. The slide valve is operated by a piston located in a hydraulic cylinder and actuated by high-pressure oil from both sides. When the compressor is started, the slide valve is fully open and the compressor is unloaded. To increase capacity, a solenoid valve on the hydraulic line opens, moving the piston in the direction of increasing capacity. To increase part-load efficiency, the slide valve is designed to consist of two parts, one traditional slide valve for capacity regulation and other for built-in volume adjustment.

Single-screw compressors are a newer design (early 1960s) compared to twin-screw compressors, and are manufactured in the range of capacity from 100 kW to 4 MW. The compressor screw is cylindrical with helical grooves mated with two star wheels (gate rotors) rotating in opposite direction from each other. Each tooth acts as the piston in the rotating "cylinder" formed by the screw flute and cylindrical main-rotor casting.

As compression occurs concurrently in both halves of the compressor, radial forces are oppositely directed, resulting in negligible net-radial loads on the rotor bearings (unlike twin-screw compressors), but there are some loads on the star wheel shafts.

Scroll compressors are currently used in relatively small-size installations, predominantly for residential air conditioning (up to 50 kW). They are recognized for low-noise operation. Two scrolls (freestanding, involute spirals bounded on one side by a flat plate) facing each other form a closed volume while one moves in a controlled orbit around a fixed point on the other fixed scroll.

The suction gas which enters from the periphery is trapped by the scrolls. The closed volumes move radially inward until the discharge port is reached, when vapor is pressed out. The orbiting scroll is driven by a short-throw crank mechanism. As in screw compressors, internal leakage should be kept low, and it occurs in gaps between cylindrical surfaces and between the tips of the involute and the opposing scroll baseplate.

Similar to the screw compressor, the scroll compressor is a constant-volume-ratio machine. Losses occur when operating conditions of the compressor do not match the built-in volume ratio (see Fig. 11-84).

Vane compressors are used in small, hermetic units, but sometimes as booster compressors in industrial applications. Two basic types are the *fixed* (roller) or single-vane type and the *rotating* or multiple-vane type. In the single-vane type, the rotor (called *roller*) is eccentrically placed in the cylinder so these two are always in contact. The contact line makes the first separation between the suction and discharge chambers while the vane (spring-loaded divider) makes the second. In the multiple-vane compressors, the rotor and the cylinder are not in the contact. The rotor has two or more sliding vanes which are held against the cylinder by centrifugal force. In the vane rotary compressors, no suction valves are needed. Since the gas enters the compressor continuously, gas pulsations are at a minimum. Vane compressors have a high volumetric efficiency because of the small clearance volume and consequent low reexpansion losses. Rotary vane compressors have a low weight-to-displacement ratio, which makes them suitable for transport applications.

Centrifugal Compressors These are sometimes called *turbocompressors* and mostly serve refrigeration systems in the capacity range of 200 to 10,000 kW. The main component is a spinning impeller wheel, backward-curved, which imparts energy to the gas being compressed. Some of the kinetic energy converts into pressure in a volute. Refrigerating centrifugal compressors are predominantly multistage, compared to other turbocompressors, which produce high-pressure ratios.

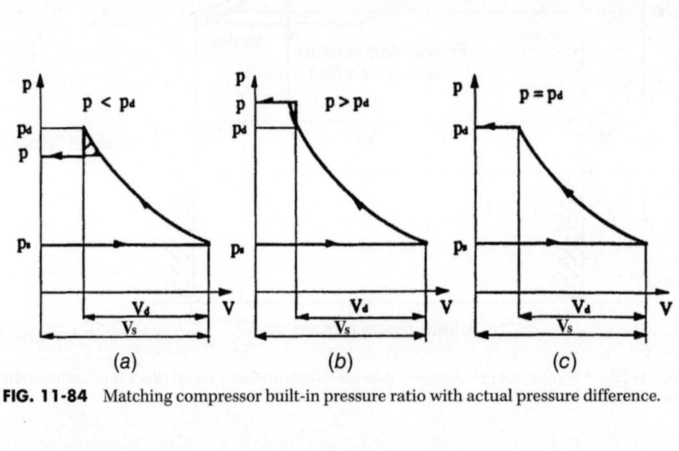

FIG. 11-84 Matching compressor built-in pressure ratio with actual pressure difference.

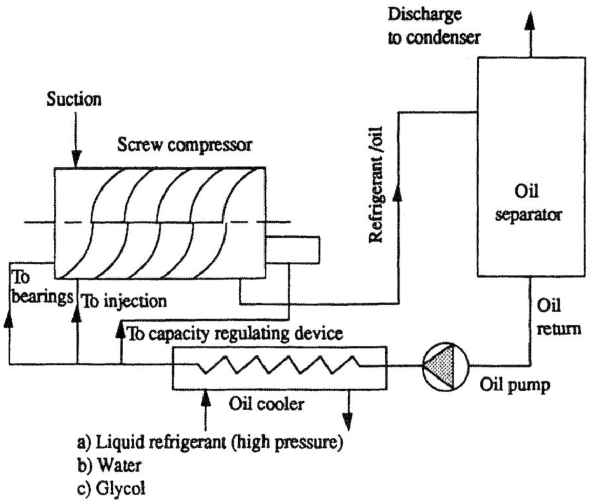

FIG. 11-85 Oil cooling in a screw compressor.

The torque T (N·m) the impeller ideally imparts to the gas is

$$T = m\,(u_{\text{tang.out}}\, r_{\text{out}} - u_{\text{tang.in}}\, r_{\text{in}}) \tag{11-97}$$

where m (kg/s) = mass flow rate
r_{out} (m) = radius of exit of impeller
r_{in} (m) = radius of entrance of impeller
$u_{tang.out}$ (m/s) = tangential velocity of refrigerant leaving impeller
$u_{tang.in}$ (m/s) = tangential velocity of refrigerant entering impeller

When refrigerant enters essentially radially, $u_{tang.in} = 0$ and torque becomes

$$T = m \times u_{\text{tang.out}} \times r_{\text{out}} \tag{11-98}$$

The power P (in watts) is the product of torque and rotative speed ω [1/s] and so is

$$P = T \times \omega = m \times u_{\text{tang.out}} \times r_{\text{out}} \times \omega \tag{11-99}$$

which for $u_{tang.ou} = r_{out} \times \omega$ becomes

$$P = m \times u^2_{\text{tang.out}} \tag{11-100}$$

or for isentropic compression

$$P = m \times \Delta h \tag{11-101}$$

The performance of a centrifugal compressor (discharge to suction-pressure ratio versus the flow rate) for different speeds is shown in Fig. 11-86.

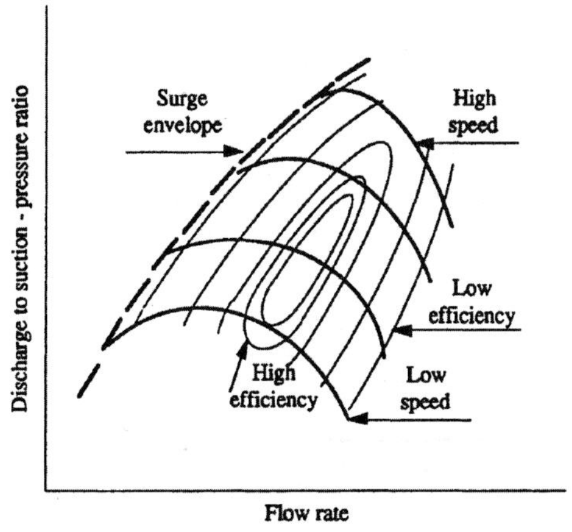

FIG. 11-86 Performance of the centrifugal compressor.

Lines of constant efficiencies show the maximum efficiency. Unstable operation sequence, called *surging*, occurs when compressors fail to operate in the range left of the surge envelope. It is characterized by noise and wide fluctuations of load on the compressor and the motor. The period of the cycle is usually 2 to 5 s, depending upon the size of the installation.

The capacity could be controlled by (1) adjusting the prerotation vanes at the impeller inlet, (2) varying the speed, (3) varying the condenser pressure, and (4) bypassing discharge gas. The first two methods are predominantly used.

Condensers These are heat exchangers that convert refrigerant vapor to a liquid. Heat is transferred in three main phases: (1) desuperheating, (2) condensing, and (3) subcooling. In reality condensation occurs even in the superheated region, and subcooling occurs in the condensation region. Three main types of refrigeration condensers are air-cooled, water-cooled, and evaporative.

Air-cooled condensers are used mostly in air conditioning and for smaller refrigeration capacities. The main advantage is the availability of cooling medium (air), but heat-transfer rates for the air side are far below values when water is used as a cooling medium. Condensation always occurs inside tubes, while the air side uses extended surface (fins).

The most common types of water-cooled refrigerant condensers are (1) shell-and-tube, (2) shell-and-coil, (3) tube-in-tube, and (4) brazed-plate. *Shell-and-tube* condensers are built up to 30-MW capacity. Cooling water flows through the tubes in a single-pass or multipass circuit. Fixed-tubesheet and straight-tube construction are common. Horizontal layout is typical, but sometimes vertical is used. Heat-transfer coefficients for the vertical types are lower due to poor condensate drainage, but less water of lower purity can be utilized. Condensation always occurs on the tubes, and often the lower portion of the shell is used as a receiver. In *shell-and-coil* condensers, water circulates through one or more continuous or assembled coils contained within the shell while refrigerant condenses outside. The tubes cannot be mechanically cleaned or replaced. *Tube-in-tube* condensers could be found in versions where condensation occurs either in the inner tube or in the annulus. Condensing coefficients are more difficult to predict, especially in the cases where tubes are formed in spiral. Mechanical cleaning is more complicated, sometimes impossible, and tubes are not replaceable. *Brazed-plate* condensers are constructed of plates brazed together to make up an assembly of separate channels. The plates are typically stainless steel, wave-style corrugated, enabling high heat-transfer rates. Performance calculation is difficult, with very few correlations available. The main advantage is the highest performance/volume (mass) ratio and the lowest refrigerant charge. The last mentioned advantage seems to be the most important feature for many applications where minimization of charge inventory is crucial.

Evaporative condensers (Fig. 11-87) are widely used due to lower condensing temperatures than in the air-cooled condensers and also lower than the water-cooled condenser combined with the cooling tower. Water demands are far lower than for water-cooled condensers. The chemical industry uses shell-and-tube condensers widely, although the use of air-cooled condensing equipment and evaporative condensers is on the increase.

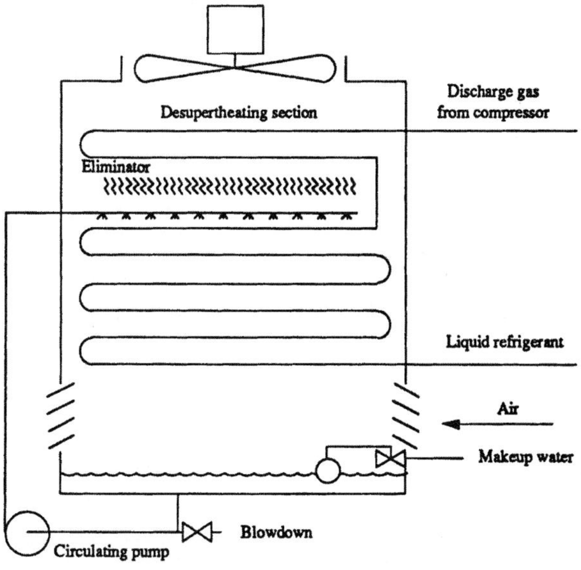

FIG. 11-87 Evaporative condenser with desuperheating coil.

Generally, cooling water is of a lower quality than normal, having also higher mud and silt content. Sometimes even replaceable copper tubes in shell-and-tube heat exchangers are required. It is advisable to use cupronickel instead of copper tubes (when water is high in chlorides) and to use conservative water-side velocities (less than 2 m/s for copper tubes).

Evaporative condensers are used quite extensively. In most cases, commercial evaporative condensers are not totally suitable for chemical plants due to the hostile atmosphere, which usually abounds in vapor and dusts that can cause either chemical (corrosion) or mechanical problems (plugging of spray nozzles).

Air-cooled condensers are similar to evaporative ones in that the service dictates the use of either more expensive alloys in the tube construction or conventional materials of greater wall thickness.

Heat rejected in the condenser Q_{Cd} consists of heat absorbed in the evaporator Q_{Eevap} and energy W supplied by the compressor:

$$Q_{Cd} = Q_{Eevap} + W \tag{11-102}$$

For the actual systems, compressor work will be higher than for ideal systems for the isentropic efficiency and other losses. In the case of hermetic or accessible compressors where an electric motor is cooled by the refrigerant, condenser capacity should be

$$Q_{Cd} = Q_{Eevap} + P_{EM} \tag{11-103}$$

It is common that compressor manufacturers provide data for the ratio of the heat rejected at the condenser to the refrigeration capacity, as shown in Fig. 11-88. The solid line represents data for the open compressors while the dotted line represents the hermetic and accessible compressors. The difference between solid and dotted line is due to all losses (mechanical and electrical in the electric motor). Condenser design is based on the value

$$Q_{Cd} = Q_{Eevap} \times \text{heat rejection ratio} \tag{11-104}$$

Thermal and mechanical design of heat exchangers (condensers and evaporators) is presented earlier in this section.

Evaporators These are heat exchangers where refrigerant is evaporated while cooling the product, fluid, or body. Refrigerant could be in direct contact with the body that is being cooled, or some other medium could be used as secondary fluid. Mostly air, water, and antifreeze are fluids that are cooled. Design is strongly influenced by the application. Evaporators for air cooling will have in-tube evaporation of the refrigerant, while liquid chillers could have refrigerant evaporation inside or outside the tube. The heat-transfer coefficient for evaporation inside the tube (versus length or quality) is shown in Fig. 11-89. Fundamentals of the heat transfer in evaporators, as well as design aspects, are presented in Sec. 11. We point out only some specific aspects of refrigeration applications.

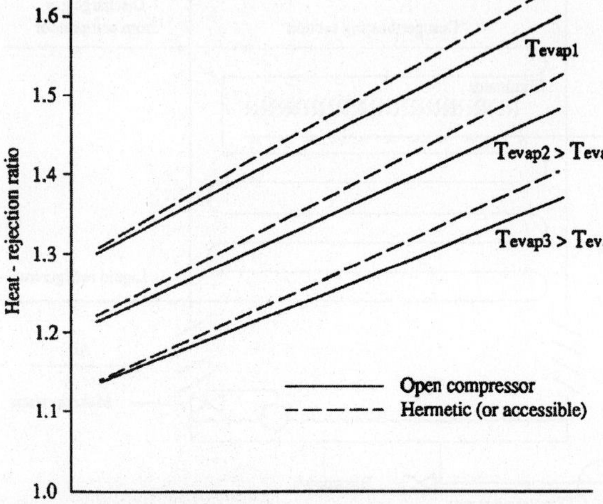

FIG. 11-88 Typical values of the heat rejection ratio of the heat rejected at the condenser to the refrigerating capacity.

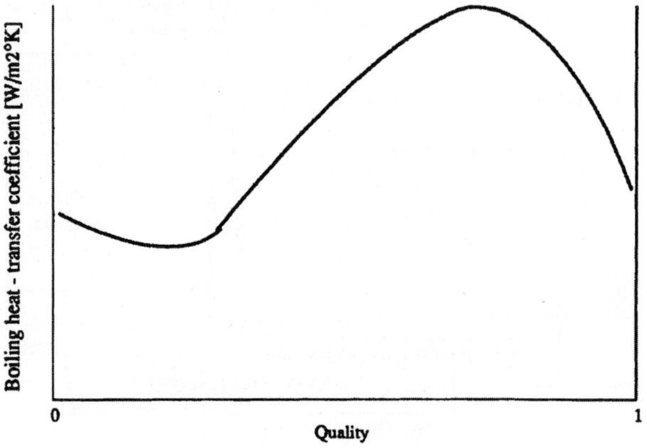

FIG. 11-89 Heat-transfer coefficient for boiling inside the tube.

Refrigeration evaporators could be classified according to the method of feed as either direct (dry) expansion or flooded (liquid overfeed). In dry expansion the evaporator's outlet is dry or slightly superheated vapor. This limits the liquid feed to the amount that can be completely vaporized by the time it reaches the end of the evaporator. In the liquid overfeed evaporator, the amount of liquid refrigerant circulating exceeds the amount evaporated by the circulation number. Decision on the type of the system to be used is one of the first in the design process. A direct-expansion evaporator is generally applied in smaller systems where compact design and low first costs are crucial. Control of the refrigerant mass flow is then obtained by either a thermoexpansion valve or a capillary tube. Figure 11-89 suggests that the evaporator surface is the most effective in the regions with quality that is neither low nor high. In dry-expansion evaporators, inlet qualities are 10 to 20 percent, but when controlled by the thermoexpansion valve, vapor at the outlet is not only dry, but even superheated.

In recirculating systems, saturated liquid ($x = 0$) is entering the evaporator. Either the pump or gravity will deliver more refrigerant liquid than will evaporate, so outlet quality could be lower than 1. The ratio of refrigerant flow rate supplied to the evaporator overflow rate of refrigerant vaporized is the circulation ratio n. When n increases, the coefficient of heat transfer will increase due to the wetted outlet of the evaporator and the increased velocity at the inlet (Fig. 11-90). In the range of $n = 2$ to 4, the overall U value for air cooler increases roughly by 20 to 30 percent compared to the direct-expansion case. Circulation rates higher than 4 are not efficient.

The price for an increase in heat-transfer characteristics is a more complex system with more auxiliary equipment: low-pressure receivers, refrigerant pumps, valves, and controls. Liquid refrigerant is predominantly pumped by mechanical pumps; however, sometimes gas at condensing pressure is used for pumping, in the variety of concepts.

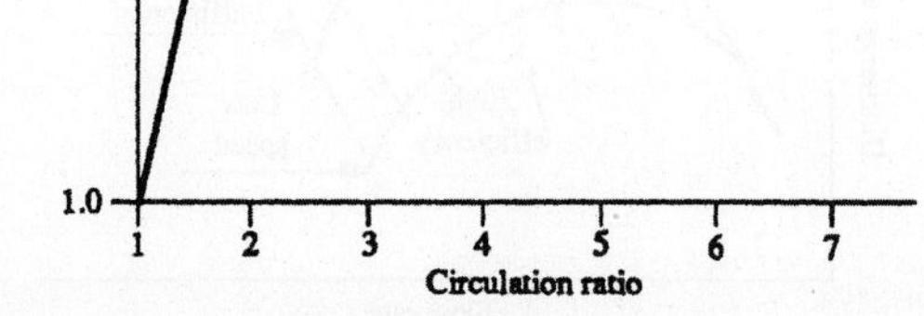

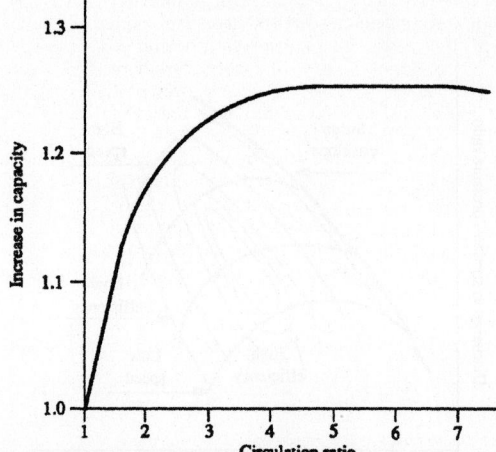

FIG. 11-90 Effect of circulation ratio on the overall heat-transfer coefficient of an air-cooling coil.

The important characteristics of the refrigeration evaporators is the presence of the oil. The system is contaminated with oil in the compressor, in spite of reasonably efficient oil separators. Some systems will recirculate oil, when miscible with refrigerant, returning it to the compressor crankcase. This is found mostly in the systems using halocarbon refrigerants. Although oils that are miscible with ammonia exist, immiscible oils are predominantly used. This inhibits the ammonia systems from recirculating the oil. In systems with oil recirculation when halocarbons are used, special consideration should be given to proper sizing and layout of the pipes. Proper pipeline configuration, slopes, and velocities (to ensure oil circulation under all operating loads) are essential for good system operation. When refrigerant is lighter than the oil in systems with no oil recirculation, oil will be at the bottom of every volume with a top outlet. Then oil must be drained periodically to avoid decreasing the performance of the equipment.

It is essential for proper design to have the data for refrigerant–oil miscibility under all operating conditions. Some refrigerant–oil combinations will always be miscible, others always immiscible, but still others will have both characteristics, depending on the temperatures and pressures applied. Defrosting is the important issue for evaporators which are cooling air below freezing. Defrosting is done periodically, actuated predominantly by time relays, but other frost indicators are used (temperature, visual, or pressure-drop sensors). Defrost technique is determined mostly by fluids available and tolerable complexity of the system. Defrosting is done by the following mechanisms when the system is off:

- Hot (or cool) refrigerant gas (the predominant method in industrial applications)
- Water (defrosting from the outside, unlike hot-gas defrost)
- Air (only when room temperature is above freezing)
- Electricity (for small systems where hot-gas defrost will be too complex and water is not available)
- Combinations of above

System Analysis Design calculations are made on the basis of the close to the highest refrigeration load; however, the system operates at the design conditions very seldom. The purpose of regulating devices is to adjust the system performance to cooling demands by decreasing the effect or performance of some component. Refrigeration systems have inherent self-regulating control which the engineer can rely on to a certain extent. When the refrigeration load starts to decrease, less refrigerant will evaporate. This causes a drop in evaporation temperature (as long as compressor capacity is unchanged) due to the imbalance in vapor being taken by the compressor and produced by evaporation in the evaporator. With a drop in evaporation pressure, the compressor capacity will decrease due to (1) lower vapor density (lower mass flow for the same volumetric flow rate) and (2) a decrease in volumetric efficiency. However, when the evaporation temperature drops, for the unchanged temperature of the medium being cooled, the evaporator capacity will increase due to an increase in the mean-temperature difference between refrigerant and cooled medium, causing a positive effect (increase) on the cooling load. With a decrease in the evaporation temperature, the heat rejection factor will increase, causing an increase in heat rejected to the condenser, but the refrigerant mass flow rate will decrease because of the compressor characteristics. These will have an opposite effect on the condenser load.

Even a simplified analysis demonstrates the necessity for better understanding of system performance under different operating conditions. Two methods could be used for more accurate analysis. The traditional method of refrigeration-system analysis is through determination of balance points, whereas in recent years, system analysis is performed by system simulation or mathematical modeling, using mathematical (equation-solving) rather than graphical (intersection of two curves) procedures. Systems with a small number of components such as the vapor compression refrigeration system could be analyzed both ways. Graphical presentation, better suited for understanding trends, is not appropriate for more-complex systems, more detailed component description, and frequent change of parameters. There are a variety of different mathematical models tailored to fit specific systems, refrigerants, available resources, demands, and complexity. Although limited in its applications, graphical representation is valuable as the starting tool and for clear understanding of the system performance.

Refrigeration capacity q_e and power P curves for the reciprocating compressor are shown in Fig. 11-91. They are functions of temperatures of evaporation and condensation:

$$q_e = q_e(t_{evap}, t_{cd}) \qquad (11\text{-}105a)$$

and

$$P = P(t_{evap}, t_{cd}) \qquad (11\text{-}105b)$$

where q_e (kW) = refrigerating capacity
P (kW) = power required by the compressor
t_{evap} (°C) = evaporating temperature
t_{cd} (°C) = condensing temperature

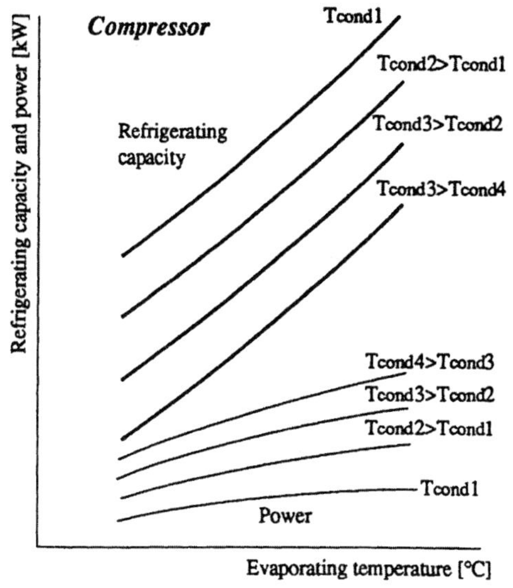

FIG. 11-91 Refrigerating capacity and power requirement for the reciprocating compressor.

A more detailed description of compressor performance is shown in the subsection on refrigeration compressors.

Condenser performance, shown in Fig. 11-92, could be simplified as

$$q_{cd} = F(t_{cd} - t_{amb}) \qquad (11\text{-}105c)$$

where q_{cd} (kW) = capacity of condenser
F (kW/°C) = capacity of condenser per unit inlet temperature difference ($F = U \times A$)
t_{amb} (°C) = ambient temperature (or temperature of condenser cooling medium)

In this analysis F will be constant, but it could be described more accurately as a function of parameters influencing heat transfer in the condenser (temperature, pressure, flow rate, fluid thermodynamic, and thermophysical characteristics, etc.).

Condenser performance should be expressed as an *evaporating effect* to enable matching with compressor and evaporator performance. The condenser evaporating effect is the refrigeration capacity of an evaporator served by a particular condenser. It is the function of the cycle, evaporating

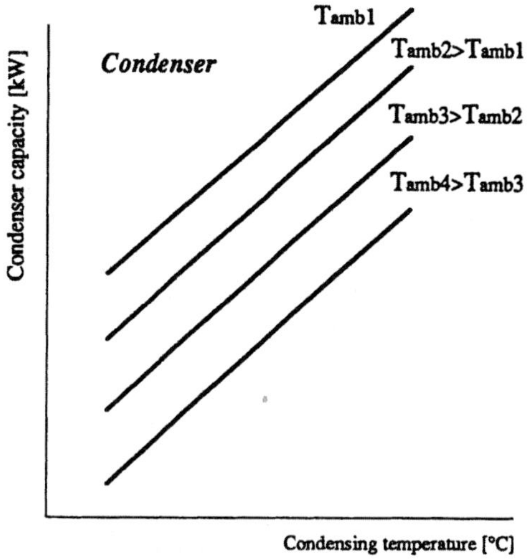

FIG. 11-92 Condenser performance.

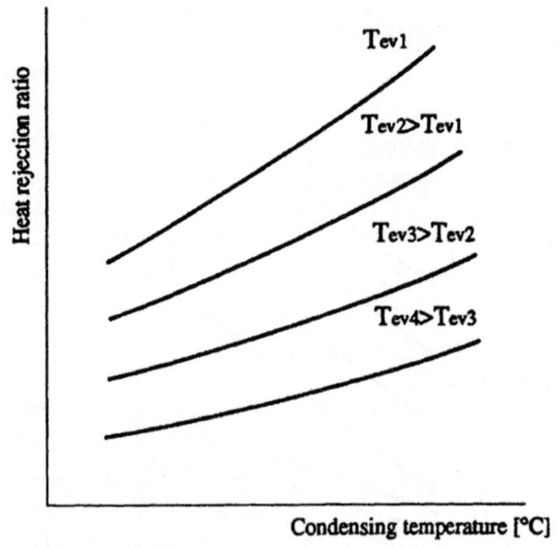

FIG. 11-93 Heat rejection ratio.

temperature, and the compressor. The evaporating effect could be calculated from the heat-rejection ratio q_{Cd}/q_e:

$$q_e = \frac{q_{Cd}}{\text{heat-rejection ratio}} \tag{11-105d}$$

The heat rejection rate is presented in Fig. 11-93 (or Fig. 11-88).

Finally, the evaporating effect of the condenser is shown in Fig. 11-94.

The performance of the condensing unit (compressor and condenser) subsystem could be developed as shown in Fig. 11-95 by superimposing two graphs, one for compressor performance and the other for condenser evaporating effect.

Evaporator performance could be simplified as

$$q_e = F_{evap}(t_{amb} - t_{evap}) \tag{11-106}$$

where q_e (kW) = evaporator capacity

F_e (kW/°C) = evaporator capacity per unit inlet temperature difference

t_{amb} (°C) = ambient temperature (or temperature of cooled body or fluid).

The diagram of the evaporator performance is shown in Fig. 11-96. The character of the curvature of the lines (variable heat-transfer rate) indicates that the evaporator is cooling air. Influences of the flow rate of cooled fluid are also shown in this diagram; i.e., higher flow rate will increase heat transfer. The same effect could be shown in the condenser performance curve. It is omitted only for the reasons of simplicity.

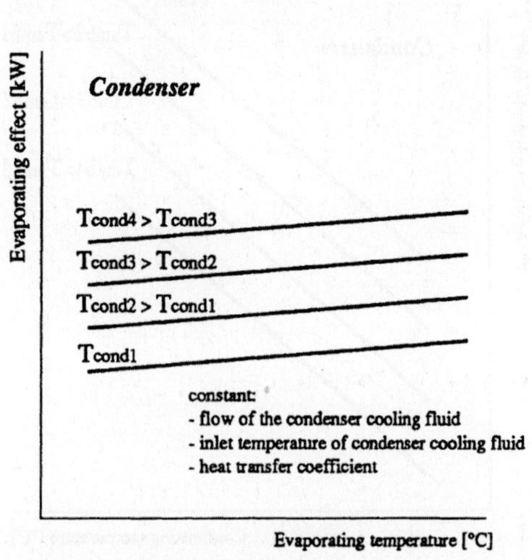

FIG. 11-94 Condenser evaporating effect.

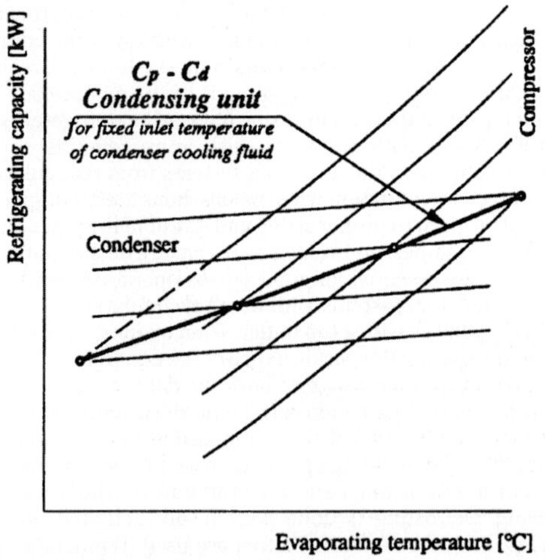

FIG. 11-95 Balance points of compressor and condenser determine performance of condensing unit for fixed temperature of condenser cooling fluid (flow rate and heat-transfer coefficient are constant).

The performance of the complete system could be predicted by superimposing the diagrams for the condensing unit and the evaporator, as shown in Fig. 11-97. Point 1 reveals a balance for constant flows and inlet temperatures of chilled fluid and fluid for condenser cooling. When this point is transferred in the diagram for the condensing unit in Fig. 11-94 or 11-95, the condensing temperature could be determined. When the temperature of entering fluid in the evaporator t_{amb1} is lowered to t_{amb2}, the new operating conditions will be determined by the state at point 2. Evaporation temperature drops from t_{evap1} to t_{evap2}. If the evaporation temperature is unchanged, the same reduction of inlet temperature could be achieved by reducing the capacity of the condensing unit from Cp to Cp^*. The new operating point 3 shows reduction in capacity for Δ due to the reduction in the compressor or the condenser capacity.

Mathematical modeling is essentially the same process, but the description of the component performance is generally much more complex and detailed. This approach enables a user to vary more parameters more easily, look into various possibilities for intervention, and predict the response of the system from different influences. Equation solving does not necessarily have to be done by successive substitution or iteration as this procedure could suggest.

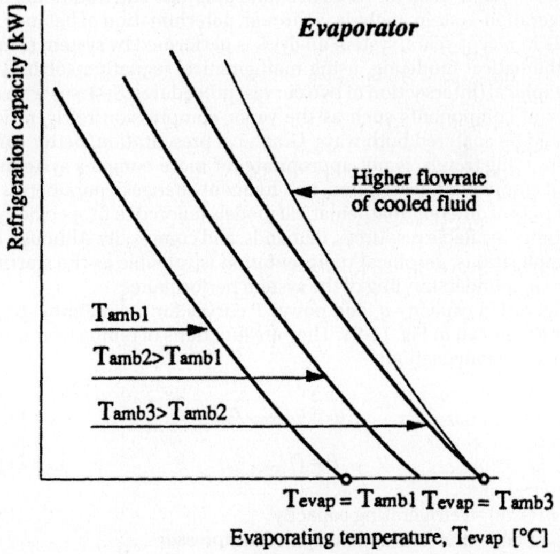

FIG. 11-96 Refrigerating capacity of evaporator.

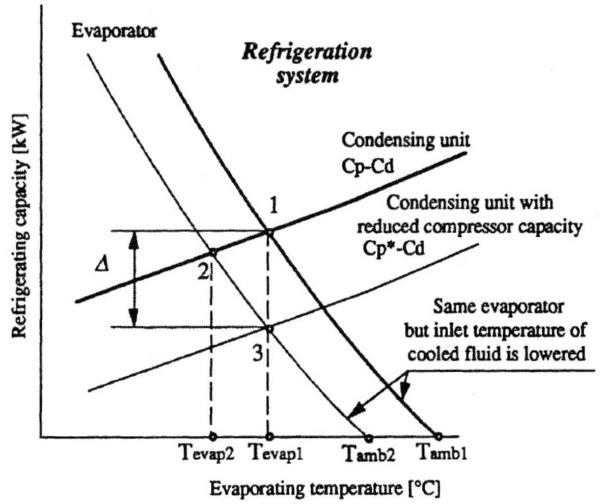

FIG. 11-97 Performance of complete refrigeration system (1), when there is reduction in heat load (2), and when for the same ambient (or inlet in evaporator) evaporation temperature is maintained constant by reducing capacity of compressor/condenser part (3).

System, Equipment, and Refrigerant Selection There is no universal rule that can be used to decide which system, equipment type, or refrigerant is the most appropriate for a given application. A number of variables influence the final-design decision:

- Refrigeration load
- Type of installation
- Temperature level of medium to be cooled
- Condensing media characteristics: type (water, air, etc.), temperature level, available quantities
- Energy source for driving the refrigeration unit (electricity, natural gas, steam, waste heat)
- Location and space available (urban areas, sensitive equipment around, limited space, etc.)
- Funds available (i.e., initial versus run-cost ratio)
- Safety requirements (explosive environment, aggressive fluids, etc.)
- Other demands (compatibility with existing systems, type of load, compactness, level of automatization, operating life, possibility to use process fluid as refrigerant)

Generally, vapor *compression systems* are considered first. They can be used for almost every task. Whenever it is possible, prefabricated elements or complete units are recommended. Reciprocating compressors are widely used for lower rates, more uneven heat loads (when frequent and wider range of capacity reduction is required). They ask for more space and have higher maintenance costs than centrifugal compressors, but are often the most economical in first costs. Centrifugal compressors are considered for huge capacities, when the evaporating temperature is not too low. Screw compressors are considered first when space in the machine room is limited, when the system has operating long hours, and when periods between service should be longer.

Direct expansions are more appropriate for smaller systems which should be compact and where there is just one or few evaporators. Overfeed (recirculation) systems should be considered for all applications where first cost for additional equipment (surge drums, low-pressure receivers, refrigerant pumps, and accessories) is lower than the savings for the evaporator surface.

Choice of refrigerant is complex and not straightforward. For industrial applications, advantages of ammonia (thermodynamic and economic) overcome drawbacks which are mostly related to low-toxicity refrigerants and panics created by accidental leaks when used in urban areas. Halocarbons have many advantages (not toxic, not explosive, odorless, etc.), but environmental issues and slightly inferior thermodynamic and thermophysical properties compared to ammonia or hydrocarbons as well as rising prices are giving the opportunity to use other options. When this text was written, the ozone depletion issue was not resolved, R22 was still used but facing phase-out, and R134a was considered to be the best alternative for CFCs and HCFCs, having similar characteristics to the already banned R12. Very often, fluid to be cooled is used as a refrigerant in the chemical industry. Use of secondary refrigerants in combination with the ammonia central-refrigeration unit is becoming a viable alternative in many applications.

Absorption systems will be considered when there is low-cost, low-pressure steam or waste heat available and the evaporation temperature and refrigeration load are relatively high. Typical application range is for water chilling at 7°C to 10°C, and capacities from 300 kW to 5 MW in a single unit. The main drawback is the difficulty in maintaining a tight system with the highly corrosive lithium bromide and an operating pressure in the evaporator and the absorber below atmospheric.

Ejector (steam-jet) refrigeration systems are used for similar applications, when the chilled-water outlet temperature is relatively high, when relatively cool condensing water and cheap steam at 7 bar are available, and for similar high duties (0.3 to 5 MW). Even though these systems usually have low first and maintenance costs, there are not many steam-jet systems running.

OTHER REFRIGERATION SYSTEMS APPLIED IN THE INDUSTRY

Absorption Refrigeration Systems Two main absorption systems are used in industrial application: lithium bromide–water and ammonia–water. Lithium bromide–water systems are limited to evaporation temperatures above freezing because water is used as the refrigerant, while the refrigerant in an ammonia–water system is ammonia and consequently can be applied for the lower-temperature requirements.

The single-effect indirect-fired lithium bromide cycle is shown in Fig. 11-98. The machine consists of five major components:

The *evaporator* is the heat exchanger where refrigerant (water) evaporates (being sprayed over the tubes) due to low pressure in the vessel. Evaporation chills water flow inside the tubes that bring heat from the external system to be cooled.

The *absorber* is a component where strong absorber solution is used to absorb the water vapor flashed in the evaporator. A solution pump sprays the lithium bromide over the absorber tube section. Cool water is passing through the tubes taking the refrigeration load, heat of dilution, heat to cool condensed water, and sensible heat for solution cooling.

The *heat exchanger* is used to improve the efficiency of the cycle, reducing consumption of steam and condenser water.

The *generator* is a component where heat brought to a system in a tube section is used to restore the solution concentration by boiling off the water vapor absorbed in the absorber.

The *condenser* is an element where water vapor, boiled in the generator, is condensed, preparing pure water (refrigerant) for discharge to an evaporator.

Heat supplied to the generator is boiling weak (dilute) absorbent solution on the outside of the tubes. Evaporated water is condensed on the outside of the condenser tubes. Water utilized to cool the condenser is usually cooled in the cooling tower. Both the condenser and generator are located in the same vessel, being at the absolute pressure of about 6 kPa. The water condensate passes through a liquid trap and enters the evaporator. Refrigerant (water) boils on the evaporator tubes and cools the water flow that brings the refrigeration load. Refrigerant that is not evaporated flows to the recirculation pump to be sprayed over the evaporator tubes. Solution with high water concentration that enters the generator increases in concentration as water evaporates. The resulting strong, absorbent solution (solution with low water concentration) leaves the generator on its way to the heat exchanger. There the stream of high water concentration that flows to the generator cools the stream of solution with low water concentration that flows to the second vessel. The solution with low water concentration is distributed over the absorber tubes. The absorber and evaporator are located in the same vessel, so the refrigerant evaporated on the evaporator tubes is readily absorbed into the absorbent solution. The pressure in the second vessel during the operation is 7 kPa (absolute). Heats of absorption and dilution are removed by cooling water (usually from the cooling tower). The resulting solution with high water concentration is pumped through the heat exchanger to the generator, completing the cycle. The heat exchanger increases the efficiency of the system by preheating, that is, reducing the amount of heat that must be added to the high water solution before it begins to evaporate in the generator.

The absorption machine operation is analyzed with the use of a lithium bromide–water equilibrium diagram, as shown in Fig. 11-99. Vapor pressure is plotted versus the mass concentration of lithium bromide in the solution. The corresponding saturation temperature for a given vapor pressure is shown on the left-hand side of the diagram. The line in the lower right corner of the diagram is the crystallization line. It indicates the point at which the solution will begin to change from liquid to solid, and this is the limit of the cycle. If the solution becomes overconcentrated, the absorption cycle will be interrupted owing to solidification, and capacity will not be restored until the unit is desolidified. This normally requires the addition of heat to the outside of the solution heat exchanger and the solution pump.

The diagram in Fig. 11-100 presents enthalpy data for LiBr–water solutions. It is needed for the thermal calculation of the cycle. Enthalpies for

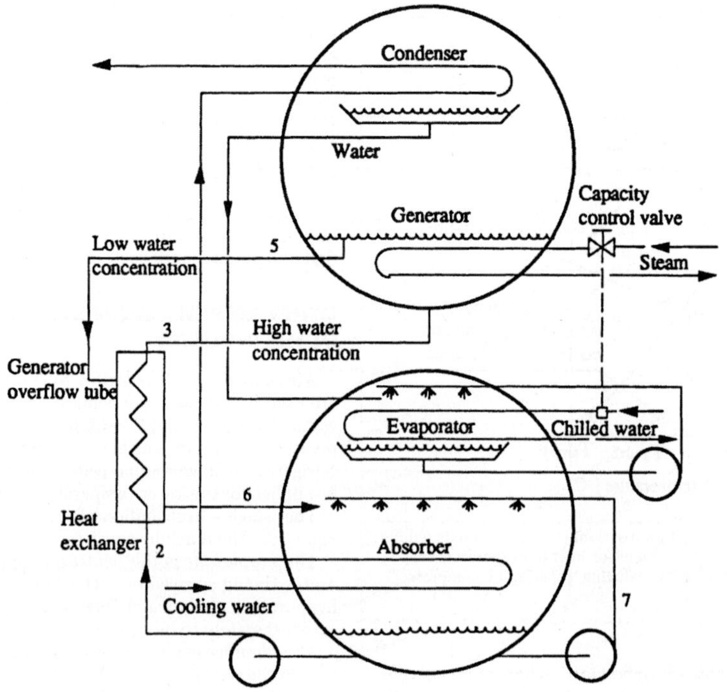

FIG. 11-98 Two-shell lithium bromide–water cycle chiller.

water and water vapor can be determined from the table of the properties of water. The data in Fig. 11-100 are applicable to saturated or subcooled solutions and are based on a zero enthalpy of liquid water at 0°C and a zero enthalpy of solid LiBr at 25°C. Since the zero enthalpy for the water in the solution is the same as that in conventional tables of properties of water, the water property tables can be used in conjunction with the diagram in Fig. 11-99.

The coefficient of performance of the absorption cycle is defined on the same principle as for the mechanical refrigeration

$$COP_{abs} = \frac{\text{useful effect}}{\text{heat input}} = \frac{\text{refrigeration rate}}{\text{heat input at generator}}$$

but note that here the denominator for COP_{abs} is heat while for the mechanical refrigeration cycle it is work. Since these two forms of energy are not equal, COP_{abs} is not as low (0.6 to 0.8) as it appears compared to COP for mechanical system (2.5 to 3.5).

The double-effect absorption unit is shown in Fig. 11-101. All major components and the operation of the double-effect absorption machine are similar to those for the single-effect machine. The primary generator, located in vessel 1, is using an external heat source to evaporate water from dilute-absorbent (high water concentration) solution. Water vapor readily flows to generator II where it is condensed on the tubes. The absorbent (LiBr) intermediate solution from generator I will pass through the heat exchanger on the way to generator II where it is heated by the condensing water vapor. The throttling valve reduces pressure from vessel 1 (about 103 kPa absolute)

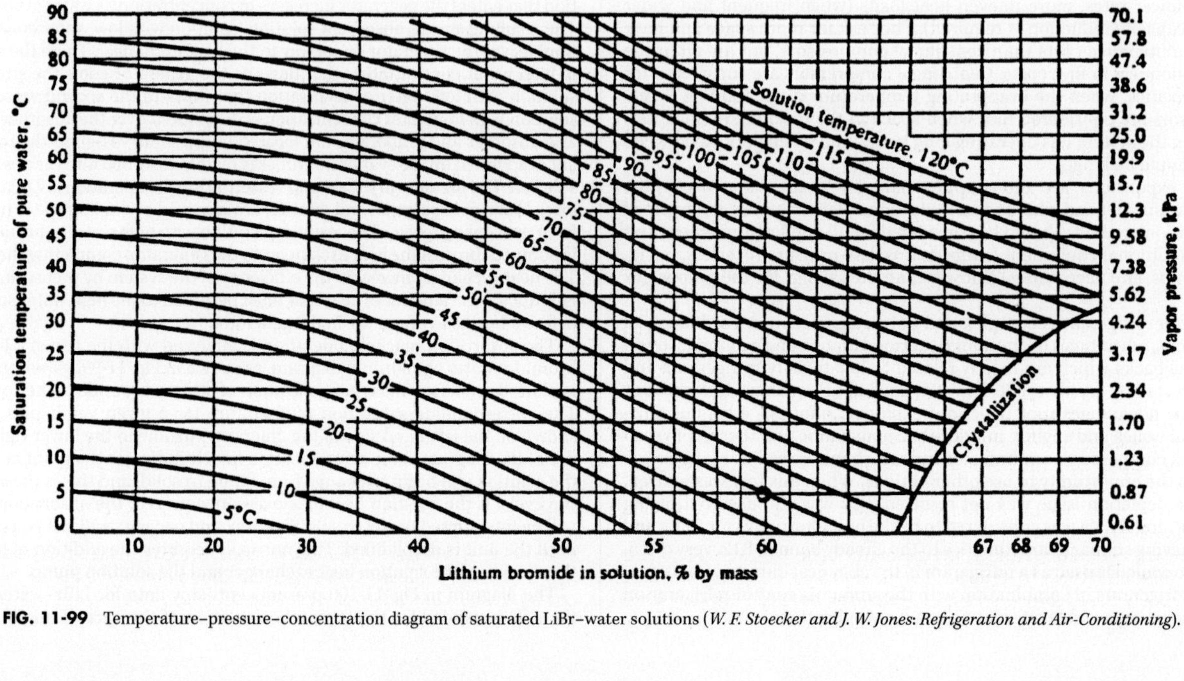

FIG. 11-99 Temperature–pressure–concentration diagram of saturated LiBr–water solutions (*W. F. Stoecker and J. W. Jones: Refrigeration and Air-Conditioning*).

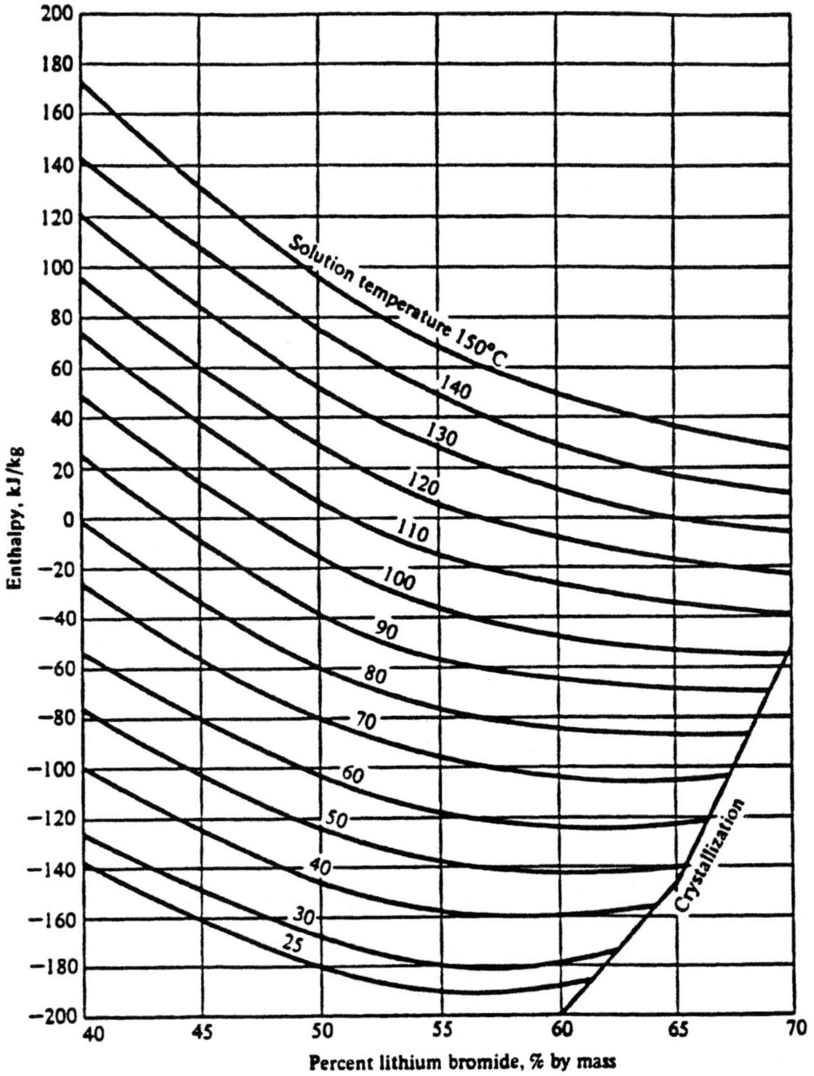

FIG. 11-100 Enthalpy of LiBr–water solutions (*W. F. Stoecker and J. W. Jones: Refrigeration and Air-Conditioning*).

to that of vessel 2. Following the reduction of pressure, some water in the solution flashes to vapor, which is liquefied at the condenser. In the high-temperature heat exchanger, the intermediate solution heats the weak (high water concentration) solution stream coming from the low-temperature heat exchanger. In the low-temperature heat exchanger, strong solution is being cooled before entering the absorber. The absorber is at the same pressure as the evaporator. The double-effect absorption units achieve higher COPs than the single-stage ones.

The ammonia-water absorption system was extensively used until the 1950s when the LiBr-water combination became popular. Figure 11-102 shows a simplified ammonia-water absorption cycle. The refrigerant is ammonia, and the absorbent is dilute aqueous solution of ammonia. Ammonia-water systems differ from water–lithium bromide equipment to accommodate major differences: Water (here absorbent) is also volatile, so the regeneration of weak water solution to strong water solution is a fractional distillation. Different refrigerant (ammonia) causes different, much higher pressures: about 1100 to 2100 kPa absolute in condenser.

Ammonia vapor from the evaporator and the weak water solution from the generator are producing strong water solution in the absorber. Strong water solution is then separated in the rectifier, producing (1) ammonia with some water vapor content and (2) very strong water solution at the bottom, in the generator. Heat in the generator vaporizes ammonia, and the weak solution returns to absorber. On its way to the absorber, the weak solution stream passes through the heat exchanger, heating the strong solution from the absorber on the way to the rectifier. The other stream, mostly ammonia vapor but with some water vapor content, flows to the condenser. To remove water as much as possible, the vapor from the rectifier passes through the analyzer where it is additionally cooled. The remaining water escaped from the analyzer passes as liquid through the condenser and the evaporator to the absorber.

Ammonia-water units can be arranged for single-stage or cascaded two-stage operation. The advantage of two-stage operation is that it creates the possibility of utilizing only part of the heat on the higher-temperature level and the rest on the lower-temperature level, but the price is increased for first cost and heat required.

Ammonia-water and lithium bromide–water systems operate under comparable COP. The ammonia-water system is capable of achieving evaporating temperatures below 0°C because the refrigerant is ammonia. Water as the refrigerant limits evaporating temperatures to no lower than freezing, better to 3°C. Advantage of the lithium bromide–water system is that it requires less equipment and operates at lower pressures. But this is also a drawback, because pressures are below atmospheric, causing air infiltration in the system which must be purged periodically. Due to corrosion problems, special inhibitors must be used in the lithium bromide–water system. The infiltration of air in the ammonia-water system is also possible, but when evaporating temperature is below −33°C. This can result in formation of corrosive ammonium carbonate.

Further Readings: ASHRAE Handbook, *Refrigeration Systems and Applications,* 1994; Bogart, M., *Ammonia Absorption Refrigeration in Industrial Processes,* Gulf Publishing Co. Houston, Tex., 1981; Stoecker, W. F., and Jones, J. W., *Refrigeration and Air-Conditioning,* McGraw-Hill, New York, 1982.

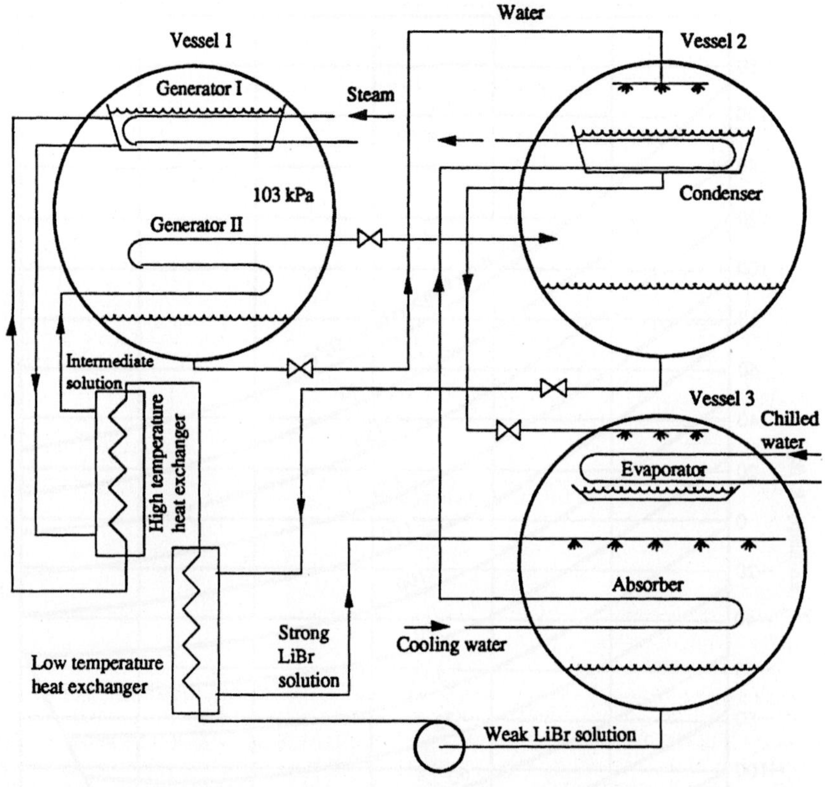

FIG. 11-101 Double-effect absorption unit.

Steam-Jet (Ejector) Systems These systems substitute an ejector for a mechanical compressor in a vapor compression system. Since the refrigerant is water, maintaining temperatures lower than that of the environment requires that the pressure of water in the evaporator be below atmospheric. A typical arrangement for the steam-jet refrigeration cycle is shown in Fig. 11-103.

Main Components These are the main components of steam-jet refrigeration systems:

1. *Primary steam ejector.* This kinetic device utilizes the momentum of a high-velocity jet to entrain and accelerate a slower-moving medium into which it is directed. High-pressure steam is delivered to the ejector nozzle. The steam expands while flowing through the nozzle where the velocity increases rapidly. The velocity of steam leaving the nozzle is around 1200 m/s. Because of this high velocity, flash vapor from the tank is continually aspired into the moving steam. The mixture of steam and flash vapor then enters the diffuser section where the velocity is gradually reduced

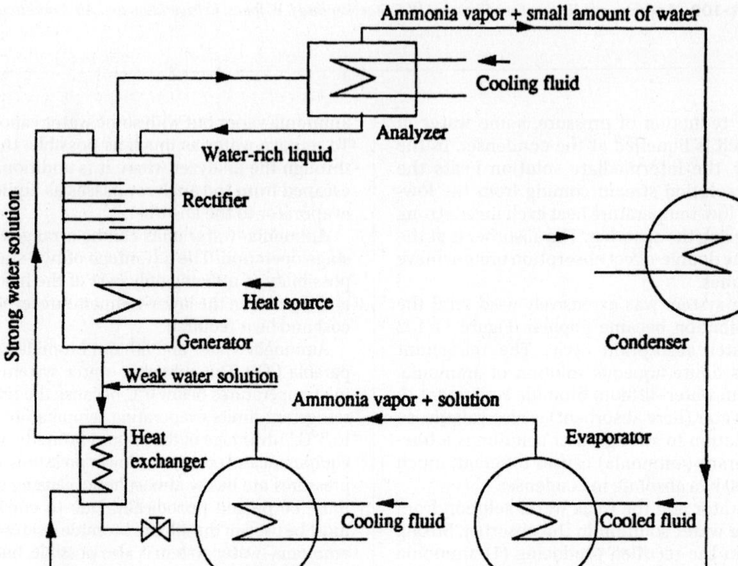

FIG. 11-102 Simplified ammonia–water absorption cycle.

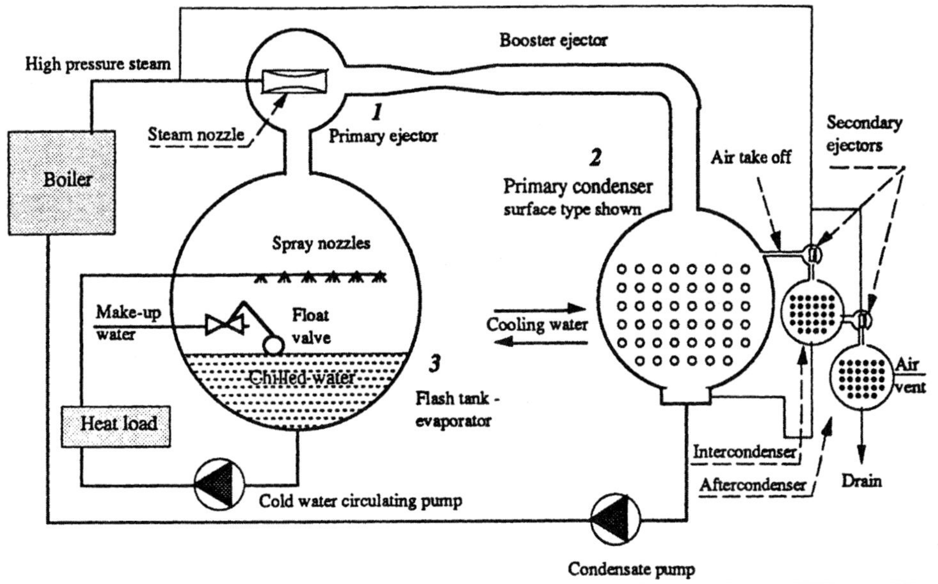

FIG. 11-103 Ejector (steam-jet) refrigeration cycle (with surface-type condenser).

because of increasing cross-sectional area. The energy of the high-velocity steam compresses the vapor during its passage through the diffuser, raising its temperature above the temperature of the condenser cooling water.

2. *Condenser.* This is the component of the system where the vapor condenses and the heat is rejected. The rate of heat rejected is

$$Q_{\text{cond}} = (W_s + W_w)\, hfg \qquad (11\text{-}107)$$

where Q_{cond} = heat rejection (kW)
W_s = primary booster steam rate (kg/s)
W_w = flash vapor rate (kg/s)
hfg = latent heat (kJ/kg)

The condenser design, surface area, and condenser cooling water quantity should be based on the highest cooling water temperature likely to be encountered. If the inlet cooling water temperature becomes hotter than the design value, the primary booster (ejector) may cease functioning because of the increase in condenser pressure.

Two types of condensers could be used: the surface condenser (shown in Fig. 11-103) and the barometric or jet condenser (Fig. 11-104). The surface condenser is of shell-and-tube design with water flowing through the tubes and steam condensed on the outside surface. In the jet condenser, condensing water and the steam being condensed are mixed directly, and no tubes are provided. The jet condenser can be barometric or a low-level type. The barometric condenser requires a height of ~10 m above the level of the water in the hot well. A tailpipe of this length is needed so that condenser water and condensate can drain by gravity. In the low-level jet type, the tailpipe is eliminated, and it becomes necessary to remove the condenser water and condensate by pumping from the condenser to the hot well. The main advantages of the jet condenser are low maintenance with the absence of tubes and the fact that condenser water of varying degrees of cleanliness may be used.

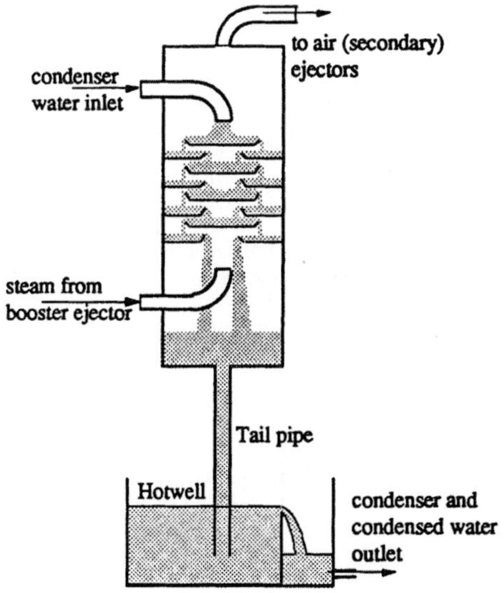

FIG. 11-104 Barometric condenser for steam-jet system.

3. *Flash tank.* This is the evaporator of the ejector system and is usually a large-volume vessel where large water surface area is needed for efficient evaporative cooling action. Warm water returning from the process is sprayed into the flash chamber through nozzles (sometimes cascades are used for maximizing the contact surface, since they are less susceptible to carryover problems), and the cooled effluent is pumped from the bottom of the flash tank.

When the steam supply to one ejector of a group is closed, some means must be provided for preventing the pressure in the condenser and flash tank from equalizing through that ejector. A compartmental flash tank is frequently used for such purposes. With this arrangement, partitions are provided so that each booster is operating on its own flash tank. When the steam is shut off to any booster, the valve to the inlet spray water to that compartment also is closed.

A float valve is provided to control the supply of makeup water to replace the water vapor that has flashed off. The flash tank should be insulated.

Applications The steam-jet refrigeration is suited for the following:

1. It is suited to processes where direct vaporization is used for concentration or drying of heat-sensitive foods and chemicals and where, besides elimination of the heat exchanger, preservation of the product quality is an important advantage.

2. It enables the use of hard water or even seawater for heat rejection, e.g., for absorption of gases (CO_2, SO_2, ClO_2, etc.) in chilled water (desorption is provided simultaneously with chilling) when a direct-contact barometric condenser is used.

Despite being simple, rugged, reliable, low cost, and vibration-free and requiring low maintenance, steam-jet systems are not widely accepted in water chilling for air conditioning due to characteristics of the cycle.

Factors Affecting Capacity Ejector (steam-jet) units become attractive when cooling relatively high-temperature chilled water with a source of about 7 bar gauge waste steam and relatively cool condensing water. The factors involved with steam-jet capacity include the following:

1. *Steam pressure.* The main boosters can operate on steam pressures from as low as 0.15 bar up to 7 bar gauge. The quantity of steam required increases rapidly as the steam pressure drops (Fig. 11-105). The best steam rates are obtained with about 7 bar. Above this pressure the change

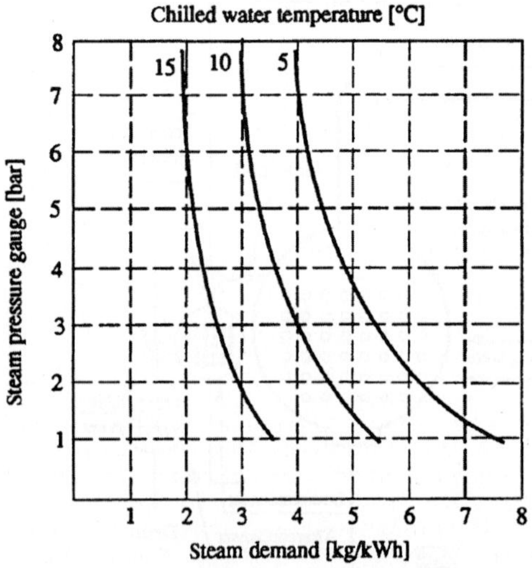

FIG. 11-105 Effect of steam pressure on steam demand at 38°C condenser temperature. (*ASHRAE 1983 Equipment Handbook*).

in quantity of steam required is practically negligible. Ejectors must be designed for the highest available steam pressure, to take advantage of the lower steam consumption for various steam-inlet pressures.

The secondary ejector systems used for removing air require steam pressures of 2.5 bar or greater. When the available steam pressure is lower than this, an electrically driven vacuum pump is used for either the final secondary ejector or the entire secondary group. The secondary ejectors normally require 0.2 to 0.3 kg/h of steam per kilowatt of refrigeration capacity.

2. *Condenser water temperature.* In comparison with other vapor compression systems, steam-jet machines require relatively large water quantities for condensation. The higher the inlet water temperature, the higher the water requirements (Fig. 11-106). The condensing water temperature has an important effect on steam rate per refrigeration effect, rapidly decreasing with colder condenser cooling water. Figure 11-107 presents data on steam rate versus condenser water inlet for given chilled-water outlet temperatures and steam pressure.

3. *Chilled-water temperature.* As the chilled-water outlet temperature decreases, the ratio of the steam/refrigeration effect decreases, thus increasing condensing temperatures and/or increasing the condensing-water requirements.

Unlike other refrigeration systems, the chilled-water flow rate is of no particular importance in steam-jet system design, because there is, due to direct heat exchange, no influence of evaporator tube velocities and related temperature differences on heat-transfer rates. Widely varying return chilled-water temperatures have little effect on steam-jet equipment.

Multistage Systems The majority of steam-jet systems being currently installed are multistage. Up to five-stage systems are in commercial operation.

Capacity Control The simplest way to regulate the capacity of most steam vacuum refrigeration systems is to furnish several primary boosters

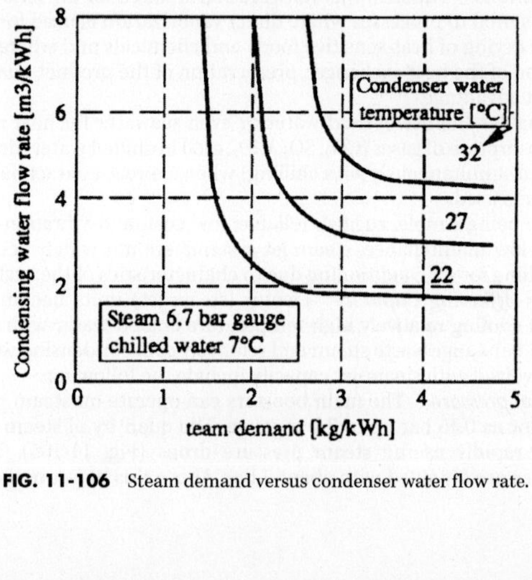

FIG. 11-106 Steam demand versus condenser water flow rate.

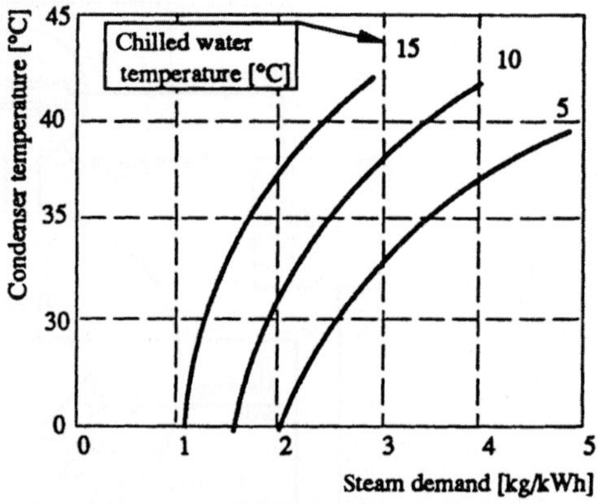

FIG. 11-107 Steam demand versus chilled-water temperature for typical steam-jet system. (*ASHRAE 1983 Equipment Handbook*).

in parallel and operate only those required to handle the heat load. It is not uncommon to have as many as four main boosters on larger units for capacity variation. A simple automatic on-off type of control may be used for this purpose. By sensing the chilled-water temperature leaving the flash tank, a controller can turn steam on and off to each ejector as required.

Additionally, two other control systems that will regulate steam flow or condenser-water flow to the machine are available. As the condenser-water temperature decreases during various periods of the year, the absolute condenser pressure will decrease. This will permit the ejectors to operate on less steam because of the reduced discharge pressure. Either the steam flow or the condenser water quantities can be reduced to lower operating costs at other than design periods. The arrangement selected depends on cost considerations between the two flow quantities. Some systems have been arranged for a combination of the two, automatically reducing steam flow down to a point, followed by a reduction in condenser-water flow. For maximum operating efficiency, automatic control systems are usually justifiable in keeping operating cost to a minimum without excessive operator attention. In general, steam savings of about 10 percent of rated booster flow are realized for each 2.5°C reduction in condensing-water temperature below the design point.

In some cases, with relatively cold inlet condenser water it has been possible to adjust automatically the steam inlet pressure in response to chilled-water outlet temperatures. In general, however, this type of control is not possible because of the differences in temperature between the flash tank and the condenser. Under usual conditions of warm condenser-water temperatures, the main ejectors must compress water vapor over a relatively high ratio, requiring an ejector with entirely different operating characteristics. In most cases, when the ejector steam pressure is throttled, the capacity of the jet remains almost constant until the steam pressure is reduced to a point at which there is a sharp capacity decrease. At this point, the ejectors are unstable, and the capacity is severely curtailed. With a sufficient increase in steam pressure, the ejectors will once again become stable and operate at their design capacity. In effect, steam jets have a vapor-handling capacity fixed by the pressure at the suction inlet. In order for the ejector to operate along its characteristic pumping curve, it requires a certain minimum steam flow rate which is fixed for any particular pressure in the condenser. (For further information on the design of ejectors, see Sec. 6.)

Further Reading and Reference: *ASHRAE 1983 Equipment Handbook;* Spencer, E., "New Development in Steam Vacuum Refrigeration," *ASHRAE Trans.* 67: 339 (1961).

Refrigerants A refrigerant is a body or substance that acts as a cooling agent by absorbing heat from another body or substance which has to be cooled. Primary refrigerants are those that are used in the refrigeration systems, where they alternately vaporize and condense as they absorb and give off heat, respectively. Secondary refrigerants are heat-transfer fluids or heat carriers. Refrigerant pairs in absorption systems are ammonia–water and lithium bromide–water, while steam (water) is used as a refrigerant in ejector systems. Refrigerants used in the mechanical refrigeration systems are far more numerous.

A list of the most significant refrigerants is presented in the *ASHRAE Handbook Fundamentals.* More data are shown in Sec. 2, "Physical and Chemical Data." Because of the rapid changes in the refrigerant issue, readers are advised to consult the most recent data and publications at the time of application.

The first refrigerants were natural: air, ammonia, CO_2, SO_2, and so on. Fast expansion of refrigeration in the second and third quarters of the 20th

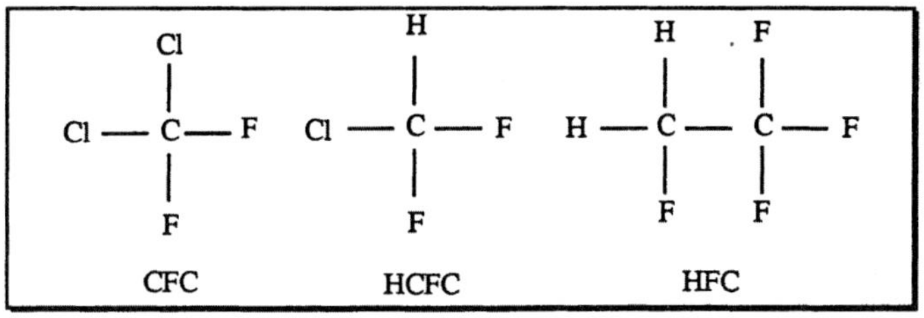

FIG. 11-108 Halocarbon refrigerants.

century is marked by the new refrigerants, chlorofluorocarbons (CFCs) and hydrochlorofluorocarbons (HCFCs). They are halocarbons that contain one or more of the three halogens chlorine, fluorine, and bromine (Fig. 11-108). These refrigerants introduced many advantageous qualities compared to most of the existing refrigerants: they are odorless, nonflammable, nonexplosive, compatible with the most engineering materials, nontoxic, and have a reasonably high COP.

In the last decade, the refrigerant issue is extensively discussed because of the accepted hypothesis that the chlorine and bromine atoms from halocarbons released to the environment were using up ozone in the stratosphere, depleting it especially above the polar regions. The Montreal Protocol and later agreements ban use of certain CFCs and halon compounds. Presently, all CFCs are out of production and the production, import, and use of HCFCs have been banned in the United States since 2015, except for HCFC-22 and HCFC-142b. A complete ban on all HCFCs in the United States is scheduled for 2030.

Chemical companies are trying to develop safe and efficient refrigerant for the refrigeration industry and application, but uncertainty in CFC and HCFC substitutes is still high. When this text was written, HFCs were a promising solution. That is true especially for R134a which seems to be the best alternative for R12. Substitutes for R22 and R502 are still under debate. Numerous ecologists and chemists are for an extended ban on HFCs as well, mostly due to significant use of CFCs in the production of HFCs. Extensive research is ongoing to find new refrigerants. Many projects are aimed to design and study refrigerant mixtures, both azeotropic (mixture which behaves physically as a single, pure compound) and zeotropic having desirable qualities for the processes with temperature glides in the evaporator and the condenser.

Ammonia (R717) is the single natural refrigerant being used extensively (beside halocarbons). It is significant in industrial applications for its excellent thermodynamic and thermophysical characteristics. Many engineers are considering ammonia as a CFC substitute for various applications. Significant work is being done on reducing the refrigerant inventory and consequently problems related to leaks of this fluid with strong odor. There is growing interest in hydrocarbons in some countries, particularly in Europe. Indirect cooling (secondary refrigeration) is under reconsideration for many applications.

Because of the vibrant refrigerant issue it will be a challenge for every engineer to find the best solution for the particular application, but basic principles are the same. Good refrigerant should have these characteristics:

- Safe: nontoxic, nonflammable, and nonexplosive
- Environmentally friendly
- Compatible with materials normally used in refrigeration: oils, metals, elastomers, etc.
- Desirable thermodynamic and thermophysical characteristics:
  - High latent heat
  - Low specific volume of vapor
  - Low compression ratio
  - Low viscosity
  - Reasonably low pressures for operating temperatures
  - Low specific heat of liquid
  - High specific heat of vapor
  - High conductivity and other heat-transfer related characteristics
  - Reasonably low compressor discharge temperatures
  - Easily detected if leaking
  - High dielectric constant
  - Good stability

Secondary Refrigerants (Antifreezes or Brines) These are mostly liquids used for transporting heat energy from the remote heat source (process heat exchanger) to the evaporator of the refrigeration system. Antifreezes or brines do not change state in the process, but there are examples where some secondary refrigerants are either changing state themselves, or just particles which are carried in them.

Indirect refrigeration systems are more prevalent in the chemical industry than in the food industry, commercial refrigeration, or comfort air conditioning. This is even more evident in the cases where a large amount of heat is to be removed or where a low temperature level is involved. The advantage of an indirect system is centralization of refrigeration equipment, which is especially important for relocation of refrigeration equipment in a nonhazardous area, for both people and equipment.

Salt Brines The typical curve of freezing point is shown in Fig. 11-109. Brine of concentration x (water concentration is $1 - x$) will not solidify at 0°C (freezing temperature for water, point A). When the temperature drops to B, the first crystal of ice is formed. As the temperature decreases to point C, ice crystals continue to form and their mixture with the brine solution forms the slush. At point C there will be part ice in the mixture $l_2/(l_1 + l_2)$ and liquid (brine) $l_1/(l_1 + l_2)$. At point D there is mixture of m_1 parts eutectic brine solution D_1 [concentration $m_1/(m_1 + m_2)$], and m_2 parts of ice [concentration $m_2/(m_1 + m_2)$]. Cooling the mixture below D solidifies the entire solution at the eutectic temperature. The eutectic temperature is the lowest temperature that can be reached with no solidification.

It is obvious that further strengthening of brine has no effect, and can cause a different reaction—salt sometimes freezes out in the installations where the concentration is too high.

Sodium chloride, an ordinary salt (NaCl), is the least expensive per volume of any brine available. It can be used in contact with food and in open systems because of its low toxicity. Heat-transfer coefficients are relatively high. However, its drawbacks are that it has a relatively high freezing point and is highly corrosive (requires inhibitors, thus must be checked on a regular schedule).

Calcium chloride ($CaCl_2$) is similar to NaCl. It is the second-lowest-cost brine, with a somewhat lower freezing point (used for temperatures as low as −37°C). It is highly corrosive and not appropriate for direct contact with food. Heat-transfer coefficients are rapidly reduced at temperatures below −20°C. The presence of magnesium salts in either sodium or calcium chloride is undesirable because they tend to form sludge. Air and carbon dioxide are contaminants, and excessive aeration of the brine should be prevented through the use of closed systems. Oxygen, required for corrosion, normally comes from the atmosphere and dissolves in the brine solution. Dilute brines dissolve oxygen more readily and are generally more corrosive than concentrated brines. It is believed that even a closed brine system will not prevent the infiltration of oxygen.

To adjust an alkaline condition to pH 7.0 to 8.5, use caustic soda (to correct up to pH 7.0) or sodium dichromate (to reduce excessive alkalinity below pH 8.5). Such slightly alkaline brines are generally less corrosive than neutral or acid ones, although with high alkalinity the activity may increase.

If the untreated brine has the proper pH value, the acidifying effect of the dichromate may be neutralized by adding commercial flake caustic soda (76 percent pure) in quantity that corresponds to 27 percent of sodium dichromate used. Caustic soda must be thoroughly dissolved in warm water before it is added to the brine.

Recommended inhibitor (sodium dichromate) concentrations are 2 kg/m^3 of $CaCl_2$ and 3.2 kg/m^3 of NaCl brine. Sodium dichromate when dissolved in water or brine makes the solution acid. Steel, iron, copper, or red brass can be used with brine circulating systems. Calcium chloride systems are generally equipped with all-iron-and-steel pumps and valves to prevent electrolysis in the event of acidity. Copper and red brass tubing is used for calcium chloride evaporators. Sodium chloride systems are using all-iron or all-bronze pumps.

Organic Compounds (Inhibited Glycols) *Ethylene glycol* is colorless, practically odorless, and completely miscible with water. Advantages are low volatility and relatively low corrosivity when properly inhibited. Main drawbacks are relatively low heat-transfer coefficients at lower temperatures due to high viscosities (even higher than for propylene glycol). It is somewhat toxic, but less harmful than methanol–water solutions. It is not appropriate for the food industry and should not stand in open containers.

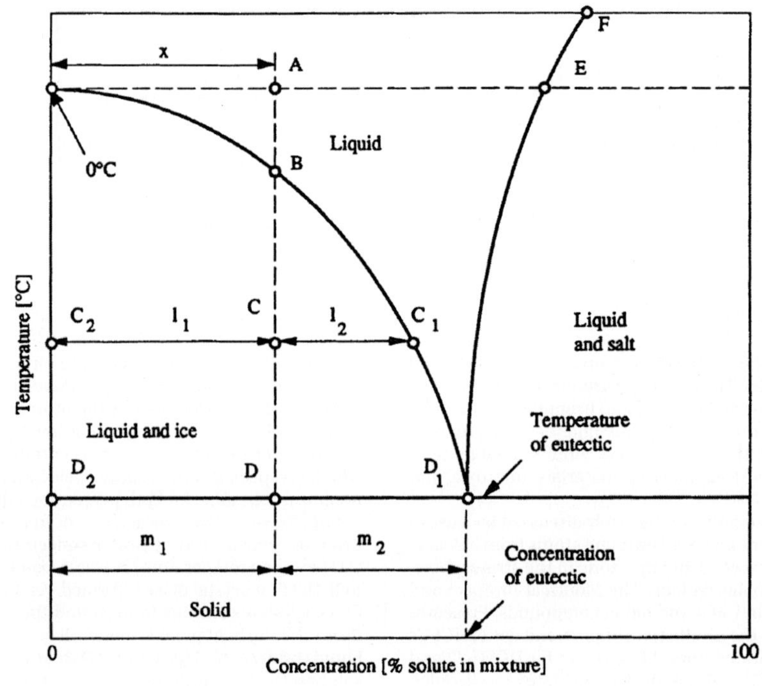

FIG. 11-109 Phase diagram of the brine.

Preferably waters that are classified as soft and are low in chloride and sulfate ions should be used for preparation of ethylene glycol solution.

Pure ethylene glycol freezes at −12.7°C. Exact composition and temperature for the eutectic point are unknown, since solutions in this region turn to viscous, glassy mass that makes it difficult to determine the true freezing point. For the concentrations lower than eutectic, ice forms on freezing, while on the concentrated, solid glycol separates from the solution.

Ethylene glycol normally has pH of 8.8 to 9.2 and should not be used below pH 7.5. Addition of more inhibitor cannot restore the solution to its original condition. Once inhibitor has been depleted, it is recommended that the old glycol be removed from the system and the new charge be installed.

Propylene glycol is very similar to ethylene glycol, but it is not toxic and is used in direct contact with food. It is more expensive and, having higher viscosity, shows lower heat-transfer coefficients.

Methanol-water is an alcohol-base compound. It is less expensive than other organic compounds and, due to lower viscosity, has better heat-transfer and pressure drop characteristics. It is used up to −35°C. Disadvantages are that (1) it is considered more toxic than ethylene glycol and thus more suitable for outdoor applications and (2) it is flammable and could be assumed to be a potential fire hazard.

For ethylene glycol systems, copper tubing is often used (up to 3 in), while pumps, cooler tubes, or coils are made of iron, steel, brass, copper, or aluminum. Galvanized tubes should not be used in ethylene glycol systems because of reaction of the inhibitor with the zinc.

Methanolewater solutions are compatible with most materials but in sufficient concentration will badly corrode aluminum.

Ethanol-water is a solution of denatured grain alcohol. Its main advantage is that it is nontoxic and thus is widely used in the food and chemical industry. By using corrosion inhibiters it could be made noncorrosive for brine service. It is more expensive than methanol–water and has somewhat lower heat-transfer coefficients. As an alcohol derivate it is flammable.

Secondary refrigerants shown below, listed under their generic names, are sold under different trade names. Some other secondary refrigerants appropriate for various refrigeration applications will be listed under their trade names. More data could be obtained from the manufacturer.

Syltherm XLT (Dow Corning Corporation). A silicone polymer (Dimethyl Polysiloxane); recommended temperature range −70°C to 250°C; odorless; low in acute oral toxicity; noncorrosive toward metals and alloys commonly found in heat transfer systems.

Syltherm 800 (Dow Corning Corporation). A silicone polymer (Dimethyl Polysiloxane); recommended temperature range −40°C to 400°C; similar to Syltherm XLT, more appropriate for somewhat higher temperatures; flash point is 160°C.

D-limonene (Florida Chemicals). A compound of optically active terpene ($C_{10}H_{16}$) derived as an extract from orange and lemon oils; limited data show very low viscosity at low temperatures—only 1 cP at −50°C; natural substance having questionable stability.

Therminol D-12 (Monsanto). A synthetic hydrocarbon; clear liquid; recommended range −40°C to 250°C; not appropriate for contact with food; precautions against ignitions and fires should be taken with this product; could be found under trade names Santotherm or Gilotherm.

Therminol LT (Monsanto). Akylbenzene, synthetic aromatic ($C_{10}H_{14}$); recommended range −70°C to −180°C; not appropriate for contact with food; precautions against ignitions and fire should be taken when dealing with this product.

Dowtherm J (Dow Corning Corporation). A mixture of isomers of an alkylated aromatic; recommended temperature range −70°C to 300°C; noncorrosive toward steel, common metals and alloys; combustible material; flash point 58°C; low toxic; prolonged and repeated exposure to vapors should be limited to 10 ppm for daily exposures of 8 h.

Dowtherm Q (Dow Corning Corporation). A mixture of dyphenylehane and alkylated aromatics; recommended temperature range −30°C to 330°C; combustible material; flash point 120°C; considered low toxic, similar to Dowtherm J.

Safety in Refrigeration Systems This is of paramount importance and should be considered at every stage of installation.

The design engineer should have safety as the primary concern by choosing a suitable system and refrigerant: selecting components, choosing materials and thicknesses of vessels, pipes, and relief valves of pressure vessels, proper venting of machine rooms, and arranging the equipment for convenient access for service and maintenance (piping arrangements, valve location, machine room layout, etc.). She or he should conform to the stipulation of the safety codes, which is also important for the purposes of professional liability.

During construction and installation, the installer's good decisions and judgments are crucial for safety, because design documentation never specifies all details. This is especially important when there is reconstruction or repair while the facility has been charged.

During operation the plant is in the hands of the operating personnel. They should be properly trained and familiar with the installation. Very often, accidents are caused by an improper practice, such as making an attempt to repair when proper preparation has not been made. Operators should be trained in first-aid procedures and how to respond to emergencies.

Most frequently needed standards and codes are listed below, and the reader can find comments in W. F. Stoecker, *Industrial Refrigeration,* vol. 2, chap. 12, Business News Publishing Co., Troy, Mich., 1995; ASHRAE Handbook,

Refrigeration System and Applications, 1994, chap. 51. These are some important standards and codes on safety that a refrigeration engineer should consult: ANSI/ASHRAE Standard 15-92, Safety Code for Mechanical Refrigeration, ASHRAE, Atlanta Ga., 1992; ANSI/ASHRAE Standard 34-92, Number Designation of Refrigerants, ASHRAE, Atlanta, Ga., 1992; ANSI/ASME Boiler and Pressure Vessel Code, ASME, New York, 1989; ANSI/ASME Code for Pressure Piping, B31, B31.5-1987, ASME, New York, 1987; ANSI/IIAR 2—1984, Equipment, Design and Installation of Ammonia Mechanical Refrigeration Systems, IIAR, Chicago, 1984; IIAR Minimum Safety Criteria for a Safe Ammonia Refrigeration Systems, Bulletin 109; IIAR, IIAR Start-up, Inspection, and Maintenance of Ammonia Mechanical Refrigeration Systems, Bulletin 110, Chicago, 1988; IIAR Recommended Procedures in Event of Ammonia Spills, Bulletin 106, IIAR, Chicago, 1977; A Guide to Good Practices for the Operation of an Ammonia Refrigeration System, IIAR Bulletin R1, 1983.

EVAPORATORS

GENERAL REFERENCES: Badger and Banchero, *Introduction to Chemical Engineering,* McGraw-Hill, New York, 1955. Standiford, *Chem. Eng.* **70:** 158–176 (Dec. 9, 1963). *Testing Procedure for Evaporators,* American Institute of Chemical Engineers, 1979. *Upgrading Evaporators to Reduce Energy Consumption,* ERDA Technical Information Center, Oak Ridge, Tenn., 1977.

PRIMARY DESIGN PROBLEMS

Heat Transfer This is the most important single factor in evaporator design, since the heating surface represents the largest part of evaporator cost. Other things being equal, the type of evaporator selected is the one having the highest heat-transfer cost coefficient under desired operating conditions in terms of joules per second per kelvin (British thermal units per hour per degree Fahrenheit) per dollar of installed cost. When power is required to induce circulation past the heating surface, the coefficient must be even higher to offset the cost of power for circulation.

Vapor-Liquid Separation This design problem may be important for a number of reasons. The most important is usually prevention of entrainment because of value of product lost, pollution, contamination of the condensed vapor, or fouling or corrosion of the surfaces on which the vapor is condensed. Vapor-liquid separation in the vapor head may also be important when spray forms deposits on the walls, when vortices increase the head requirements of circulating pumps, and when short-circuiting allows vapor or unflashed liquid to be carried back to the circulating pump and heating element. Evaporator performance is rated on the basis of *steam economy*—kilograms of solvent evaporated per kilogram of steam used. Heat is required (1) to raise the feed from its initial temperature to the boiling temperature, (2) to provide the minimum thermodynamic energy to separate liquid solvent from the feed, and (3) to vaporize the solvent. The first can be changed appreciably by reducing the boiling temperature or by heat interchange between the feed and the residual product and/or condensate. The greatest increase in steam economy is achieved by reusing the vaporized solvent. This is done in a *multiple-effect evaporator* by using the vapor from one effect as the heating medium for another effect in which boiling takes place at a lower temperature and pressure. Another method of increasing the utilization of energy is to employ a *thermocompression* evaporator, in which the vapor is compressed so that it will condense at a temperature high enough to permit its use as the heating medium in the same evaporator.

Selection Problems Aside from heat-transfer considerations, the selection of type of evaporator best suited for a particular service is governed by the characteristics of the feed and product. Points that must be considered include crystallization, salting and scaling, product quality, corrosion, and foaming. In the case of a *crystallizing evaporator,* the desirability of producing crystals of a definite uniform size usually limits the choice to evaporators having a positive means of circulation. *Salting,* which is the growth on body and heating-surface walls of a material having a solubility that increases with an increase in temperature, is frequently encountered in crystallizing evaporators. It can be reduced or eliminated by keeping the evaporating liquid in close or frequent contact with a large surface area of crystallized solid. *Scaling* is the deposition and growth on body walls, and especially on heating surfaces, of a material undergoing an irreversible chemical reaction in the evaporator or having a solubility that decreases with an increase in temperature. Scaling can be reduced or eliminated in the same general manner as salting. Both salting and scaling liquids are usually best handled in evaporators that do not depend on boiling to induce circulation. *Fouling* is the formation of deposits other than salt or scale and may be due to corrosion, solid matter entering with the feed, or deposits formed by the condensing vapor.

Product Quality Considerations of product quality may require low holdup time and low-temperature operation to avoid thermal degradation. The low holdup time eliminates some types of evaporators, and other types are also eliminated because of poor heat-transfer characteristics at low temperature. Product quality may dictate special materials of construction to avoid metallic contamination or a catalytic effect on decomposition of the product. *Corrosion* may also influence evaporator selection, since the advantages of evaporators having high heat-transfer coefficients are more apparent when expensive materials of construction are indicated. Corrosion and erosion are frequently more severe in evaporators than in other types of equipment because of the high liquid and vapor velocities used, the frequent presence of solids in suspension, and the necessary concentration differences.

EVAPORATOR TYPES AND APPLICATIONS

Evaporators may be classified as follows:

1. Heating medium is separated from evaporating liquid by tubular heating surfaces.
2. Heating medium is confined by coils, jackets, double walls, flat plates, etc.
3. Heating medium is brought into direct contact with the evaporating liquid.
4. Heating is done by solar radiation.

By far the largest number of industrial evaporators employ tubular heating surfaces. Circulation of liquid past the heating surface may be induced by boiling or by mechanical means. In the latter case, boiling may or may not occur at the heating surface.

Forced-Circulation Evaporators (Fig. 11-110*a, b, c*) Although it may not be the most economical for many uses, the forced-circulation (FC) evaporator is suitable for the widest variety of evaporator applications. The use of a pump to ensure circulation past the heating surface makes it possible to separate the functions of heat transfer, vapor-liquid separation, and crystallization. The pump withdraws liquor from the flash chamber and forces it through the heating element back to the flash chamber. Circulation is maintained regardless of the evaporation rate; so this type of evaporator is well suited to *crystallizing operation,* in which solids must be maintained in suspension at all times. The liquid velocity past the heating surface is limited only by the pumping power needed or available and by accelerated corrosion and erosion at the higher velocities. *Tube velocities* normally range from a minimum of about 1.2 m/s (4 ft/s) in salt evaporators with copper or brass tubes and liquid containing 5 percent or more solids up to about 3 m/s (10 ft/s) in caustic evaporators having nickel tubes and liquid containing only a small amount of solids. Even higher velocities can be used when corrosion is not accelerated by erosion.

Highest heat-transfer coefficients are obtained in FC evaporators when the liquid is allowed to boil in the tubes, as in the type shown in Fig. 11-110*a*. The heating element projects into the vapor head, and the liquid level is maintained near and usually slightly below the top tube sheet. This type of FC evaporator is not well suited to salting solutions because boiling in the tubes increases the chances of salt deposit on the walls and the sudden flashing at the tube exits promotes excessive nucleation and production of fine crystals. Consequently, this type of evaporator is seldom used except when there are headroom limitations or when the liquid forms neither salt nor scale.

Swirl Flow Evaporators One of the most significant problems in the thermal design of once-through, tube-side evaporators is the poor predictability of the loss of ΔT upon reaching the critical heat flux condition. This situation may occur through flashing due to a high wall temperature or due to process needs to evaporate most of, if not all, the liquid entering the evaporator. It is the result of sensible heating of the vapor phase which accumulates at the heat-transfer surface, dries out the tube wall, and blocks the transfer of heat to the remaining liquid.

In some cases, even with correctly predicted heat-transfer coefficients, the unexpected ΔT loss can reduce the actual performance of the evaporator by as much as 200 percent below the predicted performance. The best approach is to maintain a high level of mixing of the phases through the heat exchanger near the heat-transfer surface.

The use of swirl flow—whereby a rotational vortex is imparted to the boiling fluid to centrifuge the liquid droplets out to the tube wall—has proved to be the most reliable means to correct for and eliminate this loss of ΔT. The

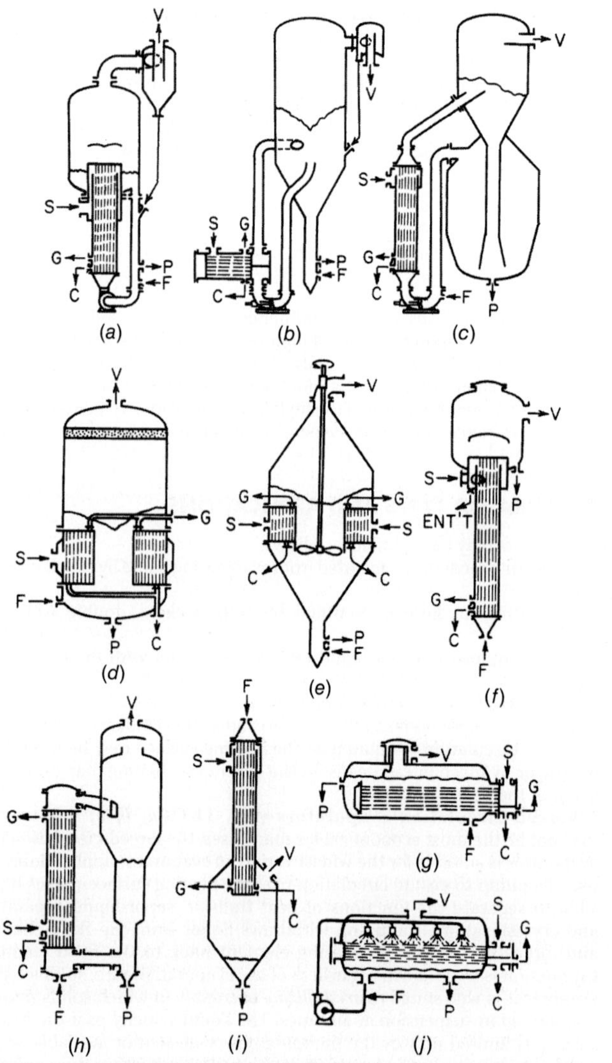

FIG. 11-110 Evaporator types. (*a*) Forced circulation. (*b*) Submerged-tube forced circulation. (*c*) Oslo-type crystallizer. (*d*) Short-tube vertical. (*e*) Propeller calandria. (*f*) Long-tube vertical. (*g*) Recirculating long-tube vertical. (*h*) Falling film. (*i, j*) Horizontal-tube evaporators. C = condensate; F = feed; G = vent; P = product; S = steam; V = vapor; ENT'T = separated entrainment outlet.

use of this technique almost always corrects the design to operate as well as or better than predicted. Also, the use of swirl flow eliminates the need to choose between horizontal and vertical orientation for most two-phase velocities. Both orientations work about the same in swirl flow.

Many commercially viable methods of inducing swirl flow inside tubes are available in the form of either swirl flow tube inserts (twisted tapes, helical cores, spiral wire inserts) or special tube configurations (twisted tube, internal spiral fins). All are designed to impart a natural swirl component to the flow inside the tubes. Each has been proved to solve the problem of tube-side vaporization at high vapor qualities up to and including complete tube-side vaporization.

By far the largest number of forced-circulation evaporators is of the submerged-tube type, as shown in Fig. 11-110*b*. The heating element is placed far enough below the liquid level or return line to the flash chamber to prevent boiling in the tubes. Preferably, the hydrostatic head should be sufficient to prevent boiling even in a tube that is plugged (and hence at steam temperature), since this prevents salting of the entire tube. Evaporators of this type sometimes have horizontal heating elements (usually two-pass), but the vertical single-pass heating element is used whenever sufficient headroom is available. The vertical element usually has a lower friction loss and is easier to clean or retube than a horizontal heater. The submerged-tube forced-circulation evaporator is relatively immune to salting in the tubes, since no supersaturation is generated by evaporation in the tubes. The tendency toward scale formation is also reduced, since supersaturation in the heating element is generated only by a controlled amount of heating and not by both heating and evaporation.

The type of *vapor* head used with the FC evaporator is chosen to suit the product characteristics and may range from a simple centrifugal separator to the crystallizing chambers shown in Fig. 11-110*b* and *c*. Figure 11-110*b* shows a type frequently used for common salt. It is designed to circulate a slurry of crystals throughout the system. Figure 11-110*c* shows a submerged-tube FC evaporator in which heating, flashing, and crystallization are completely separated. The crystallizing solids are maintained as a fluidized bed in the chamber below the vapor head, and little or no solids circulate through the heater and flash chamber. This type is well adapted to growing coarse crystals, but the crystals usually approach a spherical shape, and careful design is required to avoid production of tines in the flash chamber.

In a submerged-tube FC evaporator, all heat is imparted as sensible heat, resulting in a temperature rise of the circulating liquor that reduces the overall temperature difference available for heat transfer. Temperature rise, tube proportions, tube velocity, and head requirements on the circulating pump all influence the selection of circulation rate. Head requirements are frequently difficult to estimate since they consist not only of the usual friction, entrance and contraction, and elevation losses when the return to the flash chamber is above the liquid level, but also of increased friction losses due to flashing in the return line and vortex losses in the flash chamber. Circulation is sometimes limited by vapor in the pump suction line. This may be drawn in as a result of inadequate vapor-liquid separation or may come from vortices near the pump suction connection to the body or may be formed in the line itself by short-circuiting from heater outlet to pump inlet of liquor that has not flashed completely to equilibrium at the pressure in the vapor head.

Advantages of forced-circulation evaporators:

1. High heat-transfer coefficients
2. Positive circulation
3. Relative freedom from salting, scaling, and fouling

Disadvantages of forced-circulation evaporators:

1. High cost
2. Power required for circulating pump
3. Relatively high holdup or residence time

Best applications of forced-circulation evaporators:

1. Crystalline product
2. Corrosive solutions
3. Viscous solutions

Frequent difficulties with forced-circulation evaporators:

1. Plugging of tube inlets by salt deposits detached from walls of equipment
2. Poor circulation due to higher than expected head losses
3. Salting due to boiling in tubes
4. Corrosion-erosion

Short-Tube Vertical Evaporators (Fig. 11-110*d*) This is one of the earliest types still in widespread commercial use. Its principal use at present is in the evaporation of cane-sugar juice. Circulation past the heating surface is induced by boiling in the tubes, which are usually 50.8 to 76.2 mm (2 to 3 in) in diameter by 1.2 to 1.8 m (4 to 6 ft) long. The body is a vertical cylinder, usually of cast iron, and the tubes are expanded into horizontal tube sheets that span the body diameter. The circulation rate through the tubes is many times the feed rate; so there must be a return passage from above the top tube sheet to below the bottom tube sheet. Most commonly used is a central well or *downtake*, as shown in Fig. 11-110*d*. So that friction losses through the downtake do not appreciably impede circulation up through the tubes, the area of the downtake should be of the same order of magnitude as the combined cross-sectional area of the tubes. This results in a downtake almost half of the diameter of the tube sheet.

Circulation and heat transfer in this type of evaporator are strongly affected by the "liquid level." Highest heat-transfer coefficients are achieved when the level, as indicated by an external gauge glass, is only about halfway up the tubes. Slight reductions in level below the optimum result in incomplete wetting of the tube walls with a consequent increased tendency to foul and a rapid reduction in capacity. When this type of evaporator is used with a liquid that can deposit salt or scale, it is customary to operate with the liquid level appreciably higher than the optimum and usually appreciably above the top tube sheet.

Circulation in the standard short-tube vertical evaporator is dependent entirely on boiling, and when boiling stops, any solids present settle out of suspension. Consequently, this type is seldom used as a crystallizing evaporator. By installing a propeller in the downtake, this objection can be overcome. Such an evaporator, usually called a *propeller calandria*, is illustrated in Fig. 11-110*e*. The propeller is usually placed as low as possible to reduce cavitation and is shrouded by an extension of the downtake well. The use of the propeller can sometimes double the capacity of a short-tube vertical evaporator. The evaporator shown in Fig. 11-110*e* includes an elutriation leg for salt manufacture similar to that used on the FC evaporator of Fig. 11-110*b*. The shape of the bottom will, of course, depend on the particular application and on whether the propeller is driven from above or below. To avoid salting when the evaporator is used for crystallizing solutions, the liquid level must be kept appreciably above the top tube sheet.

Advantages of short-tube vertical evaporators:

1. High heat-transfer coefficients at high temperature differences
2. Low headroom
3. Easy mechanical descaling
4. Relatively inexpensive

Disadvantages of short-tube vertical evaporators:

1. Poor heat transfer at low temperature differences and low temperature
2. High floor space and weight
3. Relatively high holdup
4. Poor heat transfer with viscous liquids

Best applications of short-tube vertical evaporators:

1. Clear liquids
2. Crystalline product if propeller is used
3. Relatively noncorrosive liquids, since body is large and expensive if built of materials other than mild steel or cast iron
4. Mild scaling solutions requiring mechanical cleaning, since tubes are short and large in diameter

Long-Tube Vertical Evaporators (Fig. 11-110*f, h, i*) More total evaporation is accomplished in this type than in all the others combined because it is normally the *cheapest per unit of capacity*. The *long-tube vertical* (LTV) evaporator consists of a simple one-pass vertical shell-and-tube heat exchanger discharging into a relatively small vapor head. Normally, no liquid level is maintained in the vapor head, and the residence time of liquor is only a few seconds. The tubes are usually about 50.8 mm (2 in) in diameter but may be smaller than 25.4 mm (1 in). Tube length may vary from less than 6 to 10.7 m (20 to 35 ft) in the rising-film version and to as great as 20 m (65 ft) in the falling-film version. The evaporator is usually operated single-pass, concentrating from the feed to discharge density in just the time that it takes the liquid and evolved vapor to pass through a tube. An extreme case is the caustic high concentrator, producing a substantially anhydrous product at 370°C (700°F) from an inlet feed of 50 percent NaOH at 149°C (300°F) in one pass up 22-mm- (8/8-in-) OD nickel tubes 6 m (20 ft) long. The largest use of LTV evaporators is for concentrating black liquor in the pulp and paper industry. Because of the long tubes and relatively high heat-transfer coefficients, it is possible to achieve higher single-unit capacities in this type of evaporator than in any other.

The LTV evaporator shown in Fig. 11-110*f* is typical of those commonly used, especially for black liquor. Feed enters at the bottom of the tube and starts to boil partway up the tube, and the mixture of liquid and vapor leaving at the top at high velocity impinges against a deflector placed above the tube sheet. This deflector is effective both as a primary separator and as a foam breaker.

In many cases, as when the ratio of feed to evaporation or the ratio of feed to heating surface is low, it is desirable to provide for *recirculation of product* through the evaporator. This can be done in the type shown in Fig. 11-110*f* by adding a pipe connection between the product line and the feed line. Higher recirculation rates can be achieved in the type shown in Fig. 11-110*h*, which is used widely for condensed milk. By extending the enlarged portion of the vapor head still lower to provide storage space for liquor, this type can be used as a batch evaporator.

Liquid temperatures in the tubes of an LTV evaporator are far from uniform and are difficult to predict. At the lower end, the liquid is usually not boiling, and the liquor picks up heat as sensible heat. Since entering liquid velocities are usually very low, true heat-transfer coefficients are low in this nonboiling zone. At some point up the tube, the liquid starts to boil, and from that point on the liquid temperature decreases because of the reduction in static, friction, and acceleration heads until the vapor-liquid mixture reaches the top of the tubes at substantially the vapor-head temperature. Thus the true temperature difference in the boiling zone is always less than the total temperature difference as measured from steam and vapor-head temperatures.

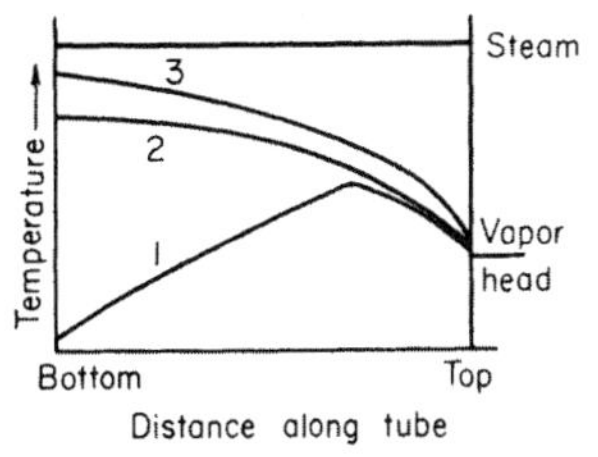

FIG. 11-111 Temperature variations in a long-tube vertical evaporator.

Although the true heat-transfer coefficients in the boiling zone are quite high, they are partially offset by the reduced temperature difference. The point in the tubes at which boiling starts and at which the maximum temperature is reached is sensitive to operating conditions, such as feed properties, feed temperature, feed rate, and heat flux. Figure 11-111 shows typical variations in liquid temperature in tubes of an LTV evaporator operating at a constant terminal temperature difference. Curve 1 shows the normal case in which the feed is not boiling at the tube inlet. Curve 2 gives an indication of the temperature difference lost when the feed enters at the boiling point. Curve 3 is for exactly the same conditions as curve 2 except that the feed contained 0.01 percent Teepol to reduce surface tension [Coulson and Mehta, *Trans. Inst. Chem. Eng.* **31:** 208 (1953)]. The surface-active agent yields a more intimate mixture of vapor and liquid, with the result that liquid is accelerated to a velocity more nearly approaching the vapor velocity, thereby increasing the pressure drop in the tube. Although the surface-active agent caused an increase of more than 100 percent in the true heat-transfer coefficient, this was more than offset by the reduced temperature difference so that the net result was a reduction in evaporator capacity. This sensitivity of the LTV evaporator to changes in operating conditions is less pronounced at high than at low temperature differences and temperature levels.

The *falling-film* version of the LTV evaporator (Fig. 11-110*i*) eliminates these problems of hydrostatic head. Liquid is fed to the tops of the tubes and flows down the walls as a film. Vapor-liquid separation usually takes place at the bottom, although some evaporators of this type are arranged for vapor to rise through the tube countercurrently to the liquid. The pressure drop through the tubes is usually very small, and the boiling-liquid temperature is substantially the same as the vapor-head temperature. The falling-film evaporator is widely used for concentrating *heat-sensitive materials*, such as fruit juices, because the holdup time is very small, the liquid is not overheated during passage through the evaporator, and heat-transfer coefficients are high even at low boiling temperatures.

The principal problem with the falling-film LTV evaporator is *feed distribution* to the tubes. It is essential that all tube surfaces be wetted continually. This usually requires recirculation of the liquid unless the ratio of feed to evaporation is quite high. An alternative to the simple recirculation system of Fig. 11-110*i* is sometimes used when the feed undergoes an appreciable concentration change and the product is viscous and/or has a high boiling point rise. The feed chamber and vapor head are divided into a number of liquor compartments, and separate pumps are used to pass the liquor through the various banks of tubes in series, all in parallel as to steam and vapor pressures. The actual distribution of feed to the individual tubes of a falling-film evaporator may be accomplished by orifices at the inlet to each tube, by a perforated plate above the tube sheet, or by one or more spray nozzles.

Both rising- and falling-film LTV evaporators are generally unsuited to salting or severely scaling liquids. However, both are widely used for black liquor, which presents a mild scaling problem, and also are used to carry solutions beyond saturation with respect to a crystallizing salt. In the latter case, deposits can usually be removed quickly by increasing the feed rate or reducing the steam rate in order to make the product unsaturated for a short time. The falling-film evaporator is not generally suited to liquids containing solids because of difficulty in plugging the feed distributors. However, it has been applied to the evaporation of saline waters saturated with $CaSO_4$ and containing 5 to 10 percent $CaSO_4$ seeds in suspension for scale prevention (Anderson, ASME Pap. 76-WA/Pwr-5, 1976).

Because of their simplicity of construction, compactness, and generally high heat-transfer coefficients, LTV evaporators are well suited to service with corrosive liquids. An example is the reconcentration of rayon spin-bath liquor, which is highly acid. These evaporators employ impervious graphite tubes, lead, rubber-covered or impervious graphite tube sheets, and rubber-lined vapor heads. Polished stainless-steel LTV evaporators are widely used for food products. The latter evaporators are usually similar to that shown in Fig. 11-110*h*, in which the heating element is at one side of the vapor head to permit easy access to the tubes for cleaning.

Advantages of long-tube vertical evaporators:
1. Low cost
2. Large heating surface in one body
3. Low holdup
4. Small floor space
5. Good heat-transfer coefficients at reasonable temperature differences (rising film)
6. Good heat-transfer coefficients at all temperature differences (falling film)

Disadvantages of long-tube vertical evaporators:
1. High headroom
2. Generally unsuitable for salting and severely scaling liquids
3. Poor heat-transfer coefficients of rising-film version at low temperature differences
4. Recirculation usually required for falling-film version

Best applications of long-tube vertical evaporators:
1. Clear liquids
2. Foaming liquids
3. Corrosive solutions
4. Large evaporation loads
5. High temperature differences—rising film, low temperature differences—falling film
6. Low-temperature operation—falling film
7. Vapor compression operation—falling film

Frequent difficulties with long-tube vertical evaporators:
1. Sensitivity of rising-film units to changes in operating conditions
2. Poor feed distribution to falling-film units

Horizontal-Tube Evaporators (Fig. 11-110*j*) In these types the steam is inside and the liquor outside the tubes. The submerged-tube version of Fig. 11-110*j* is seldom used except for the preparation of boiler feedwater. Low entrainment loss is the primary aim: the horizontal cylindrical shell yields a large disengagement area per unit of vessel volume. Special versions use deformed tubes between restrained tube sheets that crack off much of a scale deposit when sprayed with cold water. By showering liquor over the tubes in the version of Fig. 11-110*f*, hydrostatic head losses are eliminated and heat-transfer performance is improved to that of the falling-film tubular type of Fig. 11-110*i*. Originally called the *Lillie*, this evaporator is now also called the *spray-film* or simply the *horizontal-tube evaporator*. Liquid distribution over the tubes is accomplished by sprays or perforated plates above the topmost tubes. Maintaining this distribution through the bundle to avoid overconcentrating the liquor is a problem unique to this type of evaporator. It is now used primarily for seawater evaporation.

Advantages of horizontal-tube evaporators:
1. Very low headroom
2. Large vapor-liquid disengaging area—submerged-tube type
3. Relatively low cost in small-capacity straight-tube type
4. Good heat-transfer coefficients
5. Easy semiautomatic descaling—bent-tube type

Disadvantages of horizontal-tube evaporators:
1. Unsuitable for salting liquids
2. Unsuitable for scaling liquids—straight-tube type
3. High cost—bent-tube type
4. Maintaining liquid distribution—film type

Best applications of horizontal-tube evaporators:
1. Limited headroom
2. Small capacity
3. Nonscaling nonsalting liquids—straight-tube type
4. Severely scaling liquids—bent-tube type

Miscellaneous Forms of Heating Surface Special evaporator designs are sometimes indicated when heat loads are small, special product characteristics are desired, or the product is especially difficult to handle. *Jacketed kettles*, frequently with agitators, are used when the product is very viscous, batches are small, intimate mixing is required, and/or ease of cleaning is an important factor. Evaporators with steam in coiled *tubes* may be used for small capacities with scaling liquids in designs that permit "cold shocking," or complete withdrawal of the coil from the shell for manual scale removal. Other designs for scaling liquids employ flat-plate heat exchangers, since in general a scale deposit can be removed more easily from a flat plate than from a curved surface. One such design, the *channel-switching evaporator*, alternates the duty of either side of the heating surface periodically from boiling liquid to condensing vapor so that scale formed when the surface is in contact with boiling liquid is dissolved when the surface is next in contact with condensing vapor.

Agitated thin-film evaporators employ a heating surface consisting of one large-diameter tube that may be either straight or tapered, horizontal or vertical. Liquid is spread on the tube wall by a rotating assembly of blades that either maintain a close clearance from the wall or actually ride on the film of liquid on the wall. The expensive construction limits application to the most difficult materials. High agitation [on the order of 12 m/s (40 ft/s) rotor-tip speed] and power intensities of 2 to 20 kW/m^2 (0.25 to 2.5 hp/ft^2) permit handling extremely viscous materials. Residence times of only a few seconds permit concentration of heat-sensitive materials at temperatures and temperature differences higher than in other types [Mutzenberg, Parker, and Fischer, *Chem. Eng.* **72:** 175–190 (Sept. 13, 1965)]. High feed-to-product ratios can be handled without recirculation.

Economic and process considerations usually dictate that agitated thin-film evaporators be operated in single-effect mode. Very high temperature differences can then be used: many are heated with Dowtherm or other high-temperature media. This enables one to achieve reasonable capacities in spite of the relatively low heat-transfer coefficients and the small surface that can be provided in a single tube [to about 20 m^2 (200 ft^2)]. The structural need for wall thicknesses of 6 to 13 mm (¼ to ½ in) is a major reason for the relatively low heat-transfer coefficients when evaporating waterlike materials.

Evaporators without Heating Surfaces The *submerged-combustion* evaporator makes use of combustion gases bubbling through the liquid as the means of heat transfer. It consists simply of a tank to hold the liquid, a burner and gas distributor that can be lowered into the liquid, and a combustion control system. Since there are no heating surfaces on which scale can deposit, this evaporator is well suited to use with severely scaling liquids. The ease of constructing the tank and burner of special alloys or nonmetallic materials makes practical the handling of highly corrosive solutions. However, since the vapor is mixed with large quantities of noncondensible gases, it is impossible to reuse the heat in this vapor, and installations are usually limited to areas of low fuel cost. One difficulty frequently encountered in the use of submerged-combustion evaporators is a high entrainment loss. Also, these evaporators cannot be used when control of crystal size is important.

Disk or *cascade evaporators* are used in the pulp and paper industry to recover heat and entrained chemicals from boiler stack gases and to effect a final concentration of the black liquor before it is burned in the boiler. These evaporators consist of a horizontal shaft on which are mounted disks perpendicular to the shaft or bars parallel to the shaft. The assembly is partially immersed in the thick black liquor so that films of liquor are carried into the hot-gas stream as the shaft rotates.

Some forms of *flash evaporators* require no heating surface. An example is a recrystallizing process for separating salts having normal solubility curves from salts having inverse solubility curves, as in separating sodium chloride from calcium sulfate [Richards, *Chem. Eng.* **59**(3): 140 (1952)]. A suspension of raw solid feed in a recirculating brine stream is heated by direct steam injection. The increased temperature and dilution by the steam dissolve the salt having the normal solubility curve. The other salt remains undissolved and is separated from the hot solution before it is flashed to a lower temperature. The cooling and loss of water on flashing cause recrystallization of the salt having the normal solubility curve, which is separated from the brine before the brine is mixed with more solid feed for recycling to the heater. This system can be operated as a multiple effect by flashing down to the lower temperature in stages and using flash vapor from all but the last stage to heat the recycle brine by direct injection. In this process no net evaporation occurs from the total system, and the process cannot be used to concentrate solutions unless heating surfaces are added.

UTILIZATION OF TEMPERATURE DIFFERENCE

Temperature difference is the driving force for evaporator operation and usually is limited, as by compression ratio in vapor compression evaporators and by available steam pressure and heat-sink temperature in single- and multiple-effect evaporators. A fundamental objective of evaporator design is to make as much of this total temperature difference available for heat transfer as is economically justifiable. Some losses in temperature difference, such as those due to *boiling point rise* (BPR), are unavoidable. However, even these can be minimized, such as by passing the liquor through effects or through different sections of a single effect in series so that only a portion of the heating surface is in contact with the strongest liquor.

Figure 11-112 shows approximate BPR losses for a number of process liquids. A correlation for concentrated solutions of many inorganic salts

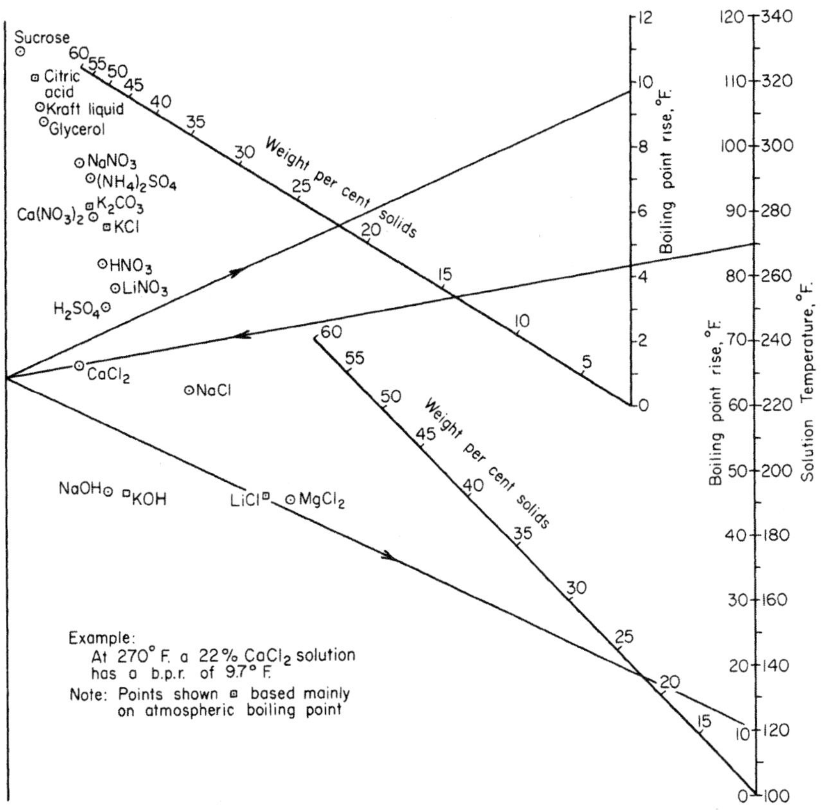

FIG. 11-112 Boiling-point rise of aqueous solutions. °C = 5/9 (°F − 32).

at the atmospheric pressure boiling point [Meranda and Furter, *J. Ch. and E. Data* **22:** 315–7 (1977)] is

$$BPR = 104.9N_2^{1.14} \tag{11-108}$$

where N_2 is the mole fraction of salts in solution. Correction to other pressures, when heats of solution are small, can be based on a constant ratio of vapor pressure of the solution to that of water at the same temperature.

The principal reducible loss in ΔT is that due to friction and to entrance and exit losses in vapor piping and entrainment separators. Pressure-drop losses here correspond to a reduction in condensing temperature of the vapor and hence a loss in available ΔT. These losses become most critical at the low-temperature end of the evaporator, both because of both the increasing specific volume of the vapor and the reduced slope of the vapor-pressure curve. Sizing of vapor lines is part of the economic optimization of the evaporator, extra costs of larger vapor lines being balanced against savings in ΔT, which correspond to savings in heating-surface requirements. Note that entrance and exit losses in vapor lines usually exceed by severalfold the straight-pipe friction losses, so they cannot be ignored.

VAPOR-LIQUID SEPARATION

Product losses in evaporator vapor may result from foaming, splashing, or entrainment. Primary separation of liquid from vapor is accomplished in the vapor head by making the horizontal plan area large enough that most of the entrained droplets can settle out against the rising flow of vapor. Allowable velocities are governed by the Souders-Brown equation $V = k\sqrt{(\rho_l - \rho_v)/\rho_v}$, in which k depends on the size distribution of droplets and the decontamination factor F desired. For most evaporators and for F between 100 and 10,000, $k \cong 0.245/(F - 50)^{0.4}$ (Standiford, *Chemical Engineers' Handbook,* 4th ed., McGraw-Hill, New York, 1963, pp. 11–35). Higher values of k (to about 0.15) can be tolerated in the falling-film evaporator, where most of the entrainment separation occurs in the tubes, the vapor is scrubbed by liquor leaving the tubes, and the vapor must reverse direction to reach the outlet.

Foaming losses usually result from the presence in the evaporating liquid of colloids or of surface-tension depressants and finely divided solids. Antifoam agents are often effective. Other means of combatting foam include the use of steam jets impinging on the foam surface, the removal of product at the surface layer, where the foaming agents seem to concentrate, and operation at a very low liquid level so that hot surfaces can break the foam. Impingement at high velocity against a baffle tends to break the foam mechanically, and this is the reason that the long-tube vertical, forced-circulation, and agitated-film evaporators are particularly effective with foaming liquids. Operating at lower temperatures and/or higher dissolved solids concentrations may also reduce foaming tendencies.

Splashing losses are usually insignificant if a reasonable height has been provided between the liquid level and the top of the vapor head. The height required depends on the violence of boiling. Heights of 2.4 to 3.6 m (8 to 12 ft) or more are provided in short-tube vertical evaporators, in which the liquid and vapor leaving the tubes are projected upward. Less height is required in forced-circulation evaporators, in which the liquid is given a centrifugal motion or is projected downward as by a baffle. The same is true of long-tube vertical evaporators, in which the rising vapor-liquid mixture is projected against a baffle.

Entrainment losses by flashing are frequently encountered in an evaporator. If the feed is above the boiling point and is introduced above or only a short distance below the liquid level, then entrainment losses may be excessive. This can occur in a short-tube-type evaporator if the feed is introduced at only one point below the lower tube sheet (Kerr, *Louisiana Agric. Expt. Stn. Bull.* 149, 1915). The same difficulty may be encountered in forced-circulation evaporators having too high a temperature rise through the heating element and thus too wide a flashing range as the circulating liquid enters the body. Poor vacuum control, especially during start-up, can cause the generation of far more vapor than the evaporator was designed to handle, with a consequent increase in entrainment.

Entrainment separators are frequently used to reduce product losses. There are a number of specialized designs available, practically all of which rely on a change in direction of the vapor flow when the vapor is traveling at high velocity. Typical separators are shown in Fig. 11-110, although not necessarily with the type of evaporator with which they may be used. The most common separator is the cyclone, which may have either a top or a bottom outlet, as shown in Fig. 11-110*a* and *b,* or may even be wrapped around the heating element of the next effect, as shown in Fig. 11-110*f.* The separation efficiency of a cyclone increases with an increase in inlet velocity, although at the cost of some pressure drop, which means a loss in available temperature difference. Pressure drop in a cyclone is from 10 to 16 velocity heads [Lawrence, *Chem. Eng. Prog.* **48:** 241 (1952)], based on the velocity in the inlet pipe. Such cyclones can be sized in the same manner as a cyclone dust collector [using velocities of about 30 m/s (100 ft/s) at atmospheric pressure] although sizes may be increased somewhat in order to reduce losses in available temperature difference.

Knitted wire mesh serves as an effective entrainment separator when it cannot be easily fouled by solids in the liquor. The mesh is available in woven metal wire of most alloys and is installed as a blanket across the top of the evaporator (Fig. 11-110*d*) or in a monitor of reduced diameter atop the vapor head. These separators have low-pressure drops, usually on the order of 13 mm (½ in) of water, and collection efficiency is above 99.8 percent in the range of vapor velocities from 2.5 to 6 m/s (8 to 20 ft/s) [Carpenter and Othmer, *Am. Inst. Chem. Eng. J.* **1:** 549 (1955)]. Chevron (hook-and-vane) type of separators is also used because of the higher-allowable velocities or because of the reduced tendency to foul with solids suspended in the entrained liquid.

EVAPORATOR ARRANGEMENT

Single-Effect Evaporators Single-effect evaporators are used when the required capacity is small, steam is cheap, the material is so corrosive that very expensive materials of construction are required, or the vapor is so contaminated that it cannot be reused. Single-effect evaporators may be operated in batch, semibatch, or continuous-batch modes or continuously. Strictly speaking, *batch evaporators* are ones in which filling, evaporating, and emptying are consecutive steps. This method of operation is rarely used since it requires that the body be large enough to hold the entire charge of feed and the heating element be placed low enough so as not to be uncovered when the volume is reduced to that of the product. The more usual method of operation is *semibatch,* in which feed is continually added to maintain a constant level until the entire charge reaches final density. *Continuous-batch* evaporators usually have a continuous feed and, over at least part of the cycle, a continuous discharge. One method of operation is to circulate from a storage tank to the evaporator and back until the entire tank is up to desired concentration and then to finish in batches. *Continuous evaporators* have essentially continuous feed and discharge, and concentrations of both feed and product remain substantially constant.

Thermocompression The simplest means of reducing the energy requirements of evaporation is to compress the vapor from a single-effect evaporator so that the vapor can be used as the heating medium in the same evaporator. The compression may be accomplished by mechanical means or by a steam jet. To keep the compressor cost and power requirements within reason, the evaporator must work with a fairly narrow temperature difference, usually from about 5.5°C to 11°C (10°F to 20°F). This means that a large evaporator heating surface is needed, which usually makes the vapor compression evaporator more expensive in first cost than a multiple-effect evaporator. However, total installation costs may be reduced when purchased power is the energy source, since the need for boiler and heat sink is eliminated. Substantial savings in operating cost are realized when electric or mechanical power is available at a low cost relative to low-pressure steam, when only high-pressure steam is available to operate the evaporator, or when the cost of providing cooling water or other heat sink for a multiple-effect evaporator is high.

Mechanical thermocompression may employ reciprocating, rotary positive-displacement, centrifugal, or axial-flow compressors. Positive-displacement compressors are impractical for all but the smallest capacities, such as portable seawater evaporators. Axial-flow compressors can be built for capacities of more than 472 m^3/s (1×10^6 ft^3/min). Centrifugal compressors are usually cheapest for the intermediate-capacity ranges that are normally encountered. In all cases, great care must be taken to keep entrainment at a minimum, since the vapor becomes superheated on compression and any liquid present will evaporate, leaving the dissolved solids behind. In some cases, a vapor-scrubbing tower may be installed to protect the compressor. A mechanical recompression evaporator usually requires more heat than is available from the compressed vapor. Some of this extra heat can be obtained by preheating the feed with the condensate and, if possible, with the product. Rather extensive heat-exchange systems with close approach temperatures are usually justified, especially if the evaporator is operated at high temperature to reduce the volume of vapor to be compressed. When the product is a solid, an elutriation leg such as that shown in Fig. 11-110*b* is advantageous, since it cools the product almost to the feed temperature. The remaining heat needed to maintain the evaporator in operation must be obtained from outside sources.

While theoretical compressor power requirements are reduced slightly by going to lower evaporating temperatures, the volume of vapor to be compressed and hence compressor size and cost increase so rapidly that low-temperature operation is more expensive than high-temperature operation. The requirement of low temperature for fruit juice concentration has led to the development of an evaporator employing a *secondary fluid,* usually Freon or ammonia. In this evaporator, the vapor is condensed in an exchanger cooled by boiling Freon. The Freon, at a much higher vapor density than the water vapor, is then compressed to serve as the heating medium for the evaporator. This system requires that the latent heat be transferred through two surfaces instead of one, but the savings in compressor size and cost are enough to justify the extra cost of heating surface or the cost of compressing through a wider temperature range.

Steam-jet thermocompression is advantageous when steam is available at a pressure appreciably higher than can be used in the evaporator. The steam jet then serves as a reducing valve while doing some useful work. The efficiency of a steam jet is quite low and falls off rapidly when the jet is not used at the vapor flow rate and terminal pressure conditions for which it was designed. Consequently, multiple jets are used when wide variations in evaporation rate are expected. Because of the low first cost and the ability to handle large volumes of vapor, steam-jet thermocompressors are used to increase the economy of evaporators that must operate at low temperatures and hence cannot be operated in multiple effect. The steam-jet thermocompression evaporator has a heat input larger than that needed to balance the system, and some heat must be rejected. This is usually done by venting some of the vapor at the suction of the compressor.

Multiple-Effect Evaporation Multiple-effect evaporation is the principal means in use for economizing on energy consumption. Most such evaporators operate on a continuous basis, although for a few difficult materials a continuous-batch cycle may be employed. In a multiple-effect evaporator, steam from an outside source is condensed in the heating element of the first effect. If the feed to the effect is at a temperature near the boiling point in the first effect, 1 kg of steam will evaporate almost 1 kg of water. The first effect operates at (but is not controlled at) a boiling temperature high enough that the evaporated water can serve as the heating medium of the second effect. Here almost another kilogram of water is evaporated, and this may go to a condenser if the evaporator is a double-effect or may be used as the heating medium of the third effect. This method may be repeated for any number of effects. Large evaporators having 6 and 7 effects are common in the pulp and paper industry, and evaporators having as many as 17 effects have been built. As a first approximation, the *steam economy* of a multiple-effect evaporator will increase in proportion to the number of effects and usually will be somewhat less numerically than the number of effects.

The increased steam economy of a multiple-effect evaporator is gained at the expense of evaporator first cost. The total heat-transfer surface will increase substantially in proportion to the number of effects in the evaporator. This is only an approximation since going from one to two effects means that about one-half of the heat transfer is at a higher temperature level, where heat-transfer coefficients are generally higher. On the other hand, operating at lower temperature differences reduces the heat-transfer coefficient for many types of evaporator. If the material has an appreciable boiling-point elevation, this will also lower the available temperature difference. The only accurate means of predicting the changes in steam economy and surface requirements with changes in the number of effects is by detailed heat and material balances together with an analysis of the effect of changes in operating conditions on heat-transfer performance.

The approximate temperature distribution in a multiple-effect evaporator is under the control of the designer, but once built, the evaporator establishes its own equilibrium. Basically, the effects are a number of series resistances to heat transfer, each resistance being approximately proportional to $1/U_nA_n$. The total available temperature drop is divided between the effects in proportion to their resistances. If one effect starts to scale, its temperature drop will increase at the expense of the temperature drops across the other effects. This provides a convenient means of detecting a drop in the heat-transfer coefficient in an effect of an operating evaporator. If the steam pressure and final vacuum do not change, the temperature in the effect that is scaling will decrease and the temperature in the preceding effect will increase.

The feed to a multiple-effect evaporator is usually transferred from one effect to another in series so that the ultimate product concentration is reached in only one effect of the evaporator. In backward-feed operation, the raw feed enters the last (coldest) effect, the discharge from this effect becomes the feed to the next-to-the-last effect, and so on until product is discharged from the first effect. This method of operation is advantageous when the feed is cold, since much less liquid must be heated to the higher temperature existing in the early effects. It is also used when the product is so viscous that high temperatures are needed to keep the viscosity low enough to give reasonable heat-transfer coefficients. When product viscosity is high but a hot product is not needed, the liquid from the first effect is sometimes flashed to a lower temperature in one or more stages and the flash vapor added to the vapor from one or more later effects of the evaporator.

In *forward-feed* operation, raw feed is introduced in the first effect and passed from effect to effect parallel to the steam flow. Product is withdrawn from the last effect. This method of operation is advantageous when the feed is hot or when the concentrated product would be damaged or would deposit scale at high temperature. Forward feed simplifies operation when liquor can be transferred by pressure difference alone, thus eliminating all intermediate liquor pumps. When the feed is cold, forward feed gives a low steam economy since an appreciable part of the prime steam is needed to heat the feed to the boiling point and thus accomplishes no evaporation. If forward feed is necessary and feed is cold, steam economy can be improved markedly by preheating the feed in stages with vapor bled from intermediate effects of the evaporator. This usually represents little increase in total heating surface or cost since the feed must be heated in any event and shell-and-tube heat exchangers are generally less expensive per unit of surface area than evaporator heating surface.

Mixed-feed operation is used only for special applications, as when liquor at an intermediate concentration and a certain temperature is desired for additional processing.

Parallel feed involves the introduction of raw feed and the withdrawal of product at each effect of the evaporator. It is used primarily when the feed is substantially saturated and the product is a solid. An example is the evaporation of brine to make common salt. Evaporators of the types shown in Fig. 11-110*b* or *e* are used, and the product is withdrawn as a slurry. In this case, parallel feed is desirable because the feed washes impurities from the salt leaving the body.

Heat recovery systems are frequently incorporated in an evaporator to increase the steam economy. Ideally, product and evaporator condensate should leave the system at a temperature as low as possible. Also, heat should be recovered from these streams by exchange with feed or evaporating liquid at the highest possible temperature. This would normally require separate liquid-liquid heat exchangers, which add greatly to the complexity of the evaporator and are justifiable only in large plants. Normally, the loss in thermodynamic availability due to flashing is tolerated since the flash vapor can then be used directly in the evaporator effects. The most commonly used is a *condensate flash* system in which the condensate from each effect but the first (which normally must be returned to the boiler) is flashed in successive stages to the pressure in the heating element of each succeeding effect of the evaporator. Product flash tanks may also be used in a backward- or mixed-feed evaporator. In a forward-feed evaporator, the principal means of heat recovery may be by use of *feed preheaters* heated by vapor bled from each effect of the evaporator. In this case, condensate may be either flashed as before or used in a separate set of exchangers to accomplish some of the feed preheating. A feed preheated by last-effect vapor may also materially reduce condenser water requirements.

Seawater Evaporators The production of potable water from saline water represents a large and growing field of application for evaporators. Extensive work done in this field to 1972 was summarized in the annual *Saline Water Conversion Reports* of the Office of Saline Water, U.S. Department of the Interior. *Steam economies* on the order of 10 kg evaporation/kg steam are usually justified because (1) unit production capacities are high, (2) fixed charges are low on capital used for public works (i.e., they use long amortization periods and have low interest rates, with no other return on investment considered), (3) heat-transfer performance is comparable with that of pure water, and (4) properly treated seawater causes little deterioration due to scaling or fouling.

Figure 11-113*a* shows a *multiple-effect* (falling-film) flow sheet as used for seawater. Twelve effects are needed for a steam economy of 10. Seawater is used to condense last-effect vapor, and a portion is then treated to prevent scaling and corrosion. Treatment usually consists of acidification to break down bicarbonates, followed by deaeration, which also removes the carbon dioxide generated. The treated seawater is then heated to successively higher temperatures by a portion of the vapor from each effect and finally is fed to the evaporating surface of the first effect. The vapor generated therein and the partially concentrated liquid are passed to the second effect, and so on until the last effect. The feed rate is adjusted relative to the steam rate so that the residual liquid from the last effect can carry away all the salts in solution, in a volume about one-third of that of the feed. Condensate formed in each effect but the first is flashed down to the following effects in sequence and constitutes the product of the evaporator.

As the feed-to-steam ratio is increased in the flow sheet of Fig. 11-113*a*, a point is reached where all the vapor is needed to preheat the feed and none is available for the evaporator tubes. This limiting case is the *multistage flash evaporator*, shown in its simplest form in Fig. 11-113*b*. Seawater is treated as before and then pumped through a number of feed heaters in series. It is

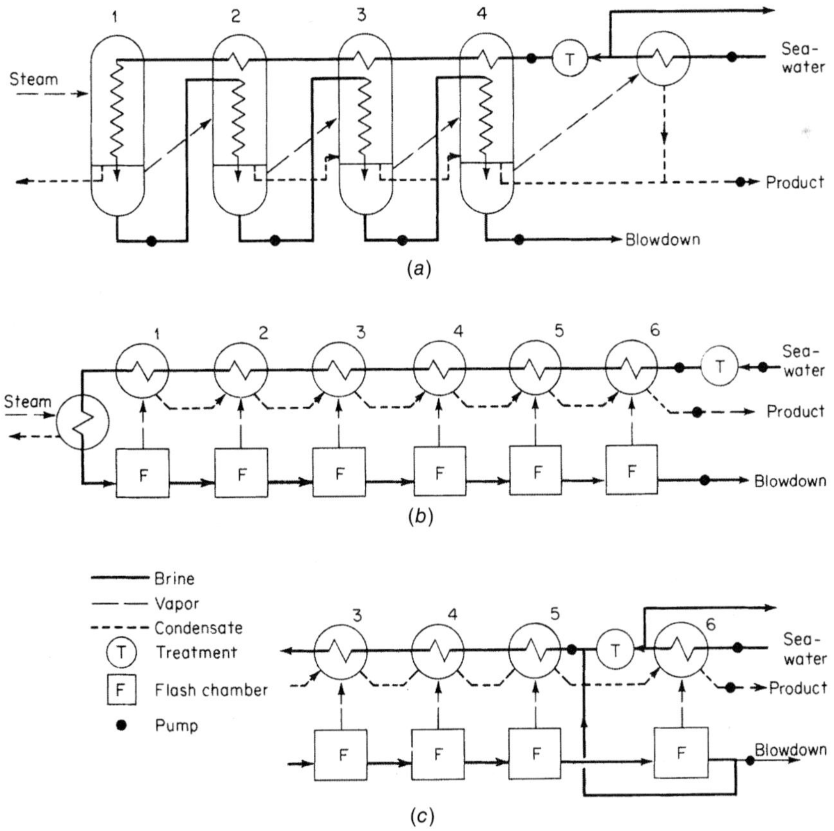

FIG. 11-113 Flow sheets for seawater evaporators. (*a*) Multiple effect (falling film). (*b*) Multistage flash (once-through). (*c*) Multistage flash (recirculating).

given a final boost in temperature with prime steam in a *brine heater* before it is flashed down in series to provide the vapor needed by the feed heaters. The amount of steam required depends on the approach-temperature difference in the feed heaters and the flash range per stage. Condensate from the feed heaters is flashed down in the same manner as the brine.

Since the flow being heated is identical to the total flow being flashed, the temperature rise in each heater is equal to the flash range in each flasher. This temperature difference represents a loss from the temperature difference available for heat transfer. There are thus two ways of increasing the steam economy of such plants: increasing the heating surface and increasing the number of stages. Whereas the number of effects in a multiple-effect plant will be about 20 percent greater than the steam economy, the number of stages in a flash plant will be 3 to 4 times the steam economy. However, a large number of stages can be provided in a single vessel by means of internal bulkheads. The heat-exchanger tubing is placed in the same vessel, and the tubes usually are continuous through a number of stages. This requires ferrules or special close tube-hole clearances where the tubes pass through the internal bulkheads. In a plant for a steam economy of 10, the ratio of flow rate to heating surface is usually such that the seawater must pass through about 152 m of 19-mm (500 ft of ¾-in) tubing before it reaches the brine heater. This places a limitation on the physical arrangement of the vessels.

Inasmuch as it requires a flash range of about 61°C (110°F) to produce 1 kg of flash vapor for every 10 kg of seawater, the multistage flash evaporator requires handling a large volume of seawater relative to the product. In the flow sheet of Fig. 11-113*b* all this seawater must be deaerated and treated for scale prevention. In addition, the last-stage vacuum varies with the ambient seawater temperature, and ejector equipment must be sized for the worst condition. These difficulties can be eliminated by using the *recirculating multistage flash* flow sheet of Fig. 11-113*c*. The last few stages, called the *reject stages,* are cooled by a flow of seawater that can be varied to maintain a reasonable last-stage vacuum. A small portion of the last-stage brine is blown down to carry away the dissolved salts, and the balance is recirculated to the *heat recovery stages.* This arrangement requires a much smaller makeup of fresh seawater and hence a lower treatment cost.

The multistage flash evaporator is similar to a multiple-effect forced-circulation evaporator, but with all the forced-circulation heaters in series. This has the advantage of requiring only one large-volume forced-circulation pump, but the sensible heating and short-circuiting losses in available temperature differences remain. A disadvantage of the flash evaporator is that the liquid throughout the system is at almost the discharge concentration. This has limited its industrial use to solutions in which no great concentration differences are required between feed and product and where the liquid can be heated through wide temperature ranges without scaling. A partial remedy is to arrange several multistage flash evaporators in series, the heat rejection section of one being the brine heater of the next. This permits independent control of concentration but eliminates the principal advantage of the flash evaporator, which is the small number of pumps and vessels required. An unusual feature of the flash evaporator is that fouling of the heating surfaces reduces primarily the steam economy rather than the capacity of the evaporator. Capacity is not affected until the heat rejection stages can no longer handle the increased flashing resulting from the increased heat input.

EVAPORATOR CALCULATIONS

Single-Effect Evaporators The heat requirements of a single-effect continuous evaporator can be calculated by the usual methods of stoichiometry. If enthalpy data or specific heat and heat-of-solution data are not available, the heat requirement can be estimated as the sum of the heat needed to raise the feed from feed to product temperature and the heat required to evaporate the water. The latent heat of water is taken at the vapor-head pressure instead of at the product temperature in order to compensate partially for any heat of solution. If sufficient vapor pressure data are available for the solution, methods are available to calculate the true latent heat from the slope of the Dühring line [Othmer, *Ind. Eng. Chem.* **32:** 841 (1940)].

The heat requirements in batch evaporation are the same as those in continuous evaporation except that the temperature (and sometimes pressure) of the vapor changes during the course of the cycle. Since the enthalpy of water vapor changes but little relative to temperature, the difference between continuous and batch heat requirements is almost always negligible. More important usually is the effect of variation of fluid properties, such as viscosity and boiling point rise, on heat transfer. These can only be estimated by a step-by-step calculation.

In selecting the *boiling temperature,* consideration must be given to the effect of temperature on heat-transfer characteristics of the type of evaporator to be used. Some evaporators show a marked drop in coefficient at low temperature—more than enough to offset any gain in available temperature difference. The condenser cooling-water temperature and cost must also be considered.

Thermocompression Evaporators Thermocompression evaporator calculations [Pridgeon, *Chem. Metall. Eng.* **28:** 1109 (1923); Peter, *Chimia (Switzerland)* **3:** 114 (1949); Petzold, *Chem. Ing. Tech.* **22:** 147 (1950); and Weimer, Dolf, and Austin, *Chem. Eng. Prog.* **76**(11): 78 (1980)] are much the same as single-effect calculations with the added complication that the heat supplied to the evaporator from compressed vapor and other sources must exactly balance the heat requirements. Some knowledge of compressor efficiency is also required. Large axial-flow machines on the order of 236 m^3/s (500,000 ft^3/min) capacity may have efficiencies of 80 to 85 percent. Efficiency drops to about 75 percent for a 14 m^3/s (30,000 ft^3/min) centrifugal compressor. Steam-jet compressors have thermodynamic efficiencies on the order of only 25 to 30 percent.

Flash Evaporators The calculation of a heat and material balance on a flash evaporator is relatively easy once it is understood that the temperature rise in each heater and the temperature drop in each flasher must all be substantially equal. The steam economy *E,* kg evaporation/kg of 1055-kJ steam (lb/lb of 1000-Btu steam) may be approximated from

$$E=\left(1-\frac{\Delta T}{1250}\right)\frac{\Delta T}{Y+R+\Delta T/N} \tag{11-109}$$

where ΔT is the total temperature drop between feed to the first flasher and discharge from the last flasher, °C; N is the number of flash stages; Y is the approach between vapor temperature from the first flasher and liquid leaving the heater in which this vapor is condensed, °C (the approach is usually substantially constant for all stages); and R°C is the sum of the boiling-point rise and the short-circuiting loss in the first flash stage. The expression for the mean effective temperature difference Δt available for heat transfer then becomes

$$\Delta t=\frac{\Delta T}{N\ln\dfrac{1-\Delta T/1250-RE/\Delta T}{1-\Delta T/1250-RE/\Delta T-E/N}} \tag{11-110}$$

Multiple-Effect Evaporators A number of approximate methods have been published for estimating performance and heating-surface requirements of a multiple-effect evaporator [Coates and Pressburg, *Chem. Eng.* **67**(6): 157 (1960); Coates, *Chem. Eng. Prog.* **45:** 25 (1949); and Ray and Carnahan, *Trans. Am. Inst. Chem. Eng.* **41:** 253 (1945)]. However, because of the wide variety of methods of feeding and the added complication of feed heaters and condensate flash systems, the only certain way to determine performance is by detailed heat and material balances. Algebraic solutions may be used, but if more than a few effects are involved, trial-and-error methods are usually quicker. These frequently involve trial-and-error within trial-and-error solutions. Usually, if condensate flash systems or feed heaters are involved, it is best to start at the first effect. The basic steps in the calculation are then as follows:

1. Estimate temperature distribution in the evaporator, taking into account boiling-point elevations. If all heating surfaces are to be equal, the temperature drop across each effect will be approximately inversely proportional to the heat-transfer coefficient in that effect.
2. Determine total evaporation required, and estimate steam consumption for the number of effects chosen.
3. From assumed feed temperature (forward feed) or feed flow (backward feed) to the first effect and assumed steam flow, calculate evaporation in the first effect. Repeat for each succeeding effect, checking intermediate assumptions as the calculation proceeds. Heat input from condensate flash can be incorporated easily since the condensate flow from the preceding effects will have already been determined.
4. The result of the calculation will be a feed to or a product discharge from the last effect that may not agree with actual requirements. The calculation must then be repeated with a new assumption of steam flow to the first effect.
5. These calculations should yield liquor concentrations in each effect that enable a revised estimate of boiling-point rises. They also give the quantity of heat that must be transferred in each effect. From the heat loads, assumed temperature differences, and heat-transfer coefficients, the heating-surface requirements can be determined. If the distribution of heating surface is not as desired, the entire calculation may need to be repeated with revised estimates of the temperature in each effect.
6. If sufficient data are available, heat-transfer coefficients under the proposed operating conditions can be calculated in greater detail and surface requirements readjusted.

Such calculations require considerable judgment to avoid repetitive trials but are usually well worth the effort. Sample calculations are given in the American Institute of Chemical Engineers' *Testing Procedure for Evaporators* and by Badger and Banchero, *Introduction to Chemical Engineering*, McGraw-Hill, New York, 1955. These balances may be done by computer, but programming time frequently exceeds the time needed to do them manually, especially when variations in flow sheet are to be investigated. The MASSBAL program of SACDA, London, Ont., provides a considerable degree of flexibility in this regard. Another program, not specific to evaporators, is ASPEN PLUS by Aspen Tech., Cambridge, Mass. Many such programs include simplifying assumptions and approximations that are not explicitly stated and can lead to erroneous results.

Optimization The primary purpose of evaporator design is to enable production of the necessary amount of satisfactory product at the lowest total cost. This requires economic-balance calculations that may include a great number of variables. Among the possible variables are the following:

1. Initial steam pressure versus cost or availability.
2. Final vacuum versus water temperature, water cost, heat-transfer performance, and product quality.
3. Number of effects versus steam, water, and pump power cost.
4. Distribution of heating surface between effects versus evaporator cost.
5. Type of evaporator versus cost and continuity of operation.
6. Materials of construction versus product quality, tube life, evaporator life, and evaporator cost.
7. Corrosion, erosion, and power consumption versus tube velocity.
8. Downtime for retubing and repairs.
9. Operating-labor and maintenance requirements.
10. Method of feeding and use of heat recovery systems.
11. Size of recovery heat exchangers.
12. Possible withdrawal of steam from an intermediate effect for use elsewhere.
13. Entrainment separation requirements.

The type of evaporator to be used and the materials of construction are generally selected on the basis of past experience with the material to be concentrated. The method of feeding can usually be decided on the basis of known feed temperature and the properties of feed and product. However, few of the listed variables are completely independent. For instance, if a large number of effects are to be used, with a consequent low temperature drop per effect, it is impractical to use a natural-circulation evaporator. If expensive materials of construction are desirable, it may be found that the forced-circulation evaporator is the cheapest and that only a few effects are justifiable.

The variable having the greatest influence on total cost is the number of effects in the evaporator. An economic balance can establish the optimum number where the number is not limited by such factors as viscosity, corrosiveness, freezing point, boiling-point rise, or thermal sensitivity. Under present U.S. conditions, savings in steam and water costs justify the extra capital, maintenance, and power costs of about seven effects in large commercial installations when the properties of the fluid are favorable, as in black-liquor evaporation. Under government financing conditions, as for plants to supply freshwater from seawater, evaporators containing from 12 to 30 or more effects can be justified.

As a general rule, the optimum number of effects increases with an increase in steam cost or plant size. Larger plants favor more effects, partly because they make it easier to install heat recovery systems that increase the steam economy attainable with a given number of effects. Such recovery systems usually do not increase the total surface needed, but do require that the heating surface be distributed between a greater number of pieces of equipment.

The most common evaporator design is based on the use of the same heating surface in each effect. This is by no means essential since few evaporators are "standard" or involve the use of the same patterns. In fact, there is no reason why all effects in an evaporator must be of the same type. For instance, the cheapest salt evaporator might use propeller calandrias for the early effects and forced-circulation effects at the low-temperature end, where their higher cost per unit area is more than offset by higher heat-transfer coefficients.

Bonilla [*Trans. Am. Inst. Chem. Eng.* **41:** 529 (1945)] developed a simplified method for distributing the heating surface in a multiple-effect evaporator to achieve minimum cost. If the cost of the evaporator per unit area of heating surface is constant throughout, then minimum cost and area will be achieved if the ratio of area to temperature difference $A/\Delta T$ is the same for all effects. If the cost per unit area z varies, as when different tube materials or evaporator types are used, then $zA/\Delta T$ should be the same for all effects.

EVAPORATOR ACCESSORIES

Condensers The vapor from the last effect of an evaporator is usually removed by a condenser. *Surface condensers* are employed when mixing of condensate with condenser cooling water is not desired. For the most part, they are shell-and-tube condensers with vapor on the shell side and a multipass flow of cooling water on the tube side. Heat loads, temperature differences, sizes, and costs are usually of the same order of magnitude as for another effect of the evaporator. Surface condensers use more cooling water and are so much more expensive that they are never used when a direct-contact condenser is suitable.

The most common type of direct-contact condenser is the countercurrent *barometric condenser*, in which vapor is condensed by rising against a rain of cooling water. The condenser is set high enough that water can discharge by gravity from the vacuum in the condenser. Such condensers are inexpensive and are economical on water consumption. They can usually be relied on to maintain a vacuum corresponding to a saturated-vapor temperature within 2.8°C (5°F) of the water temperature leaving the condenser [How, *Chem. Eng.* **63**(2): 174 (1956)]. The ratio of water consumption to vapor condensed can be determined from the following equation:

$$\frac{\text{Water flow}}{\text{Vapor flow}} = \frac{H_\upsilon - h_2}{h_2 - h_1} \tag{11-111}$$

where H_υ = vapor enthalpy and h_1 and h_2 = water enthalpies entering and leaving the condenser. Another type of direct-contact condenser is the *jet* or *wet condenser*, which makes use of high-velocity jets of water both to condense the vapor and to force noncondensible gases out the tailpipe. This type of condenser is frequently placed below barometric height and requires a pump to remove the mixture of water and gases. Jet condensers usually require more water than the more common barometric-type condensers and cannot be throttled easily to conserve water when operating at low evaporation rates.

Vent Systems Noncondensible gases may be present in the evaporator vapor as a result of leakage, air dissolved in the feed, or decomposition reactions in the feed. When the vapor is condensed in the succeeding effect, the noncondensibles increase in concentration and impede heat transfer. This occurs partially because of the reduced partial pressure of vapor in the mixture but mainly because the vapor flow toward the heating surface creates a film of poorly conducting gas at the interface. (See Thermal Design of Condensers, Multicomponent Condensers earlier in this section for means of estimating the effect of noncondensible gases on the steam-film coefficient.) The most important means of reducing the influence of noncondensibles on heat transfer is by properly channeling them past the heating surface. A positive vapor flow path from inlet to vent outlet should be provided, and the path should preferably be tapered to avoid pockets of low velocity where noncondensibles can be trapped. Excessive clearances and low-resistance channels that could bypass vapor directly from the inlet to the vent should be avoided [Standiford, *Chem. Eng. Prog.* **75**: 59–62 (July 1979)].

In any event, noncondensible gases should be vented well before their concentration reaches 10 percent. Since gas concentrations are difficult to measure, the usual practice is to overvent. This means that an appreciable amount of vapor can be lost.

To help conserve steam economy, venting is usually done from the steam chest of one effect to the steam chest of the next. In this way, excess vapor in one vent does useful evaporation at a steam economy only about 1 less than the overall steam economy. Only when there are large amounts of noncondensible gases present, as in beet sugar evaporation, is it desirable to pass the vents directly to the condenser to avoid serious losses in heat-transfer rates. In such cases, it can be worthwhile to recover heat from the vents in separate heat exchangers, which preheat the entering feed.

The noncondensible gases eventually reach the condenser (unless vented from an effect above atmospheric pressure to the atmosphere or to auxiliary vent condensers). These gases will be supplemented by air dissolved in the condenser water and by carbon dioxide given off on decomposition of bicarbonates in the water if a barometric condenser is used. These gases may be removed by the use of a water-jet-type condenser but are usually removed by a separate vacuum pump.

The vacuum pump is usually of the steam-jet type if high-pressure steam is available. If high-pressure steam is not available, more expensive mechanical pumps may be used. These may be either a water-ring (Hytor) type or a reciprocating pump.

The primary source of noncondensible gases usually is air dissolved in the condenser water. Figure 11-114 shows the dissolved-gas content of freshwater and seawater, calculated as equivalent air. The lower curve for seawater includes only dissolved oxygen and nitrogen. The upper curve includes carbon dioxide that can be evolved by complete breakdown of bicarbonate in seawater. Breakdown of bicarbonates is usually not appreciable in a condenser but may go almost to completion in a seawater evaporator. The large increase in gas volume as a result of possible bicarbonate breakdown is illustrative of the uncertainties involved in sizing vacuum systems.

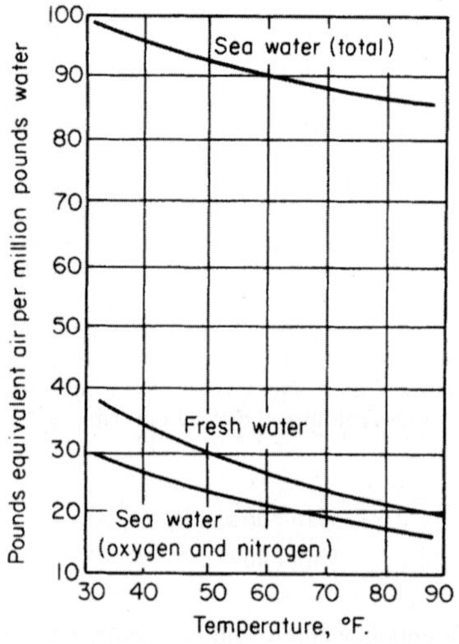

FIG. 11-114 Gas content of water saturated at atmospheric pressure. °C = 5/9 (°F − 32).

By far the largest load on the vacuum pump is water vapor carried with the noncondensible gases. Standard power-plant practice assumes that the mixture leaving a surface condenser will have been cooled 4.2°C (7.5°F) below the saturation temperature of the vapor. This usually corresponds to about 2.5 kg water vapor/kg air. One advantage of the countercurrent barometric condenser is that it can cool the gases almost to the temperature of the incoming water and thus reduce the amount of water vapor carried with the air.

In some cases, as with pulp mill liquors, the evaporator vapors contain constituents more volatile than water, such as methanol and sulfur compounds. Special precautions may be necessary to minimize the effects of these compounds on heat transfer, corrosion, and condensate quality. They can include removing most of the condensate countercurrent to the vapor entering an evaporator heating element, channeling vapor and condensate flow to concentrate most of the "foul" constituents into the last fraction of vapor condensed (and keeping this condensate separate from the rest of the condensate), and flashing the warm evaporator feed to a lower pressure to remove much of the foul constituents in only a small amount of flash vapor. In all such cases, special care is needed to properly channel vapor flow past the heating surfaces so there is a positive flow from steam inlet to vent outlet with no pockets, where foul constituents or noncondensibles can accumulate.

Salt Removal When an evaporator is used to make a crystalline product, a number of means are available for concentrating and removing the salt from the system. The simplest is to provide settling space in the evaporator itself. This is done in the types shown in Fig. 11-110*b*, *c*, and *e* by providing a relatively quiescent zone in which the salt can settle. Sufficiently high slurry densities can usually be achieved in this manner to reach the limit of pumpability. The evaporators are usually placed above barometric height so that the slurry can be discharged intermittently on a short time cycle. This permits the use of high velocities in large lines that have little tendency to plug.

If the amount of salts crystallized is on the order of 1 ton/h or less, a salt trap may be used. This is simply a receiver that is connected to the bottom of the evaporator and is closed off from the evaporator periodically for emptying. Such traps are useful when insufficient headroom is available for gravity removal of the solids. However, traps require a great deal of labor, give frequent trouble with the shutoff valves, and also can upset evaporator operation completely if a trap is reconnected to the evaporator without first displacing all air with feed liquor.

EVAPORATOR OPERATION

The two principal elements of evaporator control are *evaporation rate* and *product concentration.* Evaporation rate in single- and multiple-effect evaporators is usually achieved by steam flow control. Conventional-control instrumentation is used (see Sec. 22), with the added precaution that the pressure drop across the meter and control valve, which reduces temperature difference available for heat transfer, not be excessive when maximum capacity is desired. Capacity control of thermocompression evaporators depends on the type of compressor; positive-displacement compressors can utilize speed control or variations in operating pressure level. Centrifugal machines normally utilize adjustable inlet-guide vanes. Steam jets may have an adjustable spindle in the high-pressure orifice or be arranged as multiple jets that can individually be cut out of the system.

Product concentration can be controlled by any property of the solution that can be measured with the requisite accuracy and reliability. The preferred method is to impose control on the rate of product withdrawal. Feed rates to the evaporator effects are then controlled by their levels. When level control is impossible, as with the rising-film LTV, product concentration is used to control the feed rate—frequently by ratioing of feed to steam with the ratio reset by product concentration, sometimes also by feed concentration. Other controls that may be needed include vacuum control of the last effect (usually by air bleed to the condenser) and temperature-level control of thermocompression evaporators (usually by adding makeup heat or by venting excess vapor, or both, as feed or weather conditions vary). For more control details, see N. Lior, ed., *Measurement and Control in Water Desalination,* Elsevier Science Publ. Co., New York, 1986, pp. 241–305.

Control of an evaporator requires more than proper instrumentation. Operator logs should reflect changes in basic characteristics, as by use of *pseudo heat-transfer* coefficients, which can detect obstructions to heat flow, hence to capacity. These are merely the ratio of any convenient measure of heat flow to the temperature drop across each effect. *Dilution* by wash and seal water should be monitored since it absorbs evaporative capacity. Detailed tests, routine measurements, and operating problems are covered more fully in AICE, *Testing Procedure for Evaporators* and by Standiford [*Chem. Eng. Prog.* **58**(11): 80 (1962)].

By far the best application of computers to evaporators is for working up operators' data into the basic performance parameters such as heat-transfer coefficients, steam economy, and dilution.

Section 12

Psychrometry, Evaporative Cooling, and Solids Drying

John P. Hecht, Ph.D. *Technical Section Head, Drying and Particle Processing, The Procter & Gamble Company; Member, American Institute of Chemical Engineers (Section Editor, Evaporative Cooling, Solids-Drying Fundamentals, Drying Equipment)*

Wayne E. Beimesch, Ph.D. *Technical Associate Director (Retired), Corporate Engineering, The Procter & Gamble Company (Drying Equipment, Operation and Troubleshooting)*

Karin Nordström Dyvelkov, Ph.D. *GEA Process Engineering A/S Denmark (Drying Equipment, Fluidized Bed Dryers, Spray Dryers)*

Ian C. Kemp, M.A. (Cantab) *Scientific Leader, GlaxoSmithKline; Fellow, Institution of Chemical Engineers; Associate Member, Institution of Mechanical Engineers (Psychrometry, Solids-Drying Fundamentals, Freeze Dryers)*

Tim Langrish, D. Phil. *School of Chemical and Biomolecular Engineering, The University of Sydney, Australia (Solids-Drying Fundamentals, Cascading Rotary Dryers)*

(Francis) Lee Smith, Ph.D., M. Eng. *Principal, Wilcrest Consulting Associates, LLC, Katy, Texas; Partner and General Manager, Albutran USA, LLC, Katy, Texas (Evaporative Cooling)*

Jason A. Stamper, M. Eng. *Technology Leader, Drying and Particle Processing, The Procter & Gamble Company; Member, Institute for Liquid Atomization and Spray Systems (Drying Equipment, Fluidized Bed Dryers, Spray Dryers)*

PSYCHROMETRY

EVAPORATIVE COOLING

Nomenclature and Units

Symbol	Definition	SI units	U.S. Customary System units
A	Area	m^2	ft^2
a_w	Water activity	—	—
a_w^{vapor}	Activity of water in vapor phase	—	—
a_w^{solid}	Activity of water in solid phase	—	—
C	Concentration	kg/m^3	lb/ft^3
C_P	Specific heat capacity at constant pressure	$J/(kg \cdot K)$	$Btu/(lb \cdot °F)$
C_w	Concentration of water in solid	kg/m^3	lb/ft^3
$D(u)$	Diffusion coefficient of water in a solid or liquid as a function of moisture content	m^2/s	ft^2/s
$D_{water/air}$	Diffusion coefficient of water in air	m^2/s	ft^2/s
D	Diameter (particle)	m	in
E	Power	W	Btu/h
F	Solids or liquid mass flow rate	kg/s	lb/h
F	Mass flux of water at surface	$kg/(m^2 \cdot s)$	$lb/(ft^2 \cdot s)$
F	Relative drying rate	—	—
G	Gas mass flow rate	kg/s	lb/h
g	Acceleration due to gravity, 9.81 m/s^2 (32.2 ft/s^2)	m/s^2	ft/s^2
H	Enthalpy of a pure substance	J/kg	Btu/lb
ΔH_{vap}	Heat of vaporization	J/kg	Btu/lb
h	Heat-transfer coefficient	$W/(m^2 \cdot K)$	$Btu/(ft^2 \cdot h \cdot °F)$
I	Humid enthalpy (dry substance and associated moisture or vapor)	J/kg	Btu/lb
J	Mass flux (of evaporating liquid)	$kg/(m^2 \cdot s)$	$lb/(ft^2 \cdot h)$
K	Mass-transfer coefficient	m/s	$lb/(ft^2 \cdot h \cdot atm)$
k_{air}	Thermal conductivity of air	$W/(m \cdot K)$	$Btu/(ft \cdot h \cdot °F)$
k_c	Mass-transfer coefficient for a concentration driving force	m/s	ft^2/s
k_p	Mass-transfer coefficient for a partial pressure driving force	$kg/(m^2 \cdot s)$	$lbm/(ft^3 \cdot s)$
L	Length; length of drying layer	m	ft
M	Molecular weight	kg/mol	lb/mol
m	Mass	kg	lb
m_{solids}	Mass of dry solids	kg	lb
N	Specific drying rate ($-dX/dt$)	1/s	1/s
N	Rotational speed (drum, impeller, etc.)	1/s	rpm
P	Total pressure	$kg/(m \cdot s^2)$	lb/in^2
P_w^{bulk}	Partial pressure of water vapor in air far from the drying material	$kg/(m \cdot s^2)$	lb/in^2
$P_w^{surface}$	Partial pressure of water vapor in air at the solid interface	$kg/(m \cdot s^2)$	lb/in^2

Symbol	Definition	SI units	U.S. Customary System units
P	Partial pressure/vapor pressure of component	kg/(m·s^2)	lb/in^2
p^{sat}_{pure}	Pure component vapor pressure	kg/(m·s^2)	lb/in^2
p_w, air	Partial pressure of water vapor in air	kg/(m·s^2)	lb/in^2
Q	Heat-transfer rate	W	Btu/h
Q	Heat flux	W/m^2	Btu/(ft^2·h)
R	Universal gas constant, 8314 J/(kmol·K)	J/(mol·K)	Btu/(mol·°F)
r	Droplet radius	m	ft
r	Radius; radial coordinate	m	ft
RH	Relative humidity	—	—
S	Percentage saturation	—	—
S	Solid-fixed coordinate	Depends on geometry	
T	Absolute temperature	K	°R
T, t	Temperature	°C	°F
T	Time	s	h
U	Velocity	m/s	ft/s
U	Mass of water/mass of dry solid	—	—
V	Volume	m^3	ft^3
V	Air velocity	m/s	ft/s
υ	Specific volume	m^3/kg	ft^3/lb
$\upsilon_{droplet}$	Droplet volume	m^3	ft^3
W	Wet-basis moisture content	—	—
$w_{avg\ dry\text{-}basis}$	Average wet-basis moisture content	—	—
X	Solids moisture content (dry basis)	—	—
Y	Mass ratio	—	—
Z	Distance coordinate	m	ft
	Dimensionless groups		
Ar	Archimedes number, $(gd_P{}^3\rho g/\mu^2)(\rho_P - \rho_G)$	—	—
Bi	Biot number, hL/κ	—	—
Gr	Grashof number, $L^3\rho^2\beta g\,\Delta T/\mu^2$	—	—
Nu	Nusselt number, hd_P/κ	—	—
Pr	Prandtl number, $\mu C_P/\kappa$	—	—

Symbol	Definition	SI units	U.S. Customary System units
Re	Reynolds number, $\rho\, d_P\, U/\mu$	—	—
Sc	Schmidt number, $\mu/\rho D$	—	—
Sh	Sherwood number, $k_Y\, d_P/D$	—	—
Le	Lewis = Sc/Pr	—	—
	Greek letters		
α	Slope	—	—
β	Psychrometric ratio	—	—
ε	Voidage (void fraction)	—	—
ζ	Dimensionless distance	—	—
η	Efficiency	—	—
θ	Dimensionless time	—	—
κ	Thermal conductivity	W/(m·K)	Btu/(ft·h·°F)
λ	Latent heat of evaporation	J/kg	Btu/lb
μ	Absolute viscosity	kg/(m·s)	lb/(ft·s)
μ_{air}	Viscosity of air	kg/(m·s)	lb/(ft·s)
ρ	Density	kg/m^3	lb/ft^3
ρ_{air}	Air density	kg/m^3	lb/ft^3
ρ_s	Mass concentration of solids	kg/m^3	lb/ft^3
ρ^o_s	Density of dry solids	kg/m^3	lb/ft^3
ρ^o_w	Density of pure water	kg/m^3	lb/ft^3
τ	Residence time of solids	s	h
Φ	Characteristic (dimensionless) moisture content	—	—
Ψ	Relative humidity	%	%
	Subscripts		
g	gas		
v	vapor		
as	adiabatic saturation		
o	reference		
s	saturation		
wb	wet bulb		
dew	dew point		
liq	liquid		

PSYCHROMETRY

General References: *ASHRAE 2002 Handbook: Fundamentals,* SI Edition, American Society of Heating, Refrigeration, and Air-Conditioning Engineers, Atlanta, Ga., 2002, chap. 6. "Psychometrics," chap. 19.2, "Sorbents and Desiccants." *Aspen Process Manual* (Internet knowledge base), Aspen Technology, 2000 onward. *Humidity and Dewpoint,* British Standard BS 1339 (rev.). Humidity and dew point, Pt. 1 (2002); Terms, definitions and formulae, Pt. 2 (2005); Psychrometric calculations and tables (including spreadsheet), Pt. 3 (2004); Guide to humidity measurement. British Standards Institution, Gunnersbury, UK. Cook and DuMont, *Process Drying Practice,* McGraw-Hill, New York, 1991, chap. 6. Keey, *Drying of Loose and Particulate Materials,* Hemisphere, New York, 1992. Poling, Prausnitz, and O'Connell, *The Properties of Gases and Liquids,* 5th ed., McGraw-Hill, New York, 2000. Earlier editions: 1st/2d eds., Reid and Sherwood (1958/1966); 3d ed., Reid, Prausnitz, and Sherwood (1977); 4th ed., Reid, Prausnitz, and Poling (1986). Soininen, "A Perspectively Transformed Psychrometric Chart and Its Application to Drying Calculations," *Drying Technol.* **4**(2): 295–305 (1986). Sonntag, "Important New Values of the Physical Constants of 1986, Vapor Pressure Formulations Based on the ITS-90, and Psychrometer Formulae," *Zeitschrift für Meteorologie* **40**(5): 340–344 (1990). Treybal, *Mass-Transfer Operations,* 3d ed., McGraw-Hill, New York, 1980. Wexler, *Humidity and Moisture,* vol. 1, Reinhold, New York, 1965.

Psychrometry is concerned with the determination of the properties of gas-vapor mixtures. These are important in calculations for humidification and dehumidification, particularly in cooling towers, air-conditioning systems, and dryers. The first two cases involve the air–water vapor system at near-ambient conditions, but dryers normally operate at elevated temperatures and may also use elevated or subatmospheric pressures and other gas-solvent systems.

TERMINOLOGY

Terminology and nomenclature pertinent to psychrometry are given below. There is often considerable confusion between the dry and wet basis, and between mass, molar, and volumetric quantities, in both definitions and calculations. Dry- and wet-basis humidities are similar at ambient conditions but can differ significantly at elevated humidities, e.g., in dryer exhaust streams. Complete interconversion formulas between four key humidity parameters are given in Table 12-1 for the air-water system and in Table 12-2 for a general gas-vapor system.

Definitions related to humidity, vapor pressure, saturation, and volume are as follows; the most useful are absolute humidity, vapor pressure, and relative humidity.

Absolute Humidity Y Mass of water (or solvent) vapor carried by unit mass of dry air (or other carrier gas). It is also known as the *humidity ratio, mixing ratio, mass ratio,* or *dry-basis humidity.* Preferred units are lb/lb or kg/kg, but g/kg and gr/lb are often used, as are ppm$_w$ and ppb$_w$ (parts per million/billion by weight); ppm$_w$ = $10^6 Y$, ppb$_w$ = $10^9 Y$.

Specific Humidity Y_W Mass of vapor per unit mass of gas-vapor mixture. Also known as *mass fraction* or *wet-basis humidity,* and is used much more rarely than dry-basis absolute humidity. $Y_W = Y/(1 + Y)$; $Y = Y_W/(1 - Y_W)$.

Mole Ratio z Number of moles of vapor per mole of gas (dry basis), mol/mol; $z = (M_g/M_\upsilon)Y$, where M_υ = molecular weight of vapor and M_g = molecular weight of gas. It may also be expressed as ppm$_v$ and ppb$_v$ (parts per million/billion by volume); ppm$_v$ = $10^6\, z$, ppb$_v$ = $10^9\, z$.

TABLE 12-1 Interconversion Formulas for Air-Water System, to Three Significant Figures

T = temperature in **kelvins** (K); P = total pressure in **pascals** (Pa or N/m^2)

Convert from:	Y (or ppm$_w$)*	y	p	Y_υ
Convert to: Absolute humidity (mixing ratio) Y (kg · kg^{-1})	1	$Y=\frac{0.622Y}{1-Y}$	$Y=\frac{0.622p}{P-p}$	$Y=\frac{0.622}{0.002167P/(Y_\upsilon T)-1}$
Mole fraction y (mol · mol^{-1})	$y=\frac{Y}{0.622+Y}$	1	$y=\frac{p}{P}$	$y=\frac{461.5Y_\upsilon T}{P}$
Vapor pressure p (Pa)	$p=\frac{PY}{0.622+Y}$	$p=yP$	1	$p=461.5Y_\upsilon T$
Volumetric humidity Y_υ (kg · m^{-3})	$Y_\upsilon=\frac{0.002167PY}{(0.622+Y)T}$	$Y_\upsilon=\frac{0.002167yP}{T}$	$Y_\upsilon=\frac{0.002167p}{T}$	1

Mole Fraction y Number of moles of vapor per mole of gas-vapor mixture (wet basis); $y=z/(1+z)$; $z=y/(1-y)$. If a mixture contains m_υ kg and n_υ mol of vapor (e.g., water) and m_g kg and n_g mol of noncondensible gas (e.g., air), with $m_\upsilon=n_\upsilon M_\upsilon$ and $m_g=n_g M_g$, then the four quantities above are defined by

$$Y=\frac{m_\upsilon}{m_g}\qquad Y_w=\frac{m_\upsilon}{m_g+m_\upsilon}\qquad z=\frac{n_\upsilon}{n_g}\qquad y=\frac{n_\upsilon}{n_g+n_\upsilon}$$

Volumetric Humidity Y_υ Mass of vapor per unit volume of gas-vapor mixture. It is sometimes, confusingly, called the absolute humidity, but it is really a vapor concentration; preferred units are kg/m^3 or lb/ft^3, but g/m^3 and gr/ft^3 are also used. It is inconvenient for calculations because it depends on temperature and pressure and on the units system; absolute humidity Y is always preferable for heat and mass balances. It is proportional to the specific humidity (wet basis); $Y_\upsilon=Y_W\rho_g$, where ρ_g is the *humid gas density* (mass of gas-vapor mixture per unit volume, wet basis). Also

$$Y_\upsilon=\frac{M_\upsilon P n_\upsilon}{RT(n_g+n_\upsilon)}$$

Vapor Pressure p Partial pressure of vapor in gas-vapor mixture, which is proportional to the mole fraction of vapor; $p=yP$, where P = total pressure, in the same units as p (Pa, N/m^2, bar, atm, or psi). Hence

$$p=\frac{n_\upsilon}{n_g+n_\upsilon}P$$

Saturation Vapor Pressure p_s Pressure exerted by pure vapor at a given temperature. When the vapor partial pressure p in the gas-vapor mixture at a given temperature equals the saturation vapor pressure p_s at the same temperature, the air is *saturated* and the absolute humidity is designated the *saturation humidity* Y_s.

Relative Humidity RH or Ψ The partial pressure of vapor divided by the saturation vapor pressure at the given temperature, usually expressed as a percentage. Thus RH = $100p/p_s$.

Percentage Absolute Humidity (Percentage Saturation) S Ratio of absolute humidity to saturation humidity, given by $S=100Y/Y_s=100p(P-p_s)/[p_s(P-p)]$. It is used much less commonly than relative humidity.

Dew Point T_{dew}, or Saturation Temperature The temperature at which a given mixture of water vapor and air becomes saturated on cooling; i.e., the temperature at which water exerts a vapor pressure equal to the partial pressure of water vapor in the given mixture.

Humid Volume υ Volume in cubic meters (cubic feet) of 1 kg (1 lb) of dry air and the water vapor it contains.

Saturated Volume v_s Humid volume when the air is saturated.

Terms related to heat balances are as follows:

Humid Heat C_s Heat capacity of unit mass of dry air and the moisture it contains. $C_s=C_{Pg}+C_{Pv}Y$, where C_{Pg} and C_{Pv} are the heat capacities of dry air and water vapor, respectively, and both are assumed constant. For approximate engineering calculations at near-ambient temperatures, in SI units, $C_s=1+1.9Y$ kJ/(kg · K) and in USCS units, $C_s=0.24+0.45Y$ (Btu/(lb · °F).

Humid Enthalpy H Heat content at a given temperature T of unit mass of dry air and the moisture it contains, relative to a datum temperature T_0, usually 0°C. As water is liquid at 0°C, the humid enthalpy also contains a term for the latent heat of water. If heat capacity is invariant with temperature, $H=(C_{Pg}+C_{Pv}Y)(T-T_0)+\lambda_0 Y$, where λ_0 is the latent heat of water at 0°C, 2501 kJ/kg (1075 Btu/lb). In practice, for accurate calculations, it is often easier to obtain the vapor enthalpy H_υ from steam tables, when $H=H_g+H_v=C_{Pg}T+H_\upsilon$.

Adiabatic Saturation Temperature T_{as} Final temperature reached by a small quantity of vapor-gas mixture into which water is evaporating. It is sometimes called the *thermodynamic wet-bulb temperature.*

Wet-Bulb Temperature T_{wb} Dynamic equilibrium temperature attained by a liquid surface from which water is evaporating into a flowing airstream when the rate of heat transfer to the surface by convection equals the rate of mass transfer away from the surface. It is very close to the adiabatic saturation temperature for the air-water system, but not for most other vapor-gas systems; see later.

CALCULATION FORMULAS

Table 12-1 gives formulas for conversion between absolute humidity, mole fraction, vapor pressure, and volumetric humidity for the air-water system, and Table 12-2 does likewise for a general gas-vapor system. Where relationships are not included in the definitions, they are given below.

In USCS units, the formulas are the same except for the volumetric humidity Y_υ. Because of the danger of confusion with pressure units, it is recommended that in both Tables 12-1 and 12-2, Y_υ be calculated in SI units and then converted.

Volumetric humidity is also related to absolute humidity and humid gas density by

$$Y_\upsilon=Y_W\rho_g=\frac{Y}{1+Y}\rho_g \tag{12-1}$$

TABLE 12-2 Interconversion Formulas for a General Gas-Vapor System

M_g, M_υ = molal mass of gas and vapor, respectively; R = 8314 J/(kmol · K); T = temperature in **kelvins** (K); P = total pressure in **pascals** (Pa or N/m^2)

Convert from:	Y (or ppm$_w$)*	y	p	Y_υ
Convert to: Absolute humidity (mixing ratio) Y (kg · kg^{-1})	1	$Y=\frac{M_\upsilon y}{M_g(1-Y)}$	$Y=\frac{pM_\upsilon}{(P-p)M_g}$	$Y=\frac{M_\upsilon}{M_g(PM_\upsilon/(Y_\upsilon RT)-1}$
Mole fraction y (mol · mol^{-1})	$y=\frac{Y}{M_\upsilon/M_g+Y}$	1	$y=\frac{p}{P}$	$y=\frac{Y_\upsilon RT}{PM_\upsilon}$
Vapor pressure p (Pa)	$p=\frac{PY}{M_\upsilon/M_g+Y}$	$p=yP$	1	$p=\frac{Y_\upsilon RT}{M_\upsilon}$
Volumetric humidity Y_υ (kg · m^{-3})	$Y_\upsilon=\frac{M_\upsilon}{RT}\frac{PY}{M_\upsilon/M_g+Y}$	$Y_\upsilon=\frac{M_\upsilon yP}{RT}$	$Y_\upsilon=\frac{M_\upsilon p}{RT}$	1

Two further useful formulas are as follows:

Parameter	General vapor-gas system	Air-water system, SI units, to 3 significant figures	Eq. no.
Density of humid gas (moist air) ρ_g, kg/m^3	$\rho_g = \frac{M_g}{RT}\left(P - p + \frac{M_\upsilon}{M_g}p\right)$	$\rho_g = \frac{P - 0.378p}{287.1T}$	(12-2)
Humid volume υ per unit mass of dry air, m^3/kg	$\upsilon = \frac{RT}{M_g(P-p)} = \frac{RT}{P} \times \left(\frac{1}{M_g} + \frac{Y}{M_\upsilon}\right)$	$\upsilon = \frac{461.5T}{P}(0.622 + Y)$	(12-3)

From Eq. (12-2), the density of dry air at 0°C (273.15 K) and 1 atm (101,325 Pa) is 1.292 kg/m^3 (0.08065 lb/ft^3). Note that the density of moist air is always lower than that of dry air.

Equation (12-3) gives the humid volume of dry air at 0°C (273.15 K) and 1 atm as 0.774 m^3/kg (12.4 ft^3/lb). For moist air, the humid volume is not the reciprocal of humid gas density; $\upsilon = (1 + Y)/\rho_g$.

The *saturation vapor pressure* of water is given by Sonntag (1990) in Pa (N/m^2) at absolute temperature T (K).

Over water:

$$\ln p_s = -6096.9385T^{-1} + 21.2409642 - 2.711193 \times 10^{-2}T + 1.673952 \times 10^{-5}T^2 + 2.433502 \ln T \qquad (12\text{-}4a)$$

Over ice:

$$\ln p_s = -6024.5282T^{-1} + 29.32707 + 1.0613868 \times 10^{-2}T - 1.3198825 \times 10^{-5}T^2 - 0.49382577 \ln T \qquad (12\text{-}4b)$$

Simpler equations for saturation vapor pressure are the Antoine equation and Magnus formula. These are slightly less accurate, but easier to calculate and also easily reversible to give T in terms of p. For the Antoine equation, given below, coefficients for numerous other solvent-gas systems are given in Poling, Prausnitz, and O'Connell, *The Properties of Gases and Liquids*, 5th ed., McGraw-Hill, New York, 2000.

$$\ln p_s = C_0 - \frac{C_1}{T - C_2} \qquad T = \frac{C_1}{C_0 - \ln ps} + C_2 \qquad (12\text{-}5)$$

Values for Antoine coefficients for the air-water system are given in Table 12-3. The standard values give vapor pressure within 0.1 percent of steam tables over the range 50 to 100°C, but an error of nearly 3 percent at 0°C. The alternative coefficients give a close fit at 0 and 100°C and an error of less than 1.2 percent over the intervening range.

The Sonntag equation strictly only applies to water vapor with no other gases present (i.e., in a partial vacuum). The vapor pressure of a gas mixture, e.g., water vapor in air, is given by multiplying the pure liquid vapor pressure by an enhancement factor f, for which various equations are available (see British Standard BS 1339 Part 1, 2002). However, the correction is typically less than 0.5 percent, except at elevated pressures, and it is therefore usually neglected for engineering calculations.

RELATIONSHIP BETWEEN WET-BULB AND ADIABATIC SATURATION TEMPERATURES

If a stream of air is intimately mixed with a quantity of water in an adiabatic system, the temperature of the air will drop and its humidity will increase. If the equilibration time or the number of transfer units approaches infinity, the air-water mixture will reach saturation. The *adiabatic saturation temperature* T_{as} is given by a heat balance between the initial unsaturated vapor-gas mixture and the final saturated mixture at thermal equilibrium:

$$C_s(T - T_{as}) = \lambda_{as}(Y_{as} - Y) \qquad (12\text{-}6)$$

This equation has to be reversed and solved iteratively to obtain Y_{as} (absolute humidity at adiabatic saturation) and hence T_{as} (the calculation is divergent in the opposite direction). Approximate direct formulas are available from various sources, e.g., British Standard BS 1339 (2002) and Liley [*IJMEE* **21**(2), 1993]. The latent heat of evaporation evaluated at the adiabatic saturation temperature is λ_{as}, which may be obtained from steam tables; humid heat C_s is evaluated at initial humidity Y. On a psychrometric chart, the adiabatic saturation process almost exactly follows a *constant-enthalpy line*, as the sensible heat given up by the gas-vapor mixture exactly balances the latent heat of the liquid that evaporates back into the mixture. The only difference is due to the sensible heat added to the water to take it from the datum temperature to T_{as}. The adiabatic saturation line differs from the constant-enthalpy line as follows, where C_{PL} is the specific heat capacity of the liquid:

$$H_{as} - H = C_{PL}T_{as}(Y_{as} - Y) \qquad (12\text{-}7)$$

Equation (12-7) is useful for calculating the adiabatic saturation line for a given T_{as} and gives an alternative iterative method for finding T_{as}, given T and Y; compared with Eq. (12-6), it is slightly more accurate and converges faster, but the calculation is more cumbersome.

The *wet-bulb temperature* is the temperature attained by a fully wetted surface, such as the wick of a wet-bulb thermometer or a droplet or wet particle undergoing drying, in contact with a flowing unsaturated gas stream. It is regulated by the rates of vapor-phase heat and mass transfer to and from the wet bulb. Assuming mass transfer is controlled by diffusion effects and heat transfer is purely convective,

$$h(T - T_{wb}) = k_y \lambda_{wb}(Y_{wb} - Y) \qquad (12\text{-}8)$$

where k_y is the corrected mass-transfer coefficient [kg/(m$^2 \cdot$ s)], h is the heat-transfer coefficient [kW/(m$^2 \cdot$ K)], Y_{wb} is the saturation mixing ratio at t_{wb}, and λ_{wb} is the latent heat (kJ/kg) evaluated at T_{wb}. Again, this equation must be solved iteratively to obtain T_{wb} and Y_{wb}.

In practice, for any practical psychrometer or wetted droplet or particle, there is significant extra heat transfer from radiation. For an Assmann psychrometer at near-ambient conditions, this is approximately 10 percent. This means that any measured real value of T_{wb} is slightly higher than the "pure convective" value in the definition. It is often more convenient to obtain wet-bulb conditions from adiabatic saturation conditions (which are much easier to calculate) by the following formula:

$$\frac{T - T_{wb}}{Y_{wb} - Y} = \frac{T - T_{as}}{Y_{as} - Y}\beta \qquad (12\text{-}9)$$

where the psychrometric ratio $\beta = \overline{C_s}k_y/h$ and $\overline{C_s}$ is the mean value of the humid heat over the range from T_{as} to T.

The advantage of using β is that it is approximately constant over normal ranges of temperature and pressure for any given pair of vapor and gas values. This avoids having to estimate values of heat- and mass-transfer coefficients h and k_y from uncertain correlations. For the air-water system, considering convective heat transfer alone, $\beta \sim 1.1$. In practice, there is an additional contribution from radiation, and β is very close to 1. As a result, the wet-bulb and adiabatic saturation temperatures differ by less than 1°C for the air-water system at near-ambient conditions (0 to 100°C, $Y < 0.1$ kg/kg) and can be taken as equal for normal calculation purposes. Indeed, typically the T_{wb} measured by a practical psychrometer or at a wetted solid surface is closer to T_{as} than to the "pure convective" value of T_{wb}.

However, for nearly all other vapor-gas systems, particularly for organic solvents, $\beta < 1$, and hence $T_{wb} > T_{as}$. This is illustrated in Fig. 12-5. The surface (wet-bulb) temperature can change as drying progresses, whereas in the air-water system it stays constant. For these systems the psychrometric ratio may be obtained by determining h/k_y from heat- and mass-transfer analogies such as the Chilton-Colburn analogy. The basic form of the equation is

$$\beta = \left(\frac{Sc}{Pr}\right)^n = Le^{-n} \qquad (12\text{-}10)$$

TABLE 12-3 Alternative Set of Values for Antoine Coefficients for Air-Water Systems

		C_0	C_1	C_2		C_0	C_1	C_2
Standard values	p in Pa	23.1963	3816.44	46.13 K	p in mmHg	18.3036	3816.44	46.13 K
Alternative values	p in Pa	23.19	3830	44.83 K	p in mmHg	18.3	3830	44.87 K

where Sc is the Schmidt number for mass-transfer properties, Pr is the Prandtl number for heat-transfer properties, and Le is the Lewis number $\kappa/(C_s\rho_g D)$, where κ is the gas thermal conductivity and D is the diffusion coefficient for the vapor through the gas. Experimental and theoretical values of the exponent n range from 0.56 [Bedingfield and Drew, *Ind. Eng. Chem.* **42:** 1164 (1950)] to $\frac{2}{3}$ [Chilton and Colburn, *Ind. Eng. Chem.* **26:** 1183 (1934)]. A detailed discussion is given by Keey (1992). Values of β for any system can be estimated from the specific heats, diffusion coefficients, and other data given in Sec. 2. See the example below.

For calculation of wet-bulb (and adiabatic saturation) conditions, the most commonly used formula in industry is the *psychrometer equation.* This is a simple linear formula that gives vapor pressure directly if the wet-bulb temperature is known, and it is therefore ideal for calculating humidity from a wet-bulb measurement using a psychrometer, although the calculation of wet-bulb temperature from humidity still requires an iteration

$$p = p_{wb} - AP(T - T_{wb}) \tag{12-11}$$

where A is the psychrometer coefficient. For the air-water system, the following formulas based on equations given by Sonntag [*Zeitschrift für Meteorologie* **40**(5): 340–344 (1990)] may be used to give A for T_{wb} up to 30°C; they are based on extensive experimental data for Assmann psychrometers.

Over water (wet-bulb temperature):

$$A = 6.5 \times 10^{-4}(1 + 0.000944T_{wb}) \tag{12-12a}$$

Over ice (ice-bulb temperature):

$$A_i = 5.72 \times 10^{-4} \tag{12-12b}$$

For other vapor-gas systems, A is given by

$$A = \frac{M_g C_s}{M_v \beta \lambda_{wb}} \tag{12-13}$$

Here β is the psychrometric coefficient for the system. As a cross-check, for the air-water system at 20°C wet-bulb temperature, 50°C dry-bulb temperature, and absolute humidity 0.002 kg/kg, $C_s = 1.006 + 1.9 \times 0.002 = 1.01$ kJ/(kg·K) and $\lambda_{wb} = 2454$ kJ/kg. Since $M_g = 28.97$ kg/kmol and $M_v = 18.02$ kg/kmol, Eq. (12-12) gives A as $6.617 \times 10^{-4}/\beta$, compared with Sonntag's value of 6.653×10^{-4} at this temperature, giving a value for the psychrometric coefficient β of 0.995; that is, $\beta \approx 1$, as expected for the air-water system.

PSYCHROMETRIC CHARTS

Psychrometric charts are plots of humidity, temperature, enthalpy, and other useful parameters of a gas-vapor mixture. They are helpful for rapid estimates of conditions and for visualization of process operations such as humidification and drying. They apply to a given system at a given pressure, the most common, of course, being air-water at atmospheric pressure. There are four types, of which the Grosvenor and Mollier types are most widely used.

The *Grosvenor chart* plots temperature (abscissa) versus humidity (ordinate). Standard charts produced by ASHRAE and other groups, or by computer programs, are usually of this type. The saturation line is a curve from bottom left to top right, and curves for constant relative humidity are approximately parallel to this. Lines from top left to bottom right may be of either constant wet-bulb temperature or constant enthalpy, depending on the chart. The two are not quite identical, so if only one is shown, correction factors are required for the other parameter. Examples are shown in Figs. 12-1 (SI units) and 12-2 (USCS units). An additional chart for a wider temperature range in USCS units is given in Perry's 8th Edition (Fig. 12-2*b*).

The *Bowen chart* is a plot of enthalpy (abscissa) versus humidity (ordinate). It is convenient to be able to read enthalpy directly, especially for near-adiabatic convective drying where the operating line approximately follows a line of constant enthalpy. However, it is very difficult to read accurately because the key information is compressed in a narrow band near the saturation line. See Cook and DuMont, *Process Drying Practice,* McGraw-Hill, New York, 1991, chap. 6.

The *Mollier chart* plots humidity (abscissa) versus enthalpy (lines sloping diagonally from top left to bottom right). Lines of constant temperature are shallow curves at a small slope to the horizontal. The chart is nonorthogonal (no horizontal lines) and hence a little difficult to plot and interpret initially. However, the area of greatest interest is expanded, and they are therefore easy to read accurately. They tend to cover a wider temperature range than Grosvenor charts, so are useful for dryer calculations. The slope of the enthalpy lines is normally $-1/\lambda$, where λ is the latent heat of evaporation. Adiabatic saturation lines are not quite parallel to constant-enthalpy lines and are slightly curved; the deviation increases as humidity increases. Figure 12-3 shows an example.

The *Salen-Soininen* perspective transformed *chart* is a triangular plot. It is tricky to plot and read, but covers a much wider range of humidity than do the other types of chart (up to 2 kg/kg) and is thus very effective for high-humidity mixtures and calculations near the boiling point, e.g., in pulp and paper drying. See Soininen, *Drying Technol.* **4**(2): 295–305 (1986).

Figure 12-4 shows a psychrometric chart for combustion products in air. The thermodynamic properties of moist air are given in Table 12-1. Figure 12-4 shows a number of useful additional relationships, e.g., specific volume and latent heat variation with temperature. Accurate figures should always be obtained from physical properties tables or by calculation using the formulas given earlier, and these charts should only be used as a quick check for verification. Figure 12-5 shows a psychrometric chart for an organic solvent system.

In the past, psychrometric charts have been used to perform quite precise calculations. To do this, additive corrections are often required for enthalpy of added water or ice, and for variations in barometric pressure from the standard level (101,325 Pa, 14.696 lbf/in^2, 760 mmHg, 29.921 inHg). It is preferable to use formulas, which give an accurate figure at any set of conditions. Psychrometric charts and tables can be used as a rough cross-check that the result has been calculated correctly. Table 12-4 gives values of saturation humidity, specific volume, enthalpy, and entropy of saturated moist air at selected conditions. Below the freezing point, these become virtually identical to the values for dry air, as saturation humidity is very low. For pressure corrections, an altitude increase of approximately 275 m (900 ft) gives a pressure decrease of 0.034 bar (1 inHg). For a recorded wet-bulb temperature of 10°C (50°F), this gives an increase in humidity of 0.00027 kg/kg (1.9 gr/lb) and the enthalpy increases by 0.68 kJ/kg (0.29 Btu/lb). This correction increases roughly proportionately for further changes in pressure, but climbs sharply as wet-bulb temperature is increased; when T_{wb} reaches 38°C (100°F), $\Delta Y = 0.0016$ kg/kg (11.2 gr/lb) and $\Delta H = 4.12$ kJ/kg (1.77 Btu/lb). Equivalent, more detailed tables in SI units can be found in the ASHRAE Handbook.

Examples Illustrating Use of Psychrometric Charts In these examples the following nomenclature is used:

t = dry-bulb temperature, °F
t_w = wet-bulb temperature, °F
t_d = dew point temperature, °F
H = moisture content, lb water/lb dry air
ΔH = moisture added to or rejected from airstream, lb water/lb dry air
h' = enthalpy at saturation, Btu/lb dry air
D = enthalpy deviation, Btu/lb dry air
$h = h' + D$ = true enthalpy, Btu/lb dry air
h_w = enthalpy of water added to or rejected from system, Btu/lb dry air
q_a = heat added to system, Btu/lb dry air
q_r = heat removed from system, Btu/lb dry air

Subscripts 1, 2, 3, etc., indicate entering and subsequent states.

Example 12-1 Determination of Moist Air Properties Find the properties of moist air when the dry-bulb temperature is 80°F and the wet-bulb temperature is 67°F.

Solution Read directly from Fig. 12-2 (Fig. 12-6*a* shows the solution diagrammatically).

Moisture content H = 78 gr/lb dry air
= 0.011 lb water/lb dry air
Enthalpy at saturation h' = 31.6 Btu/lb dry air
Enthalpy deviation D = −0.1 Btu/lb dry air
True enthalpy h = 31.5 Btu/lb dry air
Specific volume υ = 13.8 ft^3/lb dry air
Relative humidity = 51 percent
Dew point t_d = 60.3°F

Example 12-2 Air Heating Air is heated by a steam coil from 30°F dry-bulb temperature and 80 percent relative humidity to 75°F dry-bulb temperature. Find the relative humidity, wet-bulb temperature, and dew point of the heated air. Determine the quantity of heat added per pound of dry air.

Solution Reading directly from the psychrometric chart (Fig. 12-2),

Relative humidity = 15 percent
Wet-bulb temperature = 51.5°F
Dew point = 25.2°F

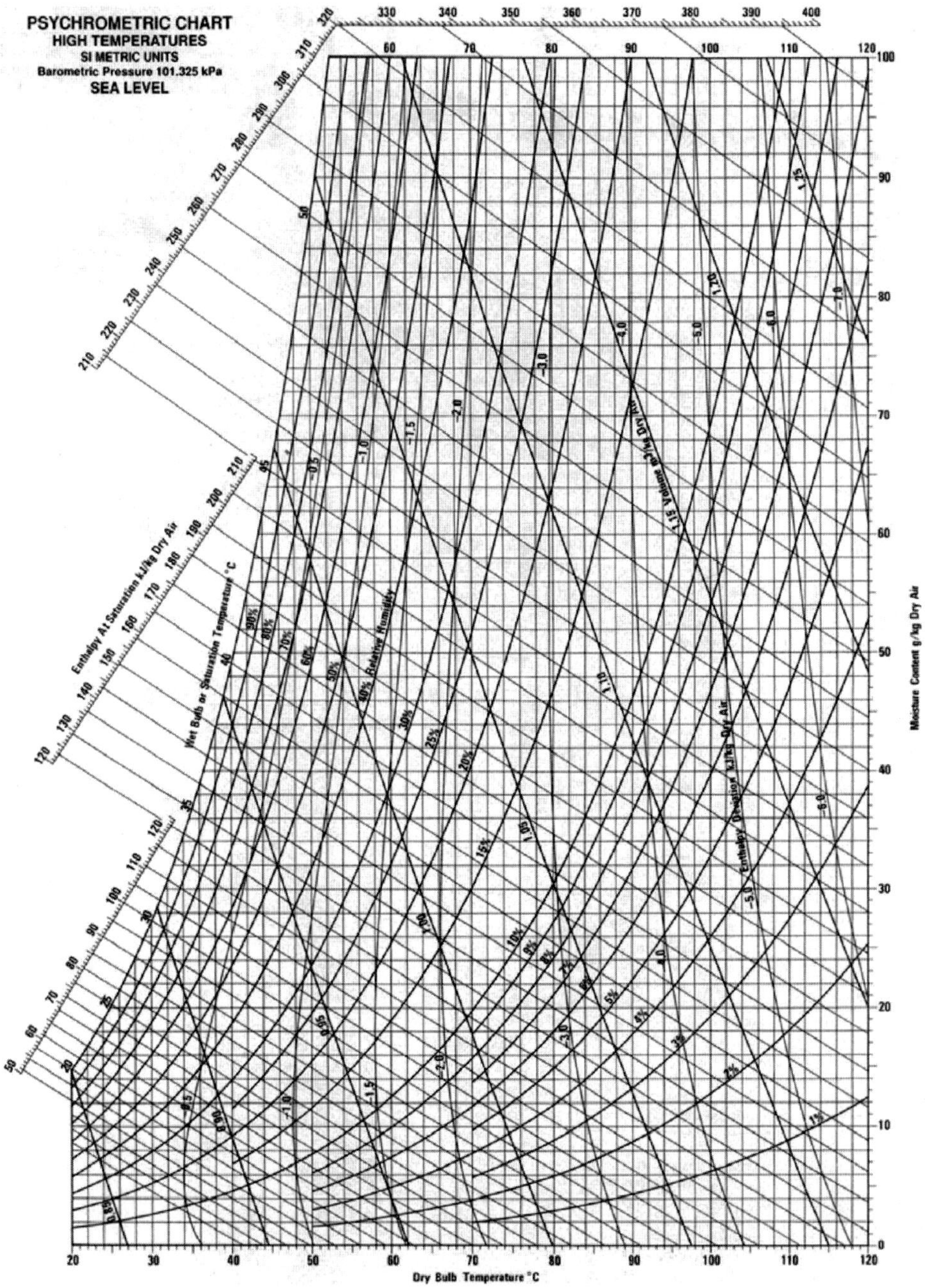

FIG. 12-1 Grosvenor psychrometric chart for the air-water system at standard atmospheric pressure, 101,325 Pa, SI units. (*Courtesy Carrier Corporation.*)

The enthalpy of the inlet air is obtained from Fig. 12-2 as $h_1 = h_1' + D_1 = 10.1 + 0.06 = 10.16$ Btu/lb dry air; at the exit, $h_2 = h_2' + D_2 = 21.1 - 0.1 = 21$ Btu/lb dry air. The heat added equals the enthalpy difference, or

$$q_a = \Delta h = h_2 - h_1 = 21 - 10.16 = 10.84 \text{ Btu/lb dry air}$$

If the enthalpy deviation is ignored, the heat added q_a is $\Delta h = 21.1 - 10.1 = 11$ Btu/lb dry air, or the result is 1.5 percent high. Figure 12-6*b* shows the heating path on the psychrometric chart.

Example 12-3 Evaporative Cooling Air at 95°F dry-bulb temperature and 70°F wet-bulb temperature contacts a water spray, where its relative humidity is increased to 90 percent. The spray water is recirculated; makeup water enters at 70°F. Determine the exit dry-bulb temperature, wet-bulb temperature, change in enthalpy of the air, and quantity of moisture added per pound of dry air.

Solution Figure 12-6*c* shows the path on a psychrometric chart. The leaving dry-bulb temperature is obtained directly from Fig. 12-2 as 72.2°F. Since the spray water enters at the wet-bulb temperature of 70°F and there is no heat added to or removed from it, this is by definition an adiabatic process and there will be no change in wet-bulb temperature. The only change in enthalpy is that from the heat content of the makeup water. This can be demonstrated as follows:

$$\text{Inlet moisture } H_1 = 70 \text{ gr/lb dry air}$$

$$\text{Exit moisture } H_2 = 107 \text{ gr/lb dry air}$$

$$\Delta H = 37 \text{ gr/lb dry air}$$

$$\text{Inlet enthalpy } h_1 = h_1' + D_1 = 34.1 - 0.22 = 33.88 \text{ Btu/lb dry air}$$

$$\text{Exit enthalpy } h_2 = h_2' + D_2 = 34.1 - 0.02 = 34.08 \text{ Btu/lb dry air}$$

$$\text{Enthalpy of added water } h_w = 0.2 \text{ Btu/lb dry air (from small diagram, 37 gr at 70°F)}$$

Then

$$q_a = h_2 - h_1 + h_w = 34.08 - 33.88 + 0.2 = 0$$

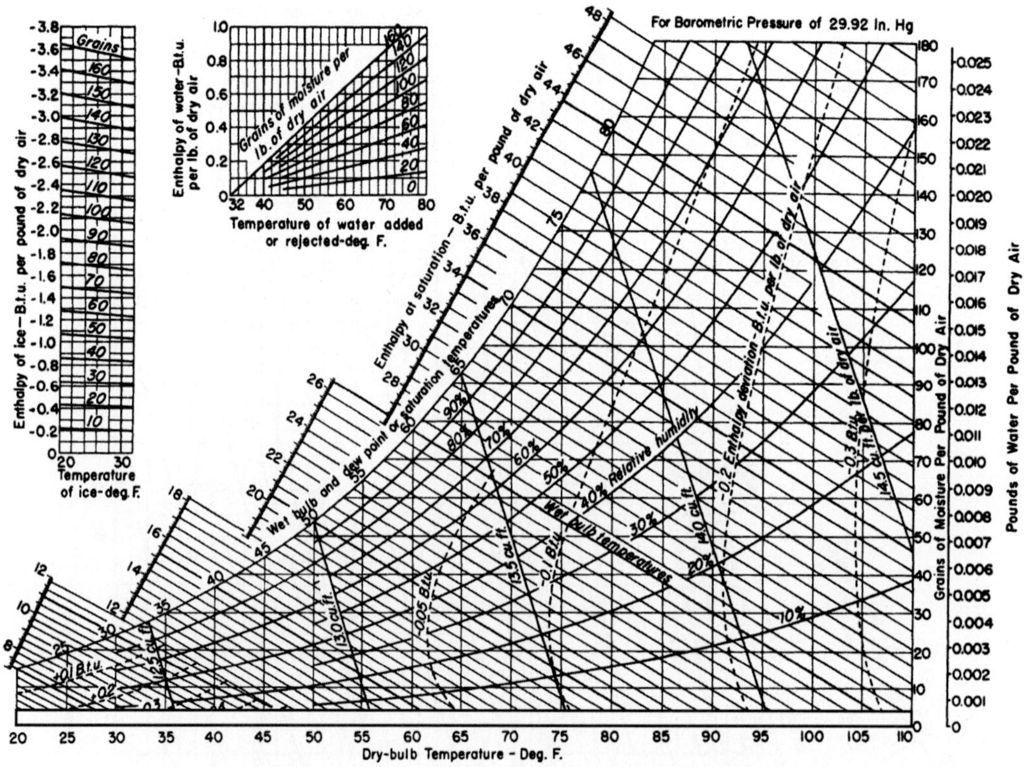

FIG. 12-2 Grosvenor psychrometric chart (medium temperature) for the air-water system at standard atmospheric pressure, 29.92 inHg, USCS units. (*Courtesy Carrier Corporation.*)

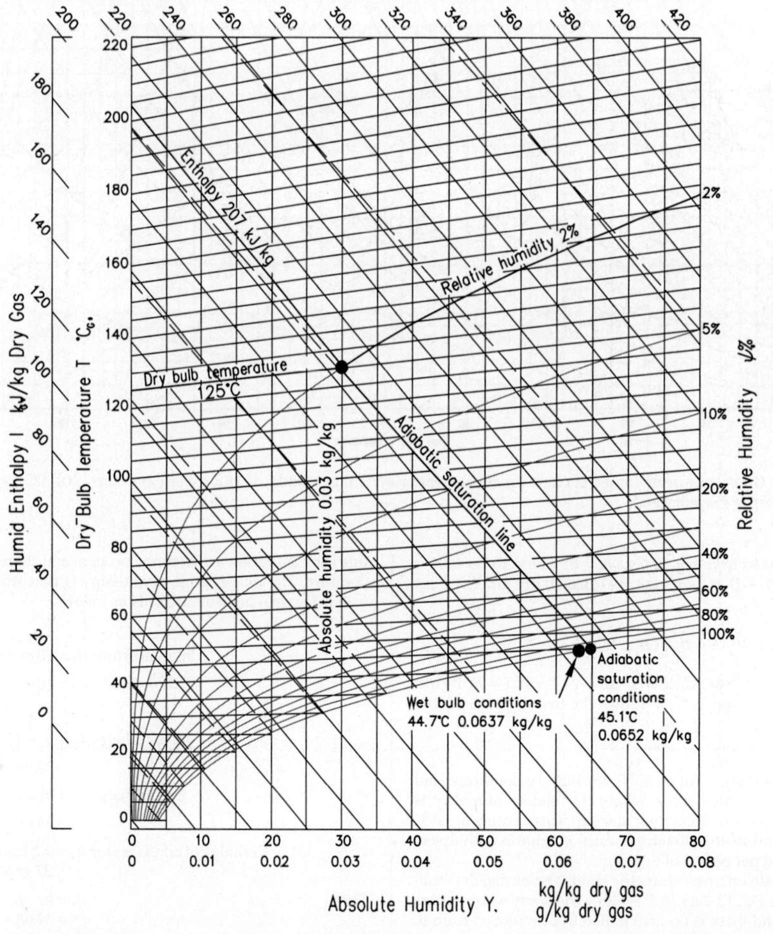

FIG. 12-3 Mollier psychrometric chart for the air-water system at standard atmospheric pressure, 101,325 Pa SI units, plots humidity (abscissa) against enthalpy (lines sloping diagonally from top left to bottom right). (*Source: Aspen Technology.*)

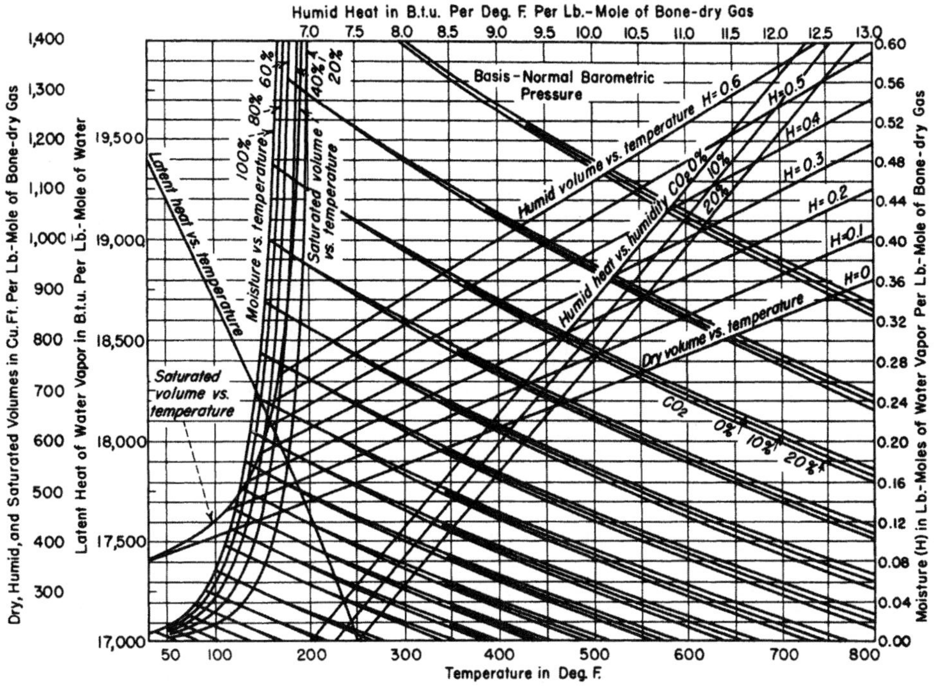

FIG. 12-4 Grosvenor psychrometric chart for air and flue gases at high temperatures, molar units [Hatta, *Chem. Metall. Eng.* 37: 64 (1930)].

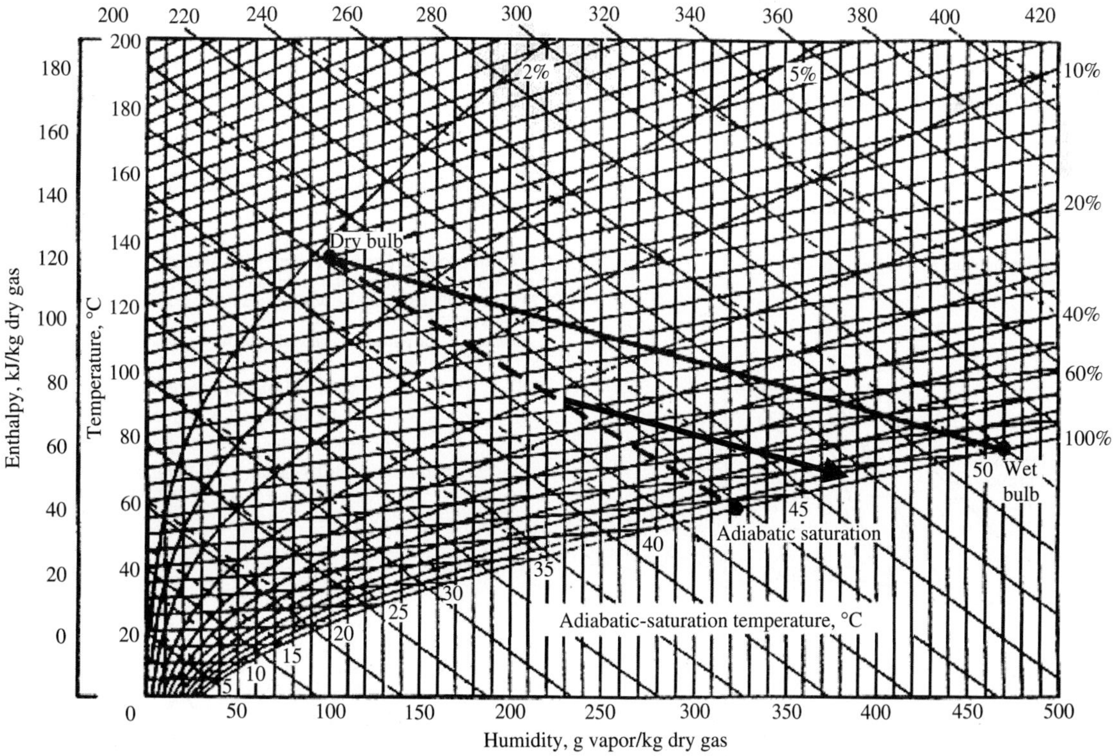

FIG. 12-5 Mollier chart showing changes in T_{wb} during an adiabatic saturation process for an organic system (nitrogen-toluene).

TABLE 12-4 Thermodynamic Properties of Saturated Air (USCS units, at standard pressure, 29.921 inHg)

Temp. T, °F	Saturation humidity H_s	Volume, ft³/lb dry air υ_a	υ_{as}	υ_s	Enthalpy, Btu/lb dry air h_a	h_{as}	h_s	Entropy, Btu/(°F·lb dry air) s_a	s_{as}	s_s	Condensed water: Enthalpy, Btu/lb h_w	Condensed water: Entropy, Btu/(lb·°F) s_w	Condensed water: Vapor pressure, inHg p_s	Temp. T, °F
−150	6.932×10^{-9}	7.775	.000	7.775	36.088	.000	36.088	0.09508	.00000	0.09508	218.77	0.4800	3.301×10^{-6}	−150
−100	9.772×10^{-7}	9.046	.000	9.046	24.037	.001	24.036	0.05897	.00000	0.05897	201.23	0.4277	4.666×10^{-5}	−100
−50	4.163×10^{-5}	10.313	.001	10.314	12.012	.043	11.969	0.02766	.00012	0.02754	181.29	0.3758	1.991×10^{-3}	−50
0	7.872×10^{-4}	11.578	.015	11.593	0.000	.835	0.835	0.00000	.00192	0.00192	158.93	0.3244	0.037645×10^{-2}	0
10	1.315×10^{-3}	11.831	.025	11.856	2.402	1.401	3.803	.00518	.00314	.00832	154.17	0.3141	0.062858	10
20	2.152×10^{-3}	12.084	.042	12.126	4.804	2.302	7.106	.01023	.00504	.01527	149.31	0.3039	0.10272	20
30	3.454×10^{-3}	12.338	.068	12.406	7.206	3.709	10.915	.01519	.00796	.02315	144.36	0.2936	0.16452	30
32	3.788×10^{-3}	12.388	.075	12.463	7.686	4.072	11.758	.01617	.00870	.02487	143.36	0.2916	0.18035	32
32*	3.788×10^{-3}	12.388	.075	12.463	7.686	4.072	11.758	.01617	.00870	.02487	0.04	0.0000	0.18037	32*
40	5.213×10^{-3}	12.590	.105	12.695	9.608	5.622	15.230	.02005	.01183	.03188	8.09	.0162	.24767	40
50	7.658×10^{-3}	12.843	.158	13.001	12.010	8.291	20.301	.02481	.01711	.04192	18.11	.0361	.36240	50
60	1.108×10^{-2}	13.096	.233	13.329	14.413	12.05	26.46	.02948	.02441	.05389	28.12	.0555	.52159	60
70	1.582×10^{-2}	13.348	.339	13.687	16.816	17.27	34.09	.03405	.03437	.06842	38.11	.0746	.73915	70
80	2.233×10^{-2}	13.601	0.486	14.087	19.221	24.47	43.69	0.03854	0.04784	0.08638	48.10	0.0933	1.0323	80
90	3.118×10^{-2}	13.853	.692	14.545	21.625	34.31	55.93	.04295	.06596	.10890	58.08	.1116	1.4219	90
100	4.319×10^{-2}	14.106	.975	15.081	24.029	47.70	71.73	.04729	.09016	.13745	68.06	.1296	1.9333	100
110	5.944×10^{-2}	14.359	1.365	15.724	26.434	65.91	92.34	.05155	.1226	.1742	78.03	.1472	2.5966	110
120	8.149×10^{-2}	14.611	1.905	16.516	28.841	90.70	119.54	.05573	.1659	.2216	88.01	.1646	3.4474	120
130	0.1116	14.864	2.652	17.516	31.248	124.7	155.9	.05985	.2245	.2844	98.00	.1817	4.5272	130
140	0.1534	15.117	3.702	18.819	33.655	172.0	205.7	.06390	.3047	.3686	107.99	.1985	5.8838	140
150	0.2125	15.369	5.211	20.580	36.063	239.2	275.3	.06787	.4169	.4848	117.99	.2150	7.5722	150
160	0.2990	15.622	7.446	23.068	38.472	337.8	376.3	.07179	.5793	.6511	128.00	.2313	9.6556	160
170	0.4327	15.874	10.938	26.812	40.882	490.6	531.5	.07565	.8273	.9030	138.01	.2473	12.203	170
180	0.6578	16.127	16.870	32.997	43.292	748.5	791.8	.07946	1.240	1.319	148.03	.2631	15.294	180
190	1.099	16.379	28.580	44.959	45.704	1255	1301	.08320	2.039	2.122	158.07	.2786	19.017	190
200	2.295	16.632	60.510	77.142	48.119	2629	2677	.08689	4.179	4.266	168.11	.2940	23.468	200

NOTE: Compiled by John A. Goff and S. Gratch. See also Keenan and Kaye. *Thermodynamic Properties of Air*, Wiley, New York, 1945. Enthalpy of dry air taken as zero at 0°F. Enthalpy of liquid water taken as zero at 32°F.

To convert British thermal units per pound to joules per kilogram, multiply by 2326; to convert British thermal units per pound dry air-degree Fahrenheit to joules per kilogram-kelvin, multiply by 4186.8; and to convert cubic feet per pound to cubic meters per kilogram, multiply by 0.0624.

*Entrapolated to represent metastable equilibrium with undercooled liquid.

Example 12-4 Cooling and Dehumidification Find the cooling load per pound of dry air resulting from infiltration of room air at 80°F dry-bulb temperature and 67°F wet-bulb temperature into a cooler maintained at 30°F dry-bulb and 28°F wet-bulb temperatures, where moisture freezes on the coil, which is maintained at 20°F.

Solution The path followed on a psychrometric chart is shown in Fig. 12-6*d*.

$$\text{Inlet enthalpy } h_1 = h_1' + D_1 = 31.62 - 0.1$$
$$= 31.52 \text{ Btu/lb dry air}$$
$$\text{Exit enthalpy } h_2 = h_2' + D_2 = 10.1 + 0.06$$
$$= 10.16 \text{ Btu/lb dry air}$$
$$\text{Inlet moisture } H_1 = 78 \text{ gr/lb dry air}$$

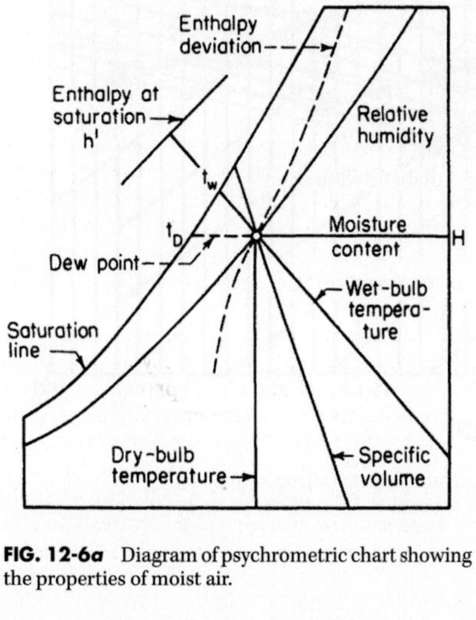

FIG. 12-6*a* Diagram of psychrometric chart showing the properties of moist air.

$$\text{Exit moisture } H_2 = 19 \text{ gr/lb dry air}$$
$$\text{Moisture rejected } \Delta H = 59 \text{ gr/lb dry air}$$
Enthalpy of rejected moisture = −1.26 Btu/lb dry air (from small diagram of Fig. 12-2)
$$\text{Cooling load } q_r = 31.52 - 10.16 + 1.26$$
$$= 22.62 \text{ Btu/lb dry air}$$

Note that if the enthalpy deviations were ignored, the calculated cooling load would be about 5 percent low.

Example 12-5 Cooling Tower Determine water consumption and the amount of heat dissipated per 1000 ft³/min of entering air at 90°F dry-bulb temperature and 70°F wet-bulb temperature when the air leaves saturated at 110°F and the makeup water is at 75°F.

Solution The path followed is shown in Fig. 12-6*e*.

$$\text{Exit moisture } H_2 = 416 \text{ gr/lb dry air}$$
$$\text{Inlet moisture } H_1 = 78 \text{ gr/lb dry air}$$
$$\text{Moisture added } \Delta H = 338 \text{ gr/lb dry air}$$
Enthalpy of added moisture $h_w = 2.1$ Btu/lb dry air (from small diagram of Fig. 12-2)

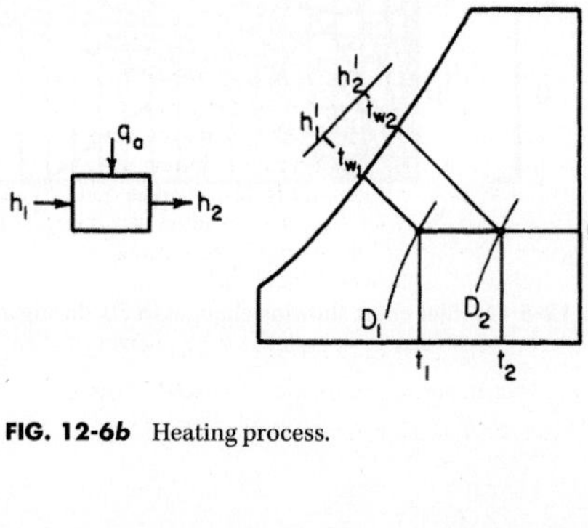

FIG. 12-6*b* Heating process.

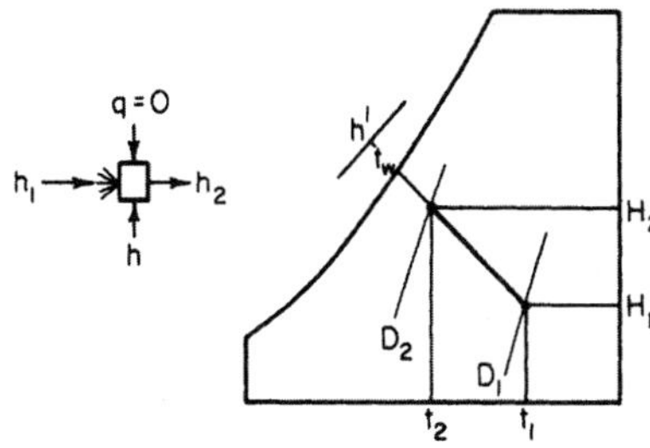

FIG. 12-6c Spray or evaporative cooling.

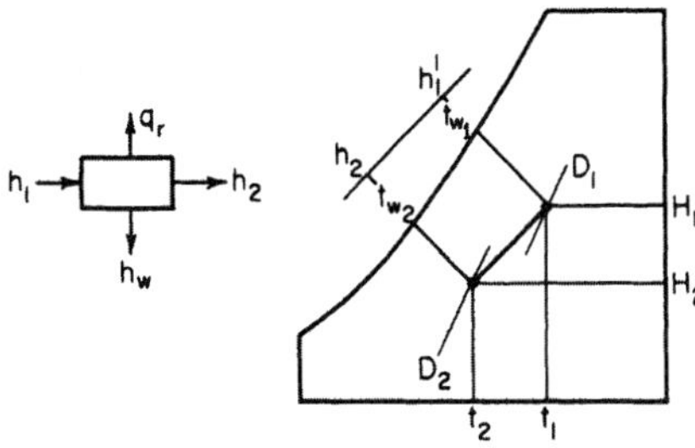

FIG. 12-6d Cooling and dehumidifying process.

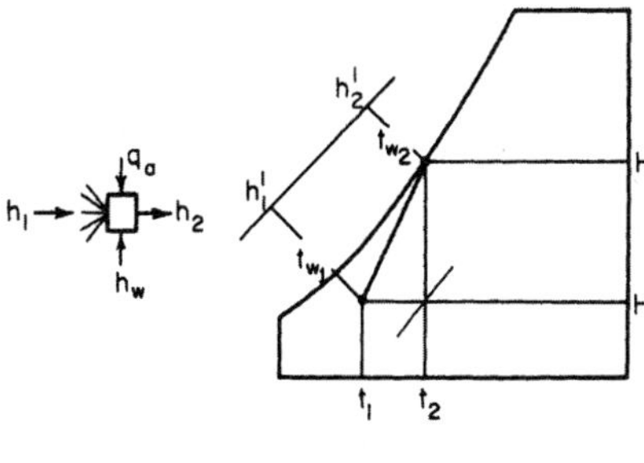

FIG. 12-6e Cooling tower.

If greater precision is desired, h_w can be calculated as

$$h_w = (338/7000)(1)(75 - 32)$$
$$= 2.08 \text{ Btu/lb dry air}$$
$$\text{Enthalpy of inlet air } h_1 = h_1' + D_1 = 34.1 - 0.18$$
$$= 33.92 \text{ Btu/lb dry air}$$
$$\text{Enthalpy of exit air } h_2 = h_2' + D_2 = 92.34 + 0$$
$$= 92.34 \text{ Btu/lb dry air}$$
$$\text{Heat dissipated} = h_2 - h_1 - h_w$$
$$= 92.34 - 33.92 - 2.08$$
$$= 56.34 \text{ Btu/lb dry air}$$
$$\text{Specific volume of inlet air} = 14.1 \text{ ft}^3\text{/lb dry air}$$
$$\text{Total heat dissipated} = \frac{(1000)(56.34)}{14.1} = 3990 \text{ Btu/min}$$

Example 12-6 Recirculating Dryer A dryer is removing 100 lb water/h from the material being dried. The air entering the dryer has a dry-bulb temperature of 180°F and a wet-bulb temperature of 110°F. The air leaves the dryer at 140°F. A portion of the air is recirculated after mixing with room air having a dry-bulb temperature of 75°F and a relative humidity of 60 percent. Determine the quantity of air required, recirculation rate, and load on the preheater if it is assumed that the system is adiabatic. Neglect the heat up of the feed and of the conveying equipment.

Solution The path followed is shown in Fig. 12-6*f*.

$$\text{Humidity of room air } H_1 = 0.0113 \text{ lb/lb dry air}$$
$$\text{Humidity of air entering dryer } H_3 = 0.0418 \text{ lb/lb dry air}$$
$$\text{Humidity of air leaving dryer } H_4 = 0.0518 \text{ lb/lb dry air}$$

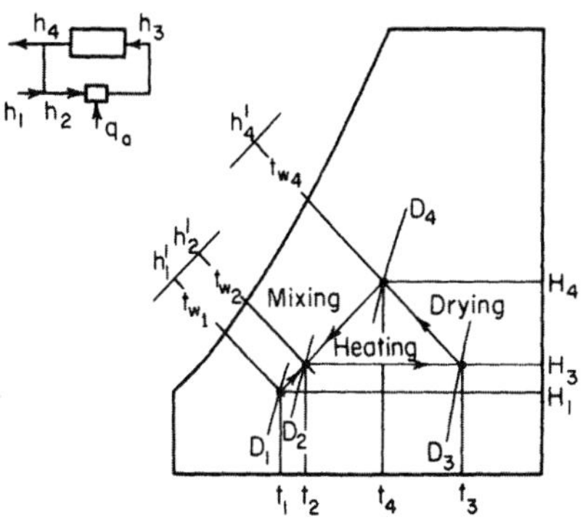

FIG. 12-6f Drying process with recirculation.

$$\text{Enthalpy of room air } h_1 = 30.2 - 0.3 = 29.9 \text{ Btu/lb dry air}$$
$$\text{Enthalpy of entering air } h_3 = 92.5 - 1.3 = 91.2 \text{ Btu/lb dry air}$$
$$\text{Enthalpy of leaving air } h_4 = 92.5 - 0.55 = 91.95 \text{ Btu/lb dry air}$$

Quantity of air required is 100/(0.0518 − 0.0418) = 10,000 lb dry air/h. At the dryer inlet the specific volume is 17.1 ft^3/lb dry air. Air volume is (10,000)(17.1)/60 = 2850 ft^3/min. Fraction exhausted is

$$\frac{X}{W_a} = \frac{0.0518 - 0.0418}{0.0518 - 0.0113} = 0.247$$

where X = quantity of fresh air and W_a = total airflow. Thus 75.3 percent of the air is recirculated. Load on the preheater is obtained from an enthalpy balance

$$q_a = 10{,}000(91.2) - 2470(29.9) - 7530(91.95) = 146{,}000 \text{ Btu/h}$$

PSYCHROMETRIC CALCULATIONS

Table 12-5 gives the steps required to perform the most common humidity calculations, using the formulas given earlier.

Methods (i) to (iii) are used to find the humidity and dew point from temperature readings from a wet- and dry-bulb psychrometer.

Method (iv) is used to find the absolute humidity and dew point from a relative humidity measurement at a given temperature.

Methods (v) and (vi) give the adiabatic saturation and wet-bulb temperatures from absolute humidity (or relative humidity) at a given temperature.

Method (vii) gives the absolute and relative humidity from a dew point measurement.

Method (viii) allows the calculation of all the main parameters if the absolute humidity is known, e.g., from a mass balance on a process plant.

Method (ix) converts the volumetric form of absolute humidity to the mass form (mixing ratio).

Method (x) allows the dew point to be corrected for pressure. The basis is that the mole fraction $y = p/P$ is the same for a given mixture composition at all values of total pressure P. In particular, the dew point measured in a compressed air duct can be converted to the dew point at atmospheric pressure, from which the humidity can be calculated. It is necessary to check that the temperature change associated with compression or expansion does not bring the dry-bulb temperature to a point where condensation can occur. Also, at these elevated pressures, it is strongly advisable to apply the enhancement factor (see British Standard BS1339 Part 1).

Psychrometric Software and Tables As an alternative to using charts or individual calculations, lookup tables have been published for many years for common psychrometric conversions, e.g., to find relative humidity given the dry-bulb and wet-bulb temperatures. These were often very extensive. To give precise coverage of T_{wb} in 1°C or 0.1°C steps, a complete table would be needed for each individual dry-bulb temperature.

Software is available that will perform calculations of humidity parameters for any point value, and for plotting psychrometric charts. Moreover, British Standard BS 1339 Part 2 (2006) provides functions as macros that can be embedded into any Excel-compatible spreadsheet. Users can therefore generate their own tables for any desired combination of parameters as well as perform point calculations. Hence, the need for published lookup tables has been eliminated. However, this software, like the previous lookup tables, is only valid for the air-water system. For other vapor-gas systems, the equations given in previous sections must be used.

TABLE 12-5 Calculation Methods for Various Humidity Parameters

	Known	Required	Method
i.	T, T_{wb}	Y	Find saturation vapor pressure p_{wb} at wet-bulb temperature T_{wb} from Eq. (12-4). Find actual vapor pressure p at dry-bulb temperature T from psychrometer equation (12-11). Find mixing ratio Y by conversion from p (Table 12-1).
ii.	T, T_{wb}	T_{dp}, d_{υ}	Find p if necessary by method (i) above. Find dew point T_{dp} from Eq. (12-4) by calculating the T corresponding to p [iteration required; Antoine equation (12-5) gives a first estimate]. Calculate volumetric humidity Y_{υ}, using Eq. (12-1).
iii.	T, T_{wb}	%RH (ψ)	Use method (i) to find p. Find saturation vapor pressure p_s at T from Eq. (12-4). Now relative humidity %RH = $100p/p_s$.
iv.	T, %RH	Y, d_{υ}	Find saturation vapor pressure p_s at T from Eq. (12-4). Actual vapor pressure $p = p_s$(%RH/100). Convert to Y (Table 12-1). Find Y_{υ} from Eq. (12-1).
v.	T, %RH (or T, Y)	T_{as}	Use method (iv) to find p and Y. Make an initial estimate of T_{as}, say, using a psychrometric chart. Calculate Y_{as} from Eq. (12-6). Find p from Table 12-1 and T_{as} from Antoine equation (12-5). Repeat until iteration converges (e.g., using spreadsheet). Alternative method: Evaluate enthalpy H_{est} at these conditions and H at initial conditions. Find H_{as} from Eq. (12-7) and compare with H_{est}. Make new estimate of Y_{as} which would give H_{est} equal to H_{as}. Find p from Table 12-1 and T_{as} from Antoine equation (12-5). Reevaluate H_{as} from Eq. (12-7) and iterate to refine value of Y_{as}.
vi.	T, %RH (or T, Y)	T_{wb}	Use method (iv) to find p and Y. Make an initial estimate of T_{wb}, e.g., using a psychrometric chart, or (for air-water system) by estimating adiabatic saturation temperature T_{as}. Find p_{wb} from psychrometer equation (12-11). Calculate new value of T_{wb} corresponding to p_{wb} by reversing Eq. (12-4) or using the Antoine equation (12-5). Repeat last two steps to solve iteratively for T_{wb} (computer program is preferable method).
vii.	T, T_{dp}	Y, %RH	Find saturation vapor pressure at dew point T_{dp} from Eq. (12-4); this is the actual vapor pressure p. Find Y from Table 12-1. Find saturation vapor pressure p_s at dry-bulb temperature T from Eq. (12-4). Now %RH = $100p/p_s$.
viii.	T, Y	T_{dp}, d_{υ}, %RH, T_{wb}	Find p by conversion from Y (Table 12-1). Then use method (ii), (iii), or (v) as appropriate.
ix.	T, Y_{υ}	Y	Find specific humidity Y_W from Eqs. (12-2) and (12-1). Convert to absolute humidity Y using $Y = Y_W/(1 - Y_W)$.
x.	T_{dp} at P_1 (elevated)	T_{dp} at P_2 (ambient)	Find vapor pressure p_1 at T_{dp} and P_1 from Eq. (12-4). Convert to vapor pressure p_2 at new pressure P_2 by the formula $p_2 = p_1P_2/P_1$. Find new dew point T_{dp} from Eq. (12-4) by calculating the T corresponding to p_2 [iteration required as in (ii)].

Software may be effectively used to draw psychrometric charts or perform calculations. A wide variety of other psychrometric software may be found on the Internet, but quality varies considerably; the source and basis of the calculation methods should be carefully checked before using the results. In particular, most methods only apply for the air-water system at moderate temperatures (below 100°C). For high-temperature dryer calculations, only software stated as suitable for this range should be used.

Reliable sources include the following:

1. The American Society of Agricultural and Biological Engineers (ASABE): http://www.asabe.org. Psychrometric data in chart and equation form in both SI and USCS units. Charts for temperature ranges of −35 to 600°F in USCS units and −10 to 120°C in SI units. Equations and calculation procedures. Air-water system and Grosvenor (temperature-humidity) charts only.
2. The American Society of Heating, Refrigerating and Air-Conditioning Engineers (ASHRAE): http://www.ashrae.org. Psychrometric Analysis CD with energy calculations and creation of custom charts at virtually any altitude or pressure. Detailed scientific basis given in ASHRAE Handbook. Air-water system and Grosvenor charts only.
3. Carrier Corporation, a United Technologies Company: http://www.training.carrier.com. PSYCH+, computerized psychrometric chart and instructional guide, including design of air-conditioning processes and/or cycles. Printed psychrometric charts also supplied. Air-water system and Grosvenor charts only.
4. Linric Company: http://www.linric.com. PsycPro generates custom psychrometric charts in English (USCS) or metric (SI) units, based on ASHRAE formulas. Air-water system and Grosvenor charts only.
5. Aspen Technology: http://www.aspentech.com. PSYCHIC, one of the Process Tools in the Aspen Engineering Suite, generates customized psychrometric charts. Mollier and Bowen enthalpy-humidity charts are produced in addition to Grosvenor. Any gas-vapor system can be handled as well as air-water; data supplied for common organic solvents. It can draw operating lines and spot points, as shown in Fig. 12-7.
6. British Standards Institution: http://www.bsigroup.com. British Standard BS 1339 Part 2 is a spreadsheet-based software program providing functions based on the latest internationally agreed standards. It calculates all key psychrometric parameters and can produce a wide range of psychrometric tables. Users can embed the functions in their own spreadsheets to do psychrometric calculations. Air-water system only (although BS 1339 Part 1 text gives full calculation methods for other gas-vapor systems). SI (metric) units. It does not plot psychrometric charts.
7. Akton Associates: http://www.aktonassoc.com. Akton provides digital versions of psychrometry charts.

PSYCHROMETRIC CALCULATIONS—WORKED EXAMPLES

Example 12-7 Determination of Moist Air Properties An air-water mixture is found from the heat and mass balance to be at 60°C (333 K) and 0.025 kg/kg (25 g/kg) absolute humidity. Calculate the other main parameters for the mixture. Take atmospheric pressure as 101,325 Pa.

Method: Consult item (vi) in Table 12-5 for the calculation methodology.

From the initial terminology section, specific humidity Y_W = 0.02439 kg/kg, mole ratio z = 0.0402 kmol/kmol, mole fraction y = 0.03864 kmol/kmol.

From Table 12-1, vapor pressure p = 3915 Pa (0.03915 bar) and volumetric humidity Y_{υ} = 0.02547 kg/m³. Dew point is given by the temperature corresponding to p at saturation. From the reversed Antoine equation (12-5), T_{dp} = 3830/(23.19 − ln 3915) + 44.83 = 301.58 K = 28.43°C.

Relative humidity is the ratio of actual vapor pressure to saturation vapor pressure at dry-bulb temperature. From the Antoine equation (12-5), p_s = exp [23.19 − 3830/(333.15 − 44.83)] = 20,053 Pa (new coefficients), or p_s = exp [23.1963 − 3816.44/(333.15 − 46.13)] = 19,921 Pa (old coefficients).

From Sonntag equation (12-4), p_s = 19,948 Pa; the difference from the Antoine is less than 0.5 percent. Relative humidity = 100 × 3915/19,948 = 19.6 percent. From a psychrometric chart, e.g., Fig. 12-1, a humidity of 0.025 kg/kg at T = 60°C lies very close to the adiabatic saturation line for 35°C. Hence a good first estimate for T_{as} and T_{wb} will be 35°C. Refining the estimate of T_{wb} by using the psychrometer equation and iterating gives

$$p_{wb} = 3915 + 6.46 \times 10^{-4}\,(1.033)(101{,}325)\,(60 - 35) = 5605 \text{ Pa}$$

From the Antoine equation,

$$T_{wb} = 3830/(23.19 - \ln 5605) + 44.83 = 307.9 \text{ K} = 34.75°\text{C}$$

Second iteration:

$$p_{wb} = 3915 + 6.46 \times 10^{-4}(1.033)(101{,}325)(60 - 34.75) = 5622 \text{ Pa}$$

$$T_{wb} = 307.96 \text{ K} = 34.81°\text{C}$$

To a sensible level of precision, T_{wb} = 34.8°C.

From Table 12-1, Y_{wb} = 5622 × 0.622/(101,325 − 5622) = 0.0365(4) kg/kg.

Enthalpy of original hot air is approximately given by $H = (C_{Pg} + C_{P\upsilon}Y)(T - T_0) + \lambda_0 Y$ (1 + 1.9 × 0.025) × 60 + 2501 × 0.025 = 62.85 + 62.5 = 125.35 kJ/kg. A more accurate calculation can be obtained from steam tables; C_{Pg} = 1.005 kJ/(kg·K) over this range, H_{υ} at 60°C = 2608.8 kJ/kg, H = 60.3 + 65.22 = 125.52 kJ/kg.

Calculation (v), method 1: If T_{as} = 34.8, from Eq. (12-6), with C_s = 1 + 1.9 × 0.025 = 1.048 kJ/(kg·K), then λ_{as} = 2419 kJ/kg (steam tables), Y_{as} = 0.025 + 1.048/2419 (60 − 34.8) = 0.0359(2) kg/kg. From Table 12-1, p = 5530 Pa. From the Antoine equation (12-5), T_{as} = 3830/(23.19 − ln 5530) + 44.83 = 307.65 K = 34.52°C. Repeat until iteration converges (e.g., using spreadsheet). Final value T_{as} = 34.57°C, Y_{as} = 0.0360 kg/kg.

Enthalpy check: From Eq. (12-7), $H_{as} - H$ = 4.1868 × 34.57 × (0.036 − 0.025) = 1.59 kJ/kg. So H_{as} = 127.11 kJ/kg. Compare H_{as} calculated from enthalpies; H_g at 34.57°C = 2564 kJ/kg, H_{est} = 34.90 + 92.29 = 127.19 kJ/kg. The iteration has converged successfully.

Note that T_{as} is 0.2°C lower than T_{wb} and Y_{as} is 0.0005 kg/kg lower than Y_{wb}, both negligible differences.

Example 12-8 Calculation of Humidity and Wet-Bulb Condition A dryer exhaust which can be taken as an air-water mixture at 70°C (343.15 K) is measured to have a relative humidity of 25 percent. Calculate the humidity parameters and wet-bulb conditions for the mixture. Pressure is 1 bar (100,000 Pa).

Method: Consult item (v) in Table 12-5 for the calculation methodology.

From the Antoine equation (12-5), using standard coefficients (which give a better fit in this temperature range), p_s = exp[23.1963 − 3816.44/(343.15 − 46.13)] = 31,170 Pa. Actual vapor pressure p = 25 percent of 31,170 = 7792 Pa (0.078 bar).

From Table 12-1, absolute humidity Y = 0.05256 kg/kg and volumetric humidity Y_{υ} = 0.0492 kg/m³. From the terminology section, mole fraction y = 0.0779 kmol/kmol, mole ratio z = 0.0845 kmol/kmol, specific humidity Y_W = 0.04994 kg/kg.

$$\text{Dew point } T_{dp} = 3816.44/(23.1963 - \ln 7792) + 46.13 = 314.22 \text{ K} = 41.07°\text{C}$$

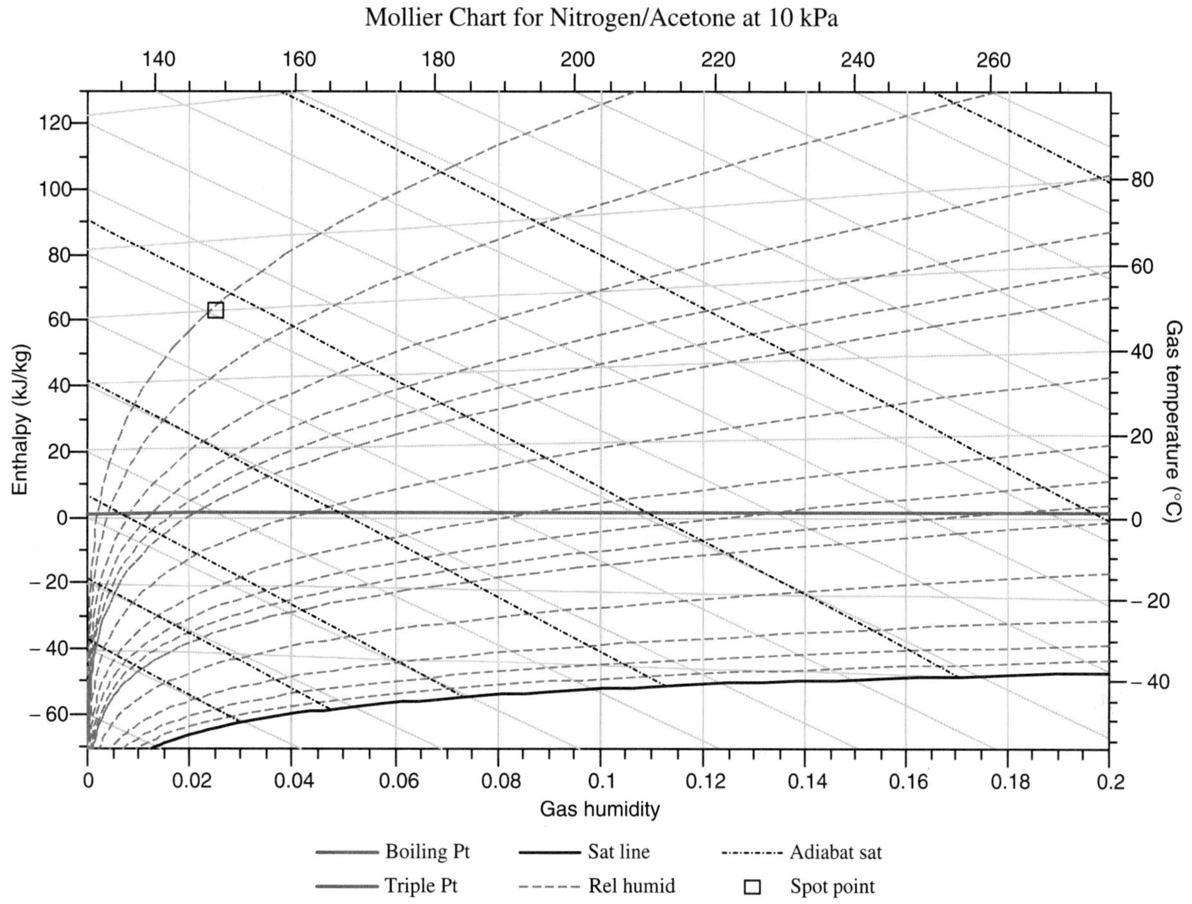

FIG. 12-7 Mollier psychrometric chart (from PSYCHIC software program) showing determination of adiabatic saturation temperature plots of humidity (abscissa) against enthalpy (lines sloping diagonally from top left to bottom right). (*Courtesy AspenTech.*)

From the psychrometric chart, a humidity of 0.0526 kg/kg at $T = 70°C$ falls just below the adiabatic saturation line for 45°C. Estimate T_{as} and T_{wb} as 45°C. Refining the estimate of T_{wb} by using the psychrometer equation and iterating gives

$$p_{wb} = 7792 + 6.46 \times 10^{-4}\,(1.0425)(10^5)(70 - 45) = 9476$$

From the Antoine equation,

$$T_{wb} = 3816.44/(23.1963 - \ln 9476) + 46.13 = 317.96\text{ K} = 44.81°C$$

Second iteration (taking $T_{wb} = 44.8$):

$$p_{wb} = 9489 T_{wb} = 317.99\text{ K} = 44.84°C$$

The iteration has converged.

Example 12-9 Calculation of Psychrometric Properties of Acetone/Nitrogen Mixture A mixture of nitrogen N_2 and acetone CH_3COCH_3 is found from the heat and mass balance to be at 60°C (333 K) and 0.025 kg/kg (25 g/kg) absolute humidity (same conditions as in Example 12-7). Calculate the other main parameters for the mixture. The system is under vacuum at 100 mbar (0.1 bar, 10,000 Pa).

Additional data for acetone and nitrogen are obtained from *The Properties of Gases and Liquids* (Prausnitz et al.). Molecular weight (molal mass) M_g for nitrogen = 28.01 kg/kmol; for acetone $M_v = 58.08$ kg/kmol. Antoine coefficients for acetone are 16.6513, 2940.46, and 35.93, with p_s in mmHg and T in K. Specific heat capacity of nitrogen is approximately 1.014 kJ/(kg·K). Latent heat of acetone is 501.1 kJ/kg at the boiling point. The psychrometric ratio for the nitrogen-acetone system is not given, but the diffusion coefficient D can be roughly evaluated as 1.34×10^{-5}, compared to 2.20×10^{-5} for water in air. As the psychrometric ratio is linked to $D^{2/3}$, it can be estimated as 0.72, which is in line with tabulated values for similar organic solvents (e.g., propanol).

Method: Consult item (vi) in Table 12-5 for the calculation methodology.

From the terminology, specific humidity $Y_W = 0.02439$ kg/kg, the same as in Example 12-7. Mole ratio $z = 0.0121$ kmol/kmol, mole fraction $y = 0.01191$ kmol/kmol—lower than in Example 12-7 because molecular weights are different.

From the Antoine equation (12-5),

$$\ln p_s = C_0 - \frac{C_1}{T - C_2} = 16.6513 - \frac{2940.46}{T - 35.93}$$

Since $T = 60°C$, $\ln p_s = 6.758$, $p_s = 861.0$ mmHg. Hence $p_s = 1.148$ bar $= 1.148 \times 10^5$ Pa. The saturation vapor pressure is higher than atmospheric pressure; this means that acetone at 60°C must be above its normal boiling point. Check: T_{bp} for acetone = 56.5°C.

Vapor pressure $p = yP = 0.01191 \times 10{,}000 = 119.1$ Pa (0.001191 bar)—much lower than before because of the reduced total pressure. This is 0.89 mmHg. Volumetric humidity $Y_v = 0.0025$ kg/m³—again substantially lower than at 1 atm.

Dew point is the temperature where p_s equals p'. From the reversed Antoine equation (12-5),

$$T = \frac{C_1}{C_0 - \ln p_s} + C_2$$

so

$$T_{dp} = \frac{2940}{16.6513 - \ln 0.89} + 35.93 = 211.27\text{ K} = -61.88°C$$

This very low dew point is due to the low boiling point of acetone and the low concentration.

Relative humidity is the ratio of actual vapor pressure to saturation vapor pressure at dry-bulb temperature. So $p = 119.1$ Pa, $p_s = 1.148 \times 10^5$ Pa, RH = 0.104 percent—again very low.

A special psychrometric chart would need to be constructed for the acetone-nitrogen system to get first estimates (this can be done using PSYCHIC, as shown in Fig. 12-7). A humidity of 0.025 kg/kg at $T = 60°C$ lies just below the adiabatic saturation line for −40°C. The wet-bulb temperature will not be the same as T_{as} for this system; since the psychrometric ratio β is less than 1, T_{wb} should be significantly above T_{as}. However, let us assume no good first estimate is available and simply take T_{wb} to be 0°C initially.

When using the psychrometer equation, we will need to use Eq. (12-13) to obtain the value of the psychrometer coefficient. Using the tabulated values above, we obtain $A = 0.00135$, about double the value for air-water. We must remember that the estimate will be very rough because of the uncertainty in the value of β. Refining the estimate of T_{wb} by using the psychrometer equation and iterating gives

$$p_{wb} = 119.1 + 1.35 \times 10^{-3}\,(10^4)\,(60 - 0) = 932.3 \text{ Pa} = 7.0 \text{ mmHg}$$

From the Antoine equation,

$$T_{wb} = 2940/(16.6513 - \ln 7) + 35.93 = 235.84 \text{ K} = -37.3°\text{C}$$

Second iteration:

$$p_{wb} = 119.1 + 1.35 \times 10^{-3}\,(10^4)\,(60 + 37.3) = 1433 \text{ Pa} = 10.7 \text{ mmHg}$$

$$T_{wb} = 241.85 \text{ K} = -31.3°\text{C}$$

Third iteration:

$$p_{wb} = 119.1 + 1.35 \times 10^{-3}\,(10^4)\,(60 + 31.3) = 1352 \text{ Pa} = 10.1 \text{ mmHg}$$

$$T_{wb} = 241.0 \text{ K} = -32.1°\text{C}$$

The iteration has converged successfully, despite the poor initial guess. The wet-bulb temperature is −32°C; given the levels of error in the calculation, it will be meaningless to express this to any greater level of precision.

In a similar way, the adiabatic saturation temperature can be calculated from Eq. (12-6) by taking the first guess as −40°C and assuming the humid heat to be 1.05 kJ/(kg · K) including the vapor:

$$Y_{as} = Y + \frac{C_s}{\lambda_{as}}\,(T - T_{as})$$

$$= 0.025 + \left(\frac{1.05}{501.1}\right)(60 + 40) = 0.235 \text{ kg/kg}$$

From Table 12-2,

$$p_{as} = 1018 \text{ Pa} = 7.63 \text{ mmHg}$$

From the Antoine equation,

$$T_{as} = 237.05 \text{ K} = -36.1°\text{C}$$

Second iteration:

$$Y_{as} = 0.025 + (1.05/501.1)(60 + 36.1) = 0.226 \text{ kg/kg} \quad p_{as} = 984 \text{ Pa} = 7.38 \text{ mmHg}$$

From the Antoine equation,

$$T_{as} = 236.6 \text{ K} = -36.6°\text{C}$$

This has converged. A more accurate figure could be obtained with more refined estimates for C_s and λ_{wb}.

MEASUREMENT OF HUMIDITY

Hygrometers *Electric hygrometers* have been the fastest-growing form of humidity measurement in recent years and are now the most commonly used sensors for process measurement. They measure the electrical resistance, capacitance, or impedance of a film of moisture-absorbing materials exposed to the gas. A wide variety of sensing elements are used. Normally, relative humidity is measured, with a corresponding temperature measurement and conversion to absolute humidity.

Mechanical hygrometers utilizing materials such as human hair, wood fiber, and plastics have been used to measure humidity. These methods rely on a change in dimension with humidity. They are not suitable for process use.

Other hygrometric techniques in process and laboratory use include electrolytic and piezoelectric hygrometers, infrared and mass spectroscopy, and vapor pressure measurement, e.g., by a Pirani gauge.

The *gravimetric method* is accepted as the most accurate humidity-measuring technique. In this method a known quantity of gas is passed over a moisture-absorbing chemical such as phosphorus pentoxide, and the increase in weight is determined. It is mainly used for calibrating standards and measurements of gases with SO_x present.

Dew Point Method The dew point of wet air is measured directly by observing the temperature at which moisture begins to form on an artificially cooled, polished surface.

Optical dew point hygrometers employing this method are often used as a fundamental technique for determining humidity. Uncertainties in temperature measurement of the polished surface, gradients across the surface, and the appearance or disappearance of fog have been much reduced in modern instruments. Automatic mirror cooling, e.g., Peltier thermoelectric, is more accurate and reliable than older methods using evaporation of a low-boiling solvent such as ether, or external coolants (e.g., vaporization of solid carbon dioxide or liquid air, or water cooling). Contamination effects have also been reduced or compensated for, but regular recalibration is still required, at least once a year.

Wet-Bulb/Dry-Bulb Method In the past, probably the most commonly used method for determining the humidity of a gas stream was the measurement of wet- and dry-bulb temperatures. The wet-bulb temperature is measured by contacting the air with a thermometer whose bulb is covered by a wick saturated with water. If the process is adiabatic, the thermometer bulb attains the wet-bulb temperature. When the wet- and dry-bulb temperatures are known, the humidity is readily obtained from charts such as Figs. 12-1 through 12-4. To obtain reliable information, care must be exercised to ensure that the wet-bulb thermometer remains wet and that radiation to the bulb is minimized. The latter is accomplished by making the relative velocity between wick and gas stream high [a velocity of 4.6 m/s (15 ft/s) is usually adequate for commonly used thermometers] or by the use of radiation shielding. In the *Assmann psychrometer,* the air is drawn past the bulbs by a motor-driven fan. Making sure that the wick remains wet is a mechanical problem, and the method used depends to a large extent on the particular arrangement. Again, as with the dew point method, errors associated with the measurement of temperature can cause difficulty.

For measurement of atmospheric humidities, the *sling* or *whirling psychrometer* was widely used in the past to give a quick and cheap, but inaccurate, estimate. A wet- and dry-bulb thermometer is mounted in a sling which is whirled manually to give the desired gas velocity across the bulb.

In addition to the mercury-in-glass thermometer, other temperature-sensing elements may be used for psychrometers. These include resistance thermometers, thermocouples, bimetal thermometers, and thermistors.

EVAPORATIVE COOLING

General References: ASHRAE is the American Society of Heating, Refrigeration and Air Conditioning Engineers: www.ashrae.org; *ASHRAE Handbook of Fundamentals,* "Climatic Design Information," chap. 14, ASHRAE, Atlanta, Ga., 2013. Cooling Technology Institute: www.cti.org. ASHRAE and CTI are both professional organizations and both websites contain technical resources and contacts for engineers.

INTRODUCTION

Evaporative cooling, using recirculated cooling water systems, is the method most widely used throughout the process industries for employing water to remove process waste heat, rejecting that waste heat into the environment. Maintenance considerations (water-side fouling control), through control of makeup water quality and control of cooling water chemistry, form one reason for this preference. Environmental considerations—by minimizing consumption of potable water, minimizing the generation and release of contaminated cooling water, and controlling the release into the environment of chemicals from leaking heat exchangers—form the second major reason.

Local ambient climatic conditions, particularly the maximum summer wet-bulb temperature, determine the design of the evaporative equipment. Typically, the wet-bulb temperature used for design is the 0.4 percent value, as listed in the *ASHRAE Handbook of Fundamentals,* equivalent to 35-h exceedance per year on average.

The first subsection below presents the classic *cooling tower* (CT), the evaporative cooling technology most widely used today. The second subsection presents the *wet surface air cooler* (WSAC), a more recent technology, combining within one piece of equipment the functions of cooling tower, circulated cooling water system, and heat exchange tube bundle. The most common application for WSACs is in the direct cooling of process streams. However, the closed-circuit cooling tower, employing WSACs for cooling the circulated cooling water (replacing the CT), is an important alternative WSAC application, presented at the end of this section.

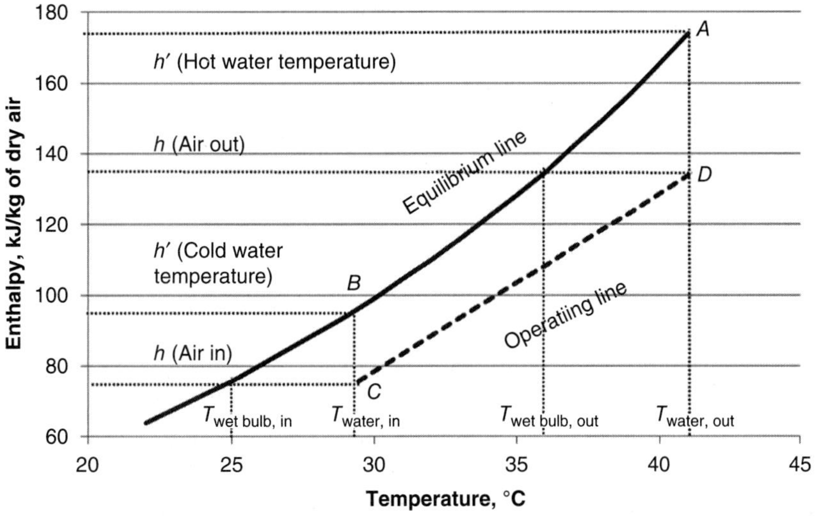

FIG. 12-8 Cooling-tower process heat balance and solution to Example 12-10.

To minimize the total annualized costs for evaporative cooling is a complex engineering task in itself, separate from classic process design. The evaluation and the selection of the best option for process cooling impact many aspects of how the overall project will be optimally designed (utilities supply, reaction and separations design, pinch analyses, 3D process layout, plot plan, etc.). Therefore, evaluation and selection of the evaporative cooling technology system should be performed at the start of the project design cycle, during conceptual engineering (Sec. 9, Process Economics, Value Improving Practices), when the potential to influence project costs is at a maximum value (Sec. 9, VIP Fig. 9-26). The relative savings achievable for selection of the optimum heat rejection technology option can frequently exceed 25 percent, for the installed cost for the technology alone.

PRINCIPLES

The processes of cooling water are among the oldest known. Usually water is cooled by exposing its surface to air. Some of the processes are slow, such as the cooling of water on the surface of a pond; others are comparatively fast, such as the spraying of water into air. These processes all involve the exposure of water surface to air in varying degrees.

The heat-transfer process involves (1) latent heat transfer owing to vaporization of a small portion of the water and (2) sensible heat transfer owing to the difference in temperatures of water and air. Approximately 80 percent of this heat transfer is due to latent heat and 20 percent to sensible heat.

COOLING TOWERS

GENERAL REFERENCES: Hensley, *Cooling Tower Fundamentals,* 2d ed., Marley Cooling Technologies,* Bridgewater, N.J., 1998. McAdams, *Heat Transmission,* 3d ed., McGraw-Hill, New York, 1954, pp. 356–365. Extensive information can be found online at the following websites: www.cti.org; www.ashrae.org; www.marleyct.com; www.spxcooling.com.

Process Description A cooling tower is a simultaneous heat- and mass-transfer device that cools a hot process water stream directly by evaporation into ambient air. The water is pumped up to the top of the tower and sprayed into flowing ambient air. The tower contains a packing (called *fill*) to increase the surface area of contact of the water with the air as it falls to the cool water collection basin. The fill is commonly made from wood slats or PVC.

Theoretical possible heat removal per unit mass of air circulated in a cooling tower depends on the temperature and moisture content of air. An indication of the moisture content of the air is its wet-bulb temperature. Ideally, then, the wet-bulb temperature is the lowest theoretical temperature to which the water can be cooled. Practically, the cold water temperature approaches but does not equal the air wet-bulb temperature in a cooling tower; this is so because it is impossible to contact all the water with fresh air as the water drops through the wetted fill surface to the basin. The magnitude of the approach to the wet-bulb temperature is dependent on the tower design. Important factors are air-to-water contact time, amount of fill surface, and breakup of water into droplets. In actual practice, cooling towers are seldom designed for approaches closer than 2.8°C (5°F).

Cooling Tower Theory The most generally accepted theory of the cooling tower heat-transfer process is that developed by Merkel [Merkel, *Z. Ver. Dtsch. Ing. Forsch.,* no. 275 (1925)]. The theory is developed using the same approach as the HTU-NTU model for mass or heat transfer in packed columns. Mass and energy balances are constructed within a differential vertical increment and then integrated over the height of the tower. The Merkel equation combines the mass- and heat-transfer processes to arrive at one driving force—enthalpy to capture the simultaneous heat- and mass-transfer processes in one equation. Both the Chilton-Colburn analogy and the fact that the Lewis number is near unity (see Psychrometry subsection) for air-water systems are used to derive the equation; this treatment is only valid for air-water systems. See Wankat, *Equilibrium Staged Separations,* Elsevier, 1988, pp. 674–688 for a lucid and detailed description of this equation and a worked example.

In the integrated form, the Merkel equation is given by

$$\frac{Ka\,V}{L} = \int_{T_2}^{T_1} \frac{C_L\,dT}{h' - h} \tag{12-14}$$

where K = mass-transfer coefficient, kg water/(m^2 · s); a = contact area, m^2/m^3 tower volume; V = active cooling volume, m^3/m^2 of plan area; L = water rate, kg/(m^2 · s); C_L = heat capacity of water, J/(kg · °C); h' = enthalpy of saturated air at water temperature, J/kg; h = enthalpy of airstream, J/kg; and T_1 and T_2 = entering and leaving water temperatures, °C.

The right-hand side of Eq. (12-14) is entirely in terms of air and water properties and is independent of tower dimensions. The left-hand side is the "tower characteristic," KaV/L, which can be determined by integration. To predict tower performance, it is necessary to know the required tower characteristics for fixed ambient and water conditions.

Figure 12-8 illustrates water and air relationships and the driving potential which exist in a counterflow tower, where air flows parallel but opposite in direction to water flow. An understanding of this diagram is important in visualizing the cooling tower process and evaluating the integral in the Merkel equation.

Figure 12-8 is an enthalpy-temperature diagram, containing two lines: an equilibrium line and an operating line. The equilibrium line is shown by *AB*. This line represents the enthalpy of saturated water vapor. It can be plotted using the definition of humid enthalpy in the Psychrometry subsection.

$$h' = (C_{p,g} + C_{p,y} \cdot Y)(T - T_0) + \lambda_0 Y \tag{12-15}$$

If we choose a reference temperature of 0°C, then λ_0 = 2501 kJ/kg. The absolute humidity at saturation is found first by calculating the vapor

*The contributions of Ken Mortensen and coworkers of Marley Cooling Technologies, Overland Park, Kansas, toward the review and update of this subsection are gratefully acknowledged.

pressure (at T) using Eq. (12-5) and then by using the following relationship from Table 12-1:

$$Y = \frac{0.622 \cdot p}{P - p} \tag{12-16}$$

The operating line describes the enthalpy of the air moving through the tower and is given by line CD. The highest temperature is that of the water entering (point D), and the lowest is that of the water leaving (point C). This difference is called the *range*. The difference between the wet-bulb temperature of the air entering and the temperature of the water exiting is called the *approach*. The operating line is given by

$$h - h_{\text{wet bulb, air in}} = \left(\frac{L}{G} \cdot C_{p,\text{water}}\right) \cdot (T - T_{\text{water, out}}) \tag{12-17}$$

The enthalpy of the wet-bulb temperature is on the equilibrium line, and it is found by using the same procedure as outlined above. Mechanical draft cooling towers normally are designed for L/G ratios ranging from 0.75 to 1.50; accordingly, the values of KaV/L vary from 0.50 to 2.50.

The tower characteristic contains the kinetic information in the design, which is affected by the nature of the fill and the velocity of the air. A useful discussion on optimization of cooling towers is found in Picardo, J. R., *Energy Conversion and Management* **57:** 167–172 (2012). Some practical guidelines on the height and cross-sectional area are given below.

Example 12-10 Calculation of Mass-Transfer Coefficient Group Determine the theoretically required KaV/L value for a cooling duty from 41°C inlet water, 29.4°C outlet water, 25°C ambient wet-bulb temperature, and an L/G ratio of 1.2.

We first evaluate the equilibrium and operating lines over the temperature range of interest. These are plotted as Fig. 12-8. The enthalpy h of the air at the wet-bulb temperature equals 74.75 kJ/kg, using Eqs. (12-5), (12-15), and (12-16). Numerical integration of the Merkel equation using a spreadsheet gives $KaV/L = 2.32$.

Cooling Tower Equipment The airflow in a cooling tower is driven by fans or by natural convection. When fans are used, it is called a *mechanical draft tower.* Two types are in use today, the forced-draft and the induced-draft towers. In the forced-draft tower, the fan is mounted at the base, and air is forced in at the bottom and discharged at low velocity through the top. This arrangement has the advantage of locating the fan and drive outside the tower, where it is convenient for inspection, maintenance, and repairs. Since the equipment is out of the hot, humid top area of the tower, the fan is not subjected to corrosive conditions. However, because of the low exit-air velocity, the forced-draft tower is subjected to excessive recirculation of the humid exhaust vapors back into the air intakes. Since the wet-bulb temperature of the exhaust air is considerably higher than the wet-bulb temperature of the ambient air, there is a decrease in performance evidenced by an increase in cold (leaving) water temperature.

The induced-draft tower is the most common type used in the United States. It is further classified into counterflow and cross-flow design, depending on the relative flow directions of water and air. Thermodynamically, the counterflow arrangement is more efficient, since the coldest water contacts the coldest air, thus obtaining maximum enthalpy potential. The greater the cooling ranges and the more difficult the approaches, the more distinct are the advantages of the counterflow type. The cross-flow tower manufacturer may effectively reduce the tower characteristic at very low approaches by increasing the air quantity to give a lower L/G ratio. The increase in airflow is not necessarily achieved by increasing the air velocity, but primarily by lengthening the tower to increase the airflow cross-sectional area. It appears then that the cross-flow fill can be made progressively longer in the direction perpendicular to the airflow and shorter in the direction of the airflow until it almost loses its inherent potential-difference disadvantage. However, as this is done, fan power consumption increases.

Ultimately, the economic choice between counterflow and cross-flow is determined by the effectiveness of the fill, design conditions, water quality, and costs of tower manufacture.

Performance of a given type of cooling tower is governed by the ratio of the weights of air to water and the time of contact between water and air. In commercial practice, the variation in the ratio of air to water is first obtained by keeping the air velocity constant at about 1148 m/(min · m^2) of active tower area [350 ft/(min · ft^2 of active tower area)] and varying the water concentration, L/(min · m^2 of ground area) [gal/(min · ft^2 of tower area)]. As a secondary operation, air velocity is varied to make the tower accommodate the cooling requirement.

The time of contact between water and air is governed largely by the time required for the water to discharge from the nozzles and fall through the tower to the basin. The time of contact is therefore obtained in a given type of unit by varying the height of the tower. Should the time of contact be insufficient, no amount of increase in the ratio of air to water will produce the desired cooling. It is therefore necessary to maintain a certain minimum height of cooling tower. When a wide approach of 8 to 11°C (15 to 20°F) to the wet-bulb temperature and a 13.9 to 19.4°C (25 to 35°F) cooling range are required, a relatively low cooling tower will suffice. A tower in which the water travels 4.6 to 6.1 m (15 to 20 ft) from the distributing system to the basin is sufficient. When a moderate approach and a cooling range of 13.9 to 19.4°C (25 to 35°F) are required, a tower in which the water travels 7.6 to 9.1 m (25 to 30 ft) is adequate. Where a close approach of 4.4°C (8°F) with a 13.9 to 19.4°C (25 to 35°F) cooling range is required, a tower is required in which the water travels from 10.7 to 12.2 m (35 to 40 ft). It is usually not economical to design a cooling tower with an approach of less than 2.8°C (5°F).

The cooling performance of any tower containing a given depth of fill varies with the water concentration. It has been found that maximum contact and performance are obtained with a tower having a water concentration of 80 to 200 L/(min · m^2 of ground area) [2 to 5 gal/(min · ft^2 of ground area)]. Thus the problem of calculating the size of a cooling tower becomes one of determining the proper concentration of water required to obtain the desired results. Once the necessary water concentration has been established, the tower area can be calculated by dividing the liters per minute circulated by the water concentration in liters per minute per square meter. The required tower size then is a function of the following:

1. Cooling range (hot water temperature minus cold water temperature)
2. Approach to wet-bulb temperature (cold water temperature minus wet-bulb temperature)
3. Quantity of water to be cooled
4. Wet-bulb temperature
5. Air velocity through the cell
6. Tower height

These considerations in combination with the Markel equation can help engineers with basic conceptual designs that can be used in conjunction with vendors for process design.

Cooling Tower Operation: Water Makeup It is the open nature of evaporative cooling systems, bringing in external air and water continuously, that determines the unique water problems these systems exhibit. Cooling towers (1) concentrate solids by the mechanisms described above and (2) wash air. The result is a buildup of dissolved solids, suspended contaminants, organics, bacteria, and their food sources in the circulating cooling water. These unique evaporative water system problems must be specifically addressed to maintain cooling equipment in good working order.

Makeup requirements for a cooling tower consist of the sum of evaporation loss, drift loss, and blowdown. Therefore,

$$W_m = W_e + W_d + W_b \tag{12-18}$$

where W_m = makeup water, W_e = evaporation loss, W_d = drift loss, and W_b = blowdown (consistent units: m^3/h or gal/min).

Evaporation loss can be estimated by

$$W_e = 0.00085 W_c (T_1 - T_2) \tag{12-19}$$

where W_c = circulating water flow, m^3/h or gal/min, at tower inlet and $T_1 - T_2$ = inlet water temperature minus outlet water temperature, °F. The 0.00085 evaporation constant is a good rule-of-thumb value. The actual evaporation rate will vary by season and climate.

Drift loss can be estimated by

$$W_d = 0.0002 W_c \tag{12-20}$$

Drift is entrained water in the tower discharge vapors. Drift loss is a function of the drift eliminator design and is typically less than 0.02 percent of the water supplied to the tower given the new developments in eliminator design.

Blowdown discards a portion of the concentrated circulating water due to the evaporation process in order to lower the system solids concentration. The amount of blowdown can be calculated according to the number of cycles of concentration required to limit scale formation. The *cycles of concentration* are the ratio of dissolved solids in the recirculating water to dissolved solids in the makeup water. Since chlorides remain soluble on concentration, cycles of concentration are best expressed as the ratio of the chloride contents of the circulating and makeup waters. Thus, the blowdown quantities required are determined from

$$\text{Cycles of concentration} = \frac{W_e + W_b + W_d}{W_b + W_d} \tag{12-21}$$

or

$$W_b = \frac{W_e - (\text{cycles} - 1)\, W_d}{\text{cycles} - 1} \tag{12-22}$$

TABLE 12-6 Blowdown (Percent)

Range, °F	2X	3X	4X	5X	6X
10	0.83	0.41	0.26	0.19	0.15
15	1.26	0.62	0.41	0.30	0.24
20	1.68	0.83	0.55	0.41	0.32
25	2.11	1.04	0.69	0.51	0.41
30	2.53	1.26	0.83	0.62	0.49

Cycles of concentration involved with cooling tower operation normally range from 3 to 5 cycles. For water qualities where operating water concentrations must be below 3 to control scaling, blowdown quantities will be large. The addition of acid or scale-inhibiting chemicals can limit scale formation at higher cycle levels and will allow substantially reduced water usage for blowdown.

The blowdown equation (12-22) translates to calculated percentages of the cooling system circulating water flow exiting to drain, as listed in Table 12-6. The blowdown percentage is based on the cycles targeted and the cooling range. The range is the difference between the system hot water and cold water temperatures.

Example 12-11 Calculation of Makeup Water Determine the amount of makeup required for a cooling tower with the following conditions:

Inlet water flow, m^3/h (gal/min)	2270 (10,000)
Inlet water temperature, °C (°F)	37.77 (100)
Outlet water temperature, °C (°F)	29.44 (85)
Drift loss, percent	0.02
Concentration cycles	5

Evaporation loss [using Eq. (12-19)]:

$$W_e,\ m^3/h = 0.00085 \times 2270 \times (37.77 - 29.44) \times (1.8°F/°C) = 28.9$$
$$W_e,\ gal/min = 127.5$$

Drift loss

$$W_d,\ m^3/h = 2270 \times 0.0002 = 0.45$$
$$W_d,\ gal/min = 2$$

Blowdown

$$W_b,\ m^3/h = 6.8$$
$$W_b,\ gal/min = 29.9$$

Makeup

$$W_m,\ m^3/h = 28.9 + 0.45 + 6.8 = 36.2$$
$$W_m,\ gal/min = 159.4$$

Fans and Pumps The fan and pump power requirements are important considerations in system design since they impact cost and performance. The power requirement of the fan depends on the configuration and the pressure drop/air velocity characteristics of the fill. The power requirement and pressure rating on the pump depend on the tower height and how the incoming water is distributed over the fill.

Fogging and Plume Abatement A phenomenon that occurs in cooling tower operation is fogging, which produces a highly visible plume and possible icing hazards. Fogging results from mixing warm, highly saturated tower discharge air with cooler ambient air that lacks the capacity to absorb all the moisture as vapor. While in the past visible plumes have not been considered undesirable, properly locating towers to minimize possible sources of complaints has now received the necessary attention. In some instances, high fan stacks have been used to reduce ground fog. Although tall stacks minimize the ground effects of plumes, they can do nothing about water vapor saturation or visibility (which can be a safety matter). The persistence of plumes is much greater in periods of low ambient temperatures.

Special care must be taken regarding the placement of cooling towers relative to other buildings to ensure a fresh air supply.

Environmental aspects have caused public awareness and concern over any visible plume, although many laypersons misconstrue cooling tower discharge as harmful. This has resulted in a new development for plume abatement known as a *wet-dry cooling tower configuration.* Reducing the relative humidity or moisture content of the tower discharge stream will reduce the frequency of plume formation. Figure 12-9 shows a "parallel path" arrangement that has been demonstrated to be technically sound but at substantially increased tower investment. Ambient air travels in parallel streams through the top dry-surface section and the evaporative section. Both sections benefit thermally by receiving cooler ambient air with the wet and dry airstreams mixing after leaving their respective sections. Water flow is arranged in series, flowing first to the dry coil section and then to the evaporation fill section. A "series path" airflow arrangement, in which dry coil sections can be located before or after the air traverses the evaporative section, also can be used. However, series-path airflow has the disadvantage of water impingement, which could result in coil scaling and restricted airflow.

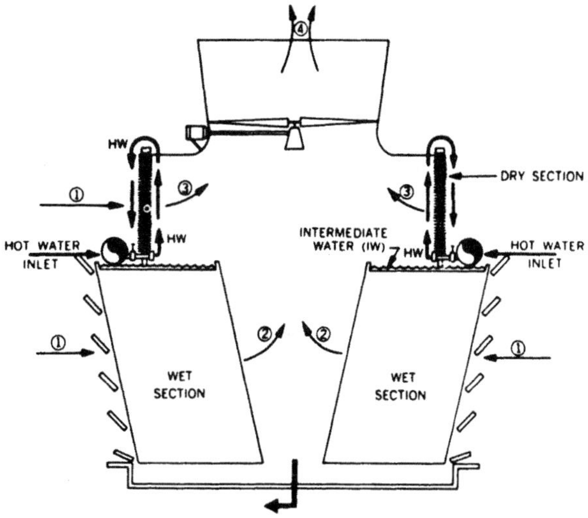

FIG. 12-9 Parallel-path cooling-tower arrangement for plume abatement. (*Marley Co.*)

Natural Draft Towers, Cooling Ponds, and Spray Ponds Natural draft towers are primarily suited to very large cooling water quantities, and the reinforced concrete structures used are as large as 80 m (260 ft) in diameter and 105 m (340 ft) high.

When large ground areas are available, large cooling ponds offer a satisfactory method of removing heat from water. A pond may be constructed at a relatively small investment by pushing up earth in an earth dike 2 to 3 m (6 to 9 ft) high.

Spray ponds provide an arrangement for lowering the temperature of water by evaporative cooling and in so doing greatly reduce the cooling area required in comparison with a cooling pond.

Natural draft towers, cooling ponds, and spray ponds are infrequently used in new construction today in the chemical processing industry. Additional information may be found in the 7th edition of Perry's Handbook.

WET SURFACE AIR COOLERS (WSACs)

GENERAL REFERENCES: Kals, "Wet Surface Aircoolers," *Chem. Engg.* July 1971; Kals, "Wet Surface Aircoolers: Characteristics and Usefulness," AIChE-ASME Heat Transfer Conference, Denver, CO., August 6–9, 1972; Elliott and Kals, "Air Cooled Condensers," *Power,* January 1990; Kals, "Air Cooled Heat Exchangers: Conventional and Unconventional," *Hydrocarbon Processing,* August 1994; Hutton, "Properly Apply Closed Circuit Evaporative Cooling," *Chem. Engg. Progress,* October 1996; Hutton, "Improved Plant Performance through Evaporative Steam Condensing," ASME 1998 International Joint Power Conference, Baltimore, Md., August 23–26, 1998; http://www.niagarablower.com/; http://www.baltimoreaircoil.com.

Principles Rejection of waste process heat through a cooling tower (CT) requires transferring the heat in two devices in series, using two different methods of heat transfer. This requires two temperature driving forces in series: first, sensible heat transfer from the process stream across the *heat exchanger* (HX) into the cooling water, and, second, sensible and latent heat transfer from the cooling water to atmosphere across the CT. Rejecting process heat with a wet surface air cooler transfers the waste heat in a single device by using a single-unit operation. The single required temperature driving force is lower because the WSAC does not require the use of cooling water sensible heat to transfer heat from the process stream to the atmosphere. A WSAC tube cross section (Fig. 12-10) shows the characteristic external tube surface having a continuous flowing film of evaporating water, which cascades through the WSAC tube bundle.

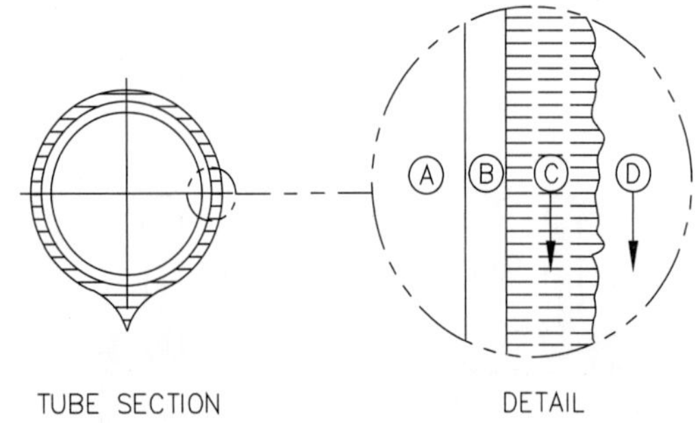

FIG. 12-10 WSAC tube cross section. Using a small *T*, heat flows from (A) the process stream, through (B) the tube, through (C) the flowing film of evaporating water, into (D) flowing ambient air.

Consequently, process streams can be economically cooled to temperatures much closer to the ambient wet-bulb temperature, as low as to within 2.2°C (4°F), depending on the process requirements and economics for the specific application.

Wet Surface Air Cooler Basics The theory and principles for the design of WSACs are a combination of those known for evaporative cooling tower design and for HX design. However, the design practices for engineering WSAC equipment remain a largely proprietary, technical art, and the details are not presented here. Any evaluation of the specifics and economics for a particular application requires direct consultation with a reputable vendor.

Because ambient air is contacted with evaporating water within a WSAC, from a distance a WSAC has a similar appearance to a CT (Fig. 12-11). Economically optimal plot plan locations for WSACs can vary: integrated into, or with, the process structure, remote to it, in a pipe rack, etc.

In the WSAC the evaporative cooling occurs on the wetted surface of the tube bundle. The wetting of the tube bundle is performed by recirculating water the short vertical distance from the WSAC collection basin, through the spray nozzles, and onto the top of the bundle (Fig. 12-12). The tube bundle is completely deluged with this cascading flow of water. Using water application rates between 12 and 24 $(m^3/h)/m^2$ (5 and 10 gpm/ft^2), the tubes have a continuous, flowing external water film, minimizing the potential for water-side biological fouling, sediment deposition, etc. Process inlet temperatures are limited to a maximum of about 85°C (185°F), to prevent external water-side mineral scaling. However, higher process inlet temperatures can be accepted by incorporating bundles of dry, air-cooled finned tubing within the WSAC unit, to reduce the temperature of the process stream to an acceptable level before it enters the wetted evaporative tube bundles.

FIG. 12-11 Overhead view of a single-cell WSAC.

FIG. 12-12 Nozzles spraying onto wetted tube bundle in a WSAC unit.

The WSAC combines within one piece of equipment the functions of cooling tower, circulated cooling water system, and water-cooled HX. In the basic WSAC configuration (Fig. 12-13), ambient air is drawn in and down through the tube bundle. This airflow is cocurrent with the evaporating water flow, recirculated from the WSAC collection basin sump to be sprayed over the tube bundles. This downward cocurrent flow pattern minimizes the generation of water mist (drift). At the bottom of the WSAC, the air changes direction through 180°, disengaging entrained fine water droplets. Drift eliminators can be added to meet very low drift requirements. Because heat is extracted from the tube surfaces by water latent heat (and not sensible heat), only about 75 percent as much circulating water is required in comparison to an equivalent CT-cooling water heat exchange application.

The differential head of the circulation water pump is relatively small, since dynamic losses are modest (short vertical pipe and a low ΔP spray nozzle) and the hydraulic head is small, only about 6 m (20 ft) from the basin to the elevation of the spray header. Combined, the pumping energy demand is about 35 percent that for an equivalent CT application. The capital cost for this complete water system is also relatively small. The pumps and motors are smaller, the piping has a smaller diameter and is much shorter, and the

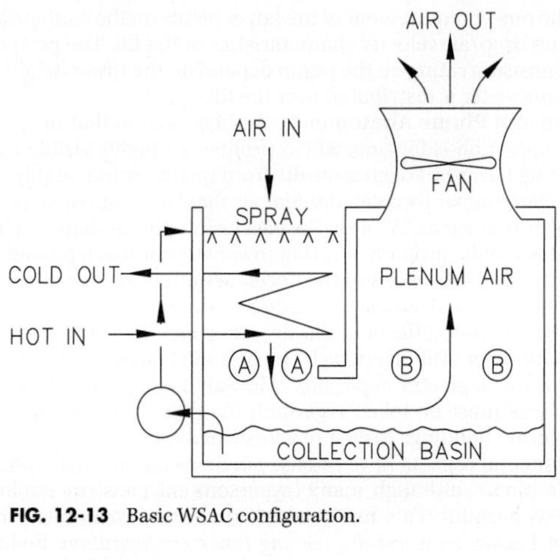

FIG. 12-13 Basic WSAC configuration.

required piping structural support is almost negligible, compared to an equivalent CT application. WSAC fan horsepower is typically about 25 percent less than that for an equivalent CT.

A WSAC is inherently less sensitive to water-side fouling. This is so because the deluge rate prevents the adhesion of waterborne material which can cause fouling within a HX. A WSAC can accept relatively contaminated makeup water, such as CT blowdown, treated sewage plant effluent, etc. WSACs can endure more cycles of concentration without fouling than can a CT application. This higher practical operating concentration reduces the relative volume for the evaporative cooling blowdown, and therefore it also reduces the relative volume of required makeup water. For facilities designed for zero liquid discharge, the higher practical WSAC blowdown concentration reduces the size and the operating costs for the downstream water treatment system. Since a hot process stream provides the unit with a heat source, a WSAC has intrinsic freeze protection while operating.

Common WSAC Applications and Configurations Employment of a WSAC can reduce process system operating costs that are not specific to the WSAC unit itself. A common WSAC application is condensation of compressed gas (Fig. 12-14). A compressed gas can be condensed in a WSAC at a lower pressure, by condensing at a temperature closer to the ambient wet-bulb temperature, typically 5.5°C (10°F) above the wet-bulb temperature. This reduced condensation pressure reduces costs, by reducing the gas compressor motor operating horsepower. Consequently, WSACs are widely applied for condensing refrigerant gases, for HVAC, process chillers, ice makers, gas-turbine inlet air cooling, chillers, etc. WSACs are also used directly to condense lower-molecular-weight hydrocarbon streams, such as ethane, ethylene, propylene, and LPG. A related WSAC application is the cooling of compressed gases (CO_2, N_2, methane, LNG, etc.), which directly reduces gas compressor operating costs (inlet and interstage cooling) and indirectly reduces downstream condensing costs (after cooling the compressed gas to reduce the downstream refrigeration load).

For combined-cycle electric power generation, employment of a WSAC increases steam turbine efficiency. Steam turbine exhaust can be condensed at a lower pressure (higher vacuum) by condensing at a temperature closer to the ambient wet-bulb temperature, typically 15°C (27°F) above the wet-bulb temperature. This reduced condensation pressure results in a lower turbine discharge pressure, increasing electricity generation by increasing output shaft power (Fig. 12-15). Due to standard WSAC configurations, a second cost advantage is gained at the turbine itself. The steam turbine can be placed at grade, rather than being mounted on an elevated platform, by venting horizontally into the WSAC, rather than venting downward to condensers located below the platform elevation, as is common for conventional water-cooled vacuum steam condensers.

A WSAC can eliminate chilled water use, for process cooling applications with required temperatures close to and just above the ambient wet-bulb temperature, typically about 3.0 to 5.5°C (5 to 10°F) above the wet-bulb temperature. This WSAC application can eliminate both chiller capital and operating costs. In such an application, either the necessary process temperature is below the practical CT water supply temperature, or they are so close to it that the use of CT water is uneconomical (a low-HX log-mean temperature difference).

WSACs can be designed to simultaneously cool several process streams in parallel separate tube bundles within a single cell of a WSAC (Fig. 12-16). Often one of the streams is closed-circuit cooling water to be used for remote cooling applications. These might be applications not compatible with a WSAC (rotating seals, bearings, cooling jackets, internal reactor cooling coils, etc.) or merely numerous, small process streams in small HXs.

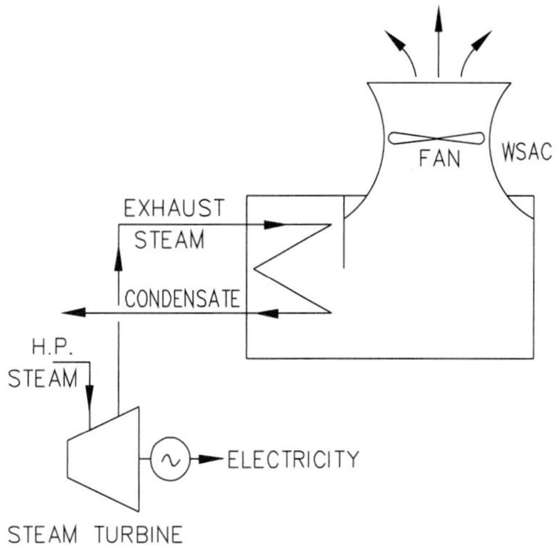

FIG. 12-15 WSAC configuration with electricity generation. A lower steam condensing pressure increases the turbine horsepower extracted.

WSAC for Closed-Circuit Cooling Systems A closed-circuit cooling system as defined by the Cooling Technology Institute (CTI) (www.cti.org) employs a closed loop of circulated fluid (typically water) remotely as a cooling medium. By definition, this medium is cooled by water evaporation involving no direct fluid contact between the air and the enclosed circulated cooling medium. Applied in this manner, a WSAC can be used as the evaporative device to cool the circulated cooling medium, used remotely to cool process streams. This configuration completely isolates the cooling water (and the hot process streams) from the environment (Fig. 12-17).

The closed circuit permits complete control of the cooling water chemistry, which permits minimizing the cost for water-side materials of construction and eliminating water-side fouling of, and fouling heat-transfer resistance in, the heat exchangers (or jackets, reactor coils, etc.). Elimination of water-side fouling is particularly helpful for high-temperature cooling applications, especially where heat recovery may otherwise be impractical (quench oils, low-density polyethylene reactor cooling, etc.).

Closed-circuit cooling minimizes circulation pumping horsepower, which must overcome only dynamic pumping losses. This results through recovery of the returning circulated cooling water hydraulic head. A closed-circuit system can be designed for operation at elevated pressures, to guarantee that any process heat-transfer leak will be into the process. Such high-pressure operation is economical, since the system overpressure is not lost during return flow to the circulation pump.

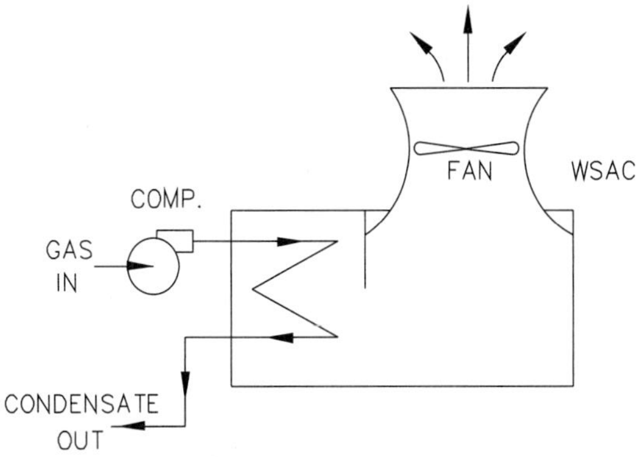

FIG. 12-14 WSAC configuration for condensing a compressed gas. A lower condensing pressure reduces compressor operating horsepower.

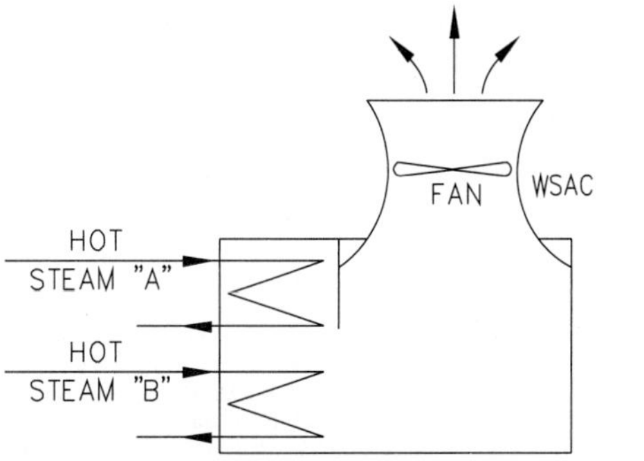

FIG. 12-16 WSAC configuration with parallel streams.

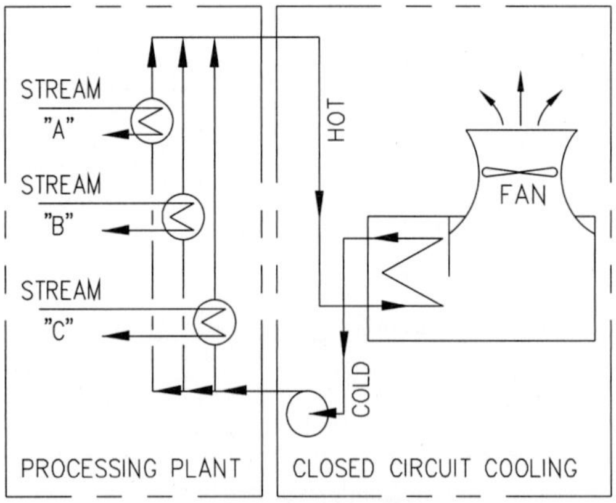

FIG. 12-17 WSAC configuration with no direct fluid contact.

Closed-circuit cooling splits the water chemistry needs into two isolated systems: the evaporating section, exposed to the environment, and the circulated cooling section, isolated from the environment. Typically, this split reduces total water chemistry costs and water-related operations and maintenance problems. However, the split permits the effective use of a low-quality or contaminated makeup water for evaporative cooling, or a water source having severe seasonal quality problems, such as high sediment loadings.

If highly saline water is used for the evaporative cooling, a reduced flow of makeup saline water would need to be supplied to the WSAC. This reduction results from using latent cooling rather than sensible cooling to reject the waste heat. This consequence reduces the substantial capital investment required for the saline water supply and return systems (canal structures) and pump stations, and the saline supply pumping horsepower. (When saline water is used as the evaporative medium, special attention is paid to materials of construction and spray water chemical treatment due to the aggravated corrosion and scaling tendencies of this water.)

Water Conservation Applications—"Wet-Dry" Cooling A modified and hybridized form of a WSAC can be used to provide what is called *wet-dry* cooling for water conservation applications (Fig. 12-18). A hybridized combination of air-cooled dry finned tubes, standard wetted bare tubes, and wet deck surface area permits the WSAC to operate without water in cold weather, reducing water consumption by about 75 percent of the total for an equivalent CT application.

Under design conditions of maximum summer wet-bulb temperature, the unit operates with spray water deluging the wetted tube bundle. The exiting water then flows down into and through the wet deck surface, where the water is cooled adiabatically to about the wet-bulb temperature and then to the sump.

As the wet-bulb temperature drops, the process load is shifted from the wetted tubes to the dry finned tubes. By bypassing the process stream around the wetted tubes, cooling water evaporation (consumption) is proportionally reduced.

When the wet-bulb temperature drops to the switch point, the process bypassing has reached 100 percent. This switch point wet-bulb temperature is at or above 5°C (41°F). As the ambient temperature drops further, adiabatic evaporative cooling continues to be used, to lower the dry-bulb temperature to below the switch point temperature. This guarantees that the entire cooling load can be cooled in the dry finned tube bundle.

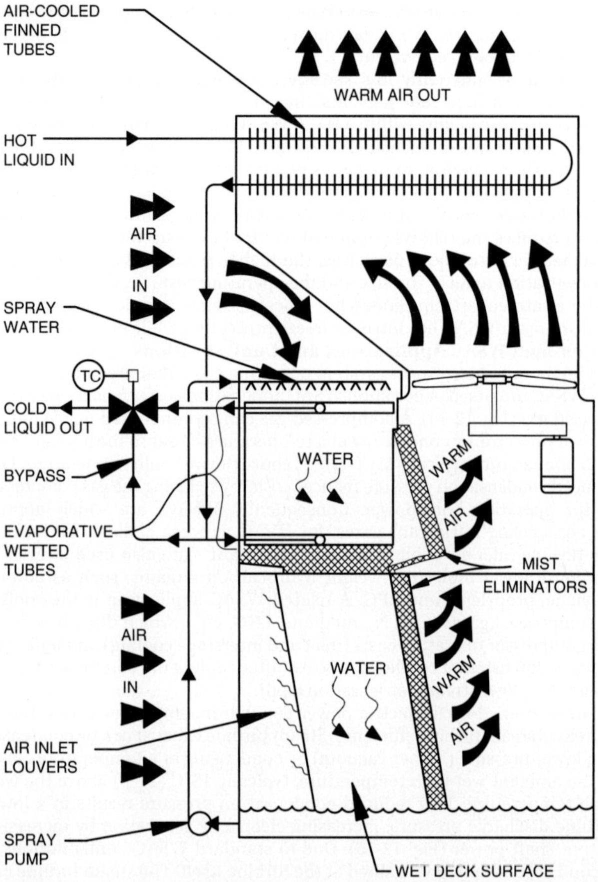

FIG. 12-18 As seasonal ambient temperatures drop, the "wet-dry" configuration for a WSAC progressively shifts the cooling load from evaporative to convective cooling.

The use of water is discontinued after ambient dry-bulb temperatures fall below the switch point temperature, since the entire process load can be cooled using only cold fresh ambient air. By using this three-step load-shifting practice, total wet-dry cooling water consumption is about 25 percent of that consumption total experienced with an equivalent CT application.

Wet-dry cooling permits significant reduction of water consumption, which is useful where makeup water supplies are limited or where water treatment costs for blowdown are high. Because a WSAC (unlike a CT) has a heat source (the hot process stream), wet-dry cooling avoids various cold-weather-related CT problems. Fogging and persistent plume formation can be minimized or eliminated during colder weather. Freezing and icing problems can be eliminated by designing a wet-dry system for water-free operation during freezing weather, typically below 5°C (41°F). In the arctic, or regions of extreme cold, elimination of freezing fog conditions is realized by not evaporating any water during freezing weather.

SOLIDS-DRYING FUNDAMENTALS

General References: Cook and DuMont, *Process Drying Practice,* McGraw-Hill, New York, 1991. *Drying Technology—An International Journal,* Taylor and Francis, New York. Hall, *Dictionary of Drying,* Marcel Dekker, New York, 1979. Keey, *Introduction to Industrial Drying Operations,* Pergamon, New York, 1978. Keey, *Drying of Loose and Particulate Materials,* Hemisphere, New York, 1992. Masters, *Spray Drying Handbook,* Wiley, New York, 1990. Mujumdar, *Handbook of Industrial Drying,* Marcel Dekker, New York, 1987. Strumillo and Kudra, *Drying: Principles, Application and Design,* Gordon and Breach, New York, 1986. van't Land, *Industrial Drying Equipment,* Marcel Dekker, New York, 1991. Tsotsas and Mujumdar, eds., *Modern Drying Technology* (vols. 1 to 6), Wiley, New York, 2011. *Aspen Process Manual* (Internet knowledge base), Aspen Technology, Boston, 2000 onward.

INTRODUCTION

Drying is the process by which volatile materials, usually water, are evaporated from a material to yield a solid product. Drying is a heat- and mass-transfer process. Heat is necessary to evaporate water. The latent heat of vaporization of water is about 2500 J/g, which means that the drying process requires a significant amount of energy. Simultaneously, the evaporating material must leave the drying material by diffusion and/or convection.

Heat transfer and mass transfer are not the only concerns when one is designing or operating a dryer. The product quality (color, particle density, hardness, texture, flavor, etc.) is also very strongly dependent on the drying conditions and the physical and chemical transformations occurring in the dryer.

Understanding and designing a drying process involves measurement and/or calculation of the following:

1. Mass and energy balances
2. Thermodynamics
3. Mass- and heat-transfer rates
4. Product quality considerations

The subsection below explains how these factors are measured and calculated and how the information is used in engineering practice.

TERMINOLOGY

Generally accepted terminology and definitions are given alphabetically in the following paragraphs.

Absolute humidity is the mass ratio of water vapor (or other solvent mass) to dry air.

Activity is the ratio of the fugacity of a component in a system relative to the standard-state fugacity. In a drying system, it is the ratio of the vapor pressure of a solvent (e.g., water) in a mixture to the pure solvent vapor pressure at the same temperature. Boiling occurs when the vapor pressure of a component in a liquid exceeds the ambient total pressure.

Bound moisture in a solid is that liquid which exerts a vapor pressure less than that of the pure liquid at the given temperature. Liquid may become bound by retention in small capillaries, by solution in cell or fiber walls, by homogeneous solution throughout the solid, by chemical or physical adsorption on solid surfaces, and by hydration of solids.

Capillary flow is the flow of liquid through the interstices and over the surface of a solid, caused by liquid-solid molecular attraction.

Constant-rate period (unhindered) is that drying period during which the rate of water removal per unit of drying surface is constant, assuming the driving force is also constant.

Convection is heat or mass transport by bulk flow.

Critical moisture content is the average moisture content when the constant-rate period ends, assuming the driving force is also constant.

Diffusion is the molecular process by which molecules, moving randomly due to thermal energy, migrate from regions of high chemical potential (usually concentration) to regions of lower chemical potential.

Dry basis expresses the moisture content of wet solid as kilograms of water per kilogram of bone-dry solid.

Equilibrium moisture content is the limiting moisture to which a given material can be dried under specific conditions of air temperature and humidity.

Evaporation is the transformation of material from a liquid state to a vapor state.

Falling-rate period (hindered drying) is a drying period during which the instantaneous drying rate continually decreases.

Free moisture content is that liquid which is removable at a given temperature and humidity. It may include bound and unbound moisture.

Hygroscopic material is material that may contain bound moisture.

Initial moisture distribution refers to the moisture distribution throughout a solid at the start of drying.

Latent heat of vaporization is the specific enthalpy change associated with evaporation.

Moisture content of a solid is usually expressed as moisture quantity per unit weight of the dry or wet solid.

Moisture gradient refers to the distribution of water in a solid at a given moment in the drying process.

Nonhygroscopic material is material that can contain no bound moisture.

Permeability is the resistance of a material to bulk or convective, pressure-driven flow of a fluid through it.

Relative humidity is the partial pressure of water vapor divided by the vapor pressure of pure water at a given temperature. In other words, the relative humidity describes how close the air is to saturation.

Sensible heat is the energy required to increase the temperature of a material without changing the phase.

Unaccomplished moisture change is the ratio of the free moisture present at any time to that initially present.

Unbound moisture in a hygroscopic material is that moisture in excess of the equilibrium moisture content corresponding to saturation humidity. All water in a nonhygroscopic material is unbound water.

Vapor pressure is the partial pressure of a substance in the gas phase that is in equilibrium with a liquid or solid phase of the pure component.

Wet basis expresses the moisture in a material as a percentage of the weight of the wet solid. Use of a dry-weight basis is recommended since the percentage change of moisture is constant for all moisture levels. When the wet-weight basis is used to express moisture content, a 2 or 3 percent change at high moisture contents (above 70 percent) actually represents a 15 to 20 percent change in evaporative load. See Fig. 12-19 for the relationship between the dry- and wet-weight bases.

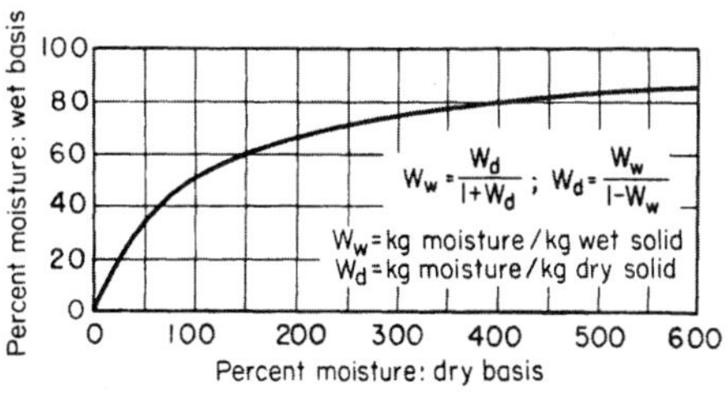

FIG. 12-19 Relationship between wet-weight and dry-weight bases.

THERMODYNAMICS

The thermodynamic driving force for evaporation is the difference in chemical potential or water activity between the drying material and the gas phase. Although drying of water is discussed in this subsection, the same concepts apply analogously for solvent drying.

For a pure water drop, the driving force for drying is the difference between the vapor pressure of water and the partial pressure of water in the gas phase. The rate of drying is proportional to this driving force; please see the discussion on drying kinetics later in this section.

$$\text{Rate} \propto \left(p_{\text{pure}}^{\text{sat}} - p_{w,\text{air}} \right) \tag{12-23}$$

The *activity* of water in the gas phase is defined as the ratio of the partial pressure of water to the vapor pressure of pure water, which is also related to the definition of *relative humidity*.

$$a_w^{\text{vapor}} = \frac{p_w}{p_{\text{pure}}^{\text{sat}}} = \frac{\%\text{RH}}{100} \tag{12-24}$$

The *activity* of water in a mixture or solid is defined as the ratio of the vapor pressure of water in the mixture to that of a reference, usually the vapor pressure of pure water. In solids drying or drying of solutions, the vapor pressure (or water activity) is lower than that for pure water. Therefore, the water activity value equals 1 for pure water and is less than 1 when binding is occurring. This is caused by thermodynamic interactions between the water and the drying material. In many standard drying references, this is called *bound water.*

$$a_w^{\text{solid}} = \frac{p_{\text{mixture}}^{\text{sat}}}{p_{\text{pure}}^{\text{sat}}} \tag{12-25}$$

When a solid sample is placed into a humid environment, water will transfer from the solid to the air or vice versa until equilibrium is established. At thermodynamic equilibrium, the water activity is equal in both phases:

$$a_w^{\text{vapor}} = a_w^{\text{solid}} = a_w \tag{12-26}$$

Sorption isotherms quantify how tightly water is bound to a solid. This is a result of chemical interactions, such as hydrogen bonding, between the solid and the water. The goal of obtaining a sorption isotherm for a given solid is to measure the equilibrium relationship between the percentage of water in the sample and the vapor pressure of the mixture. The sorption isotherm describes how dry a product can get if contacted with humid air for an infinite amount of time. An example of a sorption isotherm is shown in Fig. 12-20. In the sample isotherm, a feed material dried with 50 percent relative humidity air ($a_w = 0.5$) will approach a moisture content of 10 percent on a dry basis. Likewise, a material kept in a sealed container will create a headspace humidity according to the isotherm; a 7 percent moisture sample in the example below will create a 20 percent relative humidity ($a_w = 0.2$) headspace in a sample jar or package.

Strictly speaking, the equilibrium moisture content of the sample in a given environment should be independent of the initial condition of

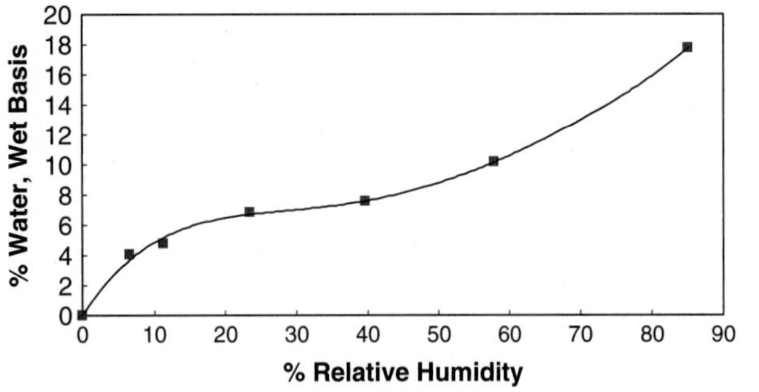

FIG. 12-20 Example of a sorption isotherm (coffee at 22°C).

the sample. However, in some cases the sorption isotherm of an initially wet sample (sometimes called a *desorption isotherm*) is different from that of an identical, but initially dry sample. This is called *hysteresis* and can be caused by irreversible changes in the sample during wetting or drying, micropore geometry in the sample, and other factors. Paper products are notorious for isotherm hysteresis. Most materials show little or no hysteresis.

Sorption isotherms cannot generally be predicted from theory. They need to be measured experimentally. The simplest method of measuring a sorption isotherm is to generate a series of controlled-humidity environments by using saturated salt solutions, allow a solid sample to equilibrate in each environment, and then analyze the solid for moisture content.

The basic apparatus is shown in Fig. 12-21, and a table of salts is shown in Table 12-7. It is important to keep each chamber sealed and to be sure that crystals are visible in the salt solution to ensure that the liquid is saturated. Additionally, the solid should be ground into a powder to facilitate mass transfer. Equilibration can take 2 to 3 weeks. Successive moisture measurements should be used to ensure that the sample has equilibrated, i.e., achieved a steady value. Care must be taken when measuring the moisture content of a sample; this is described later in this section.

Another common method of measuring a sorption isotherm is to use a dynamic vapor sorption device. This machine measures the weight change of a sample when exposed to humidity-controlled air. A series of humidity points are programmed into the unit, and it automatically delivers the proper humidity to the sample and monitors the weight. When the weight is stable, an equilibrium point is noted and the air humidity is changed to reflect the next setting in the series. When one is using this device, it is critical to measure and record the starting moisture of the sample, since the results are often reported as a percentage of change rather than a percentage of moisture.

There are several advantages to the dynamic vapor sorption device. First, any humidity value can be dialed in, whereas salt solutions are not available for every humidity value and some are quite toxic. Second, since the weight is monitored as a function of time, it is clear when equilibrium is reached; however, this can be a slow process, so care must be taken to ensure equilibrium is actually achieved. The dynamic devices also give the sorption/desorption rates, although these can easily be misused (see the drying kinetics subsection later). The salt solution method, however, is significantly less expensive to buy and maintain. Samples created in salt solution chambers can also be qualitatively assessed for physical characteristics such as stickiness, flowability, or deliquescence.

An excellent reference on all aspects of sorption isotherms is that by Bell and Labuza, *Moisture Sorption,* 2d ed., American Association of Cereal Chemists, St. Paul, Minnesota, 2000.

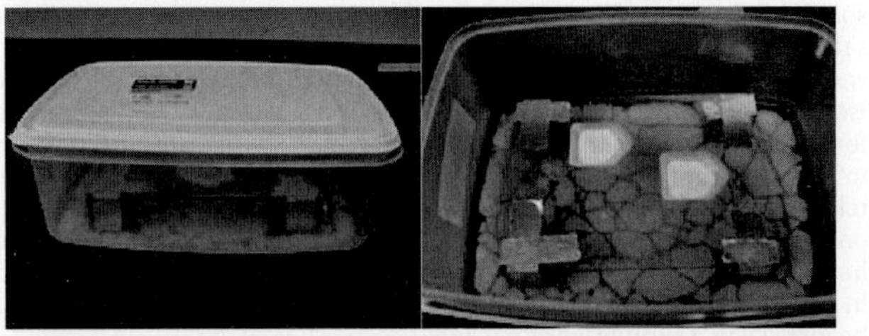

FIG. 12-21 Sorption isotherm apparatus. A saturated salt solution is in the bottom of the sealed chamber; samples sit on a tray in the headspace.

TABLE 12-7 Maintenance of Constant Humidity

Saturated salt solution		% Relative humidity at 25°C
Lithium bromide	LiBr	6.37
Lithium chloride	LiCl	11.30
Potassium acetate	$KC_2H_3O_2$	22.51
Sodium iodide	NaI	38.17
Sodium bromide	NaBr	57.57
Sodium chloride	NaCl	75.29
Potassium chloride	KCl	84.34

MECHANISMS OF MOISTURE TRANSPORT WITHIN SOLIDS

Drying requires moisture to travel to the surface of a material. There are several mechanisms by which this can occur:

1. *Diffusion of moisture through solids.* Diffusion is a molecular process, brought about by random wanderings of individual molecules. If all the water molecules in a material are free to migrate, they tend to diffuse from a region of high moisture concentration to one of lower moisture concentration, thereby reducing the moisture gradient and equalizing the concentration of moisture.
2. *Convection of moisture within a liquid or slurry.* If a flowable solution is drying into a solid, then liquid motion within the material brings wetter material to the surface.
3. *Evaporation of moisture within a solid and gas transport out of the solid by diffusion and/or convection.* Evaporation can occur within a solid if it is boiling or porous. Subsequently vapor must move out of the sample.
4. *Capillary flow of moisture in porous media.* The reduction of liquid pressure within small pores due to surface tension forces causes liquid to flow in porous media by capillary action.

DRYING KINETICS

This subsection discusses the rate of drying. The kinetics of drying dictate the size of industrial drying equipment, which directly affects the capital and operating costs of a process involving drying. The rate of drying can also influence the quality of a dried product since other simultaneous phenomena, such as heat transfer, shrinkage, microstructure development, and chemical reactions, are often affected by the moisture and temperature history of the material.

The most classical drying kinetics problem is that of a pure water drop drying in air, as shown in Example 12-12.

Example 12-12 Drying of a Pure Water Drop See Marshall, *Atomization & Spray Drying,* 1986. Calculate the time to dry a drop of water, given the air temperature and relative humidity as a function of drop size.

Solution Assume that the drop is drying at the wet-bulb temperature. Begin with an energy balance [Eq. (12-27)]

$$\text{Mass flux} = \frac{h\,(T_{\text{air}} - T_{\text{drop}})}{\Delta H_{\text{vap}}} \tag{12-27}$$

Next, a mass balance is performed on the drop. The change in mass equals the flux times the surface area.

$$\frac{\rho\, dV_{\text{droplet}}}{dt} = -A \times \text{mass flux} \tag{12-28}$$

Evaluating the area and volume for a sphere gives

$$\rho \cdot 4\pi R^2 \frac{dR}{dt} = -4\pi R^2 \times \text{mass flux} \tag{12-29}$$

Combining Eqs. (12-28) and (12-29) and simplifying give

$$\rho \frac{dR}{dt} = \frac{-h\,(T_{\text{air}} - T_{\text{drop}})}{\Delta H_{\text{vap}}} \tag{12-30}$$

A standard correlation for heat transfer to a sphere is given by Ranz and Marshall, "Evaporation from Drops," *Chem. Eng. Prog.* **48**(3): 141–146 and **48**(4): 173–180 (1952), as

$$\text{Nu} = \frac{h(2R)}{k_{\text{air}}} = 2 + 0.6\,\text{Re}^{0.5}\text{Pr}^{0.33} \tag{12-31}$$

For small drop sizes or for stagnant conditions, the Nusselt number has a limiting value of 2.

$$\text{Nu} = \frac{h(2R)}{k_{\text{air}}} = 2 \tag{12-32}$$

$$h = \frac{k_{\text{air}}}{R} \tag{12-33}$$

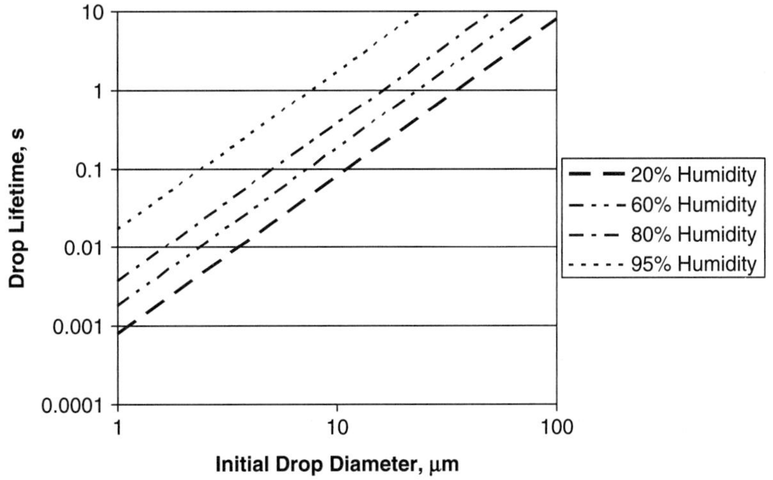

FIG. 12-22 Drying time of pure water drops as function of relative humidity at 25°C.

Insertion into Eq. (12-30) gives

$$R\frac{dR}{dt} = \frac{k_{air}(T_{air} - T_{drop})}{\rho\, \Delta H_{vap}} \tag{12-34}$$

Integration yields

$$\frac{R_0^2}{2} - \frac{R^2}{2} = \frac{k_{air}(T_{air} - T_{drop}) \cdot t}{\rho \Delta H_{vap}} \tag{12-35}$$

where R_0 = initial drop radius, m.

Now the total lifetime of a drop can be calculated from Eq. (12-35) by setting $R = 0$:

$$t = \frac{\rho\, \Delta H_{vap}\, R_0^2}{2k_{air}(T_{air} - T_{drop})} \tag{12-36}$$

The effects of drop size and air temperature are readily apparent from Eq. (12-36). The temperature of the drop is the wet-bulb temperature and can be obtained from a psychrometric chart, as described in the previous subsection. Sample results are plotted in Fig. 12-22.

The above solution for drying of a pure water drop cannot be used to predict the drying rates of drops containing solids. Drops containing solids will not shrink uniformly and will develop internal concentration gradients (falling-rate period) in most cases.

Drying Curves and Periods of Drying The most basic and essential kinetic information on drying of solid materials is a *drying curve.* A drying curve describes the drying kinetics and how they change during drying. The drying curve is affected by the material properties, size or thickness of the drying material, and drying conditions. In this subsection, the general characteristics of drying curves and their uses are described. Experimental techniques to obtain drying curves are discussed in the Experimental Methods subsection.

Several representations of a typical drying curve are shown in Fig. 12-23. The top plot, Fig. 12-23*a,* is the moisture content (dry basis) as a function of time. The middle plot, Fig. 12-23*b,* is the drying rate as a function of time, the derivative of the top plot. The bottom plot, Fig. 12-23*c,* is the drying rate as affected by the average moisture content of the drying material. Since the material loses moisture as time passes, the progression of time in this bottom plot is from right to left.

Some salient features of the drying curve show the different periods of drying. These are common periods, but not all occur in every drying process. The first period of drying is called the *induction* period. This period occurs when material is being heated early in drying. The second period of drying is called the *constant-rate* period. During this period, the surface remains wet enough to maintain the vapor pressure of water on the surface. Once the surface dries sufficiently, the drying rate decreases and the falling-rate period occurs. This period can also be referred to as *hindered drying.*

Figure 12-23 shows the transition between constant- and falling-rate periods of drying occurring at the critical point. The *critical point* refers to the average moisture content of a material at this transition. This is a useful concept, but the critical point can depend on drying conditions.

The subsections below show examples of drying curves and the phenomena that give rise to different common types.

Introduction to Internal and External Mass-Transfer Control—Drying of a Slab The concepts in drying kinetics are best illustrated with a simple example—air drying of a slab. Consider a thick slab of homogeneous wet material, as shown in Fig. 12-24. In this particular example, the slab is dried on an insulating surface under constant conditions. The heat for drying is carried to the surface with hot air, and air carries water vapor

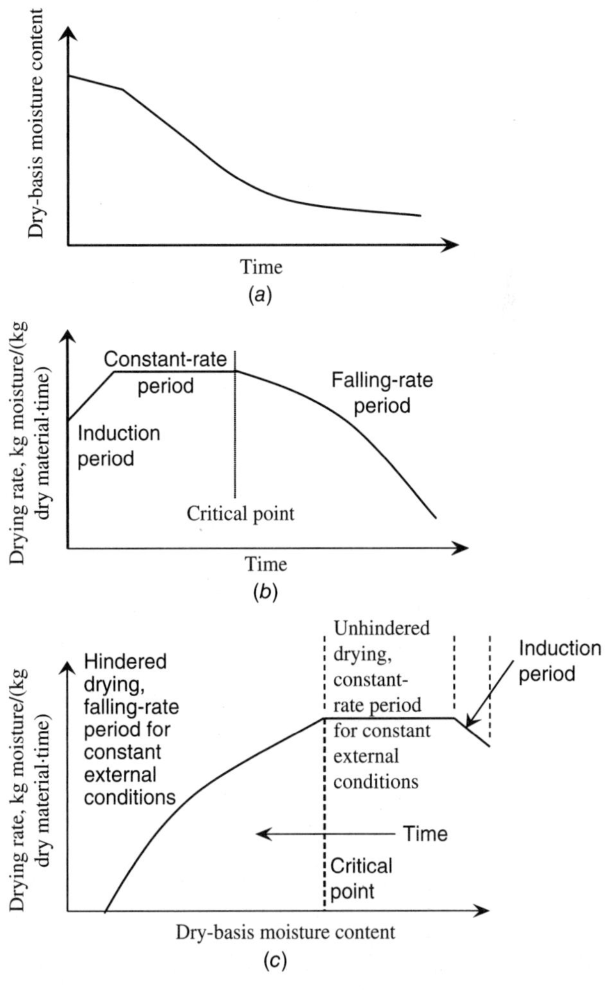

FIG. 12-23 Several common representations of a typical drying curve.

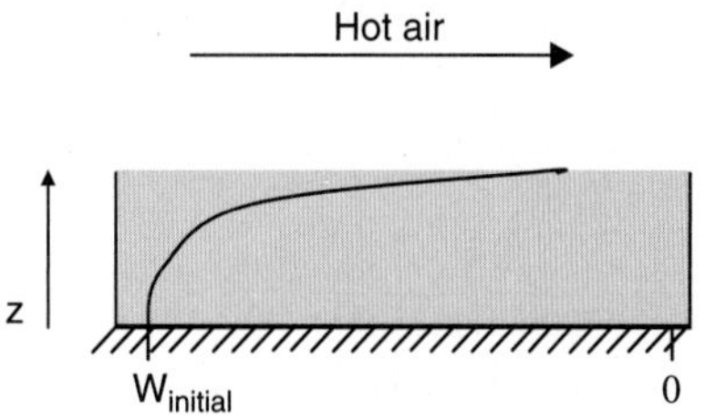

FIG. 12-24 Drying of a slab.

from the surface. At the same time, a moisture gradient forms within the slab, with a dry surface and a wet interior. The curved line is the representation of the gradient. At the bottom of the slab ($z = 0$), the material is wet and the moisture content is drier at the surface.

The following processes must occur to dry the slab:

1. Heat transfer from the air to the surface of the slab
2. Mass transfer of water vapor from the surface of the slab to the bulk air
3. Mass transfer of moisture from the interior of the slab to the surface of the slab

Depending on the drying conditions, thickness, and physical properties of the slab, any of the above steps can be rate-limiting. Figure 12-25 shows two examples of rate-limiting cases. Example 12-14 shows how to compute these from physical property data and how the same material can exhibit different drying curves.

The top example shows the situation of *external rate control.* In this situation, the heat transfer to the surface and/or the mass transfer from the surface to the vapor phase is slower than mass transfer to the surface from the bulk of the drying material. In this limiting case, the moisture gradient in the material is minimal, and the rate of drying will be constant as long as the average moisture content remains high enough to maintain a high water activity (see the subsection on thermodynamics for a discussion of the relationship between moisture content and water vapor pressure). External rate control leads to the observation of a constant-rate period drying curve.

The bottom example shows the opposite situation: *internal rate control.* In the case of heating from the top, internal control refers to a slow rate of mass transfer from the bulk of the material to the surface of the material. Diffusion, convection, and capillary action (in the case of porous media) are possible mechanisms for mass transfer of moisture to the surface of the slab. In the internal rate control situation, moisture is removed from the surface by the air faster than moisture is transported to the surface. This regime is caused by relatively thick layers or high values of the mass- and heat-transfer coefficients in the air. Internal rate control leads to the observation of a falling-rate period drying curve.

Generally speaking, drying curves show both behaviors. When drying begins, the surface is often wet enough to maintain a constant-rate period and is therefore externally controlled. But as the material dries, the mass-transfer rate of moisture to the surface often slows, causing the rate to decrease since the lower moisture content on the surface causes a lower water vapor pressure. However, some materials begin dry enough that there is no observable constant-rate period.

Note that falling-rate periods do sometimes occur when drying is externally controlled. The drying rate depends on the water activity at the surface of the material, and the rate, by itself, is not a measure of an internal moisture gradient. The observation of falling-rate periods with external rate control is more likely during drying of powder, where moisture can be tightly bound but the distance for diffusion to the drying surface is small.

Concept of a Characteristic Drying Rate Curve In 1958, van Meel observed that the drying rate curves, during the falling-rate period, for a specific material often show the same shape (Fig. 12-26), so that a single characteristic drying curve can be drawn for the material being dried. Strictly speaking, the concept should only apply to materials of the same specific size (surface area to material ratio) and thickness, but Keey (1992) shows evidence that it applies over a somewhat wider range with reasonable accuracy. In the absence of experimental data, a linear falling-rate curve is often a reasonable first guess for the form of the characteristic function (good approximation for milk powder, fair for ion-exchange resin and silica gel). At each volume-averaged free moisture content, it is assumed that there is a corresponding specific drying rate relative to the unhindered drying rate in the first drying period that is independent of the external drying conditions. *Volume-averaged* means averaging over the volume (distance cubed for a sphere) rather than just the distance. The *relative drying rate* is defined as

$$f = \frac{N}{N_m} \tag{12-37}$$

where N is the drying rate, N_m is the rate in the constant-rate period, and the characteristic moisture content becomes

$$\Phi = \frac{\overline{X} - X_e}{X_{cr} - X_e} \tag{12-38}$$

where $\overline{X}$ is the volume-averaged moisture content, X_{cr} is the moisture content at the critical point, and X_e is that at equilibrium. Thus, the drying curve is normalized to pass through the point (1, 1) at the critical point of transition in drying behavior and the point (0, 0) at equilibrium.

This representation leads to a simple lumped-parameter expression for the drying rate in the falling-rate period, namely,

$$N = fN_m = f[k\phi_m(Y_W - Y_G)] \tag{12-39}$$

Here k is the external mass-transfer coefficient, ϕ_m is the humidity-potential coefficient (corrects for the humidity not being a strictly true representation of the driving force; close to unity most of the time), Y_W is the humidity above a fully wetted surface, and Y_G is the bulk-gas humidity. Equation (12-39) has been used extensively as the basis for understanding the behavior of industrial drying plants owing to its simplicity and the separation of the parameters

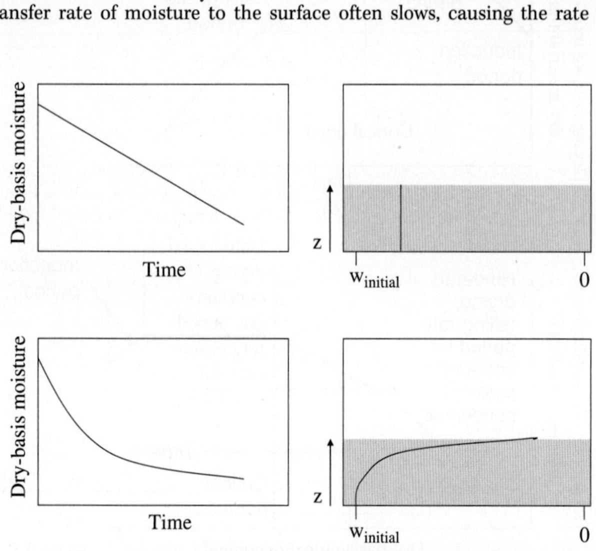

FIG. 12-25 Drying curves and corresponding moisture gradients for situations involving external heat- and mass-transfer control and internal mass-transfer control.

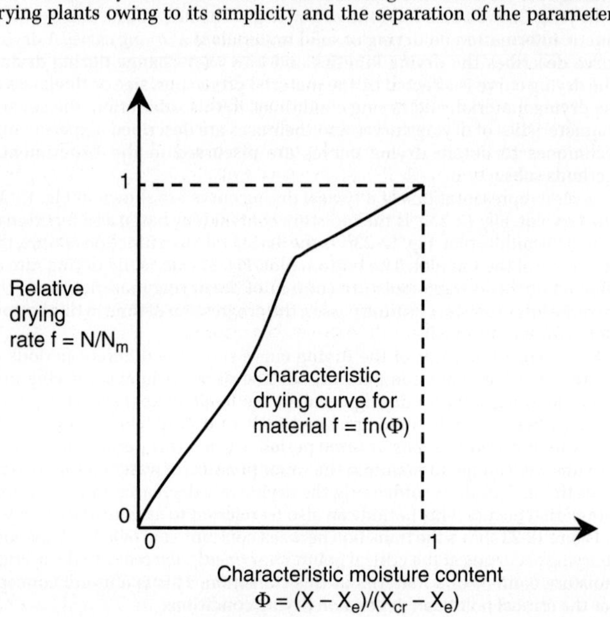

FIG. 12-26 Characteristic drying curve.

that influence the drying process: the material itself f, the design of the dryer k, and the process conditions $\phi_m(Y_W - Y_G)f$.

Example 12-13 Characteristic Drying Curve Application Suppose (with nonhygroscopic solids, X_e = 0 kg/kg) that we have a linear falling-rate curve, with a maximum drying rate N_m of 0.5 kg moisture/(kg dry solids · s) from an initial moisture content of 1 kg moisture/kg dry solids. If the drying conditions around the sample are constant, what is the time required to dry the material to a moisture content of 0.2 kg moisture/kg dry solids?

The linear falling-rate drying curve is given by this relationship:

$$N = -\frac{dX}{dt} = X \cdot 0.5 \frac{\text{kg}}{\text{kg}} \frac{1}{\text{s}} \tag{12-40}$$

where N = the drying rate, kg/(kg · s) and X = average dry basis moisture content. Rearranging and integrating Eq. (12-40) gives

$$\int_1^{0.2} \frac{dX}{X} = \ln\left(\frac{0.2}{1}\right) = -0.5 \cdot \frac{\text{kg}}{\text{kg}} \frac{1}{\text{s}} \cdot t \tag{12-41}$$

$$t = 3.21 \text{ s}$$

The characteristic drying curve, however, is clearly a gross approximation. A common drying curve will be found only if the volume-averaged moisture content reflects the moistness of the surface in some fixed way. An additional worked example using a linear falling-rate drying curve is in the Continuous Agitated Dryer subsection.

For example, in the drying of impermeable timbers, for which the surface moisture content reaches equilibrium quickly, there is unlikely to be any significant connection between the volume-averaged and the surface moisture contents, so the concept is unlikely to apply. While the concept might not be expected to apply to the same material with different thickness, Pang finds that it applies for different thicknesses in the drying of softwood timber (Keey, 1992), and its applicability appears to be wider than the theory might suggest. A paper by Kemp and Oakley (2002) explains that many of the errors in the assumptions in this method often cancel out, meaning that the concept has wide applicability.

DRYER MODELING, DESIGN, AND SCALE-UP

General Principles Models and calculations on dryers can be categorized in terms of (1) the level of complexity used and (2) the purpose or type of calculation (design, or scale-up). A fully structured approach to dryer modeling can be developed from these principles, as described below and in greater detail by Kemp and Oakley (2002).

In this section, we cover the principles and refer the reader to specific examples relevant for each type of dryer in the Drying Equipment subsection.

Levels of Dryer Modeling Modeling can be carried out at four different levels, depending on the amount of data available and the level of detail and precision required in the answer.

Level 1. *Heat and mass balances.* These balances give information on the material and energy flows to and from the dryer, but do not address the kinetics or required size of the equipment.

Level 2. *Scoping.* Approximate or scoping calculations give rough sizes and throughputs (mass flow rates) for dryers, using simple data and making some simplifying assumptions. Either heat-transfer control or first-order drying kinetics is assumed.

Level 3. *Scaling.* Scaling calculations give overall dimensions and performance figures for dryers by scaling up drying curves from small-scale or pilot-plant experiments.

Level 4. *Detailed.* Rigorous or detailed methods aim to track the temperature and drying history of the solids and find local conditions inside the dryer. Naturally, these methods use more complex modeling techniques with many more parameters and require many more input data.

Types of Dryer Calculations The user may wish to design a new dryer, dry a different formulation, or improve the performance of an existing dryer. Three types of calculations are possible:

- Design of a new dryer to perform a given duty, using information from the process flowsheet and physical properties databanks
- Performance calculations for an existing dryer at a new set of operating conditions or to dry a different material
- Scale-up from laboratory-scale or pilot-plant experiments to a full-scale dryer

Solids drying is very difficult to model reliably, particularly in the falling-rate period which usually has the main effect on determining the overall drying time. Falling-rate drying kinetics depend strongly on the internal moisture transport within a solid. This is highly dependent on the internal structure, which in turn varies with the upstream process, the solids formation step, and often between individual batches. Hence, many key drying parameters within solids (e.g., diffusion coefficients) cannot be predicted from theory alone, or obtained from physical property databanks; practical measurements are required. Because of this, experimental work is almost always necessary to design a dryer accurately, and scale-up calculations are more reliable than design based only on thermodynamic data. The experiments are used to verify the theoretical model and find the difficult-to-measure parameters; the full-scale dryer can then be modeled more realistically.

Heat and Mass Balance The heat and mass balance on a generic continuous dryer is shown schematically in Fig. 12-27. In this case, mass flows and moisture contents are given on a dry basis.

The *mass balance* is usually performed on the principal solvent and gives the evaporation rate E (kg/s). In a contact or vacuum dryer, this is approximately equal to the exhaust vapor flow, apart from any noncondensibles. In a convective dryer, this gives the increased outlet humidity of the exhaust. For a continuous dryer at steady-state operating conditions,

$$E = F(X_I - X_O) = G(Y_O - Y_I) \tag{12-42}$$

This assumes that the dry gas flow G and dry solids flow F do not change between dryer inlet and outlet. Mass balances can also be performed on the overall gas and solids flows to allow for features such as air leaks and solids entrainment in the exhaust gas stream.

In a design calculation (including scale-up), the required solids flow rate, inlet moisture content X_I, and outlet moisture X_O are normally specified, and the evaporation rate and outlet gas flow are calculated. In a performance calculation, this is normally reversed; the evaporation rate under new operating conditions is found, and the new solids throughput or outlet moisture content is back-calculated.

For a batch dryer with a dry mass m of solids, a mass balance only gives a snapshot at one point during the drying cycle and an instantaneous drying rate, given by

$$E = m\left(\frac{-dX}{dt}\right) = G\,(Y_O - Y_I) \tag{12-43}$$

The *heat balance* on a continuous dryer takes the generic form

$$GI_{GI} + FI_{SI} + Q_{\text{in}} = GI_{GO} + FI_{SO} + Q_{wl} \tag{12-44}$$

Here I is the enthalpy (kJ/kg dry material) of the solids or gas plus their associated moisture. Enthalpy of the gas includes the latent heat term for the vapor. Expanding the enthalpy terms gives

$$\begin{aligned} &G(C_{sI}T_{GI} + \lambda Y_I) + F(C_{PS} + X_I C_{PL})T_{SI} + Q_{\text{in}} \\ &= G(C_{sO}T_{GO} + \lambda Y_O) + F(C_{PS} + X_O C_{PL})T_{SO} + Q_{wl} \end{aligned} \tag{12-45}$$

Here C_s is the humid heat $C_{PG} + YC_{PY}$. In convective dryers, the left-hand side is dominated by the sensible heat of the hot inlet gas $GC_{sI}T_{GI}$; in contact dryers, the heat input from the jacket Q_{in} is dominant. In both cases, the largest single term on the right-hand side is the latent heat of the vapor $G\lambda Y_O$. Other terms are normally below 10 percent. *This shows why the operating line of a convective dryer on a psychrometric chart is roughly parallel to a constant-enthalpy line.*

The corresponding equation for a batch dryer is

$$GI_{GI} + Q_{\text{in}} = GI_{GO} + m\frac{dI_s}{dt} + Q_{wl} \tag{12-46}$$

Further information on heat and mass balances, including practical advice on industrial dryers, is given later in this section. Worked examples are shown in the continuous band and tunnel dryer, pneumatic conveying dryer, and spray dryer subsections.

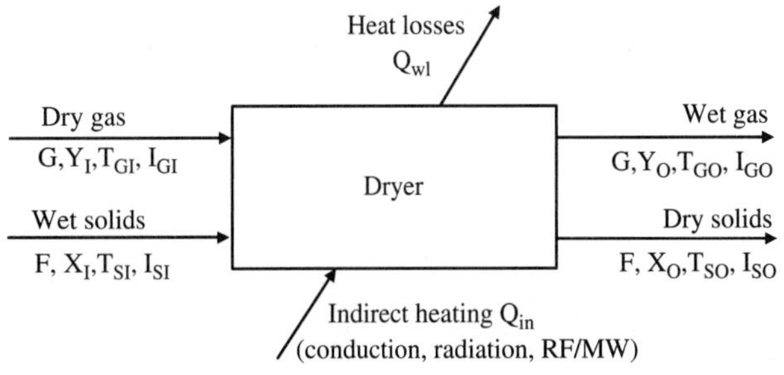

FIG. 12-27 Heat and material flows around a continuous dryer.

Scoping Design Calculations In scoping calculations, some approximate dryer dimensions and drying times are obtained based mainly on a heat and mass balance, without measuring a drying curve or other experimental drying data. They allow the cross-sectional area of convective dryers and the volume of batch dryers to be estimated quite accurately, but are less effective for other calculations and can yield overoptimistic results.

Some examples of scoping calculations are shown later in the batch agitated dryer, spray drying, drum dryer, and sheet dryer subsections.

Scaling Models These models use experimental data from drying kinetics tests in a laboratory, pilot-plant or full-scale dryer, and are thus more accurate and reliable than methods based only on estimated drying kinetics. They treat the dryer as a complete unit, with drying rates and air velocities averaged over the dryer volume, except that, if desired, the dryer can be subdivided into a small number of sections. These methods are used for layer dryers (tray, oven, horizontal-flow band, and vertical-flow plate types) and for a simple estimate of fluidized-bed dryer performance. For batch dryers, they can be used for scale-up by refining the scoping design calculation.

The basic principle is to take an experimental drying curve and perform two transformations: (1) from test operating conditions to full-scale operating conditions and (2) from test dimensions to full-scale dryer dimensions. If the operating conditions of the test (e.g., temperature, gas velocity, agitation rate) are the same as those for the full-scale plant, then the first correction is not required.

Scaling models are the main design method traditionally used by dryer manufacturers. Pilot-plant test results are scaled to a new set of conditions on a dryer with greater airflow or surface area by empirical rules, generally based on the external driving forces (temperature, vapor pressure, or humidity driving forces). By implication, therefore, a characteristic drying curve concept is again being used, scaling the external heat and mass transfer and assuming that the internal mass transfer changes in proportion. A good example is the set of rules described under Fluidized-Bed Dryers, which include the effects of temperature, gas velocity, and bed depth on drying time in the initial test and the full-scale dryer. A worked example is shown in the fluidized-bed drying subsection.

Specific Drying Rate Concept An intuitive, useful method for scale-up of layer dryers from experimental data has been developed and reported by C. Moyers [*Drying Technol.* **12**(1 & 2): 393–417 (1994)]. The method defines a *specific drying rate* (SDR) as

$$\text{SDR} = \frac{m_S}{A_c} = \frac{\rho_s z}{(1+X_1)\tau} \tag{12-47}$$

where ρ_s is the bulk density of the dry solids, z is the layer thickness, X_1 is the initial moisture content, τ is the drying time, and A_c is the surface area of contact.

The method assumes that the SDR is constant between scales. The article presents practical examples using laboratory data for continuous rotating shelf dryers, plate dryers, and continuous paddle dryers.

Detailed or Rigorous Models These models aim to predict local conditions within the dryer and the transient condition of the particles and gas in terms of temperature, moisture content, velocity, etc. Naturally, they require many more input data on the dryer equipment and material properties as well as computational tools (hardware and software) to solve the equations. There are many published models of this type in the academic literature. They give the possibility of more-detailed results, but the potential cumulative errors are also greater. Two types are discussed here: incremental models and *computational fluid dynamics* (CFD) models.

Incremental Model The one-dimensional incremental model is a key analysis tool for several types of dryers. A set of simultaneous equations is solved at a given location (Fig. 12-28), and the simulation moves along the dryer axis in a series of steps or increments, hence the name. A spreadsheet or other computer program is needed, and any number (sometimes thousands) of increments may be used.

Examples of incremental models are shown later in the pneumatic conveying drying, sheet drying, and electromagnetic drying subsections.

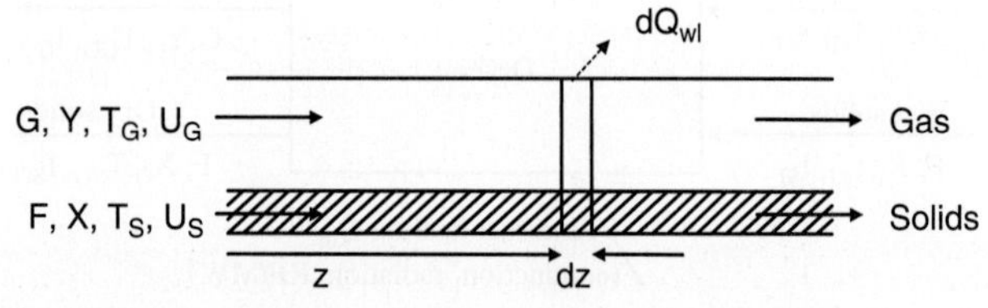

FIG. 12-28 Principle of the incremental model.

Increments may be stated in terms of time (dt), length (dz), or moisture content (dX). A set of six simultaneous equations is then solved, and ancillary calculations are also required, e.g., to give local values of gas and solids properties. The generic set of equations (for a time increment Δt) is as follows:

Heat transfer to particle:

$$Q_P = h_{PG}A_P(T_G - T_S) \tag{12-48}$$

Mass transfer from particle:

$$\frac{-dX}{dt} = \text{function}\,(X, Y, T_P, T_G, h_{PG}, A_P) \tag{12-49}$$

Mass balance on moisture:

$$G\,\Delta Y = -F\,\Delta X = F\,\frac{-dX}{dt}\,\Delta t \tag{12-50}$$

Heat balance on particle:

$$\Delta T_S = \frac{Q_P\,\Delta t - \lambda_{ev} m_P\,\Delta X}{m_P(C_{PS} + C_{PL}X)} \tag{12-51}$$

Heat balance for increment:

$$-\Delta T_G = \frac{F(C_{PS} + C_{PL}X)\,\Delta T_S + G(\lambda_0 + C_{PY}T_G)\,\Delta Y + \Delta Q_{WL}}{G(C_{PG} + C_{PY}Y)} \tag{12-52}$$

Particle transport:

$$\Delta z = U_S\,\Delta t \tag{12-53}$$

The mass and heat balance equations are the same for any type of dryer, but the particle transport equation is completely different, and the heat- and mass-transfer correlations are also somewhat different as they depend on the environment of the particle in the gas (i.e., single isolated particles, agglomerates, clusters, layers, fluidized beds, or packed beds). The mass-transfer rate from the particle is regulated by the drying kinetics and is thus obviously material-dependent (at least in falling-rate drying).

The model is effective and appropriate for dryers where both solids and gas are approximately in axial plug flow, such as pneumatic conveying and cascading rotary dryers. However, it runs into difficulties where there is recirculation or radial flow.

The incremental model is also useful for measuring variations in local conditions such as temperature, solids moisture content, and humidity along the axis of a dryer (e.g., plug-flow fluidized bed), through a vertical layer (e.g., tray or band dryers), or during a batch drying cycle (using time increments, not length).

Any fundamental mathematical model of drying contains mass and energy balances, constitutive equations for mass- and heat-transfer rates, and physical properties. Table 12-8 shows the differential mass balance equations that can be used for common geometries and solved as incremental models. Note there are two sets of differential mass balances—one including shrinkage and one not including shrinkage. When moisture leaves a drying material, the material can either shrink, or develop porosity, or both.

The equations in Table 12-8 are insufficient on their own. Some algebraic relationships are needed to formulate a complete problem, as illustrated in Example 12-14. Equations for the mass- and heat-transfer coefficients are also needed for the boundary conditions presented in Table 12-8. These require the physical properties of the air, the object geometry, and the Reynolds number.

Analytical solutions exist for the equations on the left-hand side of Table 12-8 in the special case of a constant diffusion coefficient. These can be found in Crank, J., *Mathematics of Diffusion,* 2d ed., Oxford University Press, Oxford, UK, 1975. However, the values of the water-solid diffusion coefficient often vary 1 to 3 orders of magnitude with the local moisture content, and so use of these analytical solutions is not recommended. Some data and some theories are available in the literature on the variation of a moisture (or solvent)/solid diffusion coefficient with moisture level; see, e.g., Zielinski, J. M., and Duda, J. L., *AIChE Journal* **38**(3): 405–413 (1992).

Example 12-14 shows the solution for a problem using numerical modeling. This example shows some of the important qualitative characteristics of drying.

Example 12-14 Air Drying of a Thin Layer of Paste Simulate the drying kinetics of 100 μm of paste initially containing 50 percent moisture (wet-basis) with dry air at 60°C, 0 percent relative humidity air at velocities of 0.1, 1.0, or 10 m/s. The diffusion coefficient of water in the material depends on the local moisture content. The length of the layer in the airflow direction is 2.54 cm (Fig. 12-29).

TABLE 12-8 Mass-Balance Equations for Drying Modeling When Diffusion Is Mass-Transfer Mechanism of Moisture Transport

Case	Mass balance without shrinkage	Mass balance with shrinkage
Slab geometry	$\frac{\partial C_w}{\partial t}=\frac{\partial}{\partial z}\left[\mathcal{D}(w)\frac{\partial C_w}{\partial z}\right]$	$\frac{\partial u}{\partial t}=\frac{\partial}{\partial s}\left[\mathcal{D}(w)\frac{\partial u}{\partial s}\right]\quad \frac{\partial s}{\partial z}=\rho_s$
	At top surface, $-\mathcal{D}(w)\frac{\partial C_w}{\partial z_{\text{top surface}}}=k_p\frac{p_w^{\text{bulk}}-p_w^{\text{surface}}}{P-p_w^{\text{surface}}}$	At top surface, $-\mathcal{D}(w)\frac{\partial u}{\partial s_{\text{top surface}}}=k_p\frac{p_w^{\text{bulk}}-p_w^{\text{surface}}}{P-p_w^{\text{surface}}}$
	At bottom surface, $\frac{\partial C_w}{\partial z_{\text{bottom surface}}}=0$	At bottom surface, $\frac{\partial u}{\partial s_{\text{bottom surface}}}=0$
Cylindrical geometry	$\frac{\partial C_w}{\partial t}=\frac{1}{r}\frac{\partial}{\partial r}\left[r\mathcal{D}(w)\frac{\partial C_w}{\partial r}\right]$	$\frac{\partial u}{\partial t}=\frac{\partial}{\partial s}\left[\rho_s^2\mathcal{D}(w)\frac{\partial u}{\partial s}\right]\quad \frac{\partial s}{\partial z}=r\rho_s$
	At surface, $-\mathcal{D}(w)\frac{\partial C_w}{\partial r_{\text{surface}}}=k_p\frac{p_w^{\text{bulk}}-p_w^{\text{surface}}}{P-p_w^{\text{surface}}}$	At surface, $-\mathcal{D}(w)r\frac{\partial uC_w}{\partial s_{\text{surface}}}=k_p\frac{p_w^{\text{bulk}}-p_w^{\text{surface}}}{P-p_w^{\text{surface}}}$
	At center, $\frac{\partial C_w}{\partial r_{\text{center}}}=0$	At center, $\frac{\partial u}{\partial s_{\text{surface}}}=0$
Spherical geometry	$\frac{\partial C_w}{\partial t}=\frac{1}{r^2}\frac{\partial}{\partial r}\left[r^2\mathcal{D}(w)\frac{\partial C_w}{\partial r}\right]$	$\frac{\partial u}{\partial t}=\frac{\partial}{\partial s}\left[\rho_s^4\mathcal{D}(w)\frac{\partial u}{\partial s}\right]\quad \frac{\partial s}{\partial z}=r^2\rho_s$
	At surface, $-\mathcal{D}(w)\frac{\partial C_w}{\partial r_{\text{surface}}}=k_p\frac{p_w^{\text{bulk}}-p_w^{\text{surface}}}{P-p_w^{\text{surface}}}$	At surface, $-\mathcal{D}(w)r^2\frac{\partial u}{\partial s_{\text{surface}}}=k_p\frac{p_w^{\text{bulk}}-p_w^{\text{surface}}}{P-p_w^{\text{surface}}}$
	At center, $\frac{\partial C_w}{\partial r_{\text{center}}}=0$	At center, $\frac{\partial u}{\partial s_{\text{bottom surface}}}=0$

The variable μ is the dry-basis moisture content. The equations that include shrinkage are taken from Van der Lijn, doctoral thesis, Wageningen (1976).

Physical Property Data Sorption isotherm data fit well to the following equation:

$$w=3.10\left(\frac{\%\text{RH}}{100}\right)^5-6.21\left(\frac{\%\text{RH}}{100}\right)^4+4.74\left(\frac{\%\text{RH}}{100}\right)^3-1.70\left(\frac{\%\text{RH}}{100}\right)^2+0.378\left(\frac{\%\text{RH}}{100}\right) \tag{12-54}$$

$$\text{Solid density}=1150\ \text{kg/m}^3$$
$$\text{Heat of vaporization}=2450\ \text{J/g}$$
$$\text{Solid heat capacity}=2.5\ \text{J/(g}\cdot\text{K)}$$
$$\text{Water heat capacity}=4.184\ \text{J/(g}\cdot\text{K)}$$

Solution The full numerical model needs to include shrinkage since the material is 50 percent water initially and the thickness will decrease from 100 to 46.5 mm during drying. Assuming the layer is viscous enough to resist convection in the liquid, diffusion is the dominant liquid-phase transport mechanism.

Table 12-8 gives the mass balance equation:

$$\frac{\partial u}{\partial t}=\frac{\partial}{\partial s}\left[\mathcal{D}(u)\frac{\partial u}{\partial s}\right]\qquad \frac{\partial s}{\partial z}=\rho_s$$

At top surface,

$$-\mathcal{D}(u)\frac{\partial u}{\partial s}=k_p\left(\frac{p_w^{\text{bulk}}-p_w^{\text{surface}}}{P-p_w^{\text{surface}}}\right) \tag{12-55}$$

At bottom surface,

$$\frac{\partial u}{\partial s_{\text{top surface}}}=0 \tag{12-56}$$

The temperature is assumed to be uniform through the thickness of the layer.

$$(C_{p,\text{solids}}+C_{p,\text{water}}\cdot w_{\text{avg,dry-basis}})\cdot m_{\text{solids}}\cdot\frac{dT_{\text{layer}}}{dt}=[h\cdot(T_{\text{air}}-T_{\text{layer}})-F\cdot\Delta H_{\text{vap}}]\cdot A \tag{12-57}$$

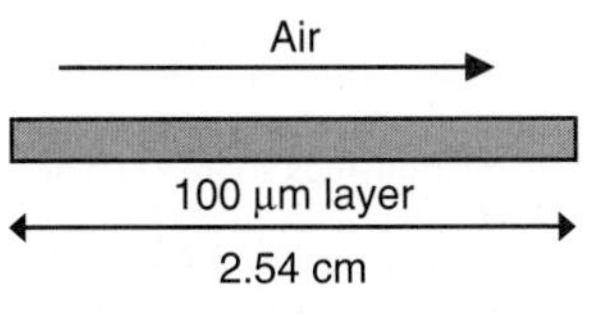

FIG. 12-29 Drying of a thin layer of paste.

Mass- and heat-transfer coefficients are given by

$$\text{Nu}=\frac{hL}{k_{\text{air}}}=0.664\ \text{Re}^{0.5}\ \text{Pr}^{0.333} \tag{12-58}$$

$$\text{Sh}=\frac{k_cL}{D_{\text{air/water}}}=0.664\ \text{Re}^{0.5}\ \text{Sc}^{0.333} \tag{12-59}$$

$$k_p=k_c\cdot\rho_{\text{air}} \tag{12-60}$$

The Reynolds number uses the length of the layer L in the airflow direction:

$$\text{Re}=\frac{VL\rho_{\text{air}}}{\mu_{\text{air}}} \tag{12-61}$$

where V = air velocity.

The Prandtl and Schmidt numbers, Pr and Sc, respectively, for air are given by

$$\text{Pr}=\frac{C_{p,\text{air}}\mu_{\text{air}}}{k_{\text{air}}}=0.70\qquad \text{Sc}=\frac{\mu_{\text{air}}}{\rho_{\text{air}}D_{\text{air/water}}}=0.73 \tag{12-62}$$

The following algebraic equations are also needed:

$$\frac{1}{\rho}=\frac{w}{\rho_w^{\circ}}+\frac{1-w}{\rho_s^{\circ}}\ \text{density of wet material (assumes volume additivity)} \tag{12-63}$$

$$C_w=w\cdot\rho\quad\text{concentration of water} \tag{12-64}$$

$$\rho_s=(1-w)\rho\quad\text{concentration of solids} \tag{12-65}$$

$$\frac{\%\text{RH}}{100}=\frac{P_{w,\text{surface}}}{P_{w,\text{sat}}}\quad\text{definition of relative humidity}$$

The Antoine equation for vapor pressure of water is

$$\ln(P_{w,\text{sat}})=23.1963-\frac{3816.44}{(T+273.15)-46.13} \tag{12-66}$$

Equation (12-67) is the dependence of diffusion coefficient on moisture content for maltodextrin from Raderer, M., et al., *Chemical Engineering Journal* **86:** 185–191 (2002). For dry-basis moisture contents of 1, 0.5, and 0.25, the values of $\mathcal{D}$ are 1.88×10^{-10}, 6.28×10^{-11}, and 1.11×10^{-11} m^2/s, respectively.

$$\mathcal{D}(u)=\exp\left(-\frac{35.8+215\cdot u}{1+10.2\cdot u}\right) \tag{12-67}$$

Result: The results of simulations for air velocities of 0.1, 1.0, and 10 m/s are shown in Fig. 12-30. The top plot shows the average moisture content of the layer as a function of time, the middle plot shows the drying rate as a function of time, and the bottom plot shows the moisture gradient in each layer after 60 s of drying.

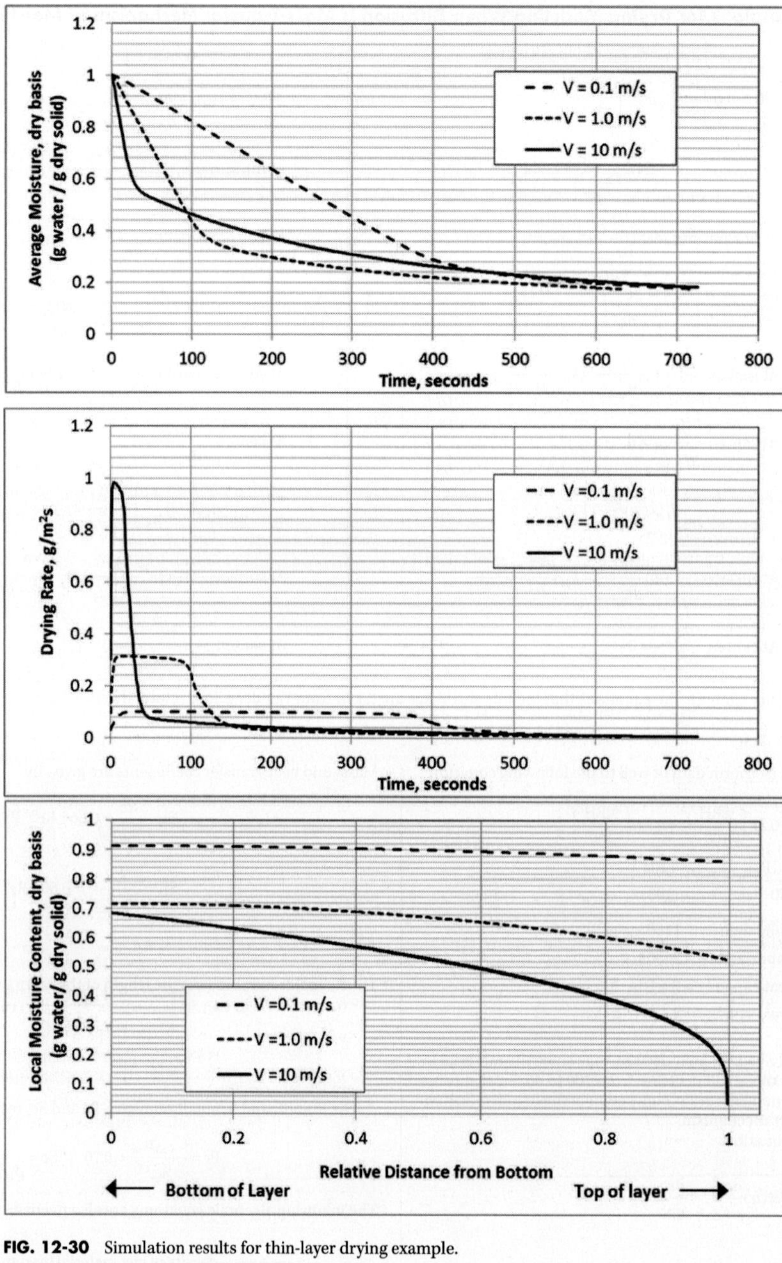

FIG. 12-30 Simulation results for thin-layer drying example.

These results illustrate the relationships between the external air conditions, drying rate, and moisture gradient. In each case, drying begins in a constant-rate period and then moves to a falling-rate period. The drying rates in the constant-rate period are controlled by the air velocity, which affects the external heat- and mass-transfer coefficients [Eqs. (12-58) and (12-59)]. The surface dries, reaching a critical moisture content, and the falling-rate period begins in each case. In the falling-rate period, the rate-limiting step is internal diffusion so the rate becomes independent of air velocity. The plot of the internal moisture gradient at 60 s (bottom plot) illustrates that the falling-rate period has begun for the 10 m/s case, but not yet for the 1.0 and 0.1 m/s cases. The equation set in this example was solved by using a differential algebraic equation solver called gPROMS from Process Systems Enterprises (www.pse.com). It can also be solved with other software and programming languages such as FORTRAN. Example 12-14 is too complicated to be done on a spreadsheet.

Computational Fluid Dynamics CFD provides a very detailed and accurate model of the gas phase, including three-dimensional effects and swirl. Where localized flow patterns have a major effect on the overall performance of a dryer and the particle history, CFD can yield immense improvements in modeling and in the understanding of physical phenomena. Conversely, where the system is well mixed or drying is dominated by falling-rate kinetics and local conditions are unimportant, CFD modeling will give little or no advantage over conventional methods, but will incur a vastly greater cost in computing time.

CFD has been extensively applied in recent years to spray dryers (Langrish and Fletcher, 2001), but it has also been useful for other local three-dimensional swirling flows, e.g., around the feed point of pneumatic conveying dryers (Kemp et al., 1991), and for other cases where airflows affect drying significantly, e.g., local overdrying and warping in timber stacks (Langrish, 1999). See Jamaleddine, T., and Ray, M., *Drying Technology* **28:** 120–154 (2010), for a comprehensive review on how CFD has been used for a wide variety of drying problems.

CFD software packages specialize in their treatment of the turbulence and other details of airflows. However, the differential equations that describe transport within the solid material in the dryer are usually greatly simplified into algebraic relationships so they can be called as user-defined functions within the solver.

Usually, one cannot simultaneously employ a detailed treatment of the equipment and the product at the same time. The engineer must decide which level of detail will enable the designs or business decisions.

TABLE 12-9 Sample Techniques for Various Dryer Types

Dryer type	Sampling method
Fluid bed dryer	Sampling cup (see Fig. 12-31)
Sheet dryer	Collect at end of dryer. Increase speed to change the drying time.
Tray dryer	Record initial moisture and mass of tray with time.
Indirect dryer	Decrease residence time with higher flow rate and sample at exit.
Spray dryer	Residence time of product is difficult to determine and change. Special probes have been developed to sample partially dried powder in different places within the dryer (ref. Langrish).

EXPERIMENTAL METHODS

Lab-, pilot-, and plant-scale experiments all play important roles in drying research. Lab-scale experiments are often necessary to study product characteristics and physical properties; pilot-scale experiments are often used in proof-of-concept process tests and to generate larger quantities of sample material; and plant-scale experiments are often needed to diagnose processing problems and to start or change a full-scale process. Quite often, however, plant data are difficult to obtain since plants are not generally designed to facilitate experimental measurements.

Measurement of Drying Curves Measuring and using experimental drying curves can be difficult. Typically, this is a three-step process. The first step is to collect samples at different times of drying, the second step is to analyze each sample for moisture, and the third step is to interpret the data to make process decisions.

Solid sample collection techniques depend on the type of dryer. Since a drying curve is the moisture content as a function of time, it must be possible to obtain material before the drying process is complete. There are several important considerations when sampling material for a drying curve:

1. The sampling process needs to be fast relative to the drying process. Drying occurring during or after sampling can produce misleading results. Samples must be sealed prior to analysis. Plastic bags do not provide a sufficient seal.

2. In heterogeneous samples, the sample must be large enough to accurately represent the composition of the mixture.

Table 12-9 outlines some sampling techniques for various dryer types.

Moisture measurement techniques are critical to the successful collection and interpretation of drying data. The key message of this subsection is that the moisture value almost certainly depends on the measurement technique and that it is essential to have a *consistent technique* when measuring moisture. Table 12-10 compares and contrasts some different techniques for moisture measurement.

The most common method is gravimetric ("loss on drying"). A sample is weighed in a sample pan or tray and placed into an oven or heater at some high temperature for a given length of time. The sample is weighed again after drying. The difference in weight is then assumed to be due to the complete evaporation of water from the sample. The sample size, temperature, and drying time are all important factors. A very large or thick sample may not dry completely in the given time; a very small sample may not accurately represent the composition of a heterogeneous sample. A low temperature can fail to completely dry the sample, and a temperature that is too high can burn the sample, causing an artificially high loss of mass.

Usually solid samples are collected as described, but in some experiments, it is more convenient to measure the change in humidity of the air due to drying. This technique requires a good mass balance of the system and is more common in lab-scale equipment than pilot- or plant-scale equipment.

Performing a Mass and Energy Balance on a Large Industrial Dryer Measuring a mass and energy balance on a large dryer is often necessary to understand how well the system is operating and how much additional capacity may be available. This exercise can also be used to detect and debug gross problems, such as leaks and product buildup.

There are four steps to this process.

1. Draw a sketch of the overall process including all the flows of mass into and out of the system. Look for places where air can leak into or out of the system. There is no substitute for physically walking around the equipment to get this information.

2. Decide on the envelope for the mass and energy balance. Some dryer systems have hot-air recycle loops and/or combustion or steam heating systems. It is not always necessary to include these to understand the dryer operation.

3. Decide on places to measure airflows and temperatures and to take feed and product samples. Drying systems and other process equipment are frequently not equipped for such measurements; the system may need minor modification, such as the installation of ports into pipes for pitot tubes or humidity probes. These ports must not leak when a probe is in place.

4. Take the appropriate measurements and calculate the mass and energy balances. In continuous operations, these measurements should be taken when the process is at a steady-state condition; data from different locations should be coordinated to be collected within a narrow time window. Care should be taken to seal samples effectively and analyze them quickly, ideally during the course of the experiment.

The measurements are inlet and outlet temperatures, humidities, and flow rates of the air inlets and outlets as well as the moisture and temperature of the feed and dry solids. The following are methods for each of the measurements:

Airflow Rate This is often the most difficult to measure. Fan curves are frequently available for blowers but are not always reliable. A small pitot tube can be used (see Sec. 22, Waste Management, in this text) to measure the local velocity. The best location for use of a pitot tube is in a straight section of pipe. Measurements at multiple positions in the cross section of the pipe or duct are advisable, particularly in laminar flow or near elbows and other flow disruptions.

Air Temperature A simple thermocouple can be used in most cases, but in some cases special care must be taken to ensure that wet or sticky material does not build up on the thermocouple. A wet thermocouple will yield a low temperature from evaporative cooling.

Air Humidity Humidity probes need to be calibrated before use, and the absolute humidity needs (or both the relative humidity and temperature need) to be recorded. If the probe temperature is below the dew point of the air in the process, then condensation on the probe will occur until the probe heats.

Feed and Exit Solids Rate These are generally known, particularly for a unit in production. Liquids can be measured by using a bucket and stopwatch. Solids can be measured in a variety of ways.

TABLE 12-10 Moisture Determination Techniques

Method	Principle	Advantages	Disadvantages
Gravimetric (loss on drying)	Water evaporates when sample is held at a high temperature. Difference in mass is recorded.	Simple technique. No extensive calibration methods are needed. Lab equipment is commonly available.	Method is slow. Measurement time is several minutes to overnight (depending on material and accuracy). Generally not suitable for process control. Does not differentiate between water and other volatile substances.
IR/NIR	Absorption of infrared radiation by water is measured.	Fast method. Suitable for very thin layers or small particles.	Only surface moisture is detected. Extensive calibration is needed.
RF/microwave	Absorption of RF or microwave energy is measured.	Fast method. Suitable for large particles.	Extensive calibration is needed.
Equilibrium relative humidity (ERH)	The equilibrium relative humidity headspace above sample in a closed chamber is measured. Sorption isotherm is used to determine moisture.	Relatively quick method. Useful particularly if a final moisture specification is in terms of water activity (to retard microorganism growth).	May give misleading results since the surface of the material will equilibrate with the air. Large particles with moisture gradients can give falsely low readings. Measurement of relative humidity can be imprecise.
Karl Fischer titration	Chemical titration that is water-specific. Material can be either added directly to a solvent or heated in an oven, with the headspace purged and bubbled through solvent.	Specific to water only and very precise. Units can be purchased with an autosampler. Measurement takes only a few minutes.	Equipment is expensive and requires solvents. Minimal calibration required. Sample size is small, which may pose a problem for heterogeneous mixtures.

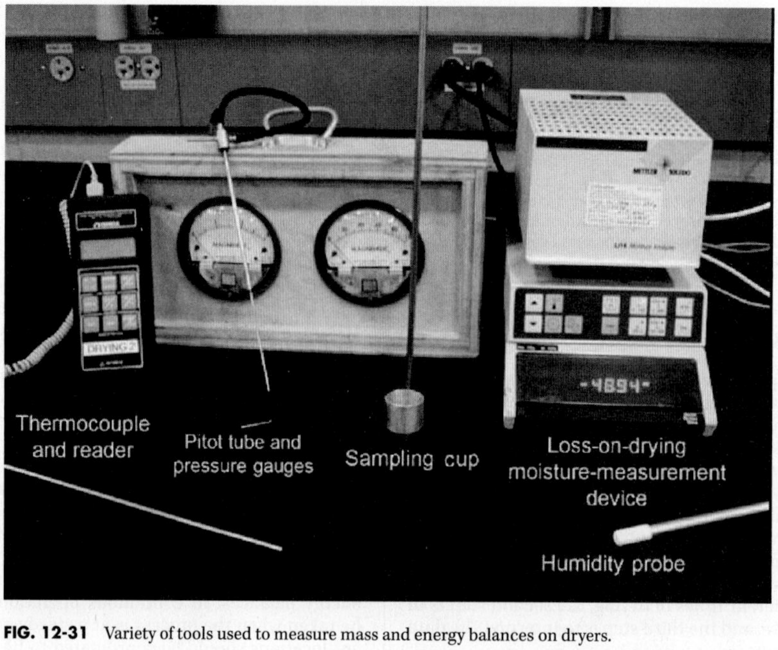

FIG. 12-31 Variety of tools used to measure mass and energy balances on dryers.

Feed and Exit Solids Moisture Content These need to be measured by using an appropriate technique, as described above. Use the same method for both the feed and exit solids. Don't rely on formula sheets for feed moisture information.

Figure 12-31 shows some common tools used in these measurements.

DRYING OF NONAQUEOUS SOLVENTS

Practical Considerations Removal of nonaqueous solvents from a material presents several practical challenges. First, solvents are often flammable and require drying either in an inert environment, such as superheated steam or nitrogen, or in a gas phase comprised solely of solvent vapor. The latter will occur in indirect or vacuum drying equipment. Second, the solvent vapor must be collected in an environmentally acceptable manner.

An additional practical consideration is the remaining solvent content that is acceptable in the final product. Failure to remove all the solvent can lead to problems such as toxicity of the final solid or can cause the headspace of packages, such as drums, to accumulate solvent vapor.

Physical Properties The physical properties that are important in solvent drying are the same as those for an aqueous system. The vapor pressure of a solvent is the most important property since it provides the thermodynamic driving force for drying. Acetone (BP 57°C), for example, can be removed from a solid at atmospheric pressure readily by boiling, but glycerol (BP 200°C) will dry only very slowly. Like water, a solvent may become bound to the solid and have a lower vapor pressure. This effect should be considered when one is designing a solvent-drying process.

Diffusion of nonaqueous solvents through a material can be slow. The diffusion coefficient is directly related to the size of the diffusing molecule, so molecules larger than water typically have diffusion coefficients that have a much lower value. This phenomenon is known as *selective diffusion.* Large diffusing molecules can become kinetically trapped in the solid matrix. Solvents with a lower molecular weight will often evaporate from a material faster than a solvent with a higher molecular weight, even if the vapor pressure of the larger molecule is higher. Some encapsulation methods rely on selective diffusion; an example is instant coffee production using spray drying, where volatile flavor and aroma components are retained in particles more than water, even though they are more volatile than water, as shown in Fig. 12-32.

PRODUCT QUALITY CONSIDERATIONS

Overview The drying operation usually has a very strong influence on final product quality and product performance measures. And the final product quality strongly influences the value of the product. Generally, a specific particle or unit size, a specific density, a specific color, and a specific target moisture are desired. Naturally every product is somewhat different, but these are usually the first things we need to get right.

Target Moisture This seems obvious, but it's very important to determine the right moisture target before we address other drying basics. Does biological activity determine the target, flowability of the powder, shelf life, etc.? Sometimes a very small (1 to 2 percent) change in the target moisture will have a profound impact on the size of dryer required. This is especially true for difficult-to-dry products with flat falling-rate drying characteristics. Therefore, spend the time necessary to get clear on what really determines the moisture target. And as noted earlier in this subsection, care should be taken to define a moisture measurement method since results are often sensitive to the method.

Particle Size Generally a customer or consumer wants a very specific particle size—and the narrower the distribution, the better. No one wants lumps or dust. The problem is that some attrition and sometimes agglomeration occur during the drying operation. We may start out with the right particle size, but we must be sure the dryer we've selected will not adversely affect particle size to the extent that it becomes a problem. Some dryers will

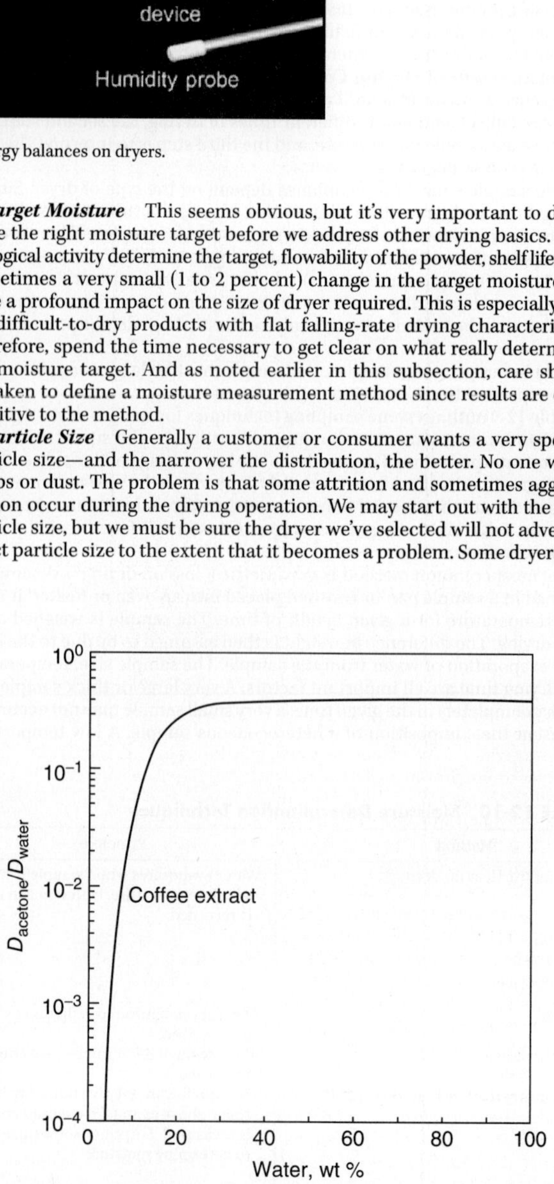

FIG. 12-32 The ratio of the diffusion coefficients of acetone to water in instant coffee as a function of moisture content (taken from Thijssen et al., *De Ingenieur,* JRG, 80, Nr. 47 (1968)]. Acetone has a much higher vapor pressure than water, but is selectively retained in coffee during drying.

treat particles more gently than others. Particle size is also important from a segregation standpoint. See Sec. 18, Liquid-Solid Operations and Equipment. Fine particles can also increase the risk of fire or explosion.

Density Customers and consumers are generally very interested in getting the product density they have specified or expect. If the product is a consumer product and is going into a box, then the density needs to be correct to fill the box to the appropriate level. If density is important, then product shrinkage during drying can be an important harmful transformation to consider. This is particularly important for biological products for which shrinkage can be very high. This is why freeze drying can be the preferred dryer for many of these materials.

Solubility Many dried products are rewet either during use by the consumer or by a customer during subsequent processing. Shrinkage can again be a very harmful transformation. Often shrinkage is a virtually irreversible transformation that creates an unacceptable product morphology. Case hardening is a phenomenon that occurs when the outside of the particle or product initially shrinks to form a very hard and dense skin that does not easily rewet. A common cause is capillary collapse, discussed along with shrinkage below.

Flowability If we're considering particles, powders, and other products that are intended to flow, then this is a very important consideration. These materials need to easily flow from bins, from hoppers, and out of boxes for consumer products. Powder flowability is a measurable characteristic using rotational shear cells (Peschl, Freeman) or translational shear cells (Jenike) in which the powder is consolidated under various normal loads; then the shear force is measured, enabling a complete yield locus curve to be constructed. This can be done at various powder moistures to create a curve of flowability versus moisture content. Some minimal value is necessary to ensure free flow. Additional information on these devices and this measure can be found in Sec. 21, Solids Processing and Particle Technology.

Color Product color is usually a very important product quality attribute, and a change in color can be caused by several different transformations.

Transformations Affecting Product Quality Drying, as with any other unit operation, has both productive and harmful transformations that occur. The primary productive transformation is water removal of course, but there are many harmful transformations that can occur and adversely affect product quality. The most common of these harmful transformations includes product shrinkage; attrition or agglomeration; loss of flavor, aroma, and nutritional value; browning reactions; discoloration; stickiness; and flowability problems. These issues were discussed briefly above, but are worth a more in-depth review.

Shrinkage Shrinkage is a particularly important transformation with several possible mechanisms to consider. It's usually especially problematic with food and other biological materials, but is a very broadly occurring phenomenon. Shrinkage generally affects solubility, wettability, texture and morphology, and absorbency. It can be observed when drying lumber when it induces stress cracking and during the drying of coffee beans prior to roasting. Tissue, towel, and other paper products undergo some shrinkage during drying. And many chemical products shrink as water evaporates, creating voids and capillaries prone to collapse as additional water evaporates. As we consider capillary collapse, there are several mechanisms worth mentioning.

Surface tension. The capillary suction created by a receding liquid meniscus can be extremely high.

Plasticization. An evaporating solvent which is also a plasticizer of polymer solute product will lead to greater levels of collapse and shrinkage.

Electric charge effects. The van der Waals and electrostatic forces can also be a strong driver of collapse and shrinkage.

Surface Tension These effects are very common and worthy of a few more comments. Capillary suction created by a receding liquid meniscus can create very high pressures for collapse. The quantitative expression for the pressure differential across a liquid-fluid interface was first derived by Laplace in 1806. The meniscus, which reflects the differential, is affected by the surface tension of the fluid. Higher surface tensions create greater forces for collapse. These strong capillary suction pressures can easily collapse a pore. We can reduce these suction pressures by using low-surface-tension fluids or by adding surfactants, in the case of water, which will also significantly reduce surface tension (from 72 to 30 dyn/cm).

The collapse can also be reduced with some dryer types. Freeze drying and heat pump drying can substantially reduce collapse, but the capital cost of these dryers is sometimes prohibitive. At the other extreme, dryers that rapidly flash off the moisture can reduce collapse. This mechanism can also be affected by particle size such that the drying is primarily boundary-layer-controlled. When the particle size becomes sufficiently small, moisture can diffuse to the surface at a rate sufficient to keep the surface wetted. This has been observed in a gel-forming food material when the particle size reached 150 to 200 μm (Genskow, "Considerations in Drying Consumer Products," *Proceedings International Drying Symposium,* Versailles, France, 1988).

Biochemical Degradation Biochemical degradation is another harmful transformation that occurs with most biological products. There are four key reactions to consider: lipid oxidation, Maillard browning, protein denaturation, and various enzyme reactions. These reactions are both heat- and moisture-dependent such that control of temperature and moisture profiles can be very important during drying.

Lipid oxidation Lipid oxidation is normally observed as a product discoloration and can be exacerbated with excessive levels of bleach. It is catalyzed by metal ions, enzymes, and pigments. Acidic compounds can be used to complex the metal ions. Synthetic antioxidants such as butylated hydroxytoluene (BHT) and butylated hydroxyanisole (BHA) can be added to the product, but are limited and coming under increased scrutiny due to toxicology concerns. It may be preferable to use natural antioxidants such as lecithin or vitamin E or to dry under vacuum or in an inert (nitrogen, steam) atmosphere.

Protein denaturation Normally protein denaturation is observed as an increase in viscosity and a decrease in wettability. It is temperature-sensitive, generally occurring between 40 and 80°C. A common drying process scheme is to dry thermally and under wet-bulb drying conditions without overheating and then vacuum, heat-pump, or freeze-dry to the target moisture.

Enzyme reactions Enzymatic browning is caused by the enzyme polyphenol oxidase which causes phenols to oxidize to orthoquinones. The enzyme is active between pH 5 and 7. A viable process scheme again is to dry under vacuum or in an inert (nitrogen, steam) atmosphere.

Maillard browning reaction This nonenzymatic reaction is observed as a product discoloration, which in some products creates an attractive coloration. The reaction is temperature-sensitive, and normally the rate passes through a maximum and then falls as the product becomes drier. The reaction can be minimized by minimizing the drying temperature, reducing the pH to acidic, or adding an inhibitor such as sulfur dioxide or metabisulfate. A viable process scheme again is to dry thermally and under wet-bulb drying conditions without overheating and then vacuum, heat-pump, or freeze-dry to the target moisture.

Some of the above reactions can be minimized by reducing the particle size and using a monodisperse particle size distribution. The small particle size will better enable wet-bulb drying, and the monodisperse size will reduce overheating of the smallest particles.

Stickiness, Lumping, and Caking These are not characteristics we generally want in our products. They generally connote poor product quality, but can be a desirable transformation if we are trying to enlarge particle size through agglomeration. Stickiness, lumping, and caking are phenomena that are dependent on product moisture and product temperature. The most general description of this phenomenon is created by measuring the cohesion (particle to particle) of powders, as described below. A related measure is *adhesion*—particle-to-wall interactions. Finally, the sticky point is a special case for materials that undergo glass transitions.

The sticky point can be determined by using a method developed by Lazar and later by Downton [Downton, Flores-Luna, and King, "Mechanism of Stickiness in Hygroscopic, Amorphous Powders," *I&EC Fundamentals* **21:** 447 (1982)]. In the simplest method, a sample of the product, at a specific moisture, is placed in a closed tube which is suspended in a water bath. A small stirrer is used to monitor the torque needed to "stir" the product. The water bath temperature is slowly increased until the torque increases. This torque increase indicates a sticky point temperature for that specific moisture. The test is repeated with other product moistures until the entire stickiness curve is determined. A typical curve is shown in Fig. 12-33.

As noted, a sticky point mechanism is a glass transition—the transition when a material changes from the glassy state to the rubbery liquid state. Glass transitions are well documented in food science (Levine and Slade). Roos and Karel ["Plasticizing Effect of Water on Thermal Behavior and Crystallization of Amorphous Food Models," *J. Food Sci.* **56**(1): 38–43 (1991)] have demonstrated that for these types of products, the glass transition temperature follows the sticky point curve within about 2°C. This makes it straightforward to measure the stickiness curve by using a *differential scanning calorimeter* (DSC). Somewhat surprisingly, even materials that are not undergoing glass transitions exhibit this behavior, as demonstrated with the detergent stickiness curve above.

Lumping and caking can be measured by using the rotational shear cells (Peschl, Freeman) or translational shear cells (Jenike) noted above for measuring flowability. The powder is consolidated under various normal loads, and then the shear force is measured, enabling a complete yield locus curve to be constructed. This can be done at various powder moistures to create a curve of cake strength versus moisture content. Slurries and dry solids are free-flowing, and there is a cohesion/adhesion peak at an intermediate moisture content, typically when voids between particles are largely full of liquid. A variety of other test methods for handling properties and flowability are available.

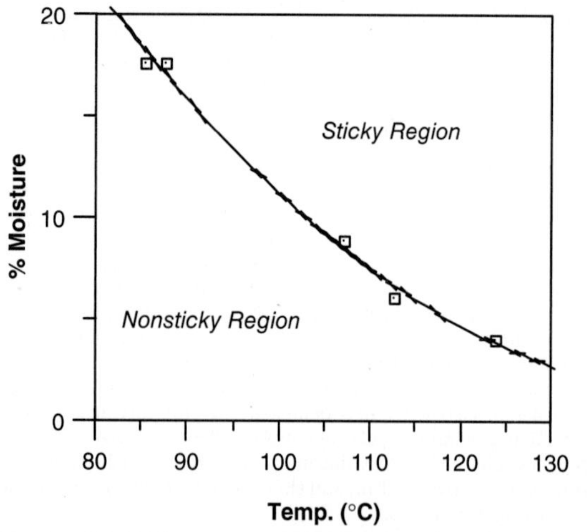

FIG. 12-33 Detergent stickiness curve.

Product quality was addressed quite comprehensively by Evangelos Tsotsas at the 2d Nordic Drying Conference [Tsotsas, "Product Quality in Drying—Luck, Trial, Experience, or Science?" 2d Nordic Drying Conference, Copenhagen, Denmark, 2003]. Tsotsas notes that 31 percent of the papers at the 12th International Drying Symposium refer to product quality. The top five were color (12 percent), absence of chemical degradation (10 percent), absence of mechanical damage (9 percent), bulk density (8 percent), and mechanical properties (7 percent). All these properties are reasonably straightforward to measure. They are physical properties, and we are familiar with them for the most part. However, down the list at a rank of 20 with only 2 percent of the papers dealing with it, we have sensory properties.

This is the dilemma—sensory properties should rank very high, but they don't because we lack the tools to measure them effectively. For the most part, these quality measures are subjective rather than objective, and frequently they require direct testing with consumers to determine the efficacy of a particular product attribute. So the issue is really a lack of physical measurement tools that directly assess the performance measures important to the consumer of the product. The lack of objective performance measures and unknown mechanistic equations also makes mathematical modeling very difficult for addressing quality problems.

The good news is that there has been a shift from the macro to the meso and now to the micro scale in drying science. We have some very powerful analytical tools to help us understand the transformations that are occurring at the meso scale and micro scale.

ADDITIONAL READINGS

Keey, *Drying of Loose and Particulate Materials,* Hemisphere, New York, 1992.

Keey, Langrish, and Walker, *Kiln Drying of Lumber,* Springer-Verlag, Heidelberg, 2000.

Keey and Suzuki, "On the Characteristic Drying Curve," *Int. J. Heat Mass Transfer* **17:** 1455–1464 (1974).

Kemp and Oakley, "Modeling of Particulate Drying in Theory and Practice," *Drying Technol.* **20**(9): 1699–1750 (2002).

Kock et al., "Design, Numerical Simulation and Experimental Testing of a Modified Probe for Measuring Temperatures and Humidities in Two-Phase Flow," *Chem. Eng. J.* **76**(1): 49–60 (2000).

Liou and Bruin, "An Approximate Method for the Nonlinear Diffusion Problem with a Power Relation between the Diffusion Coefficient and Concentration. 1. Computation of Desorption Times," *Int. J. Heat Mass Transfer* **25:** 1209–1220 (1982a).

Liou and Bruin, "An Approximate Method for the Nonlinear Diffusion Problem with a Power Relation between the Diffusion Coefficient and Concentration. 2. Computation of the Concentration Profile," *Int. J. Heat Mass Transfer* **25:** 1221–1229 (1982b).

Marshall, "Atomization and Spray Drying," *AIChE Symposium Series,* no. 2, p. 89 (1986).

Oliver and Clarke, "Some Experiments in Packed-Bed Drying," *Proc. Inst. Mech. Engrs.* **187:** 515–521 (1973).

Perré and Turner, "The Use of Macroscopic Equations to Simulate Heat and Mass Transfer in Porous Media," in Turner and Mujumdar, eds., *Mathematical Modeling and Numerical Techniques in Drying Technology,* Marcel Dekker, New York, 1996, pp. 83–156.

Ranz and Marshall, "Evaporation from Drops," *Chem. Eng. Prog.* **48**(3): 141–146 and **48**(4): 173–180 (1952).

Schlünder, "On the Mechanism of the Constant Drying Rate Period and Its Relevance to Diffusion Controlled Catalytic Gas Phase Reactions," *Chem. Eng. Sci.* **43:** 2685–2688 (1988).

Schoeber and Thijssen, "A Short-cut Method for the Calculation of Drying Rates for Slabs with Concentration-Dependent Diffusion Coefficient," *AIChE. Symposium Series* **73**(163): 12–24 (1975).

Sherwood, "The Drying of Solids," *Ind. And Eng. Chem.* **21**(1): 12–16 (1929).

Suzuki et al., "Mass Transfer from a Discontinuous Source," *Proc. PACHEC '72,* Kyoto, Japan, **3:** 267–276 (1972).

Thijssen and Coumans, "Short-cut Calculation of Non-isothermal Drying Rates of Shrinking and Non-shrinking Particles Containing an Expanding Gas Phase," *Proc. 4th Int. Drying Symp., IDS '84,* Kyoto, Japan, **1:** 22–30 (1984).

Thijssen and Rulkens, "Retention of Aromas in Drying Food Liquids," *De Ingenieur, JRG,* **80**(47) (1968).

Van der Lijn, "Simulation of Heat and Mass Transfer in Spray Drying," doctoral thesis, Wageningen, 1976.

Van Meel, "Adiabatic Convection Batch Drying with Recirculation of Air," *Chem. Eng. Sci.* **9:** 36–44 (1958).

Viollez and Suarez, "Drying of Shrinking Bodies," *AIChE J.* **31:** 1566–1568 (1985).

Waananan, Litchfield, and Okos, "Classification of Drying Models for Porous Solids," *Drying Technol.* **11**(1): 1–40 (1993).

SOLIDS-DRYING EQUIPMENT—GENERAL ASPECTS

General References: Cook and DuMont, *Process Drying Practice,* McGraw-Hill, New York, 1991. *Drying Technology—An International Journal,* Taylor and Francis, New York, 1982 onward. Hall, *Dictionary of Drying,* Marcel Dekker, New York, 1979. Keey, *Introduction to Industrial Drying Operations,* Pergamon, New York, 1978. Mujumdar, ed., *Handbook of Industrial Drying,* Marcel Dekker, New York, 1995. Van't Land, *Industrial Drying Equipment,* Marcel Dekker, New York, 1991. Tsotsas and Mujumdar, eds., *Modern Drying Technology* (vols. 1–6), Wiley, 2011. *Aspen Process Manual* (Internet knowledge base), Aspen Technology, Boston, 2000 onward. Nonhebel and Moss, *Drying of Solids in the Chemical Industry,* CRC Press, Cleveland, Ohio, 1971.

CLASSIFICATION AND SELECTION OF DRYERS

Drying equipment may be classified in several ways. Effective classification is vital in selection of the most appropriate dryer for the task and in understanding the key principles on which it operates. The main drying-process attributes are as follows:

1. Form of feed and product—particulate (solid or liquid feed), sheet, slab
2. Mode of operation—batch or continuous
3. Mode of heat transfer—convective (direct), conductive (indirect), radiative, or dielectric
4. Condition of solids—static bed, moving bed, fluidized or dispersed
5. Gas-solids contacting—parallel flow, perpendicular flow, or through-circulation
6. Gas flow pattern—cross-flow, cocurrent, or countercurrent

Other important features of the drying system are the type of carrier gas (air, inert gas, or superheated steam/solvent), use of gas or solids recycle, type of heating (indirect or direct-fired), and operating pressure (atmospheric or vacuum).

However, in the selection of a group of dryers for preliminary consideration in a given drying problem, the most important factor is often category 1, the form, handling characteristics, and physical properties of the wet material.

Table 12-11 shows the major categories of drying equipment, organized by feed type. This section compares these types in general terms. Each dryer type listed in Table 12-11 is discussed in greater detail in the Drying Equipment subsection.

Description of Dryer Classification and Selection Criteria

1. *Form of Feed and Product* Dryers are specifically designed for particular feed and product forms; dryers handling films, sheets, slabs, and bulky artifacts form a clear subset. Most dryers are for particulate products, but the feed may range from a solution or slurry (free-flowing liquid) through a sticky paste to wet filter cakes, powders, or granules (again relatively free-flowing). The ability to successfully mechanically handle the feed and product is a key factor in dryer selection (see Table 12-11).

The drying kinetics also depend strongly on solids properties, particularly particle size and porosity. The surface area/volume ratio and the internal pore structure control the extent to which an operation is diffusion-limited, i.e., diffusion into and out of the pores of a given solids particle, not through the voids among separate particles.

2. *Mode of Operation* Batch dryers are typically used for low throughputs (under 50 kg/h), for long drying times, or where the overall process is predominantly batch. Continuous dryers dominate for high throughputs, high evaporation rates, and where the rest of the process is continuous. Dryers that are inherently continuous can be operated in semibatch mode (e.g., small-scale spray dryers) and vice versa.

3. *Mode of Heat Transfer*

Direct (convective) dryers These are the general operating characteristics of direct dryers.

a. Direct contacting of hot gases with the solids is employed for solids heating and vapor removal. Note, in some cases, hot exhaust gas from combustion containing CO and CO_2 should not contact the product and this exhaust gas heats fresh air that contacts the product using a separate heat exchanger; these are still considered "direct" dryers.

TABLE 12-11 Classification of Commercial Dryers Based on Feed Materials Handled

Type of dryer	Liquids	Slurries	Pastes and sludges	Free-flowing	Granular, crystalline, or fibrous solids	Large solids, special forms and shapes	Continuous sheets	Discontinuous sheets
	True and colloidal solutions; emulsions. Examples: inorganic salt solutions, extracts, milk, blood, waste liquors, rubber latex	Pumpable suspensions. Examples: pigment slurries, soap and detergents, calcium carbonate, bentonite, clay slip, lead concentrates	Examples: filter-press cakes, sedimentation sludges, centrifuged solids, starch, detergents	100-mesh (150-μm) or less. Relatively free-flowing in wet state. Dusty when dry. Examples: centrifuged precipitates	Larger than 100-mesh (150 μm). Examples: rayon staple, salt crystals, sand, ores, potato strips, synthetic rubber	Examples: pottery, brick, rayon cakes, shotgun shells, hats, painted objects, rayon skeins, lumber, pet food, croutons	Examples: paper, impregnated fabrics, cloth, cellophane, plastic sheets	Examples: veneer, wallboard, photographic prints, leather, foam rubber sheets
Batch tray dryers Material is loaded onto a static tray and dried without agitation. Includes vacuum drying.	Not applicable	Not applicable	Suitable only if material can be preformed. Suited to batch operation.	Not applicable	Usually not suited for materials smaller than 30-mesh (0.5 mm). Suited to small capacities.	Primarily useful for small objects.	Not applicable	Not applicable
Continuous tray and gravity dryers Material dries on trays and moves between successive trays by gravity.	Not applicable	Not applicable	Suitable for small-scale and large-scale production.	See comments under Pastes and Sludges. Vertical turbo-tray applicable	Suitable for large particles (>500 μm) or small discrete objects.	Not suitable for large or fragile objects.	Not applicable	Not applicable
Continuous band and tunnel dryers Material is loaded on a continuous band or belt and is dried by convection, conduction, or radiant heat.	Not applicable	Only crystal filter dryer or centrifuge dryer may be suitable.	Suitable for materials that can be preformed. Will handle large capacities	Not applicable	Usually not suited for materials smaller than 30-mesh (0.5 mm). Material does not tumble or mix.	Suited to a wide variety of shapes and forms. Can be used to convey materials through heated zones.	Not applicable	Special designs are required. Suited to veneers.
Batch agitated and rotating dryers Batch dryer where material is moved to enhance mass- and heat-transfer rates	Atmospheric or vacuum. Suitable for small batches. Easily cleaned. Solvents can be recovered. Material agitated while dried.	See comments under Liquids.	See comments under Liquids.	See comments under Liquids.	Suitable for small batches. Easily cleaned. Material is agitated during drying, causing some degradation and/or balling up.	Not applicable	Not applicable	Not applicable
Continuous agitated and rotating dryers Material is mechanically agitated or turned over during drying.	Applicable with dry-product recirculation.	Applicable with dry-product recirculation.	Suitable only if product does not stick to walls and does not dust. Recirculation of product may prevent sticking. Generally requires recirculation of dry product. Little dusting occurs.	Suitable for most materials and especially for high capacities, provided dusting is not too severe. Chief advantage is low dust loss. Well suited to most materials and capacities, particularly those requiring drying at steam temperature	Suitable for most materials, especially for high capacities. Dusting or crystal abrasion will limit its use. Low dust loss. Material must not stick or be temperature-sensitive.	Not applicable	Not applicable	Not applicable
Fluidized- and spouted-bed dryers Particulate material is suspended/fluidized and simultaneously dried by upward-moving air.	Applicable only as fluid-bed granulator with inert bed or dry-solids recirculator.	See comments under Liquids.	See comments under Liquids.	Fluidized bed suitable, if not too dusty. Internal coils can supplement heating, especially for fine powders. Suitable for high capacities.	Fluidized bed suitable for crystals, granules, and very short fibers. Suitable for high capacities. Suitable for spouted beds for granules over 800 μm.	Only applicable for spouted beds if objects are conveyable in gas stream.	Not applicable	Not applicable
Pneumatic conveying dryers Particulate material is suspended in hot air (or superheated steam) and conveyed.	See comments under Slurries.	Can be used only if product is recirculated (backmixed) to make feed suitable for handling.	Usually requires recirculation of dry product to make suitable feed. Well suited to high capacities. Disintegration usually required.	Suitable for materials that are easily suspended in a gas stream and lose moisture readily. Well suited to high capacities.	Suitable for materials conveyable in a gas stream. Well suited to high capacities. Only surface moisture usually removed. Product may suffer physical degradation.	Not applicable	Not applicable	Not applicable

TABLE 12-11 Classification of Commercial Dryers Based on Feed Materials Handled *(Continued)*

Type of dryer	Liquids	Slurries	Pastes and sludges	Free-flowing	Granular, crystalline, or fibrous solids	Large solids, special forms and shapes	Continuous sheets	Discontinuous sheets
Spray dryers A liquid or slurry feed is atomized/sprayed, and the resulting droplets dry into particles.	Suited for large capacities. Product is usually powdery, spherical, and free-flowing. High temperatures can sometimes be used with heat-sensitive materials. Products generally have low bulk density.	See comments under Liquids. Pressure-nozzle atomizers subject to erosion.	Requires special pumping equipment to feed the atomizer. See comments under Liquids.	Not applicable unless feed is pumpable.	Not applicable	Not applicable	Not applicable	Not applicable
Drum and thin-film dryers Liquid or paste feed is distributed on the outside of a hot, slowly rotating drum.	Single, double, or twin. Atmospheric or vacuum operation. Product flaky and usually dusty. Maintenance costs may be high.	See comments under Liquids. Twin-drum dryers are widely used.	Can be used only when paste or sludge can be made to flow. See comments under Liquids.	Not applicable	Not applicable	Not applicable	Not applicable	Not applicable
Sheet dryers Discrete or continuous solid sheets are dried by blowing air or steam through them or over them. Heat can be added by multiple methods.	Not applicable	Not applicable	Not applicable	Not applicable	Not applicable	Not applicable	See description for applications. Web can be controlled with tension.	See description for applications. Items being dried need to be positioned or held in place.
Freeze dryers Air is removed from the system. Energy must be supplied via conduction or radiation.	Expensive. Usually used only for high-value products such as pharmaceuticals; products which are heat-sensitive and readily oxidized.	See comments under Liquids.	See comments under Liquids.	See comments under Liquids.	Expensive. Usually used on pharmaceuticals and related products which cannot be dried successfully by other means. Applicable to fine chemicals.	See comments under Granular Solids.	Applicable in special cases such as emulsion-coated films.	See comments under Granular Solids.
Electromagnetic drying (infrared, radiofrequency, microwave) Batch or continuous operation. Electromagnetic energy added to material.	Can be used in combination with other dryers such as drum. Infrared: only for thin films. Microwave/radio-frequency: expensive, specialty applications.	See comments under Liquids.	See comments under Liquids.	Infrared: only for thin layers Radiofrequency/microwave	Primarily suited to drying surface moisture. Not suited for thick layers.	Especially suited for drying and baking paint and enamels.	Infrared. Useful for laboratory work or in conjunction with other methods. Radiofrequency/microwave: successful for foam rubber.	See comments under Continuous Sheets.

b. Drying temperatures may range up to 750°C, the limiting temperature for most common structural metals. At higher temperatures, radiation becomes an important heat-transfer mechanism.

c. At gas temperatures below the boiling point, the vapor content of gas influences the rate of drying and the final moisture content of the solid. With gas temperatures above the boiling point throughout, the vapor content of the gas has only a slight retarding effect on the drying rate and final moisture content. Thus, superheated vapors of the liquid being removed (e.g., steam) can be used for drying.

d. For low-temperature drying, dehumidification of the drying air may be required when atmospheric humidities are excessively high.

e. Efficiency increases with an increase in the inlet gas temperature for a constant exhaust temperature.

f. Because large amounts of gas are required to supply all the heat for drying, dust recovery or volatile organic compound (VOC) equipment may be very large and expensive, especially when drying very small particles.

Indirect (contact or conductive) dryers These differ from direct dryers with respect to heat transfer and vapor removal:

a. Heat is transferred to the wet material by conduction through a solid retaining wall, usually metallic.

b. Surface temperatures may range from below freezing in the case of freeze dryers to above 500°C in the case of indirect dryers heated by combustion products.

c. Indirect dryers are suited to drying under reduced pressures and inert atmospheres, to permit the recovery of solvents and to prevent the occurrence of explosive mixtures or the oxidation of easily decomposed materials.

d. Indirect dryers using condensing fluids as the heating medium are generally economical from the standpoint of heat consumption, since they furnish heat only in accordance with the demand made by the material being dried.

e. Dust recovery and dusty or hazardous materials can be handled more satisfactorily in indirect dryers than in direct dryers.

Electromagnetic (Infrared, Radiofrequency, Microwave) These dryers use energy in the form of electromagnetic radiation.

a. Infrared dryers depend on the transfer of radiant energy to evaporate moisture. The radiant energy is supplied electrically by infrared lamps, by electric resistance elements, or by incandescent refractories heated by gas. The last method has the added advantage of convection heating. Infrared heating is not widely used in the chemical industries for the removal of moisture. Its principal use is in baking or drying paint films (curing) and in heating thin layers of materials. It is sometimes used to give supplementary heating on the initial rolls of paper machines (cylinder dryers).

b. Dielectric dryers (radiofrequency or microwave) have not yet found a wide field of application, but are increasingly used. Their fundamental characteristic of generating heat within the solid indicates potentialities for drying massive geometric objects such as wood, sponge-rubber shapes, and ceramics, and for reduced moisture gradients in layers of solids. Power costs are generally much higher than the fuel costs of conventional methods; a small amount of dielectric heating (2 to 5 percent) may be combined with thermal heating to maximize the benefit at minimum operating cost. The high capital costs of these dryers must be balanced against product quality and process improvements. See more in the Drying Equipment part of this subsection.

4. *Condition of Solids* In solids-gas contacting equipment, the solids bed can exist in any of the following four conditions:

Static This is a dense bed of solids in which each particle rests upon another at essentially the settled bulk density of the solids phase. Specifically, *there is no relative motion among solids particles* (Fig. 12-34).

Moving This is a slightly expanded bed of solids in which the particles are separated only enough to flow one over another. Usually the flow is downward under the force of gravity (Fig. 12-35), but upward motion by mechanical lifting or agitation may also occur within the process vessel (Fig. 12-36). In some cases, lifting of the solids is accomplished in separate equipment, and solids flow in the presence of the gas phase is downward only. The latter is a moving bed as usually defined in the petroleum industry. In this definition, *solids motion is achieved by either mechanical agitation or gravity force.*

Fluidized This is an expanded condition in which the solids particles are supported by drag forces caused by the gas phase passing through the interstices among the particles at some critical velocity. The superficial gas velocity upward is less than the terminal settling velocity of the solids particles; the gas velocity is not sufficient to entrain and convey continuously all the solids. Specifically, the solids phase and the gas phase are intermixed and *together behave as a boiling fluid* (Fig. 12-37).

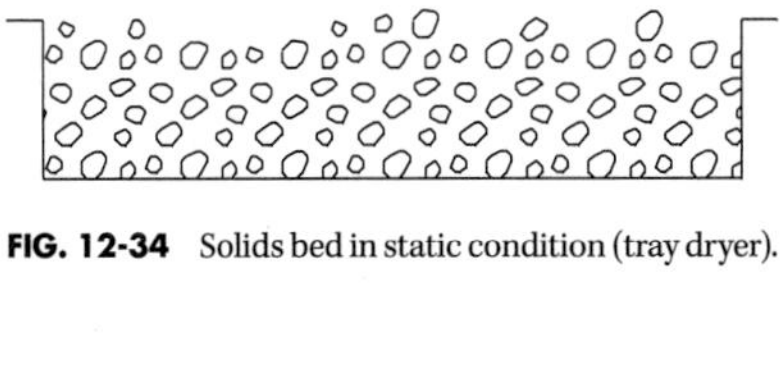

FIG. 12-34 Solids bed in static condition (tray dryer).

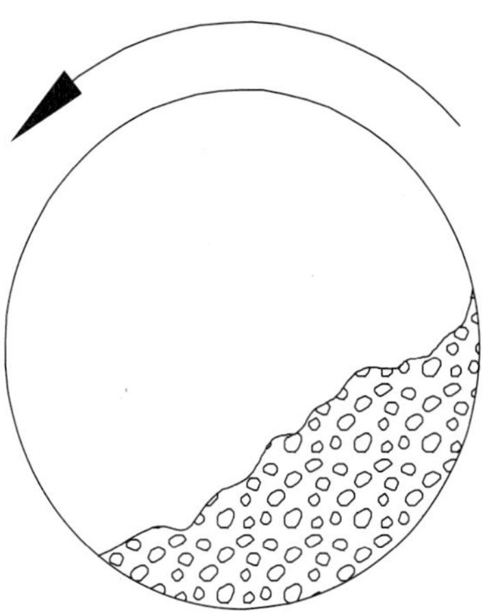

FIG. 12-35 Horizontal moving bed.

Dispersed or dilute This is a fully expanded condition in which the solids particles are so widely separated that they exert essentially no influence upon one another. Specifically, the solids phase is so fully dispersed in the gas that the *density of the suspension is essentially that of the gas phase alone* (Fig. 12-38). Commonly, this situation exists when the gas velocity at all points in the system exceeds the terminal settling velocity of the solids and the particles can be lifted and continuously conveyed by the gas; however, this is not always true. Cascading rotary dryers, countercurrent-flow spray dryers, and gravity settling chambers are three exceptions in which the gas velocity is insufficient to entrain the solids completely.

Cascading (direct) rotary dryers with lifters illustrate all four types of flow in a single dryer. Particles sitting in the lifters (flights) are a static bed. When they are in the rolling bed at the bottom of the dryer, or rolling off the top of the lifters, they form a moving bed. They form a falling curtain which is initially dense (fluidized) but then spreads out and becomes dispersed.

Dryers where the solid forms the continuous phase (static and moving beds) are called *layer dryers,* while those where the gas forms the continuous phase (fluidized and dispersed solids) are classified as *dispersion dryers.* Gas-particle heat and mass transfer is much faster in dispersion dryers, and so these are often favored where high drying rates, short drying times, or high solids throughput is required. Layer dryers are very suitable for slow-drying materials requiring a long residence time.

Because heat transfer and mass transfer in a gas-solids-contacting operation take place at the solids' surfaces, maximum process efficiency can be expected with a maximum exposure of solids surface to the gas phase, together with thorough mixing of gas and solids. Both are important. Within any arrangement of particulate solids, gas is present in the voids among the particles and contacts all surfaces except at the points of particle contact. When the solids are fluidized or dispersed, the gas moves past them rapidly, and external heat- and mass-transfer rates are high. When the solids bed is in a static or slightly moving condition, however, gas within the voids is cut off from the main body of the gas phase and can easily become saturated, so that local drying rates fall to low or zero values. Some transfer of energy

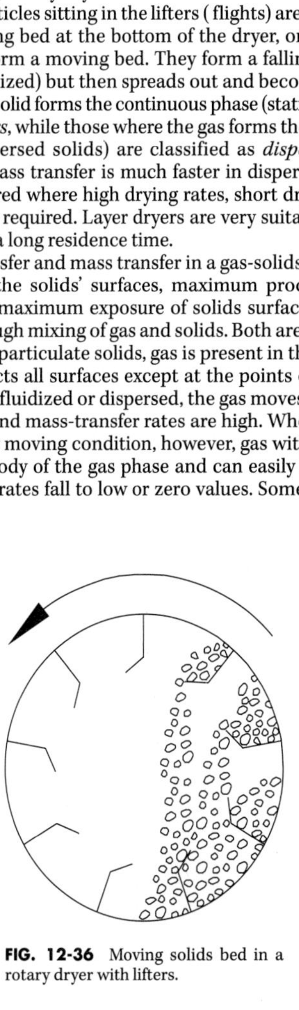

FIG. 12-36 Moving solids bed in a rotary dryer with lifters.

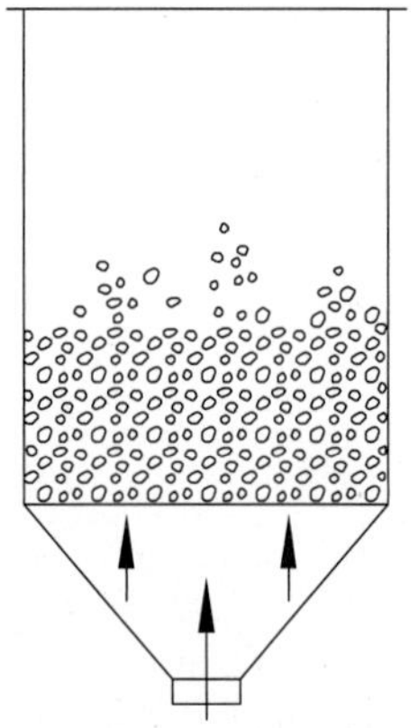

FIG. 12-37 Fluidized solids bed.

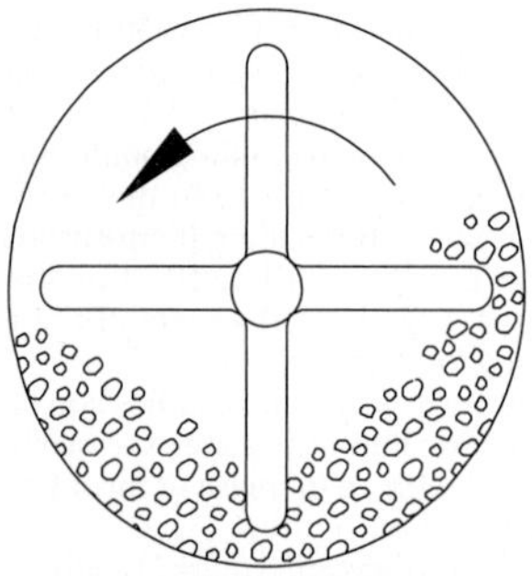

FIG. 12-39 Paddle dryer.

and mass may occur by diffusion, but it is usually insignificant. The problem can be much reduced by using through-circulation of gas instead of cross-circulation, or by agitating and mixing the solids.

Solids Agitation and Mixing There are four alternatives:

1. No agitation, e.g., tray and band dryers. This is desirable for friable materials. However, drying rates can be extremely low, particularly for cross-circulation and vacuum drying.
2. Mechanical agitation, e.g., vertical pan and paddle dryers. This improves mixing and drying rates, but may give attrition depending on agitator speed; and solids may stick to the agitator. These are illustrated in Fig. 12-39.
3. Vessel rotation, e.g., double-cone and rotary dryers. Mixing and heat transfer are better than for static dryers but may be less than for mechanical agitation. Formation of balls and lumps may be a problem.
4. Airborne mixing, e.g., fluidized beds and flash and spray dryers. Generally there is excellent mixing and mass transfer, but feed must be dispersible and entrainment and gas cleaning are higher.

Solids transport In continuous dryers, the solids must be moved through the dryer. These are the main methods of doing this:

1. Gravity flow (usually vertical), e.g., turbo-tray, plate and moving-bed dryers, and rotary dryers (due to the slope)
2. Mechanical conveying (usually horizontal), e.g., band, tunnel, and paddle dryers
3. Airborne transport, e.g., fluidized beds and flash and spray dryers
4. Vibration

Solids flow pattern For most continuous dryers, the solids are basically in plug flow; backmixing is low for nonagitated dryers but can be extensive for mechanical, rotary, or airborne agitation. Exceptions are well-mixed fluidized beds, fluid-bed granulators, and spouted beds (well-mixed) and spray and spray/fluidized-bed units (complex flow patterns).

5. *Gas-Solids Contacting* Where there is a significant gas flow, it may contact a bed of solids in the following ways:

a. Parallel flow or cross-circulation. The direction of gas flow is parallel to the surface of the solids phase. Contacting is primarily at the interface between phases, with possibly some penetration of gas into the voids among the solids near the surface. The solids bed is usually in a static condition (Fig. 12-40).

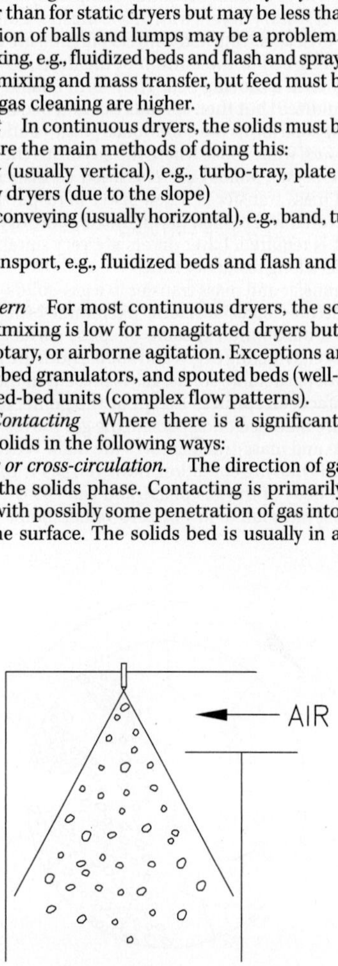

FIG. 12-38 Solids in a dilute condition near the top of a spray dryer.

b. Perpendicular flow or *impingement.* The direction of gas flow is normal to the phase interface. The gas impinges on the solids bed. Again the solids bed is usually in a static condition (Fig. 12-41). This most commonly occurs when the solids are a continuous sheet, film, or slab.

c. Through circulation. The gas penetrates and flows through interstices among the solids, circulating more or less freely around the individual particles (Fig. 12-42). This may occur when solids are in static, moving, fluidized, or dilute conditions.

6. *Gas Flow Pattern in Dryer* Where there is a significant gas flow, it may be in cross-flow, cocurrent, or countercurrent flow compared with the direction of solids movement.

a. Cocurrent gas flow. The gas phase and solids particles both flow in the same direction (Fig. 12-43).

b. Countercurrent gas flow. The direction of gas flow is exactly opposite to the direction of solids movement.

c. Cross-flow of gas. The direction of gas flow is at a right angle to that of solids movement, across the solids bed (Fig. 12-44).

The difference between these is shown most clearly in the gas and solids temperature profiles along the dryer. For cross-flow dryers, all solids particles are exposed to the same gas temperature, and the solids temperature approaches the gas temperature near the end of drying (Fig. 12-45). In cocurrent dryers, the gas temperature falls throughout the dryer, and the final solids temperature is much lower than that for the cross-flow dryer (Fig. 12-46). Hence cocurrent dryers are very suitable for drying heat-sensitive materials, although it is possible to get a solids temperature peak inside the dryer. Conversely, countercurrent dryers give the most even temperature gradient throughout the dryer, but the exiting solids come into contact with the hottest, driest gas (Fig. 12-47). These can be used to heat-treat the solids or to give low final moisture content (minimizing the local equilibrium moisture content) but are obviously unsuitable for thermally sensitive solids.

SELECTION OF DRYING EQUIPMENT

Dryer Selection Considerations Dryer selection is a challenging task. These are some important considerations:

- Batch dryers are almost invariably used for mean throughputs below 50 kg/h, and continuous dryers are generally used above 1 ton/h; in the intervening range, either may be suitable.
- Liquid or slurry feeds, large objects, or continuous sheets and films require completely different equipment from particulate feeds.
- Particles and powders below 1 mm are effectively dried in dispersion or contact dryers, but most through-circulation units are unsuitable. Conversely, for particles of several millimeters or above, through-circulation dryers, rotary dryers, and spouted beds are very suitable.
- Through-circulation and dispersion convective dryers (including fluidized-bed, rotary, and pneumatic types) and agitated or rotary contact dryers generally give better drying rates than nonagitated cross-circulated or contact tray dryers.
- Nonagitated dryers (including through-circulation) may be preferable for fragile particles where it is desired to avoid attrition.

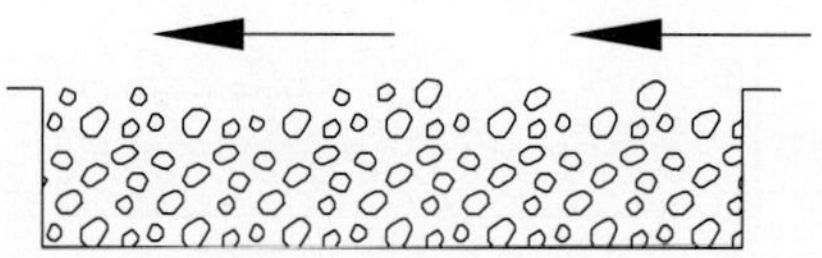

FIG. 12-40 Parallel gas flow over a static bed of solids.

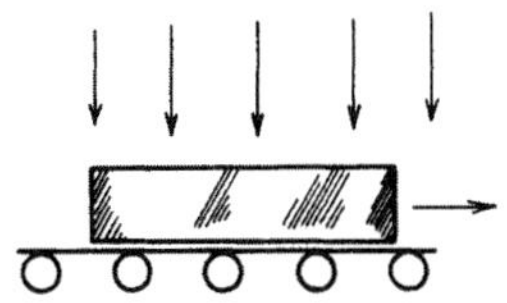

FIG. 12-41 Circulating gas impinging on a large solid object in perpendicular flow, in a roller-conveyor dryer.

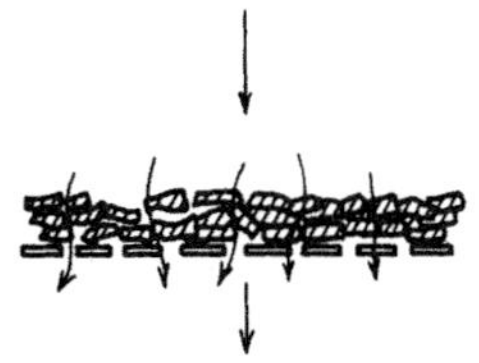

FIG. 12-42 Gas passing through a bed of preformed solids, in through-circulation on a perforated-band dryer.

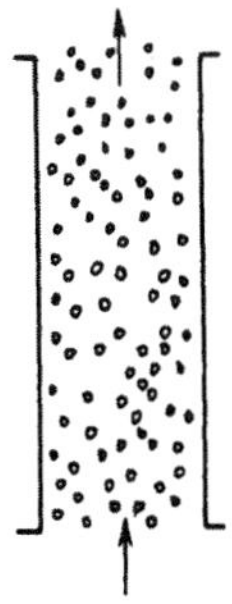

FIG. 12-43 Cocurrent gas-solids flow in a vertical-lift dilute-phase pneumatic conveyor dryer.

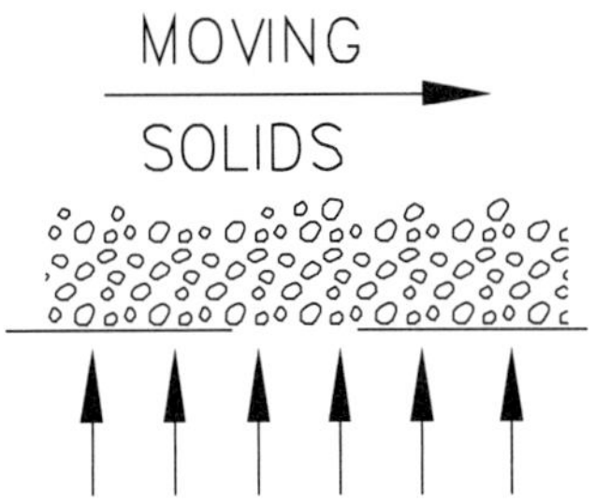

FIG 12-44 Cross-flow of gas and solids in a fluid-bed or band dryer.

- For organic solvents or solids which are highly flammable, are toxic, or decompose easily, contact dryers are often preferable to convective dryers, as containment is better and environmental emissions are easier to control. If a convective dryer is used, a closed-cycle system using an inert carrier gas (e.g., nitrogen) is often required.
- Cocurrent, vacuum, and freeze dryers can be particularly suitable for heat-sensitive materials.

A detailed methodology for dryer selection, including the use of a rule-based expert system, has been described by Kemp [*Drying Technol.* **13**(5–7): 1563–1578 (1995) and **17**(7 and 8): 1667–1680 (1999)].

A simpler step-by-step procedure is given here.

1. *Initial selection of dryers.* Select those dryers which appear best suited to handling the wet material and the dry product, which fit into the continuity of the process as a whole, and which will produce a product of the desired physical properties. This preliminary selection can be made with the aid of Table 12-11, which classifies the various types of dryers on the basis of the materials handled.
2. *Initial comparison of dryers.* The dryers so selected should be evaluated approximately from available cost and performance data. From this evaluation, those dryers that appear to be uneconomical or unsuitable from the standpoint of performance should be eliminated from further consideration.
3. *Drying tests.* Drying tests should be conducted in those dryers still under consideration. These tests will determine the optimum operating conditions and the product characteristics and will form the basis for firm quotations from equipment vendors.
4. *Final selection of dryer.* From the results of the drying tests and quotations, the final selection of the most suitable dryer can be made.

These are the important factors to consider in the preliminary selection of a dryer:

1. Properties of the material being handled
   a. Physical characteristics when wet (stickiness, cohesiveness, adhesiveness, flowability)
   b. Physical characteristics when dry
   c. Corrosiveness
   d. Toxicity
   e. Flammability
   f. Particle size
   g. Abrasiveness
2. Drying characteristics of the material
   a. Type of moisture (bound, unbound, or both)
   b. Initial moisture content (maximum and range)
   c. Final moisture content (maximum and range)
   d. Permissible drying temperature
   e. Probable drying time for different dryers
   f. Level of nonwater volatiles
3. Flow of material to and from the dryer
   a. Quantity to be handled per hour (or batch size and frequency)
   b. Continuous or batch operation
   c. Process prior to drying
   d. Process subsequent to drying

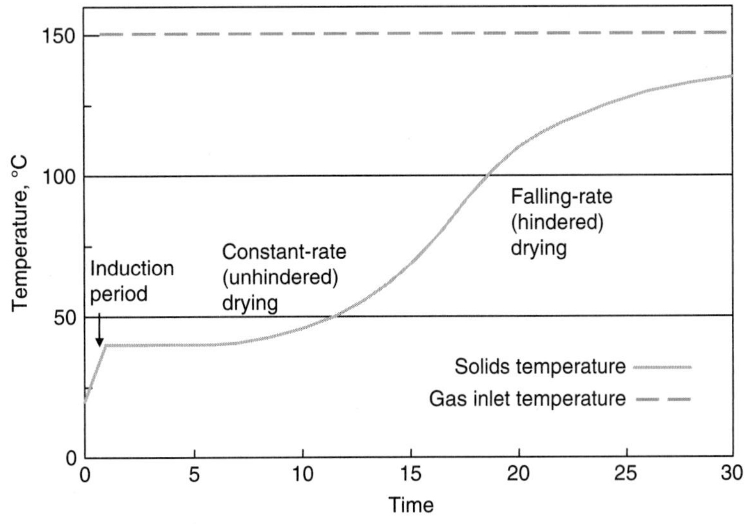

FIG. 12-45 Temperature profiles along a continuous plug-flow dryer for cross-flow of gas and solids. (*Aspen Technology Inc.*)

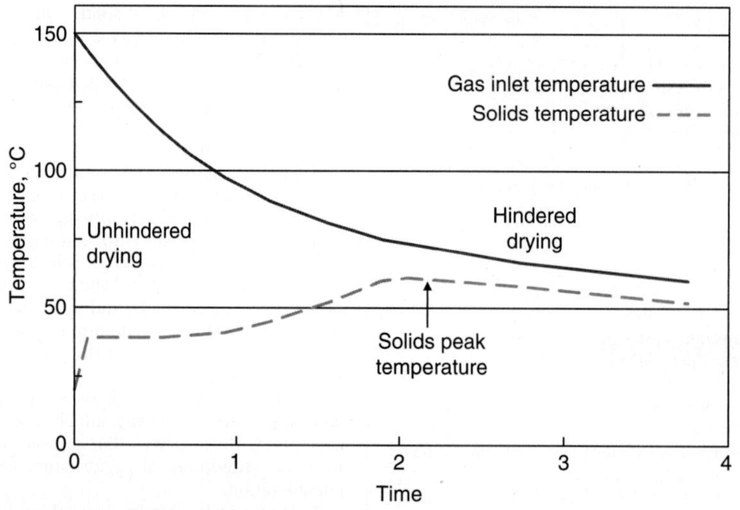

FIG. 12-46 Temperature profiles along a continuous plug-flow dryer for cocurrent flow of gas and solids. (*Aspen Technology Inc.*)

4. Product quality
   a. Shrinkage
   b. Contamination
   c. Uniformity of final moisture content
   d. Decomposition of product
   e. Overdrying
   f. Particle size distribution (if applicable)
   g. Product temperature
   h. Bulk density
5. Recovery and environmental considerations
   a. Dust recovery
   b. Solvent recovery
6. Facilities available at site of proposed installation
   a. Space
   b. Flow rate, temperature, humidity, and cleanliness of air
   c. Available fuels
   d. Available electric power
   e. Permissible noise, vibration, dust, or heat losses
   f. Source of wet feed
   g. Exhaust-gas outlets/permissible VOC discharge levels

Following preliminary selection of suitable types of dryers, a rough evaluation of the size and cost should be made to eliminate those which are obviously uneconomical. Information for this evaluation can be obtained from material presented under discussion of the various dryer types. When data are inadequate, usually preliminary cost and performance data can be obtained from the equipment manufacturer. In comparing dryer performance, the factors in the preceding list which affect dryer performance should be properly weighed.

DRYER DESCRIPTIONS

Batch Tray Dryers

Examples and Synonyms Direct heat tray dryer, batch through-circulation dryer, vacuum shelf dryer, through-circulation drying room, vacuum oven, vacuum-shelf dryer.

Description A tray or compartment dryer is an enclosed, insulated housing in which solids are placed upon tiers of trays in the case of particulate solids or stacked in piles or upon shelves in the case of large objects. Heat transfer may be *direct* from gas to solids by circulation of large volumes of hot gas or *indirect* by use of heated shelves, radiator coils,

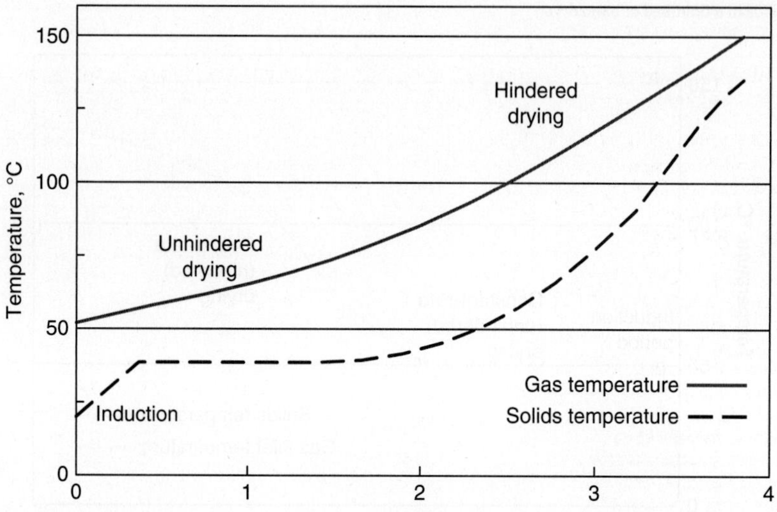

FIG. 12-47 Temperature profiles along a continuous plug-flow dryer for countercurrent flow of gas and solids. (*Aspen Technology Inc.*)

or refractory walls inside the housing. In indirect-heat units, excepting vacuum-shelf equipment, circulation of a small quantity of gas is usually necessary to sweep moisture vapor from the compartment and prevent gas saturation and condensation. Compartment units are employed for the heating and drying of lumber, ceramics, sheet materials (supported on poles), painted and metal objects, and all forms of particulate solids.

Field of Application Because of the high labor requirements usually associated with loading or unloading the compartments, batch compartment equipment is rarely economical except in the following situations:

1. A long heating cycle is necessary because the size of the solid objects or permissible heating temperature requires a long holdup for internal diffusion of heat or moisture. This case may apply when the cycle will exceed 12 to 24 h.

2. The production of several different products requires strict batch identity and thorough cleaning of equipment between batches. This is a situation existing in many small, multiproduct plants, e.g., for pharmaceuticals or specialty chemicals.

3. The quantity of material to be processed does not justify investment in more expensive, continuous equipment. This case would apply in many pharmaceutical drying operations.

Further, because of the nature of solids-gas contacting, which is usually by parallel flow and rarely by through-circulation, heat transfer and mass transfer are comparatively inefficient. For this reason, use of tray and compartment equipment is restricted primarily to ordinary drying and heat-treating operations.

Auxiliary Equipment If noxious gases, fumes, or dust is given off during the operation, dust or fume recovery equipment will be necessary in the exhaust gas system. Condensers are employed for the recovery of valuable solvents from dryers. To minimize heat losses, thorough insulation of the compartment with brick or other insulating compounds is necessary. Vacuum-shelf dryers require auxiliary stream jets or other vacuum-producing devices, intercondensers for vapor removal, and occasionally wet scrubbers or (heated) bag-type dust collectors.

Uniform depth of loading in dryers and furnaces handling particulate solids is essential to consistent operation, minimum heating cycles, or control of final moisture. After a tray has been loaded, the bed should be leveled to a uniform depth. Special preform devices, noodle extruders, pelletizers, etc., are employed occasionally for preparing pastes and filter cakes so that screen bottom trays can be used and the advantages of through-circulation approached.

Control of tray and compartment equipment is usually maintained by control of the circulating air temperature (and humidity) and rarely by the solids temperature. On vacuum units, control of the absolute pressure and heating-medium temperature is utilized. In direct dryers, cycle controllers are frequently employed to vary the air temperature or velocity across the solids during the cycle; e.g., high air temperatures may be employed during a constant-rate drying period while the solids surface remains close to the air wet-bulb temperature. During the falling-rate periods, this temperature may be reduced to prevent case hardening or other degrading effects caused by overheating the solids surfaces. In addition, higher air velocities may be employed during early drying stages to improve heat transfer; however, after surface drying has been completed, this velocity may need to be reduced to prevent dusting. Two-speed circulating fans are employed commonly for this purpose.

Direct-Heat Tray Dryers Satisfactory operation of tray-type dryers depends on maintaining a constant temperature and a uniform air velocity over all the material being dried.

Circulation of air at velocities of 1 to 10 m/s is desirable to improve the surface heat-transfer coefficient and to eliminate stagnant air pockets. Proper airflow in tray dryers depends on sufficient fan capacity, on the design of ductwork to modify sudden changes in direction, and on properly placed baffles. *Nonuniform airflow is one of the most serious problems in the operation of tray dryers.*

Tray dryers may be of the tray-truck or the stationary-tray type. In the former, the trays are loaded on trucks which are pushed into the dryer; in the latter, the trays are loaded directly into stationary racks within the dryer. Trays may be square or rectangular, with 0.5 to 1 m^2 per tray, and may be fabricated from any material compatible with corrosion and temperature conditions. When the trays are stacked in the truck, there should be a clearance of not less than 4 cm between the material in one tray and the bottom of the tray immediately above. When material characteristics and handling permit, the trays should have screen bottoms for additional drying area. Metal trays are preferable to nonmetallic trays, since they conduct heat more readily. Tray loadings range usually from 1 to 10 cm deep.

Steam is the usual heating medium, and a standard heater arrangement consists of a main heater before the circulating fan. When steam is not available or the drying load is small, electric heat can be used. For temperatures above 450 K, products of combustion can be used, or indirect-fired air heaters.

Air is circulated by propeller or centrifugal fans; the fan is usually mounted within or directly above the dryer. Total pressure drop through the trays, heaters, and ductwork is usually in the range of 2.5 to 5 cm of water. Air recirculation is generally on the order of 80 to 95 percent except during the initial drying stage of rapid evaporation. Fresh air is drawn in by the circulating fan, frequently through dust filters. In most installations, air is exhausted by a separate small exhaust fan with a damper to control air recirculation rates.

Prediction of heat- and mass-transfer coefficients in direct heat tray dryers In convection phenomena, heat-transfer coefficients depend on the geometry of the system, the gas velocity past the evaporating surface, and the physical properties of the drying gas. In estimating drying rates, the use of heat-transfer coefficients is preferred because they are usually more reliable than mass-transfer coefficients. In calculating mass-transfer coefficients from drying experiments, the partial pressure at the surface is usually inferred from the measured or calculated temperature of the evaporating surface. Small errors in temperature have negligible effect on the heat-transfer coefficient but introduce relatively large errors in the partial pressure and hence in the mass-transfer coefficient.

For many cases in drying, the heat-transfer coefficient is proportional to U_g^n, where U_g is an appropriate local gas velocity. For flow parallel to plane plates, the exponent n has been reported to range from 0.35 to 0.8. The differences in exponent have been attributed to differences in flow pattern in the space above the evaporating surface, particularly whether it is laminar or turbulent, and whether the length is sufficient to allow fully developed flow. In the absence of applicable specific data, the heat-transfer coefficient for the parallel-flow case can be taken, for estimating purposes, as

$$h = \frac{8.8 J^{0.8}}{D_c^{0.2}} \qquad (12\text{-}68)$$

where h is the heat-transfer coefficient, W/($m^2 \cdot$ K); J is the gas mass flux, kg/($m^2 \cdot$ s); and D_c is a characteristic dimension of the system. The experimental data have been weighted in favor of an exponent of 0.8 in conformity with the usual Colburn j factor, and average values of the properties of air at 370 K have been incorporated. Typical values are in the range 10 to 50 W/($m^2 \cdot$ K).

Experimental data for drying from flat surfaces have been correlated by using the equivalent diameter of the flow channel or the length of the evaporating surface as the characteristic length dimension in the Reynolds number. However, the validity of one versus the other has not been established. The proper equivalent diameter probably depends at least on the geometry of the system, the roughness of the surface, and the flow conditions upstream of the evaporating surface. For airflow impinging normally to the surface from slots and nozzles, the heat-transfer coefficient can be obtained from the well-known Martin correlation: Martin, "Heat and Mass Transfer Between Impinging Gas Jets and Solid Surfaces," *Advances in Heat Transfer,* vol. 13, Academic Press, 1977, pp. 1–66. This correlation uses relevant geometric properties such as the diameter of the holes, the spacing between the slots/nozzles, and the distance between the slots/nozzles and the sheet. See Example 12-23 for calculation of the heat-transfer coefficient from an array of jets.

Most efficient performance is obtained with plates having open areas equal to 2 to 3 percent of the total heat-transfer area. The plate should be located at a distance equal to four to six hole (or equivalent) diameters from the heat-transfer surface.

As with many drying calculations, the most reliable design method is to perform experimental tests and to scale up. By measuring performance on a single tray with similar layer depth, air velocity, and temperature, the specific drying rate (SDR) concept as described in the Solids Drying Fundamentals subsection can be applied to give the total area and number of trays required for the full-scale dryer.

Performance data for direct heat tray dryers A standard two-truck dryer is illustrated in Fig. 12-48. Adjustable baffles or a perforated distribution plate is normally employed to develop 0.3 to 1.3 cm of water pressure drop at the wall through which air enters the truck enclosure. This will enhance the uniformity of air distribution, from top to bottom, among the trays.

Performance data on some typical tray and compartment dryers are tabulated in Table 12-12. These indicate that an overall rate of evaporation of 0.0025 to 0.025 kg water/(s·m^2) of tray area may be expected from tray and tray-truck dryers. The thermal efficiency of this type of dryer will vary from 20 to 50 percent, depending on the drying temperature used and the humidity of the exhaust air. In drying to very low moisture contents under temperature restrictions, the thermal efficiency may be on the order of 10 percent. Maintenance will run from 3 to 5 percent of the installed cost per year.

Batch Through-Circulation Dryers These may be either of shallow bed or deep bed type. In the first type of batch through-circulation dryer, heated air passes through a stationary permeable bed of the wet material placed on removable screen-bottom trays suitably supported in the dryer. This type is similar to a standard tray dryer *except that hot air passes through the wet solid*

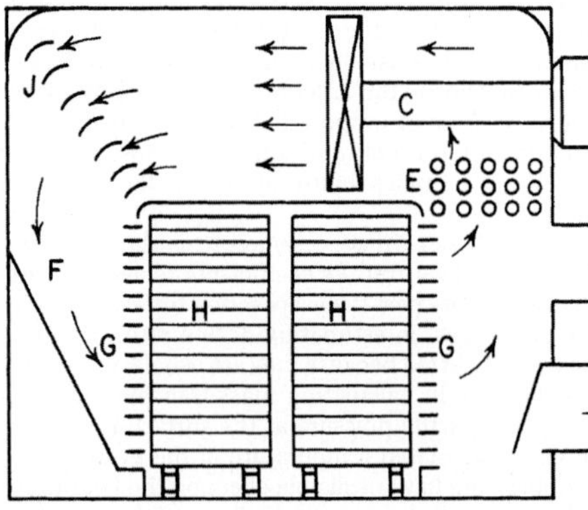

FIG. 12-48 Double-truck tray dryer. (*A*) Air inlet duct. (*B*) Air exhaust duct with damper. (*C*) Adjustable-pitch fan, 1 to 15 hp. (*D*) Fan motor. (*E*) Fin heaters. (*F*) Plenum chamber. (*G*) Adjustable air blast nozzles. (*H*) Trucks and trays.

instead of across it. The pressure drop through the bed of material can be estimated using the Ergun equation [Ergun, S., "Fluid Flow through Packed Columns," *Chem. Eng. Progress* **48** (1952)] and does not usually exceed about 2 cm of water. In the second type, deep perforated-bottom trays are placed on top of plenum chambers in a closed-circuit hot air circulating system. In some food-drying plants, the material is placed in finishing bins with perforated bottoms; heated air passes up through the material and is removed from the top of the bin, reheated, and recirculated. The latter types involve a pressure drop through the bed of material of 1 to 8 cm of water at relatively low air rates. Table 12-13 gives performance data on three applications of batch through-circulation dryers. Batch through-circulation dryers are restricted in application to granular materials (particle size typically 1 mm or greater) that permit free flow-through circulation of air. Drying times are usually much shorter than in parallel-flow tray dryers. Design methods are included in the subsection Continuous Through-Circulation Dryers.

Vacuum-Shelf Dryers Vacuum-shelf dryers are indirectly heated batch dryers consisting of a vacuum-tight chamber usually constructed of cast iron or steel plate; heated, supporting shelves within the chamber; a vacuum source; and usually a condenser. One or two doors are provided, depending on the size of the chamber. The doors are sealed with resilient gaskets of rubber or similar material.

Hollow shelves of flat steel plate are fastened permanently inside the vacuum chamber and are connected in parallel to inlet and outlet headers. The heating medium, entering through one header and passing through the hollow shelves to the exit header, is generally steam, ranging in pressure from 700 kPa gauge to subatmospheric pressure for low-temperature operations. Low temperatures can be provided by circulating hot water, and high temperatures can be obtained by circulating hot oil or other heat-transfer liquids. Some small dryers employ electrically heated shelves. The material to be dried is placed in pans or trays on the heated shelves. The trays are generally of metal to ensure good heat transfer between the shelf and the tray.

Vacuum-shelf dryers may vary in size from 1 to 24 shelves, the largest chambers having overall dimensions of 6 m wide, 3 m long, and 2.5 m high.

TABLE 12-13 Performance Data for Batch Through-Circulation Dryers*

Kind of material	Granular polymer	Vegetable	Vegetable seeds
Capacity, kg product/h	122	42.5	27.7
Number of trays	16	24	24
Tray spacing, cm	43	43	43
Tray size, cm	91.4 × 104	91.4 × 104	85 × 98
Depth of loading, cm	7.0	6	4
Physical form of product	Crumbs	0.6-cm diced cubes	Washed seeds
Initial moisture content, % dry basis	11.1	669.0	100.0
Final moisture content, % dry basis	0.1	5.0	9.9
Air temperature, °C	88	77 dry-bulb	36
Air velocity, superficial, m/s	1.0	0.6 to 1.0	1.0
Tray loading, kg product/m²	16.1	5.2	6.7
Drying time, h	2.0	8.5	5.5
Overall drying rate, kg water evaporated/(h·m²)	0.89	11.86	1.14
Steam consumption, kg/kg water evaporated	4.0	2.42	6.8
Installed power, kW	7.5	19	19

*Courtesy of Wolverine Proctor & Schwartz, Inc.

Vacuum is applied to the chamber, and vapor is removed through a large pipe which is connected to the chamber in such a manner that if the vacuum is broken suddenly, the in-rushing air will not greatly disturb the bed of material being dried. This line leads to a condenser where moisture or solvent that has been vaporized is condensed. The noncondensible exhaust gas goes to the vacuum source, which may be a wet or dry vacuum pump or a steam-jet ejector.

Vacuum-shelf dryers are used extensively for drying pharmaceuticals, temperature-sensitive or easily oxidizable materials, and materials so valuable that labor cost is insignificant. They are particularly useful for handling small batches of materials wet with toxic or valuable solvents. Recovery of the solvent is easily accomplished without danger of passing through an explosive range. Dusty materials may be dried with negligible dust loss. Hygroscopic materials may be completely dried at temperatures below that required in atmospheric dryers. The equipment is employed also for freeze-drying processes, for metallizing-furnace operations, and for the manufacture of semiconductor parts in controlled atmospheres. All these latter processes demand much lower operating pressures than do ordinary drying operations.

Design methods for vacuum-shelf dryers Heat is transferred to the wet material by conduction through the shelf and bottom of the tray and by radiation from the shelf above. The critical moisture content will not be necessarily the same as for atmospheric tray drying, as the heat-transfer mechanisms are different.

During the constant-rate period, moisture is rapidly removed. Often 50 percent of the moisture will evaporate in the first hour of a 6- to 8-h cycle. The drying time has been found to be proportional to between the first and second powers of the depth of loading. Vacuum-shelf dryers operate in the range of 1 to 25 mmHg pressure. For size-estimating purposes, a heat-transfer coefficient of 20 J/(m²·s·K) may be used. The area employed in this case should be the shelf area in direct contact with the trays. For the same

TABLE 12-12 Manufacturer's Performance Data for Tray and Tray-Truck Dryers*

Material	Color	Chrome yellow	Toluidine red	Half-finished Titone	Color
Type of dryer	2-truck	16-tray dryer	16-tray	3-truck	2-truck
Capacity, kg product/h	11.2	16.1	1.9	56.7	4.8
Number of trays	80	16	16	180	120
Tray spacing, cm	10	10	10	7.5	9
Tray size, cm	60 × 75 × 4	65 × 100 × 2.2	65 × 100 × 2	60 × 70 × 3.8	60 × 70 × 2.5
Depth of loading, cm	2.5 to 5	3	3.5	3	
Initial moisture, % bone-dry basis	207	46	220	223	116
Final moisture, % bone-dry basis	4.5	0.25	0.1	25	0.5
Air temperature, °C	85–74	100	50	95	99
Loading, kg product/m²	10.0	33.7	7.8	14.9	9.28
Drying time, h	33	21	41	20	96
Air velocity, m/s	1.0	2.3	2.3	3.0	2.5
Drying, kg water evaporated/(h·m²)	0.59	65	0.41	1.17	0.11
Steam consumption, kg/kg water evaporated	2.5	3.0	—	2.75	
Total installed power, kW	1.5	0.75	0.75	2.25	1.5

*Courtesy of Wolverine Proctor & Schwartz, Inc.

TABLE 12-14 Standard Vacuum-Shelf Dryers*

Shelf area, m^2	Floor space, m^2	Weight average, kg	Pump capacity, m^3/s	Pump motor, kW	Condenser area, m^2	Price/m^2 (1995) Carbon steel	Price/m^2 (1995) 304 stainless steel
0.4–1.1	4.5	540	0.024	1.12	1	$110	$170
1.1–2.2	4.5	680	0.024	1.12	1	75	110
2.2–5.0	4.6	1130	0.038	1.49	4	45	65
5.0–6.7	5.0	1630	0.038	1.49	4	36	65
6.7–14.9	6.4	3900	0.071	2.24	9	27	45
16.7–21.1	6.9	5220	0.071	2.24	9	22	36

*Stokes Vacuum, Inc.

reason, the shelves should be kept free from scale and rust. Air vents should be installed on steam-heated shelves to vent noncondensible gases. The heating medium should not be applied to the shelves until after the air has been evacuated from the chamber, to reduce the possibility of the material's overheating or boiling at the start of drying. Case hardening (formation of hard external layer) can sometimes be avoided by retarding the rate of drying in the early part of the cycle.

Some performance data for vacuum-shelf dryers are given in Table 12-14. The thermal efficiency of a vacuum-shelf dryer is usually on the order of 60 to 80 percent. Table 12-15 gives operating data for one organic and two inorganic compounds.

Continuous Tray and Gravity Dryers

Examples and Synonyms Turbo-tray dryer, plate dryer, moving-bed dryer, gravity dryer.

Description Continuous tray dryers are equivalent to batch tray dryers, but with the solids moving between trays by a combination of mechanical movement and gravity. Gravity (moving-bed) dryers are normally through-circulation convective dryers with no internal trays where the solids gradually descend by gravity. In all these types, the net movement of solids is vertically downward.

Turbo-Tray Dryers The turbo-tray dryer (also known as rotating tray, rotating shelf, or Wyssmont Turbo-Dryer) is a continuous dryer consisting of a stack of rotating annular shelves in the center of which turbo-type fans revolve to circulate the air over the shelves. Wet material enters through the roof, falling onto the top shelf as it rotates beneath the feed opening. After completing one rotation, the material is wiped by a stationary wiper through radial slots onto the shelf below, where it is spread into a uniform pile by a stationary leveler. The action is repeated on each shelf, with transfers occurring once in each revolution. From the last shelf, material is discharged through the bottom of the dryer (Fig. 12-49).

The rate at which each fan circulates air can be varied by changing the pitch of the fan blades. In final drying stages, in which diffusion controls or the product is light and powdery, the circulation rate is considerably lower than in the initial stage, in which high evaporation rates prevail. In the majority of applications, air flows through the dryer upward in counterflow to the material. In special cases, required drying conditions dictate that airflow be cocurrent or both countercurrent and cocurrent with the exhaust leaving at some level between solids inlet and discharge. A separate cold-air-supply fan is provided if the product is to be cooled before being discharged.

By virtue of its vertical construction, the turbo-type tray dryer has a stack effect, the resulting draft being frequently sufficient to operate the dryer with natural draft. Pressure at all points within the dryer is maintained close to atmospheric. Most of the roof area is used as a breeching, lowering the exhaust velocity to settle dust back into the dryer.

Heaters can be located in the space between the trays and the dryer housing, where they are not in direct contact with the product, and thermal efficiencies up to 3500 kJ/kg (1500 Btu/lb) of water evaporated can be obtained by reheating the air within the dryer. For materials which have a tendency to foul internal heating surfaces, an external heating system is employed.

The turbo-tray dryer can handle materials from thick slurries [1 million N·s/m^2 (100,000 cP) and over] to fine powders. Filter-press cakes are granulated before feeding. Thixotropic materials are fed directly from a rotary filter by scoring the cake as it leaves the drum. Pastes can be extruded onto the top shelf and subjected to a hot blast of air to make them firm and free-flowing after one rotation.

The turbo-tray dryer is manufactured in sizes from package units 2 m in height and 1.5 m in diameter to large outdoor installations 20 m in height and 11 m in diameter. Tray areas range from 1 m^2 up to about 2000 m^2. The number of shelves in a tray rotor varies according to space available and the minimum rate of transfer required, from as few as 12 shelves to as many as 58 in the largest units. Standard construction permits operating temperatures up to 615 K, and high-temperature heaters permit operation at temperatures up to 925 K.

TABLE 12-15 Performance Data for Vacuum-Shelf Dryers

Material	Sulfur black	Calcium carbonate	Calcium phosphate
Loading, kg dry material/m^2	25	17	33
Steam pressure, kPa gauge	410	410	205
Vacuum, mmHg	685–710	685–710	685–710
Initial moisture content, % (wet basis)	50	50.3	30.6
Final moisture content, % (wet basis)	1	1.15	4.3
Drying time, h	8	7	6
Evaporation rate, kg/(s·m^2)	8.9×10^{-4}	7.9×10^{-4}	6.6×10^{-4}

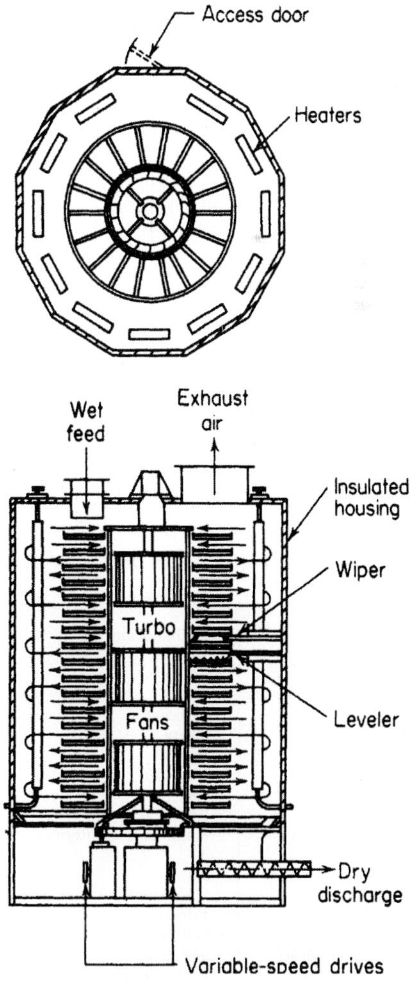

FIG. 12-49 Turbo-Dryer. (*Wyssmont Company, Inc.*)

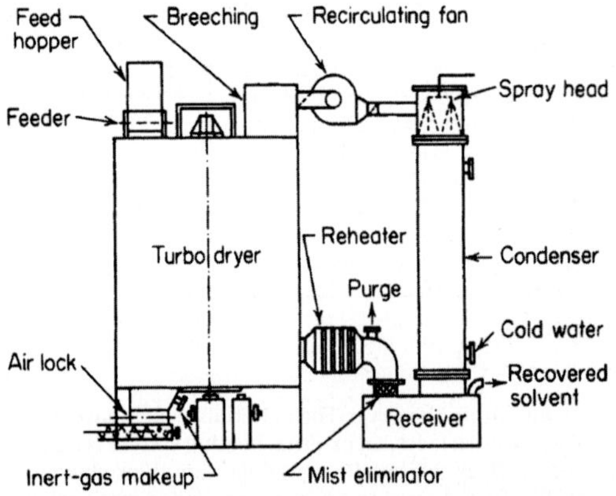

FIG. 12-50 Turbo-Dryer in closed circuit for continuous drying with solvent recovery. (*Wyssmont Company, Inc.*)

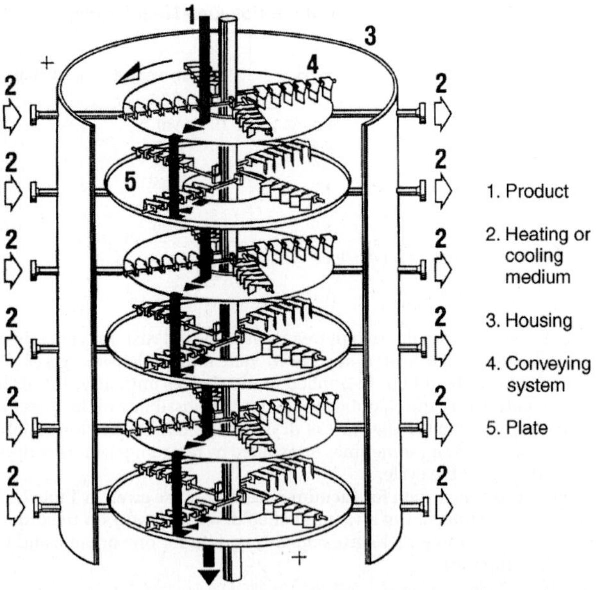

FIG. 12-51 Indirect heat continuous plate dryer for atmospheric, gastight, or full-vacuum operation. (*Krauss Maffei.*)

Design methods for turbo-tray dryers The heat- and mass-transfer mechanisms are similar to those in batch tray dryers, except that constant turning over and mixing of the solids significantly improve drying rates. Design usually must be based on previous installations or pilot tests by the manufacturer; apparent heat-transfer coefficients are typically 30 to 60 J/(m^2·s·K) for dry solids and 60 to 120 J/(m^2·s·K) for wet solids. Turbo-tray dryers have been employed successfully for the drying and cooling of calcium hypochlorite, urea crystals, calcium chloride flakes, and sodium chloride crystals. The Wyssmont "closed-circuit" system, as shown in Fig. 12-50, consists of the turbo-tray dryer with or without internal heaters, recirculation fan, condenser with receiver and mist eliminators, and reheater. Feed and discharge are through a sealed wet feeder and lock, respectively. This method is used for continuous drying without leakage of fumes, vapors, or dust to the atmosphere. A unified approach for scaling up dryers, such as turbo-tray, plate, conveyor, or any other dryer type that forms a defined layer of solids next to a heating source, is the SDR method described by Moyers [*Drying Technol.* **12**(1 & 2): 393–417 (1994)].

Performance and cost data for turbo-tray dryers Performance data for four applications of closed-circuit drying are included in Table 12-16. Operating, labor, and maintenance costs compare favorably with those of direct heat rotating equipment.

Plate Dryers The plate dryer is an *indirectly heated,* fully continuous dryer available for three modes of operation: atmospheric, gastight, or full vacuum. The dryer is of vertical design, with horizontal, heated plates mounted inside the housing. The plates are heated by hot water, steam, or thermal oil, with operating temperatures up to 320°C possible. The product enters at the top and is conveyed through the dryer by a product transport system consisting of a central-rotating shaft with arms and plows. (See dryer schematic, Fig. 12-51.) The thin product layer [approximately 12- mm (0.5-in) depth] on the surface of the plates, coupled with frequent product turnover by the conveying system, results in short retention times (approximately 5 to 40 min), true plug flow of the material, and uniform drying. The vapors are removed from the dryer by a small amount of heated purge gas or by vacuum. The material of construction of the plates and housing is normally stainless steel, with special metallurgies also available. The drive unit is located at the bottom of the dryer and supports the central-rotating shaft. Typical speed of the dryer is 1 to 7 rpm.

The plate dryer may vary in size from 5 to 35 vertically stacked plates with a heat-exchange area between 3.8 and 175 m^2. Depending upon the loose-bulk density of the material and the overall retention time, the plate dryer can process up to 5000 kg/h of wet product.

The plate dryer is limited in its scope of applications only in the consistency of the feed material (the products must be friable, be free-flowing, and not undergo phase changes) and drying temperatures up to 320°C. Applications include specialty chemicals, pharmaceuticals, foods, polymers, pigments, etc. Initial moisture or volatile level can be as high as 65 percent, and the unit is often used as a final dryer to take materials to a bone-dry state, if necessary. The plate dryer can also be used for heat treatment, removal of waters of hydration (bound moisture), solvent removal, and a product cooler.

The atmospheric plate dryer is a dust-tight system. The dryer housing is an octagonal, panel construction, with operating pressure in the range of ±0.5 kPa gauge. An exhaust air fan draws the purge air through the housing for removal of the vapors from the drying process. The purge air velocity through the dryer is in the range of 0.1 to 0.15 m/s, resulting in minimal

TABLE 12-16 Turbo-Dryer Performance Data in Wyssmont Closed-Circuit Operations*

Material dried	Antioxidant	Water-soluble polymer	Antibiotic filter cake	Petroleum coke
Dried product, kg/h	500	85	2400	227
Volatiles composition	Methanol and water	Xylene and water	Alcohol and water	Methanol
Feed volatiles, % wet basis	10	20	30	30
Product volatiles, % wet basis	0.5	4.8	3.5	0.2
Evaporation rate, kg/h	53	16	910	302
Type of heating system	External	External	External	External
Heating medium	Steam	Steam	Steam	Steam
Drying medium	Inert gas	Inert gas	Inert gas	Inert gas
Heat consumption, J/kg	0.56×10^6	2.2×10^6	1.42×10^6	1.74×10^6
Power, dryer, kW	1.8	0.75	12.4	6.4
Power, recirculation fan, kW	5.6	5.6	37.5	15
Materials of construction	Stainless-steel interior	Stainless-steel interior	Stainless-steel interior	Carbon steel
Dryer height, m	4.4	3.2	7.6	6.5
Dryer diameter, m	2.9	1.8	6.0	4.5
Recovery system	Shell-and-tube condenser	Shell-and-tube condenser	Direct-contact condenser	Shell-and-tube condenser
Condenser cooling medium	Brine	Chilled water	Tower water	Chilled water
Location	Outdoor	Indoor	Indoor	Indoor
Approximate cost of dryer (2004)	$300,000	$175,000	$600,000	$300,000
Dryer assembly	Packaged unit	Packaged unit	Field-erected unit	Field-erected unit

*Courtesy of Wyssmont Company, Inc.

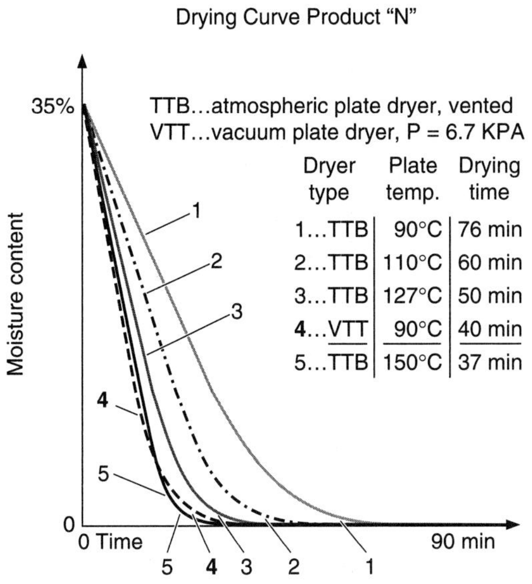

FIG. 12-52 Plate dryer drying curves demonstrating impact of elevated temperature and/or operation under vacuum. (*Krauss Maffei.*)

dusting and small dust filters for the exhaust air. The air temperature is normally equal to the plate temperature. The gastight plate dryer, together with the components of the gas recirculation system, forms a closed system. The dryer housing is semicylindrical and is rated for a nominal pressure of 5 kPa gauge. The flow rate of the recirculating purge gas must be sufficient to absorb the vapors generated from the drying process. The gas temperature must be adjusted according to the specific product characteristics and the type of volatile. After condensation of the volatiles, the purge gas (typically nitrogen) is recirculated back to the dryer via a blower and heat exchanger. Solvents such as methanol, toluene, and acetone are normally evaporated and recovered in the gastight system.

The vacuum plate dryer is provided as part of a closed system. The vacuum dryer has a cylindrical housing and is rated for full-vacuum operation (typical pressure range of 3 to 27 kPa absolute). The exhaust vapor is evacuated by a vacuum pump and is passed through a condenser for solvent recovery. There is no purge gas system required for operation under vacuum. Of special note in the vacuum-drying system are the vacuum feed and discharge locks, which allow for continuous operation of the plate dryer under full vacuum.

Comparison Data—Plate Dryers Comparative studies have been done on products under both atmospheric and vacuum drying conditions. See Fig. 12-52. These curves demonstrate (1) the improvement in drying achieved with elevated temperature and (2) the impact to the drying process obtained with vacuum operation. Note that curve 4 at 90°C, pressure at 6.7 kPa absolute, is comparable to the atmospheric curve at 150°C. Also the comparative atmospheric curve at 90°C requires 90 percent more drying time than the vacuum condition. The dramatic improvement with the use of vacuum is important to note for heat-sensitive materials.

The above drying curves have been generated via testing on a plate dryer simulator. The test unit duplicates the physical setup of the production dryer; therefore, linear scale-up from the test data can be made to the full-scale dryer. Because of the thin product layer on each plate, drying in the unit closely follows the normal type of drying curve in which the constant-rate period (steady evolution of moisture or volatiles) is followed by the falling-rate period of the drying process. This results in higher heat-transfer coefficients and specific drying capacities on the upper plates of the dryer compared to the lower plates. The average specific drying capacity for the plate dryer is in the range of 2 to 20 kg/($m^2 \cdot h$) (based on final dry product). Performance data for typical applications are shown in Table 12-17.

Gravity or Moving-Bed Dryers A body of solids in which the particles, consisting of granules, pellets, beads, or briquettes, flow downward by gravity at substantially their normal settled bulk density through a vessel in contact with gases is defined frequently as a *moving-bed* or *tower dryer.* Moving-bed equipment is frequently used for grain drying and plastic pellet drying, and it also finds application in blast furnaces, shaft furnaces, and petroleum refining. Gravity beds are also employed for the cooling and drying of extruded pellets and briquettes from size enlargement processes.

A gravity dryer consists of a stationary vertical, usually cylindrical housing with openings for the introduction of solids (at the top) and removal of solids (at the bottom), as shown schematically in Fig. 12-53. Gas flow is through the solids bed and may be cocurrent or countercurrent and, in some instances, cross-flow. By definition, the rate of gas flow upward must be less than that required for fluidization.

Fields of Application One of the major advantages of the gravity-bed technique is that it lends itself well to true intimate countercurrent contacting of solids and gases. This provides for efficient heat transfer and mass transfer. Gravity-bed contacting also permits the use of the solid as a heat-transfer medium, as in pebble heaters.

Gravity vessels are applicable to coarse granular free-flowing solids which are comparatively dust-free. The solids must possess physical properties in size and surface characteristics so that they will not stick together, bridge, or segregate during passage through the vessel. The presence of significant quantities of fines or dust will close the passages among the larger particles through which the gas must penetrate, increasing pressure drop. Fines may also segregate near the sides of the bed or in other areas where gas velocities are low, ultimately completely sealing off these portions of the vessel. The high efficiency of gas-solids contacting in gravity beds is due to the uniform distribution of gas throughout the solids bed; hence choice of feed and its preparation are important factors to successful operation. Preforming techniques such as pelleting and briquetting are employed frequently for the preparation of suitable feed materials.

Gravity vessels are suitable for low-, medium-, and high-temperature operation; in the last case, the housing will be lined completely with refractory brick. Dust recovery equipment is minimized in this type of operation

TABLE 12-17 Plate Dryer Performance for Three Applications*

Product	Plastic additive	Pigment	Foodstuff
Volatiles	Methanol	Water	Water
Production rate, dry	362 kg/hr	133 kg/hr	2030 kg/hr
Inlet volatiles content	30%	25%	4%
Final volatiles content	0.1%	0.5%	0.7%
Evaporative rate	155 kg/hr	44 kg/hr	70 kg/hr
Heating medium	Hot water	Steam	Hot water
Drying temperature	70°C	150°C	90°C
Dryer pressure	11 kPa abs	Atmospheric	Atmospheric
Air velocity	NA	0.1 m/sec	0.2 m/sec
Drying time, min	24	23	48
Heat consumption, kcal/kg dry product	350	480	100
Power, dryer drive	3 kW	1.5 kW	7.5 kW
Material of construction	SS 316L/316Ti	SS 316L/316Ti	SS 316L/316Ti
Dryer height	5 m	2.6 m	8.2 m
Dryer footprint	2.6 m diameter	2.2 m by 3.0 m	3.5 m by 4.5 m
Location	Outdoors	Indoors	Indoors
Dryer assembly	Fully assembled	Fully assembled	Fully assembled
Power, exhaust fan	NA	2.5 kW	15 kW
Power, vacuum pump	20 kW	NA	NA

*Krauss Maffei

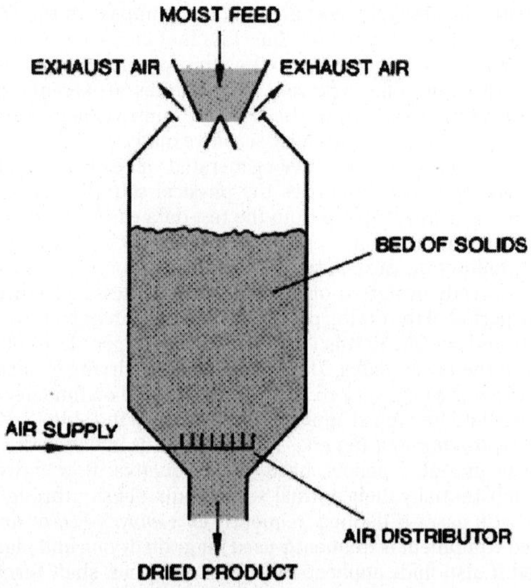

FIG. 12-53 Moving-bed gravity dryer.

since the bed actually performs as a dust collector itself, and dust in the bed will not, in a successful application, exist in large quantities.

Other advantages of gravity beds include flexibility in gas and solids flow rates and capacities, variable retention times from minutes to several hours, space economy, ease of start-up and shutdown, the potentially large number of contacting stages, and ease of control by using the inlet and exit gas temperatures.

Maintenance of a uniform rate of solids movement downward over the entire cross-section of the bed is one of the most critical operating problems encountered. For this reason, gravity beds are designed to be as high and narrow as practical. In a vessel of large cross-section, discharge through a conical bottom and center outlet will usually result in some degree of "ratholing" through the center of the bed. Flow through the center will be rapid while essentially stagnant pockets are left around the sides. To overcome this problem, multiple outlets may be provided in the center and around the periphery; table unloaders, rotating plows, wide moving grates, and multiple-screw unloaders are employed; insertion of inverted cone baffles in the lower section of the bed, spaced so that flushing at the center is retarded, is also a successful method for improving uniformity of solids movement.

Continuous Band and Tunnel Dryers

Examples and Synonyms Ceramic tunnel kilns, moving truck dryers, trolleys, atmospheric belt/band dryers, vibrating bed dryers, vacuum belt dryers, vibrating tray dryers, conveyor dryers, continuous-tunnel, belt-conveyor, or screen-conveyor (band) dryers.

Description Continuous-tunnel dryers are batch truck or tray compartments, operated in series. The solids to be processed are placed in trays or on trucks which move progressively through the tunnel in contact with hot gases. In high-temperature operations, radiation from walls and refractory lining may be significant also. Operation is semicontinuous; when the tunnel is filled, one truck is removed from the discharge end as each new truck is fed into the inlet end.

Applications of tunnel equipment are essentially the same as those for batch tray and compartment units previously described, namely, practically all forms of particulate solids and large solid objects. Auxiliary equipment and the special design considerations discussed for batch trays and compartments apply also to tunnel equipment. For size-estimating purposes, tray and truck tunnels and furnaces can be treated in the same manner as discussed for batch equipment.

Belt-conveyor and *screen-conveyor (band) dryers* are truly continuous in operation, carrying a layer of solids on an endless conveyor.

Conveyor dryers are more suitable than (multiple) batch compartments for large-quantity production, usually giving investment and installation savings. Belt and screen conveyors which are truly continuous represent major labor savings over batch operations, but require additional investment for automatic feeding and unloading devices.

Airflow can be totally cocurrent, countercurrent, or a combination of both. In addition, cross-flow designs are employed frequently, with the heating air flowing back and forth across the belts in series. Reheat coils may be installed after each cross-flow pass to maintain constant-temperature operation; large propeller-type circulating fans are installed at each stage, and air may be introduced or exhausted at any desirable points. Conveyor dryers possess the maximum flexibility for any combination of airflow and temperature staging. Contact drying is also possible, usually under vacuum, with the bands resting on heating plates (vacuum band dryer).

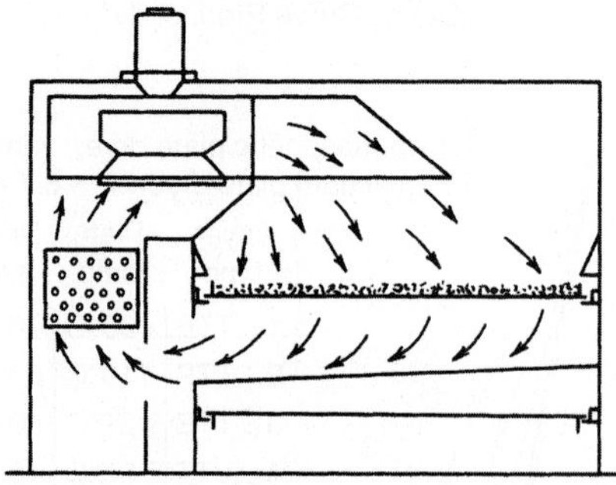

FIG. 12-54 Section view of a continuous through-circulation conveyor dryer. (*Proctor & Schwartz, Inc.*)

Ceramic tunnel kilns handling large irregular-shaped objects must be equipped for precise control of temperature and humidity conditions to prevent cracking and condensation on the product. The internal mechanisms that cause cracking when drying clay and ceramics have been studied extensively. Information on ceramic tunnel kiln operation and design is reported fully in publications such as *The American Ceramic Society Bulletin, Ceramic Industry,* and *Transactions of the British Ceramic Society.*

Continuous Through-Circulation Band Dryers Continuous through-circulation dryers operate on the principle of blowing hot air through a permeable bed of wet material passing continuously through the dryer. Drying rates are high because of the large area of contact and short distance of travel for the internal moisture.

The most widely used type is the horizontal conveyor dryer (also called *perforated band* or *conveying-screen* dryer), in which wet material is conveyed as a layer 2 to 15 cm deep (sometimes up to 1 m), on a horizontal mesh screen, belt, or perforated apron, while heated air is blown either upward or downward through the bed of material. This dryer consists usually of a number of individual sections, complete with fan and heating coils, arranged in series to form a housing or tunnel through which the conveying screen travels. As shown in the sectional view in Fig. 12-54, the air circulates through the wet material and is reheated before reentering the bed. It is not uncommon to circulate the hot gas upward in the wet end and downward in the dry end, as shown in Fig. 12-55. A portion of the air is exhausted continuously by one or more exhaust fans, not shown in the sketch, which handle air from several sections. Since each section can be operated independently, extremely flexible operation is possible, with high temperatures usually at the wet end, followed by lower temperatures; in some cases, a unit with cooled or specially humidified air is employed for final conditioning. The maximum pressure drop that can be taken through the bed of solids without developing leaks or air bypassing is roughly 50 mm of water.

Example 12-15 Mass and Energy Balance on a Dryer with Partially Recycled Air A continuous through-air dryer is producing 648 kg/h of a coarse granular product, as illustrated in Fig. 12-56. The material enters the dryer at 40 percent moisture and exits at 10 percent moisture, on wet basis. The airflow rate is 4750 kg/h, and the ambient temperature and absolute humidity are 22°C and 0.01 kg/kg, respectively. Measurements are being taken on the system to assess performance. To check the measurements against expected values, calculate the relative humidity and dew point of the exhaust air if 85 percent of the mass flow of air exiting the dryer is recycled. Neglect heat losses and sensible heating of the solids. Use the following physical properties:

$$\Delta H_{vap} = 2257 \text{ kJ/kg};\ C_{p,air} = 1.0 \text{ J/(g·K)};\ C_{p,\text{water vapor}} = 1.9 \text{ J/(g·K)}$$

We start by calculating the dry solids flow rate into the process, which equals 648(1 − 0.4) = 389 kg/h. The dry-basis moisture content of the feed and product are then calculated to be $X_{product} = 0.111$ g water/g dry material and $X_{feed} = 0.667$ g water/g dry material. The relationship from the Solids Drying Fundamentals subsection was used: $X = w/(1 - w)$, where X is the dry-basis moisture content and w is the wet-basis moisture content.

The drying rate of the system equals the dry mass flow rate times the difference in the dry-basis moisture contents:

$$\text{Drying rate} = \dot{m}_{dry}(X_{feed} - X_{product}) = 216 \text{ kg/h}$$

Now a series of steady-state mass balance equations can be written. Dry airflow rates (kg/h) are denoted by a, and water vapor flow rates (kg/h) are denoted by w. Subscripts refer to streams.

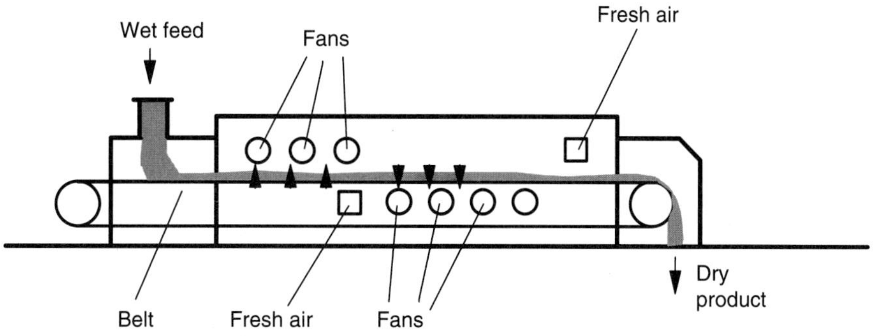

FIG. 12-55 Longitudinal view of a continuous through-circulation conveyor dryer with intermediate airflow reversal.

Air mass balances:

$$a_4 = a_5 = a_6 \quad a_4 = a_1 + a_2 \quad a_5 = a_2 + a_3 \quad \frac{a_2}{a_5} = \text{recycle ratio}$$

Water vapor mass balances:

$$w_5 = w_4 + \text{drying rate} \quad w_4 = w_1 + w_2 \quad w_5 = w_2 + w_3 \quad w_4 = w_6 \quad \frac{w_2}{w_5} = \text{recycle ratio}$$

Energy balances:

$$a_4 Cp_4 (T_4 - T_5) = \dot{m} \cdot \Delta H_{\text{vap}} \quad a_4 Cp_4 (T_4 - T_6) = a_6 Cp_6 T_6 = a_1 Cp_1 T_1 + a_2 Cp_2 T_2$$
$$T_2 = T_3 = T_5$$

The equations above can be solved on a spreadsheet iteratively. The results of the calculation are shown in Table 12-18.

This type of calculation is very helpful to understanding performance of an industrial dryer. Once it is set up, different scenarios can be explored, such as changing the recycle rate. However, it is important to note that changing the conditions in the system can also affect the drying rate. For example, an increase of the recycle rate would increase the air velocity but also increase the humidity in the dryer. Using a drying kinetics model, such as the "characteristic curve" method (if that model is appropriate for the material), along with this model can help to optimize the system.

Through-circulation drying requires that the wet material be in a state of granular or pelleted subdivision so that hot air may be readily blown through it. Many materials meet this requirement without special preparation. Others require special and often elaborate pretreatment to render them suitable for through-circulation drying. The process of converting a wet solid to a form suitable for through-circulation of air is called *preforming*, and often the success or failure of this contacting method depends on the preforming step. Fibrous, flaky, and coarse granular materials are usually amenable to drying without preforming. They can be loaded directly onto the conveying screen by suitable spreading feeders of the oscillating-belt or vibrating type or by spiked drums or belts feeding from bins. When materials must be preformed, several methods are available, depending on the physical state of the wet solid.

1. Relatively dry materials such as centrifuge cakes can sometimes be granulated to give a suitably porous bed on the conveying screen.
2. Pasty materials can often be preformed by extrusion to form spaghetti-like pieces, about 6 mm in diameter and several centimeters long.
3. Wet pastes that cannot be granulated or extruded may be predried and preformed on a steam-heated finned drum. Preforming on a finned drum may be desirable also in that some predrying is accomplished.
4. Thixotropic filter cakes from rotary vacuum filters that cannot be preformed by any of the above methods can often be scored by knives on the filter, the scored cake discharging in pieces suitable for through-circulation drying.
5. Material that shrinks markedly during drying is often reloaded during the drying cycle to 2 to 6 times the original loading depth. This is usually done after a degree of shrinkage which, by opening the bed, has destroyed the effectiveness of contact between the air and solids.
6. In a few cases, powders have been pelleted or formed in briquettes to eliminate dustiness and permit drying by through-circulation. Table 12-19 gives a list of materials classified by preforming methods suitable for through-circulation drying.

Steam-heated air is the usual heat-transfer medium employed in these dryers, although combustion gases may also be used. Temperatures above 600 K are not usually feasible because of the problems of lubricating the conveyor, chain, and roller drives. Recirculation of air is in the range of 60 to 90 percent of the flow through the bed. Conveyors may be made of wire-mesh screen or perforated-steel plate. The minimum practical screen opening size is about 30-mesh (0.5 mm). Multiple bands in series may be used.

Vacuum band dryers utilize heating by conduction and are a continuous equivalent of vacuum tray (shelf) dryers, with the moving bands resting on heating plates. Drying is usually relatively slow, and it is common to find several bands stacked above one another, with material falling to the next band and flowing in opposite directions on each pass, to reduce dryer length and give some product turnover.

Design Methods for Continuous Band Dryers A scoping calculation is a good starting point for designing a new system or for understanding an existing system. The required solids throughput F and the inlet and outlet moisture content X_I and X_O are known, as is the ambient humidity Y_I. If the inlet gas temperature T_{GI} is chosen, the outlet gas conditions (temperature T_{GO} and humidity Y_O) can be found, either by calculation or (more simply and quickly) by using the constant-enthalpy lines on a psychrometric chart.

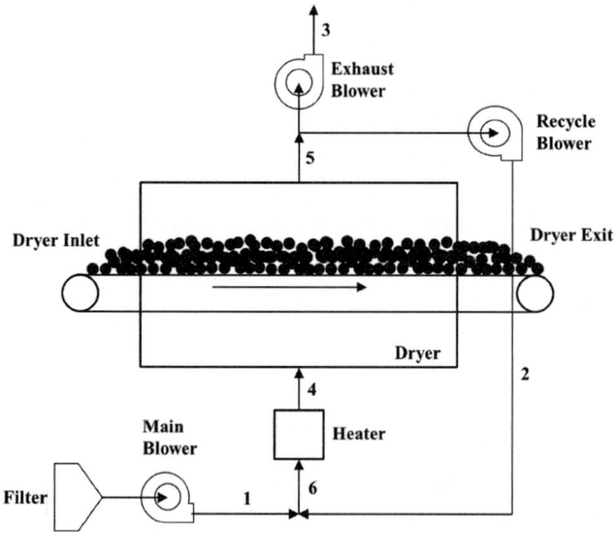

FIG. 12-56 Drying of a course granular material with partially recycled air.

TABLE 12-18 Results From Mass- and Energy-Balance Calculation for Dryer with Recycle

Stream	Absolute humidity, kg/kg	Temperature, °C	% Relative humidity	Dew point, °C
1	0.010	22.0	57.5	13.3
2	0.051	70.5	26.2	42.4
3	0.051	70.5	26.2	42.4
4	0.045	84.7	12.7	39.8
5	0.051	70.5	26.2	42.4
6	0.045	63.8	30.7	39.8

TABLE 12-19 Methods of Preforming Some Materials for Through-Circulation Drying

No preforming required	Scored on filter	Granulation	Extrusion	Finned drum	Flaking on chilled drum	Briquetting and squeezing
Cellulose acetate	Starch	Kaolin	Calcium carbonate	Lithopone	Soap flakes	Soda ash
Silica gel	Aluminum hydrate	Cryolite	White lead	Zinc yellow		Cornstarch
Scoured wool		Lead arsenate	Lithopone	Calcium carbonate		Synthetic rubber
Sawdust		Cornstarch	Titanium dioxide	Magnesium carbonate		
Rayon waste		Cellulose acetate	Magnesium carbonate			
Fluorspar		Dye intermediates	Aluminum stearate			
Tapioca			Zinc stearate			
Breakfast food						
Asbestos fiber						
Cotton linters						
Rayon staple						

However, it may be necessary to allow for heat losses and sensible heating of solids, which typically reduce the useful enthalpy of the inlet gas by 10 to 20 percent. Also, if tightly bound moisture is being removed, the heat of wetting to break the bonds should be allowed for. The gas mass flow rate G can now be calculated, as it is the only unknown in the mass balance on the solvent [Eq. (12-69)]. For through-circulation and dispersion dryers, the cross-sectional area A is given by

$$A = \frac{G}{\rho_{GI} U_G} = \frac{F(X_I - X_O)}{\rho_{GI} U_G (Y_O - X_I)} \tag{12-69}$$

The linear dimensions of a rectangular bed can then be calculated. The result is usually accurate to within 10 percent, and it can be further improved by better estimates of velocity and heat losses.

The method gives no information about solids residence time or dryer length. A minimum drying time t_{min} can be calculated by evaluating the maximum (unhindered) drying rate N_{cr}, assuming gas-phase heat-transfer control and estimating a gas-to-solids heat-transfer coefficient. The simple Eq. (12-70) then applies:

$$t_{min} = \frac{X_I - X_O}{N_{cr}} \tag{12-70}$$

Alternatively, it may be assumed that first-order falling-rate kinetics apply throughout the drying process, and one can scale the estimated drying time by using Eq. (12-70).

To correct from a calculated constant-rate (unhindered) drying time t_{CR} to first-order falling-rate kinetics, the following equation is used, where X_1 is the initial, X_2 the final, and X_E the equilibrium moisture content (all must be dry-basis):

$$\frac{t_{FR}}{t_{CR}} = \frac{X_1 - X_E}{X_1 - X_2} \ln\left(\frac{X_1 - X_E}{X_2 - X_E}\right) \tag{12-71}$$

However, these crude methods can give serious underestimates of the required drying time, and it is much better to measure the drying time experimentally and apply scaling methods. Example 12-16 applies this method to a batch rotary dryer.

A more detailed mathematical method of a through-circulation dryer has been developed by Thygeson [*Am. Inst. Chem. Eng. J.* **16**(5): 749 (1970)]. Rigorous modeling is possible with a two-dimensional incremental model, with steps both horizontally along the belt and vertically through the layer; nonuniformity of the layer across the belt could also be allowed for, if desired. Heat-transfer coefficients are typically in the range of 100 to 200 W/(m^2·K), and the relationship $h_c = 12(\rho_g U_g/d_p)^{0.5}$ may be used for a first estimate, where ρ_g is gas density (kg/m^3); U_g, local gas velocity (m/s); and d_p, particle diameter (m). For 5-mm particles and air at 1 m/s, 80°C, and 1 kg/m^3 [mass flux 1 kg/(m^2·s)] this gives h_c = 170 W/(m^2·K).

In actual practice, design of a continuous through-circulation dryer is best based upon data taken in pilot-plant tests. Loading and distribution of solids on the screen are rarely as nearly uniform in commercial installations as in test dryers; 50 to 100 percent may be added to the test drying time for commercial design.

Performance and Cost Data for Continuous Band and Tunnel Dryers Experimental performance data are given in Table 12-20 for numerous common materials. Performance data from several commercial through-circulation conveyor dryers are given in Table 12-21. Labor requirements vary depending on the time required for feed adjustments, inspection, etc. These dryers may consume as little as 1.1 kg of steam/kg of water evaporated, but 1.4 to 2 is a more common range. Thermal efficiency is a function of final moisture required and percent air recirculation.

Conveying-screen dryers are fabricated with conveyor widths from 0.3- to 4.4-m sections 1.6 to 2.5 m long. Each section consists of a sheet-metal enclosure, insulated sidewalls and roof, heating coils, a circulating fan, inlet air distributor baffles, a fines catch pan under the conveyor, and a conveyor screen (Fig. 12-54). Cabinet and auxiliary equipment fabrication is of aluminized steel or stainless-steel materials. Prices do not include temperature controllers, motor starters, preform equipment, or auxiliary feed and discharge conveyors.

Table 12-22 gives approximate purchase costs for equipment with type 304 stainless-steel hinged conveyor screens and includes steam-coil heaters, fans, motors, and a variable-speed conveyor drive. Cabinet and auxiliary equipment fabrication is of aluminized steel or stainless-steel materials. Prices do not include temperature controllers, motor starters, preform equipment, or auxiliary feed and discharge conveyors. These may add \$75,000 to \$160,000 to the dryer purchase cost (2005 costs).

Batch Agitated and Rotating Dryers

Examples and Synonyms Pan dryers, spherical and conical dryers, side-screw, Nauta, turbosphere, batch paddle, ploughshare.

Description An *agitated dryer* is defined as one on which the housing enclosing the process is stationary while solids movement is accomplished by an internal mechanical agitator. A rotary dryer is one in which the outer housing rotates. Many forms are in use, including batch and continuous versions. The batch forms are almost invariably heated by conduction with operation under vacuum. Vacuum is used in conjunction with drying or other chemical operations when low solids temperatures must be maintained because heat will cause damage to the product or change its nature; when air combines with the product as it is heated, causing oxidation or an explosive condition; when solvent recovery is required; and when materials must be dried to extremely low moisture levels. Vertical agitated pan, spherical, and conical dryers are mechanically agitated; tumbler or double-cone dryers have a rotating shell. All these types are typically used for the drying of solvent or water-wet, free-flowing powders in small batch sizes of 1000 L or less, as frequently found in the pharmaceutical, specialty chemical, and fine chemicals industries. Corrosion resistance and cleanability are often important, and common materials of construction include SS 304 and 316 as well as Hastelloy. The batch nature of operation is of value in the pharmaceutical industry to maintain batch identification. In addition to pharmaceutical materials, the conical mixer dryer is used to dry polymers, additives, inorganic salts, and many other specialty chemicals. As the size increases, the ratio of jacket heat-transfer surface area to volume falls, extending drying times. For larger batches, horizontal agitated pan dryers are more common, but there is substantial overlap of operating ranges. Drying times may be reduced for all types by heating the internal agitator, but this increases complexity and cost.

Mechanical versus rotary agitation Agitated dryers are applicable to processing solids that are relatively free-flowing and granular when discharged as product. Materials that are not free-flowing in their feed condition can be treated by recycle methods as described in the subsection Continuous Rotary Dryers. In general, agitated dryers have applications similar to those of rotating vessels. Their chief advantages compared with the latter are twofold: (1) Large-diameter rotary seals are not required at the solids and gas feed and exit points because the housing is stationary, and for this reason gas leakage problems are minimized. Rotary seals are required only at the points of entrance of the mechanical agitator shaft. (2) Use of a mechanical agitator for solids mixing introduces shear forces which are helpful for breaking up lumps and agglomerates. Balling and pelleting of sticky solids, an occasional occurrence in rotating vessels, can be prevented by special agitator design. The problems concerning dusting of fine particles in direct-heat units are identical to those discussed in subsection Continuous Rotary Dryers.

Heated Agitators For all agitated dryers, in addition to the jacket heated area, heating the agitator with the same medium as the jacket (hot water,

TABLE 12-20 Experimental Through-Circulation Drying Data for Miscellaneous Materials

Material	Physical form	Moisture contents, kg/kg dry solid			Inlet-air tempera-ture, K	Depth of bed, cm	Loading, kg product/m^2	Air velocity, m/s × 10^1	Experimental drying time, s × 10^{-2}
		Initial	Critical	Final					
Alumina hydrate	Briquettes	0.105	0.06	0.00	453	6.4	60.0	6.0	18.0
Alumina hydrate	Scored filter cake	9.60	4.50	1.15	333	3.8	1.6	11.0	90.0
Alumina hydrate	Scored filter cake	5.56	2.25	0.42	333	7.0	4.6	11.0	108.0
Aluminum stearate	0.7-cm extrusions	4.20	2.60	0.003	350	7.6	6.5	13.0	36.0
Asbestos fiber	Flakes from squeeze rolls	0.47	0.11	0.008	410	7.6	13.6	9.0	5.6
Asbestos fiber	Flakes from squeeze rolls	0.46	0.10	0.0	410	5.1	6.3	9.0	3.6
Asbestos fiber	Flakes from squeeze rolls	0.46	0.075	0.0	410	3.8	4.5	11.0	2.7
Calcium carbonate	Preformed on finned drum	0.85	0.30	0.003	410	3.8	16.0	11.5	12.0
Calcium carbonate	Preformed on finned drum	0.84	0.35	0.0	410	8.9	25.7	11.7	18.0
Calcium carbonate	Extruded	1.69	0.98	0.255	410	1.3	4.9	14.3	9.0
Calcium carbonate	Extruded	1.41	0.45	0.05	410	1.9	5.8	10.2	12.0
Calcium stearate	Extruded	2.74	0.90	0.0026	350	7.6	8.8	5.6	57.0
Calcium stearate	Extruded	2.76	0.90	0.007	350	5.1	5.9	6.0	42.0
Calcium stearate	Extruded	2.52	1.00	0.0	350	3.8	4.4	10.2	24.0
Cellulose acetate	Granulated	1.14	0.40	0.09	400	1.3	1.4	12.7	1.8
Cellulose acetate	Granulated	1.09	0.35	0.0027	400	1.9	2.7	8.6	7.2
Cellulose acetate	Granulated	1.09	0.30	0.0041	400	2.5	4.1	5.6	10.8
Cellulose acetate	Granulated	1.10	0.45	0.004	400	3.8	6.1	5.1	18.0
Clay	Granulated	0.277	0.175	0.0	375	7.0	46.2	10.2	19.2
Clay	1.5-cm extrusions	0.28	0.18	0.0	375	12.7	100.0	10.7	43.8
Cryolite	Granulated	0.456	0.25	0.0026	380	5.1	34.2	9.1	24.0
Fluorspar	Pellets	0.13	0.066	0.0	425	5.1	51.4	11.6	7.8
Lead arsenate	Granulated	1.23	0.45	0.043	405	5.1	18.1	11.6	18.0
Lead arsenate	Granulated	1.25	0.55	0.054	405	6.4	22.0	10.2	24.0
Lead arsenate	Extruded	1.34	0.64	0.024	405	5.1	18.1	9.4	36.0
Lead arsenate	Extruded	1.31	0.60	0.0006	405	8.4	26.9	9.2	42.0
Kaolin	Formed on finned drum	0.28	0.17	0.0009	375	7.6	44.0	9.2	21.0
Kaolin	Formed on finned drum	0.297	0.20	0.005	375	11.4	56.3	12.2	15.0
Kaolin	Extruded	0.443	0.20	0.008	375	7.0	45.0	10.16	18.0
Kaolin	Extruded	0.36	0.14	0.0033	400	9.6	40.6	15.2	12.0
Kaolin	Extruded	0.36	0.21	0.0037	400	19.0	80.7	10.6	30.0
Lithopone (finished)	Extruded	0.35	0.065	0.0004	408	8.2	63.6	10.2	18.0
Lithopone (crude)	Extruded	0.67	0.26	0.0007	400	7.6	41.1	9.1	51.0
Lithopone	Extruded	0.72	0.28	0.0013	400	5.7	28.9	11.7	18.0
Magnesium carbonate	Extruded	2.57	0.87	0.001	415	7.6	11.0	11.4	17.4
Magnesium carbonate	Formed on finned drum	2.23	1.44	0.0019	418	7.6	13.2	8.6	24.0
Mercuric oxide	Extruded	0.163	0.07	0.004	365	3.8	66.5	11.2	24.0
Silica gel	Granular	4.51	1.85	0.15	400	3.8–0.6	3.2	8.6	15.0
Silica gel	Granular	4.49	1.50	0.215	340	3.8–0.6	3.4	9.1	63.0
Silica gel	Granular	4.50	1.60	0.218	325	3.8–0.6	3.5	9.1	66.0
Soda salt	Extruded	0.36	0.24	0.008	410	3.8	22.8	5.1	51.0
Starch (potato)	Scored filter cake	0.866	0.55	0.069	400	7.0	26.3	10.2	27.0
Starch (potato)	Scored filter cake	0.857	0.42	0.082	400	5.1	17.7	9.4	15.0
Starch (corn)	Scored filter cake	0.776	0.48	0.084	345	7.0	26.4	7.4	54.0
Starch (corn)	Scored filter cake	0.78	0.56	0.098	380	7.0	27.4	7.6	24.0
Starch (corn)	Scored filter cake	0.76	0.30	0.10	345	1.9	7.7	6.7	15.0
Titanium dioxide	Extruded	1.2	0.60	0.10	425	3.0	6.8	13.7	6.3
Titanium dioxide	Extruded	1.07	0.65	0.29	425	8.2	16.0	8.6	6.0
White lead	Formed on finned drum	0.238	0.07	0.001	355	6.4	76.8	11.2	30.0
White lead	Extruded	0.49	0.17	0.0	365	3.8	33.8	10.2	27.0
Zinc stearate	Extruded	4.63	1.50	0.005	360	4.4	4.2	8.6	36.0

steam, or thermal oil) will increase the heat-exchange area. This is usually accomplished via rotary joints. Obviously, heating the screw or agitator will mean shorter batch drying times, which yields higher productivity and better product quality owing to shorter exposure to the drying temperature, but capital and maintenance costs will be increased. In pan and conical dryers, the area is increased only modestly, by 15 to 30 percent; but in horizontal pan and paddle dryers, the opportunity is much greater and indeed the majority of the heat may be supplied through the agitator.

Also the mechanical power input of the agitator can be a significant additional heat source, and microwave assistance has also been used in filter dryers and conical dryers to shorten drying times (and is feasible in other types).

Vacuum processing All these types of dryer usually operate under vacuum, especially when drying heat-sensitive materials or when removing flammable organic solvents rather than water. The heating medium is hot water, steam, or thermal oil, with most applications in the temperature range of 50 to 150°C and pressures in the range of 3 to 30 kPa absolute. The vapors generated during the drying process are evacuated by a vacuum pump and passed through a condenser for recovery of the solvent. A dust filter is normally mounted over the vapor discharge line as it leaves the dryer, thus allowing any entrapped dust to be pulsed back into the process area. Standard cloth-type dust filters are available, along with sintered metal filters.

In vacuum processing and drying, a major objective is to create a large temperature-driving force between the jacket and the product. To accomplish this purpose at fairly low jacket temperatures, it is necessary to reduce the internal process pressure so that the liquid being removed will boil at a lower vapor pressure. It is not always economical, however, to reduce the internal pressure to extremely low levels because of the large vapor volumes thereby created. It is necessary to compromise on operating pressure, considering leakage, condensation problems, and the size of the vapor lines and pumping system. Very few vacuum dryers operate below 5 mmHg pressure on a commercial scale. Air in-leakage through gasket surfaces will be in the range of 0.2 kg/(h · lin m of gasketed surface) under these conditions. To keep vapor partial pressure and solids temperature low without pulling excessively high vacuum, a nitrogen bleed may be introduced, particularly in the later stages of drying. The vapor and solids surface temperatures then fall below the vapor boiling point, toward the wet-bulb temperature.

Vertical Agitated Dryers This classification includes vertical pan dryers, filter dryers, and spherical and conical dryers.

Vertical pan dryer The basic vertical pan dryer consists of a short, squat vertical cylinder (Fig. 12-57 and Table 12-23) with an outer heating jacket and an internal rotating agitator, again with the axis vertical, which mixes the solid and sweeps the base of the pan. Heat is supplied by circulation of hot water, steam, or thermal fluid through the jacket; it may also be used for cooling at the end of the batch cycle, using cooling water or refrigerant. The agitator is usually a plain set of solid blades, but may be a ribbon-type screw or internally heated blades. Product is discharged from a door at the lower

TABLE 12-21 Performance Data for Continuous Through-Circulation Dryer*

	Kind of material					
	Inorganic pigment	Cornstarch	Fiber staple	Charcoal briquettes	Gelatin	Inorganic chemical
Capacity, kg dry product/h	712	4536	1724 Stage A, Stage B,	5443	295	862
Approximate dryer area, m^2	22.11	66.42	57.04 35.12	52.02	104.05	30.19
Depth of loading, cm	3	4		16	5	4
Air temperature, °C	120	115 to 140	130 to 100 100	135 to 120	32 to 52	121 to 82
Loading, kg product/m^2	18.8	27.3	3.5 3.3	182.0	9.1	33
Type of conveyor, mm	1.59 by 6.35 slots	1.19 by 4.76 slots	2.57-diameter holes, perforated plate	8.5 × 8.5 mesh screen	4.23 × 4.23 mesh screen	1.59 × 6.35 slot
Preforming method or feed	Rolling extruder	Filtered and scored	Fiber feed	Pressed	Extrusion	Rolling extruder
Type and size of preformed particle, mm	6.35-diameter extrusions	Scored filter cake	Cut fiber	64 × 51 × 25	2-diameter extrusions	6.35-diameter extrusions
Initial moisture content, % bone-dry basis	120	85.2	110	37.3	300	111.2
Final moisture content, % bone-dry basis	0.5	13.6	9	5.3	11.1	1.0
Drying time, min	35	24	11	105	192	70
Drying rate, kg water evaporated/(h · m^2)	38.39	42.97	17.09	22.95	9.91	31.25
Air velocity (superficial), m/s	1.27	1.12	0.66	1.12	1.27	1.27
Heat source per kg water evaporated, steam kg/kg gas (m^3/kg)	Gas 0.11	Steam 2.0	Steam 1.73	Waste heat	Steam 2.83	Gas 0.13
Installed power, kW	29.8	119.3	194.0	82.06	179.0	41.03

*Courtesy of Wolverine Proctor & Schwartz, Inc.

side of the wall. Sticky materials may adhere to the agitator or be difficult to discharge.

Filter dryer The basic Nutsche filter dryer is like a vertical pan dryer, but with the bottom heated plate replaced by a filter plate (see also Sec. 18, Liquid-Solid Operations and Equipment). Hence, a slurry can be fed in and filtered, and the wet cake dried in situ. These units are especially popular in the pharmaceutical industry, as containment is good and a difficult wet-solids transfer operation is eliminated by carrying out both filtration and drying in the same vessel. Drying times tend to be longer than for vertical pan dryers, as the bottom plate is no longer heated. Some types (e.g., Mitchell Thermovac, Krauss-Maffei TNT) invert the unit between the filtration and drying stages to avoid this problem.

These are popular in the pharmaceutical and specialty chemicals industries as two operations are performed in the same piece of equipment without intermediate solids transfer, and containment is good.

Spherical dryer Sometimes called the *turbosphere,* this is another agitated dryer with a vertical axis mixing shaft, but rotation is typically faster than in the vertical pan unit, giving improved mixing and heat transfer. The dryer chamber is spherical, with solids discharge through a door or valve near the bottom.

Conical mixer dryer This is a vertically oriented conical vessel with an internally mounted rotating screw. Figure 12-58 shows a schematic of a typical conical mixer dryer. The screw rotates about its own axis (speeds up to 100 rpm) and around the interior of the vessel (speeds up to 0.4 rpm). Because it rotates around the full circumference of the vessel, the screw provides a self-cleaning effect for the heated vessel walls, as well as effective agitation; it may also be internally heated. Either top-drive (via an internal rotating arm) or bottom-drive (via a universal joint) may be used; the former is more common. The screw is cantilevered in the vessel and requires no additional support (even in vessel sizes up to 20-m^3 operating volume). Cleaning of the dryer is facilitated with *clean-in-place* (CIP) systems that can be used for cleaning, and/or the vessel can be completely flooded with water or solvents. The dryer makes maximum use of the product-heated areas—the filling volume of the vessel (up to the knuckle of the dished head) is the usable product loading. In some applications, microwaves have been used to provide additional energy input and shorten drying times.

TABLE 12-22 Conveyor-Screen-Dryer Costs*

Length	2.4-m-wide conveyor	3.0-m-wide conveyor
7.5 m	\$8600/m^2	\$7110/m^2
15 m	\$6700/m^2	\$5600/m^2
22.5 m	\$6200/m^2	\$5150/m^2
30 m	\$5900/m^2	\$4950/m^2

*National Drying Machinery Company, 1996.

Horizontal Pan Dryer This dryer consists of a stationary cylindrical shell, mounted horizontally, in which a set of agitator blades mounted on a revolving central shaft stir the solids being treated. These dryers tend to be used for larger batches than vertical agitated or batch rotating dryers. Heat is supplied by circulation of hot water, steam, or other heat-transfer fluids through the jacket surrounding the shell and, in larger units, through the hollow central shaft. The agitator can take many different forms, including simple paddles, ploughshare-type blades, a single discontinuous spiral, or a double continuous spiral. The outer blades are set as closely as possible to the wall without touching, usually leaving a gap of 0.3 to 0.6 cm. Modern units occasionally employ spring-loaded shell scrapers mounted on the blades. The dryer is charged through a port at the top and emptied through one or more discharge nozzles at the bottom. Vacuum is applied and maintained by any of the conventional methods, i.e., steam jets, vacuum pumps, etc.

A similar type, the batch indirect rotary dryer, consists of a rotating horizontal cylindrical shell, suitably jacketed. Vacuum is applied to this unit through hollow trunnions with suitable packing glands. Rotary glands must be used also for admitting and removing the heating medium from the jacket. The inside of the shell may have lifting bars, welded longitudinally, to assist agitation of the solids. Continuous rotation is needed while emptying the solids, and a circular dust hood is frequently necessary to enclose the discharge-nozzle turning circle and prevent serious dust losses to the atmosphere during unloading. A typical horizontal pan vacuum dryer is illustrated in Fig. 12-59.

Tumbler or Double-Cone Dryers These are rotating batch vacuum dryers, as shown in Fig. 12-60. Some types are an offset cylinder, but a double-cone shape is more common. They are very common in the pharmaceutical and fine chemicals industries. The gentle rotation can give less attrition than in some mechanically agitated dryers; on the other hand, formation of lumps and balls is more likely. The sloping walls of the cones permit more rapid emptying of solids when the dryer is in a stationary position, compared to a horizontal cylinder, which requires continuous rotation during emptying to convey product to the discharge nozzles. Several new designs of the double-cone type employ internal tubes or plate coils to provide additional heating surface.

On all rotating dryers, the vapor outlet tube is stationary; it enters the shell through a rotating gland and is fitted with an elbow and an upward extension so that the vapor inlet, usually protected by a felt dust filter, will be near the top of the shell at all times.

Design, Scale-Up, and Performance Like all batch dryers, agitated and rotating dryers are primarily sized to physically contain the required batch volume. Note that the nominal capacity of most dryers is significantly lower than their total internal volume, because of the headspace needed for mechanical drives, inlet ports, suction lines, dust filters, etc. Care must be taken to determine whether a stated percentage fill is based on nominal capacity or geometric volume. Vacuum dryers are usually filled to 50 to 65 percent of their total shell volume.

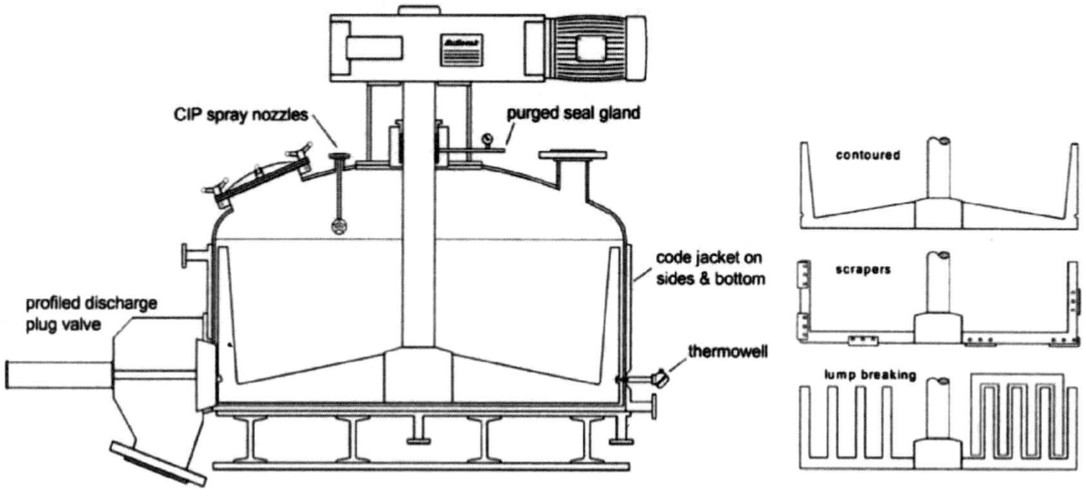

FIG. 12-57 Vertical pan dryer. (*Buflovak Inc.*)

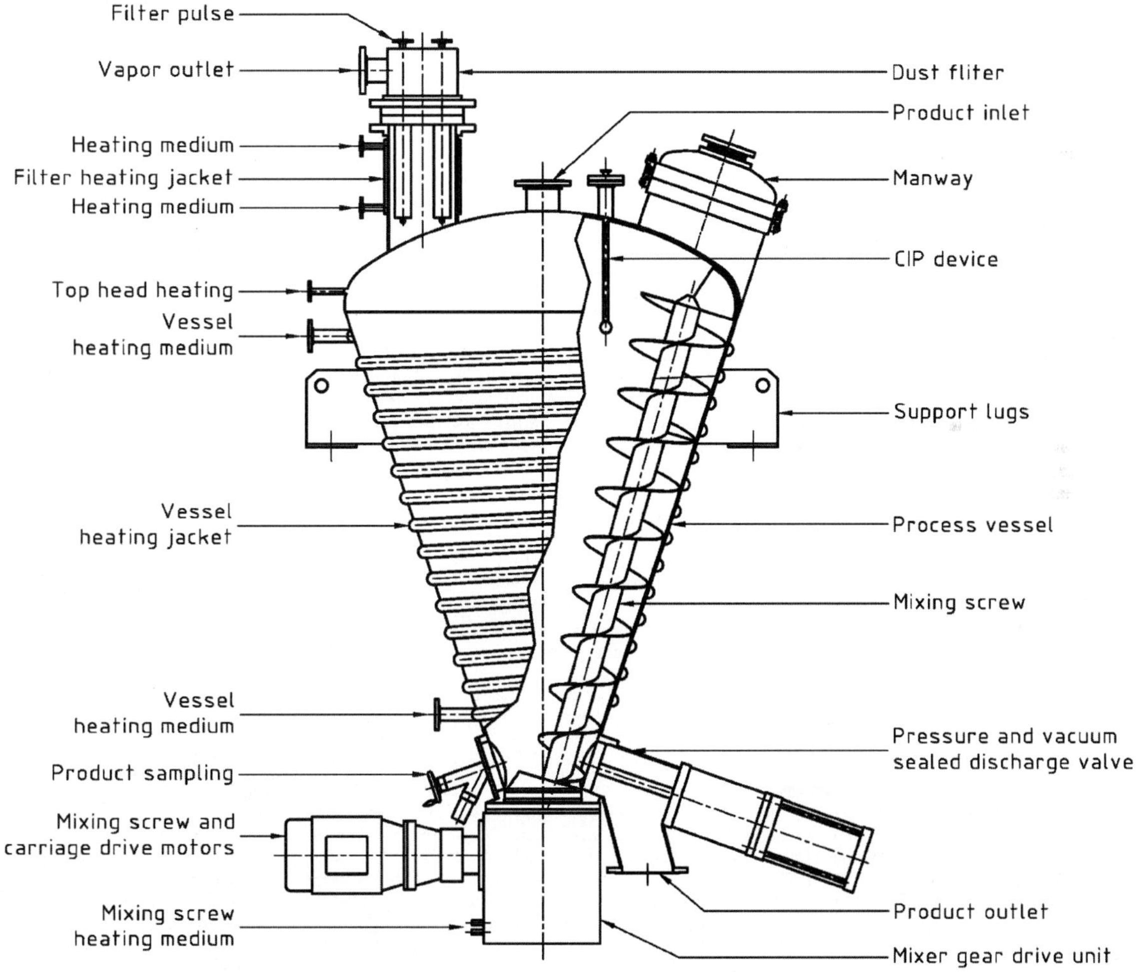

FIG. 12-58 Bottom-drive conical mixer dryer. (*Krauss Maffei.*)

TABLE 12-23 Dimensions of Vertical Pan Dryers (*Buflovak Inc.*)

					Jacketed area, ft^2				
I.D, ft	Product depth, ft	Working volume, ft^3	USG	Jacketed height, ft	Cylinder wall	Bottom	Total	Discharge door, in	
3	0.75	5.3	40	1.0	9	7	16	5	8
4	1	12.6	94	2.0	25	13	38	6	8
5	1	19.6	147	2.0	31	20	51	8	9
6	1	28.3	212	2.0	38	28	66	8	9
8	1	50.3	377	2.0	50	50	101	8	9
10	1.5	117.8	884	3.0	94	79	173	12	12

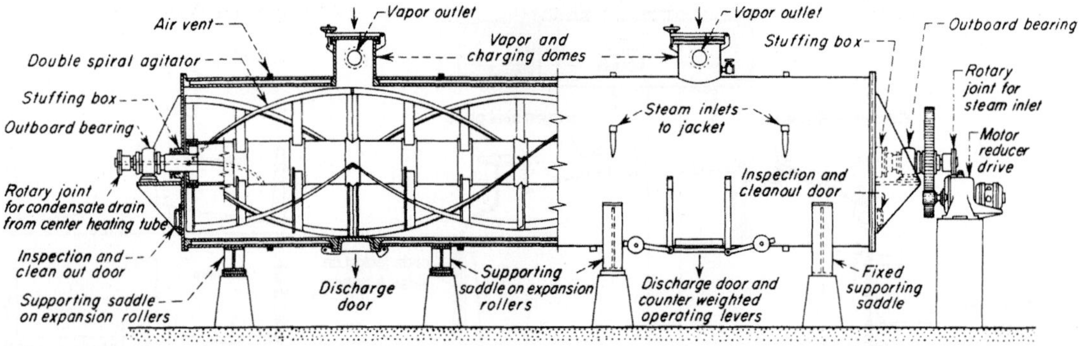

FIG. 12-59 A typical horizontal pan vacuum dryer. (*Blaw-Knox Food & Chemical Equipment, Inc.*)

The standard scoping calculation methods for batch conduction drying apply. The rate of heat transfer from the heating medium through the dryer wall to the solids can be expressed by the usual formula

$$Q = hA\,\Delta T_m \tag{12-72}$$

where Q = heat flux, J/s (Btu/h); h = overall heat-transfer coefficient, J/(m^2·s·K) [Btu/(h·ft^2 jacket area·°F)]; A = total jacket area, m^2 (ft^2); and ΔT_m = log-mean temperature driving force from heating medium to the solids, K (°F).

The overall heat-transfer rate is almost entirely dependent upon the film coefficient between the inner jacket wall and the solids, which depends on the dryer type and agitation rate and, to a large extent, on the solids characteristics. Overall heat-transfer coefficients may range from 30 to 200 J/(m^2·s·K), based upon total area if the dryer walls are kept reasonably clean. Heat-transfer coefficients as low as 5 or 10 J/(m^2·s·K) may be encountered if caking on the walls occurs.

For estimating purposes without tests, a reasonable coefficient for ordinary drying, and without taking the product to absolute dryness, may be assumed as h = 50 J/(m^2·s·K) for mechanically agitated dryers (although higher figures have been quoted for conical and spherical dryers) and 35 J/(m^2·s·K) for rotating units. The true heat-transfer coefficient is usually higher, but this conservative assumption makes some allowance for the slowing down of drying during the falling-rate period. However, if at all possible, it is always preferable to do pilot-plant tests to establish the drying time of the actual material. Drying trials are conducted in small pilot dryers (50- to 100-L batch units) to determine material handling and drying retention times. Variables such as drying temperature, vacuum level, and screw speed are analyzed during the test trials. Scale-up to larger units is done based upon the area/volume ratio of the pilot unit versus the production dryer. In most applications, the overall drying time in the production models is in the range of 2 to 24 h.

Agitator or rotation speeds range from 3 to 8 rpm. Faster speeds yield a slight improvement in heat transfer but consume more power and in some cases, particularly in rotating units, can cause more "balling up" and other stickiness-related problems.

In all these dryers, the surface area tends to be proportional to the square of the diameter D^2, and the volume to diameter cubed D^3. Hence the area/volume ratio falls as the diameter increases, and drying times increase. It can be shown that the ratio of drying times in the production and pilot-plant dryers is proportional to the cube root of the ratio of batch volumes. However, if the agitator of the production unit is heated, the drying time increase can be reduced or reversed. Table 12-24 gives basic geometric relationships for agitated and rotating batch dryers, which can be used for approximate size estimation or (with great caution) for extrapolating drying times obtained from one dryer type to another. Note that these do not allow for nominal capacity or partial solids fill. For the paddle (horizontal pan) dryer with heated agitator, R is the ratio of the heat transferred through the agitator to that through the walls, which is proportional to the factor hA for each case.

In the absence of experimental data, the following method may be used for scoping calculations. For constant-rate drying, the drying time t_{CR} can be calculated from

$$t_{CR} = \frac{m_s(X_{initial} - X_{final})\lambda_{ev}}{h_{ws}\Delta T_m A_s} \tag{12-73}$$

where $X_{initial}$ = initial dry-basis moisture content; X_{final} = final dry-basis moisture content; λ_{ev} = latent heat of vaporization; A_s = surface area; ΔT_m = log-mean temperature difference.

To correct from a calculated constant-rate (unhindered) drying time t_{CR} to first-order falling-rate kinetics, the following equation is used, where X_1 is the initial, X_2 the final, and X_E the equilibrium moisture content (all must be dry-basis):

$$\frac{t_{FR}}{t_{CR}} = \frac{X_1 - X_E}{X_1 - X_2} \ln\left(\frac{X_1 - X_E}{X_2 - X_E}\right) \tag{12-74}$$

Note that $t_{FR} \geq t_{CR}$. Likewise, to convert to a two-stage drying process with constant-rate drying down to X_{cr} and first-order falling-rate drying beyond, the equation is

$$\frac{t_{2S}}{t_{CR}} = \frac{X_1 - X_{cr}}{X_1 - X_2} + \frac{X_{cr} - X_E}{X_1 - X_2} \ln\left(\frac{X_{cr} - X_E}{X_2 - X_E}\right) \tag{12-75}$$

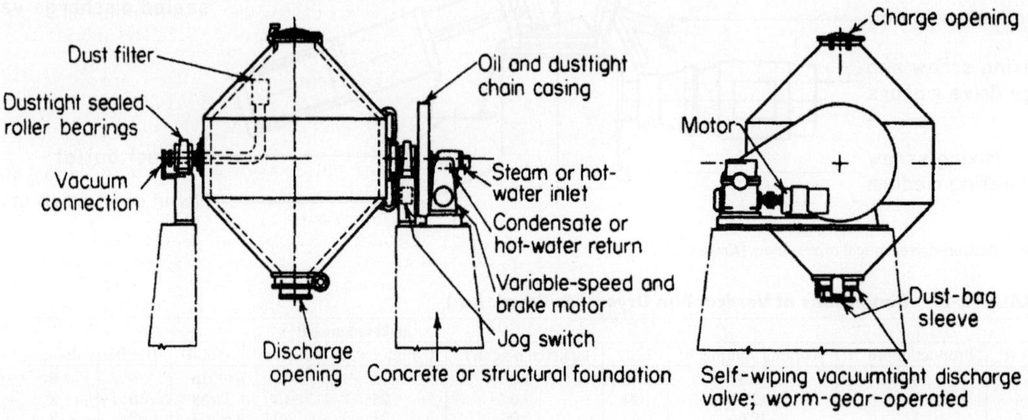

FIG. 12-60 Rotating (double-cone) vacuum dryer. (*Stokes Vacuum, Inc.*)

TABLE 12-24 Calculation of Key Dimensions for Various Batch Contact Dryers (Fig. 12-61 Shows the Geometries)

Dryer type	Volume as $f(D)$	Typical L/D	Diameter as $f(V)$	Surface area as $f(D)$	Ratio A/V
Tumbler/double-cone	$V=\frac{\pi D^3}{12}\left(\frac{L}{D}\right)$	1.5	$D=\left[\frac{12V}{\pi(L/D)}\right]^{1/3}$	$A=\frac{\pi D^2}{2}\left[\left(\frac{L}{D}\right)^2+1\right]^{1/2}$	$\frac{A}{V}=\frac{6}{D}\left[1+\left(\frac{L}{D}\right)^2\right]^{1/2}$
Vertical pan	$V=\frac{\pi D^3}{4}\left(\frac{L}{D}\right)$	0.5	$D=\left[\frac{4V}{\pi(L/D)}\right]^{1/3}$	$A=\pi D^2\left(\frac{L}{D}+\frac{1}{4}\right)$	$\frac{A}{V}=\frac{4}{D}\left(1+\frac{D}{4L}\right)$
Spherical	$V=\frac{\pi D^3}{6}\left(\frac{L}{D}\right)$	1	$D=\left[\frac{6V}{\pi(L/D)}\right]^{1/3}$	$A=\pi D^2\left(\frac{L}{D}\right)$	$\frac{A}{V}=\frac{6}{D}$
Filter dryer	$V=\frac{\pi D^3}{4}\left(\frac{L}{D}\right)$	0.5	$D=\left[\frac{4V}{\pi(L/D)}\right]^{1/3}$	$A=\pi D^2\left(\frac{L}{D}\right)$	$\frac{A}{V}=\frac{4}{D}$
Conical agitated	$V=\frac{\pi D^3}{12}\left(\frac{L}{D}\right)$	1.5	$D=\left[\frac{12V}{\pi(L/D)}\right]^{1/3}$	$A=\frac{\pi D^2}{2}\left[\left(\frac{L}{D}\right)^2+\frac{1}{4}\right]^{1/2}$	$\frac{A}{V}=\frac{6}{D}\left[1+\frac{1}{4}\left(\frac{D}{L}\right)^2\right]^{1/2}$
Paddle (horizontal agitated)	$V=\frac{\pi D^3}{4}\left(\frac{L}{D}\right)$	5	$D=\left[\frac{4V}{\pi(L/D)}\right]^{1/3}$	$A=\pi D^2\left(\frac{L}{D}\right)$	$\frac{A}{V}=\frac{4}{D}$
Paddle, heated agitator	$V=\frac{\pi D^3}{4}\left(\frac{L}{D}\right)$	5	$D=\left[\frac{4V}{\pi(L/D)}\right]^{1/3}$	$A=\pi D^2\left(\frac{L}{D}\right)(1+R)$	$\frac{A}{V}=\frac{4}{D}(1+R)$

Example 12-16 Calculations for Batch Dryer For a 10-m^3 batch of material containing 5000 kg of dry solids and 30 percent moisture (dry basis), estimate the size of vacuum dryers required to contain the batch at 50 percent volumetric fill. Jacket temperature is 200°C, applied pressure is 100 mbar (0.1 bar), and the solvent is water (take latent heat as 2400 kJ/kg). Assuming the heat-transfer coefficient based on the total surface area to be 50 W/(m^2·K) for all types, calculate the time to dry to 5 percent for (*a*) unhindered (constant-rate) drying throughout, (*b*) first-order falling-rate (hindered) drying throughout, (*c*) the case where experiment shows the actual drying time for a conical dryer to be 12.5 h and other cases are scaled accordingly. Take $R = 5$ with the heated agitator. Assume the material is nonhygroscopic (equilibrium moisture content $X_E = 0$).

Solution The dryer volume V must be 20 m^3, and the diameter is calculated from column 4 of Table 12-24, assuming the default L/D ratios. Table 12-25 gives the results. Water at 100 mbar boils at 46°C so take ΔT as 200 − 46 = 154°C. Then Q is found from Eq. (12-72). For constant-rate drying throughout, drying time t_{CR} = evaporation rate/heat input rate and was given by

$$t_{CR}=\frac{m_s(X_O-X_I)\lambda_{ev}}{h_{ws}\Delta T_{ws}A_s}=\frac{5000(0.3-0.05)(2400)}{0.05(154A_s)} \tag{12-76}$$

This gives t_{CR} as 389,610/A_s s or 108.23/A_s h. Values for A_s and calculated times for the various dryer types are given in Table 12-25.

For falling-rate drying throughout, time t_{FR} is given by Eq. (12-77); the multiplying factor for drying time is 1.2 ln 6 = 2.15 for all dryer types.

$$\frac{t_{FR}}{t_{CR}}=\left(\frac{X_1-X_E}{X_1-X_2}\right)\ln\left(\frac{X_1-X_E}{X_2-X_E}\right)=\frac{0.3}{0.25}\ln\frac{0.3}{0.05} \tag{12-77}$$

If the material showed a critical moisture content, the calculation could be split into two sections for constant-rate and falling-rate drying. Likewise, the experimental drying time t_{expt} for the conical dryer is 12.5 h which is a factor of 3.94 greater than the constant-rate drying time. A very rough estimate of drying times for the other dryer types has been made by applying the same scaling factor (3.94) to their constant-rate drying times. Two major sources of error are possible: (1) The drying kinetics could differ between dryers; and (2) if the estimated heat-transfer coefficient for either the base case or the new dryer type is in error, then the scaling factor will be wrong. All drying times have been shown in hours, as this is more convenient than seconds.

The paddle with heated agitator has the shortest drying time, and the filter dryer the longest (because the bottom plate is unheated). Other types are fairly comparable. The spherical dryer would usually have a higher heat-transfer coefficient and shorter drying time than shown.

An excellent model for a variety of agitated vacuum dryers has been developed and validated against experimental data. See Schlünder, E. and Mollekopf, N., "Vacuum Contact Drying of Free Flowing Mechanically Agitated Particulate Material," *Chemical Engineering and Processing: Process Intensification* **18**(2): 93–111 (March–April 1984).

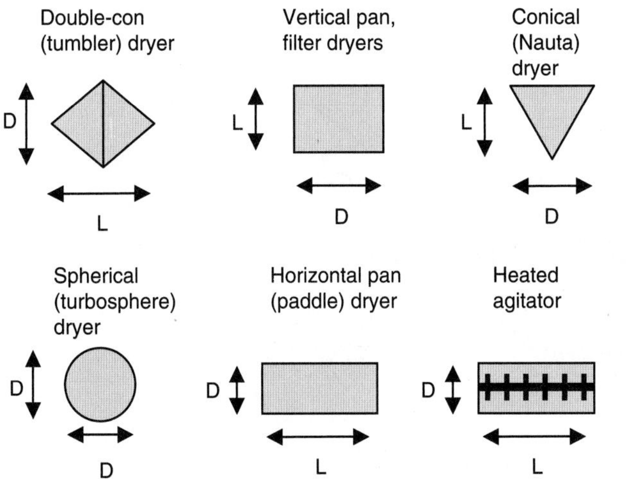

FIG. 12-61 Basic geometries for batch dryer calculations.

Performance Data for Batch Vacuum Rotary Dryers Typical performance data for horizontal pan vacuum dryers are given in Table 12-26. Size and cost data for rotary agitator units are given in Table 12-27. Data for double-cone rotating units are shown in Table 12-28.

Continuous Agitated and Rotary Dryers

Examples and Synonyms Disk, Porcupine, Nara, Solidaire, Forberg, steam tube, paddle dryers, continuous rotary dryers, rotary kiln, steam-tube dryer, rotary calciner, Roto-Louvre dryer.

Continuous Agitated Dryers: Description Continuous agitated dryers, often known as *paddle* or *horizontal agitated dryers*, consist of one or more horizontally mounted shells with internal mechanical agitators, which may take many different forms. They are a continuous equivalent of the horizontal pan dryer and are similar in construction, but usually of larger dimensions. They have many similarities to continuous indirect rotary dryers and are sometimes classified as rotary dryers, but this is a misnomer because the outer shell does not rotate, although in some types there is an inner shell which does. Frequently, the internal agitator is heated, and a wide variety of designs exist. Often two intermeshing agitators are used. There are important variants with high-speed agitator rotation and supplementary convective heating by hot air.

The basic differences are in the type of agitator, with the two key factors being the heat-transfer area and solids handling/stickiness characteristics. Unfortunately, the types giving the highest specific surface area (multiple tubes and coils) are often also the ones most susceptible to fouling and blockage and most difficult to clean. Figure 12-62 illustrates a number of different agitator types.

Paddle Dryers Product trials are conducted in small pilot dryers (8- to 60-L batch or continuous units) to determine material handling and process retention times. Variables such as drying temperature, pressure level, and shaft speed are analyzed during the test trials. For initial design purposes, the heat-transfer coefficient for paddle dryers is typically in the range of 10 W/(m^2·K) (light, free-flowing powders) up to 150 W/(m^2·K) (dilute slurries). However, it is preferable to scale up from the test results, fitting the data to estimate the heat-transfer coefficient and scaling up on the basis of total area of heat-transfer surfaces, including heated agitators. Typical length/diameter ratios are between 5 and 8, similar to rotary dryers and greater than some batch horizontal pan dryers.

TABLE 12-25 Comparative Dimensions and Drying Times for Various Batch Contact Dryers

Dryer type	h, kW/(m²·K)	L/D	D, m	L, m	A, m²	t_{CR}, h	t_{FR}, h	t_{expt}, h
Tumbler/double-cone	0.05	1.5	3.71	5.56	38.91	2.78	5.98	11.0
Vertical pan	0.05	0.5	3.71	1.85	32.37	3.34	7.19	13.2
Spherical	0.05	1	3.37	3.37	35.63	3.04	6.54	12.0
Filter dryer	0.05	0.5	3.71	1.85	21.58	5.01	10.77	19.8
Conical agitated	0.05	1.5	3.71	5.56	34.12	3.17	6.82	12.5
Paddle (horizontal agitated)	0.05	5	1.72	8.60	46.50	2.33	5.01	9.2
Paddle, heated agitator	0.05	5	1.72	8.60	278.99	0.39	0.83	1.52

TABLE 12-26 Performance Data of Vacuum Rotary Dryer*

Material	Diameter × length, m	Initial moisture, % dry basis	Batch Steam pressure, Pa × 10³	Final Agitator speed, r/min	dry weight, kg	moisture, % dry basis	Pa × 10³	Time, h	Evaporation, kg/(h·m²)
Cellulose acetate	1.5 × 9.1	87.5	97	5.25	610	6	90–91	7	1.5
Starch	1.5 × 9.1	45–48	103	4	3630	12	88–91	4.75	7.3
Sulfur black	1.5 × 9.1	50	207	4	3180	1	91	6	4.4
Fuller's earth/mineral spirit	0.9 × 3.0	50	345	6	450	2	95	8	5.4

*Stokes Vacuum, Inc.

TABLE 12-27 Standard Rotary Vacuum Dryer*

Diameter, m	Length, m	Heating surface, m²	Working capacity, m³†	Agitator speed, r/min	Drive, kW	Weight, kg	Purchase price (1995) Carbon steel	Purchase price (1995) Stainless steel (304)
0.46	0.49	0.836	0.028	7½	1.12	540	$ 43,000	$ 53,000
0.61	1.8	3.72	0.283	7½	1.12	1680	105,000	130,000
0.91	3.0	10.2	0.991	6	3.73	3860	145,000	180,000
0.91	4.6	15.3	1.42	6	3.73	5530	180,000	205,000
1.2	6.1	29.2	3.57	6	7.46	11,340	270,000	380,000
1.5	7.6	48.1	6.94	6	18.7	15,880	305,000	440,000
1.5	9.1	57.7	8.33	6	22.4	19,050	330,000	465,000

*Stokes Vacuum, Inc. Prices include shell, 50-lb/in²-gauge jacket, agitator, drive, and motor; auxiliary dust collectors, condensers.

†Loading with product level on or around the agitator shaft.

TABLE 12-28 Standard (Double-Cone) Rotating Vacuum Dryer*

Working capacity, m³	Total volume, m³	Heating surface, m²	Drive, kW	Floor space, m²	Weight, kg	Purchase cost (1995) Carbon steel	Purchase cost (1995) Stainless steel
0.085	0.130	1.11	.373	2.60	730	$ 32,400	$ 38,000
0.283	0.436	2.79	.560	2.97	910	37,800	43,000
0.708	1.09	5.30	1.49	5.57	1810	50,400	57,000
1.42	2.18	8.45	3.73	7.15	2040	97,200	106,000
2.83	4.36	13.9	7.46	13.9	3860	198,000	216,000
4.25	6.51	17.5	11.2	14.9	5440	225,000	243,000
7.08	10.5	*38.7	11.2	15.8	9070	324,000	351,000
9.20	13.9	*46.7	11.2	20.4	9980	358,000	387,000
11.3	16.0	*56.0	11.2	26.0	10,890	378,000	441,000

*Stokes Vacuum, Inc. Price includes dryer, 15-lb/in² jacket, drive with motor, internal filter, and trunnion supports for concrete or steel foundations. Horsepower is established on 65 percent volume loading of material with a bulk density of 50 lb/ft³. Models of 250 ft³, 325 ft³, and 400 ft³ have extended surface area.

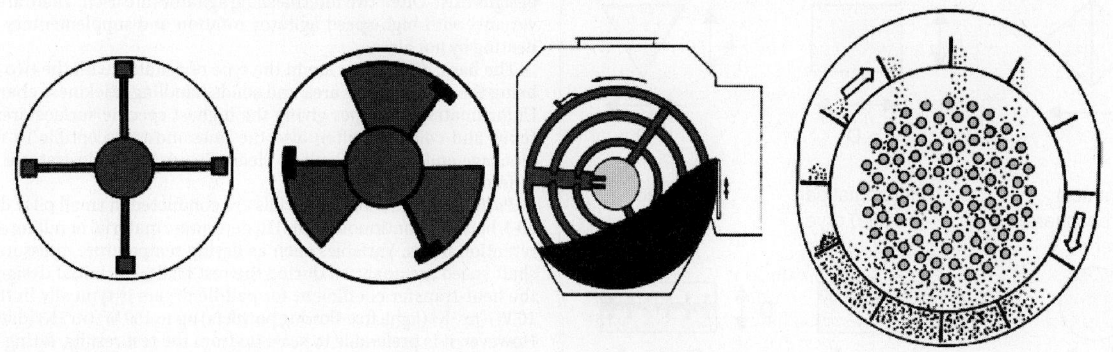

FIG. 12-62 Typical agitator designs for paddle (horizontal agitated) dryers. (*a*) Simple unheated agitator. (*b*) Heated cut-flight agitator. (*c*) Multicoil unit. (*d*) Tube bundle.

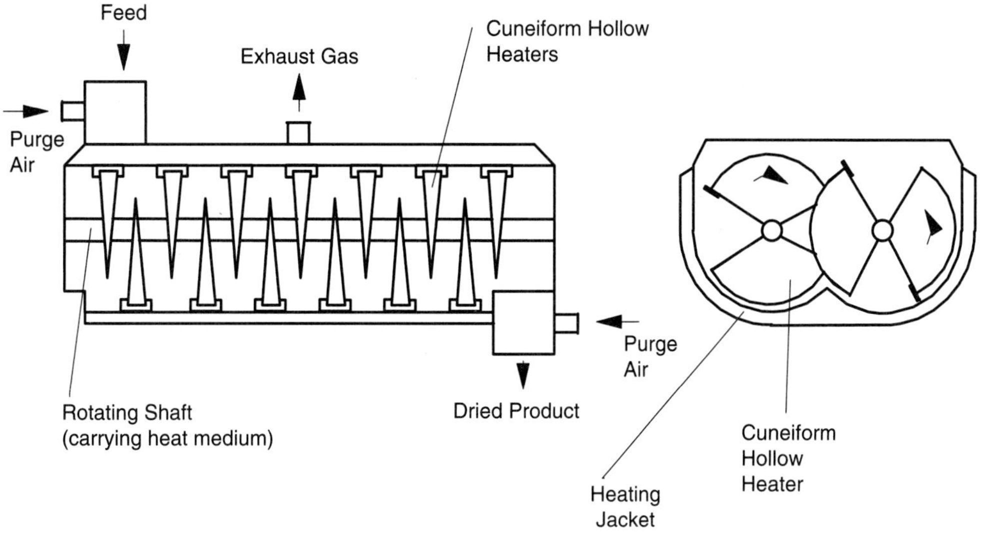

FIG. 12-63 Nara twin-shaft paddle dryer.

The most common problem with paddle dryers (and with their closely related cousins, steam-tube and indirect rotary dryers) is the buildup of sticky deposits on the surface of the agitator or outer jacket. This leads, first, to reduced heat-transfer coefficients and slower drying and, second, to blockages and stalling of the rotor. Also thermal decomposition and loss of product quality can result. The problem is usually most acute at the feed end of the dryer, where the material is wettest and stickiest. A wide variety of different agitator designs have been devised to try to reduce stickiness problems and enhance cleanability while providing a high heat-transfer area. Many designs incorporate a high-torque drive combined with rugged shaft construction to prevent rotor stall during processing, and stationary mixing elements are installed in the process housing which continually clean the heat-exchange surfaces of the rotor to minimize any crust buildup and ensure an optimum heat-transfer coefficient at all times. Another alternative is to use two parallel intermeshing shafts, as in the Nara paddle dryer (Fig. 12-63). Suitably designed continuous paddle and batch horizontal pan dryers can handle a wide range of product consistencies (dilute slurries, pastes, friable powders) and can be used for processes such as reactions, mixing, drying, cooling, melting, sublimation, distilling, and vaporizing.

Continuous Rotary Dryers: Description A rotary dryer consists of a cylinder that rotates on suitable bearings and that is usually slightly inclined to the horizontal. The cylinder length may range from 4 to more than 10 times the diameter, which may vary from less than 0.3 to more than 3 m. Solids fed into one end of the drum are carried through it by gravity, with rolling, bouncing and sliding, and drag caused by the airflow either retarding or enhancing the movement, depending on whether the dryer is cocurrent or countercurrent. It is possible to classify rotary dryers into direct-fired, where heat is transferred to the solids by direct exchange between the gas and the solids, and indirect, where the heating medium is separated from physical contact with the solids by a metal wall or tube. Many rotary dryers contain flights or lifters, which are attached to the inside of the drum and which cascade the solids through the gas as the drum rotates.

For handling large quantities of granular solids, a cascading rotary dryer is often the equipment of choice. If the material is not naturally free-flowing, recycling of a portion of the final dry product may be used to precondition the feed, either in an external mixer or directly inside the drum. Hanging link chains and/or scrapper chains are also used for sticky feed materials.

Their operating characteristics when performing heat- and mass-transfer operations make them suitable for the accomplishment of drying, chemical reactions, solvent recovery, thermal decompositions, mixing, sintering, and agglomeration of solids. The specific types included are the following:

Direct cascading rotary dryer (cooler). This is usually a bare metal cylinder but with internal flights (shelves) which lift the material and drop it through the airflow. It is suitable for low- and medium-temperature operations, the operating temperature being limited primarily by the strength characteristics of the metal employed in fabrication.

Direct rotary dryer (cooler). As above but without internal flights.

Direct rotary kiln. This is a metal cylinder lined on the interior with insulating block and/or refractory brick. It is suitable for high-temperature operations.

Indirect steam-tube dryer. This is a bare metal cylinder provided with one or more rows of metal tubes installed longitudinally in the shell. It is suitable for operation up to available steam temperatures or in processes requiring water cooling of the tubes.

Indirect rotary calciner. This is a bare metal cylinder surrounded on the outside by a fired or electrically heated furnace. It is suitable for operation at medium temperatures up to the maximum that can be tolerated by the metal wall of the cylinder, usually 650 to 700 K for carbon steel and 800 to 1025 K for stainless steel.

Direct Roto-Louvre dryer. This is one of the more important special types, differing from the direct rotary unit in that true *through-circulation* of gas through the solids bed is provided. Like the direct rotary, it is suitable for low- and medium-temperature operation.

Direct heat rotary dryer. The direct heat units are generally the simplest and most economical in operation and construction, when the solids and gas can be permitted to be in contact. The required gas flow rate can be obtained from a heat and mass balance. The bed cross-sectional area is found from a scoping design calculation (a typical gas velocity is 3 m/s for cocurrent and 2 m/s for countercurrent units). Length is normally between 5 and 10 times the drum diameter (an L/D value of 8 can be used for initial estimation) or can be calculated by using an incremental model (see Example 12-17).

A typical schematic diagram of a rotary dryer is shown in Fig. 12-64, while Fig. 12-65 shows typical lifting flight designs.

Residence Time, Standard Configuration The residence time in a rotary dryer τ represents the average time that particles are present in the equipment, so it must match the required drying time.

The calculation of the residence time of material in the dryer is complex since the holdup depends on the design of the flights and material properties, such as the angle of repose. The flow of the material in the equipment to calculate the residence times has been the subject of a number of historical papers, including those by Sullivan et al. (U.S. Bureau of Mines Tech. Paper 384), Friedman and Marshall equation [*Chem. Eng. Progr.* **45**(8): 482 (1949)], Saeman and Mitchell [*Chem. Eng. Progr.* **50**(9): 467 (1954)], and Schofield and Glikin [*Trans. IChemE* **40:** 183 (1962)].

The most complete analysis of particle motion in rotary dryers is given by Matchett and Baker [*J. Sep. Proc. Technol.* **8:** 11 (1987)]. They considered both the airborne phase (particles falling through air) and the dense phase (particles in the flights or the rolling bed at the bottom). Typically, particles spend 90 to 95 percent of the time in the dense phase, but the majority of the drying takes place in the airborne phase. In the direction parallel to the dryer axis, most particle movement occurs through four mechanisms: by gravity and air drag in the airborne phase, and by bouncing, sliding, and rolling in the dense phase.

The combined particle velocity in the airborne phase is U_{P1}, which is the sum of the gravitational and air drag components for cocurrent dryers and the difference between them for countercurrent dryers. The dense-phase velocity, arising from bouncing, sliding, and rolling, is denoted U_{P2}.

Papadakis et al. [*Dry. Tech.* **12**(1&2): 259–277 (1994)] rearranged the Matchett and Baker model from its original "parallel" form into a more

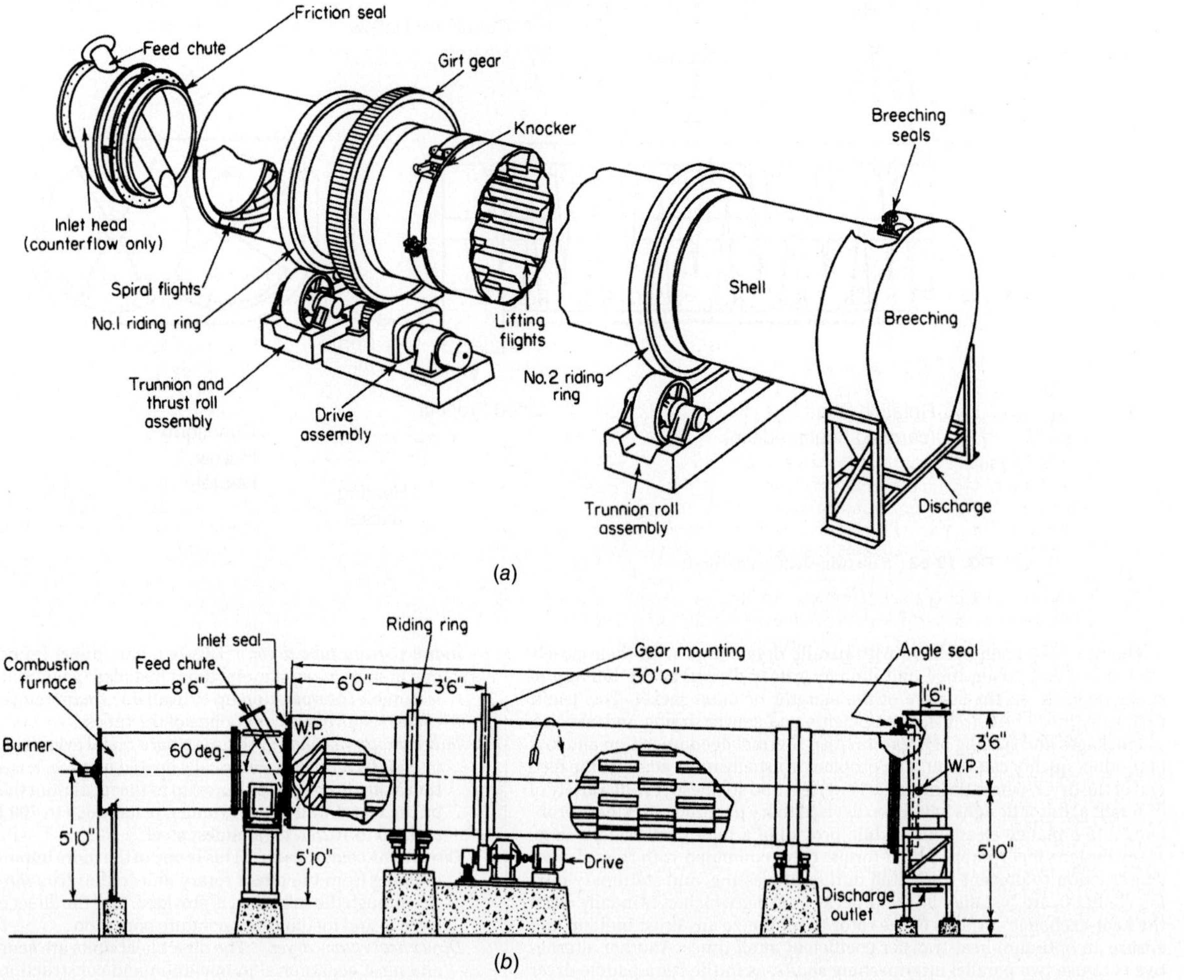

FIG. 12-64 Component arrangement (*a*) and elevation (*b*) of countercurrent direct-heat rotary dryer. (*Air Preheater Company, Raymond & Bartlett Snow Products.*)

computationally convenient "series" form. The sum of the calculated residence times in the airborne and dense phases, τ_G and τ_S, respectively, is the total solids residence time. The dryer length is simply the sum of the distances traveled in the two phases.

$$\tau = \tau_G + \tau_S \qquad (12\text{-}78)$$

$$L = \tau_G U_{P1} + \tau_S U_{P2} \qquad (12\text{-}79)$$

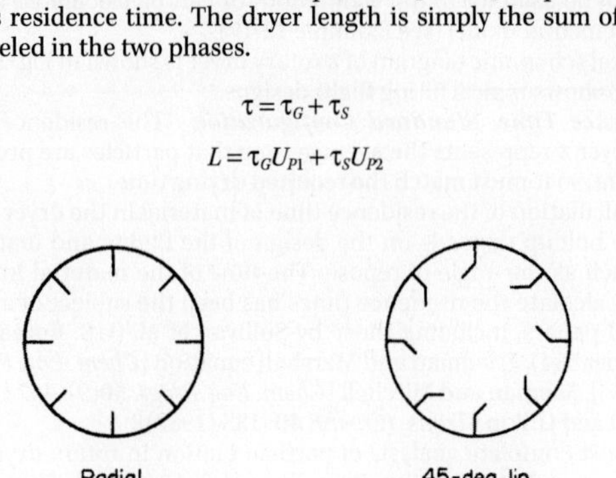

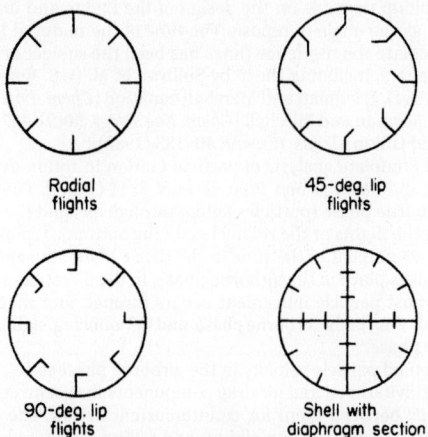

FIG. 12-65 Typical lifting flight designs.

For airborne phase motion, the velocity is affected by gravity and air drag.

$$U_{P1} = U_{P1}^{O} + U_{P1}^{d} \qquad (12\text{-}80)$$

The velocity U_{P1}^{O} due to the gravitational component is most conveniently expressed as

$$U_{P1}^{O} = \sqrt{\frac{gD}{2}\tan\alpha}\, K_{\text{fall}} \sqrt{\frac{D_e}{D\cos\alpha}} = \sqrt{\frac{gD}{2}\tan\alpha}\, K_K \qquad (12\text{-}81)$$

where K_{fall} is a parameter that allows for particles falling from a number of positions, with different times of flight and lifting times, and is generally between 0.7 and 1; and the effective diameter (internal diameter between lips of flights) is D_e.

The contribution of air drag on the velocity of falling can be calculated using Eqs. (12-82) and (12-83). The value is positive for concurrent airflow and negative for countercurrent airflow. For Reynolds numbers up to 220, where $\text{Re} = \dfrac{U_{G,\text{super}} d_P \rho_g}{\mu_g}$,

$$U_{P1}^{d} = 7.45 \times 10^{-4}\, \text{Re}^{2.2} \frac{\mu U_{G\,\text{super}} t_a^*}{\rho_P d_p^2} \qquad (12\text{-}82)$$

Above this Reynolds number, the following equation was recommended by Matchett and Baker (1987):

$$U_{P1}^{d} = 125 \frac{\mu U_{G\,\text{super}} t_a^*}{\rho_P d_p^2} \qquad (12\text{-}83)$$

The variable t_a^* is the average time of flight of a particle in the airborne phase when the dryer is at the "design loaded," i.e., if the powder fills and does not overflow the flights; see Fig. 12-65. If the dryer has more powder than this (overloaded), then the flights will not be able to carry all of it up as far and so the time of flight will be lower.

The average time of flight of the particles can be estimated from

$$t_f = \left(\frac{2D}{g\cos\alpha}\right)^{1/2} K_{\text{fall}} \qquad (12\text{-}84)$$

Here, since we are designing the dryer, $t_f = t_a^*$.

Bouncing, rolling, and sliding are not so easily analyzed theoretically. Matchett and Baker (cited above) suggested that the dense-phase velocity could be characterized in terms of a dimensionless dense-phase velocity number a, using Eq. (12-85). Values of a are in the range of 1 to 4.

$$U_{P2} = a \cdot N \cdot D \cdot \tan\alpha \qquad (12\text{-}85)$$

Other workers suggested that, in underloaded and design-loaded dryers, bouncing was a significant transport mechanism, whereas for overloaded dryers, rolling was important. Bouncing mechanisms can depend on the airborne phase velocity U_{P1}, since this affects the angle at which the particles hit the bottom of the dryer and the distance they move forward. Rolling mechanisms would be expected to depend on the depth of the bottom bed, and hence on the difference between the actual holdup and the design-loaded holdup.

As an example of the typical numbers involved, Matchett and Baker [*J. Sep. Proc. Technol.* **9:** 5 (1988)] used their correlations to assess the data of Saeman and Mitchell for an industrial rotary dryer with D = 1.83 m and L = 10.67 m, with a slope of 4°, 0.067 m/m. For a typical run with U_G = 0.98 m/s and N = 0.08 r/s, they calculated that U_{P1}° = 0.140 m/s, U_{P1}^{d} = −0.023 m/s, U_{P_1} = 0.117 m/s, and U_{P_2} = −0.02 m/s. The dryer modeled was countercurrent and therefore had a greater slope and lower gas velocity than those of a cocurrent unit; for the latter, U_{P1}° would be lower and U_{P1}^{d} positive and larger. The ratio τ_S/τ_G is approximately 12 in this case, so that the distance traveled in dense-phase motion would be about twice that in the airborne phase.

Kemp and Oakley [*Dry. Tech.* **20**(9): 1699 (2002)] showed that the ratio τ_G/τ_S can be found by comparing the average time of flight from the top of the dryer to the bottom t_f to the average time required for the particles to be lifted by the flights t_d. They derived the following equation:

$$\frac{\tau_S}{\tau_G} = \frac{t_d}{t_f} = \frac{K_{fl}}{N}\sqrt{\frac{g}{D}} \qquad (12\text{-}86)$$

Here all the unknowns have been rolled into a single dimensionless parameter K_{fl}, given by

$$K_{fl} = \frac{\theta}{\pi\sqrt{2}\sin\theta}\sqrt{\frac{D}{D_e}} \qquad (12\text{-}87)$$

Here D_e is the effective diameter (internal diameter between lips of flights), and the solids are carried in the flights for an angle 2θ, on average, before falling. Kemp and Oakley concluded that K_{fl} can be taken to be 0.4 to a first (and good) approximation. For overloaded dryers with a large rolling bed, K_{fl} will increase. The form of Eq. (12-86) is very convenient for design purposes since it does not require D_e, which is unknown until a decision has been made on the type and geometry of the flights.

The model of Matchett and Baker has been shown by Kemp (Proc. IDS 2004, B, 790) to be similar in form to that proposed by Saeman and Mitchell:

$$\tau = \frac{1.1L}{ND\left[\tan\alpha \cdot (K_K/K_{fl}\sqrt{2}+a) + (1/K_{fl})\sqrt{\frac{1}{gD}\cdot U_{P1}^{d}}\right]} \qquad (12\text{-}88)$$

In Eq. (12-88), $K_K/(K_{fl}\sqrt{2})$ will typically be on the order of unity, and reported values of a are in the range of 1 to 4. The airborne gravity component is usually smaller than the dense-phase motion but is not negligible.

Heat- and Mass-Transfer Estimates Many rotary dryer studies have correlated heat- and mass-transfer data in terms of an overall volumetric heat-transfer coefficient $U_\upsilon a$ [W/(m³·K)], defined by

$$Q = U_\upsilon a \cdot V_{\text{dryer}} \cdot \Delta T_m \qquad (12\text{-}89)$$

Here Q is the overall rate of heat transfer between the gas and the solids (W), V_{dryer} is the dryer volume (m³), and ΔT_m is an average temperature driving force (K). When one is calculating the average temperature driving force, it is important to distinguish between the case of heat transfer with dry particles, where the change in the particle temperature is proportional to the change in the gas temperature, and the case of drying particles, where the particle temperature does not change so significantly. Where the particles are dry, the average temperature difference is the logarithmic mean of the temperature differences between the gas and the solids at the inlet and outlet of the dryer. The volumetric heat-transfer coefficient itself consists of a heat-transfer coefficient U_υ, based on the effective area of contact between the gas and the solids, and the ratio a of this area to the dryer volume. Thus, this procedure eliminates the need to specify where most of the heat transfer occurs (e.g., to material in the air, on the flights, or in the rolling bed). Empirical correlations are of the form

$$U_\upsilon a = \frac{K' U_{G\,\text{super}}^{n}}{D} \qquad (12\text{-}90)$$

where K' depends on the solids properties, the flight geometry, the rotational speed, and the dryer holdup. In Eq. (12-90), a is the surface area per unit volume of the powder. Table 12-29 gives the values of n chosen by various authors, and Table 12-30 gives references and conditions for a few published studies. The most accepted value for the exponent n is 0.67; however, this is not universally true. This is not surprising considering the complicated particle flow mechanics in the equipment. Experimental data on the materials used are preferred.

An alternative procedure is the use of a conventional film heat-transfer coefficient h_f [W/(m²·K)]

$$Q = h_f \cdot A_s \cdot \Delta T \qquad (12\text{-}91)$$

Here Q is the local heat-transfer rate (W), A_s is the total surface area of all the particles (m²), and ΔT is the temperature difference between the gas and the solids (K). The method has the advantages that h_f can be determined by relatively simple tests (or calculated from appropriate correlations in the literature), variations in operating conditions can be allowed for, and analogies between heat and mass transfer allow the film coefficients for these processes to be related. However, the area for heat transfer must be estimated under the complex conditions of gas-solids interaction present in particle cascades. Schofield and Glikin (1962) estimated this area to be the surface area of particles per unit mass $6/(\rho_P d_P)$, multiplied by the fraction of solids in the drum that are cascading through the gas at any moment, which was estimated as the fraction of time spent by particles cascading through the gas:

$$A_S = \frac{6}{\rho_P d_P}\frac{t_f}{t_f + t_d} \qquad (12\text{-}92)$$

Schofield and Glikin estimated the heat-transfer coefficient by using the correlation given by McAdams (1954), which correlates data for gas-to-particle heat transfer in air to about 20 percent over a range of Reynolds numbers (Re_P, defined in the previous subsection) between 17 and 70,000:

$$\text{Nu}_P = 0.33 \cdot \text{Re}_P^{1/2} \qquad (12\text{-}93)$$

TABLE 12-29 Values of the Index *n* in Correlations for the Volumetric Heat-Transfer Coefficient (after Baker, 1983)

Author(s)	Exponent n
Saeman and Mitchell (1954)	0
Friedman and Marshall (1949)	0.16
Aiken and Polsak (1982)	0.37
Miller et al. (1942)	0.46–0.60
McCormick (1962)	0.67
Myklestad (1963)	0.80

TABLE 12-30 Summary of the Predictions Using the Correlations for the Volumetric Heat-Transfer Coefficients of Various Authors (after Baker, 1983)

	$U_\upsilon a$, W/(m³·K)	
Author(s)	$U_{G\text{super}_1}$ = 1 m/s	$U_{G\text{super}_1}$ = 3 m/s
Miller et al. (1942)		
Commercial data	248	516
Pilot-scale data	82	184
Friedman and Marshall (1949)	67	138
Saeman and Mitchell (1954)	495–1155	1032–2410
Myklestad (1963)	423	1019

Here the particle Nusselt number is Nu_P, where $Nu_P = h_f d_P/k_G$, and k_G is the thermal conductivity of the gas [W/(m·K)]. They stated that the heat-transfer rates predicted by this procedure were much larger than those measured on an industrial cooler, which is probably due to the particles on the inside of the cascades not experiencing the full gas velocity. Kamke and Wilson (1986) used a similar approach to model the drying of wood chips, but used the Ranz-Marshall (1952) equation to predict the heat-transfer coefficient:

$$Nu = 2 + 0.6 \cdot Re_P^{1/2} \cdot Pr_G^{1/3} \tag{12-94}$$

where Pr_G is the Prandtl number of the gas.

Drying Time Estimates Sometimes, virtually all the drying takes place in the airborne phase. Under such circumstances, the airborne-phase residence time τ_G and the drying time are virtually the same, and the required drying time can be estimated from equivalent times in drying kinetics experiments, e.g., using a thin-layer test (Langrish, D.Phil. thesis, 1988).

Example 12-17 Sizing of a Cascading Rotary Dryer The average gas velocity passing through a cocurrent, adiabatic, cascading rotary dryer is 4 m/s. The particles moving through the dryer have an average diameter of 5 mm (Sauter mean diameter), a solids density of 600 kg/m³, and a shape factor of 0.75. The particles enter with a moisture content of 0.50 kg/kg (dry basis) and leave with a moisture content of 0.15 kg/kg (dry basis). The drying rate may be assumed to decrease linearly with average moisture content, with no unhindered (constant-rate) drying period. In addition, let us assume that the solids are nonhygroscopic (so that the equilibrium moisture content is zero; hygroscopic means that the equilibrium moisture content is nonzero).

The inlet humidity is 0.10 kg/kg (dry basis) due to the use of a direct-fired burner, and the ratio of the flow rates of dry solids to dry gas is unity ($F/G = 1$). The gas temperature at the inlet to the dryer is 800°C, and the gas may be assumed to behave as a pure water vapor/air mixture.

Suppose that this dryer has a slope α of 4° and a diameter D of 1.5 m, operating at a rotational speed N of 0.04 r/s.

What residence time is required to dry the solid material to the target moisture content? How long does the dryer need to be?

Data:

$U = 4$ m/s	$X_I = 0.50$ kg/kg	$F/G = 1$
$d_{particle} = 0.005$ m	$X_O = 0.15$ kg/kg	$T_{GI} = 800$°C
$\rho_P = 600$ kg/m³	$X_{cr} = 0.50$ kg/kg	$Y_I = 0.10$ kg/kg
$\alpha_P = 0.75$	$X_e = 0.0$ kg/kg	

Solution

Application of concept of characteristic drying curve: A linear falling rate curve implies the following equation for the drying kinetics [see Solids Drying Fundamentals subsection, Eqs. (12-29) and (12-30)]:

$f = \Phi$ assumption of linear drying kinetics

where f is the drying rate relative to the initial drying rate

$$f = \frac{N}{N_{initial}} \quad (12\text{-}37)$$

Since the material begins drying in the falling-rate period, the critical moisture content can be taken as the initial moisture content. The equilibrium moisture content is zero since the material is not hygroscopic.

$$\Phi = \frac{\bar{X} - X_{eq}}{X_{cr} - X_{eq}} = \frac{\bar{X}}{0.5} \tag{12-95}$$

Application of mass balances (theory): A mass balance around the inlet and any section of the dryer is shown in Fig. 12-66.

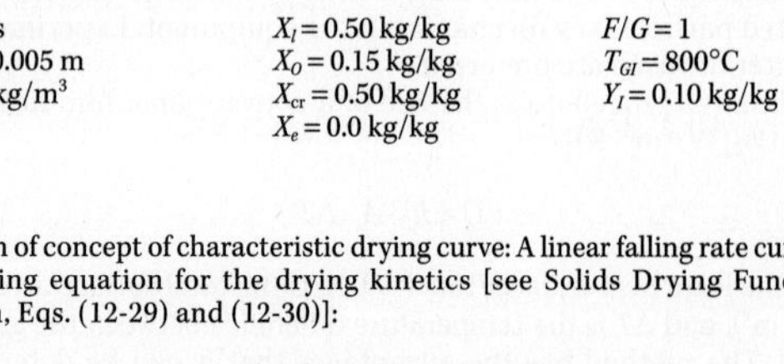

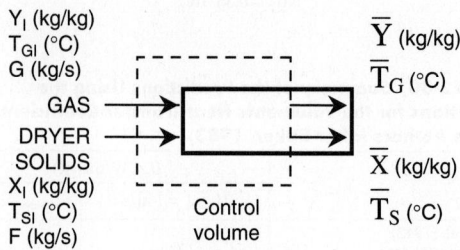

FIG. 12-66 Mass balance around a typical section of a cocurrent dryer.

The essential idea is to calculate the average gas humidity $\bar{Y}$ at each average moisture content $\bar{X}$.

A differential mass balance on the air at any position in the bed is

$$F \cdot dX = -G \cdot dy \tag{12-96}$$

$$-\frac{dY}{dX} = \frac{F}{G} = \frac{\bar{Y} - Y_I}{X_I - \bar{X}} \tag{12-97}$$

where Y = gas humidity, kg moisture/kg dry gas
X = solids moisture content, kg moisture/kg dry solids
W_G = flow rate of dry gas, kg dry gas/s
W_S = flow rate of dry solids, kg dry solids/s

Application of mass balances: Plugging in the numbers gives the relationship between absolute humidity and moisture in the solids at any position.

$$\frac{\bar{Y} - 0.1}{0.5 - \bar{X}} = 1 = \frac{F}{G} \tag{12-98}$$

$$\bar{Y} = 0.6 - \bar{X} \tag{12-99}$$

$$Y_O = 0.6 - 0.15 = 0.45 \text{ kg/kg} \tag{12-100}$$

From the Mollier chart:

$$T_{wb} = 79°C \quad Y_s^* = 0.48 \text{ kg/kg}$$

For the whole dryer,

$$\bar{Y} = 0.275 \text{ kg/kg}$$

The mass balance information is important, but not the entire answer to the question. Now the residence time can be calculated from the kinetics.

Application of concept of characteristic drying curve to estimating drying rates in practice (theory): The overall (required) change in moisture content is divided into a number of intervals of size ΔX, and the problem is solved using a spreadsheet; note that ΔX is difference in the dry-basis moisture content, not distance. The sizes of the intervals need not be the same and should be finer where the fastest moisture content change occurs. For the sake of simplicity, this example will use intervals of uniform size. Then the application of the concept of a characteristic drying curve gives the following outcomes:

$$-\frac{d\bar{X}}{dt} = \text{drying rate} = \frac{f \cdot k \cdot \phi}{\rho_p} \frac{A_p}{V_p} (Y_s^* - \bar{Y}) \tag{12-101}$$

where f = relative drying rate in interval (dimensionless)
$\bar{Y}$ = average humidity in interval, kg/kg
ϕ = humidity potential coefficient, close to unity
ρ_p = density of dry solids, kg/m³
Y_s^* = humidity at saturation, from the adiabatic saturation contour on Mollier chart (Fig. 12-67)

$$\frac{A_p}{V_p} = \frac{\text{particle surface area}}{\text{particle volume}} = \frac{6}{\phi_p \cdot d_p} \tag{12-102}$$

where $d_{particle}$ = Sauter-mean particle diameter for mixture (volume-surface diameter), m
ϕ_p = particle shape factor, unity for spheres (dimensionless)
k = mass-transfer coefficient, kg/(m²·s), obtained from heat-transfer coefficient (often easier to obtain) using the Chilton-Colburn analogy

$$k \cdot \phi = \frac{\beta}{C_{PY}} h \tag{12-103}$$

where β = psychrometric ratio, close to unity for air/water vapor system
C_{PY} = humid heat capacity = $C_{PG} + YC_P = 1050 + 0.275 \times 2000 = 1600$ J/(kg·K)
C_{PG} = specific heat capacity of dry gas (air) = 1050 J/(kg·K)
C_{PV} = specific heat capacity of water vapor = 2000 J/(kg·K)
h = heat-transfer coefficient, W/(m²·K) = $\frac{\lambda_G}{d_{particle}} \cdot Nu_p$
λ_G = thermal conductivity of gas = 0.02 W/(m·K)
Nu_P = particle Nusselt number = $2 + 0.6\, Re_P^{0.5}\, Pr^{0.33}$
Pr = Prandtl number for air = 0.7
Re_P = particle Reynolds number = $\frac{U \cdot d_{particle}}{\nu}$

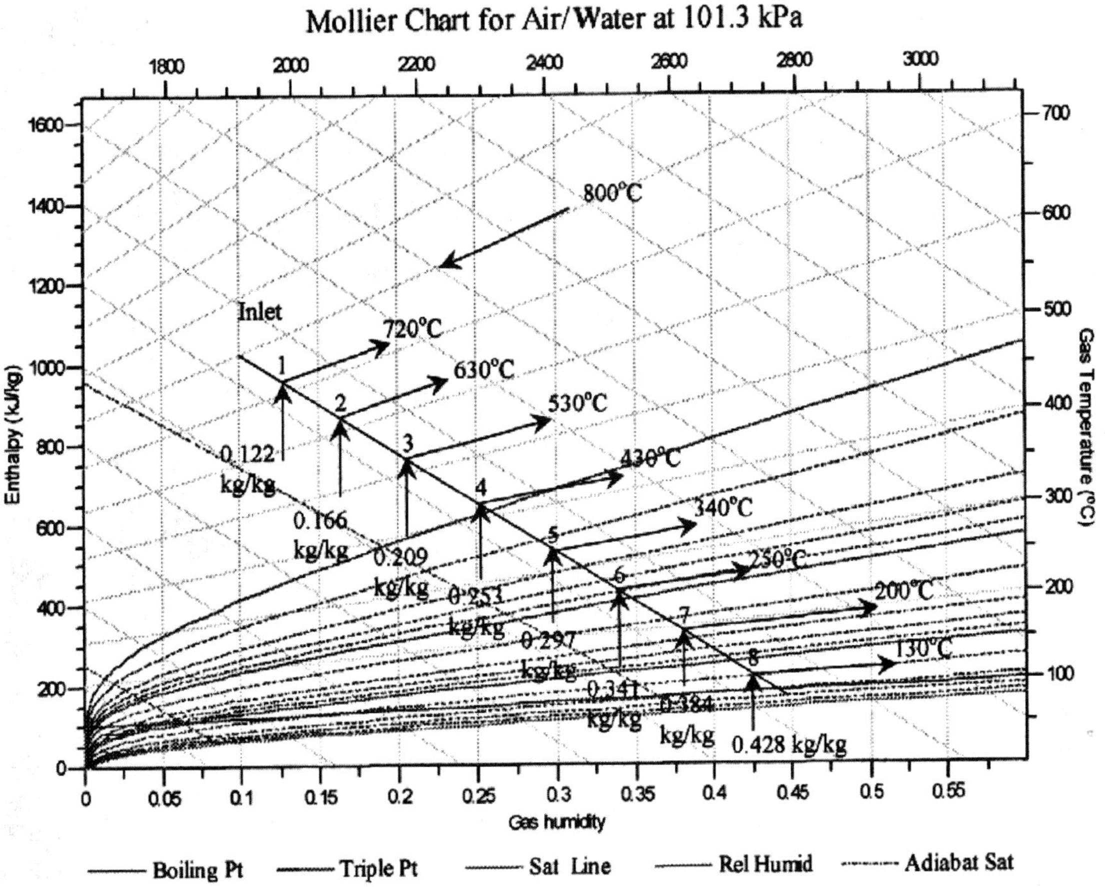

FIG. 12-67 Enthalpy humidity chart used to generate the results in Table 12-31 plots humidity (abscissa) versus enthalpy (lines sloping diagonally from top left to bottom right).

U = relative velocity between gas and particles; in cascading rotary dryers, this is almost constant throughout the dryer and close to the superficial gas velocity $U_{G\,super}$ (equals 4 m/s in this example)

ν = kinematic viscosity of gas at average T_G in dryer = 15×10^{-6} m²/s

We might do a more accurate calculation by finding the gas properties at the conditions for each interval.

Application of concept of characteristic drying curve to estimating drying rates in practice

From the relationships above, $Re_p = \frac{U \cdot d_p}{\nu} = \frac{4 \cdot 0.005}{15 \times 10^{-6}} = 1333; Nu_p = 21.5; h = \frac{0.02}{0.005} \cdot 21.5 = 86 \frac{W}{m^2 \cdot K}; k = \frac{86}{1600} = 0.054 \frac{kg}{m^2 \cdot s}; \frac{A_p}{V_p} = \frac{6}{0.75 \cdot 0.005} = 1600\ m^{-1}$.

Plugging into Eq. (12-98) gives 0.

$$\frac{d\bar{X}}{dt} = f \frac{0.054 \cdot 1600}{600}(0.48 - \bar{Y}) = 0.14 \cdot f \cdot (0.48 - \bar{Y}) \tag{12-104}$$

As stated prior to this example, most of the drying in a cascading rotary dryer occurs while the particles are falling through the air. In this example, we will assume that is when all the drying occurs. We will also assume that while a particle is falling, the temperature can be calculated using a quasi-steady-state approximation. This means that we equate the heat transfer to the particle by convection to the evaporative cooling caused by drying, and we neglect the energy accumulation term.

$$h(T_G - T_S) = f \cdot N_w \cdot \Delta H_{vap} \tag{12-105}$$

where N_w is the maximum (unhindered) drying rate. For completely unhindered drying, f = 1 and T_S (the temperature of the solid) equals the wet-bulb temperature T_W. Plugging those into Eq. (12-105) and taking the ratio give

$$f = \frac{T_G - T_S}{T_G - T_W} \tag{12-106}$$

Now that the relationships are defined, we can perform the incremental calculation using a spreadsheet. This is shown in Table 12-31. We begin with column 2, where we set the increments of dry-basis moisture X from the inlet to the exit of the dryer (which are known from the problem statement). The number of increments used is arbitrary; a higher number will give a more precise solution. Column 3 is the average of X in each increment of column 1. Column 4 is the gas-phase composition Y, as calculated from Eq. (12-98).

Column 5 is the gas temperature, which is obtained from the Mollier diagram in Fig. 12-68. Mollier diagrams are explained in the Psychrometry subsection. The gas temperature cools due to the evaporation of water; the line on the diagram is an adiabatic saturation line. The first point to mark on the diagram, indicated by a star, is the inlet air condition (absolute humidity = 0.1 g/g and 800°C). The wet-bulb temperature of this air is 78°C, obtained by following the adiabatic saturation line to the saturation curve at the bottom and reading the temperature. If calculations are preferred to using the diagram, Eq. (12-6) may be used for the gas temperature and the procedure described in Table 12-5vi can be used to calculate the wet-bulb temperature.

The values of f in column 6 are calculated from the average solids moisture content, Eq. (12-94); the temperature of the solids in column 7 is calculated from Eq. (12-103); and the drying rate is calculated from Eq. (12-103). The required drying time for the increment of ΔX in column 2 is calculated using the drying rate in column 10 (rightmost column). The drying times for all increments add to 50.82 s.

The value of 50.82 s is the time required for the particles when they are falling through the air. However, most of the residence time of the particles in the dryer is spent slowly rotating on the flights in a dense phase. The next steps in this analysis are to estimate the ratio of the time falling in the airborne phase to that of the time in dense phase and then the time in the dense phase per unit length of the dryer.

To estimate the time of the particles in the dense phase, we can use Eq. (12-87).

$$\frac{\tau_S}{\tau_G} = \frac{t_d}{t_f} = \frac{K_{fl}}{N}\sqrt{\frac{g}{D}} = \frac{0.4}{0.04\ s^{-1}}\sqrt{\frac{9.81\ m/s^2}{1.5\ m}} = 25.6 \tag{12-107}$$

$$\tau_S = 25.6 \cdot 50.8 = 1300\ s$$

The total required residence time is therefore $\tau_S + \tau_G = \tau = 1350.8$ s.

TABLE 12-31 Variation in Process Conditions for the Example of a Cocurrent Cascading Rotary Dryer

Interval	X_p, kg/kg	$\bar{X}$, kg/kg	$\bar{Y}$, kg/kg	$T_{\bar{G}}$, °C	T_{wb}, °C	$f = \Phi$	$T_{\bar{S}}$, °C	$d\bar{X}/dt$, kg/(kg·s)	Δt_p, s
1	0.500	0.478	0.122	720	79.0	0.956	107	0.04656	0.94
2	0.456	0.434	0.166	630	78.5	0.869	151	0.03683	1.19
3	0.412	0.391	0.209	530	78.5	0.781	177	0.02820	1.55
4	0.369	0.347	0.253	430	78.0	0.694	186	0.02067	2.12
5	0.325	0.303	0.297	340	78.0	0.606	181	0.01424	3.07
6	0.281	0.259	0.341	250	78.0	0.519	161	0.00892	4.91
7	0.238	0.216	0.384	200	78.0	0.431	147	0.00470	9.32
8	0.194	0.172	0.428	130	78.0	0.344	112	0.00158	27.73
out	0.150	**Total required gas-phase residence time (s) = 50.82 s (summation of last column)**							

The length of dryer per second of residence time can be estimated by using Eq. (12-89).

First we need to calculate the particle velocity due to air drag while falling. Since $\text{Re}_p = 1300$, we use Eqs. (12-83) and (12-84).

$$U_{P1}^d = 125 \cdot \frac{\mu \cdot U_{G,\text{super}} t_a^*}{\rho_p d_p^2} = 125 \cdot \frac{1.8 \times 10^{-5} \frac{\text{kg}}{\text{m} \cdot \text{s}} \cdot \frac{4\ \text{m}}{\text{s}} \cdot 0.554\ \text{s}}{\left(600 \frac{\text{kg}}{\text{m}^3}\right)(0.005\ \text{m})^2} = 0.332\ \text{m/s}$$

$$t_a^* = t_f = \left(\frac{2D}{g \cos \alpha}\right)^{1/2} K_{\text{fall}} = \left(\frac{2 \cdot 1.5}{9.81 \cos 4°}\right)^{1/2} (1.0) = 0.554\ \text{s}$$

We can apply Eq. (12-88), using $K_K/(K_{fl}\sqrt{2}) \approx 1$, $K_{fl} \approx 0.4$, $K_{\text{fall}} \approx 1$, and take a to equal 2.5 (within the range of 1 to 4).

$$\frac{\tau}{L} = \frac{1.1}{(0.04\ \text{s}^{-1})(1.5\ \text{m})} \times \left[\tan 4° \cdot (1 + 2.5) + \frac{1}{0.4} \sqrt{\frac{1}{9.81 \frac{\text{m}}{\text{s}^2} \cdot 1.5\ \text{m}}} \cdot 0.332\ \text{m/s}\right] = 30\ \text{s/m}$$

Since = 1300 s, L = 1350.2/30 = 45 m. The dryer length/diameter ratio is therefore 45 m/1.5 m = 30, which is significantly larger than the recommended ratio of between 5:1 and 10:1. The remedy would then be to use a larger dryer diameter and repeat these calculations. The larger dryer diameter would decrease the gas velocity, slowing the particle velocity along the drum, increasing the residence time per unit length, and hence decreasing the required drum length, to give a more normal length/diameter ratio.

Performance and Cost Data for Direct Heat Rotary Dryers Table 12-32 gives estimating-price data for direct rotary dryers employing steam-heated air. Higher-temperature operations requiring combustion chambers and fuel burners will cost more. The total installed cost of rotary dryers including instrumentation, auxiliaries, allocated building space, etc., will run from 150 to 300 percent of the purchase cost. Simple erection costs average 10 to 20 percent of the purchase cost.

Operating costs will include 5 to 10 percent of one worker's time, plus power and fuel required. Yearly maintenance costs will range from 5 to 10 percent of total installed costs. Total power for fans, dryer drive, and feed and product conveyors will be in the range of $0.5D^2$ to $1.0D^2$. Thermal efficiency of a high-temperature direct heat rotary dryer will range from 55 to 75 percent and, with steam-heated air, from 30 to 55 percent.

Table 12-32 gives some performance data for some cocurrent rotary dryers. A representative list of materials dried in direct heat rotary dryers is given in Table 12-33.

Indirect Heat Rotary Steam-Tube Dryers Probably the most common type of indirect heat rotary dryer is the steam-tube dryer (Fig. 12-69). Steam-heated tubes running the full length of the cylinder are fastened symmetrically in one, two, or three concentric rows inside the cylinder and rotate with it. Tubes may be simple pipe with condensate draining by gravity into the discharge manifold or bayonet type. Bayonet-type tubes are also employed when units are used as water-tube coolers. When one is handling sticky materials, one row of tubes is preferred. These are occasionally shielded at the feed end of the dryer to prevent buildup of solids behind them. Lifting flights are usually inserted behind the tubes to promote solids agitation.

Wet feed enters the dryer through a chute or screw feeder. The product discharges through peripheral openings in the shell in ordinary dryers. These openings also serve to admit purge air to sweep moisture or other evolved gases from the shell. In practically all cases, gas flow is countercurrent to solids flow. To retain a deep bed of material within the dryer, normally a 10 to 20 percent fill level, the discharge openings are supplied with removable chutes extending radially into the dryer. These, on removal, permit complete emptying of the dryer.

Steam is admitted to the tubes through a revolving steam joint into the steam side of the manifold. Condensate is removed continuously, by gravity through the steam joint to a condensate receiver and by means of lifters in the condensate side of the manifold. By employing simple tubes, noncondensibles are continuously vented at the other ends of the tubes through

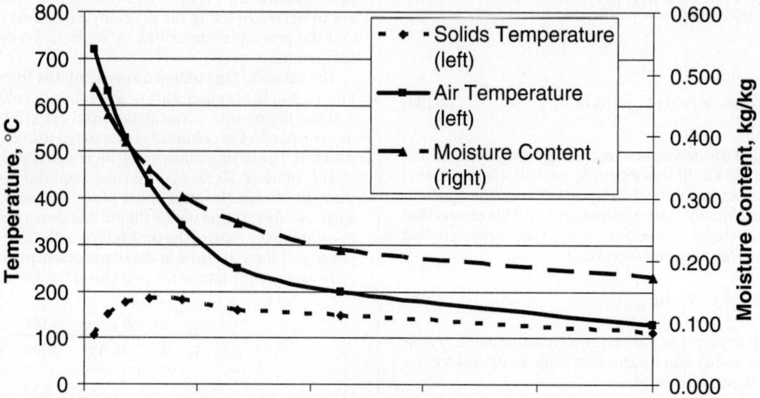

FIG. 12-68 Typical variation of process conditions through a cocurrent cascading rotary dryer.

TABLE 12-32 Warm-Air Direct-Heat Cocurrent Rotary Dryers: Typical Performance Data*

Dryer size, m × m	1.219 × 7.62	1.372 × 7.621	1.524 × 9.144	1.839 × 10.668	2.134 × 12.192	2.438 × 13.716	3.048 × 16.767
Evaporation, kg/h	136.1	181.4	226.8	317.5	408.2	544.3	861.8
Work, 10^8 J/h	3.61	4.60	5.70	8.23	1.12	1.46	2.28
Steam, kg/h at kg/m^2 gauge	317.5	408.2	521.6	725.7	997.9	131.5	2041
Discharge, kg/h	408	522	685	953	1270	1633	2586
Exhaust velocity, m/min	70	70	70	70	70	70	70
Exhaust volume, m^3/min	63.7	80.7	100.5	144.4	196.8	257.7	399.3
Exhaust fan, kW	3.7	3.7	5.6	7.5	11.2	18.6	22.4
Dryer drive, kW	2.2	5.6	5.6	7.5	14.9	18.6	37.3
Shipping weight, kg	7700	10,900	14,500	19,100	35,800	39,900	59,900
Price, FOB Chicago	$158,000	$168,466	$173,066	$204,400	$241,066	$298,933	$393,333

*Courtesy of Swenson Process Equipment Inc.
NOTE:
Material: heat-sensitive solid
Maximum solids temperature: 65°C
Feed conditions: 25 percent moisture, 27°C
Product conditions: 0.5 percent moisture, 65°C
Inlet-air temperature: 165°C
Exit-air temperature: 71°C
Assumed pressure drop in system: 200 mm
System includes finned air heaters, transition piece, dryer, drive, product collector, duct, and fan.
Prices are for carbon steel construction and include entire dryer system (November, 1994).
For 304 stainless-steel fabrication, multiply the prices given by 1.5.

Sarco-type vent valves mounted on an auxiliary manifold ring, also revolving with the cylinder.

Vapors (from drying) are removed at the feed end of the dryer to the atmosphere through a natural-draft stack and settling chamber or wet scrubber. When employed in simple drying operations with 3.5×10^5 to 10×10^5 Pa steam, draft is controlled by a damper to admit only sufficient outside air to sweep moisture from the cylinder, discharging the air at 340 to 365 K and 80 to 90 percent saturation. In this way, shell gas velocities and dusting are minimized. When used for solvent recovery or other processes requiring a sealed system, sweep gas is recirculated through a scrubber-gas cooler and blower.

Steam-tube dryers are used for the continuous drying, heating, or cooling of granular or powdery solids which cannot be exposed to ordinary atmospheric or combustion gases. They are especially suitable for fine, dusty particles because of the low gas velocities required for purging of the cylinder. Tube sticking is avoided or reduced by employing recycle, shell knockers, etc., as previously described; tube scaling by sticky solids is one of the major hazards to efficient operation. The dryers are suitable for drying, solvent recovery, and chemical reactions. Steam-tube units have found effective employment in soda ash production, replacing more expensive indirect heat rotary calciners.

TABLE 12-33 Representative Materials Dried in Direct-Heat Rotary Dryers*

Material dried	Moisture content, % (wet basis) Initial	Final	Heat efficiency %
High-temperature:			
Sand	10	0.5	61
Stone	6	0.5	65
Fluorspar	6	0.5	59
Sodium chloride (vacuum salt)	3	0.04	70–80
Sodium sulfate	6	0.1	60
Ilmenite ore	6	0.2	60–65
Medium-temperature:			
Copperas	7	1 (moles)	55
Ammonium sulfate	3	0.10	50–60
Cellulose acetate	60	0.5	51
Sodium chloride (grainer salt)	25	0.06	35
Cast-iron borings	6	0.5	50–60
Styrene	5	0.1	45
Low-temperature:			
Oxalic acid	5	0.2	29
Vinyl resins	30	1	50–55
Ammonium nitrate prills	4	0.25	30–35
Urea prills	2	0.2	20–30
Urea crystals	3	0.1	50–55

*Taken from *Chem. Eng.*, June 19, 1967, p. 190, Table III.

Design methods for indirect heat rotary steam-tube dryers Heat-transfer coefficients in steam-tube dryers range from 30 to 85 W/(m^2·K). Coefficients will increase with increasing steam temperature because of increased heat transfer by radiation. In units carrying saturated steam at 420 to 450 K, the heat flux will range from 6300 W/m^2 for difficult-to-dry and organic solids to 1890 to 3790 W/m^2 for finely divided inorganic materials. The effect of steam pressure on heat-transfer rates up to 8.6×10^5 Pa is illustrated in Fig. 12-70.

Performance and cost data for indirect heat rotary steam-tube dryers Table 12-34 contains data for a number of standard sizes of steam-tube dryers. Prices tabulated are for ordinary carbon-steel construction. Installed costs will run from 150 to 300 percent of the purchase cost.

The thermal efficiency of steam-tube units will range from 70 to 90 percent, if a well-insulated cylinder is assumed. This does not allow for boiler efficiency, however, and is therefore not directly comparable with direct heat units such as the direct heat rotary dryer or indirect heat calciner.

Operating costs for these dryers include 5 to 10 percent of one person's time. Maintenance will average 5 to 10 percent of the total installed cost per year. Table 12-35 outlines typical performance data from three drying applications in steam-tube dryers.

Indirect Rotary Calciners and Kilns These large-scale rotary processors are used for very high temperature operations. Operation is similar to that of rotary dryers. For additional information, refer to Perry's 7th Edition, pages 12-56 to 12-58.

Indirect Heat Calciners Indirect heat rotary calciners, either batch or continuous, are employed for heat treating and drying at higher temperatures than can be obtained in steam-heated rotating equipment. They generally require a minimum flow of gas to purge the cylinder, to reduce dusting, and are suitable for gas-sealed operation with oxidizing, inert, or reducing atmospheres. Indirect calciners are widely utilized, and some examples of specific applications are as follows:

1. Activating charcoal
2. Reducing mineral high oxides to low oxides
3. Drying and devolatilizing contaminated soils and sludges
4. Calcination of alumina oxide–based catalysts
5. Drying and removal of sulfur from cobalt, copper, and nickel
6. Reduction of metal oxides in a hydrogen atmosphere
7. Oxidizing and "burning off" of organic impurities
8. Calcination of ferrites

This unit consists essentially of a cylindrical retort, rotating within a stationary insulation-lined furnace. The latter is arranged so that fuel combustion occurs within the annular ring between the retort and the furnace.

To prevent sliding of solids over the smooth interior of the shell, agitating flights running longitudinally along the inside wall are frequently provided. These normally do not shower the solids as in a direct heat vessel, but merely prevent sliding so that the bed will turn over and constantly expose new surface for heat and mass transfer. To prevent scaling of the shell interior by sticky solids, cylinder scraper and knocker arrangements are occasionally employed. For example, a scraper chain is fairly common practice in soda ash calciners, while knockers are frequently utilized on metallic oxide calciners.

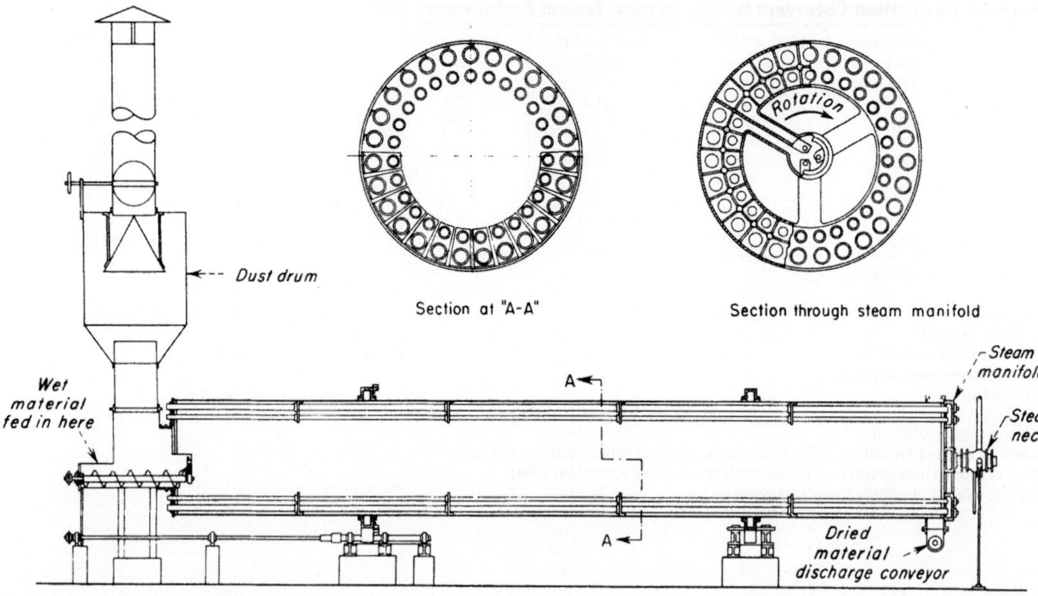

FIG. 12-69 Steam-tube rotary dryer.

In general, the temperature range of operation for indirect heat calciners can vary over a wide range, from 475 K at the low end to approximately 1475 K at the high end. All types of carbon steel, stainless, and alloy construction are used, depending upon the temperature, process, and corrosion requirements.

Design methods for calciners In indirect heat calciners, heat transfer is primarily by radiation from the cylinder wall to the solids bed. The thermal efficiency ranges from 30 to 65 percent. By utilization of the furnace exhaust gases for preheated combustion air, steam production, or heat for other process steps, the thermal efficiency can be increased considerably. The limiting factors in heat transmission lie in the conductivity and radiation constants of the shell metal and solids bed. If the characteristics of these are known, equipment may be accurately sized by employing the Stefan-Boltzmann radiation equation. Apparent heat-transfer coefficients will range from 17 $W/(m^2 \cdot K)$ in low-temperature operations to 85 $W/(m^2 \cdot K)$ in high-temperature processes.

Cost data for calciners Power, operating, and maintenance costs are similar to those previously outlined for direct and indirect heat rotary dryers. Estimating purchase costs for preassembled and frame-mounted rotary calciners with carbon-steel and type 316 stainless-steel cylinders are given in Table 12-36 together with size, weight, and motor requirements. Sale price includes the cylinder, ordinary angle seals, furnace, drive, feed conveyor, burners, and controls. Installed cost may be estimated, not including

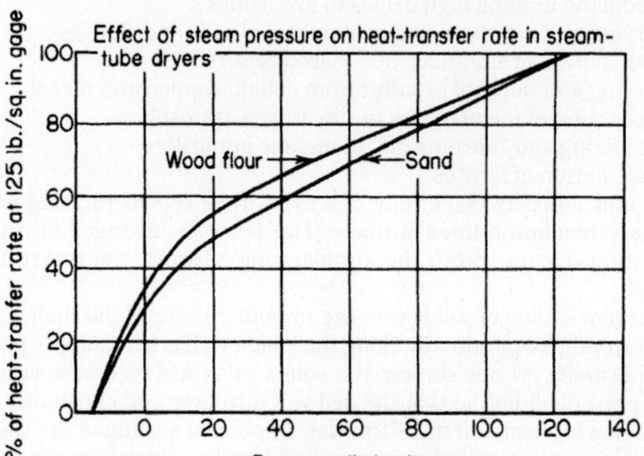

FIG. 12-70 Effect of steam pressure on the heat-transfer rate in steam-tube dryers.

building or foundation costs, at up to 50 percent of the purchase cost. A layout of a typical continuous calciner with an extended cooler section is illustrated in Fig. 12-71.

Direct Heat Roto-Louvre Dryer One of the more important special types of rotating equipment is the Roto-Louvre dryer. As illustrated in Fig. 12-72, hot air (or cooling air) is blown through louvres in a double-wall rotating cylinder and up through the bed of solids. The latter moves continuously through the cylinder as it rotates. Constant turnover of the bed ensures uniform gas contacting for heat and mass transfer. The annular gas passage behind the louvres is partitioned so that contacting air enters the cylinder only beneath the solids bed. The number of louvres covered at any one time is roughly 30 percent. Because air circulates through the bed, fillage levels of 13 to 15 percent or greater are employed.

Roto-Louvre dryers range in size from 0.8 to 3.6 m in diameter and from 2.5 to 11 m long. The largest unit is reported capable of evaporating 5500 kg/h of water. Hot gases from 400 to 865 K may be employed. Because gas flow is through the bed of solids, high pressure drop, from 7 to 50 cm of water, may be encountered within the shell. For this reason, both a pressure inlet fan and an exhaust fan are provided in most applications to maintain the static pressure within the equipment as close as possible to atmospheric.

Roto-Louvre dryers are suitable for processing coarse granular solids which do not offer high resistance to airflow, do not require intimate gas contacting, and do not contain significant quantities of dust.

Heat transfer and mass transfer from the gas to the surface of the solids are extremely efficient; hence the equipment size required for a given duty is frequently less than that required when an ordinary direct heat rotary vessel with lifting flights is used. Purchase price savings are partially balanced, however, by the more complex construction of the Roto-Louvre unit. A Roto-Louvre dryer will have a capacity roughly 1.5 times that of a single-shell rotary dryer of the same size under equivalent operating conditions. Because of the cross-flow method of heat exchange, the average *temperature* is not a simple function of inlet and outlet temperatures. Three applications of Roto-Louvre dryers are outlined in Table 12-37. Installation, operating, power, and maintenance costs will be similar to those experienced with ordinary direct heat rotary dryers. *Thermal efficiency* will range from 30 to 70 percent.

Additional Readings

Aiken and Polsak, "A Model for Rotary Dryer Computation," in Mujumdar, ed., *Drying '82,* Hemisphere, New York, 1982, pp. 32–35.

Baker, "Cascading Rotary Dryers," chap. 1 in Mujumdar, ed., *Advances in Drying,* vol. 2, Hemisphere, New York, 1983, pp. 1–51.

Friedman and Mahall, "Studies in Rotary Drying. Part 1. Holdup and Dusting. Part 2. Heat and Mass Transfer," *Chem. Eng. Progr.* **45:** 482–493, 573–588 (1949).

Hirosue and Shinohara, "Volumetric Heat Transfer Coefficient and Pressure Drop in Rotary Dryers and Coolers," *1st Int. Symp. on Drying* **8** (1978).

Kamke and Wilson, "Computer Simulation of a Rotary Dryer. Part 1. Retention Time. Part 2. Heat and Mass Transfer," *AIChE J.* **32:** 263–275 (1986).

TABLE 12-34 Standard Steam-Tube Dryers*

Size, diameter × length, m	Tubes		m^2 of free area	Dryer speed, r/min	Motor size, hp	Shipping weight, kg	Estimated price
	No. OD (mm)	No. OD (mm)					
0.965 × 4.572	14 (114)		21.4	6	2.2	5,500	$152,400
0.965 × 6.096	14 (114)		29.3	6	2.2	5,900	165,100
0.965 × 7.620	14 (114)		36.7	6	3.7	6,500	175,260
0.965 × 9.144	14 (114)		44.6	6	3.7	6,900	184,150
0.965 × 10.668	14 (114)		52.0	6	3.7	7,500	196,850
1.372 × 6.096	18 (114)	18 (63.5)	58.1	4.4	3.7	10,200	203,200
1.372 × 7.620	18 (114)	18 (63.5)	73.4	4.4	3.7	11,100	215,900
1.372 × 9.144	18 (114)	18 (63.5)	88.7	5	5.6	12,100	228,600
1.372 × 10.668	18 (114)	18 (63.5)	104	5	5.6	13,100	243,840
1.372 × 12.192	18 (114)	18 (63.5)	119	5	5.6	14,200	260,350
1.372 × 13.716	18 (114)	18 (63.5)	135	5.5	7.5	15,000	273,050
1.829 × 7.62	27 (114)	27 (76.2)	118	4	5.6	19,300	241,300
1.829 × 9.144	27 (114)	27 (76.2)	143	4	5.6	20,600	254,000
1.829 × 10.668	27 (114)	27 (76.2)	167	4	7.5	22,100	266,700
1.829 × 12.192	27 (114)	27 (76.2)	192	4	7.5	23,800	278,400
1.829 × 13.716	27 (114)	27 (76.2)	217	4	11.2	25,700	292,100
1.829 × 15.240	27 (114)	27 (76.2)	242	4	11.2	27,500	304,800
1.829 × 16.764	27 (114)	27 (76.2)	266	4	14.9	29,300	317,500
1.829 × 18.288	27 (114)	27 (76.2)	291	4	14.9	30,700	330,200
2.438 × 12.192	90 (114)		394	3	11.2	49,900	546,100
2.438 × 15.240	90 (114)		492	3	14.9	56,300	647,700
2.438 × 18.288	90 (114)		590	3	14.9	63,500	736,600
2.438 × 21.336	90 (114)		689	3	22.4	69,900	838,200
2.438 × 24.387	90 (114)		786	3	29.8	75,300	927,100

*Courtesy of Swenson Process Inc. (prices from November, 1994). Carbon steel fabrication; multiply by 1.75 for 304 stainless steel.

TABLE 12-35 Steam-Tube Dryer Performance Data

	Class 1	Class 2	Class 3
Class of materials handled	High-moisture organic, distillers' grains, brewers' grains, citrus pulp	Pigment filter cakes, blanc fixe, barium carbonate, precipitated chalk	Finely divided inorganic solids, water-ground mica, water-ground silica, flotation concentrates
Description of class	Wet feed is granular and damp but not sticky or muddy and dries to granular meal	Wet feed is pasty, muddy, or sloppy; product is mostly hard pellets	Wet feed is crumbly and friable; product is powder with very few lumps
Normal moisture content of wet feed, % dry basis	233	100	54
Normal moisture content of product, % dry basis	11	0.15	0.5
Normal temperature of wet feed, K	310–320	280–290	280–290
Normal temperature of product, K	350–355	380–410	365–375
Evaporation per product, kg	2	1	0.53
Heat load per lb product, kJ	2250	1190	625
Steam pressure normally used, kPa gauge	860	860	860
Heating surface required per kg product, m^2	0.34	0.4	0.072
Steam consumption per kg product, kg	3.33	1.72	0.85

TABLE 12-36 Indirect-Heat Rotary Calciners: Sizes and Purchase Costs*

Diameter, ft	Overall cylinder length	Heated cylinder length	Cylinder drive motor hp	Approximate Shipping weight, lb	Approximate sale price in carbon steel construction†	Approximate sale price in No. 316 stainless construction
4	40 ft	30 ft	7.5	50,000	$275,000	$325,000
5	45 ft	35 ft	10	60,000	375,000	425,000
6	50 ft	40 ft	20	75,000	475,000	550,000
7	60 ft	50 ft	30	90,000	550,000	675,000

*Courtesy of ABB Raymone (Bartlett-SnowTM).
†Prices for November, 1994.

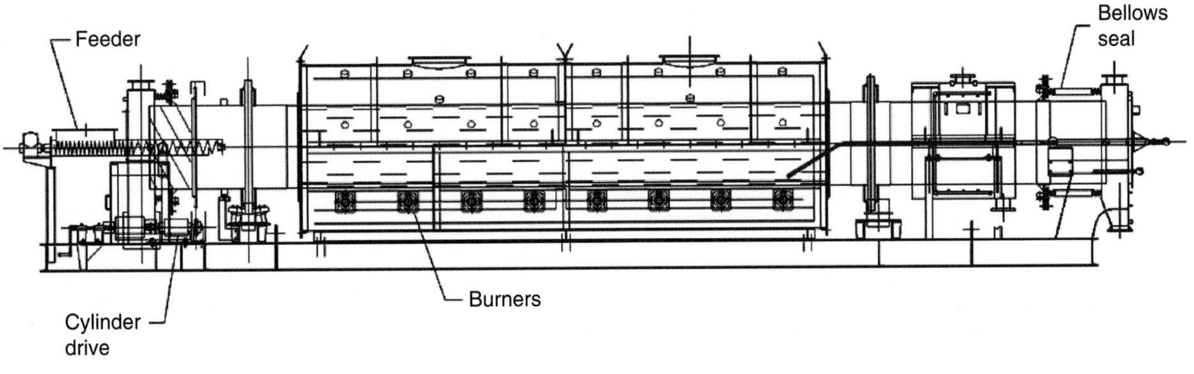

FIG. 12-71 Gas-fired rotary calciner with integral cooler. (*Air Preheater Company, Raymond & Bartlett Snow Products.*)

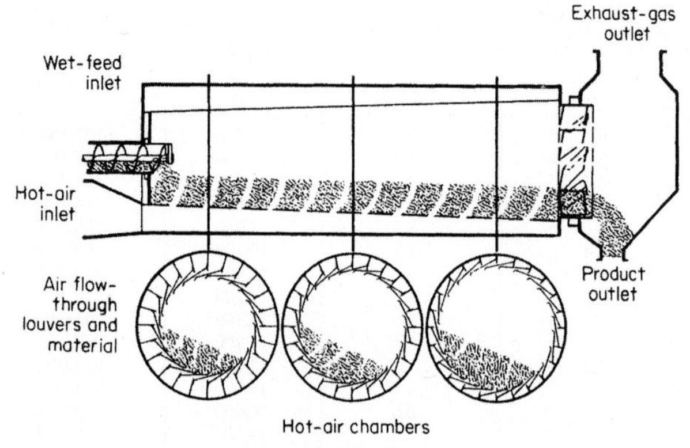

FIG. 12-72 FMC Link-Belt Roto-Louvre dryer.

Kemp, "Comparison of Particle Motion Correlations for Cascading Rotary Dryers," *Drying 2004—Proceedings of the 14th International Drying Symposium* (IDS 2004), São Paulo, Brazil, Aug. 22–25, 2004, vol. B., pp. 790–797.

Kemp and Oakley, "Modeling of Particulate Drying in Theory and Practice," *Drying Technol.* **20**(9): 1699–1750 (2002).

Langrish, "The Mathematical Modeling of Cascading Rotary Dryers," DPhil thesis, University of Oxford, 1988.

Matchett and Baker, "Particle Residence Times in Cascading Rotary Dryers. Part 1—Derivation of the Two-Stream Model," *J. Separ. Proc. Technol.* **8:** 11–17 (1987).

Matchett and Baker, "Particle Residence Times in Cascading Rotary Dryers. Part 2—Application of the Two-Stream Model to Experimental and Industrial Data," *J. Separ. Proc. Technol.* **9:** 5 (1988).

McCormick, "Gas Velocity Effects on Heat Transfer in Direct Heat Rotary Dryers," *Chem. Eng. Progr.* **58:** 57–61 (1962).

Miller, Smith, and Schuette, "Factors Influencing the Operation of Rotary Dryers. Part 2. The Rotary Dryer as a Heat Exchanger," *Trans. AIChE* **38:** 841–864 (1942).

Myklestad, "Heat and Mass Transfer in Rotary Dryers," *Chem. Eng. Progr. Symp. Series* **59:** 129–137 (1963).

Papadakis et al., "Scale-up of Rotary Dryers," *Drying Technol.* **12**(1&2): 259–278 (1994).

Ranz and Marshall, "Evaporation from Drops, Part 1," *Chem. Eng. Progr.* **48:** 123–142, 251–257 (1952).

Saeman and Mitchell, "Analysis of Rotary Dryer Performance," *Chem. Eng. Progr.* **50**(9): 467–475 (1954).

Schofield and Glikin, "Rotary Dryers and Coolers for Granular Fertilisers," *Trans. IChemE* **40:** 183–190 (1962).

Sullivan, Maier, and Ralston, "Passage of Solid Particles through Rotary Cylindrical Kilns," U.S. Bureau of Mines Tech. Paper, 384, 44 (1927).

Fluidized-Bed and Spouted-Bed Dryers

Examples and synonyms Fluid beds, fluidized beds, spouted beds, vibrating fluidized beds, vibro-fluidized bed.

Description A fluidized bed is a deep layer of particles supported by both a distributor plate (containing numerous small holes) and the fluidizing gas. The bed has many properties of a liquid; the particles seek their own level, assume the shape of the vessel they are in, and exhibit buoyancy effects.

The basic principles of fluidized-bed technology are thoroughly described in Sec. 17, Gas-Solid Operations and Equipment. The technology has several advantages. These include no moving parts, rapid heat and mass transfer between gas and particles, rapid heat transfer between the gas/particle bed and immersed objects, intense mixing, and continuous or batch operation. These advantages allow fluidized beds to be used as both dryers and coolers.

As described in Sec. 17, the process parameter of the highest importance is the fluidizing gas velocity in the fluidized bed, also referred to as the superficial gas velocity. This velocity is of nominal character since the flow field will be disturbed and distorted by the presence of the solid phase and the turbulent fluctuations created by the gas/solid interaction.

Proper design and operation of a fluidized-bed dryer requires consideration of fluidization and drying characteristics of a material, the fluidization velocity, the particle size distribution, the design of the gas distributor plate, the operating conditions, and the mode of operation.

Fluidization characteristics have been investigated by Geldart, resulting in the well-known Geldart diagram (Fig. 12-73). The Geldart diagram shows that particulate material can be handled successfully in a fluidized bed only if it is not too fine or too coarse with a mean particle size between 20 μm and 10 mm. Fluidized beds are best suited for flowable particles that are regular in shape and not too sticky. Needle- or leaflike shaped particles should be considered as nonfluidizable.

The total drying time needed to reach the final moisture and the heat sensitivity of the material is an important parameter for design of an industrial plant. Small batch fluidized-bed tests can measure a drying curve as shown in Fig. 12-74. Figure 12-74*a* shows two drying curves for the same material. The curves differ based on the bed loading. The drying curves clearly show that the moisture is rapidly evaporating while the material is maintained at a low temperature. This particular material does not have a constant-rate period, evidenced by the decline in drying rate and rise of temperature with time. This is indicative of the drying of the surface of the particles; as they dry, the driving force for evaporation decreases. A moisture sorption isotherm, described in the Solids Drying Fundamentals subsection, is how this decrease can be quantified. The falling rate, by itself, does not mean that the rate is limited by internal mass transfer within each particle.

Additional drying curves can be measured to determine whether the drying rate of a material is internally or externally limited. Internally limited materials are slow to dry with moisture that is tightly bound and unable to move to the evaporating surface of the individual particles quickly enough; changes in superficial velocity and bed depth do not influence drying. Externally limited materials are influenced by the external drying conditions at which they are dried including both drying temperature and superficial gas velocity.

The drying curve data presented in Fig. 12-74*a* were normalized as shown in Fig. 12-74*b*. This suggests externally limited drying behavior. In this particular material, the rate is falling due to the dryness of the particles but not due to slow moisture transport through each particle. Rate-limiting steps in drying processes are described further in the Solids-Drying Fundamentals subsection; specifically, see Fig. 12-25.

See Table 12-38 to learn how to increase the throughput by using the air velocity or bed depth, depending on whether the drying rate is controlled externally or internally.

For production, increasing gas velocity is beneficial for externally limited materials, giving reduced drying time and either a higher throughput or a smaller bed area, but gives no real benefit for internally limited materials; likewise, increasing bed depth is beneficial for internally limited materials, giving either a higher throughput or enabling use of a smaller bed area with the same drying time but not for externally limited materials. However, using unnecessarily high gas velocity or an unnecessarily deep bed can increase the pressure drop, operating costs, elutriation, attrition, and backmixing.

TABLE 12-37 Manufacturer's Performance Data for FMC Link-Belt Roto-Louvre Dryer*

Material dried	Ammonium sulfate	Foundry sand	Metallurgical coke
Dryer diameter	2 ft 7 in	6 ft 4 in	10 ft 3 in
Dryer length	10 ft	24 ft	30 ft
Moisture in feed, % wet basis	2.0	6.0	18.0
Moisture in product, % wet basis	0.1	0.5	0.5
Production rate, lb/h	2500	32,000	38,000
Evaporation rate, lb/h	50	2130	8110
Type of fuel	Steam	Gas	Oil
Fuel consumption	255 lb/h	4630 ft^3/h	115 gal/h
Calorific value of fuel	837 Btu/lb	1000 Btu/ft^3	150,000 Btu/gal
Efficiency, Btu, supplied per lb evaporation	4370	2170	2135
Total power required, hp	4	41	78

*Courtesy of Material Handling Systems Division, FMC Corp. To convert British thermal units to kilojoules, multiply by 1.06; to convert horsepower to kilowatts, multiply by 0.746.

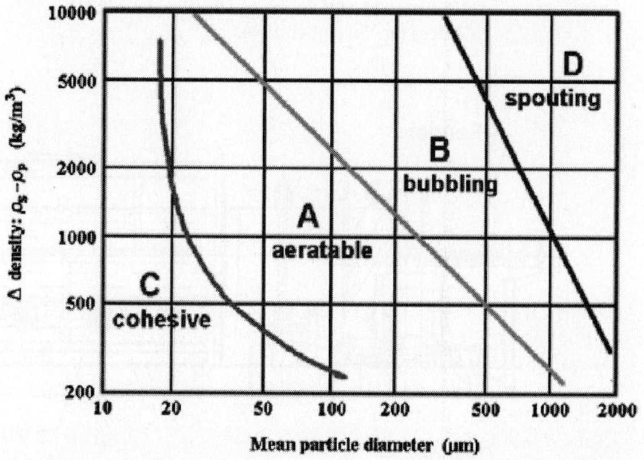

FIG. 12-73 Geldart diagram.

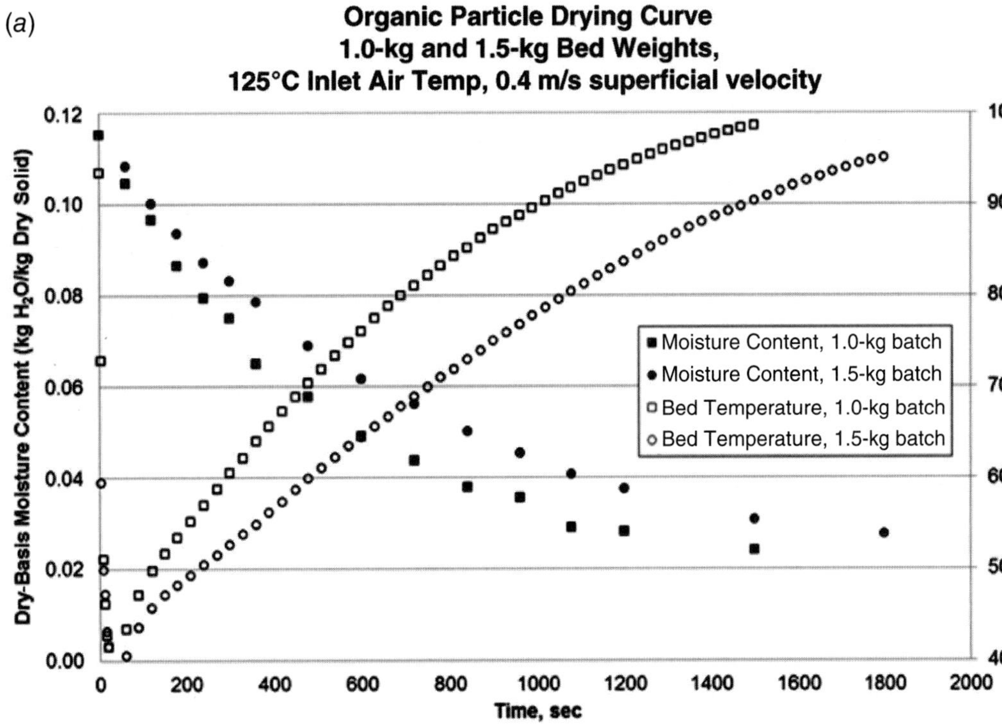

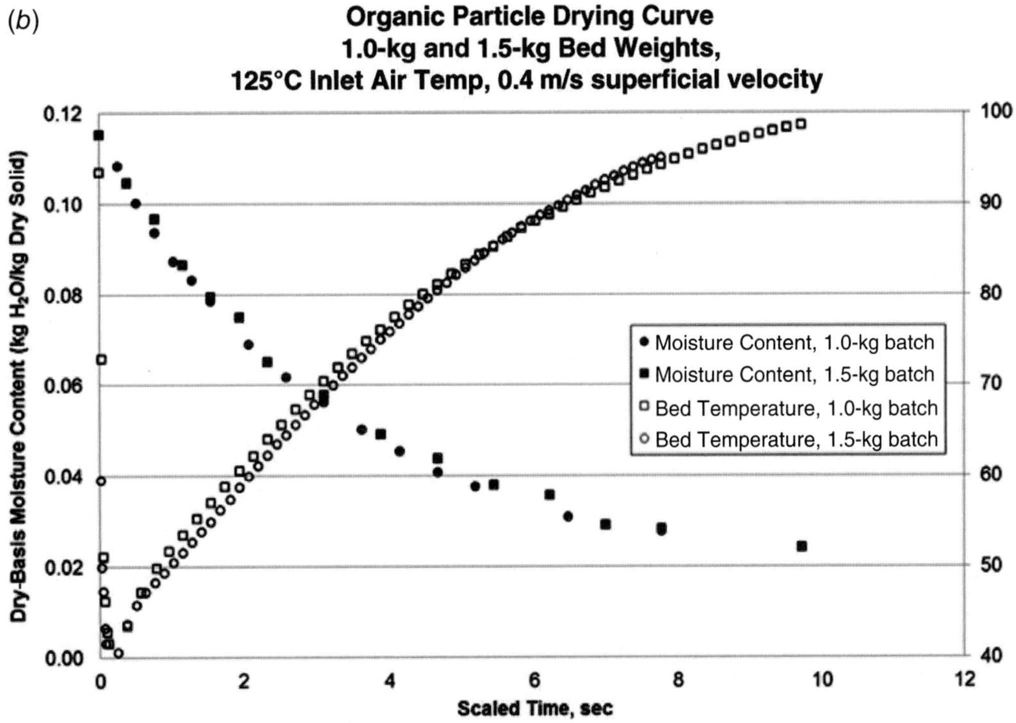

FIG. 12-74 Drying curve of organic material.

The fluidization velocity is of major importance, as indicated in the introduction. Each material will have individual requirements for the gas velocity and pressure drop to provide good fluidization. An investigation of the relationship between fluidization velocity and bed pressure drop for a given material is called a *fluidization curve*. An example is shown in Fig. 12-75. The results are illustrative and intended to give a clear picture of the relationship. The minimum fluidization velocity can be estimated from the Wen and Yu correlation [*AIChE J.* **12**(3): 610–612 (1966)] given in Sec. 19.

At a superficial velocity below the value required for minimum fluidization, the pressure drop over the bed will increase proportionally with the velocity. Above a critical velocity, the pressure drop corresponds to the weight of the fluidized mass of material. This is referred to as the *minimum fluidization*; the bed is said to be in an incipiently fluidized state. A further increase in the superficial velocity will result in little or no increase in the pressure drop. The particle layer now behaves as a liquid, and the bed volume expands considerably. At even higher gas velocities the motion will be

TABLE 12-38 Comparison of Internally and Externally Limited Drying

	Internally limited drying	
Batch	Air velocity	2X air velocity ≈ no change in drying time
	Bed depth	2X bed depth ≈ no change in drying time
Continuous	Air velocity	2X air velocity ≈ no change in production rate
	Bed depth	2X bed depth ≈ production rate increases
	Externally limited drying	
Batch	Air velocity	2X air velocity ≈ 1/2 drying time
	Bed depth	2X bed depth ≈ 2X drying time
Continuous	Air velocity	2X air velocity ≈ 2X production rate
	Bed depth	2X bed depth ≈ no change in production rate

stronger, and the excess gas flow will tend to appear as bubbles. In this state the particle layer will undergo vigorous mixing, while still appearing as a dense layer of fluidlike material or a boiling liquid. A further increase in the superficial velocity will result in the solid phase being entrained by the gas flow and will appear as lean phase pneumatic transport. Accordingly, the pressure drop falls to zero. Figure 17-4 provides examples of these fluidization regimes.

However, as a general recommendation, a value between the critical value and the value where the pressure drop falls off will be right. A first choice could be a factor of 2 to 5 times the minimum fluidization velocity. The fluidizing velocity value that will serve a drying task best cannot be derived exactly from the diagram and must be verified through experimentation.

There are additional consequences to using a high air velocity in fluidized beds such as particle elutriation and attrition. Given a particle size distribution in the bed, the finer particles will have a lower terminal velocity and will be selectively carried out of the bed, i.e., elutriated. See Geldart et al., "Entrainment of FCC from Fluidized Beds—a New Correlation for Elutriation Rate Constant $K^*_{i\infty}$," *Powder Technol.* **95:** 240–247 (1998), for more information on this transformation.

The design of the gas distributor plate is important for several reasons. First, the plate serves as a manifold, distributing fluidization and drying gas evenly and preventing dead spots in the bed. This requires an even pattern of orifices in the plate and a sufficient pressure drop over the plate. As a general rule, the pressure drop across the gas distributor should equal at least one-half of the pressure drop across the powder bed with the following range (of pressure drops across the plate) of 500 to 2500 Pa. The estimation of the pressure drop in design situations may be difficult except for the case of the traditional perforated sheet with cylindrical holes perpendicular to the plane of the plate, as shown in Fig. 12-76. For this type of plate, see McAllister et al., "Perforated-Plate Performance," *Chem. Eng. Sci.* **9:** 25–35 (1958). A calculation using this formula will show that a plate giving a required pressure drop of 1500 Pa and a typical fluidizing velocity of 0.35 m/s will need an open area of roughly 1 percent. Provided by a plate of 1-mm thickness and 1-mm-diameter holes, this requires approximately 12,732 holes per square meter.

However, this type of plate is being replaced in most fluid-bed applications because of its inherent disadvantages, which are caused by the manufacture of the plate, i.e., punching holes of a smaller diameter than the

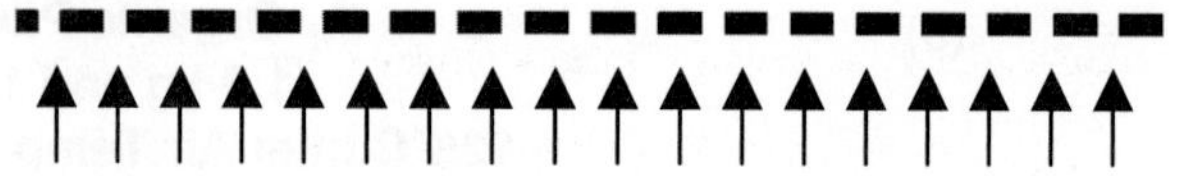

FIG. 12-76 Traditional perforated plate for fluid-bed application.

FIG. 12-77 Conidur plate for fluid-bed application. (*Hein, Lehmann Trenn- und Fördertechnik GmbH.*)

FIG. 12-78 Gill Plate for fluid-bed application. (*GEA*)

thickness of the plate itself. The result is that the plates are weak and are prone to sifting finer particles. The perpendicular flow pattern also means that the plate does not provide a transport capacity for lumps of powder along the plane of the plate.

This transport capacity is provided by so-called gill-type plates of which there are two distinct categories. One category is the type where plates are punched in a very fine regular pattern, not only to provide holes or orifices but also to deform the plate so that each orifice acquires a shape suited for acceleration of the gas flow in magnitude and direction. An example of this type is shown in Fig. 12-77, representing the Conidur type of plate.

The particular feature of Conidur sheets is the specific hole shape which creates a directional airflow to help in discharging the product and influences the retention time in the fluid bed. The special method of manufacturing Conidur sheets enables finishing of fine perforations in sheets with an initial thickness many times over the hole width. Perforations of only 100 μm in an initial sheet thickness of 0.7 mm are possible. With holes this small 1 m^2 of plate may comprise several hundred thousand individual orifices.

The capacity of contributing to the transport of powder in the plane of the plate due to the horizontal component of the gas velocity is also the present for the second category of plates of the gill type. Figure 12-78 shows an example.

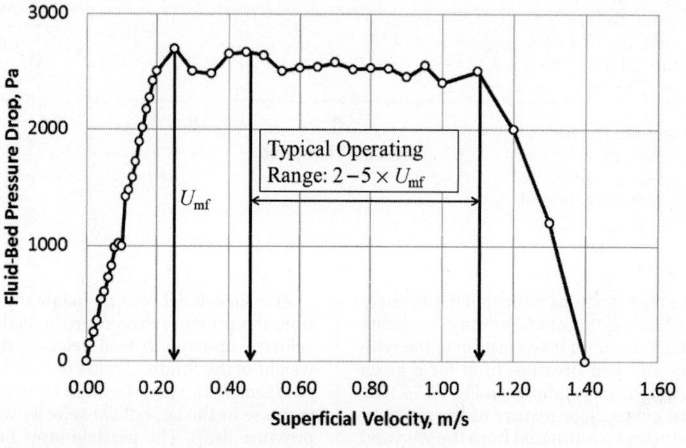

FIG. 12-75 Fluid-bed pressure drop versus fluidizing velocity. (*revised, GEA*)

FIG. 12-79 Non-sifting Gill Plate. (*Patented by GEA*)

In this type of plate, the holes or orifices are large, and the number of gills per square meter is just a few thousand. The gas flow through each of the gills has a strong component parallel to the plate, providing powder transport capacity as well as a cleaning effect. The gills are punched individually or in groups and can be oriented individually to provide a possibility of articulating the horizontal transport effect.

In certain applications in the food and pharmaceutical industries, the nonsifting property of a fluid-bed plate is particularly appreciated. This property of a gill-type plate can be enhanced as illustrated in Fig. 12-79, where the hole after punching is additionally deformed so that the gill overlaps the orifice.

The final type of fluid-bed plate mentioned here is the so bubble plate type. Illustrated in Fig. 12-80, in principle it is a gill-type plate. The orifice is cut out of the plate, and the bubble is subsequently pressed so that the orifice is oriented in a predominantly horizontal direction. A fluid-bed plate will typically have only 1600 holes per m^2. By this technology a combination of three key features is established. The plate is nonsifting, it has directional transport capacity that can be articulated through individual orientation of bubbles, and it is totally free of cracks that may compromise sanitary aspects of the installation.

The operating conditions of a fluid bed are, to a high degree, dictated by the properties of the material to be dried. For most products, the temperature is of primary importance, since the fluidized state results in very high heat-transfer rates so that heat sensitivity may restrict temperature and thereby prolong process time.

To achieve the most favorable combination of conditions to carry out a fluid-bed drying process, it is necessary to consider the different modes of fluid-bed drying available.

Industrial Fluid-Bed Drying The first major distinction between fluid-bed types is the choice of mode: batch or continuous.

Batch fluid beds may appear in several forms. The process chamber has a perforated plate or screen in the bottom and a drying gas outlet at the top, usually fitted with an internal filter. The drying gas enters the fluid bed through a plenum chamber below the perforated plate and leaves through the filter arrangement. The batch of material is enclosed in the process chamber for the duration of the process.

Figure 12-81 shows a sketch of a typical batch fluid-bed dryer as used in the food and pharmaceutical industries. The process chamber is conical in order to create a freeboard velocity in the upper part of the chamber that is lower than the fluidizing velocity just above the plate. The enclosed product batch is prevented from escaping the process chamber and will therefore allow a freer choice of fluidizing velocity than is the case in a continuous fluid bed, as described later.

Continuous fluid beds may be even more varied than batch fluid beds. The main distinction between continuous fluid beds will be according to the solids flow pattern in the dryer. The continuous fluid bed will have an inlet point for moist granular materials to be dried and an outlet for the dried material. If the moist material is immediately fluidizable, it can be introduced directly onto the plate and led through the bed in a plug-flow pattern that will enhance control of product residence time and temperature control. If the moist granular material is too sticky or cohesive due to surface moisture and requires a certain degree of drying before fluidization, it can be handled by a backmix fluid bed, to be described later.

Continuous plug-flow beds are designed to lead the solids flow along a distinct path through the bed. Baffles will be arranged to prevent or limit solids mixing in the horizontal direction. Thereby the residence time distribution of the solids becomes narrow. The bed may be of cylindrical or rectangular shape.

The temperature and moisture contents of the solids will vary along the path of solids through the bed and thereby enable the solids to come close to equilibrium with the drying gas. A typical plug-flow fluid bed is shown in Fig. 12-82.

Continuous plug-flow beds of stationary as well as vibrating type may benefit strongly from use of the gill-type fluid-bed plates with the capacity for controlling the movement of powder along the plate and around bends and corners created by baffles. Proper use of these means may make it possible to optimize the combination of fluidization velocity, bed layer height, and powder residence time.

FIG. 12-80 Bubble Plate. (*Patented by GEA.*)

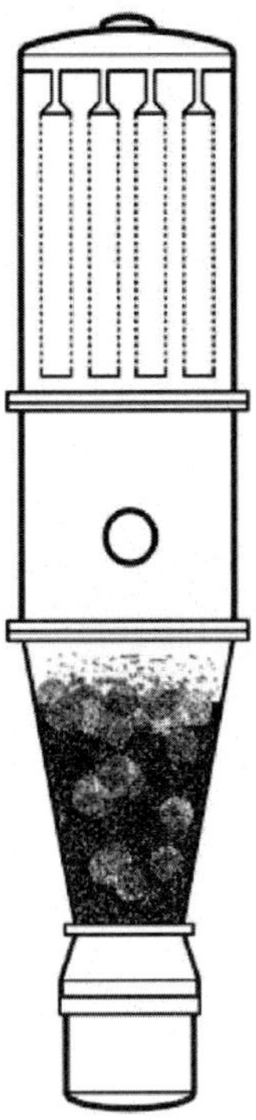

FIG. 12-81 Batch-type fluid bed. (*Aeromatic-Fielder.*)

Continuous backmix beds are used in particular when the moist granular material needs a certain degree of drying before it can fluidize. By distributing the material over the surface of an operating fluid bed arranged for total solids mixing, also called *backmix flow*, it will be absorbed by the dryer material in the bed, and lumping as well as sticking to the chamber surfaces will be avoided. The distribution of the feed can be arranged in different ways, among which a rotary thrower is an obvious choice. A typical backmix fluid bed is shown in Fig. 12-83. Backmix fluid beds can be of box-shaped design or cylindrical.

The whole mass of material in the backmix fluid bed will be totally mixed, and all powder particles in the bed will experience the same air temperature regardless of their position on the drying curve illustrated in Fig. 12-74*a*. The residence time distribution becomes very wide, and part of the material may get a very long residence time while another part may get a very short time.

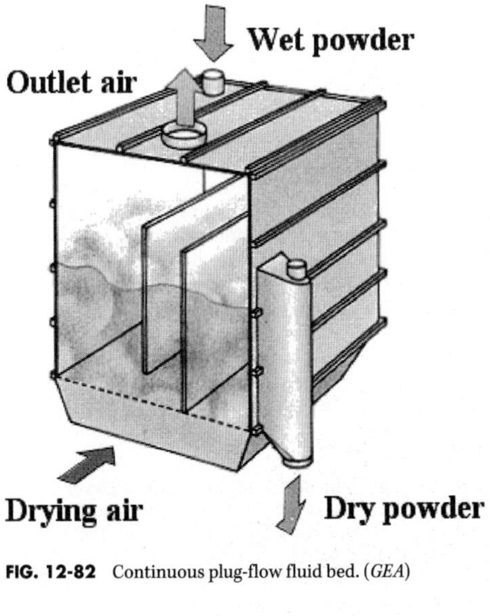

FIG. 12-82 Continuous plug-flow fluid bed. (*GEA*)

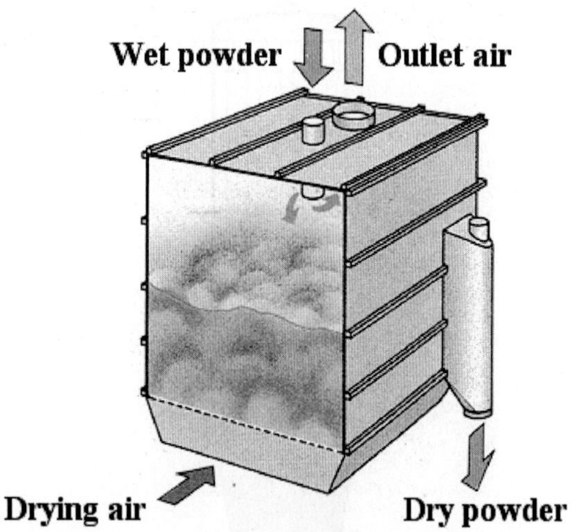

FIG. 12-83 Continuous backmix fluid bed. (*GEA*)

Continuous-contact fluid beds are common in the chemical industry as the solution to the problem arising from materials requiring low fluidizing air temperature due to heat sensitivity and high energy input to complete the drying operation.

The main feature of the contact fluid bed is the presence of heating panels, which are plate or tube structures submerged in the fluidized-bed layer and heated internally by an energy source such as steam, water, or oil. The fluidized state of the bed provides very high heat-transfer rates between the fluidizing gas, the fluidized material, and any objects submerged in the bed. The result is that a very significant portion of the required energy input can be provided by the heating panels without risk of overheating the material. The fluidized state of the bed ensures that the material in the bed will flow with little restriction around the heating panels.

Design and Scale-Up of Fluid Beds When fluid-bed technology can be applied to drying of granular products, significant advantages compared to other drying processes can be observed. Design variables such as fluidizing velocity, critical moisture content for fluidization, and residence time required for drying to the specified residual moisture must, however, be established by experimental or pilot test before design steps can be taken. Reliable and highly integrated fluid-bed systems of either batch or continuous type can be designed, but only by using a combination of such pilot tests and industrial experience.

For scale-up based on an experimentally recorded batch drying curve, including performance mode calculations and altering operating conditions, Kemp and Oakley (2002) showed that the drying time for a given range of moisture content ΔX scales according to the following relationships:

Externally limited (fast drying material):
$$Z = \frac{\Delta\tau_2}{\Delta\tau_1} = \frac{(m_B/A)_2 G_1 (T_{GI} - T_{wb})_1}{(m_B/A)_1 G_2 (T_{GI} - T_{wb})_2} \quad (12\text{-}108)$$

Internally limited (slow drying material):
$$Z = \frac{\Delta\tau_2}{\Delta\tau_1} = \frac{(T_{GI} - T_{wb})_1}{(T_{GI} - T_{wb})_2} \quad (12\text{-}109)$$

Here 1 denotes experimental or original conditions and 2 denotes full-scale or new conditions; Z is the normalization factor; G is gas mass flux; m_B/A is bed mass per unit area, proportional to bed depth z. This method can be used to scale a batch drying curve section by section. Almost always, one of these two simplified limiting cases applies, known as externally limited and internally limited normalization.

For a typical pilot-plant experiment, the fluidization velocity and temperature driving forces are similar to those of the full-size bed, but the bed diameter or depth can be much less. Hence, for externally limited normalization, the m_B/A term dominates; Z can be much greater than 1.

Example 12-18 Scaling a Batch Fluidized-Bed Dryer An experimental batch drying curve has been measured at 100°C, and the drying time was 30 min. The bed diameter was 0.15 m with a bed mass of 1.0 kg. Assume that temperature driving forces are proportional to $T - T_{wb}$, with $T_{wb} = 30$°C and that the air mass flux = 0.55 kg/(m^2·s). Assuming a scaling normalization factor of 2.5, calculate the bed area for a new dryer that can produce 1000 kg. The new bed will operate at 150°C with $T_{wb} = 38$°C and an air mass flux $G_2 = 0.75$ kg/(m^2·s).

$$A = \frac{(1000\text{ kg})(0.55\text{ kg/m}^2\cdot\text{s})(150° - 38°)}{(1\text{ kg}/0.017\text{ m}^2)(0.75\text{ kg/m}^2\cdot\text{s})(100° - 30°)(2.5)} = 9.03\text{ m}^2$$

More complicated mathematical models exist for design and scaling of fluid-bed dryers, namely, those described by Tsotsas et al., "Experimental Investigation and Modelling of Continuous Fluidized Bed Drying under Steady-State and Dynamic Conditions," *Chem. Eng. Sci.* **57:** 5021–5038 (2002).

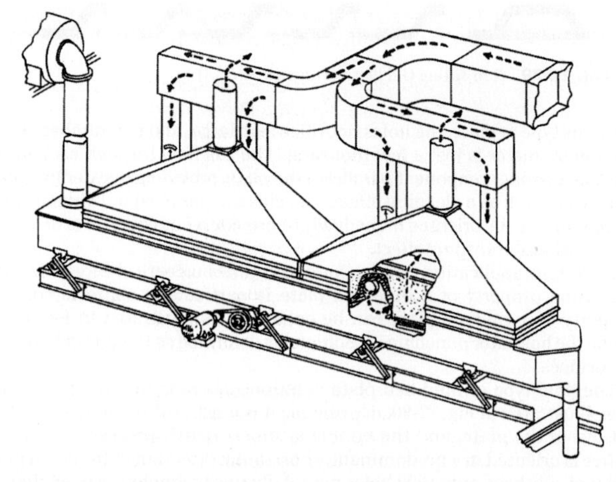

FIG. 12-84 Vibrating conveyor dryer. (*Carrier Vibrating Equipment, Inc.*)

Vibrating Fluidized-Bed Dryers Information on vibrating conveyors and their mechanical construction is given in Sec. 21, Solids Processing and Particle Technology. The vibrating conveyor dryer is a modified form of fluidized-bed equipment, in which fluidization is maintained by a combination of pneumatic and mechanical forces. The heating gas is introduced into a plenum beneath the conveying deck through ducts and flexible hose connections and passes up through a screen, perforated, or slotted conveying deck, through the fluidized bed of solids, and into an exhaust hood (Fig. 12-84). If ambient air is employed for cooling, the sides of the plenum may be open and a simple exhaust system used; however, because the gas distribution plate may be designed for several inches of water pressure drop to ensure a uniform velocity distribution through the bed of solids, a combination pressure-blower exhaust-fan system is desirable to balance the pressure above the deck with the outside atmosphere and prevent gas in-leakage or blowing at the solids feed and exit points.

Units are fabricated in widths from 0.3 to 1.5 m. Lengths are variable from 3 to 50 m; however, most commercial units will not exceed a length of 10 to 16 m per section. Power required for the vibrating drive will be approximately 0.4 kW/m^2 of deck.

Capacity is primarily limited by the air velocity that can be used without excessive dust entrainment. Table 12-39 shows limiting air velocities suitable for various solids particles. Usually, the equipment is satisfactory for particles larger than 150 μm.

When a stationary vessel is employed for fluidization, all solids being treated must be fluidized; nonfluidizable fractions fall to the bottom of the bed and may eventually block the gas distributor. The addition of mechanical vibration to a fluidized system offers the following advantages:

1. Equipment can handle nonfluidizable solids fractions.
2. Prescreening or sizing of the feed is less critical than in a stationary fluidized bed.
3. Air channeling at the incipient fluidization velocity is reduced.

TABLE 12-39 Estimating Maximum Superficial Air Velocities Through Vibrating-Conveyor Screens*

	Velocity, m/s	
Mesh size	2.0 specific gravity	1.0 specific gravity
200	0.22	0.13
100	0.69	0.38
50	1.4	0.89
30	2.6	1.8
20	3.2	2.5
10	6.9	4.6
5	11.4	7.9

*Carrier Vibrating Equipment, Inc.

4. Fluidization may be accomplished with lower pressures and gas velocities.

5. Vibrating conveyor dryers are suitable for free-flowing solids containing mainly surface moisture.

Retention is limited by conveying speeds which range from 0.02 to 0.12 m/s. Bed depth rarely exceeds 7 cm, although units are fabricated to carry 30- to 46-cm-deep beds; these also employ plate and pipe coils suspended in the bed to provide additional heat-transfer area. Vibrating dryers are not suitable for fibrous materials which mat or for sticky solids which may ball or adhere to the deck.

For estimating purposes for direct heat drying applications, it can be assumed that the average exit gas temperature leaving the solids bed will approach the final solids discharge temperature on an ordinary unit carrying a 5- to 15-cm-deep bed. Calculation of the heat load and selection of an inlet air temperature and superficial velocity (Table 12-39) will then permit approximate sizing, provided an approximation of the minimum required retention time can be made.

Vibrating conveyors employing direct contacting of solids with hot, humid air have also been used for the agglomeration of fine powders, chiefly for the preparation of agglomerated water-dispersible food products. Control of inlet air temperature and dew point permits the uniform addition of small quantities of liquids to solids by condensation on the cool incoming-particle surfaces. The wetting section of the conveyor is followed immediately by a warm-air drying section and particle screening.

Spouted Beds The spouted-bed technique was developed primarily for solids too coarse to be handled in fluidized beds, typically classified as type D on the Geldart diagram.

Although their applications overlap, the methods of gas-solids mixing are completely different. A schematic view of a spouted bed is given in Fig. 12-85. Mixing and gas-solids contacting are achieved first in a fluid "spout," flowing upward through the center of a loosely packed bed of solids. Particles are entrained by the fluid and conveyed to the top of the bed. They then flow downward in the surrounding annulus as in an ordinary gravity bed, countercurrently to gas flow. The mechanisms of gas flow and solids flow in spouted beds were first described by Mathur and Gishler [*Am. Inst. Chem. Eng. J.* **1**(2): 157–164 (1955)]. Drying studies have been carried out by Cowan [*Eng. J.* **41:** 5, 60–64 (1958)], and a theoretical equation for predicting the minimum fluid velocity necessary to initiate spouting was developed by Madonna and Lama [*Am. Inst. Chem. Eng. J.* **4**(4): 497 (1958)]. Investigations to determine maximum spoutable depths and to develop theoretical relationships based on vessel geometry and operating variables have been carried out by Lefroy [*Trans. Inst. Chem. Eng.* **47**(5): T120–128 (1969)] and Reddy [*Can. J. Chem. Eng.* **46**(5): 329–334 (1968)]. Information on the scale-up of spouted beds is provided by Passos, Mujumdar, and Massarani [*Drying Technol.* **12**(1–2): 351–391 (1994)].

Gas flow in a spouted bed is partially through the spout and partially through the annulus. About 30 percent of the gas entering the system immediately diffuses into the downward-flowing annulus. Near the top of the bed, the quantity in the annulus approaches 66 percent of the total gas flow; the gas flow through the annulus at any point in the bed equals that which would flow through a loosely packed solids bed under the same conditions of pressure drop. Solids flow in the annulus is both downward and slightly inward. As the fluid spout rises in the bed, it entrains more and more particles, losing velocity and gas into the annulus. The volume of solids displaced by the spout is roughly 6 percent of the total bed.

On the basis of experimental studies, Mathur and Gishler derived an empirical correlation to describe the minimum fluid flow necessary for spouting, in 3- to 12-in-diameter columns:

$$u = \frac{D_p}{D_c}\left(\frac{D_o}{D_c}\right)^{1/3}\left[\frac{2gL(\rho_s - \rho_f)}{\rho_f}\right]^{0.5} \tag{12-110}$$

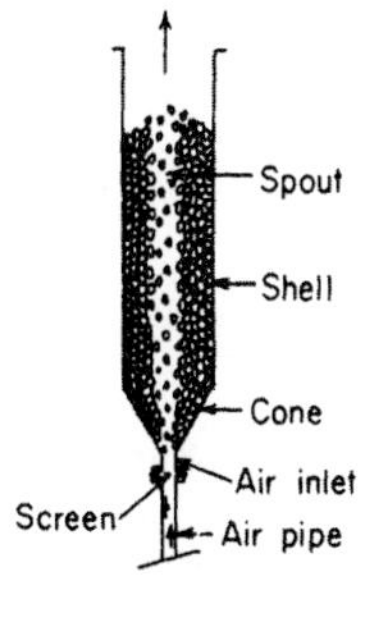

FIG. 12-85 Schematic diagram of spouted bed. Schematic diagram of spouted bed. [Mathur and Gishler, *Am. Inst. Chem. Eng. J.* **1:** 2, 15 (1955).]

where u = superficial fluid velocity through the bed; D_p = particle diameter; D_c = column (or bed) diameter; D_o = fluid inlet orifice diameter; L = bed height; ρ_s = absolute solids density; ρ_f = fluid density; and g = gravity acceleration. The inlet orifice diameter, air rate, bed diameter, and bed depth were all found to be critical and interdependent:

1. In a given-diameter bed, deeper beds can be spouted as the gas inlet orifice size is decreased.
2. Increasing bed diameter increases spoutable depth.
3. As indicated by Eq. (12-110), the superficial fluid velocity required for spouting increases with bed depth and orifice diameter and decreases as the bed diameter is increased.

Additional Reading

Davidson and Harrison, *Fluidized Particles,* Cambridge University Press, Cambridge, UK, 1963.

Geldart, *Powder Technol.* **6:** 201–205 (1972).

Geldart, *Powder Technol.* **7:** 286–292 (1973).

Grace, "Fluidized-Bed Hydrodynamics," chap. 8.1 in *Handbook of Multiphase Systems,* McGraw-Hill, New York, 1982.

Gupta and Mujumdar, "Recent Developments in Fluidized Bed Drying," chap. 5 in Mujumdar, ed., *Advances in Drying,* vol. 2, Hemisphere, Washington, D.C., 1983, p. 155.

Kemp and Oakley, "Modeling of Particulate Drying in Theory and Practice," *Drying Technol.* **20**(9): 1699–1750 (2002).

Kunii and Levenspiel, *Fluidization Engineering,* 2d ed., Butterworth-Heinemann, Stoneham, Mass., 1991.

McAllister et al., "Perforated-Plate Performance," *Chem. Eng. Sci.* **9:** 25–35 (1958).

Poersch, *Aufbereitungs-Technik* **4:** 205–218 (1983).

Richardson, "Incipient Fluidization and Particulate Systems," chap. 2 in Davidson and Harrison, eds., *Fluidization,* Academic Press, London, 1972.

Romankows, "Drying," chap. 12 in Davidson and Harrison, eds., *Fluidization,* Academic Press, London, 1972.

Vanacek, Drbohlar, and Markvard, *Fluidized Bed Drying,* Leonard Hill, London, 1965.

Pneumatic Conveying Dryers

Synonyms and Examples Flash dryer, spin flash dryer, ring dryer.

Description Pneumatic conveyor dryers comprise a long tube or duct carrying a gas at high velocity, a fan to propel the gas, a suitable feeder for addition and dispersion of particulate solids in the gas stream, and a cyclone collector or other separation equipment for final recovery of solids from the gas.

Pneumatic conveying dryers simultaneously dry and convey particles by using high-velocity hot air. The quantity and velocity of the gas phase are sufficient to lift and convey the solids against the forces of gravity and friction. These systems are sometimes incorrectly called *flash dryers* when in fact the moisture is not actually "flashed" off. (True flash dryers are sometimes used for soap drying to describe moisture removal when pressure is quickly reduced.) Pneumatic systems may be distinguished by two characteristics:

1. Retention of a given solids particle in the system is on average very short, usually no more than a few seconds. This means that any process conducted in a pneumatic system cannot be internally controlled (diffusion-controlled). The solids particles must be so small that heat transfer and mass transfer from the interiors are essentially instantaneous.
2. On an energy content basis, the system is balanced at all times; i.e., there is sufficient energy in the gas (or solids) present in the system at any time to complete the work on all the solids (or gas) present at the same time. This is significant in that there is no lag in response to control changes or in starting up and shutting down the system; no partially processed residual solids or gas need be retained between runs.

It is for these reasons that pneumatic equipment is especially suitable for processing heat-sensitive, easily oxidized, explosive, or flammable materials which cannot be exposed to process conditions for extended periods.

The solids feeder may be of any type: Screw feeders, venturi sections, high-speed grinders, and dispersion mills are employed. For pneumatic conveyors, selection of the correct feeder to obtain thorough initial dispersion of solids in the gas is of major importance. For example, by employing an air-swept hammer mill in a drying operation, 65 to 95 percent of the total heat may be transferred within the mill itself if all the drying gas is passed through it. Fans may be of the induced-draft or the forced-draft type. The former is usually preferred because the system can then be operated under a slight negative pressure. Dust and hot gas will not be blown out through leaks in the equipment. Cyclone separators are preferred for low investment. If maximum recovery of dust or noxious fumes is required, the cyclone may be followed by a wet scrubber or bag collector.

Pneumatic conveyors are suitable for materials which are granular and free-flowing when dispersed in the gas stream, so they do not stick on the conveyor walls or agglomerate. Sticky materials such as filter cakes may be dispersed and partially dried by an air-swept disintegrator in many cases. Otherwise, dry product may be recycled and mixed with fresh feed, and then the two dispersed are together in a disintegrator. Coarse material containing internal moisture may be subjected to fine grinding in a hammer mill.

The main requirement in all applications is that the operation be instantaneously completed; internal diffusion of moisture must not be limiting in drying operations, and particle sizes must be small enough that the thermal conductivity of the solids does not control during heating and cooling operations. Pneumatic conveyors are rarely suitable for abrasive solids. Pneumatic conveying can result in significant particle size reduction, particularly when crystalline or other friable materials are being handled. This may or may not be desirable but must be recognized if the system is selected. The action is similar to that of a fluid-energy grinder.

Pneumatic conveyors may be single-stage or multistage. The former is employed for evaporation of small quantities of surface moisture. Multistage installations are used for difficult drying processes, e.g., drying heat-sensitive products containing large quantities of moisture and drying materials initially containing internal as well as surface moisture.

Typical single- and two-stage drying systems are illustrated in Figs. 12-86, 12-87, and 12-88. Figure 12-86 illustrates the flow diagram of a single-stage dryer with a paddle mixer, a screw conveyor followed by a rotary disperser for introduction of the feed into the airstream at the throat of a venturi section. The drying takes place in the drying column after which the dry product is collected in a cyclone. A diverter introduces the option of recycling part of the product into the mixer in order to handle somewhat sticky products. The environmental requirements are met with a wet scrubber in the exhaust stream.

Figure 12-87 illustrates a two-stage dryer where the initial feed material is dried in a flash dryer by using the spent drying air from the second stage. This semidried product is then introduced into the second-stage flash dryer for contact with the hottest air. This concept is in use in the pulp and paper industry. Its use is limited to materials that are dry enough on the surface after the first stage to avoid plugging of the first-stage cyclone. The main advantage of the two-stage concept is the heat economy which is improved considerably over that of the single-stage concept.

Figure 12-88 is an elevation view of an actual single-stage dryer. It employs an integral coarse-fraction classifier to separate undried particles for recycle.

Several typical products dried in pneumatic conveyors are described in Table 12-40.

Design methods for pneumatic conveyor dryers Depending upon the temperature sensitivity of the product, inlet air temperatures between 125 and 750°C are employed. With a heat-sensitive solid, a high initial moisture content should permit use of a high inlet air temperature. Evaporation of surface moisture takes place at essentially the wet-bulb air temperature. Until this has been completed, by which time the air will have cooled significantly, the surface-moisture film prevents the solids temperature from exceeding the wet-bulb temperature of the air. Pneumatic conveyors are used for solids having initial moisture contents ranging from 3 to 90 percent, wet basis. The air quantity required and solids-to-gas loading are fixed by

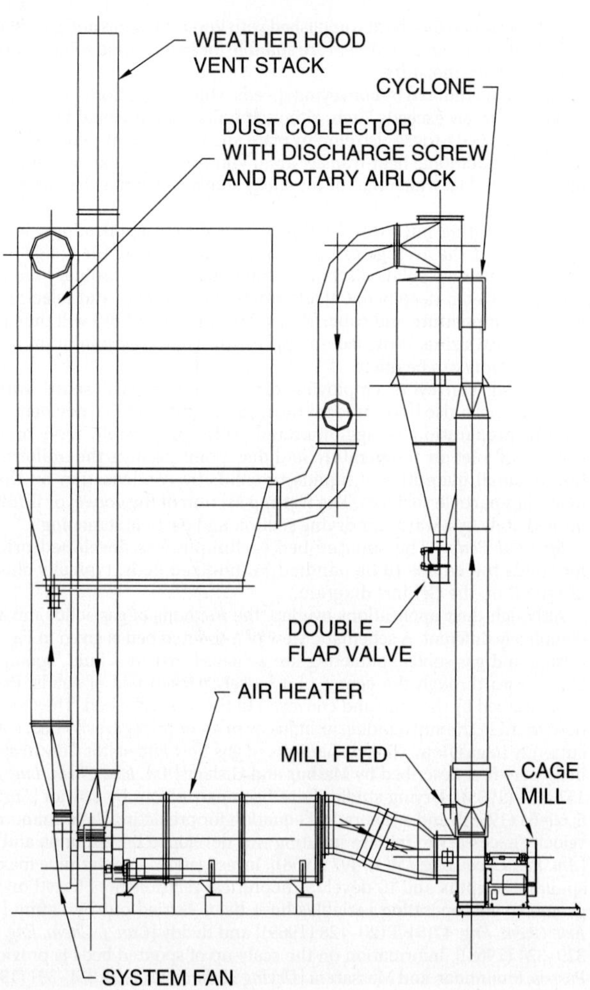

FIG. 12-86 Flow diagram of single-stage flash dryer. (*Air Preheater Company, Raymond & Bartlett Snow Products.*)

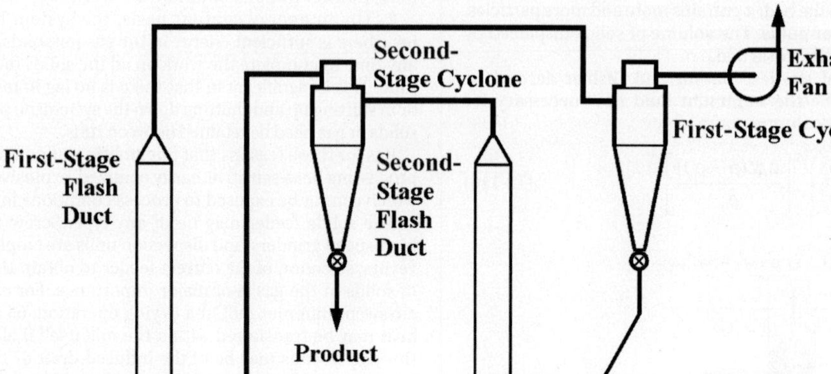

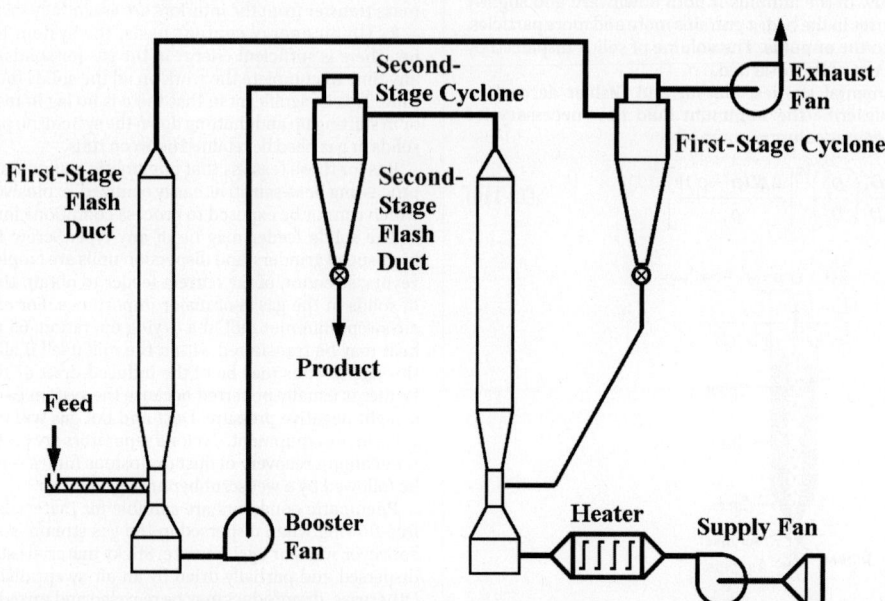

FIG. 12-87 Flow diagram of countercurrent two-stage flash dryer. (*GEA*)

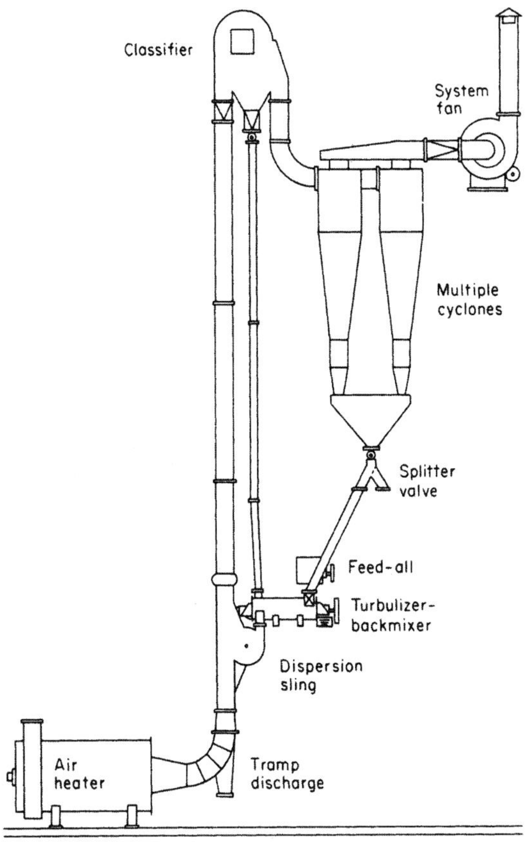

FIG. 12-88 Flow diagram of Strong Scott flash dryer with integral coarse-fraction classifier. (*Bepex Corp.*)

the moisture load, the inlet air temperature, and frequently the exit air humidity. See Example 12-19 for a calculation of the mass and energy balance of a pneumatic conveying dryer. The gas velocity in the conveying duct must be sufficient to convey the largest particle. This may be calculated accurately by methods given in Sec. 17, Gas-Solids Operations and Equipment. For estimating purposes, a velocity of 25 m/s, calculated at the exit air temperature, is frequently employed. The exit solids temperature will approach the exit gas dry-bulb temperature.

Observation of operating conveyors indicates that the solids are rarely uniformly dispersed in the gas phase. With infrequent exceptions, the particles move in a streaklike pattern, following a streamline along the duct wall where the flow velocity is at a minimum. Complete or even partial diffusion in the gas phase is rarely experienced even with low-specific-gravity particles. Air velocities may approach 20 to 30 m/s. It is doubtful, however, that even finer and lighter materials reach more than 80 percent of this speed, while heavier and larger fractions may travel at much slower rates [Fischer, *Mech. Eng.* **81**(11): 67–69 (1959)]. Very little information and few operating data have been published on pneumatic conveyor dryers which would permit a true theoretical basis for design.

Therefore, firm design always requires pilot tests. It is believed, however, that the significant velocity effect in a pneumatic conveyor is the difference in velocities between gas and solids, which is strongly linked to heat- and mass-transfer coefficients and is the reason why a major part of the total drying actually occurs in the feed input section.

See Mills, D., *Pneumatic Conveying Design Guide*, 3d ed., Elsevier, Amsterdam, Netherlands, 2015 for comprehensive information on pneumatic conveying systems.

For estimating purposes, the conveyor cross-section is fixed by the assumed air velocity and quantity. The standard scoping design method is used, obtaining the required gas flow rate from a heat and mass balance, and the duct cross-sectional area and diameter from the gas velocity (if unknown, a typical value is 20 m/s). An incremental model may be used to predict drying conditions along the duct. However, several parameters are hard to obtain, and conditions change rapidly near the feed point. Hence, for reliable estimates of drying time and duct length, pilot-plant tests should always be used. A conveyor length larger than 50 duct diameters is rarely required. The length of the full-scale dryer should always be somewhat larger than required in pilot-plant tests, because wall effects are higher in small-diameter ducts. This gives greater relative velocity (and thus higher heat transfer) and lower particle velocity in the pilot-plant dryer, both effects giving a shorter length than the full-scale dryer for a given amount of drying. If desired, the length difference on scale-up can be predicted by using the incremental model and the pilot-plant data to back-calculate the uncertain parameters; see Kemp, *Drying Technol.* **12**(1&2): 279–297 (1994) and Kemp and Oakley (2002).

An alternative method of estimating dryer size very roughly is to estimate a volumetric heat-transfer coefficient [typical values are around 2000 J/($m^3 \cdot s \cdot K$)] and thus calculate dryer volume.

Pressure drop in the system may be computed by methods described in Sec. 6, Fluid and Particle Dynamics. To prevent excessive leakage into or out of the system, which may have a total pressure drop of 2000 to 4000 Pa, rotary air locks or screw feeders are employed at the solids inlet and discharge.

Ring Dryers The ring dryer is a development of flash, or pneumatic conveyor, drying technology, designed to increase the versatility of application of this technology and overcome many of its limitations.

One of the great advantages of flash drying is the very short retention time, typically no more than a few seconds. However, in a conventional flash dryer, residence time is fixed, and this limits its application to materials in which the drying mechanism is not diffusion-controlled and where a range of moisture within the final product is acceptable. The ring dryer offers two advantages over the flash dryer. First, residence time is controlled by the use of an adjustable internal classifier that allows fine particles, which dry quickly, to leave while larger particles, which dry slowly, have an extended residence time within the system. Second, the combination of the classifier with an internal mill can allow simultaneous grinding and drying with control of product particle size and moisture. Available with a range of different feed systems to handle a variety of applications, the ring dryer provides wide versatility.

The essential difference between a conventional flash dryer and the ring dryer is the manifold centrifugal classifier. The manifold provides classification of the product about to leave the dryer by using differential centrifugal force.

TABLE 12-40 Typical Products Dried in Pneumatic Conveyor Dryers (Barr-Rosin)

Material	Initial moisture, wet basis, %	Final moisture, wet basis, %	Plant configuration
Expandable polystyrene beads	3	0.1	Single-stage flash
Coal fines	23	1.0	Single-stage flash
Polycarbonate resin	25	10	Single-stage flash
Potato starch	42	20	Single-stage flash
Aspirin	22	0.1	Single-stage flash
Melamine	20	0.05	Single-stage flash
Com gluten meal	60	10	Feed-type ring dryer
Maize fiber	60	18	Feed-type ring dryer
Distillers dried grains (DDGs)	65	10	Feed-type ring dryer
Vital wheat gluten	70	7	Full-ring dryer
Casein	50	10	Full-ring dryer
Tricalcium phosphate	30	0.5	Full-ring dryer
Zeolite	45	20	Full-ring dryer
Orange peels	82	10	Full-ring dryer
Modified com starch	40	10	P-type ring dryer
Methylcellulose	45	25	P-type ring dryer

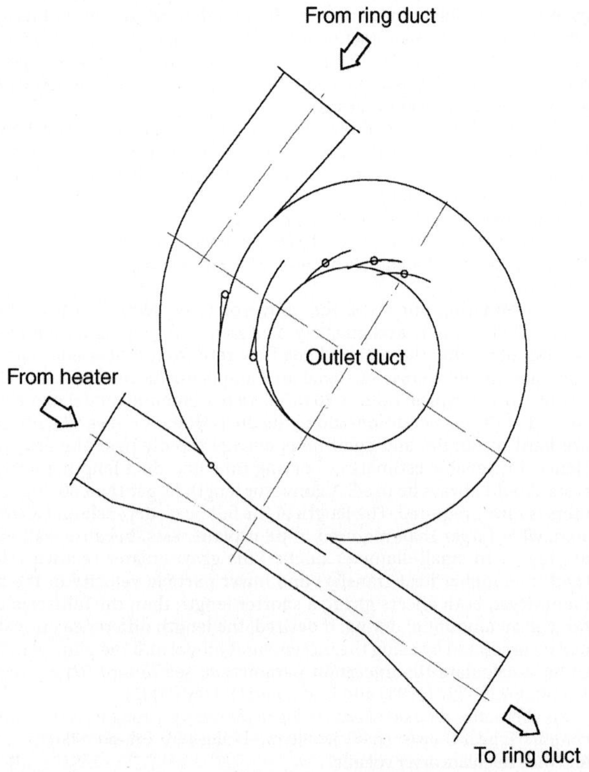

FIG. 12-89 Full manifold classifier for ring dryer. (*Barr-Rosin.*)

The manifold, as shown in Fig. 12-89, uses the centrifugal effect of an airstream passing around the curve to concentrate the product into a moving layer, with the dense material on the outside and the light material on the inside.

This enables the adjustable splitter blades within the manifold classifier to segregate the denser, wetter material and return it for a further circuit of drying. Fine, dried material is allowed to leave the dryer with the exhaust air and to pass to the product collection system. This selective extension of residence time ensures a more evenly dried material than is possible from a conventional flash dryer. Many materials that have traditionally been regarded as difficult to dry can be processed to the required moisture content in a ring dryer. The recycle requirements of products in different applications can vary substantially depending upon the scale of operation, ease of drying, and finished-product specification. The location of reintroduction of undried material back into the drying medium has a significant impact upon the dryer performance and final-product characteristics.

The full ring dryer is the most versatile configuration of the ring dryer. See Fig. 12-90. It incorporates a multistage classifier which allows much higher recycle rates than the single-stage manifold. This configuration usually incorporates a disintegrator which provides adjustable amounts of product grinding depending upon the speed and manifold setting. For sensitive or fine materials, the disintegrator can be omitted. Alternative feed locations are available to suit the material sensitivity and the final-product requirements. The full ring configuration gives a very high degree of control of both residence time and particle size, and it is used for a wide variety of applications from small production rates of pharmaceutical and fine chemicals to large production rates of food products, bulk chemicals, and minerals.

Other ring dryer configurations are available, including ones that have a cyclone within the loop to enable readdition of larger particles to remix or redispersion with the feed.

An important element in optimizing the performance of a flash or ring dryer is the degree of dispersion at the feed point. Maximizing the product surface area in this region of highest evaporative driving force is a key objective in the design of this type of dryer. Ring dryers are fed using equipment similar to that of conventional flash dryers. Ring dryers with vertical configuration are normally fed by a flooded screw and a disperser which propels the wet feed into a high-velocity venturi, in which the bulk of the evaporation takes place. The full ring dryer normally employs an air-swept disperser or mill within the drying circuit to provide screenless grinding when required. Together with the manifold classifier this ensures a product with a uniform particle size. For liquid, slurry, or pasty feed materials, backmixing of the feed with a portion of the dry product will be carried out to produce a conditioned friable material. This further increases the versatility of the ring dryer, allowing it to handle sludge and slurry feeds with ease.

The air velocity required and air/solids ratio are determined by the evaporative load, the air inlet temperature, and the exhaust air humidity. Too high an exhaust air humidity would prevent complete drying, so then a higher air inlet temperature and air/solids ratio would be required. The air velocity within the dryer must be sufficient to convey the largest particle, or agglomerate. The air/solids ratio must be high enough to convey both the product and backmix, together with internal recycle from the manifold. For estimating purposes, a velocity of 25 m/s, calculated at dryer exhaust conditions, is appropriate for both pneumatic conveyor and ring dryers.

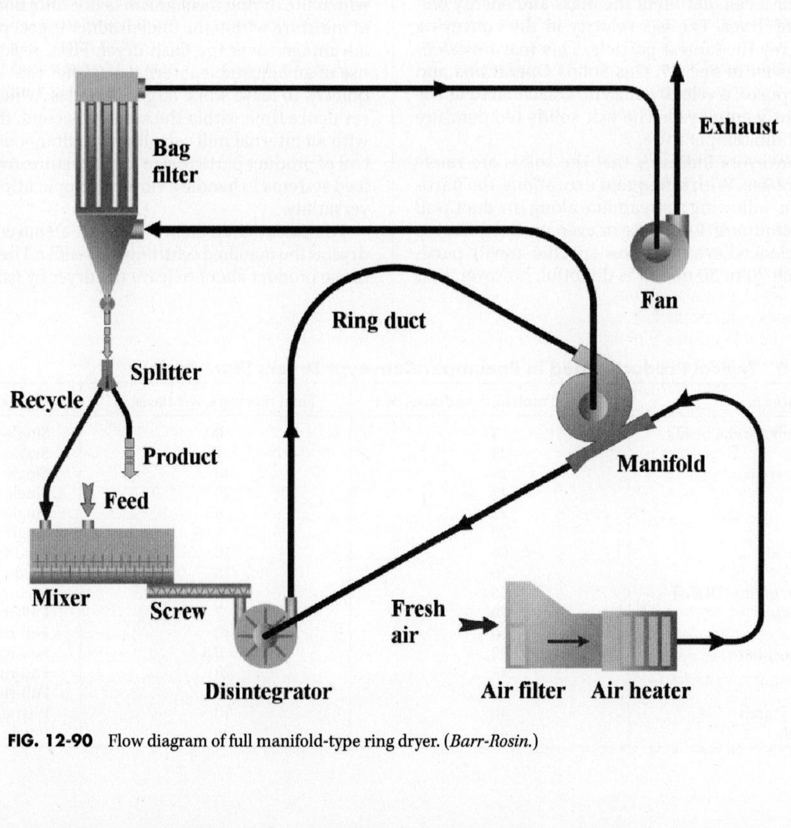

FIG. 12-90 Flow diagram of full manifold-type ring dryer. (*Barr-Rosin.*)

Example 12-19 Mass and Energy Balance for a Pneumatic Conveying Dryer Calculate the exit temperature and relative humidity of a pneumatic conveying system with an inlet air temperature, mass flow rate, and absolute humidity of 700°C, 9 kg/s, and 0.01 g/g, respectively. The feed material has a flow rate of 6.12 kg/s and wet-basis moisture content of 35 percent. Use the following physical properties: $C_{p,\text{air}} = 1.0\ \text{J/(g}\cdot\text{K)}$, $C_{p,\text{dry solids}} = \text{J/(g}\cdot\text{K)}$, $C_{p,\text{liquid water}} = 4.18\ \text{J/(g}\cdot\text{K)}$, $C_{p,\text{water vapor}} = 2.0\ \text{J/(g}\cdot\text{K)}$, and heat of vaporization of water (at 100°C) = 2257 J/g.

Solution First, assume all the water in the feed evaporates. If all the water in the feed evaporates, then the mass flow of water vapor exiting the dryer will equal the water vapor entering the dryer (0.01 · 9 kg/s) plus the water evaporated (0.35 · 6.12 kg/s), which equals 2.24 kg/s. The dry air feed rate is (1 − 0.01) · 9 kg/s, so the absolute humidity in the exhaust is 2.24/[(1 − 0.01) · 9] = 0.239 kg water vapor/kg dry air.

The energy balance can be calculated on a spreadsheet, by setting the sum of the enthalpy terms to zero and solving iteratively for T_{out}. The last three terms calculate the change of enthalpy of water entering as a liquid (within the solid) in the feed at one temperature and exiting as a vapor at a different temperature. Enthalpy is a state function, so we can choose any convenient calculation path. Since we have the heat of vaporization at 100°C, it is convenient to sum the energy terms for the heating of the liquid water to the evaporation temperature of 100°C, vaporize it at that temperature, then account for the enthalpy difference between water vaper at 100°C and the exit temperature. The solution for T_{out} is 87°C in this case.

$$
\begin{aligned}
&G_{\text{dry air}} C_{p,\text{ air}} (T_{\text{air in}} - T_{\text{out}}) + G_{\text{water vapor out}} C_{p,\text{ water vapor}} (T_{\text{air in}} - T_{\text{out}}) \\
&\quad + F_{\text{solids in}} C_{p,\text{ solids}} (T_{\text{solids in}} - T_{\text{out}}) \\
&\quad + F_{\text{liquid water out}} C_{p,\text{ liquid water}} (T_{\text{solids in}} - T_{\text{evap}}) - G_{\text{evaporated}} \cdot \Delta H_{\text{vap}} \\
&\quad + G_{\text{evaporated}} C_{p,\text{ water vapor}} (T_{\text{evap}} - T_{\text{out}}) = 0
\end{aligned}
$$

$$
\begin{aligned}
&(9.44)(1.0 - 0.01)(700 - 87) + (9.44)(0.01)(2.0)(700 - 87) \\
&\quad + (3.98)(1.46)(21 - 87) + (2.14)(4.18)(21 - 100) - (2.14)(2257) \\
&\quad + (2.14)(2.0)(100 - 87) = 0
\end{aligned}
$$

Air at a temperature of 87°C and an absolute humidity of 0.239 kg/kg has a relative humidity of 45 percent. The *relative humidity* is the ratio of the partial pressure to the vapor pressure of pure water at the exhaust temperature. Table 12-1 can be used to calculate the partial pressure of water and Eq. (12-5) to calculate the vapor pressure of pure water.

An exhaust relative humidity of 45 percent may be too high for the product. To make this judgment, a moisture sorption isotherm is needed. If this product were to have an isotherm such as the one shown in Fig. 12-20, then the moisture of the product would be at least 8 percent exiting the dryer. To target a specific moisture content of the exiting material, the isotherm can be used to select a maximum exit relative humidity and the calculation above can be repeated. An isotherm equation could be included in the calculation algorithm, the assumption of complete evaporation can also be changed, and an estimate of heat losses to the environment can be included if more-exact calculations are needed.

Agitated Flash Dryers Agitated flash dryers produce fine powders from feeds with high solids contents, in the form of filter cakes, pastes, or thick, viscous liquids. Many continuous dryers are unable to dry highly viscous feeds. Spray dryers require a pumpable feed. Conventional flash dryers often require backmixing of dry product to the feed in order to fluidize. Other drying methods for viscous pastes and filter cakes are well known, such as contact, drum, band, and tray dryers. They all require long processing time, large floor space, high maintenance, and aftertreatment such as milling.

The agitated flash dryer offers a number of process advantages, such as ability to dry pastes, sludges, and filter cakes to a homogeneous, fine powder in a single-unit operation; continuous operation; compact layout; effective heat- and mass-transfer short drying times; negligible heat loss and high thermal efficiency; and easy access and cleanability.

The agitated flash dryer (Fig. 12-91) consists of four major components: feed system, drying chamber, heater, and exhaust air system. Wet feed enters the feed tank, which has a slow-rotating impeller to break up large particles. The level in the feed tank is maintained by a level controller. The feed is metered at a constant rate into the drying chamber via a screw conveyor mounted under the feed tank. If the feed is shear-thinning and can be pumped, the screw feeder can be replaced by a positive displacement pump.

The drying chamber is the heart of the system consisting of three important components: air disperser, rotating disintegrator, and drying section. Hot, drying air enters the air disperser tangentially and is introduced into the drying chamber as a swirling airflow. The swirling airflow is established by a guide-vane arrangement. The rotating disintegrator is mounted at the base of the drying chamber. The feed, exposed to the hot, swirling airflow and the agitation of the rotating disintegrator, is broken up and dried. The fine, dry particles exit with the exhaust air and are collected in the bag filter. The speed of the rotating disintegrator controls the particle size. The outlet air temperature controls the product moisture content.

The drying air is heated either directly or indirectly, depending upon the feed material, powder properties, and available fuel source. The heat sensitivity of the product determines the drying air temperature. The highest possible value is used to optimize thermal efficiency. A bag filter is usually recommended for collecting the fine particles produced. The exhaust fan maintains a slight vacuum in the dryer, to prevent powder leakage into the surroundings. The appropriate process system is selected according to the feed and powder characteristics, available heating source, energy utilization, and operational health and safety requirements.

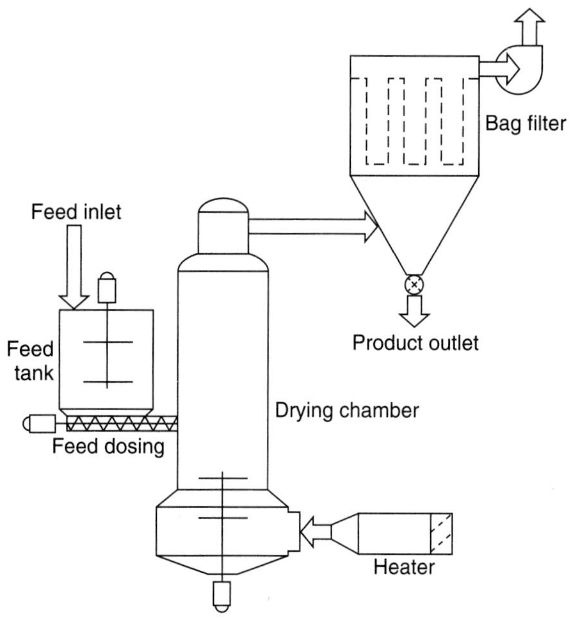

FIG. 12-91 Agitated flash dryer with open cycle. (*GEA*)

Open systems use atmospheric air for drying. In cases where products pose a potential for dust explosion, plants are provided with pressure relief or suppression systems. For recycle systems, the drying system medium is recycled, and the evaporated solvent is recovered as condensate. There are two alternative designs. In the self-inertizing mode, oxygen content is held below 5 percent by combustion control at the heater. This is recommended for products with serious dust explosion hazards. In the inert mode, nitrogen is the drying gas. This is used when an organic solvent is evaporated or product oxidation during drying must be prevented.

Design methods The size of the agitated flash dryer is based on the evaporation rate required. The operating temperatures are product-specific. Once established, they determine the airflow requirements. The drying chamber is designed based on air velocity (approximately 3 to 4 m/s) and residence time (product-specific).

Spray Dryers Spray drying is a process for the transformation of a pumpable liquid feed into dried particulates in a single operation. The process comprises atomization of the feed followed by intense contact with hot air. The dry particulate product is formed while the spray droplets are still suspended in the hot drying air. The process is concluded by the separation and recovery of the product from the drying air.

Industrial Applications Thousands of products are spray-dried. The most common products may include agrochemicals, catalysts, ceramics, chemicals, detergents, dyestuffs and pigments, foodstuffs, pharmaceuticals, and waste products. A few examples are shown in Table 12-41.

For each of these product groups and any other product, successful drying depends on the proper selection of a plant concept and operational parameters, in particular inlet and outlet temperatures and the atomization method. The air temperatures are traditionally established through experiments and test work. The inlet temperatures reflect the heat sensitivity of the different products, and the outlet temperatures the willingness of the products to release moisture. The percentage of moisture in the feed is an indication of feed viscosity and other properties that influence the pumpability atomization behavior.

A spray-drying plant comprises six process stages, as shown in Table 12-42.

Preatomization Spray drying may require a number of operations prior to the drying process. These operations are meant to ensure optimal atomization processes for the given feedstock. Elements for preatomization include but are not limited to low- and/or high-pressure pumping, high-shear mixing and/or in-line milling, viscosity modifications, and preheating. These processes can be achieved by any number of unit operations or equipment.

TABLE 12-41 Some Products That Have Been Successfully Spray-Dried

Product	Air temperature, K		Water in feed, %	Air/evap. ratio, kg/kg	Air/prod. ratio, kg/kg
	In	Out			
Yeast	500	335	86	15.4	96.2
Zinc sulfate	600	380	55	12.4	15.2
Lignin	475	365	63	24.3	41.4
Aluminum hydroxide	590	325	93	9.7	128.4
Silica gel	590	350	95	10.9	206.5
Magnesium carbonate	590	320	92	9.5	108.7
Tannin extract	440	340	46	26.4	22.5
Coffee extract	420	355	70	40.6	94.8
Detergent	505	395	50	25.4	25.4
Manganese sulfate	590	415	50	16.3	16.3
Aluminum sulfate	415	350	70	40.5	94.4
Urea resin	535	355	60	14.8	22.1
Sodium sulfide	500	340	50	16.5	16.5
Pigment	515	335	73	14.4	39

Atomization Stage Atomization creates very a large surface area which enables rapid evaporation from the droplets in the spray. For example, atomization of 1 L of water into a uniform spray of 100-μm droplets results in approximately 1.9×10^9 individual particles with a combined surface area of 60 m^2.

Atomization is the primary means to create the final particle attributes; the droplet size distribution from the nozzle directly affects the particle size distribution of the dry powder. Atomization fundamentals are discussed further in Sec. 14 of this text and in the Lefebvre and Lipp references at the end of this section. In this section, we focus on the most important elements of atomization for spray-drying applications.

Choice of atomizer system The choice of atomizer system for a specific spray-drying operation depends upon

1. The particle size distribution required in the final dried product
2. The physical and chemical properties of the feed liquid (e.g., rheology)
3. The shape of the drying chamber

In this subsection, we introduce and compare the three most common atomization methods for spray drying: rotary atomization, hydraulic nozzles (also called pressure nozzles), and two-fluid nozzles (also called pneumatic nozzles). Table 12-43 shows a comparison of these three methods.

Other specialized atomizers such as ultrasonic nozzles may be used; their use is severely limited based on high operating costs, low individual production rates, and inability to handle viscous feedstocks.

TABLE 12-42 Stages of Spray Drying

Process stages for spray drying	Methods
1. Preatomization	Pumping In-line milling Viscosity modification Preheating
2. Atomization	Rotary atomization Pressure nozzle atomization Two-fluid nozzle atomization
3. Spray/hot air interaction	Cocurrent flow Countercurrent flow Mixed flow
4. Evaporation	Drying Particle morphology
5. Post-treatment	Fluid-bed dryer Sieve Cooling Agglomeration
6. Recovery	Drying chamber Dry collector Wet collector

Rotary Atomizer The liquid feed is supplied to the atomizer by gravity or hydraulic pressure. A liquid distributor system leads the feed to the inner part of a rotating wheel. Since the wheel is mounted on a spindle supported by bearings in the atomizer structure, the liquid distributor is usually formed as an annular gap or a ring of holes of various shapes or orifices concentric with the spindle and wheel. The liquid is forced to follow the wheel either by friction or by contact with internal vanes in the wheel. Due to the high centrifugal forces acting on the liquid, it moves rapidly toward the rim of the wheel, where it is ejected as a film or a series of jets or ligaments (see Fig. 12-92).

By interaction with the surrounding air, the liquid breaks up to form a spray of droplets of varying size. The spray pattern is virtually horizontal with a spray angle said to be 180°. The mean droplet size of the spray depends strongly on the atomizer wheel speed and to a much lesser degree on the feed rate and the feed physical properties such as viscosity. More details about spray characteristics such as droplet size distribution are given below.

As indicated earlier, the atomizer wheel speed is the important parameter influencing the spray droplet size and thus the particle size of the final product. Also important for the atomization process is the selection of a wheel capable of handling a specific liquid feed with characteristic properties such as abrasiveness, high viscosity/nonnewtonian behavior. A highly abrasive feedstock can quickly erode a wheel if proper materials are not used.

The most common design of atomizer wheel has radial vanes. This wheel type is widely used in the chemical industry and is virtually blockage-free and simple to operate, even at very high speed. For high-capacity applications, the number and height of the vanes may be increased to maintain limited liquid-film thickness conditions on each vane.

Wheels with radial vanes have one important drawback, i.e., their capacity for pumping large amounts of air through the wheel. This so-called air pumping effect causes unwanted product aeration, resulting in powders of low bulk density for some sensitive spray-dried products.

Unwanted air pumping effect and product aeration can be reduced through careful wheel design involving change of the shape of the vanes that may appear forward-curved, as seen in Fig. 12-93*a*. This wheel type is used widely in the dairy industry to produce powders of high bulk density. The powder bulk density may increase as much as 15 percent when a curved vane wheel is replacing a radial vane wheel of standard design.

Another way of reducing the air pumping effect is to reduce the space between the vanes so that the liquid feed takes up a larger fraction of the available cross-sectional area. This feature is used with the so-called bushing wheels shown in Fig. 12-93*b*. This wheel combines two important design aspects. The air pumping effect is reduced by reducing the flow area to a number of circular orifices, each 5 to 10 mm in diameter.

Table 12-44 gives the main operational parameters for three typical atomizers covering the wide range of capacity and size.

The Niro F1000 atomizer is one of the largest rotary atomizers offered to industry today. It has a capacity up to 200 ton/h in one single atomizer. As indicated above, the atomizer wheel speed is the important parameter influencing the spray droplet size. The wheel speed U also determines the power consumption P_s of the atomizer; see Table 12-44 for calculations and estimates for various atomizers. The capacity limit of an atomizer is normally its maximum power rating.

Since the atomizer wheel peripheral speed is proportional to the rotational speed, the maximum feed rate that can be handled by a rotary atomizer declines with the square of the rotational speed. The maximum feed rates indicated in Table 12-44 are therefore not available at the higher end of the speed ranges.

The rotary atomizer has one distinct advantage over other means of atomization. The degree or fineness of atomization achieved at a given speed is only slightly affected by changes in the feed rate. In other words, the rotary atomizer has a large turndown capability.

Hydraulic pressure nozzle In hydraulic pressure nozzles, the liquid is fed to the nozzle under pressure. In the nozzle orifice, the pressure energy is converted to kinetic energy. The internal parts of the nozzle are normally designed to apply a certain amount of swirl to the feed flow so that it issues from the orifice as a high-speed film in the form of a cone with a desired vertex angle (see Fig. 12-94). This film disintegrates readily into droplets due to instabilities. The vertex or spray angle is normally on the order of 50° to 80°, a much narrower spray pattern than is seen with rotary atomizers. This means that spray drying chamber designs for pressure nozzle atomization differ substantially from designs used with rotary atomizers. The droplet size distribution produced by a pressure nozzle atomizer varies inversely with the pressure and to some degree with the feed rate and viscosity. The capacity of a pressure nozzle varies with the square root of the pressure. To obtain a certain droplet size, the pressure nozzle must operate very close to the design pressure and feed rate. This implies that the pressure nozzle has very little turndown capability.

TABLE 12-43 Comparison of Rotary, Hydraulic, and Two-Fluid Atomizers

Atomizer	Rotary	Hydraulic (Pressure)	Two-fluid
Advantages	High feed rates Can tolerate abrasive materials Negligible blocking Low-pressure feed system Simple size adjustment Narrow size distribution Minimal agglomeration Large turndown	Simple construction Low cost Low energy consumption Spray characteristic can be changed by nozzle design Dryer chambers generally smaller than those for rotary	Can atomize highly viscous materials Control of size, pattern, and capacity during operation Low-pressure feed system Dryer chambers generally smaller than those for rotary Can produce sprays of high homogeneity and small size
Disadvantages	Higher energy consumption than pressure nozzles Cannot tolerate highly viscous fluids Requires a large chamber diameter	Control of size, pattern, and capacity during operation not possible Swirl nozzles not suitable for suspensions due to phase separation Tendency to clog Wear can greatly affect nozzle performance Requires high-pressure pumps Low turndown	Higher cost than pressure nozzles Must be operated within design parameters for both liquid and air feeds
Cost considerations	Cost is in the capital investment of the atomizer wheel and motor	Cost is in the high-pressure pumping of the feed	Cost is in the requirements for large air compressors
Key considerations	Can be limited as the wheel periphery speed required for fine atomization puts the wheel under extreme tensile stress Due to high capacity of the atomizers, typically only one is required in most dryers	Suited for coarse atomization as pressures above 300 bar are not practical Since capacity of an individual nozzle is limited, dryer may require multiple nozzles	Droplet size tends to vary inversely with the ratio of gas to liquid The capacity is not linked to performance, so some turndown is possible Since capacity of an individual nozzle is limited, dryer may require multiple nozzles
Power consumption	5–11 kWh/ton	up to 3 kWh/ton	up to 25 kWh/ton
Typical droplet size	40–100 μm	50–250 μm	5–500 μm
Typical spray angle	180°	50°–80°	10°–60°

Hydraulic pressure nozzles cannot combine the capability for fine atomization with high feed capacity in one single unit. Many spray dryer applications, where pressure nozzles are applied, require multinozzle systems with the consequence that start-up, operational control, and shutdown procedures become more complicated.

Two-fluid nozzle atomization In two-fluid nozzle atomizers, the liquid feed is fed to the nozzle under marginal or no pressure conditions. An additional flow of gas, normally air, is fed to the nozzle under pressure. Near the nozzle orifice, internally or externally, the two fluids (feed and pressurized gas) are mixed and the pressure energy is converted to kinetic energy, as shown in Fig. 12-95. The flow of feed disintegrates into droplets during the interaction with the high-speed gas flow which may have sonic velocity.

The spray angle obtained with two-fluid nozzles is normally on the order of 10° to 60°. The spray pattern may be narrow and is related to the spread of a free jet of gas. Spray-drying chamber designs for two-fluid nozzle atomization are very specialized according to the application.

The droplet size produced by a two-fluid nozzle atomizer varies inversely with the ratio of gas to liquid and with the pressure of the atomization gas. The capacity of a two-fluid nozzle is not linked to its atomization performance. Therefore, two-fluid nozzles can be attributed with some turndown capability.

Two-fluid nozzles share with pressure nozzles the lack of high feed capacity combined with fine atomization in one single unit. Many spray dryer applications with two-fluid nozzle atomization have a very high number of individual nozzles. The main advantage of two-fluid nozzles is the capability to achieve very fine atomization.

Table 12-45 shows several relationships between liquid properties and spray qualities.

In general, spray drying operation parameters are experience and pilot-scale testing. Droplet size is critical for all spray-drying operations. When droplet size data are unavailable for a spray, the scientific literature contains numerous empirical relationships that can be used to make predictions of the droplet sizes in a spray. If any difference between the atomization means mentioned here were to be pointed out, it would be the tendency for two-fluid nozzles to have the wider particle size distribution and narrower pressure nozzles with rotary atomizers in between.

Spray/Hot Air Contact Atomization is first and most important process stage in spray drying. The final result of the process does, however, depend to a very large degree on the second stage, the spray/hot air contact. This stage influences the quality of the product. In general terms, three possible forms can be defined. These are depicted in Fig. 12-96 as cocurrent, countercurrent, and mixed flow.

Different drying chamber forms and different methods of hot air introduction accompany the different flow pattern forms and are selected according to

- Required particle size in product specification
- Required particle form
- Temperature or heat sensitivity of the dried particle

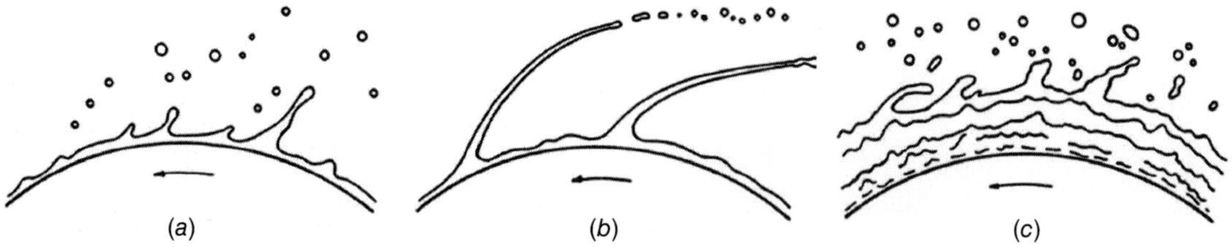

FIG. 12-92 The regimes of droplet formation in rotary disk atomizer: (*a*) drop regime; (*b*) ligament regime; (*c*) sheet regime. (*Reprinted with permission from Bayvel and Orzochowski, Liquid Atomization, Taylor and Francis, Washington D.C., 1993.*)

TABLE 12-44 Operational Parameters of Rotary Atomizers (GEA)

Rotary atomizer design					
Atomizer type		F1.5X/FX1	F100	F160	F1000
Nominal power rating	kW	1.5	75	160	1100
Maximum feed rate	ton/h	0.3	20	30	140
Atomizer wheel diameter	mm	100	210	210/240	350
Minimum speed	rpm	10,000	7400	6000	5800
Maximum speed	rpm	30,000	18,200	18,600	11,500
Typical periphery velocity U	m/s	141	160	165	161
Typical specific power P_s ($P_s = U^2/3600$)	kWh/ton	5.52	7.11	7.56	7.20

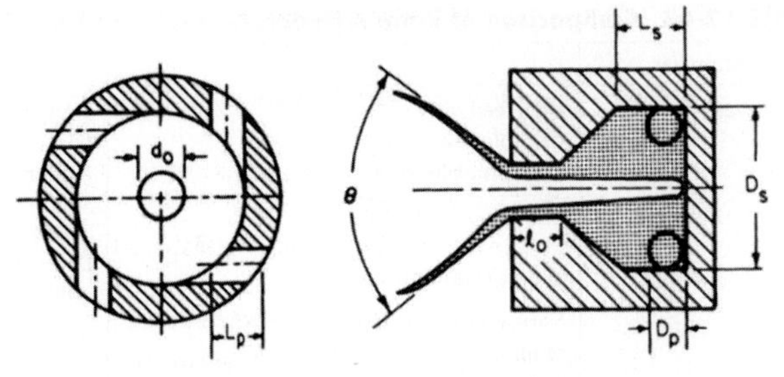

FIG. 12-94 Schematic view of a simplex swirl atomizer. (*Reprinted with permission from Lefebvre, A.H., Atomization and Sprays, Taylor and Francis, Washington D.C., 1989.*)

Figure 12-96*a* shows a cocurrent cone-based tall form chamber with roof gas disperser. This chamber design is used primarily with pressure nozzle atomization to produce powders of large particle sizes with a minimum of agglomeration. The chamber can be equipped with an oversize cone section to maximize powder discharge from the chamber bottom. This type of dryer is often used for dyestuffs, baby foods, detergents, and instant coffee powder.

Figure 12-96*b* shows a standard cocurrent cone-based chamber with roof gas disperser. The chamber can have either single- or two-point discharge and can be equipped with rotary or nozzle atomization. Fine or moderately coarse powders can be produced. This type of dryer finds application in dairy, food, chemical, pharmaceutical, agrochemical, and polymer industries. A version of Fig. 12-96*b* with a flat-based cocurrent chamber can be used with limited building height.

Figure 12-96*c* shows a countercurrent flow chamber with pressure nozzle atomization. This design is in limited use because it cannot produce heat-sensitive products. Detergent powder is the main application; see Huntington, D. L., *Drying Technol.* **22**(6): 1261–1287 (2004), for a large-scale example.

Figure 12-96*d* shows a high-temperature chamber with the hot gas distributor arranged internally on the centerline of the chamber. The atomizer is rotary. Inlet temperature in the range of 600 to 1000°C can be utilized in the drying of non-heat-sensitive products in the chemical and mining industries. Kaolin and mineral flotation concentrates are typical examples.

Figure 12-96*e* shows a mixed-flow chamber with pressure nozzle atomization arranged in *fountain* mode. This design is ideal for producing a coarse product in a limited-size low-cost drying chamber. This type of dryer is used extensively for ceramic products. Powder removal requires a sweeping suction device. One of few advantages is ease of access for manual cleaning. These are widely used in production of flavoring materials.

The layouts in Fig. 12-96 can be augmented with an integrated fluid-bed chamber in the spray dryer. The final stage of the drying process is accomplished in a fluid bed located in the lower cone of the chamber. This type of operation allows lower outlet temperatures to be used, leading to fewer temperature effects on the powder and higher energy efficiency. Similarly, the layouts can be modified with an integrated belt chamber where product is sprayed onto a moving belt, which also acts as the air exhaust filter. It is highly suitable for slowly crystallizing and high-fat products. Previous operational difficulties derived from hygienic problems on the belt have been overcome, and the integrated belt dryer is now moving the limits of products that can be dried by spray drying.

In general terms, selection of chamber design and flow pattern form follows these guidelines:

- Use cocurrent spray drying for heat-sensitive products of fine as well as coarse particle size, where the final product temperature must be kept lower than the dryer outlet temperature.

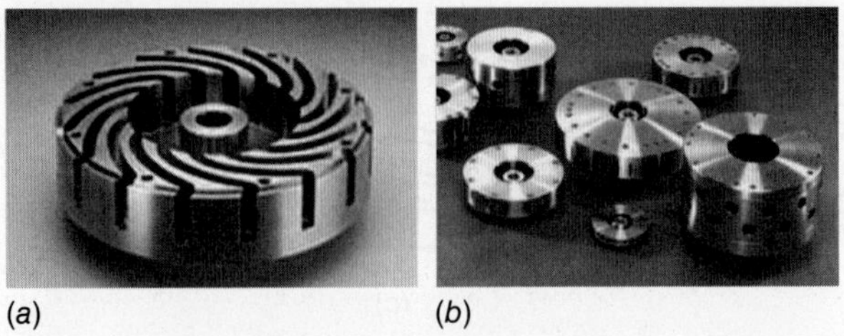

(*a*) (*b*)

FIG. 12-93 Rotary atomizers: (*a*) forward-curved vane; (*b*) vaned and bushing-type rotary wheels. (*GEA*)

- Use countercurrent spray drying for products which are not heat-sensitive, but may require some degree of heat treatment to obtain a special characteristic, i.e., porosity or bulk density. In this case the final powder temperature may be higher than the dryer outlet temperature.
- Use mixed-flow spray drying when a coarse product is required and the product can withstand short time exposure to heat without adverse effects on dried product quality.
- Dryers with rotary atomizers have a wider diameter to accommodate the spray pattern without wall buildup.

Evaporation Stage Evaporation takes place from a moisture film that establishes on the droplet surface. The droplet surface temperature is kept low and close to the adiabatic saturation temperature of the drying air. As the temperature of the drying air drops off and the solids content of the droplet/particle increases, the evaporation rate is reduced. The drying chamber design must provide a sufficient residence time in suspended condition for the particle to enable completion of the moisture removal.

During the evaporation stage, the atomized spray droplet size distribution may undergo changes as droplets shrink, expand, collapse, fracture, or agglomerate. The quality of a spray-dried product is often strongly dependent on its morphological characteristics. Attributes include particle size, density, ability to dissolve, fragility, retention of trace volatile components (aroma), etc.

Typical morphological changes that may occur are outlined in Fig. 21-175 of Perry's 8th ed. and in Fig. 12-97 [Walton and Mumford, "The Morphology of Spray-Dried Particles—The Effects of Process Variables upon the Morphology of Spray-dried Particles," *Trans IChemE* **77**(Part A): 442–460 (1999)]. See also numerous articles from C. J. King.

These morphological transformations can be difficult to predict a priori and require experimentation to determine final particle properties. A number of experimental studies have been conducted on single droplets to better understand the mechanisms [Hecht, J. P., and King, C. J., *Ind. Eng. Chem. Res.* **39:** 1766–1774 (2000)].

Post-treatment Manipulating powder properties including moisture content, particle size, density, morphology, and dispersibility can be done with a variety of unit operations. These include fluid-bed dryers and agglomerators, coaters, sieves, granulator, presses, etc. The treatment will depend on the final uses of the product dried. Post-treatment may take place prior to or after dry product recovery.

Product Recovery Product recovery is the last stage of the spray-drying process. Two distinct systems are used:

- In two-point discharge, primary discharge of a coarse powder fraction is achieved by gravity from the base of the drying chamber. The fine fraction is recovered by secondary equipment downstream of the chamber air exit.
- In single-point discharge, total recovery of dry product is accomplished in the dryer separation equipment.

Collection of powder from an airstream is a large subject area of its own. In spray drying, dry collection of powder in a nondestructive way is achieved by use of cyclones, filters with textile bags or metallic cartridges, and electrostatic precipitators or a combination thereof.

With the current emphasis on environmental protection, many spray dryers are equipped with additional means to collect even the finest fraction. This collection is often destructive to the powder. Equipment in use includes wet scrubbers, bag or other kinds of filters, and in a few cases incinerators.

Industrial Designs and Systems Thousands of different products are processed in spray dryers representing a wide range of feed and product properties as well as drying conditions. The flexibility of the spray-drying concept, which is the main reason for this wide application, is described by the following systems.

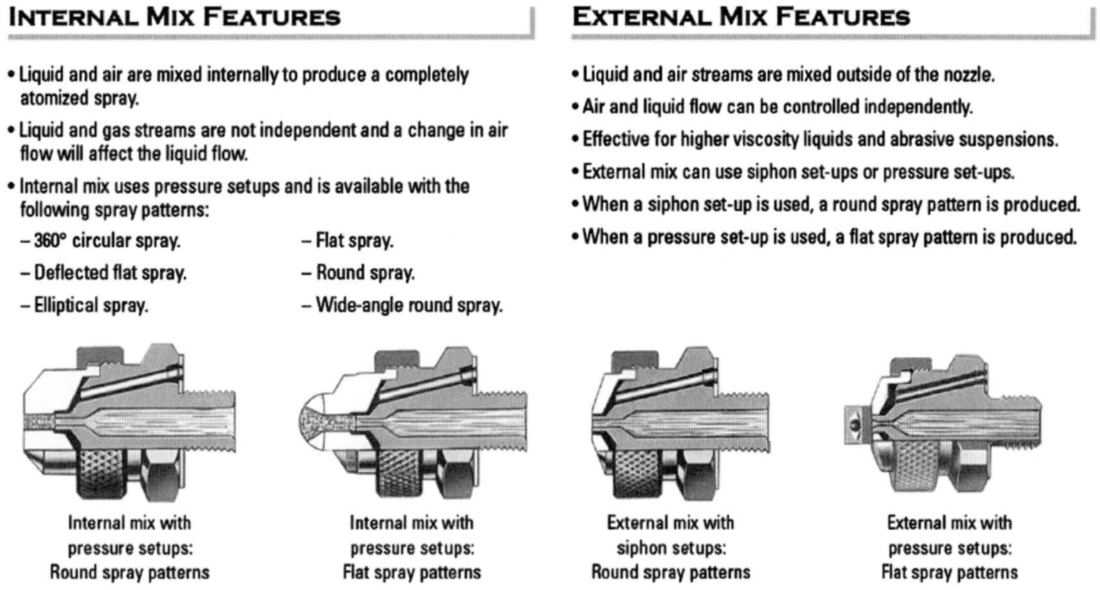
INTERNAL MIX FEATURES
• Liquid and air are mixed internally to produce a completely atomized spray.
• Liquid and gas streams are not independent and a change in air flow will affect the liquid flow.
• Internal mix uses pressure setups and is available with the following spray patterns:
– 360° circular spray.
– Flat spray.
– Deflected flat spray.
– Round spray.
– Elliptical spray.
– Wide-angle round spray.
EXTERNAL MIX FEATURES
• Liquid and air streams are mixed outside of the nozzle.
• Air and liquid flow can be controlled independently.
• Effective for higher viscosity liquids and abrasive suspensions.
• External mix can use siphon set-ups or pressure set-ups.
• When a siphon set-up is used, a round spray pattern is produced.
• When a pressure set-up is used, a flat spray pattern is produced.
Internal mix with pressure setups: Round spray patterns
Internal mix with pressure setups: Flat spray patterns
External mix with siphon setups: Round spray patterns
External mix with pressure setups: Flat spray patterns

FIG. 12-95 Schematic view of two-fluid atomizers. (*Spraying Systems Co.*)

Plant Layouts All the above-mentioned chamber layouts can be used in open-cycle, partial-recycle, or closed-cycle layouts. The selection is based on the needs of operation, feed, drying gas, solvent and powder specification, and environmental considerations.

An open-cycle layout is by far the most common in industrial spray drying. The open layout involves intake of drying air from the atmosphere and discharge of exhaust air to the atmosphere. Drying air can be supplemented by a waste heat source to reduce overall fuel consumption. The heater may be direct, i.e., natural gas burner, or indirect by steam-heated heat exchanger or other heat recovery systems. An example of an open-cycle layout is shown in Fig. 12-98.

A closed-cycle layout is used for drying inflammable or toxic solvent feedstocks or gases. The closed-cycle layout ensures complete solvent recovery and prevents explosion and fire risks. The reason for the use of a solvent system is often to avoid oxidation/degradation of the dried product. Consequently closed-cycle plants are gastight installations operating with an inert drying medium, usually nitrogen. These plants operate at a slight gauge pressure to prevent inward leakage of air.

Partial recycle is used in a plant type applied for products of moderate sensitivity toward oxygen. The atmospheric drying air is heated in a direct fuel-burning heater. Part of the exhaust air, depleted of its oxygen content by the combustion, is dried by using a condenser and recycled to the heater. This type of plant is also designated self-inertizing.

As a consequence, the amount of drying air or gas required for drying one unit of feed or product varies considerably. A quick scoping estimate of the size of an industrial spray dryer can be made on this basis. The required evaporation rate or product rate can be multiplied by the relevant ratio to give the mass flow rate of the drying gas. The next step would be to calculate the size of a spray-drying chamber to allow the drying gas at outlet conditions for a given residence time.

TABLE 12-45 Properties of Fluids and How They Influence Atomization (Spraying Systems)

	Increase with specific gravity	Increase in viscosity	Increase in fluid temperature	Increase in surface tension
Pattern quality	Negligible	Deteriorates	Improves	Negligible
Capacity	Decreases	—	—	No effect
Spray angle	Negligible	Decreases	Increases	Decreases
Drop size	Negligible	Increases	Decreases	Increases
Drop velocity	Decreases	Decreases	Increases	Negligible
Impact	Negligible	Decreases	Increases	Negligible
Wear	Negligible	Decreases	—	No effect

Example 12-20 Scoping Exercise for Size of Spray Dryer Estimate the size of a zinc sulfate spray dryer with cylindrical chamber with diameter D, height H equal to D, and a 60° conical bottom. The dryer has an evaporative capacity of 2.0 ton/h and requires a drying gas flow rate of 8.45 kg/s with a residence time of 25 s. The outlet gas density is 0.89 kg/m³.

The dryer has a nominal geometric volume V_{chamber} (cylinder on top of cone) of

$$V_{\text{chamber}} = \frac{\pi}{4} D^2 \times \left(H + \frac{1}{3} \cdot \frac{\sqrt{3}}{2} D \right) = 1.01 \times D^3$$

Based on the drying gas flow rate, outlet gas density, and residence time, the required chamber volume is

$$V_{\text{chamber}} = (8.45 \text{ kg/s})/(0.89 \text{ kg/m}^3) \times 25 \text{ s} = 237 \text{ m}^3$$

The chamber size now becomes

$$D = \sqrt[3]{\frac{237}{1.01}} = 6.2 \text{ m}$$

The selection of the plant concept involves the dryer modes illustrated in Fig. 12-96. For different products a range of plant concepts are available to secure successful drying at the lowest cost. These concepts are illustrated in Fig. 12-99.

Figure 12-99 shows a traditional spray dryer layout with a cone-based chamber and roof gas disperser. The chamber has two-point discharge and rotary atomization. The powder leaving the chamber bottom as well as the fines collected by the cyclone is conveyed pneumatically to a conveying cyclone from which the product discharges. A bag filter serves as the common air pollution control system.

Figure 12-99 also shows closed-cycle spray dryer layout used to dry certain products with a nonaqueous solvent in an inert gas flow. The background for this may be product sensitivity to water and oxygen or severe explosion risk. Typical products can be tungsten carbide or pharmaceuticals.

Figure 12-99 also shows an integrated fluid-bed chamber layout of the type used to produce agglomerated product. The drying process is accomplished in several stages, the first being a spray dryer with atomization. The second stage is an integrated static fluid bed located in the lower cone of the chamber. The final stages are completed in external fluid beds of the vibrating type. This type of operation allows lower outlet temperatures to be used, leading to fewer temperature effects on the powder and higher energy efficiency. The chamber has a mixed-flow concept with air entering and exiting at the top of the chamber. This chamber is ideal for heat-sensitive, sticky products. It can be used with pressure nozzle as well as rotary atomization. An important feature is the return of fine particles to the chamber to enhance the agglomeration effect. Many products have been made feasible for spray drying by the development of this concept, which was initially

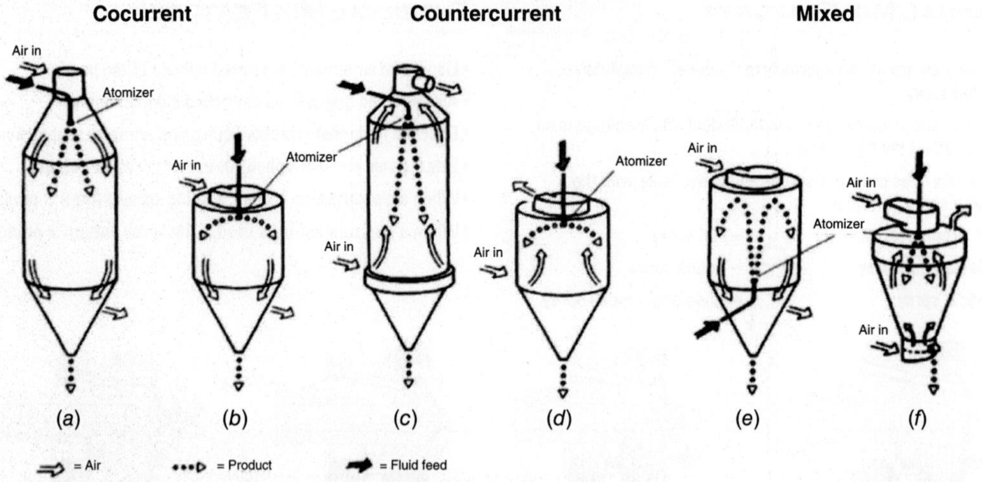

FIG. 12-96 Different forms of spray/hot air contact. (*revised, GEA*)

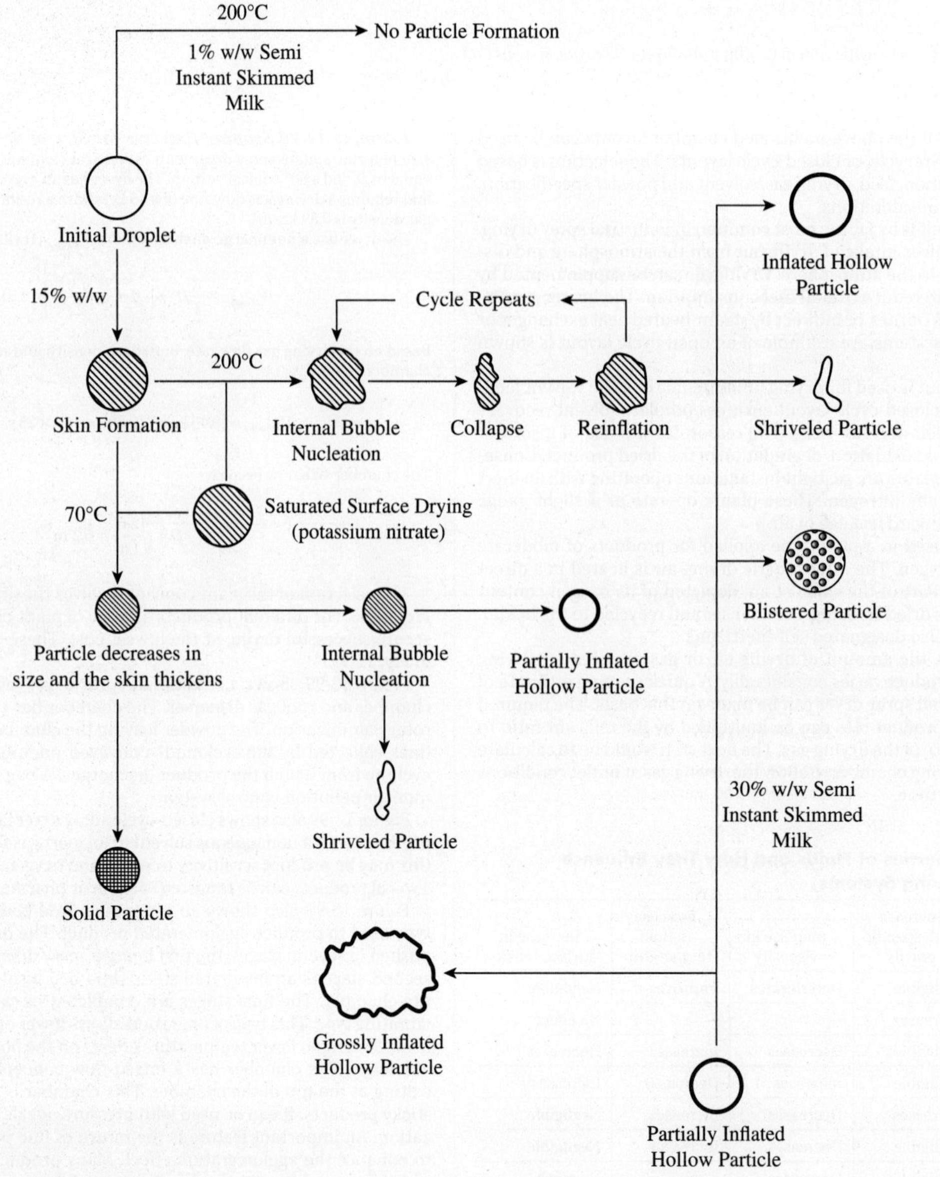

FIG. 12-97 Description of possible particle morphologies. (*Reprinted with permission from Walton and Mumford, Trans IChemE, 77(Part A): 442–460, 1999.*)

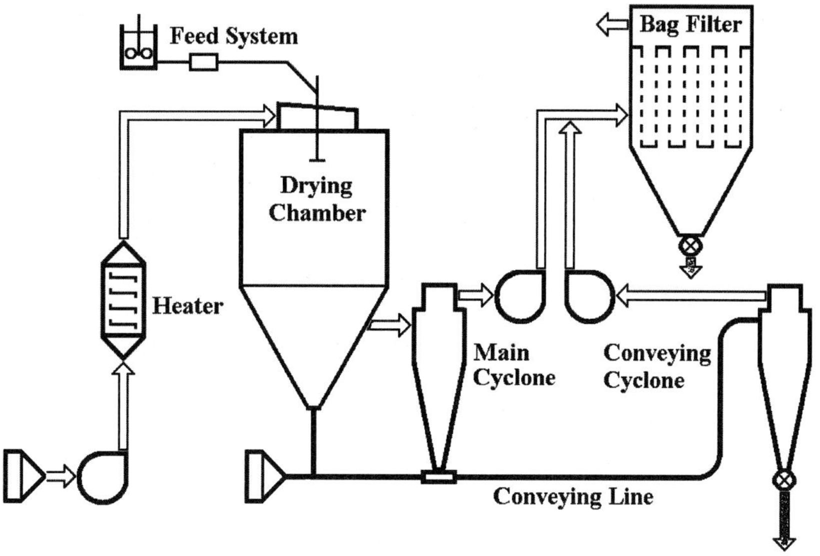

FIG. 12-98 Spray dryer with rotary atomizer and pneumatic powder conveying. (*GEA*)

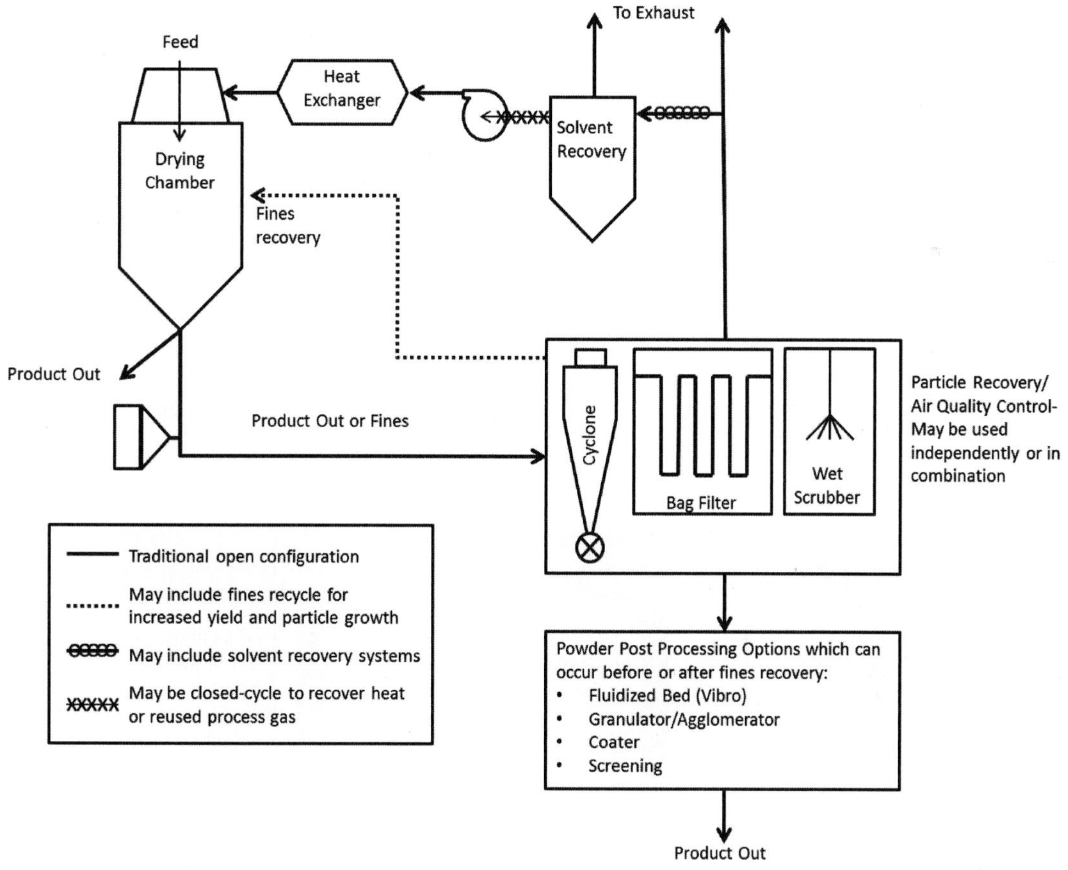

FIG. 12-99 Spray dryer layout with multiple possible configurations.

aimed at the food and dairy industry. Recent applications have, however, included dyestuffs, agrochemicals, polymers, and detergents.

Spray Dryer Modeling Modeling of spray dryers is a unique challenge due to the vast differences in the length scales [dryer (1.0 to 10 μ in diameter) compared to droplets (100 μm in diameter)], billions of particles, one or more sprays, and multiple, complex micro-scale transformations. Making measurements inside a spray dryer to develop and validate models is notoriously difficult. *Computational fluid dynamics* (CFD) can be used to examine the complex airflow patterns within many dryers and also to track particle trajectories. Other modeling techniques have examined the drying kinetics of various droplet types. Invariably, efforts to date have simplified either the air patterns in the dryer or the drop-drying dynamics. A recent notable effort to combine these scales is the EDECAD Project [Verdurmen et al., "Agglomeration in Spray Drying Installations (The EDECAD Project): Stickiness Measurements and Simulation Results," *Drying Technol.* **24:** 721–726 (2006)].

Drynetics is a commercially available method to incorporate experimental single-drop drying into CFD software. Experiments can be conducted on individual droplets of a feed to determine their drying properties. The results are then transferred to the CFD software with the help of appropriate mathematical models, making it possible to simulate the drying process accurately. This modeling workflow is seen in Fig. 12-100.

Drying kinetics as well as morphology formation during drying of a single particle can be found experimentally by using an apparatus such as the *drying kinetics analyzer* (DKA) which is based on the principle of ultrasonic levitation. In an ultrasonic levitator a small particle may be held constant against gravity due to the forces of an ultrasonic field between the so-called transmitter and the so-called reflector. While the levitated particle is drying, it may be monitored with a camera to record the morphology development and by an infrared device to record the development in particle temperature. A levitator can be encapsulated in a drying chamber so that the drying gas temperature and humidity may be set arbitrarily. If the drying gas is injected through small holes in the reflector below the particle, as in the DKA, the relative velocity between the gas and droplets in a spray dryer may be simulated. Further, equipment such as the DKA may also be used to analyze the solidification of a melted particle (Coolnetics) which is relevant, e.g., for congealing processes. Here a melt or a solid particle can be inserted into the ultrasonic field. If a solid particle is inserted, it may subsequently be melted, e.g., using laser light or by infrared radiation.

Example 12-21 Mass and Energy Balance on a Spray Dryer A pilot-scale spray dryer has a nominal evaporative rate of 1.0 kg water/h. The dryer is typically operated with an inlet air temperature of 200°C. The dryer is operated with an inlet airflow of 26.4 kg/h. The feedstock has a moisture content of 70 percent (wet basis) and is fed in to the dryer at 25°C. The powder exits the dryer at 6 percent moisture and the same temperature as the exiting air. The ambient air conditions are 22°C and 55 percent relative humidity.

1. Calculate the relative humidity in the exhaust airstream.
2. Calculate the exit air and product temperature.
3. Estimate the increase in production if the moisture content of the feedstock is decreased from 70 to 65 percent.

Solution The mass balance is given by the following equations:

$$F_{\text{bone dry solids in}} = F_{\text{bone dry solids out}}$$
$$F_{\text{liquid water in}} = F_{\text{liquid water out}} + F_{\text{evaporated}}$$
$$G_{\text{dry air in}} = G_{\text{dry air out}}$$
$$G_{\text{water vapor in}} + F_{\text{evaporated}} = G_{\text{water vapor out}}$$

The wet-basis moisture content of the incoming feedstock and outgoing powder are given by

$$w_{\text{in}} = \frac{F_{\text{liquid water in}}}{F_{\text{liquid water in}} + F_{\text{bone dry solids in}}}$$
$$w_{\text{out}} = \frac{F_{\text{liquid water out}}}{F_{\text{liquid water out}} + F_{\text{bone dry solids out}}}$$

The relationship between the total airflow and the absolute humidity is given by

$$G_{\text{dry air}} = G_{\text{air}} \frac{1}{1+Y} = 10 \text{ kg/h} \frac{1}{1+0.009} = 9.91 \text{ kg/h}$$

The absolute humidity of each air stream is given by

$$Y_{\text{in}} = \frac{G_{\text{water vapor in}}}{G_{\text{dry air in}}}$$
$$Y_{\text{out}} = \frac{G_{\text{water vapor out}}}{G_{\text{dry air out}}}$$

The mass flow rate of bone-dry solids into the dryer can be calculated from the evaporation rate and the incoming and exit moisture contents:

$$F_{\text{evap}} = F_{\text{liquid water in}} - F_{\text{liquid water out}}$$
$$F_{\text{evap}} = \frac{w_{\text{in}}}{1-w_{\text{in}}} F_{\text{bone dry solids in}} - \frac{w_{\text{out}}}{1-w_{\text{out}}} F_{\text{bone dry solids out}}$$

This can be simplified as follows:

$$F_{\text{solids}} = \frac{F_{\text{evaporated}}}{\dfrac{w_{\text{in}}}{1-w_{\text{in}}} - \dfrac{w_{\text{out}}}{1-w_{\text{out}}}} = \frac{1.0 \text{ kg / h}}{\dfrac{0.7}{0.3} - \dfrac{0.06}{0.94}} = 0.44 \text{ kg / h}$$
$$F_{\text{liquid water in}} = \frac{w_{\text{in}}}{1-w_{\text{in}}} F_{\text{solids}} = \frac{0.7}{0.3} 0.44 \text{ kg / h} = 1.03 \text{ kg / h}$$
$$F_{\text{liquid water out}} = \frac{w_{\text{out}}}{1-w_{\text{out}}} F_{\text{solids}} = \frac{0.06}{0.94} 0.44 \text{ kg / h} = 0.03 \text{ kg / h}$$

Since the dryer heats ambient air, the absolute humidity can be determined from the psychrometric chart. The mass flow rate of dry air and water vapor can be calculated from the overall airflow rate and the absolute humidity of the incoming air:

$$G_{\text{water vapor in}} = G_{\text{air}} \frac{1}{1+Y_{\text{in}}} Y_{\text{in}} = 26.16 \text{ kg/h} \cdot 0.009 = 0.24 \text{ kg/h}$$
$$G_{\text{water vapor out}} = G_{\text{water vapor in}} + F_{\text{evap}} = 0.24 \text{ kg/h} + 1.0 \text{ kg/h} = 1.24 \text{ kg/h}$$
$$Y_{\text{out}} = \frac{G_{\text{water vapor out}}}{G_{\text{dry air out}}} = \frac{1.24 \text{ kg/h}}{26.16 \text{ kg/h}} = 0.047$$

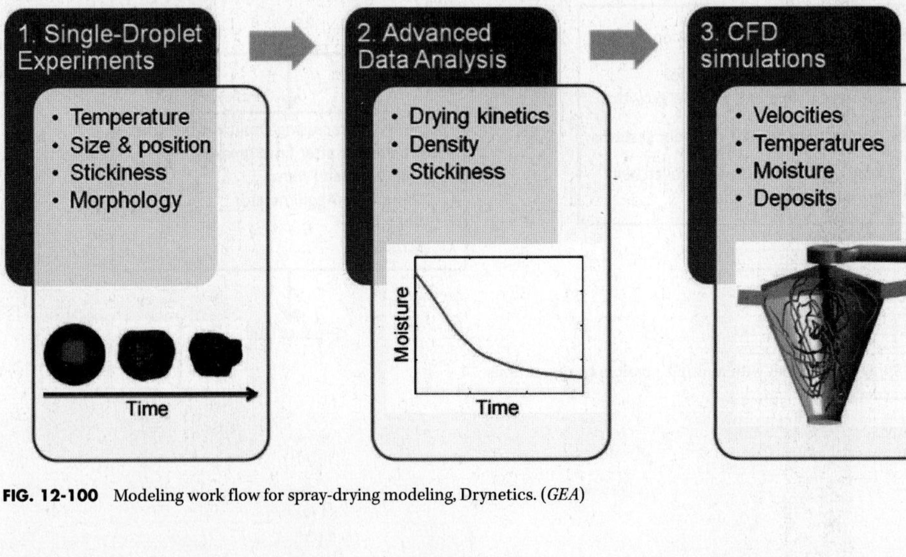

FIG. 12-100 Modeling work flow for spray-drying modeling, Drynetics. (*GEA*)

Next an energy balance must be used to estimate the outgoing air and product temperature:

$$H_{\text{dry air in}} + H_{\text{water vapor in}} + H_{\text{bone dry solids in}} + H_{\text{liquid water in}}$$
$$= H_{\text{dry air out}} + H_{\text{water vapor out}} + H_{\text{bone dry solids out}} + H_{\text{liquid water out}}$$
$$+ \text{heat losses to surroundings}$$

Heat losses to the surrounding environment can be difficult to calculate and can be neglected for a first approximation. This assumption is more valid for larger systems than smaller systems and is neglected in this example.

The equation above was rearranged in terms of enthalpy differences:

$$\Delta H_{\text{dry air}} + \Delta H_{\text{water vapor}} + \Delta H_{\text{solids}} + \Delta H_{\text{liquid water}} + \Delta H_{\text{evap}} = 0$$

Since enthalpy is a state function, the path for determination must be stated. There are several paths that would yield equivalent results. For this example, it is assumed that the evaporation is occurring at the inlet temperature and that the water vapor is being heated from the inlet temperature to the outlet temperature. The terms of the equation can be evaluated by using

$$\Delta H_{\text{dry air}} = G_{\text{dry air}} C_{p,\text{ air}} (T_{\text{air in}} - T_{\text{air out}})$$
$$\Delta H_{\text{water vapor}} = G_{\text{water vapor out}} C_{p,\text{ water vapor}} (T_{\text{air in}} - T_{\text{air out}})$$
$$\Delta H_{\text{solids}} = F_{\text{solids}} C_{p,\text{ solids}} (T_{\text{solids in}} - T_{\text{solids out}})$$
$$\Delta H_{\text{liquid water}} = F_{\text{liquid water out}} C_{p,\text{ liquid water}} (T_{\text{solids in}} - T_{\text{solids out}})$$
$$\Delta H_{\text{evap}} = -G_{\text{evaporated}} \cdot \Delta H_{\text{vap}} + G_{\text{evap}} C_{p,\text{ water vapor}} (T_{\text{evap}} - T_{\text{out}})$$

Since the powder and air exit the dryer at the same temperature, $T_{\text{air out}} = T_{\text{solids out}} = T_{\text{out}}$, the relationships above can be rewritten as

$$G_{\text{dry air}} C_{p,\text{ air}} (T_{\text{air in}} - T_{\text{out}}) + G_{\text{water vapor out}} C_{p,\text{ water vapor}} (T_{\text{air in}} - T_{\text{out}})$$
$$+ F_{\text{solids in}} C_{p,\text{ solids}} (T_{\text{solids in}} - T_{\text{out}})$$
$$+ F_{\text{liquid water out}} C_{p,\text{ liquid water}} (T_{\text{solids in}} - T_{\text{out}}) - G_{\text{evap}} \cdot \Delta H_{\text{vap}}$$
$$+ G_{\text{evap}} C_{p,\text{ water vapor}} (T_{\text{evap}} - T_{\text{out}}) = 0$$

From the steam tables ΔH_{vap} at 25°C = 2442 kJ/kg, h_l = 105 kJ/kg, and at 200°C (superheated, low pressure) h_g = 2880 kJ/kg.

$$(26.16\text{ kg/h})(1\text{ kJ/kg}\cdot{}^\circ\text{C})(200^\circ\text{C} - T_{\text{out}}) + (1.24\text{ kg/h})(2\text{ kJ/kg}\cdot{}^\circ\text{C})(200^\circ\text{C} - T_{\text{out}})$$
$$+ (0.44\text{ kg/h})\left[2.5\frac{\text{kJ}}{\text{kg}\cdot{}^\circ\text{C}}\right](25^\circ\text{C} - T_{\text{out}})$$
$$+ (1.03\text{ kg/h})(4.18\text{ kJ/kg}\cdot{}^\circ\text{C})(25^\circ\text{C} - T_{\text{out}}) - 1.0\text{ kg/h}\cdot\frac{2775\text{ kJ}}{\text{kg}}$$
$$+ (1.0\text{ kg/h})(2.5\text{ kJ/kg}\cdot{}^\circ\text{C})(100^\circ\text{C} - T_{\text{out}}) = 0$$
$$T_{\text{out}} = 99.9^\circ\text{C}$$

With the absolute humidity defined and the outlet temperature calculated, the exit relative humidity was determined from the psychrometric chart as 7.1 percent.

Assuming the drying kinetics and the nominal evaporation rate remain the same with a decrease in the inlet moisture content (there may be a slight adjustment to the energy balance):

$$F_{\text{out}} = \left(1 + \frac{w_{\text{out}}}{1 - w_{\text{out}}}\right) \frac{F_{\text{evap}}}{\dfrac{w_{\text{in}}}{1 - w_{\text{in}}} - \dfrac{w_{\text{out}}}{1 - w_{\text{out}}}} = \left(1 + \frac{0.06}{0.94}\right) \frac{1.0\text{ kg/h}}{\dfrac{0.7}{0.3} - \dfrac{0.06}{0.94}} = 0.47\text{ kg/h}$$

$$F_{\text{out}} = \left(1 + \frac{w_{\text{out}}}{1 - w_{\text{out}}}\right) \frac{F_{\text{evap}}}{\dfrac{w_{\text{in}}}{1 - w_{\text{in}}} - \dfrac{w_{\text{out}}}{1 - w_{\text{out}}}} = \left(1 + \frac{0.06}{0.94}\right) \frac{1.0\text{ kg/h}}{\dfrac{0.65}{0.35} - \dfrac{0.06}{0.94}} = 0.59\text{ kg/h}$$

This suggests the throughput can increase by 27 percent assuming the inlet moisture content can be decreased by 5 percent.

Additional Readings

Bayvel and Orzechowski, *Liquid Atomization*, Taylor & Francis, New York, 1993.

Brask, A., T. Ullum, A. Thybo, and S. K. Andersen, *High-Temperature Ultrasonic Levitator for Investigating Drying Kinetics of Single Droplets.* 6th International Conference on Multiphase Flow, Leipzig, Germany, 2007.

Geng Wang et al., "An Experimental Investigation of Air-Assist Non-Swirl Atomizer Sprays," *Atomisation and Spray Technol.* **3:** 13–36 (1987).

Lefebvre, *Atomization and Sprays*, Hemisphere, New York, 1989.

Lipp, C., *Practical Spray Technology*, Lake Innovation, LLC, Lake Jackson, Texas (2013).

Marshall, "Atomization and Spray Drying," *Chem. Eng. Prog. Mng. Series*, **50**(2) (1954).

Masters, *Spray Drying in Practice*, SprayDryConsult International ApS, Denmark, 2002.

Ullum, T., J. Sloth, A. Brask, and M. Wahlberg, *CFD Simulation of a Spray Dryer Using an Empirical Drying Model.* 16th International Drying Symposium, Hyderabad, India, 2008, pp. 301, 308.

Walzel, P., "Zerstäuben von Flüssigkeiten," *Chem.-Ing.-Tech.* 62 (1990) Nr. 12, S. 983–994.

Nuzzo, M., A. Millqvist-Fureby, J. Sloth, and B. Bergenstahl, "Surface Composition and Morphology of Particles Dried Individually and by Spray Drying," *Drying Technol.* **33**(6) (2015).

Drum and Thin-Film Dryers

Synonyms and Examples Drum dryer, film drum dryer, thin-film dryer (note: this term is used by the paper industry for heated cylinder dryers—these are covered in Sheet Drying).

Description Drum (Film-Drum) Dryers A film of liquid or paste is spread onto the outer surface of a rotating, internally heated drum. Heat transfer occurs by conduction. At the end of the revolution the dry product is removed by a doctor's knife. The material can be in the form of powder, flakes, or chips and typically is 100 to 300 μm thick. Drum dryers cannot handle feedstocks that do not adhere to metal, products that dry to a glazed film, or thermoplastics. The drum is heated normally by condensing steam or in vacuum drum dryers by hot water. Figure 12-101 shows three of the many possible forms. The dip feed system is the simplest and most common arrangement, but is not suitable for viscous or pasty materials. The nip feed system is usually employed on double-drum dryers, especially for viscous materials, but it cannot handle lumpy or abrasive solids. With nip feed systems, the fluid is exposed to the hot surface of the drum, possibly causing the liquid to boil. This may change the fluid rheology due to sudden evaporation and liquid heating or degrade the fluid. Lumpy or abrasive solids are usually applied by roller, and this is also effective for sticky and pasty materials. Spray and splash devices are used for feeding heat-sensitive, low-viscosity materials. Vacuum drum dryers are simply conventional units encased in a vacuum chamber with a suitable air lock for product discharge. Air impingement is also used as a secondary heat source on drum and can dryers, as shown in Fig. 12-102. Impingement or other additional air can be used to purge saturated vapor away from the drums to aid in drying.

In drum drying, the moist material covers a hot surface which supplies the heat required for the drying process.

Let us consider a moist material lying on a hot flat plate of infinite extent. Figure 12-103 illustrates the temperature profile for the fall in temperature from T_H in the heating fluid to T_G in the surrounding air. It is assumed that the temperatures remain steady, unhindered drying takes place, and there is no air gap between the material being dried and the heating surface.

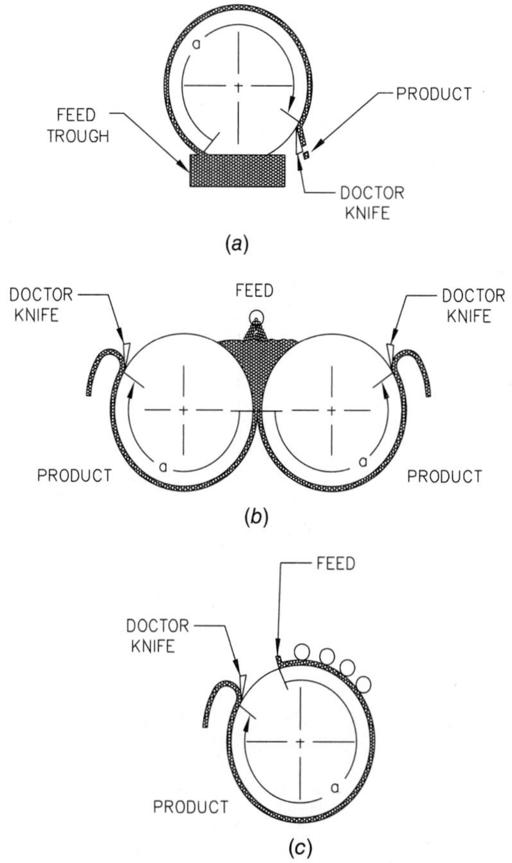

FIG. 12-101 Main types of drum dryers. (*a*) Dip; (*b*) nip; (*c*) roller.

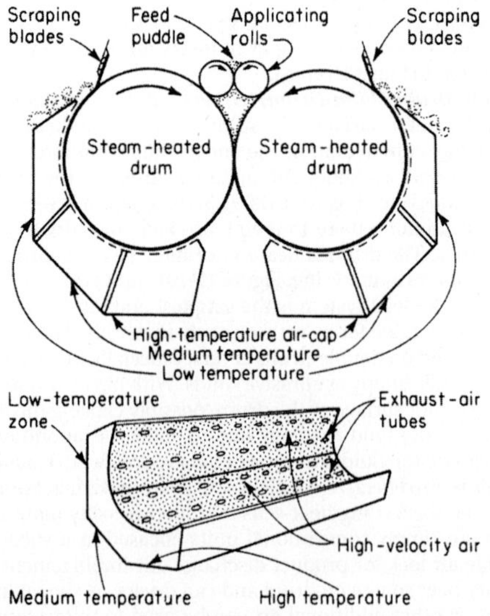

FIG. 12-102 Example of the use of air impingement in drying as a secondary heat source on a double-drum dryer. (*Sloan, C.E., et. al, Chem. Eng.,* **197**, *June 19, 1967*)

The heat conducted through the wall and material is dissipated by evaporation of moisture and convection from the moist surface to the surrounding air. A heat balance yields

$$U(T_H - T_S) = N_W \Delta H_{VS} + h_C(T_S - T_G) \tag{12-111}$$

where U is the overall heat-transfer coefficient. This coefficient is found from the reciprocal law of summing resistances in series:

$$\frac{1}{U} = \frac{1}{h_H} + \frac{b_B}{\lambda_B} + \frac{b_s}{\lambda_s} \tag{12-112}$$

in which h_H is the heat-transfer coefficient for convection inside the heating fluid. If condensing steam is used, this coefficient is very large normally and the corresponding resistance $1/h_H$ is negligible. Rearrangement of Eq. (12-111) yields an expression for the maximum drying rate

$$N_W = \frac{U(T_H - T_S) - h_C(T_S - T_G)}{\Delta H_{VS}} \tag{12-113}$$

Equation (12-113), as it stands, would give an overestimate of the maximum drying rate for the case of contact drying over heated rolls, when there are significant heat losses from the ends of the drum and only part of the drum's surface can be used for drying. In the roller drying arrangements shown in Fig. 12-101, only a fraction a of the drum's periphery is available from the point of pickup to the point where the solids are peeled off.

Let q_E be the heat loss per unit area from the ends. The ratio of the end areas to cylindrical surface, from a drum of diameter D and length L, is $2(1/4 \cdot \pi D^2)/\pi DL$ or $D/2L$. Equation (12-113) for the maximum drying rate under roller drying conditions thus becomes

$$N_W = \frac{aU(T_H - T_S) - h_C(T_S - T_G) - Dq_E/2L}{\Delta H_{VS}} \tag{12-114}$$

The total evaporation from the drum is $N_W a(\pi DL)$. Equation (12-114) could be refined further, as it neglects the effect caused by the small portion of the drum's surface being covered by the slurry in the feed trough, as well as thermal conduction through the axial shaft to the bearing mounts. The use of Eq. (12-114) to estimate the maximum drying rate is illustrated in Example 12-22.

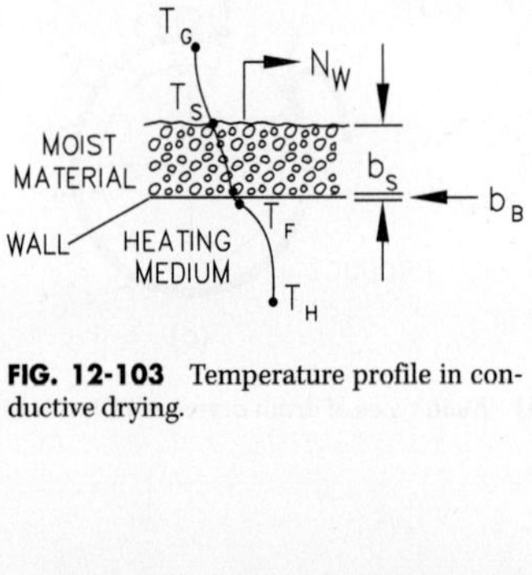

FIG. 12-103 Temperature profile in conductive drying.

Example 12-22 Heat-Transfer Calculations on a Drum Dryer A single rotating drum of 1.250-m diameter and 3 m wide is internally heated by saturated steam at 0.27 MPa. As the drum rotates, a film of slurry 0.05 mm thick is picked up and dried. The dry product is removed by a knife, as shown in Fig. 12-101*a*. About three-quarters of the drum's surface is available for evaporating moisture. Estimate the maximum drying rate when the outside air temperature T_G is 15°C and the surface temperature is 90°C; and compare the effectiveness of the unit with a dryer without end effects and in which all the surface could be used for drying.

Data:

Heat-transfer coefficient $h_C = 50\ \text{W/(m}^2\cdot\text{K)}$
Thickness of cylinder wall $b_B = 10$ mm
Thermal conductivity of wall $\lambda_B = 40\ \text{W/(m}\cdot\text{K)}$
Thermal conductivity of slurry film $\lambda_s = 0.10\ \text{W/(m}\cdot\text{K)}$
Film transfer coefficient for condensing steam $h_H = 2.5\ \text{kW/(m}^2\cdot\text{K)}$
Overall heat-transfer coefficient U: The thermal resistances are as follows:

Steam side	$1/2.5 = 0.40\ \text{m}^2\cdot\text{K/kW}$
Wall	$0.01/0.04 = 0.25\ \text{m}^2\cdot\text{K/kW}$
Film side	$5.0 \times 10^{-5}/0.1 \times 10^{-3} = 0.5\ \text{m}^2\cdot\text{K/kW}$

∴ Overall resistance $= 0.40 + 0.25 + 0.5 = 1.15\ \text{m}^2\cdot\text{K/kW}$

$$U = 1/1.15 = 0.870\ \text{kW/(m}^2\cdot\text{K)}$$

Wall temperature T_B: At 0.27 MPa, the steam temperature is 130°C. If it is assumed that the temperature drops between the steam and the film surface are directionally proportional to the respective thermal resistances, it follows that

$$\frac{T_H - T_B}{T_H - T_S} = \frac{0.40 + 0.25}{1.15} = 0.565$$

$$\therefore T_B = T_H - 0.565(T_H - T_S)$$
$$= 130 - 0.565\,(130 - 90)$$
$$= 107.4°\text{C}$$

Heat losses from ends q_E: For an emissivity ~1 and an air temperature of 15°C with a drum temperature of 107.4°C, one finds [see Eq. (12-120)]

$$q_E = 798\ \text{W/(m}^2\cdot\text{s)}$$

Maximum drying rate N_W: From Eq. (12-114),

$$N_W = \frac{aU(T_H - T_S) - h_C(T_S - T_G) - Dq_E/2L}{\Delta H_{VS}}$$
$$= \frac{0.75 \times 0.870(130 - 90) - 0.05(90 - 15) - (1.25 \times 0.798)/6}{2382}$$
$$= 0.0093\ \text{kg/(m}^2 \times \text{s)}$$

The ideal maximum rate is given by Eq. (12-113) for an endless surface:

$$N_N = \frac{U(T_H - T_S) - h_C(T_S - T_G)}{\Delta H_{VS}}$$
$$= \frac{0.870(130 - 90) - 0.05(90 - 15)}{2382}$$
$$= 0.0130\ \text{kg/(m}^2\cdot\text{s)}$$

Therefore the effectiveness of the dryer is $0.0093/0.0130 = 0.714$.

The predicted thermal efficiency η is

$$\eta = 1 - \frac{h_C(T_S - T_G) + Dq_E/2L}{aU(T_H - T_S)}$$
$$= 1 - \frac{0.05(90 - 15) + (1.25 \times 0.798)/6}{0.75 \times 0.869(130 - 90)}$$
$$= 0.850$$

These estimates may be compared with the range of values found in practice, as shown in Table 12-46 (Nonhebel and Moss, *Drying of Solids in the Chemical Industry,* Butterworths, London, 1971, p. 168).

The typical performance is somewhat less than the estimated maximum evaporative capacity, although values as high as 25 $\text{g/(m}^2\cdot\text{s)}$ have been reported. As the solids dry out, the thermal resistance of the film increases and the evaporation falls off accordingly. Heat losses through the bearing of the drum shaft have been neglected, but the effect of radiation is accounted for in the value of h_C taken. In the case of drying organic pastes, the heat losses have been determined to be 2.5 kW/m^2 over the whole surface, compared with 1.75 kW/m^2 estimated here for the cylindrical surface. The inside surface of the drum has been assumed to be clean, and scale would reduce the heat transfer markedly.

For constant hygrothermal conditions, the base temperature T_B is directly proportional to the thickness of the material over the hot surface. When the wet-bulb

TABLE 12-46 Drum Dryer Operating Information

	This estimate	Typical range
Specific evaporation, g/(m²·s)	9.3	7–11
Thermal efficiency	0.85	0.4–0.7

temperature is high and the layer of material is thick enough, the temperature T_B will reach the boiling point of the moisture. Under these conditions, a mixed vapor-air layer interposes between the material and the heating surface. This is known as the *Leidenfrost effect*, and the phenomenon causes a greatly increased thermal resistance to heat transfer to hinder drying.

Thin-Film Dryers Evaporation and drying take place in a single unit, normally a vertical chamber with a vertical rotating agitator which almost touches the internal surface. The feed is distributed in a thin layer over the heated inner wall and may go through liquid, slurry, paste, and wet solid forms before emerging at the bottom as a dry solid. These dryers are based on wiped-film or scraped-surface (Luwa-type) evaporators and can handle viscous materials and deal with the "cohesion peak" experienced by many materials at intermediate moisture contents. They also offer good containment. Disadvantages are complexity, limited throughput, and the need for careful maintenance. Continuous or semibatch operation is possible. A typical unit is illustrated in Fig. 12-104.

Sheet Dryers

Synonyms and Examples Cylinder dryer, drum dryer (note: this term is used by the paper industry, not to be confused with the drum dryers for pastes in this text), stenter dryers, and tenter dryers.

Description The construction of dryers where both the feed and the product are in the form of a sheet, web, or film is markedly different from that for dryers used in handling particulate materials. The main users are the paper and textile industries. Almost invariably the material is formed into a very long sheet (often hundreds or thousands of meters long) which is dried in a continuous process. The sheet is wound onto a bobbin at the exit from the dryer; this may be several meters in diameter and several meters wide. Alternatively, the sheet may be chopped into shorter sections.

The heat-transfer calculations [Eqs. (12-114) through (12-115)] used in the Drum Drying subsection are directly applicable for sheet drying when a sheet is in contact with a roller.

Cylinder Dryers and Paper Machines The most common type of dryer in papermaking is the cylinder dryer (Fig. 12-105), which is a contact dryer. The paper web is taken on a convoluted path during which it wraps around the surface of cylinders that are internally heated by steam or hot water. In papermaking, the sheet must be kept taut, and a large number of cylinders are used, with only short distances between them and additional small unheated rollers to maintain the tension. Normally, a continuous sheet of felt is also used to hold the paper onto the cylinders, and this also becomes damp and is dried on a separate cylinder.

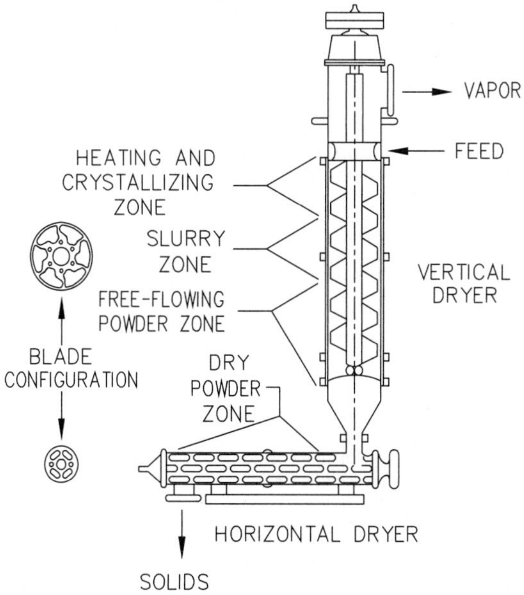

FIG. 12-104 Continuous thin-film dryer.

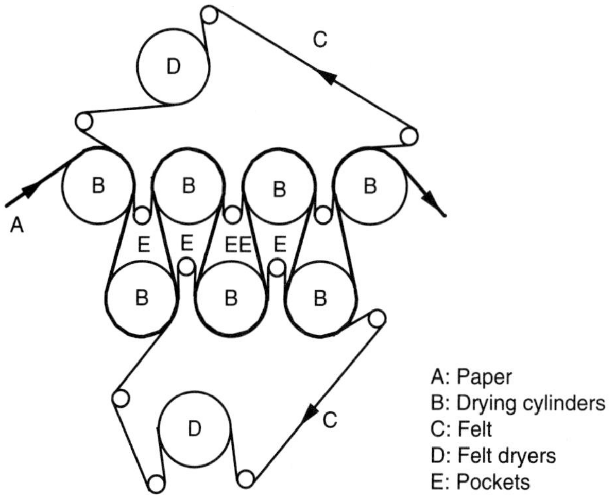

FIG. 12-105 Cylinder dryer (paper machine).

Most of the heating is conductive, through contact with the drums. However, infrared assistance is frequently used in the early stages of modern paper machines. This gets the paper sheet up to the wet-bulb temperature more rapidly, evaporates greater surface moisture, and enables reduction of the number of cylinders for a given throughput. Hot air jets (jet foil dryer) may also be used to supplement heating at the start of the machine. Infrared and dielectric heating may also be used in the later stages to assist the drying of the interior of the sheet.

Although paper is the most common application, multicylinder dryers can also be used for polymer films and other sheet-type feeds.

Convective dryers may be used as well in papermaking. In the Yankee dryer (Fig. 12-106), high-velocity hot airstreams impinging on the web surface give heating by cross-convection. The "Yankees" are barbs holding the web in place. Normally the cylinder is also internally heated, giving additional conduction heating of the lower bed surface. In the rotary through-dryer (Fig. 12-107), the drum surface is perforated and hot air passes from the outside to the center of the drum, so that it is a through-circulation convective dryer.

Another approach to drying of sheets has been to suspend or "float" the web in a stream of hot gas, using the Coanda effect, as illustrated in Fig. 12-108. Air is blown from both sides, and the web passes through as an almost flat sheet (with a slight "ripple"). The drying time is reduced because the heat transfer from the impinging hot air jets is faster than that from stagnant hot air in a conventional oven. It is essential to control the tension of the web very accurately. The technique is particularly useful for drying coated paper, as the expensive surface coating can stick to cylinder dryers.

Stenters (Tenters) and Textile Dryers These are the basic type of dryer used for sheets or webs in the textile industry. The sheet is held by its edges by clips (clip stenter) or pins (pin stenter), which not only suspend

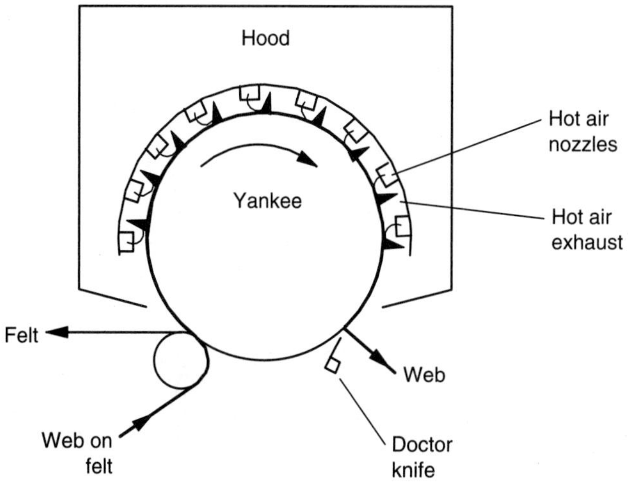

FIG. 12-106 Yankee dryer.

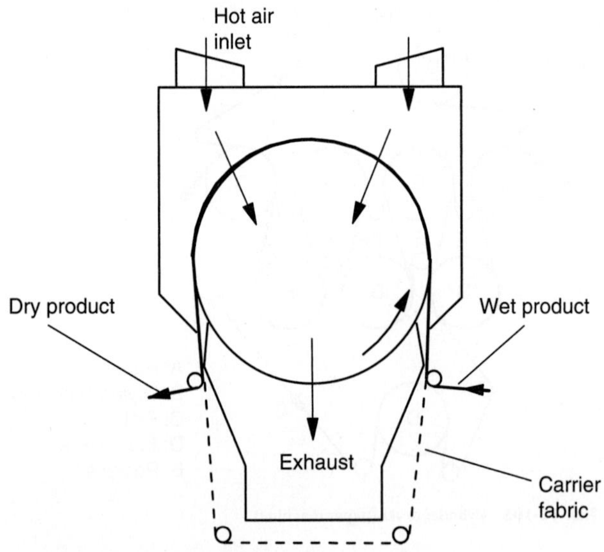

FIG. 12-107 Rotary through-dryer.

the sheet but also keep it taut and regulate its width—a vital consideration in textile drying. Drying is by convection; hot air is introduced from one or both sides, passes over the surface of the sheet, and permeates through it. Infrared panels may also be used to supply additional heat. A schematic diagram of the unit is shown in Fig. 12-109. A typical unit is 1.4 m wide and handles 2 to 4 tons/h of material.

Heavy-duty textiles with thick webs may need a long residence time, and the web can be led up and down in "festoons" to reduce dryer length. Substantial improvements in drying rates have been obtained with radio-frequency heating assistance.

Air impingement dryers as in Fig. 12-108 may also be used for textiles.

Example 12-23 Impinging Air Drying of Sheets Estimate the dryer length needed to dry a continuous thin polyethylene sheet moving at 0.1 m/s using impinging air from a wet-basis moisture of 40 to 10 percent. Assume that moisture is evenly distributed on the top surface of the sheet and that the sheet nonhygroscopic (i.e., there is no bound water). Ambient air at 22°C and 50 percent relative humidity is heated to 120°C. The impinging air dryer has an array of jets that are 7 cm apart from one another, 1 cm in diameter, and 5 cm above the sheet. The air velocity through each jet is 10 m/s. Estimate the dryer size reductions possible if (a) the impinging air were predried, using a dessicant wheel, to 10 percent relative humidity, (b) the inlet moisture content was reduced to 35 percent, and (c) the air temperature was increased to 130°C.

Physical properties: Pr = 0.71; $D_{water/air} = 2.7 \times 10^{-5}$ m²/s; kinematic viscosity of air = 2.2×10^{-5} m²/s; Sc = 0.64 (= 2.2/2.7); k_{air} = 0.03 W/(m · K); ΔH_{vap} = 2450 kJ/kg. Both belt and polyethylene sheet are 1 kg/m² with a specific heat of 2 J/(g · K) and both enter the dryer at 22°C.

Solution For this example, we will create an incremental calculation using a spreadsheet.

We will make use of a well-known correlation for prediction of heat- and mass-transfer coefficients for air impinging on a surface from arrays of holes (jets). This correlation uses relevant geometric properties such as the diameter of the holes, the distance between the holes, and the distance between the holes and the sheet [Martin, "Heat and Mass Transfer Between Impinging Gas Jets and Solid Surfaces," *Advances in Heat Transfer*, vol. 13, Academic Press, Cambridge, Mass., 1977, pp. 1–66].

$$\frac{\text{Sh}}{\text{Sc}^{0.42}} = \frac{\text{Nu}}{\text{Pr}^{0.42}} = \left[1 + \left(\frac{H/D}{0.6/\sqrt{f}}\right)^6\right]^{-0.05} \times \sqrt{f}\,\frac{1 - 2.2\sqrt{f}}{1 + 0.2(H/D - 6)\sqrt{f}} \times \text{Re}^{2/3} \qquad (12\text{-}115)$$

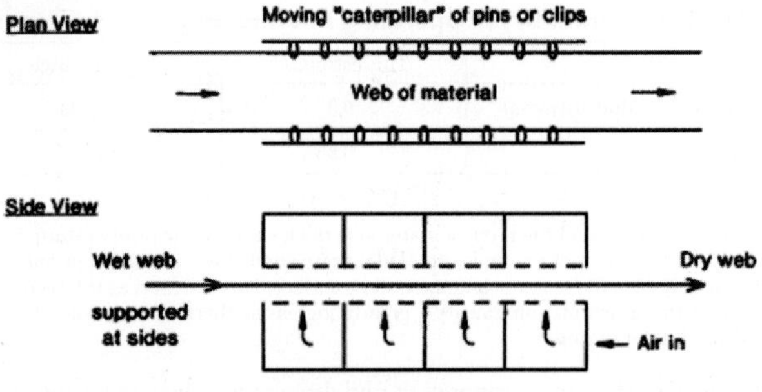

FIG. 12-109 Stenter or tenter for textile drying.

where D = inside diameter of nozzle, m

$$f = \frac{\pi}{2\sqrt{3}}\left(\frac{D}{L_D}\right)^2$$

H = distance from nozzle to sheet, m
L_D = average distance between nozzles, m
Nu = Nusselt number
Pr = Prandtl number
Re = Reynolds number = $\frac{wD}{\nu}$
Sc = Schmidt number
Sh = Sherwood number
w = velocity of air at nozzle exit, m/s
ν = kinematic viscosity of air, m²/s

The heat- and mass-transfer coefficients were then calculated from the definitions of the Nusselt and Sherwood numbers.

$$\text{Nu} = \frac{hD}{k_{th}}$$

where k_{th} = thermal conductivity of air, W/(m · K)
D = diameter of holes in air bars, m

$$\text{Sh} = \frac{k_m^* D}{\text{diff}}$$

where diff = diffusion coefficient of water vapor in air, m²/s.

Plugging in the values of H = 5 cm, D = 1 cm, L = 7 cm, and the values of Pr, Sc, $D_{water/air}$ and k_{air} gives heat- and mass-transfer coefficients of 74 W/(m² · K) and 0.0705 m/s, respectively.

Now we write an energy balance equation for the temperature of the sheet, which we assume to be a single value through the thickness of the sheet (it is the same as in Example 12-14):

$$(C_{p,\text{solids}} + C_{p,\text{water}} \cdot w_{\text{avg,dry-basis}}) \cdot \frac{m_{\text{sheet}}}{A} \cdot \frac{dT_{\text{sheet}}}{dt} = h \cdot (T_{\text{air}} - T_{\text{sheet}}) - F \cdot \Delta H_{\text{vap}} \qquad (12\text{-}116)$$

We also write a mass balance:

$$\frac{dm_w}{dt}\frac{1}{A} = k_c \cdot (C_w^{\text{surface}} - C_w^{\text{bulk}}) \qquad (12\text{-}117)$$

The specific heat values ($C_{p,\text{solids}}$ and $C_{p,\text{water}}$), the basis weight of the sheet ($\frac{m_{\text{sheet}}}{A}$), the heat- and mass-transfer coefficients (h and k_c), heat of vaporization (ΔH_{vap}), and the bulk water concentration in the air (C_w^{bulk}) are all constant. Values of the sheet temperature, the drying flux (F), the water concentration in the air immediately adjacent to the sheet (C_w^{surface}), and the mass loading of water ($\frac{m_{\text{water}}}{A}$) on the sheet all change with time. We convert the time into distance from the dryer feed point by using the sheet velocity.

The concentration of water vapor (also called the *volumetric humidity*) immediately adjacent to the wet sheet is calculated using equations from the psychrometry section. Specifically, Eq. (12-5) is used to calculate the vapor pressure of the water at the sheet temperature, and Table 12-1 is used to calculate the volumetric humidity from the vapor pressure and sheet temperature.

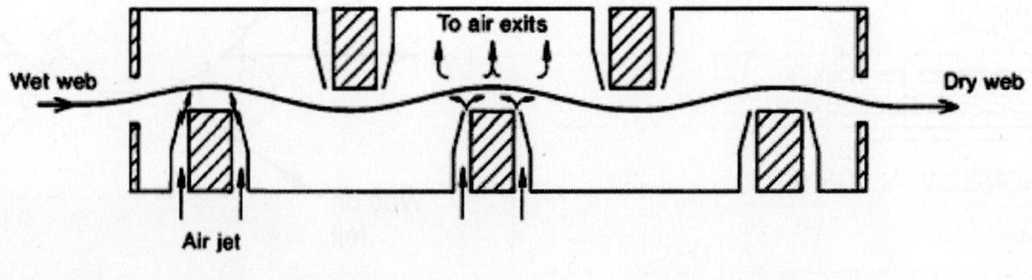

FIG. 12-108 Air flotation (impingement) dryer.

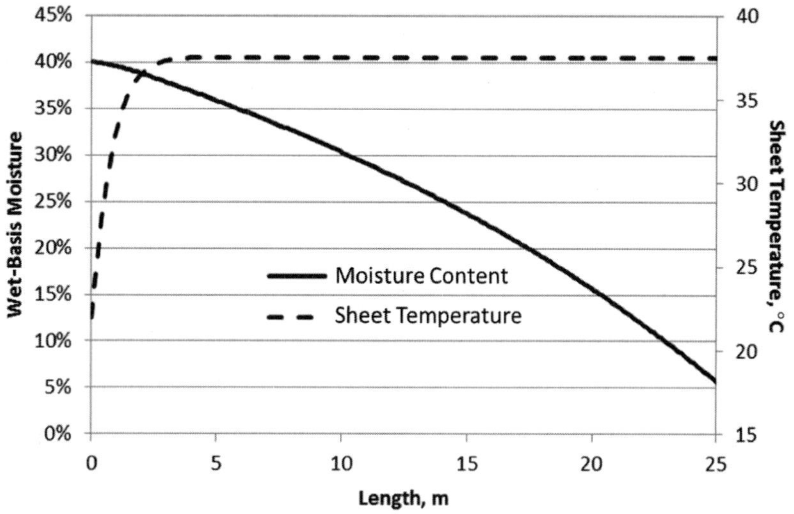

FIG. 12-110 Simulation results for sheet drying example.

The two equations above were solved in stepwise explicit manner with a spreadsheet. Each row represents a small time step. The temperature from the previous time step was used to calculate all the changing quantities in the new time step.

Results from this calculation are shown in Fig. 12-110. For the base conditions, the results show that a dryer length of 23.0 m is needed.

In Table 12-47, the results for all the cases are shown.

The results show us the relative sensitivity of the process to some typical methods to increase the drying rate and therefore reduce the dryer size. The biggest handle is the reduction of the initial moisture content. This is a nonlinear effect with wet-basis moisture content. In practice, this is often accomplished by a mechanical dewatering process upstream of thermal drying. The dry-basis moisture content is 0.667 for the base case and 0.35/0.65 = 0.538. So a reduction from 40 percent to 35 percent on a wet basis equates to a drying load reduction of nearly 20 percent,

$$(1 - 0.538/0.667) \times 100 \text{ percent.}$$

The heat-transfer calculations [Eqs. (12-110) through (12-113)] used in the Drum Drying subsection are directly applicable for heating/drying of sheets in contact with rollers.

Freeze Dryers* In freeze drying (lyophilization), the feed material is frozen and ice sublimes directly to vapor. This gives gentle drying, preserves heat-sensitive materials, and preserves the solid structure without shrinkage and deformation. The process must operate at temperatures and vapor partial pressures below the triple point (0.006 bar and 0.01°C for water), normally by operating at high vacuum (vacuum freeze drying—see Fig. 12-111), although atmospheric freeze drying with highly dehumidified air is sometimes possible. Because of the low driving forces, freeze drying is generally an expensive option in both equipment and operating cost, with typical process times of hours or days.

Applications Freeze drying is mainly used for high-value products where the gain in product quality justifies the high cost, particularly in the food, beverage, and pharmaceutical industries. The major advantages are as follows:

- Preservation of original flavor, aroma, color, shape, and texture (or development of special food textures and flavor effects)
- Retention of original distribution of soluble substances such as sugars, salts, and acids, which can migrate to the product surface in conventional drying
- Negligible shrinkage, resulting in excellent and near-instantaneous rehydration characteristics
- Negligible product loss
- Minimal risk of cross-contamination

Freeze drying is used for selected vegetables, fruits, meat, fish, and beverage products, such as instant coffee (flavor and aroma retention), strawberries (color preservation), and chives (shape preservation). For pharmaceuticals, solutions of sterile products may be dried in glass vials, or blocks or slabs of material may be dried on trays.

The freeze-drying process Industrial freeze drying is carried out in three steps:

1. Freezing of the feed material
2. Primary drying, i.e., sublimation drying of the main ice content, corresponding to the constant-rate period in conventional drying
3. Secondary drying, i.e., desorption drying of the internal or bound moisture or hydrates, corresponding to the falling-rate period in conventional drying

Unlike many conventional processes, the primary drying period is usually the longest. The secondary drying period is relatively short and may run at a higher temperature.

Freezing The freezing methods applied for solid products are all conventional freezing methods such as blast freezing, individual quick freezing (IQF), or similar. To ensure good stability of the final product during storage, a product temperature of −20 to −30°C should be achieved to ensure that more than 95 percent of the free water is frozen. Freezing may be performed within the dryer or externally. External freezing can debottleneck a dryer being used for repeated cycles, e.g., in the food industry where 2 to 3 batches may be run per day.

TABLE 12-47 Results for Example 12-23: Impinging Air Drying of Sheets

	Cases			
	1	2	3	4
Air temperature, °C	120	130	120	120
Relative humidity of air (before heating), percent	40	40	10	40
Initial wet-basis moisture content, percent	40	40	40	35
Dryer length needed, m	23	20.9	22	17.8

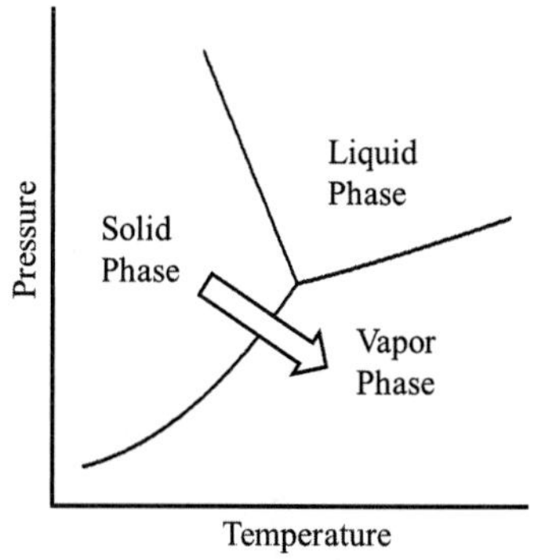

FIG. 12-111 Phase diagram for freeze drying.

*Special thanks are due to Prof. A. Basseri, University of Turin, Italy, for his input to this subsection.

The free water freezes to pure ice crystals, leaving the soluble substances as high concentrates or even crystallized. Solid products maintain their natural cell structure, as long as the ice crystals are small enough to avoid damaging the cells. For liquid feeds (with no cell structure) the product structure is formed by the freezing process as an intercrystalline matrix of the concentrated product around the ice crystals.

Freezing rate is a key parameter; small ice crystals are obtained by quick freezing, while slow freezing gives larger ice crystals. The structure of the matrix can affect the freeze-drying performance as well as the appearance, mechanical strength, and solubility rate. Small ice crystals lead to light color (high surface reflection of light) and a good mechanical strength of the freeze-dried product, but give diffusion restrictions for vapor transport inside the product (particularly for solutions) and hence slower drying. Large ice crystals lead to the opposite results. An optimum may be achieved by initial fast freezing followed by annealing at a higher temperature, allowing crystal growth and a more porous structure and significantly reducing primary drying time, as shown in Table 12-48.

Thus the freezing method must be carefully adapted to the quality criteria of the finished product. Common methods include

- Drum freezing, by which a thin slab of 1.5 to 3 mm is frozen within 1.5 to 3 min
- Belt freezing, by which a slab of 6 to10 mm passing through different freezing zones is frozen during 10 to 20 min
- Shelf freezing in situ in the dryer, particularly for pharmaceuticals
- Foaming, used to influence the structure and mainly to control the density of the freeze-dried product

Batch Freeze Dryers Freeze dryers are normally multishelf units with the product in trays or multiple glass vials. The main components are (1) the vacuum chamber, heating plates, and vapor traps, all built into the freeze dryer and (2) the external systems, such as the transport system for the product trays, the deicing system, and the support systems for supply of heat, vacuum, and refrigeration.

In some units, the trays are carried in tray trolleys suspended in an overhead rail system for easy transport and quick loading and unloading, as illustrated in Fig. 12-112. Heat transfer is by conduction from the shelf below and radiation from the shelves above and below. As driving forces are low, a large heating surface is desirable. The heating plates should be at uniform temperatures, not exceeding 2°C to 3°C across the dryer, so the distribution of the heating medium (usually silicone thermal oil) to the heating plates and the flow rate inside the plates are very important factors.

If the product has been frozen externally, the operation vacuum should be reached quickly (within 10 min) to avoid the risk of product melting, and the heating plates are cooled to approximately 25°C. When the operating vacuum is achieved, the heating plate temperature is raised quickly to the desired operating temperature. Important features of a modern freeze-drying plant include a built-in vapor trap, allowing a large opening for vapor flow to the condenser, and a *continuous deicing* (CDI) system, reducing the ice layer on the condenser to a maximum of 6 to 8 mm. At 1 mbar pressure, the vapor flow rate is typically about 1 $m^3/(s \times m^2$ of tray area).

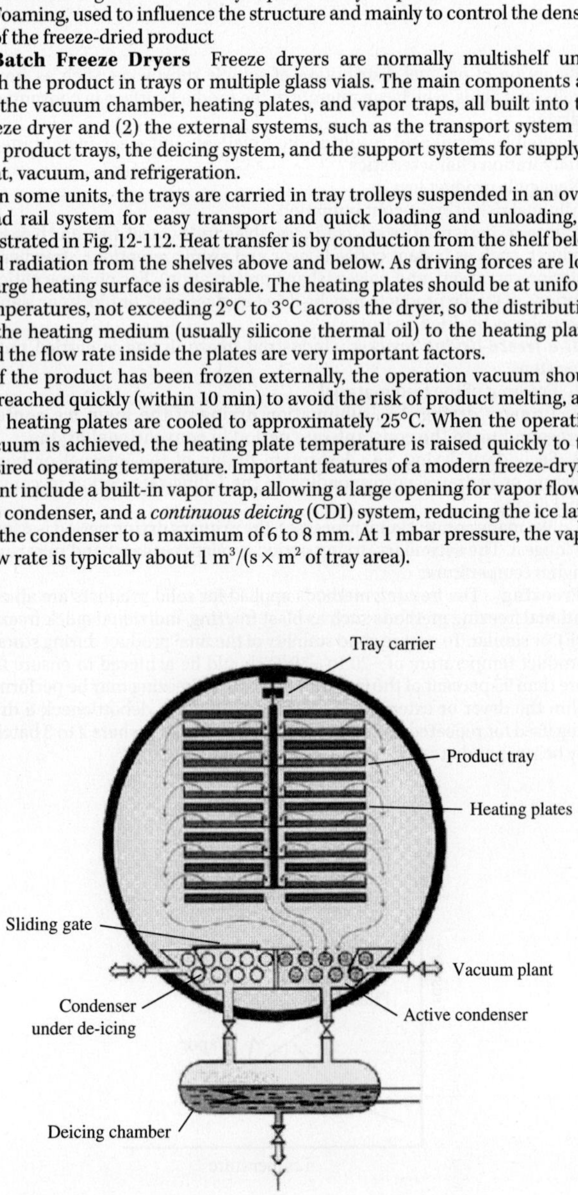

FIG. 12-112 Cross-section of RAY batch freeze dryer. (*GEA*)

Cycle Operating Conditions Low shelf temperatures give slower drying and increase the refrigeration load; about 75 percent of the energy costs relate to the refrigeration plant, and if the set temperature is 10°C lower than optimum, its energy consumption will increase by approximately 50 percent. However, although the product is kept cool by the sublimation, it must be kept below the temperature where product "collapse" or "meltback" occurs. This value can be measured at small scale with a freeze-drying microscope, and it can be chosen as an upper limit for the shelf temperature in primary drying. A control strategy is generally based on applied pressure and shelf temperature. In secondary drying, there is no danger of meltback, and higher shelf temperatures may be usable. Final moisture is typically 2 to 3 percent. Typical vacuum levels are 0.4 to 1.3 mbar absolute (40 to 130 Pa) for foods and 10 to 20 Pa for pharmaceuticals. Table 12-48 shows a typical cycle for a pharmaceutical product.

Continuous Freeze Drying The utility requirements for batch freeze dryers vary considerably over the cycle. During sublimation drying, the requirements are 2 to 2.5 times the average requirement, and the external systems must be designed accordingly. To overcome this peak load and to meet the market request for high unit capacities, continuous freeze dryer designs have been developed and implemented for coffee drying. These require vacuum locks for product both entering and leaving the dryer. As the tray stacks move through the freeze dryer, they can pass through different temperature zones to give a heating profile, selected so that overheating of dry surfaces is avoided.

Design Methods The size of the freeze-drying plant is based on the batch size and average sublimation capacity required as well as on the product type and form. The evaporation temperature of the refrigeration plant depends on the required vacuum. At 1 mbar it will be –35 to –40°C depending on the vapor trap performance. Sample data are shown in Table 12-49.

There is extensive specialized literature on freeze drying, and hands-on courses are available, e.g., from Biopharma.

Electromagnetic Drying Methods

Examples and Synonyms Infrared, radiofrequency, microwave, electromagnetic heating, dielectric heating

Description Electromagnetic drying methods employ radiation (infrared, radiofrequency, and microwave) to produce heating. In the instances of radiofrequency and microwave heating, electromagnetic radiation is absorbed by dipolar liquids, such as water or liquids containing dissolved salts. In infrared heating, heat radiates from an extremely hot element.

Generally, dielectric methods heat volumetrically and are suitable for thicker materials whereas infrared energy is considered a surface-heating method.

Since these are heating methods, drying systems also need to enable moisture removal either by using air or heating the material above the boiling temperature. These heating methods are often used in conjunction with hot air.

Dielectric Methods (Radiofrequency and Microwave) Schiffmann (1995) states that dielectric/radiofrequency heating operates in the range of 1 to 100 MHz, while microwave frequencies range from 300 MHz to 300 GHz. The electromagnetic spectrum shown in Fig. 12-113 illustrates the relative differences in wavelength and frequency between radio waves and microwaves. All electromagnetic waves are characterized by both wavelength and frequency. An example of a simplified electromagnetic wave is shown in Fig. 12-114. An electromagnetic wave is a combination of an electric component E and a magnetic component H. Note that E and H are perpendicular to each other and both are perpendicular to the direction of travel. The devices used for generating microwaves are called *magnetrons* and *klystrons* whereas the devices used to generate dielectric frequencies are referred to as *oscillators* and *triodes* or *tetrodes*.

Water molecules are dipolar (i.e., they have an asymmetric charge center) and are normally randomly oriented. The rapidly changing polarity of a microwave or radiofrequency field attempts to pull these dipoles into alignment with the field. As the field changes polarity, the dipoles return to a

TABLE 12-48 Example Pharmaceuticals Freeze-Drying Cycles, with and Without Annealing During Freezing

	No annealing		With annealing	
	Shelf temperature, °C	Time, h	Shelf temperature, °C	Time, h
Freezing	−40	6	−30, −5, −40	10
Primary drying	0	55	0	40
Secondary drying	25	2	25	2
Total	—	63	—	52

TABLE 12-49 Freeze Dryer Performance Date, Niro Ray and Conrad Types

	Tray area, m^2	Typical sublimation capacity: Flat tray, kg/h	Typical sublimation capacity: Ribbed tray, kg/h	Electricity consumption, kWh/kg, sublimated	Steam consumption, kg/kg sublimated
RAY Batch Plant—1 mbar					
RAY 75	68	68	100	1.1	2.2
RAY 100	91	91	136	1.1	2.2
RAY 125*	114	114	170	1.1	2.2
CONRAD Continuous Plant—1 mbar					
CONRAD 300	240	240	360	1.0	2.0
CONRAD 400	320	320	480	1.0	2.0
CONRAD 500*	400	400	600	1.0	2.0

*Other sizes available.

random orientation before being pulled the other way. This buildup and decay of the field, and the resulting stress on the molecules, causes a conversion of electric field energy to stored potential energy, then to random kinetic or thermal energy. Hence dipolar molecules such as water absorb energy in these frequency ranges. The power developed per unit volume P_υ by this mechanism is

$$P_\upsilon = kE^2 f\varepsilon' \tan\delta = kE^2 f\varepsilon'' \qquad (12\text{-}118)$$

where k is a dielectric constant, depending on the units of measurement, E is the electric field strength (V/m^3), f is the frequency, ε' is the relative dielectric constant or relative permeability, tan δ is the loss tangent or dissipation factor, and ε'' is the loss factor.

The field strength and the frequency are dependent on the equipment, while the dielectric constant, dissipation factor, and loss factor are material-dependent. The electric field strength is also dependent on the location of the material within the microwave/radiofrequency cavity [Turner and Ferguson, "A Study of the Power Density Distribution Generated during the Combined Microwave and Convective Drying of Softwood," *Drying Technol.* **12**(5–7): 1411–1430 (1995)], which is one reason why domestic microwave ovens have rotating turntables (so that the food is exposed to a range of microwave intensities). This mechanism is the major one for the generation of heat within materials by these electromagnetic fields.

There is also a heating effect due to ionic conduction. The water inside a material may contain ions such as sodium, chloride, and hydroxyl; these ions are accelerated and decelerated by the changing electric field. The collisions that occur as a result of the rapid accelerations and decelerations lead to an increase in the random kinetic (thermal) energy of the material. This type of heating is not significantly dependent on either temperature or frequency. The power developed per unit volume P_υ from this mechanism is

$$P_\upsilon = E^2 qn\mu \qquad (12\text{-}119)$$

where q is the amount of electric charge on each of the ions, n is the charge density (ions/m^3), and μ is the level of mobility of the ions.

Schiffmann (1995) indicates that the dielectric constant of water is more than an order of magnitude higher than that of most underlying materials, and the overall dielectric constant of most materials is usually nearly proportional to moisture content up to the critical moisture content, often around 20 to 30 percent. Hence microwave and radiofrequency methods preferentially heat and dry wetter areas in most materials, a process which tends to give more-uniform final moisture contents. The dielectric constant of air is very low compared with that of water, so lower density usually means lower heating rates. For water and other small molecules, the effect of increasing temperature is to decrease the heating rate slightly, hence leading to a self-limiting effect.

Other effects (frequency, conductivity, specific heat capacity, etc.) are discussed by Schiffmann (1995), but are less relevant because the range of available frequencies (which do not interfere with radio transmissions) is small (2.45 GHz, 910 MHz). Higher frequencies lead to lower penetration depths into a material than lower frequencies do. Sometimes the 2.45-GHz frequency has a penetration depth as low as 2.5 cm (1 in). For in-depth heating (*volumetric heating*), radio frequencies with lower frequencies and longer wavelengths are often used. Note that not all frequencies are available to use in all geographies as designated by the International Telecommunication Union, as seen in Table 12-50.

Also note that microwave and radiofrequency generators are often used in conjunction with other dryer types to enhance drying rates, especially in thicker materials. Achieving uniform heating is challenging when using these electromagnetic heating methods.

Infrared Methods Infrared (IR) radiation is commonly used in the dehydration of coated films and to even out the moisture content profiles in the drying of paper and boards. The mode of heating is essentially on the material surface, and IR sources are relatively inexpensive compared with dielectric sources.

The heat flux obtainable from an IR source is given by

$$q = F\alpha\varepsilon\left(T^4_{\text{source}} - T^4_{\text{drying material}}\right) \qquad (12\text{-}120)$$

where q = heat flux, W/m^2; α = Stefan-Boltzmann constant = 5.67×10^{-8} W/(m$^2\cdot$K^4); ε = emissivity (0 to 1); F = view factor; and T = absolute temperature of the source or drying material.

The emissivity is a property of the material. The limiting value is 1 (blackbody); shiny surfaces have a low value of emissivity. The view factor is a fractional value that depends on the geometric orientation of the source with respect to the heating object.

It is very important to recognize the T^4 dependence on the heat flux. IR sources need to be very hot to give appreciable heat fluxes. Therefore, IR sources should not be used with flammable materials. Improperly designed IR systems can also overheat materials and equipment.

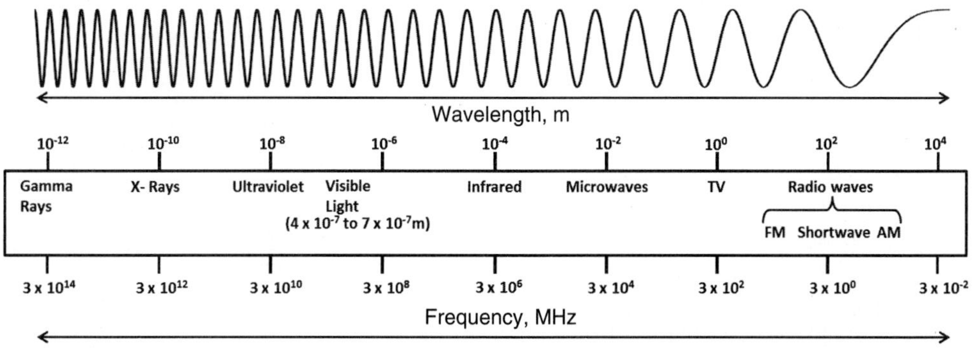

FIG. 12-113 Electromagnetic spectrum.

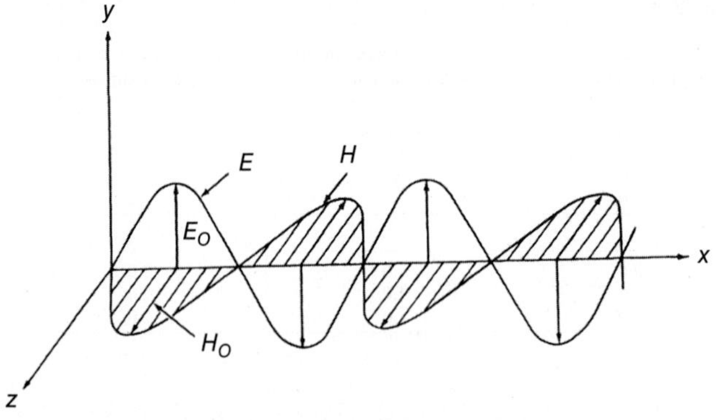

FIG. 12-114 Illustration of a simple EM wave. Here E and H represent the electrical and magnetic components of the wave, respectively. E_0 and H_0 show their respective amplitudes. (*Reprinted with permission from Mujumdar, A.S., Handbook of Industrial Drying, 2nd ed., Marcel Dekker Inc., New York, 1995.*)

Example 12-24 Sheet Drying with Convection and Infrared In the same sheet drying problem as in Example 12-23, the belt has been removed, and now the sheet to be dried will be suspended in air using tensioning rollers (see Fig. 12-104). The same impinging air heat-transfer process occurs at the bottom of the web, but now flat infrared heat panels will be installed along the length of the dryer above the sheet.

Calculate the dryer length needed to dry the material from the inlet at 40 percent wet basis to the outlet at 10 percent dry basis without infrared and with the infrared on with panel temperatures at 300, 500, and 700°C. Assume the emissivity and view factors are unity.

We will use the energy and mass balance equations as before, except we will include the infrared term. We will assume that the web is thin enough that we can neglect temperature gradients through its thickness.

$$(C_{p,\text{solids}} + C_{p,\text{water}} \cdot w_{\text{avg,dry-basis}}) \cdot \frac{m_{\text{sheet}}}{A} \cdot \frac{dT_{\text{sheet}}}{dt} = h \cdot (T_{\text{air}} - T_{\text{sheet}}) + F_{\text{View}}\sigma\varepsilon\left(T^4_{\text{source}} - T^4\right) - F \cdot \Delta H_{\text{vap}} \qquad (12\text{-}121)$$

We also write a mass balance:

$$\frac{dm_w}{dt}\frac{1}{A} = k_c \cdot (C_w^{\text{surface}} - C_w^{\text{bulk}}) \qquad (12\text{-}122)$$

Results are shown in Table 12-51.

The calculation indicates that the drying rate can be greatly accelerated by the infrared. However, there are some idealizations such as the view factor and emissivity value. Running infrared panels can introduce problems since they apply heat to anything they

TABLE 12-50 Frequency Designation by the International Telecommunication Union (Schiffmann, 1995)

Center frequency (MHz)	Frequency range (MHz)	Maximum radiation limit*
6.780	6.765–6.795	Under consideration
13.560	13.553–13.567	Unrestricted
27.120	26.957–27.283	Unrestricted
40.680	40.660–40.700	Unrestricted
433.920	433.050–434.790	Under consideration
915.000	902–928	Unrestricted
2450	2400–2500	Unrestricted
5800	5725–5875	Unrestricted
24,125	24,000–24,250	Under consideration
61,250	61,000–61,500	Under consideration
122,500	122,000–123,000	Under consideration
245,000	240,000–246,000	Under consideration

*The term *unrestricted* applies to the fundamental and all other frequency components falling within the designated band. Special measures to achieve compatibility may be necessary where other equipment satisfying immunity requirements is placed close to ISM equipment.

TABLE 12-51 Results for Example 12-24: Sheet Drying with Convection and Infrared

Distance, m	IR panel temperature, °C
23.0	None
13.4	300
6.4	500
3.0	700

"see" and they need to be extremely hot. The panels must be shut off if the line stops, or else the material may be damaged or pose a fire risk. Infrared heating needs special considerations if the process is expected to create volatile or flammable materials.

Additional Reading

Schiffman, R, "Microwave and Dielectric Drying," chap. 11 in Mujumdar, A., *Handbook of Industrial Drying*, 2d ed., Marcel Dekker, New York, 1995, pp. 345–372.

OPERATION AND TROUBLESHOOTING

Troubleshooting Dryer troubleshooting is not extensively covered in the literature, but a systematic approach has been proposed in Kemp and Gardiner, "An Outline Method for Troubleshooting and Problem-Solving in Dryers," *Drying Technol.* **19**(8): 1875–1890 (2001). The main steps of the algorithm are as follows:

- *Problem definition*—definition of the dryer problem to be solved
- *Data gathering*—collection of relevant information, e.g., plant operating data
- *Data analysis*, e.g., heat and mass balance, *and identification of the cause of the problem*
- *Conclusions and actions*—selection and implementation of a solution in terms of changes to process conditions, equipment, or operating procedures
- *Performance auditing*—monitoring to ensure that the problem was permanently solved

The algorithm might also be considered as a "plant doctor." The doctor collects data, or symptoms, and makes a diagnosis of the cause(s) of the problem. Then alternative solutions, or treatments, are considered and a suitable choice is made. The results of the treatment are reviewed (i.e., the process is monitored) to ensure that the "patient" has returned to full health. See Fig. 12-115.

The algorithm is an excellent example of the "divergent-convergent" (brainstorming) method of problem solving. It is important to list all possible causes and solutions, no matter how ridiculous they may initially seem; there may actually be some truth in them, or they may lead to a new and better idea.

Problem Categorization In the problem definition stage, it is extremely useful to categorize the problem, as the different broad groups require different types of solution. Five main categories of dryer problems can be identified:

1. Drying process performance (outlet moisture content too high, throughput too low)
2. Materials handling (dried material too sticky to get out of dryer, causing blockage)
3. Product quality (too many fines in product, bulk density too low/high, discoloration, etc.)
4. Mechanical breakdown (catastrophic sudden failure)
5. Safety, health, and environmental issues (drying air temperature too high, buildup of material in dryer, etc.)

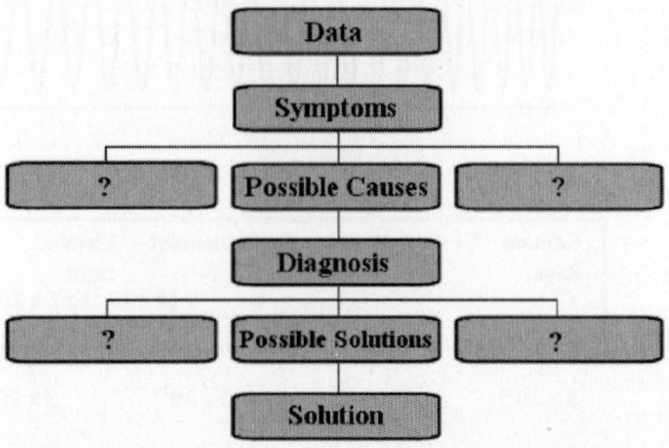

FIG. 12-115 Schematic diagram of algorithm for dryer troubleshooting.

Experience suggests that the majority of problems are of the first three types, and these are about equally split over a range of industries and dryer types. Ideally, unforeseen safety, health, and environmental issues will be rare, as these will have been identified in the safety case before the dryer is installed or during commissioning. Likewise, major breakdowns should be largely avoided by a planned maintenance program.

Drying Performance Problems Performance problems can be further categorized as

1. Heat and mass balance deficiencies (not enough heat input to do the evaporation)
2. Drying kinetics (drying too slowly, or solids residence time in dryer is too short)
3. Equilibrium moisture limitations (reaching a limiting value, or regaining moisture in storage)

For the heat and mass balance, the main factors are

- Solids throughput
- Inlet and outlet moisture content
- Temperatures and heat supply rate
- Leaks and heat losses

As well as problem-solving, these techniques can be used for performance improvement and debottlenecking.

Drying kinetics, which are affected by temperature, particle size, and structure, are limited by external heat and mass transfer to and from the particle surface in the early stages; but internal moisture transport is the main parameter at lower moisture.

Equilibrium moisture content increases with higher relative humidity, or with lower temperature. Problems that depend on the season of the year, or vary between day and night (both suggesting a dependence on ambient temperature and humidity), are often related to equilibrium moisture content.

Materials Handling Problems The vast majority of handling problems in a dryer concern sticky feedstocks. Blockages can be worse than performance problems as they can close down a plant completely, without warning. Most stickiness, adhesion, caking, and agglomeration problems are due to *mobile liquid bridges* (surface moisture holding particles together). These are extensively described in particle technology textbooks. Unfortunately, these forces tend to be at a maximum when the solid forms the continuous phases and surface moisture is present, which is the situation for most filter and centrifuge cakes at discharge. By comparison, slurries (where the liquid forms the continuous phase) and dry solids (where all surface moisture has been eliminated) are relatively free-flowing and incur fewer problems.

Other sources of problems include electrostatics (most marked with fine and dry powders) and *immobile liquid bridges*, the so-called "sticky-point phenomenon." This latter is sharply temperature-dependent, with only a weak dependence on moisture content, in contrast to mobile liquid bridges. It occurs for only a small proportion of materials, but is particularly noticeable in amorphous powders and foods and is often linked to the glass transition temperature.

Product Quality Problems (These do not include the moisture level of the main solvent.) Many dryer problems either concern product quality or cannot be solved without considering the effect of any changes on product quality. Thus it is a primary consideration in most troubleshooting, although product quality measurements are specific to the particular product, and it is difficult to generalize. However, typical properties may include color, taste (not easily quantifiable), bulk density, viscosity of a paste or dispersion, dispersibility, or rate of solution. Others are more concerned with particle size, size distribution (e.g., coarse or fine fraction), or powder handling properties such as rate of flow through a standard orifice. These property measurements are nearly always made off-line, either by the operator or by the laboratory, and many are very difficult to characterize in a rigorous quantitative manner. (See also the Fundamentals subsection.)

Storage problems, very common in industry, result if the product from a dryer is free-flowing when packaged, but has caked and formed solid lumps by the time it is received by the customer. Sometimes the entire internal contents of a bag or drum have welded together into a huge lump, making it impossible to discharge.

Depending on the situation, there are at least three different possible causes:

1. *Equilibrium moisture content:* hygroscopic material is absorbing moisture from the air on cooling.
2. *Incomplete drying:* product is continuing to lose moisture in storage.
3. *Psychrometry:* humid air is cooling and reaching its dew point.

The three types of problem have some similarities and common features, but the solution to each one is different. Therefore, it is essential to understand which mechanism is actually in play.

Option 1: The material is hygroscopic and is absorbing moisture back from the air in storage, where the cool air has a higher relative humidity than the hot dryer exhaust. *Solution:* Pack and seal the solids immediately on discharge in tough impermeable bags (usually double- or triple-lined to reduce the possibility of tears and pinholes), and minimize the ullage (airspace above the solids in the bags) so that the amount of moisture that can be absorbed is too low to cause any significant problem. Dehumidifying the air to the storage area is also possible, but often very expensive.

Option 2: The particles are emerging with some residual moisture, and they continue to dry after being stored or bagged. As the air and solids cool, the moisture in the air comes out as dew and condenses on the surface of the solids, causing caking by mobile liquid bridges. *Solution:* If the material is meeting its moisture content specification, cool the product more effectively before storage, to stop the drying process. If the outlet material is wetter than stated in the specification, alter dryer operating conditions or install a postdryer.

Option 3: Warm, wet air is getting into the storage area or the bags, either because the atmosphere is warm with a high relative humidity (especially in the tropics) or because dryer exhaust air has been allowed to enter. As in option 2, when the temperature falls, the air goes below its dew point and condensation occurs on the walls of the storage area or inside the bags, or on the surface of the solids, leading to caking. *Solution:* Avoid high-humidity air in the storage area. Ensure the dryer exhaust is discharged a long way away. If the ambient air humidity is high, consider cooling the air supply to storage to bring it below its dew point and reduce its absolute humidity.

Dryer Operation

Start-Up Considerations It is important to start up the heating system before introducing product into the dryer. This will minimize condensation and subsequent product buildup on dryer walls. It is also important to minimize off-quality production by not overdrying or underdrying during the start-up period. Proper control system design can aid in this regard. The dryer turndown ratio is also an important consideration during start-up. Normally the dryer is started up at the lowest end of the turndown ratio, and it is necessary to match heat input with capacity load.

Shutdown Considerations The sequence for dryer shutdown is also very important and depends on the type of dryer. The sequence must be thoroughly thought through to prevent significant off-quality product or a safety hazard. The outlet temperature during shutdown is a key operating variable to follow.

Energy Considerations The first consideration is to minimize the moisture content of the dryer feed, e.g., with dewatering equipment, and to establish as high an outlet product moisture target as possible. Other energy considerations vary widely by dryer type. In general, heating with gas, fuel oil, and steam is significantly more economical than heating with electricity. Hence radiofrequency (RF), microwave, and infrared drying is energy-intensive. Direct heating is more efficient than indirect in most situations. Sometimes air recycle (direct or indirect) can be effective in reducing energy consumption. And generally operating at high inlet temperatures is more economical.

Recycle In almost all situations, the process system must be able to accommodate product recycle. The question is how to handle it most effectively, considering the product quality, equipment size, and energy.

Improvement Considerations *The first consideration is to evaluate mass and energy balances to identify problem areas.* See the Experimental Methods part of this subsection for guidance on how to conduct a mass and energy balance on an industrial dryer. This will identify air leaks and excessive equipment heat losses and will enable determination of overall energy efficiency.

A simplified heat balance will show what might need to be done to debottleneck a convective (hot gas) dryer, i.e., increase its production rate F.

$$F(X_I - X_O)\lambda_{\text{evap}} \approx GC_{PG}(T_{GI} - T_{GO}) - Q_{wl}$$

Before proceeding along this line, however, it is necessary to establish that the dryer is genuinely heat and mass balance–limited. If the system is controlled by kinetics or equilibria, changing the parameters may have undesirable side effects, e.g., increasing the product moisture content.

The major alternatives are then as follows (assuming gas specific heat capacity C_{PG} and latent heat of evaporation λ_{evap} are fixed):

1. Increase the gas flow rate G, as it usually increases pressure drop, so new fans and gas cleaning equipment may be required.
2. Increase the inlet gas temperature T_{GI} which is usually limited by risk of thermal damage to product.
3. Decrease the outlet gas temperature T_{GO}. But note that this increases NTUs, outlet humidity, and relative humidity and reduces both the temperature and humidity driving forces. Hence it may require a longer drying time and a larger dryer, and it may also increase equilibrium and outlet moistures.
4. Reduce inlet moisture content X_I, say, by dewatering by gas blowing, centrifuging, vacuum or pressure filtration, or a predryer.
5. Reduce heat losses Q_{wl} by insulation, removing leaks, etc.

Dryer Safety This subsection discusses some of the key considerations in dryer safety. General safety considerations are discussed in Sec. 23, Process Safety, and should be referred to for additional guidance.

Fires, explosions, and, to a lesser extent, runaway decompositions are the primary hazards associated with drying operations. The outbreak of fire is a result of ignition which may or may not be followed by an explosion. A hazardous situation is possible if

1. The product is combustible.
2. The product is wetted by a flammable solvent.
3. The dryer is direct-fired.

An explosion can be caused by dust or flammable vapors, both of which are fires that rapidly propagate, causing a pressure rise in a confined space.

Dust Explosions Dispersion dryers can be more hazardous than layer-type dryers if one is drying a solid combustible material which is then dispersed in air, particularly if the product is a fine particle size. If this finely dispersed product is then exposed to an ignition source, an explosion can result. The following conditions (van't Land, *Industrial Drying Equipment*, Marcel Dekker, New York, 1991) will be conducive to fire and explosion hazard:

1. Small particle sizes, generally less than 75 μm, which are capable of propagating a flame
2. Dust concentrations within explosive limits, generally 10 to 60 g/m^3
3. Ignition source energy of 10 to 1000 mJ or as low as 5 mJ for highly explosive dust sources
4. Atmosphere supporting combustion

Since most product and hence dust compositions vary widely, it is generally necessary to do quantitative testing in approved test equipment.

Flammable Vapor Explosions This can be a problem for products wetted by flammable solvents if the solvent concentration exceeds 0.2 percent v/v in the vapor phase. The ignition energy of vapor-air mixtures is lower (<1 mJ) than that of dust-air suspensions. Many of these values are available in the literature, but testing may sometimes be required.

Ignition Sources There are many possible sources of an ignition, and they need to be identified and addressed by both designers and operators. A few of the most common ignition sources are

1. Spontaneous combustion
2. Electrostatic discharge
3. Electric or frictional sparks
4. Incandescent solid particles from heating system

Safety hazards must be addressed with proper dryer design specifications. The following are a few key considerations in dryer design.

Inert system design The dryer atmosphere is commonly made inert with nitrogen, but superheated steam or self-inertized systems are also possible. Self-inertized systems are not feasible for flammable solvent systems. These systems must be operated with a small overpressure to ensure no oxygen ingress. And continuous on-line oxygen concentration monitoring is required to ensure that oxygen levels remain well below the explosion hazard limit.

Relief venting Relief vents that are properly sized relieve and direct dryer explosions to protect the dryer and personnel if an explosion does occur. Normally they are simple pop-out panels with a minimum length of ducting to direct the explosion away from personnel or other equipment.

Suppression systems Suppression systems typically use an inert gas such as carbon dioxide to minimize the explosive peak pressure rise and fire damage. The dryer operating pressure must be properly monitored to detect the initial pressure rise followed by shutdown of the dryer operating systems and activation of the suppression system.

Clean design Care should be taken in the design of both the dryer and dryer ancillary equipment (cyclones, filters, etc.) to eliminate ledges, crevices, and other obstructions that can lead to dust and product buildup. Smooth drying equipment walls will minimize deposits. This can go a long way in prevention. No system is perfect, of course, and a routine cleaning schedule is also recommended.

Start-up and shutdown Start-up and shutdown situations must be carefully considered in designing a dryer system. These situations can create higher than normal dust and solvent concentrations. This coupled with elevated temperatures can create a hazard well beyond that of normal continuous operation.

Environmental Considerations Environmental considerations are continuing to be an increasingly important aspect of dryer design and operation as environmental regulations are tightened. The primary environmental problems associated with drying are particulate and *volatile organic compound* (VOC) emissions. Noise can be an issue with certain dryer types.

Environmental Regulations These vary by country, and it is necessary to know the specific regulations in the country in which the dryer will be installed. It is also useful to have some knowledge of the direction of regulations so that the environmental control system is not obsolete by the time it becomes operational.

Particulate emission problems can span a wide range of hazards. Generally, there are limits on both toxic and nontoxic particles in terms of annual and peak emissions limits. Particles can present toxic, bacterial, viral, and other hazards to human, animal, and plant life.

Likewise, VOC emissions can span a wide range of hazards and issues from toxic gases to smelly gases.

Environmental Control Systems We should consider environmental hazards before the drying operation is even addressed. The focus should be on minimizing the hazards created in the upstream processing operations. After potential emissions are minimized, these hazards must be dealt with during dryer system design and then subsequently with proper operational and maintenance procedures.

Particle Emission Control Equipment The four most common methods of particulate emissions control are as follows:

1. *Cyclone separators* The advantage of cyclones is they have relatively low capital and operating costs. The primary disadvantage is that they become increasingly ineffective as the particle size decreases. As a general rule of thumb, we can say that they are 100 percent efficient with particles larger than 20 μm and 0 percent efficient with particles smaller than 1 μm. Cyclones can also be effective precleaning devices to reduce the load on downstream bag filters.
2. *Scrubbers* The more general classification is wet dedusters, the most common of which is the wet scrubber. The advantage of wet scrubbers is that they can remove fine particles that the cyclone does not collect. The disadvantages are that they are more costly than cyclones and they can turn air contamination into water contamination, which may then require additional cleanup before the cleaning water is put into the sewer.
3. *Bag filters* The advantages of filters are that they can remove very fine particles; and bag technologies continue to improve and enable ever-smaller particles to be removed without excessive pressure drops or buildup. The primary disadvantages are higher cost relative to cyclones and greater maintenance costs, especially if frequent bag replacement is necessary.
4. *Electrostatic precipitators* The capital cost of these systems is relatively high, and maintenance is critical to effective operation.

VOC Control Equipment The four most prevalent equipment controls are

1. *Scrubbers* Similar considerations as above apply.
2. *Absorbers* These systems use a high-surface-area absorbent, such as activated carbon, to remove the VOC absorbate.
3. *Condensers* These systems are generally only feasible for recovering solvents from nonaqueous wetted products.
4. *Thermal and catalytic incinerators* These can be quite effective and are generally a low capital and operating cost solution, except in countries with high energy costs.

Noise Noise analysis and abatement is a very specialized area. Generally, the issue with dryers is associated with the fans, particularly for systems requiring fans that develop very high pressures. Noise is a very big issue that needs to be addressed with pulse combustion dryers, and it can be an issue with very large dryers such as rotary dryers and kilns.

Additional considerations regarding environmental control and waste management are addressed in Sec. 22, Waste Management, and Sec. 23, Process Safety.

Control and Instrumentation The purpose of the control and instrumentation system is to provide a system that enables the process to produce the product at the desired moisture target and to meet other quality control targets discussed earlier (density, particle size, color, solubility, etc.). This segment discusses key considerations for dryer control and instrumentation. Additional more-detailed information can be found in Sec. 8, Process Control.

Proper control of product quality starts with the dryer selection and design. Sometimes two-stage or multistage systems are required to meet product quality targets. Multistage systems enable us to better control temperature and moisture profiles during drying. Assuming the proper dryer design has been selected, we must then design the control and instrumentation system to meet all product quality targets.

Manual versus Automatic Control Dryers can be controlled either manually or automatically. Generally, lab-, pilot-, and small-scale production units are controlled manually. These operations are usually batch systems, and manual operation provides lower cost and greater flexibility. The preferred mode for large-scale, continuous dryers is automatic.

Key Control Variables Product moisture and product temperature are key control variables. Ideally both moisture and temperature measurement are done on-line, but frequently moisture measurement is done off-line and temperature (or exhaust air temperature) becomes the primary control variable. And generally the inlet temperature will control the rate of production, and the outlet temperature will control the product moisture and other product quality targets.

Common Control Schemes Two relatively simple, but common control schemes in many dryer systems (Fig. 12-116) are as follows:

1. The outlet air temperature is controlled by feed rate regulation with the inlet temperature controlled by gas heater regulation.
2. The outlet air temperature is controlled by heater regulation with the feed rate held constant.

Alternatively, product temperatures can replace air temperatures with the advantage of better control and the disadvantage of greater maintenance of the product temperature sensors.

Other Instrumentation and Control

Pressure Pressure and equipment pressure drops are important to proper dryer operation. Most dryers are operated under vacuum. This prevents dusting to the environment, but excess leakage in decreases dryer efficiency. Pressure drops are especially important for stable fluid-bed operation.

Air (gas) flow rate Obviously gas flows are another important parameter for proper dryer operation. Pitot tubes are useful when a system has no permanent gas flow sensors. Averaging pitot tubes work well in permanent installations. The devices work best in straight sections of ductwork which are sometimes difficult to find and make accurate measurement a challenge.

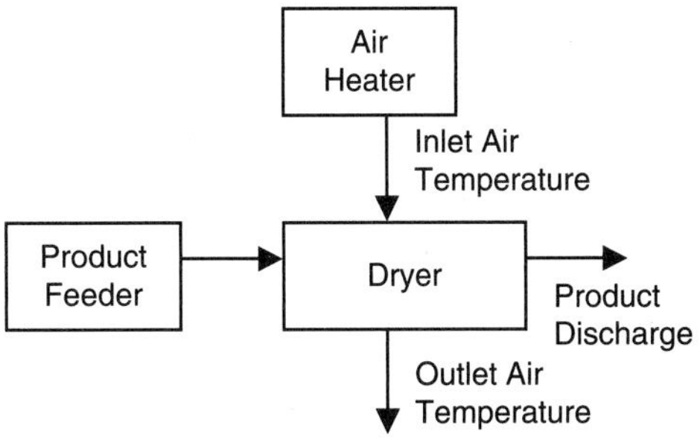

FIG. 12-116 Typical dryer system.

Product feed rate It's important to know that product feed rates and feed rate changes are sometimes used to control finished-product moistures. Weigh belt feeders are common for powdered products, and there is a wide variety of equipment available for liquid feeds. Momentum devices are inexpensive but less accurate.

Humidity The simplest method is sometimes the best. Wet- and dry-bulb temperature measurement to get air humidity is simple and works well for the occasional gas humidity measurement. The problem with permanent humidity measurement equipment is the difficulty of getting sensors robust enough to cope with a hot, humid, and sometimes dusty environment. If these are used, be careful about placement and inspection to ensure that product does not accumulate on the sensor.

Interlocks Interlocks are another important feature of a well-designed control and instrumentation system. Interlocks are intended to prevent damage to the dryer system or to personnel, especially during the critical periods of start-up and shutdown. The following are a few key interlocks to consider in a typical dryer system.

Drying chamber damage This type of damage can occur when the chamber is subjected to significant vacuum when the exhaust fans are started up before the supply fans.

Personnel injury This interlock is to prevent injury due to entering the dryer during operation, but more typically to prevent dryer start-up with personnel in the main chamber or inlet or exhaust air ductwork on large dryers. This typically involves microswitches on access doors coupled with proper door lock devices and tags.

Assurance of proper start-up and shutdown These interlocks ensure, e.g., that the hot air system is started up before the product feed system and that the feed system is shut down before the hot air system.

Heater system There are a host of important heater system interlocks to prevent major damage to the entire drying system. Additional details can be found in Sec. 23, Process Safety.

Section 13

Distillation

Michael F. Doherty, Ph.D. *Professor of Chemical Engineering, University of California—Santa Barbara (Section Editor)*

Zbigniew T. Fidkowski, Ph.D. *Process Engineer, Evonik Industries (Distillation Systems, Batch Distillation)*

M. F. Malone, Ph.D. *Professor of Chemical Engineering and Vice-Chancellor for Research and Engagement, University of Massachusetts—Amherst (Batch Distillation, Enhanced Distillation)*

Ross Taylor, Ph.D. *Distinguished Professor of Chemical Engineering, Clarkson University (Simulation of Distillation Processes)*

Certain portions of this section draw heavily on the work of J. D. Seader, Jeffrey J. Siirola, and Scott D. Barnicki, authors of this section in the 7th edition.

DEGREES OF FREEDOM AND DESIGN VARIABLES

DISTILLATION SYSTEMS

ENHANCED DISTILLATION

PETROLEUM AND COMPLEX-MIXTURE DISTILLATION

BATCH DISTILLATION

Nomenclature and Units

Symbol	Definition	SI units	U.S. Customary System units
A	Absorption factor		
A	Area	m^2	ft^2
C	Number of chemical species		
D	Distillate flow rate	kg·mol/s	lb·mol/h
D	Diffusion coefficient	m^2/s	ft^2/h
E	Efficiency		
E	Energy flux	kW/m^2	$Btu/(ft^2 \cdot h)$
$\mathcal{E}$	Energy transfer rate	kW	Btu/h
F	Feed flow rate	kg·mol/s	lb·mol/h
H	Column height	m	ft
H	Enthalpy	J/(kg·mol)	Btu/(lb·mol)
H	Liquid holdup	kg·mol	lb·mol
$\mathbb{H}$	Height of a transfer unit	m	ft
K	Vapor-liquid equilibrium ratio (K value)		
K_D	Chemical equilibrium constant for dimerization		
K_d	Liquid-liquid distribution ratio		
L	Liquid flow rate	kg·mol/s	lb·mol/h
N	Number of equilibrium stages		
N_c	Number of relationships		
N_i	Number of design variables		
N_{min}	Minimum number of equilibrium stages		
N_p	Number of phases		
N_r	Number of repetition variables		
N_o	Number of variables		
$\mathcal{N}$	Rate of mass transfer	kg·mol/s	lb·mol/h
N	Molar flux	$kg \cdot mol/(m^2 \cdot s)$	$lb \cdot mol/(ft^2 \cdot h)$
$\mathbb{N}$	Number of transfer units		
P	Pressure	Pa	psia
P^{sat}	Vapor pressure	Pa	psia
Q	Heat-transfer rate	kW	Btu/h
Q_c	Condenser duty	kW	Btu/h
Q_r	Reboiler duty	kW	Btu/h
R	External-reflux ratio		
R_{min}	Minimum-reflux ratio		
S	Sidestream flow rate	kg·mol/s	lb·mol/h
S	Stripping factor		
S	Vapor-sidestream ratio		
Sc	Schmidt number		
T	Temperature	K	°R
U	Liquid-sidestream rate	kg·mol/s	lb·mol/h
V	Vapor flow rate	kg·mol/s	lb·mol/h
W	Vapor-sidestream rate	kg·mol/s	lb·mol/h
X	Relative mole fraction in liquid phase		
Y	Relative mole fraction in vapor phase		
a	Activity		
a	Area	m^2	ft^2
b	Component flow rate in bottoms	kg·mol/s	lb·mol/h
c	Number of chemical species		
c	Molar density	$kg \cdot mol/m^3$	$lb \cdot mol/ft^3$
d	Component flow rate in distillate	kg·mol/s	lb·mol/h
d	Mass-transfer driving force		
e	Rate of heat transfer	kW	Btu/h
f	Component flow rate in feed	kg·mol/s	lb·mol/h
f	Fugacity	Pa	psia

Symbol	Definition	SI units	U.S. Customary System units
h	Height	m	ft
h	Heat-transfer coefficient	kW/m^2	$Btu/(ft^2 \cdot h)$
κ	Mass-transfer coefficient	m/s	ft/h
l	Component flow rate in liquid	kg·mol/s	lb·mol/h
p	Pressure	kPa	psia
q	Measure of thermal condition of feed		
q	Heat flux	kW/m^2	$Btu/ ft^2 \cdot h)$
q_c	Condenser duty	kW	Btu/h
q_r	Reboiler duty	kW	Btu/h
r	Sidestream ratio		
s	Liquid-sidestream ratio		
t	Time	s	H
u	Velocity	m/s	ft/h
v	Component flow rate in vapor	kg·mol/s	lb·mol/h
w	Weight fraction		
x	Mole fraction in liquid		
y	Mole fraction in vapor		
z	Mole fraction in feed		
	Greek Symbols		
α	Relative volatility		
γ	Activity coefficient		
ε	TBK efficiency		
ξ	Dimensionless time		
ρ	Density	kg/m^3	lb/ft^3
μ	Viscosity	N/m^2	
σ	Surface tension	N/m	
θ	Time for batch distillation	s	h
Θ	Parameter in Underwood equations		
Φ	Fugacity coefficient of pure component		
$\hat{\Phi}$	Fugacity coefficient in mixture		
Φ_A	Fraction of a component in feed vapor that is not absorbed		
Φ_S	Fraction of a component in entering liquid that is not stripped		
Ψ	Factor in Gilliland correlation		
	Subscripts and Superscripts		
EQ	Equilibrium		
f	Froth		
hk	Heavy key		
i	Component index		
j	Stage index		
L	Liquid		
lk	Light key		
MV	Murphree vapor		
o	Overall		
s	Superficial		
t	Mixture or total		
V	Vapor		
*	Equilibrium composition		
	Acronyms		
HETP	Height equivalent to a theoretical plate		
n.b.p.	Normal boiling point (1-atm pressure)		
NTU	Number of transfer units		

General References: Billet, *Distillation Engineering,* Chemical Publishing, New York, 1979; Doherty and Malone, *Conceptual Design of Distillation Systems,* McGraw-Hill, New York, 2001; Fair and Bolles, "Modern Design of Distillation Columns," *Chem. Eng.* **75**(9): 156 (Apr. 22, 1968); Fredenslund, Gmehling, and Rasmussen, *Vapor-Liquid Equilibria Using UNIFAC, A Group Contribution Method,* Elsevier, Amsterdam, 1977; Friday and Smith, "An Analysis of the Equilibrium Stage Separation Problem—Formulation and Convergence," *AIChE J.* **10:** 698 (1964); Gorak and Sorensen, *Distillation: Fundamentals and Principles,* Elsevier (2014); Gorak and Olujic, *Distillation: Equipment and Processes,* Elsevier (2014); Gorak and Schoenmakers, *Distillation: Operation and Applications,* Elsevier (2014); Hengstebeck, *Distillation—Principles and Design Procedures,* Reinhold, New York, 1961; Henley and Seader, *Equilibrium-Stage Separation Operations in Chemical Engineering,* Wiley, New York, 1981; Hoffman, *Azeotropic and Extractive Distillation,* Wiley, New York, 1964; Holland, *Fundamentals and Modeling of Separation Processes,* Prentice-Hall, Englewood Cliffs, N.J., 1975; Holland, *Fundamentals of Multicomponent Distillation,* McGraw-Hill, New York, 1981; King, *Separation Processes,* 2d ed., McGraw-Hill, New York, 1980; Kister, *Distillation Design,* McGraw-Hill, New York, 1992; Kister, *Distillation Operation,* McGraw-Hill, New York, 1990; Robinson and Gilliland, *Elements of Fractional Distillation,* 4th ed., McGraw-Hill, New York, 1950; Rousseau, ed., *Handbook of Separation Process Technology,* Wiley-Interscience, New York, 1987; Seader, J. D., "The B. C. (Before Computers) and A. D. of Equilibrium-Stage Operations," *Chem. Eng. Educ.* **14**(2): 88–103 (1985); Seader, "The Rate-based Approach for Modeling Staged Separations," *Chem. Eng. Progress,* **85**(10): 41–49 (1989); Smith, *Design of Equilibrium Stage Processes,* McGraw-Hill, New York, 1963; Seader and Henley, *Separation Process Principles,* Wiley, New York, 1998; Taylor and Krishna, *Multicomponent Mass Transfer,* Wiley, New York, 1993; Treybal, *Mass Transfer Operations,* 3d ed., McGraw-Hill, New York, 1980; *Ullmann's Encyclopedia of Industrial Chemistry,* vol. **B3,** VCH, Weinheim, 1988; Van Winkle, *Distillation,* McGraw-Hill, New York, 1967.

INTRODUCTION TO DISTILLATION OPERATIONS

GENERAL PRINCIPLES

Separation operations achieve their objective by the creation of two or more coexisting zones that differ in temperature, pressure, composition, and/or phase state. Each molecular species in the mixture to be separated responds in a unique way to differing environments offered by these zones. Consequently, as the system moves toward equilibrium, each species establishes a different concentration in each zone, and this results in a separation between the species.

The separation operation called *distillation* uses vapor and liquid phases at essentially the same temperature and pressure for the coexisting zones. Various kinds of devices such as *random* or *structured packings* and *plates* or *trays* are used to bring the two phases into intimate contact. Trays are stacked one above the other and enclosed in a cylindrical shell to form a *column.* Packings are also generally contained in a cylindrical shell between hold-down and support plates. The column may be operated continuously or in batch mode, depending on a number of factors, such as scale and flexibility of operations and the solids content of feed. A typical tray-type continuous distillation column plus major external accessories is shown schematically in Fig. 13-1.

The *feed* material, which is to be separated into fractions, is introduced at one or more points along the column shell. Because of the difference in density between vapor and liquid phases, liquid runs down the column, cascading from tray to tray, while vapor flows up the column, contacting liquid at each tray.

Liquid reaching the bottom of the column is partially vaporized in a heated *reboiler* to provide *boil-up,* which is sent back up the column. The remainder of the bottom liquid is withdrawn as *bottoms,* or bottom product. Vapor reaching the top of the column is cooled and condensed to liquid in the *overhead condenser.* Part of this liquid is returned to the column as *reflux* to provide liquid overflow. The remainder of the overhead stream is withdrawn as *distillate,* or overhead product. In some cases only part of the vapor is condensed so that a vapor distillate can be withdrawn.

This overall flow pattern in a distillation column provides countercurrent contacting of vapor and liquid streams on all the trays through the column. Vapor and liquid phases on a given tray approach thermal, pressure, and composition equilibria to an extent dependent on the efficiency of the contacting tray.

The *lighter* (lower-boiling temperature) components tend to concentrate in the vapor phase, while the *heavier* (higher-boiling temperature) components concentrate in the liquid phase. The result is a vapor phase that becomes richer in light components as it passes up the column and a liquid phase that becomes richer in heavy components as it cascades downward. The overall separation achieved between the distillate and the bottoms depends primarily on the *relative volatilities* of the components, the number of contacting trays in each column section, and the ratio of the liquid-phase flow rate to the vapor-phase flow rate in each section.

If the feed is introduced at one point along the column shell, the column is divided into an upper section, which is often called the *rectifying* section, and a lower section, which is often referred to as the *stripping* section. In *multiple-feed* columns and in columns from which a liquid or vapor *sidestream* is withdrawn, there are more than two column sections between the two end-product streams. The notion of a column section is a useful concept for finding alternative *systems* (or *sequences*) of columns for separating multicomponent mixtures, as described in the subsection Distillation Systems.

All separation operations require energy input in the form of heat or work. In the conventional distillation operation, as typified in Fig. 13-1, energy needed to separate the species is added in the form of heat to the reboiler at the bottom of the column, where the temperature is highest. Heat is also removed from a condenser at the top of the column, where the temperature is lowest. This often results in a large energy-input requirement and low overall thermodynamic efficiency, especially if the heat removed in the condenser is wasted. Complex distillation operations that offer higher thermodynamic efficiency and lower energy-input requirements have been developed and are also discussed in the subsection Distillation Systems.

Batch distillation is preferred for small feed flows or seasonal production, which is carried out intermittently in "batch campaigns." In this mode, the feed is charged to a still that provides vapor to a column where the separation occurs. Vapor leaving the top of the column is condensed to provide liquid reflux back to the column as well as a distillate stream containing the product. Under normal operation, this is the only stream leaving the device. In addition to the batch rectifier just described,

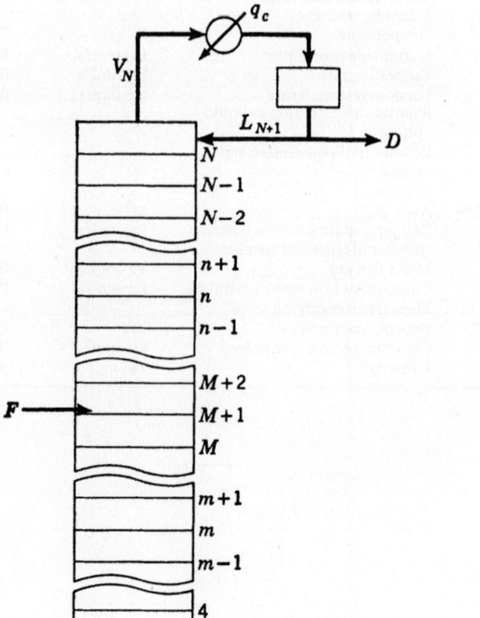

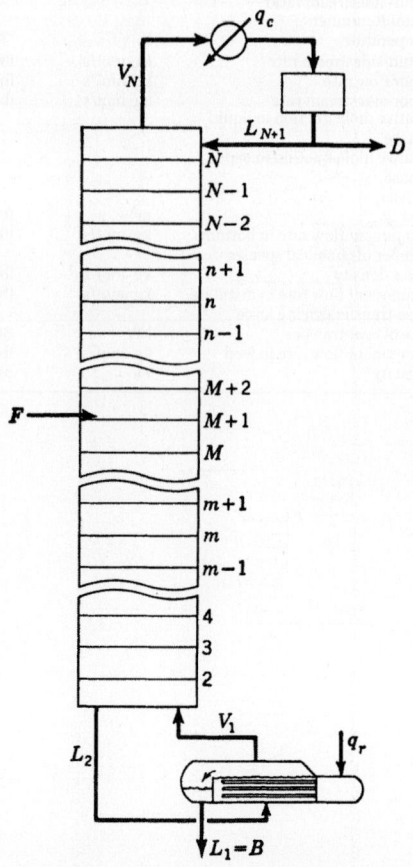

FIG. 13-1 Schematic diagram and nomenclature for a simple continuous distillation column with one feed, a total overhead condenser, and a partial reboiler.

other batch configurations are possible as discussed in the subsection Batch Distillation. Many of the concepts and methods discussed for continuous distillation are useful for developing models and design methods for batch distillation.

EQUILIBRIUM AND NONEQUILIBRIUM-STAGE CONCEPTS

The transfer processes taking place in an actual distillation column are a complicated interplay between the thermodynamic phase equilibrium properties of the mixture, the rates of intra- and interphase mass and energy transport, and multiphase flows. Simplifications are needed to develop tractable models. The landmark concept of the *equilibrium-stage model* was developed by Sorel in 1893; in this model, the liquid in each stage is considered to be well mixed, and the vapor and liquid streams leaving each stage are in thermodynamic equilibrium with each other. This is needed so that thermodynamic phase equilibrium relations can be used to determine the temperature and composition of the equilibrium streams at a given pressure. A hypothetical column composed of equilibrium stages (instead of actual contact trays) is designed to accomplish the separation specified for the actual column. The number of hypothetical equilibrium stages required is then converted to a number of actual trays by means of *tray efficiencies,* which describe the extent to which the performance of an actual contact tray duplicates the performance of an equilibrium stage. Alternatively and preferably, tray inefficiencies can be accounted for by using rate-based models that are described later in this section.

When we use the equilibrium-stage concept, we separate the design of a distillation column into three major steps: (1) Thermodynamic data and methods needed to predict equilibrium-phase compositions are assembled. (2) The number of equilibrium stages and the energy input required to accomplish a specified separation, or the separation that will be accomplished in a given number of equilibrium stages for a given energy input, are calculated. (3) The number of equilibrium stages is converted to an equivalent number of actual contact trays or height of packing, and the column diameter is determined. Much of the third step is eliminated if a rate-based model is used. This section deals primarily with equilibrium and rate-based models of distillation. Section 4 covers the first step, but a summary of methods and some useful data are included in this section. Section 14 covers equipment design.

RELATED SEPARATION OPERATIONS

The simple and complex distillation operations just described all have two things in common: (1) Both rectifying and stripping sections are provided so that a separation can be achieved between two components that are adjacent in volatility; and (2) the separation is effected only by the addition and removal of energy and not by the addition of any mass separating agent (MSA) such as in liquid-liquid extraction.

Sometimes, alternative single- or multiple-stage vapor-liquid separation operations, of the types shown in Fig. 13-2, may be more suitable than distillation for the specified task.

A single-stage flash, as shown in Fig. 13-2*a,* may be appropriate if (1) the relative volatility between the two components to be separated is very large; (2) the recovery of only one component in one of the two product streams is to be achieved, without regard to the separation of the other components; or (3) only a partial separation is to be made. A common example is the separation of light gases such as hydrogen and methane from aromatics. The desired temperature and pressure of a flash may be established by the use of heat exchangers, a valve, a compressor, or a pump upstream of the vessel, used to separate the product vapor and liquid phases. Depending on the original condition of the feed, it may be partially condensed or partially vaporized in a so-called flash operation.

If the recovery of only one component is required rather than a sharp separation between two components of adjacent volatility, their absorption or stripping in a single section of stages may be sufficient. If the feed is vapor at separation conditions, absorption is used either with a liquid MSA absorbent of relatively low volatility, as in Fig. 13-2*b,* or with reflux produced by an overhead partial condenser, as in Fig. 13-2*c*. The choice usually depends on the ease of partially condensing the overhead vapor or of recovering and recycling the absorbent. If the feed is liquid at separation conditions, stripping is used, either with an externally supplied vapor stripping agent of relatively high volatility, as shown in Fig. 13-2*d,* or with boil-up produced by a partial reboiler, as in Fig. 13-2*e*. The choice depends on the ease of partially reboiling the bottoms or of recovering and recycling the stripping agent.

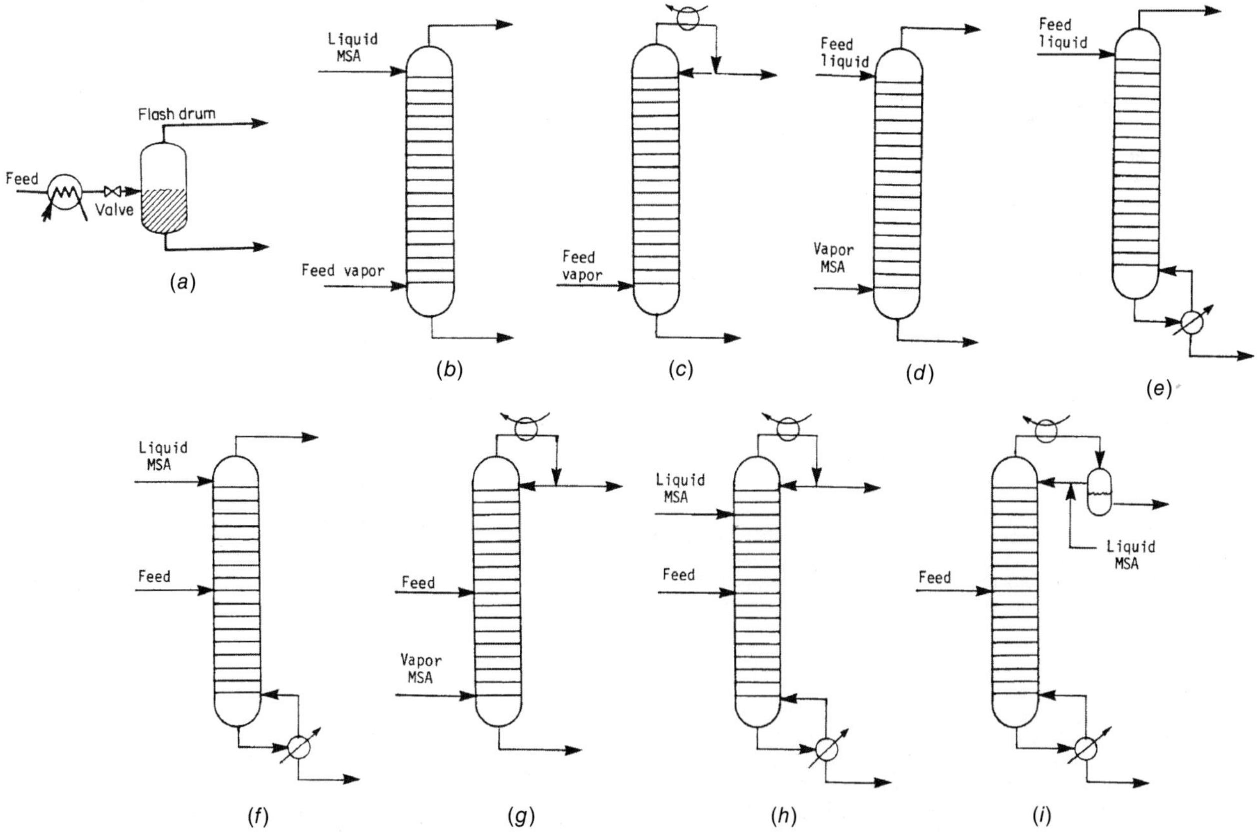

FIG. 13-2 Separation operations related to distillation. (*a*) Flash vaporization or partial condensation. (*b*) Absorption. (*c*) Rectifier. (*d*) Stripping. (*e*) Reboiled stripping. (*f*) Reboiled absorption. (*g*) Refluxed stripping. (*h*) Extractive distillation. (*i*) Azeotropic distillation.

If a relatively sharp separation is needed between two components of adjacent volatility, but either an undesirably low temperature is required to produce reflux at the column operating pressure or an undesirably high temperature is required to produce boil-up, then refluxed stripping, as shown in Fig. 13-2*g*, or reboiled absorption, as shown in Fig. 13-2*f*, may be used. In either case, the choice of MSA follows the same consideration given for simple absorption and stripping.

When the volatility difference between the two components to be separated is so small that a very large number of stages would be required, then extractive distillation, as shown in Fig. 13-2*h*, should be considered. Here, an MSA is selected that increases the volatility difference enough to reduce the stage requirement to a reasonable number. Usually, the MSA is a polar compound of low volatility that leaves in the bottoms, from which it is recovered and recycled. It is introduced in an appreciable amount near the top stage of the column so as to affect the volatility difference over most of the stages. Some reflux to the top stage is used to minimize the MSA content in the distillate. An alternative to extractive distillation is azeotropic distillation, which is shown in Fig. 13-2*i* in just one of its many modes. In a common mode, an MSA that forms a heterogeneous minimum-boiling azeotrope with one or more components of the feed is used. The azeotrope is taken overhead, and the MSA-rich phase is decanted and returned to the top of the column as reflux.

Many other multistaged configurations are possible. One important variation of a stripper, shown in Fig. 13-2*d*, is a refluxed stripper, in which an overhead condenser is added. Such a configuration is sometimes used to steam-strip sour water containing NH_3, H_2O, phenol, and HCN.

All the separation operations shown in Fig. 13-2, as well as the simple and complex distillation operations described earlier, are referred to here as distillation-type separations because they have much in common with respect to calculations of (1) thermodynamic properties, (2) vapor-liquid equilibrium stages, and (3) column sizing. In fact, as will be evident from the remaining treatment of this section, the trend is toward single generalized digital computer program packages that compute many or all distillation-type separation operations.

This section also includes a treatment of distillation-type separations from a rate-based point of view that uses principles of mass- and heat-transfer rates. Section 14 also presents details of that subject as applied to absorption and stripping.

THERMODYNAMIC DATA AND MODELS

Reliable thermodynamic data are essential for the accurate design or analysis of distillation columns. The failure of equipment to perform at specified levels is often attributable, at least in part, to the lack of such data.

This subsection summarizes and presents examples of phase equilibrium data currently available to the designer. The thermodynamic concepts used are presented in Sec. 4 Thermodynamics.

PHASE EQUILIBRIUM DATA

For a binary mixture, pressure and temperature fix the equilibrium vapor and liquid compositions. Thus, experimental data are often presented in the form of tables of vapor mole fraction y and liquid mole fraction x for one constituent over a range of temperature T for a fixed pressure P or over a range of pressure for a fixed temperature. Compilations of such data may be found in Hala, Wichterle, Polak, and Boublik (*Vapour-Liquid Equilibrium Data at Normal Pressures,* Pergamon, Oxford, 1968); Hirata, Ohe, and Nagahama (*Computer Aided Data Book of Vapor-Liquid Equilibria,* Elsevier, Amsterdam, 1975); Wichterle, Linek, and Hala (*Vapor-Liquid Equilibrium Data Bibliography,* Elsevier, Amsterdam, 1973, Supplement I, 1976, Supplement II, 1979); Ohe (*Vapor-Liquid Equilibrium Data,* Elsevier, Amsterdam, 1989); Ohe (*Vapor-Liquid Equilibrium Data at High Pressure,* Elsevier, Amsterdam, 1990); Walas (*Phase Equilibria in Chemical Engineering,* Butterworth, Boston, 1985); and, particularly, Gmehling and Onken [*Vapor-Liquid Equilibrium Data Collection,* DECHEMA Chemistry Data ser., vol. 1 (parts 1–10), Frankfurt, 1977]. Extensive databases of phase equilibrium measurements are readily available in most process simulators, together with models for correlating, interpolating, and extrapolating (care is needed here) the data. Many of these simulators also provide graphical display of the data for easy visualization and interpretation.

For application to distillation (a nearly isobaric process), binary-mixture data are often plotted, for a fixed pressure, as y versus x, with a line of 45° slope included for reference, and as T versus y and x, as shown in Figs. 13-3 to 13-8. In some binary systems, one of the components is more volatile than the other over the entire composition range. This is the case in Figs. 13-3 and 13-4 for the benzene-toluene system at pressures of both 101.3 and 202.6 kPa (1 and 2 atm), where benzene is more volatile than toluene.

For other binary systems, one of the components is more volatile over only a part of the composition range. Two systems of this type, ethyl acetate–ethanol and chloroform-acetone, are shown in Figs. 13-5 to 13-7. Figure 13-5 shows that chloroform is less volatile than acetone below a concentration of 66 mol% chloroform and that ethyl acetate is more volatile than ethanol below a concentration of 53 mol% ethyl acetate. Above these concentrations, volatility is reversed. Such mixtures are known as azeotropic mixtures, and the composition in which the reversal occurs, which is the composition in which vapor and liquid compositions are equal, is the azeotropic composition, or azeotrope. The azeotropic liquid may be homogeneous or heterogeneous (two immiscible liquid phases). Non-azeotrope-forming mixtures such as benzene and toluene in Figs. 13-3 and 13-4 can be separated by simple distillation into two essentially pure products. By contrast, simple distillation of azeotropic mixtures will at best yield the azeotrope and one essentially pure species. The distillate and bottoms products obtained depend on the feed composition and whether a minimum-boiling azeotrope is formed as with the ethyl acetate–ethanol mixture in Fig. 13-6 or a maximum-boiling azeotrope is formed as with the chloroform-acetone mixture in Fig. 13-7. For example, if a mixture of 30 mol% chloroform and 70 mol% acetone is fed to a simple distillation column, such as that shown in Fig. 13-1, operating at 101.3 kPa (1 atm), the distillate could approach pure acetone and the bottoms could approach the maximum-boiling azeotrope.

An example of heterogeneous-azeotrope formation is shown in Fig. 13-8 for the water–normal butanol system at 101.3 kPa. At liquid compositions between 0 and 3 mol% butanol and between 40 and 100 mol% butanol, the liquid phase is homogeneous. Phase splitting into two separate liquid phases (one with 3 mol% butanol and the other with 40 mol% butanol) occurs for any overall liquid composition between 3 and 40 mol% butanol. A minimum-boiling heterogeneous azeotrope occurs at 92°C (198°F) when the vapor composition is equal to the overall composition of the two co-existing equilibrium liquid phases at 25 mol% butanol.

For mixtures containing more than two species, an additional degree of freedom is available for each additional component. Thus, for a four-component system, the equilibrium vapor and liquid compositions are fixed only if the pressure, temperature, and mole fractions of two components are set. Representation of multicomponent vapor-liquid equilibrium data in tabular or graphical form of the type shown earlier for binary systems is

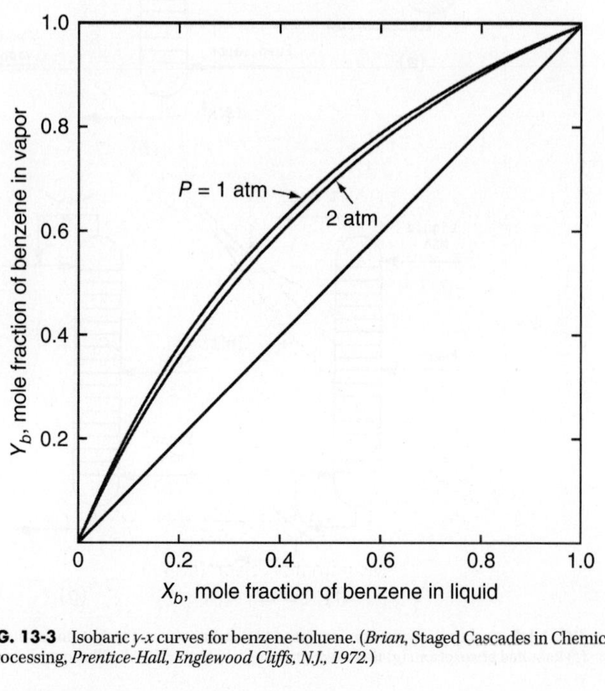

FIG. 13-3 Isobaric y-x curves for benzene-toluene. (*Brian,* Staged Cascades in Chemical Processing, *Prentice-Hall, Englewood Cliffs, N.J., 1972.*)

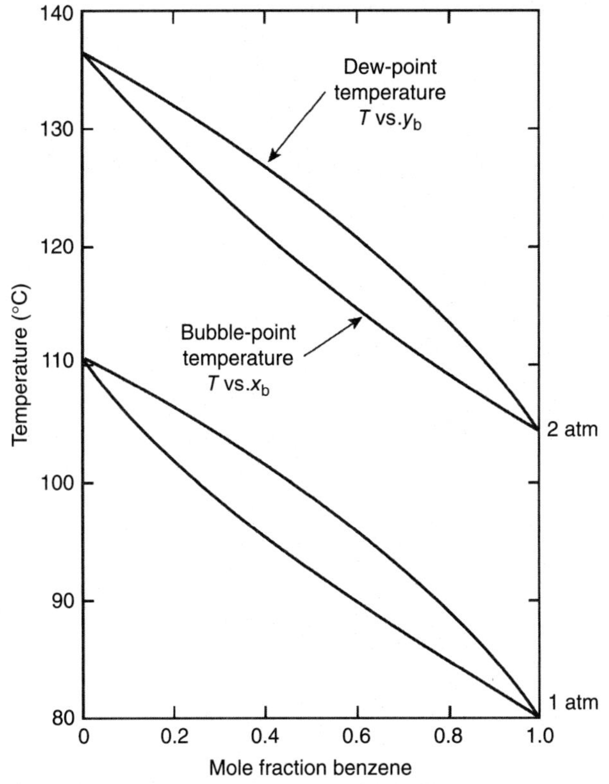

FIG. 13-4 Isobaric vapor-liquid equilibrium curves for benzene-toluene. (*Brian,* Staged Cascades in Chemical Processing, *Prentice-Hall, Englewood Cliffs, N.J., 1972.*)

either difficult or impossible. Instead, such data, as well as binary-system data, are commonly represented in terms of *K* values (vapor-liquid equilibrium ratios), which are defined by

$$K_i = \frac{y_i}{x_i} \tag{13-1}$$

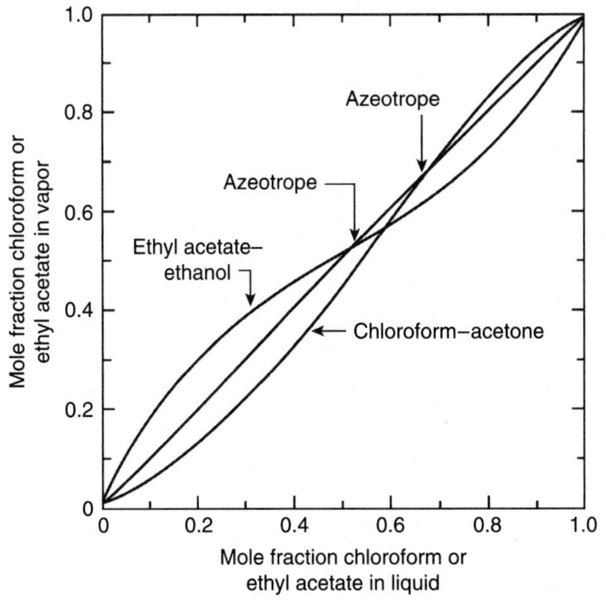

FIG. 13-5 Vapor-liquid equilibria for the ethyl acetate–ethanol and chloroform-acetone systems at 101.3 kPa (1 atm).

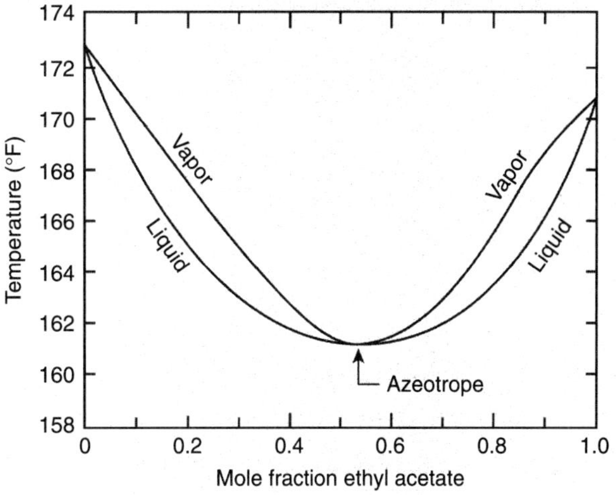

FIG. 13-6 Liquid boiling points and vapor condensation temperatures for minimum-boiling azeotrope mixtures of ethyl acetate and ethanol at 101.3-kPa (1-atm) total pressure.

and are correlated empirically or theoretically in terms of temperature, pressure, and phase compositions in the form of tables, graphs, and equations. The *K* values are widely used in multicomponent distillation calculations, and the ratio of the *K* values of two species, called the relative volatility,

$$\alpha_{ij} = \frac{K_i}{K_j} \tag{13-2}$$

is a convenient index of the relative ease or difficulty of separating components *i* and *j* by distillation. Rarely is distillation used on a large scale if the relative volatility is less than 1.05, with *i* more volatile than *j*.

GRAPHICAL *K* VALUE CORRELATIONS

As discussed in Sec. 4, the *K* value of a species is a complex function of temperature, pressure, and equilibrium vapor- and liquid-phase compositions. However, for mixtures of compounds of similar molecular structure and size, the *K* value depends mainly on temperature and pressure. For example, several major graphical *K* value correlations are available for light-hydrocarbon systems. The easiest to use are the DePriester charts [*Chem. Eng. Prog. Symp. Ser. 7,* **49:** 1 (1953)], which cover 12 hydrocarbons (methane, ethylene,

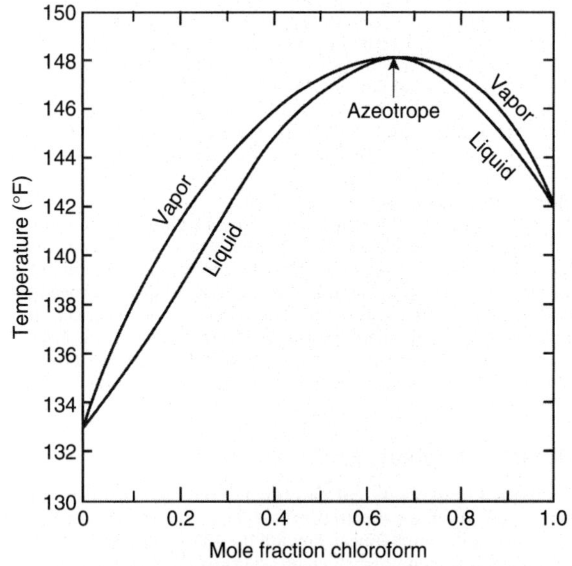

FIG. 13-7 Liquid boiling points and vapor condensation temperatures for maximum-boiling azeotrope mixtures of chloroform and acetone at 101.3-kPa (1-atm) total pressure.

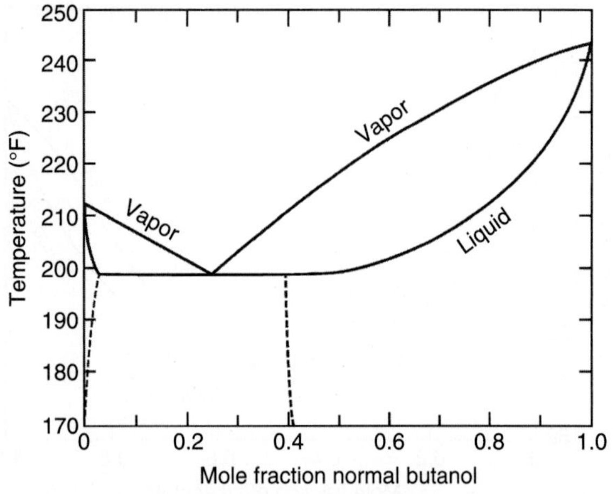

FIG. 13-8 Vapor-liquid equilibrium data for an *n*-butanol–water system at 101.3 kPa (1 atm); phase splitting and heterogeneous-azeotrope formation.

ethane, propylene, propane, isobutane, isobutylene, *n*-butane, isopentane, *n*-pentane, *n*-hexane, and *n*-heptane). These charts are a simplification of the Kellogg charts (*Liquid-Vapor Equilibria in Mixtures of Light Hydrocarbons, MWK Equilibrium Constants, Polyco Data,* 1950) and include additional experimental data. The Kellogg charts, and hence the DePriester charts, are based primarily on the Benedict-Webb-Rubin equation of state [*Chem. Eng. Prog.* **47:** 419 (1951); **47:** 449 (1951)], which can represent both the liquid and the vapor phases and can predict *K* values quite accurately when the equation constants are available for the components in question.

A trial-and-error procedure is required with any *K* value correlation that takes into account the effect of composition. One cannot calculate *K* values until phase compositions are known, and those cannot be known until the *K* values are available to calculate them. For *K* as a function of *T* and *P* only, the DePriester charts provide good starting values for the iteration. These nomographs are shown in Fig. 2-10*a* and *b*. The SI versions of these charts were developed by Dadyburjor [*Chem. Eng. Prog.* **74**(4): 85 (1978)].

The Kellogg and DePriester charts and their subsequent extensions and generalizations use the molar average boiling points of the liquid and vapor phases to represent the composition effect. An alternative measure of composition is the convergence pressure of the system, which is defined as that pressure at which the *K* values for all the components in an isothermal mixture converge to unity. It is analogous to the critical point for a pure component in the sense that the two phases become indistinguishable. The behavior of a complex mixture of hydrocarbons for a convergence pressure of 34.5 MPa (5000 psia) is illustrated in Fig. 13-9.

Two major graphical correlations based on convergence pressure as the third parameter (besides temperature and pressure) are the charts published by the Gas Processors Association (GPA, *Engineering Data Book,* 9th ed., Tulsa, Okla., 1981) and the charts of the American Petroleum Institute (API, *Technical Data Book—Petroleum Refining,* New York, 1966) based on the procedures from Hadden and Grayson [*Hydro-carbon Process., Pet. Refiner* **40**(9): 207 (1961)]. The former uses the method proposed by Hadden [*Chem. Eng. Prog. Symp. Ser. 7,* **49:** 53 (1953)] for the prediction of convergence pressure as a function of composition.

The GPA convergence pressure charts are primarily for alkane and alkene systems, but they include charts for nitrogen, carbon dioxide, and hydrogen sulfide. The charts may not be valid when appreciable amounts of naphthenes or aromatics are present; the API charts use special procedures for such cases. Useful extensions of the convergence pressure concept to more varied mixtures include the nomographs of Winn [*Chem. Eng. Prog. Symp. Ser. 2,* **48:** 121 (1952)], Hadden and Grayson [*Hydro-carbon Process., Pet. Refiner* **40**(9): 207 (1961)], and Cajander, Hipkin, and Lenoir [*J. Chem. Eng. Data* **5:** 251 (1960)].

ANALYTICAL *K* VALUE CORRELATIONS

The widespread availability and use of digital computers for distillation calculations have given impetus to the development of analytical expressions for *K* values. McWilliams [*Chem. Eng.* **80**(25): 138 (1973)] presents a regression equation and accompanying regression coefficients that represent the DePriester charts of Fig. 2-10. Regression equations and coefficients for various versions of the GPA convergence pressure charts are available from the GPA.

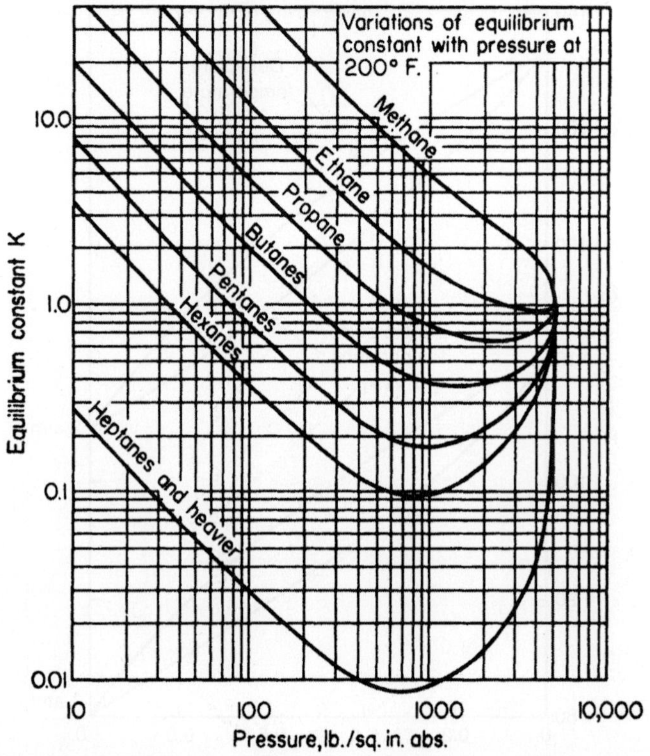

FIG. 13-9 Typical variation of *K* values with total pressure at constant temperature for a complex mixture. Light hydrocarbons in admixture with crude oil. [*Katz and Hachmuth,* Ind. Eng. Chem., ***29:*** *1072 (1937).*]

Preferred analytical correlations are less empirical and most often are theoretically based on one of two exact thermodynamic formulations, as derived in Sec. 4. When a single pressure-volume-temperature (*P-V-T*) equation of state is applicable to both vapor and liquid phases, the formulation used is

$$K_i = \frac{\hat{\Phi}_i^L}{\hat{\Phi}_i^V} \tag{13-3}$$

where the mixture fugacity coefficients $\hat{\Phi}_i^L$ for the liquid and $\hat{\Phi}_i^V$ for the vapor are derived by classical thermodynamics from the *P-V-T* expression. Consistent equations for enthalpy can be similarly derived.

Until recently, equations of state that have been successfully applied to Eq. (13-3) have been restricted to mixtures of nonpolar compounds, namely, hydrocarbons and light gases. These equations include those of Benedict-Webb-Rubin (BWR), Soave (SRK) [*Chem. Eng. Sci.* **27:** 1197 (1972)], who extended the remarkable Redlich-Kwong equation, and Peng-Robinson (PR) [*Ind. Eng. Chem. Fundam.* **15:** 59 (1976)]. The SRK and PR equations belong to a family of so-called cubic equations of state. The Starling extension of the BWR equation (*Fluid Thermodynamic Properties for Light Petroleum Systems,* Gulf, Houston, 1973) predicts *K* values and enthalpies of the normal paraffins up through *n*-octane, as well as isobutane, isopentane, ethylene, propylene, nitrogen, carbon dioxide, and hydrogen sulfide, including the cryogenic region. Computer programs for *K* values derived from the SRK, PR, and other equations of state are widely available in all computer-aided process design and simulation programs. The ability of the SRK correlation to predict *K* values even when the pressure approaches the convergence pressure is shown for a multicomponent system in Fig. 13-10. Similar results are achieved with the PR correlation. The Wong-Sandler mixing rules for cubic equations of state now permit such equations to be extended to mixtures of organic chemicals, as shown in a reformulated version by Orbey and Sandler [*AIChE J.* **41**(3): 683–690 (1995)].

An alternative *K* value formulation that has received wide application to mixtures containing polar and/or nonpolar compounds is

$$K_i = \frac{\gamma_i^L \Phi_i^L}{\hat{\Phi}_i^V} \tag{13-4}$$

where different equations of state may be used to predict the pure-component liquid fugacity coefficient Φ_i^L and the vapor-mixture fugacity

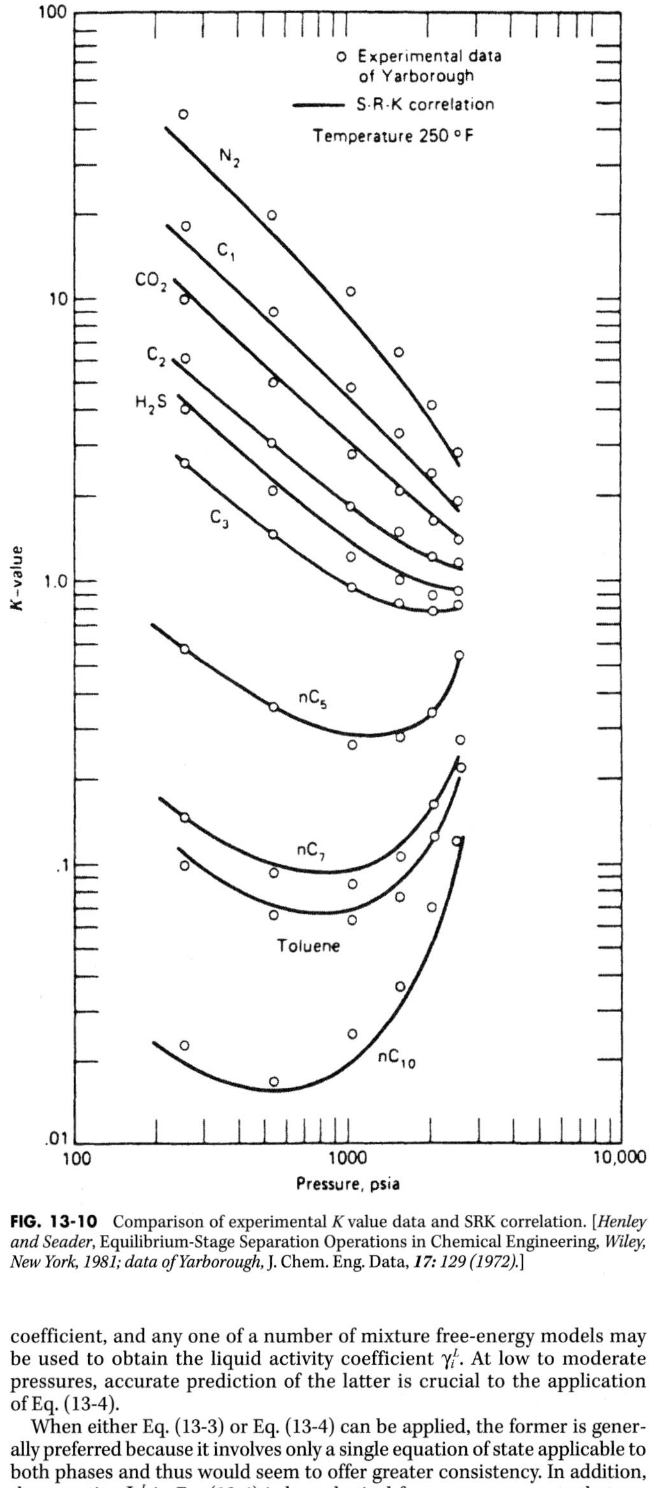

FIG. 13-10 Comparison of experimental K value data and SRK correlation. [*Henley and Seader*, Equilibrium-Stage Separation Operations in Chemical Engineering, *Wiley, New York, 1981; data of Yarborough,* J. Chem. Eng. Data, ***17:** 129 (1972).*]

coefficient, and any one of a number of mixture free-energy models may be used to obtain the liquid activity coefficient γ_i^L. At low to moderate pressures, accurate prediction of the latter is crucial to the application of Eq. (13-4).

When either Eq. (13-3) or Eq. (13-4) can be applied, the former is generally preferred because it involves only a single equation of state applicable to both phases and thus would seem to offer greater consistency. In addition, the quantity Φ_i^L in Eq. (13-4) is hypothetical for any components that are supercritical. In that case, a modification of Eq. (13-4) that uses Henry's law is sometimes applied.

For mixtures of hydrocarbons and light gases, Chao and Seader (CS) [*AIChE J.* **7:** 598 (1961)] applied Eq. (13-4) by using an empirical expression for Φ_i^L based on the generalized corresponding-states P-V-T correlation of Pitzer et al., the Redlich-Kwong equation of state for $\hat{\Phi}_i^V$, and the regular solution theory of Scatchard and Hildebrand for γ_i^L. The predictive ability of the last-named theory is exhibited in Fig. 13-11 for the heptane-toluene system at 101.3 kPa (1 atm). Five pure-component constants for each species (T_v, P_v, ω, δ, and v^L) are required to use the CS method which, when applied within the restrictions discussed by Lenoir and Koppany [*Hydrocarbon Process.* **46**(11): 249 (1967)], gives good results. Revised

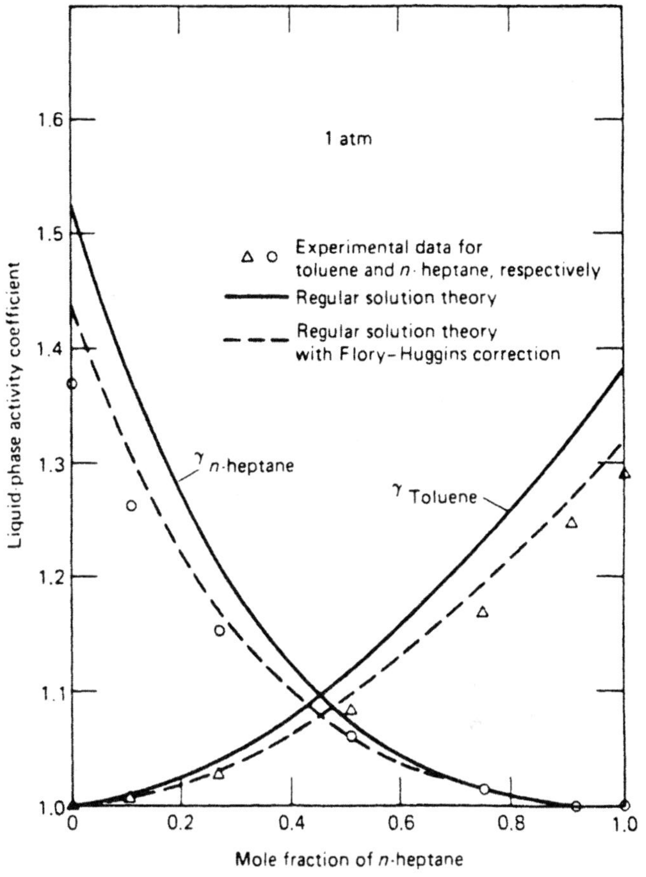

FIG. 13-11 Liquid-phase activity coefficients for an n-heptane–toluene system at 101.3 kPa (1 atm). [*Henley and Seader,* Equilibrium-Stage Separation Operations in Chemical Engineering, *Wiley, New York, 1981; data of Yerazunis et al.,* AIChE J., ***10:** 660 (1964).*]

coefficients of Grayson and Streed (GS) (Paper 20-P07, Sixth World Pet. Conf. Frankfurt, June 1963) for the Φ_i^L expression permit the application of the CS correlation to higher temperatures and pressures and give improved predictions for hydrogen. Jin, Greenkorn, and Chao [*AIChE J.* **41:** 1602 (1995)] present a revised correlation for the standard-state liquid fugacity of hydrogen, applicable from 200 to 730 K.

For mixtures containing polar substances, more complex predictive equations for γ_i^L that involve binary-interaction parameters for each pair of components in the mixture are required for use in Eq. (13-4), as discussed in Sec. 4. Four popular expressions are the Wilson, NRTL, UNIFAC, and UNIQUAC equations. The preferred expressions for representing activity coefficients are the NRTL and UNIQUAC equations. Extensive listings of binary-interaction parameters for use in all but the UNIFAC equation are given by Gmehling and Onken [*Vapor-Liquid Equilibrium Data Collection,* DECHEMA Chemistry Data ser., vol. 1 (parts 1–10), Frankfurt, 1977]. They obtained the parameters for binary systems at 101.3 kPa (1 atm) from best fits of the experimental T-y-x equilibrium data by setting Φ_i^V and Φ_i^L to their ideal-gas, ideal-solution limits of 1.0 and P^{sat}/P, respectively, with the vapor pressure P^{sat} given by a three-constant Antoine equation, whose values they tabulate.

The Wilson equation is particularly useful for systems that are highly nonideal but do not undergo phase splitting, as exemplified by the ethanol-hexane system, whose activity coefficients are shown in Fig. 13-12.

Carboxylic acids (e.g., formic acid and acetic acid) tend to dimerize in the vapor phase according to the chemical equilibrium expression

$$K_D = \frac{P_D}{P_M^2} = 10^{A+B/T} \tag{13-5}$$

where K_D is the chemical equilibrium constant for dimerization, P_D and P_M are partial pressures of dimer and monomer, respectively, in torr, and T is in Kelvin. Values of A and B for the first four normal aliphatic acids are

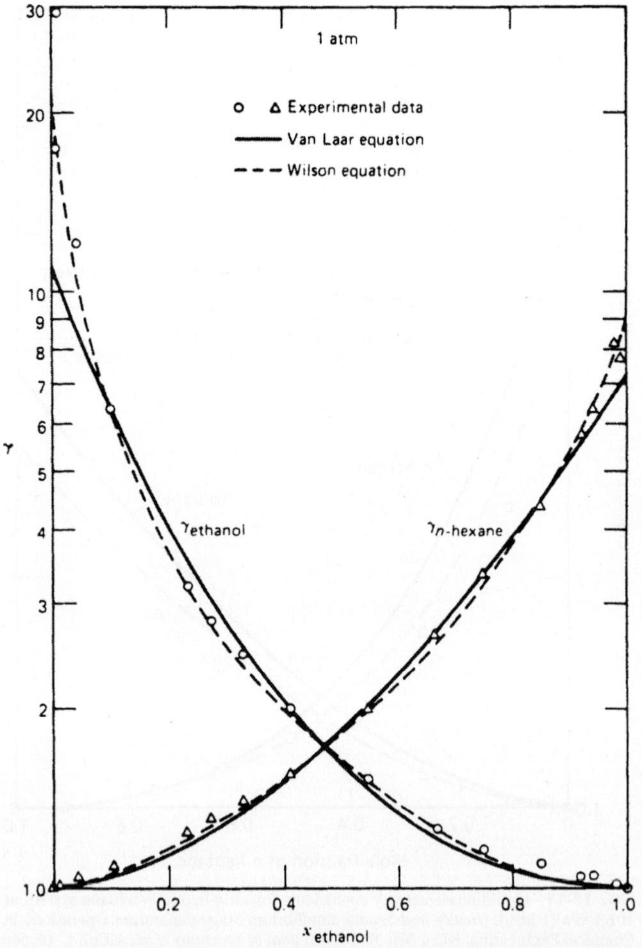

FIG. 13-12 Liquid-phase activity coefficients for an ethanol–*n*-hexane system. [*Henley and Seader,* Equilibrium-Stage Separation Operations in Chemical Engineering, *Wiley, New York, 1981; data of Sinor and Weber,* J. Chem. Eng. Data, *5: 243–247 (1960).*]

	A	*B*
Formic acid	−10.743	3083
Acetic acid	−10.421	3166
n-Propionic acid	−10.843	3316
n-Butyric acid	−10.100	3040

As shown by Marek and Standart [*Collect. Czech. Chem. Commun.* **19:** 1074 (1954)], it is preferable to correlate and use liquid-phase activity coefficients for the dimerizing component by considering separately the partial pressures of the monomer and dimer. For example, for a binary system of components 1 and 2, when only compound 1 dimerizes in the vapor phase, the following equations apply if an ideal gas is assumed:

$$P_1 = P_D + P_M \tag{13-6}$$

$$y_1 = \frac{P_M + 2P_D}{P} \tag{13-7}$$

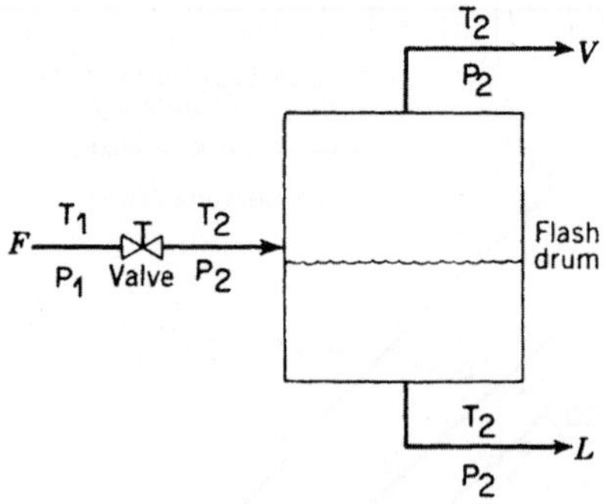

FIG. 13-13 Equilibrium flash separator.

These equations when combined with Eq. (13-5) lead to the following equations for liquid-phase activity coefficients in terms of measurable quantities:

$$\gamma_1 = \frac{Py_1}{P_1^{sat}x_1}\left\{\frac{1+(1+4K_DP_1^{sat})^{0.5}}{1+[1+4K_DPy_1(2-y_1)]^{0.5}}\right\} \tag{13-8}$$

$$\gamma_2 = \frac{Py_1}{P_2^{sat}x_2}\left(\frac{2(1-y_1+[1+4K_DPy_1(2-y_1)]^{0.5}}{(2-y_1)\{1+[1+4K_DPy_1(2-y_1)]^{0.5}\}}\right) \tag{13-9}$$

Detailed procedures, including computer programs for evaluating binary-interaction parameters from experimental data and then using these parameters to predict *K* values and phase equilibria, are given in terms of the UNIQUAC equation by Prausnitz et al. (*Computer Calculations for Multicomponent Vapor-Liquid and Liquid-Liquid Equilibria,* Prentice-Hall, Englewood Cliffs, N.J., 1980) and in terms of the UNIFAC group contribution method by Fredenslund, Gmehling, and Rasmussen (*Vapor-Liquid Equilibria Using UNIFAC,* Elsevier, Amsterdam, 1980). Both use the method of Hayden and O'Connell [*Ind. Eng. Chem. Process Des. Dev.* **14:** 209 (1975)] to compute $\hat{\Phi}_i^V$ in Eq. (13-4). When the system temperature is greater than the critical temperature of one or more components in the mixture, Prausnitz et al. use a Henry's law constant $H_{i,M}$ in place of the product $\gamma_i^L \Phi_i^L$ in Eq. (13-4). Otherwise Φ_i^L is evaluated from vapor pressure data with a Poynting saturated-vapor fugacity correction. When the total pressure is less than about 202.6 kPa (2 atm) and all components in the mixture have a critical temperature that is greater than the system temperature, then $\Phi_i^L = P_i^{sat}/P$ and $\Phi_i^V = 1.0$. Equation (13-4) then reduces to

$$K_i = \frac{\gamma_i^L P_i^{sat}}{P} \tag{13-10}$$

which is referred to as a modified Raoult's law *K* value. If, furthermore, the liquid phase is ideal, then $\gamma_i^L = 1.0$ and

$$K_i = \frac{P_i^{sat}}{P} \tag{13-11}$$

which is referred to as a Raoult's law *K* value that is dependent solely on the vapor pressure P_i^{sat} of the component in the mixture. The UNIFAC method is being periodically updated with new group contributions; for example, see Gmehling et al. [*Ind. Eng. Chem. Res.* **42:** 183 (2003)].

SINGLE-STAGE EQUILIBRIUM FLASH CALCULATIONS

The simplest continuous distillation process is the adiabatic single-stage equilibrium flash process pictured in Fig. 13-13. Feed temperature and the pressure drop across the valve are adjusted to vaporize the feed to the desired extent, while the drum provides disengaging space to allow the vapor to separate from the liquid. The expansion across the valve is at constant enthalpy, and this fact can be used to calculate T_2 (or T_1 to give a desired T_2).

A degrees-of-freedom analysis indicates that the variables subject to the designer's control are $C + 3$ in number. The most common way to use these is to specify the feed rate, composition, and pressure ($C + 1$ variables) plus the drum temperature T_2 and pressure P_2. This operation will give one point on the *equilibrium flash curve* shown in Fig. 13-14. This curve shows the relation at constant pressure between the fraction V/F of the feed flashed and the drum temperature. The temperature at $V/F = 0.0$ when the first bubble

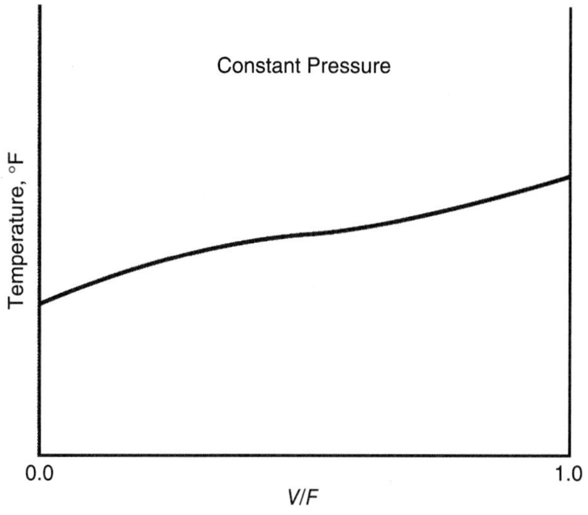

FIG. 13-14 Equilibrium flash curve.

of vapor is about to form (saturated liquid) is the *bubble point* temperature of the feed mixture, and the value at $V/F = 1.0$ when the first droplet of liquid is about to form (saturated vapor) is the *dew point* temperature.

BUBBLE POINT AND DEW POINT

For a given drum pressure and feed composition, the bubble and dew point temperatures bracket the temperature range of the equilibrium flash. At the bubble point temperature, the total vapor pressure exerted by the mixture becomes equal to the confining drum pressure, and it follows that $\sum y_i = 1.0$ in the bubble formed. Since $y_i = K_i x_i$ and since the x_i's still equal the feed compositions (denoted by z_i), calculation of the bubble point temperature involves a trial-and-error search for the temperature which, at the specified pressure, makes $\sum K_i z_i = 1.0$. If instead the temperature is specified, one can find the bubble point pressure that satisfies this relationship.

At the dew point temperature, y_i still equals z_i, and the relationship $\sum x_i = \sum z_i / K_i = 1.0$ must be satisfied. As in the case of the bubble point, a trial-and-error search for the dew point temperature at a specified pressure is involved. Or, if the temperature is specified, the dew point pressure can be calculated.

ISOTHERMAL FLASH

The calculation for a point on the flash curve that is intermediate between the bubble point and the dew point is referred to as an isothermal flash calculation because T_2 is specified. Except for an ideal binary mixture, procedures for calculating an isothermal flash are iterative. A popular and recommended method is the following, due to Rachford and Rice [*J. Pet. Technol.* 4(10): sec. 1, p. 19, and sec. 2, p. 3 (1952)]. The component mole balance ($Fz_i = Vy_i + Lx_i$), phase distribution relation ($K_i = y_i/x_i$), and total mole balance ($F = V + L$) can be combined to give

$$x_i = \frac{z_i}{1 + (V/F)(K_i - 1)} \qquad (13\text{-}12)$$

$$y_i = \frac{K_i z_i}{1 + (V/F)(K_i - 1)} \qquad (13\text{-}13)$$

Since $\sum x_i - \sum y_i = 0$,

$$f\left(\frac{V}{F}\right) = \sum_i \frac{z_i(1 - K_i)}{1 + (V/F)(K_i - 1)} = 0 \qquad (13\text{-}14)$$

Equation (13-14) is solved iteratively for V/F, followed by the calculation of values of x_i and y_i from Eqs. (13-12) and (13-13) and L from the total mole balance. Any one of a number of numerical root-finding procedures such as the Newton-Raphson, secant, false-position, or bisection method can be used to solve Eq. (13-14). Values of K_i are constants if they are independent of liquid and vapor compositions. Then the resulting calculations are straightforward. Otherwise, the K_i values must be periodically updated for composition effects, perhaps after each iteration, using prorated values of x_i and y_i from Eqs. (13-12) and (13-13). Generally the iterations are continued until the change in the absolute value of V/F is sufficiently small and until the absolute value of the residual $f(V/F)$ is close to zero. When converged, $\sum x_i$ and $\sum y_i$ will each be very close to a value of 1, and, if desired, T_1 can be computed from an energy balance around the valve if no heat exchanger is used. Alternatively, if T_1 is fixed, as mentioned earlier, a heat exchanger must be added before, after, or in place of the valve with the required heat duty being calculated from an energy balance. The limits of applicability of Eqs. (13-12) to (13-14) are the bubble point, at which $V = 0$ and $x_i = z_i$, and the dew point, at which $L = 0$ and $y_i = z_i$. At these limits, Eq. (13-14) reduces to the bubble point equation

$$\sum_i K_i x_i = 1 \qquad (13\text{-}15)$$

and the dew point equation, respectively,

$$\sum_i \frac{y_i}{K_i} = 1 \qquad (13\text{-}16)$$

For a *binary feed,* specification of the flash drum temperature and pressure fixes the equilibrium-phase compositions, which are related to the K values by

$$x_1 = \frac{1 - K_2}{K_1 - K_2} \quad \text{and} \quad y_1 = \frac{K_1 K_2 - K_1}{K_2 - K_1}$$

The mole balance can be rearranged to

$$\frac{V}{F} = \frac{z_1(K_1 - K_2)/(1 - K_2) - 1}{K_1 - 1}$$

If K_1 and K_2 are functions of temperature and pressure only (ideal solutions), the flash curve can be calculated directly without iteration.

ADIABATIC FLASH

In Fig. 13-13, if P_2 and the feed-stream conditions (that is, F, z_i, T_1, P_1) are known, then the calculation of T_2, V, L, y_i, and x_i is referred to as an adiabatic flash. In addition to Eqs. (13-12) to (13-14) and the total mole balance, the following energy balance around both the valve and the flash drum combined must be included:

$$H^F F = H^V V + H^L L \qquad (13\text{-}17)$$

By taking a basis of $F = 1.0$ mol and eliminating L with the total mole balance, Eq. (13-17) becomes

$$f_2\{V, T_2\} = H^F - V(H^V - H^L) - H^L = 0 \qquad (13\text{-}18)$$

With T_2 now unknown, Eq. (13-14) becomes

$$f_1\{V, T_2\} = \sum_i \frac{z_i(1 - K_i)}{1 + V(K_i - 1)} = 0 \qquad (13\text{-}19)$$

A number of iterative procedures have been developed for solving Eqs. (13-18) and (13-19) simultaneously for V and T_2. Frequently, and especially if the feed contains components of a narrow range of volatility, convergence is rapid for a tearing method in which a value of T_2 is assumed, Eq. (13-19) is solved iteratively by the isothermal flash procedure, and, using that value of V, Eq. (13-18) is solved iteratively for a new approximation of T_2, which is then used to initiate the next cycle until T_2 and V converge. However, if the feed contains components of a wide range of volatility, it may be best to invert the sequence and assume a value for V, solve Eq. (13-19) for T_2, solve Eq. (13-18) for V, and then repeat the cycle. If the K values and/or enthalpies are sensitive to the unknown phase compositions, it may be necessary to solve Eqs. (13-18) and (13-19) simultaneously by a Newton or other suitable iterative technique. Alternatively, the two-tier method of Boston and Britt [*Comput. Chem. Eng.* **2:** 109 (1978)], which is also suitable for difficult isothermal flash calculations, may be applied.

Other Flash Specifications Flash drum specifications in addition to (P_2, T_2) and (P_2, adiabatic) are possible but must be applied with care, as discussed by Michelsen [*Comp. Chem. Eng.* **17:** 431 (1993)]. Most computer-aided process design and simulation programs permit a wide variety of flash specifications.

Three-Phase Flash Single-stage equilibrium flash calculations become considerably more complex when an additional liquid phase can form, as in mixtures of water with hydrocarbons, water with ethers, and water with higher alcohols (containing four or more carbon atoms). Procedures for computing such situations are referred to as three-phase flash methods, which are given for the general case by Henley and Rosen (*Material and Energy Balance Computations,* Wiley, New York, 1968, chap. 8). When the two liquid phases are almost mutually insoluble, they can be considered separately, and relatively simple procedures apply, as discussed by Smith (*Design of Equilibrium Stage Processes,* McGraw-Hill, New York, 1963). Condensation of such mixtures may result in one liquid phase being formed before the other. Computer-aided process design and simulation programs all contain a Gibbs free-energy routine that can compute a three-phase flash by minimization of the Gibbs free energy. Many important and subtle aspects of three-phase flash calculations are discussed by Michelsen [*Fluid Phase Equil.* **9**(1): 21 (1982)], McDonald and Floudas [*AIChE J.* **41:** 1798 (1995)], and Wasylkiewicz et al. [*Ind. Eng. Chem. Research* **35:** 1395 (1996)].

Complex Mixtures Feed analyses in terms of component compositions are usually not available for complex hydrocarbon mixtures with a final normal boiling point above about 38°C (100°F) (*n*-pentane). One method of handling such a feed is to break it down into pseudocomponents (narrow-boiling fractions) and then estimate the mole fraction and *K* value for each such component. Edmister [*Ind. Eng. Chem.* **47:** 1685 (1955)] and Maxwell (*Data Book on Hydrocarbons,* Van Nostrand, Princeton, N.J., 1958) give charts that are useful for this estimation. Once *K* values are available, the calculation proceeds as described above for multicomponent mixtures. Another approach to complex mixtures is to obtain an American Society for Testing and Materials (ASTM) or true-boiling point (TBP) curve for the mixture and then use empirical correlations to construct the atmospheric-pressure equilibrium flash vaporization (EFV) curve, which can then be corrected to the desired operating pressure. A discussion of this method and the necessary charts is presented in a later subsection Petroleum and Complex-Mixture Distillation.

GRAPHICAL METHODS FOR BINARY DISTILLATION

Multistage distillation under continuous, steady-state operating conditions is widely used in practice to separate a variety of mixtures. Table 13-1, taken from the study of Mix, Dweck, Weinberg, and Armstrong [*AIChE Symp. Ser.* **76**(192): 10 (1980)], lists key components along with typical stage requirements to perform the separation for 27 industrial distillation processes. The design of multistage columns can be accomplished by graphical techniques when the feed mixture contains only two components. The *y-x* diagram method developed by McCabe and Thiele [*Ind. Eng. Chem.* **17:** 605 (1925)] uses only phase equilibrium and mole balance relationships. The method assumes an adiabatic column (no heat losses through the column walls) and constant latent heat for the binary mixture at all compositions (which requires, among other things, equal latent heat for both components). The method is exact only for those systems in which the energy effects on vapor and liquid rates leaving the stages are negligible. However, the approach is simple and gives a useful first estimate of the column design, which can be refined by using the enthalpy composition diagram method of Ponchon [*Tech. Mod.* **13**(20): 55 (1921)] and Savarit [*Arts Metiers* **75**(18): 65, 142, 178, 241, 266, 307 (1922)]. This approach uses the energy balance in addition to mole balance and phase equilibrium relationships and is rigorous when enough calorimetric data are available to construct the diagram without assumptions.

With the widespread availability of computers, the preferred approach to design is equation-based since it provides answers rapidly and repeatedly without the tedium of redrawing graphs. Such an approach is especially useful for sensitivity analysis, which gives insight into how a design changes under variations or uncertainty in design parameters such as thermodynamic properties; feed flow rate, composition, temperature, and pressure; and desired product compositions. Nevertheless, diagrams are useful for quick approximations, for interpreting the results of equation-based methods, and for demonstrating the effect of various design variables. The *x-y* diagram is the most convenient for these purposes, and its use is developed in detail here. The use of the enthalpy composition diagram is given by Smith (*Design of Equilibrium Stage Processes,* McGraw-Hill, New York, 1963) and Henley and Seader (*Equilibrium-Stage Separation Operations in Chemical Engineering,* Wiley, New York, 1981). An approximate equation-based approach based on the enthalpy composition diagram was proposed by Peters [*Ind. Eng. Chem.* **14:** 476 (1922)] with additional aspects developed later by others. Doherty and Malone (*Conceptual Design of Distillation Systems,* McGraw-Hill, New York, 2001, app. A) describe this method for binary mixtures and extend it to multicomponent systems. The approach is exact when the enthalpy composition surfaces are linear.

TABLE 13-1 Key Components and Typical Number of (Real) Stages Required to Perform the Separation for Distillation Processes of Industrial Importance

Key components	Typical number of trays
Hydrocarbon systems	
Ethylene-ethane	73
Propylene-propane	138
Propyne–1–3-butadiene	40
1–3 Butadiene-vinyl acetylene	130
Benzene-toluene	34, 53
Benzene-ethyl benzene	20
Benzene–diethyl benzene	50
Toluene–ethyl benzene	28
Toluene-xylenes	45
Ethyl benzene–styrene	34
o-Xylene-*m*-xylene	130
Organic systems	
Methonol-formaldehyde	23
Dichloroethane-trichloroethane	30
Acetic acid–acetic anhydride	50
Acetic anhydride–ethylene diacetate	32
Vinyl acetate–ethyl acetate	90
Ethylene glycol–diethylene glycol	16
Cumene-phenol	38
Phenol-acetophenone	39, 54
Aqueous systems	
HCN-water	15
Acetic acid–water	40
Methanol-water	60
Ethanol-water	60
Isopropanol-water	12
Vinyl acetate–water	35
Ethylene oxide–water	50
Ethylene glycol–water	16

PHASE EQUILIBRIUM DIAGRAMS

Three types of binary phase equilibrium curves are shown in Fig. 13-15. The *y-x* diagram is almost always plotted for the component that is the more volatile (denoted by the subscript 1) in the region where distillation is to

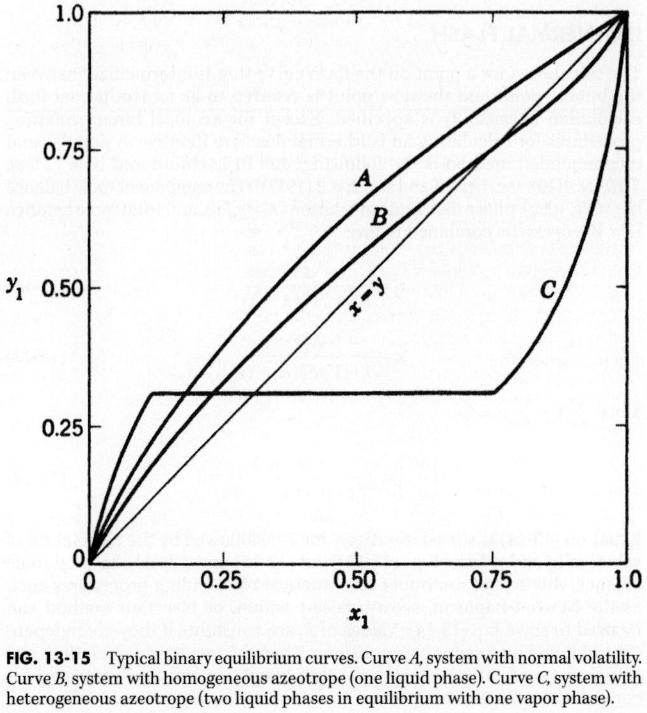

FIG. 13-15 Typical binary equilibrium curves. Curve *A*, system with normal volatility. Curve *B*, system with homogeneous azeotrope (one liquid phase). Curve *C*, system with heterogeneous azeotrope (two liquid phases in equilibrium with one vapor phase).

take place. Curve A shows the common case in which component 1 remains more volatile over the entire composition range. Curve B is typical of many systems (e.g., ethanol-water) in which the component that is more volatile at low values of x_1 becomes less volatile than the other component at high values of x_1. The vapor and liquid compositions are identical for the homogeneous azeotrope where curve B crosses the 45° diagonal (that is, $x_1 = y_1$). A heterogeneous azeotrope is formed by curve C, in which there are two equilibrium liquid phases and one equilibrium vapor phase.

An azeotrope limits the separation that can be obtained between components by simple distillation. For the system described by curve B, the maximum overhead-product concentration that could be obtained from a feed with $z_1 = 0.25$ is the azeotropic composition. Similarly, a feed with $x_1 = 0.9$ could produce a bottom-product composition no lower than the azeotrope.

The phase rule permits only two variables to be specified arbitrarily in a binary two-phase mixture at equilibrium. Consequently, the curves in Fig. 13-15 can be plotted at either constant temperature or constant pressure, but not both. The latter is more common. The y-x diagram can be plotted in mole, weight, or volume fractions. The units used later for the phase flow rates must, of course, agree with those used for the equilibrium data. Mole fractions, which are almost always used, are applied here.

It is sometimes permissible to assume constant *relative volatility* to approximate the equilibrium curve quickly. Then by applying Eq. (13-2) to components 1 and 2,

$$\alpha = \frac{K_1}{K_2} = \frac{y_1 x_2}{x_1 y_2}$$

which can be rewritten as (using $x_2 = 1 - x_1$ and $y_2 = 1 - y_1$)

$$y_1 = \frac{x_1 \alpha}{1 + (\alpha - 1)x_1} \tag{13-20}$$

With a constant value for α, this equation provides a simple, approximate expression for representing the equilibrium $y = x$ diagram. Doherty and Malone (*Conceptual Design of Distillation Systems*, McGraw-Hill, New York, 2001, sec. 2.3) discuss this approximation in greater detail and give a selection of binary mixtures for which the approximation is reasonable. At a constant pressure of 1 atm, these include benzene + toluene, $\alpha = 2.34$; benzene + p-xylene, $\alpha = 4.82$; and hexane + p-xylene, $\alpha = 7.00$.

MCCABE-THIELE METHOD

Operating Lines The McCabe-Thiele method is based on representation of the material balance equations as operating lines on the y-x diagram. The lines are made straight by the assumption of *constant molar overflow*, which eliminates the need for an energy balance. The liquid-phase flow rate is assumed to be constant from tray to tray in each section of the column between addition (feed) and withdrawal (product) points. If the liquid rate is constant, the vapor rate must also be constant.

The constant-molar-overflow assumption rests on several underlying thermodynamic assumptions. The most important one is equal molar heats of vaporization for the two components. The other assumptions are adiabatic operation (no heat leaks) and no heat of mixing or sensible heat effects. These assumptions are most closely approximated for close-boiling isomers. The result of these assumptions on the calculation method can be illustrated with Fig. 13-16, which shows two material balance envelopes cutting through the top section (above the top feed stream or sidestream) of the column. If the liquid flow rate L_{n+1} is assumed to be identical to L_{n-1}, then $V_n = V_{n-2}$ and the component material balance for both envelopes 1 and 2 can be represented by

$$y_n = \left(\frac{L}{V}\right) x_{n+1} + \frac{Dx_D}{V} \tag{13-21}$$

where y and x have a stage subscript n or $n + 1$, but L and V need to be identified only with the section of the column to which they apply. Equation (13-21) has the analytical form of a straight line where L/V is the slope and Dx_D/V is the y intercept at $x = 0$.

The effect of a sidestream withdrawal point is illustrated by Fig. 13-17. The material balance equation for the column section below the sidestream is

$$y_n = \frac{L'}{V'} x_{n+1} + \frac{Dx_D + Sx_S}{V'} \tag{13-22}$$

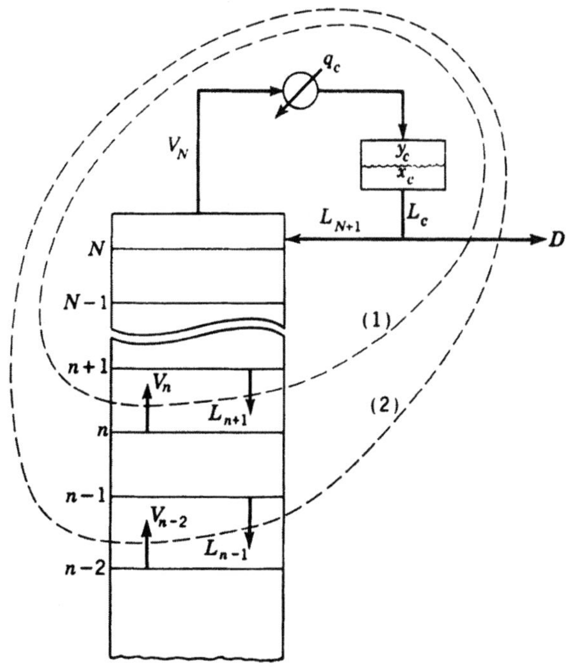

FIG. 13-16 Two material balance envelopes in the top section of a distillation column.

where the primes designate the L and V below the sidestream. Since the sidestream must be a saturated phase, $V = V'$ if a liquid sidestream is withdrawn and $L = L'$ if it is a vapor.

If the sidestream in Fig. 13-17 is a feed (not necessarily a saturated liquid or vapor), the balance for the section below the feed becomes

$$y_n = \frac{L'}{V'} x_{n+1} + \frac{Dx_D - Fz_F}{V'} \tag{13-23}$$

Similar equations can be written for the bottom section of the column. For the envelope shown in Fig. 13-18,

$$y_m = \frac{L'}{V'} x_{m+1} - \frac{Bx_B}{V'} \tag{13-24}$$

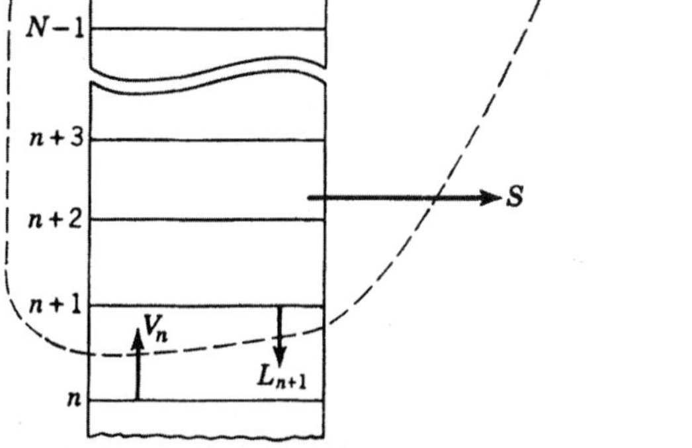

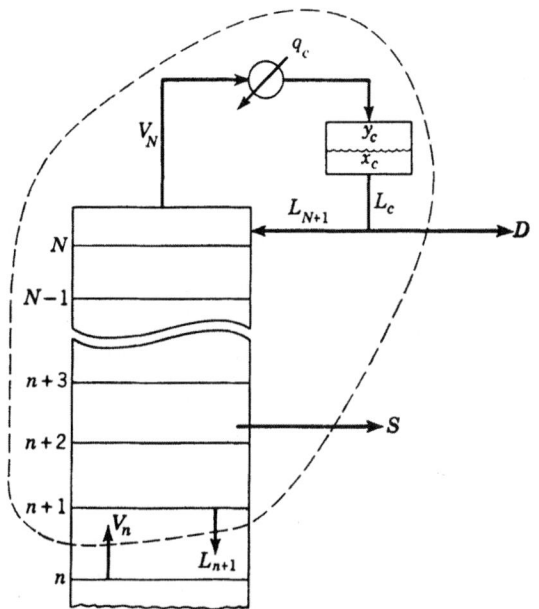

FIG. 13-17 Material balance envelope that contains two external streams D and S, where S represents a sidestream product withdrawn above the feed plate.

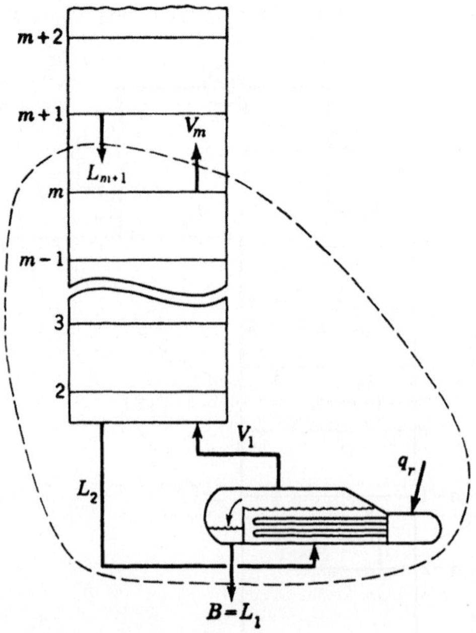

FIG. 13-18 Material balance envelope around the bottom end of the column. The partial reboiler is equilibrium stage 1.

where the subscript m is used to identify the stage number in the bottom section.

Equations such as (13-21) through (13-24), when plotted on the y-x diagram, furnish a set of *operating lines.* A point on an operating line represents two *passing streams,* and the operating line itself is the locus of all possible pairs of passing streams within the column section to which the line applies.

An operating line can be located on the y-x diagram if (1) two points on the line are known or (2) one point and the slope are known. The known points on an operating line are usually its intersection with the y-x diagonal and/or its intersection with another operating line.

The slope L/V of the operating line is termed the *internal reflux ratio.* This ratio in the operating line equation for the top section of the column [see Eq. (13-21)] is related to the *external reflux ratio* $R = L_{N+1}/D$ by

$$\frac{L}{V} = \frac{L_{N+1}}{V_N} = \frac{RD}{(1+R)D} = \frac{R}{1+R} \tag{13-25}$$

when the reflux stream L_{N+1} is a saturated liquid.

Thermal Condition of the Feed The slope of the operating line changes whenever a feed stream or a sidestream is passed. To calculate this change, it is convenient to introduce a quantity q, which is defined by the following equations for a feed stream F:

$$L' = L + qF \tag{13-26}$$

$$V = V' + (1-q)F \tag{13-27}$$

The primes denote the streams below the stage to which the feed is introduced. The value of q is a measure of the thermal condition of the feed and represents the moles of saturated liquid formed in the feed stage per mole of feed. The value of q for a particular feed can be estimated from

$$q = \frac{\text{energy to convert 1 mol of feed to saturated vapor}}{\text{molar heat of vaporization}}$$

It takes on the following values for various thermal conditions of the feed:

Subcooled liquid feed: $q > 1$
Saturated liquid feed: $q = 1$
Partially flashed feed: $0 < q < 1$
Saturated vapor feed: $q = 0$
Superheated vapor feed: $q < 0$

Equations analogous to (13-26) and (13-27) can be written for a sidestream, but the value of q will be either 1 or 0, depending on whether the sidestream is taken from the liquid or the vapor stream.

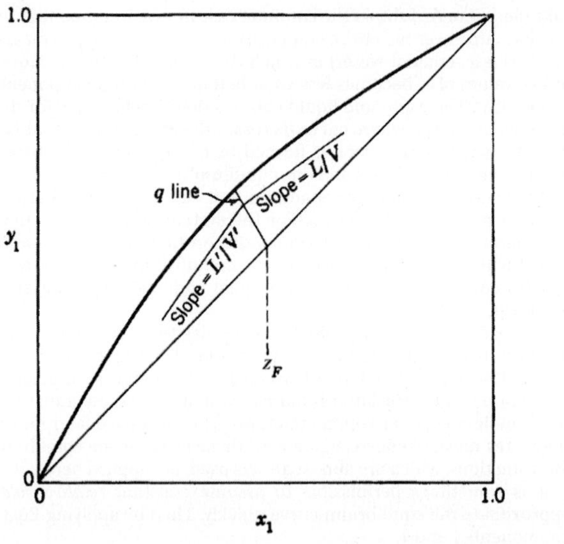

FIG. 13-19 Typical intersection of the two operating lines at the q line for a feed stage. The q line shown is for a partially flashed feed.

The quantity q can be used to derive the "q line equation" for a feed stream or a sidestream. The q line is the locus of all points of intersection of the two operating lines, which meet at the feed stream or sidestream stage. This intersection must occur along that section of the q line between the equilibrium curve and the $y = x$ diagonal. At the point of intersection, the same y, x point must satisfy both the operating line equation above the feed stream (or sidestream) stage and the one below the feed stream (or sidestream) stage. Subtracting one equation from the other gives for a feed stage

$$(V - V')y = (L - L')x + Fz_F$$

which, when combined with Eqs. (13-26) and (13-27), gives the q line equation

$$y = \frac{q}{q-1}x - \frac{z_F}{q-1} \tag{13-28}$$

A q line construction for a partially flashed feed is given in Fig. 13-19. It is easily shown that the q line must intersect the diagonal at z_F. The slope of the q line is $q/(q-1)$. All five q line cases are shown in Fig. 13-20. Note that when $q = 1$, the q line has infinite slope and is vertical.

The derivation of Eq. (13-28) assumes a single-feed column and no sidestream. However, the same result is obtained for other column configurations. Typical q line constructions for sidestream stages are shown in Fig. 13-21. Note that the q line for a sidestream must always intersect the diagonal at the composition (y_S or x_S) of the sidestream. Figure 13-21 also shows the intersections of the operating lines with the diagonal construction line. The top operating line must always intersect the diagonal at

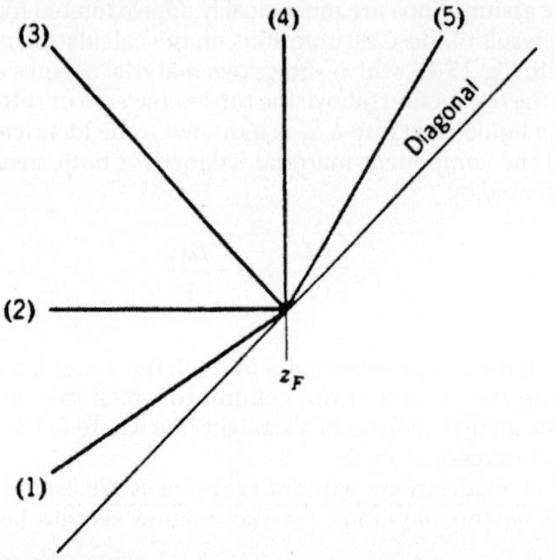

FIG. 13-20 All five cases of q lines: (1) superheated vapor feed, (2) saturated vapor feed, (3) partially vaporized feed, (4) saturated liquid feed, and (5) subcooled liquid feed. Slope of q line is $q/(q-1)$.

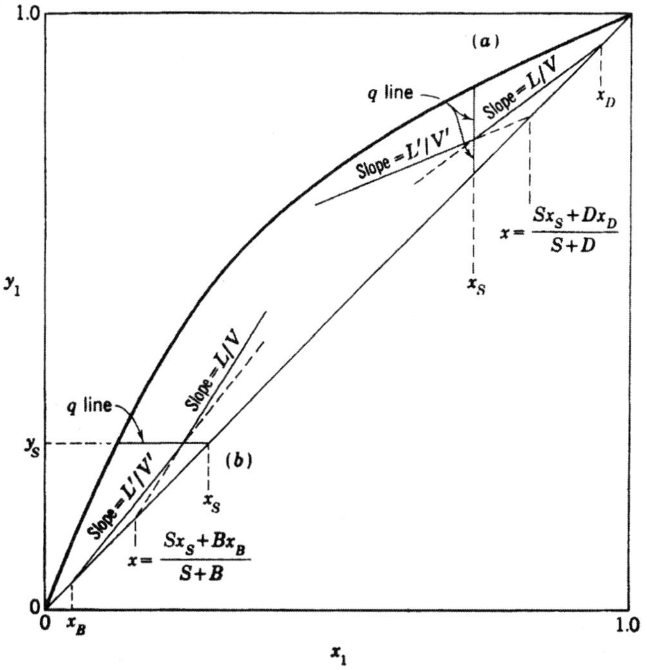

FIG. 13-21 Typical construction for a sidestream showing the intersection of the two operating lines with the q line and with the x-y diagonal. (a) Liquid sidestream near the top of the column. (b) Vapor sidestream near the bottom of the column.

the overhead-product composition x_D. This can be shown by substituting $y = x$ in Eq. (13-21) and using $V - L = D$ to reduce the resulting equation to $x = x_D$. Similarly (except for columns in which open steam is introduced at the bottom), the bottom operating line must always intersect the diagonal at the bottom-product composition x_B.

Equilibrium-Stage Construction The use of the equilibrium curve and the operating lines to "step off" equilibrium stages is illustrated in Fig. 13-22. The plotted portions of the equilibrium curve (curved) and the operating line (straight) cover the composition range existing in the column section shown in the lower right-hand corner of the figure. If y_n and x_n represent the compositions (in terms of the more volatile component) of the equilibrium vapor and liquid leaving stage n, then point (y_n, x_n) on the equilibrium curve must represent the equilibrium stage n. The operating line is the locus for compositions of all possible pairs of passing streams within the section, and therefore a horizontal line (dashed) at y_n must pass through the point (y_n, x_{n+1}) on the operating line since y_n and x_{n+1} represent passing streams. Likewise, a vertical line (dashed) at x_n must intersect the operating line at point (y_{n-1}, x_n). The equilibrium stages above and below stage n can be located by a vertical line through (y_n, x_{n+1}) to find (y_{n+1}, x_{n+1}) and a horizontal line through (y_{n-1}, x_n) to find (y_{n-1}, x_{n-1}). This procedure can be repeated by alternating the use of equilibrium and operating lines upward or downward through the column to find the total number of equilibrium stages.

Total Column Construction The graphical construction for an entire column is shown in Fig. 13-23. The process, pictured in the lower right-hand corner of the diagram, is an existing column with a number of actual trays equivalent to eight equilibrium stages. A partial reboiler (equivalent to an equilibrium stage) and a total condenser are used. This column configuration has $C + 2N + 9$ design variables (degrees of freedom), which must be specified to define one unique operation [see subsection Degrees of Freedom and Design Variables, especially Fig. 13-59 and Eq. (13-109)]. These may be used as follows as the basis for a graphical solution:

Specifications	Degrees of freedom
Stage pressures (including reboiler)	N
Condenser pressure	1
Stage heat leaks (except reboiler)	$N - 1$
Pressure and heat leak in reflux divider	2
Feed stream	$C + 2$
Feed-stage location	1
Total number of stages N	1
One overhead purity	1
Reflux temperature	1
Eternal reflux ratio	1
	$C + 2N + 9$

Pressures can be specified at any level below the safe working pressure of the column. The condenser pressure will be set at 275.8 kPa (40 psia), and all pressure drops within the column will be neglected. The equilibrium curve in Fig. 13-23 represents data at that pressure. All heat leaks will be assumed to be zero. The feed composition is 40 mol% of the more volatile component 1, and the feed rate is 0.126 kg·mol/s (1000 lb·mol/h) of saturated liquid ($q = 1$). The feed-stage location is fixed at stage 4, and the total number of stages is fixed at eight.

The overhead purity is specified as $x_D = 0.95$. The reflux temperature is the bubble point temperature (saturated reflux), and the external reflux ratio is set at $R = 4.5$.

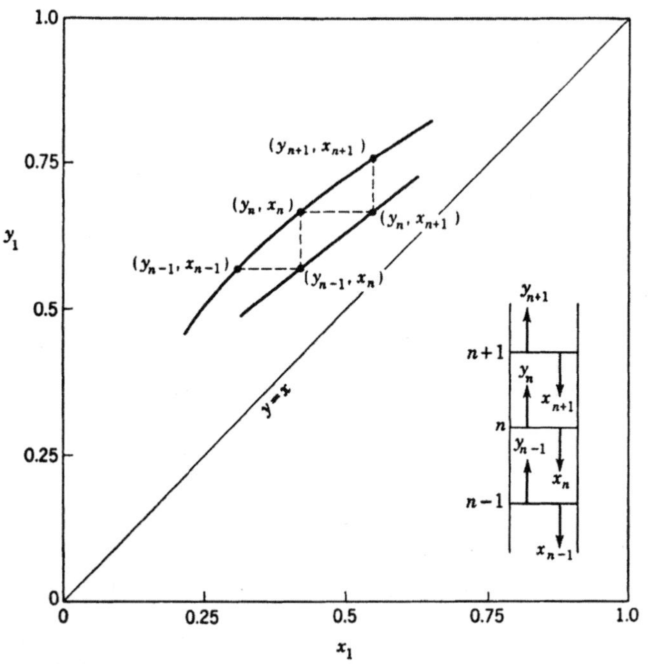

FIG. 13-22 Illustration of how equilibrium stages can be located on the x-y diagram through the alternating use of the equilibrium curve and the operating line.

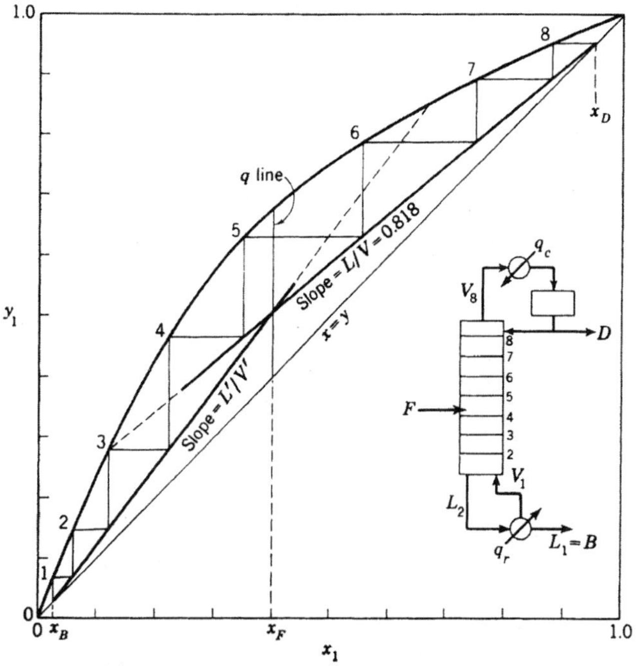

FIG. 13-23 Construction for a column with a bubble point feed, a total condenser, and a partial reboiler.

We need answers to the following two questions: First, what bottom-product composition x_B will the column produce under these specifications? Second, what is the value of the top vapor rate V_N in this operation, and will it exceed the maximum vapor rate capacity for this column, which is assumed to be 0.252 kg·mol/s (2000 lb·mol/h) at the top-tray conditions?

The solution is started by using Eq. (13-25) to convert the external reflux ratio of 4.5 to an internal reflux ratio of $L/V = 0.818$. The distillate composition $x_D = 0.95$ is then located on the diagonal, and the upper operating line is drawn as shown in Fig. 13-23.

If the x_B value were known, the bottom operating line could be immediately drawn from the x_B value on the diagonal up to its required intersection point with the upper operating line on the feed q line. In this problem, since the number of stages is fixed, the value of x_B that gives a lower operating line that will require exactly eight stages must be found by trial and error. An x_B value is assumed, and the resulting lower operating line is drawn. The stages can be stepped off by starting from either x_B or x_D; x_B was used in this case.

Note that the lower operating line is used until the fourth stage is passed, at which time the construction switches to the upper operating line. This is necessary because the vapor and liquid streams passing each other between the fourth and fifth stages must fall on the upper line.

The x_B that requires exactly eight equilibrium stages is $x_B = 0.026$. An overall component balance gives $D = 0.051$ kg·mol/s (405 lb·mol/h). Then

$$V_N = V_B = L_{N+1} + D = D(R+1) = 0.051(4.5 + 1.0)$$
$$= 0.280 \text{ kg}\cdot\text{mol/s (2230 lb}\cdot\text{mol/h)}$$

which exceeds the column capacity of 0.252 kg·mol/s (2007 lb·mol/h). This means that the column cannot provide an overhead-product yield of 40.5 percent at 95 percent purity. Either the purity specification must be reduced, or we must be satisfied with a lower yield. If the distillate specification ($x_D = 0.95$) is retained, the reflux rate must be reduced. This will cause the upper operating line to pivot upward around its fixed point of $x_D = 0.95$ on the diagonal. The new intersection of the upper line with the q line will lie closer to the equilibrium curve. The x_B value must then move upward along the diagonal because the eight stages will not "reach" as far as before. The higher x_B composition will reduce the recovery of component 1 in the 95 percent overhead product.

Another entire column with a partially vaporized feed, a liquid sidestream rate equal to D withdrawn from the second stage from the top, and a total condenser is shown in Fig. 13-24. The specified compositions are $z_F = 0.40$, $x_B = 0.05$, and $x_D = 0.95$. The specified L/V ratio in the top section is 0.818. These specifications permit the top operating line to be located and the two top stages stepped off to determine the liquid sidestream composition $x_s = 0.746$. The operating line below the sidestream must intersect the

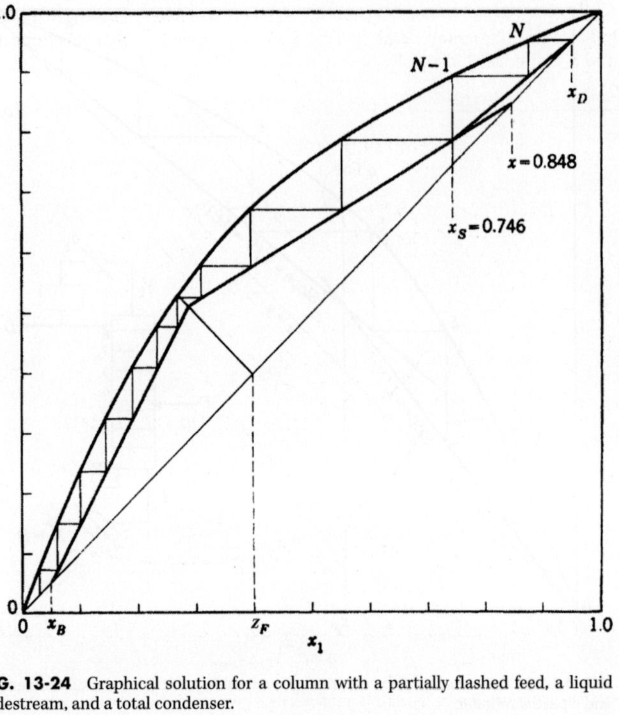

FIG. 13-24 Graphical solution for a column with a partially flashed feed, a liquid sidestream, and a total condenser.

diagonal at the "blend" of the sidestream and the overhead stream. Since S was specified to be equal to D in rate, the intersection point is

$$x = \frac{(1.0)(0.746) + (1.0)(0.95)}{1.0 + 1.0} = 0.848$$

This point plus the point of intersection of the two operating lines on the sidestream q line (vertical at $x_s = 0.746$) permits the location of the middle operating line. (The slope of the middle operating line could also have been used.) The lower operating line must run from the specified x_B value on the diagonal to the required point of intersection on the feed q line. The stages are stepped off from the top down in this case. The sixth stage from the top is the feed stage, and a total of about 11.4 stages are required to reach the specified $x_B = 0.05$.

Fractional equilibrium stages have meaning. The 11.4 will be divided by a tray efficiency, and the rounding up to an integral number of actual trays should be done after that division. For example, if the average tray efficiency for the process modeled in Fig. 13-24 is 80 percent, then the number of actual trays required is 11.4/0.8 = 14.3, which is rounded up to 15.

Feed-Stage Location The *optimum* feed-stage location is that location which, with a given set of other operating specifications, will result in the widest separation between x_D and x_B for a given number of stages. Or, if the number of stages is not specified, the optimum feed location is the one that requires the lowest number of stages to accomplish a specified separation between x_D and x_B. Either of these criteria will always be satisfied if the operating line farthest from the equilibrium curve is used in each step, as in Fig. 13-23.

It can be seen from Fig. 13-23 that the optimum feed location would have been the fifth tray for that operation. If a new column were being designed, that should be the designer's choice. However, when an existing column is being modeled, the feed stage on the diagram should correspond as closely as possible to the actual feed tray in the column. It can be seen that a badly mislocated feed (a feed that requires one to remain with an operating line until it closely approaches the equilibrium curve) can be very wasteful insofar as the effectiveness of the stages is concerned.

Minimum Stages A column operating at total reflux is represented in Fig. 13-25*a*. Enough material has been charged to the column to fill the reboiler, the trays, and the overhead condensate drum to their working levels. The column is then operated with no feed and with all the condensed overhead stream returned as reflux ($L_{N+1} = V_N$ and $D = 0$). Also all the liquid reaching the reboiler is vaporized and returned to the column as vapor. Since F, D, and B are all zero, $L_{n+1} = V_n$ at all points in the column. With a slope of unity ($L/V = 1.0$), the operating line must coincide with the diagonal throughout the column. Total reflux operation gives the minimum number of stages required to effect a specified separation between x_B and x_D.

Minimum Reflux The minimum reflux ratio is defined as that ratio which if decreased by an infinitesimal amount would require an infinite number of stages to accomplish a specified separation between two components. The concept has meaning only if a separation between two components is specified and the number of stages is not specified. Figure 13-25*b* illustrates the minimum reflux condition. As the reflux ratio is reduced, the two operating lines swing upward, pivoting around the specified x_B and x_D values, until one or both touch the equilibrium curve. For equilibrium curves shaped like the one shown, the contact occurs at the feed q line, resulting in a *feed pinch point*. Often an equilibrium curve will dip down closer to the diagonal at higher compositions. In such cases, the upper operating line may make contact before its intersection point on the q line reaches the equilibrium curve, resulting in a *tangent pinch point*. Wherever the contact appears, the intersection of the operating line with the equilibrium curve produces a pinch point that contains a very large number of stages, and a zone of constant composition is formed (see Doherty and Malone, *Conceptual Design of Distillation Systems*, McGraw-Hill, New York, 2001, chap. 3, and Sec. 4.6 for additional information).

Intermediate Reboilers and Condensers When a large temperature difference exists between the ends of the column due to a wide boiling point difference between the components, intermediate reboilers or condensers may be used to add heat at a lower temperature or remove heat at a higher temperature, respectively. (A distillation column of this type is shown in *Perry's Chemical Engineers' Handbook*, 7th ed., 1986, Fig. 13-2*a*.) A column operating with an intermediate reboiler and an intermediate condenser in addition to a regular reboiler and a condenser is illustrated with the solid lines in Fig. 13-26. The dashed lines correspond to simple distillation with only a bottoms reboiler and an overhead condenser. Total boiling and condensing heat loads are the same for both columns. As shown by Kayihan [*AIChE Symp. Ser.* **76**(192): 1 (1980)], the addition of intermediate reboilers and intermediate condensers increases thermodynamic efficiency but requires additional stages, as is clear from the positions of the operating lines in Fig. 13-26.

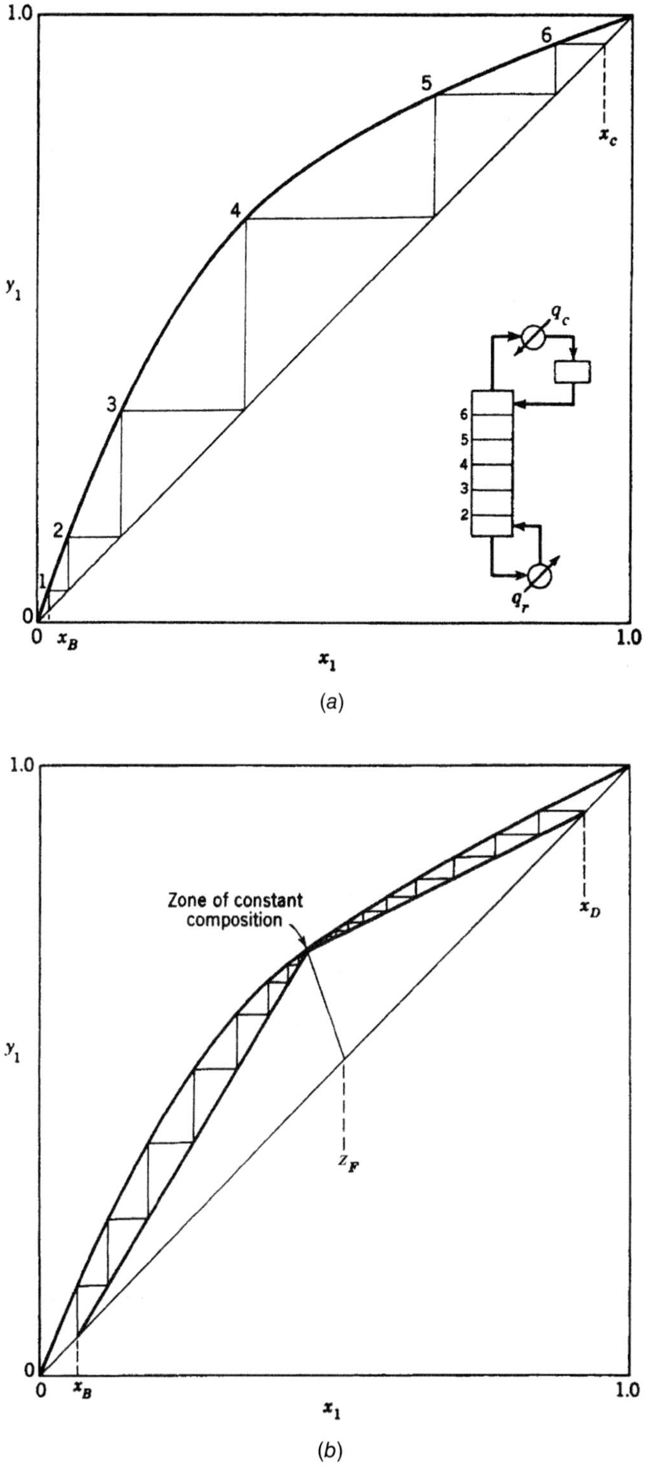

FIG. 13-25 McCabe-Thiele diagrams for limiting cases. (*a*) Minimum stages for a column operating at total reflux with no feeds or products. (*b*) Minimum reflux for a binary system of normal volatility.

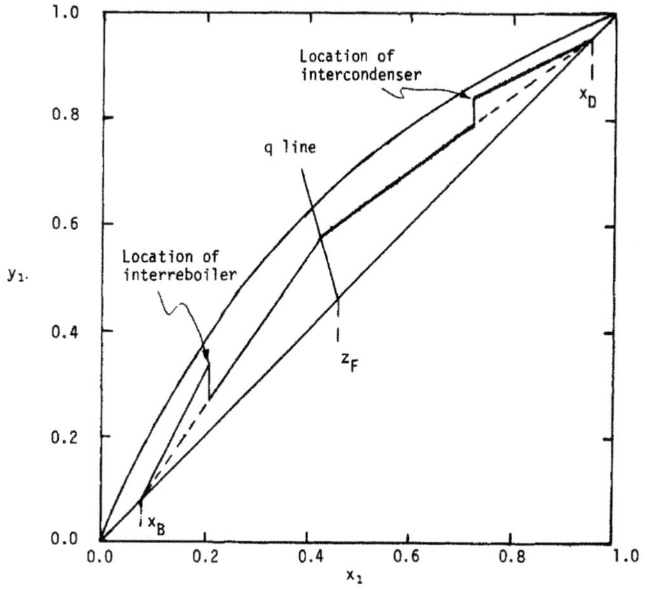

FIG. 13-26 McCabe-Thiele diagram for columns with and without an intermediate reboiler and an intermediate condenser.

Optimum Reflux Ratio The general effect of the operating reflux ratio on fixed costs, operating costs, and the sum of these is shown in Fig. 13-27. In ordinary situations, the minimum on the total cost curve will generally occur at an operating reflux ratio in the interval 1.1 to 2 times the minimum value. Generally, the total cost curve rises slowly from its minimum value as the operating reflux ratio increases, and very steeply as the operating reflux ratio decreases. In the absence of a detailed cost analysis for the specific separation of interest, it is recommended to select operating reflux ratios closer to 1.5 to 2.0 times the minimum value (see Doherty and Malone 2001, chap. 6, for additional discussion).

Difficult Separations Some binary separations may pose special problems because of extreme purity requirements for one or both products or because of a relative volatility close to 1. The *y-x* diagram is convenient for stepping off stages at extreme purities if it is plotted on log-log paper. However, such cases are best treated by equation-based design methods.

Equation-Based Design Methods Exact design equations have been developed for mixtures with constant relative volatility. Minimum stages can be computed with the Fenske equation, minimum reflux from the Underwood equation, and the total number of stages in each section of the column from either the Smoker equation [*Trans. Am. Inst. Chem. Eng.* **34:** 165 (1938); the derivation of the equation is shown, and its use is illustrated by Smith 1963] or Underwood's method. A detailed treatment of these approaches is given in Doherty and Malone (2001, chap. 3). Equation-based methods have also been developed for nonconstant relative volatility mixtures (including nonideal and azeotropic mixtures) by Julka and Doherty [*Chem. Eng. Sci.* **45:** 1801 (1990); *Chem. Eng. Sci.* **48:** 1367 (1993)], and Fidkowski et al. [*AIChE J.* **37:** 1761 (1991)]. Also see Doherty and Malone (2001, chap. 4).

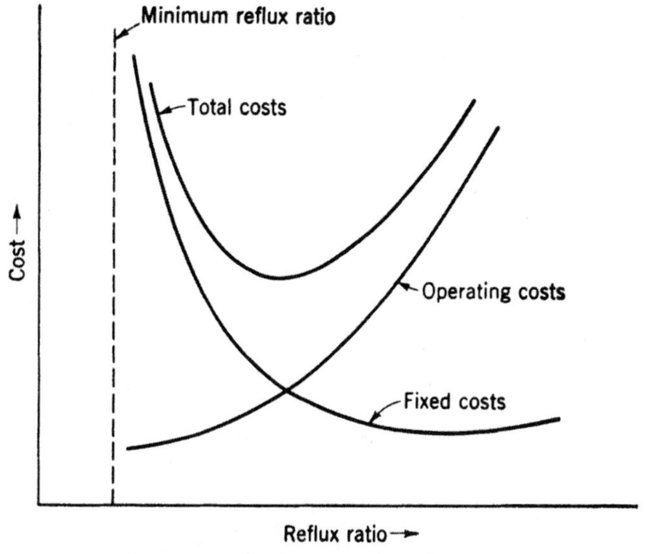

FIG. 13-27 Location of the optimum reflux for a given feed and specified separation.

Stage Efficiency The use of the *Murphree plate efficiency* is particularly convenient on *y-x* diagrams. The Murphree efficiency is defined for the vapor phase as

$$E^{MV} = \frac{y_n - y_{n-1}}{y_n^* - y_{n-1}} \tag{13-29}$$

where y_n^* is the composition of the vapor that would be in equilibrium with the liquid leaving stage n and is the value read from the equilibrium curve. The y_{n-1} and y_n are the actual (nonequilibrium) values for vapor streams leaving the $n-1$ and n stages, respectively. Note that for the y_{n-1} and y_n values, we assume that the vapor streams are completely mixed and uniform in composition. An analogous efficiency can be defined for the liquid phase.

The application of a 50 percent Murphree vapor-phase efficiency on a *y-x* diagram is illustrated in Fig. 13-28. A pseudoequilibrium curve is drawn halfway (on a vertical line) between the operating lines and the true equilibrium curve. The true equilibrium curve is used for the first stage (the partial reboiler is assumed to be an equilibrium stage), but for all other stages the vapor leaving each stage is assumed to approach the equilibrium value y_n^* only 50 percent of the way. Consequently, the steps in Fig. 13-28 represent actual trays.

In general, the application of a constant efficiency to each stage as in Fig. 13-28 will not give the same answer as obtained when the number of equilibrium stages (obtained by using the true equilibrium curve) is divided by the same efficiency factor.

The prediction and use of stage efficiencies are described in detail in Sec. 14. Alternative approaches based on mass-transfer rates are preferred, as described in the subsection Nonequilibrium Modeling.

Miscellaneous Operations The *y-x* diagrams for several other column configurations have not been presented here. The omitted items are *partial condensers, rectifying columns* (feed introduced to the bottom stage), *stripping columns* (feed introduced to the top stage), total reflux in the top section but not in the bottom section, multiple feeds, and the introduction of *open steam* to the bottom stage to eliminate the reboiler. These configurations are discussed in Smith (1963) and Henley and Seader (1981), who also describe the more rigorous Ponchon-Savarit method, which is not covered here.

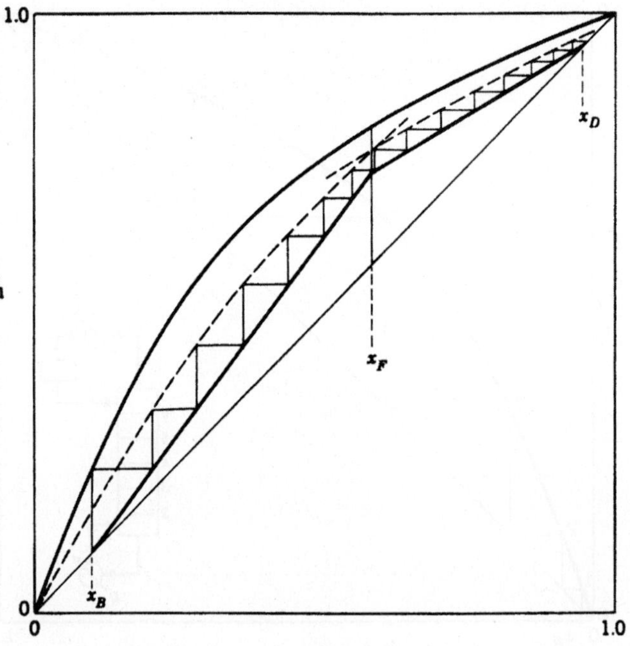

FIG. 13-28 Application of a 50 percent Murphree vapor-phase efficiency to each stage (excluding the reboiler) in the column. Each step in the diagram corresponds to an actual stage.

APPROXIMATE MULTICOMPONENT DISTILLATION METHODS

Some approximate calculation methods for the solution of multicomponent, multistage separation problems continue to serve useful purposes even though computers are available to provide more rigorous solutions. The available phase equilibrium and enthalpy data may not be accurate enough to justify the longer rigorous methods. Or in extensive design and optimization studies, a large number of cases can be worked quickly and cheaply by an approximate method to define roughly the optimum specifications, which can then be investigated more exactly with a rigorous method.

Two approximate multicomponent shortcut methods for simple distillation are the Smith-Brinkley (SB) method, which is based on an analytical solution of the finite-difference equations that can be written for staged separation processes when stages and interstage flow rates are known or assumed, and the Fenske-Underwood-Gilliland (FUG) method, which combines Fenske's total reflux equation and Underwood's minimum reflux equation with a graphical correlation by Gilliland that relates actual column performance to total and minimum reflux conditions for a specified separation between two key components. Thus, the SB and FUG methods are rating and design methods, respectively. Both methods work best when mixtures are nearly ideal.

The SB method is not presented here, but it is presented in detail in the 6th edition of *Perry's Chemical Engineers' Handbook.* Extensions of the SB method to nonideal mixtures and complex configurations are developed by Eckert and Hlavacek [*Chem. Eng. Sci.* **33:** 77 (1978)] and Eckert [*Chem. Eng. Sci.* **37:** 425 (1982)], respectively, but are not discussed here. However, the approximate and very useful method of Kremser [*Nat. Pet. News* **22**(21): 43 (1930)] for application to absorbers and strippers is discussed at the end of this subsection.

FENSKE-UNDERWOOD-GILLILAND (FUG) SHORTCUT METHOD

In this approach, Fenske's equation [*Ind. Eng. Chem.* **24:** 482 (1932)] is used to calculate $N_{\min}$, which is the number of plates required to make a specified separation at total reflux, that is, the minimum value of N. Underwood's equations [*J. Inst. Pet.* **31:** 111 (1945); **32:** 598 (1946); **32:** 614 (1946); and *Chem. Eng. Prog.* **44:** 603 (1948)] are used to estimate the minimum reflux ratio $R_{\min}$. The empirical correlation of Gilliland [*Ind. Eng. Chem.* **32:** 1220 (1940)] shown in Fig. 13-29 then uses these values to give N for any specified R, or R for any specified N. Limitations of the Gilliland correlation are discussed by Henley and Seader (*Equilibrium-Stage Separation Operations in Chemical Engineering,* Wiley, New York, 1981). The following equation, developed by Molokanov et al. [*Int. Chem. Eng.* **12**(2): 209 (1972)], satisfies the endpoints and fits the Gilliland curve reasonably well:

$$\frac{N - N_{\min}}{N+1} = 1 - \exp\left(\frac{1+54.4\Psi}{11+117.2\Psi} \times \frac{\Psi - 1}{\Psi^{0.5}}\right) \tag{13-30}$$

where $\Psi = (R - R_{\min})/(R+1)$.

The Fenske total reflux equation can be written as

$$\left(\frac{x_i}{x_r}\right)_D = (\alpha_i)^{N_{\min}} \left(\frac{x_i}{x_r}\right)_B \tag{13-31}$$

or as

$$N_{\min} = \frac{\log[(Dx_D/Bx_B)_i(Bx_B/Dx_D)_r]}{\log \alpha_i} \tag{13-32}$$

where i is any component and r is an arbitrarily selected reference component in the definition of relative volatilities

$$\alpha_i = \frac{K_i}{K_r} = \frac{y_i x_r}{y_r x_i} \tag{13-33}$$

The particular value of α_i used in Eqs. (13-31) and (13-32) is the effective value calculated from Eq. (13-34) defined in terms of values for each stage in the column by

$$\alpha^N = \alpha_N \alpha_{N-1} \cdots \alpha_2 \alpha_1 \tag{13-34}$$

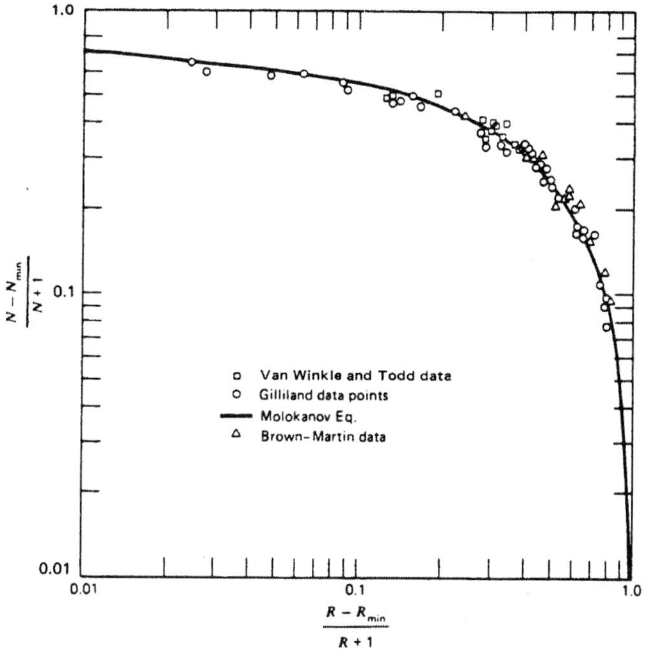

FIG. 13-29 Comparison of rigorous calculations with Gilliland correlation. [*Henley and Seader,* Equilibrium-Stage Separation Operations in Chemical Engineering, *Wiley, New York, 1981; data of Van Winkle and Todd,* Chem. Eng., ***78(21)****: 136 (Sept. 20, 1971); data of Gilliland,* Elements of Fractional Distillation, *4th ed., McGraw-Hill, New York, 1950; data of Brown and Martin,* Trans. Am. Inst. Chem. Eng., ***35****: 679 (1939).*]

Equations (13-31) and (13-32) are exact relationships between the splits obtained for components i and r in a column at total reflux. However, the value of α_i must always be estimated, and this is where the approximation enters. It is usually estimated from

$$\alpha = (\alpha_{top}\alpha_{bottom})^{1/2} \qquad (13\text{-}35)$$

or

$$\alpha = (\alpha_{top}\alpha_{middle}\alpha_{bottom})^{1/3} \qquad (13\text{-}36)$$

As a side note, the separation that will be accomplished in a column with a known number of equilibrium stages can often be reasonably well estimated by specifying the split of one component (designated as the reference component r), setting N_{min} equal to 40 to 60 percent of the number of equilibrium stages (not actual trays), and then using Eq. (13-32) to estimate the splits of all the other components. This is an iterative calculation because the component splits must first be arbitrarily assumed to give end compositions that can be used to give initial end-temperature estimates. The α_{top} and α_{bottom} values corresponding to these end temperatures are used in Eq. (13-5) to give α_i values for each component. The iteration is continued until the α_i values do not change from trial to trial.

The Underwood minimum reflux equations of main interest are those that apply when some of the components do not appear in either the distillate or the bottom products at minimum reflux. These equations are

$$\sum_i \frac{\alpha_i (x_{iD})_{min}}{\alpha_i - \Theta} = R_{min} + 1 \qquad (13\text{-}37)$$

and

$$\sum_i \frac{\alpha_i z_{iF}}{\alpha_i - \Theta} = 1 - q \qquad (13\text{-}38)$$

The relative volatilities α_i are defined by Eq. (13-33), R_{min} is the minimum reflux ratio, and q describes the thermal condition of the feed (1 for a saturated liquid feed and 0 for a saturated vapor feed). The z_{iF} values are available from the given feed composition. The Θ is the common root for the top section equations and the bottom section equations developed by Underwood for a column at minimum reflux with separate zones of constant composition in each section. The common root value must fall between α_{hk} and α_{lk}, where hk and lk stand for *heavy key* and *light key*, respectively. The *key components* are the ones the designer wants to separate. In the butane-pentane splitter problem in Example 13-1, the light key is n-C_4 and the heavy key is i-C_5.

The α_i values in Eqs. (13-37) and (13-38) are effective values obtained from Eq. (13-35) or Eq. (13-36). Once these values are available, Θ can be calculated in a straightforward iteration from Eq. (13-38). Since the α-Θ difference can be small, Θ should be determined to four decimal places to avoid numerical difficulties.

The $(x_{iD})_{min}$ values in Eq. (13-37) are minimum reflux values, that is, the overhead composition that would be produced by the column operating at the minimum reflux with an infinite number of stages. When the light key and the heavy key are adjacent in relative volatility and the specified split between them is sharp, or when the relative volatilities of the other components are not close to those of the two keys, only the two keys will distribute at minimum reflux and the $(x_{iD})_{min}$ values are easily determined. This is often the case, and it is the only one considered here. Other cases in which some of or all of the nonkey components distribute between distillate and bottom products are discussed in detail by Henley and Seader (1981).

The FUG method is convenient for new column design with the following specifications:

1. A value for R/R_{min}
2. Desired split on the reference component (usually chosen as the heavy key)
3. Desired split on one other component (usually the light key)

However, the total number of equilibrium stages N, N/N_{min}, or the external reflux ratio can be substituted for one of these three specifications. Note that the feed location is automatically specified as the optimum one; this is assumed in the Underwood equations. The assumption of saturated liquid reflux is also inherent in the Fenske and Underwood equations (i.e., the reflux is *not* subcooled). An important limitation on the Underwood equations is the assumption of constant molar overflow. As discussed by Henley and Seader (1981), this assumption can lead to a prediction of the minimum reflux that is considerably lower than the actual value. No such assumption is inherent in the Fenske equation. An exact calculation technique for minimum reflux is given by Tavana and Hansen [*Ind. Eng. Chem. Process Des. Dev.* **18:** 154 (1979)]. Approximate explicit expressions for minimum reflux for various types of splits in three- and four-component mixtures were developed by Glinos and Malone [*Ind. Eng. Chem. Process Des. Dev.* **23:** 764 (1984)] as well as lumping rules for applying their expressions to mixtures containing more than four components. These expressions are fairly accurate (usually within 5 percent of the exact value for R_{min}) and are extremely convenient for using in the FUG method since they remove the tedious calculation of R_{min} via the Underwood equations. A computer program for the FUG method is given by Chang [*Hydrocarbon Process.* **60**(8): 79 (1980)]. The method is best applied to mixtures that form ideal or nearly ideal solutions, and it should not be used for strongly nonideal or azeotropic mixtures.

Example 13-1 Application of FUG Method A large butane-pentane splitter is to be shut down for repairs. Some of its feed will be diverted temporarily to an available smaller column, which has only 11 trays plus a partial reboiler. The feed enters on the middle tray. Past experience with similar feeds indicates that the 11 trays plus the reboiler are roughly equivalent to 10 equilibrium stages and that the column has a maximum top vapor capacity of 1.75 times the feed rate on a mole basis. The column will operate at a condenser pressure of 827.4 kPa (120 psia). The feed will be at its bubble point (q = 1.0) at the feed tray conditions and has the following composition on the basis of 0.0126 kg·mol/s (100 lb·mol/h):

Component	Fx_{iF}
C_3	5
i-C_4	15
n-C_4	25
i-C_5	20
n-C_5	35
	100

The original column normally has less than 7 mol% i-C_5 in the overhead and less than 3 mol% n-C_4 in the bottom product when operating at a distillate rate of D/F = 0.489. Can these product purities be produced on the smaller column at D/F = 0.489?

Pressure drops in the column will be neglected, and the K values will be read at 827 kPa (120 psia) in both column sections from the DePriester nomograph in Fig. 2-10*b*. When constant molar overflow is assumed in each section, the rates in pound moles per hour in the upper and lower sections are as follows:

Top section	Bottom section
$D=(0.489)(100)=48.9$	$B=100-48.9=51.1$
$V=(1.75)(100)=175$	$V'=V=175$
$L=175-48.9=126.1$	$L'=L+F=226.1$
$\frac{V}{L}=1.388$	$\frac{V'}{L'}=0.7739$

$$\frac{L}{L'}=\frac{126.1}{226.1}=0.5577$$

$$R=\frac{126.1}{48.9}=2.579$$

NOTE: To convert pound-moles per hour to kilogram-moles per second, multiply by 1.26×10^{-4}.

Application of the FUG method is demonstrated on the splitter. Specifications necessary to model the existing column include these:

1. $N=10$, total number of equilibrium stages
2. Optimum feed location (which may or may not reflect the actual location)
3. Maximum V/F at the top tray of 1.75
4. Split on one component given in the following paragraphs

The solution starts with an assumed arbitrary split of all the components to give estimates of top and bottom compositions that can be used to get initial end temperatures. The α_i's evaluated at these temperatures are averaged with the α at the feed-stage temperature (assumed to be the bubble point of the feed) by using Eq. (13-36). The initial assumption for the split on i-C_5 is $Dx_D/Bx_B=3.15/16.85$. As mentioned earlier, N_{min} usually ranges from $0.4N$ to $0.6N$, and the initial N_{min} value assumed here will be $(0.6)(10)=6.0$. Equation (13-32) can be rewritten as

$$\left(\frac{Dx_D}{Fx_F-Dx_D}\right)=\alpha_i^{6.0}\left(\frac{3.15}{16.85}\right)=\alpha_i^{6.0}(0.1869)$$

or

$$Dx_{iD}=\frac{0.1869\alpha_i^{6.0}}{1+0.1869\alpha_i^{6.0}}Fx_{iF}$$

The evaluation of this equation for each component is as follows:

Component	α_i	$\alpha_i^{6.0}$	$0.1869\,\alpha_i^{6.0}$	Fx_{iF}	Dx_{iD}	Bx_{iB}
C_3	5.00			5	5.0	0.0
i-C_4	2.63	330	61.7	15	14.8	0.2
n-C_4	2.01	66	12.3	25	25.1	1.9
i-C_5	1.00	1.00	0.187	20	3.15	16.85
n-C_5	0.843	0.36	0.0672	35	2.20	32.80
				100	48.25	51.75

The end temperatures corresponding to these product compositions are 344 K (159°F) and 386 K (236°F). These temperatures plus the feed bubble point temperature of 358 K (185°F) provide a new set of α_i's that vary only slightly from those used earlier. Consequently, the $D=48.25$ value is not expected to vary greatly, and it will be used to estimate a new i-C_5 split. The desired distillate composition for i-C_5 is 7 percent, so it will be assumed that $Dx_D=(0.07)(48.25)=3.4$ for i-C_5 and that the split on that component will be 3.4/16.6. The results obtained with the new α_i's and the new i-C_5 split are as follows:

Component	$\alpha_i^{6.0}$	$0.2048\,\alpha_i^{6.0}$	Fx_{iF}	Dx_{iD}	Bx_{iB}	x_{iD}	x_{iB}
C_3			5	5.0	0.0	0.102	0.000
i-C_4	322	65.9	15	14.8	0.2	0.301	0.004
n-C_4	68	13.9	25	23.3	1.7	0.473	0.033
i-C_5	1.00	0.205	20	3.4	16.6	0.069	0.327
n-C_5	0.415	0.085	35	2.7	32.3	0.055	0.636
			100	49.2	50.8	1.000	1.000

The calculated i-C_5 composition in the overhead stream is 6.9 percent, which is close enough to the target value of 7.0 for now.

Table 13-2 shows subsequent calculations using the Underwood minimum reflux equations. The α and x_D values in Table 13-2 are those from the Fenske total reflux calculation. As noted earlier, the x_D values should be those at minimum reflux. This inconsistency may reduce the accuracy of the Underwood method, but to be useful, a shortcut method must be fast, and it has not been shown that a more rigorous estimation of x_D values results in an overall improvement in accuracy. The calculated R_{min} is 0.9426. The actual reflux assumed is obtained from the specified maximum top vapor rate of 0.022 kg·mol/s [175 lb·(mol/h)] and the calculated D of 49.2 (from the Fenske equation).

$$L_{N+1}=V_N-D$$

$$R=\frac{V_N}{D}-1=\frac{175}{49.2}-1=2.557$$

The values of $R_{min}=0.9426$, $R=2.557$, and $N=10$ are now used with the Gilliland correlation in Fig. 13-29 or Eq. (13-30) to check the initially assumed value of 6.0 for N_{min}. Equation (13-30) gives $N_{min}=6.95$, which differs from the assumed value.

Repetition of the calculations with $N_{min}=7$ gives $R=2.519$, $R_{min}=0.9782$, and a calculated check value of $N_{min}=6.85$, which is close enough. The final product compositions and the α values used are as follows:

Component	α_i	Dx_{iD}	Bx_{iB}	x_{iD}	x_{iB}
C_3	4.98	5.00	0	0.1004	0.0
i-C_4	2.61	14.91	0.09	0.2996	0.0017
n-C_4	2.02	24.16	0.84	0.4852	0.0168
i-C_5	1.00	3.48	16.52	0.0700	0.3283
n-C_5	0.851	2.23	32.87	0.0448	0.6532
		49.78	50.32	1.0000	1.0000

These results indicate that the 7 percent composition of i-C_5 in D and the 3 percent composition of i-C_4 in B obtained in the original column can also be obtained with the smaller column. These results disagree somewhat with the answers obtained from a rigorous computer solution, as shown in the following comparison. However, given the approximations that went into the FUG method, the agreement is good.

	x_{iD}		x_{iB}	
Component	Rigorous	FUG	Rigorous	FUG
C_3	0.102	0.100	0.0	0.0
i-C_4	0.299	0.300	0.006	0.002
n-C_4	0.473	0.485	0.037	0.017
i-C_5	0.073	0.070	0.322	0.328
n-C_5	0.053	0.045	0.635	0.653
	1.000	1.000	1.000	1.000

TABLE 13-2 Application of Underwood Equations

				$\theta=1.36$		$\theta=1.365$					
Component	x_F	α	αx_F	$\alpha-\theta$	$\frac{\alpha x_F}{\alpha-\theta}$	$\alpha-\theta$	$\frac{\alpha x_F}{\alpha-\theta}$	x_D	αx_D	$\alpha-\theta$	$\frac{\alpha x_D}{\alpha-\theta}$
C_3	0.05	4.99	0.2495	3.63	0.0687	3.625	0.0688	0.102	0.5090	3.6253	0.1404
i-C_4	0.15	2.62	0.3930	1.26	0.3119	1.255	0.3131	0.301	0.7886	1.2553	0.6282
n-C_4	0.25	2.02	0.5050	0.66	0.7651	0.655	0.7710	0.473	0.9555	0.6553	1.4581
i-C_5	0.20	1.00	0.2000	−0.36	−0.5556	−0.365	−0.5479	0.069	0.0690	−0.3647	−0.1892
n-C_5	0.35	0.864	0.3024	−0.496	−0.6097	−0.501	−0.6036	0.055	0.0475	−0.5007	−0.0949
	1.00				−0.0196		+0.0014	1.000			$1.9426=R_m+1$

Interpolation gives $\theta=1.3647$.

KREMSER EQUATION

Starting with the classical method of Kremser [*Nat. Pet. News* **22**(21): 43 (1930)], approximate methods of increasing complexity have been developed to calculate the behavior of groups of equilibrium stages for a countercurrent cascade, such as those used in simple absorbers and strippers of the type shown in Fig. 13-2*b* and *d*. However, none of these methods can adequately account for stage temperatures that are considerably higher or lower than the two entering stream temperatures for absorption and stripping, respectively, when appreciable composition changes occur. Therefore, only the simplest form of the Kremser method is presented here. Fortunately, rigorous computer methods described later can be applied when accurate results are required. The Kremser method is most useful for making preliminary estimates of absorbent and stripping agent flow rates or equilibrium-stage requirements. The method can also be used to quickly extrapolate the results of a rigorous solution to a different number of equilibrium stages.

Consider the general adiabatic countercurrent cascade of Fig. 13-30 where v and ℓ are molar component flow rates. Regardless of whether the cascade is an absorber or a stripper, components in the entering vapor will tend to be absorbed and components in the entering liquid will tend to be stripped. If more moles are stripped than absorbed, the cascade is a stripper; otherwise, the cascade is an absorber. The Kremser method is general and applies to either case. Application of component material balance and phase equilibrium equations successively to stages 1 through $N-1$, 1 through $N-2$, etc., as shown by Henley and Seader (1981), leads to the following equations originally derived by Kremser. For each component i,

$$(v_i)_N = (v_i)_0(\Phi_i)_A = (l_i)_{N+1}[1-(\Phi_i)_S] \tag{13-39}$$

where

$$(\Phi_i)_A = \frac{(A_i)_e - 1}{(A_i)_e^{N+1} - 1} \tag{13-40}$$

is the fraction of component i in the entering vapor that is not absorbed,

$$(\Phi_i)_S = \frac{(S_i)_e - 1}{(S_i)_e^{N+1} - 1} \tag{13-41}$$

is the fraction of component i in the entering liquid that is not stripped,

$$(A_i)_e = \left(\frac{L}{K_i V}\right)_e \tag{13-42}$$

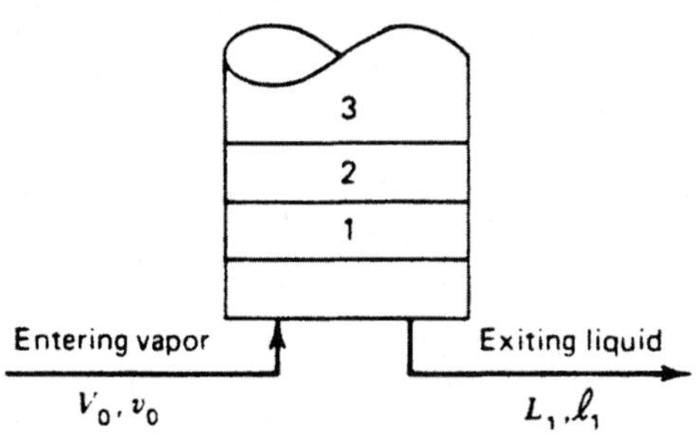

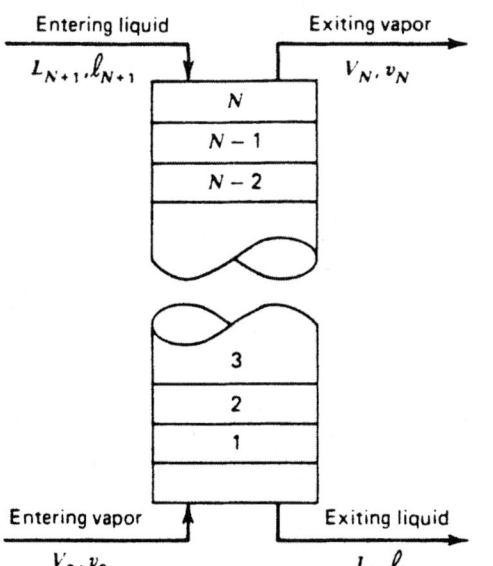

FIG. 13-30 General adiabatic countercurrent cascade for simple absorption or stripping.

is the effective or average absorption factor for component i, and

$$(S_i)_e = \frac{1}{(A_i)_e} \tag{13-43}$$

is the effective or average stripping factor for component i. When the entering streams are at the same temperature and pressure and negligible absorption and stripping occur, effective component absorption and stripping factors are determined simply by entering stream conditions. Thus, if K values are composition-independent, then

$$(A_i)_e = \frac{1}{(S_i)_e} = \frac{L_{N+1}}{K_i(T_{N+1}, P_{N+1})V_0} \tag{13-44}$$

When entering stream temperatures differ or moderate to appreciable absorption or stripping occurs, the values of A_i and S_i should be based on the effective average values of L, V, and K_i in the cascade. However, even then, Eq. (13-44) with T_{N+1} replaced by $(T_{N+1} + T_0)/2$ may be able to give a first-order approximation of $(A_i)_e$. In the case of an absorber, $L_{N+1} < L_e$ and $V_0 > V_e$ will be compensated to some extent by $K_i\{(T_{N+1} + T_0)/2, P\} < K_i\{T_e, P\}$. A similar compensation, but in opposite directions, will occur in the case of a stripper. Equations (13-40) and (13-41) are plotted in Fig. 13-31.

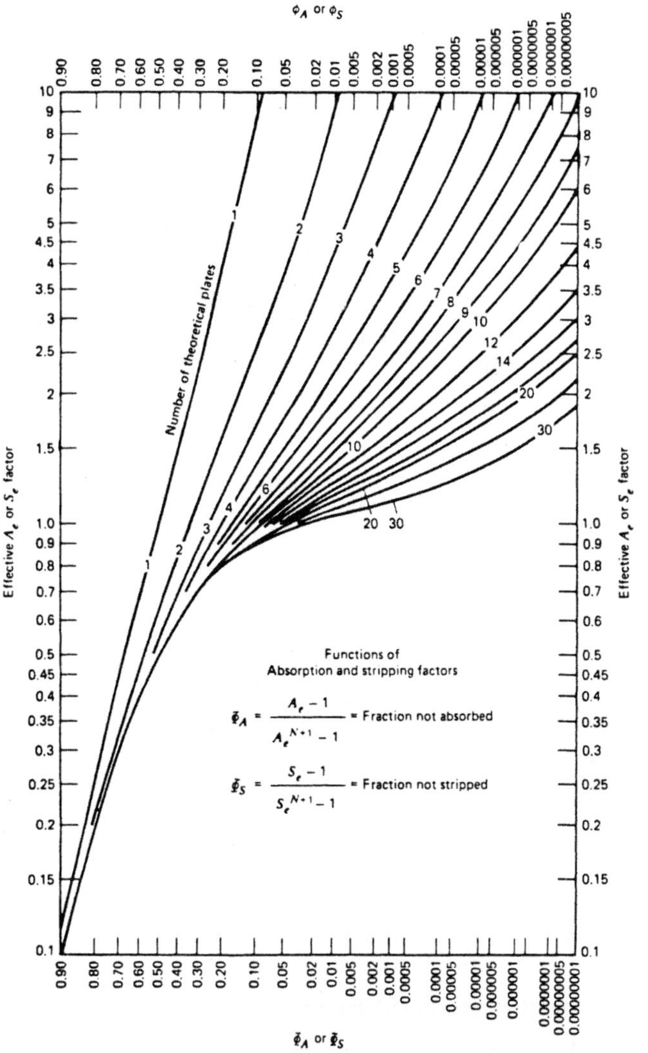

FIG. 13-31 Absorption and stripping factors. [*W. C. Edmister,* AIChE J., *3: 165–171 (1957).*]

Components having large values of A_e or S_e absorb or strip, respectively, to a large extent. Corresponding values of Φ_A and Φ_S approach a value of 1 and are almost independent of the number of equilibrium stages.

An estimate of the minimum absorbent flow rate for a specified amount of absorption from the entering gas of some key component K for a cascade with an infinite number of equilibrium stages is obtained from Eq. (13-40) as

$$(L_{N+1})_{\min} = K_K V_0 [1 - (\Phi_K)_A] \tag{13-45}$$

The corresponding estimate of minimum stripping agent flow rate for a stripper is obtained as

$$(V_0)_{\min} = \frac{L_{N+1}[1-(\Phi_K)_S]}{K_K} \tag{13-46}$$

Example 13-2 Calculation of Kremser Method For the simple absorber specified in Fig. 13-32, a rigorous calculation procedure as described below gives the results in Table 13-3. Values of Φ were computed from component product flow rates, and corresponding effective absorption and stripping factors were obtained by iterative calculations in using Eqs. (13-40) and (13-41) with $N = 6$. Use the Kremser method to estimate component product rates if N is doubled to a value of 12.

Assume that values of A_e and S_e will not change with a change in N. Application of Eqs. (13-40), (13-41), and (13-39) gives the results in the last four columns of Table 13-3. Because of its small value of A_e, the extent of absorption of C_1 is unchanged. For the other components, somewhat increased amounts of absorption occur. The degree of stripping of the absorber oil is essentially unchanged. Overall, only an additional 0.5 percent of absorption occurs. The greatest increase in absorption occurs for n-C_4, to the extent of about 4 percent.

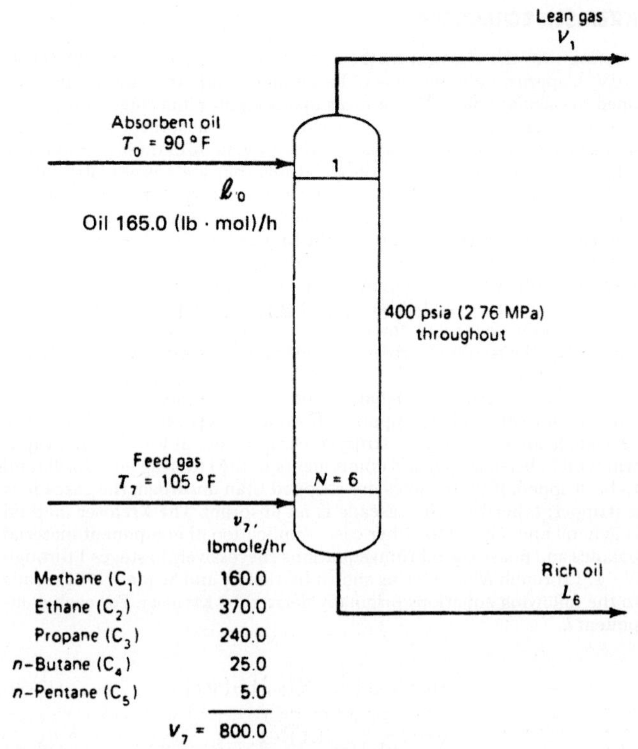

FIG. 13-32 Specifications for the absorber example.

TABLE 13-3 Results of Calculations for Simple Absorber of Fig. 13-32

	$N = 6$ (rigorous method)				$N = 12$ (Kremser method)					
	(lb · mol)/h						(lb · mol)/h			
Component	$(v_i)_6$	$(\ell_i)_1$	$(\Phi_i)_A$	$(\Phi_i)_S$	$(A_i)_e$	$(S_i)_e$	$(v_i)_{12}$	$(\ell_i)_1$	$(\Phi_i)_A$	$(\Phi_i)_S$
C_1	147.64	12.36	0.9228		0.0772		147.64	12.36	0.9228	
C_2	276.03	94.97	0.7460		0.2541		275.98	94.02	0.7459	
C_3	105.42	134.58	0.4393		0.5692		103.46	136.54	0.4311	
nC_4	1.15	23.85	0.0460		1.3693		0.16	24.84	0.0063	
nC_5	0.0015	4.9985	0.0003		3.6		0	5.0	0.0	
Absorber oil	0.05	164.95		0.9997		0.0003	0.05	164.95		0.9997
Totals	530.29	435.71					527.29	437.71		

NOTE: To convert pound-moles per hour to kilogram-moles per hour, multiply by 0.454.

SIMULATION OF DISTILLATION PROCESSES

Chemical engineers have been solving distillation problems by using the *equilibrium-stage model* since 1893 when Sorel outlined the concept to describe the distillation of alcohol. Since that time, it has been used to model a wide variety of distillation-like processes, including simple distillation (single-feed, two-product columns), complex distillation (multiple-feed, multiple-product columns), extractive and azeotropic distillation, petroleum distillation, absorption, liquid-liquid extraction, stripping, and supercritical extraction.

Real distillation processes, however, nearly always operate away from equilibrium. In recent years it has become possible to simulate distillation and absorption as the mass-transfer rate-based operations that they really are, using what have become known as *nonequilibrium* (NEQ) or *rate-based* models [Taylor et al., *CEP* (2003, July 28, pp. 28–39)].

EQUILIBRIUM-STAGE MODELING

A schematic diagram of an equilibrium stage is shown in Fig. 13-33*a*. Vapor from the stage below and liquid from a stage above are brought into contact on the stage together with any fresh or recycle feeds. The vapor and liquid streams leaving the stage are assumed to be in equilibrium with each other. A complete separation process is modeled as a sequence of these *equilibrium stages*, as shown in Fig. 13-33*b*.

The MESH Equations (the 2*c* + 3 Formulation) The equations that model equilibrium stages often are referred to as the MESH equations. The M equations are the material balance equations, E stands for equilibrium equations, S stands for mole fraction summation equations, and H refers to the heat or enthalpy balance equations.

There are two types of material balance: the total material balance

$$V_{j+1} + L_{j-1} + F_j - (1 + r_j^V)V_j - (1 + r_j^L)L_j = 0 \tag{13-47}$$

and the component material balance

$$V_{j+1}y_{i,j+1} + L_{j-1}x_{i,j-1} + F_j z_{i,j} - (1 + r_j^V)V_j y_{i,j} - (1 + r_j^L)L_j x_{i,j} = 0 \tag{13-48}$$

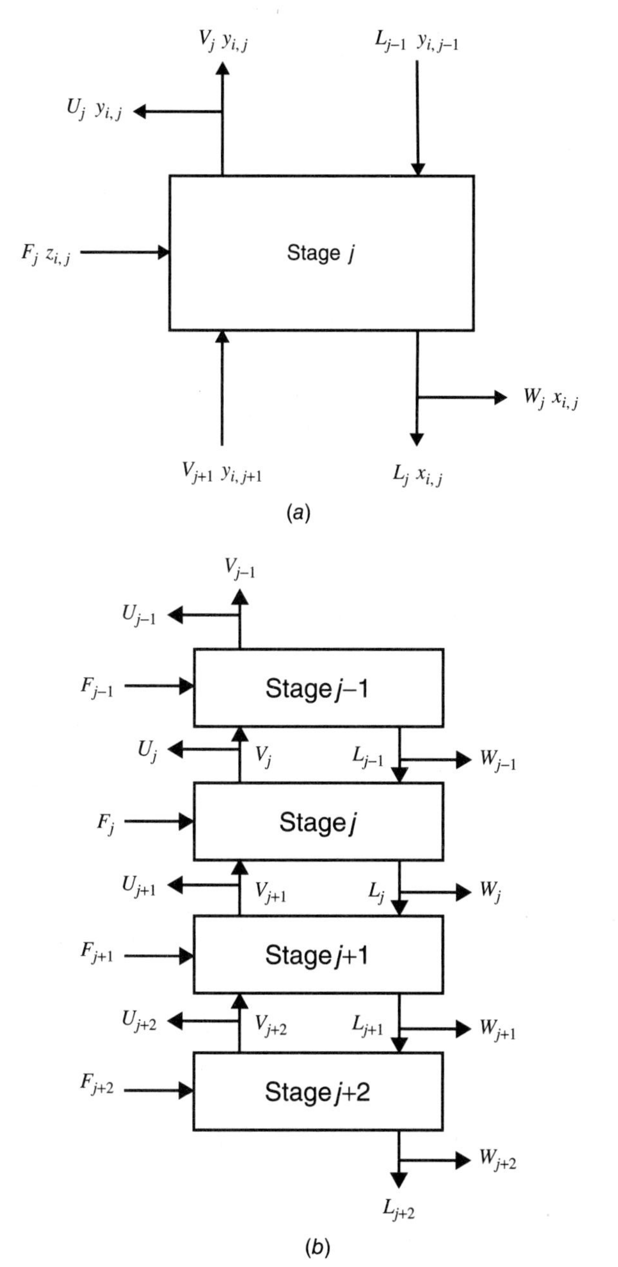

FIG. 13-33 (*a*) The equilibrium stage. (*b*) Multistage column.

In these material balance equations, r_j is the ratio of sidestream flow to interstage flow:

$$r_j^V = \frac{U_j}{V_j} \qquad r_j^L = \frac{W_j}{L_j} \tag{13-49}$$

Mole fractions must be forced to sum to unity, thus

$$\sum_{i=1}^{c} x_{i,j} = 1 \quad \sum_{i=1}^{c} y_{i,j} = 1 \quad \sum_{i=1}^{c} z_{i,j} = 1 \tag{13-50}$$

The enthalpy balance is given by

$$V_{j+1}H_{j+1}^V + L_{j-1}H_{j-1}^L + F_jH_j^F - (1+r_j^V)V_jH_j^V - (1+r_j^L)L_jH_j^L - Q_j = 0 \tag{13-51}$$

The superscripted H's are the enthalpies of the appropriate phase.

TABLE 13-4 Equations for an Equilibrium Stage

Equation	Equation no.	Number
Total mass balance	(13-47)	1
Component mass balances	(13-48)	c
Total energy balance	(13-51)	1
Mole fraction summation equations	(13-50)	2
Equilibrium equations	(13-52)	c
		$2c + 4$

To complete the model, it is usual to add equations that relate the compositions of the two streams leaving the stage. In the standard model of a distillation column, we assume that these two streams are in equilibrium with each other. Thus, the mole fractions in the exiting streams are related by the familiar equations of phase equilibrium:

$$y_{i,j} = K_{i,j}x_{i,j} \tag{13-52}$$

The $K_{i,j}$ are the equilibrium ratios or K values for species i on stage j.

Degrees-of-Freedom Analysis and Problem Formulation Table 13-4 summarizes the equations for a single equilibrium stage (in a sequence of stages). There are $2c + 4$ equations per stage, of which only $2c + 3$ are independent. Thus, one of these equations must be ignored in a (computer-based) method to solve the equations. In some methods we disregard the total material balance; an alternative is to combine the vapor and liquid mole fraction summation equations, as was done for flash calculations by Rachford and Rice [*J. Pet. Technol.* **4**(10): sec. 1, p. 19, and sec. 2, p. 3 (1952)]. The mole fraction summation equations for the feed are omitted here, as are the mole fraction summation equations for interstage vapor and liquid traffic; the latter "belong" to the equation set for adjacent stages (see, however, the subsection Degrees of Freedom, in which this topic is revisited in greater detail).

The quantities for stage j that appear in these equations are summarized in Table 13-5. The total number of variables appearing in these equations is $3c + 10$. Note that the K values and enthalpies that also appear in the MESH equations are not included in the table of variables, nor are equations for their estimation included in the list of equations. Thermodynamic properties are functions of temperature, pressure, and composition, quantities that do appear in the table of variables.

The $2c + 3$ unknown variables normally determined by solving the MESH equations are the c vapor mole fractions $y_{i,j}$, the c liquid mole fractions $x_{i,j}$, the stage temperature T_j, and the vapor and liquid flow V_j and L_j. The remaining variables, $c + 7$ in number (the difference between the number of variables and the number of independent equations), that need to be specified are the stage pressure P_j, the feed flow rate, $c - 1$ mole fractions in the feed (the last is then computed directly from the feed mole fraction summation equation, which was not included in the equations for stage j in Table 13-4), temperature and pressure of the feed, the stage heat duty Q_j, and the sidestream flows U_j and W_j. It is important to recognize that the other flows and composition variables appearing in the MESH equations are associated with the equivalent equations for adjacent stages.

TABLE 13-5 Variables for an Equilibrium Stage

Variable	Symbol	Number
Vapor flow rate	V_j	1
Liquid flow rate	L_j	1
Feed flow rate	F_j	1
Vapor sidestream flow	W_j	1
Liquid sidestream flow	U_j	1
Vapor-phase mole fractions	$y_{i,j}$	c
Liquid-phase mole fractions	$x_{i,j}$	c
Feed composition	$z_{i,j}$	c
Stage temperature	T_j	1
Stage pressure	P_j	1
Feed temperature		1
Feed pressure		1
Stage heat duty	Q_j	1
		$3c + 10$

For a column of N stages, we must solve $N(2c + 3)$ equations. The following table shows how we may need to solve hundreds or even thousands of equations.

c	N	$N(2c+3)$
2	10	70
3	20	180
5	50	650
10	30	690
40	100	8300

The 2c + 1 Formulation An alternative form of the MESH equations is used in many algorithms. In this variation, we make use of the component flow defined by

$$v_{ij} = V_j y_{ij} \qquad l_{ij} = L_j x_{ij} \qquad f_{ij} = F_j z_{ij} \tag{13-53}$$

In terms of the component flow, the component material balance becomes

$$v_{i,j+1} + l_{i,j-1} + f_{ij} - (1 + r_j^V) v_{i,j} - (1 + r_j^L) l_{i,j} = 0 \tag{13-54}$$

Since the total vapor and liquid flow rates are, by definition, the sum of the component flow rates of the respective phases, the summation equations and total mass balance equation are satisfied automatically. Thus, the number of equations and variables per stage have been reduced by 2 to 2c + 1.

Condenser and Reboiler The MESH equations given above apply to all the interior stages of a column. In addition, any reboiler and condenser must be considered. On an actual plant, a total condenser may be followed by a reflux accumulator and a stream divider. The accumulator does not add any equations to a steady-state model (but it is important to consider in an un-steady-state model), but the stream splitter is a separate unit with its own temperature, pressure, and heat loss, and it is modeled by the appropriate balance equations (see the subsection on Degrees of Freedom, where we address this topic in greater detail). In practice the condenser and reflux splitter often are modeled as a single combined unit, and the MESH equations described previously may be used to model these stages with only some minor modifications. For example, for a total condenser *and* the reflux stream splitter at the top of a distillation column, the liquid distillate is U_1 and the reflux ratio $R = 1/r_1^L$. For a partial condenser (with no stream splitter needed), the vapor product is V_1 and the reflux ratio is $R = L_1/V_1$. A condenser/splitter device that provides both vapor and liquid products is given by a combination of these two units. Finally, for a partial reboiler at the base of a column, the bottoms flow rate is L_N. Note that an equilibrium stage with a sidestream is considered here to be a single unit in essentially the same way.

In computer-based methods for solving the MESH equations, it is common to replace the energy balance of the condenser (with or without the associated stream splitter) and reboiler with a specification equation. Possible specifications include

- The flow rate of the distillate/bottoms product stream
- The mole fraction of a component in either the distillate or bottoms product stream
- Component flow rate in either the distillate or bottoms product stream
- A reflux/reboil ratio or rate
- The temperature of the condenser or reboiler
- The heat duty to the condenser or reboiler

If the condenser and/or reboiler heat duties are not specified, it is possible to calculate them from the energy balances after all the other model equations have been solved.

For a total condenser, the vapor composition used in the equilibrium relations is that determined during a bubble point calculation based on the actual pressure and liquid compositions found in the condenser. These vapor mole fractions are not used in the component mass balances since there is no vapor stream from a total condenser. It often happens that the temperature of the reflux stream is below the bubble point temperature of the condensed liquid (subcooled condenser). In such cases it is necessary to specify either the actual temperature of the reflux stream or the difference in temperature between the reflux stream and the bubble point of the condensate.

Solution of the MESH Equations We may identify several classes of methods of solving the *equilibrium*-stage model equations:

- *Graphical methods.* These methods were developed before modern computer methods became widely adopted. Some graphical methods retain their value for a variety of reasons and were discussed at length earlier in this section.
- *Approximate methods.* In these, a great many simplifying assumptions are made to obtain solutions to the model equations. These methods were the subject of the immediately preceding subsection.
- *Computer-based methods.* From the late 1950s to the early 1990s, hardly a year passed without the publication of at least one (and usually many more than one) new algorithm for solving these equations [Seader, "The B. C. (Before Computers) and A. D. of Equilibrium-Stage Operations," *Chem. Eng. Educ.* **14**(2): 88–103 (1985)]. One could make the case that it was the equilibrium model that brought computing into chemical engineering in the first place! One of the incentives for this activity has always been a desire to solve problems with which existing methods have trouble. The evolution of algorithms for solving the stage model equations has been influenced by, among other things, the availability (or lack) of sufficient computer storage and speed, the development of mathematical techniques that can be exploited, the complexity of physical property (K value and enthalpy) correlations, and the form of the model equations being solved. We continue with a brief discussion of computer-based methods. Many computer-based methods are discussed at length in a number of textbooks [see, e.g., Holland, *Fundamentals of Multicomponent Distillation,* McGraw-Hill, New York, 1981; King, *Separation Processes,* 2d ed., McGraw-Hill, New York, 1980; Seader and Henley, *Separation Process Principles,* Wiley, New York, 1998; Haas (in Kister 1992)]. Seader (1985) has written an interesting and elegant history of equilibrium-stage simulation. Other reviews of simulation methods can be found in this chapter in the 7th edition of this handbook.

Examples In what follows we illustrate possible column specifications by considering four examples from Seader (*Perry's Chemical Engineers' Handbook,* 7th ed., 1997, pp. 13-46–13-49). They are a simple distillation column, a more complicated distillation column, an absorber, and a reboiled stripper. A simultaneous convergence method was used for the calculations reported below. The computer program that was used for these exercises automatically generated initial estimates of all the unknown variables (flows, temperatures, and mole fractions). In most cases the results here differ only very slightly from those obtained by Seader (almost certainly due to small differences in physical property constants).

Example 13-3 Simple Distillation Column Compute stage temperatures, interstage vapor and liquid flow rates and compositions, and reboiler and condenser duties for the butane-pentane splitter studied in Example 13-1. The specifications for this problem are summarized next and in Fig. 13-34.

The specifications made in this case are:

Variable	Number	Value
Number of stages	1	11
Feed stage location	1	6
Component flows in feed	$c = 5$	5, 15, 25, 20, 35 lb·mol/h
Feed pressure	1	120 psia
Feed vapor fraction	1	0
Pressure on each stage including condenser and reboiler	$N = 11$	$P_j = 120$ psia
Heat duty on each stage except reboilers and condensers	$N - 2 = 9$	$Q_j = 0$
Vapor flow to condenser (replaces heat duty of reboiler)	1	$V_2 = 175$ lb·mol/h
Distillate flow rate (replaces heat duty of condenser)	1	$D = 48.9$ lb·mol/h
Total	31	

In addition, we have assumed that the pressure of the reflux divider is the same as the pressure of the condenser, the heat loss from the reflux divider is zero, and the reflux temperature is the boiling point of the condensed overhead vapor.

The Peng-Robinson equation of state was used to estimate K values and enthalpy departures [as opposed to the De Priester charts used in Example 13-1 and by Seader (1977)].

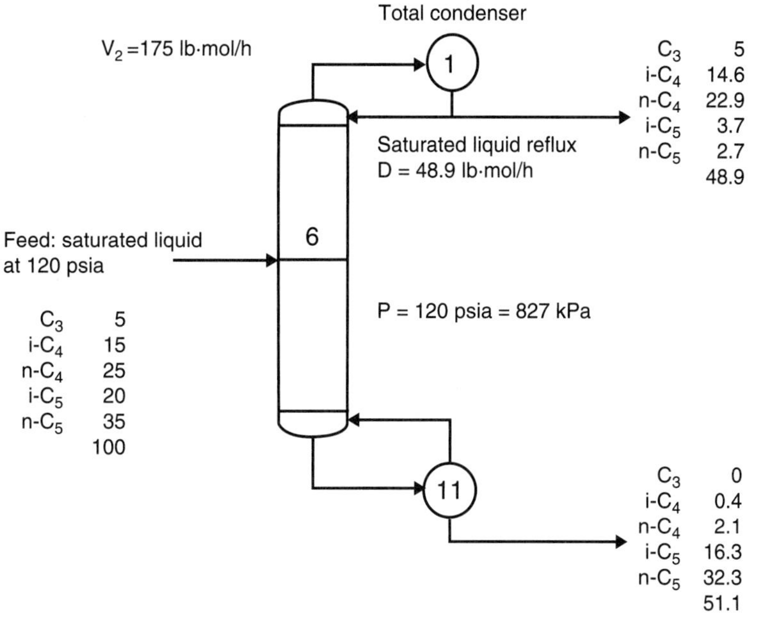

FIG. 13-34 Specifications and calculated product stream flows for butane-pentane splitter. Flows are in pound-moles per hour.

With 11 stages and 5 components, the equilibrium-stage model has 143 equations to be solved for 143 variables (the unknown flow rates, temperatures, and mole fractions). Convergence of the computer algorithm was obtained in just four iterations. Computed product flows are shown in Fig. 13-34.

A pseudobinary McCabe-Thiele diagram for this multicomponent system is shown in Fig. 13-35. For systems with more than two components, these diagrams can only be computed from the results of a computer simulation. The axes are defined by the relative mole fractions:

$$X = \frac{x_{lk}}{x_{lk} + x_{hk}} \qquad Y = \frac{y_{lk}}{y_{lk} + y_{hk}}$$

where the subscripts *lk* and *hk* refer to light and heavy key, respectively. The lines in the diagram have similar significance as would be expected from our knowledge of McCabe-Thiele diagrams for binary systems discussed earlier in this section; the triangles correspond to equilibrium stages. The operating lines are not straight because of heat effects and because the feed is not in the best location.

The fact that the staircase of triangles visible in Fig. 13-35 fails to come close to the corners of the diagram where $X = Y = 1$ and $X = Y = 0$ shows that the separation is not especially sharp. It is worth asking what can be done to improve the separation obtained with this column. The parameters that have a significant effect on the separation are the numbers of stages in the sections above and below the feed, the reflux ratio, and a product flow rate (or reflux flow). Figure 13-36 shows how the mole fraction of *i*-pentane in the overhead and of *n*-butane in the bottom product changes with the reflux ratio. For the base case considered above, the reflux ratio is 2.58 (calculated from the results of the simulation). It is clear that increasing the reflux ratio has the desired effect of improving product purity. This improvement in purity is, however, accompanied by an increase in both the operating cost, indicated in Fig. 13-36 by the increase in reboiler duty, and capital cost, because a larger column would be needed to accommodate the increased internal flow. Note, however, that the curves that represent the mole fractions of the keys in the overhead and bottoms appear to flatten, showing that product purity will not increase indefinitely

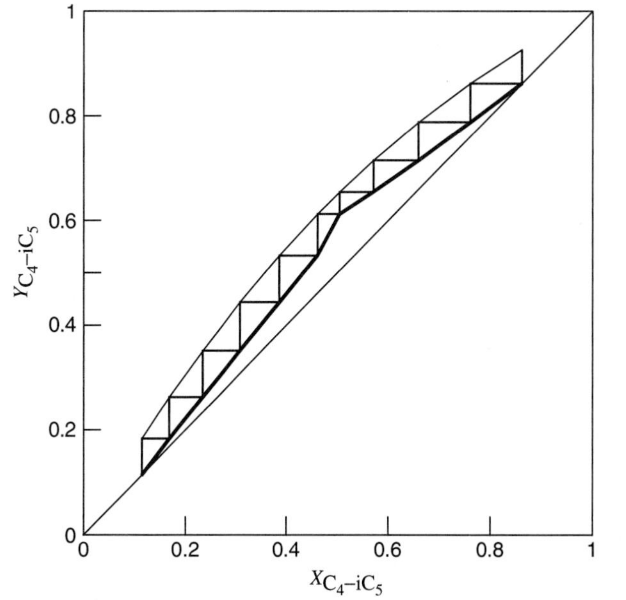

FIG. 13-35 Multicomponent McCabe-Thiele diagram for butane-pentane splitter in Fig. 13-34.

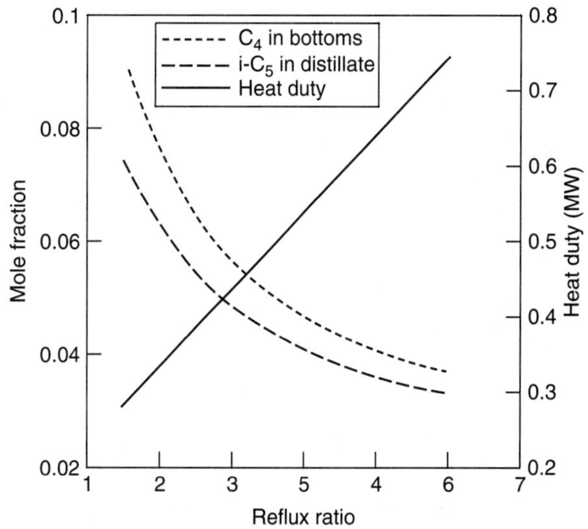

FIG. 13-36 Product mole fractions and reboiler heat duty as a function of the reflux ratio for butane-pentane splitter in Fig. 13-34.

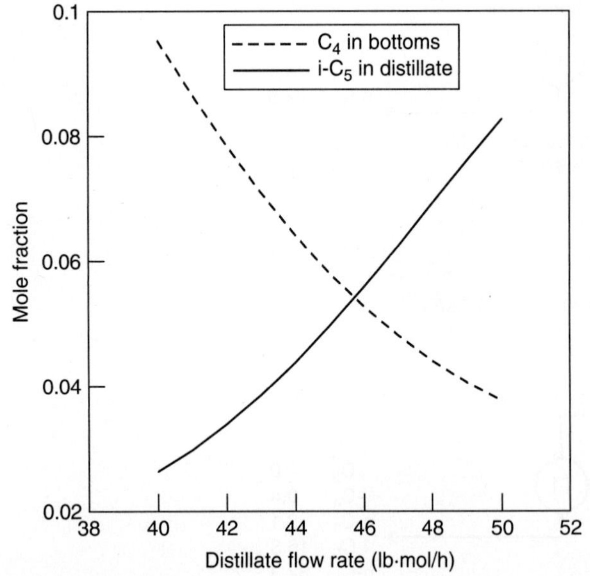

FIG. 13-37 Product mole fraction duty as a function of the distillate rate for butane-pentane splitter in Fig. 13-34.

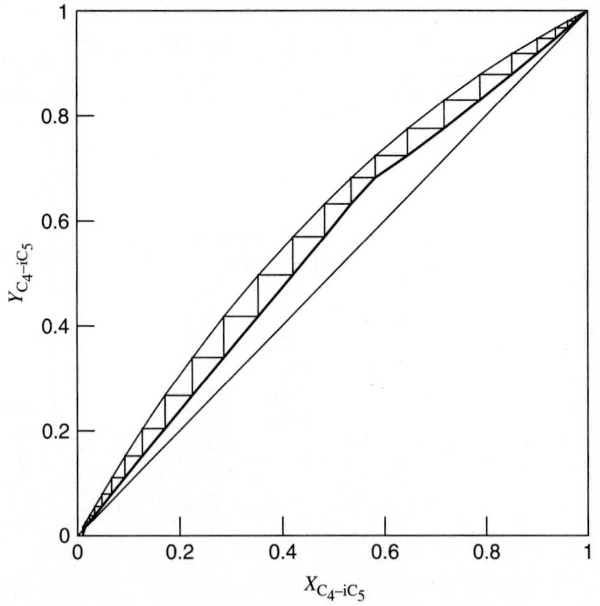

FIG. 13-38 Multicomponent McCabe-Thiele diagram for butane-pentane splitter after optimization to improve product purities.

as the reflux ratio increases. Further improvement in product purity can best be made by changing a different specification.

Figure 13-37 shows the tradeoff in product purities when we change the specified distillate flow rate, maintaining all other specifications at the values specified in the base case. At lower distillate flow rates, the mole fraction of heavy key (i-C_5) in the distillate is small, and at higher distillate flow rates the mole fraction of light key (n-C_4) in the bottoms is small. Both cannot be small simultaneously. From this result we see that the "best" overall product purities are obtained when the distillate rate is in the vicinity of 45 lb · mol/h. On reflection, this should not come as a surprise; the flow rate of the light key (n-butane) and all components with a higher volatility is 45 lb · mol/h. However, even with the distillate flow rate set to 45 lb · mol/h there remains room for improvement in the separation.

The other key design specifications here are the total number of stages and the location of the feed stage. In most cases, increasing the number of stages will improve the separation. On increasing the number of stages to 26, with the feed to stage 12, increasing the overhead vapor flow to 195 lb · mol/h, and decreasing the distillate rate to 45 lb · mol/h, we obtain the following products:

Mole flows, lb · mol/h	Feed	Top	Bottom
Propane	5.00	5.00	0.00
Isobutane	15.00	15.00	0.00
n-Butane	25.00	24.81	0.19
Isopentane	20.00	0.16	19.84
n-Pentane	35.00	0.03	34.97
Total	100.00	45.00	55.00

The McCabe-Thiele diagram for this configuration, shown in Fig. 13-38, shows that the product purities have improved significantly.

The temperature and liquid phase composition profiles for this final case are shown in Fig. 13-39. The temperature increases from top to bottom in the column. This is normally the case in distillation columns (exceptions may occur with cold feeds or feeds with boiling points significantly lower than that of the mixture on stages above the feed stage). The composition profiles also are as expected. The components more volatile than the light key (n-butane) are concentrated above the feed; those compounds less volatile than the heavy key (i-pentane) are concentrated below the feed. The mole fractions of the two keys exhibit maxima, the light key above the feed stage and the heavy key below the feed stage. The decrease in the mole fraction of light key over the top few stages is necessary to accommodate the increase in the composition of the lighter compounds. Similar arguments pertain to the decrease in the mole fraction of the heavy key over the stages toward the bottom of the column.

Flow profiles are shown in Fig. 13-39c. Note the step change in the liquid flow rate around the feed stage. Had the feed been partially vaporized, we would have observed changes in both vapor and liquid flows around the feed stage, and a saturated vapor feed would significantly change only the vapor flow profile. The slight (in this case) curvature in the flow profiles is due to enthalpy changes.

Example 13-4 Light Hydrocarbon Distillation Compute stage temperatures, interstage vapor and liquid flow rates and compositions, and reboiler and condenser duties for the light hydrocarbon distillation column shown in Fig. 13-40. How might the separation be improved?

This more complicated example features a partial condenser (with vapor product) and a vapor sidestream withdrawn from the 13th stage. The SRK equation of state may be used for estimating the K values and enthalpy departures for thermodynamic properties.

The specifications made in this case are summarized as follows:

Variable	Number	Value
Number of stages	1	17
Feed stage location	1	9
Component flows in feed	$c = 5$	3, 20, 37, 35, 5 lb · mol/h
Feed pressure	1	260 psia
Feed vapor fraction	1	0
Pressure on each stage including condenser and reboiler	$N = 17$	$P_j = 250$ psia
Heat duty on each stage except reboilers and condensers	$N - 2 = 15$	$Q_j = 0$
Reflux rate (replaces heat duty of reboiler)	1	$L_1 = 150$ lb · mol/h
Distillate flow rate (replaces heat duty of condenser)	1	$D = 23$ lb · mol/h
Sidestream stage	1	13
Sidestream flow rate and phase	2	37 lb · mol/h vapor
Total	46	

As in Example 13-3, we have assumed that the pressure of the reflux divider is the same as the pressure of the condenser, the heat loss from the reflux divider is zero, and the reflux temperature is the boiling point of the condensed overhead vapor.

The specifications were selected to obtain three products: a vapor distillate rich in C_2 and C_3, a vapor sidestream rich in n-C_4, and a bottoms rich in n-C_5 and n-C_6, as summarized in the following table:

Mole flows, lb · mol/h	Feed	Top	Bottom	Sidestream
Ethane	3.00	3.00	0.0	0.0
Propane	20.00	18.3	0.0	1.6
n-Butane	37.00	1.7	9.6	25.7
n-Pentane	35.00	0.0	25.8	9.2
n-Hexane	5.00	0.0	4.5	0.5
Total molar flow	100.00	23.0	40.0	37.0

Convergence of the simultaneous convergence method was obtained in five iterations.

Further improvement in the purity of the sidestream as well as of the other two products could be obtained by increasing the reflux flow rate (or reflux ratio) and the number of stages in each section of the column. If, for example, we increase the number

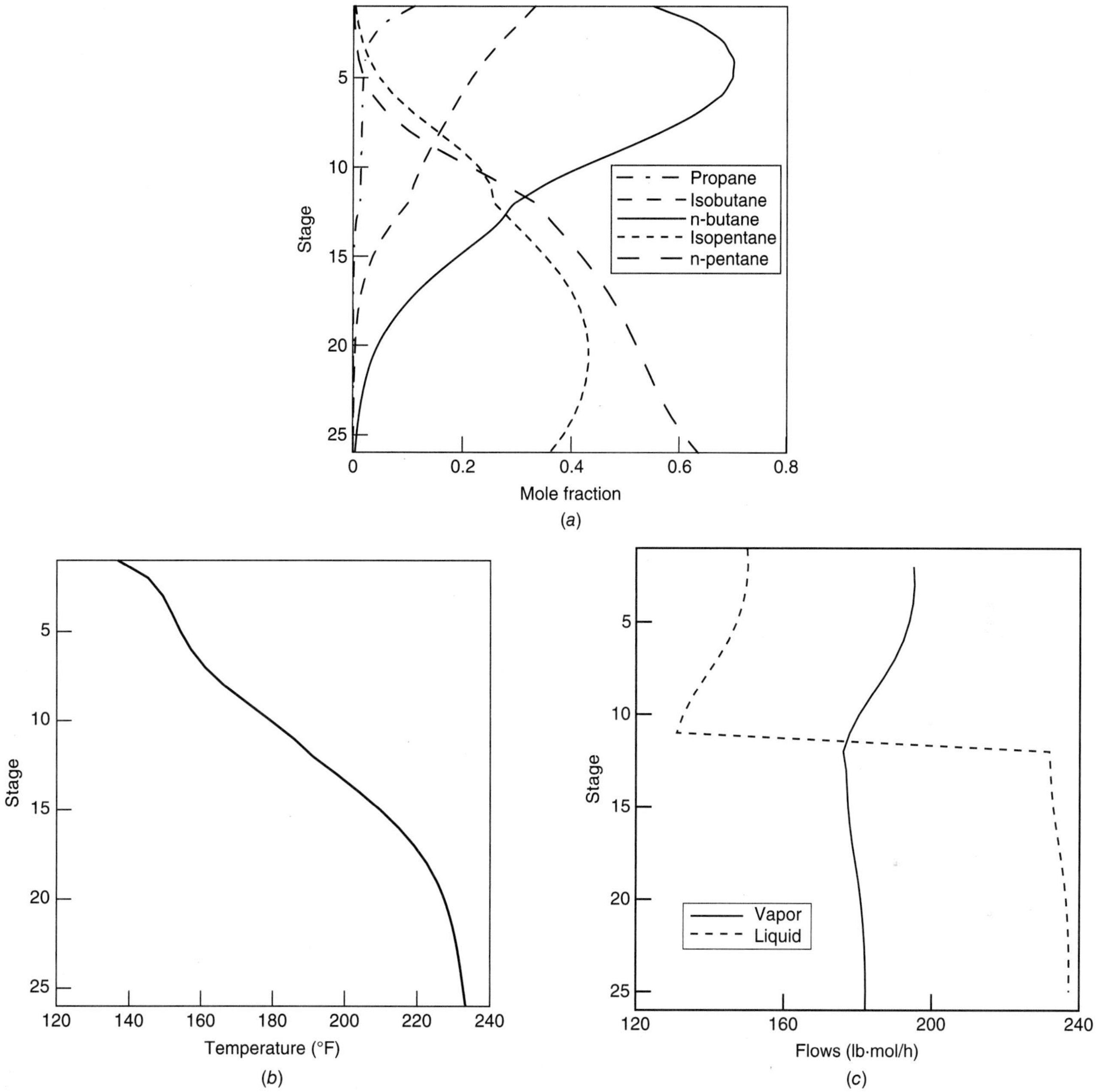

FIG. 13-39 (*a*) Composition, (*b*) temperature, and (*c*) flow profiles in butane-pentane splitter.

of stages to 25 (including condenser and reboiler in this total), with the feed to stage 7 and the sidestream removed from stage 17, we obtain the following:

Mole flows, lb · mol/h	Feed	Top	Bottom	Sidestream
Ethane	3.00	3.00	0.00	0.00
Propane	20.00	19.46	0.00	0.54
n-Butane	37.00	0.54	7.46	29.00
n-Pentane	35.00	0.00	27.93	7.07
n-Hexane	5.00	0.00	4.61	0.39
Total	100.0	23.0	40.0	37.00

The McCabe-Thiele diagram for this design, showing that the feed is to the optimum stage, is shown in Fig. 13-41. The flow profiles are shown in Fig. 13-42; note the step changes due to both the feed and the sidestream. As was the case in Example 13-3, the curvature in the flow profiles is due to enthalpy changes.

Example 13-5 Absorber Compute stage temperatures and interstage vapor and liquid flow rates and compositions for the absorber specifications shown in Fig. 13-43. Note that a second absorber oil feed is used in addition to the main absorber oil and that heat is withdrawn from the seventh theoretical stage. The oil may be taken to be *n*-dodecane.

The specifications made in this case are summarized as follows:

Variable	Number	Value(s)
Number of stages	1	8
Location of feed 1	1	1
Component flows in feed 1	$c = 6$	0, 0, 0, 0, 0, 250 lb · mol/h
Pressure of feed 1	1	400 psia
Temperature of feed 1	1	80°F
Location of feed 2	1	4
Component flows in feed 2	$c = 6$	13, 3, 4, 5, 5, 135 lb · mol/h
Pressure of feed 2	1	400 psia
Temperature of feed 2	1	80°F
Location of feed 3	1	8
Component flows in feed 3	$c = 6$	360, 40, 25, 15, 15, 10 lb · mol/h
Pressure of feed 3	1	400 psia
Temperature of feed 3	1	80°F
Pressure on each stage	$N = 8$	$P_j = 400$ psia; $j = 1, 2, \ldots, 8$
Heat duty on each stage	$N = 8$	$Q_j = 0, j = 1, 2, \ldots, 6, 8; Q_7 = 150{,}000$ Btu/h
Total	44	

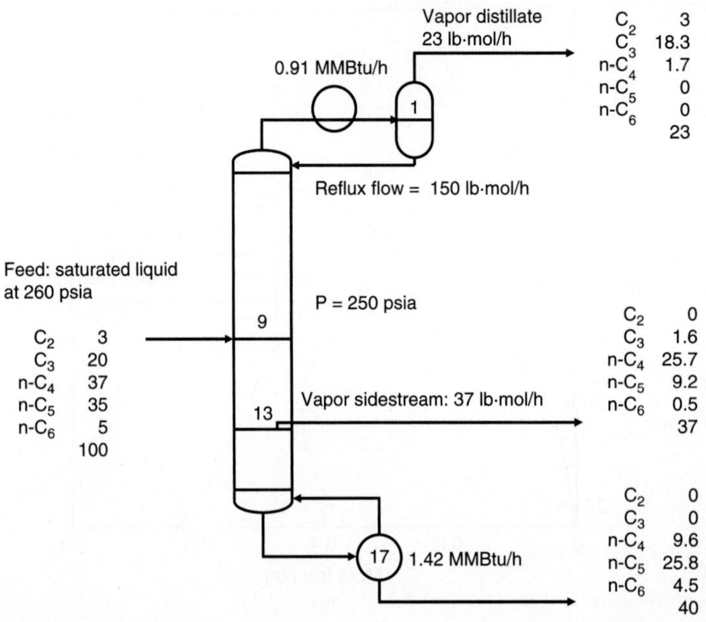

FIG. 13-40 Specifications and calculated product stream flows and heat duties for light hydrocarbon still. Flows are in pound-moles per hour.

The simultaneous solution method solved this example in four iterations. The Peng-Robinson equation of state was used to estimate K values and enthalpy departures and by Seader (*Perry's Chemical Engineers' Handbook,* 7th ed. p. 13-47), who solved this problem by using the SR method.

The computed product flows are summarized as follows:

Mole flows, lb · mol/h	Lean oil	Secondary oil	Rich gas	Lean gas	Rich oil
Methane	0.0	13.0	360.0	303.7	69.3
Ethane	0.0	3.0	40.0	10.7	32.3
Propane	0.0	4.0	25.0	0.2	28.8
n-Butane	0.0	4.0	15.0	0.0	19.0
n-Pentane	0.0	5.0	10.0	0.0	15.0
n-Dodecane	250.0	135.0	0.0	0.0	385.0
Total molar flow	250.0	164.0	450.0	314.6	549.4

The energy withdrawn from stage 7 has the effect of slightly increasing the absorption of the more volatile species in the rich oil leaving the bottom of the column. Temperature and flow profiles are shown in Fig. 13-44. The temperature profile shows the rise in temperature toward the bottom of the column that is typical of gas absorption processes. The bottom of the column is where the bulk of the absorption takes place, and the temperature rise is a measure of the heat of absorption. The liquid flow profile exhibits a step change due to the secondary oil feed at the midpoint of the column.

Example 13-6 Reboiled Stripper Compute the stage temperatures and interstage vapor and liquid flow rates and compositions and reboiler heat duty for the reboiled stripper shown in Fig. 13-45. Thermodynamic properties may be estimated by using the Peng-Robinson equation of state.

The specifications made in this case are summarized in Fig. 13-45 and in the following table. The specified bottoms rate is equivalent to removing most of the n-C_5 and n-C_6 in the bottoms.

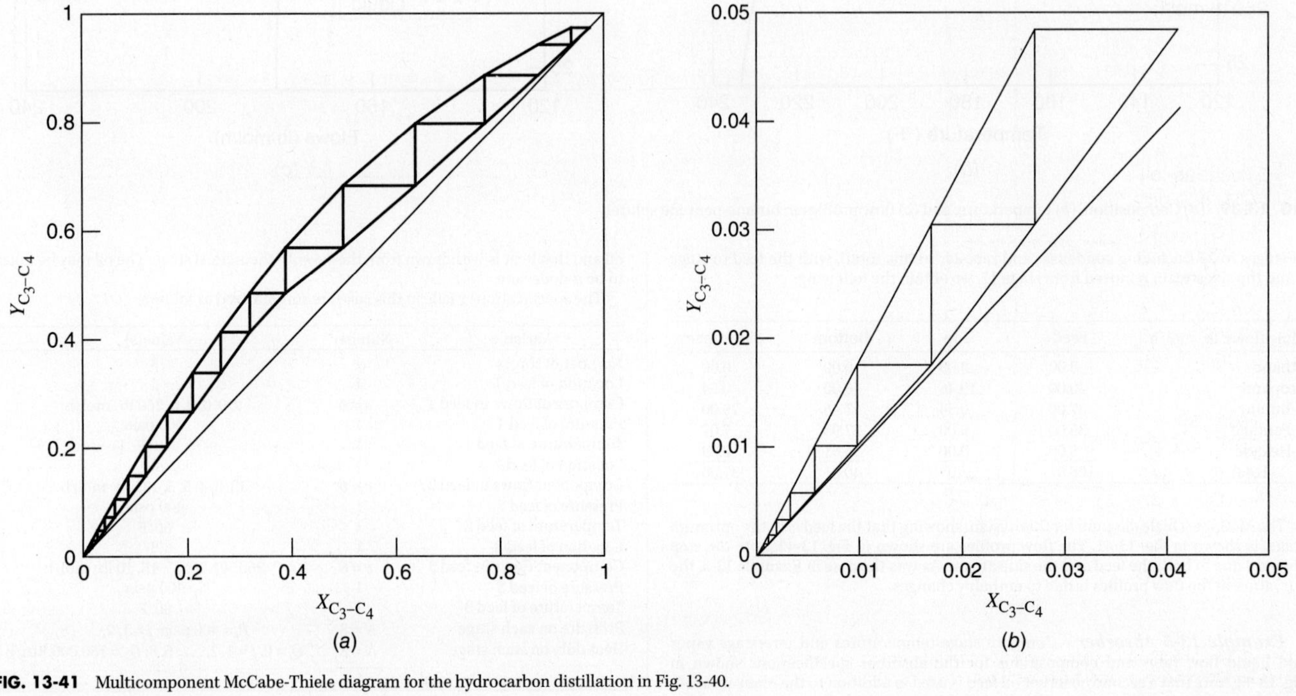

FIG. 13-41 Multicomponent McCabe-Thiele diagram for the hydrocarbon distillation in Fig. 13-40.

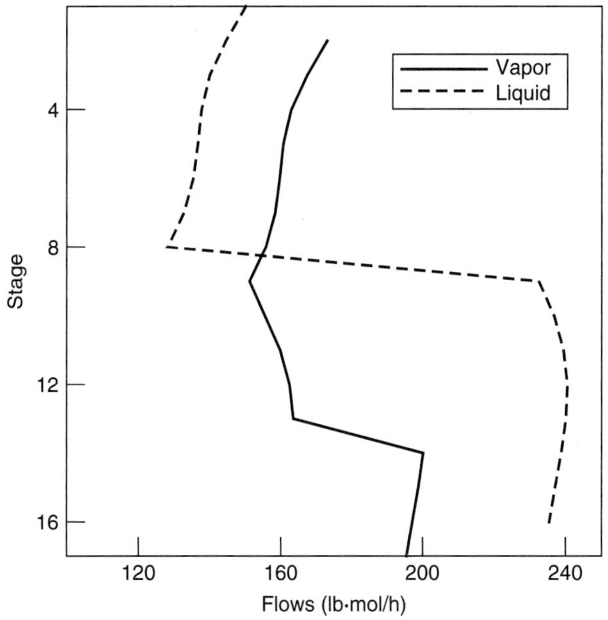

FIG. 13-42 Flow profiles in hydrocarbon distillation in Fig. 13-40.

Variable	Number	Value(s)
Number of stages	1	8
Component flows in feed	$c = 7$	0.22, 59.51, 73.57, 153.2, 173.2, 58.22, 33.63 lb · mol/h
Pressure of feed	1	150 psia
Temperature of feed	1	40.8°F
Pressure on each stage	$N = 8$	$P_j = 150$ psia; $j = 1, \ldots, 8$
Heat duty on each stage (except reboiler)	$N - 1 = 7$	$Q_j = 0, j = 1, \ldots, 7$
Bottom product flow rate	1	99.33 lb · mol/h
Total	26	

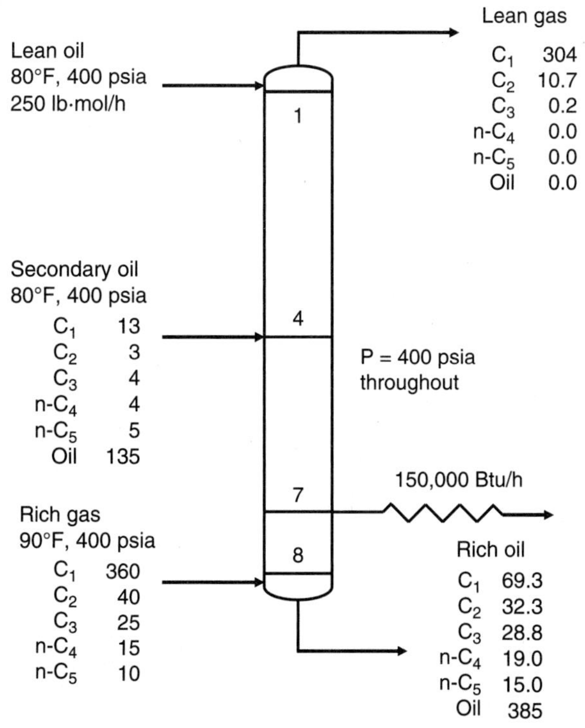

FIG. 13-43 Specifications and calculated product stream flows and heat duties for absorber. Flows are in pound-moles per hour.

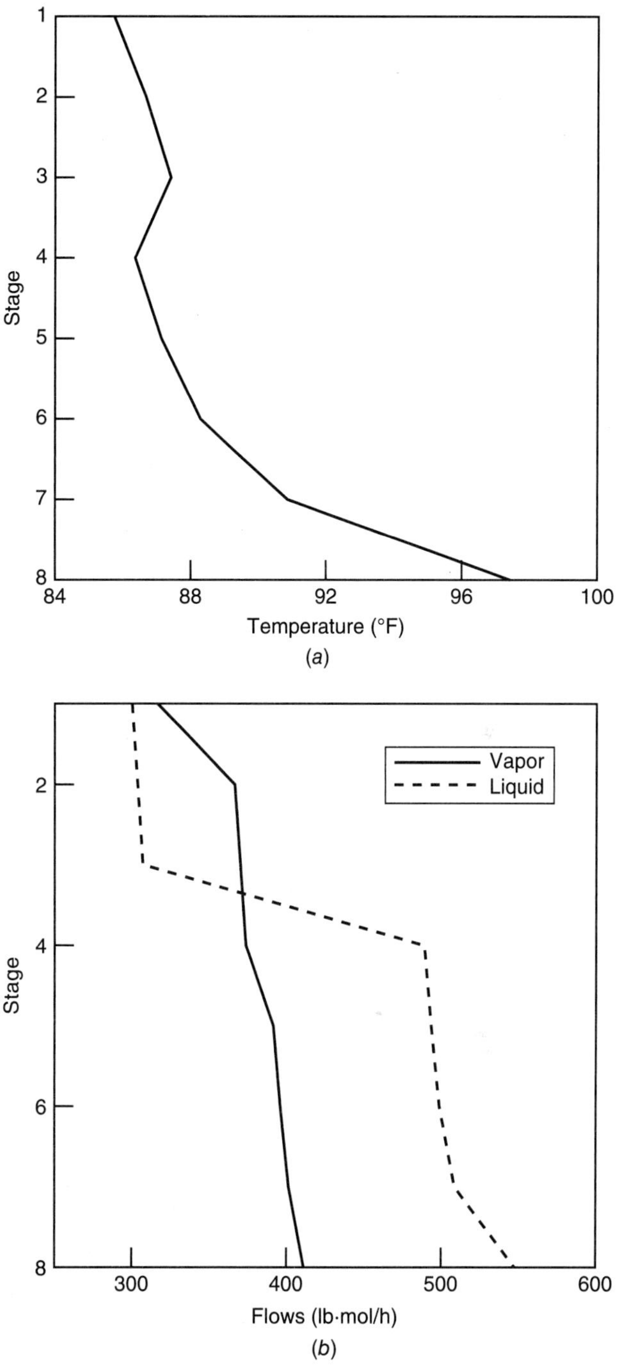

FIG. 13-44 (*a*) Temperature and (*b*) flow profiles in absorber in Fig. 13-43.

The computed product flows and reboiler duty are shown alongside the specifications in Fig. 13-45 and in the following table:

Mole flows, lb · mol/h	Feed	Overhead	Bottoms
Nitrogen	0.22	0.22	0.0
Methane	59.51	59.51	0.0
Ethane	73.57	73.57	0.0
Propane	153.22	153.17	0.05
n-Butane	173.22	148.10	25.12
n-Pentane	58.22	15.08	43.14
n-Hexane	33.63	2.61	31.02
Total	551.59	452.26	99.33

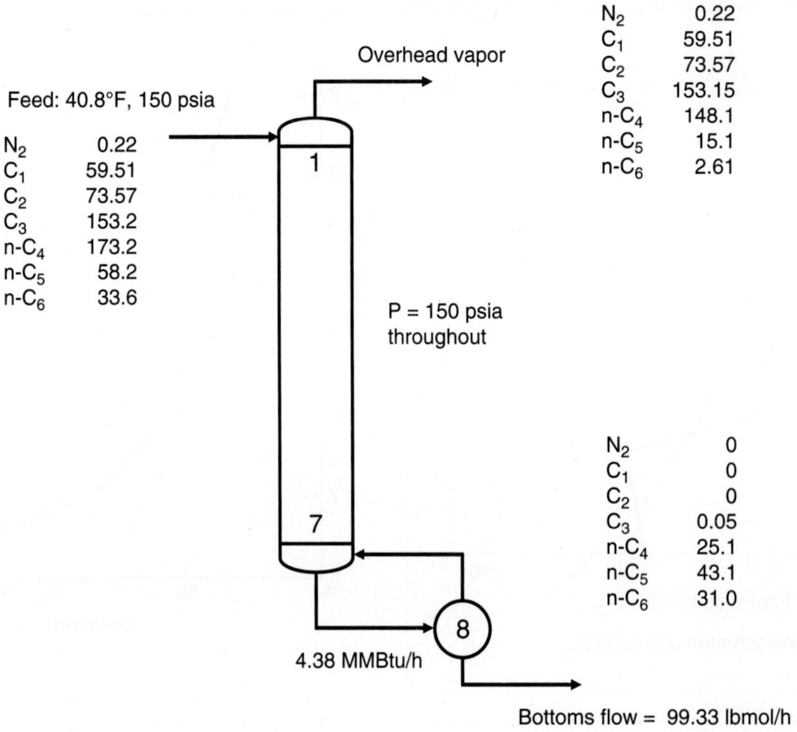

FIG. 13-45 Specifications and calculated product stream flows and reboiler heat duty for a reboiled stripper. Flows are in pound-moles per hour.

Computed temperature, flow rates, and vapor-phase mole fraction profiles, shown in Fig. 13-46, are not of the shapes that might have been expected. Vapor and liquid flow rates for n-C_4 change dramatically from stage to stage.

Example 13-7 An Industrial i-Butane/n-Butane Fractionator Klemola and Ilme [*Ind. Eng. Chem.* **35:** 4579 (1996)] and Ilme (Ph.D. thesis, University of Lapeenranta, Finland, 1997) report data from an industrial *i*-butane/*n*-butane fractionator that is used here as the basis for this example.

The column has 74 valve trays, the design details of which can be found in Example 13-11. The feed was introduced onto tray 37.

To properly model an existing column, it is necessary to know all feed and product conditions (flow rate, temperature, pressure, and composition). Flows, temperatures, and pressures often are available from standard instrumentation. It may be necessary to obtain additional samples of these streams to determine their composition. Such sampling should be scheduled as part of a plant trial to ensure that measured data are consistent. Ideally, multiple sets of plant measurements should be obtained at different operating conditions, and care should be taken to ensure that operating data are obtained at steady state since a steady-state model can only be used to describe a column at steady state. Measurements should be taken over a time interval longer than the residence time in the column and time-averaged to avoid a possible mismatch between feed and product data. Condenser and reboiler heat duties should be known (or available from the appropriate energy balance) whenever possible.

The measured compositions and flow rates of the feed and products for the C_4 splitter are summarized in the following table:

Measured Feed and Product Flows and Compositions (Mass %) for *i*-Butane/*n*-Butane Fractionator (Ilme 1997)

Species	Feed	Top	Bottom
Propane	1.50	5.30	0.00
Isobutane	29.4	93.5	0.30
n-Butane	67.7	0.20	98.1
C_4 olefins	0.50	1.00	0.20
Neopentane	0.10	0.00	0.20
Isopentane	0.80	0.00	1.10
n-Pentane	0.10	0.00	0.10
Total flow, kg/h	26,234	8011	17,887

Other measured parameters are as follows:

Other Details of the *i*-Butane/*n*-Butane Fractionator

Reflux flow rate, kg/h	92,838
Reflux temperature, °C	18.5
Column top pressure, kPa	658.6
Pressure drop per tray, kPa	0.47
Feed pressure, kPa	892.67
Boiler duty, MW	10.24

Rarely, and this is a case in point, are plant data in exact material balance, and it will be necessary to reconcile errors in such measurements before continuing. The feed and product compositions, as adjusted by Ilme so that they satisfy material balance constraints, are provided next. Note how the C_4 olefins are assigned to isobutene and 1-butene.

Adjusted Feed and Product Compositions (Mass %) and Flows for *i*-Butane/*n*-Butane Fractionator (Ilme 1997)

Species	Feed	Top	Bottom
Propane	1.54	4.94	0.00
Isobutane	29.5	94.2	0.3
n-Butane	67.7	0.20	98.1
Isobutene	0.13	0.23	0.08
1-Butene	0.20	0.41	0.10
Neopentane	0.11	0.00	0.17
Isopentane	0.77	0.00	1.12
n-Pentane	0.08	0.00	0.11
Total flow, kg/h	26,122	8123	17,999

To proceed with building a model of this column, we specify the number of stages equal to the number of trays plus condenser and reboiler (N = 76). The common arrangement of locating the actual feed between stages may require modeling as two separate feeds, the liquid portion to the stage below and the vapor portion to the stage above. In this illustration, the feed is (assumed to be) saturated liquid, and we provide just a single feed to stage 38.

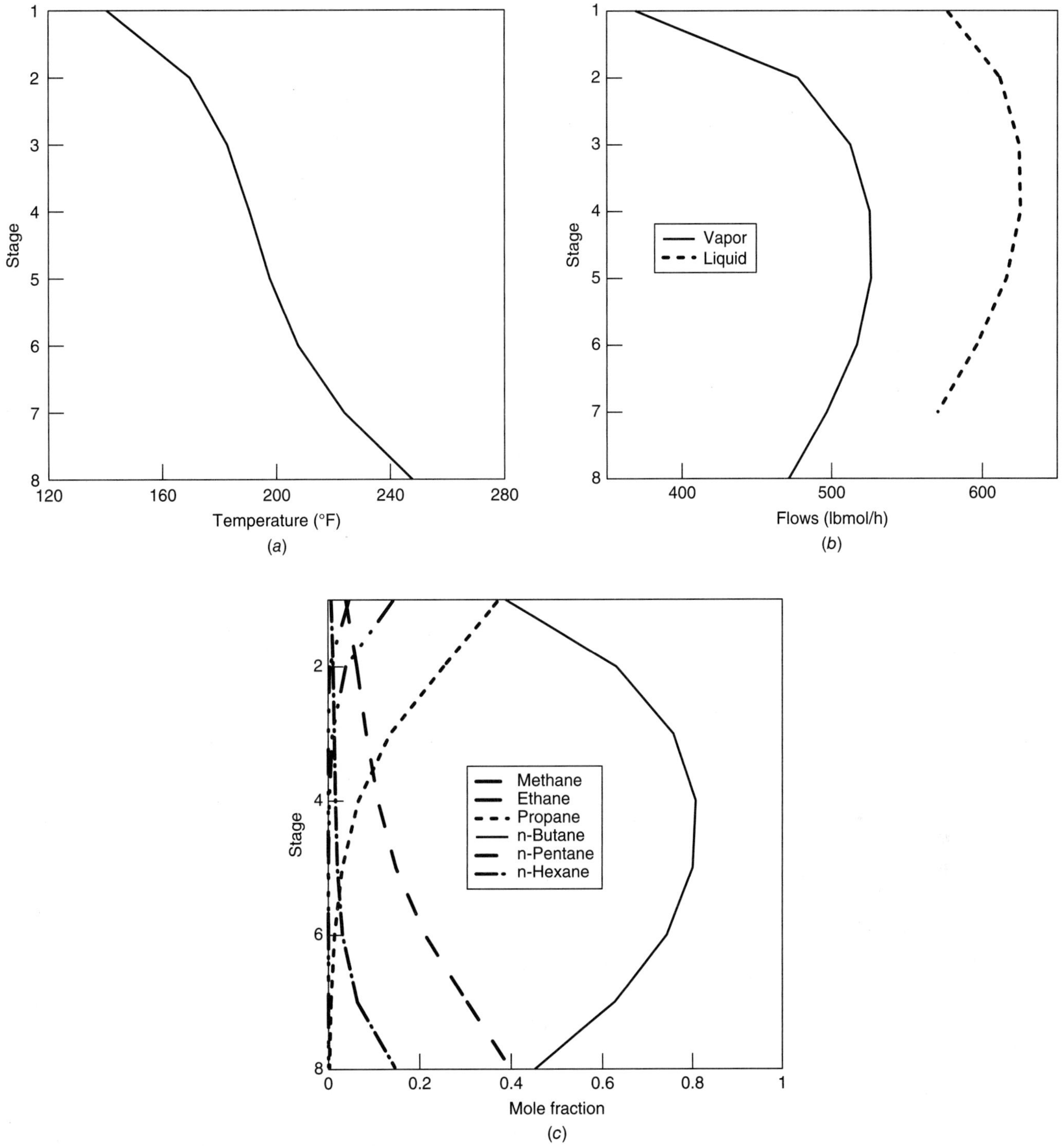

FIG. 13-46 (*a*) Temperature, (*b*) flow, and (*c*) vapor mole fraction profiles in reboiled stripper in Fig. 13-45.

Upon computing the bubble point of the overhead product, we find that the measured reflux temperature is well below the estimated boiling point. Thus, we choose the subcooled condenser model. The steady-state concept of the "subcooled" condenser often does not exist in practice. Instead, the condenser is in vapor-liquid equilibrium, with the vapor augmented by a blanket of noncondensable gas (which has the effect of lowering the dew point of the overhead vapor). The subcooled condenser is a convenient work-around for steady-state models (as is needed here), but not for dynamic models. We assume a partial reboiler.

The pressure of the top stage is specified (at 658.6 kPa). The pressures of all trays below the top tray can then be fixed from the knowledge of the per-tray pressure drop (0.47 kPa). The pressure of the condenser is not known here. Thus, in the absence of further information, we make the condenser pressure equal to the top tray pressure (knowing that in practice it will be lower).

It is advisable to use the plant set points in building a model. For example, it is possible that a column simulation might involve the specification of the reflux ratio and bottoms flow rate because such specifications are (relatively) easy to converge. It is quite likely that the column may be controlled by using the temperature at some specific location (e.g., the temperature of tray 48). This specification should be used in building the model. In this case, we do not know the set points used for controlling the column, but we do have enough information to allow us to compute the reflux ratio from plant flow data ($R = 11.6$). Finally, the bottom product flow is specified as equal to the adjusted value reported above (17,999 kg/h).

The specifications made to model this column are summarized as follows:

Variable	Number	Value
Number of stages	1	$N = 76$
Feed stage location	1	39
Component flows in feed	$c = 8$	See previous table
Feed pressure	1	120 psia
Feed vapor fraction	1	0
Pressure at the top of the column	1	658.6 kPa
Pressure drop per stage	$N - 1 = 75$	0.47 kPa
Heat duty on each stage except reboilers and condensers	$N - 2 = 74$	$Q_j = 0$
Reflux ratio (replaces heat duty of condenser)	1	$R = 11.588$
Bottoms flow rate (replaces heat duty of reboiler)	1	$B = 17{,}999$ kg/h
Temperature of reflux	1	291.65 K
Total	165	

Finally, we must select appropriate methods of estimating thermodynamic properties. Ilme (1997) used the SRK equation of state to model this column, whereas Klemola and Ilme [*Ind. Eng. Chem.* **35:** 4579 (1996)] had earlier used the UNIFAC model for liquid-phase activity coefficients, the Antoine equation for vapor pressures, and the SRK equation for vapor-phase fugacities only. For this exercise we used the Peng-Robinson equation of state. Computed product compositions and flow rates are shown in the following table:

Specified Feed and Computed Product Compositions (Mass %) and Flows for *i*-Butane/*n*-Butane Fractionator (Ilme 1997)

Compound	Feed	Top	Bottom
Propane	1.54	4.95	0.00
Isobutane	29.49	93.67	0.53
n-Butane	67.68	0.73	97.89
Isobutene	0.13	0.29	0.06
1-Butene	0.20	0.36	0.13
Neopentane	0.11	0.00	0.16
Isopentane	0.77	0.00	1.12
n-Pentane	0.08	0.00	0.12
Total flow, kg/h	26,122	8123.01	17,999

The agreement with the adjusted material balance (tabulated previously) appears to be quite good, and to a first approximation it seems that we have a good model of the column.

Note that although this column is distilling a mixture containing at least eight identifiable compounds, only two are present in significant amounts, and therefore this is essentially a binary separation. It is usually relatively straightforward to match product compositions in processes involving only two different species simply by adjusting the number of equilibrium stages. We return to this point later.

Efficiencies In actual operation, the trays of a distillation column rarely, if ever, operate at equilibrium despite attempts to approach this condition by proper design and choice of operating conditions. The usual way of dealing with departures from equilibrium in multistage towers is through the use of stage and/or overall efficiencies.

The overall column efficiency is defined by

$$E_o = \frac{N_{\text{EQ}}}{N_{\text{actual}}} \tag{13-55}$$

where N_{EQ} is the number of equilibrium stages.

There are many different definitions of stage (or tray) efficiency, with that of Murphree [*Ind. Eng. Chem.* **17:** 747–750, 960–964 (1925)] being by far the most widely used in separation process calculations:

$$E_{i,J}^{MV} = \frac{y_{i,j} - y_{i,j+1}}{y_{i,j}^* - y_{i,j+1}} \tag{13-56}$$

Here $E_{i,J}^{MV}$ is the Murphree vapor efficiency for component i on stage j, and $y_{i,J}^*$ is the composition of the vapor in equilibrium with the liquid. Other types of efficiency include that of Hausen [*Chemie Ingr. Tech.* **25:** 595 (1953)], vaporization (see Holland, 1975), and generalized Hausen [Standart, *Chem. Eng. Sci.* **20:** 611 (1965)]. There is by no means a consensus on which is best. Arguments for and against various types are presented by, among others, Standart [*Chem. Eng. Sci.* **20:** 611 (1965); **26:** 985 (1971)], Holland and McMahon [*Chem. Eng. Sci.* **25:** 431 (1972)], and Medina et al. [*Chem. Eng. Sci.* **33:** 331 (1978); **34:** 1105 (1979)]. Possibly the most soundly based definition, the generalized Hausen efficiency of Standart (1965), is never used in industrial practice. Seader [*Chem. Eng. Progress* **85**(10): 41 (1989)] summarizes the shortcomings of efficiencies.

The Murphree (and Hausen) efficiencies of both components in a binary mixture are equal; although they cannot be less than 0, they may be greater than 1. A table of typical values of Murphree tray efficiency can be found in Sec. 14. Also described in Sec. 14 are methods for estimating Murphree efficiencies when they are not known.

For multicomponent systems (those with more than two components), there are $c - 1$ independent component efficiencies, and there are sound theoretical reasons as well as experimental evidence for not assuming the individual component efficiencies to be alike; indeed, they may take values between plus and minus infinity. Component efficiencies are more likely to differ for strongly nonideal mixtures. While models exist for estimating efficiencies in multicomponent systems [see chapter 13 in Taylor and Krishna (*Multicomponent Mass Transfer,* Wiley, New York, 1993) for a review of the literature], they are not widely used and have not (yet) been included in any of the more widely used commercial simulation programs.

The fact that component efficiencies in multicomponent systems are unbounded means that the arithmetic average of the component Murphree efficiencies is useless as a measure of the performance of a multicomponent distillation process. Taylor, Baur, and Krishna [*AIChE J.* **50:** 3134 (2004)] proposed the following efficiency for multicomponent systems:

$$\varepsilon_j = \frac{\sqrt{\sum_{i=1}^{c} (\Delta y_{i,j})^2}}{\sqrt{\sum_{i=1}^{c} (\Delta y_{i,j}^*)^2}} \tag{13-57}$$

This efficiency has a simple and appealing physical significance; it is the ratio of the length of the actual composition profile (in mole fraction space) to the length of the composition profile given by the equilibrium-stage model. The Taylor-Baur-Krishna (TBK) efficiency has just one value per stage regardless of the number of components in the mixture; it can never be negative. For binary systems in tray columns, the TBK average efficiency simplifies to the Murphree efficiency [Taylor, Baur, and Krishna (2004)].

Murphree efficiencies are easily incorporated within simultaneous convergence algorithms (something that is not always easy, or even possible, with some tearing methods). As an aside, note that vaporization efficiencies are very easily incorporated in all computer algorithms, a fact that has helped to prolong the use of these quantities in industrial practice despite the lack of any convenient way to relate them to the fundamental processes of heat and mass transfer. Unfortunately, it is not at all easy to include the more fundamentally sound TBK efficiencies in a computer method for equilibrium-stage simulations.

Efficiencies are often used to fit actual operating data, along with the number of equilibrium stages in each section of the column (between feed and product takeoff points). The maximum number of these efficiencies is the number of independent efficiencies per stage ($c - 1$) times the number of stages—potentially a very large number, indeed. This many adjustable parameters may lead to a model that fits the data very well, but has no predictive ability (i.e., cannot describe how the column will behave when something changes). At the other extreme, the overall efficiency defined by Eq. (13-55) is just a single parameter that can improve the robustness of the model and speed of convergence, but it may be difficult to match actual temperature or composition profiles because there is unlikely to be a one-to-one correspondence between the model stages and actual trays. A compromise often used in practice is to use just one value for all components and all stages in a single section of a column. Efficiencies should not be used to model condensers and reboilers; it is usually safe to assume that they are equilibrium devices. It is also unwise to employ Murphree efficiencies for trays with a vapor product since any Murphree efficiency less than 1 will necessarily lead to the prediction of a subcooled vapor.

Example 13-8 The Industrial i-Butane/n-Butane Fractionator (Again) With the material on efficiencies in mind, we return to the model of the C_4 splitter that we developed in Example 13-7.

It is possible to estimate the overall efficiency for a column such as this one simply by adjusting the number of equilibrium stages in each section of the column that is needed to match the mass fractions of *i*-butane in the distillate and *n*-butane in the bottoms. Using the SRK equation of state for estimating thermodynamic properties, Ilme (1997) found that 82 equilibrium stages (plus condenser and reboiler) and the feed to stage 38 were required. This corresponds to an overall column efficiency of 82/74 = 111 percent. Klemola and Ilme (1996) used the UNIFAC model for liquid-phase activity coefficients, the Antoine equation for vapor pressures, and the SRK equation for vapor-phase fugacities only and found that 88 ideal stages were needed; this corresponded to an overall efficiency of

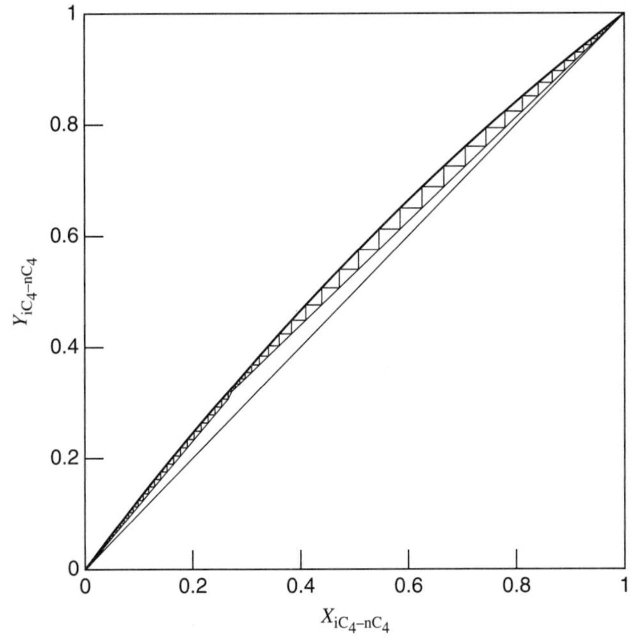

FIG. 13-47 McCabe-Thiele diagram for C_4 splitter.

119 percent. With the Peng-Robinson equation of state for the estimation of thermodynamic properties, we find that 84 stages are needed (while maintaining the feed to the center stage as is the case here); the overall column efficiency for this model is 114 percent. The differences between these efficiencies are not large in this case, but the important point here is that efficiencies—all types—depend on the choice of model used to estimate the thermodynamic properties. Caution must therefore be exercised when one is using efficiencies determined in this way to predict column performance.

As an alternative to varying the number of stages, we may prefer to maintain a one-to-one correspondence between the number of stages and the number of actual trays, 74 in this case (plus condenser and reboiler), with the feed to tray 38. Using the Peng-Robinson equation of state and a Murphree stage efficiency of 116 percent, we find the product mass fractions that are in excellent agreement with the plant data. The McCabe-Thiele (Hengstebeck) diagram for this case, assembled from the results of the simulation, is shown in Fig. 13-47. Composition profiles computed from this model are shown in Fig. 13-48. Note that the mole fractions are shown on a logarithmic axis so that all the composition profiles can easily be seen.

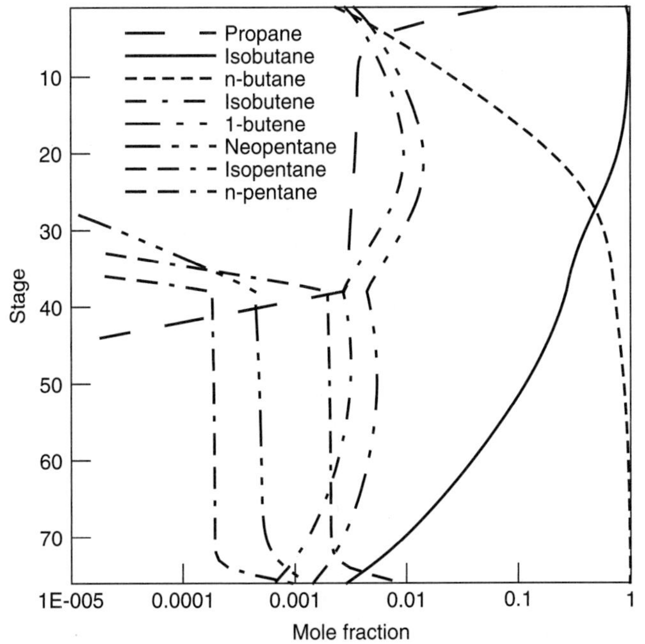

FIG. 13-48 Liquid-phase mole fraction profiles for *i*-butane/*n*-butane fractionator.

It must be remembered that this is essentially a binary separation and that it is usually relatively straightforward to match product compositions in processes involving only two different species. In other cases involving a greater number of species with significant concentrations, it will likely be necessary to vary both the number of stages and the component efficiencies to match plant data. We do not recommend adjusting thermodynamic model parameters to fit plant data since this can have unfortunate consequences on the prediction of product distributions, process temperatures, and pressures.

The performance of a packed column often is expressed in terms of the height equivalent to a theoretical plate (HETP) for packed columns. The HETP is related to the height of packing H by

$$\text{HETP} = \frac{H}{N_{\text{EQ}}} \tag{13-58}$$

In this case N_{EQ} is the number of equilibrium stages (theoretical plates) needed to accomplish the separation that is possible in a real packed column of height H.

Example 13-9 HETP of a Packed Absorber McDaniel, Bassyoni, and Holland [*Chem. Eng. Sci.* **25:** 633 (1970)] presented the results of field tests on a packed absorber in a gas plant. The packed section was 23 ft in height, and the column was 3 ft in diameter and filled with 2-in metallic Pall rings. The measured feeds are summarized in the following table.

Stream	Lean oil	Rich gas
Pressure, psia	807	807
Temperature, °F	2.9	0
Mole flows, lb · mol/h		
Carbon dioxide	0.0	14.1
Nitrogen	0.0	5.5
Methane	0.0	2655.8
Ethane	0.0	199.9
Propane	0.0	83.2
Isobutane	0.0	19.1
n-Butane	0.0	10.9
Isopentane	0.2	3.5
n-Pentane	0.2	1.5
n-Hexane	4.5	0.4
n-Heptane	17.2	0.2
n-Octane	54.5	0.1
n-Nonane	50.5	0.0
n-Decane	61.7	0.0
Total molar flow	188.9	2994.2

Determine the number of equilibrium stages needed to match the lean gas product flow for this column (2721.1 lb · mol/h, 93.6 percent methane).

As a first step, we choose an appropriate thermodynamic model that will be used to estimate K values and enthalpies. Either the Chao-Seader method or the Peng-Robinson equation of state could be considered for this system. It turns out not to be possible to match the plant data with the Chao-Seader method since even one equilibrium stage overpredicts the separation by a very significant amount. It is not even possible to match the exit flows by using the Chao-Seader model combined with a stage efficiency as low as 0.00001. With the Peng-Robinson equation of state, however, it is possible to obtain reasonable agreement with the measured overall product flows by using precisely one equilibrium stage! This suggests that the HETP for this column is 23 ft, a value much higher than any HETP ever published by a packing vendor. In this case the reason for the unrealistic estimate of the HETP has nothing to do with the packing; it is the extreme sensitivity of the simulation to the thermodynamic model, emphasizing the need for caution when one is using efficiencies and HETPs to model some absorption (and distillation) processes.

Using a Simulator to Solve Distillation Problems Computer-based methods for solving distillation (and related) column simulation problems now are reasonably reliable. Nevertheless, at times such methods fail to converge. The principal cause of convergence failures is generally a poor initial estimate of the variables being computed. In the discussion that follows, we identify some of the reasons that a simulation might be difficult to converge, along with suggestions on what might make the problem more amenable to solution. The key idea is to modify the problem that is difficult to solve as posed into one that is easy to solve—essentially to provide an improved initial estimate that is more likely to lead to convergence. Note that most simulators allow a calculation to be restarted from an older converged solution. The solution to the "easy" problem may then be used as a starting point for the more difficult problem. By doing this we are using a form of continuation (albeit executed manually, at least in part) to solve those problems with which the algorithm at hand may have trouble.

As a rule, the degree of difficulty increases with increasing nonideality. Simultaneous convergence methods are often recommended for simulating strongly nonideal systems (as opposed to tearing or inside-out methods), but even SC methods can experience difficulties with strongly nonideal systems. A possible remedy is to make the system "less nonideal." It is likely that an activity coefficient model is part of the model used to describe the thermodynamics of these systems, and the source of the convergence difficulties often encountered with such systems. First solving an equivalent ideal system that entirely omits the activity coefficient model (i.e., using Raoult's law) may provide a converged solution that may be an adequate starting point for the nonideal system of interest. However, since many simulators use ideal solution thermodynamic models in any self-initialization method, this technique may not provide enough help.

A measure of the nonideality of the system is given by the magnitude of the interaction parameters for the activity coefficient model. It is possible, therefore, to lessen the degree of nonideality by reducing the interaction parameters enough to make the problem easy to solve. The parameters may then be increased in size in a series of steps until the desired values are reached, each time using the solution converged by using the previous set of parameter values as the starting point. It is essential that the parameters return to their correct values in the final step because intermediate solutions have no meaning, serving merely as an aid to convergence. Using the stage efficiency as a continuation parameter also is useful for such cases, provided that the simulation uses standard specifications (more on this topic below).

The most strongly nonideal systems are those that may exhibit two liquid phases. We have avoided a detailed discussion of such systems in this section because special algorithms are needed for these cases; see, however, the section on azeotropic distillation and, for example, chapter 8 of Doherty and Malone (*Conceptual Design of Distillation Systems*, McGraw-Hill, New York, 2001) for entry points to the literature.

Large heat effects can lead to convergence difficulties. For such systems it is the enthalpies that are the source of the nonlinearity that leads to convergence failures. It is generally not straightforward to modify enthalpies in a simulator because no adjustable parameters exert their influence over the enthalpy in a way comparable to that of the interaction parameters in the activity coefficient model. The use of a constant-enthalpy model in distillation calculations, if available, will lead to constant molar flows from stage to stage (within each separate section of the column), a condition often approached in many real distillation (but not absorption) columns. Thus, if the simulator includes a constant-enthalpy model, then this can be used to obtain a converged solution that may provide a good starting point for the problem with a more realistic enthalpy model.

High pressure adds to the difficulties of converging simulation models. It is likely that an equation of state will be used to estimate fugacity coefficients and enthalpy departures in such systems. Mixtures become increasingly nonideal as the pressure is raised. In some cases the column may operate close to the critical point at which the densities of both phases approach each other. In other cases the iterations may take the estimates of temperature and composition into regions where the equation of state can provide only one mathematically real root for density or compressibility. Occurrences of this behavior often are a source of convergence difficulties. For such systems we suggest reducing the pressure until the problem becomes easy to solve. A converged solution obtained in this way may be used as the starting point for subsequent calculations at increasingly higher pressures (up to that desired).

Very large numbers of stages can pose their own kind of convergence difficulty. A possible remedy is to reduce the number of stages until a converged solution can be obtained. This solution can then be used as the starting point for a problem with more stages. Interpolation will have to be used to estimate the values of the flows, temperatures, and mole fractions for any added stages, something not available in all programs.

Nonstandard specifications are very likely to be the source of convergence difficulties. It is all too easy to specify a desired product purity or component flow rate that simply cannot be attained with the specified column configuration. There is always (at least) one solution if the reflux ratio and bottoms flow rate are specified (the so-called standard specifications), which is likely to converge easily. Other specifications that can cause difficulties for similar reasons include specifying temperatures and compositions anywhere in the column and specifying condenser or reboiler heat duties. A way to circumvent this kind of difficulty is first to obtain a converged solution for a case involving standard specifications. Once the behavior of the column is understood, it will be possible to make sensible nonstandard specifications, again using an old converged result as a starting point.

Columns in which temperature and/or compositions change over a wide range in a limited number of stages pose their own difficulties. Some highly nonideal systems exhibit this kind of behavior (see Example 13-11). For cases such as this, it is wise to limit per-iteration changes to temperature and composition. Most modern computer methods will do this as a matter or course, and problems with this cause are not the source of convergence difficulties that they once were.

This is by no means an exhaustive list of the reasons that computer-based simulations fail. Indeed, in many cases it is a combination of more than one of the factors that leads to difficulty. In those cases it may be necessary to combine several of the strategies we have outlined to solve the simulation problem. Often, however, there is no substitute for trial and error. Haas (chap. 4 in Kister 1992) offers some additional insight on using simulators to solve distillation column models.

Example 13-10 Multiple Steady States in Distillation This example is one of the most famous in the literature on distillation column modeling, having been studied, in one form or another, by many investigators, including Shewchuk [Ph.D. thesis in *Chem. Eng.* Appendix A1, University of Cambridge (1974)], Magnussen et al. [*I. Chem. E. Symp. Series* **56** (1979)], Prokopakis and Seider [*AIChE J.* **29:** 49 (1983)], Venkataraman and Lucia [*Comput. Chem. Eng.* **12:** 55 (1988)], and Rovaglio and Doherty [*AIChE J.* **36:** 39 (1990)]. The column simulated here is adapted from the work of Prokopakis and Seider and shown in Fig. 13-49. The ethanol-benzene-water ternary

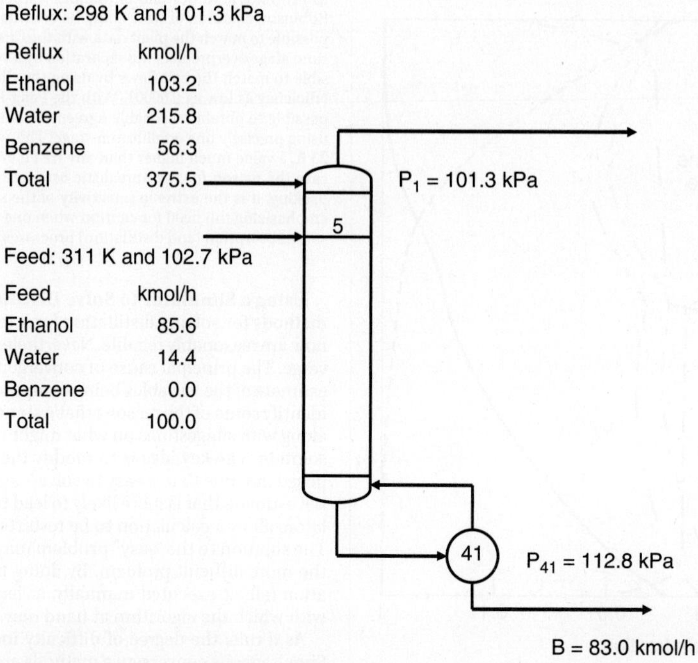

FIG. 13-49 Azeotropic distillation tower for distillation of an ethanol-water mixture using benzene as a mass separating agent. [*After Prokopakis and Seider,* AIChE J. ***29:** 49 (1983).*]

system actually splits into two liquid phases when the overhead vapor is condensed and cooled below its bubble point. One liquid phase is sent to a second column, and the other is returned to the main column as (cold) reflux. Here, in common with others, this column is modeled by ignoring the condenser and decanter. Reflux is simulated by a feed of appropriate composition, temperature, and pressure to the top of the column. The UNIQUAC method was used for estimating the activity coefficients, with parameters given by Prokopakis and Seider. The numerical results are very sensitive to the choice of activity coefficient model and associated parameters; qualitatively, however, the behavior illustrated here is typical of many systems.

This system is considered difficult because convergence of the MESH equations can be difficult to obtain with any algorithm. In fact, for the specifications considered here, there are no less than three solutions; the composition profiles are shown in Fig. 13-50. The goal of the distillation is to recover high-purity ethanol in the bottom stream from the column by using benzene as a mass separating agent. The low-purity profile in Fig. 13-50*a*, containing a large amount of water in the bottom product, is easily obtained from an ideal solution starting point (but with severe restrictions on the maximum allowed temperature change per iteration). The intermediate profile in Fig. 13-50*b* is rather more difficult to obtain. We were able to find it by using, as a starting point, a profile that had been converged at a stage efficiency of 0.7. The high-purity solution in Fig. 13-50*c*, containing very little water in the bottom product, is also easily obtained from an initial profile calculated by assuming that the stage efficiency is quite low (0.3). Multiple solutions for this column have been reported by many authors. In fact, with the parameters used here, the three solutions exist over a narrow range of ethanol feed flows. Taylor, Achuthan, and Lucia [*Comput. Chem. Eng.* **20:** 93 (1996)] found complex-valued solutions to the MESH equations for values outside this range.

Multiple steady-state solutions of the MESH equations have been found for many systems, and the literature on this topic is quite extensive. An introduction to the literature is provided by Bekiaris, Guttinger, and Morari [*AIChE J.* **46:** 955 (2000)]. Chavez, Seader, and Wayburn [*Ind. Eng. Chem. Fundam.* **25:** 566 (1986)] used homotopy methods to find multiple solutions for some systems of interlinked columns. Parametric continuation has been used to detect multiple solutions of the MESH

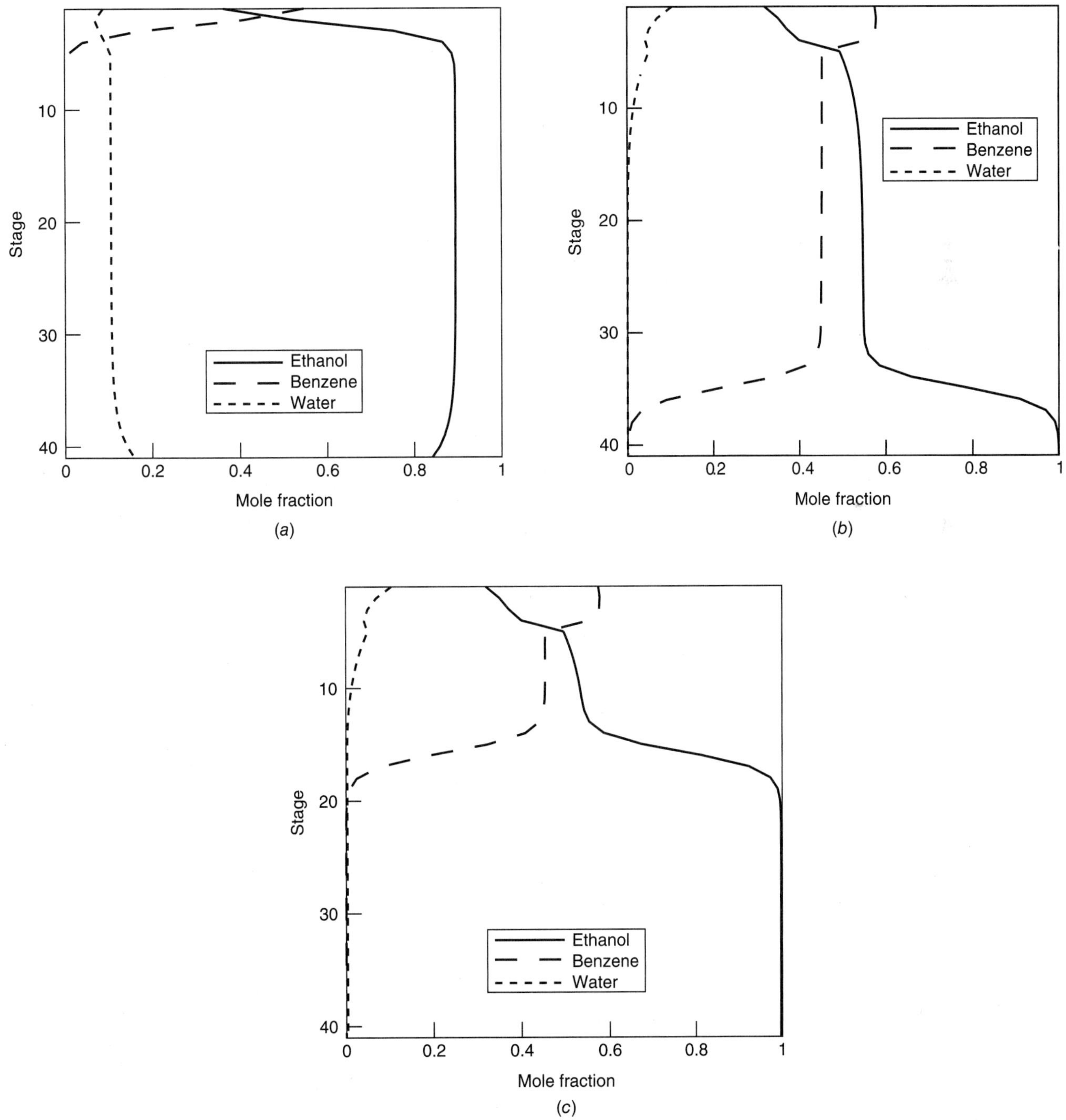

FIG. 13-50 Liquid-phase mole fraction profiles for ethanol-benzene-water distillation. (*a*) Low-purity profile. (*b*) Intermediate-purity profile. (*c*) High-purity profile.

equations [Ellis et al., *Comput. Chem. Eng.* **10:** 433 (1986); Kovach and Seider, *Comput. Chem. Eng.* **11:** 593 (1987); Burton, Ph.D. thesis in Chem. Eng., Cambridge University, 1986]. That real distillation columns can possess multiple steady states has been confirmed by the experimental work of Kienle et al. [*Chem. Eng. Sci.* **50:** 2691 (1995)], Køggersbol et al. [*Comput. Chem. Eng.* **20:** S835 (1996)], Gaubert et al. [*Ind. Eng. Chem. Res.* **40:** 2914 (2001)], and others.

NONEQUILIBRIUM MODELING

Although the widely used equilibrium-stage models for distillation, described previously, have proved to be adequate for binary and close-boiling, ideal and near-ideal multicomponent vapor/liquid mixtures, their deficiencies for general multicomponent mixtures have long been recognized. Even Murphree [*Ind. Eng. Chem.* **17:** 747–750, 960–964 (1925)], who formulated the widely used plate efficiencies that carry his name, pointed out clearly their deficiencies for multicomponent mixtures and when efficiencies are small. Walter and Sherwood [*Ind. Eng. Chem.* **33:** 493 (1941)] showed that experimentally measured efficiencies could cover an enormous range, with some values less than 10 percent. Toor [*AIChE J.* **3:** 198 (1957)] predicted that the Murphree vapor efficiencies in some multicomponent systems could cover the entire range of values from minus infinity to plus infinity, a result that was verified experimentally by others.

In recent years, a new approach to the modeling of distillation and absorption processes has become available: the *nonequilibrium* or *rate-based* models. These models treat these classical separation processes as the mass-transfer-rate–governed processes that they really are, and they avoid entirely the (a priori) use of concepts such as efficiency and HETP [Krishnamurthy and Taylor, *AIChE J.* **31:** 449–465 (1985); Taylor, Kooijman, and Hung, *Comput. Chem. Eng.* **18:** 205–217 (1994)].

A schematic diagram of a nonequilibrium (NEQ) stage is given in Fig. 13-51. This NEQ stage may represent (part of) the two phases on a tray or within a section of a packed column. The wavy line in the middle of the box represents the phase interface. This illustration is intended only to aid in understanding the basic principles of nonequilibrium modeling; the actual flow patterns and the shape of the phase boundary are very complicated and depend on, among other things, the equipment design, the column operation, and the physical properties of the system.

In a nonequilibrium model, separate balance equations are written for each distinct phase. The material balances for each species in the vapor and liquid phases on an arbitrary stage j are

$$(1+r_j^V)V_j y_{i,j} - V_{j+1} y_{i,j+1} - f_{j+1}^V + \mathcal{N}_{i,j}^V = 0 \qquad (13\text{-}59)$$

$$(1+r_j^L)L_j x_{i,j} - L_{j-1} x_{i,j+1} - f_{j+1}^L + \mathcal{N}_{i,j}^L = 0 \qquad (13\text{-}60)$$

where r_j^V and r_j^L are the ratios of sidestream to interstage flows and are defined by Eqs. (13-49); $f_{i,j}^p$ is the external feed flow rate of species i in phase p to stage j; and $\mathcal{N}_{i,j}$ is the rate of mass transfer across the phase interface (with units of moles per second or equivalent). Formally, we may write

$$\mathcal{N}_{i,j}^p = \int N_{i,j}^p \, da_j \qquad (13\text{-}61)$$

where $N_{i,j}^p$ is the molar flux in phase p [with units of mol/(m$^2 \cdot$ s) or equivalent] at a particular point in the two-phase dispersion, and da_j is the portion of interfacial area through which that flux passes. A material balance around the interface yields

$$\mathcal{N}_{i,j}^V = \mathcal{N}_{i,j}^L \qquad (13\text{-}62)$$

The sum of the phase and interface balances yields the component material balance for the stage as a whole, the equation used in the equilibrium-stage model.

The energy balance is treated in a similar way, split into two parts—one for each phase, each part containing a term for the rate of energy transfer across the phase interface.

$$(1+r_j^V)V_j H_j^V - V_{j+1} H_{j+1}^V - F_j^V H_j^{VF} + \mathcal{E}_j^V + Q_j^V = 0 \qquad (13\text{-}63)$$

$$(1+r_j^L)L_j H_j^L - L_{j-1} H_{j-1}^L - F_j^L H_j^{LF} + \mathcal{E}_j^L + Q_j^L = 0 \qquad (13\text{-}64)$$

where $\mathcal{E}_j^p$ is the rate of energy transfer across the phase interface in phase p and is defined by

$$\mathcal{E}_j^p = \int_a E_j^p \, da \qquad (13\text{-}65)$$

where a is the interfacial area and E_j^p is the energy flux across the interface from/to phase p. An energy balance at the phase interface yields

$$\mathcal{E}_j^V = \mathcal{E}_j^L \qquad (13\text{-}66)$$

In addition, we need summation equations for the mole fractions in the vapor and liquid phases.

A review of early applications of NEQ models is available in chapter 14 of Taylor and Krishna (*Multicomponent Mass Transfer*, Wiley, New York, 1993).

It is worth emphasizing that Eqs. (13-59) to (13-66) hold regardless of the models used to calculate the interphase transport rates $\mathcal{N}_{i,j}$ and $\mathcal{E}_j^p$. With a mechanistic model of sufficient complexity it is possible, at least in principle, to account for mass transfer from bubbles in the froth on a tray as well as to entrained droplets in a spray, as well as transport between the phases flowing over and through the elements of packing in a packed column. However, a completely comprehensive model for estimating mass-transfer rates in all the possible flow regimes does not exist at present, and simpler approaches are used.

The simplest approach is to say that the molar fluxes at a vapor-liquid interface may be expressed as

$$N_i^V = c_t^V k_i^V (y_i^V - y_i^I) + y_i^V N_t^V \qquad (13\text{-}67)$$

with a similar expression for the liquid phase.

$$N_i^L = c_t^L k_i^L (x_i^I - x_i^L) + x_i^L N_t^L \qquad (13\text{-}68)$$

In these equations, c_t^V and c_t^L are the molar densities of the superscripted phases, y_i^V is the mole fraction in the bulk vapor phase, x_i^L is the mole fraction in the bulk liquid phase, and x_i^I and y_i^I are the mole fractions of species i at the phase interface. Also N_t^p is the total molar flux in phase p, and k_i^V and k_i^L are the mass-transfer coefficients for the vapor and liquid phases (with units of velocity), respectively. Methods for estimating mass-transfer coefficients in distillation processes are discussed briefly below and at greater length in Sec. 5 of this handbook.

The second term on the right-hand sides of Eqs. (13-67) and (13-68) is not often important in distillation (its neglect is equivalent to the assumption of equimolar counterflows in the column), but it can be quite significant in gas absorption.

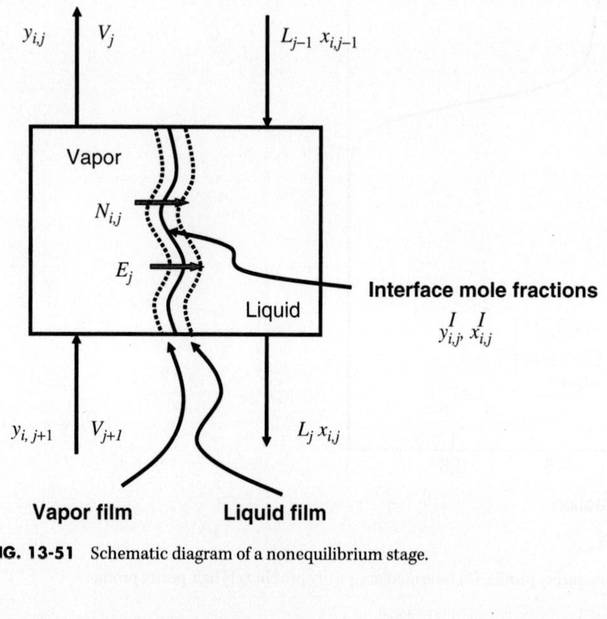

FIG. 13-51 Schematic diagram of a nonequilibrium stage.

The energy fluxes are related by

$$E^V = q^V + \sum_{i=1}^{c} N_i^V \bar{H}_i^V = E = q^L + \sum_{i=1}^{c} N_i^L \bar{H}_i^L = E^L \quad (13\text{-}69)$$

with $$q^V = h^V(T^V - T^I) \qquad q^L = h^L(T^I - T^L) \quad (13\text{-}70)$$

where h^V and h^L are the heat-transfer coefficients in the vapor and liquid phases, respectively.

The inclusion in the model of the mass and energy transport equations introduces the mole fractions and temperature at the interface. It is common in almost all treatments of mass transfer across a phase boundary to assume that the mole fractions in the vapor and liquid phases at the interface are in equilibrium with each other. We may, therefore, use the very familiar equations from phase equilibrium thermodynamics to relate the interface mole fractions

$$y_i^I = K_i x_i^I \quad (13\text{-}71)$$

where the superscript I denotes the interface compositions and K_i is the vapor-liquid equilibrium ratio (or K value) for component i. In equilibrium-stage calculations, the equilibrium equations are used to relate the composition of the streams leaving the stage, and the K values are evaluated at the composition of the two exiting streams and the stage temperature (usually assumed to be the same for both phases). In nonequilibrium models, the K values are evaluated at the interface composition and temperature by using exactly the same thermodynamic property models as those used in equilibrium-stage simulations. The interface composition and temperature must, therefore, be computed during a nonequilibrium column simulation. Strictly speaking, the composition and temperature at the interface vary with position in two-phase dispersion that exists on a tray or within the confines of a packed bed. In most NEQ models, the interface state is assumed to be uniform on the stage; thus, the model of a single stage includes one set of mass-transfer, heat-transfer, and phase equilibrium equations.

In equilibrium-stage models, the compositions of the leaving streams are related through the assumption that they are in equilibrium (or by the use of an efficiency equation). It is important to recognize that efficiencies are not used in a nonequilibrium model; they may, however, be calculated from the results obtained by solving the model equations.

Degrees of Freedom Table 13-6 summarizes the equations for a single nonequilibrium stage. There are $6c + 5$ independent equations per stage. As with the equilibrium-stage model discussed previously, we have not included the feed mole fraction summation equation, or those for the vapor and liquid streams coming from adjacent stages.

The variables appearing in these equations are summarized in Table 13-7. It is important to recognize that we have not included mass- and heat-transfer coefficients in the table of variables. These quantities are considered analogous to the thermodynamic properties of the equilibrium-stage model and are functions of other variables (as discussed in greater detail below).

The $6c + 5$ variables for each stage determined during the solution of the nonequilibrium model equations are the vapor and liquid flow V_j and L_j, respectively; the bulk vapor mole fractions, $y_{i,j}$ (c in number); the bulk liquid mole fractions $x_{i,j}$ (c); the vapor and liquid temperatures T_j^V and, T_j^L, respectively; the interface mole fractions and temperature $y_{i,j}^I$ (c), $x_{i,j}^I$ (c), and T_j^I; and the mass-transfer rates $\mathcal{N}_{i,j}^V$ (c) and $\mathcal{N}_{i,j}^L$ (c). Note that the interface material balance, Eq. (13-62), means that only one set of mass-transfer rates really needs to be counted in the set of variables for this stage, say, $\mathcal{N}_{i,j}$ (this reduces the number of variables being computed). The remaining variables, $2c + 7$ in number, that need to be specified, are identified in Table 13-7. It is important to recognize that the other flows and composition variables appearing in the nonequilibrium-stage model equations are associated with the equivalent equations for adjacent stages. Although it appears that the number of degrees of freedom is higher for this more complicated model, this is misleading. The additional variables that are specified here take into account that there is one additional heat duty (one per phase) and that the feed is split into vapor and liquid fractions. In practice, the overall feed flow, pressure, temperature (or vapor fraction), and composition would be specified, and the vapor and liquid component flows in the feed determined from an adiabatic flash.

TABLE 13-6 Equations for a Nonequilibrium Stage

Equation	Reference	Number
Vapor phase component balance	(13-59)	c
Liquid phase component balance	(13-60)	c
Interface material balance	(13-62)	c
Vapor bulk mole fraction summation equation		1
Liquid bulk mole fraction summation equation		1
Vapor phase energy balance	(13-63)	1
Liquid phase energy balance	(13-64)	1
Equilibrium at the interface	(13-71)	c
Mass transfer in the vapor phase	(13-67)	$c - 1$
Mass transfer in the liquid phase	(13-68)	$c - 1$
Energy balance at phase interface	(13-69)	1
Summation equation for vapor mole fractions at phase interface		1
Summation equation for liquid mole fractions at phase interface		1
Summation equations for bulk vapor and liquid mole fractions		2
	Total	$6c + 7$

TABLE 13-7 Variables for a Nonequilibrium Stage

Variable	Symbol	Number	Specified
Vapor and liquid flow rates	V_j, L_j	2	
Sidestream flow rates	U_j, W_j	2	Yes
Vapor-phase bulk composition	$y_{i,j}$	c	
Liquid-phase bulk composition	$x_{i,j}$	c	
Vapor interface composition	$y_{i,j}^I$	c	
Liquid interface composition	$x_{i,j}^I$	c	
Component feed flow rates	$f_{i,j}^V, f_{i,j}^L$	$2c$	Yes
Feed pressure and temperature		2	Yes
Vapor-phase temperature	T_j^V	1	
Liquid-phase temperature	T_j^L	1	
Interface temperature	T_j^I	1	
Stage pressure	P_j	1	Yes
Heat loss from vapor phase	Q_j^V	1	Yes
Heat loss from liquid phase	Q_j^L	1	Yes
Mass-transfer rates in the vapor phase	$\mathcal{N}V_{i,j}$	c	
Mass-transfer rates in the liquid phase	$\mathcal{N}L_{i,j}$	c	
	Total	$8c + 12$	$2c + 7$

Physical Properties The only physical properties needed for an equilibrium-stage simulation are those needed to estimate K values and enthalpies; these same properties are needed for nonequilibrium models as well. Enthalpies are required for the energy balance equations; vapor-liquid equilibrium ratios are needed for the calculation of driving forces for mass and heat transfer. The need for mass- (and heat-) transfer coefficients means that nonequilibrium models are rather more demanding of physical property data than are equilibrium-stage models. These coefficients may depend on a number of other physical properties, as summarized in Table 13-8.

Methods for estimating physical and transport properties are described in Secs. 4 and 5 of this handbook (see also Poling et al., *The Properties of Gases and Liquids*, McGraw-Hill, 5th ed., 2001).

Flow Models In a real column the composition of the vapor and liquid phases changes due to flow across a tray or over and around packing. Thus, the bulk phase mole fractions that appear in the rate equations (13-67) and (13-68) vary with position and should not automatically be assumed to be equal to the average exit mole fractions that appear in the material balance equations. In practice, we assume a flow pattern for the vapor and

TABLE 13-8 Physical Property Needs of Equilibrium and Nonequilibrium Models

Property	EQ model	NEQ model	Used for
K values	Yes	Yes	Driving forces
Enthalpy	Yes	Yes	Energy balances
Activity coefficient	Yes	Yes	K values, enthalpies
Fugacity coefficients	Yes	Yes	K values, enthalpies
Vapor pressure	Yes	Yes	K values
Heat capacity	Yes	Yes	Enthalpies, heat-transfer coefficient
Mass-transfer coefficients		Yes	Mass-transfer rate equations
Heat-transfer coefficients		Yes	Energy-transfer rate equation
Density		Yes	Mass-transfer coefficients
Diffusion coefficients		Yes	Mass-transfer coefficients
Viscosity		Yes	Mass-transfer coefficients
Surface tension		Yes	Mass-transfer coefficients
Thermal conductivity		Yes	Heat-transfer coefficients

liquid phases, and this allows us to determine appropriate average mole fractions for use in the rate equations. There are three flow models in general use: mixed flow, plug flow, and dispersion flow (Lockett, *Distillation Tray Fundamentals*, Cambridge University Press, Cambridge, England, 1986). A flow model needs to be identified for each phase. If both phases are assumed to be well mixed, then the average mole fractions are indeed equal to the mole fractions in the exit streams. This is the simplest (and an often used) approach that leads to the most conservative simulation (lowest mass-transfer rates, tallest column); at the opposite extreme is plug flow. The most realistic model is dispersion flow (see Lockett, *Distillation Tray Fundamentals*, Cambridge University Press, Cambridge, England, 1986), but this model is not included in most computer implementations of NEQ models because it is quite complicated. For further discussion of the importance of flow models and the equations used to estimate average compositions, see Taylor and Krishna (1993) and Kooijman and Taylor [*Chem. Eng. J.* **57:** 177 (1995)].

Mass-Transfer Coefficients Mass-transfer coefficients (and the equally important interfacial area, a parameter with which they often are combined) may be computed from empirical correlations or theoretical models.

The mass-transfer performance of trays often is expressed by way of a dimensionless group called the number of transfer units [see Lockett (1986), Kister (1992), and Sec. 14 for additional background]. These dimensionless numbers are defined by

$$\mathbb{N}^V = k^V a' t_V = \frac{k^V a h_f}{u_s^V} \tag{13-72}$$

$$\mathbb{N}^L = k^L \bar{a} t_L = \frac{k^L a h_f Z}{Q_L/W} \tag{13-73}$$

where h_f = froth height, m
Z = liquid flow path length, m
W = weir length, m
$Q_L = \dfrac{L}{c_t^L}$ = volumetric liquid flow rate, m³/s
$u_s = \dfrac{V}{c_t^V A_{bub}}$ = superficial vapor velocity, m/s
A_{bub} = bubbling area of tray, m²
h_L = clear liquid height on tray, m
a' = interfacial area per unit volume of vapor, m²/m³
$\bar{a}$ = interfacial area per unit volume of liquid, m²/m³

These areas are related to the interfacial area per unit volume of froth a by

$$a' = \frac{1}{1-a} \qquad \bar{a} = \frac{a}{\alpha} \tag{13-74}$$

where $\alpha = h_L/h_f$ is the relative froth density. Also t_V and t_L are the vapor- and liquid-phase residence times, defined by

$$t_V = (1-\alpha) h_f u_s \tag{13-75}$$

$$t_L = \frac{Z}{u_L} = \frac{h_L Z W}{Q_L} \tag{13-76}$$

The *AIChE Bubble Tray Design Manual*, published in 1958 (see also Gerster et al., *Tray Efficiencies in Distillation Columns*, AIChE, New York, 1958), presented the first comprehensive procedure for estimating the numbers of transfer units in distillation. For many years this work represented the only such procedure available in the open literature; the work of organizations such as Fractionation Research Incorporated (FRI) was available only to member companies. Other comprehensive procedures for trays appeared in the 1980s [Zuiderweg, *Chem. Eng. Sci.* **37:** 1441 (1982); Chan and Fair, *Ind. Eng. Chem. Proc. Des. Dev.* **23:** 814, 820 (1984)]. Readers are referred to Kister (1992), Lockett (1986), Klemola and Ilme (1996), and Sec. 14 of this handbook for summaries and references to what is available in the open literature.

Example 13-11 Mass-Transfer Coefficient in a Tray Column Consider again the C_4 splitter that formed the basis of Examples 7 and 8. The key design parameters for the valve trays are given in the following table (from Klemola and Ilme 1996):

Column height	51.8 m	Downcomer area center	0.86 m²
Column diameter	2.9 m	Tray spacing	0.6 m
Number of trays	74	Hole diameter	39 mm
Weir length, side	1.859 m	Total hole area	0.922 m²
Weir length, center	2.885 m	Outlet weir height	51 mm
Liquid flow path length	0.967 m per pass	Tray thickness	2 mm
Active area	4.9 m²	Number of valves per tray	772
Downcomer area, side	0.86 m²	Free fractional hole area	18.82%

Estimate the mass-transfer coefficients for tray 7, where the flow and physical properties are estimated to be summarized as:

	Gas/vapor	Liquid
Flow, mol/s	590	550
Density, kg/m³	16.8	520
Viscosity, N/m²	8.6×10^{-6}	1.35×10^{-4}
Molecular weight, kmol/kg	58.0	58.0
Diffusivity, m²/s	800×10^{-9}	1.0×10^{-9}
Surface tension, N/m	0.014	

We use the AIChE correlation to illustrate the general approach, noting that the correlation was not developed specifically for valve trays (few methods were). In this model, the number of transfer units is given by

$$\mathbb{N}^V = \frac{0.776 + 4.57 h_w - 0.238 F_s + 104.8 Q_L/W}{\sqrt{\mathrm{Sc}^V}} \tag{13-77}$$

$$\mathbb{N}^L = 19{,}700\sqrt{D^L}\,(0.4F_s + 0.17)t_L \tag{13-78}$$

In the preceding expressions, h_w is the weir height (m). The vapor-phase Schmidt number Sc^V is defined by $\mathrm{Sc}^V = \mu^V/(\rho^V D^V)$, which here takes the value $\mathrm{Sc}^V = 0.640$.

The superficial velocity is computed next from $u_s^V = V/(c_t^V A_{bub}) = 0.42$ m/s. Here F_s is the so-called F factor and is $F_s = u_s^V\sqrt{\rho^V} = 1.7\ (\mathrm{kg/ms})^{1/2}$. The volumetric liquid flow is $Q_L = L/c_t^L = 0.061$ m³/s.

The froth height on the tray is estimated (by using the methods in Sec. 14 of this handbook—see also Lockett, *Distillation Tray Fundamentals*, Cambridge University Press, Cambridge, UK, 1986; Kister 1992) to be $h_f = 0.143$ m. The liquid-phase residence time is $t_L = h_L Z W_l/Q_L = 4.67$ s.

The number of transfer units follows for the vapor phase from Eq. (13-77) as $\mathbb{N}^V = 2.45$ and for the liquid phase from Eq. (13-78) as $\mathbb{N}^L = 2.33$. The products of the vapor- and liquid-phase mass-transfer coefficients and the interfacial area follow directly from the second parts of the same equations. Note that it is not possible with these correlations to separate the mass-transfer coefficient from the interfacial area. In practice this is not a concern because it is the mass-transfer rates that are needed rather than the fluxes, and the product suffices for NEQ model computations.

The diffusivities used in this example were for the light key–heavy key pair of components. For systems similar to this one, the diffusion coefficients of all binary pairs in the mixture would be expected to have similar values. This will not be the case for mixtures of components that differ sharply in their fundamental properties (e.g., size, polarity). For these more highly nonideal mixtures, it is necessary to estimate the mass-transfer coefficients for each of the binary pairs.

The number of transfer units for packed columns is defined by

$$\mathbb{N}^V = \frac{k^V a' H}{u_V} \tag{13-79}$$

$$\mathbb{N}^L = \frac{k^L a' H}{u_L} \tag{13-80}$$

where $u_V = V/(c_t^V A_c)$ and $u_L = L/(c_t^L A_c)$ are the superficial vapor and liquid velocities, with A_c the cross-sectional area of the column; a' is the interfacial area per unit volume. The height of a transfer unit (HTU) is defined as

$$\mathbb{H}^V = \frac{H}{\mathbb{N}^V} = \frac{u_V}{k^V a'} \tag{13-81}$$

$$\mathbb{H}^L = \frac{H}{\mathbb{N}^L} = \frac{u_L}{k^L a'} \tag{13-82}$$

Methods of estimating numbers and heights of transfer units and mass-transfer coefficients and interfacial areas in packed columns are reviewed by Ponter and Au Yeung (in *Handbook of Heat and Mass Transfer*, Gulf Pub., Houston, 1986), by Wang et al. [*Ind. Eng. Chem. Res.* **44:** 8715 (2005)], and in Sec. 5 of this handbook; one such method is illustrated next.

Example 13-12 Mass-Transfer Coefficients in a Packed Column Estimate the mass-transfer coefficients at the top of the packed gas absorber in Example 13-9. The column has 23 ft of 2-in metallic Pall rings and is 3 ft in diameter. The specific surface area of this packing is 112 m^2/m^3. The flows and physical properties are estimated to be as follows:

	Gas/vapor	Liquid
Flow, mol/s	332	37
Temperature, K	267	266
Density, kg/m^3	51.2	705
Viscosity, N/m^2	1.17×10^{-5}	2.43×10^{-4}
Molecular weight, kmol/kg	16.8	87.9
Diffusivity, m^2/s	100×10^{-9}	4.6×10^{-9}
Surface tension, N/m	0.0085	

We will use the well-known correlation of Onda et al. (see Sec. 14) to estimate the mass-transfer coefficients. The vapor-phase coefficient is given by

$$\frac{k^V}{a_p D^V} = A\left(\frac{\rho^V u^V}{\mu^V a_p}\right)^{0.7}\left(\frac{\mu^V}{\rho^V D^V}\right)^{0.33}(a_p d_p)^{-2} \tag{13-83}$$

where d_p is the nominal packing size (2 in = 0.0508 m), and a_p is the specific surface area of the packing. Also, A is a constant that has the value 2 if nominal packing size is less than 0.012 m, otherwise, $A = 5.23$.

The liquid-phase mass-transfer coefficient is given by

$$k^L = 0.0051\left(\frac{\rho^L}{\mu^L g}\right)^{-1/3}\left(\frac{u_L}{a'\mu^L}\right)^{2/3}\left(\frac{\mu^L}{\rho^L D^L}\right)^{-1/2}(a_p d_p)^{0.4} \tag{13-84}$$

Finally, the interfacial area per unit volume is given by

$$a' = a_p\left\{1-\exp\left[-1.45\left(\frac{\sigma_c}{\sigma^L}\right)^{0.75}\left(\frac{\rho^L u_L}{a_p \mu_L}\right)^{0.1}\left(\frac{a_p u_L^2}{g}\right)^{-0.05}\left(\frac{u_L^2 \rho^L}{a_p \sigma^L}\right)^{0.2}\right]\right\} \tag{13-85}$$

The vapor and liquid velocities are calculated to be $u_V = V/(c_t^V A_t) = 0.42$ m/s and $u_L = L/(c_t^L A_t) = 0.0017$ m/s. Substituting these values into Eqs. (13-83) to (13-85) gives $k^V = 0.0021$ m/s, $a' = 96.2$ m^2/m^3, and $k^L = 3.17 \times 10^{-4}$ m/s.

Solving the NEQ Model Equations In general, a nonequilibrium model of a column has many more equations than does an equivalent equilibrium-stage model. Nevertheless, we use essentially the same computational approaches to solve the nonequilibrium model equations: simultaneous convergence [Krishnamurthy and Taylor, *AIChE J.* **31:** 449–465 (1985)] and continuation methods [Powers et al., *Comput. Chem. Eng.* **12:** 1229 (1988)]. Convergence of a nonequilibrium model is likely to be slower than that of the equilibrium model because of the greater number of model equations and the associated overhead in evaluating a greater number of physical properties. Finally, we note that the strategies we have outlined for helping to converge equilibrium-stage simulation may prove equally useful when simulating distillation operations using the nonequilibrium models described here.

Equipment Design As we have already seen, the estimation of mass-transfer coefficients and interfacial areas from empirical correlations nearly always requires us to know something about the column design. At the very least we need to know the diameter and type of column internals (although usually we need to know more than that since most empirical correlations for mass-transfer coefficients have some dependency on equipment design parameters, such as the weir height of trays). This need for more or less complete equipment design details suggests that nonequilibrium models cannot be used in preliminary process design (before any actual equipment design has been carried out). This is not true, however. Column design methods are available in the literature as well in most process simulation programs, and it is straightforward to carry out equipment sizing calculations at the same time as the stage equations are being solved (Taylor et al., 1994). This does not add significantly to the difficulty of the calculation, while providing the very significant advantage of allowing nonequilibrium or rate-based models to be used at all stages of process simulation.

Example 13-13 A Nonequilibrium Model of a C4 Splitter Consider again the C_4 splitter that formed the basis of Examples 7, 8, and 11.

In this example we do not need to guess how many stages to use in each section of the column. The real column had 74 valve trays; the model column includes 74 model trays with the feed to tray 38 [plus a (subcooled) condenser and a reboiler, both of which are modeled as equilibrium stages, as described previously]. All operating specifications are the same as for the corresponding equilibrium-stage model and are given in Examples 7 and 8. It is necessary to choose models that allow for the estimation of the rates of interphase mass transfer; that means selecting vapor and liquid flow models and correlations to estimate the mass-transfer coefficients in each phase. In this case, the AIChE correlations were used. It is known that this method is more conservative than others (i.e., the predicted efficiencies are lower). The importance of the flow model is clear from the simulation results tabulated below. The predicted component Murphree efficiencies computed with Eq. (13-56) vary more widely from stage to stage and from component to component than might be expected for a system such as this. The TBK efficiency, on the other hand, does not change by more than a few percentage points over the height of the column; the value in the following table is an average of that computed for each tray from the simulation results using Eq. (13-57).

Vapor flow model	Liquid flow model	i-C_4 in distillate, %	n-C_4 in bottoms, %	TBK efficiency, %
Mixed	Mixed	90.2	96.3	63
Plug	Mixed	92.2	97.2	78
Plug	Dispersion	93.9	98.0	106

Internal vapor and liquid composition data rarely are available, but such data are the best possible for model discrimination and validation. It is often relatively easy to match even a simple model only to product compositions. In the absence of composition profiles, the internal temperature profile can often be as useful, provided that it is known to which phase a measured temperature pertains. The following table compares the few available measured tray temperatures with those computed during the simulation. The agreement is quite good.

	Temperature, °C	
Tray	Measured	Predicted
9	47.5	48.6
65	62.2	62.5
74	63.2	63.1

A portion of the McCabe-Thiele diagram for the simulation involving plug flow of vapor and dispersion flow of the liquid is shown in Fig. 13-52. For a nonequilibrium column, these diagrams can only be constructed from the results of a computer simulation. Note that the triangles that represent the stages extend beyond the curve that represents the equilibrium line; this is so because the efficiencies are greater than 100 percent.

In this case, the converged composition and temperature profiles have the same shape as those obtained with the equilibrium-stage model (with specified efficiency), and therefore they are not shown. The reason for the similarity is that this is basically a binary separation of very similar compounds. The important point here is that, unlike the equilibrium-stage model simulations, the nonequilibrium model predicted how the column would perform; *no parameters were adjusted to provide a better fit to the plant data.* That is not to say, of course, that NEQ models cannot be used to fit plant data. In principle, the mass-transfer coefficients and interfacial area (or parameters in the equations used to estimate them) can be tuned to help the model better fit plant data.

MAXWELL-STEFAN APPROACH

Strictly speaking, Eqs. (13-67) and (13-68) are valid only for describing mass transfer in binary systems under conditions where the rates of mass transfer are low. Most industrial distillation and absorption processes, however, involve more than two different chemical species. The most fundamentally sound way to model mass transfer in multicomponent systems is to use the Maxwell-Stefan (MS) approach (Taylor and Krishna, *Multicomponent Mass Transfer*, Wiley, New York, 1993).

The MS equation for diffusion in a mixture with any number of different species can be written as

$$d_i = -\sum_{k=1}^{c}\frac{x_i x_k (u_i - u_k)}{Đ_{ik}} \tag{13-86}$$

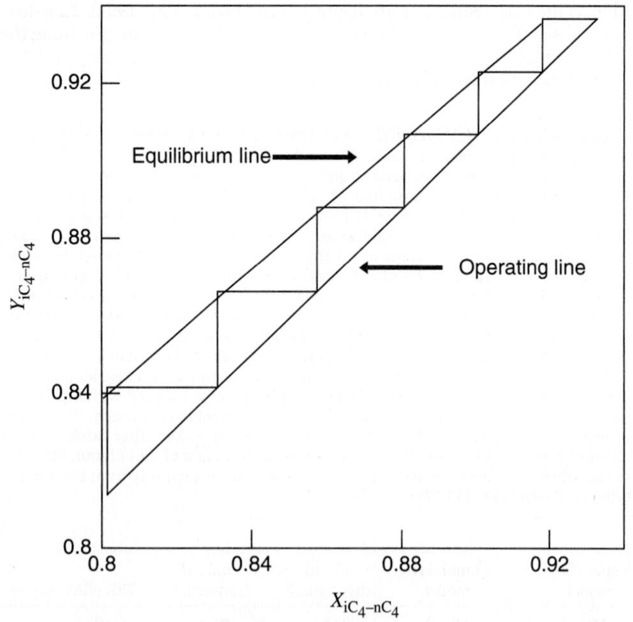

FIG. 13-52 Expanded view near upper right corner of McCabe-Thiele diagram for C_4 splitter.

where $Đ_{ik}$ is the Maxwell-Stefan diffusion coefficient for the binary *i-k* pair of components. Methods for estimating these coefficients are discussed by Taylor and Krishna (1993) (see also Sec. 5 of this handbook and Poling et al., *The Properties of Gases and Liquids*, McGraw-Hill, 5th ed., 2001).

In Eq. (13-86), d_i is termed the generalized driving force. For an ideal gas mixture, the driving force is related to the partial pressure gradient and the mole fraction gradient as follows:

$$d_i = \frac{1}{P}\frac{dp_i}{dz} = \frac{dx_i}{dz} \tag{13-87}$$

For a nonideal fluid, the driving force is related to the chemical potential gradient

$$d_i = \frac{x_i}{RT}\frac{d\mu_i}{dz} \tag{13-88}$$

Equation (13-86) for ideal gas mixtures may be derived by using nothing more complicated than Newton's second law: *The sum of the forces acting on the molecules of a particular species is directly proportional to the rate of change of momentum.* The rate of change of momentum between different species is proportional to the concentrations (mole fractions) of the different species and to their relative velocity (see also Taylor and Krishna 1993 for a more complete derivation). Equation (13-86) is more familiar in the form

$$d_i = -\sum_{k=1}^{c} \frac{x_i N_k - x_k N_i}{c_t Đ_{ik}} \tag{13-89}$$

where we have replaced the velocities with the molar fluxes $N_i = c_t u_i$ (see Sec. 5). Only $c - 1$ of Eq. (13-89) are independent; the mole fraction of the last component is obtained by the mole fraction summation equations for both phases.

Solving the MS equations can be quite involved (see Taylor and Krishna 1993). Most often, a simple film model solution of Eq. (13-89) is used, leading to a simple difference approximation to the MS equations

$$\Delta x_i = -\sum_{k=1}^{c} \frac{\bar{x}_i N_k - \bar{x}_k N_i}{c_t \kappa_{i,k}} \tag{13-90}$$

where $\kappa_{i,k}$ is the mass-transfer coefficient for the binary *i-k* pair of components. The Maxwell-Stefan mass-transfer coefficients can be estimated from existing correlations for mass-transfer coefficients using the binary MS diffusion coefficients.

For a nonideal fluid, the driving force is related to the chemical potential gradient

$$d_i = \frac{x_i}{RT}\frac{d\mu_i}{dz} \tag{13-91}$$

The difference approximation of this expression is somewhat more involved because we have to include the derivative of the activity (or fugacity) coefficient in the approximation. If, as we often assume (not always with justification), the resistance to mass transfer in the liquid phase is negligible, then the MS equations for the liquid phase can safely be replaced by

$$\Delta x_i = x_i^I - x_i^L = 0 \tag{13-92}$$

The use of the MS equations in place of the simpler Eqs. (13-67) and (13-68) does not change the number of independent model equations or the number of degrees of freedom.

Example 13-14 The Need for Rigorous Maxwell-Stefan-Based NEQ Models Design a distillation column to separate a feed of 20 mol/s methanol, 10 mol/s isopropanol, and 20 mol/s water. The bottom product is to contain no more than 0.5 mol% methanol, and the distillate is to contain at least 99 mol% methanol but no more than 50 ppm water.

As a first step, we try to design the column by using the equilibrium-stage model. Following Doherty and Malone (2001), the NRTL model was used for the activity coefficients and the Antoine equation for the vapor pressures. Doherty and Malone estimate the minimum reflux as 5; we used a value 50 percent higher in this example and specified the bottoms product rate at 30 mol/s; this choice provides a consistent basis for the comparison of different models. The number of stages and the location of the feed were varied until a column configuration was obtained that met the desired product purity: 80 total stages (including total condenser and partial reboiler) with the feed to stage 16.

Efficiencies of alcohol-water and alcohol-alcohol systems obtained experimentally in sieve tray columns varied from 60 to 100 percent (Sec. 14 in the seventh edition of this handbook). After specifying an average efficiency of 80 percent, we find that 99 total stages with the feed to stage 21 were needed to get the distillate product below 50 ppm water.

If we use the nonequilibrium model to design a sieve tray column, we find that a column with 84 trays (plus condenser and reboiler) and with the feed to tray 21 (stage 22) will produce an overhead product of the desired purity. The reflux ratio and bottoms flows were maintained at the values employed for the equilibrium-stage design. The AIChE method was used for estimating the mass-transfer coefficient–interfacial area products, and the vapor and liquid phases were assumed to be in plug flow. The pressure was assumed to be constant in the column (an assumption that would need to be relaxed at a later stage of the design exercise). The computer simulation also provided a preliminary tray design; that for the trays above the feed is summarized in the following table:

Column diameter, m	1.76
Total tray area, m²	2.43
Number of flow passes	2
Tray spacing, m	0.6
Liquid flow path length, m	0.75
Active area, % total	91.4
Total hole area, % active	14
Downcomer area, % total	4.3
Hole diameter, mm	5
Hole pitch, mm	12
Weir type	Segmental
Combined weir length, m	1.55
Weir height, mm	50

To converge the nonequilibrium model at the specified reflux ratio, it was necessary first to solve the problem at a much lower reflux ratio $R = 2$ and then increase R in steps until the desired value of 7.5 was reached.

The liquid composition, flow, and temperature profiles are shown in Fig. 13-53. In this system, the vapor and liquid temperatures estimated by the rate-based model are quite close (as often is the case in distillation operations).

The McCabe-Thiele diagram for this column is shown in Fig. 13-54. Note that in this case the triangles that represent the stages do not touch the equilibrium line. The length of the vertical section of each step in Fig. 13-54 is a measure of the efficiency of that tray. The component Murphree efficiencies calculated from the simulation results and Eq. (13-56), as well as the TBK average efficiency defined by Eq. (13-57), are shown in Fig. 13-55. The efficiency of methanol in the stripping section is seen to be around 80 percent, that of isopropanol is approximately 75 percent, while that of water is close to 90 percent in the bulk of the column before falling off on the bottom few trays. All

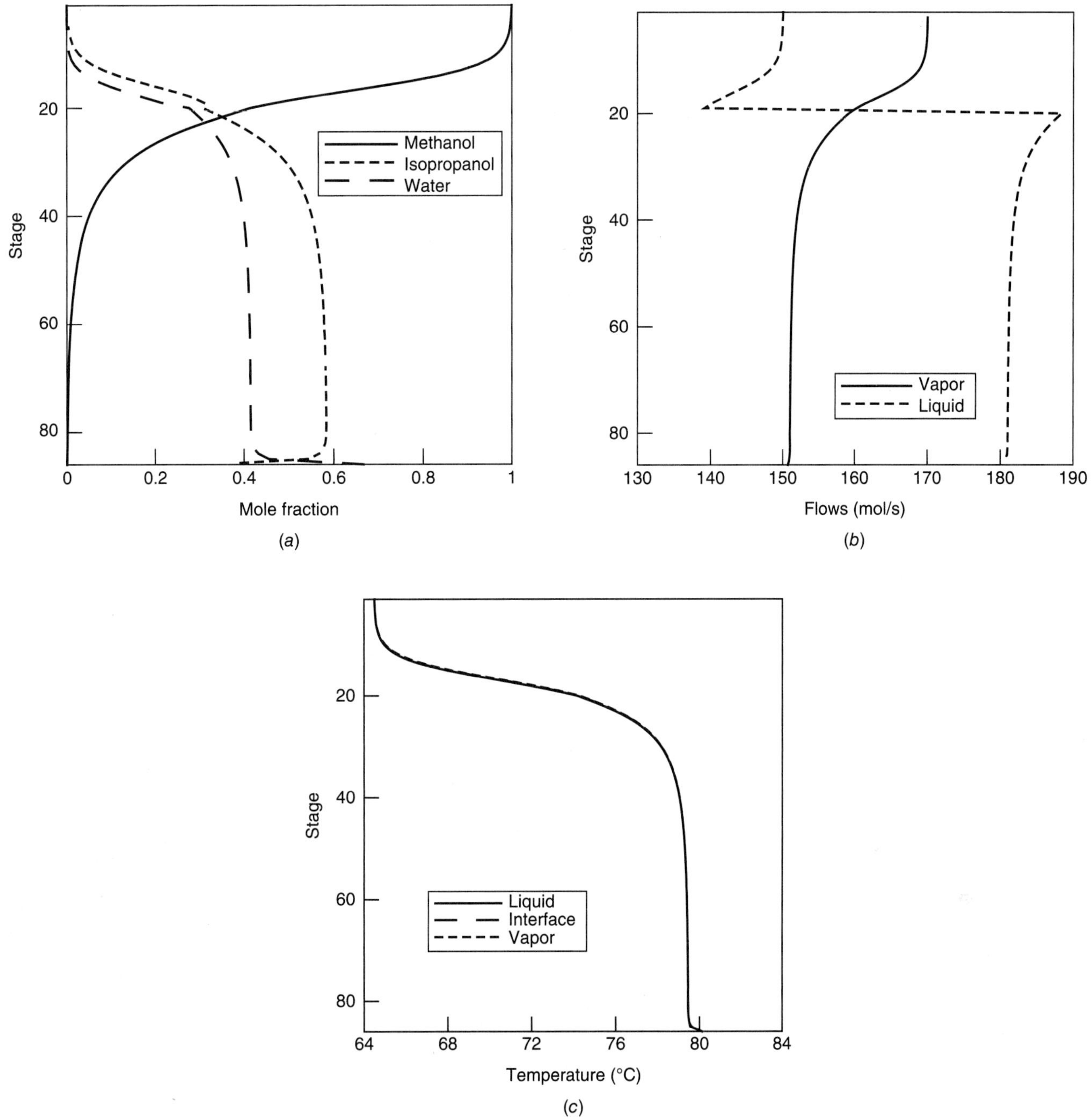

FIG. 13-53 Composition, flow, and temperature profiles in nonideal distillation process.

component efficiencies are found to be lower in the rectifying section. The TBK average efficiency, also shown in Fig. 13-55, is close to the Murphree efficiency of methanol and varies from 60 percent in the top of the column to 78 percent. Thus, the constant value of 80 percent used above appears to be appropriate, and yet the column designed with the constant-efficiency model required no less than 99 stages (97 trays)!

With 84 trays as opposed to 78 equilibrium stages (not counting condenser and reboiler in either case), we find an overall efficiency of 93 percent, a figure that is quite at odds with the values of the individual component efficiencies seen in Fig. 13-55. How, then, is it possible that the nonequilibrium model suggests that the column needs only six trays more than the number of equilibrium stages? It is because the efficiency of water is so much higher than that of the alcohols that a column design can produce high-purity methanol while producing the 50-ppm methanol bottom product in so few extra stages. Note that nonequilibrium models will not always lead to a design with fewer trays than might be suggested by a constant-efficiency model; it is just as likely for the mass-transfer-rate–based model to predict that more stages will be needed. It all depends on the differences between the component efficiencies.

Individual component efficiencies can vary as much as they do in this example only when the diffusion coefficients of the three binary pairs that exist in this system differ significantly. For ideal or nearly ideal systems, all models lead to essentially the same results. This example demonstrates the importance of mass-transfer models for nonideal systems, especially when trace components are a concern. For further discussion of this example, see Doherty and Malone (2001) and Baur et al. [*AIChE J.* **51:** 854 (2005)]. It is worth noting that there is extensive experimental evidence for mass-transfer effects for this system, and it is known that nonequilibrium models accurately describe the behavior of this system, whereas equilibrium models (and equal-efficiency models) sometimes predict completely erroneous product compositions [Pelkonen et al., *Ind. Eng. Chem. Res.* **36:** 5392 (1997) and *Chem. Eng. Process* **40:** 235 (2001); Baur et al., *Trans. I. Chem. E.* **77:** 561 (1999)].

Nonequilibrium models should be preferred to equilibrium models when efficiencies are unknown, cannot be reliably predicted, and are low or highly variable in nonideal systems and in processes where trace components are a concern.

There is a rapidly growing body of literature on nonequilibrium modeling of distillation and absorption processes. An extended bibliography is available at www.chemsep.org/publications. A brief review of other applications follows.

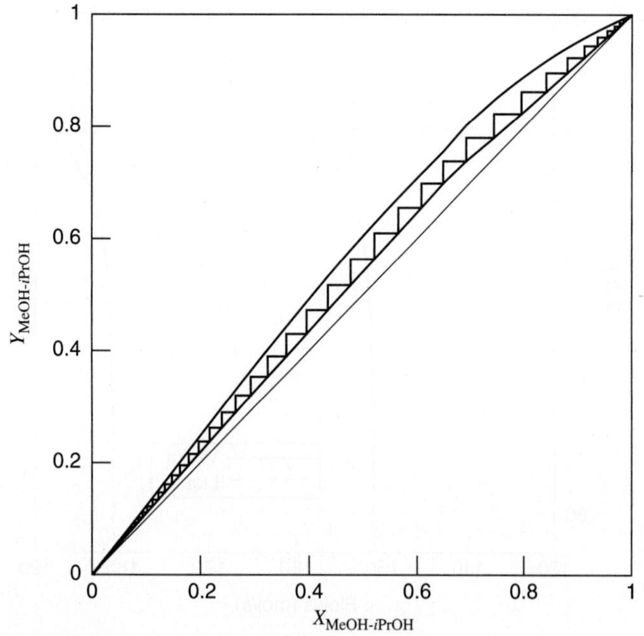

FIG. 13-54 McCabe-Thiele diagram for nonideal distillation column.

Simulation methods currently in use for three-phase systems and systems involving chemical reaction use the equilibrium-stage model (Doherty and Malone, 2001). Three-phase distillation remains rather poorly understood compared to conventional distillation operations that involve just a single liquid phase. It is important to be able to correctly predict the location of the stages where a second liquid phase can form (e.g., to determine the appropriate location for a sidestream decanter). The limited experimental data available suggest that efficiencies are low and highly variable, commonly between 25 and 50 percent. Clearly, a model based on the assumption of equilibrium on every stage cannot hope to be able to *predict* column performance. Cairns and Furzer [*Ind. Eng. Chem. Res.* **29:** 5392 (1997)] explicitly warn against incorporating Murphree efficiencies into the equlilibrium-stage model for three-phase systems, although Müller and Marquardt [*Ind. Eng. Chem. Res.* **36:** 5410 (1997)] find that an efficiency modified EQ stage model is perfectly adequate for their column for the dehydration of ethanol using cyclohexane.

It is possible to develop nonequilibrium models for systems with more than two phases, as shown by Lao and Taylor [*Ind. Eng. Chem. Res.* **33:** 2367 (1994)], Eckert and Vaněk [*Comput. Chem. Eng.* **25:** 603 (2001)], and Higler et al. [*Comput. Chem. Eng.* **28:** 2021 (2004)]. Experimental work that can be used to evaluate these models is scarce; see, however, Cairns and Furzer (1997) and Springer et al. [*Chem. Eng. Res. Design* **81:** 413 (2003)].

There is now an extensive literature on using nonequilibrium models for reactive distillation; see, for example, Taylor and Krishna [*Chem. Eng. Sci.* **55:** 6139 (2000)], Sundmacher and Kienle (*Reactive Distillation: Status and Future Directions*, Wiley-VCH, New York, 2003), Noeres et al. [*Chem. Eng. Processing* **42:** 157 (2003)], and Klöcker et al. [*Chem. Eng. Processing* **44:** 617 (2005)]. Gas absorption accompanied by chemical reaction for a long time has been modeled by using mass-transfer rate-based concepts; see, for example, Cornelisse et al. [*Chem. Eng. Sci.* **35:** 1245 (1980)], Pacheco and Rochelle [*Ind. Eng. Chem. Res.* **37:** 4107 (1998)], and a review by Kenig et al. [*Chem. Eng. Technol.* **26:** 631 (2003)]. For such systems, the chemical reaction influences the efficiencies to such an extent that the concept loses its meaning.

Even at steady state, efficiencies vary from component to component and with position in a column. Thus, if the column is not at steady state, efficiencies also must vary with time as a result of changes to flow rates and composition inside the column. Thus, equilibrium-stage models with efficiencies should not be used to model the dynamic behavior of distillation and absorption columns. Nonequilibrium models for studying column dynamics are described by Kooijman and Taylor [*AIChE J.* **41:** 1852 (1995)], Baur et al. [*Chem. Eng. Sci.* **56:** 2085 (2001)], Gunaseelan and Wankat [*Ind. Eng. Chem. Res.* **41:** 5775 (2002)], Peng et al. [*Chem. Eng. Sci.* **58:** 2671 (2003)], and Kenig et al. [*Chem. Eng. Sci.* **54:** 5195 (1999)].

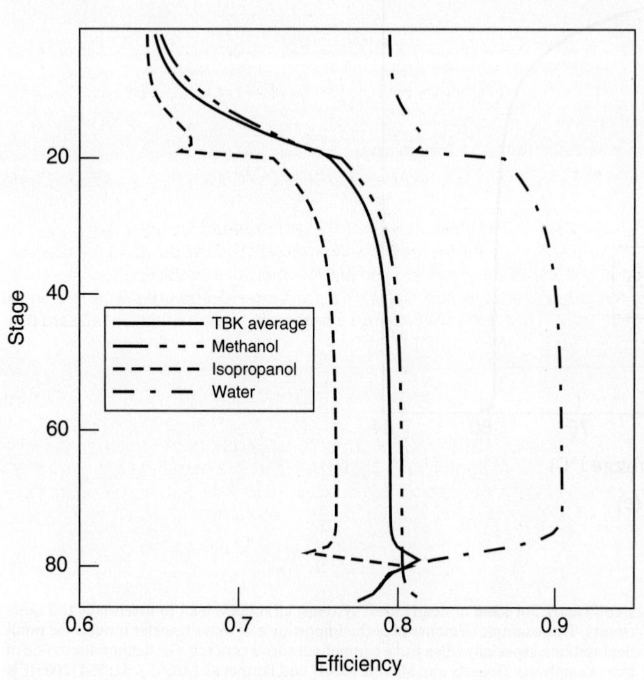

FIG. 13-55 Component Murphree efficiencies and TBK average efficiency [defined by Eq. (13-57)] for nonideal distillation.

SOFTWARE FOR DISTILLATION COLUMN SIMULATIONS

Computer software for equilibrium-stage and nonequilibrium column models is available from a number of suppliers. Many other models have been implemented primarily for research purposes and are not available commercially.

In Table 13-9 we list several suppliers of column simulation models that are available commercially, without in any way claiming that this list is exhaustive or providing an endorsement of any particular package. We recommend that users interested in any of these (or other) packages carry out an independent evaluation that focuses on the ability of the package to tackle the simulation problems of direct interest. The simulations described in this subsection were carried out with *ChemSep*.

TABLE 13-9 Selected List of Suppliers of Column Simulation Software

Supplier	Website	EQ model	NEQ model
Aspen Tech	www.aspentech.com	Yes	Yes
Bryan Research & Engineering	www.bre.com	Yes	
ChemSep	www.chemsep.com	Yes	Yes
Chemstations	www.chemstations.net	Yes	Yes
Deerhaven Technical Software	www.deerhaventech.com	Yes	
Honeywell	www.honeywell.com	Yes	Yes
Process Systems Engineering	www.psenterprise.com	Yes	Yes
ProSim	www.prosim.net	Yes	
SimSci-ESSCOR	www.simsci-esscor.com	Yes	Yes
VMG	virtualmaterials.com	Yes	

DEGREES OF FREEDOM AND DESIGN VARIABLES

DEFINITIONS

In the models described in previous subsections, we have determined the degrees of freedom on a case-by-case basis. We now develop a general approach to the topic.

For separation processes, a design solution is possible if the number of independent equations equals the number of unknowns

$$N_i = N_v - N_c \tag{13-93}$$

where N_v is the total number of variables (unknowns) involved in the process under consideration, N_c is the number of restricting relationships among the unknowns (independent equations), and N_i is the degrees of freedom that must be specified for there to be exactly the same number of unknowns as there are independent equations in the model. The concept of degrees of freedom in this context is similar to the same concept that appears in the Gibbs phase rule. The degrees of freedom is the number of design variables that must be specified to define one unique operation (solution) of the process.

The variables N_i with which the designer of a separation process must be concerned are

1. Stream concentrations (e.g., mole fractions)
2. Temperatures
3. Pressures
4. Stream flow rates
5. Repetition variables N_r

The first three are intensive variables. The fourth is an extensive variable that is not considered in the usual phase rule analysis. The fifth is neither an intensive nor an extensive variable but is a single degree of freedom that the designer uses in specifying how often a particular element is repeated in a unit. For example, a distillation column section is composed of a series of equilibrium stages, and when the designer specifies the number of stages that the section contains, he or she uses the single degree of freedom represented by the repetition variable ($N_r = 1.0$). If the distillation column contains more than one section (such as above and below a feed stage), the number of stages in each section must be specified, and as many repetition variables exist as there are sections, that is, $N_r = 2$.

The various restricting relationships N_c can be classified as

1. Inherent
2. Mass balance
3. Energy balance
4. Phase distribution
5. Chemical equilibrium

The inherent restrictions are usually the result of definitions and take the form of identities. For example, the concept of the equilibrium stage involves the inherent restrictions that $T^V = T^L$ and $P^V = P^L$ where the superscripts V and L refer to the equilibrium exit streams.

The mass balance restrictions are the C balances written for the C components present in the system. (Since we will deal with only nonreactive mixtures, each chemical compound present is a phase rule component.) An alternative is to write $C - 1$ component balances and one overall mass balance.

The phase distribution restrictions reflect the requirement that $f_i^V = f_i^L$ at equilibrium where f is the fugacity. This may be expressed by Eq. (13-1). In vapor-liquid systems, it should always be recognized that all components appear in both phases to some extent, and there will be such a restriction for each component in the system. In vapor-liquid-liquid systems, each component will have three such restrictions, but only two are independent. In general, when all components exist in all phases, the number of restricting relationships due to the distribution phenomenon will be $C(N_p - 1)$, where N_p is the number of phases present.

For the analysis here, the forms in which the restricting relationships are expressed are unimportant. Only the number of such restrictions is important.

ANALYSIS OF ELEMENTS

An *element* is defined as part of a more complex *unit*. The unit may be all or only part of an operation or the entire *process*. Our strategy will be to analyze all elements that appear in a separation process and to determine the number of design variables associated with each. The appropriate elements can then be quickly combined to form the desired units, and the various units can be combined to form the entire process. Of course, allowance must be made for the connecting streams (*interstreams*) whose variables are counted twice when elements or units are joined.

The simplest element is a *single homogeneous stream*. The variables necessary to define it are

	N_v^e
Compositions	$C-1$
Temperature	1
Pressure	1
Flow rate	1
	$C+2$

There are no restricting relationships when the stream is considered only at a point. Henley and Seader (*Equilibrium-Stage Separation Operations in Chemical Engineering,* Wiley, New York, 1981) count all C compositions as variables, but then have to include as a restriction

$$\sum_i x_i = 1 \quad \text{or} \quad \sum_i y_i = 1 \tag{13-94}$$

A stream divider simply splits a stream into two or more streams of the same composition. Consider Fig. 13-56, which shows the division of the condensed overhead liquid L_c into distillate D and reflux L_{N+1}. The divider is permitted to operate nonadiabatically if desired.

Three mass streams and one possible "energy stream" are involved, so

$$N_v^e = 3(C+2)+1 = 3C+7 \tag{13-95}$$

Each mass stream contributes $C+2$ variables, but an energy stream has only its rate q as a variable. The independent restrictions are as follows:

	N_c^e
T and P identities between L_{N+1} and D	2
Composition identities between L_{N+1} and D	$C-1$
Mass balances	C
Energy balance	1
	$2C+2$

The number of design variables for the element is given by

$$N_i^e = N_v^e - N_c^e$$
$$= (3C+7) - (2C+2) = C+5 \tag{13-96}$$

Specification of the feed stream L_c ($C + 2$ variables), the ratio L_{N+1}/D, the "heat leak" q, and the pressure of either stream leaving the divider uses these design variables and defines one unique operation of the divider.

A simple equilibrium stage (no feed or sidestreams) is depicted in Fig. 13-57. Four mass streams and a heat leak (or heat addition) stream provide the following number of variables:

$$N_v^e = 4(C+2)+1 = 4C+9 \tag{13-97}$$

Vapor and liquid streams V_n and L_n, respectively, are in equilibrium with each other by definition and therefore are at the same T and P. These two inherent identities when added to C-component balances, one energy balance, and the C phase distribution relationships give

$$N_v^e = 2C+3 \tag{13-98}$$

Then

$$N_i^e = N_v^e - N_c^e \tag{13-99}$$
$$= (4C+9) - (2C+3) = 2C+6 \tag{13-100}$$

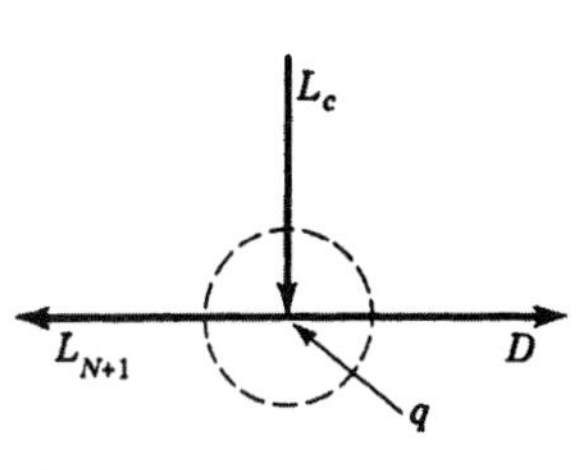

FIG. 13-56 Stream divider.

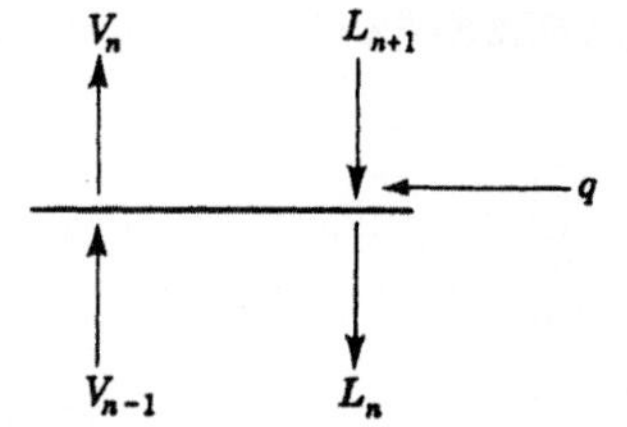

FIG. 13-57 Simple equilibrium stage.

These design variables can be used as follows:

Specifications	N_i^e
Specification of L_{n+1} stream	$C+2$
Specification of V_{n-1} stream	$C+2$
Pressure of either leaving stream	1
Heat leak q	1
	$2C+6$

The results of the analyses for all the various elements commonly encountered in distillation processes are summarized in Table 13-10. Details of the analyses are given by Smith (*Design of Equilibrium Stage Processes,* McGraw-Hill, New York, 1967) and in a somewhat different form by Henley and Seader (1981).

ANALYSIS OF UNITS

A *unit* is defined as a combination of elements and may or may not constitute the entire process. By definition,

$$N_v^u = N_r + \sum_i N_i^e \tag{13-101}$$

and

$$N_i^u = N_v^u - N_c^u \tag{13-102}$$

where N_c^u refers to *new* restricting relationships (identities) that may arise when elements are combined. Here N_c^u does not include any of the restrictions considered in calculating the N_i^e's for the various elements. It includes only the stream identities that exist in each interstream between two elements. The interstream variables ($C + 2$) were counted in each of the two elements when their respective N_i^e's were calculated. Therefore, $C + 2$ new restricting relationships must be counted for each interstream in the combination of elements to prevent redundancy.

The simple absorber column shown in Fig. 13-58 is analyzed here to illustrate the procedure. This unit consists of a series of simple equilibrium stages of the type in Fig. 13-57. Specification of the number of stages N uses the single repetition variable and

$$N_c^u = N_r + \sum_i N_i^e = 1 + N(2C+6) \tag{13-103}$$

since $N_i^e = 2C + 6$ for a simple equilibrium stage in Table 13-10. There are $2(N - 1)$ interstreams, and therefore $2(N - 1)(C + 2)$ new identities (not previously counted) come into existence when elements are combined.

TABLE 13-10 Design Variables N_i^e for Various Elements

Element	N_e^v	N_e^c	N_e^f
Homogeneous stream	$C+2$	0	$C+2$
Stream divider	$3C+7$	$2C+2$	$C+5$
Stream mixer	$3C+7$	$C+1$	$2C+6$
Pump	$2C+5$	$C+1$	$C+4$
Heater	$2C+5$	$C+1$	$C+4$
Cooler	$2C+5$	$C+1$	$C+4$
Total condenser	$2C+5$	$C+1$	$C+4$
Total reboiler	$2C+5$	$C+1$	$C+4$
Partial condenser	$3C+7$	$2C+3$	$C+4$
Partial reboiler	$3C+7$	$2C+3$	$C+4$
Simple equilibrium stage	$4C+9$	$2C+3$	$2C+6$
Feed stage	$5C+11$	$2C+3$	$3C+8$
Sidestream stage	$5C+11$	$3C+4$	$2C+7$
Adiabatic equilibrium flash	$3C+6$	$2C+3$	$C+3$
Nonadiabatic equilibrium flash	$3C+7$	$2C+3$	$C+4$

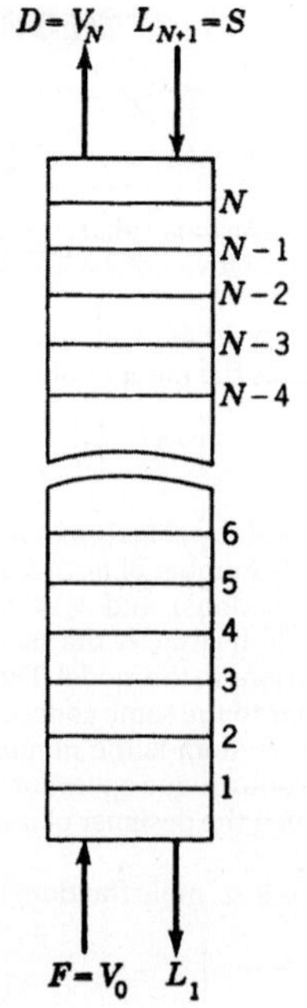

FIG. 13-58 Simple absorption column.

Subtraction of these restrictions from N_v^u gives N_i^u, the design variables that must be specified.

$$N_i^u = N_v^u - N_c^u = N_r + \sum_i N_i^e - N_c^u \tag{13-104}$$

$$= [1 + N(2C+6)] - 2[(N-1)(C+2)] \tag{13-105}$$

$$= 2C + 2N + 5 \tag{13-106}$$

These might be used as follows:

Specifications	N_i^u
Two feed streams	$2C+4$
Number of stages N	1
Pressure of either stream leaving each stage	N
Heat leak for each stage	N
	$2C+2N+5$

A more complex unit is shown in Fig. 13-59, which is a schematic diagram of a distillation column with one feed, a total condenser, and a partial reboiler. Dotted lines encircle the six connected elements (or units) that constitute the distillation operation. The variables N_v^u that must be considered in the analysis of the entire process are just the sum of the N_i^e's for these six elements since here $N_r = 0$. Using Table 13-10, we get the following:

Element (or unit)	$N_v^u = \sum_i N_i^e$
Total condenser	$C+4$
Reflux divider	$C+5$
$N-(M+1)$ equilibrium stages	$2C+2(N-M-1)+5$
Feed stage	$3C+8$
$M-1$ equilibrium stages	$2C+2(M-1)+5$
Partial reboiler	$C+4$
	$10C+2N+27$

Here, the two units of $N-(M+1)$ and $M-1$ stages are treated just as elements. Nine interstreams are created by the combination of elements, so

$$N_c^u = 9(C+2) = 9C + 18 \tag{13-107}$$

The number of design variables is

$$N_i^u = C + 2N + 9\, N_v^u = (10C + 2N + 27) - (9C + 18) \tag{13-108}$$

$$= C + 2N + 9 \tag{13-109}$$

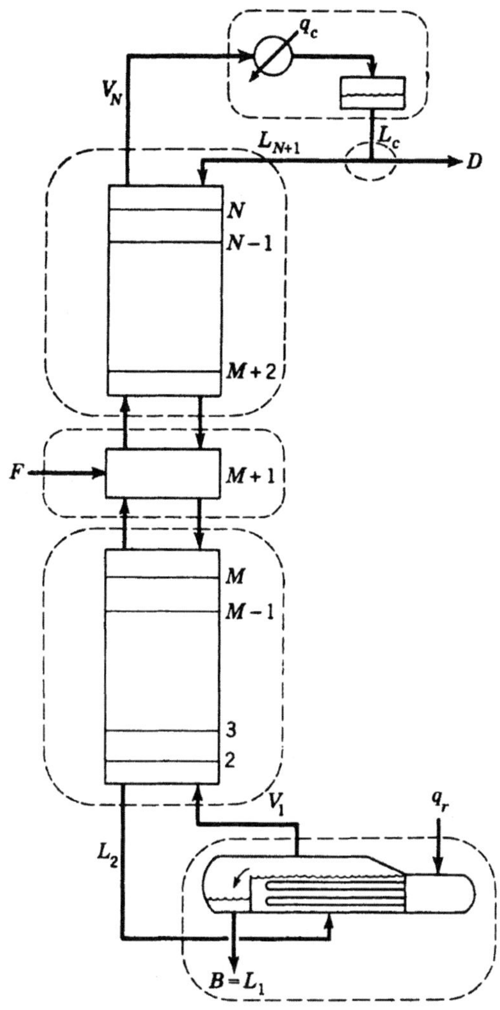

FIG. 13-59 Distillation column with one feed, a total condenser, and a partial reboiler.

One set of specifications that is particularly convenient for computer solutions is the following:

Specifications	N_i^u
Pressure of either stream leaving each stage (including reboiler)	N
Pressure of stream leaving condenser	1
Pressure of either stream leaving reflux divider	1
Heat leak for each stage (excluding reboiler)	$N - 1$
Heat leak for reflux divider	1
Feed stream	$C + 2$
Reflux temperature	1
Total number of stages N	1
Number of stages below feed M	1
Distillate rate D/F	1
Reflux ratio L_{N+1}/D	1
	$C + 2N + 9$

Other specifications often used in place of one or more of the last four listed are the fractional recovery of one component in either D or B and/or the composition of one component in either D or B.

OTHER UNITS AND COMPLEX PROCESSES

In Table 13-11, the number of design variables is summarized for several distillation-type separation operations, most of which are shown in Fig. 13-2. For columns not shown in Figs. 13-1 or 13-2 that involve additional feeds or sidestreams, add $C + 3$ degrees of freedom for each additional feed ($C + 2$ to define the feed and 1 to designate the feed stage) and 2 degrees of freedom for each sidestream (1 for the sidestream flow rate and 1 to designate the sidestream-stage location). Any number of elements or units can be combined to form complex processes. No new rules beyond those developed earlier are necessary for the analysis. Further examples are given in Henley and Seader (1981). An alternative method for determining the degrees of freedom for equipment and processes is given by Pham [*Chem. Eng. Sci.* **49:** 2507 (1994)].

TABLE 13-11 Design Variables N_i^u for Separation Units

Unit	N_i^u*
Distillation (partial reboiler-total condenser)	$C + 2N + 9$
Distillation (partial reboiler-partial condenser)	$C + 2N + 6$
Absorption	$2C + 2N + 5$
Rectification (partial condenser)	$C + 2N + 3$
Stripping	$2C + 2N + 5$
Reboiled stripping (partial reboiler)	$C + 2N + 3$
Reboiled absorption (partial reboiler)	$2C + 2N + 6$
Refluxed stripping (total condenser)	$2C + 2N + 9$
Extractive distillation (partial reboiler–total condenser)	$2C + 2N + 12$

*N includes reboiler, but not condenser.

DISTILLATION SYSTEMS

Distillation systems for the separation of nonazeotropic mixtures are discussed in this subsection. Many of the results extend also to azeotropic mixtures when the desired splits do not try to break azeotropes or cross a distillation boundary.

Whenever we wish to separate a mixture into multiple products, various combinatorial possibilities of column arrangements are feasible, and the optimal (usually the least expensive) column configurations are sought. For example, there are at least two possible ways of separating a ternary mixture of components A, B, and C into pure-product streams (where A is the most volatile and C the least volatile component):

1. *Direct split,* where component A is separated from BC first (A/BC split) and then mixture BC is distilled to separate B from C
2. *Indirect split,* where C is removed first (AB/C) and then mixture AB is distilled

The difference in energy required for these splits can be assessed by simply comparing the total minimum vapor flows (summed over all the columns) for each column sequence. Example calculations were performed for a mixture with relative volatilities $\alpha_A = 4$, $\alpha_B = 2$, and $\alpha_C = 1$ containing 90 percent of A, 5 percent of B, and 5 percent of C in the feed stream. The indirect split requires 58 percent more energy than the direct split, assuming that columns are connected by a liquid stream in both cases. Moreover, the system using more energy requires bigger heat exchangers and larger column diameters, which increase the capital investment costs. Therefore, the direct split configuration would clearly be a better choice in this case.

One of the most important factors that determines the column configuration is the formulation (or goals) of the separation task with respect to the total flow sheet. Although a mixture may consist of C components, it does not mean that all C products are necessary. The components contained in streams recycled into the process (e.g., unreacted reactants recycled to the reactor) usually do not have to be separated from each other. Also, separation of streams that are later mixed (blended) should be avoided, if possible. The separation system needs to be optimized together with the entire plant, either simultaneously or in a hierarchical approach, as described by Douglas (*The Conceptual Design of Chemical Processes,* McGraw-Hill, New York, 1988).

Heuristic methods have been widely used for the synthesis of distillation sequences to avoid lengthy calculations. Many heuristics are intuitive, such as "Remove corrosive components first," "Remove the most plentiful components first," or "Remove the lightest component first." Since heuristics are just rules of thumb, they sometimes conflict with one another, and they may provide wrong answers even if they are not in conflict among themselves [Malone, Glinos, Marques, and Douglas, *AIChE J.* **31:** 683 (1985)]. More exact, algorithmic synthesis methods and cost optimization should be used in practice. The total energy of separation has been identified as the major component of the total cost. Energy-saving distillation schemes for light hydrocarbons with high levels of flexibility and operability were discussed by Petterson and Wells [*Chem. Eng.* **84**(20): 78, 1977]. Tedder and Rudd presented a three-part paper [*AIChE J.* **24:** 303 (1978)] in which they compared eight distillation systems for the separation of ternary mixtures and determined the regions of economic optimality (with respect to feed compositions and relative volatilities). They proposed several heuristics for comparing various configurations. They evaluated several rank-order functions that allow for a comparison of various configurations without detailed cost calculations. Interestingly, one of the rank-order functions used successfully to compare various systems was the total minimum vapor flow. Minimum vapor flow will also be used to compare various distillation systems presented here. Finally, in the third part of their paper, Tedder and Rudd proposed a design method for various column networks.

In other approaches, Gomez-Munoz and Seader [*Comp. & Chem. Eng.* **9:** 311 (1985)] proposed an optimization algorithm based on maximum thermodynamic efficiency. However, Smith and Linhoff [*Chem. Eng. Res. Des.* **66:** 195 (1988)] pointed out the importance of a simultaneous design of the separation network and the rest of the process. They developed a pinch method based on the temperature–heat duty diagram to accomplish this task. Nishida, Stephanopoulos, and Westerberg [*AIChE J.* **27:** 321 (1981)] presented a comprehensive review of process synthesis, including distillation. Westerberg [*Comp. & Chem. Eng.* **9:** 421 (1985)] discussed methods for the synthesis of distillation systems that include sharp splits, nonsharp splits, thermal linking, and heat integration. Agrawal [*AIChE J.* **49:** 379 (2003)] presented a method for systematic generation of distillation configurations, including conventional and complex columns. Sorting through the alternatives and selecting the low-cost systems can be quite tedious, and it is best done with computer-aided optimization strategies such as those proposed by Caballero and Grossmann [*Ind. Eng. Chem. Res.* **40:** 2260 (2001)]. Giridhar and Agrawal [*Comp. & Chem. Eng.* **34:** 73 (2010)] divided possible column configurations (for the separation of zeotropic *C*-component mixtures into almost pure components) into two groups: (1) basic configurations, having $(C-1)$ columns; (2) nonbasic configurations, having more than $(C-1)$ columns, where the same product is made from more than one place. They calculated minimum energy requirements and observed that basic configurations always contain the column sequence with the lowest heat duty. Therefore they concluded that nonbasic configurations are not the most energy efficient and should be eliminated from the search space, thus reducing the number of possible column sequences. More recently, Shah and Agrawal [*AIChE J.* **56:** 1759 (2010)] developed a simple-to-use matrix method for generating all basic configurations. They applied it to identify more than 70 new column configurations that can potentially have lower heat duty than existing petroleum crude distillation.

POSSIBLE CONFIGURATIONS OF DISTILLATION COLUMNS

This subsection describes one possible way of generating the feasible configurations of distillation columns for separation of mixtures that do not form azeotropes. Components are named A, B, C, D, . . . , and they are listed in the order of decreasing volatility (or increasing boiling temperature). We limit our considerations to splits where the most volatile (lightest) component and the least volatile (heaviest) component do not distribute between the top and bottom products. For simplicity, we consider only separations where final products are relatively pure components. Systems containing simultaneously simple and complex distillation columns are considered. Simple columns are the conventional columns with one feed stream and two product streams; complex columns have multiple feeds and/or multiple product streams.

The combinatorial possibilities for separating a three-component mixture ABC into two product streams in which the most volatile component and the least volatile component do not distribute between the top and bottom products are

1. A/BC—Top product A is separated from bottom product BC.
2. AB/BC—Component B distributes between both product streams.
3. AB/C—Top product AB is separated from bottoms C.

These separations are also often referred to as splits; sharp splits (none of the components distribute) in cases 1 and 3, and a nonsharp split in case 2. When we add binary separations, the column configuration corresponding to split 1 is known as the direct split; the column configuration corresponding to split 2 is called the prefractionator system, and configuration 3 is called the indirect split (see Fig. 13-60). In the prefractionator system, the binary columns separating components AB and BC may be stacked together, forming one column with three products A, B, and C as indicated by the dashed envelope in Fig. 13-60*b*. These systems are called here the *basic column configurations*. Basic column configurations are configurations in which the types of interconnecting streams are not defined. (Note that this is a different definition of basic configuration than the one used by Giridhar and Agrawal [*Comp. & Chem. Eng.* **34:** 73 (2010)].) The arrows on the flow sheets symbolize the net material flow, but the types of streams connecting the columns (liquid, vapor, two-phase, multiple streams) are not specified. Reboilers and condensers are deliberately not shown in Fig. 13-60 because for some types of interconnecting streams they are not necessary.

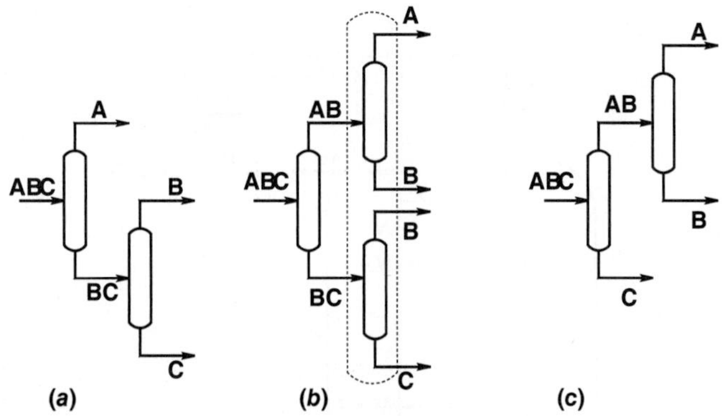

FIG. 13-60 Basic column configurations. (*a*) Direct split. (*b*) Prefractionator system. (*c*) Indirect split.

Symbolic-network representations of these separation systems, called *state-task networks* (STNs), are shown in Fig. 13-61. In this representation, the states (feeds, intermediate mixtures, and products) are represented by the nodes (ABC, AB, BC, A, B, C) in the network, and the tasks (separations) are depicted as lines (1, 2, . . . , 6) connecting the nodes, where arrows denote the net flow of material. This STN representation was used by Sargent to represent distillation systems [*Comp. & Chem. Eng.* **22:** 31 (1998)] and has been widely used ever since. Originally STNs were introduced by Kondili, Pantelides, and Sargent [*Comp. & Chem. Eng.* **17:** 211 (1993)] for representing batch processes.

Interconnecting streams may be liquids, vapors, or two-phase mixtures. The total energy of separation can be minimized for the direct split if the two columns have a liquid connection; for the indirect split, it is minimized if the columns have a vapor connection, and for the prefractionator system if the top connection is vapor and the bottom connection is liquid. See Fidkowski and Krolikowski [*AIChE J.* **33:** 643 (1987); **36:** 1275 (1990)].

A complete direct split configuration, including reboilers and condensers, is shown in Fig. 13-62*a*. The numbers used to represent the column sections in this figure correspond to the numbers used to represent the tasks in the STN (Fig. 13-61*a*). By eliminating the reboiler from the first column in Fig. 13-62*a* and supplying the boil-up from the second column, we obtain the side rectifier arrangement, shown in Fig. 13-62*b*. An alternative side rectifier arrangement is seen more clearly in Fig. 13-62*c*, where the stripping section of the binary BC column (section 4) has been moved and lumped with the first column. Note that all these three configurations can be represented by one state-task network, shown in Fig. 13-61*a*. In each configuration in Fig. 13-62, feed ABC is separated in section 1 to get component A and in section 2 to get mixture BC. Then mixture BC is further separated in section 3 to give component B and in section 4 to produce C. Therefore, all these three systems are topologically equivalent to the same basic column configuration represented by the STN in Fig. 13-61*a* and the column arrangement in Fig. 13-60*a*. However, if we take into account reboilers and condensers, we see that only the two side rectifier configurations (Fig. 13-62*b* and *c*) are topologically equivalent. Side stripper configurations can be obtained from the indirect split in an analogous way (Fig. 13-63*a*, *b*, and *c*).

By eliminating the reboiler and condenser in the prefractionator column in Fig. 13-64*a* (the column containing sections 1 and 2), we obtain a *thermally coupled system*, also known as a Petlyuk system, shown in Fig. 13-64*b* [Petlyuk, Platonov, and Slavinskii, *Int. Chem. Eng.* **5:** 555 (1965)]. Side stripper, side rectifier, and Petlyuk systems can also be built as divided wall columns, as explained in detail in the following subsection on thermally coupled systems.

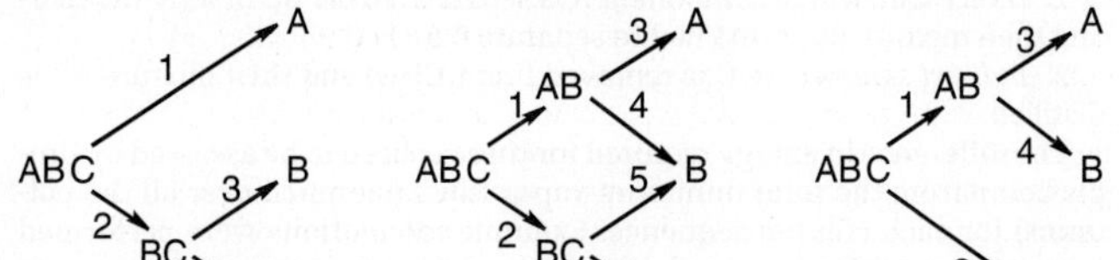

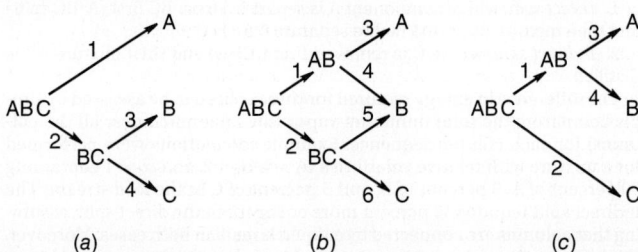

FIG. 13-61 State-task networks. (*a*) Direct split. (*b*) Prefractionator system. (*c*) Indirect split.

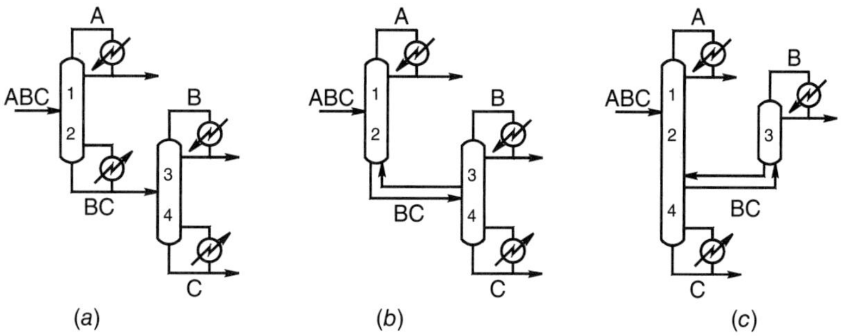

FIG. 13-62 (*a*) Direct split configuration. (*b*), (*c*) Side rectifier configurations.

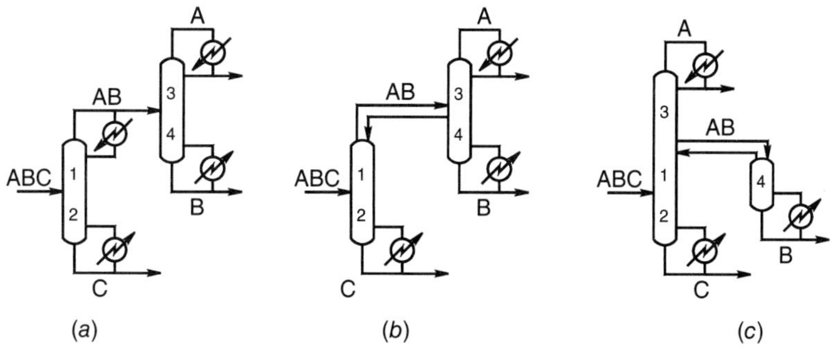

FIG. 13-63 (*a*) Indirect split configuration. (*b*), (*c*) Side stripper configurations.

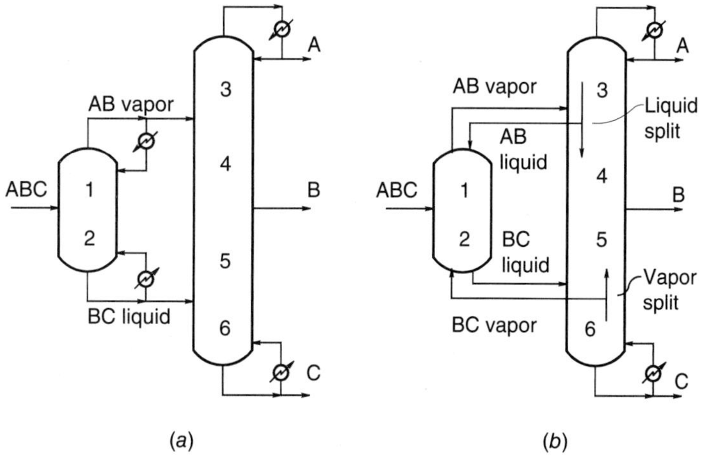

FIG. 13-64 (*a*) Prefractionator system. (*b*) Thermally coupled system.

There are many other possible ternary column systems that have different interconnecting streams between the columns. For example, if we eliminate only the reboiler (or only the condenser) from the prefractionator column in Fig. 13-64*a*, we obtain a partially thermally coupled system [Agrawal and Fidkowski, *AIChE J.* **45:** 485 (1999), U.S. Patent 5,970,742]. In other instances, one may significantly increase the thermodynamic efficiency of the direct split and the indirect split if a portion of the interconnecting stream is vaporized and fed to the second column below the liquid connection [Agrawal and Fidkowski, *Ind. Eng. Chem. Res.* **38:** 2065 (1999)].

There are six possible types of splits for a quaternary mixture. The ternary mixtures resulting from these splits may be separated in one of the three possible ternary splits, as described previously. Table 13-12 summarizes the resulting number of possible basic column configurations. This number does not account for the various possible types of interconnecting streams.

This method of generating various column configurations [Fidkowski, *AIChE J.* **52:** 2098 (2006)] is very similar to the methods used previously for conventional systems; see, for example, Rathore, Van Wormer, and Powers [*AIChE J.* **20:** 491 (1974); **20:** 940 (1974)]. The only difference is that in addition to sharp splits, the nonsharp splits are included here, which leads to unconventional systems. Similar methods were proposed by Rong et al. [Rong, Kraslawski, and Turunen, *Ind. Eng. Chem. Res.* **42:** 1204 (2003)] to generate all the possible quaternary thermally coupled configurations.

The basic column configurations for all 22 quaternary distillation systems are shown in Fig. 13-65. Five of these configurations consist of only sharp splits; hence each species appears in only one product stream. Seventeen of the configurations have at least one nonsharp split, which results in each distributing component appearing in product streams from two different locations. Again, the interconnecting streams could be liquid,

TABLE 13-12 Number of Basic Column Configurations for Separation of a Four-Component Mixture

Split	Distillate/bottoms	Number of configurations
1	A/BCD	3
2	AB/BCD	3
3	AB/CD	1
4	ABC/BCD	$3 \times 3 = 9$
5	ABC/CD	3
6	ABC/D	3
	Total	22

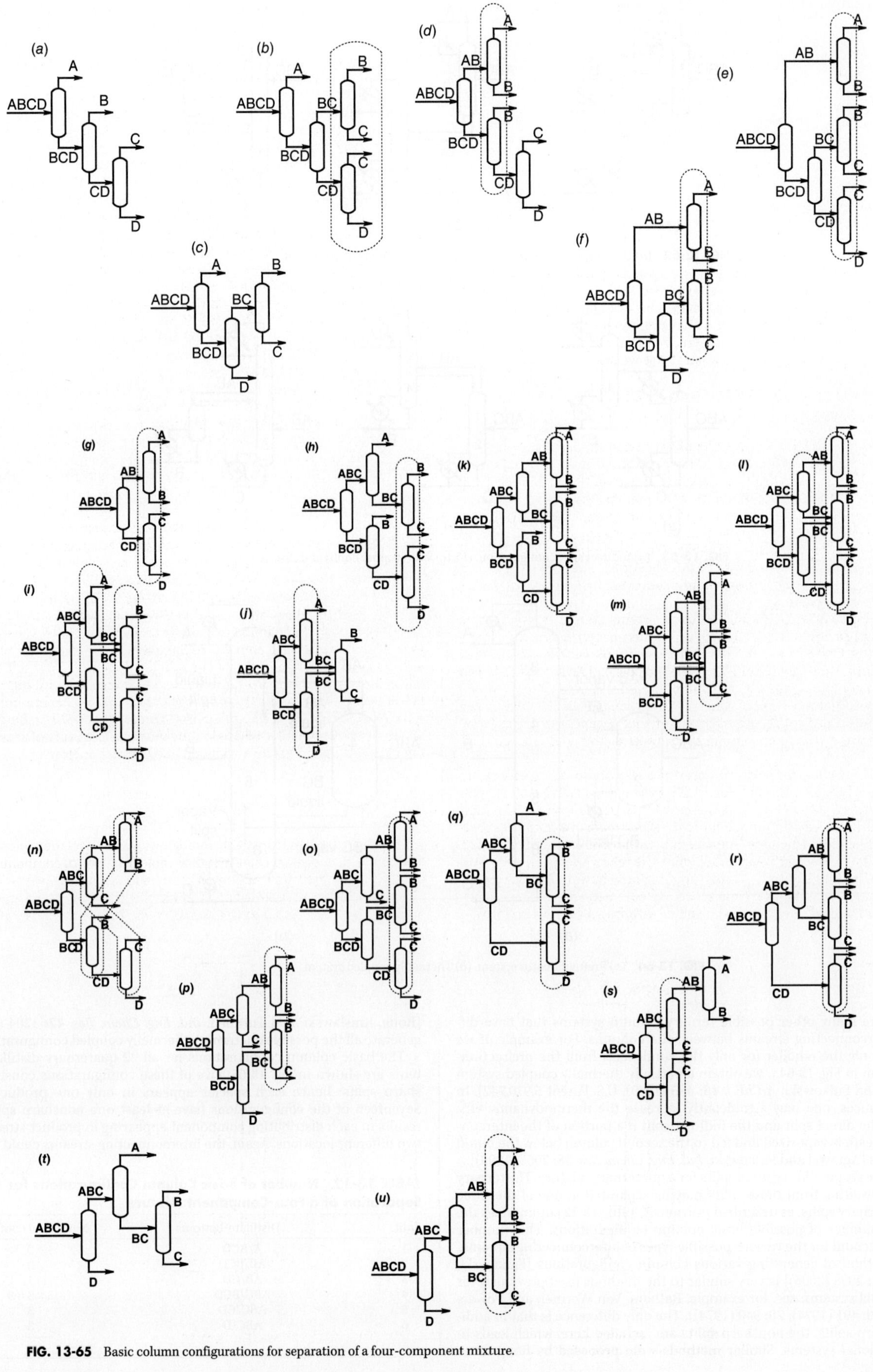

FIG. 13-65 Basic column configurations for separation of a four-component mixture.

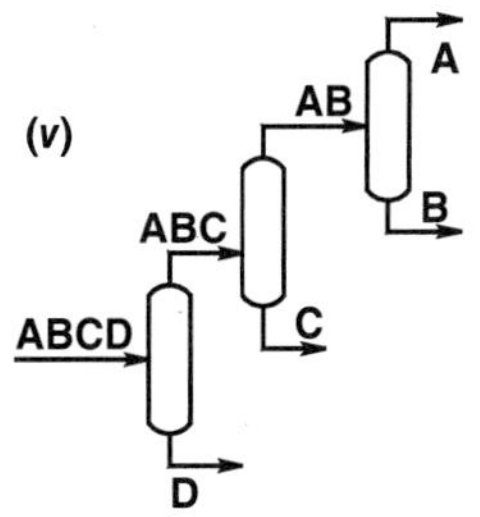

FIG. 13-65 (*Continued*)

vapor, two-phase, or two-way liquid and vapor, as in thermally linked columns. There might also be several alternatives for column stacking, which increases the number of possible configurations even further. Subsequently, one can analyze possible splits for a five-component mixture, and so on. These splits create quaternary and ternary products that can be further separated by one of the column configurations we have discussed.

THERMALLY COUPLED SYSTEMS AND DIVIDING WALL COLUMNS

In recent years there has been significant interest in thermally coupled systems and dividing wall columns for ternary mixtures. In this subsection we discuss such column arrangements, their energy requirements, design and optimization methods, controllability and operability, experimental and industrial experience, and extension to more than three components.

Two columns are thermally coupled if a vapor (liquid) stream is sent from the first column to the second column and then a return liquid (vapor) stream is implemented between the same locations. These streams, when introduced at the top or bottom of a column, provide (at least partial) reflux or boil-up to this column.

The development of thermally coupled systems started with attempts to find energy-saving schemes for the separation of ternary mixtures into three products. One of the first industrial applications was the side rectifier configuration for air separation. The side stripper configuration followed naturally. By combining the two, we obtain the fully thermally coupled system of Petlyuk, Platonov, and Slavinskii [*Int. Chem. Eng.* **5:** 555 (1965)]; see Fig. 13-64*b*. It consists of the prefractionator that accepts the ternary feed stream followed by the main column that produces the products (product column).

The dividing wall column was invented as a way of producing three pure products from a single column [Monro, U.S. Patent 2,134,882 (1933); Wright, U.S. Patent 2,471,134 (1949)]. In some cases it is possible to achieve high purity of the intermediate component in the sidestream of a distillation column. This is possible when the sidestream is withdrawn above the feed as a liquid or below the feed as a vapor and when the relative volatilities of components differ significantly. In many applications, however, a sidestream is contaminated to an appreciable extent by either the light or the heavy component. For example, if a sidestream is withdrawn from the rectifying section, it must contain not only the intermediate component but also some of the most volatile component. This contamination problem can be eliminated by adding a dividing wall that prevents the most volatile component (and the heaviest component) from entering the zone where the intermediate component is withdrawn; see Fig. 13-66.

The dividing wall column is topologically equivalent to the fully thermally coupled system (Fig. 13-64*b*). The prefractionator and the main product column are built in one shell, separated by a dividing wall (Fig. 13-66*c*). Similarly, the side rectifier (Fig. 13-62*c*) and side stripper (Fig. 13-63*c*) configurations can be built in one shell as dividing wall columns [Agrawal, *Ind. Eng. Chem. Res.* **40:** 4258 (2001)]. The corresponding dividing wall columns are shown in Fig. 13-66*a* and *b*, respectively. The corresponding sections in the thermally coupled systems and the dividing wall columns have exactly the same numbers and perform exactly the same separation tasks, indicating that the column arrangements are topologically equivalent.

Various design, simulation, and evaluation methods have been developed for the distillation systems shown in Figs. 13-62 to 13-64. These include those by Stupin and Lockhart [*Chem. Eng. Prog.* **68**(10): 71 (1972)]; Fidkowski and Krolikowski [*AIChE J.* **32:** 537 (1986)]; Nikolaides and Malone [*Ind. Eng. Chem. Res.* **27:** 811 (1988)]; Rudd [Distillation Supplement to *The Chemical Engineer*, S14 (1992)]; Triantafyllou and Smith [*Trans. Inst. Chem. Eng.* **70**(Part A): 118 (1992)]; Finn [*Gas Sep. Purif.* **10:** 169 (1996)]; Annakou and Mizsey [*Ind. Eng. Chem. Eng.* **35:** 1877 (1996)]; Hernandez and Jimenez [*Comp. & Chem. Eng.* **23:** 1005 (1999)]; Dunnebier and Pantelides [*Ind. Eng. Chem. Res.* **38:** 162 (1999)]; Watzdorf, Bausa, and Marquardt [*AIChE J.* **45:** 1615 (1999)]; and Kim [*J. Chem. Eng. Japan*, **34:** 236 (2001)]. The fully thermally coupled system uses less energy than any other ternary column configuration [Fidkowski and Krolikowski, *AIChE J.* **33:** 643–653 (1987)]. The energy savings may be on the order of 30 to 50 percent, depending on the feed composition and volatilities of the components. Similar energy savings are possible in partially thermally coupled columns, where only one connection (top or bottom) between the columns is thermally coupled and the other is just a single liquid or vapor stream together with an associated condenser or reboiler, respectively [Agrawal and Fidkowski, *AIChE J.* **45:** 485 (1999)]. Fidkowski and Krolikowski [*AIChE J.* **32:** 537 (1986)] solved analytically the optimization problem for the minimum vapor boil-up rate from the reboiler in the main column for the fully thermally coupled system (shown in Fig. 13-64*b*), assuming constant molar overflow and constant relative volatilities among the components. The solution depends on the splits of vapor and liquid between the main product column and the prefractionator [or between both sides of the dividing wall (Fig. 13-66*c*)]. The minimum vapor flow for each column in the system is shown in Fig. 13-67 as a function of β, where the parameter β is defined as the fractional recovery of the distributing component B in the top product of the prefractionator

$$\beta = \frac{V_1 y_B - L_1 x_B}{f_B} \tag{13-110}$$

where V_1 and L_1 are the vapor and liquid flows at the top of the prefractionator (section 1), y_B and x_B are corresponding mole fractions of component B, and f_B denotes the molar flow of component B in the feed stream. The lower line in Fig. 13-67 is the minimum vapor flow in the first column, the prefractionator. At $\beta = 0$, it corresponds to the A/BC split and at $\beta = 1$ to the AB/C split.

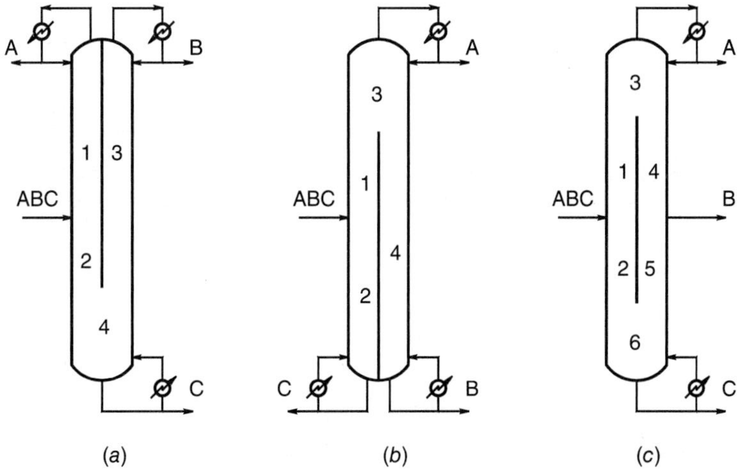

FIG. 13-66 Dividing wall columns equivalent to (*a*) side rectifier configuration, (*b*) side stripper configuration, (*c*) thermally coupled system.

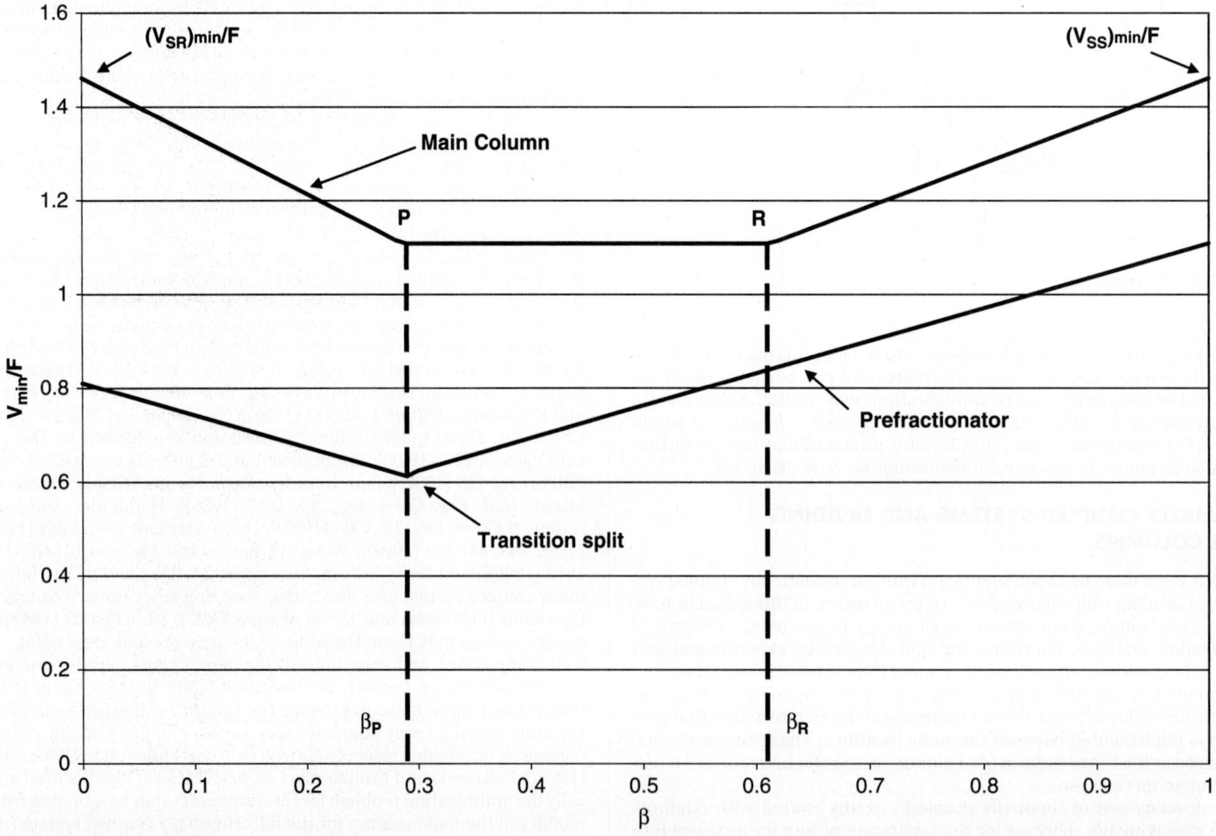

FIG. 13-67 Minimum vapor flows in the thermally coupled system; bottom curve, minimum vapor flow in prefractionator; top curve, minimum vapor flow from the reboiler of the main column (total minimum vapor flow for entire system) for $\alpha_A = 6.25$, $\alpha_B = 2.5$, $\alpha_C = 1.0$ and feed mole fractions $z_A = 0.33$, $z_B = 0.33$, and $z_C = 0.34$.

There is a minimum in the minimum vapor flow in the prefractionator for $\beta = \beta_P$, the so-called transition split; β_P can be calculated as

$$\beta_P = \frac{\alpha_B - \alpha_C}{\alpha_A - \alpha_C} \tag{13-111}$$

where the α's are relative to any reference component.

The transition split divides direct-type splits from indirect-type splits as discussed by Doherty and Malone (*Conceptual Design of Distillation Systems*, 2001, chaps. 4 and 5); also see Fidkowski, Doherty, and Malone [*AIChE J.* **39:** 1301(1993)]. The upper line in Fig. 13-67 is the minimum vapor flow leaving the reboiler of the main column, which also corresponds to the minimum vapor flow for the entire system since all the vapor for the total system is generated by this reboiler. For $\beta = 0$, the minimum vapor flow for the entire thermally coupled system (i.e., main column) becomes equal to the minimum vapor flow for the side rectifier system (i.e., main column of the side-rectifier system; see Fig. 13-62*b* or *c*) $(V_{SR})_{\min}$; for $\beta = 1$ it is equal to the minimum vapor flow of the entire side stripper system $(V_{SS})_{\min}$ (which is the sum of the vapor flows from both reboilers in this system; see Fig. 13-63*b* or *c*). Coincidentally, the values of these two minimum vapor flows are always the same: $(V_{SR})_{\min} = (V_{SS})_{\min}$. For $\beta = \beta_R$ the main column is pinched at both feed locations; that is, the minimum vapor flows for separations A/B and B/C are equal.

The minimum vapor flow for the entire thermally coupled system is flat over a wide range of β: $\beta_P \le \beta \le \beta_R$. This is the reason why dividing wall columns usually work well without tight control of the vapor or liquid split between both sides of the partition. The optimally designed fully thermally coupled system should operate with a fractional recovery of B in the top product of the prefractionator placed somewhere between points *P* and *R*. The transition split *P* is located at one end of the optimal section *PR*, and it is not a recommended design point for normal operation because process disturbances may move the operating point outside the optimal section *PR* shown in Fig. 13-67.

Although invented long ago, dividing wall columns and fully thermally coupled distillation systems were not implemented in practice until the late 1980s. The major objections concerned controllability and operability. More recently, however, several papers have shown that control of these systems is possible; see Hernandez and Jimenez [*Ind. Eng. Chem. Res.* **38:** 3957 (1999)]; Mutalib, Zeglam, and Smith [*Trans. Inst. Chem. Eng.* **76:** 319 (1998)]; and Halvorsen and Skogestad [*Comp. Chem. Eng.* **21,** Suppl., S249 (1997)]. A major control problem was attributed to an inability to set the vapor split between the main column and prefractionator due to conflicting pressure drop requirements between sections 6 and 2, as well as sections 1 and 3 in Fig. 13-64*b*. For the BC vapor to flow from section 6 to section 2, the pressure in the main column at the top of section 6 must be *higher* than the pressure in the prefractionator at the bottom of section 2. But the pressure at the bottom of section 3 in the main column must be *lower* than the pressure at the top of section 1 in the prefractionator, or else the AB vapor stream will not flow from the prefractionator to the main column. Liquid split control is easier to realize in practice by using liquid collectors, overflows, or pumps. How much vapor flows straight up the main column and how much vapor splits off to the prefractionator depends on the pressure drops in the middle sections of the main column (sections 4 and 5 in Fig. 13-64*b*, between the interconnecting streams) and in the prefractionator. These pressure drops depend on the height of these sections, the type of packing or stages, and liquid flows. Therefore, these pressure drops cannot be easily controlled in the configuration shown in Fig. 13-64*b;* moreover, they may even be such that vapor flows in the wrong direction.

Agrawal and Fidkowski [*AIChE J.* **44:** 2565 (1998); U.S. Patent 6,106,674] proposed robustly operable two-column configurations that cleverly overcome this design and control problem. One is shown in Fig. 13-68. This new configuration is topologically equivalent to the original thermally coupled configuration in Fig. 13-64*b* and retains its energy advantage. In the new Agrawal and Fidkowski configuration, column section 6 is just shifted from the product column to the bottom of the prefractionator; thus the two interconnecting vapor streams flow in the *same direction.* The first column (with sections 1, 2, and 6) in Fig. 13-68 operates at a slightly higher pressure than the second column, and the relative flows of the vapor streams can be changed by using a valve on one of them, as shown in the figure. The Agrawal and Fidkowski thermally coupled systems are expected to have higher investment cost than dividing wall columns, and the same investment cost as a Petlyuk thermally coupled system. However, for certain feed compositions and volatilities, the

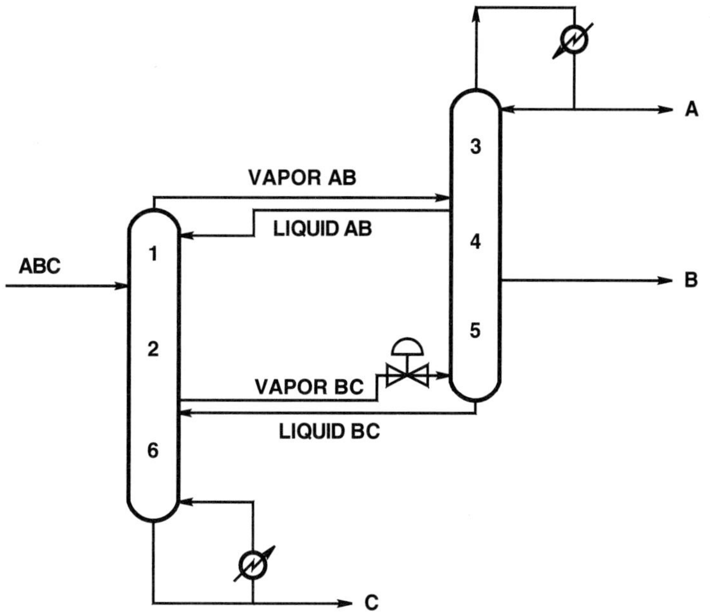

FIG. 13-68 Agrawal and Fidkowski thermally coupled system (topologically equivalent to the Petlyuk system shown in Fig. 13-64*b*).

energy optimum in Fig. 13-67 may be narrow (i.e., the interval *PR* may be short), and dividing wall columns may not be able to operate at the optimum. On the other hand, the Agrawal and Fidkowski configurations can operate at the optimum because of better control. Also, the Agrawal and Fidkowski configurations are useful in cases where high-purity products (especially B) are required. This is so because the two columns are built in separate shells that may contain more stages in one of the shells than could be accommodated in the corresponding side of the dividing wall column.

Experimental tests of dividing wall columns were carried out by Mutalib, Zeglam, and Smith [*Trans. Inst. Chem. Eng.* **76:** 319 (1998)]. Today there are about 60 dividing wall columns in operation; 42 are owned by BASF [Parkinson, *CEP* (2005, July), p. 10].

Thermally coupled systems can also be devised for multicomponent mixtures. Sargent and Gaminibandara (in *Optimization in Action*, ed. L. W. C. Dixon, Academic Press, London, 1976, p. 267) presented a natural extension of the Petlyuk column sequence to multicomponent systems. Agrawal [*Ind. Eng. Chem. Res.* **35:** 1059 (1996); *Trans. Inst. Chem. Eng.* **78:** 454 (2000)] presented a method for generating an even more complete superstructure from which all the known column configurations (including thermally coupled systems) can be derived. Fidkowski and Agrawal [*AIChE J.* **47:** 2713 (2001)] presented a method for calculating the minimum vapor flows in multicomponent thermally coupled systems. They analyzed the quaternary fully thermally coupled system in detail (there are many equivalent column configurations, all with the minimum number of column sections, which is 10, as well as one reboiler and one condenser; one of the configurations is shown in Fig. 13-69). They showed that one of the optimum solutions (with the minimum value of the total minimum vapor flow rate from the single reboiler in the system) occurs when the quaternary feed column (far left column in Fig. 13-69) and both ternary columns (the two middle columns in Fig. 13-69) perform transition splits. They also concluded that the optimized quaternary fully coupled system always requires less energy than the five sharp-split conventional systems (where each column performs a sharp split and has one feed, two products, one reboiler, and one condenser). The basic configurations for these five sharp-split systems are shown in Fig. 13-65*a, c, g, t,* and *v*. Selecting the best thermally coupled column configuration can be tedious without computer-aided tools

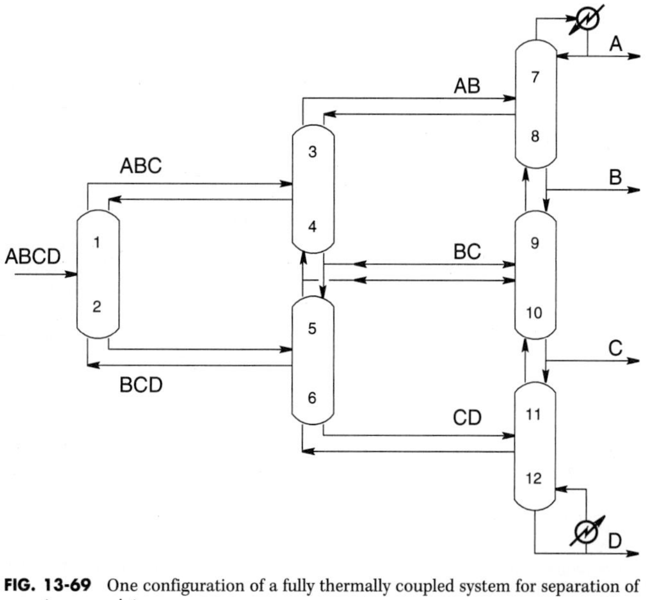

FIG. 13-69 One configuration of a fully thermally coupled system for separation of a quaternary mixture.

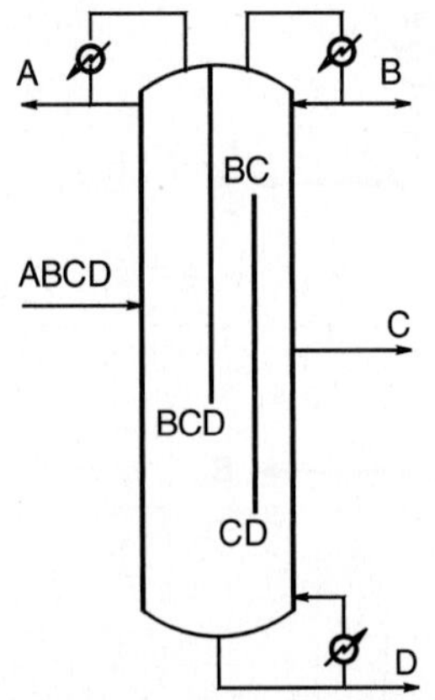

FIG. 13-70 One possible dividing wall column for separation of a quaternary mixture.

such as the disjunctive programming approach developed by Caballero and Grossmann [*Ind. Eng. Chem. Res.* **40:** 2260 (2001)].

All the multicomponent thermally coupled configurations have a corresponding dividing wall column equivalent. Keibel [*Chem. Eng. Technol.* **10:** 92 (1987)] has shown examples of columns with multiple dividing walls, separating three, four, and six components. Agrawal [*Ind. Eng. Chem. Res.* **40:** 4258 (2001)] presented several examples of quaternary columns with partitions and multiple reboilers and condensers. One of these examples is shown in Fig. 13-70.

THERMODYNAMIC EFFICIENCY

Thermodynamic efficiency can be a useful figure of merit (in place of total cost, or total vapor rate) for comparing alternative column configurations. This is especially true for cryogenic distillations where very low temperatures are necessary and where highly efficient "cold box" designs are needed to achieve them. The thermodynamic efficiency of thermally coupled and other distillation systems for the separation of ternary mixtures was analyzed by Agrawal and Fidkowski [*Ind. Eng. Chem. Res.* **37:** 3444 (1998)]. Feed composition regions for column configurations with the highest thermodynamic efficiency are shown in Fig. 13-71. Often, the efficiency of the direct split or the indirect split is better than the efficiency of thermally coupled systems. This is primarily due to the ability of these configurations to accept

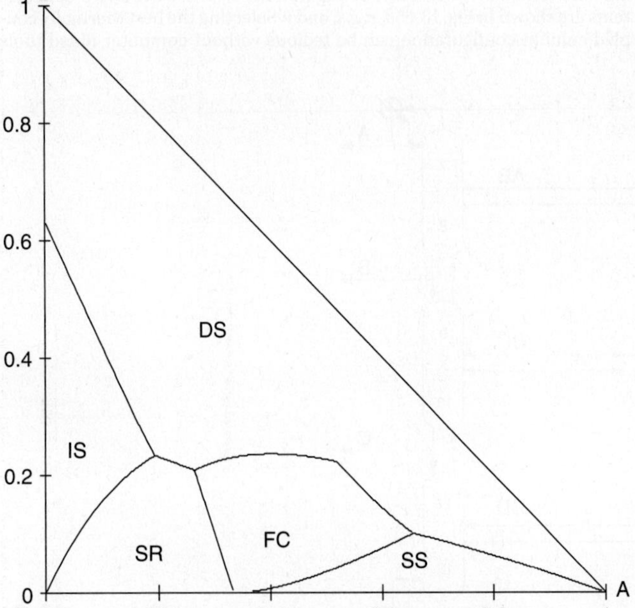

FIG. 13-71 Feed composition regions of column configurations with highest thermodynamic efficiency; DS—direct split, IS—indirect split, SR—side rectifier, SS—side stripper, FC—fully thermally coupled. Example for $\alpha_A = 4.0$, $\alpha_B = 2.0$, and $\alpha_C = 1.0$.

or reject heat at the intermediate boiling or condensing temperatures of the binary submixtures. The fully thermally coupled system can only accept heat at the temperature of the highest-boiling component (boiling point of C) and reject heat at the lowest temperature (condensation temperature of A). This conclusion gave rise to new, more thermodynamically efficient thermally coupled configurations, as discussed by Agrawal and Fidkowski [*Ind. Eng. Chem. Res.* **38:** 2065 (1999); U.S. Patent 6,116,051].

HEAT INTEGRATION

In this subsection we describe heat pumps, the multieffect distillation of binary mixtures, the synthesis of multicomponent distillation systems with heat integration, and multieffect distillation for thermally coupled configurations.

Two columns are heat-integrated when they exchange heat indirectly through a heat exchanger. This is different than in thermally linked configurations, where the heat exchange between columns is direct, through the material stream connecting the columns. The objective of heat integration in distillation systems is to save energy. Heat integration is realized by matching heat sources (usually condensers) with heat sinks (usually reboilers). The other heat exchangers considered for heat integration might be feed preheaters and product coolers. Typical examples of heat integration schemes are heat transfer from a condenser to a reboiler, or heat exchange between a (hot) column feed and a reboiler. If the temperature of the heat source (condenser) is higher than the temperature of the heat sink (reboiler), the opportunity for the match is straightforward. If the condenser temperature is too low, one may increase the condensing pressure or use a heat pump.

For example, the reboiler and condenser from the same column may be heat-integrated by using a heat pump, as discussed by Null [*Chem. Eng. Prog.* **72**(7): 58 (1976)]. A heat pump may use an external fluid that is vaporized in the condenser, then compressed and condensed in the column reboiler (Fig. 13-72*a*). Another heat pump (Fig. 13-72*b*) uses column overhead vapor, which is compressed and condensed in the reboiler and then returned to the top of the column as reflux. A third possibility is to use the column bottoms, which is let down in pressure and vaporized in the condenser, then compressed and fed back to the bottom of the column as boil-up (Fig. 13-72*c*).

The entire rectifying section can be pressurized, and the heat can be transferred between any desired stages of the rectifying and stripping sections. This is called *secondary reflux and vaporization* (SRV) distillation. It reduces the consumption of both hot and cold utilities and reduces the sizes of the reboiler and condenser. However, capital cost is increased by additional intermediate heat exchangers; moreover, since the process is more thermodynamically reversible, it requires more stages to achieve the desired separation.

In multiple-column systems, possibilities for heat integration may be created by increasing the pressure in one of the columns, to increase the temperature of the condenser. When the temperature of the condenser becomes higher than the temperature of some other column reboiler, it is possible to heat-integrate these streams via a heat exchanger and to reuse the heat rejected from the condenser. However, there are several drawbacks to this procedure. The required heat-transfer area in the integrated exchangers needs to be increased due to smaller temperature differences between the process streams than would normally exist when hot and cold external utilities are used to provide boil-up and condensation. Higher-pressure columns need hotter external heating utilities that are more expensive. Separation in higher-pressure columns is more difficult (because relative volatilities tend to decrease with increasing pressure), and more stages and energy may be required. Finally, higher-pressure column shells and piping may be more expensive, although this is not always certain, since the overall dimensions decrease with pressure.

One of the first industrial applications of heat-integrated distillation was a double column for air separation, developed by Linde in the beginning of the 20th century (Fig. 13-73). Air is compressed (typically to about 6 bar) and fed in the bottom of the high-pressure column. Nitrogen is condensed at the top of the high-pressure column. The heat of condensation provides boil-up to the low-pressure column by vaporizing oxygen in the sump located above the nitrogen condenser. Another industrial example of heat integration occurs in the large-scale methanol-water separation in the production of methanol from synthesis gas; see Siirola [*Adv. Chem. Eng.* **23:** 1 (1996)]. Several alternative designs were evaluated, and the regions of superior cost were developed on volatility-feed composition diagrams. Three heat-integrated designs were better than a single column:

1. *Split feed configuration*, in which the feed is split between the high-pressure and the low-pressure columns
2. *Sloppy first bottom split, second column pressurized*, in which the bottom from the first column is further separated in the high-pressure column
3. *Sloppy first bottom split, first column pressurized*, in which the bottom from the first, high-pressure column is further separated in the low-pressure column

These heat-integrated systems were economically advantageous because of the large feed flow rate. The last configuration was built.

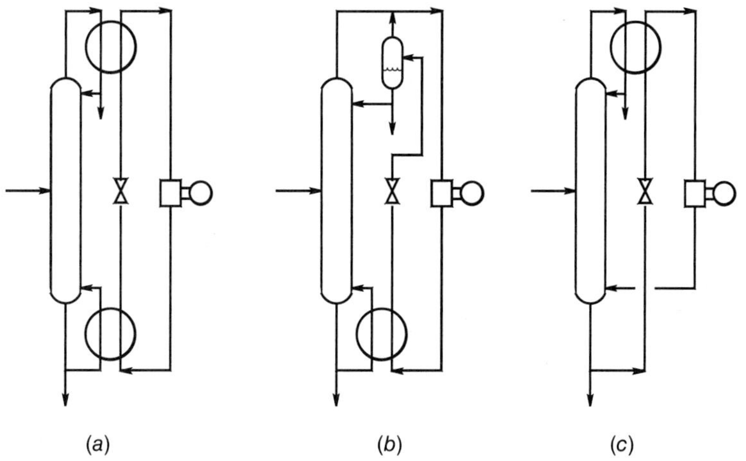

FIG. 13-72 Heat pumps transfer heat from the top to the bottom of the column. (*a*) External fluid heat pump. (*b*) Heat pump using column overhead. (*c*) Heat pump using column bottoms.

Design heuristics and computer-aided design methods for heat-integrated distillation systems have been developed in many publications. Wankat [*Ind. Eng. Chem. Res.* **32:** 894 (1993)] analyzed 23 different multieffect distillation systems obtained by dividing a single column into two or three columns and operating them at various pressures or compositions, to obtain the temperatures necessary for heat integration. He developed heuristics to devise a feasible system. Rathore, Van Wormer, and Powers [*AIChE J.* **20:** 491 (1974) and **20:** 940 (1974)] presented an algorithmic synthesis method for multicomponent separation systems with energy integration. The five-component separation problem was solved by decomposing it into all the possible sharp splits and then examining all the possible integrations of heat exchangers. Thermally coupled systems, however, were not considered. Umeda, Niida, and Shiriko [*AIChE J.* **25:** 423 (1979)] published a thermodynamic approach to heat integration in distillation systems based on pinching heat source and heat sink curves on the temperature–heat duty diagram. Linhoff and coworkers (*A User Guide on Process Integration for the Efficient Use of Energy*, Institution of Chemical Engineers, Rugby, UK, 1982) developed the very successful *pinch technology* approach, commonly used for heat exchanger synthesis and heat integration of entire plants. But, most successfully, Andrechowich and Westerberg [*AIChE J.* **31:** 1461 (1985)] presented a simple conceptual design approach wherein temperature-enthalpy diagrams are used to select the best column stacking. This design method is the recommended starting point for heat integration studies.

Even thermally coupled systems can be heat-integrated, as discovered by Agrawal [*AIChE J.* **46:** 2211 (2000)]. To achieve heat integration without compressors, the vapor connections between the columns must be eliminated. This is shown in Fig. 13-74*a*. The configuration is constructed by extending the prefractionator to a full column with a reboiler and condenser, and adding sections 3x and 6x (analogous to sections 3 and 6 in the product column). By balancing heat duties, only liquid interconnecting streams remain, and the configuration is still thermodynamically equivalent to the original thermally coupled system (Fig. 13-64*b*). The new thermally coupled system with liquid connections only is easier to control than the original thermally coupled system because there is no need to control the internal split of the vapor flow. On a side note, Ramapriya et al. [*AIChE J.* **60:** 2949 (2014)] also developed dividing wall column schemes equivalent to this thermally coupled system with liquid connections only. It is now possible to heat-integrate the columns by pressurizing one of them and allowing heat transfer from the high-pressure condenser to the low-pressure reboiler (Fig. 13-74*b*). The thermally coupled system and dividing wall column are equally energy efficient. Therefore, the thermally coupled system with heat integration is even more energy efficient than the dividing wall column.

There is a significant cost associated with heat integration (e.g., for heat pump compressors, taller columns with more stages, thicker walls of high-pressure equipment, higher cost of high-temperature utilities), which is why such configurations are not widely used. The operational flexibility of a heat-integrated system also becomes more constrained. However, heat-integrated systems can be quite economical for some applications. Typical applications where heat-integrated systems are preferred include

- Cryogenic separation, where very low-temperature cooling utilities do not exist and need to be created, which is very expensive. Therefore, it makes sense to reuse the expensive refrigeration.
- Very large-scale processes, with large feed and product flow rates. Because of the large feed rate, multiple columns may be required to process a given feed, anyway. These cases have the potential to save a lot of energy due to heat integration. Also, pressurizing some equipment in these large-scale processes allows for decreased equipment sizes or increased production rates at a fixed equipment size.

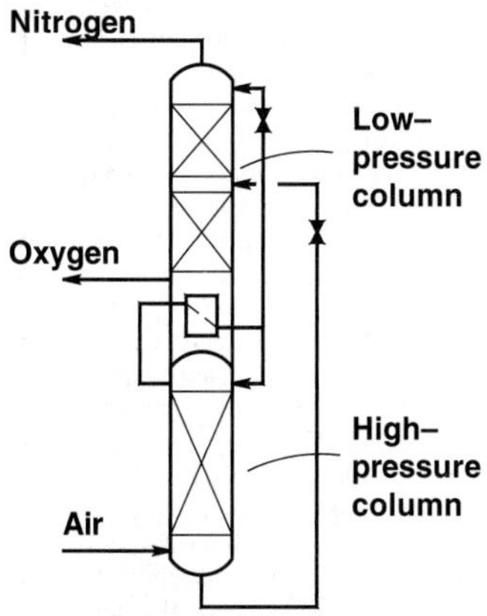

FIG. 13-73 Double-column arrangement for air separation.

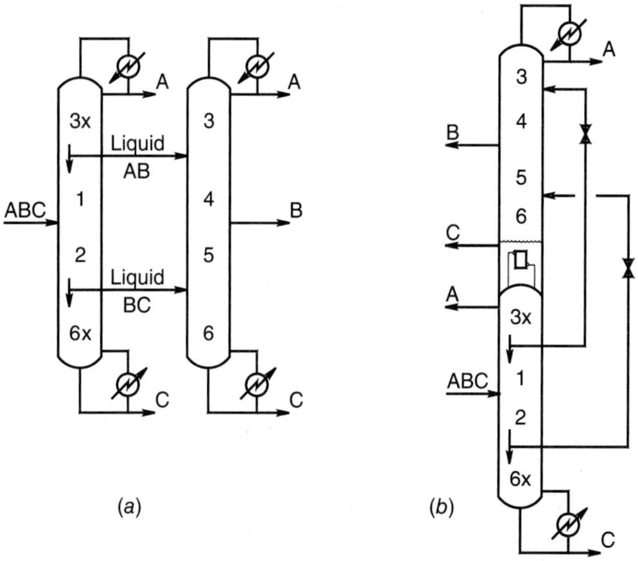

FIG. 13-74 (*a*) Configuration equivalent to thermally coupled system, but without vapor interconnecting streams. (*b*) Heat-integrated thermally coupled system.

IMBALANCED FEEDS

In many practical cases, feed compositions are far from equimolar, with some components present in very small amount (e.g., less than 2 percent). In these cases the top or bottom product flow rate is less than 2 percent of the feed flow rate. The design for imbalanced feeds may include

- Various column diameters
- Intermittent pumping of heavies from the reboiler
- Intermittent firing up a specially devoted column to purify the dirty product made at a small rate
- In a continuous operation, an overrefluxed section (to keep the column diameter the same over the height of the column)
- Producing an impure product to be purified later or discarded

ENHANCED DISTILLATION

In distillation operations, separation results from differences in vapor- and liquid-phase compositions arising from the partial vaporization of a liquid mixture or the partial condensation of a vapor mixture. The vapor phase becomes enriched in the more volatile components, while the liquid phase is depleted of those same components. In many situations, however, the change in composition between the vapor and liquid phases in equilibrium becomes small (so-called pinched condition), and a large number of successive partial vaporizations and partial condensations are required to achieve the desired separation. In some cases, the vapor and liquid phases may have identical compositions because of the formation of an azeotrope, and no separation by simple distillation is possible.

Several enhanced distillation-based separation techniques have been developed for close-boiling or low-relative-volatility systems, and for systems that exhibit azeotropic behavior. All of these special techniques are ultimately based on the same differences between the vapor and liquid compositions as in ordinary distillation, but they rely on some additional mechanism to further modify the vapor-liquid behavior of the key components. These enhanced techniques can be classified according to their effect on the relationship between the vapor and liquid compositions:

1. *Azeotropic distillation and pressure-swing distillation.* Methods that cause or exploit azeotrope formation or behavior to change the boiling characteristics and separability of the mixture.

2. *Extractive distillation and salt distillation.* Methods that primarily modify liquid-phase behavior to change the relative volatility of the components of the mixture.

3. *Reactive distillation.* Methods that use chemical reaction to modify the composition of the mixture or, alternatively, use existing vapor-liquid differences between reaction products and reactants to enhance the performance of a reaction.

AZEOTROPY

At low to moderate pressure ranges typical of most industrial applications, the fundamental composition relationship between the vapor and liquid phases in equilibrium can be expressed as a function of the total system pressure, the vapor pressure of each pure component, and the liquid-phase activity coefficient of each component *i* in the mixture:

$$y_i P = x_i \gamma_i P_i^{sat} \tag{13-112}$$

In systems that exhibit ideal liquid-phase behavior, the activity coefficients γ_i are equal to unity, and Eq. (13-112) simplifies to Raoult's law. For nonideal liquid-phase behavior, a system is said to show negative deviations from Raoult's law if $\gamma_i < 1$ and, conversely, positive deviations from Raoult's law if $\gamma_i > 1$. In sufficiently nonideal systems, the deviations may be so large that the temperature-composition phase diagrams exhibit extrema, as shown in Figs. 13-6, 13-7, and 13-8. At such maxima or minima, the equilibrium vapor and liquid compositions are identical. Thus,

$$y_i = x_i \qquad \text{for all } i = 1, \cdots, c \tag{13-113}$$

and the system is said to form an azeotrope (from the Greek word meaning "to boil unchanged"). Azeotropic systems show a minimum in the T versus x, y diagram when the deviations from Raoult's law are positive (Fig. 13-6) and a maximum in the T versus x, y diagram when the deviations from Raoult's law are negative (Fig. 13-7). If, at these two conditions, a single liquid phase is in equilibrium with the vapor phase, the azeotrope is homogeneous. If multiple-liquid-phase behavior is exhibited at the azeotropic condition, the azeotrope is heterogeneous. For heterogeneous azeotropes, the vapor-phase composition is equal to the *overall* composition of the two (or more) liquid phases (Fig. 13-8). These conditions are consequences of the general definition of an azeotrope in any kind of mixture (i.e., homogeneous, heterogeneous, reactive, or in any combination), which is as follows:

At fixed and constant pressure, an azeotropic state (normally called "an azeotrope") is one in which mass transfer occurs between a vapor and one or more co-existing equilibrium liquid phases in a closed system while the composition of each phase remains constant, but not necessarily equal. Moreover, the temperature of all co-existing equilibrium phases is identical and constant during the entire vaporization or condensation process. As a consequence, at homogeneous azeotropes (in which there is only one equilibrium liquid phase) there is a stationary point (i.e., a minimum, maximum, or saddle, see Figs. 13-6 and 13-7) in the boiling temperature surface where the vapor composition is equal to the liquid composition (by Gibbs-Konovalov theory). At heterogeneous azeotropes (in which there are multiple equilibrium liquid phases), the composition of the vapor is equal to the *overall* composition of the two or more liquid phases. Heterogeneous azeotropes do not have a stationary point in the boiling temperature (pressure) surface at the azeotropic point (see Fig. 13-8), and maximum boiling (minimum pressure) heterogeneous azeotropes cannot exist.

Azeotropes of all types have one degree of freedom, which is why only one thermodynamic state variable (e.g., pressure) is allowed to be fixed. Temperature could also be selected as the variable to fix (in place of pressure), in which case the pressure of all co-existing equilibrium phases is identical and constant during the azeotropic transformation. In principle, a different variable (other than P or T) could be fixed, but this is normally not considered.

For brevity, an azeotrope is often defined to be a state whereby at constant pressure (or constant temperature) the mixture can be completely vaporized or condensed without change in the temperature (pressure) or composition of each co-existing equilibrium phase. It is understood by those skilled in the art that this short-hand definition carries with it all the additional detail cited above.

The property shared by homogeneous and heterogeneous azeotropes (i.e., that the overall liquid composition is equal to the vapor composition) provides a means for finding azeotropes experimentally and computationally.

For additional reading see Prigogine and Defay, *Chemical Thermodynamics*, 4th ed., Longmans Green and Co., London, 1967; Rowlinson, *Liquids and Liquid Mixtures*, 2d ed., Butterworths, London, 1969; Doherty and Malone, *Conceptual Design of Distillation Systems*, McGraw-Hill, 2001, chaps. 5, 8, app. C.

Mixtures with only small deviations from Raoult's law (i.e., ideal or nearly ideal mixtures) may form an azeotrope but only if the saturated vapor pressure curves of the two pure components cross each other (such a point is called a Bancroft point). In such a situation, the azeotrope occurs at the temperature and pressure where the curves cross, and perhaps also in the vicinity close to the Bancroft point [e.g., cyclohexane (n.b.p. 80.7°C) and benzene (n.b.p. 80.1°C) form an almost ideal mixture, yet they exhibit a minimum-boiling azeotrope with roughly equal proportions of each component]. As the boiling point difference between the components increases, the composition of the azeotrope shifts closer to one of the pure components (toward the lower-boiling pure component for minimum-boiling azeotropes, and toward the higher-boiling pure component for maximum-boiling azeotropes). For example, the minimum-boiling azeotrope between methanol (n.b.p. 64.5°C) and toluene (n.b.p. 110.6°C) occurs at about 90 mol% methanol, and the minimum-boiling azeotrope between methyl acetate (n.b.p. 56.9°C) and water (n.b.p. 100°C) occurs at about 90 mol% methyl acetate. Mixtures of components whose boiling points differ by more than about 50°C generally do not exhibit azeotropes distinguishable from the pure components, even if large deviations from Raoult's law are present. As a qualitative guide to liquid-phase activity coefficient behavior, Robbins [*Chem. Eng. Prog.* **76**(10): 58 (1980)] developed a matrix of chemical families, shown in Table 13-13, that indicates expected deviations from Raoult's law.

The formation of two liquid phases within some boiling temperature range is generally an indication that the system will also exhibit a minimum-boiling azeotrope, since two liquid phases may form when deviations from Raoult's law are large and positive. The fact that immiscibility does occur, however, does not guarantee that the azeotrope will be heterogeneous; the azeotropic composition may not necessarily fall within the composition range of the two-liquid phase region, as is the case for the methyl acetate–water and tetrahydrofuran-water systems. Since large positive deviations

TABLE 13-13 Solute-Solvent Group Interactions

Solute class	Group	Solvent class 1	2	3	4	5	6	7	8	9	10	11	12
	H-donor												
1	Phenol	0	0	–	0	–	–	–	–	–	–	–	–
2	Acid, thiol	0	0	–	0	–	–	0	0	0	0	–	–
3	Alcohol, water	–	–	0	+	+	0	–	–	–	–	–	–
4	Active-H on multihalo paraffin	0	0	+	0	–	–	–	–	–	–	0	–
	H-acceptor												
5	Ketone, amide with no H on N, sulfone, phosphine oxide	–	–	+	–	0	+	–	–	–	+	–	–
6	Tertamine	–	–	0	–	+	0	–	–	0	+	0	0
7	Secamine	–	0	–	–	+	+	0	0	0	0	0	–
8	Priamine, ammonia, amide with 2H on N	–	0	–	–	+	+	0	0	–	+	–	–
9	Ether, oxide, sulfoxide	–	0	+	–	+	0	0	–	0	+	0	–
10	Ester, aldehyde, carbonate, phosphate, nitrate, nitrite, nitrile, intramolecular bonding, e.g., *o*-nitro phenol	–	0	+	–	+	+	0	–	–	0	–	–
11	Aromatic, olefin, halogen aromatic, multihalo paraffin without active H, monohalo paraffin	+	+	+	0	+	0	0	–	0	+	0	0
	Non-H-bonding												
12	Paraffin, carbon disulfide	+	+	+	+	+	0	+	+	+	+	0	0

SOURCE: Robbins, L. A., *Chem. Eng. Prog.*, **76**(10), 58–61 (1980), by permission.

from Raoult's law are required for liquid-liquid phase splitting, maximum-boiling azeotropes ($\gamma_i < 1$) are never heterogeneous.

Additional general information on the thermodynamics of phase equilibria and azeotropy is available in Malesinski (*Azeotropy and Other Theoretical Problems of Vapour-Liquid Equilibrium,* Interscience, London, 1965), Swietoslawski (*Azeotropy and Polyazeotropy,* Pergamon, London, 1963), Van Winkle (*Distillation,* McGraw-Hill, New York, 1967), Smith and Van Ness (*Introduction to Chemical Engineering Thermodynamics,* McGraw-Hill, New York, 1975), Wizniak [*Chem. Eng. Sci.* **38:** 969 (1983)], and Walas (*Phase Equilibria in Chemical Engineering,* Butterworths, Boston, 1985). Horsley (*Azeotropic Data-III,* American Chemical Society, Washington, 1983) compiled an extensive list of experimental azeotropic boiling point and composition data for binary and some multicomponent mixtures. Another source for azeotropic data and activity coefficient model parameters is the multivolume *Vapor-Liquid Equilibrium Data Collection* (DECHEMA, Frankfurt, 1977), a compendium of published experimental VLE data. Most of the data have been tested for thermodynamic consistency and have been fitted to the Wilson, UNIQUAC, Van Laar, Margules, and NRTL equations. An extensive two-volume compilation of azeotropic data for 18,800 systems involving 1700 compounds, entitled *Azeotropic Data* by Gmehling et al., was published in 1994 by VCH Publishers, Deerfield Beach, Fla. A computational method for determining the temperatures and compositions of all homogeneous azeotropes of a multicomponent mixture, from liquid-phase activity coefficient correlations, is given by Fidkowski, Malone, and Doherty [*Computers and Chem. Eng.* **17:** 1141 (1993)]. The method was generalized to determine all homogeneous and heterogeneous azeotropes by Wasylkiewicz, Doherty, and Malone [*Ind. Eng. Chem. Res.* **38:** 4901 (1999)].

RESIDUE CURVE MAPS AND DISTILLATION REGION DIAGRAMS

The simplest form of distillation involves boiling a multicomponent liquid mixture in an open evaporation from a single-stage batch still. As the liquid is boiled, the vapor generated is removed from contact with the liquid as soon as it is formed. Because the vapor is richer in the more volatile components than the liquid, the composition and boiling temperature of the liquid remaining in the still change continuously over time and move progressively toward less volatile compositions and higher temperatures until the last drop is vaporized. This last composition may be a pure-component species, or a maximum-boiling azeotrope, and it may depend on the initial composition of the mixture charged to the still.

The trajectory of liquid compositions starting from some initial composition is called a residue curve, and the collection of all such curves for a given mixture is called a residue curve map. Arrows are usually added to these curves, pointing in the direction of increasing time, which corresponds to increasing temperature, and decreasing volatility. If the liquid is well mixed and the vaporization is slow, such that the escaping vapor is in phase equilibrium with the residual liquid, then residue curve maps contain exactly the same information as the corresponding phase equilibrium diagram for the mixture, but they represent it in a way that is much more useful for understanding distillation systems. Composition changes taking place in simple batch distillation can be described mathematically by the following ordinary differential equation

$$\frac{dx_i}{d\xi} = x_i - y_i \qquad \text{for all } i = 1, \cdots, c \tag{13-114}$$

where ξ is a dimensionless nonlinear time scale. Normally, y_i and x_i are related by an isobaric VLE model. Integrating these equations forward in time leads to the less volatile final compositions; integrating them backward in time leads to the more volatile compositions that would produce a residue curve passing through the specified initial composition. A *residue curve map* (RCM) is generated by varying the initial composition and integrating Eq. (13-114) both forward and backward in time [Doherty and Perkins, *Chem. Eng. Sci.* **33:** 281 (1978); Doherty and Malone 2001, chap. 5]. Unlike a binary y-x plot, relative-volatility information is not represented on an RCM. Therefore, it is difficult to determine the ease of separation from a residue curve map alone. The steady states of Eq. (13-114) are the constant-composition trajectories corresponding to $dx_i/d\xi = 0$ for all $i = 1, \ldots, c$. The steady states therefore correspond to *all* the pure components and *all* the azeotropes in the mixture.

Residue curve maps can be constructed for mixtures of any number of components, but they can be pictured graphically only up to four components. For binary mixtures, a T vs. x, y diagram or a y-x diagram suffices; the system is simple enough that vapor-phase information can be included with liquid-phase information without confusion. For ternary mixtures, liquid-phase compositions are plotted on a triangular diagram, similar to that used in liquid-liquid extraction. Four-component systems can be plotted in a three-dimensional tetrahedron. The vertices of the triangular diagram or tetrahedron represent the pure components. Any binary, ternary, and quaternary azeotropes are placed at the appropriate compositions on the edges or the interior of the triangle and tetrahedron.

The simplest form of ternary RCM, as exemplified for the ideal normal-paraffin system of pentane-hexane-heptane, is illustrated in Fig. 13-75*a*, using a right-triangle diagram. Maps for all other nonazeotropic ternary mixtures are qualitatively similar. Each of the infinite number of possible residue curves originates at the pentane vertex, travels toward and then

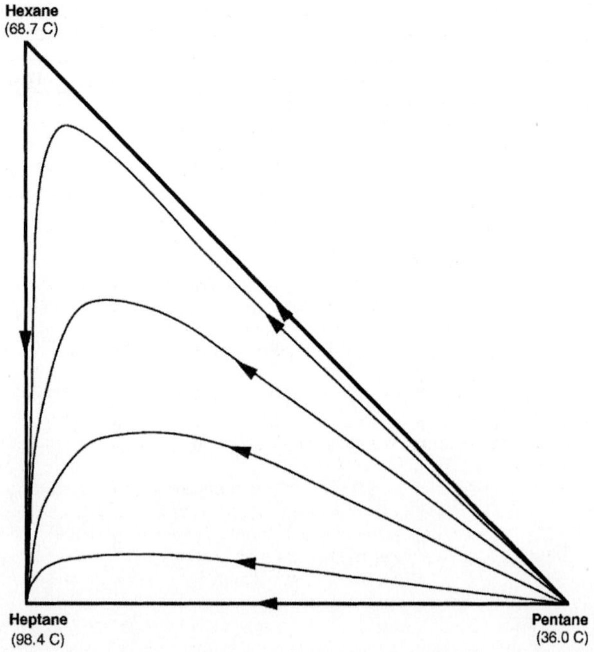

FIG. 13-75*a* Residue curve map: Nonazeotropic pentane-hexane-heptane system at 1 atm.

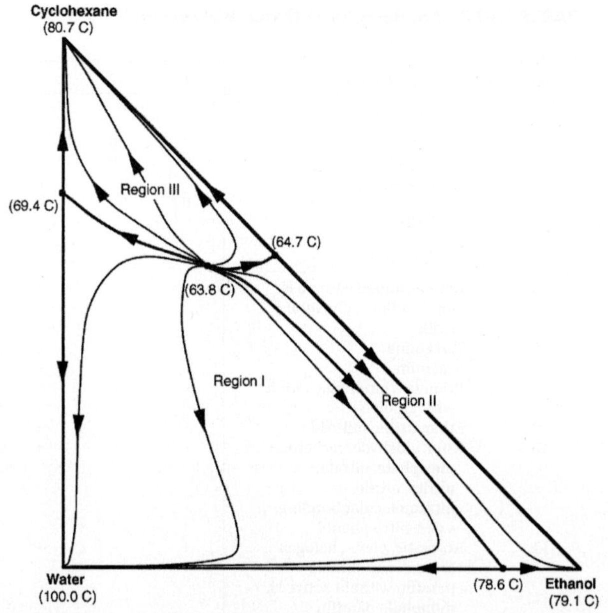

FIG. 13-75*c* Residue curve map: Ethanol-cyclohexane-water system at 1 atm containing four minimum-boiling azeotropes (three binary and one ternary) and three distillation regions.

away from the hexane vertex, and terminates at the heptane vertex. The family of all residue curves that originate at one composition and terminate at another composition defines a *distillation region.* Systems that do not involve azeotropes have only one region—the entire composition space. However, for many systems, not all residue curves originate or terminate at the same two compositions. Such systems will have more than one distillation region. The residue curve that divides two distillation regions in which adjacent residue curves originate from different compositions or terminate at different compositions is called a *simple batch distillation boundary* or *separatrix.* Distillation boundaries are related to the existence of azeotropes. In the composition space for a binary system, the distillation boundary is a point (the azeotropic composition). For three components, the distillation boundary is a curve; for four components, the boundary is a surface; and so on.

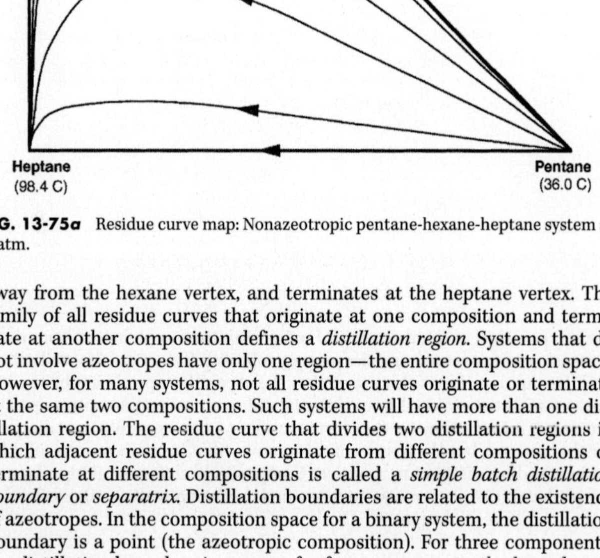

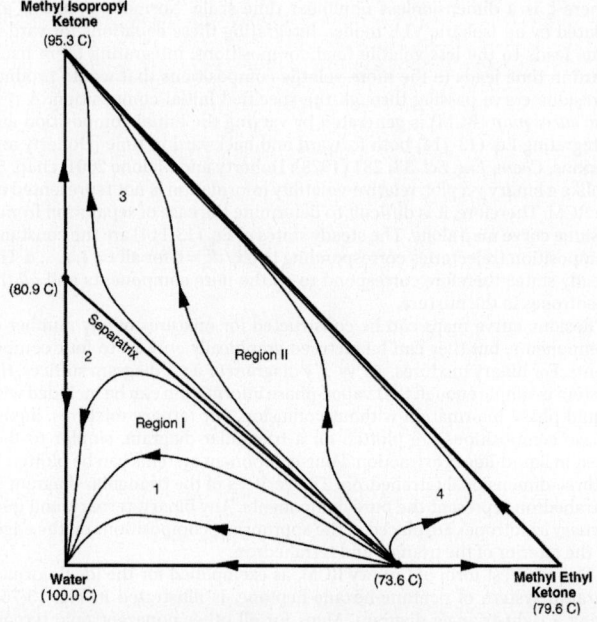

FIG. 13-75*b* Residue curve map: MEK-MIPK-water system at 1 atm containing two minumum-boiling binary azeotropes.

The boundaries of the composition diagram (e.g., the edges of a composition triangle) also form region boundaries since they divide physically realistic residue curves with positive compositions from unrealistic curves with negative compositions. All pure components and azeotropes in a system lie on region boundaries. Within each region, the most volatile composition on the boundary (either a pure component or a minimum-boiling azeotrope, and the origin of all residue curves in that region) is called the *low-boiling node.* The least volatile composition on the boundary (either a pure component or a maximum-boiling azeotrope, and the terminus of all residue curves in that region) is called the *high-boiling node.* All other pure components and azeotropes are called *intermediate-boiling saddles.* Adjacent regions may share some (but not all) nodes and saddles. Pure components and azeotropes are labeled as nodes and saddles as a result of the boiling points of all the components and azeotropes in a system. If one species is removed, the labeling of all remaining pure components and azeotropes, particularly those that were saddles, may change. Distillation boundaries always originate or terminate at saddle azeotropes, but never at pure component saddles—distillation boundaries can be calculated by using the method proposed by Lucia and Taylor [*AIChE J.* **52:** 582 (2006)]. Ternary saddle azeotropes are particularly interesting because they are more difficult to detect experimentally (being neither minimum-boiling nor maximum-boiling). However, their presence in a mixture implies the existence of distillation boundaries, which may have an important impact on the design of a separation system. The first ternary saddle azeotrope to be detected experimentally was reported by Ewell and Welch [*Ind. Eng. Chem.* **37:** 1224 (1945)], and a particularly comprehensive set of experimental residue curves were reported by Bushmakin and Kish [*J. Appl. Chem. USSR (Engl. Trans.)* **30:** 205 (1957)] for a ternary mixture with a ternary saddle azeotrope (reproduced as Fig. 5.9 in Doherty and Malone 2001). More ternary saddle azeotropes are reported in Gmehling et al. (*Azeotropic Data,* 1994).

Both methylethylketone (MEK) and methylisopropylketone (MIPK) form minimum-boiling homogeneous azeotropes with water (Fig. 13-75*b*). In this ternary system, a distillation boundary connects the binary azeotropes and divides the RCM into two distillation regions, I and II. The high-boiling node of region I is pure water, while the low-boiling node is the MEK-water azeotrope. In region II, the high- and low-boiling nodes are MIPK and the MEK-water azeotrope, respectively. These two regions, however, have a different number of saddles—one in region I and two in region II. This leads to region I having three sides, while region II has four sides. The more complicated cyclohexane-ethanol-water system (Fig. 13-75*c*) has three boundaries and three regions, all of which are four-sided and share the ternary azeotrope as the low-boiling node.

The liquid composition profiles in continuous staged or packed distillation columns operating at infinite reflux and boil-up are closely approximated by simple distillation residue curves [Van Dongen and Doherty, *Ind. Eng. Chem. Fundam.* **24:** 454 (1985)]. Residue curves are also indicative of many aspects of the general behavior of continuous columns operating at more practical reflux ratios. For example, to a first approximation, the stage-to-stage liquid compositions (along with the distillate and bottoms compositions) of a single-feed, two-product, continuous distillation column lie on the same residue curve. Therefore, for systems with distillation boundaries and multiple regions, distillation composition profiles are constrained to lie in specific regions. The precise boundaries of these distillation regions are a function of reflux ratio, but they are closely approximated by the RCM distillation boundaries. If an RCM distillation boundary exists in a system, a corresponding continuous distillation boundary will also exist. Both types of boundaries correspond exactly at all pure components and azeotropes.

Residue curves can be constructed from experimental data, or they can be calculated by integrating Eq. (13-114) if equation-of-state or activity-coefficient expressions are available (e.g., Wilson binary-interaction parameters, UNIFAC groups). However, considerable information on system behavior can still be deduced from a simple qualitative sketch of the RCM distillation boundaries based only on pure-component and azeotrope boiling point data and approximate azeotrope compositions. Rules for constructing such qualitative *distillation region diagrams* (DRDs) are given by Foucher et al. [*Ind. Eng. Chem. Res.* **30:** 760, 2364 (1991)]. For ternary systems containing no more than one ternary azeotrope and no more than one binary azeotrope between each pair of components, 125 such DRDs are mathematically possible [Matsuyama and Nishimura, *J. Chem. Eng. Japan* **10:** 181 (1977); Doherty and Caldarola, *Ind. Eng. Chem. Fundam.* **24:** 474 (1985); Peterson and Partin, *Ind. Eng. Chem. Res.* **36:** 1799 (1997)], although only a dozen or so represent most systems commonly encountered in practice.

Figure 13-76 illustrates all the 125 possible DRDs for ternary systems [see Peterson and Partin, *Ind. Eng. Chem. Res.* **36:** 1799 (1997)]. Azeotropes are schematically depicted generally to have equimolar composition, distillation boundaries are shown as straight lines, and the arrows on the distillation boundaries indicate increasing temperature. These DRDs are indexed in Table 13-14 according to a temperature profile sequence of position numbers, defined in a keyed triangular diagram at the bottom of the table, arranged by increasing boiling point. Positions 1, 3, and 5 are the pure components in order of decreasing volatility. Positions 2, 4, and 6 are binary azeotropes at the positions shown in the keyed triangle, and position 7 is the ternary azeotrope. Azeotrope position numbers are deleted from the temperature profile if the corresponding azeotrope is known not to exist. Note that not every conceivable temperature profile corresponds to a thermodynamically consistent system, and such combinations have been excluded from the index. As is evident from the index, some DRDs are consistent with more than one temperature profile. Also, some temperature profiles are consistent with more than one DRD. In such cases, the correct diagram for a system must be determined from residue curves obtained from experimental or calculated data.

Schematic DRDs are particularly useful in determining the implications of possibly unknown ternary saddle azeotropes by postulating position 7 at interior positions in the temperature profile. Also note that some combinations of binary azeotropes require the existence of a ternary saddle azeotrope. As an example, consider the system acetone (56.4°C), chloroform (61.2°C), and methanol (64.7°C) at 1 atm pressure. Methanol forms minimum-boiling azeotropes with both acetone (54.6°C) and chloroform (53.5°C), and acetone forms a maximum-boiling azeotrope (64.5°C) with chloroform. Experimentally there are no data for maximum- or minimum-boiling ternary azeotropes for this mixture. Assuming no ternary azeotrope, the temperature profile for this system is 461325, which from Table 13-14 is consistent with DRD 040 and DRD 042. However, Table 13-14 also indicates that the pure-component and binary azeotrope data are consistent with three temperature profiles involving a ternary saddle azeotrope, namely, 4671325, 4617325, and 4613725. All three of these temperature profiles correspond to DRD 107. Calculated residue curve trajectories for the acetone-chloroform-methanol system at 1 atm pressure, as shown in Fig. 13-77, show the existence of a ternary saddle azeotrope and DRD 107 as the correct approximation of the distillation regions. Ewell and Welch [*Ind. Eng. Chem.* **37:** 1224 (1945)] confirmed experimentally such a ternary saddle at 57.5°C.

APPLICATIONS OF RCM AND DRD

Residue curve maps and distillation region diagrams are very powerful tools for understanding all types of batch and continuous distillation operations, particularly when combined with other information such as liquid-liquid binodal curves. Applications include

1. *System visualization.* Location of distillation boundaries, azeotropes, distillation regions, feasible products, and liquid-liquid regions.
2. *Evaluation of laboratory data.* Location and confirmation of saddle ternary azeotropes and a thermodynamic consistency check of data.
3. *Process synthesis.* Concept development, construction of flow sheets for new processes, and redesign or modification of existing process flow sheets.
4. *Process modeling.* Identification of infeasible or problematic column specifications that could cause simulation convergence difficulties or failure, and determination of initial estimates of column parameters, including feed-stage location, number of stages in the stripping and enriching sections, reflux ratio, and product compositions.
5. *Control analysis/design.* Analysis of column balances and profiles to aid in control system design and operation.
6. *Process troubleshooting.* Analysis of separation system operation and malfunction, examination of composition profiles, and tracking of trace impurities with implications for corrosion and process specifications.

Material balances for mixing or continuous separation operations at steady state are represented graphically on triangular composition diagrams such as residue curve maps or distillation region diagrams by straight lines connecting pertinent compositions. The straight lines are exact representations of the compositions due to the lever rule. Overall flow rates are found by the inverse-lever-arm rule. Distillation material balance lines are governed by two constraints:

1. The bottoms, distillate, and overall feed compositions must lie on the same straight line.
2. The bottoms and distillate compositions must lie (to a very close approximation) on the same residue curve.

Since residue curves do not cross simple batch distillation boundaries, the distillate and bottoms compositions must be in the same distillation region with the mass balance line intersecting a residue curve in two places. Mass balance lines for mixing and for other separations not involving vapor-liquid equilibria, such as extraction and decantation, are of course not limited by distillation boundaries.

For a given multicomponent mixture, a single-feed, two-product distillation column (simple column) can be designed with enough stages, reflux, and material balance control to produce separations ranging from the *direct-split* mode of operation (low-boiling node taken as distillate) to the *indirect-split* mode (high-boiling node taken as bottoms). The bow-tie-shaped set of reachable product compositions for a simple distillation column is roughly bounded by the (straight) material balance lines that connect the feed composition to the sharpest direct separation and the sharpest indirect separation possible (see Fig. 13-78). A more accurate approximation involves replacing two of the straight-line segments of the bow tie with the residue curve through the feed composition [Stichlmair and Herguijuela, *AIChE J.* **38:** 1523 (1992)]. The exact shape of the reachable product composition regions involves replacing two of the straight-line segments of the bow tie with a locus of pinch points, as explained by Wahnschafft et al. [*Ind. Eng. Chem. Res.* **31:** 2345 (1992)] and Fidkowski, Doherty, and Malone [*AIChE J.* **39:** 1303 (1993)]. Since residue curves are deflected by saddles, it is generally not possible to obtain a saddle product (pure component or azeotrope) from a simple distillation column.

Consider the recovery of MIPK from an MEK-MIPK-water mixture. The approximate bow tie regions of product compositions for three different feeds are shown in Fig 13-78. From feed F3, which is situated in a different distillation region than the desired product, pure MIPK cannot be obtained at all. Feed F1 is more favorable, with the upper edge of the bow tie region along the MEK-MIPK (water-free) face of the composition triangle and part of the lower edge along the MEK-water (MIPK-free) face. There are conditions under which both the water in the MIPK bottoms product can be driven to low levels (high-product purity) and MIPK in the distillate can be driven to low levels (high-product recovery), although achieving such an operation depends on having an adequate number of stages and reflux ratio.

Although feed F2 lies in the same distillation region as F1, the bow tie region for feed F2 is significantly different than that for F1, with the upper edge along the water-MIPK (MEK-free) face of the triangle and the lower edge along the distillation boundary. From this feed it is not possible to simultaneously achieve a high-purity MIPK specification while obtaining high MIPK recovery. If the column is operated to get a high purity of MIPK, then the material balance line runs into the distillation boundary. Alternatively, if the column is operated to obtain a high recovery of MIPK (by removing the MEK-water azeotrope as distillate), the material balance requires the bottoms to lie on the water-MIPK face of the triangle.

The number of saddles in a particular distillation region can have significant impact on column profile behavior, process stability, and convergence behavior in process simulation of the system. Referring to the MIPK-MEK-water system in Fig. 13-75*b*, region I contains one saddle (MIPK-water azeotrope), while region II contains two saddles (pure MEK and the MIPK-water azeotrope). These are three- and four-sided regions, respectively. In a three-sided region, all residue curves track toward the solitary saddle. However, in a four- (or more) sided region with saddles on either side of a node, some residue curves will tend to track toward one saddle, while others track toward another saddle. For example, residue curve 1 in region I originates from the MEK-water azeotrope low-boiling node and travels first toward the single

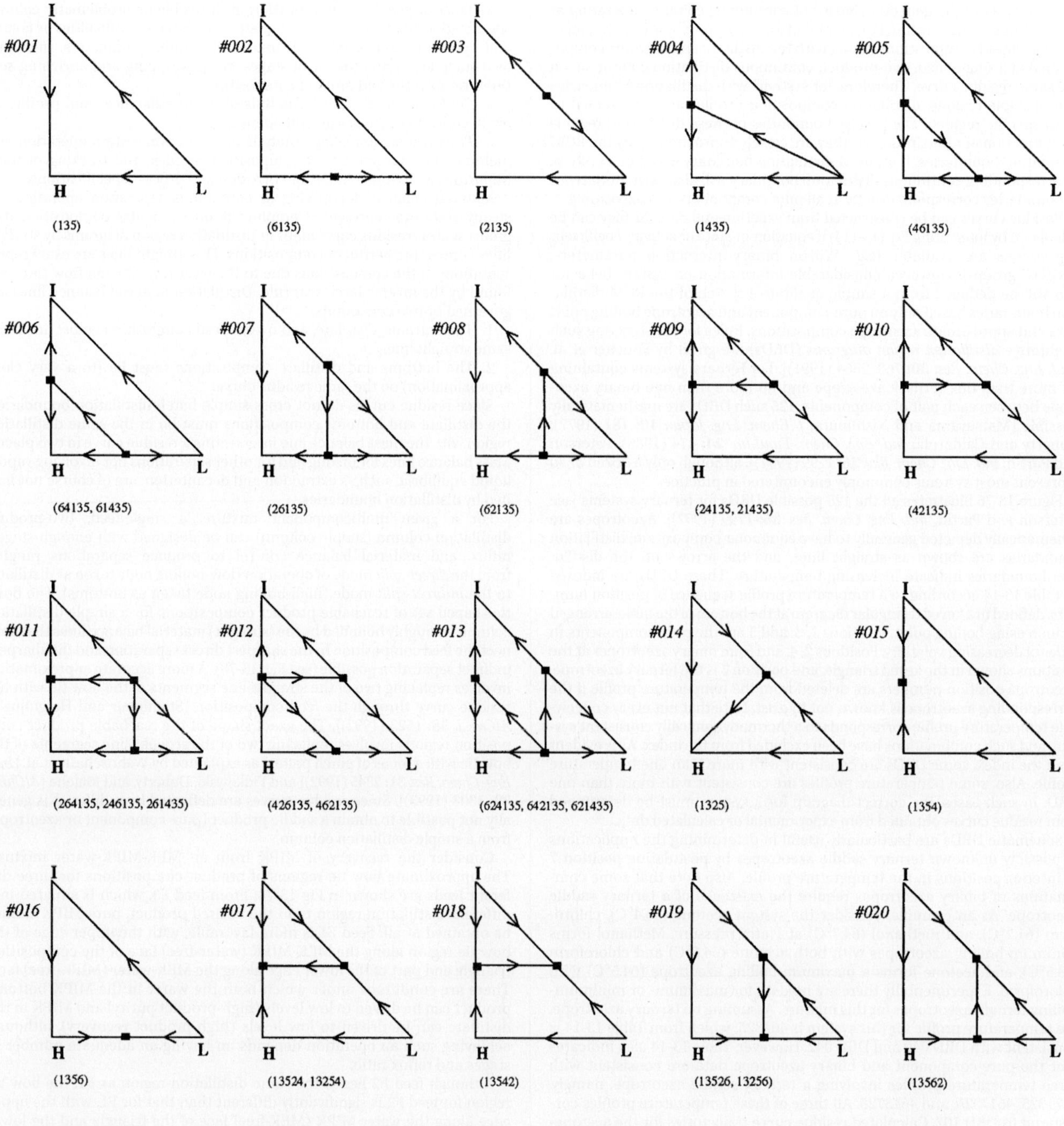

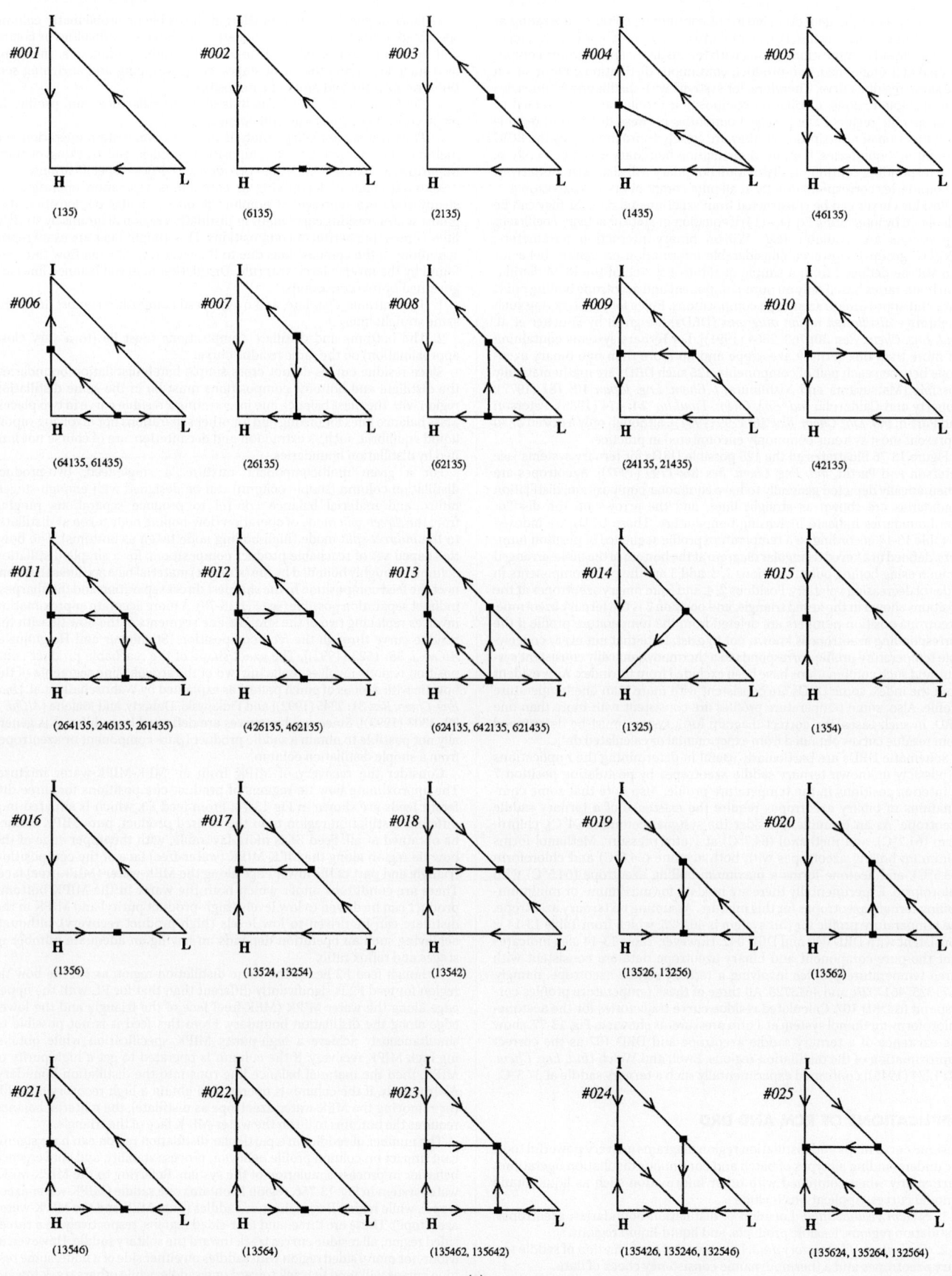

FIG. 13-76 Distillation region diagrams for ternary mixtures.

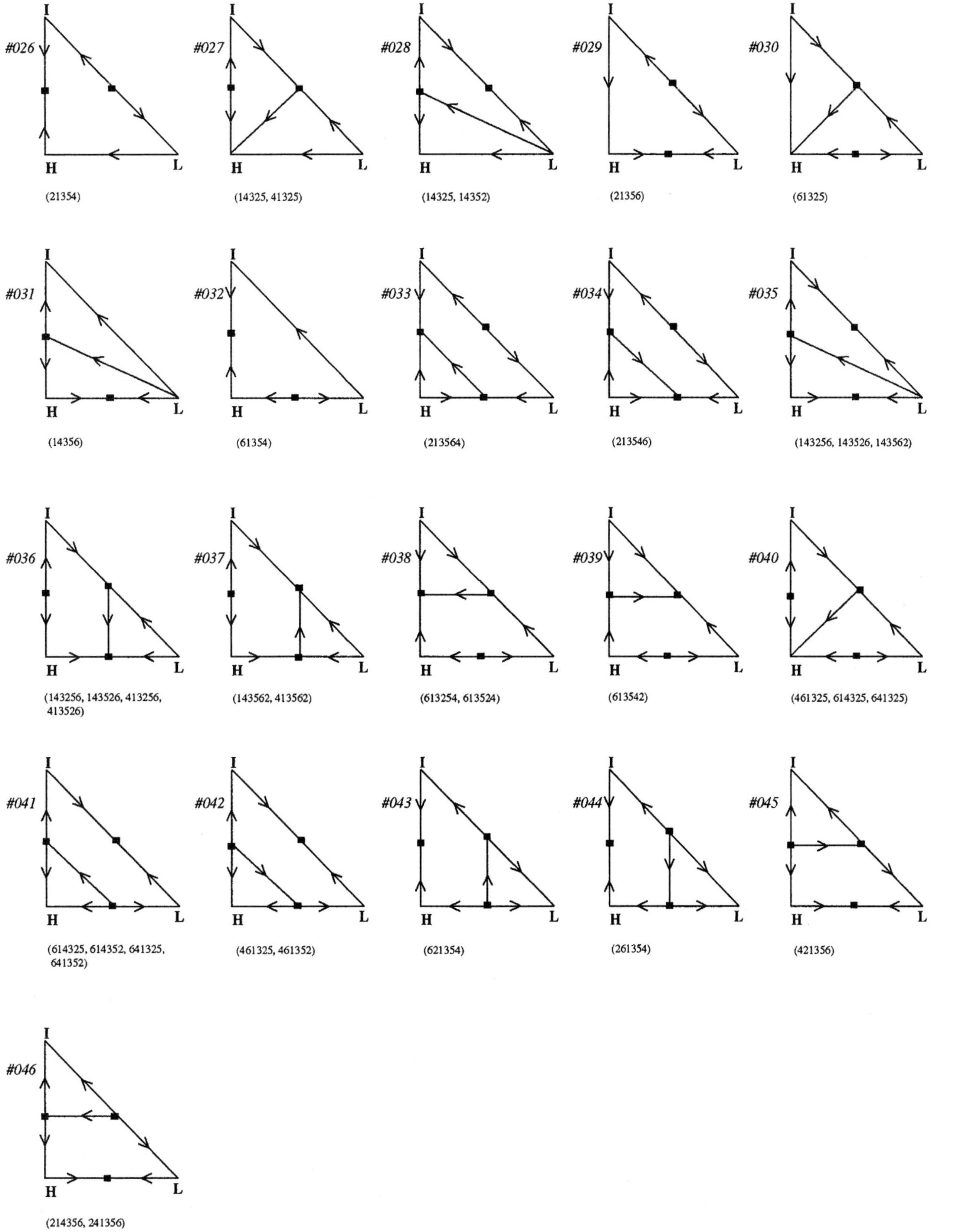
#026
I
H
L
(21354)
#027
(14325, 41325)
#028
(14325, 14352)
#029
(21356)
#030
(61325)
#031
(14356)
#032
(61354)
#033
(213564)
#034
(213546)
#035
(143256, 143526, 143562)
#036
(143256, 143526, 413256, 413526)
#037
(143562, 413562)
#038
(613254, 613524)
#039
(613542)
#040
(461325, 614325, 641325)
#041
(614325, 614352, 641325, 641352)
#042
(461325, 461352)
#043
(621354)
#044
(261354)
#045
(421356)
#046
(214356, 241356)

(*b*)

FIG. 13-76 (*Continued*)

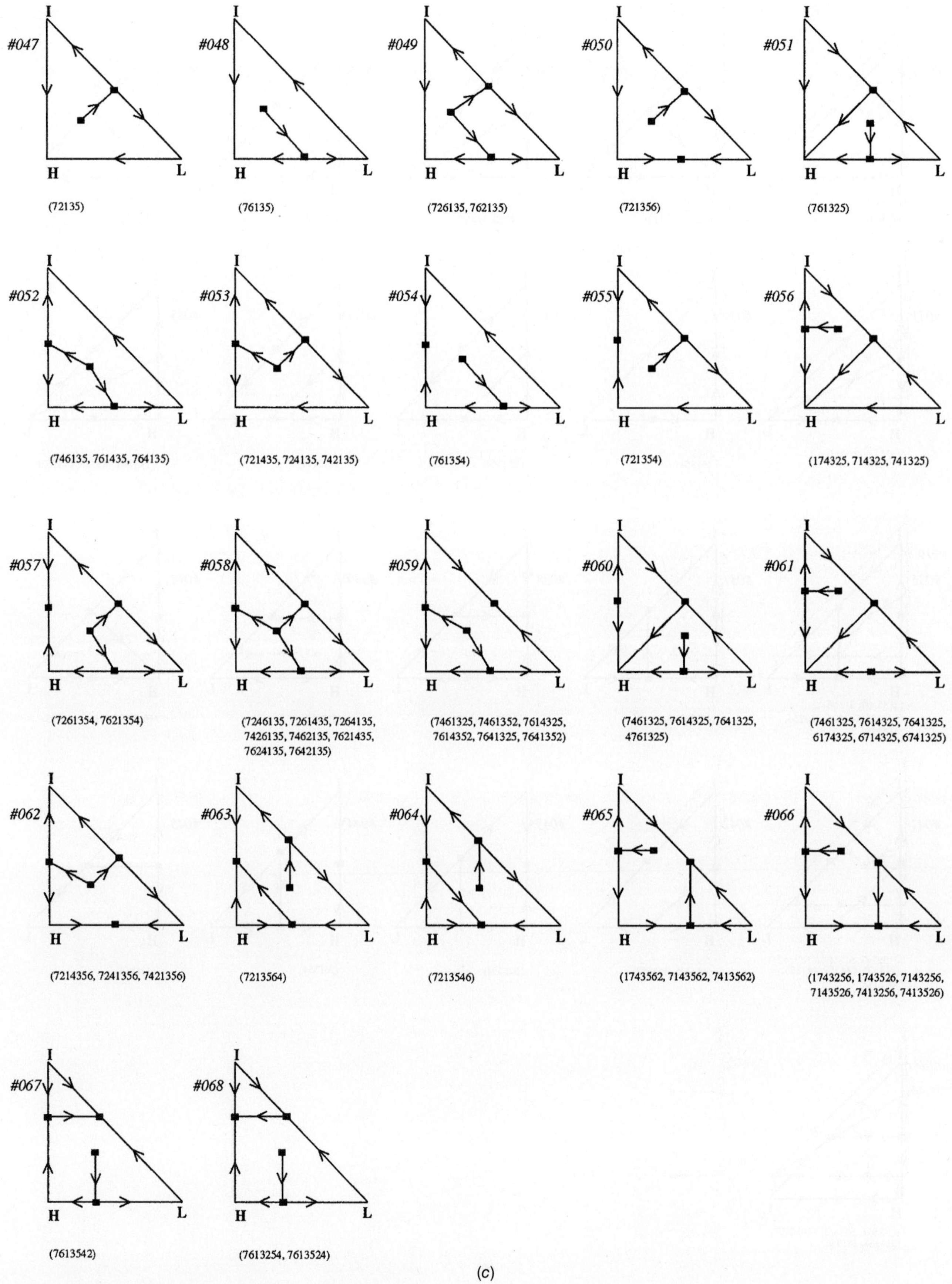

(c)

FIG. 13-76 (*Continued*)

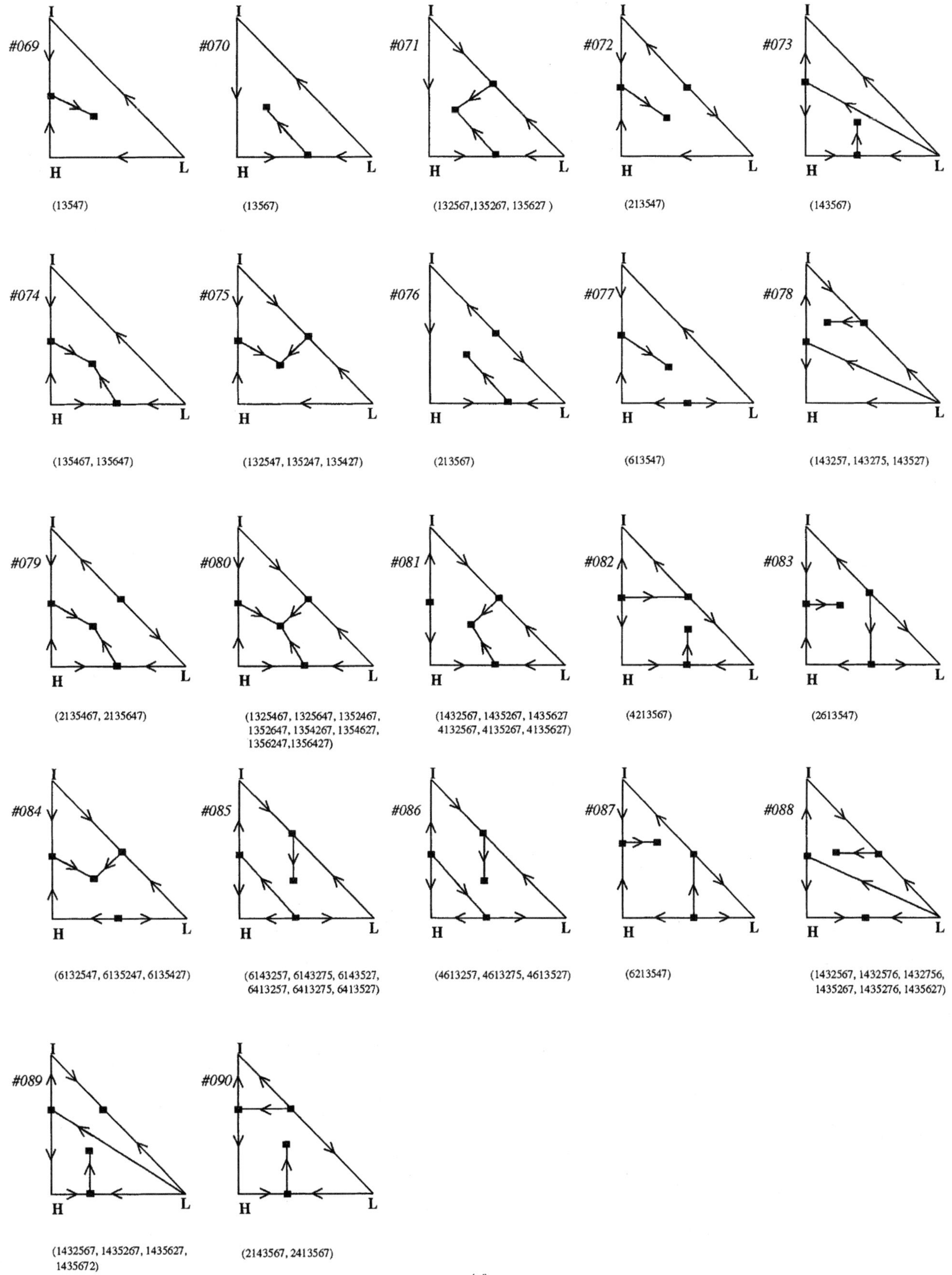

(*d*)

FIG. 13-76 (*Continued*)

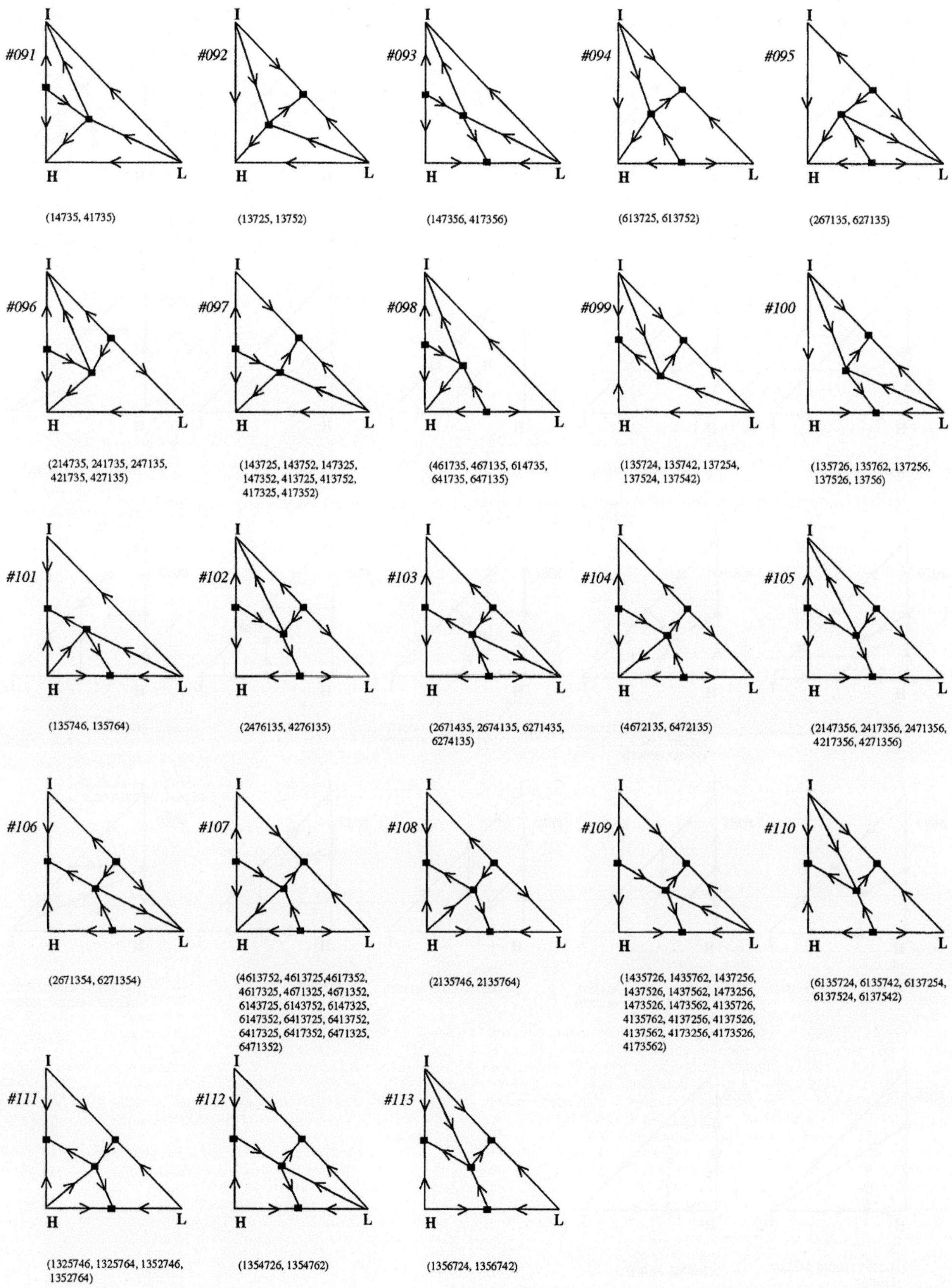

(e)

FIG. 13-76 *(Continued)*

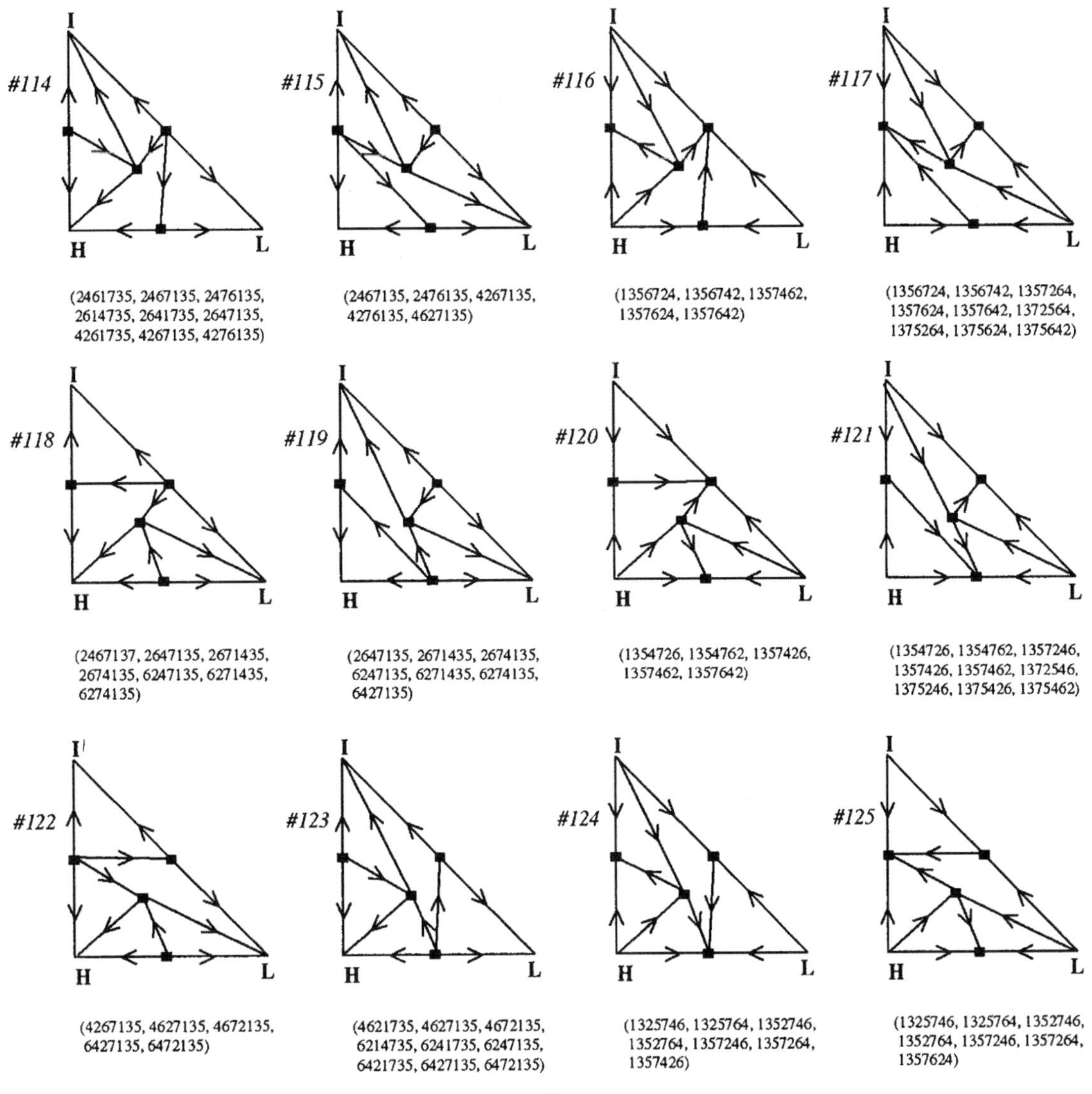

(*f*)

FIG. 13-76 (*Continued*)

saddle of the region (MIPK-water azeotrope) before ending at the water high-boiling node. Likewise, residue curve 2 and all other residue curves in region I follow the same general path.

In region II, residue curve 3 originates from the MEK-water azeotrope, travels toward the MIPK-water saddle azeotrope, and ends at pure MIPK. However, residue curve 4 follows a completely different path, traveling toward the pure MEK saddle before ending at pure MIPK. Some multicomponent columns have been designed for operation in four-sided regions with the feed composition adjusted so that both the high-boiling and low-boiling nodes can be obtained simultaneously as products. However, small perturbations in feed composition or reflux can result in feasible operation on many different residue curves that originate and terminate at these product compositions. Multiple steady states and composition profiles that shift dramatically from tracking toward one saddle to the other are possible [Kovach and Seider, *AIChE J.* **33:** 1300 (1987); Pham, Ryan, and Doherty, *AIChE J.* **35:** 1585 (1989)]. Consider a column operating in region II of the MIPK-MEK-water diagram. Figure 13-79 shows the composition and temperature profiles for the column operating at three different sets of operating conditions and two feed locations, as given in Table 13-15. The desired product specification is 97 mol% MIPK, no more than 3 mol% MEK, and less than 10 ppm residual water. For case A (Fig. 13-79*a*), the column profile tracks up the water-free side of the diagram. A pinched zone (i.e., section of little change in tray temperature and composition) occurs above the feed between the feed tray (tray 4) and tray 18. The temperature remains constant at about 93°C throughout the pinch zone. Product specifications are met.

When the feed composition becomes slightly enriched in water, as with case B, the column profile changes drastically (Fig. 13-79*b*). At the same reflux and boil-up, the column no longer meets specifications. The MIPK product is too lean in MIPK and too rich in water. The profile now tracks generally up the left side of region II. Note also the dramatic change in the temperature profile. A pinched zone still exists above the feed between trays 4 and 18, but the tray temperature in the zone has dropped to 80°C (from 93°C). Most of the trays are required to move through the vicinity of the saddle. Typically, pinches (if they exist) occur close to saddles and nodes.

In case C (Fig. 13-79*c*), increasing the boil-up ratio to 6 brings the MIPK product back within specifications, but the production rate and recovery have dropped off. In addition, the profile has switched back to the right side of the region, and the temperatures on trays in the pinched zone (trays 4 through 18) are back to 93°C. Such a drastic fluctuation in tray temperature with a relatively minor adjustment of the manipulated variable (boil-up in this case) can make control difficult. This is especially true if the control strategy involves maintaining a constant temperature on one of the trays between trays 4 and 18. If a tray is selected that exhibits wide temperature swings, the control system may have a difficult time compensating for disturbances. Such columns are also often difficult to model with a process simulator. Design algorithms often rely on

TABLE 13-14 Temperature Profile—DRD # Table*

Temp. Profile	DRD #	Temp. Profile	DRD #	Temp. Profile	DRD #	Temp. Profile	DRD #	Temp. Profile	DRD #	Temp. Profile	DRD #
135	001	137524	099	624135	013	1357462	120	2671435	103	6247135	118
1325	014	137526	100	627135	095		121		119		119
1354	015	137542	099	641325	041		116		118		123
1356	016	137562	100		040	1357624	117	2674135	118	6271354	106
1435	004	143256	035	641352	041		116		119	6271435	118
2135	003		036	641735	098		125		103		103
6135	002	143257	078	642135	013	1357642	120	4132567	081		119
13254	017	143275	078	647135	098		117	4135267	081	6274135	103
13256	019	143526	036	714325	056		116	4135627	081		118
13524	017		035	721354	055	1372546	121	4135726	109		119
13526	019	143527	078	721356	050	1372564	117	4135762	109	6413257	085
13542	018	143562	035	721435	053	1375246	121	4137256	109	6413275	085
13546	021		037	724135	053	1375264	117	4137526	109	6413527	085
13547	069	143567	073	726135	049	1375426	121	4137562	109	6413725	107
13562	020	143725	097	741325	056	1375462	121	4173256	109	6413752	107
13564	022	143752	097	742135	053	1375624	117	4173526	109	6417325	107
13567	070	147325	097	746135	052	1375642	117	4173562	109	6417352	107
13725	092	147352	097	761325	051	1432567	089	4213567	082	6421735	123
13752	092	147356	093	761354	054		088	4217356	105	6427135	122
14325	028	174325	056	761435	052		081	4261735	114		123
027		213546	034	762135	049	1432576	088	4267135	122		119
14352	028	213547	072	764135	052	1432756	088		115	6471325	107
14356	031	213564	033	1325467	080	1435267	081		114	6471352	107
14735	091	213567	076	1325647	080		089	4271356	105	6472135	104
21354	026	214356	046	1325746	111		088	4276135	114		122
21356	029	214735	096		124	1435276	088		102		123
21435	009	241356	046		125	1435627	089		115	6714325	061
24135	009	241735	096	1325764	111		081	4613257	086	6741325	061
26135	007	246135	011		124		088	4613275	086	7143256	066
41325	027	247135	096		125	1435672	089	4613527	086	7143526	066
41735	091	261354	044	1352467	080	1435726	109	4613725	107	7143562	065
42135	010	261435	011	1352647	080	1435762	109	4613752	107	7213546	064
46135	005	264135	011	1352746	125	1437256	109	4617325	107	7213564	063
61325	030	267135	095		111	1437526	109	4617352	107	7214356	062
61354	032	413256	036		124	1437562	109	4621735	123	7241356	062
61435	006	413526	036	1352764	124	1473256	109	4627135	123	7246135	058
62135	008	413562	037		111	1473526	109		115	7261354	057
64135	006	413725	097		125	1473562	109		122	7261435	058
72135	047	413752	097	1354267	080	1743256	066	4671325	107	7264135	058
76135	048	417325	097	1354627	080	1743526	066	4671352	107	7413256	066
132546	024	417352	097	1354726	120	1743562	065	4672135	104	7413526	066
132547	075	417356	093		121	2135467	079		122	7413562	065
132564	025	421356	045		112	2135647	079		123	7421356	062
132567	071	421735	096	1354762	121	2135746	108	4761325	060	7426135	058
135246	024	426135	012		120	2135764	108	6132547	084	7461325	060
135247	075	427135	096		112	2143567	090	6135247	084		059
135264	025	461325	040	1356247	080	2147356	105	6135427	084		061
135267	071		042	1356427	080	2413567	090	6135724	110	7461352	059
135426	024	461352	042	1356724	113	2417356	105	6135742	110	7462135	058
135427	075	461735	098		117	2461735	114	6137254	110	7613254	068
135462	023	462135	012		116	2467135	114	6137524	110	7613524	068
135467	074	467135	098	1356742	116		115	6137542	110	7613542	067
135624	025	613254	038		117		118	6143257	085	7614325	061
135627	071	613524	038		113	2471356	105	6143275	085		060
135642	023	613542	039	1357246	121	2476135	115	6143527	085		059
135647	074	613547	077		125		114	6143725	107	7614352	059
135724	099	613725	094		124		102	6143752	107	7621354	057
135726	100	613752	094	1357264	117	2613547	083	6147325	107	7621435	058
135742	099	614325	041		125	2614735	114	6147352	107	7624135	058
135746	101		040		124	2641735	114	6174325	061	7641325	061
135762	100	614735	098	1357426	121	2647135	118	6213547	087		060
135764	101	621354	043		120		119	6214735	123		059
137254	099	614352	041		124		114	6214735	123	7641352	059
137256	100	621435	013			2671354	106	6241735	123	7642135	058

Ternary DRD table lookup procedure:

1. Classify a system by writing down each position number in ascending order of boiling points.
   - A position number is not written down if there is no azeotrope at that position.
   - The resulting sequence of numbers is known as the *temperature profile.*
   - Each temperature profile will have a minimum of three numbers and a maximum of seven numbers.
   - List multiple temperature profiles when you have incomplete azeotropic data.
   - All seven position numbers are shown on the diagram.
2. Using the table, look up the temperature profile(s) to find the corresponding DRD #.

*Table 13-14 and Fig. 13-76 developed by Eric J. Peterson, Eastman Chemical Co.

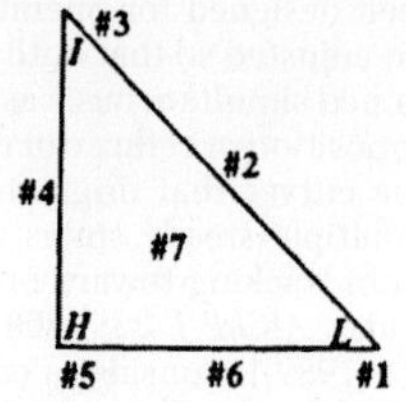

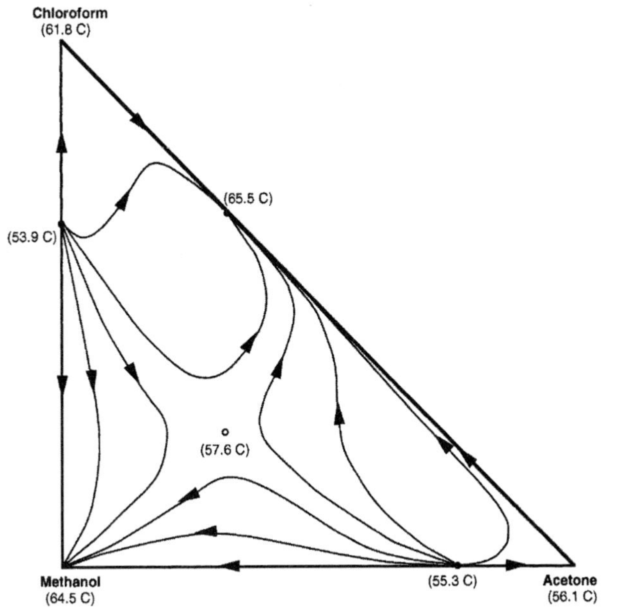

FIG. 13-77 Residue curves for acetone-chloroform-methanol system at 1 atm pressure suggesting a ternary saddle azeotrope.

perturbation of a variable (such as reflux or reboil) while checking for convergence of column heat and material balances. In situations where the column profile is altered drastically by minor changes in the perturbed variable, the simulator may be close to a feasible solution, but successive iterations may appear to be very far apart. The convergence routine may continue to oscillate between column profiles and never reach a solution. Likewise, when an attempt is made to design a column to obtain product compositions in different distillation regions, the simulation will never converge.

AZEOTROPIC DISTILLATION

The term *azeotropic distillation* has been applied to a broad class of fractional distillation-based separation techniques when specific azeotropic behavior is exploited to effect a separation. The agent that causes the

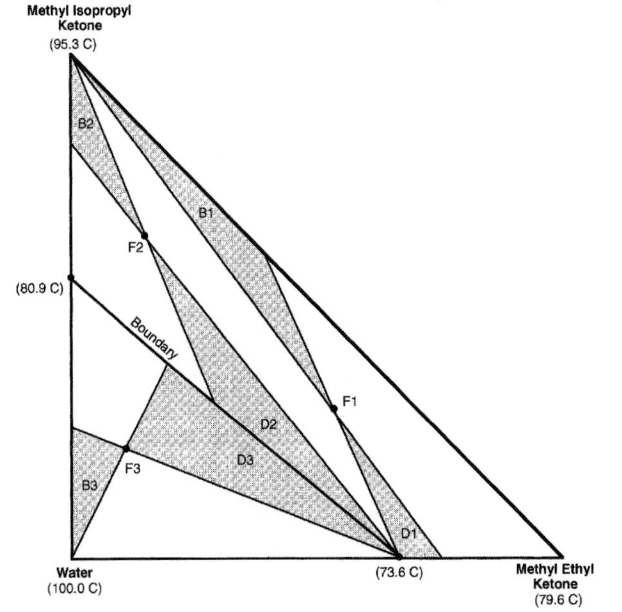

FIG. 13-78 MEK-MIPK-water system. Approximate product composition regions for a simple distillation column.

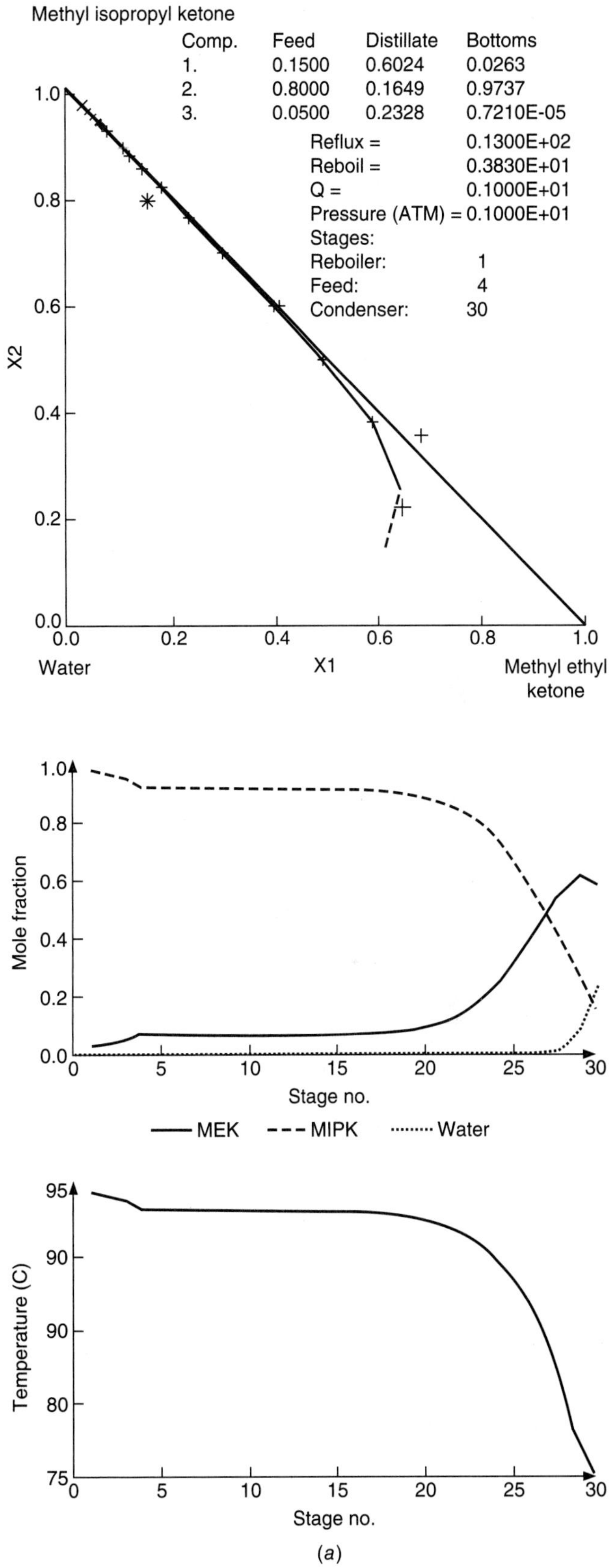

FIG. 13-79 Sensitivity of composition and temperature profiles for MEK-MIPK-water system at 1 atm.

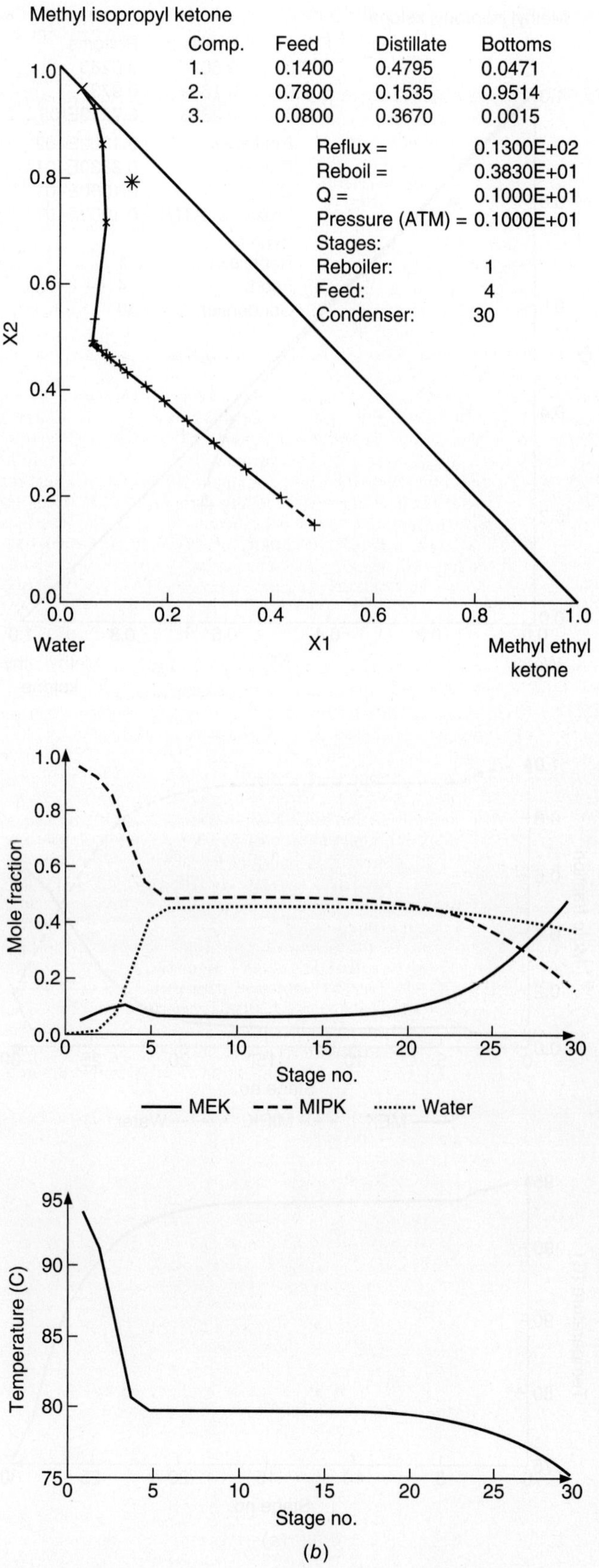

FIG. 13-79 (*Continued*)

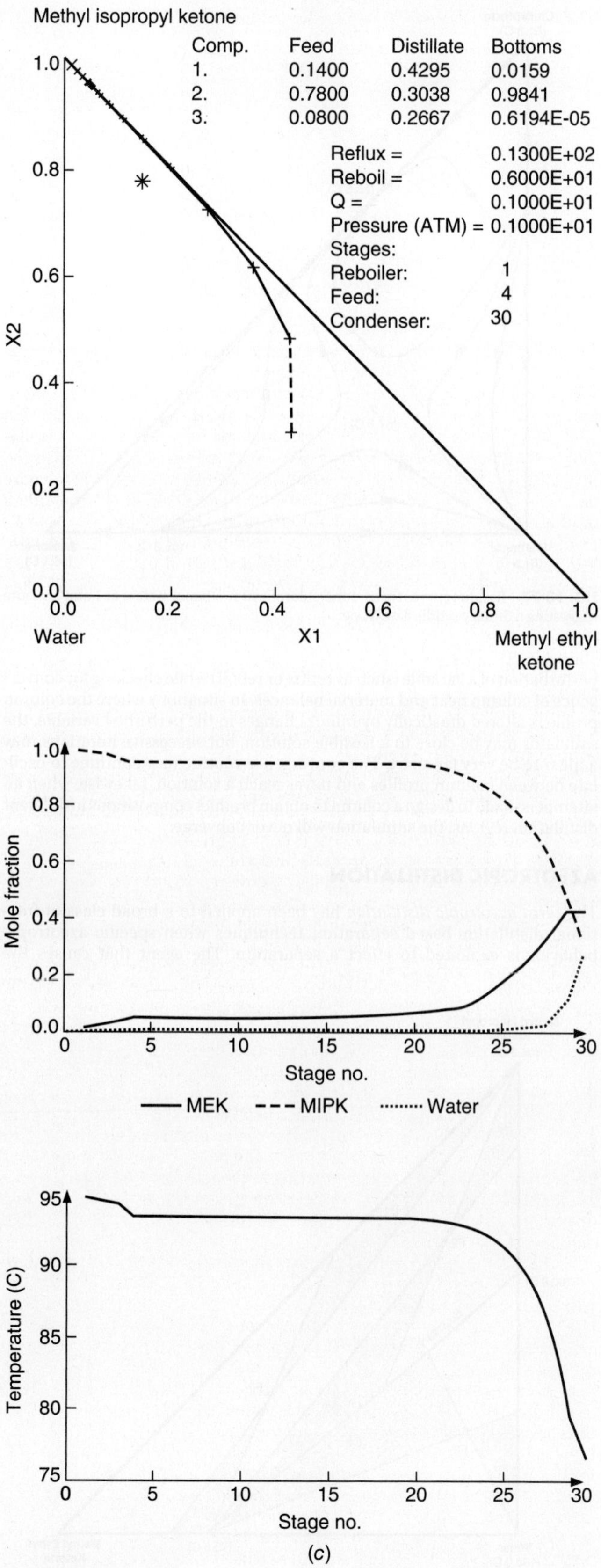

FIG. 13-79 (*Continued*)

TABLE 13-15 Sets of Operating Conditions for Fig. 13-79

Case	Reflux ratio	Reboil ratio		Feed composition	Distillate composition	Bottoms composition
A	13	3.8	MEK	0.15	0.60	0.03
			MIPK	0.80	0.16	0.97
			water	0.05	0.24	7 ppm
B	13	3.8	MEK	0.14	0.48	0.05
			MIPK	0.78	0.15	0.95
			water	0.08	0.37	20,000 ppm
C	13	6	MEK	0.14	0.43	0.02
			MIPK	0.78	0.30	0.98
			water	0.08	0.27	6.5 ppm

specific azeotropic behavior, often called the *entrainer*, may already be present in the feed mixture (a self-entraining mixture) or may be an added mass separation agent. Azeotropic distillation techniques are used throughout the petrochemical and chemical processing industries for the separation of close-boiling, pinched, or azeotropic systems for which simple distillation is either too expensive or impossible. With an azeotropic feed mixture, the presence of the azeotroping agent results in the formation of a more favorable azeotropic pattern for the desired separation. For a close-boiling or pinched feed mixture, the azeotroping agent changes the dimensionality of the system and allows separation to occur along a less pinched path. Within the general heading of azeotropic distillation techniques, several approaches have been followed in devising azeotropic distillation flow sheets, including:

1. Choosing an entrainer to give a residue curve map with specific distillation regions and node temperatures
2. Exploiting changes in azeotropic composition with total system pressure
3. Exploiting the curvature of distillation region boundaries
4. Choosing an entrainer to cause azeotrope formation in combination with liquid-liquid immiscibility

The first three of these are solely VLE-based approaches, involving a series of simple distillation column operations and recycles. The final approach relies on distillation (VLE), but it also exploits another physical phenomenon, liquid-liquid phase formation (phase splitting), to assist in entrainer recovery. This approach is the most powerful and versatile. Examples of industrial uses of azeotropic distillation grouped by method are given in Table 13-16.

The choice of the appropriate azeotropic distillation method and the resulting flow sheet for the separation of a particular mixture are strong functions of the separation objective. For example, it may be desirable to recover all constituents of the original feed mixture as pure components, or only some as pure components and others as azeotropic mixtures suitable for recycle. Not every objective may be obtainable by azeotropic distillation for a given mixture and portfolio of candidate entrainers.

Exploiting Homogeneous Azeotropes Homogeneous azeotropic distillation refers to a flow sheet structure in which azeotrope formation is exploited or avoided in order to accomplish the desired separation in one or more distillation columns. Either the azeotropes in the system do not exhibit two-liquid-phase behavior, or the liquid-phase behavior is not or cannot be exploited in the separation sequence. The structure of a particular sequence will depend on the geometry of the residue curve map or distillation region diagram for the feed mixture-entrainer system. Two approaches are possible:

1. Selection of an entrainer such that the desired products all lie within the *same* distillation region (the products may be pure components or azeotropic mixtures)
2. Selection of an entrainer such that some type of distillation boundary-crossing mechanism is employed to separate desired products that lie in *different* regions.

As mentioned previously, ternary mixtures can be represented by 125 different residue curve maps or distillation region diagrams. However, feasible distillation sequences using the first approach can be developed for breaking homogeneous binary azeotropes by the addition of a third component only for those more restricted systems that do not have a distillation boundary connected to the azeotrope and for which one of the original components is a node. For example, from Fig. 13-76 the following eight residue curve maps are suitable for breaking homogeneous minimum-boiling azeotropes: DRD 002, 027, 030, 040, 051, 056, 060, and 061 as collected in Fig. 13-80. To produce the necessary distillation region diagrams, an entrainer must be found that is either: (1) an intermediate boiler that forms no azeotropes (DRD 002), or (2) lowest-boiling or intermediate-boiling and forms a maximum-boiling azeotrope with the lower-boiling original component (A). In these cases, the entrainer may also optionally form a minimum-boiling azeotrope with the higher boiling of the original components or a minimum-boiling ternary azeotrope. In all cases, after the addition of the entrainer, the higher-boiling original component (B) is a high-boiling node and is removed as bottoms product from a first column operated in the indirect-split mode with the lower-boiling original component recovered as distillate in a second column; see the flow sheet in Fig. 13-80.

The seven residue curve maps suitable for breaking homogeneous maximum-boiling binary azeotropes (DRD 028, 031, 035, 073, 078, 088, 089) are shown in Fig. 13-81. In this case, the entrainer must form a minimum-boiling azeotrope with the higher-boiling original component and either a maximum-boiling azeotrope or no azeotrope with the lower-boiling original component. In all cases, after the addition of the entrainer, the lower-boiling original component is a low-boiling node and is removed as distillate from a first column operated in the direct-split mode with the higher-boiling original component recovered as bottoms product in a second column; see the flow sheet in Fig. 13-81.

The restrictions on the boiling point and azeotrope formation of the entrainer act as efficient screening criteria for entrainer selection. Entrainers that do not show appropriate boiling point characteristics can be discarded without detailed analysis. However, the entrainers in Fig. 13-80 do suffer from serious drawbacks that limit their practical application. DRD 002 requires that the entrainer be an intermediate-boiling component that forms no azeotropes. Unfortunately, these are often difficult criteria to meet because any intermediate boiler will be closer-boiling to both of the original components and, therefore, will be more likely to be at least pinched or even form azeotropes. The remaining feasible distillation region diagrams require that the entrainer form a maximum-boiling azeotrope with the lower-boiling original component. Because maximum-boiling azeotropes are relatively rare, finding a suitable entrainer may be difficult.

For example, the dehydration of organics that form homogeneous azeotropes with water is a common industrial problem. It is extremely difficult to find an intermediate-boiling entrainer that also does not form an azeotrope with water. Furthermore, the resulting separation is likely to be close-boiling or pinched throughout most of the column, requiring a large number of stages. For example, consider the separation of valeric acid (187.0°C) and water. This system exhibits a minimum-boiling azeotrope (99.8°C) with a composition and boiling point close to those of pure water. Ignoring for the moment potentially severe corrosion problems, formic acid (100.7°C), which is an intermediate boiler and which forms a maximum-boiling azeotrope with water (107.1°C), is a candidate entrainer (DRD 030, Fig. 13-82*a*). In the conceptual sequence shown in Fig. 13-82*b*, a recycle of the formic acid–water maximum-boiling azeotrope is added to the original valeric acid–water feed, which may be of any composition. Using the indirect-split mode of operation, the high-boiling node valeric acid is removed in high purity and high recovery as bottoms in a first column, which by mass balance produces a formic acid–water distillate. This binary mixture is fed to a second column that produces pure water as distillate and the formic acid–water azeotrope as bottoms for recycle to the first column. The inventory of formic acid is an important optimization variable in this theoretically feasible but difficult separation scheme.

Exploiting Pressure Sensitivity Breaking a homogeneous azeotrope that is part of a distillation boundary (i.e., the desired products lie in different distillation regions on either side of the boundary) requires that the boundary be "crossed" by the separation system. This may be done by mixing some external stream with the original feed stream in one region such that the resulting composition is in another region for further processing. However, the external stream must be completely regenerated, and mass balance must be preserved. For example, it is not possible to break a homogeneous binary azeotrope simply by adding one of the products to cross the azeotropic composition.

The composition of many azeotropes varies with the system pressure (Horsley, *Azeotropic Data-III*, American Chemical Society, Washington, 1983; Gmehling et al., *Azeotropic Data*, VCH Publishers, Deerfield Beach, Fla., 1994). This effect can be exploited to separate azeotropic mixtures by so-called pressure-swing distillation if at some pressure the azeotrope simply disappears, such as does the ethanol-water azeotrope at pressures below 11.5 kPa. However, pressure sensitivity can still be exploited if the azeotropic composition and related distillation boundary change sufficiently over a moderate change in total system pressure. A composition in one distillation region at one pressure could be in a different region at a different pressure. A two-column sequence for separating a binary maximum-boiling azeotrope is shown in Fig. 13-83 for a system in which the azeotropic composition at pressure P1 is richer in component B than the azeotropic composition at pressure P2. The first column, operating at pressure P1, is fed a mixture of fresh feed plus recycle stream from the

TABLE 13-16 Examples of Azeotropic Distillation

System	Type	Entrainer(s)	Remark
	Exploitation of homogeneous azeotropes		
No known industrial examples			
	Exploitation of pressure sensitivity		
THF-water	Minimum-boiling azeotrope	None	Alternative to extractive distillation
Methyl acetate-methanol	Minimum-boiling azeotrope	None	Element of recovery system for alternative to production of methyl acetate by reactive distillation; alternative to azeotropic, extractive distillation
Alcohol-ketone systems	Minimum-boiling azeotropes	None	
Ethanol-water	Minimum-boiling azeotrope	None	Alternative to extractive distillation, salt extractive distillation, heterogeneous azeotropic distillation; must reduce pressure to less than 11.5 kPa for azeotrope to disappear
	Exploitation of boundary curvature		
Hydrochloric acid-water	Maximum-boiling azeotrope	Sulfuric acid	Alternative to salt extractive distillation
Nitric acid-water	Maximum-boiling azeotrope	Sulfuric acid	Alternative to salt extractive distillation
	Exploitation of azeotropy and liquid phase immiscibility		
Ethanol-water	Minimum-boiling azeotrope	Cyclohexane, benzene, heptane, hexane, toluene, gasolene, diethyl ether	Alternative to extractive distillation, pressure-swing distillation
Acetic acid-water	Pinched system	Ethyl acetate, propyl acetate, diethyl ether, dichloroethane, butyl acetate	
Butanol-water	Minimum-boiling azeotrope	Self-entraining	
Acetic acid-water-vinyl acetate	Pinched, azeotropic system	Self-entraining	
Methyl acetate-methanol	Minimum-boiling azeotrope	Toluene, methyl isobutyl ketone	Element of recovery system for alternative to production of methyl acetate by reactive distillation; alternative to extractive pressure-swing distillation
Diethoxymethanol-water-ethanol	Minimum-boiling azeotropes	Self-entraining	
Pyridine-water	Minimum-boiling azeotrope	Benzene	Alternative to extractive distillation
Hydrocarbon-water	Minimum-boiling azeotrope	Self-entraining	

second column such that the overall composition lies on the A-rich side of the azeotropic composition at P1. Pure component A is recovered as distillate, and a mixture near the azeotropic composition is produced as bottoms. The pressure of this bottoms stream is changed to P2 and fed to the second column. This feed is on the B-rich side of the azeotropic composition at P2. Pure component B is now recovered as the distillate, and the azeotropic bottoms composition is recycled to the first column. An analogous flow sheet can be used for separating binary homogeneous minimum-boiling azeotropes. In this case the pure components are recovered as bottoms in both columns, and the distillate from each column is recycled to the other column.

For pressure-swing distillation to be practical, the azeotropic composition must vary at least 5 percent (preferably 10 percent or more) over a moderate pressure range (not more than 10 atm between the two pressures). With a very large pressure range, refrigeration may be required for condensation of the low-pressure distillate, or an impractically high reboiler temperature may result in the high-pressure column. The smaller the variation of azeotrope composition over the pressure range, the larger the recycle flow rates between the two columns. In particular, for minimum-boiling azeotropes, the pressure-swing distillation approach requires high energy usage and high capital costs (large-diameter columns) because both recycled azeotropic compositions must be taken overhead. Moreover, one lobe of an azeotropic VLE diagram is often pinched regardless of pressure; therefore, one of the columns will require a large number of stages to produce the corresponding pure-component product.

General information on pressure-swing distillation can be found in Van Winkle (*Distillation,* McGraw-Hill, New York, 1967), Wankat (*Equilibrium-Staged Separations,* Elsevier, New York, 1988), and Knapp and Doherty [*Ind. Eng. Chem. Res.* **31:** 346 (1992)]. Only a relatively small fraction of azeotropes are sufficiently pressure-sensitive for a pressure-swing process to be economical. Some applications include the minimum-boiling azeotrope tetrahydrofuran-water [Tanabe et al.; U.S. Patent 4,093,633 (1978)], and maximum-boiling azeotropes of hydrogen chloride–water and formic acid–water (Horsley, *Azeotropic Data-III,* American Chemical Society, Washington, 1983). Since distillation boundaries move with pressure-sensitive azeotropes, the pressure-swing principle can also be used for overcoming distillation boundaries in multicomponent azeotropic mixtures.

Exploiting Boundary Curvature A second approach to boundary crossing exploits boundary curvature to produce compositions in different distillation regions. When distillation boundaries exhibit extreme curvature, it may be possible to design a column such that the distillate and bottoms compositions are on the same residue curve in one distillation region, while the feed composition (which is not required to lie on the column composition profile) is in another distillation region. For such a column to meet material balance constraints (i.e., bottom, distillate, feed compositions on a straight line), the feed must be located in a region where the boundary is concave.

As an example, Van Dongen (Ph.D. thesis, University of Massachusetts, 1983) considered the separation of a methanol–methyl acetate mixture, which forms a homogeneous azeotrope, using *n*-hexane as an entrainer. The distillation boundaries for this system (Fig. 13-84*a*) are somewhat curved. A separation sequence that exploits this boundary curvature is shown in Fig. 13-84*b*. Recycled methanol–methyl acetate binary azeotrope and methanol–methyl acetate–hexane ternary azeotrope are added to the original feed F0 to produce a net feed composition F1 for column C1 designed to lie on a line between pure methanol and the curved part of the boundary between regions I and II. Column C1 is operated in the indirect-split mode, producing the high-boiling node methanol as a bottoms product, and by mass balance, a distillate near the curved boundary. The distillate, although in region I, becomes feed F2 to column C2, which is operated in the direct-split mode entirely in region II, producing the low-boiling node ternary azeotrope as distillate and, by mass balance, a methanol–methyl acetate mixture as bottoms (B2). This bottoms mixture is on the opposite side of the methanol–methyl acetate azeotrope from the original feed F0. The bottoms product from C2 is finally fed to binary distillation column C3, which produces pure methyl acetate as bottoms product (B3) and the methanol–methyl acetate azeotrope as distillate (D3). The distillates from columns C2 and C3 are recycled to column C1. The distillate and bottoms compositions for column C2 lie on the same residue curve, and the composition profile lies entirely within region II, even though its feed composition is in region I.

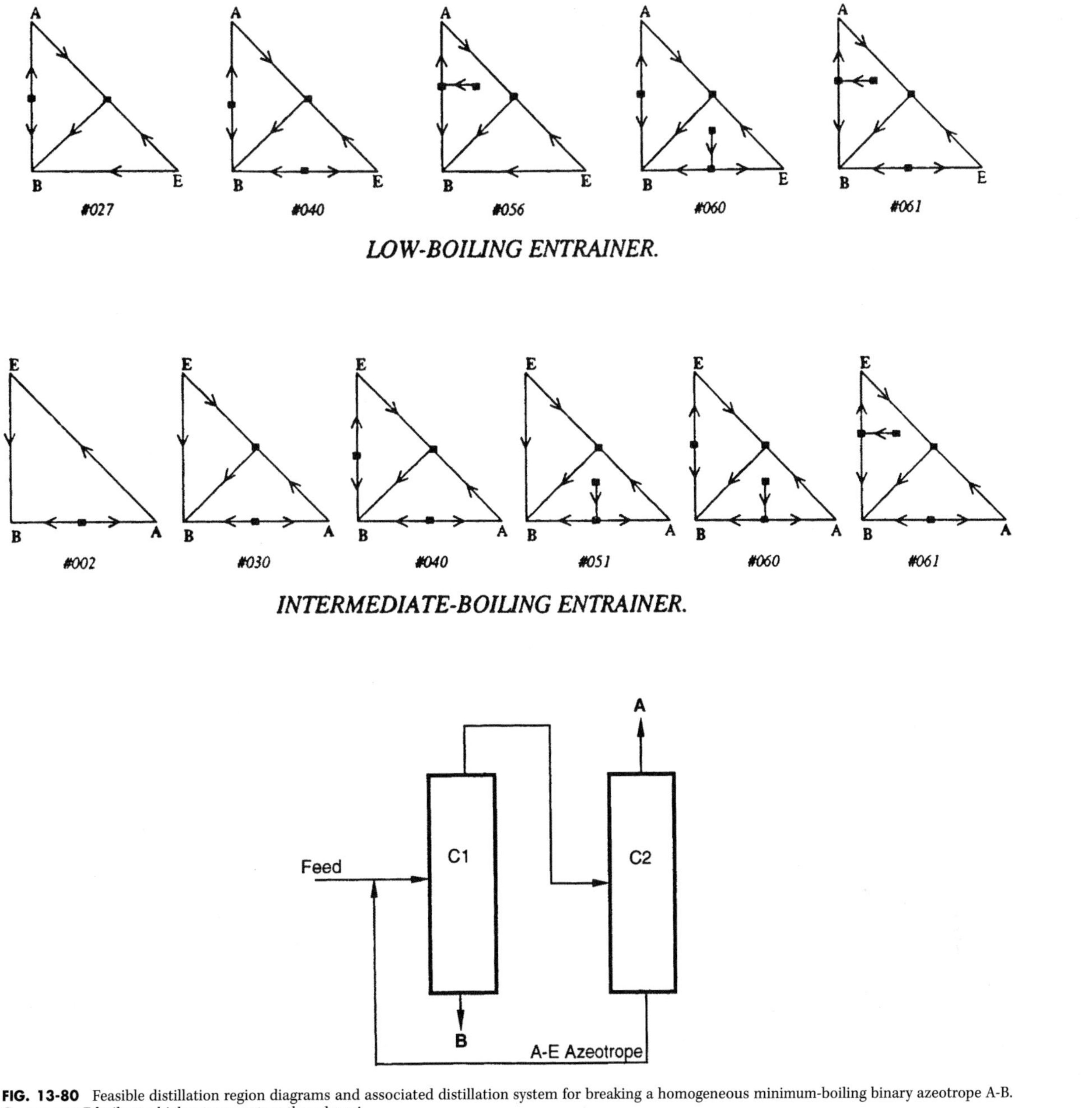

FIG. 13-80 Feasible distillation region diagrams and associated distillation system for breaking a homogeneous minimum-boiling binary azeotrope A-B. Component B boils at a higher temperature than does A.

Additional information on exploiting boundary curvature, including the useful concept of a *pitchfork distillation boundary*, can be found in Doherty and Malone (*Conceptual Design of Distillation Systems*, McGraw-Hill, 2001, sec. 5.4).

Exploiting boundary curvature for breaking azeotropes is very similar to exploiting pressure sensitivity from a mass balance point of view, and it suffers from the same disadvantages. These separation schemes have large recycle flows, and in the case of minimum-boiling azeotropes, the recycle streams are distillates. However, in the case of maximum-boiling azeotropes, these recycles are bottoms products, and the economics are improved. One such application, illustrated in Fig. 13-85, is the separation of the maximum-boiling nitric acid–water azeotrope by adding sulfuric acid. Recycled sulfuric acid is added to a nitric acid–water mixture near the azeotropic composition to produce a net feed F1 in region II. The first column, operated in the direct-split mode, produces a nitric acid distillate and a bottoms product, by mass balance, near the distillation boundary. In this case, sulfuric acid associates with water so strongly and the distillation boundary is so curved and nearly tangent to the water–sulfuric acid edge of the composition diagram that the second column operating in the indirect-split mode in region I, producing sulfuric acid as bottoms product, also produces a distillate close enough to the water specification that a third column is not required (Thiemann et al., in *Ullmann's Encyclopedia of Industrial Chemistry,* 5th ed., vol. A17, VCH Verlagsgesellschaft mbH, Weinheim, 1991).

Exploiting Azeotropy and Liquid-Phase Immiscibility One powerful and versatile separation approach exploits several physical phenomena simultaneously, including nonideal vapor-liquid behavior, where possible, and liquid-liquid behavior to bypass difficult distillation separations. For example, the overall separation of close-boiling mixtures can be made easier by the addition of an entrainer that introduces liquid-liquid immiscibility and forms a heterogeneous minimum-boiling azeotrope with one (generally the lower-boiling) of the key components. Two-liquid-phase formation provides a means of breaking this azeotrope, thus simplifying the entrainer recovery and recycle process. Moreover, since liquid-liquid tie lines are unaffected by distillation boundaries (and the separate liquid phases are often located in different distillation regions), liquid-liquid phase splitting is a powerful mechanism for crossing distillation boundaries. The phase separator is usually a simple decanter, but sometimes a multistage extractor is substituted. The decanter or extractor can also be replaced by some other non-VLE-based separation

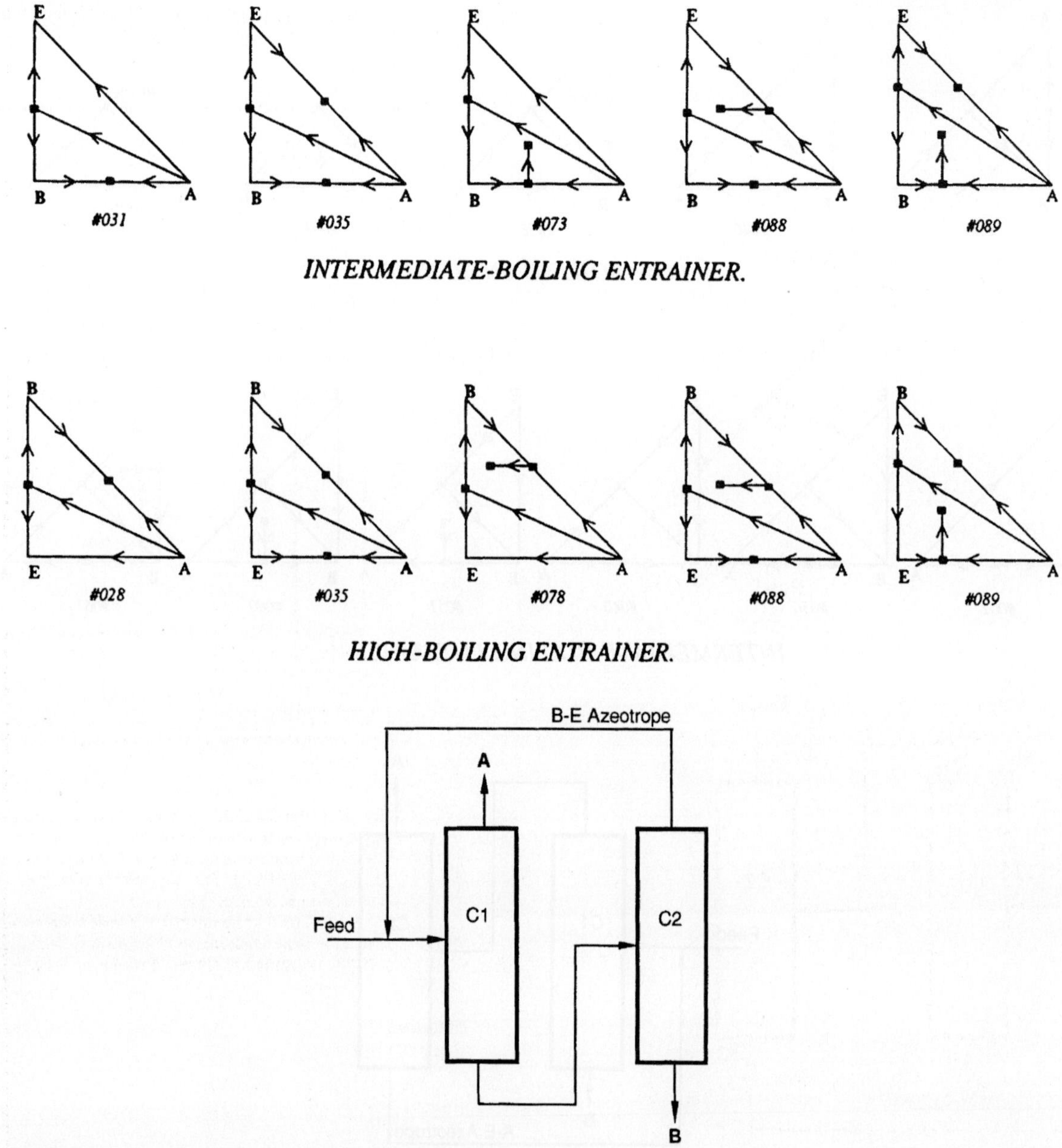

FIG. 13-81 Feasible distillation region diagrams and associated distillation system for breaking a homogeneous maximum-boiling binary azeotrope A-B. Component B boils at a higher temperature than does A.

technique, such as membrane permeation, chromatography, adsorption, or crystallization. Also, sequences may include additional separation operations (distillations or other methods) for preconcentration of the feed mixture, entrainer recovery, and final-product purification.

The simplest case of combining VLE and LLE is the separation of a binary heterogeneous azeotropic mixture. One example is the dehydration of 1-butanol, a self-entraining system, in which butanol (117.7°C) and water form a minimum-boiling heterogeneous azeotrope (93.0°C). As shown in Fig. 13-86, the fresh feed may be added to either column C1 or C2, depending on whether the feed is on the organic-rich side or the water-rich side of the azeotrope. The feed may also be added into the decanter directly if it does not move the overall composition of the decanter outside of the two-liquid phase region. Column C1 produces anhydrous butanol as a bottoms product and a composition close to the butanol-water azeotrope as the distillate. After condensation, the azeotrope rapidly phase-separates in the decanter. The upper layer, consisting of 78 wt% butanol, is refluxed totally to column C1 for further butanol recovery. The water layer, consisting of 92 wt% water, is fed to column C2. This column produces pure water as a bottoms product and, again, a composition close to the azeotrope as distillate for recycle to the decanter. Sparged steam may be used in C2, saving the cost of a reboiler. A similar flow sheet can be used for dehydration of hydrocarbons and other species that are largely immiscible with water.

A second example of the use of liquid-liquid immiscibilities in an azeotropic distillation sequence is the separation of the ethanol-water minimum-boiling homogeneous azeotrope. For this separation, a number of entrainers have been proposed, which are usually chosen to be immiscible with water and form a ternary minimum-boiling (preferably heterogeneous) azeotrope with ethanol and water (and, therefore, usually also binary minimum-boiling azeotropes with both ethanol and water). All such systems correspond to DRD 058, although the labeling of the vertices depends on whether the entrainer is lower-boiling than ethanol, intermediate-boiling, or higher-boiling than water. The residue curve map for the case of cyclohexane as entrainer was illustrated in Fig. 13-75*c*. One three-column distillation sequence is shown in Fig. 13-87. Other two-, three-, or four-column sequences have been described by Knapp and Doherty (*Kirk-Othmer Encyclopedia of Chemical Technology*, 5th ed., vol. 8, Wiley, New York, 2004, p. 786).

Fresh aqueous ethanol feed is first preconcentrated to nearly the azeotropic composition in column C3, while producing a water bottoms product.

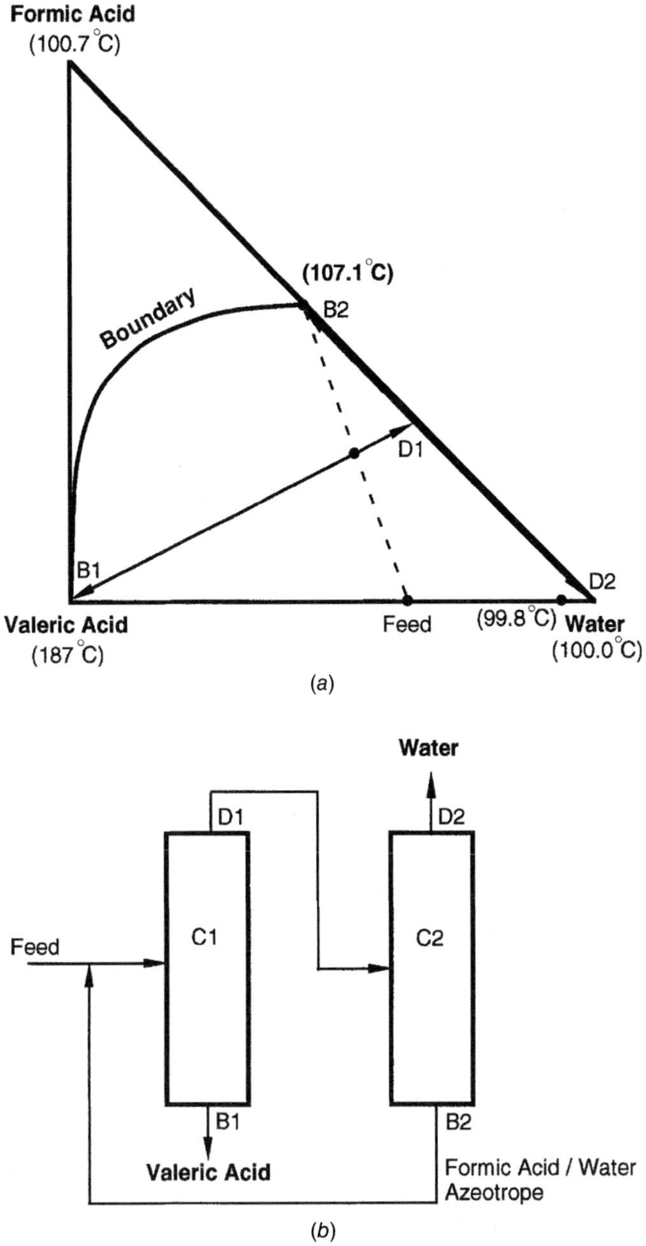

FIG. 13-82 Valeric acid–water separation with formic acid. (*a*) Mass balances on distillation region diagram. (*b*) Conceptual sequence.

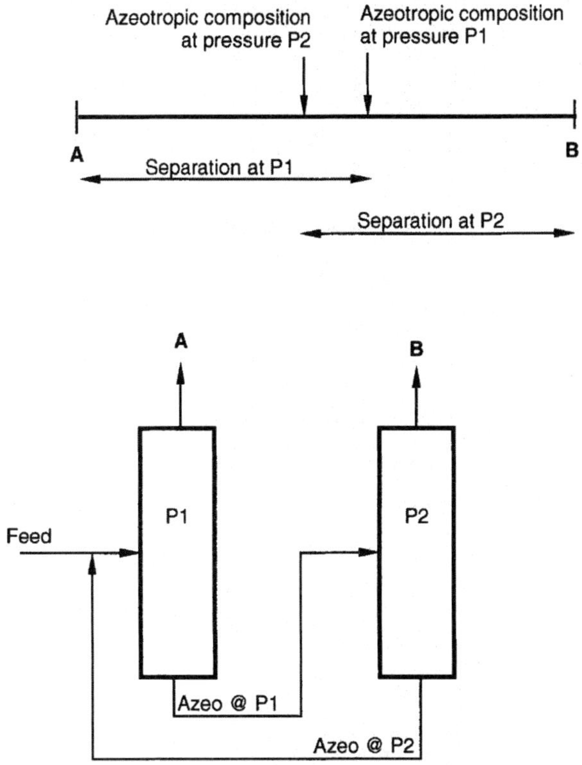

FIG. 13-83 Conceptual sequence for separating maximum-boiling binary azeotrope with pressure-swing distillation.

The distillate from C3 is sent to column C1, which is refluxed with the entire organic (entrainer-rich) layer, recycled from a decanter. Mixing of these two streams is the key to this sequence as it allows the overall feed composition to cross the distillation boundary into region II. Column C1 is operated to recover pure high-boiling node ethanol as a bottoms product and to produce a distillate close to the ternary azeotrope. If the ternary azeotrope is heterogeneous (as it is in this case), it is sent to the decanter for phase separation. If the ternary azeotrope is homogeneous (as it is in the alternative case of ethyl acetate as the entrainer), the distillate is first mixed with water before being sent to the decanter. The inventory of entrainer is adjusted to allow column C1 to operate essentially between two nodes, although such practice, as discussed previously, is relatively susceptible to instabilities from minor feed or reflux perturbations. Refluxing a fraction of the water-rich decanter layer results in an additional degree of freedom to mitigate against variability in the feed composition. The remaining portion of the water layer from the decanter is stripped of residual cyclohexane in column C2, which may be operated either in the direct-split mode (producing low-boiling node ternary azeotrope as distillate and, by mass balance, an ethanol-water bottoms for recycle to C3) or in the indirect-split mode (producing high-boiling node water as bottoms and, by mass balance, a ternary distillate near the distillation boundary). (The distillate may be recycled to the decanter, the top of column C1, or the C1 feed.) The indirect-split mode alternatives are discussed in greater detail by Knapp and Doherty (*Kirk-Othmer Encyclopedia of Chemical Technology,* 5th ed., vol. 8, Wiley, New York, 2004, p. 786).

Design and Operation of Azeotropic Distillation Columns Simulation and design of azeotropic distillation columns are a difficult computational problem, but one that is readily handled, in most cases, by widely

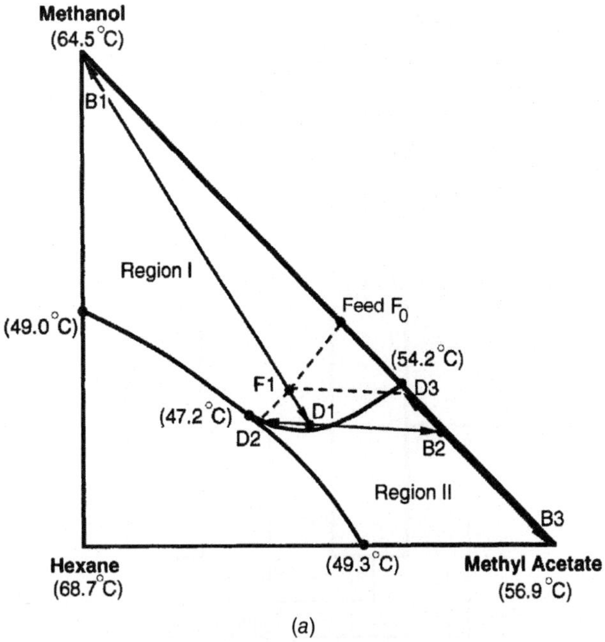

FIG. 13-84 Separation of methanol–methyl acetate by exploitation of distillation boundary curvature.

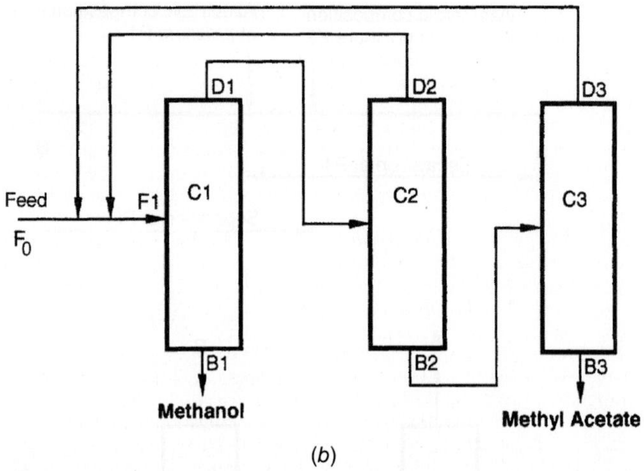

FIG. 13-84 (*Continued*)

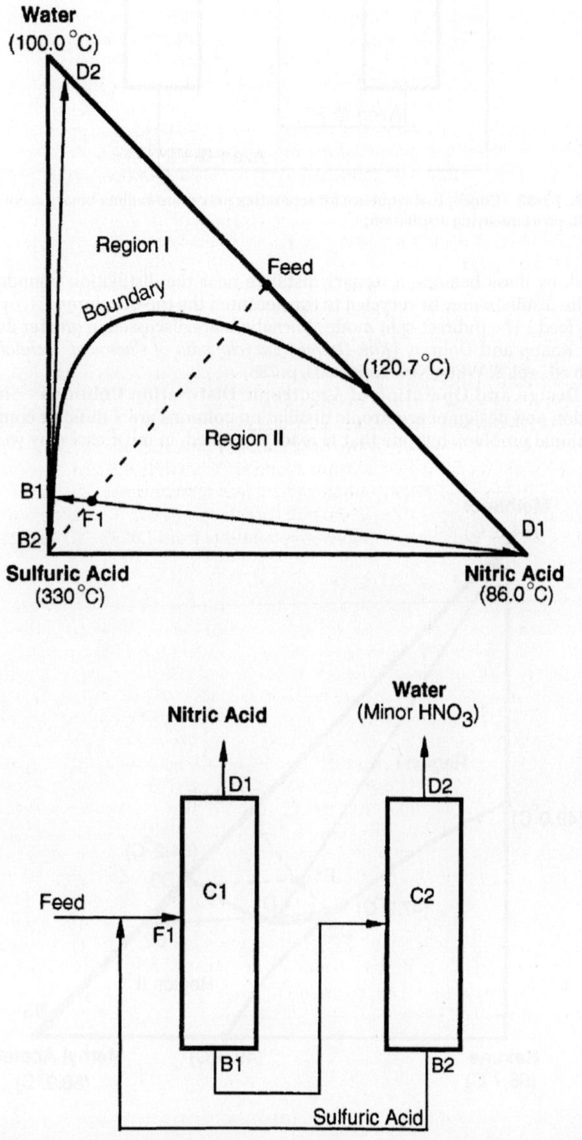

FIG. 13-85 Separation of nitric acid–water system with sulfuric acid in a two-column sequence exploiting extreme boundary curvature.

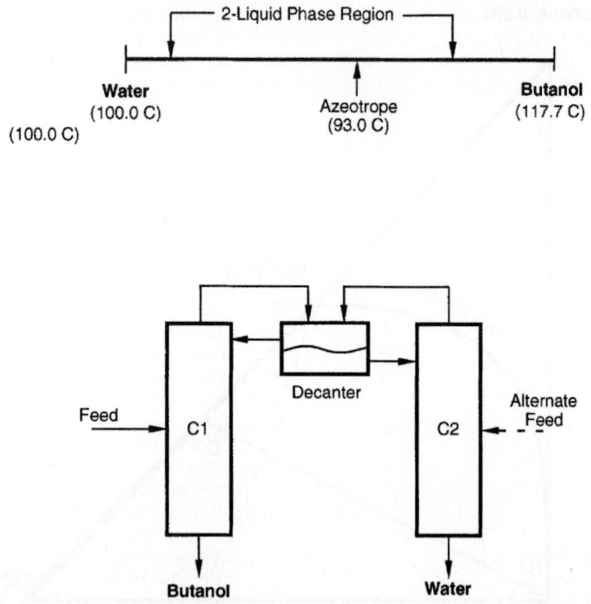

FIG. 13-86 Separation of butanol-water with heterogeneous azeotropic distillation.

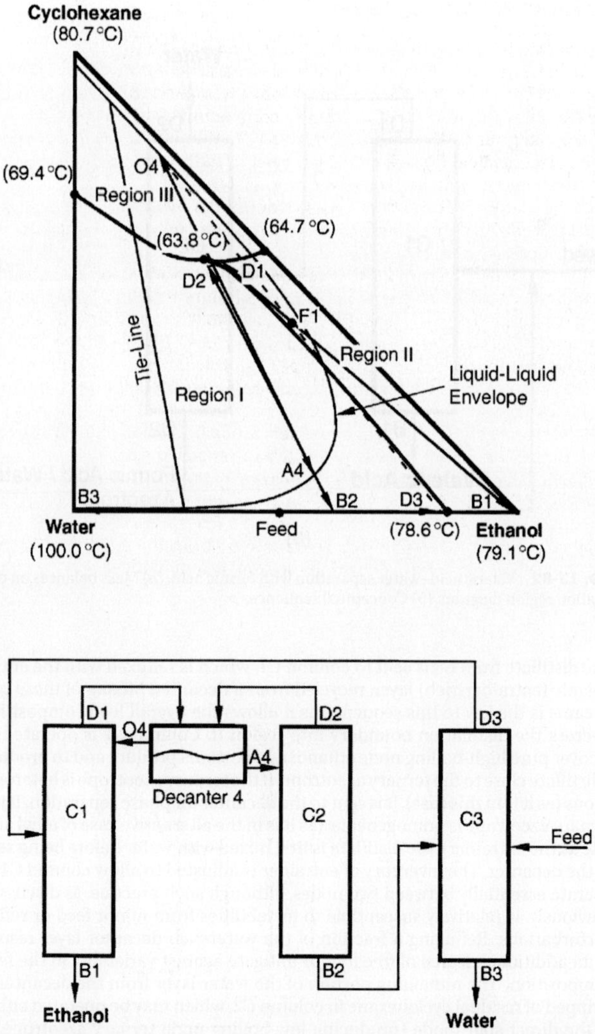

FIG. 13-87 Three-column sequence for ethanol dehydration with cyclohexane (operating column C2 in the direct-split mode).

available commercial computer process simulation packages [Glasscock and Hale, *Chem. Eng.* **101**(11): 82 (1994)]. Most simulators are capable of modeling the steady-state and dynamic behavior of both homogeneous azeotropic distillation systems and those systems involving two-liquid phase behavior within the column, if accurate thermodynamic data and activity coefficient or equation-of-state models are available. However, VLE and VLLE estimated or extrapolated from binary data or predicted from such methods as UNIFAC may not be able to accurately locate boundaries and predict the extent of liquid immiscibilities. Moreover, different activity coefficient models fit to the same experimental data often give very different results for the shape of distillation boundaries and liquid-liquid regions. Therefore, the design of separation schemes relying on boundary curvature should not be attempted unless accurate, reliable experimental equilibrium data are available.

Two liquid phases can occur within a column in the distillation of heterogeneous systems. Older references, e.g., Robinson and Gilliland (*Elements of Fractional Distillation,* McGraw-Hill, New York, 1950), state that the presence of two liquid phases in a column should be avoided as much as possible because performance may be reduced. However, subsequent studies indicate that problems with two-phase flow have been overstated [Herron et al., *AIChE J.* **34:** 1267 (1988); Harrison, *Chem. Eng. Prog.* **86**(11): 80 (1990)]. Based on case history data and experimental evidence, there is no reason to expect unusual capacity or pressure-drop limitations, and standard correlations for these parameters should give acceptable results. Because of the intense nature of the gas-liquid-liquid mixing on trays, mass-transfer efficiencies are relatively unaffected by liquid-liquid phase behavior. The falling-film nature of gas-liquid-liquid contact in packing, however, makes that situation more uncertain. Reduced efficiencies may be expected in systems where one of the keys distributes between the phases.

EXTRACTIVE DISTILLATION

Extractive distillation is a partial vaporization process in the presence of a miscible, high-boiling, nonvolatile mass separation agent, normally called the *solvent,* which is added to an azeotropic or nonazeotropic feed mixture to alter the volatilities of the key components without the formation of any additional azeotropes. Extractive distillation is used throughout the petrochemical and chemical processing industries for the separation of close-boiling, pinched, or azeotropic systems for which simple single-feed distillation is either too expensive or impossible. It can also be used to obtain products that are residue curve saddles, a task not generally possible with single-feed distillation.

Figure 13-88 illustrates the classical implementation of an extractive distillation process for the separation of a binary mixture. The configuration consists of a double-feed extractive column (C1) and a solvent recovery column (C2). The components A and B may have a low relative volatility or form a minimum-boiling azeotrope. The solvent is introduced into the extractive column at a high concentration a few stages below the condenser, but above the primary-feed stage. Since the solvent is chosen to be nonvolatile, it remains at a relatively high concentration in the liquid phase throughout the sections of the column below the solvent-feed stage.

One of the components, A (not necessarily the most volatile species of the original mixture), is withdrawn as an essentially pure distillate stream. Because the solvent is nonvolatile, at most a few stages above the solvent-feed stage are sufficient to rectify the solvent from the distillate. The bottoms product, consisting of B and the solvent, is sent to the recovery column.

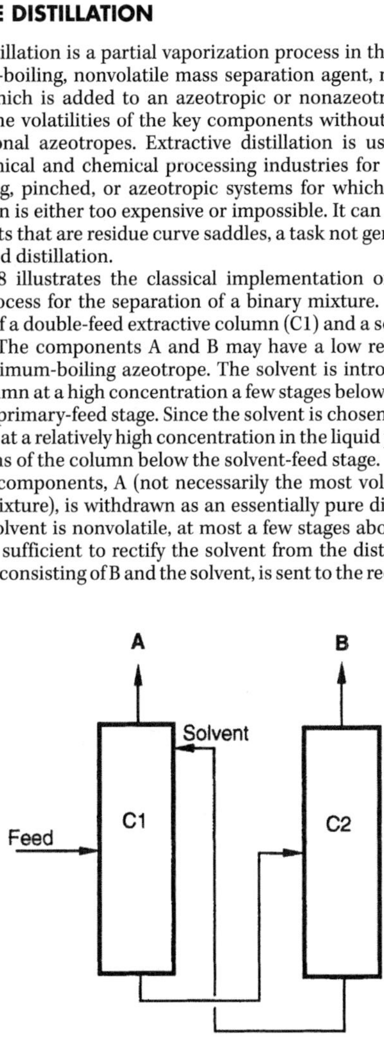

FIG. 13-88 Typical extractive distillation sequence. Component A is less associated with the solvent.

The distillate from the recovery column is pure B, and the solvent-bottoms product is recycled to the extractive column.

Extractive distillation works by the exploitation of the selective solvent-induced enhancements or moderations of the liquid-phase nonidealities of the original components to be separated. The solvent selectively alters the activity coefficients of the components being separated. To do this, a high concentration of solvent is necessary. Several features are essential:

1. The solvent must be chosen to affect the liquid-phase behavior of the key components differently; otherwise, no enhancement in separability will occur.
2. The solvent must be higher-boiling than the key components of the separation and must be relatively nonvolatile in the extractive column, in order to remain largely in the liquid phase.
3. The solvent should not form additional azeotropes with the components in the mixture to be separated.
4. The extractive column *must* be a double-feed column, with the solvent feed above the primary feed. The column must have an extractive section (middle section) between the rectifying section and the stripping section.

As a consequence of these restrictions, separation of binary mixtures by extractive distillation corresponds to only two possible three-component distillation region diagrams, depending on whether the binary mixture is pinched or close-boiling (DRD 001), or forms a minimum-boiling azeotrope (DRD 003). The addition of high-boiling solvents can also facilitate the breaking of maximum-boiling azeotropes (DRD 014)—for example, splitting the nitric acid–water azeotrope with sulfuric acid. However, as explained in the subsection on azeotropic distillation, this type of separation might be better characterized as exploiting extreme boundary curvature rather than extractive distillation because the important liquid-phase activity coefficient modification occurs in the bottom of the column. Although many references show sulfuric acid being introduced high in the column, in fact two separate feeds are not required.

Examples of industrial uses of extractive distillation grouped by distillation region diagram type are given in Table 13-17. Achievable product compositions in double-feed extractive distillation columns are very different from the bow tie regions for single-feed columns. For a given solvent, only one of the pure components in the original binary mixture can be obtained as distillate from the extractive column (the higher-boiling of which is a saddle for close-boiling systems, and both of which are saddles for minimum-boiling azeotropic systems). However, different solvents are capable of selecting either A or B as distillate (but not both). Simple tests are available for determining which component is the distillate, as discussed later in this section.

Extractive distillation is generally only applicable to systems in which the components to be separated contain one or more different functional groups. Extractive distillation is usually uneconomical for separating stereoisomers, homologs, or structural isomers containing the same functional groups, unless the differences in structure also contribute to significantly different polarity, dipole moment, or hydrophobic character. One such example is the separation of ethanol from isopropanol, where the addition of methyl benzoate raises the relative volatility from 1.09 to 1.27 [Berg et al., *Chem. Eng. Comm.* **66:** 1 (1988)].

Solvent Effects in Extractive Distillation In the ordinary distillation of ideal or nonazeotropic mixtures, the component with the lowest pure-component boiling point is always recovered primarily in the distillate, while the highest boiler is recovered primarily in the bottoms. The situation is not as straightforward for an extractive distillation operation. With some solvents, the key component with the lower pure-component boiling point in the original mixture will be recovered in the distillate as in ordinary distillation. For another solvent, the expected order may be reversed, and the component with the higher pure-component boiling point will be recovered in the distillate. The possibility that the expected relative volatility may be reversed by the addition of solvent is entirely a function of the way the solvent interacts with and modifies the activity coefficients and, thus, the volatility of the components in the mixture.

In normal applications of extractive distillation (i.e., pinched, close-boiling, or azeotropic systems), the relative volatilities between the light and heavy key components will be unity or close to unity. Assuming an ideal vapor phase and subcritical components, the relative volatility between the light and heavy keys of the desired separation can be written as the product of the ratios of the pure-component vapor pressures and activity coefficients whether the solvent is present or not:

$$\alpha_{L,H} = \left(\frac{P_L^{\text{sat}}}{P_H^{\text{sat}}}\right)\left(\frac{\gamma_L}{\gamma_H}\right) \tag{13-115}$$

where L and H denote the lower-boiling and higher-boiling key pure component, respectively.

TABLE 13-17 Examples of Extractive Distillation, Salt Extractive Distillation

System	Type	Solvent(s)	Remark
Ethanol-water	Minimum-boiling azeotrope	Ethylene glycol, acetate salts for salt process	Alternative to azeotropic distillation, pressure swing distillation
Benzene-cyclohexane	Minimum-boiling azeotrope	Aniline	
Ethyl acetate-ethanol	Minimum-boiling azeotrope	Higher esters or alcohols, aromatics	Process similar for other alcohol-ester systems
THF-water	Minimum-boiling azeotrope	Propylene glycol	Alternative to pressure swing distillation
Acetone-methanol	Minimum-boiling azeotrope	Water, aniline, ethylene glycol	
Isoprene-pentane	Minimum-boiling azeotrope	Furfural, DMF, acetonitrile	
Pyridine-water	Minimum-boiling azeotrope	Bisphenol	
Methyl acetate-methanol	Minimum-boiling azeotrope	Ethylene glycol monomethyl ether	Element of recovery system for alternative to production of methyl acetate by reactive distillation; alternative to azeotropic, pressure, swing distillation
C4 alkenes/C4 alkanes/ C4 dienes	Close-boiling and minimum-boiling azeotropes	Furfural, DMF, acetonitrile, *n*-methylpyrolidone	
C5 alkenes/C5 alkanes/ C5 dienes	Close-boiling and minimum-boiling azeotropes	Furfural, DMF, acetonitrile, *n*-methylpyrolidone	
Heptane isomers-cyclohexane	Close-boiling	Aniline, phenol	
Heptane isomers-toluene	Close-boiling and minimum-boiling azeotropes	Aniline, phenol	
Vinyl acetate-ethyl acetate	Close-boiling	Phenol, aromatics	Alternative to simple distillation
Propane-propylene	Close-boiling	Acrylonitrile	Alternative to simple distillation, adsorption
Ethanol-isopropanol	Close-boiling	Methyl benzoate	Alternative to simple distillation
Hydrochloric acid-water	Maximum-boiling azeotrope	Sulfuric acid, calcium chloride for salt process	Sulfuric acid process relies heavily on boundary curvature
Nitric acid-water	Maximum-boiling azeotrope	Sulfuric acid, magnesium nitrate for salt process	Sulfuric acid process relies heavily on boundary curvature

The addition of the solvent has an indirect effect on the vapor-pressure ratio. Because the solvent is high-boiling and is generally added at a relatively high molar ratio to the primary-feed mixture, the temperature of an extractive distillation process tends to increase over that of a simple distillation of the original mixture (unless the system pressure is lowered). The result is a corresponding increase in the vapor pressure of both key components. However, the rise in operating temperature generally *does not* result in a significant modification of the relative volatility; the ratio of vapor pressures often remains approximately constant, unless the slopes of the vapor-pressure curves differ significantly. The ratio of the vapor pressures typically remains greater than unity, following the "natural" volatility of the system.

Since activity coefficients have a strong dependence on composition, the effect of the solvent on the activity coefficients is generally more pronounced. However, the magnitude and direction of change are highly dependent on the solvent concentration as well as on the liquid-phase interactions between the solvent and the key components. The solvent acts to lessen the nonidealities of the key component whose liquid-phase behavior is similar to that of the solvent, while enhancing the nonideal behavior of the dissimilar key. The solvent and the key component that show most similar liquid-phase behavior tend to exhibit weak molecular interactions. These components form an ideal or nearly ideal liquid solution. The activity coefficient of this key approaches unity, or may even show negative deviations from Raoult's law if solvating or complexing interactions occur. On the other hand, the dissimilar key and the solvent demonstrate unfavorable molecular interactions, and the activity coefficient of this key increases. The positive deviations from Raoult's law are further enhanced by the diluting effect of the high-solvent concentration, and the value of the activity coefficient of this key may approach the infinite dilution value, often a very large number.

The natural relative volatility of the system is enhanced when the activity coefficient of the lower-boiling pure component is increased by the solvent addition (γ_L/γ_H increases and $P_L^{sat}/P_H^{sat} > 1$). In this case, the lower-boiling pure component will be recovered in the distillate as expected. For the higher-boiling pure component to be recovered in the distillate, the addition of the solvent must decrease the ratio γ_L/γ_H such that the product of γ_L/γ_H and P_L^{sat}/P_H^{sat} (that is, α_{LH}) in the presence of the solvent is less than unity. Generally, the latter is more difficult to achieve and requires higher solvent-to-feed ratios. It is normally better to select a solvent that forces the lower-boiling component overhead.

The effect of solvent concentration on the activity coefficients of the key components is shown in Fig. 13-89 for the methanol-acetone system with either water or methylisopropylketone (MIPK) as a solvent. For an initial feed mixture of 50 mol% methanol and 50 mol% acetone (no solvent present), the ratio of the activity coefficients of methanol and

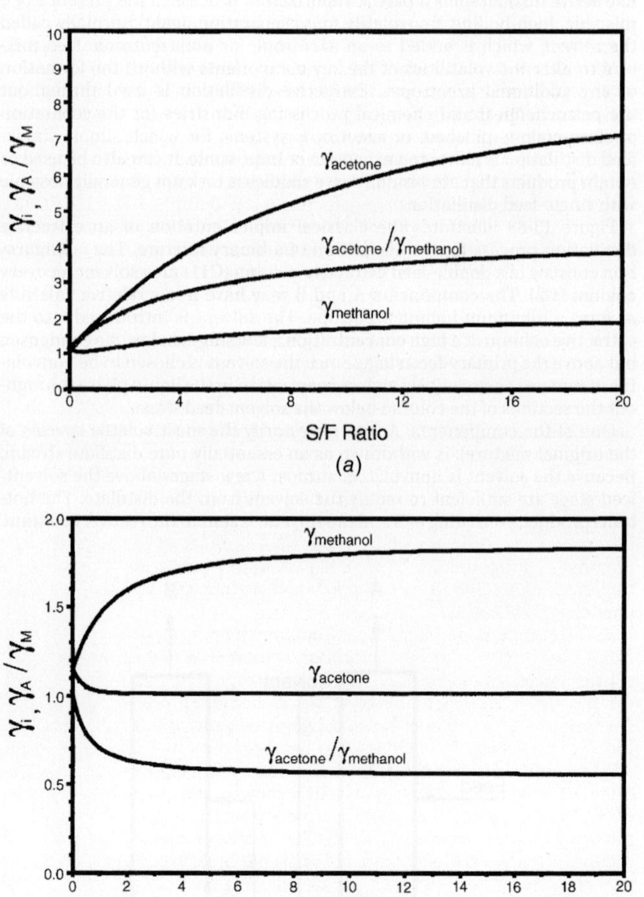

FIG. 13-89 Effect of solvent concentration on activity coefficients for acetone-methanol system. (*a*) Water solvent. (*b*) MIPK solvent.

acetone is close to unity. With water as the solvent, the activity coefficient of the similar key (methanol) rises slightly as the solvent concentration increases, while the coefficient of acetone approaches the relatively large infinite dilution value. With methylisopropylketone as the solvent, acetone is the similar key, and its activity coefficient drops toward unity as the solvent concentration increases, while the activity coefficient of the methanol increases.

Several methods are available for determining whether the lower- or higher-boiling pure component will be recovered in the distillate. For a series of solvent concentrations, the binary *y-x* phase diagram for the low-boiling and high-boiling keys can be plotted on a solvent-free basis. At a particular solvent concentration (dependent on the selected solvent and keys), the azeotropic point in the binary plot disappears at one of the pure-component corners. The component corresponding to the corner where the azeotrope disappears is recovered in the distillate (Knapp and Doherty, *Kirk-Othmer Encyclopedia of Chemical Technology,* 5th ed., vol. 8, Wiley, New York, 2004, p. 786). LaRoche et al. [*Can. J. Chem. Eng.* **69:** 1302 (1991)] present a related method in which the $\alpha_{LH} = 1$ line is plotted on the ternary composition diagram. If this line intersects the lower-boiling pure component + solvent binary face, then the lower-boiling component will be recovered in the distillate, and vice versa if the $\alpha_{LH} = 1$ line intersects the higher-boiling pure component + solvent face. A very simple method, if an accurate residue curve map is available, is to examine the shape and inflection of the residue curves as they approach the pure solvent vertex. Whichever solvent-key component face the residue curves predominantly tend toward as they approach the solvent vertex is the key component that will be recovered in the bottoms with the solvent (see property 6, p. 193, in Doherty and Malone, *Conceptual Design of Distillation Systems,* McGraw-Hill, 2001). In Fig. 13-90*a,* all residue curves approaching the water (solvent) vertex are inflected toward the methanol-water face, with the result that methanol will be recovered in the bottoms and acetone in the distillate. Alternatively, with MIPK as the solvent, all residue curves show inflection toward the acetone-MIPK face (Fig. 13-90*b*), indicating that acetone will be recovered in the bottoms and methanol in the distillate.

Extractive Distillation Design and Optimization Extractive distillation column composition profiles have a very characteristic shape on a ternary diagram. The composition profile for the separation of methanol-acetone with water is given in Fig. 13-91. Stripping and rectifying profiles start at the bottoms and distillate compositions, respectively, track generally along the faces of the composition triangle, and then turn toward the high-boiling (solvent) node and low-boiling node, respectively. For a feasible single-feed design, these profiles must cross at some point. However, in an extractive distillation they cannot cross. The extractive section profile acts at the bridge between these two sections. Most of the key component separation occurs in this section in the presence of high solvent composition.

The variable that has the most significant impact on the economics of an extractive distillation is the solvent-to-feed flow rate ratio *S/F*. For close-boiling or pinched nonazeotropic mixtures, no minimum-solvent flow rate is required to effect the separation because the separation is always theoretically possible (if not economical) in the absence of the solvent. However, the extent of enhancement of the relative volatility is largely determined by the solvent composition in the lower column sections and hence the *S/F* ratio. The relative volatility tends to increase as the *S/F* ratio increases. Thus, a given separation can be accomplished in fewer equilibrium stages. As an illustration, the total number of theoretical stages required as a function of *S/F* ratio is plotted in Fig. 13-92*a* for the separation of the nonazeotropic mixture of vinyl acetate and ethyl acetate using phenol as the solvent.

For the separation of a minimum-boiling binary azeotrope by extractive distillation, there is clearly a minimum-solvent flow rate below which the separation is impossible (due to the azeotrope). For azeotropic separations, the number of equilibrium stages is infinite at or below $(S/F)_{min}$ and decreases rapidly with increasing solvent feed flow, and then may asymptote, or rise slowly. The relationship between the total number of stages and the *S/F* ratio for a given purity and recovery for the azeotropic acetone-methanol system with water as solvent is shown in Fig. 13-92*b*. A rough idea of $(S/F)_{min}$ can be determined from a pseudobinary diagram or by plotting the $\alpha_{LH} = 1$ line on a ternary diagram. The solvent composition at which the azeotrope disappears in a corner of the pseudobinary diagram is an indication of $(S/F)_{min}$ [LaRoche et al., *Can. J. Chem. Eng.* **69:** 1302 (1991)]. An exact method for calculating $(S/F)_{min}$ is given by Knapp and Doherty [*AIChE J.* **40:** 243 (1994)]. Typically, operating *S/F* ratios for economically acceptable solvents are between 2 and 5. Higher *S/F* ratios tend to increase the diameter of both the extractive column and the solvent recovery columns, but they tend to reduce the required number of equilibrium stages and minimum reflux ratio. Moreover, higher *S/F* ratios lead to higher reboiler temperatures,

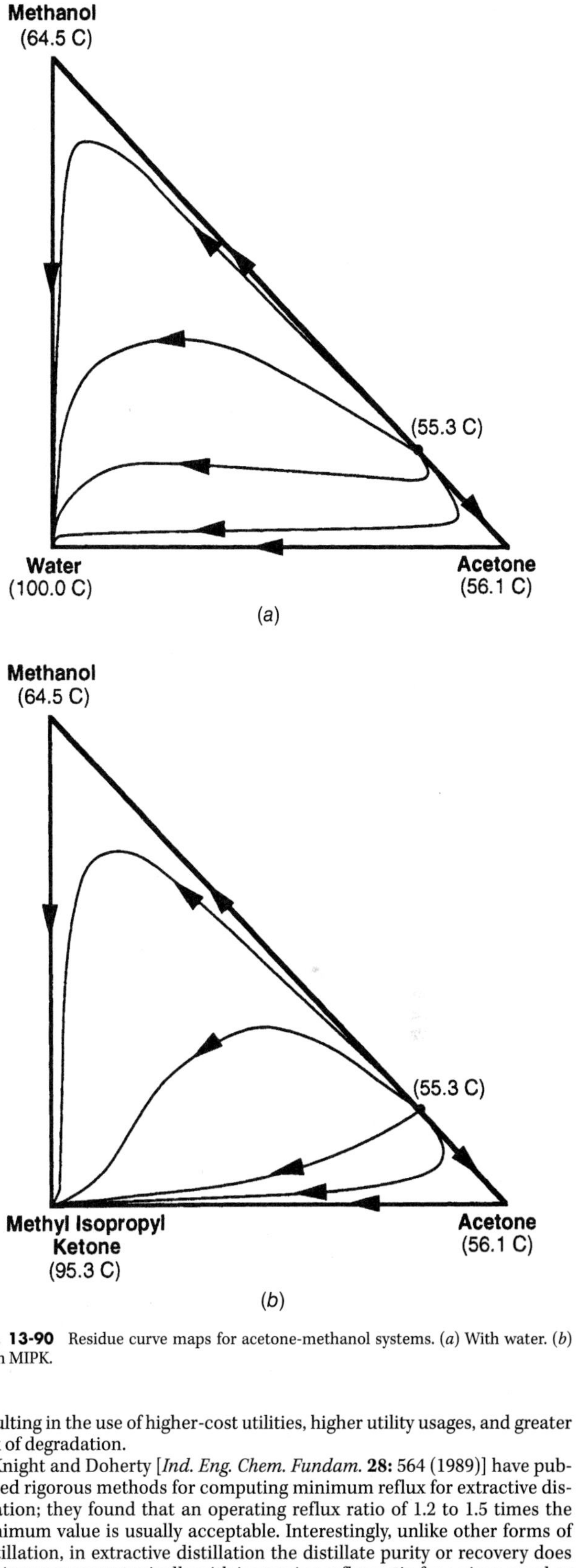

FIG. 13-90 Residue curve maps for acetone-methanol systems. (*a*) With water. (*b*) With MIPK.

resulting in the use of higher-cost utilities, higher utility usages, and greater risk of degradation.

Knight and Doherty [*Ind. Eng. Chem. Fundam.* **28:** 564 (1989)] have published rigorous methods for computing minimum reflux for extractive distillation; they found that an operating reflux ratio of 1.2 to 1.5 times the minimum value is usually acceptable. Interestingly, unlike other forms of distillation, in extractive distillation the distillate purity or recovery does not increase monotonically with increasing reflux ratio for a given number of stages. Above a maximum reflux ratio, the separation can no longer be achieved, and the distillate purity actually decreases for a given number of stages [LaRoche et al., *AIChE J.* **38:** 1309 (1992)]. The difference between R_{min} and R_{max} increases as the *S/F* ratio increases. Large amounts of reflux lower the solvent composition in the upper section of the column, degrading

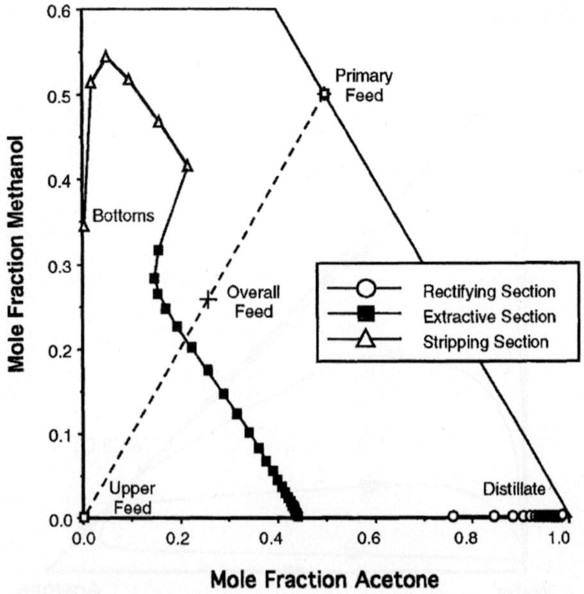

FIG. 13-91 Extractive distillation column composition profile for the separation of acetone-methanol with water.

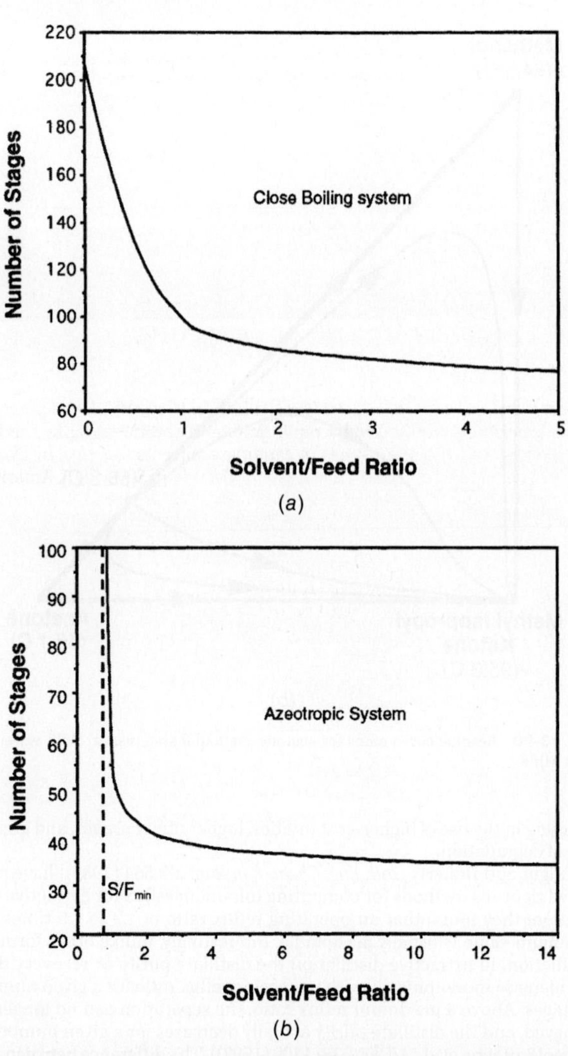

FIG. 13-92 Number of theoretical stages versus solvent-to-feed ratio for extractive distillation. (*a*) Close-boiling vinyl acetate–ethyl acetate system with phenol solvent. (*b*) Azeotropic acetone-methanol system with water solvent.

rather than enhancing column performance. Because the reflux ratio goes through a maximum, the conventional control strategy of increasing reflux to maintain purity can be detrimental rather than beneficial. However, R_{max} generally occurs at impractically high reflux ratios and is typically not a major concern.

The thermal quality of the solvent feed has no effect on the value of S/F_{min}, but it does affect the minimum reflux to some extent, especially as the S/F ratio increases. The maximum reflux ratio R_{max} occurs at higher values of the reflux ratio as the upper-feed quality decreases; a subcooled upper feed provides additional refluxing capacity, and less external reflux is required for the same separation. It is also sometimes advantageous to introduce the primary feed to the extractive distillation column as a vapor to help maintain a higher solvent composition on the feed tray and the trays immediately below.

Robinson and Gilliland (*Elements of Fractional Distillation,* McGraw-Hill, New York, 1950), Smith (*Design of Equilibrium Stage Processes,* McGraw-Hill, New York, 1963), Van Winkle (*Distillation,* McGraw-Hill, New York, 1967), and Walas (*Chemical Process Equipment,* Butterworths, Boston, 1988) discuss rigorous stage-to-stage design techniques as well as shortcut and graphical methods for determining minimum stages, $(S/F)_{min}$, minimum reflux, and the optimum locations of the solvent and primary feed points. Knapp and Doherty [*AIChE J.* **40:** 243 (1994)] have published column design methods based on geometric arguments and fixed-point analysis that are capable of calculating $(S/F)_{min}$, as well as the minimum and maximum reflux ratios. Most commercial simulators can solve multiple-feed extractive distillation heat and material balances, but they do not include straightforward techniques for calculating $(S/F)_{min}$ or the minimum and maximum reflux ratios.

Solvent Screening and Selection Choosing an effective solvent can have the most profound effect on the economics of an extractive distillation process. The approach most often adopted is to first generate a short list of potential solvents by using simple qualitative screening and selection methods. Experimental verification is best undertaken only after a list of promising candidate solvents has been generated and some chance at economic viability has been demonstrated via preliminary process modeling.

Solvent selection and screening approaches can be divided into two levels of analysis. The first level focuses on the identification of functional groups or chemical families that are likely to give favorable solvent–key component molecular interactions. The second level of analysis identifies and compares individual candidate solvents. The various methods of analysis are described briefly and illustrated with an example of choosing a solvent for the methanol-acetone separation.

First Level: Broad Screening by Functional Group or Chemical Family

Homologous series. Select candidate solvents from the high-boiling homologous series of both light and heavy key components. Favor homologs of the heavy key because this tends to enhance the natural relative volatility of the system. Homologous components tend to form ideal solutions and are unlikely to form azeotropes [Scheibel, *Chem. Eng. Prog.* **44**(12): 927 (1948)].

Robbins chart. Select candidate solvents from groups in the Robbins chart (Table 13-13) that tend to give positive (or no) deviations from Raoult's law for the key component desired in the distillate and negative (or no) deviations for the other key.

Hydrogen-bonding characteristics. Select candidate solvents from groups that are likely to cause the formation of hydrogen bonds with the key component to be removed in the bottoms, or disruption of hydrogen bonds with the key to be removed in the distillate. The formation and disruption of hydrogen bonds are often associated with strong negative and positive deviations, respectively, from Raoult's law. Several authors have developed charts indicating expected hydrogen bonding interactions between families of compounds [Ewell et al., *Ind. Eng. Chem.* **36:** 871 (1944); Gilmont et al., *Ind. Eng. Chem.* **53:** 223 (1961); Berg, *Chem. Eng. Prog.* **65**(9): 52 (1969)]. Table 13-18 presents a hydrogen bonding classification of chemical families and a summary of deviations from Raoult's law.

Polarity characteristics. Select candidate solvents from chemical groups that tend to show higher polarity than one key component or lower polarity than the other key. Polarity effects are often cited as a factor in causing deviations from Raoult's law [Hopkins and Fritsch, *Chem. Eng. Prog.* **51**(8): (1954); Carlson et al., *Ind. Eng. Chem.* **46:** 350 (1954); Prausnitz and Anderson, *AIChE J.* **7:** 96 (1961)]. The general trend in polarity based on the functional group of a molecule is given in Table 13-19. The chart is best for molecules of similar size. A more quantitative measure of the polarity of a molecule is the polarity contribution to the three-term Hansen solubility parameter. A tabulation of calculated three-term solubility parameters is provided by Barton (*CRC Handbook of Solubility Parameters and Other Cohesion Parameters,* CRC Press, Boca Raton, Fla., 1991), along with a group contribution method for calculating the three-term solubility parameters of compounds not listed in the reference.

TABLE 13-18 Hydrogen Bonding Classification of Chemical Families

Class	Chemical family			
H-Bonding, Strongly Associative (HBSA)	Water Primary amides Secondary amides	Polyacids Dicarboxylic acids Monohydroxy acids	Polyphenols Oximes Hydroxylamines	Amino alcohols Polyols
H-Bond Acceptor-Donor (HBAD)	Phenols Aromatic acids Aromatic amines Alpha H nitriles	Imines Monocarboxylic acids Other monoacids Peracids	Alpha H nitros Azines Primary amines Secondary amines	*n*-alcohols Other alcohols Ether alcohols
H-Bond Acceptor (HBA)	Acyl chlorides Acyl fluorides Hetero nitrogen aromatics Hetero oxygen aromatics	Tertiary amides Tertiary amines Other nitriles Other nitros Isocyanates Peroxides	Aldehydes Anhydrides Cyclo ketones Aliphatic ketones Esters Ethers	Aromatic esters Aromatic nitriles Aromatic ethers Sulfones Sulfolanes
π-Bonding Acceptor (π-HBA)	Alkynes Alkenes	Aromatics Unsaturated esters		
H-Bond Donor (HBD)	Inorganic acids Active H chlorides	Active H fluorides Active H iodides	Active H bromides	
Non-Bonding (NB)	Paraffins Nonactive H chlorides	Nonactive H fluorides Sulfides	Nonactive H iodides Disulfides	Nonactive H bromides Thiols

Deviations from Raoult's Law

H-Bonding classes	Type of deviations	Comments
HBSA + NB HBAD + NB	Alway positive dev., HBSA + NB often limited miscibility	H-bonds broken by interactions
HBA + HBD	Always negative dev.	H-bonds formed by interactions
HBSA + HBD HBAD + HBD	Always positive deviations, HBSA + HBD often limited miscibility	H-bonds broken and formed; dissociation of HBSA or HBAD liquid most important effect
HBSA + HBSA HBSA + HBAD HBSA + HBA HBAD + HBAD HBAD + HBA	Usually positive deviations; some give maximum-boiling azeotropes	H-bonds broken and formed
HBA + HBA HBA + NB HBD + HBD HBD + NB NB + NB	Ideal, quasi-ideal systems; always positive or no deviations; azeotropes, if any, minimum-boiling	No H-bonding involved

NOTE: π-HBA is *enhanced* version of HBA.

Second Level: Identification of Individual Candidate Solvents

Boiling point characteristics. Select only candidate solvents that boil at least 30°C to 40°C above the key components to ensure that the solvent is relatively nonvolatile and remains largely in the liquid phase. With this boiling point difference, the solvent should also not form azeotropes with the other components.

Selectivity at infinite dilution. Rank candidate solvents according to their selectivity at infinite dilution. The selectivity at infinite dilution is defined as the ratio of the activity coefficients at infinite dilution of the two key components in the solvent. Since solvent effects tend to increase as solvent concentration increases, the infinite-dilution selectivity gives an upper bound on the efficacy of a solvent. Infinite-dilution activity coefficients can be predicted by using such methods as UNIFAC, ASOG, MOSCED (Reid et al., *Properties of Gases and Liquids,* 4th ed., McGraw-Hill, New York, 1987). They can be found experimentally by using a rapid gas-liquid chromatography method based on relative retention times in candidate solvents (Tassios, in *Extractive and Azeotropic Distillation,* Advances in Chemistry Series 115, American Chemical Society, Washington, 1972), and they can be correlated to bubble point data [Kojima and Ochi, *J. Chem. Eng. Japan* **7**(2): 71 (1974)]. DECHEMA (*Vapor-Liquid Equilibrium Data Collection,* Frankfort, 1977) has also published a compilation of experimental infinite-dilution activity coefficients.

TABLE 13-19 Relative Polarities of Functional Groups

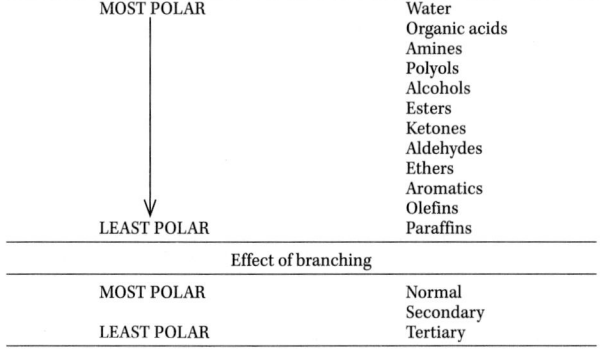

MOST POLAR	Water
↓	Organic acids
	Amines
	Polyols
	Alcohols
	Esters
	Ketones
	Aldehydes
	Ethers
	Aromatics
	Olefins
LEAST POLAR	Paraffins
Effect of branching	
MOST POLAR	Normal
	Secondary
LEAST POLAR	Tertiary

Experimental measurement of relative volatility. Rank candidate solvents by the increase in relative volatility caused by the addition of the solvent. One technique is to experimentally measure the relative volatility of a fixed-composition, key component + solvent mixture (often a 1/1 ratio of each key, with a 1/1 to 3/1 solvent/key ratio) for various solvents [Carlson et al., *Ind. Eng. Chem.* **46:** 350 (1954)]. The Othmer equilibrium still is the apparatus of choice for these measurements [Zudkevitch, *Chem. Eng. Comm.* **116:** 41 (1992)].

At atmospheric pressure, methanol and acetone boil at 64.5°C and 56.1°C, respectively, and they form a minimum-boiling azeotrope at 55.3°C. The natural volatility of the system is acetone > methanol, so the favored solvents most likely will be those that cause acetone to be recovered in the distillate. However, for the purposes of the example, a solvent that reverses the natural volatility will also be identified. First, by examining the polarity of ketones and alcohols (Table 13-19), solvents favored for the recovery of methanol in the bottoms would come from groups more polar than methanol, such as acids, water, and polyols. Turning to the Robbins chart (Table 13-13), we see that favorable groups are amines, alcohols, polyols, and water since these show expected positive deviations for acetone and zero or negative deviations for methanol. For reversing the natural volatility, solvents should be chosen that are less polar than acetone, such as ethers, hydrocarbons, and aromatics. Unfortunately, both ethers and hydrocarbons are expected to give positive deviations for both acetone and methanol, so they should be discarded. Halohydrocarbons and ketones are expected to give positive

TABLE 13-20 Comparison of Candidate Solvents for Methanol/Acetone Extractive Distillation

Solvent	Boiling pt. (°C)	Azeotrope formation	$\gamma^{\infty}_{Acetone}$	γ^{∞}_{MeOH}	$\gamma^{\infty}_{Acetone}/\gamma^{\infty}_{MeOH}$
MEK	79.6	With MeOH	1.01	1.88	0.537
MIPK	102.0	No	1.01	1.89	0.534
MIBK	115.9	No	1.06	2.05	0.517
Ethanol	78.3	No	1.85	1.04	1.78
1-Propanol	97.2	No	1.90	1.20	1.58
1-Butanol	117.8	No	1.93	1.33	1.45
Water	100.0	No	11.77	2.34	5.03
EG	197.2	No	3.71	1.25	2.97

$\gamma^{\infty}_{Acetone} = 1.79$ (in MeOH)
$\gamma^{\infty}_{MeOH} = 1.81$ (in acetone)

deviations for methanol and either negative or no deviations for acetone. The other qualitative indicators show that both homologous series (ketones and alcohols) look promising. Thus, after discounting halohydrocarbons for environmental reasons, the best solvents will probably come from alcohols, polyols, and water for recovering methanol in the bottoms and ketones for recovering acetone in the bottoms. Table 13-20 shows the boiling points and experimental or estimated infinite-dilution activity coefficients for several candidate solvents from the aforementioned groups. Methylethylketone boils too low, as does ethanol, and it also forms an azeotrope with methanol. These two candidates can be discarded. Other members of the homologous series, along with water and ethylene glycol, have acceptable boiling points (at least 30°C higher than those of the keys). Of these, water (the solvent used industrially) clearly has the largest effect on the activity coefficients, followed by ethylene glycol. Although inferior to water or ethylene glycol, both MIPK and MIBK would probably be acceptable for reversing the natural volatility of the system.

Extractive Distillation by Salt Effects A second method of modifying the liquid-phase behavior (and thus the relative volatility) of a mixture to effect a separation is by the addition of a nonvolatile, soluble, ionic salt. The process is analogous to extractive distillation with a high-boiling liquid. In the simplest case, for the separation of a binary mixture, the salt is fed at the top of the column by dissolving it in the hot reflux stream before introduction into the column. To function effectively, the salt must be adequately soluble in both components throughout the range of compositions encountered in the column. Since the salt is completely nonvolatile, it remains in the liquid phase on each tray and alters the relative volatility throughout the length of the column. No rectification section is needed above the salt feed. The bottoms product is recovered from the salt solution by evaporation or drying, and the salt is recycled. The ions of a salt are typically capable of causing much larger and more selective effects on liquid-phase behavior than the molecules of a liquid solvent. As a result, salt-to-feed ratios of less than 0.1 are typical.

The use of a salting agent presents a number of problems not associated with a liquid solvent, such as the difficulty of transporting and metering a solid or saturated salt solution, slow mixing or dissolution rate of the salt, limits to solubility in the feed components, and the potential for corrosion. However, in the limited number of systems for which an effective salt can be found, the energy usage, equipment size, capital investment, and ultimate separation cost can be significantly reduced compared to that for extractive distillation using a liquid solvent [Furter, *Chem. Eng. Commun.* **116:** 35 (1992)]. Applications of salt extractive distillation include acetate salts to produce absolute ethanol, magnesium nitrate for the production of concentrated nitric acid as an alternative to the sulfuric acid solvent process, and calcium chloride to produce anhydrous hydrogen chloride. Other examples are noted by Furter [*Can. J. Chem. Eng.* **55:** 229 (1977)].

One problem limiting the consideration of salt extractive distillation is the fact that the performance and solubility of a salt in a particular system are difficult to predict without experimental data. Some recent advances have been made in modeling the VLE behavior of organic aqueous salt solutions using modified UNIFAC, NRTL, UNIQUAC, and other approaches [Kumar, *Sep. Sci. Tech.* **28**(1): 799 (1993)].

REACTIVE DISTILLATION

Reactive distillation is a process in which chemical reaction and distillation are carried out simultaneously within a fractional distillation apparatus. Reactive distillation may be advantageous for liquid-phase reaction systems when the reaction must be carried out with a large excess of one or more of the reactants, when a reaction can be driven to completion by the removal of one or more of the products as they are formed, or when the product recovery or by-product recycle scheme is complicated or made infeasible by azeotrope formation.

For consecutive reactions in which the desired product is formed in an intermediate step, excess reactant can be used to suppress additional series reactions by keeping the intermediate-species concentration low. A reactive distillation can achieve the same result by removing the desired intermediate from the reaction zone as it is formed. Similarly, if the equilibrium constant of a reversible reaction is small, high conversions of one reactant can be achieved by the use of a large excess of the other reactant. Alternatively, by Le Chatelier's principle, the reaction can be driven to completion by the removal of one or more of the products as they are formed. Typically, reactants can be kept much closer to stoichiometric proportions in a reactive distillation.

When a reaction mixture exhibits azeotropes, the recovery of products and recycle of excess reagents can be quite complicated and expensive. Reactive distillation can provide a means of breaking azeotropes by altering or eliminating the condition for azeotrope formation in the reaction zone through the combined effects of vaporization-condensation and consumption-production of the species in the mixture. Alternatively, a reaction may be used to convert the species to components that are more easily distilled. In each of these situations, the conversion and selectivity often can be improved markedly, with much lower reactant inventories and recycle rates, and much simpler recovery schemes. The capital savings can be quite dramatic. A list of applications of reactive distillation appearing in the literature is given in Table 13-21. Additional industrial applications are described by Sharma and Mahajani (chap. 1 in Sundmacher and Kienle, eds., *Reactive Distillation*, Wiley-VCH, New York, 2003).

TABLE 13-21 Applications of Reactive Distillation

Process	Reaction type	Reference
Methyl acetate from methanol and acetic acid	Esterification	Agreda et al., *Chem. Eng. Prog.* **86**(2): 40 (1990)
General process for ester formation	Esterification	Simons, "Esterification" in *Encyclopedia of Chemical Processing and Design*, Vol 19, Dekker, New York, 1983
Diphenyl carbonate from dimethyl carbonate and phenol	Esterification	Oyevaar et al., U.S. Patent 6,093,842 (2000)
Dibutyl phthalate from butanol and phthalic acid	Esterification	Berman et al., *Ind. Eng. Chem.* **40:** 2139 (1948)
Ethyl acetate from ethanol and butyl acetate	Transesterification	Davies and Jeffreys, *Trans. Inst. Chem. Eng.* **51:** 275 (1973)
Recovery of acetic acid and methanol from methyl acetate by-product of vinyl acetate production	Hydrolysis	Fuchigami, *J. Chem. Eng. Jap.* **23:** 354 (1990)
Nylon 6,6 prepolymer from adipic acid and hexamethylenediamine	Amidation	Jaswal and Pugi, U.S. Patent 3,900,450 (1975)
MTBE from isobutene and methanol	Etherification	DeGarmo et al., *Chem. Eng. Prog.* **88**(3): 43 (1992)
TAME from pentenes and methanol	Etherification	Brockwell et al., *Hyd. Proc.* **70**(9): 133 (1991)
Separation of close boiling 3- and 4-picoline by complexation with organic acids	Acid-base	Duprat and Gau, *Can. J. Chem. Eng.* **69:** 1320 (1991)
Separation of close-boiling meta and para xylenes by formation of tert-butyl meta-xyxlene	Transalkylation	Saito et al., *J. Chem. Eng. Jap.* **4:** 37 (1971)
Cumene from propylene and benzene	Alkylation	Shoemaker and Jones, *Hyd. Proc.* **67**(6): 57 (1987)
General process for the alkylation of aromatics with olefins	Alkylation	Crossland, U.S. Patent 5,043,506 (1991)
Production of specific higher and lower alkenes from butenes	Diproportionation	Jung et al., U.S. Patent 4,709,115 (1987)
4-Nitrochlorobenzene from chlorobenzene and nitric acid	Nitration	Belson, *Ind. Eng. Chem. Res.* **29:** 1562 (1990)
Production of methylal and high purity formaldehyde		Masamoto and Matsuzaki, *J. Chem. Eng. Jap.* **27:** 1 (1994)

Although reactive distillation has many potential applications, it is not appropriate for all situations. Since it is in essence a distillation process, it has the same range of applicability as other distillation operations. Distillation-based equipment is not designed to effectively handle solids, supercritical components (where no separate vapor and liquid phases exist), gas-phase reactions, or high-temperature or high-pressure reactions such as hydrogenation, steam reforming, gasification, and hydrodealkylation.

Simulation, Modeling, and Design Feasibility Because reaction and separation phenomena are closely coupled in a reactive distillation process, simulation and design are significantly more complex than those of sequential reaction and separation processes. In spite of the complexity, however, most commercial computer process modeling packages offer reliable and flexible routines for simulating steady-state reactive distillation columns, with either equilibrium or kinetically controlled reaction models [Venkataraman et al., *Chem. Eng. Prog.* **86**(6): 45 (1990)]. As with other enhanced distillation processes, the results are very sensitive to the thermodynamic models chosen and the accuracy of the VLE data used to generate model parameters. Of equal if not greater significance is the accuracy of data and models for reaction rate as a function of catalyst concentration, temperature, and composition. Very different conclusions can be drawn about the feasibility of a reactive distillation if the reaction is assumed to reach chemical equilibrium on each stage of the column or if the reaction is assumed to be kinetically controlled [Barbosa and Doherty, *Chem. Eng. Sci.* **43:** 541 (1988); Chadda, Malone, and Doherty, *AIChE J.* **47:** 590 (2001)]. Tray holdup and stage requirements are two important variables directly affected by the reaction time relative to the residence time inside the column. Unlike distillation without reaction, product feasibility can be quite sensitive to changes in tray holdup and production rate.

When an equilibrium reaction occurs in a vapor-liquid system, the phase compositions depend not only on the relative volatility of the components in the mixture, but also on the consumption (and production) of species. Thus, the condition for azeotropy in a nonreactive system ($y_i = x_i$ for all i) no longer holds true in a reactive system and must be modified to include reaction stoichiometry:

$$\frac{y_1 - x_1}{\nu_1 - \nu_T x_1} = \frac{y_i - x_i}{\nu_i - \nu_T x_i} \quad \text{for all} \quad i = 1, \ldots, c \tag{13-116}$$

where $$\nu_T = \sum_{i-1}^{c} \nu_i$$

and ν_i represents the stoichiometric coefficient of component i (negative for reactants, positive for products).

Phase compositions that satisfy Eq. (13-116) are stationary points on a phase diagram and have been labeled *reactive azeotropes* by Barbosa and Doherty [*Chem. Eng. Sci.* **43:** 529 (1988)]. At a reactive azeotrope, the mass exchange between the vapor and liquid phases and the generation (or consumption) of each species are balanced such that the composition of neither phase changes. Reactive azeotropes show the same distillation properties as ordinary azeotropes and therefore affect the achievable products. Reactive azeotropes are predicted to exist in many reacting mixtures, and they have been confirmed experimentally in the reactive boiling mixture of acetic acid + isopropanol + isopropyl acetate + water [Song et al., *Nature* **388:** 561 (1997); Huang et al., *Chem. Eng. Sci.* **60:** 3363 (2005)].

Reactive azeotropes are not easily visualized in conventional *y-x* coordinates but become apparent upon a transformation of coordinates, which depends on the number of reactions, the order of each reaction (for example, $A + B \rightleftharpoons C$ or $A + B \rightleftharpoons C + D$), and the presence of nonreacting components. The general vector-matrix form of the transform for c reacting components, with R reactions, and I nonreacting components, has been derived by Ung and Doherty [*Chem. Eng. Sci.* **50:** 23 (1995)]. For the transformed mole fraction of component i in the liquid phase X_i, they give

$$X_i = \left[\frac{x_i - \mathbf{v}_i^T (\mathbf{\nu}_{\text{Ref}})^{-1} \mathbf{x}_{\text{Ref}}}{1 - \mathbf{v}_{\text{TOT}}^T (\mathbf{\nu}_{\text{Ref}})^{-1} \mathbf{x}_{\text{Ref}}}\right] \quad i = 1 \cdots c - R \tag{13-117}$$

where $\mathbf{v}_i^T$ = row vector of stoichiometric coefficients of component i for each reaction
$\mathbf{\nu}_{\text{Ref}}$ = square matrix of stoichiometric coefficients for R reference components in R reactions
$\mathbf{x}_{\text{Ref}}$ = column vector of mole fractions for R reference components in liquid phase
$\mathbf{v}_{\text{TOT}}^T$ = row vector composed of the sum of stoichiometric coefficients for each reaction

An equation identical to (13-117) defines the transformed mole fraction of component i in the vapor phase Y_i, where the terms in x are replaced by terms in y.

The transformed variables describe the system composition with or without reaction and sum to unity as do x_i and y_i. The condition for reactive azeotropy becomes $X_i = Y_i$. Barbosa and Doherty have shown that phase diagrams and distillation diagrams constructed by using the transformed composition coordinates have the same properties as phase and distillation diagrams for nonreactive systems and similarly can be used to help design feasibility and operability studies [*Chem. Eng. Sci.* **43:** 529, 541, 1523, and 2377 (1988)]. Residue curve maps in transformed coordinates for the reactive system methanol–acetic acid–methyl acetate–water are shown in Fig. 13-93. Note that the nonreactive azeotrope between water and methyl acetate has disappeared, while the methyl acetate–methanol azeotrope remains intact. Only those azeotropes containing all the reactants or products will be altered by the reaction (water and methyl acetate can back-react to form acetic acid and methanol, whereas methanol and methyl acetate

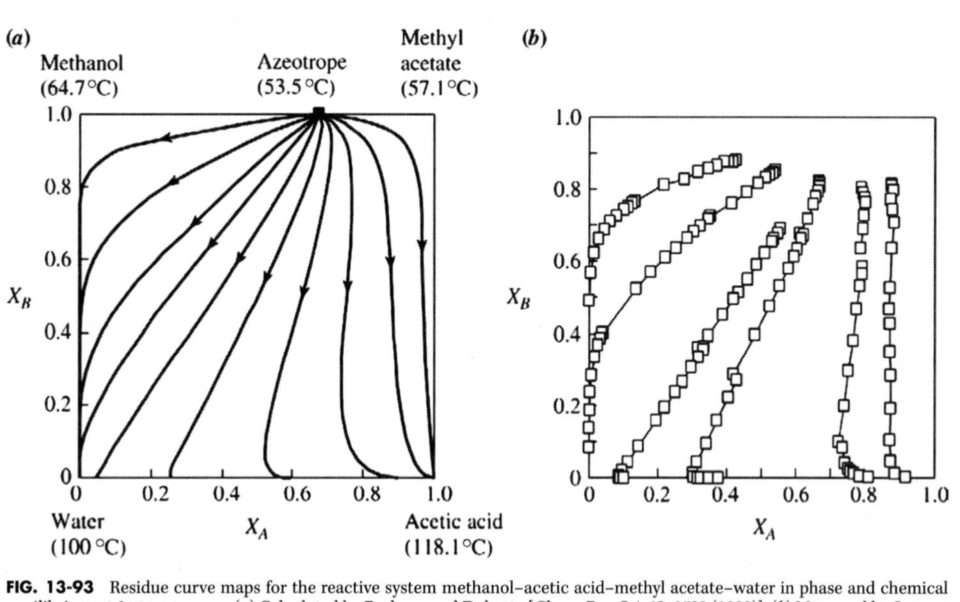

FIG. 13-93 Residue curve maps for the reactive system methanol–acetic acid–methyl acetate–water in phase and chemical equilibrium at 1 atm pressure. (*a*) Calculated by Barbosa and Doherty [*Chem. Eng. Sci.* **43:** 1523 (1988)]. (*b*) Measured by Song et al. [*Ind. Eng. Chem. Res.* **37:** 1917 (1998)].

cannot further react in the absence of either water or acetic acid). This reactive system consists of only one distillation region in which the methanol–methyl acetate azeotrope is the low-boiling node and acetic acid is the high-boiling node.

The situation becomes more complicated when the reaction is kinetically controlled and does not come to complete chemical equilibrium under the conditions of temperature, liquid holdup, and rate of vaporization in the column reactor. Venimadhavan et al. [*AIChE J.* **40:** 1814 (1994); **45:** 546 (1999)] and Rev [*Ind. Eng. Chem. Res.* **33:** 2174 (1994)] show that the concept of a reactive azeotrope generalizes to the concept of a *reactive fixed point*, whose existence and location are a function of approach to equilibrium as well as the evaporation rate [see also Frey and Stichlmair, *Trans IChemE* **77,** Part A, 613 (1999); Chadda, Malone, and Doherty, *AIChE J.* **47:** 590 (2001); Chiplunkar et al., *AIChE J.* **51:** 464 (2005)]. In the limit of simultaneous phase and reaction equilibrium, a reactive fixed point becomes identical to the thermodynamic concept of a reactive azeotrope.

These ideas have been extended to reacting systems with (1) multiple chemical reactions [Ung and Doherty, *Ind. Eng. Chem. Res.* **34:** 2555, 3195 (1995)], (2) multiple liquid phases [Ung and Doherty, *Chem. Eng. Sci.* **50:** 3201 (1995); Qi, Kolah, and Sundmacher, *Chem. Eng. Sci.* **57:** 163 (2002); Qi and Sundmacher, *Comp. Chem. Eng.* **26:** 1459 (2002)], (3) membrane separations [Huang et al., *Chem. Eng. Sci.* **59:** 2863 (2004)], (4) finite rates of vapor-liquid mass transfer [Baur, Taylor, and Krishna, *Chem. Eng. Sci.* **56:** 2085 (2001); Nisoli, Doherty, and Malone, *AIChE J.* **50:** 1795 (2004)], (5) column design and multiple steady-states (Güttinger, Dorn, and Morari, *Ind. Eng. Chem. Res.* **36:** 794 (1997); Hauan, Hertzberg, and Lien, *Comput. Chem. Eng.* **21:** 1117 (1997); Sneesby et al., *Ind. Eng. Chem. Res.* **36:** 1855 (1997); Bessling et al., *Chem. Eng. Technol.* **21:** 393 (1998); Okasinski and Doherty, *Ind. Eng. Chem. Res.* **37:** 2821 (1998); Sneesby, Tade, and Smith, *Trans. IChemE.* **76,** Part A, 525 (1998); Güttinger and Morari, *Ind. Eng. Chem. Res.* **38:** 1633, 1649 (1999); Higler, Taylor, and Krishna, *Chem. Eng. Sci.* **54:** 1389 (1999); Mohl et al., *Chem. Eng. Sci.* **54:** 1029 (1999); Chen et al., *Comput. Chem. Eng.* **26:** 81 (2002)]. Much useful information is available in Taylor and Krishna [*Chem. Eng. Sci.* **55:** 5183 (2000)] and Sundmacher and Kienle (*Reactive Distillation*, Wiley-VCH, New York, 2003).

Mechanical Design and Implementation Issues The choice of catalyst has a significant impact on the mechanical design and operation of the reactive column. The catalyst must allow the reaction to occur at reasonable rates at the relatively low temperatures and pressures common in distillation operations (typically less than 10 atm and between 50°C and 250°C). The selection of a homogeneous catalyst, such as a high-boiling mineral acid, allows the use of more traditional tray designs and internals (albeit designed with exotic materials of construction to avoid corrosion, and allowance for high-liquid holdups to achieve suitable reaction contact times). With a homogeneous catalyst, lifetime is not a problem, as it is added (and withdrawn) continuously. Alternatively, heterogeneous solid catalysts require either complicated mechanical means for continuous replenishment or relatively long lifetimes to avoid constant maintenance. As with other multiphase reactors, the use of a solid catalyst adds resistance to mass transfer from the bulk liquid (or vapor) to the catalyst surface, which may be the limiting resistance. The catalyst containment system must be designed to ensure adequate liquid-solid contacting and to minimize bypassing. A number of specialized column internal designs, catalyst containment methods, and catalyst replenishment systems have been proposed for both homogeneous and heterogeneous catalysts. A partial list of these methods is given in Table 13-22; see also the useful ideas presented by Krishna [*Chem. Eng. Sci.* **57:** 1491 (2002); and chap. 7 in Sundmacher and Kienle, eds., *Reactive Distillation*, Wiley-VCH, New York, 2003].

Heat management is another important consideration in the implementation of a reactive distillation process. Conventional reactors for highly exothermic or endothermic reactions are often designed as modified shell-and-tube heat exchangers for efficient heat transfer. However, a trayed or packed distillation column is a rather poor mechanical design for the management of the heat of reaction. Although heat can be removed or added in the condenser or reboiler easily, the only mechanism for heat transfer in the column proper is through vaporization (or condensation). For highly exothermic reactions, a large excess of reactants may be required as a heat sink, necessitating high reflux rates and larger-diameter columns to return the vaporized reactants back to the reaction zone. Often a prereactor of conventional design is used to accomplish most of the reaction and heat removal before feeding to the reactive column for final conversion, as exemplified in most processes for the production of tertiary amyl methyl ether (TAME) [Brockwell et al., *Hyd. Proc.* **70**(9): 133 (1991)]. Highly endothermic reactions may require intermediate reboilers. None of these heat management issues preclude the use of reactive distillation, but they must be taken into account during the design phase. A comparison of heat of reaction and average heat of vaporization data for a system, as in Fig. 13-94, gives some indication of potential heat imbalances [Sundmacher, Rihko, and Hoffmann, *Chem. Eng. Comm.* **127:** 151 (1994)]. The heat-neutral systems [$-\Delta H_{react} \approx \Delta H_{vap}$ (avg)] such as methyl acetate and other esters can be accomplished in one reactive column, whereas the MTBE and TAME processes, with higher heats of reaction than that of vaporization, often include an additional prereactor. One exception is the catalytic distillation process for cumene production, which is accomplished without a prereactor. Three moles of benzene reactant are vaporized (and refluxed) for every mole of cumene produced. The relatively high heat of reaction is advantageous in this case as it reduces the overall

TABLE 13-22 Catalyst Systems for Reactive Distillation

Description	Application	Reference
	Homogeneous catalysis	
Liquid-phase mineral-acid catalyst added to column or reboiler	Esterifications Dibutyl phlalate Methyl acetate	Keyes, *Ind. Eng. Chem.*, **24,** 1096 (1932) Berman et al., *Ind. Eng. Chem.*, **40,** 2139 (1948) Agreda et al., U.S. Patent 4,435,595 (1984)
	Heterogeneous catalysis	
Catalyst-resin beads placed in cloth bags attached to fiberglass strip. Strip wound around helical stainless steel mesh spacer	Etherifications Cumene	Smith et al., U.S. Patent 4,443,559 (1981) Shoemaker and Jones, *Hyd.* **57**(6), 57 (1987)
Ion exchange resin beads used as column packing	Hydrolysis of methyl acetate	Fuchigami, *J. Chem. Eng. Jap.*, **23,** 354 (1990)
Molecular sieves placed in bags or porous containers	Alkylation of aromatics	Crossland, U.S. Patent 5,043,506 (1991)
Ion exchange resins formed into Raschig rings	MTBE	Flato and Hoffman, *Chem. Eng. Tech.*, **15,** 193 (1992)
Granular catalyst resin loaded in corrugated sheet casings	Dimethyl acetals of formaldehyde	Zhang et al., Chinese Patent 1,065,412 (1992)
Trays modified to hold catalyst bed	MTBE	Sanfilippo et al., Eur. Pat. Appl. EP 470,625 (1992)
Distillation trays constructed of porous catalytically active material and reinforcing resins	None specified	Wang et al., Chinese Patent 1,060,228 (1992)
Method described for removing or replacing catalyst on trays as a liquid slurry	None specified	Jones, U.S. Patent, 5,133,942 (1992)
Catalyst bed placed in downcomer, designed to prevent vapor flow through bed	Etherifications, alkylations	Asselineau, Eur. Pat. Appl. EP 547,939 (1993)
Slotted plate for catalyst support designed with openings for vapor flow	None specified	Evans and Stark, Eur. Pat. Appl. EP 571,163 (1993)
Ion exchanger fibers (reinforced ion exchange polymer) used as solid-acid catalyst	Hydrolysis of methyl acetate	Hirata et al., Jap. Patent 05,212,290 (1993)
High-liquid holdup trays designed with catalyst bed extending below tray level, perforated for vapor-liquid contact	None specified	Yeoman et al., Int. Pat. Appl., WO 9408679 (1994)
Catalyst bed placed in downcomer, in-line withdrawal/addition system	None specified	Carland, U.S. Patent, 5,308,451 (1994)

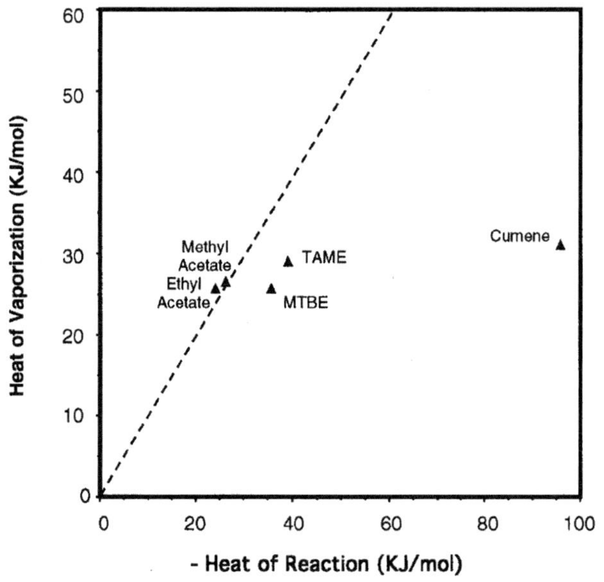

FIG. 13-94 Similarity of heats of reaction and vaporization for compounds made by reactive distillation.

heat duty of the process by about 30 percent [Shoemaker and Jones, *Hyd. Proc.* **57**(6): 57 (1987)].

Distillation columns with multiple conventional side reactors were first suggested by Schoenmakers and Buehler [*German Chem. Eng.* **5:** 292 (1982)] and have the potential to accommodate gas-phase reactions, highly exo- or endothermic reactions, catalyst deactivation, and operating conditions outside the normal range suitable for distillation (e.g., short contact times, high temperature and pressure); see Krishna (chap. 7 in Sundmacher and Kienle, eds., *Reactive Distillation*, Wiley-VCH, New York, 2003). This process concept has been applied commercially to produce ethyl acetate from ethanol [Gadewar et al., U.S. Patent 9,079,851 B2 (2015)].

Process Applications The production of esters from alcohols and carboxylic acids illustrates many of the principles of reactive distillation as applied to equilibrium-limited systems. The true thermodynamic equilibrium constants for esterification reactions are usually in the range of 5 to 20. Large excesses of alcohols must be used to obtain acceptable yields, resulting in large recycle flow rates. In a reactive distillation scheme, the reaction is driven to completion by removal of the water of esterification. The method used for removal of the water depends on the boiling points, compositions, and liquid-phase behavior of any azeotropes formed between the products and reactants and largely dictates the structure of the reactive distillation flow sheet.

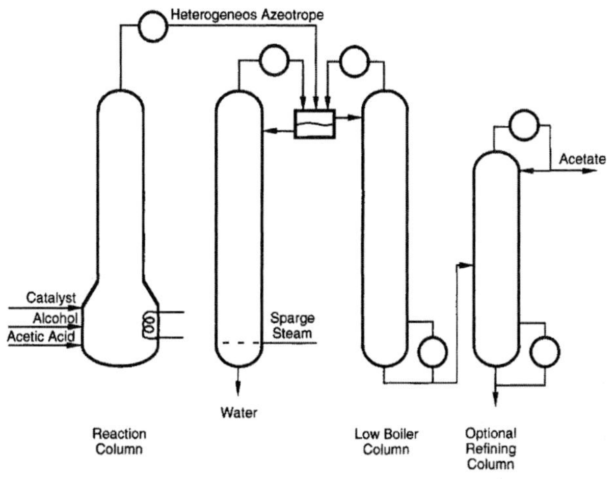

FIG. 13-95 Flow sheet for making esters that form a heterogeneous minimum-boiling azeotrope with water.

When the ester forms a binary low-boiling azeotrope with water or a ternary alcohol-ester-water azeotrope and that azeotrope is heterogeneous (or can be moved into the two-liquid phase region), the flow sheet illustrated in Fig. 13-95 can be used. Such a flow sheet works for the production of ethyl acetate and higher homologs. In this process scheme, acetic acid and the alcohol are continuously fed to the reboiler of the esterification column (reflux not shown in the column) along with a homogeneous strong-acid catalyst. Since the catalyst is largely nonvolatile, the reboiler acts as the primary reaction section. The alcohol is usually fed in slight excess to ensure complete reaction of the acid and to compensate for alcohol losses through distillation of the water-ester-(alcohol) azeotrope. The esterification column is operated so that the low-boiling, water-laden azeotrope is taken as the distillation product. Upon cooling, the distillate separates into two liquid phases. The aqueous layer is steam-stripped, with the organics recycled to the decanter or reactor. The ester layer from the decanter contains some water and possibly alcohol. Part of this layer may be refluxed to the esterification column (reflux not shown in the figure). The remainder is fed to a low-boiler column where the water-ester and alcohol-ester azeotropes are removed overhead and recycled to the decanter or reactor. The dry, alcohol-free ester is then optionally taken overhead in a final refining column. Additional literature on the application of reactive distillation to ester production includes papers by Hanika, Kolena, and Smejkal [*Chem. Eng. Sci.* **54:** 5205 (1999)], Schwarzer and Hoffmann [*Chem. Eng. Technol.* **25:** 975 (2002)], Steinigeweg and Gmehling [*Ind. Eng. Chem. Res.* **41:** 5483 (2002)], and Omata, Dimian, and Bliek [*Chem. Eng. Sci.* **58:** 3159, 3175 (2003)].

Methyl acetate cannot be produced in high purity by using the simple esterification scheme just described. The methyl acetate–methanol–water system does not exhibit a ternary minimum-boiling azeotrope, the methyl acetate–methanol azeotrope is lower-boiling than the water–methyl acetate azeotrope, a distillation boundary extends between these two binary azeotropes, and the heterogeneous region does not include either azeotrope, nor does it cross the distillation boundary. Consequently, the water of esterification cannot be removed effectively, and methyl acetate cannot be separated from the methanol and water azeotropes by a simple decantation in the same manner as that previously outlined. Conventional sequential reaction-separation processes rely on large excesses of acetic acid to drive the reaction to higher conversion to methyl acetate, necessitating a capital- and energy-intensive acetic acid–water separation and large recycle streams. The crude methyl acetate product, contaminated with water and methanol, can be purified by a number of enhanced distillation techniques, such as pressure-swing distillation (Harrison, U.S. Patent 2,704,271, 1955), extractive distillation with ethylene glycol monomethylether as the solvent (Kumerle, German Patent 1,070,165, 1959), or azeotropic distillation with an aromatic or ketone entrainer (Yeomans, Eur. Patent Appl. 060717 and 060719, 1982). The end result is a capital- and energy-intensive process typically requiring multiple reactors and distillation columns.

The reactive distillation process in Fig. 13-96 provides a mechanism for overcoming both the limitations on conversion due to chemical equilibrium and the difficulties in purification imposed by the water–methyl acetate and methanol–methyl acetate azeotropes [Agreda and Partin, U.S. Patent 4,435,595, 1984; Agreda, Partin, and Heise, *Chem. Eng. Prog.* **86**(2): 40 (1990)]. Conceptually, this flow sheet can be thought of as four heat-integrated distillation columns (one of which is also a reactor) stacked on top of each other. The primary reaction zone consists of a series of countercurrent flashing stages in the middle of the column. Adequate residence time for the reaction is provided by high-liquid-holdup bubble cap trays with specially designed downcomer sumps to further increase tray holdup. A nonvolatile homogeneous catalyst is fed at the top of the reactive section and exits with the underflow water by-product. The extractive distillation section, immediately above the reactive section, is critical in achieving high methyl acetate purity. As shown in Fig. 13-93, simultaneous reaction and distillation eliminates the water–methyl acetate azeotrope (and the distillation boundary of the nonreactive system). However, pure methyl acetate remains a saddle in the reactive system and cannot be obtained as a pure component by simple reactive distillation. The acetic acid feed acts as a solvent in an extractive-distillation section placed above the reaction section, breaking the methanol-methyl acetate azeotrope, and yielding a pure methyl acetate distillate product. The uppermost rectification stages serve to remove any acetic acid from the methyl acetate product, and the bottommost stripping section removes any methanol and methyl acetate from the water by-product. The countercurrent flow of the reactants results in high local excesses at each end of the reactive section, even though the overall feed to the reactive column is stoichiometric. Therefore, the large excess of acetic acid at the top of the reactive section prevents methanol from reaching the distillate; similarly, methanol at the bottom of the reactive section keeps acetic acid from the water bottoms. Temperature and composition profiles for this reactive extractive distillation column are shown in Fig. 13-97*a* and *b*, respectively.

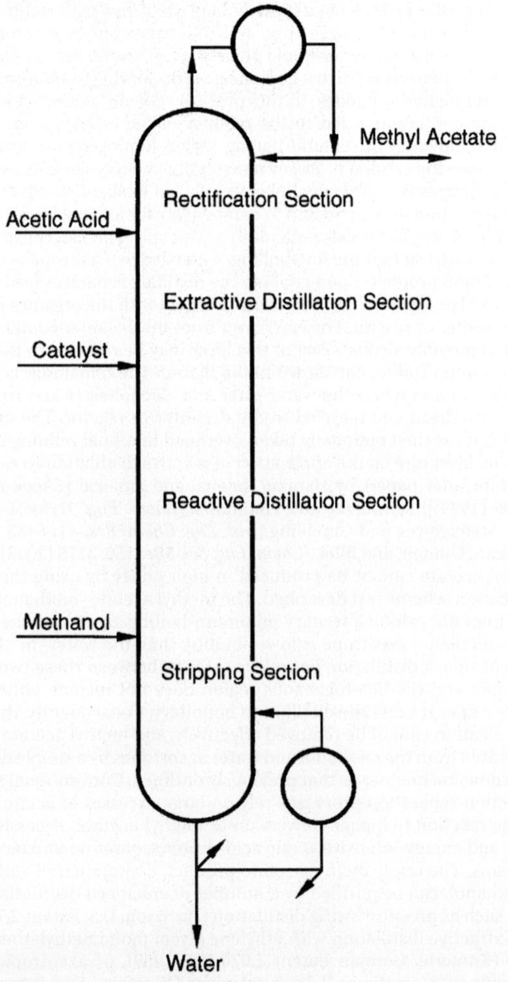

FIG. 13-96 Integrated reactive extractive distillation column for the production of methyl acetate.

Much has been written about this reactive distillation scheme, including works by Bessling et al. [*Chem. Eng. Tech.* **21:** 393 (1998)], Song et al. [*Ind. Eng. Chem. Res.* **37:** 1917 (1998)], Huss et al. [*Comput. Chem. Eng.* **27:** 1855 (2003)], Siirola ("An Industrial Perspective on Process Synthesis," in *Foundations of Computer-Aided Process Design*, ed. Biegler and Doherty, AIChE, New York, 1995, pp. 222–233), and Krishna (chap. 7 in *Reactive Distillation*, ed. Sundmacher and Kienle, Wiley-VCH, New York, 2003).

SYNTHESIS OF MULTICOMPONENT SEPARATION SYSTEMS

The sequencing of distillation columns and other types of equipment for the separation of multicomponent mixtures has received much attention in recent years. Although one separator of complex design can sometimes be devised to produce more than two products, more often a sequence of two-product separators is preferable. Often, the sequence includes simple distillation columns. A summary of sequencing methods, prior to 1977, that can lead to optimal or near-optimal designs is given by Henley and Seader (1981). Methods for distillation column sequencing are reviewed by Modi and Westerberg [*Ind. Eng. Chem. Res.* **31:** 839 (1992)], who also present a more generally applicable method based on a marginal price that is the change in price of a separation operation when the separation is carried out in the absence of nonkey components. The synthesis of sequences that consider a wide range of separation operations in a knowledge-based approach is given by Barnicki and Fair for liquid mixtures [*Ind. Eng. Chem. Res.* **29:** 421 (1990)] and for gas/vapor mixtures [*Ind. Eng. Chem. Res.* **31:** 1679 (1992)]. The problem decomposition approach of Wahnschafft, Le Rudulier, and Westerberg [*Ind. Eng. Chem. Res.* **32:** 1121 (1993)] is directed to the synthesis of complex separation sequences that involve nonsharp splits and recycle, including azeotropic distillation. The method was applied by using a computer-aided separation process designer called *SPLIT.* The approach developed by Ryll, Blagov, and Hasse [*Chem. Eng Sci.* **84:** 315 (2012)] for the synthesis of multicomponent azeotropic distillation systems is especially notable. An expert system, called *EXSEP,* for the synthesis of solvent-based separation trains is presented by Brunet and Liu [*Ind. Eng. Chem. Res.* **32:** 315 (1993)]. The use of ternary composition diagrams and residue curve maps is reviewed and evaluated for application to the synthesis of complex separation sequences by Fien and Liu [*Ind. Eng. Chem. Res.* **33:** 2506 (1994)]. In recent years, many optimization-based process synthesis schemes have been proposed for distillation systems; see the review by Chen and Grossmann [*Annu. Rev. Chem. Biomol. Eng.* **8** (2017)].

Synthesis schemes for reactive distillation have been proposed by Ismail, Proios, and Pistikopoulos [*AIChE J.* **47:** 629 (2001)], Jackson and Grossmann [*Comput. Chem. Eng.* **25:** 1661 (2001)], Schembecker and Tlatlik [*Chem. Eng. Process.* **42:** 179 (2003)], and Burri and Manousiouthakis [*Comput. Chem. Eng.* **28:** 2509 (2004)].

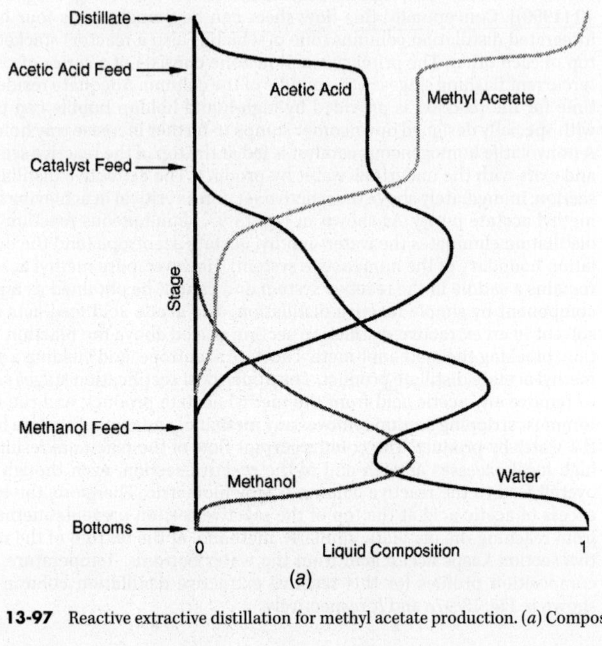

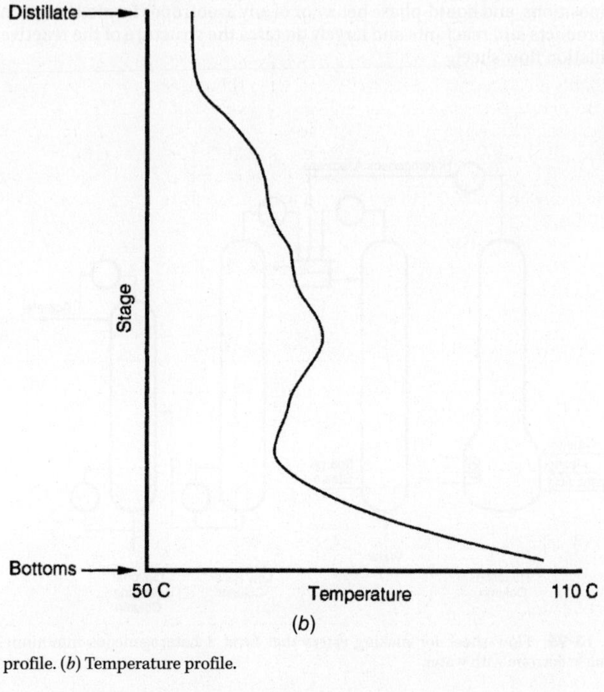

FIG. 13-97 Reactive extractive distillation for methyl acetate production. (*a*) Composition profile. (*b*) Temperature profile.

PETROLEUM AND COMPLEX-MIXTURE DISTILLATION

Although the principles of multicomponent distillation apply to petroleum, synthetic crude oil, and other complex mixtures, this subject warrants special consideration for the following reasons:

1. Such feedstocks are of exceedingly complex composition, consisting of, in the case of petroleum, many different types of hydrocarbons and perhaps of inorganic and other organic compounds. The number of carbon atoms in the components may range from 1 to more than 50, so the compounds may exhibit atmospheric-pressure boiling points from −162°C (−259°F) to more than 538°C (1000°F). In a given boiling range, the number of different compounds that exhibit only small differences in volatility multiplies rapidly with increasing boiling point. For example, 16 of the 18 octane isomers boil within a range of only 12°C (22°F).

2. Products from the distillation of complex mixtures are in themselves complex mixtures. The character and yields of these products vary widely, depending on the source of the feedstock. Even crude oils from the same locality may exhibit marked variations.

3. The scale of petroleum-distillation operations is generally large, and as discussed in detail by Nelson (*Petroleum Refinery Engineering,* 4th ed., McGraw-Hill, New York, 1958) and Watkins (*Petroleum Refinery Distillation,* 2d ed., Gulf, Houston, 1979), such operations are common in several petroleum refinery processes, including atmospheric distillation of crude oil, vacuum distillation of bottoms residuum obtained from atmospheric distillation, main fractionation of gaseous effluent from catalytic cracking of various petroleum fractions, and main fractionation of effluent from thermal coking of various petroleum fractions. These distillation operations are conducted in large pieces of equipment that can consume large quantities of energy. Therefore, the optimization of design and operation is very important and often leads to a relatively complex equipment configuration.

CHARACTERIZATION OF PETROLEUM AND PETROLEUM FRACTIONS

Although much progress has been made in identifying the chemical species in petroleum, it is generally sufficient for the purposes of designing and analyzing the operation of distillation plants to characterize petroleum and petroleum fractions by gravity, laboratory distillation curves, component analysis of light ends, and hydrocarbon-type analysis of middle and heavy ends. From such data, as discussed in the *Technical Data Book—Petroleum Refining* [American Petroleum Institute (API), Washington], five different average boiling points and an index of paraffinicity can be determined. These are then used to predict the physical properties of complex mixtures by a number of well-accepted correlations, whose use will be explained in detail and illustrated with examples. Many other characterizing properties or attributes such as sulfur content, pour point, water and sediment content, salt content, metals content, Reid vapor pressure, Saybolt Universal viscosity, aniline point, octane number, freezing point, cloud point, smoke point, diesel index, refractive index, cetane index, neutralization number, wax content, carbon content, and penetration are generally measured for a crude oil or certain of its fractions, according to well-specified ASTM tests. But these attributes are of much less interest here, even though feedstocks and products may be required to meet certain specified values of the attributes.

The gravity of a crude oil or petroleum fraction is generally measured by the ASTM D287 test or the equivalent ASTM D1298 test, and it may be reported as specific gravity (SG) 60/60°F [measured at 60°F (15.6°C) and referred to water at 60°F (15.6°C)] or, more commonly, as API gravity, which is defined as

$$\text{API gravity} = 141.5/(\text{SG } 60/60°\text{F}) - 131.5 \qquad (13\text{-}118)$$

Water thus has an API gravity of 10.0, and most crude oils and petroleum fractions have values of API gravity in the range of 10 to 80. Light hydrocarbons (*n*-pentane and lighter) have values of API gravity ranging upward from 92.8.

The volatility of crude oil and petroleum fractions is characterized in terms of one or more laboratory distillation tests that are summarized in Table 13-23. The ASTM D86 and D 1160 tests are reasonably rapid batch laboratory distillations involving the equivalent of approximately one equilibrium stage and no reflux except for that caused by heat losses. Apparatus typical of the D 86 test is shown in Fig. 13-98 and consists of a heated 100-mL or 125-mL Engler flask containing a calibrated thermometer of suitable range to measure the temperature of the vapor at the inlet to the condensing tube, an inclined brass condenser in a cooling bath using a suitable coolant, and a graduated cylinder for collecting the distillate. A stem correction is not applied to the temperature reading. Related tests using similar apparatus are the D 216 test for natural gasoline and the Engler distillation.

In the widely used ASTM D86 test, 100 mL of sample is charged to the flask and heated at a sufficient rate to produce the first drop of distillate from the lower end of the condenser tube in 5 to 15 min, depending on the nature of the sample. The temperature of the vapor at that instant is recorded as the initial boiling point (IBP). Heating is continued at a rate such that the time from the IBP to 5 vol% recovered of the sample in the cylinder is 60 to 75 s. Again, vapor temperature is recorded. Then successive vapor temperatures are recorded for 10 to 90 percent recovered in intervals of 10, and at 95 percent recovered, with the heating rate adjusted so that 4 to 5 mL is collected per minute. At 95 percent recovered, the burner flame is increased if necessary to achieve a maximum vapor temperature, referred to as the endpoint (EP) in 3 to 5 additional min. The percent recovery is reported as the maximum percent recovered in the cylinder. Any residue remaining in the flask is reported as percent residue, and percent loss is reported as the difference between 100 mL and the sum of the percent recovery and percent residue. If the atmosphere test pressure P is other than 101.3 kPa (760 torr), temperature readings may be adjusted to that pressure by the Sidney Young equation, which for degrees Fahrenheit is

$$T_{760} = T_P + 0.00012(760 - P)(460 + T_P) \qquad (13\text{-}119)$$

Another pressure correction for percent loss can also be applied, as described in the ASTM test method.

The results of a typical ASTM distillation test for an automotive gasoline are given in Table 13-24, in which temperatures have already been corrected to a pressure of 101.3 kPa (760 torr). It is generally assumed that percent loss corresponds to volatile noncondensables that are distilled off at the beginning of the test. In that case, the percent recovered values in Table 13-24 do not correspond to percent evaporated values, which are of greater scientific value. Therefore, it is common to adjust the reported temperatures according to a linear interpolation procedure given in the ASTM test method to obtain corrected temperatures in terms of percent evaporated at the standard intervals as included in Table 13-24. In the example, the corrections are not large because the loss is only 1.5 vol%.

Although most crude petroleum can be heated to 600°F (316°C) without noticeable cracking, when ASTM temperatures exceed 475°F (246°C), fumes may be evolved, indicating decomposition, which may cause thermometer readings to be low. In that case, the following correction attributed to S. T. Hadden may be applied:

$$\Delta T_{corr} = 10^{-1.587 + 0.004735T} \qquad (13\text{-}120)$$

where T = measured temperature, °F
ΔT_{corr} = correction to be added to T, °F

TABLE 13-23 Laboratory Distillation Tests

Test name	Reference	Main applicability
ASTM (atmospheric)	ASTM D 86	Petroleum fractions or products, including gasolines, turbine fuels, naphthas, kerosines, gas oils, distillate fuel oils, and solvents that do not tend to decompose when vaporized at 760 mmHg
ASTM [vacuum, often 10 torr (1.3 kPa)]	ASTM D 1160	Heavy petroleum fractions or products that tend to decompose in the ASTM D 86 test but can be partially or completely vaporized at a maximum liquid temperature of 750°F (400°C) at pressures down to 1 torr (0.13 kPa)
TBP [atmospheric or 10 torr (1.3 kPa)]	Nelson,* ASTM D 2892	Crude oil and petroleum fractions
Simulated TBP (gas chromatography)	ASTM D 2887	Crude oil and petroleum fractions
EFV (atmospheric, superatmospheric, or subatmospheric)	Nelson†	Crude oil and petroleum fractions

*Nelson, *Petroleum Refinery Engineering,* 4th ed., McGraw-Hill, New York, 1958, pp. 95–99.
†Ibid., pp. 104–105.

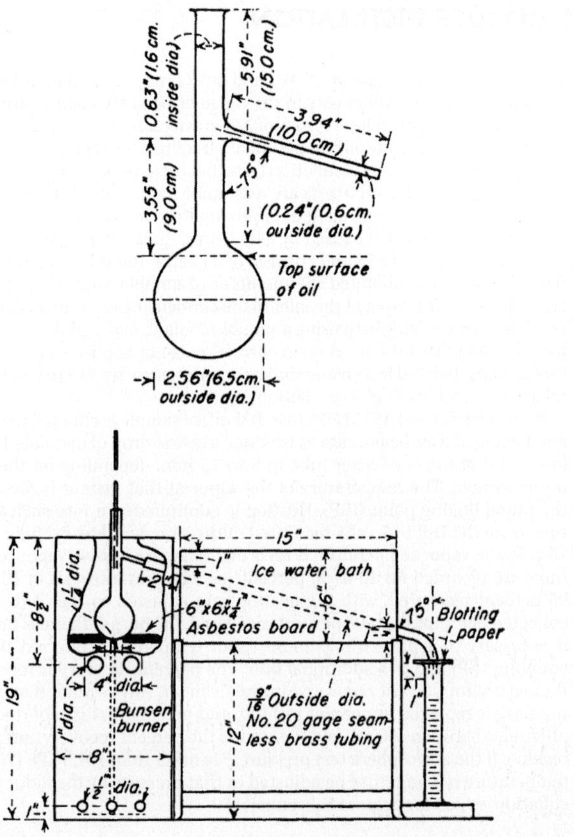

FIG. 13-98 ASTM distillation apparatus; detail of distilling flask is shown in the upper figure.

At 500°F and 600°F (260°C and 316°C), the corrections are 6°F and 18°F (3.3°C and 10°C), respectively.

As discussed by Nelson (*Petroleum Refinery Engineering*, 4th ed., McGraw-Hill, New York, 1958), virtually no fractionation occurs in an ASTM distillation. Thus, components in the mixture do distill one by one in the order of their boiling points but as mixtures of successively higher boiling points. The IBP, EP, and intermediate points have little theoretical significance, and, in fact, components boiling below the IBP and above the EP are present in the sample. Nevertheless, because ASTM distillations are quickly conducted, have been successfully automated, require only a small sample, and are quite reproducible, they are widely used for comparison and as a basis for specifications on a large number of petroleum intermediates and products, including many solvents and fuels. Typical ASTM curves for several such products are shown in Fig. 13-99.

Data from a *true boiling point* (TBP) distillation test provide a much better theoretical basis for characterization. If the sample contains compounds that have moderate differences in boiling points such as in a light gasoline containing light hydrocarbons (e.g., isobutane, *n*-butane, isopentane), a plot of overhead vapor distillate temperature versus percent distilled in a TBP test would appear in the form of steps as in Fig. 13-100. However, if the sample has a higher average boiling range when the number of close-boiling isomers increases, the steps become indistinct, and a TBP curve such as that in Fig. 13-101 results. Because the degree of separation for a TBP distillation test is much higher than that for an ASTM distillation test, the IBP is lower and the EP is higher for the TBP method than for the ASTM method, as shown in Fig. 13-101.

A standard TBP laboratory distillation test method has not been well accepted. Instead, as discussed by Nelson (1958, pp. 95–99), batch distillation equipment that can achieve a good degree of fractionation is usually considered suitable. In general, TBP distillations are conducted in columns with 15 to 100 theoretical stages at reflux ratios of 5 or greater. Thus, the new ASTM D2892 test method, which involves a column with 14 to 17 theoretical

TABLE 13-24 Typical ASTM D 86 Test Results for Automobile Gasoline Pressure, 760 torr (101.3 kPa)

Percent recovered basis (as measured)			Percent evaporated basis (as corrected)		
Percent recovered	*T*, °F	Percent evaporated	Percent evaporated	*T*, °F	Percent recovered
0(IBP)	98	1.5	1.5	98	(IBP)
5	114	6.5	5	109	3.5
10	120	11.5	10	118	8.5
20	150	21.5	20	146	18.5
30	171	31.5	30	168	28.5
40	193	41.5	40	190	38.5
50	215	51.5	50	212	48.5
60	243	61.5	60	239	58.5
70	268	71.5	70	264	68.5
80	300	81.5	80	295	78.5
90	340	91.5	90	334	88.5
95	368	96.5	95	360	93.5
EP	408			408	(EP)

NOTE: Percent recovery = 97.5; percent residue = 1.0; percent loss = 1.5. To convert degrees Fahrenheit to degrees Celsius, °C = (°F − 32)/1.8.

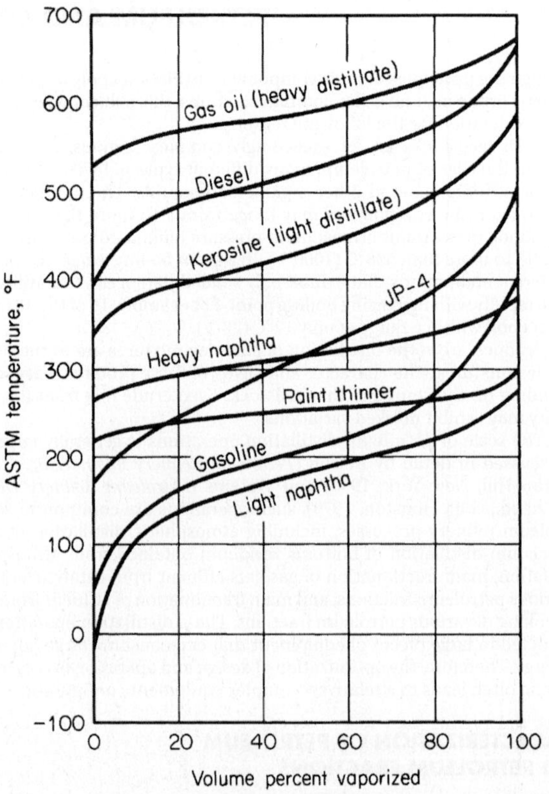

FIG. 13-99 Representative ASTM D86 distillation curves.

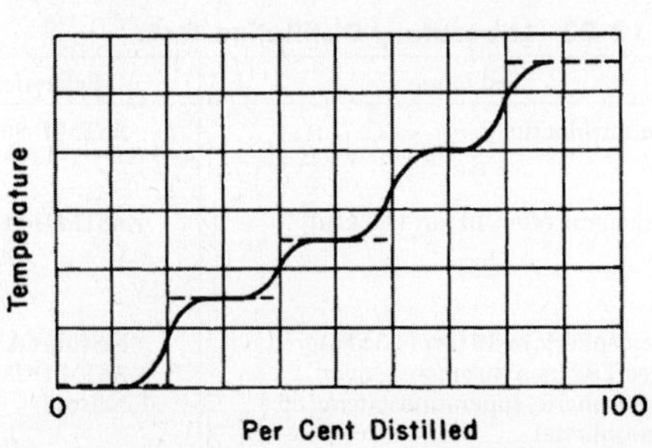

FIG. 13-100 Variation of boiling temperature with percent distilled in TBP distillation of light hydrocarbons.

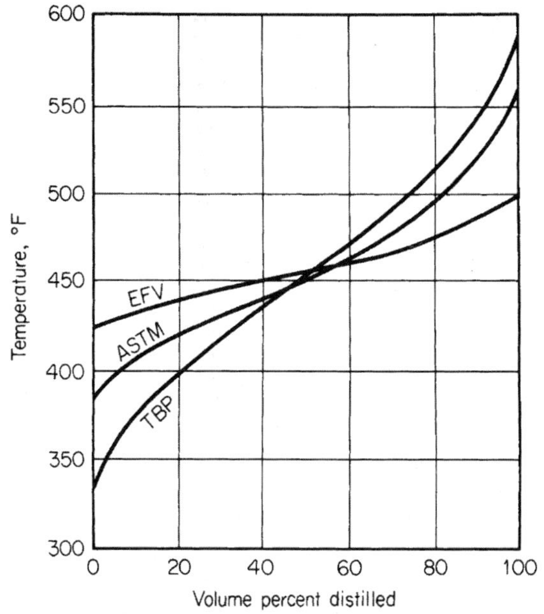

FIG. 13-101 Comparison of ASTM, TBP, and EFV distillation curves for kerosine.

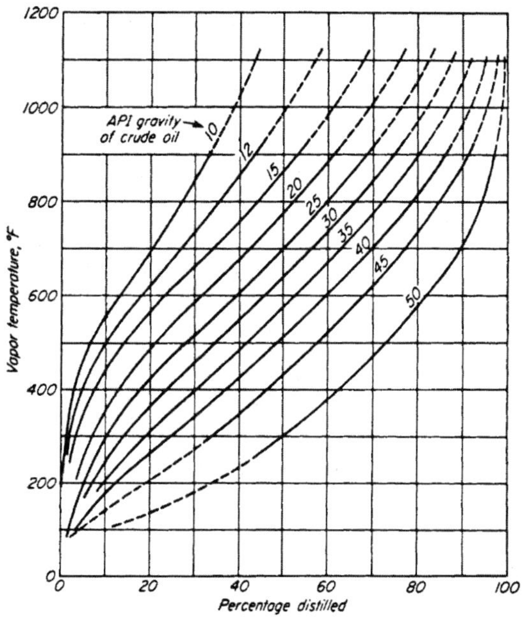

FIG. 13-102 Average true-boiling-point distillation curves of crude oils. *(From W. E. Edmister,* Applied Hydrocarbon Thermodynamics, *vol. 1, 1st ed., 1961 Gulf Publishing Company, Houston, Texas, Used with permission. All rights reserved.)*

stages and a reflux ratio of 5, essentially meets the minimum requirements. Distillate may be collected at a constant or a variable rate. Operation may be at 101.3-kPa (760-torr) pressure or at a vacuum at the top of the column as low as 0.067 kPa (0.5 torr) for high-boiling fractions, with 1.3 kPa (10 torr) being common. Results from vacuum operation are extrapolated to 101.3 kPa (760 torr) by the vapor-pressure correlation of Maxwell and Bonner [*Ind. Eng. Chem.* **49:** 1187 (1957)], which is given in great detail in the API *Technical Data Book—Petroleum Refining* (1966) and in the ASTM D2892 test method. It includes a correction for the nature of the sample (paraffin, olefin, naphthene, and aromatic content) in terms of the UOP characterization factor, UOP-K, as given by

$$\text{UOP-K} = \frac{(T_B)^{1/3}}{\text{SG}} \tag{13-121}$$

where T_B is the mean average boiling point in degrees Rankine, which is the arithmetic average of the molal average boiling point and the cubic volumetric average boiling point. Values of UOP-K for *n*-hexane, 1-hexene, cyclohexene, and benzene are 12.82, 12.49, 10.99, and 9.73, respectively. Thus, paraffins with their lower values of specific gravity tend to have high values, and aromatics tend to have low values of UOP-K. A movement toward an international TBP standard is discussed by Vercier and Mouton [*Oil Gas J.* **77**(38): 121 (1979)].

A crude oil assay always includes a whole crude API gravity and a TBP curve. As discussed by Nelson (1958, pp. 89–90) and as shown in Fig. 13-102, a reasonably consistent correlation (based on more than 350 distillation curves) exists between whole crude API gravity and the TBP distillation curve at 101.3 kPa (760 torr). Exceptions not correlated by Fig. 13-102 are highly paraffinic or naphthenic crude oils.

An alternative to TBP distillation is simulated distillation by gas chromatography. As described by Green, Schmauch, and Worman [*Anal. Chem.* **36:** 1512 (1965)] and Worman and Green [*Anal. Chem.* **37:** 1620 (1965)], the method is equivalent to a 100-theoretical-plate TBP distillation; is very rapid, reproducible, and easily automated; requires only a small microliter sample; and can better define initial and final boiling points. The ASTM D2887 standard test method is based on such a simulated distillation and is applicable to samples having a boiling range greater than 55°C (100°F) for temperature determinations as high as 538°C (1000°F). Typically, the test is conducted with a gas chromatograph having a thermal conductivity detector, a programmed temperature capability, helium or hydrogen carrier gas, and column packing of silicone gum rubber on a crushed firebrick or diatomaceous earth support.

It is important to note that simulated distillation does not always separate hydrocarbons in the order of their boiling points. For example, high-boiling multiple-ring-type compounds may be eluted earlier than normal paraffins (used as the calibration standard) of the same boiling point. Gas chromatography is also used in the ASTM D2427 test method to determine quantitatively ethane through pentane hydrocarbons.

A third fundamental type of laboratory distillation, which is the most tedious to perform of the three types of laboratory distillations, is equilibrium flash vaporization (EFV), for which no standard test exists. The sample is heated in such a manner that the total vapor produced remains in contact with the total remaining liquid until the desired temperature is reached at a set pressure. The volume percent vaporized at these conditions is recorded. To determine the complete flash curve, a series of runs at a fixed pressure is conducted over a range of temperatures sufficient to cover the range of vaporization from 0 to 100 percent. As seen in Fig. 13-101, the component separation achieved by an EFV distillation is much less than that achieved by the ASTM or TBP distillation tests. The initial and final EFV points are the bubble point and the dew point, respectively, of the sample. If desired, EFV curves can be established at a series of pressures.

Because of the time and expense involved in conducting laboratory distillation tests of all three basic types, it has become increasingly common to use empirical correlations to estimate the other two distillation curves when the ASTM, TBP, or EFV curve is available. Preferred correlations given in the API *Technical Data Book—Petroleum Refining* (1966) are based on the work of Edmister and Pollock [*Chem. Eng. Prog.* **44:** 905 (1948)], Edmister and Okamoto [*Pet. Refiner* **38**(8): 117 (1959); **38**(9): 271 (1959)], Maxwell (*Data Book on Hydrocarbons,* Van Nostrand, Princeton, N.J., 1950), and Chu and Staffel [*J. Inst. Pet.* **41:** 92 (1955)]. Because of the lack of sufficiently precise and consistent data on which to develop the correlations, they are, at best, first approximations and should be used with caution. Also, they do not apply to mixtures containing only a few components of widely different boiling points. Perhaps the most useful correlation of the group is Fig. 13-103 for converting between ASTM D86 and TBP distillations of petroleum fractions at 101.3 kPa (760 torr). The ASTM D2889 test method, which presents a standard method for calculating EFV curves from the results of an ASTM D86 test for a petroleum fraction having a 10 to 90 vol% boiling range of less than 55°C (100°F), is also quite useful.

APPLICATIONS OF PETROLEUM DISTILLATION

Typical equipment configurations for the distillation of crude oil and other complex hydrocarbon mixtures in a crude unit, a catalytic cracking unit, and a delayed coking unit of a petroleum refinery are shown in Figs. 13-104, 13-105, and 13-106. The initial separation of crude oil into fractions is conducted in two main columns, shown in Fig. 13-104. In the first column, called the atmospheric tower or topping still, partially vaporized crude oil, from which water, sediment, and salt have been removed, is mainly rectified, at a feed tray pressure of no more than about

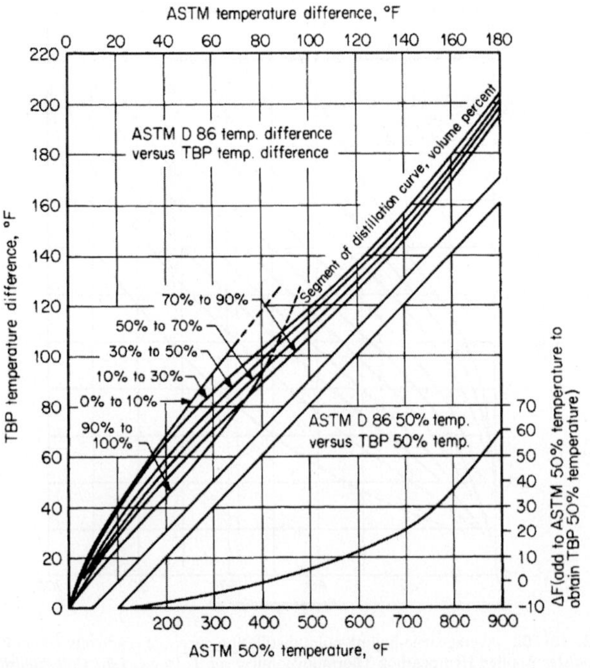

FIG. 13-103 Relationship between ASTM and TBP distillation curves. *(From W. C. Edmister,* Applied Hydrocarbon Thermodynamics, *vol. 1, 1st ed., 1961 Gulf Publishing Company, Houston, Tex. Used with permission. All rights reserved.)*

276 kPa (40 psia), to yield a noncondensable light-hydrocarbon gas, a light naphtha, a heavy naphtha, a light distillate (kerosine), a heavy distillate (diesel oil), and a bottoms residual of components whose TBP exceeds approximately 427°C (800°F). Alternatively, other fractions, shown in Fig. 13-99, may be withdrawn. To control the IBP of the ASTM D86 curves, each of the sidestreams of the atmospheric tower and the vacuum and main fractionators of Figs. 13-104, 13-105, and 13-106 may be sent to sidecut strippers, which use a partial reboiler or steam stripping. Additional stripping by steam is commonly used in the bottom of the atmospheric tower as well as in the vacuum tower and other main fractionators.

Additional distillate in the TBP range of approximately 427°C to 593°C (800°F to 1100°F) is recovered from bottoms residuum of the atmospheric tower by rectification in a vacuum tower, also shown in Fig. 13-104, at the minimum practical overhead condenser pressure, which is typically 1.3 kPa (10 torr). The use of special low-pressure-drop trays or column packing permits the feed tray pressure to be approximately 5.3 to 6.7 kPa (40 to 50 torr) to obtain the maximum degree of vaporization. Vacuum towers may be designed or operated to produce several different products, including heavy distillates, gas-oil feedstocks for catalytic cracking, lubricating oils, bunker fuel, and bottoms residua of asphalt (5 to 8 API gravity) or pitch (0 to 5 API gravity). The catalytic cracking process of Fig. 13-105 produces a superheated vapor at approximately 538°C (1000°F) and 172 to 207 kPa (25 to 30 psia) of a TBP range that covers hydrogen to compounds with normal boiling points above 482°C (900°F). This gas is sent directly to a main fractionator for rectification to obtain products that are typically gas and naphtha [204°C (400°F) ASTM EP approximately], which are often fractionated further to produce relatively pure light hydrocarbons and gasoline; a light cycle oil [typically 204°C to 371°C (400°F to 700°F) ASTM D86 range], which may be used for heating oil, hydrocracked, or recycled to the catalytic cracker; an intermediate cycle oil [typically 371°C to 482°C (700°F to 900°F) ASTM D86 range], which is generally recycled to the catalytic cracker to extinction; and a heavy gas oil or bottom slurry oil.

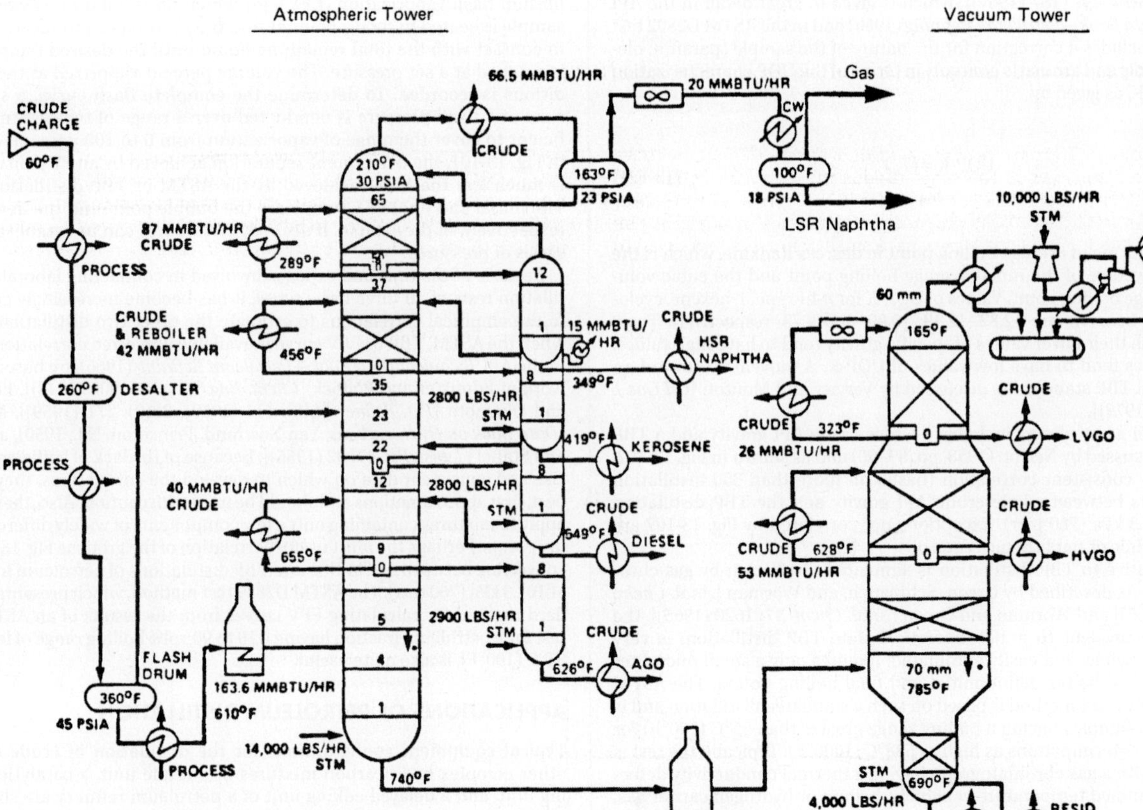

FIG. 13-104 Crude unit with atmospheric and vacuum towers. [*Kleinschrodt and Hammer,* Chem. Eng. Prog. 79*(7): 33 (1983).*]

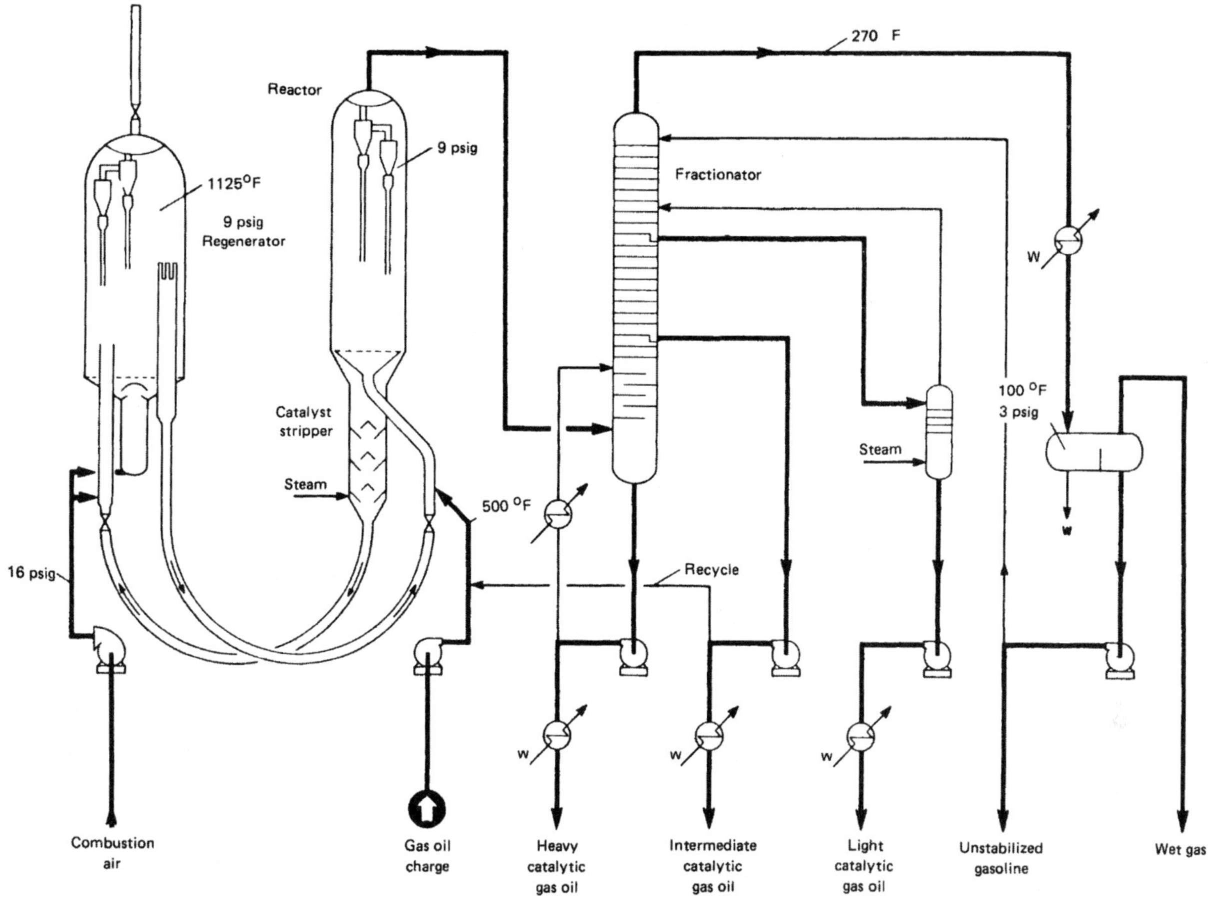

FIG. 13-105 Catalytic cracking unit. [*New Horizons*, Lummus Co., New York (1954)].

Vacuum-column bottoms, bottoms residuum from the main fractionation of a catalytic cracker, and other residua can be further processed at approximately 510°C (950°F) and 448 kPa (65 psia) in a delayed-coker unit, as shown in Fig. 13-106, to produce petroleum coke and gas of TBP range that covers methane (with perhaps a small amount of hydrogen) to compounds with normal boiling points that may exceed 649°C (1200°F). The gas is sent directly to a main fractionator that is similar to the type used in conjunction with a catalytic cracker, except that in the delayed-coking operation the liquid to be coked first enters into and passes down through the bottom trays of the main fractionator to be preheated by

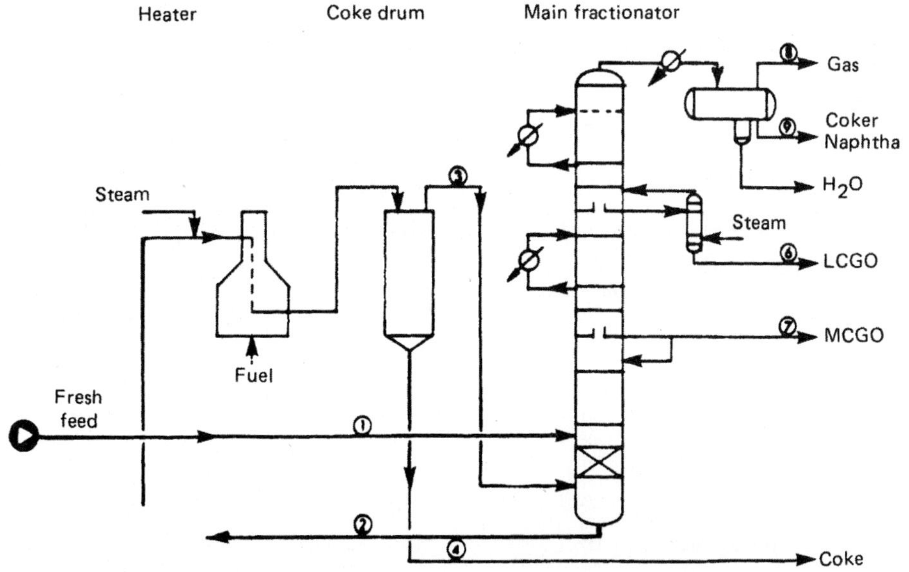

FIG. 13-106 Delayed-coking unit. (Watkins, *Petroleum Refinery Distillation*, 2d ed., Gulf, Houston, Tex., 1979).

and to scrub coker vapor of entrained coke particles and condensables for recycling to the delayed coker. Products produced from the main fractionator are similar to those produced in a catalytic cracking unit, except for more unsaturated cyclic compounds, and include gas and coker naphtha, which are further processed to separate out light hydrocarbons and a coker naphtha that generally needs hydrotreating, and light and heavy coker gas oils, both of which may require hydrocracking to become suitable blending stocks.

DESIGN PROCEDURES

Two general procedures are available for designing fractionators that process petroleum, synthetic crude oils, and complex mixtures. The first, which was originally developed for crude units by Packie [*Trans. Am. Inst. Chem. Eng. J.* **37:** 51 (1941)], extended to main fractionators by Houghland, Lemieux, and Schreiner [*Proc. API,* sec. III, *Refining,* 385 (1954)], and further elaborated and described in great detail by Watkins (*Petroleum Refinery Distillation,* 2d ed., Gulf, Houston, 1979), uses material and energy balances, with empirical correlations to establish tray requirements, and is essentially a hand calculation procedure that is a valuable learning experience and is suitable for preliminary designs. Also, when backed by sufficient experience from previous designs, this procedure is adequate for final design.

In the second procedure, which is best applied with a digital computer, the complex mixture being distilled is represented by actual components at the light end and by perhaps 30 pseudocomponents (e.g., petroleum fractions) over the remaining portion of the TBP distillation curve for the column feed. Each of the pseudocomponents is characterized by a TBP range, an average normal boiling point, an average API gravity, and an average molecular weight. Rigorous material balance, energy balance, and phase equilibrium calculations are then made by an appropriate equation-tearing method, as shown by Cecchetti et al. [*Hydrocarbon Process.* **42**(9): 159 (1963)] or a simultaneous-correction procedure as shown, for example, by Goldstein and Stanfield [*Ind. Eng. Chem. Process Des. Dev.* **9:** 78 (1970)] and Hess et al. [*Hydrocarbon Process.* **56**(5): 241 (1977)]. Highly developed procedures of the latter type, suitable for preliminary or final design, are included in most computer-aided steady-state process design and simulation programs as a special case of interlinked distillation, wherein the crude tower or fractionator is converged simultaneously with the sidecut stripper columns.

Regardless of the procedure used, certain initial steps must be taken for the determination or specification of certain product properties and yields based on the TBP distillation curve of the column feed, the method of providing column reflux, the column-operating pressure, the type of condenser, and the type of sidecut stripper and stripping requirements. These steps are developed and illustrated with several detailed examples by Watkins (1979). Only one example, modified from one given by Watkins, is considered briefly here to indicate the approach taken during the initial steps.

For the atmospheric tower shown in Fig. 13-107, suppose distillation specifications are as follows:

- Feed: 50,000 bbl (at 42 U.S. gal each) per stream day (BPSD) of 31.6 API crude oil.
- Measured light-ends analysis of feed:

Component	Volume percent of crude oil
Ethane	0.04
Propane	0.37
Isobutane	0.27
n-Butane	0.89
Isopentane	0.77
n-Pentane	1.13
	3.47

- Measured TBP and API gravity of feed, computed atmospheric pressure EFV (from API *Technical Data Book*), and molecular weight of feed:

Volume percent vaporized	TBP, °F	EFV, °F	°API	Molecular weight
0	−130	179		
5	148	275	75.0	91
10	213	317	61.3	106
20	327	394	50.0	137
30	430	468	41.8	177
40	534	544	36.9	223
50	639	619	30.7	273
60	747	696	26.3	327
70	867	777	22.7	392
80	1013	866	19.1	480

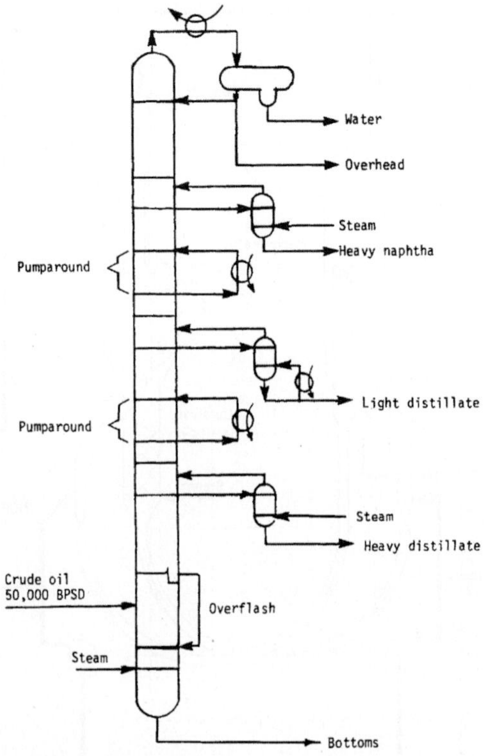

FIG. 13-107 Crude atmospheric tower.

- Product specifications:

	ASTM D86, °F		
Desired cut	5%	50%	95%
Overhead (OV)			253
Heavy naphtha (HN)	278	314	363
Light distillate (LD)	398	453	536
Heavy distillate (HD)	546	589	
Bottoms (B)			

NOTE: To convert degrees Fahrenheit to degrees Celsius, °C = (°F − 32)/1.8.

- TBP cut point between the heavy distillate and the bottoms = 650°F
- Percent overflash = 2 vol% of feed
- Furnace outlet temperature = 343°C (650°F) maximum
- Overhead temperature in reflux drum = 49°C (120°F) minimum

From the product specifications, distillate yields are computed as follows: From Fig. 13-103 and the ASTM D86 50 percent temperatures, TBP 50 percent temperatures of the three intermediate cuts are obtained as 155°C, 236°C, and 316°C (311°F, 456°F, and 600°F) for the HN, LD, and HD, respectively. The TBP cut points, corresponding volume fractions of crude oil, and flow rates of the four distillates are readily obtained by starting from the specified 343°C (650°F) cut point as follows, where CP is the cut point and T is the TBP temperature (°F):

$$CP_{HD,B} = 650°F$$
$$CP_{HD,B} - T_{HD50} = 650 - 600 = 50°F$$
$$CP_{LD,HD} = T_{HD50} - 50 = 600 - 50 = 550°F$$
$$CP_{LD,HD} - T_{LD50} = 550 - 456 = 94°F$$
$$CP_{HN,LD} = T_{LD50} - 94 = 456 - 94 = 362°F$$
$$CP_{HN,LD} - T_{HN50} = 362 - 311 = 51°F$$
$$CP_{OV,HN} = T_{HN50} - 51 = 311 - 51 = 260°F$$

These cut points are shown as vertical lines on the crude oil TBP plot of Fig. 13-108, from which the following volume fractions and flow rates of product cuts are readily obtained:

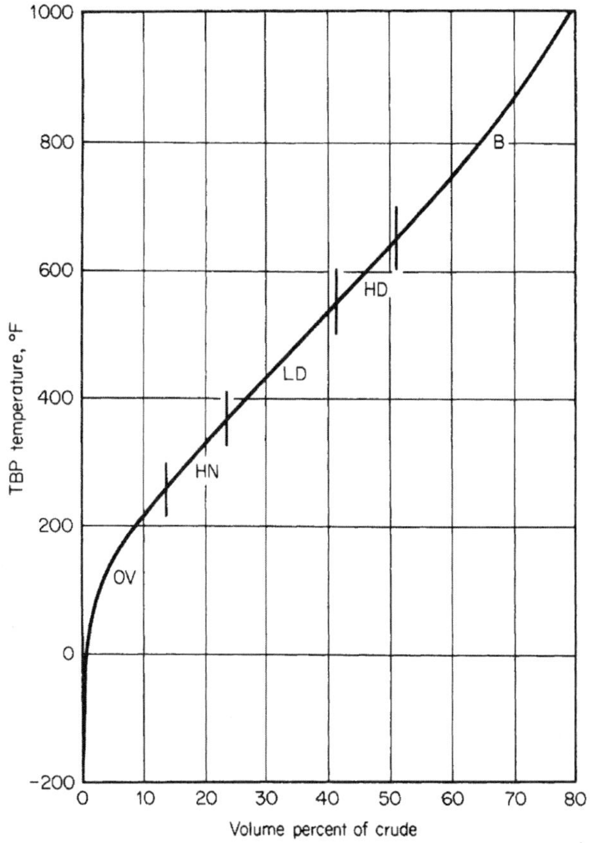

FIG. 13-108 Example of crude oil TBP cut points.

Desired cut	Volume percent of crude oil	BPSD
Overhead (OV)	13.4	6,700
Heavy naphtha (HN)	10.3	5,150
Light distillate (LD)	17.4	8,700
Heavy distillate (HD)	10.0	5,000
Bottoms (B)	48.9	24,450
	100.0	50,000

As shown in Fig. 13-109, methods of providing column reflux include (*a*) conventional top-tray reflux, (*b*) pump-back reflux from sidecut strippers, and (*c*) pump-around reflux. The latter two methods essentially function as intercondenser schemes that reduce the top-tray reflux requirement. As shown in Fig. 13-110 for the example being considered, the internal-reflux flow rate decreases rapidly from the top tray to the feed-flash zone for case *a*. The other two cases, particularly case *c*, result in better balancing of the column-reflux traffic. Because of this and the opportunity provided to recover energy at a moderate- to high-temperature level, pump-around reflux is the most commonly used technique. However, not indicated in Fig. 13-110 is the fact that in cases *b* and *c* the smaller quantity of reflux present in the upper portion of the column increases the tray requirements. Furthermore, the pump-around circuits, which extend over three trays each, are believed to be equivalent for mass-transfer purposes to only one tray each. Representative tray requirements for the three cases are included in Fig. 13-109. In case *c*, heat-transfer rates associated with the two pump-around circuits account for approximately 40 percent of the total heat removed in the overhead condenser and from the two pump-around circuits combined.

Bottoms and three sidecut strippers remove light ends from products and may use steam or reboilers. In Fig. 13-109 a reboiled stripper is used on the light distillate, which is the largest sidecut withdrawn. Steam-stripping rates in sidecut strippers and at the bottom of the atmospheric column may vary from 0.45 to 4.5 kg (1 to 10 lb) of steam per barrel of stripped liquid, depending on the fraction of stripper feed liquid that is vaporized.

Column pressure at the reflux drum is established to totally condense the overhead vapor or some fraction thereof. Flash-zone pressure is approximately 69 kPa (10 psia) higher. Crude oil feed temperature at flash-zone pressure must be sufficient to vaporize the total distillates plus the overflash, which is necessary to provide reflux between the lowest sidestream-product

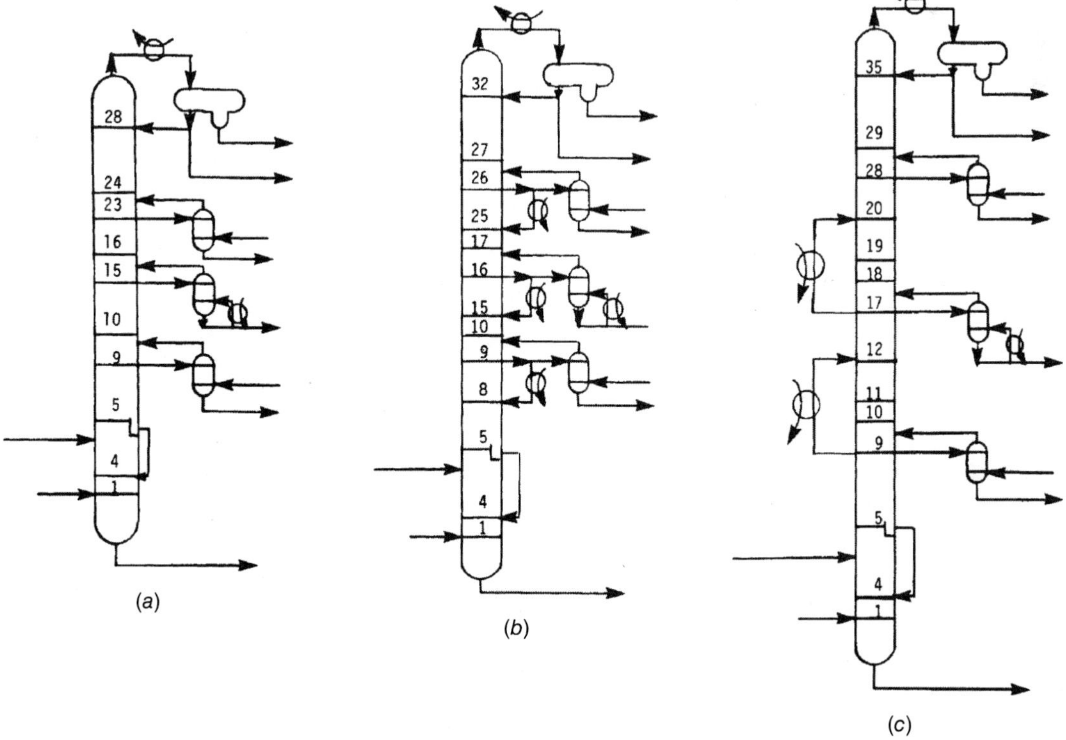

FIG. 13-109 Methods of providing reflux to crude units. (*a*) Top reflux. (*b*) Pump-back reflux. (*c*) Pump-around reflux.

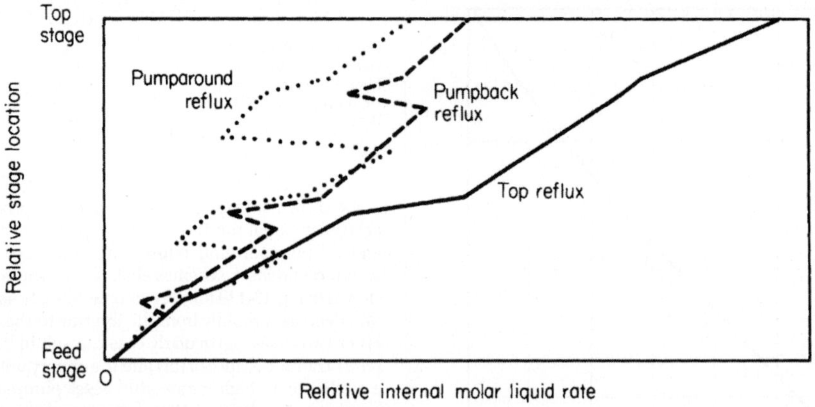

FIG. 13-110 Comparison of internal reflux rates for three methods of providing reflux.

drawoff tray and the flash zone. Calculations are made by using the crude oil EFV curve corrected for pressure. For the example being considered, percent vaporized at the flash zone must be 53.1 percent of the feed.

Tray requirements depend on internal reflux ratios and ASTM 5-95 gaps or overlaps and may be estimated by the correlation of Packie [*Trans. Am. Inst. Chem. Eng. J.* **37:** 51 (1941)] for crude units and the correlation of Houghland, Lemieux, and Schreiner [*Proc. API,* sec. III, *Refining,* 385 (1954)] for main fractionators.

Example 13-15 Simulation Calculation of an Atmospheric Tower The ability of a rigorous calculation procedure to simulate the operation of an atmospheric tower with its accompanying sidecut strippers may be illustrated by comparing commercial-test data from an actual operation with results computed with the REFINE program of ChemShare Corporation, Houston, Texas. (See also the DESIGN II program from WinSim, Inc., Sugar Land, Texas; http://www.winsim.com.) The tower configuration and plant operating conditions are shown in Fig. 13-111.

Light-component analysis and the TBP and API gravity for the feed are given in Table 13-25. Representation of this feed by pseudocomponents is given in Table 13-26 based on 16.7°C (30°F) cuts from 82°C to 366°C (180°F to 690°F), followed by 41.7°C (75°F) and then 55.6°C (100°F) cuts. Actual tray numbers are shown in Fig. 13-111. Corresponding theoretical-stage numbers, which were determined by trial and error to obtain a reasonable match of computed- and measured-product TBP distillation curves, are shown in parentheses. Overall tray efficiency appears to be approximately 70 percent for the tower and 25 to 50 percent for the sidecut strippers.

Results of rigorous calculations and comparison to plant data, when possible, are shown in Figs. 13-112, 13-113, and 13-114. Plant temperatures are in good agreement with computed values in Fig. 13-112. Computed sidestream-product TBP distillation curves are in reasonably good agreement with values converted from plant ASTM distillations, as shown in Fig. 13-113. Exceptions are the initial points of all four cuts and the higher-boiling end of the heavy-distillate curve. This would seem to indicate that more theoretical stripping stages should be added and that either the percent vaporization of the tower feed in the simulation is too high or the internal reflux rate at the lower drawoff tray is too low. The liquid-rate profile in the tower is shown in Fig. 13-114. The use of two or three pump-around circuits instead of one would result in a better traffic pattern than that shown.

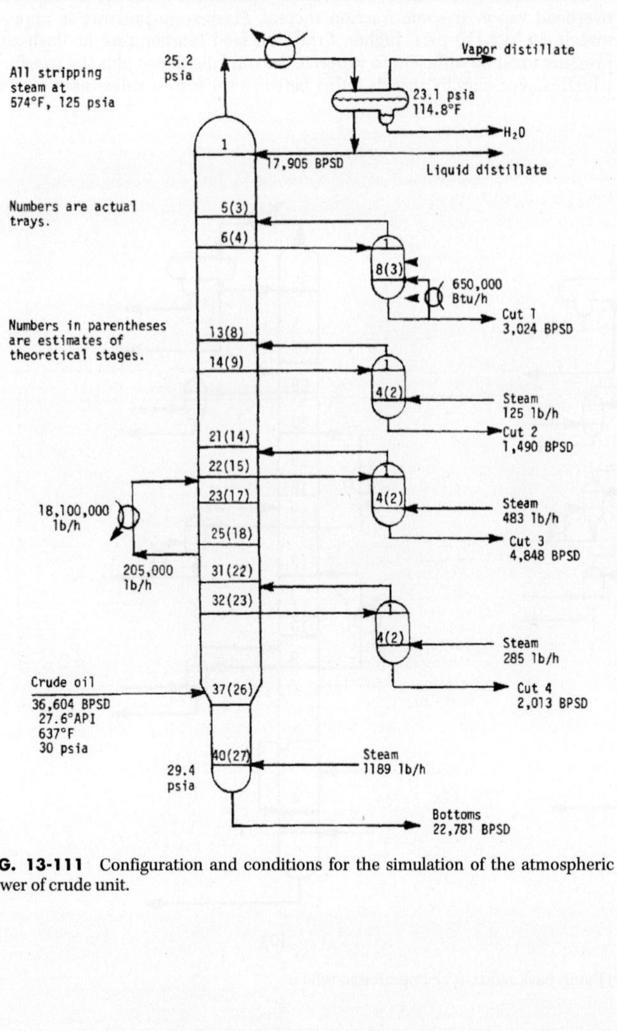

FIG. 13-111 Configuration and conditions for the simulation of the atmospheric tower of crude unit.

TABLE 13-25 Light-Component Analysis and TBP Distillation of Feed for the Atmospheric Crude Tower of Fig. 13-114

Light-component analysis		
Component		Volume percent
Methane		0.073
Ethane		0.388
Propane		0.618
n-Butane		0.817
n-Pentane		2.05
TBP distillation of feed		
API gravity	TBP, °F	Volume percent
80	−160	0.1
70	155	5
57.5	242	10
45	377	20
36	499	30
29	609	40
26.5	707	50
23	805	60
20.5	907	70
17	1054	80
10	1210	90
−4	1303	95
−22	1467	100

NOTE: To convert degrees Fahrenheit to degrees Celsius, °C = (°F − 32)/1.8.

TABLE 13-26 Pseudo-Component Representation of Feed for the Atmospheric Crude Tower of Fig. 13-114

No.	Component name	Molecular weight	Specific gravity	API gravity	(lb · mol)/h
1	Water	18.02	1.0000	10.0	.00
2	Methane	16.04	.3005	339.5	7.30
3	Ethane	30.07	.3561	265.8	24.54
4	Propane	44.09	.5072	147.5	37.97
5	*n*-Butane	58.12	.5840	110.8	43.84
6	*n*-Pentane	72.15	.6308	92.8	95.72
7	131 ABP	83.70	.6906	73.4	74.31
8	180 ABP	95.03	.7152	66.3	66.99
9	210 ABP	102.23	.7309	62.1	65.83
10	240 ABP	109.78	.7479	57.7	70.59
11	270 ABP	118.52	.7591	54.9	76.02
12	300 ABP	127.69	.7706	52.1	71.62
13	330 ABP	137.30	.7824	49.4	67.63
14	360 ABP	147.33	.7946	46.6	64.01
15	390 ABP	157.97	.8061	44.0	66.58
16	420 ABP	169.37	.8164	41.8	63.30
17	450 ABP	181.24	.8269	39.6	59.92
18	480 ABP	193.59	.8378	37.4	56.84
19	510 ABP	206.52	.8483	35.3	59.05
20	540 ABP	220.18	.8581	33.4	56.77
21	570 ABP	234.31	.8682	31.5	53.97
22	600 ABP	248.30	.8804	29.2	52.91
23	630 ABP	265.43	.8846	28.5	54.49
24	660 ABP	283.37	.8888	27.7	51.28
25	690 ABP	302.14	.8931	26.9	48.33
26	742 ABP	335.94	.9028	25.2	109.84
27	817 ABP	387.54	.9177	22.7	94.26
28	892 ABP	446.02	.9288	20.8	74.10
29	967 ABP	509.43	.9398	19.1	50.27
30	1055 ABP	588.46	.9531	17.0	57.12
31	1155 ABP	665.13	.9829	12.5	50.59
32	1255 ABP	668.15	1.0658	1.3	45.85
33	1355 ABP	643.79	1.1618	−9.7	29.39
34	1436 ABP	597.05	1.2533	−18.6	21.19
		246.90	.8887	27.7	1922.43

NOTE: To convert (lb · mol)/h to (kg · mol)/h, multiply by 0.454.

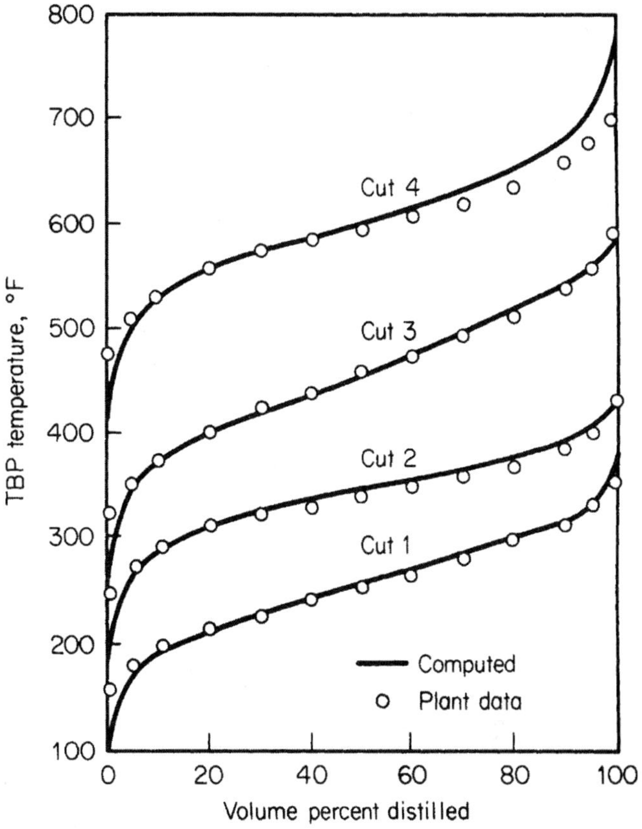

FIG. 13-113 Comparison of computed TBP curves with plant data for the example of Fig. 13-111.

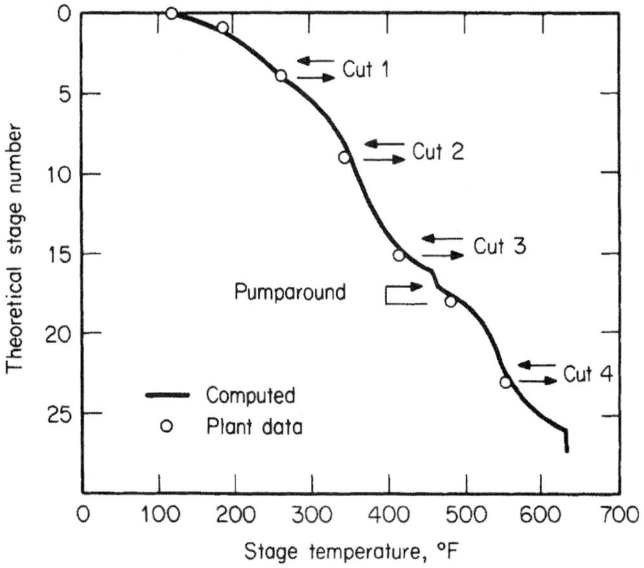

FIG. 13-112 Comparison of computed stage temperatures with plant data for the example of Fig. 13-111.

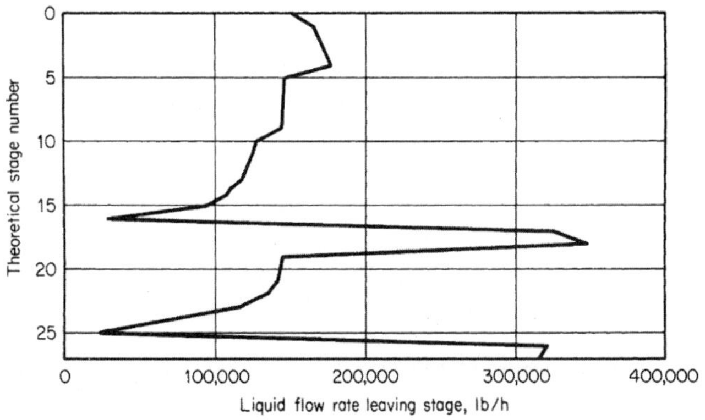

FIG. 13-114 Liquid rate profile for the example of Fig. 13-111.

BATCH DISTILLATION

Batch distillation, which is the process of separating a specific quantity (the charge) of a liquid mixture into products, is used extensively in the laboratory and in small production units that may have to serve for many mixtures. When there are c components in the feed, one batch column will often suffice where $c - 1$ simple continuous distillation columns would be required.

Many larger installations also feature a batch still. The material to be separated may be high in solids content, or it might contain tars or resins that would plug or foul a continuous unit. The use of a batch unit can keep solids separated and permit convenient removal at the termination of the process.

SIMPLE BATCH DISTILLATION

The simplest form of batch distillation consists of a heated vessel (pot or boiler), a condenser, and one or more receiving tanks. No trays or packing is provided. Feed is charged into the vessel and brought to boiling. Vapors are condensed and collected in a receiver. No reflux is returned. The rate of vaporization is sometimes limited to prevent "bumping" the charge and to avoid overloading the condenser, but other controls are minimal. This process is often referred to as a Rayleigh distillation.

If we represent the moles of vapor with V, the moles of liquid in the pot with H, the mole fraction of the more volatile component in this liquid with x, and the mole fraction of the same component in the vapor with y, a material balance yields

$$-y\,dV = d(Hx) \tag{13-122}$$

Since $dV = -dH$, substitution and expansion give

$$y\,dH = H\,dx + x\,dH \tag{13-123}$$

Rearranging and integrating give

$$\ln\frac{H_i}{H_f} = \int_{x_f}^{x_i}\frac{dx}{y-x} \tag{13-124}$$

where subscript i represents the initial condition and f the final condition of the liquid in the pot. The integration limits have been reversed to obtain a positive integral. Equation (13-124) is equivalent to an integrated form of the defining expression for residue curves in Eq. (13-114), with appropriate substitutions for the variable ξ (see below).

If phase equilibrium is assumed between liquid and vapor, the right-hand side of Eq. (13-124) may be evaluated from the area under a curve of $1/(y-x)$ versus x between the limits x_i and x_f. If the mixture is a binary system for which the relative volatility α can be approximated as a constant over the range considered, then the VLE relationship

$$y = \frac{\alpha x}{1+(\alpha-1)x} \tag{13-125}$$

can be substituted into Eq. (13-124) and a direct integration can be made:

$$\ln\left(\frac{H_f}{H_i}\right) = \frac{1}{\alpha-1}\ln\left[\frac{x_f/(1-x_f)}{x_i/(1-x_i)}\right] + \ln\left(\frac{1-x_i}{1-x_f}\right) \tag{13-126}$$

For any two components A and B of a multicomponent mixture, if constant α values can be assumed for all pairs of components, then $dH_A/dH_B = y_A/y_B = \alpha_{A,B}(x_A/x_B)$. When this is integrated, we obtain

$$\ln\left(\frac{H_{A,f}}{H_{A,i}}\right) = \alpha_{A,B}\ln\left(\frac{H_{B,f}}{H_{B,i}}\right) \tag{13-127}$$

where $H_{A,i}$ and $H_{A,f}$ are the moles of component A in the pot before and after distillation and $H_{B,i}$ and $H_{B,f}$ are the corresponding moles of component B. Mixtures that cannot be accurately described by using a constant relative volatility require some form of numerical or graphical integration for the solution of Eq. (13-124).

As an example, consider the distillation of an ethanol-water mixture at 101.3 kPa (1 atm). The initial charge is 100 mol of liquid containing 18 mol% ethanol, and the mixture must be reduced to a maximum ethanol concentration in the still of 6 mol%. By using equilibrium data interpolated from Gmehling and Onken [*Vapor-Liquid Equilibrium Data Collection*, DECHEMA Chemistry Data Ser., vol. 1, Part 1, Frankfurt (1977)], we get the following:

x	y	$y-x$	$1/(y-x)$
0.18	0.517	0.337	2.97
0.16	0.502	0.342	2.91
0.14	0.485	0.345	2.90
0.12	0.464	0.344	2.90
0.10	0.438	0.338	2.97
0.08	0.405	0.325	3.08
0.06	0.353	0.293	3.41

The area under a curve of $1/(y-x)$ versus x between $x = 0.06$ and 0.18 is $0.358 = \ln(H_i/H_f)$, so that $H_f = 100/1.43 = 70.0$ mol. The liquid remaining consists of (70.0)(0.06) = 4.2 mol of ethanol and 65.8 mol of water. By material balance, the total accumulated distillate must contain 18.0 − 4.2 = 13.8 mol of alcohol and 82.0 − 65.8 = 16.2 mol of water. The total distillate is 30 mol, and the average distillate composition is 13.8/30 = 0.46 mole fraction ethanol. The time, rate of heating, and vapor rate required to carry out the process are related by the energy balance and operating policy, which can be considered separately.

Graphical solutions of models lend significant insight, but there are many cases where such solutions are not possible or where repeated solutions are desired for different conditions. Progress in computer-based models, ranging from specialized simulation software to more general-purpose tools, now permits rapid solutions for most models. It is a simple matter to find a numerical solution to this model using a general-purpose computational tool such as Matlab or Mathematica.

The simple batch still provides only one theoretical plate of separation. Its use is usually restricted to laboratory work or preliminary manufacturing in which the products will be held for additional separation at a later time, when most of the volatile component must be removed from the batch before it is processed further, for separation of the batch from heavy undesired components.

BATCH DISTILLATION WITH RECTIFICATION

To obtain products with a narrow composition range, a batch rectifying still is commonly used. The *batch rectifier* consists of a pot (or reboiler) as in simple distillation, plus a rectifying column, a condenser, some means of accumulating and splitting off a portion of the condensed vapor (distillate) for reflux, and one or more product receivers (Fig. 13-115).

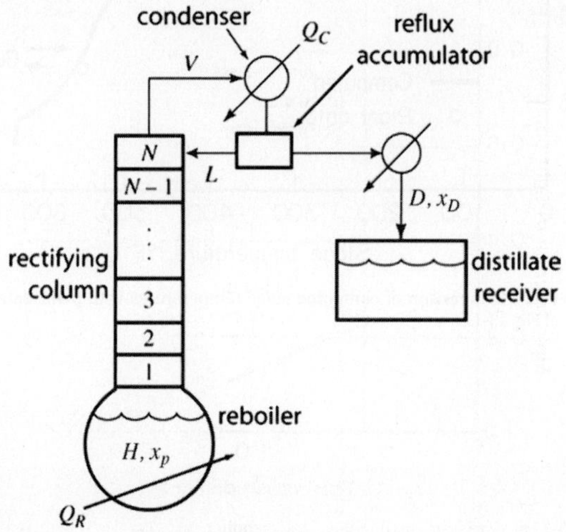

FIG. 13-115 Schematic of a batch rectifier.

The temperature of the distillate is controlled near the bubble point, and reflux is returned at or near the upper column temperature to permit a true indication of reflux quantity and to improve the column operation. A heat exchanger is used to subcool the remainder of the distillate, which is sent to a product receiver. The column may operate at an elevated pressure or at vacuum, in which case appropriate additional devices must be included to obtain the desired pressure. Equipment design methods for batch still components, except for the pot, typically follow the same principles as those presented for continuous distillation under the assumption of conditions close to a steady state (but see the comments below on the effects of holdup). The design should be checked for each mixture if several mixtures are to be processed. The design should be checked at more than one point for each mixture, since the compositions in the pot and in the column change as the distillation proceeds. The pot design is based on the batch size and the vaporization rate, which are related to the time and rate of heating and cooling available. For existing equipment, the pot size will determine the size of the batch or at least a range of feasible sizes H_i.

In operation, a batch of liquid is charged to the pot, and the system is first brought to steady state under total reflux. A portion of the overhead condensate is then continuously withdrawn in accordance with the established reflux policy. "Cuts" are made by switching to alternate receivers, at which time the operating conditions, such as reflux rate, may also be changed. The entire column operates as an enriching or rectifying section. As time proceeds, the composition of the liquid in the pot becomes less rich in the more volatile components, and distillation of a cut is stopped when the accumulated distillate attains the desired average composition or temperature.

OPERATING METHODS

A batch distillation can be operated in several ways:

1. *Constant reflux ratio, varying overhead composition.* The reflux ratio is set at a predetermined value at which it is maintained for the entire run. Since the pot liquid composition is changing, the instantaneous composition of the distillate also changes. The progress of the distillate and pot compositions in a particular binary separation is illustrated in Fig. 13-116. The variation of the distillate composition for a multicomponent batch distillation is shown in Fig. 13-117 (these distillate product cuts have relatively low purity). The shapes of the curves are functions of volatility, reflux ratio, and number of theoretical plates. The distillation is continued until the average distillate composition is at the desired value. In the case of a binary mixture, the overhead is then typically diverted to another receiver, and an intermediate or "slop" cut is withdrawn until the remaining pot liquid meets the required specification. The intermediate cut is usually added to the next batch, which can therefore have a somewhat different composition from the previous batch. For a multicomponent mixture, two or more intermediate cuts may be taken between the product cuts. It is preferable to limit the size of the intermediate cuts as far as is practical because they reduce the total amount of fresh feed that can be processed.

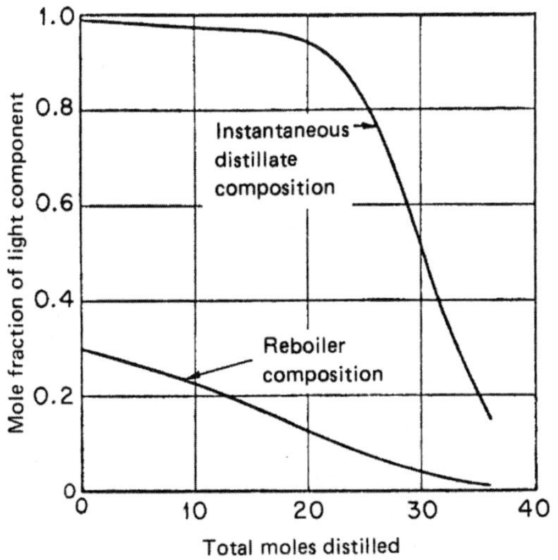

FIG. 13-116 Variation in distillate and reboiler compositions with the amount distilled in binary batch distillation at a constant reflux ratio.

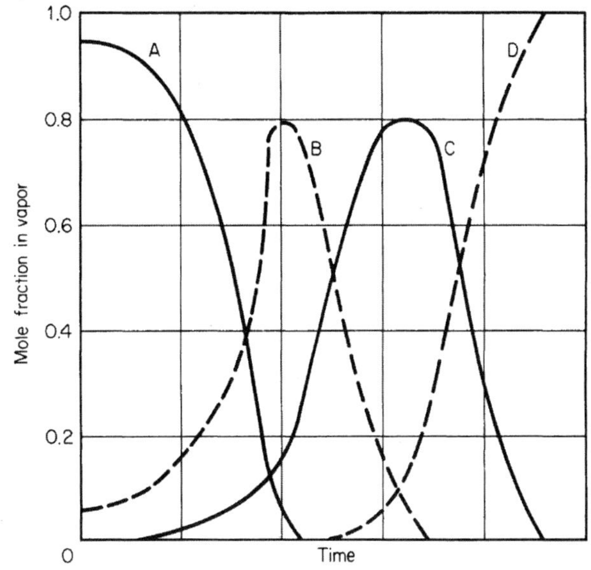

FIG. 13-117 Distillate composition for a batch distillation of a four-component mixture at a constant reflux ratio.

2. *Constant overhead composition, varying reflux.* If we wish to maintain a constant overhead composition in the case of a binary mixture, the amount of reflux returned to the column must be constantly increased throughout the run. As time proceeds, the pot is gradually depleted of the lighter component. The increase in reflux is typically gradual at first and more rapid near the end of a cut. Finally, a point is reached at which there is little of the lighter component remaining in the pot, and the reflux ratio has attained a very high value. The receivers are then changed, the reflux is reduced, and an intermediate cut is taken as before. This technique can also be extended to a multicomponent mixture.

3. *Other methods.* Instead of fixing reflux ratio, distillation may be run at constant reflux flow, or constant takeoff flow. Since boil-up usually diminishes during a batch, fixing reflux flow results in gradually increasing reflux ratio, whereas fixing distillate flow would have the opposite effect. Operating at constant reflux flow ensures that the column receives sufficient reflux to keep it wet.

4. A cycling procedure can also be used for the column operation. The unit operates at total reflux until a steady state is established. The distillate is then taken as total drawoff for a short time, after which the column is returned to total reflux operation. This cycle is repeated throughout the course of distillation. An alternative scheme is to interrupt vapor flow to the column periodically by the use of a solenoid-operated butterfly valve in the vapor line from the pot. In both cases, the equations needed to describe the system are complex, as shown by Schrodt et al. [*Chem. Eng. Sci.* **22:** 759 (1967)]. Several investigators have also proposed that batch distillation be programmed to attain *time optimization* by proper variation of the reflux ratio. A comprehensive discussion was first presented by Coward [*Chem. Eng. Sci.* **22:** 503 (1967)] and reviewed and updated by Kim and Diwekar [*Rev. Chem. Eng.* **17:** 111 (2001)]. Typical control instrumentation is described by Block [*Chem. Eng.* **74:** 147 (Jan. 16, 1967)].

5. More complex operations may involve the withdrawal of sidestreams, provision for intercondensers, the addition of feeds to trays, and periodic feed additions to the pot.

APPROXIMATE CALCULATION PROCEDURES FOR BINARY MIXTURES

Useful intuition is provided by an analysis for a binary mixture based on the McCabe-Thiele graphical method. In addition to the usual assumptions of an adiabatic column and constant molar overflow on the trays, the following procedure assumes that the holdup of liquid on the trays, in the column, and in the condenser is negligible compared to the holdup in the pot. (The effects of holdup can be significant and are discussed in a later subsection.)

As a first step, the minimum reflux ratio should be determined. Point D in Fig. 13-118 represents the desired distillate composition and is located on the diagonal since a total condenser is assumed and $x_D = y_D$. Point F represents the initial composition in the pot x_{pi} and for the vapor entering the

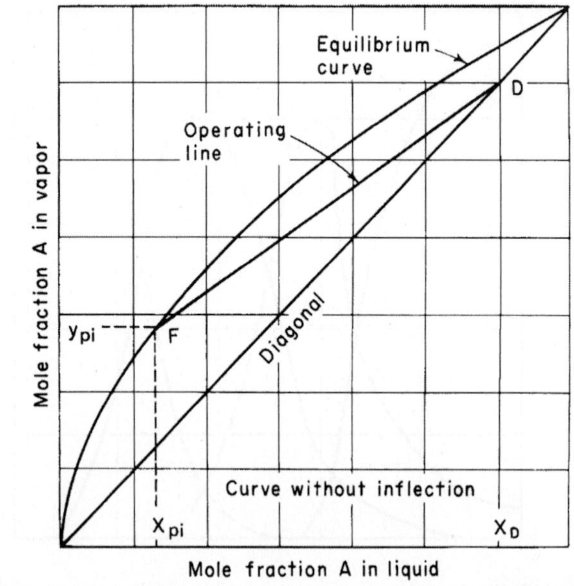

FIG. 13-118 Determination of the minimum reflux for a relatively ideal equilibrium curve.

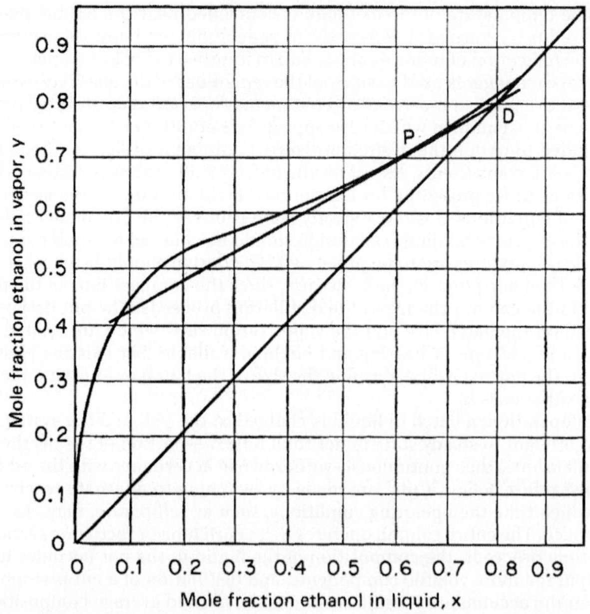

FIG. 13-119 Determination of minimum reflux for an equilibrium curve with an inflection point.

bottom of the rectifying column y_{pi}. The minimum internal reflux is found from the slope of the line *DF*

$$\left(\frac{L}{V}\right)_{\min} = \frac{y_D - y_{pi}}{x_D - x_{pi}} \tag{13-128}$$

where *L* is the liquid flow rate and *V* is the vapor rate, both in moles per hour. Since $V = L + D$ (where *D* is distillate rate) and the external reflux ratio *R* is defined as $R = L/D$,

$$\frac{L}{V} = \frac{R}{R+1} \tag{13-129}$$

or

$$R_{\min} = \frac{(L/V)_{\min}}{1-(L/V)_{\min}} \tag{13-130}$$

The condition of minimum reflux for an equilibrium curve with an inflection point *P* is shown in Fig. 13-119. In this case the minimum internal reflux is

$$\left(\frac{L}{V}\right)_{\min} = \frac{y_D - y_P}{x_D - x_P} \tag{13-131}$$

The operating reflux ratio is usually 1.5 but may be as much as 10 times the minimum. By using the ethanol-water equilibrium curve for 101.3-kPa (1-atm) pressure shown in Fig. 13-119 but extending the line to a convenient point for readability, $(L/V)_{\min} = (0.800 - 0.695)/(0.800 - 0.600) = 0.52$ and $R_{\min} = 1.083$.

Batch Rectification at Constant Reflux Ratio Using an analysis similar to the simple batch still, Smoker and Rose [*Trans. Am. Inst. Chem. Eng.* **36:** 285 (1940)] developed the following equation:

$$\ln \frac{H_i}{H_f} = \int_{x_{pf}}^{x_{pi}} \frac{dx_p}{x_D - x_p} \tag{13-132}$$

An overall material balance on the light component gives the average or accumulated distillate composition $x_{D,\text{avg}}$.

$$x_{D,\text{avg}} = \frac{H_i x_{pi} - H_f x_{pf}}{H_i - H_f} \tag{13-133}$$

If the integral on the right side of Eq. (13-132) is denoted by ξ, the time θ for distillation can be found by

$$\theta = (R+1)\frac{H_i(e^{\xi}-1)}{Ve^{\xi}} \tag{13-134}$$

An alternative equation is

$$\theta = \frac{R+1}{V}(H_i - H_f) \tag{13-135}$$

Development of these equations is given by Block [*Chem. Eng.* **68:** 88 (Feb. 6, 1961)]. The calculation process is illustrated schematically in Fig. 13-120. Operating lines are drawn with the same slope but intersecting the 45° line at different points. The number of theoretical plates

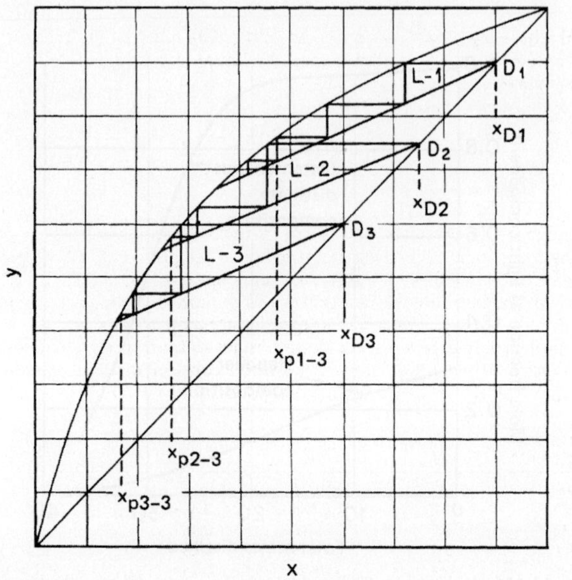

FIG. 13-120 Graphical method for constant-reflux operation.

under consideration is stepped off to find the corresponding bottoms composition (i.e., still pot composition) for each distillate composition. In Fig. 13-120, operating line $L-1$ with slope L/V drawn from point D_1 where the distillate composition is x_{D1} and the pot composition is $x_{p1\text{-}3}$ for three theoretical plates, x_{D2} has a corresponding pot composition of $x_{p2\text{-}3}$, etc. By using these pairs of distillate and pot compositions, the right-hand side of Eq. (13-132) can be evaluated, and $x_{D,\text{avg}}$ can be found from Eq. (13-133). An iterative calculation is required to find the value of H_f that corresponds to a specified $x_{D,\text{avg}}$.

To illustrate the use of these equations, consider a charge of 520 mol of an ethanol-water mixture containing 18 mol% ethanol to be distilled at 101.3 kPa (1 atm). Suppose that the vaporization rate is 75 mol/h, and the product specification is 80 mol% ethanol. Let $L/V = 0.75$, corresponding to a reflux ratio $R = 3.0$. If the column section has six theoretical plates and the pot provides an additional seventh, find how many moles of product will be obtained, what the composition of the pot residue will be, and the amount of time that the distillation will take.

Using the vapor-liquid equilibrium data, plot a y-x diagram. Draw a number of operating lines at a slope of 0.75. Note the composition at the 45° intersection, and step off seven stages on each to find the equilibrium value of the bottoms pot composition. Some of the results are tabulated in the following table:

x_D	x_p	$x_D - x_p$	$1/(x_D - x_p)$
0.800	0.323	0.477	2.097
0.795	0.245	0.550	1.820
0.790	0.210	0.580	1.725
0.785	0.180	0.605	1.654
0.780	0.107	0.673	1.487
0.775	0.041	0.734	1.362

By using an iterative procedure, integrating between x_{pi} of 0.18 and various lower limits, we find that $x_{D,\text{avg}} = 0.80$ when $x_{pf} = 0.04$, at which time the value of the integral $= 0.205 = \ln(H_i/H_f)$, so that $H_f = 424$ mol. The product collected $= H_i - H_f = 520 - 424 = 96$ mol. From Eq. (13-134),

$$\theta = \frac{(4)(520)(e^{0.205} - 1)}{75(e^{0.205})} = 5.2\,\text{h} \tag{13-136}$$

Batch Rectification at Constant Distillate Composition Bogart [*Trans. Am. Inst. Chem. Eng.* **33:** 139 (1937)] developed the following equation for constant distillate composition with the column holdup assumed to be negligible:

$$\theta = \frac{H_i(x_D - x_{pi})}{V} \int_{x_{pf}}^{x_{pi}} \frac{dx_p}{(1 - L/V)(x_D - x_p)^2} \tag{13-137}$$

and where the terms are defined as before. The quantity distilled can then be found by material balance once the initial and final pot compositions are known.

$$H_i - H_f = \frac{H_i(x_{pi} - x_{pf})}{x_D - x_{pf}} \tag{13-138}$$

A schematic example is shown in Fig. 13-121. The distillate composition is held constant by increasing the reflux as the pot composition becomes more dilute. Operating lines with varying slopes $(= L/V)$ are drawn from the known distillate composition, and the given number of stages is stepped off to find the corresponding bottoms (still pot) compositions.

As an example, consider the same ethanol-water mixture used previously to illustrate constant reflux, but now with a constant distillate composition of $x_D = 0.90$. The following table is compiled:

L/V	R	x_p	$x_D - x_p$	$1/(1 - L/V)(x_D - x_p)^2$
0.600	1.50	0.654	0.147	115.7
0.700	2.33	0.453	0.348	27.5
0.750	3.00	0.318	0.483	17.2
0.800	4.00	0.143	0.658	11.5
0.850	5.67	0.054	0.747	11.9
0.900	9.00	0.021	0.780	16.4

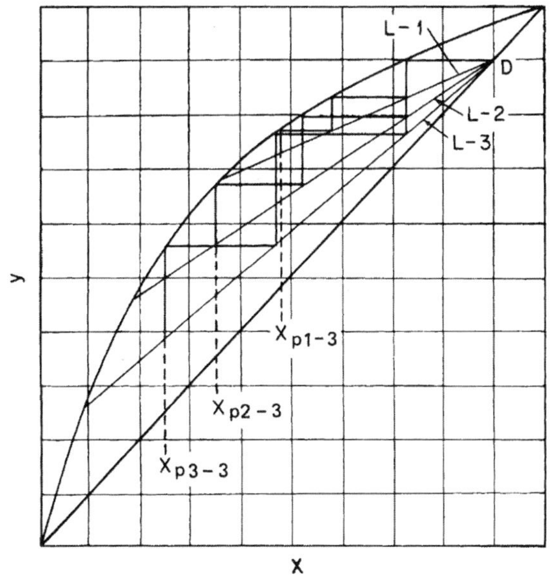

FIG. 13-121 Schematic of constant distillate composition operation.

If the right-hand side of Eq. (13-137) is integrated by using a limit for x_{pf} of 0.04, the value of the integral is 1.615, and the time is

$$\theta = \frac{(520)(0.800 - 0.180)(1.615)}{75} = 7.0\,\text{h} \tag{13-139}$$

The quantity distilled can be found from Eq. (13-140):

$$H_i - H_f = \frac{(520)(0.180 - 0.040)}{0.800 - 0.040} = 96\,\text{mol} \tag{13-140}$$

EFFECTS OF COLUMN HOLDUP

When the holdup of liquid on the trays and in the condenser and reflux accumulator is significant compared with the holdup in the pot, the distillate composition at constant reflux ratio changes with time at a different rate than when the column holdup is negligible because of two separate effects.

First, with an appreciable column holdup, the composition of the charge to the pot will be higher in the light component than the pot composition at the start of the distillation. The reason is that before product takeoff begins, the column holdup must be supplied, and due to the rectification, its average composition is higher in the lighter component than that of the liquid charged as feed to the pot. Thus, when overhead takeoff begins, the pot composition is lower than it would be if there were negligible column holdup, and the separation is more difficult than expected based on the composition of the feed. The second effect of column holdup is to slow the rate of exchange of the components; the holdup exerts an inertial effect, which prevents compositions from changing as rapidly as they would otherwise, and the degree of separation is usually improved.

Both these effects occur at the same time and change in importance during the course of distillation. Although a number of studies were made and approximate methods developed for predicting the effect of liquid holdup during the 1950s and 1960s (summarized in the 6th edition of *Perry's Chemical Engineers' Handbook*), it is now simpler to use simulation methods to determine the effect of holdup on a case-by-case basis.

As an example, consider a batch rectifier fed with a 1 to 1 mixture of ethanol and n-propanol. The rectifier has eight theoretical stages in the column and is operated at a reflux ratio of 19. The distillate and pot compositions are shown in Fig. 13-122 for various values of the holdups.

In Fig. 13-122a, the holdup on each stage is 0.01 percent of the initial pot holdup, and in the reflux accumulator it is 0.1 percent of the initial pot holdup (for a total of 0.108 percent). Because this model calculation does not begin with a total reflux period, there is a very small initial distillate cut

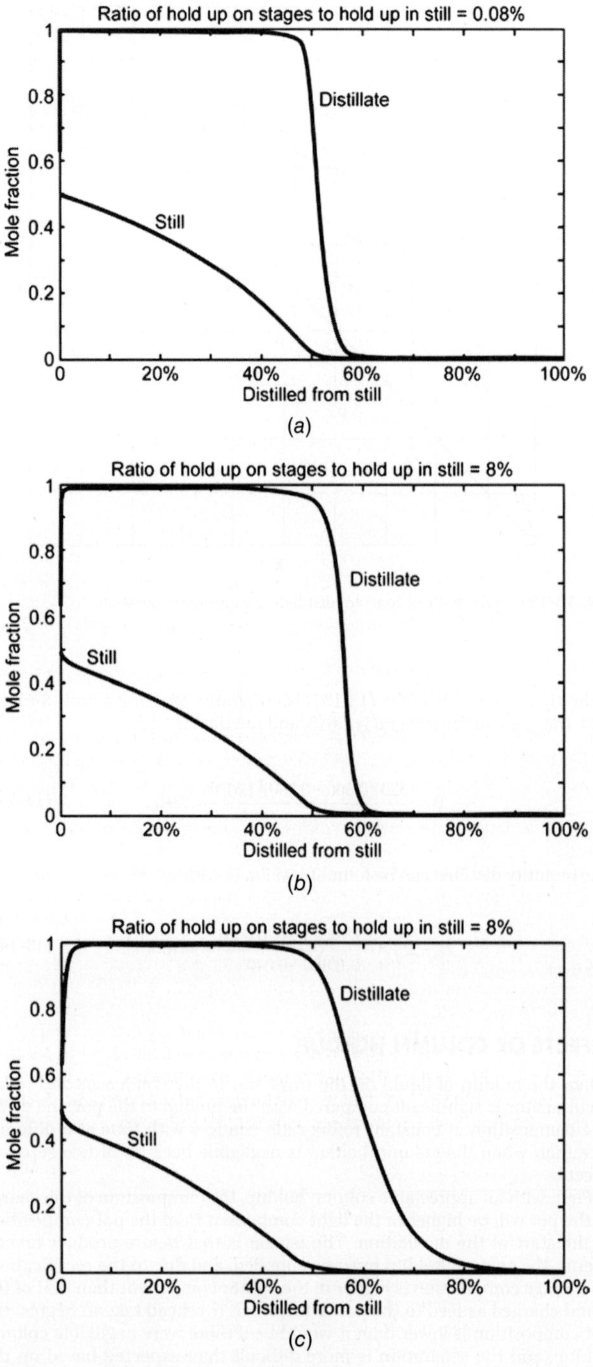

FIG. 13-122 Effects of holdup on batch rectifier.

with relatively low ethanol purity. This is followed by a high-purity distillate cut. An intermediate cut of approximately 10 percent of the initial batch size can be collected, leaving the pot with a high purity of *n*-propanol. The column holdup for the case shown in Fig. 13-122*b* is 1 percent of the initial batch size on each stage, while the reflux accumulator holdup remains small at 0.1 percent (for a total of 8.1 percent). In this case, both the first low-purity cut and the intermediate cut are somewhat larger for the same purity specifications. These effects are substantially larger when the reflux accumulator has a more significant holdup, as shown in Fig. 13-122*c*, corresponding to a holdup of 1 percent on each stage and 5 percent in the reflux accumulator (for a total of 13 percent). Similar effects are found for multicomponent mixtures. The impact of column and condenser holdup is most important when a high-purity cut is desired for a component that is present in relatively small amounts in the feed.

SHORTCUT METHODS FOR MULTICOMPONENT BATCH RECTIFICATION

For preliminary studies of batch rectification of multicomponent mixtures, shortcut methods that assume constant molar overflow and negligible vapor and liquid holdup are useful in some cases (see the preceding discussion about the effects of holdup). The method of Diwekar and Madhaven [*Ind. Eng. Chem. Res.* **30:** 713 (1991)] can be used for constant reflux or constant overhead rate. The method of Sundaram and Evans [*Ind. Eng. Chem. Res.* **32:** 511 (1993)] applies only to the case of constant reflux, but it is easy to implement. Both methods employ the Fenske-Underwood-Gilliland (FUG) shortcut procedure at successive time steps. Thus, batch rectification is treated as a sequence of continuous, steady-state rectifications.

CALCULATION METHODS AND SIMULATION

Model predictions such as those shown in Fig. 13-122 are relatively straightforward to obtain by using modern simulation models and software tools. As discussed in earlier editions of this handbook, such models and algorithms for their solutions have been the subject of study since the early 1960s. Detailed calculation procedures for binary and multicomponent batch distillation were initially focused on binary mixtures of constant relative volatility. For example, Huckaba and Danly [*AIChE J.* **6:** 335 (1960)] developed a simulation model that incorporated more details than can be included in the simple analytical models we have described. They assumed constant-mass tray holdups, adiabatic tray operation, and linear enthalpy relationships, but they did include energy balances around each tray, and they incorporated the use of nonequilibrium trays by means of specified tray efficiencies. Experimental data were provided to validate the simulation. Meadows ["Multicomponent Batch-Distillation Calculations on a Digital Computer," *Chem. Eng. Prog. Symp. Ser.* **46**(59): 48–55 (1963)] presented a multicomponent batch distillation model that included equations for energy, material, and volume balances around theoretical trays. The assumptions made were perfect mixing on each tray, negligible vapor holdup, adiabatic operation, and constant-volume tray holdup. Distefano [*AIChE. J.* **14:** 190 (1968)] extended the model and developed a procedure that was used to simulate several commercial batch distillation columns successfully. Boston et al. (in *Foundations of Computer-Aided Chemical Process Design,* vol. 2, ed. Mah and Seider, American Institute of Chemical Engineers, New York, 1981, p. 203) further extended the model, provided a variety of practical sets of specifications, and used modern numerical procedures and equation formulations to efficiently handle the nonlinear and often stiff nature of the multicomponent batch distillation problem.

It is important to note that in using computer-aided models for batch distillation, the various assumptions of the model can have a significant impact on the accuracy of the results; for example, see the previous discussion of the effects of holdup. Uncertainties in the physical and chemical parameters in the models can be addressed most effectively by a combination of sensitivity calculations using simulation tools, along with comparison to data. The mathematical treatment of stiffness in the model equations can also be very important, and there is often a substantial advantage in using simulation tools that take special account of this stiffness. (See the 7th edition of *Perry's Chemical Engineers' Handbook* for a more detailed discussion of this aspect.)

The availability of detailed models and solution methods has enabled many new studies of complex, mixtures, configurations, and operating and control strategies for batch distillation.

SEMIBATCH DISTILLATION

A typical example of this process is a distillation in which feed is supplied continuously and distillate is withdrawn continuously. This is applicable for mixtures with significant amounts of volatile component. At first, a small portion of the pot is charged and heated. Then the feed is supplied at such a rate that the low-boiling component instantaneously flashes off and is withdrawn as distillate. At the end of charging, there is a full pot containing mostly the less volatile component (usually the product). This component may be withdrawn as bottom product, or it may be distilled to purify it from heavier contaminants.

Another example of a semibatch process is constant-level distillation. In this application, one solvent is replaced with another in the presence of a heavy nonvolatile product, as may be encountered in pharmaceutical production. One option for switching solvents is to use simple distillation repeatedly. Initially, a portion of the first solvent is removed by boiling. Then the second solvent is added, and a simple distillation removes more of the first solvent along with some of the second. Repetition of the latter step can be used to reduce the concentration of the first solvent to very small levels.

Gentilcore [*Chem. Eng. Progr.* **98**(1): 56 (2002)] describes an alternative strategy of "constant-level" batch distillation, where the replacement solvent is added at a rate to keep the volume of liquid in the pot constant. For simple distillation without rectification, the analog of Eq. (13-124) is

$$\frac{S}{H}=\int_{x_f}^{x_i}\frac{dx}{y} \tag{13-141}$$

and the analog of Eq. (13-126) is

$$\frac{S}{H}=\frac{1}{\alpha}\ln\frac{x_i}{x_f}+\frac{\alpha-1}{\alpha}(x_i-x_f) \tag{13-142}$$

where the mole fractions refer to the compositions of the original solvent, and S is the amount of the second solvent added to the batch. The amount of solute, a nonvolatile heavy product, is small compared to the size of the batch (alternatively, the analysis can be done on a solute-free basis). The second solvent is assumed to be pure, and the rate of addition is manipulated to keep a constant level in the pot. Compared to the repeated application of simple distillation, this semibatch operation can typically reduce solvent use by one-half or more, depending on the volatility and the desired compositions. This is also a more efficient use of equipment at the expense of a somewhat more complex operation.

An example provided by Gentilcore shows a 60 percent savings in the use of replacement solvent.

INDUSTRIAL OPERATING PRACTICES

Batch columns often purify products coming out of a reactor. The reactor and column usually work in campaigns. A campaign is devoted to making one product (although sometimes two products are made simultaneously, such as isomers), and a campaign may last from a few days to several weeks. After the campaign is over, the equipment is cleaned and another campaign to make different products begins.

The objective of industrial batch distillation is normally to maximize the production rate, provided that purity and yield constraints are satisfied. Usually one or two products are separated from a multicomponent mixture. A product can be obtained as a distillate or as a bottom product.

The simplest batch distillation is the removal of volatile impurities from the final (bottom) product. This can be treated as pseudobinary distillation, with the light impurities being the first pseudocomponent (in reality, the impurities are always multicomponent mixtures) and the product being the second component.

Another typical case (pseudoternary system) is a separation of a desired product from light (more volatile) impurities and heavy (less volatile) impurities. The process involves the following steps: charging the column, heating and degassing, establishing reflux, low-boils cut, front cut, product cut, after-cut, pumping out heavies, and possibly washing the pot. These steps will be discussed next.

Charging must be as fast as possible to maximize the production rate. Therefore, the batch must be charged as soon as possible after the previous batch is finished (if it is safe to do so). We need to ensure that the feed pump works at a sufficiently high rate. If not, the pump rotor or the entire pump may need to be replaced. The piping should be inspected and understood; for example, there may be a manual valve left partially closed on the way to the pot. Also, we may consider feeding the column that is under vacuum, to increase the feeding rate. Finally, we should be charging as much as possible, but without overcharging. Partial charge is a waste of equipment capacity, but charging too much may result in damaging column internals when distillation starts.

One should not wait until charging is complete to start the heating step. The heat exchanger may be internal to the pot (a submerged tube bundle) or external (forced circulation reboiler or thermosyphon). The heat should be turned on as soon as the internal bundle is covered with liquid or as soon as it is possible to continuously operate a circulation pump without it going dry. To reduce heating time, the heat to the reboiler should be maximized.

After reflux is established, the column is usually kept at total reflux for some time. The objective of this step is to establish a concentration profile in the column, with more volatile components closer to the top and heavier components situated lower in the column. What often interferes with establishing the desired column profile is material that may be left from the previous batch in the top accumulator. This material is usually the heavy component. It will be washed down by total reflux, but it will take time for it to travel down the column. Heat duty for the total reflux step should be lowered, and column pressure drop must be kept below maximum, to keep the column away from the flood point.

After the steady state in the column is reached, we may begin taking off distillate. The purity of the distillate will be changing during the process, so the material will be directed to various tanks as so-called cuts. First, the most volatile components will be taken off as the "low-boils cut." This material is usually disposed of as waste. When the product starts to appear in the distillate, but it is still below the purity specification, we switch the distillate to a different tank. This material is referred to as "front cut" or "slop cut." This cut will have to be reprocessed. When the purity of the product in the distillate is high enough, we may start collecting the "product cut." However, if this is the first distillation in the campaign, we need to make sure that the line to the pure tank and the tank itself are clean and dry (they are usually cleaned with water before the campaign). So part of the first product cut is used to flash the line and the tank, and then it is pumped away to the front-cut tank, to be reprocessed. Sometimes the flash needs to be repeated. Then we begin taking the product cut and continue until the purity of the distillate starts decreasing (thus taking the product purity out of specification). So distillate purity needs to be either measured or inferred from the temperature at the top of the column or inferred from the distillate amount. That is usually determined empirically by trial and error.

If the purity of the distillate is below the specification and it can no longer be blended with the existing product taken so far, the material is switched to the "after-cut tank." If there are no more tanks available, it can be directed to the front-cut tank, although mixing materials of different compositions is never thermodynamically efficient. This material should be reprocessed.

Finally, product content in the distillate is very small or boil-up stops because the pot becomes almost empty or the residual heavy material does not boil anymore. That is the end of distillation. The pot may need to be pumped out or (to save time) new material is charged on top of the left-over residue of heavy impurities. This practice cannot be continued over too many batches because eventually, when the amount of heavy impurities increases, distillation time and yield losses will increase. Therefore, from time to time the pot needs to be cleaned—usually washed with water.

To maximize the yield of the batch, the column holdup (containing valuable product) should be recovered. A part of this holdup resides in the top accumulator. Therefore, at the end of the batch, this accumulator should be emptied, if possible into the product tank. It is a very simple step, but it is often overlooked in industrial practice. Also, the piping from the accumulator to the tank should be blown, e.g., with nitrogen, to recover the product.

The second part of the holdup is the liquid that resides on the packing or trays. After heat is turned off, boil-up vanishes, and this liquid falls down to the pot and mixes with the heavy residue. If the pot needs to be washed, the column holdup is lost, which may be a significant amount. A simple valve on the liquid return to the pot, closed after the batch is over, prevents these losses.

Slop cuts need to be reprocessed, at extra expense of time and energy. Common strategies are either to mix the slop cuts with the next-batch charge or to collect them and distill separately. In some cases, when separation is very easy or purity specification is not stringent or the amount of impurity is small, some slop cuts may not be needed.

Parameters used to optimize the column are: reflux ratio, heat duty, column pressure, and switch points between cuts. These parameters may change between cuts or even during a single cut. Typically, computer recipes contain all this detailed information.

The reflux ratio for slop cuts should be optimized. High reflux ratios reduce the sizes of slop cuts, which give less material to reprocess. However, high reflux ratios also increase the time of the cuts.

For the product cut, the reflux ratio should have the smallest value that still provides the desired purity of the cut. Certain cuts, with easily separable components, may proceed at total takeoff (zero reflux); at other times, total reflux is used intermittently with total takeoff.

Heat duty should be maximized to maximize distillation rate, provided that the column does not flood (the column pressure drop needs to be monitored).

The column pressure is chosen so that we could easily boil the mixture in the pot and condense the distillate. Lower pressure usually makes separation easier. However, too low a pressure may cause column flooding. It is not uncommon to reduce the pressure as distillation proceeds and heavier components need to be boiled off.

To summarize, batch distillation is one of the oldest, most widely approved chemical engineering processes. It seems to be quite simple because there is only one column to control. It provides great flexibility because many different materials can be distilled in the same equipment. However, there are many disadvantages of batch distillation compared to continuous distillation. One is that cleaning between campaigns or even in one campaign (blowing the lines) takes time and energy and creates waste streams. Another disadvantage is related to slop cuts: these are cuts that need to be reprocessed using additional time and energy. Thermodynamic inefficiencies are also related to holdups; for example, the top accumulator collects liquid from the condenser. This liquid (at constant reflux) changes its composition in time, so condensates of various compositions are mixed together over time. Finally, the batch nature of the process creates additional problems, either in the form of mistakes (product inadvertently sent to a wrong tank) or idle times when one process has to wait for another.

ALTERNATIVE EQUIPMENT CONFIGURATIONS

The batch rectifier shown schematically in Fig. 13-115 is by far the most common configuration of equipment. Several alternative special-purpose configurations have been studied and offer potential advantages in particular applications. Also see Doherty and Malone (*Conceptual Design of Distillation Systems,* McGraw-Hill, 2001, pp. 407–409, 417–419).

For instance, a simple batch distillation can be combined with a stripping column to give the batch stripper shown in Fig. 13-123. The pot holds the batch charge and provides liquid reflux to the stripping section. The reboiler provides vapor to the column and has relatively small holdup. The product stream B in the bottom is concentrated in the higher-boiling compound, and the pot gradually becomes more concentrated in the lighter component. Multiple "cuts" can be taken as products, and the reboil rate either can be constant or can be adjusted by analogy with the reflux ratio in the batch rectifier.

For mixtures containing large concentrations of a heavy component, the batch stripper can be advantageous.

The more complex "middle vessel" column combines aspects of both the batch rectifier and the batch stripper, as shown in Fig. 13-124. The middle vessel arrangement was described qualitatively by Robinson and Gilliland (*Elements of Fractional Distillation,* McGraw-Hill, New York, 1950, p. 388) and analyzed by Bortolini and Guirase [*Quad. Ing. Chim. Ital.* **6:** 150 (1970)]. This configuration requires more equipment and is more complex, but it can produce both distillate and bottoms product cuts simultaneously. Barolo and Botteon [*AIChE J.* **43:** 2601 (1997)] pointed out that the middle vessel configuration at total reflux and reboil and with the appropriate collection equipment for distillate and bottoms products (not shown in Fig. 13-124) can concentrate a ternary mixture into its three pure fractions. This and analogous configurations for mixtures with more components have been studied by Hasebe et al. [*J. Chem. Eng. Japan* **29:** 1000 (1996); *Computers Chem. Eng.* **23:** 523 (1999)] and experimentally by Wittgens and Skogestad [*IChemE Symp Ser.* **142:** 239 (1997).]

The batch stripper and the middle vessel configurations offer the ability to make separations for certain azeotropic mixtures that are not possible or that cannot be done efficiently in the batch rectifier.

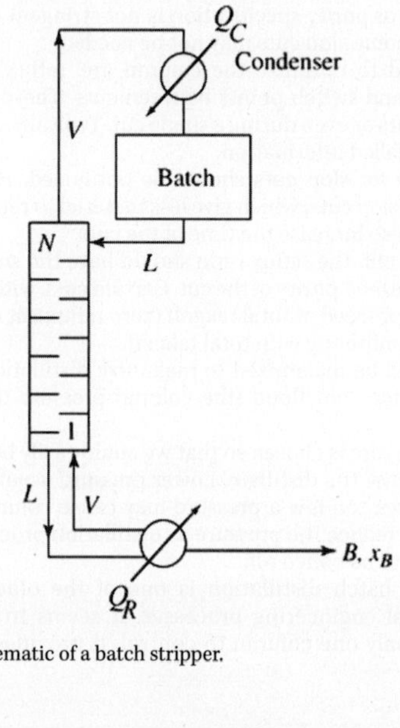

FIG. 13-123 Schematic of a batch stripper.

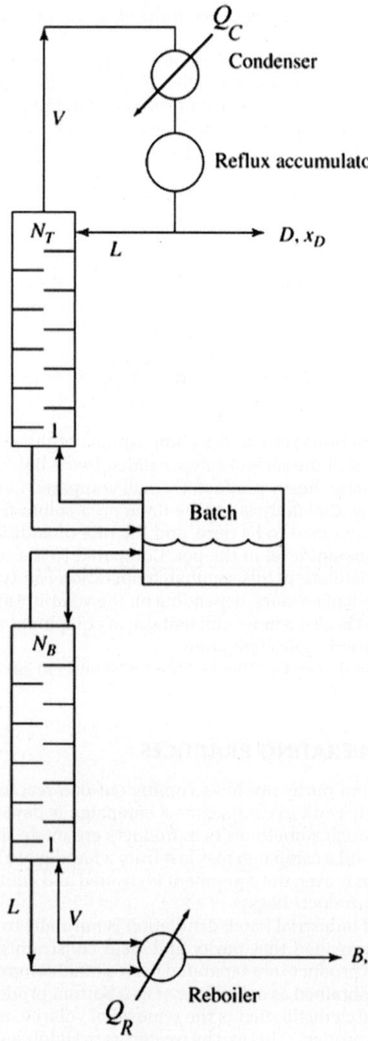

FIG. 13-124 Middle vessel batch distillation.

BATCH DISTILLATION OF AZEOTROPIC MIXTURES

Although azeotropic distillation is covered in an earlier subsection, it is appropriate to consider the application of residue curve maps to batch distillation here. (See the subsection Enhanced Distillation for a discussion of residue curve maps.) An essential point is that the sequence, number, and limiting composition of each cut from a batch distillation depend on the form of the residue curve map and the composition of the initial charge to the still. As with continuous distillation operation, the set of reachable products (cuts) for a given charge to a batch distillation is constrained by the residue curve–map distillation boundaries. Furthermore, some pure components can be produced as products from the batch stripper but not the batch rectifier, and vice versa. Doherty and Malone (*Conceptual Design of Distillation Systems,* chap. 9, McGraw-Hill, New York, 2001) give more details, but the main points are the following.

In the batch rectifier, the limiting cuts, obtainable with a sufficiently large number of stages and reflux, begin with the low-boiling node that defines the distillation region containing the feed composition. For the batch stripper, the first limiting cut is the high-boiling node. In either case, the subsequent cuts depend on the structure of the residue curve map.

For the batch rectifier, as the low-boiling component or azeotrope is removed, the still composition moves along a straight material balance line through the initial feed composition and the low-boiling node, and away from the initial composition, until it reaches the edge of the composition triangle or a distillation boundary. The path then follows the edge or distillation boundary to the high-boiling node of the region.

As an example, consider the residue curve map structure shown in Fig. 13-125 for a mixture of methanol, methyl propionate, and water at a

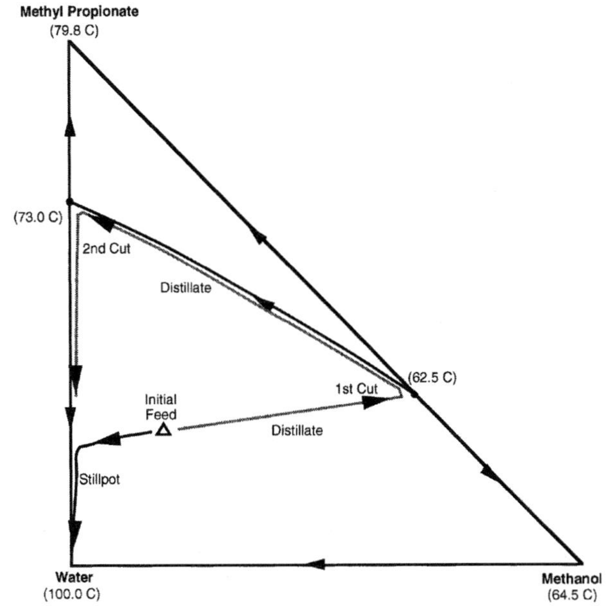

FIG. 13-125 Residue curve map and batch rectifier paths for methanol, methyl propionate, and water.

pressure of 1 atm. There are two minimum-boiling binary azeotropes joined by a distillation boundary that separates the compositions into two distillation regions. Feeds in the upper and lower regions will have different distillate products. For the sample feed shown, and with a sufficient number of theoretical stages and reflux, the distillate will approach the low-boiling azeotrope of methanol and methyl propionate at 62.5°C. The still pot composition changes along the straight-line segment as shown until it is nearly free of methanol. At that point, the distillate composition changes along the distillation boundary to a composition for the second cut at or near the methyl propionate–water azeotrope. The still pot composition eventually approaches pure water.

For the same feed, a batch stripper can be used to remove a bottoms product that approaches pure water. The pot composition (overhead) will contain all three components near the point of intersection of the distillation boundary with a straight line extended from the water vertex through the feed composition.

For this mixture, it is not possible to isolate the pure components in a batch rectifier or batch stripper. The use of additional equipment such as a decanter to exploit liquid-liquid phase behavior or the addition of a fourth component or chemical reactions can sometimes be used to effect the separation.

The product cuts for azeotropic mixtures are also sensitive to the curvature of the distillation boundaries; see Doherty and Malone (*Conceptual Design of Distillation Systems,* McGraw-Hill, New York, 2001; pp. 403–404) and additional references there.

Equipment for Distillation, Gas Absorption, Phase Dispersion, and Phase Separation

Henry Z. Kister, M.E., C.Eng., C.Sc. *Senior Fellow and Director of Fractionation Technology, Fluor Corporation; Member, National Academy of Engineering (NAE); Fellow, American Institute of Chemical Engineers; Fellow, Institution of Chemical Engineers (U.K.); Member, Institute of Energy (Section Editor, Equipment for Distillation and Gas Absorption)*

Paul M. Mathias, Ph.D. *Senior Fellow and Technical Director, Fluor Corporation; Fellow, American Institute of Chemical Engineers (Design of Gas Absorption Systems)*

Daniel E. Steinmeyer, P.E., M.S. *Distinguished Science Fellow, Monsanto Company (Retired); Fellow, American Institute of Chemical Engineers; Member, American Chemical Society (Phase Dispersion, Liquid in Gas Systems)*

W. Roy Penney, Ph.D., P.E. *Professor Emeritus, Department of Chemical Engineering, University of Arkansas; Fellow, American Institute of Chemical Engineers (Gas-in-Liquid Dispersions)*

Valerie S. Monical, B.S. *Fellow, Ascend Performance Materials, Inc. (Phase Separation)*

James R. Fair, Ph.D., P.E. *(Deceased) Professor of Chemical Engineering, University of Texas; Fellow, American Institute of Chemical Engineers; Member, American Chemical Society, American Society for Engineering Education, National Society of Professional Engineers (Section Editor of the 7th edition and major contributor to the 5th, 6th, and 7th editions)*

INTRODUCTION

DESIGN OF GAS ABSORPTION SYSTEMS

EQUIPMENT FOR DISTILLATION AND GAS ABSORPTION: TRAY COLUMNS

EQUIPMENT FOR DISTILLATION AND GAS ABSORPTION: PACKED COLUMNS

OTHER TOPICS FOR DISTILLATION AND GAS ABSORPTION EQUIPMENT

PHASE DISPERSION

PHASE SEPARATION

Nomenclature

Symbol	Definition	SI units	U.S. Customary System units
a, a_e	Effective interfacial area	m^2/m^3	ft^2/ft^3
a_p	Packing surface area per unit volume	m^2/m^3	ft^2/ft^3
A	Absorption factor $L_M/(mG_M)$	-/-	-/-
A	Cross-sectional area	m^2	ft^2
A_a	Active area, same as bubbling area	m^2	ft^2
A_B	Bubbling (active) area	m^2	ft^2
A_D	Downcomer area (straight vertical downcomer)	m^2	ft^2
A_{da}	Downcomer apron area	m^2	ft^2
A_{DB}	Area at bottom of downcomer	m^2	ft^2
A_{DT}	Area at top of downcomer	m^2	ft^2
A_e, A'	Effective absorption factor (Edmister)	-/-	-/-
A_f	Fractional hole area	-/-	-/-
A_h	Hole area	m^2	ft^2
A_N	Net (free) area	m^2	ft^2
A_S	Slot area	m^2	ft^2
A_{SO}	Open slot area	m^2	ft^2
A_T	Tower cross-sectional area	m^2	ft^2
c	Concentration	$kg \cdot mol/m^3$	$lb \cdot mol/ft^3$
c'	Stokes-Cunningham correction factor for terminal settling velocity	-/-	-/-
C	C-factor for gas loading, Eq. (14-77)	m/s	ft/s
C_1, C_2	Parameters in system limit equation	m/s	ft/s
C_3, C_4	Constants in Robbins' packing pressure drop correlation	-/-	-/-
CAF	Flood C-factor, Eq. (14-88)	m/s	ft/s
CAF_0	Uncorrected flood C-factor, Fig. 14-31	—	ft/s
C_d	Coefficient in clear liquid height correlation, Eq. (14-116)	-/-	-/-
C_G	Gas C-factor; same as C	m/s	ft/s
C_L	Liquid loading factor, Eq. (14-144)	m/s	ft/s
C_{LG}	A constant in packing pressure drop correlation, Eq. (14-143)	$(m/s)^{0.5}$	$(ft/s)^{0.5}$
CP	Capacity parameter (packed towers), Eq. (14-140)		
C_{SB}, C_{sb}	C-factor at entrainment flood, Eq. (14-80)	m/s	ft/s
C_{sbf}	Capacity parameter corrected for surface tension	m/s	ft/s
C_v, C_V	Discharge coefficient, Fig. 14-36	-/-	-/-
C_w	A constant in weep rate equation, Eq. (14-123)	-/-	-/-
C_{XY}	Coefficient in Eq. (14-159) reflecting angle of inclination	-/-	-/-
d	Diameter	m	ft
d_b	Bubble diameter	m	ft
d_h, d_H	Hole diameter	mm	in
d_o	Orifice diameter	m	ft
d_{pc}	Cut size of a particle collected in a device, 50% mass efficiency	μm	ft
d_{psd}	Mass median size particle in the pollutant gas	μm	ft
d_{pa50}	Aerodynamic diameter of a real median size particle	μm	ft
d_w	Weir diameter, circular weirs	mm	in
D	Diffusion coefficient	m^2/s	ft^2/s
D	Tube diameter (wetted-wall columns)	m	ft
D_{32}	Sauter mean diameter	m	ft
D_g	Diffusion coefficient	m^2/s	ft^2/h
D_p	Packing particle diameter	m	ft
D_T	Tower diameter	m	ft
D_{tube}	Tube inside diameter	m	ft
D_{vm}	Volume mean diameter	m	ft
e	Absolute entrainment of liquid	$kg \cdot mol/h$	$lb \cdot mol/h$
e	Entrainment, mass liquid/mass gas	kg/kg	lb/lb
E	Plate or stage efficiency, fractional	-/-	-/-
E	Power dissipation per mass	W	Btu/lb
E_a	Murphree tray efficiency, with entrainment, gas concentrations, fractional	-/-	-/-
E_g	Point efficiency, gas phase only, fractional	-/-	-/-
E_{OC}	Overall column efficiency, fractional	-/-	-/-
E_{OG}	Overall point efficiency, gas concentrations, fractional	-/-	-/-
E_{mv}, E_{MV}	Murphree tray efficiency, gas concentrations, fractional	-/-	-/-

Symbol	Definition	SI units	U.S. Customary System units
E_s	Entrainment, kg entrained liquid per kg gas upflow	kg/kg	lb/lb
f	Fractional approach to flood	-/-	-/-
f	Liquid maldistribution fraction	-/-	-/-
f_{max}	Maximum value of f above which separation cannot be achieved	-/-	-/-
f_w	Weep fraction, Eq. (14-121)	-/-	-/-
F	Fraction of volume occupied by liquid phase, system limit correlation, Eq. (14-171)	-/-	-/-
F	F-factor for gas loading Eq. (14-76)	$m/s(kg/m^3)^{0.5}$	$ft/s(lb/ft^3)^{0.5}$
F_{LG}	Flow parameter, Eq. (14-89) and Eq. (14-141)	-/-	-/-
F_p	Packing factor	m^{-1}	ft^{-1}
F_{pd}	Dry packing factor	m^{-1}	ft^{-1}
FPL	Flow path length	m	ft
Fr	Froude number, clear liquid height correlation, Eq. (14-120)	-/-	-/-
Fr_h	Hole Froude number, Eq. (14-114)	-/-	-/-
F_w	Weir constriction correction factor, Fig. 14-39	-/-	-/-
g	Gravitational constant	m/s^2	ft/s^2
g_c	Conversion factor	$1.0\ kg \cdot m/(N \cdot s^2)$	$32.2\ lb \cdot ft/(lbf \cdot s^2)$
G	Gas phase mass velocity	$kg/(s \cdot m^2)$	$lb/(hr \cdot ft^2)$
G_f	Gas loading factor in Robbins' packing pressure drop correlation	$kg/(s \cdot m^2)$	$lb/(h \cdot ft^2)$
G_M	Gas phase molar velocity	$kg \cdot mol/(s \cdot m^2)$	$lb \cdot mol/(h \cdot ft^2)$
GPM	Liquid flow rate	—	gpm
h	Pressure head	mm	in
h'_{dc}	Froth height in downcomer	mm	in
h'_L	Pressure drop through aerated mass on tray	mm	in
h_c, h_C	Clear liquid height on tray	mm	in
h_{cl}	Clearance under downcomer	mm	in
h_{ct}	Clear liquid height at spray to froth transition	mm	in
h_d	Dry pressure drop across tray	mm	in
h_{da}	Head loss due to liquid flow under downcomer apron	mm	in
h_{dc}	Clear liquid height in downcomer	mm	in
h_{ds}	Calculated clear liquid height, Eq. (14-108)	mm	in
h_f	Height of froth	mm	in
h_{fow}	Froth height over the weir, Eq. (14-117)	mm	in
h_{hg}	Hydraulic gradient	mm	in
h_{Lo}	Packing holdup in preloading regime, fractional	-/-	-/-
h_{Lt}	Clear liquid height at froth to spray transition, corrected for effect of weir height, Eq. (14-96)	mm	in
h_{ow}	Height of crest over weir	mm	in
h_T	Height of contacting	m	ft
h_t	Total pressure drop across tray	mm	in
h_w, h_W	Weir height	mm	in
H	Height of a transfer unit	m	ft
H	Henry's law constant	kPa/mol fraction	atm/mol fraction
H'	Henry's law constant	$kPa/(kmol \cdot m^3)$	$psi/(lb \cdot mol \cdot ft^3)$
H_G	Height of a gas phase transfer unit	m	ft
H_L	Height of a liquid phase transfer unit	m	ft
H_{OG}	Height of an overall transfer unit, gas phase concentrations	m	ft
H_{OL}	Height of an overall transfer unit, liquid phase concentrations	m	ft
H'	Henry's law coefficient	kPa/mol frac	atm/mol frac
HETP	Height equivalent to a theoretical plate or stage	m	ft
J^*_G	Dimensionless gas velocity, weep correlation, Eq. (14-124)	-/-	-/-
J^*_L	Dimensionless liquid velocity, weep correlation, Eq. (14-125)	-/-	-/-
k	Individual phase mass transfer coefficient	$kmol/(s \cdot m^2 \cdot mol\ frac)$	$lb \cdot mol/(s \cdot ft^2 \cdot mol\ frac)$
k_1	First order reaction velocity constant	1/s	1/s
k_2	Second-order reaction velocity constant	$m^3/(s \cdot kmol)$	$ft^3/(h \cdot lb \cdot mol)$
k_g	Gas mass-transfer coefficient, wetted-wall columns [see Eq. (14-172) for unique units]		

Nomenclature *(Continued)*

Symbol	Definition	SI units	U.S. Customary System units
k_G	Gas phase mass-transfer coefficient	kmol/(s · m² · mol frac)	lb · mol/(s · ft² · mol frac)
k_L	Liquid phase mass-transfer coefficient	kmol/(s · m² · mol mol frac)	lb · mol/(s · ft² · mol frac)
K	Constant in trays dry pressure drop equation	mm · s²/m²	in · s²/ft²
K	Vapor–liquid equilibrium ratio	-/-	-/-
K_C	Dry pressure drop constant, all valves closed	mm · s²/m²	in · s²/ft²
K_D	Orifice discharge coefficient, liquid distributor	-/-	-/-
K_g	Overall mass-transfer coefficient	kg · mol/ (s · m² · atm)	lb · mol/ (h · ft² · atm)
K_O	Dry pressure drop constant, all valves open	mm · s²/m²	in · s²/ft²
K_{OG}, K_G	Overall mass-transfer coefficient, gas concentrations	kmol/ (s · m² · mol frac)	lb · mol/ (s · ft² · mol frac)
K_{OL}	Overall mass-transfer coefficient, liquid concentrations	kmol/ (s · m² · mol frac)	lb.mol/ (s · ft² · mol frac)
L	Liquid mass velocity	kg/(m² · s)	lb/ft² · h
L_f	Liquid loading factor in Robbins' packing pressure drop correlation	kg/(s · m²)	lb/(h · ft²)
L_m	Molar liquid downflow rate	kg · mol/h	lb · mol/h
L_M	Liquid molar mass velocity	kmol/(m² · s)	lb · mol/(ft² · h)
L_S	Liquid velocity, based on superficial tower area	m/s	ft/s
L_w	Weir length	m	in
m	An empirical constant based on Wallis' countercurrent flow limitation equation, Eqs. (14-123) and (14-143)	-/-	-/-
m	Slope of equilibrium curve = dy^*/dx	-/-	-/-
M	Molecular weight	kg/kmol	lb/(lb · mol)
n	Parameter in spray regime clear liquid height correlation, Eq. (14-84)	mm	In
n_A	Rate of solute transfer	kmol/s	lb · mol/s
n_D	Number of holes in orifice distributor	-/-	-/-
N_a	Number of actual trays	-/-	-/-
N_A, N_t	Number of theoretical stages	-/-	-/-
N_{OG}	Number of overall gas-transfer units	-/-	-/-
N_p	Number of tray passes	-/-	-/-
p	Hole pitch (center-to-center hole spacing)	mm	in
p	Partial pressure	kPa	atm
P_{BM}	Logarithmic mean partial pressure of inert gas	kPa	atm
P, p_T	Total pressure	kPa	atm
P^0	Vapor pressure	kpa	atm
Q, q	Volumetric flow rate of liquid	m³/s	ft³/s
Q'	Liquid flow per serration of serrated weir	m · ³/s	ft³/s
Q_D	Downcomer liquid load, Eq. (14-79)	m/s	ft/s
Q_L	Weir load, Eq. (14-78)	m³/(h · m)	gpm/in
Q_{MW}	Minimum wetting rate	m³/(h · m²)	gpm/ft²
R	Reflux flow rate	kg · mol/h	lb · mol/h
R	Gas constant		
R_h	Hydraulic radius	m	ft
R_{vw}	Ratio of valve weight with legs to valve weight without legs, Table (14-11)	-/-	-/-
Re_G	Gas Reynolds number (defined on page 14-63)	-/-	-/-
Re_L	Liquid Reynolds number, Eq. 14-165	-/-	-/-
S	Length of corrugation side, structured packing	m	ft
S	Stripping factor mG_M/L_M	-/-	-/-
S	Tray spacing	mm	in
S_e, S'	Effective stripping factor (Edmister)	-/-	-/-
SF	Derating (system) factor, Table 14-9	-/-	-/-
t_t	Tray thickness	mm	in
t_v	Valve thickness	mm	in
T	Absolute temperature	K	°R
TS	Tray spacing; same as S	mm	in
U,u	Linear velocity of gas	m/s	ft/s
U_a	Velocity of gas through active area	m/s	ft/s
U_a^*	Gas velocity through active area at froth to spray transition	m/s	ft/s
U_h, u_h	Gas hole velocity	m/s	ft/s
U_L, u_L	Liquid superficial velocity based on tower cross-sectional area	m/s	ft/s
U_n	Velocity of gas through net area	m/s	ft/s
U_{nf}	Gas velocity through net area at flood	-/-	-/-
U_t	Superficial velocity of gas	m/s	ft/s
v_H	Horizontal velocity in trough	m/s	ft/s
V	Linear velocity	m/s	ft/s
V	Molar vapor flow rate	kg · mol/s	lb · mol/h
W	Weep rate	m³/s	gpm
x	Mole fraction, liquid phase (note 1)	-/-	-/-
x'	Mole fraction, liquid phase, column 1 (note 1)		
x''	Mole fraction, liquid phase, column 2 (note 1)		
x^*, x°	Liquid mole fraction at equilibrium (note 1)	-/-	-/-
y	Mole fraction, gas or vapor phase (note 1)	-/-	-/-
y'	Mole fraction, vapor phase, column 1 (note 1)		
y''	Mole fraction, vapor phase, column 2 (note 1)		
y^*, y°	Gas mole fraction at equilibrium (note 1)		
Z	Characteristic length in weep rate equation, Eq. (14-126)	m	ft
Z_p	Total packed height	m	ft
	Greek symbols		
α	Relative volatility	-/-	-/-
β	Tray aeration factor, Fig. 14-38	-/-	-/-
ε	Void fraction	-/-	-/-
ϕ	Contact angle	deg	deg
ϕ	Relative froth density	-/-	-/-
γ	Activity coefficient	-/-	-/-
Γ	Flow rate per length	kg/(s · m)	lb/(s · ft)
δ	Effective film thickness	m	ft
η	Collection eficiency, fractional	-/-	-/-
η	Factor used in froth density correlation, Eq. (14-118)	-/-	-/-
λ	Stripping factor = $m/(L_M/G_M)$	-/-	-/-
μ	Absolute viscosity	Pa · s	cP or lb/(ft · s)
μm	Micrometers	m	-/-
ν	Kinematic viscosity	m²/s	cS
π	3.1416. . . .	-/-	-/-
θ	Residence time	s	s
θ	Angle of serration in serrated weir	deg	deg
ρ	Density	kg/m³	lb/ft³
ρ_M	Valve metal density	kg/m³	lb/ft³
σ	Surface tension	mN/m	dyn/cm
χ	Parameter used in entrainment correlation, Eq. (14-95)	-/-	-/-
ψ	Fractional entrainment, moles liquid entrained per mole liquid downflow	k · mol/ k · mol	lb · mol/ lb · mol
Φ	Fractional approach to entrainment flood	-/-	-/-
ΔP	Pressure drop per length of packed bed	mmH₂O/m	inH₂O/ft
$\Delta\rho$	$\rho_L - \rho_G$	kg/m³	lb/ft³
	Subscripts		
A	Species A		
AB	Species A diffusing through species B		
B	Species B		
B	Based on the bubbling area		
d	Dry		
da	Downcomer apron		
dc	Downcomer		
dry	Uncorrected for entrainment and weeping		
e	Effective value		
f	Froth		
Fl	Flood		
flood	At flood		
G, g	Gas or vapor		
h	Based on hole area (or slot area)		
H_2O	Water		
i	Interface value		
L, l	Liquid		
m	Mean		

Nomenclature *(Continued)*

Symbol	Definition	SI units	U.S. Customary System units
min	Minimum		
MOC	At maximum operational capacity		
	Subscripts		
n, N	On stage n		
N	At the inlet nozzle		
NF, nf	Based on net area at flood		
p	Particle		
S	Superficial		
t	Total		
ult	At system limit (ultimate capacity)		

Symbol	Definition	SI units	U.S. Customary System units
V, v	Vapor		
w	Water		
1	Tower bottom		
2	Tower top		
	Dimensionless groups		
N_{Fr}	Froude number = $(U_L^2)/(Sg)$		
N_{Re}	Reynolds number = $(D_{tube}U_{ge}\rho_G)/(\mu_G)$		
N_{Sc}	Schmidt number = $\mu/(\rho D)$		
N_{We}	Weber number = $(U_L^2\rho_L S)/(\sigma g_c)$		

NOTE: 1. Unless otherwise specified, refers to concentration of more volatile component (distillation) or solute (absorption).

General References: American Institute of Chemical Engineers, *AIChE Equipment Testing Procedure—Trayed and Packed Columns: A Guide to Performance Evaluation,* 3d ed., Wiley, New York, 2014; Astarita, G., *Mass Transfer with Chemical Reaction,* Elsevier, New York, 1967; Astarita, G., D. W. Savage, and A. Bisio, *Gas Treating with Chemical Solvents,* Wiley, New York, 1983; Billet, R., *Distillation Engineering,* Chemical Publishing Co., New York, 1979; Billet, R., *Packed Column Analysis and Design,* Ruhr University, Bochum, Germany, 1989; Chattopadhyay, P., *Distillation Engineering Handbook,* Tata McGraw-Hill Education Private Ltd, New Delhi, 2012; Danckwerts, P. V., *Gas-Liquid Reactions,* McGraw-Hill, New York, 1970; Gorak, A., and Olujic, Z., eds., *Distillation Equipment and Processes,* Elsevier, New York, 2014; Gorak, A., and Schoenmakers H., eds., *Distillation Operation and Applications,* Elsevier, New York, 2014; *Distillation and Absorption 1987,* Institution of Chemical Engineers, Rugby, UK, 1987; *Distillation and Absorption 1992,* Institution of Chemical Engineers, Rugby, UK, 1992; *Distillation and Absorption 1997,* Institution of Chemical Engineers. Rugby, UK, 1997; *Distillation and Absorption 2002,* Institution of Chemical Engineers, Rugby, UK, 2002; *Distillation and Absorption 2006,* Institution of Chemical Engineers, Rugby, UK, 2006; *Distillation and Absorption 2010,* Einhoven University of Technology, The Netherlands, 2010. *Distillation and Absorption 2014,* EFCE, DECHEMA e.V., Frankfurt am Main, Germany, 2014; *Distillation Topical Conference Proceedings,* AIChE Spring Meetings (separate *Proceedings* book for each topical conference), Houston, Tex., March 1999; Houston, Tex., April 22–26, 2001; New Orleans, La., March 10–14, 2002; New Orleans, La., March 30–April 3, 2003; Atlanta, Ga., April 10–13, 2005; Tampa, Fla., 2009; Chicago, Ill., 2011; San Antonio, Tex., 2013; Hines, A. L., and R. N. Maddox, *Mass Transfer—Fundamentals and Applications,* Prentice Hall, Englewood Cliffs, N.J., 1985; Hobler, T., *Mass Transfer and Absorbers,* Pergamon Press, Oxford, UK, 1966; Kister, H. Z., *Distillation Operation,* McGraw-Hill, New York, 1990; Kister, H. Z., *Distillation Design,* McGraw-Hill, New York, 1992; Kister, H. Z., and G. Nalven, eds., *Distillation and Other Industrial Separations,* reprints from CEP, AIChE, New York, 1998; Kister, H. Z., *Distillation Troubleshooting,* Wiley, New York, 2006; Kister Distillation Symposium Proceedings, *Topical Conference Proceedings* (separate Proceedings book for each topical conference), Austin, Tex., 2015; San Antonio, Tex., 2017; Kohl, A. L., and R. B. Nielsen, *Gas Purification,* 5th ed., Gulf Publishing, Houston, 1997; Lockett, M. J., *Distillation Tray Fundamentals,* Cambridge University Press, Cambridge, UK, 1986; Maćkowiak, J., *Fluid Dynamics of Packed Columns,* Springer-Verlag, Berlin Heidelberg, 2010; Schweitzer, P. A., ed., *Handbook of Separation Techniques for Chemical Engineers,* 3d ed., McGraw-Hill, New York, 1997; Sherwood, T. K., R. L. Pigford, C. R. Wilke, *Mass Transfer,* McGraw-Hill, New York, 1975; Stichlmair, J., and J. R. Fair, *Distillation Principles and Practices,* Wiley, New York, 1998; Strigle, R. F., Jr., *Packed Tower Design and Applications,* 2d ed., Gulf Publishing, Houston, 1994; Treybal, R. E., *Mass Transfer Operations,* 3d ed, McGraw-Hill, New York, 1980.

INTRODUCTION

DEFINITIONS

Gas absorption is a unit operation in which soluble components of a gas mixture are dissolved in a liquid. The inverse operation, called stripping or desorption, is used when we wish to transfer volatile components from a liquid mixture into a gas. Both absorption and stripping, in common with distillation (Sec. 13), make use of special equipment for bringing gas and liquid phases into intimate contact. This section is concerned with the design of gas–liquid contacting equipment, as well as with the design of absorption and stripping processes.

EQUIPMENT

Absorption, stripping, and distillation operations are usually carried out in vertical, cylindrical columns or towers in which devices such as plates or packing elements are placed. The gas and liquid normally flow countercurrently, and the devices serve to provide the contacting and development of interfacial surface through which mass transfer takes place. Background material on this mass transfer process is given in Sec. 5.

DESIGN PROCEDURES

The procedures to be followed in specifying the principal dimensions of gas absorption and distillation equipment are described in this section and are supported by several worked-out examples. The experimental data required for executing the designs are keyed to appropriate references or to other sections of this handbook.

For absorption, stripping, and distillation, there are three main steps involved in design:

1. *Data on the gas–liquid or vapor–liquid equilibrium for the system at hand.* If absorption, stripping, and distillation operations are considered equilibrium-limited processes, which was the usual approach in the past but is less so now due to the availability of commercial software, these data are critical for determining the maximum possible separation. In some cases, the operations are considered rate-based (see later in this section and also Sec. 13), but they require knowledge of equilibrium at the phase interface. Other data required include physical properties such as viscosity and density and thermodynamic properties such as enthalpy. Section 2 deals with sources of such data.

2. *Information on the liquid- and gas-handling capacity of the contacting device chosen for the particular separation problem.* Such information includes pressure drop characteristics of the device, in order that an optimum balance between capital cost (column cross section and height) and energy requirements might be achieved. Capacity and pressure drop characteristics of the available devices are covered later in Sec. 14.

3. *Determination of the required height of contacting zone for the separation to be made as a function of properties of the fluid mixtures and mass-transfer efficiency of the contacting device.* This determination involves the calculation of mass-transfer parameters such as heights of transfer units and tray efficiencies as well as equilibrium or rate parameters such as theoretical stages or numbers of transfer units. An additional consideration for systems in which chemical reaction occurs is the provision of adequate residence time for desired reactions to occur, or minimal residence time to prevent undesired reactions from occurring. For equilibrium-based operations, the parameters for required height are covered in the present section, but guidance is also provided for the use of commercial software.

DATA SOURCES IN THE HANDBOOK

Sources of data for the analysis or design of absorbers, strippers, and distillation columns are manifold, and a detailed listing of them is outside the scope of this section. Some key sources within the handbook are shown in Table 14-1.

EQUILIBRIUM DATA

Finding reliable gas–liquid and vapor–liquid equilibrium data usually is the most time-consuming task associated with the design of absorbers and other gas–liquid contactors, and yet it may be the most important task at hand. For gas solubility, an important data source is the set of volumes edited by Kertes et al., *Solubility Data Series,* published by Pergamon Press (1979 ff.). In the introduction to each volume, there is an excellent discussion and definition of the various methods by which gas solubility data have been reported, such as the Bunsen coefficient, the Kuenen coefficient, the Ostwalt coefficient, the absorption coefficient, and the Henry's law coefficient. The fifth edition of *The Properties of Gases and Liquids* by Poling, Prausnitz, and O'Connell (McGraw-Hill, New York, 2000) provides data and recommended estimation methods for gas solubility as well as the broader area of vapor–liquid equilibrium. Online databases for vapor–liquid equilibrium are increasingly available, but they may entail a fee. DETHERM on the web (http://i-systems.dechema.de/detherm/) is a comprehensive source of data. NIST-TDE [Frenkel et al., *J. Chem. Inf. Model* **45:** 816–838 (2005); http://trc.nist.gov/tde.html] also provides a comprehensive source of data, and in addition it is available in process-simulation software tools, such as those from Aspen Technology, Inc.

TABLE 14-1 Directory to Key Data for Absorption and Gas-Liquid Contactor Design

Type of data	Section
Phase equilibrium data	
Gas solubilities	2
Pure component vapor pressures	2
Equilibrium *K* values	13
Thermal data	
Heats of solution	2
Specific heats	2
Latent heats of vaporization	2
Transport property data	
Diffusion coefficients	
Liquids	2
Gases	2
Viscosities	
Liquids	2
Gases	2
Densities	
Liquids	2
Gases	2
Surface tensions	2
Packed tower data	
Pressure drop and flooding	14
Mass transfer coefficients	5
HTU, physical absorption	5
HTU with chemical reaction	14
Height equivalent to a theoretical plate (HETP)	
Plate tower data	
Pressure drop and flooding	14
Plate efficiencies	14
Costs of gas–liquid contacting equipment	14

DESIGN OF GAS ABSORPTION SYSTEMS

GENERAL DESIGN PROCEDURE

The design engineer usually must determine (1) the best solvent; (2) the best gas velocity through the absorber, or, equivalently, the vessel diameter; (3) the height of the vessel and its internal members, which is the height and type of packing or the number of contacting trays; (4) the optimum solvent circulation rate through the absorber and stripper; (5) temperatures of streams entering and leaving the absorber and stripper, and the quantity of heat to be removed to account for the heat of solution and other thermal effects; (6) pressures at which the absorber and stripper will operate; and (7) the mechanical design of the absorber and stripper vessels (predominantly columns or towers), including flow distributors and packing supports. This subsection covers these aspects.

The problem presented to the designer of a gas absorption system usually specifies the following quantities: (1) gas flow rate; (2) gas composition of the component or components to be absorbed; (3) operating pressure and allowable pressure drop across the absorber; (4) minimum recovery of one or more of the solutes; and, possibly, (5) the solvent to be employed. Items 3, 4, and 5 may be subject to economic considerations and therefore may be left to the designer. For a determination of the number of variables that must be specified to fix a unique solution for the absorber design, one may use the same phase-rule approach described in Sec. 13 for distillation systems.

Recovery and recycle of the solvent, occasionally by chemical means but more often by stripping, is almost always required and is considered an integral part of the absorption system process design. A more complete solvent-stripping operation normally will result in a less costly absorber because of a lower concentration of residual solute in the regenerated (lean) solvent, but this may increase the overall cost of the entire absorption system. A more detailed discussion of these and other economic considerations is presented later in this section.

The design calculations presented in this subsection are relatively simple and usually can be done by using a calculator or spreadsheet. In many cases, the calculations are explained through design diagrams. Most engineers today will perform rigorous, detailed calculations using process simulators. The design procedures presented here are intended to complement the rigorous computerized calculations by presenting approximate estimates and insight into the essential elements of absorption and stripping operations. These relatively simple design procedures are especially useful for understanding trends and for checking the results from commercial simulators.

SELECTION OF SOLVENT AND NATURE OF SOLVENTS

When a choice is possible, preference is given to solvents with high solubilities for the target solute and high selectivity for the target solute over the other species in the gas mixture. A high solubility reduces the flow rate of liquid to be circulated. The solvent should have the advantages of low volatility, low cost, low corrosive tendencies, high stability, low viscosity, low tendency to foam, and low flammability. Since the exit gas normally leaves saturated with solvent, solvent loss can be costly and can cause environmental problems. The choice of the solvent is a key factor in the economic analysis of the process and in its compliance with environmental regulations.

Typically, a solvent that is chemically similar to the target solute or that reacts with it will provide high solubility. Water is often used for polar and acidic solutes (e.g., HCl), oils for light hydrocarbons, and special chemical solvents for acid gases such as CO_2, SO_2, and H_2S. Solvents are classified as physical and chemical. A chemical solvent forms complexes or chemical compounds with the solute, while physical solvents have only weaker interactions with the solute. Physical and chemical solvents are compared by examining the solubility of CO_2 in propylene carbonate (representative physical solvent) and aqueous monoethanolamine (MEA; representative chemical solvent).

Figures 14-1 and 14-2 present correlations (based on data) for the solubility of CO_2 in the two representative physical and chemical solvents, each at two temperatures: 40°C and 100°C. The propylene carbonate data are from Zubchenko et al. [*Zhur. Priklad. Khim.* **44:** 2044–2047 (1971)], and the MEA data are from Jou, Mather, and Otto [*Can. J. Chem. Eng.* **73:** 140–147 (1995)]. The two figures have the same content, but Fig. 14-2 focuses on the low-pressure region by converting both composition and pressure to the logarithm scale. Examination of the two sets of data reveals the differences between physical and chemical solvents, which are summarized in the following table:

Characteristic	Physical solvent	Chemical solvent
Solubility variation with pressure	Relatively linear	Highly nonlinear
Low-pressure solubility	Low	High
High-pressure solubility	Continues to increase	Levels off
Heat of solution—related to variation of solubility with temperature at fixed pressure	Relatively low and approximately constant with solute loading	Relatively high and decreases somewhat with increased solute loading
Typical value for heat of solution	12 kJ/gmol for CO_2 in propylene carbonate 40°C	85 kJ/gmol for CO_2 in 30 wt% MEA at 40°C and low CO_2 loading

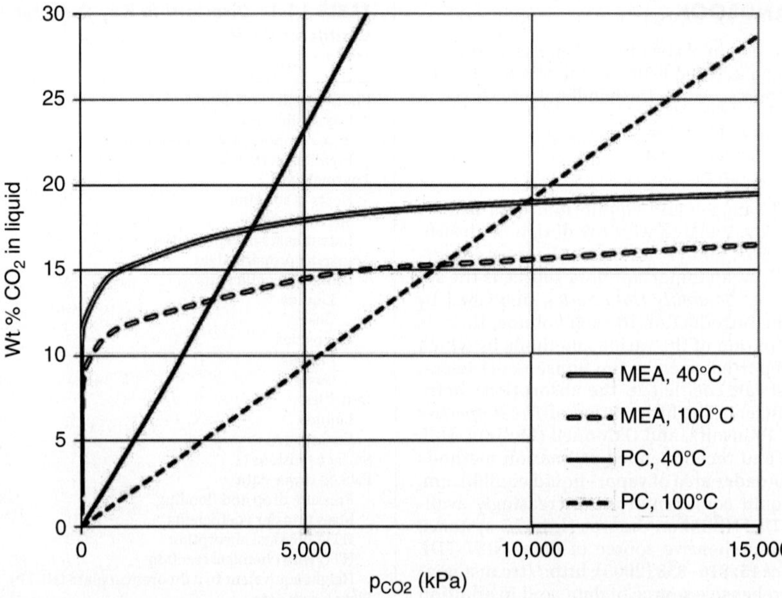

FIG. 14-1 Solubility of CO_2 in 30 wt% MEA and propylene carbonate. Linear scale.

Chemical solvents are usually preferred when the solute must be reduced to very low levels, when high selectivity is needed, and when the solute partial pressure is low. However, the strong absorption at low solute partial pressures and the high heat of solution are disadvantages for stripping. For chemical solvents, the strong nonlinearity of the absorption makes it necessary that accurate absorption data for the conditions of interest be available.

SELECTION OF SOLUBILITY DATA

Solubility values are necessary for design because they determine the liquid rate necessary for complete or economical solute recovery. Equilibrium data generally will be found in one of three forms: (1) solubility data expressed either as weight or mole percent or as Henry's law coefficients, (2) pure-component vapor pressures, or (3) equilibrium distribution coefficients (*K* values). Data for specific systems may be found in Sec. 2, and Sec. 4 provides a discussion of Henry's law coefficients; additional references to sources of data are presented in this section.

To define completely the solubility of gas in a liquid, it is generally necessary to state the temperature, equilibrium partial pressure of the solute gas in the gas phase, and the concentration of the solute gas in the liquid phase. Strictly speaking, the total pressure of the system should also be identified, but for low pressures (less than about 507 kPa or 5 atm), the solubility for a particular partial pressure of the solute will be relatively independent of the total pressure.

For many physical systems, the equilibrium relationship between solute partial pressure and liquid-phase concentration is given by Henry's law:

$$p_A = Hx_A \tag{14-1}$$

or

$$p_A = H'c_A \tag{14-2}$$

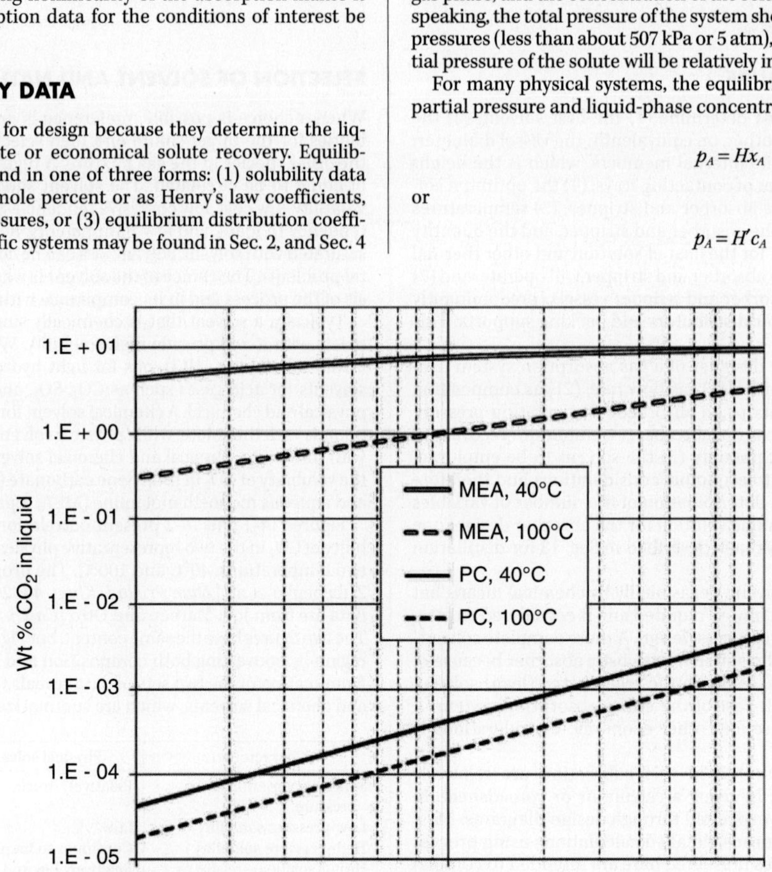

FIG. 14-2 Solubility of CO_2 in 30 wt% MEA and propylene carbonate. Logarithm scale and focus on low-pressure region.

where H is the Henry's law coefficient expressed in kPa per mole fraction solute in liquid and H' is the Henry's law coefficient expressed in kPa·m^3/kmol. Section 4 discusses conversions between H and H' and other variants of Henry's law coefficients.

Figure 14-1 indicates that Henry's law is valid to a good approximation for the solubility of CO_2 in propylene carbonate. In general, Henry's law is a reasonable approximation for physical solvents. If Henry's law holds, the solubility is defined by knowing (or estimating) the value of the constant H (or H').

Note that the assumption of Henry's law will lead to incorrect results for the solubility of chemical systems such as CO_2-MEA (Figs. 14-1 and 14-2) and HCl-H_2O. Solubility modeling for chemical systems requires the use of a *speciation model,* as described later in this section and also in Sec. 4.

For quite a number of physically absorbed gases, Henry's law holds very well when the partial pressure of the solute is less than about 101 kPa (1 atm). For partial pressures above 101 kPa, H may be independent of the partial pressure (Fig. 14-1), but this needs to be verified for the system of interest. The variation of H with temperature is a strongly nonlinear function of temperature, as discussed by Poling, Prausnitz, and O'Connell (*The Properties of Gases and Liquids,* 5th ed., McGraw-Hill, New York, 2000) and Smith and Harvey [*Chem. Eng. Progress* **103:** 33 (2007)]. One should consult these references and the discussion in Sec. 4 when temperature and pressure extrapolations of Henry's law data are needed. Further discussion of Henry's law is presented in Sec. 4.

The use of Henry's law constants is illustrated by the following example.

Example 14-1 Gas Solubility We wish to find out how much hydrogen can be dissolved in 100 weights of water from a gas mixture when the total pressure is 101.3 kPa (760 torr; 1 atm), the partial pressure of the H_2 is 26.7 kPa (200 torr), and the temperature is 25°C. For partial pressures up to about 100 kPa, the value of H is given in Sec. 3 as 7.17×10^6 kPa (7.08×10^4 atm) at 25°C. According to Henry's law,

$$x_{H2} = p_{H2}/H_{H2} = 26.7/7.08 \times 10^6 = 3.72 \times 10^{-6}$$

The mole fraction x is the ratio of the number of moles of H_2 in solution to the total moles of all constituents contained. To calculate the weights of H_2 per 100 weights of H_2O, one can use the following formula, where the subscripts A and w correspond to the solute (hydrogen) and solvent (water):

$$\left(\frac{x_A}{1-x_A}\right)\frac{M_A}{M_W}100 = \left(\frac{3.72 \times 10^{-6}}{1-3.72 \times 10^{-6}}\right)\frac{2.02}{18.02}100$$

$$= 4.17 \times 10^{-5} \text{ weights } H_2/100 \text{ weights } H_2O$$

$$= 0.42 \text{ parts per million weight or } 0.42 \text{ ppmw}$$

Pure-component vapor pressure can be used for predicting solubilities for systems in which *Raoult's law* is valid. For such systems, $p_A = p_A^0 x_A$, where p_A^0 is the pure-component vapor pressure of the solute and p_A is its partial pressure. Extreme care should be exercised when using pure-component vapor pressures to predict gas absorption behavior. Both vapor-phase and liquid-phase nonidealities can cause significant deviations from Raoult's law, and this is often the reason particular solvents are used, that is, because they have special affinity for particular solutes. Poling, Prausnitz, and O'Connell (*The Properties of Gases and Liquids,* 5th ed., McGraw-Hill, New York, 2000) provide an excellent discussion of the conditions where Raoult's law is valid. Vapor-pressure data are available in Sec. 2 for a variety of materials.

Whenever data are available for a given system under similar conditions of temperature, pressure, and composition, *equilibrium distribution coefficients* ($K = y/x$) provide a much more reliable tool for predicting vapor-liquid distributions. Detailed discussions of equilibrium K values are presented in Secs. 4 and 13.

CALCULATION OF LIQUID-TO-GAS RATIO

The minimum possible liquid rate is readily calculated from the composition of the entering gas and the solubility of the solute in the exit liquor, with equilibrium being assumed. It may be necessary to estimate the temperature of the exit liquid based on the heat of solution of the solute gas. Values of latent heat and specific heat and values of heats of solution (at infinite dilution) are given in Sec. 2.

The actual liquid-to-gas ratio (solvent circulation rate) normally will be greater than the minimum by as much as 25 to 100 percent, and the estimated factor may be arrived at by economic considerations as well as judgment and experience. For example, in some packed-tower applications involving very soluble gases or vacuum operation, the minimum quantity of solvent needed to dissolve the solute may be insufficient to keep the packing surface thoroughly wet, leading to poor distribution of the liquid stream.

When the solute concentration in the inlet gas is low and when a significant fraction of the solute is absorbed (this often the case), the approximation

$$y_1 G_M = x_1 L_M = (y_1^\circ/m)L_M \tag{14-3}$$

leads to the conclusion that the ratio mG_M/L_M represents the fractional approach of the exit liquid to saturation with the inlet gas,

$$mG_M/L_M = y_1^\circ/y_1 \tag{14-4}$$

Optimization of the liquid-to-gas ratio in terms of total annual costs often suggests that the molar liquid-to-gas ratio L_M/G_M should be about 1.2 to 1.5 times the theoretical minimum corresponding to equilibrium at the rich end of the tower (infinite height or number of trays), provided flooding is not a problem. This, for example, would be an alternative to assuming that $L_M/G_M \approx m/0.7$.

When the exit-liquor temperature rises because of the heat of absorption of the solute, the value of m changes through the tower, and the liquid-to-gas ratio must be chosen to give reasonable values of m_1G_M/L_M and m_2G_M/L_M, where the subscripts 1 and 2 refer to the bottom and top of the absorber, respectively. For this case, the value of m_2G_M/L_M will be taken to be somewhat less than 0.7, so the value of m_1G_M/L_M will not approach unity too closely. This rule-of-thumb approach is useful only when the solute concentration is low and heat effects are negligible.

When the solute has a large heat of solution or when the feed gas contains high concentrations of the solute, one should consider the use of internal cooling coils or intermediate liquid withdrawal and cooling to remove the heat of absorption.

SELECTION OF EQUIPMENT

Trays and packings (both random and structured) have been extensively used for gas absorption; structured packings are seeing increasing usage, particularly for applications requiring low pressure drop and high surface area. Compared to trays, packings have the advantages of availability in low-cost, corrosion-resistant materials (such as plastics and ceramics), low pressure drop (which can be an advantage when the tower is in the suction of a fan or compressor), easy and economic adaptability to small-diameter (less than 0.6-m or 2-ft) columns, and excellent handling of foams. Trays are much better for handling solids and fouling applications, offer greater residence time for slow absorption reactions, can better handle high L/G ratios and intermediate cooling, give better liquid turndown, and are more robust and less prone to reliability issues such as those resulting from poor distribution. Details on the operating characteristics of tray and packed towers are given later in this subsection.

COLUMN DIAMETER AND PRESSURE DROP

Flooding determines the minimum possible diameter of the absorber column, and the usual design is for 60 to 80 percent of the flooding velocity. In near-atmospheric applications, pressure drop usually needs to be minimized to reduce the cost of energy for compression of the feed gas. For systems having a significant tendency to foam, the maximum allowable velocity will be lower than the estimated flooding velocity. Methods for predicting flooding velocities and pressure drops are given later in this section.

COMPUTATION OF TOWER HEIGHT

The required height of a gas absorption or stripping tower for physical solvents depends on (1) the phase equilibria involved; (2) the specified degree of removal of the solute from the gas; and (3) the mass-transfer efficiency of the device. These three considerations apply to both tray and packed towers. Items 1 and 2 dictate the required number of theoretical stages (tray tower) or transfer units (packed tower). Item 3 is derived from the tray efficiency and spacing (tray tower) or from the height of one transfer unit (packed tower). Solute removal specifications are usually derived from economic considerations.

For tray towers, the approximate design methods described in this subsection may be used in estimating the number of theoretical stages, and the tray efficiencies and spacings for the tower can be specified on the basis of the information given later. Considerations involved in the rigorous design of theoretical stages for tray towers are treated in Sec. 13.

For packed towers, the continuous differential nature of the contact between gas and liquid leads to a design procedure involving the solution of differential equations, as described in the next subsection. Note that the design procedures discussed in this section are not applicable to reboiled absorbers, which should be designed according to the procedures described in Sec. 13.

Caution is advised in distinguishing between systems involving pure physical absorption and those in which chemical reactions can significantly affect design procedures. Chemical systems require additional procedures, as described later in this section.

SELECTION OF STRIPPER OPERATING CONDITIONS

Stripping involves the removal of one or more components from the solvent through the application of heat or contacting it with a gas such as steam, nitrogen, or air. The operating conditions chosen for stripping normally result in a low solubility of solute (i.e., a high value of m), so the ratio mG_M/L_M will be larger than unity. A value of 1.4 may be used for rule-of-thumb calculations involving pure physical absorption. For tray-tower calculations, the stripping factor $S = KG_M/L_M$, where $K = y^0/x$ usually is specified for each tray.

When the solvent from an absorption operation must be regenerated for recycling to the absorber, one may employ a "pressure-swing" or "temperature-swing" concept, or a combination of the two, in specifying the stripping operation. In pressure-swing operation, the temperature of the stripper is about the same as that of the absorber, but the stripping pressure is much lower. In temperature-swing operation, the pressures are about equal, but the stripping temperature is much higher than the absorption temperature.

In pressure-swing operation, a portion of the gas may be "sprung" from the liquid by the use of a flash drum upstream of the stripper feed point. This type of operation has been discussed by Burrows and Preece [*Trans. Inst. Chem. Eng.* **32:** 99 (1954)] and by Langley and Haselden [*Inst. Chem. Eng. Symp. Ser. (London)*, no. 28 (1968)]. If the flashing of the liquid takes place inside the stripping tower, this effect must be accounted for in the design of the upper section in order to avoid overloading and flooding near the top of the tower.

Often the rate at which residual absorbed gas can be driven from the liquid in a stripping tower is limited by the rate of a chemical reaction, in which case the liquid-phase residence time (and hence the tower liquid holdup) becomes the most important design factor. Thus, many stripper regenerators are designed on the basis of liquid holdup rather than on the basis of mass-transfer rate.

Approximate design equations applicable only to the case of pure physical desorption are developed later in this subsection for both packed and tray stripping towers. A more rigorous approach using distillation concepts may be found in Sec. 13. A brief discussion of desorption with chemical reaction is given in the subsection Absorption with Chemical Reaction.

DESIGN OF ABSORBER-STRIPPER SYSTEMS

The solute-rich liquor leaving a gas absorber normally is distilled or stripped to regenerate the solvent for recirculation back to the absorber, as depicted in Fig. 14-3. The conditions selected for the absorption step (e.g., temperature, pressure, L_M/G_M) will affect the design of the stripping tower, and conversely, a selection of stripping conditions will affect the absorber design. The choice of optimum operating conditions for an absorber-stripper system therefore involves a combination of economic factors and practical judgments as to the operability of the system within the context of the overall process flow sheet. In Fig. 14-3, the stripping vapor is provided by a reboiler; alternately, an extraneous stripping gas may be used.

An appropriate procedure for executing the design of an absorber-stripper system is to set up a carefully selected series of design cases and then evaluate the equipment costs, the operating costs, and the operability of each case. Equipment costs are discussed briefly in the subsection Column Costs later in this chapter.

IMPORTANCE OF DESIGN DIAGRAMS

One of the first things a designer should do is to lay out a carefully constructed equilibrium curve $y^0 = F(x)$ on an xy diagram, as shown in Fig. 14-4. A horizontal line corresponding to the inlet-gas composition y_1 is then the locus of feasible outlet-liquor compositions, and a vertical line corresponding to the inlet-solvent-liquor composition x_2 is the locus of outlet-gas compositions. These lines are indicated as $y = y_1$ and $x = x_2$, respectively, on Fig. 14-4.

For gas absorption, the region of feasible operating lines lies above the equilibrium curve; for stripping, the feasible region for operating lines lies below the equilibrium curve. These feasible regions are bounded by the equilibrium curve and by the lines $x = x_2$ and $y = y_1$. By inspection, one should be able to visualize those operating lines that are feasible and those that would lead to "pinch points" within the tower. Also, it is possible to determine if a particular proposed design for solute recovery falls within the feasible envelope.

Once the design recovery for an absorber has been established, the operating line can be constructed by first locating the point x_2, y_2 on the diagram. The intersection of the horizontal line corresponding to the inlet gas composition y_1 with the equilibrium curve $y^0 = F(x)$ defines the theoretical minimum liquid-to-gas ratio for systems in which there are no intermediate pinch points. This operating line that connects this point with the point x_2, y_2 corresponds to the minimum value of L_M/G_M. The actual design value of L_M/G_M should normally be around 1.2 to 1.5 times this minimum value. Thus, the actual design operating line for a gas absorber will pass through the point x_2, y_2 and will intersect the line $y = y_1$ to the left of the equilibrium curve.

For stripping, one begins by using the design specification to locate the point x_1, y_1; then the intersection of the vertical line $x = x_2$ with the equilibrium curve $y^0 = F(x)$ defines the theoretical minimum gas-to-liquid ratio. The actual value of G_M/L_M is chosen to be about 20 to 50 percent higher than this minimum, so the actual design operating line will intersect the line $x = x_2$ at a point somewhat below the equilibrium curve.

PACKED-TOWER DESIGN

Methods for estimating the height of the active section of counterflow differential contactors such as packed towers, spray towers, and falling-film absorbers are based on rate expressions representing mass transfer at a

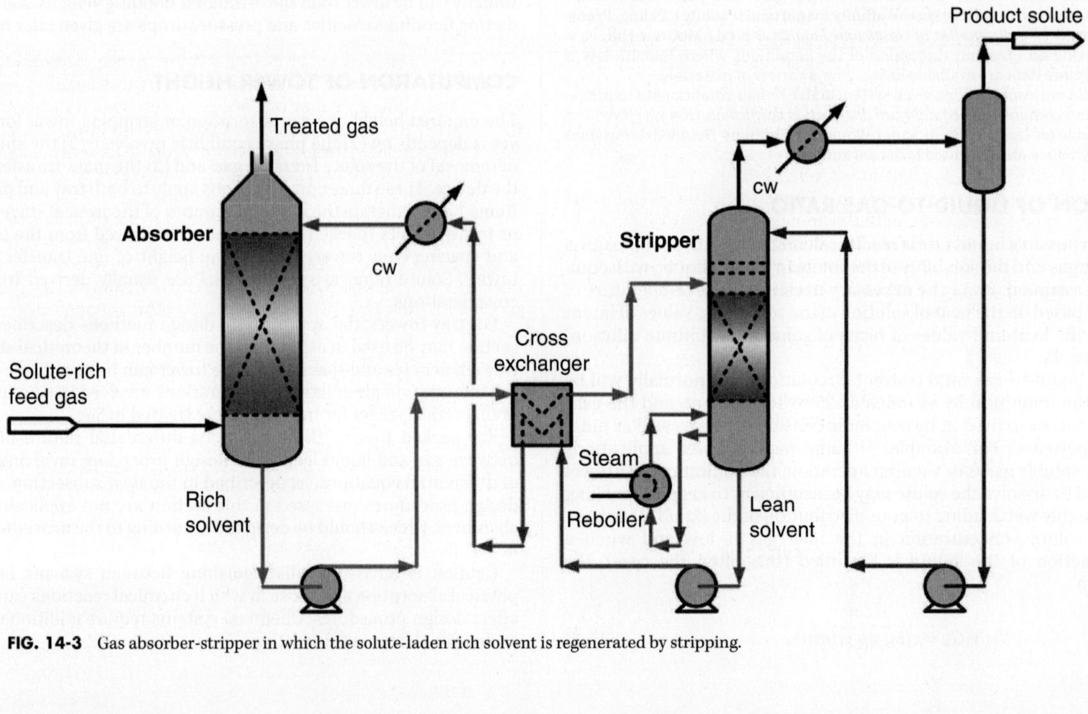

FIG. 14-3 Gas absorber-stripper in which the solute-laden rich solvent is regenerated by stripping.

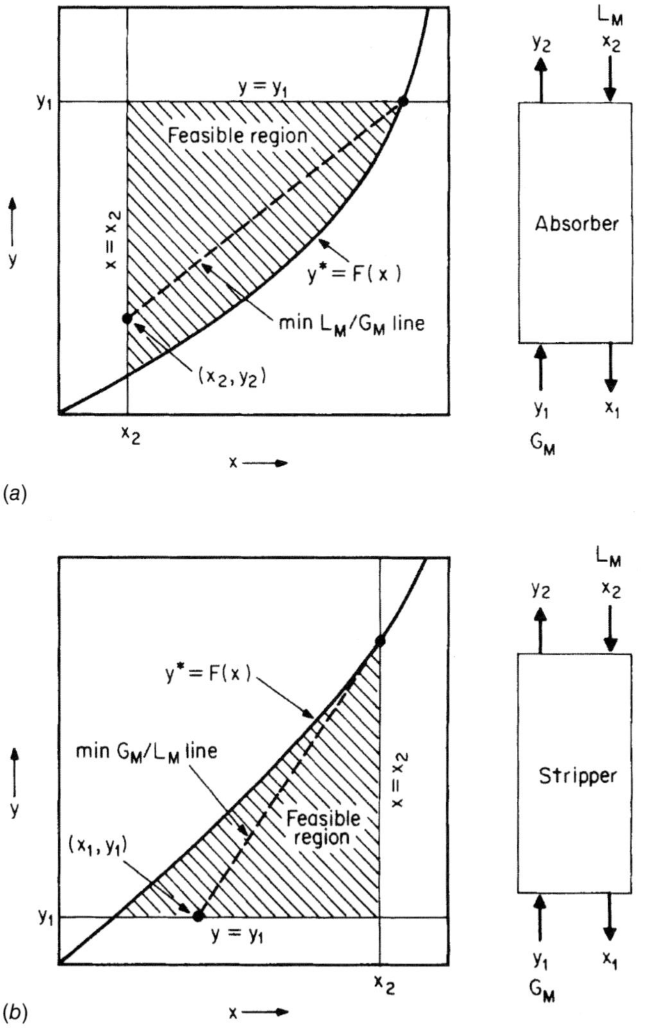

FIG. 14-4 Design diagrams for (*a*) absorption and (*b*) stripping.

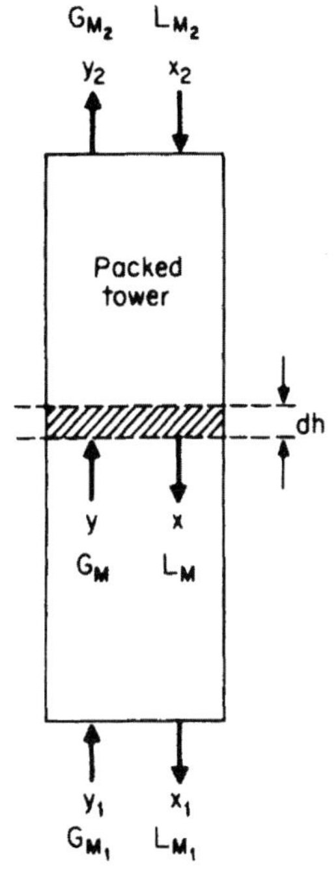

FIG. 14-5 Nomenclature for material balances in a packed-tower absorber or stripper.

point on the gas–liquid interface and on material balances representing the changes in bulk composition in the two phases that flow past each other. The rate expressions are based on the interphase mass-transfer principles described in Sec. 5. A combination of such expressions leads to an integral expression for the number of transfer units or to equations related closely to the number of theoretical stages. The paragraphs that follow set forth convenient methods for using such equations, first in a general case and then for cases in which simplifying assumptions are valid.

Use of Mass-Transfer-Rate Expression Figure 14-5 shows a section of a packed absorption tower together with the nomenclature that will be used in developing the equations that follow. In a differential section dh, we can equate the rate at which solute is lost from the gas phase to the rate at which it is transferred through the gas phase to the interface as follows:

$$-d(G_M y) = -G_M\,dy - y\,dG_M = N_A a\,dh \tag{14-5}$$

In Eq. (14-5), G_M is the gas-phase molar velocity [kmol/(s · m²)], N_A is the mass-transfer flux [kmol/(s · m²)], and a is the effective interfacial area (m²/m³).

When only one component is transferred,

$$dG_M = -N_A a\,dh \tag{14-6}$$

Substitution of this relation into Eq. (14-5) and rearranging yield

$$dh = -\frac{G_M\,d_y}{N_A a(1-y)} \tag{14-7}$$

For this derivation we use the gas-phase rate expression $N_A = k_G(y - y_i)$ and integrate over the tower to obtain

$$h_{T l} = \int_{y_2}^{y_1} \frac{G_M\,dy}{k_G a(1-y)(y-y_i)} \tag{14-8}$$

Multiplying and dividing by y_{BM} place Eq. (14-8) into the $H_G N_G$ format

$$h_T = \int_{y_2}^{y_1} \left(\frac{G_M}{k_G a y_{BM}}\right) \frac{y_{BM}\,dy}{(1-y)(y-y_i)}$$

$$= H_{G,\text{av}} \int_{y_2}^{y_1} \frac{y_{BM}\,dy}{(1-y)(y-y_i)} = H_{G,\text{av}} N_G \tag{14-9}$$

The general expression given by Eq. (14-8) is more complex than normally is required, but it must be used when the mass-transfer coefficient varies from point to point, as may be the case when the gas is not dilute or when the gas velocity varies as the gas dissolves. The values of y_i to be used in Eq. (14-8) depend on the local liquid composition x_i and on the temperature. This dependency is best represented by using the operating and equilibrium lines as discussed later.

Example 14-2 illustrates the use of Eq. (14-8) for scrubbing chlorine from air with aqueous caustic solution. For this case, one can make the simplifying assumption that y_i, the interfacial partial pressure of chlorine over the caustic solution, is zero due to the rapid and complete reaction of the chlorine after it dissolves. We note that the feed gas is not dilute.

Example 14-2 Packed Height Requirement Let us compute the height of packing needed to reduce the chlorine concentration of a chlorine–air mixture containing 0.503 mole-fraction chlorine to 0.0403 mole fraction. The inlet gas flow rate is of 0.537 kg/(s · m²), or 396 lb/(h · ft²). On the basis of test data described by Sherwood and Pigford (*Absorption and Extraction,* McGraw-Hill, 1952, p. 121) the value of $k_G a y_{BM}$ at a gas velocity equal to that at the bottom of the packing is equal to 0.1175 kmol/(s · m³), or 26.4 lb · mol/(h · ft³). The equilibrium back pressure y_i can be assumed to be negligible.

Solution. By assuming that the mass-transfer coefficient varies as the 0.8 power of the local gas mass velocity, we can derive the following relation:

$$\hat{K}_G a = k_G a y_{BM} = 0.1175\left[\frac{71y + 29(1-y)}{71y_1 + 29(1-y_1)}\left(\frac{1-y_1}{1-y}\right)\right]^{0.8}$$

where 71 and 29 are the molecular weights of chlorine and air, respectively. Noting that the inert-gas (air) mass velocity is given by $G'_M = G_M(1-y) = 5.34 \times 10^{-3}$ kmol/(s · m²), or 3.94 lb · mol/(h · ft²), and introducing these expressions into the integral gives

$$h_T = 1.82 \int_{0.0403}^{0.503} \left[\frac{1-y}{29.+42y}\right]^{0.8} \frac{dy}{(1-y)^2 \ln[1/(1-y)]}$$

This definite integral can be evaluated numerically by the use of Simpson's rule to obtain $h_T = 0.303$ m (0.99 ft).

Note that if the exit mole fraction of chlorine is lowered to 0.0203, the required column height will rise by about 28 percent.

Use of Operating Curve Often it is not possible to assume that $y_i = 0$ as in Example 14-2, due to diffusional resistance in the liquid phase or to the accumulation of solute in the liquid stream. When the backpressure cannot be neglected, it is necessary to supplement the equations with a material balance representing the operating line or curve. In view of the countercurrent flows into and from the differential section of packing shown in Fig. 14-5, a steady-state material balance leads to the following equivalent relations:

$$d(G_M y) = d(L_M x) \tag{14-10}$$

$$G'_M \frac{dy}{(1-y)^2} = L'_M \frac{dx}{(1-x)^2} \tag{14-11}$$

where L'_M = molar mass velocity of the inert-liquid component and G'_M = molar mass velocity of the inert gas; L_M, L'_M, G_M, and G'_M are superficial velocities based on the total tower cross section.

Equation (14-11) is the differential equation of the operating curve, and its integral around the upper portion of the packing is the equation for the operating curve.

$$G'_M \left[\frac{y}{1-y} - \frac{y_2}{1-y_2}\right] = L'_M \left[\frac{x}{1-x} - \frac{x_2}{1-x_2}\right] \tag{14-12}$$

For dilute solutions in which the mole fractions of x and y are small, the total molar flows G_M and L_M will be nearly constant, and the operating-curve equation is

$$G_M(y - y_2) = L_M(x - x_2) \tag{14-13}$$

This equation gives the relation between the bulk compositions of the gas and liquid streams at each height in the tower for conditions in which the operating curve can be approximated as a straight line.

Figure 14-6 shows the relationship between the operating curve and the equilibrium curve $y_i = F(x_i)$ for a typical example involving solvent recovery, where y_i and x_i are the interfacial compositions (assumed to be in equilibrium). Once y is known as a function of x along the operating curve, y_i can be found at corresponding points on the equilibrium curve by

$$(y - y_i)/(x_i - x) = k_L/k_G = L_M H_G / G_M H_L \tag{14-14}$$

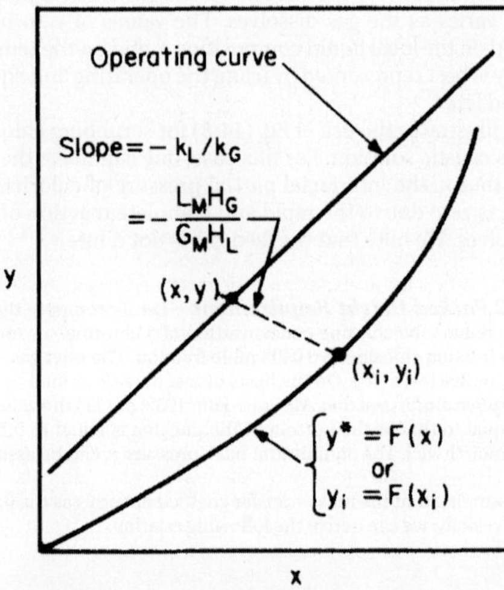

FIG. 14-6 Relationship between equilibrium curve and operating curve in a packed absorber; computation of interfacial compositions.

where L_M = molar liquid mass velocity, G_M = molar gas mass velocity, H_L = height of one transfer unit based on liquid-phase resistance, and H_G = height of one transfer unit based on gas-phase resistance. Using this equation, the integral in Eq. (14-8) can be evaluated.

Calculation of Transfer Units In the general case, the equations described here must be used to calculate the height of packing required for a given separation. However, if the local mass-transfer coefficient $k_G a y_{BM}$ is approximately proportional to the first power of the local gas velocity G_M, then the height of one gas-phase transfer unit, defined as $H_G = G_M / k_G a y_{BM}$, will be constant in Eq. (14-9). Similar considerations lead to an assumption that the height of one overall gas-phase transfer unit H_{OG} may be taken as constant. The height of packing required is then calculated according to the relation

$$h_T = H_G N_G = H_{OG} N_{OG} \tag{14-15}$$

where N_G = number of gas-phase transfer units and N_{OG} = number of overall gas-phase transfer units. When H_G and H_{OG} are not constant, it may be valid to use averaged values between the top and bottom of the tower and the relation

$$h_T = H_{G,av} N_G = H_{OG,av} N_{OG} \tag{14-16}$$

In these equations, the terms N_G and N_{OG} are defined by Eqs. (14-17) and (14-18).

$$N_G = \int_{y_2}^{y_1} \frac{y_{BM}\, dy}{(1-y)(y-y_i)} \tag{14-17}$$

$$N_G = \int_{y_0}^{y_1} \frac{y^0_{BM}\, dy}{(1-y)(y-y^0)} \tag{14-18}$$

Equation (14-18) is the more useful one in practice. It requires either actual experimental H_{OG} data or values estimated by combining individual measurements of H_G and H_L by Eq. (14-19). Correlations for H_G, H_L, and H_{OG} in nonreacting systems are presented in Sec. 5.

$$H_{OG} = \frac{y_{BM}}{y^0_{BM}} H_G + \frac{mG_M}{L_M} \frac{x_{BM}}{y^0_{BM}} H_L \tag{14-19a}$$

$$H_{OL} = \frac{x_{BM}}{x^0_{BM}} H_L + \frac{L_M}{mG_M} \frac{y_{BM}}{x^0_{BM}} H_G \tag{14-19b}$$

On occasion, the changes in gas flow and in the mole fraction of inert gas can be neglected so that inclusion of terms such as $1-y$ and y^0_{BM} can be approximated, as is shown below.

One such simplification was suggested by Wiegand [*Trans. Am. Inst. Chem. Eng.* **36:** 679 (1940)], who pointed out that the logarithmic-mean mole fraction of inert gas y^0_{BM} (or y_{BM}) is often very nearly equal to the arithmetic mean. Thus, substitution of the relation

$$\frac{y^0_{BM}}{(1-y)} = \frac{(1-y^0)+(1-y)}{2(1-y)} = \frac{y-y^0}{2(1-y)} + 1 \tag{14-20}$$

into the equations presented previously leads to the simplified forms

$$N_G = \frac{1}{2}\ln\left[\frac{1-y_2}{1-y_1}\right] + \int_{y_2}^{y_1} \frac{dy}{y-y_1} \tag{14-21}$$

$$N_{OG} = \frac{1}{2}\ln\frac{1-y_2}{1-y_1} + \int_{y2}^{y1} \frac{dy}{y-y^0} \tag{14-22}$$

The second (integral) terms represent the numbers of transfer units for an infinitely dilute gas. The first terms, usually only a small correction, give the effect of a finite level of gas concentration.

The procedure for applying Eqs. (14-21) and (14-22) involves two steps: (1) evaluation of the integrals and (2) addition of the correction corresponding to the first (logarithmic) term. The discussion that follows deals only with the evaluation of the integral term (first step).

The simplest possible case occurs when (1) both the operating and equilibrium lines are straight (i.e., the solutions are dilute); (2) Henry's law is valid ($y^0/x = y_i/x_i = m$); and (3) absorption heat effects are negligible. Under these conditions, the integral term in Eq. (14-21) may be computed by Colburn's equation [*Trans. Am. Inst. Chem. Eng.* **35:** 211 (1939)]:

$$N_{OG} = \frac{1}{1 - mG_M/L_M} \ln\left[\left(1 - \frac{mG_M}{L_M}\right)\left(\frac{y_1 - mx_2}{y_2 - mx_2}\right) + \frac{mG_M}{L_M}\right] \qquad (14\text{-}23)$$

Figure 14-7 is a plot of Eq. (14-23) from which the value of N_{OG} can be read directly as a function of mG_M/L_M and the ratio of concentrations. This plot and Eq. (14-23) are equivalent to the use of a logarithmic mean of terminal driving forces, but they are more convenient because one does not need to compute the exit-liquor concentration x_1.

In many practical situations involving nearly complete cleanup of the gas, an approximate result can be obtained from the equations just presented even when the simplifications are not valid, that is, solutions are concentrated and heat effects occur. In such cases the driving forces in the upper part of the tower are very much smaller than those at the bottom, and the value of mG_M/L_M used in the equations should be the ratio of the operating line L_M/G_M in the low-concentration region near the top of the tower.

Another approach is to divide the tower arbitrarily into a lean section (near the top) where approximate methods are valid, and to deal with the rich section separately. If the heat effects in the rich section are appreciable, consideration should be given to installing cooling units near the bottom of the tower. In any event, a design diagram showing the operating and equilibrium curves should be prepared to check the applicability of any simplified procedure. Figure 14-10, presented in Example 14-6, is one such diagram for an adiabatic absorption tower.

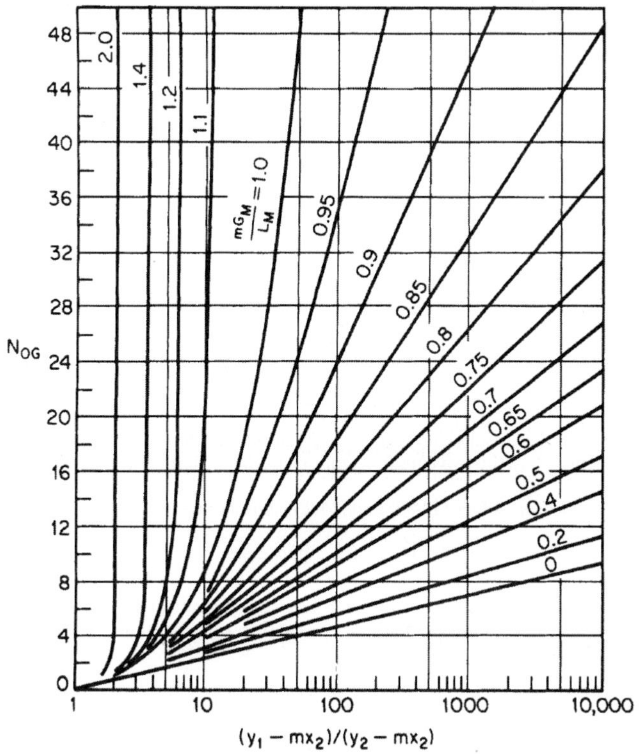

FIG. 14-7 Number of overall gas-phase mass-transfer units in a packed absorption tower for constant mG_M/L_M; solution of Eq. (14-23). (*From Sherwood and Pigford,* Absorption and Extraction, *McGraw-Hill, New York, 1952.*)

Stripping Equations Stripping or desorption involves the removal of a volatile component from the liquid stream by contact with an inert gas such as nitrogen or steam or the application of heat. Here the change in concentration of the liquid stream is of prime importance, and it is more convenient to formulate the rate equation analogous to Eq. (14-6) in terms of the liquid composition x. This leads to the following equations defining the number of transfer units and the height of transfer units based on liquid-phase resistance:

$$h_T = H_L \int_{x_2}^{x_1} \frac{x_{BM}\,dx}{(1-x)(x_i - x)} = H_L N_L \qquad (14\text{-}24)$$

$$h_T = H_{OL} \int_{x_2}^{x_1} \frac{x^0_{BM}\,dx}{(1-x)(x^0 - x)} = H_{OL} N_{OL} \qquad (14\text{-}25)$$

where, as before, subscripts 1 and 2 refer to the bottom and top of the tower, respectively (see Fig. 14-5).

In situations where one cannot assume that H_L and H_{OL} are constant, these terms need to be incorporated inside the integrals in Eqs. (14-24) and (14-25), and the integrals must be evaluated numerically (using Simpson's rule, for example). In the normal case involving stripping without chemical reactions, the liquid-phase resistance will dominate, making it preferable to use Eq. (14-25) together with the approximation $H_L \approx H_{OL}$.

The Weigand approximations of these integrals, in which arithmetic means are substituted for the logarithmic means (x_{BM} and x^0_{BM}), are

$$N_L = \frac{1}{2} \ln \frac{1 - x_1}{1 - x_2} + \int_{x_1}^{x_2} \frac{dx}{x - x_i} \qquad (14\text{-}26)$$

$$N_{OL} = \frac{1}{2} \ln \frac{1 - x_1}{1 - x_2} + \int_{x_1}^{x_2} \frac{dx}{x - x^\circ} \qquad (14\text{-}27)$$

In these equations, the first term is a correction for finite liquid-phase concentrations, and the integral term represents the numbers of transfer units required for dilute solutions. In most practical stripper applications, the first (logarithmic) term is relatively small.

For dilute solutions in which both the operating and the equilibrium lines are straight and in which heat effects can be neglected, the integral term in Eq. (14-27) is

$$N_{OL} = \frac{1}{1 - \dfrac{L_m}{mG_m}} \ln\left[\left(1 - \frac{L_m}{mG_m}\right)\left(\frac{x_2 - y_1/m}{x_1 - y_1/m}\right) + \frac{L_m}{mG_m}\right] \qquad (14\text{-}28)$$

This equation is analogous to Eq. (14-23). Thus, Fig. 14-7 is applicable if the concentration ratio $(x_2 - y_1/m)/(x_1 - y_1/m)$ is substituted for the abscissa and the parameter on the curves is identified as L_M/mG_M.

Example 14-3 Air Stripping of VOCs from Water A 0.45-m-diameter packed column was used by Dvorak et al. [*Environ. Sci. Tech.* **20:** 945 (1996)] for removing trichloroethylene (TCE) from wastewater by stripping with atmospheric air. The column was packed with 25-mm Pall rings, fabricated from polypropylene, to a height of 3.0 m. The TCE concentration in the entering water was 38 parts per million by weight (ppmw). A molar ratio of entering water to entering air was kept at 23.7. What degree of removal was to be expected? The temperatures of water and air were 20°C. Pressure was atmospheric.

Solution. For TCE in water, the Henry's law coefficient may be taken as 417 atm/mf at 20°C; note that this value of 417 atm/mf at 20°C has a high uncertainty, about 20 to 30 percent. In this low-concentration region, the coefficient is constant and equal to the slope of the equilibrium line m. The solubility of TCE in water, based on $H = 417$ atm, is 2390 ppm. Because of this low solubility, the entire resistance to mass transfer resides in the liquid phase. Thus, Eq. (14-25) may be used to obtain N_{OL}, the number of overall liquid phase transfer units.

In the equation, the ratio $x^0_{BM}/(1 - x)$ is unity because of the very dilute solution. It is necessary to have a value of H_L for the packing used, at given flow rates of liquid and gas. Methods for estimating H_L may be found in Sec. 5. Dvorak et al. found $H_{OL} = 0.8$ m. Then, for $h_T = 3.0$ m, $N_L = N_{OL} = 3.0/0.8 = 3.75$ transfer units.

Transfer units may be calculated from Eq. (14-25), replacing mole fractions with ppm concentrations, and since the operating and equilibrium lines are straight,

$$N_{OL} = \ln\left(\frac{38}{(\text{ppm})_{\text{exit}}}\right) = 3.75$$

Solving, $(\text{ppm})_{exit} = 0.9$. Thus, the stripped water would contain 0.9 parts per million of TCE. If the column height is doubled, to 4.5 m, the stripped water would contain 0.021 parts per million of TCE, a concentration reduction by a factor of 43.

Use of HTU and K_Ga Data In estimating the size of a commercial gas absorber or liquid stripper, it is desirable to have data on the overall mass-transfer coefficients (or heights of transfer units) for the system of interest and at the desired conditions of temperature, pressure, solute concentration, and fluid velocities. Such data is best obtained in an apparatus of pilot-plant or semiworks size to avoid the complexities of scale-up. Within the packing category, there are both random and ordered (structured) packing elements. Physical characteristics of these devices will be described later.

When no K_Ga or HTU data are available, their values may be estimated by means of a generalized model. A summary of useful models is given in Sec. 5. The values obtained may then be combined by the use of Eq. (14-19) to obtain values of H_{OG} and H_{OL}. This simple procedure is not valid when the rate of absorption is limited by chemical reaction.

Use of HETP Data for Absorber Design Distillation design methods (see Sec. 13) normally involve determination of the number of theoretical equilibrium stages N. Thus, when packed towers are employed in distillation applications, it is common practice to rate the efficiency of tower packings in terms of the height of packing equivalent to one theoretical stage (HETP).

The HETP of a packed-tower section, valid for either distillation or dilute-gas absorption and stripping systems in which constant molal overflow can be assumed and in which no chemical reactions occur, is related to the height of one overall gas-phase mass-transfer unit H_{OG} by the equation

$$\text{HETP} = H_{OG}\frac{\ln\left(\dfrac{mG_M}{L_M}\right)}{\left(\dfrac{mG_M}{L_M}-1\right)} \tag{14-29}$$

For gas absorption systems in which the inlet gas is concentrated, the corrected equation is

$$\text{HETP} = \left(\frac{y^0_{BM}}{1-y}\right)_{av} H_{OG}\frac{\ln\left(\dfrac{mG_M}{L_M}\right)}{\left(\dfrac{mG_M}{L_M}-1\right)} \tag{14-30}$$

where the correction term $y^0_{BM}/(1-y)$ is averaged over each individual theoretical stage. The equilibrium compositions corresponding to each theoretical stage may be estimated by the methods described in the next subsection, Tray-Tower Design. These compositions are used in conjunction with the local values of the gas and liquid flow rates and the equilibrium slope m to obtain values for H_G, H_L, and H_{OG} corresponding to the conditions on each theoretical stage, and the local values of the HETP are then computed by Eq. (14-30). The total height of packing required for the separation is the summation of the individual HETPs computed for each theoretical stage.

TRAY-TOWER DESIGN

The design of a tray tower for gas absorption and gas-stripping operations involves many of the same principles employed in distillation calculations, such as the determination of the number of theoretical trays needed to achieve a specified composition change (see Sec. 13). Distillation differs from absorption because it involves the separation of components based on the distribution of the various substances between a vapor phase and a liquid phase when all components are present in both phases. In distillation, the new phase is generated from the original phase by the vaporization or condensation of the volatile components, and the separation is achieved by introducing reflux to the top of the tower.

In gas absorption, the new phase consists of a relatively nonvolatile solvent (absorption) or a relatively insoluble gas (stripping), and normally no reflux is involved. This section discusses some of the considerations peculiar to gas absorption calculations for tray towers and some of the approximate design methods that can be applied (when simplifying assumptions are valid).

Graphical Design Procedure Construction of design diagrams (xy curves showing the equilibrium and operating curves) should be an integral part of any design involving the distribution of a single solute between a solvent and an inert gas. The number of theoretical trays can be stepped off rigorously, provided the curvatures of the operating and equilibrium lines are correctly represented in the diagram. The procedure is valid even though a solvent is present in the liquid phase and an inert gas is present in the vapor phase.

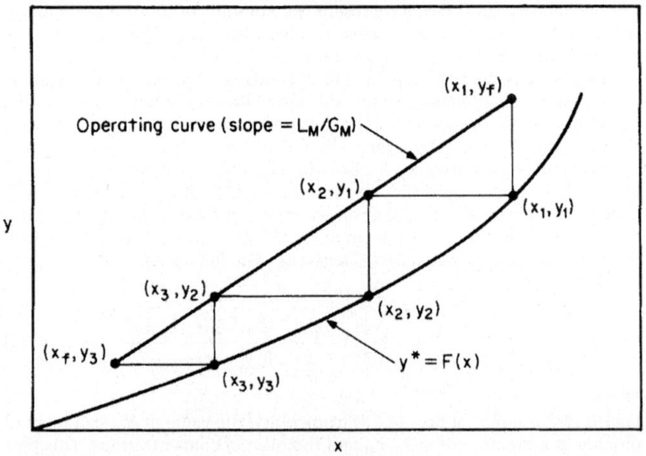

FIG. 14-8 Graphical method for a three-theoretical-plate gas-absorption tower with inlet-liquor composition x_f and inlet-gas composition y_f.

Figure 14-8 illustrates the graphical method for a three-theoretical-stage system. Note that in gas absorption the operating line is above the equilibrium curve, whereas in distillation this does not happen. In gas stripping, the operating line will be below the equilibrium curve.

On Fig. 14-8, note that the stepping-off procedure begins on the operating line. The starting point x_f, y_3 represents the compositions of the entering lean wash liquor and of the gas exiting from the top of the tower, as defined by the design specifications. After three steps, one reaches the point x_1, y_f representing the compositions of the solute-rich feed gas y_f and of the solute-rich liquor leaving the bottom of the tower x_1.

Algebraic Method for Dilute Gases By assuming that the operating and equilibrium curves are straight lines and that heat effects are negligible, Souders and Brown [*Ind. Eng. Chem.* **24:** 519 (1932)] developed the following equation:

$$(y_1 - y_2)/(y_1 - y_2^\circ) = (A^{N+1} - A)/(A^{N+1} - 1) \tag{14-31}$$

where N = number of theoretical trays, y_1 = mole fraction of solute in the entering gas, y_2 = mole fraction of solute in the leaving gas, $y_2^\circ = mx_2$ = mole fraction of solute in equilibrium with the incoming solvent (zero for a pure solvent), and A = absorption factor = L_M/mG_M. Note that the absorption factor is the reciprocal of the expression given in Eq. (14-4) for packed columns.

Note that for the limiting case of $A = 1$, the solution is given by

$$(y_1 - y_2)/(y_1 - y_2^\circ) = N/(N+1) \tag{14-32}$$

Although Eq. (14-31) is convenient for computing the composition of the exit gas as a function of the number of theoretical stages, an alternative equation derived by Colburn [*Trans. Am. Inst. Chem. Eng.* **35:** 211 (1939)] is more useful when the number of theoretical plates is the unknown:

$$N = \frac{\ln[(1-A^{-1})(y_1 - y_2^\circ)/(y_2 - y_2^\circ) + A^{-1}]}{\ln(A)} \tag{14-33}$$

The numerical results obtained by using either Eq. (14-31) or Eq. (14-33) are identical. Thus, the two equations may be used interchangeably as the need arises.

Comparison of Eqs. (14-33) and (14-23) shows that

$$N_{OG}/N = \ln(A)/(1-A^{-1}) \tag{14-34}$$

thus revealing the close relationship between theoretical stages in a plate tower and mass-transfer units in a packed tower. Equations (14-23) and (14-33) are related to each other by virtue of the relation

$$h_T = H_{OG}N_{OG} = (\text{HETP})N \tag{14-35}$$

Algebraic Method for Concentrated Gases When the feed gas is concentrated, the absorption factor, which is defined in general as $A = L_M/KG_M$ and where $K = y^0/x$, can vary throughout the tower due to changes in temperature and composition. An approximate solution to this problem can be obtained by substituting the "effective" adsorption factors A_e and A' derived by Edmister [*Ind. Eng. Chem.* **35:** 837 (1943)] into the equation

$$\frac{y_1 - y_2}{y_1} = \left[1 - \frac{1}{A'}\frac{(L_M x)_2}{(G_M y)_1}\right]\frac{A_e^{N+1} - A_e}{A_e^{N+1} - 1} \tag{14-36}$$

where subscripts 1 and 2 refer to the bottom and top of the tower, respectively, and the absorption factors are defined by the equations

$$A_e = \sqrt{A_1(A_2 + 1) + 0.25} - 0.5 \tag{14-37}$$

$$A' = A_1(A_2 + 1)/(A_1 + 1) \tag{14-38}$$

This procedure has been applied to the absorption of C_5 and lighter hydrocarbon vapors into a lean oil, for example.

Stripping Equations When the liquid feed is dilute and the operating and equilibrium curves are straight lines, the stripping equations analogous to Eqs. (14-31) and (14-33) are

$$(x_2 - x_1)/(x_2 - x_1^0) = (S^{N+1} - S)/(S^{N+1} - 1) \tag{14-39}$$

where $x_1^0 = y_1/m$; $S = mG_M/L_M = A^{-1}$; and

$$N = \frac{\ln\left[(1 - A)\frac{(x_2 - x_1^0)}{(x_1 - x_1^0)} + A\right]}{\ln(S)} \tag{14-40}$$

For systems in which the concentrations are large and the stripping factor S may vary along the tower, the following Edmister equations [*Ind. Eng. Chem.* **35:** 837 (1943)] are applicable:

$$\frac{x_2 - x_1}{x_2} = \left[1 - \frac{1}{S'}\frac{(G_M y)_1}{(L_M x)_2}\right]\frac{S_e^{N+1} - S_e}{S_e^{N+1} - 1} \tag{14-41}$$

where

$$S_e = \sqrt{S_2(S_1 + 1) + 0.25} - 0.5 \tag{14-42}$$

$$S' = S_2(S_1 + 1)/(S_2 + 1) \tag{14-43}$$

and the subscripts 1 and 2 refer to the bottom and top of the tower, respectively.

Equations (14-37) and (14-42) represent two different ways of obtaining an effective factor S, and a value of A_e obtained by taking the reciprocal of S_e from Eq. (14-42) will not check exactly with a value of A_e derived by substituting $A_1 = 1/S_1$ and $A_2 = 1/S_2$ into Eq. (14-37). Regardless of this fact, the equations generally give reasonable results for approximate design calculations.

It should be noted that throughout this section the subscripts 1 and 2 refer to the bottom and to the top of the apparatus, respectively, regardless of whether it is an absorber or a stripper. This has been done to maintain internal consistency among all the equations and to prevent the confusion created in some derivations in which the numbering system for an absorber is different from the numbering system for a stripper.

Tray Efficiencies in Tray Absorbers and Strippers Computations of the theoretical trays N assume that the liquid on each tray is completely mixed and that the vapor leaving the tray is in equilibrium with the liquid. In practice, complete equilibrium cannot exist since interphase mass transfer requires a finite driving force. This leads to the definition of an overall tray efficiency

$$E = N_{\text{theoretical}}/N_{\text{actual}} \tag{14-44}$$

which can be correlated with the system design variables.

Mass-transfer theory indicates that for trays of a given design, the factors that have the biggest influence on E in absorption and stripping towers are the physical properties of the fluids and the dimensionless ratio mG_M/L_M. Systems in which mass transfer is gas-film-controlled may be expected to have efficiencies as high as 50 to 100 percent, whereas tray efficiencies as low as 1 percent have been reported for the absorption of low-solubility (large-m) gases into solvents of high viscosity.

The fluid properties of interest are represented by the Schmidt numbers of the gas and liquid phases. For gases, the Schmidt numbers are normally close to unity and independent of temperature and pressure. Thus, gas-phase mass-transfer coefficients are relatively independent of the system.

By contrast, the liquid-phase Schmidt numbers range from about 10^2 to 10^4 and depend strongly on temperature. The temperature dependence of the liquid-phase Schmidt number derives primarily from the strong dependence of the liquid viscosity on temperature.

Consideration of the preceding discussion in connection with the relationship between mass-transfer coefficients (see Sec. 5)

$$1/K_G = 1/k_G + m/k_L \tag{14-45}$$

indicates that the variations in the overall resistance to mass transfer in absorbers and strippers are related primarily to variations in the liquid-phase viscosity μ and the slope m. O'Connell [*Trans. Am. Inst. Chem. Eng.* **42:** 741 (1946)] used the preceding findings and correlated the tray efficiency in terms of the liquid viscosity and the gas solubility. The O'Connell correlation for absorbers (Fig. 14-9) has Henry's law constant in lb · mol/(atm · ft³), the pressure in atmospheres, and the liquid viscosity in centipoise.

The best procedure for making tray efficiency corrections (which can be quite significant, as seen in Fig. 14-9) is to use experimental data from a prototype system that is large enough to be representative of the actual commercial tower.

Example 14-4 Actual Trays for Steam Stripping The number of actual trays required for steam-stripping an acetone-rich liquor containing 0.573 mol% acetone in water is to be estimated. The design overhead recovery of acetone is 99.88 percent, leaving 18.5 ppm weight of acetone in the stripper bottoms. The design operating temperature and pressure are 101.3 kPa and 94°C respectively, the average liquid-phase viscosity is 0.30 cP, and the average value of $K = y°/x$ for these conditions is 33.

By choosing a value of $mG_M/L_M = S = A^{-1} = 1.4$ and noting that the stripping medium is pure steam (i.e., $x_1^0 = 0$), the number of theoretical trays according to Eq. (14-40) is

$$N = \frac{\ln[(1 - 0.714)(1000) + 0.714]}{\ln(1.4)} = 16.8$$

The O'Connell parameter for gas absorbers is $\rho_L/KM\mu_L$, where ρ_L is the liquid density, lb/ft³; μ_L is the liquid viscosity, cP; M is the molecular weight of the liquid; and $K = y°/x$. For the present design

$$\rho_L/KM\mu_L = 60.1/(33 \times 18 \times 0.30) = 0.337$$

and according to the O'Connell graph for absorbers (Fig. 14-9), the overall tray efficiency for this case is estimated to be 30 percent. Thus, the required number of actual trays is 16.8/0.3 = 56 trays.

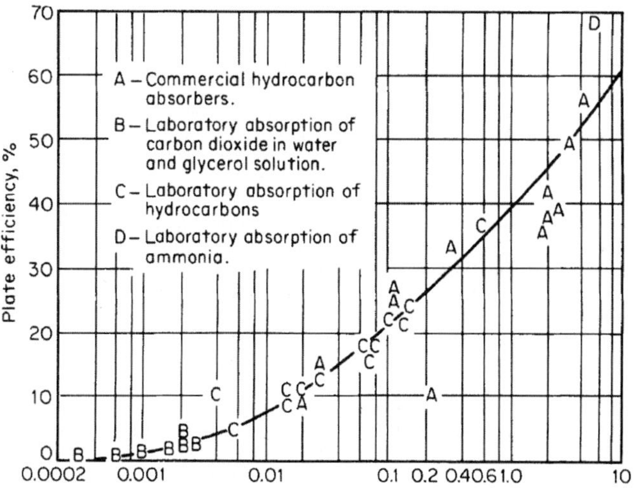

FIG. 14-9 O'Connell correlation for overall column efficiency E_{0c} for absorption. H is in (atm · ft³)/lb · mol, P is in atm, μ is in cP. To convert $P/(H\mu)$ in lb · mol/(ft³ · cP) to kg · mol/(m³ · Pa · s), multiply by 1.60 * 10⁴. [*O'Connell,* Trans. Am. Chem. Eng. *42: 741 (1946).*]

HEAT EFFECTS IN GAS ABSORPTION

Overview One of the most important considerations involved in designing gas absorption towers is to determine whether temperatures will vary along the height of the tower due to heat effects; note that the solute solubility often depends strongly on temperature. The simplified design procedures described earlier in this section become more complicated when heat effects cannot be neglected. Our purpose in this section is to help readers to understand and design gas absorption towers where heat effects are significant and cannot be ignored.

Heat effects that cause temperatures to vary from point to point in a gas absorber include (1) the heat of solution of the solute (including heat of condensation, heat of mixing, and heat of reaction); (2) the heat of vaporization or condensation of the solvent; (3) the exchange of sensible heat between the gas and liquid phases; and (4) the loss of sensible heat from the fluids to internal or external coils.

There are a number of systems where heat effects definitely cannot be ignored. Examples include the absorption of ammonia in water, dehumidification of air using concentrated H_2SO_4, absorption of HCl in water, absorption of SO_3 in H_2SO_4, and absorption of CO_2 in chemical solvents like alkanolamines. Even for systems where the heat effects are mild, they may not be negligible; an example is the absorption of acetone in water.

Thorough and knowledgeable discussions of the problems involved in gas absorption with significant heat effects have been presented by Coggan and Bourne [*Trans. Inst. Chem. Eng.* **47:** T96, T160 (1969)]; Bourn, von Stockar, and Coggan [*Ind. Eng. Chem. Proc. Des. Dev.* **13:** 115, 124 (1974)]; and von Stockar and Wilke [*Ind. Eng. Chem. Fundam.* **16:** 89 (1977)]. The first two of these references discuss tray-tower absorbers and include experimental studies of the absorption of ammonia in water. The third reference discusses the design of packed-tower absorbers and includes a shortcut design method based on a semitheoretical correlation of rigorous design calculations. All these authors demonstrate that when the solvent is volatile, the temperature inside an absorber can go through a maximum. They note that the least expensive and most common of solvents—water—is capable of exhibiting this "hot-spot" behavior.

Several approaches may be used in modeling absorption with heat effects, depending on the job at hand: (1) treat the process as isothermal by assuming a particular temperature, then add a safety factor; (2) employ the classical adiabatic method, which assumes that the heat of solution manifests itself only as sensible heat in the liquid phase and that the solvent vaporization is negligible; (3) use semitheoretical shortcut methods derived from rigorous calculations; and (4) employ rigorous methods available from a process simulator.

While simpler methods are useful for understanding the key effects involved, rigorous methods (usually using commercial software) are recommended for final designs. This subsection also discusses the range of safety factors that are required if simpler methods are used.

Effects of Operating Variables Conditions that give rise to significant heat effects are (1) an appreciable heat of solution and/or (2) absorption of large amounts of solute in the liquid phase. The second condition is favored when the solute concentration in the inlet gas is large, when the liquid flow rate is relatively low (small L_M/G_M), when the solubility of the solute in the liquid is high, and/or when the operating pressure is high.

If the solute-rich gas entering the bottom of an absorber tower is cold, the liquid phase may be cooled somewhat by transfer of sensible heat to the gas. A much stronger cooling effect can occur when the solute is volatile and the entering gas is not saturated with respect to the solvent. It is possible to experience a condition in which solvent is being evaporated near the bottom of the tower and condensed near the top. Under these conditions a pinch point may develop in which the operating and equilibrium curves approach each other at a point inside the tower.

In the references previously cited, the authors discuss the influence of operating variables upon the performance of towers when large heat effects are involved. Some key observations are as follows:

Operating Pressure Raising the pressure may increase the separation effectiveness considerably. Calculations for the absorption of methanol in water from water-saturated air showed that doubling the pressure doubles the allowable concentration of methanol in the feed gas while still achieving the required concentration specification in the off gas.

Temperature of Lean Solvent The temperature of the entering (lean) solvent has surprisingly little influence upon the temperature profile in an absorber since any temperature changes are usually caused by the heat of solution or the solvent vaporization. In these cases, the temperature profile in the liquid phase is usually dictated solely by the internal heat effects.

Temperature and Humidity of the Rich Gas Cooling and consequent dehumidification of the feed gas to an absorption tower can be very beneficial. A high humidity (or relative saturation with the solvent) limits the capacity of the gas to take up latent heat and hence is unfavorable to absorption. Thus dehumidification of the inlet gas is worth considering in the design of absorbers with large heat effects.

Liquid-to-Gas Ratio The L/G ratio can have a significant influence on the development of temperature profiles in gas absorbers. High L/G ratios tend to result in less strongly developed temperature profiles due to the increased heat capacity of the liquid phase. As the L/G ratio is increased, the operating line moves away from the equilibrium line, and more solute is absorbed per stage or packing segment. However, there is a compensating effect; since more heat is liberated in each stage or packing segment, the temperatures will rise, which causes the equilibrium line to shift up. As the L/G ratio is decreased, the concentration of solute tends to build up in the upper part of the absorber, and the point of highest temperature tends to move upward in the tower until finally the maximum temperature occurs at the top of the tower. Of course, the capacity of the liquid to absorb solute falls progressively as L/G is reduced.

Number of Stages or Packing Height When the heat effects combine to produce an extended zone in the tower where little absorption takes place (i.e., a pinch zone), the addition of trays or packing height will have no useful effect on separation efficiency. In this case, increases in absorption may be obtained by increasing solvent flow, introducing strategically placed coolers, cooling and dehumidifying the inlet gas, or raising the tower pressure.

Equipment Considerations When the solute has a large heat of solution and the feed gas contains a high concentration of solute, as in the absorption of HCl in water, the effects of heat release during absorption may be so pronounced that the installation of heat-transfer surface to remove the heat of absorption may be as important as providing sufficient interfacial area for the mass-transfer process itself. The added heat-transfer area may consist of internal cooling coils on the trays, or the liquid may be withdrawn from the tower, cooled in an external heat exchanger, and then returned to the tower.

In most cases the rate of heat liberation is largest near the bottom of the tower, where most of the solute absorption occurs, so cooling surfaces or intercoolers are required only at the lower part of the column. Coggan and Bourne [*Trans. Inst. Chem. Eng.* **47:** T96, T160 (1969)] found, however, that the optimal position for a single interstage cooler does not necessarily coincide with the position of the maximum temperature of the center of the pinch. They found that in a 12-tray tower, two strategically placed interstage coolers tripled the allowable ammonia feed concentration for a given off-gas specification. For a case involving methanol absorption, it was found that greater separation was possible in a 12-stage column with two intercoolers than in a simple column with 100 stages and no intercoolers.

In the case of HCl absorption, a shell-and-tube heat exchanger often is used as a cooled wetted-wall vertical-column absorber so that the exothermic heat of reaction can be removed continuously as it is released into a liquid film.

Installation of heat-exchange equipment to precool and dehumidify the feed gas to an absorber also deserves consideration, in order to take advantage of the cooling effects created by vaporization of solvent in the lower sections of the tower.

Classical Isothermal Design Method When the feed gas is sufficiently dilute, the exact design solution may be approximated by the isothermal one over the broad range of L/G ratios, since heat effects are generally less important when washing dilute-gas mixtures. The problem, however, is one of defining the term *sufficiently dilute* for each case. For a new absorption duty, the assumption of isothermal operation must be subjected to verification by the use of a rigorous design procedure.

When heat-exchange surface is being provided in the design of an absorber, the isothermal design procedure can be rendered valid by virtue of the exchanger design specification. With ample surface area and a close approach, isothermal operation can be guaranteed.

For preliminary screening and feasibility studies or for rough estimates, one may wish to employ a version of the isothermal design method that assumes that the liquid temperatures in the tower are everywhere equal to the inlet-liquid temperature. In their analysis of packed-tower designs, von Stockar and Wilke [*Ind. Eng. Chem. Fundam.* **16:** 89 (1977)] showed that the isothermal method tended to underestimate the required height of packing by a factor of as much as 1.5 to 2. Thus, for rough estimates, one may wish to assume that the absorber temperature is equal to the inlet-liquid temperature and then apply a design factor to the result.

Another case in which the constant-temperature method is used involves the direct application of experimental K_Ga values obtained at the desired conditions of inlet temperatures, operating pressures, flow rates, and feed-stream compositions. The assumption here is that, regardless of any temperature profiles that may exist within the actual tower, the procedure of "working the problem in reverse" will yield a correct result. One should, however, be cautious about extrapolating such data from the original basis and take care to use compatible equilibrium data.

Classical Adiabatic Design Method The classical adiabatic design method assumes that the heat of solution serves only to heat up the liquid stream, and there is no vaporization of the solvent. This assumption makes it feasible to relate increases in the liquid-phase temperature to the solute concentration x by a simple enthalpy balance. The equilibrium curve can then be adjusted to account for the corresponding temperature rise on an xy diagram. The adjusted equilibrium curve will be concave upward as the concentration increases, tending to decrease the driving forces near the bottom of the tower, as illustrated in Fig. 14-10 in Example 14-6.

Colburn [*Trans. Am. Inst. Chem. Eng.* **35:** 211 (1939)] has shown that when the equilibrium line is straight near the origin but curved slightly at its upper end, N_{OG} can be computed approximately by assuming that the equilibrium curve is a parabolic arc of slope m_2 near the origin and passing through the point x_1, K_1x_1 at the upper end. The Colburn equation for this case is

$$N_{OG} = \frac{1}{1 - m_2G_M/L_M} \times \ln\left[\frac{(1-m_2G_M/L_M)^2}{1-K_1-G_M/L_M}\left(\frac{y_1 - mx_2x_2}{y_2 - m_2x_2}\right) + \frac{m_2G_M}{L_M}\right] \quad (14\text{-}46)$$

Comparison by von Stockar and Wilke [*Ind. Eng. Chem. Fundam.* **16:** 89 (1977)] between the rigorous and the classical adiabatic design methods for packed towers indicates that the simple adiabatic design methods underestimate packing heights by as much as a factor of 1.25 to 1.5. Thus, when using the classical adiabatic method, one should probably apply a design safety factor.

A slight variation of the preceding method accounts for increases in the solvent content of the gas stream between the inlet and the outlet of the tower and assumes that the evaporation of solvent tends to cool the liquid. This procedure offsets a part of the temperature rise that would have been predicted with no solvent evaporation and leads to the prediction of a shorter tower.

Rigorous Design Methods A detailed discussion of rigorous methods for the design of packed and tray absorbers when large heat effects are involved is beyond the scope of this subsection. In principle, material and energy balances may be executed under the same constraints as for rigorous distillation calculations (see Sec. 13). Further discussion on this subject is given in the subsection Absorption with Chemical Reaction.

Direct Comparison of Design Methods The following problem, originally presented by Sherwood, Pigford, and Wilke (*Mass Transfer*, McGraw-Hill, New York, 1975, p. 616), was employed by von Stockar and Wilke [*Ind. Eng. Chem. Fundam.* **16:** 89 (1977)] as the basis for a direct comparison between the isothermal, adiabatic, semitheoretical shortcut, and rigorous design methods for estimating the height of packed towers.

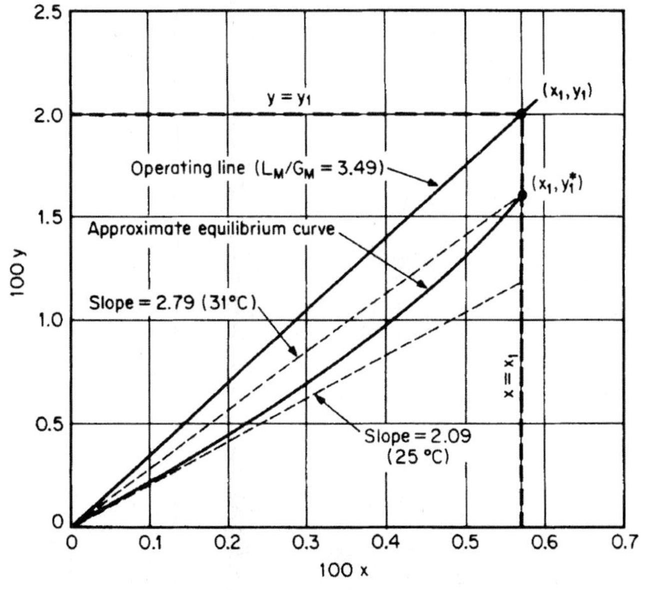

FIG. 14-10 Design diagram for adiabatic absorption of acetone in water, Example 14-6.

Example 14-5 Packed Absorber, Acetone into Water Inlet gas to an absorber consists of a mixture of 6 mol% acetone in air saturated with water vapor at 15°C and 101.3 kPa (1 atm). The scrubbing liquor is pure water at 15°C, and the inlet gas and liquid rates are given as 0.080 and 0.190 kmol/s, respectively. The liquid rate corresponds to 20 percent over the theoretical minimum as calculated by assuming a value of x_1 corresponding to complete equilibrium between the exit liquor and the incoming gas. H_G and H_L are given as 0.42 and 0.30 m, respectively, and the acetone equilibrium data at 15°C are $p_A^0 = 19.7$ kPa (147.4 torr), $\gamma_A = 6.46$, and $m_A = 6.46 \times 19.7/101.3 = 1.26$. The heat of solution of acetone is 7656 cal/gmol (32.05 kJ/gmol), and the heat of vaporization of solvent (water) is 10,755 cal/gmol (45.03 kJ/gmol). The problem calls for determining the height of packing required to achieve a 90 percent recovery of the acetone.

The following table compares the results obtained by von Stockar and Wilke [*Ind. Eng. Chem. Fundam.* **16:** 89 (1977)] for the various design methods:

Design method used	N_{OG}	Packed height, m	Design safety factor
Rigorous	5.56	3.63	1.00
Shortcut rigorous	5.56	3.73	0.97
Classical adiabatic	4.01	2.38	1.53
Classical isothermal	3.30	1.96	1.85

It should be clear from this example that there is considerable room for error when approximate design methods are employed in situations involving large heat effects, even for a case in which the solute concentration in the inlet gas is only 6 mol%.

Example 14-6 Solvent Rate for Absorption Let us consider the absorption of acetone from air at atmospheric pressure into a stream of pure water fed to the top of a packed absorber at 25°C. The inlet gas at 35°C contains 2 percent by volume of acetone and is 70 percent saturated with water vapor (4 percent H_2O by volume). The mole-fraction acetone in the exit gas is to be reduced to 1/400 of the inlet value, or 50 ppmv. For 100 kmol of feed-gas mixture, how many kilomoles of freshwater should be fed to provide a positive-driving force throughout the packing? How many transfer units will be needed according to the classical adiabatic method? What is the estimated height of packing required if $H_{OG} = 0.70$ m?

The latent heats at 25°C are 7656 kcal/kmol for acetone and 10,490 kcal/kmol for water, and the differential heat of solution of acetone vapor in pure water is given as 2500 kcal/kmol. The specific heat of air is 7.0 kcal/(kmol · K).

Acetone solubilities are defined by the equation

$$K = y^\circ/x = \gamma_a p_a/p_T \quad (14\text{-}47)$$

where the vapor pressure of pure acetone, p_A^0, in mm Hg (torr), with the temperature T in Kelvin, is given by (Sherwood et al., *Mass Transfer*, McGraw-Hill, New York, 1975, p. 537):

$$p_A^0 = \exp\,(18.1594 - 3794.06/T) \quad (14\text{–}48)$$

and the liquid-phase-activity coefficient may be approximated for low concentrations ($x \le 0.01$) by the equation

$$\gamma_a = 6.5 \exp\,(2.0803 - 601.2/T) \quad (14\text{-}49)$$

Typical values of acetone solubility as a function of temperature at a total pressure of 760 mm Hg are shown in the following table:

t, °C	25	30	35	40
γ_a	6.92	7.16	7.40	7.63
p_a, mm Hg	229	283	346	422
$K = \gamma_a p_a^0/760$	2.09	2.66	3.37	4.23

For dry gas and liquid water at 25°C, the following enthalpies are computed for the inlet- and exit-gas streams (basis, 100 kmol of gas entering):

Entering gas:

Acetone	$2(2500 + 7656) = 20{,}312$ kcal
Water vapor	$4(10{,}490) = 41{,}960$
Sensible heat	$(100)(7.0)(35 - 25) = 7{,}000$
	69,272 kcal

Exit gas (assumed saturated with water at 25°C):

Acetone	$(2/400)(94/100)(2500) = 12$ kcal
Water vapor	$94\left(\frac{23.7}{760 - 23.7}\right)(10{,}490) = 31{,}600$
	31,612 kcal

Enthalpy change of liquid = 69,272 − 31,612 = 37,660 kcal/100 kmol gas. Thus, $\Delta t = t_1 - t_2 = 37{,}660/18L_M$, and the relation between L_M/G_M and the liquid-phase temperature rise is

$$L_M/G_M = (37{,}660)/(18)(100)\,\Delta t = 20.92/\Delta t$$

The following table summarizes the critical values for various assumed temperature rises:

Δt, °C	L_M/G_M	K_1	K_1G_M/L_M	m_2G_M/L_M
0		2.09	0.	0.
2	10.46	2.31	0.221	0.200
3	6.97	2.42	0.347	0.300
4	5.23	2.54	0.486	0.400
5	4.18	2.66	0.636	0.500
6	3.49	2.79	0.799	0.599
7	2.99	2.93	0.980	0.699

Evidently a temperature rise of 7°C would not be a safe design because the equilibrium line nearly touches the operating line near the bottom of the tower, creating a pinch. A temperature rise of 6°C appears to give an operable design, and for this case $L_M = 349$ kmol per 100 kmol of feed gas.

The design diagram for this case is shown in Fig. 14-10, in which the equilibrium curve is drawn so that the slope at the origin m_2 is equal to 2.09 and passes through the point $x_1 = 0.02/3.49 = 0.00573$ at $y_1^* = 0.00573 \times 2.79 = 0.0160$.

The number of transfer units can be calculated from the adiabatic design equation, Eq. (14-46):

$$N_{OG} = \frac{1}{1-0.599}\ln\left[\frac{(1-0.599)^2}{(1-0.799)}(400)+0.599\right] = 14.4$$

The estimated height of tower packing by assuming $H_{OG} = 0.70$ m and a design safety factor of 1.5 is

$$h_T = (14.4)(0.7)(1.5) = 15.1 \text{ m } (49.6 \text{ ft})$$

For this tower, one should consider the use of two or more shorter packed sections instead of one long section.

Another point to be noted is that this calculation can be done more easily today by using a process simulator. However, the details are presented here to help the reader gain familiarity with the key assumptions and results. The results of the approximate calculation in Example 14-6 are remarkably accurate. A rigorous simulation by a modern simulator using 14 theoretical stages results in a temperature change of about 6°C and an acetone concentration in the exit air of 6 ppm by volume.

MULTICOMPONENT SYSTEMS

When no chemical reactions are involved in the absorption of more than one soluble component from an insoluble gas, the design conditions (temperature, pressure, liquid-to-gas ratio) are normally determined by the volatility or physical solubility of the least soluble component for which the recovery is specified.

The more volatile (i.e., less soluble) components will only be partially absorbed even for an infinite number of trays or transfer units. This can be seen in Fig. 14-11, in which the asymptotes become vertical for values of mG_M/L_M greater than unity. If the amount of volatile component in the fresh solvent is negligible, then the limiting value of y_1/y_2 for each of the highly volatile components is

$$y_1/y_2 = S/(S-1) \tag{14-50}$$

where $S = mG_M/L_M$ and the subscripts 1 and 2 refer to the bottom and top of the tower, respectively.

When the gas stream is dilute, absorption of each constituent can be considered separately as if the other components were absent. The following example illustrates the use of this principle.

Example 14-7 Multicomponent Absorption, Dilute Case Air entering a tower contains 1 percent acetaldehyde and 2 percent acetone. The liquid-to-gas ratio for optimum acetone recovery is $L_M/G_M = 3.1$ mol/mol when the fresh-solvent temperature is 31.5°C. The value of y^o/x for acetaldehyde has been measured as 50 at the boiling point of a dilute solution, 93.5°C. What will the percentage recovery of acetaldehyde be under conditions of optimal acetone recovery?

Solution. If the heat of solution is neglected, y^o/x at 31.5°C is equal to 50(1200/7300) = 8.2, where the factor in parentheses is the ratio of pure-acetaldehyde vapor pressures at 31.5 and 93.5°C, respectively. Since L_M/G_M is equal to 3.1, the value of S for the aldehyde is $S = mG_M/L_M = 8.2/3.1 = 2.64$, and $y_1/y_2 = S/(S-1) = 2.64/1.64 = 1.61$. The acetaldehyde recovery is therefore equal to 100 × 0.61/1.61 = 38 percent recovery.

In concentrated systems the change in gas and liquid flow rates within the tower and the heat effects accompanying the absorption of all components must be considered. A trial-and-error calculation from one theoretical stage to the next usually is required if accurate results are to be obtained, and in such cases calculation procedures similar to those described in Sec. 13 normally are used. A computer procedure for multicomponent adiabatic absorber design has been described by Feintuch and Treybal [*Ind. Eng. Chem. Process Des. Dev.* **17:** 505 (1978)]. Also see Holland, *Fundamentals and Modeling of Separation Processes,* Prentice Hall, Englewood Cliffs, N.J., 1975.

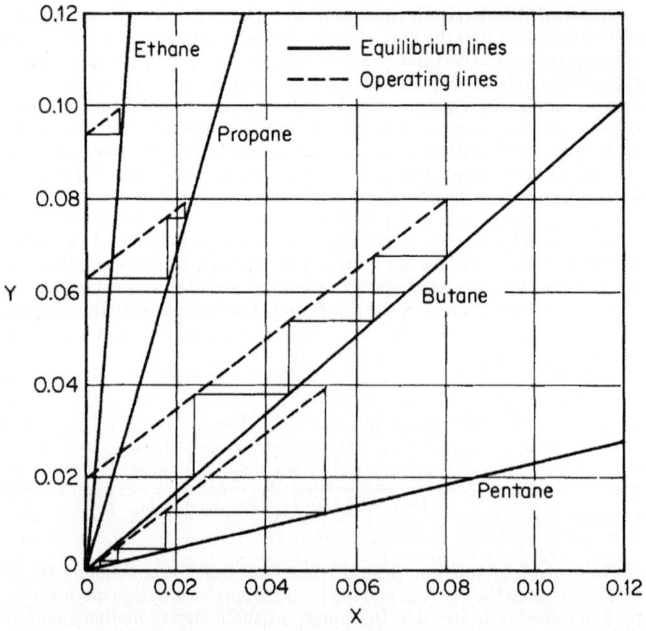

FIG. 14-11 Graphical design method for multicomponent systems; absorption of butane and heavier components in a solute-free lean oil.

In concentrated systems, the changes in the gas and liquid flow rates within the tower and the heat effects accompanying the absorption of all components must be considered. A trial-and-error calculation from one theoretical stage to the next is usually required if accurate and reliable results are to be obtained, and in such cases calculation procedures similar to those described in Sec. 13 must be used.

When two or more gases are absorbed in systems involving chemical reactions, the system is much more complex. This topic is discussed later in the subsection Absorption with Chemical Reaction.

Graphical Design Method for Dilute Systems The following notation for multicomponent absorption systems has been adapted from Sherwood, Pigford, and Wilke (*Mass Transfer,* McGraw-Hill, New York, 1975, p. 415):

L_M^S = moles of solvent per unit time
G_M^0 = moles of rich feed gas to be treated per unit time
X = moles of one solute per mole of solute-free solvent fed to top of tower
Y = moles of one solute in gas phase per mole of rich feed gas

Subscripts 1 and 2 refer to the bottom and the top of the tower, respectively, and the material balance for any one component may be written as

$$L_M^S (X - X_2) = G_M^0 (Y - Y_2) \tag{14-51}$$

or else as

$$L_M^S (X_1 - X) = G_M^0 (Y_1 - Y) \tag{14-52}$$

For the special case of absorption from lean gases with relatively large amounts of solvent, the equilibrium lines are defined for each component by the relation

$$Y^0 = K'X \tag{14-53}$$

Thus, the equilibrium line for each component passes through the origin with slope K', where

$$K' = K(G_M/G_M^0)/(L_M/L_M^S) \tag{14-54}$$

and $K = y^o/x$. When the system is sufficiently dilute, $K' = K$.

The liquid-to-gas ratio is chosen on the basis of the least soluble component in the feed gas that must be absorbed completely. Each component will then have its own operating line with slope equal to L_M^S/G_M^0 (i.e., the operating lines for the various components will be parallel).

A typical diagram for the complete absorption of pentane and heavier components is shown in Fig. 14-11. The oil used as solvent is assumed to be solute-free (i.e., $X_2 = 0$), and the "key component," butane, was identified as that component absorbed in appreciable amounts whose equilibrium line is most nearly parallel to the operating lines (i.e., the K value for butane is approximately equal to L_M^S/G_M^0).

In Fig. 14-11, the composition of the gas with respect to components more volatile than butane will approach equilibrium with the liquid phase at the bottom of the tower. The gas compositions of the components less volatile (heavier) than butane will approach equilibrium with the oil entering the tower, and since $X_2 = 0$, the components heavier than butane will be completely absorbed.

Four theoretical trays have been stepped off for the key component (butane) on Fig. 14-11, and are seen to give a recovery of 75 percent of the butane. The operating lines for the other components have been drawn with the same slope and placed so as to give approximately the same number of theoretical trays. Figure 14-11 shows that equilibrium is easily achieved in fewer than four theoretical trays and that for the heavier components nearly complete recovery is obtained in four theoretical trays. The diagram also shows that absorption of the light components takes place in the upper part of the tower, and the final recovery of the heavier components takes place in the lower section of the tower.

Algebraic Design Method for Dilute Systems The design method just described can be performed algebraically by using the following modified version of the Kremser formula:

$$\frac{Y_1 - Y_2}{Y_1 - mX_2} = \frac{(A^0)^{N+1} - A^0}{(A^0)^{N+1} - 1} \qquad (14\text{-}55)$$

where for dilute gas absorption $A^0 = L_M^S/mG_M^0$ and $m \approx K = y^0/x$.

The left side of Eq. (14-55) represents the efficiency of absorption of any one component of the feed gas mixture. If the solvent is solute-free so that $X_2 = 0$, the left side is equal to the fractional absorption of the component from the rich feed gas. When the number of theoretical trays N and the liquid and gas feed rates L_M^S and G_M^0 have been fixed, the fractional absorption of each component may be computed directly, and the operating lines need not be placed by trial and error as in the graphical method described previously.

According to Eq. (14-55), when A^0 is less than unity and N is large,

$$(Y_1 - Y_2)/(Y_1 - mX_2) = A^0 \qquad (14\text{-}56)$$

Equation (14-56) may be used to estimate the fractional absorption of more volatile components when A^0 of the component is greater than A^0 of the key component by a factor of 3 or more.

When A^0 is much larger than unity and N is large, the right side of Eq. (14-55) becomes equal to unity. This signifies that the gas will leave the top of the tower in equilibrium with the incoming oil, and when $X_2 = 0$, it corresponds to complete absorption of the component in question. Thus, the least volatile components may be assumed to be at equilibrium with the lean oil at the top of the tower.

When $A^0 = 1$, the right side of Eq. (14-56) simplifies as follows:

$$(Y_1 - Y_2)/(Y_1 - mX_2) = N/(N+1) \qquad (14\text{-}57)$$

For systems in which the absorption factor A^0 for each component is not constant throughout the tower, an effective absorption factor for use in the equations just presented can be estimated by the Edmister formula

$$A_e^0 = \sqrt{A_1^0(A_2^0 + 1) + 0.25} - 0.5 \qquad (14\text{-}58)$$

This procedure is a reasonable approximation only when no pinch points exist within the tower and when the absorption factors vary in a regular manner between the bottom and the top of the tower.

Example 14-8 Multicomponent Absorption, Concentrated Case A hydrocarbon feed gas is to be treated in an existing four-theoretical-tray absorber to remove butane and heavier components. The recovery specification for the key component, butane, is 75 percent. The composition of the exit gas from the absorber and the required liquid-to-gas ratio are to be estimated. The feed-gas composition and the equilibrium K values for each component at the temperature of the (solute-free) lean oil are presented in the following table:

Component	Mole %	K value
Methane	68.0	74.14
Ethane	10.0	12.00
Propane	8.0	3.43
Butane	8.0	0.83
Pentane	4.0	0.23
C_6 plus	2.0	0.065

For $N = 4$ and $Y_2/Y_1 = 0.25$, the value of A^0 for butane is found to be equal to 0.89 from Eq. (14-55) by using a trial-and-error method. The values of A^0 for the other components are then proportional to the ratios of their K values to that of butane. For example, $A^0 = 0.89(0.833/12.0) = 0.062$ for ethane. The values of A^0 for each of the other components and the exit-gas composition as computed from Eq. (14-55) are shown in the following table:

Component	A^0	Y_2, mol/100 mols feed	Exit gas mol%	
			Example 14-8	Rigorous
Methane	0.010	67.3	79.1	79.5
Ethane	0.062	9.4	11.1	11.0
Propane	0.216	6.3	7.4	7.1
Butane	0.890	2.0	2.4	2.4
Pentane	3.182	0.027	0.03	0.05
C_6 Plus	11.406	0.0012	0.0014	0.0004

The molar liquid-to-gas ratio required for this separation is computed as $L_M^s/G_M^0 = A^0 \times K = 0.89 \times 0.833 = 0.74$.

We note that this example is the analytical solution to the graphical design problem shown in Fig. 14-11, which therefore is the design diagram for this system.

The results in the last column in the preceding table have been obtained from a rigorous model in which the solvent rate has been calculated to obtain 75 percent recovery of butane. The excellent agreement with the simple model in Example 14-8 demonstrates that the simple procedure captures the essential characteristics of the absorption process.

The simplified design calculations presented in this subsection are intended to reveal the key features of gas absorption in multicomponent systems. It is expected that rigorous computations, based on the methods presented in Sec. 13, will be used in design practice. Nevertheless, it is valuable to study these simplified design methods and examples because they provide insight into the key aspects of multicomponent absorption.

ABSORPTION WITH CHEMICAL REACTION

Introduction Many present-day commercial gas absorption processes involve systems in which chemical reactions take place in the liquid phase; an example of the absorption of CO_2 by MEA was presented earlier in this section. These reactions greatly increase the capacity of the solvent and enhance the rate of absorption when compared to physical absorption systems. In addition, the selectivity of reacting solutes is greatly increased over that of nonreacting solutes. For example, MEA has a strong selectivity for CO_2 compared to chemically inert solutes such as CH_4, CO, O_2, or N_2. Note that the design procedures presented here are theoretically and practically related to biofiltration, which is discussed in Sec. 22, Waste Management.

A necessary prerequisite to understanding the subject of absorption with chemical reaction is the development of a thorough understanding of the principles involved in physical absorption, as discussed earlier in this section and in Sec. 5. Excellent references on the subject of absorption with chemical reactions are the books by Dankwerts (*Gas-Liquid Reactions*, McGraw-Hill, New York, 1970), Astarita et al. (*Gas Treating with Chemical Solvents*, Wiley, New York, 1983), and Kohl and Nielsen (*Gas Purification*, Gulf Publishing Company, Houston, 1997).

Recommended Overall Design Strategy When one is considering the design of a gas absorption system involving chemical reactions, the following procedure is recommended:

1. Consider the possibility that the physical design methods described earlier in this section may be applicable.

2. Determine whether commercial design overall K_Ga values are available for use in conjunction with the traditional design method, being careful to note whether the conditions under which the K_Ga data were obtained are essentially the same as for the new design. Contact the various tower-packing vendors for information about whether K_Ga data are available for your system and conditions.

3. Consider the possibility of scaling up the design of a new system from experimental data obtained in a laboratory bench-scale or small pilot-plant unit.

4. Consider the possibility of developing for the new system a rigorous, theoretically based design procedure that will be valid over a wide range of design conditions. Note that commercial software is readily available today to develop a rigorous model in a relatively small amount of time. These topics are further discussed in the subsections that follow.

Dominant Effects in Absorption with Chemical Reaction When the solute is absorbing into a solution containing a reagent that chemically reacts with it, diffusion and reaction effects become closely coupled. It is thus important for the design engineer to understand the key effects. Figure 14-12 shows the concentration profiles that occur when solute *A* undergoes an irreversible second-order reaction with component *B*, dissolved in the liquid, to give product *C*.

$$A + bB \rightarrow cC \tag{14-59}$$

The rate equation is

$$r_A = -k_2 C_A C_B \tag{14-60}$$

Figure 14-12 shows that the fast reaction takes place entirely in the liquid film. In such instances, the dominant mass-transfer mechanism is physical

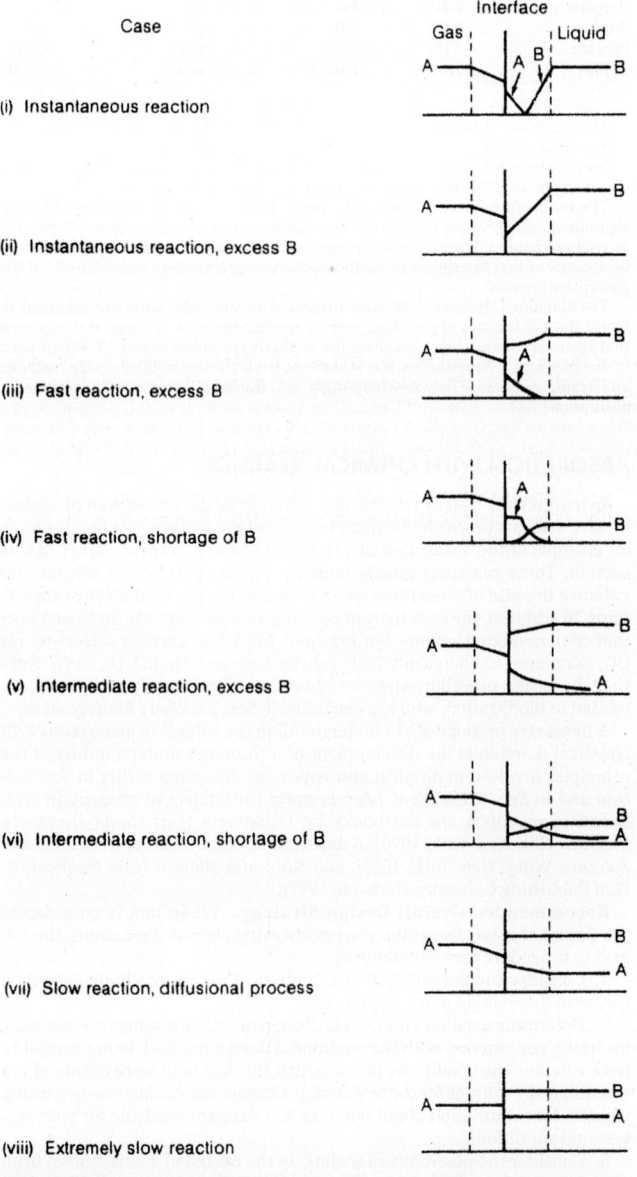

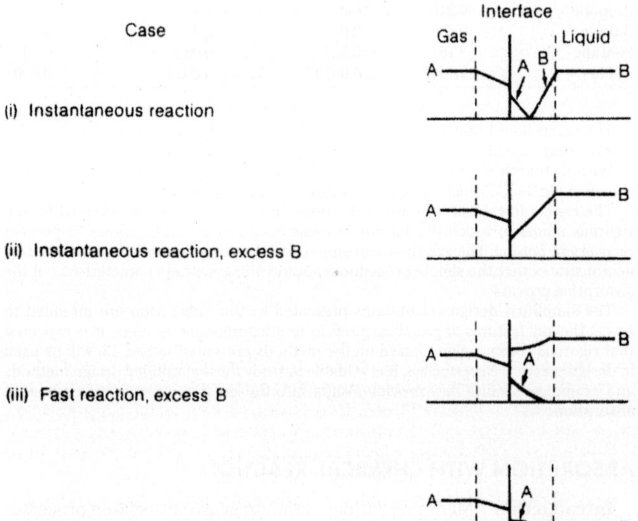

FIG. 14-12 Vapor- and liquid-phase concentration profiles near an interface for absorption with chemical reaction.

absorption, and physical design methods are applicable, but the resistance to mass transfer in the liquid phase is lower due to the reaction. On the other extreme, a slow reaction occurs in the bulk of the liquid, and its rate has little dependence on the resistance to diffusion in either the gas or the liquid films. Here the mass-transfer mechanism is that of chemical reaction, and holdup in the bulk liquid is the determining factor.

The Hatta number is a dimensionless group used to characterize the importance of the speed of reaction relative to the diffusion rate.

$$N_{\text{Ha}} = \frac{\sqrt{D_A k_2 C_{B0}}}{k_L^0} \tag{14-61}$$

As the Hatta number increases, the effective liquid-phase mass-transfer coefficient increases. Figure 14-13, which was first developed by Van Krevelen and Hoftyzer [*Rec. Trav. Chim.* **67:** 563 (1948)] and later refined by Perry and Pigford and by Brian et al. [*AIChE J.* **7:** 226 (1961)], shows how the enhancement (defined as the ratio of the effective liquid-phase mass-transfer coefficient to its physical equivalent $\phi = k_L/k_L^0$) increases with N_{Ha} for a second-order, irreversible reaction of the kind defined by Eqs. (14-60) and (14-61). The various curves in Fig. 14-13 were developed based on penetration theory, and they depend on the parameter $\phi_\infty - 1$, which is related to the diffusion coefficients and reaction coefficients, as shown below.

$$\phi_\infty = \sqrt{\frac{D_A}{D_B}} + \sqrt{\frac{D_A}{D_B}}\left(\frac{C_B}{C_A b}\right) \tag{14-62}$$

For design purposes, the entire set of curves in Fig. 14-13 may be represented by the following two equations:

For, $N_{\text{Ha}} \geq 2$:

$$k_L/k_L^0 = 1 + (\phi_\infty - 1)\{1 - \exp[-(N_{\text{Ha}} - 1)/(\phi_\infty - 1)]\} \tag{14-63}$$

For, $N_{\text{Ha}} \leq 2$:

$$k_L/k_L^0 = 1 + (\phi_\infty - 1)\{1 - \exp[-(\phi_\infty - 1)^{-1}]\}\exp[1 - 2/N_{\text{Ha}}] \tag{14-64}$$

Equation (14-64) was originally reported by Porter [*Trans. Inst. Chem. Eng.* **44:** T25 (1966)], and Eq. (14-64) was derived by William M. Edwards, the author of the sixth edition of this handbook.

The Van Krevelen-Hoftyzer (Fig. 14-13) relationship was tested by Nijsing et al. [*Chem. Eng. Sci.* **10:** 88 (1959)] for the second-order system in which CO_2 reacts with either NaOH or KOH solutions. Nijsing's results are shown in Fig. 14-14 and can be seen to be in excellent agreement with the second-order-reaction theory. Indeed, these experimental data are well described by Eqs. (14-62) and (14-63) when values of $b = 2$ and $D_A/D_B = 0.64$ are employed in the equations.

Applicability of Physical Design Methods Physical design models such as the classical isothermal design method or the classical adiabatic design method may be applicable for systems in which chemical reactions are either extremely fast or extremely slow, or when chemical equilibrium is achieved between the gas and liquid phases.

If the chemical reaction is extremely fast and irreversible, the rate of absorption may in some cases be completely governed by gas-phase resistance. For practical design purposes, one may assume, for example, that this gas-phase mass-transfer-limited condition will exist when the ratio y_i/y is less than 0.05 everywhere in the apparatus.

From the basic mass-transfer flux relationship for species *A* (Sec. 5),

$$N_A = k_G(y - y_i) = k_L(x_i - x) \tag{14-65}$$

one can readily show that this condition on y_i/y requires that the ratio x/x_i be negligibly small (i.e., a fast reaction) and that the ratio $mk_G/k_L = mk_G/k_L^0\phi$ be less than 0.05 everywhere in the apparatus. The ratio $mk_G/k_L^0\phi$ will be small if the equilibrium backpressure of the solute over the liquid is small (i.e., small *m* or high reactant solubility), or the reaction enhancement factor $\phi = k_L/k_L^0$ is very large, or both. The reaction enhancement factor ϕ will be large for all extremely fast pseudo-first-order reactions and will be large for extremely fast second-order irreversible reaction systems in which there is sufficiently large excess of liquid reagent.

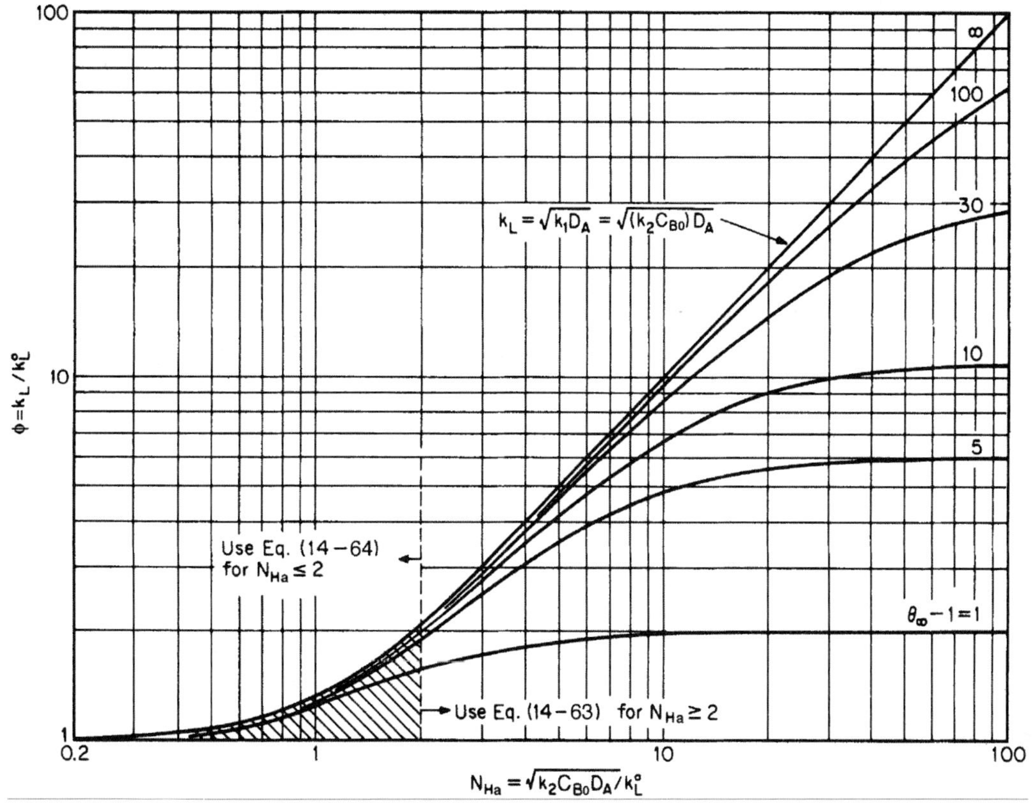

FIG. 14-13 Influence of irreversible chemical reactions on the liquid-phase mass-transfer coefficient k_L. [*Adapted from Van Krevelen and Hoftyzer*, Rec. Trav. Chim. *67: 563 (1948).*]

Figure 14-12, case (ii), illustrates the gas-film and liquid-film concentration profiles one might find in an extremely fast (gas-phase mass-transfer-limited), second-order irreversible reaction system. The solid curve for reagent B represents the case in which there is a large excess of bulk liquid reagent B^0. Figure 14-12, case (iv), represents the case in which the bulk concentration B^0 is not large enough to prevent the depletion of B near the liquid interface.

Whenever these conditions on the ratio y_i/y apply, the design can be based on the physical rate coefficient k_G or on the height of one gas-phase mass-transfer unit H_G. The gas-phase mass-transfer-limited condition is approximately valid for the following systems: absorption of NH_3 into water or acidic solutions, absorption of H_2O into concentrated sulfuric acid, absorption of SO_2 into alkali solutions, absorption of H_2S from a gas stream into a strong alkali solution, absorption of HCl into water or alkaline solutions, or absorption of Cl_2 into strong alkali solutions.

When the liquid-phase reactions are extremely slow, the gas-phase resistance can be neglected, and one can assume that the rate of reaction has a

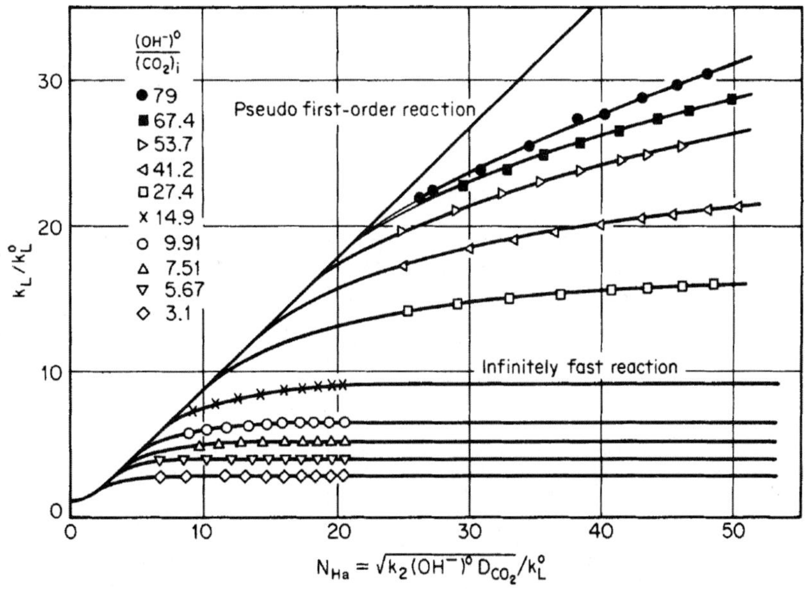

FIG. 14-14 Experimental values of k_L/k_L^0 for absorption of CO_2 into NaOH solutions at 20°C. [*Data of Nijsing et al.*, Chem. Eng. *Sci. 10: 88 (1959).*]

predominant effect upon the rate of absorption. In this case the differential rate of transfer is given by the equation

$$dn_A = R_A f_H S\, dh = (k_L^0 a/\rho_L)(c_i - c) S\, dh \qquad (14\text{-}66)$$

where n_A = rate of solute transfer, R_A = volumetric reaction rate (function of c and T), f_H = fractional liquid volume holdup in tower or apparatus, S = tower cross-sectional area, h = vertical distance, k_L^0 = liquid-phase mass-transfer coefficient for pure physical absorption, a = effective interfacial mass-transfer area per unit volume of tower or apparatus, ρ_L = average molar density of liquid phase, c_i = solute concentration in liquid at gas–liquid interface, and c = solute concentration in bulk liquid.

Although the right side of Eq. (14-66) remains valid even when chemical reactions are extremely slow, the mass-transfer driving force may become increasingly small, until finally $c \approx c_i$. For extremely slow first-order irreversible reactions, the following rate expression can be derived from Eq. (14-66):

$$R_A = k_1 c = k_1 c_i/(1 + k_1 \rho_L f_H/k_L^0 a) \qquad (14\text{-}67)$$

where k_1 = first-order reaction rate coefficient.

For dilute systems in countercurrent absorption towers in which the equilibrium curve is a straight line (i.e., $y_i = mx_i$), the differential relation of Eq. (14-66) is formulated as

$$dn_A = -G_M S\, dy = k_1 c f_H S\, dh \qquad (14\text{-}68)$$

where G_M = molar gas-phase mass velocity and y = gas-phase solute mole fraction.

Substitution of Eq. (14-67) into Eq. (14-68) and integration lead to the following relation for an extremely slow first-order reaction in an absorption tower:

$$y_2 = y_1 \exp\left[-\frac{k_1 \rho_L f_H h_T/(mG_m)}{1 + k_1 \rho_L f_H/(k_L^0 a)}\right] \qquad (14\text{-}69)$$

In Eq. (14-69), subscripts 1 and 2 refer to the bottom and top of the tower, respectively.

As discussed above, the Hatta number N_{Ha} usually is used as the criterion for determining whether a reaction can be considered extremely slow. A reasonable criterion for slow reactions is

$$N_{Ha} = \sqrt{k_1 D_A}/k_L^0 \le 0.3 \qquad (14\text{-}70)$$

where D_A = liquid-phase diffusion coefficient of the solute in the solvent. Figure 14-12, cases (vii) and (viii), illustrates the concentration profiles in the gas and liquid films for the case of an extremely slow chemical reaction.

Note that when the second term in the denominator of the exponential in Eq. (14-69) is very small, the liquid holdup in the tower can have a significant influence upon the rate of absorption if an extremely slow chemical reaction is involved.

When chemical equilibrium is achieved quickly throughout the liquid phase, the problem becomes one of properly defining the physical and chemical equilibria for the system. It is sometimes possible to design a tray-type absorber by assuming chemical equilibrium relationships in conjunction with a stage efficiency factor, as is done in distillation calculations. Rivas and Prausnitz [*AIChE J.* **25:** 975 (1979)] have presented an excellent discussion and example of the correct procedures to be followed for systems involving chemical equilibria.

Traditional Design Method The traditional procedure for designing packed-tower gas absorption systems involving chemical reactions makes use of overall mass-transfer coefficients as defined by the equation

$$K_G a = n_A/(h_T S p_T \Delta y_{1m}^0) \qquad (14\text{-}71)$$

where $K_G a$ = overall volumetric mass-transfer coefficient, n_A = rate of solute transfer from the gas to the liquid phase, h_T = total height of tower packing, S = tower cross-sectional area, p_T = total system pressure, and Δy_{1m}^0 is defined by the equation

$$\Delta y_{1m}^0 = \frac{(y-y^0)_1 - (y-y^0)_2}{\ln\left[\dfrac{(y-y^0)_1}{(y-y^0)_2}\right]} \qquad (14\text{-}72)$$

in which subscripts 1 and 2 refer to the bottom and top of the absorption tower, respectively, y = mole-fraction solute in the gas phase, and y^0 = gas-phase solute mole fraction in equilibrium with bulk-liquid-phase solute concentration x. When the equilibrium line is straight, $y^\circ = mx$.

The traditional design method normally makes use of overall $K_G a$ values even when resistance to transfer lies predominantly in the liquid phase. For example, the CO_2-NaOH system that is most commonly used for comparing $K_G a$ values of various tower packings is a liquid-phase-controlled system. When the liquid phase is controlling, extrapolation to different concentration ranges or operating conditions is not recommended since changes in the reaction mechanism can cause k_L to vary unexpectedly, and the overall $K_G a$ do not capture such effects.

Overall $K_G a$ data may be obtained from tower-packing vendors for many of the established commercial gas absorption processes. Such data often are based either on tests in large-diameter test units or on actual commercial operating data. Since application to untried operating conditions is not recommended, the preferred procedure for applying the traditional design method is equivalent to duplicating a previously successful commercial installation. When this is not possible, a commercial demonstration at the new operating conditions may be required, or else one could consider using some of the more rigorous methods described later.

While the traditional design method is reported here because it has been used extensively in the past, it should be used with extreme caution. In addition to the lack of an explicit liquid-phase resistance term, the method has other limitations. Equation (14-71) assumes that the system is dilute ($y_{BM} \approx 1$) and that the operating and equilibrium lines are straight, which are weak assumptions for reacting systems. Also, Eq. (14-65) is strictly valid only for the temperature and solute partial pressure at which the original test was done, even though the total pressure p_T appears in the denominator.

In using Eq. (14-71), therefore, it should be understood that the numerical values of $K_G a$ will be a complex function of pressure, temperature, the type and size of packing employed, the liquid and gas mass flow rates, and the system composition (e.g., the degree of conversion of the liquid-phase reactant).

Figure 14-15 illustrates the influence of system composition and degree of reactant conversion upon the numerical values of $K_G a$ for the absorption of CO_2 into sodium hydroxide at constant conditions of temperature, pressure, and type of packing. An excellent experimental study of the influence of operating variables upon overall $K_G a$ values is that of Field et al. (*Pilot-Plant Studies of the Hot Carbonate Process for Removing Carbon Dioxide and Hydrogen Sulfide*, U.S. Bureau of Mines Bulletin 597, 1962).

Table 14-2 illustrates the observed variations in $K_G a$ values for different packing types and sizes for the CO_2-NaOH system at a 25 percent reactant conversion for two different liquid flow rates. The lower rate of 2.7 kg/(s · m^2)

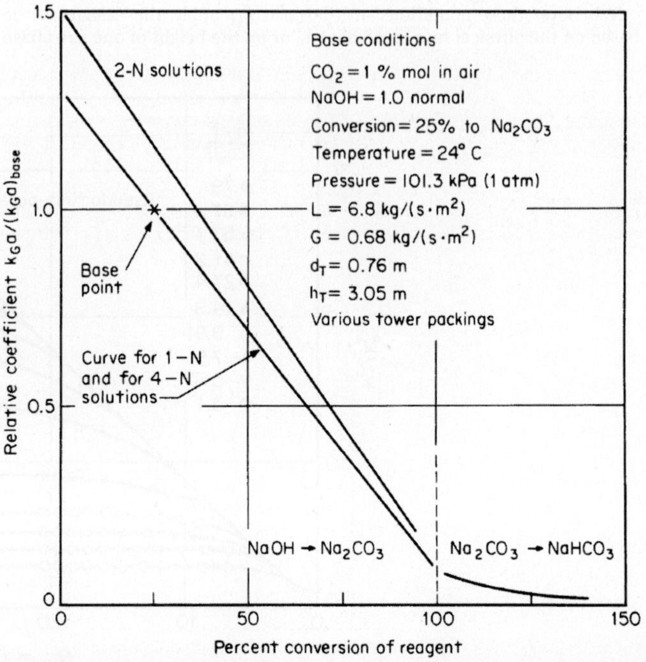

FIG. 14-15 Effects of reagent-concentration and reagent-conversion levels upon the relative values of $K_G a$ in the CO_2-NaOH-H_2O system. [*Adapted from Eckert et al.,* Ind. Eng. Chem. *59(2): 41 (1967).*]

TABLE 14-2 Typical Effects of Packing Type, Size, and Liquid Rate on K_Ga in a Chemically Reacting System, K_Ga, kmol/(h·m³)

	L = 2.7 kg/(s·m²)				L = 13.6 kg/(s·m²)			
Packing size, mm	25	38	50	75–90	25	38	50	75–90
Berl-saddle ceramic	30	24	21		45	38	32	
Raschig-ring ceramic	27	24	21		42	34	30	
Raschig-ring metal	29	24	19		45	35	27	
Pall-ring plastic	29	27	26*	16	45	42	38*	24
Pall-ring metal	37	32	27	21*	56	51	43	27*
Intalox-saddle ceramic	34	27	22	16*	56	43	34	26*
Super-Intalox ceramic	37*		26*		59*		40*	
Intalox-saddle plastic	40*		24*	16*	56*		37*	26*
Intalox-saddle metal	43*	35*	30*	24*	66*	58*	48*	37*
Hy-Pak metal	35	32*	27*	18*	54	50*	42*	27*

Data courtesy of the Norton Company.

Operating conditions: CO_2, 1 percent mole in air; NaOH, 4 percent weight (1 normal); 25 percent conversion to sodium carbonate; temperature, 24°C (75°F); pressure, 98.6 kPa (0.97 atm); gas rate = 0.68 kg/(s·m²) = 0.59 m/s = 500 lb/(h·ft²) = 1.92 ft/s except for values with asterisks, which were run at 1.22 kg/(s·m²) = 1.05 m/s = 900 lb/(h·ft²) = 3.46 ft/s superficial velocity; packed height, 3.05 m (10 ft); tower diameter, 0.76 m (2.5 ft). To convert table values to units of (lb·mol)/(h·ft³), multiply by 0.0624.

or 2000 lb/(h·ft²) is equivalent to 4 U.S. gal/(min·ft²) and is typical of the liquid rates employed in fume scrubbers. The higher rate of 13.6 kg/(s·m²) or 10,000 lb/(h·ft²) is equivalent to 20 U.S. gal/(min·ft²) and is more typical of absorption towers such as those used in CO_2 removal systems, for example. We also note that two gas velocities are represented in the table, corresponding to superficial velocities of 0.59 and 1.05 m/s (1.94 and 3.44 ft/s).

Table 14-3 presents a typical range of K_Ga values for chemically reacting systems. The first two entries in the table represent systems that can be designed by the use of purely physical design methods because they are completely gas-phase mass transfer limited. To ensure a negligible liquid-phase resistance in these two tests, the HCl was absorbed into a solution maintained at less than 8 wt% HCl, and the NH_3 was absorbed into a water solution maintained below pH 7 by the addition of acid. The last two entries in Table 14-3 represent liquid-phase mass-transfer-limited systems.

Scaling Up from Laboratory Data Laboratory experimental techniques offer an efficient and cost-effective route to develop commercial absorption designs. For example, Ouwerkerk (*Hydrocarbon Process.*, April 1978, pp. 89–94) revealed that both laboratory and small-scale pilot plant data were employed as the basis for the design of an 8.5-m (28-ft) diameter commercial Shell Claus off-gas treating (SCOT) tray-type absorber. Ouwerkerk claimed that the cost of developing comprehensive design procedures can be minimized, especially in the development of a new process, by the use of these modern techniques.

In a 1966 paper that is considered a classic, Dankwerts and Gillham [*Trans. Inst. Chem. Eng.* **44:** T42 (1966)] showed that data taken in a small stirred-cell laboratory apparatus could be used in the design of a packed-tower absorber when chemical reactions are involved. They showed that if the packed-tower mass-transfer coefficient in the absence of reaction (k_L^0) can be reproduced in the laboratory unit, then the rate of absorption in the laboratory apparatus will respond to chemical reactions in the same way as in the packed column, even though the means of agitating the liquid in the two systems may be quite different.

According to this method, it is not necessary to investigate the kinetics of the chemical reactions in detail, nor is it necessary to determine the solubilities or diffusivities of the various reactants in their unreacted forms. To use the method for scaling up, one must independently obtain data on the values of the interfacial area per unit volume a and the physical mass-transfer coefficient k_L^0 for the commercial packed tower. Once these data have been measured and tabulated, they can be used directly for scaling up the experimental laboratory data for any new chemically reacting system.

Dankwerts and Gillham did not investigate the influence of the gas-phase resistance in their study (for some processes, gas-phase resistance may be neglected). However, in 1975 Dankwerts and Alper [*Trans. Inst. Chem. Eng.* **53:** T42 (1975)] showed that by placing a stirrer in the gas space of the stirred-cell laboratory absorber, the gas-phase mass-transfer coefficient k_G in the laboratory unit could be made identical to that in a packed-tower absorber. When this was done, laboratory data for chemically reacting systems having a significant gas-side resistance could successfully be scaled up to predict the performance of a commercial packed-tower absorber.

If it is assumed that the values for k_G, k_L^0, and a have been measured for the commercial tower packing to be used, the procedure for using the laboratory stirred-cell reactor is as follows:

1. The gas-phase and liquid-phase stirring rates are adjusted to produce the same values of k_G and k_L^0 as will exist in the commercial tower.
2. For the reaction system under consideration, experiments are made at a series of bulk-liquid and bulk-gas compositions representing the compositions to be expected at different levels in the commercial absorber (on the basis of material balance).
3. The ratios of $r_A(c_i, B^0)$ are measured at each pair of gas and liquid compositions.

For the dilute-gas systems, one form of the equation to be solved in conjunction with these experiments is

$$h_T = \frac{G_M}{a} \int_{y2}^{y1} \frac{dy}{r_A} \qquad (14\text{-}73)$$

where h_T = height of commercial tower packing, G_M = molar gas-phase mass velocity, a = effective mass-transfer area per unit volume in the commercial

TABLE 14-3 Typical K_Ga Values for Various Chemically Reacting Systems, kmol/(h·m³)

Gas-phase reactant	Liquid-phase reactant	K_Ga	Special conditions
HCl	H_2O	353	Gas-phase limited
NH_3	H_2O	337	Gas-phase limited
Cl_2	NaOH	272	8% weight solution
SO_2	Na_2CO_3	224	11% weight solution
HF	H_2O	152	
Br_2	NaOH	131	5% weight solution
HCN	H_2O	114	
HCHO	H_2O	114	Physical absorption
HBr	H_2O	98	
H_2S	NaOH	96	4% weight solution
SO_2	H_2O	59	
CO_2	NaOH	38	4% weight solution
Cl_2	H_2O	8	Liquid-phase limited

Data courtesy of the Norton Company.

Operating conditions (see text): 38-mm ceramic Intalox saddles; solute gases, 0.5–1.0 percent mole; reagent conversions = 33 percent; pressure, 101 kPa (1 atm); temperature, 16–24°C; gas rate = 1.3 kg/(s·m²) = 1.1 m/s; liquid rates = 3.4 to 6.8 kg/(s·m²); packed height, 3.05 m; tower diameter, 0.76 m. Multiply table values by 0.0624 to convert to (lb·mol)/(h·ft³).

tower, y = mole fraction solute in the gas phase, and r_A = experimentally determined rate of absorption per unit of exposed interfacial area.

By using the series of experimentally measured rates of absorption, Eq. (14-73) can be integrated numerically to determine the height of packing required in the commercial tower.

A number of different types of experimental laboratory units could be used to develop design data for chemically reacting systems. Charpentier [*ACS Symp. Ser.* **72:** 223–261 (1978)] has summarized the state of the art with respect to methods of scaling up laboratory data and has tabulated typical values of the mass-transfer coefficients, interfacial areas, and contact times to be found in various commercial gas absorbers, as well as in currently available laboratory units.

The laboratory units that have been used to date for these experiments were designed to operate at a total system pressure of about 101 kPa (1 atm) and at near-ambient temperatures. In practical situations, it may become necessary to design a laboratory absorption unit that can be operated either under vacuum or at elevated pressure and over a range of temperatures in order to apply the Dankwerts method.

It would be desirable to reinterpret existing data for commercial tower packings to extract the individual values of the interfacial area a and the mass-transfer coefficients k_G and k_L^0 to facilitate a more general usage of methods for scaling up from laboratory experiments. Some progress has already been made, as described later in this section. In the absence of such data, it is necessary to operate a pilot plant or a commercial absorber to obtain k_G, k_L^0, and a as described by Ouwerkerk (*Hydrocarbon Process.*, April 1978, pp. 89–94).

Modern techniques use rigorous computer-based modeling methods to extract fundamental parameters from laboratory-scale measurements and then apply them to the design of commercial absorption towers. These techniques are covered next.

Rigorous Computer-Based Absorber Design While the techniques described earlier in this section are very useful for understanding the key effects in commercial absorbers, current design methods used in industrial practice for chemically reactive systems are increasingly based on rigorous computerized methods, and these are commercially available from software vendors. The advantages of the rigorous methods are as follows: (1) Approximations do not have to be made for special cases (e.g., fast chemical reactions or mass-transfer resistance dominated by the gas or liquid phase), and all effects can be simultaneously modeled. (2) Fundamental quantities such as kinetic parameters and mass-transfer coefficients can be extracted from laboratory and pilot-scale equipment, and applied to commercial absorber towers. (3) Integrated models can be developed for an entire absorption process flow sheet (e.g., the absorber-stripper system with heat integration presented in Fig. 14-3), and consequently the entire system may be optimized.

Computer programs for chemically reacting systems are available from several vendors. The specific approaches used to model the chemically reacting absorption system are slightly different among the different vendors. The general approach used and the benefits obtained are identified by considering a specific example of broad current interest: removal of CO_2 from flue gases discharged by a power plant using aqueous monoethanolamine (MEA), as presented by Zhang et al. [*Ind. Eng. Chem. Res.* **48:** 9233 (2009)].

The development and application of a rigorous model for a chemically reactive system typically involves five steps: (1) development of a thermodynamic model to describe the physical and chemical equilibrium; (2) adoption and use of a modeling framework to describe the mass transfer and chemical reactions; (3) parameterization of the mass-transfer and kinetic models based on laboratory, pilot-plant, or commercial-plant data; (4) parameterization of the hydraulic models to estimate operating features such as pressure drop and holdup; and (5) use of the integrated model to optimize the process and perform equipment design.

Development of Thermodynamic Model for Physical and Chemical Equilibrium The first and perhaps most important step in the development of the thermodynamic model is the *speciation*, or representation of the set of chemical reactions. For CO_2 absorption in aqueous MEA solutions, the set of reactions is

$$CO_2 + MEA + H_2O \leftrightarrow MEACOO^- + H_3O^+ \quad (14\text{-}74a)$$

$$CO_2 + OH^- \leftrightarrow HCO_3^- \quad (14\text{-}74b)$$

$$HCO_3^- + H_2O \leftrightarrow CO_3^{2-} + H_3O^+ \quad (14\text{-}74c)$$

$$MEAH^+ + H_2O \leftrightarrow MEA + H_3O^+ \quad (14\text{-}74d)$$

$$2H_2O \leftrightarrow H_3O^+ + OH^- \quad (14\text{-}74e)$$

In addition, a model is needed that can describe the nonideality of a system containing molecular and ionic species. Zhang et al. (2009) adopted the model developed by Hilliard in his Ph.D. thesis [chemical engineering, University of Texas (2008)]. The combination of the speciation set of reactions [Eqs. (14-74a) to (14-74e)] and the nonideality model is capable of representing the solubility data, such as those presented in Figs. 14-1 and 14-2, to good accuracy. In addition, the model accurately and correctly represents the actual species present in the aqueous phase, which is important for faithful description of the chemical kinetics and species mass transfer across the interface. Finally, the thermodynamic model facilitates accurate modeling of the heat effects, such as those discussed in Example 14-6. The Gibbs-Helmholtz equation provides the rigorous relationship between vapor-liquid equilibrium and calorimetric data, Mathias and O'Connell [*I&EC Res.*, **51:** 5090 (2012)].

Wang et al. [*J. Geochem. Explor.* **101:** 112 (2009)], Chen and Song [*AIChE J.* **50:** 1928 (2004)], and Zhang, Que, and Chen [*Fluid Phase Equil.* **311:** 67 (2011)] have provided comprehensive discussions of speciation and electrolyte thermodynamic models.

Adoption and Use of Modeling Framework The rate of species generation and diffusion by chemical reaction can be described by film theory, penetration theory, or a combination of the two. The most popular description is in terms of a two-film theory, which is diagrammed in Fig. 14-16 for absorption. Accordingly, there exists a stable interface separating the gas and the liquid. A certain distance from the interface, large fluid motions exist, and these distribute the material rapidly and equally so that no concentration gradients develop; these regions are referred to as the "bulk" vapor and liquid in Fig. 14-16. Next to the interface, however, there are regions in which the fluid motion is slow; in these regions, termed *films* and denoted by lengths δ_V and δ_L, material is transferred by diffusion alone. At the gas–liquid interface, material is transferred instantaneously, so the gas and liquid are in physical equilibrium at the interface. This means that y_{AI} and x_{AI} in Fig. 14-16 satisfy the equality of fugacity relationships. The rate of diffusion in absorption is therefore the rate of diffusion in the gas and liquid films adjacent to the interface. The model framework is completed by including terms for species generation (chemical equilibrium *and* chemical kinetics) in the gas and liquid film and bulk regions. Taylor, Krishna, and Kooijman (*Chem. Eng. Progress*, July 2003, p. 28) have provided an excellent discussion of rate-based models; these authors emphasize that the diffusion flux for multicomponent systems must be based on the Maxwell-Stefan approach. The book by Taylor and Krishna (*Multicomponent Mass Transfer*, Wiley, New York, 1993) provides a detailed discussion of the Maxwell-Stefan approach.

Parameterization of Mass-Transfer, Hydraulic, and Kinetic Models The mass-transfer and chemical kinetic rates required in the rigorous model are typically obtained from the literature, but they must be carefully evaluated, and fine-tuning through pilot-plant and commercial data is highly recommended.

Mass-transfer coefficient models for the vapor and liquid coefficients are of the general form

$$k_{i,j}^L = a\rho_L f(D_{i,j}^m, \mu_L, \rho_L, a, \text{internal characteristics}) \quad (14\text{-}75a)$$

$$k_{i,j}^V = a\rho_V f(D_{i,j}^m, \mu_V, \rho_V, a, \text{internal characteristics}) \quad (14\text{-}75b)$$

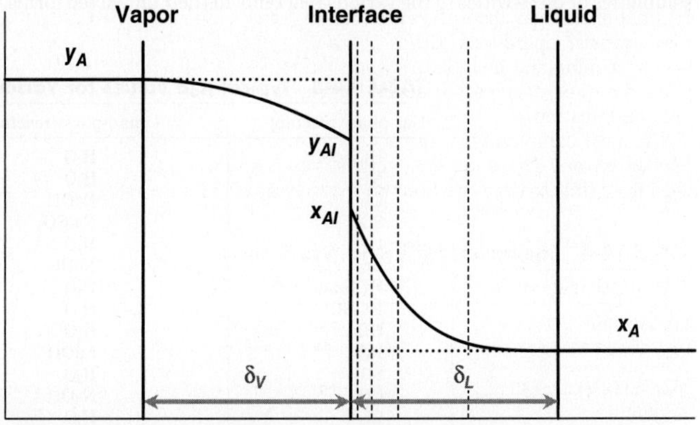

FIG. 14-16 Concentration profiles in the vapor and liquid phases near the interface. The two regions in the extreme left and right represent the bulk vapor and liquid, with mole fractions y_A and x_A, respectively. The two inside regions represent the vapor and liquid films, with thicknesses δ_V and δ_L, respectively. The liquid film has been discretized into five segments (vertical dashed lines) with thinner (finer) segments close to the interface.

where a = effective interfacial area per unit volume, $D^m_{i,j}$ are the Stefan-Maxwell diffusion coefficients, P = pressure, ρ = molar density, and μ = viscosity. The functions in Eqs. (14-75*a*) and (14-75*b*) are correlations that depend on the column internals. Popular correlations in the literature are those by Onda at al. [*J. Chem.. Eng. Jap.* **1:** 56 (1968)] for random packing, Bravo and Fair [*Ind. Eng. Chem. Proc. Des. Dev.* **21:** 162 (1982)] for structured packing, Chan and Fair [*Ind. Eng. Chem. Proc. Des. Dev.* **23:** 814 (1984)] for sieve trays, Scheffe and Weiland [*Ind. Eng. Chem. Res.* **26:** 228 (1987)] for valve trays, and Hughmark [*AIChE J.* **17:** 1295 (1971)] for bubble-cap trays. These references provide correlations of heat transfer, mass transfer, interfacial area, liquid holdup, and pressure drop, among other factors.

Kinetic models are usually developed by replacing a subset of the speciation reactions by kinetically-limited reversible reactions. For example, Zhang et al. (2009) replaced equilibrium reactions (14-74*a*) and (14-74*b*) with kinetically reversible reactions, and they retained the remaining three reactions as very fast and hence effectively at equilibrium. The kinetic constants were tuned using kinetic data, here from Aboudheir [Ph.D. thesis, chemical engineering, University of Regina (2002)]. It is very important that the kinetic models asymptote to the equilibrium limit for large residence times (e.g., very tall columns), as has been demonstrated by Mathias and Gilmartin [*Energy Procedia* **63:** 1171 (2014)].

There are many details that require attention in the rigorous models. The concentration variation in the liquid film is highly nonlinear, and hence the liquid film must be discretized, with more clustered segments in the vicinity of the interface, as is depicted by the vertical dashed lines in Fig. 14-16. The segment height should be chosen to be small enough so that the model is effectively simulating a continuous distribution. In both discretization and segment height, tests should be done to confirm that increased fine graining does not change the results—that is, that the asymptotic limit has been reached. Another choice that must be made is the *flow model*, which determines the bulk properties based on those of adjacent segments; for example, RateSep from AspenTech offers four flow models (*Mixed, CounterCurrent, Vplug*, and *Vplug-Pavg*), and these flow models have been described and analyzed by Zhang et al. (2009).

It is highly recommended that the mass-transfer correlations be tested and improved by using laboratory, pilot-plant, or commercial data for the specific application. Zhang et al. (2009) show that the model they developed accurately describes the University of Texas pilot plant data for measured results such as fraction of CO_2 removal and temperature profiles.

There are many examples in the technical literature where detailed process models have provided good results similar to those of Zhang et al. (2009). However, users of commercial software should perform due diligence to ensure that the models provide reliable and accurate results for the particular application of interest to them. Luo et al. [*Energy Procedia* **1:** 1249 (2009)] performed a systematic test of several commercial and in-house codes against 16 sets of data from four different pilot plants. Their key conclusions are as follows: "Basically all the simulators are capable of giving reasonable predictions on overall performance, i.e., CO_2 absorption rate. The reboiler duties are less well predicted, as well as concentration and temperature profiles. For the reboiler temperature there is very much scatter."

Commercial software generally provides for correction factors to adjust generalized correlations to the particular application. Users of commercial codes are again urged to perform due diligence and to apply reasonable correction factors (say, within ± 20%) when reliable pilot-plant or commercial data are available.

Deployment of Rigorous Model for Process Optimization and Equipment Design It is usually valuable to develop an integrated model for the absorption-stripping system, including the cross exchanger (see Fig. 14-3). Øi and Kvam (*Energy Procedia* **63:** 1186 (2014)] used various models to simulate entire flow sheets. They found that even though the models gave different absolute results, they predicted the same trends when flow sheet improvements (e.g., vapor compression or split flows) were adopted.

In this subsection, we have used the example of CO_2 removal from flue gases using aqueous alkanolamines to demonstrate the development and application of a rigorous model for a chemically reactive system. Modern software enables rigorous description of complex chemically reactive systems, but it is very important to carefully evaluate the models and to tune them using experimental data. A favorable result is that integrated models may be relied upon to predict trends accurately, and hence their best value may be to efficiently evaluate process improvements.

Use of Literature for Specific Systems A large body of experimental data obtained in bench-scale laboratory units and in small-diameter packed towers has been published since the early 1940s. One might wish to consider using such data for a particular chemically reacting system as the basis for scaling up to a commercial design. Extreme caution is recommended in interpreting such data for the purpose of developing commercial designs because extrapolating the data can lead to serious errors. Extrapolation to temperatures, pressures, or liquid-phase reagent conversions different from those that were used by the original investigators definitely should be regarded with caution. As noted earlier in this subsection, rigorous models are recommended to perform these extrapolations.

The General References at the beginning of this subsection can be an excellent source of information on specific chemically reacting systems. *Gas-Liquid Reactions* by Dankwerts (McGraw-Hill, New York, 1970) contains a tabulation of references to specific chemically reactive systems. *Gas Treating with Chemical Solvents* by Astarita et al. (Wiley, New York, 1983) deals with the absorption of acid gases and includes an extensive listing of patents. *Gas Purification* by Kohl and Nielsen (Gulf Publishing, Houston, 1997) provides a practical description of techniques and processes in widespread use and typically also sufficient design and operating data for specific applications.

In searching for data on a particular system, we recommend a computerized search of *Chemical Abstracts, Engineering Index*, and *National Technical Information Service* (NTIS) databases. Modern search engines such as *Google Scholar* will also rapidly produce much potentially valuable information.

The experimental data for the system CO_2-NaOH-Na_2CO_3 are unusually comprehensive and well known as the result of the work of many experimenters. A serious study of the data and theory for this system therefore is recommended as the basis for developing a good understanding of the kind and quality of experimental information needed for design purposes.

EQUIPMENT FOR DISTILLATION AND GAS ABSORPTION: TRAY COLUMNS

Distillation and gas absorption are the prime and most common gas–liquid mass-transfer operations. Other operations that are often performed in similar equipment include stripping (often considered part of distillation), direct-contact heat transfer, flashing, washing, humidification, and dehumidification.

The most common types of contactors by far used for these are tray and packed towers. These are the focus of this subsection. Other contactors used from time to time and their applications are listed in Table 14-4.

In this subsection, the terms *gas* and *vapor* are used interchangeably. *Vapor* is more precise for distillation, where the gas phase is at equilibrium. Also, the terms *tower* and *column* are used interchangeably.

A cross-flow tray (Fig. 14-17) consists of the *bubbling area* and the *downcomer*. Liquid descending the downcomer from the tray above enters the bubbling area. Here, the liquid contacts gas ascending through the tray perforations, forming froth or spray. An *outlet weir* on the downstream side of the bubbling area helps maintain liquid level on the tray. Froth

TABLE 14-4 Equipment for Liquid–Gas Systems

Equipment designation	Mode of flow	Gross mechanism	Continuous phase	Primary process applications
Tray column	Cross-flow, countercurrent	Integral	Liquid and/or gas	Distillation, absorption, stripping, DCHT, washing
Packed column	Countercurrent, cocurrent	Differential	Liquid and/or gas	Distillation, absorption, stripping, humidification, dehumidification, DCHT, washing
Wetted-wall (falling-film) column	Countercurrent, cocurrent	Differential	Liquid and/or gas	Distillation, absorption, stripping, evaporation
Spray chamber	Cocurrent, cross-flow, countercurrent	Differential	Gas	Absorption, stripping, humidification, dehumidification
Agitated vessel	Complete mixing	Integral	Liquid	Absorption
Line mixer	Cocurrent	Differential	Liquid or gas	Absorption, stripping

DCHT = direct contact heat transfer.

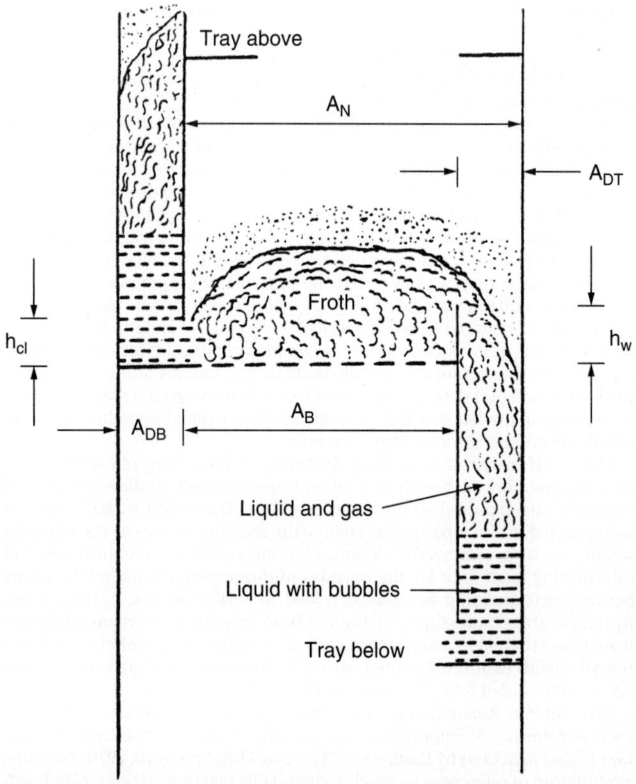

FIG. 14-17 Schematic of a tray operating in the froth regime. (*Based on H. Z. Kister,* Distillation Design, *copyright © 1992 by McGraw-Hill; reprinted by permission.*)

overflowing the weir enters the outlet downcomer. Here, gas disengages from the liquid, and the liquid descends to the tray below. The bubbling area can be fitted with many types of tray hardware. The most common types by far are:

Sieve trays (Fig. 14-18*a*) are perforated plates. The velocity of upflowing gas keeps the liquid from descending through the perforations (weeping). At low gas velocities, liquid weeps through the perforations, bypassing part of the tray and reducing tray efficiency. Because of this, sieve trays have relatively poor turndown.

Fixed valve trays (Fig. 14-18*b*) have the perforations covered by a fixed cover, often a section of the tray floor pushed up. Their performance is similar to that of sieve trays, although it has been argued (Hebert and Sandford, *Chem. Eng. Progr.*, May 2016, p. 34) that the horizontal vapor deflection in valve trays reduces the potential for entrainment.

Moving valve trays (Fig. 14-18*c*) have the perforations covered by movable disks (valves). Each valve rises as the gas velocity increases. The upper limit of the rise is controlled by restricting legs on the bottom of the valve (Fig. 14-18*c*) or by a cage structure around the valve. As the gas velocity falls, some valves close completely, preventing weeping. This gives the moving valve tray good turndown.

Table 14-5 is a general comparison of the three main tray types, assuming proper design, installation, and operation. Sieve and valve trays are comparable in capacity, efficiency, entrainment, and pressure drop. The turndown of moving valve trays is much better than that of sieve and fixed valve trays. Sieve trays are least expensive; valve trays cost only slightly more. Maintenance, fouling tendency, and effects of corrosion are least troublesome in fixed valve and sieve trays (provided the perforations or fixed valves are large enough) and most troublesome with moving valve trays.

Fixed valve and sieve trays prevail when fouling or corrosion is expected, or if turndown is unimportant. Moving valve trays prevail when high turndown is required. The energy saved, even during short turndown periods, usually justifies their small additional costs.

An excellent updated detailed comparison of the common tray types that reflects the current industry trend to shift to fixed valve trays was presented by Hebert and Sandford (*Chem. Eng. Progr.*, May 2016, p. 34). Caution is required in comparing tray types on capacity and efficiency because hole sizes and open area often affect these more than the tray type.

DEFINITIONS

Tray Area Definitions Some of these are illustrated in Fig. 14-17.

Total tower cross-section area A_T The inside cross-section area of the empty tower (without trays or downcomers).

Net area A_N *(also called free area)* The total tower cross-section area A_T minus the area at the top of the downcomer A_{DT}. The net area represents the smallest area available for vapor flow in the intertray spacing.

Bubbling area A_B *(also called active area)* The total tower cross-section area minus the sum of downcomer top area A_{DT}, downcomer seal area A_{DB}, and any other nonperforated areas on the tray. The bubbling area represents the area available for vapor flow just above the tray floor.

Hole area A_h The total area of the perforations on the tray. The hole area is the smallest area available for vapor passage on a sieve tray.

Slot area A_S The total (for all open valves) vertical curtain area through which vapor passes in a horizontal direction as it leaves the valves. It is a function of the narrowest opening of each valve and the number of valves that are open. The slot area is normally the smallest area available for vapor flow on a valve tray.

Open slot area A_{SO} The slot area when all valves are open.

Fractional hole area A_f The ratio of hole area to bubbling area (sieve trays) or slot area to bubbling area (valve trays).

Vapor and Liquid Load Definitions *F-factor* F This is the square root of the kinetic energy of the gas, defined by Eq. (14-76). The velocity in Eq. (14-76) is usually (not always) based on the tower cross-sectional area A_T, the net area A_N, or the bubbling area A_B. The user should beware of any data for which the area basis is not clearly specified.

$$F = u\sqrt{\rho_G} \tag{14-76}$$

C-factor C The *C*-factor, defined in Eq. (14-77), is the best gas load term for comparing capacities of systems of different physical properties. It has the same units as velocity (m/s or ft/s) and is directly related to droplet entrainment. As with the *F*-factor, the user should beware of any data for which the area basis is not clearly specified.

$$C = u\sqrt{\frac{\rho_G}{\rho_L - \rho_G}} \tag{14-77}$$

Weir load For trays (as distinct from downcomers), liquid load is normally defined as

$$Q_L = \frac{\text{volume of liquid}}{\text{length of outlet weir}} = \frac{Q}{L_w} \tag{14-78}$$

This definition describes the flux of liquid horizontally across the tray. Units often used are $m^3/(h \cdot m)$, $m^3/(s \cdot m)$, gpm/in, and gpm/ft.

Downcomer liquid load For downcomer design, the liquid load is usually defined as the liquid velocity at the downcomer entrance (m/s or ft/s):

$$Q_D = \frac{\text{volume of liquid}}{\text{downcomer entrance area}} = \frac{Q}{A_{DT}} \tag{14-79}$$

FLOW REGIMES ON TRAYS

Three main flow regimes exist on industrial distillation trays. These regimes may all occur on the same tray under different liquid and gas flow rates (Fig. 14-19). An excellent discussion of the fundamentals and modeling of these flow regimes was presented by Lockett (*Distillation Tray Fundamentals,* Cambridge University Press, Cambridge, 1986). An excellent overview of these as well as less common flow regimes was given by Prince (*PACE,* June 1975, p. 31; July 1975, p. 18).

*Froth regime (or mixed regime; Fig. 14-20*a*).* This is the most common operating regime in distillation practice. Each perforation bubbles vigorously. The bubbles circulate rapidly through the liquid, are of nonuniform sizes and shapes, and travel at varying velocities. The froth surface is mobile and not level and is generally covered by droplets. Bubbles are formed at the tray perforations and are swept away by the froth.

As gas load increases in the froth regime, jetting begins to replace bubbling in some holes. The fraction of holes that is jetting increases with gas velocity. When jetting becomes the dominant mechanism, the dispersion

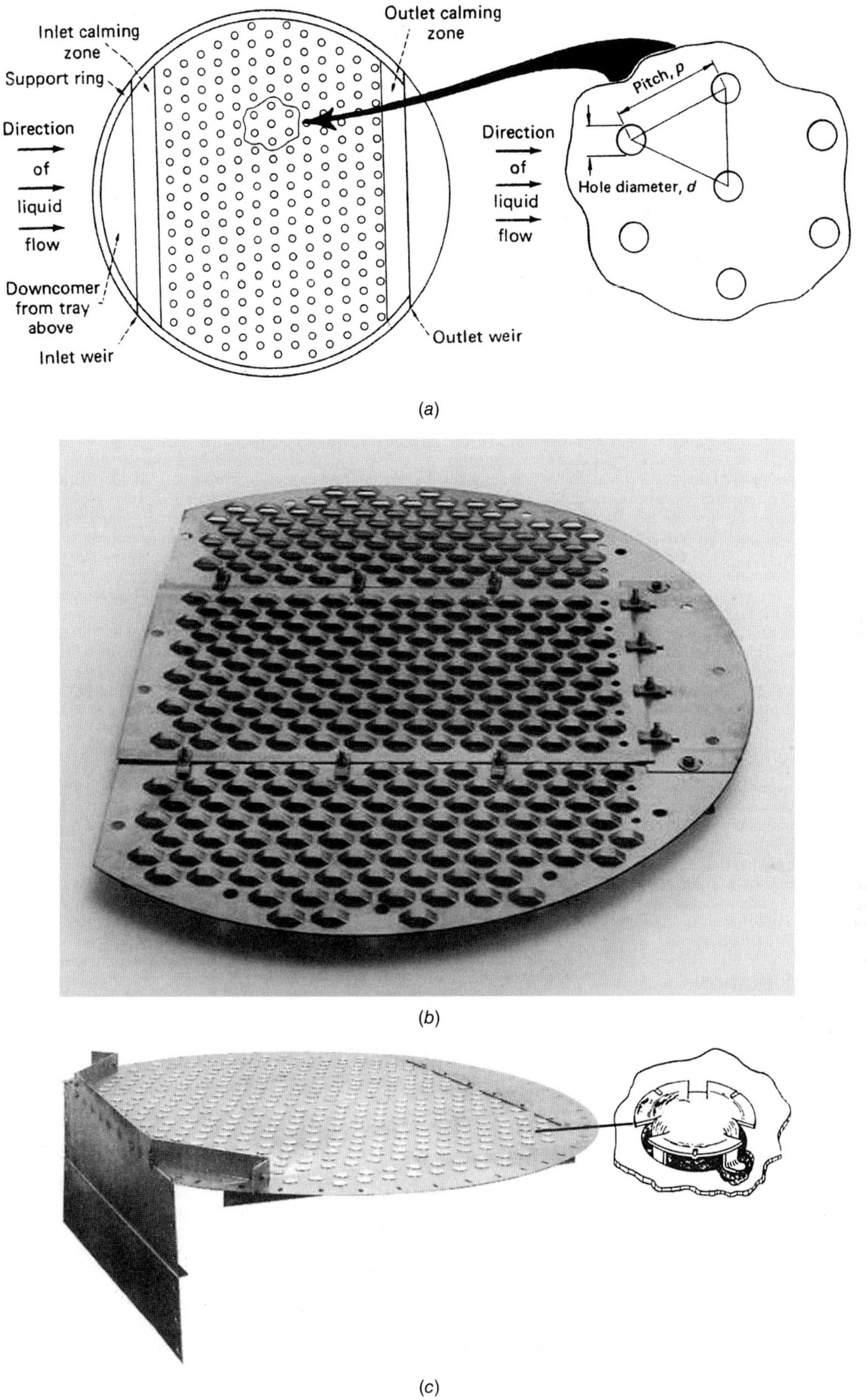

FIG. 14-18 Common tray types. (*a*) Sieve. (*b*) Fixed valve. (*c*) Moving valve with legs. [*Part a, from Henry Z. Kister,* Chem. Eng., *September 8, 1980; reprinted courtesy of* Chemical Engineering. *Part b, Courtesy of Sulzer Chemtech and Fractionation Research Inc. (FRI). Part c, courtesy of Koch-Glitsch LP.*]

TABLE 14-5 Comparison of the Common Tray Types

	Sieve trays	Fixed valve tray	Moving valve tray
Capacity	High	High	High to very high
Efficiency	High	High	High
Turndown	About 2:1. Not generally suitable for operation under variable loads	About 2.5:1. Not generally suitable for operation under variable loads	About 4:1 to 5:1. Some special designs achieve 8:1 or more
Entrainment	Moderate	Moderate	Moderate
Pressure drop	Moderate	Moderate	Slightly higher
Cost	Low	Low	About 20 percent higher
Maintenance	Low	Low	Moderate
Fouling tendency	Low to very low	Low to very low	Moderate
Effects of corrosion	Low	Very low	Moderate
Main applications	(1) Most columns when turndown is not critical (2) High fouling and corrosion potential	(1) Most columns when turndown is not critical (2) High fouling and corrosion potential	(1) Most columns (2) Services where turndown is important

changes from froth to spray. Prado et al. [*Chemical Engineering Progr.* **83**(3): 32 (1987)] showed that the transition from froth to spray takes place gradually as jetting replaces bubbling in 45 to 70 percent of the tray holes.

Emulsion regime (Fig. 14-20b). At high liquid loads and relatively low gas loads, the high-velocity liquid bends the swarms of gas bubbles leaving the orifices and tears them off, so most of the gas becomes emulsified as small bubbles within the liquid. The mixture behaves as a uniform two-phase fluid, which obeys the Francis weir formula [see the subsection Pressure Drop and Eq. (14-109) (Hofhuis and Zuiderweg, *IChemE Symp. Ser.* **56:** 2.2/1 (1979); Zuiderweg, *Int. Chem. Eng.* **26**(1): 1 (1986)]. In industrial practice, the emulsion regime is the most common in high-pressure and high-liquid-rate operation.

Spray regime (or drop regime, Fig. 14-20c). At high gas velocities and low liquid loads, the liquid pool on the tray floor is shallow and easily atomized by the high-velocity gas. The dispersion becomes a turbulent cloud of liquid droplets of various sizes that reside at high elevations above the tray and follow free trajectories. Some droplets are entrained to the tray above, while others fall back into the liquid pools and become reatomized. In contrast to the liquid-continuous froth and emulsion regimes, the phases are reversed in the spray regime: here the gas is the continuous phase, while the liquid is the dispersed phase.

The spray regime often occurs where gas velocities are high and liquid loads are low (e.g., vacuum and rectifying sections at low liquid loads).

Three-layered structure. Van Sinderen, Wijn, and Zanting [*Trans. IChemE.* **81:** Part A, p. 94 (January 2003)] postulate a tray dispersion consisting of a bottom liquid-rich layer where jets/bubbles form; an intermediate liquid-continuous froth layer where bubbles erupt, generating drops; and a top gas-continuous layer of drops. The intermediate layer that dampens the bubbles and jets disappears at low liquid rates, and the drop layer approaches the tray floor, similar to the classic spray regime.

PRIMARY TRAY CONSIDERATIONS

Number of Passes Tray liquid may be split into two or more flow passes to reduce tray liquid load Q_L (Fig. 14-21). Each pass carries $1/N_p$ fraction of the total liquid load (e.g., ¼ in four-pass trays). Liquid in each pass reverses direction on alternate trays. Two-pass trays have perfect symmetry with full remixing in the center downcomers. Four-pass trays are symmetric along the centerline, but the side and central passes are nonsymmetric. Also, the center and off-center downcomers only partially remix the liquid, allowing any maldistribution to propagate. Maldistribution can cause major loss of efficiency and capacity in four-pass trays. Three-pass trays are even more prone to maldistribution due to their complete nonsymmetry. Most designers avoid three-pass trays altogether, jumping from two to four passes. Good practices for liquid and vapor balancing and for avoiding maldistribution in multipass trays were described by Pilling (*Chemical Engineering Progr.*, June 2005, p. 22), Bolles [*AIChE J.* **22**(1): 153 (1976)], and Kister (*Distillation Operation,* McGraw-Hill, New York, 1990).

Common design practice is to minimize the number of passes, resorting to a larger number only when the liquid load exceeds 100 to 140 $m^3/(h \cdot m)$ (11 to 15 gpm/in) of outlet weir length (Davies and Gordon, *Petro/Chem Eng.*, December 1961, p. 228). Trays smaller than 1.5 m (5 ft) in diameter seldom use more than a single pass; those with 1.5- to 3-m (5- to 10-ft) diameters seldom use more than two passes. In high liquid services with towers larger than 4 m (13 ft) in diameter four pass trays are common, and in some "mega towers" (larger than 7-m, or 23-ft in diameter) 6 or 8 pass trays are used.

Tray Spacing Taller spacing between successive trays raises capacity, leading to a smaller tower diameter, but it also raises tower height. There is an economic tradeoff between tower height and diameter. As long as the tradeoff exists, tray spacing has little effect on tower economics and is set to provide adequate access. In towers larger than 1.5 m (5 ft) in diameter, tray spacing is typically 600 mm (24 in), large enough to permit a worker to crawl between trays. In very large towers (>6-m or 20-ft diameter), tray spacings of 750 mm (30 in) are often used. In chemical towers (as distinct from petrochemical, refinery, and gas plants), 450 mm (18 in) has been a popular tray spacing. With towers smaller than 1.5 m (5 ft), tower walls are reachable from the manways, there is no need to crawl, and it becomes difficult to support thin and tall columns, so smaller tray spacing (typically 380 to 450 mm or 15 to 18 in) is favored. Towers taller than 50 m (160 ft) also favor smaller tray spacings (400 to 450 mm or 16 to 18 in). Finally, cryogenic towers enclosed in cold boxes favor very small spacings, as small as 150 to 200 mm (6 to 8 in), to minimize the size of the cold box.

More detailed considerations for setting tray spacing were discussed by Kister (*Distillation Operation,* McGraw-Hill, New York, 1990) and Mukherjee (*Chem. Eng.*, September 2005, p. 53).

Outlet Weir The outlet weir should maintain a liquid level on the tray high enough to provide sufficient gas–liquid contact without causing excessive pressure drop, downcomer backup, or a capacity limitation. Weir heights are usually set at 40 to 80 mm (1.5 to 3 in). In this range, weir heights have little effect on distillation efficiency [Van Winkle, *Distillation,* McGraw-Hill, New York, 1967; Kreis and Raab, *IChemE Symp. Ser.* **56:** 3.2/63 (1979); Kister, *Chem. Eng. Prog.*, June 2008, p. 39]. In operations where long residence times are necessary (e.g., chemical reaction, absorption, stripping, dust and mist scrubbing), taller weirs do improve efficiency, weirs 80 to 100 mm (3 to 4 in) are more common (Lockett, *Distillation Tray Fundamentals,* Cambridge University Press, Cambridge, UK, 1986), and success has been reported with up to 300 mm (12 in) weirs (Flowers, Evans, and Payne, Distillation Topical Conference, AIChE Spring Meeting, April 2018, Orlando, Florida).

Adjustable weirs (Fig. 14-22*a*) are used to provide additional flexibility. They are uncommon with conventional trays, but they are used with some proprietary trays. Swept-back weirs (Fig. 14-22*b*) are used to extend the effective length of side weirs, either to help balance liquid flows to nonsymmetric tray passes or to reduce the tray liquid loads. Picket fence weirs (Fig. 14-22*c*) are used to shorten the effective length of a weir, either to help balance multipass trays' liquid flows (they are used in center and

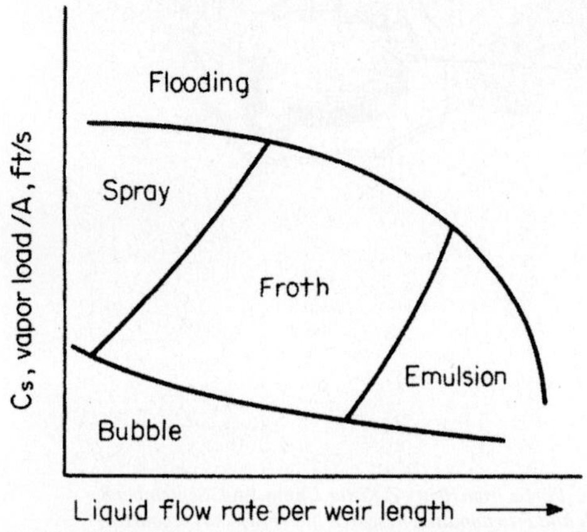

FIG. 14-19 The flow regime likely to exist on a distillation tray as a function of vapor and liquid loads. (*From H. Z. Kister,* Distillation Design, *copyright ©1992 by McGraw-Hill; reprinted by permission.*)

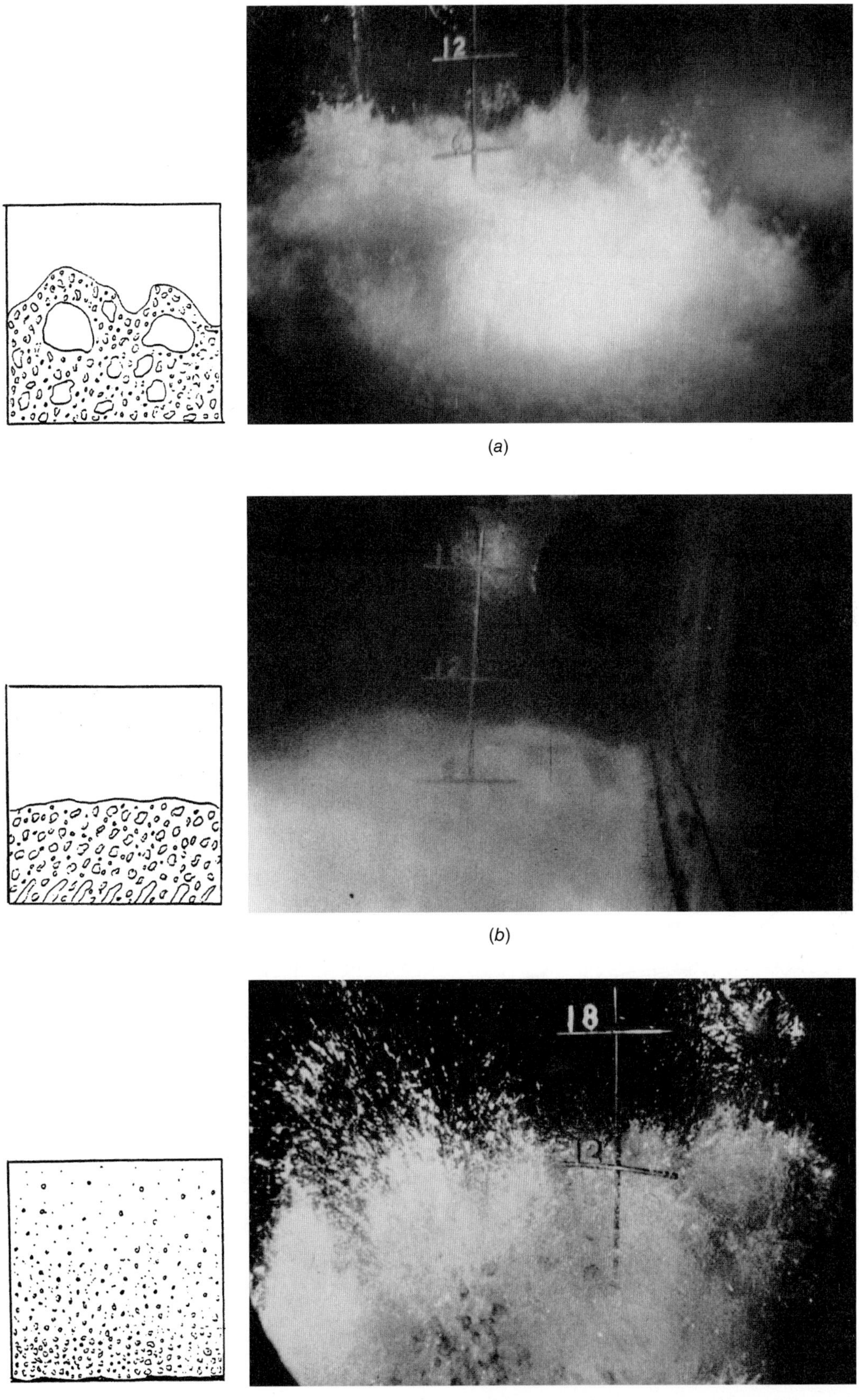

FIG. 14-20 Distillation flow regimes: schematics and photos. (*a*) Froth. (*b*) Emulsion. (*c*) Spray. [*Schematics from H. Z. Kister,* Distillation Design, *copyright © 1992 by McGraw-Hill, Inc.; reprinted by permission. Photographs courtesy of Fractionation Research Inc. (FRI).*]

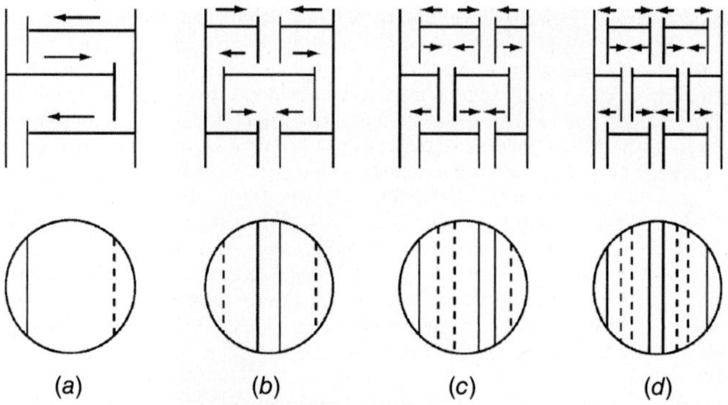

FIG. 14-21 Flow passes on trays. (*a*) Single-pass. (*b*) Two-pass. (*c*) Three-pass. (*d*) Four-pass.

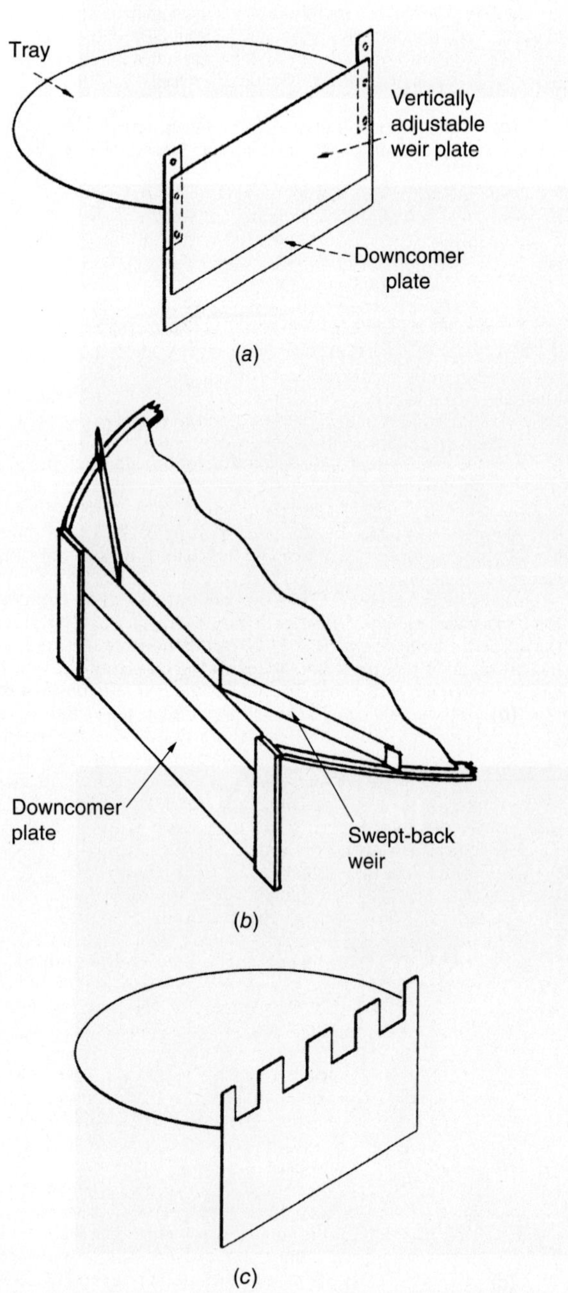

FIG. 14-22 Unique outlet weir types. (*a*) Adjustable. (*b*) Swept back. (*c*) Picket fence. (*Parts a, c, from H. Z. Kister,* Distillation Operation, *copyright © 1990 by McGraw-Hill; reprinted by permission. Part b, courtesy of Koch-Glitsch LP.*)

off-center weirs) or to raise tray liquid load and prevent drying in low-liquid-load services. To be effective, the pickets (or "weir blocks") need to be tall, typically around 300 to 400 mm (12 to 16 in) above the top of the weir. Excessive picketing (>70%) of the outlet weir length may induce excessive entrainment and premature flooding and should generally be avoided. An excellent discussion of weir picketing practices was provided by Summers and Sloley (*Hydroc. Proc.,* January 2007, p. 67).

Downcomers A downcomer is the drainpipe of the tray. It conducts liquid from one tray to the tray below. The fluid entering the downcomer is far from pure liquid; it is essentially the froth on the tray, typically 20 to 30 percent liquid by volume, with the balance being gas. Due to the density difference, most of this gas disengages in the downcomer and vents back to the tray from the downcomer entrance. Some gas bubbles usually remain in the liquid even at the bottom of the downcomer, ending on the tray below [Lockett and Gharani, *IChemE Symp. Ser.* **56:** 2.3/43 (1979)].

The straight, segmental vertical downcomer (Fig. 14-23*a*) is the most common downcomer geometry. It is simple and inexpensive and gives good use of the tower area for downflow. Circular downcomers (downpipes) (Fig. 14-23*b*) are cheaper, but they use tower area poorly and are only suitable for very low liquid loads. Sloped downcomers (Fig. 14-23*c, d*) improve

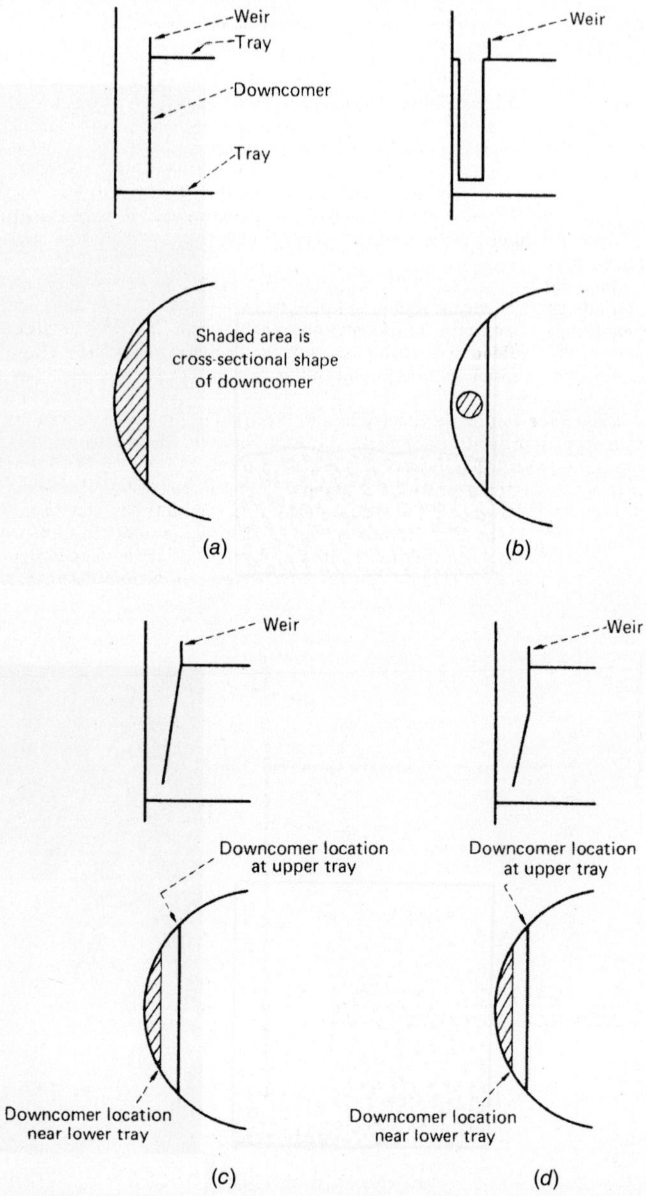

FIG. 14-23 Common downcomer types. (*a*) Segmental. (*b*) Circular. (*c, d*) Sloped. (*From Henry Z. Kister, Chem. Eng., December 29, 1980; reprinted courtesy of* Chemical Engineering.)

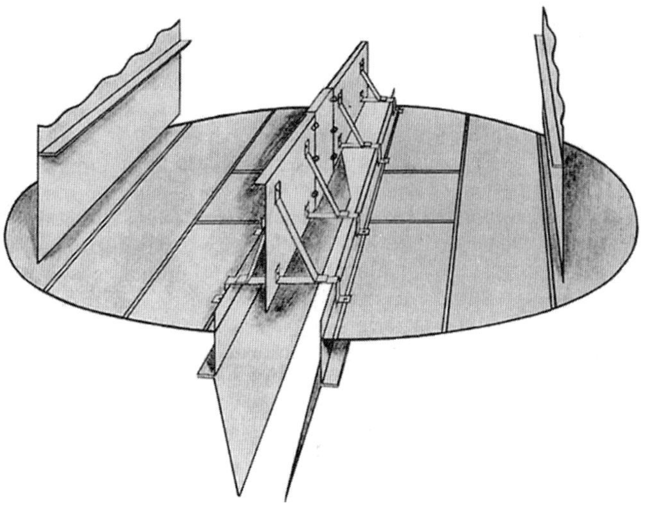

FIG. 14-24 Antijump baffle. (*Reprinted courtesy of Koch-Glitsch LP.*)

tower area use for downflow. They provide sufficient area and volume for gas–liquid disengagement at the top of the downcomer, gradually narrowing as the gas disengages, minimizing the loss of bubbling area at the foot of the downcomer. Sloped downcomers are invaluable when large downcomers are required such as at high liquid loads, high pressures, and foaming systems. In conventional trays, typical ratios of downcomer top to bottom areas are 1.5 to 2, and higher ratios are used in some high-capacity trays (see later discussion).

Antijump baffles (Fig. 14-24) are sometimes installed just above center and off-center downcomers of multipass trays to prevent liquid from one pass skipping across the downcomer onto the next pass. Such liquid jump adds to the liquid load on each pass, leading to premature flooding. These baffles are essential with proprietary trays that induce forward push (see later discussion).

Clearance Under the Downcomer Restricting the downcomer bottom opening prevents gas from the tray from rising up the downcomer and interfering with its liquid descent (*downcomer unsealing*). A common design practice makes the downcomer clearance 13 mm (0.5 in) lower than the outlet weir height (Fig. 14-25) to ensure submergence at all times (Davies and Gordon, *Petro/Chem Eng.,* November 1961, p. 250). This practice is sound in the froth and emulsion regimes, where tray dispersions are liquid continuous, but it is ineffective in the spray regime, where tray dispersions are gas

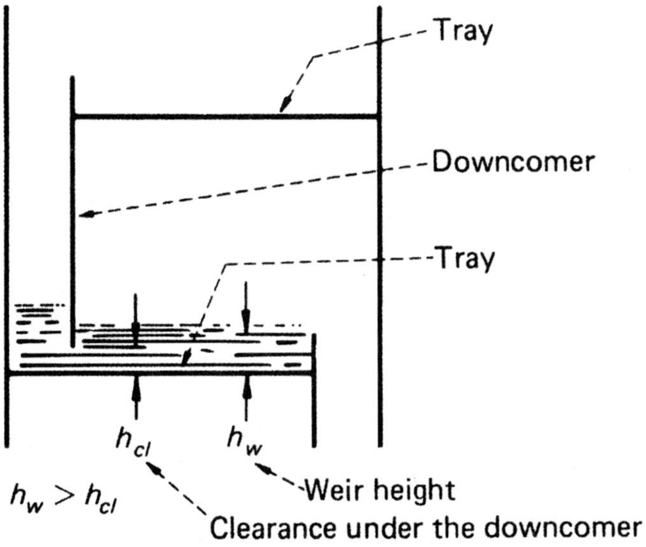

FIG. 14-25 A common design practice of ensuring a positive downcomer seal. (*From Henry Z. Kister, Chem. Eng., December 29, 1980; reprinted courtesy of* Chemical Engineering.)

continuous and there is no submergence. Also, this practice can be unnecessarily restrictive at high liquid loads where high crests over the weirs sufficiently protect the downcomers from gas rise. Generally, downcomer clearances in the spray regime need to be smaller, while those in the emulsion regime can be larger, than those set by the above practice. Seal pans and inlet weirs are devices sometimes used to help with downcomer sealing while keeping downcomer clearances large. Details are in Kister's book (*Distillation Operation,* McGraw-Hill, New York, 1990).

Hole Sizes Small holes slightly enhance tray capacity when limited by entrainment flood. Reducing sieve hole diameters from 13 to 5 mm (½ to $^3/_{16}$ in) at a fixed hole area typically enhances capacity by 3 to 8 percent, more at low liquid loads. Small holes are effective for reducing entrainment and enhancing capacity in the spray regime [Q_L < 20 $m^3/(h \cdot m)$ of weir]. Hole diameter has only a small effect on pressure drop, tray efficiency, and turndown.

On the debit side, the plugging tendency increases exponentially as hole diameters diminish. Smaller holes are also more prone to corrosion. While 5-mm ($^3/_{16}$-in) holes easily plug even with scale and rust, 13-mm (½-in) holes are quite robust and are therefore very common. The small holes are only used in clean, noncorrosive services. Holes smaller than 5 mm are usually avoided because they require drilling (larger holes are punched), which is much more expensive. For highly fouling services, 19- to 25-mm (¾- to 1-in) holes are preferred.

Similar considerations apply to fixed valves. Small fixed valves have a slight capacity advantage, but they are far more prone to plugging than larger fixed valves.

For round moving valves, a common orifice size is 39 mm (117/32 in). The float opening is usually of the order of 8 to 10 mm (0.3 to 0.4 in).

In recent years there has been a trend toward minivalves, both fixed and moving. These are smaller and therefore give a slight capacity advantage while being more prone to plugging.

Fractional Hole Area Typical sieve and fixed valve tray hole areas are 8 to 12 percent of the bubbling areas. Smaller fractional hole areas bring about a capacity reduction when limited by entrainment or downcomer backup flood or by excessive pressure drop. At above 12 percent of the bubbling areas, the capacity gains from higher hole areas become marginal, while weeping, and at high liquid loads also channeling, escalate.

Typical open-slot areas for moving valve trays are 14 to 15 percent of the bubbling area. Here the higher hole areas can be afforded due to the high turndown of the valves.

Moving valves can have a sharp or a smooth ("venturi") orifice. The venturi valves have one-half the dry pressure drop of the sharp-orifice valves, but they are far more prone to weeping and channeling than the sharp-orifice valves. Sharp orifices are almost always preferred.

Multipass Balancing There are two balancing philosophies: equal bubbling areas and equal flow path lengths. Equal bubbling areas means that all active area panels on Fig. 14-21*d* are of the same area, and each panel has the same hole (or open-slot) area. In a four-pass tray, one-quarter of the gas flows through each panel. To equalize the L/G ratio on each panel, the liquid needs to be split equally to each panel. Since the center weirs are longer than the side weirs, more liquid tends to flow toward the center weir. To equalize, side weirs are often swept back (Fig. 14-22*b*), while center weirs often contain picket fences (Fig. 14-22*c*).

The alternative philosophy (equal flow path lengths) provides more bubbling and perforation areas in the central panels of Fig. 14-21*d* and less in the side panels. To equalize the L/G ratio, less liquid needs to flow toward the sides, which is readily achieved because the center weirs are naturally longer than the side weirs. Usually there is no need for swept-back weirs, and only minimal picket fencing is required at the center weir.

Equal flow path panels are easier to fabricate and are cheaper, while equal bubbling areas have a robustness and reliability advantage due to the ease of equally splitting the fluids. The author had good experience with both when they were well designed. Pass balancing is discussed in detail by Pilling (*Chem. Eng. Prog.,* June 2005, p. 22) and by Jaguste and Kelkar (*Hydroc. Proc.,* March 2006, p. 85).

Channel Baffles Excessive weeping is a major issue in services that handle large liquid loads with small vapor loads. To prevent weeping, the hole area can be reduced, but when it is reduced below about 5 percent of the bubbling area vapor–liquid contact suffers. For these cases, channel baffles (or downsizing baffles) can be used (Fig. 14-26) to block out a large fraction of the active area. The only vapor–liquid contact area is between the baffles, with the regions to the sides of the baffles blanked off. Their common design strategy is to largely reduce the bubbling area with only a small reduction in weir length (to accommodate the high weir load). The full flow path length is preserved (Fig. 14-26) to permit good contacting and efficiency. Summers and Chambers (Kister Distillation Symposium Proceedings, AIChE Spring Meeting, Austin, Tex., 2015) provide an excellent description of the good practices for channel baffles.

FIG. 14-26 A tray with channel baffles. The only active area is between the baffles. The area to the sides of the baffles is blanked. (*Courtesy of Sulzer Chemtech AG.*)

TRAY CAPACITY ENHANCEMENT

High-capacity trays evolved from conventional trays by including one or more capacity enhancement features such as those discussed below. These features enhance not only the capacity but usually also the complexity and cost. These features have varying impact on the efficiency, turndown, plugging resistance, pressure drop, and reliability of the trays.

Truncated Downcomers/Forward Push Trays Truncated downcomers/forward push trays include the Nye™ Tray, Maxfrac™ (Fig. 14-27*a*), Triton™, and MVGT™. In all these, the downcomer from the tray above terminates about 100 to 150 mm (4 to 6 in) above the tray floor. Liquid from the downcomer issues via holes or slots, directed downward or in the direction of liquid flow. The tray floor under each downcomer is equipped with fixed valves or side perforations. Gas issuing in this region, typically 10 to 20 percent of the total tray gas, is deflected horizontally in the direction of liquid flow by the downcomer floor. This horizontal gas flow pushes liquid droplets toward the tower wall directly above the outlet downcomer. The tower wall catches this liquid and directs it downward into the downcomer. This deentrains the gas space. In multipass trays, antijump baffles (Fig. 14-24), typically 300 mm or taller, are installed above center and off-center downcomers to catch the liquid and prevent its jumping from pass to pass. The rest of the tray features are similar to those of conventional trays. The tray floor may contain fixed valves, moving valves, or sieve holes.

Trays from this family are proprietary, and they have been extensively used in the last three decades with great success. Compared to equivalent conventional trays, the truncated downcomer/forward push trays give about 8 to 12 percent more gas-handling capacity at much the same efficiency.

High Top-to-Bottom Downcomer Area and Forward Push Sloping downcomers from top to bottom raises the available tray bubbling area and, therefore, the gas-handling capacity (see the subsection Downcomers). As long as the ratio of top to bottom areas is not excessive, sloping does not lower downcomer capacity. Downcomer choke flood restricts the downcomer entrance, not exit, because there is much less gas at the downcomer bottom. However, a high top-to-bottom area ratio makes the downcomer bottom a very short chord, which makes distribution of liquid to the tray below difficult. To permit high top-to-bottom area ratios, some trays use a special structure (Fig. 14-27*b*) to change the downcomer shape from segmental to semiarc or multichordal. This high ratio of top to bottom areas, combined with forward push (above) imparted by bubblers and directional fixed or moving valves, and sometimes directional baffles, is used in trays such as Superfrac™ III (Fig. 14-27*b*) and IV and V-Grid Plus™. When the downcomer inlet areas are large, these trays typically gain 15 to 20 percent capacity compared to equivalent conventional trays at much the same efficiency. Trays from this family are proprietary, and they have been used successfully for about two decades.

Large Number of Truncated Downcomers These include the MD™ (Fig. 14-27*c*) and Hi-Fi™ trays. The large number of downcomers raises the total weir length, moving tray operation toward the peak capacity point of 20 to 30 $m^3/(h \cdot m)$ (2 to 3 gpm/in) of outlet weir (see Fig. 14-30). The truncated downcomers extend about halfway to the tray below, discharging their liquid via holes or slots at the downcomer floor. The area directly under the downcomers is perforated or valved, and there is enough open height between the tray floor and the bottom of the downcomer for this perforated or valved area to be effective in enhancing the tray bubbling area.

Trays from this family are proprietary and have been successfully used for about five decades. Their strength is in high-liquid-load services where reducing weir loads provides major capacity gains. Compared to conventional trays, they can gain as much as 20 to 30 percent capacity but at an efficiency loss. The efficiency loss is of the order of 10 to 20 percent due to the large reduction in flow path length (see the subsection Tray Efficiency). When using these trays, the separation is maintained by either using more trays (typically at shorter spacing) or raising reflux and boilup. This lowers the net capacity gains to 10 to 20 percent above conventional trays. In some variations, forward push slots and antijump baffles are incorporated to enhance the capacity by another 10 percent.

Radial Trays These include the Slit™ tray and feature radial flow of liquid. In the efficiency-maximizing A variation (Fig. 14-27*d*), a multipipe distributor conducts liquid from each center downcomer to the periphery of the tray below, so liquid flow is from periphery to center on each tray. The capacity-maximizing B variation has central and peripheral (ring) downcomers on alternate trays, with liquid flow alternating from center-to-periphery to periphery-to-center on successive trays. The trays are arranged at small spacings (typically, 200 to 250 mm, or 8 to 10 in) and contain small fixed valves. Slit trays are used in chemical and pharmaceutical low-liquid-rate applications [<40 $m^3/(h \cdot m)$ or 4 gpm/in of outlet weir], typically at pressures ranging from moderate vacuum to slight superatmospheric.

Centrifugal Force Deentrainment These trays use a contact step similar to that in conventional trays, followed by a separation step that disentrains the tray dispersion by using centrifugal force. Separation of entrained liquid before the next tray allows very high gas velocities, as high as 50 percent above the system limit (see the subsection System Limit: The Ultimate Capacity of Fractionators), to be achieved. The capacity of these trays can be over 50 percent above that of conventional trays. The efficiency of these trays can be 20 percent less than that of conventional trays due to their typical very short flow paths (see the subsection Tray Efficiency).

These trays include the Ultrfrac™, the ConSep™ (Fig. 14-27*e*), and the Swirl Tube™ trays. This technology has been sporadically used in Eastern Europe for quite some time. It is now making inroads into distillation in the rest of the world, and it looks very promising.

OTHER TRAY TYPES

Bubble-Cap Trays (Fig. 14-28*a*) These are flat perforated plates with risers (chimneylike pipes) around the holes, and caps in the form of inverted cups over the risers. The caps are usually (but not always) equipped with slots through which some of the gas comes out, and may be round or rectangular. Liquid and froth are trapped on the tray to a depth at least equal to the riser or weir height, giving the bubble-cap tray a unique ability to operate at very low gas and liquid rates.

The bubble-cap tray was the workhorse of distillation before the 1960s. It was superseded by the much cheaper (as much as 10 times) sieve and valve trays. Compared to the bubble-cap trays, sieve and valve trays also offer slightly higher capacity and efficiency and lower entrainment and pressure drop, and they are less prone to corrosion and fouling. Today, bubble-cap trays are only used in special applications where liquid or gas rates are very low. A large amount of information on bubble-cap trays is documented in several texts (e.g., Bolles in B. D. Smith, *Design of Equilibrium Stage Processes,* McGraw-Hill, New York, 1963; Bolles, *Pet. Proc.,* February 1956, p. 65; March 1956, p. 82; April 1956, p. 72; May 1956, p. 109; Ludwig, *Applied Process Design for Chemical and Petrochemical Plants,* 2d ed., vol. 2, Gulf Publishing, Houston, 1979).

Dual-Flow Trays These are sieve trays with no downcomers (Fig. 14-28*b*). Liquid continuously weeps through the holes, hence their low efficiency. At peak loads they are typically 5 to 10 percent less efficient than sieve or valve trays, but as the gas rate is reduced, the efficiency gap rapidly widens, giving poor turndown. The absence of downcomers gives dual-flow trays more area, and therefore greater capacity, less entrainment, and less pressure drop, than conventional trays. Their pressure drop is further reduced by their large fractional hole area (typically 18 to 30 percent of the tower area). However, this low pressure drop also renders dual-flow trays prone to gas and liquid maldistribution.

In general, gas and liquid flows pulsate, with a particular perforation passing both gas and liquid intermittently, but seldom simultaneously. In

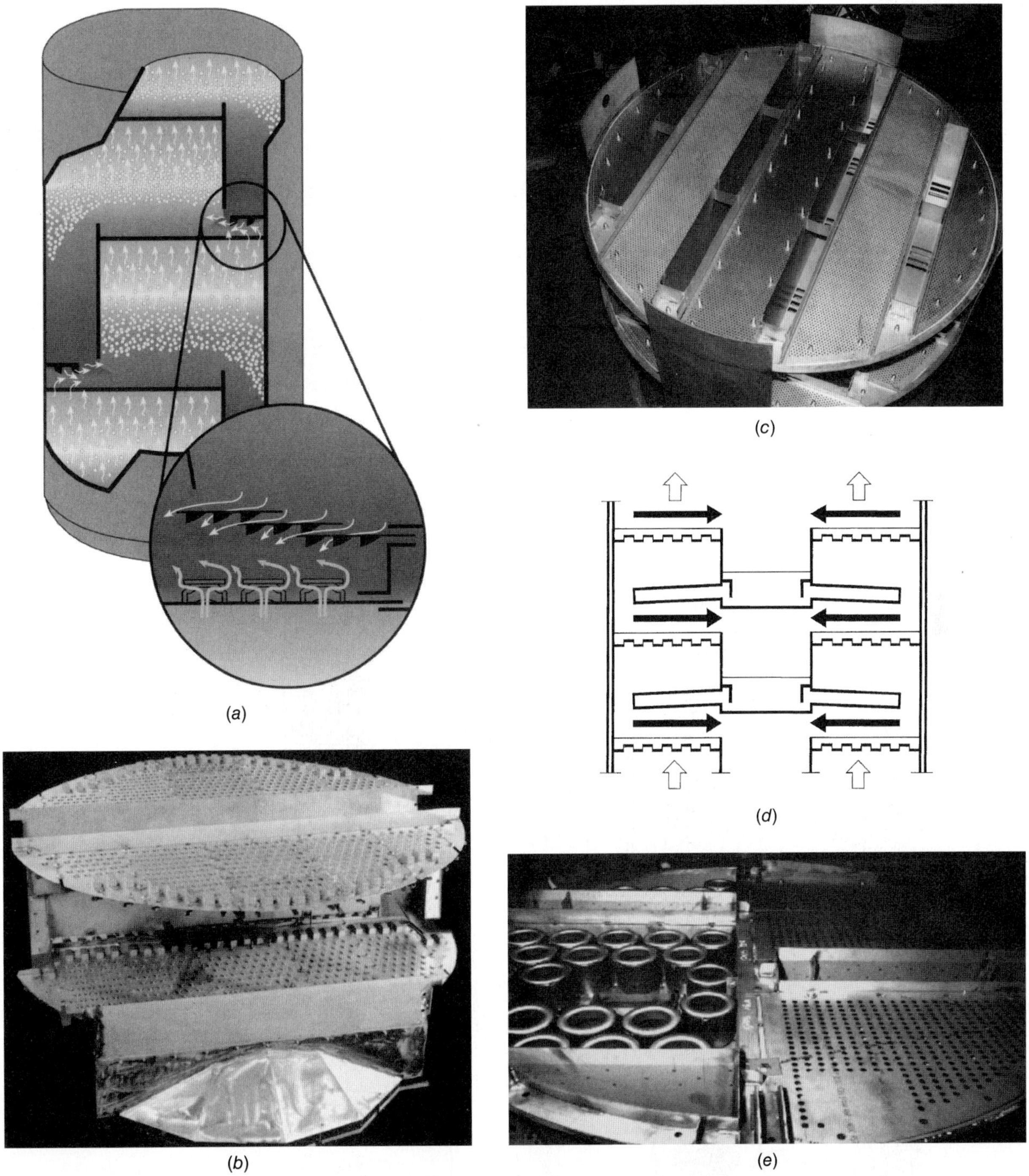

FIG. 14-27 Tray capacity enhancement. (*a*) Truncated downcomer/forward-push principle illustrated with a schematic of the Maxfrac™ tray. (*b*) High top-to-bottom area ratio illustrated with a two-pass Superfrac™ tray. Note the baffle in the front side downcomer that changes the side downcomer shape from segmental to multichordal. Also note the bubble promoters on the side of the upper tray and in the center of the lower tray, which give forward push to the tray liquid. (*c*) Top view of an MD™ tray with four downcomers. The decks are perforated. The holes in the downcomer lead the liquid to the active area of the tray below, which is rotated 90°. (*d*) Schematic of the Slit™ tray, type A, showing distribution pipes. Heavy arrows depict liquid movement; open arrows, gas movement. (*e*) The ConSep™ tray. The right-hand side shows sieve panels. On the left-hand side, these sieve panels were removed to show the contact cyclones that catch the liquid from the tray below. (*Parts a, b, courtesy of Koch-Glitsch LP; part c, courtesy of UOP LLC; part d, courtesy of Kühni AG; part e, courtesy of Sulzer Chemtech Ltd. and Shell Global Solutions International BV.*)

large-diameter (>2.5 m, or 8 ft) dual-flow trays, the pulsations sometimes develop into sloshing, instability, and vibrations. The Ripple Tray™ is a proprietary variation in which the tray floor is corrugated to minimize this instability.

With large holes (16 to 25 mm), dual-flow trays are some of the most fouling-resistant and corrosion-resistant devices in the industry. This defines their main application: highly fouling services, slurries, and corrosive services. Dual-flow trays are also the least expensive and easiest to install and maintain.

A wealth of information for the design and rating of dual-flow trays, much of it originating from FRI data, was published by Garcia and Fair [*Ind. Eng. Chem. Res.* **41:**1632 (2002)].

Baffle Trays Baffle trays ("shed decks," "shower decks") (Fig. 14-29*a*) are solid half-circle plates, sloped slightly in the direction of outlet flow,

FIG. 14-28 Other trays. (*a*) Bubble-cap tray. (*b*) Dual-flow tray. [*Part a, courtesy of Koch-Glitsch LP; part b, courtesy of Fractionation Research Inc. (FRI).*]

with or without weirs at the end. Gas contacts the liquid as it showers from the plate. This contact is inefficient, typically giving 30 to 40 percent of the efficiency of conventional trays. This limits their application mainly to heat-transfer and scrubbing services. The capacity is high and pressure drop is low due to the high open area (typically 50 percent of the tower cross-sectional area). Since there is not much that can plug up, the baffle trays are perhaps the most fouling-resistant devices in the industry, and their main application is in extremely fouling services. To be effective in these services, their liquid rate needs to exceed 20 $m^3/(h \cdot m)$ (2 gpm/in) of outlet weir and dead spots formed due to poor support design eliminated (Kister, *Distillation Troubleshooting,* Wiley, New York, 2006).

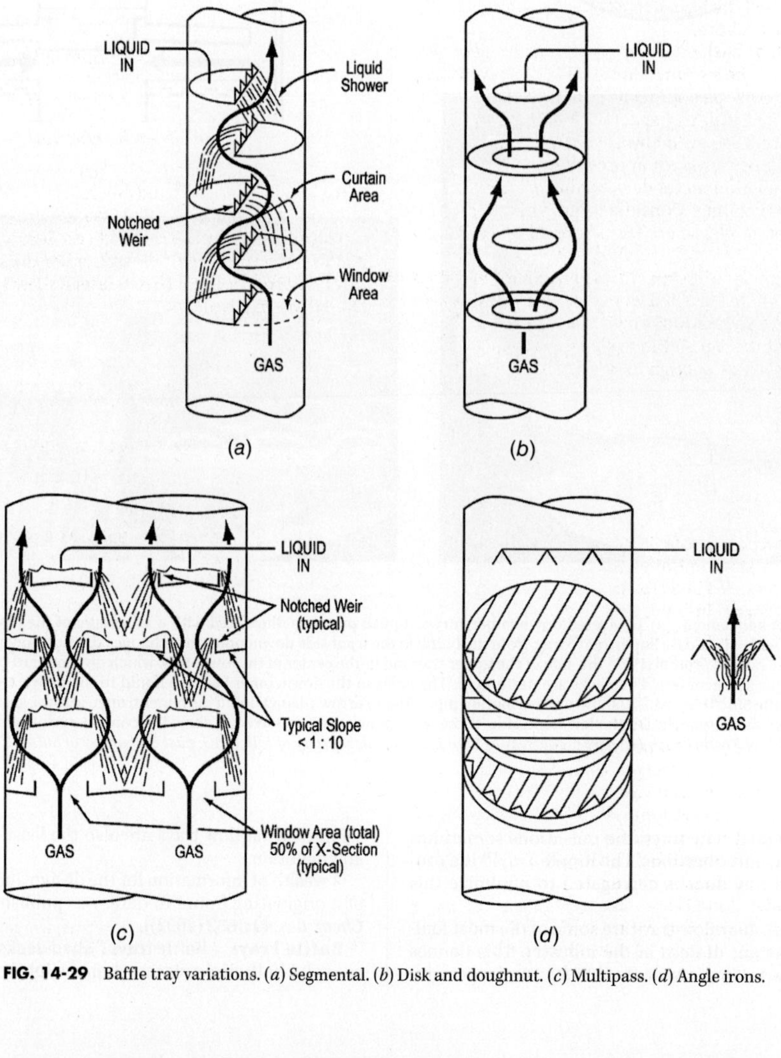

FIG. 14-29 Baffle tray variations. (*a*) Segmental. (*b*) Disk and doughnut. (*c*) Multipass. (*d*) Angle irons.

There are several geometric variations. The disk and doughnut trays (Fig. 14-29*b*) replace the half-circle segmental plates with alternate plates shaped as disks and doughnuts, each occupying about 50 percent of the tower cross-sectional area. In large towers, multipass baffle trays (Fig. 14-29*c*) are used. Another variation uses angle irons, with one layer oriented at 90° to the one below (Fig. 14-29*d*). Multipass baffle trays, as well as angle irons, require good liquid (and to a lesser extent, also good gas) distribution, as has been demonstrated from field heat-transfer measurements [Kister and Schwartz, *Oil & Gas J.,* p. 50 (May 20, 2002)]. Excellent overviews of the fundamentals and design of baffle trays were given by Fair and Lemieux (Fair, *Hydro. Proc.,* May 1993, p. 75; Lemieux, *Hydroc. Proc.,* September 1983, p. 106). Mass-transfer efficiency data with baffle trays by Fractionation Research Inc. (FRI) have been released and presented together with their correlation (Fair, paper presented at the AIChE Annual Meeting, San Francisco, November 2003). Kister and Olsson (*Chem. Eng. Prog.,* July 2011, p. 22) analyzed published test data from FRI and others, as well as plant data, to derive an improved baffle tray flood correlation based on the liquid velocity through the windows.

FLOODING

Flooding is by far the most common upper capacity limit of a distillation tray. The column diameter is set to ensure that the column can achieve the required throughput without flooding. Towers are usually designed to operate at 80 to 90 percent of the flood limit.

Flooding is an excessive accumulation of liquid inside a column. Flood symptoms include a rapid rise in pressure drop (the accumulating liquid increases the liquid head on the trays), liquid carryover from the column top, reduction in bottom flow rate (the accumulating liquid does not reach the tower bottom), and instability (accumulation is non-steady-state). This liquid accumulation is generally induced by one of the following mechanisms.

Entrainment (Jet) Flooding Froth or spray height rises with gas velocity. As the froth or spray approaches the tray above, some of the liquid is aspirated into the tray above as entrainment. Upon a further increase in gas flow rate, massive entrainment of the froth or spray begins, causing liquid accumulation and flood on the tray above.

Entrainment flooding can be subclassified into *spray entrainment flooding* (common) and *froth entrainment flooding* (uncommon). Froth entrainment flooding occurs when the froth envelope approaches the tray above, and it is therefore only encountered with small tray spacings (<450 mm or 18 in) in the froth regime. At larger (and often even lower) tray spacing, the froth breaks into spray well before the froth envelope approaches the tray above.

The entrainment flooding prediction methods described here are based primarily on spray entrainment flooding. Considerations unique to froth entrainment flooding can be found elsewhere (Kister, *Distillation Design,* McGraw-Hill, New York, 1992).

Spray Entrainment Flooding Prediction Most entrainment flooding prediction methods derive from the original work of Souders and Brown [*Ind. Eng. Chem.* **26**(1): 98 (1934)]. Souders and Brown theoretically analyzed entrainment flooding in terms of droplet settling velocity. Flooding occurs when the upward vapor velocity is high enough to suspend a liquid droplet, giving

$$C_{SB} = u_{S,\text{flood}}\sqrt{\frac{\rho_G}{\rho_L - \rho_G}} \tag{14-80}$$

The Souders and Brown constant C_{SB} is the *C*-factor [Eq. (14-77)] at the entrainment flood point. Most modern entrainment flooding correlations retain the Souders and Brown equation (14-80) as the basis, but they depart from the notion that C_{SB} is a constant. Instead, they express C_{SB} as a weak function of several variables, which differ from one correlation to another. Depending on the correlation, C_{SB} and $u_{S,\text{flood}}$ are based either on the net area A_N or on the bubbling area A_B.

The constant C_{SB} is roughly proportional to the tray spacing to a power of 0.5 to 0.6 (Kister, *Distillation Design,* McGraw-Hill, New York, 1992). Figure 14-30 demonstrates the effect of liquid rate and fractional hole area on C_{SB}. As liquid load increases, C_{SB} first increases, then peaks, and finally declines. Some interpret the peak as the transition from the froth to the spray regime [Porter and Jenkins, *I. Chem. E. Symp. Ser.* **56,** Summary Paper, London (1979)]. C_{SB} increases slightly with fractional hole area at lower liquid rates, but there is little effect of fractional hole area on C_{SB} at high liquid rates. C_{SB} slightly increases as hole diameter is reduced.

For sieve trays, the entrainment flood point can be predicted by using the method by Kister and Haas [*Chem. Eng. Progr.* **86**(9): 63 (1990)]. The method is said to reproduce a large database of measured flood points to within ±15 percent. $C_{SB,\text{flood}}$ is based on the net area. The equation is

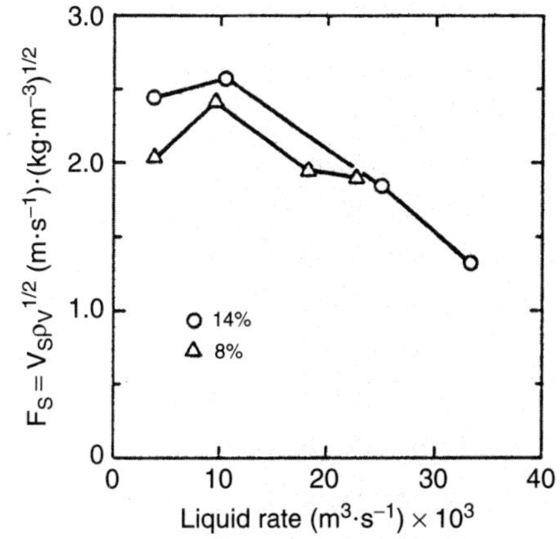

FIG. 14-30 Effect of liquid rate and fractional hole area on flood capacity. FRI sieve tray test data, cyclohexane/*n*-heptane, 165 kPa (24 psia), D_T = 1.2 m (4 ft), S = 610 mm (24 in), h_w = 51 mm (2 in), d_H = 12.7 mm (0.5 in), straight downcomers, A_D/A_T = 0.13. (*From T. Yanagi and M. Sakata,* Ind. Eng. Chem. Proc. Des. Dev. *21: 712; copyright © 1982, American Chemical Society, reprinted by permission.*)

$$C_{SB,\text{flood}} = 0.0277(d_h^2\,\sigma/\rho_L)^{0.125}(\rho_G/\rho_L)^{0.1}(\text{TS}/h_{ct})^{0.5} \tag{14-81}$$

where d_h = hole diameter, mm
σ = surface tension, mN/m (dyn/cm)
ρ_G, ρ_L = vapor and liquid densities, kg/m^3
TS = tray spacing, mm
h_{ct} = clear liquid height at the froth-to-spray transition, mm; obtained from:

$$h_{ct} = h_{ct,H_2O}\,(996/\rho_L)^{0.5(1-n)} \tag{14-82}$$

$$h_{ct,H_2O} = \frac{0.497\,A_f^{-0.791}d_h^{0.833}}{1+0.013\,Q_L^{-0.59}\,A_f^{-1.79}} \tag{14-83}$$

$$n = 0.00091 d_h/A_f \tag{14-84}$$

In Eq. (14-83), Q_L = m^3 liquid downflow/(h · m weir length) and A_f = fractional hole area based on active ("bubbling") area; for instance, $A_f = A_h/A_a$.

The Kister and Haas method can also be applied to valve trays, but the additional approximations reduce its data prediction accuracy for valve trays to within ±20 percent. For valve trays, adaptations of Eqs. (14-81) to (14-84) are required:

$$d_h = \frac{4 \times (\text{area of opening of one fully open valve})}{\text{wetted perimeter of opening of one fully open valve}} \tag{14-85}$$

$$A_f = \frac{\text{no. of valves} \times (\text{area of opening of one fully open valve})}{\text{active (bubbling) area}} \tag{14-86}$$

A correlation for valve tray entrainment flooding that has gained respect and popularity throughout the industry is the Glitsch "Equation 13" (Glitsch, Inc., *Ballast Tray Design Manual,* 6th ed., 1993; available from Koch-Glitsch, Wichita, Kans.). This equation has been applied successfully for valve trays from different manufacturers, as well as for sieve trays with large fractional hole areas (12 to 15 percent). With tray spacings of 600 mm and higher, its flood prediction accuracy for valve trays has generally been within ±10 percent in the author's experience. The Glitsch correlation is

$$\frac{\%\text{flood}}{100} = \frac{C_B}{\text{CAF}} + 1.359\frac{Q\ \text{FPL}}{A_B\ \text{CAF}} \tag{14-87}$$

where

$$\text{CAF} = 0.3048\ \text{CAF}_0\ \text{SF} \tag{14-88}$$

C_B is the operating *C*-factor based on the bubbling area, m/s; Q is the liquid flow rate, m^3/s; A_B is the bubbling area, m^2; FPL is the flow path length, m, that is, the horizontal distance between the inlet downcomer and the

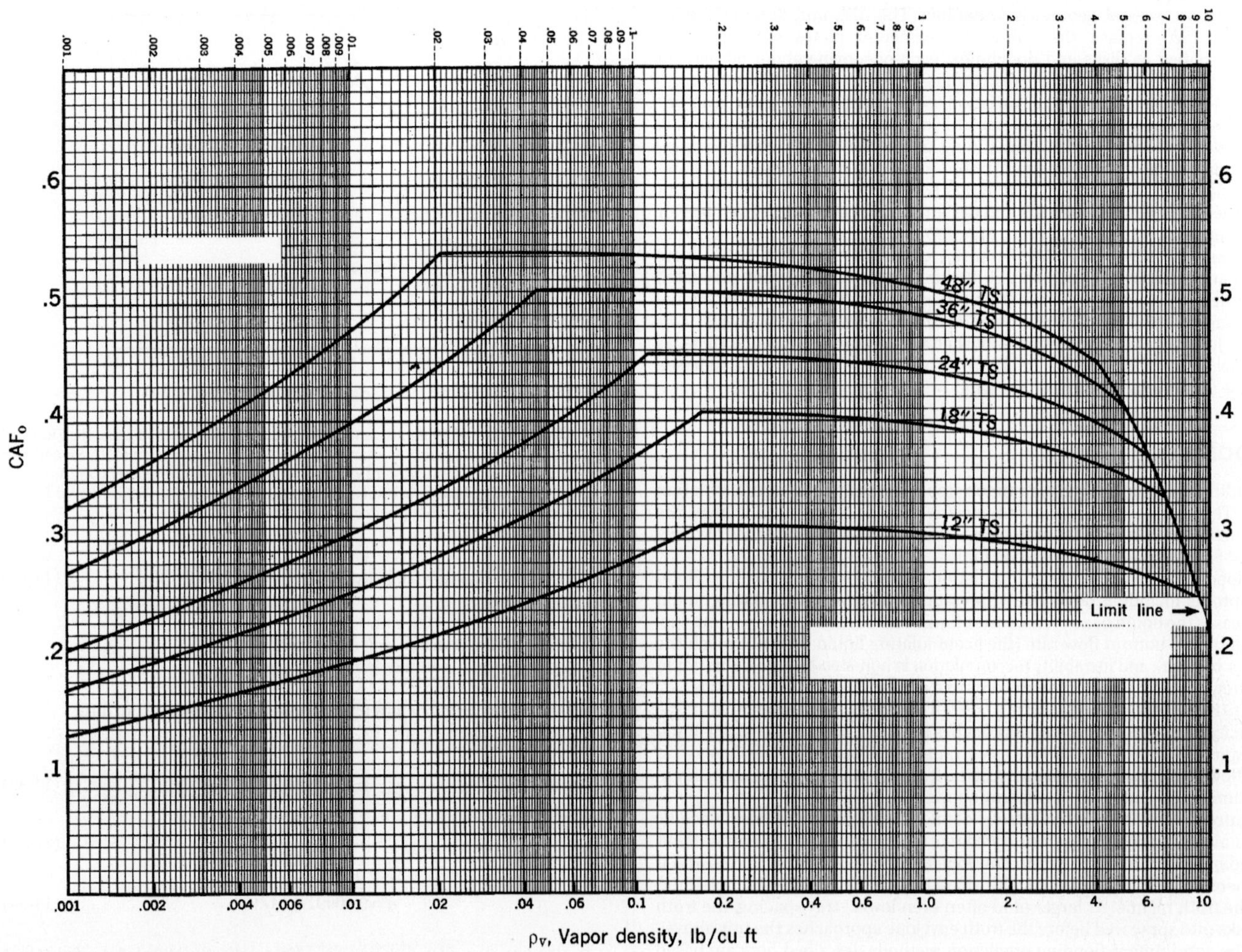

FIG. 14-31 Flood capacity of moving valve trays. (*Courtesy of Koch-Glitch LP.*)

outlet weir. The flow path length becomes shorter as the number of passes increases. CAF_0 and CAF are the flood *C*-factors. CAF_0 is obtained from Fig. 14-31 in English units (ft/s). Equation (14-88) converts CAF_0 to the metric CAF (m/s) and corrects it by using a system factor SF. Values of SF are given in Table 14-9.

The Fair correlation [*Pet/Chem Eng.* **33**(10): 45 (September 1961)] for decades has been the standard of the industry for entrainment flood prediction. It uses a plot (Fig. 14-32) of surface-tension-corrected Souders and Brown flood factor C_{SB} against the dimensionless flow parameter. The flow parameter represents a ratio of liquid to vapor kinetic energies:

$$F_{LG} = \frac{L}{G}\left(\frac{\rho_G}{\rho_L}\right)^{0.5} \qquad (14\text{-}89)$$

Low values of F_{LG} indicate vacuum operation; high values indicate operation at higher pressures or at high liquid/vapor loadings. The liquid-to-gas ratio L/G is based on mass flow rates. For multipass trays, the ratio needs to be divided by the number of passes. The strength of the correlation is at the lower flow parameters. At higher flow parameters (high L/G ratios, high pressures, emulsion flow), Fig. 14-32 gives excessively conservative predictions, with the low values of C_{sbf} to the right likely to result from downcomer flow restrictions rather than excessive entrainment. The curves may be expressed in equation form as [Lygeros and Magoulas, *Hydrocarbon Proc.* **65**(12): 43 (1986)]:

$$C_{sbf} = 0.0105 + 8.127(10^{-4})(TS^{0.755})\exp[-1.463\, F_{LG}^{\;0.842}] \qquad (14\text{-}90)$$

where TS = plate spacing, mm.

Figure 14-32 or Eq. (14-90) may be used for sieve, valve, or bubble-cap trays. The value of the capacity parameter (ordinate term in Fig. 14-32) may be used to calculate the maximum allowable vapor velocity through the net area of the plate:

$$U_{nf} = C_{sbf}\left(\frac{\sigma}{20}\right)^{0.2}\left(\frac{\rho_L-\rho_g}{\rho_g}\right)^{0.5} \qquad (14\text{-}91)$$

where U_{nf} = gas velocity through net area at flood, m/s
C_{sbf} = flood capacity parameter corrected for surface tension, m/s
σ = liquid surface tension, mN/m (dyn/cm)
ρ_L = liquid density, kg/m³
ρ_G = gas density, kg/m³

The application of the correlation is subject to the following restrictions:

1. System is low or nonfoaming.
2. Weir height is less than 15 percent of tray spacing.
3. Sieve-tray perforations are 13 mm (½ in) or less in diameter.
4. Ratio of slot (bubble cap), perforation (sieve), or full valve opening (valve plate) area A_h to active area A_a is 0.1 or greater. Otherwise the value of U_{nf} obtained from Fig. 14-32 should be corrected:

A_h/A_a	$U_{nf}/U_{nf,\text{Fig. 14-31}}$
0.10	1.00
0.08	0.90
0.06	0.80

where A_h = total slot, perforated, or open-valve area on tray.

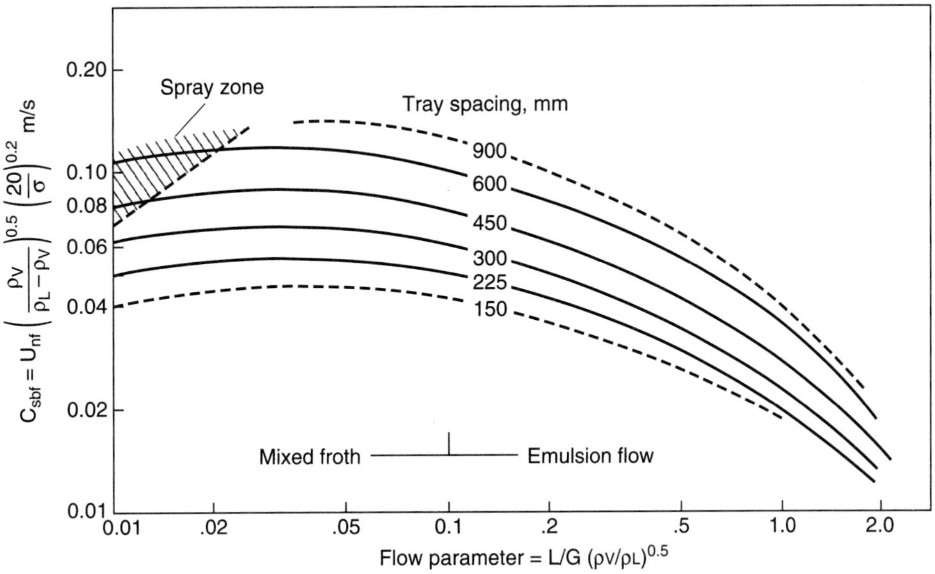

FIG. 14-32 Fair's entrainment flooding correlation for columns with cross-flow trays (sieve, valve, bubble-cap). [*Fair, Pet/Chem Eng* **33***(10): 45 (September 1961).*]

Example 14-9 Flooding of a Distillation Tray An available sieve tray column of 2.5 m diameter is being considered for an ethylbenzene/styrene separation. An evaluation of loading at the top tray will be made. Key dimensions of the single-pass tray are:

Column cross section, m^2	4.91
Downcomer area, m^2	0.25
Net area, m^2	4.66
Active area, m^2	4.41
Hole area, m^2	0.617
Hole diameter, mm	4.76
Weir length, m	1.50
Weir height, mm	38
Tray spacing, mm	500

Conditions and properties at the top tray are:

Temperature,°C	78
Pressure, torr	100
Vapor flow, kg/h	25,500
Vapor density, kg/m^3	0.481
Liquid flow, kg/h	22,000
Liquid density, kg/m^3	841
Surface tension, mN/m	25

Solution. The method of Kister and Haas gives:

$$Q_L = \frac{22{,}000}{841 \times 1.50} = 17.44\ \text{m}^3/\text{h}\cdot\text{m weir}$$

$$A_f = \frac{0.617}{4.41} = 0.14$$

By Eq. (14-83), $h_{ct,\text{H2O}} = 7.98$ mm

Eq. (14-84): $n = 0.0309$

Eq. (14-82): $h_{ct} = 8.66$ mm

Finally, by Eq. (14-81),

$$C_{SB,\text{flood}} = 0.0277[(4.76^2)(25/841)]^{0.125} \times (0.481/841)^{0.1}(500/8.66)^{0.5}$$

$$= 0.0947\ \text{m/s}$$

Alternatively, applying the Fair correlation:

The flow parameter $F_{LG} = 0.021$ [Eq. (14-89)]. From Fig. 14-31, $C_{sbf} = 0.095$ m/s. Then, correcting for surface tension, based on the net area,

$$C_{sbf} = 0.095(25/20)^{0.2} = 0.0993\ \text{m/s}$$

about 5 percent higher than the answer obtained from Kister and Haas.

For the design condition, the C-factor based on the net area is

$$C = \frac{25{,}500}{3600(0.481)(4.66)}\sqrt{\frac{0.481}{841-0.481}} = 0.0756\ \text{m/s}$$

or about 80 percent of flood. The proposed column is entirely adequate for the service required.

System Limit (Ultimate Capacity) This limit is discussed later under the subsection System Limit: The Ultimate Capacity of Fractionators.

Downcomer Backup Flooding Aerated liquid backs up in the downcomer because of tray pressure drop, liquid height on the tray, and frictional losses in the downcomer apron (Fig. 14-33). All these increase with increasing liquid rate. Tray pressure drop also increases as the gas rate rises. When the backup of aerated liquid exceeds the tray spacing, liquid accumulates on the tray above, causing downcomer backup flooding.

Downcomer backup is calculated from the pressure balance

$$h_{dc} = h_t + h_w + h_{ow} + h_{da} + h_{hg} \tag{14-92}$$

where h_{dc} = clear liquid height in downcomer, mm liquid
h_t = total pressure drop across the tray, mm liquid
h_w = height of weir at tray outlet, mm liquid
h_{ow} = height of crest over weir, mm liquid
h_{da} = head loss due to liquid flow under downcomer apron, mm liquid
h_{hg} = liquid gradient across tray, mm liquid

The heights of head losses in Eq. (14-92) should be in consistent units, e.g., millimeters or inches of liquid under operating conditions on the tray.

As noted, h_{dc} is calculated in terms of equivalent clear liquid. Actually, the liquid in the downcomer is aerated and actual backup is

$$h'_{dc} = \frac{h_{dc}}{\phi_{dc}} \tag{14-93}$$

where ϕ_{dc} is an average *relative* froth density (ratio of froth density to liquid density) in the downcomer. Design must not permit h'_{dc} to exceed the value of tray spacing plus weir height; otherwise, flooding can be precipitated.

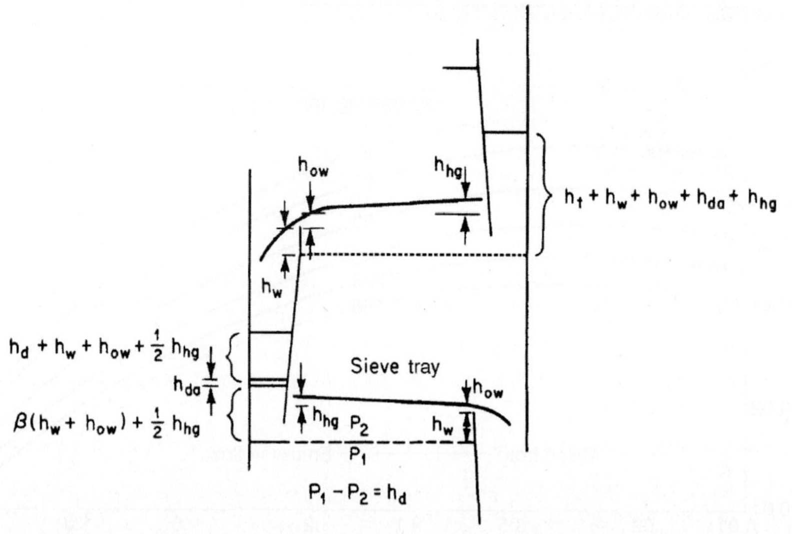

FIG. 14-33 Pressure-drop contributions for trays. h_d = pressure drop through cap or sieve, equivalent height of tray liquid; h_w = height of weir; h_{ow} = weir crest; h_{hg} = hydraulic gradient; h_{da} = loss under downcomer.

The value of ϕ_{dc} depends on the tendency for gas and liquid to disengage (froth to collapse) in the downcomer. For cases favoring rapid bubble rise (low gas density, low liquid viscosity, low system foamability), collapse is rapid, and fairly clear liquid fills the bottom of the downcomer (Fig. 14-17). For such cases, it is usual practice to employ a higher value of ϕ_{dc}. For cases favoring slow bubble rise (high gas density, high liquid viscosity, high system foamability), lower values of ϕ_{dc} should be used. As the critical point is approached in high-pressure distillations and absorptions, special precautions with downcomer sizing are mandatory. Table 14-6 lists values of ϕ_{dc} commonly used by the industry.

Downcomer Choke Flooding This is also called downcomer entrance flood or downcomer velocity flood. A downcomer must be sufficiently large to transport all the liquid downflow. Excessive friction losses in the downcomer entrance, and/or excessive flow rate of gas venting from the downcomer in counterflow, will impede liquid downflow, initiating liquid accumulation (termed downcomer choke flooding) on the tray above. The prime design parameter is the downcomer top area. Farther down the downcomer, gas disengages from the liquid, and the volumes of aerated liquid downflow and vented gas upflow are greatly reduced. With sloped downcomers, the downcomer bottom area is normally set at 50 to 60 percent of the top downcomer area. This taper is small enough to keep the downcomer top area the prime choke variable.

There is no satisfactory published correlation for downcomer choke. The best that can be done in the absence of data or correlation is to apply the criteria for maximum velocity of clear liquid at the downcomer entrance. Kister (*Distillation Operation,* McGraw-Hill, New York, 1990) surveyed the multitude of published criteria for maximum downcomer velocity and incorporated them into a single set of guidelines (Table 14-7). Some values, marked with an asterisk (*), were revised to reflect the author's recent experiences. The values given in Table 14-7 are not conservative. For a conservative design, multiply the values from Table 14-7 by a safety factor of 0.75. For very highly foaming systems, where antifoam application is undesirable, there are benefits to reducing downcomer design velocities down to 0.1 to 0.15 ft/s.

Another criterion sometimes used is to provide sufficient residence time in the downcomer to allow adequate disengagement of gas from the descending liquid, so that the liquid is relatively gas-free by the time it enters the tray below. Inadequate removal of gas from the liquid may choke the downcomer. Kister (*Distillation Operation,* McGraw-Hill, New York, 1990) reviewed various published criteria for downcomer residence times and recommended those by Bolles (private communication, 1977) and Erbar and Maddox (Maddox, *Process Engineer's Absorption Pocket Handbook,* Gulf Publishing, Houston, 1985). Both sets of guidelines are similar and are summarized in Table 14-8. The residence times in Table 14-8 are *apparent residence times,* defined as the ratio of the total downcomer volume to the clear liquid flow in the downcomer.

As a segmental downcomer becomes smaller, its width decreases faster than its length, turning the downcomer into a long, narrow slot. This geometry increases the resistance to liquid downflow and to the upflow of disengaging gas. Small downcomers are also extremely sensitive to foaming, fouling, construction tolerances, and the introduction of debris. Generally, segmental downcomers smaller than 5 percent of the column

TABLE 14-6 Criteria for Downcomer Aeration Factors

Foaming tendency	Bolles' criterion*		Glitsch's criterion†		Fair et al.'s criterion‡	
	Examples	ϕ_{dc}	Examples	ϕ_{dc}	Examples	ϕ_{dc}
Low	Low-molecular-weight hydrocarbons§ and alcohols	0.6	$\rho_G < 1.0$ lb/ft³	0.6	Rapid bubble rise systems, such as low gas density, low liquid viscosity	0.5
Moderate	Distillation of medium-molecular-weight hydrocarbons	0.5	$1.0 < \rho_G < 3.0$ lb/ft³	0.5		
High	Mineral oil absorbers	0.4	$\rho_G > 3.0$ lb/ft³	0.4		
Very high	Amines, glycols	0.3			Slow bubble rise systems, such as high gas density, high liquid viscosity, foaming systems	0.2–0.3

*"Distillation Theory and Practice—an Intensive Course," University of New South Wales/University of Sydney, August 9–11, 1977.
†Glitsch, Inc., *Ballast Tray Design Manual,* 6th ed., 1993; available from Koch-Glitsch LP, Wichita, Kans.
‡R. H. Perry and D. W. Green (eds.), *Perry's Chemical Engineers' Handbook,* 7th ed., McGraw-Hill, 1997.
§The author believes that low-molecular-weight hydrocarbons refer to light hydrocarbons at near atmospheric pressure or under vacuum. The foam stability of light-hydrocarbon distillation at medium and high pressure is best inferred from the Glitsch criterion.

To convert from lb/ft³, to kg/m³, multiply by 16.0.

SOURCE: From H. Z. Kister, *Distillation Design,* copyright © 1992 by McGraw-Hill, Inc.; reprinted by permission.

TABLE 14-7 Maximum Downcomer Velocities

		Clear liquid velocity in downcomer, ft/s		
Foaming tendency	Example	18-in spacing	24-in spacing	30-in spacing
Low	Low-pressure (<100 psia) light hydrocarbons, stabilizers, air-water simulators	0.4–0.5	0.5*	0.5*
Medium	Oil systems, crude oil distillation, absorbers, midpressure (100–300 psia) hydrocarbons	0.3–0.4	0.4–0.5	0.4–0.5*
High	Amines, glycerine, glycols, high-pressure (>300 psia) light hydrocarbons	0.2–0.25	0.2–0.25	0.2–0.3

*Revised from previous versions.
To convert from ft/s to m/s, multiply by 0.3048; from in to mm, multiply by 25.4; from psia to bar, multiply by 0.0689.
SOURCE: From H. Z. Kister, *Distillation Operation*, copyright 1990 by McGraw-Hill, Inc.; reprinted by permission.

cross-sectional area should be avoided. Additional discussion of small downcomers is available (Kister, *Distillation Operation,* McGraw-Hill, New York, 1990).

Derating ("System") Factors With certain systems, traditional flooding equations consistently give optimistic predictions. To allow for this discrepancy, an empirical derating or system factor (SF < 1.0) is applied. To obtain the actual or derated flood load, the flood gas load (entrainment flooding) or flood liquid load (downcomer choke) obtained from the traditional equations is multiplied by the derating factor. In the case of downcomer backup flood, the froth height from the traditional flood equation is divided by the derating factor.

Derating factors are vaguely related to the foaming tendency, but they are also applied to nonfoaming systems where standard flooding equations consistently predict too high. Sometimes, derating factors are used solely as overdesign factors. Brierley (*Chem. Eng. Prog.,* July 1994, p. 68) states that some derating factors actually evolved from plant misoperation or from misinterpretation of plant data. Kister (*Distillation Operation,* McGraw-Hill, New York, 1990) compiled the derating factors found in the literature into Table 14-9.

The application of derating factors is fraught with inconsistent practices and confusion. Caution is required. The following need to be carefully specified:

1. The flooding mechanism to which the derating factor applies (entrainment, downcomer backup, downcomer choke, or all these) must be specified.
2. Avoiding double derating. For instance, the values in Table 14-9 may apply with Eq. (14-81) because Eq. (14-81) does not take foaminess into account. However, they will double-derate a flood calculation that is made with a correlation or criteria that already take foaminess into account, such as the criteria for downcomer choke in Tables 14-7 and 14-8. Similarly, two different factors from Table 14-9 may apply to a single system; only one should be used.
3. Derating factors vary from source to source, and they may depend on the correlation used as well as the system. For instance, some caustic wash applications have a track record of foaming more severely than other caustic wash applications (see note in Table 14-9). The derating factors in Table 14-9 are a useful guide, but they are far from absolute.

ENTRAINMENT

Entrainment (Fig. 14-34) is liquid transported by the gas to the tray above. As the lower tray liquid is richer with the less-volatile components, entrainment counteracts the mass-transfer process, reducing tray efficiency. At times entrainment may transport nonvolatile impurities upward to contaminate the tower overhead product, or it may damage rotating machinery located in the path of the overhead gas.

TABLE 14-8 Recommended Minimum Residence Time in the Downcomer

Foaming tendency	Example	Residence time, s
Low	Low-molecular-weight hydrocarbons,* alcohols	3
Medium	Medium-molecular-weight hydrocarbons	4
High	Mineral oil absorbers	5
Very high	Amines and glycols	7

*The author believes that low-molecular-weight hydrocarbons refer to light hydrocarbons at atmospheric conditions or under vacuum. The foaming tendency of light-hydrocarbon distillation at medium pressure [>7 bar (100 psia)] is medium; at high pressure [>21 bar (300 psia)], it is high.
SOURCE: W. L. Bolles (Monsanto Company), private communication, 1977.

Effect of Gas Velocity Entrainment increases with gas velocity to a high power. Generally, smaller powers, indicative of a relatively gradual change, are typical of low-pressure systems. Higher powers, which indicate a steep change, are typical of high-pressure systems.

Due to the steep change of entrainment with gas velocity at high pressure, the gas velocity at which entrainment becomes significant tends to coincide with the flood point. At low pressure, the rate of change of entrainment with gas velocity is much slower, and entrainment can be significant even if the tray is operating well below the flood point. For this reason, excessive entrainment is a common problem in low-pressure and vacuum systems, but it is seldom troublesome with high-pressure systems. If encountered at high pressure, entrainment usually indicates flooding or abnormality.

Effect of Liquid Rate As the liquid rate is raised at constant gas rate, entrainment first diminishes, then passes through a minimum, and finally increases [Sakata and Yanagi, *I. Chem. E. Symp. Ser.* **56:** 3.2/21 (1979); Porter and Jenkins, *I. Chem. E. Symp. Ser.* **56:** Summary Paper (1979); Friend, Lemieux, and Schreiner, *Chem. Eng.,* October 31, 1960, p. 101]. The entrainment minima coincide with the maxima in plots of entrainment flood *F*-factor against liquid load (Fig. 14-30). At the low liquid loads (spray regime), an increase in liquid load suppresses atomization, drop formation, and consequently entrainment. At higher liquid loads, an increase in liquid load reduces the effective tray spacing, thereby increasing entrainment. The entrainment minima have been interpreted by many workers as the tray dispersion change from predominantly spray to the froth regime [Porter and Jenkins, *I. Chem. E. Symp. Ser.* **56:** Summary Paper (1979); Kister and Haas, *I. Chem. E. Symp. Ser.* **104:** A483 (1987)].

Effect of Other Variables Entrainment diminishes with higher tray spacing and increases with hole diameter [Kister and Haas, *I. Chem. E. Symp. Ser.* **104:** A483 (1987); *Ind. Eng. Chem. Res.* **27:** 2331 (1988); Lemieux and Scotti, *Chem. Eng. Prog.* **65**(3): 52 (1969)]. The hole diameter effect is large in the spray regime but small in the froth regime. In the spray regime, entrainment also increases as the fractional hole area is lowered, but this variable has little effect in the froth regime [Yanagi and Sakata, *Ind. Eng. Chem. Proc. Des. Dev.* **21:** 712 (1982); and Kister and Haas, *I. Chem. E. Symp. Ser.* **104:** A483 (1987)].

Entrainment Prediction For spray regime entrainment, the Kister and Haas correlation was shown to give good predictions to a wide commercial and pilot-scale data bank [*I. Chem. E. Symp. Ser.* **104:** A483 (1987)]. The correlation is

$$E_S = 4.742^{(10/\sqrt{\sigma})^{1.64}} \chi^{(10/\sqrt{\sigma})} \tag{14-94}$$

where

$$\chi = 872 \left(\frac{u_B h_{Lt}}{\sqrt{d_H S}} \right)^4 \left(\frac{\rho_G}{Q_L \rho_L} \right) \left(\frac{\rho_L - \rho_G}{\sigma} \right)^{0.25} \tag{14-95}$$

and

$$h_{Lt} = \frac{h_{ct}}{1 + 0.00262 h_w} \tag{14-96}$$

The terms in Eqs. (14-94) through (14-96) are in the metric units described in the Nomenclature table at the beginning of this section.

The recommended range of application of the correlation is given in Table 14-10. The clear liquid height at the froth-to-spray transition h_{ct} is calculated using the corrected Jeronimo and Sawistowski [*Trans. Inst. Chem. Engnrs.* **51:** 265 (1973)] correlation as per Eqs. (14-82) to (14-84).

For decades, the Fair correlation [*Pet/Chem. Eng.* **33**(10): 45 (September 1961)] has been used for entrainment prediction. In the spray regime, the Kister and Haas correlation was shown to be more accurate [Koziol and Mackowiak, *Chem. Eng. Process.* **27:** 145 (1990)]. In the froth regime, the Kister and Haas correlation does not apply, and Fair's correlation remains the standard of the industry. Fair's correlation (Fig. 14-35) predicts entrainment in

TABLE 14-9 Derating ("System") Factors

System	Factor	Reference	Notes
Nonfoaming regular systems	1.0	1–4	
High pressure (ρ_G >1.8 lb/ft^3)	$2.94/\rho_G^{0.32}$	2	Do not double-derate.
Low-foaming			
Depropanizers	0.9	4	
H_2S strippers	0.9	3, 4	
	0.85	2	
Fluorine systems (freon, BF_3)	0.9	1, 4	
Hot carbonate regenerators	0.9	2, 4	
Moderate-foaming			
Deethanizers			
Absorbing type, top section	0.85	1–4	
Absorbing type, bottom section	1.0	3	
	0.85	1, 2, 4	
Refrigerated type, top section	0.85	4	
	0.8	3	
Refrigerated type, bottom section	1.0	1, 3	
	0.85	4	
Demethanizers			
Absorbing type, top section	0.85	1–3	
Absorbing type, bottom section	1.0	3	
	0.85	1, 2	
Refrigerated type, top section	0.8	3	
Refrigerated type, bottom section	1.0	3	
Oil absorbers			
Above 0°F	0.85	1–4	Ref. 2 proposes these for "absorbers" rather than "oil absorbers."
Below 0°F	0.95	3	
	0.85	1, 4	
	0.8	2	
Crude towers	1.0	3	
	0.85	4	
Crude vacuum towers	1.0	3	
	0.85	2	
Furfural refining towers	0.85	2	
	0.8	4	
Sulfolane systems	1.0	3	
	0.85	4	
Amine regenerators	0.85	1–4	
Glycol regenerators	0.85	1, 4	
	0.8	3	
	0.65	2	
Hot carbonate absorbers	0.85	1, 4	
Caustic wash	0.65	2	The author suspects that this low factor refers only to some caustic wash applications but not to others.
Heavy-foaming			
Amine absorbers	0.8	2	
	0.75	3, 4	
	0.73	1	
Glycol contactors	0.73	1	Ref. 2 recommends 0.65 for glycol contactors in glycol synthesis gas service, 0.5 for others.
	0.65	3, 4	
	0.50	2	
Sour water strippers	0.5–0.7	3	
	0.6	2	
Oil reclaimer	0.7	2	
MEK units	0.6	1, 4	
Stable foam			
Caustic regenerators	0.6	2	
	0.3	1, 4	
Alcohol synthesis absorbers	0.35	2, 4	

References:

1. Glitsch, Inc., *Ballast Tray Design Manual*, Bulletin 4900, 6th ed., 1993. Available from Koch-Glitsch, Wichita, Kans.
2. Koch Engineering Co., Inc., *Design Manual—Flexitray*, Bulletin 960-1, Wichita, Kans., 1982.
3. Nutter Engineering, *Float Valve Design Manual*, 1976. Available from Sulzer ChemTech, Tulsa, Okla.
4. M. J. Lockett, *Distillation Tray Fundamentals*, Cambridge University Press, Cambridge, England, 1986.

To convert lb/ft^3 to kg/m^3, multiply by 16.0.

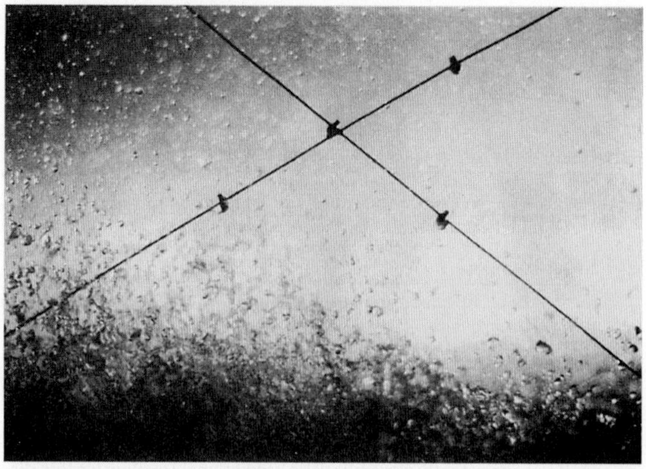

FIG. 14-34 Entrainment. [*Reprinted courtesy of Fractionation Research Inc. (FRI).*]

terms of the flow parameter [Eq. (14-89)] and the ratio of gas velocity to entrainment flooding gas velocity. The ordinate values Ψ are fractions of gross liquid downflow, defined as follows:

$$\psi = \frac{e}{L_m + e} \tag{14-97}$$

where e = absolute entrainment of liquid, mol/time
L_m = liquid downflow rate without entrainment, mol/time

Figure 14-35 also accepts the validity of the Colburn equation [*Ind. Eng. Chem.* **28:** 526 (1936)] for the effect of entrainment on efficiency:

$$\frac{E_a}{E_{mv}} = \frac{1}{1 + E_{mv}[\psi/(1-\psi)]} \tag{14-98}$$

where E_{mv} = Murphree vapor efficiency [see Eq. (14-134)]
E_a = Murphree vapor efficiency, corrected for recycle of liquid entrainment

The Colburn equation is based on complete mixing on the tray. For incomplete mixing, such as liquid approaching plug flow on the tray, Rahman and Lockett [*I. Chem. E. Symp. Ser.* **61:** 111 (1981)] and Lockett et al. [*Chem. Eng. Sci.* **38:** 661 (1983)] have provided corrections.

Fair (paper presented at the AIChE Annual Meeting, Chicago, Ill., November 1996) correlated data for efficiency reduction due to the rise of entrainment near entrainment flood, getting

$$\ln \psi = A + B\Phi + C\Phi^2 \tag{14-99}$$

where Φ is the fractional approach to entrainment flood and A, B, and C are constants given by

	A	*B*	*C*
Highest likely efficiency loss	−3.1898	−4.7413	7.5312
Median (most likely) efficiency loss	−3.2108	−8.9049	11.6291
Lowest likely efficiency loss	4.0992	−29.9141	25.3343

TABLE 14-10 Recommended Range of Application for the Kister and Haas Spray Regime Entrainment Correlation

Flow regime	Spray only
Pressure	20–1200 kPa (3–180 psia)
Gas velocity	0.4–5 m/s (1.3–15 ft/s)
Liquid flow rate	3–40 m^3/(m·h) (0.5–4.5 gpm/in)
Gas density	0.5–30 kg/m^3 (0.03–2 lb/ft^3)
Liquid density	450–1500 kg/m^3 (30–90 lb/ft^3)
Surface tension	5–80 mN/m
Liquid viscosity	0.05–2 cP
Tray spacing	400–900 mm (15–36 in)
Hole diameter	3–15 mm (0.125–0.75 in)
Fractional hole area	0.07–0.16
Weir height	10–80 mm (0.5–3 in

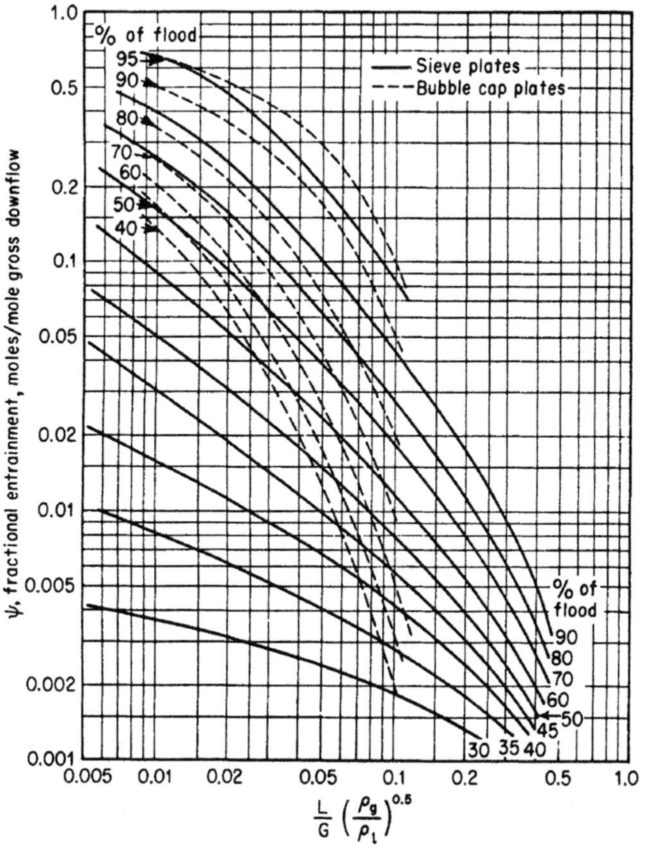

FIG. 14-35 Entrainment correlation. L/G = liquid/gas mass ratio; ρ_l and ρ_g = liquid and gas densities. [*Fair*, Pet./Chem. Eng. ***33(10)****: 45 (September 1961).*]

Either the Kister and Haas or the Fair method can be used to evaluate Φ. The correlation has been tested with sieve trays in the flow parameter range of 0.024 to 0.087.

Example 14-10 Entrainment Effect on Tray Efficiency For the column in Example 14-9, estimate the efficiency loss should the operation be pushed from the design 80 percent of flood to 90 percent of flood. The midrange dry Murphree tray efficiency is 70 percent.

Solution The vapor and liquid densities and L/V ratio remain unchanged from Example 14-9, and so is the flow parameter (calculated 0.021 in Example 14-9). At 80 and 90 percent of flood, respectively, Fig. 14-35 gives $\psi = 0.15$ and 0.24. The respective efficiency reductions are calculated from Eq. (14-98),

$$\frac{E_a}{E_{mv}} = \frac{1}{1 + 0.70[0.15/(1-0.15)]} = 0.89$$

$$\frac{E_a}{E_{mv}} = \frac{1}{1 + 0.70[0.24/(1-0.24)]} = 0.82$$

signifying an efficiency loss from 62 to 57 percent.

Alternatively, at 80 and 90 percent of entrainment flood, the median value of ψ from Eq. (14-99) is

$$\ln \psi = -3.2108 - 8.9049(0.80) + 11.6291(0.80^2)$$

giving $\psi = 0.056$, and from Eq. (14-98), $E_o/E_{mv} = 0.96$

$$\ln \psi = -3.2108 - 8.9049(0.90) + 11.6291(0.90^2)$$

giving $\psi = 0.164$, and from Eq. (14-98), $E_o/E_{mv} = 0.88$

signifying an efficiency reduction from 67 to 62 percent.

PRESSURE DROP

In vacuum distillation, excessive pressure drop causes excessive bottom temperatures which, in turn, increase degradation, polymerization, coking, and fouling, and also load up the column, vacuum system, and reboiler. In the suction of a compressor, excessive pressure drop increases the compressor size and energy usage. Such services try to minimize tray pressure drop. Methods for estimating pressure drops are similar for most conventional trays. The total pressure drop across a tray is given by

$$h_t = h_d + h'_L \tag{14-100}$$

where h_t = total pressure drop, mm liquid
h_d = pressure drop across the dispersion unit (dry hole for sieve trays; dry valve for valve trays), mm liquid
h'_L = pressure drop through aerated mass over and around the disperser, mm liquid

It is convenient and consistent to relate all of these pressure-drop terms to the height of equivalent clear liquid (deaerated basis) on the tray, in either millimeters or inches of liquid.

Pressure drop across the disperser is calculated by variations of the standard orifice equation:

$$h_d = K\left(\frac{\rho_G}{\rho_L}\right)U_h^2 \tag{14-101}$$

where U_h = linear gas velocity through slots (valve trays) or perforations (sieve tray), m/s.

For sieve trays, $K = 50.8/C_v^2$. Values of C_v are taken from Fig. 14-36. Values from Fig. 14-36 may be calculated from

$$C_v = 0.74(A_h/A_a) + \exp[0.29(t_t/d_h) - 0.56] \tag{14-102a}$$

Recent work by Summers (*Chem. Eng.*, June 2009, p. 36) and Summers and Cai (Chem. Eng., August 2017, p. 38) shows that the dry pressure drop data fit can be improved by using the alternative equation

$$C_v = (d_h/p)^{0.1}\,[0.997 - 0.34/\{1.0 + (4.925\, t_t/d_h)^{3.582}\}] \tag{14-102b}$$

where t_t is the tray thickness, mm, and p is the center-to-center hole pitch, mm.

For Sulzer's fixed valve trays, Summers and van Sinderen (*Distillation 2001: Topical Conference Proceedings,* AIChE Spring National Meeting, Houston, April 22–26, 2001, p. 444) provided the following equation for K:

$$K = 58 + 386A_f \quad \text{for MVG fixed valves} \tag{14-103a}$$

$$K = 58 + 461A_f \quad \text{for SVG and LVG fixed valves} \tag{14-103b}$$

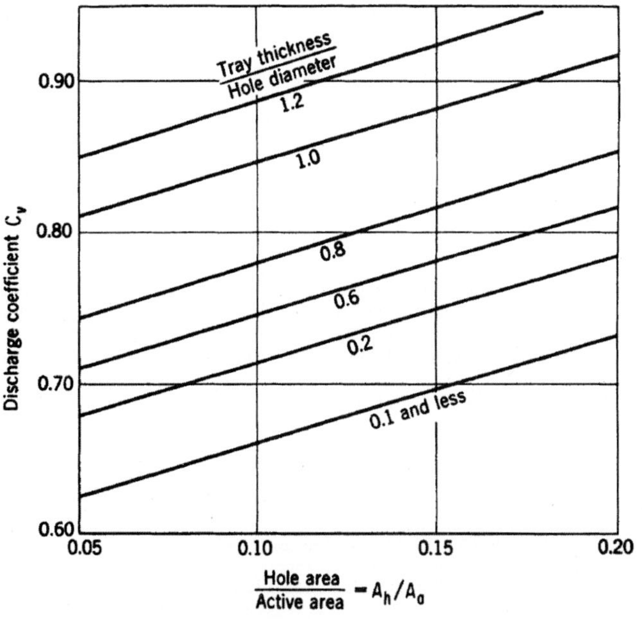

FIG. 14-36 Discharge coefficients for gas flow, sieve trays. [*Liebson, Kelley, and Bullington, Pet. Refiner* ***36****(3): 288 (1957).*]

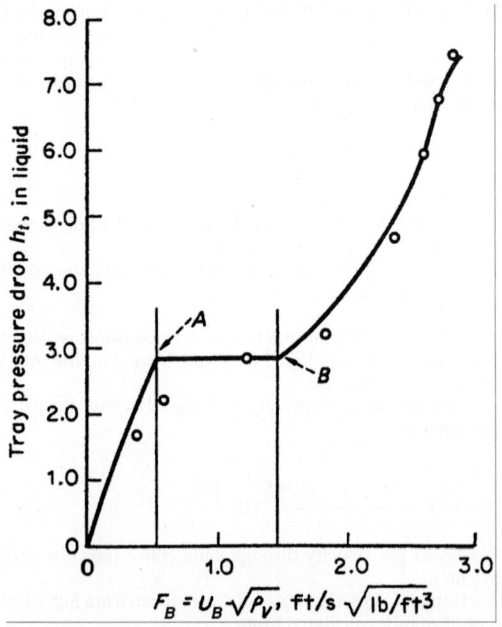

FIG. 14-37 Typical moving valve tray pressure-drop profile. (*From G. F. Klein, Chem. Eng., May 3, 1982, p. 81; reprinted courtesy of* Chemical Engineering.)

Figure 14-37 illustrates the pressure drop of a typical moving valve tray as a function of gas velocity. At low velocities, all valves are closed. Gas rises through the crevices between the valves and the tray deck, with increasing pressure drop as the gas velocity rises. Once point *A*, the closed balance point (CBP), is reached, some valves begin to open. Upon further increase in gas velocity, more valves open until point *B*, the open balance point (OBP), is reached. Between points *A* and *B*, gas flow area increases with gas velocity, keeping pressure drop constant. Further increases in gas velocity increase pressure drop similar to that in a sieve tray.

The term *K* in Eq. (14-101) depends on valve slot area, orifice geometry, deck thickness, and the type, shape, and weight of the valves. These are best obtained from the manufacturer's literature, but they can also be calculated from Bolles' [*Chem. Eng. Prog.* **72**(9): 43 (1976)], Lockett's (*Distillation Tray Fundamentals,* Cambridge University Press, Cambridge, UK, 1986), and Klein's (*Chem. Eng.,* May 3, 1982, p. 81) methods.

For valve trays, Klein gives the following values for *K* (in $s^2 \cdot mm/m^2$) in Eq. (14-101), when based on the total hole area (not slot area):

	Sharp orifice	Venturi valve
All valves open (K_O)	$254.5\,(2.64/t_t)^{0.5}$	122
All valves closed (K_C)	1683	841

The velocity at which the valves start to open (point *A*) is given by

$$U_{h,\text{closed}} = 1.14[t_v(R_{vw}/K_C)(\rho_M/\rho_G)]^{0.5} \tag{14-104}$$

where $U_{h,\text{closed}}$ = hole area at point *A*, m/s; t_v = valve thickness, mm; R_{vw} = ratio of valve weight with legs to valve weight without legs, given in Table 14-11; K_C = orifice coefficient with all valves closed (see above), $s^2 \cdot mm/m^2$; ρ_M = valve metal density, kg/m³ (about 8000 kg/m³ for steel); ρ_G = gas density, kg/m³.

The velocity at which all the valves are open $U_{h,\text{open}}$ can be calculated from

$$U_{h,\text{open}} = U_{h,\text{closed}}(K_C/K_O)^{0.5} \tag{14-105}$$

TABLE 14-11 R_{VW} Values for Eq. (14-104)

Valve type	Sharp	Venturi
Three-leg	1.23	1.29
Four-leg	1.34	1.45
Cages (no legs)	1.00	1.00

Pressure drop through the aerated liquid [h'_L, in Eq. (14-100)] is calculated by

$$h'_L = \beta h_{ds} \tag{14-106}$$

where β = aeration factor, dimensionless
h_{ds} = calculated height of clear liquid, mm (dynamic seal)

The aeration factor β has been determined from Fig. 14-38 for valve and sieve trays. For sieve trays, values of β in the figure may be calculated from

$$\beta = 0.0825 \ln \frac{Q}{L_w} - 0.269 \ln F_h + 1.679 \tag{14-107}$$

where L_w = weir length, m
F_h = *F*-factor for flow through holes, $F_h = U_h\rho_G^{0.5}$, m/s (kg/m³)$^{0.5}$

For sieve and valve trays,

$$h_{ds} = h_w + h_{ow} + 0.5h_{hg} \tag{14-108}$$

where h_w = weir height, mm
h_{ow} = height of crest over weir, mm clear liquid
h_{hg} = hydraulic gradient across tray, mm clear liquid

The value of weir crest h_{ow} may be calculated from the Francis weir equation and its modifications for various weir types. For a segmental weir and for height in millimeters of clear liquid,

$$h_{ow} = 664\left(\frac{Q}{L_w}\right)^{2/3} \tag{14-109}$$

where Q = liquid flow, m³/s
L_w = weir length, m

For serrated weirs,

$$h_{ow} = 851\left(\frac{Q'}{\tan\theta/2}\right)^{0.4} \tag{14-110}$$

where Q' = liquid flow, m³/s per serration
θ = angle of serration, deg

For circular weirs,

$$h_{ow} = 44{,}300\left(\frac{Q}{d_w}\right)^{0.704} \tag{14-111}$$

where q = liquid flow, m³/s
d_w = weir diameter, mm

For most sieve and valve trays, the hydraulic gradient is small and can be dropped from Eq. (14-108). Some calculation methods are available and are detailed in previous editions of this handbook. A rule of thumb for hydraulic gradients on sieve and valve trays which gives reasonable fit (albeit with large scatter) to FRI and industrial gamma scan data (Kister, *The Distillation Topical Conference*, AIChE Spring Meeting, Houston, Tex., April 2012) is 17 mm/m (0.2 in/ft) of flow path length in the froth and emulsion regimes (Q_L >54 m³/h · m or >6 gpm/in of outlet weir length), and zero in the spray regime (Q_L <18 m³/h · m or <2 gpm/in of outlet weir). For weir loads between 18 and 54 m³/h · m the rule recommends linear interpolation in weir loads.

As noted, the weir crest h_{ow} is calculated on an equivalent clear-liquid basis. A more realistic approach is to recognize that in general a froth or spray flows over the outlet weir (settling can occur upstream of the weir if a large "calming zone" with no dispersers is used). Bennett et al. [*AIChE J.* **29:** 434 (1983)] allowed for froth overflow in a comprehensive study of pressure drop across sieve trays; their correlation for residual pressure drop h'_L in Eq. (14-100) is presented in detail in the seventh edition of this handbook, including a worked example. Although more difficult to use, the method of Bennett et al. was recommended when determination of pressure drop is of critical importance.

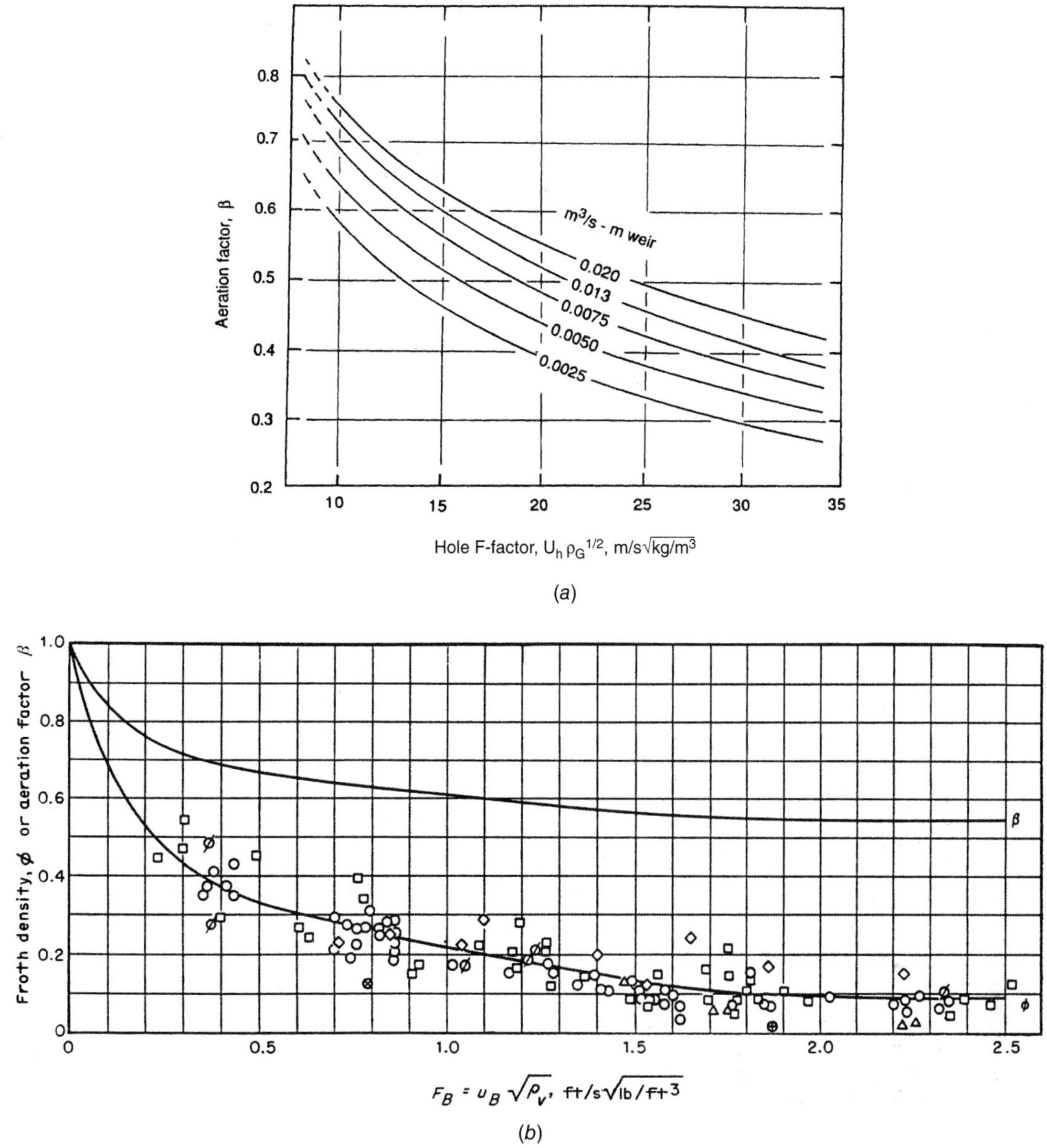

FIG. 14-38 Aeration factor for pressure drop calculation. (*a*) Sieve trays. [*Bolles and Fair,* Encyclopedia of Chemical Processing and Design, *vols. 16, 86. J. M. McKetta (ed.), Marcel Dekker, New York, 1982.*] (*b*) Valve trays. (*From G. F. Klein,* Chem. Eng., *May 3, 1982, p. 81; reprinted courtesy of* Chemical Engineering.)

Example 14-11 Pressure Drop, Sieve Tray For the conditions of Example 14-9, estimate the pressure drop for flow across one tray. The thickness of the tray metal is 2 mm and the pitch is 12.4 mm. The superficial *F*-factor is 2.08 m/s(kg/m³)$^{1/2}$.

Solution Equations (14-100), (14-106), and (14-107), where $h_t = h_d + \beta(h_w + h_{ow})$, are used. For $F_S = 2.08$, $F_B = 2.32$ and $F_H = 16.55$. From Example 14-9, $L_w = 1.50$ m and $h_w = 38$ mm. For a liquid rate of 22,000 kg/hr, $Q = 7.27(10^{-3})$ m³/s, and $Q/L_w = 4.8(10^{-3})$. By Eq. (14-107) or Fig. 14-38, $\beta = 0.48$. From Eq. (14-102a) or Fig. 14-36, $C_v = 0.75$. Then, by Eq. (14-101), $h_d = 29.0$ mm liquid. Alternatively, from Eq. (14-102b), $C_v = 0.88$, giving $h_d = 21.0$ mm liquid. For conservative estimate, select $h_d = 29.0$ mm liquid. Using Eq. (14-109), $h_{ow} = 18.9$ mm. Finally, $h_t = h_d + \beta(h_w + h_{ow}) = 29.0 + 0.48(38 + 18.9) = 56.4$ mm liquid.

When straight or serrated segmental weirs are used in a column of circular cross section, a correction may be needed for the distorted pattern of flow at the ends of the weirs, depending on liquid flow rate. The correction factor F_w from Fig. 14-39 is used directly in Eq. (14-109). Even when circular downcomers are used, they are often fed by the overflow from a segmental weir.

Loss Under Downcomer The head loss under the downcomer apron, as millimeters of liquid, may be estimated from

$$h_{da} = 165.2\left(\frac{Q}{A_{da}}\right)^2 \qquad (14\text{-}112)$$

where Q = volumetric flow of liquid, m³/s and A_{da} = most restrictive (minimum) area of flow under the downcomer apron, m². Equation (14-112) was derived from the orifice equation with an orifice coefficient of 0.6. Although the loss under the downcomer is small, the clearance is significant from the aspect of tray stability and liquid distribution.

The term A_{da} should be taken as the most restrictive area for liquid flow in the downcomer outlet. Usually, this is the area under the downcomer apron (i.e., the downcomer clearance times the length of the segmental downcomer), but not always. For instance, if an inlet weir is used and the area between the segmental downcomer and the inlet weir is smaller than the area under the downcomer apron, the smaller area should be used.

OTHER HYDRAULIC LIMITS

Weeping Weeping is liquid descending through the tray perforations, short-circuiting the contact zone, which lowers tray efficiency. At the tray floor, liquid static head that acts to push liquid down the perforations is counteracted by the gas pressure drop that acts to hold the liquid on the tray. When the static head overcomes the gas pressure drop, weeping occurs.

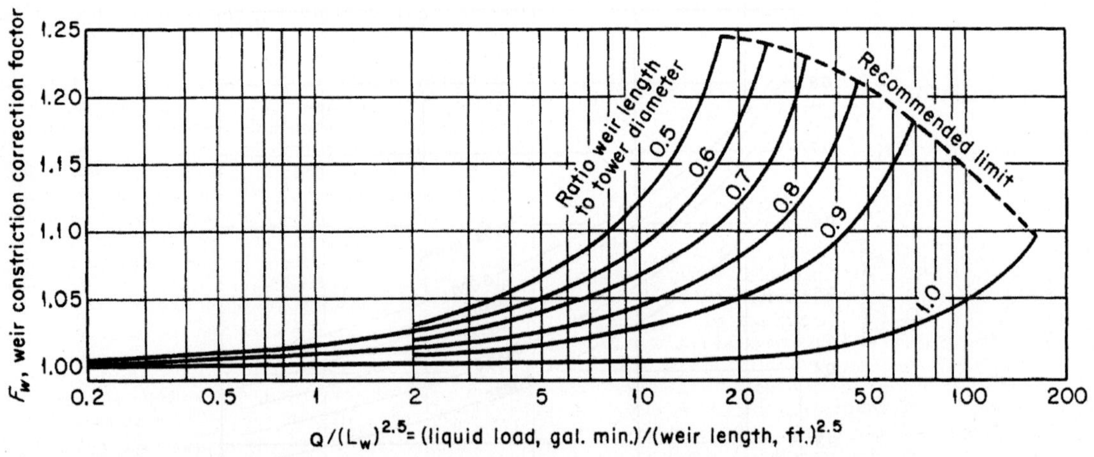

FIG. 14-39 Correction for effective weir length. To convert gallons per minute to cubic meters per second, multiply by 6.309×10^{-5}; to convert feet to meters, multiply by 0.3048. [*Bolles, Pet. Refiner* ***25:*** *613 (1946).*]

Some weeping usually takes place under all conditions due to sloshing and oscillation of the tray liquid. Generally, this weeping is too small to appreciably affect tray efficiency. The *weep point* is the gas velocity at which weeping first becomes noticeable. At this point, little efficiency is lost. As gas velocity is reduced below the weep point, the weep rate increases. When the weep rate becomes large enough to significantly reduce tray efficiency, the lower tray operating limit is reached.

The main factor that affects weeping is the fractional hole area. The larger it is, the smaller the gas pressure drop and the greater the weeping tendency. Larger liquid rates and taller outlet weirs increase liquid heads and therefore weeping. Hole diameter has a complex effect on weeping, detailed by Lockett and Banik [*Ind. Eng. Chem. Proc. Des. Dev.* **25:** 561 (1986)].

Tests by Lockett and Banik show that weeping is often nonuniform, with some hydraulic conditions favoring weeping from the tray inlet and others from the tray outlet. Weeping from the tray inlet is particularly detrimental to tray efficiency because the weeping liquid bypasses two trays.

Weep Rate Prediction Lockett and Banik [*Ind. Eng. Chem. Proc. Des. Dev.* **25:** 561 (1986)] and Hsieh and McNulty (*Chem. Eng. Progr.,* July 1993, p. 71) proposed correlations for predicting weep rates from sieve trays. Colwell and O'Bara (Paper presented at the AIChE Meeting, Houston, April 1989) recommended the Lockett and Banik correlation for low pressures (<1100 kPa or 165 psia), and the Hsieh and McNulty correlation for high pressures (>1100 kPa or 165 psia). They also corrected the Lockett and Banik correlation to improve its accuracy near the weep point.

The *Lockett and Banik correlation* (as corrected by Colwell and O'Bara) is

$$\frac{W}{A_h} = \frac{29.45}{\sqrt{\mathrm{Fr}_h}} - 44.18 \qquad \mathrm{Fr}_h < 0.2 \tag{14-113a}$$

$$\frac{W}{A_h} = \frac{1.841}{\mathrm{Fr}_h^{1.533}} \qquad \mathrm{Fr}_h > 0.2 \tag{14-113b}$$

where

$$\mathrm{Fr}_h = 0.373 \frac{u_h^2}{h_c} \frac{\rho_v}{\rho_L - \rho_v} \tag{14-114}$$

Equations (14-113) and (14-114) use English units: W is the weep rate, gpm; A_h is the hole area, ft^2; u_h is the hole velocity, ft/s; and h_c is the clear liquid height, in. Colwell's [*Ind. Eng. Chem. Proc. Des. Dev.* **20**(2): 298 (1981)] method below has been recommended for obtaining the clear liquid height h_c in Eq. (14-114).

$$h_c = \phi_f \left\{ 0.527 \left(\frac{Q_L(1-f_w)}{C_d \phi_f} \right)^{2/3} + h_w \right\} \tag{14-115}$$

where ϕ_f is given by Eq. (14-119) and C_d is given by Eq. (14-116),

$$\left.\begin{aligned} C_d &= 0.61 + 0.08\frac{h_{fow}}{h_w} & \frac{h_{fow}}{h_w} < 8.135 \\ C_d &= 1.06\left(1+\frac{h_w}{h_{fow}}\right)^{1.5} & \frac{h_{fow}}{h_w} > 8.135 \end{aligned}\right\} \tag{14-116}$$

$$h_{fow} = h_f - h_w \tag{14-117}$$

where h_f is given by Eq. (14-122). The froth density ϕ_f is calculated from

$$\eta = 12.6\mathrm{Fr}^{0.4}\left(\frac{A_B}{A_h}\right)^{0.25} \tag{14-118}$$

$$\phi_f = \frac{1}{\eta + 1} \tag{14-119}$$

$$\mathrm{Fr} = 0.37 \frac{\rho_v u_B^2}{h_c(\rho_L - \rho_v)} \tag{14-120}$$

The term f_w in Eq. (14-115) is the ratio of weep rate from the tray to the total liquid flow entering the tray, calculated as follows:

$$f_w = W/\mathrm{GPM} \tag{14-121}$$

Some trial and error is required in this calculation because the clear liquid height h_c and the froth density ϕ_f depend on each other, and the weep fraction f_w depends on the clear liquid height h_c. Clear liquid height is related to froth height and froth density by

$$h_c = \phi_f h_f \tag{14-122}$$

The terms in Eqs. (14-115) to (14-122) are in English units and are explained in the Nomenclature table.

With large-diameter trays and low liquid loads, a small ratio of W/A_h corresponds to a large fractional weep. Under these conditions, the Lockett and Banik correlation is inaccurate. The correlation is unsuitable for trays with very small (<3-mm or $\frac{1}{8}$-in) holes. The correlation appears to fit most data points to an accuracy of ±15 to ±30 percent.

The *Hsieh and McNulty correlation* (*Chem. Eng. Progr.,* July 1993, p. 71) is

$$\sqrt{J_G^*} + m\sqrt{J_L^*} = C_w \tag{14-123}$$

where

$$J_G^* = u_h \left[\frac{\rho_G}{gZ(\rho_L - \rho_G)} \right]^{0.5} \tag{14-124}$$

and

$$J_L^* = \frac{W}{448.83 A_h}\left[\frac{\rho_L}{gZ(\rho_L - \rho_G)}\right]^{0.5} \quad (14\text{-}125)$$

$$Z = h_C^{1.5}/(12 d_H^{0.5}) \quad (14\text{-}126)$$

The terms in Eqs. (14-123) to (14-126) are in English units and are explained in the Nomenclature. For sieve trays, $m = 1.94$ and $C_w = 0.79$. Note that the constants are a slight revision of those presented in the original paper (C. L. Hsieh, private communication, 1991). Clear liquid height is calculated from Colwell's correlation [Eqs. (14-115) to (14-122)]. The Hsieh and McNulty correlation applies to trays with 9 percent and larger fractional hole area. For trays with smaller hole area, Hsieh and McNulty expect the weeping rate to be smaller than predicted.

Weeping from Valve Trays An analysis of weeping from valve trays [Bolles, *Chem. Eng. Progs.* **72**(9): 43 (1976)] showed that in a well-designed valve tray, the weep point is below the gas load at which the valves open; and throughout the valve opening process, the operating point remains above the weep point. In contrast, if the tray contains too many valves, or the valves are too light, excessive valve opening occurs before the gas pressure drop is high enough to counter weeping. In this case, weeping could be troublesome.

Weep point correlations for valve trays were presented by Bolles [*Chem. Eng. Progs.* **72**(9): 43 (1976)] and by Klein (*Chem. Eng.*, Sept. 17, 1984, p. 128). Hsieh and McNulty (*Chem. Eng. Progr.*, July 1993, p. 71) gave a complex extension of their weep rate correlation to valve trays.

Dumping As gas velocity is lowered below the weep point, the fraction of liquid weeping increases until all the liquid fed to the tray weeps through the holes and none reaches the downcomer. This is the *dump point*, or the *seal point*. The dump point is well below the range of acceptable operation of distillation trays. Below the dump point, tray efficiency is slashed, and mass transfer is extremely poor. Operation below the dump point can be accompanied by severe hydraulic instability due to unsealing of downcomers.

Extensive studies on dumping were reported by Prince and Chan [*Trans. Inst. Chem. Engr.* **43:** T49 (1965)]. The Chan and Prince dump-point correlation was recommended and is presented in detail elsewhere (Kister, *Distillation Design*, McGraw-Hill, 1992). Alternatively, the dump point can be predicted by setting the weep rate equal to 100 percent of the liquid entering the tray in the appropriate weep correlation.

Stability at Low Vapor Rates Summers (*Distillation and Absorption 2010*, Einhoven University of Technology, The Netherlands, p. 611) analyzed weep data to show that the following criterion for stable tray operation at turndown keeps the weep rate well below 30 percent:

$$\eta = (h_d/h'_L)^{0.5} > 0.6 \quad (14\text{-}126a)$$

where η is the stability factor, dimensionless; h_d is the dry pressure drop, mm of liquid; and h'_L is the clear liquid height, mm liquid. Although the clear liquid heights used in Summers' derivation were calculated from pressure drop data using Eq. (14-100), he later recommended to the author (private communication, 2017) that the Colwell correlation [Eq. (14-115)] be used for the clear liquid height h'_L calculation. The stability factor is proportional to the hole Froude number, which represents the ratio of the upward vapor inertial forces keeping the liquid on the tray to the downward gravity forces pulling the liquid down the holes.

For two-pass and multipass trays, an additional instability may result from vapor channeling through one panel and not another. Summers proposes an additional preliminary criterion for such trays, but this criterion needs data verification. Olsson and Kister (*Kister Distillation Symposium*, AIChE Spring Meeting, March 26–30, San Antonio, Tex., 2017) identified a multiplicity of stable hydraulic steady states for two-pass and multipass moving valve trays operating in the flat region of Fig. 14-37, some with severely channeled vapor through one of the panels, with a likely efficiency and turndown loss. Their findings need data verification.

Turndown The turndown ratio is the ratio of the normal operating (or design) gas throughput to the minimum allowable gas throughout. The minimum allowable throughput is usually set by excessive weeping, while normal operating throughput is a safe margin away from the relevant flooding limit.

Sieve and fixed valve trays have a poor turndown ratio (about 2:1). Their turndown can be improved by blanking some rows of tray holes, which reduces the tendency to weep, but this will also reduce the tray's maximum capacity. Turndown of *moving valve trays* is normally between about 4:1 and 5:1. Special valve designs can achieve even better turndown ratios, between 6:1 and 10:1, and even more. Turndown can also be enhanced by blanking strips (which require valve removal) or valve leg crimping. Sloley and Fleming (*Chem. Eng. Progr.*, March 1994, p. 39) stress that correct implementation of turndown enhancement is central to achieving a desired turndown. When poorly implemented, turndown may be restricted by poor vapor–liquid contact rather than by weeping.

Vapor Channeling All the correlations in this section assume an evenly distributed tray vapor. When the vapor preferentially channels through a tray region, premature entrainment flood and excessive entrainment take place due to a high vapor velocity in that region. At the same time, other regions become vapor-deficient and tend to weep, which lowers tray efficiency.

Work by Davies [*Pet. Ref.* **29**(8): 93 and **29**(9): 121 (1950)] based on bubble-cap tray studies suggests that the vapor pressure drop of the tray (the *dry pressure drop*) counteracts channeling. The higher the dry tray pressure drop, the greater the tendency for vapor to spread uniformly over the bubbling area. If the dry tray pressure drop is too small compared with the channeling potential, channeling prevails.

Perhaps the most common vapor channeling mechanism is *vapor cross-flow channeling* (VCFC, Fig. 14-40). The hydraulic gradient on the tray induces preferential vapor rise at the outlet and middle of the tray and a vapor-deficient region near the tray inlet. The resulting high vapor velocities near the tray outlet step up entrainment, while the low vapor velocities near the tray inlet induce weeping. Interaction between adjacent trays (Fig. 14-40) accelerates both the outlet entrainment and the inlet weeping. The net result is excessive entrainment and premature flooding at the tray middle and outlet, simultaneous with weeping from the tray inlet, accompanied by a loss of efficiency and turndown.

VCFC takes place when the following four conditions exist simultaneously [Kister, Larson, and Madsen, *Chem. Eng. Progr.*, November 1992, p. 86; Kister, *The Chemical Engineer* **544:** 18 (1993); Kister, Clancy-Jundt, and Miller., *PTQ* **Q1:** 39 (2015)]:

1. Absolute pressure <800 kPa (120 psia).
2. High liquid rates [>50 $m^3/(m \cdot h)$ or 6 gpm/in of outlet weir].
3. High ratio (>2:1) of flow path length to tray spacing.
4. Low dry tray pressure drop. On sieve and fixed valve trays, this means high (>11 and 12 percent, respectively) fractional hole area. On moving valve trays, this means venturi valves (smooth orifices) or long-legged valves (>15 percent slot area). On all trays, the channeling tendency and severity escalate rapidly as the dry pressure drop diminishes (e.g., as fractional hole area increases).

Hartman (*Distillation 2001: Topical Conference Proceedings*, AIChE Spring National Meeting, Houston, Tex., April 22–26, 2001, p. 108) reports VCFC even with conventional valve trays (14 percent slot area) at a very high ratio (3.6:1) of flow path length to tray spacing and tray truss obstruction.

VCFC is usually avoided by limiting fractional hole areas, avoiding venturi valves, and using forward-push devices. Resitarits and Pappademos (Paper presented at the AIChE Annual Meeting, Reno, Nevada, November 2001) cited tray inlet inactivity as a contributor to VCFC, and they advocate inlet forward-push devices to counter it.

Downcomer Unsealing When a downcomer loses its liquid seal, gas rises through it and interferes with liquid descent, leading to capacity bottlenecks, poor separation, instability, and inability to start up.

On conventional trays, at weir loads exceeding 20 to 30 $m^3/h \cdot m$ of outlet weir length, the outlet weir generates a frothy pool (Fig. 14-20*a*, *b*). This liquid pool will seal the downcomer as long as the downcomer clearance is lower than the liquid head. In contrast, at low weir loads (<20 to 30 $m^3/h \cdot m$ of outlet weir length), and especially at high gas velocities, the gas atomizes

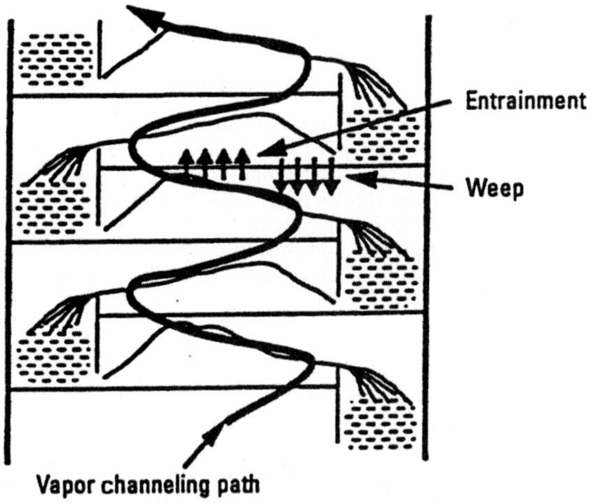

FIG. 14-40 Vapor cross-flow channeling. Note entrainment near the tray middle and outlet, and weep near the tray inlet. (*Kister, H. Z., K. F. Larson, and P. Madsen,* Chem. Eng. Prog., *Nov. 1992, p. 86; reproduced with permission.*)

the tray liquid, forming a dispersion known as the spray regime or drop regime. In this regime, there is no frothy pool, and the liquid resides as atomized drops in the inter-tray space (Fig. 14-20*c*). The continuous gas phase reaches the downcomer outlet. The absence of static seal preventing gas rise up the downcomer renders the downcomers prone to seal loss.

Downcomer unsealing is common at start-ups and/or with low weir loads. With little liquid on the trays, the gas ascends through both the trays and downcomers, following the path of least resistance. A large downcomer clearance gives little resistance, allowing a large amount of gas into the downcomer. The high gas velocity through the downcomer impedes liquid descent. Unable to descend, the liquid accumulates on the tray above, flooding it. The flood may either propagate up the tower or intermittently flood a few trays and then dump, generating instability. Detailed modeling and troubleshooting guidelines were presented by Kister and Mohamed (*Gas Processing*, January/February 2015, p. 39, and March/April 2015, p. 49). Keeping downcomer clearances at 25 mm or less in clean services (up to 40 mm in fouling services), judicious use of inlet weirs, and switching to well-designed bubble-cap trays have been effectively applied to prevent downcomer unsealing. It is important to emphasize that these measures should only be considered at low weir loads (<20 to 30 $m^3/h \cdot m$ of outlet weir length). At weir loads exceeding 30 $m^3/h \cdot m$ of outlet weir length, a liquid pool forms on the tray, rendering these measures unnecessary, even detrimental.

Downcomer seal loss is also encountered in many high-capacity trays, in which maximizing capacity leads to downcomers that terminate in the inter-tray vapor space, well above the floor of the tray below. Terminating in the vapor space, the liquid pool on the tray does not reach their outlets, so vapor can easily break into these downcomers. Bernard (*Distillation Topical Conference Proceedings*, AIChE Spring Meeting, Chicago, Ill., March 2011) demonstrated that a model derived for low weir loads can be extended to evaluate unsealing in high-capacity trays, and to devise a successful start-up strategy.

Downcomer seal loss can also be caused by tray damage or by tray plugging. With a plugged tray, the pressure drop through the active area rises, incurring high resistance to gas flow, making going up a downcomer the path of lesser resistance. When the tray plugging is extensive, gas can break into downcomers, especially when there are multiple downcomers.

TRANSITION BETWEEN FLOW REGIMES

Froth–Spray Froth–spray transition has been investigated for sieve trays using a variety of techniques. The gradual nature of this transition bred a multitude of criteria for defining it and made its correlation difficult. Excellent overviews were given by Lockett (*Distillation Tray Fundamentals*, Cambridge University Press, Cambridge, UK, 1986) and Prado, Johnson, and Fair [*Chem. Eng. Progr.* **83**(3): 32 (1987)]. Porter and Jenkins [*I. Chem. E. Symp. Ser.* **56:** Summary Paper (1979)] presented a simple correlation for the froth-to-spray transition.

$$F_{LG}\frac{N_p A_B}{L_w} = 0.0191 \tag{14-127}$$

The terms of this equation are in English units and are explained in the Nomenclature. This correlation is based on the premise that froth-to-spray transition occurs when the entrainment versus liquid load relationship passes through a minimum (see the subsection Entrainment). Alternatively, it was argued that the minimum represents a transition from the froth regime to a partially developed spray region (Kister, Pinczewski, and Fell, paper presented in the 90th National AIChE Meeting, Houston, April 1981). If this alternative argument is valid, then when the correlation predicts froth, it is highly unlikely that the column operates in the spray regime; but when it predicts spray, the column may still be operating in the froth regime. Entrainment studies by Ohe (*Distillation 2005: Topical Conference Proceedings*, AIChE Spring National Meeting, Atlanta, April 10–13, 2005, p. 283) argue that the entrainment minima represent minimum liquid residence times on the tray and are unrelated to the froth–spray transition.

A second correlation is by Pinczeweski and Fell [*Ind. Eng. Chem. Proc. Des. Dev.* **21:** 774 (1982)]

$$u_B\sqrt{\rho_G} = 2.25\left(\frac{Q_L\sqrt{\rho_L}}{100}\right)^n \tag{14-128}$$

The terms of Eq. (14-128) are in English units and are explained in the Nomenclature. The exponent n is calculated from Eq. (14-84). Equation (14-128) is based on transition data obtained from orifice jetting measurements for the air–water system and on entrainment minimum data for some hydrocarbon systems.

Summers and Sloley (*Hydroc. Proc.*, January 2007, p. 67) extended a correlation by Lockett (*Distillation Tray Fundamentals*, Cambridge University Press, Cambridge, UK, 1986) to define a spray factor given by

$$\text{Spray factor} = Kh_c\rho_L^{0.5}/(d_h u_h \rho_V^{0.5}) \tag{14-129}$$

where $K = 1$ for sieve trays, 2.5 for movable or fixed valves, h_c is the clear liquid height, mm, calculated using the Colwell correlation Eq. (14-115), u_h is the hole velocity, m/s, and other symbols are as described above.

According to Summers and Sloley, spray factors exceeding 2.78 in Eq. (14-129) signify operation definitely outside of the spray regime. Below 2.78, the lower the spray factor, the higher the potential for the operation to be in the spray regime, with vapor lifting liquid off the tray deck (or blowing the liquid off the tray). The equation indicates that devices where the vapor enters the tray horizontally are much less prone to liquid lifting off the tray. Compared to the two earlier correlations, Eq. (14-129) shows a much stronger dependence on the fractional hole area.

Froth–Emulsion Froth–emulsion transition occurs [Hofhuis and Zuiderweg, *I. Chem. E. Symp. Ser.* **56:** 2, 2/1 (1979)] when the aerated mass begins to obey the Francis weir formula. Using this criterion, the latest version of this transition correlation is

$$F_{LG}\frac{N_p A_B}{L_w h_c} = 0.0208 \tag{14-130}$$

The terms of this equation are in English units and are explained in the Nomenclature; h_c is calculated from the Hofhuis and Zuiderweg equation.

$$h_c = 2.08\left(F_{LG}\frac{N_p A_B}{L_w}p\right)^{0.25} h_w^{0.5} \tag{14-131}$$

An inspection of the experimental data correlated shows that this, too, is a gradual transition, which occurs over a range of values rather than at a sharp point.

TRAY EFFICIENCY

Definitions

Overall Column Efficiency This is the ratio of the number of theoretical stages to the number of actual stages

$$E_{OC} = N_t/N_a \tag{14-132}$$

Since tray efficiencies vary from one section to another, it is best to apply Eq. (14-132) separately for the rectifying and stripping sections. In practice, efficiency data and prediction methods are often too crude to give a good breakdown between the efficiencies of different sections, and so Eq. (14-132) is applied over the entire column.

Point Efficiency This is defined by Eq. (14-133) (Fig. 14-41*a*):

$$E_{OG} = \left(\frac{y_n - y_{n-1}}{y_n^* - y_{n-1}}\right)_{\text{point}} \tag{14-133}$$

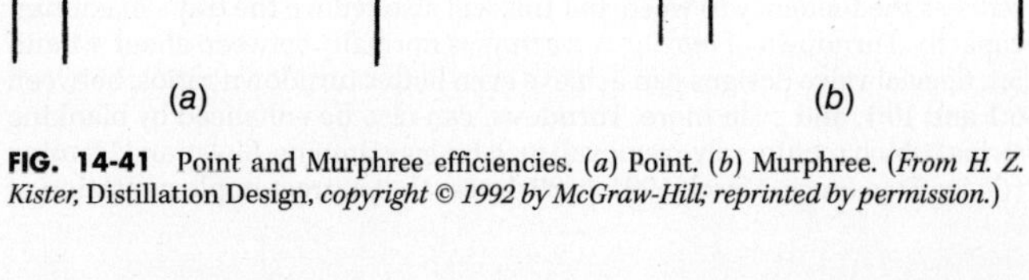

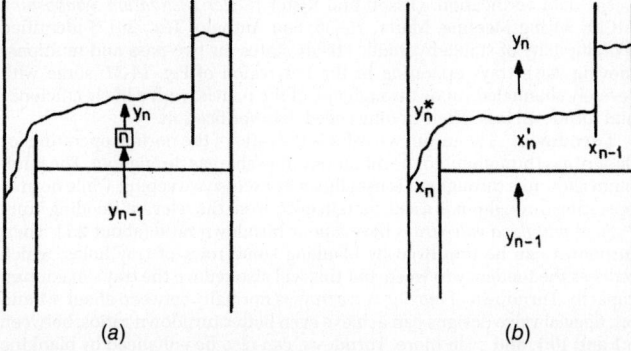

FIG. 14-41 Point and Murphree efficiencies. (*a*) Point. (*b*) Murphree. (*From H. Z. Kister,* Distillation Design, *copyright © 1992 by McGraw-Hill; reprinted by permission.*)

where y_n^* is the composition of vapor in equilibrium with the liquid at point n. The term y_n is actual vapor composition at that point. The point efficiency is the ratio of the change of composition at a point to the change that would occur on a theoretical stage. As the vapor composition at a given point cannot exceed the equilibrium composition, fractional point efficiencies are always lower than 1. If there is a composition gradient on the tray, point efficiency will vary between points on the tray.

Murphree Tray Efficiency [*Ind. Eng. Chem.* **17:** 747 (1925)] This is the same as point efficiency, except that it applies to the entire tray instead of to a single point (Fig. 14-41*b*):

$$E_{MV} = \left(\frac{y_n - y_{n-1}}{y_n^* - y_{n-1}}\right)_{\text{tray}} \tag{14-134}$$

If both liquid and vapor are perfectly mixed, liquid and vapor compositions on the tray are uniform, and the Murphree tray efficiency will coincide with the point efficiency at any point on the tray. In practice, a concentration gradient exists in the liquid, and x_n at the tray outlet is lower than x'_n on the tray (see Fig. 14-41*b*). This frequently lowers y_n^* relative to y_n, thus enhancing tray efficiency [Eq. (14-134)] compared with point efficiency. The value of y_n^* may even drop below y_n. In this case, E_{MV} exceeds 100 percent [Eq. (14-134)].

Overall column efficiency can be calculated from the Murphree tray efficiency by using the relationship developed by Lewis [*Ind. Eng. Chem.* **28:** 399 (1936)].

$$E_{OC} = \frac{\ln[1+E_{MV}(\lambda-1)]}{\ln \lambda} \tag{14-135}$$

where

$$\lambda = m\frac{G_M}{L_M} \tag{14-136}$$

Equation (14-135) is based on the assumption of constant molar overflow and a constant value of E_{MV} from tray to tray. It needs to be applied separately to each section of the column (rectifying and stripping) because G_M/L_M, and therefore λ, varies from section to section. Where molar overflow or Murphree efficiencies vary throughout a section of column, the section needs to be divided into subsections small enough to render the variations negligible.

The point and Murphree efficiency definitions just given are expressed in terms of vapor concentrations. Analogous definitions can be made in terms of liquid concentrations. Further discussion is elsewhere (Lockett, *Distillation Tray Fundamentals,* Cambridge University Press, Cambridge, UK, 1986).

Fundamentals Figure 14-42 shows the sequence of steps for converting phase resistances to a tray efficiency. Gas and liquid film resistances are added to give the point efficiency. Had both vapor and liquid on the tray been perfectly mixed, the Murphree tray efficiency would have equaled the point efficiency. Since the phases are not perfectly mixed, a model of the vapor and liquid mixing patterns is needed for converting point efficiency to tray efficiency. Liquid mixing patterns are plug flow, backmixing, and stagnant zones, while vapor mixing patterns are perfect mixing and plug flow.

Lewis [*Ind. Eng. Chem.* **28:** 399 (1936)] was the first to derive quantitative relationships between the Murphree and the point efficiencies. He derived three mixing cases, assuming plug flow of liquid in all. The Lewis cases give the maximum achievable tray efficiency. In practice, efficiency is lower due to liquid and vapor nonuniformities and liquid mixing.

Most tray efficiency models are based on Lewis case 1 with vapor perfectly mixed between trays. For case 1, Lewis derived the following relationship:

$$E_{MV,\text{dry}} = \frac{\exp(\lambda E_{OG})-1}{\lambda} \tag{14-137}$$

The "dry" Murphree efficiency calculated thus far takes into account the vapor and liquid resistances and the vapor–liquid contact patterns, but it is uncorrected for the effects of entrainment and weeping. This correction converts the dry efficiency to a "wet" or actual Murphree tray efficiency. Colburn [Eq. (14-98), under Entrainment] incorporated the effect of entrainment on efficiency, assuming perfect mixing of liquid on the tray.

Factors Affecting Tray Efficiency Following is a summary based on the industry's experience. A detailed discussion of the fundamentals is found in Lockett's book (*Distillation Tray Fundamentals,* Cambridge University Press, Cambridge, UK, 1986). A detailed discussion of the reported

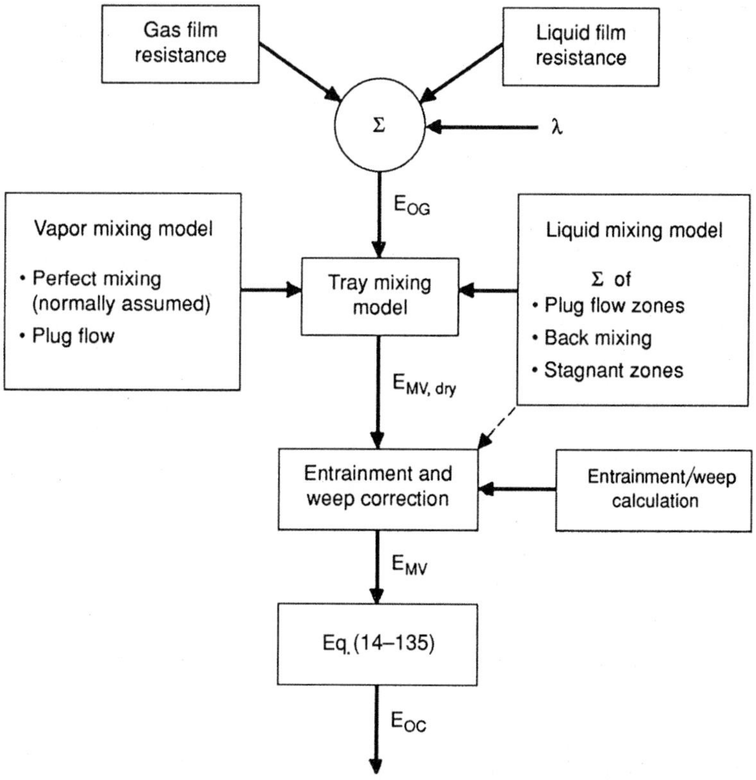

FIG. 14-42 Sequence of steps for theoretical prediction of tray efficiency. (*From H. Z. Kister,* Distillation Design, *copyright © 1992 by McGraw-Hill; reprinted by permission.*)

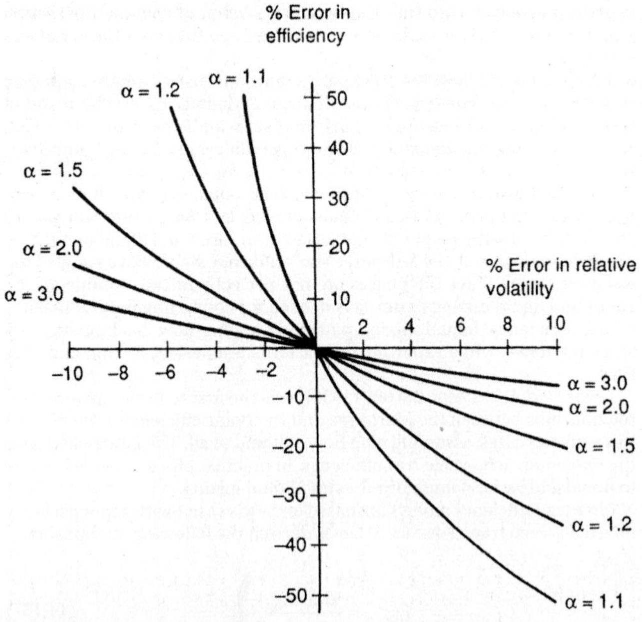

FIG. 14-43 Direct effect of errors in relative volatility on error in tray efficiency. (*From H. Z. Kister,* Distillation Design, *copyright © 1992 by McGraw-Hill; reprinted by permission.*)

experience, and the basis of statements made in this section, are in Kister's book (*Distillation Design,* McGraw-Hill, New York, 1992) and later paper (*Chem. Eng. Progr.*, June 2008, p. 39). One useful finding both by Kister and by Hennigan (*Distillation 2011: Topical Conference Proceedings,* AIChE Spring National Meeting, Chicago, Ill., March 2011, p. 151) is that for hydraulically sound tray designs, tray geometry has only a small influence on tray efficiency.

Errors in Vapor–Liquid Equilibrium (VLE) Errors in relative volatility are the most underrated factor affecting tray efficiency. Figure 14-43 shows the direct effect of the errors [Deibele and Brandt, *Chem. Ing. Tech.* **57**(5): 439 (1985); Roy P. and G. K. Hobson, *I. Chem. E. Symp. Ser.* **104:** A273 (1987)]. At very low relative volatilities ($\alpha < 1.2$), small errors in VLE have a huge impact on tray efficiency. For instance, at $\alpha = 1.1$, a −3 percent error gives a tray efficiency 40 to 50 percent higher than its true value (Fig. 14-43). Since VLE errors are seldom lower than ±2 to 3 percent, tray efficiencies of low-volatility systems become meaningless unless accompanied by VLE data. Likewise, comparing efficiencies derived for a low-volatility system by different sources is misleading unless one is using identical VLE.

Figure 14-43 shows that errors in relative volatility are a problem only at low relative volatilities; for $\alpha > 1.5$ to 2.0, VLE errors have negligible direct impact on tray efficiency.

Most efficiency data reported in the literature are obtained at total reflux, and there are no indirect VLE effects. For measurements at finite reflux ratios, the indirect effects below compound the direct effect of Fig. 14-43. Consider a case where $\alpha_{\text{apparent}} < \alpha_{\text{true}}$ and test data at a finite reflux are analyzed to calculate tray efficiency. Due to the volatility difference, $R_{\text{min,apparent}} > R_{\text{min,true}}$. Since the test was conducted at a fixed reflux flow rate, $(R/R_{\text{min}})_{\text{apparent}} < (R/R_{\text{min}})_{\text{true}}$. A calculation based on the apparent R/R_{min} will give more theoretical stages than a calculation based on the true R/R_{min}. This means a higher apparent efficiency than the true value.

The indirect effects add to those of Fig. 14-43, widening the gap between true and apparent efficiency. The indirect effects exponentially escalate as minimum reflux is approached. Small errors in VLE or reflux ratio measurement (this includes column material balance as well as reflux rate) alter R/R_{min}. Near minimum reflux, even small R/R_{min} errors induce huge errors in the number of stages, and therefore in tray efficiency. Efficiency data obtained near minimum reflux are therefore meaningless and potentially misleading.

Accuracy of Mass Balance and Reflux/Reboil Measurements Errors in these variables affect efficiencies derived from test data. This includes errors in physical properties (e.g., densities, latent heats) that are used for evaluating the molar reflux ratio. When the measured apparent molar reflux ratio is lower than the true reflux ratio, a calculation based on the measured reflux ratio will give a false high tray efficiency. These errors escalate exponentially as minimum reflux is approached.

Liquid Flow Patterns on Large Trays The most popular theoretical models (below) postulate that liquid crosses the tray in plug flow with superimposed backmixing, and that the vapor is perfectly mixed. Increasing tray diameter promotes liquid plug flow and suppresses backmixing.

The presence of stagnant zones on large-diameter distillation trays is well established, but the associated efficiency loss is poorly understood; in some cases, significant efficiency losses, presumably due to stagnant zones, were reported [Weiler, Kirkpatrick, and Lockett, *Chem. Eng. Progr.* **77**(1): 63 (1981)], while in other cases, no efficiency difference was observed [Yanagi and Scott, *Chem. Eng. Progr.* **69**(10): 75 (1973)]. Several techniques are available for eliminating stagnant regions (see Kister, *Distillation Design,* McGraw-Hill, New York, 1992, for some), but their effectiveness for improving tray efficiency is uncertain.

Weir Height Taller weirs raise the liquid level on the tray in the froth and emulsion regimes. This increases interfacial area and vapor contact time, which should theoretically enhance efficiency. In the spray regime, weir height affects neither liquid level nor efficiency. In distillation systems, the improvement of tray efficiency due to taller weirs is small, often marginal.

Length of Liquid Flow Path Longer liquid flow paths enhance the liquid–vapor contact time and the significance of liquid plug flow, and therefore raise efficiency. Typically, doubling the flow path length (such as going from two-pass to one-pass trays at a constant tower diameter) raises tray efficiency by 5 to 15 percent.

Fractional Hole Area Efficiency increases with a reduction in fractional hole area. Tests by Yanagi and Sakata [*Ind. Eng. Chem. Proc. Des. Dev.* **21:** 712 (1982)] in commercial-scale towers showed a 5 to 10 percent increase in tray efficiency when fractional hole area was lowered from 14 to 8 percent (Fig. 14-44).

Hole Diameter The jury is out on the effect of hole diameter on tray efficiency. There is, however, a consensus that the effect of hole diameter on efficiency is small, often negligible.

Tray Spacing Commercial-scale test data (Kister, *Chem. Eng. Progr.*, June 2008, p. 39) showed little effect of tray spacing on tray efficiency.

Vapor–Liquid Loads and Reflux Ratio Vapor and liquid loads, as well as the reflux ratio, have a small effect on tray efficiency (Fig. 14-44) as long as no capacity or hydraulic limits (flood, weep, channeling, etc.) are violated.

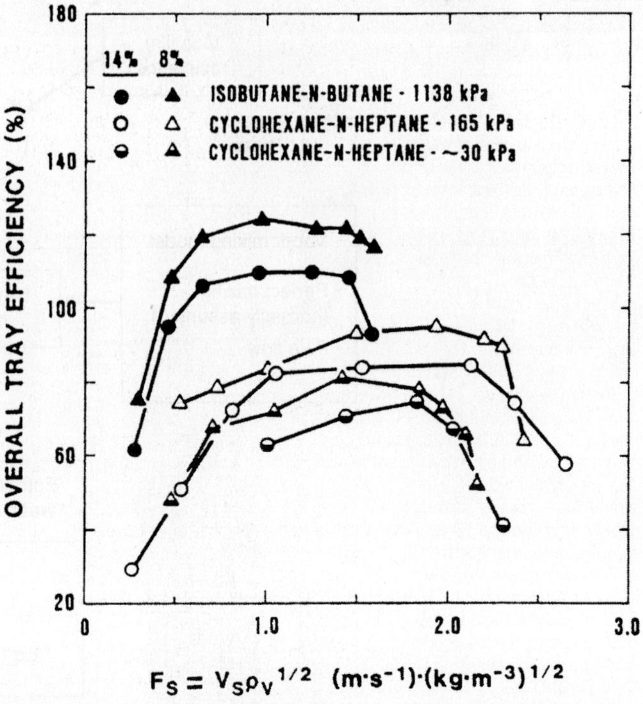

FIG. 14-44 Efficiency reduction when fractional hole area is increased, also showing little effect of vapor and liquid loads on efficiency in the normal operating range (between excessive weeping and excessive entrainment). Also shown is the small increase in efficiency with pressure. FRI data, total reflux, D_T = 1.2 m, S = 610 mm, h_w = 50.8 mm, d_H = 12.7 mm. (*Reprinted with permission from T. Yanagi and M. Sakata,* Ind. Eng. Chem. Proc. Des. Dev. *21: 712; copyright © 1982, American Chemical Society.*)

Viscosity, Relative Volatility Efficiency increases as liquid viscosity and relative volatility diminish. These effects are reflected in the O'Connell correlation, to be described later.

Two Liquid Phases Herron, Kruelskie, and Fair's [*AIChE J.* **34**(8): 1267 (1988)] oil–water tests in a pilot-scale tower showed that gas agitation on the trays caused the two liquid phases to behave as a homogenous liquid that followed general correlations for pressure drop, liquid holdup, froth height, downcomer backup, and entrainment with no foaming or unusual efficiency trends. They concluded that when immiscible liquids are both present in significant proportions, designers need not fear unusual hydraulic or mass-transfer behavior. The Davies, Ali, and Porter [*AIChE J.* **33**(1): 161 (1987)] *n*-hexane/1-propanol/water tests in an Oldershaw column showed foaming only on those plates with liquid composition near that of the one-phase/two-phase transition. Mortaheb, Kosuge, and Asano [*Chem. Eng. J.* **88:** 59 (2002)] concurred with Herron et al. on tray efficiency but noted that the second liquid phase raised the mass-transfer resistance in the liquid.

Surface Tension There is uncertainty about the effect of surface tension on tray efficiency. Often, it is difficult to divorce the surface tension effects from those of other physical properties.

Pressure Tray efficiency slightly increases with pressure (Fig. 14-44), reflecting the rise of efficiency with a reduction in liquid viscosity and in relative volatility, which generally accompany a distillation pressure increase.

At pressures exceeding 10 to 20 bar (150 to 300 psia), and especially at high liquid rates, vapor entrainment into the downcomer liquid becomes important, and tray efficiency decreases with further increases in pressure [Zuiderweg, *Int. Chem. Eng.* **26**(1): 1 (1986)].

Maldistribution Maldistribution can cause major efficiency reduction in multipass trays (>two passes). Further discussion appears in the subsection Number of Passes.

OBTAINING TRAY EFFICIENCY

Efficiency prediction methods are listed below in decreasing order of reliability. An excellent paper by Hennigan (*Distillation 2011: Topical Conference Proceedings,* AIChE Spring National Meeting, Chicago, Ill., March, 2011, p. 151) offers methods to critically review efficiency numbers and to manage the uncertainties involved.

Rigorous Testing Rigorous testing of a plant column is generally the most reliable method of obtaining tray efficiency. Test procedures can be found elsewhere (AIChE Equipment Testing Procedures Committee, *AIChE Equipment Testing Procedure—Trayed and Packed Columns,* 3d ed., Wiley, New York, 2014; Cai, "Column Performance Testing Procedures," chap. 3 in Gorak and Schoenmakers, *Distillation Operation and Application,* Academic Press, New York, 2014).

Scale-Up from an Existing Commercial Column As long as data are for the same system under similar process conditions, loadings, and operating regimes, data obtained in one column directly extend to another. Fractional hole area and the number of tray passes have a small but significant effect on efficiency, and any changes in these parameters need to be allowed for during scale-up. The empirical information in the subsection Factors Affecting Tray Efficiency can be used to estimate the magnitude of the changes on efficiency.

Scale-Up from Existing Commercial Column to Different Process Conditions During scale-up, test data are analyzed by computer simulation. The number of theoretical stages is varied until the simulated product compositions and temperature profile match the test data. Tray efficiency is determined by the ratio of theoretical stages to actual trays. In this procedure, errors in VLE are offset by compensating errors in tray efficiency. For instance, if the relative volatility calculated by the simulation is too high, fewer stages will be needed to match the measured data, that is, "apparent" tray efficiency will be lower. Scale-up will be good as long as the VLE and efficiency errors continue to offset each other equally. This requires that process conditions (feed composition, feed temperature, reflux ratio, etc.) remain unchanged during scale-up.

When process conditions change, the VLE and efficiency errors no longer offset each other equally. If the true relative volatility is higher than simulated, then the scale-up will be conservative. If the true relative volatility is lower than simulated, the scale-up will be optimistic. A detailed discussion is found in Kister, *Distillation Design,* McGraw-Hill, New York, 1992.

Experience Factors These are tabulations of efficiencies previously measured for various systems. Tray efficiency is insensitive to tray geometry (above), so in the absence of hydraulic anomalies and issues with VLE data, efficiencies measured in one tower are extensible to others distilling the same system. A small allowance to variations in tray geometry as discussed above is in order. Caution is required with mixed aqueous-organic systems, where concentration may have a marked effect on physical properties, relative volatility, and efficiency. Table 14-12 shows typical tray efficiencies reported in the literature.

Vital, Grossel, and Olsen [*Hydroc. Proc.* **63**(11): 147 (1984)] and Garcia and Fair [*Ind. Eng. Chem. Res.* **39:** 1809 (2000)] present an extensive tabulation of tray efficiency data collected from the published literature.

The *GPSA Engineering Data Book* (10th ed., Gas Processors Association, 1987) and Kaes (*Refinery Process Modeling—A Practical Guide to Steady State Modeling of Petroleum Processes Using Commercial Simulators,* Athens Printing Co., Athens, Ga., 2000) tabulate typical efficiencies in gas plant and refinery columns, respectively. Pilling (Paper presented at the 4th Topical Conference on Separations Science and Technology, November 1999, available from Sulzer Chemtech, Tulsa, Okla.) tabulated more typical efficiencies. Similar information is often available from simulation guide manuals. The quality and reliability of efficiencies from these sources vary and are generally lower than the reliability of actual measured data.

Scale-Up from a Pilot- or Bench-Scale Column This is a very common scale-up. No reduction in efficiency on scale-up is expected as long as several precautions are observed. These precautions, generally relevant to pilot- or bench-scale columns, are spelled out with specific reference to the Oldershaw column.

Scale-Up from Oldershaw Columns One laboratory-scale device that found wide application in efficiency scale-up is the Oldershaw column [Fig. 14-45, Oldershaw, *Ind. Eng. Chem. Anal. Ed.* **13:** 265 (1941)]. This glass column is available from a number of laboratory supply houses. Typical column diameters are 25 to 100 mm (1 to 4 in), with tray spacing the same as the column diameter. Metal Oldershaw columns were available in the past but nowadays are scarce.

Fair, Null, and Bolles [*Ind. Eng. Chem. Process Des. Dev.* **22:** 53 (1983)] found that efficiency measurements in Oldershaw columns closely approach the point efficiencies [Eq. (14-133)] measured in commercial sieve-tray columns (Fig. 14-46) providing (1) the systems being distilled are the same, (2) comparison is made at the same relative approach to the flood point, (3) operation is at total reflux, and (4) a standard Oldershaw device is used in the laboratory experimentation.

A mixing model can be used to convert the Oldershaw point efficiencies to overall column efficiencies. This enhances the commercial column efficiency estimates. A conservative approach suggested by Fair et al. is to apply the Oldershaw column efficiency as the estimate for the overall column efficiency of the commercial column, taking no credit for the greater plug-flow character upon scale-up. The author prefers this conservative approach, considering the poor reliability of mixing models.

Previous work with Oldershaw columns [Ellis, Barker, and Contractor, *Trans. Instn. Chem. Engnrs.* **38:** 21 (1960)] presents an additional note of caution. Cellular (i.e., wall-supported) foam may form in pilot or Oldershaw columns, but is rare in commercial columns. For a given system, higher Oldershaw column efficiencies were measured under cellular foam conditions than under froth conditions. For this reason, Gerster [*Chem. Eng. Progr.* **59**(3): 35 (1963)] warned that when cellular foam can form, scale-up from an Oldershaw column may be dangerous. The conclusions presented by Fair et al. do not extend to Oldershaw columns operating in the cellular foam regime. Cellular foam can be identified by lower pilot column capacity compared to a standard mixture that is visualized not to form cellular foam.

Heat losses are a major issue in pilot and Oldershaw columns and can lead to optimistic scale-up. Special precautions are needed to keep these at a minimum. Vacuum jackets with viewing ports are commonly used.

Uses of Oldershaw columns for less conventional systems and applications were described by Fair, Reeves, and Seibert (Topical Conference on Distillation, AIChE Spring Meeting, New Orleans, March 10–14, 2002, p. 27). The applications described include scale-up in the absence of good VLE, steam stripping efficiencies, individual component efficiencies in multicomponent distillation, determining component behavior in azeotropic separation, and foam testing.

Empirical Efficiency Prediction An empirical correlation that has been the standard of the industry for distillation tray efficiency prediction for seven decades is the O'Connell plot [*Trans. Am. Inst. Chem. Eng.* **42:** 741 (1946)], Fig. 14-47. O'Connell extended an earlier correlation by Drickamer and Bradford [*Trans. Am. Inst. Chem. Eng.* **39:** 319 (1943)] of tray efficiency as a function of liquid viscosity for petroleum cuts. O'Connell's extension added relative volatility to the *x*-axis, making it suitable to a wide range of chemical and refinery systems.

Lockett (*Distillation Tray Fundamentals,* Cambridge University Press, Cambridge, UK, 1986) noted some theoretical sense in O'Connell's correlation. Higher viscosity usually implies lower liquid diffusivity and therefore greater liquid-phase resistance and lower efficiency. Higher relative volatility increases the significance of the liquid-phase resistance, thus reducing efficiency. Chen and Chuang [*Ind. Eng. Chem. Res.* **34**(9): 3078 (1995)]

TABLE 14-12 Representative Tray Efficiencies

Tray	System	Column diameter, ft	Tray spacing, in	Pressure, psia	Efficiency, %	% hole (slot) area	Ref.
Sieve	Methanol-water	3.2	15.7	14.7	70–90	10.8	2
	Ethanol-water	2.5	14	14.7	75–85	10.4	1
	Methanol-water	3.2	15.7	14.7	90–100	4.8	2
	Ethylbenzene-styrene	2.6	19.7	1.9	70	12.3	5
	Benzene-toluene	1.5	15.7	14.7	60–75	8	10
	Methanol-*n* propanol-sec butanol	6.0	18	18	64*		6
	Mixed xylene + C_8-C_{10} paraffins and naphthenes	13.0	21	25	86*		4
	Cyclohexane-*n*-heptane	4.0	24	5	60–70	14	9
				24	80	14	9
		4.0	24	5	70–80	8	8
				24	90	8	8
	Isobutane-*n*-butane	4.0	24	165	110	14	9
		4.0	24	165	120	8	8
		4.0	24	300	110	8	8
		4.0	24	400	100	8	8
	n-Heptane-toluene	1.5	15.7	14.7	60–75	8	10
	Methanol-water	2.0	13.6	14.7	68–72	10	11
	Isopropanol-water	2.0	13.6	14.7	59–63		11
	Toluene-methylcyclohexane	2.0	13.6	14.7	70–82		11
	Toluene stripping from water	4	24	14.7	31–42	8	13
Valve	Methanol-water	3.2	15.7	14.7	70–80	14.7	2
	Ethanol-water	2.5	14	14.7	75–85		1
	Ethyl benzene-styrene	2.6	19.7	1.9	75–85		3
	Cyclohexane-*n*-heptane	4.0	24	24	70–96*		7
	Isobutane-*n*-butane	4.0	24	165	108–121*		7
	Cyclohexane-n-heptane	4.0	24	24	77–93†	14.7	12
				5	57–86†	14.7	12
	Isobutane-*n*-butane	4.0	24	165	110–123†	14.7	12
	C_3-C_4 splitter	5.6	24	212	65–67‡	12	14

References:

1. Kirschbaum, *Distillier-Rektifiziertechnik*, 4th ed., Springer-Verlag, Berlin and Heidelberg, 1969.
2. Kastanek and Standart, *Sep. Sci.* **2**, 439 (1967).
3. Billet and Raichle, *Chem. Ing. Tech.*, 38, 825 (1966); **40**, 377 (1968).
4. AIChE Research Committee, *Tray Efficiency in Distillation Columns*, final report, University of Delaware, Newark, 1958.
5. Billet R., IChemE., *Symp. Ser.* **32**, p. 4:42 (1969).
6. Mayfield et al., *Ind. Eng. Chem.*, **44**, 2238 (1952).
7. Fractionation Research, Inc. "Report of Tests of Nutter Type B Float Valve Tray," July 2, 1964 from Sulzer Chem Tech.
8. Sakata and Yanagi, *IChemE., Eng. Symp.* Ser., no. 56, 3.2/21 (1979).
9. Yanagi and Sakata, *Ind. Eng. Chem. Process Des. Dec.*, **21**, 712 (1982).
10. Zuiderweg and Van der Meer, *Chem. Tech.* (Leipzig), **24**, 10 (1972).
11. Korchinsky, *Trans. I. Chem. E.*, **72**, Part A, 472 (1994).
12. Glitsch, Inc. "Glitsch Ballast Trays," Bulletin 159/160 (FRI Topical Report 15, 1958). Available from Koch-Glitsch LP, Wichita, Kans.
13. Kunesh et al., Paper presented at the AIChE Spring National Meeting, Atlanta, Ga., 1994.
14. Remesat, Chuang, and Svrcek, *Trans. I Chem. E.*, Vol. 83, Part A, p. 508, May 2005.

Notes:

*Rectangular Sulzer BDP valves.

†Glitsch V-1 round valves (Koch-Glitsch).

‡Two-pass trays, short path length.

To convert feet to meters, multiply by 0.3048; to convert inches to centimeters, multiply by 2.54; and to convert psia to kilopascals, multiply by 6.895.

showed that the O'Connell correlation can be derived from theory if one assumes that distillation mass transfer is liquid-film controlled. Recently, Duss and Taylor (*Kister Distillation Symposium*, AIChE Spring Meeting, San Antonio, Tex., March 26–30, 2017, p. 285) successfully showed that the O'Connell Correlation can be derived from first principles by assuming that the number of transfer units in the vapor equals the number of transfer units in the liquid. For each phase, the transfer unit equals the mass transfer coefficient times the interfacial area per unit volume in the froth, times the residence time in the froth. They also proposed a slightly modified O'Connell equation that shows promise.

Lockett expresses the O'Connell plot in equation form:

$$E_{OC} = 0.492(\mu_L \alpha)^{-0.245} \tag{14-138}$$

The viscosity is in cP, and E_{OC} is fractional. The volatility and viscosity are evaluated at the average arithmetic temperature between the column top and bottom temperatures. The relative volatility is between the key components.

The O'Connell correlation was based on data for bubble-cap trays. For sieve and valve trays, its predictions are likely to be slightly conservative.

Schon (*Distillation 2011: Topical Conference Proceedings*, AIChE Spring National Meeting, Chicago, Ill., March 2011) reports success with a modified O'Connell analysis (MOCA) in which empirical correction factors are added to the slope and intercept in Eq. (14-138) and are individually fitted for each stage to match the simulation of test data.

Theoretical Efficiency Prediction Theoretical tray efficiency prediction is based on the two-film theory and the sequence of steps in Fig. 14-42. Almost all methods evolved from the AIChE model (AIChE Research Committee, *Bubble Tray Design Manual*, American Institute of Chemical Engineers, New York, 1958). This model was developed over five years in the late 1950s in three universities. Since then, several aspects of the AIChE model have been criticized, corrected, and modified. Reviews are given by Lockett (*Distillation Tray Fundamentals*, Cambridge University Press, Cambridge, UK, 1986) and Chan and Fair [*Ind. Eng. Chem. Proc. Des. Dev.* **23:** 814 (1984)]. An improved version of the AIChE model, which alleviated several of its shortcomings, updated its hydraulic and mass-transfer relationships, and generally gave good predictions when tested against a wide data bank, was produced by Chan and Fair.

The Chan and Fair correlation is considered the most reliable fundamental correlation for tray efficiency, but even this correlation has been unable to rectify several theoretical and practical limitations inherited

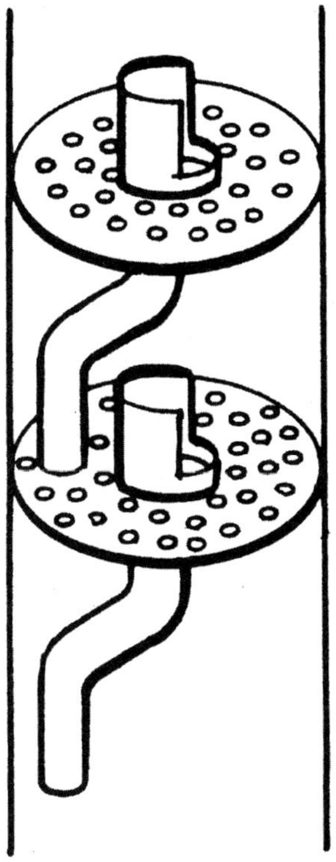

FIG. 14-45 An Oldershaw column. (*From H. Z. Kister,* Distillation Design, *copyright © 1992 by McGraw-Hill; reprinted by permission.*)

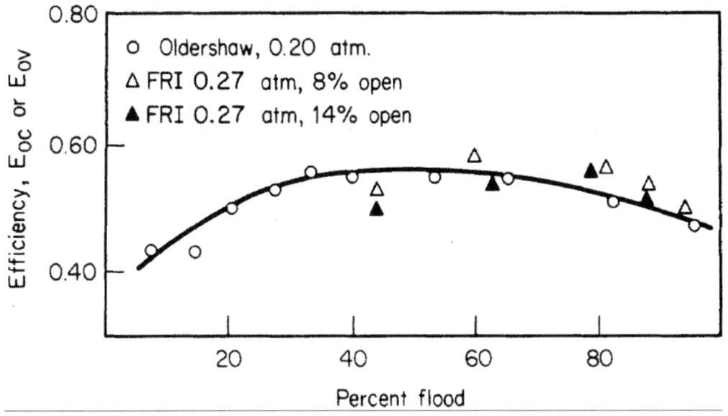

FIG. 14-46 Overall column efficiency of 25-mm Oldershaw column compared with point efficiency of 1.22-m-diameter-sieve sieve-plate column of Fractionation Research, Inc. System = cyclohexane-*n*-heptane. [*Fair, Null, and Bolles,* Ind. Eng. Chem. Process Des. Dev. ***22:*** *53 (1982).*]

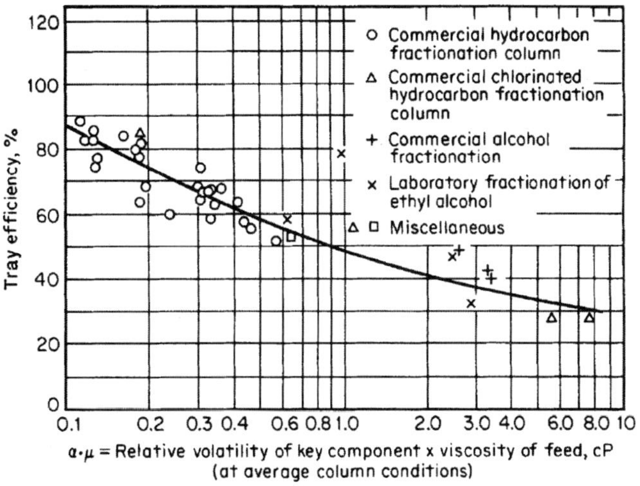

FIG. 14-47 O'Connell correlation for overall column efficiency E_{OC} for distillation. To convert centipoises to pascal-seconds, multiply by 10^{-3}. [*O'Connell,* Trans. Am. Inst. Chem. Eng. ***42:*** 741 *(1946).*]

from the AIChE correlation (see Kister, *Distillation Design,* McGraw-Hill, New York, 1992). Garcia and Fair [*Ind. Eng. Chem. Res.* **39:** 1818, (2000)] proposed a more fundamental and accurate model that is also more complicated to apply.

The prime issue that appears to plague fundamental tray efficiency methods is their tendency to predict efficiencies of 80 to 100 percent for distillation columns larger than 1.2 m (4 ft) in diameter. In the real world, most columns run closer to 60 percent efficiency. Cai and Chen (*Distillation 2003: Topical Conference Proceedings,* AIChE Spring National Meeting, New Orleans, La., March 30–April 3, 2003) show that published eddy diffusivity models, which are based on small-column work, severely underestimate liquid backmixing and overestimate plug flow in commercial-scale columns, leading to optimistic efficiency predictions. Which other limitations (if any) in the theoretical methods contribute to the mismatch, and to what degree, is unknown. For this reason, the author would not recommend any currently published theoretical tray efficiency correlation for obtaining design efficiencies.

Example 14-12 Estimating Tray Efficiency For the column in Example 14-9, estimate the tray efficiency, given that the relative volatility near the feed point is 1.3 and the viscosity is 0.25 cP.

Solution Table 14-12 presents measurements by Billet (*Packed Column Analysis and Design,* Ruhr University, Bochum, Germany, 1989) for ethylbenzene-styrene under similar pressure with sieve and valve trays. The column diameter and tray spacing in Billet's tests were close to those in Example 14-9. Since both have single-pass trays, the flow path lengths are similar. The fractional hole area (14 percent in Example 14-9) is close to that in Table 14-12 (12.3 percent for the tested sieve trays, 14 to 15 percent for standard valve trays). So the values in Table 14-12 should be directly applicable, that is, 70 to 85 percent. A conservative estimate would be 70 percent. The actual efficiency should be about 5 to 10 percent higher.

Alternatively, using Eq. (14-138) or Fig. 14-47, $E_{OC} = 0.492(0.25 \times 1.3)^{-0.245} = 0.65$ or 65 percent. As stated, the O'Connell correlation tends to be slightly conservative. This confirms that 70 percent will be a good estimate.

EQUIPMENT FOR DISTILLATION AND GAS ABSORPTION: PACKED COLUMNS

Packings are generally divided into three classes:

1. *Random or dumped packings* (Figs. 14-48 and 14-49) are discrete pieces of packing, of a specific geometric shape, that are "dumped" or randomly packed into the column shell.
2. *Structured or systematically arranged packings* (Fig. 14-50) are crimped layers of corrugated sheets (usually) or wire mesh. Sections of these packings are stacked in the column.
3. *Grids.* These are also systematically arranged packings, but instead of wire mesh or corrugated sheets, these use an open-lattice structure.

Random and structured packings are common in commercial practice. The application of grids is limited primarily to heat-transfer and wash services and/or where a high fouling resistance is required. Grids are discussed in detail elsewhere (Kister, *Distillation Design,* McGraw-Hill, New York, 1992).

Figure 14-51 is an illustrative cutaway of a packed tower, depicting typical internals. This tower has a structured-packed top bed and a random-packed bottom bed. Each bed rests on a support grid or plate. The lower bed has a holddown grid at its top to restrict packing uplift. Liquid to each of the beds is supplied by a liquid distributor. An intermediate distributor, termed a *redistributor,* is used to introduce feed and/or to remix liquid at regular height intervals. The intermediate distributor in Fig. 14-51 is not self-collecting, so a chevron collector is used to collect the liquid from the bed above. An internal pipe passes this liquid to the distributor below. The collected liquid is mixed with the fresh feed (not shown) before entering the distributor. The reboiler return enters behind a baffle above the bottom sump.

As illustrated, the packing needs to be interrupted and a distributor added at each point where a feed enters or a product leaves. A simple

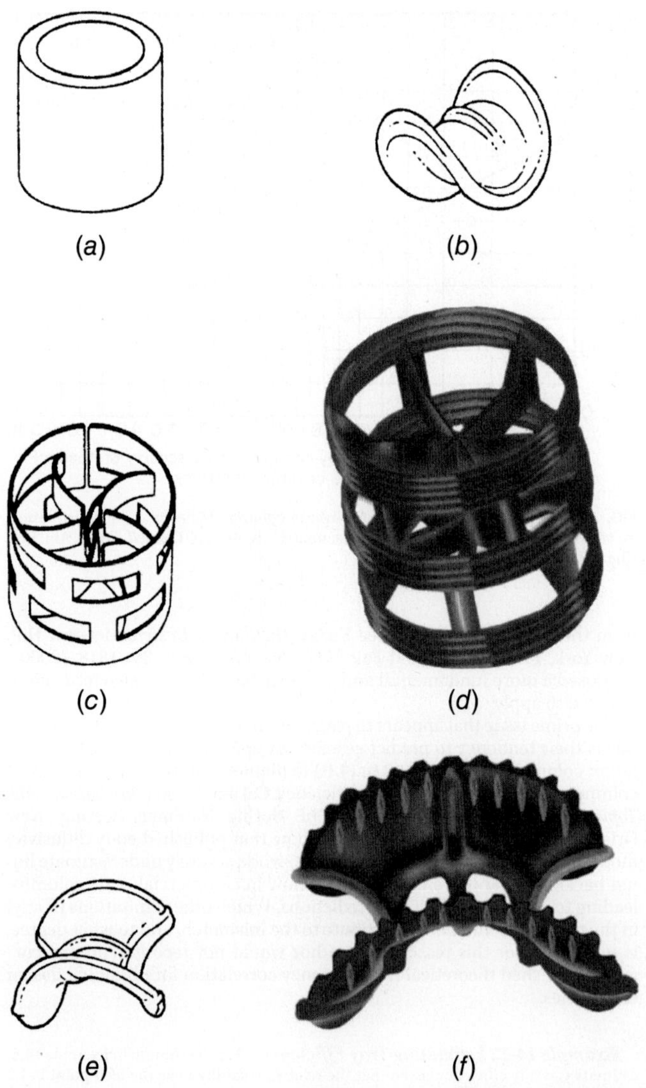

FIG. 14-48 Common first- and second-generation random packings. (*a*) Raschig ring (metal, plastic, ceramic). (*b*) Berl saddle (ceramic). (*c*) Pall ring (metal). (*d*) Pall ring (plastic). (*e*) Intalox saddle (ceramic). (*f*) Super Intalox saddle (plastic). (*Parts d, f, courtesy of Koch-Glitsch LP.*)

distillation tower with a single feed will have a minimum of two beds, a rectifying bed and a stripping bed.

PACKING OBJECTIVES

The objective of any packing is to maximize efficiency for a given capacity, at an economic cost. To achieve these goals, packings are shaped to

1. *Maximize the specific surface area, that is, the surface area per unit volume.* This maximizes vapor–liquid contact area, and, therefore, efficiency. A corollary is that efficiency generally increases as the random packing size is decreased or as the space between structured packing layers is decreased.

2. *Spread the surface area uniformly.* This improves vapor–liquid contact, and, therefore, efficiency. For instance, a Raschig ring (Fig. 14-48*a*) and a Pall® ring (Fig. 14-48*c*) of an identical size have identical surface areas per unit volume, but the Pall® ring has a superior spread of surface area and therefore gives much better efficiency.

3. *Maximize the void space per unit column volume.* This minimizes resistance to gas upflow, thereby enhancing packing capacity. A corollary is that capacity increases with random packing size or with the space between structured packing layers. Comparing with the first objective, a tradeoff exists; the ideal size of packing is a compromise between maximizing efficiency and maximizing capacity.

4. *Minimize friction.* This favors an open shape that has good aerodynamic characteristics.

5. *Minimize cost.* Packing costs, as well as the requirements for packing supports and column foundations, generally rise with the weight per unit volume of packing. A corollary is that packings become cheaper as the size increases (random packing) and as the space between layers increases (structured packing).

RANDOM PACKINGS

Historically, there were four generations of evolution in random packings. The first generation (1907 to the 1950s) produced two basic simple shapes—the Raschig ring and the Berl saddle (Fig. 14-48*a, b*) that became the ancestors of modern random packings. These packings have been superseded by more modern packing and are seldom used in modern distillation practice.

The second generation (late 1950s to the early 1970s) produced two popular geometries—the Pall® ring, which evolved from the Raschig ring, and the Intalox® saddle (Fig. 14-48*c–f*), which evolved from the Berl saddle. BASF developed the Pall® ring by cutting windows in the Raschig ring and bending the window tongues inward. This opened up the ring, lowering the aerodynamic resistance and dramatically enhancing capacity. The bent tongues improved area distribution around the particle, giving also better efficiency. These improvements made the first-generation Raschig rings obsolete for distillation. Berl saddles (ceramics) are still used due to their good breakage resistance. The second-generation packings are still popular and extensively used in modern distillation practice. The Pall® ring is still a standard random packing up to now in many applications and serves as a benchmark packing for comparison.

The third generation (the mid-1970s to the late 1990s) has produced a multitude of popular geometries, most of which evolved from the Pall® ring and Intalox® saddle, featuring a more open ("lattice") structure to reduce pressure drop and slightly enhance capacity while retaining the same surface area per unit volume and therefore the efficiency. The IMTP® (Figure 14-49*a*) and similar shapes have been the dominant metal random packing in the last two decades. The CMR® (Figure 14-49*b*) and similar shapes are another popular third-generation packing both in metal and plastic. A more comprehensive description of the various packings is given elsewhere (Kister, *Distillation Design*, McGraw-Hill, New York, 1992). The third generation of packing was a significant, yet not large, improvement over the second generation.

The fourth generation (late 1990s to date) opened the lattice structures of the third-generation packings even more (Fig. 14-49*c–e*), providing relatively small additional pressure drop and capacity improvements, compared to the third-generation packings of the same efficiency. At present, both the third- and fourth-generation packings are popular.

STRUCTURED PACKINGS

Structured packings have been around since as early as the 1940s. The very early structured packings, such as Panapak, never became popular, and they are seldom used nowadays.

The first generation of modern structured packings began in the late 1950s with high-efficiency wire-mesh packings such as Goodloe®, Hyperfil®, and the Sulzer® (wire-mesh) packings. By the early 1970s, these packings had made substantial inroads into vacuum distillation, where their low pressure drop per theoretical stage is a major advantage. In these services, they are extensively used today. Their high cost, high sensitivity to solids, and low capacity hindered their application outside vacuum distillation.

The corrugated-sheet packing, first introduced by Sulzer in the late 1970s, started a second generation of structured packings. With a high capacity, lower cost, and lower sensitivity to solids, while still retaining a high efficiency, these corrugated-sheet packings became competitive with conventional internals, especially for revamps. The 1980s saw an accelerated rise in the popularity of structured packings, to the point of their becoming one of the most popular column internals in use today.

Corrugated structured packings are fabricated from thin, corrugated (crimped) metal sheets, arranged parallel to one another. The corrugated sheets are assembled into an element (Figs. 14-50*a, c* and 14-51). The sheets in each element are arranged at a fixed angle to the vertical. Table 14-14 contains geometric data for several corrugated packings.

Geometry (Fig. 14-52) The crimp size defines the opening between adjacent corrugated layers. Smaller B, h, and S yield narrower openings, more sheets (and, therefore, greater surface area) per unit volume, and more efficient packing, but higher resistance to gas upflow, lower capacity, and enhanced sensitivity to plugging and fouling.

The corrugations spread gas and liquid flow through a single element in a series of parallel planes. To spread the gas and liquid uniformly in all radial planes, adjacent elements are rotated so that sheets of one element are at a

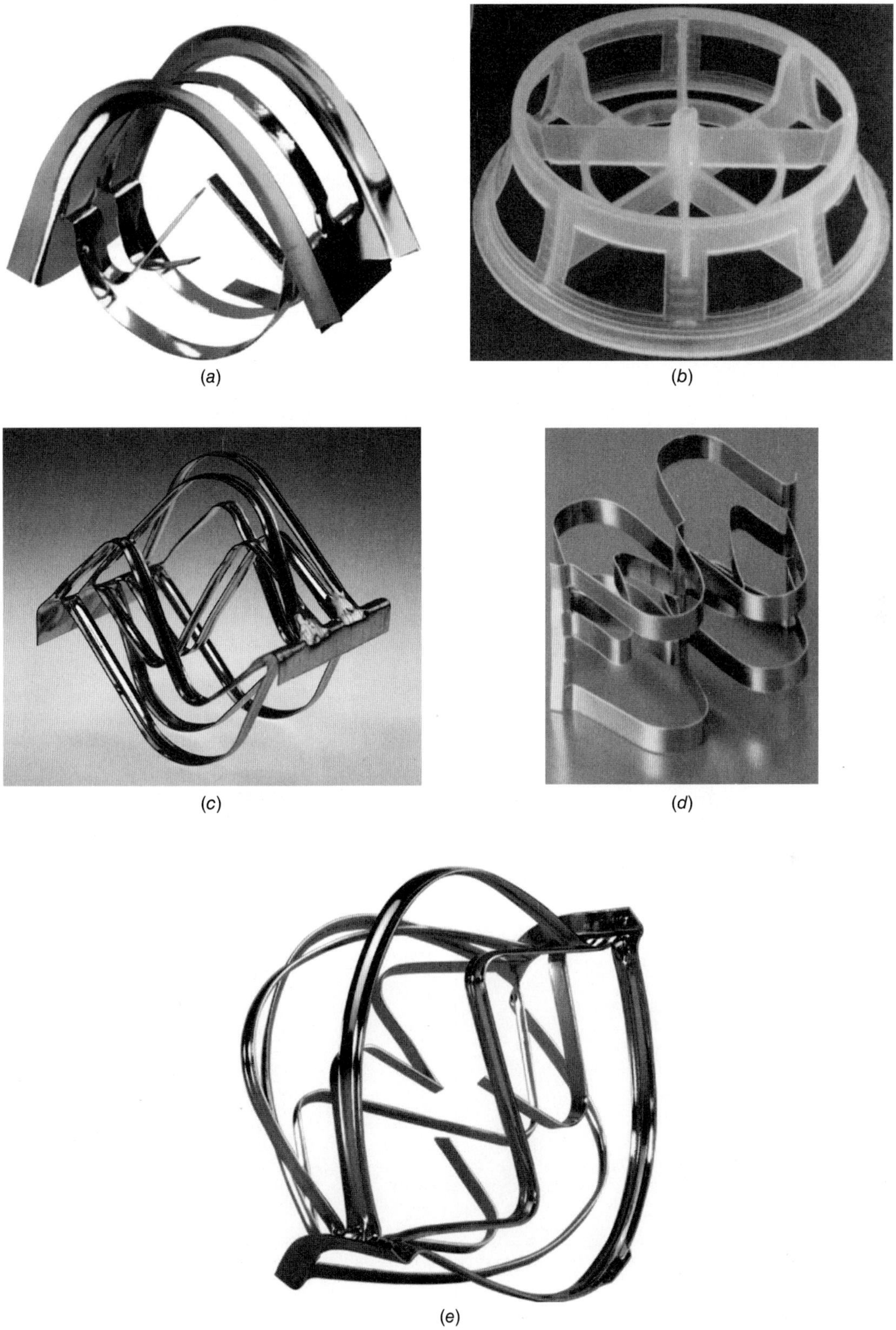

FIG. 14-49 Common third- and fourth-generation random packings. (*a*) Intalox Metal Tower Packing (IMTP). (*b*) Cascade Mini-Ring (CMR) (plastic). (*c*) NexRing (metal). (*d*) Raschig Super-Ring (metal). (*e*) Intalox Ultra Random Packing (metal). (*Parts a, b, and e courtesy of Koch-Glitsch LP; part c, courtesy of Sulzer Chemtech AG; part d, courtesy of Rashig AG.*)

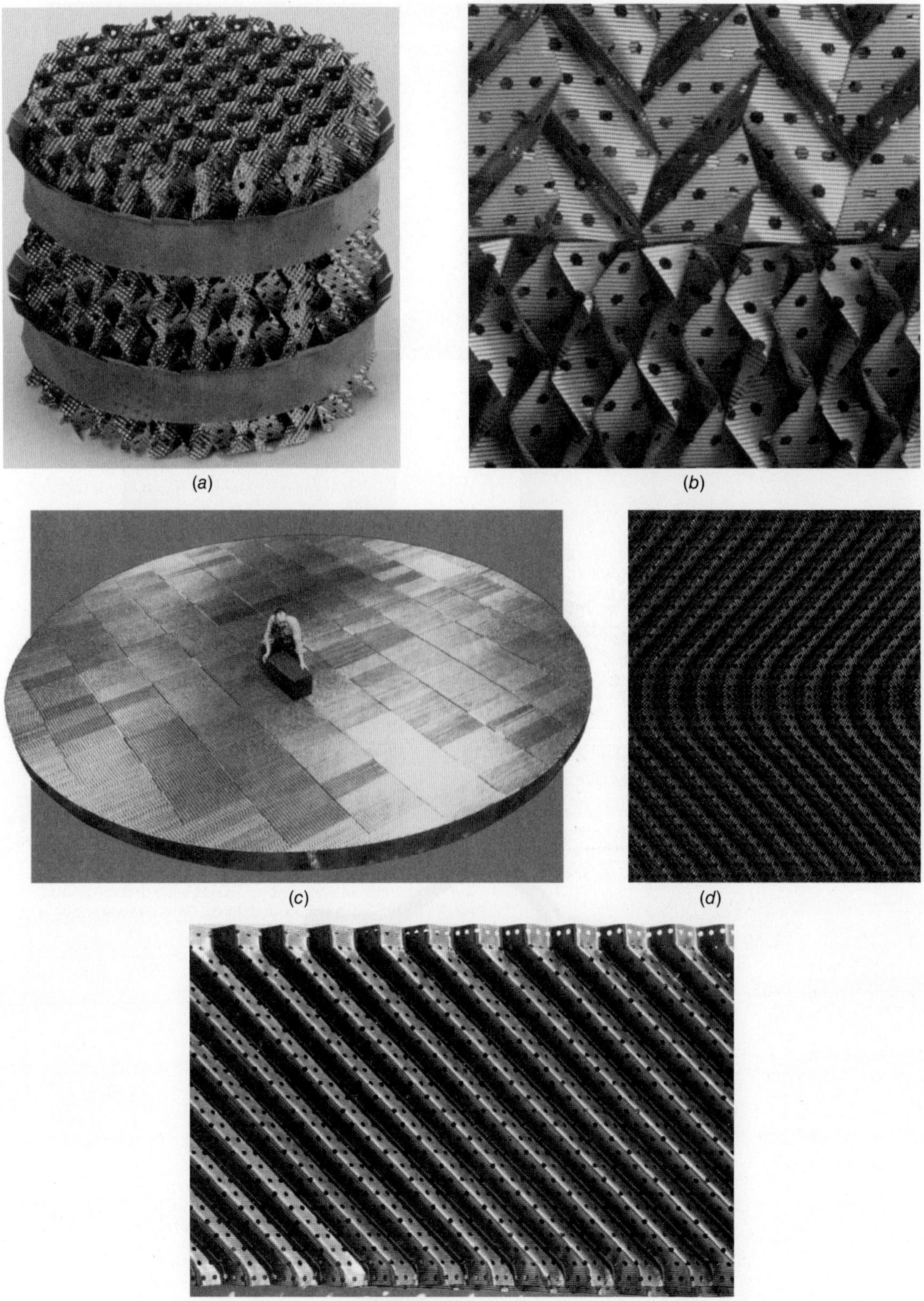

FIG. 14-50 Common structured packings. (*a*) A small element of Mellapak™ showing embossed surface, holes, and corrugated-sheet arrangement. (*b*) A closeup of the surface of Flexipac™ showing grooved surface and holes. (*c*) Fitting structured packing elements to a large-diameter tower. (*d*) Mellapak Plus™, a third-generation structured packing, showing a 45° inclination angle in the element and near-vertical inclination at the element-to-element transition. Note that in the tower, the successive layers will be oriented 90° to each other as in part b. (*e*) A sheet of Flexipac HC™, a third-generation structured packing showing a 45° inclination angle in the element and near-vertical inclination at the element-to-element top and bottom transitions. (*Parts a, d, courtesy of Sulzer Chemtech; parts b, c, e, courtesy of Koch-Glitsch LP.*)

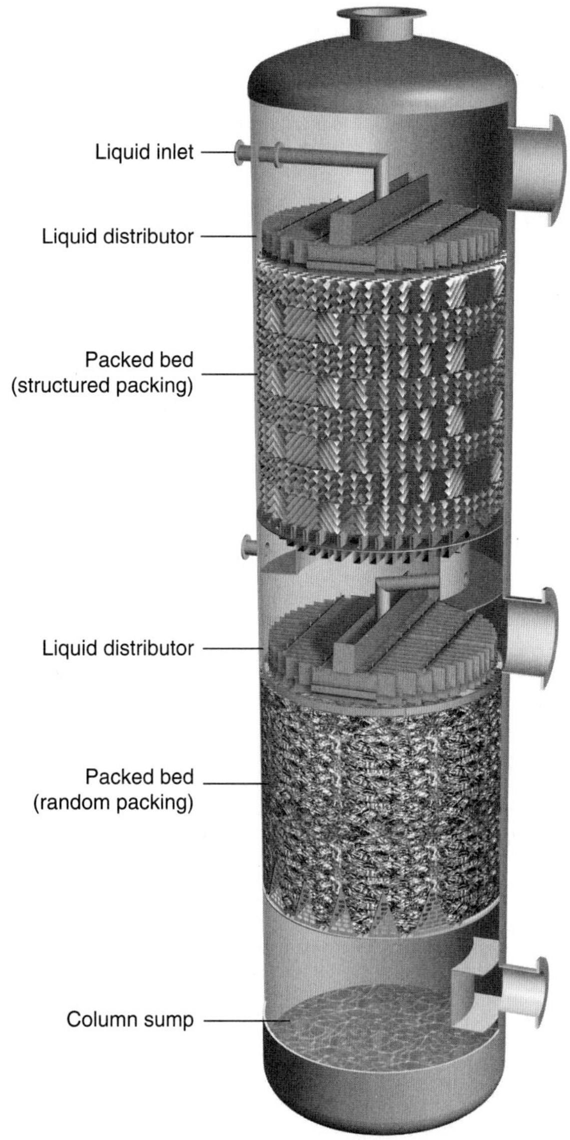

FIG. 14-51 Illustrative cutaway of a packed tower, depicting an upper bed of structured packing and a lower bed of random packing. (*Courtesy of Sulzer Chemtech.*)

fixed angle to the layer below (Fig. 14-51). For good spread, element height is relatively short (typically 200 to 300 mm, 8 to 12 in), and the angle of rotation is around 90°.

The surfaces of a few structured packings (especially those used in highly fouling environments) are smooth. Most structured packings have a roughened or enhanced surface that assists the lateral spread of liquid, promotes film turbulence, and enhances the area available for mass transfer. Texturing commonly employed is embossing and grooving (Fig. 14-50*a, b*).

The surfaces of most (but not all) structured packings contain holes that serve as communication channels between the upper and lower surfaces of each sheet. If the holes are too small, or nonexistent, both sides of a sheet will be wet only at low liquid rates. At high liquid rates, *sheeting* or *blanking* will cause liquid to run down the top surface with little liquid wetting the bottom surface [Chen and Chuang, *Hydroc. Proc.* **68**(2): 37 (1989)], which may lower efficiency. Usually, but not always, the holes are circular (Fig. 14-50*a, b*), about 4 mm in diameter. Olujic et al. (*Distillation 2003: Topical Conference Proceedings*, AIChE Spring National Meeting, New Orleans, La., 2003, p. 523) showed that the hole diameter has a complex effect, strongly dependent on packing size, on both capacity and efficiency.

Inclination Angle In each element, corrugated sheets are most commonly inclined at about 45° to the vertical (typically indicated by the letter Y following the packing size). This angle is large enough for good drainage of liquid, avoiding stagnant pockets and regions of liquid accumulation, and small enough to prevent gas from bypassing the metal surfaces. In some packings, the inclination angle to the vertical is steepened to 30° (typically indicated by the letter X following the packing size). This improves drainage, and therefore capacity, but at the expense of reduced gas–liquid contact, and therefore efficiency.

A recent development followed the realization that liquid drainage was restricted at the element-to-element transition rather than inside elements [Lockett, Victor, and Billingham, *IChemE Symp. Ser.* **152:** 400 (2006)]. This means that the liquid accumulation leading to flood initiates at the transition region. A third generation of structured packing started, often referred to as *high-performance packings,* in which the main body of each element has layers inclined at 45°, but the ends of each element are vertical or almost vertical to permit drainage at this end region (Fig. 14-50*d, e*; but keep in mind that successive elements are rotated 90° rather than continuous, as shown in Fig. 14-50*d*). These high-performance packings offer greater capacity and lower pressure drop compared to equivalent 45° inclined packings, with efficiency the same with some (Pilling and Haas, *Topical Conference Proceedings,* AIChE Spring Meeting, New Orleans, March 10–14, 2002, p. 132; McNulty and Sommerfeldt in "Distillation: Horizons for the New Millennium," *Topical Conference Proceedings,* AIChE Spring Meeting, Houston, Tex., March 1999, p. 89) and lower with others [Olujic et al., *Chem. Eng. and Proc.* **42:** 55 (2003)].

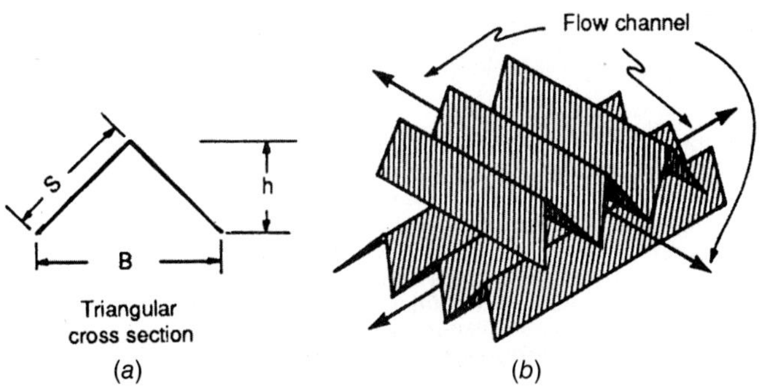

FIG. 14-52 Crimp geometry in structured packings. (*a*) Flow channel cross section. (*b*) Flow channel arrangement. (*From J. R. Fair and J. L. Bravo, Chem. Eng. Progr., Jan. 1990, p. 19; reproduced courtesy of the American Institute of Chemical Engineers.*)

PACKED-COLUMN FLOOD AND PRESSURE DROP

Pressure drop of a gas flowing upward through a packing countercurrently to liquid flow is characterized graphically in Fig. 14-53. At very low liquid rates, the effective open cross section of the packing is not appreciably different from that of dry packing, and pressure drop is due to flow through a series of variable openings in the bed. Thus, pressure drop is proportional

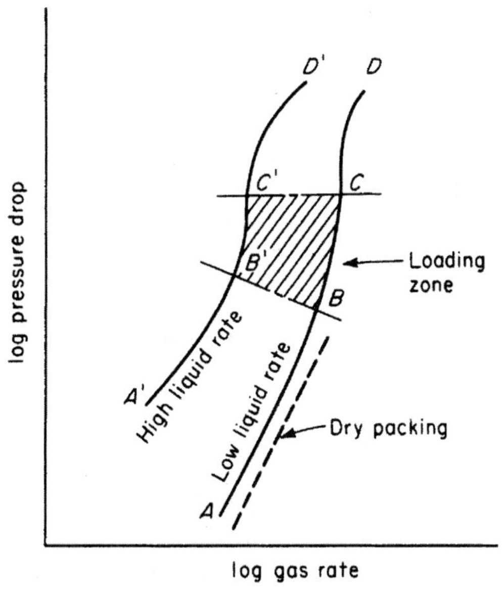

FIG. 14-53 Pressure-drop characteristics of packed columns.

approximately to the square of the gas velocity, as indicated in the region *AB*. At higher liquid rates, the effective open cross section is smaller because of the presence of liquid (region *A′B′*). The pressure drop is higher, but still proportional to the square of the gas velocity.

At higher gas rates, a portion of the energy of the gas stream is used to support an increasing quantity of liquid in the column. For all liquid rates, a zone is reached where pressure drop is proportional to a gas flow rate to a power distinctly higher than 2; this zone is called the *loading zone.* The increase in pressure drop is due to the liquid accumulation in the packing voids (region *BC* or *B′C′*).

As the liquid holdup increases, the effective orifice diameter may become so small that the liquid surface becomes continuous across the cross section of the column. Column instability occurs concomitantly with a rising continuous-phase liquid body in the column. Pressure drop shoots up with only a slight change in gas rate (condition *C* or *C′*). The phenomenon is called *flooding* and is analogous to entrainment flooding in a tray column.

Alternatively, a phase inversion occurs, and gas bubbles through the liquid. The column is not unstable and can be brought back to gas-phase continuous operation by merely reducing the gas rate. A stable operating condition beyond flooding (region *CD* or *C′D′*) may form with the liquid as the continuous phase and the gas as the dispersed phase [Lerner and Grove, *Ind. Eng. Chem.* **43:** 216 (1951); Teller, *Chem. Eng.* **61**(9): 168 (1954); Leung et al., *Ind. Eng. Chem. Fund.* **14**(1): 63 (1975); Buchanan, *Ind. Eng. Chem. Fund.* **15**(1): 87 (1976)].

For total-reflux distillation in packed columns, regions of loading and flooding are identified by their effects on mass-transfer efficiency, as shown in Fig. 14-54. Gas and liquid rate increase together, and a point is reached at which liquid accumulates rapidly (point *B*) and effective surface for mass transfer decreases rapidly.

Flood-Point Definition In 1966, Silvey and Keller [*Chem. Eng. Progr.* **62**(1): 68 (1966)] listed 10 different flood point definitions that have been used by different literature sources. A later survey (Kister and Gill, *Proceedings of Chemeca 92*, Canberra, Australia, 1992, p. 185-2) listed twice that many. As Silvey and Keller pointed out, the existence of so many definitions puts into question what constitutes flooding in a packed tower, and at what gas rate it occurs. Symptoms used to identify flood in these definitions include the appearance of liquid on top of the bed, excessive entrainment, a sharp rise in pressure drop, a sharp rise in liquid holdup, and a sharp drop in efficiency. The survey of Kister and Gill suggests that most flood point definitions describe the point of flooding initiation (*incipient flooding*; point *C* or *C′* on Figs. 14-53 and 14-54). The different incipient flooding definitions gave surprisingly little scatter of flood point data (for a given packing under similar operating conditions). It follows that any definition describing flooding initiation should be satisfactory.

The author believes that due to the variations in the predominant symptom with the system and the packing, the use of multiple symptoms is most appropriate. The author prefers the following definition by Fair and Bravo [*Chem. Eng. Symp. Ser.* **104:** A183 (1987)]: "A region of rapidly increasing pressure drop with simultaneous loss of mass-transfer efficiency. Heavy entrainment is also recognized as a symptom of this region." An almost identical definition was presented earlier by Billet (*Distillation Engineering,* Chem. Publishing Co., New York, 1979).

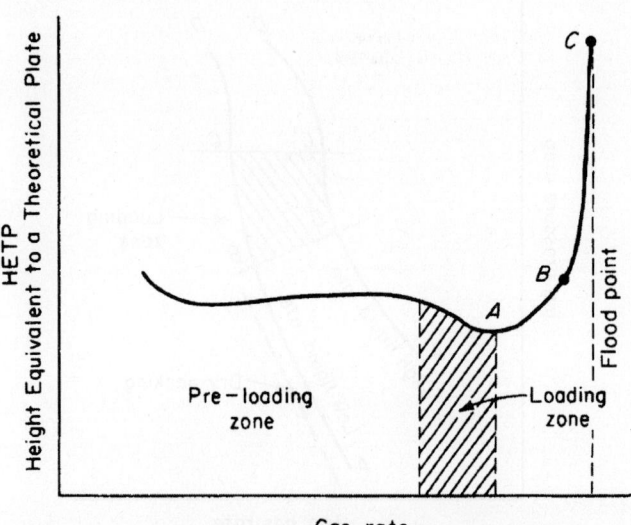

FIG. 14-54 Efficiency characteristics of packed columns (total-reflux distillation.)

The maximum useful capacity (MUC, often also referred to as maximum efficient capacity, or maximum operational capacity) is defined as the "maximum vapor rate that provides normal efficiency of a packing when approaching flood" (i.e., point B in Fig. 14-54) (Strigle, *Packed Tower Design and Applications,* 2d ed., Gulf Publishing, Houston, Tex., 1994; Cai, chap. 3 in Gorak and Schoenmakers *Distillation Operation and Application*, Academic Press, New York, 2014). The MUC is clear-cut in Fig. 14-54. On the other hand, locating the MUC in other cases is difficult and leaves a lot of room for subjectivity.

In most cases [Kister and Gill, *Chem. Eng. Progr.* **87**(2): 32 (1991)], the velocity at which MUC is reached is related to the flood point velocity by

$$u_{S,MUC} = 0.95\ u_{S,Fl} \tag{14-139}$$

Flood and Pressure Drop Prediction The first generalized correlation of packed-column flood points was developed by Sherwood, Shipley, and Holloway [*Ind. Eng. Chem.* **30:** 768 (1938)] on the basis of laboratory measurements primarily on the air–water system with random packing. Later work with air and liquids other than water led to modifications of the Sherwood correlation, first by Leva [*Chem. Eng. Progr. Symp. Ser.* **50**(1): 51 (1954)], who also introduced the pressure drop curves, and later in a series of papers by Eckert. The generalized flooding–pressure drop chart by Eckert [*Chem. Eng. Progr.* **66**(3): 39 (1970)], included in previous editions of this handbook, was modified and simplified by Strigle (*Packed Tower Design and Applications,* 2d ed., Gulf Publishing, Houston, Tex., 1994) (Fig. 14-55). It is often called the generalized pressure drop correlation (GPDC). The ordinate is a capacity parameter [Eq. (14-140)] related to the Souders-Brown coefficient used for tray columns.

$$\mathrm{CP} = C_s F_p^{0.5} \nu^{0.05} = U_s \left(\frac{\rho_G}{\rho_L - \rho_G}\right)^{0.50} F_p^{0.5} \nu^{0.05} \tag{14-140}$$

where U_S = superficial gas velocity, ft/s
ρ_G, ρ_L = gas and liquid densities, lb/ft³ or kg/m³
F_p = packing factor, ft⁻¹
ν = kinematic viscosity of liquid, cS
C_S = *C*-factor, Eq. (14-77), based on tower superficial cross-sectional area, ft/s
CP = capacity factor, dimensional [units consistent with Eq. (14-140) and its symbols]

The abscissa scale term is the same flow parameter used for trays (dimensionless):

$$F_{LG} = \frac{L}{G}\left(\frac{\rho_G}{\rho_L}\right)^{0.5} \tag{14-141}$$

For structured packing, Kister and Gill [*Chem. Eng. Symp. Ser.* **128:** A109 (1992)] noticed a much steeper rise of pressure drop with flow parameter than that predicted from Fig. 14-55, and they presented a modified chart (Fig. 14-56).

The GPDC charts in Figs. 14-55 and 14-56 do not contain specific flood curves. Both Strigle and Kister and Gill recommend calculating the flood point from the flood pressure drop, ΔP_{flood} (inch of water per foot of packings) given by the Kister and Gill equation

$$\Delta P_{\text{flood}} = 0.12 F_p^{0.7} \tag{14-142}$$

Equation (14-142) permits finding the pressure drop curve in Fig. 14-55 or 14-56 at which incipient flooding occurs.

For low-capacity random packings, such as the small first-generation packings and those smaller than 1-in diameter (F_p > 60 ft⁻¹), calculated flood pressure drops are well in excess of the upper pressure drop curve in Fig. 14-55. For these packings only, the original Eckert flood correlation [*Chem. Eng. Prog.* **66**(3): 39 (1970)] found in pre-1997 editions of this handbook and other major distillation texts is suitable.

The packing factor F_p is empirically determined for each packing type and size. Values of F_p, together with general dimensional data for individual packings, are given for random packings in Table 14-13 (to go with Fig. 14-55) and for structured packings in Table 14-14 (to go with Fig. 14-56).

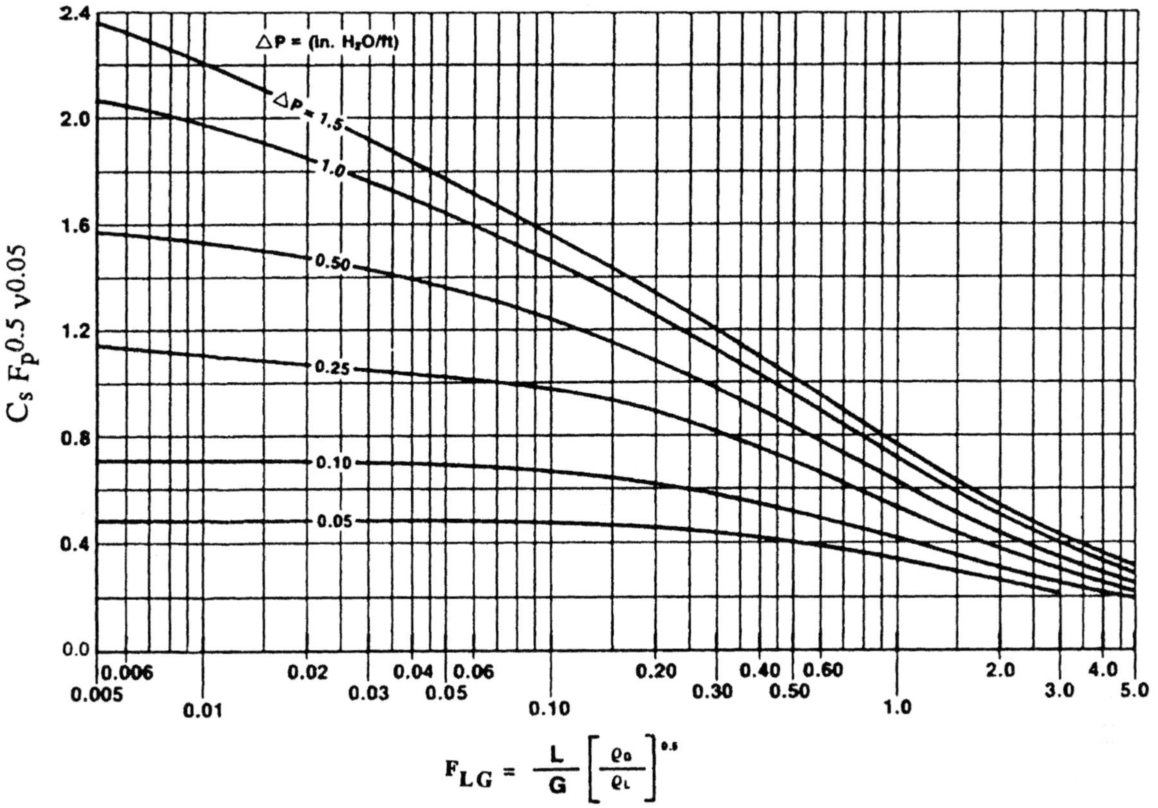

FIG. 14-55 Generalized pressure drop correlation of Eckert as modified by Strigle. To convert inches H_2O to mm H_2O, multiply by 83.31. (*From* Packed Tower Design and Applications by *Ralph E. Strigle, Jr. Copyright © 1994 by Gulf Publishing Co., Houston, Texas. Used with permission. All rights reserved.*)

Packing flood and pressure drop correlations should always be used with caution. Kister and Gill [*Chem. Eng. Progr.* **87**(2): 32 (1991)] showed that deviations from the GPDC predictions tend to be systematic and not random. To avoid regions in which the systematic deviations lead to poor prediction, they superimposed experimental data points for each individual packing on the curves of the GPDC. Figure 14-57 is an example. This method requires a single chart for each packing type and size. It provides the highest possible accuracy as it interpolates measured data and identifies uncertain regions. A set of charts is in Chapter 10 of Kister's book (*Distillation Design,* McGraw-Hill, New York, 1992) with updates in Kister, Larson, and Gill, (Paper presented at the AIChE Spring National Meeting, Houston, Tex., March 19–23, 1995) and in Kister, Scherffius, Afshar, and Abkar (*Distillation 2007: Topical Conference Proceedings,* 2007 AIChE Spring National Meeting, Houston, Tex., April 22–27, 2007, p.445). The latter reference also discusses correct and incorrect applications of those interpolation charts.

There are many alternative methods for flood and pressure drop prediction. The Billet and Schultes [*IChemE Symp. Ser.* **104:** A171, B255 (1987);

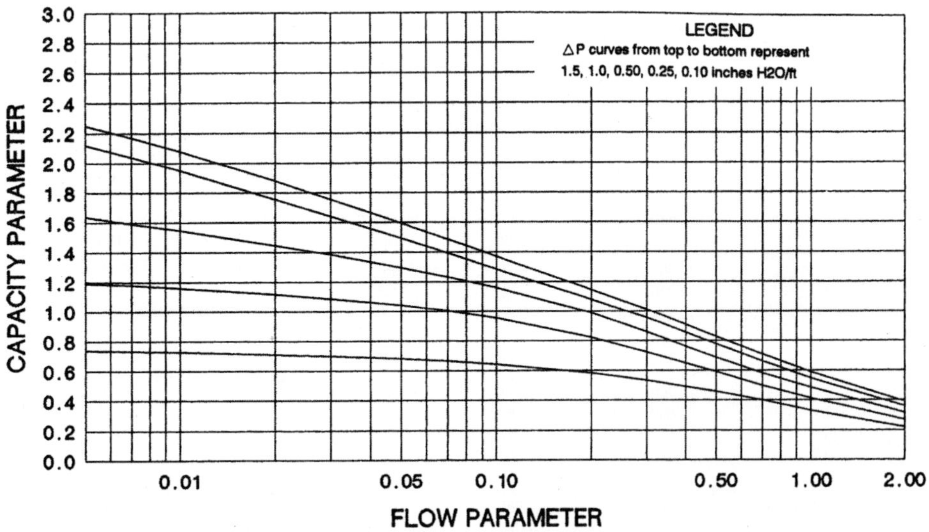

FIG. 14-56 The Kister and Gill GPDC (SP) chart for structured packings only. Abscissa and ordinate same as in Fig. 14-55. (*From Kister, H. Z., and D. R. Gill,* IChemE Symp. Ser. *128: p. A109, 1992. Reprinted courtesy of IChemE.*)

TABLE 14-13 Characteristics of Random Packings

Name	Size, mm, or no. (#)	Bed density,* kg/m³	Area, m²/m³	% voids	Packing factor m⁻¹ Normal F_p†	Packing factor m⁻¹ Dry F_{pd}‡	Vendor
			Metals				
Pall rings (also Flexi-rings, Ballast rings, P-rings)	16	510	360	92	256	262	Various
	25	325	205	94	183	174	
	38	208	130	95	131	91	
	50	198	105	96	89	79	
	90	135	66	97	59	46	
Metal Intalox (IMTP) [also I-rings, AHPP, RSMR, MSR]§	25	224	207	97	134	141	Koch-Glitsch [Sulzer, AMACS, RVT, Montz]§
	40	153	151	97	79	85	
	50	166	98	98	59	56	
	70	141	60	98	39	—	
Nutter rings	#0.7	177	226	98	—	128	Sulzer
	#1	179	168	98	98	89	
	#1.5	181	124	98	79	66	
	#2	144	96	98	59	56	
	#2.5	121	83	98	52	49	
	#3.0	133	66	98	43	36	
Raschig Super-ring	#0.3	340	315	96	—	—	Raschig
	#0.5	275	250	98	—	—	
	#0.7	185	175	98	—	—	
	#1	220	160	98	82	—	
	#1.5	170	115	98	59	—	
	#2	155	98	99	49	—	
	#3	150	80	98	36	—	
Cascade mini-rings (CMR)	#1	389	250	96	131	102	Koch-Glitsch
	#1.5	285	190	96	95	—	
	#2	234	151	97	72	79	
	#2.5	195	121	97	62	—	
	#3	160	103	98	46	43	
	#4	125	71	98	33	32	
	#5	108	50	98	26¶	—	
Fleximax	#300	—	141	98	85	—	Koch-Glitsch
	#400	—	85	98	56	—	
Jaeger Tripacks (Top-Pak)	#1	223	118	96	85	—	Raschig
	#2	170	75	98	46	—	
VSP	25	352	206	98	105¶	—	Raschig
	50	296	112	96	69	—	
Ralu-rings	25	310	215	98	157¶	—	Raschig
	38	260	135	97	92¶	—	
	50	200	105	98	66¶	—	
Hiflow	25	298	203	96	—	—	RVT
	50	175	92	98	52	—	
Hy-Pak, K-Pak,	#1	262	174	97	148	—	Koch-Glitsch,
	#1.5	180	118	98	95	—	
	#2	161	92	98	85	—	
	#3	181	69	98	52	—	
Bialecki rings	25	368	223	95	—	—	
	50	264	112	97	—	—	
Raschig rings (1/16 in wall)	19	1500	245	80	722	—	Various
	25	1140	185	86	472	492	
	50	590	95	92	187	223	
	75	400	66	95	105	—	
			Ceramics				
Berl saddles	6	900	900	60	—	2950	Various
	13	865	465	62	790	900	
	25	720	250	68	360	308	
	38	640	150	71	215	154	
	50	625	105	72	150	102	
Intalox, Flexi-saddles, Torus-saddles, Novalox	6	864	984	65	—	2720	Various
	13	736	623	71	660	613	
	25	672	256	73	302	308	
	50	608	118	76	131	121	
	75	576	92	79	72	66	
Raschig rings	6	960	710	62	—	5250	Various
	13	880	370	64	1900	1705	
	25	670	190	74	587	492	
	50	660	92	74	213	230	
	75	590	62	75	121	—	
Pall ring	25	620	220	75	350	—	Raschig
	38	540	164	78	180	—	
	50	550	121	78	142	—	
	80	520	82	78	85¶	—	

TABLE 14-13 Characteristics of Random Packings *(Continued)*

Name	Size, mm, or no. (#)	Bed density,* kg/m³	Area, m²/m³	% voids	Packing factor m⁻¹		Vendor
					Normal F_p [†]	Dry F_{pd} [‡]	
			Ceramics				
Hiflow rings	38	409	108	83	121	—	RVT
	50	405	89	82	95	—	
	70	333	52	86	49	—	
			Plastics				
Pall rings	15	95	350	87	320	348	Various
	25	71	206	90	180	180	
	40	70	131	91	131	131	
	50	60	102	92	85	82	
	90	43	85	95	56	39	
Super Intalox,	25	83	207	90	131	131	Various
Flexi-saddles,	50	60	108	93	92	85	
Super-Torus, Novalox	75	48	89	96	59	46	
Cascade mini-rings	#1A	53	185	97	98	92	Koch-Glitsch
(CMR)	#2A	46	115	97	59	—	
	#3A	40	74	98	39	33	
Raschig Super-ring	#0.6	62	205	96	105[¶]	—	Raschig
	#2	55	100	96	49	—	
Ralu-ring	15	80	320	94	230[¶]	—	Raschig
	25	56	190	94	135	—	
	38	65	150	95	80	—	
	50	60	110	95	55	—	
	90	40	75	96	38	—	
	125	30	60	97	30[¶]	—	
Snowflake	—	51	92	97	43	—	Koch-Glitsch
Nor-Pac	25	72	180	92	102	—	NSW
	38	61	144	93	69	—	
	50	53	102	94	46	—	
Tri-Packs	#1	67	158	93	53[¶]	—	Raschig
(Hacketten)	#2	53	125	95	39[¶]	43	
Ralu-Flow	#1	55	165	95	67[¶]	—	Raschig
	#2	54	100	95	38[¶]	—	
Hiflow	25	63	192	92	138	—	RVT
	50	59	110	93	66	—	
	90	34	65	97	30	—	
Envipac	33	58	192	94	—	—	ENVIMAC
	60	39	110	96	—	—	
	90	40	65	96	—	—	
Lanpac	90	67	148	93	46	—	Lantec
Impac	#3	83	213	91	49	—	
Tellerettes	25	112	180	87	—	131	Ceilcote
	50	59	125	93	—	—	

*Values are approximate and may not be consistent. Actual number depends on wall thickness and material.

†Packing factor F_p from Kister, *Distillation Design*, McGraw-Hill, 1992; Kister, Larson, and Gill, paper presented at the Houston AIChE meeting, March 19–23, 1995; Strigle, *Packed Tower Design and Applications*, 2d ed., Gulf Publishing Co., Houston, Tex., 1994; Kister et al., in *Distillation 2007: Topical Conference Proceedings*, 2007 AIChE Spring National Meeting, Houston, Tex.

‡Dry packing factor F_{pd} from Robbins, *Chem. Eng. Progr.*, **87**(1), 19 (1990).

§The bracketed packings are similar to, but not the same as, the Koch-Glitsch IMTP. Some of them have holes that the IMTP do not have, and others have sizes that are different and are made by different dies.

¶Packing factor supplied by packing vendor.

Vendors: Koch-Glitsch LLP, Wichita, Kansas; Raschig GmbH, Ludwigshafen/Rhein, Germany; Sulzer Chemtech Ltd., Winterthur, Switzerland; RVT Process Equipment GmbH, Steinwiesen, Germany; AMACS, 14211 Industry Street, Houston Texas, 77053; Julius Montz GmbH, Hilden, Germany; ENVIMAC Engineering GmbH, Oberhausen, Germany; Ceilcote Co., Berea, Ohio; NSW Corp., Roanoke, Virginia; Lantec Products Inc., Agoura Hills, California.

TABLE 14-14 Characteristics of Structured Packings

Name	Size or number	Area, m^2/m^3	% voids*	Packing factor, m^{-1} Normal F_p†	Packing factor, m^{-1} Dry F_{pd}‡	Vendor
		Metals, corrugated sheets				
Mellapak	125Y	125	99	33		Sulzer
	170Y	170	99	39		
	2Y	223	99	46		
	250Y	250	98	66		
	350Y	350	98	75		
	500Y	500	98	112		
	750Y	750	97			
	125X	125	99	16		
	170X	170	99	20		
	2X	223	99	23		
	250X	250	98	26		
	350X	350	98			
	500X	500	98	82		
Mellapak Plus	202Y		99			Sulzer
	252Y	250	98	39		
	352Y		98			
	452Y	350	98	69		
	752Y	500	98	131		
Flexipac	700Y	710	96			Koch-Glitsch
	500Y	495	97			
	1Y	420	98	98	(105)	
	350Y	350	98			
	1.6Y	290	98	59		
	250Y	250	99			
	2Y	220	99	49	(36)	
	2.5Y	150	99			
	3.5Y	80	99	30	(15)	
	4Y	55	99	23	(10.5)	
	1X	420	98	52		
	350X	350	98			
	1.6X	290	98	33		
	250X	250	99			
	2X	220	99	23		
	2.5X	150	99			
	3X	110	99	16		
	3.5X	80	99			
	4X	55	99			
Flexipac High-capacity	700	710	96	223		Koch-Glitsch
	500Z	495	97			
	1Y	420	98	82		
	350Y	350	98			
	1.6Y	290	99	56		
	250Y	250	99			
	2Y	220	99	43		
Intalox	1T	310	98	66		Koch-Glitsch
	1.5T	250	99			
	2T	215	99	56		
	3T	170	99	43		
	4T	135	99			
	5T	90	99			
	5TX	90	99			
Super-Pak	250	250	98	39§		Raschig
	350	350	98	69§		
Ralu-Pak	250YC	250	98	66		Raschig
Rhombopac	4M	151				Sulzer
	6M	230		59		
	9M	351				
Max-Pak	0.5-in	229	98	39		Raschig
Montz-Pak	B1-100Y	100	99			Montz
	B1-125Y	125	99			
	B1-200Y	200	98			
	B1-250Y	250	97	66		
	B1-300Y	300	97			
	B1-350Y	350	96			
	B1-500Y	500	95			
	B1-500X	500	96			
	BSH-250¶	250	95			
	BSH-500¶	500	91			
	B1-250M	250		43		
	B1-350M	350				
	B1-500M	500				
		Wire Mesh				
Sulzer	AX	250	95			Sulzer
	BX	492	90	69	(52.5)	
	CY	700	85			
	BX Plus					
Wire gauze	BX	495	93			Koch-Glitsch
Montz-Pak	A3-500	500	91			Montz
Goodloe	765	1010	96			Koch-Glitsch
	773	1920	95			
	779	2640	92			
Hyperfil	2300	2300	93.6	394/230§,‖	460	Knit Mesh
	1900	1900	94.8	312/180§,‖		
	1400	1400	96.0	180/131§,‖		
		Ceramic				
Flexeramic	28	260	66	131		Koch-Glitsch
	48	160	77	79		
	88	100	86	49		
		Plastic				
Mellapak	125Y	125				Sulzer
	250Y	250	96	72		
	125X	125				
	250X	250				
Montz-Pak	C2-200	200	95			Montz
Ralu-Pak	30/160	160	92			Raschig
Multifil plastics	P1500	1500	88.5			Knit Mesh

*% voids vary with material thickness and values may not be consistent.
†Packing factors from Kister, *Distillation Design*, McGraw-Hill, 1992; Kister, Larson, and Gill, paper presented at the Houston AIChE Meeting, March 19–23, 1995; and Kister et al., in *Distillation 2007: Proceedings of Topical Conference*, AIChE Spring Meeting, Houston, Tex., April 22–26, 2007.
‡Dry packing factors from Robbins, *Chem. Eng. Prog.*, p. 87, May 1991.
§These packing factors supplied by the packing vendor.
¶These are expanded metal packings.
‖First figure is for hydrocarbon service, second figure for aqueous service.
Vendors: Sulzer Chemtech Ltd., Winterthur, Switzerland; Koch-Glitsch LLP, Wichita, Kansas; Raschig GmbH, Ludwigshafen/Rhein, Germany; Julius Montz GmbH, Hilden, Germany; Knit Mesh/Enhanced Separation Technologies, Houston, Texas; Kuhni Ltd., Allschwil, Switzerland.

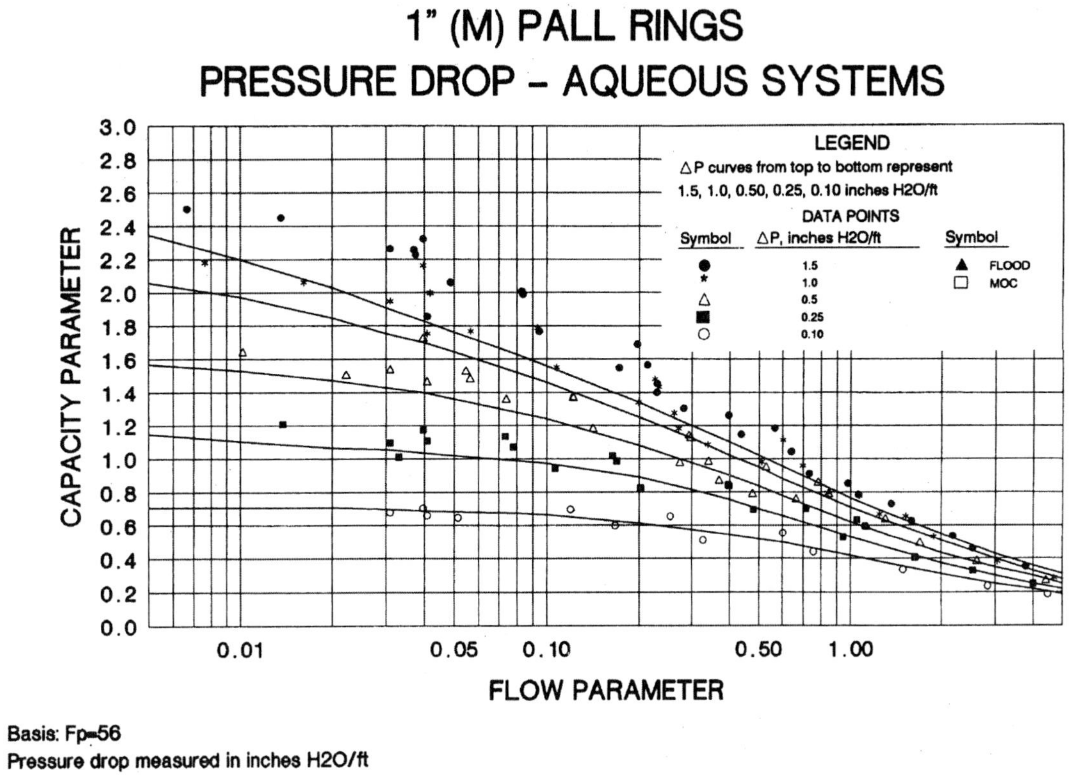

(*a*)

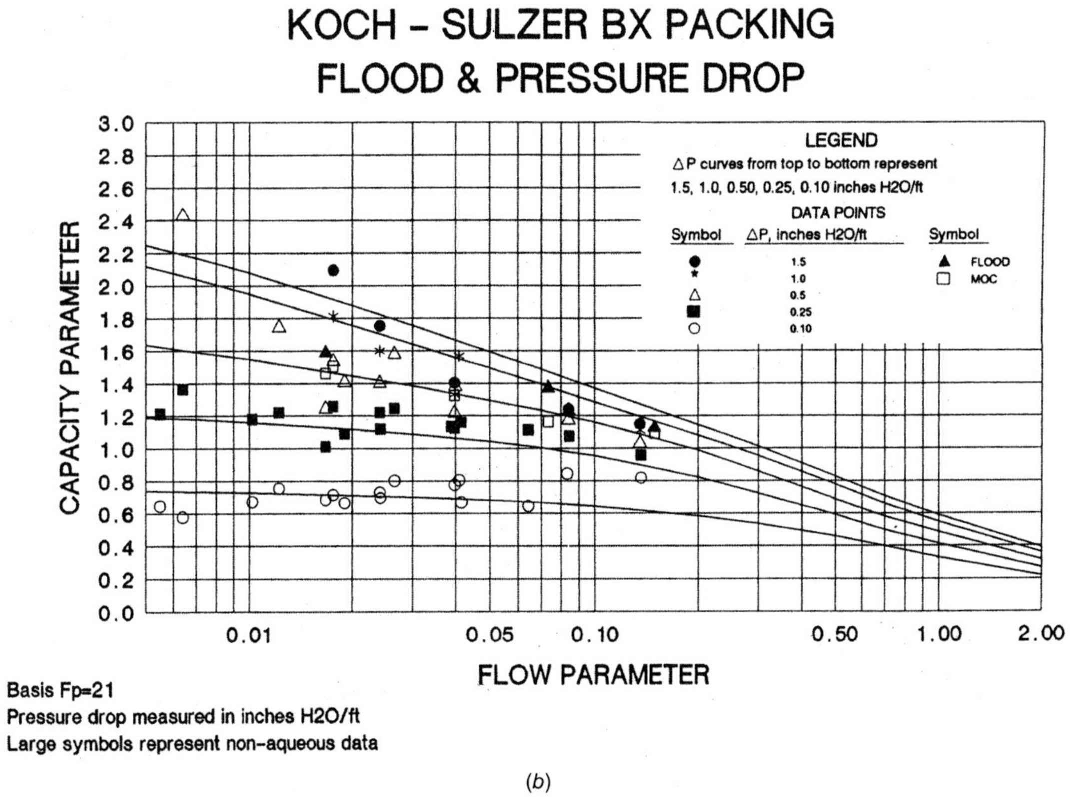

(*b*)

FIG. 14-57 Superimposing experimental pressure-drop data for a given packing generates a GPDC interpolation chart for this packing. (*a*) A random packing; chart is based on Eckert's GPDC, Fig. 14-55. (*b*) A structured packing; chart is based on Kister and Gill's GPDC (SP), Fig. 14-56. (*From Kister, H. Z.,* Distillation Design, *copyright © McGraw-Hill, 1992; used with permission.*)

Trans. IChemE. **77:** Part A, p. 498 (September 1999)] and the Maćkowiak (*Fluid Dynamics of Packed Columns*, Springer/Heidelberg, New York, 2010) correlations are versions of the GPDC that take the liquid holdup into account. The Eiden and Bechtel correlation [*IChemE Symp. Ser.* **142:** 757 (1997)] is a version of the GPDC in which accuracy is improved by using constants representative of packing shape instead of packing factors. The Lockett and Billingham correlation [*IChemE Symp. Ser.* **152:** 400 (2006)] uses a Wallis correlation

$$C_G^{0.5} + mC_L^{0.5} = C_{LG} \tag{14-143}$$

where

$$C_L = u_L[\rho_L/(\rho_L - \rho_G)]^{0.5} \tag{14-144}$$

and was shown to work well for high-surface-area (> 400 m^2/m^3) structured packings. Here C_G is the gas *C*-factor, Eq. (14-77), based on the tower superficial cross-sectional area, and m and C_{LG} are constants, available from the cited reference for some packing.

A drawback of most of these correlations (except that of Eiden and Bechtel) is the unavailability of constants for many, often most, of the modern popular packings.

The preceding methods apply to nonfoaming systems. Foaming systems can be handled either by applying additional derating (system) factors to the flood correlation (see Table 14-9) or by limiting the calculated pressure drop to 0.25 in of water per foot of packing (Hausch, in *Distillation Tools for the Practicing Engineer*, Topical Conference Proceedings, AIChE Spring Meeting, New Orleans, March 10–14, 2002, p. 119).

Derating ("System") Factors A review of some design criteria by Kooijman and Taylor (chap. 5 in Gorak and Schoenmakers, *Distillation Operation and Application*, Academic Press, New York, 2014) argued that the system factors in Table 14-9 can be extended to random packings by raising the relevant derating factor to the power of 0.7, or for a conservative design to a power of 0.8. The author prefers the more cautious approach of applying the same system factors as in Table 14-9. Comments in the tray section regarding the application of derating factors also extend to random packings.

Pressure Drop The GPDC discussed above (Figs. 14-55 and 14-56) and the Kister and Gill interpolation charts provide popular methods for calculating packing pressure drops. An alternative popular method that is particularly suitable for lower liquid loads was presented by Robbins.

For gas flow through dry packings, pressure drop may be estimated by the use of an orifice equation. For irrigated packings, pressure drop increases because of the presence of liquid, which effectively decreases the available cross section for gas flow (Fig. 14-53). Robbins (*Chem. Eng. Progr.*, May 1991, p. 87) uses the approach of correcting the dry pressure drop for the presence of liquid. The total pressure drop then becomes

$$\Delta P_t = \Delta P_d + \Delta P_L \tag{14-145}$$

where ΔP_t = total pressure drop, inches H_2O per foot of packing

$$\Delta P_d = \text{dry pressure drop} = C_3 G_f^2 10^{(C_4 L_f)} \tag{14-146}$$

$$\Delta P_L = \text{pressure drop due to liquid presence}$$

$$= 0.4[L_f/20{,}000]^{0.1}[C_3 G_f^2 10^{(C_4 L_f)}]^4 \tag{14-147}$$

$$G_f = \text{gas loading factor} = 986 F_s (F_{pd}/20)^{0.5} \tag{14-148}$$

$$L_f = \text{liquid loading factor} = L(62.4/\rho_L)(F_{pd}/20)^{0.5}\mu_L^{0.1} \tag{14-149}$$

The term F_{pd} is a dry packing factor, specific for a given packing type and size. Values of F_{pd} are given in Tables 14-13 and 14-14. For operating pressures above atmospheric, and for certain packing sizes, L_f and G_f are calculated differently:

$$G_f = 986 F_s (F_{pd}/20)^{0.5} 10^{0.3\rho_G} \tag{14-150}$$

$$L_f = L(62.4/\rho_L)(F_{pd}/20)^{0.5}\mu_L^{0.2} \quad F_{pd} > 200 \tag{14-151a}$$

$$L_f = L(62.4/\rho_L)(20/F_{pd})^{0.5}\mu_L^{0.1} \quad F_{pd} < 15 \tag{14-151b}$$

The Robbins equations require careful attention to dimensions. However, the use of the equations has been simplified through the introduction of Fig. 14-58. The terms L_f and G_f are evaluated, and the ΔP_L is obtained directly from the chart. Basic nomenclature for the Robbins method follows:

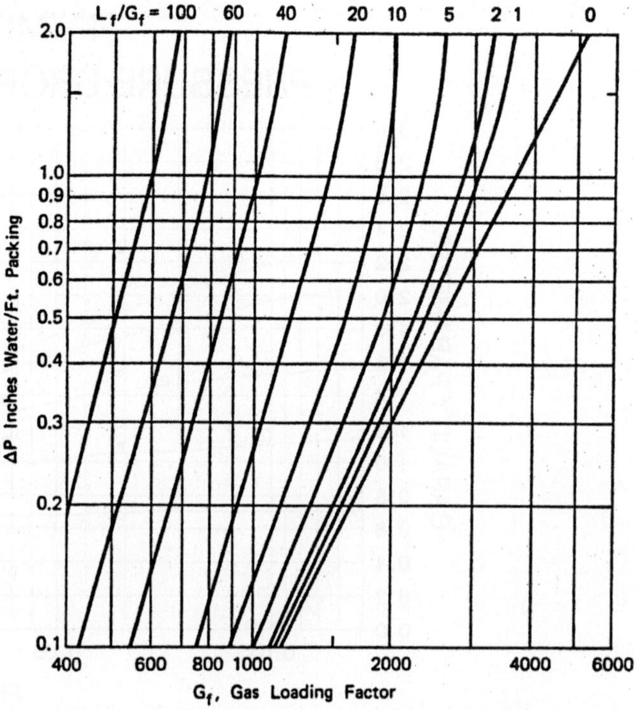

FIG. 14-58 The Robbins generalized pressure-drop correlation. (*From L. A. Robbins. Chem* Eng. Progr., *May 1991, p. 87, reprinted courtesy of the American Institute of Chemical Engineers.*)

$C_3 = 7.4(10)^{-8}$
$C_4 = 2.7(10)^{-5}$
F_{pd} = dry packing factor, ft^{-1}
F_s = superficial *F*-factor for gas, $U_t\rho_G^{0.5}$, $ft/s(lb/ft^3)^{0.5}$
G = gas mass velocity, $lb/hr \cdot ft^2$
G_f = gas loading factor, $lb/hr \cdot ft^2$
L = liquid mass velocity, $lb/hr \cdot ft^2$
L_f = liquid loading factor, $lb/hr \cdot ft^2$
ΔP = pressure drop, inches H_2O/ft packing (× 83.3 = mm H_2O/m packing)
ρ_G = gas density, lb/ft^3
ρ_L = liquid density, lb/ft^3
μ_L = liquid viscosity, cP

The Robbins correlation applies near atmospheric pressure and under vacuum, but it is not suitable above 3 bar absolute. For high (>0.3) flow parameters [Eq. (14-141)], the correlation has only been tested with air–water data.

For flood and MOC predictions, Robbins recommends his pressure drop method together with Eqs. (14-142) (flood) and (14-139) (MUC).

The GPDC and Robbins correlations are empirical. Fundamental correlations are also available. Most of these use the channel model, which attributes the pressure drop to the resistance to flow in a multitude of parallel channels. The channels may have bends, expansions, and contractions. Popular applications of this approach are the Rocha et al. correlation [Rocha, Bravo, and Fair, *Ind. Eng. Chem. Res.* **32:** 641 (1993)] for structured packing and the Maćkowiak (*Fluid Dynamics of Packed Columns*, Springer-Verlag, Berlin Heidelberg, 2010) and Billet (*Packed Column Analysis and Design*, Ruhr University, Bochum, Germany, 1989) methods. Stichlmair et al. (*Distillation Principles and Practices*, Wiley, New York, 1998; *Gas Sep. Purif.* **3:** March 1989, p. 19) present alternative correlations using the particle model, which attributes packing pressure drop to friction losses due to drag of a particle. This is similar to the Ergun model for single-phase flow [*Chem. Eng. Prog.* **48**(2): 89 (1952)].

Duss (chap. 4 in Gorak, A., and Z. Olujic, eds., *Distillation Equipment and Processes,* Elsevier, New York, 2014; AIChE *Distillation Topical Conference Proceedings,* San Antonio, Tex., April 2013) notes that at very low gas Reynolds numbers, experienced in deep vacuum distillation, typically below 3 to 10 mbar, the gas flow regime changes to laminar, causing the friction factor to rise. This is accounted for in the fundamental models described previously, but it is often unaccounted for by empirical and many vendor methods that are based on turbulent flow. When applying an empirical or vendor correlation for structured packings under deep vacuum, Duss recommends checking the gas Reynolds number

$$Re_G = 1000\, F \rho_G^{0.5} D_p / \mu_G$$

where Re_G is the gas Reynolds number, dimensionless; F is the F-factor, defined by Eq. (14-76), m/s (kg/m^3)$^{0.5}$; D_p is the packing hydraulic diameter, m; μ_G is the gas viscosity, cP, and other symbols are as described above. For structured packings, Duss uses $D_p = 4/a_p$. Values of a_p are listed in Table 14-14. Maćkowiak (*Fluid Dynamics of Packed Columns,* Springer/Heidelberg, New York, 2010) states that random packings 25 mm and larger are always operated in the turbulent range. Duss gives the critical gas Reynolds numbers for changing from turbulent flow to laminar flow at 300 and 500 for structured packings with corrugation angles of 45° and 30° from the vertical, respectively.

Example 14-13 Packed-Column Pressure Drop Air and water are flowing countercurrently through a bed of 2-in metal Pall rings. The air mass velocity is 2.03 kg/s · m^2 (1500 lbs/hr · ft^2), and the liquid mass velocity is 12.20 kg/s · m^2 (9000 lbs/hr · ft^2). Calculate the pressure drop by the generalized pressure drop (GPDC, Fig. 14-55) and the Robbins methods. Properties: $\rho_G = 0.074$ lbs/ft^3; $\rho_L = 62.4$ lbs/ft^3, $\mu_L = 1.0$ cP, $\nu = 1.0$ cS. The packing factor $F_p = 27$ ft^{-1}. For Robbins, $F_{pd} = 24$ ft^{-1}. The flow parameter $F_{LG} = L/G\,(\rho_G/\rho_L)^{0.5} = (9000/1500)\,(0.074/62.4)^{0.5} = 0.207$. The F-factor $= F_s = U_t\rho_G^{0.5} = G/(\rho_G^{0.5}3600) = 1500/[(0.074)^{0.5}\,(3600)] = 1.53$ ft/s(lb/ft^3)$^{0.5}$.

Using the GPDC method, the capacity parameter [by Eq. (14-140)] $= U_t[\rho_G/(\rho_L - \rho_G)]^{0.5} F_p^{0.5}\,\nu^{0.05}$, which is roughly equivalent to

$$\frac{F_S}{\rho_L^{0.5}} F_p^{0.5} \nu^{0.05} = \frac{1.53}{62.4^{0.5}} 27^{0.5}\,(1.0)$$
$$= 1.01$$

Referring to Fig. 14-55, the intersection of the capacity parameter and the flow parameter lines gives a pressure drop of 0.38 in H_2O/ft packing.

Using the Robbins method, $G_f = 986F_s\,(F_{pd}/20)^{0.5} = 986(1.53)(24/20)^{0.5} = 1653$. $L_f = L\,(62.4/\rho_L)(F_{pd}/20)^{0.5}\,\mu^{0.1} = 9000\,(1.0)(1.095)(1.0) = 9859$. $L_f/G_f = 5.96$.

From Fig. 14-58, pressure drop = 0.40 in H_2O/ft packing.

Example 14-14 Does the Reynolds Number Matter? For the column in Example 14-13, would the pressure drop depend on Reynolds number if the packing used is a structured packing of 250 m^2/m^3?

Converting to metric, the F-factor is F (metric) $= 1.221 \times 1.53 = 1.87$ m/s (kg/m^3)$^{0.5}$, and the gas density is $0.074 \times 16.018 = 1.19$ kg/m^3. From Duss's criterion, the hydraulic diameter is $4/250 = 0.016$ m. Using a viscosity of 0.018 cP gives $Re_G = 1000 \times 1.87 \times 1.19^{0.5} \times 0.016/0.0181 = 1800$, well above the critical Reynolds number of 300.

PACKING EFFICIENCY

HETP versus Fundamental Mass Transfer The two-film model gives the following transfer unit relationship:

$$H_{OG} = H_G + \lambda H_L \tag{14-152}$$

where H_{OG} = height of an overall transfer unit, gas concentration basis, m
H_G = height of a gas-phase transfer unit, m
H_L = height of a liquid-phase transfer unit, m
$\lambda = m/(L_M/G_M)$ = slope of equilibrium line/slope of operating line

In design practice, a less rigorous parameter, HETP, is used to express packing efficiency. The HETP is the height of packed bed required to achieve a theoretical stage. The terms H_{OG} and HETP may be related under certain conditions:

$$\text{HETP} = H_{OG}\left[\frac{\ln\lambda}{(\lambda - 1)}\right] \tag{14-153}$$

and since

$$Z_p = (H_{OG})(N_{OG}) = (\text{HETP})(N_t) \tag{14-154}$$

$$N_{OG} = N_t[\ln \lambda/(\lambda - 1)] \tag{14-155}$$

Equations (14-153) and (14-155) have been developed for binary mixture separations, and they hold for cases where the operating line and equilibrium line are straight. Thus, when there is curvature, the equations should be used for sections of the column where linearity can be assumed. When the equilibrium line and operating line have the same slope, HETP $= H_{OG}$ and $N_{OG} = N_t$ (theoretical stages).

An alternative parameter popular in Europe is the NTSM (number of theoretical stages per meter), which is simply the reciprocal of the HETP.

Factors Affecting HETP: An Overview Generally, packing efficiency increases (HETP decreases) when the following occur:

- Packing surface area per unit volume increases. Efficiency increases as the particle size decreases (random packing, Fig. 14-59) or as the channel size narrows (structured packing, Fig. 14-60).
- The packing surface is better distributed around a random packing element.

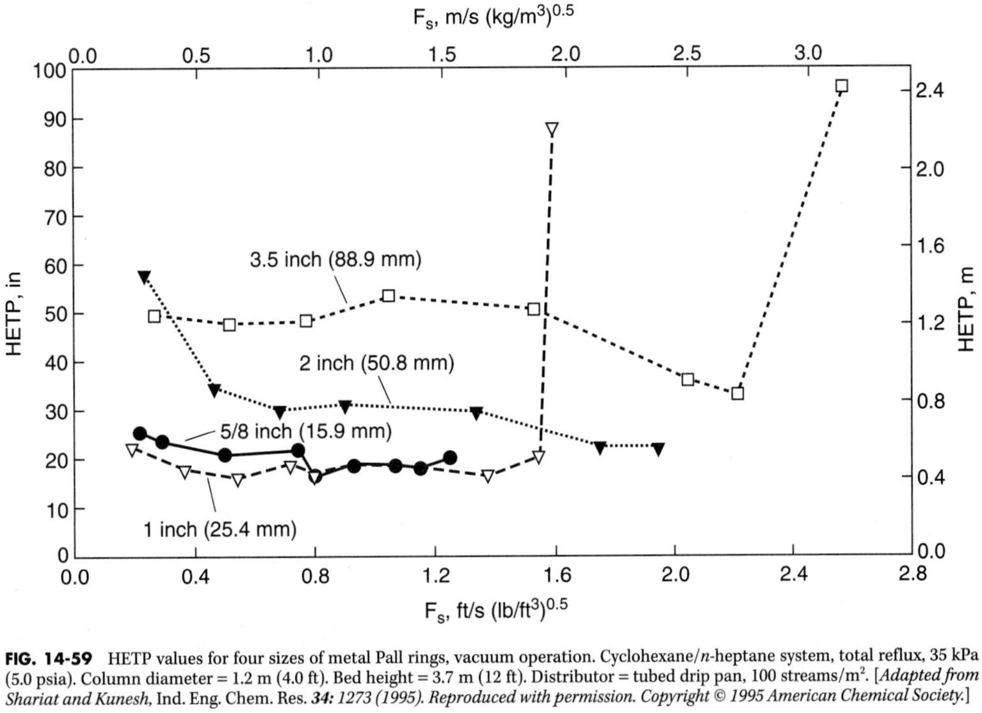

FIG. 14-59 HETP values for four sizes of metal Pall rings, vacuum operation. Cyclohexane/n-heptane system, total reflux, 35 kPa (5.0 psia). Column diameter = 1.2 m (4.0 ft). Bed height = 3.7 m (12 ft). Distributor = tubed drip pan, 100 streams/m^2. [*Adapted from Shariat and Kunesh,* Ind. Eng. Chem. Res. *34: 1273 (1995). Reproduced with permission. Copyright © 1995 American Chemical Society.*]

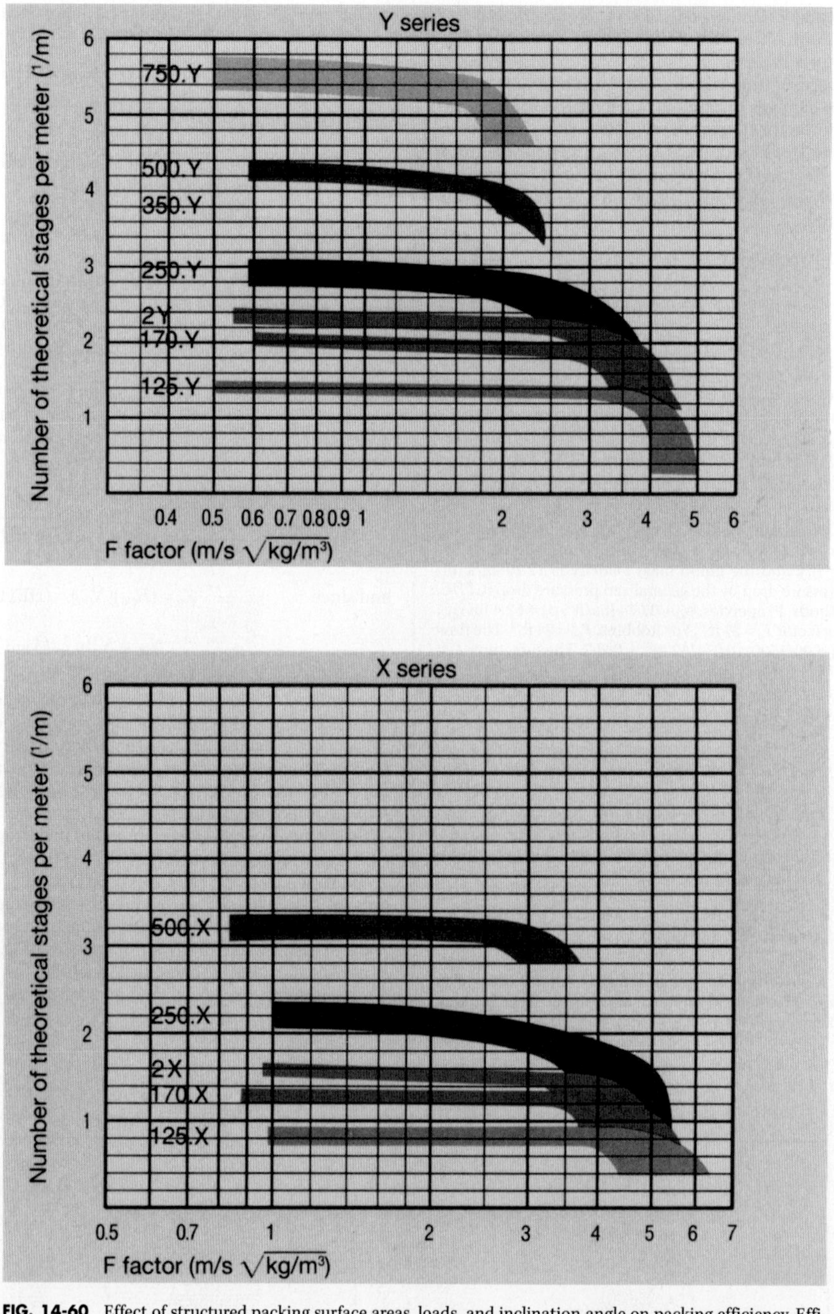

FIG. 14-60 Effect of structured packing surface areas, loads, and inclination angle on packing efficiency. Efficiency expressed as number of theoretical stages per meter, the reciprocal of HETP. Sulzer data, chlorobenzene–ethylbenzene, 100 mbar, at total reflux; 250-mm-diameter test column. (*Reprinted courtesy of Sulzer Chemtech.*)

- Y-structured packings (45° inclination) give better efficiencies than X-structured packings (60° inclination to the horizontal) of the same surface areas (Fig. 14-60).
- High-performance structured packings (45° inclination) give much the same efficiencies as the Y-structured packings.
- For constant *L/V* operation in the preloading regime, generally liquid and vapor loads have little effect on random and most corrugated sheet structured packings (Figs. 14-59 and 14-60). HETP increases with loadings in some wire-mesh structured packing.
- Liquid and vapor are well distributed. Both liquid and vapor maldistribution have a major detrimental effect on packing efficiency.
- Other. These include *L/V* ratio (lambda), pressure, and physical properties. These come into play in some systems and situations, as discussed below.

HETP Prediction HETP can be predicted from mass-transfer models, rules of thumb, and data interpolation.

Mass-Transfer Models Development of a reliable mass-transfer model for packing HETP prediction has been inhibited by a lack of understanding of the complex two-phase flow that prevails in packings, by the shortage of commercial-scale efficiency data for the newer packings, and by difficulty in quantifying the surface generation in modern packings. Bennett and Ludwig (*Chem. Eng. Prog.*, April 1994, p. 72) point out that the abundant air–water data cannot be reliably used for assessing real system mass-transfer resistance due to variations in turbulence, transport properties, and interfacial areas. More important, the success and reliability of rules of thumb for predicting packing efficiency made it difficult for mass-transfer models to compete.

For random packings, the Bravo and Fair correlation [*Ind. Eng. Chem. Proc. Des. Dev.* **21:** 162 (1982)] has been one of the most popular theoretical correlations. It was shown (e.g., McDougall, *Chem SA*, October 1985, p. 255) to be better than other theoretical correlations, yet it produced large discrepancies when compared to test data [Shariat and Kunesh, *Ind. Eng. Chem. Res.* **34**(4): 1273 (1995)]. For structured packings, the Bravo, Fair, and Rocha correlation [*Chem. Eng. Progr.* **86**(1): 19 (1990); *Ind. Eng. Chem. Res.* **35:** 1660 (1996)] is one of the most popular theoretical correlations. This correlation is based on the two-film theory. Interfacial areas are calculated from the packing geometry and an empirical wetting parameter.

Alternate popular theoretical correlations for random packings, structured packings, or both [e.g., Billet and Schultes, *Trans. IChemE* **77:** Part A, p. 498 (September 1999); Maćkowiak, *Chem. Eng. Res. Des.* **99:** 28 (2015)] are also available.

Rules of Thumb Since in most circumstances packing HETP is sensitive to only few variables, and due to the unreliability of even the best mass-transfer model, it has been the author's experience that rules of thumb for HETP are more accurate and more reliable than mass-transfer models. A similar conclusion was reached by Porter and Jenkins (*IChemE Symp. Ser.* **56,** Summary paper, London, 1979).

The majority of published random packing rules of thumb closely agree with one another. They are based on second-, third-, and fourth-generation random packings and should not be applied to the obsolete first-generation packings. Porter and Jenkins's (*IChemE Symp. Ser.* **56,** Summary paper, London, 1979), Frank's (*Chem. Eng.*, March 14, 1977, p. 40), Harrison and France's (*Chem. Eng.*, April 1989, p. 121), Chen's (*Chem. Eng.*, March 5, 1984, p. 40), and Walas' (*Chem. Eng.*, March 16, 1987, p. 75) general rules of thumb are practically the same, have been successfully tested against an extensive data bank, and are slightly conservative, and therefore suitable for design.

For small-diameter columns, the rules of thumb presented by Frank (*Chem. Eng.*, March 14, 1977, p. 40), Ludwig (*Applied Process Design for Chemical and Petrochemical Plants,* vol. 2, 2d ed., Gulf Publishing, Houston, Tex., 1979), and Vital et al. [*Hydrocarbon Processing* **63**(12): 75 (1984)] are identical. The author believes that for small columns, the more conservative value predicted from either the Porter and Jenkins or the Frank-Ludwig-Vital rule should be selected for design. Summarizing:

$$\text{HETP} = 18D_P \tag{14-156}$$

$$\text{HETP} > D_T \quad \text{for } D_T < 0.67\text{ m} \tag{14-157}$$

where D_P and D_T are the packing and tower diameters, m, respectively, and the HETP is in meters. In high-vacuum columns (<100 mbar), and when $\lambda = mG_M/L_M$ is outside the range of 0.5 to 2.0, the preceding rules may be optimistic (see below).

The preceding rules of thumb were based on experience with Pall rings. The packing diameter may be difficult to establish for some of the modern packings, especially those of saddle or flat shape. For these, Kister and Larson (in Schweitzer, *Handbook of Separation Techniques for Chemical Engineers,* 3d ed., McGraw-Hill, 1997) extended Eq. (14-156) by expressing the packing diameter in terms of the more fundamental surface area per unit volume a_P, m^2/m^3. For Pall rings, it can be shown that

$$a_P = 5.2/D_P \tag{14-158}$$

and Eq. (14-156) becomes

$$\text{HETP} = 93/a_P \tag{14-159}$$

Harrison and France (*Chem. Eng.*, April 1989, p. 121) presented the only published rule of thumb for structured packings efficiency as a function of packing crimp. Kister and Larson reexpressed it in terms of the surface area per unit volume to accommodate a wider range of packing geometries. The final expression is

$$\text{HETP} = 100C_{XY}/a_P + 0.10 \tag{14-160}$$

Specific surface areas are listed in Tables 14-13 and 14-14.

The preceding rules of thumb apply to organic and hydrocarbon systems, whose surface tensions are relatively low ($\sigma < 25$ mN/m). For higher surface tensions, the liquid does not adhere well to the packing surfaces (*underwetting*), causing higher HETPs. In a water-rich system (σ = 70 mN/m or so), HETPs obtained from Eqs. (14-156), (14-159), and (14-160) need to be doubled. For intermediate surface tension systems (some amines and glycols, whose surface tension at column conditions is 40 to 50 mN/m), HETPs obtained from Eqs. (14-156), (14-159), and (14-160) need to be multiplied by 1.5.

For random packings, Eqs. (14-156) and (14-159) apply for packings of 25-mm diameter and larger. For smaller packings, use of the a_p at 25 mm often gives a slightly conservative HETP estimate. For structured packing, C_{XY} in Eq. (14-160) reflects the effect of the inclination angle (Fig. 14-60). $C_{XY} = 1$ for Y-type, S-type, or high-performance packings, and $C_{XY} = 1.45$ for the larger (<300 m^2/m^3) X-type packings. There are insufficient data to determine C_{XY} for high specific area X-type packings, but Fig. 14-60 suggests it is somewhat lower than 1.45.

Compared to experimental data, the preceding rules of thumb are slightly conservative. Since packing data are usually measured with perfect distribution, a slight conservative bias is necessary to extend these data to the good, yet imperfect, distributors used in the real world. For poor distributors, these rules of thumb will usually predict well below the HETPs measured in the field.

Lockett (*Chem. Eng. Progr.*, January 1998, p. 60) simplified the fundamental Bravo-Fair-Rocha correlation [*Ind. Eng. Chem. Res.* **35:** 1660 (1996)] to derive an alternative rule of thumb for structured packing efficiency. This rule of thumb predicts HETPs under perfect distribution conditions. Lockett recommends caution when applying this rule of thumb for aqueous systems because it does not predict the effects of underwetting.

Service-Oriented Rules of Thumb Strigle (*Packed Tower Design and Applications,* 2d ed., Gulf Publishing, Houston, Tex., 1994) proposed a multitude of rules of thumb as a function of the service, column pressure, and physical properties. These rules are based on the extensive experience of Strigle and the Norton Company (now merged with Koch-Glitsch LP).

Data Interpolation Interpolation of experimental HETP data is the most reliable means of obtaining design HETP values. This is hardly surprising in an area where our understanding of the theory is so poor that rules of thumb can do better than theoretical models. The author believes that it is best to derive HETP from experimental data, and to check it against a rule of thumb.

Eckert [*Chem. Eng. Progr.* **59**(5): 76 (1963)], Chen (*Chem. Eng.*, March 5, 1984, p. 40), and Vital et al. [*Hydroc. Proc.* **63**(12): 75 (1984)] tabulated experimental HETP data for various random packings. Kister (*Distillation Design,* McGraw-Hill, 1992) extended these tabulations and included published HETP data and a detailed procedure for interpolating such HETP data. A prerequisite to any interpolation of packing data is thorough familiarity with the factors that affect HETP. Overlooking any of the factors listed can easily lead to poor interpolation and grossly incorrect design. In particular, it is imperative to recognize that the quality of distribution in pilot towers is generally superior to the quality of distribution in commercial towers.

Underwetting Laboratory- and pilot-scale distillation experiments with systems that exhibit large differences in surface tension along the column such as methanol-water showed a sharp drop in efficiency at the high-surface-tension end of the column [Ponter et al., *Trans. Instn. Chem. Engineers [London]* 45: T345 (1967)]. There appeared to be a critical methanol composition below which performance deteriorated rapidly. The poor performance at the low-methanol-concentration end appeared independent of the type and size of packing. Visual observations with disk columns attributed these effects to underwetting.

Underwetting is a packing surface phenomenon, which breaks up liquid film. The tendency of the liquid film to break (the degree of wetting) is expressed by the contact angle (Fig. 14-61). A contact angle of 0° indicates perfect wetting; an angle of 180° indicates no wetting. Mersmann and Deixler [*Chem. Ing. Tech.* **58**(1): 19 (1986)] provide a preliminary chart for estimating contact angles. The contact angle depends on both the surface and the liquid and is a strong function of composition. In systems with large surface tension gradients, both contact angles and minimum wetting rates may vary rapidly with changes of composition or surface tension. Extensive studies by Ponter et al. [*Trans. Instn. Chem. Engineers [London]* **45:** T345 (1967); also, Ponter and Au-Yeung, *Chem. Ing. Tech.* **56**(9): 701 (1984)] and experiences reported by Niggemann et al. (chap. 7, in Gorak and Schoenmakers, *Distillation Operation and Application,* Academic Press, New York, 2014) showed that

- Underwetting is most significant in aqueous-organic systems, and it tends to occur at the high-surface-tension (aqueous) end of the composition range. Liquid viscosity may also have an effect.
- Underwetting may be alleviated by changing the material and surface roughness of the packing.

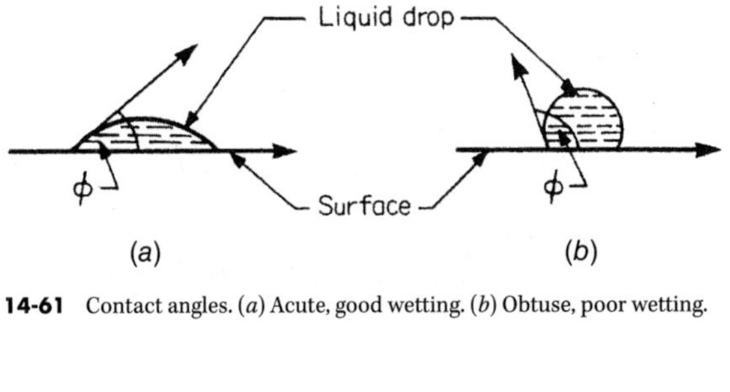

FIG. 14-61 Contact angles. (*a*) Acute, good wetting. (*b*) Obtuse, poor wetting.

- In systems susceptible to underwetting, column efficiency can sometimes (but not always) be improved by the addition of small amounts of surfactants, or it may deteriorate in service due to deposition that increases the contact angle (e.g., hydrophobic deposits in an aqueous system).
- The flow rates needed to wet a surface are higher than those at which dewetting occurs.

Effect of Lambda Most packed-column efficiency testing has been at total reflux. Some tests for both random and structured packings [Billet, *Packed Towers Analysis and Design*, Ruhr University, Bochum, Germany, 1989; Meier, Hunkeler, and Stocker, *IChemE Symp. Ser.* 56: 3.3/1 (1979); Eckert and Walter, *Hydroc. Proc.* **43**(2): 107 (1964)] suggest that efficiencies at finite reflux are similar to those at total reflux when lambda ($\lambda = mG_M/L_M$, which is the ratio of the slope of the equilibrium curve to the slope of the operating line) ranges between 0.5 and 2.0. This range is typical for most distillation systems.

Koshy and Rukovena [*Hydroc. Proc.* **65**(5): 64 (1986)], experimenting with methanol-water and water-DMF using #25 IMTP packing in a pilot-scale column, observed a sharp efficiency drop when λ was greater than 2 or lower than 0.5. The efficiency loss escalated as λ deviated more from this range. Koshy and Rukovena recognized that surface tension gradients and underwetting may have influenced some of their findings, but they argued that the lambda effect is the major cause for the efficiency differences observed in their tests. High-relative-volatility systems are those most likely to be affected by λ because at low volatility, λ ranges from 0.5 to 2. Strigle (*Packed Tower Design and Applications*, 2d ed., Gulf Publishing, Houston, Tex., 1994) quantified the lambda effect on HETP using the following equation:

$$\text{Actual HETP/standard HETP} = 1 + 0.278[\text{ABS}(\ln \lambda)^3] \qquad (14\text{-}161)$$

For $0.5 < \lambda < 2$, Eq. (14-161) gives a ratio of less than 1.1; that is, it has little influence on HETP.

Pressure Generally, pressure has little effect on HETP of both random and structured packing, at least above 100 mbar abs. At deep vacuum (<100 mbar), there are data to suggest that efficiency decreases as pressure is lowered for random packings [Zelvinski, Titov, and Shalygin, *Khim Tekhnol. Topl. Masel.* **12**(10) (1966)], but most of these data can also be explained by poor wetting or maldistribution. When reducing the pressure steeply lowers temperatures, rapid liquid viscosity escalation can reduce efficiency [Bratmöller and Scholl, *Chem. Eng. Res. Design* **99:** 75 (2015)].

At high pressures (>15 to 20 bar), and high liquid rates, best characterized as high flow parameters, above 0.2 [Eq. (14-141)], structured packing efficiency diminishes as pressure is raised (Kister, Larson, and Yanagi, *Chem. Eng. Progr.*, February 1994, p. 23; Chambers, McCarley, Cai, and Vennavelli, *Kister Distillation Symposium Proceedings*, AIChE Spring Meeting, Austin, Tex., 2015, p. 364).

With structured packings (only), FRI's high-pressure (10 to 30 bar; flow parameters >0.25) distillation tests measured maxima, termed *humps* in the HETP versus load plot, typically at 65 to 90 percent of flood [Fitz, Shariat, and Kunesh, *IChemE Symp. Ser.* **142:** 829 (1997); Cai et al., *Trans IChemE* **81:** Part A, p. 85 (2003); Chambers, McCarley, Cai, and Vennavelli, *Kister Distillation Symposium Proceedings*, AIChE Spring Meeting, Austin, Tex., 2015, p. 364]. These humps (Fig. 14-62) were not observed with lower-pressure distillation (flow parameters <0.2), and they appeared to intensify with higher pressure. The humps did not always occur; some tests at different distributor positioning and with larger packing showed no humps.

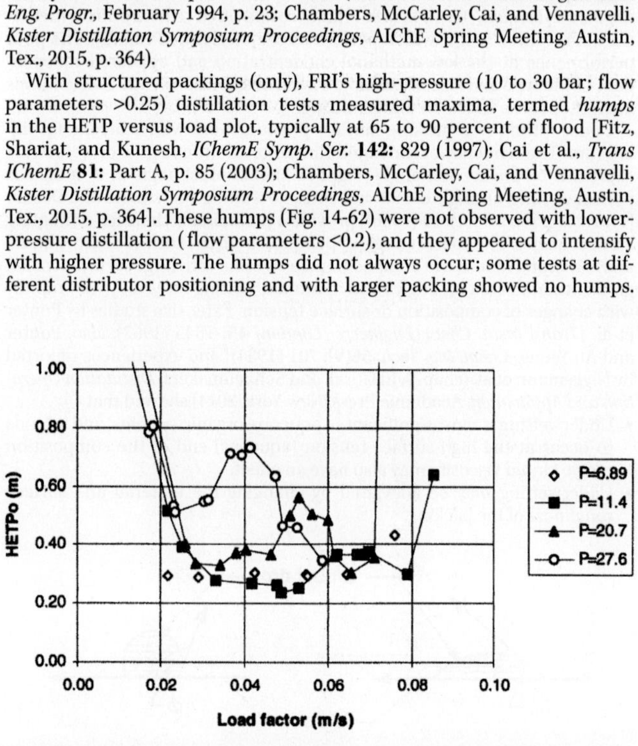

FIG. 14-62 HETP data as measured in the FRI column for the iC_4/nC_4 system at different pressures (bara), showing efficiency humps. (*From J. L. Nooijen, K. A. Kusters, and J. J. B. Pek,* IChemE Symp. Ser. *142: p. 885, 1997. Reprinted courtesy of IChemE.*)

Zuiderweg et al. [*IChemE Symp. Ser.* **142:** 865 (1997); *Trans. IChemE* **81:** Part A, p. 85 (January 2003)] and Nooijen et al. [*IChemE Symp. Ser.* **142:** 885 (1997)] explain the humps by two-phase backmixing. At the high liquid loads in high-pressure distillation, flow unevenness prematurely floods some of the packing channels, carrying vapor bubbles downward and recirculating liquid upward. Chambers et al. applied observations from Kerst, Judat, and Schlünder's experiments with falling films [*Chem. Eng. Sci.* **55:** 4189 (2000)] to shed more light on the humps. As pressure increases, the smooth liquid films on the packing progressively turn to wavy, then to troughs, crests, coarse drops, and finally fine drops. The backmixing of the liquid drops reduces the efficiency leading to the hump formation. In addition, as liquid flow rate and pressure are raised, the falling liquid chokes an increasing proportion of the corrugation channels, forcing vapor to ascend through the less loaded channels, reducing contact and capacity.

Physical Properties Data presented by a number of workers [e.g., Vital, Grossel, and Olsen, *Hydroc. Proc.* **63**(12): 75 (1984)] suggest that, generally, random packing HETP is relatively insensitive to system properties for nonaqueous systems. A survey of data in Chapter 11 of Kister's *Distillation Design* (McGraw-Hill, New York, 1992) leads to a similar conclusion for structured packings. For water-rich systems, packing HETPs tend to be much higher than for nonaqueous systems due to their high lambda or surface underwetting, as discussed previously. High hydrogen concentrations (>30 percent or so in the gas) have also led to low packing efficiencies (Kister et al., *Proc. 4th Ethylene Producers Conference,* AIChE, New Orleans, La., p. 283, 1992), possibly due to the fast-moving hydrogen molecule dragging heavier molecules with it as it diffuses from a liquid film into the vapor.

Errors in VLE and Reflux Ratios These affect packing HETP in the same way as they affect tray efficiency. The discussions and derivation earlier in this subsection apply equally to tray and packed towers.

Comparison of Various Packing Efficiencies for Absorption and Stripping In past editions of this handbook, extensive data on absorption/stripping systems were given. Emphasis was given to the following systems:

Carbon dioxide–1 N caustic	Liquid phase chemical reaction controlling
Ammonia–air–water	Liquid and gas phases contributing; chemical reaction contributing
Air–water	Gas phase controlling
Sulfur dioxide–air–water	Liquid and gas phase controlling
Carbon dioxide–air–water	Liquid phase controlling

The reader may refer to the data in the 5th edition of this handbook.

Summary In the preloading regime, packing size, type, and distribution affect HETP. With aqueous-organic systems, HETP may be sensitive to underwetting and composition. A lambda value ($\lambda = mG_M/L_M$) outside the range of 0.5 to 2.0 causes HETP to rise, and so does a high hydrogen concentration. HETP of structured packings may also be affected by pressure (at high pressure) and vapor and liquid loads.

MALDISTRIBUTION AND ITS EFFECTS ON PACKING EFFICIENCY

Modeling and Prediction Maldistribution may drastically reduce packing efficiency. HETP may increase by a factor of as high as 2 or 3 due to maldistribution. Shariat and Kunesh [*Ind. Eng. Chem. Res.* **34**(4): 1273 (1995)] provide a good demonstration.

Early models [Mullins, *Ind. Chem. Mfr.* **33:** 408 (1957); Manning and Cannon, *Ind. Eng. Chem.* **49**(3): 347 (1957)] expressed the effect of liquid maldistribution on packing efficiency in terms of a simple channeling model. A portion of the liquid bypasses the bed, undergoing negligible mass transfer, and then rejoins and contaminates the rest of the liquid. Huber et al. [*Chem. Ing. Tech.* **39:** 797 (1967); *Chem. Eng. Sci.* **21:** 819 (1966)] and Zuiderweg et al. [*IChemE Symp. Ser.* **104:** A217 (1987)] replaced the simple bypassing with variations in the local L/V ratios. The overirrigated parts have a high L/V ratio, the underirrigated parts a low L/V ratio. Regions with low L/V ratios experience pinching and, therefore, produce poor separation.

Huber et al. [*Chem. Ing. Tech.* **39:** 797 (1967)] and Yuan and Spiegel [*Chem. Ing. Tech.* **54:** 774 (1982)] added lateral mixing to the model. Lateral deflection of liquid by the packing particles tends to homogenize the liquid, thus counteracting the channeling and pinching effect.

A third factor is the nonuniformity of the flow profile through the packing. This nonuniformity was observed as far back as 1935 [Baker, Chilton, and Vernon, *Trans. Instn. Chem. Engrs.* **31:** 296 (1935)] and was first modeled by Cihla and Schmidt [*Coll. Czech. Chem. Commun.* **22:** 896 (1957)]. Hoek (Ph.D. thesis, The University of Delft, The Netherlands, 1983) combined all three factors into a single model, leading to the zone-stage model described next.

The Zone-Stage Model Zuiderweg, Kunesh, et al. [*IChemE Symp. Ser.* **104:** A217, A233 (1987)] extended Hoek's work combining the effects of local L/V ratio, lateral mixing, and flow profile into a model describing the effect of liquid maldistribution on packing efficiency. This work was performed at Fractionation Research Inc. (FRI) and at The University of Delft in the Netherlands. The model postulates that, in the absence of maldistribution, there is a "basic" (or "true" or "inherent") HETP that is a function of the packing and the system only. This HETP can be inferred from data for small towers, in which lateral mixing is strong enough to offset any pinching. For a given initial liquid distribution, the model uses a diffusion-type equation to characterize the splitting and recombining of liquid streams in the horizontal and vertical directions. The mass transfer is then calculated by integrating the liquid flow distribution at each elevation and the basic HETP. Zuiderweg et al. successfully applied the model to predict measured effects of maldistribution on packing efficiency. However, this model is difficult to use and has not gained industrywide acceptance.

Empirical Prediction Moore and Rukovena [*Chemical Plants and Processing* (European edition), August 1987, p. 11] proposed the empirical correlation in Fig. 14-63 for efficiency loss due to liquid maldistribution in packed towers containing Pall® rings or IMTP® packing. This correlation was shown to work well for several case studies (Fig. 14-63), is simple to use, and is valuable, at least as a preliminary guide.

To quantify the quality of liquid irrigation, the correlation uses the distribution quality rating index. Typical indexes are 10 to 70 percent for most standard commercial distributors, 75 to 90 percent for intermediate-quality distributors, and over 90 percent for high-performance distributors. Moore and Rukovena present a method for calculating a distribution-quality rating index from distributor geometry. Their method is described in detail in their paper as well as in Kister's book (*Distillation Operation,* McGraw-Hill, New York, 1990). For structured packing, the Moore and Rukovena method does not credit the liquid spreading performed by the top layer of packings, and therefore it may penalize distributors that irrigate the packings in the form of a drip line like in Fig. 14-68*e* (Spiegel, *Chemical Engineering and Processing,* 2006, p. 1011). Spiegel proposed a "wetting index," defined as the fraction of area wetted by liquid after the first layer of packing to the tower cross-sectional area, which accounts for the liquid spread at the top layer of packing. Unfortunately, generic guidelines for the compilation and application of the wetting index method have not been published.

Maximum Liquid Maldistribution Fraction f_{max} To characterize the sensitivity of packed beds to maldistribution, Lockett and Billingham [*Trans. IChemE* **80:** Part A, p. 373, (May 2002); *Trans. IChemE* **81:** Part A, p. 134 (January 2003)] modeled maldistribution as two parallel columns, one receiving more liquid $(1+f)L$, the other receiving less $(1-f)L$. The vapor was assumed to be equally split (Fig. 14-64) without lateral mixing. Because of the different L/V ratios, the overall separation is less than is obtained at uniform distribution. A typical calculated result (Fig. 14-65) shows the effective number of stages from the combined two-column system decreasing as the maldistribution fraction f increases. Figure 14-65*a* shows that the decrease is minimal in short beds (e.g., 10 theoretical stages) or when the maldistribution fraction is small. Figure 14-65*a* shows that there is a limiting fraction f_{max} that characterizes the maximum maldistribution that still permits achieving the required separation. Physically, f_{max} represents the maldistribution fraction at which one of the two parallel columns in the model becomes pinched. Figure 14-65*b* highlights the steep drop in packing

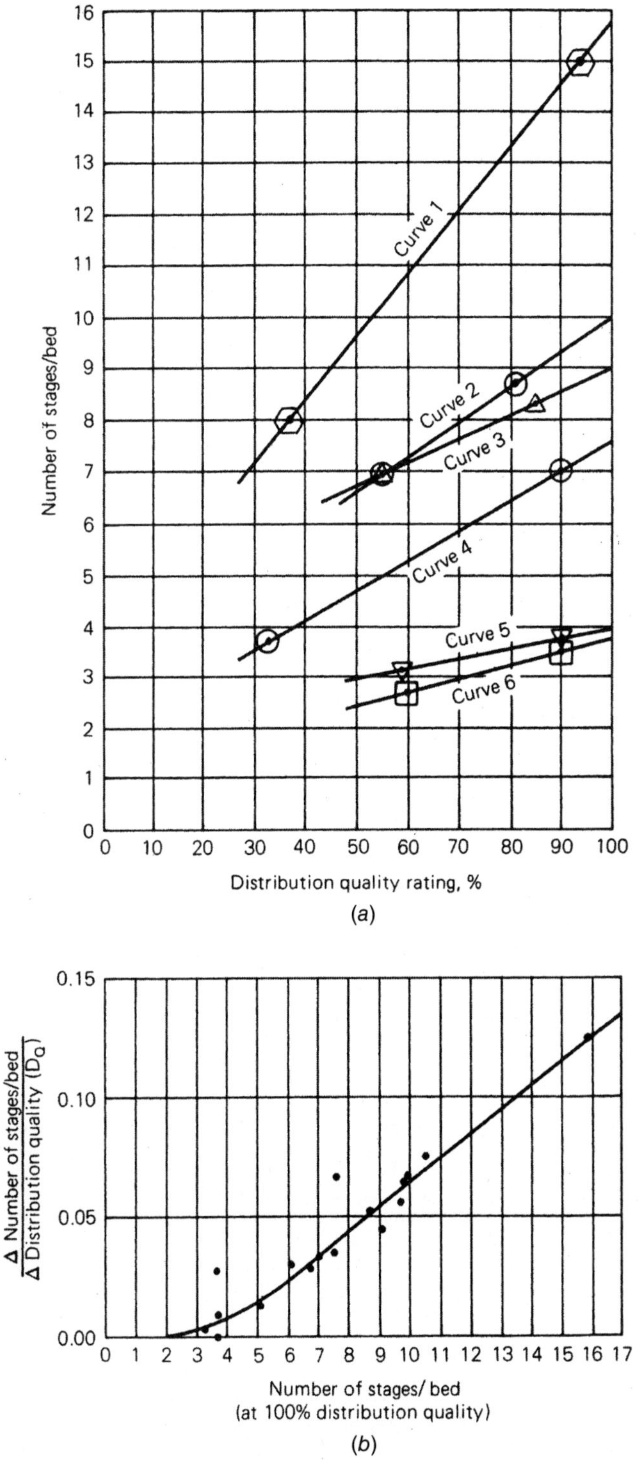

FIG. 14-63 Effect of irrigation quality on packing efficiency. (*a*) Case histories demonstrating efficiency enhancement with higher distribution quality rating. (*b*) Correlation of the effect of irrigation quality on packing efficiency. (*From F. Moore and F. Rukovena,* Chemical Plants and Processing, *Europe edition, Aug. 1987; reprinted courtesy of* Chemical Plants and Processing.)

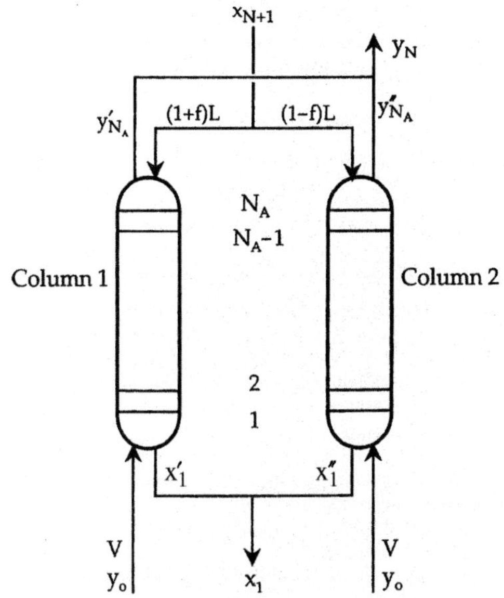

FIG. 14-64 Parallel-columns model. (*From Lockett and Billingham,* Trans. IChemE *80: Part A, p. 373, May 2002; reprinted courtesy of IChemE.*)

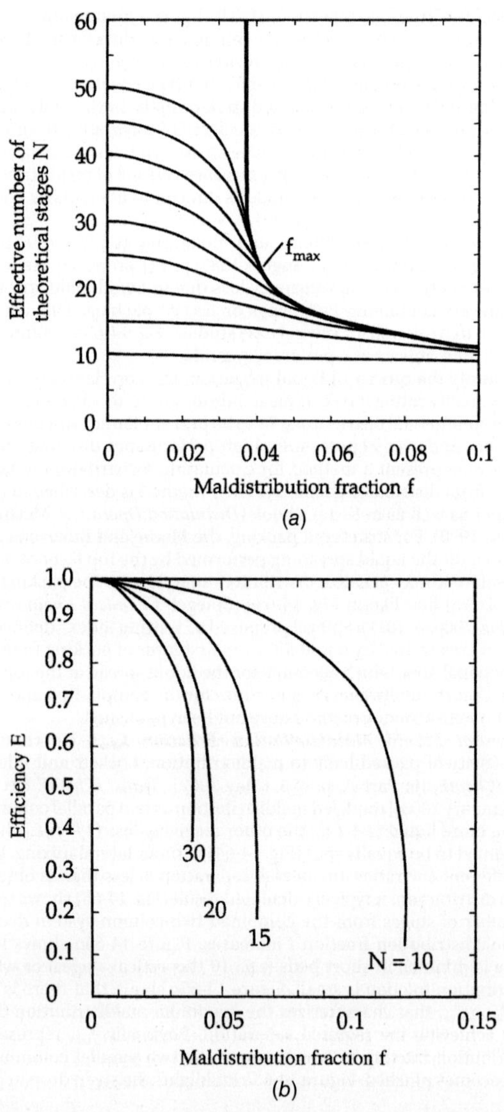

FIG. 14-65 Loss of efficiency due to maldistribution as a function of maldistribution fraction f and the number of stages per bed for a given case study. (*a*) f_{max} and reduction in number of stages. (*b*) Showing larger efficiency loss at higher number of stages per bed and higher f. The steep drops indicate pinching associated with f_{max}. (*From Lockett and Billingham,* Trans. IChemE *80: Part A, p. 373, May 2002; reprinted courtesy of IChemE.*)

efficiency upon the onset of this pinch. Billingham and Lockett derived the following equation for f_{max} in a binary system:

$$f_{max} = \frac{y_{N+1} - y_N}{y_N - y_\circ} + \frac{x_1 - x_\circ}{x_{N+1} - x_\circ} - \left(\frac{y_{N+1} - y_N}{y_N - y_\circ}\right)\left(\frac{x_1 - x_\circ}{x_{N+1} - x_\circ}\right) \qquad (14\text{-}162)$$

This equation can be used to calculate f_{max} directly without the need for a parallel column model. Billingham and Lockett show that the various terms in Eq. (14-162) can be readily calculated from the output of a steady-state computer simulation. Multicomponent systems are represented as binary mixtures, either by lumping components together to form a binary mixture of pseudolight and pseudoheavy components, or by normalizing the mole fractions of the two key components. Once f_{max} is calculated, Billingham and Lockett propose the following guidelines:

- f_{max} <0.05, extremely sensitive to maldistribution. The required separation will probably not be achieved.
- 0.05 <f_{max} <0.10, sensitive to maldistribution, but separation can probably be achieved.
- 0.10 <f_{max} <0.20, not particularly sensitive to maldistribution.
- f_{max} >0.20 insensitive to maldistribution.

Figure 14-65*b* shows that shortening the bed can increase f_{max}. Relative volatility and L/V ratio also affect f_{max}. The bed length and L/V ratio can often be adjusted to render the bed less sensitive to maldistribution.

Implications of Maldistribution to Packing Design Practice These are discussed at length with extensive literature citation in Kister's book *Distillation Operation,* McGraw-Hill, New York, 1990. Following are the highlights:

1. Three factors appear to establish the effect of maldistribution on efficiency:

a. Pinching. Regional changes in L/V ratio cause regional composition pinches.

b. Lateral mixing. Packing particles deflect both liquid and vapor laterally. This promotes mixing of vapor and liquid and counteracts the pinching effect.

c. Liquid nonuniformity. Liquid flows unevenly through the packing and tends to concentrate at the wall.

2. At small tower-to-packing diameter ratios (D_T/D_p < 10), the lateral mixing cancels out the pinching effect, and a greater degree of maldistribution can be tolerated without a serious efficiency loss. At high ratios (D_T/D_p > 40), the lateral mixing becomes too small to offset the pinching effect. The effects of maldistribution on efficiency are therefore most severe in large-diameter columns and small-diameter packings.

A good design practice is to seek a packing size that gives a D_T/D_p between 10 and 40. This is often impractical, and higher ratios are common. When D_T/D_p exceeds 100, avoiding efficiency loss due to maldistribution is challenging. Either ratios exceeding 100 should be avoided, or a special allowance should be made for loss of efficiency due to maldistribution.

3. A large-diameter packed bed cannot return to health from any severe maldistribution [Olujic and Jansen, *Chem. Eng. Res. Des.* **99:** 2 (2015)].

4. Wall flow effects become large when D_T/D_p falls below about 10. Packing diameter should be selected so that D_T/D_p exceeds 10.

Commercial structured packings come with integral wall wipers at regular vertical intervals. These redirect liquid running down the wall into the packings and prevent gas from bypassing the bed at the wall region, and they need to be adequately stretched out during packing installation. Experiments by Olujic and Jansen [*Chem. Eng. Res. Des.* **99:** 2 (2015)] showed that the amount of wall flow differs to a certain extent, depending on the number and effectiveness of the wall wipers used.

5. Columns containing fewer than five theoretical stages per bed are relatively insensitive to liquid maldistribution. With 10 or more stages per bed, efficiency can be extremely sensitive to maldistribution (Strigle, *Packed Tower Design and Applications,* 2d ed., Gulf Publishing, Houston, Tex., 1994) (Fig. 14-66). Beds consisting of small packings or structured packings, which develop more theoretical stages per bed, are therefore more sensitive to maldistribution than equal-depth beds of larger packings. This is clearly demonstrated by FRI's experiments [Shariat and Kunesh, *Ind. Eng. Chem. Res.* **34**(4): 1273 (1995)]. Lockett and Billingham [*Trans. IChemE* **81:** Part A, p. 131 (January 2003)] concur with these comments when their procedure (above) indicates high sensitivity to maldistribution, but they allow a higher number of stages per bed when the sensitivity is low.

6. Maldistribution tends to be a greater problem at low liquid flow rates than at high liquid flow rates [Zuiderweg, Hoek, and Lahm, *IChemE Symp. Ser.* **104:** A217 (1987)]. The tendency to pinch and to spread unevenly is generally higher at the lower liquid flow rates.

7. A packed column has reasonable tolerance for a uniform or smooth variation in liquid distribution and for a variation that is totally random (*small-scale maldistribution*). The impact of discontinuities or zonal flow (*large-scale maldistribution*) is much more severe [Zuiderweg et al. 1987; Kunesh, *Chem. Eng.,* December 7, 1987, p. 101; Kunesh, Lahm, and Yanagi, *Ind. Eng. Chem. Res.* **26**(9): 1845 (1987); Olujic and Jansen, *Chem. Eng. Res. Des.* **99:** 2 (2015)]. This is so because the local pinching of small-scale maldistribution is evened out by the lateral mixing and therefore causes few ill effects. In contrast, the lateral mixing either is powerless to rectify a large-scale maldistribution or takes considerable bed length to do so (meanwhile, efficiency is lost).

Figure 14-65 shows HETPs measured in tests that simulate various types of maldistribution in FRI's 1.2-m column containing a 3.6-m bed of 1-in Pall® rings. The y axis is the ratio of measured HETP in the maldistribution tests to the HETP obtained with an excellent distributor. Analogous measurements with structured packing were reported by Fitz, King, and Kunesh [*Trans. IChemE* **77:** Part A, p. 482 (1999)]. Generally, the response of the structured packings resembled that of the Pall® rings, except as noted below.

Figure 14-66*a* shows virtually no loss of efficiency when a distributor uniformly tilts, such that the ratio of highest to lowest flow is 1.25 (i.e., a 1.25 tilt). In contrast, an 11 percent chordal blank of a level distributor causes packing HETP to rise by 50 percent.

Figure 14-66*b* compares continuous tilts with ratios of highest to lowest flow of 1.25 and 1.5 to a situation where one-half of the distributor passes

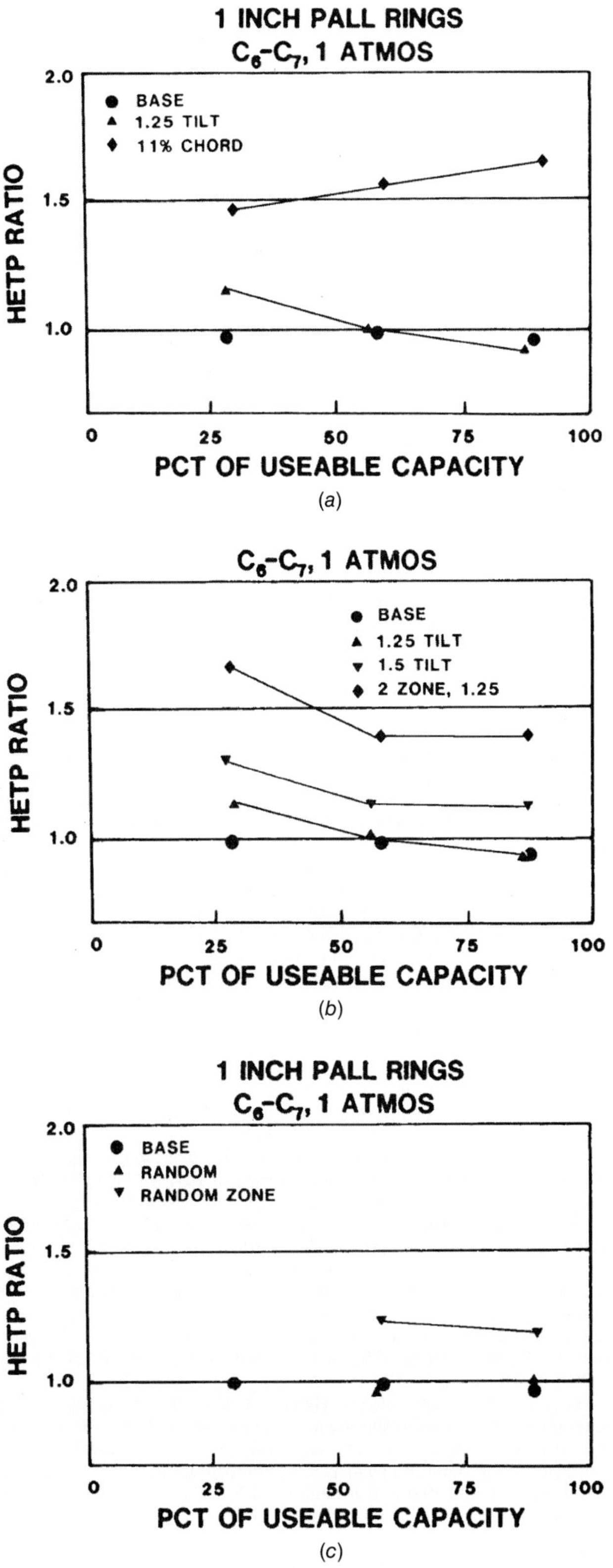

FIG. 14-66 Comparing the effects of "small-scale" and "large-scale" maldistribution on packing HETP. (*a*) Comparing the effect of a simulated continuous tilt (max/min flow ratio = 1.25) with the simulated effect of blanking a chordal area equal to 11 percent of the tower area. (*b*) Comparing the effects of simulated continuous tilts (max/min flow ratios of 1.25 and 1.5) with the effects of a situation where one-half of the distributor passes 25 percent more liquid than the other half. (*c*) Comparing the effects of random maldistribution with those of zonal maldistribution. (*Reprinted with permission from J. G. Kunesh, L. Lahm, and T. Yahagi,* Ind. Eng. Chem. Res. ***26:** p. 1845; copyright © 1987, American Chemical Society.*)

25 percent more liquid than the other half. The latter ("zonal") situation causes a much greater rise in HETP than a "uniform" maldistribution with twice as much variation from maximum to minimum.

Figure 14-66*c* shows results of tests in which flows from individual distributor drip points were varied in a gaussian pattern (maximum/mean = 2). When the pattern was randomly assigned, there was no efficiency loss. When the variations above the mean were assigned to a "high zone," and those below the mean to a "low zone," HETP rose by about 20 percent.

With structured packing, both random and zonal maldistribution caused about the same loss of efficiency at the same degree of maldistribution. Olujic and Jansen [*Chem. Eng. Res. Des.* **99:** 2 (2015)] provide a detailed survey on large-scale maldistribution tests with structured packings. With large surface area structured packings, even a relatively small degree of zonal maldistribution can be detrimental to HETP. In tests by Pavlenko et al. [*Theor. Found. Chem. Eng.* **43:** 1 (2009)], chordal blanking of as little as 10 percent of a tower cross section area, which gave no visible deterioration of liquid distribution, reduced bed efficiency by one-third in 450 m^2/m^3 structured packings.

8. A packed bed appears to have a "natural distribution," which is an inherent and stable property of the packings. An initial distribution that is better than natural will rapidly degrade to it, and one that is worse will finally achieve it, but sometimes at a slow rate. If the rate is extremely slow, recovery from a maldistributed pattern may not be observed in practice (Zuiderweg et al. 1987). Even though the volumetric distribution improves along the bed, the concentration profile could have already been damaged, and pinching occurs (Bonilla, *Chem. Eng. Prog.*, March 1993, p. 47).

9. Liquid maldistribution lowers packing turndown. The 2-in Pall rings curve in Fig. 14-59 shows HETP rise upon reaching the distributor turndown limit.

10. The major source of gas maldistribution is undersized gas inlet and reboiler return nozzles, leading to the entry of high-velocity gas jets into the tower. These jets persist through low-pressure-drop devices such as packings. Installing gas distributors and improving gas distributor designs, even inlet baffles, have alleviated many of these problems. Vapor distribution is most troublesome in large-diameter columns. Strigle (*Packed Tower Design and Applications,* 2d ed., Gulf Publishing, Houston, Tex., 1994) recommends considering a gas distributing device whenever the gas nozzle *F*-factor ($F_N = u_N \rho_G^{0.5}$) exceeds 27 m/s $(kg/m^3)^{0.5}$, or the kinetic energy of the inlet gas exceeds eight times the pressure drop through the first foot of packing, or the pressure drop through the bed is less than 0.65 mbar/m. Gas maldistribution is best tackled at the source by paying attention to the gas inlet arrangements.

11. A poor initial liquid maldistribution may cause gas maldistribution in the loading region, that is, at high gas rates [Stoter, Olujic, and de Graauw, *IChemE Symp. Ser.* **128:** A201 (1992); Kouri and Sohlo, *IChemE Symp. Ser.* **104:** B193 (1987)]. At worst, initial liquid maldistribution may induce local flooding, which would channel the gas. The segregation tends to persist down the bed. Outside the loading region, the influence of the liquid flow on gas maldistribution is small or negligible. Similarly, in high-gas-velocity situations, the liquid distribution pattern in the bottom structured packing layers is significantly influenced by a strongly maldistributed inlet gas flow [Olujic et al., *Chem. Eng. and Processing* **43:** 465 (2004)]. Duss [*IChemE Symp. Ser.* **152:** 418 (2006)] suggests that high liquid loads such as those experienced in high-pressure distillation also increase the susceptibility to gas maldistribution.

12. The effect of gas maldistribution on packing performance is riddled with unexplained mysteries. FRI's (Cai, paper presented at the AIChE Annual Meeting, Reno, Nev., 2001) commercial-scale tests show little effect of gas maldistribution on both random and structured packing efficiencies. The Cai et al. [*Trans IChemE* **81:** Part A, p. 85 (2003)] distillation tests in a 1.2-m-diameter tower showed that blocking the central 50 percent or the chordal 30 percent of the tower cross-sectional area beneath a 1.7-m-tall bed of 250 m^2/m^3 structured packing had no effect on packing efficiency, pressure drop, or capacity. The blocking did not permit gas passage but allowed the collection of the descending liquid. Simulator tests with similar blocking with packing heights ranging from 0.8 to 2.4 m [Olujic et al., *Chemical Engineering and Processing* **43:** 465 (2004); *Distillation 2003: Topical Conference Proceedings,* AIChE Spring National Meeting, New Orleans, La., AIChE, 2003, p. 567] differed, showing that a 50 percent chordal blank raised pressure drop, gave a poorer gas pattern, and prematurely loaded the packing. They explain the difference by the ability of liquid to drain undisturbed from the gas in the blocked segment in the FRI tests. Olujic et al. found that while gas maldistribution generated by collectors and by central blockage of 50 percent of the cross-sectional areas was smoothed after two to three layers of structured packing, a chordal blockage of 30 to 50 percent of cross-sectional area generated maldistribution that penetrated deeply into the bed.

13. Computational fluid dynamics (CFD) has been an effective tool for analyzing the effects of gas inlet geometry on gas maldistribution in packed beds. Using CFD, Wehrli et al. [*Trans. IChemE* **81:** Part A, p. 116 (January 2003)]

found that a very simple device such as the V-baffle (see Fig. 14-69*b*) gives much better distribution than a bare nozzle, while a more sophisticated vane device such as a Schoepentoeter (see Fig. 14-70*c*) is even better. Implications of the gas inlet geometry to gas distribution in refinery vacuum towers was studied by Vaidyanathan et al. (*Distillation 2001: Topical Conference Proceedings*, AIChE Spring National Meeting, Houston, Tex., April 22–26, 2001, p. 287); Paladino et al. (*Distillation 2003: Topical Conference Proceedings*, AIChE Spring National Meeting, New Orleans, La., 2003, p. 241); Torres et al. (*Distillation 2003: Topical Conference Proceedings,* AIChE Spring National Meeting, New Orleans, La., 2003, p. 284); Waintraub et al. (*Distillation 2005: Topical Conference Proceedings*, AIChE Spring National Meeting, Atlanta, Ga., 2005, p. 79); and Wehrli et al. (*IChemE* Symp. Ser 152, London, 2006). Vaidyanathan et al., Torres et al., and Stemich and Spiegel [*Chem. Eng. Res. Des.* **89:** 1392 (2011)] examined the effect of the geometry of a chimney tray (e.g., see Fig. 14-71) above the inlet on gas distribution and liquid entrainment. Paladino et al. demonstrated that the presence of liquid in the feed affects the gas velocity profile and must be accounted for in modeling. Paladino et al. and Waintraub et al. used their two-fluid model to study the velocity distributions and entrainment generated by different designs of vapor horns (e.g., see Fig. 14-70). Wehrli et al. produced pilot-scale data simulating a vacuum tower inlet, which can be used in CFD model validation. Ali et al. [*Trans. IChemE* **81:** Part A, p. 108 (January 2003)] found that the gas velocity profile obtained using a commercial CFD package compared well to those measured in a 1.4-m simulator equipped with structured packing together with commercial distributors and collectors. Their CFD model effectively pointed them to a collector design that minimizes gas maldistribution.

14. Liquid maldistribution has little effect on pressure drop in the preloading regime, while tending to lower pressure drop in the loading regime [Olujic and Jansen, *Chem. Eng. Res. Des.* **99:** 2 (2015)].

PACKED-TOWER SCALE-UP

Process versus Equipment Scale-Up The superb article on packing scale-up, Schoenmakers and Spiegel (chap. 10 in Gorak, A., and Olujic, Z., eds., *Distillation Equipment and Processes*, Elsevier, New York, 2014) and the classic book by Krell (*Handbook of Laboratory Distillation,* 2d ed., Elsevier, New York, 1982) guide many of the recommendations that follow, with full endorsement by the author. *Process scale-up*, which tests the feasibility of a proposed process, identifies azeotropes, reactions, component accumulation, fouling, and corrosive potential, compares separation sequences, and looks for unexpected reactions, can be distinguished from *equipment scale-up*, which focuses on the number of separation stages needed in order to avoid tower oversizing or undersizing. The demarcation between the two is not sharp and varies in different literature references.

Process Scale-Up (Miniplants) A miniplant generates a bench-scale plant as equivalent as possible to the intended production plant. Usually the column diameters are about 50 mm. It is essential to correctly set up the process steps, recycles, purges, side draws, and residence times that influence reactions, product decompositions, and fouling. Schoenmakers and Spiegel emphasize that steady state may be difficult to reach unless the miniplant operates around the clock, and that bench-scale batch distillation tests are unsuitable for generating process scale-up data for continuous distillation. Sampling, analyses, and controls are also very important.

Minimizing Heat Losses Due to the high area-to-volume ratio, heat losses are much greater in bench-scale columns than in production plants. Good insulation, heating the column wall close to the internal temperature, and avoiding column diameters smaller than 50 mm are key.

Residence Times Bench-scale columns usually have more residence time than plant columns, and this has a major impact on reactions and decomposition. Using packings rather than trays, adding glass balls to reduce the vessel volume, and side heating can help. Feed and drawoff piping need to correctly mimic the intended plant column because fast reactions may occur in the vapor or liquid.

Packings for Process Scale-Up Here the emphasis is on packings that will do the job. The diameter and height effects, which are primary for equipment scale-up, are secondary for process scale-up. Laboratory packings are discussed in Krell's book and also in some vendor brochures (e.g., Cannon Instrument Co., *Pro-Pak Protruded Metal Distillation Packing* Bulletin, State College, Pa.; Sulzer Chemtech, *Structured Packings Bulletin,* Winterthur, Switzerland).

Calibration and Precautions This step tests the tower using a standard system and is a prerequisite for both process and equipment scale-up. The standard system should be nonhazardous, of stable components, nonfoaming, and nonfouling, and its equilibrium, thermal, and transport properties should be well known. A constant relative volatility over the concentration range, such as that of the chlorobenzene-ethylbenzene system [Ottenbacher et al., *Chem. Eng. Res. Des.* **89:** 1427 (2011)], is desirable. Schoenmakers and Spiegel name chlorobenzene-ethylbenzene above 50 mbar and cis-trans decalin at lower pressures. Deibele, Goedecke, and Schoenmakers [*IChemE Symp. Ser.* **142:** 1021 (1997)], and Gann et al. [*Chem. Ing. Tech.* **64**(1): 6 (1992)] provide an extensive list of factors that can affect this scale-up, including test mixture, contaminants, packing pretreatment, column structure, packing installation, snug fit at the wall, column insulation, vacuum tightness, measurement and control, liquid distribution, reflux subcooling, prewetting, sampling, analysis, adjusting the number of stages to avoid pinches and analysis issues, evaluation procedure, and more. Data from laboratory columns can be particularly sensitive to some of these factors.

Equipment Scale-Up The considerations that follow are mostly important for equipment scale-up purposes.

Diameter For random packings, there are many reports [Billet, *Distillation Engineering,* Chem Publishing Co., New York, 1979; Chen, *Chem. Eng.*, March 5, 1984, p. 40; Zuiderweg, Hoek, and Lahm, *IChemE Symp. Ser.* **104:** A217 (1987)] of an increase in HETP with column diameter. Billet and Mackowiak's (Billet, *Packed Column Analysis and Design,* Ruhr University, Bochum, Germany, 1989) scale-up chart for Pall® rings implies that efficiency decreases as column diameter increases.

Practically all sources explain the increase of HETP with column diameter in terms of enhanced maldistribution or issues with the scale-up procedure. Bench-scale and pilot columns seldom operate at column-to-packing diameter ratios (D_T/D_p) larger than 20; under these conditions, lateral mixing effectively offsets a loss of efficiency due to maldistribution pinch. In contrast, industrial-scale columns usually operate at D_T/D_p ratios of 30 to 100; under these conditions, lateral mixing is far less effective for offsetting maldistribution pinch.

To increase D_T/D_p, it may appear attractive to perform the bench-scale tests using a smaller packing size than will be used in the prototype. Deibele, Goedecke, and Schoenmakers [*IChemE Symp. Ser.* **142:** 1021 (1997)], Goedecke and Alig (Paper presented at the AIChE Spring National Meeting, Atlanta, Ga., April 1994), and Gann et al. [*Chem. Ing. Tech.* **64**(1): 6 (1992)] studied scale-up from 50- to 75-mm-diameter packed columns directly to industrial columns. Goedecke and Alig show that for wire-mesh structured packing, bench-scale efficiency tends to be better than large-column efficiency, while for corrugated-sheet structured packing, the opposite is true, possibly due to excessive wall flow. For some packings, variation of efficiency with loads at bench scale completely differs from its variation in larger columns. For one structured packing, Kuhni Rombopak 9M (this packing is not currently available), there was little load effect, and there was good consistency between data obtained from different sources—at least for one test mixture. Deibele et al. present an excellent set of practical guidelines to improve scale-up reliability. So, it appears that great caution is required when scaling up packing data from bench-scale columns.

Height Experimental data for random packings show that HETP slightly increases with bed depth [Billet, *Distillation Engineering,* Chemical Publishing Co., New York, 1979; *Packed Tower Analysis and Design,* Ruhr University, Bochum, Germany, 1989; Eckert and Walter, *Hydrocarbon Processing* **43**(2): 107 (1964)].

For structured packing, some tests with Mellapak 250Y [Meier, Hunkeler, and Stöcker, *IChemE Symp. Ser.* **56:** 3, 3/1 (1979)] showed no effect of bed height on packing efficiency, while others [Cai et al., *Trans IChemE* **81:** Part A, p. 89 (January 2003)] did show a significant effect.

The effect of bed depth on packing HETP is attributed to liquid maldistribution. Zuiderweg et al. [*IChemE Symp. Ser.* **104:** A217 (1987)] suggest that the uneven irrigation generates an uneven concentration profile and localized pinching near the bottom of the beds. The tests by Martin, Bravo, and Fair (Paper presented at the National AIChE Meeting, New Orleans, La., 1988) confirm that the problem area is near the bottom. According to the zone-stage and Lockett and Billingham models discussed previously, as well as the empirical correlation by Moore and Rukovena (Fig. 14-63), the more stages per bed, the greater is the rise in HETP with bed depth. The presence and extent of maldistribution play an important role in determining the bed-depth effect.

As the bed depth increases, end effects (i.e., mass transfer in the region of liquid introduction and in the region where liquid drips from the packing supports) become less important. Such end effects tend to lower the HETP observed in short columns, such as pilot-plant columns.

In summary, bed depth may significantly influence HETP. This adds uncertainty to scale-up. Shallow test beds should be avoided. Most investigators use beds at least 1.5 m tall, and often more than 3 m tall. The FRI sampling technique (discussed in the subsection Sampling that follows) can detect maldistribution along the bed height. Ottenbacher et al. [*Chem. Eng. Res. Des.* **89:** 1427 (2011)] recommend making bed heights at least 4 to 6 m and to apply efficiencies from shorter test beds with care because they may be too optimistic.

Loadings For many random and corrugated-sheet structured packings, HETP is independent of vapor and liquid loadings (Figs. 14-59 and 14-60). For wire-mesh and some corrugated-sheet structured packings, HETP changes with gas and liquid loads.

Wu and Chen [*IChemE Symp. Ser.* **104:** B225 (1987)] recommend pilot testing over the entire range between the expected minimum and maximum operating rates and taking the highest measured HETP as the basis for scale-up. The author concurs. With structured packings, the load effect may be due to liquid rather than gas loads, and the pilot tests should cover the range of liquid loads (i.e., m/s based on column cross section) that is expected in the prototype.

Wetting For operation at low liquid loads, the onset of minimum wetting can adversely affect scale-up, particularly with random packings and aqueous systems. Scale-up reliability at low liquid loads can be improved by pilot testing at the composition range expected in the prototype, and by using identical packing materials and surface treatment in the pilot tests and in the prototype.

Underwetting At the aqueous end of aqueous-organic columns, underwetting is important. Rapid changes of concentration profiles and physical properties in organic-water separations complicate scale-up [Eiden and Kaiser, *IChemE Symp. Ser.* **142:** 757 (1997); also Schoenmakers and Spiegel, chap. 10 in Gorak, A., and Olujic, Z., eds., *Distillation Equipment and Processes,* Elsevier, New York, 2014]. Near the onset of underwetting, HETP becomes strongly dependent on composition, packing material and surface roughness, and the presence of surfactants. Scale-up reliability can be enhanced by pilot testing at the composition range expected in the prototype and by using identical packing material and surface treatment in the pilot tests and in the prototype.

Preflooding For one structured packing test with an aqueous system, Billet (*Packed Column Analysis and Design,* Ruhr University, Bochum, Germany, 1989) measured higher efficiency for a preflooded bed than for a non-preflooded bed. Presumably, the preflooding improved either wetting or distribution. Billet and Schoenmakers and Spiegel recommend preflooding the packing, both in the prototype and in the pilot column, to ensure maximum efficiency.

Sampling Fractionation Research Inc. (FRI) developed a sampling technique that eliminates the influence of "end effects" and detects a maldistributed composition profile. This technique [Silvey and Keller, *IChemE Symp. Ser.* **32:** 4:18 (1969)] samples the bed at frequent intervals, typically every 0.6 m or so. HETP is determined from a plot of these interbed samples rather than from the top and bottom compositions.

It is imperative that the interbed samplers catch representative samples, which are an average through the bed cross section. Caution is required when the liquid is highly aerated and turbulent (e.g., above 1300 kPa or above 1 m/min). The author highly recommends the FRI sampling technique for all other conditions.

Aging Billet (*Packed Column Analysis and Design,* Ruhr University, Bochum, Germany, 1989) showed that for some plastic packings in aqueous systems, the efficiency after one week's operation was almost double the efficiency of new packings. Little further change was observed after one week. Billet explains the phenomenon by improved wetting. He recommends that data for plastic packings should only be used for scale-up after being in operation for an adequately long period.

DISTRIBUTORS

Liquid Distributors A liquid distributor (or redistributor) should be used in any location in a packed column where an external liquid stream is introduced. Liquid redistributors are also used between packed beds to avoid excessive bed lengths that may impair packing efficiency or mechanical strength. It is best to have the packing supplier also supply the distributor, with the user critically reviewing the design. The user must provide the supplier with concise information about the plugging, corrosive, and foaming tendencies of the service as well as the range of liquid flow rates that it needs to handle and the physical properties of the liquid.

Olsson (*Chem. Eng. Progr.,* October 1999, p. 57) discussed the key for successful distributor design and operation. He states that it is critical to correctly evaluate the fouling potential of the service and to design for it (e.g., preventing small holes, filtering the feed); to avoid gas entry into liquid distributors (e.g., no flashing feed into a liquid distributor); to systematically check the irrigation pattern using a method such as the circle analysis of Moore and Rukovena [*Chem. Plants and Process* (European ed.), August 1987, p. 11; described in detail in Kister's *Distillation Operation,* McGraw-Hill, New York, 1990]; to water-test any liquid distributor (major suppliers have dedicated test stands, which provide qualitative observations as well as random sampling of liquid rates issuing from various irrigation points from which a statistical coefficient of variation CV is derived; CV < 5% is good); to ensure correct entry of a feed into a liquid distributor; and to thoroughly inspect a distributor. Kister [*Trans. IChemE* **81:** Part A, p. 5 (January 2003)] found that between 80 and 90 percent of the distributor failures reported in the literature in the last 50 years could have been prevented if users and suppliers had followed Olsson's measures.

A minimum of 40 irrigation points per square meter has been recommended, with 60 to 100 per square meter being ideal [Strigle, *Packed Tower Design and Applications,* 2d ed., Gulf Publishing, Houston, Tex., 1994; Kister, *Distillation Operation,* McGraw-Hill, New York, 1990]. Commercial-scale tests with both random and structured packings showed no improvement in packing efficiency by increasing the irrigation point density above 40 per square meter [Fitz, King, and Kunesh, *Trans. IChemE* **77:** Part A, p. 482 (1999); Olujic and Jansen, *Chem. Eng. Res. Des.* **99:** 2 (2015)]. So going to larger numbers of irrigation points per square meter provides little improvement while leading to smaller holes, which increases the plugging tendency. With high-surface-area structured packings, a larger irrigation point density was shown to help packing efficiency (Olujic and Jansen 2015), and the recommended irrigation point densities per square meter are >120 for 350 m^2/m^3 packings, >160 for 500 m^2/m^3 packings, and >200 for larger areas [Kooijman and Taylor, chap. 5 in Gorak and Schoenmakers, *Distillation Operation and Application,* Academic Press, New York, 2014].

In orifice-type distributors, which are the most common type, the head-flow relationship is given by the Torricelli orifice equation

$$Q = 3.96 \times 10^{-4} K_D n_D d_h^2 h^{0.5} \qquad (14\text{-}163)$$

where Q is the liquid flow rate, m^3/h; K_D is the orifice discharge coefficient, with a recommended value of 0.707 (Chen, *Chem. Eng.,* March 5, 1984, p. 40); n_D is the number of holes; d_h is the hole diameter, mm; and h is the liquid head, mm. Equation (14-163) shows that at a given Q, increasing n leads to either smaller d or smaller h.

Figures 14-67 and 14-68 show common distributor types used for distillation and absorption. An excellent detailed discussion of the types of distributors and their pros and cons was given by Bonilla (*Chem. Eng. Progr.,* March 1993, p. 47). The *perforated pipe* (or *ladder pipe*) distributor (Fig. 14-67*a*) has holes on the underside of the pipes. It is inexpensive, provides a large open area for vapor flow, and does not rely on gravity. On the debit side, it is typically designed for high-velocity heads, 500 to 1000 mm of water, which is 5 to 10 times more than gravity distributors, requiring [per Eq. (14-163)] either fewer irrigation points or the use of plugging-prone smaller holes. The high hole velocities make it prone to corrosion and erosion. These disadvantages make it relatively unpopular. A gravity variation of this distributor uses a liquid drum above the distributor that gravity-feeds it.

Spray distributors (Fig. 14-67*b*) are pipe headers with spray nozzles fitted on the underside. The spray nozzles are typically wide-angle (often 120°) full-cone. Spray distributors are unpopular in distillation but are common in heat transfer, in washing and scrubbing services (especially in refinery towers), and in small-diameter towers where a single spray nozzle can be used. They are inexpensive and offer a large open area for vapor flow and a robustness for handling of fouling fluids when correctly designed, and the sprays themselves contribute to mass and heat transfer. On the debit side, the spray cones often generate regions of over- and underirrigation, the sprays may not be homogeneous, and the spray nozzles are prone to corrosion, erosion, and damage. With highly subcooled liquids, the spray angle may collapse when pushed at high pressure drops (above 100 to 150 kPa) (Fractionation Research Inc., *A Spray Collapse Study,* motion picture 919, Stillwater, Okla., 1985). The design and spray pattern are highly empirical. Sprays also generate significant entrainment to the section above [Trompiz and Fair, *Ind. Eng. Chem, Res.* **39**(6): 1797 (2000)].

Orifice pan distributors (Fig. 14-68*a*) and *orifice tunnel* distributors (Fig. 14-68*b*) have floor holes for liquid flow and circular (Fig. 14-68*a*) or rectangular (Fig. 14-68*b*) risers for vapor passages. When they are used as redistributors, a hat is installed above each riser to prevent liquid from the bed above from dripping into the risers. Unlike the ladder pipe and spray distributors that operate by pressure drop, orifice distributors operate by gravity and therefore use a much smaller liquid head, typically 100 to 150 mm at maximum rates. Using Eq. (14-163), the lower head translates to either more distributions points (n_D), which helps irrigation quality, or larger hole diameters, which resists plugging. However, the low liquid velocities, large residence times, and open pans (or troughs) make them more prone to plugging than the pressure distributors. Leakage through cracks at the support ring may generate maldistribution, and it is problematic especially when thermal expansion to process temperatures is significant. A good hole pattern and avoidance of oversized risers are essential. Orifice distributors are self-collecting, a unique advantage for redistributors. Orifice distributors are one of the most popular types and are favored whenever the liquid loads are high enough to afford hole diameters large enough to resist plugging (>12 mm).

Orifice trough (or *orifice channel*) distributors (Fig. 14-68*c–f*) are some of the most popular types. The trough construction does away with the multitude of joints in the orifice pans, making them far more leak-resistant, a major advantage in large towers and low-liquid-rate applications. Liquid

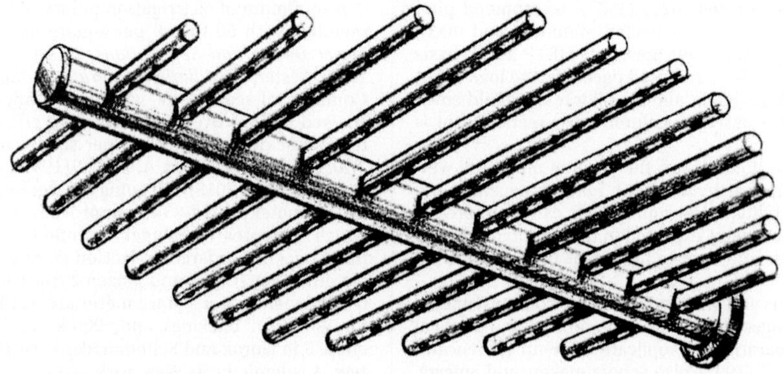

(a)

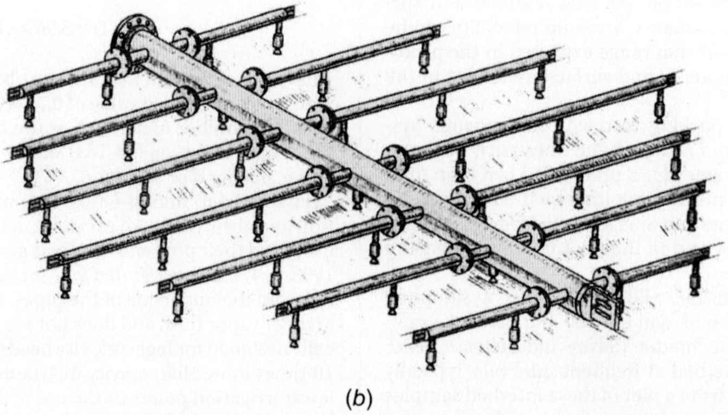

(b)

FIG. 14-67 Pressure liquid distributors. (*a*) Ladder pipe. (*b*) Spray. (*Courtesy of Koch-Glitsch LP.*)

from a central parting box (Fig. 14-68*c, e*) or middle channel (Fig. 14-68*d*) is metered into each trough. The troughs can have floor holes, but elevating the holes above the floor (Fig. 14-68*c–g*) is preferred because it enhances plugging resistance. Tubes (Fig. 14-68*c, d, f*) or baffles (Fig. 14-68*e*) direct the liquid issuing from the elevated holes downward onto the packings. Orifice trough distributors are not self-collecting. When used for redistribution, they require a liquid collector to be installed above them.

Turndown of orifice distributors is constrained to about 2:1 by Eq. (14-163). For example, a 100-mm liquid head at the design drops to 25 mm when the liquid rate is halved. Lower heads give poor irrigation and high sensitivity to levelness. Turndown is often enhanced by using two rows of side tubes (in the Fig. 14-68*c* type) or of side holes (in the Fig. 14-68*d* or *e* types). *Perforated drip tubes* (as in Fig. 14-68*d*) are popular in either orifice trough or orifice pan distributors. The lower, smaller hole is active at low liquid rates, with the larger upper hole becoming active at higher liquid rates. The use of perforated drip tubes is not recommended when the vapor dew point is much higher than the liquid bubble point because liquid may boil in the tubes, causing dryout underneath [Kister, Stupin, and Oude Lenferink, *IChemE Symp. Ser.* **152:** 409 (2006)].

A popular type of the orifice trough distributor is the *splash plate* distributor (Fig. 14-68*e*). The splash plates spread the issuing liquid over their lengths, making it possible to reduce the number of irrigation points. This is a special advantage with small liquid rates because fewer irrigation points (at a given head) translate to larger, more fouling-resistant hole diameters [Eq. (14-163)]. Lack of the drip tubes eliminates the possible in-tube boiling issue mentioned previously.

Multistage orifice trough distributors (Fig. 14-68*f*) also try to provide good irrigation at low liquid rates without resorting to plugging-prone small holes. The primary stage uses fewer irrigation points. Liquid from the primary stage is further split at the secondary stage. The secondary stage is small, so leveling and small flow variations are not of great concern. The secondary stage may use the same or a different liquid splitting principle from that of the primary stage. Even short layers of structured packings have been used as a secondary distribution stage.

Notched trough distributors (Fig. 14-68*g*) consist of parallel troughs with side V notches. These distributors obey the triangular notch equation instead of the orifice equation, which makes the flow proportional to $h^{2.5}$ [instead of $h^{0.5}$ in Eq. (14-163)]. This high power renders the distributor highly sensitive to out-of-levelness and hydraulic gradients and makes it difficult to incorporate a large number of distribution points. Since the liquid issues sideways, it is difficult to predict where the liquid will hit the packings. Baffles are sometimes used to direct the liquid downward. Overall, the quality of distribution is inferior to that of orifice distributors, making notched-trough distributors unpopular. Their strength is their insensitivity to fouling and corrosive environments and their ability to handle high liquid rates at good turndown.

With any trough distributor, and especially those with V notches, excessive hydraulic gradients must be avoided. This is often achieved by using more parting boxes.

The hydraulic gradient is highest where the liquid enters the troughs, approaching zero at the end of the trough. The hydraulic gradient (between entry point and trough end) can be calculated from [Moore and Rukovena, *Chemical Plants and Processing* (European ed.), August 1987, p. 11]

$$h_{hg} = 51\, v_H^2 \qquad (14\text{-}164)$$

where h_{hg} is the hydraulic gradient head, mm, and v_H is the horizontal velocity in the troughs, m/s.

Flashing Feed and Vapor Distributors When the feed or reflux is a flashing feed, the vapor must be separated out of the liquid before the liquid enters a liquid distributor. At low velocities (only), this can be achieved by a *bare nozzle* (Fig. 14-69*a*). A *V baffle* (Fig. 14-69*b*) is sometimes installed as a primitive flashing feed or vapor distributor.

For better vapor-liquid separation and distribution, with smaller-diameter towers (<1.5 m), a *flash chamber* (Fig. 14-69*c*) separates the liquid from the vapor, with the collected liquid descending via downpipes to a liquid distributor below. The flash chamber can be peripheral (Fig. 14-69*c*) or central. In larger towers, *gallery distributors* (Fig. 14-69*d*) are preferred. The flashing feed enters the peripheral section of the upper plate (the gallery) where vapor disengages and flows up, with liquid descending through holes (Fig. 14-69*d*) or down pipes onto the liquid distributor below. Alternatively, an external knockout pot is sometimes used to give separate vapor and liquid feeds.

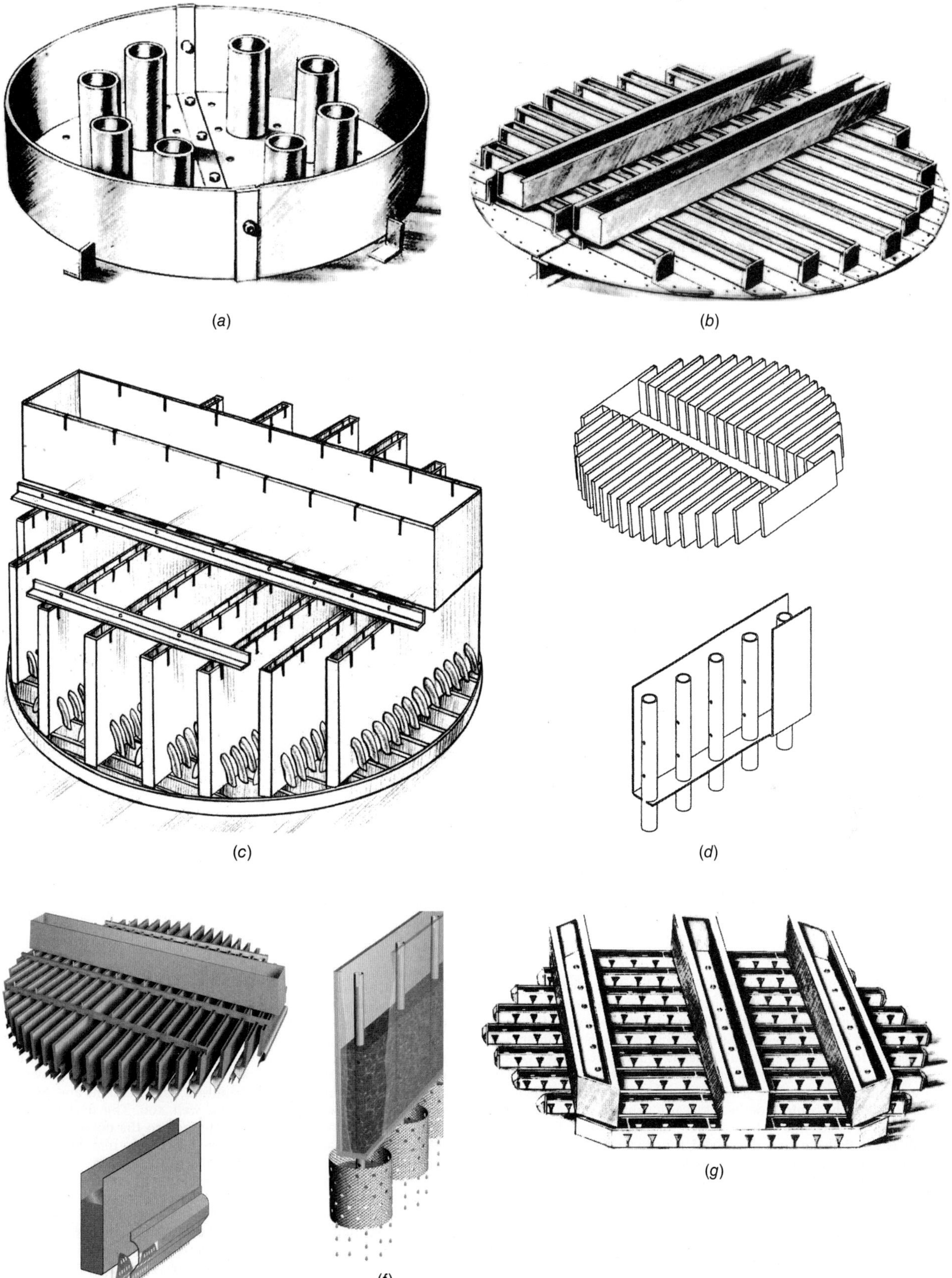

FIG. 14-68 Gravity liquid distributors. (*a*) Orifice pan. (*b*) Orifice tunnel. (*c*) Orifice tube, using external drip tubes. (*d*) Orifice trough, using internal drip tubes. (*e*) Splash plate orifice trough. (*f*) Two-stage orifice trough. (*g*) Notched trough. (*Parts a–c, g, courtesy of Koch-Glitsch LP; parts d–f, courtesy of Sulzer Chemtech.*)

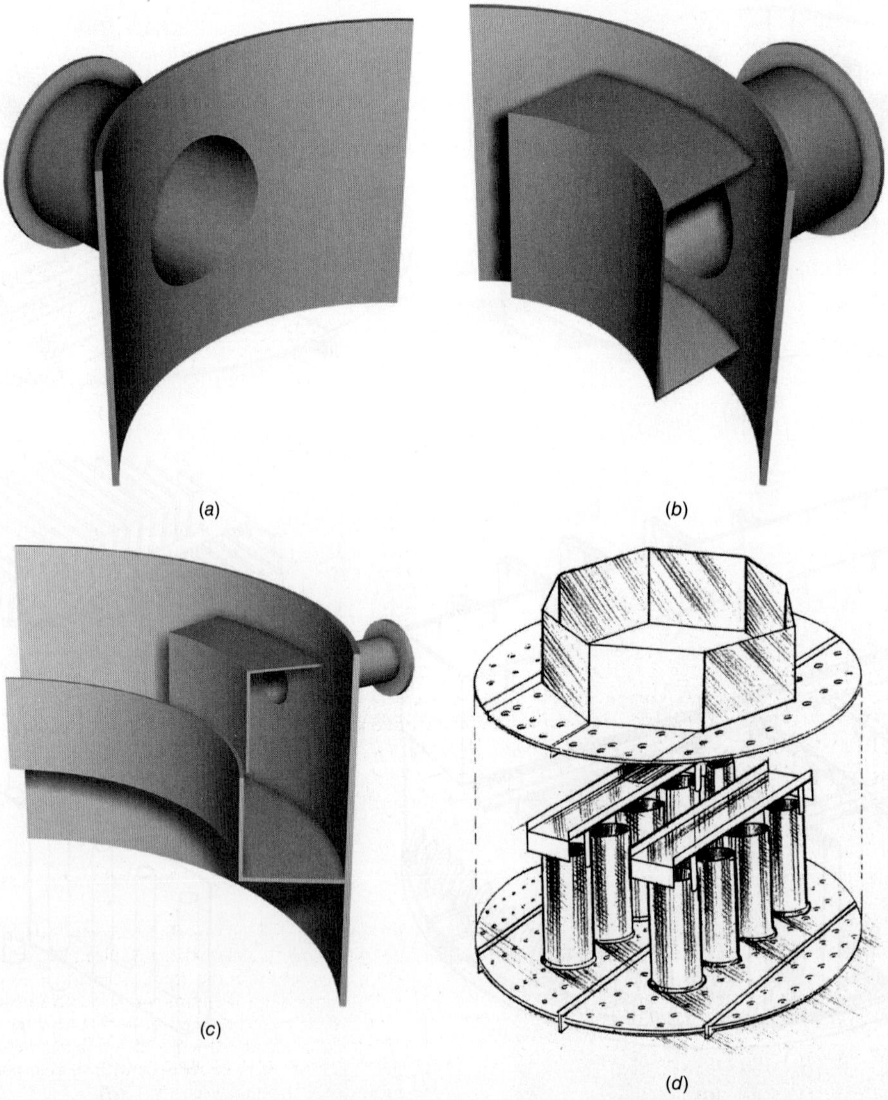

FIG. 14-69 Flashing feed and vapor distributors. (*a*) Bare nozzle. (*b*) Rounded V baffle. (*c*) Peripheral flash box—the box extends right around the tower wall, with the collected liquid descending via downpipes to a liquid distributor below. (*d*) Gallery distributor—the feed enters the gallery area (upper plate). (*Parts a–c, courtesy of Sulzer Chemtech; part d, courtesy of Koch-Glitsch LP.*)

The *vapor horn* (Fig. 14-70*a*) is unique for high-velocity feeds in which vapor is the continuous phase with liquid present as suspended drops in the feed vapor. This is common when the feed makes up the bulk of the vapor traffic in the tower section above. Typical examples are feeds to refinery vacuum and crude towers and rich solution feeds to hot carbonate regenerators. A tangential helical baffle or vapor horn, covered at the top, open at the bottom, and spiraling downward, is used at the feed entry. This baffle forces the vapor to follow the contour of the vessel as it expands and decreases in velocity. Liquid droplets, due to their higher mass, tend to collide with the tower wall, which deflects them downward, thus reducing upward entrainment. Large forces, generated by hurricane-force winds and vapor flashing, are absorbed by the entire tower wall rather than by a small area. A wear plate is required at the tower wall. Some designs have vane openings on the inside wall.

Internal vanes to knock down the liquid and to deflect it downward may (Fig. 14-70*a*) or may not be included in the vapor horn design. If there are no vanes, the top of the horn typically goes down about ½ of the nozzle diameter for 180° horns and ¾ of nozzle diameter for 270° horns. The idea is that the vapor flow projection at 360° practically clears the inlet. The vapor horn with internal vanes is typically closed at the end and is not sloped.

Alternatively, multivane triangular diffusers (Fig. 14-70*b, c*) such as the Schoepentoeter™ have been successful for high-velocity vapor-rich feeds. These are used with radial (as distinct from tangential) nozzles. The vanes knock out the liquid and direct it downward while the vapor expands to the tower diameter. A recent development, the Schoepentoeter Plus,™ adds catching rims at the vane outlets to reduce liquid reentrainment.

Pilot-scale tests by Fan et al. [*IChemE Symp. Ser.* **142:** 899 (1997)] compared vapor distribution and entrainment from sparger, vapor horns, and multivane triangular diffusers. Vapor horns gave the best overall performance considering vapor distribution, entrainment, and pressure drop, with multivane distributors doing well, too. The designs of the inlets compared, however, were not optimized so the comparison could have reflected deviations from optimum rather than real differences between the devices.

Low-velocity vapor-only feeds often enter via bare nozzles or V baffles. At higher velocities, perforated vapor spargers are used. At high velocities, vapor horns and multivane triangular diffusers are often preferred. Alternatively or additionally, a vapor distributor may be mounted above the feed. The vapor distributor is a chimney tray (Fig. 14-71) where liquid is collected on the deck and flows via downcomers or is drawn out while vapor passes through the chimneys. To be effective as a vapor distributor, the pressure drop through the chimneys needs to be high enough to counter the maldistributed vapor. It was recommended to make the pressure drop through the chimneys at least equal to the velocity head at the tower inlet nozzle (Strigle, *Random Packings and Packed Towers,* Gulf Publishing, Houston, Tex., 1987), with common pressure drops ranging from 25 to 200 mm water.

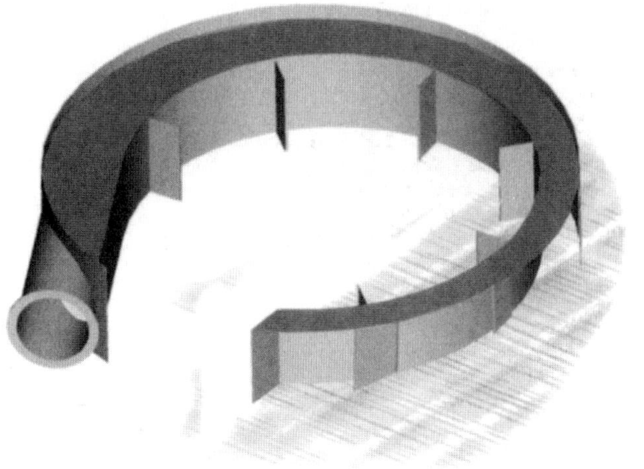

(a)

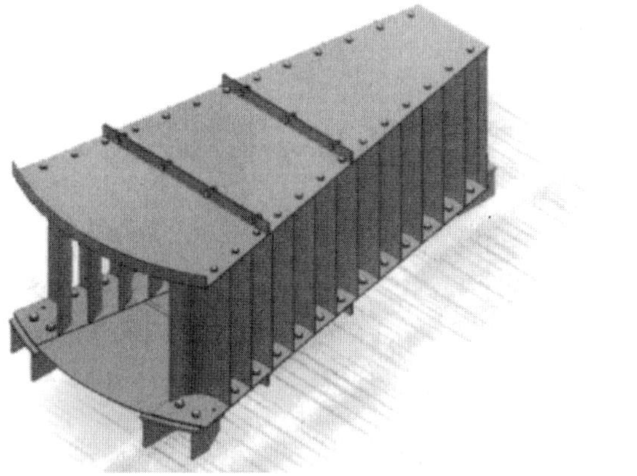

(b)

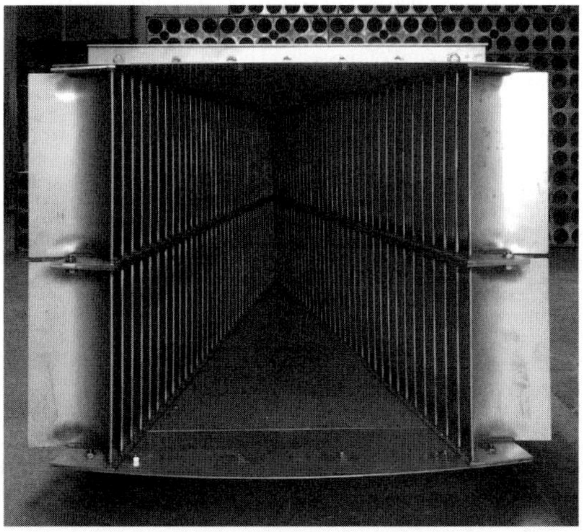

(c)

FIG. 14-70 High-velocity flashing feed and vapor distributors. (*a*) Vapor horn. (*b*) Radial vane distributor. (*c*) Schoepentoeter. (*Parts a, b, courtesy of Koch-Glitsch LP; part c, courtesy of Sulzer Chemtech.*)

OTHER PACKING CONSIDERATIONS

Liquid Holdup Liquid holdup is the liquid present in the void spaces of the packing. Reasonable liquid holdup is necessary for good mass transfer and efficient tower operation, but beyond that, it should be kept low. High holdup increases tower pressure drop, the weight of the packing, the support load at the bottom of the packing and tower, and the tower drainage time. Most important, when distilling thermally unstable materials, excessive holdup raises product degradation and fouling, and with hazardous chemicals, it increases undesirable inventories.

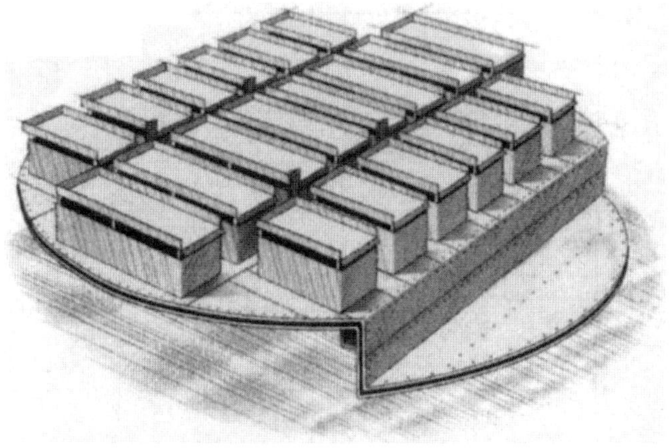

FIG. 14-71 Chimney tray vapor distributor. (*Reprinted courtesy of Koch-Glitsch LP.*)

The effect of liquid and gas rates on the operating holdup is shown in Figs. 14-72 and 14-73. In the preloading regime, holdup is essentially independent of gas velocity, but it is a strong function of liquid flow rate and packing size. Smaller packings and high liquid rates tend to have greater holdup. Liquid holdup significantly increases with liquid viscosity above 2.5 cP, with only a slight effect below 2.5 cP [Zakeri, Einbu, and Svenden, *Chem. Eng. Res. Des.* **90:** 585 (2012); Bratmöller and Scholl, *Chem. Eng. Res. Design* **99:** 75 (2015); Alix and Raynal, *Chem. Eng. Res. Des.* **86:** 585 (2008)].

Maćkowiak (*Fluid Dynamics of Packed Columns*, Springer/Heidelberg, New York, 2010) characterizes holdup behavior according to the dimensionless liquid Reynolds number Re_L given by

$$Re_L = 1000\, u_L\, \rho_L / a_p\, \mu_L \qquad (14\text{-}165)$$

where u_L is the liquid superficial velocity, m/s, a_p is the surface area per unit volume, m^2/m^3 (values listed in Tables 14-13 and 14-14), ρ_L is the liquid density, kg/m^3, and μ_L is the liquid viscosity, cP. Reynolds numbers less than 1 indicate laminar flow; numbers above 10 indicate turbulent flow.

In the far most common turbulent flow regime, there is little viscosity effect and a smaller dependence on liquid velocity. In laminar flow, liquid holdup increases more strongly with smaller packing size and significantly increases with liquid viscosity.

Holdup can be estimated by using Buchanan's correlation [*Ind. Eng. Chem. Fund.* **6:** 400 (1967)], as recommended in previous editions of this handbook. More recent correlations by Billet and Schultes [*Trans.IChemE* **77:** Part A, p. 498 (1999)], by Maćkowiak (*Fluid Dynamics of Packed Columns*, Springer/Heidelberg, New York, 2010), and by Mersmann and Deixler [*Chem. Ing. Tech.*

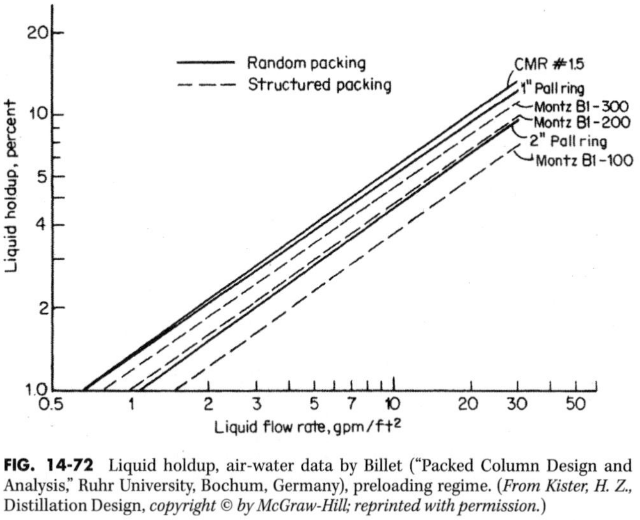

FIG. 14-72 Liquid holdup, air-water data by Billet ("Packed Column Design and Analysis," Ruhr University, Bochum, Germany), preloading regime. (*From Kister, H. Z.,* Distillation Design, *copyright © by McGraw-Hill; reprinted with permission.*)

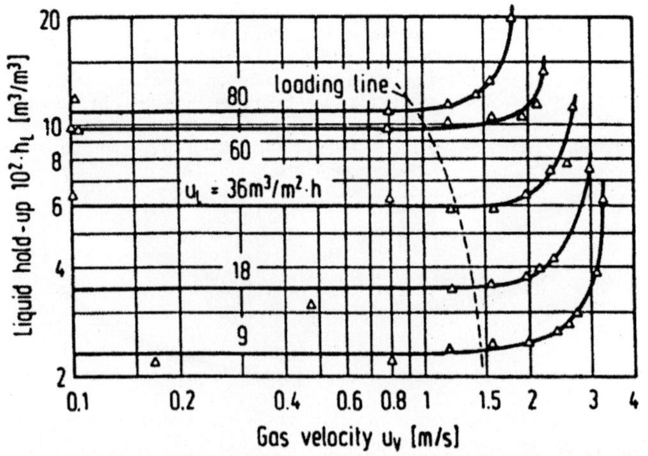

FIG. 14-73 Effect of liquid and gas rates on the operating holdup of modern random packings (25-mm NorPac®). [*From R. Billet and M. Schultes,* IChemE Symp. Ser. ***104:*** *p. A159, 1987. Reprinted courtesy of the Institution of Chemical Engineers (UK).*]

58(1): 19 (1986)] apply to modern random packings as well as to some structured packings.

Stichlmair and Fair (*Distillation Principles and Practice*, Wiley-VCH, New York, 1998) show that liquid holdup is primarily a function of the liquid Froude number, the specific surface area of the packing, and physical properties. They recommend the correlation by Engel, Stichlmair, and Geipel [*Ind. Eng. Chem. Symp. Ser.* **142:** 939 (1997)].

$$h_{Lo} = 0.93\left(\frac{U_L^2 a_p}{g}\right)^{1/6}\left(\frac{\mu_L^2 a_p^3}{\rho_L^2 g}\right)^{1/10}\left(\frac{\sigma a_p^2}{1000\rho_L g}\right)^{1/8} \tag{14-166}$$

where h_{Lo} = liquid holdup, fractional
U_L = liquid superficial velocity, m/s
a_p = packing specific surface area, m^2/m^3
g = acceleration due to gravity, m/s
μ_L = liquid viscosity, kg/(m · s)
σ = surface tension, mN/m

The Engel, Stichlmair, and Geipel correlation applies only in the preloading regime. The packing geometry is represented in the correlation solely by the readily available specific surface area (Tables 14-13 and 14-14).

Minimum Wetting Rate The minimum wetting rate (MWR) is the lower stability limit of packings. Below this liquid load, the liquid film on the packing surfaces breaks up and dewetting occurs. The area available for mass transfer diminishes, and efficiency drops.

Hiller et al. [*Chem. Eng. Res. Des.* **99:** 220 (2015)] note a hysteresis between wetting and dewetting, reflected by the industrial practice of starting up low-liquid-load columns at higher liquid rates to achieve initial wetting.

Schmidt [*IChemE Symp. Ser.* **56:** 3.1/1 (1979)] described the MWR in terms of a force balance at a dry patch along the path of a falling liquid film. While the gravity and viscous forces resist dewetting, the surface tension and vapor shear forces tend to dewet the falling film. The MWR therefore rises with an increase in surface tension and liquid density and with a decrease in liquid viscosity. Large packing sizes and poor surface wetting characteristics also contribute to higher MWR.

Schmidt presented a fundamental correlation to predict minimum wetting for Raschig and Pall® rings. More popular have been the Glitsch rules of thumb [Table 14-15; Glitsch, Inc. (now Koch Glitsch), Bulletin 345, Dallas, Tex., 1986] for CMR® random packings with packing surface areas around 200 m^2/m^3. To extend these rules to other random packings, Kister (*Distillation Design*, McGraw-Hill, New York, 1992) applied Schmidt's model to give

$$Q_{MW} \approx (Q_{MW} \text{ from Table 14-15}) \times (200/a_p)^{0.5} \tag{14-167}$$

The Glitsch brochure did not state the surface tension bases of Table 14-15. The author believes that they conservatively apply to organic and hydrocarbon systems (σ < 25 mN/m). For water (σ = 70 mN/m), the author believes that double the values from Table 14-15 is reasonable.

A more recent rule by Stichlmair and Fair (*Distillation Principles and Practice*, Wiley, New York, 1998) recommends higher MWR values than in Table 14-15, 2 and 10 $m^3/(h \cdot m^2)$ for organic and aqueous systems, respectively.

TABLE 14-15 Glitsch's Rule of Thumb for Minimum Wetting

Basis: CMR with a_p > 150 m^2/m^3

Material	Minimum wetting rate, $m^3/(m^2 \cdot h)$
Unglazed ceramic (chemical stoneware)	0.5
Oxidized metal (carbon steel, copper)	0.7
Surface-treated metal (etched stainless steel)	1.0
Glazed ceramic	2.0
Glass	2.5
Bright metal (stainless steel, tantalum, other alloys)	3.0
PVC-CPVC	3.5
Polypropylene	4.0
Fluoropolymers (PTFE type)	5.0

Some surface treatments of the packing (e.g., oxidizing, sandblasting, etching) can substantially reduce the MWR. Chuang and Miller [*Can. J. Chem. Eng.* **66**(6): 377 (1988)] tested a metallic random packing with an aqueous system at low liquid rates [about 0.4 $m^3/(h \cdot m^2)$]. They used two alternative techniques for oxidizing the packing surfaces. The packings oxidized with the more effective technique gave a column efficiency twice as high as those oxidized by the others.

Superior wetting characteristics characterize structured packings. Satisfactory performance was reported down to 0.25 and 0.12 $m^3/(h \cdot m^2)$ in corrugated-sheet and wire-mesh metal structured packings, respectively, with organic systems. FRI's commercial-scale tests (Fitz and Kunesh, paper presented at the AIChE Annual Meeting, Chicago, November 1996) demonstrated good efficiencies for a hydrocarbon system in Mellapak® 250Y down to 0.5 $m^3/(h \cdot m^2)$ (no lower limit reached) with good liquid and vapor distribution. Hiller et al. [*Chem. Eng. Res. Des.* **99:** 220 (2015)] plotted operating points for 16 different industrial columns, successfully separating both organic and aqueous temperature-sensitive mixtures. About half of these were in the range of 0.2 to 1 $m^3/(h \cdot m^2)$. The *F*-factor [Eq. (14-76)] range for most was around 1.5 $Pa^{0.5}$. On this basis, Niggemann, Rix, and Meier (chap. 7 in Gorak and Schoenmakers, *Distillation Operation and Application*, Academic Press, New York, 2014) propose a minimum of 0.2 $m^3/(h \cdot m^2)$ and 1.0 $m^3/(h \cdot m^2)$ for organic and aqueous systems, respectively, with structured packings. They emphasize that the success of designs at such low liquid loads depends on achieving good liquid distribution.

Two Liquid Phases Two liquid phases often occur in heterogeneous distillation, in steam stripping, and in solvent recovery. Harrison (*Chem. Eng. Progr.*, November 1990, p. 80) and Meier et al. [*IChemE Symp. Ser.* **152:** 267 (2006)] saw no reason to expect unusual capacity or pressure drop limitations due to the presence of the second liquid phase, suggesting that standard correlations for these should apply.

In contrast, both expressed uncertainty in predicting packing efficiency. Harrison presented two case studies: In one, adding water to two water-insoluble organics had no effect on HETP; in another, a key component was soluble in both liquid phases, and HETP was about 50 percent above normal. Harrison argued that a second liquid phase leads to lower efficiency only when it impairs diffusion of the key species.

Extensive work by Wozny, Repke, et al. [*Ind. Eng. Chem. Res.* **49:** 822 (2010); *Chem. Eng. Res. Des.* **85**(A1): 50 (2007)] explained their findings and the earlier ones. Water tends to form rivulets or drops, while organics tend to flow as closed films on the packing. Both liquid phases overlay each other, causing strong interactions that are difficult to predict. Individual effective areas for the two liquid phases may be increased or decreased, depending on the flow characteristics on the packing. For example, one liquid phase may overflow the other and shield it from contact with the gas. Overall, the presence of a second liquid phase may increase, decrease, or not change the mass transfer efficiency. Meier et al. recommend obtaining efficiencies by scaling up laboratory-scale data using a similar type of packing.

Both Harrison and Meier et al. emphasize adequately distributing each liquid phase to the packing. Harrison noted that a well-designed ladder pipe distributor can maintain high velocities and low residence times that provide good mixing. With trough distributors that separate the phases and then distribute each to the packing, a light-to-heavy phase maldistribution may occur, especially when the phase ratio and separation vary. Meier et al. noted the existence of a cloudy two-liquid layer between the clear light and heavy liquid and recommend an additional middle distribution point for this layer. They also noticed that phase separation unevenness can have a large influence on the phase ratio irrigated to the packing.

High Viscosity and Surface Tension Bravo (Paper presented at the AIChE Spring National Meeting, Houston, Tex., 1995) studied a system that had 425-cP viscosity, 350 mN/m surface tension, and a high foaming tendency. He found that efficiencies were liquid-phase-controlled and could be estimated from theoretical HTU models. Capacity was less than predicted by conventional methods, which do not account for the high viscosity. Design equations for orifice distributors extended well to the system once the orifice coefficient was calculated as a function of the low Reynolds number and the surface tension head was taken into account.

Böcker and Ronge [*Chem. Eng. Technol.* **28**(1) (2005)] verified Bravo's findings. Experimenting with chlorobenzene/ethylbenzene stripping in a 50-mm column, they raised the viscosity from 0.26 cP to 25 cP and 260 cP by adding up to 8 percent of polybutadiene to the feed. At the higher viscosities, liquid drainage deteriorated, which lowered flood points and raised pressure drops, especially at high liquid loads. Packing efficiency declined as liquid viscosity was raised, reaching 20 percent of the nonviscous efficiency at 260 cP, due to poorer wetting of the packing. Bratmöller and Scholl [*Chem. Eng. Res. Design* **99:** 75 (2015)] confirmed declining packing efficiency with higher liquid viscosity for a distillation system in a 50-mm column with structured packings at pressures from 950 to 20 mbar, over which there was a sixfold increase in liquid viscosity.

OTHER TOPICS FOR DISTILLATION AND GAS ABSORPTION EQUIPMENT

COMPARING TRAYS AND PACKINGS

Most separations can be performed either with trays or with packings. The following factors represent economic pros and cons that favor each and may be overridden. For instance, column complexity is a factor favoring trays, but gas plant demethanizers that often use one or more interreboilers are traditionally packed.

Factors Favoring Packings

Vacuum systems. Packing pressure drop is much lower than that of trays because the packing open area approaches the tower cross-sectional area, while the tray's open area is only 8 to 15 percent of the tower cross-sectional area. Also, the tray liquid head, which incurs substantial pressure drop (typically about 50 mm of the liquid per tray), is absent in packing. Typically, tray pressure drop is on the order of 10 mbar per theoretical stage, compared to 3 to 4 mbar per theoretical stage with random packings and about one-half of that with structured packings.

Consider a vacuum column with 10 theoretical stages, operating at 70-mbar top pressure. The bottom pressure will be 170 mbar with trays, but only 90 to 110 mbar with packings. The packed tower will have a much better relative volatility in the lower parts, thus reducing reflux and reboil requirements and bottom temperature. These translate to less product degradation, greater capacity, and smaller energy consumption, giving packings a major advantage.

Lower-pressure-drop applications. When the gas is moved by a fan through the tower, or when the tower is in the suction of a compressor, the smaller packing pressure drop is often a controlling consideration. This is particularly true for towers operating close to atmospheric pressure. Here excessive pressure drop in the tower increases the size of the fan or compressor (new plant), bottlenecks them (existing plant), and largely increases power consumption. Due to the compression ratio, pressure drop at the compressor discharge is far less important and seldom a controlling consideration.

Revamps. The pressure drop advantage is invaluable in vacuum column revamps and can be translated to a capacity gain, an energy gain, a separation improvement, or various combinations of these benefits. Likewise, for towers in the suction of compressors or when gas is moved through the tower by a fan, replacing trays with packings reduces the compression ratio or fan pressure drop and helps debottleneck the compressor or fan.

Packings also offer an easy tradeoff between capacity and separation. In the loaded sections of the tower, larger packings can overcome capacity bottlenecks at the expense of loss in separation. The separation loss can often be regained by retrofitting with smaller packings in sections of the tower that are not highly loaded. In tray towers, changing tray spacing gives similar results, but it is more difficult to do.

Foaming (and emulsion). The low gas and liquid velocities in packing suppress foam formation. The large open area of the larger random packing promotes foam dispersal. Both attributes make random packing excellent for handling foams. In many cases, recurrent foaming was alleviated by replacing trays with random packing, especially when tray downcomers were poorly designed.

Switching from trays to structured packing can aggravate foaming. While the low gas and liquid velocities help, the solid walls restrict the lateral movement of foams and give support to the foams.

Small-diameter columns. Columns with diameters less than 1 m (3 ft) are difficult to access from inside to install and maintain the trays. "Cartridge" trays or an oversized diameter are often used. Either option is expensive. Cartridge trays also run into problems with sealing to the tower wall, matching tower to tray hardware (Sands, *Chem. Eng.*, April 2006, p. 86) and being easy to dislodge. Packing is normally a cheaper and more desirable alternative.

Corrosive systems. The practical range of packing materials is wider. Ceramic and plastic packings are cheap and effective. Trays can be manufactured in nonmetals, but packing is usually a cheaper and more desirable alternative.

Low liquid holdup. Packings have lower liquid holdup than do trays. This is often advantageous for reducing polymerization, degradation, or the inventory of hazardous materials.

Batch distillation. Because of the smaller liquid holdup of packing, a higher percentage of the liquid can be recovered as top product.

Factors Favoring Trays

Solids. Trays handle solids much more easily than packing. Both gas and liquid velocities on trays are often an order of magnitude higher than through packing, providing a sweeping action that keeps tray openings clear. Solids tend to accumulate in packing voids. There are fewer locations on trays where solids can be deposited. Plugging in liquid distributors has been a common trouble spot. Cleaning trays is much easier than cleaning packings.

Not all trays are fouling-resistant. Floats on moving valve trays tend to "stick" to deposits on the tray deck. Fouling-resistant trays have large sieve holes or large fixed valves, and these should be used when plugging and fouling are the primary considerations.

There is much that can be done to alleviate plugging with random packing. Large, open packing with minimal pockets offers good plugging resistance. Distributors that resist plugging have large holes (>13-mm diameter). Such large holes are readily applied with high liquid flow rates, but they are often not practical for small liquid flow rates.

Maldistribution. The sensitivity of packing to liquid and gas maldistribution has been a common cause of failures in packed towers. Maldistribution issues are most severe in large-diameter towers, long beds, small liquid flow rates, and smaller packing. Structured packing is generally more prone to maldistribution than random packing. While good distributor design, water testing, and inspection can eliminate most maldistribution issues, it only takes a few small details that fall through the cracks to turn success into failure. Due to maldistribution, there are far more failures experienced with packing than in trays, and it takes more trials "to get it right" than with trays. This makes trays more robust.

Complex towers. Interreboilers, intercondensers, cooling coils, and side drawoffs are more easily incorporated in trays than in packed towers. In packed towers, every complexity requires additional distribution and/or liquid collection equipment.

Feed composition variation. One way of allowing for design uncertainties and feedstock variation is by installing alternate feed points. In packed towers, every alternate feed point requires expensive distribution equipment.

Performance prediction. Due to their sensitivity to maldistribution, there is greater uncertainty in predicting packed column performance.

Chemical reaction, absorption. Here the much higher liquid holdup on trays provides greater residence time for absorption or chemical reaction than does packing.

Turndown. Moving valve and bubble-cap trays normally give better turndown than packings. Unless very expensive distributors are used, packed tower turndown is usually limited by distributor turndown.

Weight. Tray towers usually weigh less than packed towers, saving on the cost of foundations, supports, and column shell.

Trays versus Random Packings The following factors generally favor trays compared to random packings, but not compared to structured packings.

Low liquid rates. With the aid of picket-fence weirs, splash baffles, reverse-flow trays, and bubble-cap trays, low liquid rates can be handled better in trays. Random packings suffer from liquid dewetting and maldistribution sensitivity at low liquid rates.

Process surges. Random packings are usually more troublesome than trays in services prone to process surges (e.g., those caused by slugs of water entering a hot oil tower, relief valve lifting, compressor surges, or instability of liquid seal loops). Structured packings are usually less troublesome than trays in such services.

Trays versus Structured Packings The following factors generally favor trays compared to structured packings, but not compared to random packings.

Packing fires. The thin sheets of structured packing (typically 0.1 mm) poorly dissipate heat away from hot spots. Also, cleaning, cooling, and washing of pyrophoric deposits can be difficult, especially when distributors or packing plug up. Many incidents of packing fires during turnarounds (while towers with structured packings were open to atmosphere) have been reported. Most of these fires were initiated by pyrophoric deposits, hot work (e.g., welding) above the packing, opening the tower while hot organics were still present, and packing metallurgy that was not fire-resistant. Detailed discussion can be found in Fractionation Research Inc. (FRI) Design Practices Committee, "Causes and Prevention of Packing Fires," *Chem. Eng.*, July 2007.

Materials of construction. Due to the thin sheets of structured packings (an order of magnitude thinner than trays), their materials of construction need to have better resistance to oxidation or corrosion. For a service in which carbon steel is usually satisfactory with trays, stainless steel is usually required with structured packings.

Column wall inspection. Due to their snug fit, structured packings are easily damaged during removal. This makes it difficult to inspect the column wall (e.g., for corrosion).

Washing and purging. Thorough removal of residual liquid, wash water, air, or process gas trapped in structured packings at start-up and shutdown is more difficult than with trays. Inadequate removal of these fluids may be hazardous.

High liquid rates. Multipass trays effectively lower the liquid load "seen" by each part of the tray. A similar trick cannot be applied with packings. The capacity of structured packings tends to rapidly fall off at high liquid rates.

Capacity and Efficiency Comparison Kister et al. [*Chem. Eng. Progr.* **90**(2): 23 (1994)] reported a study of the relative capacity and efficiency of conventional trays, modern random packings, and conventional structured packings. They found that, for each device optimally designed for the design requirements, a rough guide could be developed on the basis of flow parameter $L/G\,(\rho_G/\rho_L)^{0.5}$ (abscissa in Figs. 14-32, 14-55, and 14-56) and the following tentative conclusions could be drawn:

Flow Parameter 0.02–0.1

1. Trays and random packings have much the same efficiency and capacity.
2. Structured packing efficiency is about 1.5 times that of trays or random packing.
3. At a parameter of 0.02, the structured packing has a 1.3–1.4 capacity advantage over random packing and trays. This advantage disappears as the parameter approaches 0.1.

Flow Parameter 0.1–0.3

1. Trays and random packings have about the same efficiency and capacity.
2. Structured packing has about the same capacity as trays and random packings.
3. The efficiency advantage of structured packing over random packings and trays decreases from 1.5 to 1.2 as the parameter increases from 0.1 to 0.3.

Flow Parameter 0.3–0.5

1. The loss of capacity of structured packing is greatest in this range.
2. The random packing appears to have the highest capacity and efficiency with conventional trays just slightly behind. Structured packing has the least capacity and efficiency.

Experience indicates that the use of structured packings has capacity/efficiency disadvantages in the higher-pressure (higher-flow-parameter) region.

Zuiderweg and Nutter [*IChemE Symp. Ser.* **128:** A481 (1992)] explain the loss of capacity/efficiency by a large degree of backmixing and vapor recycle at high flow parameters, promoted by the solid walls of the corrugated packing layers.

SYSTEM LIMIT: THE ULTIMATE CAPACITY OF FRACTIONATORS

Liquid drops of various sizes form in the gas–liquid contact zones of tray or packed towers. Small drops are easily entrained upward, but their volume is usually too small to initiate excessive liquid accumulation (flooding). When the gas velocity is high enough to initiate a massive carryover of the larger drops to the tray above, or upward in a packed bed, liquid accumulation (entrainment flooding) takes place. This flood can be alleviated by increasing the tray spacing or using more hole areas on trays or by using larger, more open packings.

Upon further increase of gas velocity, a limit is reached when the superficial gas velocity in the gas–liquid contact zone exceeds the settling velocity of large liquid drops. At gas velocities higher than this, ascending gas lifts and carries over much of the tray or packing liquid, causing the tower to flood. This flood is termed *system limit* or *ultimate capacity*. This flood cannot be debottlenecked by improving packing size or shape, tray hole area, or tray spacing. The system limit gas velocity is a function only of physical properties and liquid flow rate. Once this limit is reached, the liquid will be blown upward. This is analogous to spraying water against a strong wind and getting drenched (Yanagai, *Chem. Eng.*, November 1990, p. 120). The system limit represents the ultimate capacity of the vast majority of existing trays and packings. In some applications, where very open packings (or trays) are used, such as in refinery vacuum towers, the system limit is the actual capacity limit.

The original work of Souders and Brown [*Ind. Eng. Chem.* **26**(1): 98 (1934), Eq. (14-80)] related the capacity of fractionators due to entrainment flooding to the settling velocity of drops. The concept of system limit was advanced by Fractionation Research Inc. (FRI), whose measurements and model have recently been published (Fitz and Kunesh, *Distillation 2001: Proceedings of Topical Conference,* AIChE Spring National Meeting, Houston, Tex., 2001; Stupin, FRI Topical Report 34, 1965, available through Special Collection Section, Oklahoma State University Library, Stillwater, Okla.). Figure 14-74 is a plot of FRI system limit data (most derived from tests with dual-flow trays with 29 percent hole area and 1.2- to 2.4-m tray spacing) against liquid superficial velocity for a variety of systems (Stupin 1965). The data show a constant-slope linear dependence of the system limit *C*-factor on the liquid load. There was a shortage of data at low liquid loads. Later data (Fig. 14-75) showed that as the liquid load was reduced, the system limit $C_{s,\text{ult}}$ stopped increasing and reached a limiting value. Based on this observation, Stupin and Kister [*Trans. IChemE* **81:** Part A, p. 136 (January 2003)] empirically revised the earlier Stupin/FRI correlation to give

$$C_1 = 0.445(1-F)\left(\frac{\sigma}{\Delta\rho}\right)^{0.25} - 1.4L_S \tag{14-168}$$

$$C_2 = 0.356(1-F)\left(\frac{\sigma}{\Delta\rho}\right)^{0.25} \tag{14-169}$$

$$C_{s,\text{ult}} = \text{smaller of } C_1 \text{ and } C_2 \tag{14-170}$$

where

$$F = \frac{1}{1 + 1.4(\Delta\rho/\rho_G)^{1/2}} \tag{14-171}$$

In Eqs. (14-168) through (14-171), $C_{s,\text{ult}}$ is the system limit *C*-factor based on the tower superficial area [see Eq. (14-77) for *C*-factor definition]; L_S is the liquid superficial velocity, m/s; σ is the surface tension, mN/m; $\Delta\rho$ is the difference between the liquid and gas densities, kg/m^3; and ρ_G is the gas density, kg/m^3.

Stupin and Kister [*Trans. IChemE* **81:** Part A, p. 136 (January 2003)] relate the flattening of the curve in Fig. 14-75 at low liquid loads to the formation of more, smaller, easier-to-entrain liquid drops when the liquid load is lowered beyond the limiting liquid load. It follows that devices that can restrict the formation of smaller drops may be able to approach the system limit capacity predicted by Stupin's original equation [Eq. (14-168)] even at low liquid loads.

The only devices capable of debottlenecking a tray system-limit device are those that introduce a new force that helps disentrain the vapor space. Devices that use centrifugal force (see the subsection Centrifugal Force Deentrainment) are beginning to make inroads into commercial distillation and have achieved capacities as high as 50 percent above the system limit. Even the horizontal vapor push (see the subsection Truncated Downcomers/Forward-Push Trays) can help settle the entrained drops, but to a much lesser extent. It is unknown whether the horizontal push alone can achieve capacities exceeding the system limit.

WETTED-WALL COLUMNS

Wetted-wall or falling-film columns are vertical tubes with liquid flowing as a thin film down a tube wall and vapor ascending through the hollow center. Because of the ease of modeling, wetted wall columns are popular as laboratory equipment, for example, to measure mass-transfer coefficients. In industry, wetted wall columns found application in services where high-heat-transfer-rate requirements are concomitant with the absorption process. Large areas of open surface are available for heat transfer for a given rate of mass transfer in this type of equipment because of the low mass-transfer rate inherent in wetted-wall equipment. In addition, this type of equipment lends itself to annular-type cooling devices.

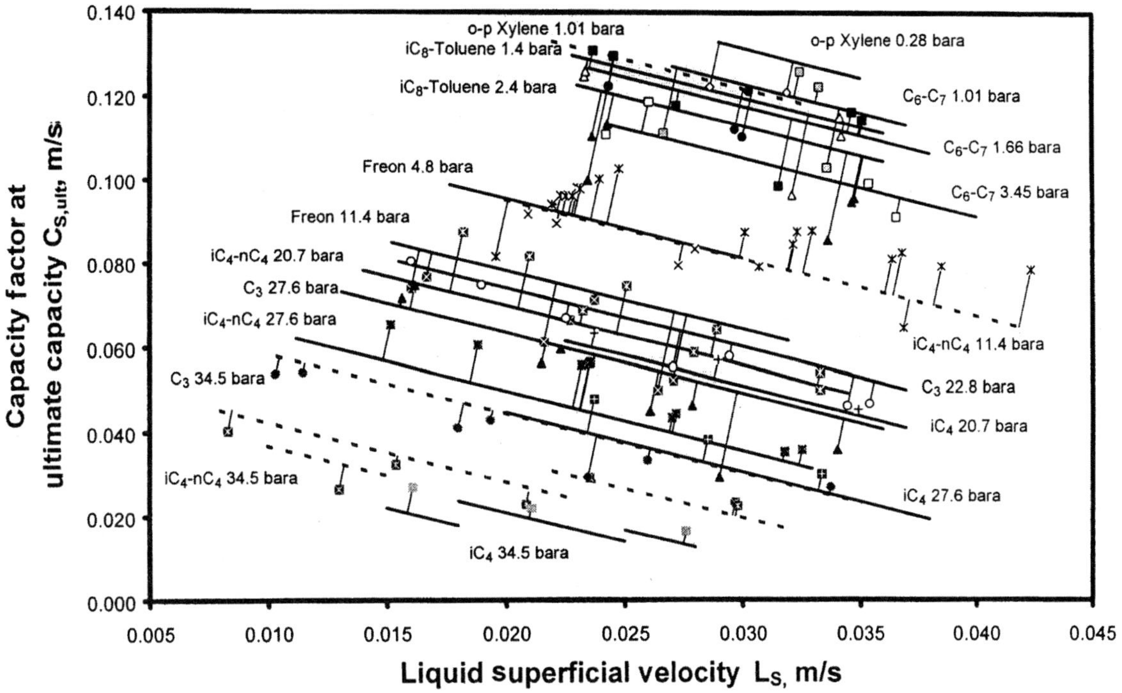

FIG. 14-74 Effect of liquid rate on ultimate capacity at higher liquid rates. (*From Stupin, W. J., and H. Z. Kister,* Trans. IChemE, *vol. 81, Part A, p. 136, January 2003. Reprinted courtesy of IChemE.*)

The classic experimental work by Gilliland and Sherwood [*Ind. Eng. Chem.* **26:** 516 (1934)] used a falling-film column to study mass transfer for the vaporization of pure liquids in air streams for streamline flow. They obtained a correlation between the Sherwood number, Reynolds number, and Schmidt number.

$$\frac{k_g D_{\text{tube}}}{D_g}\frac{P_{BM}}{P} = 0.023 N_{\text{Re}}^{0.83} N_{\text{Sc}}^{0.44} \tag{14-172}$$

where D_g = diffusion coefficient, ft²/h
D_{tube} = inside diameter of tube, ft
k_g = mass-transfer coefficient, gas phase, lb · mol/(h · ft²) (lb · mol/ft³)
P_{BM} = logarithmic mean partial pressure of inert gas, atm
P = total pressure, atm
N_{Re} = Reynolds number, gas phase
N_{Sc} = Schmidt number, gas phase

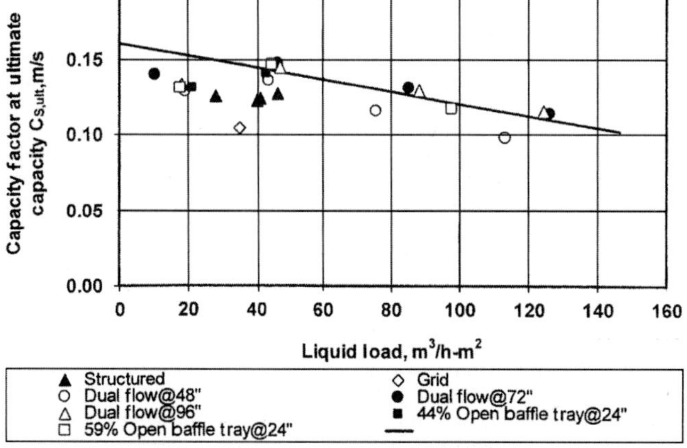

FIG. 14-75 Comparison of original ultimate capacity correlation to test data, C_6/C_7, 1.66 bar. (*From Stupin, W. J., and H. Z. Kister,* Trans. IChemE, *vol. 81, Part A, p. 136, January 2003. Reprinted courtesy of IChemE.*)

Note that the group on the left side of Eq. (14-172) is dimensionless. When turbulence promoters are used at the inlet-gas section, an improvement in gas mass-transfer coefficient for the absorption of water vapor by sulfuric acid was observed by Greenewalt [*Ind. Eng. Chem.* **18:** 1291 (1926)]. A falling off of the rate of mass transfer below that indicated in Eq. (14-172) was observed by Cogan and Cogan (thesis, Massachusetts Institute of Technology, 1932) when a calming zone preceded the gas inlet in ammonia absorption.

Considerable work on wetted wall columns was reported in previous editions of *Perry's Handbook*. Many published contributions have been added since [for example, Nielsen, Kiil, Thomsen, and Dam-Johansen, *Chem. Eng. Sci.* **53**(3): 495 (1998); Erasmus and Nieuwoudt, *Ind. Eng. Chem. Res.* **40**(10): 2310 (2001); Spedding, *The Chemical Engineering Journal* **37**(3): 165 (1988); Strumillo and Porter, *AIChE J.* **11**(6): 1139 (1965)]. Due to the bulkiness of this work compared to the relatively low industrial usage of wetted wall columns, this work will not be addressed in detail in this edition of *Perry's Handbook*, and the reader is referred to the earlier editions and to the update articles.

COLUMN COSTS

The estimation of column costs for preliminary process evaluations requires consideration not only of the basic type of internals but also of their effect on overall system cost. For a distillation system, for example, the overall system can include the vessel (column), attendant structures, supports, and foundations; auxiliaries such as reboiler, condenser, feed heater, and control instruments; and connecting piping. The choice of internals influences all these costs, but other factors influence them as well. A complete optimization of the system requires a full-process simulation model that can cover all pertinent variables influencing economics.

The cost estimation method presented here follows the guidelines of Peters, Timmerhaus, and West (*Plant Design and Economics for Chemical Engineers*, 5th ed., McGraw-Hill Education, 2003). Alternative methods were presented by Erwin (*Industrial Chemical Process Design*, 2d ed., McGraw-Hill, New York, 2014), and Towler and Sinnott (*Chemical Engineering Design—Principles, Practice, and Economics of Plant and Process Design*, 2d ed., Elsevier, Amsterdam, 2013).

Cost of Internals Purchased costs of trays may be estimated from Fig. 14-76, with corrections for tray material taken from Table 14-16. For two-pass trays, the cost is 15 to 20 percent higher. Figure 14-77 provides similar information for random packings. Note that for Figs. 14-76 and 14-77,

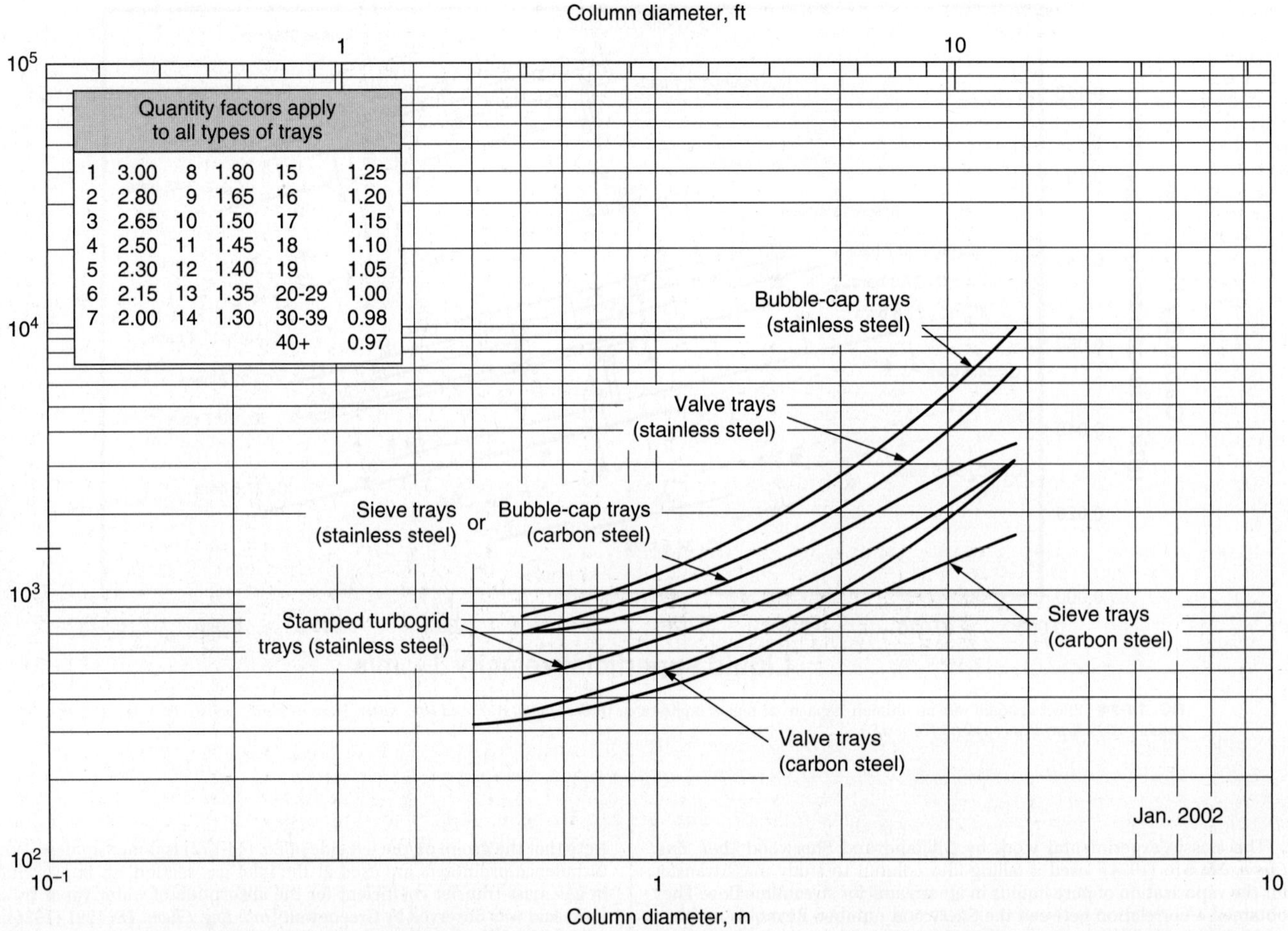

FIG. 14-76 Purchased cost of trays in tray columns. Price includes tray decks, downcomers, and structural steel parts. (*Peters, Timmerhaus and West,* Plant Design and Economics for Chemical Engineers, *5th Ed., McGraw-Hill Education, 2003. Reprinted with permission.*)

TABLE 14-16 Relative Fabricated Cost for Metals Used in Tray-Tower Construction*

Materials of construction	Relative cost per ft² of tray area (based on carbon steel = 1)
Sheet-metal trays	
Steel	1
4–6% chrome—½ moly alloy steel	2.1
11–13% chrome type 410 alloy steel	2.6
Red brass	3
Stainless steel type 304	4.2
Stainless steel type 347	5.1
Monel	7.0
Stainless steel type 316	5.5
Inconel	8.2
Cast-iron trays	2.8

*Peters and Timmerhaus, *Plant Design and Economics for Chemical Engineers*, 4th ed., McGraw-Hill, New York, 1991. To convert cost per square foot to cost per square meter, multiply by 10.76.

the effective cost date is January 2002, with the Marshall and Swift cost index being taken as 1167. Table 14-17 is based on 1990 cost data, but it only provides the relative costs between different materials.

In early 2017, costs of 0.2-mm-thick 410 SS corrugated sheet structured packings for a bed about 20 ft tall were on the order of $40 to $45 per cubic foot for Y packing of 125 m^2/m^3, rising to $55 to $60 per cubic foot and $70 to $75 per cubic foot for third-generation (high-performance) structured packings of 250 m^2/m^3 and 350 m^2/m^3, respectively, of the same material and thickness. Reducing the sheet thickness to the less-fire-resistant 0.1 mm in the same material reduces the costs to an order of $40 to $45 per cubic foot and $50 to $55 per cubic foot for third-generation (high-performance) packings of 250 m^2/m^3 and 350 m^2/m^3, respectively. See Table 14-14 for common packings of these surface areas. For third- and fourth-generation random packings, 410 SS and carbon steel, respectively, costs were on the order of $100 to $110 per cubic foot and $85 to $95 per cubic foot for 25-mm packings, declining to $60 to $65 and $55 to $60 per cubic foot, respectively, for 50-mm packings. These figures allow for distributors, collectors, and supports, but the need for special distributors and redistributors can double the cost of packings on a volumetric basis.

It should be recognized that, because of the requirement for distributors, redistributors, collectors, and holddowns, and because of competition, there are likely to be large variations in these costs from tower to tower and from supplier to supplier. Also, packings sold in very large quantities carry discounts. So the values in Figs. 14-76 and 14-77, as well as the numbers just cited, are far from reliable and are only suitable for a very preliminary indication, not for evaluation. The supplier should always be contacted for a reliable estimate.

Cost of Columns The fabricated cost of the vessel, including heads, skirt, nozzles, and ladderways, is usually estimated on the basis of weight. Figure 14-78 provides cost data for the shell and heads, and Fig. 14-79 provides cost data for connections. For very approximate estimates of complete columns, including internals, Fig. 14-80*a* and *b* may be used. As for Figs. 14-76 and 14-77, the effective cost date for Figs. 14-78 through 14-80 is January 2002, when the cost index was 1167.

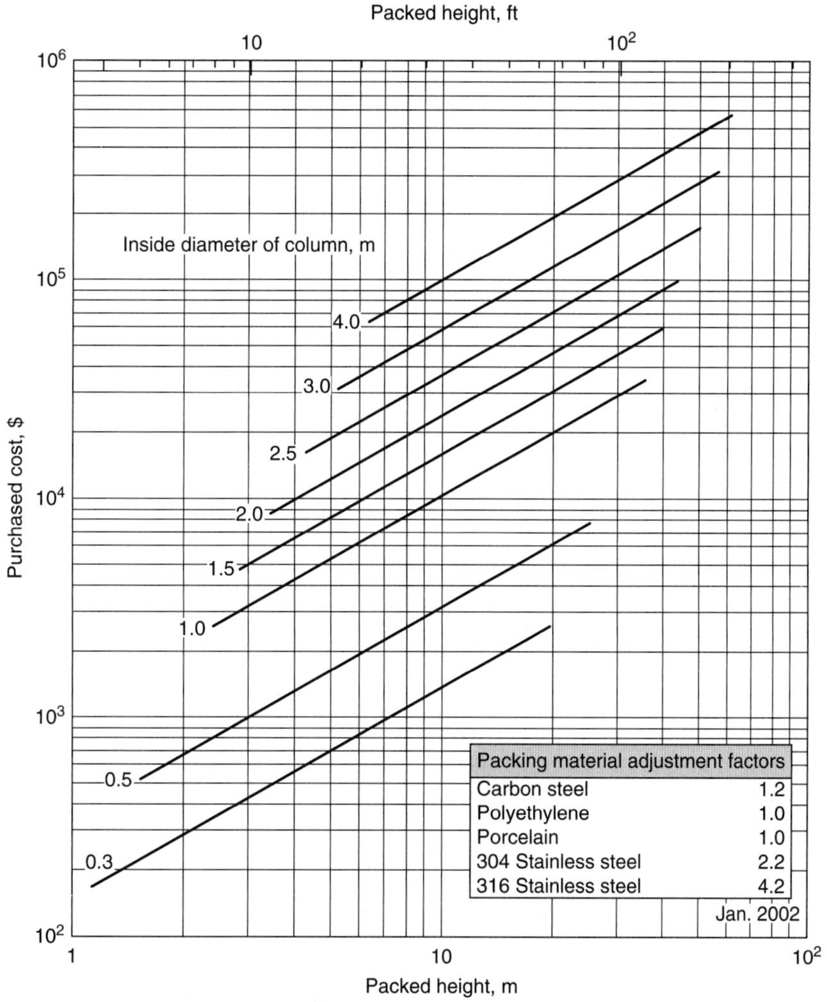

FIG. 14-77 Purchased cost of random packings (price includes packing supports and distributors). (*Peters, Timmerhaus and West,* Plant Design and Economics for Chemical Engineers, *5th Ed., McGraw-Hill Education, 2003. Reprinted with permission.*)

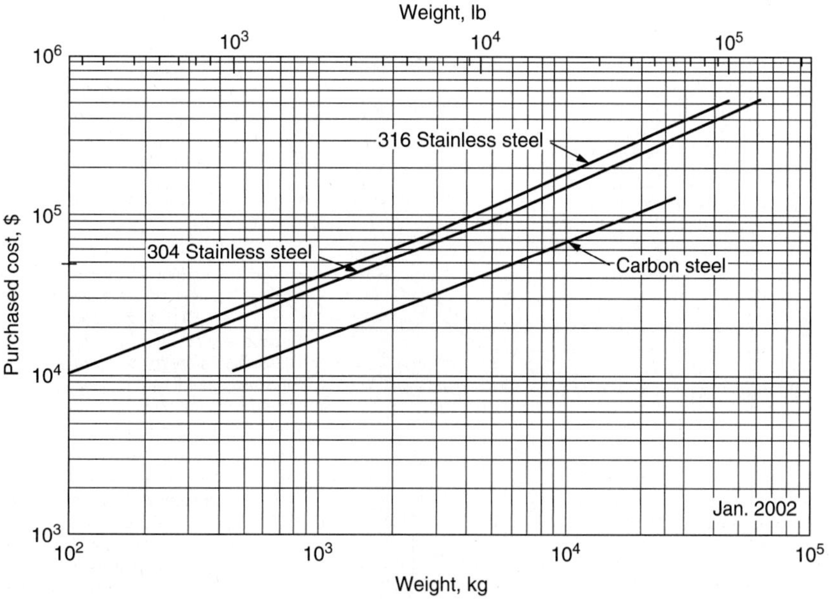

FIG. 14-78 Purchased cost of columns. Costs are for shell with two heads and skirt, but without trays, packings, or connections. (*Peters, Timmerhaus and West,* Plant Design and Economics for Chemical Engineers, *5th Ed., McGraw-Hill Education, 2003. Reprinted with permission.*)

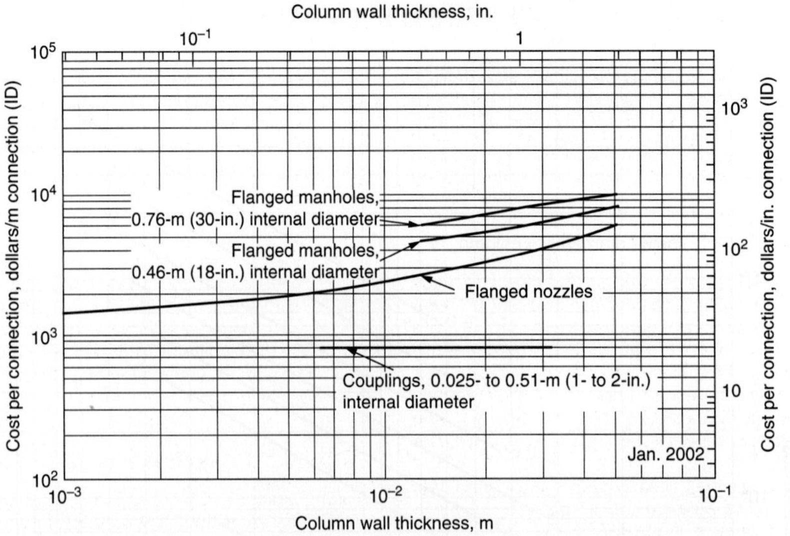

FIG. 14-79 Approximate installed cost of steel-tower connections. Values apply to 136-kg (300-lb) connections. Multiply costs by 0.9 for 68 kg (150-lb) connections and by 1.2 for 272-kg (600-lb) connections. (*Peters, Timmerhaus and West,* Plant Design and Economics for Chemical Engineers, *5th Ed., McGraw-Hill Education, 2003. Reprinted with permission.*)

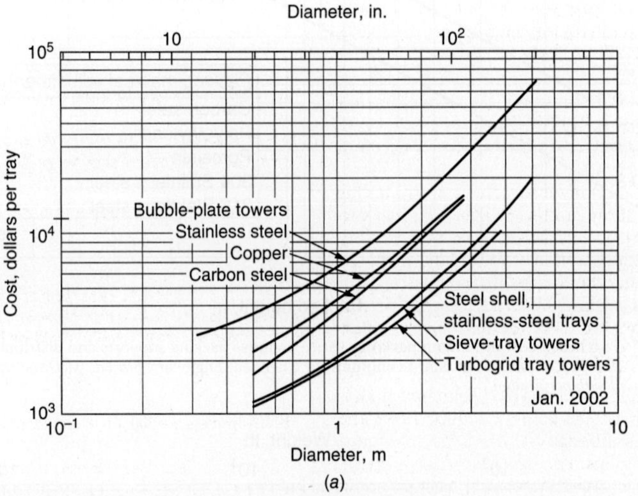

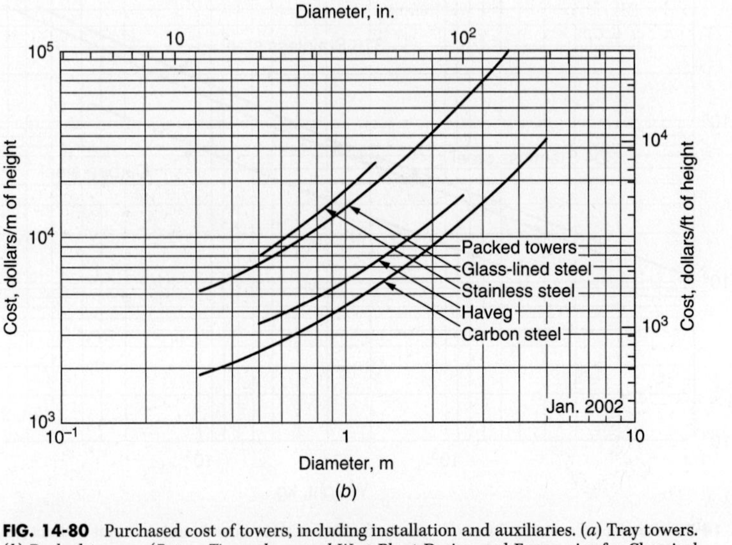

FIG. 14-80 Purchased cost of towers, including installation and auxiliaries. (*a*) Tray towers. (*b*) Packed towers. (*Peters, Timmerhaus and West,* Plant Design and Economics for Chemical Engineers, *5th Ed., McGraw-Hill Education, 2003. Reprinted with permission.*)

PHASE DISPERSION

General References: For an overall discussion of gas–liquid breakup, see Brodkey, *The Phenomena of Fluid Motions,* Addison-Wesley, Reading, Mass., 1967. For a discussion of atomization devices and how they work, see Lipp, *Practical Spray Technology*, Lake Innovation LLC, Lake Junaluska, N.C., 2012. See also Lefebvre, *Atomization and Sprays,* Hemisphere, New York, 1989. For a discussion of how power input controls drop size in high-energy gas–liquid contact, see Steinmeyer [*Chem. Engr. Prog.* **91**(7): 72–80 (1995)].

BASICS OF INTERFACIAL CONTACTORS

Steady-State Systems: Bubbles and Droplets Bubbles are made by injecting vapor below the liquid surface. In contrast, droplets are commonly made by atomizing nozzles that inject liquid into a vapor. Bubble and droplet systems are fundamentally different, mainly because of the enormous difference in the density of the injected phase. There are situations where each is preferred. Bubble systems tend to have much higher interfacial area, as shown by comparing Example 14-17 to Examples 14-15 and 14-16. Because of their greater area, bubble systems will usually give a closer approach to equilibrium.

However, droplet systems can enable much higher energy input (via gas-phase pressure drop in cocurrent systems) and, as a result, can dominate applications where a quick quench is needed. See Examples 14-22 and 14-23. Conversely, droplet systems can also be designed for very low pressure drop, which is advantageous in applications such as vacuum condensers.

Unstable Systems: Froths and Hollow-Cone Atomizing Nozzles We usually think of interfacial contact as a steady-state system of raining droplets or rising bubbles, but some of the most efficient interfacial contactors take advantage of unstable interfacial geometry. The most common is the distillation tray, which operates with a wild mix of bubbles, jets, films, and droplets. The mix is often described as *froth.* Gas pressure drop provides the energy to create the froth.

A variant on the froth contact is the reverse jet contactor (see Example 14-23), which can be considered an upside-down distillation tray, operated above the flooding velocity in cocurrent flow of gas and liquid. It is limited to one stage.

An entirely different unstable contactor involves the thin expanding liquid film produced by a hollow-cone spray nozzle. Because of fresh surface and the thinness of the film, this can give very high transfer for liquid-limited systems. Two applications are direct contact condensation and the removal of volatile components from a high-boiling residual liquid.

Surface Tension Makes Liquid Sheets and Liquid Columns Unstable Surface tension is the energy required to make an increment of interfacial surface. A sheet or column of liquid has greater surface than a sphere, hence surface tension converts sheets and columns to droplets. See Fig. 14-81.

There are many different atomizers, but the underlying principle of all is the same—to first generate a flat sheet or a liquid column. Liquid sheets and columns are unstable; a small surface disturbance on either will propagate, and the liquid will reshape itself into droplets. The key property in controlling this process is surface tension. Surface tension gets a high exponent in all the atomization correlations.

Little Droplets and Bubbles versus Big Droplets and Bubbles—Coalescence versus Breakup When big drops are subjected to shear forces, as in falling rain, the droplets are distorted; and if the distortions are great enough, the big droplets break into little ones. This is why raindrops never exceed a certain size. A variant on this is breakup in highly turbulent systems such as that in high-velocity quench systems or pneumatic nozzles. Here the droplets are distorted by the energy of the turbulent eddies.

But little droplets and bubbles have greater surface per unit of liquid than big ones do. Hence little droplets tend to coalesce into big ones and will grow larger if given enough quiet time.

While the primary difficulty is estimating the interfacial area due to the unstable interface, a secondary problem is that freshly made, unstable surface gives higher transfer than older, more stable surface.

Empirical Design Tempered by Operating Data The net of these is that interfacial area is difficult to predict, and interfacial contactors are difficult to design. Prediction methods are given next, but they should always be tempered by operating experience.

INTERFACIAL AREA—IMPACT OF DROPLET OR BUBBLE SIZE

Transfer is aided by increased interfacial area. Interfacial area per unit volume a_D of a single droplet or bubble is inversely proportional to the diameter of the droplet or bubble D.

$$a_D = 6/D \qquad (14\text{-}173a)$$

To estimate the total interfacial area in a given volume, the a_D value is multiplied by the fractional holdup of dispersed phase in the total volume.

$$a = a_D(\Phi_D) \qquad (14\text{-}173b)$$

where a = interfacial area/volume and Φ_D = fraction of volume in dispersed phase = holdup.

Fractional holdup in a continuous process depends on the velocities of the two phases, as if they were flowing by themselves.

$$\Phi_D = \text{(dispersed phase volume)/(volume of dispersed and continuous phases)}$$

Example 14-15 Interfacial Area for Droplets/Gas in Cocurrent Flow For equal mass flow of gas and liquid and with gas density 0.001 of liquid density, the gas velocity in cocurrent flow will be 1000 times the liquid velocity. This sets Φ_D.

$$\Phi_D = 1/(1 + 1000) = 0.00099$$

If the droplets are 500 μm in diameter, Eqs. (14-173*a*) and (14-173*b*) give

$$a = (6/0.0005)(0.00099) = 12 \text{ m}^2/\text{m}^3$$

If the droplets are 100 μm in diameter, Eqs. (14-173*a*) and (14-173*b*) give

$$a = (6/0.0001)(0.00099) = 60 \text{ m}^2/\text{m}^3$$

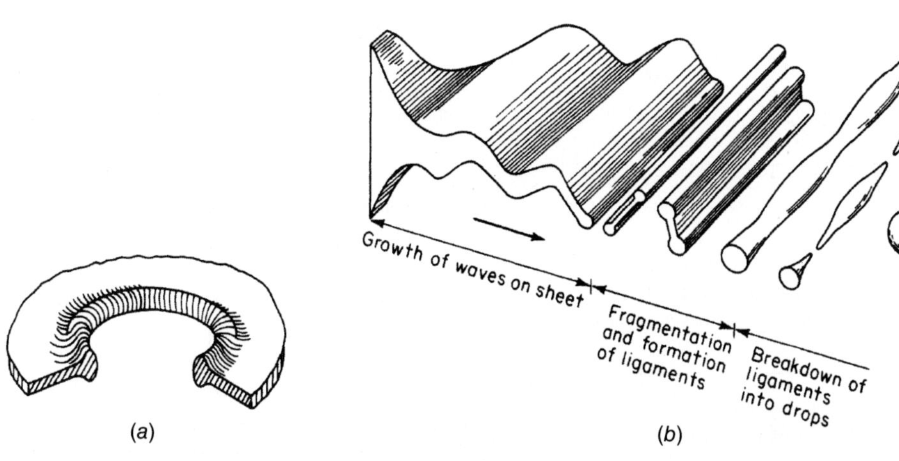

FIG. 14-81 Sheet breakup. (*a*) By perforation. [*After Fraser et al.,* Am. Inst. Chem. Eng. J. *8(5): 672 (1962).*] (*b*) By sinusoidal wave growth. [*After Dombrowski and Johns,* Chem. Eng. Sci. ***18****: 203 (1963).*]

Example 14-16 Interfacial Area for Droplets Falling in a Vessel Droplet systems rarely exceed a Φ_D value of 0.01. At this low level, Φ_D in a low-velocity countercurrent contactor can be approximated by Eq. (14-174).

$$\Phi_D = U_L/(U_t - U_G) \tag{14-174}$$

where U_L = liquid superficial velocity
U_t = terminal velocity of droplet
U_G = gas superficial velocity

With a gas superficial velocity of 1.5 m/s, for equal mass flow of gas and liquid, with gas density 0.001 of liquid density, and with 500-μm-diameter droplets falling at a vessel settling of 2.5 m/s, Eq. (14-174) gives a fractional holdup of liquid of

$$\Phi_D = (0.001)1.5/(2.5 - 1.5) = 0.0015$$

Equations (14-173*a*) and (14-173*b*) then give

$$a = (6/0.0005)(0.0015) = 18 \text{ m}^2/\text{m}^3$$

Example 14-17 Interfacial Area for Bubbles Rising in a Vessel For bubble systems (gases dispersed in liquids), fractional holdup can approach 0.5, as shown by Fig. 14-82. However, before reaching this holdup, the bubble systems shift to an unstable mix of bubbles and vapor jets. Hence an exact comparison to Example 14-15 isn't possible because at the 1.5 m/s velocity of Example 14-15, the system becomes a froth. But at about one-fifth the velocity of Example 14-15, an estimate of interfacial area is possible.

If the bubble size is 10,000 μm and fractional holdup is 0.4, Eqs. (14-173*a*) and (14-173*b*) give an interfacial area of

$$a = (6/0.01)(0.4) = 240 \text{ m}^2/\text{m}^3$$

Measured interfacial area in distillation trays is consistent with this high value.

Note the much higher interfacial area than in the droplet systems of Examples 14-15 and 14-16. The higher interfacial area when the gas is dispersed explains why bubbling and froth systems often give better performance than droplet systems. The big difference in interfacial area stems from the much larger volume per unit of mass of gas, that is, the lower density of the gas than the liquid.

RATE MEASURES, TRANSFER UNITS, APPROACH TO EQUILIBRIUM, AND BYPASSING

What Controls Mass/Heat Transfer: Liquid or Gas Transfer or Bypassing Either the gas side or the liquid side of the interface can be controlling.

Liquid-Controlled In fractionation systems with high viscosity or component relative volatility that greatly exceeds 1, the liquid side will be controlling. This is clearly illustrated by Fig. 14-47, which shows a sharp decline in efficiency with either a rise in liquid viscosity or a rise in *component relative volatility.*

Note that *high component relative volatility* means the same thing as *sparingly soluble.* Oxygen dissolving in a fermentation reactor is an example of a system being liquid controlled due to a *sparingly soluble gas.* Another application that is liquid controlled is the removal of *high relative volatility components* from residual oil.

Still another case where liquid controls is in condensing a pure vapor, as in Example 14-24, or absorbing a pure gas, as in Example 14-25.

Gas-Controlled The gas side dominates in gas cooling applications. An example is the quenching of a furnace effluent with a vaporizing liquid. In this application, the liquid is nearly uniform in temperature. Restated, the reduction in driving force across the liquid side of the interface is unimportant.

Other applications that are gas-side-controlled include removal of a component such as NH_3 from a gas by using an acidic liquid, or removing a component such as SO_2 from a gas with a basic liquid. See Examples 14-20 through 14-23.

Bypassing-Controlled Trayed or packed columns operate with countercurrent flow and can achieve many equilibrium stages in series by good distribution of gas and liquid and careful control of details. Other devices such as sprays are vulnerable to bypassing and are limited to one equilibrium stage.

Rate Measures for Interfacial Processes The terminology used for reporting rate data can be confusing. Normally rate data are reported on a volumetric basis with transfer rate and effective area combined. For example, k_La denotes mass-transfer data per unit volume. The subscript L means it is referenced to the molar concentration difference between the interface and the bulk liquid. This is commonly used on data involving a sparingly soluble (high relative volatility) component. Note that the lowercase k means the data deal only with the resistance in the liquid phase.

Less commonly, data are given as k_Ga. The subscript G means it is referenced to the molar concentration difference between the interface and the gas. This might be used for data on absorbing a gas such as NH_3 by a highly acidic liquid. Note that k_Ga only deals with the resistance in the gas phase.

When one is dealing with direct contact heat transfer, the corresponding terms are h_La and h_Ga. Here the driving force is the temperature difference. The subscript L means that we are dealing with a liquid-limited process such as condensing a pure liquid. How to convert k_La data to an h_La value is illustrated by Example 14-24.

There are ways to combine the liquid and gas resistance to get an overall transfer rate such as K_Ga (as denoted by the uppercase K). However, data are rarely reported in this form.

Approach to Equilibrium Although rate measures such as k_Ga and h_Ga are often cited in the literature, they are often not as useful to designers as the simpler concept of approach to equilibrium. *Approach to equilibrium* compares the transfer between liquid and gas phases to the best possible that could be achieved in a single backmixed equilibrium stage.

Approach to equilibrium is easy to understand and easy to apply. Examples 14-18 through 14-23 illustrate its use.

Example 14-18 Approach to Equilibrium—Perfectly Mixed, Complete Exchange This would be approximated by a very long pipeline contactor where an acidic aqueous stream is injected to cool the gas and remove NH_3.

If the adiabatic saturation temperature of the gas is 70°C, at the exit of the contactor, the gas would be cooled to 70°C.

Similarly, at the exit of the contactor, the NH_3 in the gas would be zero, regardless of the initial concentration.

Example 14-19 Approach to Equilibrium—Complete Exchange but with 10 Percent Gas Bypassing A spray column is used, and an acidic liquid rains down on the gas of Example 14-18. If the initial NH_3 is 1000 ppm and 10 percent of the gas bypasses, the NH_3 in the exit gas would be

$$0.1(1000) = 100 \text{ ppm}$$

Similarly, if the gas enters at 120°C, at the exit we will find 10 percent of the differential above the adiabatic saturation temperature. For an adiabatic saturation temperature of 70°C, the exit gas temperature will be

$$70 + 0.1(120 - 70) = 75°\text{C}$$

Approach to Equilibrium—Finite Contactor with No Bypassing When there is no bypassing, the measure that sets the approach is the ratio of change to driving force. This ratio is called the *number of transfer units* N_G. It is dimensionless. For heat-transfer applications, it can be envisioned as a conventional heat exchanger where a vaporizing liquid cools a gas:

$$\text{No. of gas-phase transfer units} = \frac{T_{G,\text{out}} - T_{G,\text{in}}}{(T_G - T_L)_{\text{av}}} = N_G \tag{14-175}$$

where T_G = gas temperature and T_L = liquid temperature. The number of transfer units N_G can also be calculated as the capability for change divided by the thermal capacitance of the flowing streams.

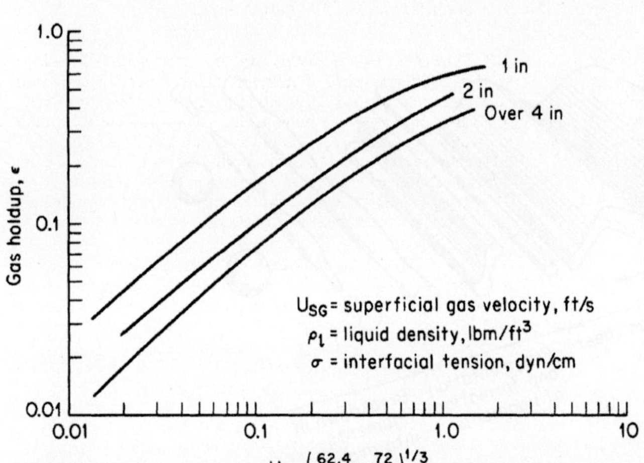

FIG. 14-82 Gas holdup correlation. [Ind. Eng. Chem. Process Des. Dev. *6: 218 (1967).*]

$$N_G = \frac{(\text{system volume})(h_G a)}{(\text{volumetric flow rate})\rho_G c_G} = \frac{(\text{gas contact time})(h_G a)}{\rho_G c_G} \quad (14\text{-}176)$$

where a = interfacial area per unit volume
h_G = heat-transfer coefficient from interface to gas
ρ_G = gas density
c_G = gas specific heat

Note that in this calculation, performance and properties all refer to the gas, which is appropriate when dealing with a gas-limited transfer process.

This leads to a way to estimate the approach to equilibrium.

$$E = 1 - e^{-N_G} \quad (14\text{-}177)$$

where E = "approach to equilibrium" fractional removal of NH_3 or fractional approach to adiabatic liquid temperature
N_G = number of transfer units calculated relative to gas flow

Example 14-20 Finite Exchange, No Bypassing, Short Contactor A short cocurrent horizontal pipeline contactor gives 86 percent removal of NH_3. There is no bypassing because of the highly turbulent gas flow and injection of liquid into the center of the pipe. What would we expect the exit gas temperature to be?

Equation (14-177) says that the back-calculated N_G is 2:

$$N_G = -\ln(1 - 0.86) = 2$$

For diffusing gases of similar molecular weight, the properties that control heat transfer follow the same rules as those that control mass transfer. As a result, the NH_3 scrubbing and gas cooling processes achieve similar approaches to equilibrium.

For an entry temperature of 120°C and an adiabatic saturation temperature of 70°C, the expected outlet temperature would be

$$70 + (1 - 0.86)(120 - 70) = 77°\text{C}$$

This looks like a powerful concept, but its value is limited due to uncertainty in estimating $h_G a$. Both h_G and a are difficult to estimate due to dependence on power dissipation, as discussed below. The primary value of the N_G concept lies in estimating an expected change from baseline data as in the comparison of Example 14-20 with Example 14-21.

Example 14-21 A Contactor That Is Twice as Long, No Bypassing If we double the length of the pipeline contactor, double the effective contact area, and double the number of transfer units to 4, what do we expect for performance?

For $N_G = 4$,

$$E = 1 - e^{-4} = 0.982$$

The NH_3 in the exit gas would be expected to drop to

$$(1 - 0.982)(1000) = 18 \text{ ppm}$$

and the expected outlet temperature would be

$$70 + (1 - 0.982)(120 - 70) = 70.9°\text{C}$$

If we double the length again, we increase the number of transfer units to 8 and achieve an approach of

$$E = 1 - e^{-8} = 0.9997$$

The outlet temperature would be

$$70 + (1 - 0.9997)(120 - 70) = 70.015°\text{C}$$

Similarly, the NH_3 in the exit gas would be

$$(1 - 0.9997)(1000) = 0.3 \text{ ppm}$$

Note that this approximates the exit condition of Example 14-18.

Transfer Coefficient—Impact of Droplet Size The transfer coefficients increase as the size of droplets decreases. This is so because the transfer process is easier if it only has to move mass or heat a shorter distance (i.e., as the bubble or droplet gets smaller).

In the limiting case of quiescent small bubbles or droplets, the transfer coefficients vary inversely with average bubble or droplet diameter. For example, in heat transfer from a droplet interface to a gas, the minimum value is

$$h_{G,\min} = \text{heat transfer coefficient from interface to gas} = 2k_G/D \quad (14\text{-}178)$$

where k_G = gas thermal conductivity and D = droplet diameter.

IMPORTANCE OF TURBULENCE

The designer usually has control over the size of a droplet. As discussed next, several of the correlations show that droplet diameter varies with turbulent energy dissipation. For example, Eqs. (14-185) and (14-196) suggest that in droplet systems

$$D \propto \{1/(\text{gas velocity})\}^{1.2}$$

and hence from Eq. (14-173)

$$a \propto 1/D \propto (\text{gas velocity})^{1.2} \quad (14\text{-}179)$$

However, just looking at the impact of velocity on droplet size underestimates the velocity impact because turbulence gives higher transfer than Eq. (14-178) predicts. Transfer coefficients increase as the mixing adjacent to the surface increases. This mixing depends on the energy dissipated into the phases. To a first approximation this transfer from droplets increases with local power dissipation raised to the 0.2 power.

$$h_{G,\text{turbulent}} \propto (\text{power dissipated})^{0.2}$$

and since power dissipation per unit volume increases with velocity^3,

$$h_{G,\text{turbulent}} \propto (\text{velocity})^{0.6} \quad (14\text{-}180)$$

The combined effect on interfacial area and on the transfer coefficient is that the effective transfer increases greatly with gas velocity. From Eqs. (14-173) and (14-180)

$$h_G a_{\text{turbulent}} \propto (\text{velocity})^{1.8} \quad (14\text{-}181)$$

For quenching operations, this means that even though residence time is cut as gas velocity goes up, the effective approach to equilibrium increases. Since the volume for a given length of pipe falls with velocity^{-1}, the expected number of transfer units N_G in a given length of pipe increases with velocity$^{0.8}$.

$$N_{G,\text{turbulent}} \propto (h_G a_{\text{turbulent}})(\text{volume}) \propto (\text{velocity})^{0.8} \quad (14\text{-}182)$$

See Example 14-22.

EXAMPLES OF CONTACTORS

High-Velocity Pipeline Contactors High-velocity cocurrent flow can give greater power input than any other approach. This is critical when extremely high rates of reaction quenching are needed.

Example 14-22 Doubling the Velocity in a Horizontal Pipeline Contactor—Impact on Effective Heat Transfer Velocity in pipeline quench systems often exceeds 62 m/s (200 ft/s). Note that this is far above the flooding velocity in distillation packing, distillation trays, or gas-sparged reactors. There are few data available to validate performance even though liquid injection into high-velocity gas streams has been historically used in quenching reactor effluent systems. However, the designer knows the directional impact of parameters as given by Eq. (14-182).

For example, if a 10-ft length of pipe gives a 90 percent approach to equilibrium in a quench operation, Eq. (14-177) says that the back-calculated N_G is 2.303:

$$N_G - \ln(1 - 0.9) = 2.303$$

Equation (14-177) says if we double velocity but retain the same length, we expect an increase of N_G to 4.0.

$$N_G = 2.303(2)^{0.8} = 4$$

and

$$E = 1 - e^{-4} = 0.982$$

Restated, the approach to equilibrium rises from 90 percent to greater than 98 percent even though the contact time is cut in half.

Vertical Reverse Jet Contactor A surprisingly effective modification of the liquid injection quench concept is to inject the liquid countercurrent *upward* into a gas flowing *downward*, with the gas velocity at some multiple of the flooding velocity defined by Eq. (14-198). The reverse jet contactor can be envisioned as an upside-down distillation tray. For large gas volumes, multiple injection nozzles are used. One advantage of this configuration is that it minimizes the chance of liquid or gas bypassing. Another advantage is that it operates in the froth region, which generates greater area per unit volume than the higher-velocity cocurrent pipeline quench.

The concept was first outlined in U.S. Patent 3,803,805 (1974) and was amplified in U.S. Patent 6,339,169 (2002). The 1974 patent presents data that clarify that the key power input is from the gas stream.

A more recent article discusses use of the reverse jet in refinery off-gas scrubbing for the removal of both SO_2 and small particles [*Hydrocarbon Processing* **84**(9): 99–106 (2005)]. This article cites downward gas velocities in the range of 10 to 37 m/s and notes gas pressure drop in the range of 6 to 20 in of water. Removals of SO_2 and fine particles were both close to 99 percent. The froth produced by the contactor reverses direction, flows down, and is largely disengaged in a vessel mounted below.

Example 14-23 The Reverse Jet Contactor, U.S. Patent 6,339,169 This patent deals with rapid cooling and removal of NH_3 from gas exiting an acrylonitrile reactor. Liquid is injected upward. The claims suggest downward-flowing gas velocity is between 20 and 25 m/s.

Gas cooling is reported to be largely complete in 0.1 s. NH_3 removal at the exit of the contactor is reported to be greater than 99 percent. The gas is cooled by water vaporizing from the injected liquid, with total water circulated being in the range of 100 times that evaporated.

Since the gas cooling and NH_3 scrubbing move in parallel, they would be expected to achieve nearly the same approach to equilibrium as long as the pH of all the liquid stays below a key threshold. The great excess of liquid enables this.

The key is high froth interfacial area per unit volume.

Simple Spray Towers The other extreme to the pipeline and reverse jet contactors is an open vessel where spray is injected down into upflowing gas to form a rain of liquid. The advantage of simple spray towers is that they give low gas pressure drop and also tend to be nonfouling.

Even though gas velocity is well below flooding velocity, the finer droplets of the spray will be entrained. Note the wide spectrum of particle sizes shown by Fig. 14-83. However, as shown by Examples 14-24 and 14-25, they can be extremely effective in liquid-limited systems.

Bypassing Limits Spray Tower Performance in Gas Cooling As shown by Example 14-19, only modest performance is achieved in gas-limited systems. The modest efficiency is due to gas bypassing. Tall spray towers are not effective countercurrent devices. Even with nominally falling droplets, there is a great deal of backmixing because there is no stabilizing pressure drop as there would be in a column filled with packing or trays. A packet of droplets weighs more than a gas-filled space. The result is that the volume that is filled with the most droplets moves down relative to all other volumes. Similarly, the gas volume that has the fewest droplets moves up more quickly than other volumes. This generates bypassing of liquid and gas. The flows are driven by the rain of droplets themselves. Anything less than perfect distribution of liquid and gas will induce a dodging action between the flowing streams. Most designers limit expectations for spray contactors to some fraction of a single equilibrium stage regardless of height.

One approach that has been employed to get better distribution in spray systems is to mount a single large-capacity nozzle in the center of the vessel with radial discharge of large droplets. The droplets are discharged with enough velocity to penetrate to the vessel walls.

Spray Towers in Liquid-Limited Systems—Hollow-Cone Atomizing Nozzles If we follow an element of liquid leaving a hollow-cone hydraulic spray nozzle, the sequence is a rapidly thinning cone followed by wave development, followed by shedding of ligaments, followed by breakage of the ligaments into droplets. See Fig. 14-81. The sequence gives high transfer for liquid-limited systems. This results from the thin sheet of the hollow cone as well as the creation of fresh surface in the breakup process.

Devolatilizers Devolatilization systems are liquid-limited due to the combination of high liquid viscosity and removal of a component with high relative volatility. Simpson and Lynn [*AIChE J.* **23**(5): 666–673 (1977)] reported oxygen stripping from water at 98 percent complete, in less than 1 ft of contact. The concept has been employed for residual devolatilization in refineries.

Spray Towers as Direct Contact Condensers Similarly, spray contactors can be highly effective for direct contact condensers, which are also liquid-limited. The high transfer rate in the initial formation of sprays is key. Kunesh [*Ind. Engr. Chem. Res.* **32**: 2387–2389 (1993)] reported a 97 percent approach to equilibrium in a hydrocarbon system in the 6-in space below the discharge of a row of hollow-cone spray nozzles.

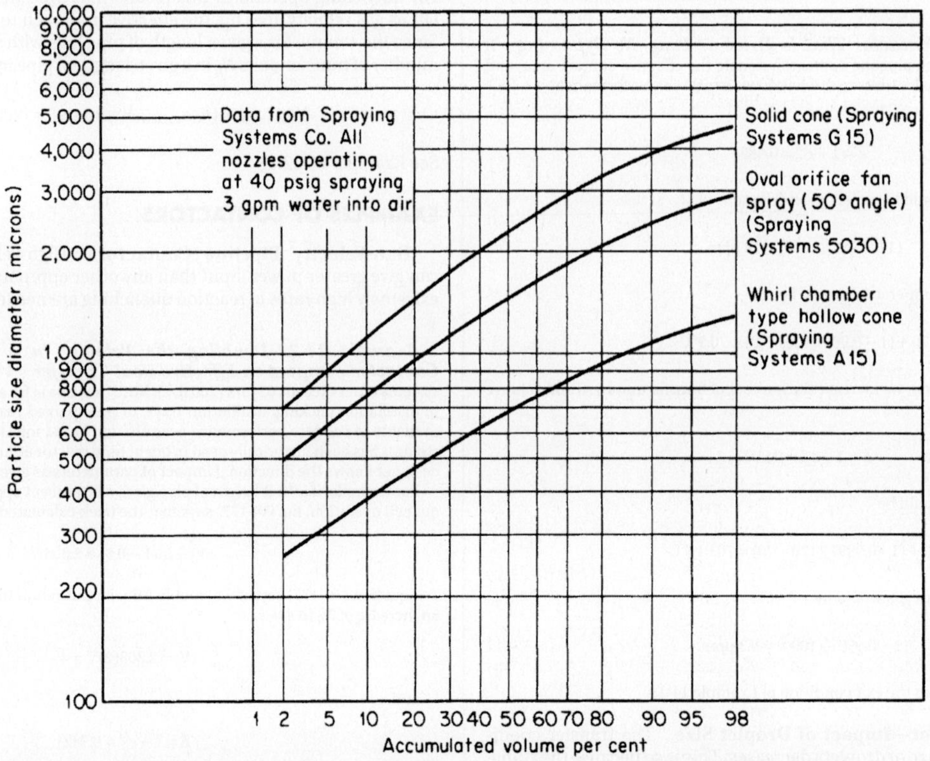

FIG. 14-83 Droplet-size distribution for three different types of nozzles. To convert pounds per square inch gauge to kilopascals, multiply by 6.89; to convert gallons per minute to cubic meters per hour, multiply by 0.227. (*Spraying Systems Inc.*)

Other results on heat transfer in a large spray condenser are given by Waintraub et al. ("Removing Packings from Heat Transfer Sections of Vacuum Towers," AIChE 2005 Spring National Meeting, *Proceedings of Topical Conference*, Apr. 10, 2005, Atlanta, Ga., p. 79). The paper highlights the importance of good gas and liquid distribution.

Converting Liquid Mass-Transfer Data to Direct Contact Heat Transfer Liquid-limited performance measures are much more commonly given for mass transfer than for heat transfer. Often mass-transfer data are reported as k_La with units of h^{-1}. This can be converted to h_La with units of Btu/(h · °F · ft³) by Eq. (14-183).

$$h_La = 187(k_La)(c_L)(\rho_L)(\mu_L/T)^{0.5} \tag{14-183}$$

where μ_L = liquid, cP
T = temperature, °R
ρ_L = liquid density, lb/ft³
c_L = liquid specific heat, Btu/(lb · °F)

The calculation of transfer units for heat transfer is relatively simple. For a liquid,

$$N_L = \frac{(\text{liquid contact time})(h_La)}{\rho_L c_L} \tag{14-184}$$

where ρ_L = liquid density and c_L = liquid specific heat. [See parallel gas expression, Eq. (14-176).]

Unlike gases, the liquid properties that control mass and heat transfer differ greatly. The key term is *diffusivity*, which for liquids drops with viscosity.

The resulting values for h_La and N_L can be surprisingly large when a pure vapor such as steam is condensed. See Example 14-24.

Example 14-24 Estimating Direct Contact Condensing Performance Based on k_La ***Mass-Transfer Data*** If an aqueous system at 560°R gives a k_La of 60 h^{-1}, what does Eq. (14-183) predict for h_La in a direct contact steam condenser?

For an aqueous system

$$\mu_L = 1 \text{ cP}$$
$$\rho_L = 62 \text{ lb/ft}^3$$
$$c_L = 1 \text{ Btu/(lb}\cdot°\text{F)}$$

and Eq. (14-183) predicts

$$h_La = 187(60)(1)(62)(1/560)^{0.5} = 29{,}400 \text{ Btu/(h}\cdot°\text{F}\cdot\text{ft}^3)$$

When a pure gas such as HCl is absorbed by low-viscosity liquid such as water, simple spray systems can also be highly effective. See Example 14-25.

Example 14-25 HCl Vent Absorber (Kister, *Distillation Troubleshooting*, Wiley, New York, 2006, p. 95) A 6-in-diameter, 8-ft-tall packed bed was giving major problems due to failure of the packing support. Water was the scrubbing fluid.

The liquid distributors were replaced with carefully positioned spray nozzles, and the packing was removed. HCl in the vent was removed to a level one-fortieth of the original design.

LIQUID-IN-GAS DISPERSIONS

Liquid Breakup into Droplets There are four basic mechanisms for breakup of liquid into droplets:

- Droplets in a field of high turbulence (i.e., high power dissipation per unit mass)
- Simple jets at low velocity
- Expanding sheets of liquid at relatively low velocity
- Droplets in a steady field of high relative velocity

These mechanisms coexist, and the one that gives the smallest drop size will control. The four mechanisms follow distinctly different velocity dependencies:

1. *Breakup in a highly turbulent field* $(1/\text{velocity})^{1.2}$. This appears to be the dominant breakup process in distillation trays in the spray regime, pneumatic atomizers, and high-velocity pipeline contactors.
2. *Breakup of a low-velocity liquid jet* $(1/\text{velocity})^0$. This governs in special applications such as prilling towers and is often an intermediate step in liquid breakup processes.
3. *Breakup of a sheet of liquid* $(1/\text{velocity})^{0.67}$. This governs drop size in most hydraulic spray nozzles.
4. *Single-droplet breakup at very high velocity* $(1/\text{velocity})^2$. This governs drop size in free fall as well as breakup when droplets impinge on solid surfaces.

Droplet Breakup—High Turbulence This is the dominant breakup mechanism for many process applications. Breakup results from local variations in turbulent pressure that distort the droplet shape. Hinze [*Am. Inst. Chem. Eng. J.* **1:** 289–295 (1953)] applied turbulence theory to obtain the form of Eq. (14-185) and took liquid–liquid data to define the coefficient:

$$D_{max} = 0.725(\sigma/\rho_G)^{0.6}/E^{0.4} \tag{14-185}$$

where E = (power dissipated)/mass — length²/time³
σ = surface tension — mass/time²
ρ_G = gas density — mass/length³

Note that D_{max} comes out with units of length. Since E typically varies with (gas velocity)³, this results in drop size dependence with $(1/\text{velocity})^{1.2}$.

The theoretical requirement for the use of Eq. (14-185) is that the microscale of turbulence << D_{max}. This is satisfied in most gas systems. For example, in three cases,

	(microscale of turbulence)/D_{max}
Distillation tray in spray regime	0.007
Pipeline @ 40 m/s and atmospheric pressure	0.012
Two-fluid atomizer using 100 m/s air	0.03

Many applications involve a three-step process with high velocity first tearing wave crests away from liquid sheets, followed by the breakup of ligaments into large droplets, followed by the breakup of the large droplets. The prediction of final droplet size based on power/mass works surprisingly well, as shown by Eqs. (14-193), (14-196), (14-197), and (14-198).

Liquid-Column Breakup Because of increased pressure at points of reduced diameter, the liquid column is inherently unstable. As a result, it breaks into small drops with no external energy input. Ideally, it forms a series of uniform drops with the size of the drops set by the fastest-growing wave. This yields a dominant droplet diameter about 1.9 times the initial diameter of the jet, as shown by Fig. 14-84. As shown, the actual breakup is quite close to prediction, although smaller satellite drops are also formed.

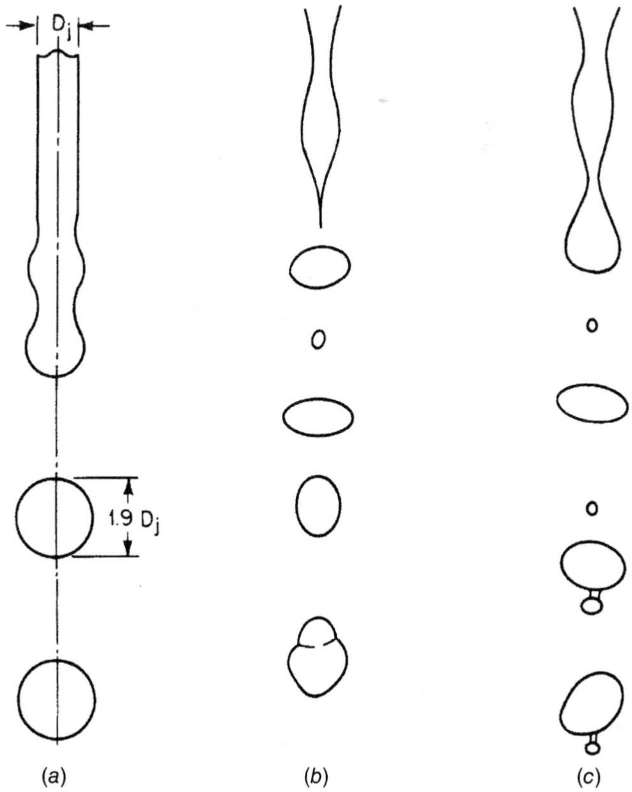

FIG. 14-84 (*a*) Idealized jet breakup suggesting uniform drop diameter and no satellites. (*b*) and (*c*) Actual breakup of a water jet as shown by high-speed photographs. [*From W. R. Marshall, "Atomization and Spray Drying," Chem. Eng. Prog. Monogr. Ser., no. 2 (1954).*]

The prime advantage of this type of breakup is the greater uniformity of drop size.

For high-viscosity liquids, the drops are larger, as shown by Eq. (14-186):

$$D = 1.9 D_j \left[1 + \frac{3\mu_\ell}{(\sigma \rho_\ell D_j)^{1/2}} \right] \tag{14-186}$$

where D = diameter of droplet
D_j = diameter of jet
μ_ℓ = viscosity of liquid
ρ_ℓ = density of liquid
σ = surface tension of liquid

These units are dimensionally consistent; any set of consistent units can be used.

As the velocity of the jet is increased, the breakup process changes and ultimately becomes a mix of various competing effects, such as the capture of small drops by bigger ones in the slowing jet and the "turbulent breakup" of the bigger drops. The high-velocity jet is occasionally used in process applications because of the very narrow spray angle (5°–20°) and the high penetration into a gas it can give. The focused stream also aids erosion of a surface.

Liquid-Sheet Breakup The basic principle of most hydraulic atomizers is to form a thin sheet that breaks by a variety of mechanisms to form ligaments of liquid, which in turn yield chains of droplets. See Fig. 14-81.

For a typical nozzle, the drop size varies with $1/(\text{pressure drop})^{1/3}$. When $(\text{velocity})^2$ is substituted for pressure drop, droplet size is seen to vary with $(\text{velocity})^{-2/3}$.

Isolated Droplet Breakup—in a Velocity Field Much effort has focused on defining the conditions under which an isolated drop will break in a velocity field. The criterion for the largest stable drop size is the ratio of aerodynamic forces to surface-tension forces defined by the Weber number, N_{We} (dimensionless):

$$N_{We\,crit} = \text{constant} = [\rho_G (\text{velocity})^2 (D_{max})/\sigma] \tag{14-187}$$

For low-viscosity fluids, $N_{We\,crit}$ commonly ranges from 10 to 20, with the larger value for a free-fall condition and the smaller for a sudden acceleration. High liquid viscosity also increases $N_{We\,crit}$.

Droplet breakup via impingement appears to follow a similar relationship, but many fewer data are available. This type of breakup can result from impingement on equipment walls or compressor blades. In general, there is less tendency to shatter on wetted surfaces.

Droplet Size Distribution Instead of the single droplet size implied by the preceding discussion, a spectrum of droplet sizes is produced. The most common ways to characterize this spectrum are as follows:

- *Volume median (mass median)* D_{vm}. This has no fundamental meaning but is easy to determine since it is at the midpoint of a cumulative-volume plot.
- *Sauter mean* D_{32}. This has the same ratio of surface to volume as the total drop population. It is typically 70 to 90 percent of D_{vm}. It is often used in transport processes and is used here to characterize drop size.
- *Maximum* D_{max}. This is the largest particle in the population. It is typically 3 to 4 times D_{32} in turbulent breakup processes, per Walzel [*International Chemical Engineering* **33:** 46 (1993)]. It is the size directly calculated from the power/mass relationship. Thus D_{32} is estimated from D_{max} by

$$D_{32} = 0.3 D_{max} \tag{14-188}$$

and D_{vm} is estimated from it by

$$D_{vm} = 0.4 D_{max} \tag{14-189}$$

However, any average drop size is fictitious, and none is completely satisfactory. For example, there is no way in which the high surface and transfer coefficients in small drops can be made available to the larger drops. Hence, a process calculation based on a given droplet size describes only what happens to that size and gives at best an approximation to the total mass.

There are a variety of ways to describe the droplet population. Figures 14-83 and 14-85 illustrate one of the most common methods, the plot of cumulative volume versus droplet size on log-normal graph paper. This satisfies the restraint of not extrapolating to a negative drop size. Its other advantages are that it is easy to plot, the results are easy to visualize, and it yields a nearly straight line at lower drop sizes.

Cumulative volume over the range of 1 to 50 percent can also be shown to vary approximately as D^2. This is equivalent to finding that the number

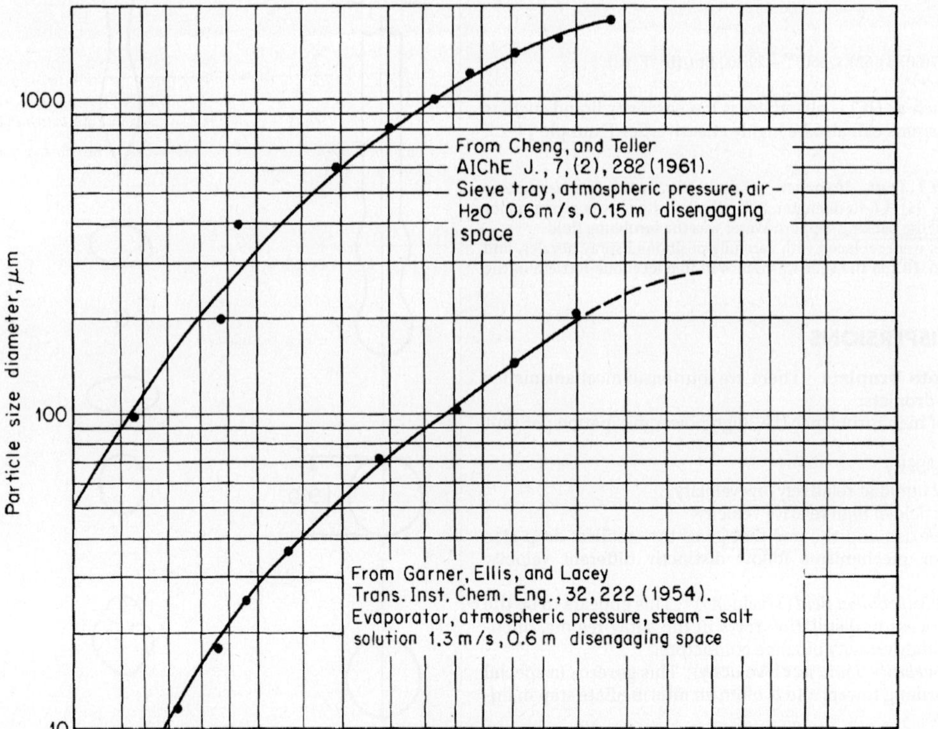

FIG. 14-85 Entrainment droplet-size distribution. To convert meters per second to feet per second, multiply by 3.28; to convert meters to feet, multiply by 3.28.

of droplets of a given size is inversely proportional to the droplet area or the surface energy of the droplet.

Atomizers The common need to disperse a liquid into a gas has spawned a large variety of mechanical devices. The different designs emphasize different advantages, such as freedom from plugging, pattern of spray, small droplet size, uniformity of spray, high turndown ratio, and low power consumption.

As shown in Table 14-17, most atomizers fall into three categories:

1. Pressure nozzles (hydraulic)
2. Two-fluid nozzles (pneumatic)
3. Rotary devices (spinning cups, disks, or vaned wheels)

These share certain features, such as relatively low efficiency and low cost relative to most process equipment. The energy required to produce the increase in area is typically less than 0.1 percent of the total energy consumption. This is so because atomization is a secondary process resulting from high interfacial shear or turbulence. As droplet sizes decrease, this efficiency decreases.

Other types are available that use sonic energy (from gas streams), ultrasonic energy (electronic), and electrostatic energy, but they are less commonly used in process industries. See Table 14-17 for a summary of the advantages and disadvantages of the different types of units. An expanded discussion is given by Lipp (*Practical Spray Technology*, Lake Innovation LLC, Lake Junaluska, N.C., 2012). Special requirements such as size uniformity in prilling towers can dictate still other approaches to dispersion. Here plates are drilled with many holes to develop nearly uniform columns.

Commonly, the most important feature of a nozzle is the size of droplet it produces. Since the heat or mass transfer that a given dispersion can produce is often proportional to $(1/D_d)^2$, fine drops are usually favored. At the other extreme, drops that are too fine will not settle, and a concern is the amount of liquid that will be entrained from a given spray operation. For example, if sprays are used to contact atmospheric air flowing at 1.5 m/s, then drops smaller than 350 mm [terminal velocity = 1.5 m/s (4.92 ft/s)] will be entrained. Even for the relative coarse spray of the hollow-cone nozzle shown in Fig. 14-83, 7.5 percent of the total liquid mass will be entrained.

Hydraulic (Pressure) Nozzles Manufacturers' data such as those shown by Fig. 14-83 are available for most nozzles for the air-water system. In Fig. 14-83, note the much coarser solid-cone spray. The coarseness results from the less uniform discharge.

Effect of Physical Properties on Drop Size Because of the extreme variety of available geometries, no attempt to encompass this variable is made here. The suggested predictive route starts with air-water droplet size data from the manufacturer at the chosen flow rate. This drop size is then corrected by Eq. (14-190) for different viscosity and surface tension:

$$\frac{D_{vm,\,\text{system}}}{D_{vm,\,\text{water}}} = \left(\frac{\sigma_{\text{system}}}{73}\right)^{0.25}\left(\frac{\mu_\ell}{1.0}\right)^{0.25} \tag{14-190}$$

where D_{vm} = volume median droplet diameter
σ = surface tension, mN/m (dyn/cm)
μ_ℓ = liquid viscosity, mPa·s (cP)

The exponential dependencies in Eq. (14-190) represent averages of values reported by a number of studies, with particular weight given to Lefebvre

TABLE 14-17 Atomizer Summary

Types of atomizer	Design features	Advantages	Disadvantages
Pressure.	Flow $\alpha(\Delta P/\rho_\ell)^{1/2}$. Only source of energy is from fluid being atomized.	Simplicity and low cost.	Limited tolerance for solids; uncertain spray with high-viscosity liquids; susceptible to erosion. Need for special designs (e.g., bypass) to achieve turndown.
1. Hollow cone.	Liquid leaves as conical sheet as a result of centrifugal motion of liquid. Air core extends into nozzle.	High atomization efficiency.	Concentrated spray pattern at cone boundaries.
a. Whirl chamber (see Fig. 14-86*a*).	Centrifugal motion developed by tangential inlet in chamber upstream of orifice.	Minimum opportunity for plugging.	
b. Grooved core.	Centrifugal motion developed by inserts in chamber.	Smaller spray angle than 1*a* and ability to handle flows smaller than 1*a*.	
2. Solid cone (see Fig. 14-86*b*).	Similar to hollow cone but with insert to provide even distribution.	More uniform spatial pattern than hollow cone.	Coarser drops for comparable flows and pressure drops. Failure to yield same pattern with different fluids.
3. Fan (flat) spray.	Liquid leaves as a flat sheet or flattened ellipse.	Flat pattern is useful for coating surfaces and for injection into streams.	Small clearances.
a. Oval or rectangular orifice (see Fig. 14-86*c*). Numerous variants on cavity and groove exist.	Combination of cavity and orifice produces two streams that impinge within the nozzle.		
b. Deflector (see Fig. 14-86*d*).	Liquid from plain circular orifice impinges on curved deflector.	Minimal plugging.	Coarser drops.
c. Impinging jets (see Fig. 14-86*e*).	Two jets collide outside nozzle and produce a sheet perpendicular to their plane.	Different liquids are isolated until they mix outside of orifice. Can produce a flat circular sheet when jets impinge at 180°.	Extreme care needed to align jets.
4. Nozzles with wider range of turndown.			
a. Spill (bypass) (see Fig. 14-86*f*).	A portion of the liquid is recirculated after going through the swirl chamber.	Achieves uniform hollow cone atomization pattern with very high turndown (50:1).	Waste of energy in bypass stream. Added piping for spill flow.
b. Poppet (see Fig. 14-86*g*).	Conical sheet is developed by flow between orifice and poppet. Increased pressure causes poppet to move out and increase flow area.	Simplest control over broad range.	Difficult to maintain proper clearances.
Two-fluid (see Fig. 14-86*h*).	Gas impinges coaxially and supplies energy for breakup.	High velocities can be achieved at lower pressures because the gas is the high-velocity stream. Liquid-flow passages can be large, and hence plugging can be minimized.	Because gas is also accelerated, efficiency is inherently lower than pressure nozzles.
Sonic.	Gas generates an intense sound field into which liquid is directed.	Similar to two-fluid but with greater tolerance for solids.	Similar to two-fluid.
Rotary wheels (see Fig. 14-86*i*) disks, and cups.	Liquid is fed to a rotating surface and spreads in a uniform film. Flat disks, disks with vanes, and bowl-shaped cups are used. Liquid is thrown out at 90° to the axis.	The velocity that determines drop size is independent of flow. Hence these can handle a wide range of rates. They can also tolerate very viscous materials as well as slurries. Can achieve very high capacity in a single unit; does not require a high-pressure pump.	Mechanical complexity of rotating equipment. Radial discharge.
Ultrasound.	Liquid is fed over a surface vibrating at a frequency > 20 kHz.	Fine atomization, small size, and low injection velocity.	Low flow rate and need for ultrasound generator.

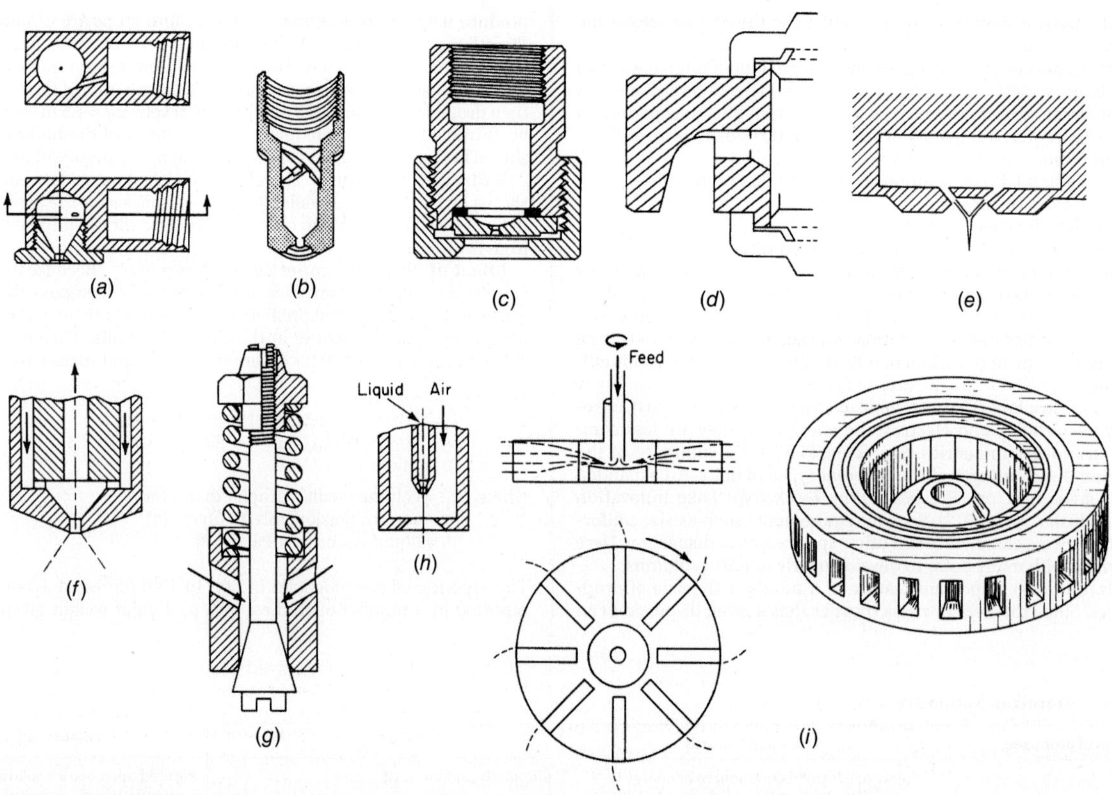

FIG. 14-86 Characteristic spray nozzles. (*a*) Whirl-chamber hollow cone. (*b*) Solid cone. (*c*) Oval-orifice fan. (*d*) Deflector jet. (*e*) Impinging jet. (*f*) Bypass. (*g*) Poppet. (*h*) Two-fluid. (*i*) Vaned rotating wheel.

(*Atomization and Sprays*, Hemisphere, New York, 1989). Since viscosity can vary over a much broader range than surface tension can, viscosity has much greater leverage over drop size. For example, it is common to find an oil with 1000 times the viscosity of water, while most liquids fall within a factor of 3 of their surface tension. Liquid density is generally even closer to that of water, and since the data are not clear that a liquid density correction is needed, none is shown in Eq. (14-190). Vapor density also has an impact on drop size, but the impact is complex, involving conflicts of a number of effects, and vapor density is commonly omitted in atomizer drop size correlations.

Effect of Pressure Drop and Nozzle Size For a nozzle with a developed pattern, the average drop size can be estimated to fall with rising ΔP (pressure drop) by Eq. (14-191):

$$\frac{D_1}{D_2} = \left(\frac{\Delta P_2}{\Delta P_1}\right)^{1/3} \tag{14-191}$$

For similar nozzles and constant ΔP, the drop size will increase with nozzle size as indicated by Eq. (14-192):

$$\frac{D_1}{D_2} = \left(\frac{\text{orifice diameter}_1}{\text{orifice diameter}_2}\right)^{1/2} \tag{14-192}$$

Once again, these relationships are averages of a number of reported values and are intended as rough guides.

The normal operating regime is well below the turbulent breakup velocity. However, the data of Kennedy [*J. of Engineering for Gas Turbines and Power* **108:** 191 (1986)] at very high pressure drop in large nozzles shows a shift to a higher dependence on pressure drop. These data suggest that turbulent droplet breakup can also be governing with hydraulic spray nozzles, although this is unusual.

Spray Angle A shift to a smaller-angle nozzle gives slightly larger drops for a given type of nozzle because of the reduced tendency of the sheet to thin. Dietrich [*Proc. 1st Int. Conf. Liq. Atomization Spray Systems*, Tokyo, 1978] shows the following:

Angle	25°	50°	65°	80°	95°
D_{vm}, μm	1459	1226	988	808	771

In calculating the impact point of spray, one should recognize that the spray angle closes in as the spray moves away from the nozzle. This is caused by a loss of momentum of the spray to the gas.

At some low flow, pressure nozzles do not develop their normal pattern but tend to approach solid streams. The required flow to achieve the normal pattern increases with viscosity.

Two-Fluid (Pneumatic) Atomizers This general category includes such diverse applications as venturi atomizers and reactor-effluent quench systems, in addition to two-fluid spray nozzles. Depending on the manner in which the two fluids meet, several of the breakup mechanisms may be applicable, but the final one is high-level turbulent rupture.

As shown by Table 14-18, empirical correlations for two-fluid atomization show a dependence on high gas velocity to supply atomizing energy, usually to a power dependence close to that for turbulent breakup. In addition, the correlations show a dependence on the ratio of gas to liquid and system dimension.

Further differences from hydraulic nozzles (controlled by sheet and ligament breakup) are the stronger increase in drop size with increasing surface tension and decreasing gas density.

TABLE 14-18 Exponential Dependence of Drop Size on Different Parameters in Two-Fluid Atomization

	Relative velocity	Surface tension	Gas density	$1 + L/G$	Atomizer dimension
Jasuja (empirical for small nozzle)	−0.9	0.45	−0.45	0.5	0.55
El-Shanawany and Lefebvre (empirical for small nozzle)	−1.2	0.6	−0.7	1	0.40
Tatterson, Dallman, and Hanratty (pipe flow)	−1	0.5	−0.5		0.5
Power/mass	−1.2	0.6	−0.6	0.4	0.4

The similarity of these dependencies to Eq. (14-185) led to a reformulation with two added terms that arise naturally from the theory of power dissipation per unit mass. The result is Eq. (14-193), which is labeled power/mass in Table 14-18.

$$D_{32} = 0.29\left(\frac{\sigma}{\rho_G}\right)^{0.6} (1/\text{velocity})^{1.2}\left(1+\frac{L}{G}\right)^{0.4} (D_{\text{nozzle}})^{0.4} \qquad (14\text{-}193)$$

where σ = surface tension
ρ_G = gas density
L/G = mass ratio of liquid flow to gas flow
D_{nozzle} = diameter of the air discharge

This is remarkably similar to the empirical two-fluid atomizer relationships of El-Shanawany and Lefebvre [*J. Energy* **4:** 184 (1980)] and Jasuja [*Trans. Am. Soc. Mech. Engr.* **103:** 514 (1981)]. For example, El-Shanawany and Lefebvre give a relationship for a prefilming atomizer:

$$D_{32} = 0.0711(\sigma/\rho_G)^{0.6}(1/\text{velocity})^{1.2}(1 + L/G)(D_{\text{nozzle}})^{0.4}(\rho_L/\rho_G)^{0.1} + 0.015[(\mu_L)^2/(\sigma \times \rho_L)]^{0.5}(D_{\text{nozzle}})^{0.5}(1 + L/G) \qquad (14\text{-}194)$$

where μ_L is liquid viscosity.

According to Jasuja,

$$D_{32} = 0.17(\sigma/\rho_G)^{0.45}(1/\text{velocity})^{0.9}(1 + L/G)^{0.5}(D_{\text{nozzle}})^{0.55} + \text{viscosity term} \qquad (14\text{-}195)$$

[Equations (14-193), (14-194), and (14-195) are dimensionally consistent; any set of consistent units on the right-hand side yields the droplet size in units of length on the left-hand side.]

The second, additive term carrying the viscosity impact in Eq. (14-194) is small at viscosities around 1 cP but can become controlling as viscosity increases. For example, for air at atmospheric pressure atomizing water, with nozzle conditions

$$D_{\text{nozzle}} = 0.076 \text{ m (3 in)}$$
$$\text{Velocity} = 100 \text{ m/s (328 ft/s)}$$
$$L/G = 1$$

El-Shanaway measured 70 μm, and his Eq. (14-194) predicted 76 μm. The power/mass correlation [Eq. (14-193)] predicts 102 μm. The agreement between both correlations and the measurement is much better than is normally achieved.

Rotary Atomizers For rotating wheels, vaneless disks, and cups, there are three regimes of operation. At low rates, the liquid is shed directly as drops from the rim. At intermediate rates, the liquid leaves the rim as threads, and at the highest rate, the liquid extends from the edge as a thin sheet that breaks down in a manner similar to a fan or hollow-cone spray nozzle. As noted in Table 14-17, rotary devices have many unique advantages, such as the ability to handle high viscosity and slurries and to produce small droplets without high pressures. The prime applications are in spray drying. See Masters (*Spray Drying Handbook*, Wiley, New York, 1991) for more details.

Pipeline Contactors The correlation for droplet diameter based on power/mass is similar to that for two-fluid nozzles. The dimensionless correlation is

$$D_{32} = 0.8(\sigma/\rho_G)^{0.6}(1/\text{velocity})^{1.2}(D_{\text{pipe}})^{0.4} \qquad (14\text{-}196)$$

(The relation is dimensionally consistent; any set of consistent units on the right-hand side yields the droplet size in units of length on the left-hand side.)

The relationship is similar to the empirical correlation of Tatterson, Dallman, and Hanratty [*Am. Inst. Chem. Eng. J.* **23**(1): 68 (1977)]

$$D_{32} \sim \left(\frac{\sigma}{\rho_G}\right)^{0.5} (1/\text{velocity})^1 (D_{\text{pipe}})^{0.5}$$

Predictions from Eq. (14-196) align well with the Tatterson data. For example, for a velocity of 43 m/s (140 ft/s) in a 0.05-m (1.8-in) equivalent diameter channel, Eq. (14-196) predicts D_{32} of 490 μm, compared to the measured 460 to 480 μm.

Entrainment Due to Gas Bubbling/Jetting Through a Liquid Entrainment generally limits the capacity of distillation trays and is commonly a concern in vaporizers and evaporators. Fortunately, it is readily controllable by simple inertial entrainment capture devices such as wire-mesh pads in gravity separators.

In distillation towers, entrainment lowers the tray efficiency, and 1 lb of entrainment per 10 lb of liquid is sometimes taken as the limit for acceptable performance. However, the impact of entrainment on distillation efficiency depends on the relative volatility of the component being considered. Entrainment has a minor impact on close separations when the difference between vapor and liquid concentrations is small, but this factor can be dominant for systems where the liquid concentration is much higher than the vapor in equilibrium with it (i.e., when a component of the liquid has a very low volatility, as in an absorber).

As shown by Fig. 14-85, entrainment droplet sizes span a broad range. The reason for the much larger drop sizes of the upper curve is the short disengaging space. For this curve, over 99 percent of the entrainment has a terminal velocity greater than the vapor velocity. For contrast, in the lower curve the terminal velocity of the largest particle reported is the same as the vapor velocity. For the settling velocity to limit the maximum drop size entrained, at least 0.8 m (30 in) of disengaging space is usually required. Note that even for the lower curve, less than 10 percent of the entrainment is in drops of less than 50 μm. The coarseness results from the relatively low power dissipation per mass on distillation trays. This means that it is relatively easy to remove by a device such as a wire-mesh pad. Over 50 percent is typically captured by the underside of the next-higher tray or by a turn in the piping leaving an evaporator. Conversely, though small on a mass basis, the smaller drops are extremely numerous. On a number basis, more than one-half of the drops in the lower curve are under 5 μm. These can serve as nuclei for fog condensation in downstream equipment.

Entrainment E is inherent in the bubbling process and can stem from a variety of sources, as shown by Fig. 14-87. However, the biggest practical problem is entrainment generated by the kinetic energy of the flowing vapor rather than the bubbling process. As the vapor velocity approaches the flooding limit [Eq. (14-197)], the entrainment rises approximately with $(\text{velocity})^8$.

Pinczewski and Fell [*Trans. Inst. Chem Eng.* **55:** 46 (1977)] show that the velocity at which vapor jets onto the tray sets the droplet size, rather than the superficial tray velocity. The power/mass correlation predicts an average drop size close to that measured by Pinczewski and Fell. Combination of this prediction with the estimated fraction of the droplets entrained gave a relationship for entrainment, Eq. (14-197). The dependence of entrainment with the eighth power of velocity even approximates the observed velocity dependence, as flooding is approached.

$$E = \frac{\text{constant }(\text{velocity})^8(\rho_G)^4}{(A_f)^3(\rho_L - \rho_G)^{2.5}\sigma^{1.5}} \qquad (14\text{-}197)$$

(Here E is the mass of entrainment per mass of vapor and A_f is the fractional open area on the tray.)

When flooding is defined as the condition that gives E of 1, the flood velocity is estimated by Eq. (14-198).

$$U_{\text{flood}} = \frac{1.25(A_f)^{3/8}[g(\rho_L - \rho_G)]^{0.3125}\sigma^{0.1875}}{\rho_G^{0.5}} \qquad (14\text{-}198)$$

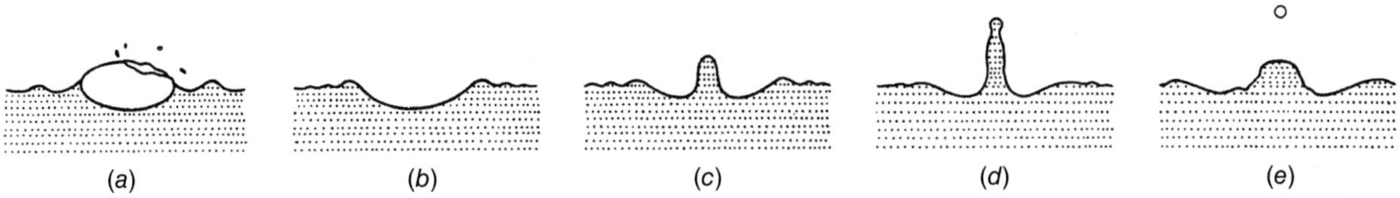

FIG. 14-87 Mechanism of the burst of an air bubble on the surface of water. [*Newitt, Dombrowski, and Knellman,* Trans. Inst. Chem. Eng. ***32:** 244 (1954).*]

The relationship is dimensionally consistent; any set of consistent units on the right-hand side yields velocity units on the left-hand side. It is similar in form to Eq. (14-168) and provides a conceptual framework for understanding the ultimate distillation column capacity concept.

"Upper Limit" Flooding in Vertical Tubes If, instead of a gas jet being injected into a liquid as in distillation, the liquid runs down the walls and the gas moves up the center of the tube, then higher velocities can be achieved than those shown by Eq. (14-198). This application is important in the design of vertical condensers.

Maharudrayya and Jayanti [*AIChE J.* **48**(2): 212–220 (2002)] show that peak pressure drop in a 25-mm vertical tube occurs at a value close to that predicted by Eq. (14-198). At this velocity, about 20 percent of the injected liquid is being entrained out of the top of the tube. However, the condition where essentially all liquid was entrained didn't occur until a velocity more than twice the value estimated from Eq. (14-198).

The higher velocities at modest entrainment observed by Maharudrayya and Jayanti were obtained with special smooth entry of gas (and exit of liquid) at the bottom of the tube. Hewitt (*Handbook of Heat Exchanger Design,* Begell House, 1992, pp. 2.3.2–2.3.23) suggests that these values should be derated by at least 35 percent for more typical sharp heat exchanger tube entry. Similar to the smooth entry effect, other data suggest that countercurrent capacity can be increased by providing an extension of the tube below the tube sheet, with the bottom of the extension cut on a steep angle (>60°) to the horizontal. The tapered extension facilitates the drainage of liquid.

An extensive data bank correlated by Diehl and Koppany [*Chem. Eng. Prog. Symp. Ser.* **65:** 77–83 (1965)] also gave higher allowable entry velocities than Eq. (14-198). Diehl and Koppany's correlation [Eq. (14-199)] is dimensional and appears to give a much higher dependence on σ than the more recent work. However, for many fluids, $\sigma^{0.5}$ is essentially the same as the combination $\sigma^{0.1875}(\rho_L - \rho_g)^{0.3125}$ that appears in Eq. (14-198). Hence Eq. (14-199) gives a similar physical property dependence.

$$U_f = F_1 F_2 (\sigma/\rho_g)^{0.5} \tag{14-199}$$

where U_f = flooding gas velocity, m/s
$F_1 = 1.22$ when $3.2d_i/\sigma > 1.0$
$= 1.22\,(3.2\,d_i/\sigma)^{0.4}$ when $3.2d_i/\sigma < 1.0$
$F_2 = (G/L)^{0.25}$
G/L = gas–liquid mass ratio
d_i = inside diameter of column, mm
σ = surface tension, mN/m (dyn/cm)
ρ_g = gas density, kg/m^3

The primary reason for citing Eq. (14-199) is the large successful experience base in practical applications. Note that the reduction in allowable gas velocity for small diameters given by the F_1 factor is conceptually the same as the effect of using smaller-diameter packing in distillation. Note also that over the range of G/L between 1 and 0.1, the Maharudrayya and Jayanti data show a similar reduction in allowable gas rate to the F_2 factor in Eq. (14-199). The phenomenon behind this is that a thicker liquid film on the tube wall is more easily entrained.

While the limiting phenomenon of upper-limit flooding in a vertical pipe is similar to *ultimate capacity* in distillation, there is a distinct difference. *Upper limit* in a vertical pipe applies to a design where a conscious effort should be made to minimize gas–liquid contact. Carried to extremes, it would involve separate tubes for liquid flowing down and vapor going up. In contrast, *ultimate capacity* in a distillation column corresponds to the condition where effective mass transfer disappears due to high entrainment. One could force more vapor up through the contactor, but fractionation would be poor.

Fog Condensation—The Other Way to Make Little Droplets For a variety of reasons, a gas or vapor can become supersaturated with a condensible component. Surface tension and mass transfer impose barriers on immediate condensation, so the growth of fog particles lags behind what equilibrium predicts. Droplets formed by fog condensation are usually much finer (0.1 to 10 μm) than those formed by mechanical breakup and hence more difficult to collect. Sometimes fog can be a serious problem, as in the atmospheric discharge of a valuable or a hazardous material. More commonly, fog is a curiosity rather than a dominating element in chemical processing.

Fog particles grow because of excess saturation in the gas. Usually this means that the gas is supersaturated (i.e., it is below its dew point). Sometimes, fog can also grow on soluble foreign nuclei at partial pressures below saturation. Increased saturation can occur through a variety of routes:

1. Mixing of two saturated streams at different temperatures. This is commonly seen in the plume from a stack. Since vapor pressure is an exponential function of temperature, the resultant mixture of two saturated streams will be supersaturated at the mixed temperature. Uneven flow patterns and cooling in heat exchangers make this route to supersaturation difficult to prevent.
2. Increased partial pressure due to reaction. An example is the reaction of SO_3 and H_2O to yield H_2SO_4, which has much lower vapor pressure than its components.
3. Isentropic expansion (cooling) of a gas, as in a steam nozzle.
4. Cooling of a gas containing a condensible vapor. Here the problem is that the gas cools faster than condensible vapor can be removed by mass transfer.

These mechanisms can be observed in many common situations. For example, fog via mixing can be seen in the discharge of breath on a cold day. Fog via adiabatic expansion can be seen in the low-pressure area over the wing of an airplane landing on a humid summer day. Fog via condensation can be seen in the exhaust from an automobile air conditioner (if you follow closely enough behind another car to pick up the ions or NO molecules needed for nucleation). All these occur at a very low supersaturation and appear to be keyed to an abundance of foreign nuclei. All these fogs also quickly dissipate as heat or unsaturated gas is added.

The supersaturation in condensers arises for two reasons. First, the condensible vapor is generally of higher molecular weight than the noncondensible gas. This means that the molecular diffusivity of the vapor will be much less than the thermal diffusivity of the gas. Restated, the ratio of N_{Sc}/N_{Pr} is greater than 1. The result is that a condenser yields more heat-transfer units $dT_g/(T_g - T_i)$ than mass-transfer units $dY_g/(Y_g - Y_i)$. Second, both transfer processes derive their driving force from the temperature difference between the gas T_g and the interface T_i. Each incremental decrease in interface temperature yields the same relative increase in temperature driving force. However, the interface vapor pressure can only approach the limit of 0. Because of this, for equal molecular and thermal diffusivities, a saturated mixture will supersaturate when cooled. The tendency to supersaturate generally increases with increased molecular weight of the condensible, increased temperature differences, and reduced initial superheating. To evaluate whether a given condensing step yields fog requires rigorous treatment of the coupled heat-transfer and mass-transfer processes through the entire condensation. Steinmeyer [*Chem. Eng. Prog.* **68**(7): 64 (1972)] illustrates this, showing the impact of foreign-nuclei concentration on calculated fog formation. See Table 14-19. Note the relatively large particles generated for cases 1 and 2 for 10,000 foreign nuclei per cubic centimeter. These are large enough to be fairly easily collected. There have been very few documented problems with industrial condensers despite the fact that most calculate to generate supersaturation along the condensing path. The explanation appears to be a limited supply of foreign nuclei.

Ryan et al. [*Chem. Eng. Progr.* **90**(8): 83 (1994)] show that separate mass- and heat-transfer rate modeling of an HCl absorber predicts 2 percent fog in the vapor. The impact is equivalent to lowering the stage efficiency to 20 percent.

Spontaneous (Homogeneous) Nucleation This process is quite difficult because of the energy barrier associated with creation of the interfacial area. It can be treated as a kinetic process with the rate a very steep function of the supersaturation ratio (S = partial pressure of condensible per vapor pressure at gas temperature). For water, an increase in S from 3.4 to

TABLE 14-19 Simulation of Three Heat Exchangers with Varying Foreign Nuclei

	1	2	3
Weight fraction, noncondensable			
Inlet	0.51	0.42	0.02
Outlet	0.80	0.80	0.32
Molecular weight			
Inert	28	29	29
Condensable	86	99	210
Temperature difference between gas and liquid interface, K			
Inlet	14	24	67
Outlet	4	10	4
Percent of liquid that leaves unit as fog Nuclei concentration in inlet particles/cm^3			
100	0.05	1.1	2.2
1,000	0.44	5.6	3.9
10,000	3.2	9.8	4.9
100,000	9.6	11.4	5.1
1,000,000	13.3	11.6	
10,000,000	14.7		
∞	14.7	11.8	5.1
Fog particle size based on 10,000 nuclei/cm^3 at inlet, μm	28	25	4

TABLE 14-20 Experimental Critical Supersaturation Ratios

	Temperature, K	S_{crit}
H_2O	264	4.91
C_2H_5OH	275	2.13
CH_3OH	264	3.55
C_6H_6	253	5.32
CCl_4	247	6.5
$CHCl_3$	258	3.73
C_6H_5Cl	250	9.5

3.9 causes a 10,000-fold increase in the nucleation rate. As a result, below a critical supersaturation (S_{crit}), homogeneous nucleation is slow enough to be ignored. Generally, S_{crit} is defined as that which limits nucleation to one particle produced per cubic centimeter per second. It can be estimated roughly by traditional theory (*Theory of Fog Condensation,* Israel Program for Scientific Translations, Jerusalem, 1967) using the following equation:

$$S_{crit} = \exp\left[0.56\frac{M}{\rho_l}\left(\frac{\sigma}{T}\right)^{3/2}\right] \tag{14-200}$$

where σ = surface tension, mN/m (dyn/cm)
ρ_l = liquid density, g/cm^3
T = temperature, K
M = molecular weight of condensible

Table 14-20 shows typical experimental values of S_{crit} taken from the work of Russel [*J. Chem. Phys.* **50:** 1809 (1969)]. Since the critical supersaturation ratio for homogeneous nucleation is typically greater than 3, it is not often reached in process equipment. However, fog formation is typically found in steam turbines. Gyarmathy [*Proc. Inst. Mech. E., Part A: J. Power and Energy* **219**(A6): 511–521 (2005)] reports fog in the range 3.5 to 5 percent of total steam flow, with average fog diameter in the range of 0.1 to 0.2 μm.

Growth on Foreign Nuclei As previously noted, foreign nuclei are often present in abundance and permit fog formation at much lower supersaturation. For example,

1. *Solids.* Surveys have shown that air contains thousands of particles per cubic centimeter in the 0.1-μm to 1-μm range suitable for nuclei. The sources range from ocean-generated salt spray to combustion processes. The concentration is highest in large cities and industrial regions. When the foreign nuclei are soluble in the fog, nucleation occurs at S values very close to 1.0. This is the mechanism controlling atmospheric water condensation. Even when not soluble, a foreign particle is an effective nucleus if wet by the liquid. Thus, a 1-μm insoluble particle with zero contact angle requires an S of only 1.001 to serve as a condensation site for water.
2. *Ions.* Amelin (*Theory of Fog Condensation,* Israel Program for Scientific Translations, Jerusalem, 1967) reports that ordinary air contains even higher concentrations of ions. These ions also reduce the required critical supersaturation, but by only about 10 to 20 percent, unless multiple charges are present.
3. *Entrained liquids.* The production of small droplets is inherent in the bubbling process, as shown by Fig. 14-86. Values range from near zero to 10,000/cm^3 of vapor, depending on how the vapor breaks through the liquid and on the opportunity for evaporation of the small drops after entrainment.

As a result of these mechanisms, most process streams contain enough foreign nuclei to cause some fogging. While fogging has been reported in only a relatively low percentage of process partial condensers, it is rarely looked for and volunteers its presence only when yield losses or pollution is intolerable.

Drop Size Distribution Monodisperse (nearly uniform droplet size) fogs can be grown by providing a long retention time for growth. However, industrial fogs usually show a broad distribution, as in Fig. 14-88. Note also that for this set of data, the sizes are several orders of magnitude smaller than those shown earlier for entrainment and atomizers.

The result, as discussed in a later subsection, is a demand for different removal devices for the small particles.

While generally fog formation is a nuisance, it can occasionally be useful because of the high surface area generated by the fine drops. An example is insecticide application.

GAS-IN-LIQUID DISPERSIONS

General References: Design methods for agitated vessels are presented by Penney in Couper et al., *Chemical Process Equipment, Selection and Design,* 3d ed., chap. 10, Elsevier, New York, 2005. A comprehensive review of all industrial fluid mixing technology is given by Paul et al., *Handbook of Industrial Mixing,* Wiley, Hoboken, N.J., 2004 and by Kresta et al., *Advances in Industrial Mixing: A Companion to the Handbook of Industrial Mixing,* Wiley, Hoboken, N.J., 2015. Comprehensive treatments of bubbles or foams are given by Akers, *Foams: Symposium 1975,* Academic Press, New York, 1973; and Exerowa and Kruglyakov, *Foam and Foam Films,* Elsevier, New York, 1998. The formation of bubbles is comprehensively treated by Clift et al., *Bubbles, Drops and Particles,* Academic Press, New York, 1978, and the literature is reviewed by Kulkarni and Joshi [*Ind. Eng. Chem. Res.* **44:** 5873 (2005)]. Design methods for unit operation in bubble columns and stirred vessels are covered by Lemoine and Morsi [*Chem. Eng. J.* **114:** 9 (2005)].

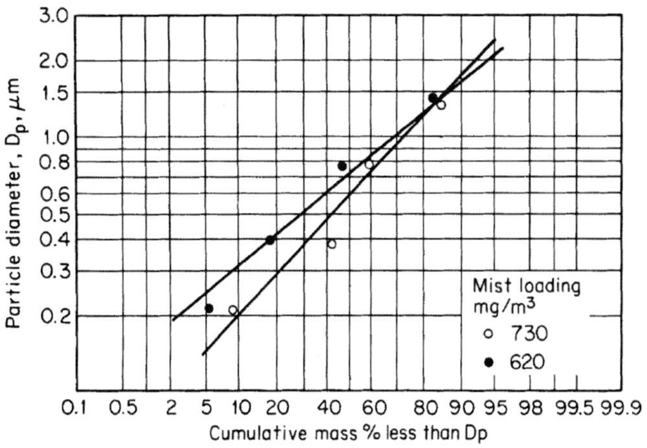

FIG 14-88 Particle-size distribution and mist loading from absorption tower in an H_2SO_4 plant [*Gillispie and Johnstone,* Chem. Eng. Prog. *51(2): 74 (1955).*]

A review of foam rheology is given by Herzhaft [*Oil & Gas Sci. & Technol.* **54:** 587 (1999)] and Heller and Kuntamukkula [*Ind. Eng. Chem. Res.* **26:** 318 (1987)]. The influence of surface-active agents on bubbles and foams is summarized in selected passages from Schwartz and Perry, *Surface Active Agents,* vol. 1, Interscience, New York, 1949; and from Schwartz, Perry, and Berch, *Surface Active Agents and Detergents,* vol. 2, Interscience, New York, 1958.

A review of foam stability also is given by de Vries and Meded [*Rubber Sticht. Delft.* No. 328, 1957]. Foam-separation methodology is discussed by Aguoyo and Lemlich [*Ind. Eng. Chem. Process Des. Dev.* **13:** 153 (1974)] and Lemlich [*Ind. Eng Chem.* **60:** 16 (1968)].

Prior references, which can be useful, are given in previous editions of this handbook; the most helpful is the eighth edition, Sec. 14, McGraw-Hill, New York, 2008.

Objectives of Gas Dispersion The dispersion of gas as bubbles in a liquid or in a plastic mass is effected for one of the following purposes: (1) gas–liquid contacting (to promote absorption or stripping, with or without chemical reaction), (2) agitation of the liquid phase, or (3) foam or froth production. Gas-in-liquid dispersions also may be produced or encountered inadvertently, sometimes undesirably.

Gas–Liquid Contacting Usually this is accomplished with conventional columns or with spray absorbers (see preceding subsection, Liquid-in-Gas Dispersions). For systems containing solids or tar likely to plug columns, absorptions accomplished by strongly exothermic reactions, or treatments involving a readily soluble gas or a condensable vapor, however, bubble columns or agitated vessels may be used to advantage.

Agitation Agitation by a stream of gas bubbles (often air) rising through a liquid is often employed when the extra expense of mechanical agitation is not justified. Gas spargers may be used for simple blending operations involving a liquid of low volatility or for applications where agitator shaft sealing is difficult.

Foam Production This is important in froth-flotation separations; in the manufacture of cellular elastomers, plastics, and glass; and in certain special applications (e.g., food products, fire extinguishers). Unwanted foam can occur in process columns, in agitated vessels, and in reactors in which a gaseous product is formed; it must be avoided, destroyed, or controlled. Berkman and Egloff (*Emulsions and Foams,* Reinhold, New York, 1941, pp. 112–152) have mentioned that foam is produced only in systems that have the proper combination of interfacial tension, viscosity, volatility, and concentration of solute or suspended solids. From the standpoint of gas comminution, foam production requires the creation of small bubbles in a liquid capable of sustaining foam.

Theory of Bubble and Foam Formation Foam is a group of bubbles separated from one another by thin films, the aggregation having a finite static life. Although nontechnical dictionaries do not distinguish between foam and froth, a technical distinction is often made. A highly concentrated dispersion of bubbles in a liquid is considered a froth even if its static life is substantially nil (i.e., it must be dynamically maintained). Thus, all foams are also froths, whereas the reverse is not true. The thin walls of bubbles comprising a foam are called *laminae* or *lamellae.*

Bubbles in a liquid originate from one of three general sources: (1) They may be formed by de-supersaturation of a solution of the gas or by the decomposition of a component in the liquid, (2) they may be introduced directly into the liquid by a bubbler or sparger or by mechanical entrainment, and (3) they may result from the disintegration of larger bubbles already in the liquid.

Generation Spontaneous generation of gas bubbles within a homogeneous liquid is theoretically impossible (Bikerman, *Foams: Theory and Industrial Applications,* Reinhold, New York, 1953, p. 10). The appearance of a bubble requires a gas nucleus as a void in the liquid. The nucleus may be in the form of a small bubble or of a solid carrying adsorbed gas, examples of the latter being dust particles, boiling chips, and a solid wall. A void can result from cavitation, mechanically or acoustically induced. Basu, Warrier, and Dhir [*J. Heat Transfer* **124:** 717 (2002)] have reviewed boiling nucleation, and Blander and Katz [*AIChE J.* **21:** 833 (1975)] have thoroughly reviewed bubble nucleation in liquids.

In a 58-page paper, with 271 references cited, Kulkarni and Joshi [*Ind. Eng. Chem. Res.* **44:** 5873 (2005)] have reviewed bubble formation and rise.

Bubble Formation, Bubble Diameter, and Bubble Rise Velocity

Formation at a Single Orifice The formation of bubbles at an orifice or capillary immersed in a liquid has been the subject of much study. The paper by Wilkinson and Van Dierendonck [*Chem. Eng. Sci.* **49:** 1429 (1994)] is an excellent starting point to review pertinent literature.

There are three regimes of bubble production (Silberman in *Proceedings of the Fifth Midwestern Conference on Fluid Mechanics,* Univ. of Michigan Press, Ann Arbor, 1957, pp. 263–284): (1) single-bubble, (2) intermediate, and (3) jet.

Single-Bubble Regime (for $Re_g = V_g d_o \rho_g / \mu_l < 100$, where V_g = gas velocity m/s, d_o = orifice diameter, m, μ_l = liquid viscosity, kg/m · s, and ρ_g = gas density, kg/m^3) Bubbles are produced one at a time, their size being influenced primarily by orifice diameter, d_o, interfacial tension, σ, and the liquid density, ρ_l.

Intermediate Regime ($100 < Re_g < 2000$) As the gas flow through a submerged orifice increases beyond the limit of the single-bubble regime, the frequency of bubble formation increases more slowly, and the bubbles begin to grow in size. Between the two regimes there may indeed be a range of gas rates over which the bubble size decreases with increasing rate, owing to the establishment of liquid currents that nip the bubbles off prematurely. The net result can be the occurrence of a minimum bubble diameter at some particular gas rate [Mater, *U.S. Bur. Mines Bull.* 260 (1927) and Bikerman, *Foams: Theory and Industrial Applications,* Reinhold, New York, 1953, p. 4]. At the upper portion of this region, the frequency becomes very nearly constant with respect to gas rate, and the bubble size correspondingly increases with gas rate. Bubble size is affected primarily by d_o, μ_l, ρ_l, and Q (the gas flow rate, m^3/s).

Kulkarni and Joshi (2005, pp. 5878–5880) have listed 17 references that contain predictive correlations for bubble size from a single submerged orifice in a liquid and 22 references (pp. 5886–5888) for correlations of a more general nature. One of those correlations, by Gaddis and Vogelpohl [*Chem. Eng. Sci.* **41:** 97 (1986)], "shows a very good match with experimental data," according to Kulkarni and Joshi (p. 5891).

$$d_b/d_o = [(6\sigma/\rho_l g d_o^2)^{4/3} + (81\mu_l Q/\pi g d_o^4) + (135Q^2/4\pi^2 g d_o^5)^{4/5}]^{1/4} \qquad (14\text{-}201)$$

where d_b = bubble diameter, d_o = orifice diameter, and g = local gravity, m/s^2

For conditions approaching constant pressure at the orifice entrance, which probably simulates most industrial applications, there is no independently verified predictive method. For air at near atmospheric pressure sparged into relatively inviscid liquids (11 ~ 100 cP), the correlation of Kumar et al. [*Can. J. Chem. Eng.* **54:** 503 (1976)] fits experimental data well. Their correlation is presented here as Fig. 14-89.

Jet Regime With further rate increases, turbulence occurs at the orifice, and the gas stream approaches the appearance of a continuous jet that breaks up 8 to 10 cm above the orifice. Actually, the stream consists of large, closely spaced, irregular bubbles with a rapid swirling motion. These bubbles disintegrate into a cloud of smaller ones of random size distribution between 0.025 cm or smaller and about 1.25 cm, with a mean size for air and water of about 0.4 cm (Leibson et al., *AIChE J.* **2:** 300–308 [1956]). According to Kulkarni and Joshi (2005, p. 5890), jetting begins when the Weber number exceeds 4

$$We = \rho_g d_o V_g^2/\sigma > 4 \qquad (14\text{-}202)$$

There are many contradictory reports about the jet regime, and theory, although helpful (see, for example, Silberman 1957), is as yet unable to describe the phenomena observed. The correlation of Kumar et al. (Fig. 14-89) is recommended.

Formation at Multiple Orifices The coverage in the eighth edition of this handbook could be helpful.

Jamialhanadi et al. [*Trans. IChemE* **79A:** 523 (2001)] have proposed a unified correlation for the estimation of average bubble size. Kulkarni and Joshi [*Chem. Eng. Res. Dev.* **89:** 1972 (2011)] have recommended this correlation for the prediction of bubble size from spargers in bubble columns.

$$d_b/d_o = [5/Bo^{1.08} + 9.26Fr^{0.36}/Ga^{0.39} + 2.14Fr^{0.51}]^{1/3} \qquad (14\text{-}203)$$

where Bo = Bond number = $\rho_l d_o^2 g/\sigma$; Fr = Froude number = $V_o^2/g d_o$; Ga = Galileo number = $\rho_l^2 d_o^3 g/\mu_l^2$, V_o = orifice velocity, m/s. For most practical spargers, the gas velocity is high; thus, inertial forces will dominate, so the third term in Eq. (14-203) will dominate. Hence, the bubble diameter is proportional to $V_g^{0.34}$, which is counter to the decrease in bubble size with increasing V_g predicted by Fig. 14-89 for $Re_g > 2000$. The correlation of Jamialhanadi et al. is thus suspect for $Re_g > 2000$. Use the result from Fig. 14-89 for Re > 2000.

Critical Weep Point Spargers must to be designed to avoid weeping into the sparger header. Kulkarni and Joshi (2011) have addressed this issue, and they recommend the correlation of Thorat et al. [*Chem. Eng. Tech.* **24**(8): 815 (2001)] for sieve trays.

$$V_c^2[\rho_g/(\rho_l - \rho_g)]/(g d_o) = 0.37 + 140 H_l (\Delta X/d_o)^{-1.6}(t/d_o)^{0.75} \qquad (14\text{-}204)$$

where V_c = critical gas velocity to avoid weeping, m/s; ΔX = distance between holes, m; H_l = static liquid head above the sparger, m; t = sparger plate thickness, m.

The correlation of Kulkarni et al. [*Chem. Eng. Res. Dev.* **87**(12): 1612 (2009)] is recommended for pipe spargers.

$$V_c^2 = [0.44(\rho_l - \rho_g) d_o g/\rho_g](L/d_o)^{-0.12}(\Delta X/d_o)^{-0.145}(H_l/d_o)^{0.67} \qquad (14\text{-}205)$$

where L = length of the sparger pipe, m.

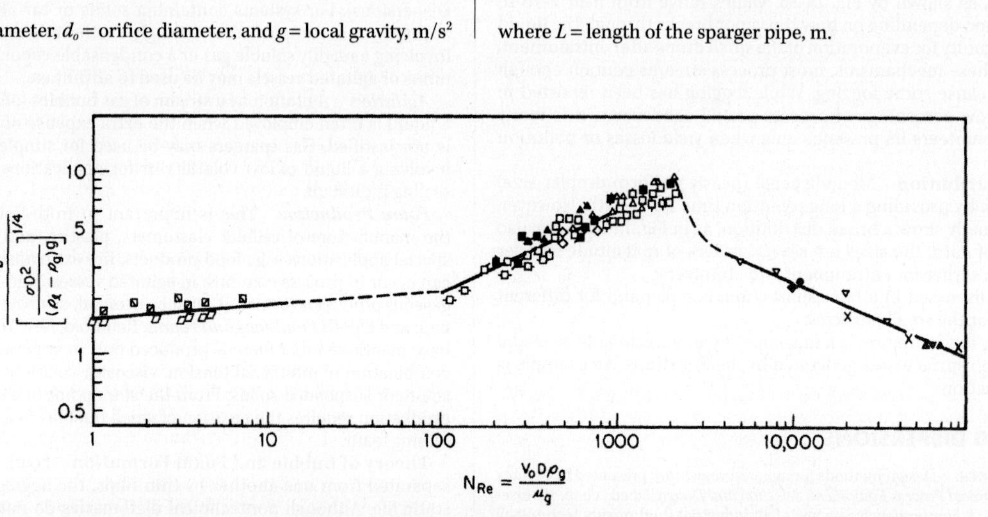

FIG. 14-89 Bubble-diameter correlation for air sparged in relatively inviscid liquids. D_b = bubble diameter, D = orifice diameter, V_o = gas velocity through the sparging orifice, ρ_g = gas density, and μ_g = gas viscosity. [*From* Can. J. Chem. Eng. *54: 503 (1976)*].

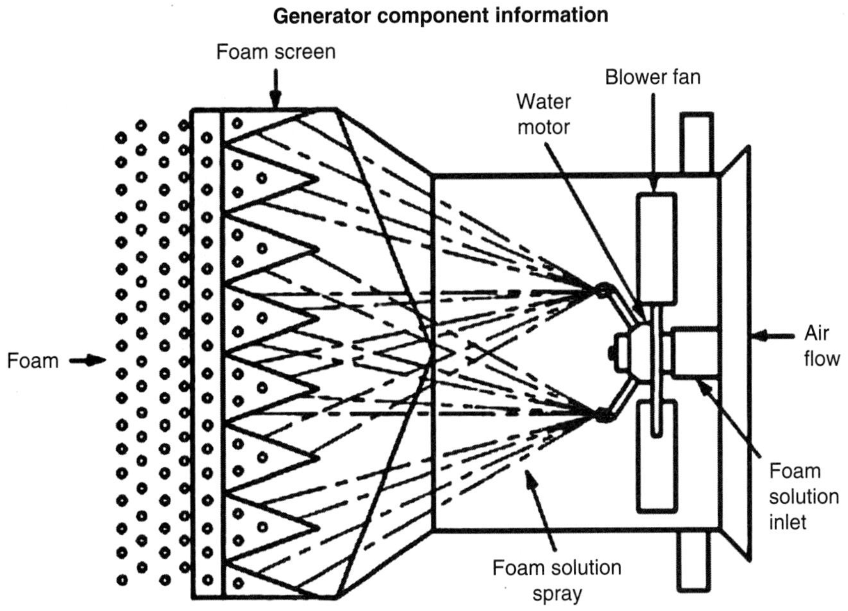

FIG. 14-90 Details of the Ansul high-expansion foam generators. (https://www.ansul.com/en/us/DocMedia/F-93137.pdf.)

The prediction of bubble size and critical weeping velocity for a submerged sparger involves complex phenomena that cannot be modeled accurately; thus, a correlational approach using the appropriate dimensionless parameters must be taken. No correlation covers accurately the wide range of conditions for which spargers are used. The results obtained by using the recommended correlations must be used cautiously and, normally, with experimental verifications prior to scale-up from the laboratory or pilot plant to the full-scale plant.

Entrainment and Mechanical Disintegration Gas can be entrained into a liquid by a solid or a stream of liquid falling from the gas phase into the liquid, by surface ripples or waves, or by a vortex in a partially baffled mechanically agitated vessel. The disintegration of sparged gas is often accomplished by mechanical agitation. Quantitative correlations for gas entrainment by liquid jets and in agitated vessels will be given later.

Foams The excellent review by Exerowa and Kruglyakov (*Foam and Foam Films,* Elsevier, New York, 1998) covers the literature pertinent to foams. A foam is formed when bubbles rise to the surface of a liquid and persist for a while without coalescence with one another or without rupture into the vapor space. Gravitational force and interfacial tension forces favor the coalescence and ultimate disappearance of bubbles. The viscosity of the liquid in a film opposes the drainage of the film and its displacement by the approach of coalescing bubbles. The higher the viscosity, the slower will be the film-thinning process; furthermore, if viscosity increases as the film grows thinner, the process becomes self-retarding.

If the liquid laminae of a foam system can be converted to impermeable solid membranes, the film viscosity can be regarded as having become infinite, and the resulting solid foam will be permanent. Likewise, if the laminae are composed of a gingham plastic or a thixotrope, the foam will be permanently stable for bubbles whose buoyancy does not permit exceeding the yield stress. For other non-Newtonian fluids, however, and for all Newtonian ones, no matter how viscous, the viscosity can only delay but never prevent foam disappearance. Foam stability is keyed to the existence of a surface skin of low interfacial tension immediately overlying a solution bulk of higher tension, latent until it is exposed by rupture of the superficial layer [Maragoni, *Nuovo Cimento* **2**(5–6): 239 (1871)]. Such a phenomenon of surface elasticity, resulting from concentration differences between bulk and the surface of the liquid, accounts for the ability of bubbles to be penetrated by missiles without damage. It is conceivable that films below a certain thickness no longer carry any bulk of solution and hence have no capacity to close surface ruptures, thus becoming vulnerable to mechanical damage that will destroy them. The Maragoni phenomenon is consistent also with the observation that neither pure liquids nor saturated solutions will sustain a foam, since neither extreme will allow the necessary differences in concentration between surface and bulk of solution.

The specific ability of certain finely divided, insoluble solids to stabilize foam has long been known (Berkman and Egloff, *Emulsions and Foams,* Reinhold, New York, 1941, p. 133; and Bikerman, *Foams: Theory and Industrial Applications,* Reinhold, New York, 1953, chap. 11). Bartsch [*Kolloidchem. Beih* **20:** 1 (1925)] found that the presence of fine galena greatly extended the life of air foam in aqueous isoamyl alcohol, and the finer the solids, the greater the stability.

The production of foams is a well-established technology, covered extensively by YouTube videos and on the Web. An example of the technology available on the Web is the JET-X high-expansion foam generators (https://www.ansul.com/en/us/DocMedia/F-93137.pdf), for which the design details are included in Fig. 14-90.

Characteristics of Dispersion

Dispersion Characteristics The chief characteristics of gas-in-liquid dispersions, like those of liquid-in-gas suspensions, are heterogeneity and instability.

The rate of rise of bubbles has been discussed by Clift, Grace, and Weber, *Bubbles, Drops and Particles,* Academic Press, New York, 1978; Benfratello, *Energ. Elettr.* **30:** 80 (1953); Haberman and Morton, *Report 802: David W. Taylor Model Basin,* Washington, September 1953; and Kulkarni and Joshi, *Ind. Eng. Chem. Res.* **44:** 5873–5931 (2005), in which reference they summarize 16 correlations for predicting the rise velocity of bubbles; however, none are accurate over the entire range of practical bubble sizes.

Figure 14-91 presents a graph of bubble rise velocities versus mean bubble size for a variety of pure liquids, liquid mixtures, and contaminated liquids. The shapes of these curves vary greatly depending on the purity and viscosity of the liquids.

Small bubbles (below 0.2 mm in diameter) are essentially rigid spheres and rise at terminal velocities that place them clearly in the laminar-flow region; hence their rising velocity may be calculated from Stokes' law [$V_b = 2g(\rho_l - \rho_g)D_b^2/18\mu_l$]. As bubble size increases to about 2 mm, the spherical shape is retained, and the Reynolds number is still so small (<10) that Stokes' law should be nearly obeyed. The Reynolds number Re is given by $\mathrm{Re} = V_b D_b/\mu_l$ where V_b = bubble velocity, m/s; D_b = bubble diameter, m; μ_l = liquid viscosity, kg/m · s.

As bubble size increases, two effects set in, however, that change the deviation from Stokes' law. At about Re = 100, a wobble begins that can develop into a helical path if the bubbles are not liberated too closely to one another [Houghton, McLean, and Ritchie, *Chem. Eng. Sci.* **7:** 40 (1957); and Houghton, Ritchie, and Thomson, *Chem. Eng. Sci.* **7:** 111 (1957)]. Furthermore, for bubbles in the range of 1 mm and larger (until distortion becomes serious), internal circulation can set in [Garner and Hammerton, *Chem. Eng. Sci.* **3**, (1954)], and, according to a theoretical analysis by Hadamard and Rybczynski (https://en.wikipedia.org/wiki/Hadamard%E2%80%93Rybczynski_equation), the rise velocity of a freely circulating bubble is 3/2 that of a rigid sphere for a low-viscosity gas and a much higher viscosity liquid. Redfield and Houghton [*Chem. Eng. Sci.* **20:** 131 (1965)] have found that CO_2 bubbles rising in pure water agree with the theoretical solution for a circulating drop below Re = 1. Extremely small quantities of impurities can retard or stop this

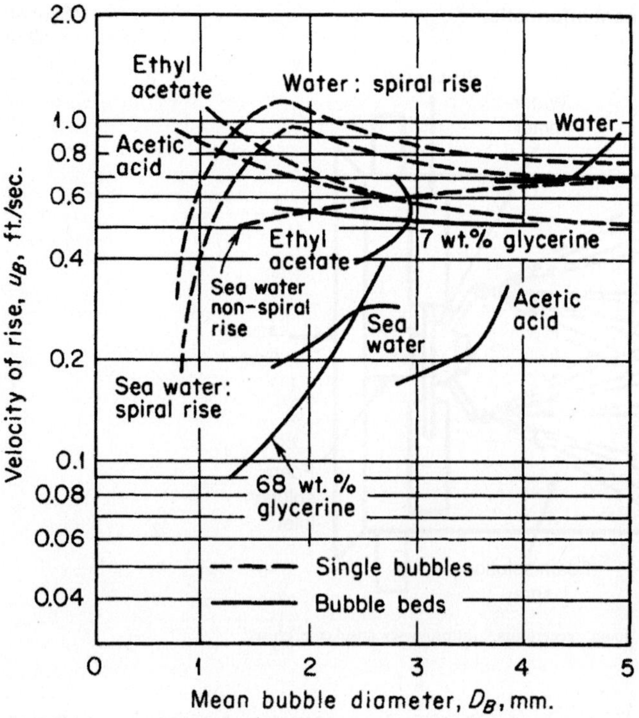

FIG. 14-91 Velocity of rising bubbles, singly and in clouds. To convert feet per second to meters per second, multiply by 0.305. [*From* Chem. Eng. Sci. *7: 48 (1957).*]

internal circulation as explained in *Bubble and Foam Chemistry*, Cambridge University Press, 2016, p. 114.

Above diameters of about 2 mm, bubbles begin to change to ellipsoids, and above 1 cm they become lens-shaped, according to Davies and Taylor [*Proc. Roy. Soc.* (London) **A200:** 379 (1950)] and Kulkarni and Joshi 2005, p. 5911. The rise velocity in thin liquids for the size range 1 mm < D_B < 20 is 20 to 30 cm/s, as indicated by Fig. 14-89.

Above a Reynolds number of the order of magnitude of 1000, bubbles assume a helmet shape, with a flat bottom. After bubbles become large enough to depart from Stokes' law at their terminal velocity, behavior is generally complicated and erratic, and the reported data scatter considerably. The rise can be slowed, furthermore, by a wall effect if the diameter of the container is not greater than 10 times the diameter of the bubbles, as shown by Uno and Kintner [*AIChE J.* **2:** 420 (1956); and Collins, *J. Fluid Mech.* **28**(1): 97 (1967)]. Work has been done to predict the rise velocity of large bubbles [Rippin and Davidson, *Chem. Eng. Sci.* **22:** 217 (1967); Mendelson, *AIChE J.* **13:** 250 (1967); Lehrer, *J. Chem. Eng. Japan* **9:** 237 (1976); and Lehrer, *AIChE J.* **26:** 170 (1980)]. The works of Lehrer present correlations for rise velocities for a wide range of system properties for bubbles outside the Stokes' law regime.

Most practical applications occur where bubbles do not act as rigid spheres; instead, they oscillate and deform. Although the predictive methods for rise velocity are not very accurate outside the rigid sphere regime, that does not deter the use of bubble rise velocity for designing suitable processing equipment. The most widely used application of bubble rise velocity is to design bottom vessel outlets and horizontal vessels to separate gas from liquid when a gas dispersion exists within a system. From Fig. 14-89 we see that the maximum rise velocity, even for pure liquids, is about 30 cm/s.

When bubbles are produced in clouds, as by a porous disperser, their behavior during rising is further complicated by interaction among themselves. In addition to the tendency for small bubbles to coalesce and large ones to disintegrate, there are two opposing influences on the rate of rise of bubbles of any particular size: (1) A "chimney effect" can develop in which a massive current upward appears at the axis of the bubble stream, leading to increased net bubble velocity; and (2) the proximity of the bubbles to one another can result in a hindered-settling condition, leading to reduced average bubble velocity. Figure 14-91 shows the data for clouds of bubbles compared with their single-bubble data for pure water and seawater and of Peebles and Garber [*Chem. Eng. Progr.* **49:** 88 (1953)] for acetic acid and ethyl acetate. The bubble clouds were produced with a sintered-glass plate of mean pore size (inferred from air wet-permeability data) of 81 μm.

The difference between the curves for pure water and seawater again illustrates the significance of small concentrations of solute with respect

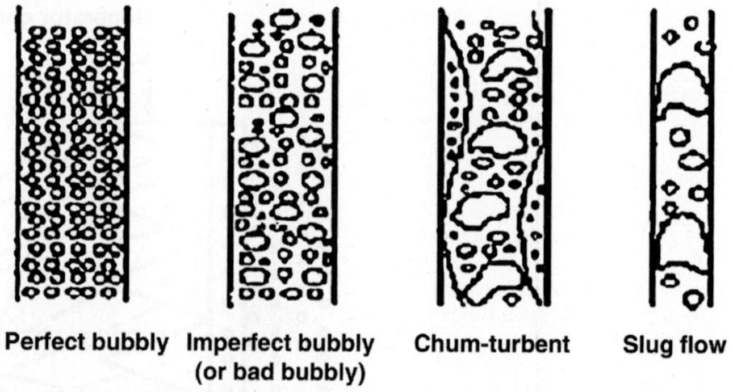

FIG. 14-92 Visual representations of flow regimes in bubble columns. [*Kantarci et al.,* Process Biochemistry *40(7): (2005), p. 2269.*]

to bubble behavior. In commercial bubble columns and agitated vessels, coalescence and breakup are so rapid and violent that the rise velocity of a single bubble is meaningless. The average rise velocity can, however, be readily calculated from holdup correlations that will be given later.

Methods of Gas Dispersion The problem of dispersing a gas in a liquid may be attacked in several ways: (1) The gas bubbles of the desired size or which grow to the desired size may be introduced directly into the liquid; (2) a volatile liquid may be vaporized by either decreasing the system pressure or increasing its temperature; (3) a chemical reaction may produce a gas; or (4) a massive bubble or stream of gas is disintegrated by fluid shear, by, perhaps, mechanical agitation or turbulence in the liquid. Bubble columns and mechanically agitated columns are most commonly used for dispersing gas into liquid in commercial-scale equipment.

Bubble Columns The technical literature covering bubble column reactors was reviewed by Kantarci et al. [*Process Biochemistry* **40**(7): 2263 (2005)] and Chen [*Biotechnology and Genetic Engineering Reviews* **8:** 379–396 (November 1990)]; similarly, Middleton and Smith (*Handbook of Mixing Equipment*, Wiley, Hoboken, N.J., 2004, chap. 11, p. 585) have summarized the technical literature for gas–liquid mixing in mechanically agitated vessels. Gas/liquid processing vessels can be very large, as evidenced by the 265,000 L algae reactors at the Alltech Hendersonville, Ky., plant. [https://www.youtube.com/watch?v=a58JJH9pxOU]

Flow Regimes in Bubble Columns Assuming uniform dispersion at the tank bottom, the likely flow regimes and flow regime visualizations are presented in Figs. 14-92 and 14-93. In larger columns, bubbly flow exists up to a superficial gas velocity of 0.05 m/s; from 0.05 to 0.07 m/s the regime transitions from bubbly to churn-turbulent.

Spargers: Simple Bubblers Gas dispersed in a liquid contained in a tank can be introduced through an open-end standpipe, horizontal perforated pipes, concentric ring pipes, or a perforated plate at the bottom of the tank.

Perforated-pipe or -plate spargers usually have orifices 1 to 10 mm in diameter [Kulkarni et al., *Chem. Eng. Res. & Des.* **87:** 1612 (2009); **89:** 1972 (2011)]. Effective design methods to minimize maldistribution are presented in the eighth edition of this handbook, p. 6-32, by Senecal [*Ind. Eng. Chem.*

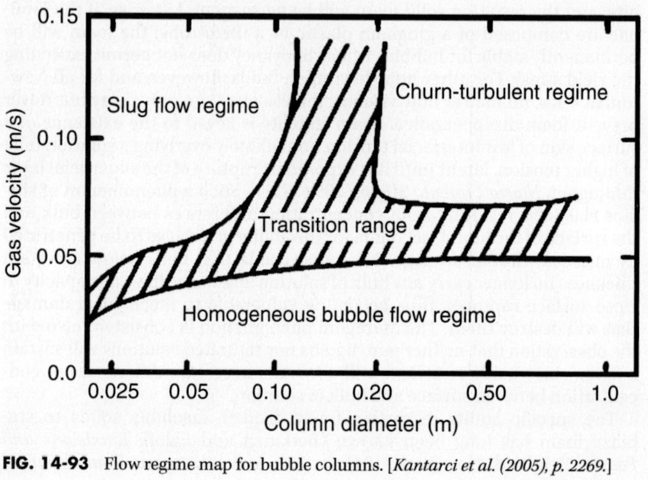

FIG. 14-93 Flow regime map for bubble columns. [*Kantarci et al. (2005), p. 2269.*]

49: 993–997 (1957)] and Kulkarni et al. 2011. For turbulent flow conditions into the sparger, the following simpler relationship will allow conservative design of a perforated-pipe sparger for a given degree of maldistribution provided the number of holes exceeds about 5 and the length/diameter <300.

$$V_h/V_0 = [C_o V_h/(\Delta V_h/V_h)]^{(1/2)} \tag{14-206}$$

where V_h = hole velocity, m/s; V_0 = velocity entering the sparger pipe, m/s; ΔV_h = difference between maximum hole velocity and minimum hole velocity, m/s; $(\Delta V_h/V_h)$ = fractional maldistribution of flow along the sparger, and C_o = the orifice coefficient for the holes. C_o for holes in a pipe wall varies from 0.6 to 0.8, depending on the ratio of wall thickness to hole diameter (t/D_h): C_o = 0.6, 0.65, 0.7, 0.75, 0.8 for (t/D_h) = 0.1, 0.34, 0.45, 0.66, 2; respectively. For example, for a thick wall pipe and small holes, C = 0.8; thus, for a fractional maldistribution of 0.05,V_h/V_0 = 4.

Simple spargers are used as agitators for large tanks, principally in the cement and oil industries. Kauffman [*Chem. Metall. Eng.* **37:** 178–180 (1930)] reported the following air rates for various degrees of agitation in a tank containing 2.7 m (9 ft) of *liquid:*

Degree of agitation	Air rate (cfm/ft²)	Air rate ([m³/min]/m²)
Moderate	0.65 [0.011 ft/s]	0.2
Complete	1.3 [0.022 ft/s]	0.4
Violent	3.1 [0.05 ft/s]	0.95

NOTE: For a liquid depth of 0.9 m (3 ft), Kauffman recommended that the listed rates be doubled.

Tank Blending Using Gas Sparging Lehrer [*Ind. Eng. Chem. Process Des. Dev.* 7: 226 (1968)] conducted liquid-blending tests with air sparging in a 0.61-m-diameter by 0.61-m-tall vessel and found that an air volume equal to about one-half of the vessel volume (for vessel A6, Fig. 9 in Lehrer, at t_b = 10 s, v_g = 0.1 fts, Q = 0.314 cfs, V_g = 3.14 ft³, and V_g/V_v = 3.14/6.54 = 0.48) gave thorough blending of inviscid liquids of equal viscosities. Using an analogy to mechanically agitated vessels in which equal tank turnovers give equal blend times, one would expect this criterion to be applicable to other vessel sizes. Liquids of unequal density and higher viscosities will require somewhat more air. This design criterion is substantiated by an analysis of the data presented by El Ezzi and Najmuldeen [*J. Eng. Res. & Application* **4**(4): 286 (2014)] for a gas lift vessel. Using a conservative design criterion of one tank volume of gas to blend a vessel is normally very practical; for example, for a 3-m-diameter × 6-m-tall vessel (42 m³ = 11,000 gal), the total gas volume required for blending is 42 m³. A gas flow of 4.2 m³/min (150 cfm) will blend this vessel in 10 minutes.

Gas spargers are usually operated at orifice Reynolds numbers in excess of 6000 in order to obtain small bubbles, which increase the interfacial area and, consequently, mass transfer. In the "turbulent regime" (superficial gas velocity > about 5 cm/s), sparger design is not critical because a balance between coalescence and breakup is established very quickly according to Towell et al. [*AIChE Symp. Ser. No. 10,* 97 (1965)]. However, a reasonably uniform orifice distribution over the column cross section is desirable, and according to Fair [*Chem. Eng.* **74:** 67 (July 3, 1967); 207 (July 17, 1967)] the orifice velocity should be less than 75 to 90 m/s.

Porous Septa In the quiescent regime, porous plates, tubes, disks, or other shapes that are made by bonding or sintering together carefully sized particles of carbon, ceramic, polymer, or metal are often used for gas dispersion, particularly in foam fractionators. Many companies supply such septa, including Mott (http://www.mottcorp.com/products) and GKN (http://www.gkn-filters.de/products_en/spargers/). The resulting septa may be used as spargers to produce much smaller bubbles than will result from a simple bubbler. Figure 14-94 shows a comparison of the bubbles emitted by a perforated-pipe sparger [0.16 cm orifices] and a porous carbon septum (120 μm pores). The air permability of porous spagers are given by the Mott Corp. in its *Porous Metal Design Guidebook*, p. 10 (https://mottcorp.com/sites/default/files/Porous%20Metal%20Design%20Guidebook%20Published%20by%20MPIF.pdf)

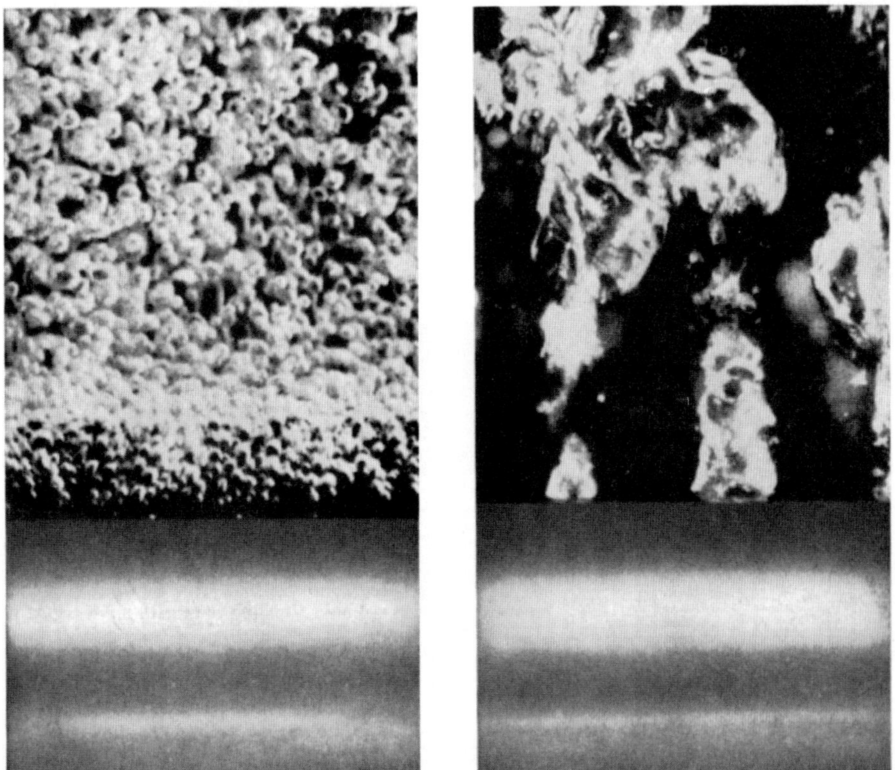

(a) (b)

FIG. 14-94 Comparison of bubbles from a porous septum and from a perforated-pipe sparger. Air in water at 70°F. (*a*) Grade 25 porous-carbon diffuser operating under a pressure differential of 13.7 in of water. (*b*) Karbate pipe perforated with 1⁄16-in holes on 1-in centers. To convert inches to centimeters, multiply by 2.54; °C = 5⁄9 (°F – 32). (*National Carbon Co.*)

An Inyo website (http://www.inyoprocess.com/images/Porous_inj/porous_sparger_sizing.pdf) states, "When an injector is sized correctly, a bubble size of 200–6000 micron will be achieved in clear water. . . . If the velocity is too high, the smaller bubbles will tend to coalesce into larger bubbles." A working limit of 0.05 m/s (10 fpm) is recommended to avoid serious coalescence. Tubes and slabs of porous material are used as porous spargers. Roe [*Sewage Works J.* **18:** 878 (1945)] claimed that silicon carbide tubes are superior to horizontal plates, principally because of the wiping action of the liquid circulating past the tube; he reported respective maximum capacities of 0.015 to 0.025 m/s for a horizontal tube and a horizontal plate of the same material (unspecified grade).

Porous septa produce small bubbles that provide high mass transfer rates. The primary resistance to mass transfer is normally in the liquid phase, for which the limiting Sherwood number ($k_l D/d_b$) is 2 (https://en.wikipedia.org/wiki/Sherwood_number), where k_l = liquid phase mass-transfer coefficient, m/s; D = solute diffusivity, m²/s, d_b = bubble diameter, m.

Precipitation and Generation Methods For a thorough understanding of the phenomena involved, bubble nucleation should be considered. A discussion of nucleation phenomena is beyond the scope of this handbook, but a starting point with recent references is Deng, Lee, and Cheng, *J. Micromech. Microeng.* **15:** 564 (2005), and Jones, Evans, and Galvin, *Adv. Colloid and Interface Sci.* **80:** 27 (1999).

Agitated Vessels Agitated vessels are used extensively for gas–liquid dispersion, especially for the purposes of mass transfer. There are three modes of impellers commonly used for gas introduction. Those three arrangements are shown schematically in Fig. 14-95. The surface aeration impeller is often a four-blade pitched-blade impeller operating about one-half impeller diameter submerged below the free liquid surface. The gas-inducing impeller is a hollow-shaft design; these impellers are always of a special design, depending on the manufacturer; Ekato's gas-inducing impellers are shown and explained in http://www.ekato.com/fileadmin/user_upload/Documents/PresseCenter/Titel_Chemical_Engineering.pdf.

Chemineer (http://www.chemineer.com/products/chemineer/impellers.html) sells four impellers that are used for gas dispersion: a six-blade disk, CD-6, and the BT-6 for use as the bottom impeller just above the ring-type perforated pipe sparger, and normally a Maxiflow W for installation above the bottom impeller. Other manufacturers sell impellers that are competitive with those offered by Chemineer.

Gas Dispersion—Vessel Headspace Boerma and Lankester [*Chem. Eng. Sci.* **23:** 799 (1968)] measured the surface aeration of a nine-bladed disk-type turbine (*Note*: A well-designed four-bladed pitched-blade turbine will give equal or better performance). In a fully baffled vessel, the optimum depth to obtain maximum gas dispersion was 15 to 50 percent of the impeller diameter. In a vessel with baffles extending only halfway to the liquid surface, the optimum impeller submergence increased with agitator speed because of the vortex formed. At optimum depth, the following correlation is recommended for larger vessels:

$$Q = 0.00015(N/10)^{2.5}(D/0.1)^{4.5} \qquad (14\text{-}207)$$

where Q = rate of gas dispersed into the liquid, m³/s, N = impeller speed, rps, and D = impeller diameter, m. Thus, a typical 3-m-diameter plant-size vessel, with four-blade pitched impeller (D = 1 m) operating at 2 rps will give gas dispersion from the headspace into the batch of $Q = 0.00015(2/10)^{2.5}(2/0.1)^{4.5}$ = 1.9 m³/s (4000 cfm). There is another means of estimating the gas entraining capability of this impeller, and that is based on the pumping capability of the impeller in conjunction with a reasonable estimate of gas holdup. The impeller pumping number ($N_Q = Q/ND^3$) for a four-bladed pitched-blade impeller is 0.75 (Couper et al., *Chemical Process Equipment Selection and Design*, Elsevier, New York, 2010, p. 285); thus, the impeller pumping capability is $0.75(2)3^3$ = 13.5 m³/s. A reasonable estimate of the maximum gas fraction within the pumped liquid is 0.15; thus, the maximum expected gas entraining capability, at the optimum impeller depth, is 0.15(13.5) = 2 m³/s, which agrees reasonably well with the 1.9 m³/s obtained by correlating the experimental results of Borema and Lankester ["The Occurrence of Minimum Stirring Rates in Gas-Liquid Reactors," *Chem. Eng. Sci.* **23:** 799–801 (1968)].

Borema and Lankester (1968) recommended headspace gas dispersion with partial baffling for fatty oils hydrogenation in stirred reactors: "The hydrogen in the head-space of the closed reactor can again be brought into contact with the liquid by a stirrer under the liquid level, . . ." and Chemineer (http://www.chemineer.com/hydrogenator.html) has this to say about using upper impellers in hydrogenation reactors, "It is important to re-incorporate the hydrogen into the batch to maintain the reaction and improve hydrogen utilization. This is very easily and efficiently accomplished by using a 4-bladed pitched impeller (P-4), or impellers, located above the BT-6 impeller. The P-4 impeller:

- Provides increased mixing intensity and draws a vortex for the reincorporation of the hydrogen gas from the vapor space,
- With proper design can handle varying liquid levels,
- Does not plug or foul as compared to self inducing impellers."

However, Middleton et al. in Paul et al. (*Handbook of Industrial Mixing*, Wiley, Hoboken, N.J., 2004) say, "The simplest self-inducer for an agitated vessel is an impeller located near the surface, sometimes with the upper part of the baffles removed so as to encourage the formation of a vortex. This is, however, a sensitive and unstable arrangement. It is better, although probably more expensive, to use a self-inducing impeller system in which gas is drawn down a hollow shaft to the low-pressure region behind the blades of a suitable, often shrouded impeller.... Various proprietary designs are available, such as the Ekato gasjet (http://www.ekato.com/products/product/show/Product/impellers/gasjet/); video included), Prasair AGR (http://citeseerx.ist.psu.edu/viewdoc/download?doi=10.1.1.624.6642&rep=rep1&type=pdf) and the Frings Friborator (http://www.frings.com/FRIBORATOR-Aerator.224+M52087573ab0.0.html)."

In many hydrogenation reactors, the impeller just underneath the free surface has, without any doubt, performed admirably; consequently, one must consider this fact very carefully when selecting a self-inducing impeller [e.g., inspired by the work of Borema and Lankester (op. cit.), in the early 1970s the capacity of a hydrogenation reactor in the Monsanto L-DOPA process (https://www.google.com/?gws_rd=ssl#q=l-dopa+monsanto+process) was more than doubled by adding a four-bladed pitched impeller just below the free liquid surface in a glass-lined reactor].

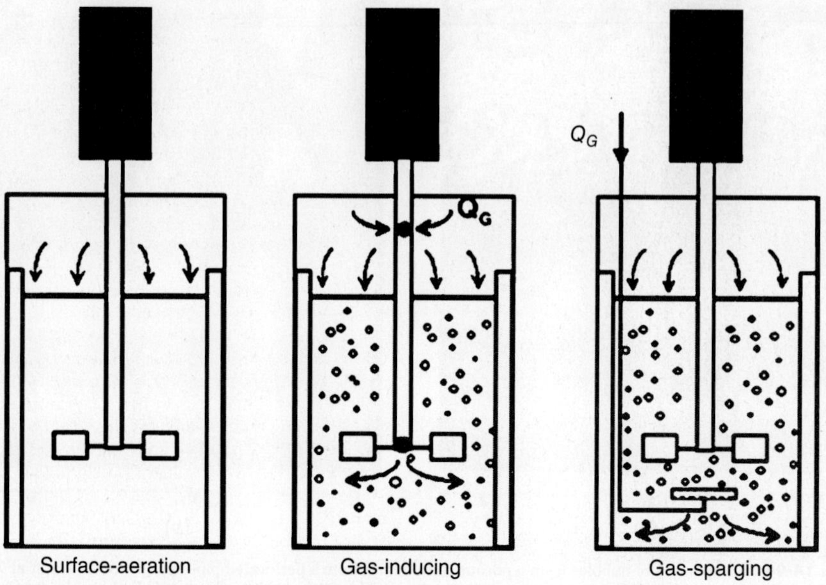

FIG 14-95 Operating modes of gas–liquid reactors. [*Lemoine*, Chem. Eng. J. ***114:** 9 (2005).*]

FIG. 14-96 The Aqua-Jet surface aerator in operation. (http://www.aqua-aerobic.com/index.cfm/products-systems/aeration-mixing/aqua-jet/.)

Critical Speed for Gas Entrainment The critical speed for air entrainment to commence has been studied by Bhattacharya, Hebert, and Kresta [*IChemE* **85**(A5): 654 (2007)] and reviewed by Patwardhan and Joshi (*Can. J. Chem. Eng.* **76:** 339 (1998)]. This parameter is of marginal practical interest because practical operation (1–10 kW/kg) is far beyond the power input level where gas entrainment starts.

Mass-Transfer Coefficient The calculation of mass transfer rates is most important; correlations for $k_L a$ have been developed by Leguay et al. (*10th European Conference on Mixing*, Elsevier, Amsterdam, 2000, p. 189) and Conway et al. "Gas-Liquid-Solid Operation of a Vortex-Ingesting Stirred Tank Reactor," available at (http://citeseerx.ist.psu.edu/viewdoc/download?doi=10.1.1.624.6642&rep=rep1&type=pdf), and the literature has been reviewed by Patwardhan and Joshi [*Ind. Eng. Chem. Res.* **38:** 49–80 (1999)] and Lemoine et al. [*Chem. Eng. and Processing* **42:** 621 (2003)].

Surface Aerators Gas dispersion through the free surface by mechanical aerators is commonplace in aerobic waste-treatment lagoons. Surface aerators are generally of four types: (1) small-diameter high-speed (normally motor-speed) propellers operating in draft tubes (Fig. 14-96), (2) large-diameter slow-speed turbines operating just below the free surface of the liquid (Fig. 14-97), often pontoon-mounted, (3) aeration jets (http://www.wateronline.com/doc/kynar-pvdf-jet-aerator-for-improved-wear-resistance-0001), and (4) fine bubble spargers.

FIG 14-97 The Voltas pontoon-mounted turbine-type surface aerator in operation. (http://www.voltas.com/voltas_water/images/pdf/aerator.pdf.)

TABLE 14-21 Performance Parameters of the Voltas Pontoon-Mounted Turbine-Type Surface Aerator*

Blade immersion (mm)	Power at aerator output shaft (kW)	Oxygenation capacity at 20°C (kg/h)	Oxygenation capacity efficiency (std.) (kg/kWh)
216	5.7	15.60	2.76
241	6.6	19.50	2.96
267	7.7	20.00	2.60
292	9.5	26.80	2.83
318	10.6	29.30	2.75
345	13.7	39.60	2.90
368	15.7	44.00	2.81
381	16.8	52.20	3.10

*http://www.voltas.com/voltas_water/images/pdf/aerator.pdf.

The performance parameters of the Voltas pontoon-mounted turbine-type surface aerator are given in Table 14-21. Note that the oxygen transfer efficiency varies from about 2.75 to 3.1 (kg/hr)/kW. Stenstrom and Rosso ("Aeration," table, University of California–Los Angeles, 2010, p. 55) give lower transfer efficiencies than those given by Voltas; process guarantees must be used to ensure that the machinery will perform as the manufacturer specifies.

Equipment Selection Ideally, the selection of equipment to produce a gas-in-liquid dispersion should be made on the basis of a complete economic analysis. The design engineer and especially the pilot-plant engineer seldom have enough information or time to do a complete economic analysis. In the following discussion, some guidelines are given as to what equipment is feasible and what equipment might prove most economical.

For producing foam for foam-separation processes, perforated-plate or porous-plate spargers are normally used. Mechanical agitators are often not effective in the light foams needed in foam fractionation. Dissolved-air flotation, based on the release of a pressurized flow in which oxygen was dissolved, has been shown to be effective for particulate removal when sparged air failed because the bubbles formed upon precipitation are smaller—down to 80 μm—than bubbles possible with sparging, typically 1000 μm [Grieves and Ettelt, *AIChE J.* **13:** 1167 (1967)]. Mechanically agitated surface aerators such as the WEMCO-Fagergren flotation unit (http://www.flsmidth.com/en-US/Industries/Categories/Products/Flotation/WEMCOFlotation/WEMCOFlotation) are used extensively for ore flotation.

To produce foam in batch processes, mechanical agitators are used almost exclusively. The gas can either be introduced through the free surface by the entraining action of the impeller or alternatively sparged beneath the impeller. In such a batch operation, the liquid level gradually rises as the foam is generated; thus, squatty impellers such as turbines are rapidly covered with foam and must almost always be sparged from below. Tall impellers such as wire whips (https://www.google.com/search?q=wire+whip+agitators&biw=1920&bih=934&source=lnms&tbm=isch&sa=X&ved=0ahUKEwjWjJPbourRAhXJsFQKHQSZDXMQ_AUIBygC) are especially well suited to entrain gas from the vapor space. Intermeshing wire whips are standard kitchen utensils for producing foamed meringues, consisting of air, vegetable oil, and egg whites. For a new application, generally some experimentation with different impellers is necessary in order to get the desired fine final bubble size without frothing over initially. For producing foams continually, an aspirating venturi nozzle and restrictions in pipes such as baffles and metal gauzes are generally most economical.

For gas absorption, the equipment possibilities are generally packed columns; plate distillation towers, possibly with mechanical agitation on every plate; deep-bed contactors (bubble columns or sparged lagoons); and mechanically agitated vessels or lagoons. Packed towers and plate distillation columns are covered earlier in this section. Generally these devices are used when a relatively large number of stages (more than two or three) is required to achieve the desired result practically.

The volumetric mass-transfer coefficients and heights of transfer units for bubble columns and packed towers have been compared for absorption of CO_2 into water by Houghton et al. [*Chem. Eng. Sci.* **7:** 26 (1957)]. The bubble column will tolerate much higher vapor velocities, and in the overlapping region (superficial gas velocities of 0.009 to 0.018 m/s), the bubble column has about three times higher mass-transfer coefficient and about three times greater height of transfer unit. The liquid in a bubble column is, for practical purposes, quite well mixed; thus, chemical reactions and component separations requiring significant plug flow of the liquid cannot be effected with bubble columns. Bubble columns and agitated vessels are the ideal equipment for processes in which the fraction of gas absorbed need not be great, possibly the gas can be recycled, and the liquid can or should be well mixed. The gas phase in bubble columns is not nearly so well back-mixed as the liquid, and often plug flow of the gas is a logical assumption, but in agitated vessels the gas phase is also well mixed.

The choice of a bubble column or an agitated vessel depends primarily on the solubility of the gas in the liquid, the corrosiveness of the liquid (often a gas compressor can be made of inexpensive material, whereas a mechanical agitator may have to be made of exotic, expensive materials), and the rate of chemical reaction as compared with the mass-transfer rate. Bubble columns and agitated vessels are seldom used for gas absorption except in chemical reactors. As a general rule, if the overall reaction rate is five times greater than the mass-transfer rate in a simple bubble column, a mechanical agitator will be most economical unless the mechanical agitator would have to be made from considerably more expensive material than the gas compressor.

In bubble columns and simply sparged lagoons, selecting the sparger is an important consideration. In the turbulent regime (superficial gas velocity greater than 0.046 to 0.06 m/s), inexpensive perforated-pipe spargers should be used. Often the holes must be placed on the pipe bottom in order to make the sparger free-draining. In the quiescent regime, porous septa will often give considerably higher overall mass-transfer coefficients than perforated plates or pipes because of the formation of small bubbles. Chain and coworkers (*First International Symposium on Chemical Microbiology,* World Health Organization, Monograph Ser. 10, Geneva, 1952) claimed that porous disks are about twice as effective as open-pipe and ring spargers for the air oxidation of sodium sulfite. Eckenfelder [*Chem. Eng. Progr.* **52**(7): 290 (1956)] has compared the oxygen-transfer capabilities of various devices on the basis of the operating power required to absorb a given quantity of O_2. The installed cost of the various pieces of equipment probably would not vary sufficiently to warrant being included in an economic analysis. Surface mechanical aerators are not included in this comparison. Of the units compared, it appears that porous tubes give the most efficient power usage. Kalinske (*Advances in Biological Waste Treatment,* Pergamon Press, New York, 1963, pp. 157–168) compared submerged sparged aerators with mechanical surface aerators. He summarized this comparison in *Water Sewage Works* **33** (January 1968). He indicated that surface aerators are significantly more efficient than subsurface aeration, both for oxygen absorption and for gas-stripping operations.

Zlokarnik and Mann (paper at Mixing Conference, Rindge, New Hampshire, August 1975) found the opposite of Kalinske, that is, subsurface diffusers, subsurface sparged turbines, and surface aerators compared approximately 4:2:1, respectively, in terms of their O_2 transfer efficiency. However, Zlokarnik [*Adv. Biochem. Eng.* **11:** 157 (1979)] later indicated that the scale-up correlation used earlier might be somewhat inaccurate. When all available information is considered, it appears that with near-optimum design, any of the aeration systems (diffusers, submerged turbines, or surface impellers) should give a transfer efficiency of at least 2.25 kg O_2/kWh. Thus, the final selection should probably be made primarily on the basis of operational reliability, maintenance, and capital costs.

Mass Transfer Mass transfer in plate and packed gas–liquid contactors has been covered previously in this section. Attention here will be limited to deep-bed contactors (bubble columns and agitated vessels). Theory underlying mass transfer between phases is discussed in Sec. 5 of this handbook.

To design deep-bed contactors for mass-transfer operations, one must have, in general, predictive methods for the following design parameters:

- Flooding (for both columns and agitator impellers)
- Agitator power requirements
- Gas-phase and liquid-phase mass-transfer coefficients
- Interfacial area
- Interface resistance
- Mean concentration driving force for mass transfer

In most cases, available methods are incomplete or unreliable, and some supporting experimental work is necessary, followed by scale-up. The methods given here should allow theoretical feasibility studies, help minimize experimentation, and permit a measure of optimization in final design.

Sparged Impellers The most common method of introducing gas into an agitated vessel is through a ring-type perforated pipe sparger.

Flooding of Agitator Impellers Impeller flooding correlations for six-blade disk (6BD) Rushton turbines and six-blade disk Smith turbines (Chemineer designation: CD-6) are presented by Bakker, Myers, and Smith [*Chem. Eng.* **101:** 98 (December 1994)]. The Bakker et al. correlation is

$$Q/ND^3 = C_{FL} N_{Fr} (D/T)^{3.5} \tag{14-208}$$

where C_{FL} = 30 (experimentally determined dimensionless constant) for a 6BD impeller and C_{FL} = 70 for a concave blade CD-6 impeller and N_{Fr} = Froude number = N^2D/g; Q = gas flow rate at flooding, m³/s; N = impeller speed, rps; D = impeller diameter, m; and T = tank diameter, m. Note that the CD-6 impeller will handle 70/30 = 2.33 times the gas a 6BD will handle, without flooding, at the same N and D; this is the great advantage of the CD-6, along with lower power decrease as the gas flow rate increases.

TABLE 14-22 Constants in Eq. (14-209) and Impeller Power Numbers

Impeller type	a	b	c	d	N_p
6BD	0.72	0.72	24	0.25	3.2
CD-6	0.12	0.44	12	0.37	5.5

Gassed Impeller Power Bakker et al. 1994 have given a gassed power correlation for 6BD and CD-6 impellers, which can be seen at (http://www.chemineer.com/products/chemineer/impellers.html).

$$P_g / P_u = [1-(b-a\mu)]\, N_{Fr}^d \tanh(cN_A) \tag{14-209}$$

where P_g = gassed power, W; P_u = ungassed power, W; $N_A = Q/ND^3$; μ = batch viscosity, Pa-s (i.e., kg/m · s); and the constants of Eq. (14-209) are given in Table 14-22.

As mentioned previously, the CD-6 experiences much less power decrease with increased gassing compared to the 6BD. For example, at $N_A = 0.15$, P_g/P_u = 0.7 for the CD-6 and 0.5 for the CD-6.

The ungassed power is calculated by

$$P_u = N_p \rho_l / N^3 D^5 \tag{14-210}$$

where ρ_l = liquid density, kg/m³; N = impeller speed, rps; D = impeller diameter, m, and the impeller power numbers, N_p, are given in Table 14-22.

Bakker et al. 1994 have given correlations for gas holdup in agitated vessels. The Bakker et al. correlation is

$$\varepsilon = C_\varepsilon (P_g / V)^A \upsilon_{sg}^B \tag{14-211}$$

where ε = fractional gas holdup, dimensionless; C_ε = 0.16, experimentally determined parameter; A = 0.33, B = 0.67; V = batch volume, m³; υ_{sg} = superficial gas velocity = $Q/[(\pi/4)T^2]$, m/s; Q = gas volumetric rate, m³/s; T = tank diameter, m. Equation (14-211) applies for both the 6BD and the CD-6.

Interfacial Area This consideration in agitated vessels has been reviewed and summarized by Tatterson (*Fluid Mixing and Gas Dispersion in Agitated Tanks,* McGraw-Hill, 1991, pp. 477, 486). Predictive methods for interfacial area are not presented here because correlations are given for the overall volumetric mass-transfer coefficient, liquid phase controlling mass transfer.

Overall Mass-Transfer Coefficient In systems with relatively sparing soluble gases, where the gas-phase resistance is negligible, the mass-transfer rate can be determined by using the concept of an overall volumetric mass-transfer coefficient k_La as follows:

$$M_s = k_La(C_s^* - C_{s,b})V_l \tag{14-212}$$

where M_s = solute molar mass-transfer rate, kg · mol/s; k_La = overall mass-transfer coefficient, 1/s; C_s^* = solute concentration in equilibrium with the liquid phase, kg·mol/m³; and $C_{s,b}$ = solute concentration in bulk of liquid, kg·mol/m³; V_l = liquid volume, m³.

Bakker et al. [*Chem. Eng.* **101:** 98 (December 1994)] have given a correlation for k_La for aqueous systems in the absence of significant surface active agents.

$$k_La = C_{kLa} (P_g / V)^a \upsilon_{sg}^b \tag{14-213}$$

where C_{kLa} = 0.015, an experimentally determined parameter. Equation (14-213) applies for both the 6BD and the CD-6.

Gas-Inducing and Gas-Sparging Impellers The fundamentals of the operation and design of these impellers are presented by Bao, Ye and Jong (*Bioreactor Engineering Research and Industrial Application II,* Springer-Verlag, Berlin, 2016), Patwardhan and Joshi [*Ind. Eng. Chem. Res.* **38:** 49 (1999)], Lemoine et al. [*Chem. Eng. and Processing* **42:** 621 (2003)], Lemoine and Morsi [*Chem. Eng. J.* **114:** 9 (2005)], Patwardhan and Joshi [*Can. J. Chem. Eng.* **76:** 339 (1998)], and Saravanan et al. [*Ind. Eng. Chem. Res.* **33:** 2226 (1994)]. There are many different designs as evidenced by the six styles shown schematically on page 56 of Patwardhan and Joshi 1999. Only two styles will be covered here: (1) a six-blade disk impeller operating near the free surface in a baffled vessel and the WEMCO flotation machine. http://

TABLE 14-23 Design Parameters for Four WEMCO Machines*

WEMCO flotation machine "Design Scale-Up"

Design parameters	Cell size designation			
	#144	#164	#190	#225
Cell volume (cu. ft.)	500	1000	1500	3000
Fourth Area (sq. ft.)	99	122	168	224
Air/pump mixing volume (cu. ft.)	15	31	51	105
Impeller speed (RPM)	192	185	164	140
Air transfer* (ACFM)	240	370	560	870
Liquid circulation* (CFM)	970	1620	1870	3370
Mechanism power* (HP)	26.7	48.3	82.7	160
Air escape velocity* (Ft/min)	2.42	3.03	3.33	3.88
Specific circulation intensity (1/min.)	1.94	1.62	1.25	1.12
Air/pulp mixing residence time (sec.)	0.72	0.92	1.26	1.49
Mechanism power number (P/pn^3D^5)	0.019	0.019	0.022	0.024
Pulp circulation velocity (Ft/min)	263	330	280	334
Air capacity number (Q_a/ND^3)	0.123	0.128	0.138	0.135
Froude Number (N^2D/g)	27.2	29.1	26.7	23.9

*http://www.flsmidth.com/en-US/Industries/Categories/Products/Flotation/WEMCOFlotation/WEMCOFlotation.

www.flsmidth.com/~/media/PDF%20Files/Liquid-Solid%20Separation/Flotation/Wemco11brochure.ashx

Six-Blade Disk Impeller At the optimum depth of submergence, the gas induction rate is estimated on the basis of a gas holdup of 15 percent going through the impeller region. The pumping number for a 6BD impeller is 0.72 (http://www.postmixing.com/mixing%20forum/impellers/impellers.htm); thus, the gas pumping number is $(0.15)0.72 = 0.11 = Q_g/ND^3$. The ratio of gassed power to ungassed power (P_g/P_{ug}) is assumed to be the same as that for a 6BD impeller operating above a ring-type sparger.

The mass transfer correlation of Wu [*Chem. Eng. Sci.* **50:** 2801–2811 (1995)] is recommended

$$k_L a = 0.0634(P/V_l)^{0.65} \qquad (14\text{-}214)$$

where P = impeller power, W; V = batch volume, m^3; and $k_L a$ = volumetric mass-transfer coefficient, 1/s.

The gas holdup within the batch is calculated by Eq. (14-211).

WEMCO Machine The design method is developed from the information presented in Table 14-23 from the WEMCO Flotation Technology Brochure (http://www.flsmidth.com/en-US/Industries/Categories/Products/Flotation/WEMCOFlotation/WEMCOFlotation).

Consider the #164 machine, which has cell volume of 1000 ft^3 (6500 gal; 28.4 m^3) (Fig. 14-98). The air capacity number (Q_a/ND^3) is given as 0.128; the liquid circulation is given as 1620 cfm, so the liquid capacity number is $[1620/\{185(3.33)^3\}] = 0.017$. The power number is given as 0.019, and $\text{Fr} = N^2D/g = 3.08^2(3.33/3.28)/9.81 = 0.98$; thus, the 29.1 in the table is incorrect. The air capacity number is 0.128; thus, the air induced is $Q_a = 0.128(185)3.33^3 = 874$ cfm, which is higher than the value of 370 in Table 14-23.

FIG 14-98 A photograph of a Model 164 WEMCO flotation machine. (*Note*: from the photo and the brochure, the impeller diameter is about 40 in.) (http://www.flsmidth.com/en-US/Industries/Categories/Products/Flotation/WEMCOFlotation/WEMCOFlotation.)

Interfacial Phenomena These can significantly affect overall mass transfer. Deckwer (*Bubble Column Reactors*, Wiley, Hoboken, N.J., 1992) has covered the effect of surfactants on mass transfer in bubble columns. In fermentation reactors, small quantities of surface-active agents (especially antifoaming agents) can drastically reduce overall oxygen transfer, and in aerobic mechanically aerated waste-treatment lagoons, overall oxygen transfer has been found to be from 0.5 to 3 times that for pure water from tests with typical sewage streams.

One cannot quantitatively predict the effect of the various interfacial phenomena; thus, these phenomena will not be covered in detail here. The following literature gives a good general review of the effects of interfacial phenomena on mass transfer: Goodridge and Robb, *Ind. Eng. Chem. Fund.* **4:** 49 (1965); Calderbank, *Chem. Eng. (London)* CE 205 (1967); Gal-Or et al., *Ind. Eng. Chem.* **61**(2): 22 (1969); Kintner, *Adv. Chem. Eng.* **4** (1963); Resnick and Gal-Or, *Adv. Chem. Eng.* **7** (1968), p. 295; Valentin, *Absorption in Gas-Liquid Dispersions*, E. & F. N. Spon, London, 1967; Elenkov, *Theor. Found Chem. Eng.* **1**(1): 117 (1967); and *Ind. Eng. Chem. Ann. Rev. Mass Transfer* **60**(1): 67 (1968); **60**(12): 53 (1968); **62**(2): 41 (1970). In the following outline, the effects of the various interfacial phenomena on the factors that influence overall mass transfer are given. Possible effects of interfacial phenomena are as follows:

1. Effect on continuous-phase mass-transfer coefficient
   a. Impurities concentrate at interface. Bubble motion produces circumferential surface-tension gradients that retard circulation and vibration, thereby decreasing the mass-transfer coefficient.
   b. Large concentration gradients and large heat effects (very soluble gases) can cause interfacial turbulence (the Marangoni effect), which increases the mass-transfer coefficient.
2. Effect on interfacial area
   a. Impurities will lower static surface tension and give smaller bubbles.
   b. Surfactants can electrically charge the bubble surface (produce ionic bubbles) and retard coalescence (soap stabilization of an oil/water emulsion is an excellent example of this phenomenon), thereby increasing the interfacial area.
   c. Large concentration gradients and large heat effects can cause bubble breakup.
3. Effect on mean mass-transfer driving force
   a. Relatively insoluble impurities concentrate at the interface, giving an interfacial resistance. This phenomenon has been used in retarding evaporation from water reservoirs.
   b. The axial concentration variation can be changed by changes in coalescence. The mean driving force for mass transfer is therefore changed.

Gas Holdup (ε) in Bubble Columns Bubble column gas holdup literature has been reviewed by Kantarci, Borak, and Ulgen [*Process Biochemistry* **40**(7): 2263 (2005)], Tsuchiya and Nakanishi [*Chem. Eng. Sci.* **47**(13/14): 3347 (1992)], and Sotelo et al. [*Int. Chem. Eng.* **34**(1): 82–90 (1994)].

With coalescing systems, holdup may be estimated from Hughmark's work [*Ind. Eng Chem. Process Des. Dev.* **6:** 218–220 (1967)] (see Fig. 14-82). For noncoalescing systems, with considerably smaller bubbles, ε can be as great as 0.6 at $U_{sg} = 0.05$ m/s, according to Mersmann [*Ger. Chem. Eng.* **1:** 1 (1978)].

A simple application of the continuity relationship determines that the gas holdup is given by the following equation, given by Joshi and Sharma [*Trans. IChemE* **57:** 244 (1979), eq. 30]:

$$\varepsilon = U_{sg}/U_b \qquad (14\text{-}215)$$

where U_{sg} = superficial gas velocity, m/s, and U_b = volume average vertical bubble rise velocity, m/s. This relationship cannot be used directly with the bubble rise velocities previously discussed because of liquid circulation within the column, which increases the effective bubble rise velocity and lowers the gas holdup; Joshi and Sharma (1979) have addressed the consequences of liquid circulation on gas holdup. An empirical equation suggested by Mashelkar [*Br. Chem. Eng.* **15:** 1297 (1970)] fits Hughmark's data [*Ind. Eng Chem. Process Des. Dev.* **6:** 218–220 (1967)] very well for column diameters >4 in (0.1 m).

$$\varepsilon = U_{sg}/(0.3 + 2U_{sg}) \qquad (14\text{-}216)$$

Kantarci [*Process Biochemistry* **40**(7): 2263 (2005)] lists 20 correlations for gas holdup in bubble columns; a thorough study would include an evaluation of each correlation; however, due to (1) sparger variations and (2) the effect of surfactants on holdup, an initial evaluation need go no further than the use of the works mentioned previously. Only an experimental study will determine accurately the holdup in real systems with varying sparger and varying interfacial effects.

Liquid-phase mass-transfer coefficients in bubble columns have been reviewed by Kantarci et al. (2005). Data by Ozturk, Schumpe, and Deckwer [*AIChE J.* **33:** 1473–1480 (1987)] are presented in Figs. 14-99 and 14-100; these

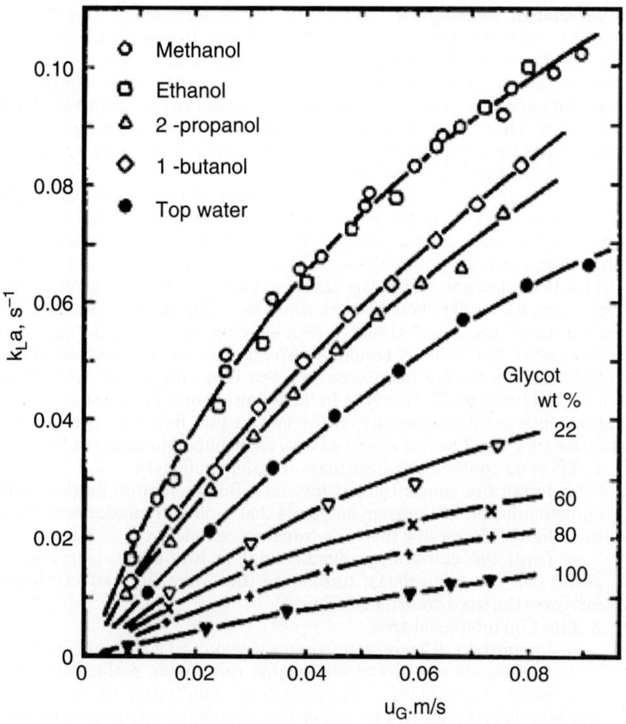

FIG 14-99 Volumetric mass-transfer coefficients in alchols and glycol solutions. [*Ozturk, Schumpe, and Deckwer,* AIChE J. *33: 1477 (1987).*]

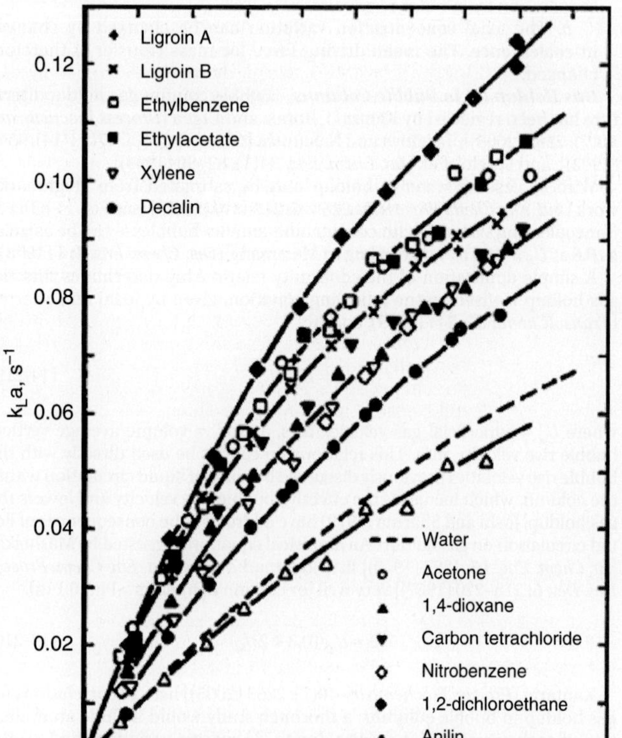

FIG 14-100 Volumetric-mass transfer coefficients in various organic liquids. [*Ozturk, Schumpe, and Deckwer,* AIChE J. *33: 1477 (1987).*]

data can be used directly to estimate the k_La for various systems. These data were used to develop a correlation for the volumetric mass-transfer coefficient

$$\mathrm{Sh}_b = 0.62\mathrm{Sc}^{0.3}\mathrm{Bo}^{0.33}\mathrm{Ga}^{0.29}\mathrm{Fr}^{0.68}(\rho_g/\rho_l)^{0.04} \tag{14-217}$$

Sh_b = Sherwood number $(k_La)d_b^2/D_l$; Sc = Schmidt number $(\mu_l/\rho_l D_l)$; Bo = Bond number $(g\rho_l d_b^2/\sigma)$; Ga = Galileo number = $(g\rho_l d_b^3/\mu_l^2)$; Fr = Froude number = $U_g/(gd_b)^{1/2}$ and k_La = volumetric mass-transfer coefficient, s^{-1}; d_b = bubble diameter = 0.003 m; D_l = diffusivity of the solute in the solvent, m^2/s; U_g = mean superficial gas velocity; μ_l = liquid viscosity, kg/m · s; g = local gravity, m/s^2; σ = interfacial tension, N/m. Based on the study of Qucker and Deckwar [*Ger. Chem. Eng.* **4:** 363 (1981)], the bubble diameter was assumed to have a constant value of 3 mm.

The following dependencies are implied:

$$k_La \; \alpha \; D_l^{0.5}\rho_l^{0.37}\mu_l^{-0.08}\sigma^{-0.33}\rho_g^{0.04}d_b^{-0.81}U_g^{0.68} \tag{14-218}$$

As mentioned earlier, surfactants and ionic solutions significantly affect mass transfer. Normally, surface effects retard coalescence and thus increase the mass transfer. For example, Hikata et al. [*Chem. Eng. J.* **22:** 61–69 (1981)] have studied the effect of KCl on mass transfer in water. As KCl concentration increased, the mass transfer increased up to about 35 percent at an ionic strength of 6 gm/L. Other investigators have found similar increases for liquid mixtures.

Axial Dispersion Backmixing in bubble columns has been extensively studied. Wiemann and Mewes [*Ind. Eng. Chem. Res.* **44:** 4959 (2005)] and Wild et al. [*Int. J. Chemical Reactor Eng.* 1: R7 (2003)] give a long list of references pertaining to backmixing in bubble columns. An excellent review article by Shah et al. [*AIChE J.* **24:** 369 (1978)] has summarized the literature prior to 1978. Works by Konig et al. [*Ger. Chem. Eng.* **1:** 199 (1978)], Lucke et al. [*Trans. Inst. Chem. Eng.* **58:** 228 (1980)], Riquarts [*Ger. Chem. Eng.* **4:** 18 (1981)], Mersmann [*Ger. Chem. Eng.* **1:** 1 (1978)], Deckwer (*Bubble Column Reactors*, Wiley, Hoboken, N.J., 1992), Yang et al. [*Chem. Eng. Sci.* **47**(9–11): 2859 (1992)], and Garcia-Calvo and Leton [*Chem. Eng. Sci.* **49**(21): 3643 (1994)] are particularly useful references.

Axial dispersion occurs in both the liquid and the gas phases. The degree of axial dispersion is affected by vessel diameter, vessel internals, gas superficial velocity, and surface-active agents that retard coalescence. For systems with coalescence-retarding surfactants, the initial bubble size produced by the gas sparger is also significant. The gas and liquid physical properties have only a slight effect on the degree of axial dispersion, except that liquid viscosity becomes important as the flow regime becomes laminar. With pure liquids, in the absence of coalescence-inhibiting, surface-active agents, the nature of the sparger has little effect on the axial dispersion, and experimental results are reasonably well correlated by the dispersion model. For the liquid phase, the correlation recommended by Deckwer et al. [*Can. J. Chem. Eng.* **58:** 190 (1980)], after the original work by Baird and Rice [*Chem. Eng. J.* **9:** 171(1975)] is as follows:

$$\frac{E_L}{(DU_G)} = 0.35\left(\frac{gD}{U_G^2}\right)^{1/3} \tag{14-219}$$

where E_L = liquid-phase axial dispersion coefficient, m^2/s; U_G = superficial velocity of the gas phase, m/s, D = vessel diameter, m; and g = local gravity, m/s^2.

The recommended correlation for the gas-phase axial-dispersion coefficient is given by Field and Davidson ["Axial Dispersion in Bubble Columns," *Trans. Inst. Chem. Eng.* **58:** 228–236 (1980)]:

$$E_G = 56.4\, D^{1.33}\left(\frac{U_G}{\varepsilon}\right)^{3.56} \tag{14-220}$$

where E_G = gas-phase axial-dispersion coefficient, m^2/s; D = vessel diameter, m; U_G = superficial gas velocity, m/s; and ε = fractional gas holdup, volume fraction.

The correlations given in the preceding paragraphs are applicable to vertical cylindrical vessels with pure liquids without coalescence inhibitors. For other vessel geometries such as columns of rectangular cross section, packed columns, and coiled tubes, the work of Shah et al. [*AIChE J.* **24:** 369 (1978)] should be consulted. For systems containing coalescence-inhibiting surfactants, axial dispersion can be vastly different from that in systems in which coalescence is negligible. Konig et al. [*Ger. Chem. Eng.* **1:** 199 (1978)] have well demonstrated the effects of surfactants and sparger type by conducting tests with weak alcohol solutions using three different porous spargers. With pure water, the sparger—and, consequently, initial bubble size—had little effect on backmixing because coalescence produced a dynamic-equilibrium bubble

TABLE 14-24 Comparison of Gas–Liquid Processing Options for Mass Transfer

	$T = Z = 3$ m, $V = 21.2$ m³ (5600 gal)				
System	Bubble column	Bubble column	Surface aeration impeller 6BD	Gas inducing impeller WEMCO[1]	Gas sparged impeller CD-6
Impeller style	None	None	6BD	WEMCO[1]	CD-6
Impeller diameter, D (m)			1.03	0.76	1.03
Impeller speed, N (rpm) [rps]			150 [2.5]	185 [3.1]	215 [3.6]
P/P_0 (ratio; gassed:ungassed)			0.36	?	0.52
Impeller power number ($P/\rho N^3 D^5$)			5	0.019[2]	3.2
Impeller power, P_i (kW)			32	35	87
Impeller specific power, $[P_i/V_l]$ (kW/m³)			1.3	1.42	5
Gas superficial velocity, v_s (m/s)	0.01	0.08	0.043	0.0033	0.11
Gas flow, Q_g (m³/s)	0.07	0.57	0.298	0.023[6]	0.75
% Impeller flood			28	?	10
N_a (aeration number) $[Q_g/ND^3]$			0.11	0.128[6]	0.2
Fr (Froude number) $[N^2D/g]$			0.65	0.98[5]	1.34
Gas holdup, ε	0.01	0.2	0.12	?	0.24
k_La (1/s)	0.0012	0.06	0.093[4]	0.023[3]	0.624
Gas compressor power, P_c (kW)	10	34	0	0	104
Total power, $P_t = P_i + P_c$ (kW)	10	34	38	35	191
Total specific power, $[P_t/V_l]$ (kW/m³)	0.47	1.6	1.8	1.65	11.2
$(k_La)/[P_t/V_l]$ (m³/kW s)	0.0026	0.038	0.05	0.014	0.056

[1]Data taken from http://www.flsmidth.com/~/media/PDF%20Files/Liquid-Solid%20Separation/Flotation/Wemco11brochure.ashx
[2]The impeller diameter is about 40" to require 35 hp (at the impeller) at 185 rpm.
[3]$k_La = 0.0159(P_i/V)^{0.86} = 0.0159(1.34)^{0.86} = 0.023$ [Sawant et al. (op. cit.)].
[4]$k_La = 0.0634(P_i/V)^{0.65} = 0.0634(1.8)^{0.65} = 0.086$ [*Wu* {CES, **50: 2801** (1995)}].
[5]With $D = 40$" (1.02 m) $N = 185$ rpm (3.08 rps), Fr $\neq$ 29.1 as the WEB table shows.
[6]The WEB given induced gas is 370 cfm; however an air capacity number of 0.128 gives $Q_g = 0.128(3.08)(1.02)^3 = 0.418$ cms = 826 cfm; something is not consistent.

size not far above the sparger. With surfactants, the average bubble size was smaller than the dynamic-equilibrium bubble size. Small bubbles produced minimal backmixing up to $\varepsilon \approx 0.40$; however, above $\varepsilon \approx 0.40$, backmixing increased very rapidly as U_G increased. The rapid increase in backmixing as ε exceeds 0.40 was postulated to occur indirectly because a bubble carries upward with it a volume of liquid equal to about 70 percent of the bubble volume, and, for $\varepsilon \approx 0.40$, the bubbles carry so much liquid upward that steady, uniform bubble rise can no longer be maintained and an oscillating, slugging flow develops, which produces fluctuating pressure at the gas distributor and the formation of large eddies. The large eddies greatly increase backmixing. For the air-alcohol-water system, the minimum bubble size to prevent unsteady conditions was about 1, 1.5, and 2 mm for U_G = 1, 3, and 5 cm/s, respectively. Any smaller bubble size produced increased backmixing. The results of Konig et al. (1978) clearly indicate that the interaction of surfactants and sparger can be very complex; thus, one should proceed very cautiously in designing systems for which surfactants significantly retard coalescence. Caution is particularly important because surfactants can produce either much more or much less backmixing than surfactant-free systems, depending on the bubble size, which, in turn, depends on the sparger used.

Table 14-24 summarizes pertinent parameters for two bubble columns—one in the single-bubble regime with v_s = 0.01 m/s and the second in the churn-turbulent regime with v_s = 0.08 m/s—and three mechanically agitated designs, a 6BD operating near the free liquid surface, a commercial surface aerator (WEMCO), and a concave blade impeller operating above a ring-style perforated pipe gas sparger. In terms of mass transfer capability, the churn-turbulent bubble column, the 6BD surface aerator, and the sparged CD-6 were about equal. The WEMCO machine lagged behind these three by a factor of 4, and the single-bubble regime and the single-bubble regime bubble column was a distant fifth. The surface aeration using a 6BD was selected because Wu [*Chem. Eng. Sci.* **50:** 2801–2811 (1995)] investigated that device. As the literature indicates, a four-blade pitched impeller is probably a more efficient machine for this application. As is evident from this evaluation of the various mass-transfer devices, there are many choices for a given process application. Reliable data and reliable design correlations are only available for a few of the possibilities. However, vendors have a storehouse of knowledge that has not been and never will be published. The truly effective process engineer uses vendor resources effectively. For example, Chemineer and other agitator manufacturers can provide design and scale-up help for a four-blade pitched impeller operating just beneath the free surface; Ekato can provide excellent help for hollow shaft and hollow impeller designs, and the FLSmidth and Co. A/S can provide assistance to define, design, and implement a WEMCO machine.

PHASE SEPARATION

Gases and liquids may be intentionally contacted as in absorption and distillation, or a mixture of phases may occur unintentionally as in vapor condensation from inadvertent cooling or liquid entrainment from a film. Regardless of the origin, it is usually desirable or necessary to separate gas–liquid dispersions. Natural separation of the liquid and gas phases occurs due to density differences; however, natural separation is generally inadequate because the process is slow and the location of the separation is uncontrolled. Separation processes are used to accelerate the natural phase separation and to control the location of the phases. Technology for phase separation has changed little in recent years; the challenge remains the correct selection and application of the existing technology to meet the needs for a particular service. Failures in phase separation can cause severe problems in plant performance from direct damage to hardware if excessive liquid reaches a turbine, to corrosion resulting from liquids in unexpected locations, to failure to meet air emission requirements.

GAS-PHASE CONTINUOUS SYSTEMS

Practical separation techniques for liquid particles in gases are discussed. Since gas-borne particulates include both liquid and solid particles, many devices used for dry-dust collection (discussed in Sec. 17 under Gas–Solids Separations) can be adapted to liquid-particle separation. Separation of liquid particulates is often desirable in chemical processes such as in counter-current-stage contacting because liquid entrainment with the gas partially reduces efficiency. Separation before entering another process step may be needed to prevent corrosion, yield loss, or equipment damage or malfunction. Separation before the atmospheric release of gases may be necessary to prevent environmental problems and for regulatory compliance.

General References: Calvert, *J. Air Pollut. Control Assoc.* **24:** 929 (1974); Calvert, *Chem. Eng.* **84**(18): 54 (1977); Calvert and Englund, eds., *Handbook of Air Pollution*

Technology, Wiley, New York, 1984; Calvert, Goldchmid, Leith, and Mehta, NTIS Publ. PB-213016, 213017, 1972; Calvert, Yung, and Leung, NTIS Publ. PB-248050, 1975; Cheremisinoff, ed., *Encyclopedia of Environmental Control Technology,* vol. 2, Gulf Publishing, Houston, 1989; *Code of Federal Regulations,* 40 (CFR 40), subchapter C—Air Programs, parts 50–99, Office of the Federal Register, Washington; Hoffman and Stein, *Gas Cyclones and Swirl Tubes: Principles, Design, and Operation,* 2d ed., Springer, New York, 2008; Katz, M.S. thesis, Pennsylvania State University, 1958; Kouba, G. E., and O. Shoham, "A Review of Gas Liquid Cylindrical Cyclone (GLCC) Technology," Production Separation Systems International Conference, Aberdeen, UK, April 23–24, 1996; Lee and Lin, *Handbook of Environmental Engineering Calculations,* 2d ed. McGraw-Hill, New York, 2007; Moen, Kolbjørn et al., U.S. Patent 9233320 B2, Jan. 12, 2016; Stern, *Air Pollution,* 3d ed., vols. 3–5, Academic Press, Orlando, Fla., 1976–77; Theodore and Buonicore, *Air Pollution Control Equipment: Selection, Design, Operation and Maintenance,* Prentice Hall, Englewood Cliffs, N.J., 1982; York, *Chem. Eng. Prog.* **50:** 421 (1954); York and Poppele, *Chem. Eng. Prog.* **59**(6): 45 (1963); Yung, Barbarika, and Calvert, *J. Air Pollut. Control Assoc.* **27:** 348 (1977).

Definitions: Mist, Fog, and Spray Little standardization has been adopted in defining gas-borne liquid particles, and this often leads to confusion in the selection, design, and operation of collection equipment. *Aerosol* applies to suspended particulate, either solid or liquid, that is slow to settle by gravity and to particles from the submicrometer range up to 10 to 20 μm. Fogs and mists are fine suspended liquid dispersions that usually result from condensation, as discussed in the subsection Liquid-in-Gas Dispersions, and they range upward in particle size from around 0.1 μm. The distinction between fogs and mists is the density of the liquid dispersed in the gas, with fog referring to a denser dispersion. Spray refers to entrained liquid droplets, as described under Liquid-in-Gas Dispersions. In such instances, size will range from the finest particles produced up to a particle whose terminal settling velocity is equal to the entraining gas velocity if some settling volume is provided. Process spray is often created unintentionally, such as by the condensation of vapors on cold duct walls and its subsequent reentrainment, or from two-phase flow in pipes, gas bubbling through liquids, and entrainment from boiling liquids. Table 14-25 lists typical ranges of particle size created by different mechanisms. Figure 14-101 compares the approximate size range of liquid particles with other particulate material and the approximate applicable size range of collection devices.

Background Much of the development of phase separation technology has been driven by air pollution and the evolving environmental regulations. Phase separation requirements are often defined by the need to meet air emissions limits; inadequate phase separation reduces the efficiency of emissions control devices. Much development was funded by the EPA and other government agencies; many of these studies were led by Calvert [Calvert and Englund, eds., *Handbook of Air Pollution Technology,* Wiley, New York, 1984; Yung, Barbarika, and Calvert, *J. Air Pollut. Control Assoc.* 27: 348 (1977)]. Several websites provide access to this information. EPA's Air Pollution Training Institute at www.apti-learn.net provides training courses on wet scrubbers, electrostatic precipitators, and other issues related to air compliance. Another useful resource is www.ntis.gov, which provides access to the National Technical Reports Library and NTIS publications, including relevant technical reports (Calvert, Goldchmid, Leith, and Mehta, NTIS Publ. PB-213016, 213017, 1972; Calvert, Yung, and Leung, NTIS Publ. PB-248050, 1975).

Gas Sampling The sampling of gases containing mists and sprays may be necessary to obtain data for collection-device design, in which case particle-size distribution, total mass loading, and gas volume, temperature, pressure, and composition may all be needed. Other reasons for sampling may be to determine equipment performance, measure yield loss, or determine compliance with regulations. Sampling of two-phase systems must consider both the distribution of the phases in the system and the potential for introduction of errors through the sampling technique.

Location of a sample probe in the process stream is critical, especially when larger particles must be sampled. Mass loading in one portion of a duct may be several times greater than in another portion as affected by flow patterns. Horizontal ducts will tend to concentrate particles toward the bottom of the duct, while vertical ducts will tend to show a higher particle concentration than the true concentration in upflow and a lower concentration in downflow. Therefore, the stream should be sampled at a number of points. The U.S. Environmental Protection Agency [*Code of Federal Regulations,* 40 (CFR 40), subchapter C—Air Programs, parts 50–99, Office of the Federal Register, Washington] has specified 8 points for ducts between 0.3 and 0.6 m (12 and 24 in) and 12 points for larger ducts, provided there are no flow disturbances for eight pipe diameters upstream and two downstream from the sampling point. When only particles smaller than 3 μm are to be sampled, the location and number of sample points are less critical since such particles remain reasonably well dispersed by Brownian motion.

TABLE 14-25 Particle Sizes Produced by Various Mechanisms

Mechanism or process	Particle-size range, μm
Liquid pressure spray nozzle	100–5000
Gas-atomizing spray nozzle	1–100
Gas bubbling through liquid or boiling liquid	20–1000
Condensation processes with fogging	0.1–30
Annular two-phase flow in pipe or duct	10–2000

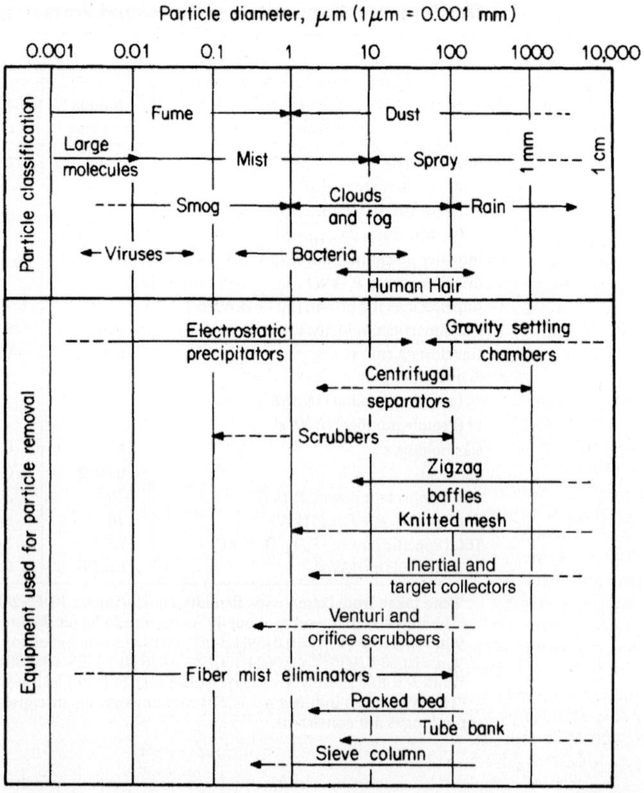

FIG. 14-101 Particle classification and useful collection equipment versus particle size.

Isokinetic sampling (velocity at the probe inlet is equal to local duct velocity) is required to get a representative sample of particles larger than 3 μm (error is small for 4- to 5-μm particles). Lower sample velocities will result in a measured concentration lower than the true concentration, while higher sample velocities will overpredict the true concentration. Sampling methods and procedures for mass loading have been developed [Calvert and Englund, eds., *Handbook of Air Pollution Technology,* Wiley, New York, 1984; Stern, *Air Pollution,* 3d ed., vols. 3–5, Academic Press, Orlando, Fla., 1976–77; *Code of Federal Regulations,* 40 (CFR 40), subchapter C—Air Programs, parts 50–99, Office of the Federal Register, Washington].

Particle Size Analysis Many particle-size-analysis methods suitable for dry-dust measurement are unsuitable for liquids because of coalescence and drainage after collection. Measurement of particle sizes in the flowing stream by using a cascade impactor is one of the better means. The impacting principle was described by Ranz and Wong [*Ind. Eng. Chem.* **44:** 1371 (1952)] and Gillespie and Johnstone [*Chem. Eng. Prog.* **51:** 75F (1955)]. An impactor designed specifically for collecting liquids was described by Brink, Kennedy, and Yu [*Am. Inst. Chem. Eng. Symp. Ser.* **70**(137): 333 (1974)].

In most cases, the design for a phase separator will be based on similar successful applications and an estimate of the range of liquid particle sizes based on the mechanism for the formation of the dispersion rather than on sample results. For new systems, predictions based on similar applications must be used.

Collection Mechanisms Mechanisms that may be used for separating liquid particles from gases are (1) gravity settling, (2) inertial (including centrifugal) impaction, (3) flow-line interception, (4) diffusional (Brownian) deposition, (5) electrostatic attraction, (6) thermal precipitation, (7) flux forces (thermophoresis, diffusiophoresis, Stefan flow), and (8) particle agglomeration (nucleation) techniques. These techniques are similar to techniques for gas–solid separations; equations and parameters for these mechanisms are given in Table 17-2. Most collection devices rarely operate with only a single mechanism, although one mechanism may so predominate that it may be referred to, for instance, as an inertial-impaction device.

Unlike solids, after collection, liquid particles coalesce and must be drained from the unit, minimizing reentrainment. Liquid coalescence allows the use of some devices like wire mesh pads, which would plug in gas–solid separations. Calvert (Calvert, Yung, and Leung, NTIS Publ. PB-248050, 1975) studied the mechanism of reentrainment in a number of liquid-particle collectors. Four types of reentrainment were typically observed: (1) transition from separated flow of gas and liquid to a two-phase region of separated-entrained flow, (2) rupture of bubbles, (3) liquid creep on the separator surface, and (4) shattering of liquid droplets and splashing. Generally, reentrainment increased with increasing gas velocity. Unfortunately, in devices collecting primarily by centrifugal and inertial impaction, primary collection efficiency increases with gas velocity; thus overall efficiency may go through a maximum as reentrainment overtakes the incremental increase in efficiency.

Design and Selection of Collection Devices The selection of phase separation equipment depends on the relative quantities of vapor and liquid, the size of the liquid particles to be removed, the separation efficiency required, the allowable pressure drop, and whether solids are also present in the system. The efficiency required may be determined by regulatory needs if a stream is being cleaned for atmospheric discharge. The particle diameter range of the dispersion combined with the efficiency will determine what the minimum particle size requiring high-efficiency removal will be. Gravity separators are the least efficient on smaller particle sizes, with centrifugal separators, impingement separators (chevrons, mesh pads, baffle separators, etc.), venturi and other scrubbers improving separation at the cost of increased pressure drop and investment. Electrostatic precipitators or fiber mist eliminators are required when the very finest particles must be removed. Because of the difficulty in predicting precisely the particle-size distribution of the liquid, experience with successful designs in similar systems should be considered, as should the experience of the equipment manufacturer. For meaningful design discussions, all of the following must be considered: operating temperature and pressure, gas flow rate and composition for normal and surge conditions, estimated liquid volume, composition, and particle-size distribution based on the process creating the dispersion, liquid and gas physical properties, available pressure drop, required efficiency, and approximate concentration, size distribution, and properties of any solids present.

Calvert and coworkers (Calvert and Englund 1984; Calvert et al. 1972; Calvert 1974; Calvert 1977; and Calvert, Yung, and Leung 1975 in General References) have suggested useful design and selection procedures for particulate-collection devices in which direct impingement and inertial impaction are the most significant mechanisms. The concept is based on the premises that the mass median aerodynamic particle diameter d_{p50} is a significant measure of the difficulty of collection of the liquid particles and that the collection device cut size d_{pc} (defined as the aerodynamic particle diameter collected with 50 percent efficiency) is a significant measure of the capability of the collection device. The aerodynamic diameter for a particle is the diameter of a spherical particle (with an arbitrarily assigned density of 1 g/cm^3) that behaves in an air stream in the same fashion as the actual particle. For airborne liquid particles, the assumption of spherical shape is reasonably accurate. For dilute aqueous particles at ambient temperatures, the actual liquid particle diameter is approximately the equivalent aerodynamic diameter. When a distribution of particle sizes that must be collected is present, the actual size distribution must be converted to a mass distribution by aerodynamic size. Often the distribution can be represented or approximated by a log-normal distribution (a straight line on a log-log plot of cumulative mass percent of particles versus diameter), which can be characterized by the mass median particle diameter d_{p50} and the standard statistical deviation of particles from the median σ_g. σ_g can be obtained from the log-log plot by $\sigma_g = D_{pa50}/D_{pe}$ at 15.87 percent $= D_{pe}$ at 84.13 percent$/D_{pa50}$.

The grade efficiency η of most collectors can be expressed as a function of the aerodynamic particle size in the form of an exponential equation. It is simpler to write the equation in terms of the particle penetration P_t (those particles not collected), where the fractional penetration $P_t = 1 - \eta$, when η is the fractional efficiency. The typical collection equation is

$$P_t = e^{(-A^a D_{pa}{}^B)} \tag{14-221}$$

where A_a and B are functions of the collection device. Calvert et al. (1975) determined that for many devices in which the primary collection mechanism is direct interception and inertial impaction—such as packed beds, knitted-mesh collectors, zigzag baffles, and target collectors such as tube banks, sieve-plate columns, and venturi scrubbers—the value of B is approximately 2.0. For cyclonic collectors, the value of B is approximately 0.67. The overall integrated penetration $\bar{P}_t$ for a device handling a distribution of particle sizes can be obtained by

$$\bar{P}_t = \int_0^W \left(\frac{dW}{W}\right) P_t \tag{14-222}$$

where (dW/W) is the mass of particles in a given narrow size distribution and P_t is the average penetration for that size range. When the particles to be collected are log-normally distributed and the collection device efficiency can be expressed by Eq. (14-221), the required overall integrated collection efficiency $\bar{P}_t$ can be related to the ratio of the device aerodynamic cut size D_{pc} to the mass median aerodynamic particle size D_{pa50}. This required ratio for a given distribution and collection is designated R_{rL}, and these relationships are illustrated graphically in Fig. 14-102. For the many devices for which B is approximately 2.0, a simplified plot (Fig. 14-103) is obtained. From these figures, by knowing the desired overall collection efficiency and particle distribution, the value of R_{rL} can be read. Substituting the mass median particle diameter gives the aerodynamic cut size required from the collection device being considered. Therefore, an experimental plot of aerodynamic cut size for each collection device versus operating parameters can be used to determine the device suitability.

Collection Equipment

Gravity Settlers Gravity can remove larger droplets. Settling or disengaging space above aerated or boiling liquids in a tank or spray zone in a tower can be very useful. If gas velocity is kept low, all particles with terminal settling velocities (see Sec. 6) above the gas velocity will eventually settle. Increasing vessel cross section in the settling zone is helpful. Terminal velocities for particles smaller than 50 μm are very low and generally not attractive for particle removal. Laminar flow of gas in long horizontal paths between trays or shelves on which the droplets settle is another effective means of using gravity. Design equations are given in Sec. 17 under the subsection Gas–Solids Separations. Settler pressure drop is very low, usually being limited to entrance and exit losses. Gravity settling in knockout pots is often used to remove bulk liquid ahead of another device such as an impingement separator, or when a gross separation is all that is needed.

Centrifugal Separation Centrifugal force can be used to enhance particle collection to several hundred times the force of gravity. The design of cyclone separators for dust removal is treated in detail in Sec. 17 under the subsection Gas–Solids Separations, and typical cyclone designs are shown in Fig. 17-55. Cyclones, if carefully designed, can be more efficient on liquids than on solids because liquids coalesce on capture and are easy to drain from the unit. However, some precautions not needed for solid cyclones are necessary to prevent reentrainment.

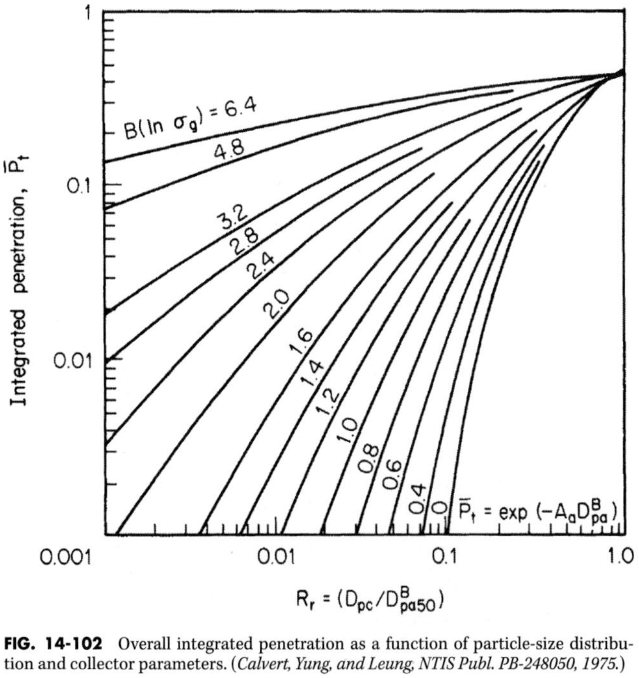

FIG. 14-102 Overall integrated penetration as a function of particle-size distribution and collector parameters. (*Calvert, Yung, and Leung, NTIS Publ. PB-248050, 1975.*)

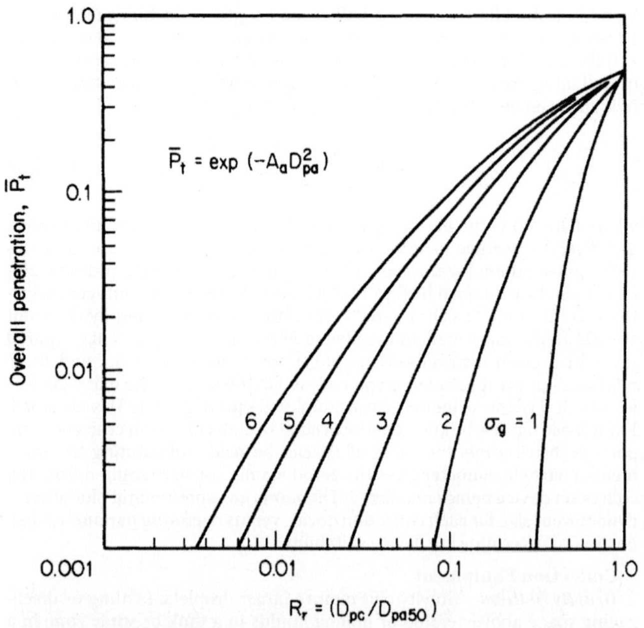

FIG. 14-103 Overall integrated penetration as a function of particle-size distribution and collector cut diameter when B = 2 in, Eq. (14-221). (*Calvert, Goldshmid, Leith, and Mehta, NTIS Publ. PB-213016, 213017, 1972.*)

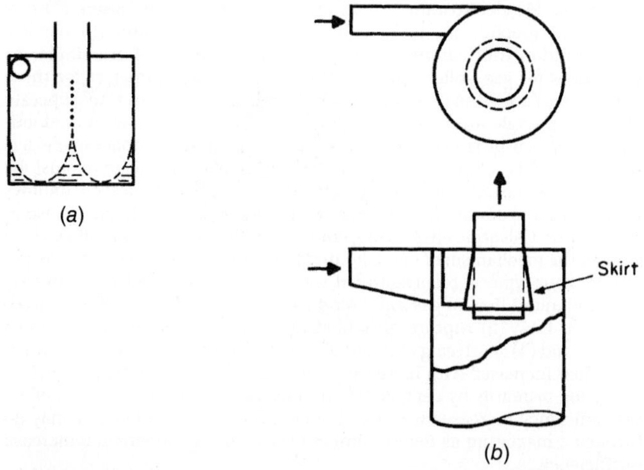

FIG. 14-104 (*a*) Liquid entrainment from the bottom of a vessel by centrifugal flow. (*Rietema and Verver,* Cyclones in Industry, *Elsevier, Amsterdam, 1961.*) (*b*) Gas-outlet skirt for liquid cyclones. (*Stern et al.,* Cyclone Dust Collectors, *American Petroleum Institute, New York, 1955.*)

Cyclone separators can be used for cases of high liquid loading or in demisting applications. Kouba and Shoham ["A Review of Gas–Liquid Cylindrical Cyclone (GLCC) Technology," Production Separation Systems International Conference, Aberdeen, UK, April 23–24, 1996] describe applications of gas-liquid cylindrical cyclones (GLCCs) in oil and gas applications, including metering, preseparation, and internal applications. Hoffman and Stein (*Gas Cyclones and Swirl Tubes: Principles, Design, and Operation*, 2d ed., Springer, New York, 2008) present a detailed analysis of demisting cyclones. Cyclone separators are more efficient than spray towers, smaller and more efficient than gravity separators, and less prone to fouling than impingement separators. Internal cyclone separators have also been used effectively in retrofit situations where existing equipment is not providing adequate phase separation.

Tests by Calvert et al. (1975) showed high primary collection efficiency on droplets down to 10 μm and in accordance with the efficiency equations of Leith and Licht [*Am. Inst. Chem. Eng. Symp. Ser.* **68**(126): 196–206 (1972)] for the specific cyclone geometry tested if entrainment is avoided. Typical entrainment points are (1) creep along the gas outlet pipe, (2) entrainment by shearing of the liquid film from the walls, and (3) vortex pickup from accumulated liquid in the bottom (Fig. 14-104*a*). Reentrainment from creep of liquid along the top of the cyclone and down the outlet pipe can be prevented by providing the outlet pipe with a flared conical skirt (Fig. 14-104*b*), which provides a point from which the liquid can drip without being caught in the outlet gas. The skirt should be slightly shorter than the gas outlet pipe but should extend below the bottom of the gas inlet. The cyclone inlet gas should not impinge on this skirt. Often the bottom edge of the skirt is V-notched or serrated. Roof skimmers or inlet raceways can also be provided to reduce liquid creep.

Reentrainment is generally reduced by lower inlet gas velocities. Calvert et al. (1975) reviewed the literature on predicting the onset of entrainment and found that of Chien and Ibele (ASME Pap. 62-WA170) to be the most reliable. Calvert applied their correlation to a liquid Reynolds number on the wall of the cyclone, $N_{Re,L} = 4Q_L/h_i\upsilon_L$, where Q_L is the volumetric liquid flow rate, cm^3/s; h_i is the cyclone inlet height, cm; and υ_L is the kinematic liquid viscosity, cm^2/s. He found that the onset of entrainment occurs at a cyclone inlet gas velocity V_{ci}, m/s, in accordance with the relationship in $V_{ci} = 6.516 - 0.2865 \ln N_{Re,L}$.

Reentrainment from the bottom of the cyclone can be prevented in several ways. If a typical long-cone dry cyclone is used and liquid is kept continually drained, vortex entrainment is unlikely. However, a vortex breaker baffle in the outlet is desirable, and perhaps a flat disk on top extending to within 2 to 5 cm (0.8 to 2 in) of the walls may be beneficial. Often liquid cyclones are built without cones and have dished bottoms. These designs require both an isolation plate to prevent the vortex from contacting the liquid level and a vortex breaker on the liquid outlet. Perforated, vertical wall baffles can also be considered to reduce bulk rotation of the liquid pool. Research continues on the optimum designs to prevent liquid reentrainment; for example, Moen et al. (U.S. Patent 9233320 B2, Jan. 12, 2016) present configurations for roof skimmers, isolation plates, and vortex breakers to improve efficiency.

As with dust cyclones, no reliable pressure-drop equations exist (see Sec. 17), although many have been published. A part of the problem is that there is no standard cyclone geometry. Calvert et al. (1975) experimentally obtained $\Delta P = 0.000513\, \rho_g (Q_g/h_i w_i)^2 (2.8 h_i w_i/d_o^2)$, where ΔP is in cm of water; ρ_g is the gas density, g/cm^3; Q_g is the gas volumetric flow rate, cm^3/s; h_i and w_i are cyclone inlet height and width, respectively, cm; and d_o is the gas outlet diameter, cm. This equation is in the same form as that proposed by Shepherd and Lapple [*Ind. Eng. Chem.* **31:** 1246 (1940)] but gives only 37 percent as much pressure drop.

Liquid cyclone efficiency can be improved somewhat by introducing a coarse spray of liquid in the cyclone inlet. Large droplets that are easily collected collide with finer particles as they sweep the gas stream in their travel to the wall. (See the subsection Wet Scrubbers regarding optimum spray size.) The most effective operation is obtained by spraying countercurrently to the gas flow in the cyclone inlet duct at liquid rates of 0.7 to 2.0 L/m^3 of gas. There are also many proprietary designs of liquid separators using centrifugal force, some of which are illustrated in Fig. 14-105. Many of these were originally developed as steam separators to remove entrained condensate. In some designs, impingement on swirl baffles aids separation.

Impingement Separation Impingement separation employs direct impact and inertial forces between particles, the gas streamlines, and target bodies to provide capture. The mechanism is discussed in Sec. 17 under the subsection Gas–Solids Separations. With liquids, droplet coalescence occurs on the target surface, and provision must be made for drainage without reentrainment. Calvert et al. (1975) studied droplet collection by impingement on targets consisting of banks of tubes, zigzag baffles, and packed and mesh beds. Figure 14-106 illustrates several types of impingement-separator designs. Methods for efficiency calculations are discussed; in practice, the equipment manufacturers' experience will most likely guide the design.

In its simplest form, an impingement separator may be nothing more than a target placed in front of a flow channel such as a disk at the end of a tube. To improve collection efficiency, the gas velocity may be increased by forming the end into a nozzle (Fig. 14-106*a*). Particle collection as a function of size may be estimated by using the target-efficiency correlation in Fig. 17-52. Since target efficiency will be low for systems with separation numbers below 5 to 10 (small particles, low gas velocities), the mist will often be subjected to a number of targets in series as in Fig. 14-106*c*, *d*, and *g*.

The overall droplet penetration is the product of penetration for each set of targets in series. For a distribution of particle sizes, an integration procedure is required to give overall collection efficiency. This target-efficiency method is suitable for predicting efficiency when the design effectively prevents the bypassing or short-circuiting of targets by the gas stream and provides adequate time to accelerate the liquid droplets to gas velocity. Katz (M.S. thesis, Pennsylvania State University, 1958) investigated a jet and

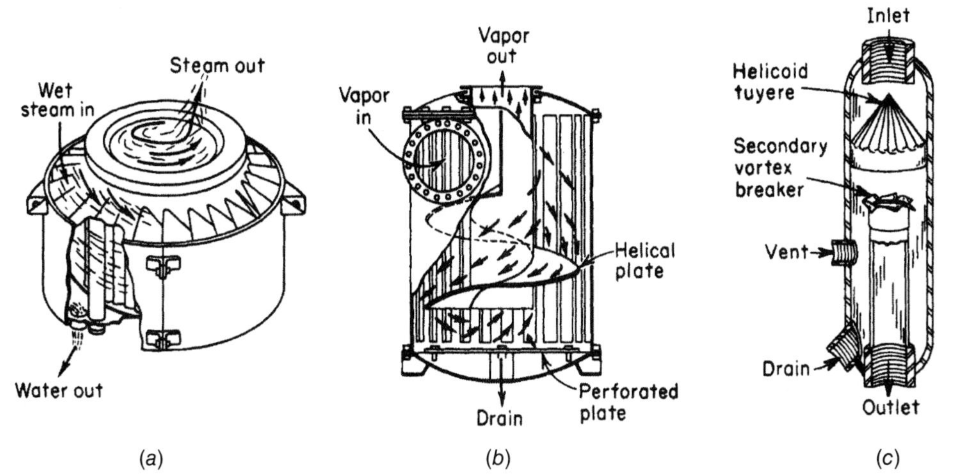

FIG. 14-105 Typical separators using impingement in addition to centrifugal force. (*a*) Hi-eF purifier. (*V. D. Anderson Co.*) (*b*) Flick separator. (*Wurster & Sanger, Inc.*) (*c*) Type RA line separator. (*Centrifix Corp., Bull. 220.*)

target-plate entrainment separator design and found the pressure drop less than would be expected to supply the kinetic energy both for droplet acceleration and gas friction. An estimate based on his results indicates that the liquid particles on the average were being accelerated to only about 60 percent of the gas velocity. The largest droplets, which are the easiest to collect, will be accelerated less than the smaller particles. This factor has a leveling effect on collection efficiency as a function of particle size, so experimental results on such devices may not show as sharp a decrease in efficiency with particle size as predicted by calculation. Katz (1958) also studied *wave-plate impingement separators* (Fig. 14-106*b*) made up of 90° formed arcs with an 11.1-mm (0.44-in) radius and a 3.8-mm (0.15-in) clearance between sheets. The pressure drop is a function of system geometry. The pressure drop for

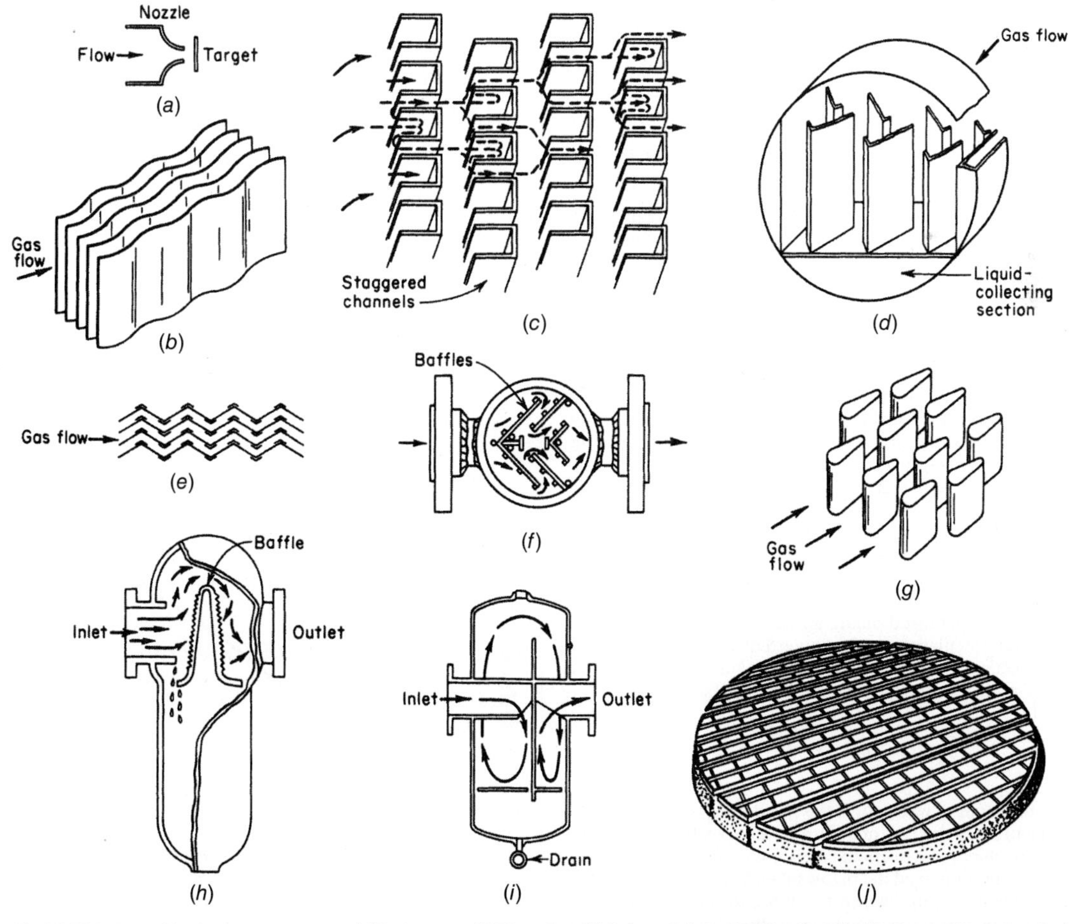

FIG. 14-106 Typical impingement separators. (*a*) Jet impactor. (*b*) Wave plate. (*c*) Staggered channels. (*Blaw-Knox Food & Chemical Equipment, Inc.*) (*d*) Vane-type mist extractor. (*Maloney-Crawford Tank and Mfg. Co.*) (*e*) Peerless line separator. (*Peerless Mfg. Co.*) (*f*) Strong separator. (*Strong Carlisle and Hammond.*) (*g*) Karbate line separator. (*Union Carbide Corporation.*) (*h*) Type E horizontal separator. (*Wright-Austin Co.*) (*i*) PL separator. (*Ingersoll Rand.*) (*j*) Wire-mesh demister. (*Otto H. York Co.*)

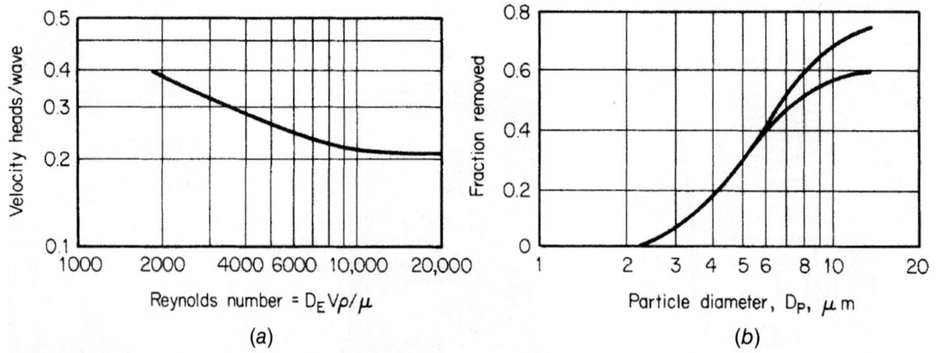

FIG. 14-107 Pressure drop and collection efficiency of a wave-plate separator. (*a*) Pressure drop. (*b*) Efficiency D_E = clearance between sheets. (*Katz, M.S. thesis, Pennsylvania State University, 1958.*)

Katz's system and collection efficiency for a separator with seven waves are shown in Fig. 14-107. Katz used the Souders-Brown expression to define a design velocity U for the gas between the waves:

$$U = K\sqrt{(\rho_l - \rho_g)/\rho_g} \tag{14-223}$$

K is 0.12 to give U in ms^{-1} (0.4 for ft/s), and ρ_l and ρ_g are liquid and gas densities in any consistent set of units. Katz found no change in efficiency at gas velocities from one-half to three times that given by the equation.

Calvert et al. (1975) investigated *zigzag baffles* of a design more like Fig. 14-106*e*. The baffles may have spaces between the changes in direction or be connected as shown. He found close to 100 percent collection for water droplets of 10 μm and larger. Some designs had high efficiencies down to 5 or 8 μm. Desirable gas velocities were 2 to 3.5 m/s (6.6 to 11.5 ft/s), with a pressure drop for a six-pass baffle of 2 to 2.5 cm (0.8 to 1.0 in) of water. On the basis of turbulent mixing, an equation was developed for predicting primary collection efficiency as a function of particle size and collector geometry:

$$\eta = 1 - \exp\left[-\frac{u_{te} n W \theta}{57.3 U_g b \tan\theta}\right] \tag{14-224}$$

where η is the fractional primary collection efficiency; u_{te} is the drop terminal centrifugal velocity in the normal direction, cm/s; U_g is the superficial gas velocity, cm/s; n is the number of rows of baffles or bends; θ is the angle of inclination of the baffle to the flow path, °; W is the width of the baffle, cm; and b is the spacing between baffles in the same row, cm. For conditions of low Reynolds number ($N_{Re,D} < 0.1$) where Stokes' law applies, Calvert obtained the value for drop terminal centrifugal velocity of $u_{te} = d_p^2 \rho_p a / 18\mu_g$, where d_p and ρ_p are the drop particle diameter, cm, and particle density, g/cm³, respectively; μ_g is the gas viscosity, P; and a is the acceleration due to centrifugal force. It is defined by the equation $a = 2U_g^2 \sin\theta / W \cos^3\theta$. For situations in which Stokes' law does not apply, Calvert recommended $u_{te} = 4d_p \rho_p a / 3C_D \rho_g$, where ρ_g is the gas density, g/cm²; and C_D is the drag coefficient from Foust et al. (*Principles of Unit Operations*, Toppan Co., Tokyo, 1959).

Calvert found that reentrainment from the baffles was affected by the gas velocity, the liquid-to-gas ratio, and the orientation of the baffles. Horizontal gas flow past vertical baffles provided the best drainage and lowest reentrainment. Safe operating regions with vertical baffles are shown in Fig. 14-108. Horizontal baffles gave the poorest drainage and the highest reentrainment, with inclined baffles intermediate in performance. Equation (14-225), developed by Calvert, predicts pressure drop across zigzag baffles. The indicated summation must be made over the number of rows of baffles present.

$$\Delta P = \sum_{i=1}^{i=n} 1.02 \times 10^{-3} f_D \rho_g \frac{U_g'^2 A_p}{2A_t} \tag{14-225}$$

ΔP is the pressure drop, cm of water; ρ_g is the gas density, g/cm³; A_p is the total projected area of an entire row of baffles in the direction of inlet gas flow, cm²; and A_t is the duct cross-sectional area, cm². The value f_D is a drag coefficient for gas flow past inclined flat plates taken from Fig. 14-109, while U_g' is the actual gas velocity, cm/s, which is related to the superficial gas velocity U_g by $U_g' = U_g/\cos\theta$. It must be noted that the angle of incidence θ for the second and successive rows of baffles is twice the angle of incidence for the first row. Most of Calvert's work was with 30° baffles, but the method correlates well with other data on 45° baffles.

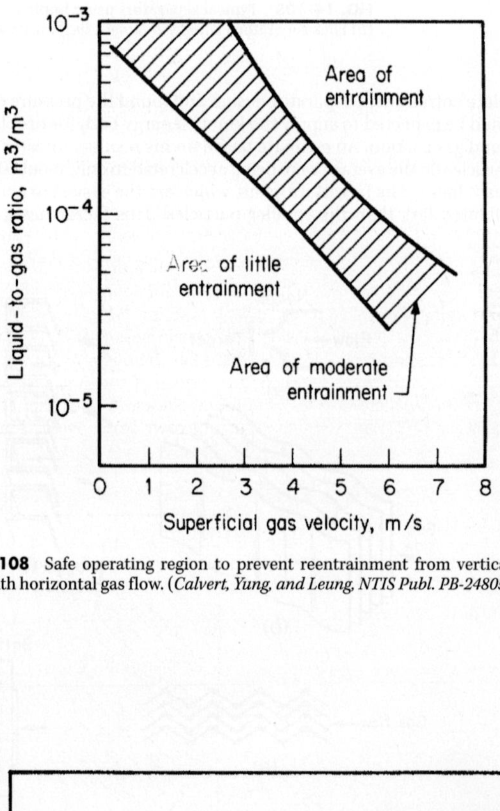

FIG. 14-108 Safe operating region to prevent reentrainment from vertical zigzag baffles with horizontal gas flow. (*Calvert, Yung, and Leung, NTIS Publ. PB-248050, 1975.*)

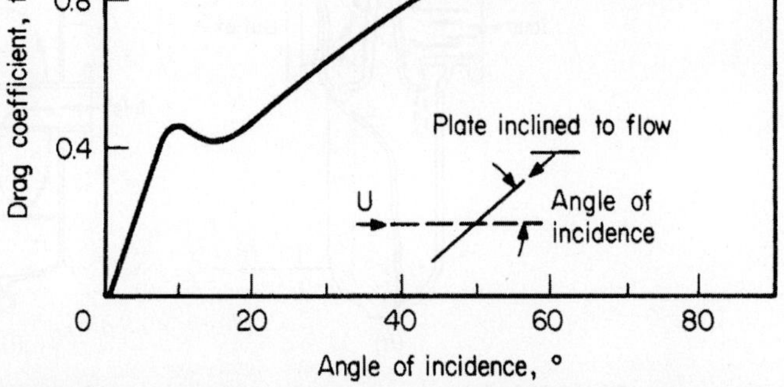

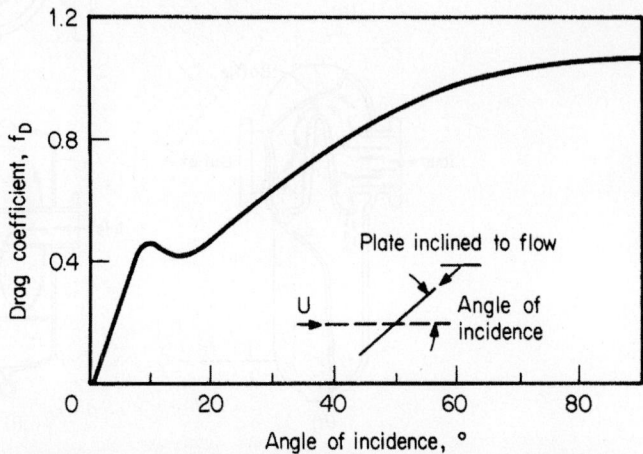

FIG. 14-109 Drag coefficient for flow past inclined flat plates for use in Eq. (14-224). [*Calvert, Yung, and Leung, NTIS Publ. PB-248050; based on Fage and Johansen,* Proc. R. Soc. (London), *116A, 170 (1927).*]

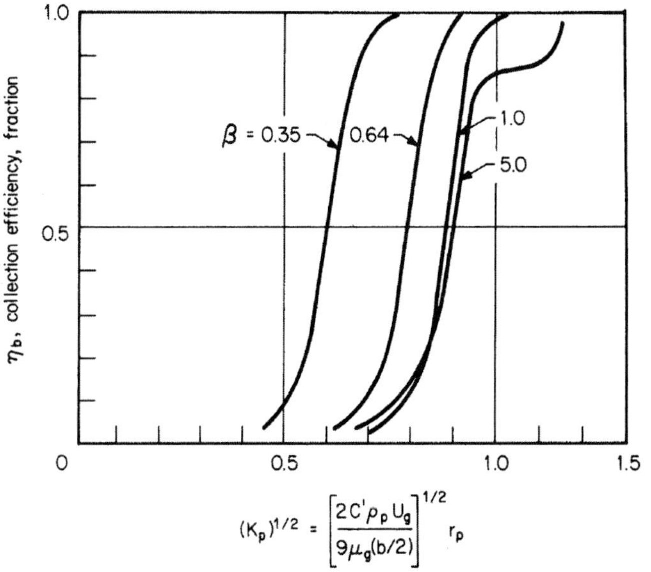

FIG. 14-110 Experimental collection efficiencies of rectangular impactors. C' is the Stokes-Cunningham correction factor; ρ_p, particle density, g/cm^3; U_g, superficial gas velocity, approaching the impactor openings, cm/s; and μ_g, gas viscosity, P. [*Calvert, Yung, and Leung, NTIS Publ. PB-248050; based on Mercer and Chow,* J. Coll. Interface Sci. *27: 75 (1968).*]

The use of multiple tube banks as a droplet collector was also studied by Calvert et al. (1975). He reported that collection efficiency for closely packed tubes follows equations for rectangular jet impaction, which can be obtained graphically from Fig. 14-110 by using a dimensionless parameter β that is based on the tube geometry; $\beta = 2l_i/b$, where b is the open distance between adjacent tubes in the row (orifice width) and l_i is the impaction length (distance between orifice and impingement plane), or approximately the distance between centerlines of successive tube rows. Note that the impaction parameter K_p is plotted to the one-half power in Fig. 14-110 and that the radius of the droplet is used rather than the diameter. Collection efficiency overall for a given size of particle is predicted for the entire tube bank by

$$\eta = 1 - (1 - \eta_b)^N \qquad (14\text{-}226)$$

where η_b is the collection efficiency for a given size of particle in one stage of a rectangular jet impactor (Fig. 14-111) and N is the number of stages in the tube bank (equal to one less than the number of rows). For widely spaced tubes, the target efficiency η_g can be calculated from Fig. 17-52 or from the impaction data of Golovin and Putnam [*Ind. Eng. Chem. Fundam.* **1:** 264 (1962)]. The efficiency of the overall tube banks for a specific particle size can then be calculated from the equation $\eta = 1 - (1 - \eta_t a'/A)^n$, where a' is the cross-sectional area of all tubes in one row, A is the total flow area, and n is the number of rows of tubes.

Calvert reported pressure drop through tube banks to be largely unaffected by liquid loading and indicated that Grimison's correlations in Sec. 6 (Tube Banks) for gas flow normal to tube banks or data for gas flow through heat-exchanger bundles can be used. However, the following equation is suggested:

$$\Delta P = 8.48 \times 10^{-3} n \rho_g U_g'^2 \qquad (14\text{-}227)$$

where ΔP is cm of water; n is the number of rows of tubes; ρ_g is the gas density, g/cm^3; and U_g' is the actual gas velocity between tubes in a row, cm/s. Calvert did find an increase in pressure drop of about 80 to 85 percent above that predicted by Eq. (14-227) in the vertical upflow of gas through tube banks due to liquid holdup at gas velocities above 4 m/s. The onset of liquid reentrainment from tube banks can be predicted from Fig. 14-111. Reentrainment occurred at much lower velocities in vertical upflow than in horizontal gas flow through vertical tube banks. While the top of the cross-hatched line of Fig. 14-111*a* predicts reentrainment above gas velocities of 3 m/s (9.8 ft/s) at high liquid loading, most of the entrainment settled to the bottom of the duct in 1 to 2 m (3.3 to 6.6 ft), and entrainment did not carry significant distances until the gas velocity exceeded 7 m/s (23 ft/s).

Packed-Bed Collectors Many different materials, including coal, coke, and broken solids, as well as normal types of tower-packing rings, saddles, and special plastic shapes, have been used over the years in packed beds to remove entrained liquids through impaction and filtration. Separators using natural materials are not available as standard commercial units but

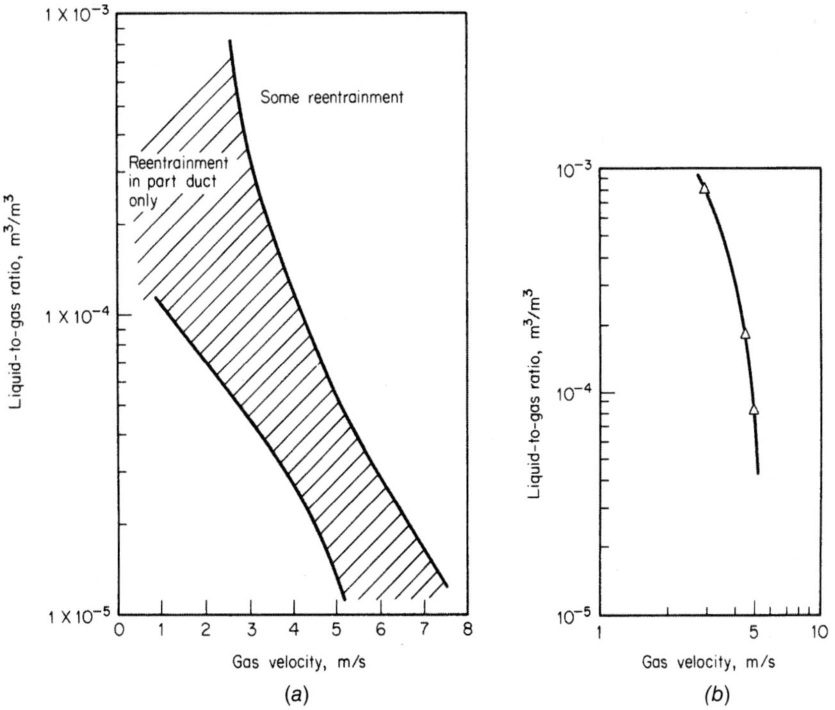

FIG. 14-111 Experimental results showing effect of gas velocity and liquid load on entrainment from (*a*) vertical tube banks with horizontal gas flow and (*b*) horizontal tube banks with upflow. To convert meters per second to feet per second, multiply by 3.281. (*Calvert, Yung, and Leung, NTIS Publ. PB-248050.*)

are designed for specific applications and have now been largely superseded by more efficient devices.

Calvert et al. (1975) generalized several studies to predict the collection efficiencies of liquid particles in any packed bed. Assumptions in the theoretical development are that the drag force on the drop is given by Stokes' law and that the number of semicircular bends to which the gas is subjected, η_1, is related to the length of the bed, Z (cm), in the direction of gas flow, the packing diameter, d_c (cm), and the gas-flow channel width, b (cm), such that $\eta_1 = Z/(d_c + b)$. The gas velocity through the channels, U_{gb} (cm/s), is inversely proportional to the bed free volume for gas flow such that $U_{gb} = U_g[1/(\varepsilon - h_b)]$, where U_g is the gas superficial velocity, cm/s, approaching the bed, ε is the bed void fraction, and h_b is the fraction of the total bed volume taken up with liquid, which can be obtained from data on liquid holdup in packed beds. The width of the semicircular channels b can be expressed as a fraction j of the diameter of the packing elements, such that $b = jd_c$. These assumptions (as modified by G. E. Goltz, personal communication) lead to an equation for predicting the penetration of a given size of liquid particle through a packed bed:

$$P_t = \exp\left[\frac{-\pi}{2(j+j^2)(\varepsilon - h_b)}\left(\frac{Z}{d_c}\right)K_p\right] \tag{14-228}$$

where

$$K_p = \frac{\rho_p d_p^2 U_g}{9\mu_g d_c} \tag{14-229}$$

Values of ρ_p and d_p are droplet density, g/cm^3, and droplet diameter, cm; μ_g is the gas viscosity, P. All other terms were defined previously. Table 14-26 gives values of j calculated from experimental data of Jackson and Calvert. Values of j for most manufactured packing appear to fall in the range from 0.16 to 0.19. The low value of 0.03 for coke may be due to the porosity of the coke itself.

Packed sections may be designed for vertical or horizontal gas flow with liquid flow either countercurrent, cocurrent, or cross-flow for horizontal gas flow. The cross-flow design can help prevent plugging in systems that contain solids. Calvert et al. (1975) tested the correlation in cross-flow packed beds, which tend to give better drainage than countercurrent beds, and has found the effect of gas-flow orientation insignificant. However, the onset of reentrainment was somewhat lower in a bed of 2.5-cm (1.0-in) Pall rings with gas upflow [6 m/s (20 ft/s)] than with horizontal cross-flow of gas. The onset of reentrainment was independent of liquid loading, and entrainment occurred at values somewhat above the flood point for packed beds as predicted by conventional correlations. In beds with more than 3 cm (1.2 in) of water pressure drop, the experimental drop with both vertical and horizontal gas flow was somewhat less than predicted by generalized packed-bed pressure-drop correlations. However, Calvert recommended these correlations for design as conservative.

Calvert's data indicate that packed beds irrigated only with the collected liquid can have collection efficiencies of 80 to 90 percent on mist particles down to 3 μm, but they have low efficiency on finer mist particles. Often, irrigated packed towers and towers with internals will be used with liquid having a wetting capability for the fine mist that must be collected. Calvert [Calvert et al. (1972); Calvert (1974)] reported on the efficiency of various gas-liquid-contacting devices for fine particles. Equation (14-228) can be used to calculate generalized design curves for collection in packed columns by finding parameters of packing size, bed length, and gas velocity that give collection efficiencies of 50 percent for various sizes of particles. Figure 14-112 illustrates such a plot for three gas velocities and two sizes of packing.

Wire-Mesh Mist Collectors Knitted mesh of varying density and voidage is widely used for entrainment separators. Its advantage is close to 100 percent removal of drops larger than 5 μm at superficial gas velocities from about 0.2 ms/s (0.6 ft/s) to 5 m/s (16.4 ft/s), depending somewhat on the design of the mesh. Pressure drop is usually no more than 2.5 cm (1 in) of water. A major disadvantage is the ease with which tars and insoluble solids plug the mesh. The separator can be made to fit vessels of any shape and can be made of any material that can be drawn into a wire. Stainless steel and plastic fibers are most common, but other metals are sometimes used. Manufacturers typically consider the design details of these separators to be proprietary. While general correlations can be useful to understand likely performance, equipment manufacturers should be consulted for expected performance in a particular application. Generally three basic types of mesh are used: (1) layers with a crimp in the same direction (each layer is actually a nested double layer); (2) layers with a crimp in alternate directions, which increases voidage, reduces sheltering, and increases target efficiency per layer, and gives a lower pressure drop per unit length; and (3) spiral-wound layers that reduce pressure drop by one-third, but fluid creep may lead to higher entrainment. The filament size can vary from about 0.15 mm (0.006 in) for fine-wire pads to 3.8 mm (0.15 in) for some plastic fibers. Typical pad thickness varies from 100 to 150 mm (4 to 6 in), but occasionally pads up to 300 mm (12 in) thick are used.

Figure 14-113 presents an early calculated estimate of mesh efficiency as a fraction of mist-particle size. Experiments by Calvert et al. (1975) confirmed the accuracy of the equation of Bradie and Dickson (*Joint Symp. Proc. Inst. Mech. Eng./Yorkshire Br. Inst. Chem. Eng.*, 1969, pp. 24–25) for primary efficiency in mesh separators:

$$\eta = 1 - \exp(-2/3)\pi a l \eta_i) \tag{14-230}$$

TABLE 14-26 Experimental Values for *j*, Channel Width in Packing as a Fraction of Packing Diameter

Packing size			
cm	in	Type of packing	j
1.27	0.5	Berl and Intalox saddles, marbles, Raschig rings	0.192
2.54	1.0	Berl and Intalox saddles, pall rings	0.190
3.8	1.5	Berl and Intalox saddles, pall rings	0.165
7.6–12.7	3–5	Coke	0.03

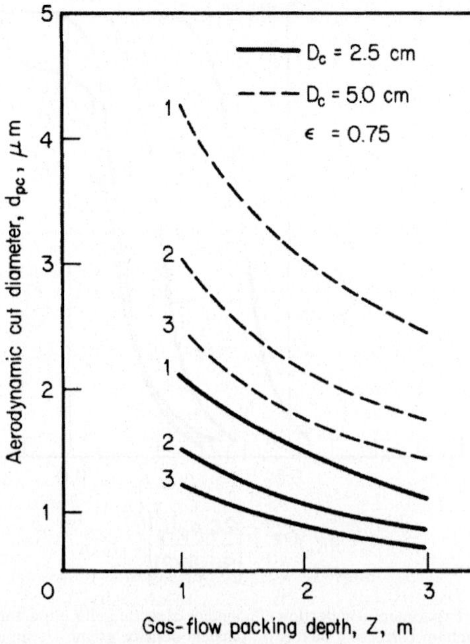

FIG. 14-112 Aerodynamic cut diameter for a typical packed-bed entrainment separator as a function of packing size, bed depth, and three gas velocities: curve 1–1.5 m/s, curve 2–3.0 m/s, and curve 3–4.5 m/s. To convert meters to feet, multiply by 3.281; to convert centimeters to inches, multiply by 0.394. [*Calvert,* J. Air Pollut. Control Assoc. *24:* 929 *(1974).*]

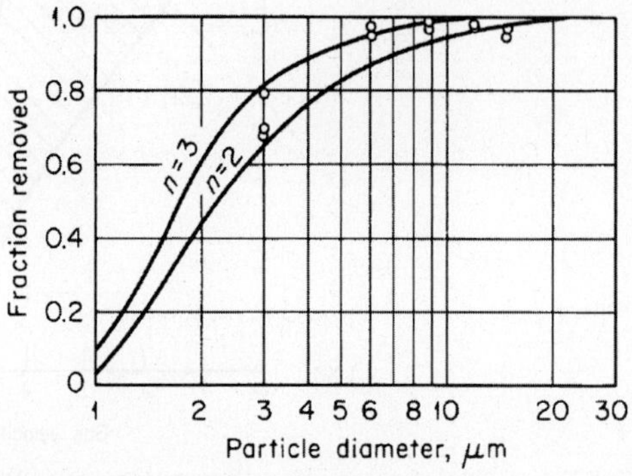

FIG. 14-113 Collection efficiency of wire-mesh separator; 6-in thickness, 98.6 percent free space, 0.006-in-diameter wire used for experiment points. Curves calculated for target area equal to 2 and 3 times the solids volume of packing. To convert inches to millimeters, multiply by 25.4.

where η is the overall collection efficiency for a given-size particle; l is the thickness of the mesh, cm, in the direction of gas flow; a is the surface area of the wires per unit volume of mesh pad, cm^2/cm^3; and η_i, the target collection efficiency for cylindrical wire, can be calculated from Fig. 17-52 or the impaction data of Golovin and Putnam [*Ind. Eng. Chem.* **1:** 264 (1962)]. The factor 2/3, introduced by Carpenter and Othmer [*Am. Inst. Chem. Eng. J.* **1:** 549 (1955)], corrects for the fact that not all the wires are perpendicular to the gas flow and gives the projected perpendicular area. If the specific mesh surface area a is not available, it can be calculated from the mesh void area ε and the mesh wire diameter d_w in cm, $a = 4(1 - \varepsilon)/d_w$.

York and Poppele [*Chem. Eng. Prog.* **59**(6): 45 (1963)] stated that factors governing maximum allowable gas velocity through the mesh are (1) gas and liquid density, (2) liquid surface tension, (3) liquid viscosity, (4) specific wire surface area, (5) entering-liquid loading, and (6) suspended-solids content. York [*Chem. Eng. Prog.* **50:** 421 (1954)] proposed the application of the Souders-Brown equation [Eq. (14-223)] for the correlation of maximum allowable gas velocity with values of K for most cases of 0.1067 m/s to give U in m/s (0.35 for ft/s). When liquid viscosity or inlet loading is high or the liquid is dirty, the value of K must be reduced. Schroeder (M.S. thesis, Newark College of Engineering, 1962) found lower values for K necessary when liquid surface tension is reduced, such as by the presence of surfactants in water. Ludwig (*Applied Process Design for Chemical and Petrochemical Plants,* 2d ed., vol. I, Gulf Publishing, Houston, 1977, p. 157) recommended reduced K values of (0.061 m/s) under vacuum at an absolute pressure of 6.77 kPa (0.98 lbf/in^2) and $K = 0.082$ m/s at 54 kPa (7.83 lbf/in^2) absolute. Most manufacturers suggest setting the design velocity at three-fourths of the maximum velocity to allow for surges in gas flow.

York and Poppele [*Chem. Eng. Prog.* **59**(6): 45 (1963)] suggested that total pressure drop through the mesh is equal to the sum of the mesh dry pressure drop plus an increment due to the presence of liquid. They considered the mesh to be equivalent to many small circular channels and used the D'Arcy formula with a modified Reynolds number to correlate friction factor f (see Fig. 14-114) for Eq. (14-231), giving dry pressure drop.

$$\Delta P_{dry} = fla\rho_g U_g^2 / 981\varepsilon^3 \qquad (14\text{-}231)$$

where ΔP is in cm of water; f is from Fig. (14-114); ρ_g is the gas density, g/cm^3; U_g is the superficial gas velocity, cm/s; and ε is the mesh porosity or void fraction; l and a are as defined in Eq. (14-230). Figure 14-114 gives data of York and Poppele for mesh crimped in the same and alternating directions and also includes the data of Satsangee, of Schurig, and of Bradie and Dickson.

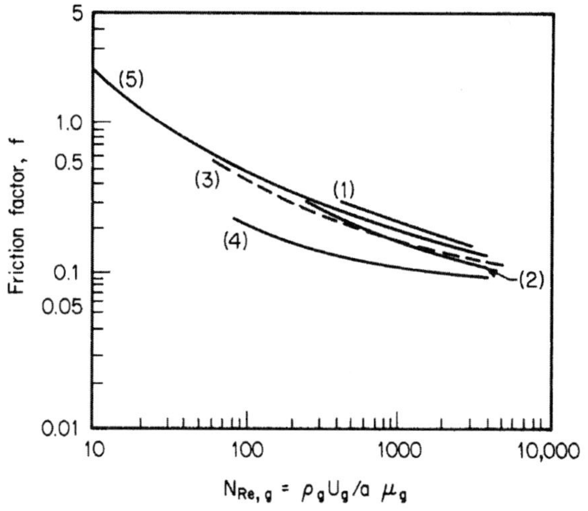

FIG. 14-114 Value of friction factor f for dry knitted mesh for Eq. (14-231). Values of York and Poppele [Chem. Eng. Prog. *50: 421 (1954)*] are given in curve 1 for mesh crimped in the alternating direction and curve 2 for mesh crimped in the same direction. Data of Bradie and Dickson (*Joint Symp. Proc. Inst. Mech. Eng./Yorkshire Br. Inst. Chem. Eng., 1969, pp. 24–25*) are given in curve 3 for layered mesh and curve 4 for spiral-wound mesh. Curve 5 is data of Satsangee (M.S. thesis, Brooklyn Polytechnic Institute, 1948) and Schurig (D.Ch.E. dissertation, Brooklyn Polytechnic Institute, 1946). (*From Calvert, Yung, and Leung, NTIS Publ. PB-248050, 1975.*)

The incremental pressure drop for wet mesh is not available for all operating conditions or for mesh of different styles. The data of York and Poppele for wet-mesh incremental pressure drop, ΔP_L in cm of water, are shown in Fig. 14-115 for parameters of liquid velocity L/A, defined as liquid volumetric flow rate, cm^3/min per unit of mesh cross-sectional area in cm^2; liquid density ρ_L is in g/cm^3.

York generally recommends the installation of the mesh horizontally with upflow of gas as in Fig. 14-106*f*; Calvert et al. (1975) tested the mesh horizontally with upflow and vertically with horizontal gas flow. He reported better drainage with the mesh vertical and somewhat higher permissible gas velocities without reentrainment, which is contrary to past practice. With horizontal flow through vertical mesh, he found collection efficiency to follow

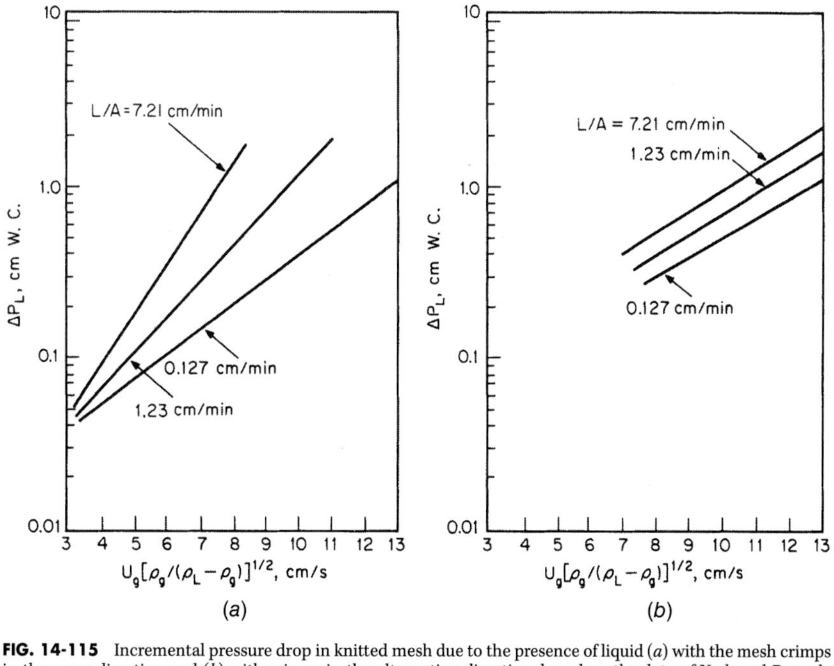

FIG. 14-115 Incremental pressure drop in knitted mesh due to the presence of liquid (a) with the mesh crimps in the same direction and (b) with crimps in the alternating direction, based on the data of York and Poppele [*Chem. Eng. Prog.* **50:** 421 (*1954*)]. To convert centimeters per minute to feet per minute, multiply by 0.0328; to convert centimeters per second to feet per second, multiply by 0.0328. (*From Calvert, Yung, and Leung, NTIS Publ. PB-248050, 1975.*)

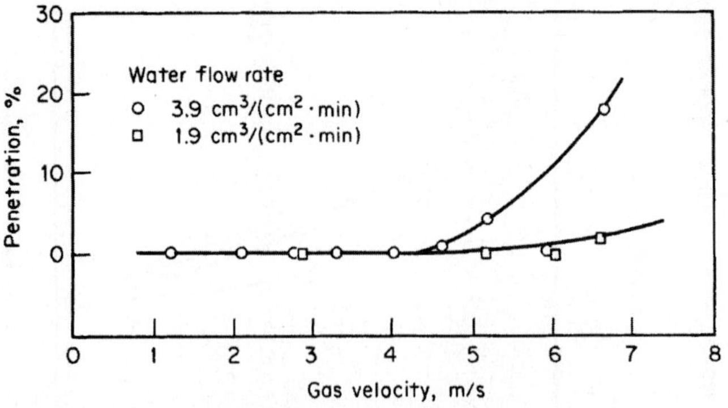

FIG. 14-116 Experimental data of Calvert with air and water in mesh with vertical upflow, showing the effect of liquid loading on efficiency and reentrainment. To convert meters per second to feet per second, multiply by 3.281; to convert cubic centimeters per square centimeter-minute to cubic feet per square foot-minute, multiply by 0.0328. (*Calvert, Yung, and Leung, NTIS Publ. PB-248050, 1975.*)

the predictions of Eq. (14-230) up to 4 m/s (13 ft/s) with air and water. Some reentrainment was encountered at higher velocities, but it did not appear serious until velocities exceeded 6.0 m/s (20 ft/s). With vertical upflow of gas, entrainment was encountered at velocities above and below 4.0 m/s (13 ft/s), depending on inlet liquid quantity (see Fig. 14-116). Figure 14-117 illustrates the onset of entrainment from mesh as a function of liquid loading and gas velocity and the safe operating area recommended by Calvert. Measurements of dry pressure drop by Calvert gave values only about one-third of those predicted from Eq. (14-231). He found the pressure drop to be highly affected by liquid load. The pressure drop of wet mesh could be correlated as a function of $U_g^{1.65}$ and parameters of liquid loading L/A, as shown in Fig. 14-118.

As indicated previously, mesh efficiency drops rapidly as particles decrease in size below 5 μm. An alternative is to use two mesh pads in series. The first mesh is made of fine wires and is operated beyond the flood point. It results in droplet coalescence, and the second mesh, using standard wire and operated below flooding, catches entrainment from the first mesh. Coalescence and flooding in the first mesh may be assisted with water sprays or irrigation. Massey [*Chem. Eng. Prog.* **53**(5): 114 (1959)] and Coykendall et al. [*J. Air Pollut. Control Assoc.* **18**: 315 (1968)] discussed

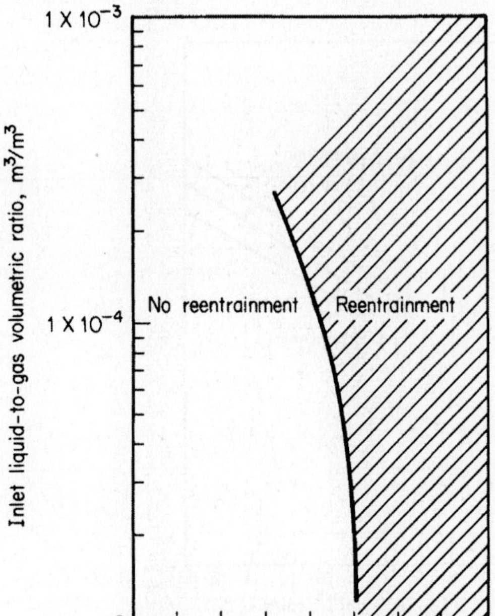

FIG. 14-117 Effect of gas and liquid rates on onset of mesh reentrainment and safe operating regions. To convert meters per second to feet per second, multiply by 3.281. (*Calvert, Yung, and Leung, NTIS Publ. PB-248050, 1975.*)

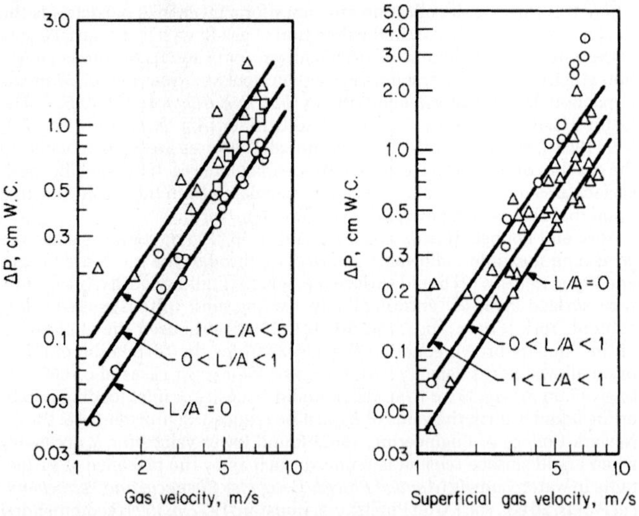

FIG. 14-118 Experimental pressure measured by Calvert as a function of gas velocity and liquid loading for (*a*) horizontal gas flow through vertical mesh and (*b*) gas upflow through horizontal mesh. Mesh thickness was 10 cm with 2.8-mm wire and void fraction of 98.2 percent, crimped in alternating directions. To convert meters per second to feet per second, multiply by 3.281; to convert centimeters to inches, multiply by 0.394. (*Calvert, Yung, and Leung, NTIS Publ. PB-248050, 1975.*)

such applications. Calvert et al. (1975) presented data on the particle size of entrained drops from mesh as a function of gas velocity, which can be used for sizing the secondary collector. A major disadvantage of this approach is high pressure drop, which can be in the range from 25 cm (10 in) of water to as high as 85 cm (33 in) of water if the mist is mainly submicrometer.

Wet Scrubbers Scrubbers have not been widely used for the collection of purely liquid particulate, probably because they are generally more complex and expensive than impaction devices of the types previously discussed. Further, scrubbers are no more efficient than the former devices for the same energy consumption. However, scrubbers of the types discussed in Sec. 17 and illustrated in Figs. 17-58 to 17-64 can be used to capture liquid particles efficiently. Their use is primarily indicated when we wish to accomplish another task at the same time, such as gas absorption or the collection of solid and liquid particulate mixtures.

Figure 14-119 gives calculated particle cut size as a function of tower height (or length) for vertical countercurrent spray towers and for horizontal-gas-flow, vertical-liquid-flow cross-current spray towers with parameters for liquid drop size. These curves are based on physical properties of standard air and water and should be used under conditions in which these are reasonable approximations. Lack of uniform liquid distribution or liquid flowing down the walls can affect the performance, requiring empirical correction factors. Many more complicated wet scrubbers use a combination of sprays or liquid atomization, cyclonic action, baffles, and targets. These combinations are not likely to be more efficient than similar devices previously discussed that operate at equivalent pressure drop. The vast majority of wet scrubbers operate at moderate pressure drop [8 to 15 cm (3 to 6 in) of water or 18 to 30 cm (7 to 12 in) of water] and cannot be expected to have high efficiency on particles smaller than 10 μm or 3 to 5 μm, respectively. Fine and submicrometer particles can be captured efficiently only in wet scrubbers that have high-energy input such as venturi scrubbers, two-phase eductor scrubbers, and flux-force-condensation scrubbers.

Venturi Scrubbers One type of venturi scrubber is illustrated in Fig. 17-58. Venturi scrubbers have been used extensively for collecting fine and submicrometer solid particulate, condensing tars and mists, and mixtures of liquids and solids. To a lesser extent, they have also been used for simultaneous gas absorption, although Lundy [*Ind. Eng. Chem.* **50**: 293 (1958)] indicated that they are generally limited to three transfer units. Venturi scrubbers use pressure drop at the throat to generate many liquid droplets. The large number of liquid droplets reduces the distance a small particle must travel to collide with a larger droplet and be captured. The collection efficiency of a venturi scrubber is highly dependent on the throat velocity or pressure drop, the liquid-to-gas ratio, and the chemical nature of wettability of the particulate. Throat velocities may range from 60 to 150 m/s (200 to 500 ft/s). Liquid injection rates are typically 0.67 to 1.4 m^3/1000 m^3 of gas. A liquid rate of 1.0 m^3 per 1000 m^3 of gas is usually close to optimum, but liquid rates as high as 2.7 m^3 (95 ft^3) have been used. Efficiency improves

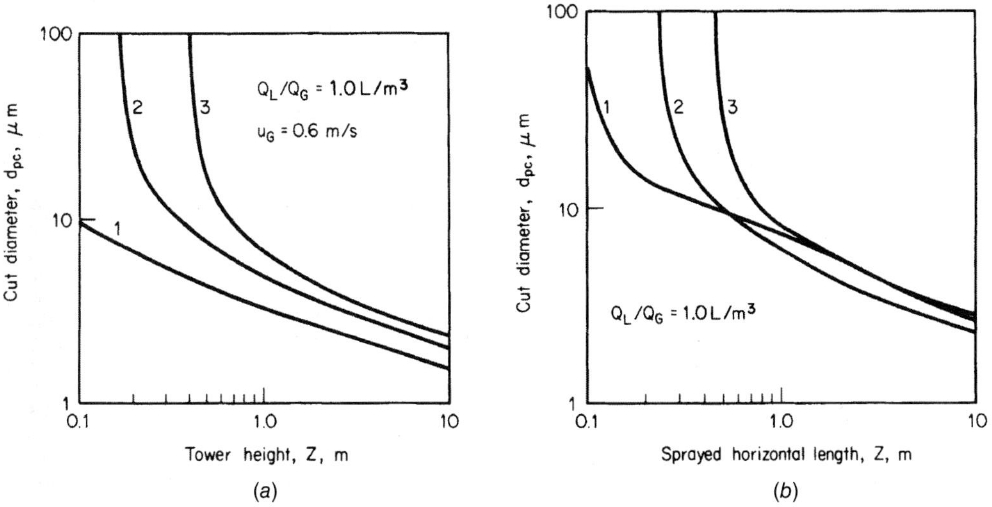

FIG. 14-119 Predicted spray-tower cut diameter as a function of sprayed length and spray droplet size for (*a*) vertical-countercurrent towers and (*b*) horizontal-cross-flow towers per Calvert [J. Air Pollut. Control Assoc. ***24:*** *929 (1974)*]. Curve 1 is for 200-μm spray droplets, curve 2 for 500-μm spray, and curve 3 for 1000-μm spray. Q_L/Q_G is the volumetric liquid-to-gas ratio, *L* liquid/m³ gas, and u_G is the superficial gas velocity in the tower. To convert liters per cubic meter to cubic feet per cubic foot, multiply by 10^{-3}.

with increased liquid rate but only at the expense of higher pressure drop and energy consumption. Pressure-drop predictions for a given efficiency are hazardous without determining the nature of the particulate and the liquid-to-gas ratio. In general, particles coarser than 1 μm can be collected efficiently with pressure drops of 25 to 50 cm of water. For appreciable collection of submicrometer particles, pressure drops of 75 to 100 cm (30 to 40 in) of water are usually required. When particles are much finer than 0.5 μm, pressure drops of 175 to 250 cm (70 to 100 in) of water have been used.

One of the problems in predicting the efficiency and required pressure drop of a venturi is the chemical nature or wettability of the particulate, which on 0.5-μm-size particles can make up to a threefold difference in the required pressure drop for its efficient collection. Calvert (Calvert et al. 1972; Calvert 1974) represented this effect with an empirical factor *f*, which is based on the hydrophobic (*f* = 0.25) or hydrophilic (*f* = 0.50) nature of the particles. Figure 14-120 gives the cut diameter of a venturi scrubber as a function of its operating parameters (throat velocity, pressure drop, and liquid-to-gas ratio) for hydrophobic particles. Figure 14-121 compares cut diameter as a function of pressure drop for an otherwise identically operating venturi on hydrophobic and hydrophilic particles. Calvert et al. (1972) give equations that can be used to construct cut-size curves similar to those of Fig. 14-121 for other values of the empirical factor *f*. Most real particles are neither completely hydrophobic nor completely hydrophilic but have *f* values lying between the two extremes. Unfortunately, no chemical-test methods have yet been devised for determining appropriate *f* values for a particulate in the laboratory, so design must be guided by experience in similar applications.

The pressure drop in a venturi scrubber is controlled by throat velocity. While some venturis have fixed throats, many are designed with variable louvers to change throat dimensions and control performance for changes in gas flow.

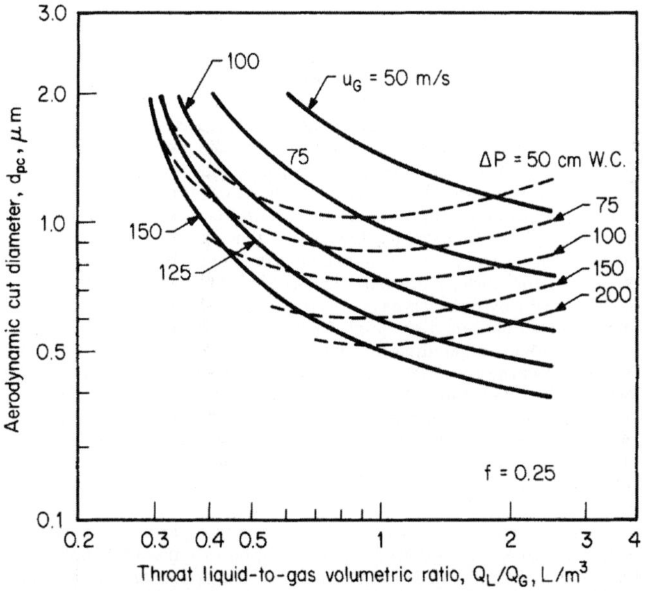

FIG. 14-120 Prediction of venturi-scrubber cut diameter for hydrophobic particles as functions of operating parameters as measured by Calvert [*Calvert, Goldshmid, Leith, and Mehta, NTIS Publ. PB-213016, 213017, 1972; and Calvert,* J. Air Pollut. Control Assoc. ***24:*** *929 (1974)*]. u_G is the superficial throat velocity, and ΔP is the pressure drop from converging to diverging section. To convert meters per second to feet per second, multiply by 3.281; to convert liters per cubic meter to cubic feet per cubic foot, multiply by 10^{-3}; and to convert centimeters to inches, multiply by 0.394.

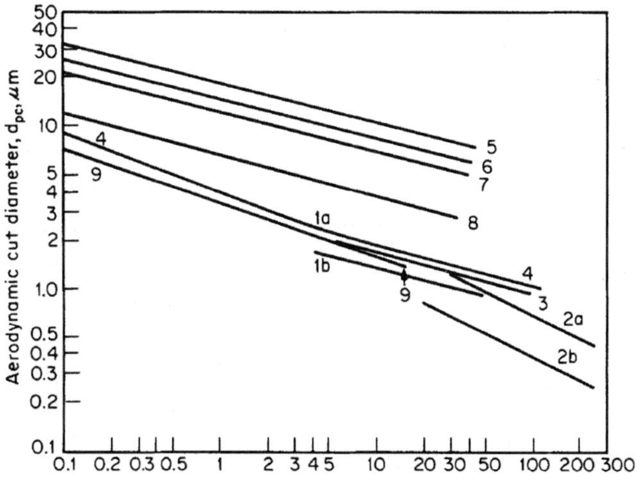

FIG. 14-121 Typical cut diameter as a function of pressure drop for various liquid-particle collectors. Curves 1*a* and *b* are single-sieve plates with froth density of 0.4 g/cm³; 1*a* has sieve holes of 0.5 cm and 1*b* holes of 0.3 cm. Curves 2*a* and *b* are for a venturi scrubber with hydrophobic particles (2*a*) and hydrophilic particles (2*b*). Curve 3 is an impingement plate, and curve 4 is a packed column with 2.5-cm-diameter packing. Curve 5 is a zigzag baffle collector with six baffles at θ = 30°. Curve 7 is for six rows of staggered tubes with 1-cm spacing between adjacent tube walls in a row. Curve 8 is similar, except that tube-wall spacing in the row is 0.3 cm. Curve 9 is for wire-mesh pads. To convert grams per cubic centimeter to pounds per cubic foot, multiply by 62.43; to convert centimeters to inches, multiply by 0.394. [*Calvert,* J. Air Pollut. Control Assoc. ***24:*** *929 (1974); and Calvert, Yung, and Leung, NTIS Publ. PB-248050, 1975.*]

Calvert (Yung, Barbarika, and Calvert 1977) critiqued the many pressure-drop equations and suggested the following simplified equation as accurate to ±10 percent:

$$\Delta P = \frac{2\rho_\ell U_g^2}{981 g_c}\left(\frac{Q_\ell}{Q_g}\right)\left[1 - x^2 + \sqrt{(x^4 - x^2)^{0.5}}\right] \qquad (14\text{-}232)$$

where

$$x = (3l_t C_{Di}\rho_g/16 d_\ell \rho_l) + 1 \qquad (14\text{-}233)$$

ΔP is the pressure drop, cm of water; ρ_ℓ and ρ_g are the density of the scrubbing liquid and gas respectively, g/cm³; U_g is the velocity of the gas at the throat inlet, cm/s; Q_ℓ/Q_g is the volumetric ratio of liquid to gas at the throat inlet, dimensionless; l_t is the length of the throat, cm; C_{Di} is the drag coefficient, dimensionless, for the mean liquid diameter, evaluated at the throat inlet; and d_l is the Sauter mean diameter, cm, for the atomized liquid. The drag coefficient C_{Di} should be evaluated by the Dickinson and Marshall [*Am. Inst. Chem. Eng. J.* **14:** 541 (1968)] correlation

$$C_{Di} = 0.22 + (24/N_{\text{Rei}})(1 + 0.15\, N_{\text{Rei}}^{0.6}) \qquad (14\text{-}234)$$

The Reynolds number, N_{Rei}, is evaluated at the throat inlet conditions as $d_\ell G_g/\mu_g$.

All venturi scrubbers must be followed by an entrainment collector for the liquid spray. These collectors will have an additional pressure drop that must be added to that of the venturi itself.

Other Scrubbers A liquid-ejector venturi (Fig. 17-59), in which high-pressure water from a jet induces the flow of gas, has been used to collect mist particles in the 1- to 2-μm range, but submicrometer particles will generally pass through an eductor. Power costs for liquid pumping are high if appreciable motive force must be imparted to the gas because jet-pump efficiency is usually less than 10 percent. Harris [*Chem. Eng. Prog.* **42**(4): 55 (1966)] described their application. Two-phase eductors have been considerably more successful on the capture of submicrometer mist particles and could be attractive in situations in which large quantities of waste thermal energy are available. However, the equivalent energy consumption is equal to that required for high-energy venturi scrubbers, and such devices are likely to be no more attractive than venturi scrubbers when the thermal energy is priced at its proper value. Sparks [*J. Air Pollut. Control Assoc.* **24:** 958 (1974)] discussed steam ejectors giving 99 percent collection of particles 0.3 to 10 μm. Energy requirements were 311,000 J/m³ (8.25 Btu/scf). Gardenier [*J. Air Pollut. Control Assoc.* **24:** 954 (1974)] operated a liquid eductor with high-pressure (6900- to 27,600-kPa) (1000- to 4000-lbf/in²) hot water heated to 200°C (392°F) that flashed into two phases as it issued from the jet. He obtained 95 to 99 percent collection of submicrometer particulate. Figure 14-122 shows the water-to-gas ratio required as a function of particle size to achieve 99 percent collection.

Effect of Gas Saturation in Scrubbing If hot unsaturated gas is introduced into a wet scrubber, spray particles will evaporate to cool and saturate the gas. The evaporating liquid molecules moving away from the target

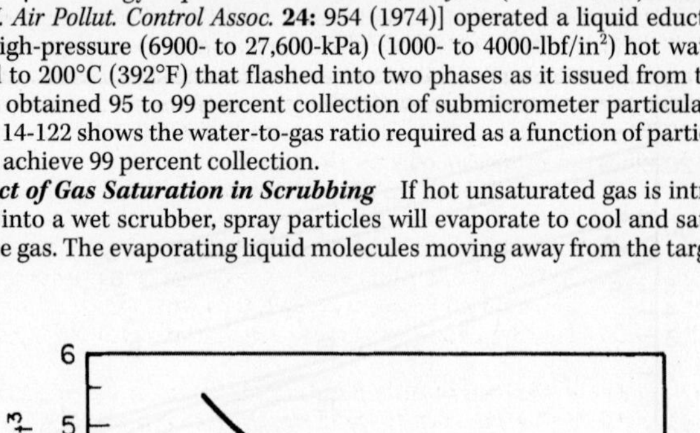

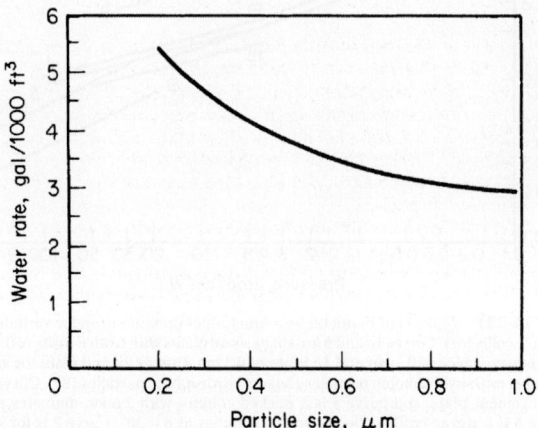

FIG. 14-122 Superheated high-pressure hot-water requirements for 99 percent collection as a function of particle size in a two-phase eductor jet scrubber. To convert gallons per 1000 cubic feet to cubic meters per 1000 cubic meters, multiply by 0.134. [*Gardenier,* J. Air Pollut. Control Assoc. *24: 954 (1974).*]

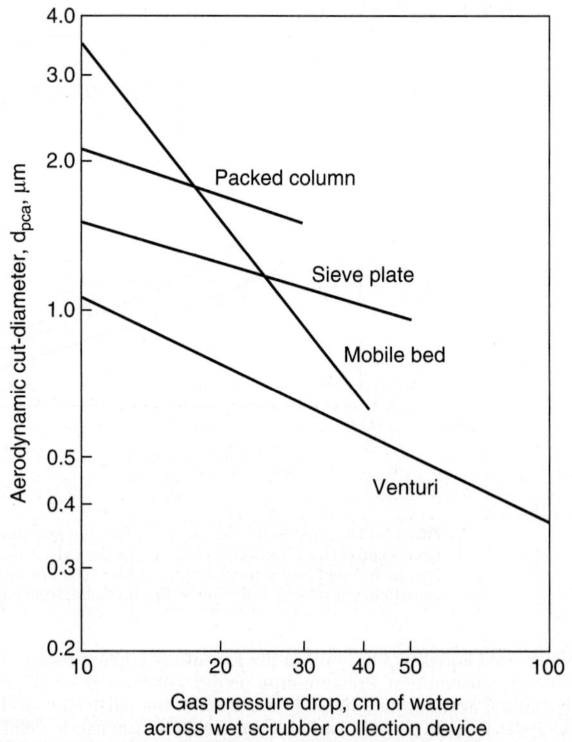

FIG. 14-123 Calvert's refined particle cut-size/power relationship for particle inertial impaction wet collectors. (*Calvert and Englund, eds.,* Handbook of Air Pollution Technology, *Wiley, New York, 1984, chap. 10, pp. 215–248, by permission.*)

droplets will repel particles that might collide with them. This results in the forces of diffusiophoresis opposing particle collection. Semrau and Witham (Air Pollut. Control Assoc. Prepr. 75-30.1) investigated temperature parameters in wet scrubbing and found a definite decrease in the efficiency of evaporative scrubbers and an enhancement of efficiency when a hot saturated gas was scrubbed with cold water rather than with recirculated hot water. Little improvement was experienced in cooling a hot saturated gas below a 50°C dew point.

Energy Requirements for Inertial-Impaction Efficiency Calvert (Calvert and Englund 1984, p. 228) combined mathematical modeling with performance tests on a variety of industrial scrubbers and obtained a refinement of the power-input/cut-size relationship as shown in Fig. 14-123. He considered these relationships sufficiently reliable to use these data as a tool for the selection of scrubber type and performance prediction. The power input for this figure is based solely on gas pressure drop across the device.

Collection of Fine Mists Inertial-impaction devices previously discussed give high efficiency on particles above 5 μm in size and often reasonable efficiency on particles down to 3 μm in size at moderate pressure drops. However, this mechanism becomes ineffective for particles smaller than 3 μm because of the particles' gas-like mobility. Only impaction devices having extremely high energy input, such as venturi scrubbers and flooded mesh pads, can give high collection efficiency on fine particles, 3 μm and smaller, including the submicrometer range. Fine particles are subjected to Brownian motion in gases, and diffusional deposition can be used for their collection. Diffusional deposition becomes highly efficient as particles become smaller, especially below 0.2 to 0.3 μm. Table 14-27 shows the typical displacement velocity of particles. Randomly oriented fiber beds having tortuous and narrow gas passages are suitable devices for using this collection mechanism. (The diffusional collection mechanism is discussed in Sec. 17 under Mechanisms of Dust Collection.) Other collection mechanisms that are efficient for fine particles are electrostatic forces and flux forces, such as thermophoresis and diffusiophoresis. Particle growth and nucleation methods are also applicable. Efficient collection of fine particles is important because particles in the range of 2.0 to around 0.2 μm are the ones that penetrate and are deposited in the lung most efficiently. Hence, particles in this range constitute the largest health hazard.

Fiber Mist Eliminators These devices are produced in various configurations and are highly efficient for fine particles. Generally, randomly oriented glass or polypropylene fibers are densely packed between

TABLE 14-27 Brownian Movement of Particles*

Particle diameter, μm	Brownian displacement of particle, μm/s
0.1	29.4
0.25	14.2
0.5	8.92
1.0	5.91
2.5	3.58
5.0	2.49
10.0	1.75

*Brink, *Can. J. Chem. Eng.,* **41,** 134 (1963). Based on spherical water particles in air at 21°C and 1 atm.

TABLE 14-28 Operating Characteristics of Various Types of Fiber Mist Eliminators as Used on Sulfuric Acid Plants*

	High efficiency	High velocity	Spray catcher
Controlling mechanism for mist collection	Brownian movement	Impaction	Impaction
Superficial velocity, m/s	0.075–0.20	2.0–2.5	2.0–2.5
Efficiency on particles greater than 3 μm, %	Essentially 100	Essentially 100	Essentially 100
Efficiency on particles 3 μm and smaller, %	95–99+	90–98	15–30
Pressure drop, cm H_2O	12–38	15–20	1.0–2.5

*Brink, Burggrabe, and Greenwell, *Chem. Eng. Prog.,* **64**(11), 82 (1968). To convert centimeters to inches, multiply by 0.394.

reinforcing screens, producing fiber beds varying in thickness, usually from 25 to 75 mm (1 to 3 in), although thicker beds can be produced. Units with efficiencies as high as 99.9 percent on fine particles have been developed (see *Chemical Engineers' Handbook,* 5th ed., p. 18-88). A combination of mechanisms interacts to provide the high overall collection efficiency: particles larger than 2 to 3 μm are collected on the fibers by inertial impaction and direct interception, while small particles are collected by Brownian diffusion. When the device is designed to use this latter mechanism as the primary means, efficiency turndown problems are eliminated as collection efficiency by diffusion increases with residence time. Pressure drop through the beds increases with velocity to the first power since the gas flow is laminar.

Three series of fiber mist eliminators are typically available. A spray-catcher series uses inertial impacting as the controlling collection mechanism and is designed for essentially 100 percent capture of droplets larger than 3 μm. The high-velocity type also uses impaction and is designed to give moderately high efficiency on particles down to 1.0 μm. Both of these types are usually produced in the form of flat panels of 25- to 50-mm (1- to 2-in) thickness. The high-efficiency type uses Brownian motion to provide high efficiency on particles less than 3 μm and is illustrated in Fig. 14-124. As mist particles are collected, they coalesce into a liquid film that wets the fibers. Liquid is moved horizontally through the bed by the gas drag force and downward by gravity. It drains down the downstream retaining screen to the bottom of the element and is returned to the process through a liquid seal. Table 14-28 gives typical operating characteristics of the three types of collectors.

Solid particulates are captured as readily as liquids in fiber beds but can rapidly plug the bed if they are insoluble. Fiber beds have often been used for mixtures of liquids and soluble solids and with soluble solids in condensing situations. Enough solvent is atomized into the gas stream entering the collector to irrigate the fiber elements and dissolve the collected particulate. Such fiber beds have been used to collect fine fumes such as ammonium nitrate and ammonium chloride smokes, and oil mists from compressed air.

Electrostatic Precipitators The principles and operation of electrical precipitators are discussed in Sec. 17 under the subsection Gas–Solids Separations. Gas–liquid electrostatic precipitation has an advantage over gas–solid precipitation because the collected liquid can readily drain. Precipitators are admirably suited to the collection of fine mists and mixtures of mists and solid particulates. They have the advantage of low pressure drop compared to venturi scrubbers or fiber mist eliminators, but they require large areas. Electrostatic precipitators can be dry or wet types, with wet precipitators using water sprays or overflowing weirs. Such precipitators operate on the principle of making all particles conductive when possible, which increases the particle migration velocity and collection efficiency. Under these conditions, particle dielectric strength becomes a much more important variable, and particles with a low dielectric constant such as condensed hydrocarbon mists become much more difficult to collect than water-wettable particles. Bakke (U.S.–U.S.S.R. Joint Work. Group Symp.: Fine Particle Control, San Francisco, 1974) developed equations for particle charge and relative collection efficiency in wet precipitators that show the effect of dielectric constant. Wet precipitators can also be used to absorb soluble gases simultaneously by adjusting the pH or the chemical composition of the liquid spray. The presence of the electric field appears to enhance absorption. Wet precipitators have found their greatest usefulness to date in handling mixtures of gaseous pollutants and submicrometer particulate (either liquid or solid, or both) such as fumes from aluminum-pot lines, carbon anode baking, fiberglass-fume control, coke-oven and metallurgical operations, chemical incineration, and phosphate-fertilizer operations. Two-stage precipitators are used increasingly for moderate-volume gas streams containing nonconductive liquid mists that will drain from the collecting plates. Their application on hydrocarbon mists has been quite successful, but careful attention must be given to fire and explosion hazards.

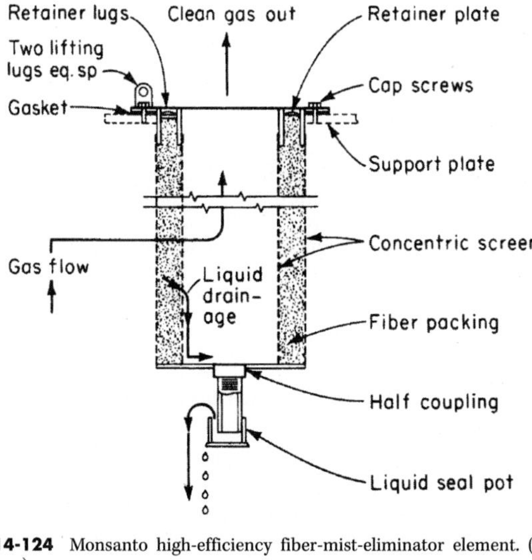

FIG. 14-124 Monsanto high-efficiency fiber-mist-eliminator element. (*Monsanto Company.*)

Electrically Augmented Collectors Collection efficiency can be enhanced by the combining of electrostatic forces with devices using other collecting mechanisms, such as impaction and diffusion. Cooper (Air Pollut. Control Assoc. Prepr. 75-02.1) evaluated the magnitude of forces operating between charged and uncharged particles and concluded that electrostatic attraction is the strongest collecting force operating on particles finer than 2 μm. Nielsen and Hill [*Ind. Eng. Chem. Fundam.* **15:** 149 (1976)] quantified these relationships, and a number of practical devices have been demonstrated. Pilat and Meyer (NTIS Publ. PB-252653, 1976) demonstrated up to 99 percent collection of fine particles in a two-stage spray tower in which the inlet particles and water spray were charged with opposite polarity. The principle has been applied to retrofitting existing spray towers to enhance collection.

Klugman and Sheppard (Air Pollut. Control Assoc. Prepr. 75-30.3) developed an ionizing wet scrubber in which the charged mist particles were collected in a grounded, irrigated cross-flow bed of Tellerette packing. Particles smaller than 1 μm have been collected with 98 percent efficiency by using two units in series. Dembinsky and Vicard (Air Pollut. Control Assoc. Prepr. 78-17.6) used an electrically augmented low-pressure [5 to 10 cm (2 to 4 in) of water] venturi scrubber to give 95 to 98 percent collection efficiency on submicrometer particles.

Particle Growth and Nucleation Fine particles may be subjected to conditions that favor the growth of particles, either through condensation or through coalescence. Saturation of a hot gas stream with water, followed by condensation on the particles acting as nuclei when the gas is cooled, can increase particle size and ease of collection. The addition of steam can produce the same results. Scrubbing of the humid gas with a cold liquid can bring diffusiophoresis into play. The introduction of cold liquid drops causes a reduction in water-vapor pressure at the surface of the cold drop. The resulting vapor-pressure gradient causes a hydrodynamic flow toward the drop known as Stefan flow, which enhances the movement of mist particles toward the spray drop. If the molecular mass of the diffusing vapor is different from the carrier gas, this density difference also produces a driving force, and the sum of these forces is known as diffusiophoresis. A mathematical description of these forces was presented by Calvert et al. (1972)

and by Sparks and Pilat [*Atmos. Environ.* **4:** 651 (1970)]. Thermal differences between the carrier gas and the cold scrubbing droplets can further enhance collection through thermophoresis. Calvert and Jhaseri [*J. Air Pollut. Control Assoc.* **24:** 946 (1974); and NTIS Publ. PB-227307, 1973] investigated condensation scrubbing in multiple-sieve plate towers.

Submicrometer droplets can be coagulated through Brownian diffusion if given ample time. The introduction of particles 50 to 100 times larger in diameter can enhance coagulation, but the addition of a broad range of particle sizes is discouraged. Increasing turbulence will aid coagulation, so fans to stir the gas or narrow, tortuous passages such as those of a packed bed can be beneficial. Sonic energy can also produce coagulation, especially the production of standing waves in the confines of long, narrow tubes. The addition of water and oil mists can sometimes aid sonic coagulation. Sulfuric acid mist [Danser, *Chem. Eng.* **57**(5): 158 (1950)] and carbon black [Stokes, *Chem. Eng. Prog.* **46:** 423 (1950)] have been successfully agglomerated with sonic energy. Sonic agglomeration is often unsuccessful because of its high energy requirements. Most sonic generators have very poor energy-transformation efficiency. Wegrzyn et al. (U.S. EPA Publ. EPA-600/7-79-004C, 1979, p. 233) reviewed acoustic agglomerators. Mednikov (*U.S.S.R. Akad. Soc.*, Moscow, 1963) suggested that the incorporation of sonic agglomeration with electrostatic precipitation could greatly reduce precipitator size.

Other Collectors Tarry particulates and other difficult-to-handle liquids have been collected on a dry, expendable phenol formaldehyde-bonded glass-fiber mat (Goldfield, *J. Air Pollut. Control Assoc.* **20:** 466 (1970)] in roll form that is advanced intermittently into a filter frame. Superficial gas velocities are 2.5 to 3.5 m/s (8.2 to 11.5 ft/s), and pressure drop is typically 41 to 46 cm (16 to 18 in) of water. Collection efficiencies of 99 percent have been obtained on submicrometer particles. Brady [*Chem. Eng. Prog.* **73**(8): 45 (1977)] discussed a cleanable modification of this approach in which the gas is passed through a reticulated foam filter that is slowly rotated and solvent-cleaned.

In collecting very fine (mainly submicron) mists of a hazardous nature where one of the collectors previously discussed has been used as the primary one (fiber-mist eliminators of the Brownian diffusion type and electrically augmented collectors are primarily recommended), there is the chance that the effluent concentration may still be too high for atmospheric release when residual concentration must be in the range of 1 to 2 μm. In such situations, secondary treatment may be needed. Removal of the residual mist by adsorption will probably be in order. See Sec. 16, Adsorption and Ion Exchange. Another possibility might be treatment of the remaining gas by membrane separation. See the subsection Membrane Separation Processes in Sec. 20.

Continuous Phase Uncertain Some situations exist, such as in two-phase gas–liquid flow, where the volume of the liquid phase may approach being equal to the volume of the vapor phase, and where it may be difficult to be sure which phase is the continuous phase. Svrcek and Monnery [*Chem. Eng. Prog.* **89**(10): 53–60 (1993)] have discussed the design of two-phase separation in a tank with gas–liquid separation in the middle, mist elimination in the top, and entrained gas-bubble removal from the liquid in the bottom. A design approach for sizing the gas–liquid disengaging space in the vessel is given using a tangential tank inlet nozzle, followed by a wire mesh mist eliminator in the top of the vessel for final separation of entrained mist from the vapor.

LIQUID-PHASE CONTINUOUS SYSTEMS

Practical separation techniques for gases dispersed in liquids are discussed in this subsection. Processes and methods for dispersing gas in liquid have been discussed earlier in this section, together with information for predicting the bubble size produced. Gas-in-liquid dispersions are also produced in chemical reactions and electrochemical cells in which a gas is liberated. Such dispersions are likely to be much finer than those produced by the dispersion of a gas. Dispersions may also be unintentionally created in the vaporization of a liquid.

General References: Adamson, *Physical Chemistry of Surfaces*, 4th ed., Wiley, New York, 1982; Akers, *Foams*, Academic Press, New York, 1976; Bikerman, *Foams*, Springer-Verlag, New York, 1973; Bikerman et al., *Foams: Theory and Industrial Applications*, Reinhold, New York, 1953; "Defoamers" and "Foams," *Encyclopedia of Chemical Technology*, 4th ed., vols. 7, 11, Wiley, New York, 1993–1994; Exerowa and Kruglyakov, *Foam and Foam Films*, Elsevier, New York, 1998; Garrett, Peter R., *Science of Defoaming—Theory, Experiment and Applications*, Taylor & Francis, Boca Raton, Fla., 2014; Sonntag and Strenge, *Coagulation and Stability of Disperse Systems*, Halsted-Wiley, New York, 1972; Wilson, ed., *Foams: Physics, Chemistry and Structure*, Springer-Verlag, London, 1989.

Types of Gas-in-Liquid Dispersions "Gas-in-liquid dispersions" describes the formation of these dispersions. For separation, an important distinction is that between unstable dispersions, which separate readily under the influence of gravity once the mixture has been removed from the influence of the dispersing force, and stable dispersions or foam. Gas–liquid contacting equipment, such as bubble towers and gas-dispersing agitators, are typical examples of equipment that produces unstable dispersions. More difficulties may result in separation when the gas is dispersed in the form of bubbles only a few micrometers in size. An example is the evolution of gas from a liquid in which it has been dissolved or released through chemical reaction such as electrolysis. Coalescence of the dispersed phase can be helpful in such circumstances.

The second type of gas-in-liquid dispersion is a stable dispersion, or foam. Separation can be extremely difficult. In many chemical processes, preventing or managing the formation of foam is important for stable operation. A pure two-component system of gas and liquid cannot produce stable dispersions. Stable foams can be produced only when an additional substance is adsorbed at the liquid-surface interface. The substance adsorbed may be in true solution but with a chemical tendency to concentrate in the interface, such as that of a surface-active agent, or it may be a finely divided solid that concentrates in the interface because it is only poorly wetted by the liquid. Surfactants and proteins are examples of soluble materials, while dust particles and extraneous dirt that includes traces of nonmiscible liquids can be examples of poorly wetted materials.

The separation of gases and liquids always involves coalescence, but enhancement of the rate of coalescence may be required only in difficult separations.

Separation of Unstable Systems The buoyancy of bubbles suspended in liquid can often be depended upon to cause the bubbles to rise to the surface and separate. This is a special case of gravity settling. The mixture is allowed to stand at rest or is moved along a flow path in laminar flow until the bubbles have surfaced. Table 14-29 shows the calculated rate of rise of air bubbles at atmospheric pressure in water at 20°C (68°F) as a function of diameter. The velocity of rise for 10-μm bubbles is very low, so long separating times are required for gas that is more finely dispersed.

For liquids other than water, the rise velocity can be approximated from Table 14-29 by multiplying by the liquid's specific gravity and the reciprocal of its viscosity (in centipoises). For bubbles larger than 100 μm, this procedure is erroneous, but the error is less than 15 percent for bubbles up to 1000 μm. More serious is the underlying assumption of Table 14-29 that the bubbles are rigid spheres. Circulation within the bubble causes notable increases in velocity in the range of 100 μm to 1 mm, and the flattening of bubbles 1 cm and larger appreciably decreases their velocity. However, in this latter size range, the velocity is so high that separation is a trivial problem.

In the design of separating chambers, static vessels or continuous-flow tanks may be used. Care must be taken to protect the flow from turbulence, which could cause backmixing of partially separated fluids or which could carry unseparated liquids rapidly to the separated-liquid outlet. Vertical baffles are sometimes used to protect rising bubbles from flow currents. Unseparated fluids should be distributed to the separating region as uniformly and with as little velocity as possible. When the bubble rise velocity is quite low, shallow tanks or flow channels should be used to minimize the residence time required.

Quite low velocity rise of bubbles due either to small bubble size or to high liquid viscosity can cause difficult situations. With low-viscosity liquids, separation-enhancing possibilities in addition to those previously enumerated are to sparge the liquid with large-diameter gas bubbles or to atomize the mixture as a spray into a tower. Large gas bubbles rising rapidly through the liquid collide with small bubbles and aid their coalescence through capture. Atomizing of the continuous phase reduces the distance that small gas bubbles must travel to reach a gas interface. Evacuation of the spray space can also be beneficial in promoting small-bubble growth and especially in promoting gas evolution when the gas has appreciable liquid solubility. Liquid heating will also reduce solubility.

Surfaces in the settling zone for bubble coalescence such as closely spaced vertical or inclined plates or tubes are beneficial. When clean, low-viscosity fluids are involved, passage of the undegassed liquid through a tightly packed pad of mesh or fine fibers at low velocity will result in efficient bubble coalescence. Problems have been experienced in degassing a water-based organic solution that has been passed through an electrolytic cell for chemical reaction in which extremely fine bubbles of hydrogen gas

TABLE 14-29 Terminal Velocity of Standard Air Bubbles Rising in Water at 20°C*

Bubble diameter, μm	10	30	50	100	200	300
Terminal velocity, mm/s	0.061	0.488	1.433	5.486	21.95	49.38

*Calculated from Stokes' law. To convert millimeters per second to feet per second, multiply by 0.003281.

are produced in the liquid within the cell. Near-total removal of hydrogen gas from the liquid is needed for process safety. This is very difficult to achieve by gravity settling alone because of the fine bubble size and the need for a coalescing surface. The use of a fine fiber medium is strongly recommended in such situations. A low-forward liquid flow through the medium is desirable to provide time for the bubbles to attach themselves to the fiber medium through Brownian diffusion. Spielman and Goren [*Ind. Eng. Chem.* **62**(10): (1970)] reviewed the literature on coalescence with porous media and reported their own experimental results [*Ind. Eng. Chem. Fundam.* **11**(1): 73 (1972)] on the coalescence of oil–water liquid emulsions. The principles are applicable to a gas-in-liquid system. Glass-fiber mats composed of 3.5-, 6-, or 12-μm-diameter fibers, varying in thickness from 1.3 to 3.3 mm, successfully coalesced and separated 1- to 7-μm oil droplets at superficial bed velocities of 0.02 to 1.5 cm/s (0.00067 to 0.049 ft/s).

In the deaeration of high-viscosity fluids such as polymers, the material is flowed in thin sheets along solid surfaces. Vacuum is applied to increase bubble size and hasten separation. The Versator (Cornell Machine Co.) degasses viscous liquids by spreading them into a thin film by centrifugal action as the liquids flow through an evacuated rotating bowl.

Separation of Foam Foams can be a severe problem in chemical-processing steps that involve gas–liquid interaction such as distillation, absorption, evaporation, chemical reaction, and particle separation and settling. It can also be a major problem in pulp and paper manufacture, oil-well drilling fluids, the production of water-based paints, the use of lubricants and hydraulic fluids, dyeing and sizing textiles, the operation of steam boilers, fermentation operations, polymerization, wet-process phosphoric acid concentration, adhesive production, and foam control in products such as detergents, waxes, printing inks, instant coffee, and glycol antifreeze.

Foam stability decreases as the liquid drainage rate increases. Drainage rate is influenced by surface viscosity, which is very temperature-sensitive. At a critical temperature, which is a function of the system, a temperature change of only a few degrees can change a slow-draining foam to a fast-draining foam. This change in drainage rate can be a factor of 100 or more; thus increasing the temperature of foam can cause its destruction. An increase in temperature may also cause liquid evaporation and lamella thinning. As the lamellae become thinner, they become more brittle and fragile. Thus, mechanical deformation or pressure changes, which cause a change in gas-bubble volume, can also cause rupture. Bendure [*Tappi* **58:** 83 (1975)] indicated 10 ways to increase foam stability: (1) increase bulk liquid viscosity, (2) increase surface viscosity, (3) maintain thick walls (higher liquid-to-gas ratio), (4) reduce liquid surface tension, (5) increase surface elasticity, (6) increase surface concentration, (7) reduce surfactant-adsorption rate, (8) prevent liquid evaporation, (9) avoid mechanical stresses, and (10) eliminate foam inhibitors. Obviously, the reverse of each of these actions, when possible, is a way to control and break foam.

Physical Defoaming Techniques Typical physical defoaming techniques include mechanical methods for producing foam stress, thermal methods involving heating or cooling, and electrical methods. Combinations of these methods may also be employed, or they may be used in conjunction with chemical defoamers. Some methods are only moderately successful when conditions are present to reform the foam, such as breaking foam on the surface of boiling liquids. In some cases it may be desirable to draw the foam off and treat it separately. Foam can always be stopped by removing the energy source creating it, but this is often impractical.

Thermal Methods Heating is often a suitable means of destroying foam. As indicated previously, raising the foam above a critical temperature (which must be determined experimentally) can greatly decrease the surface viscosity of the film and change the foam from one that is slow-draining to one that is fast-draining. Coupling such heating with a mechanical force such as a revolving paddle to cause foam deformation is often successful. Other effects of heating include the expansion of the gas in the foam bubbles, which increases strain on the lamella walls and requires them to move and flex. Evaporation of solvent may occur, causing thinning of the walls. At sufficiently high temperatures, desorption or decomposition of stabilizing substances may occur. Placing a high-temperature bank of steam coils at the maximum foam level is one control method. As the foam approaches or touches the coil, it collapses. Designers should keep in mind that the coil will often become coated with solute.

The application of radiant heat to a foam surface is another option. Depending on the situation, the radiant source may be electric lamps, Glow-bar units, or gas-fired radiant burners. Hot gases from burners will enhance film drying of the foam. Heat may also be applied by jetting or spraying hot water on the foam. This approach is a combination of methods since the jetting produces mechanical shear, and the water itself provides dilution and change in foam-film composition. Cooling can also destroy foam if it is carried to the point of freezing since the formation of solvent crystals destroys the foam structure. Less drastic cooling such as spraying a hot foam with cold water may be effective; cooling may cause foam bubbles to shrink, coupled with the effects of shear and dilution mentioned earlier. In general, moderate cooling will be less effective than heating since the surface viscosity is being modified in the direction of a more stable foam.

Mechanical Methods Static or rotating breaker bars or slowly revolving paddles are sometimes successful. Their application in conjunction with other methods is often better. As indicated in the theory of foams, they will work better if installed at a level at which the foam has had some time to age and drain. A rotating breaker works by deforming the foam, which causes rupture of the lamella walls. Rapidly moving slingers will throw the foam against the vessel wall and may cause impact on other foam outside the envelope of the slinger. In some instances, stationary bars or closely spaced plates will limit the rise of foam. The action here is primarily one of providing surface for the coalescence of the foam. The wettability of the surface, whether moving or stationary, is often important. Usually a surface not wetted by the liquid is superior, as in the case of porous media for foam coalescence. However, in both cases there are exceptions for which wettable surfaces are preferred. Shkodin [*Kolloidn. Zh.* **14:** 213 (1952)] found molasses foam to be destroyed by contact with a wax-coated rod and unaffected by a clean glass rod.

Goldberg and Rubin [*Ind. Eng. Chem. Process Des. Dev.* **6:** 195 (1967)] showed in tests with a disk spinning vertically to the foam layer that most mechanical procedures, whether centrifugation, mixing, or blowing through nozzles, consist basically of the application of shear stress. Subjecting foam to an air-jet impact can also provide a source of drying and evaporation from the film, especially if the air is heated. Other effective means of destroying bubbles are to lower a frame of metal points periodically into the foam or to shower the foam with falling solid particles.

Pressure and Acoustic Vibrations These methods for rupturing foam are really special forms of mechanical treatment. Change in pressure in the vessel containing the foam stresses the lamella walls by expanding or contracting the gas inside the foam bubbles. Oscillation of the vessel pressure subjects the foam to repeated film flexing. Parlow [*Zucker* **3:** 468 (1950)] controlled foam in sugar-syrup evaporators with high-frequency air pulses. It is by no means certain that high-frequency pulsing is needed in all cases. Lower frequency and higher amplitude could be equally beneficial. Acoustic vibration is a similar phenomenon, causing localized pressure oscillation by using sound waves. Impulses at 6 kHz have been found to break froth from coal flotation [Sun, *Min. Eng.* **3:** 865 (1958)]. Sonntag and Strenge (*Coagulation and Stability of Disperse Systems,* Halsted-Wiley, New York, 1972, p. 121) reported foam suppression with high-intensity sound waves (11 kHz, 150 dB) but indicated that the procedure was too expensive for large-scale application. The Sontrifuge (Teknika Inc., a subsidiary of Chemineer, Inc.) is a commercially available low-speed centrifuge employing sonic energy to break the foam. Walsh [*Chem. Process.* **29:** 91 (1966)], Carlson [*Pap. Trade J.* **151:** 38 (1967)], and Thorhildsen and Rich [*TAPPI* **49:** 95A (1966)] have described the unit.

Electrical Methods As colloids, most foams typically have electrical double layers of charged ions that contribute to foam stability. Accordingly, foams can be broken by the influence of an external electric field. While few commercial applications have been developed, Sonntag and Strenge (1972, p. 114) indicated that foams can be broken by passage through devices much like electrostatic precipitators for dusts. Devices similar to two-stage precipitators having closely spaced plates of opposite polarity should be especially useful. Sonntag and Strenge, in experiments with liquid–liquid emulsions, indicated that the colloid structure can be broken at a field strength on the order of 8 to 9×10^5 V/cm.

Chemical Defoaming Techniques Sonntag and Strenge (1972, p. 111) described two chemical methods for foam breaking. One method is causing the stabilizing substances to be desorbed from the interface, such as by displacement with other more surface-active but nonstabilizing compounds. The second method is to effect chemical changes in the adsorption layer, leading to a new structure. Some defoamers may act purely by mechanical means but will be discussed in this subsection since their action is generally considered to be chemical in nature. Often chemical defoamers act in more than one way.

Chemical Defoamers The addition of chemical foam breakers is the most elegant way to break a foam. Effective defoamers cause very rapid disintegration of the foam and often need be present only in parts per million. The great diversity of compounds used for defoamers and the many different systems in which they are applied make a brief and orderly discussion of their selection difficult. Compounds needed to break aqueous foams may be different from those needed for aqueous-free systems. Most defoamers are insoluble or nonmiscible in the foam continuous phase, but some work best because of their ready solubility. Lichtman (*Defoamers,* 3d ed., Wiley, New York, 1979) has presented a concise summary of the application and use of defoamers. Rubel (*Antifoaming and Defoaming Agents,* Noyes Data Corp., Park Ridge, N.J., 1972) has reviewed the extensive patent literature on defoamers. Defoamers are also discussed extensively in the General References at the beginning of this subsection.

TABLE 14-30 Major Types and Applications of Defoamers

Classification	Examples	Applications
Silicones	Dimethyl silicone, trialkyl and tetraalkyl silanes	Lubricating oils; distillation; fermentation; jam and wine making; food processing
Aliphatic acids or esters	Mostly high-molecular-weight compounds; diethyl phthalate; lauric acid	Papermaking; wood-pulp suspensions; water-based paints; food processing
Alcohols	Moderate- to high-molecular-weight monohydric and polyhydric alcohols; octyl alcohol; C-12 to C-20 alcohols; lauryl alcohol	Distillation; fermentation; papermaking; glues and adhesives
Sulfates or sulfonates	Alkali metal salts of sulfated alcohols, sulfonic acid salts; alkyl-aryl sulfonates; sodium lauryl sulfate	Nonaqueous systems; mixed aqueous and nonaqueous systems; oil-well drilling muds; spent H_3SO_4 recovery; deep-fat frying
Amines or amides	Alkyl amines (undecyloctyl and diamyl methyl amine); polyamides (acyl derivatives of piperazine)	Boiler foam; sewage foam; fermentation; dye baths
Halogenated compounds	Fluochloro hydrocarbons with 5 to 50 C atoms; chlorinated hydrocarbons	Lubrication-oil and grease distillation; vegetable-protein glues
Natural products	Vegetable oils; waxes, mineral oils plus their sulfated derivatives (including those of animal oils and fats)	Sugar extraction; glue manufacture; cutting oils
Fatty-acid soaps	Alkali, alkaline earth, and other metal soaps; sodium stearate; aluminum stearate	Gear oils; paper stock; paper sizing; glue solutions
Inorganic compounds	Monosodium phosphate mixed with boric acid and ethyl carbonate, disodium phosphate; sodium aluminate, bentonite and other solids	Distillation; instant coffee; boiler feedwater; sugar extraction
Phosphates	Alkyl-alkalene diphosphates; tributyl phosphate in isopropanol	Petroleum-oil systems; foam control in soap solutions
Hydrophobic silica	Finely divided silica in polydimethyl siloxane	Aqueous foaming systems
Sulfides or thio derivatives	Metallic derivatives of thio ethers and disulfides, usually mixed with organic phosphite esters; long-chain alkyl thienyl ketones	Lubricating oils; boiler water

One useful method of aqueous defoaming is to add a nonfoam stabilizing surfactant that is more surface-active than the stabilizing substance in the foam. Thus a foam stabilized with an ionic surfactant can be broken by the addition of a very surface-active but nonstabilizing silicone oil. The silicone displaces the foam stabilizer from the interface by virtue of its insolubility. However, it does not stabilize the foam because its foam films have poor elasticity and rupture easily.

A major requirement for a defoamer is cost-effectiveness. Accordingly, some useful characteristics are low volatility (to prevent stripping from the system before it is dispersed and does its work), ease of dispersion and strong spreading power, and surface attraction-orientation. Chemical defoamers must also be selected with regard to their possible effect on product quality and their environmental and health suitability. For instance, silicone antifoam agents are effective in textile jet dyeing but reduce the fire retardancy of the fabric. Mineral-oil defoamers in sugar evaporation have been replaced by specifically approved materials. The tendency is no longer to use a single defoamer compound but to use a formulation specially tailored for the application, comprising carriers, secondary antifoam agents, emulsifiers, and stabilizing agents in addition to the primary defoamer. Carriers, usually hydrocarbon oils or water, serve as the vehicle to support the release and spread of the primary defoamer. Secondary defoamers may provide a synergistic effect for the primary defoamer or modify its properties such as spreadability or solubility. Emulsifiers may enhance the speed of dispersion, while stabilizing agents may enhance defoamer stability or shelf life.

Hydrophobic silica defoamers work on a basis that may not be chemical at all. They are basically finely divided solid silica particles dispersed in a hydrocarbon or silicone oil that serves as a spreading vehicle. Kulkarni [*Ind. Eng. Chem. Fundam.* **16:** 472 (1977)] theorizes that this mixture defoams by the penetration of the silica particle into the bubble and the rupture of the wall. Table 14-30 lists major types of defoamers and typical applications.

Other Chemical Methods These methods rely chiefly on destroying the foam stabilizer or neutralizing its effect through methods other than displacement and are applicable when the process will permit changing the chemical environment. Foams stabilized with alkali esters can be broken by acidification since the equivalent free acids do not stabilize foam. Foams containing sulfated and sulfonated ionic detergents can be broken with the addition of fatty-acid soaps and calcium salts.

Ionic surfactants adsorb at the foam interface and orient with the charged group immersed in the lamellae and their uncharged tails pointed into the gas stream. As the film drains, the charged groups, which repel each other, tend to be moved more closely together. The repulsive force between like charges hinders drainage and stabilizes the film. The addition of a salt or an electrolyte to the foam screens the repulsive effect, permits additional drainage, and can reduce foam stability.

Foam Prevention Chemical prevention of foam differs from defoaming only in that compounds or mixtures are added to a stream prior to processing to prevent the formation of foam either during processing or during customer use. Such additives, sometimes distinguished as antifoam agents, are usually in the same chemical class of materials as defoamers. However, they are usually specifically formulated for the application. The use of antifoam is very common in chemical processing applications. Typical examples of products formulated with antifoam agents are laundry detergents (to control excess foaming), automotive antifreeze, instant coffee, and jet-aircraft fuel. An alternative to antifoam agents in some chemical processes is the removal of trace impurities such as surface-active agents before processing such as by treatment with activated carbon [Pool, *Chem. Process.* **21**(9): 56 (1958)].

Automatic Foam Control In processing materials when foam can accumulate, it is often desirable to measure the height of the foam layer continuously and to dispense defoamer automatically as required to control the foam. Other corrective action can also be taken automatically. Methods of sensing the foam level have included electrodes in which the electrical circuit is completed when the foam touches the electrode [Nelson, *Ind. Eng. Chem.* **48:** 2183 (1956); Browne, U.S. Patent 2,981,693, 1961], floats designed to rise in a foam layer (Carter, U.S. Patent 3,154,577, 1964), and change in power input required to turn a foam-breaking impeller as the foam level rises (Yamashita, U.S. Patent 3,317,435, 1967). Timers to control the duration of defoamer addition have also been used. Browne suggested the automatic addition of defoamer through a porous wick when the foam level reached the level of the wick. Foam control was also discussed by Kroll [*Ind. Eng. Chem.* **48:** 2190 (1956)].

Section 15

Liquid-Liquid Extraction and Other Liquid-Liquid Operations and Equipment

Timothy C. Frank, Ph.D. *Fellow, The Dow Chemical Company; Fellow, American Institute of Chemical Engineers (Section Editor)*

Bruce S. Holden, M.S. *Principal Research Scientist, The Dow Chemical Company; Fellow, American Institute of Chemical Engineers*

A. Frank Seibert, Ph.D., P.E. *Technical Manager, Separations Research Program, The University of Texas at Austin; Fellow, American Institute of Chemical Engineers*

INTRODUCTION AND OVERVIEW

THERMODYNAMIC BASIS FOR LIQUID-LIQUID EXTRACTION

SOLVENT SCREENING METHODS

LIQUID DENSITY, VISCOSITY, AND INTERFACIAL TENSION

LIQUID-LIQUID DISPERSION FUNDAMENTALS

PROCESS FUNDAMENTALS AND BASIC CALCULATION METHODS

The contributions of Lise Dahuron, William D. Prince, and Loren C. Wilson, coauthors of Sec. 15 in the 8th edition, and Lanny A. Robbins and Roger W. Cusack, coauthors of Sec. 15 in the 7th edition, are gratefully acknowledged.

CALCULATION PROCEDURES

LIQUID-LIQUID EXTRACTION EQUIPMENT

PROCESS CONTROL CONSIDERATIONS

LIQUID-LIQUID PHASE SEPARATION EQUIPMENT

EMERGING DEVELOPMENTS

Nomenclature

A given symbol may represent more than one property. The appropriate meaning should be apparent from the context. The equations given in Sec. 15 reflect the use of the SI or cgs system of units and not ft-lb-s units, unless otherwise noted in the text. The gravitational conversion factor g_c needed to use ft-lb-s units is not included in the equations.

Symbol	Definition	SI units	U.S. Customary System units
a	Interfacial area per unit volume	m^2/m^3	ft^2/ft^3
a_p	Specific packing surface area (area per unit volume)	m^2/m^3	ft^2/ft^3
a_w	Specific wall surface area (area per unit volume)	m^2/m^3	ft^2/ft^3
b_{ij}	NRTL model regression parameter	K	K
A	Envelope-style downcomer area	m^2	ft^2
A	Area between settled layers in a decanter	m^2	ft^2
A_{col}	Column cross-sectional area	m^2	ft^2
A_{dow}	Area for flow through a downcomer (or upcomer)	m^2	ft^2
$A_{i,j}/RT$	van Laar binary interaction parameter	Dimensionless	Dimensionless
A_o	Cross-sectional area of a single hole	m^2	in^2
C	Concentration (mass or moles per unit volume)	kg/m^3 or $kgmol/m^3$ or gmol/L	lb/ft^3 or $lbmol/ft^3$
C_A^i	Concentration of component A at the interface	kg/m^3 or $kgmol/m^3$ or gmol/L	lb/ft^3 or $lbmol/ft^3$
C^*	Concentration at equilibrium	kg/m^3 or $kgmol/m^3$ or gmol/L	lb/ft^3 or $lbmol/ft^3$
C_D	Drag coefficient	Dimensionless	Dimensionless
C_o	Initial concentration	kg/m^3 or $kgmol/m^3$ or gmol/L	lb/ft^3 or $lbmol/ft^3$
C_p	Packing constant (empirical)	Dimensionless	Dimensionless
C_t	Concentration at time t	kg/m^3 or $kgmol/m^3$ or gmol/L	lb/ft^3 or $lbmol/ft^3$
d	Drop diameter	m	in
d_C	Critical packing dimension	m	in
d_i	Diameter of an individual drop	m	in
d_m	Characteristic diameter of media in a packed bed	m	in
d_o	Orifice or nozzle diameter	m	in
d_p	Sauter mean drop diameter	m	in
d_{32}	Sauter mean drop diameter	m	in
D_{col}	Column diameter	m	in or ft
D_{eq}	Equivalent diameter giving the same area	m	in
D_h	Equivalent hydraulic diameter	m	in
$\mathcal{D}_i$	Distribution ratio for a given chemical species including all its forms (unspecified units)		
D_i	Impeller diameter or characteristic mixer diameter	m	in or ft
D_{sm}	Static mixer diameter	m	in or ft
D_t	Tank diameter	m	ft
$\mathcal{D}$	Molecular diffusion coefficient (diffusivity)	m^2/s	cm^2/s
$\mathcal{D}_{AB}$	Mutual diffusion coefficient for components A and B	m^2/s	cm^2/s
E	Mass or mass flow rate of extract phase	kg or kg/s	lb or lb/h
E'	Solvent mass or mass flow rate (in the extract phase)	kg or kg/s	lb or lb/h
E	Axial mixing coefficient (eddy diffusivity)	m^2/s	cm^2/s

Symbol	Definition	SI units	U.S. Customary System units
$\mathcal{E}^C$	Extraction factor for case C [Eq. (15-93)]	Dimensionless	Dimensionless
$\mathcal{E}_i$	Extraction factor for component i	Dimensionless	Dimensionless
$\mathcal{E}_s$	Stripping section extraction factor	Dimensionless	Dimensionless
$\mathcal{E}_w$	Washing section extraction factor	Dimensionless	Dimensionless
f_{da}	Fractional downcomer area	Dimensionless	Dimensionless
f_{ha}	Fractional hole area	Dimensionless	Dimensionless
F	Mass or mass flow rate of feed phase	kg or kg/s	lb or lb/h
F	Force	N	lbf
F'	Feed mass or mass flow rate (feed solvent only)	kg or kg/s	lb or lb/h
F_R	Solute reduction factor (ratio of inlet to outlet concentrations)	Dimensionless	Dimensionless
g	Gravitational acceleration	$9.807\ m/s^2$	$32.17\ ft/s^2$
G_{ij}	NRTL model parameter	Dimensionless	Dimensionless
h	Height of coalesced layer at a sieve tray	m	in
h	Head loss due to frictional flow	m	in
h	Height of dispersion band in batch decanter	m	in
h_i^E	Excess enthalpy of mixing	J/gmol	Btu/lbmol or cal/gmol
H	Dimensionless group defined by Eq. (15-118)	Dimensionless	Dimensionless
H	Dimension of envelope-style downcomer [Fig. 15-37]	m	in or ft
ΔH	Steady-state dispersion band height in a continuously fed decanter	m	in
HDU	Height of a dispersion unit	m	in
H_e	Height of a transfer unit due to resistance in extract phase	m	in
HETS	Height equivalent to a theoretical stage	m	in
H_{oc}	Height of an overall mass-transfer unit based on continuous phase	m	in
H_{or}	Height of an overall mass-transfer unit based on raffinate phase	m	in
H_r	Height of a transfer unit due to resistance in raffinate phase	m	in
HTU	Height of a transfer unit (phase of reference evident from the context of its use)	m	in
k	Individual mass-transfer coefficient	m/s or cm/s	ft/h
k	Mass-transfer coefficient (unspecified units)		
k_m	Membrane-side mass-transfer coefficient	m/s or cm/s	ft/h
k_o	Overall mass-transfer coefficient	m/s or cm/s	ft/h
k_c	Continuous-phase mass-transfer coefficient	m/s or cm/s	ft/h
k_d	Dispersed-phase mass-transfer coefficient	m/s or cm/s	ft/h
k_s	Setschenow constant	L/gmol	L/gmol
k_s	Shell-side mass-transfer coefficient	m/s or cm/s	ft/h
k_t	Tube-side mass-transfer coefficient	m/s or cm/s	ft/h
K	Partition ratio (unspecified units)		

Nomenclature *(Continued)*

Symbol	Definition	SI units	U.S. Customary System units
K'_s	Stripping section partition ratio (in Bancroft coordinates)	Mass ratio/mass ratio	Mass ratio/mass ratio
K'_w	Washing section partition ratio (in Bancroft coordinates)	Mass ratio/mass ratio	Mass ratio/mass ratio
K'	Partition ratio, mass ratio basis (Bancroft coordinates)	Mass ratio/mass ratio	Mass ratio/mass ratio
K''	Partition ratio, mass fraction basis	Mass fraction/mass fraction	Mass fraction/mass fraction
K^o	Partition ratio, mole fraction basis	Mole fraction/mole fraction	Mole fraction/mole fraction
K^{vol}	Partition ratio (volumetric concentration basis)	Ratio of kg/m³ or kgmol/m³ or gmol/L	Ratio of lb/ft³ or lbmol/ft³
L	Downcomer (or upcomer) length	m	in or ft
L_{fp}	Length of flow path in Eq. (15-143)	m	in or ft
m	Local slope of equilibrium line (unspecified concentration units)		
m'	Local slope of equilibrium line (in Bancroft coordinates)	Mass ratio/mass ratio	Mass ratio/mass ratio
m_{dc}	Local slope of equilibrium line for dispersed-phase concentration plotted versus continuous-phase concentration		
m_{er}	Local slope of equilibrium line for extract-phase concentration plotted versus raffinate-phase concentration		
m^{vol}	Local slope of equilibrium line (volumetric concentration basis)	Ratio of kg/m³ or kgmol/m³ or gmol/L	Ratio of lb/ft³ or lbmol/ft³ units
M	Mass or mass flow rate	kg or kg/s	lb or lb/h
MW	Molecular weight	kg/kgmol or g/gmol	lb/lbmol
N	Number of theoretical stages	Dimensionless	Dimensionless
N_A	Flux of component A (mass or mol/area/unit time)	(kg or kgmol)/(m²·s)	(lb or lbmol)/(ft²·s)
N_{holes}	Number of holes	Dimensionless	Dimensionless
N_{or}	Number of overall mass-transfer units based on the raffinate phase	Dimensionless	Dimensionless
N_s	Number of theoretical stages in stripping section	Dimensionless	Dimensionless
N_w	Number of theoretical stages in washing section	Dimensionless	Dimensionless
P	Pressure	bar or Pa	atm or lbf/in²
P	Dimensionless group defined by Eq. (15-117)	Dimensionless	Dimensionless
P	Power	W or kW	HP or ft·lbf/h
Pe	Péclet number Vb/E, where V is liquid velocity, E is axial mixing coefficient, and b is a characteristic equipment dimension	Dimensionless	Dimensionless
$P_{i,extract}$	Purity of solute i in extract (in wt%)	wt%	wt%
$P_{i,feed}$	Purity of solute i in feed (in wt%)	wt%	wt%
P_o	Power number $P/(\rho_m \omega^3 D_i^5)$	Dimensionless	Dimensionless
ΔP_{dow}	Pressure drop for flow through a downcomer (or upcomer)	bar or Pa	atm or lbf/in²
ΔP_o	Orifice pressure drop	bar or Pa	atm or lbf/in²
Q	Volumetric flow rate	m³/s	ft³/min
R	Universal gas constant	8.31 J · K/kgmol	1.99 Btu · °R/lbmol
R	Mass or mass flow rate of raffinate phase	kg or kg/s	lb or lb/h
R_A	Rate of mass transfer (moles per unit time)	kgmol/s	lbmol/h
Re	Reynolds number: for pipe flow, $Vd\rho/\mu$; for an impeller, $\rho_m \omega D_i^2/\mu_m$; for drops, $V_{so} d_p \rho_c/\mu_c$; for flow in a packed-bed coalescer, $Vd_m\rho_c/\mu$; for flow through an orifice, $V_o d_o \rho_d/\mu_d$	Dimensionless	Dimensionless
Re_{Stokes}	$\rho_c \Delta\rho g d^3{}_p/18\mu_c{}^2$	Dimensionless	Dimensionless
S	Mass or mass flow rate of solvent phase	kg or kg/s	lb or lb/h
S	Dimension of envelope-style downcomer (Fig. 15-37)	m	ft
S'	Solvent mass or mass flow rate (extraction solvent only)	kg or kg/s	lb or lb/h
S'_s	Mass flow rate of extraction solvent within stripping section	kg/s	lb/h
S'_w	Mass flow rate of extraction solvent within washing section	kg/s	lb/h
$S_{i,j}$	Separation power for separating component i from component j [defined by Eq. (15-100)]	Dimensionless	Dimensionless
S_{tip}	Impeller tip speed	m/s	ft/s
t_b	Batch mixing time	s or h	min or h
T	Temperature (absolute)	K	°R
u_t	Stokes' law terminal or settling velocity of a drop	m/s or cm/s	ft/s or ft/min
$u_{t\infty}$	Unhindered settling velocity of a single drop	m/s or cm/s	ft/s or ft/min
v	Molar volume	m³/kgmol or cm³/gmol	ft³/lbmol
V	Liquid velocity (or volumetric flow per unit area)	m/s	ft/s or ft/min
$\mathcal{V}$	Volume	m³	ft³ or gal
V_{cf}	Continuous-phase flooding velocity	m/s	ft/s or ft/min
V_{cflow}	Cross-flow velocity of continuous phase at sieve tray	m/s	ft/s or ft/min
V_{df}	Dispersed-phase flooding velocity	m/s	ft/s or ft/min
V_{drop}	Average velocity of a dispersed drop	m/s	ft/s or ft/min
V_{ic}	Interstitial velocity of continuous phase	m/s	ft/s or ft/min
$V_{o,max}$	Maximum velocity through an orifice or nozzle	m/s	ft/s or ft/min
V_s	Slip velocity	m/s	ft/s or ft/min
V_{so}	Slip velocity at low dispersed-phase flow rate	m/s	ft/s or ft/min
V_{sm}	Static mixer superficial liquid velocity (entrance velocity)	m/s	ft/s or ft/min
W	Mass or mass flow rate of wash solvent phase	kg or kg/s	lb or lb/h
W'_s	Mass flow rate of wash solvent within stripping section	kg/s	lb/h
W'_w	Mass flow rate of wash solvent within washing section	kg/s	lb/h
We	Weber number: for an impeller, $\rho_c \omega^2 D_i^3/\sigma$; for flow through an orifice or nozzle, $V_o^2 d_0 \rho_d/\sigma$; for a static mixer, $V^2{}_{sm} D_{sm} \rho_c/\sigma$	Dimensionless	Dimensionless

Nomenclature *(Continued)*

Symbol	Definition	SI units	U.S. Customary System units
x	Mole fraction solute in feed or raffinate	Mole fraction	Mole fraction
X	Concentration of solute in feed or raffinate (unspecified units)		
X''	Mass fraction solute in feed or raffinate	Mass fractions	Mass fractions
X'	Mass solute/mass feed solvent in feed or raffinate	Mass ratios	Mass ratios
X_f^B	Pseudoconcentration of solute in feed for case B [Eq. (15-90)]	Mass ratios	Mass ratios
X_f^C	Pseudoconcentration of solute in feed for case C [Eq. (15-92)]	Mass ratios	Mass ratios
$X_{i,\text{extract}}$	Concentration of solute i in extract	Mass fraction	Mass fraction
$X_{i,\text{feed}}$	Concentration of solute i in feed	Mass fraction	Mass fraction
$X_{i,\text{raffinate}}$	Concentration of solute i in raffinate	Mass fraction	Mass fraction
X_{ij}	Concentration of component i in the phase richest in j	Mass fraction	Mass fraction
y	Mole fraction solute in solvent or extract	Mole fraction	Mole fraction
Y	Concentration of solute in the solvent or extract (unspecified units)		
Y''	Mass fraction solute in solvent or extract	Mass fraction	Mass fraction
Y'	Mass solute/mass extraction solvent in solvent or extract	Mass ratio	Mass ratio
Y_s^B	Pseudoconcentration of solute in solvent for case B [Eq. (15-91)]	Mass ratio	Mass ratio
z	Dimension or direction of mass transfer	m	in or ft
z	Sieve tray spacing	m	in or ft
z	Point representing feed composition on a tie line		
z_i	Number of electronic charges on an ion	Dimensionless	Dimensionless
Z_t	Total height of extractor	m	ft
	Greek symbols		
$\alpha_{i,j}$	Separation factor for solute i with respect to solute j	Dimensionless	Dimensionless
$\alpha_{i,j}$	NRTL model parameter	Dimensionless	Dimensionless
$\gamma_{i,j}$	Activity coefficient of i dissolved in j	Dimensionless	Dimensionless
γ_i^I	Activity coefficient of component i in phase I	Dimensionless	Dimensionless
γ_i^R	Activity coefficient, residual part of UNIFAC	Dimensionless	Dimensionless
ε	Void fraction	Dimensionless	Dimensionless
ε	Fractional open area of a perforated plate	Dimensionless	Dimensionless

Symbol	Definition	SI units	U.S. Customary System units
	Greek symbols		
ζ	Tortuosity factor [Table 15-17]	Dimensionless	Dimensionless
θ	Residence time for total liquid	s	s or min
θ_i	Fraction of solute i extracted from feed	Dimensionless	Dimensionless
λ_m	Membrane thickness	mm	in
μ	Liquid viscosity	Pa·s	cP
μ_i^I's	Chemical potential of component i in phase I	J/gmol	Btu/lbmol
μ_m	Mixture mean viscosity defined in Eq. (15-162)	Pa·s	cP
μ_w	Reference viscosity (of water)	Pa·s	cP
ξ_{batch}	Efficiency of a batch experiment [Eq. (15-157)]	Dimensionless	Dimensionless
$\xi_{\text{continuous}}$	Efficiency of a continuous process [Eq. (15-158)]	Dimensionless	Dimensionless
ξ_m	Murphree stage efficiency	Dimensionless	Dimensionless
ξ_{md}	Murphree stage efficiency based on dispersed phase	Dimensionless	Dimensionless
ξ_o	Overall stage efficiency	Dimensionless	Dimensionless
$\Delta\pi$	Osmotic pressure gradient	bar or Pa	atm or lbf/in^2
ρ	Liquid density	kg/m^3	lb/ft^3
ρ_m	Mixture mean density [Eq. (15-160)]	kg/m^3	lb/ft^3
σ	Interfacial tension	N/m	dyn/cm
$\tau_{i,j}$	NRTL model parameter	Dimensionless	Dimensionless
ϕ	Volume fraction	Dimensionless	Dimensionless
ϕ_d	Volume fraction of dispersed phase (holdup)	Dimensionless	Dimensionless
$\phi_{d,\text{feed}}$	Volume fraction of dispersed phase in feed	Dimensionless	Dimensionless
ϕ_o	Initial dispersed-phase holdup in feed to a decanter	Dimensionless	Dimensionless
φ	Volume fraction of voids in a packed bed	Dimensionless	Dimensionless
χ	Parameter indicating which phase is likely to be dispersed [Eq. (15-36)]	Dimensionless	Dimensionless
ω	Impeller speed	Rotations/s	Rotations/min
	Additional subscripts		
c	Continuous phase		
d	Dispersed phase		
e	Extract phase		
f	Feed phase or flooding condition (when combined with d or c)		
i	Component i		
j	Component j		
H	Heavy liquid		
L	Light liquid		
max	Maximum value		
min	Minimum value		
o	Orifice or nozzle		
r	Raffinate phase		
s	Solvent		

General References: Seader, J. D., E. J. Henley, and D. K. Roper, *Separation Process Principles with Applications Using Process Simulators,* 4th ed., Wiley, New York, 2016; Zhang, J., B. Zhao, and B. Schreiner, *Separation Hydrometallurgy of Rare Earth Metals*, Springer, Berlin, 2016; Koch, J., and G. Shiveler, "Design Principles for Liquid-Liquid Extraction," *Chem. Eng. Prog.* **111**(11): 22-30 (2015); Leng, D. E., and R. V. Calabrese, "Immiscible Liquid-Liquid Systems," chap. 12 in *Advances in Industrial Mixing*, ed. S. M. Kresta et al., Wiley, New York, 2015, and chap. 12 in *Handbook of Industrial Mixing*, ed. E. L. Paul, V. A. Atiemo-Obeng, and S. M. Kresta, Wiley, New York, 2004; G. Wypych, *Handbook of Solvents*, ChemTec, New York, 2014; P. C. Wankat, *Separation Process Engineering*, 3d ed., Prentice-Hall, Upper Saddle River, NJ, 2012; Seibert, A. F., "Extraction and Leaching," chap. 14 in *Chemical Process Equipment: Selection and Design*, 3d ed., ed. J. R. Couper et al., Butterworth-Heinemann, Oxford, UK, 2012; Kislik, V. S., *Solvent Extraction: Classical and Novel Approaches*, Elsevier, Oxford, UK, 2012; Müller, E., et al., "Liquid-Liquid Extraction," in *Ullmann's Encyclopedia of Industrial Chemistry*, 6th ed., Wiley-VCH, New York, 2002, updated online, 2008; Aguilar, M., and J. L. Cortina, *Solvent Extraction and Liquid Membranes: Fundamentals and Applications in New Materials*, CRC Press, Boca Raton, 2010; Glatz, D. J., and W. Parker, "Enriching Liquid-Liquid Extraction," *Chem. Eng. Magazine* **111**(11): 44–48 (2004); Rydberg, J., M. Cox, C. Musikas, and G. R. Choppin, eds., *Solvent Extraction Principles and Practice*, 2d ed., Marcel Dekker, New York, 2004; Marcus, Y., and A. J. SenGupta, eds., *Ion Exchange and Solvent Extraction*, vol. 17, Marcel Dekker, New York, 2004, and earlier volumes in the series; Cheremisinoff, N. P., *Industrial Solvents Handbook*, 2d ed. Marcel Dekker, New York, 2003; Van Brunt, V., and J. S. Kanel, "Extraction with Reaction," chap. 3 in *Reactive Separation Processes*, ed. S. Kulprathipanja, Taylor & Francis, New York, 2002; Benitez, J., *Principles and Modern Applications of Mass Transfer Operations*, Wiley-VCH, New York, 2002; Bart, H.-J., *Reactive Extraction*, Springer, Berlin, 2001; Flick, E. W., *Industrial Solvents Handbook*, 5th ed., Noyes, Westwood, NJ, 1998; Robbins, L. A., "Liquid-Liquid Extraction," sec. 1.9 in *Handbook of Separation Techniques for Chemical Engineers*, 3d ed., ed. P. A. Schweitzer, McGraw-Hill, New York, 1997; Lo, T. C., "Commercial Liquid-Liquid Extraction Equipment," sec. 1.10 in *Handbook of Separation Techniques for Chemical Engineers*, 3d ed., ed. P. A. Schweitzer, McGraw-Hill, New York, 1997; Humphrey, J. L., and G. E. Keller, "Extraction," chap. 3 in *Separation Process Technology*, McGraw-Hill, New York, 1997, pp. 113–151; Cusack, R. W., and D. J. Glatz, "Apply Liquid-Liquid Extraction to Today's Problems," *Chem. Eng. Magazine* **103**(7): 94–103 (1996); *Liquid-Liquid Extraction Equipment*, J. C. Godfrey and M. J. Slater, eds., Wiley, New York, 1994; Zaslavsky, B. Y., *Aqueous Two-Phase Partitioning*, Marcel Dekker, New York, 1994; Strigle, R. F., "Liquid-Liquid Extraction," chap. 11 in *Packed Tower Design and Applications*, 2d ed., Gulf, Houston, 1994; Schügerl, K., *Solvent Extraction in Biotechnology*, Springer-Verlag, Berlin, 1994; Schügerl, K., "Liquid-Liquid Extraction (Small Molecules)," chap. 21 in *Biotechnology*, 2d ed., vol. 3, ed. G. Stephanopoulos, VCH, New York, 1993; Kelley, B. D., and T. A. Hatton, "Protein Purification by Liquid-Liquid Extraction," chap. 22 in *Biotechnology*, 2d ed., vol. 3, ed. G. Stephanopoulos, VCH, New York, 1993; Lo, T. C., and M. H. I. Baird, "Extraction, Liquid-Liquid," in *Kirk-Othmer Encyclopedia of Chemical Technology*, 4th ed., vol. 10, ed. J. I. Kroschwitz and M. Howe-Grant, Wiley, New York, 1993, pp. 125–180; Thornton, J. D., ed., *Science and Practice of Liquid-Liquid Extraction*, vol. 1, *Phase Equilibria; Mass Transfer and Interfacial Phenomena; Extractor Hydrodynamics, Selection, and Design*, and vol. 2, *Process Chemistry and Extraction Operations in the Hydrometallurgical, Nuclear, Pharmaceutical, and Food Industries*, Oxford, New York, 1992; Cusack, R. W., P. Fremeaux, and D. J. Glatz, "A Fresh Look at Liquid-Liquid Extraction," pt. 1, "Extraction Systems," *Chem. Eng. Magazine* **98**(2): 66–67 (1991); Cusack, R. W., and P. Fremeauz, pt. 2, "Inside the Extractor," *Chem. Eng. Magazine* **98**(3): 132–138, 1991; Cusack, R. W., and A. E. Karr, pt. 3, "Extractor Design and Specification," *Chem. Eng. Magazine* **98**(4): 112–120, 1991; *Methods in Enzymology*, vol. 182, *Guide to Protein Purification*, ed. M. P. Deutscher, Academic, New York, 1990; Wankat, P. C., *Equilibrium Staged Separations*, Prentice Hall, Englewood Cliffs, NJ, 1988; Blumberg, R., *Liquid-Liquid Extraction*, Academic, New York, 1988; Skelland, A. H. P., and D. W. Tedder, "Extraction—Organic Chemicals Processing," chap. 7 in *Handbook of Separation Process Technology*, ed. R. W. Rousseau, Wiley, New York, 1987; Chapman, T. W., "Extraction—Metals Processing," chap. 8 in *Handbook of Separation Process Technology*, ed. R. W. Rousseau, Wiley, New York, 1987; Novak, J. P., J. Matous, and J. Pick, *Liquid-Liquid Equilibria*, Studies in Modern Thermodynamics Series, vol. 7, Elsevier, Amsterdam, 1987; Bailes, P. J., et al., "Extraction, Liquid-Liquid" in *Encyclopedia of Chemical Processing and Design*, vol. 21, ed. J. J. McKetta and W. A. Cunningham, Marcel Dekker, New York, 1984, pp. 19–166; Lo, T. C., M. H. I. Baird, and C. Hanson, eds., *Handbook of Solvent Extraction*, Wiley, New York, 1983, and Krieger, Huntington, NY, 1991; Sørensen, J. M., and W. Arlt, *Liquid-Liquid Equilibrium Data Collection*, DECHEMA Chemistry Data Series, *Binary Systems*, vol. V, pt. 1, 1979, *Ternary Systems*, vol. V, pt. 2, 1980, *Ternary and Quaternary Systems*, vol. 5, pt. 3, 1980, Macedo, M. E. A., and P. Rasmussen, Suppl. 1, vol. V, pt. 4, 1987, DECHEMA, Frankfurt; Wisniak, J., and A. Tamir, *Liquid-Liquid Equilibrium and Extraction, a Literature Source Book*, vols. I and II, Elsevier, Amsterdam, 1980–1981, Suppl. 1, 1985; Treybal, R. E., *Mass Transfer Operations*, 3d ed., McGraw-Hill, New York, 1980; King, C. J., *Separation Processes*, 2d ed., McGraw-Hill, New York, 1980, and Dover, Mineola, N.Y., 2013; Laddha, G. S., and T. E. Degaleesan, *Transport Phenomena in Liquid Extraction*, McGraw-Hill, New York, 1978; Brian, P. L. T., *Staged Cascades in Chemical Processing*, Prentice-Hall, Englewood Cliffs, NJ, 1972; Pratt, H. R. C., *Countercurrent Separation Processes*, Elsevier, Amsterdam, 1967; Treybal, R. E., *Liquid Extraction*, 2d ed., McGraw-Hill, New York, 1963.

INTRODUCTION AND OVERVIEW

Liquid-liquid extraction is a process for separating the components of a liquid (the feed) by contact with a second liquid phase (the solvent). The process takes advantage of differences in the chemical properties of the feed components, such as differences in polarity and hydrophobic/hydrophilic character, to separate them. Stated more precisely, the transfer of components from one liquid to the other is driven by a deviation from thermodynamic equilibrium, and the equilibrium state depends on the nature of the molecular interactions between the feed components and the solvent. The potential for separating the feed components is determined by differences in these molecular interactions.

A liquid-liquid extraction process produces a solvent-rich stream called the extract that contains a portion of the feed, and an extracted-feed stream called the raffinate. A commercial process almost always includes two or more auxiliary operations in addition to the extraction operation itself. These extra operations are needed to treat the extract and raffinate streams for the purposes of isolating a desired product, recovering the solvent for recycle back to the extractor, and purging unwanted components from the process. A typical process includes two distillation operations in addition to extraction.

Liquid-liquid extraction is used to recover desired components from a crude liquid mixture or to remove unwanted contaminants. In developing a process, the project team must decide what solvent or solvent mixture to use, how to isolate product and solvent from the extract, and how to remove solvent residues from the raffinate. The team must also decide what temperature or range of temperatures should be used for the extraction, what process scheme to employ among many possibilities, and what type of equipment to use for liquid-liquid contacting and phase separation. The variety of commercial equipment options is large and includes stirred tanks and decanters (settlers), specialized mixer-settlers, a wide variety of agitated and nonagitated extraction columns or towers, and various types of centrifuges.

Because of the availability of hundreds of commercial solvents and extractants, as well as a wide variety of established process schemes and equipment options, liquid-liquid extraction is a versatile technology with a wide range of commercial applications. It is used in the processing of many commodity and specialty chemicals, including petrochemicals, coal- and wood-derived chemicals, complex organics such as pharmaceuticals and agricultural chemicals, and metals and nuclear fuel (hydrometallurgy). Liquid-liquid extraction also is an important operation in industrial wastewater treatment, food processing, and the recovery of biomolecules from fermentation broth and other forms of biomass.

HISTORICAL PERSPECTIVE

The art of solvent extraction has been practiced in one form or another since ancient times. It appears that until the 19th century, solvent extraction was mainly used to isolate desired components such as perfumes and dyes from plant solids and other natural sources [Aftalion, F., *A History of the International Chemical Industry*, 2d ed., Chemical Heritage Foundation, Philadelphia, 2001; and Taylor, F. S., *A History of Industrial Chemistry*, Abelard-Schuman, New York, 1957]. However, several early applications involving liquid-liquid contacting are described by E. Blass, T. Liebel, and

M. Haeberl ["Solvent Extraction—A Historical Review," *International Solvent Extraction Conf. (ISEC) '96 Proceedings*, Univ. of Melbourne, Melbourne, 1996], including washing oil with water to remove color.

The modern practice of liquid-liquid extraction has its roots in the middle to late 19th century when extraction became an important laboratory technique. In 1855, Adolf Fick introduced fundamental concepts of diffusion underlying mass transfer [*Ann. Phys. (Berlin)* **170**(1): 59–86 (1855)]. Later, the partition ratio concept describing how a solute partitions between two liquid phases at equilibrium was introduced by M. Berthelot and E. Jungfleisch [*Ann. Chim. Phys.* (4th Ser.) **26:** 396–407 (1872)] and further defined by W. H. Nernst [*Z. Phys. Chem.* **8**(1): 110–139 (1891)]. At about the same time, J. Willard Gibbs published his landmark treatise on chemical thermodynamics [*Trans. Conn. Acad. Arts Sci.* **3:** 108–248 (1876); **3:** 343–527 (1878)]. These and other advances were accompanied by a growing chemical industry. In 1883, T. Göring received a patent for a countercurrent extraction process using ethyl acetate solvent to recover acetic acid from "pyroligneous acid" produced by pyrolysis of wood [Ger. Patent 28064 (1883)], and in 1901, L. C. Reese received a patent for a stirred extraction column [U.S. Patent 679,575 (1901)].

With the emergence of the chemical engineering profession in the 1890s and the early 20th century, additional attention was given to process fundamentals and development of a more quantitative basis for process design. Many of the advances made in the study of distillation and absorption were readily adapted to liquid-liquid extraction, owing to its similarity as another diffusion-based operation [Sherwood, T. K., *Ind. Eng. Chem.* **33**(4): 424–429 (1941)]. Examples include the introduction of the equilibrium-stage approach to analyzing performance [Sorel, E., *La Rectification de l'Alcohol*, Gauthier-Villars, Paris, 1893], the application of mass-transfer coefficients [Lewis, W. K., *Ind. Eng. Chem.* **8**(9): 825–833 (1916); and Lewis, W. K., and W. G. Whitman, *Ind. Eng. Chem.* **16**(12): 1215–1220 (1924)], the use of graphical stagewise design methods [McCabe, W. L., and E. W. Thiele, *Ind. Eng. Chem.* **17**(6): 605–611 (1925); Evans, T. W., *Ind. Eng. Chem.* **26**(8): 860–864 (1934); and Thiele, E. W., *Ind. Eng. Chem.* **27**(4): 392–396 (1935)], countercurrent theoretical-stage calculations [Kremser, A., *National Petroleum News* **22**(21): 43–49 (1930); and Souders, M., and G. G. Brown, *Ind. Eng. Chem.* **24**(5): 519–522 (1932)], and the transfer unit concept introduced in the late 1930s by A. P. Colburn and others [*Ind. Eng. Chem.* **33**(4): 459–467 (1941)]. Additional background is given by M. J. Hampe, S. Hartland, and M. J. Slater [Chap. 2 in *Liquid-Liquid Extraction Equipment*, ed. J. C. Godfrey and M. J. Slater, Wiley, New York, 1994].

The number of commercial applications continued to grow, and by the 1930s liquid-liquid extraction had replaced various chemical treatment methods for refining mineral oil and coal tar products [Varteressian, K. A., and M. R. Fenske, *Ind. Eng. Chem.* **28**(8): 928–933 (1936)]. Extraction also was used to recover acetic acid from waste liquors generated in the production of cellulose acetate, and in various nitration and sulfonation processes [Hunter, T. G., and A. W. Nash, *The Industrial Chemist* **9**(102–104): 245–248, 263–266, 313–316 (1933)]. Here, Hunter and Nash also describe early mixer-settler equipment, mixing jets, and various extraction columns, including the spray column, baffle tray column, sieve tray column, and a packed column filled with Raschig rings or with coke breeze, a solid by-product of coal or petroleum.

Much of the liquid-liquid extraction technology in practice today was first introduced to industry during a period of vigorous innovation and growth of the chemical industry as a whole from about 1920 to 1970. This period saw the introduction of many new equipment designs, including specialized mixer-settler equipment, mechanically agitated extraction columns, and centrifugal extractors, as well as a great increase in the availability of different types of industrial solvents. A variety of alcohols, ketones, esters, and chlorinated hydrocarbons became available in large quantities beginning in the 1930s as petroleum refiners and chemical companies found ways to manufacture them inexpensively using the by-products of petroleum refining and natural gas processing. The advances of this period also included the development of fractional extraction process schemes, including work described by R. E. Cornish et al. [*Ind. Eng. Chem.* **26**(4): 397–406 (1934)] and by E. W. Thiele [*Ind. Eng. Chem.* **27**(4): 392–396 (1935)]. A well-known commercial example involving the use of extract reflux is the UDEX process for separating aromatic from aliphatic hydrocarbons, a process developed jointly by The Dow Chemical Company and Universal Oil Products in the 1940s. The early UDEX units used diethylene glycol and diglycolamine as extraction solvents. Over the years, these were supplanted by higher-boiling glycols for greater capacity and lower energy consumption in the associated distillation operations. Later, a number of specialty solvents were introduced by others, including sulfolane (2,3,4,5-tetrahydrothiophene-1,1-dioxide) and NMP (*N*-methyl-2-pyrrolidone).

The ready availability of many solvents and extractants, combined with the tremendous growth of the chemical industry, drove the development and implementation of many new industrial applications. Handbooks of chemical process technology provide a glimpse of some of these [*Kent and Riegel's Handbook of Industrial Chemistry and Biotechnology*, 11th ed., ed. J. A. Kent, Springer, Berlin, 2007; *Chemical Processing Handbook*, ed. J. J. McKetta, Marcel Dekker, New York, 1993; and Austin, G. T., *Shreve's Chemical Process Industries*, 5th ed., McGraw-Hill, New York, 1984], but many remain proprietary and are not widely known. The better-known examples include the separation of aromatics from aliphatics as mentioned previously, the extraction of phenolic compounds from coal tars and liquors, the recovery of ε-caprolactam for the production of polyamide-6 (nylon-6), the recovery of hydrogen peroxide from oxidized anthraquinone solution, many processes involving the washing of crude organic streams with alkaline or acidic solutions and water, and the detoxification of industrial wastewater prior to biotreatment using steam-strippable organic solvents. The pharmaceutical and specialty chemicals industry also began using liquid-liquid extraction in the production of new synthetic drug compounds and other complex organics. In these processes, often involving multiple batch reaction steps, liquid-liquid extraction generally is used for the recovery of intermediates or crude products prior to final isolation of a pure product by crystallization. In the mining and metals industry, specialty organophosphorus compounds and alkyl amines were developed as extractants for recovery and purification of metal ions in aqueous acid solution (hydrometallurgy), including the recovery of uranium from phosphate-rock acid leachate liquor (the PUREX technology originating with the Manhattan Project during World War II) and the purification of copper by removal of arsenic impurities. Extraction processes also were developed for bioprocessing applications. Examples include the use of amyl acetate to recover penicillin and other antibiotics from fermentation broth, the recovery of citric acid from broth using trialkylamine extractants, and the use of water-soluble polymers in aqueous two-phase extraction for the purification of proteins.

Since the 1970s, the use of supercritical or near-supercritical fluids as an alternative to using liquid solvents for extraction has received a great deal of attention in the R&D community. Some processes were developed many years before then; for example, the propane deasphalting process used to refine lubricating oils uses propane at near-supercritical conditions. This technology dates back to the 1930s [McHugh, M. A., and V. J. Krukonis, *Supercritical Fluid Processing*, 2d ed., Butterworth-Heinemann, Oxford, UK, 1993]. In recent years, the use of supercritical fluids has found a number of commercial applications and has displaced earlier liquid-liquid extraction applications, particularly for the recovery of high-value products meant for consumption by humans, including decaffeinated coffee, flavor components from citrus oils, and a variety of nutraceuticals from natural sources.

Progress continues to be made toward improving extraction technology and its application, including the introduction of improved methods for calculating solvent properties and screening candidate solvents and solvent blends, improved methods for overall process conceptualization and optimization, and improved methods for equipment design. Progress also is being made by applying the technology developed for a particular application in one industry to improve another application in another industry. For example, much can be learned by comparing equipment and practices used in organic chemical production with those used in the inorganic chemical industry (and vice versa), or by comparing practices used in commodity chemical processing with those used in the specialty chemicals industry. And new concepts offering potential for significant improvements continue to be described in the literature. (See the subsection Emerging Developments.)

USES FOR LIQUID-LIQUID EXTRACTION

For many separation applications, the use of liquid-liquid extraction is an alternative to the various distillation schemes described in Sec. 13, Distillation. In many of these cases, a distillation process is more economical largely because the extraction process requires extra operations to process the extract and raffinate streams, and these operations usually involve the use of distillation anyway. However, in certain cases the use of liquid-liquid extraction is more cost-effective than using distillation alone because it can be implemented with smaller equipment and/or lower energy consumption. In effect, a difficult distillation is exchanged for an easier one by first using extraction to transfer specific feed components into a second liquid phase (the extract), enabling easier isolation of a key component by distillation of this second liquid. Normally, distillation also will be needed to remove solvent residues from the remaining feed (the raffinate), and the solvent should be chosen to make this an easy distillation as well.

In particular, liquid-liquid extraction may be preferred when the relative volatility of key components is less than 1.3 or so, such that distillation alone requires an unusually tall distillation tower or high reflux ratios and high energy consumption. In certain cases, the distillation option may be improved by adding a solvent (extractive distillation) or an entrainer (azeotropic distillation) directly to the distillation tower instead of first using extraction (see Sec. 13). The driving force for these *enhanced distillation* processes is a function of specific molecular interactions *and* the vapor pressures of the components, whereas the driving force for liquid-liquid extraction is a function of molecular interactions alone. Which process scheme is best in terms of higher selectivity or lower solvent usage and lower energy consumption will vary depending on the specific chemical system and specific process requirements.

Extraction also may be preferred when the distillation option requires operation at pressures less than about 70 mbar (about 50 mmHg) and an unusually large-diameter distillation tower is required, or when most of the feed must be taken overhead to isolate a desired bottoms product. Extraction also may be attractive when distillation requires the use of high-pressure steam for the reboiler or refrigeration for overheads condensation [Null, H. R., *Chem. Eng. Prog.* **76**(8): 42–49 (1980)], or when the desired product is temperature sensitive and extraction can provide a gentler separation process.

Of course, liquid-liquid extraction also may be a useful option when the components of interest cannot be separated by using distillation methods simply because key components are not sufficiently volatile. An obvious example is the recovery of metal ions from aqueous solution in hydrometallurgy [Cox, M., Chap. 1 in *Science and Practice of Liquid-Liquid Extraction,* vol. 2, ed. J. D. Thornton, Oxford University Press, New York, 1992]. Another example is the use of liquid-liquid extraction employing a steam-strippable solvent to remove nonstrippable, low-volatility contaminants from wastewater [Robbins, L. A., *Chem. Eng. Prog.* **76**(10): 58–61 (1980)]. The same process scheme often provides a cost-effective alternative to direct distillation or stripping of volatile impurities when the relative volatility of the impurity with respect to water is less than about 10 [Robbins, L. A., U.S. Patent 4,236,973 (1980); Hwang, Y. L., G. E. Keller, and J. D. Olson, *Ind. Eng. Chem. Res.* **31:** 1753–1759 (1992); and Frank, T. C., et al., *Ind. Eng. Chem. Res.* **46**(11): 3774–3786 (2007)].

Liquid-liquid extraction also can be an attractive alternative to separation methods other than distillation—for example, as an alternative to crystallization from solution to remove dissolved salts from a crude organic feed. Extraction of the salt content into water eliminates the need to filter solids from the mother liquor, often a difficult or expensive operation. Extraction also may compete with process-scale chromatography, an example being the recovery of hydroxytyrosol (3,4-dihydroxy-phenylethanol), an antioxidant food additive, from olive-processing wastewaters [Guzman, J. F.-B., et al., U.S. Patent 6,849,770 (2005)]. In hydrometallurgy, extraction is an alternative to various fixed-bed ion-exchange processes.

The attractiveness of liquid-liquid extraction for a given application compared to alternative separation technologies often depends upon the concentration of solute in the feed. The recovery of acetic acid from aqueous solutions is a well-known example [Brown, W. V., *Chem. Eng. Prog.* **59**(10): 65–68 (1963)]. In this case, extraction generally is more economical than distillation when handling dilute to moderately concentrated feeds, while distillation is more economical at higher concentrations. In the treatment of water to remove trace amounts of organics, when the concentration of impurities in the feed is greater than about 20 to 50 ppm, liquid-liquid extraction may be more economical than adsorption of the impurities by using carbon beds because the latter may require frequent and costly replacement of the adsorbent [L. A. Robbins, *Chem. Eng. Prog.* **76**(10): 58–61 (1980)]. At lower concentrations of impurities, adsorption may be the more economical option because the usable lifetime of the carbon bed is longer.

Examples of cost-effective liquid-liquid extraction processes that use relatively low-boiling solvents include the recovery of acetic acid from aqueous solutions using diethyl ether or ethyl acetate [King, C. J., chap. 18.5 in *Handbook of Solvent Extraction,* ed. T. C. Lo, M. H. I. Baird, and C. Hanson, Wiley, New York, 1983, and Krieger, Huntington, NY, 1991] and the recovery of phenolic compounds from water by using methyl isobutyl ketone [Greminger, D. C., et al., *Ind. Eng. Chem. Proc. Des. Dev.* **21**(1): 51–54 (1982)]. In these processes, the solvent is recovered from the extract by distillation, and dissolved solvent is removed from the raffinate by steam stripping (Fig. 15-1). The solvent circulates through the process in a closed loop.

A well-known application of liquid-liquid extraction in petrochemical operations involves the extraction of aromatic compounds from hydrocarbon mixtures using high-boiling (low volatility) polar solvents. The production rates are very large, and a high-boiling solvent (relative to the product solute) is used to minimize energy consumption in subsequent distillations of the extract. A number of processes have been developed to recover benzene, toluene, and xylene (BTX) as feedstock for chemical manufacturing or to refine hydrocarbon fractions for use as lubricants and fuels. This general technology is described in detail in the subsection Single-Solvent Fractional Extraction with Extract Reflux under Calculation Procedures. A typical flow diagram is shown in Fig. 15-2. For smaller-scale operations, a relatively light polar solvent may be used; processes using *N,N*-dimethylformamide or acetonitrile have been developed for the removal of polynuclear aromatic and

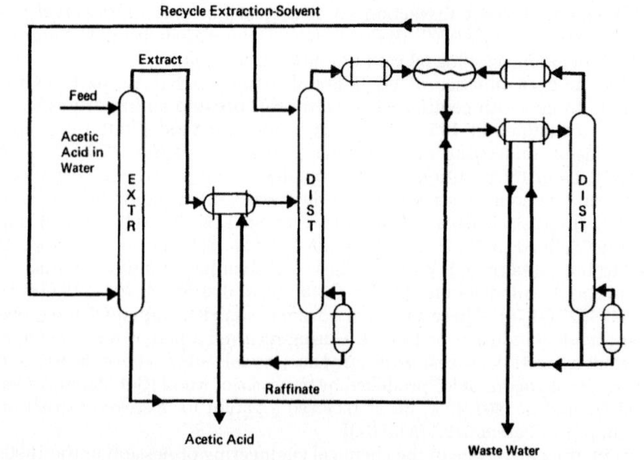

FIG. 15-1 Typical process for extraction of acetic acid from water.

sulfur-containing contaminants from used motor oils [Sherman, J. H., J. W. Hershberger, and R. T. Taylor, U.S. Patent 6,320,090 (2001)]. An alternative process uses a blend of methyl ethyl ketone + 2-propanol and small amounts of aqueous KOH [Rincón, J., P. Cañizares, and M. T. García, *Ind. Eng. Chem. Res.* **44**(20): 7854–7859 (2005)].

Liquid-liquid extraction also is used to remove CO_2, H_2S, and other acidic contaminants from liquefied petroleum gas (LPG) generated during the operation of fluid catalytic crackers and cokers in petroleum refineries and from liquefied natural gas (LNG)—a process called sweetening. The acid gases are extracted from the liquefied hydrocarbons into water by reversible reaction with various aqueous amine extractants. Typical amines are methyldiethanolamine (MDEA), diethanolamine (DEA), monoethanolamine (MEA), and diglycolamine (DGA). In a typical process (Fig. 15-3), the treated hydrocarbon liquid (the raffinate) is washed with water to remove residual amine, and the loaded aqueous amine solution (the extract) is regenerated in a stripping tower for recycle back to the extractor [Nielsen, R. B., et al., *Hydrocarbon Proc.* **76:** 49–59 (1997)]. The technology is similar to that used to scrub CO_2 and H_2S from gas streams [Oyenekan, B. A., and G. T. Rochelle, *Ind. Eng. Chem. Res.* **45**(8): 2465–2472 (2006); and Jassim, M. S., and G. T. Rochelle, *Ind. Eng. Chem. Res.* **45**(8): 2457–2464 (2006)], except that the process involves liquid-liquid contacting instead of gas-liquid contacting. Because of this, in a typical refinery, a single large stripping tower often will be used to regenerate aqueous amines coming from a variety of gas absorbers and liquid-liquid extractors (called *liquid treaters* in the industry). In certain applications, organic acids such as formic acid are present in low concentrations in the hydrocarbon feed. These contaminants will react with the amine to form heat-stable amine salts that accumulate in the extraction solvent recycle loop over time, requiring periodic purging or regeneration of the extractant [Price, J., and D. Burns, *Hydrocarb. Process.* **74:** 140–141 (1995)].

A typical extraction process used in hydrometallurgical applications is outlined in Fig. 15-4. This technology involves transferring the desired metal ion from the ore leachate liquor, an aqueous acid, into an organic solvent phase containing specialty extractants that form a reversible complex with the metal ion. The organic phase is later contacted with another aqueous solution to regenerate the solvent and transfer the metal ion into a clean solution from which it can be recovered by electrolysis or another method [Sole, K. C., A. M. Feather, and P. M. Cole, *Hydrometallurgy* **78:** 52–78 (2005); and Zhang, J., B. Zhao, and B. Schreiner, *Separation Hydrometallurgy of Rare Earth Metals,* Springer, Berlin, 2016]. In this industry, the initial extraction of metal ion from the aqueous acid leachate liquor into an organic phase is called *solvent extraction,* while the extraction of unwanted impurities from the loaded organic phase is called *washing* or *scrubbing,* and back-extraction of the metal ion of interest from the organic phase into clean aqueous solution is called *stripping.* Note that these terms have more general meanings in nonhydrometallurgical applications. (See the subsection Definitions.) Hydrometallurgy is discussed in more detail in the subsection Reaction-Enhanced Extraction under Commercial Process Schemes. A related process technology uses metals complexed with various organophosphorus compounds as recyclable homogeneous catalysts; liquid-liquid extraction is used to transfer the metal complex between the reaction phase and a separate liquid phase after reaction. Different ligands having different polarities are chosen to facilitate the use of various extraction and recycle schemes [Kanel, J. S., et al., U.S. Patents 6,294,700 (2001) and 6,303,829 (2001)].

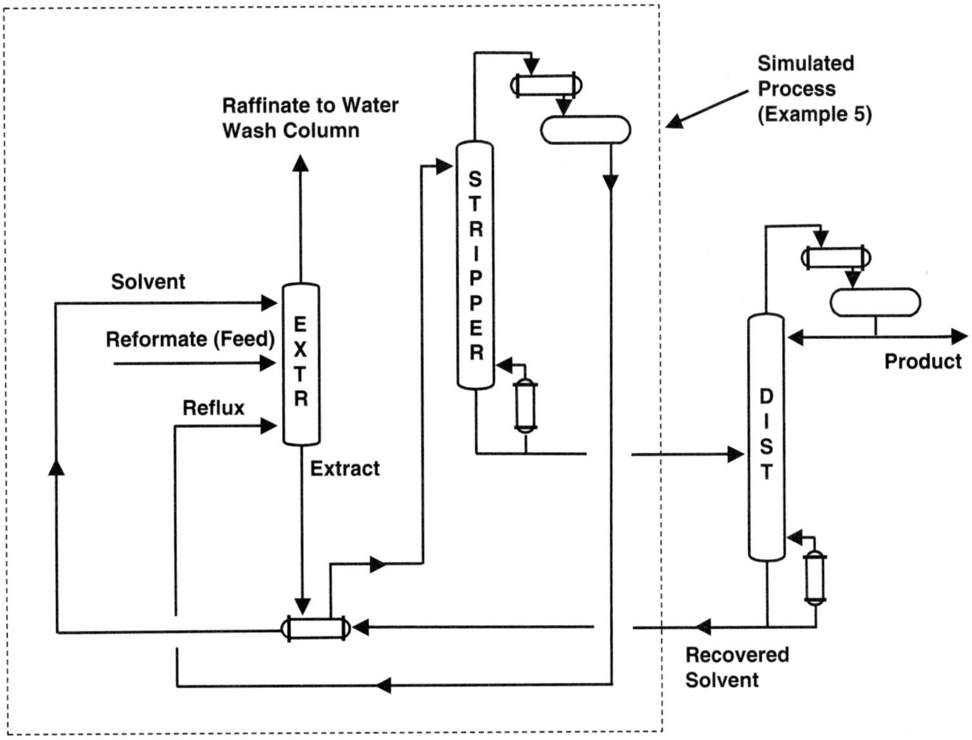

FIG. 15-2 Flow sheet of a simplified aromatic extraction process (see Example 15-5).

Bioprocessing is another category of useful liquid-liquid extraction applications. A longstanding example involves the recovery of antibiotics and other complex organics from aqueous fermentation broth by using a variety of oxygenated organic solvents such as acetates and ketones. Although some of these products are unstable at the required extraction conditions (particularly if pH must be low for favorable partitioning), short-contact-time centrifugal extractors may be used to minimize exposure. Centrifugal extractors also help overcome problems associated with the formation of emulsions between solvent and broth. In a number of applications, the whole broth can be processed without prior removal of cell debris and other solids, a practice that can significantly reduce costs. For detailed information, see "The History of Penicillin Production," A. L. Elder, ed., *Chemical Engineering Progress Symposium Series No. 100,* vol. 66, pp. 37–42, American Institute of Chemical Engineers, New York, 1970; S. W. Queener and R. W. Swartz, "Penicillins: Biosynthetic and Semisynthetic," in *Secondary Products of Metabolism, Economic Microbiology,* vol. 3, ed. A. H. Rose, Academic, New York, 1979; and T. Z. Chaung et al., *J. Chinese Inst. Chem. Eng.* **20**(3): 155–161 (1989). Another well-known commercial application of liquid-liquid extraction in bioprocessing involves the recovery of citric acid from fermentation broth with tertiary amine extractants [Baniel, A. M., R. Blumberg, and

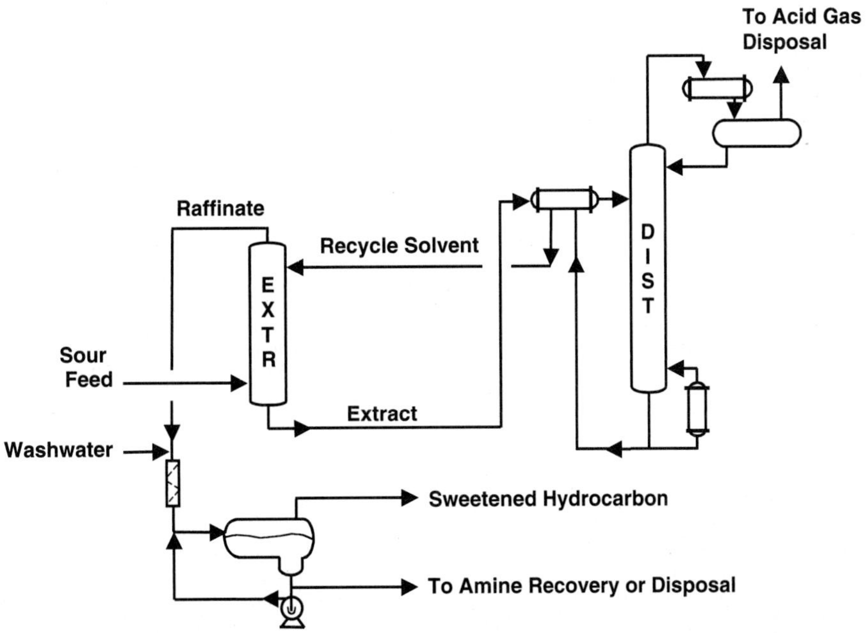

FIG. 15-3 Typical process for extracting acid gases from LPG or LNG.

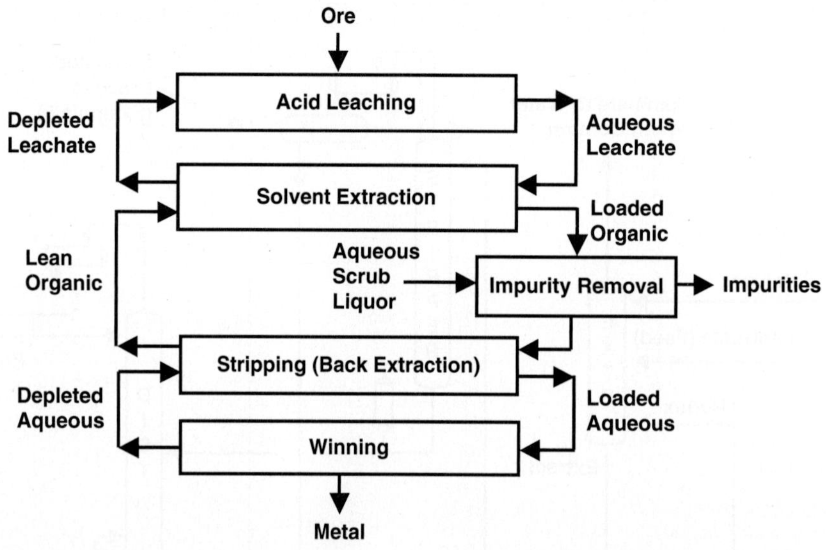

FIG. 15-4 Example process scheme used in hydrometallurgical applications. [*Taken from Cox, Chap. 1 in* Science and Practice of Liquid-Liquid Extraction, *vol. 2, Thornton, (Oxford, 1992), with permission. Copyright 1992 Oxford University Press.*]

K. Hajdu, U.S. Patent 4,275,234 (1981)]. This type of process is discussed in Reaction-Enhanced Extraction under Commercial Process Schemes. Another bioprocessing example is the application of extraction in lignocellulosic ethanol production [Zhang, J., and B. Hu, "Liquid-Liquid Extraction," chap. 3 in *Separation and Purification Technologies in Biorefineries*, ed. S. Ramaswamy, H.-J. Huang, and B. V. Ramarao, Wiley, New York, 2013].

DEFINITIONS

Extraction terms defined by the International Union of Pure and Applied Chemistry (IUPAC) generally are recommended. See N. M. Rice, H. M. N. H. Irving, and M. A. Leonard, *Pure Appl. Chem.* (IUPAC) **65**(11): 2373–2396 (1993); and J. Inczédy, *Pure Appl. Chem.* (IUPAC) **66**(12): 2501–2512 (1994). *Liquid-liquid extraction* is a process for separating components dissolved in a liquid feed by contact with a second liquid phase. *Solvent extraction* is a broader term that describes a process for separating the components of any matrix by contact with a liquid, and it includes liquid-solid extraction (leaching) as well as liquid-liquid extraction. The *feed* to a liquid-liquid extraction process is the solution that contains the components to be separated. The major liquid component (or components) in the feed can be referred to as the *feed solvent* or the *carrier solvent*. Minor components in solution often are referred to as *solutes*. The *extraction solvent* is the immiscible or partially miscible liquid added to the process to create a second liquid phase for the purpose of extracting one or more solutes from the feed. It is also called the *separating agent* and may be a mixture of several individual solvents (a *mixed solvent* or a *solvent blend*). The extraction solvent also may be a liquid comprised of an *extractant* dissolved in a liquid *diluent*. In this case, the extractant species is primarily responsible for the extraction of solute due to a relatively strong attractive interaction with the desired solute, forming a reversible adduct or molecular complex. The diluent itself does not contribute significantly to the extraction of solute, and in this respect it is not the same as a true extraction solvent, although a diluent may improve mass transfer by reducing viscosity. A *modifier* may be added to the diluent to increase the solubility of the extractant or otherwise enhance the effectiveness of the extractant. The phase leaving a liquid-liquid contactor rich in extraction solvent is called the *extract*. The *raffinate* is the liquid phase left from the feed after it is contacted by the extract phase. The word *raffinate* originally referred to a "refined product"; however, common usage has extended its meaning to describe the feed phase after extraction whether that phase is a product or not.

Industrial liquid-liquid extraction most often involves processing two immiscible or partially miscible liquids in the form of a *dispersion* of droplets of one liquid (the *dispersed phase*) suspended in the other liquid (the *continuous phase*). The dispersion will exhibit a distribution of drop diameters d_i often characterized by the volume-to-surface-area average diameter or *Sauter mean drop diameter*. The term *emulsion* generally refers to a liquid-liquid dispersion with a dispersed-phase mean drop diameter on the order of 1 μm or less.

The tension that exists between two liquid phases is called the *interfacial tension*. It is a measure of the energy or work required to increase the surface area of the liquid-liquid interface, and it affects the size of dispersed drops. Its value, in units of force per unit length or energy per unit area, reflects the compatibility of the two liquids. Systems that have low compatibility (low mutual solubility) exhibit high interfacial tension. Such a system tends to form relatively large dispersed drops and low interfacial area to minimize contact between the phases. Systems that are more compatible (with higher mutual solubility) exhibit lower interfacial tension and more easily form small dispersed droplets.

A *theoretical* or *equilibrium stage* accomplishes the effect of intimately mixing two liquid phases until equilibrium concentrations are reached, then physically separating the two phases into clear layers. The *partition ratio K* commonly is defined for a given solute as the solute concentration in the extract phase divided by that in the raffinate phase after equilibrium is attained in a single stage of contacting. A variety of concentration units are used, so it is important to determine how partition ratios have been defined in the literature for a given application. The term *partition ratio* is preferred, but it also is referred to as the *distribution constant, distribution coefficient*, or the *K* value. It is a measure of the thermodynamic potential of a solvent for extracting a given solute and can be a strong function of composition and temperature. In some cases, the partition ratio transitions between a value less than unity and a value greater than unity as a function of solute concentration. A system of this type is called a *solutrope* [Smith, A. S., *Ind. Eng. Chem.* **42**(6): 1206–1209 (1950)]. The term *distribution ratio*, designated by $\mathcal{D}_i$, is used in analytical chemistry to describe the distribution of a species that undergoes chemical reaction or dissociation, in terms of the total concentration of analyte in one phase over that in the other, regardless of its chemical form.

The *extraction factor* $\mathcal{E}$ is a dimensionless process variable that characterizes the capacity of the extract phase to carry solute relative to the feed phase. Its value largely determines the number of theoretical stages required to transfer solute from the feed to the extract. The extraction factor is analogous to the stripping factor in distillation and is the ratio of the slope of the equilibrium line to the slope of the operating line in a McCabe-Thiele type of stagewise graphical calculation. For a dilute to moderately concentrated extraction process with straight equilibrium and operating lines, $\mathcal{E}$ is constant and equal to the partition ratio for the solute of interest times the ratio of the solvent flow rate to the feed flow rate. The separation factor $\alpha_{i,j}$ measures the relative enrichment of solute *i* in the extract phase, compared to solute *j*, after one theoretical stage of extraction. It is equal to the ratio of *K* values for components *i* and *j* and is used to characterize the *selectivity* a solvent has for a given solute.

A *standard extraction* process is one in which the primary purpose is to transfer solute from the feed phase into the extract phase in a manner analogous to stripping in distillation. *Fractional extraction* refers to a process in which two or more solutes present in the feed are sharply separated from each other, one fraction leaving the extractor in the extract and the other in

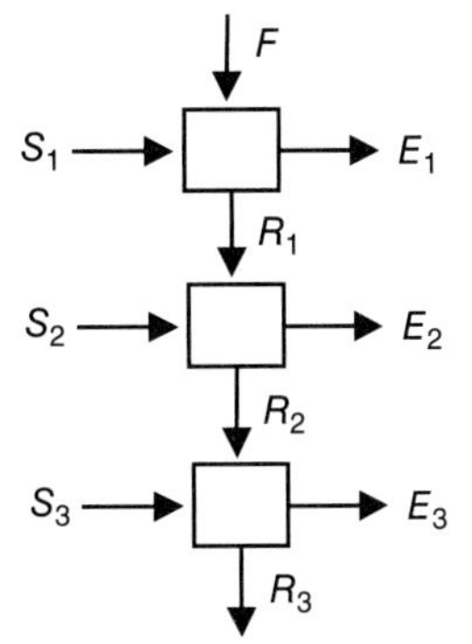

FIG. 15-5 Cross-current extraction.

the raffinate. In the chemical process industries, *stripping* generally refers to a standard extraction process or the portion of a fractional extraction process where solute transfers from the feed phase into the extract phase. *Washing* refers to the portion of a fractional extraction process (or a separate operation) where unwanted impurities transfer out of the extract phase. *Cross-current or cross-flow extraction* (Fig. 15-5) is a series of discrete stages in which the raffinate *R* from one extraction stage is contacted with additional fresh solvent *S* in each subsequent stage. *Countercurrent extraction* (Fig. 15-6) is an extraction scheme in which the extraction solvent enters the process at the end of the extraction farthest from where the feed *F* enters, and the two phases pass each other in countercurrent fashion. The objective is to transfer one or more components from the feed solution *F* into the extract *E*. Compared to cross-current operation, countercurrent operation generally allows operation with less solvent. When a *staged contactor* is used, the two phases are mixed with droplets of one phase suspended in the other, but the phases are separated before leaving each stage. A *countercurrent cascade* uses multiple staged contactors with countercurrent flow of solvent and feed streams from stage to stage. When a *differential contactor* is used, one of the phases remains dispersed as drops throughout the contactor as the phases pass each other in countercurrent fashion. The dispersed phase is then allowed to coalesce at the end of the device before being discharged. For these types of processes, *mass-transfer units* (or the related *mass-transfer coefficients*) often are used instead of theoretical stages to characterize separation performance. For a given phase, mass-transfer units are defined as the integral of the differential change in solute concentration divided by the deviation from equilibrium, between the limits of inlet and outlet solute concentrations. A single transfer unit represents the change in solute concentration equal to that achieved by a single theoretical stage when the extraction factor is equal to 1.0. It differs from a theoretical stage at other values of the extraction factor.

Flooding generally refers to excessive breakthrough or entrainment of one liquid phase into the discharge stream of the other. The flooding characteristics of an extractor limit its hydraulic capacity. Flooding can be caused by excessive flow rates within the equipment, by *phase inversion* due to accumulation and coalescence of dispersed droplets, or by the formation of stable dispersions or emulsions due to the presence of surface-active impurities or excessive agitation. The *flood point* refers to the specific total volumetric *throughput* in $(m^3/h)/m^2$ or gpm/ft^2 of cross-sectional area (or the equivalent *phase velocity* in m/s or ft/s) at which flooding begins.

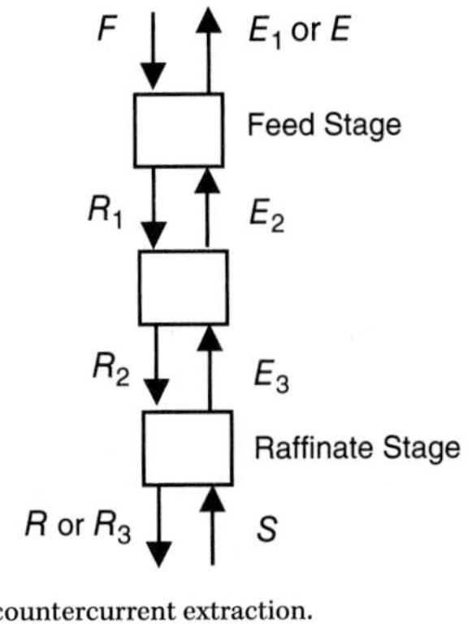

FIG. 15-6 Standard countercurrent extraction.

DESIRABLE SOLVENT PROPERTIES

Common industrial solvents generally are single-functionality organic solvents such as ketones, esters, alcohols, linear or branched aliphatic hydrocarbons, aromatic hydrocarbons, and so on; or water, which may contain acids or bases to adjust pH and may be mixed with water-soluble organic solvents. More complex solvents are sometimes used to obtain specific properties needed for a given application. These include compounds with multiple functional groups such as diols or triols, glycol ethers, and alkanol amines, as well as specialty organics such as pine-derived solvents (terpenes), citrus-derived solvents (such as limonene, a cyclic terpene), sulfolane (2,3,4,5-tetrahydrothiophene-1,1-dioxide), and NMP (*N*-methyl-2-pyrrolidone). Solvent properties have been summarized in a number of handbooks and databases, including those by N. P. Cheremisinoff, *Industrial Solvents Handbook*, 2d ed., revised and expanded, CRC Press, Boca Raton, Fla., 2003; G. Wypych, *Handbook of Solvents*, 2d ed., ChemTec, New York, 2014; A. Wypych and G. Wypych, *Databook of Solvents*, ChemTec, New York, 2014; *Solvents Database*, CD-ROM, ver. 4.0, ChemTec, New York, 2014; C. L. Yaws, *Thermodynamic and Physical Property Data*, 2d ed., Gulf, Houston, 1998; and E. W. Flick, *Industrial Solvents Handbook*, 5th ed., Noyes, Westwood, NJ, 1998. Also see D. Prat et al. [*Org. Process Res. Dev.* **17**(12): 1517–1525 (2013)] for a discussion of solvents used in pharmaceutical processing, and B. G. Cox [*Org. Proc. Res. Dev.* **19**(12): 1800–1808 (2015)] for a discussion of acids and bases in aqueous organic solvents.

Organic solvents are sometimes blended to obtain specific properties, another approach to achieving a multifunctional solvent with properties tailored for a given application. Examples are discussed by I. Escudero, J. L. Cabezas, and J. Coca [*Chem. Eng. Comm.* **173:** 135–146 (1999)] and by M. L. van Delden et al. [*Chem. Eng. Technol.* **29**(10): 1221–1226 (2006)]. As discussed earlier, a solvent also may be a liquid containing a dissolved extractant species, the extractant chosen because it forms an adduct or molecular complex with the desired solute.

In terms of desirable properties, no single solvent or solvent blend can be best in every respect. The choice of solvent often is a compromise, and the relative weighting given to the various considerations depends on the given situation. Assessments should take into account long-term sustainability and overall cost of ownership. An example case study is discussed by V. H. Shah et al. [*Ind. Eng. Chem. Res.* **55**(6): 1731–1739 (2016)]. Normally, the factors considered in choosing a solvent include the following:

1. *Loading capacity.* This property refers to the maximum concentration of solute the extract phase can hold before two liquid phases can no longer coexist or solute precipitates as a separate phase. If a specialized extractant is used, loading capacity may be determined by the point at which all the extractant in solution is completely occupied by solute and extractant solubility limits capacity. If loading capacity is low, a high solvent-to-feed ratio may be needed even if the partition ratio is high.

2. *Partition ratio* $K_i = Y_i/X_i$. Partition ratios on the order of $K_i = 10$ or higher are desired for an economical process because they allow operation with minimal amounts of solvent (more specifically, with a minimal solvent-to-feed ratio) and production of higher solute concentrations in the extract—unless the solute concentration in the feed already is high and a limitation in the solvent's loading capacity determines the required solvent-to-feed ratio. Because high partition ratios generally allow for low solvent use, smaller and less costly extraction equipment may be used, and costs for solvent recovery and recycle are lower. In principle, partition ratios less than $K_i = 1.0$ may be accommodated by using a high solvent-to-feed ratio, but usually at much higher cost.

3. *Solute selectivity.* In certain applications, it is important not only to recover a desired solute from the feed, but also to separate it from other solutes present in the feed and thereby achieve a degree of solute purification. The selectivity of a given solvent for solute *i* compared to solute *j* is characterized by the separation factor $\alpha_{i,j} = K_i/K_j$. Values must be greater than $\alpha_{i,j} = 1.0$ to achieve an increase in solute purity (on a solvent-free basis). When solvent blends are used in a commercial process, often it is because the blend provides higher selectivity, and often at the expense of a somewhat lower partition ratio. The degree of purification that can be achieved also depends on the extraction scheme chosen for the process, the amount of extraction solvent, and the number of stages employed.

4. *Mutual solubility.* Low liquid-liquid mutual solubility between feed and solvent phases is desirable because it reduces the separation requirements for removing solvents from the extract and raffinate streams. Low solubility of extraction solvent in the raffinate phase often results in high relative volatility for stripping the residual solvent in a raffinate stripper, allowing low-cost removal of solvent from the raffinate [Hwang, Y. L., G. E. Keller, and J. D. Olson, *Ind. Eng. Chem. Res.* **31**(7): 1753–1759 (1992)]. Low solubility of feed solvent in the extract phase reduces separation requirements for recovering solvent for recycle and producing a purified product solute. In some cases, if the solubility of feed solvent in the extract is high, more than one distillation operation will be required to separate the extract phase.

If mutual solubility is nil (as for aliphatic hydrocarbons dissolved in water), the need for stripping or another treatment method may be avoided as long as efficient liquid-liquid phase separation can be accomplished. However, very low mutual solubility normally is achieved at the expense of a lower partition ratio and loading capacity for extracting the desired solute. Mutual solubility also limits the solvent-to-feed ratios that can be used, because a point can be reached where the solvent stream is so large it dissolves the entire feed stream, or the solvent stream is so small it is dissolved by the feed, and these can be real limitations for systems with high mutual solubility.

5. *Stability.* The solvent should have little tendency to react with the product solute and form unwanted by-products, causing a loss in yield. Also it should not react with feed components or degrade to undesirable contaminants that cause the development of undesirable odors, colors, or tars that foul equipment over time, or cause difficulty achieving desired product purity, or accumulate in the process because they are difficult to purge.

6. *Density difference.* As a general rule, a difference in density between solvent and feed phases on the order of 0.1 to 0.3 g/mL is preferred. A value that is too low makes for poor or slow liquid-liquid phase separation and may require the use of a centrifuge. A value that is too high makes it difficult to build high dispersed-droplet population density for good mass transfer—that is, it is difficult to mix the two phases together and maintain high holdup of the dispersed phase within the extractor—but this depends on the viscosity of the continuous phase.

7. *Viscosity.* Low (waterlike) viscosity is preferred because higher viscosity generally increases mass-transfer resistance and makes liquid-liquid phase separation more difficult. Sometimes an extraction process is operated at an elevated temperature where viscosity is significantly lower for better mass-transfer performance, even when this results in a lower partition ratio. Low viscosity at ambient temperatures also facilitates the transfer of solvent from storage to processing equipment.

8. *Interfacial tension.* Preferred values for interfacial tension between the feed phase and the extraction solvent phase generally are in the range of 5 to 25 dyn/cm (1 dyn/cm is equivalent to 10^{-3} N/m). Systems with lower values easily emulsify. For systems with higher values, dispersed droplets tend to coalesce easily, resulting in low interfacial area and poor mass-transfer performance unless mechanical agitation is used.

9. *Recoverability.* The economical recovery of solvent from the extract and raffinate is critical to commercial success. Solvent physical properties should facilitate low-cost options for solvent recovery, recycle, and storage. For example, the use of relatively low-boiling organic solvents with low heats of vaporization generally allows cost-effective use of distillation and stripping for solvent recovery. Solvent properties also should enable low-cost methods for purging from the overall process impurities that may accumulate over time (both low-boiling and high-boiling impurities). One of the challenges often encountered in using a high-boiling solvent or extractant involves accumulation of high-boiling impurities in the solvent phase and difficulty in removing them from the process. Another consideration is the ease with which solvent residues can be reduced to low levels in final extract or raffinate products, particularly for food-grade products and pharmaceuticals.

10. *Freezing point.* Solvents that are liquids at all anticipated ambient temperatures are desirable because they avoid the need for freeze protection and/or thawing of frozen solvent prior to use. Sometimes an "antifreeze" additive such as water or an aliphatic hydrocarbon can be added to the solvent, or the solvent is supplied as a mixture of related compounds instead of a single pure component to suppress the freezing point.

11. *Safety and health.* Solvents with low toxicity and low potential for fire and reactive chemical hazards are preferred as inherently safe solvents. Low mammalian toxicity and low dermal absorption rate reduce the potential for injury through acute exposure. In all cases, solvents must be used with a full awareness of potential hazards and in a manner consistent with measures needed to avoid hazards and prevent injury. A detailed hazard assessment and safety review will be needed for any new process operation and for significant changes to an existing operation. For information on the safe use of solvents and their potential hazards, see Sec. 23, Safety and Handling of Hazardous Materials. Also see D. A. Crowl and J. F. Louvar, *Chemical Process Safety: Fundamentals with Applications*, 3d ed., Prentice-Hall, Upper Saddle River, NJ, 2011; C. L. Yaws, *Handbook of Chemical Compound Data for Process Safety*, Elsevier, Amsterdam, 1997; S. Mannan, *Lees' Loss Prevention in the Process Industries*, 4th ed., Butterworth-Heinemann, Oxford, UK, 2012; and *Bretherick's Handbook of Reactive Chemical Hazards*, 7th ed., ed. P. G. Urben, 2 vols., Academic Press, New York, 2007. A thorough review of the medical literature also must be conducted to ascertain chronic toxicity issues. Measures needed to avoid unsafe exposures must be incorporated into process designs and implemented in operating procedures. See D. L. Goetsch, *Occupational Safety and Health for Technologists, Engineers, and Managers*, 8th ed., Pearson, London, 2014.

12. *Environmental requirements.* The solvent must have physical or chemical properties that allow effective control of emissions in vents, wastewater, and other discharge streams. Preferred properties include low aquatic toxicity and low potential for fugitive emissions from leaks or spills. It also is desirable for a solvent to have low photoreactivity in the atmosphere and be biodegradable so it does not persist in the environment. Efficient technologies for capturing solvent vapors from vents and condensing them for recycle include activated carbon adsorption with steam regeneration or vacuum-swing regeneration [Smallwood, I. M., *Solvent Recovery Handbook*, 2d ed., Blackwell, Oxford, UK, 2002; Technical Bulletins EPA 452/B-02-001 and EPA 456/F-99-004, U.S. Environmental Protection Agency, 1999; and Pezolt, D. J., et al., *Environ. Prog.* **16**(1): 16–19 (1997)]. The optimization of a process to increase the efficiency of solvent utilization is a key aspect of waste minimization and reduction of environmental impact. An opportunity may exist to reduce solvent use through the application of countercurrent processing and other chemical engineering principles aimed at improving processing efficiencies. For a discussion of environmental issues in process design, see D. T. Allen and D. R. Shonnard, *Green Engineering: Environmentally Conscious Design of Chemical Processes*, Prentice-Hall, Upper Saddle River, NJ, 2002]. Also see Sec. 22, Waste Management.

13. *Multiple uses.* It is desirable to use as the extraction solvent a material that can serve a number of purposes in the manufacturing plant. This avoids the cost of storing and handling multiple solvents. It may be possible to use a single solvent for a number of different extraction processes practiced in the same facility, either in different equipment operated at the same time or by using the same equipment in a series of product campaigns. In other cases, the solvent used for extraction may be one of the raw materials for a reaction carried out in the same facility, or a solvent used in another operation such as crystallization.

14. *Materials of construction.* It is desirable for a solvent to allow the use of common, relatively inexpensive materials of construction at moderate temperatures and pressures. Material compatibility and potential for corrosion are discussed in Sec. 25, Materials of Construction.

15. *Availability and cost.* The solvent should be readily available at a reasonable cost. Considerations include the initial fill cost, the investment costs associated with maintaining a large solvent inventory in the plant (particularly when expensive extractants are used), as well as the cost of makeup solvent.

COMMERCIAL PROCESS SCHEMES

For the purpose of illustrating process concepts, liquid-liquid extraction schemes typically practiced in industry may be categorized into a number of general types.

Standard Extraction Also called simple extraction or single-solvent extraction, standard extraction is by far the most widely practiced type of extraction operation. It can be practiced using single-stage or multistage processing, cross-current or countercurrent flow of solvent, and batch-wise or continuous operation. Figure 15-6 illustrates the contacting stages and liquid streams associated with a typical multistage, countercurrent scheme. Standard extraction is analogous to stripping in distillation (as defined in Sec. 13) because the process involves transferring or *stripping* components from the feed phase into another phase. Note that the feed (F) enters the process where the extract stream (E) leaves the process, analogous to feeding the top of a stripping tower. And the raffinate (R) leaves where the extraction solvent (S) enters. Standard extraction is used to remove contaminants from a crude liquid feed (product purification) or to recover valuable components from the feed (product recovery). Applications can involve very dilute feeds, such as when purifying a liquid product or detoxifying a wastewater stream, or concentrated feeds, such as when recovering a crude product from a reaction mixture. In either case, standard extraction can be used to transfer a high fraction of solute from the feed phase into the extract. Note, however, that transfer of the desired solute or solutes may be accompanied by transfer of unwanted solutes. Because of this, standard extraction normally cannot achieve satisfactory solute purity in the extract stream unless the separation factor for the desired solute with respect to unwanted solutes is at least $\alpha_{i,j} = K_i/K_j = 20$ and usually much higher. This depends on the crude feed purity and the product purity specification. (See the subsection Potential for Solute Purification Using Standard Extraction under Process Fundamentals and Basic Calculation Methods.)

Fractional Extraction Fractional extraction combines solute recovery with cosolute rejection. In principle, the process can achieve high solute recovery and high solute purity even when the solute separation factor is fairly low, as low as $\alpha_{i,j} = 4$ or so (see the subsection Dual-Solvent Fractional Extraction under Calculation Procedures. Lower values of $\alpha_{i,j}$ may be considered in special cases, but this will require using an unusually large number of contacting stages.) Dual-solvent fractional extraction uses an extraction solvent (S) and a wash solvent (W) and includes a stripping

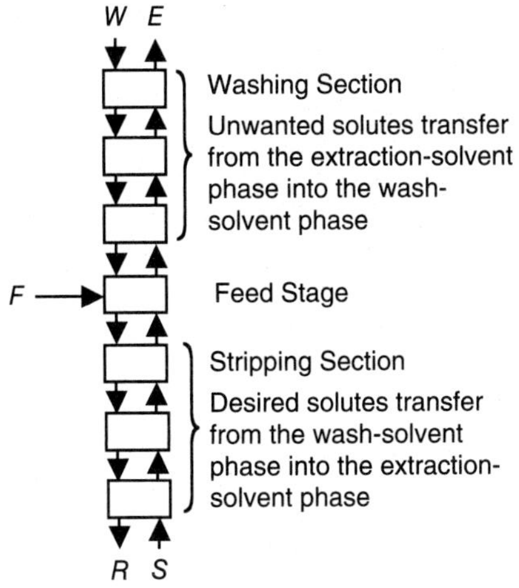

FIG. 15-7 Dual-solvent fractional extraction without reflux.

section at the raffinate end of the process (for product-solute recovery) and a washing section at the extract end of the process (for cosolute rejection and product purification) (Fig. 15-7). The feed enters the process at an intermediate stage located between the extract and raffinate ends. In this respect, the process is analogous to a middle-fed fractional distillation, although the analogy is not exact; wash solvent is added to the extract end of the process instead of returning a reflux stream. The desired solutes transfer into the extraction solvent (the extract phase) within the stripping section, and unwanted solutes transfer into the wash solvent (the raffinate phase) within the washing section. Typically, the feed stream consists of feed solutes pre-dissolved in wash solvent or extraction solvent; or, if they are liquids, they may be injected directly into the process. To maximize performance, a fractional extraction process may be operated such that the washing and stripping sections are carried out in different equipment and at different temperatures. The stripping section is sometimes called the extraction section, and the washing section is sometimes called the enriching section, the scrubbing section, or the absorbing section. A dual-solvent fractional extraction process involving reflux to the washing section is shown in Fig. 15-8.

In a special case referred to as single-solvent fractional extraction with extract reflux, the wash solvent is comprised of components that enter the overall process with the feed and return as reflux (Fig. 15-9). This is the type of extraction scheme commonly used to recover aromatic components from crude hydrocarbon mixtures using high-boiling polar solvents (as in Fig. 15-2). A reflux stream rich in light aromatics including benzene

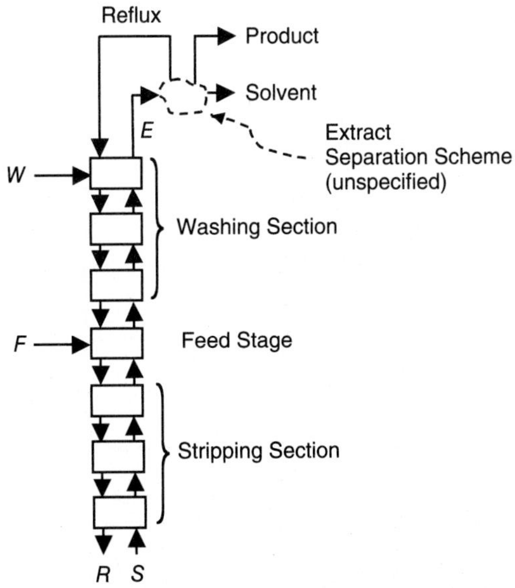

FIG. 15-8 Process concepts for dual-solvent fractional extraction with extract reflux.

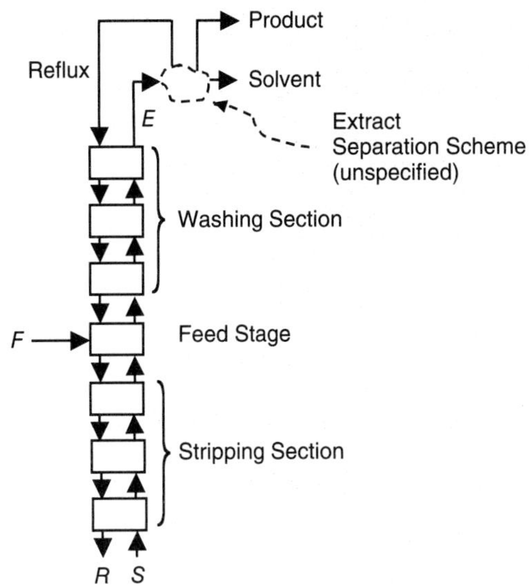

FIG. 15-9 Process concepts for single-solvent fractional extraction with extract reflux. The process flow sheet shown in Fig. 15-2 is an example of this general process scheme.

is refluxed to the washing section to serve as wash solvent. This process scheme is very similar in concept to fractional distillation. In practice, it is used only for a very limited number of chemical systems [Stevens, G. W., and H. R. C. Pratt, Chap. 6, in *Science and Practice of Liquid-Liquid Extraction*, vol. 1, ed. J. D. Thornton, Oxford University Press, Oxford, UK, 1992, pp. 379–395]. More detailed discussion is given in Single-Solvent Fractional Extraction with Extract Reflux under Calculation Procedures.

In terms of common practice, fractional extraction operations may be classified into several types: (1) standard extraction augmented by addition of a washing section utilizing a relatively small amount of feed solvent as the wash solvent; (2) full fractionation (less common); and (3) full fractionation with solute reflux (much less common). The first two categories are examples of dual-solvent fractional extraction. The third category can be practiced as dual-solvent or single-solvent fractional extraction.

In the first type of operation, a relatively small amount of feed solvent is added to a short washing section as wash solvent. (The word *short* is used here in an extraction column context, but refers in general to a relatively few theoretical stages.) This approach is useful for systems exhibiting a moderate to high solute separation factor ($\alpha_{i,j} > 20$ or so) and requiring a boost in product-solute purity. An example involves recovery of an organic solute from a dilute brine feed by using a partially miscible organic solvent. In this case, the inorganic salt present in the aqueous feed stream has some solubility in the organic solvent phase because of water that saturates that phase, and the partition ratio for transfer of salt into the organic phase is small (i.e., the partition ratio for transfer of salt into wash water is high). Adding wash water to the extract end of the process has the effect of washing a portion of the soluble salt content out of the organic extract. The reduction in salt content depends on how much wash water is added and how many washing stages or transfer units are used in the design.

The second type of fractional extraction operation involves the use of stripping and washing sections without reflux (Fig. 15-7) to separate a mixture of feed solutes with close K values. In this case, the solute separation factor is low to moderate. Normally, $\alpha_{i,j}$ must be greater than about 4 for a commercially viable process (which requires only a moderate number of contacting stages). E. G. Scheibel [*Chem. Eng. Prog.* **44**(9): 681–690 (1948); and **44**(10): 771–782 (1948)] gives several instructive examples of fractional extraction: (1) separation of ortho and para chloronitrobenzenes using heptane and 85 percent aqueous methanol as solvents ($\alpha_{para,ortho} \approx$ 1.6 to 1.8); (2) separation of ethanol and isopropanol by using water and xylene ($\alpha_{ethanol,isopropanol} \approx 2$); and (3) separation of ethanol and methyl ethyl ketone (MEK) by using water and kerosene ($\alpha_{ethanol,MEK} \approx$ 10 to 20). The first two applications demonstrate fractional extraction concepts, but a sharp separation is not achieved using a moderate number of stages because the selectivity of the solvent is too low. In these kinds of applications, fractional extraction might be combined with another separation operation to complete the separation. (See the subsection Hybrid Extraction Processes.) In Scheibel's third example, the selectivity is much higher, and nearly complete separation is achieved by using a total of about seven theoretical stages.

In another example, D. L. Venter and I. Nieuwoudt [*Ind. Eng. Chem. Res.* **37**(10): 4099–4106 (1998)] describe a dual-solvent extraction process using hexane and aqueous tetraethylene glycol to selectively recover *m*-cresol from coal pyrolysis liquors also containing *o*-toluonitrile. This process has been successfully implemented in industry. The separation factor for *m*-cresol with respect to *o*-toluonitrile varies from 5 to 70 depending upon solvent ratios and the resulting liquid compositions. The authors compare a standard extraction configuration (bringing the feed into the first stage) with a fractional extraction configuration (bringing the feed into the second stage of a seven theoretical-stage process).

Another example of the use of dual-solvent fractional extraction concepts involves the recovery of ε-caprolactam monomer (for nylon-6 production) from a two-liquid-phase reaction mixture containing ammonium sulfate plus smaller amounts of other impurities, using water and benzene as solvents [Simons, A. J. F., and N. F. Haasen, chap. 18.4 in *Handbook of Solvent Extraction*, ed. T. C. Lo, M. H. I. Baird, and C. Hanson, Wiley, New York, 1983, and Krieger, Huntington, NY, 1991]. In this application, the separation factor for caprolactam with respect to ammonium sulfate is high because the salt greatly favors partitioning into water; however, separation factors with respect to the other impurities are smaller. V. Alessi et al. [*Chem. Eng. Technol.* **20:** 445–454 (1997)] describe two process schemes used in industry. These are outlined in Fig. 15-10. The simpler scheme (Fig. 15-10*a*) is a straightforward dual-solvent fractional extraction process that isolates caprolactam (CPL) in a benzene extract stream and ammonium sulfate (AS) in the aqueous raffinate. The feed stage is comprised of mixer M1 and settler S1, and separate extraction columns are used for the washing and stripping sections. In Fig. 15-10*a*, these are denoted by C1 and C2, respectively. Minor impurity components also present in the feed must exit the process in either the extract or the raffinate. The more complex scheme (Fig. 15-10*b*) eliminates addition of benzene to the feed stage and adds a back-extraction section at the extract end of the process (denoted by C4) to extract CPL from the benzene phase leaving the washing section. Also, a separate fractional extractor (denoted as C1 in Fig. 15-10*b*) is added between the original stripping and washing sections in order to treat the benzene phase leaving the stripping section and to recover the CPL content of the CPL-rich aqueous stream leaving the feed stage. In the C1 extractor, the CPL transfers into the benzene stream that ultimately enters the upper washing section, leaving hydrophilic impurities in an aqueous purge stream that exits at the bottom. The resulting process scheme includes two purge streams for rejecting minor impurities: a stream rich in high-boiling organic impurities leaving the bottom of the benzene distillation tower, and the aqueous stream rich in hydrophilic impurities leaving the bottom of the C1 extractor. This sophisticated design separates the feed into four streams instead of just two, allowing separate removal of two impurity fractions to increase the purity of the two main products. The caprolactam is made to transfer into either an aqueous or a benzene-rich stream as desired, by judicious choice of solvent-to-feed ratio at the various sections in the process.

A dual-solvent fractional extraction process can provide a powerful separation scheme, and some authors suggest that fractional extraction is not used as much as it could be. In many cases, instead of using full fractional extraction, standard extraction is used to recover solute from a crude feed; and if the solvent-to-feed ratio is less than 1.0, concentrate the solute in a smaller solute-bearing stream. Another operation such as crystallization, adsorption, or process chromatography is then used downstream for solute purification. Perhaps fractional extraction schemes should be evaluated more often as alternative processing schemes that may have advantages.

The third type of fractional extraction operation involves refluxing a portion of the extract stream back to the extract end (washing section) of the process. As mentioned earlier, this process can be practiced as a dual-solvent process (Fig. 15-8) or as a single-solvent process (Figs. 15-2 and 15-9). The process for extracting aromatics from aliphatics in petrochemical operations is one of the best known commercial examples. But fractional extraction processes employing extract reflux also have been developed in other industries, including selected applications in hydrometallurgy [Xie, F., et al., *Minerals Engineering* **56:** 10–28 (2014)]. Unlike in distillation, however, the use of reflux is not common. The reflux consists of a portion of the extract stream from which a significant amount of solvent has been removed. Injection of this solvent-lean, concentrated extract back into the washing section increases the total amount of solute and the amount of raffinate phase present in that section of the extractor. This can boost separation performance by allowing the process to operate at a more favorable location within the phase diagram, resulting in a reduction in the number of theoretical stages or transfer units needed within the washing section. This also allows the process to boost the concentration of solute in the extract phase above that in equilibrium with the feed phase. The increased amount of solute present within the process may require the use of extra solvent to avoid approaching the plait point at the feed stage (the composition at which only a single liquid

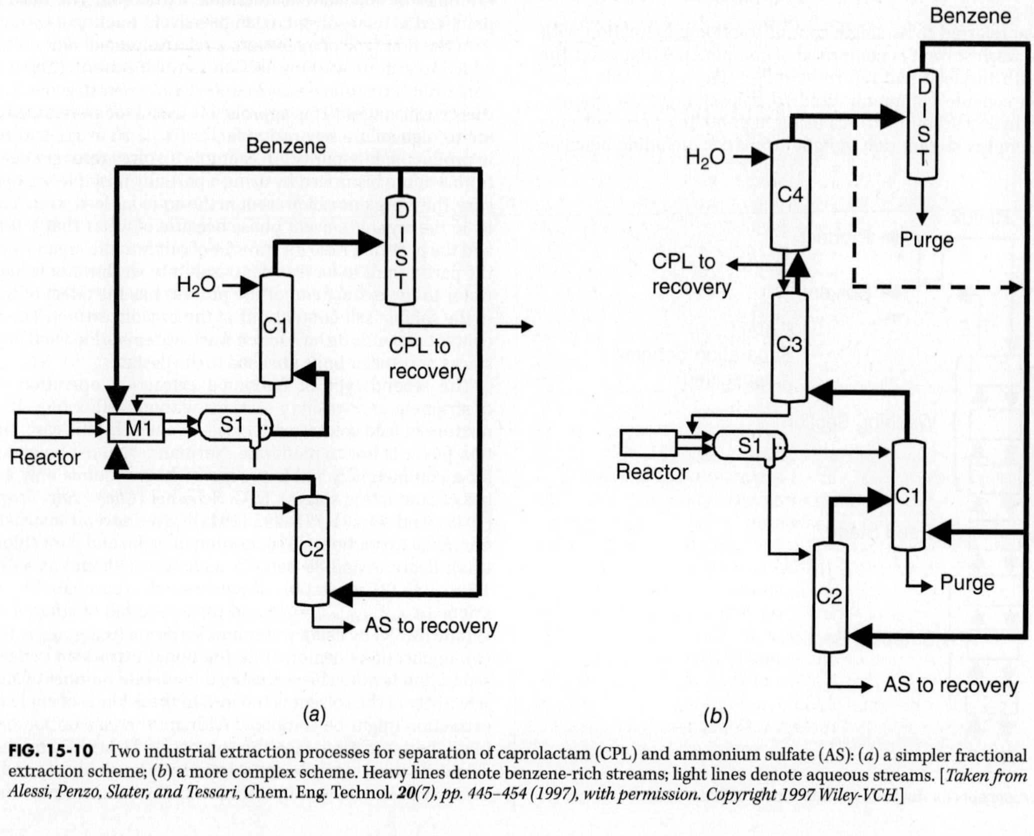

FIG. 15-10 Two industrial extraction processes for separation of caprolactam (CPL) and ammonium sulfate (AS): (*a*) a simpler fractional extraction scheme; (*b*) a more complex scheme. Heavy lines denote benzene-rich streams; light lines denote aqueous streams. [*Taken from Alessi, Penzo, Slater, and Tessari,* Chem. Eng. Technol. ***20****(7), pp. 445–454 (1997), with permission. Copyright 1997 Wiley-VCH.*]

phase can exist at equilibrium). Because of this, using reflux may involve a tradeoff between a reduction in the number of theoretical stages and an increase in the total liquid traffic within the process equipment, requiring larger-capacity equipment and increasing the cost of solvent recovery and recycle. This tradeoff is discussed by E. G. Scheibel with regard to extraction column design [*Ind. Eng. Chem.* **47**(11): 2290–2293 (1955)]. The potential benefit that can be derived from the use of extract reflux is greatest for applications utilizing solvents with a low solute separation factor and low partition ratios (as in the example illustrated in Fig. 15-2). In these cases, reflux serves to reduce the number of required theoretical stages or transfer units to a practical number (normally on the order of 10 or so) or to reduce the solvent-to-feed ratio required for the desired separation.

The fractional extraction schemes just described are typical of those practiced in industry. A related kind of process employs a second solvent in a separate extraction operation to wash the raffinate produced in an upstream extraction operation. This process scheme is particularly useful when the wash solvent is only slightly soluble in the raffinate and can easily be removed. An example is the use of water to remove residual amine solvent from the treated hydrocarbon stream in an acid-gas extraction process (Fig. 15-3).

A fourth type of fractional extraction operation involves the use of reflux at both ends of a dual-solvent process—that is, reflux to the raffinate end of the process (the stripping section) as well as reflux to the extract end of the process (the washing section). E. G. Scheibel discusses several potential flow sheets of this type [*Chem. Eng. Prog.* **62**(9): 76–81 (1966)]. Although rare, selected applications are described by F. Xie et al. for processing of rare earth elements in hydrometallurgy [*Minerals Engineering* **56:** 10–28 (2014)]. In this case, the number of contacting stages required by the separation is unusually high (with some applications requiring hundreds of mixer-settler stages), and dual reflux helps to minimize this number. In the special case of *single-solvent* fractional extraction with extract reflux, A. H. P. Skelland [*Ind. Eng. Chem.* **53**(10): 799–800 (1961)] has pointed out that the addition of raffinate reflux is not effective from a strictly thermodynamic point of view as it cannot reduce the required number of theoretical stages in this special case.

Dissociative Extraction This process scheme normally involves partitioning of weak organic acids or bases between water and an organic solvent phase. Whether the solute partitions mainly into one phase or the other depends upon whether it is in its neutral state or its charged ionic state and the ability of each phase to solvate that form of the solute. In general, water interacts much more strongly with the charged species, and the ionic form will strongly favor partitioning into the aqueous phase. The nonionic form generally will favor partitioning into the organic phase.

The pK_a is the pH at which 50 percent of the solute is in the dissociated (ionized) state. It is a function of solute concentration and normally is reported for dilute conditions. For an organic acid (RCOOH) dissolved in aqueous solution, the amount of solute in the dissociated state relative to that in the nondissociated state is $[RCOO^-]/[RCOOH] = 10^{pH-pK_a}$. Extraction of an organic acid out of an organic feed into an aqueous phase is greatly facilitated by operating at a pH above the acid's pK_a value because most of the acid will be deprotonated to yield the dissociated form ($RCOO^-$). On the other hand, partitioning of the organic acid from an aqueous feed into an organic solvent is favored by operating at a pH below its pK_a to ensure that most of the acid is in the protonated (nondissociated) form. Another example involves extraction of a weak base, such as a compound with amine functionality (RNH_2), out of an organic phase into water at a pH below the pK_a. This will protonate or neutralize most of the base, yielding the ionized form (RNH_3^+), and will favor extraction into water. It follows that extracting an organic base out of an aqueous feed into an organic solvent is favored by operating at a pH above its pK_a because this yields most of the solute in the free base (nonionized) form. For weak bases, $pK_a = 14 - pK_b$, and the relative amount of solute in the dissociated state in the aqueous phase is given by 10^{pK_a-pH}. As a rule, to obtain the maximum partition ratio for an extraction, the pH should be maintained about 2 pH units from the solute's pK_a value to obtain essentially complete dissociation or nondissociation, as appropriate for the extraction. In a typical continuous extraction process, the pH of the aqueous stream leaving the process is controlled at a constant pH set point by injection of acid or base at the opposite end of the process, and a pH gradient exists within the process. The pH set point may be adjusted to optimize performance. The effect of pH on the partition ratio is discussed in Effect of pH for Ionizable Organic Solutes under Thermodynamic Basis for Liquid-Liquid Extraction. Determination of the optimum pH for the extraction of compounds with multiple ionizable groups and thus multiple pK_a values is discussed by L. S. Crocker, Y. Wang, and J. A. McCauley [*Org. Process Res. Dev.* **5**(1): 77–79 (2001)].

In fractional dissociative extraction, a sharp separation of feed solutes is achieved by taking advantage of a difference in their pK_a values. If the difference in pK_a is sufficient, controlling pH at a specific value can yield high K values for one solute fraction and very low K values for another fraction, thus allowing a sharp separation. For example, a mixture of two organic bases can be separated by contacting the mixture with an aqueous acid containing less than the stoichiometric amount of acid needed to neutralize (ionize) both bases. The stronger of the two bases reacts with the acid to yield the dissociated form in the aqueous phase, while the other base remains nondissociated in a separate organic phase. Buffer compounds may be used to control pH within a desired range for improved separation results [Ma, G., and A. Jha, *Org. Process Res. Dev.* **9**(6): 847–852 (2005)]. Buffers are discussed by D. D. Perrin and B. Dempsey [*Buffers for pH and Metal Ion Control*, Chapman and Hall, London, 1979]. For additional discussion, see M. W. T. Pratt, chap. 21 in *Handbook of Solvent Extraction*, ed. T. C. Lo, M. H. I. Baird, and C. Hanson, Wiley, New York, 1983, and Krieger, Huntington, NY, 1991, and M. M. Anwar, A. S. Arif, and D. W. Pritchard, *Solvent Ext. Ion Exch.* **16:** 931 (1998).

pH-Swing Extraction A pH-swing extraction process uses dissociative extraction concepts to recover and purify ionizable organic solutes in a forward- and back-extraction scheme, each extraction operation carried out at a different pH. For example, in the forward extraction, the desired solute may be in its nonionized state so it can be extracted out of a crude aqueous feed into an organic solvent. The extract stream from this operation is then fed to a separate extraction operation where the solute is ionized by readjustment of pH and back-extracted into clean water. This scheme can achieve both high recovery and high purity if the impurity solutes are not ionizable or have pK_a values that differ greatly from those of the desired solute. A pH-swing extraction scheme commonly is used for recovery and purification of antibiotics and other complex organic solutes with some ionizable functionality.

Reaction-Enhanced Extraction This scheme involves enhancement of the partition ratio for extraction through the use of a reactive extractant that forms a reversible adduct or molecular complex with the desired solute. For a discussion of process fundamentals, see C. J. King, chap. 15 in *Handbook of Separation Process Technology*, ed. R. W. Rousseau, Wiley, New York, 1987, and H.-J. Bart, *Reactive Extraction*, Springer, Berlin, 2001. Because reactive extractants form strong specific interactions with the solute molecule, they can provide much higher partition ratios and generally are more selective than conventional physical solvents. For extraction of solute from an aqueous feed solution, the extractant compound often is dissolved in a diluent liquid such as kerosene or another high-boiling hydrocarbon, and the resulting molecular complex resides in the organic phase. Well-known examples include recovery of metals from acid leachate solutions in hydrometallurgy (Fig. 15-4) and extraction of carboxylic acids from aqueous fermentation broth, as discussed below. The same principle may be applied in reverse to extract compounds from an organic feed, the extractant being dissolved in aqueous solution and the resulting complex residing mainly in the aqueous phase. A commercial example of this kind is the extraction of dissolved acid gases (primarily CO_2 and H_2S) from liquefied hydrocarbons using aqueous alkanolamines (Fig. 15-3), an application that is closely related to the removal of acid gases from vapor-phase hydrocarbons by absorption into the same or similar fluids [R. B. Nielsen et al., *Hydrocarbon Proc.* **76:** 49–59 (1997)]. Another example involves extraction of terpenyl amine from organic solution using aqueous acetic acid [R. Schulz et al., *Ind. Eng. Chem. Res.* **55**(19): 5763–5769 (2016)].

Although there are many successful commercial applications, it is important to note that the use of high-boiling extractants can present severe difficulties whenever high-boiling impurities are present. A number of commercial processes have failed because there was no economical option for purging high-boiling contaminants that accumulated in the solvent phase over time, so care must be taken to address this possibility when developing a new application. The advantages and disadvantages of using high-boiling solvents or extractants versus low-boiling solvents are discussed by C. J. King in the context of acetic acid recovery [chap. 18.5 in *Handbook of Solvent Extraction*, ed. T. C. Lo, M. H. I. Baird, and C. Hanson, Wiley, New York, 1983, and Krieger, Huntington, NY, 1991]. Also see the discussion by J. Price and D. Burns regarding accumulation of high-boiling impurities in hydrocarbon sweetening operations [*Hydrocarb. Process.* **74:** 140–141 (1995)].

Reviews of reactive extractants used in hydrometallurgy are given by M. Cox [chap. 1 in *Science and Practice of Liquid-Liquid Extraction*, vol. 2, ed. J. D. Thornton, Oxford University Press, Oxford, UK, 1992, pp. 1–27], by A. M. Wilson et al., *Chem. Soc. Rev.* **43:** 123–134 (2014); and by J. Zhang, B. Zhao, and B. Schreiner [*Separation Hydrometallurgy of Rare Earth Metals*, Springer, Berlin, 2016]. Also see *Solvent Extraction Principles and Practice*, 2d ed., ed. J. Rydberg et al., Marcel Dekker, New York, 2004; and V. S. Kislik, *Solvent Extraction: Classical and Novel Approaches*, Elsevier, Oxford, UK, 2012. Extractants used in hydrometallurgy generally are classified according to the mechanism of solute-solvent interaction in solution: (1) cationic (carboxylic and organophosphorus acids);

(2) anionic (primary amines and quaternary amines); (3) chelating (compounds such as hydroxyoximes that form multiple bonds to a central metal ion); (4) ion-pair-forming (such as trialklyamines) and (5) solvating (nonionic compounds including tri-n-butyl phosphate). Blends of the different types also are sometimes used (called synergistic extraction).

Another well-known class of applications involves the formation of ion-pair interactions between a carboxylic acid dissolved in an aqueous feed and alkylamine extractants such as trioctylamine dissolved in a hydrocarbon diluent, as discussed by R. Wennersten [*J. Chem. Technol. Biotechnol.* **33B:** 85–94 (1983)], by C. J. King and others [*Ind. Eng. Chem. Res.* **29**(7): 1319–1338 (1990); and *Chemtech* **22:** 285 (1992)], and by A. Schunk and G. Maurer [*Ind. Eng. Chem. Res.* **44**(23): 8837–8851 (2005)]. Extractants also may be used to facilitate the extraction of other ionizable organic solutes, including certain antibiotics [R. A. Pai, M. F. Doherty, and M. F. Malone, *AIChE J.* **48**(3): 514–526 (2002)]. Sometimes mixing extractants with promoter compounds (called modifiers) provides synergistic effects that dramatically enhance the partition ratio. An example is discussed by M. Atanassova and I. Dukov [*Sep. Purif. Technol.* **40:** 171–176 (2004)]. Also see the discussion of combined physical (hydrogen-bonding) and reaction-enhanced extraction by S. C. Lee [*Biotechnol. Prog.* **22**(3): 731–736 (2006)].

The chemistry involved in the removal of acid gases from hydrocarbons using aqueous alkanolamines is discussed by P. V. Danckwerts [*Chem. Eng. Sci.* **34:** 443–446 (1979)] and by G. S. Hwang et al. [*Phys. Chem. Chem. Phys.* **17:** 831–839 (2015)].

Extractive Reaction This scheme combines reaction and separation in the same unit operation for the purpose of facilitating a desired reaction to produce a desired product. For a discussion of process fundamentals, see K. D. Samant and K. M. Ng, *AIChE J.* **44**(6): 1363–1381 (1998). To avoid confusion, the term *extractive reaction* is recommended for this type of process, while the term *reaction-enhanced extraction* is recommended for a process involving formation of reversible solute-extractant complexes and enhanced partition ratios for the purpose of facilitating a desired separation. The term *reactive extraction* is a more general term commonly used for both types of processes.

In general, extractive reaction involves conducting a reaction in the presence of two liquid phases and taking advantage of differences in the partitioning of reactants, products, and homogeneous catalyst (if used) between the two liquids to improve reaction performance. The second liquid phase either is deliberately added to the system or it naturally forms during the course of the reaction. The classes of reactions that can benefit from an extractive reaction scheme include chemical-equilibrium-limited reactions (such as esterifications, transesterifications, and hydrolysis reactions), where it is important to remove a product or coproduct from the reaction zone to drive conversion, and consecutive or sequential reactions (such as nitrations, sulfonations, and alkylations), where the goal may be to produce only the mono- or difunctional product and minimize the formation of subsequent addition products. For additional discussion, see H. J. Gorissen, *Chem Eng. Sci.* **58:** 809–814 (2003); and V. Van Brunt and J. S. Kanel, chap. 3 in *Reactive Separation Processes*, ed. S. Kulprathipanja, Taylor & Francis, Abingdon, UK, 2002, pp. 51–92.

The manufacture of fatty acid methyl esters (FAME) for use as biodiesel fuel by transesterification of triglyceride oils and greases provides an example of a chemical-equilibrium-limited extractive reaction [Kiss, A. A., *Process Intensification Technologies for Biodiesel Production—Reactive Separation Processes*, Springer, Berlin, 2014; and Van Gerpen, J., *Fuel Process Technol.* **86:** 1097–1107 (2005)]. Low-grade triglycerides are reacted with methanol to produce FAME plus glycerol as a by-product. Because glycerol is only partially miscible with the feed and the FAME product, it transfers from the reaction zone into a separate glycerol-rich liquid phase. Excess methanol normally is needed to obtain complete conversion, but periodic removal of the glycerol-rich phase helps minimize the required amount. In another example, M. Minotti, M. F. Doherty, and M. F. Malone [*Ind. Eng. Chem. Res.* **37**(12): 4748–4755 (1998)] studied the esterification of aqueous acetic acid by reaction with butanol in an extractive reaction process involving the extraction of the butyl acetate product into a separate butanol-rich phase. The authors concluded that cocurrent processing is preferred over countercurrent processing in this case. Their general conclusions likely apply to other applications involving extraction of a reaction product out of the reaction phase to drive conversion. The cocurrent scheme is equivalent to a series of two-liquid-phase stirred-tank reactors approaching the performance of a plug-flow reactor. C. Rohde, R. Marr, and M. Siebenhofer [Paper No. 232f, AIChE Annual Meeting, Austin, Tex., 2004] studied the esterification of acetic acid with methanol to produce methyl acetate. Their extractive reaction scheme involves selective transfer of methyl acetate into a high-boiling solvent such as n-nonane.

An example of a sequential-reaction extractive reaction is the manufacture of 2,4-dinitrotoluene as a precursor to 2,4-diaminotoluene and toluene diisocyanate (TDI) based polyurethanes. In traditional processes, liquid-phase nitration of toluene is conducted using concentrated nitric and sulfuric acids, which form a separate liquid phase. Toluene transfers into the acid phase, where it reacts with nitronium ion, and the reaction product transfers back into the organic phase [*Nitration—Recent Laboratory and Industrial Developments*, ed. L. F. Albright, R. V. C. Carr, and R. J. Schmitt, *ACS Symposium Series*, vol. 623, American Chemical Society, Washington, 1996)]. In these processes, careful control of liquid-liquid contacting conditions is required to obtain high yield of the desired product and to minimize the formation of impurities. A similar process is used for nitration of benzene to mononitrobenzene, a precursor to aniline used in the manufacture of many products, including methylenediphenylisocyanate (MDI) for polyurethanes [Quadros, P. A., M. S. Reis, and C. M. S. G. Baptista, *Ind. Eng. Chem. Res.* **44**(25): 9414–9421 (2005)].

Another category of extractive reaction involves the extraction of a product solute during microbial fermentation (biological reaction) to avoid microbe inhibition effects, allowing an increase in fermenter productivity. An example involving the production of ethanol is discussed by C. Weilnhammer and E. Blass [*Chem. Eng. Technol.* **17:** 365–373 (1994)], and an example involving the production of propionic acid is discussed by Z. Gu, B. A. Glatz, and C. E. Glatz [*Biotechnol. and Bioeng.* **57**(4): 454–461 (1998)].

Temperature-Swing Extraction Temperature-swing processes take advantage of a change in K value with temperature. An extraction example is the commercial process used to recover citric acid from whole fermentation broth by using trioctylamine (TOA) extractant [Baniel, A. M., R. Blumberg, and K. Hajdu, U.S. Patent 4,275,234 (1981); Wennersten, R., *J. Chem. Biotechnol.* **33B:** 85–94 (1983); and Pazouki, M., and T. Panda, *Bioprocess Eng.* **19:** 435–439 (1998)]. This process involves a forward reaction-enhanced extraction carried out at 20 to 30°C in which citric acid transfers from the aqueous phase into the extract phase. Relatively pure citric acid is subsequently recovered by back extraction into clean water at 80 to 100°C, also liberating the TOA extractant for recycle. This temperature-swing process is feasible because partitioning of citric acid into the organic phase is favored at the lower temperature but not at 80 to 100°C.

Partition ratios can be particularly sensitive to temperature when solute-solvent interactions in one or both phases involve specific attractive interactions such as formation of ion-pair bonds (as in trialkyamine–carboxylic acid interactions) or hydrogen bonds, or when mutual solubility between feed and extraction solvent involves hydrogen bonding. An interesting example is the extraction of citric acid from water with 1-butoxy-2-propanol (common name propylene glycol n-butyl ether) as solvent (Fig. 15-11). This example illustrates how important it can be when developing and optimizing an extraction operation to understand how K varies with temperature, regardless of whether a temperature-swing process is contemplated. Of course, changes in other properties such as mutual solubility and viscosity also must be considered. For additional discussion, see the subsection Temperature Effect under Thermodynamic Basis for Liquid-Liquid Extraction.

Reversed Micellar Extraction This scheme involves use of microscopic water-in-oil micelles formed by surfactants and suspended within a hydrophobic organic solvent to isolate proteins from an aqueous feed. The micelles essentially are microdroplets of water having dimensions on the order of the protein to be isolated. These stabilized water droplets provide a compatible environment for the protein, allowing its recovery from a crude aqueous feed without significant loss of protein activity [Ayala, G. A., et al., *Biotechnol. and Bioeng.* **39:** 806–814 (1992); and Bordier, C., *J. Biolog. Chem.* **256**(4): 1604–1607 (February 1981)]. Also see the discussion of ultrafiltration membranes for concentrating micelles in Membrane-Based Coalescers under Liquid-Liquid Phase Separation Equipment.

Aqueous Two-Phase Extraction Also called aqueous biphasic extraction, this technique generally involves the use of two incompatible water-miscible polymers [normally polyethylene glycol (PEG) and dextran, a starch-based polymer], or a water-miscible polymer and a salt (such as PEG and Na_2SO_4), to form two immiscible aqueous phases each containing 75+ percent water. This technology provides mild conditions for recovery of proteins and other biomolecules from broth or other aqueous feeds with minimal loss of activity (Walter, H., and G. Johansson, eds., *Aqueous Two Phase Systems, Methods in Enzymology*, vol. 228, Academic Press, New York, 1994; Zaslavsky, B. Y., *Aqueous Two-Phase Partitioning*, Marcel Dekker, New York, 1994; and Blanch, H. W., and D. S. Clark, Chap. 6 in *Biochemical Engineering*, Marcel Dekker, New York, 1997, pp. 474–482). The effect of salts on the liquid-liquid phase equilibrium of polyethylene glycol + water mixtures has been extensively studied [Salabat, A., *Fluid Phase Equilibr.* **187–188:** 489–498 (2001)]. A typical phase diagram, for PEG 6000 + Na_2SO_4 + water, is shown in Fig. 15-12. The hydraulic characteristics of the aqueous two-phase system PEG 4000 + Na_2SO_4 + water in a countercurrent sieve plate column have been reported by A. Hamidi et al. [*J. Chem. Technol. Biotechnol.* **74:** 244–249 (1999)].

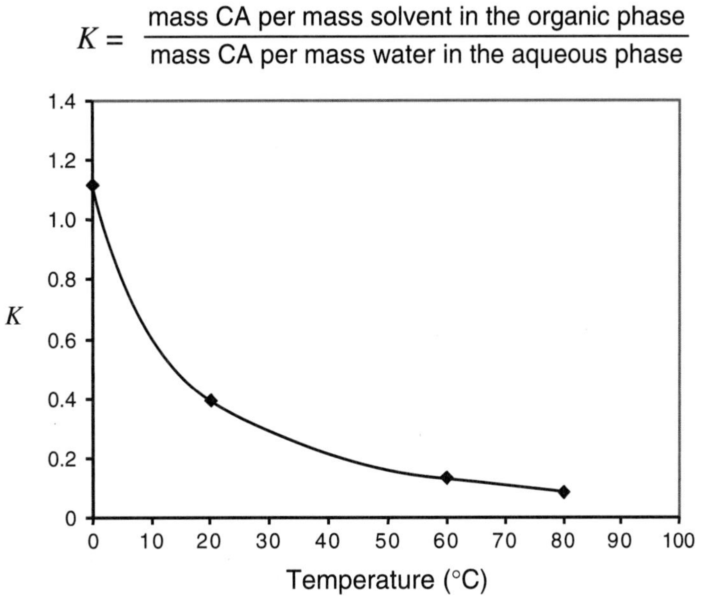

FIG. 15-11 Partition ratio as a function of temperature for recovery of citric acid (CA) from water using 1-butoxy-2-propanol (propylene glycol *n*-butyl ether). (*Data generated by The Dow Chemical Company.*)

Two immiscible aqueous phases also may be formed by using two incompatible salts. An example is the system formed by using the hydrophilic organic salt 1-butyl-3-methylimidazolium chloride and a water-structuring (kosmotropic) salt such as K_3PO_4 [K. E. Gutowski et al., *J. Am. Chem. Soc.* **125:** 6632 (2003)].

Enantioselective Extraction The isolation of specific enantiomers from racemic mixtures is of increasing importance in the production of pharmaceutical active ingredients, flavor and aroma chemicals, agricultural chemicals, and other biologically active products. Although most published studies involve use of chromatography or crystallization, a number of liquid-liquid extraction applications involving specialized chiral extractants have been reported. Selectivity generally is not high, so multistage processing is required. This subject is reviewed by B. Schuur et al. [*Org. Biomol. Chem.* **9:** 36–51 (2011)]. Recent studies are reported by R. Lavie [*Ind. Eng. Chem. Res.* **50**(22): 12750–12756 (2011)], by Z. Ren et al. [*J. Chem. Eng. Data* **59**(8): 2517–2522 (2014)], by B. Schuur et al. [*Chirality* **27:** 123–130 (2015)], and by Y. Wang et al. [*Org. Process Res. Dev.* **19**(9): 1082–1087 (2015)].

Hybrid Extraction Processes Hybrid processes employ an extraction operation in close association with another unit operation. A hybrid scheme is employed because an individual unit operation cannot achieve all the separation goals, or because the hybrid process is more economical. Common examples include the following.

Extraction-Distillation An example involves the use of extraction to break the methanol + dichloromethane azeotrope. The near-azeotropic overheads from a distillation tower can be fed to an extractor where water is used to extract the methanol content and generate nearly methanol-free dichloromethane (saturated with roughly 2000 ppm water). A related type of extraction-distillation operation involves closely coupling extraction with the distillate or bottoms stream produced by a distillation tower, such that the distillation specification for that stream can be relaxed. This approach has been used to facilitate distillation of aqueous acetic acid to produce acetic acid as a bottoms product, taking a mixture of acetic acid and water overhead [R. G. Gualy et al., U.S. Patent 5,492,603 (1996)]. The distillate is sent to an extraction tower to recover the acetic acid content

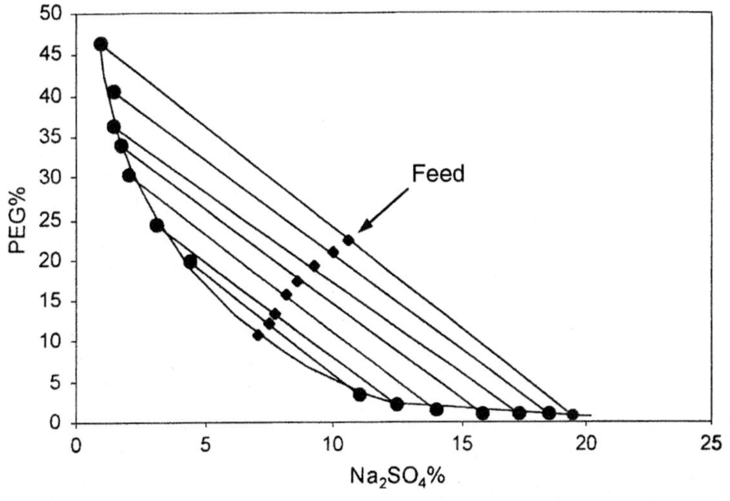

FIG. 15-12 Equilibrium phase diagram for PEG 6000 + Na_2SO_4 + water at 25°C. [*Reprinted from Salabat,* Fluid Phase Equilibr. ***187–188,*** *pp. 489–498 (2001), with permission. Copyright 2001 Elsevier B. V.*]

for recycle back to the process. The hybrid process allows operation with lower energy consumption compared to distillation alone because it allows the distillation tower to operate with a reduced requirement for recovering acetic acid in the bottoms stream, which permits relaxation of the minimum concentration of acetic acid allowed in the distillate. Another type of hybrid process involves combining liquid-liquid extraction with azeotropic or extractive distillation of the extract (Skelland, A. H. P., and D. W. Tedder, chap. 7 in *Handbook of Separation Process Technology*, ed. R. W. Rousseau, Wiley, New York, 1987, pp. 449–453). The solvent serves both as the extraction solvent for the upstream liquid-liquid extraction operation and as the entrainer for a subsequent azeotropic distillation or as the distillation solvent for a subsequent extractive distillation. (For a detailed discussion of azeotropic and extraction distillation concepts, see Sec. 13, Distillation.) The solvent-to-feed ratio must be optimized with regard to both the liquid-liquid extraction operation and the downstream distillation operation. An example is the use of ethyl acetate to extract acetic acid from an aqueous feed, followed by azeotropic distillation of the extract to produce a dry acetic acid bottoms product and an ethyl acetate + water overheads stream. In this example, ethyl acetate serves as the extraction solvent in the extractor and as the entrainer for removing water overhead in the distillation tower. Another hybrid extraction-distillation design, also employing a relatively low-boiling solvent, is described by Y.-C. Chen et al. [*Ind. Eng. Chem. Res.* **54**(31): 7715–7727 (2015)]. Examples involving extractive distillation and high-boiling solvents can be seen in the various processes used to recover aromatics from aliphatic hydrocarbons. See the subsection Single-Solvent Fractional Extraction with Extract Reflux under Calculation Procedures.

Extraction-Crystallization Extraction often is used in association with a crystallization operation. In the pharmaceutical and specialty chemical industries, extraction is used to recover a product compound (or remove impurities) from a crude reaction mixture, with subsequent crystallization of the product from the extract (or from the pre-extracted reaction mixture). In many of these applications, the product needs to be delivered as a pure crystalline solid, so crystallization is a necessary operation. (For a detailed discussion of crystallization operations, see Sec. 18, Liquid-Solid Operations and Equipment.) The desired solute can sometimes be crystallized directly from the reaction mixture with sufficient purity and yield, thus avoiding the cost of the extraction operation; however, direct crystallization generally is more difficult because of higher impurity concentrations. In cases where direct crystallization is feasible, deciding whether to use extraction prior to crystallization or crystallization alone involves consideration of a number of tradeoffs and ultimately depends on the relative robustness and economics of each approach [Anderson, N. G., *Org. Process Res. Dev.* **8**(2): 260–265 (2004)]. A well-known example of extraction-crystallization is the recovery of penicillin from fermentation broth by using a pH-swing forward and back extraction scheme followed by final purification using crystallization [Queener, S., and R. Swartz, "Penicillins: Biosynthetic and Semisynthetic," in *Secondary Products of Metabolism, Economic Microbiology*, vol. 3, ed. A. H. Rose, Academic, New York, 1979]. Extraction is used for solute recovery and initial purification, followed by crystallization for final purification and isolation as a crystalline solid. Another category of extraction-crystallization processes involves the use of extraction to recover solute from the spent mother liquor leaving a crystallization operation. In yet another example, K. Maeda et al. [*Ind. Eng. Chem. Res.* **38**(6): 2428–2433 (1999)] describe a crystallization-extraction hybrid process for separating fatty acids (lauric and myristic acids). In comparing these process options, the potential uses of extraction should include efficient countercurrent processing schemes, because these may significantly reduce solvent usage and cost.

Neutralization-Extraction A common example of neutralization-extraction involves neutralization of residual acidity (or basicity) in a crude organic feed by injection of an aqueous base (or aqueous acid) combined with washing the resulting salts into water. The neutralization and washing operations may be combined within a single extraction column as illustrated in Fig. 15-13. Also see the discussion by K. L. A. Koolen [*Design of Simple and Robust Process Plants*, Wiley-VCH, Weinheim, 2001, pp. 159–161].

Reaction-Extraction This technique involves chemical modification of solutes in solution in order to more easily extract them in a subsequent extraction operation. Applications generally involve modification of impurity compounds to facilitate purification of a desired product. An example is the oxygenation of sulfur-containing aromatic impurities present in fuel oil by using H_2O_2 and acetic acid, followed by liquid-liquid extraction into an aqueous acetonitrile solution [Y. Shiraishi and T. Hirai, *Energy and Fuels* **18**(1): 37–40 (2004); and Y. Shiraishi et al., *Ind. Eng. Chem. Res.* **41:** 4362–4375 (2002)]. Another example involves esterification of aromatic alcohol impurities to facilitate their separation from apolar hydrocarbons by using an aqueous extractant solution [B. Kuzmanović et al., *Ind. Eng. Chem. Res.* **43**(23): 7572–7580 (2004)]. Another type of reaction-extraction hybrid process involves closely coupled reaction-extraction steps in batchwise, multistep processing of specialty chemicals [McConvey, I. F., and P. Nancarrow, chap. 10 in *Pharmaceutical Process Development*, ed. A. J. Blacker and M. T. Williams, RSC Publishing, Cambridge, UK, 2011]. Also see the discussion of OATS processing in Phase Transition Extraction and Tunable Solvents under Emerging Developments.

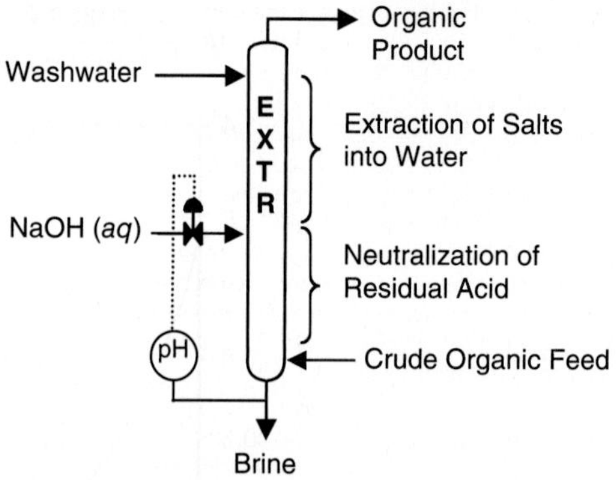

FIG. 15-13 Example of neutralization-extraction hybrid process implemented in an extraction column.

Reverse Osmosis–Extraction In certain applications, reverse osmosis (RO) or nanofiltration membranes may be used to reduce the volume of an aqueous stream and increase the solute concentration, in order to reduce the size of downstream extraction and solvent recovery equipment. R. W. Wytcherley, J. C. Gentry, and R. G. Gualy [U.S. Patents 5,492,625 (1996) and 5,624,566 (1997)] describe such a process for carboxylic acid solutes. Water is forced through the membrane when the operating pressure drop exceeds the natural osmotic pressure difference generated by the concentration gradient:

$$\text{Flux} = \frac{\mathcal{P}}{\lambda_m}(\Delta P - \Delta\pi) \qquad (15\text{-}1)$$

where $\mathcal{P}$ is a permeability coefficient for water, λ_m is the membrane thickness, ΔP is the operating pressure drop, and $\Delta\pi$ is the osmotic pressure gradient, a function of solute concentration on each side of the membrane. Normally the solute also will permeate the membrane to a small extent. The maximum possible concentration of solute in the concentrate is limited by that corresponding to an osmotic pressure of about 70 bar (about 1000 psig), as this is the maximum pressure rating of commercially available membrane modules (typical). For acetic acid, this maximum concentration is about 25 wt%. Depending upon whether the particular organic permeate of interest can swell or degrade the membrane material, the concentration achieved in practice may need to be reduced well below this limit to avoid excessive membrane deterioration. In general, a membrane preconcentrator is considered for feeds containing on the order of 3 wt% solute or less. In these cases, a more moderate membrane operating pressure may be used, and the preconcentrator can provide a significant reduction in the volume of feed entering the extraction process. The stream entering the membrane module normally must be carefully prefiltered to avoid fouling the membrane. The modeling of mass transfer through RO membranes, with an emphasis on cases involving solute-membrane interactions, is discussed by H. Mehdizadeh, Kh. Molaiee-Nejad, and Y. C. Chong [*J. Membrane Sci.* **267:** 27–40 (2005)]. Most pressure-driven membrane-based separations of this type have been developed for aqueous feeds; however, specialized solvent resistant nanofiltration membranes also are available and may be used to concentrate non-aqueous feeds through permeation of small-molecule solvents. For a detailed review, see P. Marchetti et al., *Chemical Reviews* **114:** 10735–10806 (2014).

Liquid-Solid Extraction (Leaching) Extraction of solubles from porous solids is a form of solvent extraction that has much in common with liquid-liquid extraction [Prabhudesai, R. K., "Leaching," Sec. 5.1 in *Handbook of Separation Techniques for Chemical Engineers*, ed. P. A. Schweitzer, McGraw-Hill, New York, 1997, pp. 5-3 to 5-31]. The main differences come from the need to handle solids and the fact that mass transfer of soluble components out of porous solids generally is much slower than mass transfer between liquids. Because of this, different types of contacting equipment operating at longer residence times often are required. Washing of nonporous solids is a related operation that generally exhibits faster mass-transfer rates compared to leaching. On the other hand, purification of nonporous solids or crystals by removal of impurities that reside within the bulk solid phase often is not economical or even feasible by using these methods, because the rate of mass transfer of impurities through the bulk solid is extremely slow. Liquid-solid extraction is covered in Sec. 18, Liquid-Solid Operations and Equipment.

Liquid-Liquid Partitioning of Fine Solids This process involves separation of small-particle solids suspended in a feed liquid, by contact with a second liquid phase. L. A. Robbins describes such a process for removing ash from pulverized coal [U.S. Patent 4,575,418 (1986)]. The process involves slurrying pulverized coal fines into a hydrocarbon liquid and contacting the resulting slurry with water. The coal slurry is cleaned by preferential transfer of ash particles into the aqueous phase. The process takes advantage of differences in surface-wetting properties to separate the different types of solid particles present in the feed.

Supercritical Fluid Extraction This process generally involves the use of CO_2 or light hydrocarbons to extract components from liquids or porous solids [Brunner, G., *Gas Extraction: An Introduction to Fundamentals of Supercritical Fluids and the Application to Separation Processes*, Springer-Verlag, Berlin, 1994; Brunner, G., ed., *Supercritical Fluids as Solvents and Reaction Media*, Elsevier, Amsterdam, 2004; and McHugh, M. A., and V. Krukonis, *Supercritical Fluid Extraction*, 2d ed., Butterworth-Heinemann, Oxford, UK, 1993]. Supercritical fluid extraction differs from liquid-liquid or liquid-solid extraction in that the operation is carried out at supercritical (or near-supercritical) conditions where the extraction fluid exhibits physical and transport properties that are inbetween those of liquid and vapor phases (intermediate density, viscosity, and solute diffusivity). Most applications involve the use of CO_2 (critical pressure = 73.8 bar at 31°C) or propane (critical pressure = 42.5 bar at 97°C). Other supercritical fluids and their critical-point properties are discussed by B. E. Poling, J. M. Prausnitz, and J. P. O'Connell [*The Properties of Gas and Liquids*, 5th ed., McGraw-Hill, New York, 2001].

Supercritical CO_2 extraction often is considered for extracting high-value soluble components from natural materials or for purifying low-volume specialty chemicals [Reverchon, E., and I. De Marco, *J. Supercrit. Fluids* **38:** 146–166 (2006)]. For products derived from natural materials, this can involve initial processing of solids followed by further processing of a crude liquid extract. Applications include decaffeination of coffee and recovery of desired compounds from plant- and animal-derived feeds, including recovery of flavor and fragrance components, neutraceuticals, and other active ingredients. An example is the use of supercritical CO_2 fractional extraction to remove terpenes from cold-pressed bergamot oil [Kondo, M., et al., *Ind. Eng. Chem. Res.* **39**(12): 4745–4748 (2000)]. A nonfood example involves the removal of unreacted dodecanol from nonionic surfactant mixtures and fractionation of the surfactant mixture based on polymer chain length [Eckert, C. A., et al., *Ind. Eng. Chem. Res.* **31**(4): 1105–1110 (1992)]. In these applications, process advantages may be obtained because solvent residues are easily removed or are nontoxic, the process can be operated at mild temperatures that avoid product degradation, the product is easily recovered from the extract fluid, or the solute separation factor and product purity can be adjusted by making small changes in the operating temperature and pressure. Although the loading capacity of supercritical CO_2 typically is low, addition of cosolvents such as methanol, ethanol, or tributylphosphate can dramatically boost capacity and enhance selectivity [Brennecke, J. F., and C. A. Eckert, *AIChE J.* **35**(9): 1409–1427 (1989)].

For processing liquid feeds, some supercritical fluid extraction processes use packed columns, in which the liquid feed phase wets the packing and flows through the column in film flow, with the supercritical fluid forming the continuous phase. In other applications, sieve trays give improved performance [Seibert, A. F., and D. G. Moosberg, *Sep. Sci. Technol.* **23:** 2049 (1988)]. In a number of these applications, concentrated solute is added back to the column as reflux to boost separation power (a form of single-solvent fractional extraction). Supercritical fluid extraction requires high-pressure equipment and may involve a high-pressure compressor. These requirements add considerable capital and operating cost. In certain cases, pumps can be used instead of compressors, to bring down the cost. The separators are run slightly below the critical point at slightly elevated pressure and reduced temperature to ensure the material is in the liquid state so it can be pumped. As a rule, supercritical fluid extraction is considerably more expensive than liquid-liquid extraction, so when the required separation can be accomplished by using a liquid solvent, liquid-liquid extraction often is more cost-effective.

Although most commercial applications of supercritical fluid extraction involve processing of high-value, low-volume products, a notable exception is the propane deasphalting process used to refine lubricating oils. This is a large-scale, commodity chemical process dating back to the 1930s. In this process and more recent versions, lube oils are extracted into propane at near-supercritical conditions. The extract phase is depressurized or cooled in stages to isolate various fractions. Compared to operation at lower pressures, operation at near-supercritical conditions minimizes the required pressure or temperature change, so the process can be more efficient. This technology also has been used for decolorization of tallow obtained from rendered animal fat [Moore, E. B., *J. Am. Oil Chem. Soc.* **27**(3): 75–80 (1950)]. For further discussion of supercritical fluid separation processes, see G. Brunner, *Annu. Rev. Chem. Biomol. Eng.* **1:** 321–342 (2010); J. Fernandes, R. Ruivo, and P. Simões, *AIChE J.* **53**(4): 825–837 (2007); and F. Gironi and M. Maschietti, *Chem. Eng. Sci.* **61:** 5114–5126 (2006).

KEY CONSIDERATIONS IN THE DESIGN OF AN EXTRACTION OPERATION

Successful approaches to designing an extraction process begin with an appreciation of the fundamentals (basic phase equilibrium and mass-transfer principles) and generally rely on both experimental studies and mathematical models or process simulations to define the commercial technology. Small-scale experiments using representative feed usually are needed to accurately quantify physical properties and phase equilibrium. Additionally, it is common practice in industry to perform miniplant or pilot-plant tests to accurately characterize the mass-transfer capabilities of the required equipment as a function of throughput [Glatz, D. J., B. C. Cross, and T. D. Lightfoot, *Chem. Eng. Prog.* **114**(2): 24–29 (2018)]. In many cases, mass-transfer resistance changes with increasing scale of operation, so an ability to accurately scale up the data also is needed. The required scale-up know-how often comes from experience operating commercial equipment of various sizes or from running pilot-scale equipment of sufficient size to develop and validate a scale-up correlation. Mathematical models are used as a framework for planning and analyzing the experiments, for correlating the data, and for estimating performance at untested conditions by extrapolation. Increasingly, designers and researchers are utilizing computational fluid dynamics (CFD) software or other simulation tools as an aid to scale-up.

Typical steps in the work process for designing and implementing an extraction operation include the following:

1. Outline the design basis, including specification of feed composition, required solute recovery or removal, product purity, and production rate.
2. Search the published literature (including patents) for information relevant to the application.
3. For dilute feeds, consider options for preconcentrating the feed to reduce the volumes of feed and solvent that must be handled by the extraction operation. Consider evaporation or distillation of a low-boiling feed solvent or the use of reverse-osmosis/nanofiltration membranes to concentrate the feed. (See the subsection Hybrid Extraction Processes under Commercial Process Schemes.)
4. Generate a list of candidate solvents based on chemical knowledge and experience. Consider solvents similar to those used in analogous applications. Use one or more of the methods described in Solvent Screening Methods to identify additional candidates. Include consideration of solvent blends and extractants.
5. Estimate key physical properties and review desirable solvent properties. Give careful consideration to safety, industrial hygiene, and environmental requirements. Use this preliminary information to trim the list of candidate solvents to a manageable size. (See the subsection Desirable Solvent Properties.)
6. Measure partition ratios for selected solvents at representative conditions.
7. Evaluate the potential for trace chemistry under extraction and solvent recovery conditions to determine whether solutes and candidate solvents are likely to degrade or react to produce unwanted impurities. For example, it is well known that penicillin G easily degrades at commercial extraction conditions, and short contact time is required for good results. Also under certain conditions acetate solvents may hydrolyze to form alcohols, certain alcohols and ethers can form peroxides, sulfur-containing solvents may degrade at elevated regeneration temperatures to form acids, chlorinated solvents may hydrolyze at elevated temperatures to form trace HCl with severe corrosion implications, and so on. In other cases, leakage of air into the process may cause formation of trace oxidation products. Understanding the potential for trace chemistry, the fate of potential impurities (i.e., where they go in the process), their possible effects on the process (including impact on product purity and interfacial tension), and devising means to avoid or successfully deal with impurities often are critical to a successful process design. Laboratory tests designed to probe the stability of feed and solvent mixtures may be needed.
8. Characterize mass-transfer difficulty in terms of the required number of theoretical stages or transfer units as a function of the solvent-to-feed ratio. Keep in mind that there will be a limit to the number of theoretical stages that can be achieved. For most cost-effective extraction operations, this limit will be in the range of 3 to 10 theoretical stages, although some can achieve more, depending upon the chemical system, type of equipment, and flow rate (throughput).

9. Estimate the cost of the proposed extraction operation relative to alternative separation technologies, such as extractive distillation, adsorption, and crystallization. Explore other options if they appear less expensive or offer other advantages.

10. If technical and economic feasibility look good, determine accurate values of physical properties and phase equilibria, particularly liquid densities, mutual solubilities (miscibility), viscosities, interfacial tension, and K values (at feed, extract, and raffinate ends of the proposed process), as well as data needed to evaluate solvent recycle options. Search available literature and databases. Assess data quality and generate additional data as needed. Develop the appropriate data correlations. Be careful to check the results of process simulation programs and other estimation methods by comparison with actual experimental data. Finalize the choice of solvent.

11. Outline an overall process flow sheet and material balance, including solvent recovery and recycle. This should be done with the aid of process simulation software. In the flow sheet, include methods needed for controlling emissions and managing wastes. Carefully consider the possibility that impurities may accumulate in the recycled solvent, and devise methods for purging these impurities. For general guidelines, see W. D. Seider, J. D. Seader, D. R. Lewin, and S. Widagdo, *Product and Process Design Principles: Synthesis, Analysis, and Evaluation*, 3d ed., Wiley, New York, 2009; and R. Turton, R. C. Bailie, W. B. Whiting, J. A. Shaeiwitz, and D. Bhattacharyya, *Analysis, Synthesis, and Design of Chemical Processes*, 4th ed., Prentice-Hall, Upper Saddle River, NJ, 2012.

12. In some cases, especially with multiple solutes and complex phase equilibria, it may be useful to perform laboratory batch experiments to simulate a continuous, countercurrent, multistage process. These experiments can be used to test/verify calculation results and determine the correct distribution of components. For additional information, see R. E. Treybal, chap. 9 in *Liquid Extraction*, 2d ed., McGraw-Hill, New York, 1963, pp. 359–393; and M. H. I. Baird and T. C. Lo, chap. 17.1 in *Handbook of Solvent Extraction*, ed. T. C. Lo, M. H. I. Baird, and C. Hanson, Wiley, New York, 1983, and Krieger, Huntington, NY, 1991.

13. Identify useful equipment options for liquid-liquid contacting and liquid-liquid phase separation, estimate approximate equipment size, and outline preliminary design specifications. (See the subsection Extractor Selection under Liquid-Liquid Extraction Equipment.) Where appropriate, consult with equipment vendors. Using small-scale experiments, determine whether sludgelike materials are likely to accumulate at the liquid-liquid interface (called formation of a rag layer). If so, it will be important to identify equipment options that can tolerate a rag layer and allow the rag to be drained or otherwise purged periodically.

14. For the most promising equipment option, run miniplant or pilot-plant tests over a range of operating conditions. Use representative feed, including all anticipated impurities, because even small concentrations of surface-active components can dramatically affect interfacial behavior. Whenever possible, the miniplant tests should be conducted by using actual material from the manufacturing plant, and they should include solvent recycle to evaluate the effects of impurity accumulation or possible solvent degradation. Run the miniplant long enough that the solvent encounters numerous cycles so that recycle effects can be seen. If difficulties arise, consider alternative solvents or process options for purging accumulated impurities.

15. Analyze miniplant data and update the preliminary design. Carefully evaluate loss of solvent to the raffinate, and devise methods to minimize losses as needed. Consult equipment vendors or other specialists regarding recommended scale-up methods.

16. Specify the final material balance for the overall process and carry out detailed equipment design calculations. Try to add some flexibility (depending on the cost) to allow for some adjustment of the process equipment during operation, to compensate for uncertainties in the design.

17. Install and start up the equipment in the manufacturing plant.

18. Troubleshoot and improve the operation as needed. Once a unit is operational, carefully measure the material balance and characterize mass-transfer performance. If performance does not meet expectations, look for defects in the equipment installation. If none are found, revisit the scale-up methodology and its assumptions.

LABORATORY PRACTICES

An equilibrium or theoretical stage in liquid-liquid extraction, as defined earlier, is routinely used in laboratory procedures. A feed solution is contacted with a solvent to remove one or more of the solutes from the feed. This can be carried out in a separating funnel or, preferably, in an agitated vessel that can produce droplets about 1 mm in diameter. After agitation has stopped and the phases separate, the two clear liquid layers are isolated by decantation. The partition ratio can then be determined directly by measuring the concentration of solute in the extract and raffinate layers. (Additional discussion is given in Liquid-Liquid Equilibrium Experimental Methods under Thermodynamic Basis for Liquid-Liquid Extraction.) When an appropriate analytical method is available only for the feed phase, the partition ratio can be determined by measuring the solute concentration in the feed and raffinate phases and calculating the partition ratio from the material balance. For the case of zero initial concentration of solute in the extraction solvent (before extraction), the partition ratio expressed in terms of mass fractions is given by

$$K'' = \frac{Y''_e}{X''_r} = \frac{M_r}{M_e}\left[\frac{M_f X''_f}{M_r X''_r} - 1\right] \tag{15-2}$$

where K'' = mass fraction solute in extract divided by that in raffinate
M_f = total mass of feed added to vial
M_s = total mass of extraction solvent before extraction
M_r = mass of raffinate phase after extraction
M_e = mass of extract phase after extraction
X''_f = mass fraction solute in feed prior to extraction
X''_r = mass fraction solute in raffinate, at equilibrium
Y''_e = mass fraction solute in extract, at equilibrium

For systems with low mutual solubility between phases, $K'' \approx (M_f/M_s)(X''_f/X''_r - 1)$. While this approach can provide useful results, laboratory procedures that include actual analysis of solute concentration in both the extract and raffinate layers plus measurement of feed and solvent weight before and after extraction are preferred because they allow calculation of the component material balances (a check of solute accountability). If the material balances are poor, the resulting K values are uncertain, and the procedures and analytical methods will need careful review and improvement.

After a single stage of liquid-liquid contact, the phase remaining from the feed solution (the raffinate) can be contacted with another quantity of fresh extraction solvent. This cross-current (or cross-flow) extraction scheme is an excellent laboratory procedure because the extract and raffinate phases can be analyzed after each stage to generate equilibrium data for a range of solute concentrations. Also, the feasibility of solute removal to low levels can be demonstrated (or shown to be problematic because of the presence of "extractable" and "nonextractable" forms of a given species). The number of cross-current treatments needed for a given separation, assuming a constant K value, can be estimated from

$$N = \frac{\ln\left(\dfrac{X_{\text{in}} - Y_{\text{in}}/K}{X_{\text{out}} - Y_{\text{in}}/K}\right)}{\ln(KS^*/F + 1)} \tag{15-3}$$

where F is the amount of feed, the feed and solvent are presaturated, and equal amounts of solvent (denoted by S^*) are used for each treatment [Treybal, R. E., *Liquid Extraction*, 2d ed., McGraw-Hill, New York, 1963, pp. 209–216]. The total amount of solvent is $N \times S^*$. The variable Y_{in} is the concentration of solute in the fresh solvent, normally equal to zero. Equation (15-3) is written in a general form without specifying the units. Any consistent system of units may be used. (See the subsection Process Fundamentals and Basic Calculation Methods.)

A cross-current scheme, although convenient for laboratory practice, is not generally cost effective for large commercial processes because solvent usage is high and the solute concentration in the combined extract is low. A number of batchwise countercurrent laboratory techniques have been developed and can be used to demonstrate countercurrent performance. (See item 12 in the previous subsection, Key Considerations in the Design of an Extraction Operation.) Several equipment vendors also make available continuously fed laboratory-scale extraction equipment. Examples include small-scale mixer-settler extraction batteries offered by Normag, MEAB, Rousselet-Robatel, Schott/QVF, and Sulzer. Small-diameter extraction columns also may be used, such as the 5/8-in (16-mm) diameter reciprocating-plate agitated column offered by Koch Modular Process Systems, and a 60-mm-diameter Kühni rotary-impeller agitated column offered by Sulzer. For a discussion of mixer-settler studies in the laboratory, see K. Benz et al. [*Chem. Eng. Technol.* **24**(1): 11–17 (2001)].

For additional discussion of laboratory techniques, see the subsections Liquid-Liquid Equilibrium Experimental Methods under Thermodynamic Basis for Liquid-Liquid Extraction and High-Throughput Experimental Methods under Solvent Screening Methods.

THERMODYNAMIC BASIS FOR LIQUID-LIQUID EXTRACTION

General References: See Sec. 4, Thermodynamics, as well as Elliot, J. R., and C. T. Lira, *Introduction to Chemical Engineering Thermodynamics*, 2d ed. Prentice-Hall, Upper Saddle River, NJ, 2012; Sandler, S. I., *Chemical, Biochemical, and Engineering Thermodynamics*, Wiley, New York, 2006; Smith, J. M., H. C. Van Ness, and M. M. Abbott, *Introduction to Chemical Engineering Thermodynamics*, 7th ed., McGraw-Hill, New York, 2005; Rydberg, J., M. Cox, C. Musikas, and G. R. Choppin, eds., *Solvent Extraction Principles and Practice*, 2d ed., Marcel Dekker, New York, 2004; Schwarzenbach, R. P., P. M. Gschwend, and D. M. Imboden, *Environmental Organic Chemistry*, 2d ed., Wiley-VCH, New York, 2002; Prausnitz, J. M., R. N. Lichtenthaler, and E. Gomez de Azevedo, *Molecular Thermodynamics of Fluid-Phase Equilibria*, 3d ed., Prentice-Hall, Upper Saddle River, NJ, 1999; Bolz, A., et al., *Pure Appl. Chem.* (IUPAC) **70:** 2233–2257 (1998); Grant, D. J. W., and T. Higuchi, *Solubility Behavior of Organic Compounds*, Techniques of Chemistry Series, vol. 21, Wiley, New York, 1990; Abbott, M. M., and J. M. Prausnitz, "Phase Equilibria," in *Handbook of Separation Process Technology*, ed. R. W. Rousseau, Wiley, New York, 1987, pp. 3–59; Novak, P. J., J. Matous, and J. Pick, *Liquid-Liquid Equilibria*, Studies in Modern Thermodynamics Series, vol. 7, Elsevier, Amsterdam, 1987; Walas, S. M., *Phase Equilibria in Chemical Engineering*, Butterworth-Heinemann, Oxford, UK, 1985; and Rowlinson, J. S., and F. L. Swinton, *Liquids and Liquid Mixtures*, 3d ed., Butterworth, Oxford, UK, 1982.

ACTIVITY COEFFICIENTS AND THE PARTITION RATIO

Phase Equilibrium Two phases are at equilibrium when the total Gibbs energy for the system is at a minimum. This criterion can be restated as follows: Two nonreacting phases are at equilibrium when the chemical potential of each distributed component is the same in each phase; that is, for equilibrium between two phases I and II containing n components

$$\mu_i^{\mathrm{I}} = \mu_i^{\mathrm{II}} \quad i = 1, 2, \ldots, n \tag{15-4}$$

For two phases at the same temperature and pressure, component activities are the same in each phase, and Eq. (15-4) can be expressed in terms of mole fractions and activity coefficients, giving

$$y_i \gamma_i^{\mathrm{I}} = x_i \gamma_i^{\mathrm{II}} \quad i = 1, 2, \ldots, n \tag{15-5}$$

where y_i and x_i represent mole fractions of component i in phases I and II, respectively. The equilibrium partition ratio, in units of mole fraction, is then given by

$$K_i^o = \frac{y_i}{x_i} = \frac{\gamma_i^{\text{raffinate}}}{\gamma_i^{\text{extract}}} \tag{15-6}$$

where y_i is the mole fraction in the extract phase and x_i is the mole fraction in the raffinate. Note that, in general, activity coefficients and K_i^o are functions of temperature and composition. For ionic compounds that dissociate in solution, the species that form and the extent of dissociation in each phase also must be taken into account. Similarly, for extractions involving adduct formation, molecular complexation, or other chemical reactions, the reaction stoichiometry is an important factor. For discussion of these special cases, see G. R. Choppin, chap. 3, and J. Rydberg et al., chap. 4, in *Solvent Extraction Principles and Practice*, 2d ed., ed. J. Rydberg et al., Marcel Dekker, New York, 2004; H.-J. Bart, *Reactive Extraction*, Springer, Berlin, 2001; F. Xie et al., *Minerals Engineering* **56:** 10–28 (2014); and R. Schulz et al., *Ind. Eng. Chem. Res.* **55**(19): 5763–5769 (2016).

The activity coefficient for a given solute is a measure of the nonideality of solute-solvent interactions in solution. In this context, the solvent is either the feed solvent or the extraction solvent depending upon which phase is considered, and the composition of the "solvent" includes all components present in that phase. For an ideal solution, activity coefficients are unity. For solute-solvent interactions that are repulsive relative to solvent-solvent interactions, γ_i is greater than 1. This is said to correspond to a positive deviation from ideal solution behavior. For attractive interactions, γ_i is less than 1, corresponding to a negative deviation. Activity coefficients often are reported for binary pairs in the limit of very dilute conditions (infinite dilution) because this represents the interaction of solute completely surrounded by solvent molecules, and this normally gives the largest value of the activity coefficient (denoted as γ_i^∞). Normally, useful approximations of the activity coefficients at more concentrated conditions can be obtained by extrapolation from infinite dilution using an appropriate activity coefficient correlation equation. (See Sec. 4, Thermodynamics.) Extrapolation in the reverse direction, that is, from finite concentration to infinite dilution, often does not provide reliable results.

In units of mass fraction, the partition ratio for a nonreacting/nondissociating solute is given by

$$K_i''(\text{mass frac. basis}) = \frac{Y_i''}{X_i''}$$

$$= K_i^o\,(\text{mole frac. basis}) \times \left[\frac{y_i(\mathrm{MW}_i - \mathrm{MW}_{\text{raffinate}}) + \mathrm{MW}_{\text{raffinate}}}{x_i(\mathrm{MW}_i - \mathrm{MW}_{\text{extract}}) + \mathrm{MW}_{\text{extract}}}\right] \tag{15-7}$$

Here, the notation MW refers to the molecular weight of solute i and the effective average molecular weights of the extract and raffinate phases, as indicated by the subscripts. For dilute systems, $K_i'' \approx K_i^o \times (\mathrm{MW}_{\text{raffinate}}/\mathrm{MW}_{\text{extract}})$. For theoretical stage or transfer unit calculations, often it is useful to express the partition ratio in terms of mass ratio coordinates introduced by W. D. Bancroft [*Phys. Rev.* **3**(1): 21–33; **3**(2): 114–136; and **3**(3): 193–209 (1895)]:

$$K_i' = \frac{Y_i'}{X_i'} = \frac{M_{\text{solute}}/M_{\text{extraction solvent}} \text{ in extract phase}}{M_{\text{solute}}/M_{\text{feed solvent}} \text{ in raffinate phase}} \tag{15-8}$$

Partition ratios also may be expressed on a volumetric basis. In that case,

$$K_i^{\text{vol}}\,(\text{mole/vol. basis}) = K_i'' \frac{\rho_{\text{extract}}}{\rho_{\text{raffinate}}} \tag{15-9}$$

$$K_i^{\text{vol}}\,(\text{mole/vol. basis}) = K_i^o \left(\frac{\rho_{\text{extract}}}{\rho_{\text{raffinate}}}\right)\left(\frac{\mathrm{MW}_{\text{raffinate}}}{\mathrm{MW}_{\text{extract}}}\right) \tag{15-10}$$

Extraction Factor The extraction factor is defined by

$$\mathcal{E}_i = m_i \frac{S}{F} \tag{15-11}$$

where $m_i = dY_i/dX_i$, the slope of the equilibrium line, and F and S are the flow rates of the feed phase and the extraction-solvent phase, respectively. On a McCabe-Thiele type of diagram, $\mathcal{E}$ is the slope of the equilibrium line divided by the slope of the operating line F/S. (See McCabe-Thiele Type of Graphical Method under Process Fundamentals and Basic Calculation Methods.) For dilute systems with straight equilibrium lines, the slope of the equilibrium line is equal to the partition ratio, $m_i = K_i$.

To illustrate the significance of the extraction factor, consider an application where K_i, S, and F are constant (or nearly so) and the extraction solvent entering the process contains no solute. When $\mathcal{E}_i = 1$, the extract stream has just enough capacity to carry all the solute present in the feed:

$$SY_{i,\text{extract}} = FX_{i,\text{feed}} \quad \text{at } \mathcal{E}_i = 1 \text{ and equilibrium conditions} \tag{15-12}$$

At $\mathcal{E}_i < 1.0$, the extract's capacity to carry solute is less than this amount, and the maximum fraction that can be extracted θ_i is numerically equal to the extraction factor:

$$(\theta_i)_{\max} = \mathcal{E}_i \quad \text{when } \mathcal{E}_i < 1.0 \tag{15-13}$$

At $\mathcal{E}_i > 1.0$, the extract phase has more than sufficient carrying capacity (in principle), and the actual amount extracted depends on the extraction scheme, number of contacting stages, and mass-transfer resistance. Even a solute for which $m_i < 1.0$ (or $K_i < 1.0$) can, in principle, be extracted to a very high degree by adjusting S/F so that $\mathcal{E}_i > 1$.

Thus, the extraction factor characterizes the relative capacity of the extract phase to carry solute present in the feed phase. Its value is a major factor determining the required number of theoretical stages or transfer units. (For further discussion, see the subsection Extraction Factor and General Performance Trends.) In general, the value of the extraction factor can vary at each point along the equilibrium curve, although in many cases it is nearly constant. Many commercial extraction processes are designed to operate with an average or overall extraction factor in the range of 1.3 to 5. Exceptions include applications where the partition ratio is very large and the solvent-to-feed ratio is set by hydraulic considerations.

Because the extraction factor is a dimensionless variable, its value should be independent of the units used in Eq. (15-11), as long as they are consistently applied. Engineering calculations often are carried out by using mole fraction, mass fraction, or mass ratio units (Bancroft coordinates). The flow rates S and F then need to be expressed in terms of total molar flow rates, total mass flow rates, or solute-free mass flow rates, respectively. In the design of extraction equipment, volume-based units often are used. Then the appropriate concentration units are mass or mole per unit volume, and flow rates are expressed in terms of the volumetric flow rate of each phase.

Separation Factor The separation factor in extraction is analogous to relative volatility in distillation. It is a dimensionless factor that measures the relative enrichment of a given component in the extract phase after one theoretical stage of extraction. For cosolutes i and j,

$$\alpha_{i,j} = \frac{(Y_i/Y_j)_{\text{extract}}}{(X_i/X_j)_{\text{raffinate}}} = \frac{(Y_i)_{\text{extract}}/(X_i)_{\text{raffinate}}}{(Y_j)_{\text{extract}}/(X_j)_{\text{raffinate}}} = \frac{K_i}{K_j} \tag{15-14}$$

The enrichment of solute i with respect to solute j can be further increased with the use of multiple contacting stages. The solute separation factor $\alpha_{i,j}$ is used to characterize the selectivity a solvent has for extracting a desired solute from a feed containing other solutes. It can be calculated by using any consistent units. As in distillation, $\alpha_{i,j}$ must be greater than 1.0 to achieve an increase in product-solute purity (on a solvent-free basis). In practice, if solute purity is an important requirement of a given application, $\alpha_{i,j}$ must be greater than 20 for standard extraction (at least) and greater than about 4 for fractional extraction, in order to have sufficient separation power using a moderate number of contacting stages. (See the subsection Potential for Solute Purification Using Standard Extraction in Process Fundamentals and Basic Calculation Methods and Dual-Solvent Fractional Extraction in Calculation Procedures.)

The separation factor also can be evaluated for solute i with respect to the feed solvent denoted as component f. The value of $\alpha_{i,f}$ must be greater than 1.0 if the proposed separation is to be feasible, that is, in order to be able to enrich solute i in a separate extract phase. Note that the feed may still be separated if $\alpha_{i,f} < 1.0$, but this would have to involve concentrating solute i in the feed phase by preferential transfer of component f into the extract phase. Although $\alpha_{i,f} > 1.0$ represents a minimum theoretical requirement for enriching solute i in a separate extract phase, most commercial extraction processes operate with values of $\alpha_{i,f}$ on the order of 20 or higher. There are exceptions to this rule, such as the UDEX process and similar processes involving extraction of aromatics from aliphatic hydrocarbons. In these applications, $\alpha_{i,f}$ can be as low as 10 and sometimes even lower. Applications such as these involve particularly difficult design challenges because of low solute partition ratios and high mutual solubility between phases. (For more detailed discussion of these kinds of systems, see the subsection Single-Solvent Fractional Extraction with Extract Reflux in Fractional Extraction Calculations.)

Minimum and Maximum Solvent-to-Feed Ratios Normally, it is possible to quickly estimate the physical constraints on solvent usage for a standard extraction application in terms of minimum and maximum solvent-to-feed ratios. As discussed earlier, the minimum theoretical amount of solvent needed to transfer a high fraction of solute i is the amount corresponding to $\mathcal{E}_i = 1$. In practice, the minimum practical extraction factor is about 1.3, because at lower values the required number of theoretical stages increases dramatically. This gives a minimum solvent-to-feed ratio for a practical process equal to

$$\left(\frac{S}{F}\right)_{\min} \approx \frac{1.3}{K_i} \tag{15-15}$$

Note that this minimum is achievable only if a sufficient number of contacting stages or transfer units can be used. (For additional discussion, see the subsection Extraction Factor and General Performance Trends.) It is also achievable only if the amount of solvent added to the feed is greater than the solubility limit in the feed phase (including solute); otherwise, only one liquid phase can exist. In certain cases involving fairly high mutual solubilities, this can be an important consideration when running a process using minimal solvent—because if the process operates close to the solubility limit, an upset in the solvent-to-feed ratio may cause the solvent phase to disappear. The maximum possible solvent-to-feed ratio is obtained when the amount of extraction solvent is so large that it dissolves the feed phase. Assuming the feed entering the process does not contain extraction solvent,

$$\left(\frac{S}{F}\right)_{\max} = \frac{1}{Y^{\text{SAT}}_{\text{feed solvent}}} \tag{15-16}$$

where $Y^{\text{SAT}}_{\text{feed solvent}}$ denotes the concentration of feed solvent in the extract phase at equilibrium.

If an application proves to be technically feasible, the choice of solvent-to-feed ratio is determined by identifying the most cost-effective ratio between the minimum and maximum limits. For most applications, the maximum solvent-to-feed ratio will be much larger than the ratio chosen for the commercial process; however, the maximum ratio can be a real constraint when dealing with applications exhibiting high mutual solubility, especially for systems that involve high solute concentrations. Solvent ratios are further constrained for a fractional extraction scheme, as discussed in Fractional Extraction Calculations.

Temperature Effect The effect of temperature on the value of the partition ratio can vary greatly from one system to another. This depends on how the activity coefficients of the components in each phase are affected by changes in temperature, including any effects due to changes in mutual solubility. For a given phase, the Gibbs-Helmholtz equation indicates that

$$\left[\frac{\partial \ln \gamma_i^\infty}{\partial(1/T)}\right]_{p,x} = -\frac{h_i^{E,\infty}}{R} \tag{15-17}$$

where γ_i^∞ is the activity coefficient for solute i at infinite dilution and h^E_i is the excess enthalpy of mixing relative to ideal solution behavior [Atik, Z., D. Gruber, M. Krummen, and J. Gmehling, *J. Chem. Eng. Data* **49**(5): 1429–1432 (2004)]. Calculating the temperature dependence of K from these basic principles is not a common practice, in part because in many cases the required enthalpy of mixing data are not available (or they are difficult to predict with sufficient accuracy), and changes in mutual solubility complicate the analysis. The temperature dependence of activity coefficients for various classes of compounds is discussed in Sec. 4, Thermodynamics, and by T. C. Frank, S. G. Arturo, and B. S. Holden, *AIChE J.* **60**(10): 3675–3690 (2014).

Systems with specific interactions between solute and solvent, such as hydrogen bonds or ion-pair bonds, often are particularly sensitive to changes in temperature because the specific interactions are strongly temperature-dependent. In general, hydrogen bonding and ion-pair formation are disrupted by increasing temperature (increasing molecular motion), and this can dominate the overall temperature dependence of the partition ratio. An example of a temperature-sensitive hydrogen bonding system is toluene + diethylamine + water [Morello, V. S., and R. B. Beckmann, *Ind. Eng. Chem.* **42:** 1079–1087 (1950)]. The partition ratio for transfer of diethylamine from water into toluene increases with increasing temperature (on a weight percent basis, $K = 0.7$ at 20°C and $K = 2.8$ at 58°C). For further discussion of the temperature dependence of K for this type of system, see T. C. Frank et al., *Ind. Eng. Chem. Res.* **46**(11): 3774–3786 (2007). An example of a temperature-sensitive system involving ion-pair formation is the commercial process used to recover citric acid from fermentation broth using trioctylamine (TOA) extractant [Pazouki, M., and T. Panda, *Bioprocess Engineering* **19:** 435–439 (1998)]. In this case, the partition ratio for transfer of citric acid into the TOA phase decreases with increasing temperature [Canari, R., and A. M. Eyal, *Ind. Eng. Chem. Res.* **43:** 7608–7617 (2004)]. Also see the discussion of Temperature-Swing Extraction in the subsection Commercial Process Schemes.

Salting-Out and Salting-In Effects for Nonionic Solutes It is well known that the presence of an inorganic salt can significantly affect the solubility of a nonionic (nonelectrolyte) organic solute dissolved in water. In most cases the inorganic salt reduces the organic solute's solubility (salting-out effect). Here, the salt increases the ionic strength of the aqueous solution, and this results in an increase in the organic solute's activity coefficient. As a result, certain solutes that are not easily extracted from water may be quite easily extracted from brine, depending upon the type of solute and the salt. In principle, the deliberate addition of a salt to an aqueous feed is an option for enhancing partition ratios and reducing the mutual solubility of the two liquid phases; however, this approach complicates the overall process and normally is not cost-effective. Difficulties include the added complexity and costs associated with recovery and recycle of the salt in the overall process, or disposal of the brine after extraction and the need to purchase makeup salt. The potential use of NaCl to enhance the extraction of ethanol from fermentation broth is discussed by V. Gomis et al. [*Ind. Eng. Chem. Res.* **37**(2): 599–603 (1998)].

When an aqueous feed contains a salt, the effect of the dissolved salt on the partition ratio for a given organic solute may be estimated by using an expression introduced by J. Setschenow [*Z. Physik. Chem.* **4:** 117–128 (1889)] and commonly written in the form

$$\log \frac{\gamma_{i,\text{brine}}}{\gamma_{i,\text{water}}} = k_s C_{\text{salt}} \tag{15-18}$$

where C_{salt} is the concentration of salt in the aqueous phase in units of gmol/L and k_s is the Setschenow constant. Equation (15-18) generally is valid for dilute organic solute concentrations and low to moderate salt concentrations.

In many cases, the salt has no appreciable effect on the activity coefficient in the organic phase because the salt solubility in that phase is low or negligible. Then

$$\log \frac{K_{i,\text{brine}}}{K_{i,\text{water}}} \approx \log \frac{\gamma_{i,\text{brine}}}{\gamma_{i,\text{water}}} = k_s\, C_{\text{salt}} \tag{15-19}$$

for extraction from the aqueous phase into an organic phase. For aromatic solutes dissolved in NaCl brine at room temperature, typical values of k_s fall within the range of 0.2 to 0.3 L/gmol. In general, k_s is found to vary with salt composition (i.e., with the type of salt) and increase with increasing organic-solute molar volume. I. Kojima and S. S. Davis [*Int. J. Pharm.* **20**(1–2): 203–207 (1984)] showed that partition ratio data for the extraction of phenol dissolved in NaCl brine (at low concentration) using CCl_4 solvent is well fit by a Setschenow equation for salt concentrations up to 4 gmol/L (about 20 wt% NaCl). Experimental values and methods for estimating Setschenow constants are discussed by N. Ni and S. H. Yalkowski [*Int. J. Pharm.* **254**(2): 167–172 (2003)] and by W.-H. Xie, W.-Y. Shiu, and D. MacKay [*Marine Environ. Res.* **44:** 429–444 (1997)]. For a detailed review of general principles, see A. M. Hyde et al., *Org Process Res. Dev.* **21:** 1355–1370 (2017).

Salts with large ions (such as tetramethylammonium chloride and sodium toluene sulfonate) may cause a "salting in" or "hydrotropic" effect whereby the salt increases the solubility of an organic solute in water, apparently by disordering the structure of associated water molecules in solution or by forming specific interactions with the organic solute. M. Agrawal and V. G. Gaikar [*Sep. Technol.* **2:** 79–84 (1992)] discuss the use of hydrotropic salts to facilitate extraction processes. For additional discussion, see E. Ruckenstein and I. Shulgin, *Ind. Eng. Chem. Res.* **41**(18): 4674–4680 (2002); and M. Akia and F. Feyzi, *AIChE J.* **52**(1): 333–341 (2006).

Effect of pH for Ionizable Organic Solutes The distribution of weak acids and bases between organic and aqueous phases is dramatically affected by the pH of the aqueous phase relative to the pK_a of the solute. As discussed earlier, the pK_a is the pH at which 50 percent of the solute is in the ionized state. (See the subsection Dissociative Extraction in Commercial Process Schemes.) For a weak organic acid (RCOOH) that dissociates into $RCOO^-$ and H^+, the overall partition ratio for extraction into an organic phase depends upon the extent of dissociation such that

$$K_{\text{weak acid}} = K_{\text{nonionized}} \div \left(1 + \frac{[\text{RCOO}^-]_{\text{aq}}}{[\text{RCOOH}]_{\text{aq}}}\right) \tag{15-20}$$

where $K_{\text{weak acid}} = [\text{RCOOH}]_{\text{org}}/([\text{RCOO}^-]_{\text{aq}} + [\text{RCOOH}]_{\text{aq}})$ is the partition ratio for both ionized and nonionized forms of the acid, and $K_{\text{nonionized}} = [\text{RCOOH}]_{\text{org}}/[\text{RCOOH}]_{\text{aq}}$ is the partition ratio for the nonionized form alone [R. E. Treybal, *Liquid Extraction*, 2d ed., McGraw-Hill, New York, 1963, pp. 38–40]. Equation (15-20) can be rewritten in terms of the pK_a for a weak acid or weak base:

$$K_{\text{weak acid}} = K_{\text{nonionized}} \div (1 + 10^{\text{pH}-\text{p}K_a}) \tag{15-21}$$

and

$$K_{\text{weak base}} = K_{\text{nonionized}} \div (1 + 10^{\text{p}K_a-\text{pH}}) \tag{15-22}$$

For weak bases, pK_a = 14 − pK_b. Appropriate values for $K_{\text{nonionized}}$ may be obtained by measuring the partition ratio at sufficiently low pH (for acids) or high pH (for bases) to ensure the solute is in its nonionized form (normally at a pH at least 2 units from the pK_a value). In Eqs. (15-21) and (15-22), it is assumed that concentrations are dilute, that dissociation occurs only in the aqueous phase, and that the acid does not associate (dimerize) in the organic phase. The effect of pH on the partition ratio for extraction of penicillin G, a complex organic containing a carboxylic acid group, is illustrated in Fig. 15-14. For a discussion of the effect of pH on the extraction of carboxylic acids with tertiary amines, see S. T. Yang, S. A. White, and S. T. Hsu, *Ind. Eng. Chem. Res.* **30**(6): 1335–1342 (1991). Another example is discussed by D. C. Greminger et al. [*Ind. Eng. Chem. Proc. Des. Dev.* **21**(1): 51–54 (1982)]; they present partition ratio data for various phenolic compounds as a function of pH.

For compounds with multiple ionizable groups, such as amino acids, the effect of pH on partitioning behavior is more complex. Amino acid partitioning is discussed by K. Schügerl [*Solvent Extraction in Biotechnology*, Springer-Verlag, Berlin, 1994] and by M. T. Gude et al. [*Ind. Eng. Chem. Res.* **35:** 4700–4712 (1996)]. Amino acids are zwitterionic (dipolar) molecules with both acid and base functionality; the pK_a values corresponding to acidic RCOOH groups generally are between 2 and 6, and pK_a values for basic RNH_3^+ amino groups are between 9 and 10 [Fuchs, D., et al., *Ind. Eng. Chem. Res.* **45**(19): 6578–6584 (2006)]. For amino acids, proteins (complex polymers of amino acids), and other such compounds containing both acid

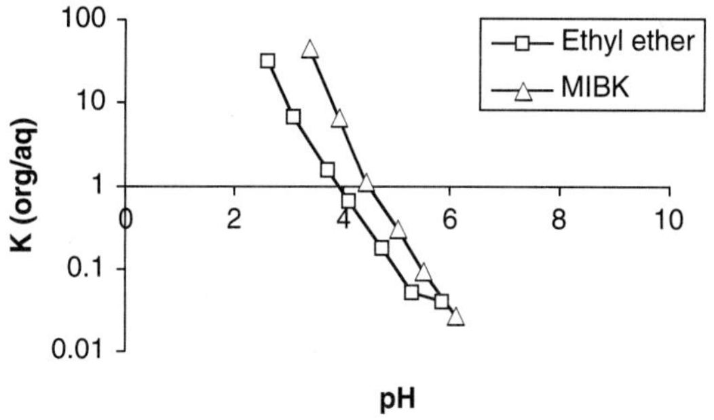

FIG. 15-14 The effect of pH on the partition ratio for extraction of penicillin G (pK_a = 2.75) from broth using an oxygenated organic solvent. The partition ratio is expressed in units of g/L in the organic phase over that in the aqueous phase. [*Data from R. L. Feder, M.S. thesis (Polytechnic Institute of Brooklyn, 1947).*]

and base functionality, aqueous solubility is lowest at the pH corresponding to the compound's isoelectric point (IEP). The IEP is the pH at which all negative charges are balanced by all positive charges and the molecule has zero net charge [van Holde, K. E., W. C. Johnson, and P. S. Ho, *Principles of Physical Biochemistry*, Prentice-Hall, Upper Saddle River, NJ, 1998]. Aqueous solubility is higher at a pH away from the IEP due to a net charge on the molecule, allowing the formation of an ionic species in solution. Above the IEP, acid groups become deprotonated, yielding a net negative charge. Below the IEP, amine groups become protonated, yielding a net positive charge. The IEP of amino acids varies widely depending on the specific acid/base functional groups. Most proteins exhibit an IEP in the range of pH 5 to 8. The well-known chelating agent ethylenediaminetetraacetic acid (EDTA) provides another example. Minimum EDTA solubility occurs at pH 2 to 3 (the IEP), which can vary somewhat depending on electrolyte concentration (ionic strength) [Battaglia, G., et al., *J. Chem. Eng. Data* **53**(2): 363–367 (2008)]. At higher pH, the carboxylic acid groups begin to deprotonate, yielding various ionized acid species ($EDTA^{1-}$ to $EDTA^{4-}$), so solubility in water is higher than the minimum. At strong acid conditions (pH < 2), the fully protonated diamine is formed (H_6EDTA^{2+}), and solubility again is higher than the minimum.

For general discussions of organic acid and base ionic equilibria, see J. N. Butler, *Ionic Equilibrium: Solubility and pH Calculations*, Wiley, New York, 1998; and M. B. Smith, *March's Advanced Organic Chemistry: Reactions, Mechanisms, and Structure*, 7th ed., chap. 8, Wiley, New York, 2013. The dissociation of inorganic salts is discussed in the book edited by D. D. Perrin [*Ionization Constants of Inorganic Acids and Bases in Aqueous Solution*, vol. 29, Pergamon, Oxford, UK, 1982]. Compilations of pK_a values are given in several handbooks [Jencks, W. P., and J. Regenstein, "Ionization Constants of Acids and Bases," in *Handbook of Biochemistry and Molecular Biology; Physical and Chemical Data*, vol. 1, 3d ed., ed. G. D. Fasman, CRC Press, Boca Raton, Fla., 1976, pp. 305–351; and *CRC Handbook of Chemistry and Physics*, 95th ed., ed. W. M. Haynes, CRC Press, Boca Raton, Fla., 2014–2015]. Also see D. D. Perrin, B. Dempsey, and E. P. Serjeant, *pK$_a$ Prediction for Organic Acids and Bases*, Chapman and Hall, London, 1981.

PHASE DIAGRAMS

Phase diagrams are used to display liquid-liquid equilibrium data across a wide composition range. Consider the binary system of water + 2-butoxyethanol (common name ethylene glycol *n*-butyl ether) plotted in Fig. 15-15. This system exhibits both an upper critical solution temperature (UCST), also called the upper consolute temperature, and a lower critical solution temperature (LCST), or lower consolute temperature. The mixture is only partially miscible at temperatures between 48°C (the LCST) and 130°C (the UCST). Most mixtures tend to become more soluble in each other as the temperature increases; that is, they exhibit UCST behavior. The presence of an LCST in the phase diagram is less common. Mixtures that exhibit LCST behavior include hydrogen-bonding mixtures such as an amine, a ketone, or an etheric alcohol plus water. Numerous water + glycol ether mixtures behave in this way [S. P. Christensen et al., *J. Chem. Eng. Data* **50**(3): 869–877 (2005)]. For these systems, hydrogen bonding leads to complete miscibility below the LCST. As temperature increases, hydrogen bonding is disrupted by increasing thermal (kinetic) energy, and hydrophobic interactions begin to dominate, leading to partial miscibility at temperatures above the LCST. The ethylene glycol + triethylamine system shown in Fig. 15-16 is another example.

Most of the ternary or pseudoternary systems used in extraction are of two types: one binary pair has limited miscibility (termed a type I system), or two

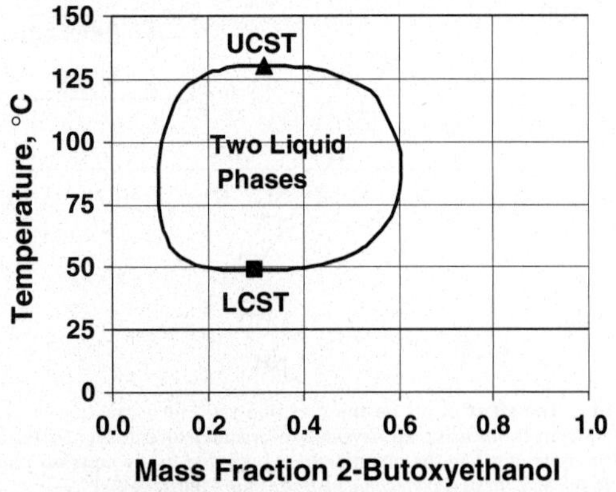

FIG. 15-15 Temperature-composition diagram for water + 2-butoxyethanol (ethylene glycol *n*-butyl ether). [*Reprinted from Christensen, Donate, Frank, LaTulip, and Wilson,* J. Chem. Eng. Data ***50****(3), pp. 869–877 (2005), with permission. Copyright 2005 American Chemical Society.*]

binary pairs have limited miscibility (a type II system). The water + acetic acid + methyl isobutyl ketone (MIBK) system shown in Fig. 15-17 is a type I system where only one of the binary pairs, water + MIBK, exhibits partial miscibility. The heptane + toluene + sulfolane system is another example of a type I system. In this case, only the heptane + sulfolane binary is partially miscible (Fig. 15-18). For a type II system, the solute has limited solubility in one of the liquids. An example of a type II system is MIBK + phenol + water (Fig. 15-19), where MIBK + water and phenol + water are only partially miscible. Some systems form more complicated phase diagrams. For example, the system water + dodecane + 2-butoxyethanol can form three liquid phases in equilibrium at 25°C [B.-J. Lin and L.-J. Chen, *J. Chem. Eng. Data* **47**(4): 992–996 (2002)]. Complex systems such as this rarely are encountered in extraction applications; however, S. Shen, Z. Chang, and H. Liu [*Sep. Purif. Technol.* **49**(3): 217–222 (2006)] describe a single-stage, three-liquid-phase extraction process for transferring phenol and *p*-nitrophenol from wastewater in separate phases. In this process, the three-phase system consists of ethylene oxide–propylene oxide copolymer + ammonium sulfate + water + an oxygenated organic solvent such as butyl acetate or 2-octanol.

For ternary systems, a three-dimensional plot is required to represent the effects of both composition and temperature on the phase behavior. Normally, ternary phase data are plotted on isothermal, two-dimensional triangular diagrams. These can be right-triangle plots, as in Fig. 15-17, or equilateral-triangle plots, as in Figs. 15-18 and 15-19. In Fig. 15-18, the line delineating the region where two liquid phases form is called the binodal locus. The lines connecting equilibrium compositions for each phase are called tie lines, as illustrated by lines *ab* and *cd*. The tie lines converge on the plait point, the point on the bimodal locus where both liquid phases attain the same composition and the tie line length goes to zero. To calculate the relative amounts of the liquid phases, the lever rule is used. For the total feed composition *z*, the fraction of phase 1 with the composition *e* is equal to the ratio of the lengths of the line segments given by *fz*/*ez* in Fig. 15-18. Data often are plotted on a mass fraction basis when differences in the molecular weights of the components are large. Plotting the phase diagram on a mole basis tends to compress the data into a small region, and details are hidden by the scale. This often is the case for systems involving water, for example.

An extraction application normally involves more than three components, including the key solute, the feed solvent, and extraction solvent (or solvent blend), plus impurity solutes. Usually, the minor impurity components do not have a major impact on the phase equilibrium. Phase equilibrium data for multicomponent systems may be represented by using an appropriate activity coefficient correlation. (See the subsection Data Correlation Equations.) However, for many dilute and moderately concentrated feeds, process design calculations are carried out as if the system were a ternary system comprised only of a single solute plus the feed solvent and extraction solvent (a pseudoternary). Partition ratios are determined for major and minor solutes by using a representative feed, and solute transfer calculations are carried out using solute *K* values as if they were completely independent of one another. This approach often is satisfactory, but its validity should be checked with a few key experiments. For industrial mixtures containing numerous impurities, a mass fraction or mass ratio basis often is used to avoid difficulties accounting for impurities of unknown structure and molecular weight.

LIQUID-LIQUID EQUILIBRIUM EXPERIMENTAL METHODS

General References: Chap. 8, "Liquid-Liquid Equilibrium," in *Measurement of the Thermodynamic Properties of Multiple Phases*, Experimental Thermodynamics, vol. VII, ed. R. D. Weir and T. W. de Loos, Elsevier, Amsterdam, 2005; Chap. 3, "Liquid-Liquid Equilibrium Measurements," in J. D. Raal and A. L. Mühlbauer, *Phase Equilibria: Measurements and Computation*, Taylor & Francis, Abingdon, UK, 1998; Newsham, D. M. T., chap. 1 in *Science and Practice of Liquid-Liquid Extraction*, vol. 1, ed. J. D. Thornton, Oxford University Press, Oxford, UK, 1992; and Novak, J. P., J. Matous, and J. Pick, *Liquid-Liquid Equilibria, Studies in Modern Thermodynamics Series*, vol. 7, Elsevier, Amsterdam, 1987, pp. 266–282.

Three general types of experimental methods commonly are used to generate liquid-liquid equilibrium data: (1) titration with visual observation of liquid clarity or turbidity (cloud point detection); (2) visual observation of clarity or turbidity for known compositions as a function of temperature; and (3) direct analysis of equilibrated liquids typically using GC or LC methods. In the titration method, one compound is slowly titrated into a known mass of

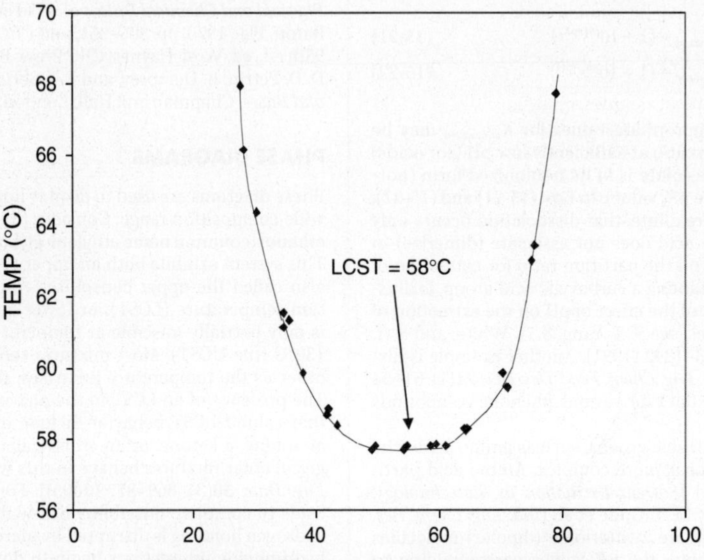

FIG. 15-16 Temperature-composition diagram for ethylene glycol + triethylamine. [*Data taken from Sørensen and Arlt,* Liquid-Liquid Equilibrium Data Collection, *DECHEMA, Binary Systems, vol. V,* pt. 1, 1979.]

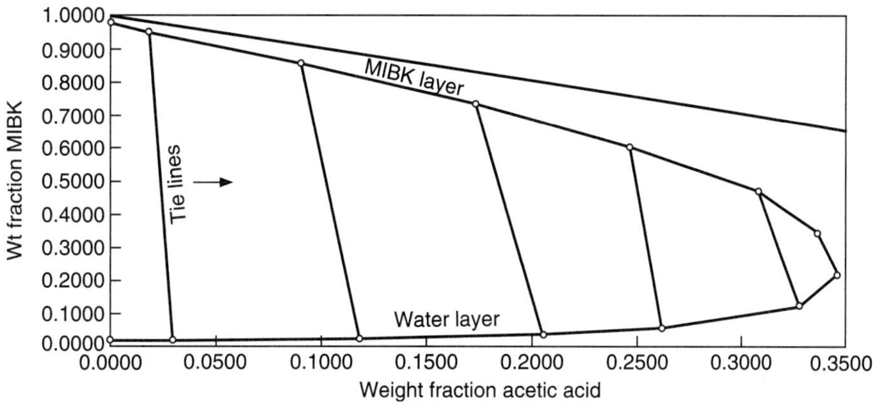

FIG. 15-17 Water + acetic acid + methyl isobutyl ketone at 25°C, a type I system.

the second compound during mixing. The titration is terminated when the mixture becomes cloudy, indicating that a second liquid phase has formed. A tie line may be determined by titrating the second compound into the first at the same temperature. This method is reasonably accurate for binary systems composed of pure materials. Because the method is visual, a trace impurity in the "titrant" that is less soluble in the second compound may cause cloudiness at a lower concentration than if pure materials were used. This method has poor precision for sparingly soluble systems.

In the second method, several mixtures of known composition are formulated and placed in glass vials or ampoules. These are placed in a bath or oven and heated or cooled until two phases become one, or vice versa. In this way, the phase boundaries of a binary system may be determined. Again, impurities in the starting materials may affect the results, and this method does not work well for sparingly soluble systems or for systems that develop significant pressure.

To obtain tie-line data for systems that involve three or more significant components, or for systems that cannot be handled in open containers, both phases must be sampled and analyzed. This generally requires the greatest effort but gives the most accurate results and can be used over the widest range of solubilities, temperatures, and pressures. This method also may be used on multicomponent systems, which are more likely to be encountered in an industrial process. For this method, an appropriate glass vessel or autoclave is selected, based on the temperature, pressure, and compounds in the mixture. It is best to either place the vessel in an oven or submerge it in a bath to ensure there are no cold or hot spots. The mixture is introduced, thermostatted, and thoroughly mixed, and the phases are allowed to separate fully. Samples are then carefully withdrawn through lines that have the minimum dead volume feasible. The sampling should be done isothermally; otherwise the collected sample may not be exactly the same as what was in the equilibrated vessel. Adding a carefully chosen, nonreactive diluent to the sample container will prevent phase splitting, and this can be an important step to ensure accuracy in the subsequent sample workup and analysis. Take sufficient purges and at least three samples from each phase. Use the appropriate analytical method and analyze a calibration standard along with the samples. Try to minimize the time between sampling and analysis.

Rydberg and others describe automated equipment for generating tie-line data, including an apparatus called AKUFVE offered by MEAB [J. Rydberg et al., chap. 4 in *Solvent Extraction Principles and Practice*, 2d ed., ed. J. Rydberg et al., Marcel Dekker, New York, 2004, pp. 193–197]. The AKUFVE apparatus employs a stirred cell, a centrifuge for phase separation, and online instrumentation for rapid generation of data. As an alternative, B. Kuzmanović et al. [*J. Chem. Eng. Data* **48:** 1237–1244 (2003)] describe a fully automated workstation for rapid measurement of liquid-liquid equilibrium using robotics for automated sampling. Reviews of phase equilibria measurement at high pressure are available elsewhere [J. M. S. Fonseca, R. Dohrn, and S. Peper, *Fluid Phase Equilibr.* **300:** 1–69 (2011); and R. Dohrn, J. M. S. Fonseca, and S. Peper, *Annu. Rev. Chem. Biomol. Eng.* **3:** 343–367 (2012)].

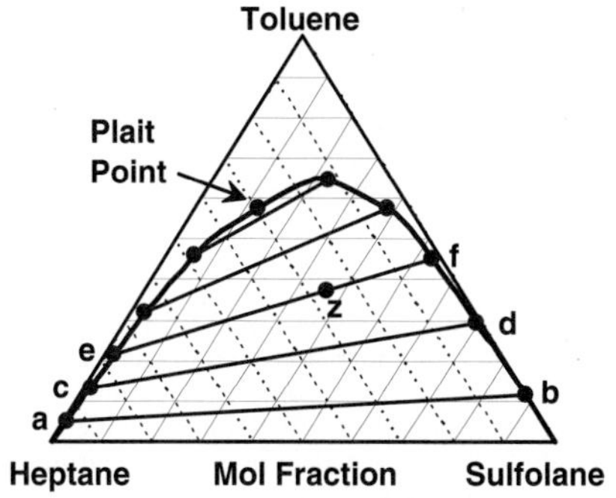

FIG. 15-18 Heptane + toluene + sulfolane at 25°C, a type I system. [*Data taken from De Fre and Verhoeye,* J. Appl. Chem. Biotechnol. ***26:** 1-19 (1976).*]

DATA CORRELATION EQUATIONS

Tie-Line Correlations Useful correlations of ternary data may be obtained by using the methods of D. B. Hand [*J. Phys. Chem.* **34**(9): 1961–2000 (1930)] and D. F. Othmer and P. E. Tobias [*Ind. Eng. Chem.* **34**(6): 693–696 (1942)]. Hand observed that plotting the equilibrium line in terms of mass ratio units on a log-log scale often gave a straight line. This relationship commonly is expressed as

$$\log \frac{X_{23}}{X_{33}} = a + b \log \frac{X_{21}}{X_{11}} \tag{15-23}$$

where X_{ij} represents the mass fraction of component i dissolved in the phase richest in component j, and a and b are empirical constants. Subscript 2 denotes the solute, while subscripts 1 and 3 denote feed solvent and extraction solvent, respectively. An equivalent expression can be written by using the Bancroft coordinate notation introduced earlier: $Y' = cX'^b$, where $c = 10^a$. Othmer and Tobias proposed a similar correlation:

$$\log \frac{1 - X_{33}}{X_{33}} = d + e \log \frac{1 - X_{11}}{X_{11}} \tag{15-24}$$

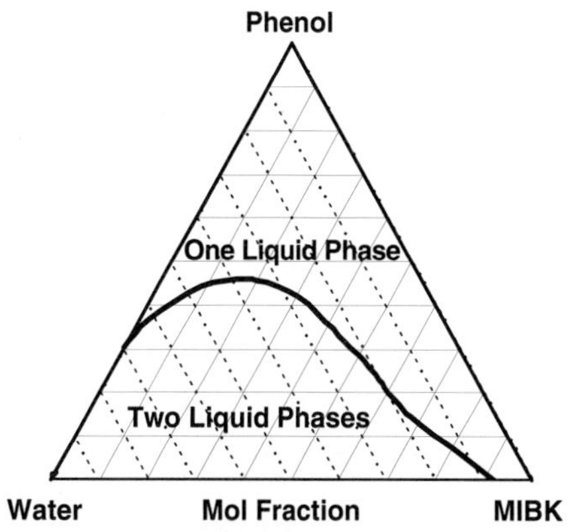

FIG. 15-19 Methyl isobutyl ketone + phenol + water at 30°C, a type II system. [*Data taken from Narashimhan, Reddy, and Chari,* J. Chem. Eng. Data ***7**, p. 457 (1962).*]

where d and e are constants. Equations (15-23) and (15-24) may be used to check the consistency of tie-line data, as discussed by A. M. Awwad et al. [*J. Chem. Eng. Data* **50**(3): 788–791 (2005)] and by I. Kirbaslar et al. [*Braz. J. Chem. Eng.* **17**(2): 191–197 (2000)].

A useful diagram is obtained by plotting the solute equilibrium line on log-log scales as X_{23}/X_{33} versus X_{21}/X_{11} [from Eq. (15-23)] along with a second plot consisting of X_{23}/X_{33} versus X_{23}/X_{13} and X_{21}/X_{31} versus X_{21}/X_{11}. This second plot is termed the limiting solubility curve. The plait point may easily be found from the intersection of the solute equilibrium line with this curve, as shown by R. E. Treybal, L. D. Weber, and J. F. Daley [*Ind. Eng. Chem.* **38**(8): 817–821 (1946)]. This type of diagram also is helpful for interpolation and limited extrapolation when equilibrium data are scarce. An example diagram is shown in Fig. 15-20 for the water + acetic acid + methyl isobutyl ketone (MIBK) system. For additional discussion of various correlation methods, see G. S. Laddha and T. E. Degaleesan, chap. 2 in *Transport Phenomena in Liquid Extraction*, McGraw-Hill, New York, 1978.

Thermodynamic Models The thermodynamic theories and equations used to model phase equilibria are reviewed in Sec. 4, Thermodynamics. These equations provide a framework for data that can help minimize the required number of experiments. An accurate liquid-liquid equilibrium (LLE) model is particularly useful for applications involving concentrated feeds where partition ratios and mutual solubility between phases are significant functions of solute concentration. Sometimes it is difficult to model LLE behavior across the entire composition range with a high degree of accuracy, depending upon the chemical system. In that case, it is best to focus on the composition range specific to the particular application at hand—to ensure the model accurately represents the data in that region of the phase diagram for accurate design calculations. Such a model can be a powerful tool for extractor design or when used with process simulation software to conceptualize, evaluate, and optimize process options. However, whether a complete LLE model is needed will depend on the application. For dilute applications where partition ratios do not vary much with composition, it may be satisfactory to characterize equilibrium in terms of partition ratios measured over the range of anticipated feed and raffinate compositions. Also, when partition ratios are always very large, on the order of 100 or larger, as can occur when washing salts from an organic phase into water, a continuous extractor is likely to operate far from equilibrium. In this case, a precise equilibrium model may not be needed because the extraction factor always is very large, and mass-transfer rates dominate performance. (See the subsection Rate-Based Calculations under Process Fundamentals and Basic Calculation Methods.)

LLE models for nonionic systems generally are developed by using either the NRTL or UNIQUAC correlation equations (see Sec. 4). Most commercial simulation software packages include these models and allow regression of data to determine model parameters. One should refer to the process simulator's operating manual for specific details. Not all simulation software will use exactly the same equation format and parameter definitions, so parameters reported in the literature may not be appropriate for direct input to the program but need to be converted to the appropriate form. In theory, activity coefficient data from binary or ternary vapor-liquid equilibria can be used for calculating liquid-liquid equilibria. While this may provide a reasonable starting point, in practice at least some of the binary parameters will need to be determined from liquid-liquid tie-line data to obtain an accurate model [D. S. Lafyatis et al., *Ind. Eng. Chem. Res.* **28**(5): 585–590 (1989)]. Detailed discussion of the application and use of NRTL and UNIQUAC is given by S. M. Walas [*Phase Equilibria in Chemical Engineering*, Butterworth-Heinemann, Oxford, UK, 1985]. The application of NRTL in the design of a liquid-liquid extraction process is discussed by D. L. Venter and I. Nieuwoudt [*Ind. Eng. Chem. Res.* **37**(10): 4099–4106 (1998)], by R. van Grieken et al. [*Ind. Eng. Chem. Res.* **44**(21): 8106–8112 (2005)], by B. Coto et al. [*Chem. Eng. Sci.* **61**: 8028–8039 (2006)]; and by Z. Li et al. [*J. Chem. Eng. Data* **59**(8): 2485–2489 (2014)]. The use of the NRTL model also is discussed in Example 15-5 under Single-Solvent Fractional Extraction with Extract Reflux in Calculation Procedures.

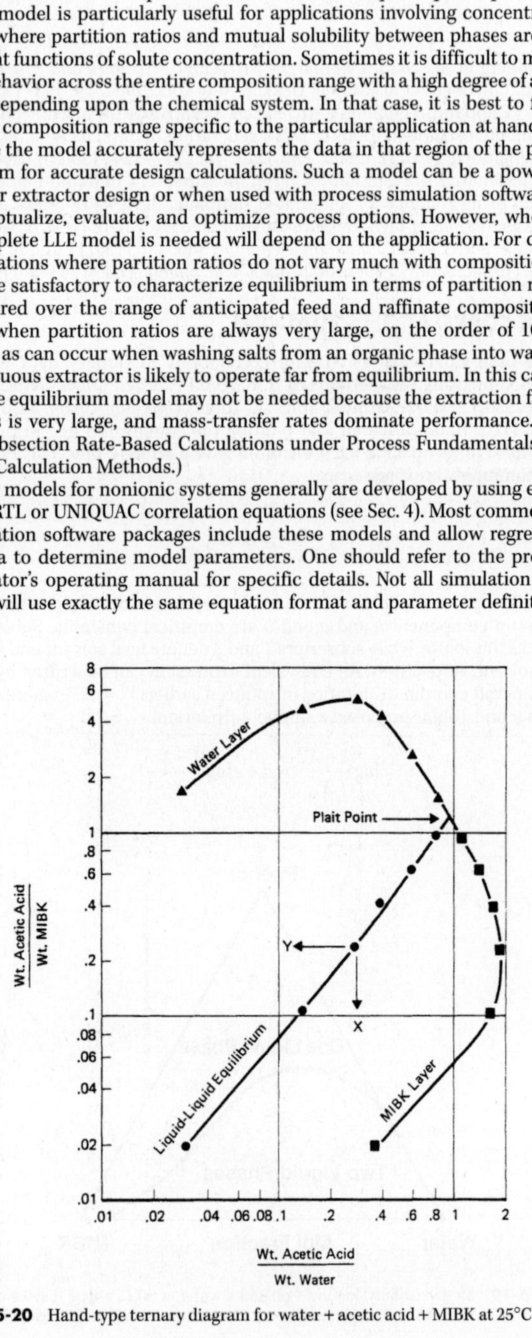

FIG. 15-20 Hand-type ternary diagram for water + acetic acid + MIBK at 25°C.

Although the NRTL or UNIQUAC equations generally are recommended for nonionic systems, a number of alternative approaches have been introduced. An example is the statistical associating fluid theory (SAFT) equation of state introduced by W. G. Chapman et al. [*Ind. Eng. Chem. Res.* **29**(8): 1709–1721 (1990)]. M.-L. Yu and Y.-P. Chen discuss the application of SAFT to correlate data for 41 binary and 8 ternary liquid-liquid systems [*Fluid Phase Equilibr.* **94**: 149–165 (1994)]. The SAFT equation often is used to correlate LLE data for polymer-solvent systems [P. K. Jog et al., *Ind. Eng. Chem. Res.* **41**(5): 887–891 (2002)]. Other methods are used to describe the behavior of ionic species (electrolytes), as discussed in Sec. 4.

Data Quality Normally it is not possible to evaluate LLE data for thermodynamic consistency [J. M. Sørensen and W. Arlt, *Liquid-Liquid Equilibrium Data Collection*, Binary Systems, vol. V, pt. 1, DECHEMA, 1979, p. 12]. The thermodynamic consistency test for VLE data involves calculating an independently measured variable from the others (usually the vapor composition from the temperature, pressure, and liquid composition) and comparing the measurement with the calculated value. Because LLE data are only very weakly affected by change in pressure, this method is not feasible for LLE. However, if the data were produced by equilibration and analysis of both phases, then at least the data can be checked to determine how well the material balance closes. This can be done by plotting the total feed composition used in the experiments along with the measured tie-line compositions on a ternary diagram. The feed composition should lie on the tie line. For very low solute concentrations, this plot may be unrevealing. Alternatively, a plot of Y_i''/Z_i'' versus X_i''/Z_i'' (where Y_i'' is the mass fraction of component i in the extract phase, X_i'' is the mass fraction of component i in the raffinate phase, and Z_i'' is the mass fraction of component i in the total feed) should give a smooth line that passes through the point (1, 1). The tie-line data also may be checked for consistency by plotting the data in the form of a Hand plot or Othmer-Tobias plot, as described in Tie-Line Correlations, and looking for outliers. Another approach is to plot the partition ratio as a function of solute concentration and look for data points that deviate significantly from otherwise smooth trends.

TABLE OF SELECTED PARTITION RATIO DATA

Table 15-1 summarizes typical partition ratio data for selected systems.

PHASE EQUILIBRIUM DATA SOURCES

A comprehensive collection of phase equilibrium data (including vapor-liquid, liquid-liquid, and solid-liquid data), known as the Dortmund Data Bank, includes LLE measurements as well as NRTL and UNIQUAC fitted parameters. The data bank also includes a compilation of infinite-dilution activity coefficients. The LLE collection is available as a series of books [J. M. Sørensen and W. Arlt, *Chemistry Data Series: Liquid-Liquid Equilibrium Data Collection, Binary Systems*, vol. V, pts. 1–4, DECHEMA, Frankfurt, 1979–1980] and as an online database that includes retrieval and modeling software. Another online database is Infotherm from Wiley. Other sources of thermodynamic data include the online IUPAC-NIST Solubility Data Series distributed by the Office of Data and Informatics of the National Institute of Standards and Technology (NIST) and compilations prepared by the Thermodynamics Research Center (TRC) in Boulder, Colo., a part of the Physical and Chemical Properties Division of NIST. An older but still useful data collection is that of H. Stephen and T. Stephen [*Solubilities of Inorganic and Organic Compounds*, vol. 1, pts. 1 and 2, Pergamon, Oxford, UK, 1960]. Also, a useful database of activity coefficients is included in the supporting information submitted with the article by M. J. Lazzaroni et al. [*Ind. Eng. Chem. Res.* **44**(11): 4075–4083 (2005)] and available from the publisher. A listing of the original sources is included. Additional sources of data are discussed by A. Skrzecz [*Pure Appl. Chem.* (IUPAC), **69**(5): 943–950 (1997)].

TABLE 15-1 Selected Partition Ratio Data

Partition ratios are listed in units of weight percent solute in the extract divided by weight percent solute in the raffinate, generally for the lowest solute concentrations given in the cited reference. The partition ratio tends to be greatest at low solute concentrations. Consult the original references for more information about a specific system.

Solute	Feed solvent	Extraction solvent	Temp. (°C)	*K* (wt % basis)	Reference
Ethanol	Cyclohexane	Ethanolamine	25	2.79	1
Acetone	Ethylene glycol	Amyl acetate	31	1.84	2
Acetone	Ethylene glycol	Ethyl acetate	31	1.85	2
Acetone	Ethylene glycol	Butyl acetate	31	1.94	2
Trilinolein	Furfural	Heptane	30	47.5	3
o-Xylene	Heptane	Tetraethylene glycol	20	0.15	4
o-Xylene	Heptane	Tetraethylene glycol	30	0.15	4
o-Xylene	Heptane	Tetraethylene glycol	40	0.16	4
Toluene	Heptane	Sulfolane	25	0.34	5
Toluene	Heptane	Sulfolane	50	0.36	5
Toluene	Heptane	Sulfolane	75	0.31	5
Toluene	Heptane	Sulfolane	100	0.33	5
Toluene	Hexane	Sulfolane	25	0.34	6
Xylene	Hexane	Sulfolane	25	0.30	6
Toluene	*n*-Hexane	Sulfolane	25	0.34	6
Xylene	*n*-Hexane	Sulfolane	25	0.30	6
Toluene	*n*-Octane	Sulfolane	25	0.35	6
Xylene	*n*-Octane	Sulfolane	25	0.25	6
Toluene	Octane	Sulfolane	25	0.35	6
Xylene	Octane	Sulfolane	25	0.25	6
1,2-Dimethoxyethane	Water	Dodecane	25	0.46	7
1,4-Dioxane	Water	Ethyl acetate	30	1.29	8
1-Butanol	Water	Benzonitrile	25	3.01	9
1-Butanol	Water	Ethyl acetate	40	5.48	10
1-Butanol	Water	Methyl *t*-butyl ether	25	7.95	11
1-Heptene	Water	1-Propanol	25	3.95	12
1-Octanol	Water	Methyl *t*-butyl ether	25	10.9	13
1-Propanol	Water	1-Heptene	25	1.36	12
1-Propanol	Water	Butyraldehyde	25	4.14	14
1-Propanol	Water	Cyclohexane	25	0.34	15
1-Propanol	Water	Di-isobutyl ketone	25	0.93	14
1-Propanol	Water	Methyl *tert*-butyl ether	25	3.79	11
2,3-Butanediol	Water	2,4-Dimethylphenol	40	1.89	16
2,3-Butanediol	Water	2-Butoxyethanol	70	1.79	17
2,3-Dichloropropene	Water	Epichlorohydrin	20	181	18
2,3-Dichloropropene	Water	Epichlorohydrin	77	69.5	18
2-Butoxyethanol	Water	Decane	22	0.45	19
2-Methoxyethanol	Water	Cyclohexanone	70	0.54	20
2-Methyl-1-propanol	Water	Benzene	25	1.18	21
2-Methyl-1-propanol	Water	Toluene	25	0.88	21
2-Propanol	Water	1-Methylcyclohexanol	20	3.66	22
2-Propanol	Water	2,2,4-Trimethylpentane	20	0.045	23
2-Propanol	Water	Carbon tetrachloride	20	1.41	24
2-Propanol	Water	Dichloromethane	20	3.56	22
2-Propanol	Water	Di-isopropyl ether	25	0.41	25
2-Propanol	Water	Di-isopropyl ether	25	0.98	26
3-Cyanopyridine	Water	Benzene	30	1.55	27
Acetaldehyde	Water	Furfural	16	0.97	28
Acetaldehyde	Water	1-Pentanol	18	1.43	28
Acetic acid	Water	1-Butanol	27	1.61	29
Acetic acid	Water	1-Hexene	25	0.0073	30
Acetic acid	Water	1-Octanol	20	0.56	31
Acetic acid	Water	20 vol % Trioctylamine + 20 vol % 1-Decanol + 60 vol % dodecane	20	0.61	32
Acetic acid	Water	2-Butanone	25	1.20	33
Acetic acid	Water	2-Ethyl-1-hexanol	20	0.58	34
Acetic acid	Water	2-Pentanol	25	1.35	35
Acetic acid	Water	2-Pentanone	25	1.00	30
Acetic acid	Water	4-Heptanone	25	0.30	30
Acetic acid	Water	70 vol % Tributylphosphate + 30 vol % dodecane	20	0.31	36
Acetic acid	Water	Cyclohexanol	27	1.33	29
Acetic acid	Water	Diethyl phthalate	20	0.22	37
Acetic acid	Water	Di-isopropyl carbinol	25	0.80	38
Acetic acid	Water	Dimethyl phthalate	20	0.34	37
Acetic acid	Water	Di-*n*-butyl ketone	25	0.38	39
Acetic acid	Water	Ethyl acetate	30	0.91	40
Acetic acid	Water	Isopropyl ether	20	0.25	41
Acetic acid	Water	Methyl cyclohexanone	25	0.93	38
Acetic acid	Water	Methylisobutyl ketone	25	0.66	42
Acetic acid	Water	Methylisobutyl ketone	25	0.76	38
Acetic acid	Water	Toluene	25	0.06	43
Acetone	Water	1-Octanol	25	0.81	44
Acetone	Water	1-Pentanol	25	4.11	44

TABLE 15-1 Selected Partition Ratio Data *(Continued)*

Partition ratios are listed in units of weight percent solute in the extract divided by weight percent solute in the raffinate, generally for the lowest solute concentrations given in the cited reference. The partition ratio tends to be greatest at low solute concentrations. Consult the original references for more information about a specific system.

Solute	Feed solvent	Extraction solvent	Temp. (°C)	K (wt % basis)	Reference
Acetone	Water	1-Pentanol	30	1.14	44
Acetone	Water	2-Octanol	30	0.66	44
Acetone	Water	Chloroform	25	1.83	45
Acetone	Water	Chloroform	25	1.72	46
Acetone	Water	Dibutyl ether	25	1.94	38
Acetone	Water	Diethyl ether	30	1.00	47
Acetone	Water	Ethyl acetate	30	1.50	48
Acetone	Water	Ethyl butyrate	30	1.28	48
Acetone	Water	Methyl acetate	30	1.15	48
Acetone	Water	Methylisobutyl ketone	25	1.91	38
Acetone	Water	Hexane	25	0.34	49
Acetone	Water	Toluene	25	0.84	38
Acrylic acid	Water	89.6 wt % Kerosene/10.4 wt % trialkylphosphine oxide (C7–C9)	25	6.50	50
Aniline	Water	Methylcyclohexane	25	2.05	51
Aniline	Water	Methylcyclohexane	50	3.41	51
Aniline	Water	Heptane	25	1.43	51
Aniline	Water	Heptane	50	2.20	51
Aniline	Water	Toluene	25	12.9	52
Benzoic acid	Water	87.4 wt % Kerosene/ 12.6 wt % tributylphosphate	25	36.0	53
Benzoic acid	Water	89.6 wt % Kerosene/10.4 wt % trialkylphosphine oxide (C7–C9)	25	1.30	50
Butyric acid	Water	20 vol % Trioctylamine + 20 vol % 1-decanol + 60 vol % dodecane	20	6.16	36
Butyric acid	Water	70 vol % Tributylphosphate + 30 vol % dodecane	20	2.51	36
Butyric acid	Water	Methyl butyrate	30	6.75	54
Citric acid	Water	25 wt % Tri-isooctylamine + 75 wt % Chloroform	25	14.1	55
Citric acid	Water	26 wt % Tri-isooctylamine + 75 wt % 1-Octanol	25	41.5	55
Epichlorohydrin	Water	2,3-Dichloropropene	20	11.4	56
Epichlorohydrin	Water	2,3-Dichloropropene	77	13.4	56
Ethanol	Water	1-Octanol	25	0.66	57
Ethanol	Water	1-Octene	25	0.036	58
Ethanol	Water	2,2,4-Trimethylpentane	5	0.027	59
Ethanol	Water	2,2,4-Trimethylpentane	40	0.041	59
Ethanol	Water	3-Heptanol	25	0.78	60
Ethanol	Water	1-Butanol	20	3.00	61
Ethanol	Water	Di-*n*-propyl ketone	25	0.59	38
Ethanol	Water	1-Hexanol	28	1.00	62
Ethanol	Water	2-Octanol	28	0.83	62
Ethyl acetate	Water	1-Butanol	40	11.1	10
Ethylene glycol	Water	Furfural	25	0.32	64
Formic acid	Water	20 vol % Trioctylamine + 20 vol % 1-decanol + 60 vol % dodecane	20	1.77	36
Formic acid	Water	70 vol % Tributylphosphate + 30 vol % dodecane	20	0.37	36
Formic acid	Water	Methyisobutyl carbinol	30	1.22	65
Furfural	Water	Toluene	25	5.64	66
Glycolic acid	Water	89.6 wt % Kerosene/10.4 wt % trialkylphosphine oxide (C7–C9)	25	0.29	67
Glyoxylic acid	Water	89.6 wt % Kerosene/10.4 wt % trialkylphosphine oxide (C7–C9)	25	0.067	67
Lactic acid	Water	20 vol % Trioctylamine + 20 vol % 1-decanol + 60 vol % dodecane	20	0.65	36
Lactic acid	Water	25 wt % Tri-isooctylamine + 75 wt % chloroform	25	19.2	55
Lactic acid	Water	26 wt % Tri-isooctylamine + 75 wt % 1-octanol	25	25.9	55
Lactic acid	Water	70 vol % Tributylphosphate + 30 vol % dodecane	20	0.14	36
Lactic acid	Water	*iso*-Amyl alcohol	25	0.35	68
Malic acid	Water	25 wt % Tri-isooctylamine + 75 wt % chloroform	25	30.7	55
Malic acid	Water	25 wt % Tri-isooctylamine + 75 wt % 1-octanol	25	59.0	55
Methanol	Water	1-Octanol	25	0.28	57
Methanol	Water	Ethyl acetate	0	0.059	69
Methanol	Water	Ethyl acetate	20	0.24	69
Methanol	Water	1-Butanol	0	0.60	70
Methanol	Water	1-Hexanol	28	0.57	71
Methanol	Water	*p*-Cresol	35	0.31	72

TABLE 15-1 Selected Partition Ratio Data *(Continued)*

Partition ratios are listed in units of weight percent solute in the extract divided by weight percent solute in the raffinate, generally for the lowest solute concentrations given in the cited reference. The partition ratio tends to be greatest at low solute concentrations. Consult the original references for more information about a specific system.

Solute	Feed solvent	Extraction solvent	Temp. (°C)	K (wt % basis)	Reference
Methanol	Water	Phenol	25	1.33	72
Methyl *t*-butyl ether	Water	1-Octanol	25	2.61	13
Methyl *t*-amyl ether	Water	2,2,4-Trimethylpentane	25	131	73
Methylethyl ketone	Water	1,1,2-Trichloroethane	25	3.44	74
Methylethyl ketone	Water	Hexane	25	1.78	75
1-Propanol	Water	Ethyl acetate	20	1.54	69
1-Propanol	Water	Heptane	38	0.54	76
p-Cresol	Water	Methylnaphthalene	35	9.89	72
Phenol	Water	Methyl isobutyl ketone	30	39.8	77
Phenol	Water	Methylnaphthalene	25	7.06	78
Phosphoric acid	Water	4-Methyl-2-pentanone	25	0.0012	79
Propionic acid	Water	20 vol % Trioctylamine + 20 vol % 1-decanol + 60 vol % dodecane	20	2.13	36
Propionic acid	Water	70 vol % Tributylphosphate + 30 vol % dodecane	20	1.02	36
Propionic acid	Water	Ethyl acetate	30	2.77	80
Propionic acid	Water	Toluene	31	0.52	81
Pyridine	Water	Chlorobenzene	25	2.10	82
Pyridine	Water	Toluene	25	1.90	83
Pyridine	Water	Xylene	25	1.26	83
t-Butanol	Water	Ethyl acetate	20	1.74	69
Tetrahydrofuran	Water	1-Octanol	20	3.31	84

References:

1. Harris, R.A., et al., *J. Chem. Eng. Data* **47**(4): 781–787 (2002).
2. Rao, M. R., and C. V. Rao, *J. Sci. Ind. Res.* **14B:** 204 (1955).
3. Chueh, P. L., and S. W. Briggs, *J. Chem. Eng. Data* **9**(2): 207–212 (1964).
4. Darwish, N. A., et al., *J. Chem. Eng. Data* **48**(6): 1614–1619 (2003).
5. De Fre, Thesis, University of Gent, 1976.
6. Dortmund Databank, private communication (2006).
7. Dortmund Databank, private communication (2006).
8. Komatsu, H., and H. Yamamoto, *Kagaku Kogaku Ronbunshu* **22**(2): 378–384 (1996).
9. Grande, M. C., et al., *J. Chem. Eng. Data* **41**(4): 926–928 (1996).
10. Dortmund Databank, private communication (2006).
11. Letcher, T. M., S. Ravindran, and S. E. Radloff, *Fluid Phase Equilibr.* **69:** 251–260 (1991).
12. Letcher, T. M., et al., *J. Chem. Eng. Data* **39**(2): 320–323 (1994).
13. Arce, A., et al., *J. Chem. Thermodyn* **28:** 3–6 (1996).
14. Letcher, T. M., et al., *J. Chem. Eng. Data* **41**(4): 707–712 (1996).
15. Plačkov and Štern, *Fluid Phase Equilibr.* **71:** 189–209 (1992).
16. Escudero, I., J. L. Cabezas, and J. Coca, *J. Chem. Eng. Data* **39**(4): 834–839 (1994).
17. Escudero, I., J. L. Cabezas, and J. Coca, *J. Chem. Eng. Data* **41**(6): 1383–1387 (1996).
18. Dortmund Databank, private communication (2006).
19. Dortmund Databank, private communication (2006).
20. Hauschild, T., and H. Knapp, *J. Solution Chem.*, **23**(3): 363–377 (1994).
21. Stephenson, R. M., *J. Chem. Eng. Data* **37**(1): 80–95 (1992).
22. Sayar, A. A., *J. Chem. Eng. Data* **36**(1): 61–65 (1991).
23. Arda, N., and A. A. Sayar, *Fluid Phase Equil,* **73:** 129–138 (1992).
24. Denzler, C. G., *J. Phys. Chem.* **49:** 358 (1945).
25. Frere, F. J., *Ind. Eng. Chem.* **41**(10): 2365–2367 (1949).
26. Letcher, T. M., S. Ravindran, and S. E. Radloff, *Fluid Phase Equilibr.* **71:** 177–188 (1992).
27. Dortmund Databank, private communication (2006).
28. Othmer, D. F., and P. E. Tobias, *Ind. Eng. Chem.* **34**(6): 690–692 (1942).
29. Skrzec, A. E., and N. F. Murphy, *Ind. Eng. Chem.* **46**(10): 2245–2247 (1954).
30. Nakahara, Masamoto, and Arai, *Kagaku Kogaku Ronbunshu* **19**(4): 663–668 (1993).
31. Ratkovics, F., et al., *J. Chem. Thermodyn.* **23:** 859–865 (1991).
32. Morales, A. F., et al., *J. Chem. Eng. Data* **48**(4): 874–886 (2003).
33. Gomis, V., et al., *Fluid Phase Equilibr.* **106:** 203–211 (1995).
34. Ratkovics, F., et al., *J. Chem. Thermodyn.* **23:** 859–865 (1991).
35. Al-Muhtaseb, S. A., and M. A. Fahim, *Fluid Phase Equilibr.* **123:** 189–203 (1996).
36. Morales, A. F., et al., *J. Chem. Eng. Data***48**(4): 874–886 (2003).
37. Dramur, U., and B. Tatli, *J. Chem. Eng. Data* **38**(1): 23–25 (1993).
38. Othmer, D. F., R. E. White, and E. Treuger, *Ind. Eng. Chem.* **33**(10): 1240–1248 (1941).
39. Othmer, D. F., R. E. White, and E. Treuger, *Ind. Eng. Chem.* **33**(10): 1240–1248 (1941).
40. Eaglesfield, P., B. K. Kelly, and J. F. Short, *Ind. Chemist* **29:** 147, 243 (1953).
41. Elgin, J., and F. M. Browning, *Trans. Am. Inst. Chem. Engrs.* **31:** 639 (1935).
42. Othmer, D. F., R. E. White, and E. Treuger, *Ind. Eng. Chem.* **33**(10): 1240–1248 (1941).
43. Woodman, *J. Phys. Chem.* **30**(9): 1283–1286 (1926).
44. Tiryaki, A., G. Gürüz, and H. Orbey, *Fluid Phase Equilibr.* **94:** 267–280 (1994).
45. Hand, D. B., *J. Phys. Chem.* **34:** 1961–2000 (1930).
46. Bancroft, W. D., and S. S. Hubard, *J. Am. Chem. Soc.* **64:** 347 (1942).
47. Krishnamurty, V. V., P. S. Murti, and C. V. Rao, *J. Sci. Ind. Res.* **12B:** 583 (1953).
48. Venkataratnam, A., R. J. Rao, and C. V. Rao, *Chem. Eng. Sci.* **7:** 102 (1957).
49. Treybal, R. E., and O. J. Vondrak, *Ind. Eng. Chem.* **41**(8): 1761–1763 (1949).
50. Li, Y., et al., *J. Chem. Eng. Data* **48:** 621–624 (2003).
51. Griswold, G., J-N. Chew, and M. E. Klecka, *Ind. Eng. Chem.* **42**(6): 1246–1251 (1950).
52. Smith, J.C., and R. E. Drexel, *Ind. Eng. Chem.* **37**(6): 601–602 (1945).
53. Mei, F., W. Qin, and Y. Dai, *J. Chem. Eng. Data* **47**(4): 941–943 (2002).
54. Murty, N. S., V. Subrahmanyam, and P. D. Murty, *J. Chem. Eng. Data* **11**(3): 335–338 (1996).
55. Dortmund Databank, private communication (2006).
56. Zhang and Liu, *J. Chem. Ind. Eng. (China)* **46**(3): 365–369 (1995).
57. Arce, A., et al., *J. Chem. Eng. Data* **39**(2): 378–380 (1994).
58. Purwanto et al., *J. Chem. Eng. Data* **41**(6): 1414–1417 (1996).
59. Wagner, G., and S. I. Sandler, *J. Chem. Eng. Data* **40**(5): 1119–1123 (1995).
60. Oualline, C. M., and M. Van Winkle, *Ind. Eng. Chem.* **44**(7): 1668–1670 (1952).
61. Drouillon, F. J., *Chim Phys. Phys.-Chim. Biol.* **22:** 149 (1925).
62. Krishnamurty, V. V., and C. V. Rao, *Trans. Indian Inst. Chem. Engrs.* **6:** 153 (1954).
63. De Andrade and D'Avila, private communication to DDB, pp. 1–7 (1991).
64. Conway, J. B., and J. J. Norton, *Ind. Eng. Chem.* **43**(6): 1433–1435 (1951).
65. Rao, M. R., M. Ramamurty, and C. V. Rao, *Chem. Eng. Sci.* **8**(3–4): 265–270 (1958).
66. Knight, *Trans. Am. Inst. Chem. Engrs.* **39:** 439 (1943).
67. Li, Y., et at., *J. Chem. Eng. Data* **48**(3): 621–624 (2003).
68. Weiser, R. B., and C. J. Geankoplis, *Ind. Eng. Chem.* **47**(4): 858–863 (1955).
69. Beech, D. G., and S. Glasstone, *J. Chem. Soc.*, p. 67 (1938).
70. Mueller, A. J., L. I. Pugsley, and J. B. Ferguson, *J. Phys. Chem.* **35**(5): 1314–1327 (1931).
71. Krishnamurty, V. V., and C. V. Rao, *J. Sci. Ind. Res.* **14B:** 614 (1955).
72. Prutton, C. F., T. J. Walsh, and A. M. Desai, *Ind. Eng. Chem.* **42**(6): 1210–1217 (1950).
73. Peschke, N., and S. I. Sandler, *J. Chem. Eng. Data* **40**(1): 315–320 (1995).
74. Newman, M., C. B. Hayworth, and R. E. Treybal, *Ind. Eng. Chem.* **41**(9): 2039–2043 (1949).
75. Treybal, R. E., and O. J. Vondrak, *Ind. Eng. Chem.* **41**(8): 1761–1763 (1949).
76. McCants, F. J., J. H. Jones, and W. H. Hopson, *Ind. Eng. Chem.* **45**(2): 454–456 (1953).
77. Narasimhan, K. S., C. C. Reddy, and K. S. Chari, *J. Chem. Eng. Data* **7**(4): 457–460 (1962).
78. Prutton, C. F., T. J. Walsh, and A. M. Desai, *Ind. Eng. Chem.* **42**(6): 1210–1217 (1950).
79. Feki, M., et al., *Can. J. Chem. Eng.* **72**(5): 939–944 (1994).
80. Rao, M. R., and C. V. Rao, *J. Sci. Ind. Res.* **14B:** 444 (1955).
81. Rao, M. R., and C. V. Rao, *J. Appl. Chem.* **6:** 270 (1956).
82. Peake, J. S., and K. E. Thompson, *Ind. Eng. Chem.* **44**(10): 2439–2441 (1952).
83. Vrieus, G. N., and E. C. Medcalf, *Ind. Eng. Chem.* **45**(5): 1098–1104 (1953).
84. Şenol, A., G. Alptekin, and A. A. Sayar, *J. Chem. Thermodyn.* **27:** 525–529 (1995).

RECOMMENDED MODEL SYSTEMS

To facilitate the study and comparison of various types of extraction equipment, H.-J. Bart et al. [chap. 3 in J. C. Godfrey and M. J. Slater, *Liquid-Liquid Extraction Equipment*, Wiley, New York, 1994] recommend several model systems. These include (1) water + acetone + toluene (high interfacial tension); (2) water + acetone + butyl acetate (moderate interfacial tension); and (3) water + succinic acid + *n*-butanol (low interfacial tension). All have solute partition ratios near $K = 1.0$. T. Mišek, R. Berger, and J. Schröter [*Standard Test Systems for Liquid Extraction*, The Instn. of Chemical Engineers, London, 1985] summarize phase equilibrium, viscosities, densities, diffusion coefficients, and interfacial tensions for these systems. Note that methyl isobutyl ketone + acetic acid + water was replaced with the water + acetone + butyl acetate system because of concerns over acetic acid dimerization and Marangoni instabilities. (See the subsection Liquid-Liquid Dispersion Fundamentals.) For test systems with a partition ratio near $K = 10$, Bart et al. recommend (1) water + methyl isopropyl ketone + toluene (high interfacial tension) and (2) water + methyl isopropyl ketone + butyl acetate (medium interfacial tension) and give references to data sources. Bart et al. also recommend a number of systems involving reactive extractants.

SOLVENT SCREENING METHODS

A variety of methods may be used to estimate solvent properties as an aid to identifying useful solvents for a new application. Many of these methods focus on thermodynamic properties; a favorable partition ratio and low mutual solubility often are necessary for an economical extraction process, so ranking candidates according to thermodynamic properties provides a useful initial screen of the more promising candidates. Keep in mind, however, that other factors also must be taken into account when selecting a solvent, as discussed in Desirable Solvent Properties under Introduction and Overview. When using the following methods, also note that the level of uncertainty may be fairly high. The uncertainty depends upon how closely the chemical system of interest resembles the systems used to develop the method. See Sec. 4 for more discussion of the thermodynamic basis of these methods.

USE OF ACTIVITY COEFFICIENTS AND RELATED DATA

Compilations of infinite-dilution activity coefficients, when available for the solute of interest, may be used to rank candidate solvents. Partition ratios at finite concentrations can be estimated from these data by extrapolation from infinite dilution using a suitable correlation equation such as NRTL. Examples of these kinds of calculations are given by S. M. Walas [*Phase Equilibria in Chemical Engineering*, Butterworth-Heinemann, Oxford, UK, 1985]. Most activity coefficients available in the literature are for small organic molecules and are derived from vapor-liquid equilibrium measurements or azeotropic composition data.

Partition ratios at infinite dilution can be calculated directly from the ratio of infinite-dilution activity coefficients for solute dissolved in the extraction solvent and in the feed solution, often providing a reasonable estimate of the partition ratio for dilute concentrations. Infinite-dilution activity coefficients may be reported in terms of a van Laar binary interaction parameter

$$\ln \gamma_{i,j}^{\infty} = \frac{A_{i,j}}{RT} \tag{15-25}$$

$$K_i^o = \frac{\gamma_i^{\infty}}{\gamma_i^{*,\infty}} = \frac{\exp(A^{i,j}/RT)}{\exp(A*^{i,j}/RT)} \tag{15-26}$$

where * denotes the extraction solvent phase [Smallwood, *Solvent Recovery Handbook*, 2d ed., Blackwell, 2002]. For example, the partition ratio for transferring acetone from water into benzene at 25°C and dilute conditions may be estimated as follows: For acetone dissolved in benzene $A_{ij}/RT = 0.47$, and for acetone dissolved in water $A_{ij}/RT = 2.29$. Then $K_i^o = e^{2.29}/e^{0.47} = 9.87/1.6 = 6.17$ (mol/mol) ≡ 1.4 (wt/wt). S. W. Briggs and E. W. Comings [*Ind. Eng. Chem.* **35**(4): 411–417 (1943)] report experimental values that range between 1.06 and 1.39 (wt/wt).

For screening candidate solvents, comparing the magnitude of the activity coefficient for the solute of interest dissolved in the solvent phase often is a good way to rank solvents. A smaller value of $\gamma_{i,\text{solvent}}$ indicates a higher K value. Solubility data available for a given solute dissolved in a range of solvents also can be used to rank solvents; higher solubility in a candidate solvent indicates a more attractive interaction (a lower activity coefficient) and therefore a higher partition ratio.

ROBBINS' CHART OF SOLUTE-SOLVENT INTERACTIONS

When available data are not sufficient (the most common situation), Robbins' chart of functional group interactions (Table 15-2) is a useful guide to ranking general classes of solvents. It is based on an evaluation of hydrogen bonding and electron donor-acceptor interactions for 900 binary systems [Robbins, L. A., *Chem. Eng. Prog.* **76**(10): 58–61 (1980)]. The chart includes 12 general classes of functional groups, divided into three main types: hydrogen-bond donors, hydrogen-bond acceptors, and non-hydrogen-bonding groups. Compounds representative of each class include (1) phenol, (2) acetic acid, (3) pentanol, (4) dichloromethane, (5) methyl isobutyl ketone, (6) triethylamine, (7) diethylamine, (8) *n*-propylamine, (9) ethyl ether, (10) ethyl acetate, (11) toluene, and (12) hexane. Robbins' chart is applicable to any process where liquid-phase activity coefficients are important, including liquid-liquid extraction, extractive distillation, azeotropic distillation, and crystallization from solution. The activity coefficient in the liquid phase is common to all these separation processes.

Here we discuss Robbins' original method. A modified version is given in Sec. 4, Thermodynamics. Robbins' chart predicts positive, negative, or zero deviations from ideal behavior for functional group interactions. For example, consider an application involving extraction of acetone from water into chloroform solvent. Acetone contains a ketone carbonyl group which is a hydrogen acceptor and a member of solute class 5 according to Table 15-2. Chloroform contains a hydrogen donor group (solvent class 4). The solute class 5 and solvent class 4 interaction in Table 15-2 is shown to give a negative deviation from ideal behavior. This indicates an attractive interaction which enhances the liquid-liquid partition ratio. Other classes of solvents shown in Table 15-2 that yield a negative deviation with a ketone (class 5) are classes 1 and 2 (phenolics and acids). Other ketones (solvent class 5) are shown to be compatible with acetone (solute class 5) and tend to give activity coefficients near 1.0, that is, nearly ideal behavior. The solvent classes 6 through 12 tend to provide repulsive interactions between these groups and acetone, and so they are not likely to exhibit partition ratios for ketones as high as the other solvent groups do.

Most of the classes in Table 15-2 are self-explanatory, but some can use additional definition. Class 4 includes halogenated solvents that have highly active hydrogens as described by R. H. Ewell, Harrison, and Berg [*Ind. Eng. Chem.* **36**(10): 871–875 (1944)]. These are molecules that have two or three halogen atoms on the same carbon as a hydrogen atom, such as dichloromethane, trichloromethane, 1,1-dichloroethane, and 1,1,2,2-tetrachloroethane. Class 4 also includes molecules that have one halogen on the same carbon atom as a hydrogen atom and one or more halogen atoms on an adjacent carbon atom, such as 1,2-dichloroethane and 1,1,2-trichloroethane. Apparently, the halogens interact intramolecularly to leave the hydrogen atom highly active. Monohalogen paraffins such as methyl chloride and ethyl chloride are in class 11 along with multihalogen paraffins and olefins without active hydrogen, such as carbon tetrachloride and perchloroethylene. Chlorinated benzenes are also in class 11 because they do not have halogens on the same carbon as a hydrogen atom. Intramolecular bonding on aromatics is another fascinating interaction which gives a net result that behaves much as does an ester group, class 10. Examples of this include *o*-nitrophenol and *o*-hydroxybenzaldehyde (salicylaldehyde). The intramolecular hydrogen bonding is so strong between the hydrogen donor group (phenol) and the hydrogen acceptor group (nitrate or aldehyde) that the molecule acts as an ester. One result is its low solubility in hot water. By contrast, the *para* derivative is highly soluble in hot water.

ACTIVITY COEFFICIENT PREDICTION METHODS

Robbins' chart provides a useful qualitative indication of interactions between classes of compounds but does not give quantitative differences within each class. For this, methods designed to calculate activity coefficients or related properties are used. The thermodynamic basis for these methods is discussed in Sec. 4, Thermodynamics. Perhaps the most widely used method is the group contribution method known by the name universal quasichemical functional group activity coefficients (UNIFAC). For a review of its current status, see J. Gmehling, D. Constantinescu, and B. Schmid, *Ann. Rev. Chem. Biomol. Eng.* **6**: 267–292 (2015). The use of UNIFAC

TABLE 15-2 Robbins' Chart of Solute-Solvent Interactions*

Solute class		Solvent class 1	2	3	4	5	6	7	8	9	10	11	12
	H donor groups												
1	Phenol	0	0	–	0	–	–	–	–	–	–	+	+
2	Acid, thiol	0	0	–	0	–	–	0	0	0	0	+	+
3	Alcohol, water	–	–	0	+	+	0	–	–	+	+	+	+
4	Active H on multihalogen paraffin	0	0	+	0	–	–	–	–	–	–	0	+
	H acceptor groups												
5	Ketone, amide with no H on N, sulfone, phosphine oxide	–	–	+	–	0	+	+	+	+	+	+	+
6	Tertiary amine	–	–	0	–	+	0	+	+	0	+	0	0
7	Secondary amine	–	0	–	–	+	+	0	0	0	0	0	+
8	Primary amine, ammonia, amide with 2H on N	–	0	–	–	+	+	0	0	+	+	+	+
9	Ether, oxide, sulfoxide	–	0	+	–	+	0	0	+	0	+	0	+
10	Ester, aldehyde, carbonate, phosphate, nitrate, nitrite, nitrile, intramolecular bonding, e.g., *o*-nitrophenol	–	0	+	–	+	+	0	+	+	0	+	+
11	Aromatic, olefin, halogen aromatic, multihalogen paraffin without active H, monohalogen paraffin	+	+	+	0	+	0	0	+	0	+	0	0
	Non-H-bonding groups												
12	Paraffin, carbon disulfide	+	+	+	+	+	0	+	+	+	+	0	0

*From Robbins, L. A., *Chem. Eng. Prog.*, **76**(10), pp. 58–61 (1980), by permission. Copyright 1980 AIChE.

for estimating LLE is discussed by J. Gmehling and A. Schedemann [*Ind. Eng. Chem. Res.* **53**(45): 17794-17805 (2014)], P. A. Gupte and R. P. Danner [*Ind. Eng. Chem. Res.* **26**(10): 2036–2042 (1987)] and by H. H. Hooper, S. Michel, and J. M. Prausnitz [*Ind. Eng. Chem. Res.* **27**(11): 2182–2187 (1988)]. S. D. Birajdar et al. [*J. Chem. Eng. Data* **59**(8): 2456–2463 (2014)] discuss the use of UNIFAC for screening candidate solvents, and G. R. Vakili-Nezhand, H. Modarress, and G. A. Mansoori [*Chem. Eng. Technol.* **22**(10): 847–852 (1999)] discuss its use for representing a complex feed containing a large number of components for which available LLE data are incomplete.

Methods based on regular solution theory include the Hansen solubility parameter model [C. M. Hansen, *Hansen Solubility Parameters: A User's Handbook*, 2d ed., CRC Press, Boca Raton, Fla., 2007] and the modified separation of cohesive energy density (MOSCED) model [E. R. Thomas and C. A. Eckert, *Ind. Eng. Chem. Proc. Des. Dev.* **23**(2): 194–209 (1984); and M. J. Lazzaroni et al., *Ind. Eng. Chem. Res.* **44**(11): 4075–4083 (2005)]. Unlike Hansen's model, MOSCED includes two parameters to represent hydrogen bonding (for both proton donor and acceptor capabilities). Example applications of these models to calculate activity coefficients for the purpose of screening candidate solvents are given by T. C. Frank et al. [*Chem. Eng. Prog.* **95**(12): 41–61 (1999)] for the Hansen model, and by I. Escudero, J. L. Cabezas, and J. Coca [*Chem. Eng. Comm.* **173**(1): 135–146 (1999)] and T. C. Frank et al. [*Ind. Eng. Chem. Res.* **46**(13): 4621–4625 (2007)] for MOSCED. In practice, these methods normally involve regression of some phase equilibrium data to determine parameter values, although estimates can be made. For example, Hansen describes methods for estimating Hansen solubility parameters [C. M. Hansen, ibid.], and R. T. Ley et al. [*Ind. Eng. Chem. Res.* **55**(18): 5415–5430 (2016)] discuss methods for calculating MOSCED parameters in the absence of data.

A method that is used in a fashion similar to how the Hansen and MOSCED models are applied is based on a modified NRTL framework and called NRTL-SAC (for segment activity coefficient). See C.-C. Chen and Y. Song, *Ind. Eng. Chem. Res.* **43**(26): 8354–8362 (2004); **44**(23): 8909–8921 (2005); and E. Sheikholeslamzadeh and S. Rohani, *Ind. Eng. Chem. Res.* **55**(18): 464–473 (2012). A method developed by P. Meyer and G. Maurer [*Ind. Eng. Chem. Res.* **34**(1): 373–381 (1995)] uses the linear solvation energy relationships (LSER) model and Kamlet-Taft solvatochromic parameters [Taft, R. W., et al., *Nature* **313:** 384 (1985); and R. W. Taft et al., *J. Pharma Sci.* **74:** 807–814 (1985)] to estimate infinite-dilution partition ratios. The SPACE model employs LSER concepts to calculate infinite-dilution activity coefficients [Hait, M. J., et al., *Ind. Eng. Chem. Res.* **32**(11): 2905–2914 (1993)]. For a discussion of LSER methods and molecular descriptors in general, see M. H. Abraham, A. Ibrahim, and A. M. Zissimos, *J. Chromatogr. A*, **1037:** 29–47 (2004).

The conductor-like screening model (COSMO) introduced by A. Klamt is based on calculation of molecular electron density profiles [Klamt, *From Quantum Chemistry to Fluid Phase Thermodynamics and Drug Design*, Elsevier, Amsterdam, 2005; and Klamt et al., *Annu. Rev. Chem. Biomol. Eng.* **1:** 101–122 (2010)]. The Klamt model is called COSMO-RS (for realistic solvation). A similar model is COSMO-SAC (segment activity coefficient) published by S. T. Lin and S. I. Sandler [*Ind. Eng. Chem. Res.* **41**(5): 899–913, 2332 (2002)]. Databases of electron density profiles (called sigma profiles) have been developed. For example, see E. Mullins et al., *Ind. Eng. Chem. Res.* **45**(12): 4389–4415 (2006), and E. Paulechka et al., *J. Chem. Eng. Data* **60**(12): 3554–3561 (2015). The application of COSMOS-RS to predict liquid-liquid equilibria is discussed by T. Banerjee et al. [*Ind. Eng. Chem. Res.* **46**(4): 1292–1304 (2007)], by L.-Y. Garcia-Chavez et al. [*Sep. Purif. Technol.* **97:** 2–10 (2012)], and by S. D. Birajdar et al. [*J. Chem. Eng. Data* **59**(8): 2456–2463 (2014)].

METHODS USED TO ASSESS LIQUID-LIQUID MISCIBILITY

In evaluating potential solvents, it is important to determine whether a given candidate will exhibit sufficiently limited miscibility with the feed liquid. Mutual solubility data for organic-solvent + water mixtures often are listed somewhere in the literature and can be obtained through a literature search. (See the subsection Phase Equilibrium Data Sources under Thermodynamic Basis for Liquid-Liquid Extraction.) However, data often are not available for pairs of organic solvents and for multicomponent mixtures showing the effect of dissolved solutes. In these cases, estimates can provide useful guidance. Note, however, that the available estimation methods normally provide limited accuracy, so it is best to measure these properties for the more promising candidates.

Phase splitting behavior can be inferred from activity coefficients. In general, partial miscibility will not occur whenever the infinite-dilution activity coefficients of the components in solution are less than 7. This is a reliable rule, but it depends upon the quality of the activity coefficient data or estimates. If γ^∞ for any one of the components is greater than 7, then partial miscibility may occur at some finite composition. The criterion $\gamma_i^\infty > 7$ often is cited as a general rule indicating a partially miscible system, but there are many exceptions. For detailed discussion, see J. M. Prausnitz, R. N. Lichtenthaler, and E. Gomez de Azevedo, *Molecular Thermodynamics of Fluid-Phase Equilibria*, 3d ed., Prentice-Hall, Upper Saddle River, NJ, 1999. Solubility parameters also can be used to assess miscibility [*Handbook of Solubility Parameters and Other Cohesion Parameters*, 2d ed., ed. A. M. F. Barton, CRC, Boca Raton, Fla., 1991].

As a complementary alternative, N. B. Godfrey's data-based method [*Chemtech* **2**(6): 359–363 (1972)] provides a quick way of qualitatively assessing whether an organic-solvent pair of interest is likely to exhibit partial miscibility at near-ambient temperatures. Godfrey assigned miscibility numbers to approximately 400 organic solvents (Table 15-3) by observing their miscibility in a series of 31 standard solvents (Table 15-4). He then showed that the general miscibility behavior of a given solvent pair can be

TABLE 15-3 Godfrey Miscibility Numbers

Acetal	23	3-Chloro-1,2-propanediol	4
Acetic acid	14	1-Chloro-2-propanol	14
Acetic anhydride	12, 19	Chlorobenzene	21
Acetol	8	1-Chlorobutane	23
Acetol acetate	10	1-Chlorodecane	27
Acetol formate	9, 17	Chloroform	19
Acetone	15, 17	1-Chloronaphthalene	22
Acetonitrile	11, 17	3-Chlorophenetole	15, 20
Acetophenone	15, 18	2-Chlorophenol	16
N-Acetylmorpholine	11	2-Chloropropane	23
Acrylonitrile	14, 18	2-Chlorotoluene	20
Adiponitrile	8, 19	Coconut oil	29
Allyl alcohol	14	*p*-Cresol	14
Allyl ether	22	4-Cyano-2,2-dimethylbutyraldehyde	11, 18
2-Allyloxyethanol	13	Cyclohexane	28
2-Aminoethanol	2	Cyclohexanecarboxylic acid	16
2-(2-Aminoethoxy) ethanol	2	Cyclohexanol	16
Aminoethylethanolamine	5	Cyclohexanone	17
1-(2-Aminoethyl) piperazine	12	Cyclohexene	26
1-Amino-2-propanol	6	Cyclooctane	29
Aniline	12	Cyclooctene	27
Anisole	20	*p*-Cymene	25
Benzaldehyde	15, 19	Decalin	29
Benzene	21	Decane	29
Benzonitrile	15, 19	1-Decanol	18
Benzyl alcohol	13	1-Decene	29
Benzyl benzoate	15, 21	Diacetone alcohol	14
Bicyclohexyl	29	Diallyl adipate	21
Bis(2-butoxyethyl) ether	23	1,2-Dibromobutane	22
Bis(2-chloroethyl) ether	20	1,4-Dibromobutane	21
Bis(2-chloroisopropyl) ether	20	Dibromoethane	19
Bis(2-ethoxyethyl) ether	15	1,2-Dibromopropane	21
Bis(2-hydroxyethyl) thiodipropionate	5	1,2-Dibutoxyethane	25
Bis(2-hydroxypropyl) maleate	6	*N,N*-Dibutylacetamide	17
Bis(2-methoxyethyl) ether	15, 17	Dibutyl ether	26
Bis(2-methoxyethyl) phthalate	11, 19	Dibutyl maleate	22
Bromobenzene	21	Dibutyl phthalate	22
1-Bromobutane	23	1,3-Dichloro-2-propanol	12
Bromocyclohexane	25	Dichloroacetic acid	13
1-Bromodecane	27	1,2-Dichlorobenzene	21
1-Bromododecane	27	1,4-Dichlorobutane	20
Bromoethane	21	1,1-Dichloroethane	20
1-Bromohexane	24	1,2-Dichloroethane	20
1-Bromo-3-methylbutane	24	*cis*-1,2-Dichloroethylene	20
1-Bromooctane	26	*trans*-1,2-Dichloroethylene	21
2-Bromooctane	26	Dichloromethane	20
1-Bromotetradecane	29	1,2-Dichloropropane	20
1,2-Butanediol	6	1,3-Dichloropropane	20
1,3-Butanediol	4	Dicyclopentadiene	26
1,4-Butanediol	3	Didecyl phthalate	26
2,3-Butanediol	12, 17	Diethanolamine	1
1-Butanol	15	Diethoxydimethylsilane	26
2-Butanol	16	*N,N*-Diethylacetamide	14
t-Butanol	16	Diethyl adipate	19
2-Buten-1-ol	15	Diethyl carbonate	21
2-Buten-1,4-diol	3	Diethyl ketone	18
2-Butoxyethanol	16	Diethyl oxalate	14, 20
2-(2-Butoxyethoxy) ethanol	15	Diethyl phthalate	13, 20
Butyl acetate	22	Diethyl sulfate	12, 21
Butyl formate	19	Diethylene glycol	5
Butyl methacrylate	23	Diethylene glycol diacetate	12, 19
Butyl oleate	26	Diethylenetriamine	9
Butyl sulfide	26	Diethyl ether	23
Butylaldoxime	15	2,5-Dihydrofuran	17
Butyric acid	16	Di-isobutyl ketone	23
Butyric anhydride	21	Di-isopropyl ketone	23
Butyrolactone	10	Di-isopropylbenzene	25
Butyronitrile	14, 19	1,2-Dimethoxyethane	17
Carbon disulfide	26	*N,N*-Dimethylacetamide	13
Carbon tetrachloride	24	*N,N*-Dimethylacetoacetamide	10
Castor oil	25	2-Dimethylaminoethanol	14
1-Chlorobutane	23	Dimethyl carbonate	14, 19
2-Chloroethanol	11	Dimethylformamide	12

TABLE 15-3 Godfrey Miscibility Numbers *(Continued)*

Dimethyl maleate	12, 19	2,5-Hexanedione	12, 17
Dimethyl malonate	11, 19	1,2,6-Hexanetriol	2
Dimethyl phthalate	12, 19	1-Hexanol	17
1,4-Dimethylpiperazine	16	Hexanoic acid	17
2,5-Dimethylpyrazine	16	1-Hexene	27
Dimethyl sebacate	22	2-Hydroxyethyl carbamate	2
2,4-Dimethylsulfonate	12, 17	2-Hydroxyethylformamide	1
Dimethyl sulfoxide	9	2-Hydroxyethylmethacrylate	12
Dioctyl phthalate	24	1-(2-Hydroxyethoxy)-2-propanol	8
1,4-Dioxane	17	2-Hydroxypropyl carbamate	3
1,4-Dioxene	15, 19	Hydroxypropyl methacrylate	14, 17
Dipentene	26	Iodobenzene	22
Dipentyl ether	26	Iodoethane	22
Diphenyl ether	22	Iodomethane	21
Diphenyl methane	21	Isoamylbenzene	25
Dipropyl sulfone	12, 17	Isobromobutane	23
Dipropylene glycol	11	2-Isobutoxyethanol	15, 17
Dodecane	29	Isobutyl acetate	21
1-Dodecanol	18	Isobutyl isobutyrate	23
1-Dodecene	29	Isobutanol	15
Epichlorohydrin	14, 19	Isophorone	18
Epoxyethylbenzene	15, 19	Isoprene	25
Ethanesulfonic acid	5	Isopropenyl acetate	19
Ethanol	14	Isopropyl acetate	19
2-Ethoxyethanol	14	Isopropyl ether	26
2-(2-Ethoxy) ethanol	13	Isopropylbenzene	24
2-Ethoxyethylacetate	15, 19	Kerosene	30
Ethyl acetate	19	2-Mercaptoethanol	9
Ethyl acetoacetate	13, 19	Mesityl oxide	18
Ethyl benzene	24	Mesitylene	24
Ethyl benzoate	21	Methacrylonitrile	15, 19
2-Ethylbutanol	17	Methanesulfonic acid	4
Ethyl butyrate	22	Methanol	12
Ethylene carbonate	6, 17	5-Methoxazolidinone	7
Ethylenediamine	9	Methoxyacetic acid	8
Ethylene glycol	2	Methoxyacetonitrile acetamide	11, 19
Ethylene glycol bis(methoxyacetate)	9, 17	3-Methoxybutanol	14
Ethylene glycol diacetate	12, 19	2-Methoxyethanol	13
Ethylene glycol diformate	8, 17	2-(2-Methoxyethoxy) ethanol	12
Ethylene monobicarbonate	10, 19	2-Methoxyethyl acetate	14, 17
Ethylformamide	9	2-Methoxyethyl methoxyacetate	15
Ethyl formate	15, 19	1-[2-Methoxy-1-methylethoxy]-2-propanol	15
2-Ethyl-1,3-hexanediol	14, 17	3-Methoxy-1,2-propanediol	5
2-Ethylhexanol	17	1-Methoxy-2-propanol	15
Ethyl hexoate	23	3-Methoxypropionitrile	11, 17
Ethyl lactate	14	3-Methoxypropylamine	15
N-Ethylmorpholine	16	3-Methoxypropylformamide	10
Ethyl orthoformate	23	Methyl acetate	15, 17
Ethyl propionate	21	Methylal	19
2-Ethylthioethanol	13	2-Methylaminoethanol	11
Ethyl trichloroacetate	21	2-Methyl-1-butene	27
Fluorobenzene	20	2-Methyl-2-butene	26
1-Fluoronaphthalene	21	Methylchloroacetate	13, 19
Formamide	3	Methylcyanoacetate	8, 17
Formic acid	5	Methylcyclohexane	29
N-Formylmorpholine	10	1-Methylcyclohexene	27
Furan	20	Methylcyclopentane	28
Furfural	11, 17	Methyl ethyl ketone	17
Furfuryl alcohol	11	Methyl formate	14, 19
Glycerol (glycerin)	1	2,2′-Methyliminodiethanol	8
Glycerol carbonate	3	Methyl isoamyl ketone	19
Glycidyl phenyl ether	13, 19	Methyl isobutyl ketone	19
Heptane	29	Methyl methacrylate	20
1-Heptanol	17	Methyl methoxyacetate	13
3-Heptanone	22	*N*-Methylmorpholine	16
4-Heptanone	23	1-Methylnaphthalene	22
1-Heptene	28	Methyl oleate	26
Hexachlorobutadiene	26	2-Methylpentane	29
Hexadecane	30	3-Methylpentane	29
1-Hexadecene	29	4-Methyl-2-pentanol	17
Hexamethylphosphoramide	15	2-Methyl-2,4-pentanediol	14
Hexane	29	4-Methyl-1-pentene	28
2,5-Hexanediol	5	*cis*-4-Methyl-2-pentene	27

TABLE 15-3 Godfrey Miscibility Numbers *(Continued)*

N-Methyl-2-pyrrolidinone	13	Styrene	22
Methyl stearate	26	Sulfolane	9, 17
α-Methylstyrene	23	1,1,2,2-Tetrabromoethane	11, 19
3-Methylsulfolane	10, 17	1,1,2,2-Tetrachloroethane	19
Mineral spirits	29	Tetrachloroethylene	25
Morpholine	14	Tetradecane	30
Nitrobenzene	14, 20	1-Tetradecene	29
Nitroethane	13, 20	Tetraethyl orthosilicate	23
Nitromethane	10, 19	Tetraethylene glycol	7
2-Nitropropane	15, 20	Tetraethylenepentamine	9
1-Nonanol	17	Tetrahydrofuran	17
Nonylphenol	17	Tetrahydrofurfuryl alcohol	13
1-Octadecene	30	Tetrahydrothophene	21
1,7-Octadiene	27	Tetralin	24
Octane	29	Tetramethylsilane	29
1-Octanethiol	26	Tetramethylurea	15
1-Octanol	17	Tetrapropylene	29
2-Octanol	17	1,1-Thiodi-2-propanol	8
2-Octanone	22	2,2′-Thiodiethanol	4
1-Octene	28	3,3′-Thiodipropionitrile	6, 19
cis-2-Octene	27	Thiophene	20
trans-2-Octene	28	Toluene	23
3,3′-Oxydipropionitrile	6	Triacetin	11, 19
Paraldehyde	15, 19	Tributylphosphate	18
polyethylene glycol PEG-200	7	Tributylamine	28
polyethylene glycol PEG-300	8	1,2,4-Trichlorobenzene	24
polyethylene glycol PEG-600	8	1,1,1-Trichloroethane	22
1,3-Pentadiene	25	1,1,2-Trichloroethane	19
Pentaethylene glycol	7	Trichloroethylene	20
Pentaethylenehexamine	9	1,1,2-Trichloro-2,2,2-trifluoroethane	27
Pentafluoroethanol	9	1,2,3-Trichloropropane	20
1,5-Pentanediol	3	Tricresyl phosphate	21
2,4-Pentanedione	12, 18	Triethanolamine	2
1-Pentanol	17	Triethyl phosphate	14
t-Pentanol	16	Triethylamine	26
Petrolatum (C14–C16 alkanes)	31	Triethylbenzene	25
Phenetole	20	Triethylene glycol	6
2-Phenoxyethanol	12	Triethylene glycol monobutyl ether	14
1-Phenoxy-2-propanol	13, 17	Triethylene glycol monomethyl ether	13
Phenyl acetate	23	Triethylenetetramine	9
Phenylacetonitrile	12, 19	Triisobutylene	29
N-Phenylethanolamine	10	Trimethyl borate	16
2-Picoline	16	Trimethyl nitrilotripropionate	12
polypropylene glycol PPG-1000	14, 23	Trimethyl phosphate	10
polypropylene glycol PPG-400	14	2,4,4-Trimethyl-1-pentane	27
Propanediamine	11, 11	2,4,4-Trimethyl-2-pentane	27
1,2-Propanediol	4	Trimethylboroxin	12, 17
1,3-Propanediol	3	2,2,4-Trimethylpentene	29
Propanesulfone	7, 19	Tripropylamine	26
1-Propanol	15	Tripropylene glycol	12
2-Propanol	15	Vinyl acetate	20
Propionic acid	15	Vinyl butyrate	22
Propionitrile	13, 17	4-Vinylcyclohexene	26
Propyl acetate	19	Naphtha	29
Propylene carbonate	9, 17	*m*-Xylene	23
Propylene oxide	17	*o*-Xylene	23
Pyridine	16	*p*-Xylene	24
2-Pyrrolidinone	10		

Reprinted from N. B. Godfrey, *CHEMTECH*, 2(6), pp. 359–363 (June, 1972), with permission. Published 1972 by the American Chemical Society.

predicted by comparing their miscibility numbers. Godfrey's rules, slightly modified, are summarized as follows:

1. If $\Delta \le 12$, where Δ is the difference in miscibility numbers, the solvents are likely to be miscible in all proportions at 25°C.
2. If $13 \le \Delta \le 15$, the solvents may be only partially miscible with an upper critical solution temperature (UCST) between 25 and 50°C. This is a borderline case. If the binary mixture is miscible, then adding a relatively small amount of water likely will induce phase splitting.
3. If $\Delta = 16$, the solvents are likely to exhibit a UCST between 25 and 75°C.
4. If $\Delta \ge 17$, the solvents are likely to exhibit a UCST above 75°C.

About 15 percent of the solvents in Table 15-3 have dual miscibility numbers A and B because the appropriate difference in miscibility numbers depends upon which end of the hydrophobic-lipophilic scale is being considered. If one of the solvents has dual miscibility numbers A and B and the other has a single miscibility number C, then Δ should be calculated as follows:

5. If $C > B$, then the solvent having miscibility number C is somewhat more lipophilic than the solvent having numbers A and B. At this end of the lipophilicity scale, the number A characterizes the solvent's miscibility behavior. Apply rules 1 through 3, using $\Delta = C - A$.
6. If $C < A$, then the solvent having miscibility number C is somewhat less lipophilic than the solvent with numbers A and B. At this end of the lipophilicity scale, the number B characterizes the solvent's miscibility behavior. Apply rules 1 through 3, using $\Delta = B - C$.

TABLE 15-4 Godfrey Standard Solvents

Miscibility number	Solvent	
1	Glycerol ("glycerin")	Hydrophilic end of scale
2	1,2-Ethanediol ("ethylene glycol")	↓
3	1,4-Butanediol	
4	2,2′-Thiodiethanol	
5	Diethylene glycol	
6	Triethylene glycol	(decreasing hydrophilicity)
7	Tetraethylene glycol	(increasing lipophilicity)
8	Methoxyacetic acid	↓
9	Dimethylsulfoxide	
10	*N*-Formylmorpholine	
11	Furfuryl alcohol	
12	2-(2-Methoxyethoxy) ethanol ("diethylene glycol methyl ether")	
13	2-Methoxyethanol ("ethylene glycol methyl ether")	↓
14	2-Ethoxyethanol ("ethylene glycol ethyl ether")	
15	2-(2-Butoxyethoxy) ethanol ("diethylene glycol *n*-butyl ether")	
16	2-Butoxyethanol ("ethylene glycol *n*-butyl ether")	
17	1,4-Dioxane	
18	3-Pentanone	
19	1,1,2,2-Tetrachloroethane	↓
20	1,2-Dichloroethane	
21	Chlorobenzene	
22	1,2-Dibromobutane	
23	1-Bromobutane	
24	1-Bromo-3-methylbutane	
25	*sec*-Amylbenzene	↓
26	4-Vinylcyclohexene	
27	1-Methylcyclohexene	
28	Cyclohexane	
29	Heptane	
30	Tetradecane	
31	Petrolatum (C_{14}–C_{16} alkanes)	Lipophilic end of scale

Reprinted from N. B. Godfrey, *CHEMTECH*, **2**(6), pp. 359–363 (June, 1972), with permission. Published 1972 by the American Chemical Society.

7. If $A \le C \le B$, then evaluate $\Delta = C - A$ and $\Delta = B - C$ and use the larger of the D values in applying rules 1 through 3. Such a mixture is likely to be miscible in all proportions at 25°C.

8. If both members of a solvent pair have dual miscibility numbers, then the pair is likely to be miscible in all proportions at 25°C.

If a compound of interest is not listed in Table 15-3 or 15-4, a compound of the same type or class may help to gauge its miscibility behavior. In cases where Godfrey's rules indicate that partial miscibility is likely, whether phase splitting actually occurs depends upon the composition of the mixture and the temperature. The composition may be close to but still outside the two-liquid-phase region on a temperature-composition diagram.

Godfrey's method is a useful guide for compounds that exhibit behavior similar to the 31 standard solvents used to define miscibility numbers. The method deals with the common situation in which a mixture exhibits a UCST; that is, solubility tends to increase with increasing temperature. Exceptions to Godfrey's rules include binary mixtures that form unusually strong hydrogen-bonding interactions. Normally, mixtures of this type are completely miscible, or they exhibit a lower critical solution temperature (LCST). Examples include ethylene glycol + triethylamine (Fig. 15-16) and glycerin + ethylbenzylamine (UCST = 280°C and LCST = 49°C) [Sørenson, J. M., and W. Arlt, *Liquid-Liquid Equilibrium Data Collection*, vol. V, pt. 1, DECHEMA, Frankfurt, 1979]. As mentioned earlier, it is not unusual for mixtures of water and amines or water and glycol ethers to exhibit LCST behavior. (See the subsection Phase Diagrams under Thermodynamic Basis for Liquid-Liquid Extraction.) This is a reason why Godfrey's method does not include water.

Sometimes the mutual solubility of a solvent pair of interest can easily be decreased by adding a third component. For example, it is common practice to add water to a solvent system containing a water-miscible organic solvent (the polar phase) and a hydrophobic organic solvent (the nonpolar phase). A typical example is the solvent system (methanol + water) + dichloromethane. An anhydrous mixture of methanol and dichloromethane is completely miscible, but adding water causes phase splitting. Adjusting the amount of water added to the polar phase also may be used to alter the K values for the extraction, density difference, and interfacial tension. Table 15-5 lists some common examples of solvent systems of this type. These systems are common candidates for fractional extractions.

COMPUTER-AIDED MOLECULAR DESIGN

Many specialized computer programs have been written specifically to identify candidate solvents with properties that best match those needed for a particular application by weighing various considerations of the kind outlined in the subsection Desirable Solvent Properties in addition to the partition ratio. The goal is to determine the optimal solvent structure that best meets the specified set of performance factors [Brignole, E. A., S. Botini, and R. Gani, *Fluid Phase Equilibr.* **29:** 125–132 (1986); and K. G. Joback and G. Stephanopoulos, *Proc. FOCAPD* **11:** 631 (1989)]. Recent studies that include reviews of previous work are given by N. D. Austin, N. V. Sahinidis, and D. W. Trahan [*Chem. Eng. Sci.* **159:** 93–105 (2017) and

TABLE 15-5 Common Solvent Systems Involving a Water-Miscible Organic Solvent and Addition of Water to Control Properties

Polar component	Nonpolar component
Methanol	*n*-Hexane, *n*-heptane, other alkanes, dichloromethane
Acetonitrile	*n*-Hexane, *n*-heptane, other alkanes, dichloromethane
Ethylene glycol, diethylene glycol, triethylene glycol, tetraethylene glycol, and propylene glycol analogs	*n*-Hexane, *n*-heptane, other alkanes, dichloromethane, amyl acetate, toluene, xylene
Ethylene glycol mono methyl ether and other glycol ethers	*n*-Hexane, *n*-heptane, other alkanes, and dichloromethane

Chem. Eng. Res. Des. **116:** 2–26 (2016)]; A. I. Papadopoulos and P. Linke [*AIChE J.* **52**(3): 1057–1070 (2006)]; A. T. Karunanithi, L. E. K. Achenie, and R. Gani [*Ind. Eng. Chem. Res.* **44**(13): 4785–4797 (2005)]; and M. Cismondi and E. A. Brignole [*Ind. Eng. Chem. Res.* **43**(3): 784–790 (2004)]. A variety of creative search strategies have been employed, including the use of stochastic algorithms to account for uncertainty [Kim, K.-J., and U. M. Diwekar, *Ind. Eng. Chem. Res.* **41**(5): 1285–1296 (2002)], the use of quantum chemistry methods for property estimation [A. Lehnamm and C. D. Maranas, *Ind. Eng. Chem. Res.* **43**(13): 3419–3432 (2004)], and the application of a genetic theory of evolution (survival of the fittest) [Nieuwoudt, I., Paper No. 233a, AIChE National Meeting, Austin, 2004; and Van Dyk, B., and I. Nieuwoudt, *Ind. Eng. Chem. Res.* **39**(5): 1423–1429 (2000)]. Similar programs have been written to facilitate the identification of alternative solvents or solvent blends as replacements for a given solvent, by attempting to identify compounds that match the physical properties of the solvent the user wishes to replace. An example is the PARIS program developed by the U.S. Environmental Protection Agency [Cabezas, H., P. F. Harten, and M. R. Green, *Chem. Eng. Magazine* **107**(3): 109 (March 2000).

HIGH-THROUGHPUT EXPERIMENTAL METHODS

In addition to the methods already described, solvents and extraction conditions may be screened by using rapid automated experimental methods and automated sample analysis. High-throughput liquid-liquid extraction methods are reviewed by D. A. Wells [*Progress in Pharmaceutical and Biomedical Analysis*, vol. 5, Elsevier, Amsterdam, 2003]. An example involving automated liquid chromatography is described by R. D. Bolden et al. [*J. Chromatogr. B.* **772:** 1–10 (2002)]. A gas chromatography method that includes automated calculation of partition ratios and mutual solubility, plus automated correlation of data with thermodynamic models, is described by D. Dechambre et al. [*Fluid Phase Equilibr.* **362:** 328–334 (2014)]. Another approach called single-drop microextraction is reviewed by Y. Yan et al. [*J. Chromatogr. A.* **1368:** 1–17 (2014)]. For a review of high-throughput methods in general, see K. Murray, ed., *Principles and Practice of High Throughput Screening*, Blackwell, Oxford, UK, 2005. The automated methods described in the subsection Liquid-Liquid Equilibrium Experimental Methods under Thermodynamic Basis for Liquid-Liquid Extraction also may be useful for screening solvents.

LIQUID DENSITY, VISCOSITY, AND INTERFACIAL TENSION

General References: See Sec. 2, Physical and Chemical Data; Rosen, M. J., *Surfactants and Interfacial Phenomena*, 4th ed., Wiley, New York, 2012; Hartland, S., *Surface and Interfacial Tension: Measurement, Theory, and Applications*, Marcel Dekker, New York, 2004; and Poling, B. E., J. M. Prausnitz, and J. P. O'Connell, *The Properties of Gases and Liquids*, 5th ed., McGraw-Hill, New York, 2000.

The utility of liquid-liquid extraction as a separation tool depends upon both phase equilibria and transport properties. The most important physical properties that influence transport properties are liquid-liquid interfacial tension, liquid density, and viscosity. These properties influence solute diffusion and the formation and coalescence of drops, and so are critical factors affecting the performance of liquid-liquid contactors and phase separators.

DENSITY AND VISCOSITY

Many handbooks contain an extensive compilation of liquid density data. These same sources often include liquid viscosity data, although fewer experimental data may be available for a particular compound. Available data compilations include those by G. Wypych, *Handbook of Solvents*, ChemTec, Toronto, 2014; A. Wypych and G. Wypych, *Solvents Database*, CD-ROM, ChemTec, Toronto, 2014; C. L. Yaws, *Thermodynamic and Physical Property Data*, 2d ed., Gulf, Houston, 1998; and E. W. Flick, *Industrial Solvents Handbook*, 5th ed., Noyes, Westwood, NJ, 1998. In addition, viscosity data for C_1–C_{28} organic compounds have been compiled by C. L. Yaws in *Handbook of Viscosity*, vols. 1–3, Elsevier, Amsterdam, 1994. Density and viscosity data also are available from the Thermodynamics Research Center at the National Institute of Standards and Technology (Boulder, Colo.) and from the DIPPR physical property databank of AIChE.

Methods for estimating density and viscosity are reviewed by B. E. Poling, J. M. Prausnitz, and J. P. O'Connell [*The Properties of Gases and Liquids*, 5th ed., McGraw-Hill, New York, 2000]. However, it is best to measure density and viscosity in the laboratory whenever possible. The methods used to measure viscosity are described in numerous books, including *Measurement of Transport Properties of Fluids*, vol. 3, ed. W. A. Wakeham, A. Nagashima, and J. V. Sengers, Blackwell, Oxford, UK, 1991; and G. E. Leblanc, R. A. Secco, and M. Kostic, "Viscosity Measurement," chap. 30 in *Measurement, Instrumentation, and Sensors Handbook*, ed. Webster, CRC Press, Boca Raton, Fla., 1999. The Stabinger method allows simultaneous measurement of viscosity and density [American Society for Testing and Materials, ASTM D7042-04 (2005)].

INTERFACIAL TENSION

Typical values of interfacial tension are listed in Tables 15-6 and 15-7. Refer to the references listed in these tables for the full data sets and for data on other mixtures. Table 15-6 shows typical values for organic + water binary mixtures. Table 15-7 shows the strong effect of the addition of a third component. Also, R. E. Treybal's classic plot of interfacial tension versus mutual solubility is given in Fig. 15-21. This information can be helpful in assessing whether interfacial tension is likely to be low, moderate, or high for a new application. However, for design purposes, interfacial tension should be measured by using representative feed and solvent because even small amounts of surface-active impurities can significantly impact the result.

Methods used to measure interfacial tension are reviewed by J. Drelich, Ch. Fang, and C. L. White ["Measurement of Interfacial Tension in Fluid-Fluid Systems," in *Encyclopedia of Surface and Colloid Science*, Marcel Dekker, New York, 2002, pp. 3152–3166]. Also see D. Megias-Alguacil, P. Fischer, and E. J. Windhab, *Chem. Eng. Sci.* **61:** 1386–1394 (2006). One class of methods derives interfacial tension values from measurement of the shape, contact angle, or volume of a drop suspended in a second liquid. These methods include the pendant drop method (a drop of heavy liquid hangs from a vertically mounted capillary tube immersed in the light liquid), the sessile drop method (a drop of heavy liquid lies on a plate immersed in the light liquid), and the spinning drop method (a drop of one liquid is suspended in a rotating tube filled with the second liquid). The sessile drop method is particularly useful for following the change in interfacial tension when surfactants or macromolecules accumulate at the surface of the drop. The spinning drop method is well suited to measuring low interfacial tensions. Another class of methods derives interfacial tension values from measurement of the force required to detach a ring of wire (Du Noüy's method), or a plate of glass or platinum foil (the Wilhelmy method), from the liquid-liquid interface. The ring or plate must be extremely clean. For the commonly used ring-pull method, the wire is usually flamed before the experiment and must be kept very horizontal and located exactly at the interface of the two liquids.

TABLE 15-6 Typical Interfacial Tensions for Different Classes of Organic + Water Binary Mixtures at 20 to 25°C

Class of organic compounds	Interfacial tension, dyn/cm
Alkanes (C_5–C_{12})	45–53
Halogenated alkanes (C_1–C_4)	30–40
Halogenated aromatics (single ring)	35–40
Aromatics (single ring)	30–40
Mononitro aromatics (single ring)	25–28
Ethers (C_4–C_6)	10–30
Esters (C_4–C_6)	10–20
Ketones (C_4–C_8)	5–15
Organic acids (C_5–C_{12})	3–15
Aniline	6–7
Alcohols (C_4–C_8)	2–8

References:
1. Demond, A. H., and A. S. Lindner, *Environ. Sci. Technol.* **27**(12): 2318–2331 (1993).
2. Fu, J., B. Li, and Z. Wang, *Chem. Eng. Sci.* **41**(10): 2673–2679 (1986).
3. Backes, H. M., et al., *Chem. Eng. Sci.* **45**(1): 275–286 (1990).

TABLE 15-7 Example Interfacial-Tension Data for Selected Ternary Mixtures

Component 1	Component 2 in phase 1, wt %	Component 3 in phase 1, wt %	Component 2 in phase 2, wt %	Component 3 in phase 2, wt %	Interfacial tension, dyn/cm
Water	Benzene	Ethanol	Benzene	Ethanol	At 25°C
	0.2	10.8	98.6	1.2	17.2
	3.6	43.7	91.3	7.9	1.99
	21.2	52.0	79.3	18.0	0.04
Water	Benzene	Acetone	Benzene	Acetone	At 30°C
	0.1	1.9	98.1	1.8	25.9
	0.2	10.3	91.2	8.6	16.1
	0.6	23.6	81.9	17.8	9.5
	2.7	45.5	68.2	30.9	3.8
Water	Benzene	Acetic acid	Benzene	Acetic acid	At 25°C
	0.3	17.2	98.6	1.3	17.3
	1.1	45.1	92.2	7.5	7.0
	7.9	64.7	77.0	21.9	2.0
Water	Hexane	Ethanol	Hexane	Ethanol	At 20°C
	0.1	32.5	99.5	0.5	9.82
	8.2	73.0	93.9	6.0	1.5
	30.0	64.0	86.2	13.2	0.096
Hexane	Methyl ethyl ketone	Water	Methyl ethyl ketone	Water	At 25°C
	0.4	99.6	0.59	0.01	40.1
	11.7	88.3	35.56	0.09	9.0
	24.5	75.5	89.88	9.97	1.1

References:
1. Sada, E., S. Kito, and M. Yamashita, *J. Chem. Eng. Data* **20**(4): 376–377 (1975).
2. Pliskin, I., and R. E. Treybal, *J. Chem. Eng. Data* **11**(1): 49–52 (1966).
3. Paul, G. W., and L. E. Marc De Chazal, *J. Chem. Eng. Data* **12**(1): 105–107 (1967).
4. Ross, S., and R. E. Patterson, *J. Chem. Eng. Data* **24**(2): 111–115 (1979).
5. Backes, H. M., et al., *Chem. Eng. Sci.* **45**(1), pp. 275–286 (1990).

For an initial assessment, an approximate value for the interfacial tension may be obtained, at least in principle, from knowledge of the maximum size of drops that can persist in a dispersion at equilibrium and without agitation. For example, if it is possible to determine drop size from a photograph of the dispersion of interest at quiescent conditions, then an estimate of interfacial tension may be obtained from the balance between interfacial tension and buoyancy forces

$$\sigma \approx d_{\max}^2 \, \Delta \rho g \tag{15-27}$$

where $d_{\max}$ is the maximum drop diameter. Antonov's rule states that interfacial tension between two liquids is approximately equal to the difference in their liquid-air surface tensions measured at the same conditions. For an organic + water system,

$$\sigma \approx \left| \sigma_{w(o)} - \sigma_{o(w)} \right| \tag{15-28}$$

where $\sigma_{w(o)}$ represents the surface tension of the water saturated with the organic and $\sigma_{o(w)}$ represents the surface tension of organic saturated with water.

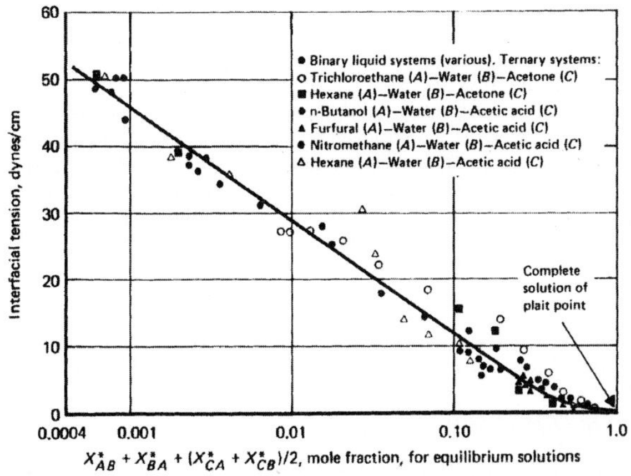

FIG. 15-21 Correlation of interfacial tension with mutual solubility for binary and ternary two-liquid-phase mixtures. [*Reprinted from Treybal,* Liquid Extraction, *2d ed., McGraw-Hill, New York, 1963. Copyright 1963 McGraw-Hill, Inc.*]

Measurements of interfacial tension are not always feasible, and calculation methods are sometimes used. The results are least reliable for interfacial tensions below about 10 dyn/cm (10^{-2} N/m). A commonly used empirical correlation of interfacial tension and mutual solubilities is given by D. J. Donahue and F. E. Bartell [*J. Phys. Chem.* **56:** 480–484 (1952)]:

$$\sigma = -3.33 - 7.21 \ln (x_1'' + x_2') \tag{15-29}$$

where σ = interfacial tension, dyn/cm (10^{-3} N/m)
x_1'' = mole fraction solubility of organic in aqueous phase
x_2' = mole fraction solubility of water in organic phase

R. E. Treybal [*Liquid Extraction*, 2d ed., McGraw-Hill, New York, 1963] modified Eq. (15-29) to expand its application to ternary systems:

$$\sigma = -5.0 - 7.355 \ln [x_1'' + x_2' + 0.5 (x_3' + x_3'')] \tag{15-30}$$

where σ = interfacial tension, dyn/cm (10^{-3} N/m)
x_3'' = mole fraction solute in aqueous phase
x_3' = mole fraction solute in organic phase

The results are plotted in Fig. 15-21. Also, J. Fu, B. Li, and C. Wang [*Chem. Eng. Sci.* **41**(10): 2673–2679 (1986)] derived a relationship for ternary mixtures:

$$\sigma = \frac{0.9414 RTz}{(A_o \exp z)(x_1'' q_1 + x_2' q_2 + x_{3r} q_3)} \tag{15-31}$$

$$z = -\ln (x_1'' + x_2' + x_{3r}) \tag{15-32}$$

where σ = interfacial tension, dyn/cm (10^{-3} N/m)
R = ideal gas law constant (8.314×10^7 dyn cm/gmol-K)
T = absolute temperature (K)
x_2'' = solubility of extract phase in raffinate phase (mole fraction)
x_2' = solubility of raffinate phase in extract phase (mole fraction)
x_{3r} = mole fraction of solute 3 in bulk phase richest in solute 3
A_o = van der Waals area of standard segment (2.5×10^9 cm^2/gmol)
q_i = van der Waals surface area ratio, usually calculated from UNIQUAC

Newer methods involve modeling of fundamental intermolecular interactions underlying both mutual solubility and interfacial tension. See the methods of B. Li and J. Fu [*Fluid Phase Equilibr.* **81:** 129–152 (1992)]; P. Wang and A. Anderko [*Ind. Eng. Chem. Res.* **52**(20): 6822–6840 (2013)]; and M. P. Andersson et al. [*J. Chem. Theory Comput.* **10**(8): 3401–3408 (2014)].

LIQUID-LIQUID DISPERSION FUNDAMENTALS

General References: Leng, D. E., and R. V. Calabrese, "Immiscible Liquid-Liquid Systems," chap. 12 in *Advances in Industrial Mixing*, ed. S. M. Kresta et al., Wiley, New York, 2015 and chap. 12 in *Handbook of Industrial Mixing*, ed. E. L. Paul, V. A. Atiemo-Obeng, and S. M. Kresta, Wiley, New York, 2004; Becher, P., *Emulsions: Theory and Practice*, 3d ed., American Chemical Society, Washington, DC, 2001; Binks, B. P., *Modern Aspects of Emulsion Science*, Royal Society of Chemistry, London, 1998; Adamson, A. W., and A. P. Gast, *Physical Chemistry of Surfaces*, 6th ed., Wiley, New York, 1997; *Liquid-Liquid Extraction Equipment*, ed. J. C. Godfrey and M. J. Slater, Wiley, New York, 1994; *Encyclopedia of Emulsion Technology*, vols. 1–4, ed. P. Becher, Marcel Dekker, New York, 1983; and Laddha, G. S., and T. E. Degaleesan, chap. 4 in *Handbook of Solvent Extraction*, ed. T. C. Lo, M. H. I. Baird, and C. Hanson, Wiley, New York, 1983, and Krieger, Huntington, NY, 1991.

HOLDUP, SAUTER MEAN DIAMETER, AND INTERFACIAL AREA

Most liquid-liquid extractors are designed to generate drops of one liquid suspended in the other rather than liquid films. The volume fraction of the dispersed phase (or holdup) within the extractor is defined as

$$\phi_d = \frac{\text{volume of dispersed phase}}{\text{total contacting volume}} \tag{15-33}$$

where the total contacting volume is the volume within the extractor minus the volume of any internals such as impellers, packing, or trays. A distribution of drop sizes will be present. The Sauter mean drop diameter d_{32} represents a volume to surface-area average diameter

$$d_{32} = \frac{\sum_{i=1}^{n} N_i d_i^3}{\sum_{i=1}^{n} N_i d_i^2} \tag{15-34}$$

where N_i is the number of drops with diameter d_i. The Sauter mean diameter often is used in the analysis and modeling of extractor performance because it is directly related to holdup and interfacial area (assuming spherical drops). It is calculated from the total dispersed volume divided by total interfacial area, and often it is expressed in the form

$$d_{32} = \frac{6\varepsilon\phi_d}{a} \tag{15-35}$$

where a is interfacial area per unit volume and ε is the void fraction within the extractor, that is, the fraction of internal volume not occupied by any packing, trays, and so on. In the remainder of Sec. 15, the Sauter mean diameter is denoted simply by d_p. In the design of extraction equipment, Eq. (15-35) often is used to calculate interfacial area from estimates of drop size and holdup.

Much less is known about the actual distribution of drop sizes existing within liquid-liquid extractors, particularly at high holdup and as a function of agitation intensity (if agitation is used) and location within the extractor. For a review, see A. Kumar and S. Hartland, chap. 17 in *Liquid-Liquid Extraction Equipment*, ed. J. C. Godfrey and M. J. Slater, Wiley, New York, 1994. Experimental methods used to measure drop size distribution include the use of a high-speed video camera [Ribeiro, M. M. M., et al., *Chem. Eng. J.* **97:** 173–182 (2004)], real-time optical measurements [Ritter, J., and M. Kraume, *Chem. Eng. Technol.* **23**(7): 579–581 (2000)], and phase-Doppler anemometry [Lohner, H., K. Bauckhage, and E. H. Schombacher, *Chem. Eng. Technol.* **21**(4): 337–341 (1998); and Willie, M., G. Langer, and U. Werner, *Chem. Eng. Technol.* **24**(5): 475–479 (2001)].

FACTORS AFFECTING WHICH PHASE IS DISPERSED

Consider mixing a batch of two liquid phases in a stirred tank. The minority phase generally will be the dispersed phase whenever the ratio of minority to majority volume fractions, or phase ratio, is less than about 0.5 (equivalent to a dispersed-phase volume fraction or holdup less than 0.33). For phase ratios between 0.5 and about 2, a region called the ambivalent range, the phase that becomes dispersed is determined in large part by the protocol used to create the dispersion. For example, pouring liquid A into a stirred tank already containing liquid B will tend to create a dispersion of A suspended in B, as long as agitation is maintained. When more of the dispersed-phase material is added to the system, the population density of dispersed drops will increase and eventually reach a point where the drops are so close together they coalesce and the phases become inverted, that is, the formerly dispersed phase becomes the continuous phase. In the ambivalent range, a sudden increase in the agitation intensity also can trigger phase inversion by increasing the number of drop-to-drop collisions. Once phase inversion occurs, it is not easily reversed because the new condition corresponds to a more stable configuration.

This phase behavior may be roughly correlated in terms of light and heavy phase properties including relative density and viscosity as follows:

$$\chi = \frac{\phi_L}{\phi_H}\left(\frac{\rho_L\mu_H}{\rho_H\mu_L}\right)^{0.3} = \frac{\phi_L}{1-\phi_L}\left(\frac{\rho_L\mu_H}{\rho_H\mu_L}\right)^{0.3} \tag{15-36}$$

where $\chi < 0.3$ light phase always dispersed
$\chi = 0.3$–0.5 light phase probably dispersed
$\chi = 0.5$–2.0 either phase can be dispersed, and phase inversion may occur
$\chi = 2.0$–3.3 heavy phase probably dispersed
$\chi > 3.3$ heavy phase always dispersed

The symbol ϕ denotes the volume fraction of light (L) and heavy (H) phases existing within the vessel. Equation (15-36) is taken from the expression recommended by W. B. Hooper [Sec. 1.11 in *Handbook of Separation Techniques for Chemical Engineers,* 3d ed., ed. P. A. Schweitzer, McGraw-Hill, New York, 1997] and L. J. Jacobs and W. R. Penney [chap. 3 in *Handbook of Separation Process Technology*, ed. R. W. Rousseau, Wiley, New York, 1987] for design of continuous decanters. It is based on the dispersed-phase data of A. H. Selker and C. A. Sleicher [*Can. J. Chem. Eng.* **43:** 298–301 (1965)].

Equation (15-36) should apply to continuously fed extraction columns and other continuous extractors as well as batch vessels. The equation is expressed here in terms of volume fractions ϕ_L/ϕ_H existing within the vessel, not volumetric flow rates of each phase entering the vessel Q_L/Q_H. The ratio of volume fractions within a continuously fed vessel can be very different from Q_L/Q_H—primarily because buoyancy allows the dispersed-phase drops to travel rapidly through the continuous phase relative to the average dispersed-phase superficial velocity. For example, a continuously fed extraction column can be designed to operate with either phase being the dispersed phase, with the main liquid-liquid interface controlled at the top of the column (for a light-phase dispersed system) or at the bottom (for a heavy-phase dispersed system). As the dispersed-to-continuous phase ratio within the column is increased, through either changes in operating variables or changes in the design of the internals, a point may be reached where the population density or holdup of dispersed drops is too large and phase inversion occurs. In the absence of stabilizing surfactants, the point of phase inversion should correspond roughly to the same general phase-ratio rules given in Eq. (15-36), with the exact conditions at which phase inversion occurs depending upon agitation intensity (if used) and the geometry of any internals (baffles, packing, trays, and so on). Certain extractors such as sieve tray columns often are designed to disperse the majority flowing phase. In extreme cases, the ratio Q_d/Q_c (where d and c represent dispersed and continuous phases) may be as high as 50, and the continuous phase may be nearly stagnant with a superficial velocity as low as 0.02 cm/s; yet the phase ratio within the extractor can be controlled within the guidelines needed to avoid phase inversion [approximated by Eq. (15-36)].

Dispersion stability also can be affected by the presence of fine solids or gas bubbles as well as surfactants. For additional discussion of factors affecting which phase is dispersed, see M. A. Norato, L. L. Tsouris, and C. Tavlarides, *Can. J. Chem. Eng.* **76:** 486–494 (1998); and A. W. Pacek et al., *AIChE J.* **40**(12): 1940–1949 (1994). For a given application, the precise conditions that lead to phase inversion must be determined by experiment. For organic + water dispersions, experimental determination may be facilitated by measuring the conductivity of the mixture. Conductivity normally will be significantly higher when water is in the continuous phase [Gilchrist, A., et al., *Chem. Eng. Sci.* **44**(10): 2381–2384 (1989)]. Another method involves monitoring the dynamics of phase inversion by using a stereo microscope and video camera [Pacek, A. W., et al., *AIChE J.* **40**(12): 1940–1949 (1994)].

SIZE OF DISPERSED DROPS

In nonagitated (static) extractors, drops are formed by flow through small holes in sieve plates or inlet distributor pipes. The maximum size of drops issuing from the holes is determined not by the hole size but primarily by the balance between buoyancy and interfacial tension forces acting on the stream or jet emerging from the hole. Neglecting any viscosity effects (i.e., assuming low dispersed-phase viscosity), the maximum drop size is proportional to the square root of interfacial tension σ divided by density difference Δρ:

$$d_{max} = (\text{const})\sqrt{\frac{\sigma}{\Delta\rho\, g}} \quad \text{for static extractors} \tag{15-37}$$

The proportionality constant typically is close to unity [Seibert, A. F., and J. R. Fair, *Ind. Eng. Chem. Res.* **27**(3): 470–481 (1988)]. Note that Eq. (15-37) indicates the maximum stable drop diameter and not the Sauter mean diameter, although the two are proportionally related and may be close in value. Smaller drops may be formed at the distributor due to jetting of the inlet liquid through the distributor holes or by mechanical pulsation of the liquid inside the distributor [Koch, J., and A. Vogelpohl, *Chem. Eng. Technol.* **24**(12): 1245–1248 (2001)]. In static extractors, hydrodynamic stresses within the main body of the extractor away from the distributor are small and normally not sufficient to cause significant drop breakage as drops flow through the extractor, although small drops may collide and coalesce into larger drops. Some authors report a small amount of drop breakage in packed columns due to collisions with packing materials [Mao, Z.-Q., J. C. Godfrey, and M. J. Slater, *Chem. Eng. Technol.* **18:** 33–40 (1995)]. Additional discussion is given in the subsection Static Extraction Columns under Liquid-Liquid Extraction Equipment.

In agitated extractors, drop size is determined by the equilibrium established between drop breakage and coalescence rates occurring within the extractor. Breakage is due to turbulent stresses caused by the agitator, so it is mainly confined to the vicinity of the agitator. Drop coalescence, however, can happen anywhere in the vessel where drops can come into close proximity with one another. Dispersed drops will begin to break into smaller droplets when turbulent stresses exceed the stabilizing forces of interfacial tension and liquid viscosity. A. N. Kolmogorov [*Dokl. Akad. Nauk* **66:** 825–828 (1949)] and J. O. Hinze [*AIChE J.* **1**(3): 289–295 (1955)] developed expressions for the maximum size of drops in an agitated liquid-liquid dispersion. Their results can be expressed as follows:

$$d_{max} = (\text{const})\left(\frac{\sigma}{\rho_c}\right)^{3/5}\left(\frac{P}{\mathcal{V}}\right)^{-2/5} \quad \text{for agitated extractors} \tag{15-38}$$

where $P/\mathcal{V}$ is the rate of mechanical energy dissipation (or power P) input to the dispersion per unit volume $\mathcal{V}$. Equation (15-38) assumes dispersed-phase holdup is low. It also assumes viscous forces that resist breakage can be neglected, a valid assumption for water and typical low- to moderate-viscosity organic solvents. Wang and Calabrese discuss how to determine when viscous resistance to breakage becomes important and show that this depends upon interfacial tension as well as dispersed-phase viscosity [Wang, C. Y., and R. V. Calabrese, *AIChE J.* **32**(4): 667–676 (1986)]. Equation (15-38) can be restated as

$$\frac{d_{max}}{D_i} \infty \mathrm{We}^{-3/5} \tag{15-39}$$

where We is a dimensionless Weber number (disruptive shear stress/cohesive interfacial tension) and D_i is a characteristic diameter. For applications involving the use of rotating impellers, D_i is the impeller diameter and the appropriate Weber number is $\mathrm{We} = \rho_c \omega^2 D_i^3/\sigma$, where ω is the impeller speed (in rotations per unit time). For static mixers, $D_i = D_{sm}$ and $\mathrm{We} = \rho_c V_{sm}^2 D_{sm}/\sigma$, where D_{sm} is the static mixer pipe diameter and V_{sm} is the superficial liquid velocity (entrance velocity). A variety of drop size models derived for various mixers and operating conditions have been tabulated by D. E. Leng and R. V. Calabrese [chap. 12 in *Handbook of Industrial Mixing*, ed. E. L. Paul, V. A. Atiemo-Obeng, and S. M. Kresta, Wiley, New York, 2004, pp. 669–675]. Also see additional discussion by Leng and Calabrese ["Immiscible Liquid-Liquid Systems," chap. 12 in *Advances in Industrial Mixing*, ed. S. M. Kresta et al., Wiley, New York, 2015] and by M. I. I. Z. Abidin, A. A. A. Raman, and M. I. M. Nor [*AIChE J.* **61**(4): 1129–1145 (2015)].

Equation (15-39) represents a limiting operating regime where the rate of drop breakage dominates performance and the coalescence rate can be neglected. Drop coalescence requires that two drops collide, and the coalescence rate increases with increasing holdup due to greater opportunity for drop-drop collisions. For agitated systems with fast coalescence at high holdup, that is, when drop coalescence dominates, drop size appears best correlated by an expression of the form $d_p/D \propto \mathrm{We}^{-n}$, where n varies between 0.35 and 0.45 [A. W. Pacek, C. C. Man, and A. W. Nienow, *Chem. Eng. Sci.* **53**(11): 2005–2011 (1998); and M. Kraume, A. Gabler, and K. Schulze, *Chem. Eng. Technol.* **27**(3): 330–334 (2004)].

When two drops first come into contact in the process of coalescing, a film of continuous phase becomes trapped between them. The film is compressed at the point of encounter until it drains away and the two drops can merge. Decreasing the viscosity of the continuous phase, by heating or by addition of a low-viscosity diluent, may promote drop coalescence by increasing the rate of film drainage. Surface-active impurities or surfactants, when present, also can affect the coalescence rate by accumulating at the surface of the drop. Surfactants tend to stabilize the film and reduce coalescence rates. Fine solid particles that are wetted by the continuous phase tend to slow film drainage, also reducing the rate of drop coalescence.

A number of semiempirical drop size data correlations have been developed for different types of extractors (static and agitated), including a term for holdup. See A. Kumar and S. Hartland, *Ind. Eng. Chem. Res.* **35**(8): 2682–2695 (1996); and A. Kumar and S. Hartland, chap. 17 in *Liquid-Liquid Extraction Equipment*, ed. J. C. Godfrey and M. J. Slater, Wiley, New York, 1994. These equations predict a characteristic drop size. They do not provide information about the drop size distribution or the minimum drop size. For discussion of minimum drop size, see G. Zhou and S. M. Kresta, *Chem. Eng. Sci.* **53**(11): 2063–2079 (1998).

STABILITY OF LIQUID-LIQUID DISPERSIONS

In designing a liquid-liquid extraction process, normally the goal is to generate an unstable dispersion that provides reasonably high interfacial area for good mass transfer during extraction and yet is easily broken to allow rapid liquid-liquid phase separation after extraction. Given enough time, most dispersions will break on standing. Often this process occurs in two distinct periods. The first is a relatively short initial period or primary break during which an interface forms between two liquid layers, one or both of which remain cloudy or turbid. This is followed by a longer period or secondary break during which the liquid layers become clarified. During the primary break, the larger drops migrate to the interface where they accumulate and begin to coalesce. If the coalescence rate is relatively slow compared to the rate at which drops rise or fall to the interface, then a layer of coalescing drops or *dispersion band* will form at the interface. The initial interface can form within a few minutes or less for drop sizes on the order of 100 to 1000 μm (0.1 to 1 mm), as in a water + toluene system, for example. When the drop size distribution in the feed dispersion is wide, smaller droplets remain suspended in one or both phases. Longer residence times are then required to break this secondary dispersion. In extreme cases, the secondary dispersion can take days or even longer to break.

When a dispersion requires a long time to break, the presence of surfactant-like impurities may be a contributing factor. Surfactants are molecules with a hydrophobic end (such as a long hydrocarbon chain) and a hydrophilic end (such as an ionic group or oxygen-containing short chain). Surfactants stabilize droplets by forming an adsorbed film at the interface and by introducing electrical repulsions between drops [Tcholakova, S., N. D. Denkov, and T. Danner, *Langmuir* **20**(18): 7444–7458 (2004)]. Both effects can interfere with drop coalescence. Surfactants also decrease the interfacial tension of the system. As more surfactant is introduced into a solution, the concentration of free surfactant molecules in the bulk liquid increases and reaches a plateau called the critical micelle concentration. At this point, any excess molecules begin forming aggregates with other surfactant molecules at the interface of the two liquids to minimize interaction with the continuous phase. The dispersed phase is then trapped inside the micelles. As more surfactant is added to the mixture, more micelles can form, and in most cases the droplets become smaller to maximize interfacial area. In theory, the maximum volume fraction of the dispersed phase should be limited to 0.74 due to the close packing density of spheres; but in practice much higher values are possible when the micelles change to other structures of different geometries such as a mix of small drops among larger ones and nonspherical shapes.

Emulsions are broken by changing conditions to promote drop coalescence, either by disrupting the film formed at the interface between adjacent drops or by interfering with the electrical forces that stabilize the drops. Water droplets usually are positively charged while oil droplets are negatively charged. Physical techniques used to break emulsions include heating (including application of microwave radiation), freezing and thawing, adsorption of surface-active compounds, filtration of fine particles that stabilize films between drops, and application of an electric field. Heating can be particularly effective for nonionic surfactants, because heating disrupts hydrogen bonding interactions that contribute to micelle stability. Chemical techniques include adding a salt to alter the charges around drops, changing the pH of the system, and adding a de-emulsifier compound (or even another type of surfactant) to interact with and alter the surfactant layer. Ionic surfactants are particularly sensitive to change in pH. Additives include bases and acids, aluminum or ferric salts, chelating agents, charged polymers (polyamines or polyacrylates), polyalcohols, silicone oils, various fatty acid esters and fatty alcohols, as well as adsorbents such as clay and lime. For further discussion, see V. N. Rajaković and D. Skala, *Sep. Purif. Technol.* **49**(2): 192–196 (2006); and G. R. Alther, *Chem. Eng. Magazine* **104**(3): 82–88 (1998). Chemical additives need to be used in sufficiently small concentrations so as not to interfere with other operations in the overall process or product quality. General information is available in L. L. Schramm, *Emulsions, Foams, and Suspensions*, Wiley-VCH, New York, 2005; P. Becher, *Emulsions: Theory and Practice*, 3d ed., American Chemical Society, Washington, DC, 2001; and B. P. Binks, *Modern Aspects of Emulsion Science*, Royal Society of Chemistry, London, 1998.

EFFECT OF SOLID-SURFACE WETTABILITY

The stability of a dispersion also may depend upon the surface properties of the container or equipment used to process the dispersion, because the walls of the vessel, or more importantly, the surfaces of any internal structures, may promote drop coalescence. In a liquid-liquid extractor or a liquid-liquid phase separator, the wetting of a solid surface by a liquid is a function of the interfacial tensions of both the liquid-solid and the liquid-liquid interfaces. For dispersed drops with low liquid-solid interfacial tension, the drops tend to spread out into films when in contact with the solid surface. In general, an aqueous liquid will tend to wet a metal or ceramic surface better than an organic liquid will, and an organic liquid will tend to wet a polymer surface better than an aqueous liquid will. However, there are many exceptions. R. F. Strigle [*Packed Tower Design and Applications*, 2d ed., chap. 11, Gulf, Houston, 1994] indicates that for packed extractors, metal packings may be wetted by either an aqueous or an organic solvent depending upon the initial exposure of the metal surface (whether the unit is started up filled with the aqueous phase or the organic phase). In general, however, metals tend to be preferentially wetted by an aqueous phase. Also, it is not uncommon for materials of construction to acquire different surface properties after aging in service due to adsorption of impurities, corrosion, or fouling. This aging effect often is observed for polymer materials. Small-scale lab tests are recommended to determine these wetting effects. For detailed discussion of wettability and its characterization, see *Contact Angle, Wettability, and Adhesion*, vols. 1–3, ed. K. L. Mittal, Wiley, New York, 2013; and J. C. Berg, ed., *Wettability*, Marcel Dekker, New York, 1993. Recently, new experimental polymer materials having specific surface roughness characteristics that allow design of specific wettability characteristics have been reported [Kota, A. K., et al., *Nat. Commun.* **3**(8): 1025 (2012); and G. Kwon et al., *MRS Commun.* **5**(3): 475–494 (2015)]. In any new application, the potential for change in wettability due to aging or fouling in service will need careful evaluation.

In liquid-liquid extraction equipment, the internals generally should be preferentially wetted by the continuous phase in order to maintain dispersed-phase drops with a high population density (high holdup). If the dispersed phase preferentially wets the internals, then drops may coalescence on contact with these surfaces, and this can result in loss of interfacial area for mass transfer and even in the formation of rivulets that flow along the internals. In an agitated extractor, this tendency may be mitigated somewhat, if needed, by increasing the agitation intensity.

MARANGONI INSTABILITIES

Numerous studies have shown that mass transfer of solute from one phase to the other can alter the behavior of a liquid-liquid dispersion because of interfacial tension gradients that form along the surface of a dispersed drop. These gradients can induce interfacial turbulence and circulation within drops, resulting in enhanced mass-transfer rates. For background on these phenomena, known as Maranoni instabilities, see C. V. Sternling and L. E. Scriven, *AIChE J.* **5**(4): 514–523 (1959); L. E. Scriven and C. V. Sternling, *Nature* **187:** 186–188 (1960); and Z.-S. Mao and J. Chen, *Chem. Eng. Sci.* **59:** 1815–1828 (2004).

The direction of mass transfer can alter the magnitude of Marangoni instabilities, affecting the rate of drop-drop coalescence and the resulting drop size. For example, A. F. Seibert and J. R. Fair [*Ind. Eng. Chem. Res.* **27**(3): 470–481 (1988)] showed that mass transfer out of a drop into the continuous phase can promote coalescence and production of larger dispersed drops. A. Kumar and S. Hartland [*Ind. Eng. Chem. Res.* **35**(8): 2682–2695 (1996)] suggest that transfer of solute from the dispersed to the continuous phase ($d \rightarrow c$) tends to produce larger drops because the concentration of transferring solute in the draining film between two approaching drops is higher than that in the surrounding continuous liquid. This lowers the local interfacial tension and accelerates drainage, thus promoting drop coalescence. For mass transfer in the opposite direction ($c \rightarrow d$), smaller drops tend to form because the solute concentration in the draining film between drops is relatively low. The magnitude of these effects depends upon system properties, the surface activity of the transferring solute, and the degree of mass transfer. Unless the solute is unusually surface-active, the effect will be small. For more information, see C. Gourdon, G. Casamatta, and G. Muratet, chap. 7 in *Liquid-Liquid Extraction Equipment*, ed. J. C. Godfrey and M. J. Slater, Wiley, New York, 1994; E. S. Perez de Oritz, chap. 3, "Marangoni Phenomena," in *Science and Practice of Liquid-Liquid Extraction*, vol. 1, ed. J. D. Thornton, Oxford University Press, Oxford, UK, 1992; and A. Grahn, *Chem. Eng. Sci.* **61:** 3586–3592 (2006).

PROCESS FUNDAMENTALS AND BASIC CALCULATION METHODS

General References: See Sec. 5, Heat and Mass Transfer, as well as J. D. Seader, E. J. Henley, and D. K. Roper, *Separation Process Principles with Applications Using Process Simulators*, 4th ed., Wiley, New York, 2016; Wankat, P. C., *Separation Process Engineering*, 3d ed., Prentice-Hall, Upper Saddle River, NJ, 2012; Godfrey, J. C., and M. J. Slater, *Liquid-Liquid Extraction Equipment*, Wiley, New York, 1994; Thornton, J. D., ed., *Science and Practice of Liquid-Liquid Extraction*, vol. 1, Oxford University Press, Oxford, UK, 1992; Wankat, P. C., *Equilibrium Staged Separations*, Prentice-Hall, Englewood Cliffs, NJ, 1988; Kirwan, D. J., chap. 2 in *Handbook of Separation Process Technology*, ed. Rousseau, Wiley, New York, 1987; Skelland, A. H. P., and D. W. Tedder, chap. 7 in *Handbook of Separation Process Technology*, ed. R. W. Rousseau, Wiley, New York, 1987; Lo, T. C., M. H. I. Baird, and C. Hanson, eds., *Handbook of Solvent Extraction*, Wiley, New York, 1983, and Krieger, Huntington, NY, 1991; King, C. J., *Separation Processes*, 2d ed., McGraw-Hill, New York, 1980, and Dover, 2013; Brian, P. L. T., *Staged Cascades in Chemical Processing*, Prentice-Hall, Englewood Cliffs, NJ, 1972; Geankoplis, C. J., *Mass Transport Phenomena*, Holt, Rinehart and Winston, New York, 1972; and Treybal, R. E., *Liquid Extraction*, 2d ed., McGraw-Hill, New York, 1963.

The fundamental mechanisms for solute mass transfer in liquid-liquid extraction involve molecular diffusion from one liquid to the other driven by a deviation from equilibrium plus convective (or advective) mass transfer due to bulk flow of the two liquids. When a liquid feed is contacted with a liquid solvent, solute transfers from the interior of the feed across a liquid-liquid interface into the interior of the solvent. This occurs while solute moves with the bulk flow of the two liquids. Given sufficient contacting time, transfer of solute from one liquid to the other will continue until the solute's chemical potential is the same in both liquids and equilibrium is achieved. The bulk flow of liquids determines residence time in the equipment and thus the amount of time available for solute diffusion. It also can affect properties of the dispersion that impact mass-transfer resistance (including drop size distribution, population density or holdup, and interfacial area).

The calculation methods used to quantify extraction processes generally involve either the calculation of the number of theoretical stages, with the application of an operating efficiency to reflect mass-transfer resistance, or calculations based on consideration of mass-transfer rates using expressions related in some way to molecular diffusion, interfacial area, and bulk flow rates. One must carefully consider flow rates, even when using the theoretical stage approach, because flow rates can have a dramatic impact on efficiency (by affecting residence time, etc.). Theoretical-stage calculations commonly are used to characterize separation difficulty regardless of the type of extractor to be used. They are also used for extractor design purposes, although for this purpose they generally should be reserved for single-stage contactors or mixer-settler cascades involving discrete stages, or for other equipment where discrete contacting zones exist, such as in a sieve tray column. Rate-based models most often are applied to differential-type contactors that lack discrete contacting stages, to staged contactors with low stage efficiencies, or to processes with extraction factors greater than about 3, indicating a mass-transfer-limited operating regime. Differential-type contactors operating at extraction factors less than 3 also may be adequately modeled with theoretical stages because these contactors operate reasonably close to equilibrium.

With either theoretical stage models or rate-based models, appropriate values for model parameters typically are determined by fitting data generated by using laboratory or pilot-plant experiments, or by analysis of the performance of large-scale commercial units. In certain cases, parameter values have been correlated as a function of physical properties and operating conditions for specific types of equipment using model systems. The reliability of the resulting correlations is generally limited to applications very similar to those used to develop the correlations. Also, most calculation methods have been developed for continuous steady-state operation. The dynamic modeling of extraction processes is discussed elsewhere [Mohanty, S., *Rev. Chem. Eng.* **16**(3): 199 (2000); Weinstein, O., R. Semiat, and D. R. Lewin, *Chem. Eng. Sci.* **53**(2): 325–339 (1998); and Steiner, L., and S. Hartland, chap. 7 in *Handbook of Solvent Extraction*, ed. T. C. Lo, M. H. I. Baird, and C. Hanson, Wiley, New York, 1983, and Krieger, Huntington, NY, 1991].

The calculation methods used for designing standard extraction operations are analogous in many respects to methods used to design absorbers and strippers in vapor-liquid and gas-liquid contacting such as those described by J. R. Ortiz-Del Castillo, et al. [*Ind. Eng. Chem. Res.* **39**(3): 731–739 (2000)] and by A. L. Kohl ["Absorption and Stripping," chap. 6 in *Handbook of Separation Process Technology*, ed. R. W. Rousseau, Wiley-Interscience, New York, 1987]. Unlike in stripping and absorption, however, liquid-liquid extraction always deals with highly nonideal systems; otherwise, only one liquid phase would exist. This nonideality contributes to difficulties in modeling and predicting phase equilibria, liquid-liquid phase behavior (hydraulics), and thus mass transfer. Also, the mass-transfer efficiency of an extractor generally is much less than that observed in distillation, stripping, or absorption equipment. For example, an overall sieve tray efficiency of 70 percent is common in distillation, but it is rare when a sieve tray extractor achieves an overall efficiency greater than 30 percent. The difference arises in part because generation of interfacial area, normally by dispersing drops of one phase in the other, generally is more difficult in liquid-liquid contactors. Unlike in distillation, the formation of liquid films often is purposely avoided; generation of dispersed droplets provides greater interfacial area for mass transfer per unit volume of extractor. (Film formation may be important in extraction applications involving centrifugal contactors or baffle tray extractors, but this is not generally the case.) In certain cases, mass-transfer rates also may be slower compared to those of gas-liquid contactors because the second phase is a liquid instead of a gas, and transport properties in that phase are less favorable. Although mass-transfer efficiency generally is lower, the specific throughput of liquid-liquid extraction equipment (in kilograms of feed processed per hour per unit volume) can be higher than is typical of vapor-liquid contactors, simply because liquids are much denser than vapors.

THEORETICAL (EQUILIBRIUM) STAGE CALCULATIONS

Calculating the number of theoretical stages is a convenient method used by process designers to evaluate separation difficulty and assess the compromise between the required equipment size (column height or the number of actual stages) and the ratio of solvent rate to feed rate required to achieve the desired separation. In any mass-transfer process, there can be an infinite number of combinations of flow rates, number of stages, and degrees of solute transfer. The optimum is governed by economic considerations. The cost of using a high solvent rate with relatively few stages should be carefully compared with the cost of using taller extraction equipment (or more equipment) capable of achieving more theoretical stages at a reduced solvent rate and operating cost. While the operating cost of an extractor is generally quite low, the operating cost for a solvent recovery distillation tower can be quite high. Another common objective for calculating the number of countercurrent theoretical stages is to evaluate the performance of liquid-liquid extraction test equipment in a pilot plant or to evaluate production equipment in an industrial plant. As mentioned earlier, most liquid-liquid extraction equipment in common use can be designed to achieve the equivalent of 1 to 8 theoretical countercurrent stages, with some designed to achieve 10 to 12 stages.

McCabe-Thiele Type of Graphical Method Graphical methods may be used to determine theoretical stages for a ternary system (solute plus feed solvent and extraction solvent) or for a pseudo-ternary with the focus placed on a key solute of interest. Although developed long ago, graphical methods are still valuable today because they help visualize the problem, clearly illustrating pinch points and other design issues not readily apparent by using other techniques. Even with computer simulations, often it is useful to plot the results for a key solute as an aid to analyzing the design. This section briefly reviews the commonly used McCabe-Thiele type of graphical method. More detailed discussions of this and other graphical methods are available elsewhere. For example, see A. F. Seibert, "Extraction and Leaching," chap. 14 in *Chemical Process Equipment: Selection and Design*, 3d ed., ed. J. R. Couper et al., Butterworth-Heinemann, Oxford, UK, 2012; P. C. Wankat, *Separation Process Engineering*, Prentice-Hall, Upper Saddle River, NJ, 2012; and C. J. King, *Separation Processes*, 2d ed., McGraw-Hill, New York, 1980, and Dover, 2013, among others. Also see instructional materials available on the Internet, such as those from the University of Colorado's LearnChemE web site. An example is the discussion of the Hunter-Nash graphical method used for liquid-liquid extractor design (www.colorado.edu/learncheme/separations/HunterHashMethodLLE.html, accessed January 15, 2018).

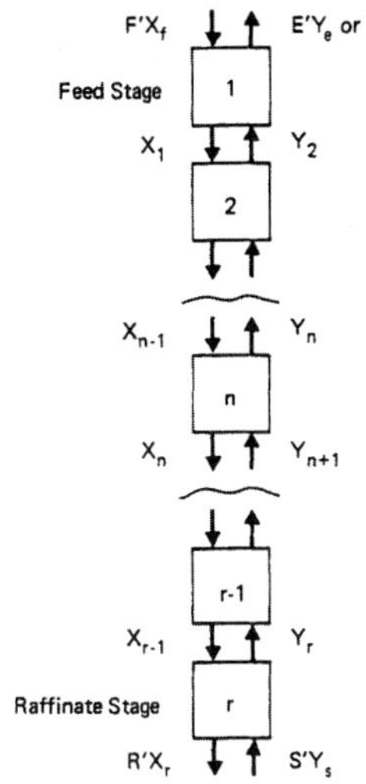

FIG. 15-22 Countercurrent extraction cascade.

In distillation calculations, the McCabe-Thiele graphical method assumes constant molar vapor and liquid flow rates and allows convenient stepwise calculation with straight operating lines and a curved equilibrium line. A similar concept can be achieved in liquid-liquid extraction by using Bancroft coordinates and expressing flow rates on a solute-free basis, that is, a constant flow rate of feed solvent F' and a constant flow rate of extraction solvent S' through the extractor [T. W. Evans, *Ind. Eng. Chem.* **26**(8): 860–864 (1934)]. The solute concentrations are then given as the mass ratio of solute to feed solvent X' and the mass ratio of solute to extraction solvent Y'. These concentrations and coordinates give a straight operating line on an X'–Y' diagram for stages 2 through $r - 1$ in Fig. 15-22. The ratio of solute-free extraction solvent to solute-free feed solvent will be constant within the extractor except at the outer stages where unsaturated feed and extraction solvent enter the process. Equilibrium data using these mass ratios have been shown to follow straight-line segments on a log-log plot (see Fig. 15-20), and they will be approximately linear over some composition range on an X'–Y' plot. When expressed in terms of Bancroft coordinates, the equilibrium line typically will curve upward at high solute concentrations, as shown in Fig. 15-23.

To illustrate the McCabe-Thiele method, consider the simplified case where feed and extraction solvents are immiscible; that is, mutual solubility is nil. Then the rate of feed solvent alone in the feed stream F' is the same

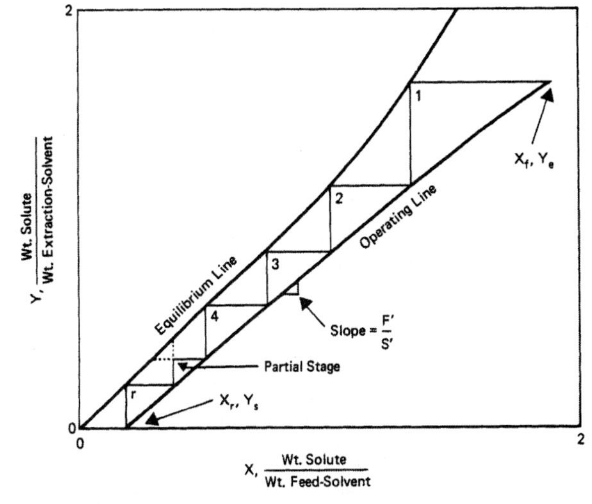

FIG. 15-23 McCabe-Thiele type of graphical stage calculation using Bancroft coordinates.

as the rate of feed solvent alone in the raffinate stream R'. In like manner, the rate of extraction solvent alone is the same in the entering stream S' as in the leaving extract stream E'. The ratio of extraction-solvent to feed-solvent flow rates is therefore $S'/F' = E'/R'$. A material balance can be written around the feed end of the extractor down to any stage n (as shown in Fig. 15-22) and then rearranged to a McCabe-Thiele type of operating line with a slope of F'/S':

$$Y'_{n+1} = \frac{F'}{S'} X'_n + \frac{E'Y'_e - F'X'_f}{S'} \tag{15-40}$$

Similarly, the same operating line can be derived from a material balance around the raffinate end of the extractor up to stage n:

$$Y'_n = \frac{F'}{S'} X'_{n-1} + \frac{S'Y'_s - R'X'_r}{S'} \tag{15-41}$$

The overall extractor material balance is given by

$$Y'_e = \frac{F'X'_f + S'Y'_s - R'X'_r}{E'} \tag{15-42}$$

The endpoints of the operating line on an X'–Y' plot (Fig. 15-23) are the points (X'_r, Y'_s) and (X'_f, Y'_e) where X' and Y' are the mass ratios for solute in the feed phase and extract phase, respectively, and subscripts f, r, s, and e denote the feed, raffinate, entering extraction solvent, and leaving extract streams. The number of theoretical stages can then be stepped off graphically as illustrated in Fig. 15-23.

Kremser-Souders-Brown Theoretical Stage Equation The Kremser-Souders-Brown (KSB) equation [A. Kremser, *Natl. Petrol. News* **22**(21): 43–49 (1930); and M. Souders and G. G. Brown, *Ind. Eng. Chem.* **24**(5): 519–522 (1932)] provides a way of calculating performance equivalent to that of a McCabe-Thiele type of graphical calculation with straight equilibrium and operating lines. In terms of Bancroft coordinates, the KSB equation may be written

$$N = \frac{\ln\left[\left(\frac{X'_f - Y'_s/m'}{X'_r - Y'_s/m'}\right)\left(1 - \frac{1}{\mathcal{E}}\right) + \frac{1}{\mathcal{E}}\right]}{\ln \mathcal{E}} \qquad \mathcal{E} = m'\frac{S'}{F'}, \mathcal{E} \neq 1 \tag{15-43}$$

where N = number of theoretical stages
X'_f = mass ratio solute to feed solvent in feed entering process (Bancroft coordinates)
X'_r = mass ratio solute to feed solvent in raffinate leaving process
Y'_s = mass ratio solute to extraction solvent in extraction solvent entering process
$\mathcal{E}$ = extraction factor
$m' = dY'/dX'$, local slope of equilibrium line in Bancroft coordinates
S' = mass flow rate of extraction solvent (solute-free basis)
F' = mass flow rate of feed solvent (solute-free units)

Solutions to Eq. (15-43) are shown graphically in Fig. 15-24. The concentration of solute in the extract leaving the process Y'_e is determined from the material balance, as in Eq. (15-42). (Note that other systems of units also may be used here, as long as they are consistently applied.)

Rearranging Eq. (15-43) yields another common form of the KSB equation:

$$\frac{X'_f - Y'_s/m'}{X'_r - Y'_s/m'} = \frac{\mathcal{E}^N - 1/\mathcal{E}}{1 - 1/\mathcal{E}} \qquad \mathcal{E} \neq 1 \tag{15-44}$$

Equations (15-43) and (15-44) can be used whenever $\mathcal{E} > 1$ or $\mathcal{E} < 1$. They cannot be used when $\mathcal{E}$ is exactly equal to unity because this would involve division by zero. When $\mathcal{E} = 1$, the number of theoretical stages is given by

$$N = \frac{X'_f - Y'_s/m'}{X'_r - Y'_s/m'} - 1 \qquad \text{for } \mathcal{E} = 1 \tag{15-45}$$

Equation (15-45) may be rewritten

$$\frac{X'_f - Y'_s/m'}{X'_r - Y'_s/m'} = N + 1 \qquad \text{for } \mathcal{E} = 1 \tag{15-46}$$

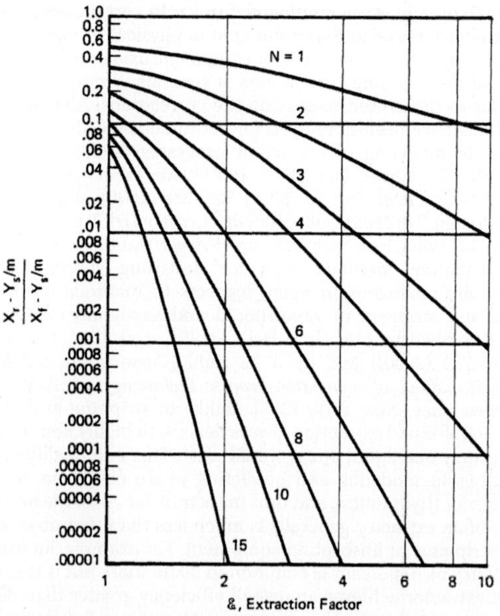

FIG. 15-24 Graphical solutions to the KSB equation [(Eq. 15-43)].

In the special case where $\mathcal{E} < 1$, the *maximum* performance potential is represented by

$$\left(\frac{X'_f - Y'_s/m'}{X'_r - Y'_s/m'}\right)_{max} \approx \frac{1}{1 - \mathcal{E}} \qquad \text{for } \mathcal{E} < 1 \text{ and large } N \tag{15-47}$$

Equation (15-47) reflects the fact that the carrying capacity of the extract stream limits performance at $\mathcal{E} < 1$, as noted in earlier discussions.

In general, Eqs. (15-43) through (15-47) (and Fig. 15-24) are valid for any concentration range in which equilibrium can be represented by a linear relationship $Y = mX + b$ (written here in general form for any system of units). For applications that involve dilute feeds, the section of the equilibrium line of interest is a straight line that extends through the origin where $Y_i = 0$ at $X_i = 0$. In this case, $b = 0$ and the slope of the equilibrium line is equal to the partition ratio ($m = K$). The KSB equation also may be used to represent a linear segment of the equilibrium curve at higher solute concentrations. In this case, the linear segment is represented by a straight line that does not extend through the origin, and m is the local slope of the equilibrium line, so $b \neq 0$ and $m \neq K$. Furthermore, a series of KSB equations may be used to model a highly curved equilibrium line by dividing the analysis into linear segments and matching concentrations where the segments meet. For equilibrium lines with moderate curvature, an approximate average slope of the equilibrium line may be obtained from the geometric mean of the slopes at low and high solute concentrations:

$$m_{\text{average}} \approx m_{\text{geometric mean}} = \sqrt{m_{\text{low}} m_{\text{high}}} \tag{15-48}$$

As we have noted, other systems of units such as mass fraction and mass flow rates or mole fraction and molar flow rates also may be used with the KSB equation; however, Bancroft coordinates and solute-free mass flow rates are recommended because then the operating line must be linear, and this normally extends the concentration range over which the KSB analysis may be used. It is important to check whether equilibrium can be adequately represented by a straight line over the concentration range of interest. The application of the KSB equation is discussed in the subsection Shortcut Calculations under Calculation Procedures. Additional discussion is given by P. C. Wankat [*Equilibrium Staged Separations*, Prentice-Hall, Englewood Cliffs, NJ, 1988] and by C. J. King [*Separation Processes*, 2d ed., McGraw-Hill, New York, 1980]. To facilitate the use of the KSB equation in computer calculations where the singularity around $\mathcal{E}$ = 1 can present difficulties, U. V. Shenoy and D. M. Fraser have proposed an alternative form of the equation [*Chem. Eng. Sci.* **58**(22): 5121–5124 (2003)].

Stage Efficiency For a multistage process, the overall stage efficiency is simply the number of theoretical stages divided by the number of actual stages times 100:

$$\xi_o(\%) = \frac{\text{theoretical stages}}{\text{actual stages}} \times 100 \qquad (15\text{-}49)$$

The fundamental stage efficiency is referred to as the Murphree stage efficiency ξ_m. The Murphree efficiency based on the dispersed phase is defined as

$$\xi_{\text{md}} = \frac{C_{d,n+1} - C_{d,n}}{C_{d,n+1} - C_d^*} \qquad (15\text{-}50)$$

where $C_{d,n+1}$ = concentration of solute i in dispersed phase at stage $n+1$
$C_{d,n}$ = concentration of solute i in dispersed phase at stage n
C_d^* = concentration of solute i in dispersed phase, at equilibrium

The overall stage efficiency is related to the Murphree stage efficiency and the extraction factor:

$$\xi_o(\%) = \frac{\ln[1 + \xi_{\text{md}}(\mathcal{E} - 1)]}{\ln \mathcal{E}} \times 100 \qquad (15\text{-}51)$$

For applications involving extraction of multiple solutes, sometimes the extraction rate and mass-transfer efficiency for each solute are significantly different. In these cases, individual efficiencies will need to be determined for each solute.

Stage efficiencies normally are determined by running miniplant tests to measure performance as a function of process variables such as feed rates, operating temperature, physical properties, impurities, and agitation (if used). A number of data correlations have been developed for various types of mixing equipment. In principle, these can be used in the estimation of mass-transfer rates and stage efficiencies, but in practice reliable design generally requires generation of miniplant data and application of mixing scale-up methods. (See the subsection Mixer-Settler Equipment under Liquid-Liquid Extraction Equipment.)

The overall efficiency of an extraction column also can be expressed as the height equivalent to a theoretical stage (HETS). This is simply the total contacting height Z_t divided by the number of theoretical stages achieved.

$$\text{HETS} = \frac{Z_t}{N} \qquad (15\text{-}52)$$

The HETS often is used to compare staged contactors with differential contactors.

RATE-BASED CALCULATIONS

This subsection reviews the basics of Fickian diffusion, the mass-transfer coefficient, and mass-transfer unit approaches to modeling extraction performance. These methods have been used for many years and continue to provide a useful basis for the design of extractors and extraction processes. Additional discussions of these and other rate-based methods are given in the books edited by J. C. Godfrey and M. J. Slater [*Liquid-Liquid Extraction Equipment*, Wiley, New York, 1994] and by J. D. Thornton [*Science and Practice of Liquid-Liquid Extraction*, vol. 1, Oxford University Press, Oxford, UK, 1992]. See R. Taylor and R. Krishna, *Multicomponent Mass Transfer*, Wiley, New York, 1993; and R. Krishna, *Ind. Eng. Chem. Res.* **55**(4): 1053–1063 (2016), for discussion of the Maxwell-Stefan approach to modeling multicomponent diffusion. For discussions of drop breakage and coalescence rates, drop size distributions, and drop population balances, see D. E. Leng and R. V. Calabrese, "Immiscible Liquid-Liquid Systems," chap. 12 in *Advances in Industrial Mixing*, ed. S. M. Kresta et al., Wiley, New York, 2015, and chap. 12 in *Handbook of Industrial Mixing*, ed. E. L. Paul, V. A. Atiemo-Obeng, and S. M. Kresta, Wiley, New York, 2004; and M. I. I. Z. Abidin, A. A. A. Raman, and M. I. H. Nor, *AIChE J.* **61**(4): 1129–1145 (2015). Also see the discussion of general approaches to analyzing dispersed-phase systems given by D. Ramkrishna, A. Sathyagal, and G. Narsimhan [*AIChE J.* **41**(1): 35–44 (1995)]. For discussions of the effect of contaminants on mass-transfer rates, see J. Saien et al., *Ind. Eng. Chem. Res.* **45**(4): 1434–1440 (2006); and A. M. Dehkordi et al., *Ind. Eng. Chem. Res.* **46**(5): 1563–1571 (2007).

Solute Diffusion and Mass-Transfer Coefficients For a binary system consisting of components A and B, the overall rate of mass transfer of component A with respect to a fixed coordinate is the sum of the rates due to diffusion and bulk flow:

$$N_A = -\mathcal{D}_{AB} \frac{\partial C_A}{\partial z} + N_A \frac{C_A}{C} \qquad (15\text{-}53)$$

where N_A = flux for component A (moles per unit area per unit time)
$\mathcal{D}_{AB}$ = mutual diffusion coefficient of A into B (area/unit time)
z = dimension or direction of mass transfer (length)
C = total concentration of A and B (mass or mole per unit volume)
C_A = concentration of A (mass or mole per unit volume)

Equation (15-53) is written for steady-state unidirectional diffusion in a quiescent liquid, assuming that the net transfer of component B is negligible. For transfer of component A across an interface or film between two liquids, it may be rewritten in the form

$$N_A = \frac{\mathcal{D}_{AB}}{\Delta z (1 - x_A)_m} \left(C_A - C_A^i\right) \qquad (15\text{-}54)$$

where $(1 - x_A)_m$ = mean mole fraction of component B
C_A^i = concentration of component A at interface
C_A = concentration of component A in bulk

For steady-state counter diffusion where $N_A + N_B = 0$, the flux equation simplifies to

$$N_A = \frac{\mathcal{D}_{AB}}{\Delta z} \left(C_A - C_A^i\right) \qquad (15\text{-}55)$$

The flux also may be written in terms of an individual mass-transfer coefficient k

$$N_A = k(C_A - C_A^i) \qquad (15\text{-}56)$$

where

$$k = \frac{\mathcal{D}_{AB}}{\Delta z (1 - x_A)_m} \qquad (15\text{-}57)$$

In Eqs. (15-53) to (15-57), the flux is expressed in terms of mass or moles per unit area per unit time, and the concentration driving force is defined in terms of mass or moles per unit volume. The units of the mass-transfer coefficients are then length per unit time. Whenever mass-transfer coefficients are reported, it is important to check how they have been defined (which may vary from our example) and how they were determined; they need to be used in the same way in any subsequent calculations. Additional discussion of mass-transfer coefficients and mass-transfer rate expressions is given in Sec. 5. Also see G. S. Laddha and T. E. Degaleesan, chap. 3 in *Transport Phenomena in Liquid Extraction*, McGraw-Hill, New York, 1978; A. H. P. Skelland, *Diffusional Mass Transfer*, Krieger, Huntington, NY, 1985; A. H. P. Skelland and D. W. Tedder, chap. 7 in *Handbook of Separation Process Technology*, ed. R. W. Rousseau, Wiley, New York, 1987; R. B. Bird, W. E. Stewart, and E. N. Lightfoot, *Transport Phenomena*, 2d ed., Wiley, New York, 2002; and E. L. Cussler, *Diffusion: Mass Transfer in Fluid Systems*, 3d ed., Cambridge, 2009. Available correlations of molecular diffusion coefficients (diffusivities) are discussed in Sec. 5 and in B. E. Poling, J. M. Prausnitz, and J. P. O'Connell, *The Properties of Gases and Liquids*, 5th ed., McGraw-Hill, New York, 2000.

Mass-Transfer Rate and Overall Mass-Transfer Coefficients In transferring from one phase to the other, a solute must overcome certain resistances: (1) movement from the bulk of the feed phase to the interface; (2) movement across the interface; and (3) movement from the interface to the bulk of the extract phase, as illustrated in Fig. 15-25. The two-film theory first used to model this process [Lewis, W. K. and W. G. Whitman, *Ind. Eng. Chem.* **16:** 1215–1220 (1924)] assumes that motion in the two phases is negligible near the interface such that the entire resistance to transfer is contained within two laminar films on each side of the interface, and mass transfer occurs by molecular diffusion through these films. The theory further invokes the following simplifying assumptions: (1) The rate of mass transfer within each phase is proportional to the difference in concentration in the bulk liquid and the interface; (2) mass-transfer resistance across the interface itself is negligible, and the phases are in equilibrium at the interface; and (3) steady-state diffusion occurs with negligible holdup of diffusing solute at the interface. Within a liquid-liquid extractor, the rate of steady-state mass transfer between the dispersed phase and the continuous phase (mass or moles per unit time per unit volume of extractor) is then expressed as

$$R_A = \frac{dC}{dt} = k_d a(C_{d,i} - C_d) = k_c a(C_c - C_{c,i}) \qquad (15\text{-}58)$$

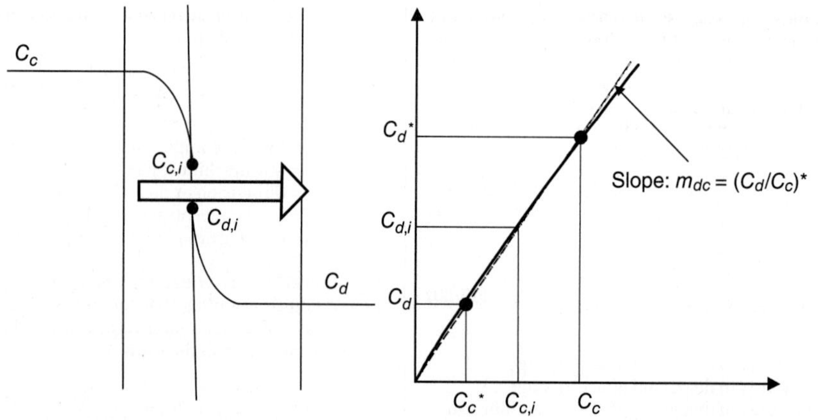

FIG. 15-25 Two-film mass transfer.

where C_i = concentration at interface (mass or moles per unit volume)
C = concentration in bulk liquid (mass or moles per unit volume)
k_c = continuous-phase mass-transfer coefficient (length per unit time)
k_d = dispersed-phase film mass-transfer coefficient (length per unit time)
a = interfacial area for mass transfer per unit volume of extractor (length^{-1})

Subscripts d and c denote the dispersed and continuous phases. The concentrations at the interface normally are not known, so the rate expression is written in terms of equilibrium concentrations assuming that the rate is proportional to the deviation from equilibrium:

$$R_A = \frac{dC}{dt} = k_{od}a(C_d^* - C_d) = k_{oc}a(C_c - C_c^*) \tag{15-59}$$

where the superscript * denotes equilibrium, and k_{oc} is an overall mass-transfer coefficient given by

$$\frac{1}{k_{oc}} = \underbrace{\frac{1}{k_c}}_{\text{Continuous phase resistance}} + \underbrace{\frac{1}{m_{dc}^{vol} k_d}}_{\text{Dispersed phase resistance}} \tag{15-60}$$

Similarly, the overall mass-transfer coefficient based on the dispersed phase is given by

$$\frac{1}{k_{od}} = \underbrace{\frac{1}{k_d}}_{\text{Dispersed phase resistance}} + \underbrace{\frac{m_{dc}^{vol}}{k_c}}_{\text{Continuous phase resistance}} \tag{15-61}$$

Assuming mass-transfer coefficients are constant over the range of conditions of interest, Eq. (15-59) may be integrated to give

$$\frac{C_c - C_c^*}{C_{c,\text{initial}} - C_c^*} = \exp(-k_{oc}a\theta) \approx \frac{C_c}{C_{c,\text{initial}}} \tag{15-62}$$

where θ is the contact time. In general, the application of mass-transfer coefficients for extractor design requires knowledge of interfacial area, normally obtained from estimates of average drop size and dispersed-phase holdup, as well as information related to contact time, normally expressed in terms of maximum phase velocities limited by flooding behavior. As we discuss later, a correction for the effect of axial mixing on scale-up also is needed for column extractors.

In Eqs. (15-60) and (15-61), $m_{dc}^{vol} = dC_d/dC_c$ is the local slope of the equilibrium line, with the equilibrium concentration of solute in the dispersed phase plotted on the ordinate (y axis), and the equilibrium concentration of solute in the continuous phase plotted on the abscissa (x axis). Note that m_{dc}^{vol} is expressed on a volumetric basis (denoted by superscript *vol*), that is, in terms of mass or mole per unit volume, because of the way the mass-transfer coefficients are defined. The mass-transfer coefficients will not necessarily be the same for each solute being extracted, so depending upon the application, mass-transfer coefficients may need to be determined for a range of different solutes. As noted earlier, other systems of units also may be used as long as they are consistently applied.

The mass-transfer coefficient in each film is expected to depend upon molecular diffusivity, and this behavior often is represented by a power-law function $k \propto \mathcal{D}^n$. For two-film theory, $n = 1$ as discussed previously [(Eq. (15-57)]. Subsequent theories introduced by R. W. Higbie [*Trans. AIChE* **31:** 365 (1935)] and by P. V. Dankwerts [*Ind. Eng. Chem.* **43:** 1460–1467 (1951)] allow for surface renewal or penetration of the stagnant film. These theories indicate a 0.5 power-law relationship. Numerous models have been developed since then, where $0.5 < n < 1.0$. The results depend upon such things as whether the dispersed drop is treated as a rigid sphere, as a sphere with internal circulation, or as oscillating drops. These theories are discussed by A. H. P. Skelland ["Interphase Mass Transfer," chap. 2 in *Science and Practice of Liquid-Liquid Extraction*, vol. 1, ed. J. D. Thornton, Oxford University Press, Oxford, UK, 1992].

Mass-transfer coefficients are functions of bulk fluid flow (convective mass transfer), as well as molecular diffusion. In the design of extraction equipment with complex flows, mass-transfer coefficients are determined by experiment and then correlated as a function of flow rates, molecular diffusivity, physical properties, and specific equipment factors such as the geometry of equipment internals and agitation intensity, if mechanical agitation is used. The application of mass-transfer coefficients also requires calculation of interfacial area, or values of $k \times a$ are correlated together. The available theories provide an approximate framework for the data. In most cases, the dominant mass-transfer resistance resides in the feed (raffinate) phase, as the slope of the equilibrium line usually is greater than unity. In that case, the overall mass-transfer coefficient based on the raffinate phase may be written

$$\frac{1}{k_{or}} = \frac{1}{k_r} + \frac{1}{m_{er}^{vol} k_e} \approx \frac{1}{k_r} \quad \text{for large } m_{er}^{vol} \tag{15-63}$$

where m_{er}^{vol} is defined by the usual convention in terms of concentration in the extract phase over that in the raffinate phase, $m_{dc}^{vol} = dC_{i,\text{extract}}/dC_{i,\text{raffinate}}$. This approximation is particularly useful when the extraction solvent is significantly less viscous than the feed liquid, so the solute diffusivity and mass-transfer coefficient in the extract phase are relatively large. For detailed discussion of mass-transfer coefficients, see A. Kumar and S. Hartland, *Trans. Inst. Chem. Eng.* Part A, **77:** 372–384 (1999).

Mass-Transfer Units The mass-transfer unit concept follows directly from mass-transfer coefficients. The choice of one or the other as a basis for analyzing a given application often is one of preference. A. P. Colburn [*Ind. Eng. Chem.* **33**(4): 450–467 (1941)] provides an early review of the relationship between the height of a transfer unit and volumetric mass-transfer coefficients ($k_{or}a$). From a differential material balance and application of the flux equations, the required contacting height of an extraction column Z_t is related to the height of a transfer unit H_{or} and the number of transfer units N_{or}

$$Z_t = \left(\frac{V_r}{k_{or}a}\right)\int_{X_{out}}^{X_{in}} \frac{dX}{X - X^*} = H_{or} \times N_{or} \tag{15-64}$$

where V_r is the velocity of the raffinate phase, a is the interfacial area per unit volume, and the superscript * denotes the equilibrium concentration. Equation (15-64) is written in terms of an overall mass-transfer coefficient based on the raffinate phase for the usual case in which the main resistance to mass transfer is in the raffinate. The transfer unit model has proved to be a convenient framework for characterizing mass-transfer performance. An advantage of this model compared to the theoretical stage model is the observation that the height of a transfer unit, H_{or}, given by the quantity $V_r/k_{or}a$, normally is less affected by changes in process operating variables compared to the height equivalent to a theoretical stage (HETS).

Thus, mass-transfer units are defined as the integral of the differential change in solute concentration divided by the deviation from equilibrium, between the limits of inlet and outlet solute concentrations:

$$N_{or} = \int_{X_{out}}^{X_{in}} \frac{dX}{X - X^*} \tag{15-65}$$

When equilibrium and operating lines are linear, Eq. (15-65) can be expressed as

$$N_{or} = \frac{\ln\left[\left(\frac{X'_f - Y'_s/m'}{X'_r - Y'_s/m'}\right)\left(1 - \frac{1}{\mathcal{E}}\right) + \frac{1}{\mathcal{E}}\right]}{1 - \frac{1}{\mathcal{E}}} \qquad \mathcal{E} = m'\frac{S'}{F'}, \mathcal{E} \neq 1 \tag{15-66}$$

where N_{or} is the number of overall mass-transfer units based on the raffinate phase. The units are the same as those used previously for the KSB equation, Eq. (15-43). Rearranging Eq. (15-66) gives

$$\frac{X'_f - Y'_s/m'}{X'_r - Y'_s/m'} = \frac{\exp[N_{or}(1 - 1/\mathcal{E})] - 1/\mathcal{E}}{1 - 1/\mathcal{E}} \tag{15-67}$$

Note that Eq. (15-66) is the same as Eq. (15-43) except in the denominator. Comparing these equations shows that the number of overall raffinate phase transfer units is related to the number of theoretical stages by

$$N_{or} = N \times \frac{\ln \mathcal{E}}{1 - 1/\mathcal{E}} \tag{15-68}$$

The difference becomes pronounced when values of the extraction factor are high. When $\mathcal{E} = 1$, the number of mass-transfer units and number of theoretical stages are the same:

$$N_{or} = N = \frac{X'_f - Y'_s/m'}{X'_r - Y'_s/m'} - 1 \qquad \text{for } \mathcal{E} = 1 \tag{15-69}$$

As with the KSB equation, in the special case where $\mathcal{E} < 1$, the maximum performance potential is represented by

$$\left(\frac{X'_f - Y'_s/m'}{X'_r - Y'_s/m'}\right)_{max} \approx \frac{1}{1 - \mathcal{E}} \qquad \text{for } \mathcal{E} < 1 \text{ and large } N_{or} \tag{15-70}$$

Equation (15-66) often is referred to as the Colburn equation. Although commonly used to represent the performance of a differential contactor, it models any steady-state, diffusion-controlled process with straight equilibrium and operating lines. As with the KSB equation, the operating line is straight even when solute concentration changes significantly as long as Bancroft coordinates are used, and both the KSB and Colburn equations can be used to model applications involving a highly curved equilibrium line by dividing the analysis into linear segments. With these approaches, these equations often can be used for applications involving high concentrations of solute.

Solutions to the Colburn equation are shown graphically in Fig. 15-26. Note the contrast to the KSB equation solutions shown in Fig. 15-24. The KSB equations are best used to model countercurrent contact devices where the separation is primarily governed by equilibrium limitations, such as extractors involving discrete stages with high stage efficiencies. The Colburn equation, on the other hand, better represents the performance of a diffusion rate-controlled contactor because performance approaches a definite limit as the extraction factor increases beyond $\mathcal{E} = 10$ or so, corresponding to a diffusion rate limitation where addition of extra solvent has little or no effect. Note that in the Colburn equation, Eq. (15-66), the extraction factor always appears as $1/\mathcal{E}$, and this is how a finite diffusion rate is taken into account. The KSB equation, Eq. (15-43), can be misleading in this regard because it predicts continued improvement as the extraction factor increases without limit.

In summary, the key relationships employed when applying the mass transfer unit model to the design of an extraction column can be expressed as follows:

$$H_{or} = \frac{V_r}{k_{or}a} = \frac{Z_t}{N_{or}} \tag{15-71}$$

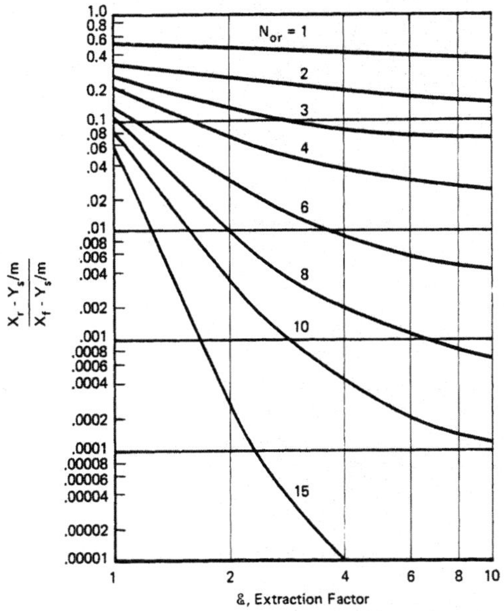

FIG. 15-26 Graphical solutions to the Colburn equation [Eq. (15-66)].

The value of H_{or} is the sum of contributions from the resistance to mass transfer in the raffinate phase (H_r) plus resistance to mass transfer in the extract phase (H_e) divided by the extraction factor $\mathcal{E}$:

$$H_{or} = H_r + \frac{H_e}{\mathcal{E}} \tag{15-72}$$

The individual transfer unit heights are given by

$$H_r = \frac{Q_r}{A_{col}\, k_r a} \tag{15-73}$$

$$H_e = \frac{Q_e}{A_{col}\, k_e a} \tag{15-74}$$

where Q = volumetric flow rate
A_{col} = column cross-sectional area
k = film mass-transfer coefficient (length per unit time)
a = interfacial mass-transfer area per unit volume of extractor

and subscripts r and e denote the raffinate and extract phases, respectively. As discussed earlier, the main resistance to mass transfer generally resides in the feed (raffinate) phase.

The lumped parameter H_{or} characterizes the efficiency of the differential contactor; higher contacting efficiency is reflected in a lower value of H_{or}. It deals directly with the ultimate design criterion, the height of the column, and reliable values often can be obtained from miniplant experiments and experience with commercial units. For processes with discrete contacting stages, mass-transfer efficiency may be expressed as the number of transfer units achieved per actual stage. For applications involving transfer of multiple solutes, the value of H_{or} or N_{or} per actual stage may differ for each solute, as discussed earlier with regard to stage efficiencies and mass-transfer coefficients.

EXTRACTION FACTOR AND GENERAL PERFORMANCE TRENDS

Because of their simplicity, the KSB equation [Eq. (15-43)] and the Colburn equation [Eq. (15-66)] are useful for illustrating a number of general trends in mass-transfer performance, in particular, helping to show how the extraction factor is related to process performance for different process configurations. For illustration, consider a simple dilute system involving immiscible liquids and zero solute concentration in the entering extraction solvent. The resulting expressions that follow are written in a general form without regard to a specific set of units.

For a single-stage batch process or a continuous extraction process that achieves one theoretical stage, the change in solute concentration expressed in terms of a solute reduction factor is

$$F_R = \frac{X_{\text{in}}}{X_{\text{out}}} = \frac{\mathcal{E} - 1/\mathcal{E}}{1 - 1/\mathcal{E}} \quad \text{for } N = 1 \tag{15-75}$$

The required solvent-to-feed ratio is then approximated by

$$\frac{S}{F} = \frac{F_R - 1}{K} \quad \text{for } N = 1 \tag{15-76}$$

After extraction, the concentration of solute in the extract, no matter what the extraction configuration, is given by

$$Y_{\text{out}} = \frac{X_{\text{in}}}{S/F}\left(1 - \frac{1}{F_R}\right) \quad \text{for } Y_{\text{in}} = 0 \tag{15-77}$$

Equation (15-77) follows from Eq. (15-42).

If the performance of a single-stage extraction is not adequate, repeated cross-current extractions can be carried out to increase solute recovery or removal. For this configuration, the reduction factor is given by

$$F_R = \left(1 + \frac{\mathcal{E}}{N}\right)^{\xi_o N} \quad \text{for cross-current operation} \tag{15-78}$$

where N is the number of repeated extractions or stages employing equal amounts of solvent, ξ_o is overall stage efficiency, and the extraction factor is expressed in terms of the total amount of solvent used by the process. Although high solute recoveries can be obtained by using cross-current processing, the required solvent usage will be high, as indicated by

$$\frac{S}{F} = \frac{N}{K}\left(F_R^{1/(\xi_o N)} - 1\right) \quad \text{for cross-current operation} \tag{15-79}$$

where S is the total amount of solvent. The concentration of solute in the combined extract will be low, as calculated by using Eq. (15-77). Comparing the results of Eqs. (15-75) and (15-76) with Eqs. (15-78) and (15-79) will show that multistage cross-current extraction yields improved performance relative to using single-stage extraction with the same total amount of solvent, but at the cost of additional contacting steps.

Compared to single-stage or cross-current processing, multistage, countercurrent processing allows a significant reduction in solvent use or an increase in separation performance. For this type of process, the reduction factor is approximated by

$$F_R = \frac{\mathcal{E}^{\xi_o N} - 1/\mathcal{E}}{1 - 1/\mathcal{E}} \quad \text{for countercurrent operation} \tag{15-80}$$

Inspection of Eqs. (15-75) and (15-80) will show how the addition of countercurrent stages magnifies the effect of the extraction factor on performance. Note that Eq. (15-80) predicts that performance will continue to improve as the value of $\mathcal{E}$ increases, approaching $F_R = \mathcal{E}^{\xi_o N}$ at high values of $\mathcal{E}$. However, stage efficiency must remain high, and this likely will require a change in some operating variable such as residence time per stage.

Multistage countercurrent processing may be practiced batchwise as well as in a continuous cascade. A batchwise countercurrent operation involves first treating a batch with extract solution as the extract leaves the process, and the last treatment is carried out by using fresh solvent as it enters the process (as in Figs. 15-6 and 15-22). A multistage, countercurrent process with discrete contacting stages (practiced either batchwise or using a continuous cascade) is well suited to applications with fairly slow rates of mass transfer because liquid-liquid contacting is carried out stagewise in separate vessels or compartments, and long residence times can be designed into each stage.

For a countercurrent extraction column with no discrete stages (or for processes operated within a diffusion-controlled regime far from equilibrium), performance is well modeled by the Colburn equation, where

$$F_R = \frac{\exp[N_{\text{or}}(1 - 1/\mathcal{E})] - 1/\mathcal{E}}{1 - 1/\mathcal{E}} \quad \text{for countercurrent operation} \tag{15-81}$$

and

$$Z_t = N_{\text{or}} \times H_{\text{or}} \tag{15-82}$$

Extraction columns are most attractive for applications with fairly fast mass transfer because residence time in the column is limited. Performance becomes mass-transfer-limited at high values of $\mathcal{E}$, approaching $F_R = \exp N_{\text{or}}$. At this point, a significant increase in performance can be achieved only by adding transfer units (column height), which corresponds to an increase in residence time for solute mass transfer.

With countercurrent processing, carried out using either a multistage cascade or an extraction column, the required solvent-to-feed ratio generally can be reduced by adding more and more stages or transfer units. As discussed in Minimum and Maximum Solvent-to-Feed Ratios, the minimum practical solvent-to-feed ratio is approximated by

$$\left(\frac{S}{F}\right)_{\text{min}} \approx \frac{1.3}{K} \quad \text{for countercurrent processing} \tag{15-83}$$

Below this value, the required number of stages or transfer units increases rapidly. At $\mathcal{E} = 1$, the number of theoretical stages and number of transfer units are equal, and

$$F_R = N + 1 = N_{\text{or}} + 1 \quad \text{for } \mathcal{E} = 1 \tag{15-84}$$

For $\mathcal{E} < 1$, the fraction of solute removed from the feed θ_i will approach a value equal to the extraction factor. In this case,

$$(F_R)_{\text{max}} = \frac{1}{1 - \mathcal{E}} \quad \text{for } \mathcal{E} < 1 \tag{15-85}$$

POTENTIAL FOR SOLUTE PURIFICATION USING STANDARD EXTRACTION

As noted earlier, the ability of a standard extraction process to isolate a desired solute from other solutes is limited. This can be illustrated by using the KSB equation [Eq. (15-43)] to calculate solute transfer for a dilute feed containing a desired solute i and an impurity solute j. On a solvent-free basis, the purity of solute i in the feed is given by

$$P_{i,\text{feed}} \text{ (in units of wt \%)} = 100\left(\frac{X''_{i,\text{feed}}}{X''_{i,\text{feed}} + X''_{j,\text{feed}}}\right) \tag{15-86}$$

Similarly, the purity of solute i in the extract is given by

$$P_{i,\text{extract}} \text{ (wt \%)} = 100\left(\frac{\theta_i X''_{i,\text{feed}}}{\theta_i X''_{i,\text{feed}} + \theta_j X''_{j,\text{feed}}}\right) \tag{15-87}$$

where θ_i is the fraction of solute extracted from the feed into the extract. By using the KSB equation to estimate θ for solutes i and j, the following expression is derived:

$$P_{i,\text{extract}} \text{ (wt \%)} = \frac{100}{1 + \left(\frac{\theta_j}{\theta_i}\right)\left(\frac{X''_{j,\text{feed}}}{X''_{i,\text{feed}}}\right)} = \frac{100}{1 + \left(\frac{\mathcal{E}_j^N - 1}{\mathcal{E}_i^N - 1}\right)\left(\frac{\mathcal{E}_i^N - 1/\mathcal{E}_i}{\mathcal{E}_j^N - 1/\mathcal{E}_j}\right)\left(\frac{X''_{j,\text{feed}}}{X''_{i,\text{feed}}}\right)} \quad \text{for } \mathcal{E} \neq 1.0 \tag{15-88}$$

Equation (15-88) assumes that no solute enters the process with the extraction solvent and that $\mathcal{E}_i$ and $\mathcal{E}_j$ are constant. An alternative expression can be written in terms of transfer units; however, the calculated results are essentially the same as a function of the number of stages or the number of transfer units because the models assume that both solute i and solute j experience the same mass-transfer resistance. Example results obtained by using Eq. (15-88) are shown in Fig. 15-27. Note that performance is not uniquely determined by a given value of $\alpha_{i,j} = K_i/K_j = \mathcal{E}_i/\mathcal{E}_j$, but depends upon the absolute value of $\mathcal{E}_i$, as well. In principle, the purity of solute i in the extract will approach a maximum value as the number of stages or transfer units approaches infinity:

$$\text{Maximum } P_{i,\text{extract}} (\%) = 100 \div \left[1 + \frac{1}{\alpha_{i,j}}\left(\frac{X''_{j,\text{feed}}}{X''_{i,\text{feed}}}\right)\right] \quad \text{in limit as } N \to \infty \tag{15-89}$$

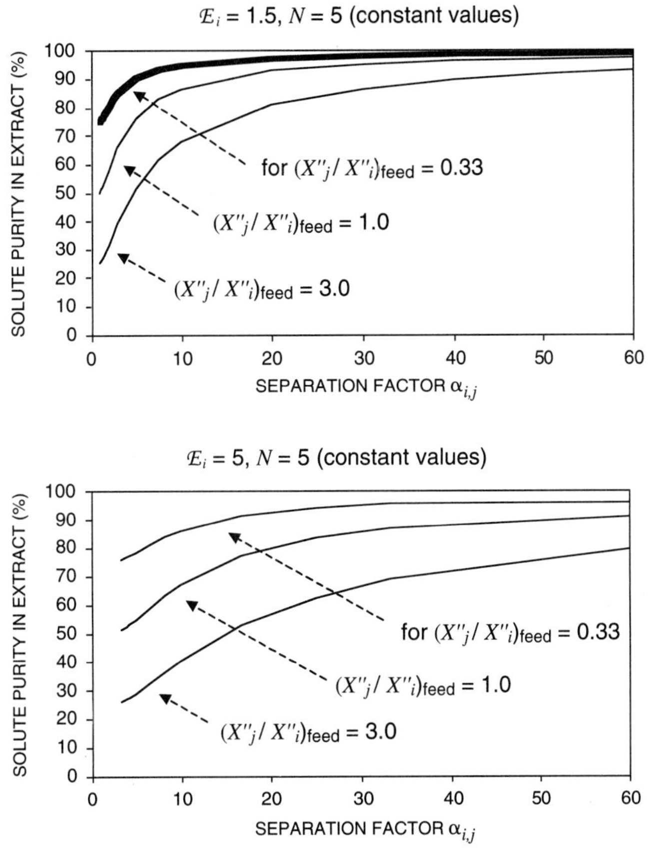

FIG. 15-27 Approximate purity of solute i in the extract ($P_{i,\text{extract}}$) versus separation factor $\alpha_{i,j}$ for standard extraction involving dilute feeds containing solutes i and j. Results obtained by using Eq. (15-88). Concentrations are in mass fraction (X'').

Of course, this theoretical maximum can never be attained in practice. Equation (15-89) follows from Eq. (15-88), noting that $\theta_j/\theta_i = 1/\alpha_{ij}$ for $N \to \infty$ as discussed by P. L. T. Brian [*Staged Cascades in Chemical Processing*, Prentice-Hall, Englewood Cliffs, NJ, 1972, p. 50]. As noted earlier, the ability to purify a desired solute is greatly enhanced by using fractional extraction (see the subsection Fractional Extraction Calculations).

CALCULATION PROCEDURES

SHORTCUT CALCULATIONS

Shortcut calculations can be quite useful to the process designer or run-plant engineer; they may be used to outline process requirements (stream and equipment sizes) early in a design project, to check the output of a process simulation program for reasonableness, to help analyze or troubleshoot a unit operating in the manufacturing plant or pilot plant, or to help explain performance trends and relationships between key process variables. In some applications involving dilute or even moderately concentrated feeds, they also may be used to specify the final design of an extraction process. In carrying out such calculations, L. A. Robbins [Sec. 1.9 in *Handbook of Separation Techniques for Chemical Engineers*, ed. P. A. Schweitzer, McGraw-Hill, New York, 1997] indicates that most liquid-liquid extraction systems can be treated as having immiscible solvents (case *A*), partially miscible solvents with a low solute concentration in the extract (case *B*), or partially miscible solvents with a high solute concentration in the extract (case *C*). These cases are illustrated in Examples 15-1 through 15-3.

Example 15-1 Shortcut Calculation, Case A Consider a 100-kg/h feed stream containing 20 wt% acetic acid in water that is to be extracted with 200 kg/h of recycle MIBK that contains 0.1 wt% acetic acid and 0.01 wt% water. The aqueous raffinate is to be extracted down to 1 wt% acetic acid. How many theoretical stages will be required and what will the extract composition be? The equilibrium data for this system are listed in Table 15-8 (in units of weight percent). The corresponding Hand plot is shown in Fig. 15-20. The Hand correlation (in mass ratio units) can be expressed as $Y' = 0.930(X')^{1.10}$, for X' between 0.03 and 0.25.

Assuming immiscible solvents, we have
$F' = 100(1 - 0.2) = 80$ kg water/h

$$X'_f = \frac{0.2}{0.8} = 0.25 \text{ kg acetic acid/kg water}$$

$$X'_r = \frac{0.01}{0.99} = 0.01 \text{ kg acetic acid/kg water}$$

$$S' = 200(1 - 0.001) = 199.8 \text{ kg MIBK/h}$$

$$Y'_s = \frac{0.2}{199.8} = 0.001 \text{ kg acetic acid/kg MIBK}$$

If we assume $R' = F'$ and $E' = S'$, we can calculate Y'_e from Eq. (15-42):

$$Y'_e = \frac{80(0.25) + 199.8(0.001) - 80(0.01)}{199.8} = 0.097 \frac{\text{kg acetic acid}}{\text{kg MIBK}}$$

Calculate $X'_1 = (0.097/0.930)^{1/1.10} = 0.128$. Then

$$m' = \frac{dY'}{dX'} = (0.930)(1.10)(X')^{0.1} \quad \text{for } X' \text{ between 0.03 and 0.25}$$

$$m'_1 = 0.833 \quad \text{at } X' = 0.128$$

$$m'_r = \frac{dY'}{dX'} = K' = 0.656 \quad \text{for } X' \text{ below 0.03}$$

TABLE 15-8 Water + Acetic Acid + Methyl Isobutyl Ketone Equilibrium Data at 25°C

Weight percent in raffinate			X'	Weight percent in extract			Y'
Water	Acetic acid	MIBK	Acetic acid	Water	Acetic acid	MIBK	Acetic acid
98.45	0	1.55	0	2.12	0	97.88	0
95.46	2.85	1.7	0.0299	2.80	1.87	95.33	0.0196
85.8	11.7	2.5	0.1364	5.4	8.9	85.7	0.1039
75.7	20.5	3.8	0.2708	9.2	17.3	73.5	0.2354
67.8	26.2	6.0	0.3864	14.5	24.6	60.9	0.4039
55.0	32.8	12.2	0.5964	22.0	30.8	47.2	0.6525
42.9	34.6	22.5	0.8065	31.0	33.6	35.4	0.9492

SOURCE: Sherwood, Evans, and Longcor, *Ind. Eng. Chem.*, **31**(9), pp. 1144–1150 (1939).

$$K'_s = 0.656 \quad \text{at } Y'_s = 0.001$$

$$\mathcal{E} = \sqrt{m'_1 m'_r}\,\frac{S'}{F'} = \frac{0.739(199.8)}{80} = 1.85$$

And N is determined from Fig. 15-24 and Eq. (15-43).

$$N = \frac{\ln\left[\left(\frac{0.25 - 0.001/0.656}{0.01 - 0.001/0.656}\right)\left(1 - \frac{1}{1.85}\right) + \frac{1}{1.85}\right]}{\ln 1.85} = 4.3 \text{ theoretical stages}$$

This result is very close to that obtained by using a McCabe-Thiele diagram (Fig. 15-23). From solubility data at $Y' = 0.1039$ kg acetic acid/kg MIBK (given in Table 15-8), the extract layer contains 5.4/85.7 = 0.0630 kg water/kg MIBK, and $Y''_e = (0.097)/(1 + 0.097 + 0.063) = 0.084$ mass fraction acetic acid in the extract.

For cases *B* and *C*, Robbins developed the concept of pseudosolute concentrations for the feed and solvent streams entering the extractor that will allow the KSB equations to be used. In case *B* the solvents are partially miscible, and the miscibility is nearly constant through the extractor. This frequently occurs when all solute concentrations are relatively low. The feed stream is assumed to dissolve extraction solvent only in the feed stage and to retain the same amount throughout the extractor. Likewise, the extraction solvent is assumed to dissolve feed solvent only in the raffinate stage. With these assumptions the primary extraction solvent rate moving through the extractor is assumed to be S', and the primary feed solvent rate is assumed to be F'. The extract rate E' is less than S', and the raffinate rate R' is less than F' because of solvent mutual solubilities.

The slope of the operating line is F'/S', just as in Eqs. (15-40) and (15-41), but only stages 2 through $r - 1$ will fall directly on the operating line. And X'_1 must be on the equilibrium line in equilibrium with Y'_e by definition. One can also calculate a pseudofeed concentration X_f^B that will fall on the operating line at $Y'_{n+1} = Y'_e$ as follows:

$$X_f^B = X'_f + \frac{S' - E'}{F'} Y'_e \tag{15-90}$$

Likewise, one knows that Y'_r will be on the equilibrium line with X'_r. One can therefore calculate a pseudoconcentration of solute in the inlet extraction solvent Y_s^B that will fall on the operating line where $X'_{n-1} = X'_r$ as follows:

$$Y_s^B = Y'_s + \frac{F' - R'}{S'} X'_r \tag{15-91}$$

For case *B*, the pseudo inlet concentration X_f^B can be used in the KSB equation with the actual value of X'_r and $\mathcal{E} = m'S'/F'$ to calculate rapidly the number of theoretical stages required. The graphical stepwise method illustrated in Fig. 15-23 also can be used. The operating line will go through points (X'_r, Y_s^B) and (X_f^B, Y'_e) with a slope of F'/S'.

Example 15-2 Shortcut Calculation, Case B Let us solve the problem in Example 15-1 by assuming case *B*. The solute (acetic acid) concentration is low enough in the extract that we may assume that the mutual solubilities of the solvents remain nearly constant. The material balance can be calculated by an iterative method.

From equilibrium data (Table 15-8) the extraction solvent (MIBK) loss in the raffinate will be about 0.016/0.984 = 0.0163 kg MIBK/kg water, and the feed solvent (water) loss in the extract will be about 5.4/85.7 = 0.0630 kg water/kg MIBK.

First iteration: Assume $R' = F' = 80$ kg water/h. Then extraction solvent in raffinate = (0.0163)(80) = 1.30 kg MIBK/h. Estimate $E' = 199.8 - 1.3 = 198.5$ kg MIBK/h. Then feed solvent in extract = (0.063)(198.5) = 12.5 kg water/h.

Second iteration: Calculate $R' = 80 - (0.063)(198.7) = 67.5$ kg water/h. And $E' = 199.8 - (0.0163)(67.5) = 198.7$ kg MIBK/h.

Third iteration: Converge $R' = 80 - (0.063)(198.7) = 67.5$ kg water/h. And Y'_e is calculated from the overall extractor material balance [Eq. (15-42)]:

$$Y'_e = \frac{80(0.25) + 199.8(0.001) - (67.5)(0.01)}{198.7} = 0.0983 \frac{\text{kg acetic acid}}{\text{kg MIBK}}$$

$$Y_e = \frac{0.0983}{1 + 0.0983 + 0.0630} = 0.0846 \text{ mass fraction acetic acid in extract}$$

From the Hand correlation of equilibrium data,

$$Y'_e = 0.930(X')^{1.10} \quad \text{for } X' \text{ between 0.03 and 0.25}$$

The raffinate composition leaving the feed (first stage) is

$$X'_1 = \left(\frac{0.0983}{0.930}\right)^{1/1.10} = 0.130$$

$$m'_1 = \frac{dY}{dX} = (0.930)(1.10)(X')^{0.1}$$

$$m'_r = \frac{dY}{dX} = K' = 0.656$$

$$m'_1 = 0.834 \quad \text{at } X'_1 = 0.13$$

$$m'_r = 0.656 \quad \text{at } X'_r = 0.01$$

$$K'_s = 0.656 \quad \text{at } Y'_s = 0.001$$

$$\mathcal{E} = \sqrt{m'_1 m'_r}\,\frac{S'}{F'} = \frac{(0.740)(199.8)}{80} = 1.85$$

And X_f^B is calculated from Eq. (15-90)

$$X_f^B = 0.25 + \frac{(199.8 - 198.7)(0.0983)}{80} = 0.251$$

and Y_s^B from Eq. (15-91):

$$Y_s^B = 0.001 + \frac{(80 - 67.5)(0.01)}{199.8} = 0.0016$$

Now N is determined from Fig. 15-24, Eq. (15-43), or the McCabe-Thiele type of plot (Fig. 15-23). For case *B*,

$$N = \frac{\ln\left[\left(\frac{0.251 - 0.0016/0.656}{0.01 - 0.0016/0.656}\right)\left(1 - \frac{1}{1.85}\right) + \frac{1}{1.85}\right]}{\ln 1.85} = 4.5 \text{ theoretical stages}$$

A less frequent situation, case *C*, can occur when the solute concentration in the extract is so high that a large amount of feed solvent is dissolved in the extract stream at the feed end of the process (at the feed stage), but a relatively small amount of feed solvent (say one-tenth as much) is dissolved by the extract stream at the raffinate end of the process (at the raffinate stage). The feed stream is assumed to dissolve the extraction solvent only in the feed stage just as in case *B*. But the extract stream is assumed to dissolve a large amount of feed solvent leaving the feed stage and a negligible amount leaving the raffinate stage. With these assumptions, the primary feed solvent rate is assumed to be R', so the slope of the operating line for case *C* is R'/S'. Again the extract rate E' is less than S', and the raffinate rate R' is less than F'.

The pseudofeed concentration for case *C*, X_f^C, can be calculated from

$$X_f^C = \frac{F'}{R'} X'_f + \frac{S' - E'}{R'} Y'_e \tag{15-92}$$

For case *C*, the value of Y'_s will fall on the operating line, and the extraction factor is given by

$$\mathcal{E}^C = \frac{m'S'}{R'} \tag{15-93}$$

On an X'–Y' diagram for case *C*, the operating line will go through points (X'_r, Y'_s) and (X_f^C, Y'_e) with a slope of R'/S' similar to Fig. 15-23. When the KSB equation is used for case *C*, use the pseudofeed concentration X_f^C from Eq. (15-92) and the extraction factor $\mathcal{E}^C$ from Eq. (15-93). The raffinate concentration X'_r and inlet solvent concentration Y'_s are used without modification. For more detailed discussion, see L. A. Robbins, sec. 1.9 in *Handbook of Separation Techniques for Chemical Engineers*, ed. P. A. Schweitzer, McGraw-Hill, New York, 1997.

Example 15-3 Number of Transfer Units Let us calculate the number of transfer units required to achieve the separation in Example 15-2. The solution to the problem is the same as in Example 15-1 except that the denominator is changed. From Eq. (15-68):

$$N_{or} = 4.5 \frac{\ln 1.85}{1 - 1/1.85} = 6.0 \text{ transfer units}$$

COMPUTER-AIDED CALCULATIONS (SIMULATIONS)

A number of process simulation programs such as Aspen Plus® and Aspen HYSYS® from AspenTech, ChemCAD® from Chemstations, ProSimPlus from ProSim, and SimSci PRO/II® from Schneider Electric, among others, can facilitate rigorous calculation of the number of theoretical stages required by a given application, provided an accurate liquid-liquid equilibrium model is employed. Some commercially available simulation packages do not include rate-based programs specifically designed for extraction process simulation; however, the equivalent number of transfer units at each stage can be calculated from knowledge of the extraction factor by using Eq. (15-68). Process simulation programs are particularly useful for concentrated systems that exhibit highly nonlinear equilibrium and operating lines, significant change in extract and raffinate flow rates within the process due to transfer of solute from one phase to the other, significant changes in the mutual solubility of the two phases as solute concentration changes, or nonisothermal operation. They also facilitate convenient calculation for complex extraction configurations such as fractional extraction with extract reflux as well as calculations involving more than three components (more than one solute). They can also facilitate process optimization by allowing rapid evaluation of numerous design cases. These programs do not provide information about mass-transfer performance in terms of stage efficiencies or extraction column height requirements, or information about the throughput and flooding characteristics of the equipment; these factors must be determined separately by using other methods. The use of simulation software to analyze extraction processes is illustrated in Examples 15-4 and 15-5.

In using simulation software, it is important to keep in mind that the reliability of the results is highly dependent upon the quality of the liquid-liquid equilibrium (LLE) model programmed into the simulation. In most cases, an experimentally validated model will be needed because UNIFAC and other estimation methods are not sufficiently accurate. It also is important to recognize, as mentioned in earlier discussions, that binary interaction parameters determined by regression of vapor-liquid equilibrium (VLE) data cannot be relied upon to accurately model the LLE behavior for the same system. On the other hand, a set of binary interaction parameters that model LLE behavior properly often will provide a reasonable VLE fit for the same system because pure-component vapor pressures often dominate the calculation of VLE.

Commercially available simulation programs often are used in a fashion similar to the classic graphical methods. When separation of specific solutes is important, the design of a new process generally focuses on determining the optimum solvent rates and number of theoretical stages needed to comply with the separation specifications according to relative K values for solutes of interest. Calculations often are made by focusing on a "soluble" key solute with a relatively high K value, and an "insoluble" key solute, expressing the design specification in terms of the maximum concentration of soluble key left in the raffinate and the maximum concentration of insoluble key contaminating the extract (analogous to "light" and "heavy" key components in distillation design). Then solutes with K values higher than that of the soluble key will go out with the extract to a greater extent, and solutes with K values less than that of the insoluble key will go out with the raffinate. If the desired separation is not feasible using a standard extraction scheme, then fractional extraction schemes should be evaluated.

For rating an existing extractor, the designer must make an estimate of the number of theoretical stages the unit can deliver and then determine the concentrations of key solutes in extract and raffinate streams as a function of the solvent-to-feed ratio, keeping in mind the fact that the number of theoretical stages a unit can deliver can vary depending upon operating conditions.

The use of process simulation software for process design is discussed by W. D. Seider, J. D. Seader, D. R. Lewin, and S. Widagdo [*Product and Process Design Principles: Synthesis, Analysis, and Evaluation*, 3d ed., Wiley, New York, 2009] and by R. Turton, R. C. Bailie, W. B. Whiting, J. A. Shaeiwitz, and D. Bhattacharyya [*Analysis, Synthesis, and Design of Chemical Processes*, 4th ed., Prentice-Hall, Upper Saddle River, NJ, 2012]. Various computational procedures for extraction simulation are discussed by L. Steiner [chap. 6 in *Liquid-Liquid Extraction Equipment*, ed. J. C. Godfrey and M. J. Slater, Wiley, New York, 1994]. In addition, a number of authors have developed specialized methods of analysis. For example, D. Sanpui, M. K. Singh, and A. Khanna [*AIChE J.* **50**(2): 368–381 (2004)] outline a computer-based approach to rate-based, nonisothermal modeling of extraction processes. B. Harjo, K. M. Ng, and C. Wibowo [*Ind. Eng. Chem. Res.* **43**(14): 3566–3576 (2004)] describe methods for visualization of high-dimensional liquid-liquid equilibrium phase diagrams as an aid to process conceptualization. This methodology can help focus the design effort by identifying specific composition regions where the design analysis will be particularly sensitive to uncertainties in the equilibrium behavior. The method of M. Minotti, M. F. Doherty, and M. F. Malone [*Ind. Eng. Chem. Res.* **35**(8): 2672–2681 (1996)] facilitates a feasibility analysis of potential solvents and process options by locating fixed points or pinches in the composition profiles determined by equilibrium and operating constraints. A. Marcilla et al. [*Ind. Eng. Chem. Res.* **38**(8): 3083–3095 (1999)] developed a method involving correlation of tie lines to calculate equilibrium compositions at each stage without iterations. To optimize the design and operating parameters of an extraction cascade, J. A. Reyes-Labarta and I. E. Grossmann [*AIChE J.* **47**(10): 2243–2252 (2001)] have proposed a calculation framework that employs nonlinear programming techniques to systematically evaluate a wide range of potential process configurations and interconnections. Focusing on another aspect of process design, R. Ravi and D. P. Rao [*Ind. Eng. Chem. Res.* **44**(26): 10016–10020 (2005)] provide an analysis of the phase rule (number of degrees of freedom) for liquid-liquid extraction processes.

Example 15-4 Extraction of Phenol from Wastewater The amount of 350 gpm (79.5 m³/h) of wastewater from a coke oven plant contains an average of 700 ppm phenol by weight that needs to be reduced to 1 ppm or less to meet environmental requirements [Karr, A. E., and S. Ramanujam, *St. Louis AIChE Symp.*, March 19, 1987]. The wastewater comes from the bottom of an ammonia stripping tower at 105°C and is to be extracted at 1.7 atm with recycle methylisobutyl ketone (MIBK) containing 5 ppm phenol. The extraction will be carried out by using a reciprocating-plate extractor (Karr column). How many theoretical stages will be required in the extractor at a solvent-to-feed ratio of 1:15, and what is the resulting extract composition?

The Aspen Plus® process simulation program is used in this example, but any of a number of process simulation programs such as those mentioned earlier also may be used for this purpose. In Aspen Plus, the EXTRACT liquid-liquid extraction unit-operation block is used to model the phenol wastewater extraction. As is typical in process simulation programs, the EXTRACT block is fundamentally a rating calculation rather than a design calculation, so the determination of the required number of stages for the separation cannot be made directly. In addition, the EXTRACT block can only handle integral numbers of theoretical stages, so the fractional number of required theoretical stages must be determined by an interpolation method.

The partition ratio for transfer of phenol from water into MIBK at 105°C is $K'' = 34$ on a mass fraction basis [Greminger, D. C., et al., *Ind. Eng. Chem. Proc. Des. Dev.* **21**(1): 51–54 (1982)]. Because the partition ratio is so high, a low solvent-to-feed ratio of 1:15 can be used and still give an extraction factor of about 2. In the EXTRACT block, a property option is available that allows the user to specify liquid-liquid K value correlations (designated as KLL Correlation in Aspen Plus) for the components involved in the extraction rather than a complete set of binary interaction parameters to define the liquid-liquid equilibria. In this example, it is time-consuming to regress a set of liquid-liquid binary interaction parameters that results in representative partition ratios, so the option of simply specifying K values directly is recommended. Because phenol will be relatively dilute in both the raffinate and extract phases, appropriate liquid-liquid K values for distribution of water and MIBK between phases at 105°C can be estimated from water–MIBK liquid-liquid equilibrium data [Řehák, M., et al., *Collect. Czech Chem. Commun.* **65:** 1471–1486 (2000)] to yield $K''_{water} = 0.0532$ and $K''_{MIBK} = 53.8$ (mass fraction basis). It is important in Aspen Plus to specify K values for all the components in the extractor in order to properly model the liquid-liquid equilibria with this approach.

The temperatures and compositions of the wastewater and solvent feed streams, as well as the wastewater feed flow rate, are specified in the problem statement. The solvent flow rate is specified as one-fifteenth of the wastewater flow rate as described above. In the EXTRACT block, the number of stages will be manually varied from 2 to 10 to observe the effect on the raffinate and extract concentrations, and it will be specified as operating adiabatically at 1.7 atm. Water is specified as the key component in the first liquid phase, and MIBK is specified as the key component in the second liquid phase. The rest of the block parameters (convergence, report, and miscellaneous block options) are allowed to remain at their default values.

The raffinate and extract concentrations resulting from successive simulation runs for 2 through 10 theoretical stages are given in Table 15-9, and the raffinate phenol concentrations are presented graphically in Fig. 15-28. Examining the results, we can see that the number of theoretical stages required to achieve the 1 ppm phenol discharge limitation falls somewhere between 7 and 8. In addition, we can see from Fig. 15-28 that the dependence of raffinate phenol concentration on the number of stages yields nearly a straight line on a semilog plot. As a result, performing a linear interpolation of the log of the raffinate concentration between 7 and 8 stages yields the number of stages required to achieve 1 ppm phenol in the raffinate:

$$N = 7 + (8-7)\left(\frac{\log 1.48 - \log 1}{\log 1.48 - \log 0.709}\right) = 7.53 \text{ theoretical stages}$$

Examining the extract phenol concentrations in Table 15-9, it is clear that they varied little for five or more stages, as is expected because nearly all the phenol contained in the wastewater feed was extracted in stages 1 through 4. As a result, the extract will contain 1.3 wt% phenol, 5.2 percent water, and 93.5 percent MIBK.

The simulation results can be checked by using a shortcut calculation—to provide confidence that the simulation is delivering a reasonable result. The KSB equation [Eq. (15-43)] can be used for this purpose with values taken from the problem specification and estimates of the phenol K' value (in Bancroft coordinates). Because phenol is always quite dilute in both the extract and raffinate phases, its K' value can be calculated from the component mass fraction K'' values according to the following approximation:

$$K'_{PhOH} \cong K''_{PhOH}\left[\frac{K''_{MIBK} - 1}{K''_{MIBK}(1 - K''_{H_2O})}\right] = 34\left[\frac{53.8 - 1}{53.8(1 - 0.0532)}\right] = 35.24$$

TABLE 15-9 Simulation Results for Extraction of Phenol from Wastewater Using MIBK (Example 15-4)

	Raffinate compositions			Extract compositions		
N	X''_{PhOH}, pmm	X''_{H_2O}, mass fraction	X''_{MIBK}, mass fraction	Y''_{PhOH}, mass fraction	Y''_{H_2O}, mass fraction	Y''_{MIBK}, mass fraction
2	101	0.98235	0.01755	0.01146	0.05226	0.93628
3	41.8	0.98237	0.01759	0.01260	0.05226	0.93514
4	17.7	0.98238	0.01760	0.01307	0.05226	0.93468
5	7.56	0.98238	0.01761	0.01326	0.05226	0.93448
6	3.29	0.98238	0.01761	0.01335	0.05225	0.93440
7	1.48	0.98238	0.01762	0.01338	0.05225	0.93436
8	0.709	0.98238	0.01762	0.01340	0.05226	0.93435
9	0.382	0.98238	0.01762	0.01340	0.05225	0.93434
10	0.242	0.98238	0.01762	0.01341	0.05225	0.93434

This value compares favorably with the value of 35.28 calculated from phenol mass ratios derived from extractor internal mole fraction profile data in the simulation output. The extraction factor [Eq. (15-11)] is then calculated with the dilute system approximation that $m_{PhOH} \cong K_{PhOH}$ and solute-free water and MIBK feed rates of 159,911 and 10,668 lb/h taken from the simulation output:

$$\mathcal{E}_{PhOH} = m_{PhOH}\frac{S}{F} \cong K''_{PhOH}\frac{S''}{F''} = K'_{PhOH}\frac{S'}{F'} = 35.24 \times \frac{10{,}668}{159{,}911} = 2.35$$

It is interesting to note that this value of the extraction factor, 2.35, is the same as those calculated on mole fraction, mass fraction, and Bancroft coordinate bases from extractor internal profile data in the simulation, a confirmation that the extraction factor is indeed independent of units as long as consistent values of m, S, and F are used. By substituting the above values into Eq. (15-43) along with concentrations taken from the problem statement and Table 15-9, the required number of stages is estimated as

$$N \approx \frac{\ln\left[\dfrac{0.0007/0.9993 - (0.000005)/(0.999995)/35.24}{0.000001/0.9824 - (0.000005)/(0.999995)/35.24}1 - 1/2.35 + 1/2.35\right]}{\ln 2.35}$$

$$= 7.18 \text{ theoretical stages}$$

The simulation result of 7.53 theoretical stages is close to this shortcut estimate, indicating that the simulation is indeed delivering reasonable results.

FRACTIONAL EXTRACTION CALCULATIONS

Dual-Solvent Fractional Extraction As discussed in the subsection Commercial Process Schemes, under Introduction and Overview, fractional extraction often may be viewed as combining product purification with product recovery by adding a washing section to the stripping section of a standard extraction process. In the stripping section, the mass transfer we focus on is the transfer of the product solute from the wash solvent into the extraction solvent. If we assume dilute conditions and use shortcut calculations for illustration, the extraction factor is given by

$$\mathcal{E}_s = K'_s \frac{S'_s}{W'_s} \tag{15-94}$$

where $\mathcal{E}_s$ = stripping section extraction factor (dimensionless)
K'_s = stripping section partition ratio, defined as equilibrium concentration of product solute in extraction solvent divided by that in wash solvent (Bancroft coordinates)
S'_s = mass flow rate of extraction solvent within stripping section (solute-free basis)
W'_s = mass flow rate of wash solvent in stripping section (solute-free basis)

The change in the concentration of product dissolved in the wash solvent, within the stripping section, can be calculated by using the KSB equation

$$\left(\frac{X'_{out}}{X'_{in}}\right)_{product} \approx \frac{1 - 1/\mathcal{E}_s}{(\mathcal{E}_s)^{N_s} - 1/\mathcal{E}_s} \tag{15-95}$$

where N_s = number of theoretical stages in stripping section
X'_{in} = concentration of product solute in wash solvent at inlet to stripping section (feed stage)
X'_{out} = concentration of product solute in wash solvent at outlet from stripping section (raffinate end of overall process)

In the washing section, we focus on transfer of impurity solute from the extraction solvent into the wash solvent. A washing extraction factor can be defined as

$$\mathcal{E}_w = \frac{1}{K'_w}\frac{W'_w}{S'_w} \tag{15-96}$$

where $\mathcal{E}_w$ = washing section extraction factor (dimensionless)
K'_w = washing section partition ratio, defined as equilibrium concentration of impurity solute in extraction solvent divided by that in wash solvent (Bancroft coordinates)
S'_w = mass flow rate of extraction solvent within washing section (solute-free basis)
W'_w = mass flow rate of wash solvent in washing section (solute-free basis)

Then the change in the concentration of impurity solute dissolved in the extraction solvent, within the washing section, is given by

$$\left(\frac{Y'_{out}}{Y'_{in}}\right)_{impurities} \approx \frac{1 - 1/\mathcal{E}_w}{(\mathcal{E}_w)^{N_w} - 1/\mathcal{E}_w} \tag{15-97}$$

where N_w = number of theoretical stages in washing section
Y'_{in} = concentration of impurity solute in extraction solvent at inlet to washing section (feed stage)
Y'_{out} = concentration of impurity solute in extraction solvent at outlet from washing section (extract end of overall process)

The ratio of extraction solvent to wash solvent in each section will be different if either solvent enters the process with the feed. Note that both K'_s and K'_w are defined as the ratio of the appropriate solute concentration in the *extraction* solvent to that in the *wash* solvent.

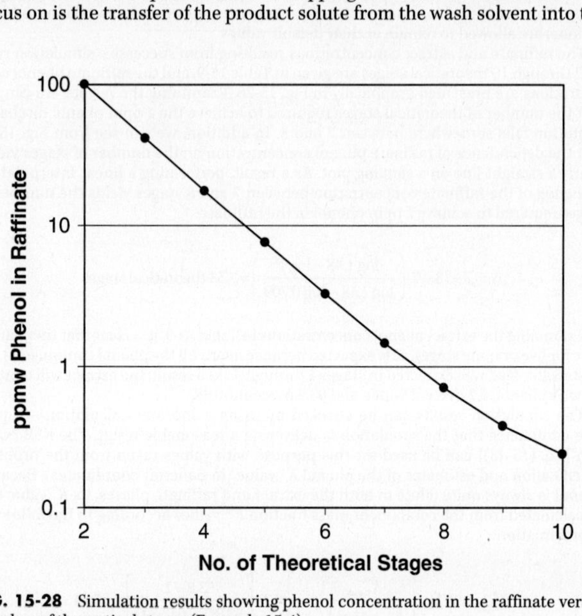

FIG. 15-28 Simulation results showing phenol concentration in the raffinate versus number of theoretical stages (Example 15-4).

The shortcut calculations just outlined illustrate the general considerations involved in analyzing a fractional extraction process. The analysis requires locating the feed stage and matching the calculations for each section with the material balance at the feed stage, an iterative procedure. B. D. Smith and W. K. Brinkley [*AIChE J.* **6**(3): 446–450 (1960)] discuss the application of the KSB equation to fractional extraction calculations including the use of reflux. Transfer unit calculations also may be used. When equilibrium and operating lines are not linear, more sophisticated calculations will be needed to take this into account. Commercially available simulation software or other computer programs often are used to carry out this procedure (see the subsection Computer-Aided Calculations). Note that with dual-solvent fractional extraction, the total solute concentration always is highest at the feed stage. This can lead to undesired behavior such as tendencies toward emulsion formation or even formation of a single liquid phase at the plait point. The minimum amounts of solvent needed to avoid these effects can be determined in laboratory tests.

Early in a project, it may be useful to consider a simplified case in which the ratio of extraction solvent to wash solvent is constant and is the same in the stripping and washing sections (i.e., the amount of solvent entering with the feed is negligible) and the extraction factors for each section are equal. For this special case, termed a *symmetric separation*, the extraction factors are

$$\mathcal{E}_s = \mathcal{E}_w = \sqrt{\alpha_{i,j}} \tag{15-98}$$

and the ratio of extraction solvent to wash solvent is given by

$$\frac{S}{W} \approx \frac{1}{\sqrt{K_s K_w}} \approx \frac{1}{\sqrt{\alpha_{i,j}}\,K_w} = \frac{\sqrt{\alpha_{i,j}}}{K_s} \tag{15-99}$$

Using these relationships, we find the number of stages required for the stripping and washing sections will be about the same, and the total number of stages required likely will be close to the minimum number—assuming symmetric separation requirements. The effects of the separation factor and the number of stages on the separation performance can be estimated by using expressions given by P. L. T. Brian [*Staged Cascades in Chemical Processing*, Prentice-Hall, Englewood Cliffs, NJ, 1972]. For a process containing two solutes *i* and *j*, with the feed entering at the middle stage, it follows from Brian's analysis that

$$S_{i,j} = \frac{Y_i}{X_i}\,\frac{1-X_i}{1-Y_i} \approx \alpha_{i,j}^{(N+1)/2} \tag{15-100}$$

where $S_{i,j}$ is termed the *separation power* of the process. Equation (15-100) is derived by assuming that the ratio of extract phase to raffinate phase within the process is constant, and that $\alpha_{i,j}$ is constant. Interestingly, Eq. (15-100) is very similar in its general form to the equation obtained by using the Fenske equation to calculate fractional distillation performance for a binary feed, assuming that the required number of theoretical stages is twice the minimum number obtained at total reflux. (See Sec. 13, Distillation.)

For a proposed symmetric separation, Eqs. (15-99) and (15-100) can be used to gauge the required flow rates, number of theoretical stages, and separation factor. For example, consider a hypothetical application with the goal of transferring 99 percent of a key solute *i* into the extract and 99 percent of an impurity solute *j* into the raffinate. For illustration, let $K_i = 2.0$ and $K_j = 0.5$, so $\alpha_{i,j} = 4$. From Eq. (15-99), the extraction-solvent-to-wash-solvent ratio should be about $S/W = 1.0$ for a symmetric separation. The number of theoretical stages is estimated by using Eq. (15-100): $S_{i,j} = 99 \times 99 = 9801$ gives $N \approx 12$ total stages for $\alpha_{i,j} = 4$. When one is evaluating candidate solvent pairs for a proposed fractional extraction process, a useful first step is to measure the equilibrium *K* values for product and impurity solutes and then assess process feasibility by using Eqs. (15-99) and (15-100). This can provide a quick way of assessing whether the measured separation factor is sufficiently large to achieve the separation goals, using a reasonable number of stages.

Single-Solvent Fractional Extraction with Extract Reflux As discussed earlier, single-solvent fractional extraction with extract reflux is practiced in the petrochemical industry to separate aromatics from crude hydrocarbon feeds. For example, a variety of extraction processes utilizing different high-boiling, polar solvents are used to separate benzene, toluene, and xylene (BTX) from aliphatic hydrocarbons and naphthenes (cycloalkanes), although processes involving extractive distillation are displacing older extraction processes. A typical hydrocarbon feed is a distillation cut containing mostly C_5 to C_9 components. Commercial extraction processes include the UDEX process (employing mainly diethylene and/or triethylene glycol), the AROSOLVAN process (*N*-methyl-2-pyrrolidone), and the Sulfolane process (2,3,4,5-tetrahydrothiophene-1,1-dioxide), among others. Although the flow diagrams for these processes differ, they all involve the use of a liquid-liquid extractor followed by a top-fed extract stripper or extractive distillation tower. A number of different processing schemes are used to isolate the aromatics and recycle the heavy solvent. Note that only high-boiling solvents are used. This is because a high-boiling (or low-volatility) solvent allows the aromatics to be taken as a distillate product and avoids the increased energy usage of boiling the recycle solvent overhead. For detailed discussion, see T. C. Lo, M. H. I. Baird, and C. Hanson, eds., *Handbook of Solvent Extraction*, chaps. 18.1 to 18.3, Wiley, New York, 1983, and Krieger, Huntington, NY, 1991; A. A. Gaile et al., *Chem. Technol. Fuels Oils* **40**(3): 131–136, and **40**(4): 215–221 (2004); and D. F. Schneider, *Chem. Eng. Prog.* **100**(7): 34–39 (2004).

Consider a process scheme involving a liquid-liquid extractor followed by a top-fed extract stripper and a product recovery distillation tower (as illustrated in Fig. 15-2). In the extractor, the feed is contacted with the polar solvent to transfer aromatics into the solvent phase. Some nonaromatics (NAs) also transfer into the solvent. The stripper is used to remove low-boiling NAs plus some aromatics from the extract to generate a reflux stream. The stripper overheads also contain some high-boiling NAs because their low solubility in the polar solvent boosts their relative volatility in the stripper. In this respect, the stripper functions as an extractive distillation tower with the high-boiling polar solvent serving as the extractive distillation solvent (serving to force high-boiling NAs overhead by increasing their activity coefficients in solution). The stripper bottoms are sent to a product recovery tower where an aromatic product cut is removed overhead and the high-boiling solvent is removed in the bottoms for return to the extractor. The stripper overheads are condensed and returned to the washing section of the extractor (the bottom of the extractor in this case) to provide reflux. As this backwash of reflux passes up through the extractor, the aromatics and a portion of the low-boiling NAs transfer back into the solvent phase, preferentially displacing high-boiling NAs which transfer out of the extract phase because of their lower solubilities in the polar solvent. Without extract reflux, the concentration of higher-boiling NAs in the extract phase would be significantly higher, and they would be difficult to completely remove in the stripper in spite of their low solubilities in the polar solvent. In this manner, low-boiling aromatics and NAs tend to build up in the extract reflux loop to provide a sort of barrier that minimizes the entry of higher-boiling NAs into the extract phase.

The use of simulation software to analyze this type of process is illustrated in Example 15-5, which considers a simplified ternary system for illustration. The simulation of an actual aromatics extraction process is more complex and can exhibit considerable difficulty converging on a solution; however, Example 15-5 illustrates the basic considerations involved in carrying out the calculations.

Example 15-5 Simplified Sulfolane Process—Extraction of Toluene from n-Heptane The amount of 40 metric tons (t) per hour (t/h) of distilled catalytic reformate from petroleum refining, containing 50 percent by weight aromatics, is to be extracted with recovered sulfolane containing 0.4 vol% aromatics in a 10-stage column contactor operating nearly adiabatically at 3 bar (gauge pressure). The extract will be fed to a 10-stage top-fed extract/paraffin stripper operating at 1 bar gauge to recover 98 percent of the aromatics with no more than 500 ppm by weight of nonaromatics. The catalytic reformate at 90°C is fed into the extractor on the fourth stage up from the bottom, and the recovered sulfolane leaving the bottom of a solvent recovery tower at 185°C is cross-exchanged with the extract stream leaving the bottom of the extractor before being fed to the top of the extractor at 105°C. Extract reflux is returned from the paraffin stripper's condenser to the bottom of the extractor with subcooling to 105°C.

1. What solvent flow and stripper reboiler duty are required to achieve the performance specifications, and what are the extract reflux rate and composition?

2. If the required aromatics recovery is increased to 99 percent, what is the effect on solvent flow and stripper reboiler duty?

In real-world commercial catalytic reformate streams, a wide range of aromatic and nonaromatic hydrocarbons must be considered, and the liquid-liquid extraction and distillation simulation becomes quite complicated. In addition, real-world applications of sulfolane extraction normally add a few percent of water to the sulfolane to reduce its pure-component freezing point of 27 to 28°C during shipping and storage [Kosters, W. C. G., chap. 18.2.3 in *Handbook of Solvent Extraction*, ed. T. C. Lo, M. H. I. Baird, and C. Hanson, Wiley, New York, 1983; Krieger, Huntington, NY, 1991]. Also, in many processes, steam is injected into the bottom of the solvent recovery tower to help strip the aromatics (i.e., the tower is both steam-stripped and reboiled), allowing operation of the tower at higher pressures without incurring (excessive) solvent thermal degradation. Another issue associated with the solvent recovery tower is the buildup of low-volatility impurities such as tars and waxes in the recycle solvent bottoms returning to the extractor. This requires continuous or periodic purging of some of the recycled solvent to avoid accumulating tars. Furthermore, in a real-world process, water may be used to wash the raffinate to recover solvent. To simplify the problem for this example, however, we model the aromatics as toluene and the NAs as *n*-heptane, consider only sulfolane as the extraction solvent, and do not include water in the calculations—to reduce the problem to a simple ternary system for illustration.

As in Example 15-4, the EXTRACT block in the Aspen Plus process simulation program (Release 27.0) is used to model this problem, but any of a number of process

TABLE 15-10 NRTL Binary Interaction Parameters for Example 15-5

Component *i*	Component *j*	b_{ij}, K
n-Heptane	Toluene	23.2040
Toluene	*n*-Heptane	−34.3180
Toluene	Sulfolane	238.952
Sulfolane	Toluene	203.243
n-Heptane	Sulfolane	1476.41
Sulfolane	*n*-Heptane	719.006

$\tau_{ij} = b_{ij}/T$(K); $\alpha_{ij} = 0.2$ for *n*-heptane + sulfolane; $\alpha_{ij} = 0.3$ for toluene + sulfolane and for *n*-heptane + sulfolane. Aspen Plus regression parameters a_{ij}, d_{ij}, e_{ij}, and f_{ij} are set to zero; $c_{ij} = \alpha_{ij}$; $\tau_{ii} = 0$; and $G_{ii} = 1$.

simulation programs such as mentioned earlier may be used for this purpose. The first task is to obtain an accurate fit of some applicable liquid-liquid equilibrium (LLE) data with an appropriate model, realizing that liquid-liquid extraction simulations are very sensitive to the quality of the LLE data fit and that estimation methods such as UNIFAC will often give misleading results for LLE that will give incorrect extractor modeling results. The NRTL liquid activity-coefficient model is utilized for this problem because it can represent a wide range of LLE systems accurately. Model parameters must be determined by regression of experimental data. The regression of the NRTL binary interaction parameters is performed within Aspen Plus in the Properties environment using the Regression run mode to ensure that the resulting parameters are consistent with the form of the NRTL model equations used within Aspen Plus.

Because the extractor operates nearly isothermally, only slightly above and below 100°C, the 100°C data of R. M. De Fre and L. A. Verhoeye [*J. Appl. Chem. Biotechnol.* **26:** 469–487 (1976)] are used as the basis for the toluene + *n*-heptane + sulfolane LLE. And because of the liquid-liquid miscibility gap for the *n*-heptane + sulfolane binary, the NRTL α_{ij} parameter for this pair is given a value of 0.2. The NRTL α_{ij} parameters for toluene + sulfolane and *n*-heptane + toluene are allowed to remain at the Aspen default value of 0.3 because of their low levels of nonideality. The temperature dependence of α_{ij} is set to zero (Aspen Plus parameter $d_{ij} = 0$). In Aspen Plus, the τ_{ij} parameter may be regressed as a function of temperature by using the expression $\tau_{ij} = a_{ij} + b_{ij}/T + e_{ij} \ln T + f_{ij}T$. In this example, all the regression parameters are set to zero except b_{ij}. The component activity coefficients are chosen as the objective function for the regression to obtain a fit that models the liquid-liquid *K* values closely, generally found to be within 5 to 10 percent in this case. The resulting b_{ij} binary parameters given in Table 15-10 are then entered into the properties section of the Aspen Plus flow sheet simulation. Pure-component properties were taken from the standard Aspen Plus pure-component databases supplied with the program.

The major unit operations in the sulfolane process usually include an extractor, paraffin stripper, solvent recovery tower, raffinate wash tower, solvent regenerator, and numerous heat exchangers; but for the purposes of this example, the simulation includes only the extractor, paraffin stripper, and extract/recovered solvent cross-exchanger—the portion of the flow sheet shown in Fig. 15-2 outlined by dotted lines. It should be recognized that the exclusion of the solvent recovery tower ignores the highly interactive behavior of the extractor, stripper, and recovery tower; but this is done here to simplify the analysis for the purposes of illustration. Note that the stripper's condenser is modeled as a separate Aspen Plus HEATER block rather than being included in the stripper block, because the Aspen Plus RADFRAC multistage distillation block used to model the stripper requires some distillate reflux if a condenser is included within the block, and generally none is required for the top-fed stripper in the sulfolane process. As a result, the stripper RADFRAC block is specified with no condenser. Also note that in the sulfolane process, the sulfolane solvent enters the top of the extractor, as it is denser than the catalytic reformate feed stream.

The 40,000 kg/h of catalytic reformate fed to stage 7 (counting from the top according to the convention in the EXTRACT block) is modeled as 50/50 *n*-heptane/toluene on a mass basis, and the residual aromatic content of the recovered sulfolane fed to the top of the extractor is 0.4 vol% toluene as given in the problem statement. As an initial guess, the sulfolane rate to the extractor was set at 120,000 kg/h or a solvent-to-feed ratio of 3.0. Depending on the feedstock, solvent-to-feed ratios can range from about 2.0 to 4.0 [Huggins, R., Paper No. 67c, AIChE Spring Meeting, Houston, March, 1997]. In the EXTRACT block, sulfolane must be specified as the key component in the first liquid phase, and *n*-heptane must be specified as the key component in the second liquid phase, because the EXTRACT block requires that the first liquid be the one exiting the bottom of the extractor. A constant-temperature profile of 105°C in the extractor is entered as an initial estimate. The rest of the block parameters (convergence, report, and miscellaneous block options) are allowed to remain at their default values.

The paraffin stripper RADFRAC block is specified with feed to the first of 10 stages, a reboiler but no condenser, a 1-bar gauge top pressure, no internal pressure drop, and a molar boil-up ratio (boil-up rate/bottoms rate) of 0.2 as an initial guess. An internal RADFRAC design specification is entered to vary the boil-up ratio from 0.10 to 0.30 to achieve a mass purity of 500 ppm *n*-heptane in the stripper bottoms on a sulfolane-free basis. To aid RADFRAC convergence, the standard algorithm was changed to Petroleum/Wide-boiling (Sum-Rates) because of the large volatility difference between the hydrocarbons and the sulfolane solvent.

A separate flow sheet Design Spec block (termed a *controller* block in some other simulators) is entered to vary the solvent feed rate to the extractor to achieve the required 98 percent toluene recovery. In addition, the extract reflux stream is called out as the flow sheet tear stream in a Wegstein convergence block to provide proper block sequencing in the simulation. (This is a numerical technique used to accelerate convergence to a solution.) Because the EXTRACT block will not execute with zero extract reflux flow to the bottom of the extractor, an initial guess is required for that stream: 10,000 kg/h of 50/50 by weight *n*-heptane/toluene at 100°C is chosen.

During simulation execution, we found that reflux tear stream convergence with the default Wegstein parameters is very oscillatory, with no convergence even with maximum iterations raised to 200. As a result, significant damping needs to be provided in the convergence block. We raised the bounds of the Wegstein *q* acceleration parameter to be between 0.75 and 1.0 for nearly full damping, after which flow sheet convergence was achieved in less than 10 iterations of every reflux tear stream loop. We also found that good initial guesses and bounds on variables needed to be set to keep the simulation from converging to an aberrant solution that was not physically valid.

With these modifications, the result is that 125,300 kg/h of sulfolane feed to the extractor is required to recover 98 percent of the toluene in the simplified reformate feed. The stage-by-stage mass fraction profile in the extractor is given in Table 15-11, from which we can see that there is very little change in concentration in either phase

TABLE 15-11 Stage Profiles for 98 Percent Recovery (Example 15-5)

	Liquid 1 profile (extract) (mass fractions)			Liquid 2 profile (raffinate) (mass fractions)		
Stage	*n*-Heptane	Toluene	Sulfolane	*n*-Heptane	Toluene	Sulfolane
1	0.02630	0.00624	0.96746	0.97239	0.01946	0.00815
2	0.02683	0.01148	0.96169	0.95558	0.03528	0.00914
3	0.02777	0.02037	0.95186	0.92780	0.06122	0.01099
4	0.02941	0.03528	0.93532	0.88328	0.10220	0.01452
5	0.03227	0.05968	0.90805	0.81532	0.16312	0.02156
6	0.03724	0.09787	0.86489	0.71912	0.24513	0.03575
7	0.04570	0.15319	0.80111	0.59709	0.33949	0.06342
8	0.04573	0.15335	0.80092	0.59676	0.33972	0.06352
9	0.04594	0.15434	0.79973	0.59482	0.34105	0.06413
10	0.04733	0.16035	0.79232	0.58333	0.34873	0.06794

	K″ values (mass fraction basis)					Volume flow (m^3/h)	
Stage	*n*-Heptane	Toluene	Sulfolane	a_{ij} Toluene/ *n*-heptane	ε Toluene	Sulfolane phase	Heptane phase
1	0.0270	0.321	118.7	11.87	2.02	110.5094	33.44022
2	0.0281	0.325	105.2	11.58	1.73	111.5296	39.63555
3	0.0299	0.333	86.6	11.12	1.73	113.3131	40.67214
4	0.0333	0.345	64.4	10.37	1.73	116.4562	42.48007
5	0.0396	0.366	42.1	9.24	1.73	122.0669	45.65451
6	0.0518	0.399	24.2	7.71	1.73	132.2865	51.29266
7	0.0765	0.451	12.6	5.90	1.66	144.7355	61.50425
8	0.0766	0.451	12.6	5.89	5.91	144.798	17.27461
9	0.0772	0.453	12.5	5.86	5.91	145.1873	17.33864
10	0.0811	0.460	11.7	5.67	5.87	146.3572	17.74118

TABLE 15-12 Stream Compositions and Conditions (Example 15-5)

	Extract	Raffinate	Reflux	Stripper bottoms
	98% Recovery—125,300 kg/h required solvent flow			
Wt. fraction *n*-heptane	0.04733	0.97239	0.57643	69 ppm
Wt. fraction toluene	0.16035	0.01946	0.41346	0.13774
Wt. fraction sulfolane	0.79232	0.00815	0.01010	0.86219
Total flow, kg/h	157,719	20,557	12,933	144,786
Temperature, °C	96.4	103.7	105.0	194.1
	99% Recovery—209,300 kg/h required solvent flow			
Wt. fraction *n*-heptane	0.03843	0.98258	0.62133	45 ppm
Wt. fraction toluene	0.10595	0.00983	0.36338	0.08900
Wt. fraction sulfolane	0.85562	0.00759	0.01530	0.91095
Total flow, kg/h	243,982	20,344	15,073	228,909
Temperature, °C	98.8	103.9	105.0	215.1

from the feed stage downward. This is so because in our simplified example we have only a single NA hydrocarbon component (*n*-heptane) to deal with, so the benefit of a backwash section in the extractor below the feed is not apparent. In a real-world profile, however, concentrations of higher-boiling NAs would decrease from the feed point to the bottom of the extractor. Also given in Table 15-11 are stage-by-stage K'' values and volumetric flows as well as the separation factor (toluene with respect to *n*-heptane) and the extraction factor profiles in the extractor. From these we can see that the separation factor for toluene with respect to *n*-heptane varies from about 6 at the bottom of the extractor to 12 at the top, and that the extraction factor is about 2 above the feed and about 6 below the feed. These separation factors are somewhat higher than the value of 4 or so normally seen in real-world aromatic extraction cases; this, too, is an artifact of the simplified ternary system used to model the process.

Another result of the simulation is that a molar boil-up ratio of 0.14 is required in the stripper to achieve the bottoms mass purity of 500 ppm *n*-heptane considering only the hydrocarbons (solvent-free basis). This boil-up ratio corresponds to a reboiler duty of 3692 kW, or roughly 6700 kg/h of 12-bar gauge steam, and it results in 12,933 kg/h of extract reflux for an extractor reflux-to-feed ratio of 0.323. Compositions and rates of the extract, raffinate, reflux, and stripper bottoms streams are given in Table 15-12.

To determine the solvent flow and other conditions required to achieve 99 percent toluene recovery, we merely need to change the specification of the recovery Design Spec block from 98 to 99 percent and reconverge the simulation. With this change and an additional 180 total reflux tear stream iterations, the result is that 209,300 kg/h of sulfolane feed to the extractor is required, 1.7 times the amount needed for 98 percent toluene recovery. A molar boil-up ratio of 0.178 is required in the stripper to maintain the bottoms mass purity of 500 ppm *n*-heptane on a solvent-free basis, even lower than that for the 98 percent recovery case. Likewise, only a slightly higher extract reflux rate is required, 15,073 kg/h, for an extractor reflux-to-feed ratio of 0.377. However, this boil-up ratio corresponds to a reboiler duty of 7022 kW, or roughly 12,800 kg/h of 12-bar gauge steam, about 90 percent higher than for the 98 percent recovery case. The much higher stripper reboiler duty required for 99 percent recovery results from the significantly greater sulfolane feed rate, indicating that the sizes of the extractor and stripper as well as the energy consumption would need to be significantly greater for that increased recovery, probably making it uneconomical in most applications with a 10-stage extractor and stripper. Compositions and rates of the extract, raffinate, reflux, and stripper bottoms streams for the 99 percent recovery case are also given in Table 15-12.

A recent study by A. F. Seibert, W. E. de Villiers, and J. L. Bravo [Paper No. 401b, AIChE Annual Meeting, San Francisco, 2016] provides experimental data on flooding capacities and mass transfer efficiencies of sieve trays for the extraction of toluene from *n*-heptane with sulfolane in a small-scale (10 cm diameter) continuous extractor at ambient temperature. Applying the performance data from this study, we can estimate preliminary values for extractor diameter, tray count, and overall length for a full-scale sieve plate extractor. A final design would require a more rigorous analysis and consultation with the extractor internals supplier.

Examining the stage-by-stage volumetric flows of the two phases in Table 15-11, we can see that the maximum combined flow occurs in the feed stage, stage 7, at a volumetric sulfolane-to-heptane phase ratio of 2.35. In the Seibert study, the flood point on the sieve trays at a phase ratio of 2.35 was measured at a continuous (heptane) phase velocity of 0.54 cm/s, so a conservative design velocity at 50 percent of that value was chosen, 0.27 cm/s. Referring to a plot of overall tray efficiency versus continuous phase velocity in the study, the average tray efficiency at that velocity is extrapolated to be 10.4 percent, which is in the lower mid-range of the measured tray efficiencies in the small-scale extractor. Applying the design phase velocity of 0.27 cm/s to the maximum volumetric heptane phase flow rate of 61.5 m^3/h in stage 7 yields an extractor cross-sectional area of 6.33 m^2 and a diameter of 2.84 m. Applying the design overall tray efficiency of 10.4 percent to the theoretical stage count of 10 results in a projected actual tray count of 96. This is quite a high tray count for a single extractor and is reported here only to illustrate the design method. In real-world applications, the tray count can be much lower because commercial-scale sieve tray extractors are designed with higher tray spacing and are operated at elevated temperatures. As a result, commercial tray efficiencies can be up to 25 percent, which is much higher than the laboratory study used as the basis for our example. (See the subsection Tray Efficiency under Sieve Tray Columns in Static Extraction Columns.)

LIQUID-LIQUID EXTRACTION EQUIPMENT

General References: Zhang, J., B. Zhao, and B. Schreiner, *Separation Hydrometallurgy of Rare Earth Metals*, Springer, Berlin, 2016; Seibert, A. F., "Extraction and Leaching," chap. 14 in *Chemical Process Equipment: Selection and Design*, 3d ed., ed. J. R. Couper et al., Butterworth-Heinemann, Oxford, UK, 2012; Robbins, L. A., sec. 1.9 in *Handbook of Separation Techniques for Chemical Engineers*, 3d ed., ed. P. A. Schweitzer, McGraw-Hill, New York, 1997; Lo, T. C., sec. 1.10 in *Handbook of Separation Techniques for Chemical Engineers*, 3d ed., ed. P. A. Schweitzer, McGraw-Hill, New York, 1997; Godfrey, J. C., and M. J. Slater, eds., *Liquid-Liquid Extraction Equipment*, Wiley, New York, 1994; *Science and Practice of Liquid-Liquid Extraction*, vol. 1, ed. J. D. Thornton, Oxford University Press, Oxford, UK, 1992; Lo, T. C., M. H. I. Baird, and C. Hanson, eds., *Handbook of Solvent Extraction*, Wiley, New York, 1983, and Krieger, Huntington, NY, 1991; Laddha, G. S., and T. E. Degaleesan, *Transport Phenomena in Liquid Extraction*, McGraw-Hill, New York, 1978; and Treybal, R. E., *Liquid Extraction*, 2d ed., McGraw-Hill, New York, 1963.

EXTRACTOR SELECTION

The common types of commercially available extraction equipment and their general features are outlined in Table 15-13. The choice of extractor type depends upon many factors, including the required number of theoretical stages or transfer units, required residence time (due to slow or fast extraction kinetics or limited solute stability), required production rate, tolerance to fouling, ease of cleaning, and availability of the required materials of construction, as well as the ability to handle high or low interfacial tension, high or low density difference, and high or low viscosities. Other factors that influence the choice of extractor include familiarity and tradition (the preferences among designers and operating companies often differ), confidence in scale-up, height constraints, and, of course, the relative capital and operating costs. The flexibility of the extractor to adjust to changes in feed properties also can be an important consideration. For example, compared to a static extractor, a mechanically agitated extractor typically provides a greater turndown ratio (ability to handle a wider range of flow rates), and agitation intensity can be adjusted in the field as needed to accommodate changes in the feed over time. Other factors that may be important include the ability to operate under pressure, to handle corrosive, highly toxic, or flammable materials, and to meet maintenance requirements, among many other possible considerations. Experience with applications similar to the current application and the use of pilot-plant testing play important roles in equipment selection. Pilot testing can address critical issues including demonstration of separation capabilities and equipment scale-up. The simplest extractor design that can meet the process requirements generally will be selected over other competing designs.

Figure 15-29 outlines the decision process recommended by L. A. Robbins [sec. 1.9 in *Handbook of Separation Techniques for Chemical Engineers*, 3d ed., ed. P. A. Schweitzer, McGraw-Hill, New York, 1997]. As an aid to decision making, Robbins recommends characterizing the feed by measuring

TABLE 15-13 Common Liquid-Liquid Extraction Equipment and Applications

Type of extractor	General features	Fields of industrial application
Static extraction columns Spray column Baffle column Packed column (random and structured packing) Sieve tray column	Deliver low to medium mass-transfer efficiency, simple construction (no internal moving parts), low capital cost, low operating and maintenance costs, best suited to systems with low to moderate interfacial tension, can handle high production rates	Petrochemical Chemical Food
Mixer-settlers Stirred-vessels with integral or external settling zones	Can deliver high stage efficiencies with long residence time, can handle high-viscosity liquids, can be adjusted in the field (good flexibility), with proper mixer-settler design can handle systems with low to high interfacial tension, can handle high production rates	Petrochemical Nuclear Fertilizer Metallurgical
Rotary-agitated columns Rotary disc contactor (RDC) Asymmetric rotating disc (ARD) contactor Oldshue-Rushton column Scheibel column Kühni column	Can deliver moderate to high efficiency (many theoretical stages possible in a single column), moderate capital cost, low operating cost, can be adjusted in the field (good flexibility), suited to low to moderate viscosity (up to several hundred centipoise), well suited to systems with moderate to high interfacial tension, can handle moderate production rates	Petrochemical Chemical Pharmaceutical Metallurgical Fertilizer Food
Reciprocating-plate column Karr column	Can deliver moderate to high efficiency (many theoretical stages possible in a single column), moderate capital cost, low operating cost, can be adjusted in the field (good flexibility), well suited to systems with low to moderate interfacial tension including mixtures with emulsifying tendencies, can handle moderate production rates	Petrochemical Chemical Pharmaceutical Metallurgical Food
Pulsed columns Packed column Sieve tray column	No internal moving parts, can deliver moderate to high efficiency, can handle moderate production rates, well suited to highly corrosive or toxic feeds requiring a hermetically sealed system	Nuclear Petrochemical Metallurgical
Centrifugal extractors	Allow short contact time for unstable solutes, minimal space requirements (minimal footprint and height), can handle systems with low density difference or tendency to easily emulsify	Petrochemical Chemical Pharmaceutical Nuclear

a flooding curve using a 1-in-diameter reciprocating-plate (Karr column) miniplant extractor. This is a plot of maximum specific throughput (close to flooding) versus agitation intensity in the Karr column. The position of the resulting curve may be used to identify the type of extractor best suited for commercial development, as illustrated in Fig. 15-30. The flooding curve reflects the liquid-liquid dispersion behavior of the system, and so it can point to options most in line with those properties. The test typically requires 40 to 200 L of feed materials (10 to 50 gal).

A number of equipment selection guides have been published. H. R. C. Pratt and C. Hanson [chap. 16 in *Handbook of Solvent Extraction*, ed. T. C. Lo, M. H. I. Baird, and C. Hanson, Wiley, New York, 1983, and Krieger, Huntington, NY, 1991] provide a detailed comparison chart for 20 equipment types considering 14 characteristics. H. R. C. Pratt and G. W. Stevens [chap. 8 in *Science and Practice of Liquid-Liquid Extraction*, vol. 1, ed. J. D. Thornton, Oxford University Press, Oxford, UK, 1992] modified the Pratt and Hanson selection guide to include solvent volatility and flammability design parameters. J. Stichlmair [*Chem. Ing. Tech.* **52**(3): 253–255 (1980)] and T. L. Holmes, A. E. Karr, and R. W. Cusack [AIChE Summer National Meeting, August, 1987] compared performance characteristics of various equipment designs in the form of a Stichlmair plot. This is a plot of typical mass-transfer efficiency versus characteristic specific throughput (for combined feed and solvent flows) for various types of extractors. Figure 15-31 represents typical performance data generated by using various small-diameter (2- to 6-in, equal to 5- to 15-cm) extractors. This type of plot is intended for use in comparing the relative performance of different extractor types and can be very helpful in this regard. It should not be used for design purposes.

Volumetric efficiency is another characteristic used to compare the different types of extractors. It can be expressed as the product of specific throughput (including feed and extraction solvent) in total volumetric flow rate per unit area (or a characteristic liquid velocity) times the number of theoretical stages achieved per unit length of extractor. It has the units of stages per unit time, or simply reciprocal time (h^{-1}). Thus, volumetric efficiency is inversely proportional to the volume of the column needed to perform a given separation. The Karr reciprocating-plate extractor provides relatively high volumetric efficiency, as it has both a high capacity per unit area and a high number of stages per meter. The Scheibel rotary-impeller column also can provide a high number of stages per meter, but the column throughput typically is less than that of a Karr column, so volumetric efficiency is less. Thus, for a given separation a Scheibel column might be somewhat shorter than a Karr column, but it will need to have a larger diameter to process the same flow rate of feed and extraction solvent. The sieve plate extractor generally exhibits moderate to high throughput, but the number of stages per meter typically is low. The Graesser raining-bucket contactor exhibits low to moderate throughput, but it is reported to have a high separating capability in certain applications.

The ability of an extractor to tolerate the presence of surface-active impurities also may be an important factor in choosing the most appropriate design. A. E. Karr, T. L. Holmes, and R. W. Cusack [*Solvent Extraction and Ion Exchange* **8**(30): 515–528 (1990)] investigated the performance of small-diameter agitated columns and found that the performance of a rotating-disk contactor (RDC) declined faster on addition of trace surface-active impurities compared to the Karr or Scheibel column. The test results indicate that care should be taken when comparing pilot tests of different types of extractors when the data were generated by using high-purity materials. The presence of surface-active impurities can lower column capacity by 20+ percent and efficiency by as much as 60 percent.

Production capacity also may be a deciding factor. Some extractors are available only in small to moderate sizes suitable for low to moderate production rates, as in specialty chemical manufacturing, while others are available in very large sizes designed to handle the very high production rates needed in the petroleum and petrochemical industries. An estimate of relative production rates (feed plus solvent) for selected extractors is given in Table 15-14. Note that the numbers are intended to represent approximate maximum values for a rough comparison. The actual values likely will vary depending upon the particular application. Keep in mind that the relative mass-transfer performance of the various designs is not represented in Table 15-14, and that very large-diameter columns are limited as to how tall they can be built.

HYDRODYNAMICS OF COLUMN EXTRACTORS

Flooding Phenomena The hydraulic capacity of a countercurrent extractor is constrained by breakthrough of one liquid phase into the discharge stream of the other, a condition called *flooding*. The point at which an extractor floods is a function of the design of the internals (as this affects the pressure drop and holdup characteristics of the extractor), the solvent-to-feed ratio and physical properties (as this affects the liquid-liquid

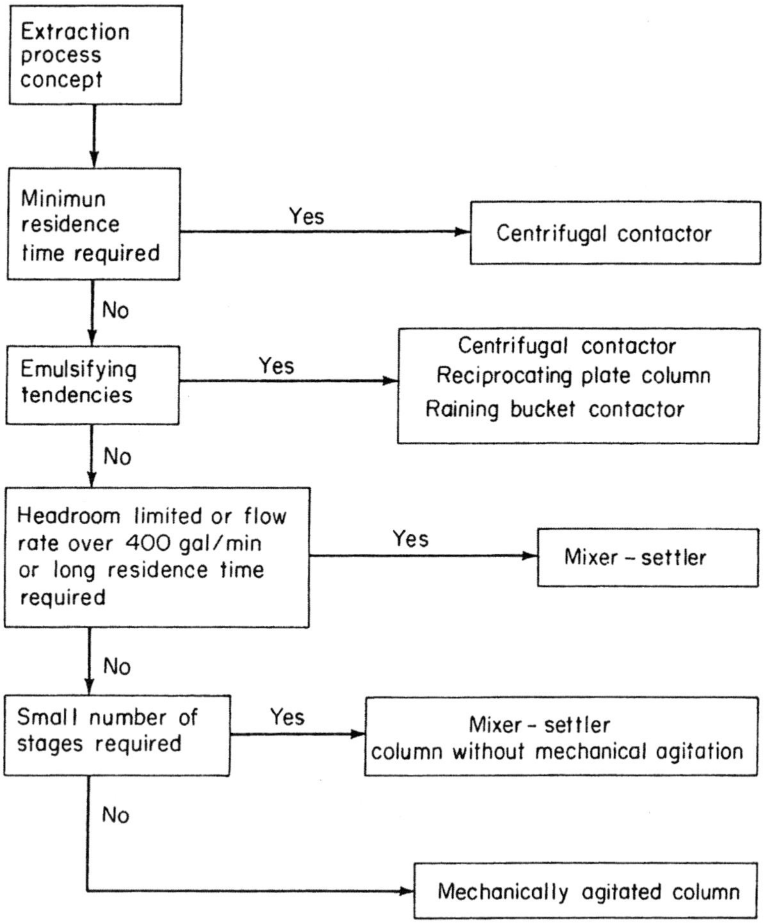

FIG. 15-29 Decision guide for extractor selection. [*Reprinted from Robbins, Sec. 1.9 in* Handbook of Separation Techniques for Chemical Engineers, *3d ed., Schweitzer, ed., McGraw-Hill, New York, 1997), with permission. Copyright 1997 McGraw-Hill, Inc.*]

dispersion behavior), the agitation intensity (if agitation is used), and the specific throughput. The latter often is expressed in terms of the volumetric flow rate per cross-sectional area; or, equivalently, in terms of liquid velocity. A plot of the maximum throughput that can be sustained just prior to flooding versus a key operating variable is called a flooding curve. Ideally, extractors are designed to operate near flooding to maximize productivity. In practice, however, many new column extractors are designed to operate at 40 to 60 percent of the predicted flood point because of uncertainties in the design, process impurity uncertainties, and to allow for future capacity increases. This practice varies from one type of extractor to another and one designer to another. In a static extraction column, countercurrent flow of the two liquid phases is maintained by virtue of the difference in their densities and the pressure drop through the equipment. Only one of the liquids may be pumped through the equipment at any desired flow rate or velocity; the maximum velocity of the other phase is then fixed by the flood point. If an attempt is made to exceed this hydraulic limit, the extractor will flood.

In extraction equipment, flooding may occur through a variety of mechanisms [Seibert, A. F., J. L. Bravo, and J. R. Fair, *ISEC '02 Proc.* **2:** 1328–1333 (2002)]:

1. Excessive flow rates of either dispersed-phase or continuous-phase, or high agitation intensity, cause dispersed-phase holdup or population density to exceed the volumetric capacity of the equipment.
2. Excessively high continuous-phase flow rate causes excessive entrainment of dispersed phase into the continuous-phase outlet.
3. Inadequate drop coalescence causes the formation of dispersion bands or layers of uncoalesced drops that entrap continuous phase between them. The continuous phase can then be entrained into the wrong outlet.
4. Operation at a high ratio of dispersed phase to continuous phase results in phase inversion. (See the subsection Liquid-Liquid Dispersion Fundamentals.)
5. Operating too close to the liquid-liquid phase boundary causes complete miscibility during an upset. This might be caused by a slight change in solvent or feed rates, by an increase in the concentration of solute in the feed, or by introduction of a surface-active impurity.
6. In sieve tray columns, excessive orifice and/or downcomer pressure drop within the extractor causes the formation of large coalesced layers that back up and overflow the trays. This might be caused by operation outside of the designed flow rates, or by fouling of the internals resulting in increased pressure drop.
7. Poor interface control allows the main liquid-liquid interface to leave the extractor. This may result from inadequate size of interface flow control valves, or operation with internals that provide inverse control responses such as those observed with sieve tray extractors. (See the subsection Process Control Considerations.)
8. Fouling and plugging of internals or the outlet flow control valves.

Accounting for Axial Mixing Differential-type column extractors are subject to axial (longitudinal) mixing, also called axial dispersion and generally referred to as backmixing. This condition refers to a departure from uniform plug flow of the swarm of dispersed drops as drops rise or fall in the column, as well as any departure from plug flow of continuous phase in the opposite direction. As a result of axial mixing, the elements of the dispersed phase and the continuous phase exhibit a distribution of residence times within the equipment, and this decreases the effective or overall concentration driving force in the contactor. Because of this effect, the actual column must be taller than simple application of an ideal, plug flow model would indicate. When one is approaching the design of a contactor, factors that may contribute to axial mixing should be considered so that measures might be taken to reduce their effects. This may involve the design of baffles to help direct the liquid traffic within the column. Also, if the transfer of solute occurs such that the continuous phase is significantly denser at the top of an extraction column than at the bottom, this may encourage circulation of continuous phase, and it may be advisable to switch the phase that is dispersed. For more information on this effect, see T. L. Holmes, A. E. Karr,

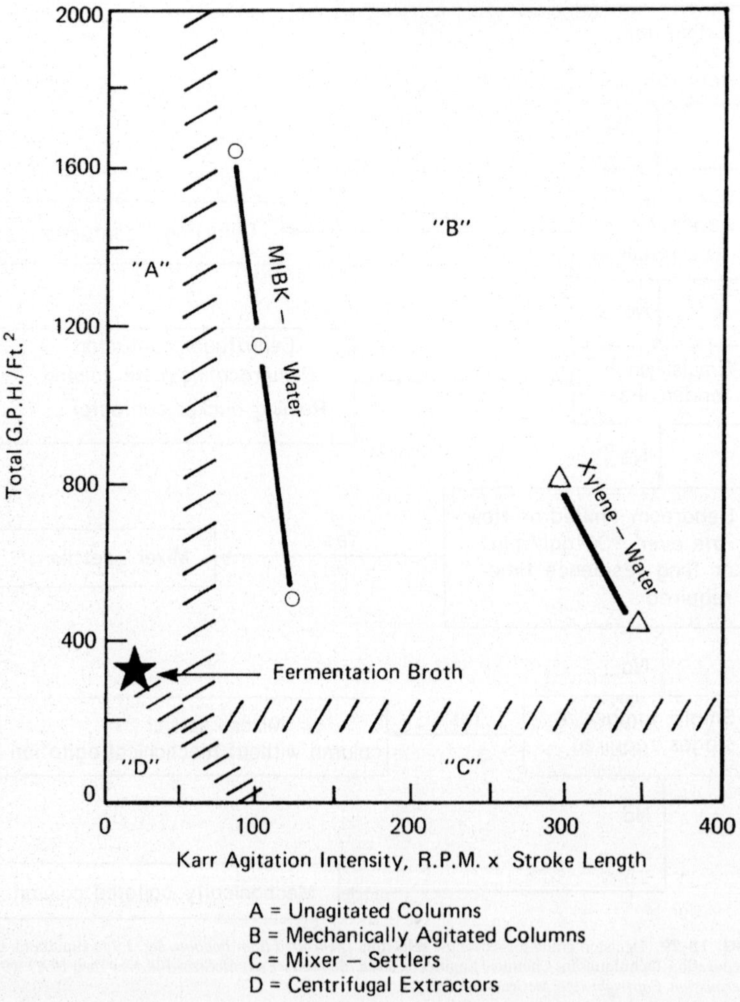

FIG. 15-30 Typical Karr column flooding characteristics. Example flooding data are shown for two applications involving MIBK + water and xylene + water (flooding occurs to the right of the indicated flooding curve). A data point for extraction of a fermentation broth is indicated by the star. Results will vary depending upon process variables including solute concentration, the presence of other solutes, and temperature. [*Reprinted from Robbins, Sec. 1.9 in* Handbook of Separation Techniques for Chemical Engineers, *3d ed., Schweitzer, ed., McGraw-Hill, New York, 1997), with permission. Copyright 1997 McGraw-Hill, Inc.*]

and M. H. I. Baird, *AIChE J.* **37**(3): 360–366 (1991); and K. Aravamudan and M. H. I. Baird, *AIChE J.* **42**(8): 2128–2140 (1996).

Axial mixing effects commonly are taken into account by using a diffusion analogy and an axial mixing coefficient E, also called the longitudinal dispersion coefficient or eddy diffusivity, to account for the spreading of the concentration profiles. At steady state, the conservation equation has the general form

$$E\frac{\partial^2 C}{\partial z^2}+V\frac{\partial C}{\partial z}+k_o a(C-C^*)=0 \qquad (15\text{-}101)$$

where V is phase velocity, k_o is an overall mass-transfer coefficient, C is solute concentration (mass or moles per unit volume), and the superscript asterisk denotes equilibrium. By using Eq. (15-101) as a foundation, the required height of extractor may be calculated from a simplified plug flow model plus application of a correction factor expressed as a function of E or a Péclet number Pe = Vb/E, where b is a characteristic equipment dimension. The required values of E must be determined by experiment. For example, calculations based on the application of mass-transfer coefficients and calculation of interfacial area from drop size and holdup correlations will need to be corrected for axial dispersion as a function of column size. A variety of axial mixing models and data correlations have been developed for various types of column extractors. For detailed discussion, see C. A. Sleicher, *AIChE J.* **5**(2): 145–149 (1959); T. Vermeulen et al., *Chem. Eng. Prog.* **62**(9): 95–102 (1966); and N. N. Li and E. N. Zeigler, *Ind. Eng. Chem.* **59**(3): 30–36 (1967). Also see the detailed discussions in G. S. Laddha and T. E. Degaleesan, *Transport Phenomena in Liquid Extraction*, McGraw-Hill, New York, 1978; H. R. C. Pratt and M. H. I. Baird, chap. 6 in *Handbook of Solvent Extraction*, ed. T. C. Lo, M. H. I. Baird, and C. Hanson, Wiley, New York, 1983, and Krieger, Huntington, NY, 1991; and J. C. Godfrey and M. J. Slater, eds., *Liquid-Liquid Extraction Equipment*, Wiley, New York, 1994. The method used by O. Becker [*Chem. Eng. Technol.* **26**(1): 35–41 (2003)] is discussed in the subsection Static Extraction Columns.

Computational fluid dynamics (CFD) simulations are beginning to be developed for certain types of extractors to better understand flow patterns. The simulation of two-liquid-phase flows around complex internals is an active research area. Examples include CFD calculations for a rotating-disk contactor by G. Modes and H.-J. Bart [*Chem. Eng. Technol.* **24**(12): 1242–1244 (2001)] and CFD calculations for a pulsed column extractor by A. Amokrane et al. [*Can. J. Chem. Eng.* **92:** 220–233 (2014)].

Liquid Distributors and Dispersers It should be recognized that the performance of a column extractor can be significantly affected by how uniformly the feed and solvent inlet streams are distributed to the cross section of the column. The requirements for distribution and redistribution vary depending upon the type of column internals (packing, trays, agitators, or baffles) and the impact of the internals on the flow of dispersed and continuous phases within the column. Important considerations in specifying a distributor include the number of holes and the hole pattern (geometric layout), hole size, number of downcomers or upcomers (if used) and their placement, the maximum to minimum flow rates the design can

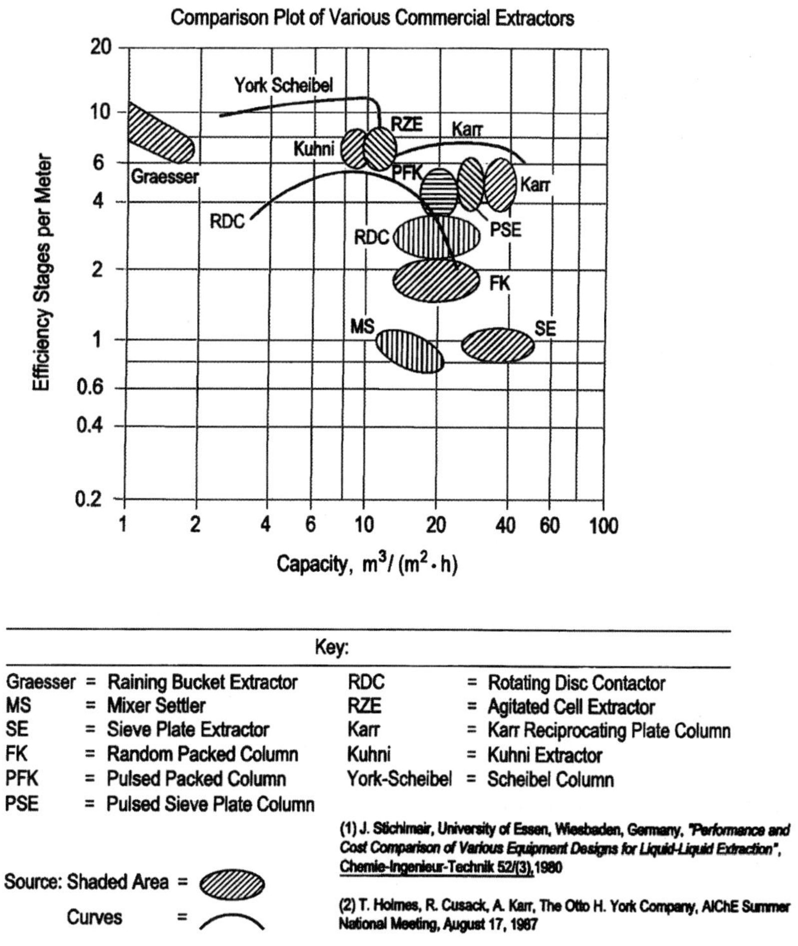

FIG. 15-31 Modified Stichlmair chart. (*Courtesy of Koch Modular Process Systems.*)

handle (turndown ratio), and resistance to fouling. Various types of liquid distributors are available, including sieve tray dispersers and ladder-type pipe distributors designed to give uniform distribution of drops across the column cross section. (See the subsection Packed Columns and Sieve Tray Columns under Liquid-Liquid Extraction Equipment for more information about these. The height of the coalesced layer on a disperser plate may be calculated by using the method described in Sieve Tray Columns.) Ring-type distributors also are used, primarily for agitated extractors. Equipment vendors should be consulted for additional information.

Typical hole sizes for distributors and dispersers are between 0.05 in (1.3 mm) and 0.25 in (6.4 mm). Small holes should be avoided in applications where the potential for plugging or fouling of the holes is a concern. For plate dispersers, the holes should be spaced no closer than about 3 hole diameters to avoid coalescence of drops emerging from adjacent holes. Design velocities for liquid exiting the holes generally are in the range of 0.5 to 1.0 ft/s (15 to 30 cm/s). Several methods have been proposed for more precisely specifying the design velocities. For detailed discussion, see A. Kumar and S. Hartland, chap. 17 in *Liquid-Liquid Extraction Equipment*, ed. J. C. Godfrey and M. J. Slater, Wiley, New York, 1994, pp. 631–635; K. Ruff, *Chem. Ing. Tech.* **50**(6): 441–443 (1978); and G. S. Laddha and T. E. Degaleesan, chap. 11 in *Transport Phenomena in Liquid Extraction*, McGraw-Hill, New York, 1978, pp. 307–310. These methods are relevant for the design of distributors/dispersers used in all types of column extractors. The liquid should issue from the hole as a jet that breaks up into drops. The jet should yield a drop size distribution that provides good interfacial area, with an average drop size smaller than the maximum given by $d_{max} = [\sigma/(\Delta\rho g)]^{0.5}$, but without creating small secondary drops that cause entrainment problems or formation of an emulsion. (See the subsection Size of Dispersed Drops in Liquid-Liquid Dispersion Fundamentals.) As a general guideline, the maximum recommended design velocity corresponds to a Weber number of about 12:

$$V_{o,\max} \approx \left(\frac{12\sigma}{d_o\rho_d}\right)^{1/2} \qquad (15\text{-}102)$$

The minimum Weber number that ensures jetting in all the holes is about 2. It is common practice to specify a Weber number between 8 and 12 for a new design. For a detailed discussion of fundamentals, see S. Homma et al., *Chem. Eng. Sci.* **61**(12): 3986–3996 (2006).

It is well established that the dispersed phase must issue cleanly from the holes. This requires that the material of the pipe or disperser plate be preferentially wetted by the continuous phase (requiring the use of plastics or plastic-coated trays in some instances), or that the dispersed phase issue from nozzles projecting beyond the surface. For plate dispersers, these may be formed by punching the holes and leaving the burr in place [Mayfield, F. D., and W. L. Church, *Ind. Eng. Chem.* **44**(9): 2253–2260 (1952)]. Once the design velocity is set, the number of holes is given by

$$N_{\text{holes}} = \frac{Q_d}{A_o V_o} \qquad (15\text{-}103)$$

where Q_d is the total volumetric flow rate of dispersed phase and A_o is the cross-sectional area of a single hole.

STATIC EXTRACTION COLUMNS

Common Features and Design Concepts Static extractors include spray-type, packed, and trayed columns often used in the petrochemical industries (Fig. 15-32). They offer the advantages of (1) availability in large diameters for high production rates, (2) simple operation with no moving parts and associated seals, (3) requirement for control of only one operating interface, and (4) relatively small required footprint compared to mixer-settler equipment. Their primary disadvantage usually is lower mass-transfer efficiency compared to that of mechanically agitated extractors. This usually limits applications to those involving low viscosities (less than about 5 cP), low to moderate interfacial tensions (typically 3 to 20 dyn/cm equal to 0.003 to 0.02 N/m), and often no more than four to six equilibrium stages. J. Koch and G. Shiveler [*Chem. Eng. Prog.* **111**(11): 22–30 (2015)] discuss important design principles to keep in mind when translating laboratory results to

TABLE 15-14 Estimated Maximum Production Rate for Selected Extractors

	Maximum* specific throughput		Maximum† diameter (typical)	Estimated maximum production rate	
Extractor type	$m^3/h/m^2$	$gal/h/ft^2$	m	m^3/h	gal/min
Mixer-settler‡	10	250	12	~1200	5200
Baffle tray column	60	1500	5	~1200	5200
Sieve plate column	50	1200	5	~1000	4300
Packed column	50	1200	5	~1000	4300
Spray column§	70	1700	4	~900	4000
Rotating disk contactor	35	850	3	~250	1100
Kühni rotating-impeller column¶	40	1000	3	~280	1200
Karr reciprocating-plate column	40	1000	3	~280	1200
Scheibel rotating-impeller column	25	600	3	~200	800
Graesser raining-bucket contactor	10	250	3	~70	300

*Typical maximum value for dispersed + continuous phase flow rates.
The actual value for a given application will depend upon physical properties and rates of extraction and phase separation and may be much lower.
†Typical value. Larger diameters may be possible.
‡Throughput and equivalent diameter are based on mixer-settler footprint. Estimates based on hydrometallurgical stripping applications.
§Larger diameters possible but not recommended due to severe backmixing.
¶Higher throughput may be achieved by increasing the column open area.

the commercial scale. Although the spray column is the least efficient static extractor in terms of mass-transfer performance, due to considerable backmixing effects, it finds use in processing feeds that would easily foul other equipment. Packed column designs can provide improved mass-transfer performance by limiting backmixing. However, packed column mass transfer can be limited by backmixing, especially with larger diameter columns and high dispersed to continuous phase flow ratios. In contrast, the sieve tray column can be designed to minimize backmixing such that the efficiency scales well with column diameter. It should be noted that mechanically agitated extractors also can experience significant backmixing and a loss of efficiency with larger diameter columns.

An understanding of the general hydraulics of a static contactor is necessary for estimating the diameter and height of the column, as this affects both capacity and mass-transfer efficiency. Accurate evaluations of characteristic drop diameter, dispersed-phase holdup, relative velocity of the two phases (called slip velocity), and flooding velocities usually are necessary. Fortunately, the relative simplicity of these devices facilitates their analysis and the approaches taken to modeling performance.

Choice of Dispersed Phase The great majority of extractor designs function by formation of dispersed drops to maximize contact area and mass transfer. Static extractors generally are designed with the majority phase dispersed in order to maximize dispersed-phase holdup and interfacial area needed for mass transfer; that is, the phase with the greatest flow rate entering the column generally is dispersed. The choice of dispersed phase also depends upon the relative viscosity of the two phases. If one phase is particularly viscous, it may be necessary to disperse that phase. Note that a few extractor designs involve formation of films or rivulets instead of drops for increased capacity and greater tolerance to fouling, but this approach is not common as it normally involves lower mass-transfer capability.

Drop Size and Dispersed-Phase Holdup Various models used to estimate the size of dispersed drops in static extractors are listed in Table 15-15. Also see the subsection Size of Dispersed Drops under Liquid-Liquid Dispersion Fundamentals. Measurements of dispersed-phase holdup within a column-type extractor often are made by stopping all flows in and out of the extractor and measuring the change in the main interface level. This technique can be prone to significant experimental error as a result of end effects, static holdup present in small laboratory packings, inaccurate measurement of the baseline interface level, inability to instantaneously stop both inlet flows and the heavy phase outlet flow, and holdup variations within a column as flooding conditions are approached. Examples of models for prediction of holdup are provided in Table 15-16. Additional models are given in J. C. Godfrey and M. J. Slater, eds., *Liquid-Liquid Extraction Equipment*, Wiley, New York, 1994. In general, an implicit calculation of the dispersed-phase holdup is usually encountered. One must be careful in evaluating the roots of these calculations, especially in the region of high dispersed-phase holdup ($\phi_d > 0.2$).

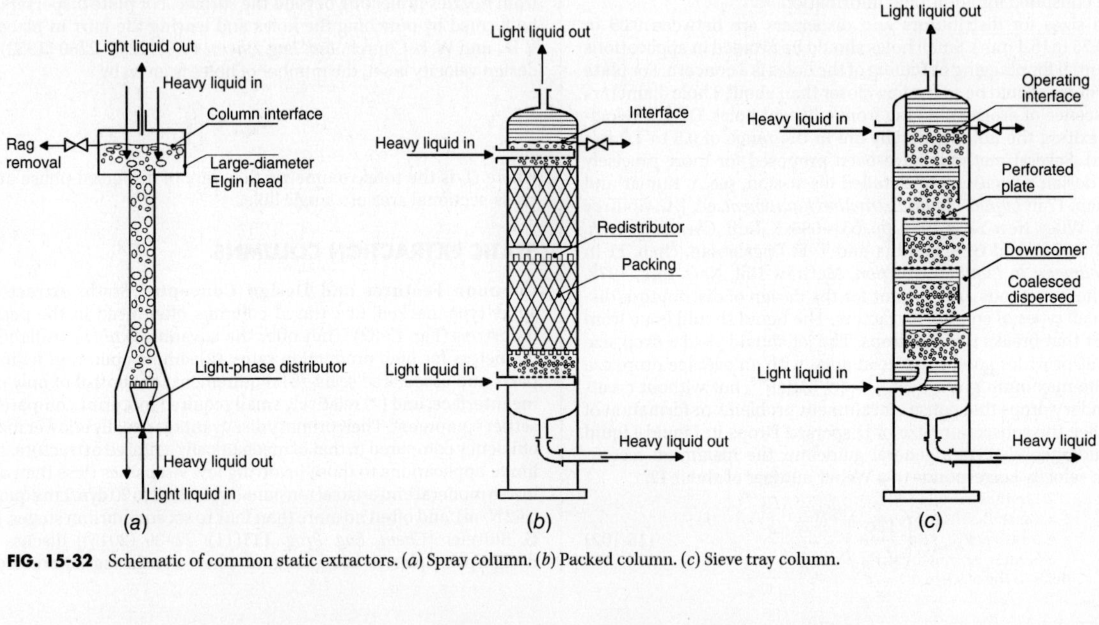

FIG. 15-32 Schematic of common static extractors. (*a*) Spray column. (*b*) Packed column. (*c*) Sieve tray column.

TABLE 15-15 Example Drop Diameter Models for Static Extractors

Example	Eq.	Comments	Ref.
$d_p = 1.15\eta\sqrt{\frac{\sigma}{\Delta\rho g}}$, $\eta = 1.0$ for no mass transfer and $c \rightarrow d$, $\eta = 1.4$ for $d \rightarrow c$	1	Spray, packing, and sieve tray	1
$d_p = 0.12 D_h \mathrm{We}_c^{-0.5} \mathrm{Re}_c^{0.15}$, developed with no mass transfer, We_c and Re_c are calculated based on slip velocity	2	SMV structured packing	2
$d_p = C_p\sqrt{\frac{\sigma}{\Delta\rho g}}$, $C_p = 1$ for $\rho_d < \rho_c$, $C_p = 0.8$ for $\rho_d > \rho_c$ developed with no mass transfer	3	Packing	3
$d_p = 1.09\sqrt{\frac{\sigma}{\Delta\rho g}}\left(1 + 700\frac{V_c\mu_c}{\sigma}\right)$, developed with no mass transfer	4	Packing	4
$d_p = 0.74 C_\psi\sqrt{\frac{\sigma}{\Delta\rho g}}\left(\frac{\Delta\rho\rho_d\sigma}{\rho_w^2\sigma_w}\right)^{-0.12}$, $C_\psi = 1$ for no mass transfer, $C_\psi = 0.84$ for $c \rightarrow d$, $C_\psi = 1.23$ for $d \rightarrow c$	5	Packing	5
$d_p = \dfrac{C_\psi}{\dfrac{1}{(6d_o\sigma/\Delta\rho g)^{1/3}} + \dfrac{1}{2.04(12\sigma/\rho_d V_o^2)}}$ $C_\psi = 1.0$ for $c \rightarrow d$ and no mass transfer, $C_\psi = 1.06$ for $d \rightarrow c$	6	Spray nozzles	5
$d_p = d_o \mathrm{E\ddot{O}}_o^{-0.35}\left[0.80 + \exp\left(-2.73\times10^{-2}\frac{\mathrm{We}_o}{\mathrm{E\ddot{O}}_o}\right)\right]$ $\mathrm{E\ddot{O}}_o = \frac{d_o^2\Delta\rho g}{\sigma}$, $\mathrm{We}_o = \frac{\rho_d d_o V_o^2}{\sigma}$	7	Perforated plate	6

References:
1. Seibert, A. F., and J. R. Fair, *Ind. Eng. Chem. Res.* **27**(3): 470–481 (1988).
2. Streiff, F. A., and S. J. Jancic, *Ger. Chem. Eng.* **7**: 178–183 (1984).
3. Billet, R., J. Mackowiak, and M. Pajak, *Chem. Eng. Process.* **19**: 39–47 (1985).
4. Lewis, J. B., I. Jones, and H. R. C. Pratt, *Trans. Instn. Chem. Engrs.* **29**: 126–148 (1951).
5. Kumar, A., and S. Hartland, *Ind. Eng. Chem. Res.* **35**(8): 2682–2695 (1996).
6. Kumar, A., and S. Hartland, Chap. 17 in *Liquid-Liquid Extraction Equipment*, J. C. Godfrey and M. J. Slater, eds. (Wiley, 1994), pp. 625–735.
Refer to the original articles for details.

TABLE 15-16 Example Holdup Models for Static Extractors

Example	Eq.	Comments	Ref.
$\phi_d = \dfrac{V_d[\cos(\pi\xi/4)]^{-2}}{\varepsilon[V_{so}\exp(-6\phi_d/\pi) - V_c/\varepsilon(1-\phi_d)]}$, $\xi = \dfrac{a_p d_p}{2}$	1	Spray, packing, and sieve tray. V_{so} is calculated by the method of Grace et al. (1976), Eqs. (15-114) to (15-118).	1
$\dfrac{V_d}{\phi_d} + \dfrac{V_c}{1-\phi_d} = \dfrac{4c_p}{3}\sqrt{\dfrac{d_p\Delta\rho g}{C_D\rho_c}}$	2	SMV structured packing. Drag coefficient, C_D is calculated by assuming a drop is a rigid sphere. Parameter c_p depends upon drop-drop and drop-packing interactions.	2
$\dfrac{V_d}{\phi_d} + \dfrac{V_c}{1-\phi_d} = \varepsilon C\left(\dfrac{4g\Delta\rho\sigma}{\rho_c^2}\right)^{0.25}\exp(-b\phi_d)$	3	Packing. Constants C and b differ for different packings. Drag coefficient = 1.	3
$\left(\dfrac{V_d}{\phi_d} + \dfrac{V_c}{1-\phi_d}\right)\left(\dfrac{a_p\rho_c}{\varepsilon^3 g\Delta\rho}\right)^{0.5} = 0.683\phi_d(1-\phi_d)$	4	Packing	4
$\phi_d = A\left\{0.27 + \left[\dfrac{\varepsilon}{g}\left(\dfrac{\rho_c}{g\sigma}\right)^{0.25}\right]^{0.78}\right\}\left[V_d\left(\dfrac{\rho_c}{g\sigma}\right)^{0.25}\right]^{0.87}\exp(B)$ $B = 3.34V_c\left(\dfrac{\rho_c}{g\sigma}\right)^{0.25}$ $A = \left(\dfrac{\Delta\rho}{\rho_c}\right)^{-0.58}\left(\dfrac{\mu_d}{\mu_w}\right)^{0.18} C\varepsilon^n\left[l\left(\dfrac{\rho_c g}{\sigma}\right)^{0.5}\right]^{-0.39}$	5	Unified model for packing, spray, Karr, pulsed perforated plate, Kühni, rotating disk. Constants C, n, and l depend upon type of contactor.	5

References:
1. Seibert, A. F., and J. R. Fair, *Ind. Eng. Chem. Res.* **27**(3): 470–481 (1988).
2. Streiff, F. A., and S. J. Jancic, *Ger. Chem. Eng.* **7**: 178–183 (1984).
3. Billet, R., J. Mackowiak, and M. Pajak, *Chem. Eng. Process.* **19**: 39–47 (1985).
4. Sitaramayya, T., and G. S. Laddha, *Chem. Eng. Sci.* **13**: 263 (1961).
5. Kumar, A., and S. Hartland Ind. Eng. *Chem. Res.* **34**: 3925–3940 (1995).
Refer to the original articles for details.

Interfacial Area The mass-transfer efficiency of most extraction devices is proportional to the area available for mass transfer (neglecting any axial mixing effects). As discussed in the subsection Liquid-Liquid Dispersion Fundamentals, for the general case where the dispersed phase travels through the column as drops, an average liquid-liquid interfacial area can be calculated from the Sauter mean drop diameter and dispersed-phase holdup:

$$a = \frac{6\varepsilon\phi_d}{d_p} \tag{15-104}$$

In most cases, the drop size distribution is not known. Various models are available to estimate the Sauter mean drop diameter and holdup for a variety of extractors (as described above).

Drop Velocity and Slip Velocity The hydraulic characteristics of a static extractor depend upon drop diameter, liquid velocities, and physical properties. The average velocity of a dispersed-phase drop (V_{drop}) and the interstitial velocity of the continuous phase V_{ic} are given by

$$V_{\text{drop}} = \frac{V_d}{\varepsilon\phi_d} \tag{15-105}$$

$$V_{ic} = \frac{V_c}{\varepsilon(1-\phi_d)} \tag{15-106}$$

where V_d = superficial velocity of dispersed phase
V_c = superficial velocity of continuous phase
ϕ_d = fraction of void volume occupied by dispersed phase
ε = void fraction of column (ε = 1.0 for sprays and sieve trays)

The slip velocity is the velocity at which a dispersed drop moves relative to the counter-flowing continuous phase, calculated by adding the magnitudes of the two phase velocities:

$$V_s = |V_{\text{drop}}| + |V_{ic}| = \frac{V_d}{\varepsilon\phi_d} + \frac{V_c}{\varepsilon(1-\phi_d)} \tag{15-107}$$

For a dispersed-phase drop of diameter d_p, slip velocity can be estimated from a balance of gravitational, buoyancy, and frictional forces:

$$F_{\text{buoyancy}} - F_{\text{gravity}} - F_{\text{drag}} = 0 \tag{15-108}$$

$$F_{\text{buoyancy}} = \rho_c \frac{\pi}{6} d_p^3 g \tag{15-109}$$

$$F_{\text{gravity}} = \rho_d \frac{\pi}{6} d_p^3 g \tag{15-110}$$

$$F_{\text{drag}} = \frac{1}{2} C_D \rho_c \frac{\pi}{4} d_p^2 V_{so}^2 \tag{15-111}$$

where V_{so} is defined as the characteristic slip velocity obtained at low dispersed-phase flow rate. Rearranging Eqs. (15-108) to (15-111) gives

$$V_{so} = \sqrt{\frac{4 \rho g d_p}{3\rho_c C_D}} \tag{15-112}$$

The slip velocity at higher holdup often is estimated from $V_s \approx V_{so}(1 - \phi_d)$. Slip velocity is a key concept in various approaches to modeling dispersed-phase holdup, flooding point, and mass transfer.

Equation (15-112) provides the basis for various methods used to predict the characteristic slip velocity. It can be difficult to use for design because of difficulty estimating the drag coefficient C_D and difficulty accounting for packing resistance or drop-drop interactions. The drag coefficient can be affected by internal circulation within the drop. For good mass transfer, it is most desirable to have circulating drops traveling through a relatively nonviscous continuous phase. Particular care should be taken in utilizing models developed primarily from studies involving small laboratory packings, because the packing resistance is particularly significant in that case. Also, many studies do not include low-interfacial tension systems even though most applications of static extractors involve low to moderate interfacial tension. Also note that surface-active impurities can reduce the characteristic drop velocity [Garner, F. H., and A. H. P. Skelland, *Ind. Eng. Chem.* **48**(1): 51–58 (1956); and Skelland, A. H. P., and C. L. Caenepeel, *AIChE J.* **18**(6): 1154–1163 (1972)], which is another reason to approach these models with care. For additional discussion, see T. Míšek, chap. 5 in *Liquid-Liquid Extraction Equipment*, ed. J. C. Godfrey and M. J. Slater, Wiley, New York, 1994.

The following method is recommended for calculating slip velocity in static extractors at low dispersed-phase holdup and very low Reynolds numbers (<2):

$$\text{If } \text{Re}_{\text{Stokes}} = \frac{\rho_c \;\; \rho g d_p^3}{18 \;\; {}_c^2} < 2, \text{ then } V_{so} = \frac{\rho g d_p^2}{18 \;\; {}_c} \qquad \text{(Stokes' law)} \tag{15-113}$$

For $\text{Re}_{\text{Stokes}} > 2$, Seibert and coworkers [Seibert, A. F., and J. L. Fair, *Ind. Eng. Chem. Res.* **27**(3): 470–481 (1988); and Seibert, A. F., B. E. Reeves, and J. R. Fair, *Ind. Eng. Chem. Res.* **29**(9): 1901–1907 (1990)] recommend the model of J. R. Grace, T. Wairegi, and T. H. Nguyen [*Trans. Inst. Chem. Eng.* **54:** 167 (1976)]. In this case, the characteristic slip velocity may be calculated from

$$V_{so} = \frac{\text{Re} \;\; {}_c}{d_p \rho_c} \tag{15-114}$$

where Re is obtained from the correlation:

$$\frac{\text{Re}}{p^{0.149}} = 0.94 H^{0.757} - 0.857 \qquad H \leq 59.3 \tag{15-115}$$

$$\frac{\text{Re}}{p^{0.149}} = 3.42 H^{0.441} - 0.857 \qquad H > 59.3 \tag{15-116}$$

And P and H are dimensionless groups defined by

$$P = \frac{\rho_c^2 \sigma^3}{{}_c^4 g \;\; \rho} \tag{15-117}$$

$$H = \frac{4 d_p^2 g \;\; \rho}{3\sigma} \left(\frac{\;\;_w}{\;\;_c}\right)^{0.14} p^{0.149} \tag{15-118}$$

and μ_w is a reference viscosity equal to 0.9 cP (9×10^{-4} Pa·s). For discussion of methods to correct slip velocity for the effect of high dispersed-phase holdup, see F. Augier, O. Masbernat, and P. Guiraud, *AIChE J.* **49**(9): 2300–2316 (2003).

Flooding Velocity Maximum flow through a countercurrent extractor is limited by the flooding velocity. See the subsection Hydrodynamics of Column Extractors for a general discussion of flooding mechanisms. Examples of published flooding models for static extractors are given in Table 15-17. Because of the many possible causes of flooding, many different approaches to modeling the flooding velocity have been taken. Some of the earlier flooding models were based on the idea that an extractor will enter a flooding condition as the slip velocity goes to zero. However, because flooding can occur in many different ways the results have not always proven reliable, so newer models were developed with other concepts in mind. Also, models developed from small-scale laboratory data can lead to problems when used for design of commercial-scale columns. For example, in packed columns a column-diameter to packing-diameter ratio of at least 8 is recommended to avoid channeling due to wall effects. This means that laboratory studies often utilize small packings with high specific packing surface areas (packing area/contacting volume). The high packing area likely provides significant resistance to drop flow, greater than that encountered in large columns containing large commercial packings. Also, many of the published laboratory data on flooding velocities were generated by using moderate to high-interfacial-tension systems. In this case, the packing surface area resistance can control the flooding mechanism. For these reasons, it is important to understand the background and basis for a given flooding model and assess whether it is appropriate for the given application.

Several published correlations of the flooding velocity (Table 15-17) have elements of the form

$$V_{cf} \propto C_1 \frac{C_2}{a_p^n} \qquad 0 < n < 1 \tag{15-119}$$

where V_{cf} is the continuous-phase velocity at which flooding occurs, a_p is the specific packing surface area, and C_1 and C_2 are empirical constants that depend upon the specific type of packing, fluid physical properties, and flow ratio. While these types of models have excellent reported fits of data, they were primarily developed by using laboratory-scale packings. Furthermore, in the limit as the packing surface area approaches zero, the predicted flooding velocity becomes infinite, an unrealistic result. Care should be taken when extrapolating such models to a larger packing size.

A. F. Seibert, B. E. Reeves, and J. R. Fair [*Ind. Eng. Chem. Res.* **29**(9): 1901–1907 (1990)] have proposed an alternative model that is derived by assuming a tightly packed arrangement of drops at flooding. It has the form

$$V_{cf} = C_1 / \left[1 + \frac{C_2}{C_3 \cos^2(C_4 a_p)} \right] \tag{15-120}$$

where parameters C_1, C_2, C_3, and C_4 are functions of system properties and flow ratio (as in Eq. 1 in Table 15-17). An advantage of this flooding model is that as the packing surface area approaches zero, a finite flooding velocity is calculated from $\cos^2(0) = 1$. For this reason, equations in the form of Eq. (15-120) can be used to predict flooding in a spray column and the ultimate capacity of a tray column. Unfortunately, very few flooding data are available for columns greater than 30 cm (12 in) in diameter. Also, many of the available flooding data have been obtained in the absence of mass transfer. With this in mind, for new designs it is recommended that flow velocities be limited to no more than 60 percent of the calculated flooding values. The final design should be refined in miniplant or pilot-plant tests using actual feed materials.

Drop Coalescence Rate The rate of drop coalescence often is assumed to be rapid (not rate-limiting) in the design of static extractors. However, this is not necessarily the case, particularly during operation at high dispersed-phase holdup and high flow ratios of dispersed phase to continuous phase. Under these conditions, a large number of drops flow through a nearly

TABLE 15-17 Example Flooding Models for Static Extractors

Example	Eq.	Comments	Ref.
$V_{cf} = \dfrac{0.178 \varepsilon V_{so}}{1 + 0.925 (V_{df}/V_{cf}) / \cos^2(\pi\zeta/4)}$ $\quad \zeta = \dfrac{a_p d_p}{2}$	1	Generalized model for spray, packing, and ultimate capacity of sieve tray. V_{so} is calculated by the method of Grace et al. (1976), Eqs. (15-114) to (15-118).	1,2
$V_{cf} = \dfrac{0.178 V_{so}}{1 + 0.925 (V_{df}/V_{cf})}$	2	Spray V_{so} is calculated by the method of Grace et al. (1976), Eqs. (15-114) to (15-118).	1
$V_{cf} = \varepsilon C \left(\dfrac{4 g \Delta\rho\sigma}{\rho_c^2} \right)^{1/4} (1 - \phi_{df})^2 [\exp(-b\phi_{df})](1 - b\phi_{df})$	3	Packing	3
$\dfrac{V_d}{V_c} = \dfrac{\phi_{df}^2}{(1 - \phi_{df})^2} \dfrac{1 + b(1 - \phi_{df})}{1 - b\phi_{df}}$	4	Constants C and b depend on packing. Drag coefficient = 1.	
$V_{cf} = \left\{ \dfrac{C[(a_p/g\varepsilon^3)(\rho_c/\Delta\rho)\sigma^{0.25}]^{-0.25}}{1 + 0.835(\rho_d/\rho_c)^{0.25}(V_{df}/V_{cf})^{0.5}} \right\}^2$	5	Packing C is a constant for each packing.	4
$V_{cf} = \left\{ \left[1 + \left(\dfrac{V_{df}}{V_{cf}} \right)^{0.5} \right]^2 \sqrt{\dfrac{a_p}{g}} \right\}^{-1} \alpha C_1 \varepsilon^{1.54} \left(\dfrac{\Delta\rho}{\rho_d} \right)^{0.41} \left[\dfrac{1}{a_p} \left(\dfrac{\Delta\rho^2 g}{\mu_c^2} \right)^{1/3} \right]^{0.3} \left(\dfrac{\mu_c}{\sqrt{\Delta\rho\sigma/a_p}} \right)^{0.15}$ $\alpha = 1$ for continuous-phase wetting, $\alpha = 1.29$ for dispersed-phase wetting	6	Packing C_1 is a constant that depends upon the type of packing.	5
$1 + 0.835 \left(\dfrac{\rho_C}{\rho_D} \right)^{1/4} \left(\dfrac{V_{cf}}{V_{df}} \right)^{1/2} = C_p \left[\dfrac{V_{cf}^2 a_p}{g\varepsilon^3} \left(\dfrac{\rho_C}{\Delta\rho} \right) \sigma^{1/4} \mu_C^{1/4} \right]^{-1/4}$ $C_p = 0.87 \dfrac{\varepsilon^{0.0068}}{a_p^{0.048}}$ for nonribbed Raschig rings $C_p = 1.2 \dfrac{\varepsilon^{0.78}}{a_p^{0.0351}}$ for Berl saddles $C_p = 1.02 \dfrac{\varepsilon^{0.0068}}{a_p^{0.048}}$ for Lessing rings	7	Packing C_p is a constant that depends upon the type of packing	6
$\dfrac{(V_{df} + V_{cf}) d_o \rho_c}{\mu_c} = 0.30137 \left(\dfrac{A_{col}}{\pi d_o^2 N_o / 4} \right)^{0.0948} A \left(\dfrac{\Delta\rho}{\rho_c} \right)^{0.1397} B \left(\dfrac{d_o^3 \rho_c^2 g}{\mu_c^2} \right)^{0.3875}$ $A = \left(\dfrac{V_{df}}{V_{cf}} \right)^{0.0593}$ $\quad B = \left(\dfrac{\sigma\rho_d d_o}{\mu_d^2} \right)^{0.0127}$ $\quad A_{col} = \dfrac{1}{1 - \dfrac{A_{\text{dow}}}{A_{\text{col}}}} \dfrac{Q_d}{f V_{df,n}}$ ϕ = fraction of flood	8	Sieve tray	7
$V_{cf} = \sqrt{\dfrac{L_{dc} - A}{B(V_{df}/V_{cf})^2 + C}}$ $\quad A = \dfrac{6\sigma}{d_p \Delta\rho g}$ $\quad B = \dfrac{1.1\rho_d}{g\Delta\rho f_{ha}^2}$ $\quad C = \dfrac{2.7\rho_c}{2g\Delta\rho f_{da}^2}$	9	Sieve tray capacity limited by coalesced layer flood	8

References:

1. Seibert, A. F., and J. R. Fair, *Ind. Eng. Chem. Res.* **27**(3): 470–481 (1988).
2. Seibert, A. F., B. E. Reeves, and J. R. Fair, *Ind. Eng. Chem. Res.* **29**(9): 1901–1907 (1990).
3. Billet, R., J. Mackowiak, and M. Pajak, *Chem. Eng. Process.* **19:** 39–47 (1985).
4. Dell, F. R., and H. R. C. Pratt, *Trans. Inst. Chem. Eng.* **29:** 89 (1951).
5. Kumar, A., and S. Hartland, *Trans. Inst. Chem. Eng.* **72**(Pt. A): 89–104 (1994).
6. Sakiadis, B. C., and A. I. Johnson, *Ind. Eng. Chem.* **46**(6): 1229-1239 (1954).
7. Rocha, J. A., et al., *Ind. Eng. Chem. Res.* **28**(12): 1873–1878 (1989).
8. Seibert, A. F., and J. R. Fair, *Ind. Eng. Chem. Res.* **32**(10): 2213–2219 (1993).

Refer to the original articles for details.

stagnant continuous phase, and these drops must coalesce at the main operating interface located at the top or bottom of the column. A. F. Seibert, J. L. Bravo, and J. R. Fair [*ISEC '02 Proc.* **2:** 1328–1333 (2002)] report that problems with coalescence are most likely when the superficial dispersed-phase velocity V_{df} is greater than about 12 percent of the characteristic slip velocity given by Eqs. (15-113) to (15-118). The rate of coalescence is influenced by the mixture's physical properties and the presence of any surface active contaminants. For these systems, miniplant tests with actual feed chemicals normally are needed to understand the rate of coalescence. If coalescence is slow, design rates will need to be reduced below those predicted by assuming rapid coalescence.

For slowly coalescing systems, the placement of coalescing material within the column at the main interface may significantly improve performance. The height of the uncoalesced layer located at the main operating interface may be reduced by adding a packing type of coalescer that is preferentially wetted by the dispersed phase. The packing may be a structured or mesh type. If plugging or fouling is a concern, a more open (lower-surface-area) structured packing may be preferred. It also may be useful to add a separate liquid-liquid phase separator outside the extractor to clarify the extract or raffinate streams. See the subsection Liquid-Liquid Phase Separation Equipment.

Mass-Transfer Coefficients As discussed in the subsection Rate-Based Calculations, the overall mass-transfer coefficient may be defined based on the dispersed phase or the continuous phase, as follows:

$$\frac{1}{k_{od}} = \frac{1}{k_d} + \frac{m_{dc}^{\text{vol}}}{k_c} \tag{15-121}$$

$$\frac{1}{k_{oc}} = \frac{1}{k_c} + \frac{1}{m_{dc}^{\text{vol}} k_d} \tag{15-122}$$

where the slope of the equilibrium line m_{dc}^{vol} is expressed in volumetric concentration units. The dispersed-phase and continuous-phase film coefficients k_d and k_c generally are functions of convection and turbulence effects, as well as molecular diffusion and the thicknesses of stagnant films at the interface between drops and the continuous phase. As mentioned earlier, normally the main resistance to mass transfer resides in the feed (raffinate) phase, so often this phase will be chosen as the basis for design calculations, whether it is the dispersed phase or the continuous phase. Examples of mass-transfer coefficient models for static extractors are given in Table 15-18. Also see A. Kumar and S. Hartland, *Trans. Inst. Chem. Eng.*,

TABLE 15-18 Example Mass-Transfer Coefficient Models

Example	Eq.	Comments	Ref.
$k_c = 0.698\left(\frac{\mathcal{D}_c}{d_p}\right)\left(\frac{d_p V_s \rho_c}{\mu_c}\right)\left(\frac{\mu_c}{\rho_c \mathcal{D}_c}\right)^{0.4}(1-\phi_d)$	1	For nonrigid drops. Spray, packing, and sieve trays.	1
$k_d = 0.023 V_s \sqrt{\frac{\rho_c \mathcal{D}_c}{\mu_c}}$	2	Model of Laddha and Degaleesan. For nonrigid drops. Spray, packing, and sieve trays.	2 1
$k_d = \frac{0.00375 V_s}{1+\mu_d/\mu_c}$	3	Model of Handlos and Baron. Approximate solution to series model. Independent of molecular diffusion. Use for large drops.	3
$\Phi = \frac{\sqrt{\mu_c/\rho_c \mathcal{D}_c}}{1+\mu_d/\mu_c}$	4	Spray, packing, and sieve trays. Use Eq. (3) if $\Phi < 6$.	1
$k_d = \frac{17.9 \mathcal{D}_d}{d_p}$	5	Laminar circulation within drops. Recommended for long contact times. For Re < 50.	4 5
$k_d = 1.14\left(\frac{\mathcal{D}_d}{d_p}\right)\left(\frac{\rho_d d_p^2 \bar{\omega}}{\mu_D}\right)^{0.56}\left(\frac{\mu_d}{\rho_d \mathcal{D}_d}\right)^{0.5}$	6	For oscillating drops. Simplified version for assumption of $\theta = 0.2$.	6
$\bar{\omega} = \frac{1}{2\pi}\left[\frac{192\sigma b}{d_p^3(3\rho_d + 2\rho_c)}\right]^{0.5} \quad b = 0.805 d_p^{0.225}, d_p \text{ in cm}$	7		
$k_c = \frac{\mathcal{D}_c}{d_p}\left[2 + 0.95\left(\frac{\rho_c V_s d_p}{\mu_c}\right)^{0.5}\left(\frac{\mu_c}{\rho_c \mathcal{D}_c}\right)^{0.33}\right]$	8	For rigid drops.	7
$k_c = 0.725\left(\frac{\mathcal{D}_c}{d_p}\right)\left(\frac{\rho_c V_s d_p}{\mu_c}\right)^{0.57}\left(\frac{\mu_c}{\rho_c \mathcal{D}_c}\right)^{0.42}(1-\phi_d)$	9	For circulating drops. Developed from correlation of spray column data.	8
$k_c = 1.4\left(\frac{\mathcal{D}_d}{d_p}\right)\left(\frac{\rho_c d_p^2 \omega}{\mu_c}\right)^{0.5}\left(\frac{\mu_c}{\rho_c \mathcal{D}_c}\right)^{0.5}$	10	For oscillating drops.	9

References:

1. Seibert, A. F., and J. R. Fair, *Ind. Eng. Chem. Res.* **27**(3): 470–481 (1988).
2. Laddha, G. S., and T. E. Degaleesan, *Transport Phenomena in Liquid Extraction* (McGraw-Hill, 1978).
3. Handlos, A. E., and T. Baron, *AIChE J.* **3:** 127 (1957).
4. Kronig, R., and J. C. Brink, *Appl. Sci. Res.* **A2:** 142 (1950).
5. Johnson, A. I., and A. E. Hamielec, *AIChE J.* **6:** 145 (1960).
6. Yamaguchi, M., T. Fujimoto, and T. Katayama, *J. Chem. Japan* **8**(5): 361–366 (1975).
7. Garner, F. H., and R. D. Suckling, *AIChE J.* **4:** 114 (1958).
8. Treybal, R. E., *Liquid Extraction* (McGraw-Hill, 1963).
9. Yamaguchi, M., S. Watanabe, and T. Katayama, *J. Chem. Japan* **8**(5): 415–417 (1975).

Refer to the original articles for details.

Part A, **77:** 372–384 (1999). The application of mass transfer coefficients requires calculation of interfacial area, normally from an estimate of drop size and holdup. See Tables 15-15 and 15-16 and Eq. (15-104). Note that the mass-transfer coefficient does not include the effect of axial mixing. This correction must be applied separately.

Axial Mixing See the subsection Accounting for Axial Mixing under Liquid-Liquid Extraction Equipment. Many approaches have been developed as discussed earlier. Becker recommends the concept of the height of a dispersion unit (HDU) to correct the height of a transfer unit for axial mixing in a spray or packed contactor [Becker, O., *Chem. Eng. Technol.* **26**(1): 35–41 (2003); *Chem. Ing. Tech.* **74:** 59–66 (2002); and Becker, O., and A. F. Seibert, *Chem. Ing. Tech.* **72:** 359–364 (2000)]. The design procedure involves first calculating $k_o a$ and then converting the result to a transfer unit height. This calculated value is then adjusted to take into account the effect of axial mixing, which varies with column diameter and height. The general analysis has the form:

$$\text{HTU} = \text{HTU}|_{\text{plug}} + \text{HDU}_o = (V/k_o a) + \text{HDU}_o \tag{15-123}$$

where

$$\text{HDU}_o = \left(p_0 + \frac{0.8}{Z_t} \frac{\mathcal{E} \ln \mathcal{E}}{\mathcal{E} - 1} \right)^{-1} \tag{15-124}$$

$$p_0 = \frac{0.1 Z_t / HTU_{\text{plug}} + 1}{0.1 Z_t / HTU_{\text{plug}} + p_1/p_2} \cdot p_1 \tag{15-125}$$

$$\frac{1}{p_1} = \frac{1}{\mathcal{E}} \text{HDU}_r + \text{HDU}_e \qquad \frac{1}{p_2} = \text{HDU}_r + \frac{1}{\mathcal{E}} \text{HDU}_e \tag{15-126}$$

HDU is related to the axial mixing coefficient E and superficial phase velocity V by

$$\text{HDU} = \frac{E}{V} \tag{15-127}$$

The parameters HDU, E, V, and k_o are for either the dispersed phase or the continuous phase, depending on the basis used to define the overall mass transfer coefficient. The subscript r denotes the raffinate phase, subscript e denotes the extract phase, and Z_t is the contacting height. For $E = 1$, the equations reduce to

$$\text{HDU}_o = \left(\frac{1}{\text{HDU}_r + \text{HDU}_e} + \frac{0.8}{Z_t} \right)^{-1} \tag{15-128}$$

The axial mixing coefficient normally is evaluated for the continuous phase. It is correlated by

$$\frac{E_c \rho_c}{\mu_c} = (C_1 \,\text{Re}_c^a + C_2 \,\text{Re}_d^b) \left(\frac{D_{\text{col}}}{100} \right)^c \tag{15-129}$$

D_{col} = column diameter, cm

where

$$\text{Re}_c = \frac{V_{ic} \rho_c}{\mu_c (a_p + a_w)} \tag{15-130}$$

$$a_w = \frac{4}{D_{\text{col}}} \tag{15-131}$$

$$\text{Re}_d = \frac{V_s d_p \rho_c}{\mu_c} \tag{15-132}$$

In Eq. (15-131), a_w is the specific wall surface (cm^2/cm^3) and a_p is the specific packing surface (cm^2/cm^3). This term is dropped for a spray column ($C_1 = 0$). The model coefficients are summarized in Table 15-19. Most of the axial mixing data available in the literature are for the continuous phase; dispersed-phase axial mixing data are rare. Becker recommends assuming $\text{HDU}_d = \text{HDU}_c$ when dispersed-phase data are not available. In Fig. 15-33, Becker presents a parity plot of HTU calculated from Eq. (15-123) based on small- and large-scale data for packed and spray columns.

Spray Columns The spray column is one of the simplest and oldest types of equipment used to contact two liquid phases in countercurrent flow. Normally it consists of an empty vertical vessel with a distributor located at one end. The distributor disperses one of the liquids into drops. These drops then rise or fall against the flow of the continuous phase, collecting at the other end of the column and finally coalescing to form a layer of clear liquid that is withdrawn from the column. Because spray columns often are used when solids are present, phases often are dispersed through pipe distributors with large holes oriented in the direction of flow. In cases where the ratio of volumetric flow rates entering the column is far from unity, the liquid entering the extractor at the smaller rate generally should be dispersed to avoid excessive backmixing. Sometimes liquid distributors are used at each end to disperse both phases, with the main liquid-liquid interface located in the middle of the column (Fig. 15-34). See the subsection Liquid Distributors and Dispersers under Liquid-Liquid Extraction Equipment.

Spray columns are inexpensive and easy to operate and provide high volumetric throughput. However, because the continuous phase flows freely through the column, backmixing effects generally are severe. As a result, spray columns rarely achieve more than one theoretical stage. Spray columns may be used when only one theoretical stage is required or when solid precipitation is prevalent and no other contacting device can be used because of plugging. Spray columns also are used for direct heat transfer between large immiscible liquid streams.

Drop Size, Holdup, and Interfacial Area Drop size is estimated by using one of the models listed in Table 15-15, and holdup is estimated from expressions given in Table 15-16. Interfacial area is then calculated by using Eq. (15-104).

Flooding Example flooding models are included in Table 15-17. A review is given by A. Kumar and S. Hartland [chap. 17 in *Liquid-Liquid Extraction Equipment*, ed. J. C. Godfrey and M. J. Slater, Wiley, New York, 1994, pp. 680–686].

Mass-Transfer Efficiency See Table 15-18. As mentioned earlier, spray columns rarely develop more than one theoretical stage due to axial mixing within the column. Nevertheless, it is necessary to determine the column height that will give this theoretical stage. S. D. Cavers [chap. 10 in *Handbook of Solvent Extraction*, ed. T. C. Lo, M. H. I. Baird, and C. Hanson, Wiley, New York, 1983, and Krieger, Huntington, NY, 1991] recommends the following equation from G. S. Laddha and T. E. Degaleesan [*Transport Phenomena in Liquid Extraction*, McGraw-Hill, New York, 1978, p. 233] to estimate the overall volumetric mass-transfer coefficient:

$$k_{oc} a = m_{dc}^{\text{vol}} k_{od} a = 0.08 \times \frac{\phi_d (1 - \phi_d)(g^3 \Delta\rho^3 / \sigma \rho_c^2)^{1/4}}{(\mu_c / \rho_c \mathcal{D}_c)^{1/2} + (1/m_{dc})(\mu_d / \rho_d \mathcal{D}_d)^{1/2}} \tag{15-133}$$

Here $\mathcal{D}_c$ and $\mathcal{D}_d$ are the solute diffusion coefficients in the continuous and dispersed phases, respectively. The height of a transfer unit can then be estimated from

$$H_{oc} = \frac{V_c}{k_{oc} a} \tag{15-134}$$

TABLE 15-19 Correlation Constants for the Becker Axial Mixing Model*

	No. of data points	C_1	a	C_2	B	c	Average relative error, %
Spray column	197	0	0	45.6	1.058	0.917	24.8
Structured packed columns and IMTP random packing	118	405.1	0.798	27.7	0.914	1.178	32.0
Structured packed columns with dual flow plates	57	284.5	0.494	35.0	0.406	0.847	18.7

*Becker, O., *Chem. Eng. Technol.*, **24**(12): 1242–1244 (2001).

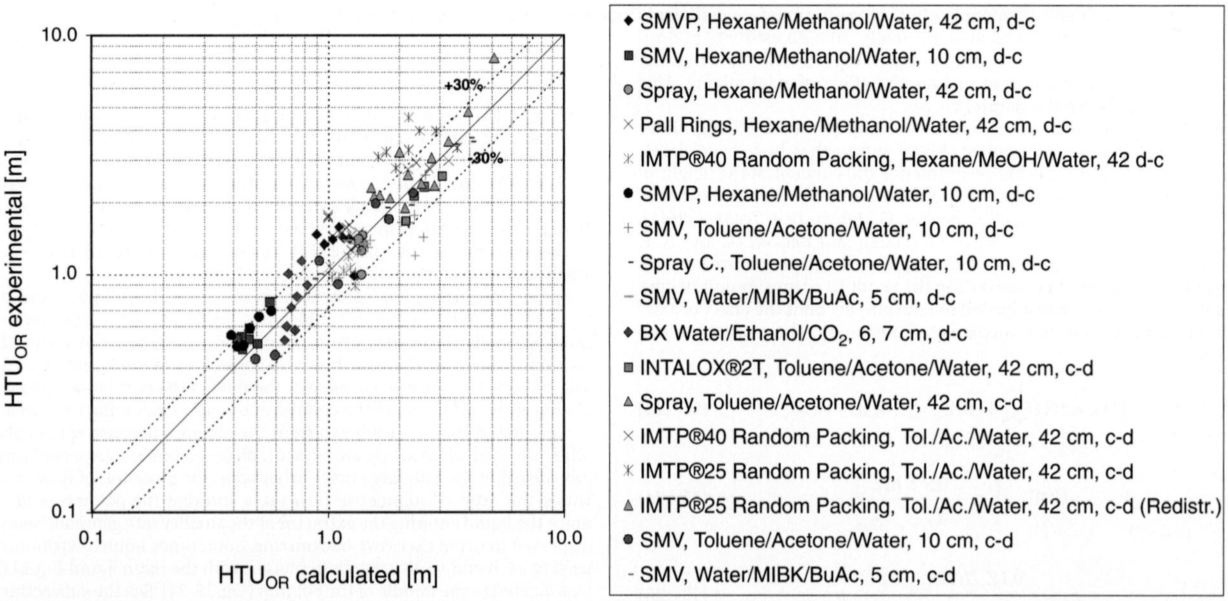

FIG. 15-33 Parity plot comparing spray and packed column results incorporating axial mixing model. [*Reprinted from Becker,* Chemie Ing. Technik 74*(1–2), pp. 59–66 (2002). Copyright 2002 Wiley-VCH.*]

where H_{oc} is the height of an overall transfer unit based on the continuous phase and V_c is the superficial velocity of the continuous phase. Equation (15-133) provides only a rough approximation. The contacting height is then calculated from the H_{oc} and the required number of transfer units.

Packed Columns Packing is used in a column extractor to provide a tortuous path for dispersed drops and to reduce axial mixing (backmixing). Packing affects interfacial area and mass transfer through its impact on the holdup and flow of drops. For further discussion of packed-column extractor design, see R. F. Strigle, chap. 11 in *Packed Tower Design and Applications*, 2d ed., Gulf, Houston, 1994; and G. W. Stevens, chap. 8 in *Liquid-Liquid Extraction Equipment*, ed. J. C. Godfrey and M. J. Slater, Wiley, New York, 1994.

The packings used for liquid-liquid extraction are essentially the same as those used in distillation and absorption service, although the distributors and dispersers and many of the associated internals are not the same. Various examples of commercially available packings used for liquid-liquid extraction service are listed in Table 15-20 (from Koch-Glitsch, Raschig, and Sulzer). Other manufacturers of packings include AMACS, GTC Technology, Kevin Enterprises, Montz, and RVT, among others. It is a good idea to consult a variety of vendors before making a selection. Illustrations of various types of packings are given in Sec. 14.

Packings are classified as either random or structured. Random packings may be wet-loaded into a column by filling the column with liquid and slowly adding the packing at the liquid surface so the packing pieces gently fall to the surface of the forming bed (typical of ceramic packings); or they may be dry-loaded by transferring them into an empty column through a chute or fabric sock while spreading to maintain uniformity (typical of metal or plastic packings). The familiar ring and saddle packings such as Raschig rings, Berl saddles, Intalox saddles, and Lessing rings are examples of ceramic packings. The more modern random packings such as Pall rings, Hy-Pak®, IMTP®, Raschig Super-Ring, and Intalox® Ultra packings are ring or saddle shapes with internal fingers and slots in the wall. These packings are more open and provide greater access to the interior surfaces for improved capacity and mass-transfer performance. Structured packings are modular assemblies placed inside the column in a specific ordered arrangement. Many are in the form of woven wire mesh or corrugated sheets arranged in layers at specific angles. For packing made from sheets, it is not clear whether surface treatments such as perforations and embossing are important in liquid-liquid extraction, so a number of smooth-surface structured packings are marketed for extraction applications (such as SMV structured packings). For best mass transfer performance, the packing should be preferentially wetted by the continuous phase. (See the subsection Effect of Solid-Surface Wettability under Liquid-Liquid Dispersion Fundamentals.) Many older packed extractors are being refurbished with newer packings and internals to achieve higher throughput while maintaining or improving separation performance. As with any packing and the associated internals, installation procedures recommended by the packing vendor need to be carefully followed to ensure the packing performs as designed. In addition to mass-transfer performance and throughput, another important consideration when choosing metal packing is the packing material and wall thickness relative to corrosion rates. The packing should have sufficient wall thickness for a reasonable service life.

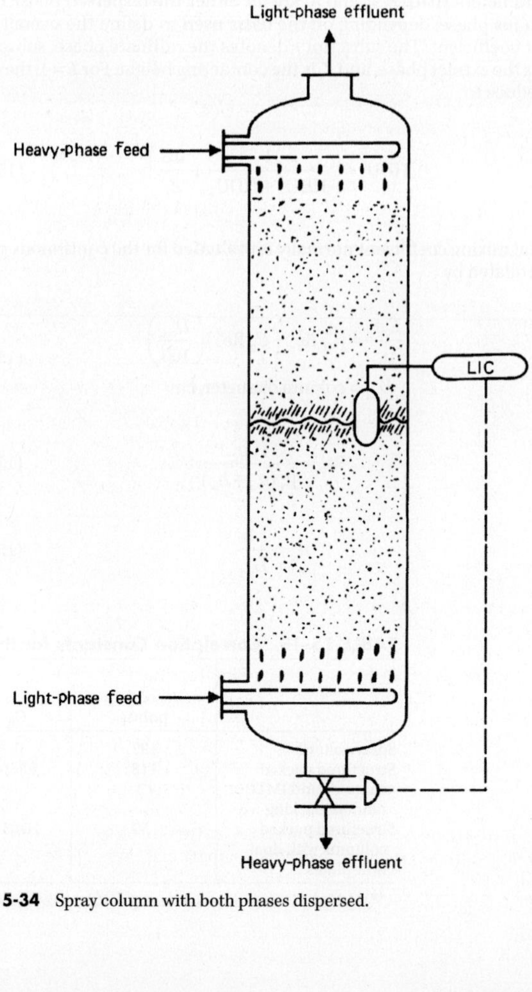

FIG. 15-34 Spray column with both phases dispersed.

TABLE 15-20 Example Random and Structured Packings Used in Packed Extractors*

Packing	Surface area† a_p, m^2/m^3	Void fraction† ε
Metal random packing		
Koch-Glitsch IMTP® 25	224	0.964
Koch-Glitsch IMTP® 40	151	0.980
Koch-Glitsch IMTP® 50	102	0.979
Koch-Glitsch IMTP® 60	84	0.983
Sulzer I-Ring #25	224	0.964
Sulzer I-Ring #40	151	0.980
Sulzer I-Ring #50	102	0.979
Nutter Ring® NR 0.7	226	0.977
Nutter Ring® NR 1	168	0.977
Nutter Ring® NR 1.5	124	0.976
Nutter Ring® NR 2	96	0.982
Nutter Ring® NR 2.5	83	0.984
HY-PAK® #1	172	0.965
HY-PAK® #1½	118	0.976
HY-PAK® #2	84	0.979
FLEXIRING® 1 in	200	0.959
FLEXIRING® 1½ in	128	0.974
FLEXIRING® 2 in	97	0.975
CMR™ 1	246	0.973
CMR™ 2	157	0.970
CMR™ 3	102	0.980
BETA RING® #1	186	0.963
BETA RING® #2	136	0.973
FLEXIMAX® 200	189	0.973
FLEXIMAX® 300	148	0.979
FLEXIMAX® 400	92	0.983
Raschig Super Rings® 0.3	315	0.960
Raschig Super Rings® 0.5	249	0.970
Raschig Super Rings® 0.6	216	0.980
Raschig Super Rings® 0.7	180	0.980
Raschig Super Rings® 1	151	0.980
Raschig Super Rings® 1.5	121	0.980
Raschig Super Rings® 2	102	0.980
Ceramic random packing		
INTALOX® Saddles 1 in	256	0.730
INTALOX® Saddles 1½ in	195	0.750
INTALOX® Saddles 2 in	118	0.760
Ceramic structured packing		
FLEXERAMIC® 28	282	0.720
FLEXERAMIC® 48	157	0.770
FLEXERAMIC® 88	102	0.850
Metal structured packing‡		
Koch-Glitsch SMV-8	417	0.978
Koch-Glitsch SMV-10	292	0.985
Koch-Glitsch SMV-16	223	0.989
Koch-Glitsch SMV-32	112	0.989
Sulzer ECP Extraction Packing SMV/SMVP 2Y	205	0.990
Sulzer ECP Extraction Packing SMV/SMVP 250.Y	256	0.988
Sulzer ECP Extraction Packing SMV/SMVP 350.Y	353	0.983
INTALOX® 2T	214	0.989
INTALOX® 3T	170	0.989
INTALOX® 4T	133	0.987
Raschig Super-Pak 250	250	0.980
Raschig Super-Pak 200	200	0.980

*Polymeric packings also are available and may be appropriate for aqueous dispersed applications, but are less often used in practice.

†Typical value for standard wall thickness. Values will vary depending upon thickness.

‡SMV structured packings are made from smooth corrugated metal sheets. They also are available with horizontal dual-flow perforated plates installed between elements (typically designated SMVP packing). These plates generally reduce backmixing and improve mass-transfer performance at the expense of a reduction in the open cross-sectional area and somewhat reduced capacity.

Liquid Distribution Good initial distribution of the dispersed phase is very important for good performance. R. F. Strigle [*Packed Tower Design and Applications*, 2d ed., chap. 11, Gulf, Houston, 1994] describes typical packed-column internals for liquid-liquid contacting. When the light phase is dispersed, a combination liquid disperser/packing support is preferred because a separate support plate can adversely affect the flow of dispersed drops. An example of a disperser plate is shown in Fig. 15-35. A ladder-type pipe distributor commonly is used to distribute the dispersed-phase feed to the initial disperser plate. Other distributor designs also are available. J. Koch and A. Vogelpohl [*Chem. Eng. Technol.* **24**(7): 695–698; **24**(8): 795–798 (2001)] discuss a sieve plate distributor design that includes a predistributor plate. Many of the concepts concerning geometric uniformity for liquid distribution in packed gas-liquid contactors [Perry, D., D. E. Nutter, and A. Hale, *Chem. Eng. Prog.* **86**(1): 30–35 (1990)] are relevant to liquid-liquid contactors as well. See the subsection Liquid Distributors and Dispersers under Liquid-Liquid Extraction Equipment.

Redistribution A. F. Seibert, B. E. Reeves, and J. R. Fair [*Ind. Eng. Chem. Res.* **29**(9): 1901–1907 (1990)] and R. R. Nemunaitis et al. [*Chem. Eng. Prog.* **67**(11): 60 (1971)] report data showing little benefit from a packed height greater than 10 ft (3 m) and recommend redistributing the dispersed phase about every 5 to 10 ft (1.5 to 3 m) to generate new droplets and constrain backmixing. A packed column often is designed with a redistributor placed between two or more packed sections. In addition, structured packings sometimes are installed with a dual-flow perforated plate (with no downcomer) between elements to reduce backmixing.

Minimum Packing Size and Drop Size For a given application there will be a minimum packing size or dimension below which random packing is too small for good extraction performance [Lewis, J. B., I. Jones, and

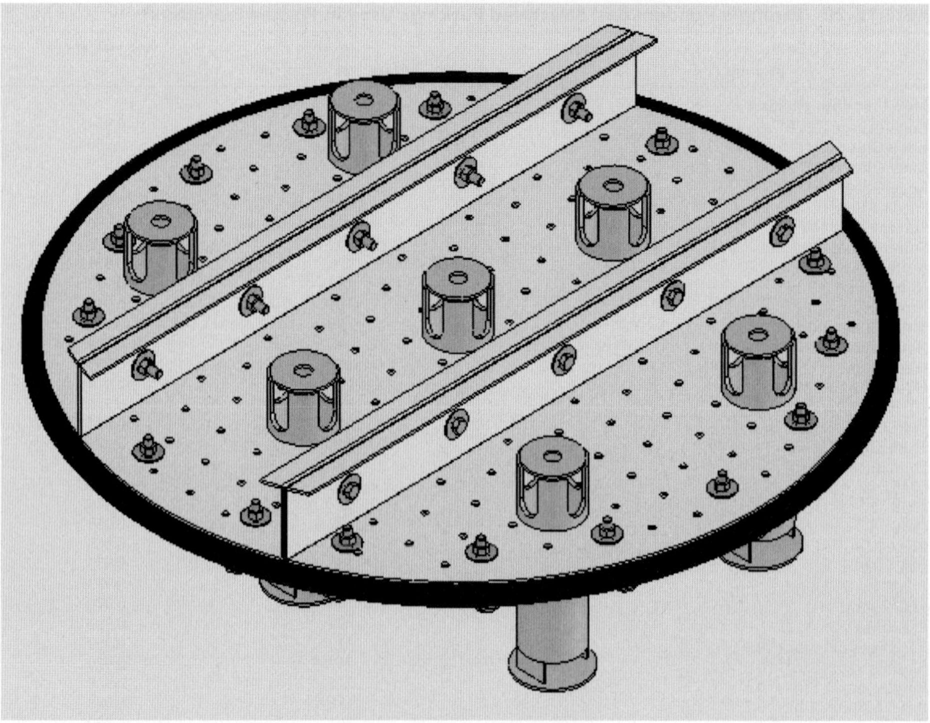

FIG. 15-35 Example of disperser plate (Sulzer model VSX). (*Courtesy of Sulzer.*)

H. R. C. Pratt, *Trans. Inst. Chem. Eng.* **29:** 126–148 (1951); R. Gayler and H. R. C. Pratt, *Trans. Inst. Chem. Eng.* **31:** 69–77 (1953); and G. S. Laddha and T. E. Degaleesan, chap. 10 in *Transport Phenomena in Liquid Extraction*, McGraw-Hill, New York, 1978, pp. 288–289]. The critical packing dimension or size has been correlated as a multiple of the maximum stable drop size obtained from Eq. (15-37):

$$d_C = 2.4\sqrt{\frac{\sigma}{\Delta\rho g}} \qquad (15\text{-}135)$$

Below this critical size, the void spaces between packing elements are too small for free flow of dispersed drops. The entry of drops into the packing is severely restricted, often resulting in formation of a coalesced layer at the entrance to the packing and flooding of the column. For packing sizes larger than d_C, the characteristic drop diameter is independent of packing size and may be estimated by using the models listed in Table 15-15. The choice of packing size above d_C generally involves a tradeoff; throughput increases with increasing packing size, while mass-transfer performance may decrease with increasing packing size due to an increase in backmixing effects. Typical random packings for commercial-scale columns are in the range of ¾ to 2 in (or about 2 to 5 cm). For small columns, the packing should be no larger than one-eighth of the column diameter to avoid channeling at the wall. This effectively restricts the size of laboratory extractors packed with random packings to no less than 4 in (10 cm) in diameter if they are intended to generate directly scalable data.

Holdup and Interfacial Area The dispersed-phase holdup in a packed-column extractor may be placed into two categories: (1) a small portion that is held in the column for extended periods (essentially permanent) and (2) a larger portion that is free to move through the packing. This is the portion that participates in transfer of solute between phases. The total is ϕ_d which here refers to the volume of dispersed phase expressed as a fraction of the void space in the packed section. H. R. C. Pratt and coworkers [*Trans. Inst. Chem. Eng.* **29:** 89–109, 110–125, 126–148 (1951); **31:** 57–68, 69–77, 78–93 (1953)] developed relationships between dispersed-phase velocity and holdup for packed columns. For standard commercial packings of 0.5 in (1.27 cm) and larger, they found that ϕ_d varies linearly with V_d up to values of $\phi_d = 0.10$ (for low values of V_d). With further increase of V_d, ϕ_d increases sharply up to a "lower transition point" resembling loading in gas-liquid contact. At still higher values of V_d an upper transition point occurs, the drops of dispersed phase tend to coalesce, and V_d can increase without a corresponding increase in ϕ_d. This regime ends in flooding. Below the upper transition point, Pratt and coworkers calculated a value for ϕ_d from Eq. (15-107) and the relationship $V_s \approx V_{so}(1 - \phi_d)$, such that

$$\frac{V_d}{\phi_d} + \frac{V_c}{1-\phi_d} = \varepsilon V_{so}(1-\phi_d) \qquad (15\text{-}136)$$

where V_{so} is the characteristic slip velocity at low dispersed-phase flow rate, which can be estimated by using Eqs. (15-113) to (15-118) or alternative methods listed in Table 15-16. Interfacial area is calculated from Eq. (15-104).

Flooding Numerous methods have been proposed for correlating flooding velocities in packed extractors as a function of the packing specific surface area and void volume. Examples are listed in Table 15-17. The earlier methods were developed by using the older-style packings such as Raschig rings and Berl saddles. For example, the well-known flooding correlation $(\sigma/\rho_c)^{0.2}(\mu_c/\Delta\rho)(a_p/\varepsilon)^{1.5}$ versus $(V_c^{1/2} + V_d^{1/2})^2\rho_c/(a_p\mu_c)$ developed by J. W. Crawford and C. R. Wilke [*Chem. Eng. Prog.* **47**(8): 423–431 (1951)] is plotted in Fig. 15-36. This is a dimensional correlation developed by using U.S. Customary System units, so the following units must be used: viscosity in lb/ft/h (equal to 2.42 times the value in cP), density in lb/ft³, interfacial tension in dyn/cm, specific packing surface area in ft²/ft³, and velocities in ft/h based on total column cross section. R. R. Nemunaitis et al. [*Chem. Eng. Prog.* **67**(11): 60–67 (1971)] modified the Crawford-Wilke correlation to include packing factors for specific types of packings (including Raschig rings, Intalox® saddles, and Pall rings). Another correlation that uses packing factors is given by B. C. Sakiadis and A. I. Johnson [*Ind. Eng. Chem.* **46**(6): 1229–1239 (1954)]. See Table 15-17 (equation 6). For this correlation, the units are viscosity in cP, interfacial tension in dyn/cm, and specific packing surface area in ft²/ft³. The generalized flooding model of A. F. Seibert, B. E. Reeves, and J. R. Fair [*Ind. Eng. Chem. Res.* **29**(9): 1901–1907 (1990)] shown in Table 15-17 (equation 1) is based on a mechanistic model validated by data for several types of packing and a range of operating scales, including data from a larger-scale column (42.5-cm inner diameter) using more modern packings: No. 25 IMTP® and No. 40 IMTP® random packings and Intalox® Structured Packing 2T.

Care must be taken in choosing the most appropriate model for a given application and in using the calculated results. A flooding correlation equation can give misleading results when data for the packing of interest was not included in the data used to develop the equation, and this is generally the case for the more modern packings. Because of significant uncertainties in the input physical properties and in the flooding prediction, new designs

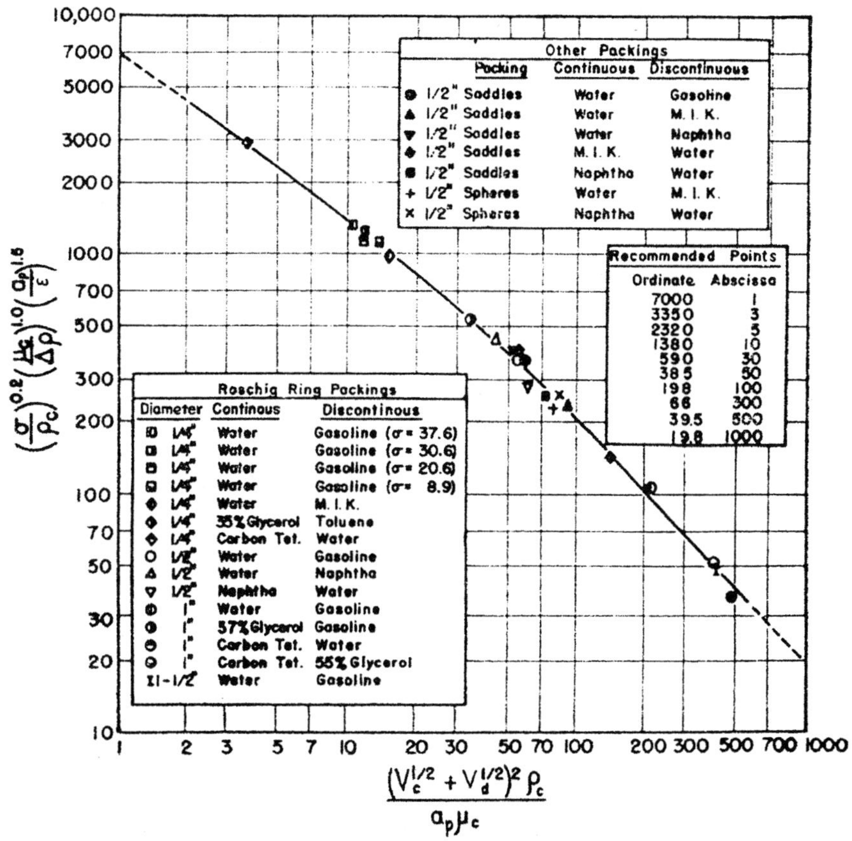

FIG. 15-36 Crawford-Wilke correlation for flooding in packed columns. Use only the units given in the text. [*Reprinted from Crawford and Wilke,* Chem. Eng. Prog. *47(8), pp. 423–431 (1951), with permission.*]

rarely are specified to operate at greater than 60 percent of the calculated flood point. If not already available, some experimental data should be generated for any new design, and in this regard, the flooding correlations may be used to scale up the pilot data to a larger packing size needed for the commercial-scale unit—by calculating the expected percentage change in capacity. This extrapolation approach also may be taken to estimate the improvement that might be achieved by retrofitting an existing commercial unit with a new packing. But again, the results should be used with caution, and consultation with packing vendors is recommended.

Packing Pressure Drop In general, the measured pressure differential across a packed extractor is mostly due to the hydrostatic head pressure. The resistance to flow caused by the packing itself normally is negligible because typical packings are large, and flooding velocities are much lower than those that would be needed to develop significant ΔP from resistance to flow between the packing elements. In some applications, solids may accumulate in the region of the packing support over time, and this may cause added pressure drop and premature flooding. For additional discussion, see G. S. Laddha and T. E. Degaleesan, chap. 10 in *Transport Phenomena in Liquid Extraction*, McGraw-Hill, New York, 1978, pp. 271–273.

Mass Transfer Table 15-21 lists typical mass-transfer performance for various packing sizes, as given by R. F. Strigle [chap. 11 in *Packed Tower Design and Applications*, 2d ed., Gulf, Houston, 1994]. The data are typical in that as the size of the packing decreases, mass-transfer performance is shown to improve (the required bed depth or packing height decreases). At the same time, the hydraulic capacity decreases, so the design problem involves finding the economic optimum for the given production rate. These guidelines are based on experience with organic aqueous systems and the use of metal slotted-ring or ceramic saddle packings, using high-performance dispersion plates for liquid distribution and redistribution between packed sections. In addition, various mass transfer coefficient models used for estimating packed column performance are included in Table 15-18. Newer methods include the calculation procedure outlined by A. F. Seibert, B. E. Reeves, and J. R. Fair [*Ind. Eng. Chem. Res.* **29**(9): 1901–1907 (1990)] and corrected for axial mixing [as in Eqs. (15-127) to (15-136)], and methods incorporating both axial mixing and dynamic behavior [Morales, C., et al., *Comput. Chem. Eng.* **31:** 1694–1701 (2007)]. While these guidelines and calculation procedures provide useful estimates, they do not replace the need for data from pilot tests and the appropriate correction for axial mixing. Table 15-22 lists selected sources of data for mass transfer in packed columns.

Packed columns also are used for applications involving mass transfer with chemical reaction. For example, C. I. Koncsag and A. Barbulescu propose a mass-transfer model for aqueous base extraction of mercaptans from gasoline [*Chem. Eng. Process. Proc. Intensif.* **47:** 1717–1725 (2008)]. They studied this mass transfer rate limited system using a 7.6 cm diameter column packed with SMV 350.Y structured packing.

Sieve Tray Columns A schematic diagram of the most common design of sieve tray column (also called a sieve plate or perforated-plate column) is shown in Fig. 15-32*c*. The light liquid is shown as the dispersed phase. The liquid flows up through the perforations of each tray and is thereby dispersed into drops that rise up through the continuous phase. The continuous liquid flows horizontally across each tray and passes to

Table 15-21 Typical Packed Extractor Performance According to Strigle

	Required bed depth for modern random packings, ft (m)		
Transfer units per bed	Nominal packing size of 1 in (2.5 cm)	Nominal packing size of 1.5 in (3.8 cm)	Nominal packing size of 2 in (5 cm)
1.5	4.4 (1.3)	5.3 (1.6)	6.2 (1.9)
2.0	7.2 (2.2)	8.6 (2.6)	10.1 (3.1)
2.5	9.9 (3.0)	11.9 (3.6)	14.0 (4.3)

Source: R. F. Strigle, chap. 11 in *Packed Tower Design and Applications*, 2d ed. (Gulf, Houston, 1994). The numbers represent typical performance achieved with good liquid distribution.

TABLE 15-22 Mass-Transfer Data for Packed Columns

System	Column diameter, in	Packing	Ref.
Water–acetic acid–ethyl acetate, cyclohexane, methylcyclohexane, ethyl acetate + benzene	1	0.25-in saddles	3
Water–acetic acid–methyl isobutyl ketone	1.95	0.23-in rings	9
	3	0.375-in plastic spheres	15
		0.375-in plastic, ceramic rings	17
		0.5-in plastic, ceramic saddles	17
Water–acetic acid–toluene	6	Montz B1-300 1-in stacked Bialecki rings	2
Water–acetone–hydrocarbon	1.88	0.25- and 0.375-in rings, 6-mm beads	19
	2–4	0.5- and 0.75-in rings	1
Water–acetone–toluene	4	0.5-in rings, ⅝-in Pall rings, IMTP® 15, SMV structured, spray	21
	16.8	IMTP® 25, IMTP® 40, Intalox® 2T structured, spray	22
	6	Montz B1-300 1-in stacked Bialecki rings	2
	4	SMV	25
	5.9, 3.2	Raschig Super Rings 0.3, 15 mm Pall Rings	12
Water–adipic acid–ethyl ether	6	0.5- and 0.75-in rings, 0.375-in spheres	7
Water–benzoic acid–carbon tetrachloride	1.95	0.25-in rings	8
Water–benzoic acid–toluene	8.7	0.5-in saddles, 0.5-in rings	20
Water–2,3 butanediol–3 methyl 1 butanol	1.6	5.5mm ceramic Raschig rings	26
Water–diethylamine–toluene	3, 4, 6	0.25- to 1-in rings	14
	3	0.375-in rings	23
Water–ethyl acetate	4	0.5-in rings	5
Water–isopar(m)	16.8	IMTP® 25, IMTP® 40, Intalox® 2T structured, spray	22
Water–kerosene	4	SMV	25
Water–mercaptans–gasoline	1.2	SMV 350Y	13
Water–methyl ethyl ketone–kerosene	18	1-in rings, 1-in saddles, 1-in Pall rings, spray	16, 4
Water–methylisobutyl-carbinol	4	0.5-in rings	24
Water–methyl ethyl ketone	4	0.5-in rings	24
Water–propionic acid–methyl isobutyl ketone	1.88	0.25- and 0.375-in rings, 6-mm beads	19
Water–propionic acid–carbon tetrachloride	4	SMV	25
Water–succinic acid–1–butanol	4	0.5-in rings, ⅝-in Pall rings, IMTP® 15, SMV structured, spray	21
	4	SMV	25
Water–toluene	6	Montz B1-300 1-in stacked Bialecki rings	2
Acetone (aq)–soybean oil, linseed oil	2	0.25-in saddles, 0.5-in rings	27
Petroleum–furfural	2	0.25-in rings	6
	1.2	0.16-in rings	18
Toluene–heptane–diethylene glycol	1.4, 2.25	Glass and brass rings	10
Sulfolane(water)–benzene/toluene/xylene-hexane	1.1	Glass rings	11

NOTE: To convert inches to centimeters, multiply by 2.54.

References:

1. Degaleesan, T. E., and G. S. Laddha, *Chem. Eng. Sci.* **21:** 199 (1966); *Indian Chem. Eng.* **8**(1): 6 (1966).
2. Billet, R., and J. Mackowiak, *Fette-Seifen-Anstrichmittel* **87:** 205–208 (1985).
3. Eaglesfield, P., B. K. Kelly, and J. F. Short, *Ind. Chem.* **29:** 147, 243 (1953).
4. Eckert, J. S., *Hydrocarbon Processing* **55**(3): 117–124 (1976).
5. Gaylor, R., and H. R. C. Pratt, *Trans. Inst. Chem. Eng. (London)* **31:** 78 (1953).
6. Garwin, L., and E. C. Barber, *Pet. Refiner* **32**(1): 144 (1953).
7. Gier, T. E., and J. O. Hougen, *Ind. Eng. Chem.* **45:** 1362 (1953).
8. Guyer, A., A. Guyer, and K. Mauli, *Helv. Chim. Acta* **38:** 790 (1955).
9. Guyer, A., A. Guyer, and K. Mauli, *Helv. Chim. Acta* **38:** 955 (1955).
10. Kishinevskii, M., and L. Mochalova, *Zh. Prikl. Khim.* **33:** 2344 (1960).
11. Habaki, H., et al., *J. Japan Petroleum Institute* **52**(4): 180–189 (2009)
12. Hlawitscka, M. W., et al., *Chem. Eng. Technol.* **38**(3): 446–454 (2015)
13. Koncsag, C. I., and A. Barbulescu, *Chem Eng & Proc* **47:** 1717–1725 (2008)
14. Liebson, I., and R. B. Beckmann, *Chem. Eng. Prog.* **49:** 405–416 (1953).
15. Moorhead, D. H., and D. M. Himmelblau, *Ind. Eng. Chem. Fundam.* **1:** 68 (1962).
16. Nemunaitis, R. R., J. S. Eckert, E. H. Foote and L. R. Rollison, *Chem. Eng. Prog.* **67**(11): 60–67 (1971).
17. Osmon, F. O., and D. M. Himmelblau, *J. Chem. Eng. Data* **6:** 551 (1961).
18. Sef and Moretu, *Nafta (Zagreb)* **5:** 125 (1954).
19. Rao, M. R., and C. V. Rao, *J. Chem. Eng. Data* **6:** 200 (1961).
20. Row, S. B., J. H. Koffolt, and J. R. Withrow, *Trans. Am, Inst. Chem.* **46:** 1229 (1954).
21. Seibert, A. F., and J. R. Fair, *Ind. Chem. Eng. Res.* **27**(3): 470 (1988).
22. Seibert, A. F., B. E. Reeves, and J. R. Fair, *Ind. Chem. Eng. Res.* **29**(9): 1901 (1990).
23. Shih, C.-K., and R. R. Kraybill, *Ind. Eng. Chem. Process. Des. Dev.* **5:** 260 (1966).
24. Smith, G. C., and R. B. Beckmann, *AIChE J.* **4:** 180 (1958).
25. Streiff, F. A., and S. J. Jancic, *Ger. Chem. Eng.* **7:** 178–183 (1984).
26. Vishwakarma, S., *Intl. J. Chem. Eng. Appl.* **1**(3): 222–224 (2010)
27. Young, C. G., and H. R. Sallans, *J. Am. Oil Chem. Soc.* **32:** 397 (1955).

the tray beneath through the downcomer. For dispersing the heavy phase, the same design may be used, but turned upside down. The trays serve to eliminate (or at least greatly reduce) the vertical recirculation of continuous phase. Mass-transfer rates may be enhanced by the repeated coalescence and redispersion into droplets of the dispersed phase at each tray, although in general the overall efficiency of a sieve tray is fairly low, on the order of 15 to 30 percent. In contrast to packed and most agitated extractors, the sieve tray efficiency remains fairly constant with increasing column diameter because of the absence of backmixing. The higher efficiencies generally are achieved for systems with low to moderate interfacial tension. As discussed earlier, the liquid entering the column at the larger volumetric flow rate generally should be dispersed to obtain satisfactory interfacial area for mass transfer.

Liquid Distribution Very good initial distribution is not as essential in a sieve tray extractor as it is in a packed extractor, because the trays provide redistribution. However, careful design is still required to prevent distributed drops from entering the bottom downcomer (or top upcomer) and to provide uniform dispersion. While the same distributors used in packed columns are applicable, simpler devices also are used. Capped pipes with holes drilled uniformly have been found to be adequate in many cases.

Drop Size, Holdup, and Interfacial Area Drop size is estimated by using one of the models listed in Table 15-15, and holdup is estimated from expressions given in Table 15-16. Interfacial area is then calculated by using Eq. (15-104).

Sieve Tray Design Perforations usually are in the range of 0.125 to 0.25 in (0.32 to 0.64 cm) in diameter, set 0.5 to 0.75 in (1.27 to 1.81 cm) apart, on square or triangular pitch. There appears to be relatively little effect of hole diameter on the mass-transfer rate, except that with systems of high interfacial tension, smaller holes will produce somewhat better mass transfer. The entire hole area is normally set at 3 to 10 percent of the column cross section, although adjustments may be needed. The velocity through the holes should be such that drops do not form slowly at the holes, but rather the dispersed phase streams through the openings as a jet that breaks up into drops at a slight distance from the tray. It is common practice to set the velocity of liquid exiting the holes to correspond to a Weber number between 4 and 12. This normally gives velocities in the range of 0.5 to 1.0 ft/s (15 to 30 cm/s). The same general guidelines used to specify hole size and velocities for plate dispersers apply to sieve tray design. See Eqs. (15-102) and (15-103) and the related discussions in the subsection Liquid Distributors and Dispersers under Liquid-Liquid Extraction Equipment.

The velocity of the continuous phase entering the downcomer (or upcomer) V_{dow}, which sets the downcomer cross-sectional area, should be set at a value lower than the terminal velocity of some arbitrarily small droplet of dispersed phase, say, $\frac{1}{32}$ or $\frac{1}{16}$ in (0.08 or 0.16 cm) in diameter; otherwise, recirculation of entrained dispersed phase around a tray will result in flooding. The terminal velocity of these small drops can be calculated by using Eqs. (15-113) to (15-118). To prevent drops from entering the downcomer, the downcomer area should be specified to achieve a somewhat higher drop velocity in the flow leaving the downcomer. Downcomer area typically is in the range of 5 to 20 percent of the total cross-sectional area, depending upon the ratio of continuous- to dispersed-phase volumetric flow rates. The downcomers should extend beyond the accumulated layer of dispersed phase on the tray, and the tray area directly opposite downcomers should be kept free of perforations.

The spacing between trays should be sufficient that (1) the "streamers" of dispersed liquid from the holes break up into drops before coalescing into the layer of liquid on the next tray; (2) the cross-flow velocity of continuous-phase liquid does not cause excessive entrainment of the dispersed phase; and (3) the column may be entered through handholes or manholes in the sides for inspection and cleaning. For systems that accumulate an interface rag, provision may be made for periodic withdrawal of the rag through the side of the column between trays. For large columns, tray spacing between 12 and 24 in (30 and 60 cm) is generally recommended.

The height of the coalesced layer at each tray is given by

$$h = \frac{\Delta P_o + \Delta P_{dow} - \phi_d g\, \Delta\rho L}{(1-\phi_d) g\, \Delta\rho} \tag{15-137}$$

where L is the downcomer length. Equation (15-137) is a slightly simplified form of the expression given by D. Mewes and W. Kunkel [*Ger. Chem. Eng.* **1:** 111–115 (1978)]. In most cases holdup is low, and Eq. (15-137) reduces to $h = (\Delta P_o + \Delta P_{dow})/(g\Delta\rho)$. The orifice pressure drop ΔP_o may be calculated by using the model of Th. Pilhofer and R. Goedl [*Chem. Ing. Tech.* **49:** 431 (1977)]:

$$\Delta P_o = \frac{1}{2}\left(1 - \frac{0.71}{\log \mathrm{Re}}\right)^{-2} \rho_d V_o^2 + 3.2\left(\frac{d_o^2 g \Delta\rho}{\sigma}\right)^{0.2} \frac{\sigma}{d_o} \tag{15-138}$$

where V_o is the velocity through the orifice, d_o is the orifice diameter, and $\mathrm{Re} = V_o d_o \rho_d/\mu_d$. The pressure drop through the downcomer ΔP_{dow} includes losses due to (1) friction in the downcomer, which should be negligible; (2) contraction and expansion upon entering and leaving the downcomer; and (3) two abrupt changes in direction. These losses total 4.5 velocity heads:

$$\Delta P_{dow} = \frac{4.5 V_{dow}^2 \rho_c}{2} \tag{15-139}$$

For large columns, the design should be specified such that the height of the coalesced layer is at least 1 in (2.5 cm) to ensure all the holes are adequately covered, and one should allow for the trays to be slightly out of level. On the other hand, the height of the coalesced layer should not be too large, as this is unproductive column height that unnecessarily increases the total column height requirement. A typical design value is about 2 in (5 cm).

Envelope-style segmental downcomers (Fig. 15-37) often are used in commercial-scale sieve tray extractors instead of circular or pipe-style downcomers. The area of an envelope downcomer is given by

$$A = \frac{H}{6S}(3H^2 + 4S^2) \tag{15-140}$$

The distance S is determined from the column diameter. The distance H is obtained from

$$S = \left[8H\left(\frac{D_{col}}{2} - \frac{H}{2}\right)\right]^{1/2} \tag{15-141}$$

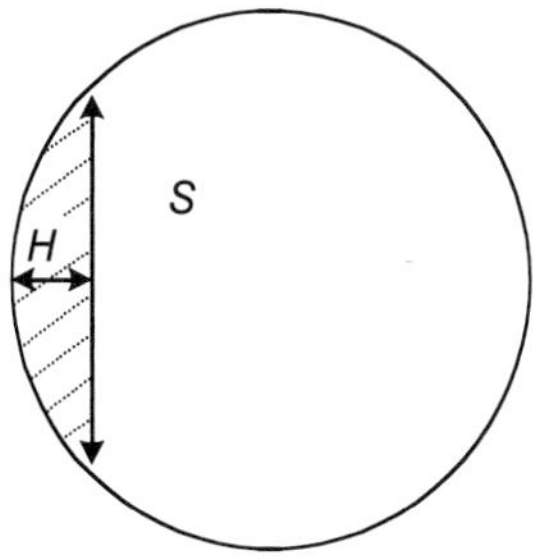

FIG. 15-37 Dimensions of an envelope-style segmental downcomer or upcomer (shaded area).

The diameter of a circular downcomer with equivalent area is given by

$$D_{eq} = \sqrt{\frac{4}{\pi}A} \tag{15-142}$$

Sieve Tray Capacity at Flooding The capacity of a sieve tray is determined by hydraulic mechanisms involved in flooding and is not completely understood, especially for larger-diameter columns. Three studies using larger equipment have been reported by J. O. Oloidi, G. Y. Jeffreys, and C. J. Mumford [*Inst. Chem. Eng. Symp. Ser.* **103:** 117–132 (1987)]; A. F. Seibert and J. R. Fair [*Ind. Eng. Chem.* **32:** 2213–2219 (1993)]; and R. B. Eldridge and J. R. Fair [*Ind. Eng. Chem.* **38:** 218–222 (1999)]. An example of sieve tray flooding data is illustrated in Fig. 15-38.

The sieve tray capacity and efficiency are strongly influenced by the height of the coalesced layer. If the height of this layer grows to the outlet of the downcomer, a sharp reduction in efficiency will result because the mass-transfer height will be significantly reduced. In this case, the downcomer area and/or total perforated area should be increased. A flooding model based on the height of the coalesced layer is given by A. F. Seibert and J. R. Fair [*Ind. Eng. Chem. Res.* **32**(10): 2213–2219 (1993)]. The result is shown in Table 15-17 (equation 9), where L_{dc} is the downcomer length, f_{ha} is the fractional hole area, and f_{da} is the fractional downcomer area.

High Cross-Flow of the Continuous Phase Miniplant tests of sieve tray extractors are often performed prior to the final design of a commercial-scale column. The design often is scaled up based on superficial velocities of the dispersed and continuous phases calculated from the volumetric flow rates and the column cross-sectional area. However, in scaling up one must be careful about the cross-flow velocity (V_{cflow}) of the continuous phase. A value may be estimated from

$$V_{c\,flow} \approx \frac{L_{fp}}{z-h} V_c \tag{15-143}$$

where L_{fp} is the length of flow path, z is the tray spacing, h is the height of coalesced layer, and V_c is the superficial continuous-phase velocity. The magnitude of the cross-flow velocity of the continuous phase can be much greater than that studied in the miniplant. Multiple downcomers or upcomers reduce the flow path length and can be utilized in new designs to reduce cross-flow velocity. Large-diameter multiple downcomer (or upcomer) trays have been reported to provide 10 to 15 percent greater capacity relative to the single-pass tray. A. F. Seibert, J. L. Bravo, and J. R. Fair [*ISEC '02 Proc.* **2:** 1328–1333 (2002)] propose a model for correcting the sieve tray capacity for high cross-flow velocity.

Mass-Transfer Data Mass-transfer data are available from the sources listed in Table 15-23. Mass-transfer performance can be expressed in terms of the number of transfer units per actual tray, or in terms of overall heights of transfer units for a given column configuration. Because sieve trays resemble and basically behave in the manner of stages, performance also can be expressed in terms of a stage efficiency, either as an overall efficiency for the entire tower or, more satisfactorily, as a Murphree efficiency for each tray. The performance of sieve trays in reactive extraction is discussed by R. S. Ettouney, M. A. El-Rifai, and A. O. Ghallab [*Chem. Eng. Process. Proc. Intensif.* **46:** 713–720 (2007)].

Tray Efficiency For low-viscosity systems (<2 cP), overall sieve tray efficiencies often are between 10 and 30 percent. One of the earliest models for predicting the overall tray efficiency was a very empirical one reported by R. E. Treybal [*Liquid Extraction*, 2d ed., McGraw-Hill, New York, 1963].

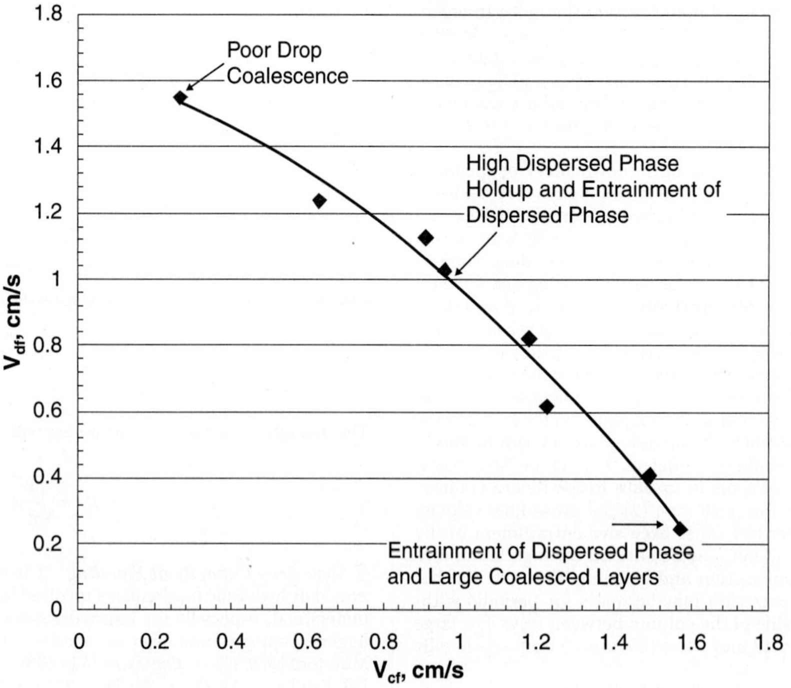

FIG. 15-38 Sieve tray flooding data. System: toluene (dispersed) + water (continuous). Tray spacing = 30.5 cm. Column diameter = 42.8 cm. [*Taken from Seibert, Bravo, and Fair,* ISEC '02 Proc. *2, pp. 1328–1333 (2002), with permission. Copyright 2002 South African Institute of Mining and Metallurgy.*]

R. Krishna Murty and C. V. Rao [*Ind. Eng. Chem. Proc. Des. Dev.* **7**(2): 166–172 (1968)] modified the Treybal model to account for hole diameter:

$$\xi_o = 0.21\left(\frac{z^{0.5}}{\sigma d_o^{0.35}}\right)\left(\frac{V_d}{V_c}\right)^{0.42} \tag{15-144}$$

where z is the tray spacing, cm; d_o is the hole diameter, cm; and σ is interfacial tension, dyn/cm. Care must be taken when using the modified Treybal equation for extractors using high dispersed to continuous flow ratios and very low interfacial tension, because in that case the model will predict an unreasonably high tray efficiency. A. F. Seibert and J. R. Fair [*Ind. Eng. Chem.* **32**(10): 2213–2219 (1993)] recommend calculating the local Murphree stage efficiency based on the dispersed phase, assuming a log mean driving force and negligible mass-transfer contribution from drop formation:

$$\xi_{md} = 1 - \exp\left[-\frac{6k_{od}\phi_d(z-h)}{d_p V_d}\right] \tag{15-145}$$

The overall tray efficiency may then be estimated by using

$$\xi_o = \frac{\ln[1+\xi_{md}(\mathcal{E}-1)]}{\ln \mathcal{E}} \tag{15-146}$$

$$\mathcal{E} = m_{dc}^{\text{vol}}\frac{V_d}{V_c} \tag{15-147}$$

Equation (15-145) assumes plug flow of the rising or falling drop population and complete mixing of the continuous phase on the tray. Also see R. B. Eldridge and J. R. Fair, *Ind. Eng. Chem. Res.* **38**: 218–222 (1999); J. A. Rocha et al., *Ind. Eng. Chem. Res.* **28**(12): 1873–1878 (1989); J. A. Rocha, J. C. Cárdenas, and J. A. García, *Ind. Eng. Chem. Res.* **28**(12): 1879–1883 (1989), and G. Shiveler, Paper No. 537a, AIChE Annual Meeting, San Francisco, 2013.

Baffle Tray Columns Baffle tray columns are similar to spray columns except that baffles are added to reduce backmixing. The baffles usually are slightly sloped to drain any solids that might settle out in the column and are designed to provide a high open area. E. J. Lemieux [*Hydrocarbon Proc.* **62**(9): 106–111 (1983)] and J. R. Fair [*Hydrocarbon Proc.* **72**(5): 75–79 (1993)] report on the performance and design of these columns for gas-liquid contacting. R. E. Treybal [*Liquid Extraction*, 2d ed., McGraw-Hill, New York, 1963] provides a brief but valuable description of a baffle tray extractor. Although no design equations or performance data are provided, Treybal indicates that commercial tray spacings should be in the range of 10 to 15 cm (4 to 6 in). Treybal also provides an interesting illustration of a baffle tray extractor in operation (Fig. 15-39). This figure shows multiple trays with a very short spacing, with the dispersed light phase moving as a layer of liquid under each tray.

Because baffle tray performance data are not widely available, the results of a pilot-scale study are summarized in Figs. 15-40 to 15-43 [Seibert, A. F., C. Lewis, and J. R. Fair, Paper No. 112a, AIChE Annual Meeting, Indianapolis, 2002]. The study was carried out using a 4.0-in (10.2-cm) diameter column set up with 5 to 30 trays. The trays were arranged in a side-to-side horizontal arrangement, as indicated in Fig. 15-39*a*. The data were generated by using the toluene (dispersed) + acetone + water (continuous) and butanol (dispersed) + succinic acid + water (continuous) systems. The effects of changes in baffle spacing and tray overlap (expressed as the percentage of total tray area covered by the next tray above or below) were measured for transfer of solute from the organic to the aqueous phase.

Hydraulic Capacity The capacity of the baffle trays at flooding was found to depend on system properties as shown in Fig. 15-40. The butanol system with its lower interfacial tension provided a much lower capacity relative to the toluene system with its higher interfacial tension. The capacity was found to be independent of tray spacing. However, capacity was strongly affected by the degree of tray overlap as shown in Fig. 15-41. A. F. Seibert, C. Lewis, and J. R. Fair have proposed a flooding model [Paper No. 112a, AIChE Annual Meeting, Indianapolis, 2002].

Baffle Tray Efficiency Baffle tray mass-transfer efficiency was observed to depend strongly on the tray spacing and system properties. In these studies, a tray spacing of about 10 cm provided a minimum HETS. The data indicate that the performance of baffle trays relative to sieve trays depends upon the interfacial tension of the system. For the high-interfacial-tension system the baffle tray performance (in terms of capacity and mass transfer) is relatively low compared to that of a sieve tray (Fig. 15-42). However, for the low-interfacial-tension system (Fig. 15-43), performance was somewhat better using 62 percent tray overlap.

TABLE 15-23 Mass-Transfer Data for Sieve Tray Columns

System	Column diameter, in	Tray spacing, in	Ref.
Benzene–acetic acid–water	1.97	3.9–6.3	27
	1.97	3.2–6.3	26
	2.2	2.8–6.3	25
	1.6 × 3.2	5.9	21
Benzene–acetone–water	3	4, 8	14
Benzene–benzoic acid–water	3	4	14
Benzene–monochloroacetic acid–water	1.97	3.9–6.3	27
Benzene–propionic acid–water	1.97	3.2–6.3	26
Butanol–acetic acid–water	3.6	10	3
Carbon tetrachloride–propionic acid–water	1.97	3.9–6.3	27
Clairsol–water	17.7	13–15	15
Ethyl acetate–acetic acid–water	2	8–24	11
Ethyl ether–acetic acid–water	8.63	4–7.2	16
Gasoline–methyl ethyl ketone–water	3.75	4.5, 6	12
Isopar(M)–water	16.8	12	22
Kerosene–acetone–water	3	4, 8	14
Kerosene–benzoic acid–water	3.63	4.75	1
Kerosene–benzoic acid–water	6	6, 12	10
Isopar-H–benzyl alcohol, methyl benzyl alcohol, acetophenone–	2 × 12	24	2
water	3	6	13
Methylisobutylcarbinol–acetic acid–water	4.18	6	6
Methyl isobutyl ketone–adipic acid–water	4.8	6–23	9
Methyl isobutyl ketone–butyric acid–water	4	6–12	18
Methyl isobutyl ketone–acetic acid–water	9.7	8–24	19, 20
Pegasol–propionic acid–water	4.8	6–11	8
Sulfolane (water)/toluene/heptane	4	8	23
Toluene–benzoic acid–water	8.75	6	17
	3.63	4.75	1
	3.56	3–9	24
	3	6	13
	2.72	9	7
	2	24	11
Toluene–diethylamine–water	4.18	6	4, 5
Toluene–water	16.8	12	22
	9.7	8–24	19
Toluene–acetone–water	16.8	12	22
	9.7	8–24	20
	4	6–12	18
2,2,4-Trimethylpentane–methyl ethyl ketone–water	3.75	4.5, 6	12

NOTE: To convert inches to centimeters, multiply by 2.54.

References:

1. Allerton, J. B., O. Strom, and R. E. Treybal, *Trans. Am. Inst. Chem. Eng.* **39:** 361 (1943).
2. Angelo, J. B., and E. N. Lightfoot, *AIChE J.* **14:** 531 (1968).
3. Gambarra, M. M. P., et al., *Latin American Applied Research* **41:** 205-209 (2011)
4. Garner, F. H., S. R. M. Ellis, and D. W. Fosbury, *Trans. Inst. Chem. Eng. (London)*, **31:** 348 (1953).
5. Garner, F. H., S. R. M. Ellis, and J. W. Hill, *AIChE J.* **1:** 185 (1955).
6. Garner, F. H., S. R. M. Ellis, and J. W. Hill, *Trans. Inst. Chem. Eng. (London)*, **34:** 223 (1956).
7. Goldberger, W. M., and R. F. Benenati, *Ind. Eng. Chem.* **51:** 641 (1959).
8. Krishna Murty, R., and C. V. Rao, *Indian J. Technol.* **5:** 205 (1967).
9. Krishna Murty, R., and C. V. Rao, *Ind. Eng. Chem. Process Des. Dev.* **7:** 166 (1968).
10. Lodh, B. B., and M. Rao, *Indian J. Technol.* **4:** 163 (1966).
11. Mayfield, F. D., and W. L. Church, *Ind. Eng. Chem.* **44:** 2253 (1952).
12. Moulton, R. W., and J. E. Walkey, *Trans. Am. Inst. Chem. Eng.* **40:** 695 (1944).
13. Murali, K., and M. R. Rao, *J. Chem. Eng. Data* **7:** 468 (1962).
14. Nandi, P., and S. B. Ghosh, *J. Indian Chem. Soc., Ind. News Ed.* **13:** 93, 103, 108 (1950).
15. Oloidi, J. O., and C. J. Mumford, *ISEC Proc.* (Munich, 1986).
16. Pyle, C., H. R. Duffey, and A. P. Colburn, *Ind. Eng. Chem.* **42:** 1042 (1950).
17. Row, S. B., J. H. Koffolt, and J. R. Withrow, *Trans. Am. Inst. Chem. Eng.* **37:** 559 (1941).
18. Rocha, J. A., J. L. Humphrey, and J. R. Fair, *Ind. Eng. Chem. Process Des.* **25:** 862 (1986).
19. Rocha, J. A., et al., *Ind. Eng. Chem. Res.* **28**(12): 1873–1878 (1989).
20. Rocha, J. A., J. C. Cardenas, and J. A. Garcia, *Ind. Eng. Chem. Res.* **28**(12): 1879–1883 (1989).
21. Shirotsuka, T., and A. Murakami, *Kagaku Kogaku* **30:** 727 (1966).
22. Seibert, A. F., and J. R. Fair, *Ind. Eng. Chem. Res.* **32**(10): 2213–2219 (1993).
23. Seibert, A. F., et al., Paper Presented at the 2016 Annual AIChE Conference in San Francisco, November 2016,
24. Treybal, R. E., and F. E. Dumoulin, *Ind. Eng. Chem.* **34:** 709 (1942).
25. Ueyama, H., and J. Kobayashi, *Bull. Univ. Osaka Prefect.* **A7:** 113 (1959).
26. Zheleznyak, A. S., *Zh. Prikl. Khim.* **40:** 689 (1967).
27. Zheleznyak, A. S., and B. I. Brounshtein, *Zh. Prikl. Khim.* **40:** 584 (1967).

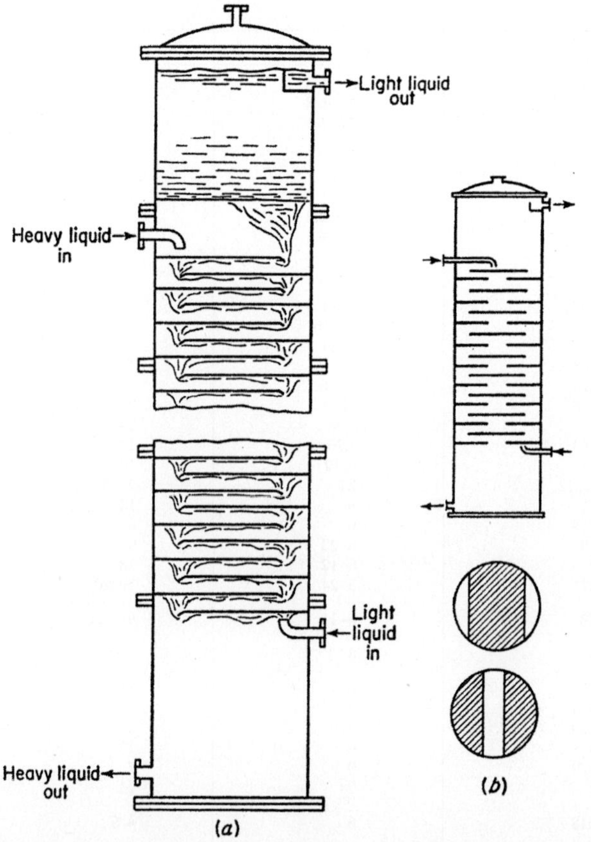

FIG. 15-39 Baffle towers. (*a*) Side-to-side flow at each tray. (*b*) Center-to-center flow (disk-and-doughnut style). (*c*) Center-to-side flow. [*Reprinted from Treybal,* Liquid Extraction, McGraw-Hill, New York, *1963), with permission. Copyright 1963 McGraw-Hill, Inc.*]

AGITATED EXTRACTION COLUMNS

In certain applications, the mass-transfer efficiency of a static extraction column is quite low, especially for systems with moderate to high interfacial tension. In these cases, efficiency may be improved by mechanically agitating the liquid-liquid dispersion within the column to better control drop size and population density (dispersed-phase holdup). Many different types of mechanically agitated extraction columns have been proposed. The more common types include various rotary-impeller columns, the reciprocating-plate column, and the rotating-disk contactor (RDC). The following is a brief review. For more detailed discussion, see J. C. Godfrey and M. J. Slater, eds., *Liquid-Liquid Extraction Equipment,* Wiley, New York, 1994; J. D. Thornton, ed., *Science and Practice of Liquid-Liquid Extraction,* vol. 1, Oxford University Press, Oxford, UK, 1992; and T. C. Lo, M. H. I. Baird, and C. Hanson, eds., *Handbook of Solvent Extraction,* Wiley, New York, 1983, and Krieger, Huntington, NY, 1991.

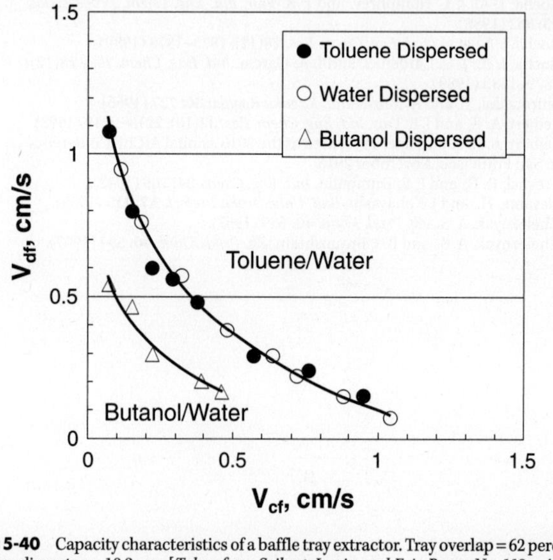

FIG. 15-40 Capacity characteristics of a baffle tray extractor. Tray overlap = 62 percent. Column diameter = 10.2 cm. [*Taken from Seibert, Lewis, and Fair, Paper No. 112a, AIChE Annual Meeting, Indianapolis (November 2002), with permission. Copyright 2002 AIChE.*]

Rotating-Impeller Columns A number of different rotating-impeller column extractors have been proposed and built over the years. Only the Scheibel and Kühni designs are reviewed here. For information about the Oldshue-Rushton design, see L. A. Robbins and R. W. Cusack, Section 15, pp. 38–40, in *Perry's Chemical Engineers' Handbook,* 7th ed., ed. R. H. Perry, D. W. Green, and J. O. Maloney, McGraw-Hill, New York, 1997; and J. Y. Oldshue, chap. 13.4 in *Handbook of Solvent Extraction,* ed. T. C. Lo, M. H. I. Baird, and C. Hanson, Wiley, New York, 1983, and Krieger, Huntington, NY, 1991.

Scheibel Extraction Column The original Scheibel column design consisted of a series of knitted-wire-mesh packed sections placed within a vertical column, with a centrally located impeller between each section and no baffles [Scheibel, E. G., and A. E. Karr, *Ind. Eng. Chem.* **42**(6): 1048–1057 (1950)]. A second-generation Scheibel design [Scheibel, *AIChE J.* **2**(1): 74–78 (1956); U.S. Patent 2,850,362 (1958)] added flat partitions or baffles to the ends of each packed section, and the impellers were surrounded by stationary shroud baffles to direct the flow of droplets discharged from the impeller tips. The new baffling arrangement improved efficiency, allowing the design of larger-diameter columns with less power input and decreased height per theoretical stage. A third design by Scheibel [U.S. Patent 3,389,970 (1968)] eliminated the wire-mesh packing and retained the use of baffles and shrouded impellers (Fig. 15-44). The packed sections were replaced by agitated sections. This design was developed because the wire-mesh packed sections were prone to fouling (plugging) and difficult to clean. A Scheibel extractor of this type is very well suited to handling mixtures with high interfacial tension and can be designed with a large number of stages. It is not as well suited for systems that tend to emulsify easily owing to the high shear rate generated by a rotating impeller. Because of its internal baffling, which controls the mixing patterns on the stages, the Scheibel column has proved to be one of the more efficient extractors in terms of height of a theoretical stage; this makes it well suited to applications that require a large number of stages or are located indoors with headroom restrictions. T. L. Holmes, A. E. Karr, and R. W. Cusack [*Solvent Extraction and Ion Exchange* **8**(3): 515–528 (1990)] have published results comparing the efficiency of the Scheibel column to that of other extractors using the system toluene + acetone + water. For additional discussion, see E. G. Scheibel, chap. 13.3 in *Handbook of Solvent Extraction,* ed. T. C. Lo, M. H. I. Baird, and C. Hanson, eds., Wiley, New York, 1983, and Krieger, Huntington, NY, 1991. A related column design called the AP column consists of alternating sections of Scheibel-type agitators and structured packing [Cusack, R. W., D. Glatz, and T. L. Holmes, *Proc. ESEC'99, Soc. Chem. Ind.,* p. 427 (2001)]. The high open area of the packing allows for higher capacity while the agitation provides increased efficiency.

As with most agitated extractors, the final design of a Scheibel column typically involves scale-up of data generated in a miniplant or pilot-plant test. The column vendor should be consulted for specific information. The key scale-up guidelines are as follows: (1) $D_t(2)/D_t(1) = [Q(2)/Q(1)]^{0.4}$; (2) $Z_t(2)/Z_t(1) = [D_t(2)/D_t(1)]^{0.70}$; (3) stage efficiency is the same for the pilot and full scale; and (4) power per unit volume is the same for each scale [Cusack, R. W., and A. E. Karr, *Chem. Eng. Magazine,* pp. 112–119 (1991)]. Industrial columns up to 10 ft (3 m) in diameter and containing 90 actual stages have been designed using the following general procedures and a 3-in (75-mm) pilot column:

1. Pilot tests usually are conducted in 3-in (75-mm-) diameter columns. The column should contain a sufficient number of stages to complete the extraction. This may require several iterations on column height.

2. The column is run over a range of throughputs $V_d + V_c$ and agitation speeds. At each condition, the concentrations of solute in extract and raffinate streams are measured after steady-state operation has been achieved (usually after three to five turnovers of column volume). At each throughput, the flood point is determined by increasing the agitation until flooding is induced. A minimum of three throughput ranges are examined in this manner. Mass-transfer performance is measured at several agitation speeds up to the flood point.

3. From the preceding mass-transfer and flooding data, the combination of specific throughput and agitation speed that gives the optimum economic performance for the required separation can be determined. This information is used to specify the specific throughput value [gal/(h · ft^3) or m^3/(h · m^3)] and agitation speed (rpm) for the commercial design. However, unlike the

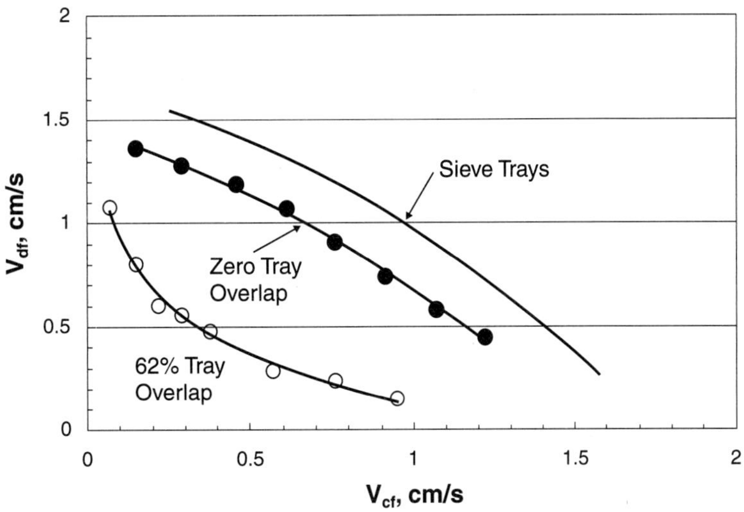

FIG. 15-41 Effect of tray overlap on baffle tray capacity. System: toluene (d) + acetone + water (c). [*Taken from Seibert, Lewis, and Fair, Paper No. 112a, AIChE Annual Meeting, Indianapolis (November 2002), with permission. Copyright 2002 AIChE.*]

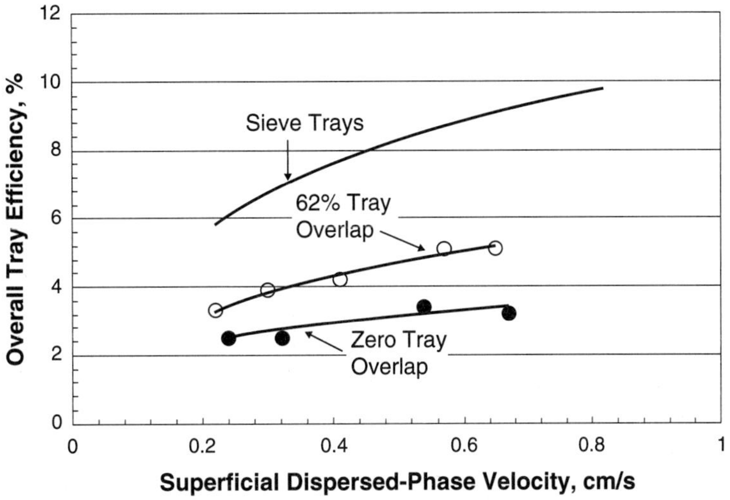

FIG. 15-42 Effect of tray overlap on baffle tray efficiency. System: toluene (*d*) + acetone + water (*c*). Tray spacing = 10.2 cm. [*Taken from Seibert, Lewis, and Fair, Paper No. 112a, AIChE Annual Meeting, Indianapolis (November 2002), with permission. Copyright 2002 AIChE.*]

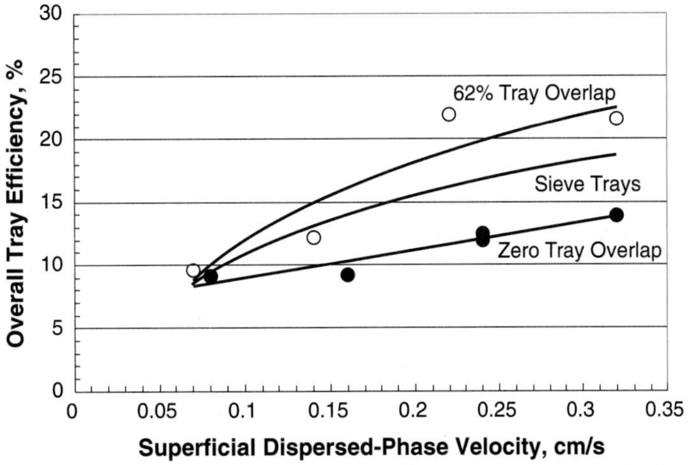

FIG. 15-43 Effect of tray overlap on baffle tray efficiency. System: *n*-butanol + succinic acid + water. Tray spacing = 10.2 cm. [*Taken from Seibert, Lewis, and Fair, Paper No. 112a, AIChE Annual Meeting, Indianapolis (November 2002), with permission. Copyright 2002 AIChE.*]

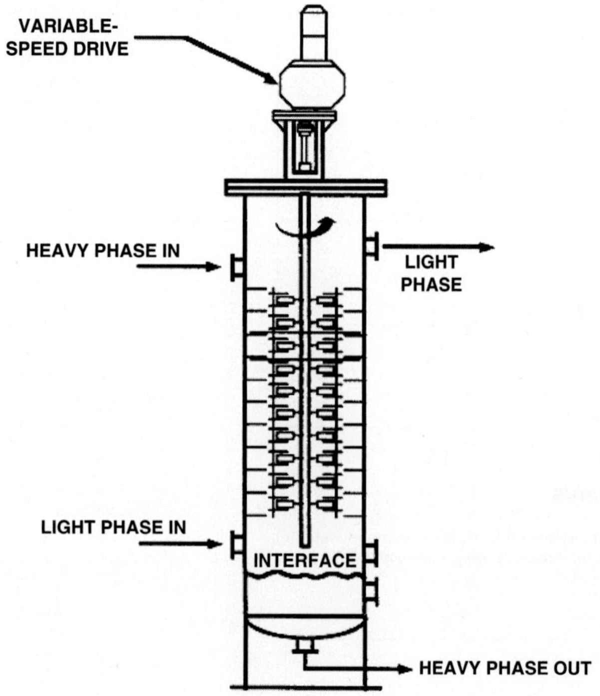

FIG. 15-44 Scheibel column extractor (third-generation design). (*Courtesy of Koch Modular Process Systems.*)

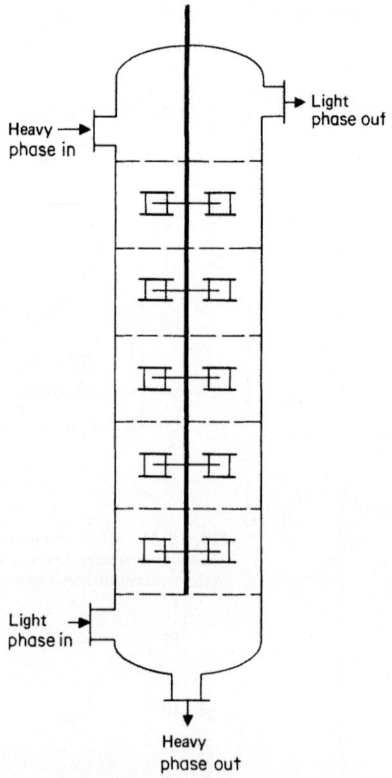

FIG. 15-45 Kühni column extractor.

RDC and Karr columns, for which the specific throughput of the scaled-up version is the same as that of the pilot column, it is a characteristic of the Scheibel column that the throughput of the scaled-up column is on the order of three to five times greater than that achieved on the 3-in-diameter pilot column. The limited throughput of the 3-in column is due to its restrictive geometry; these restrictions are removed in the scaled-up columns.

4. Once the column diameter is determined, the stage geometry can be fixed. The geometry of a stage is a complex function of the column diameter. In the 3-in pilot column, the stage height-to-diameter ratio is on the order of 1:3. On a 10-ft (3-m) diameter column, it is on the order of 1:8. The recommended ratio of height to diameter is $Z_t(2)/Z_t(1) = [D_t(2)/D_t(1)]^{0.70}$.

5. The principle of the Scheibel column scale-up procedure is to maintain the same stage efficiency. Therefore, the scaled-up column will have the same number of actual stages as the pilot column. The only difference is that the stages will be taller, to take into account the effect of axial mixing. With the agitator dimensions determined, the speed is then calculated to give the same power input per unit of throughput. Scheibel found that power input can be correlated by

$$P = 1.85\rho\omega^3 D_i^5 \tag{15-148}$$

where P is the power input per mixing stage, D_i is the impeller diameter, ρ is the average liquid density, and ω is the impeller speed (rotations per unit time).

Kühni Column Like the Scheibel column, the Kühni column uses shrouded (closed) turbine impellers as mixing elements on a central shaft (Fig. 15-45). Perforated partitions or stator plates extend over the vessel cross section to separate the extraction stages and reduce backmixing between stages. The fractional free-flow area between compartments can be adjusted by changing the free area around the rotor shaft and/or the perforations in the stator plate. As the free-flow area increases, throughput increases at the expense of increased axial mixing of the continuous phase and reduced mass-transfer performance. Throughput typically varies from about 30 m^3/(h · m^2) [750 gal/(h · ft^2)] to significantly higher values depending upon the specific design factors chosen to meet the requirements of a given application.

A. Mögli and U. Bühlmann [chap. 13.5 in *Handbook of Solvent Extraction*, ed. T. C. Lo, M. H. I. Baird, and C. Hanson, Wiley, New York, 1983, and Krieger, Huntington, NY, 1991] outline general considerations for specifying a commercial design from pilot data. The column vendor should be consulted for specific information. The scale-up procedures are based upon hydrodynamic and geometric similarity between the pilot-scale and plant-scale designs. Individual stage geometry (impeller size and free area of the stator plate) may be tailored for each stage, especially in cases where physical properties vary significantly along the column length. A. Mögli and U. Bühlmann suggest design options to maintain a somewhat uniform interfacial area along the column to minimize the impacts of axial mixing. H. R. C. Pratt and G. W. Stevens [chap. 8 in *Science and Practice of Liquid-Liquid Extraction*, vol. 1, ed. J. D. Thornton, Oxford University Press, Oxford, UK, 1992, p. 541] provide recommended scale-up factors for a Kühni column as follows: $D_i/D_t = 0.33$ to 0.5, compartment height = 0.2 to $0.3D_t$, and the fractional free area of the stator plates = 0.2 to 0.4. The minimum recommended diameter for the pilot column is 60 mm (2.4 in) for specifying columns up to 1 m in diameter and 150 mm (6 in) for specifying larger-diameter columns.

A stagewise computational procedure is proposed by A. Kumar and S. Hartland [*Ind. Eng. Chem. Res.* **38**(3): 1040–1056 (1999)] for design of a Kühni column. The procedure considers backflow of the continuous phase, with an attempt to estimate average drop size, drop size distribution, dispersed-phase holdup, flooding velocities, mass-transfer coefficients, and axial mixing. A design example for extraction of aniline from water is presented. This approach to design can be very useful for initial estimates, but as with all agitated extractors, some pilot testing is recommended for a final commercial design. Also see the discussions by M. Asadollahzadeh et al. [*Sep. Purif. Technol.* **158:** 275–285 (2016)] and L. N. Gomes et al. [*Ind. Eng. Chem. Res.* **43**(4): 1061–1070 (2004)].

Reciprocating-Plate Columns Another approach to agitating a dispersion within an extraction column is the use of reciprocating plates. This generally results in a more uniform drop size distribution because the shear forces are more evenly distributed over the entire cross section of the column. Reciprocating-plate extractors have a wide turndown range and are well suited to systems with moderate interfacial tension. They often can handle systems exhibiting a tendency to emulsify, and because of their high open-area design, they can handle slurries of solids, some containing as much as 30 percent solids by weight. Several types of reciprocating-plate extractors have been designed; design differences generally involve differences in the plate open area and plate spacing as well as the inclusion or omission of static baffles or downcomers. For detailed discussion of these designs, see T. C. Lo and J. Procházka, chap. 12 in *Handbook of Solvent Extraction*, ed. T. C. Lo, M. H. I. Baird, and C. Hanson, Wiley, New York, 1983, and Krieger, Huntington, NY, 1991; and M. H. I. Baird et al., chap. 11 in *Liquid-Liquid Extraction Equipment*, ed. J. C. Godfrey and M. J. Slater, Wiley, New York, 1994.

The Karr reciprocating-plate column (Fig. 15-46) is a popular example. It uses dual-flow plates with 50 to 60 percent open area and has no downcomers [A. E. Karr, *AIChE J.* **5**(4): 446–452 (1959); A. E. Karr and T. C. Lo, *Chem. Eng. Prog.* **72**(11): 68–70 (1976); and A. E. Karr, *AIChE J.* **31**(4): 690–692 (1985)]. Because of the high open area, a Karr column may be operated with relatively

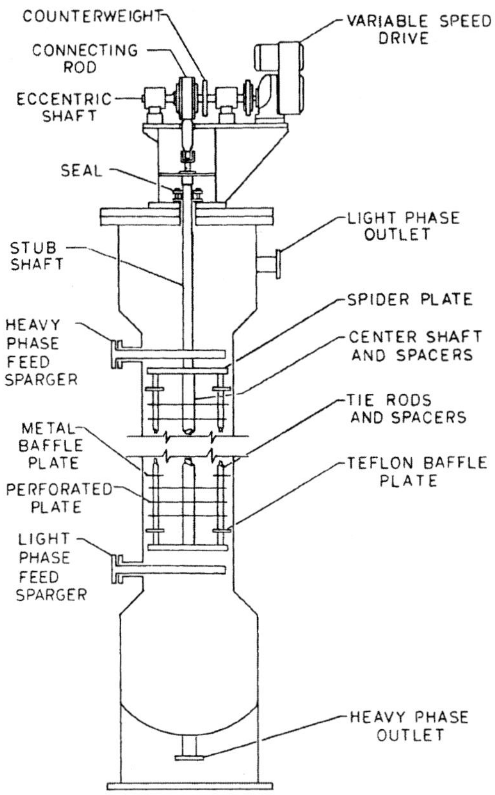

FIG. 15-46 Karr reciprocating-plate extraction column.

high throughput compared to other types of agitated columns, up to about 1000 gal/(h·ft²) [40 m³/(h·m²)] depending upon the application. The plates are mounted on a central shaft that moves up and down through a stroke length of up to 2 in (5 cm). As the diameter of the column increases, the HETS achieved by the column tends to increase due to axial mixing effects. For columns with a diameter greater than 1 ft (0.3 m), doughnut-shaped baffle plates may be added every five plates (typically) within the plate stack to minimize axial mixing. A Karr column also is well suited for corrosive systems because the plates can be fabricated from nonmetallic materials. H. R. C. Pratt and G. W. Stevens [chap. 8 in *Science and Practice of Liquid-Liquid Extraction,* vol. 1, ed. J. D. Thornton, Oxford University Press, Oxford, UK, 1992, p. 556] provide recommended geometric design and operating conditions for a Karr column as follows: reciprocation amplitude = 1 to 2 in (2.5 to 5 cm) with a 1-in amplitude being most common; reciprocation speed = 10 to 400 complete strokes (up and down) per minute; plate spacing = 2 to 6 in (5 to 15 cm); hole pitch = 0.625 to 0.75 in (1.6 to 1.9 cm); hole diameter = 0.50 to 0.625 in (1.3 to 1.6 cm); plate wall clearance = 1.25 to 2.5 in (3.2 to 6.4 cm). The plate spacing may be graduated to produce uniform drop size and population density along the length of the column, particularly for systems with high solute concentrations and depending upon how physical properties change along the column length [A. E. Karr, U.S. Patent 4,200,525 (1980)].

M. H. I. Baird et al. [chap. 11 in *Liquid-Liquid Extraction Equipment,* ed. J. C. Godfrey and M. J. Slater, Wiley, New York, 1994] discuss and summarize correlations for predicting holdup and flooding, mean drop diameter, axial mixing, mass transfer, and reciprocating-plate column performance. A. Kumar and S. Hartland [*Ind. Eng. Chem. Res.* **38**(3): 1040–1056 (1999)] present a correlation-based computational procedure for design of a Karr reciprocating-plate column, and they give an example for separation of acetone from water by using toluene. A backmixing model is described by A. Stella et al. [*Ind. Eng. Chem. Res.* **45**(19): 6555–6562 (2006)].

As with other agitated extractors, the final design of a commercial-scale Karr column is based on pilot test data. The column vendor should be consulted for specific information. The following general procedure is recommended:

1. For specifying commercial columns up to 6.5 ft (2 m) in diameter, testing in a pilot column of 1-in (25-mm) diameter is sufficient. If the anticipated scaled-up diameter is greater than 6.5 ft, then the pilot tests should be conducted in a 2-in (50-mm) diameter column. The column should be tall enough to accomplish the complete extraction. This may require several iterations on column height.

2. The column is first optimized with regard to plate spacing. The plate spacing is adjusted along the length of the column to obtain the same tendency to flood everywhere in the column. If one particular section appears to flood early, limiting the throughput, then the plate spacing should be increased in this section. This will decrease the power input into that section. Similarly, in sections that appear to be undermixed because the population of drops is low, the plate spacing should be decreased.

3. Once the plate spacing is optimized, the column is run over a range of total throughputs ($V_d + V_c$) and agitation speeds. There should be a minimum of three throughput levels and at each throughput three agitation speeds. After steady state is attained at each condition (usually three to five turnovers of column volume), samples are taken and the separation is measured. At each condition the flood point also is determined. In small-scale tests, the data used for scale-up should be collected at a point very close to flooding, say, 95 percent of flooding. Scaling these data typically results in a commercial-scale unit that operates at roughly 80 or 85 percent of flooding.

4. From the data, plots are made of volumetric efficiency and agitation speed at each throughput level. From these plots the condition that gives the maximum volumetric efficiency is selected for scale-up. For additional discussion, see T. C. Lo and J. Procházka, chap. 12 in *Handbook of Solvent Extraction,* ed. T. C. Lo, M. H. I. Baird, and C. Hanson, Wiley, New York, 1983, and Krieger, Huntington, NY, 1991.

5. For scale-up, the following parameters are kept constant: total throughput per unit area, plate spacing, and stroke length. The height and agitation speed of the scaled-up column are then calculated from the following relationships:

$$\frac{Z_t(2)}{Z_t(1)} = \left(\frac{D_{\text{col}}(2)}{D_{\text{col}}(1)}\right)^{0.38} \tag{15-149}$$

$$\frac{\text{SPM}(2)}{\text{SPM}(1)} = \left(\frac{D_{\text{col}}(1)}{D_{\text{col}}(2)}\right)^{0.14} \tag{15-150}$$

Here Z_t is the plate stack height, D_{col} is the column diameter, SPM is the reciprocating speed (complete strokes per minute), and 1 and 2 denote the pilot column and the scaled-up column, respectively.

A. E. Karr and S. Ramanujam [St. Louis AIChE Symp., March 19, 1987] propose a power per unit volume normalization factor for scale-up of the reciprocation speed if the pilot column plates have a different open area than the industrial scale plates, as follows:

$$\frac{\text{SPM}(2)}{\text{SPM}(1)} = \left(\frac{D_{\text{col}}(1)}{D_{\text{col}}(2)}\right)^{0.14} \left(\frac{\varepsilon(2)^2}{1-\varepsilon(2)^2}\right)\left(\frac{1-\varepsilon(1)^2}{\varepsilon(1)^2}\right) \tag{15-151}$$

where ε is the fractional open area of the perforated plate.

Rotating-Disk Contactor The rotary-disk contactor (RDC) is a vertical column containing an assembly of rotating disks and stationary baffles or stators. A typical design is illustrated in Fig. 15-47. The column is formed into compartments by horizontal doughnut-shaped or annular

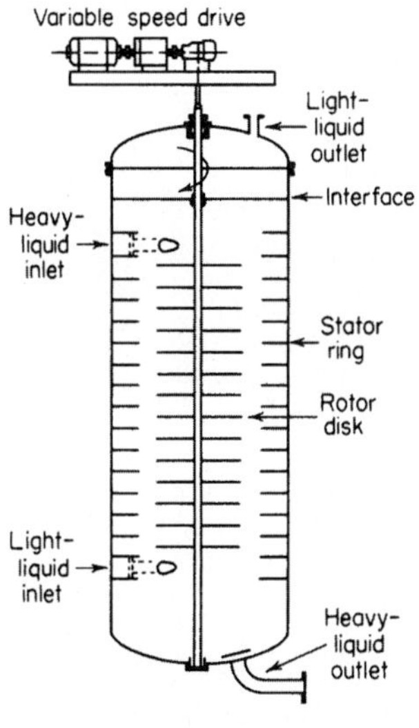

FIG. 15-47 Typical rotating-disk contactor.

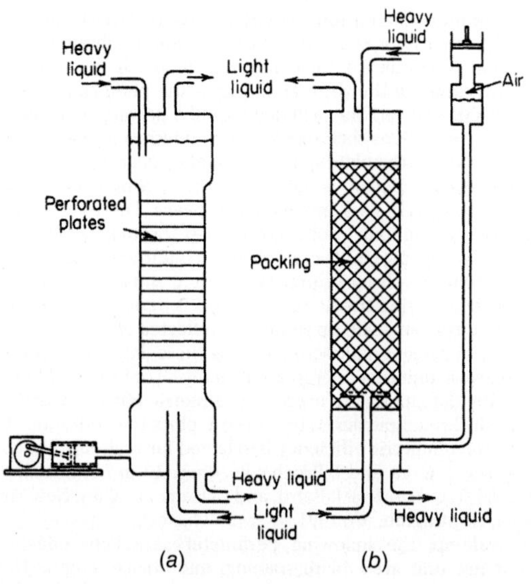

FIG. 15-48 Pulsed-liquid columns. (*a*) Sieve tray column with pump-type pulse generator. (*b*) Packed column with air pulser.

baffles, and within each compartment agitation is provided by a rotating, centrally located, horizontal disk. The rotating disk is smooth and flat and has a diameter less than that of the opening in the stationary baffles. The RDC extractor has been widely used because of its simplicity of construction, availability in relatively large diameters for high production rates, and low power consumption. For detailed reviews, see chaps. 9 and 17 in *Liquid-Liquid Extraction Equipment*, ed. J. C. Godfrey and M. J. Slater, Wiley, New York, 1994; and chaps. 13.1 and 13.2 in *Handbook of Solvent Extraction*, ed. T. C. Lo, M. H. I. Baird, and C. Hanson, Wiley, New York, 1983, and Krieger, Huntington, NY, 1991. Also see A. M. I. Al-Rahawi, *Chem. Eng. Technol.* **30**(2): 184–192 (2007); and C. Drumm and H.-J. Bart, *Chem. Eng. Technol.* **29**(11): 1297–1302 (2006).

The RDC has a moderate throughput typically in the range of 20 to 35 $m^3/(h \cdot m^2)$ [500 to 850 $gal/(h \cdot ft^2)$], and it can be turned down to 20 to 35 percent of the design rate. However, the relatively open arrangement leads to some backmixing and results in only moderate mass-transfer performance. As a consequence, some RDC columns are being replaced by more efficient extractor designs. The RDC can be used for systems with moderate viscosities up to about 100 cP and can be used for systems that tend to foul easily. The RDC also is suitable for systems with slow mass-transfer rates requiring only a few theoretical stages. An RDC can have difficulty handling feeds with emulsion formation tendencies, so it may not be suitable for some systems with low interfacial tension and low density difference.

Pulsed-Liquid Columns These are packed or tray column extractors in which a rapid reciprocating motion of relatively short amplitude is applied to the liquid contents to give improved rates of extraction (Fig. 15-48). Liquid pulsing improves the mass-transfer performance at a cost of somewhat reduced throughput. For detailed reviews of this technology, see D. H. Logsdail and M. J. Slater, chap. 11.2 in *Handbook of Solvent Extraction*, ed. T. C. Lo, M. H. I. Baird, and C. Hanson, Wiley, New York, 1983, and Krieger, Huntington, NY, 1991; H. R. C. Pratt and G. W. Stevens, chap. 8 in *Science and Practice of Liquid-Liquid Extraction*, vol. 1, ed. J. D. Thorton, Oxford University Press, Oxford, UK, 1992; and H. Haverland and M. J. Slater, chap. 10 in *Liquid-Liquid Extraction Equipment*, ed. J. C. Godfrey and M. J. Slater, Wiley, New York, 1994. Also see J. M. Bujalski et al., *Chem. Eng. Sci.* **61:** 2930–2938 (2006), for discussion of a disk and doughnut type of column extractor operated with pulsed liquid. Externally pulsing the liquid to impart mechanical agitation allows for a sealed agitated extraction column with no moving parts. This feature is important for special applications involving highly corrosive or dangerously radioactive liquids, and it is the main reason why pulsed columns commonly are applied in the extraction and separation of dissolved metals in atomic energy operations. Pulsed-liquid contactors are similar to reciprocating-plate extractors in their basic operation. However, considerably more energy generally is required to move the entire column of liquid than to move the plates. For this reason, a reciprocating-plate or other type of mechanically agitated column design generally is preferred, unless special conditions require a sealed extraction column.

Raining-Bucket Contactor (a Horizontal Column) The "raining-bucket" contactor, originally developed by the Graesser Company in the United Kingdom, consists of a horizontal column or shell, as illustrated in Fig. 15-49. The shell slowly rotates about a central axis, and during operation a main liquid-liquid interface is maintained near the centerline. The light phase is continuous in the upper half of the shell, and the heavy phase is continuous in the lower half. Buckets mounted within the shell pick up continuous phase in one half and discharge it as dispersed droplets into the other half. As a result, each phase is dispersed. The raining-bucket design is intended for systems with low density difference and low interfacial tension, that is, systems that tend to emulsify easily. It was originally developed for handling difficult settling systems in the coal-tar industry. A detailed review is given by J. Coleby [chap. 13.6 in *Handbook of Solvent Extraction*, ed. T. C. Lo, M. H. I. Baird, and C. Hanson, Wiley, New York, 1983, and Krieger, Huntington, NY, 1991]. Units currently are available through the Biotechna Company.

The rotor assembly of a raining-bucket contactor is made of a series of disks that divide the shell into a series of compartments. Each compartment contains an assembly of buckets. A small gap is maintained between the edge of the disks and the interior wall of the shell to allow for flow between compartments. The gap needs to be small to minimize backmixing. During operation, the phases are fed and removed from opposite ends of the column to produce a countercurrent flow. Throughput generally is low compared to that of other mechanically agitated extractors owing to the limited cross-sectional area available for flow. Rotational speeds are in the range of 0.25 to 40 rpm depending upon the contactor diameter and physical properties of the phases. J. Coleby [chap. 13.6 in *Handbook of Solvent Extraction*, ed. T. C. Lo, M. H. I. Baird, and C. Hanson, Wiley, New York, 1983, and Krieger, Huntington, NY, 1991] indicates that raining-bucket contactors can achieve up to 0.3 theoretical stage per compartment depending upon the application. Applications should not involve too high a viscosity in either phase. Dispersing drops in a high-viscosity continuous phase can result in slow liquid-liquid phase separation, and this can severely limit mass-transfer performance and the throughput of the extractor. Experience

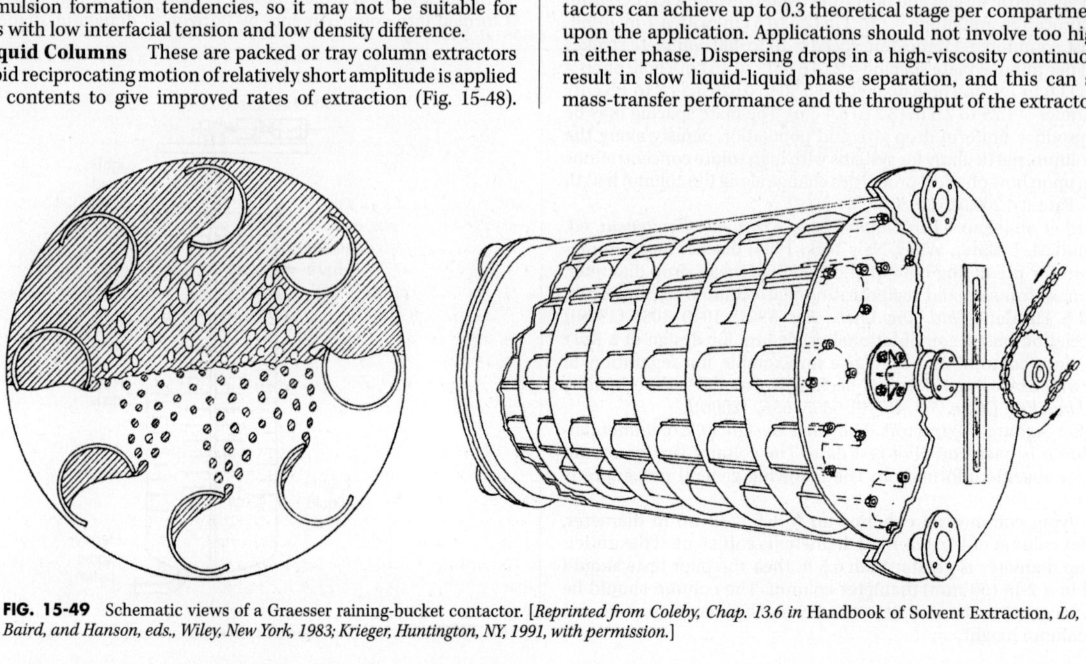

FIG. 15-49 Schematic views of a Graesser raining-bucket contactor. [*Reprinted from Coleby, Chap. 13.6 in* Handbook of Solvent Extraction, *Lo, Baird, and Hanson, eds., Wiley, New York, 1983; Krieger, Huntington, NY, 1991, with permission.*]

indicates that careful attention to this possibility is needed if viscosity is on the order of 30 cP or greater. A theoretical approach to estimating axial mixing and efficiency in a raining-bucket extractor is presented by M. Dente and G. Bozzano [*Ind. Eng. Chem. Res.* **43**(16): 4761–4767 (2004)]. A biotechnology application is described by S. Jarudilokkul, E. Paulsen, and D. C. Stuckey [*Biotechnol. Prog.* **16**(6): 1071–1078 (2000)].

MIXER-SETTLER EQUIPMENT

Mixer-settlers are used in hydrometallurgical processing for recovery of metals from aqueous acid solutions, and in multistep batchwise production of specialty chemicals, including pharmaceuticals and agricultural chemicals, among other applications. In principle, any mixer may be coupled with any settler to obtain a complete stage. The function of a single stage within the cascade is to contact the liquids so that equilibrium is closely approached (achieving a high stage efficiency), and then to separate the liquids so they can be routed to the next stage. The design must strike a balance between contacting and settling requirements; that is, the liquids should be mixed with sufficient intensity to suspend drops and facilitate good mass transfer, but not so intensely that drop sizes are too small and settling of the resulting dispersion is problematic.

A mixer-settler operation may be carried out batchwise or with a continuous feed. If batchwise operation is chosen, the same vessel used for mixing often is used for settling. Batchwise extraction in a stirred tank is a common operation in multistep, batchwise manufacture of pharmaceuticals and agricultural chemicals. Such equipment allows flexibility to accommodate batch-to-batch variability, can ensure a single batch remains isolated from other batches throughout the manufacturing process (sometimes a regulatory requirement for pharmaceuticals), and is suitable for multipurpose plants producing a variety of products in campaigns. A batchwise process may be implemented in cocurrent, cross-current, or countercurrent multistage arrangements. A countercurrent operation is carried out as in Figs. 15-6 and 15-22, by initially treating the feed batch with extract solution as the extract leaves the process. The final treatment is carried out using fresh solvent as it enters the process. A two-stage batchwise countercurrent process scheme is common practice.

Continuously operated devices may place the mixing and settling functions in separate vessels or combine them into a single, specially designed vessel with compartments for mixing and settling. Continuous mixer-settlers are particularly attractive for applications requiring several equilibrium stages and long residence times due to slow extraction kinetics, especially for applications involving the use of reactive extractants or viscous fluids. Mixing commonly is done using rotating impellers. Impeller type, shape, size, tip speed, and position within the mixing vessel may be adjusted to optimize the overall design. A static mixer may be a feasible alternative, but only if the required mass transfer can be accomplished in the short contacting time these devices allow, without generating a difficult-to-separate dispersion. Mixer-settlers may offer other advantages, including easy start-up and operation, the ability to handle very high production rates and suspended solids, and the ability to achieve high stage efficiency with proper design. For systems that accumulate rag layers (sludges) between settled liquid layers, the rag material may easily be removed at each settler. As a potential disadvantage, difficult-to-break emulsions may be formed from the shear due to mixing and pumping liquids between tanks. Mixer-settlers also generally require large floor space, and the relatively long residence time in a mixer-settler can be a disadvantage if the desired solute is degraded over time at the required extraction conditions.

Mass-Transfer Models Because the mass-transfer coefficient and interfacial area for mass transfer of solute are complex functions of fluid properties and the operational and geometric variables of a stirred-tank extractor or mixer, the approach to design normally involves scale-up of miniplant data. The mass-transfer coefficient and interfacial area are influenced by numerous factors that are difficult to precisely quantify. These include drop coalescence and breakage rates as well as complex flow patterns that exist within the vessel (a function of impeller type, vessel geometry, and power input). Nevertheless, it is instructive to review available mass-transfer coefficient and interfacial area models for the insights they can offer.

The correlation of A. H. P. Skelland and L. T. Moeti [*Ind. Eng. Chem. Res.* **29**(11): 2258–2267 (1990)] for estimating individual continuous-phase mass-transfer coefficients is given by

$$\frac{k_c d_p}{\mathcal{D}_c} = 1.237 \times 10^{-5} \left(\frac{\mu_c}{\rho_c \mathcal{D}_c}\right)^{1/3} \left(\frac{D_i \omega^2}{g}\right)^{5/12} \left(\frac{D_i}{d_p}\right)^2 \times \left(\frac{d_p}{D_t}\right)^{1/2} \left(\frac{\rho_d d_p^2 g}{\sigma}\right)^{5/4} \phi_d^{-1/2} \tag{15-152}$$

where ω is impeller speed (rotations per unit time), $\mathcal{D}_i$ is impeller diameter, D_t is tank diameter, and D_c is the solute diffusion coefficient for the continuous phase. Equation (15-152) is restricted to dispersed-phase holdup values less than $\phi_d = 0.06$. Other studies are described by H. D. Schindler and R. E. Treybal [*AIChE J.* **14**(5): 790–798 (1968)] and by R. B. Keey and J. B. Glen [*AIChE J.* **15**(6): 942–947 (1969)]. Equation (15-152) normally is used to estimate performance for applications in which the feed phase is the continuous phase and the partition ratio for transfer of solute into the extract is large. In this case, the overall resistance to mass transfer is dominated by the continuous-phase resistance (the feed or raffinate phase). Relatively little information is available about individual dispersed-phase mass-transfer coefficients. A. H. P. Skelland and Hu Xien [*Ind. Eng. Chem. Res.* **29**(3): 415–420 (1990)] offer a correlation of k_d values for batchwise extraction of solute from the dispersed phase into the continuous phase.

To use these correlation equations, it is necessary to identify which phase will be dispersed and to estimate the dispersed drop size and holdup as a function of throughput near flooding conditions—for an estimate of interfacial area. For relevant discussions, see the subsections Factors Affecting Which Phase Is Dispersed and Size of Dispersed Drops under Liquid-Liquid Dispersion Fundamentals. Holdup is a complex function of flow rates, impeller type, vessel geometry, and power input, as well as physical properties. For most impeller types, correlations for estimating holdup are not available. However, B. Weinstein and R. E. Treybal [*AIChE J.* **19**(2): 304–312; **19**(4): 851–852 (1973)] offer the following correlations for estimating holdup in a vessel agitated using a six-blade disk-style flat-blade turbine (Rushton):

For a baffled vessel with a gas-liquid surface:

$$\frac{\phi_d}{\phi_{d,\text{feed}}} = 0.764 \left(\frac{P Q_d \mu_c^2}{V_t \sigma^3}\right)^{0.300} \left(\frac{\mu_c^3}{Q_d \rho_c^2 \sigma}\right)^{0.178} \left(\frac{\rho_c}{\Delta\rho}\right)^{0.0741} \times \left(\frac{\sigma^3 \rho_c}{\mu_c^4 g}\right)^{0.276} \left(\frac{\mu_d}{\mu_c}\right)^{0.136} \tag{15-153}$$

For a liquid-full vessel without baffles:

$$\frac{\phi_d}{\phi_{d,\text{feed}}} = 3.39 \left(\frac{P Q_d \mu_c^2}{V_t \sigma^3}\right)^{0.247} \left(\frac{\mu_c^3}{Q_d \rho_c^2 \sigma}\right)^{0.427} \left(\frac{\rho_c}{\Delta\rho}\right)^{0.430} \times \left(\frac{\sigma^3 \rho_c}{\mu_c^4 g}\right)^{0.401} \left(\frac{\mu_d}{\mu_c}\right)^{0.0987} \tag{15-154}$$

Baffles are not needed if the vessel is operated full of liquid with no head space. In Eqs. (15-153) and (15-154), $\phi_{d,\text{feed}}$ is the volume fraction of the phase that ultimately becomes the dispersed phase, for the combined streams entering the vessel: $\phi_{d,\text{feed}} = Q_d/(Q_d + Q_c)$. If $\phi_d/\phi_{d,\text{feed}}$ is calculated to be greater than 1.0, it should be taken as 1.0. These equations are not applicable to other types of impellers.

When an estimate of ϕ_d is available, then $a \approx 6\varepsilon\phi_d/d_p$ [Eq. (15-104)]. If the individual mass-transfer coefficients can be estimated with reasonable accuracy, a value for the overall coefficient k_{or} can be calculated from the individual coefficients as in Eq. (15-63). The stage efficiency for a continuous process can then be estimated from

$$\xi_{mr} = 1 - \exp\left(\frac{-k_{or} a \theta}{\phi_d}\right) \tag{15-155}$$

where ξ_{mr} is the Murphree raffinate-based stage efficiency and θ is the residence time for total liquid in the vessel [Treybal, R. E., "Liquid Extractor Performance," *Chem. Eng. Prog.* **62**(9): 67–75 (1966); and G. S. Laddha and T. E. Degaleesan, *Transport Phenomena in Liquid Extraction*, McGraw-Hill, New York, 1978, p. 418]. Also see the discussion by A. H. P. Skelland and J. S. Kanel [*Ind. Eng. Chem. Res.* **31**(3): 908–920 (1992)]. These authors describe an extraction model framework that includes terms representing drop breakage and coalescence effects.

Miniplant Tests As mentioned earlier, for most liquid-liquid extraction applications involving mixer-settlers, the requirements for satisfactory performance with respect to mixing and settling are determined by using small miniplant or pilot-plant tests. For mixer design, the usual procedure is to run continuous experiments for a specific mixer geometry and type of impeller, generating performance data over a range of residence times and agitation intensities. The experimental program typically involves testing a variety of impellers and impeller locations until satisfactory results are obtained, with the ultimate goal of scaling up the miniplant design to achieve the same performance at the commercial scale. The design of settlers is discussed in the section Liquid-Liquid Separation Equipment. With careful design, most extractions require residence times in the range of

1 to 3 min. However, for reaction-enhanced extractions having relatively slow chemical kinetics compared to mass transfer, longer times in the range of 10 to 15 min are not unusual. As noted earlier, it is important to consider the time required to settle the dispersion after mixing and to determine the optimum mixing intensity that provides good mass transfer with reasonable ease of settling.

In these tests, extraction efficiency may be expressed in terms of a Murphree efficiency as

$$\xi = \frac{C_o - C_t}{C_o - C^*} \qquad (15\text{-}156)$$

where C_o is the initial concentration of solute in the feed, C_t is the concentration in the outlet for a given residence time or at time t for a batch process, and C^* is the concentration at equilibrium. Normally, the extraction efficiency is determined from continuous experiments. If batch extraction data are available for the same solvent-to-feed ratio, the efficiency of a continuous process may be estimated by fitting the batch data to a first-order rate expression

$$\xi_{\text{batch}} = 1 - \exp(-kt_b) \qquad (15\text{-}157)$$

where ξ_{batch} for the batch experiment is measured as a function of t_b, the batch mixing time [J. C. Godfrey, chap. 12 in *Liquid-Liquid Extraction Equipment*, ed. J. C. Godfrey and M. J. Slater, Wiley, New York, 1994]. The efficiency of the continuous process is calculated from the expression

$$\xi_{\text{continuous}} = \frac{k\theta}{1 + k\theta} \qquad (15\text{-}158)$$

where θ is the total liquid residence time for the continuous process. This approach is valid for most diffusion-rate-controlled processes, but may not be valid for reaction-enhanced processes in which the chemical reaction rate may be rate-limiting and not necessarily first-order.

When the ratio of phases entering a mixer-settler stage is far from unity, recycling a portion of the minority phase from the settler back to the mixer sometimes improves the settling of the dispersion by boosting the phase ratio in the settler. [See the subsection Gravity Decanters (Settlers) under Liquid-Liquid Phase Separation Equipment.] The stage efficiency also may be enhanced. For example, when the extract (solvent) is the minority phase (because K is greater than unity) and mass-transfer rates are poor, recycling the settled extract phase can boost the mass-transfer efficiency [R. E. Treybal, *Ind. Eng. Chem. Fundam.* **3**(3): 185–188 (1964)].

Liquid-Liquid Mixer Design Many different types of impellers are used for liquid-liquid extraction, including flat-blade, pitched-blade, and axial-flow turbines, marine-type propellers, and special pump-mix impellers. With pump-mix designs, the impeller serves not only to mix the fluids, but also to move the fluids through the extraction stages of a mixer-settler cascade. The agitated vessel should be baffled if the vessel is operated with a gas-liquid surface, to avoid forming a vortex. As noted earlier in reference to Eq. (15-154), baffles are not needed if the vessel is operated with the liquid full [B. Weinstein and R. E. Treybal, *AIChE J.* **19**(2): 304–312 (1973)].

The design of a liquid-liquid mixer includes specification of impeller type and rotational speed (or tip speed), the number of impellers required, the ratio of impeller diameter to vessel diameter D_i/D_t, and the location of impeller(s) and any baffles within the vessel. A single impeller generally can be used for vessels with a height-to-diameter ratio less than 1.2 and liquid density ratios within the range of $0.9 < \rho_d/\rho_c < 1.1$. Multiple impeller designs are used to improve circulation and power distribution in tall vessels. For detailed discussions of liquid-liquid mixer design, see D. E. Leng and R. V. Calabrese, chap. 12 in *Handbook of Industrial Mixing, Science and Practice*, ed. E. L. Paul, V. A. Atiemo-Obeng, and S. M. Kresta, Wiley, New York, 2004; and M. F. Edwards and M. R. Baker, chap. 7, and M. F. Edwards, M. R. Baker, and J. C. Godfrey, chap. 8, in *Mixing in the Process Industries*, 2d ed., ed. N. Harnby, M. F. Edwards, and A. W. Nienow, Butterworth-Heinemann, Oxford, UK, 1992. Also see D. Daglas and M. Stamatoudis, *Chem. Eng. Technol.* **23**(5): 437–440 (2000), for a discussion of the effect of impeller vertical position on drop size; and M. Willie, G. Langer, and U. Werner, *Chem. Eng. Technol.* **24**(5): 475–479 (2001), for a discussion of the influence of power input on drop size distribution for a variety of impeller types.

The mixing power per unit volume $P/\mathcal{V}$ is a function of impeller rotational speed ω, impeller diameter D_i, and the Power number (P_o) for the type of impeller and vessel geometry:

$$\frac{P}{\mathcal{V}} = P_o\left(\frac{\rho_m \omega^3 D_i^5}{\mathcal{V}_{\text{tank}}}\right) \qquad (15\text{-}159)$$

In Eq. (15-159), the mixture mean density is given by

$$\rho_m = \phi_d \rho_d + (1 - \phi_d)\rho_c \qquad (15\text{-}160)$$

Power numbers for different impeller types depend upon the impeller Reynolds number. Representative relationships of Power number versus Reynolds number for several types of impellers are given in Fig. 15-50. For additional information on a variety of impellers, see Sec. 6 and R. R. Hemrajani and G. B. Tatterson, chap. 6 in *Handbook of Industrial Mixing, Science and Practice*, ed. E. L. Paul, V. A. Atiemo-Obeng, and S. M. Kresta, Wiley, New York, 2004.

The power P in Eq. (15-159) does not include losses associated with the motor and drive unit. These losses can contribute as much as 30 to 40 percent to the overall power requirement. The drive supplier should be consulted for specific information. For pump-mix impellers, knowledge of the power characteristics for pumping is required in addition to that for mixing. For a discussion of these special cases, see J. C. Godfrey, chap. 12 in *Liquid-Liquid Extraction Equipment*, ed. J. C. Godfrey and M. J. Slater, Wiley, New York, 1994; and K. K. Singh et al., *Ind. Eng. Chem. Res.* **46**(7): 2180–2190 (2007).

A. H. P. Skelland and G. G. Ramsay [*Ind. Eng. Chem. Res.* **26**(1): 77–81 (1987)] correlated the minimum impeller speed needed to completely disperse one liquid in another in an agitated vessel with standard baffles as follows:

$$\frac{\omega_{\min}^2 \rho_m D_i}{g\,\Delta\rho} = C^2\left(\frac{D_t}{D_i}\right)^{2\alpha} \phi^{0.106}\left(\frac{\mu_m^2 \sigma}{D_i^5 \rho_m g^2 \Delta\rho^2}\right)^{0.084} \qquad (15\text{-}161)$$

The mixture mean density is given by Eq. (15-160), and the mixture mean viscosity is given by

$$\mu_m = \frac{\mu_c}{1 - \phi_d}\left(1 + \frac{1.5\mu_d \phi_d}{\mu_d + \mu_c}\right) \qquad (15\text{-}162)$$

The authors determined correlation constants C and α for five common types of impellers (two axial-flow impellers and three radial-flow impellers) and four impeller locations within a standard tank configuration. The specific power requirement can then be estimated by using Eq. (15-159). The power required to disperse one liquid phase into another typically is in the range of 0.2 to 0.8 kW/m^3 (1 to 4 hp/1000 gal) [Edwards, M. F., M. R. Baker, and J. C. Godfrey, chap. 8 in *Mixing in the Process Industries*, 2d ed., ed. N. Harnby, M. F. Edwards, and A. W. Nienow, Butterworth-Heinemann, Oxford, UK, 1992, p. 144].

Scale-Up Criteria It is common practice to scale up a miniplant design on the basis of equal residence time, constant power per unit volume, and geometric similarity such that the ratio D_i/D_t is held constant and the same types of impeller, tank geometry, and baffling are used. R. E. Treybal [*Chem. Eng. Prog.* **62**(9): 67–75 (1966)] indicated that in using this criterion, stage efficiency for liquid-liquid extraction is likely to increase on scale-up, so it is expected to yield a conservative design. With this approach, P/D_i^3 is constant and proportional to $P_o\omega^3 D_i^5/D_i^3 = P_o\omega^3 D_i^2$. Assuming that the Power number is independent of scale, this yields the relationship

$$\frac{\omega(2)}{\omega(1)} = \left(\frac{D_i(1)}{D_i(2)}\right)^{2/3} = \left(\frac{D_t(1)}{D_t(2)}\right)^{2/3} \qquad (15\text{-}163)$$

A. H. P. Skelland and G. G. Ramsay [*Ind. Eng. Chem. Res.* **26**(1): 77–81 (1987)] indicated that Eq. (15-163) is somewhat conservative, in general agreement with Treybal. Based on an analysis of mixing data generated at low holdup, they indicate that the exponent ⅔ may be replaced with 0.71 as a scale-up rule. Skelland and Ramsay also considered the criteria for scale-up to a tank design involving a different ratio of D_i/D_t at the large scale.

D. E. Leng and R. V. Calabrese [chap. 12 in *Handbook of Industrial Mixing: Science and Practice*, ed. E. L. Paul, V. A. Atiemo-Obeng, and S. M. Kresta, Wiley, New York, 2004, p. 732] showed that constant power per unit volume yields the following relationship if a change in drop size is desired (again, for applications with low holdup):

$$\frac{d_{\max}(1)}{d_{\max}(2)} \approx \frac{\omega(2)^{6/5} D_i(2)^{4/5}}{\omega(1)^{6/5} D_i(1)^{4/5}} \quad \text{for } \mathrm{Re} = \frac{\rho_m \omega D_i^2}{\mu_m} > 10^4 \qquad (15\text{-}164)$$

Equation (15-164) reduces to Eq. (15-163) when $d_{\max}(1)$ is set equal to $d_{\max}(2)$.

The constant power per unit volume scale-up criterion is equivalent to scaling the impeller tip speed ($S_{\text{tip}} = \pi D_i\omega$) by the ratio $S_{\text{tip}}(2)/S_{\text{tip}}(1) = [D(2)/D(1)]^{1/3}$. It follows that when the tank diameter is doubled, the impeller tip speed must increase by a factor of 1.26 to maintain constant power

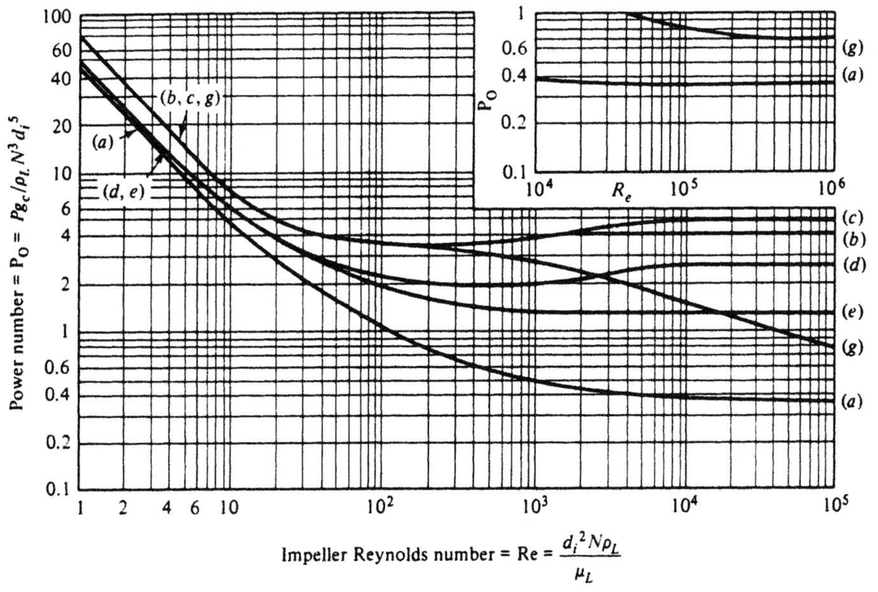

FIG. 15-50 Power for agitation impellers immersed in single-phase liquids, baffled vessels with a gas-liquid surface (except curves *c* and *g*). Curves correspond to (*a*) marine impellers; (*b*) flat-blade turbines, width = $D_i/5$; (*c*) disk flat-blade turbines (Rushton) with or without a gas-liquid surface; (*d*) curved blade turbines; (*e*) pitched blade turbines; (*g*) flat-blade turbines, no baffles, no gas-liquid interface, no vortex.

Notes on Fig. 15-50:

1. All the curves are for axial impeller shafts, with liquid depth equal to the tank diameter D_t.
2. Curves *a* to *e* are for open vessels, with a gas-liquid surface, fitted with four baffles, baffle width = $D_t/10$ to $D_t/12$. The impeller is set at a distance $C = D_i$ or greater from the bottom of the vessel.
3. Curve *a* is for marine propellers, $D_i/D_t \approx \frac{1}{3}$. The effect of changing D_i/D_t is apparently felt only at very high Reynolds numbers.
4. Curves *b* to *e* are for turbines. For disk flat-blade (Rushton) turbines, curve *c*, the effect of changing D_i/D_t is negligible in the range $0.15 < D_i/D_t < 0.50$. For open types (without the disk), curve *b*, the effect of D_i/D_t may be strong.
5. Curve *g* is for disk flat-blade turbines operated in unbaffled vessels filled with liquid and covered, so that no vortex forms. If baffles are present, the power characteristics at high Reynolds numbers are essentially the same as curve *b* for baffled open vessels, with only a slight increase in power.
6. For very deep tanks, two impellers normally are mounted on the same shaft, one above the other. For all flat-blade turbines, at a spacing of $1.5D_i$ or greater, the combined power for both will approximate that for a single turbine.

SOURCE: R. E. Treybal, *Mass-Transfer Operations*, McGraw-Hill, New York, 1980, p. 152. For more detailed information, consult Paul, Atiemo-Obeng, and Kresta, eds., *Handbook of Industrial Mixing*, Wiley, New York, 2004.

per unit volume. If the Skelland and Ramsay exponent of 0.71 is applied in Eq. (15-163) instead of ⅔, then tip speed scales as $S_{tip}(2)/S_{tip}(1) = [D(2)/D(1)]^{0.29}$ and doubling the tank diameter involves increasing the tip speed by a factor of 1.22.

W. Podgórska and J. Baldyga [*Chem. Eng. Sci.* **56:** 741–746 (2001)] presented a model of drop breakage and coalescence and compared four scale-up criteria for agitated liquid-liquid dispersions:

I. Equal power per unit volume and geometric similarity
II. Equal average circulation time and geometric similarity
III. Equal power per unit mass and equal average circulation time ($D_i/D_t \neq$ constant)
IV. Equal tip speed and geometric similarity

For slow-coalescing systems and systems at low holdup, the rate of drop breakage dominates. In this case, according to the analysis of Podgórska and Baldyga, criteria I and II yield smaller drops on scale-up, and criteria III and IV yield larger drops. For fast-coalescing systems, the rate of drop coalescence begins to dominate breakage. In this case, the authors indicate that I and III yield nearly the same drop size with scale-up, II yields much smaller drops, and IV yields larger drops. Podgórska and Baldyga recommend III for fast-coalescing systems, although they point out a limitation in terms of the maximum size of tank that this criterion will allow. See J. De Bona et al. [*Chem. Eng. J.* **296:** 112–121 (2016)] for discussion of special cases where the normal assumption of uniform solute concentration in the dispersed phase is not valid.

Based on the analyses just described, when taken together, it appears that scaling according to constant power per unit volume and geometric similarity generally will give satisfactory results, although the resulting design may not be optimal. For a new design, generally it is advisable to specify a variable-speed drive that can operate within a range of tip speeds. This provides flexibility for further adjustment and optimization of the process in the plant, and it also allows flexibility to accommodate variability in feed composition (a likely scenario in an industrial process).

Specialized Mixer-Settler Equipment As mentioned earlier, any mixer and settler can be combined to produce a stage, and the stages are in turn arranged in a multistage cascade. A great many specialized designs have been developed in an effort to reduce costs—for example, by minimizing or eliminating interstage pumping or by combining the various stages into a single vessel. With proper design, these devices generally can achieve overall stage efficiencies in excess of 80 percent, with many providing 90 to 95 percent stage efficiency. Only a few of the more commonly used types are mentioned here. For more detailed discussions, see chaps. 9.1 to 9.5 in *Handbook of Solvent Extraction*, ed. T. C. Lo, M. H. I. Baird, and C. Hanson, Wiley, New York, 1983, and Krieger, Huntington, NY, 1991; J. C. Godfrey, chap. 12 in *Liquid-Liquid Extraction Equipment*, ed. J. C. Godfrey and M. J. Slater, Wiley, New York, 1994; K. T. Hossain et al., *Ind. Eng. Chem. Process Des. Dev.* **22**(4): 553–563 (1983); N. L. Eckert and L. S. Gormely, *Chem. Eng. Res. Des.* **67:** 175–184 (1989); and K. K. Singh et al., *AIChE J.* **54**(1): 42–55 (2008).

Several pump-mix combinations have been developed by industry to simplify overall plant layout and minimize the number of pumps for greater economy. The IMI axial pump-mix and draft tube (Fig. 15-51*a*) has the pumping and mixing impellers on the same shaft. The upper part of the tank contains the draft tube and the mixing impeller. The pumping impeller for transferring the dispersion to the settler is in the lower part of the tank. There is a potential disadvantage of forming smaller and hard to separate drops when pumping a dispersion versus pumping a single phase. The Kemira design (Fig. 15-51*b*) uses a pumping impeller located near the bottom of the tank along with a mixing impeller located near the central zone of the tank. The draft tube is eliminated, and a dispersion is not pumped in this design. The Davy CMS design (Fig. 15-52) uses a pump-mix impeller in a large tank that provides both mixing and settling capability over a wide range of phase flow ratios. The dispersion occurs in the central section of the tank, and the separation occurs in the upper and lower separation zones.

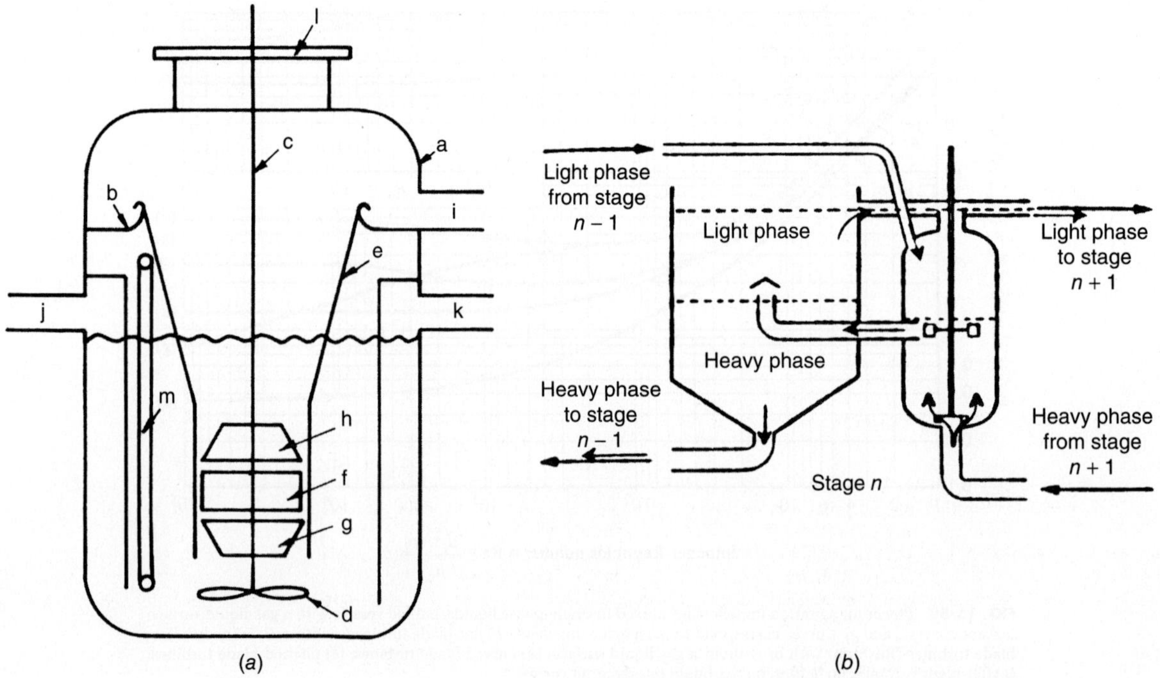

FIG. 15-51 Types of pump-mix arrangements for mixer-settler extractors. (*a*) IMI pump mix with mixing and pumping impellers (a, vessel; b, internal deck; c, shaft; d, mixing impeller; e, draft tube; f, pumping impeller; g and h, guide vanes; i, dispersion discharge; j, light-phase feed; k, heavy-phase-feed; l, mounting flange; m, sight glass). (*b*) Kemira mixer-settler. [*Figure 15-51a taken from Lo, Baird, and Hanson, eds.,* Handbook of Solvent Extraction, *Wiley, New York, 1983; Krieger, Huntington, NY, 1991), with permission. Figure 15-51b taken from Mattila,* ISEC '74 Proc., *London, 1974, with permission.*]

A compact alternating arrangement of mixers and settlers has been adopted in many of the "box-type" extractors developed originally for processing radioactive solutions. These designs are used for many other processes, with literally dozens of modifications. An example is the pump-mix mixer-settler (Fig. 15-53), in which adjacent stages have common walls [Coplan, B. V., J. K. Davidson, and E. L. Zebroski, *Chem. Eng. Prog.* **50**(8): 403–408 (1954)]. In this case, the impellers pump as well as mix by drawing the heavy liquid upward through the hollow impeller shaft and discharging it at a higher level through the hollow impeller. Rectangular tanks are not ideal for good mixing; however, the compromise in mixing and settling performance is offset by the compact and economical design.

Vertical arrangement of the stages is desirable, for then a single drive may be used for agitators and the floor space requirement of a cascade is reduced to that of a single stage. The Lurgi extractor configuration has the mixer and

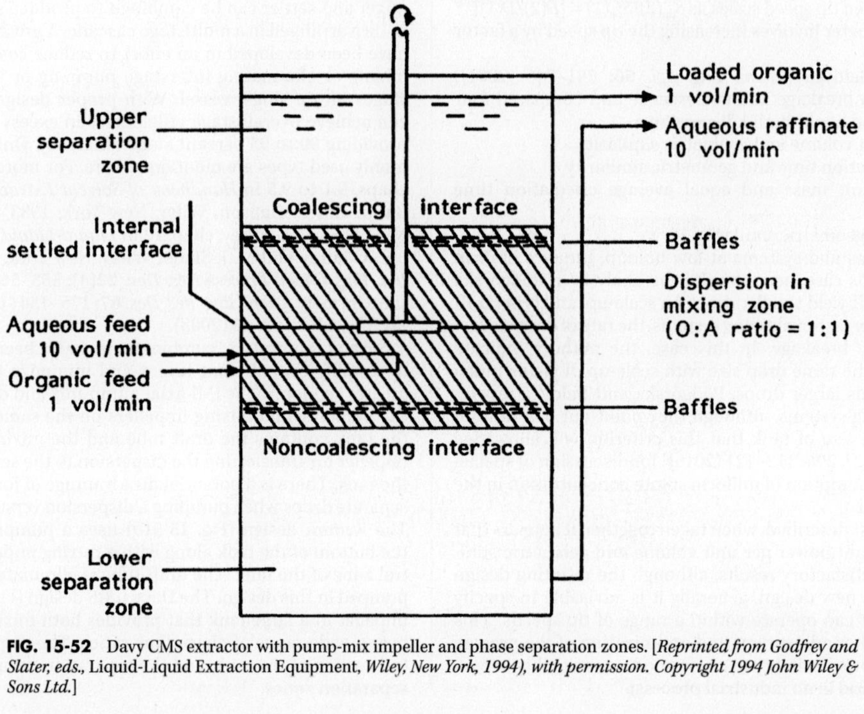

FIG. 15-52 Davy CMS extractor with pump-mix impeller and phase separation zones. [*Reprinted from Godfrey and Slater, eds.,* Liquid-Liquid Extraction Equipment, *Wiley, New York, 1994), with permission. Copyright 1994 John Wiley & Sons Ltd.*]

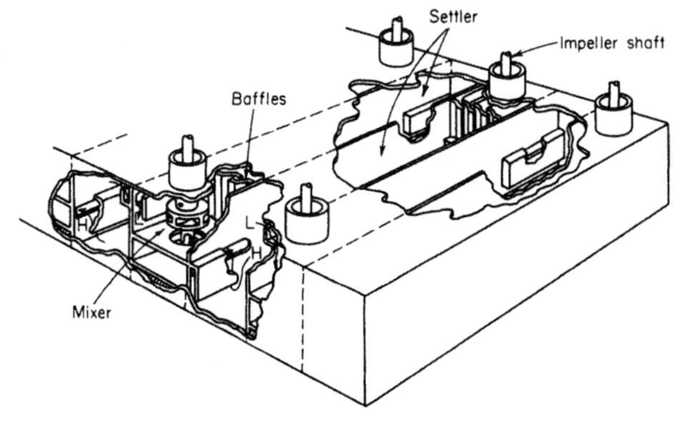

FIG. 15-53 Pump-mix box-type mixer-settler. [*Taken from Coplan, Davidson, and Zebroski,* Chem. Eng. Prog. ***50***, *p. 403 (1954), with permission.*]

settlers in separate vertical shells interconnected with piping [Guccione, E., *Chem. Eng. Magazine* **73**(4): 78–80 (1966)]. A great many other designs are known. For example, the Fenske and Long extractor [Fenske, M. R., and R. B. Long, *Chem. Eng. Prog.* **51**(4): 194–198 (1955); Long, R. B., and M. R. Fenske, *Ind. Eng. Chem.* **53**(10): 791–798 (1961); Long, R. B., *Ind. Eng. Chem. Fundam.* **1:** 152 (1962)] is a vertical stack of mixer-settler stages. This design employs a reciprocating plate at each stage to mix the two phases.

Suspended-Fiber Contactor The Merichem Fiber-Film® contactor is used in petroleum refining operations to wash hydrocarbon streams with caustic or other treating solutions [Suarez, F. J., U.S. Patent 5,997,731 (1999)]. The hydrocarbon feed and wash fluid are brought together within a vertical pipe or wash column containing fibers suspended from the top, as shown in Fig. 15-54. The two liquids flow concurrently down the column through the bed of fibers. The fibers are attached at the top of the column but not at the bottom. Liquid-liquid contacting is facilitated through capillary and surface-wetting effects. This arrangement avoids (or minimizes) the formation of small, dispersed drops, and this helps to minimize entrainment of aqueous phase into the hydrocarbon outlet. Little information about the mass-transfer performance and design requirements for this type of contactor has been published.

CENTRIFUGAL EXTRACTORS

A centrifugal extractor multiplies the force of gravity acting on two liquid phases. Centrifugal extractors can facilitate a liquid-liquid extraction process by reducing diffusion path lengths and increasing the driving force for liquid-liquid phase separation. They can achieve very high specific throughput with very low liquid residence time. A wide variety of machine types are available, ranging from relatively simple devices used primarily for phase separation or for single-stage liquid-liquid contacting with separation to more complex machines designed to provide the equivalent of multistage liquid-liquid contacting within a single unit. Some machines are designed to handle feeds containing solids such as whole fermentation broth. This section provides a brief overview with a description of several machines for illustration. More detailed descriptions of centrifuge design and performance are available from equipment vendors. For additional discussion, see R. A. Leonard, "Design Principles and Applications of Centrifugal Contactors for Solvent Extraction," chap. 10 in *Ion Exchange and Solvent Extraction: A Series of Advances* (volume 19), ed. B. A. Moyer, CRC Press, Boca Raton, Fla., 2010; U. Janoske and M. Piesche, *Chem. Eng. Technol.* **22**(3): 213–216 (1999); R. A. Leonard, D. B. Chamberlain, and C. Conner, *Sep. Sci. Tech.* **32**(1–4): 193–210 (1997); E. Blass, chap. 14 in *Liquid-Liquid Extraction Equipment*, ed. J. C. Godfrey and M. J. Slater, Wiley, New York, 1994; K. Schügerl, *Solvent Extraction in Biotechnology*, Springer-Verlag, Berlin, 1994; F. Otillinger and E. Blass, "Mass Transfer in Centrifugal Extractors," *Chem. Eng. Technol.* **11:** 312–320 (1988); and M. Hafez, chap. 15 in *Handbook of Solvent Extraction*, ed. T. C. Lo, M. H. I. Baird, and C. Hanson, Wiley, New York, 1983, and Krieger, Huntington, NY, 1991.

Centrifugal extractors can be beneficial when the liquid density difference is small, when short contact time is needed to avoid product degradation, when feed and solvent easily emulsify, or in cases where high specific throughput is needed due to limitations in available floor space or ceiling height. Centrifugal extractors also can provide flexibility in operation in cases where feed variability is high, by allowing adjustment of feed rate and rotational speed as needed to obtain satisfactory performance. Potential disadvantages generally derive from difficulties associated with maintaining high-speed rotating machinery, relatively high purchase prices compared to those of some other types of extractors, and limitations as to the number of theoretical stages that can be achieved per machine (generally less than one or up to five or six theoretical stages, depending upon throughput and the type of machine). Another consideration for some machines with close internal clearances is the potential for plugging if any solids are present in the feed; however, as previously noted, some machines are specifically designed to handle and discharge solids.

Commercial-scale centrifuges normally are continuously fed machines unless the scale of the operation is very low, as in some low-volume bioprocessing operations where very-high-g operation and long processing times are needed. A continuously fed centrifugal extractor can deliver high multiples of g at much lower residence time (given by holdup volume of the feed phase divided by volumetric feed rate) compared to a batch process. The maximum hydraulic capacity (or nominal capacity) of a continuously operated machine often is not realized in commercial applications because the feed rate needs to be turned down in order to have sufficient residence time for good extraction and phase separation performance.

In evaluating options, it generally is not possible to accurately predict performance because of the complexity of the hydrodynamics within a centrifuge. While high-g operation can promote good performance, in certain

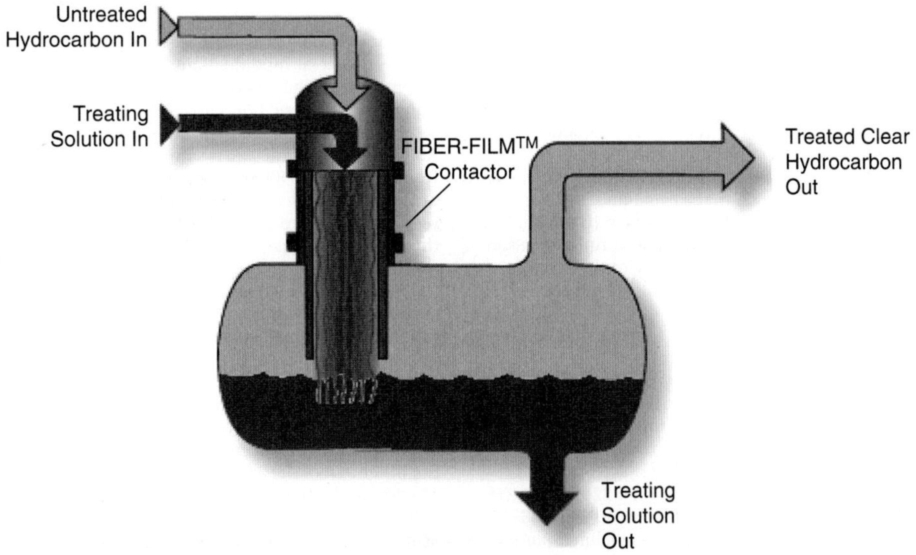

FIG. 15-54 Merichem Fiber-Film™ contactor. (*Courtesy of Merichem Chemicals and Refinery Services, LLC.*)

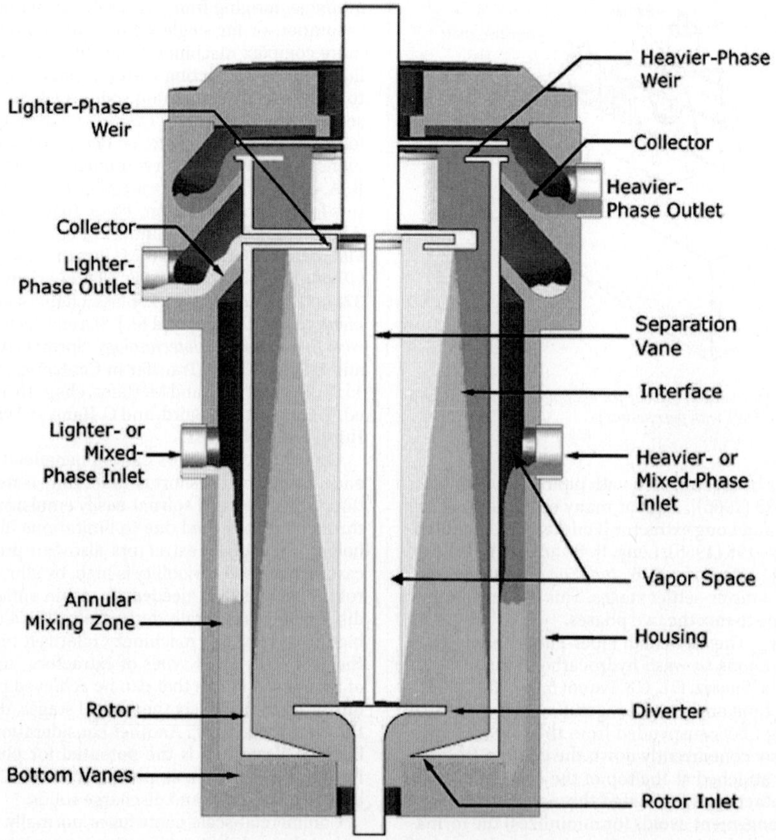

FIG. 15-55 CINC centrifugal separator. (*Courtesy of CINC Processing Equipment, Inc.*)

cases the extremely rapid acceleration generated within the machine also can promote backmixing or emulsification. Miniplant tests using small units generally are needed, and vendors often offer testing services.

Single-Stage Centrifugal Extractors The types of centrifuges used in extraction operations are quite varied. Differences include vertical versus horizontal configuration, fluid-filled versus operation with an air core, pressurized or unpressurized operation, generation of low to extremely high multiples of gravitational acceleration (500 up to 20,000 × *g* or higher), as well as differences in the liquid holdup volume, design of internals, internal clearances, and purchase price. The simpler machines, such as the CINC separator (Fig. 15-55) and the Rousselet-Robatel model BXP, have relatively large internal clearances. An air core is maintained within the machine, and liquid layers decant over internal weirs. Flow restrictions in the overflow piping need to be minimized to avoid any pressure imbalance between light- and heavy-liquid overflow lines because this can affect the location of the liquid-liquid interface and the liquid overflow/underflow split. These machines often are used for washing operations and other extraction applications with high *K* values requiring few theoretical stages. They often serve as the separator in a mixer-settler stage, such that solvent and feed are first mixed in a static mixer or a separate vessel before being fed to the centrifuge. Some mixing occurs within the centrifuge itself, so if the extraction is sufficiently fast, solvent and feed might be fed directly to the centrifuge to accomplish both mixing and phase separation. Multiple units can be connected in a countercurrent mixer-settler cascade if needed. Processes with five to seven units are typical, while processes with as many as 50 units have been reported. Multiple-unit mixer-settler processes utilizing centrifuges at each stage generally involve production of high-value, low-volume products. Stacked-disk types of machines also are available from numerous vendors and may be used in a similar extraction scheme (generally requiring some type of mixer in the feed line). These machines contain an internal stack of conical disks with a small gap between disks on the order of millimeters [Janoske, U., and M. Piesche, *Chem. Eng. Technol.* **22**(3): 213–216 (1999); and Mannweiler, K., and M. Hoare, *Bioproc. Biosystems Eng.* **8**(1–2): 19–25 (1992)]. Stacked-disk machines can be thought of as inclined-plate or lamella-type decanters operating in a centrifugal field (see the subsection Liquid-Liquid Phase Separation Equipment). They magnify the separation power by greatly reducing the distance the dispersed phase must travel before coalescing at a surface, at the expense of somewhat higher complexity and closer internal clearances.

Figure 15-55 shows a cutaway drawing of a CINC separator showing an outer annular space where solvent and feed mix before entering the interior of a rotating drum. Although this type of machine is not designed to separate solids from feeds, a clean-in-place option is offered to facilitate periodic removal of solids that accumulate in the internals. In applications in which one or more of the feed liquids is somewhat viscous, special consideration must be given to the design of the centrifuge internals such that pressure drop through the machine is not excessive. In certain applications, feed with viscosities as high as several hundred centipoise may be handled; however, special modifications to the internals are needed, and throughput must be reduced compared to that in typical operation. Maximum or nominal volumetric flow capacities for CINC machines range from 110 L/h to 136 m^3/h (0.5 to 600 gal/min) depending upon the size of the unit. The Rousselet-Robatel design is somewhat similar. These machines range in size from 50 L/h up to 80 m^3/h (0.2 to 350 gal/min). They are designed to generate only moderate centrifugal force and are generally limited to applications requiring no more than about 25,000 $g \cdot s$ [maximum *g* acceleration times the liquid residence time (in seconds) based on total volumetric flow rate and liquid holdup in the machine].

The CENTREK single-stage extractor from MEAB consists of a funnel-shaped centrifugal-bowl centrifuge mounted above a mixing tank containing a submerged stirrer. An internal "hydrolock" is used to control the position of the liquid-liquid interface in the bowl. According to the manufacturer, this is especially important for multistage, cascade operation. The unit can tolerate some amount of solids in the feed and is available in nominal capacities of 20 L/h to 20 m^3/h (0.1 to 90 gal/min).

Centrifugal Extractors Designed for Multistage Performance At the other end of the spectrum are the more complex machines designed to provide multistage or differential liquid-liquid contacting and separation within a single unit. Some machines promote the formation of very thin films or small drops for efficient liquid-liquid contacting and separation. Others provide multiple zones for mixing and separating the phases. All are designed with complex internals and close clearances. These machines

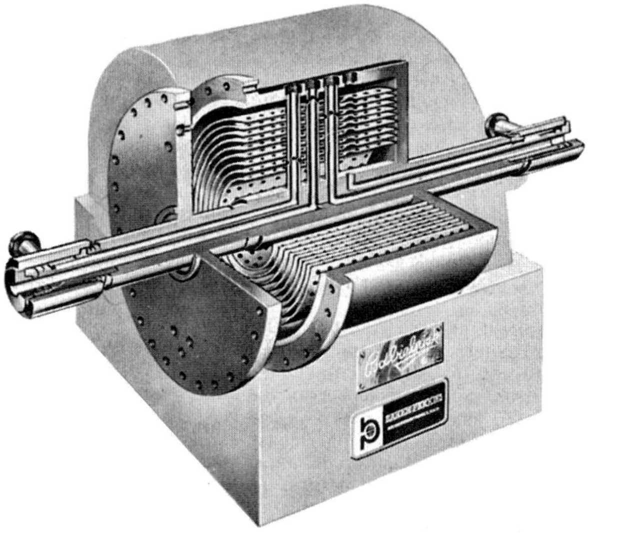

FIG. 15-56 Podbielniak centrifugal extractor. (*Courtesy of B&P Littleford.*)

typically achieve two to five theoretical stages depending upon operating conditions, with some authors claiming as many as seven or eight stages.

The classic machine of this type is the Podbielniak extractor (Fig. 15-56). The body of the extractor is a horizontal cylindrical drum containing concentric perforated cylinders. The liquids are introduced through the horizontal rotating shaft with the help of special mechanical seals; the light liquid is fed internally to the drum periphery and the heavy liquid to the axis of the drum. Rapid rotation (up to several thousand revolutions per minute, depending on size) causes radial counterflow of the liquids, which then flow out through the shaft. Materials of construction include steel, stainless steel, Hastelloy, and other corrosion-resistant alloys. The Podbielniak design provides extremely low holdup of liquid per stage, and this led to its extensive use in the extraction of antibiotics, such as penicillin and the like, for which multistage extraction and phase separation must be done rapidly to avoid chemical destruction of the product at the acidic extraction conditions [Podbielniak, W. J., H. R. Kaiser, and G. J. Ziegenhorn, chap. VI in *Chemical Engineering Progress Symposium Series No. 100*, vol. **66:** 43–50 (1970)]. Podbielniak extractors have been used in all phases of pharmaceutical manufacturing, in petroleum processing (both solvent refining and acid treating), in the extraction of uranium from ore leach liquors, and for clarification and phase separation work. F. M. Jacobsen and G. H. Beyer [*AIChE J.* **2**(3): 283–289 (1956)] describe operating characteristics and the number of theoretical stages achieved for a specific application.

The Quadronics (Liquid Dynamics) extractor is a horizontally rotated device, a variant of the Podbielniak extractor, in which either fixed or adjustable orifices may be inserted radially as a package. These permit control of the mixing intensity as the liquids pass radially through the extractor. Flow capacities, depending on machine size, range from 0.34 to 340 m^3/h (1.5 to 1500 gal/min).

The Luwesta (Centriwesta) extractor is a development from Coutor [H. Eisenlohr, *Ind. Chemist* **27:** 271 (1951)]. This centrifuge revolves about a vertical axis and contains three actual stages. It operates at 3800 rotations per minute and handles approximately 5 m^3/h (1300 gal/h) total liquid flow at 12-kW power requirement. Provision is made in the machine for the accumulation of solids separated from the liquids, for periodic removal. It is used, more extensively in Europe than in the United States, for the extraction of acetic acid, pharmaceuticals, and similar products.

The de Laval extractor contains a number of perforated cylinders revolving about a vertical shaft [Palmqvist, F. T. E., and S. Beskow, U.S. Patent 3,108,953 (1959)]. The liquids follow a spiral path about 25 m (82 ft) long, in countercurrent fashion radially, and mix when passing through the perforations. There are no published performance data.

The Rousselet-Robatel LX multistage centrifugal extractor is designed with up to seven internal mixing/separation stages. Each stage consists of a mixing chamber where the two phases are mixed by means of a stationary agitation disk mounted on a central drum. The high relative speed between the stationary disk and the rotating walls of the mixing chamber creates a liquid-liquid dispersion with high interfacial area to facilitate rapid mass transfer. The agitation disk and the mixing chamber's inlet and outlet channels form a pump that draws the two phases from the adjacent stages and transfers the dispersion to a settling chamber, where it is separated by centrifugal force. The manufacturer claims that high stage efficiencies can be achieved. Extract and raffinate phases are removed from the machine by gravity discharge, or an internal centripetal pump can be employed to discharge these streams under pressure. Nominal flow rates range from 25 L/h up to 80 m^3/h.

PROCESS CONTROL CONSIDERATIONS

GENERAL REFERENCES: Wilkinson, W. L., and J. Ingham, chap. 27.2, and Plonsky, S. P., chap. 27.3, in *Handbook of Solvent Extraction*, ed. T. C. Lo, M. H. I. Baird, and C. Hanson, Wiley, New York, 1983, and Krieger, Huntington, NY, 1991.

STEADY-STATE PROCESS CONTROL

Control of a continuous liquid-liquid extraction process generally refers to maintaining satisfactory dispersion of one phase in another for good mass-transfer performance while also maintaining the required production rate. This must be done without entering a flooding condition. It is common practice to set up a continuously fed extractor to handle a range of feed rates while maintaining other operating variables at constant preset values. These include the solvent flow rate, temperatures, and mechanical variables (if agitation or centrifugation is employed). For extraction processes that experience large swings in feed flow rate, the solvent flow rate may be manipulated to maintain a constant solvent-to-feed ratio, in order to reduce the volume of extract that needs to be processed. In this case, the extractor must be able to operate within a fairly wide range of volumetric throughput.

A common cause of upsets in operation is contamination of the feed by trace amounts of impurities that affect interfacial tension and drop coalescence, so it is important to control upstream operations to avoid contamination. Upsets or deviations from desired performance also can be caused by changes in the purity of solvent entering from solvent recovery equipment, so adequate control of closely coupled solvent recovery operations is needed to ensure good extractor performance. Periodic monitoring of the interfacial tension of light and heavy phases at the feed location (where interfacial tension is likely to be lowest due to higher solute concentration) may be useful for understanding the range of values that can be tolerated, and trends in the data may provide warning of an impending flooding or coalescence problem.

Steady-state control of a continuously fed extraction column requires maintenance of the location of the liquid-liquid interface at one end of the column. The main interface should be maintained at the top of the column when the light phase is dispersed and at the bottom of the column when the heavy phase is dispersed. If needed, extraction columns can be designed with an expanded-diameter settling zone to facilitate liquid-liquid phase separation by reducing liquid velocities and the area available for drop coalescence. If sufficient clarification of the phases cannot be achieved, then it may be necessary to add an external device such as a gravity decanter or packed coalescer. (See the subsection Liquid-Liquid Phase Separation Equipment.) Sometimes a column is built with expanded ends at both top and bottom to allow the option of operating with either phase dispersed.

The position of the main operating interface in an extraction column, whether located at the top or the bottom, generally is controlled by adjusting the outlet flow of the heavy phase; the heavy-phase outlet valve opens to lower the interface and closes to raise the interface, and the light phase is allowed to overflow the top of the column. The location of the liquid-liquid interface can be determined using a sensor such as a guided wave radar probe. Older techniques include measuring the differential pressure (if density difference is sufficiently large) or the capacitance of the liquid across the settling zone (for aqueous/organic systems). A float-based technique that rests at the position of the interface can also be used. The general interface control concept is illustrated in Fig. 15-57. O. Weinstein, R. Semiat, and D. R. Lewin [*Chem. Eng. Sci.* **53**(2): 325–339 (1997)] studied the light-phase dispersed case (with the main interface maintained at the top of the column) and recommend controlling the main interface level by manipulating the continuous-phase feed flow rate instead of the continuous-phase outlet flow rate. The authors developed a dynamic model of the hydrodynamics and mass transfer in a countercurrent liquid-liquid extraction column, and the simulation results indicate faster dynamic response using their alternative scheme.

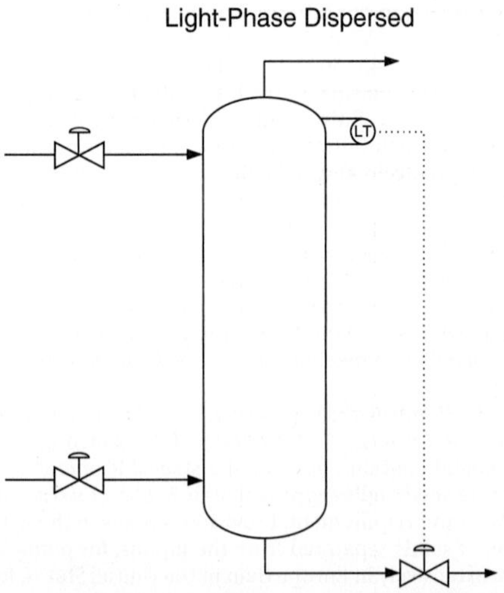

FIG. 15-57 Typical interface control for a light-phase dispersed process (with the main interface located at the top of the column). The same basic arrangement can be used for the heavy-phase dispersed case, but the level transmitter would be located differently to reflect the location of the main interface at the bottom of the column.

When a continuous extraction column begins to flood, often one of the first indications is the appearance of an interface at the wrong end of the column; so adding instrumentation that can detect such an interface (such as one or more conductivity probes when phase inversion involves the formation of a continuous aqueous phase) may help identify a flooding condition in time to take corrective action. Sometimes a rag layer will accumulate at the liquid-liquid interface, and it is necessary to provide a means for periodically draining the rag to avoid entrainment into the extract or raffinate. It may be useful to add instrumentation that can detect the rag at high positions to warn an operator before breakthrough occurs; however, often the approach taken is to drain the interface region on a predetermined schedule. Installing sensors to detect a rag layer can be problematic because they are easily fouled.

For a continuous extraction column, it is important to control the holdup of each phase within the column to obtain high interfacial area for good mass transfer. For nonagitated extraction columns, this is set by proper design of the internals and maintaining flow velocities during operation within a fairly narrow range of values needed for good performance. Agitated columns allow greater flexibility in this regard because agitation intensity can be adjusted in the plant to maintain good performance over a wider range of flow velocities and as the properties of the feed change. In industrial practice, agitation intensity normally is set at a constant rate or manually adjusted at infrequent intervals in response to a significant change in feed characteristics. Model-based control schemes offer potential for automatic adjustment of agitation intensity and other variables for faster response [Mjalli, F. S., *Chem. Eng. Sci.* **60**(1): 239–253 (2005); and Mjalli, F. S., N. M. Abdel-Jabbar, and J. P. Fletcher, *Chem. Eng. Processing* **44**(3): 531–542 and 543–555 (2005)]. Careful programming will be needed to avoid inappropriate control actions when sensors are out of calibration. Real-time measurement of dispersed-phase holdup also may be helpful; J. Chen et al. [*Ind. Eng. Chem. Res.* **41**(7): 1868–1872 (2002)] report a method for a pulsed-liquid column. They studied a system consisting of 30 percent trialkyl(C_{6-8}) phosphine oxide in kerosene + nitric acid solution, with the acid phase dispersed.

For some extraction operations, particularly fractional extractions, it may be useful to control a temperature profile across the process. In extraction columns, this is normally done by controlling the temperature of entering feed and solvent streams. Heating jackets generally are not effective because of insufficient heat-transfer area. Internal heating or cooling coils are problematic because they are difficult and expensive to install and can interfere with other column internals and liquid-liquid traffic within the column. For fractional extraction, the stripping and washing operations may be carried out in separate equipment with external heating or cooling of the streams entering the equipment.

For startup of column extractors, generally it is best to start from dilute-solute conditions to avoid unstable operation. For example, when starting a column in which the feed is the continuous phase, first fill the column with solute-lean feed liquid before starting the flow of solvent and actual feed. This way, the solvent quickly becomes dispersed, and mass transfer approaches steady state from dilute conditions, promoting faster and more stable startup.

SIEVE TRAY COLUMN INTERFACE CONTROL

Control of the main liquid-liquid interface for a sieve tray column can be counterintuitive because of complexity caused by the presence of multiple interfaces within the column. For example, if the interface level is too high, the usual control response is to allow the heavy phase to flow out the bottom of the column for a time until the desired level is reached (using the scheme outlined in Fig. 15-57). Ideally, this should lower the interface level, as shown in Fig. 15-58*a*. This is a typical response for most differential contactors such as packed or spray columns. However, for the sieve tray column, the initial response can actually be a rise in the interface level for a short time (Fig. 15-58*b*). In some cases, this can result in entrainment of heavy phase out the top of the tower. This inverse response is caused by changes in the coalesced layer heights at each tray. Neglecting any correction for dispersed-phase holdup, the height of the coalesced layer is affected by the pressure drop through the sieve holes and downcomer:

$$h \approx \frac{\Delta P_o + \Delta P_{\text{dow}}}{\Delta\rho g} = \frac{C_1 V_o^2 + C_2 V_{\text{dow}}^2}{\Delta\rho g} \tag{15-165}$$

where h is the coalesced layer height, ΔP_o is the pressure drop through perforations, ΔP_{dow} is the pressure drop through the downcomer, V_o is the average velocity through a perforation (orifice), V_{dow} is the average velocity through the downcomer, and C_1 and C_2 are constants related to tray geometry and physical properties. Tray designs often vary as to which contribution, orifice or downcomer pressure drop, controls the height of the coalesced layer. The inverse response can cause significant control problems if the downcomer pressure drop is much greater than the orifice pressure drop, and this issue should be addressed during design.

LIQUID-LIQUID PHASE SEPARATION EQUIPMENT

General References: Sinnott, R. K., *Chemical Engineering Design, Coulson and Richardson's Chemical Engineering,* vol. 6, 4th ed., Butterworth-Heinemann, Oxford, UK, 2005; Müller, E., et al., "Liquid-Liquid Extraction," in *Ullmann's Encyclopedia of Industrial Chemistry*, 6th ed., Wiley-VCH, New York, 2002, updated online, 2008; Hooper, W. B., sec. 1.11 in *Handbook of Separation Techniques for Chemical Engineers*, 3d ed., ed. P. A. Schweitzer, McGraw-Hill, New York, 1997; Hartland, S., and S. A. K. Jeelani, chap. 13 in *Liquid-Liquid Extraction Equipment*, ed. J. C. Godfrey and M. J. Slater, Wiley, New York, 1994; Monnery, W. D., and W. Y. Svrcek, *Chem. Eng. Prog.* **90**(9): 29–40 (1994); and Jacobs, J. L., and W. R. Penney, chap. 3 in *Handbook of Separation Process Technology*, ed. R. W. Rousseau, Wiley, New York, 1987.

OVERALL PROCESS CONSIDERATIONS

The ability to separate a mixture of two liquid phases is critical to the successful operation of many chemical and refining processes. Besides its obvious importance to liquid-liquid extraction and washing operations, liquid-liquid phase separation can be a critical factor in other operations, including two-liquid-phase reaction, azeotropic distillation, crude oil processing, and industrial wastewater treatment. Sometimes the required phase separation can be accomplished within the main process equipment, such as an extraction column or a batchwise, stirred-tank reactor; but in many cases a stand-alone separator is used. These include many types of gravity decanters (or settlers), filter-type coalescers, coalescers filled with granular media, centrifuges, and hydrocyclones.

The path that a liquid-liquid mixture takes through a chemical process on its way to the separator often has a dramatic impact on separation difficulty once the mixture arrives. For this reason, the first steps toward designing a decanter or other type of liquid-liquid phase separator should include a study of the overall process flow sheet to determine whether changes in upstream processing conditions can make for an easier and more robust separation. For example, if the main stream entering the separator is produced by mixing a number of smaller streams, look for opportunities to remove fine solids that contaminate the main stream by filtering solids from one or more small streams before they enter the larger stream. Also, standard centrifugal pumps are notorious for producing stable dispersions. If this type of pump is used upstream of a separator, determine whether the turbulence caused by the pump is contributing to phase separation difficulty and if so, consider using gravity or pressurized flow (if possible) or replacing a high-shear pump and piping system with a lower-shear design.

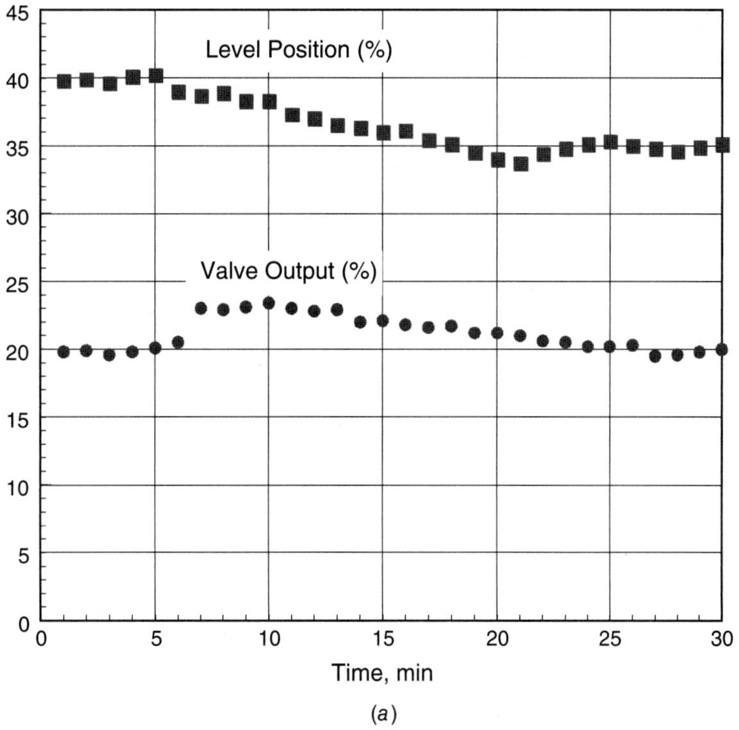

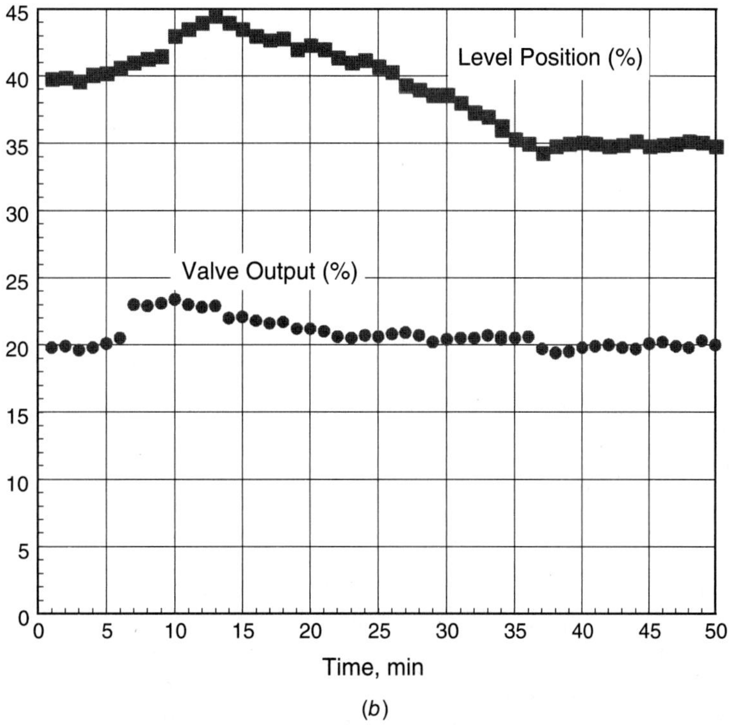

FIG. 15-58 Dynamic response to a change in heavy-phase flow rate. (*a*) Normal dynamic response to increasing outlet heavy-phase flow (packing). (*b*) Dynamic response to increasing outlet heavy-phase flow rate (sieve trays).

If a dispersion proves to be particularly difficult to separate, it may be due to the presence of some contaminant acting as a surfactant. Such contaminants may be oxidation products produced in trace amounts owing to leakage of air into the process, or they may be the products of corrosion of upstream equipment. They also may be materials that are intentionally added upstream to solve a problem there, such as cleaning agents or antifouling agents, but their presence, even in very small concentration, may cause unintended phase separation difficulties downstream.

FEED CHARACTERISTICS

Traditionally, the guidelines for selection and design of a gravity decanter or other type of separator focus on the size of dispersed drops. However, drop diameter often cannot be accurately predicted during the design of a new process, especially the size of the smaller drops in the distribution of drop sizes, and often this information is not available for an existing process because of sampling difficulties. Furthermore, knowledge of drop size alone is not

TABLE 15-24 Shake Test Characterizations

Type	Shake test observations	Interfacial tension*	Density difference*	Viscosity of each phase*	Presence of fine solids or surfactants*
I	Dispersion band collapses within 5 min with crystal-clear liquids on top and bottom	Moderate to high, 10 dyn/cm or higher	$\Delta\rho > 0.1$ g/cm^3	$\mu < 5$ cP	Negligible
II	Dispersion band collapses within 10 to 20 min with clear liquids on top and bottom	Moderate, ~10 dyn/cm	$\Delta\rho > 0.1$ g/cm^3	$\mu < 20$ cP	Negligible
III	Dispersion band collapses within 20 min but one or more phases remain cloudy	Low to moderate, 3–10 dyn/cm	$\Delta\rho > 0.05$ g/cm^3	$\mu < 100$ cP	Might be present in low concentration
IVa	Stable dispersion is formed (dispersion band does not collapse within an hour or longer)—high viscosity	Low to high	$\Delta\rho > 0.1$ g/cm^3	$\mu > 100$ cP in one of the phases	Negligible
IVb	Stable dispersion is formed—low interfacial tension	< 3 dyn/cm	$\Delta\rho > 0.1$ g/cm^3	$\mu < 100$ cP	Negligible
IVc	Stable dispersion is formed—low density difference	Low to high	$\Delta\rho < 0.05$ g/cm^3	$\mu < 100$ cP	Negligible
IVd	Stable dispersion is formed—stabilized by surface-active components or solids	Low	$\Delta\rho > 0.1$ g/cm^3	$\mu < 100$ cP	Enough surfactant/ solids to keep emulsion stable

*Typical physical properties. Behavior also depends upon the shear history of the fluid. For this test, a sample is characterized by the results of the shake test (second column), not its physical properties. Physical properties are listed only as typical values.

sufficient because it says nothing about the rate of drop coalescence. In light of this, it is recommended instead to characterize the feed material in terms of the results of simple shake tests, as indicated in Table 15-24. This basic information can be very helpful in identifying an appropriate separator type.

In Table 15-24, feed materials are classified into four main types according to the results of a shake test. Typical values of interfacial tension, density difference, and viscosity also are listed. The shake test can be as simple as shaking a representative feed by hand in a sealed graduated cylinder (about an inch in diameter) for 15 seconds. The graduated cylinder is then placed on the bench, the time is recorded, and the progress of the separation is observed. For systems with drops that coalesce quickly, a sharp interface will quickly form between two settling liquid layers, and the rate at which drops fall or rise to the interface will determine the rate of phase separation or clarification of the layers. For many other systems, however, drops will accumulate at the interface forming a dispersion band, that is, a layer of slowly coalescing drops, and the rate at which the drops coalesce determines the rate of phase separation. Whether a system is fast-coalescing or slow-coalescing is an important question that is easily answered by performing a simple shake test. Figure 15-59 illustrates the details of a batch settling profile. Once the dispersion band has disappeared, one or both of

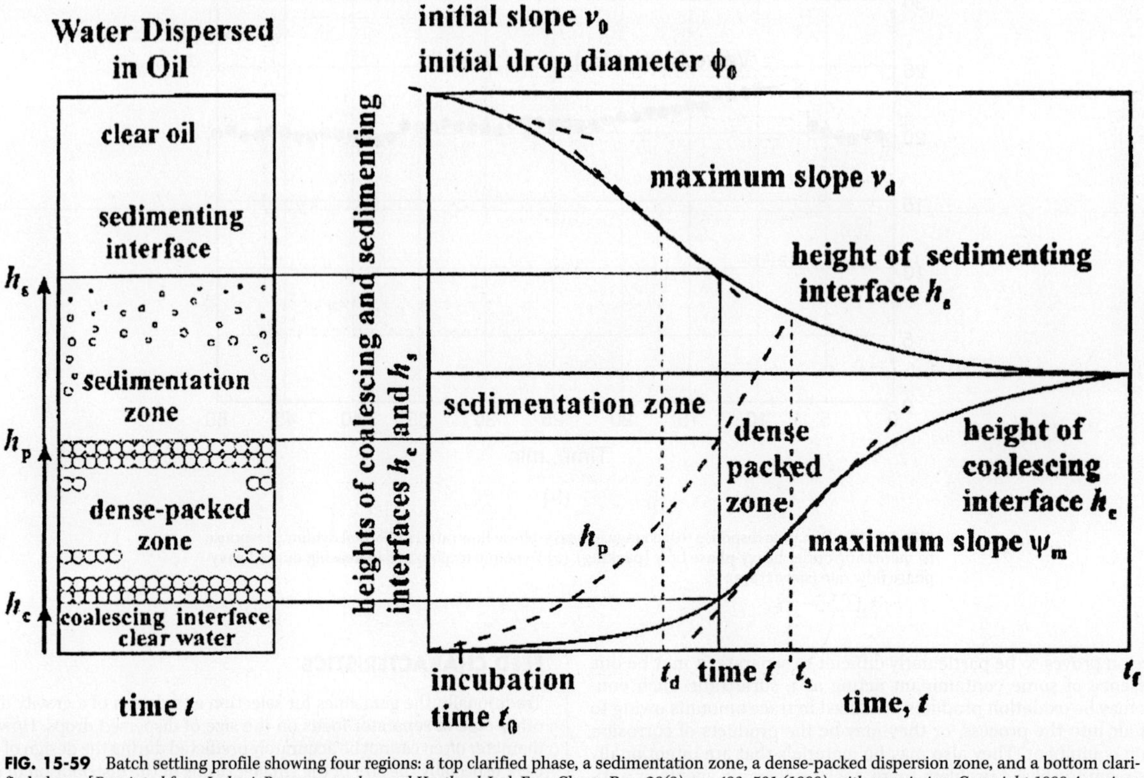

FIG. 15-59 Batch settling profile showing four regions: a top clarified phase, a sedimentation zone, a dense-packed dispersion zone, and a bottom clarified phase. [*Reprinted from Jeelani, Panoussopoulos, and Hartland,* Ind. Eng. Chem. Res. *38(2), pp. 493–501 (1999), with permission. Copyright 1999 American Chemical Society.*] Consult the original article for a detailed description.

the phases may remain cloudy. If so, this typically indicates the presence of droplets on the order of 100 μm in diameter or smaller. To reduce variability, this type of test can be implemented using a glass vessel with shaft-driven pitch-blade impeller and baffles to avoid vortexing. For additional discussion of dispersion properties, see the subsection Liquid-Liquid Dispersion Fundamentals.

GRAVITY DECANTERS (SETTLERS)

Gravity decanters or settlers are simple vessels designed to allow time for two liquid phases to settle into separate layers (Fig. 15-60). Ideally, clear top and bottom layers form above and below a sharp interface or dispersion band. The top and bottom layers serve as clarifying zones. The height of the dispersion band, if present, generally remains constant during steady-state operation, although it may vary with position. The choice of where to locate the phase boundary within the vessel depends on whether more or less height is needed in the upper or lower clarification zones to obtain the desired clarity in the discharge streams. It can also depend on whether the inventory of one particular layer within the vessel should be minimized, as when handling reactive fluids such as monomers. Gravity decanters are well suited for separating type I feeds defined in Table 15-24 and, in most cases, type II feeds as well. It is common for coalescence to be the limiting factor in the separation of type II mixtures, so the design and sizing of the decanter will differ from those of the fast-coalescing systems.

Design Considerations Gravity decanters normally are specified as horizontal vessels with a length-to-diameter ratio greater than 2 (and often greater than 4) to maximize the phase boundary (cross-sectional area) between the two settled layers. This provides more effective utilization of the vessel volume compared to vertical decanters, although vertical decanters may be more practical for low-flow applications or when space requirements limit the footprint of the vessel.

The volume fraction of the minority phase is an important parameter in the operation of a decanter. Vessels handling less than 10 to 20 percent dispersed phase typically contain a wider distribution of droplet diameters with a long tail in the small size range [Barnea, E., and J. Mizrahi, *Trans. Instn. Chem. Engrs.* **53:** 61–69 (1975)]. These decanters have a smaller capacity than when they contain more-concentrated dispersions. If one of the phases has a concentration lower than 20 percent in the feed mixture, it might be worthwhile to recycle the low-concentration phase to the feed point to boost the phase ratio within the separator vessel. Also, in certain cases increasing the operating temperature increases the drop coalescence rate. The result of either is a reduction in the dispersion band height for a given throughput, allowing an increase in the capacity of the settler. This behavior often can be attributed to a reduction in the continuous-phase viscosity.

Numerous methods are used to control the location of the interface inside the decanter. A boot or sump sometimes is included in the design to increase the path traveled by the heavy phase before exiting the vessel, to maximize the clarification zone for the light phase, or to minimize the inventory of heavy phase within the vessel. The interface can even be located inside the boot for one of these reasons. When a rag layer forms at the interface between settled layers, adding one or more nozzles in the vicinity of the interface will allow periodic draining of the rag (Fig. 15-61). Instruments such as differential pressure cells, conductance probes, or density meters are commonly used to control the location of the interface in a decanter. These instruments can be prone to fouling, and their operation can be compromised by the presence of a dispersion band or a rag layer. In that case, an alternative is to use an overflow leg or seal loop as illustrated in Figs. 15-60 and 15-61. The following expression can be used to specify the loop dimensions [Bocangel, J., *Chem. Eng. Magazine* **93**(2): 133–135 (1986); and Aerstin, F., and G. Street, *Applied Chemical Process Design*, Plenum, New York, 1982]:

$$Z_2 = \frac{(h_L + Z_1 - Z_3)\rho_L}{\rho_H} + Z_3 - h_H \qquad (15\text{-}166)$$

where Z_1, Z_2, and Z_3 are the heights shown in Fig. 15-61 and h_L and h_H are the head losses in the light- and heavy-liquid discharge piping. An overflow leg can work reasonably well, provided that the densities of the two phases and the height of the dispersion band do not change significantly in operation (as in an upset). The light phase also may be removed through a takeoff tube entering the vessel from the bottom. This design provides added flexibility by allowing adjustment of the pipe length in the field without altering the vessel itself. Care should be taken to avoid the possibility of inducing a swirling motion as liquid enters the top of the weir. Swirling motions may be avoided or minimized by adding vanes or slots at the entrance.

To allow the phases to settle and remain calm, any form of turbulence or vortexing inside the decanter should be avoided. Introduction of the feed

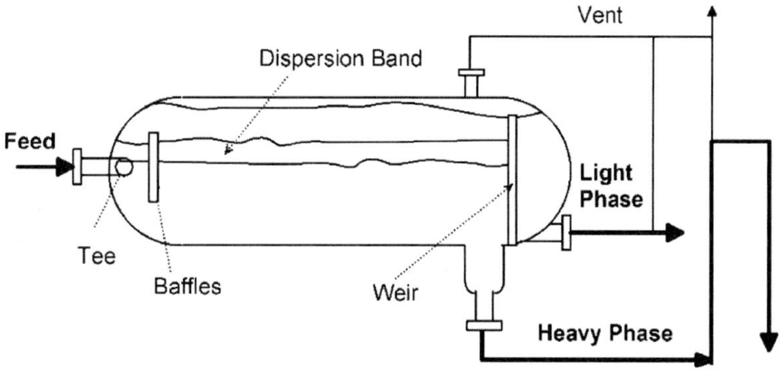

FIG. 15-60 Typical horizontal gravity decanter design.

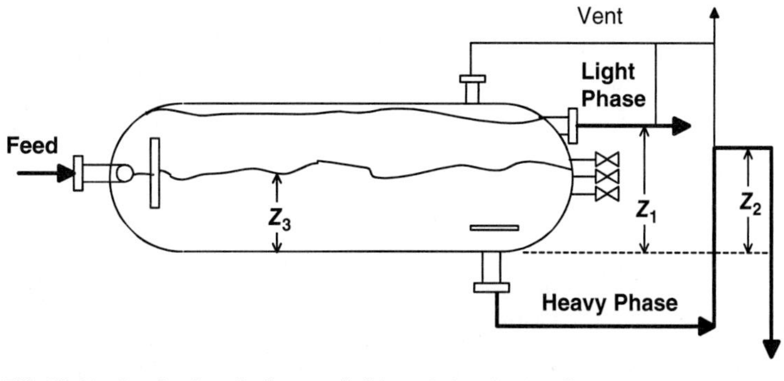

FIG. 15-61 Overflow loop for the control of the main interface in a decanter.

stream into the decanter should be located close to the interface to facilitate phase separation. Turbulence can arise from the inlet liquid entering the vessel at too high a velocity, forming a jet that disturbs the liquid layers. To counter these flow patterns, the feed into the gravity settler should enter the vessel at a velocity of less than about 1 m/s (3 ft/s) as a general rule. This can be achieved by enlarging the feed line in the last 1 to 2 m (3 to 6 ft) leading to the vessel, to slow down the feed velocity at the inlet nozzle. In addition, a quiet feed zone may be created by installing a baffle plate in front of the feed pipe or a cap at the end of the feed line, with slots machined into the side of the pipe. Some designers are now using computational fluid dynamics (CFD) methods to analyze general flow patterns as an aid to specifying decanter designs.

Vented Decanters When the liquid-liquid stream to be decanted also contains a gas or vapor, provisions for venting the decanter must be included. This often is the case when decanting overheads condensate from an azeotropic distillation tower operating under vacuum, as some amount of air leakage is virtually unavoidable, or when decanting liquids from an extractor operating at a higher pressure. A common design used for this service when the amount of gas is low is shown in Fig. 15-62. The feed enters the vessel at a point below the liquid level, so any gas must flow up through the liquid before disengaging in the vapor head space. An alternative design is illustrated in Fig. 15-63. With this design, the feed is introduced to the top of the vessel in the vapor headspace so that gases can be freely discharged and disengaged with no back-pressure. One drawback to this approach is that the feed liquids are dropped onto the light liquid surface, and significant quantities of heavy liquid may be carried over to the light liquid draw-off nozzle owing to the resulting turbulence. To mitigate this effect, a quiescent zone may be provided immediately below the top feed nozzle by means of a perforated baffle, as shown in Fig. 15-63. The baffle separates the disturbance caused by the entering feed from a calm separation zone where the two liquid phases can coalesce and disengage prior to draw-off.

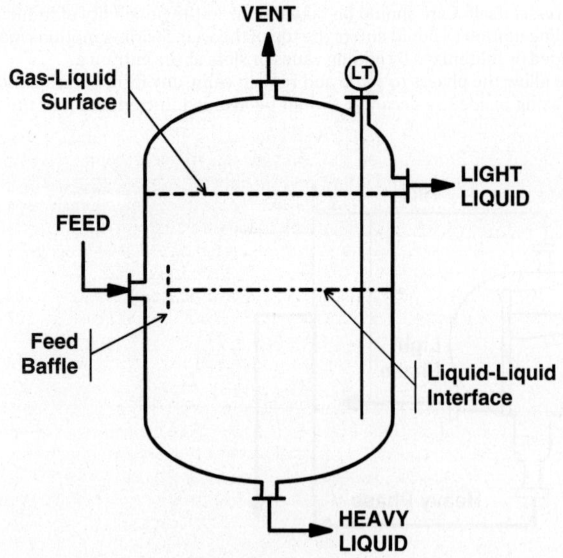

FIG. 15-62 Vertical decanter with submerged feed.

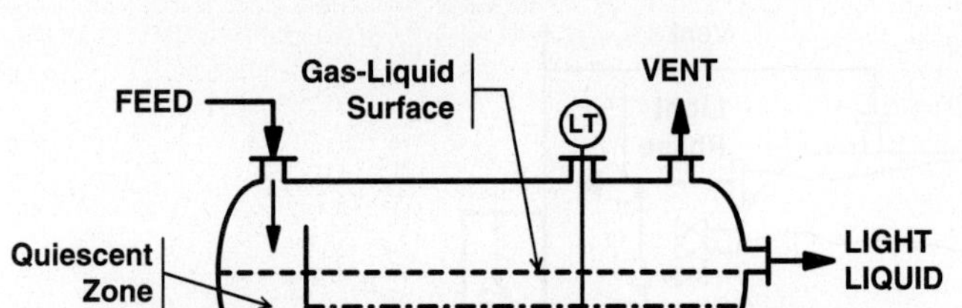

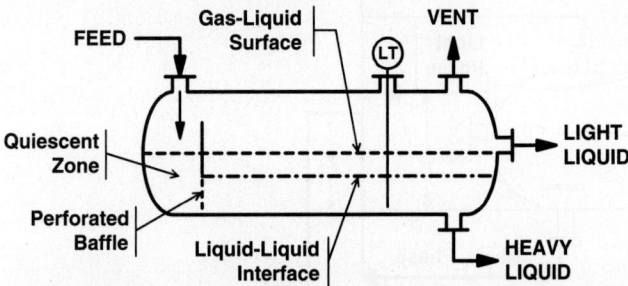

FIG. 15-63 Horizontal decanter with feed entering from the top and a baffled quiescent zone.

Decanters with Coalescing Internals Adding coalescing internals may improve decanter performance by promoting the growth of drops and may reduce the size of vessel required to handle dispersions with slow coalescence (as in type II systems in Table 15-24). A wide variety of internals have been used, including wire mesh, knitted wire or fibers, and flat or corrugated plates. When plates are used, the coalescer is sometimes referred to as a lamella-type coalescer. Plates typically are arranged in packets installed at a slight angle with respect to horizontal. The plates shorten the distance that drops must rise or fall to a coalescing surface and guide the flow of the resulting coalesced film [Menon, W., W. Rommel, and E. Blass, *Chem. Eng. Sci.* **48**(1): 159–168 (1993); and Menon, W., and E. Blass, *Chem. Eng. Technol.* **14:** 11–19 (1991)]. Arranging the plates in packets of opposite slopes promotes flow reversal, and this may lead to more frequent drop–drop collisions [Berger, R., *Int. Chem. Eng.* **29**(3): 377–387 (1989)]. The Merichem Fiber-Film® contactor described earlier in the subsection Suspended-Fiber Contactor under Mixer-Settler Equipment also may be used to promote the growth of dispersed drops in a stream feeding a gravity decanter. The dispersed phase normally must preferentially wet the coalescence media for the media to be effective. If the feed contains solids, the potential for plugging the internals should be carefully evaluated. In certain cases, it may be necessary to allow access to the vessel internals for thorough cleaning. For more information, see E. Müller et al., "Liquid-Liquid Extraction," *Ullmann's Encyclopedia of Industrial Chemistry*, 6th ed., Wiley-VCH, New York, 2002, updated online, 2008.

Sizing Methods Sizing a decanter involves quantifying the relationship between the velocity of liquid to the phase boundary between settled layers and the average height of a dispersion band formed at the boundary. For fast-coalescing systems, the height of the dispersion band is negligible. Performance is determined solely by the rate of droplet rise or fall to the interface compared with the rate of flow through the decanter. In this case, design methods based on Stokes' law may be used to size the decanter, and residence time in the vessel becomes a key factor. In many cases, however, coalescence is slow and the shake tests show a coalescence band that requires a fair amount of time to disappear. Then performance is determined by the volumetric flow rate of liquid to the boundary between the two settled layers, the boundary area available for coalescence, and the steady-state height of the dispersion band. For these systems, residence time is not a reliable guide for decanter design.

Stokes' Law Design Method This method is described by W. B. Hooper and L. L. Jacobs [sec. 1.11 in *Handbook of Separation Techniques for Chemical Engineers,* 3d ed., ed. P. A. Schweitzer, McGraw-Hill, New York, 1997]; and by L. L. Jacobs and W. R. Penney [chap. 3 in *Handbook of Separation Process Technology*, ed. R. W. Rousseau, Wiley, New York, 1987]. It assumes that the drop coalescence rate is rapid and relies on knowledge of drop size. The terminal settling velocity of a drop is computed by using Stokes' law

$$u_t = \frac{gd^2\Delta\rho}{18\mu_c} \tag{15-167}$$

where d is a characteristic minimum drop diameter. (See Sec. 6 for a detailed discussion of terminal settling velocity.) Note that which phase is continuous and which is dispersed can make a significant difference, as only the continuous-phase viscosity appears in Eq. (15-167). The decanter size is then specified such that

$$\frac{Q_c}{A} < u_t \tag{15-168}$$

where Q_c is the volumetric flow rate of the continuous phase and A is the cross-sectional area between the settled layers. This analysis assumes no effect of swirling or other deviation from quiescent flow, so a safety factor of 20 percent often is applied. Hooper as well as Jacobs and Penney both indicate that designing for a Reynolds number $\mathrm{Re} = VD_h\rho_c/\mu_c$ less than 20,000 or so should provide sufficiently quiescent conditions, where V is the continuous-phase cross-flow velocity and D_h is the hydraulic diameter of the continuous-phase layer (given by four times the flow area divided by the perimeter of the flow channel, including the interface). However, this Reynolds number criterion is a general guideline rather than a firm not-to-exceed number. Both sources cite multiple Reynolds number ranges where varying degrees of swirling effects might be encountered. In particular, decanters with large hydraulic diameters will result in higher Reynolds numbers for a given velocity, which does not necessarily mean that quiescent conditions are contra-indicated. In these cases, a high Reynolds number may only indicate that viscous drag effects due to the vessel walls are insignificant relative to inertial forces and do not influence the flow profile through the decanter.

Decanter design methods based on Stokes' law generally assume a minimum design droplet size of 150 μm, and this appears to be a reasonably conservative value for many chemical process applications. However, the size of drops created in a process is highly dependent on a number of factors such as agitation and turbulence in the liquid, and in particular on the interfacial tension of the liquid-liquid system at process temperatures. The general rule of a minimum design drop size of 150 μm is probably adequate for systems with interfacial tensions of 10 dyn/cm or more as in type I and II systems in Table 15-24, but at lower interfacial tensions a smaller drop size specification may be needed because of the ease of formation of secondary dispersions. For separating secondary dispersions, it is common to assume a design drop size of no more than 100 μm with drop sizes less than 10 μm often being typical [Speth, H., et al., *Sep. Purif. Technol.* **29:** 113–119 (2002) and Clayfield, E. J., et al., *J. Coll. Int. Sci.* **104**(2): 500–511 (1985)]. For more detailed discussion, see S. Hartland and S. A. K. Jeelani, chap. 13 in *Liquid-Liquid Extraction Equipment*, ed. J. C. Godfrey and M. J. Slater, Wiley, New York, 1994, pp. 509–516.

The method just described neglects any reduction in settling velocity due to the presence of neighboring drops at high population density (hindered settling). For best results, experimental data showing the relationship between settling velocity and initial dispersed-phase holdup should be generated. A simplified expression that neglects any drop coalescence during settling may be suitable for approximate design purposes

$$u_t \approx u_{t\infty}(1-\phi_o)\frac{\mu_c}{\mu_d} \qquad (15\text{-}169)$$

where u_t is an average settling velocity used to specify the decanter design, $u_{t\infty}$ is the velocity of an isolated drop calculated from Eq. (15-167), and ϕ_o is the initial holdup. For more detailed discussion, see M. Ishii and N. Zuber, *AIChE J.* **25:** 843–855 (1979); and P. K. Das, *Chem. Eng. Technol.* **20:** 475–477 (1997).

Design Methods for Systems with Slow Coalescence For slow-coalescing systems, simple Stokes' law calculations will not provide a reliable design. Instead, it is necessary to understand the height of the dispersion band as a function of throughput. S. A. K. Jeelani and S. Hartland [*AIChE J.* **31:** 711–720 (1985)] recommend correlating decanter performance by using an expression of the form

$$\frac{1}{Q/A} = \frac{1}{k_1 \Delta H} + \frac{1}{k_2} \qquad (15\text{-}170)$$

where ΔH is an average steady-state dispersion band height, Q is total volumetric throughput, and k_1 and k_2 are empirical constants. The general relationship between ΔH and Q/A also may be expressed in terms of a power law equation of the form

$$\Delta H \propto \left(\frac{Q}{A}\right)^a \propto \left(\frac{Q_c}{A}\right)^a \propto \left(\frac{Q_d}{A}\right)^a \qquad (15\text{-}171)$$

Equations (15-170) and (15-171) represent decanter performance for a given feed with constant properties, that is, a constant composition and phase ratio. Note that the analysis can be done in terms of total flow Q or the flow of continuous phase Q_c or dispersed phase Q_d. Typically, the value of the exponent a is greater than 2.5 [Barnea, E., and J. Mizrahi, *Trans. Inst. Chem. Eng.* **53:** 61–91 (1975); and Golob, J., and R. Modic, *Trans. Inst. Chem. Eng.* **55:** 207–211 (1977)]. The required size of a commercial-scale decanter may be determined by operating a small miniplant decanter to obtain values for the constants in Eqs. (15-170) and (15-171). Scale-up to the larger size generally follows the same relationship as long as the phase ratio and other operating variables are maintained constant. A commercial-scale decanter normally is designed for a throughput Q/A that yields a value of ΔH no larger than 15 percent of the total decanter height. Designs specifying taller dispersion bands are avoided because a sudden change in feed rate can trigger a dramatic increase in the height of the dispersion band that quickly floods the vessel. The dynamic response of ΔH has been studied by S. A. K. Jeelani and S. Hartland [*AIChE J.* **34**(2): 335–340 (1988)].

In certain cases, batch experiments may be used to size a continuous decanter [Jeelani, S. A. K., and S. Hartland, *AIChE J.* **31:** 711–720 (1985)]. In a batch experiment similar to the simple shake test described earlier, the change in the height of the dispersion band with time may follow a relationship given by

$$\frac{1}{-dh/dt} = \frac{1}{k_1 h} + \frac{1}{k_2} \qquad (15\text{-}172)$$

where h is the height of the batch dispersion band varying with time t. The constants k_1 and k_2 in Eq. (15-172) are the same as those used in the steady-state equation [Eq. (15-170)], assuming the batch test conditions (phase ratio and turbulence) are the same. Jeelani and Hartland have derived a number of models for systems with different coalescence behaviors [Jeelani, S. A. K., and S. Hartland, *Chem. Eng. Sci.* **42**(8): 1927–1938 (1987)]. The most appropriate coalescence model is determined in batch tests and then is used to estimate ΔH versus throughput Q/A for a continuous decanter. For additional information, see S. Hartland and S. A. K. Jeelani, chap. 13 in *Liquid-Liquid Extraction Equipment*, ed. J. C. Godfrey and M. J. Slater, Wiley, New York, 1994; C. Nadiv and R. Semiat, *Ind. Eng. Chem. Res.* **34**(7): 2427–2435 (1995); S. A. K. Jeelani and S. Hartland, *Ind. Eng. Chem. Res.* **37**(2): 547–554 (1998); S. A. K. Jeelani, K. Panoussopoulos, and S. Hartland, *Ind. Eng. Chem. Res.* **38**(2): 493–501 (1999); and G.-Z. Yu and Z.-S. Mao, *Chem. Eng. Technol.* **27**(4): 407–413 (2004). Development of design methods for specifying continuous decanters with coalescing internals using batch test data is a current area of research [Mungma, N., P. Chuttrakul, and A. Pfennig, *Jurnal Teknologi* **67**(4): 55–58 (2014)].

Several authors have derived correlations relating the height of the dispersion band to the density of each phase, the density difference, the viscosities, and the interfacial tension of aqueous/organic or aqueous/aqueous two-phase systems [Golob, J., and R. Modic, *Trans. Inst. Chem. Eng.* **55:** 207–211 (1977); and J. A. Asenjo et al., *Biotech. and Bioeng.* **79**(2): 217–223 (2002)]. These correlations can provide useful estimates, but the results are generally valid only for the systems similar to those used to develop the correlations, and they should be used with caution. For new applications, some experimental work will be needed for reliable design.

OTHER TYPES OF SEPARATORS

Packed Coalescers As noted earlier, adding coalescing internals to a decanter can improve decanter performance by promoting the growth of small drops. The same concept can be applied in a separate coalescer vessel to treat the stream feeding the decanter. Systems of type III or type IV (Table 15-24) in particular may benefit, that is, applications involving a need to break a secondary dispersion. Coalescers can also be effective in promoting phase separation and collection when the concentration of the dispersed phase is very low (<150 mg/L).

Coalescers typically are packed with a granular material, corrugated lamellar sheets, a mesh material (made of metal wire, fiberglass, or polymer filaments), or fine fibers of various types in woven or nonwoven composite sheets. In vertical coalescers, the typical flow configuration is upflow if the light phase is dispersed and downflow if the heavy phase is dispersed. Coalescers containing fairly large media such as beds of granules or wire mesh may be able to tolerate a feed containing some fine solids, although occasional backwashing or sparging of the media may be required to remove solids buildup and maintain coalescer pressure drop and efficiency. Coalescers containing fine granules, fine fibers, or porous membranes generally require that the feed be free of solids to avoid plugging, so prefiltration may be necessary. For more detailed information, see J. Li and Y. Gu, *Sep. Purif. Technol.* **42:** 1–13 (2005); C. Shin and G. G. Chase, *AIChE J.* **50**(2): 343–350 (2004); T. H. Wines and R. L. Brown, *Chem. Eng. Magazine* **104**(12): 104–109 (1997); and P. M. Hennessey et al., *Hydrocarbon Proc.* **74:** 107–124 (1995).

In most applications, the packing material should be wetted by the dispersed phase to some degree for best performance; however, this will depend on the size of dispersed droplets. For very fine droplets on the order of 10 μm or smaller, surface wetting is not the primary coalescence mechanism [Davies, G. A., and G. V. Jeffreys, *Filtration and Separation*, pp. 349–354 (July/August 1969)]. In these cases, the packing promotes coalescence by providing a tortuous path that holds dispersed drops in close contact, facilitating drop-drop collisions. In other cases involving larger drops, a drop interception and wettability mechanism becomes important; that is, the media provide a target for drop-solid surface collisions, and the surface becomes wetted with drops that merge together and leave the media as larger drops. In this case, an intermediate (optimum) wettability may be needed to most effectively promote the growth and dislodging of drops from the media [Shin, C., and G. G. Chase, *AIChE J.* **50**(2): 343–350 (2004)]. In general, the degree to which flow path/collision mechanisms and/or surface wettability are important for good performance depends on the drop size distribution and dispersed-phase holdup in the feed, as well as system physical properties and whether surfactants or fine particulates are present. (See the subsection Stability of Liquid-Liquid Dispersions under Liquid-Liquid Dispersion Fundamentals.) All this affects the choice of media, media size and porosity, and coalescer dimensions as a function of throughput. For a given application, some experimental work generally will be needed to sort this out and identify an effective and reliable design.

In cases where wettability is important, various types of sand, zeolites, glass fibers, and other inorganic materials may be used to facilitate coalescence of aqueous drops dispersed in organic feeds. Carbon granules, polymer beads, or polymer fibers may be useful in coalescing organic drops dispersed in water. Sparse oil dispersions in water may be effectively treated with resin bead beds at high removal efficiencies [R. T. O'Connell, Technical

Reprints, T-237 (1981) and T-246 (1983), Graver Water Systems]. The packing material should resist disarming by impurities, meaning that impurities should not become adsorbed and degrade the surface wettability characteristics over time. This can happen with charged or surfactant-like impurities; S. Paria and P. K. Yuet [*Ind. Eng. Chem. Res.* **45**(2): 712–718 (2006)] describe the adsorption of cationic surfactants at sand–water interfaces, a phenomenon that can alter surface wettability. In a few cases, the packing needs to age in service to develop its most effective surface properties. J. R. Madia et al. [*Env. Sci. Technol.* **10**(10): 1044–1046 (1976)] describe a chromatography method for screening potential media with regard to surface wettability.

For granular bed coalescers, typical granule sizes include 12 × 16 Tyler screen mesh (between 1.4 and 1 mm) and 24 × 48 Tyler mesh (0.7 to 0.3 mm). Smaller sizes sometimes are used as well. Typical bed heights range from 8 in to 4 ft (0.2 to 1.2 m), with the taller beds used with the larger granules. Layered beds may be used. For example, the front of the coalescer may contain a thin layer of fine media with low porosity and high tortuosity characteristics to facilitate drop–drop collisions of very small droplets, followed by a layer of coarser media having the wetting characteristics needed to further grow and shed larger drops.

For fine-fiber coalescers, the coalescing media normally are arranged in the form of a filter cartridge. T. H. Wines and R. L. Brown [*Chem. Eng. Magazine* **104**(12): 104–109 (1997)] describe a coalescing mechanism in which a drop (on the order of 0.2 to 50 μm) becomes adsorbed onto a fiber and then moves along the fiber with the bulk liquid flow until colliding with another adsorbed drop at the intersection where two fibers cross. Fiber diameter and wettability are important properties because they affect porosity (tortuous path) and wettable surface area. Like a packed-bed coalescer, a filter-type coalescer may be constructed in layers: an initial prefilter zone to remove particulates and minimize fouling, a primary coalescence zone where small droplets grow to larger ones, and a secondary coalescence zone with greater porosity and having surface-wetting characteristics optimized to grow the larger drops. New experimental materials having specific surface roughness characteristics that affect wettability and facilitate oil–water separation have been reported [Kota, A. K., et al., *Nat. Commun.* **3**(8): 1025 (2012); and Kwon, G., et al., *MRS Commun.* **5**(3): 475–494 (2015)].

Pressure drop, an important consideration in the design of any coalescer, depends upon media size and shape, bed height or filter thickness, and throughput. Methods for calculating pressure drop through packed beds and porous media are described in Sec. 6. For approximately spherical media, the pressure drop due to frictional losses, assuming incompressible media, may be estimated from

$$\frac{\Delta P}{L} = \frac{150(1-\varphi)^2 \mu V}{d_m^2 \varphi^3} + \frac{1.75 \rho_c V^2}{d_m \varphi^3} \qquad \mathrm{Re}_{\mathrm{particle}} = \frac{V \rho_c d_m}{\mu} \leq 10 \qquad (15\text{-}173)$$

where L is the length of the packed section, V is the superficial velocity of the total liquid flow, d_m is an equivalent spherical diameter of the media particles (given by six times the mean ratio of particle volume to particle surface area), and φ is the volume fraction of voids (flow channels) within the bed [Ergun, S., *Chem. Eng. Prog.* **48**(2): 89–94 (1952)]. Also see M. Leva, *Chem. Eng. Magazine* **56**(5): 115–117 (1949), or M. Leva, *Fluidization*, McGraw-Hill, New York, 1959. The minimum value of φ for a tightly ordered bed of uniform spherical particles is 0.26, but of course for real media this will vary depending upon the particle size distribution and particle shape. The second term in Eq. (15-173) often is neglected at $\mathrm{Re}_{\mathrm{particle}} \leq 1$. For fiber media, d_m can be thought of as a characteristic fiber dimension. For a discussion of pressure drop through fiber beds, see C. Shin and G. G. Chase, *AIChE J.* **50**(2): 343–350 (2004); and J. Li and Y. Gu, *Sep. and Purif. Tech.* **42:** 1–13 (2005). In practice, pressure drop data may be correlated by using an equation of the same form as Eq. (15-173), $\Delta P/L = aV + bV^2$, where a and b are empirically determined constants. Media and equipment suppliers generally will have some experimental data showing $\Delta P/L$ versus flow rate.

Membrane-Based Coalescers Membranes used for liquid-liquid phase separation generally are classified as microfiltration membranes (pore sizes greater than about 0.1 μm) or ultrafiltration membranes (pore sizes generally in the range of 0.01 to 0.1 μm). Most membrane-based coalescers described in the literature function by size exclusion. Microporous membranes are used for initial pretreatment of process water to remove relatively large dispersed oil drops. Ultrafiltration membranes may be used to concentrate the remaining microdroplets or micelles. In typical applications, the majority of the continuous phase flows through the pores of the membrane by pressure difference and collects on the permeate side as a clarified solution. The microdroplets or micelles are rejected and flow with the remaining continuous phase (the retentate), tangentially along the membrane surface to the retentate outlet of the membrane module [Voges, Wu, and Dalan, *Chem. Process.* **63**(4): 40–43 (2001)]. Ultrafiltration membranes can be very efficient at removing colloidal particles of an emulsion but normally will not stop dissolved oil from permeating. In some applications, fouling can occur owing to deposition of a coalesced layer that obstructs the pores. For applications involving oil-in-water dispersions and permeation of water, the rate of fouling by oil deposition may be reduced by using a hydrophilic membrane material or hydrophilic surface treatments [Munirasu, S., M. A. Haija, and F. Banat, *Proc. Saf. Environ.* **100:** 183–202 (2016); and Moslehyani, A., et al., *Jurnal Teknologi* **78**(1): 217–222 (2016)].

Selecting the membrane best suited for a given application requires laboratory testing. The membrane material must be compatible with the feed, and the module should exhibit high permeation flow while maintaining good micelle rejection. The pore size and the molecular weight cutoff reported by the manufacturer are good indicators of membrane performance potential, but other factors such as membrane/solute interaction and fouling ultimately determine practical feasibility. Key operating parameters include temperature, feed flow rate, and permeate-to-feed ratio. Scale-up consists of adding membrane modules to handle the required production rate [Eykamp, W., and J. Steen, chap. 18 in *Handbook of Separation Process Technology*, ed. R. W. Rousseau, Wiley, New York, 1987].

New experimental membranes with specific surface roughness and porosity that affect wettability and facilitate oil–water separation have been reported [Kota, A. K., et al., *Nat. Commun.* **3**(8): 1025 (2012); and Kwon, G., et al., *MRS Commun.* **5**(3): 475–494 (2015)]. These materials can be made to exhibit a variety of surface properties, including superhydrophilic/superoleophobic behavior. The authors describe processes employing both hydrophilic/oleophobic and hydrophobic porous membranes, with the separation of oil and water being driven by a difference in the capillary forces acting on the two liquid phases.

Another approach to clarifying an oil-in-water dispersion involves the use of a hydrophobic microporous membrane with intentional permeation of the oil phase through the membrane [Mercelat, A. Y. J., "Fundamental Study of Hydrophobic Microporous Membrane Contactors for the Recovery of Insoluble Oil from Oil–Water Mixtures," PhD dissertation, University of Texas at Austin, 2016]. This technology is an outcome of a program to improve the recovery of submicron oil droplets (<0.5 μm) from lysed algae slurries [Seibert, A. F., and M. Poenie, U.S. Patent appl. 2011/0174734 A1 (2011)]. With this approach employing a hydrophobic membrane, oil droplets dispersed in a continuous aqueous phase are fed on the shell side of a hollow-fiber microporous membrane contactor. The oil droplets coalesce on the outer surface of the hydrophobic hollow fibers to form an oil film. As a result, water is retained on the shell side while oil permeates the membrane and exits through the tube side. The rate of oil permeation depends on the transmembrane pressure differential and the amount of oil entering in the feed [Mercelat, ibid.].

Centrifuges A stacked-disk centrifuge or other type of centrifuge may be a cost-effective option for liquid-liquid phase separation whenever the use of a gravity decanter/coalescer proves to be impractical because rates of drop settling or coalescence are too low. This may be the case for type III and type IV systems (Table 15-24) in particular. Factors involved in specifying a centrifuge are discussed in the subsection Centrifugal Extractors under Liquid-Liquid Extraction Equipment.

Hydrocyclones Liquid-liquid hydrocyclones, like centrifuges, utilize centrifugal force to facilitate the separation of two liquid phases [Svarovsky, L., and M. T. Thew, eds., *Hydrocyclones: Analysis and Applications*, Kluwer, Dordrecht, NL, 1992; Svarovsky, L., *Hydrocyclones*, Holt, Rinehart, and Winston, New York, 1984; and Bradley, D., *The Hydrocyclone*, Pergamon, Oxford, UK, 1965]. Instead of using rotating internals, as in a centrifuge, a hydrocyclone generates centrifugal force using fluid pressure to create rotational fluid motion (Fig. 15-64). Feed enters the hydrocyclone through a tangential-entry nozzle. A primary vortex rich in the heavy phase forms along the inner wall, and a secondary vortex rich in the light phase forms near the centerline. The underflow stream (heavy phase) exits the cyclone through the apex of the cone (underflow nozzle). The overflow stream (light phase) exits through the vortex finder, a tube extending from the cylinder roof into the interior. The feed split can be adjusted by changing the relative diameters of the vortex finder and underflow nozzle. A hydrocyclone is not completely filled with liquid; an air core exists at the centerline. A commercial-scale hydrocyclone multiplies the force of gravity by a factor of 100 to 1000 or so, depending on the diameter and operating pressure. Hydrocyclones traditionally have been used for liquid-solid separations, but by adjusting their design (cone angle and length, vortex finder length, and so on) they can be applied to liquid-liquid separations [R. Mozley, *Filtration and Sep.*, pp. 474–477 (Nov./Dec. 1983)]. For example, they are widely used to remove oil from oily process water [D. Hadjiev et al., chap. 4 in *Environmental Technologies and Trends*, Springer, Berlin, 1997].

Identifying the optimal hydrocyclone geometry for a given application involves the use of flow calculations as a guide, with some experimental testing needed to refine the design [C. Gomez et al., *Soc. Petrol. Eng. J.* **7**(4): 353–361 (2002); J. Caldentey et al., *Soc. Petrol. Eng. J.* **7**(4): 362–372 (2002); and

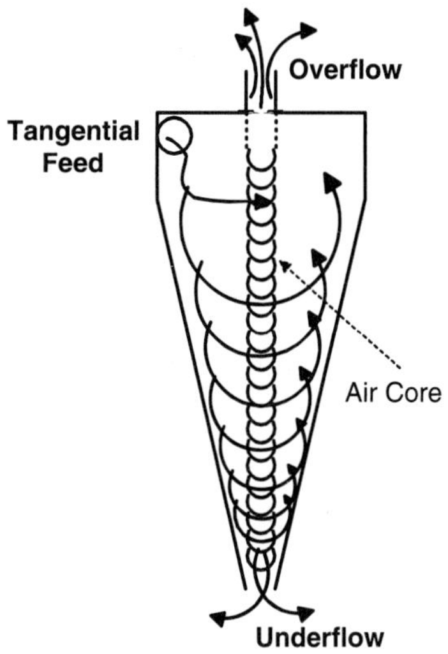

FIG. 15-64 Flow patterns in a hydrocyclone.

D. A. Colman and M. T. Thew, *Chem. Eng. Res. Des.* **61**(7): 233–240 (1983)]. The efficiency of separation will depend upon fluid viscosity, density difference, the inlet dispersed-phase drop size distribution, and interfacial tension. Because fluid flow generally is turbulent at the top of a hydrocyclone and the rotation of liquid within the device produces a high shear field, mixtures with low interfacial tension may tend to emulsify or create foam. Choosing a material of construction for the cone that is wetted by the heavy phase may improve the effectiveness of the device. In testing a specific geometry, the main operating variables are the feed pressure, the feed flow rate, and the split ratio; that is, the relative amounts of fluid exiting top and bottom, with adjustments made by changing the sizes of orifices in the underflow and overflow nozzles. If testing indicates satisfactory performance, hydrocyclones can be relatively inexpensive and simple-to-operate units (no moving parts). Because sufficient centrifugal force cannot be generated in large-diameter units, scale-up consists of connecting multiple small units in parallel. Units are sometimes placed in series to provide multiple stages of separation. Numerical simulations of hydrocyclone performance and flow profiles are described by Z. S. Bai and H.-L. Wang [*Chem. Eng. Technol.* **29**(10): 1161–1166 (2006)] and by S. Murphy et al. [*Chem. Eng. Sci.* **62:** 1619–1635 (2007)].

Electrotreaters In an electrostatic coalescer, an electric field is applied to a dispersion to induce dipoles or net charges on the suspended drops, which facilitates their migration and coalescence [Waterman, L. C., *Chem. Eng. Prog.* **61**(10): 51–57 (1965); and Yamaguchi, M., chap. 16 in *Liquid-Liquid Extraction Equipment*, ed. J. C. Godfrey and M. J. Slater, Wiley, New York, 1994]. This technology is applicable only to a nonconductive continuous phase and an aqueous dispersed phase. Once the water drops are sufficiently large, they settle to the bottom of the vessel while the clarified oil phase migrates to the top. The top and bottom zones are kept quiet and out of the electric field. In cases where inlet salt content is high, a multistage, countercurrent desalting system can be used. Units with ac or dc voltage are available.

Electrotreaters are high-voltage electrostatic devices that can arc under certain conditions. For this reason, a careful review of safety considerations is needed, especially for applications involving flammable liquids. Evaluating feasibility and generating design data normally involve close consultation with the equipment vendors. This technology is applied on a very large scale in the petroleum industry for desalting of crude oil.

EMERGING DEVELOPMENTS

MEMBRANE-BASED PROCESSES

Polymer Membranes Extraction processes employing polymer membranes are sometimes referred to as nondispersive or pertraction operations. The use of membranes in extraction offers a number of potential advantages including (1) constant well-defined mass-transfer area; (2) the ability to operate at very low solvent-to-feed ratios independent of other operating variables; (3) very low holdup of solvent and product within the extractor, thus providing low residence time similar to a centrifugal extractor; (4) dispersion-free liquid-liquid contacting that eliminates the need for liquid-liquid interface control and phase separation; (5) elimination of entrainment; (6) no requirement for a difference in density between liquid phases; and (7) linear scale-up by addition of extra modules in parallel, so performance at large scale can be determined directly from small-scale tests using a single module. This last point suggests, however, that the economy of scale may not be as large as it is for extractors that are scaled up as a single larger unit.

The most important advantages that membranes can offer to the process designer are those that overcome an inherent limitation of another type of extractor, as in the ability to handle liquids with close or even equal densities and the ability to operate at extremely low solvent-to-feed ratios. The types of applications where membrane extraction is likely to be most attractive include solids-free (or pre-filtered) feeds with moderate to high interfacial tension (to avoid breakthrough) and a solute partition ratio greater than 50 or so. In principle, $K > 50$ allows operation using a solvent-to-feed ratio of 1:25 or less (for an extraction factor of 2). The primary disadvantages of membrane-based extractors are the added mass-transfer resistance across the membrane, limited fiber-side or tube-side throughput, and concerns about plugging and fouling from insoluble solids and limited membrane life in industrial service. The useful life of a membrane module is a critical factor, as the frequency with which membrane modules must be replaced has a dramatic impact on overall cost.

For most liquid-liquid extraction applications, a porous membrane is used and extraction involves transfer through a liquid-liquid interface immobilized within the pores. One of the most promising contactors for this type of extraction is the microporous hollow-fiber (MHF) contactor (Fig. 15-65). The MHF contactor resembles a shell-and-tube heat exchanger in which the tube walls are porous and are capable of immobilizing a liquid-liquid interface within the pores. For a hydrophobic polymeric membrane which is preferentially wetted by the hydrocarbon, the hydrocarbon solvent phase usually is fed to the interior of the fiber (the fiber-bore side), while the aqueous carrier phase is fed to the shell side. In this configuration, the aqueous fluid is maintained at a higher pressure relative to the hydrocarbon phase, to immobilize the liquid-liquid interface within each pore. Care must be taken to avoid using excessive shell-side pressures, or breakthrough of the aqueous phase can occur. This breakthrough pressure is a function of the interfacial tension and pore size. Earlier versions of MHF contactors provided a parallel-flow design, but this design suffered from shell-side bypassing [Seibert, A. F., et al., *Sep. Sci. Technol.* **28**(1–3): 343 (1993)]. An improved design eliminates bypassing by incorporating a central baffle, shell-side cross-flow distribution, and uniform fiber spacing (Fig. 15-65 and Table 15-25).

As in conventional extraction, the mass transfer of solute occurs across a liquid-liquid interface. However, unlike in conventional extraction, the interface is maintained in submicron pores, and three mass-transfer resistances are present: tube-side (k_t), shell-side (k_s), and pore or membrane-side (k_m). The overall mass-transfer coefficient k_{os} based on the aqueous phase (shell side) and a hydrophobic membrane with pores filled by the hydrocarbon phase is given by

$$\frac{1}{k_{os}} = \frac{1}{m^{\text{vol}} k_t} + \frac{1}{m^{\text{vol}} k_m} + \frac{1}{k_s} \tag{15-174}$$

where m^{vol} is the local slope of the equilibrium line for the solute of interest, with the equilibrium concentration of solute in the tube-side liquid plotted on the y axis and the equilibrium concentration of solute in the shell-side liquid plotted on the x axis. Equation (15-174) assumes the tube-side fluid wets the pores.

The mass-transfer efficiencies of various MHF contactors have been studied by many researchers. L. Dahuron and E. L. Cussler [*AIChE J.* **34**(1): 130–136 (1988)] developed a membrane mass-transfer coefficient model (k_m); M.-C. Yang and E. L. Cussler [*AIChE J.* **32**(11): 1910–1916 (1986)] developed a shell-side mass-transfer coefficient model (k_s) for flow directed radially into the fibers; and R. Prasad and K. K. Sirkar [*AIChE J.* **34**(2): 177–188 (1988)] developed a tube-side mass-transfer coefficient model (k_t). Additional studies have been published by R. Prasad and K. K. Sirkar ["Membrane-Based Solvent Extraction," in *Membrane Handbook*, ed. W. S. W. Ho and K. K. Sirkar, Chapman and Hall, London, 1992]; by B. W. Reed, M. J. Semmens, and

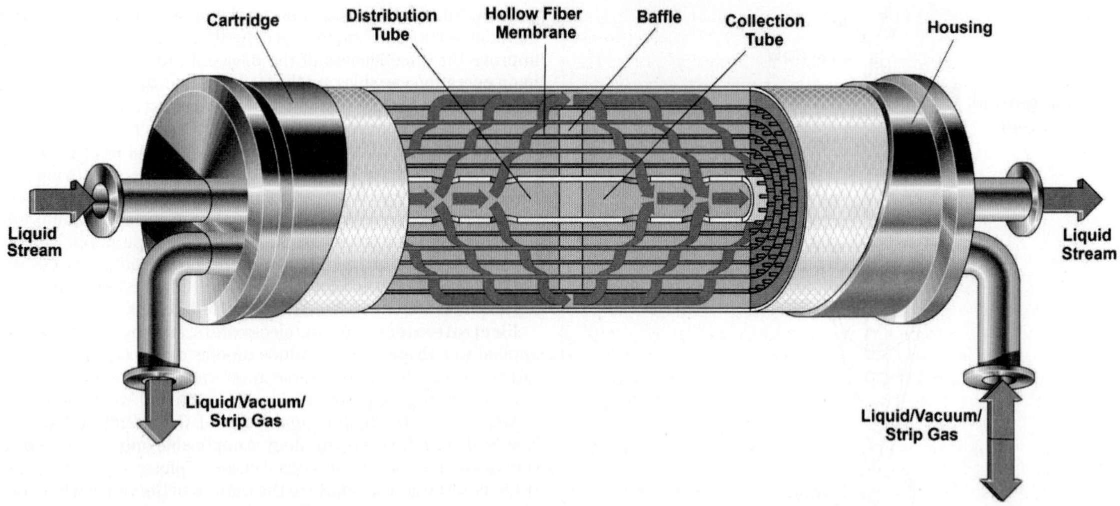

FIG. 15-65 Schematic of the 3M™ Liqui-Cel™ Membrane Contactor. (*Courtesy of 3M. 3M and Liqui-Cel are trademarks of 3M Company.*)

E. L. Cussler ["Membrane Contactors," *Membrane Separations Technology: Principles and Applications*, ed. R. D. Noble and S. A. Stern, Elsevier, Amsterdam, 1995]; by Y. Qin and J. M. S. Cabral [*AIChE J.* **43**(8): 1975–1988 (1997)]; by A. Baudot, J. Floury, and H. E. Smorenburg [*AIChE J.* **47**(8): 1780–1793 (2001)]; by M. J. González-Muñoz et al. [*J. Membrane Sci.* **213**(1–2): 181–193 (2003) and *J. Membrane Sci.* **255**(1–2): 133–140 (2005)]; by B. Saikia, N. N. Dutta, and N. N. Dass [*J. Membrane Sci.* **225**(1–2): 1–13 (2003)]; by S. Bocquet et al. [*AIChE J.* **51**(4): 1067–1079 (2005)]; and by S. Schlosser, R. Kertesz, and J. Martak [*Sep. Purif. Technol.* **41**: 237 (2005)]. For an overall review of membrane-based extraction processes, see A. K. Pabby and A. M. Sastre [*J. Membrane Sci.* **430**: 263–303 (2013)]. A review of mass-transfer correlations for hollow-fiber membrane modules is given by T. Liang and R. L. Long [*Ind. Eng. Chem. Res.* **44**(20): 7835–7843 (2005)]. T. Eksangsri, H. Habaki, and J. Kawasaki [*Sep. Purif. Technol.* **46**: 63–71 (2005)] discuss the effect of hydrophobic versus hydrophilic membranes for a specific application involving transfer of solute from an aqueous feed to an organic solvent. A. J. Karabelas and A. G. Asimakopoulou [*J. Membrane Sci.* **272**(1–2): 78–92 (2006)] discuss process and equipment design considerations. A 650 kg/h commercial application of the hollow-fiber contactor is described by W. Ratajczak et al. for extraction of phenol from hydrocarbons using aqueous alkali solution [*Przemysł Chemiczny* **86**: 262 (2007); and Š. Schlosser and J. Marták, *Membrany - Teoria i Praktyka - Zeszyt III*, pp. 123–153, Toruń, Poland, 2009, in English, download from researchgate.net, April 2016].

In general, researchers have treated MHF contactors as differential contacting devices. However, A. F. Seibert and J. R. Fair [*Sep. Sci. Technol.* **32**(1–4): 573–583 (1997)] and A. F. Seibert et al. [*ISEC '96 Proc.* **2**: 1137 (1996)] suggest that the baffled MHF contactor can be treated as a staged countercurrent contactor. Their recommendations are based on studies using a commercial-scale membrane extraction system. Also see the discussion by H. M. Yeh [*J. Membrane Sci.* **269**(1–2): 133–141 (2006)] regarding the use of internal reflux in a parallel plate membrane module to boost liquid velocities for enhanced performance.

Liquid Membranes For reviews of this subject, see *Liquid Membranes: Principles and Applications in Chemical Separations and Wastewater Treatment*, ed. V. S. Kislik, Elsevier, Amsterdam, 2010; *Solvent Extraction and Liquid Membranes*, ed. M. Aguilar and J. L. Cortina, CRC Press, Boca Raton, Fla., 2008; M. F. San Román, et al., *J. Chem. Technol. Biotechnol.* **85**(1): 2–10 (2010); and R. D. Noble and J. D. Way, eds., *Liquid Membranes: Theory and Applications*, Symposium Series No. 347, American Chemical Society, Washington, DC, 1987.

TABLE 15-25 Baffled MHF Contactor Geometric Characteristics

Baffles per module	1
Module diameter, cm	9.8
Module length, cm	71
Effective fiber length, cm	63.5
Fiber outside diameter, μm	300
Fiber inside diameter, μm	240
Porosity of fiber	0.3
Number of fibers per module	30,000
Contact area per module, cm^2	81,830
Interfacial area, cm^2/cm^3	27
Tortuosity	2.6

Reprinted from Seibert and Fair, *Sep. Sci. Technol.*, **32**(1–4), pp. 573–583 (1997), with permission. Copyright 1997 Taylor & Francis.

Emulsion liquid-membrane (ELM) extraction involves the intentional formation of an emulsion between two immiscible liquid phases followed by suspension of the emulsion in a third liquid that forms an outer continuous phase. The encapsulated liquid and the continuous phase are miscible. The liquid-membrane phase is immiscible with the other phases and normally must be stabilized by using surfactants. If the continuous phase is aqueous, the suspended phase is a water-in-oil emulsion. If the continuous phase is organic, the emulsion is the oil-in-water type. This technology differs from traditional liquid-liquid extraction processes in that it allows transfer of solute between miscible liquids by introducing an immiscible liquid membrane between them. A typical process involves first forming a stable emulsion and contacting it with the continuous phase to transfer solute between the encapsulated phase and the continuous phase, followed by steps for separating the emulsion and continuous phases and breaking the emulsion. The emulsion must be sufficiently stable to remain intact during processing, but not so stable that it cannot be broken after processing, and this may present a challenge for commercial implementation. The technology is described by J. W. Frankenfeld and N. N. Li [chap. 19 in *Handbook of Separation Process Technology*, ed. R. W. Rousseau, Wiley, New York, 1987].

Potential applications of ELM extraction include separation of aromatic and aliphatic hydrocarbons [M. Chakraborty and H.-J. Bart, *Chem. Eng. Technol.* **28**(12): 1518–1524 (2005)], separation and concentration of amino acids [M. P. Thien, T. A. Hatton, and D. I. C. Wang, *Biotech. and Bioeng.* **32**(5): 604–615 (1988)], and recovery of penicillin G from fermentation broth [K. H. Lee, S. C. Lee, and W. K. Lee, *J. Chem. Technol. Biotechnol.* **59**(4): 365–370, 371–376 (1994); S. C. Lee et al., *J. Membrane Sci.* **124**: 43–51 (1997); and S. C. Lee and S. M. Yeo, *J. Ind. Eng. Chem.* **8**(2): 114–119 (2002)]. The latter application involves transfer of the penicillin G solute ($pK_a = 2.7$) from the continuous phase (consisting of a filtered broth adjusted to a pH of about 3) into the membrane phase (typically *n*-lauryltrialkylmethyl amine extractant dissolved in kerosene) and then into the interior aqueous phase (clean water at a pH of about 8). S. C. Lee et al. [*J. Membrane Sci.* **124**: 43–51 (1997)] show that the operation can be carried out in a continuous countercurrent extraction column. The product is later obtained by separating the emulsion droplets from the continuous phase by using filtration, and this is followed by breaking the emulsion and isolating the interior aqueous phase from the amine extractant phase. A polyamine surfactant is used to stabilize the emulsion during extraction. In another study, J. Laso et al. [*Ind. Eng. Chem. Res.* **54**(12): 3218–3224 (2015)] report laboratory-scale process data for both nondispersive and ELM extraction schemes for liquid-liquid extraction of zinc from spent pickling liquor, both carried out using hollow-fiber membrane contactors.

Supported liquid-membrane (SLM) processes involve the introduction of a microporous solid membrane to serve as a support for the liquid membrane. See the review by N. M. Kocherginsky, Q. Yang, and L. Seelam [*Sep. Purif. Technol.* **53**: 171–177 (2007)]. The microporous membrane provides well-defined interfacial area and eliminates the need for a surfactant. As in the penicillin ELM application described previously, SLM applications often employ an extractant solution as the liquid-membrane phase to enable a facilitated transport mechanism. The extractant species interacts with the desired solute at the feed side and then carries the solute across the membrane to the other side, where solute transfers into a stripping solution. Such a process, whether using a surfactant-stabilized emulsion or a supported liquid membrane, allows forward and back extraction (or stripping) in a single operation. Example applications include the extraction of metal ions from aqueous solution [Ho, W. S. W., and B. Wang, *Ind. Eng. Chem. Res.* **41**(3): 381–388 (2002); L. Canet and P. Seta, *Pure Appl. Chem. (IUPAC)* **73**(12): 2039–2046 (2001)] and the recovery of aromatic acids or bases from an aqueous feed [Dastgir, M. G., et al., *Ind. Eng. Chem. Res.* **44**(20): 7659–7667 (2005)]. One of the challenges encountered in using supported liquid membranes is the difficulty in controlling trans-membrane pressure drop and maintaining the liquid membrane on the support; it may become dislodged and entrained into the flowing phases. Various approaches to stabilizing the supported liquid have been proposed. These are discussed by Dastgir et al. [ibid.].

A related technology termed thin layer extraction involves immobilizing an extractant liquid within a thin microporous solid. Solute is transferred from the feed liquid by repeatedly exposing the supported extractant first to the feed liquid and then to a recipient liquid (stripping solution) [Lavie, R., *AIChE J.* **54**(4): 957–964 (2008); U.S. patent 8,021,554 (2011); and *Ind. Eng. Chem. Res.* **53**(47): 18283–18290 (2014)].

ELECTRICALLY ENHANCED EXTRACTION

An electric field may be used to enhance the performance of an aqueous-organic liquid-liquid contactor by promoting either drop breakup or drop coalescence, depending upon the operating conditions and how the field is applied. The technology normally involves dispersing an electrically conductive phase (the aqueous phase) within a continuous nonconductive phase, applying a high-voltage electric field (either ac or dc) across the continuous phase, and taking advantage of the effect the electric field has on the shape, size, and motion of the dispersed drops. The potential advantages of this technology include more precise control of drop size and motion for improved control of mass transfer and phase separation within an extractor. Potential disadvantages include the requirement for more complex equipment, difficulties in scaling up the technology to handle large production rates, and safety hazards involved in processing flammable liquids in high-voltage equipment.

A number of different equipment configurations and operating concepts have been proposed. M. Yamaguchi [chap. 16 in *Liquid-Liquid Extraction Equipment*, ed. J. C. Godfrey and M. J. Slater, Wiley, New York, 1994] classifies the proposed equipment into three general types: perforated-plate and spray columns, mixed contactors, and liquid-film contactors. For example, M. Yamaguchi and M. Kanno [*AIChE J.* **42**(9): 2683–2686 (1996)] describe an apparatus in which a dc voltage is applied between two electrodes in the presence of a nitrogen gas interface. Aqueous drops form in the presence of the electric field, and they are first attracted to the gas-liquid interface. Once the drops contact the interface, the charge on the drops is reversed, and the drops fall back to coalesce at the bottom of the vessel. P. J. Bailes and E. H. Stitt [U.S. Patent 4,747,921 (1988)] describe a rotating-impeller extraction column containing alternating zones of high voltage (to promote dispersed drop coalescence) and high-intensity mixing (to promote redispersion of drops). In this design, the electric field serves to promote drop coalescence so that dispersed drops experience alternating drop breakup and growth as they move through the agitated column. T. C. Scott and R. M. Wham [*Ind. Eng. Chem. Res.* **28**(1): 94–97 (1989)] and T. C. Scott, D. W. DePaoli, and W. G. Sisson [*Ind. Eng. Chem. Res.* **33**(5): 1237–1244 (1994)] describe a nonagitated apparatus called an emulsion-phase contactor. This device employs an electric field to induce the formation of a stable emulsion or dispersion band, with clear organic and aqueous layers above and below. The aqueous phase is fed to the middle or top of the dispersion band; it flows down through the band and is removed from a clarified aqueous zone maintained at the bottom. The lighter organic phase is fed to the bottom; it moves up through the dispersion band and is removed from the top. The net result is countercurrent contacting with very high interfacial area and significantly improved mass transfer in terms of the number of transfer units achieved for a given contactor height.

Another approach involves electrostatically spraying aqueous solutions into a continuous organic phase to create dispersed drops within a spray column contactor [L. R. Weatherley et al., *J. Chem. Technol. Biotechnol.* **48**(4): 427–438 (1990)]. A high voltage is applied between electrodes, one connected to a nozzle where dispersed drops are formed and the other placed within the continuous organic phase. J. Petera et al. [*Chem. Eng. Sci.* **60:** 135–149 (2005)] discuss the modeling of drop size and motion within such a device. For additional discussion, see C. Tsouris et al. [*Ind. Eng. Chem. Res.* **34**(4): 1394–1403 (1995)], C. Tsouris et al. [*AIChE J.* **40**(11): 1920–1923 (1994)], G. Gneist and H.-J. Bart [*Chem. Eng. Technol.* **25**(2): 129–133, **25**(9): 899–904 (2002)], and T. Elperin and A. Fominykh [*Chem. Eng. Technol.* **29**(4): 507–511 (2006)].

PHASE TRANSITION EXTRACTION AND TUNABLE SOLVENTS

Phase transition extraction (PTE) involves transitioning between single-liquid-phase and two-liquid-phase states to facilitate a desired separation. A. Ullmann, Z. Ludmer, and R. Shinnar [*AIChE J.* **41**(3): 488–500 (1995)] showed that extraction of an antibiotic from fermentation broth into an organic solvent could be improved by transitioning across a UCST phase boundary using heating and cooling. The results showed much higher stage efficiency compared to a standard extraction technique without phase transition and much faster phase separation. The phase transition may be induced by a change in temperature or a change in composition through addition and/or removal of organic solvents or antisolvents [R. Gupta, R. Mauri, and R. Shinnar, *Ind. Eng. Chem. Res.* **35**(7): 2360–2368 (1996)]. N. Alizadeh and K. Ashtari describe a temperature-change-induced phase transition process for extracting silver(I) from aqueous solution using dinitrile solvents [*Sep. Purif. Technol.* **44:** 79–84 (2005)]. Another process that exploits a phase transition to facilitate separation and recycle of solvent after extraction utilizes ethylene oxide–propylene oxide copolymers in aqueous two-phase extraction of proteins [J. Persson et al., *J. Chem. Technol. Biotechnol.* **74:** 238–243 (1999)]. After extraction, the polymer-rich extract phase is heated above its LCST to form two layers: an aqueous layer containing the majority of protein and a polymer-rich layer that can be decanted and recycled to the extraction. An example involving use of glycol ether solvent is given by J. R. Allen et al. [*Biotechnol. Prog.* **23**(5): 1163–1170 (2007)].

Another approach termed organic aqueous tunable solvents (OATS) utilizes pressurized CO_2 to control phase splitting and tune partition ratios in organic + water mixtures [P. Pollet et al., *Acc. Chem. Res.* **43**(9): 1237–1245 (2010); and C. A. Eckert et al., *J. Phys. Chem. B*, **108**(47): 18108–18118 (2004)]. The addition of CO_2 at pressures of 3 to 80 bar sometimes causes phase splitting of a normally miscible organic + water system, yielding an organic phase rich in CO_2 (the gas-expanded phase) and an aqueous phase containing little CO_2. The CO_2 dissolved in the organic phase acts as an antisolvent such that change in the CO_2 pressure causes change in the distribution of solute between the organic and aqueous phases. This gas-expanded-liquid technique is of particular interest for improving reaction-extraction schemes used in multistage processing of pharmaceuticals and other specialty chemicals by enabling a process wherein a desired reaction is carried out in a batch reactor at conditions of complete organic + water miscibility, followed by the addition of CO_2 to the reactor to induce phase splitting for product isolation and recovery/recycle of homogeneous catalyst. After decantation of one or both liquid layers for further processing, the CO_2 can be vented and the cycle repeated.

In related studies, T. Adrian, J. Freitag, and G. Maurer [*Chem. Eng. Technol.* **23**(10): 857–860 (2000)] have demonstrated the ability to induce phase splitting in the completely miscible 1-propanol + water system by pressurization with CO_2 at pressures above 74 bar. The partition ratio for transfer of methyl anthranilate from the aqueous phase to the organic phase varied between 1 and about 13 with change in pressure and temperature. J. Lu et al. [*Ind. Eng. Chem. Res.* **43**(7): 1586–1590 (2004)] demonstrated a reduction in the lower critical solution temperature for the partially miscible THF + water system by the addition of CO_2 at more moderate pressures (on the order of 10 bar). The authors showed that the partition ratio for transfer of a water-soluble dye from the organic phase to the aqueous phase can be increased dramatically by increasing CO_2 pressure.

IONIC LIQUIDS AND DEEP EUTECTIC SOLVENTS

Ionic liquids are low-melting organic salts that are liquids at or near ambient temperature [R. D. Rogers and K. R. Seddon, *Science* **302:** (2003); and M. C. J. Moita et al., *J. Chem. Eng. Data* **57:** 2702–2709 (2012)]. A well-known example is 1-butyl-3-methylimidazolium hexafluorophosphate (commonly denoted by [BMIM][PF_6]). A new class of ionic liquid analogue known as a deep eutectic solvent is formed by mixing a hydrogen bond acceptor compound (such as choline chloride, a quaternary ammonium salt) with a metal salt or a hydrogen bond donor (such as glycerol) to form a low-melting eutectic mixture [E. L. Smith, A. P. Abbott, and K. S. Ryder, *Chem. Rev.* **114**(21): 11060–11082 (2014); and S. P. Verevkin et al., *Ind. Eng. Chem. Res.* **54**(13): 3498–3504 (2015)]. Characteristic physical properties of ionic liquids and their analogues include very low vapor pressure and high viscosity (relative to molecular organic solvents). The solvation, polarity, and miscibility behavior of the liquid can vary widely depending on composition. Both hydrophobic and hydrophilic ionic liquids are available.

The potential use of ionic liquids (and their analogues) for liquid-liquid extraction has gained considerable attention [G. Parkinson, *Chem. Eng. Prog.* **100**(9): 7–9 (2004); *Ionic Liquids IIIB: Fundamentals, Challenges, and Opportunities*, ed. R. D. Rogers and K. R. Seddon, Oxford University Press, Oxford, UK, 2005; S. T. Anjan, *Chem. Eng. Prog.* **102**(12): 30–39 (2006); and X. Han and D. W. Armstrong, *Acc. Chem. Res.* **40**(11): 1079–1086 (2007)]. Potential applications in industrial processing include olefin/paraffin separation [G. J. Belluomini, et al., *Ind. Eng. Chem. Res.* **48**(24): 11168–11174 (2009); and Y. Wang et al., *J. Chem. Eng. Data* **60**(1): 28–36 (2015)], extraction of high-boiling organics from water [D.-Z. Chen et al., *Sep. Purif. Technol.* **104:** 263–267 (2013); and L. Y. Garcia-Chavez, B. Schuur, and A. B. de Haan, *Ind. Eng. Chem. Res.* **52**(13): 4902–4910 (2013)], extraction of aromatics from aliphatic hydrocarbons [M. Larriba et al., *Ind. Eng. Chem. Res.* **52**(7): 2714–2720 (2013)], and extractive desulfurization of liquid fuels [S. A. Dharaskar et al., *Ind. Eng. Chem. Res.* **53**(51): 19845–19854 (2014)]. The use of ionic liquids to extract metal ions from aqueous solution is discussed by A. E. Visser et al. [*Sep. Sci. Technol.* **36**(5–6): 785–804 (2001)] and by K. Nakashima et al. [*Ind. Eng. Chem. Res.* **44**(12): 4368–4372 (2005)].

The possibility of switching a solvent system from ionic to nonionic states also has been investigated [P. G. Jessop et al., *Nature* **436:** 1102 (2005)]. The authors report that a 50:50 blend of 1-hexanol and 1,8-diazabicyclo-[5.4.0]-undec-7-ene (DBU) becomes ionic when CO_2 is bubbled through the solution. The CO_2 reacts to form a mixture of 1-hexylcarbonate anion and $DBUH^+$ cation, a viscous ionic liquid. The reaction can be reversed by using N_2 to strip the weakly bound CO_2 from solution. This returns the solution to its less viscous, nonionic state and provides a basis for a switchable solvent system. In another example, a change in temperature from about 20°C to 80°C is used to switch transfer of solute-carrying polymer micelles between an aqueous phase and [BMIM][PF_6] ionic liquid [Z. Bai and T. P. Lodge, *Langmuir* **26**(11): 8887–8892 (2010)]. The partitioning of micelles, which is reversible, is driven by the lower critical solution temperature (LCST) phase behavior of the polymer in water.

The challenges involved in using ionic liquids for extraction appear similar to those encountered using nonvolatile extractants dissolved in a diluent, including difficulty dealing with buildup of high-boiling impurities in the solvent phase over time. Additionally, solvent stability and recovery will need to be quite high for the process to be economical if the cost of makeup solvent is high. Potential advantages include the possibility of obtaining higher K values, allowing the use of lower solvent-to-feed ratios. Also, very low volatility allows for essentially zero fugitive emissions to the environment and may allow simpler extract and raffinate separations. For example, volatile components may easily be separated from the ionic liquid by using evaporation under vacuum instead of multistage distillation.

MINIATURIZED EXTRACTION

The implementation of liquid-liquid extraction within micro- and milli-scale devices is an active research area [Gürsel, I. V., et al., *Chem. Eng. J.* **283:** 855–868 (2016); Ciceri, D., J. M. Perera, and G. W. Stevens, *J. Chem. Technol. Biotechnol.* **89:** 771–786 (2014); and Assmann, N., A. Ładosz, and P. R. von Rohr, *Chem. Eng. Technol.* **36**(6): 921–936 (2013)]. Such work may be directed at miniaturization of chemical processes to overcome heat and mass transfer limitations inherent in large-scale equipment (process intensification) or to minimize inventories of hazardous intermediates. It also may be directed toward the development of high-throughput research tools for rapid screening of chemistry and chemical processing options, or for the development of inexpensive, portable analytical chemistry devices and sensors. Potential advantages of miniaturization derive from high surface-to-volume ratios, short diffusion path length, and the inherent portability of a small device. Miniaturization also may enable modular design approaches via plug and play of specialized devices. Potential disadvantages include difficulties mixing and separating liquids in very small channels, difficulty handling systems of high interfacial tension, and difficulty realizing reasonably large production rates when chemical production is the goal. Compared to extraction practices employed in traditional large-scale manufacturing operations, the miniaturization approach is inherently limited to low production volumes; however, production may be increased to some extent by scaling up from micro- to milli-scale devices, by continuous operation, and by using multiple devices in parallel. The scalability of such devices is discussed by A. Woitalka, S. Kuhn, and K. F. Jensen [*Chem. Eng. Sci.* **116:** 1–8 (2014)]. The application of surfactants and electric fields to facilitate mass transfer and phase separation is discussed by J. G. Kralj, M. A. Schmidt, and K. F. Jensen [*Lab Chip* **5:** 531–535 (2005)]. Miniaturized extraction columns are described by A. Holbach et al. [*Chem. Eng. Process. Process Intensif.* **80**(1): 21–28 (2014)]; and by C. Schulze et al. [*Chem. Ing. Tech.* **87**(8): 1052 (2015)].

Section 16

Adsorption and Ion Exchange

M. Douglas LeVan, Ph.D. *J. Lawrence Wilson Professor of Engineering Emeritus, Department of Chemical and Biomolecular Engineering, Vanderbilt University; Member, American Institute of Chemical Engineers, American Chemical Society, International Adsorption Society (Section Coeditor)*

Giorgio Carta, Ph.D. *Lawrence R. Quarles Professor, Department of Chemical Engineering, University of Virginia; Member, American Institute of Chemical Engineers, American Chemical Society (Section Coeditor)*

James A. Ritter, Ph.D. *L. M. Weisiger Professor of Engineering and Carolina Distinguished Professor, Department of Chemical Engineering, University of South Carolina; Member, American Institute of Chemical Engineers, American Chemical Society, International Adsorption Society (Sorption Equilibrium, Process Cycles, Equipment)*

Krista S. Walton, Ph.D. *Professor and Robert "Bud" Moeller Faculty Fellow, School of Chemical & Biomolecular Engineering, Georgia Institute of Technology; Member, American Institute of Chemical Engineers, American Chemical Society, International Adsorption Society (Adsorbents)*

DESIGN CONCEPTS

ADSORBENTS AND ION EXCHANGERS

SORPTION EQUILIBRIUM

CONSERVATION EQUATIONS

RATE AND DISPERSION FACTORS

BATCH ADSORPTION

FIXED-BED TRANSITIONS

CHROMATOGRAPHY

PROCESS CYCLES

EQUIPMENT

Nomenclature and Units

a	Specific external surface area per unit bed volume, m^2/m^3
a_v	Surface area per unit particle volume, m^2/m^3 particle
A	Surface area of solid, m^2/kg
A_s	Chromatography peak asymmetry factor (Fig. 16-27)
b	Correction factor for resistances in series (Fig. 16-12)
c	Fluid-phase concentration, mol/m^3 fluid
c_p	Pore fluid-phase concentration, mol/m^3
c^s	Fluid-phase concentration at particle surface, mol/m^3
C°_{pf}	Ideal gas heat capacity, J/(mol·K)
C_s	Heat capacity of sorbent solid, J/(kg·K)
d_p	Particle diameter, m
D	Fluid-phase diffusion coefficient, m^2/s
D_{app}	Apparent diffusion coefficient, m^2/s [Eq. (16-76)]
D_L	Axial dispersion coefficient, m^2/s [Eq. (16-79)]
D_p	Pore diffusion coefficient, m^2/s [Eqs. (16-64), (16-66), (16-67), (16-69)]
D_s	Adsorbed-phase (solid, surface, particle, or micropore) diffusion coefficient, m^2/s [Eqs. (16-70), (16-71)]
D_0	Diffusion coefficient corrected for thermodynamic driving force, m^2/s [Eq. (16-71)]
$\bar{D}$	Ionic self-diffusion coefficient, m^2/s [Eqs. (16-73), (16-74)]
F	Fractional approach to equilibrium
F_v	Volumetric flow rate, m^3/s
h	Enthalpy, J/mol; reduced height equivalent to theoretical plate [Eq. (16-183)]
htu	Reduced height equivalent to a transfer unit [Fig. (16-13)]
HETP	Height equivalent to a theoretical plate, m [Eq. (16-158)]
HTU	Height equivalent to a transfer unit, m [Eq. (16-92)]
J	Mass-transfer flux relative to molar average velocity, $mol/(m^2 \cdot s)$; *J* function [Eq. (16-148)]
k	Rate coefficient, s^{-1} [Eq. (16-83)]
k_a	Forward rate constant for reaction kinetics, $m^3/(mol \cdot s)$
k_c	Rate coefficient based on fluid-phase concentration driving force, $m^3/(kg \cdot s)$ (Table 16-12)
k_f	External mass-transfer coefficient, m/s [Eq. (16-78)]
k_n	Rate coefficient based on adsorbed-phase concentration driving force, s^{-1} (Table 16-12)
k'	Retention factor [Eq. (16-156)]
K	Isotherm parameter
K^c	Molar selectivity coefficient
K'	Rational selectivity coefficient
L	Bed length, m
m	Isotherm exponent; flow ratio in TMB or SMB systems [Eq. (16-199)]
M_r	Molecular mass, kg/kmol
M_s	Mass of adsorbent, kg
n	Adsorbed-phase concentration, mol/kg adsorbent or mol/m^3 adsorbent
n^s	Monolayer adsorption capacity, mol/kg adsorbent or mol/m^3 adsorbent; ion-exchange capacity, g-equiv/kg or g-equiv/m^3
N	Number of transfer units or number of reaction units; $k_f aL/(\varepsilon v^{ref})$ for external mass transfer; $15(1-\varepsilon)\varepsilon_p D_p L/(\varepsilon v^{ref} r_p^2)$ for pore diffusion; $15\Lambda D_s L/(\varepsilon v^{ref} r_p^2)$ for solid diffusion; $k_n \Lambda L/(\varepsilon v^{ref})$ for linear driving-force approximation; $k_a c_{ref} \Lambda L/[(1-R)\varepsilon v^{ref}]$ for reaction kinetics (Table 16-13)
N_p	Number of theoretical plates [Eq. (16-157)]
N_{Pe}	$v^{ref}L/D_L$, bed Peclet number (number of dispersion units)
p	Partial pressure, Pa; switch time, s
P	Pressure, Pa
P^s	Vapor pressure, Pa
Pe	Particle-based Peclet number, $d_p v/D_L$
Q_i	Amount of component *i* injected with feed, mol
r, R	Separation factor [Eqs. (16-30), (16-32)]; particle radial coordinate, m
r_c	Column internal radius, m
r_m	Stokes-Einstein radius of molecule, m [Eq. (16-68)]
r_p	Particle radius, m
r_{pore}	Pore radius, m
r_s	Radius of subparticles, m
$\mathfrak{R}$	Gas constant, $Pa \cdot m^3/(mol \cdot K)$
Re	Reynolds number based on particle diameter, $\rho_f d_p \varepsilon v/\mu$
Sc	Schmidt number, $\mu/(\rho_f D)$
Sh	Sherwood number, $k_f d_p/D$
t	Time, s
t_c	Cycle time, s
t_f	Feed time, s
t_R	Chromatographic retention time, s
T	Absolute temperature, K
u	Superficial velocity, m/s
u_s	Adsorbent velocity in TMB or SMB systems, $kg/(m^2 \cdot s)$
u_f	Fluid-phase internal energy, J/mol
u_s, u_{sol}	Stationary-phase and sorbent solid internal energy, J/kg
v	Interstitial velocity, m/s
V_f	Extraparticle fluid volume, m^3
W	Volume adsorbed as liquid, m^3; baseline width of chromatographic peak, s (Fig. 16-26)
x	Adsorbed-phase mole fraction; particle coordinate, m
y	Fluid-phase mole fraction
z	Bed axial coordinate, m; ionic valence

Greek Letters

α	Separation factor or selectivity
β	Scaling factor in Polanyi-based models; slope in gradient elution chromatography [Eq. (16-190)]
Δ	Peak width at half height, s (Fig. 16-26)
ε	Extraparticle void fraction of packed bed (extraparticle porosity); adsorption potential in Polanyi model, J/mol
ε_p	Intraparticle void fraction (intraparticle porosity)
ε_b	Total bed void fraction (inside and outside particles) [(Eq. 16-4)]
γ	Activity coefficient
Γ	Surface excess, mol/m^2 (Fig. 16-4)
κ	Boltzmann constant
λ	Isosteric heat of adsorption, J/mol [Eq. (16-7)]
Λ	Partition ratio [Eq. (16-125)]
Λ^∞	Ultimate fraction of solute adsorbed in batch
μ	Fluid viscosity, kg/(m·s)
μ_0	Zeroth moment, $mol \cdot s/m^3$ [Eq. (16-153)]
μ_1	First moment, s [Eq. (16-154)]
ν	Kinematic viscosity, m^2/s
Ω	Cycle-time dependent LDF coefficient [Eq. (16-91)]
ω	Parameter defined by Eq. (16-185*b*)
φ	Volume fraction or mobile-phase modulator concentration, mol/m^3
π	Spreading pressure, N/m [(Eq. (16-20)]
ψ_p, ψ_s	LDF correction factors (Table 16-12)
Ψ	Mechanism parameter for combined resistances (Fig. 16-12)
ρ	Subparticle radial coordinate, m
ρ_b	Bulk adsorbent density in a packed bed, kg/m^3
ρ_f	Fluid density, kg/m^3
ρ_p	Particle density, kg/m^3 [Eq. (16-1)]
ρ_s	Skeletal particle density, kg/m^3 [Eq. (16-2)]
σ^2	Second central moment, s^2 [Eq. (16-155)]
τ	Dimensionless time [Eq. (16-120)]
τ_1	Dimensionless time [Eq. (16-127) or (16-129)]
τ_p	Tortuosity factor [Eq. (16-65)]
ξ	Particle dimensionless radial coordinate (r/r_p)
ζ	Dimensionless bed axial coordinate (z/L)

Subscripts

a	Adsorbed phase
f	Fluid phase
i, j	Component index
tot	Total

Superscripts

-	An averaged concentration
^	A combination of averaged concentrations
*	Dimensionless concentration variable
e	Equilibrium
ref	Reference (indicates feed or initial values)
s	Saturation
SM	Service mark
TM	Trademark
0	Initial fluid concentration in batch
0′	Initial adsorbed-phase concentration in batch
∞	Final state approached in batch

General References: Adamson, A. W., *Physical Chemistry of Surfaces,* Wiley, New York, 1990; Barrer, R. M., *Zeolites and Clay Minerals as Adsorbents and Molecular Sieves,* Academic Press, New York, 1978; Breck, D. W., *Zeolite Molecular Sieves,* Wiley, New York, 1974; Carta, G., and Jungbauer, A., *Protein Chromatography—Process Development and Scale-Up,* Wiley-VCH, Weinheim, Germany, 2010; Cheremisinoff, P. N., and F. Ellerbusch, *Carbon Adsorption Handbook,* Ann Arbor Science, Ann Arbor, 1978; Crittenden, B., and W. J. Thomas, *Adsorption Technology and Design,* Butterworth-Heinemann, Oxford, UK, 1998; Do, D. D., *Adsorption Analysis: Equilibria and Kinetics,* Imperial College, London, 1998; Dorfner, K., ed., *Ion Exchangers,* W. de Gruyter, Berlin, 1991; Dyer, A., *An Introduction to Zeolite Molecular Sieves,* Wiley, New York, 1988; EPA, *Process Design Manual for Carbon Adsorption,* U.S. Envir. Protect. Agency, Cincinnati, 1973; Gembicki, S. A., A. R. Oroskar, and J. A. Johnson, "Adsorption, Liquid Separation," in *Kirk-Othmer Encyclopedia of Chemical Technology,* 4th ed., Wiley, New York, 1991; Guiochon, G., A. Felinger, D. G. Shirazi, and A. M. Katti, *Fundamentals of Preparative and Nonlinear Chromatography,* Elsevier, 2006; Gregg, S. J., and K. S. W. Sing, *Adsorption, Surface Area and Porosity,* Academic Press, New York, 1982; Helfferich, F. G., *Ion Exchange,* McGraw-Hill, New York, 1962; reprinted by University Microfilms International, Ann Arbor, Michigan; Helfferich, F. G., and G. Klein, *Multicomponent Chromatography,* Marcel Dekker, New York, 1970; Jaroniec, M., and R. Madey, *Physical Adsorption on Heterogeneous Solids,* Elsevier, New York, 1988; Kärger, J., and D. M. Ruthven, *Diffusion in Zeolites and Other Microporous Solids,* Wiley, New York, 1992; Kärger, J., D. M. Ruthven, and D. N. Theodorou, *Diffusion in Nanoporous Materials* (2 vol.), Wiley, New York, 2012; Keller, G. E., R. A. Anderson, and C. M. Yon, "Adsorption" in *Handbook of Separation Process Technology,* ed. Rousseau, R. W., Wiley-Interscience, New York, 1987; Keller, J. U., and R. Staudt, *Gas Adsorption Equilibria: Experimental Methods and Adsorption Isotherms,* Springer, New York, 2005; M. R. Ladisch, *Bioseparations Engineering: Principles, Practice, and Economics,* Wiley, New York, 2001; LeVan, M. D., and G. Carta, "Adsorption and Ion Exchange," in *Perry's Chemical Engineers' Handbook,* 8th ed., ed. D. W. Green, McGraw-Hill, New York, 2008; Rhee, H.-K., R. Aris, and N. R. Amundson, *First-Order Partial Differential Equations: Volume 1. Theory and Application of Single Equations; Volume 2. Theory and Application of Hyperbolic Systems of Quasi-Linear Equations,* Prentice-Hall, Englewood Cliffs, N.J., 1986, 1989; Rodrigues, A. E., M. D. LeVan, and D. Tondeur, eds., *Adsorption: Science and Technology,* Kluwer Academic Publishers, Dordrecht, The Netherlands, 1989; Rudzinski, W., and D. H. Everett, *Adsorption of Gases on Heterogeneous Surfaces,* Academic Press, San Diego, 1992; Ruthven, D. M., *Principles of Adsorption and Adsorption Processes,* Wiley, New York, 1984; Ruthven, D. M., S. Farooq, and K. S. Knaebel, *Pressure Swing Adsorption,* VCH Publishers, New York, 1994; Seader, J. D., and E. J. Henley, *Separation Process Principles,* 2d ed., Wiley, New York, 2006; Sherman, J. D., C. M. and Yon, "Adsorption, Gas Separation," in *Kirk-Othmer Encyclopedia of Chemical Technology,* 4th ed., Wiley, New York, 1991; Streat, M., and F. L. D. Cloete, "Ion Exchange," in *Handbook of Separation Process Technology,* ed. Rousseau, R. W., Wiley, New York, 1987; Suzuki, M., *Adsorption Engineering,* Elsevier, Amsterdam, 1990; Tien, C., *Adsorption Calculations and Modeling,* Butterworth-Heinemann, Newton, Mass., 1994; Valenzuela, D. P., and A. L. Myers, *Adsorption Equilibrium Data Handbook,* Prentice-Hall, Englewood Cliffs, N.J., 1989; Vermeulen, T., M. D. LeVan, N. K. Hiester, and G. Klein, G. "Adsorption and Ion Exchange" in *Perry's Chemical Engineers' Handbook,* 6th ed., ed. R. H. Perry and D. W. Green, McGraw-Hill, New York, 1984; Wankat, P. C., *Large-Scale Adsorption and Chromatography,* CRC Press, Boca Raton, Fla., 1986; Yang, R. T., *Adsorbents: Fundamentals and Applications,* Wiley, Hoboken, N.J., 2003; Yang, R. T., *Gas Separation by Adsorption Processes,* Butterworth, Stoneham, Mass., 1987; Young, D. M., and A. D. Crowell, *Physical Adsorption of Gases,* Butterworths, London, 1962.

DESIGN CONCEPTS

INTRODUCTION

Adsorption and ion exchange share so many common features in regard to application in batch and fixed-bed processes that they can be grouped together for a unified treatment. These processes involve the transfer and resulting distribution of one or more solutes between a fluid phase and a sorbent. The partitioning of a single solute between fluid and adsorbed phases or the selectivity of a sorbent toward multiple solutes makes it possible to separate solutes from a bulk fluid phase or from one another.

This section treats batch and fixed-bed operations and reviews process cycles and equipment. In practice, fixed-bed operation with the sorbent in granule, bead, or pellet form is the predominant way of conducting sorption separations and purifications. Therefore, fixed beds including chromatographic separations are given primary attention here with respect to both interpretation and prediction.

Adsorption involves, in general, the accumulation (or depletion) of solute molecules at an interface (including gas–liquid interfaces, as in foam fractionation, and liquid–liquid interfaces, as in detergency). Here we consider only solid adsorbents and ion exchangers and thus only gas–solid and liquid–solid interfaces, with solute distributed selectively between the fluid and solid phases. Adsorption is typically restricted to the interfacial surface; thus, highly porous solids with very large internal area per unit volume are preferred. Adsorbent surfaces are often physically and/or chemically heterogeneous, and bonding energies may vary widely from one site to another. For separations or temporary storage, we usually seek to promote physical adsorption or *physisorption,* which involves van der Waals forces, and retard chemical adsorption or *chemisorption,* which involves chemical bonding (and often dissociation, as in catalysis). The former is well suited for a regenerable process, while the latter generally destroys the capacity of the adsorbent.

Adsorbents are natural or synthetic materials of amorphous or microcrystalline structure. Those used on a large scale, in order of quantity used, are activated carbon, molecular sieves, silica gel, and activated alumina [see Keller, Anderson, and Yon (1987) in General References].

Ion exchange usually occurs throughout a swollen cross-linked charged polymer, termed a resin. In cation exchange, ions of positive charge (cations) from the fluid (usually an aqueous solution) replace dissimilar ions of the same charge initially in the resin. In anion exchange the same process occurs but involving ions of negative charge (anions). In both cases the ion exchanger contains permanently bound functional groups. Cation-exchange resins generally contain bound sulfonic acid groups; less commonly, these groups are carboxylic, phosphonic, phosphinic, and so on. Anion-exchange resins generally involve quaternary ammonium groups (strongly basic) or other amino groups (weakly basic).

Most ion exchangers in large-scale use are based on synthetic resins that are either preformed and then functionalized with ionogenic groups, as for polystyrene, or formed from ionogenic monomers (e.g., olefinic acids, amines, or phenols). Certain natural and synthetic zeolites are also used as ion exchangers.

Ion exchange may be thought of as a reversible reaction involving chemically equivalent quantities. A common example for cation exchange is the water-softening reaction

$$Ca^{++} + 2NaR \rightleftharpoons CaR_2 + 2Na^+$$

where R represents a stationary univalent anionic site in the polyelectrolyte network of the exchanger phase.

Table 16-1 classifies sorption operations by the type of interaction and the basis for the separation. In addition to the normal sorption operations of adsorption and ion exchange, some other similar separations are included. Applications are discussed in the subsection Process Cycles.

Example 16-1 Surface Area and Pore Volume of Adsorbent A simple example will show the extent of internal area in a typical granular adsorbent. A fixed bed is packed with particles of a porous adsorbent material. The bulk density of the packing is 500 kg/m^3, and the interparticle void fraction is 0.40. The intraparticle porosity is 0.50, with two-thirds of this in cylindrical pores of diameter 1.4 nm and the rest in much larger pores. Find the surface area of the adsorbent and, if solute has formed a complete monomolecular layer 0.3 nm thick inside the pores, determine the percent of the particle volume and the percent of the total bed volume filled with adsorbate.

From surface-area-to-volume ratio considerations, the internal area is practically all in the small pores. One gram of the adsorbent occupies 2 cm^3 as packed and has 0.4 cm^3 in small pores, which gives a surface area of 1150 m^2/g (or about 1 mi^2 per 5 lb or 6.3 mi^2/ft^3 of packing). Based on the area of the annular region filled with adsorbate, the solute occupies 22.5 percent of the internal pore volume and 13.5 percent of the total packed-bed volume.

DESIGN STRATEGY

The design of sorption systems is based on a few underlying principles. First, knowledge of *sorption equilibrium* is required. This equilibrium, between solutes in the fluid phase and the solute-enriched phase of the solid, supplants what in most chemical engineering separations is a fluid–fluid equilibrium. The selection of the sorbent material with an understanding of its equilibrium properties (i.e., capacity and selectivity as a function of temperature and component concentrations) is of primary importance. Second, because sorption operations take place in batch, in fixed beds, or in simulated moving beds, the processes have *dynamical character.* Such operations generally do not run at steady state, although steady state may be approached in a simulated moving bed. Fixed-bed processes often approach a periodic condition called a periodic state or cyclic steady state, with several different feed steps constituting a cycle. Thus, some knowledge of how transitions travel through a bed is required. This introduces both time and space into the analysis, in contrast to many other chemical engineering operations that can be analyzed

TABLE 16-1 Classification of Sorptive Separations

Type of interaction	Basis for separation	Examples
Adsorption	Equilibrium	Numerous purification and recovery processes for gases and liquids Activated carbon-based applications Desiccation using silica gels, aluminas, and zeolites Oxygen from air by PSA using LiX and 5A zeolites
	Rate	Nitrogen from air by PSA using carbon molecular sieve Nitrogen and methane using titanosilicate ETS-4
	Molecular sieving	Separation on *n*- and *iso*-parafins using 5A zeolite Separation of xylenes using zeolite
Ion exchange (electrostatic)	Equilibrium	Deionization Water softening Rare earth separations Recovery and separation of pharmaceuticals (e.g., amino acids, proteins)
Ligand exchange	Equilibrium	Chromatographic separation of glucose-fructose mixtures with Ca-form resins Removal of heavy metals with chelating resins Affinity chromatography
Solubility	Equilibrium	Partition chromatography
None (purely steric)	Equilibrium partitioning in pores	Size exclusion or gel permeation chromatography

at steady state with only a spatial dependence. For accurate design, it is crucial to understand fixed-bed performance in relation to adsorption equilibrium and rate behavior. Finally, many *practical aspects* must be included in design so that a process starts up and continues to perform well, and that it is not so overdesigned that it is wasteful. While these aspects are process-specific, they include an understanding of dispersive phenomena at the bed scale and, for regenerative processes, knowledge of aging characteristics of the sorbent material, with consequent changes in sorption equilibrium.

Characterization of Equilibria Phase equilibrium between fluid and sorbed phases for one or many components in adsorption or two or more species in ion exchange is usually the single most important factor affecting process performance. In most processes, it is much more important than mass and heat transfer rates; a doubling of the stoichiometric capacity of a sorbent or a significant change in the shape of an isotherm would often have a greater impact on process performance than a doubling of transfer rates.

A difference between adsorption and ion exchange with completely ionized resins is indicated in the *variance* of the systems. In adsorption, part of the solid surface or pore volume is vacant. This diminishes as the fluid-phase concentration of solute increases. In contrast, for ion exchange the sorbent has a fixed total capacity and merely exchanges solutes while conserving charge. Variance is defined as the number of independent variables in a sorption system at equilibrium—that is, variables that one can change separately and thereby control the values of all others. Thus, it also equals the difference between the total number of variables and the number of independent relations connecting them. Many cases arise in which ion exchange is accompanied by chemical reaction (neutralization or precipitation, in particular), or adsorption is accompanied by evolution of sensible heat. The concept of variance helps greatly to assure correct interpretations and predictions.

The working capacity of a sorbent depends on fluid concentrations and temperatures. Graphical depiction of sorption equilibrium for single-component adsorption or binary ion exchange (monovariance) is usually in the form of isotherms [$n_i = n_i(c_i)$ or $n_i(p_i)$ at constant T] or isosteres [$p_i = p_i(T)$ at constant n_i]. Representative forms are shown in Fig. 16-1. An important dimensionless group dependent on adsorption equilibrium is the **partition ratio** [see Eq. (16-125)], which is a measure of the relative affinities of the sorbed and fluid phases for solute.

Historically, isotherms have been classified as *favorable* (concave downward) or *unfavorable* (concave upward). These terms refer to the spreading tendencies of transitions in fixed beds. A favorable isotherm gives a compact transition, whereas an unfavorable isotherm leads to a broad one.

Example 16-2 Calculation of Variance In mixed-bed deionization of a solution of a single salt, there are eight concentration variables: two each for cation, anion, hydrogen, and hydroxide. There are six connecting relations: two for ion exchange and one for neutralization equilibrium, and two ion-exchanger and one solution electroneutrality relation. The variance is therefore 8 − 6 = 2.

Adsorbent/Ion Exchanger Selection Guidelines for sorbent selection are different for regenerative and nonregenerative systems. For a nonregenerative system, one generally wants a high capacity and a strongly favorable isotherm for a purification and additionally high selectivity for a separation. For a regenerative system, high overall capacity and selectivity are again desired, but needs for cost-effective regeneration leading to a reasonable working capacity influence what is sought after in terms of isotherm shape. For separations by pressure-swing adsorption (or vacuum pressure-swing adsorption), generally one wants a linear to slightly favorable isotherm

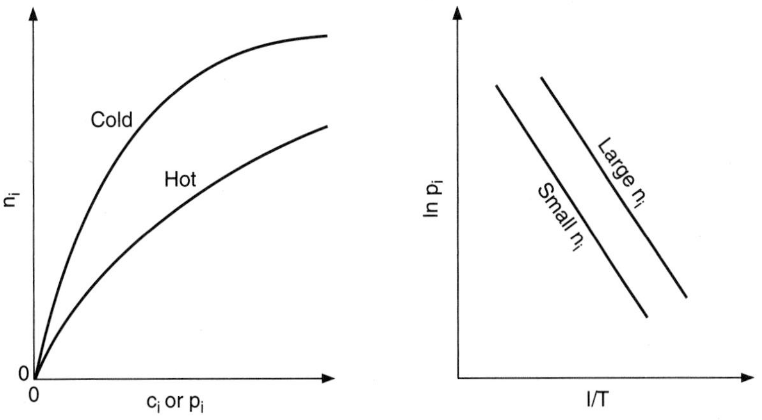

FIG. 16-1 Isotherms (*left*) and isosteres (*right*). Isosteres plotted using these coordinates are nearly straight parallel lines, with deviations caused by the dependence of the isosteric heat of adsorption on temperature and loading.

TABLE 16-2 Methods of Analysis of Fixed-Bed Transitions

Method	Purpose	Approximations
Local equilibrium theory	Shows wave character—simple waves and shocks Usually indicates best possible performance Better understanding	Mass and heat transfer very rapid Dispersion usually neglected If nonisothermal, then adiabatic
Mass-transfer zone	Design based on stoichiometry and experience	Isothermal MTZ length largely empirical Regeneration often empirical
Constant pattern and related analyses	Gives asymptotic transition shapes and upper bound on MTZ	Deep bed with fully developed transition
Full rate modeling	Accurate description of transitions Appropriate for shallow beds, with incomplete wave development General numerical solutions by finite difference or collocation methods	Various to few

(although purifications can operate economically with more strongly favorable isotherms). Temperature-swing adsorption usually operates with moderately to strongly favorable isotherms, in part because one is typically dealing with heavier solutes, and these are adsorbed rather strongly (e.g., organic solvents on activated carbon and water vapor on zeolites). Exceptions exist, however; for example, water is adsorbed on silica gel and activated alumina only moderately favorably, with some isotherms showing unfavorable sections. Equilibria for ion-exchange separations generally vary from moderately favorable to moderately unfavorable; depending on feed concentrations, alternates often exist for different steps of a regenerative cycle. Other factors in sorbent selection are mechanical and chemical stability, mass-transfer characteristics, and cost.

Fixed-Bed Behavior The number of transitions occurring in a fixed bed of initially uniform composition before it becomes saturated by a constant composition feed stream is generally equal to the variance of the system. This introductory discussion will be limited to single-transition systems.

Methods for analysis of fixed-bed transitions are shown in Table 16-2. Local equilibrium theory is based solely on stoichiometric concerns and system nonlinearities. A transition becomes a "simple wave" (a gradual transition), a "shock" (an abrupt transition), or a combination of the two. In other methods, mass-transfer resistances are incorporated.

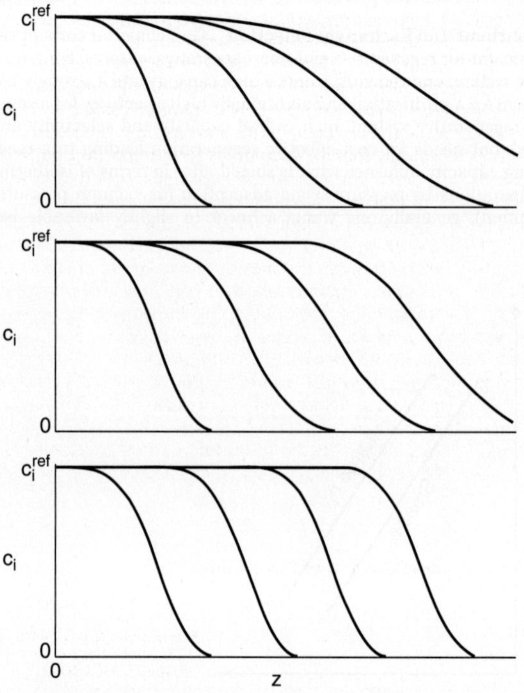

FIG. 16-2 Limiting fixed-bed behavior: simple wave for unfavorable isotherm (*top*), square-root spreading for linear isotherm (*middle*), and constant pattern for favorable isotherm (*bottom*). [From LeVan in Rodrigues et al. (eds.), Adsorption: Science and Technology, *Kluwer Academic Publishers, Dordrecht, The Netherlands, 1989; reprinted with permission.*]

The asymptotic behavior of transitions under the influence of mass-transfer resistances in long, "deep" beds is important. The three basic asymptotic forms are shown in Fig. 16-2. With an unfavorable isotherm, the breadth of the transition becomes proportional to the depth of bed it has passed through. For the linear isotherm, the breadth becomes proportional to the square root of the depth. For the favorable isotherm, the transition approaches a constant breadth called a *constant pattern.*

Design of nonregenerative sorption systems and many regenerative ones often relies on the concept of the mass-transfer zone or MTZ, which closely resembles the constant pattern [Collins, *Chem. Eng. Prog. Symp. Ser. No. 74* **63:** 31 (1974); Keller, Anderson, and Yon (1987) in General References]. The length of this zone (depicted in Fig. 16-3) together with stoichiometry can be used to predict accurately how long a bed can be used before breakthrough.

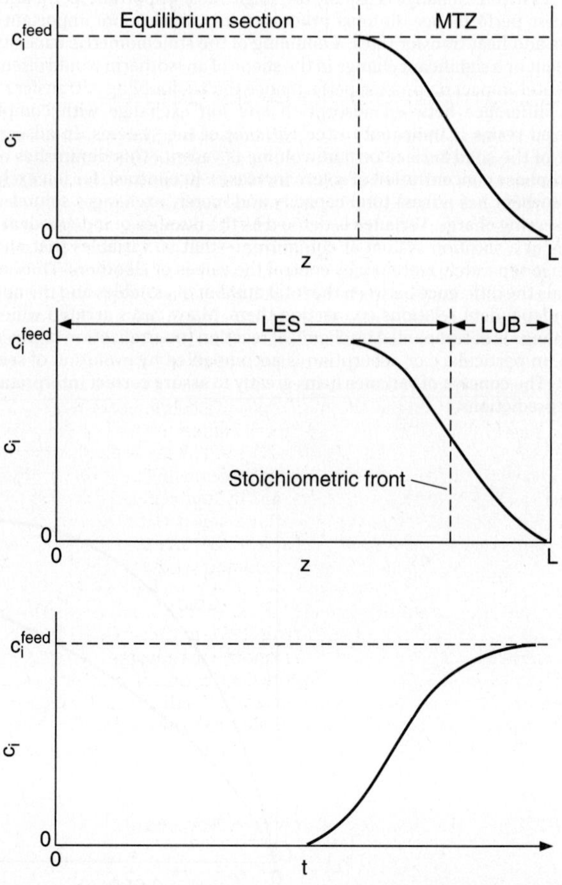

FIG. 16-3 Bed profiles (*top* and *middle*) and breakthrough curve (*bottom*). The bed profiles show the mass-transfer zone (MTZ) and equilibrium section at breakthrough. The stoichiometric front divides the MTZ into two parts with contributions to the length of equivalent equilibrium section (LES) and the length of equivalent unused bed (LUB).

Upstream of the mass-transfer zone, the adsorbent is in equilibrium with the feed. Downstream, the adsorbent is in its initial state. Within the mass-transfer zone, the fluid-phase concentration drops from the feed value to the initial, presaturation state.

Equilibrium with the feed is not attained in this region. As a result, because an adsorption bed must typically be removed from service shortly after breakthrough begins, the full capacity of the bed is not used. Obviously, the broader the mass-transfer zone is, the greater will be the extent of unused capacity. Also shown in the figure are the length of the equivalent equilibrium section (LES) and the length of the equivalent unused bed (LUB). The length of the MTZ is divided between these two.

Adsorption with strongly favorable isotherms and ion exchange between strong electrolytes can usually be carried out until most of the stoichiometric capacity of the sorbent has been utilized, corresponding to a thin MTZ. Consequently, the total capacity of the bed is practically constant regardless of the composition of the solution being treated.

The effluent concentration history is the breakthrough curve, also shown in Fig. 16-3. The effluent concentration stays at or near zero or a low residual concentration until the transition reaches the column outlet. The effluent concentration then rises until it becomes unacceptable, this time being called the breakthrough time, which is typically much shorter than the time for the feed concentration to reach the column outlet. The feed step must stop and, for a regenerative system, the regeneration step begins.

Two dimensionless variables play key roles in the analysis of single-transition systems (and some multiple-transition systems). These are the *throughput parameter* [see Eq. (16-129)] and the *number of transfer units* (see Table 16-13). The former is time made dimensionless so that it equals unity at the stoichiometric center of a breakthrough curve. The latter is, as in packed tower calculations, a measure of mass-transfer resistance.

Cycles Design methods for cycles rely on mathematical modeling (or empiricism) and often extensive pilot plant experiments. Many cycles can be easily analyzed by applying the methods previously described to the collection of steps. In some cycles, however, especially those operated with short cycle times or in shallow beds, transitions may not be fully developed, even at a periodic state, and the complexity may be compounded by multiple sorbates.

A wide variety of complex process cycles have been developed. Systems with many beds incorporating multiple sorbents, possibly in layered beds, are in use. Mathematical models constructed to analyze such cycles can be complex. With a large number of variables and nonlinear equilibria involved, it is usually not beneficial to make all variables in such models dimensionless; doing so does not help appreciably in making comparisons with other largely dissimilar systems. If dimensionless variables are used, these usually begin with a dimensionless bed length and a dimensionless time, which is often different from the throughput parameter.

Practical Aspects A number of process-specific concerns are accounted for in good design. In regenerable systems, sorbents age, losing capacity because of fouling by heavy contaminants, loss of surface area or crystallinity, oxidation, and the like. Mass-transfer resistances may increase over time. Because of particle shape, size distribution, or column packing method, dispersion may be more pronounced than would normally be expected. The humidity of an entering stream will usually affect a solvent recovery application. Safety, including the possibility of a fire, may be a concern. For gas-phase adsorption, scale-up from an isothermal laboratory column to a nonisothermal pilot plant column to a largely adiabatic process column requires careful judgment. If the MTZ concept is utilized, the length of the MTZ cannot be reliably determined solely from knowledge on other systems. Experience plays the key role in accounting for these and other such factors.

ADSORBENTS AND ION EXCHANGERS

CLASSIFICATIONS AND CHARACTERIZATIONS

Adsorbents Table 16-3 classifies common adsorbents by structural type and water adsorption characteristics. Zeolite and silicalite adsorbents take advantage of their crystalline structure and/or their molecular sieving properties. The hydrophobic (nonpolar surface) or hydrophilic (polar surface) character may vary depending on the competing adsorbate. A large number of zeolites have been identified, and these include both synthetic and naturally occurring (e.g., mordenite and chabazite) varieties.

The classifications in Table 16-3 are intended only as a rough guide. For example, a carbon molecular sieve is truly amorphous but has been manufactured to have certain structural, rate-selective properties. Similarly, the extent of hydrophobicity of an activated carbon will depend on its ash content and its level of surface oxidation.

Zeolites are crystalline aluminosilicates. Zeolitic adsorbents have had their water of hydration removed by calcination to create a structure with well-defined openings into crystalline cages. The molecular sieving properties of zeolites are based on the sizes of these openings. Two crystal types are common: type A (with openings formed by four sodalite cages) and type X or Y (with openings formed by six sodalite cages). Cations balancing charge and their locations determine the size of opening into a crystal unit cell. Nominal opening sizes for the most common synthetic zeolites are 0.3 nm for KA, 0.4 nm for NaA, 0.5 nm for CaA, and 1.0 nm for NaX. Further details, including effective molecular diameters, are widely available [Barrer (1978); Breck (1974); Ruthven (1984); and Yang (1987) in General References].

Many adsorbents, particularly the amorphous adsorbents, are characterized by their pore size distribution. The distribution of small pores is usually determined by analysis, using one of several available methods, of a cryogenic nitrogen adsorption isotherm, although other probe molecules are also used. The distribution of large pores is usually determined by mercury porisimetry [Gregg and Sing (1982) in General References].

Table 16-4 shows the IUPAC classification of pores by size. Micropores are small enough that a molecule is attracted to both of the opposing walls forming the pore. The potential energy functions for these walls superimpose to create a deep well, and strong adsorption results. Hysteresis, that is, when the adsorption and the desorption branches of an isotherm are different, is generally not observed. (However, water vapor adsorbed in the micropores of activated carbon shows a large hysteresis loop, and the desorption branch is sometimes used with the Kelvin equation to determine the pore size distribution.) Capillary condensation occurs in mesopores, and a hysteresis loop is typically found. Macropores form important paths for molecules to diffuse into a particle; for gas-phase adsorption, they do not fill with adsorbate until the gas phase becomes saturated.

New adsorbents and ion-exchange materials are constantly under development. The class of microporous crystalline materials with molecular sieving properties is extensive and includes many metal silicates and phosphates, with the zeolites (aluminosilicates) being the most broadly useful. The titanosilicate ETS-4 is currently being implemented for separation of CH_4 and N_2. Among other classes of materials, the recently developed π-complexation sorbents are particularly promising; these use weak chemical bonds instead of purely physical interaction as the basis for separation [Yang (2003) in General References]. Two other adsorbent materials of considerable research interest, but which have not been implemented for commercial separations, are carbon nanotubes and the mesoporous silicas MCM-41 and SBA-15. Carbide-derived carbons (CDCs) are a new entry into

TABLE 16-3 Classification of Common Adsorbents

	Amorphous	Structured
Hydrophobic	Activated carbon Polymers	Carbon molecular sieves Silicalite
Hydrophilic	Silica gel Activated alumina	Common zeolites: 3A (KA), 4A (NaA), 5A (CaA), 13X (NaX), Mordenite, Chabazite, etc.

TABLE 16-4 Classification of Pore Sizes

Type	Slit width* (w)	Characteristic
Micropore†	$w < 2$ nm	Superimposed wall potentials
Mesopore	2 nm $< w <$ 50 nm	Capillary condensation
Macropore	$w > 50$ nm	Effectively flat walled until $p \rightarrow P^s$

*Or pore diameter.
†Further subdivided into ultramicropores and supermicropores.
Gregg and Sing (1982) in General References; Thommes et al., *Pure Appl. Chem.* **87:** 1051 (2015).

the adsorbent field. They have surface areas ranging from 500 m^2/g to 2000 m^2/g, depending on the starting precursor and smaller pore size distributions than typical activated carbons; adsorption studies are in early stages, as CDCs have been evaluated more often as electrochemical capacitors [Simon and Gogotsi, *Nature Materials* **7:** 845 (2008)]. Polymers with intrinsic microporosity (PIMs) are also under development as adsorbents, as are many other materials targeted for specific adsorbates.

Metal-organic frameworks (MOFs) are a new class of porous materials that entered the adsorbent literature in the late 1990s. Early research focused on their potential application in natural gas storage and hydrogen storage systems (Farrusseng, *Metal-Organic Frameworks: Applications from Catalysis to Gas Storage*, Wiley-VCH, Weinheim, 2011). The materials are now widely studied for applications to CO_2 capture, natural gas upgrading, and air purification but have yet to achieve commercial success [Li et al., *Chem. Reviews* **112:** 869 (2012)].

MOFs are synthesized by the self-assembly of metal ions coordinated to organic ligands most commonly through carboxylic acid groups or nitrogens. The resulting porous networks exhibit among the highest pore volumes and surface areas of any known adsorbents. They are also highly crystalline with distinct pore sizes and shapes. More than 20,000 structures have been discovered; these possess a large diversity of chemical composition, topology, and textural properties.

Among the most important examples, MOF-5 (or IRMOF-1) is one of the earliest porous MOFs. This structure is formed by 4-coordinated Zn-O tetrahedral [Eddaoudi et al., *Science* **295:** 469 (2002)]; this material has a surface area exceeding 1500 m^2/g and pores sizes >1.2 nm but undergoes structural degradation in the presence of humidity or aqueous conditions. HKUST-1 (Cu-BTC) was the first permanently porous MOF assembled from the copper-paddlewheel building unit [Chui et al., *Science* **283:** 1148 (1999)]. The MOF possesses open metal sites (coordinatively unsaturated) that are available for interaction with adsorbate molecules. HKUST-1 has two types of pores: a larger center channel with diameter of 0.9 nm and smaller side pockets with diameter of 0.35 nm. This MOF displays improved water stability over MOF-5 and is currently produced at the kilogram scale by BASF and marketed as Basolite C300. The aluminum-based MOF MIL-53(Al), produced commercially as Basolite A100, is synthesized in water and comprised of $AlO_4(OH)_2$ corner-sharing octahedral chains connected by terephthalate groups to form a three-dimensional structure with 8.5 Å pores [Loiseau et al., *Chemistry-A European Journal* **10:** 1373 (2004)]. MIL-53(A1) is a stable structure with a surface area exceeding 1300 m^2/g. The zirconium-based MOF UiO-66 is synthesized from zirconium chloride and terephthalic acid in dimethyl formamide [Cavka et al., *J. Am. Chem. Soc.* **130:** 13850 (2008)]. UiO-66 possesses triangular-shaped pores with diameter of approximately 0.6 nm. The terephthalic acid ligand can be functionalized in a variety of ways with chemical groups such as $-NH_2$ and $-NO_2$ [Cmarik et al., *Langmuir* **28:** 15606 (2012)]. UiO-66 is one of the most robust and chemically stable MOFs in the literature.

MOFs have received intense attention in the scientific literature, but the materials have yet to find commercial application as adsorbents due to problems with chemical stability and synthesis scale-up. Of the thousands of known MOF structures, on the order of 50 structures exhibit high stability in the presence of water. Several hundred materials are "partially stable," with the remaining majority being unstable. For more details on MOF stability, a comprehensive review is available [Burtch et al., *Chemical Reviews* **114:** 10575 (2014)].

Ion Exchangers Ion exchangers are classified according to (1) their functionality and (2) the physical properties of the support matrix. Cation and anion exchangers are classified in terms of their ability to exchange positively or negatively charged species. Strongly acidic and strongly basic ion exchangers are ionized and thus are effective at nearly all pH values (e.g., pH 0–14). Weakly acidic exchangers are typically effective above pH 5. Weakly basic resins are effective below pH 9. Weakly acidic and weakly basic exchangers are often easier to regenerate, but leakage due to incomplete exchange may occur. Chelating resins containing iminodiacetic acid form specific metal complexes with metal ions with complex stability constants that follow the same order as those for EDTA. However, depending on pH, they also function as weak cation exchangers. The achievable ion-exchange capacity depends on the concentration of ionogenic groups and their availability as exchange sites. The latter is a function of the support matrix.

Polymer-based, synthetic ion exchangers known as resins are available commercially in gel type or truly porous forms. Gel-type resins are not porous in the usual sense of the word, since their structure depends on swelling in the solvent in which they are immersed. Removal of the solvent usually results in a collapse of the three-dimensional structure, and no significant surface area or pore diameter can be defined by the ordinary techniques available for truly porous materials. In their swollen state, gel-type resins approximate a true molecular-scale solution. Thus, an internal porosity ε_p can be identified only in terms of the equilibrium uptake of water or other liquid. When cross-linked polymers are used as the support matrix, the internal porosity so defined varies approximately in inverse proportion to the degree of cross-linking, and swelling is typically more pronounced in solvents with a high dielectric constant. The ion held by the exchanger also influences the resin swelling. Thus, the size of the resin particles changes during the ion-exchange process as the resin is changed from one form to another, and this effect is more dramatic for resins with a lower degree of cross-linking. The choice of degree of cross-linking is dependent on several factors, including the extent of swelling, the exchange capacity, the intraparticle diffusivity, the ease of regeneration, and the physical and chemical stability of the exchanger under chosen operating conditions. The concentration of ionogenic groups determines the capacity of the resin. Although the capacity per unit mass of dry resin is insensitive to the degree of cross-linking, except for very highly cross-linked resins, the exchange capacity per unit volume of swollen resin increases significantly with degree of cross-linking, so long as the mesh size of the polymer network allows the ions free access to functional groups within the interior of the resin. The degree of cross-linking also affects the rate of ion exchange. The intraparticle diffusivity decreases nearly exponentially with the mesh size of the matrix. As a result, resins with a lower degree of cross-linking are normally required for the exchange of bulky species, such as organic ions with molecular weight in excess of 100. The regeneration efficiency is typically greater for resins with a lower degree of cross-linking. Finally, the degree of cross-linking also affects the long-term stability of the resin. Strongly acidic and strongly basic resins are subject to irreversible oxidative degradation of the polymer and thermal and oxidative decomposition of functional groups. Generally, more highly cross-linked resins are less prone to irreversible chemical degradation, but they may be subject to osmotic breakage caused by volume changes that occur during cyclic operations. In general, experience shows that an intermediate degree of cross-linking is often preferred. However, readers are referred to manufacturers' specifications for resin stability data at different operating conditions.

Truly porous, synthetic ion exchangers are also available. These materials retain their macroporosity even after removal of the solvent and have measurable surface areas and pore size. The term *macroreticular* is commonly used for resins prepared from a phase separation technique, where the polymer matrix is prepared with the addition of a liquid that is a good solvent for the monomers, but in which the polymer is insoluble. Matrices prepared in this way usually have the appearance of a conglomerate of gel-type microspheres held together to form an interconnected porous network. Macroporous resins possessing a more continuous gellular structure interlaced with a pore network have also been obtained with different techniques and are commercially available. Since higher degrees of cross-linking are typically used to produce truly porous ion-exchange resins, these materials tend to be more stable under highly oxidative conditions, more attrition-resistant, and more resistant to breakage due to osmotic shock than their gel-type counterparts. Moreover, since their porosity does not depend entirely on swelling, they can be used in solvents with low dielectric constant where gel-type resins can be ineffective. In general, compared to gel-type resins, truly porous resins typically have somewhat lower capacities and can be more expensive. Thus, for ordinary ion-exchange applications involving small ions under nonharsh conditions, gel-type resins are usually preferred.

Adsorbents for biomacromolecules such as proteins have special properties [see Carta and Jungbauer (2010) in General References]. First, these adsorbents need to have large pore sizes. A ratio of pore radius to molecule radius larger than 5 is desirable to prevent excessive diffusional hindrance (see the subsection Intraparticle Mass Transfer). Thus, for typical proteins, pore radii need to be in excess of 10–15 nm. Second, functional groups for interactions with proteins are usually attached to a matrix backbone via a spacer arm to provide accessibility. Third, hydrophilic matrices are preferred to limit nonspecific interactions with the backbone and prevent global unfolding or denaturation of the protein. Thus, if hydrophobic supports are used, their surfaces are usually rendered hydrophilic by incorporating hydrophilic coatings such as dextran or polyvinyl alcohol. Finally, materials stable in sodium hydroxide solutions or other solutions used for clean-in-place are preferred. Support matrices for protein adsorption can be classified into the following four broad groups: hydrophilic gels, including cross-linked agarose, cross-linked dextran, cross-linked polyacrylamide, and cellulose; rigid macroporous media, including silica/ceramic, alumina, polystyrene-DVB, and polymethacrylate; composite media, including agarose-dextran and ceramic-polyacrylamide; and monoliths, including polystyrene-DVB, polymethacrylate, and silica-based materials. Monoliths for biomacromolecule adsorption are generally synthesized in place in a column, resulting in a continuous random structure of linked, generally nonporous microparticles intercalated by a network of flow-through pores. The continuous structure allows independent control of microparticle size and flow-through pore sizes, resulting in rapid adsorption and moderate pressure drop. Although the adsorption capacity for proteins is smaller than for porous-particle-based adsorbents, they offer significant advantages for rapid high-resolution separations and for adsorption of very large bioparticles such as plasmid DNA and viruses [see Josic et al., *J. Chrom. B.* **752:** 191 (2001)].

Functional groups or ligands incorporated in materials for biomacromolecule adsorption can be classified in the following four broad groups: cation exchange, including sulfopropyl (SP), methyl sulfonate (S), and carboxymethyl (CM); anion exchange, including quaternary ammonium ion (Q), quaternary aminoethyl (QAE), and diethylamino ethyl (DEAE); hydrophobic interaction, including ether, isopropyl, butyl, octyl, and phenyl ligands in order of increasing hydrophobicity; and biospecific interaction or affinity ligands, including amino acids, dyes, and proteins covalently bound to the base matrix and transition metals supported via complexing ligands. Among the affinity ligands, protein A is the most popular for the selective adsorption of antibodies. Hybrid or multimodal ligands, incorporating both charged and hydrophobic moieties, are also popular. Polystyrene-based particles and silica particles functionalized with hydrocarbon chains from C2 to C18 in length are also used for reversed-phase chromatographic separations. However, because of the limited stability of biomacromolecules, their process applications are limited. Inert, hydrophilic porous matrices and gels are used extensively for size exclusion chromatography (SEC). A useful general reference is Janson, *Protein Purification: Principles, High-Resolution Methods, and Applications*, 3d ed., Wiley, New York, 2011.

PHYSICAL PROPERTIES

Selected data on some commercially available adsorbents and ion exchangers are given in Tables 16-5 and 16-6. The purpose of the tables is twofold: to assist the engineer or scientist in identifying materials suitable for a needed application, and to supply typical physical property values.

In addition to the particulate adsorbents listed in Table 16-5, some adsorbents are available in structured form for specific applications. Monoliths, papers, and paint formulations have been developed for zeolites, with these driven by the development of adsorbent wheels, adsorptive refrigeration, and other applications requiring low pressure drop across the column. Carbon monoliths are also available, as are activated carbon fibers, created from polymeric materials and sold in the form of fabrics, mats, felts, and papers for use in various applications, including in pleated form in filters. Zeolitic and carbon membranes are also available, with the latter developed for separation by "selective surface flow" [Rao and Sircar, *J. Membrane Sci.* **85:** 253 (1993)].

Excellent sources of information on the characteristics of adsorbents or ion-exchange products for specific applications are the manufacturers themselves. Additional information on adsorbents and ion exchangers is available in many of the General References [e.g., Yang (2003)] and in several articles in *Kirk-Othmer Encyclopedia of Chemical Technology*. A comprehensive summary of commercial ion exchangers, including manufacturing methods, properties, and applications, is given by Dorfner (1991).

Several densities and void fractions are commonly defined. For adsorbents, the *bulk density* ρ_b denotes the mass of clean material per unit volume of a packed bed. The *extraparticle void fraction* ε denotes the void volume outside the adsorbent particles per unit volume of a packed bed. The *intraparticle porosity* ε_p denotes the void volume inside the adsorbent particles per unit volume of particle. Derived quantities are defined as follows. The dry *particle density* ρ_p is

$$\rho_p = \rho_b/(1-\varepsilon) \tag{16-1}$$

The *skeletal density* ρ_s of a particle (or crystalline density for a pure chemical compound) is

$$\rho_s = \rho_p/(1-\varepsilon_p) \tag{16-2}$$

For an adsorbent or ion exchanger, the *wet density* ρ_w of a particle is related to the liquid density ρ_f by

$$\rho_w = \rho_p + \varepsilon_p \rho_f \tag{16-3}$$

The *total void fraction* ε_b in a packed bed (outside and inside particles) is

$$\varepsilon_b = \varepsilon + (1-\varepsilon)\varepsilon_p \tag{16-4}$$

TABLE 16-5 Physical Properties of Adsorbents

Material and uses	Particle shape*	Size range, U.S. standard mesh†	Internal porosity, %	Bulk dry density, kg/L	Average pore diameter, nm	Surface area, km²/kg	Capacity (dry), kg/kg
Aluminas							
Low-porosity (fluoride sorbent)	G, S	8–14, etc.	40	0.70	~7	0.32	0.20
High-porosity (drying, separations)	G	Various	57	0.85	4–14	0.25–0.36	0.25–0.33
Desiccant, $CaCl_2$-coated	G	3–8, etc.	30	0.91	4.5	0.2	0.22
Activated bauxite	G	8–20, etc.	35	0.85	5		0.1–0.2
Chromatographic alumina	G, P, S	80–200, etc.	30	0.93			~0.14
Silicates and aluminosilicates							
Molecular sieves	S, C, P	Various					
Type 3A (dehydration)			~30	0.62–0.68	0.3	~0.7	0.21–0.23
Type 4A (dehydration)			~32	0.61–0.67	0.4	~0.7	0.22–0.26
Type 5A (separations)			~34	0.60–0.66	0.5	~0.7	0.23–0.28
Type 13X (purification)			~38	0.58–0.64	1.0	~0.6	0.25–0.36
Silicalite (hydrocarbons)	S, C, P	Various		0.64–0.70	0.6	~0.4	0.12–0.16
Dealuminated Y (hydrocarbons)	S, C, P	Various		0.48–0.53	0.8	0.5–0.8	0.28–0.42
Mordenite (acid drying)				0.88	0.3–0.8		0.12
Chabazite (acid drying)				0.72	0.4–0.5		0.20
LiX, LiLSX (air separation)	S	14–40		0.61	1.0		
Titanosilicate ETS-4 (N_2/CH_4)	S	Various	25–30	1.0	0.37		
Silica gel (drying, separations)	G, P	Various	38–48	0.70–0.82	2–5	0.6–0.8	0.35–0.50
Magnesium silicate (decolorizing)	G, P	Various	~33	~0.50		0.18–0.30	
Calcium silicate (fatty-acid removal)	P		75–80	~0.20		~0.1	
Clay, acid-treated (refining of petroleum, food products)	G	4–8		0.85			
Fuller's earth (same)	G, P	<200		0.80			
Diatomaceous earth	G	Various		0.44–0.50		~0.002	
Carbons							
Shell-based	G	Various	60	0.45–0.55	2	0.8–1.6	0.40
Wood-based	G	Various	~80	0.25–0.30		0.8–1.8	~0.70
Petroleum-based	G, C	Various	~80	0.45–0.55	2	0.9–1.3	0.3–0.4
Peat-based	G, C, P	Various	~55	0.30–0.50	1–4	0.8–1.6	0.5
Lignite-based	G, P	Various	70–85	0.40–0.70	3	0.4–0.7	0.3
Bituminous-coal-based	G, P	8–30, 12–40	60–80	0.40–0.60	2–4	0.9–1.2	0.4
Synthetic-polymer-based (pyrolized)	S	20–100	40–70	0.49–0.60		0.1–1.1	
Carbon molecular sieve (air separation)	C	Various	35–50	0.5–0.7	0.3–0.6		0.5–0.20
Organic polymers							
Polystyrene (removal of organics, e.g., phenol; antibiotics recovery)	S	20–60	40–60	0.64	4–20	0.3–0.7	
Polyacrylic ester (purification of pulping wastewater; antibiotics recovery)	G, S	20–60	50–55	0.65–0.70	10–25	0.15–0.4	
Phenolic & phenolic amine resin (decolorizing and deodorizing solutions)	G	16–50	45	0.42		0.08–0.12	0.45–0.55

*Shapes: C, cylindrical pellets; F, fibrous flakes; G, granules; P, powder; S, spheres.

†U.S. Standard sieve sizes (given in parentheses) correspond to the following diameters in millimeters: (3) 6.73, (4) 4.76, (8) 2.98, (12) 1.68, (14) 1.41, (16) 1.19, (20) 0.841, (30) 0.595, (40) 0.420, (50) 0.297, (60) 0.250, (80) 0.177, (200) 0.074.

TABLE 16-6 Physical Properties of Ion-Exchange Materials

Material	Shape* of particles	Bulk wet density (drained), kg/L	Moisture content (drained), % by weight	Swelling due to exchange, %	Maximum operating temperature,† °C	Operating pH range	Exchange capacity: Dry, equivalent/kg	Exchange capacity: Wet, equivalent/L
Cation exchangers: strongly acidic								
Polystyrene sulfonate								
Homogeneous (gel) resin	S				120–150	0–14		
4% cross-linked		0.75–0.85	64–70	10–12			5.0–5.5	1.2–1.6
6% cross-linked		0.76–0.86	58–65	8–10			4.8–5.4	1.3–1.8
8–10% cross-linked		0.77–0.87	48–60	6–8			4.6–5.2	1.4–1.9
12% cross-linked		0.78–0.88	44–48	5			4.4–4.9	1.5–2.0
16% cross-linked		0.79–0.89	42–46	4			4.2–4.6	1.7–2.1
20% cross-linked		0.80–0.90	40–45	3			3.9–4.2	1.8–2.0
Porous structure								
12–20% cross-linked	S	0.81	50–55	4–6	120–150	0–14	4.5–5.0	1.5–1.9
Cation exchangers: weakly acidic								
Acrylic (pK 5) or methacrylic (pK 6)								
Homogeneous (gel) resin	S	0.70–0.75	45–50	20–80	120	4–14	8.3–10	3.3–4.0
Macroporous	S	0.67–0.74	50–55	10–100	120		~8.0	2.5–3.5
Polystyrene phosphonate	G, S	0.74	50–70	<40	120	3–14	6.6	3.0
Polystyrene iminodiacetate	S	0.75	68–75	<100	75	3–14	2.9	0.7
Polystyrene amidoxime	S	~0.75	58	10	50	1–11	2.8	0.8–0.9
Zeolite (Al silicate)	G	0.85–0.95	40–45	0	60	6–8	1.4	0.75
Zirconium tungstate	G	1.15–1.25	~5	0	>150	2–10	1.2	1.0
Anion exchangers: strongly basic								
Polystyrene-based								
Trimethyl ammonium (type I)								
Homogeneous, 8% CL	S	0.70	46–50	~20	60–80	0–14	3.4–3.8	1.3–1.5
Porous, 11% CL	S	0.67	57–60	15–20	60–80	0–14	3.4	1.0
Dimethyl hydroxyethyl ammonium (type II)								
Homogeneous, 8% CL	S	0.71	~42	15–20	40–80	0–14	3.8–4.0	1.2
Porous, 10% CL	S	0.67	~55	12–15	40–80	0–14	3.8	1.1
Acrylic-based								
Homogeneous (gel)	S	0.72	~70	~15	40–80	0–14	~5.0	1.0–1.2
Porous	S	0.67	~60	~12	40–80	0–14	3.0–3.3	0.8–0.9
Anion exchangers: intermediately basic (pK 11)								
Polystyrene-based	S	0.75	~50	15–25	65	0–10	4.8	1.8
Epoxy-polyamine	S	0.72	~64	8–10	75	0–7	6.5	1.7
Anion exchangers: weakly basic (pK 9)								
Aminopolystyrene								
Homogeneous (gel)	S	0.67	~45	8–12	100	0–7	5.5	1.8
Porous	S	0.61	55–60	~25	100	0–9	4.9	1.2
Acrylic-based amine								
Homogeneous (gel)	S	0.72	~63	8–10	80	0–7	6.5	1.7
Porous	S	0.72	~68	12–15	60	0–9	5.0	1.1

*Shapes: C, cylindrical pellets; G, granules; P, powder; S, spheres.
†When two temperatures are shown, the first applies to H form for cation, or OH form for anion, exchanger; the second, to salt ion.

SORPTION EQUILIBRIUM

The quantity of a solute adsorbed can be defined conveniently in terms of moles or volume (for adsorption) or ion equivalents (for ion exchange) per unit mass or volume (dry or wet) of sorbent. Common units for adsorption are mol/(m^3 of fluid) for the fluid-phase concentration c_i and mol/(kg of clean adsorbent) for adsorbed-phase concentration n_i. For gases, partial pressure may replace concentration.

Many models have been proposed for adsorption and ion exchange equilibria. The most important factor in selecting a model from an engineering standpoint is to have an accurate mathematical description over the entire range of process conditions. It is usually fairly easy to obtain correct capacities at selected points, but isotherm shape over the entire range is often a critical concern for a regenerable process.

GENERAL CONSIDERATIONS

Forces Molecules are attracted to surfaces as the result of two types of forces: dispersion-repulsion forces (also called London or van der Waals forces) such as described by the Lennard-Jones potential for molecule–molecule interactions; and electrostatic forces, which exist as the result of a molecule or surface group having a permanent electric dipole or quadrupole moment or net electric charge.

Dispersion forces are always present and in the absence of any stronger force will determine equilibrium behavior, as with adsorption of molecules with no dipole or quadrupole moment on nonoxidized carbons and silicalite.

If a surface is polar, its resulting electric field will induce a dipole moment in a molecule with no permanent dipole and, through this polarization, increase the extent of adsorption. Similarly, a molecule with a permanent dipole moment will polarize an otherwise nonpolar surface, thereby increasing the attraction.

For a polar surface and molecules with permanent dipole moments, attraction is strong, as for water adsorption on a hydrophilic adsorbent. Similarly, for a polar surface, a molecule with a permanent quadrupole moment will be attracted more strongly than a similar molecule with a weaker moment; for example, nitrogen is adsorbed more strongly than oxygen on zeolites [Sherman and Yon (1991) in General References].

Surface Excess With a Gibbs dividing surface placed at the surface of the solid, the surface excess of component i, Γ_i (mol/m^2), is the amount of that component per unit area of solid contained in the region near the surface, above that contained at the fluid-phase concentration far from the surface. This is depicted in Fig. 16-4. The quantity adsorbed per unit mass of adsorbent is

$$n_i = \Gamma_i A \tag{16-5}$$

where A (m^2/kg) is the surface area of the solid.

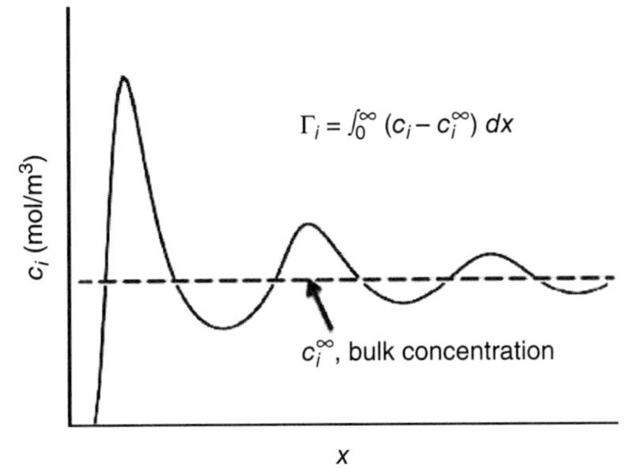

FIG. 16-4 Depiction of surface excess Γ_i. The force field of the solid concentrates component i near the surface; the concentration c_i is low at the surface because of short-range repulsive forces between adsorbate and surface.

For a porous adsorbent, the amount adsorbed in the pore structure per unit mass of adsorbent, based on surface excess, is obtained by the difference

$$n_i = n_i^{\text{tot}} - V_p c_i^\infty \tag{16-6}$$

where n_i^{tot} (mol/kg) is the total amount of component i contained within the particle's pore volume V_p (m³/kg), and c_i^∞ is the concentration outside of the particle. If n_i differs significantly from n_i^{tot} (as it will for weakly adsorbed species), then it is important to consider adsorbed-phase quantities in terms of surface excesses.

Classification of Isotherms by Shape Representative isotherms are shown in Fig. 16-5, as classified by Brunauer and coworkers. Curves that are concave downward throughout (type I) have historically been designated as "favorable," while those that are concave upward throughout (type III) are "unfavorable." Other isotherms (types II, IV, and V) have one or more inflection points. The designations favorable and unfavorable refer to fixed-bed behavior for the uptake step, with a favorable isotherm maintaining a compact wave shape. A favorable isotherm for uptake is unfavorable for the discharge step. This becomes particularly important for a regenerative process, in which a favorable isotherm may be too favorable for regeneration to occur effectively.

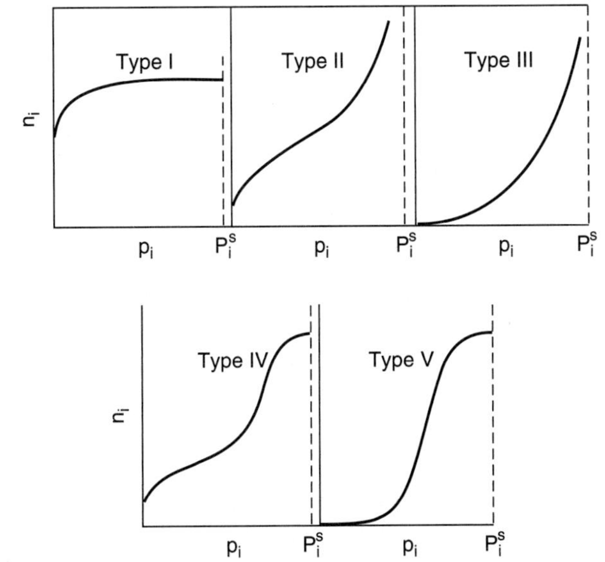

FIG. 16-5 Representative isotherm types. p_i and P_i^s are pressure and vapor pressure of the solute. [Brunauer, *J. Am. Chem. Soc.* **62:** 1723 (1940); reprinted with permission; see also Thommes et al., *Pure Appl. Chem.* **87:** 1051 (2015).]

Categorization of Equilibrium Models Historically, sorption equilibrium has been approached from different viewpoints. For adsorption, many models for flat surfaces have been used to develop explicit equations and equations of state for pure components and mixtures, and many of the resulting equations are routinely applied to porous materials. Explicit equations for pore filling have also been proposed, generally based on the Polanyi potential theory. Ion exchange adds to these approaches concepts of absorption or dissolution and exchange reactions. Statistical mechanics and molecular dynamics contribute to our understanding of all of these approaches (Steele, *The Interaction of Gases with Solid Surfaces,* Pergamon, Oxford, 1974; Nicholson and Parsonage, *Computer Simulation and the Statistical Mechanics of Adsorption,* Academic Press, New York, 1982). Mixture models are often based on the adsorbed solution theory, which uses thermodynamic equations from vapor–liquid equilibria with volume replaced by surface area and pressure replaced by a two-dimensional spreading pressure. Other approaches include lattice theories and mass-action exchange equilibrium.

Heterogeneity Adsorbents and ion exchangers can be physically and chemically heterogeneous. Although exceptions exist, solutes generally compete for the same sites. Models for adsorbent heterogeneity have been developed for both discrete and continuous distributions of energies [Ross and Olivier, *On Physical Adsorption,* Interscience, New York, 1964; Jaroniec and Madey (1988); Rudzinski and Everett (1992) in General References].

Isosteric Heat of Adsorption The most useful heat of adsorption for fixed-bed calculations is the *isosteric heat of adsorption,* which is given by the Clausius-Clapeyron type relation

$$\lambda_i = \Re T^2 \left.\frac{\partial \ln p_i}{\partial T}\right|_{n_i, n_j} \tag{16-7}$$

where the n_j can be dropped for single-component adsorption. λ_i is positive by convention. If isosteres are straight lines when plotted as $\ln p_i$ versus T^{-1} (see Fig. 16-1), then Eq. (16-7) can be integrated to give

$$\ln p_i = f(n_i) - \frac{\lambda_i}{\Re T} \tag{16-8a}$$

where $f(n_i)$ is an arbitrary function dependent only on n_i. Many other heats of adsorption have been defined, and their utility depends on the application [Ross and Olivier, *On Physical Adsorption,* Interscience, New York, 1964; Young and Crowell (1962) in General References].

From Eq. (16-8*a*), if a single isotherm is known in the pressure explicit form $p_i = p_i(n_i, T)$ and if λ_i is known at least approximately, then equilibria can be estimated over a narrow temperature range using

$$\ln \frac{p_i}{p_i^{\text{ref}}} = \frac{\lambda_i}{\Re}\left(\frac{1}{T^{\text{ref}}} - \frac{1}{T}\right) \quad (\text{const } n_i) \tag{16-8b}$$

Similarly, Eq. (16-8*b*) can be used to calculate the isosteric heat of adsorption from two isotherms.

Experiments Sorption equilibria are measured using equipment and methods classified as volumetric, gravimetric, flow-through (frontal analysis), and chromatographic. Equipment is discussed elsewhere [Yang (1987); Keller and Staudt (2005) in General References]. Heats of adsorption can be determined from isotherms measured at different temperatures or measured independently by calorimetric methods.

Dimensionless Concentration Variables Where appropriate, isotherms will be written here using the dimensionless system variables

$$c_i^* = \frac{c_i}{c_i^{\text{ref}}} \qquad n_i^* = \frac{n_i}{n_i^{\text{ref}}} \tag{16-9}$$

where the best choice of reference values depends on the operation.

In some cases, to allow for some preloading of the adsorbent, it will be more convenient to use the dimensionless transition variables

$$c_i^* = \frac{c_i - c_i'}{c_i'' - c_i'} \qquad n_i^* = \frac{n_i - n_i'}{n_i'' - n_i'} \tag{16-10}$$

where single and double primes indicate initial and final concentrations, respectively. Figure 16-6 shows n_i plotted versus c_i for a sample system. Superimposed are an upward transition (loading) and a downward transition (unloading), shown by the respective positions of (c_i, n_i) and (c_i', n_i') and (c_i'', n_i'').

SINGLE COMPONENT OR EXCHANGE

The simplest relationship between solid-phase and fluid-phase concentrations is the **linear isotherm**

$$n_i = K_i c_i \qquad \text{or} \qquad n_i = K_i' p_i \tag{16-11}$$

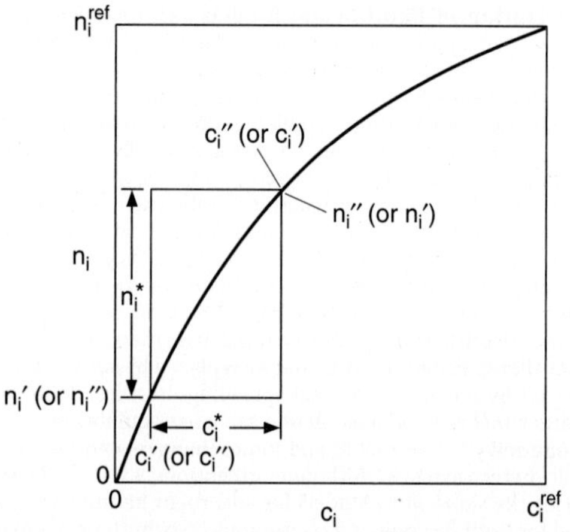

FIG. 16-6 Isotherm showing concentration variables for a transition from (c_i', n_i') to (c_i'', n_i'').

Thermodynamics requires that a linear limit be approached in the Henry's law region for all isotherm equations.

Flat-Surface Isotherm Equations The classification of isotherm equations into two broad categories for flat surfaces and pore filling reflects their origin. This distinction does not restrict equations developed for flat surfaces from being applied successfully to describe data for porous adsorbents.

The classical isotherm for a homogeneous flat surface, and most popular of all nonlinear isotherms, is the *Langmuir isotherm*

$$n_i = \frac{n_i^s K_i c_i}{1 + K_i c_i} \tag{16-12}$$

where n_i^s is the monolayer capacity approached at large concentrations and K_i is an equilibrium constant. These parameters are often determined by plotting $1/n_i$ versus $1/c_i$. The derivation of the isotherm assumes negligible interaction between adsorbed molecules. For temperature-dependent adsorption equilibrium and a constant isosteric heat of adsorption, K_i is given by

$$K_i = K_i^0 \exp(\lambda_i/\Re T) \tag{16-13}$$

where K_i^0 is a constant. Note that at low concentrations, the Langmuir isotherm approaches the linear limit $(n_i = n_i^s K_i c_i)$. The Langmuir isotherm can also be written for two or more patches of differing energies to account for surface heterogeneity, with the total amount adsorbed being the sum of the patchwise contributions. A wide variety of gas adsorption equilibria on different adsorbents can be described using only two patches [Ritter et al., *Langmuir* **27:** 4700 (2011); Bhadra et al., *Langmuir* **28:** 6935 (2012)]. This dual process Langmuir (DPL) model is given by

$$n_i = \left(\frac{n_{1,i}^s K_{1,i} p_i}{1 + K_{1,i} p_i}\right)_{\text{Site 1}} + \left(\frac{n_{2,i}^s K_{2,i} p_i}{1 + K_{2,i} p_i}\right)_{\text{Site 2}} \tag{16-14}$$

where $n_{j,i}^s$ and $K_{j,i}$ are the saturation capacity and or equilibrium constant on site j. For temperature dependence, a form of Eq. (16-13) is applied to each site. Another variation on Eq. (16-12) is the multisite Langmuir isotherm [Nitta et al., *J. Chem. Eng. Japan* **17:** 39 (1984)], which is

$$K_i p_i = \frac{n_i/n_i^s}{(1 - n_i/n_i^s)^{a_i}} \tag{16-15}$$

where a_i is a constant.

The classical isotherm for multilayer adsorption on a homogeneous, flat surface is the *BET isotherm* [Brunauer et al., *J. Am. Chem. Soc.* **60:** 309 (1938)]

$$n_i = \frac{n_i^s K_i p_i}{[1 + K_i p_i - (p_i/P_i^s)][1 - (p_i/P_i^s)]} \tag{16-16}$$

where p_i is the pressure of the adsorbable component and P_i^s is its vapor pressure. It is useful for gas–solid systems in which condensation is approached, fitting type II behavior.

For a heterogeneous flat surface, a classical isotherm is the *Freundlich isotherm*

$$n_i = K_i (c_i)^{m_i} \tag{16-17}$$

where m_i is positive and generally not an integer. The isotherm corresponds approximately to an exponential distribution of heats of adsorption. Although it lacks the required linear behavior in the Henry's law region, it can often be used to correlate data on heterogeneous adsorbents over wide ranges of concentration.

Several isotherms combine aspects of both the Langmuir and Freudlich equations. One that has been shown to be effective in describing data mathematically for heterogeneous adsorbents is the *Tóth isotherm* [Toth, *Acta Chim. Acad. Sci. Hung.* **69:** 311 (1971)]

$$n_i = \frac{n_i^s p_i}{\left[(1/K_i) + p_i^{m_i}\right]^{1/m_i}} \tag{16-18}$$

This three-parameter equation behaves linearly in the Henry's law region and reduces to the Langmuir isotherm for $m_i = 1$. Other well-known isotherms include the *Langmuir-Freundlich isotherm* or *Sips isotherm* [Sips, *J. Chem. Phys.* **16:** 490 (1948); Koble and Corrigan, *Ind. Eng. Chem.* **44:** 383 (1952)] or *loading ratio correlation* with prescribed temperature dependence [Yon and Turnock, *AIChE Symp. Ser.* **67**(117): 75 (1971)]

$$n_i = \frac{n_i^s (K_i p_i)^{m_i}}{1 + (K_i p_i)^{m_i}} \tag{16-19}$$

Equations of state are also used. From an equation of state written in terms of the two-dimensional spreading pressure π, the corresponding isotherm is easily determined, as described later for adsorption of mixtures [see Eq. (16-43)]. The two-dimensional equivalent of an ideal gas is an ideal surface gas, which is described by

$$\pi A = n_i \Re T \tag{16-20}$$

which gives the linear isotherm, Eq. (16-11). Many more complicated equations of state are available, including two-dimensional analogs of the virial equation and equations of van der Waals, Redlich-Kwong, Peng-Robinson, etc. [Adamson (1990) in General References; Patrykiejewet et al., *Chem. Eng. J.* **15:** 147 (1978); Haydel and Kobayashi, *Ind. Eng. Chem. Fundam.* **6:** 546 (1967)].

Pore-Filling Isotherm Equations Most pore-filling models are grounded in the Polanyi potential theory. In Polanyi's model, an attracting potential energy field is assumed to exist adjacent to the surface of the adsorbent and concentrates vapors there. Adsorption takes place wherever the strength of the field, independent of temperature, is great enough to compress the solute to a partial pressure greater than its vapor pressure. The last molecules to adsorb form an equipotential surface containing the adsorbed volume. The strength of this field, called the adsorption potential ε (J/mol), was defined by Polanyi to be equal to the work required to compress the solute from its partial pressure to its vapor pressure

$$\varepsilon = \Re T \ln(P_i^s/p_i) \tag{16-21}$$

The same result is obtained by considering the change in chemical potential. In the basic theory, W (m³/kg), the volume adsorbed as saturated liquid at the adsorption temperature, is plotted versus ε to give a characteristic curve. Data measured at different temperatures for the same solute-adsorbent pair should fall on this curve. Using the method, it is possible to use data measured at a single temperature to predict isotherms at other temperatures. Data for additional, homologous solutes can be collapsed into a single "correlation curve" by defining a scaling factor β, the most useful of which has been V/V^{ref}, the adsorbate molar volume as saturated liquid at the adsorption temperature divided by that for a reference compound. Thus, by plotting W versus ε/β or ε/V for data measured at various temperatures for various similar solutes and a single adsorbent, a single curve should be obtained. Variations of the theory are often used to evaluate properties for components near or above their critical points [Grant and Manes, *Ind. Eng. Chem. Fund.* **5:** 490 (1966)].

The most popular equations used to describe the shape of a characteristic curve or a correlation curve are the two-parameter *Dubinin-Radushkevich* (DR) equation

$$\frac{W}{W_0} = \exp\left[-k\left(\frac{\varepsilon}{\beta}\right)^2\right] \tag{16-22}$$

and the three-parameter *Dubinin-Astakhov* (DA) equation

$$\frac{W}{W_0} = \exp\left[-k\left(\frac{\varepsilon}{\beta}\right)^m\right] \tag{16-23}$$

where W_0 is micropore volume and m is related to the pore size distribution [Gregg and Sing (1982)]. Neither of these equations has correct limiting behavior in the Henry's law regime.

Ion Exchange The *mass-action law* describes ion exchange equilibrium in fully ionized exchanger systems as

$$z_B A + z_A \bar{B} \rightleftharpoons z_B \bar{A} + z_A B$$

where overbars indicate the ionic species in the ion exchanger and z_A and z_B are the valences of counterions A and B. The associated equilibrium relation is

$$K^c_{A,B} = K_{A,B}\left(\frac{\gamma_A}{\bar{\gamma}_A}\right)^{z_B}\left(\frac{\bar{\gamma}_B}{\gamma_B}\right)^{z_A} = \left(\frac{n_A}{c_A}\right)^{z_B}\left(\frac{c_B}{n_B}\right)^{z_A} \tag{16-24}$$

where $K^c_{A,B}$ is the apparent equilibrium constant or molar selectivity coefficient, $K_{A,B}$ is the thermodynamic equilibrium constant based on activities, and the γs are activity coefficients. Often concentrations are represented in terms of equivalent ionic fractions based on solution normality c_{tot} and ion exchanger capacity n^s as $c^*_i = z_i c_i / c_{tot}$ and $n^*_i = z_i n_i / n_{tot}$, where $c_{tot} = \Sigma z_j c_j$ and $n_{tot} = \Sigma z_j n_j = n^s$ with the summations extended to all counter-ion species. A rational selectivity coefficient is then defined as

$$K'_{A,B} = K^c_{A,B}\left(\frac{n_{tot}}{c_{tot}}\right)^{z_A - z_B} = \left(\frac{n^*_A}{c^*_A}\right)^{z_B}\left(\frac{c^*_B}{n^*_B}\right)^{z_A} \tag{16-25}$$

For the exchange of ions of equal valence ($z_A = z_B$), $K^c_{A,B}$ and $K'_{A,B}$ are coincident and, to a first approximation, independent of solution normality. When $z_A > z_B$, $K^c_{A,B}$ decreases with solution normality. Thus, ion exchangers exhibit an increasing affinity for ions of lower valence as the solution normality increases.

An alternate form of Eq. (16-25) is

$$n^*_A = K^c_{A,B}\left(\frac{n_{tot}}{c_{tot}}\right)^{(z_A - z_B)/z_B}\left(\frac{1 - n^*_A}{1 - c^*_A}\right)^{z_A/z_B} c^*_A \tag{16-26}$$

When B is in excess (that is $c^*_A, n^*_A \to 0$), this reduces to

$$n^*_A = K^c_{A,B}\left(\frac{n_{tot}}{c_{tot}}\right)^{(z_A - z_B)/z_B} c^*_A = K_A c^*_A \tag{16-27}$$

where the linear equilibrium constant $K_A = K^c_{A,B}(n_{tot}/c_{tot})^{(z_A/z_B)-1}$ decreases with solution normality c_{tot} when $z_A > z_B$ or increases when $z_A < z_B$.

Table 16-7 gives equilibrium constants $K_{A,B}$ for cross-linked cation and anion exchangers for a variety of counterions. The values given for cation exchangers are based on ion A replacing Li^+, and those given for anion exchangers are based on ion A replacing Cl^-. The selectivity for a particular ion generally increases with decreasing hydrated ion size and increasing degree of cross-linking. The selectivity coefficient for any two ions A and D can be obtained from Table 16-7 from values of $K_{A,B}$ and $K_{D,B}$ as

$$K_{A,D} = K_{A,B}/K_{D,B} \tag{16-28}$$

The values given in this table are only approximate, but they are adequate for process screening purposes with Eqs. (16-24) and (16-25). Rigorous calculations generally require that activity coefficients be accounted for. However, for the exchange between ions of the same valence at solution concentrations of 0.1 N or less, or between any ions at 0.01 N or less, the solution-phase activity coefficients will be similar enough that they can be omitted.

Models for ion exchange equilibria based on the mass-action law taking into account solution and exchanger-phase nonidealities with equations similar to those for liquid mixtures have been developed by several authors [see Smith and Woodburn, *AIChE J.* **24:** 577 (1978); Mehablia et al., *Chem. Eng. Sci.* **49:** 2277 (1994)]. Thermodynamics-based approaches are also available [Soldatov in Dorfner (1991) in General References; Novosad and Myers, *Can J. Chem. Eng.* **60:** 500 (1982); Myers and Byington in Rodrigues, ed., *Ion Exchange Science and Technology,* NATO ASI Series, No. 107, Nijhoff, Dordrecht, 1986, pp. 119–145] as well as approaches for the exchange of macromolecules [see Carta and Jungbauer (2010)] also taking into account steric-hindrance effects [Brooks and Cramer, *AIChE J.* **38:** 12 (1992)].

Example 16-3 Calculation of Useful Ion-Exchange Capacity An 8 percent cross-linked sulfonated resin is used to remove calcium from a solution containing 0.0007 mol/L Ca^{+2} and 0.01 mol/L Na^+. The total resin ion-exchange capacity is 2.0 equiv/L. Estimate the resin capacity for calcium removal for the conditions given.

From Table 16-7, $K_{Ca,Na} = 5.16/1.98 = 2.6$. Since $n_{tot} = 2$ equiv/L and $c_{tot} = 2 \times 0.0007 + 0.01 = 0.011$ equiv/L, $K'_{Ca,Na} = 2.6 \times 2/0.011 = 470$. Thus, with $C^*_{Ca} = 2 \times 0.0007/0.011 = 0.13$, Eq. (16-25) gives $470 = (n^*_{Ca}/0.13)[(1 - 0.13)/(1 - n^*_{Na})]^2$ or $n^*_{Ca} = 0.9$. The available capacity for calcium is $n_{Ca} = 0.9 \times 2.0/2 = 0.9$ mol/L.

Donnan Uptake The uptake of an electrolyte as a neutral ion pair of a salt is called Donnan uptake. It is generally negligible at low ionic concentrations. Above 0.5 g·equiv/L with strongly ionized exchangers (or at lower concentrations with those more weakly ionized), the resin's fixed ion-exchange capacity is measurably exceeded as a result of electrolyte invasion. With only one coion species Y (matching the charge sign of the fixed groups in the resin), its uptake n_Y equals the total excess uptake of the counterion.

TABLE 16-7 Equilibrium Constants for Polystyrene DVB Cation and Anion Exchangers

Strong acid sulfonated cation exchangers (Li^+ reference ion)

	Degree of cross-linking				Degree of cross-linking		
Counterion	4% DVB	8% DVB	16% DVB	Counterion	4% DVB	8% DVB	16% DVB
Li^+	1.00	1.00	1.00	Mg^{++}	2.95	3.29	3.51
H^+	1.32	1.27	1.47	Zn^{++}	3.13	3.47	3.78
Na^+	1.58	1.98	2.37	Co^{++}	3.23	3.74	3.81
NH_4^+	1.90	2.55	3.34	Cu^{++}	3.29	3.85	4.46
K^+	2.27	2.90	4.50	Cd^{++}	3.37	3.88	4.95
Rb^+	2.46	3.16	4.62	Ni^{++}	3.45	3.93	4.06
Cs^+	2.67	3.25	4.66	Ca^{++}	4.15	5.16	7.27
Ag^+	4.73	8.51	22.9	Pb^{++}	6.56	9.91	18.0
Tl^+	6.71	12.4	28.5	Ba^{++}	7.47	11.5	20.8

Strong base anion exchangers, 8% DVB (Cl^- reference ion)

Counterion	Type I resin*	Type II resin†	Counterion	Type I resin*	Type II resin†
Salicylate	32	28	Cyanide	1.6	1.3
Iodide	8.7	7.3	Chloride	1.0	1.0
Phenoxide	5.2	8.7	Hydroxide	0.05–0.07	0.65
Nitrate	3.8	3.3	Bicarbonate	0.3	0.5
Bromide	2.8	2.3	Formate	0.2	0.2
Nitrite	1.2	1.3	Acetate	0.2	0.2
Bisulfite	1.3	1.3	Fluoride	0.09	0.1
Cyanide	1.6	1.3	Sulfate	0.15	0.15

*Trimethylamine.
†Dimethyl-hydroxyethylamine.
Data from Bonner and Smith, *J. Phys. Chem.* **61:** 326 (1957) and Wheaton and Bauman, *Ind. Eng. Chem.* **43:** 1088 (1951).

Equilibrium is described by the mass-action law. For the case of a resin in *A*-form in equilibrium with a salt *AY*, the excess counterion uptake is given by [Helfferich (1962), pp. 133–147, in General References].

$$z_Y n_Y = \left[\frac{n^{s2}}{4} + (z_Y c_Y)^2 \left(\frac{\gamma_{AY}}{\bar{\gamma}_{AY}}\right)^2 \left(\frac{\bar{a}_w}{a_w}\right)^{\hat{v}_{AY}/\hat{v}_w}\right]^{1/2} - \frac{n^s}{2} \qquad (16\text{-}29)$$

where γs are activity coefficients, *a*s are water activities, and *v*s are partial molar volumes. For dilute conditions, Eq. (16-29) predicts a squared dependence of n_Y on c_Y. Thus, the electrolyte sorption isotherm has a strong positive curvature. Donnan uptake is more pronounced for resins of lower degree of cross-linking and for counterions of low valence.

Separation Factor By analogy with the mass-action case and appropriate for both adsorption and ion exchange, a *separation factor r* can be defined based on dimensionless system variables [Eq. (16-9)] by

$$r = \frac{c_i^*(1-n_i^*)}{n_i^*(1-c_i^*)} \quad \text{or} \quad r = \frac{n_B^*/c_B^*}{n_A^*/c_A^*} \qquad (16\text{-}30)$$

This term is analogous to relative volatility or its reciprocal (or to an equilibrium selectivity). Similarly, the assumption of a constant separation factor is a useful assumption in many sorptive operations. [It is constant for the Langmuir isotherm, as described below, and for mass-action equilibrium with $z_a = z_b$ in Eq. (16-24).] This gives the constant separation factor isotherm

$$n_i^* = \frac{c_i^*}{r + (1-r)c_i^*} \qquad (16\text{-}31)$$

The separation factor *r* identifies the equilibrium increase in n_i^* from 0 to 1, which accompanies an increase in c_i^* from 0 to 1. For a concentration change over only part of the isotherm, a separation factor *R* can be defined for the dimensionless transition variables [Eq. (16-10)]. This separation factor is

$$R = \frac{n_i''/c_i''}{n_i'/c_i'} = \frac{c_i^*(1-n_i^*)}{n_i^*(1-c_i^*)} \qquad (16\text{-}32)$$

and gives an equation identical to Eq. (16-31) with *R* replacing *r*.

Figure 16-7 shows constant separation factor isotherms for a range of *r* (or *R*) values. The isotherm is linear for $r = 1$, favorable for $r < 1$, rectangular (or irreversible) for $r = 0$, and unfavorable for $r > 1$. As a result of symmetry properties, if *r* is defined for adsorption of component *i* or exchange of ion *A* for *B*, then the reverse process is described by $1/r$.

The Langmuir isotherm, Eq. (16-12), corresponds to the constant separation factor isotherm with

$$r = 1/\left(1 + K_i c_i^{\text{ref}}\right) \qquad (16\text{-}33)$$

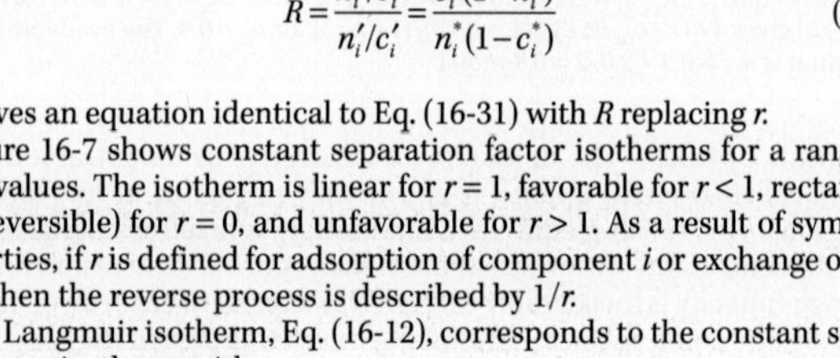

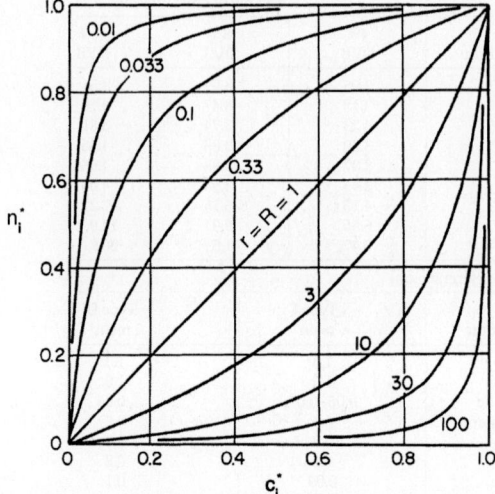

FIG. 16-7 Constant separation factor isotherm as a function of the separation factor *r* (or interchangeably *R*). Each isotherm is symmetric about the perpendicular line connecting (0,1) and (1,0). Isotherms for *r* and $1/r$ are symmetric about the 45° line.

for system variables [Eq. (16-9)] or

$$R = \frac{r + (1-r)\left(c_i'/c_i^{\text{ref}}\right)}{r + (1-r)\left(c_i''/c_i^{\text{ref}}\right)} \qquad (16\text{-}34)$$

for transition variables [Eq. (16-10)]. Vermeulen et al. (1984) give additional properties of constant separation factor isotherms.

Example 16-4 Application of Isotherms Thomas [*Ann. N.Y. Acad. Sci.* **49:** 161 (1948)] provides the following Langmuir isotherm for the adsorption of anthracene from cyclohexane onto alumina:

$$n_i = \frac{22c_i}{1 + 375c_i}$$

with n_i in mol anthracene/kg alumina and c_i in mol anthracene/L liquid.

a. What are the values of K_i and n_i^s according to Eq. (16-12)?

$$K_i = 375 \text{ L/mol}$$

$$n_i^s = \frac{22}{K_i} = 0.0587 \text{ mol/kg}$$

b. For a feed concentration of 8.11×10^{-4} mol/L, what is the value of *r*?

$$r = \frac{1}{1 + K_i(8.11 \times 10^{-4})} = 0.766 \qquad \text{[from Eq. (16-33)]}$$

c. If the alumina is presaturated with liquid containing 2.35×10^{-4} mol/L and the feed concentration is 8.11×10^{-4} mol/L, what is the value of *R*?

$$R = 0.834 \qquad \text{[from Eq. (16-32) or (16-34)]}$$

MULTICOMPONENT ADSORPTION OR ION EXCHANGE

When more than one adsorbed species or more than two ion-exchanged species interact in some manner, equilibrium becomes more complicated. Usually, thermodynamics provides a sound basis for prediction.

Adsorbed-Solution Theory The common thermodynamic approach to multicomponent adsorption treats adsorption equilibrium in a way analogous to fluid-fluid equilibrium. The theory has as its basis the Gibbs adsorption isotherm [Young and Crowell (1962) in General References], which is

$$A d\pi = \sum_i n_i d\mu_i \quad (\text{const } T) \qquad (16\text{-}35)$$

where μ_i is chemical potential. For an ideal gas ($d\mu_i = \Re T d \ln p_i$), if it is assumed that an adsorbed solution is defined with a pure-component standard state (as is common for a liquid solution), then Eq. (16-35) can be integrated to give [Rudisill and LeVan, *Chem. Eng. Sci.* **47:** 1239 (1992)]

$$p_i = \gamma_i x_i P_i^{\text{ref}}(T, \pi) \qquad (16\text{-}36)$$

where γ_i and x_i are the adsorbed-phase activity coefficient and mole fraction of component *i* and P_i^{ref} is the standard state, specified to be at the temperature and spreading pressure of the mixture.

Equation (16-36) with $\gamma_i = 1$ provides the basis for the ideal adsorbed-solution theory [Myers and Prausnitz, *AIChE J.* **11:** 121 (1965)]. The spreading pressure for a pure component is determined by integrating Eq. (16-35) for a pure component to obtain

$$\frac{\pi A}{\Re T} = \int_0^{P_i^{\text{ref}}} \frac{n_i}{p_i} dp_i \qquad (16\text{-}37)$$

where n_i is given by the pure-component isotherm. Also, since $\sum x_i = 1$, Eq. (16-36) with $\gamma_i = 1$ gives $\sum(p_i/P_i^{\text{ref}}) = 1$. With no area change on mixing for the ideal solution, the total number of moles adsorbed per unit weight of adsorbent is determined using a two-dimensional form of Amagat's law:

$$\frac{1}{n_{\text{tot}}} = \sum \frac{x_i}{n_i^{\text{ref}}} \qquad (16\text{-}38)$$

where $n_{\text{tot}} = \sum n_i$ and n_i^{ref} is given by the pure-component isotherm at P_i^{ref}. Adsorbed-phase concentrations are calculated using $n_i = x_i n_{\text{tot}}$. Generally, different values of π [or $\pi A/(\Re T)$] must be tried until one is found that satisfies Eq. (16-37) and $\sum x_i = 1$.

Example 16-5 Application of Ideal Adsorbed-Solution Theory Consider a binary adsorbed mixture for which each pure component obeys the Langmuir equation, Eq. (16-12). Let $n_1^s = 4$ mol/kg, $n_2^s = 3$ mol/kg, $K_1p_1 = K_2p_2 = 1$. Use the ideal adsorbed-solution theory to determine n_1 and n_2.

Substituting the pure component Langmuir isotherm

$$n_i = \frac{n_i^s K_i p_i}{1 + K_i p_i}$$

into Eq. (16-37) and integrating gives

$$\frac{\pi A}{\Re T} = n_i^s \ln(1 + K_i P_i^{\text{ref}})$$

which can be solved explicitly for $K_iP_i^{\text{ref}}$. Values are guessed for $\pi A/(\Re T)$, values of $K_iP_i^{\text{ref}}$ are calculated from the preceding equation, and $\Sigma x_i = \Sigma K_i p_i/(K_i P_i^{\text{ref}}) = 1$ is checked to see if it is satisfied. Trial and error gives $\pi A/(\Re T) = 3.8530$ mol/kg, $K_1P_1^{\text{ref}} = 1.6202$, $K_2P_2^{\text{ref}} = 2.6123$, and $x_1 = 0.61720$. Evaluating the pure-component isotherms at the reference pressures and using Eq. (16-38) gives $n_{\text{tot}} = 2.3475$ mol/kg, and finally $n_i = x_i n_{\text{tot}}$ gives $n_1 = 1.4489$ mol/kg and $n_2 = 0.8986$ mol/kg.

Other approaches to account for various effects have been developed. Negative deviations from Raoult's law (i.e., $\gamma_i < 1$) are often found due to adsorbent heterogeneity [e.g., Myers, *AIChE J.* **29:** 691 (1983)]. Thus, contributions include accounting for adsorbent heterogeneity [Valenzuela et al., *AIChE J.* **34:** 397 (1988)] and excluded pore-volume effects [Myers, in Rodrigues, LeVan, and Tondeur (1989) in General References]. Several activity coefficient models have been developed to account for nonideal adsorbate–adsorbate interactions, including a spreading pressure-dependent activity coefficient model [e.g., Talu and Zwiebel, *AIChE J.* **32:** 1263 (1986)] and a vacancy solution theory [Suwanayuen and Danner, *AIChE J.* **26:** 68–76 (1980)].

Langmuir-Type Relations For systems composed of solutes that individually follow Langmuir isotherms, the traditional multicomponent Langmuir equation, obtained via a kinetic derivation, is

$$n_i = \frac{n_i^s K_i p_i}{1 + \sum_j K_j p_j} \tag{16-39}$$

where j is summed over all components. This equation has been criticized on thermodynamic grounds because it does not satisfy the Gibbs adsorption isotherm unless all monolayer capacities n_i^s are equal.

To satisfy the Gibbs adsorption isotherm for unequal monolayer capacities, explicit isotherms can be obtained in the form of a series expansion [LeVan and Vermeulen, *J. Phys. Chem.* **85:** 3247 (1981)]. A two-term form is

$$n_1 = \frac{(n_1^s + n_2^s)K_1 p_1}{2(1 + K_1 p_1 + K_2 p_2)} + \frac{(n_1^s - n_2^s)K_1 p_1 K_2 p_2}{(K_1 p_1 + K_2 p_2)^2} \ln(1 + K_1 p_1 + K_2 p_2) \tag{16-40}$$

where the subscripts may be interchanged. Multicomponent forms are also available [Frey and Rodrigues, *AIChE J.* **40:** 182 (1994)].

A variation of Eq. (16-39) follows from an extension of the DPL model Eq. (16-14) to gas mixture adsorption equilibria and is given by [Ritter et al., *Langmuir* **27:** 4700 (2011)]

$$n_i = \sum_{j=1}^{2} \left(\frac{n_{j,i}^s K_{j,i} p_i}{1 + \sum_{k=1}^{N} K_{n,k} p_k} \right)_{\text{Site } j} \tag{16-41}$$

where n_i is the total amount of component i adsorbed on both sites j from an N component mixture. Since there are two different free energies for each component in the mixture, that is, one for each site, this leads to perfect positive (PP) or perfect negative (PN) correlation. When a pair of components both see site 1 as the high-energy site and site 2 as the low-energy site, then their energies correlate in a PP fashion, and $K_{n,k} = K_{j,k}$. When one component sees site 1 as the high-energy site and another component sees site 1 as the low-energy site, then their energies correlate in a PN fashion, and $K_{n,k} = K_{1,k}$ if $j = 2$ and $K_{n,k} = K_{2,k}$ if $j = 1$. This PP and PN behavior allows the DPL model to predict very nonideal behavior, including azeotropes using only single component parameters.

Example 16-6 Comparison of Binary Langmuir Isotherms Use the numerical values in Example 16-5 to evaluate the binary Langmuir isotherms given by Eqs. (16-39) and (16-40) and compare results with the exact answers given in Example 16-5.

Equation (16-39) gives $n_1 = 1.3333$ mol/kg and $n_2 = 1.0000$ mol/kg for an average deviation from the exact values of approximately 10 percent. Equation (16-40) gives $n_1 = 1.4413$ mol/kg and $n_2 = 0.8990$ mol/kg for an average deviation of about 0.6 percent.

Freundlich-Type Relations A binary Freundlich isotherm, obtained from the ideal adsorbed solution theory in loading-explicit closed form [Crittenden et al., *Environ. Sci. Technol.* **19:** 1037 (1985)], is

$$c_i = \frac{n_i}{\sum_{j=1}^{N} n_j} \left(\frac{\sum_{j=1}^{N} (n_j/m_j)}{K_i/m_i} \right)^{1/m_i} \tag{16-42}$$

Equations of State If an equation of state is specified for a multicomponent adsorbed phase of the form $\pi A/(\Re T) = f(n_1, n_2, \ldots)$, then the isotherms are determined using [Van Ness, *Ind. Eng. Chem. Fundam.* **8:** 464–473 (1969)]

$$\ln\left(\frac{k_i p_i}{n_i/M_s}\right) = \int_A^\infty \left[\left. \frac{\partial[\pi A/(\Re T)]}{\partial n_i} \right|_{T,A,n_j} - 1 \right] \frac{dA}{A} \tag{16-43}$$

where, because integration is over A, M_s is mass of adsorbent, units for n and A are mol and m^2 (rather than mol/kg and m^2/kg), and $K_ip_i/(n_i/M_s) = 1$ is the linear lower limit approached for the ideal surface gas [see Eqs. (16-12) and (16-20)].

Ion Exchange—Stoichiometry In most applications, except for some weak-electrolyte and some concentrated-solution cases, the following summations apply:

$$\sum_i z_i c_i = c_{\text{tot}} = \text{const} \qquad \sum_i z_i n_i = n_{\text{tot}} = \text{const} \tag{16-44a}$$

In equivalent-fraction terms, the sums become

$$\sum_i c_i^* = 1 \qquad \sum_i n_i^* = 1 \tag{16-44b}$$

Mass Action Here the equilibrium relations, consistent with Eq. (16-25), are

$$K'_{ij} = \left(\frac{n_i^*}{c_i^*}\right)^{z_j} \left(\frac{c_j^*}{n_j^*}\right)^{z_i} \tag{16-45}$$

For an N-species system, with N values of c^* (or n^*) known, the N values of n^* (or c^*) can be found by simultaneous solution of the $N - 1$ independent i,j combinations for Eq. (16-45) using Eq. (16-44a); one n_j^*/c_j^* is assumed, the other values can be calculated using Eq. (16-45), and the sum of the trial n^* values (or c^* values) is compared with Eq. (16-44b).

Because an N-component system has $N - 1$ independent concentrations, a three-component equilibrium can be plotted in a plane and a four-component equilibrium in a three-dimensional space. Figure 16-8 shows a triangular plot of c^* contours in equilibrium with the corresponding n^* coordinates.

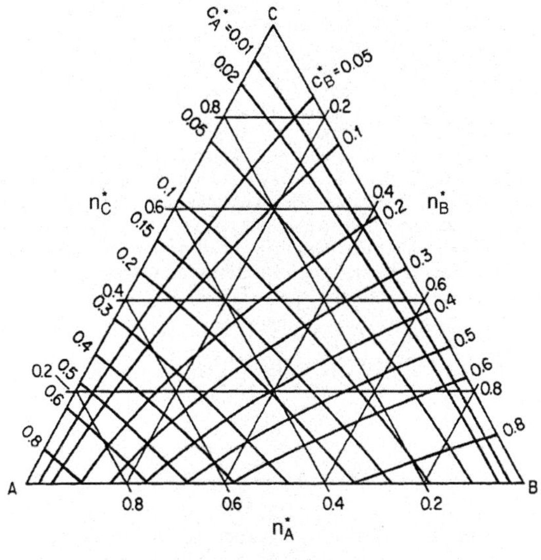

FIG. 16-8 Ideal mass-action equilibrium for three-component ion exchange with unequal valences. $K'_{A,C} = 8.06$; $K'_{B,C} = 3.87$. Duolite C-20 polystyrenesulfonate resin, with Ca as A, Mg as B, and Na as C. [Klein et al., *Ind. Eng. Chem. Fund.* **6:** 339 (1967); reprinted with permission.]

Improved models for ion-exchange equilibria are in Smith and Woodburn [*AIChE J.* **24:** 577 (1978)]; Mehablia et al. [*Chem. Eng. Sci.* **49:** 2277 (1994)]; Soldatov [in Dorfner (1991) in General References]; Novosad and Myers, [*Can J. Chem. Eng.* **60:** 500 (1982)]; and Myers and Byington [in Rodrigues, ed., *Ion Exchange: Science and Technology*, NATO ASI Series, No. 107, Nijhoff, Dordrecht, 1986, pp. 119–145].

Constant Separation-Factor Treatment If the valences of all species are equal, the separation factor α_{ij} applies, where

$$\alpha_{ij} = k_{ij} = \frac{n_i^* c_j^*}{c_i^* n_j^*} \tag{16-46}$$

For a binary system, $r = \alpha_{BA} = 1/\alpha_{AB}$. The symbol r applies primarily to the process, while α is oriented toward interactions between pairs of solute species. For each binary pair, $r_{ij} = \alpha_{ji} = 1/\alpha_{ij}$.

Equilibrium then is given explicitly by

$$n_i^* = \frac{c_i^*}{\sum_j \alpha_{ji} c_j^*} = \frac{\alpha_{iN} c_i^*}{\sum_j \alpha_{jN} c_j^*} \tag{16-47}$$

and

$$c_i^* = \frac{n_i^*}{\sum_j \alpha_{ij} n_j^*} = \frac{\alpha_{Ni} n_i^*}{\sum_j \alpha_{Nj} n_j^*} \tag{16-48}$$

For the constant separation factor case, the c^* contours in a plot like Fig. 16-8 are linear.

CONSERVATION EQUATIONS

Material balances, often an energy balance, and occasionally a momentum balance are needed to describe an adsorption process. These are written in various forms depending on the specific application and the desire for simplicity or rigor. Reasonably general material balances for various processes are given next. An energy balance is developed for a fixed bed for gas-phase application and simplified for liquid-phase application. Momentum balances for pressure drop in packed beds are given in Sec. 6.

MATERIAL BALANCES

At the nanoscale, a sorbable component exists at three locations—in a sorbed phase, in pore fluid, and in fluid outside particles. As a consequence, in material balances, time derivatives must include terms involving n_i, c_{pi} (the pore concentration), and c_i (the extraparticle concentration). Let $\bar{n}_i$ represent n_i averaged over particle volume, and let $\bar{c}_{pi}$ represent c_{pi} averaged over pore fluid volume.

For batch or stirred tank processes, in terms of the mass of adsorbent M_s (kg), extraparticle volume of fluid V_f (m^3), and volumetric flow rates F_v (m^3/s) in and out of a tank, the material balance on component i is

$$M_s \frac{d\hat{n}_i}{dt} + \frac{d(V_f c_i)}{dt} = F_{v,\text{in}}\, c_{\text{in},i} - F_{v,\text{out}}\, c_i \tag{16-49}$$

with

$$\hat{n}_i = \bar{n}_i + (\varepsilon_p/\rho_p)\bar{c}_{pi} = \bar{n}_i + [(1-\varepsilon)\varepsilon_p/\rho_b]\bar{c}_{pi} \tag{16-50}$$

where ρ_p and ρ_b are particle and bulk densities, and ε and ε_p are void fraction (extraparticle volume fraction) and particle porosity, respectively.

For a fixed-bed process, the material balance for component i is

$$\rho_b \frac{\partial \hat{n}_i}{\partial t} + \varepsilon \frac{\partial c_i}{\partial t} + \varepsilon \frac{\partial (v c_i)}{\partial z} = \varepsilon D_L \frac{\partial}{\partial z}\left(c \frac{\partial y_i}{\partial z} \right) \tag{16-51}$$

where v is interstitial fluid velocity, D_L is a Fickian axial dispersion coefficient, and $y_i = c_i/c$ is the fluid-phase mole fraction of component i. v is related to the superficial velocity u by $v = u/\varepsilon$, where u is equal to the volumetric flow rate divided by the cross-sectional area of the fixed bed. In certain applications, the group $\rho_b \hat{n}_i$ in the first term of Eq. (16-51) is replaced by $(1-\varepsilon)\hat{n}_i$, where $\hat{n}_i$ has units of mol/(m^3 of adsorbent). An alternative form, grouping together fluid-phase concentrations rather than intraparticle concentrations, is

$$\rho_b \frac{\partial \bar{n}_i}{\partial t} + \varepsilon_b \frac{\partial \hat{c}_i}{\partial t} + \varepsilon \frac{\partial (v c_i)}{\partial z} = \varepsilon D_L \frac{\partial}{\partial z}\left(c \frac{\partial y_i}{\partial z} \right) \tag{16-52}$$

where, noting Eq. (16-4), $\hat{c}_i$ is defined by

$$\varepsilon_b \hat{c}_i = \varepsilon c_i + (1-\varepsilon)\varepsilon_p \bar{c}_{pi} \tag{16-53}$$

For moving-bed processes, a term is added to Eq. (16-51) to obtain

$$\rho_b \left[\frac{\partial \hat{n}_i}{\partial t} + v_s \frac{\partial \hat{n}_i}{\partial z} \right] + \varepsilon \left[\frac{\partial c_i}{\partial t} + \frac{\partial (v c_i)}{\partial z} - D_L \frac{\partial}{\partial z}\left(c \frac{\partial y_i}{\partial z} \right) \right] = 0 \tag{16-54}$$

where v_s is the solid-phase velocity (opposite in sign to v for a countercurrent process).

ENERGY BALANCE

Many different forms of the energy balance have been used in fixed-bed adsorption studies. The form chosen for a particular study depends on the process considered (e.g., temperature-swing adsorption or pressure-swing adsorption) and on the degree of approximation that is appropriate [see Walton and LeVan, *Ind. Eng. Chem. Res.* **42:** 6938 (2003); **44:** 7474 (2005)].

An energy balance for a fixed-bed process, ignoring dispersion, is

$$\rho_b \frac{\partial u_s}{\partial t} + \frac{\partial (\varepsilon_b \hat{c} u_f)}{\partial t} + \frac{\partial (\varepsilon v c h_f)}{\partial z} = -\frac{2h_w (T - T_w)}{r_c} \tag{16-55}$$

where h_w is a heat transfer coefficient for energy transfer with the column wall and r_c is the radius of the column. The second term of Eq. (16-55) combines contributions from both pore and extraparticle fluid.

Thermodynamic paths are necessary to evaluate the enthalpy (or internal energy) of the fluid phase and the internal energy of the stationary phase. For gas-phase processes at low and modest pressures, the enthalpy departure function for pressure changes can be ignored and a reference state for each pure component chosen to be ideal gas at temperature T^{ref}, and a reference state for the stationary phase (adsorbent plus adsorbate) chosen to be adsorbate-free solid at T^{ref}. Thus, for the gas phase

$$h_f = \sum_i y_i h_{fi} = \sum_i y_i \left[h_{fi}^{\text{ref}} + \int_{T^{\text{ref}}}^{T} C_{pfi}^{\circ}\, dT \right] \tag{16-56}$$

$$u_f = h_f - P/c \tag{16-57}$$

and for the stationary phase

$$u_s = u_{\text{sol}} + n u_a \approx u_{\text{sol}} + n h_a \tag{16-58}$$

$$u_{\text{sol}} = u_{\text{sol}}^{\text{ref}} + \int_{T^{\text{ref}}}^{T} C_s\, dT \tag{16-59}$$

The enthalpy of the adsorbed phase h_a is evaluated along a path for which the gas-phase components undergo temperature change from T_{ref} to T and then are adsorbed isothermally, giving

$$h_a = \sum_i x_i h_{fi} - \left(\frac{1}{n}\right) \sum_i \int_0^{n_i} q_i^{st}(n_i, n_j, T)\, dn_i \tag{16-60}$$

The isosteric heat of adsorption q_i^{st} is composition-dependent, and the sum of integrals in Eq. (16-60) is difficult to evaluate for multicomponent adsorption if the isosteric heats indeed depend on loading. Because each isosteric heat depends on the loadings of all components, the sum must be evaluated for a path beginning with clean adsorbent and ending with the proper loadings of all components. If the isosteric heat of adsorption is constant, as is commonly assumed, then the energy balance [Eq. (16-55)] becomes

$$\left[\rho_b\left(C_s+\sum_i n_i C^\circ_{pfi}\right)+\varepsilon_b c C^\circ_{pf}\right]\frac{\partial T}{\partial t}-\rho_b\sum_i q_i^{st}\frac{\partial n_i}{\partial t}-\frac{\partial(\varepsilon_b P)}{\partial t}+\varepsilon v C^\circ_{pf}\frac{\partial T}{\partial z}=-\frac{2h_w(T-T_w)}{r_c} \tag{16-61}$$

where Eq. (16-52) with $D_L = 0$ has been used. Equation (16-61) is a popular form of the energy balance for fixed-bed adsorption calculations. Often the first summation on the left-hand side, which involves gas-phase heat capacities, is neglected, or the gas-phase heat capacities are replaced by adsorbed-phase heat capacities. It is also possible to estimate the adsorbed-phase heat capacities from thermodynamic paths given gas-phase heat capacities and temperature-dependent adsorption equilibrium [Walton and LeVan, *Ind. Eng. Chem. Res.* **44:** 178 (2005)].

Nonisothermal liquid-phase processes may be driven by changes in feed temperature or heat addition or withdrawal through a column wall. For these, heats of adsorption and pressure effects are generally of less concern. For this case, a suitable energy balance is

$$\left(\rho_b C_s+\varepsilon_b c C^\circ_{pf}\right)\frac{\partial T}{\partial t}+\varepsilon v C^\circ_{pf}\frac{\partial T}{\partial z}=-\frac{2h_w(T-T_w)}{r_c} \tag{16-62}$$

RATE AND DISPERSION FACTORS

The performance of adsorption processes results in general from the combined effects of equilibrium or thermodynamic factors and rate factors. It is convenient to consider first *thermodynamic factors.* These determine the process performance in a limit where the system behaves ideally, that is, without mass transfer and kinetic limitations and with the fluid phase in perfect plug flow. *Rate factors* determine the efficiency of the real process in relation to the ideal process performance. Rate factors include heat- and mass-transfer limitations, reaction kinetic limitations, and hydrodynamic dispersion resulting from the velocity distribution across the bed and from mixing and diffusion in the interparticle void space.

TRANSPORT AND DISPERSION MECHANISMS

Figure 16-9 depicts porous adsorbent particles in an adsorption bed with sufficient generality to illustrate the nature and location of individual transport and dispersion mechanisms. Each mechanism involves a different driving force and, in general, gives rise to a different form of mathematical result.

Intraparticle Transport Mechanisms Intraparticle transport may be limited by *pore diffusion, solid diffusion, reaction kinetics,* or by two or more of these mechanisms together.

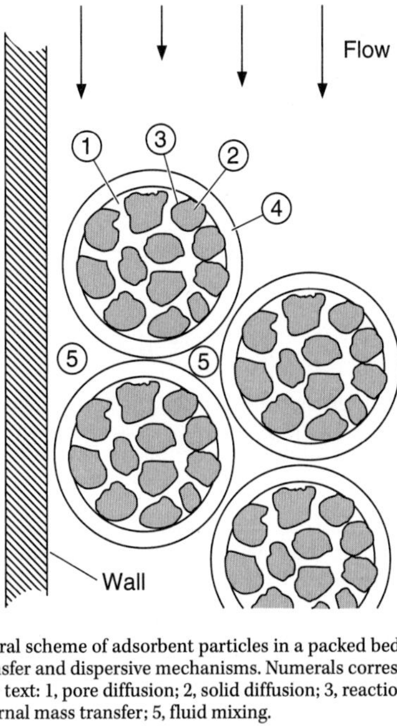

FIG. 16-9 General scheme of adsorbent particles in a packed bed showing the locations of mass transfer and dispersive mechanisms. Numerals correspond to numbered paragraphs in the text: 1, pore diffusion; 2, solid diffusion; 3, reaction kinetics at phase boundary; 4, external mass transfer; 5, fluid mixing.

1. *Pore diffusion* denotes transport in fluid-filled pores sufficiently large that the adsorbing molecules escape the force field of the adsorbent surface. Thus, this process is often referred to as *macropore diffusion.* The driving force for such a diffusion process can be approximated by the gradient in mole fraction or, if the molar concentration is constant, by the gradient in concentration of the diffusing species within the pores.

2. *Solid diffusion* denotes transport in the adsorbed phase for conditions where the diffusing molecules never escape the force field of the adsorbent surface. Solid diffusion may occur along a surface by an activated process involving jumps between adsorption sites, which is referred to as *surface diffusion,* in small pores, such as in activated carbon, which is referred to as *micropore diffusion* or *nanopore diffusion,* within a crystal, as in the case of zeolites, which is referred to as *intracrystalline diffusion,* or in gel-type ion-exchange resins, which is referred to as *gel* or *homogenous diffusion.* The common feature is that, in general, the driving force for solid diffusion can be approximated by the gradient in concentration of the species in the adsorbed state.

3. *Reaction kinetics* denotes a kinetic resistance to binding at a fluid–adsorbent interface or on a functional ligand. Rates of adsorption and desorption in porous adsorbents are generally controlled by mass transfer within the pore network rather than by the kinetics of sorption at the fluid–solid interface. Exceptions are the cases of chemisorption and affinity-adsorption systems used for biological separations, where the kinetics of bond formation can be exceedingly slow.

Intraparticle convection can also occur in packed beds when the adsorbent particles have very large and well-connected pores. Although, in general, bulk flow through the pores of the adsorbent particles is only a small fraction of the total flow, intraparticle convection can affect the transport of very slowly diffusing species such as macromolecules. The driving force for convection, in this case, is the pressure drop across each particle that is generated by the frictional resistance to flow experienced by the fluid as this flows through the packed bed [Rodrigues et al., *Chem. Eng. Sci.* **46:** 2765 (1991); Carta et al., *Sep. Technol.* **2:** 62 (1992); Frey et al., *Biotechnol. Progr.* **9:** 273 (1993); Liapis and McCoy, *J. Chromatogr.* **599:** 87 (1992)]. Intraparticle convection can also be significant when there is a total pressure difference between the center of the particle and the outside, such as is experienced in pressurization and depressurization steps of pressure-swing adsorption or when more gas is drawn into an adsorbent to equalize pressure as adsorption occurs from the gas phase within a porous particle [Lu et al., *AIChE J.* **38:** 857 (1992); Lu et al., *Gas Sep. Purif.* **6:** 89 (1992); Taqvi et al., *Adsorption* **3:** 127 (1997)].

Extraparticle Transport and Dispersion Mechanisms Extraparticle mechanisms are affected by the design of the contacting device and depend on the hydrodynamic conditions outside the particles.

4. *External mass transfer* denotes transport in the boundary layer at the external surfaces of the adsorbent particles. The driving force is the concentration difference across the boundary layer, which, in turn, is affected by the hydrodynamic conditions outside the particles.

5. *Dispersion* denotes mixing or lack of mixing in the fluid phase surrounding the adsorbent particles. Examples include the existence of a radial velocity distribution or dead zones in a packed bed column, diffusion in the extraparticle fluid, or inefficient mixing in an agitated contactor. In packed-bed adsorbers, dispersion is often described in terms of an *axial dispersion coefficient* whereby all mechanisms contributing to dispersion in the fluid phase are lumped together in a single effective coefficient.

Heat Transfer Since adsorption is generally accompanied by the evolution of heat, the rate of heat transfer between the adsorbent particles and the fluid phase may be important. In addition, heat transfer can occur across the column wall in small-diameter beds and is important in energy applications of adsorption. In gas adsorption systems, even with highly porous particles, the controlling heat transfer resistance is generally associated with extraparticle transport [Lee and Ruthven, *Can. J. Chem. Eng.* **57:** 65 (1979)], so the temperature within the particles is essentially uniform. In liquid-phase adsorption, intraparticle and extraparticle heat transfer resistances are generally comparable. However, in this case the heat capacity of the fluid phase is sufficiently high that temperature effects may be negligible except in extreme cases. General discussions of heat-transfer effects in adsorbents and adsorption beds are found in Suzuki (1990), pp. 187–208 and pp. 275–290, and in Ruthven (1984), pp. 189–198 and pp. 215–219 in General References.

INTRAPARTICLE MASS TRANSFER

Diffusional mass transfer in adsorption systems can be described phenomenologically in terms of Fick's first law:

$$J_i = -D_i(c_i)\frac{\partial c_i}{\partial x} \tag{16-63}$$

This expression can be used to describe both pore and solid diffusion so long as the driving force is expressed in terms of the appropriate concentrations. Although the driving force should be more correctly expressed in terms of chemical potentials, Eq. (16-63) provides a qualitatively and quantitatively correct representation of adsorption systems so long as the diffusivity is allowed to be a function of the adsorbate concentration. The diffusivity will be constant only for a limited number of adsorption systems.

Pore Diffusion When transport occurs through a network of fluid-filled pores inside the particles, the diffusion flux can be expressed in terms of a pore diffusion coefficient D_{pi} as:

$$J_i = -\varepsilon_p D_{pi}\frac{\partial c_{pi}}{\partial r} \tag{16-64}$$

D_{pi} is smaller than the molecular diffusivity D_i as a result of the random orientation of the pores, which gives a longer diffusion path, and the variation in the pore diameter. Both effects are commonly accounted for by a tortuosity factor τ_p such that $D_{pi} = D_i/\tau_p$. In principle, predictions of the tortuosity factor can be made if the pore structure, pore size, and shape distributions are known (see Dullien, *Porous Media: Fluid Transport and Pore Structure,* Academic Press, New York, 1979). In some cases, approximate prediction can be obtained from the following equations.

Mackie and Meares, *Proc. Roy. Soc.* **A232:** 498 (1955):

$$\tau_p = (2-\varepsilon_p)^2/\varepsilon_p \tag{16-65a}$$

Wakao and Smith, *Chem. Eng. Sci.* **17:** 825 (1962):

$$\tau_p = 1/\varepsilon_p \tag{16-65b}$$

Suzuki and Smith, *Chem. Eng. J.* **3:** 256 (1972):

$$\tau_p = \varepsilon_p + 1.5(1-\varepsilon_p) \tag{16-65c}$$

The predictive value of these equations is limited, and vastly different results are obtained from each. All of them, on the other hand, predict that τ_p increases as the porosity decreases.

For catalyst particles, Satterfield (*Heterogeneous Catalysis in Practice,* McGraw-Hill, 1980) recommends the use of a value of $\tau_p = 4$ when no other information is available, and this can be used for many adsorbents. In general, however, it is more reliable to treat the tortuosity as an empirical constant that is determined experimentally for any particular adsorbent.

For adsorbent materials, experimental tortuosity factors generally fall in the range 2–6 and generally decrease as the particle porosity is increased. Higher apparent values may be obtained when transport is affected by other resistances, while values much lower than 2 generally indicate that solid diffusion either occurs in parallel or controls the process.

Ruthven (1984) summarizes methods for the measurement of effective pore diffusivities that can be used to obtain tortuosity factors by comparison with the estimated pore diffusion coefficient of the adsorbate. Molecular diffusivities can be estimated with the methods in Sec. 6.

For gas-phase diffusion in small pores at low pressure, the molecular mean free path may be larger than the pore diameter, giving rise to Knudsen diffusion. Satterfield (*Mass Transfer in Heterogeneous Catalysis,* MIT, Cambridge, MA, 1970, p. 43) gives the following expression for the pore diffusivity:

$$D_{pi} = \frac{1}{\tau_p}\left[\frac{3}{4r_{\text{pore}}}\left(\frac{\pi M_{ri}}{2\Re T}\right)^{1/2} + \frac{1}{D_i}\right]^{-1} \tag{16-66}$$

where r_{pore} is the average pore radius, T the absolute temperature, and M_{ri} the molecular weight.

For liquid-phase diffusion of large adsorbate molecules, when the ratio $\lambda_m = r_m/r_{\text{pore}}$ of the molecule radius r_m to the pore radius is significantly greater than zero, the pore diffusivity is reduced by steric interactions with the pore wall and hydrodynamic resistance. When $\lambda_m < 0.2$, the following expressions derived by Brenner and Gaydos [*J. Coll. Int. Sci.* **58:** 312 (1977)] for a hard sphere molecule diffusing in a long cylindrical pore can be used

$$D_{pi} = \frac{D_i}{\tau_p}(1-\lambda_m)^{-2}\left[1+\frac{9}{8}\lambda_m \ln\lambda_m - 1.539\lambda_m\right] \tag{16-67}$$

r_m is the Stokes-Einstein radius of the solute that can be determined from the free diffusivity as

$$r_m = \frac{\kappa T}{6\pi\mu D_i} \tag{16-68}$$

where κ is the Boltzmann constant. When $\lambda_m > 0.2$, the centerline approximation [Anderson and Quinn, *Biophys. J.* **14:** 130, (1974)] can be used instead of Eq. (16-67)

$$D_{pi} = \frac{D_i}{\tau_p}(1-2.1044\lambda_m + 2.089\lambda_m^3 - 0.984\lambda_m^5)\times 0.865 \tag{16-69}$$

The 0.865 factor is used to match this equation to the Brenner and Gaydos expression for $\lambda_m = 0.2$. In these cases, the pore concentration c_{pi} is related to the external concentration c_i by the partition ratio $(1-\lambda_m)^2$.

Solid Diffusion In the case of pore diffusion just discussed, transport occurs within the fluid phase contained inside the particle; here the solute concentration is generally similar in magnitude to the external fluid concentration. A solute molecule transported by pore diffusion may attach to the sorbent and detach many times along its path. In other cases, attachment can be essentially permanent, but in both cases, only detached molecules undergo transport. In contrast, the following four instances illustrate cases where diffusion of adsorbate molecules occurs in their adsorbed state within phases that are distinct from the pore fluid:

1. Movement of mobile adsorbed solute molecules along pore surfaces, without detaching
2. Transport in a homogeneously dissolved state, as for a neutral molecule inside a sorbent gel or in a pore filled with a liquid that is immiscible with the external fluid
3. Ion transport in charged ion-exchange resins
4. Advance of an adsorbate molecule from one cage to another within a zeolite crystal

In these cases, the diffusion flux may be written in terms of the adsorbed solute concentration as

$$J_i = -\rho_p D_{si}\frac{\partial n_i}{\partial r} \tag{16-70}$$

The diffusion coefficient in these phases D_{si} is usually considerably smaller than that in fluid-filled pores; however, the adsorbate concentration is often much larger. Thus, the diffusion rate can be smaller or larger than can be expected for pore diffusion, depending on the magnitude of the fluid/solid partition coefficient.

Numerical values for solid diffusivities D_{si} in adsorbents are sparse and vary greatly. Moreover, they may be strongly dependent on the adsorbed phase concentration of solute. Hence, locally conducted experiments and interpretation must be used to a great extent. Summaries of available data for surface diffusivities in activated carbon and other adsorbent materials and for micropore diffusivities in zeolites are given in Ruthven (1984), Yang (1987), Suzuki (1990), and Karger and Ruthven (1992); see General References.

Surface diffusivities are generally strongly dependent on the fractional surface coverage and increase rapidly at surface coverage greater than 80 percent [see, for example, Yang et al., *AIChE J.* **19:** 1052 (1973)]. For estimation purposes, the correlation of Sladek et al. [*Ind. Eng. Chem. Fundam.* **13:** 100 (1974)] can be used to predict surface diffusivities for gas-phase adsorption on a variety of adsorbents.

Zeolite crystallite diffusivities for sorbed gases range from 10^{-7} to 10^{-14} cm^2/s. These diffusivities generally show a strong increase with the adsorbate concentration that is accounted for by the Darken thermodynamic correction factor

$$D_{si} = D_{0i} \frac{d \ln a_i}{d \ln n_i} = D_{0i} \frac{n_i / p_i}{dn_i / dp_i} \tag{16-71}$$

where D_{0i} is the corrected diffusivity and a_i is the thermodynamic activity of the species in the adsorbed phase. Corrected diffusivities D_{0i} calculated according to this equation are often found to be essentially independent of concentration. If the adsorption equilibrium isotherm obeys the Langmuir equation [Eq. (16-12)], Eq. (16-71) yields:

$$D_{si} = D_{0i} \left(1 - \frac{n_i}{n_i^s} \right)^{-1} \tag{16-72}$$

The effect of temperature on diffusivities in zeolite crystals can be expressed in terms of the Eyring equation [see Ruthven (1984) in General References].

In ion-exchange resins, diffusion is further complicated by electrical coupling effects. In a system with M counterions, diffusion rates are described by the Nernst-Planck equations [Helfferich (1962) in General References]. Assuming complete Donnan exclusion, these equations can be written as:

$$J_i = -\rho_p \frac{1}{z_i} \sum_{j=1}^{M-1} \bar{D}_{i,j} \frac{\partial z_j n_j}{\partial r} \tag{16-73}$$

with

$$\bar{D}_{i,j} = -\frac{\bar{D}_i(\bar{D}_j - \bar{D}_M) z_i^2 n_i}{\sum_{k=1}^{M} \bar{D}_k z_k^2 n_k} \tag{16-74a}$$

$$\bar{D}_{i,i} = \bar{D}_i - \frac{\bar{D}_i(\bar{D}_i - \bar{D}_M) z_i^2 n_i}{\sum_{k=1}^{M} \bar{D}_k z_k^2 n_k} \tag{16-74b}$$

which are dependent on the *ionic self diffusivities* $\bar{D}_i$ of the individual species. As a qualitative rule, ionic diffusivities of inorganic species in cross-linked polystyrene-DVB ion-exchange resins compared with those in water are 1:10 for monovalent ions, 1:100 for divalent ions, and 1:1000 for trivalent ions. Table 16-8 shows typical ionic diffusivities of inorganic ions in cation and anion-exchange resins; larger organic ions, however, can have ionic diffusivities much smaller than inorganic ions of the same valence [see, for example, Jones and Carta, *Ind. Eng. Chem. Res.* **32:** 117 (1993)].

For mixtures of unlike ions (the usual case), the apparent diffusivity will be intermediate between these values because of the electrical coupling effect. For a system with two counterions A and B, with charge z_A and z_B, Eqs. (16-73) and (16-74) reduce to:

$$J_A = -\rho_p \bar{D}_{A,B} \frac{\partial n_A}{\partial r} = -\rho_p \frac{\bar{D}_A \bar{D}_B \left[z_A^2 n_A + z_B^2 n_B \right]}{z_A^2 \bar{D}_A n_A + z_B^2 \bar{D}_B n_B} \frac{\partial n_A}{\partial r} \tag{16-75}$$

which shows that the apparent diffusivity $\bar{D}_{A,B}$ varies between $\bar{D}_A$ when the ionic fraction of species A in the resin is very small and $\bar{D}_B$ when the ionic fraction of A in the resin approaches unity, indicating that the ion present in smaller concentration has the stronger effect on the local interdiffusion rate.

Combined Pore and Solid Diffusion In porous adsorbents and ion-exchange resins, intraparticle transport can occur with pore and solid diffusion in parallel. The dominant transport process is the faster one, and this depends on the relative diffusivities and concentration gradients in the pore fluid and in the adsorbed phase. Often, equilibrium between the pore fluid and the solid phase can be assumed to exist locally at each point within a particle. In this case, the mass-transfer flux is expressed in terms of an apparent diffusivity $D_{\text{app},i}$ by:

$$J_i = -\left[\varepsilon_p D_{pi} + \rho_p D_{si} \frac{dn_i^e}{dc_i} \right] \frac{\partial c_{pi}}{\partial r} = -D_{\text{app},i}(c_{pi}) \frac{\partial c_{pi}}{\partial r} \tag{16-76}$$

where dn_i^e/dc_i is the derivative of the adsorption isotherm and it has been assumed that at equilibrium $c_{pi} = c_i$. This equation suggests that in such an adsorbent, pore and solid diffusivities can be obtained by determining the apparent diffusivity $D_{\text{app},i}$ for conditions of no adsorption $(dn_i^e/dc_i = 0)$ and for conditions of strong adsorption, where dn_i^e/dc_i is large. If the adsorption isotherm is linear over the range of experimental measurement, then

$$D_{\text{app},i} = \varepsilon_p D_{pi} + \rho_p K_i D_{si} \tag{16-77}$$

Thus, a plot of the apparent diffusivity versus the linear adsorption equilibrium constant K_i should be linear so long as D_{pi} and D_{si} remain constant.

In a particle having a *bidispersed pore structure* comprising spherical adsorptive subparticles of radius r_s forming a macroporous aggregate, separate flux equations can be written for the macroporous network in terms of Eq. (16-64) and for the subparticles themselves in terms of Eq. (16-70) if solid diffusion occurs.

EXTERNAL MASS TRANSFER

Because of the complexities encountered with a rigorous treatment of the hydrodynamics around particles in practical solid-fluid systems, mass transfer to and from the adsorbent is described in terms of a mass-transfer coefficient k_f. The flux at the particle surface is:

$$J_i = k_f (c_i - c_i^s) \tag{16-78}$$

where c_i and c_i^s are the solute concentrations in the bulk fluid and at the particle surface, respectively. k_f can be estimated from available correlations

TABLE 16-8 Self Diffusion Coefficients in Polystyrene-Divinylbenzene Ion Exchangers (Units of 10^{-7} cm^2/s)*

Temperature	0.3°C					25°C				
Cross-linking, %	4	6	8	12	16	4	6	8	12	16
Cation exchangers (sulfonated): Dowex 50										
Na^+	6.7		3.4	1.15	0.66	14.1		9.44		2.40
Cs^+			6.6		1.11			13.7		3.10
Ag^+			2.62		1.00			6.42		2.75
Zn^{2+}			0.21		0.03			0.63		0.14
La^{3+}	0.30		0.03		0.002			0.092		0.005
Anion exchangers (dimethyl hydroxyethylamine): Dowex 2										
Cl^-		1.25					3.54			
Br^-	(1.8)	1.50	0.63		0.06	(4.3)	3.87	2.04		0.26
I^-		0.35					1.33			
BrO_3^-		1.76					4.55			
WO_4^{2-}		0.60					1.80			
PO_4^{3-}		0.16					0.57			

*Data from Boyd and Soldano, *J. Am. Chem. Soc.* **75:** 6091 (1953).

TABLE 16-9 Recommended Correlations for External Mass Transfer Coefficients in Adsorption Beds ($Re = \varepsilon v \rho_f d_p/\mu$, $Sc = \mu/\rho_f D$)

Equation	Re	Phase	Ref.
$Sh = 1.15\left(\frac{Re}{\varepsilon}\right)^{0.5} Sc^{0.33}$	Re > 1	Gas/liquid	Carberry, *AIChE J.* **6:** 460 (1960)
$Sh = 2.0 + 1.1 Re^{0.6} Sc^{0.33}$	3 < Re < 10^4	Gas/liquid	Wakao and Funazkri, *Chem. Eng. Sci.* **33:** 1375 (1978)
$Sh = 1.85\left(\frac{1-\varepsilon}{\varepsilon}\right)^{0.33} Re^{0.33} Sc^{0.33}$	Re < 40	Liquid	Kataoka et al., *J. Chem. Eng. Japan* **5:** 132 (1972)
$Sh = \frac{1.09}{\varepsilon} Re^{0.33} Sc^{0.33}$	0.0015 < Re < 55	Liquid	Wilson and Geankoplis, *Ind. Eng. Chem. Fundam.* **5:** 9 (1966)
$Sh = \frac{0.25}{\varepsilon} Re^{0.69} Sc^{0.33}$	55 < Re < 1050	Liquid	Wilson and Geankoplis, *Ind. Eng. Chem. Fundam.* **5:** 9 (1966)

in terms of the Sherwood number $Sh = k_f d_p/D_i$, the Reynolds number, $Re = \varepsilon v \rho_f d_p/\mu$, and the Schmidt number $Sc = \mu/\rho_f D_i$. For packed-bed operations, the correlations in Table 16-9 are recommended. A plot of these equations is given in Fig. 16-10 for representative ranges of Re and Sc with $\varepsilon = 0.4$.

External mass-transfer coefficients for particles suspended in agitated contactors can be estimated from equations in Levins and Glastonbury [*Trans. Instn. Chem. Eng.* **50:** 132 (1972)] and Armenante and Kirwan [*Chem. Eng. Sci.* **44:** 2871 (1989)].

AXIAL DISPERSION IN PACKED BEDS

The axial dispersion coefficient in a packed bed [cf. Eq. (16-51)] lumps together all mechanisms leading to dispersion in the fluid phase, including molecular diffusion, convective mixing, and nonuniformities in the fluid velocity across the packed bed. As such, the axial dispersion coefficient is best determined experimentally for each specific system.

The effects of *flow nonuniformities*, in particular, can be severe in gas systems when the ratio of bed-to-particle diameters is small; in liquid systems when viscous fingering occurs as a result of large viscosity gradients in the adsorption bed; when very small particles (<10 μm) are used, such as in high-performance liquid chromatography systems; and in large-diameter beds. A lower bound of the axial dispersion coefficient can be estimated for well-packed beds from correlations that follow.

Neglecting flow nonuniformities, the contributions of molecular diffusion and turbulent mixing arising from stream splitting and recombination around the sorbent particles can be considered additive [Langer et al., *Int. J. Heat Mass Trans.* **21:** 751 (1978)]; thus, the axial dispersion coefficient D_L can be expressed as:

$$\frac{D_L}{D_i} = \gamma_1 + \gamma_2 \frac{d_p v}{D_i} = \gamma_1 + \gamma_2 \frac{ReSc}{\varepsilon} \qquad (16\text{-}79)$$

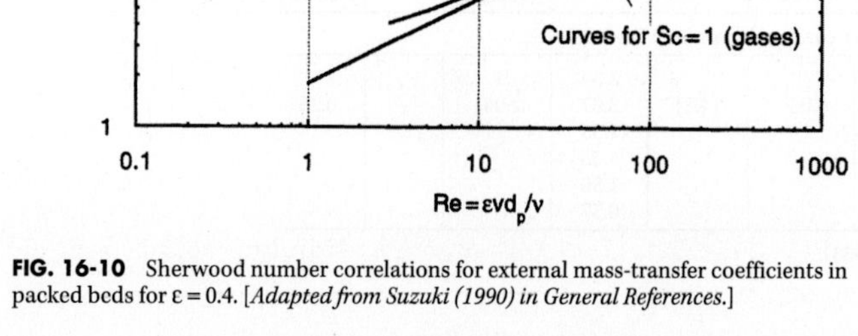

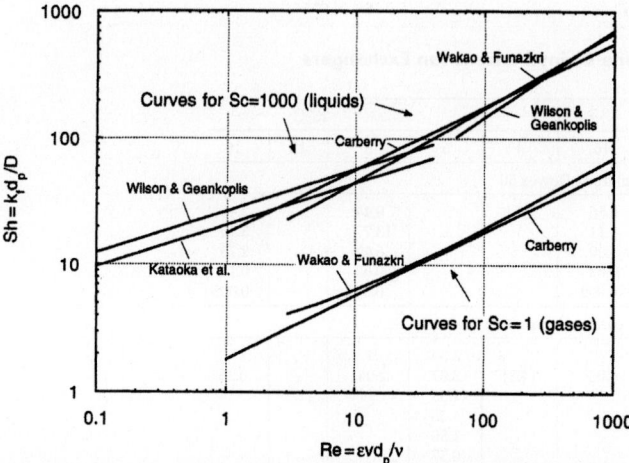

FIG. 16-10 Sherwood number correlations for external mass-transfer coefficients in packed beds for $\varepsilon = 0.4$. [*Adapted from Suzuki (1990) in General References.*]

or, in terms of a particle-based Peclet number ($Pe = d_p v/D_L$), as:

$$\frac{1}{Pe} = \frac{\gamma_1 \varepsilon}{ReSc} + \gamma_2 \qquad (16\text{-}80)$$

The first term in Eqs. (16-79) and (16-80) accounts for molecular diffusion, and the second term accounts for mixing. For the first term, Wicke [*Ber. Bunsenges* **77:** 160 (1973)] has suggested:

$$\gamma_1 = 0.45 + 0.55\varepsilon \qquad (16\text{-}81)$$

which, for typical void fractions, $\varepsilon = 0.35 - 0.45$ gives $\gamma_1 = 0.64 - 0.70$ [Ruthven (1984) in General References]. Expressions for the axial mixing term γ_2 in Eq. (16-79) are given in Table 16-10. The expression of Wakao and Funazkri includes an axial diffusion term, γ_1, that varies from 0.7 for nonporous particles to $20/\varepsilon$, depending on the intraparticle mass-transfer mechanism. For strongly adsorbed species, Wakao and Funazkri suggest that the effective axial dispersion coefficient is much larger than that predicted on the basis of nonporous, nonadsorbing particles. The Gunn expression includes a term σ_v^2 accounting for deviations from plug flow, which is defined as the dimensionless variance of the distribution of the ratio of velocity to average velocity over the cross section of the bed. The parameter values included in this equation are valid for spherical particles. Values for nonspherical particles can be found in the original reference.

Figure 16-11 compares predicted values of D_L/D_i for $\sigma_v = 0$ and $\varepsilon = 0.4$ with Sc = 1 (gases at low pressure), and Sc = 1000 (small molecules in liquids), based on the equations in Table 16-10.

Correlations for axial dispersion in beds packed with very small particles (<10 μm) that take into account the holdup of liquid in the bed are discussed by Horvath and Lin [*J. Chromatogr.* **126:** 401 (1976)].

RATE EQUATIONS

Rate equations are used to describe interphase mass transfer in batch systems, packed beds, and other contacting devices for sorptive processes and are formulated in terms of fundamental transport properties of adsorbent and adsorbate.

General Component Balance For a spherical adsorbent particle:

$$\varepsilon_p \frac{\partial c_{pi}}{\partial t} + \rho_p \frac{\partial n_i}{\partial t} = \frac{1}{r^2}\frac{\partial}{\partial r}(-r^2 N_i) \qquad (16\text{-}82)$$

For particles that have no macropores, such as gel-type ion-exchange resins, or when the adsorption capacity is high, the first term on the left side of this equation, representing solute hold up in the pore fluid, may be neglected. Ignoring bulk flow terms, the fluxes N_i and J_i are equal. In this case, coupling the component balance with the flux expressions previously introduced gives the rate equations in Table 16-11. Typical boundary conditions are also included in this table. Generally, these equations apply to particles that can be approximated as spherical and of a uniform size and properties. An appropriately chosen mean particle size must be used in these equations when dealing with adsorbents having a broad particle size distribution. The appropriate average depends on the controlling mass-transfer mechanism. For intraparticle mass-transfer mechanisms, the volume or mass-average particle size usually provides the best prediction.

TABLE 16-10 Coefficients for Axial Dispersion Correlations in Packed Beds Based on Eq. (16-79)

γ_1	γ_2	Ref.
0.73	$0.5\left(1+\frac{13\gamma_1\varepsilon}{\text{Re Sc}}\right)^{-1}$	Edwards and Richardson, *Chem. Eng. Sci.* **23:** 109 (1968)
Nonporous particles: 0.7 Porous particles: $\leq 20/\varepsilon$	0.5	Wakao and Funazkri, *Chem. Eng. Sci.* **33:** 1375 (1978)
1	$\frac{3}{4}\varepsilon+\frac{\pi^2\varepsilon(1-\varepsilon)}{6}\ln(\text{Re Sc})$	Koch and Brady, *J. Fluid Mech.* **154:** 399 (1985)
0.714	$\frac{\sigma_\upsilon^2}{2}+(1+\sigma_\upsilon^2)\left\{\gamma(1-p)^2+\gamma^2 p(1-p)^3\left[e^{-\frac{1}{\gamma p(1-p)}}-1\right]\right\}$ with $\gamma = 0.043\ \text{ReSc}/(1-\varepsilon)$ $p = 0.33\exp(-24/\text{Re}) + 0.17$	Gunn, *Chem. Eng. Sci.* **2:** 363 (1987)

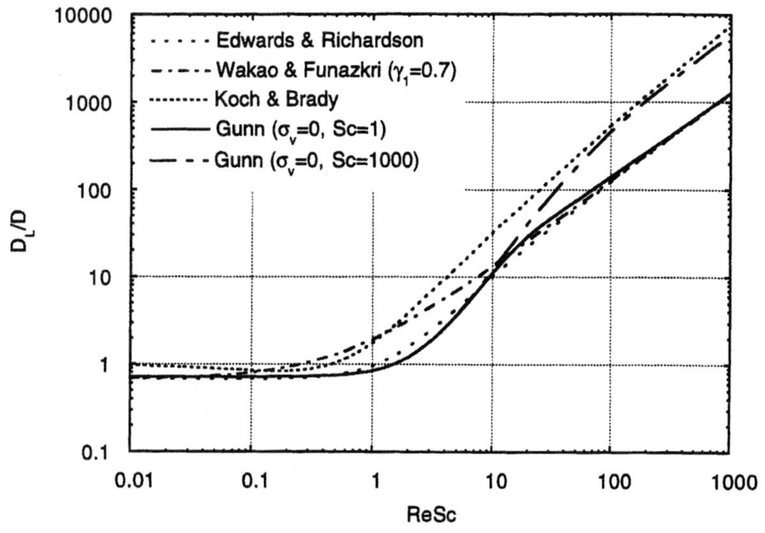

FIG. 16-11 Axial dispersion coefficient correlations for well-packed beds for $\varepsilon = 0.4$.

TABLE 16-11 Rate Equations for Description of Mass Transfer in Spherical Adsorbent Particles

Mechanism	Flux equation	Rate equation
A. Pore diffusion	16-64	$\varepsilon_p\frac{\partial c_{pi}}{\partial t}+\rho_p\frac{\partial n_i}{\partial t}=\frac{1}{r^2}\frac{\partial}{\partial r}\left(\varepsilon_p D_{pi} r^2\frac{\partial c_{pi}}{\partial r}\right)$ $(\partial c_{pi}/\partial r)_{r=0}=0$, $(\varepsilon_p D_{pi}\partial c_{pi}/\partial r)_{r=r_p}=k_f(c_i-c_{pi\vert r=r_p})$ or $(c_{pi})_{r=r_p}=c_i$ for no external resistance
B. Solid diffusion	16-70	$\frac{\partial n_i}{\partial t}=\frac{1}{r^2}\frac{\partial}{\partial r}\left(D_{si} r^2\frac{\partial n_i}{\partial t}\right)$ $(\partial n_i/\partial r)_{r=0}=0$, $(\rho_p D_{si}\partial n_i/\partial r)_{r=r_p}=k_f(c_i-c_i^s)$ or $(n_i)_{r=r_p}=n_i^e(c_i)$ for no external resistance
C. Parallel pore and solid diffusion (local equilibrium between pore and adsorbed phase)	16-76	$\left(\varepsilon_p+\rho_p\frac{dn_i^e}{dc_i}\right)\frac{\partial c_{pi}}{\partial t}=\frac{1}{r^2}\frac{\partial}{\partial r}\left[r^2\left(\varepsilon_p D_{pi}+\rho_p D_{si}\frac{dn_i^e}{dc_i}\right)\frac{\partial c_{pi}}{\partial r}\right]$ $(\partial c_{pi}/\partial r)_{r=0}=0$, $[(\varepsilon_p D_{pi}+\rho_p D_{si} dn_i^e/dc_i)\partial c_{pi}/\partial r]_{r=r_p}=k_f(c_i-c_{pi\vert r=r_p})$ or $(c_{pi})_{r=r_p}=c_i$ for no external resistance
D. Diffusion in bidispersed particles (no external resistance)	16-64 and 16-70	$\frac{\partial n_i}{\partial t}=\frac{1}{\rho^2}\frac{\partial}{\partial\rho}\left(D_{si}\rho^2\frac{\partial n_i}{\partial\rho}\right)$, $(\partial n_i/\partial\rho)_{\rho=0}=0$, $(n_i)_{\rho=r_s}=n_i^e(c_{pi})$ $\bar{n}_i(r,t)=\frac{3}{r_s^3}\int_0^{r_s}\rho^2 n_i\,d\rho$ $\varepsilon_p\frac{\partial c_{pi}}{\partial t}+\rho_p\frac{\partial\bar{n}_i}{\partial t}=\frac{1}{r^2}\frac{\partial}{\partial r}\left(\varepsilon_p D_{pi} r^2\frac{\partial c_{pi}}{\partial r}\right)$, $(\partial c_{pi}/\partial r)_{r=0}=0$, $(c_{pi})_{r=r_p}=c_i$

Linear Driving Force Approximation Simplified expressions can also be used for an approximate description of adsorption in terms of rate coefficients for both extraparticle and intraparticle mass transfer controlling. As an approximation, the rate of adsorption on a particle can be written as:

$$\frac{\partial \hat{n}_i}{\partial t} = k\, f(n_i, c_i) \tag{16-83}$$

where k is a rate coefficient, and the function $f(n_i, c_i)$ is a driving force relationship. The variables k_c and k_n are used to denote rate coefficients based on fluid-phase and adsorbed-phase concentration driving forces, respectively.

Commonly used forms of this rate equation are given in Table 16-12, where $\Lambda = \rho_b n_i^{\text{ref}} / c_i^{\text{ref}}$ is the partition ratio based on the feed concentration as a reference [cf. Eq. (16-125)]. For adsorption bed calculations with constant separation factor systems, somewhat improved predictions are obtained using correction factors ψ_s and ψ_p defined in Table 16-12.

The linear driving force (LDF) approximation is obtained when the driving force is expressed as a concentration difference. This approximation was originally developed to describe packed-bed dynamics under linear equilibrium conditions [Glueckauf, *Trans. Far. Soc.* **51:** 1540 (1955)]. This form is exact for a nonlinear isotherm only when external mass transfer is controlling. However, this expression can also be used for nonlinear systems with pore or solid diffusion mechanisms as an approximation, since it provides qualitatively correct results.

Alternate driving force approximations, item 2B in Table 16-12 for solid diffusion, and item 3B in Table 16-12 for pore diffusion, provide somewhat more accurate results in constant pattern packed-bed calculations with pore or solid diffusion controlling for constant separation factor systems.

The reaction kinetics approximation is mechanistically correct for systems where the reaction step at pore surfaces or other fluid-solid interfaces is controlling. This may occur in the case of chemisorption on porous catalysts and in affinity adsorbents that involve very slow binding steps. In these cases, the mass-transfer parameter k is replaced by a second-order reaction rate constant k_a. The corresponding driving force for a constant separation factor isotherm is given in column 4 of Table 16-12. When diffusion controls the process, it is still possible to describe the system by its apparent second-order kinetic behavior, since it usually provides a good approximation to a more complex exact form for single-transition systems (see the subsection Fixed-Bed Transitions).

Combined Intraparticle Resistances When solid diffusion and pore diffusion operate in parallel, the effective rate is the sum of these two rates. When solid diffusion predominates, mass transfer can be represented approximately in terms of the LDF approximation, replacing k_n in column 2 of Table 16-12 with

$$k_n^c = \frac{15\psi_s D_{si}}{r_p^2} + \frac{15(1-\varepsilon)\psi_p \varepsilon_p D_{pi}}{\Lambda r_p^2} \tag{16-84}$$

When pore diffusion predominates, use of column 3 in Table 16-12 is preferable, with k_n^c replacing k_n.

For particles with a *bidispersed pore structure*, the mass-transfer parameter k_n in the LDF approximation (column 2 in Table 16-12) can be approximated by the series-combination of resistances as:

$$\frac{1}{k_n^c} = \frac{1}{b_s}\left[\frac{\Lambda r_p^2}{15(1-\varepsilon)\psi_p \varepsilon_p D_{pi}} + \frac{r_s^2}{15\psi_s D_{si}}\right] \tag{16-85}$$

where b_s is a correction to the driving force that is described next. In the limiting cases where the controlling resistance is diffusion through the particle pores or diffusion within the subparticles, the rate coefficients $k_n = 15(1-\varepsilon)\psi_p\varepsilon_p D_{pi}/\Lambda r_p^2$ and $k_n = 15\psi_s D_{si}/r_s^2$ are obtained.

Overall Resistance With a linear isotherm ($R = 1$), the overall mass transfer resistance is the sum of intraparticle and extraparticle resistances. Thus, the overall LDF coefficient for use with a particle-side driving force (column 2 in Table 16-12) is:

$$\frac{1}{k_n^o} = \frac{\Lambda r_p}{3(1-\varepsilon)k_f} + \frac{1}{k_n^c} \tag{16-86}$$

or

$$\frac{1}{k_c^o} = \frac{\rho_p r_p}{3(1-\varepsilon)k_f} + \frac{\rho_b}{\Lambda k_n^c} \tag{16-87}$$

for use with a fluid-phase driving force (column 1 in Table 16-12).

In either equation, k_n^c is given by Eq. (16-84) for parallel pore and surface diffusion or by Eq. (16-85) for a bidispersed particle. For nearly linear isotherms ($0.7 < R < 1.5$), the same linear addition of resistance can be used as a good approximation to predict the adsorption behavior of packed beds, since solutions for all mechanisms are nearly identical. With a highly favorable isotherm ($R \to 0$), however, the rate at each point is controlled by the resistance that is locally greater, and the principle of additivity of resistances breaks down. For approximate calculations with intermediate values of R, an overall transport parameter for use with the LDF approximation can be

TABLE 16-12 Expressions for Rate Coefficient *k* and Driving Force Relationships for Eq. (16-83)

Mechanism	1. External film	2. Solid diffusion	3. Pore diffusion	4. Reaction kinetics
Expression for rate coefficient, k	$k_c = \frac{k_f a}{\rho_b} = \frac{3(1-\varepsilon)k_f}{\rho_b r_p}$	$k_n = \frac{15\psi_s D_{si}}{r_p^2}$	$k_n = \frac{15\psi_p (1-\varepsilon)\varepsilon_p D_{pi}}{\Lambda r_p^2}$	k_a
Expressions for driving force relationship, $f(n_i, c_i)$, and correction factors, ψ				
A. Linear driving force (LDF)	$c - c_i^e$	$n_i^e - \bar{n}_i$	$n_i^e - \bar{n}_i$	—
LDF for constant R	$c_i - \frac{Rc^{\text{ref}}\bar{n}_i/n_i^{\text{ref}}}{1-(R-1)\bar{n}_i/n_i^{\text{ref}}}$	$\frac{n^{\text{ref}}c_i/c_i^{\text{ref}}}{R+(R-1)c_i/c_i^{\text{ref}}} - \bar{n}_i$	$\frac{n^{\text{ref}}c_i/c_i^{\text{ref}}}{R+(R-1)c_i/c_i^{\text{ref}}} - \bar{n}_i$	—
ψ_s or ψ_p for constant R	—	$\frac{0.894}{1-0.106R^{0.5}}$	$\frac{0.775}{1-0.225R^{0.5}}$	—
B. Alternate driving force for constant R	—	$\frac{n_i^{e2} - \bar{n}_i^2}{2\bar{n}_i}$	$\frac{n_i^e - \bar{n}_i}{\left[1-(R-1)\bar{n}_i/n_i^{\text{ref}}\right]^{0.5}}$	—
ψ_s or ψ_p for constant R	—	$\frac{0.590}{1-0.410R^{0.5}}$	$\frac{0.548}{1-0.452R^{0.5}}$	—
C. Reaction kinetics for constant R	—	—	—	$\frac{c_i(n_i^{\text{ref}} - \bar{n}_i) - R\bar{n}_i(c_i^{\text{ref}} - c_i)}{1-R}$

References: 1A. Beaton and Furnas, *Ind. Eng. Chem.* **33:** 1500 (1941); Michaels, *Ind. Eng. Chem.* **44:** 1922 (1952)
2A,3A. Glueckauf and Coates, *J. Chem. Soc.* 1315 (1947); *Trans. Faraday Soc.* **51:** 1540 (1955); Hall et al., *Ind. Eng. Chem. Fundam.* **5:** 212 (1966)
2B. Vermeulen, *Ind. Eng. Chem.* **45:** 1664 (1953)
3B. Vermeulen and Quilici, *Ind. Eng. Chem. Fundam.* **9:** 179 (1970)
4C. Hiester and Vermeulen, *Chem. Eng. Progr.* **48:** 505 (1952)

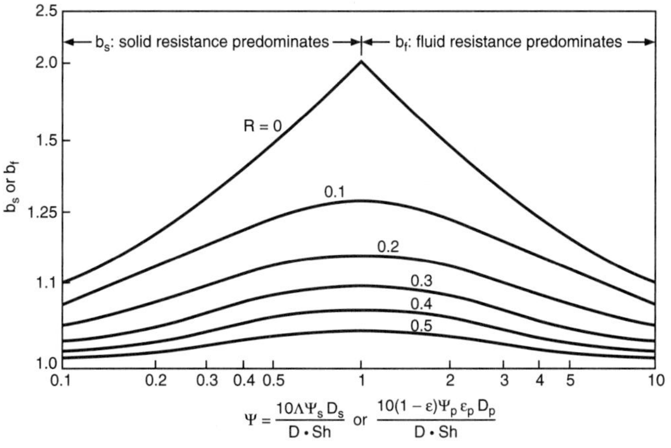

FIG. 16-12 Correction factors for addition of mass-transfer resistances, relative to effective overall solid phase or fluid phase rates, as a function of the mechanism parameter. Each curve corresponds to both b_s and b_f over its entire range.

calculated from the following relationship for solid diffusion and film resistance in series:

$$\frac{\Lambda r_p}{3(1-\varepsilon)k_f} + \frac{r_p^2}{15\psi_s D_{si}} = \frac{b_s}{k_n^o} = \frac{b_f \Lambda}{\rho_b k_c^o} \tag{16-88}$$

b_s and b_f are correction factors that are given by Fig. 16-12 as a function of the separation factor R and the mechanism parameter

$$\Psi = \frac{10\psi_s D_{si}\Lambda}{D_i}\frac{1}{\text{Sh}} \tag{16-89}$$

Axial Dispersion Effects In adsorption bed calculations, axial dispersion effects are typically accounted for by the axial diffusion–like term in the bed conservation equations [Eqs. (16-51) and (16-52)]. For nearly linear isotherms ($0.5 < R < 1.5$), the combined effects of axial dispersion and mass-transfer resistances on the adsorption behavior of packed beds can be expressed approximately in terms of an apparent rate coefficient k_c for use with a fluid-phase driving force (column 1, Table 16-12):

$$\frac{1}{k_c} = \frac{\rho_b}{\varepsilon}\frac{D_L}{v^2} + \frac{1}{k_c^o} \tag{16-90}$$

which extends the linear addition principle to combined axial dispersion and mass-transfer resistances. Even for a highly nonlinear isotherm (e.g., $R = 0.33$), the linear addition principle expressed by this equation provides a useful approximation except in the extreme case of low mass-transfer resistance and large axial dispersion, when $D_L\rho_p k_c^o / v^2\varepsilon >> 5$ [Garg and Ruthven, *Chem. Eng. Sci.* **30:** 1192 (1975)]. However, when the isotherm is irreversible ($R \rightarrow 0$), the linear addition principle breaks down, and axial dispersion has to be taken into account by explicit models (see the subsection Fixed-Bed Transitions).

Rapid Adsorption-Desorption Cycles For rapid cycles with particle diffusion controlling, when the cycle time t_c is much smaller than the time constant for intraparticle transport, the LDF approximation becomes inaccurate. The generalized expression

$$\frac{\partial \hat{n}_i}{\partial t} = \Omega k_n (n_i^e - \bar{n}_i) \tag{16-91}$$

can be used for packed-bed calculations when the parameter Ω is defined to be a function of the cycle time such that the amount of solute adsorbed and desorbed during a cycle is equal to that obtained by solution of the complete particle diffusion equations. Graphical and analytical expressions for Ω in the case of a single particle, usable for very short beds, are given by Nakao and Suzuki [*J. Chem. Eng. Japan* **16:** 114 (1983)] and Carta [*Chem. Eng. Sci.* **48:** 622 (1993)]. With equal adsorption and desorption times, $t_a = t_d = t_c/2$, Ω approaches the value $\pi^2/15$ for long cycle times and the asymptote $\Omega = 1.877/\sqrt{t_c k_n}$ for short cycle times [Alpay and Scott, *Chem. Eng. Sci.* **47:** 499 (1992)]. However, other results by Raghavan et al. [*Chem. Eng. Sci.* **41:** 2787 (1986)] indicate that a limiting constant value of Ω (larger than 1) is obtained for very short cycles, when calculations are carried out for beds of finite length.

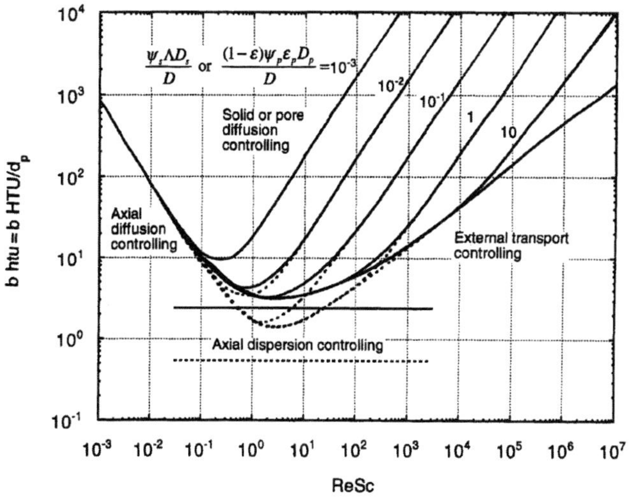

FIG. 16-13 Effect of ReSc group, distribution ratio, and diffusivity ratio on height of a transfer unit. Dotted lines for gas and solid lines for liquid-phase systems.

Determination of Controlling Rate Factor The most important physical variables determining the controlling rate factor are particle size and structure, flow rate, fluid- and solid-phase diffusivities, partition ratio, and fluid viscosity. When multiple resistances and axial dispersion can potentially affect the rate, the spreading of a concentration wave in a fixed bed can be represented approximately in terms of the single LDF rate parameter k. In customary separation-process calculations, the height of an adsorption bed can be calculated approximately as the product of the number of transfer units times the height of one fluid-phase transfer unit (HTU). The HTU is related to the LDF rate parameters k_c and k_n by:

$$\text{HTU} = \frac{\varepsilon v}{\rho_b k_c} = \frac{\varepsilon v}{\Lambda k_n} \tag{16-92}$$

Figure 16-13 is a plot of the dimensionless HTU (htu = HTU/d_p) multiplied times the correction factor b_f (between 1 and 2) as a function of the dimensionless velocity ReSc = $\varepsilon v d_p/D$ and a ratio of the controlling diffusivity to the fluid-phase diffusivity, generated on the basis of results of Vermeulen et al. (1984) using typical values of the individual physical factors likely to be found in adsorption beds. This figure can be used to determine the controlling rate factor from a knowledge of individual physical parameters. If fluid-side effects control, the dimensionless HTU is given by the bottom curve (dotted for gas and solid for liquid-phase systems). If intraparticle diffusion controls, the dimensionless HTU is given by a point above the lower envelope on the appropriate diffusional contour (through the ψ_s, the contour value depends slightly on the separation factor R). If pore and solid diffusion occur in parallel, the reciprocal of the HTU is the sum of the reciprocals of the HTU values for the two mechanisms. Near the intersections of the diffusional contours with the envelope, the dimensionless HTU is the sum of the HTU values for fluid-side and particle-side resistances.

Example 16-7 Estimation of Rate Coefficient for Gas Adsorption An adsorption bed is used to remove methane from a methane-hydrogen mixture at 10 atm (abs.) (10.1 bar) and 25°C (298 K), containing 10 mol% methane. Activated carbon particles having a mean diameter $d_p = 0.17$ cm, a surface area $A = 1.1 \times 10^7$ cm^2/g, a bulk density $\rho_b = 0.509$ g/cm^3, a particle density $\rho_p = 0.777$ g/cm^3, and a skeletal density $\rho_s = 2.178$ g/cm^3 are used as the adsorbent. Based on data of Grant et al. [*AIChE J.* **8:** 403 (1962)], adsorption equilibrium is represented by $n = 2.0 \times 10^{-3}K_A p_A/(1 + K_A p_A)$ mol/g adsorbent, with $K_A = 0.346$ atm^{-1}. Estimate the rate coefficient and determine the controlling rate factor for a superficial velocity of 30 cm/s.

1. The intraparticle void fraction is $\varepsilon_p = (0.777^{-1} - 2.178^{-1})/(0.777^{-1}) = 0.643$ and the extraparticle void fraction is $\varepsilon = (0.509^{-1} - 0.777^{-1})/(0.509^{-1}) = 0.345$. The pore radius is estimated from $r_p = 2\varepsilon_p/(A\rho_p) = 1.5 \times 10^{-7}$ cm.

2. The fluid phase diffusivity is $D = 0.0742$ cm^2/s. The pore diffusivity is estimated from Eq. (16-66) with a tortuosity factor $\tau_p = 4$; $D_p = 1.45 \times 10^{-3}$ cm^2/s.

3. The fluid-side mass transfer coefficient is estimated from Fig. 16-10. For these conditions, the kinematic viscosity is $\nu = \mu/\rho_f = 0.108$ cm^2/s, Re = 30 × 0.17/0.108 = 47, and Sc = 0.108/0.0742 = 1.5. From Fig. 16-10 or equations in Table 16-9, Sh ~ 13.

4. The isotherm parameters based on the feed concentration are $R = 1/(1 + K_A p_A) = 0.4$ and $\Lambda = 0.509 \times 7.68 \times 10^{-4}/4.09 \times 10^{-5} = 9.56$. For pore diffusion, $\psi_p = 0.961$ from item 3A in Table 16-12. Thus, $(1 - \varepsilon)\psi_p\varepsilon_p D_p/D = (1 - 0.345) \times 0.961 \times 0.643 \times 1.45 \times 10^{-3}/0.0742 = 7.9 \times 10^{-3}$. From Fig. 16-13 at ReSc = 69, b htu ~ 150. b is found from Fig. 16-12. However, since the mechanism parameter Ψ is very small, b ~ 1. Thus, $k_n = \varepsilon v/(\text{htu } d_p\Lambda) = 0.12$ s^{-1}.

This value applies to the driving force $n_i^e - \bar{n}_i$. Since pore diffusion is dominant, this value is very close to the value $k_n = 15(1-\varepsilon)\psi_p\varepsilon_p D_p/(r_p^2\Lambda) = 0.13\ s^{-1}$ obtained directly from Table 16-12. It should be noted that surface diffusion is neglected in this estimation. Its occurrence could significantly affect the overall mass transfer rate [see Suzuki (1990), pp. 70–85 in General References].

Example 16-8 Estimation of Rate Coefficient for Ion Exchange Estimate the rate coefficient for flow of a 0.01-M water solution of NaCl through a bed of cation exchange particles in hydrogen form with ε = 0.4. The superficial velocity is 0.2 cm/s and the temperature is 25°C. The particles are 600 μm in diameter, and the diffusion coefficient of sodium ion is 1.2×10^{-5} cm²/s in solution and 9.4×10^{-7} cm²/s inside the particles (cf. Table 16-8). The bulk density is 0.7 g dry resin/cm³ of bed, and the capacity of the resin is 4.9 mequiv/g dry resin. The mass action equilibrium constant is 1.5. The fluid kinematic viscosity is ν = 0.913 cm²/s.

1. Estimate the fluid-side mass-transfer coefficient; $Re = \varepsilon v d_p/\nu = 0.2 \times 0.06/0.00913 = 1.3$, $Sc = \nu/D = 0.00913/1.2 \times 10^{-5} = 761$. From Fig. 16-10 or Table 16-9, Sh ~ 23. Thus, $k_f = D\,Sh/d_p = 4.5 \times 10^{-3}$ cm/s.

2. From the equilibrium constant, $R = 1/K_{Na,H} = 0.67$. Thus, from Table 16-12, item 2A, $\psi_s = 0.979$. Using $n^{ref} = 4.9$ mequiv/g and $c^{ref} = 0.01$ mmole/cm³, $\Lambda = \rho_b n^{ref}/c^{ref} = 343$. Thus, $\psi_s D_s \Lambda/D = 26$ and the external mass-transfer resistance is controlling (cf. Fig. 16-13).

3. The rate coefficient for use with a fluid-phase driving force is $k_c = 3(1-\varepsilon)k_f/(\rho_b r_p) = 0.39$ cm³/(g·s).

Example 16-9 Estimation of Rate Coefficient for Protein Adsorption Estimate the rate coefficient for the adsorption of an antibody ($M_r = 150{,}000$) from a 1 mg/cm³ aqueous solution (μ = 1 mPa s) in a column with ε = 0.4 operated at 0.05 cm/s and 25°C. The adsorbent particles have diameter $d_p = 100$ μm, porosity $\varepsilon_p = 0.5$, and pore radius $r_{pore} = 30$ nm. The adsorption isotherm is $n = 500c/(1 + 9.0c)$, where n is in mg/cm³ particle volume and c in mg/cm³.

1. Estimate k_f. $Re = \varepsilon v d_p/\nu = 0.05 \times 100 \times 10^{-4}/0.01 = 0.05$. For a 150,000 molecular mass globular protein in water, $D \sim 4 \times 10^{-7}$ cm²/s [Tyn and Gusek, *Biotechnol. Bioeng.* **35:** 327 (1990)], $Sc = \nu/D = 25{,}000$. From Table 16-9, $Sh = (1.09/0.4)(0.05)^{0.33}(25{,}000)^{0.33} = 29$. Thus, $k_f = D\,Sh/d_p = 1.1\times10^{-3}$ cm/s.

2. Determine the controlling resistance. From the isotherm, $n^{ref} = (500 \times 1)/(1 + 9.0 \times 1) = 50$ mg/cm³ particle. Thus, $\Lambda = (1-\varepsilon)n^{ref}/c^{ref} = (1-0.4) \times 50/1 = 30$ and $R = 1/(1 + 9.0 \times 1) = 0.1$ [cf. Eqs. (16-33) and (16-34)]. From Eq. (16-68), $r_m = \kappa T/6\pi\mu D = (1.38 \times 10^{-16} \times 298)/(6\pi \times 0.01 \times 4 \times 10^{-7}) = 5.5$ nm. Thus, $\lambda_m = r_m/r_{pore} = 5.5/30 = 0.18$. From Eq. (16-67) with $\tau_p = 4$, obtain $D_p = 5.8 \times 10^{-8}$ cm²/s. From Table 16-12, item 3A, $\psi_p = 0.83$, giving $10(1-\varepsilon)\psi_p\varepsilon_p D_p/(D \times Sh) = 0.012$. Thus, from Fig. 16-13, the intraparticle pore diffusion resistance is dominant. For these conditions, from Table 16-12, $k_n = 15\psi_p(1-\varepsilon)\varepsilon_p D_p/\Lambda r_p^2 = 2.8 \times 10^{-4}\ s^{-1}$. A similar approximate result is obtained directly from Fig. 16-13 with $ReSc = 0.05 \times 25{,}000 = 1250$ and $(1-\varepsilon)\psi_p\varepsilon_p D_p/D \sim 0.04$, giving $HTU/d_p \sim 1000$. $k_n \sim 2 \times 10^{-4}\ s^{-1}$ is obtained by using the HTU value in Eq. (16-92). These calculations assume that intraparticle transport is by pore diffusion alone, which is typically limiting for protein adsorption in porous adsorbents.

BATCH ADSORPTION

In this section, the transient adsorption of a solute from a dilute solution in a constant-volume, well-mixed batch system or, equivalently, adsorption of a pure gas is considered. The solutions provided can approximate the response of a stirred vessel containing suspended adsorbent particles, or that of a very short adsorption bed. Uniform, spherical particles of radius r_p are assumed. These particles, initially of uniform adsorbate concentration, are assumed to be exposed to a step change in concentration of the external fluid.

In general, solutions are obtained by coupling the basic conservation equation for the batch system, Eq. (16-49), with the appropriate rate equation. Rate equations are summarized in Tables 16-11 and 16-12 for different controlling mechanisms.

Solutions are provided for external mass-transfer control, intraparticle diffusion control, and mixed resistances for the case of constant V_f and $F_{v,in} = F_{v,out} = 0$. The results are in terms of the fractional approach to equilibrium $F = (\hat{n}_i - \hat{n}_i^{0'})/(\hat{n}_i^\infty - \hat{n}_i^{0'})$, where $\hat{n}_i^{0'}$ and $\hat{n}_i^\infty$ are the initial and ultimate solute concentrations in the adsorbent. The solution concentration is related to the amount adsorbed by the material balance $c_i = c_i^0 - (\hat{n}_i - \hat{n}_i^{0'})M_s/V_f$.

Two general cases are considered: (1) adsorption under conditions of constant or nearly constant external solution concentration (equivalent to infinite fluid volume); and (2) adsorption in a batch with finite volume. In the latter case, the fluid concentration varies from c_i^0 to c_i^∞ when equilibrium is eventually attained. $\Lambda^\infty = (c_i^0 - c_i^\infty)/c_i^0 = M_s(\hat{n}_i^\infty - \hat{n}_i^0)/(V_f c_i^0)$ represents the fraction of adsorbate that is ultimately adsorbed and determines which general case should be considered in the analysis of experimental systems. Generally, when $\Lambda^\infty \geq 0.1$, solutions for the second case (finite batch volume) are required.

EXTERNAL MASS-TRANSFER CONTROL

The intraparticle concentration is uniform, and the rate equation is given by column 1 in Table 16-12.

For a *Langmuir isotherm* with negligible solute accumulation in the particle pores, the solution for an infinite fluid volume is:

$$(1-R)(1-n_i^{0'}/n_i^0)F - R\ln(1-F) = (3k_f t/r_p)(c_i^0/\rho_p n_i^0) \qquad (16\text{-}93)$$

where $n_i^0 = n_i^\infty = n_i^S K_i c_i^0/(1+K_i c_i^0)$ is the adsorbate concentration in the particle at equilibrium with the fluid concentration. The predicted behavior is shown in Fig. 16-14 for $n_i^{0'} = 0$. In the *irreversible limit* ($R = 0$), F increases linearly with time while in the *linear limit* ($R = 1$), $1 - F$ decreases exponentially with time.

For a finite fluid volume ($\Lambda^\infty > 0$), the fractional approach to equilibrium is given by:

$$\left[1-\frac{b'(1-R^\infty)}{2c'}\right]\frac{1}{q'}\ln\frac{(2c'F-b'-q')(-b'+q')}{(2c'F-b'+q')(-b'-q')} - \frac{1-R^\infty}{2c'}\ln(1-b'F+c'F^2) = \frac{3k_f t}{r_p}\frac{c_i^0}{\rho_p n_i^\infty} \qquad (16\text{-}94)$$

where

$$b' = \frac{1-R^\infty}{1-R^0} + \Lambda^\infty \qquad (16\text{-}95a)$$

$$c' = \Lambda^\infty(1-R^\infty) \qquad (16\text{-}95b)$$

$$q' = (b'^2 - 4c')^{0.5} \qquad (16\text{-}95c)$$

$$R^0 = \frac{1}{1+K_i c_i^o} \qquad (16\text{-}95d)$$

$$R^\infty = \frac{1}{1+K c_i^\infty} \qquad (16\text{-}95e)$$

The predicted behavior is shown in Fig. 16-15 for $R^0 = 0.5$ with different values of Λ^∞.

SOLID DIFFUSION CONTROL

For a constant diffusivity and an infinite fluid volume, the solution is:

$$F = 1 - \frac{6}{\pi^2}\sum_{n=1}^{\infty}\frac{1}{n^2}\exp\left(-\frac{n^2\pi^2 D_{si} t}{r_p^2}\right) \qquad (16\text{-}96)$$

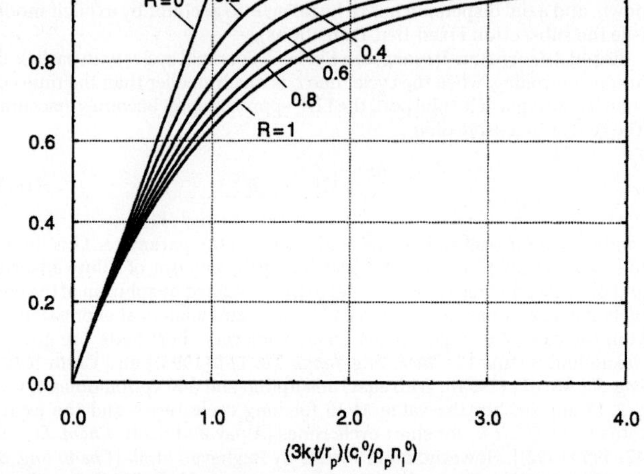

FIG. 16-14 Constant separation factor batch adsorption curves for external mass-transfer control with an infinite fluid volume and $n_i^0 = 0$.

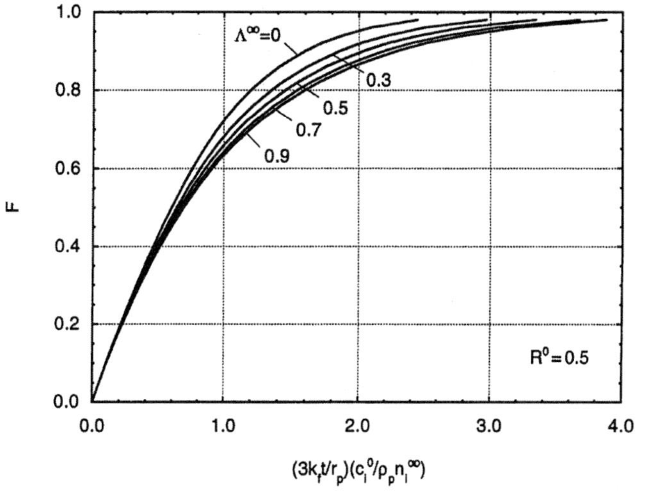

FIG. 16-15 Constant separation factor batch adsorption curves for external mass-transfer control with a finite fluid volume, $n_i^{0'}=0$ and $R^0=0.5$.

For short times, this equation does not converge rapidly. The following approximations can be used instead [Helfferich and Hwang, in Dorfner (1991), pp. 1277–1309 in General References]:

$$F=\frac{6}{r_p}\left(\frac{D_{si}t}{\pi}\right)^{0.5} \qquad F<0.2 \tag{16-97}$$

$$F=\frac{6}{r_p}\left(\frac{D_s t}{\pi}\right)^{0.5}-\frac{3D_{si}t}{r_p^2} \qquad F<0.8 \tag{16-98}$$

For values of $F>0.8$, the first term ($n=1$) in Eq. (16-96) is generally sufficient. If the controlling resistance is diffusion in the subparticles of a bidispersed adsorbent, Eq. (16-96) applies with r_s replacing r_p.

For a finite fluid volume, the solution is:

$$F=1-6\sum_{n=1}^{\infty}\frac{\exp(-p_n^2 D_{si}t/r_p^2)}{9\Lambda^\infty/(1-\Lambda^\infty)+(1-\Lambda^\infty)p_n^2} \tag{16-99}$$

where the p_ns are the positive roots of

$$\frac{\tan p_n}{p_n}=\frac{3}{3+(1/\Lambda^\infty-1)p_n^2} \tag{16-100}$$

The predicted behavior is shown in Fig. 16-16. F is calculated from Eq. (16-96) for $\Lambda^\infty=0$ and from Eq. (16-99) for $\Lambda^\infty>0$. Significant deviations from the $\Lambda^\infty=0$ curve exist for $\Lambda^\infty>0.1$.

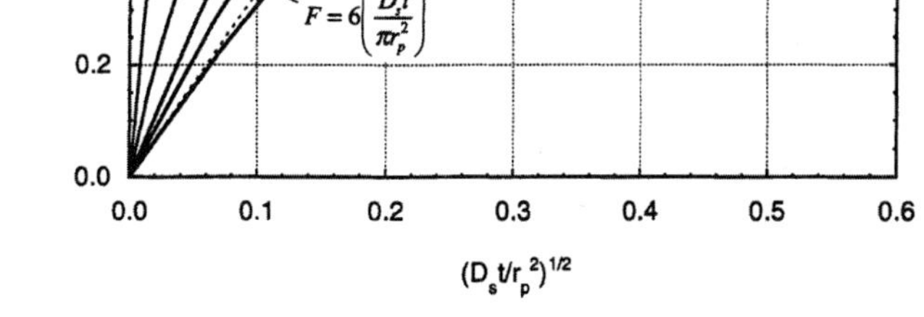

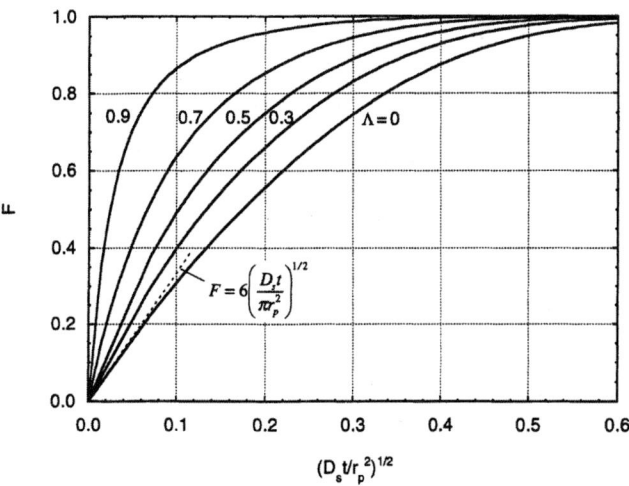

FIG. 16-16 Batch adsorption curves for solid diffusion control. The curve for $\Lambda^\infty=0$ corresponds to an infinite fluid volume. [*Adapted from Ruthven (1984) in General References, with permission.*]

For nonconstant diffusivity, a numerical solution of the conservation equations is generally required. In molecular sieve zeolites, when equilibrium is described by the Langmuir isotherm, the concentration dependence of the intracrystalline diffusivity can often be approximated by Eq. (16-72). The relevant rate equation is:

$$\frac{\partial n_i}{\partial t}=\frac{D_{0i}}{r^2}\frac{\partial}{\partial r}\left(\frac{r^2}{1-n_i/n_i^s}\frac{\partial n_i}{\partial r}\right) \tag{16-101}$$

A numerical solution of this equation for a constant surface concentration (infinite fluid volume) is given by Garg and Ruthven [*Chem. Eng. Sci.* **27:** 417 (1972)]. The solution depends on the value of $\lambda=(n_i^0-n_i^{0'})/(n_i^s-n_i^{0'})$. Because of the effect of adsorbate concentration on the effective diffusivity, for large concentration steps, adsorption is faster than desorption, while for small concentration steps, when D_s can be taken to be essentially constant, adsorption and desorption curves are mirror images of each other as predicted by Eq. (16-96); see Ruthven (1984), p. 175.

In binary ion exchange, intraparticle mass transfer is described by Eq. (16-75) and is dependent on the ionic self diffusivities of the exchanging counterions. A numerical solution of the corresponding conservation equation for spherical particles with an infinite fluid volume is given by Helfferich and Plesset [*J. Chem. Phys.* **66**(28): 418 (1958)]. The numerical results for the case of two counterions of equal valence where a resin bead, initially partially saturated with A, is completely converted to the B form, is expressed by:

$$F=\left\{1-\exp\left[\pi^2\left(f_1(\alpha')\tau_D+f_2(\alpha')\tau_D^2+f_3(\alpha')\tau_D^3\right)\right]\right\}^{1/2} \tag{16-102}$$

with

$$f_1(\alpha')=-(0.570+0.430\alpha'^{0.775})^{-1} \tag{16-103a}$$

$$f_2(\alpha')=(0.260+0.782\alpha')^{-1} \tag{16-103b}$$

$$f_3(\alpha')=-(0.165+0.177\alpha')^{-1} \tag{16-103c}$$

where $\tau_D=\bar{D}_A t/r_p^2$ and $\alpha'=1+(\bar{D}_A/\bar{D}_B-1)n_A^{0'}/n^s$ for $0.1\le\alpha'\le 10$. The predicted behavior is shown in Fig. 16-17. When $\alpha'=1$ (equal ion diffusivities or $n_A^{0'}\sim 0$), Eq. (16-102) coincides with Eq. (16-96). For $\alpha'\ne 1$, the exchange rate is faster or slower, depending on which counterion is initially present in the ion exchanger and on the initial level of saturation.

For an initially fully saturated particle, the exchange rate is faster when the faster counterion is initially in the resin, with the difference in rate becoming more important as conversion from one form to the other progresses. Helfferich (1962, pp. 270–271) gives explicit expressions for the exchange of ions of unequal valence.

PORE DIFFUSION CONTROL

The rate equation is given by item A in Table 16-11. With pore fluid and adsorbent at equilibrium at each point within the particle and for a constant diffusivity, the rate equation can be written as:

$$\frac{\partial c_{pi}}{\partial t}=\frac{\varepsilon_p D_{pi}}{\varepsilon_p+\rho_p dn_i^e/dc_i}\frac{1}{r^2}\frac{\partial}{\partial r}\left(r^2\frac{\partial c_{pi}}{\partial r}\right) \tag{16-104}$$

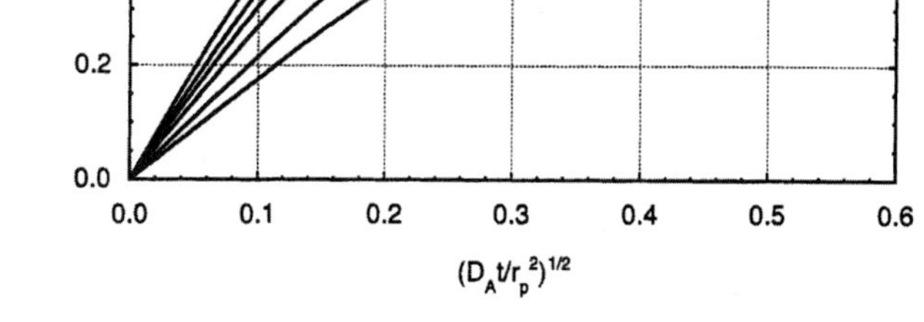

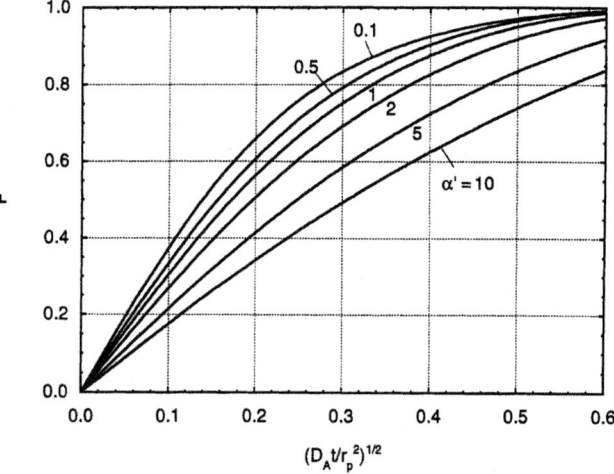

FIG. 16-17 Batch ion exchange for two equal-valence counterions. The exchanger is initially uniformly loaded with ion A in concentration $n_A^{0'}$ and is completely converted to the B form. $\alpha'=1+(\bar{D}_A/\bar{D}_B-1)n_A^{0'}/n^s$.

For a *linear isotherm* ($n_i = K_i c_i$), this equation is identical to the conservation equation for solid diffusion, except that the solid diffusivity D_{si} is replaced by the expression $\varepsilon_p D_{pi}/(\varepsilon_p + \rho_p K_i)$. Thus, Eqs. (16-96) and (16-99) can be used for pore diffusion control with infinite and finite fluid volumes simply by replacing D_{si} with this expression.

When the adsorption isotherm is nonlinear, a numerical solution is generally required. For a *Langmuir system* with negligible solute holdup in the pore fluid, item A in Table 16-11 gives:

$$\frac{\partial n_i}{\partial t} = \frac{\varepsilon_p D_{pi}}{\rho_p n_i^s K_i} \frac{1}{r^2} \frac{\partial}{\partial r}\left[\frac{r^2}{(1-n_i/n_i^s)^2} \frac{\partial n_i}{\partial r}\right] \tag{16-105}$$

This equation has the same form as that obtained for solid diffusion control with D_{si} replaced by the expression $\varepsilon_p D_{pi}/[\rho_p n_i^s K_i (1-n_i/n_i^s)^2]$. Numerical results for the case of adsorption on an initially clean particle are given in Fig. 16-18 for different values of $\lambda = n_i^0/n_i^s = 1 - R$. The uptake curves become increasingly steeper, as the nonlinearity of the isotherm, measured by the parameter λ, increases. The desorption curve shown for a particle with $n_i^{0'}/n_i^s = 0.9$ shows that for the same step in concentration, adsorption occurs much more quickly than desorption. This difference, however, becomes smaller as the value of λ is reduced, and in the linear region of the adsorption isotherm ($\lambda \to 0$), adsorption and desorption curves are mirror images. The solution in Fig. 16-18 is applicable to a nonzero initial adsorbent loading by redefining λ as $(n_i^0 - n_i^{0'})/(n_i^s - n_i^{0'})$ and the dimensionless time variable in the abscissa as $[\varepsilon_p D_{pi} t/\rho_p (1-n_i^{0'}/n_i^s)^2 n_i^s K_i r_p^2]^{1/2}$ [Ruthven (1984) in General References].

In the *irreversible limit* ($R < 0.1$), the adsorption front within the particle approaches a sharp transition separating an inner core into which the adsorbate has not yet penetrated from an outer layer in which the adsorbed phase concentration is uniform at the saturation value. The dynamics of this process is described approximately by the shrinking-core model [Yagi and Kunii, *Chem. Eng. (Japan)* **19:** 500 (1955)]. For an infinite fluid volume, the solution is:

$$\frac{\varepsilon_p D_{pi} t}{r_p^2} \frac{c_i^0}{\rho_p n_i^s} = \frac{1}{2} - \frac{1}{3} F - \frac{1}{2}(1-F)^{2/3} \tag{16-106}$$

or, in explicit form [Brauch and Schlunder, *Chem. Eng. Sci.* **30:** 540 (1975)]:

$$F = 1 - \left\{\frac{1}{2} + \cos\left[\frac{\pi}{3} + \frac{1}{3}\cos^{-1}\left(1 - \frac{12\varepsilon_p D_{pi} t}{r_p^2} \frac{c_i^0}{\rho_p n_i^s}\right)\right]\right\}^3 \tag{16-107}$$

For a finite fluid volume with $0 < \Lambda^\infty \leq 1$, the solution is [Teo and Ruthven, *Ind. Eng. Chem. Process Des. Dev.* **25:** 17 (1986)]:

$$\frac{\varepsilon_p D_{pi} t}{r_p^2} \frac{c_i^0}{\rho_p n_i^s} = I_2 - I_1 \tag{16-108}$$

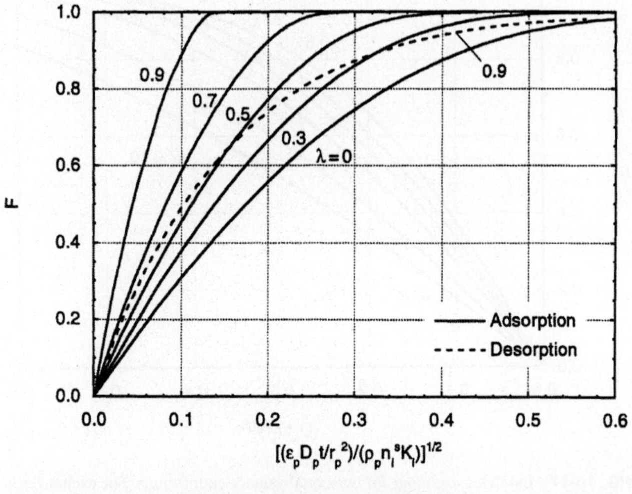

FIG. 16-18 Constant separation factor batch adsorption curves for pore diffusion control with an infinite fluid volume. λ is defined in the text.

where

$$I_1 = \frac{1}{\lambda' \Lambda^\infty \sqrt{3}}\left[\tan^{-1}\frac{2\eta - \lambda'}{\lambda'\sqrt{3}} - \tan^{-1}\frac{2-\lambda'}{\lambda'\sqrt{3}}\right] + \frac{1}{6\lambda'\Lambda^\infty} \ln\left[\frac{\lambda'^3 + \eta^3}{\lambda'^3 + 1}\left(\frac{\lambda' + 1}{\lambda' + \eta}\right)^3\right] \tag{16-109a}$$

$$I_2 = \frac{1}{3\Lambda^\infty} \ln \frac{\lambda'^3 + \eta^3}{\lambda'^3 + 1} \tag{16-109b}$$

$$\eta = (1-F)^{1/3} \tag{16-109c}$$

$$\lambda' = \left(\frac{1}{\Lambda^\infty} - 1\right)^{1/3} \tag{16-109d}$$

Solutions for $R = 0$ analogous to Eqs. (16-107) and (16-108) but accounting explicitly for a distribution of particle sizes are available in Carta and Ubiera, *AIChE J.* **49:** 3066 (2003).

COMBINED RESISTANCES

In general, exact analytic solutions are available only for the linear ($R = 1$) and irreversible limits ($R \to 0$). Intermediate cases require numerical solution or the use of approximate driving force expressions (see the subsection Rate and Dispersion Factors).

Parallel Pore and Solid Diffusion Control With a *linear isotherm*, assuming equilibrium between the pore fluid and the solid adsorbent, batch adsorption is described by Eqs. (16-96) and (16-99), replacing D_{si} with the expression $(\varepsilon_p D_{pi} + \rho_p D_{si})/(\varepsilon_p + \rho_p K_i)$.

External Mass Transfer and Intraparticle Diffusion Control With a linear isotherm, the solution for combined external mass transfer and pore diffusion control with an infinite fluid volume is (Crank, *Mathematics of Diffusion*, 2d ed., Clarendon Press, 1975):

$$F = 1 - \sum_{n=1}^{\infty} \frac{6\text{Bi}^2 \exp\left[-(p_n^2 \varepsilon_p D_{pi} t/r_p^2)/(\varepsilon_p + \rho_p K_i)\right]}{p_n^2\left[p_n^2 + \text{Bi}(\text{Bi}-1)\right]} \tag{16-110}$$

where $\text{Bi} = k_f r_p/\varepsilon_p D_{pi}$ is the *Biot number* and the p_ns are the positive roots of

$$p_n \cot p_n = 1 - \text{Bi} \tag{16-111}$$

For a finite fluid volume, the solution is:

$$F = 1 - 6\sum_{n=1}^{\infty} \frac{\exp\left[-(p_n^2 \varepsilon_p D_{pi} t/r_p^2)/(\varepsilon_p + \rho_p K_i)\right]}{\frac{9\Lambda^\infty}{1-\Lambda^\infty} + (1-\Lambda^\infty)p_n^2 - (5\Lambda^\infty + 1)\frac{p_n^2}{\text{Bi}} + (1-\Lambda^\infty)\frac{p_n^4}{\text{Bi}^2}} \tag{16-112}$$

where the p_ns are the positive roots of

$$\frac{\tan p_n}{p_n} = \frac{3 - \frac{1-\Lambda^\infty}{\Lambda^\infty}\frac{p_n^2}{\text{Bi}}}{3 + \frac{1-\Lambda^\infty}{\Lambda^\infty}\frac{(\text{Bi}-1)p_n^2}{\text{Bi}}} \tag{16-113}$$

These expressions can also be used for the case of external mass transfer and solid diffusion control by substituting D_{si} for $\varepsilon_p D_{pi}/(\varepsilon_p + \rho_p K_i)$ and $k_f r_p/(\rho_p K_i D_{si})$ for the Biot number.

In the *irreversible limit*, the solution for combined external resistance and pore diffusion with infinite fluid volume is [Yagi and Kunii, *Chem. Eng. (Japan)* **19:** 500 (1955)]:

$$\frac{\varepsilon_p D_{pi} t}{r_p^2} \frac{c_i^0}{\rho_p n_i^s} = \frac{1}{2} - \frac{1}{3}\left(1 - \frac{1}{\text{Bi}}\right)F - \frac{1}{2}(1-F)^{2/3} \tag{16-114}$$

For a finite fluid volume, the solution is [Teo and Ruthven, *Ind. Eng. Chem. Process Des. Dev.* **25:** 17 (1986)]:

$$\frac{\varepsilon_p D_{pi} t}{r_p^2} \frac{c_i^0}{\rho_p n_i^s} = \left(1 - \frac{1}{\text{Bi}}\right) I_2 - I_1 \tag{16-115}$$

where I_1 and I_2 are given by Eqs. (16-109*a*) and (16-109*b*).

Bidispersed Particles For particles of radius r_p comprising adsorptive subparticles of radius r_s that define a macropore network, conservation equations are needed to describe transport both within the macropores and within the subparticles and are given in Table 16-11, item D. Detailed equations and solutions for a linear isotherm are given in Ruthven (1984, p. 183) and Ruckenstein et al. [*Chem. Eng. Sci.* **26:** 1306 (1971)]. The solution for a linear isotherm with no external resistance and an infinite fluid volume is:

$$F = 1 - \frac{18}{\beta + 3\alpha} \sum_{m=1}^{\infty} \sum_{n=1}^{\infty} \left(\frac{n^2\pi^2}{p_{n,m}^4} \right) \times \frac{\exp(-p_{n,m}^2 D_{si} t / r_s^2)}{\alpha + \frac{\beta}{2}\left[1 + \frac{\cot p_{n,m}}{p_{n,m}}(p_{n,m}\cot p_{n,m} - 1)\right]} \tag{16-116}$$

where the $p_{n,m}$ values are the roots of the equation

$$\alpha p_{n,m}^2 - n^2\pi^2 = \beta(p_{n,m}\cot p_{n,m} - 1) \tag{16-117}$$

and

$$\alpha = \frac{D_{si}/r_s^2}{D_{pi}/r_p^2} \tag{16-118a}$$

$$\beta = \frac{3\alpha\rho_p K_i}{\varepsilon_p} \tag{16-118b}$$

In these equations, D_{si} is the diffusivity in the subparticles, and D_{pi} is the diffusivity in the pore network formed by the subparticles.

For large K_i values, the uptake curve depends only on the value of the parameter β representing the ratio of characteristic time constants for diffusion in the pores and in the subparticles. For small β values, diffusion in the subparticles is controlling, and the solution coincides with Eq. (16-96) with r_s replacing r_p. For large β values, pore diffusion is controlling, and the solution coincides with Eq. (16-96) with $\varepsilon_p D_{pi}/(\varepsilon_p + \rho_p K_i)$ replacing D_{si}.

Lee [*AIChE J.* **24:** 531 (1978)] gives the solution for batch adsorption with bidispersed particles for the case of a finite fluid volume.

MULTICOMPONENT SYSTEMS

Describing the kinetics of competitive adsorption in multicomponent systems requires, in general, a numerical solution of the relevant conservation equations. Approximate analytical solutions, neglecting solute holdup in the pore fluid, are available for the case where the isotherms are highly favorable and sharp fronts are formed within the particles when pore diffusion controls [Martin et al., *J. Chromatogr. A* **1079:** 105 (2005)].

FIXED-BED TRANSITIONS

As discussed in the subsection Design Concepts, a large fraction of adsorption and ion-exchange processes takes place in fixed beds. Two classical methods for analyzing fixed-bed transitions are described here. First, local equilibrium theory is presented. In this theory, all mass-transfer resistances are ignored to focus on the often-dominating role of isotherm shape. Second, results of constant pattern analysis are presented. This gives the maximum breadth to which a mass-transfer zone will spread for various rate mechanisms. It is therefore conservative for design purposes. Both of these methods pertain to behavior in deep beds. For shallow beds, the equations that follow must be solved for the particular case of interest.

DIMENSIONLESS SYSTEM

For both methods, Eq. (16-52) is considered, that is, the material balance for a fixed bed, written in the form

$$\rho_b \frac{\partial \bar{n}_i}{\partial t} + \varepsilon_b \frac{\partial c_i}{\partial t} + \varepsilon \frac{\partial (v c_i)}{\partial z} = \varepsilon D_L \frac{\partial}{\partial z}\left(c \frac{\partial y_i}{\partial z} \right) \tag{16-119}$$

where it has been assumed that D_L is constant and that $\hat{c}_i \approx c_i$.

Dimensionless variables can be defined for time, the axial coordinate, and velocity:

$$\tau = \frac{\varepsilon v^{\text{ref}} t}{L} \tag{16-120}$$

$$\zeta = \frac{z}{L} \tag{16-121}$$

$$v^* = \frac{v}{v^{\text{ref}}} \tag{16-122}$$

respectively, where L is the bed length, v^{ref} is the interstitial velocity at the bed inlet, and τ is equal to the number of empty bed volumes of feed passed into the bed. Accordingly, the material balance becomes

$$\rho_b \frac{\partial \bar{n}_i}{\partial \tau} + \varepsilon_b \frac{\partial c_i}{\partial \tau} + \frac{\partial (v^* c_i)}{\partial \zeta} = \frac{1}{N_{\text{Pe}}} \frac{\partial}{\partial \zeta}\left(c \frac{\partial y_i}{\partial \zeta} \right) \tag{16-123}$$

where $N_{\text{Pe}} = v^{\text{ref}} L / D_L$ is a Peclet number for the bed or a number of dispersion units. Equation (16-123) or a similar equation is often the material balance used in nonisothermal problems, in problems involving adsorption of nontrace components, and in calculations of cycles.

For a trace, isothermal system, $v^* = 1$, and using the dimensionless system variables for concentrations [Eq. (16-10)], Eq. (16-123) becomes

$$\Lambda \frac{\partial \bar{n}_i^*}{\partial \tau} + \varepsilon_b \frac{\partial c_i^*}{\partial \tau} + \frac{\partial c_i^*}{\partial \zeta} = \frac{1}{N_{\text{Pe}}} \frac{\partial^2 c_i^*}{\partial \zeta^2} \tag{16-124}$$

where Λ is the partition ratio, defined by

$$\Lambda = \rho_b n_i^{\text{ref}} / c_i^{\text{ref}} \tag{16-125}$$

This important dimensionless group is the volumetric capacity of the bed for the sorbable component divided by the concentration of the sorbable component in the feed. The stoichiometric capacity of the bed for solute is exactly equal to Λ empty bed volumes of feed (to saturate the sorbent at the feed concentration) plus a fraction of a bed volume of feed to fill the voids outside and inside the particles. Alternatively, Eq. (16-124) is also obtained using the dimensionless transition variables for concentrations [Eq. (16-11)], but now the partition ratio in the first term of Eq. (16-124) pertains to the transition and is given by

$$\Lambda = \rho_b \frac{n_i'' - n_i'}{c_i'' - c_i'} \tag{16-126}$$

Equation (16-124) is a commonly used form of material balance for a fixed-bed adsorber.

If the system under consideration involves the use of the sorbent for only a single feed step or reuse after uniform regeneration, as in many applications with activated carbons and ion exchangers, then one of two paths is often followed at this point to simplify Eq. (16-124) further. The second term on the left-hand side of the equation is often assumed to be negligibly small (usually a good assumption), and time is redefined as

$$\tau_1 = \tau / \Lambda \tag{16-127}$$

to give

$$\frac{\partial \bar{n}_i^*}{\partial \tau_1} + \frac{\partial c_i^*}{\partial \zeta} = \frac{1}{N_{\text{Pe}}} \frac{\partial^2 c_i^*}{\partial \zeta^2} \tag{16-128}$$

Alternatively, in the absence of axial dispersion, a variable of the form

$$\tau_1 = \frac{\tau - \varepsilon_b \zeta}{\Lambda} \tag{16-129}$$

can be defined to reduce Eq. (16-124) directly to

$$\frac{\partial \bar{n}_i^*}{\partial \tau_1} + \frac{\partial c_i^*}{\partial \zeta} = 0 \tag{16-130}$$

The variable τ_1 defined by Eq. (16-127) or (16-129) is a throughput parameter, equal to unity (hence, the 1 subscript) at the time when the stoichiometric center of the concentration wave leaves the bed. This important group, in essence a dimensionless time variable, determines the location of the stoichiometric center of the transition in the bed at any time.

LOCAL EQUILIBRIUM THEORY

In local equilibrium theory, fluid and sorbed phases are assumed to be in local equilibrium with one another at every axial position in the bed. Thus, because of uniform concentrations, the overbar on n_i^* is not necessary, and we have $\hat{c}_i \approx c_i$ [note Eqs. (16-52) and (16-119)].

Single-Transition System For a system described by a single material balance, Eq. (16-130) gives

$$\frac{\partial \tau_1}{\partial \zeta} = -\frac{\partial c_i^* / \partial \zeta}{\partial c_i^* / \partial \tau_1} = \frac{dn_i^*}{dc_i^*} \tag{16-131}$$

where $d\tau_1/d\zeta$ is the reciprocal of a concentration velocity. Equation (16-131) is the equation for a *simple wave* (or gradual transition or proportionate pattern). If a bed is initially uniformly saturated, then $d\tau_1/d\zeta = \tau_1/\zeta$. Thus, for the dimensionless system, the reciprocal of the velocity of a concentration is equal to the slope of the isotherm at that concentration. Furthermore, from Eq. (16-131), the depth of penetration of a given concentration into the bed is directly proportional to time, so the breadth of a simple wave increases in direct proportion to the depth of its penetration into the bed (or to time). Thus, for the simple wave, the length of the MTZ is proportional to the depth of the bed through which the wave has passed. Consideration of isotherm shape indicates that a simple wave occurs for an unfavorable dimensionless isotherm ($d^2n_i^*/dc_i^{*2} > 0$), for which low concentrations will go faster than high concentrations. Equation (16-131) also pertains to a linear isotherm, in which case the wave is called a *contact discontinuity* because it has neither a tendency to spread nor to sharpen. If a mass-transfer resistance is added to the consideration of wave character for unfavorable isotherms, the wave will still asymptotically approach the simple wave result given by Eq. (16-131).

For a favorable isotherm ($d^2n_i^*/dc_i^{*2} < 0$), Eq. (16-131) gives the impossible result that three concentrations can coexist at one point in the bed (see Example 16-10). The correct solution is a *shock* (or abrupt transition) and not a simple wave. Mathematical theory has been developed for this case to give "weak solutions" to conservation laws. The form of the solution is

$$\text{Shock speed} = \frac{\text{change in flux}}{\text{change in accumulated quantity}}$$

where the changes are jump discontinuities across the shock. The reciprocal of this equation, using Eq. (16-130), is

$$\frac{d\tau_1}{d\zeta} = \frac{\Delta n_i^*}{\Delta c_i^*} \qquad (16\text{-}132)$$

where the differences are taken across the shock.

It is also possible to have a *combined wave*, which has both gradual and abrupt parts. The general rule for an isothermal, trace system is that in passing from the initial condition to the feed point in the isotherm plane, the slope of the path must not decrease. If it does, then a shock chord is taken for part of the path. Referring to Fig. 16-19, for a transition from (0,0) to (1,1), the dashes indicate shock parts, which are connected by a simple wave part between points P_1 and P_2.

Example 16-10 Transition Types For the constant separation-factor isotherm given by Eq. (16-31), determine breakthrough curves for $r = 2$ and $r = 0.5$ for transitions from $c_i^* = 0$ to $c_i^* = 1$.

Using Eq. (16-131), we obtain

$$\frac{\tau_1}{\zeta} = \frac{r}{[r + (1-r)c_i^*]^2}$$

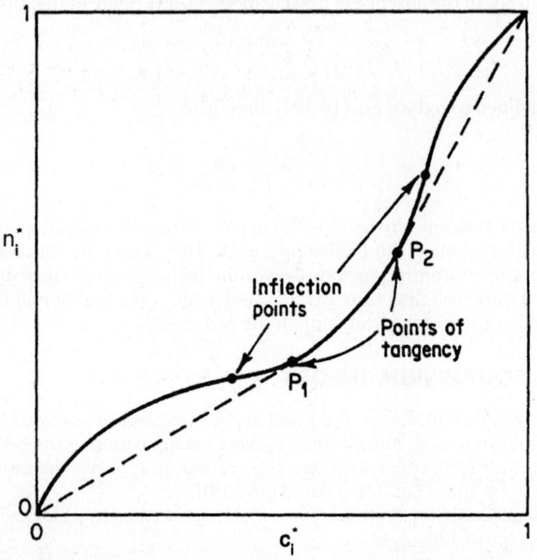

FIG. 16-19 Path in isotherm plane for a combined wave. [After Tudge, *Can. J. Phys.* **39:** 1611 (1961).]

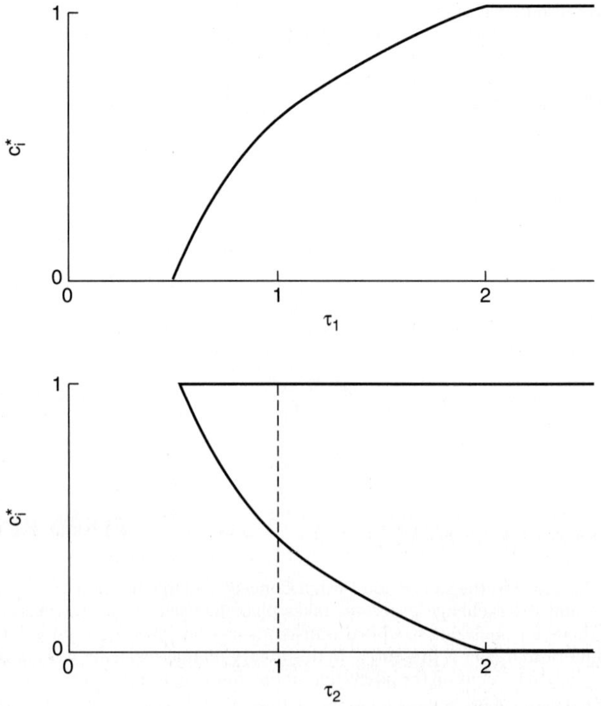

FIG. 16-20 Breakthrough curves for $r = 2$ (*top*) and $r = 0.5$ (*bottom*) for Example 16-9.

This equation, evaluated at $\zeta = 1$, is plotted for $r = 2$ and $r = 0.5$ in Fig. 16-20. Clearly, the solution for $r = 0.5$ is not physically correct. Equation (16-132), with $d\tau_1/d\zeta = \tau_1/\zeta$, is applied to this case to give the shock indicated by the dashed line. Alternatively, bed profiles can be obtained by evaluating equations at τ_1 = const.

Multiple-Transition System Local equilibrium theory for multiple transitions begins with some combination of material and energy balances, written

$$\rho_b \frac{\partial n_i}{\partial \tau} + \varepsilon_b \frac{\partial c_i}{\partial \tau} + \frac{\partial (v^* c_i)}{\partial \zeta} = 0 \qquad (i = 1,2,\ldots) \qquad (16\text{-}133)$$

$$\rho_b \frac{\partial u_s}{\partial \tau} + \varepsilon_b \frac{\partial (cu_f)}{\partial \tau} + \frac{\partial (v^* ch_f)}{\partial \zeta} = 0 \qquad (16\text{-}134)$$

which are Eq. (16-123) written with no axial dispersion and Eq. (16-55) written for an adiabatic bed.

For a simple wave, application of the method of characteristics (hodograph transformation) gives

$$\frac{d\tau}{d\zeta} = \frac{d(\rho_b n_1 + \varepsilon_b c_1)}{d(v^* c_1)} = \frac{d(\rho_b n_2 + \varepsilon_b c_2)}{d(v^* c_2)} = \cdots = \frac{d(\rho_b u_s + \varepsilon_b c u_f)}{d(v^* ch_f)} \qquad (16\text{-}135)$$

where the derivatives are taken along the path of a transition (i.e., directional derivatives).

If a simple wave is not possible on physical grounds, then it (or part of it) is replaced by a shock, given by

$$\frac{d\tau}{d\zeta} = \frac{\Delta(\rho_b n_1 + \varepsilon_b c_1)}{\Delta(v^* c_1)} = \frac{\Delta(\rho_b n_2 + \varepsilon_b c_2)}{\Delta(v^* c_2)} = \cdots = \frac{\Delta(\rho_b u_s + \varepsilon_b c u_f)}{\Delta(v^* ch_f)} \qquad (16\text{-}136)$$

Extensions When more than two conservation equations are to be solved simultaneously, matrix methods for eigenvalues and left eigenvectors are efficient [Jeffrey and Taniuti, *Nonlinear Wave Propagation,* Academic Press, New York, 1964; Jacob and Tondeur, *Chem. Eng. J.* **22:** 187 (1981), **26:** 41 (1983); Davis and LeVan, *AIChE J.* **33:** 470 (1987); Rhee, Aris, and Amundson (1986, 1989) in General References].

Nontrace isothermal systems give the "adsorption effect" (i.e., significant change in fluid velocity because of loss or gain of solute). Criteria for the existence of simple waves, contact discontinuities, and shocks are changed somewhat [Peterson and Helfferich, *J. Phys. Chem.* **69:** 1283 (1965); LeVan et al., *AIChE J.* **34:** 996 (1988); Frey, *AIChE J.* **38:** 1649 (1992)].

Local equilibrium theory also pertains to adsorption with axial dispersion, since this mechanism does not disallow the existence of equilibrium between stationary and fluid phases across the cross section of the bed [Rhee et al., *Chem. Eng. Sci.* **26:** 1571 (1971)]. This will be discussed in further detail from the standpoint of the constant pattern.

Example 16-11 Two-Component Isothermal Adsorption Two components present at low mole fractions are adsorbed isothermally from an inert fluid in an initially clean bed. The system is described by $\rho_b = 500$ kg/m^3, $\varepsilon_b = 0.7$, and the binary Langmuir isotherm

$$n_i = \frac{n_i^s K_i c_i}{1 + K_1 c_1 + K_2 c_2} \qquad (i = 1,2)$$

with $n_1^s = n_2^s = 6$ mol/kg, $K_1 = 40$ m^3/mol, and $K_2 = 20$ m^3/mol. The feed is $c_1 = c_2 = 0.5$ mol/m^3. Find the bed profile.

Using the isotherm to calculate loadings in equilibrium with the feed gives $n_1 = 3.87$ mol/kg and $n_2 = 1.94$ mol/kg. An attempt to find a simple wave solution for this problem fails because of the favorable isotherms (as in the bottom figure of Fig. 16-20). To obtain the two shocks, Eq. (16-136) is written

$$\frac{d\tau}{d\zeta} = \frac{\Delta(\rho_b n_1 + \varepsilon_b c_1)}{\Delta c_1} = \frac{\Delta(\rho_b n_2 + \varepsilon_b c_2)}{\Delta c_2}$$

The concentration of one of the components will drop to zero in the shock nearest the bed inlet. If it is component 1, then using feed values and the preceding equation, that shock would be at

$$\frac{\tau}{\zeta} = \frac{\rho_b n_1 + \varepsilon_b c_1}{c_1} \approx \frac{\rho_b n_1}{c_1} = 3870$$

Similarly, if the second component were to disappear in the first shock, $\tau/\zeta = 1940$. Material balance considerations require that the shorter distance is accepted, so component 1 disappears in the first shock.

The concentrations of component 2 on the plateau downstream of the first shock are then calculated from

$$\frac{\tau}{\zeta} = \frac{\Delta(\rho_b n_2 + \varepsilon_b c_2)}{\Delta c_2} \approx \frac{\rho_b \Delta n_2}{\Delta c_2} = 3870$$

and its pure component isotherm, giving $c_2 = 0.987$ mol/m^3 and $n_2 = 5.71$ mol/kg. The location of this shock is determined using these concentrations and

$$\frac{\tau}{\zeta} = \frac{\rho_b n_2 + \varepsilon_b c_2}{c_2} \approx \frac{\rho_b n_2}{c_2}$$

which gives $\tau/\zeta = 2890$. The bed profile is plotted in Fig. 16-21 using ζ/τ as the abscissa. This example can also be worked with the *h*-transformation and can be extended to adiabatic adsorption and thermal regeneration as shown elsewhere [LeVan and Carta (2008) in General References].

CONSTANT PATTERN BEHAVIOR FOR FAVORABLE ISOTHERMS

With a favorable isotherm and a mass-transfer resistance or axial dispersion, a transition approaches a *constant pattern*, which is an asymptotic shape beyond which the wave will not spread. The wave is said to be "self-sharpening." (If a wave is initially broader than the constant pattern, it will sharpen to approach the constant pattern.) Thus, for an initially uniformly loaded bed, the constant pattern gives the maximum breadth of the MTZ. As bed length is increased, the constant pattern will occupy an increasingly smaller fraction of the bed. (Square-root spreading for a linear isotherm gives this same qualitative result.)

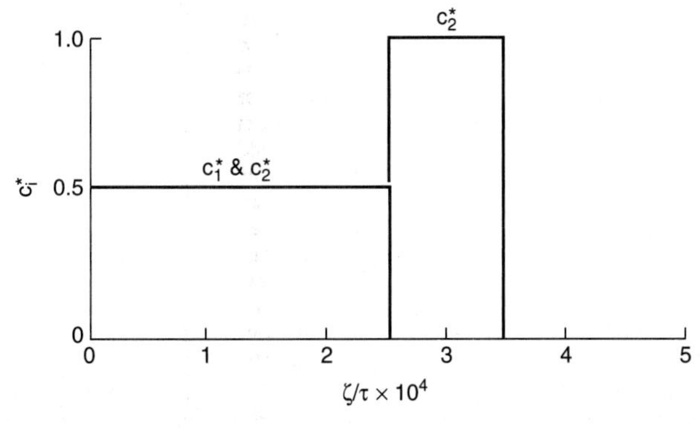

FIG. 16-21 Bed profiles for two-component isothermal adsorption, Example 16-10.

The treatment here is restricted to the Langmuir or constant separation factor isotherm, single-component adsorption, dilute systems, isothermal behavior, and mass-transfer resistances acting alone. References to extensions are given later in this discussion. Different isotherms have been considered, and the theory is well developed for general isotherms.

Asymptotic Solution Rate equations for the various mass-transfer mechanisms are written in dimensionless form in Table 16-13 in terms of a number of transfer units, $N = L/\mathrm{HTU}$, for particle-scale mass-transfer resistances, a number of reaction units for the reaction kinetics mechanism, and a number of dispersion units, N_{Pe}, for axial dispersion. For pore and solid diffusion, $\xi = r/r_p$ is a dimensionless radial coordinate, where r_p is the radius of the particle. If a particle is bidisperse, then r_p can be replaced by r_s, the radius of a subparticle. For preliminary calculations, Fig. 16-13 can be used to estimate N for use with the LDF approximation when more than one resistance is important.

In constant pattern analysis, equations are transformed into a new coordinate system that moves with the wave. Variables are changed from (ζ, τ_1) to $(\zeta - \tau_1, \tau_1)$. The new variable $\zeta - \tau_1$ is equal to zero at the stoichiometric center of the wave. Equation (16-130) for a bed with no axial dispersion, when transformed to the $(\zeta - \tau_1, \tau_1)$ coordinate system, becomes

$$-\frac{\partial \bar{n}_i^*}{\partial(\zeta - \tau_1)} + \frac{\partial \bar{n}_i^*}{\partial \tau_1} + \frac{\partial c_i^*}{\partial(\zeta - \tau_1)} = 0 \qquad (16\text{-}137)$$

The constant pattern is approached as the τ_1 dependence in this equation disappears. Thus, discarding the derivative with respect to τ_1 and integrating, using the condition that $\bar{n}_i^*$ and c_i^* approach zero as $N(\zeta - \tau_1) \to \infty$ [or approach unity as $N(\zeta - \tau_1) \to -\infty$], gives simply

$$\bar{n}_i^* = c_i^* \qquad (16\text{-}138)$$

For adsorption with axial dispersion, the material balance transforms to

$$-\frac{\partial n_i^*}{\partial(\zeta - \tau_1)} + \frac{\partial n_i^*}{\partial \tau_1} + \frac{\partial c_i^*}{\partial(\zeta - \tau_1)} = \frac{1}{N_{\mathrm{Pe}}} \frac{\partial^2 c_i^*}{\partial(\zeta - \tau_1)^2} \qquad (16\text{-}139)$$

The partial derivative with respect to τ_1 is discarded and the resulting equation integrated once to give

$$-n_i^* + c_i^* = \frac{1}{N_{\mathrm{Pe}}} \frac{dc_i^*}{d(\zeta - \tau_1)} \qquad (16\text{-}140)$$

After eliminating n_i^* or c_i^* using the adsorption isotherm, Eq. (16-140) can be integrated directly to obtain the constant pattern.

For other mechanisms, the particle-scale equation must be integrated. Equation (16-138) is used to advantage. For example, for external mass transfer acting alone, the dimensionless rate equation in Table 16-13 would be transformed into the $(\zeta - \tau_1, \tau_1)$ coordinate system and derivatives with respect to τ_1 discarded. Equation (16-138) is then used to replace c_i^* with $\bar{n}_i^*$ in the transformed equation. Furthermore, for this case there are assumed to be no gradients within the particles, so $\bar{n}_i^* = n_i^*$. After making this substitution, the transformed equation can be rearranged to

$$-\frac{dn_i^*}{n_i^* - c_i^*} = d(\zeta - \tau_1) \qquad (16\text{-}141)$$

Since n_i^* and c_i^* are related by the adsorption isotherm, Eq. (16-141) can be integrated.

The integration of Eq. (16-140) or (16-141) as an indefinite integral will give an integration constant that must be evaluated to center the transition properly. The material balance depicted in Fig. 16-22 is used. The two shaded regions must be of equal area if the stoichiometric center of the transition is located where the throughput parameter is unity. Thus,

$$\int_{-\infty}^{0} (1 - n_i^*)\, d(\zeta - \tau_1) = \int_{0}^{\infty} n_i^*\, d(\zeta - \tau_1) \qquad (16\text{-}142)$$

Integrating Eq. (16-142) by parts gives

$$\int_{0}^{1} (\zeta - \tau_1)\, dn_i^* = 0 \qquad (16\text{-}143)$$

For all mechanisms except axial dispersion, the transition can be centered just as well using c_i^* because of Eq. (16-138). For axial dispersion, the

TABLE 16-13 Constant Pattern Solutions for Constant Separation Factor Isotherm ($R < 1$)

Mechanism	N	Dimensionless rate equation[1]	Constant pattern	Refs.
Pore diffusion	$\frac{15(1-\varepsilon)\varepsilon_p D_p L}{\varepsilon v^{ref} r_p^2}$	$\frac{\partial n_i^*}{\partial \tau_1} = \frac{N}{15}\frac{1}{\xi^2}\frac{\partial}{\partial \xi}\left(\xi^2 \frac{\partial c_i^*}{\partial \xi}\right)$	Numerical for $0 < R < 1$ Analytical for $R = 0$	A B
Solid diffusion	$\frac{15\Lambda D_s L}{\varepsilon v^{ref} r_p^2}$	$\frac{\partial n_i^*}{\partial \tau_1} = \frac{N}{15}\frac{1}{\xi^2}\frac{\partial}{\partial \xi}\left(\xi^2 \frac{\partial n_i^*}{\partial \xi}\right)$	Numerical for $0 < R < 1$ Analytical for $R = 0$	C D
External mass transfer	$\frac{k_f a L}{\varepsilon v^{ref}}$	$\frac{\partial \bar{n}_i^*}{\partial \tau_1} = N(c_i^* - c_i^{e*})$	$\frac{1}{1-R}\ln\left[\frac{(1-c_i^*)^R}{c_i^*}\right] - 1 = N(\zeta - \tau_1)$	E
Linear driving force	$\frac{k_n \Lambda L}{\varepsilon v^{ref}}$	$\frac{\partial \bar{n}_i^*}{\partial \tau_1} = N(n_i^{e*} - \bar{n}_i^*)$	$\frac{1}{1-R}\ln\left[\frac{1-c_i^*}{c_i^{*R}}\right] + 1 = N(\zeta - \tau_1)$	F
Reaction kinetics	$\frac{k_a c^{ref} \Lambda L}{(1-R)\varepsilon v^{ref}}$	$\frac{\partial \bar{n}_i^*}{\partial \tau_1} = N[(1-\bar{n}_i^*)c_i^* - R\bar{n}_i^*(1-c_i^{e*})]$	$\frac{1}{1-R}\ln\left[\frac{1-c_i^*}{c_i^*}\right] = N(\zeta - \tau_1)$	G
Axial dispersion	$\frac{v^{ref} L}{D_L}$	Eq. (16-128)	$\frac{1}{1-R}\ln\left[\frac{1-c_i^*}{c_i^{*R}}\right] = N(\zeta - \tau_1)$	H

1: Dimensional rate equations are given in Tables 16-11 and 16-12.
A: Hall et al., *Ind. Eng. Chem. Fundam.* **5:** 212 (1966).
B: Hall et al., *Ind. Eng. Chem. Fundam.* **5:** 212 (1966); Cooper and Liberman, *Ind. Eng. Chem. Fundam.* **9:** 620 (1970). Analytical solution for $R = 0$ is

$$\frac{15}{2}\ln[1+(1-c_i^*)^{1/3}+(1-c_i^*)^{2/3}] - \frac{15}{\sqrt{3}}\tan^{-1}\left[\frac{2(1-c_i^*)^{1/3}+1}{\sqrt{3}}\right] + \frac{5\pi}{2\sqrt{3}} - \frac{5}{2} = N(\zeta - \tau_1)$$

C: Hall et al., *Ind. Eng. Chem. Fundam.* **5:** 212 (1966); Garg and Ruthven, *Chem. Eng. Sci.* **28:** 791, 799 (1973).
D: Hall et al., *Ind. Eng. Chem. Fundam.* **5:** 212 (1966) [see also Cooper, *Ind. Eng. Chem. Fundam.* **4:** 308 (1965); Ruthven (gen. refs.)]. Analytical solution for $R = 0$ is

$$c_i^* = 1 - \frac{6}{\pi^2}\sum_{n=0}^{\infty}\frac{1}{n^2}\exp\left\{-n^2\left[\left(\frac{\pi^2}{15}\right)N(\tau_1 - \zeta) + 0.64\right]\right\}$$

E: Michaels, *Ind. Eng. Chem.* **44:** 1922 (1952); Miura and Hashimoto, *J. Chem. Eng. Japan* **10:** 490 (1977).
F: Glueckauf Coates, *J. Chem. Soc.* **1947:** 1315 (1947); Vermeulen, *Advances in Chemical Engineering,* **2:** 147 (1958); Hall et al., *Ind. Eng. Chem. Fundam.* **5:** 212 (1966); Miura and Hashimoto, *J. Chem. Eng. Japan* **10:** 490 (1977).
G: Walter, *J. Chem. Phys.* **13:** 229 (1945); Hiester and Vermeulen, *Chem. Eng. Progress* **48:** 505 (1952).
H: Acrivos, *Chem. Eng. Sci.* **13:** 1 (1960); Coppola and LeVan, *Chem. Eng. Sci.* **36:** 967 (1981).

transition should be centered using n_i^* provided the fluid-phase accumulation term in the material balance, Eq. (16-124), can be neglected. If fluid-phase accumulation is important, then the transition for axial dispersion can be centered by taking into account the relative quantities of solute held in the fluid and adsorbed phases.

Constant pattern solutions for the individual mechanisms and constant separation factor isotherm are given in Table 16-13. The solutions all have the expected dependence on R—the more favorable the isotherm, the sharper the profile.

Figure 16-23 compares the various constant pattern solutions for $R = 0.5$. The curves are of a similar shape. The solution for reaction kinetics is perfectly symmetrical. The curves for the axial dispersion fluid-phase concentration profile and the linear driving force approximation are identical except that the latter occurs one transfer unit further down the bed. The curve for external mass transfer is exactly that for the linear driving force approximation turned upside down [i.e., rotated 180° about $c_i^* = n_i^* = 0.5$, $N(\zeta - \tau_1) = 0$]. The linear driving force approximation provides a good approximation for both pore diffusion and surface diffusion.

Because of the close similarity in shape of the profiles shown in Fig. 16-23 (as well as likely variations in parameters; e.g., concentration-dependent surface diffusion coefficient), a controlling mechanism cannot be reliably determined from transition shape. If reliable correlations are not available and rate parameters cannot be measured in independent experiments, then particle diameters, velocities, and other factors should be varied, and the observed impact should be considered in relation to the definitions of the numbers of transfer units.

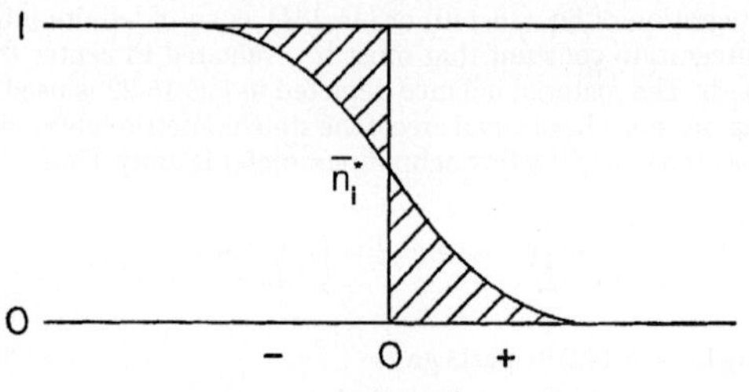

FIG. 16-22 Material balance for centering profile. [From LeVan in Rodrigues et al. (eds), *Adsorption: Science and Technology*, Kluwer Academic Publishers, Dordrecht, The Netherlands, 1989; reprinted with permission.]

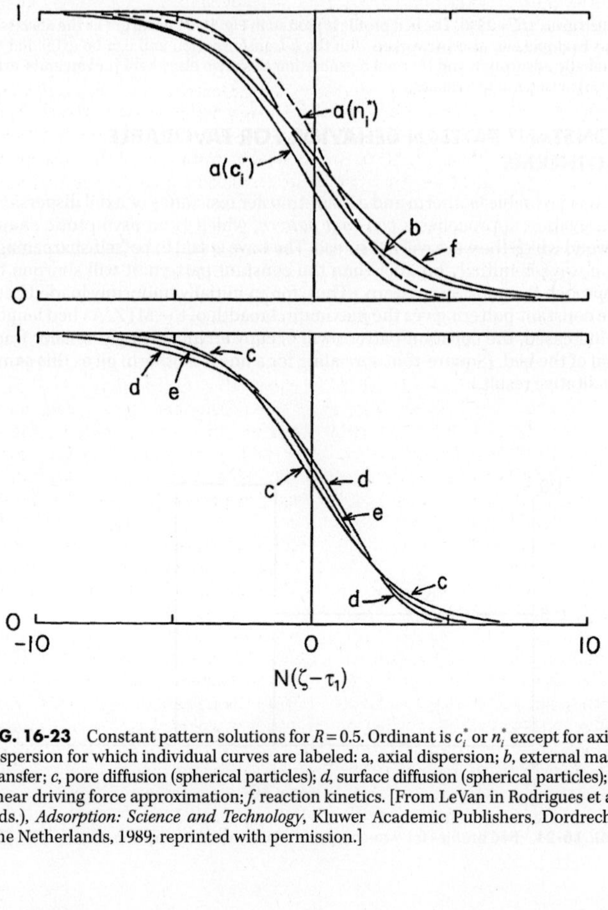

FIG. 16-23 Constant pattern solutions for $R = 0.5$. Ordinant is c_i^* or n_i^* except for axial dispersion for which individual curves are labeled: a, axial dispersion; *b*, external mass transfer; *c*, pore diffusion (spherical particles); *d*, surface diffusion (spherical particles); *e*, linear driving force approximation; *f*, reaction kinetics. [From LeVan in Rodrigues et al. (eds.), *Adsorption: Science and Technology*, Kluwer Academic Publishers, Dordrecht, The Netherlands, 1989; reprinted with permission.]

Example 16-12 Estimation of Breakthrough Time With reference to Example 16-9, determine the 10 percent breakthrough time and the column dynamic binding capacity if the column is 20 cm long.

1. From Example 16-9, $\Lambda = 30$, $R = 0.1$, and $D_p = 5.8 \times 10^{-8}$ cm^2/s. For pore diffusion control (Table 16-13, item A), $N = 15(1-\varepsilon)\varepsilon_p D_p L/(\varepsilon v r_p^2) = 4.2$. Based on the numerical solution of Hall et al. [*Ind. Eng. Chem. Fundam.* **5:** 212 (1966)], $N(1-\tau_1) \sim 1.3$ with $c_i^* = 0.1$, giving $\tau_1 = 0.69$. The corresponding time is $t = (L/\varepsilon v)(\varepsilon + \Lambda\tau_1) = (20/0.05)(0.4 + 30 \times 0.69) \sim 8400$ s. The column dynamic binding capacity ($DBC_{10\%}$) is the amount of adsorbate retained in the column per unit column volume when the outlet concentration reaches 10 percent of the feed. As a fraction of equilibrium capacity, $DBC_{10\%}$ is equal to τ_1, neglecting the adsorbate held in the pore volume and the leakage prior to 10 percent of breakthrough.

2. From Example 16-9, the rate coefficient for the linear driving-force approximation (Table 16-13, item E) is $k_n = 2.8 \times 10^{-4}$ s^{-1}. Thus, $N = k_n \Lambda L/\varepsilon v = 2.8 \times 10^{-4} \times 30 \times 20/0.05 = 3.4$ and

$$N(1-\tau_1) = \frac{1}{1-0.1}\ln\left(\frac{1-0.1}{0.1^{0.1}}\right) + 1 = 1.1$$

with $c_i^* = 0.1$, giving $\tau_1 = 0.67$. The corresponding breakthrough time $t = 8200$ s.

3. If the isotherm is approximated as irreversible ($R = 0$), from Table 16-13, item B, $N(1-\tau_1) \sim 1.03$ with $c_i^* = 0.1$, giving $\tau_1 = 0.75$ with a corresponding breakthrough time $t = 9100$ s.

Breakthrough Behavior for Axial Dispersion Breakthrough behavior for adsorption with axial dispersion in a deep bed is not adequately described by the constant pattern profile for this mechanism. Equation (16-128), the partial differential equation of the second order Fickian model, requires two boundary conditions for its solution. The constant pattern pertains to a bed of infinite depth—in obtaining the solution the downstream boundary condition $c_i^* \to 0$ as $N_{Pe}\zeta \to \infty$ is applied. Breakthrough behavior presumes the existence of a bed outlet, and a boundary condition must be applied there.

The full mathematical model for this problem is Eq. (16-128) with boundary conditions

$$c_i^* - \frac{1}{N_{Pe}}\frac{\partial c_i^*}{\partial \zeta} = 1 \quad \text{at} \quad \zeta = 0 \tag{16-144}$$

$$\frac{\partial c_i^*}{\partial \zeta} = 0 \quad \text{at} \quad \zeta = 1 \tag{16-145}$$

and an initial condition. Equation (16-144) specifies a constant flux at the bed inlet, and Eq. (16-145), the Danckwerts-type boundary condition at the bed outlet, is appropriate for fixed-bed adsorption, provided that the partition ratio is large.

The solution to this model for a deep bed indicates an increase in velocity of the fluid-phase concentration wave during breakthrough. This is most dramatic for the rectangular isotherm—the instant the bed becomes saturated, the fluid-phase profile jumps in velocity from that of the adsorption transition to that of the fluid, and a near shocklike breakthrough curve is observed [Coppola and LeVan, *Chem. Eng. Sci.* **36:** 967 (1981)].

Extensions Existence, uniqueness, and stability criteria have been developed for the constant pattern [Cooney and Lightfoot, *Ind. Eng. Chem. Fundam.* **4:** 233 (1965); Rhee et al., *Chem. Eng. Sci.* **26:** 1571 (1971); Rhee and Amundson, *Chem. Eng. Sci.* **26:** 1571 (1971), **27:** 199 (1972), **29:** 2049 (1974)].

The rectangular isotherm has received special attention. For this, many of the constant patterns are developed fully at the bed inlet, as shown for external mass transfer [Klotz, *Chem. Revs.* **39:** 241 (1946)], pore diffusion [Vermeulen, *Adv. Chem. Eng.* **2:** 147 (1958); Hall et al., *Ind. Eng. Chem. Fundam.* **5:** 212 (1966)], pore diffusion with film resistance in series [Weber and Chakraborti, *AIChE J.* **20:** 228 (1974)], the linear driving force approximation [Cooper, *Ind. Eng. Chem. Fundam.* **4:** 308 (1965)], reaction kinetics [Hiester and Vermeulen, *Chem. Eng. Progress* **48:** 505 (1952); Bohart and Adams, *J. Amer. Chem. Soc.* **42:** 523 (1920)], and axial dispersion [Coppola and LeVan, *Chem. Eng. Sci.* **38:** 991 (1983)].

Multiple mass-transfer resistances have been considered in many studies [Vermeulen, *Adv. in Chem. Eng.* **2:** 147 (1958); Vermeulen et al. (1984) and Ruthven (1984) in General References; Fleck et al., *Ind. Eng. Chem. Fundam.* **12:** 95 (1973); Yoshida et al., *Chem. Eng. Sci.* **39:** 1489 (1984)].

Treatments of constant pattern behavior have been carried out for multicomponent adsorption [Vermeulen, *Adv. in Chem. Eng.* **2:** 147 (1958); Vermeulen et al. (1984) and Ruthven (1984) in General References; Rhee and Amundson, *Chem. Eng. Sci.* **29:** 2049 (1974); Cooney and Lightfoot, *Ind. Eng. Chem. Fundam.* **5:** 25 (1966); Cooney and Strusi, *Ind. Eng. Chem. Fundam.* **11:** 123 (1972); Bradley and Sweed, *AIChE Symp. Ser. No. 152* **71:** 59 (1975)]. The behavior is such that coexisting compositions advance through the bed together at a uniform rate; this is the *coherence* concept of Helfferich and coworkers [see Helfferich (1962) and Helfferich and Klein (1970) in General References].

Nontrace systems have been considered [Sircar and Kumar, *Ind. Eng. Chem. Proc. Des. Dev.* **22:** 271 (1983)].

Constant patterns have been developed for adiabatic adsorption [Pan and Basmadjian, *Chem. Eng. Sci.* **22:** 285 (1967); Ruthven et al., *Chem. Eng. Sci.* **30:** 803 (1975); Kaguei et al., *Chem. Eng. Sci.* **42:** 2964 (1987)].

The constant pattern concept has also been extended to circumstances with nonplug flows, with various degrees of rigor, including flow profiles in tubes [Sartory, *Ind. Eng. Chem. Fundam.* **17:** 97 (1978); Tereck et al., *Ind. Eng. Chem. Res.* **26:** 1222 (1987)], wall effects [Vortmeyer and Michael, *Chem. Eng. Sci.* **40:** 2135 (1985)], channeling [LeVan and Vermeulen in *Fundamentals of Adsorption,* ed. Myers and Belfort, Engineering Foundation, New York, 1984, pp. 305–314; *AIChE Symp. Ser. No. 233* **80:** 34 (1984)], networks [Avilés and LeVan, *Chem. Eng. Sci.* **46:** 1935 (1991)], and general structures of constant cross section [Rudisill and LeVan, *Ind. Eng. Chem. Res.* **29:** 1054 (1991)].

SQUARE ROOT SPREADING FOR LINEAR ISOTHERMS

The simplest isotherm is $n_i^* = c_i^*$ corresponding to $R = 1$. For this isotherm, the rate equation for external mass transfer, the linear driving force approximation, or reaction kinetics can be combined with Eq. (16-130) to obtain

$$\frac{\partial \bar{n}_i^*}{\partial \tau_1} = -\frac{\partial c_i^*}{\partial \zeta} = N(c_i^* - \bar{n}_i^*) \tag{16-146}$$

The solution to this equation, with initial condition $n_i^* = 0$ at $\tau_1 = 0$ and boundary condition $c_i^* = 1$ at $\zeta = 0$, originally obtained for an analogous heat transfer case [Anzelius, *Z. Angew Math. Mech.* **6:** 291 (1926); Schumann, *J. Franklin Inst.* **208:** 405 (1929)], is

$$c_i^* = J(N\zeta, N\tau_1) \qquad n_i^* = 1 - J(N\tau_1, N\zeta) \tag{16-147}$$

where the J function is [Hiester and Vermeulen, *Chem. Eng. Prog.* **48:** 505 (1952)]

$$J(s,t) = 1 - \int_0^s e^{-t-\xi} I_0(2\sqrt{t\xi})\, d\xi \tag{16-148}$$

where I_0 is the modified Bessel function of the first kind of order zero. This linear isotherm result can be generalized to remove the assumption that $\hat{c}_i^* \approx c_i^*$ if the throughput parameter is redefined as

$$\tau_1 = \frac{\tau - \varepsilon\zeta}{(1-\varepsilon)(\rho_p K_i + \varepsilon_p)}$$

A series approximation suitable for practical calculations of the J-function is given by Tan, *Chem. Eng.* **84:** 158 (1977). For large argument values, J is approximated by [Vermeulen et al. (1984) in General References]

$$J(s,t) \sim \frac{1}{2}\operatorname{erfc}(\sqrt{s} - \sqrt{t}) \tag{16-149}$$

A derivation for particle-phase diffusion accompanied by fluid-side mass transfer has been carried out by Rosen [*J. Chem. Phys.* **18:** 1587 (1950); *J. Chem. Phys.* **20:** 387 (1952); *Ind. Eng. Chem.* **46:** 1590 (1954)] with the limiting form at $N > 50$:

$$c_i^* = \frac{1}{2}\operatorname{erfc}\left[\frac{\sqrt{N}}{2}(\zeta - \tau_1)\right] \tag{16-150}$$

For axial dispersion in a semi-infinite bed with a linear isotherm, the complete solution has been obtained for a constant flux inlet boundary condition [Lapidus and Amundson, *J. Phys. Chem.* **56:** 984 (1952); Brenner, *Chem. Eng. Sci.* **17:** 229 (1962); Coates and Smith, *Soc. Petrol. Engrs. J.* **4:** 73 (1964)]. For large N, the leading term is

$$c_i^* = \frac{1}{2}\operatorname{erfc}\left[\frac{\sqrt{N}}{2\sqrt{\tau_1}}(\zeta - \tau_1)\right] \tag{16-151}$$

All of these solutions are very similar and show, for large N, a wave with breadth proportional to the square root of the bed depth through which it has passed.

ANALYTICAL SOLUTION FOR REACTION KINETICS

In general, full time-dependent analytical solutions to differential equation–based models of the previously described mechanisms have not been obtained for nonlinear isotherms. Only for reaction kinetics with the constant separation factor isotherm has a full solution been found [Thomas, *J. Amer.*

Chem. Soc. **66:** 1664 (1944)]. Referred to as the *Thomas solution,* it has been extensively studied [Amundson, *J. Phys. Colloid Chem.* **54:** 812 (1950); Hiester and Vermeulen, *Chem. Eng. Progress* **48:** 505 (1952); Gilliland and Baddour, *Ind. Eng. Chem.* **45:** 330 (1953); Vermeulen, *Adv. in Chem. Eng.* **2:** 147 (1958)]. The solution to Eq. (16-130) for item 4C in Table 16-12 for the same initial and boundary conditions as Eq. (16-146) is

$$\begin{aligned} c_i^* &= \frac{J(RN\zeta, N\tau_1)}{J(RN\zeta, N\tau_1) + e^{-(R-1)N(\zeta-\tau_1)}[1 - J(N\zeta, RN\tau_1)]} \\ n_i^* &= \frac{1 - J(N\tau_1, RN\zeta)}{J(RN\zeta, N\tau_1) + e^{-(R-1)N(\zeta-\tau_1)}[1 - J(N\zeta, RN\tau_1)]} \end{aligned} \qquad (16\text{-}152)$$

The solution gives all of the expected asymptotic behaviors for large N—the proportionate pattern spreading of the simple wave if $R > 1$, the constant pattern if $R < 1$, and square root spreading for $R = 1$.

NUMERICAL METHODS AND CHARACTERIZATION OF WAVE SHAPE

For the solution of sophisticated mathematical models of adsorption cycles incorporating complex multicomponent equilibrium and rate expressions, two numerical methods are popular, including implementations in commercially available software packages. These are finite difference and weighted residual methods. The former vary in the manner in which distance variables are discretized, ranging from simple backward difference stage models (akin to the plate theory of chromatography) to more involved schemes exhibiting little numerical dispersion. The weighted residual method of orthogonal collocation is often thought to be faster computationally, but oscillations in the polynomial trial function can be a problem. The choice of best method is often the preference of the user.

For both the finite difference and weighted residual methods, a set of coupled ordinary differential equations results that are integrated forward in time using the method of lines. Various software packages implementing Gear's method are popular.

The development of mathematical models is described in several of the General References [Guiochon Felinger-Shirazi, and Katti (2006), Rhee, Aris, and Amundson (1986, 1989), Ruthven (1984), Ruthven, Farooq, and Knaebel (1994), Suzuki (1990), Tien (1994), Wankat (1986), and Yang (1987)]. See also Finlayson [*Numerical Methods for Problems with Moving Fronts,* Ravenna Park, Washington, 1992; Holland and Liapis, *Computer Methods for Solving Dynamic Separation Problems,* McGraw-Hill, New York, 1982; Villadsen and Michelsen, *Solution of Differential Equation Models by Polynomial Approximation,* Prentice-Hall, Englewood Cliffs, N.J., 1978].

For the characterization of wave shape and breakthrough curves, three methods are popular. The MTZ method [Michaels, *Ind. Eng. Chem.* **44:** 1922 (1952)] measures the breadth of a wave between two chosen concentrations (e.g., $c_i^* = 0.05$ and 0.95 or $c_i^* = 0.01$ and 0.99). Outside of a laboratory, the measurement of full breakthrough curves is rare, so the breadth of the MTZ is often estimated from an independently determined stoichiometric capacity and a measured small concentration in the "toe" of the breakthrough curve. A second method for characterizing wave shape is by the slope of the breakthrough curve at its midheight (i.e., $c_i^* = 0.5$ [Vermeulen et al. (1984)]). The use of moments of the slope of breakthrough curves is a third means for characterization. They can often be used to extract numerical values for rate coefficients for linear systems [Ruthven (1984) and Suzuki (1990) in General References; Nauman and Buffham, *Mixing in Continuous Flow Systems,* Wiley-Interscience, New York, 1983]. The method of moments is discussed further in the following part of this section.

CHROMATOGRAPHY

CLASSIFICATION

Chromatography is a sorptive separation process where a portion of a solute mixture (feed) is introduced at the inlet of a column containing a selective adsorbent (stationary phase) and separated over the length of the column by the action of a carrier fluid (mobile phase) that is continually supplied to the column following introduction of the feed. The mobile phase is generally free of the feed components, but it may contain various other species introduced to modulate the chromatographic separation. The separation occurs as a result of the different partitioning of the feed solutes between the stationary and mobile phases. Chromatography is used both in the analysis of mixtures and in preparative and process-scale applications. It can be used for both trace-level and for bulk separations both in the gas and the liquid phase.

Modes of Operation The classical modes of operation of chromatography as enunciated by Tiselius [*Kolloid Z.* **105:** 101 (1943)] are: elution chromatography, frontal analysis, and displacement development. Basic features of these techniques are illustrated in Fig. 16-24. Often, each of the different modes can be implemented with the same equipment and stationary phase. The results are, however, quite different in the three cases.

Elution Chromatography The components of the mobile phase supplied to the column after feed introduction have less affinity for the stationary phase than any of the feed solutes. Under trace conditions, the feed solutes travel through the column as bands or zones at different velocities that depend only on the composition of the mobile phase and the operating temperature and that exit from the column at different times.

Two variations of the technique exist: *isocratic elution,* when the mobile phase composition is kept constant, and *gradient elution,* when the mobile phase composition is varied during the separation. Isocratic elution is often the method of choice for analysis and in process applications when the retention characteristics of the solutes to be separated are similar and not dramatically sensitive to very small changes in operating conditions. Isocratic elution is also generally practical for systems where the equilibrium isotherm is linear or nearly linear. In all cases, isocratic elution results in a dilution of the separated products.

In gradient elution, the eluting strength of the mobile phase is gradually increased after supplying the feed to the column. In liquid chromatography, this is accomplished by changing the mobile phase composition. The gradient in the eluting strength of the mobile phase that is established in the column is used to modulate the separation, allowing control of the retention time of weakly and strongly retained components. A similar effect can be obtained in gas chromatography by modulating the column temperature. In either case, the column has to be brought back to the initial conditions before the next cycle is commenced.

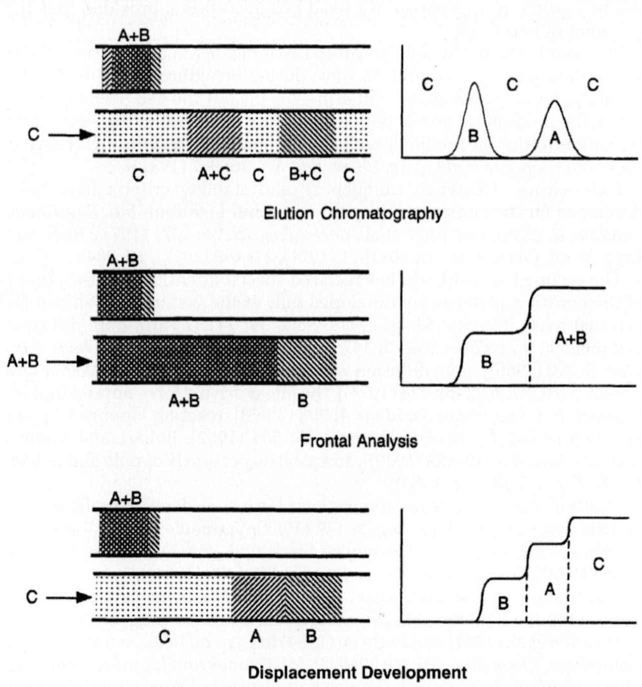

FIG. 16-24 Modes of operation of chromatography for the separation of a mixture of two components A and B. Figures on the left represent a schematic of the column with sample passing through it. Top diagrams show column at end of feed step. Figures on the right show the corresponding effluent concentrations as might be seen by a nonspecific detector. C is either the eluent or the displacer.

Generally, gradient elution is best suited for the separation of complex mixtures that contain both species that interact weakly with the stationary phase and species that interact strongly. Since the eluting strength of the mobile phase is adjusted continuously, weakly retained components of a mixture are separated in the initial phase when the relative eluting strength of the mobile phase is low, while strongly retained components are separated later in the gradient when the eluting strength is high. In addition, the technique is used to obtain reproducible chromatographic separations when the solute retention characteristics are extremely sensitive to the operating conditions, as in the case of the chromatography of biopolymers, such as proteins. These molecules are often found to transition from being very strongly retained to being completely unretained over an extremely small range of mobile phase compositions, making it difficult to obtain reproducible isocratic separations.

Frontal Analysis The feed mixture to be separated is continuously supplied to the column where the mixture components are competitively adsorbed on the stationary phase. These components are then partially separated in a series of fronts, preceded downstream by the least strongly retained species forming a pure component band, and upstream by the feed mixture. The technique is best suited for the removal of strongly adsorbed impurities present in trace amounts from an unretained or weakly adsorbed product of interest. In this case, a large amount of feed can be processed before the impurities begin to break through. When this point is reached, the bed is washed to remove any desired product from the interstitial voids, and the adsorbent is regenerated. The method can only provide a single component in pure form, but avoids product dilution completely. Multicomponent separations require a series of processing steps; either a series of frontal analysis separations, or a combination of elution and displacement separations. Example 16-11 illustrates bed concentration profiles for the frontal analysis separation of two components under local equilibrium conditions.

Displacement Development The column is partially loaded with the feed mixture as in frontal analysis, usually for conditions where all solutes of interest are strongly and competitively adsorbed on the stationary phase. The feed supply is then stopped and a mobile phase containing a component that has an affinity for the stationary phase stronger than any of the feed components, termed the *displacer*, is fed to the column. The advancement of the displacer front through the column causes displacement of the feed components from the adsorbent and their competitive readsorption downstream of the displacer front. As in frontal chromatography, the less strongly retained species tends to migrate faster down the column concentrating in a band farthest from the displacer front, while the most strongly adsorbed feed solute tends to move more slowly concentrating in a band adjacent to the displacer front. If the column is sufficiently long, all feed components eventually become distributed into a pattern of adjacent pure component bands where each upstream component acts as a displacer for each downstream species located in the band immediately downstream. When this occurs, all bands in the displacement train move at the same velocity, which is equal to the velocity of the displacer front, and the bed concentration profile is called an *isotachic pattern.*

The various operational steps of a displacement development separation are shown in Fig. 16-25. Ideally, the separated species exit the column as adjacent rectangular bands in order of increasing affinity for the stationary phase as shown in this figure. In practice, dispersion effects result in a partial mixing of adjacent bands requiring recycling of portions of the effluent that do not meet purity requirements. Following separation, the displacer has to be removed from the column with a suitable regenerant and the initial conditions of the column restored before the next cycle. Column regeneration may consume a significant portion of the cycle when removal of the displacer is difficult.

Displacement chromatography is suitable for the separation of multicomponent bulk mixtures. For dilute multicomponent mixtures, it allows a simultaneous separation and concentration. Thus, it permits the separation of compounds with extremely low separation factors without the excessive dilution that would be obtained in elution techniques.

Other modes of operation, including recycle and flow reversal schemes and continuous chromatography, are discussed in Ganetsos and Barker (*Preparative and Production Scale Chromatography,* Marcel Dekker, New York, 1993).

CHARACTERIZATION OF EXPERIMENTAL CHROMATOGRAMS

Method of Moments The first step in the analysis of chromatographic systems is often a characterization of the column response to small pulse injections of a solute under trace conditions in the Henry's law limit of the isotherm. For such conditions, the statistical moments of the response peak are used to characterize the chromatographic behavior. Such an approach is generally preferable to other descriptions of peak properties that are specific to Gaussian behavior, since the statistical moments are directly correlated to equilibrium and dispersion parameters. Useful references are Schneider and Smith [*AIChE J.* **14:** 762 (1968)], Suzuki and Smith [*Chem. Eng. Sci.* **26:** 221 (1971)], Carbonell et al. [*Chem. Eng. Sci.* **9:** 115 (1975); **16:** 221 (1978)], and Carta and Jungbauer [(2010), pp. 237–246 in General References].

The most important moments are:

$$\mu_0 = \int_0^\infty c_i\, dt \tag{16-153}$$

$$\mu_1 = \frac{1}{\mu_0}\int_0^\infty c_i t\, dt \tag{16-154}$$

$$\sigma^2 = \frac{1}{\mu_0}\int_0^\infty c_i (t-\mu_1)^2\, dt \tag{16-155}$$

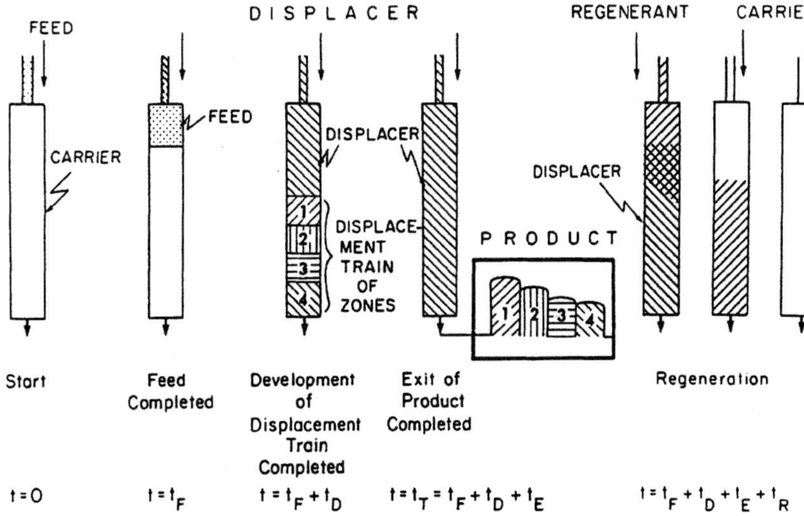

FIG. 16-25 Operational steps in displacement chromatography. The column, initially equilibrated with a carrier solvent at time 0, is loaded with feed until time t_F and supplied with displacer for a time $t_D + t_E$. Development of the displacement train occurs during the time t_D and elution of the separated products ends at time t_E. t_R is the time required to remove the displacer from the column and restore the initial conditions. Components are numbered in order of decreasing affinity for the stationary phase. [Reference: Horvath et al., *J. Chromatogr.* **218:** 365 (1981). Reprinted with permission of *J. Chromatogr.*]

where c_i is the peak profile. μ_0 represents the area, μ_1 the mean residence time, and σ^2 the variance of the response peak. Moments can be calculated by numerical integration of experimental profiles and time can be replaced by mobile phase volume in these calculations.

The *retention factor* is defined as:

$$k_i' = \frac{\mu_1 - \mu_1^0}{\mu_1^0} \tag{16-156}$$

where μ_1^0 is the first moment obtained for an unadsorbed tracer which is excluded from the stationary phase. k_i' is the equilibrium ratio of the amount of solute in the stationary phase (including any pores) and the amount in the external mobile phase. An alternative commonly used definition of the retention factor uses as a reference the first moment of an inert species that has access to all the pores.

The number of plates, N_p, and the height equivalent to a theoretical plate, HETP, are defined as measures of dispersion effects as follows:

$$N_p = \frac{L}{\text{HETP}} = \frac{\mu_1^2}{\sigma^2} \tag{16-157}$$

$$\text{HETP} = \frac{\sigma^2 L}{\mu_1^2} \tag{16-158}$$

A high number of plates and a low HETP indicate a high column efficiency.

Higher moments can also be computed and used to define the skewness of the response peak. However, difficulties often arise in such computations as a result of drifting of the detection system.

In practice, experimental peaks can be affected by extracolumn retention and dispersion factors associated with the injector, connecting tubing, and any detector. For linear chromatography conditions, the apparent response parameters are related to their corresponding true column value by

$$\mu_{1,\text{apparent}} = \mu_1 + \mu_{1,\text{injector}} + \mu_{1,\text{tubing}} + \mu_{1,\text{detector}} \tag{16-159}$$

$$\sigma^2_{\text{apparent}} = \sigma^2 + \sigma^2_{\text{injector}} + \sigma^2_{\text{tubing}} + \sigma^2_{\text{detector}} \tag{16-160}$$

Approximate Methods For certain conditions, symmetrical, Gaussian-like peaks are obtained experimentally. Such peaks may be empirically described by:

$$c_i = \frac{Q_i/F_v}{\sigma\sqrt{2\pi}} \exp\left[-\left(\frac{t-t_{Ri}}{\sigma\sqrt{2}}\right)^2\right] = \frac{Q_i/F_v}{t_{Ri}}\sqrt{\frac{N_p}{2\pi}} \exp\left[-\frac{N_p}{2}\left(\frac{t}{t_{Ri}}-1\right)^2\right] \tag{16-161}$$

where Q_i is the amount of solute injected, F_v is the volumetric flow rate, and t_{Ri} is the peak apex time. The relationships between the moments and other properties of such peaks are shown in Fig. 16-26. For such peaks, approximate calculations of the number of plates can be done with the following equations:

$$N_p = 5.54\left(\frac{t_{Ri}}{\Delta}\right)^2 \tag{16-162}$$

$$N_p = 16\left(\frac{t_{Ri}}{W}\right)^2 \tag{16-163}$$

$$N_p = 2\pi\left(\frac{c_i^{\max} t_{Ri}}{\mu_0}\right)^2 \tag{16-164}$$

where Δ is the peak width at half peak height and W is the distance between the baseline intercepts of the tangents to the inflection points of the peak.

In general, Gaussian behavior can be tested by plotting the cumulative fractional recovery $\int_0^t c_i\,dt/\mu_0$ versus time on probability-linear coordinates; if the plot is linear, Gaussian behavior is confirmed. For nearly Gaussian peaks with asymmetry factor less than about 1.6 (see below), calculations of N_p based on Eqs. (16-162) to (16-164) provide results close to those obtained with a rigorous calculation of moments. When deviations from Gaussian behavior are significant, however, large errors can be obtained using these expressions.

Tailing Peaks Tailing peaks can be obtained experimentally when the column efficiency is very low, when there are large extracolumn dispersion effects, when the stationary phase is heterogeneous (in the sense that it contains different adsorption sites), or when the adsorption equilibrium deviates from the Henry's law limit. Asymmetrical tailing peaks can sometimes be described empirically by an exponentially modified Gaussian (EMG) defined as the convolute integral of a Gaussian constituent with mean time t_G and standard deviation σ_G and an exponential decay with time constant τ_G [Grushka, *Anal. Chem.* **44:** 1733 (1972)]:

$$c_i = \frac{Q_i/F_v}{\sigma_G \tau_G \sqrt{2\pi}} \int_0^\infty \exp\left[-\left(\frac{t-t_G-t'}{\sigma_G\sqrt{2}}\right)^2 - \frac{t'}{\tau_G}\right] dt' \tag{16-165}$$

The corresponding moments are calculated directly as $\mu_1 = t_G + \tau_G$ and $\sigma^2 = \sigma_G^2 + \tau_G^2$ and the *peak skew* as:

$$\text{Peak skew} = \frac{1}{\mu_0 \sigma^3}\int_0^\infty c_i (t-\mu_1)^3\,dt \sim \frac{2(\tau_G/\sigma_G)^3}{[1+(\tau_G/\sigma_G)^2]^{3/2}} \tag{16-166}$$

For EMG peaks, peak skew increases with the ratio τ_G/σ_G. Figure 16-27 illustrates the characteristics of such a peak calculated for $\tau_G/\sigma_G = 1.5$. In general, with $\tau_G/\sigma_G > 1$ (peak skew > 0.7), a direct calculation of the moments is required to obtain a good approximation of the true value of N_p, since other methods give a large error (Yau et al., *Modern Size-Exclusion Liquid Chromatography,* Wiley, New York, 1979). Alternatively, Eq. (16-165) can be fitted to experimental peaks to determine the optimum values of t_G, σ_G, and τ_G.

In practice, the calculation of peak skew for highly tailing peaks is rendered difficult by baseline errors in the calculation of third moments. The *peak asymmetry factor*, $A_s = b/a$, at 10 percent of peak height (see Fig. 16-27) is thus often used. An approximate relationship between peak skew and A_s for tailing peaks, based on data in Yau et al. (*Modern Size-Exclusion Liquid Chromatography,* Wiley, New York, 1979) is: Peak skew ~ $[0.51 + 0.19/(A_s - 1)]^{-1}$. Values of $A_s < 1.25$ (corresponding to peak skew < 0.7) are generally desirable for an efficient chromatographic separation.

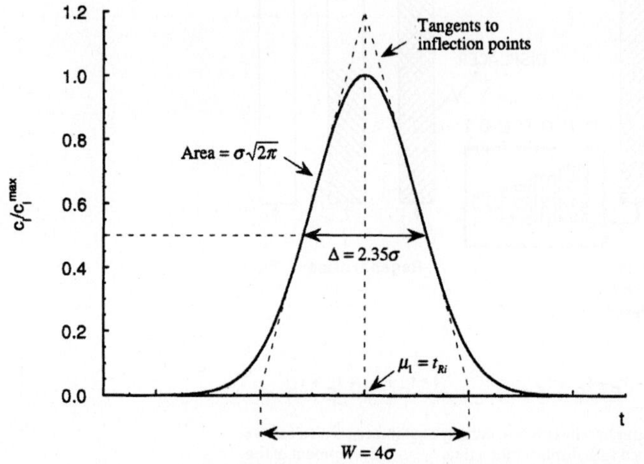

FIG. 16-26 Properties of a Gaussian peak. $c_i^{\max}$ is the peak height; t_{Ri}, the peak apex time; σ, the standard deviation; Δ, the peak width at midheight; and W, the distance between the baseline intercepts of the tangents to the peak.

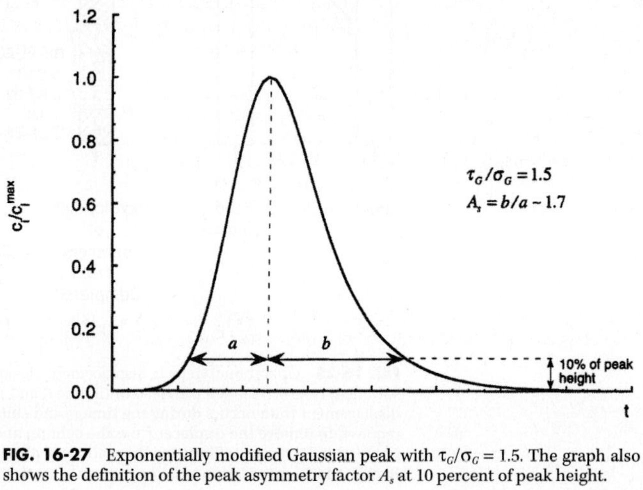

FIG. 16-27 Exponentially modified Gaussian peak with $\tau_G/\sigma_G = 1.5$. The graph also shows the definition of the peak asymmetry factor A_s at 10 percent of peak height.

Resolution The chromatographic separation of two components, A and B, under trace conditions with small feed injections can be characterized in terms of the resolution, R_s:

$$R_s = \frac{2(t_{R,A} - t_{R,B})}{W_A + W_B} \sim \frac{\Delta t_R}{4\sigma_{AB}} \tag{16-167}$$

where Δt_R is the difference in retention time of the two peaks and $\sigma_{AB} = (\sigma_A + \sigma_B)/2$ is the average of their standard deviations. Time can be replaced with mobile phase volume passed through the column. When Eq. (16-161) is applicable (that is, for nearly Gaussian peaks), the resolution for two closely spaced peaks is approximated by:

$$R_s = \frac{1}{2}\frac{\alpha-1}{\alpha+1}\frac{\bar{k}'}{1+\bar{k}'}\sqrt{N_p} \sim \frac{\alpha-1}{4}\frac{k'_A}{1+k'_A}\sqrt{N_p} \quad \text{for } \alpha \sim 1 \tag{16-168}$$

where $\alpha = k'_A/k'_B$ and $\bar{k}' = (k'_A + k'_B)/2$.

Equation (16-168) shows that the resolution is the result of the independent effects of the separation selectivity (α), column efficiency (N_p), and capacity ($\bar{k}'$). Generally, peaks are essentially completely resolved when $R_s = 1.5$ (>99.5 percent separation). In practice, values of $R_s \sim 1$, corresponding to ~98 percent separation, are often considered adequate.

The preceding equations are accurate to within about 10 percent for feed injections that do not exceed 40 percent of the final peak width. For large, rectangular feed injections, the baseline width of the response peak is approximated by:

$$W \sim 4\sigma + t_F \tag{16-169}$$

where 4σ is the baseline width obtained with a pulse injection and t_F is the duration of the actual feed injection. In this case, the resolution is defined as [see Ruthven (1984), pp. 324–331]:

$$R_s = \frac{t_{R,A} - t_{R,B} - t_F}{4\sigma_{AB}} \tag{16-170}$$

For strongly retained components ($\bar{k}' >> 1$), the number of plates required to obtain a given resolution with a finite feed injection is approximated by:

$$N_p = 4R_s^2\left(\frac{\alpha+1}{\alpha-1}\right)^2\left(1+\frac{t_F}{4\sigma_{AB}R_s}\right)^2 \tag{16-171}$$

PREDICTION OF CHROMATOGRAPHIC BEHAVIOR

The conservation equations and the rate models described in the subsection Rate and Dispersion Factors can normally be used for a quantitative description of chromatographic separations. Alternatively, *plate models* can be used for an approximate prediction, lumping together all dispersion contributions into a single parameter, the HETP or the number of plates [Sherwood et al., *Mass Transfer,* McGraw-Hill, New York, 1975, p. 576; Dondi and Guiochon, *Theoretical Advancements in Chromatography and Related Techniques,* NATO-ASI, Series C: Mathematical and Physical Sciences, vol. 383, Kluwer, Dordrecht, 1992, pp. 1–61]. Exact analytic solutions are generally available for linear isocratic elution under trace conditions [see Dondi and Guiochon, cited above, and Ruthven (1984), pp. 324–335; Suzuki (1990), pp. 224–243; and Carta and Jungbauer (2010), pp. 246–258 in General References]. Other cases generally require numerical solution [see Guiochon, Felinger-Shirazi, and Katti (2006) in General References] or approximate treatments with simplified rate models.

Isocratic Elution In the simplest case, feed with concentration c_i^F is applied to the column for a time t_F followed by the pure carrier fluid. Under trace conditions, for a linear isotherm with external mass-transfer control, the linear driving force approximation or reaction kinetics (see Table 16-12), solution of Eq. (16-146) gives the following expression for the dimensionless solute concentration at the column outlet:

$$c_i^* = J(N, N\tau_1) - J(N, N\tau_1') \tag{16-172}$$

where N is the number of transfer units given in Table 16-13 and $\tau_1 = (\varepsilon v t/L - \varepsilon)/[(1-\varepsilon)(\rho_p K_i + \varepsilon_p)]$ the throughput parameter (see Square Root Spreading for Linear Isotherms in the subsection Fixed-Bed Transitions). τ_1' represents the value of τ_1 with time measured from the end of the feed step. Thus, the column effluent profile is obtained as the difference between a breakthrough profile for a feed started at $t = 0$ and another for a feed started at $t = t_F$.

The behavior predicted by this equation is illustrated in Fig. 16-28 with $N = 80$. $\tau_F = (\varepsilon v t_F/L)/[(1-\varepsilon)(\rho_p K_i + \varepsilon_p)]$ is the dimensionless duration of the feed step and is equal to the amount of solute fed to the column divided by the sorption capacity. Thus, at $\tau_F = 1$, the column has been supplied with an amount of solute equal to the stationary phase capacity. The graph shows the transition from a case where complete saturation of the bed occurs before elution ($\tau_F = 1$) to incomplete saturation as τ_F is progressively reduced. The lower curves with $\tau_F \le 0.4$ are seen to be nearly Gaussian and centered at a dimensionless time $\tau_m \sim (1 - \tau_F/2)$. Thus, as $\tau_F \to 0$, the response curve approaches a Gaussian centered at $\tau_1 = 1$.

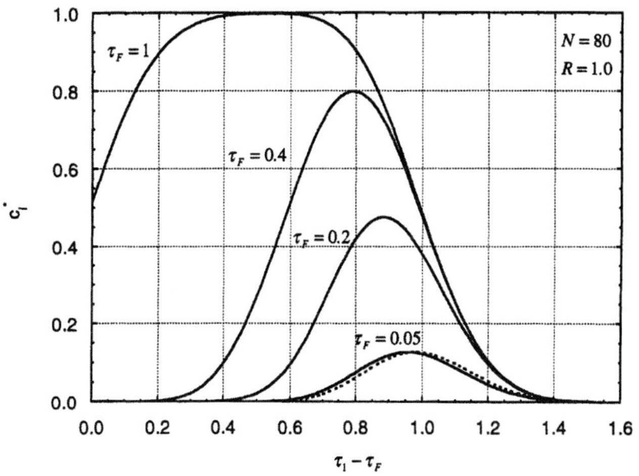

FIG. 16-28 Elution curves under trace linear equilibrium conditions for different feed loading periods and $N = 80$. Solid lines, Eq. (16-172); dashed line, Eq. (16-174) for $\tau_F = 0.05$.

When τ_F is small (<<0.4), the solution for a feed pulse represented by a Dirac's delta function at $\zeta = 0$ can be used in lieu of Eq. (16-172). In terms of the dimensionless concentration $c_i^* = c_i/c_i^F$ (Sherwood et al., *Mass Transfer,* McGraw-Hill, New York, 1975, pp. 571–577):

$$c_i^* = \frac{\tau_F N}{\sqrt{\tau_1}} e^{-N} e^{-N\tau_1} I_1(2N\sqrt{\tau_1}) \tag{16-173}$$

where I_1 is the Bessel function of the imaginary argument. When N is larger than ~5, this equation is approximated by:

$$c_i^* = \frac{\tau_F}{2\sqrt{\pi}}\sqrt{\frac{N}{\tau_1}}\frac{\exp[-N(\sqrt{\tau_1}-1)^2]}{(\tau_1)^{1/4}} \tag{16-174}$$

The behavior predicted by Eq. (16-174) is shown in Fig. 16-29 as $c_i^*/(\tau_F\sqrt{N})$ versus τ_1 for different values of N. For $N > 50$, the response peak is symmetrical, the peak apex occurs at $\tau_1 = 1$, and the dimensionless peak height is $c_i^{*\max} = \tau_F\sqrt{N/4\pi}$. A comparison of this equation with Eq. (16-172) with $N = 80$ is shown in Fig. 16-28 for $\tau_F = 0.05$.

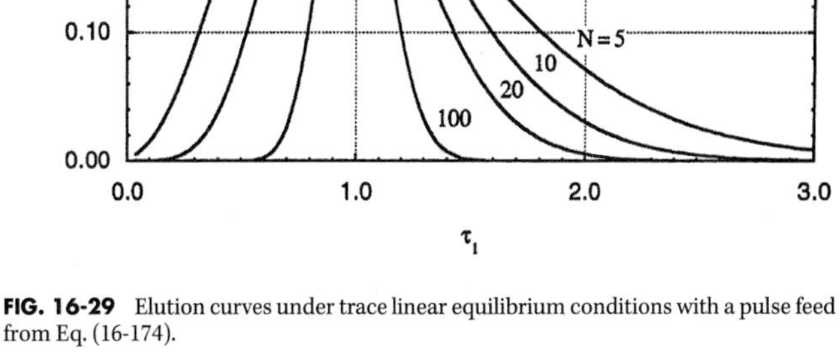

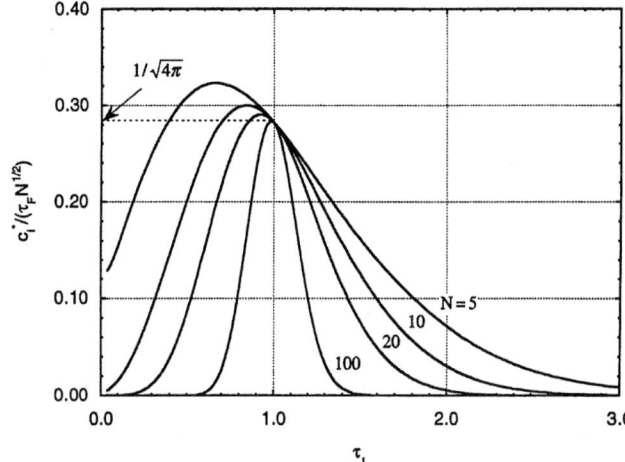

FIG. 16-29 Elution curves under trace linear equilibrium conditions with a pulse feed from Eq. (16-174).

The moments of the response peak predicted by Eq. (16-173) are

$$\mu_1 = \frac{L}{v}\left[1+\frac{1-\varepsilon}{\varepsilon}(\varepsilon_p+\rho_p K_i)\right] = \frac{L}{v}(1+k_i') \qquad (16\text{-}175)$$

$$\sigma^2 = \frac{2\mu_1^2}{N}\left(\frac{k_i'}{1+k_i'}\right)^2 \qquad (16\text{-}176)$$

where $k_i' = (1-\varepsilon)(\varepsilon_p+\rho_p K_i)/\varepsilon$. Correspondingly, the number of plates and the HETP are:

$$N_p = \frac{N}{2}\left(\frac{1+k_i'}{k_i'}\right)^2 \qquad (16\text{-}177)$$

$$\text{HETP} = \frac{2L}{N}\left(\frac{k_i'}{1+k_i'}\right)^2 \qquad (16\text{-}178)$$

Since the term $(1+k_i')/k_i'$ approaches unity for large values of k', the number of plates is about equal to one-half the number of transfer units for a strongly retained component. For these conditions, when $N_p \sim N/2$, Eqs. (16-174) and (16-161) produce the same peak retention time, peak spreading, and predict essentially the same peak profile.

In the general case of axially dispersed plug flow with bidispersed particles, the first and second moment of the pulse response are [Haynes and Sarma, *AIChE J.* **19:** 1043 (1973)]:

$$\mu_1 = \frac{L}{v}\left[1+\frac{1-\varepsilon}{\varepsilon}(\varepsilon_p+\rho_p K_i)\right] = \frac{L}{v}(1+k_i') \qquad (16\text{-}179)$$

$$\sigma^2 = \frac{2LD_L}{v^3}(1+k_i')^2 + 2\frac{L}{v}\frac{\varepsilon k_i'^2}{1-\varepsilon}\left[\frac{r_p}{3k_f}+\frac{r_p^2}{15\varepsilon_p D_{pi}}+\frac{\rho_p K_i}{(\varepsilon_p+\rho_p K_i)^2}\frac{r_s^2}{15D_{si}}\right] \qquad (16\text{-}180)$$

Correspondingly, the number of plates and the plate height are:

$$N_p = \left\{\frac{2D_L}{vL}+\frac{2\varepsilon}{1-\varepsilon}\frac{v}{L}\left(\frac{k'}{1+k_i'}\right)^2\times\left[\frac{r_p}{3k_f}+\frac{r_p^2}{15\varepsilon_p D_{pi}}+\frac{\rho_p K_i}{(\varepsilon_p+\rho_p K_i)^2}\frac{r_s^2}{15D_{si}}\right]\right\}^{-1} \qquad (16\text{-}181)$$

$$\text{HETP} = \frac{2D_L}{v}+\frac{2\varepsilon v}{1-\varepsilon}\left(\frac{k_i'}{1+k_i'}\right)^2\times\left[\frac{r_p}{3k_f}+\frac{r_p^2}{15\varepsilon_p D_{pi}}+\frac{\rho_p K_i}{(\varepsilon_p+\rho_p K_i)^2}\frac{r_s^2}{15D_{si}}\right] \qquad (16\text{-}182)$$

In dimensionless form, a *reduced HETP*, $h = \text{HETP}/d_p$, analogous to the reduced HTU (cf. Fig. 16-13), is obtained as a function of the dimensionless velocity ReSc:

$$h = \frac{b}{\text{ReSc}} + a + c\text{ReSc} \qquad (16\text{-}183)$$

where

$$b = 2\varepsilon\gamma_1 \qquad (16\text{-}184a)$$

$$a = 2\gamma_2 \qquad (16\text{-}184b)$$

$$c = \frac{1}{30}\frac{1}{1-\varepsilon}\left(\frac{k'}{1+k'}\right)^2\left[\frac{10}{\text{Sh}}\frac{\tau_p}{\varepsilon_p}+\frac{\rho_p K_i}{(\varepsilon_p+\rho_p K_i)^2}\frac{r_s^2}{D_{si}}\frac{D_{pi}}{r_p^2}\right] \qquad (16\text{-}184c)$$

Equation (16-183) is qualitatively the same as the **van Deemter equation** [van Deemter and Zuiderweg, *Chem. Eng. Sci.* **5:** 271 (1956)] and is equivalent to other empirical reduced HETP expressions such as the **Knox equation** [Knox, *J. Chromatogr. Sci.* **15:** 352 (1977)].

The Sherwood number, Sh, is estimated from Table 16-9, and the dispersion parameters γ_1 and γ_2 from Table 16-10 for well-packed columns. Typical values are $a = 1$–4 and $b = 0.5$–1. Since HETP ~ 2 HTU, Fig. 16-13 can also be used for approximate calculations.

Concentration Profiles In the general case but with a linear isotherm, the concentration profile can be found by numerical inversion of the Laplace-domain solution of Haynes and Sarma [see Lenhoff, *J. Chromatogr.* **384:** 285 (1987)] or by direct numerical solution of the conservation and rate equations. For the special case of no axial dispersion, an explicit time-domain solution is also available in the cyclic steady state for repeated injections of arbitrary duration t_F followed by an elution period t_E with cycle time $t_c = t_F + t_E$ [Carta, *Chem. Eng. Sci.* **43:** 2877 (1988)]. For the linear driving force mechanism, the solution is

$$c_i^* = \phi\frac{2}{\pi}\sum_{j=1}^{\infty}\left\{\frac{1}{j}\exp\left(-\frac{j^2N}{j^2+\omega^2}\right)\sin(j\pi\phi)\times\cos\left[\frac{2j\pi}{t_c}\left(t-\frac{t_F}{2}-\frac{L}{v}\right)-\frac{j\omega N}{j^2+\omega^2}\right]\right\} \qquad (16\text{-}185)$$

where

$$\phi = \frac{t_F}{t_c} \qquad (16\text{-}185a)$$

$$\omega = \frac{N\varepsilon v t_c/L}{2\pi(1-\varepsilon)(\rho_p K_i+\varepsilon_p)} \qquad (16\text{-}185b)$$

The average effluent concentration in a product cut between times t_1 and t_2 can be calculated directly from

$$\bar{c}_i^* = \frac{\int_{t_1}^{t_2} c_i^*\,dt}{t_2-t_1}\phi + \frac{t_c}{t_2-t_1}\frac{2}{\pi^2}\times\sum_{j=1}^{\infty}\left\{\frac{1}{j^2}\exp\left(-\frac{j^2N}{j^2+\omega^2}\right)\sin(j\pi\phi)\sin\left[j\pi\frac{t_2-t_1}{t_c}\right]\times\cos\left[\frac{2j\pi}{t_c}\left(\frac{t_1+t_2-t_F}{2}-\frac{L}{v}\right)-\frac{j\omega N}{j^2+\omega^2}\right]\right\} \qquad (16\text{-}186)$$

The same equations can also be used as a good approximation for other rate mechanisms with N calculated according to Table 16-13. When t_c is large, Eq. (16-185) describes a single feed injection and approaches the results of Eqs. (16-173) and (16-148) for small and large values of ϕ, respectively. For sample calculations, see Carta, *Chem. Eng. Sci.* **43:** 2877 (1988) and Seader and Henley (2006) in General References.

When the adsorption equilibrium is nonlinear, skewed peaks are obtained, even when N is large. For a constant separation-factor isotherm with $R < 1$ (favorable), the leading edge of the chromatographic peak is steeper than the trailing edge. When $R > 1$ (unfavorable), the opposite is true.

Figure 16-30 portrays numerically calculated chromatographic peaks for a constant separation factor system showing the effect of feed loading on the elution profile with $R = 0.5$ ($\tau_1 - \tau_F = [\varepsilon v(t - t_F)/L - \varepsilon]/\Lambda$). When the dimensionless feed time $\tau_F = 1$, the elution curve comprises a sharp leading profile reaching the feed concentration followed by a gradual decline to zero. As τ_F is reduced, breakthrough of the leading edge occurs at later times, while the trailing edge, past the peak apex, continues to follow the same profile. As the amount of feed injected approaches zero, mass-transfer resistance reduces the solute concentration to values that fall in the Henry's law limit of the isotherm, and the peak retention time gradually approaches the value predicted in the infinite dilution limit for a linear isotherm.

For high feed loads, the shape of the diffuse trailing profile and the location of the leading front can be predicted from local equilibrium theory (see the subsection Fixed-Bed Transitions). This is illustrated in Fig. 16-30 for $\tau_F = 0.4$. For the diffuse profile (a "simple wave"), Eq. (16-131) gives:

$$c_i^* = \frac{R}{1-R}\left[\left(\frac{1}{R(\tau_1-\tau_F)}\right)^{1/2}-1\right] \qquad (16\text{-}187)$$

Thus, the effluent concentration becomes zero at $\tau_1 - \tau_F = 1/R$. The position of the leading edge (a "shock front") is determined from Eq. (16-132):

$$\tau_{1s} = \tau_F + \frac{1}{R}\left[1-\sqrt{(1-R)\tau_F}\right]^2 \qquad (16\text{-}188)$$

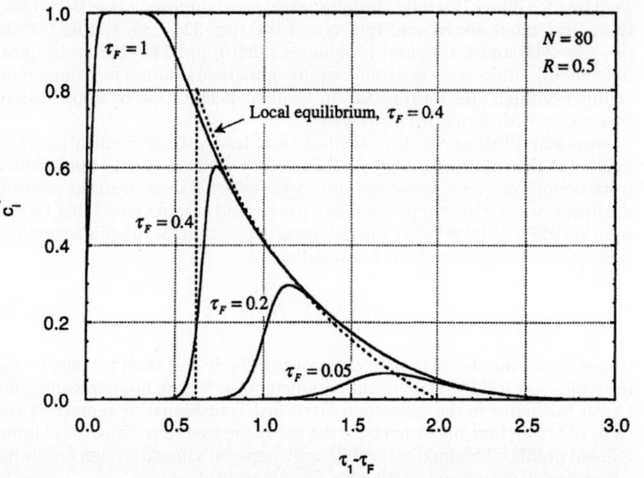

FIG. 16-30 Elution curves under trace conditions with a constant separation factor isotherm for different feed loadings and $N = 80$. Solid lines, rate model; dashed line, local equilibrium theory for $\tau_F = 0.4$.

and the peak highest concentration by [Golshan-Shirazi and Guiochon, *Anal. Chem.* **60:** 2364 (1988)]:

$$c_i^{*\max} = \frac{R}{1-R}\,\frac{\sqrt{(1-R)\tau_F}}{1-\sqrt{(1-R)\tau_F}} \tag{16-189}$$

The local equilibrium curve is in approximate agreement with the numerically calculated profiles except at very low concentrations when the isotherm becomes linear and near the peak apex. This occurs because band-spreading, in this case, is dominated by adsorption equilibrium, even if the number of transfer units is not very high. A similar treatment based on local equilibrium for a two-component mixture is given by Golshan-Shirazi and Guiochon [*J. Phys. Chem.* **93:** 4143 (1989)].

Prediction of multicomponent nonlinear chromatography accounting for rate factors requires numerical solution [see Guiochon, Felinger-Shirazi, and Katti (2006) and Numerical Methods and Characterization of Wave Shape in the subsection Fixed-Bed Transitions).

Linear Gradient Elution Analytical solutions are available for special cases under trace conditions with a linear isotherm. Other situations normally require a numerical solution. General references are Snyder (in Horvath, ed., *High Performance Liquid Chromatography: Advances and Perspectives,* vol. 1, Academic Press, 1980, p. 208), Antia and Horvath [*J. Chromatogr.* **484:** 1 (1989)], Yamamoto et al. (*Ion Exchange Chromatography of Proteins,* Marcel Dekker, 1988), and Guiochon, Felinger-Shirazi, and Katti (2006).

The most commonly used gradients are linear gradients where the starting solvent is gradually mixed with a second gradient-forming solvent at the column entrance to yield a volume fraction φ of the mobile phase modulator that increases linearly with time:

$$\varphi = \varphi_0 + \beta t \tag{16-190}$$

When the mobile phase modulator is a dilute solute in a solvent, as when the modulator is a salt, φ indicates molar concentration. As before, time can be replaced by the volume of mobile phase passed through the column.

Under *trace conditions,* the retention of the modulator in the column is independent of the presence of any solutes. The modulator concentration at the column exit is thus approximated by

$$\varphi = \varphi_0 + \beta\left[t - \frac{L}{v}(1+k'_M)\right] \tag{16-191}$$

where k'_M is the retention factor of the modulator.

For a small feed injection, the modulator concentration φ_R at which a feed solute is eluted from the column is obtained from the following integral:

$$G(\varphi_R) = \frac{\beta L}{v} = \int_{\varphi_0}^{\varphi_R} \frac{d\varphi}{k'_i(\varphi) - k'_M} \tag{16-192}$$

where $k'_i(\varphi)$ is the solute retention factor expressed as a function of the mobile phase modulator concentration. Note that the solute retention time in the column is affected by the steepness of the gradient at the column entrance.

$k'_i(\varphi)$ can be obtained experimentally from isocratic elution experiments at different φ values, or from linear gradient elution experiments where the ratio $G = \beta L/v$ is varied. In the latter case, the retention factor is obtained by differentiation of Eq. (16-192) from $k'_i(\varphi_R) = k'_M + (dG/d\varphi_R)^{-1}$.

Table 16-14 gives explicit expressions for chromatographic peak properties in isocratic elution and linear gradient elution for two cases.

In *reversed-phase chromatography* (RPC), the mobile phase modulator is typically a water-miscible organic solvent, and the stationary phase is a hydrophobic adsorbent. In this case, the logarithm of solute retention factor is commonly found to be linearly related to the volume fraction of the organic solvent.

In *ion-exchange chromatography* (IEC), the mobile phase modulator is typically a salt in aqueous solution, and the stationary phase is an ion exchanger. For dilute conditions, the solute retention factor is commonly found to be a power-law function of the salt normality [cf. Eq. (16-27) for ion-exchange equilibrium].

Band broadening is also affected by the gradient steepness. This effect is expressed in Table 16-14 by a *band compression factor C,* which is a function of the gradient steepness and of equilibrium parameters. Since $C < 1$, gradient elution yields peaks that are sharper than those that would be obtained in isocratic elution at $\varphi = \varphi_R$.

Other cases, involving an arbitrary relationship between the solute retention factor and the modulator concentration, can be handled analytically using the approaches of Frey [*Biotechnol. Bioeng.* **35:** 1055 (1990)] and Carta and Stringfield [*J. Chromatogr.* **605:** 151 (1992)].

Displacement Development A complete prediction of displacement chromatography accounting for rate factors requires a numerical solution since the adsorption equilibrium is nonlinear and intrinsically competitive. When the column efficiency is high, however, useful predictions can be obtained with the local equilibrium theory (see the subsection Fixed-Bed Transitions). For constant separation factor systems, the h-*transformation* of Helfferich and Klein (1970) or the method of Rhee et al. [*AIChE J.* **28:** 423 (1982)] can be used [see also Helfferich, *Chem. Eng. Sci.* **46:** 3320 (1991)].

TABLE 16-14 Expressions for Predictions of Chromatographic Peak Properties in Linear Gradient Elution Chromatography Under Trace Conditions with a Small Feed Injection and Inlet Gradient Described by $\varphi = \varphi_o + \beta t$ (Adapted from Refs. A and B)

Parameter	Isocratic	Gradient elution—RPC (Ref. A)	Gradient elution—IEC (Ref. B)
Dependence of retention factor on modulator concentration	—	$k' - k'_M = \alpha_s e^{-S\varphi}$	$k' - k'_M = \alpha_z \varphi^{-z}$
Retention time	$t_R = \frac{L}{\upsilon}(1+k'_o)$	$t_R = \frac{L}{\upsilon}(1+k'_M) + \frac{1}{S\beta}\ln\left[\frac{S\beta L}{\upsilon}(k'_o - k'_M)+1\right]$	$t_R = \frac{L}{\upsilon}(1+k'_M) + \frac{1}{\beta}\left[\frac{\alpha_Z \beta L}{\upsilon}(Z+1)+\varphi_o^{Z+1}\right]^{1/(Z+1)} - \frac{\varphi_o}{\beta}$
Mobile phase composition at column exit	$\varphi_R = \varphi_o$	$\varphi_R = \varphi_o + \beta\left[t_R - \frac{L}{\upsilon}(1+k'_M)\right]$	$\varphi_R = \varphi_o + \beta\left[t_R - \frac{L}{\upsilon}(1+k'_M)\right]$
Retention factor at peak elution	$k'_R = k'_o$	$k'_R = \frac{k'_o + (k'_o - k'_M)\,S\beta L/\upsilon}{1+(k'_o - k'_M)\,S\beta L/\upsilon}$	$k'_R = \alpha_Z\left[\frac{\alpha_Z \beta L}{\upsilon}(Z+1)+\varphi_o^{Z+1}\right]^{-Z/(Z+1)} + k'_M$
Peak standard deviation	$\sigma = \frac{L/\upsilon}{\sqrt{N_p}}(1+k'_o)$	$\sigma = C\frac{L/\upsilon}{\sqrt{N_{pR}}}(1+k'_R)$	$\sigma = C\frac{L/\upsilon}{\sqrt{N_{pR}}}(1+k'_R)$
Band compression factor	—	$C = \frac{(1+p+p^2/3)^{1/2}}{1+p}$ $p = \frac{(k'_o - k'_M)(1+k'_M)\,S\beta L/\upsilon}{1+k'_o}$	$C = \begin{cases} \sqrt{M'} & \text{for } M' < 0.25 \\ \frac{3.22M'}{1+3.13M'} & \text{for } 0.25 < M' < 0.25 \\ 1 & \text{for } M' > 120.25 \end{cases}$ $M' = \frac{1}{2}\,\frac{1+k'_R}{1+k'_M}\,\frac{Z+1}{Z}$

References: A. Snyder in Horvath (ed.), *High Performance Liquid Chromatography: Advances and Perspectives,* vol. 1, Academic Press, 1980, p. 208.
B. Yamamoto, *Biotechnol. Bioeng.* **48:** 444 (1995).
Solute equilibrium parameters: α_s,S for RPC and α_Z,Z for IEC
Solute retention factor for initial mobile phase: k'_o
Retention factor of mobile phase modulator: k'_M
Plate number obtained for $k' = k'_R$: N_{pR}

Analyses of displacement chromatography by the method of characteristics with non-Langmuirian systems is discussed by Antia and Horvath [*J. Chromatogr.* **556:** 199 (1991)] and Carta and Dinerman [*AIChE J.* **40:** 1618 (1994)]. Optimization studies and analyses by computer simulations are discussed by Jen and Pinto [*J. Chromatogr.* **590:** 47 (1992)], Katti and Guiochon [*J. Chromatogr.* **449:** 24 (1988)], and Phillips et al. [*J. Chromatogr.* **54:** 1 (1988)]. A detailed example of displacement chromatography based on the *h*-transformation and the preceding equations is given elsewhere [LeVan and Carta (2008)].

DESIGN FOR TRACE SOLUTE SEPARATIONS

The design objectives of the analyst and the production line engineer are generally quite different. For analysis the primary concern is typically resolution. Hence operating conditions near the minimum value of the HETP or the HTU are desirable (see Fig. 16-13).

In preparative chromatography, however, it is generally desirable to reduce capital costs by maximizing the *productivity*, or the amount of feed processed per unit column volume, subject to specified purity requirements. This reduction, however, must be balanced against operating costs that are determined mainly by the mobile phase flow rate and the pressure drop. In practice, preparative chromatography is often carried out under *overload conditions*, that is, in the nonlinear region of the adsorption isotherm. Optimization under these conditions is discussed in Guiochon, Felinger-Shirazi, and Katti (2006). General guidelines for trace-level, isocratic binary separations, in the Henry's law limit of the isotherm are:

1. The stationary phase is selected to provide the maximum selectivity. Where possible, the retention factor is adjusted (by varying the mobile phase composition, temperature, or pressure) to an optimum value that generally falls between 2 and 10. Resolution is adversely affected when $k' << 2$, while product dilution and separation time increase greatly when $k' >> 10$. When this is not possible for all feed components and large differences exist among the k'-values of the different solutes, gradient elution should be considered.
2. The average feed mixture charging rate, molar or volumetric, is fixed by the raw material supply or the demand for finished product.
3. The value of N_p required to achieve a desired resolution is determined by Eq. (16-168) or (16-171). Since $N = L/\text{HTU} \sim 2N_p = 2L/\text{HETP}$, Fig. 16-13 or Eq. (16-183) can be used to determine the range of the dimensionless velocity ReSc that maximizes N_p for a given particle diameter and column length.
4. The allowable pressure drop influences the choice of the particle size and helps determine the column length. Equations for estimating the pressure drop in packed beds are given in Sec. 6.
5. For a binary separation, the component bands may occupy only a small portion of the total column volume at any given instant. In such cases, the productivity is improved by cyclic feed injections, timed so that the most strongly retained component from an injection elutes just before the least strongly retained component from the following injection [see Fig. 10.5 in Carta and Jungbauer (2010)]. For the linear isotherm case with $\bar{k}' > 1$, when the same resolution is maintained between bands of the same injections and bands of successive injections, the cycle time t_c and the plate number requirement are:

$$t_c = 2(t_{R,A} - t_{R,B}) = \frac{2L}{v}(k'_A - k'_B) \tag{16-193}$$

$$N_p = 4R_s^2\left(\frac{\alpha+1}{\alpha-1}\right)^2(1-2\phi)^{-2} \tag{16-194}$$

where $\phi = t_F/t_c$ is the fraction of the cycle time during which feed is supplied to the column. The productivity, P = volume of feed/(time × bed volume), is:

$$\begin{aligned} P &= \frac{1}{4R_s^2}\left(\frac{\alpha-1}{\alpha+1}\right)^2 \frac{\varepsilon v}{\text{HETP}}\phi(1-2\phi)^2 \\ &= \frac{1}{4R_s^2}\left(\frac{\alpha-1}{\alpha+1}\right)^2 \frac{D_i}{d_p^2}\left(\frac{\text{ReSc}}{b/\text{ReSc}+a+c\text{ReSc}}\right)\phi(1-2\phi)^2 \end{aligned} \tag{16-195}$$

For a given resolution, P is maximized when $\phi = 1/6$ (i.e., feed is supplied for one-sixth of the cycle time), and by the use of small particle sizes. The function ReSc/(b/ReSc + a + cReSc) generally increases with ReSc, so that productivity generally increases with the mobile phase velocity. For typical columns, however, this function is within about 10 percent of its maximum value (~$1/c$) when ReSc is in the range 30–100. Thus, increasing the velocity above this range must be balanced against the costs associated with the higher pressure drop. Similar results are predicted with Eq. (16-185).

When using highly selective adsorbents for chromatographic separations, the process is often operated in a so-called load-wash-elute mode. The product of interest is selectively adsorbed in the load step, unbound impurities are removed in the wash step, and the product is recovered in the elution step. Optimization of productivity for this mode of operation is discussed in Carta and Jungbauer (2010), pp. 311–321. In general, when intraparticle mass transfer is controlling, an optimum residence time exists for the load step that maximizes productivity.

PROCESS CYCLES

GENERAL CONCEPTS

The mode of operation of an adsorption process may consist of just one step with the adsorbent removed at the end for reactivation or disposal. Such a process may be carried out in a fixed bed or agitated vessel. These are discussed in Batch Adsorption. Although there are many practical applications for which the sorbent is discarded after one use, most applications involve the removal of adsorbates from the adsorbent (i.e., regeneration). This allows the adsorbent to be reused and the adsorbates to be recovered, and the adsorption process is carried out in a cyclic fashion.

A cyclic adsorption process generally consists of repeated adsorption and desorption (regeneration) cycles carried out *in situ*. Some such applications may involve a single bed as in simple water softening, but most applications involve multiple beds running in a sequence. While one or more beds are in an adsorption mode, one or more beds are in a desorption mode, and after a certain time the roles of beds switch and the cycle continues. Cyclic adsorption processes are commonly carried out today in fixed and simulated moving beds. Regeneration is accomplished by changing a thermodynamic variable such as temperature, total pressure, partial pressure, or adsorptive selectivity. These cyclic adsorption processes are respectively called temperature-swing adsorption (TSA), pressure-swing adsorption (PSA), inert purge, and displacement purge. Ideal equilibrium cycles for each of these processes are shown in Fig. 16-31. Alternative to equilibrium-based separations are separations based on different mass transfer rates of the adsorbates in the adsorbent (i.e., kinetic separations) or size exclusion. Examples of the separation modes are given in Table 16-1.

BATCH ADSORPTION

Some applications of adsorption and ion exchange are achieved by sorbent–fluid contact in batch equipment. Batch methods are well adapted to laboratory use and have also been applied on a larger scale. In a batch run for either adsorption or ion exchange, a sorbent is added to a fluid, mixed, and separated. Batch treatment is adopted when the capacity and equilibrium of the sorbent are large enough to give nearly complete sorption in a single step, as in purifying and decoloration of laboratory preparations with carbons and clays. Batch runs are useful in the measurement of equilibrium isotherms and adsorptive diffusion rates.

Some commercial applications use the adsorbent on a throwaway basis. Reasons for using sorption nonregeneratively are usually: (1) low cost of the sorbent, (2) high value of the product, (3) very low dosage (sorbent-to-fluid ratio), and (4) difficulty in desorbing the sorbates. Magnesium perchlorate and barium oxide are used for drying, iron sponge (hydrated iron oxide on wood chips) is used to remove hydrogen sulfide, and sodium or potassium hydroxide is applied to remove sulfur compounds or carbon dioxide. In wastewater treatment, powdered activated carbon (PAC) is added to enhance biological treatment but is not regenerated; instead, it remains with the sludge. Silica gel is used as a desiccant in packaging, especially with electronics and medicines. Activated carbon is used in packaging and storage to adsorb other chemicals for preventing the tarnishing of silver, retarding the ripening or spoiling of fruits, "gettering" (scavenging) out-gassed solvents from electronic components, and removing odors. Synthetic zeolites, or blends of zeolites with silica gel, are used in dual-pane windows to adsorb water during initial dry-down and any in-leakage and to adsorb organic solvents emitted from the sealants during their cure; this prevents fogging between the sealed panes that could result from the condensation of water or the solvents [Ausikaitis, *Glass Digest* **61:** 69 (1982)]. Activated carbon is used to treat recirculated air in office buildings, apartments, and manufacturing plants using thin filter-like frames to treat the large volumes of air with low pressure drop. On a smaller scale, activated carbon filters are in kitchen hoods, air conditioners, electronic air purifiers, and faucets and refrigerators for water purification. On the smallest scale, gas masks containing carbon or carbon impregnated with promoters are used to protect individual wearers from industrial odors, toxic chemicals, and gas-warfare chemicals. Activated carbon fibers have been formed into fabrics for clothing to protect against vesicant and percutaneous chemical vapors [Macnair and Arons, in Cheremisinoff and Ellerbusch (1978) in General References] and to prevent the scent of humans from being detected by animals being hunted.

Ion exchangers are sometimes used on a throwaway basis also, with detergency being an extensive application. In the laboratory, ion exchangers are used to deionize water, purify reagents, and prepare inorganic sols. In medicine, they are used as antacids, for sodium reduction, for sustained release of drugs, in skin-care preparations, and in toxin removal.

ADSORPTION CYCLES

The maximum efficiency that a cyclic adsorption process can approach for any given set of operating conditions is given by the adsorptive loading in equilibrium with the feed. This is true whether the separation is based on equilibrium, kinetics, or molecular sieving, because it is anticipated that a faster diffusing or nonexcluded component in a fluid approaches its equilibrium loading. There are several factors that reduce the practical (or "operating") adsorption performance: mass-transfer resistance (see above), deactivation, and incomplete regeneration. The severity of regeneration influences how closely the dynamic capacity of an adsorbent resembles that of fresh, virgin material. Regeneration requires a reduction in the driving force for adsorption. This is accomplished by increasing the equilibrium driving force for the adsorbed species to desorb from the solid to the surrounding fluid.

TEMPERATURE-SWING ADSORPTION

A temperature-swing or thermal-swing adsorption (TSA) process cycle is one in which desorption takes place at a temperature much higher than adsorption. The elevation of temperature is used to shift the adsorption equilibrium and effect regeneration of the adsorbent. Figure 16-31*a* illustrates the principle of the TSA cycle. The feed fluid containing an adsorbate at a partial pressure of p_1 is passed through an adsorbent at temperature T_1. This adsorption step continues until the equilibrium loading n_1 is achieved with p_1. Next, the adsorbent temperature is raised to T_2 (heating step) so that the partial pressure in equilibrium with n_1 is increased to p_2, creating a partial pressure driving force for desorption into fluid containing less than p_2 of the adsorbate. By passing a purge fluid across the adsorbent, the adsorbate is swept away, and the equilibrium proceeds down the isotherm to some point such as p_1, n_2. (As a practical matter, in some applications, roll-up of the adsorbed-phase concentration occurs during heating such that in some regions of the bed p_2 reaches the condensation pressure of the component, causing a condensed liquid phase to form temporarily in particles [Friday and LeVan, *AIChE J.* **31:** 1322 (1985)]. Also, a heel of adsorbate is often left in the bed for an optimal process, especially for very favorable isotherms, for which it is difficult to remove trace quantities of adsorbate.) During a cooling step, the adsorbent temperature is returned to T_1. The new equilibrium p_3, n_2 represents the best-quality product that can be produced from the adsorbent at a regenerated loading of n_2 in the simplest cycle. The adsorption step is now repeated. The differential loading, $n_1 - n_2$, is the maximum loading that can be achieved for a TSA cycle operating between a feed containing p_1 at temperature T_1, regeneration at T_2, and a product containing a partial pressure p_3 of the adsorbate. The regeneration fluid will contain an average partial pressure between p_2 and p_1 and will therefore have

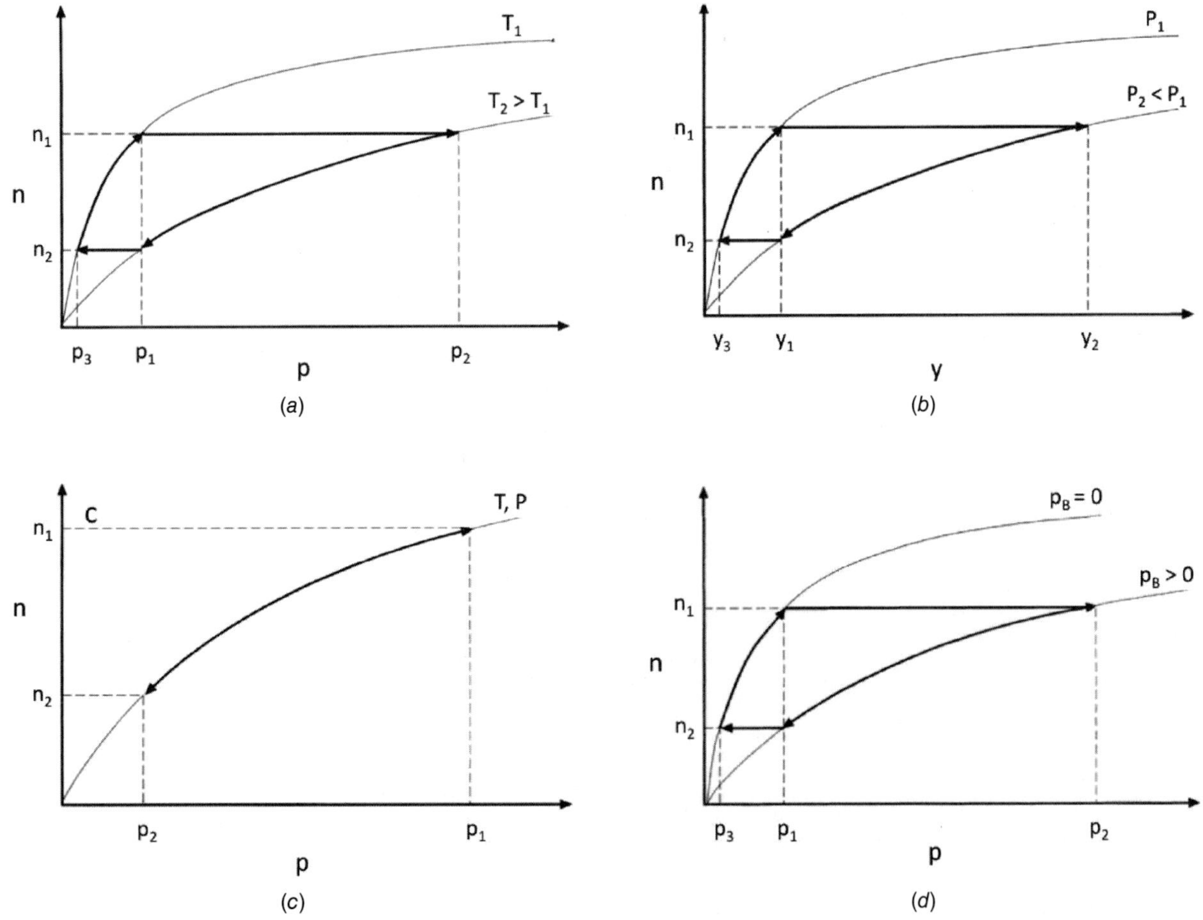

FIG. 16-31 Idealized adsorption cycles: (*a*) temperature swing, (*b*) pressure swing, (*c*) inert-purge swing, and (*d*) displacement-purge swing. (*Reprinted with permission of UOP.*)

accomplished a concentration of the adsorbate in the regenerant fluid. For liquid-phase adsorption, the partial pressure can be replaced by the fugacity of the adsorbate. Then, the entire preceding discussion is applicable whether the regeneration is by a fluid in the gas or liquid phase.

In a TSA cycle, the heating step must provide the thermal energy necessary to raise the adsorbate, adsorbent, and adsorber temperatures, to desorb the adsorbate, and to make up for heat losses. Heating is accomplished by either direct contact of the adsorbent with the heating medium (external heat exchange to a purge gas) or by indirect means (heating elements, coils, or panels inside the adsorber). Direct heating is the most commonly used, especially for stripping-limited heating. Indirect heating can be considered for stripping-limited heating, but the complexity of indirect heating limits its practicality to heating-limited regeneration where purge gas is in short supply. Microwave fields [Benchanaa et al., *Thermochim. Acta* **152:** 43 (1989)] and dielectric fields [Burkholder et al., *Ind. Eng. Chem. Fundam.* **25:** 414 (1986)] are also used to supply indirect heating.

Other Cycle Steps Besides the necessary adsorption and heating steps, TSA cycles may employ additional steps. A purge or sweep gas removes the thermally desorbed components from the adsorbent, and cooling returns it to adsorption temperature. Although the cooling is normally accomplished as a separate step after the heating, sometimes adsorption is started on a hot bed. If certain criteria are met [Basmadjian, *Can. J. Chem. Eng.* **53:** 234 (1975)], the dynamic adsorption efficiency is not significantly affected by the lack of a cooling step.

For liquid-phase adsorption cycles when the unit is to treat a product of significant value, there must be a step to remove the liquid from the adsorbent and one to displace any liquid regenerant thoroughly before filling with the valuable fluid. Because adsorbents and ion exchangers are porous, some retention of the product is unavoidable, but it needs to be minimized to maximize recovery. When regeneration is by a gas, removal and recovery are accomplished by a drain (or pressure-assisted drain) step using the gas to help displace liquid from the sorbent before heating. When the regenerant is another liquid, the feed or product can be displaced out of the adsorbent. When heating and cooling are complete, liquid feed must be introduced again to the adsorbent with a corresponding displacement of the gas or liquid regenerant. In ion exchange, these steps for draining and filling are commonly referred to as "sweetening off" and "sweetening on," respectively.

Applications Drying is the most common gas-phase application of TSA. The natural gas, chemical, and cryogenics industries all use adsorbents to dry streams. Zeolites, activated alumina, and silica gel are used for drying pipeline natural gas. Alumina and silica gel are used because they have higher equilibrium capacity and are more easily regenerated with waste-level heat [Crittenden, *Chem. Engr.* **452:** 21 (1988); Goodboy and Fleming, *Chem. Eng. Progr.* **80:** 63 (1984); Ruthven, *Chem. Eng. Progr.* **84:** 42 (1988)]. The low dewpoint that can be achieved with zeolites is especially important when drying cryogenic-process feed streams to prevent freeze-up. Zeolites dry natural gas before liquefaction to liquefied natural gas (LNG) and before ethane recovery using the cryogenic turboexpander process [Anderson in Katzer, ed., *Molecular Sieves—II, Am. Chem. Soc. Symp. Ser.* **40:** 637 (1977); Brooking and Walton, *The Chem. Engr.* **257:** 13 (1972)]. Zeolites, silica gel, and activated alumina are used to dry synthesis gas, inert gas, hydrocracker gas, rare gases, and reformer recycle H_2. Because 3A and pore-closed 4A zeolites size-selectively adsorb water but exclude hydrocarbons, they are used extensively to dry reactive streams such as cracked gas in order to prevent coke formation on the adsorbent. This molecular sieving increases the recovery of hydrocarbons by reducing the coadsorption that would otherwise cause them to be desorbed and lost with the water.

Another area of application for TSA processes is in sweetening. H_2S, mercaptans, organic sulfides and disulfides, and carbonyl sulfide all must be removed from natural gas, H_2, biogas, and refinery streams in order to prevent corrosion and catalyst poisoning. Natural gas feed to steam methane reforming is sweetened in order to protect the sulfur-sensitive, low-temperature shift catalyst. Wellhead natural gas is treated by TSA to prevent pipeline corrosion using 4A zeolites to remove sulfur compounds without the coadsorption of CO_2 that would cause shrinkage. Sweetening and drying of refinery hydrogen streams are needed to prevent poisoning of reformer catalysts. Adsorption can be used to dry and sweeten these in the same unit.

TSA processes are applied to the removal of many inorganic pollutants. CO_2 is removed from base-load and peak-shaving natural-gas liquefaction facilities using 4A zeolite in a TSA cycle. The Sulfacid and Hitachi fixed-bed processes, the Sumitomo and BF moving-bed processes, and the Westvaco fluidized-bed process all use activated carbon adsorbents to remove SO_2 from flue gases and sulfuric acid plant tail gases [Juentgen, *Carbon* **15:** 273 (1977)]. Activated carbon with a catalyst is used by the Unitaka process to remove NO_x by reacting with ammonia, and activated carbon has been used to convert NO to NO_2, which is removed by scrubbing. Mercury vapor from air and other gas streams is removed and recovered by activated carbon impregnated with elemental sulfur; the Hg is then recovered by thermal oxidation in a retort [Lovett and Cunniff, *Chem. Eng. Progr.* **70:** 43 (1974)]. Applications for HCl removal from Cl_2, chlorinated hydrocarbons, and reformer catalyst gas streams use TSA with mordenite and clinoptilolite zeolites [Dyer (1988), pp. 102–105 in General References]. Activated aluminas are also used for HCl adsorption as well as fluorine and boron-fluorine compounds from alkylation processes [Crittenden, *Chem. Engr.* **452:** 21 (1988)].

Another important application of TSA is for the purification of air before it enters cryogenic air separation units. These TSA prepurification units (PPUs) remove carbon dioxide, water vapor, and trace organic compounds, among many other compounds, to avoid plugging, prevent explosions, and minimize corrosion in the cryogenic units [Kumar et al., *Adsorption* **9:** 243 (2003)]. The TSA vessels usually contain consecutive layers of activated alumina and molecular sieve zeolite (e.g., NaX). Additional layers of other adsorbents may also be used depending on the feed components [Kumar et al., *Adsorption* **9:** 243 (2003)]. Regeneration is carried out at high temperature using a purge gas at ambient pressure, usually nitrogen in an amount equivalent to about 10 percent of the feed gas.

PRESSURE-SWING ADSORPTION

A pressure-swing adsorption (PSA) process cycle is one in which desorption takes place at a pressure much lower than adsorption. Reduction of pressure is used to shift the adsorption equilibrium and affect regeneration of the adsorbent. Figure 16-31*b* depicts a simplified PSA cycle. Feed containing adsorbate at a mole fraction of $y_1 = p_1/P_1$ is passed through an adsorbent bed at conditions T_1, P_1, and the adsorption step continues until the equilibrium loading n_1 is achieved with y_1. Next, the total pressure is reduced to P_2 during the depressurization (or blowdown) step. Now, although the partial pressure in equilibrium with n_1 is still p_1, there is a concentration driving force of $y_2 = p_1/P_2 > y_1$ for desorption into fluid containing less than y_2. By passing a fluid across the adsorbent in a purge step, adsorbate is swept away, and the equilibrium proceeds down the isotherm to some point such as y_1, n_2. (The choice of y_1 is arbitrary and need not coincide with the feed composition.) The adsorbent is then repressurized to P_1. The new equilibrium y_3, n_2 represents the best-quality (light) product that can be produced from the adsorbent at a regenerated loading of n_2. The adsorption step is now repeated. The differential loading, $n_1 - n_2$, is the maximum loading that can be achieved for a PSA cycle operating between a feed containing y_1 and a light product containing y_3 of the adsorbate. The regeneration fluid will contain an average concentration between y_2 and y_1 and will therefore have accomplished a concentration of the adsorbate in the regenerant gas, that is, in the heavy product. There is no analog for a liquid-phase PSA process cycle.

Thus, in a PSA process cycle, regeneration is achieved by a depressurization that must reduce the partial pressure of the adsorbates to allow desorption. These cycles operate at constant temperature, requiring no heating or cooling steps. Rather, they use the exothermic heat of adsorption remaining in the adsorbent to supply the energy needed for desorption. Pressure-swing cycles are classified as: (1) PSA, which, although used broadly, usually swings between a high superatmospheric and a low superatmospheric pressure; (2) VSA (vacuum-swing adsorption), which swings from a slightly superatmospheric pressure to a low subatmospheric pressure; (3) PVSA, which swings from a high superatmospheric pressure to a low subatmospheric pressure; and (4) rapid (R) PSA or rapid cycle (RC) PSA, characterized by very fast cycle times (seconds) and high bed velocities, with minimal bed pressure drop afforded in some cases by using structured adsorbents. Two other classes also exist that affect separation by forcing a significant pressure gradient along the adsorbent bed by using small adsorbent particles and very fast cycle times (seconds). These are (5) PSPP (pressure-swing parametric pumping) and (6) an older version of RPSA, better termed pressure drop (PD) PSA so as not to confuse it with its modern analog (these are discussed in the subsection Parametric Pumping). For all six classes of PSA, the broad principles nevertheless remain the same, with their performance judged by the purity and recovery of any of the desirable species in the light or heavy product and either the feed throughput—that is, the amount of feed processed per unit time per unit mass of adsorbent—or the productivity—that is, the amount of product produced per unit time per unit mass of adsorbent.

Low pressure is not as effective in totally reversing adsorption as is temperature elevation unless very high feed-to-purge-pressure ratios are applied (e.g., deep vacuum). Therefore, most PSA cycles are characterized by high residual loadings and thus low operating loadings. These low capacities at high concentrations require that cycle times be short for reasonably sized beds (usually minutes). These short cycle times are attainable because adsorbent particles respond quickly to changes in pressure.

Cycle Steps A PSA cycle may have several other steps in addition to the basic adsorption (or feed), depressurization (countercurrent), and

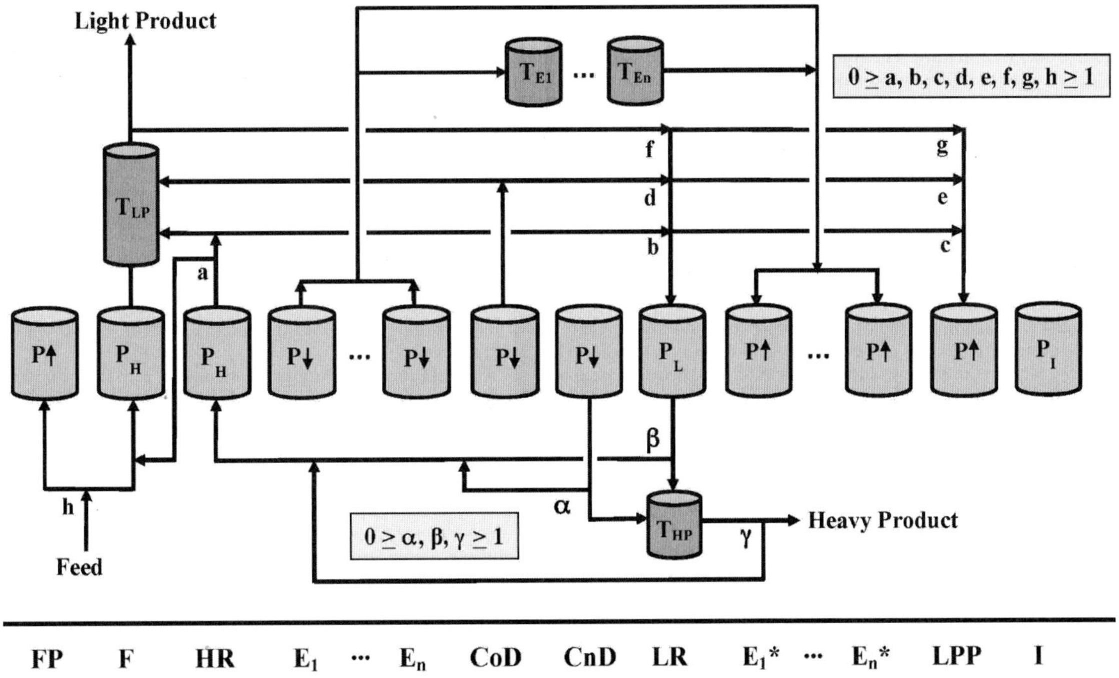

FIG. 16-32 Possible steps in a PSA cycle: feed (F), cocurrent depressurization (CoD), countercurrent depressurization (CnD), purge or light reflux (LR), rinse or heavy reflux (HR), bed-to-bed pressure equalization (E), light product or feed pressurization (LPP, FP), bed-to-tank-to-bed pressure equalization (with T), and idle (I).

repressurization steps introduced with reference to Fig. 16-31*b*. Cocurrent depressurization, purge (or light reflux), bed-to-bed pressure equalization, and rinse (or heavy reflux) steps can be added to increase separation efficiency and improve recovery of the light and/or heavy products. Thus, there are just these seven possible fundamental steps in any PSA process. Each of these steps and how they communicate with other beds undergoing other cycle steps by providing or receiving gas are shown in Fig. 16-32. Two additional but less common steps are also shown in this figure. They are a bed-to-tank-to-bed equalization step and an idle step. The former uses a tank that does not contain any adsorbent to equalize with a bed in two successive steps. The latter occurs when a bed sits idle for a certain time during the cycle, and although it performs no separative work while doing so and thus should be avoided, idle steps are required in some cases to schedule the sequence of cycle steps in each bed so they all align properly in time.

The sequence of steps from left to right in Fig. 16-32 is just one way they may be carried out in time in a PSA process. Another example would be to have the cocurrent depressurization step between two pressure equalization steps. All of these steps may be used in a PSA process, or just some of them, or just the three basic steps, depending on the application.

A PSA cycle schedule is defined by the cycle step sequence and the number of beds, with each bed running the exact same sequence but out of phase with each other. A PSA process usually contains two or more beds interconnected in some fashion, but a process with only one bed can also be used with the proper use of tanks (uncommon). When just considering the use of some or all of these seven to nine steps and the fact that modern PSA processes routinely operate with 2 to 20 interconnected beds, literally thousands of different designs or permutations exist in the development of a PSA cycle schedule [see Mehrotra et al., *Adsorption* **16:** 113 (2010)]. To further complicate the design of a PSA process, there are many choices to be made on the gas flow interconnects or couplings between the beds, as shown in Fig. 16-32. The role of each cycle step and the options that exist with it are discussed next.

The feed (F) step or adsorption step is carried out at the highest pressure (P_H) in the cycle and is usually the main light product production step. Some of the feed may be used to finish pressurizing the bed to P_H (as denoted by h). By the end of the F step, the more weakly adsorbed species are recovered as light product, but there is still a significant amount of light product held up in the bed in the interparticle and intraparticle void spaces.

A heavy refux (HR) step, bed-to-bed or bed-to-tank-to-bed pressure equalization steps, and/or a cocurrent depressurization (CoD) step can be added after the feed step and before the countercurrent depressurization (CnD) step to flush the weakly adsorbed species out of the bed cocurrently, thereby also filling the bed with the more strongly adsorbed species. These steps either recycle this gas containing some more weakly adsorbed species back to another bed or it may be taken as additional light product (as denoted by a, b, ... g, h).

The main role of the HR step is to produce a high-purity heavy product (just like in distillation) by taking heavy product gas selected from a variety of sources (as denoted by α, β, and γ), compressing it to P_H, and recycling it back to a bed just after the feed step to further concentrate it. The HR step should be thought of as a more concentrated feed step. Instead of at P_H, the HR step may also be carried out at some intermediate pressure after an E or CoD step.

Notice that "n" equalization steps are indicated in Fig. 16-32, either bed-to-bed or bed-to-tank-to-bed. Typically, for every additional equalization step, another adsorbent bed is required, possibly with the use of idle steps, unless a tank is used instead [Ebner et al., *Adsorption* **21:** 229 (2015)]. For example, Xu et al. [U.S. Patent 6,454,838 (2002)] discloses two PSA cycle schedules with four equalization steps and just six beds, in one case using idle steps and in another case without using any idle steps because an equalization tank is used.

Even without any equalization steps, the CoD step can be an important step to include. For example, in some applications, the purity of the more strongly adsorbed components has also been shown to be heavily dependent on the CoD step [Cen and Yang, *Ind. Eng. Chem. Fundam.* **25:** 758 (1986)]. CoD is optional because there is always a countercurrent one, that is, the CnD step. Skarstrom developed criteria to determine when the use of the CoD step is justified [Skarstrom, in Li, *Recent Developments in Separation Science,* vol. II, CRC Press, Boca Raton, 1975].

The next step, the CnD step, is one of the three basic cycle steps, as it constitutes the beginning of heavy product production and is thus required in all PSA processes. Once the CnD step commences, unless there is an HR step, all the weakly adsorbed species exiting the bed during this step end up in the heavy product. This reduces the light component recovery and the heavy product purity. The roles of the HR, E, and CoD steps in preventing some of the weakly adsorbed species from ending up in the heavy product via recycle are now clear, with the most efficient cycle being one that most closely matches available pressures and adsorbate concentrations to the appropriate portion of the bed at the proper point in the cycle.

The CnD step may optionally be followed by a purge or light reflux (LR) step. This step is carried out at the lowest pressure (P_L) in the PSA cycle and strips additional adsorbates from the adsorbent by decreasing their partial pressures, and it also flushes them from the voids. The main role of the LR step is to produce a high-purity light product (again, just like in distillation).

This is also a heavy product production step unless there is a HR step with certain values of α, β, and γ selected. The LR step can begin toward the end of CnD or immediately afterward. The source of the LR gas may be from the F, HR, or CoD step (as denoted by f, d, and b), with F and CoD being most common.

After the LR step, a bed begins repressurization, either via one or more equalization steps followed by light product pressurization (LPP) or feed pressurization (FP), with LPP being preferred when a high-purity light product is sought. It is worth noting that the equalization steps not only conserve gas via recycle, but they also conserve compression energy. The source of LPP can be from the F, HR, or CoD step (as denoted by g, e, and c). The source of FP is obvious (as denoted by h). FP should be followed by LPP if both are used to keep the heavier adsorbate species more toward the feed end of the bed. Once the bed is back to the feed pressure, the cycle repeats with the feed step. The example provided next serves to illustrate the foundation of a PSA cycle schedule, the associated bed interconnects or coupled steps, and how every bed undergoes the same sequence of cycle steps just out of phase with each other.

The six-bed, 16-step PSA cycle schedule with four equalization steps and three idle steps is shown in Fig. 16-33 [adapted from Xu et al., U.S. Patent 6,454,838 (2002)]. Although it looks complicated, there is much symmetry to a PSA cycle schedule. Bed 1 carries out the cycle step sequence in order by progressing step by step in time. A careful examination shows that this is the case for every bed; they are just out of phase with each other. There are also the six indicated boxes (unit blocks) within this cycle schedule, necessarily one for each bed. Within each unit block, every cycle step must be occurring somewhere in one of the beds. For example, in the leftmost unit block, the cycle step sequence is also written in order by progressing step by step in going from bed 1 to bed 2 and so on to bed 6. Notice that three idle steps are required in this PSA cycle schedule to align properly and thus accommodate all the equalization steps. This is also a continuous feed PSA cycle, with bed 1 being fed first in time, then bed 6, and so on. The feed step is the longest step in the cycle and occupies four unit time steps in each unit block. Some steps occupy only one unit time step, while others occupy two unit time steps, and one occupies three unit time steps. The number of unit time steps in a unit block is a PSA cycle schedule design choice, with a minimum number always advantageous [Mehrotra et al., *Adsorption* **17:** 337 (2011)]. This unequal step time PSA cycle schedule is common because it minimizes the number of beds for a long cycle step sequence and thus maximizes the throughput or productivity. Consider that if each cycle step were the same duration, then 16 beds would be required to run this same 16-step cycle step sequence [Reynolds et al., *Ind. Eng. Chem. Res.* **45:** 4278 (2006)]. More details on PSA cycle schedules and their creation are provided in the works of Ritter and coworkers [*Adsorption* **15:** 406 (2009); **16:** 113 (2010); **17:** 337 (2011); **21:** 229 (2015)].

Applications Major uses for PSA processes include the purification of lighter adsorbate species with a feed containing <10 vol% contaminants, as well as applications where heavy adsorbate species are present at high concentrations (>10 vol%), that is, for bulk separations. From 1960 to 1980, the major applications were air drying, hydrogen purification, large-scale oxygen production from air, nitrogen production from air using carbon molecular sieve, and small-scale medical oxygen production (even using a single-bed PSA unit). From 1980 to 1990, additional applications include helium recovery and purification, carbon dioxide recovery, natural gas purification, small-scale nitrogen production from air, linear from branched hydrocarbons (isosiv), solvent vapor recovery, and trace contaminant removal (military). From 2000 onward, new applications include hydrocarbon recovery from nitrogen, helium upgrading from dilute helium streams, ethanol dehydration at an elevated temperature, portable medical oxygen production from air using RCPSA, hydrogen upgrading using RCPSA, and biogas purification (landfill gas) using a titanium silicate molecular sieve (ETS-4). Newer applications still under development include olefin/paraffin separations, carbon dioxide capture from flue gas and from spacecraft cabin air for NASA, bulk purification of ammonia from nitrogen and hydrogen, xenon concentration from air, and even carbon monoxide isotope separation and enrichment [see Bhadra et al., *Adsorption* **19:** 11 (2013)]. Some of the more important or unique commercial and developmental applications are discussed next.

One of the earliest applications of PSA was the original Skarstrom two-bed, four-step cycle [adsorption (F), countercurrent depressurization (CnD), countercurrent purge (LR), and cocurrent repressurization (FP)] that was designed to dry an ambient air stream to less than 1 ppm H_2O [Skarstrom, in Li, *Recent Developments in Separation Science,* vol. II, CRC Press, Boca Raton, 1975]. Instrument-air dryers still use a PSA cycle similar to Skarstrom's with activated alumina or silica gel [Armond, in Townsend, *The Properties and Applications of Zeolites,* The Chemical Society, London, 1980]. With hydrocarbons being excluded by small-pore zeolites, PSA air dryers designed to work with air-brake compressors could achieve a 10 to 30 K dewpoint depression, even at high discharge air temperatures in the presence of compressor oil [Ausikaitis, in Katzer, *Molecular Sieves—II, Am. Chem. Soc. Symp. Ser.* **40:** 681 (1977)]. PSA air dryers are still ubiquitous today.

The next major application for PSA technology was hydrogen purification from a feed stream containing 60 to 70 vol% H_2. The high-purity hydrogen employed in processes such as hydrogenation, hydrocracking, and ammonia and methanol production is produced by PSA cycles with layered adsorbent beds typically containing three layers of alumina, activated carbon, and zeolites [Baksh and Terbot, U.S. Patent 6,503,299 (2003)]. Layered beds of activated carbon, zeolite, and carbon molecular sieve have also been used [Martin et al., *Adv. Cryog. Eng.* **31:** 1071 (1986); Ruthven, Farooq, and Knaebel (1994), p. 242 in General References]. The impurities to be removed include water vapor, ammonia, carbon oxides, nitrogen, oxygen, methane, and heavier hydrocarbons. This important application of PSA technology began with four-bed PSA systems routinely achieving hydrogen purities as high as 99.999 vol%. To improve recovery and productivity and also purity to 99.9999 vol%, five- to 16-bed PSA systems with complex PSA cycle schedules (e.g., see Fig. 16-33), such as the UOP Polybed™ PSA systems [Elseviers et al., *50 Years of PSA Technology for H_2 Purification,* UOP LLC (2015)], have been commercialized over the past 50 years. A change of the adsorbent inventory moved the focus from H_2 purification to the recovery of CO_2 from a very large-scale 20-bed PSA process, which began operation at an iron-making facility in Korea in 2013 [Elseviers et al., *50 Years of PSA Technology for H_2 Purification,* UOP LLC (2015)]. A unique RCPSA process for upgrading hydrogen in various refinery streams was jointly developed by Xebec Adsorption Inc. and ExxonMobil and commercialized in 2007. It uses a complex PSA

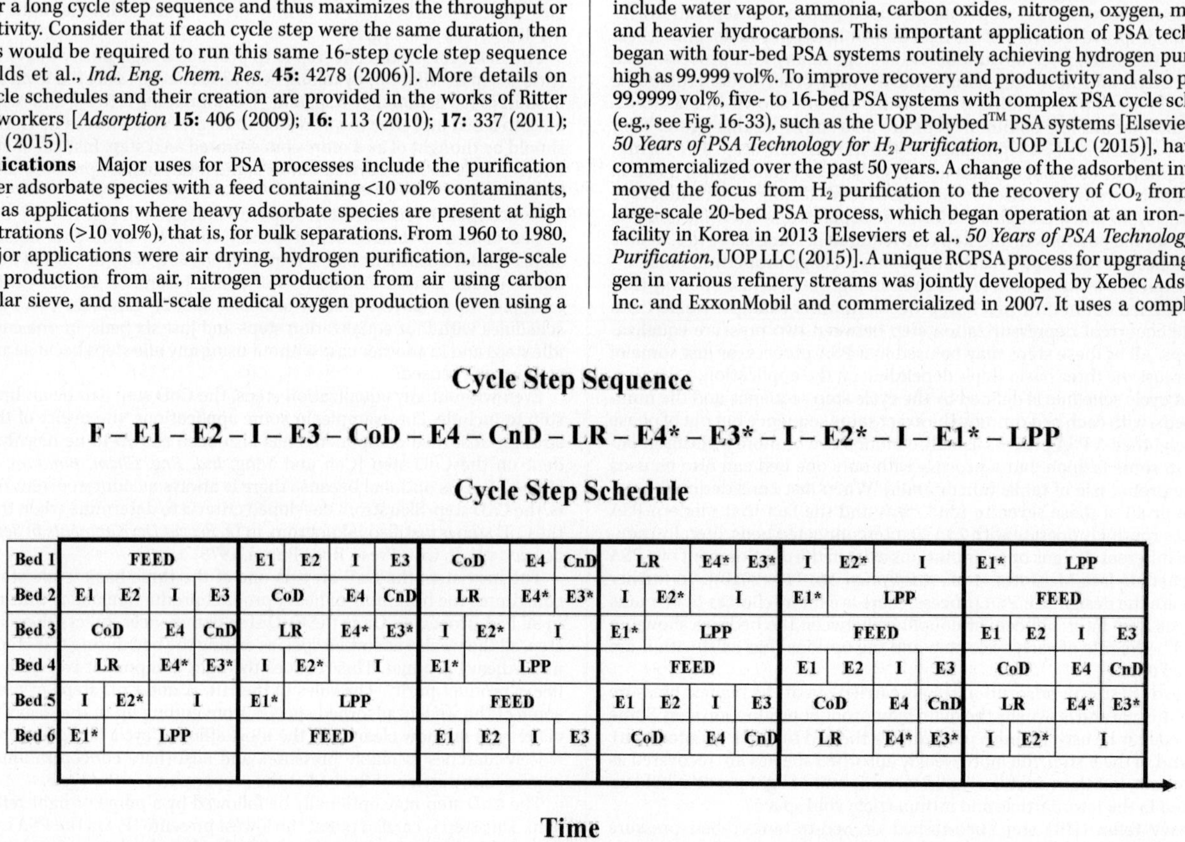
Cycle Step Sequence
F – E1 – E2 – I – E3 – CoD – E4 – CnD – LR – E4* – E3* – I – E2* – I – E1* – LPP
Cycle Step Schedule
Bed 1 FEED E1 E2 I E3 CoD E4 CnD LR E4* E3* I E2* I E1* LPP
Bed 2 E1 E2 I E3 CoD E4 CnD LR E4* E3* I E2* I E1* LPP FEED
Bed 3 CoD E4 CnD LR E4* E3* I E2* I E1* LPP FEED E1 E2 I E3
Bed 4 LR E4* E3* I E2* I E1* LPP FEED E1 E2 I E3 CoD E4 CnD
Bed 5 I E2* I E1* LPP FEED E1 E2 I E3 CoD E4 CnD LR E4* E3*
Bed 6 E1* LPP FEED E1 E2 I E3 CoD E4 CnD LR E4* E3* I E2* I
Time

FIG. 16-33 Cycle step sequence and schedule for a six-bed, 16-step PSA process with equalization idle steps (*Adapted from Xe et al., U.S. patent number 6,454,838, 2002*).

cycle schedule operating with cycle times of a few seconds or less in a rotary, five-bed system [Connor et al., U.S. Patent 6,406,523 (2002)] containing a structured adsorbent [Keefer et al., U.S. Patent 6,692,626 (2004)].

Ambient air separation was perhaps the next major type of bulk separation commercialized using PSA. PSA process cycles are used to produce oxygen, nitrogen, or both from air. Synthetic zeolites, clinoptilolite, mordenite, and carbon molecular sieves are all used in various PSA, VSA, VPSA, and RCPSA cycles. Since the mid-1980s, a LiLSX zeolite has been the preferred adsorbent for oxygen production for most large- or small-scale applications [Yang (2003) in General References]. The 85 to 95 percent purity oxygen produced is employed for electric furnace steel, waste water treating, solid waste combustion, and kilns [Martin et al., *Adv. Cryog. Eng.* **31:** 1071 (1986)]. Small medical oxygen PSA units have been used for some time for patients requiring inhalation therapy in the hospital and at home [Cassidy and Holmes, *AIChE Symp. Ser.* **80:** 68 (1984)] and for pilots on board aircraft [Tedor, Horch, and Dangieri, *SAFE J.* **12:** 4 (1982)]. Today, portable medical oxygen PSA technology based on RCPSA technology has flourished because it has made patients more ambulatory. These RCPSA units, provided by many different vendors, use cycle times less than 10 seconds, which makes them very small, lightweight, low power, and battery operated [Rao et al., *AIChE J.* **60:** 3330 (2014)]. Lower-purity oxygen (25 to 55 percent) can be produced to enhance combustion, chemical reactions, and ozone production [Sircar, in Rodrigues, LeVan, and Tondeur (1989), pp. 285–321 in General References]. The O_2 depleted product in the tail gas (heavy product) from an O_2 PSA unit, which can be tuned to range from about 10 to 15 vol%, has also been used for hypoxic training [Kotliar, U.S. Patent 5,799,652 (1998)], with variants also being proposed for hypoxic fire prevention and suspension systems [Kotliar, U.S. Patent 6,314,754 (2001)]. High-purity nitrogen (up to 99.99 percent) for inert blanketing is produced in PSA and VSA processes using zeolites and carbon molecular sieves [Kawai and Kaneko, *Gas Sep. & Purif.* **3:** 2 (1989); Ruthven et al. 1994].

Another very successful bulk separation done by PSA is the UOP IsoSiv[SM] process. The PSA process exploits the pore-size selectivity of zeolite 5A to adsorb straight-chain molecules while excluding branched and cyclic species. This PSA process separates C_5 to C_9 range hydrocarbons into a normal-hydrocarbon fraction of better than 95 percent purity, and a higher-octane isomer fraction with less than 2 percent normals [Cassidy and Holmes, *AIChE Symp. Ser.* **80:** 68 (1984)].

Methane is upgraded to natural gas pipeline quality by another PSA process. The methane is recovered from fermentation gases of landfills and wastewater purification plants and from poor-quality natural gas wells and tertiary oil recovery. Carbon dioxide is the major bulk contaminant, but the gases contain water and other undesired components such as sulfur and halogen compounds, alkanes, and aromatics [Kumar and VanSloun, *Chem. Eng. Progr.* **85:** 34–40 (1989)]. These impurities are removed by TSA using activated carbon or carbon molecular sieves, and then the CO_2 is adsorbed using a PSA cycle. The cycle can use zeolites or silica gel in an equilibrium-selective separation [Knaebel, *Adsorption* **9:** 87 (2003)] or a carbon molecular sieve in a rate-selective separation [Kapoor and Yang, *Chem. Eng. Sci.* **44:** 1723 (1989); Richter et al., *Petrochem.* **40:** 432 (1987)]. The newest commercial PSA systems that treat natural gas or landfill gas use a titanium silicate (ETS-4) adsorbent to kinetically separate CH_4 from N_2 and/or CO_2 [Butwell et al., U.S. Patent 6,197,092 (2001)]. An interesting series of PSA-TSA-PSA, PSA-PSA-PSA, or PSA-PSA processes has also been developed for this purpose [Knaebel, U.S. Patent 8,211,211 (2012)].

Another important application of PSA is for the purification of air before it enters cryogenic air separation units. PSA competes with TSA for this application, depending on many factors [Kumar et al., *Adsorption* **9:** 243 (2003)]. These PSA prepurification units (PPUs), like the TSA PPUs, remove carbon dioxide, water vapor, and trace organic compounds, among many other compounds, to avoid plugging, prevent explosions, and minimize corrosion in the cryogenic units [Kumar et al., *Adsorption* **9:** 243 (2003)]. The PSA PPU usually contains activated alumina and operates a four-step Skarstrom-like cycle using two beds or even three or more beds to produce more purified air. Regeneration is carried out at ambient pressure (P_L) using a purge gas (LR), usually nitrogen in an amount equivalent to about 40 percent of the feed gas.

In the late 1980s and throughout the 1990s, PSA technology emerged as an efficient way to concentrate and recover trace chemicals [White, *AIChE Symp. Ser.* **84:** 129 (1988); Ritter and Yang, *Ind. Eng. Chem. Res.* **30:** 1023 (1991)], solvent vapors [Robbins et al., U.S. Patent 4,857,084; Hall and Larrinaga, *Proceedings from the 16th National Industrial Energy Technology Conference*, Houston 1994] and gasoline vapors [Pezolt et al., *Environ. Prog.* **16:** 16 (1997); Liu et al., *AIChE J.* **46:** 540 (2000); Liu et al., *Sep. Purif. Technol.* **20:** 111 (2000)] from air. In each case, the focus was to produce clean air as a light product and a concentrated vapor as a heavy product using a two-bed, four-step Skarstrom-like PSA cycle.

A dual reflux PSA cycle with both stripping and enriching PSA sections and a feed location along the axial length of the column (just like a distillation column), first proposed by Wilson [U.S. Patent 4,359,328 (1982)], has been used to obtain high enrichment and recovery of both products. With the feed being delivered to a two-bed system at the low pressure of the PSA cycle, the separation is much better than that obtainable by normal PSA and suggests that it is possible to obtain two nearly pure products from a dilute feed stream [see McIntire et al., *Ind. Eng. Chem. Res.* **41:** 3499 (2002)]. When using just the enriching section in the same way, that is, with the feed at P_L, trace levels of Xe were concentrated up to 80 times with 90 percent recovery using 13X zeolite in a two-bed system with a pressure ratio of only 12.5 [Yoshida et al., *Ind. Eng. Chem. Res.* **42:** 1795 (2003)]. This Xe enrichment PSA technology was commercialized in Japan. A similar enriching PSA cycle was able to separate N_2 from air using 13X zeolite in a two-bed, four-step system [Reynolds et al., *Ind. Eng. Chem. Res.* **45:** 3256 (2006)]. The heavy product contained only 0.8 vol% O_2 with an N_2 recovery of 23.7 percent. Enriching and dual reflux PSA cycles were also used to concentrate organic vapors to the point where they condensed into a liquid without the need of any other unit operation [R. Wakasugi et al., *J. Chem. Eng. Japan* **37:** 374 (2004); *Adsorption* **11:** 561 (2005)].

One of the newest applications for PSA is CO_2 capture and concentration from a variety of feed streams where the CO_2 is the intended product. These include chemical process flue gas, where it has been commercialized by Mitsubishi Heavy Industry using a layered PSA bed containing X and A type zeolites and activated carbon [Izumi, 8th International Conference on Fundamentals of Adsorption, Sedona, AR, May 2004]; food grade CO_2 from H_2 production processes commercialized by Linde AG [Voss, *Adsorption* **11:** 527 (2005)]; as previously mentioned, recovery of CO_2 from a very large-scale 20-bed PSA process, which began operation at an iron-making facility in Korea in 2013 [Elseviers et al., *50 Years of PSA Technology for H_2 Purification*, UOP LLC (2015)]; and for coal-fired power plant flue gas, where it is still under development [Reynolds et al., *Ind. Eng. Chem. Res.* **45:** 4278 (2006)]. It has also been shown experimentally that it is possible to concentrate CO_2 in spacecraft cabins from about 4000 ppm to over 90 vol% at 82 percent recovery using a three-bed, eight-step heavy reflux PSA cycle using 13X zeolite (Erden et al., "New PSA Cycle for CO_2 Removal During Closed-Loop Human Space Exploration Missions," AIChE Annual Meeting, Salt Lake City, UT, November 2015).

PURGE/CONCENTRATION SWING ADSORPTION

A purge-swing adsorption cycle is usually considered to be one in which desorption takes place at the same temperature and total pressure as adsorption. Desorption is accomplished either by partial-pressure reduction using an inert gas purge or by adsorbate displacement with another adsorbable component. Purge cycles operate adiabatically at nearly constant inlet temperature and require no heating or cooling steps. As with PSA, they can utilize the heat of adsorption remaining in the adsorbent (if any) to supply the heat of desorption. Purge processes are classified as (1) inert or (2) displacement.

Inert Purge In inert-purge desorption cycles, *inert* refers to the fact that the purge gas is not adsorbed significantly at the cycle conditions. Inert purging desorbs the adsorbate solely by partial pressure reduction. Figure 16-31*c* depicts a simplified inert-purge swing cycle using a nonadsorbing purge fluid. The feed stream containing an adsorbate at a partial pressure of p_1 is passed through an adsorbent at temperature T and total pressure P, and the adsorption step continues until the equilibrium loading n_1 is achieved. Next, the nonadsorbing fluid is introduced to reduce the partial pressure below p_1. Therefore, there is a partial pressure driving force for desorption into the purge fluid, and the equilibrium proceeds down the isotherm to the point p_2, n_2, where p_2 represents the best-quality product that can be produced from the adsorbent at a regenerated loading of n_2. The adsorption step is now repeated, and the differential loading is $n_1 - n_2$. The regeneration fluid will contain an average partial pressure between p_2 and p_1, and the cycle will have transferred the adsorbate to a fluid from which it may be more easily separated, if desired, by means such as distillation.

Like PSA cycles, inert-purge processes are characterized by high residual loadings, low operating loadings, and short cycle times (minutes). Bulk separations of contaminants not easily separable at high concentration and of weakly adsorbed components are especially suited to inert-purge-swing adsorption. Another version of UOP's IsoSiv process employs H_2 in an inert-purge cycle for separating C_5 to C_9 naphtha by adsorbing straight-chain molecules and excluding branched and cyclic species on size selective 5A zeolite [Cassidy and Holmes, *AIChE Symp. Ser.* **80:** 68 (1984)]. Automobiles made in the United States have canisters of activated carbon to adsorb gasoline vapors lost from the fuel intake system or the gas tank; the vapors are desorbed by an inert purge of air that is drawn into the intake manifold as fuel when the engine is running [Clarke et al., *S.A.E. Trans.* **76:** 824 (1968); Johnson et al., in *Carbon Materials for Advanced Technologies*, ed. T. D. Burchell, Pergamon, New York, 1999]. UOP's Adsorptive Heat Recovery drying system has been commercialized for drying azeotropic ethanol to be blended with gasoline into gasohol; the process uses a closed loop of N_2 as the inert purge to desorb the water [Garg and Yon, *Chem. Eng. Progr.* **82:** 54 (1986)].

Displacement Purge Isothermal, isobaric regeneration of the adsorbent can also be accomplished by using a purge fluid that can adsorb. In displacement-purge stripping, *displacement* refers to the displacing action of the purge fluid caused by its ability to adsorb at the cycle conditions. Figure 16-31*d* depicts a simplified displacement-purge swing cycle using an adsorbable purge. Again, the feed stream containing an adsorbate at a partial pressure of p_1 is passed through an adsorbent bed at temperature T_1, and the adsorption step continues until the equilibrium loading n_1 is achieved. Next the displacement fluid, B, is introduced. The presence of another adsorbable species reduces the adsorptivity of the key adsorbate, A. Therefore, there exists a partial pressure driving force for desorption into the purge fluid, and the equilibrium proceeds down the isotherm to some point such as p_1, n_2 (again, arbitrary.) Next, the adsorbent is recharged with a fluid that contains no component B, shifting the effective isotherm to where the equilibrium of component A is p_3. The new equilibrium p_3, n_2 represents the best-quality product that can be produced from the adsorbent at a regenerated loading of n_2. The adsorption step is now repeated. The differential loading $(n_1 - n_2)$ is the maximum loading that can be achieved for a pressure-swing cycle operating between a feed containing y_1 and a product containing a partial pressure p_3 of the adsorbate. The regeneration fluid will contain an average partial pressure between p_2 and p_1 and will therefore have accomplished a concentration of the adsorbate in the regenerant gas.

Displacement-purge cycles are not as dependent on energy from the heat of adsorption remaining on the adsorbent, because the adsorption of purge releases most or all of the energy needed to desorb the adsorbate. It is best if the adsorbate is more selectively adsorbed than the displacement purge, so that the adsorbates can easily desorb the purge fluid during adsorption. The displacement purge must be carefully selected because it contaminates both the product stream and the recovered adsorbate and requires separation for recovery (e.g., by distillation).

Displacement-purge processes are more efficient for less selective adsorbate/adsorbent systems, while systems with high equilibrium loading of adsorbate will require more purging [Sircar and Kumar, *Ind. Eng. Chem. Proc. Des. Dev.* **24:** 358 (1985)]. Several displacement-purge-swing processes have been commercialized for the separation of branched and cyclic C_{10}–C_{18} from straight-chain molecules using the molecular-size selectivity of 5A zeolite: Exxon's Ensorb, UOP's IsoSiv, Texaco Selective Finishing (TSF), Leuna Werke's Parex, and the Shell Process [Ruthven (1984)]. All use a purge of normal paraffin or light naphtha with a carbon number of two to four less than the feed stream except for Ensorb, which uses ammonia [Yang (1987) in General References]. UOP has also developed a similar process, OlefinSiv, which separates isobutylene from normal butenes with displacement purge and a size-selective zeolite [Adler and Johnson, *Chem. Eng. Progr.* **75:** 77 (1979)]. Solvent extraction to regenerate activated carbon is another example of a displacement-purge cycle; the adsorbent is then usually steamed to remove the purge fluid [Martin and Ng, *Water Res.* **18:** 59 (1984)]. The best use of solvent regeneration is for water phase adsorption where the separation of water from carbon would use too much steam and where purge and water are easily separated, and for vapor-phase where the adsorbate is highly nonvolatile but soluble. Air Products has developed a process for separating ethanol and water on activated carbon using acetone as a displacement agent and adding a water rinse to improve the recovery of two products [Sircar, U.S. Patent 5,026,482 (1991)].

Displacement-purge forms the basis for most simulated continuous countercurrent systems (see hereafter) such as the UOP Sorbex^{SM} processes. UOP has licensed about one hundred Sorbex units for its family of processes: Parex^{SM} to separate *p*-xylene from C_8 aromatics, Molex^{SM} for *n*-paraffin from branched and cyclic hydrocarbons, Olex^{SM} for olefins from paraffin, Sarex^{SM} for fructose from dextrose plus polysaccharides, Cymex^{SM} for *p*- or *m*-cymene from cymene isomers, and Cresex^{SM} for *p*- or *m*-cresol from cresol isomers. Toray Industries' Aromax^{SM} process is another for the production of *p*-xylene [Otani, *Chem. Eng.* **80**(9): 106 (1973)]. Illinois Water Treatment [*Making Waves in Liquid Processing*, Illinois Water Treatment Company, IWT Adsep System, Rockford, IL, **6**(1): 1984] and Mitsubishi [Ishikawa, Tanabe, and Usui, U.S. Patent 4,182,633 (1980)] have also commercialized displacement-purge processes for the separation of fructose from dextrose.

ION EXCHANGE

Except in very small-scale applications, ion exchangers are used in cyclic operations involving sorption and desorption steps. A typical ion-exchange cycle used in water-treatment applications involves (*a*) *backwash*—used to remove accumulated solids obtained by an upflow of water to expand (50–80 percent expansion is typical) and fluidize the exchanger bed; (*b*) *regeneration*—a regenerant is passed slowly through the spent bed to restore the original ionic form of the exchanger; (*c*) *rinse*—water is passed through the bed to remove regenerant from the void volume and, in the case of porous exchangers, from the resin pores; (*d*) *loading*—the fresh solution to be treated is passed through the bed until leakage begins to occur. Water softening is practiced in this way with a cation exchange column in sodium form. At the low ionic strength used in the loading step, calcium and magnesium are strongly preferred over sodium, allowing nearly complete removal. Since the selectivity for divalent cations decreases sharply with ionic concentration, regeneration with a concentrated sodium chloride solution is also very efficient. Removal of sulfates from boiler feed water is done by similar means with anion exchangers in chloride form.

Many ion-exchange columns operate with downflow and are regenerated in the same direction (Fig. 16-34*a*). However, a better regeneration and lower leakage during loading can be achieved by passing the regenerant countercurrently to the loading flow. Specialized equipment is available to perform countercurrent regeneration (see Equipment in this subsection). One approach (Fig. 16-34*b*) is to apply a vacuum to remove the regenerant at the top of the bed.

Complete deionization with ion-exchange columns is the classical method of producing ultrapure water for boiler feed, in electronics manufacture, and for other general uses in the chemical and allied industries. Deionization requires use of two exchangers with opposite functionality to remove both cations and anions. These can be in separate columns, packed in adjacent layers in the same column, or, more frequently, in a mixed bed. In the latter case, the two exchangers are intimately mixed during the loading step. For regeneration, backwashing separates the usually lighter anion exchanger from the usually denser cation exchanger. The column typically has a screened distributor at the interface between the two exchangers, so that they may be separately regenerated without removing them from the column. The most common cycle (Fig. 16-35) permits sequential regeneration of the two exchangers, first with alkali flowing downward through the anion exchanger to the interface distributor and then acid flowing downward from the interface distributor through the cation exchanger. After regeneration and rinsing, the exchangers are remixed by compressed air. To alleviate the problem of intermixing of the two different exchangers and chemical penetration through the wrong one, an inert material of intermediate density can be used to provide a buffer zone between layers of cation and anion exchangers.

When recovery of the sorbed solute is of interest, the cycle is modified to include a displacement step. In the manufacture of pharmaceuticals, ion exchangers are used extensively in recovery and separation. Many of these compounds are amphoteric and are positively or negatively charged, depending on the solution pH. Thus, for example, using a cation exchanger, loading can be carried out at a low pH and displacement at a high pH. Differences in selectivity for different species can be used to carry out separations during the displacement [Carta et al., *AIChE Symp. Ser.* **84:** 54 (1988)]. Multibed cycles are also used to facilitate integration with other chemical process operations. Figure 16-36 shows a two-bed ion-exchange system using both cation and anion exchangers to treat and recover chromate from rinse water in plating operations. The cation exchanger removes trivalent chromium, while the anion exchanger removes hexavalent chromium as an anion. Regeneration of the cation exchanger with sulfuric acid produces a concentrated solution of trivalent chromium as the sulfate salt. The hexavalent chromium is eluted from the anion exchanger with sodium

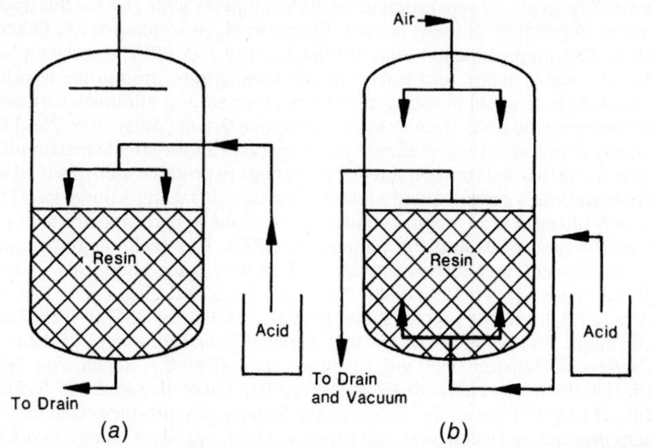

FIG. 16-34 Ion-exchanger regeneration. (*a*) Conventional. Acid is passed downflow through the cation-exchange resin bed. (*b*) Counterflow. Regenerant solution is introduced upflow with the resin bed held in place by a dry layer of resin.

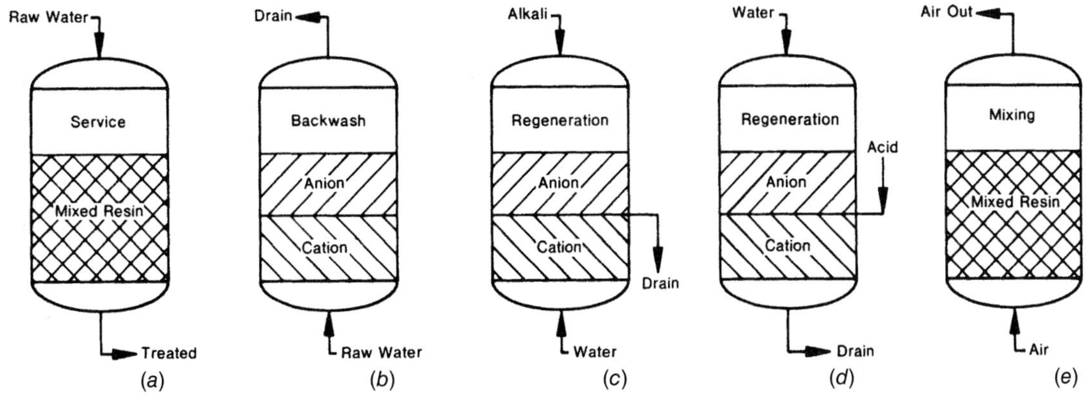

FIG. 16-35 Principles of mixed-bed ion exchange. (*a*) Service period (loading). (*b*) Backwash period. (*c*) Caustic regeneration. (*d*) Acid regeneration. (*e*) Resin mixing.

hydroxide in a concentrated solution. This solution is recycled to the plating tank by passing it through a second cation exchange column in hydrogen form to convert the sodium chromate to a dilute chromic acid solution that is concentrated by evaporation.

PARAMETRIC PUMPING

The term *parametric pumping* was coined by Wilhelm et al. [Wilhelm, Rice, and Bendelius, *Ind. Eng. Chem. Fundam.* **5:** 141 (1966)] to describe a liquid-phase adsorption process in which separation is achieved by periodically reversing not only flow but also an intensive thermodynamic property such as temperature, which influences adsorption equilibrium. Moreover, they considered the concurrent cycling of pressure, pH, and electrical and magnetic fields. A lot of research and development has been conducted on thermal, pressure, and pH-driven cycles, but to date only gas-phase pressure-swing parametric pumping has found extensive commercial acceptance.

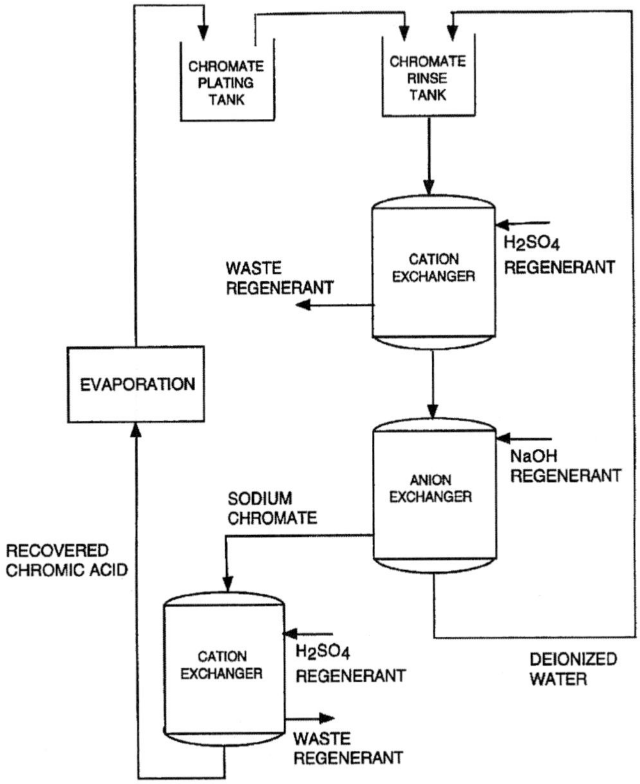

FIG. 16-36 Multicomponent ion-exchange process for chromate recovery from plating rinse water. (*Adapted from "Ion Exchange Resins for Metal Plating and Surface Finishing," Rhom and Haas Company, Philadelphia, PA, USA, December 1999.*)

Temperature Two modes of temperature parametric-pumping cycles have been defined—direct and recuperative. In direct mode, an adsorbent column is heated and cooled while the fluid feed is pumped forward and backward through the bed from reservoirs at each end. When the feed is a binary fluid, one component will concentrate in one reservoir and one in the other. In recuperative mode, the heating and cooling takes place outside the adsorbent column. Parametric pumping, thermal and pH modes, have been widely studied for separation of liquid mixtures. However, the primary success for separating gas mixtures in thermal mode has been the separation of propane/ethane on activated carbon [Jencziewski and Myers, *Ind. Eng. Chem. Fundam.* **9:** 216 (1970)] and of air/SO_2 on silica gel [Patrick et al., *Sep. Sci.* **7:** 331 (1972)]. The difficulty with applying the thermal mode to gas separation is that in a fixed volume, gas pressure increases during the hot step, which defeats the desorption purpose of this step. No thermal parametric-pumping cycle has yet been practiced commercially.

Pressure Another approach to parametric pumping is accomplished by pressure cycling of an adsorbent. An adsorbent bed is alternately pressurized with forward flow and depressurized with backward flow through the column from reservoirs at each end. Like TSA parametric pumping, one component concentrates in one reservoir and one in the other. The pressure mode of parametric pumping has been called pressure-swing parametric pumping (PSPP) and rapid pressure-swing adsorption (RPSA) or pressure drop (PD) PSA to avoid confusing it with its modern RC PSA (or RPSA) analog discussed in the PSA section. It was developed to minimize process complexity and investment at the expense of product recovery. RPSA (or PDPSA) is practiced in single-bed [Keller and Jones, in Flank, *Adsorption and Ion Exchange with Synthetic Zeolites* **135:** 275–286 (1980)] and multiple-bed (Earls and Long, U.S. Patent number 4,194,892, 1980) implementations. Adsorbers are short (about 0.3 to 1.3 m), and particle sizes are very small (about 150 to 400 mm). The total cycle time, including adsorption (F), dead (idle) time, countercurrent depressurization (CnD), and sometimes a second dead (idle) time, ranges from a few to about 30 seconds. The feature of RPSA (or PD PSA) that differentiates it from traditional PSA is the existence of axial pressure profiles throughout the cycle, much as temperature gradients are present in TSA parametric pumping. Whereas PSA processes have essentially constant pressure through the bed at any given time, the flow resistance of the very small adsorbent particles produces substantial pressure drop in the bed. These pressure dynamics are important to the attainment of separation performance. The light product end of the bed stays essentially at P_H over the entire cycle, while the feed/heavy product end of the bed cycles between P_H and P_L, thereby forcing adsorption and desorption of the heavier adsorbate species. Effectively, the light species is parametrically pumped up a pressure gradient to the light product end of the bed, while the heavier species are removed from the other end when the cycle is changing from P_H to P_L. RPSA (or PD PSA) has been commercialized for the production of oxygen and for the recovery of ethylene and chlorocarbons (the selectively adsorbed species) in an ethylene-chlorination process while purging nitrogen (the less selectively adsorbed species).

SIMULATED MOVING BED SYSTEMS

The concept of a simulated moving-bed (SMB) system was originally used in a process developed and licensed under the name UOP Sorbex process [Broughton et al., *Pet. Int. (Milan)* **23**(3): 91 (1976); **23**(5): 26 (1976)]. The basic process, used for separating a binary mixture, is illustrated in Fig. 16-37. As shown, the process employs a stack of packed beds connected to a

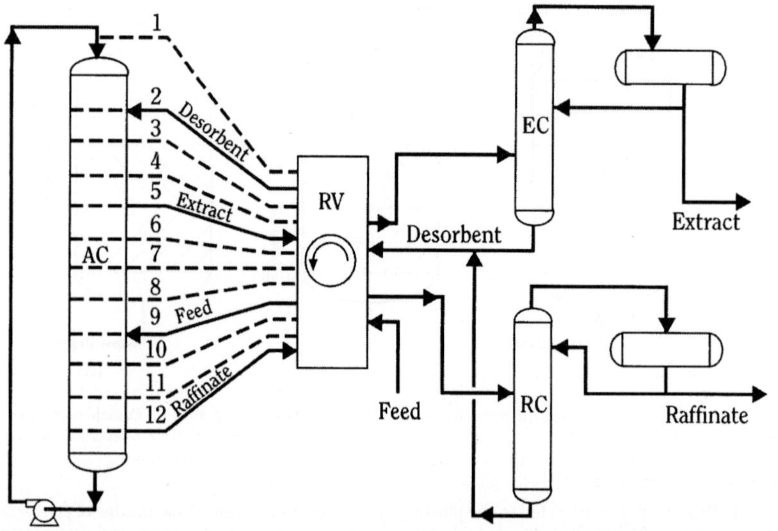

FIG. 16-37 UOP Sorbex process. (*Reprinted with permission of John Wiley & Sons, Inc. Reference: Gembicki, Oroskar, and Johnson, "Adsorption, Liquid Separation," in* Kirk-Othmer Encyclopedia of Chemical Technology, *4th ed., John Wiley & Sons, Inc., New York, 1991.*)

rotary valve (RV) that allows the introduction of feed and desorbent and the collection of extract and raffinate streams at different junctions. Feed and withdrawal points are switched periodically as shown, resulting in a periodic counterflow of adsorbent. Even with a small number of packed beds, the periodic countercurrent action closely simulates the behavior of a true countercurrent system without the complexities associated with particle flows. Distillation columns are shown in Fig. 16-37 integrated with the adsorption system to recover and recycle the desorbent. Although the Sorbex process was originally applied to hydrocarbon separations, extensive industrial applications have been developed for sugars, amino acids, and fine chemicals, especially chiral separations. Practical operation is not restricted to rotary valves. Using multiple individual valves to control the distribution of flows is often more practical, especially on a smaller scale.

The basic principle of operation is illustrated in Fig. 16-38 by reference to an equivalent true countercurrent moving-bed (TMB) system comprising four idealized moving-bed columns or "zones." The feed containing components *A* and *B* is supplied between zones II and III. The least strongly adsorbed species, *A*, is recovered between zones III and IV, while the more strongly adsorbed species, *B*, is recovered between zones I and II. The adsorbent is recirculated from the bottom of zone I to the top of zone IV. A desorbent or eluent makeup stream is added to the fluid recycled from zone IV, and the combined stream is fed to the bottom of zone I. The main purpose of each zone is as follows: Zone III adsorbs *B* while letting *A* pass through. Zone II desorbs *A* while adsorbing *B*. Zone I desorbs *B*, allowing recycle of the adsorbent. Zone IV adsorbs *A*. Proper selection of operating conditions is needed to obtain the desired separation. The ensuing analysis is based on local equilibrium and plug flow conditions and assumes linear isotherms with a nonadsorbable desorbent. In the following u^j [m^3/(m$^2 \cdot$ s)] represents the fluid-phase velocity in zone *j* and u_s [kg/(m$^2 \cdot$ s)] the adsorbent velocity, with both velocities defined based on the column cross-sectional area. Net upward transport of each component is determined by the component velocity

$$u_i^j = u^j - u_s\left(\frac{\varepsilon_p}{\rho_p} + \frac{n_i}{c_i}\right) \qquad (16\text{-}196)$$

The following inequalities must be met to obtain the desired separation:

$$u_B^{\mathrm{I}} > 0 \qquad (16\text{-}197a)$$

$$u_A^{\mathrm{II}} > 0, \quad u_B^{\mathrm{II}} < 0 \qquad (16\text{-}197b)$$

$$u_A^{\mathrm{III}} > 0, \quad u_B^{\mathrm{III}} < 0 \qquad (16\text{-}197c)$$

$$u_A^{\mathrm{IV}} < 0 \qquad (16\text{-}197d)$$

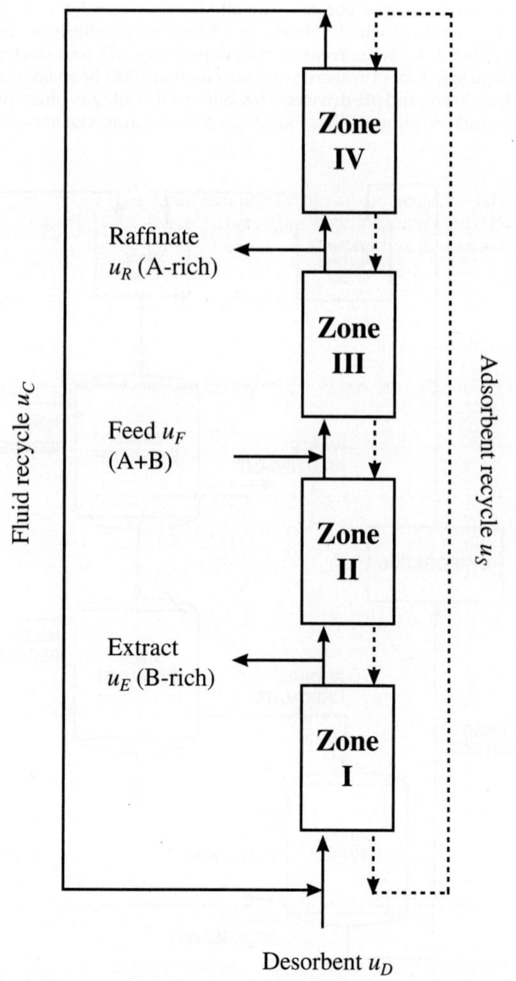

FIG. 16-38 General scheme of a true moving-bed (TMB) adsorption system for binary separations. *A* is less strongly retained than *B*.

Accordingly, A moves downward in zone IV and upward in zone III and is recovered between these two zones. Similarly, B moves upward in zone I and downward in zone II and is recovered between these two zones. Combining Eqs. (16-196) and (16-197) yields the following constraints:

$$m^{\mathrm{I}} > K_B \tag{16-198a}$$

$$K_B > m^{\mathrm{II}} > K_A \tag{16-198b}$$

$$K_B > m^{\mathrm{III}} > K_A \tag{16-198c}$$

$$m^{\mathrm{IV}} < K_A \tag{16-198d}$$

where K_i is the linear isotherm slope [cf. Eq. (16-12)] and

$$m^j = \frac{u^j - u_s \varepsilon_p / \rho_p}{u_s} \tag{16-199}$$

is a flow ratio (m^3/kg). Inequalities (16-197b) and (16-197c) determine whether separation will occur and can be represented on the $m^{\mathrm{III}} - m^{\mathrm{II}}$ plane in Fig. 16-39. Since $m^{\mathrm{III}} > m^{\mathrm{II}}$, only the region above the 45° line is valid. Values of m^{II} and m^{III} below K_A or above K_B result in incomplete separation. Thus, complete separation requires operation within the shaded triangular region. The vertex of this triangle represents the point of maximum productivity under ideal conditions. In practice, mass-transfer resistances and deviations from plug flow will result in imperfect separation even within the shaded region. As a result, operating away from the vertex and closer to the 45° line is usually needed at the expense of lower productivity. By introducing a safety margin $\beta \geq 1$ [Seader and Henley (2006) in General References], Eqs. (16-198) are transformed to

$$m^{\mathrm{I}} = K_B\beta \tag{16-200a}$$

$$m^{\mathrm{II}} = K_A\beta \tag{16-200b}$$

$$m^{\mathrm{III}} = K_B/\beta \tag{16-200c}$$

$$m^{\mathrm{IV}} = K_A/\beta \tag{16-200d}$$

$\beta = 1$ yields maximum productivity under ideal conditions, while larger values provide a more robust design up to the maximum $\beta = \sqrt{K_B/K_A}$. For a given β, external and internal flow velocities are calculated from

$$u_S = u_F/(K_B/\beta - K_A\beta) \tag{16-201a}$$

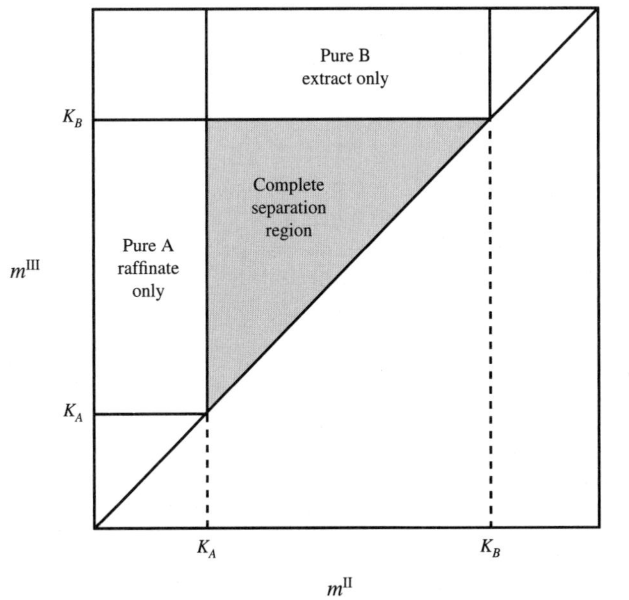

FIG. 16-39 m^{III}-m^{II} plane showing regions of complete and incomplete separation. Area below the 45° line is invalid.

$$u_E = u_S(K_B - K_A)\beta \tag{16-201b}$$

$$u_R = u_S(K_B - K_A)/\beta \tag{16-201c}$$

$$u_D = u_E + u_R - u_F \tag{16-201d}$$

$$u^{\mathrm{I}} = u_S(K_B\beta + \varepsilon_p/\rho_p) \tag{16-201e}$$

$$u^{\mathrm{II}} = u^{\mathrm{I}} - u_E \tag{16-201f}$$

$$u^{\mathrm{III}} = u^{\mathrm{II}} + u_F \tag{16-201g}$$

$$u^{\mathrm{IV}} = u^{\mathrm{III}} - u_R \tag{16-201h}$$

$$u_C = u^{\mathrm{I}} - u_D \tag{16-201i}$$

Analogous relationships are derived for SMB systems where each zone comprises a number of fixed beds operated in a merry-go-round sequence, as shown in Fig. 16-40. External flow velocities are calculated from Eqs. (16-201a) to (16-201d), replacing u_S with

$$u_{S,\mathrm{SMB}} = \frac{\rho_b L}{p} \tag{16-202}$$

where L is the length of a single bed and p is the switching period. Internal flow velocities equivalent to the TMB operation are increased from the values calculated from Eqs. (16-201e) to (16-201i) to compensate for the extraparticle fluid carried along in each bed at each switch according to

$$u^j_{\mathrm{SMB}} = u^j + \frac{\varepsilon L}{p} \tag{16-203}$$

In practice, a small number of beds in series in each zone provide a close approach to the performance of ideal, true countercurrent system. Industrial SMB systems normally use one to three beds per zone.

Complete Design and Extensions Complete design of SMB systems requires a full description of equilibrium and rate factors. For an existing SMB unit, initial stream flow rates and switching period can be selected based on Eqs. (16-201) to (16-203) so that the operating point lies within the desired separation region. Column length design requires a dynamic adsorption model including a description of mass-transfer rates, adsorption kinetics, and axial dispersion. An analytical solution for the linear isotherm with an LDF model is available in Carta and Jungbauer [(2010), pp. 327–338]

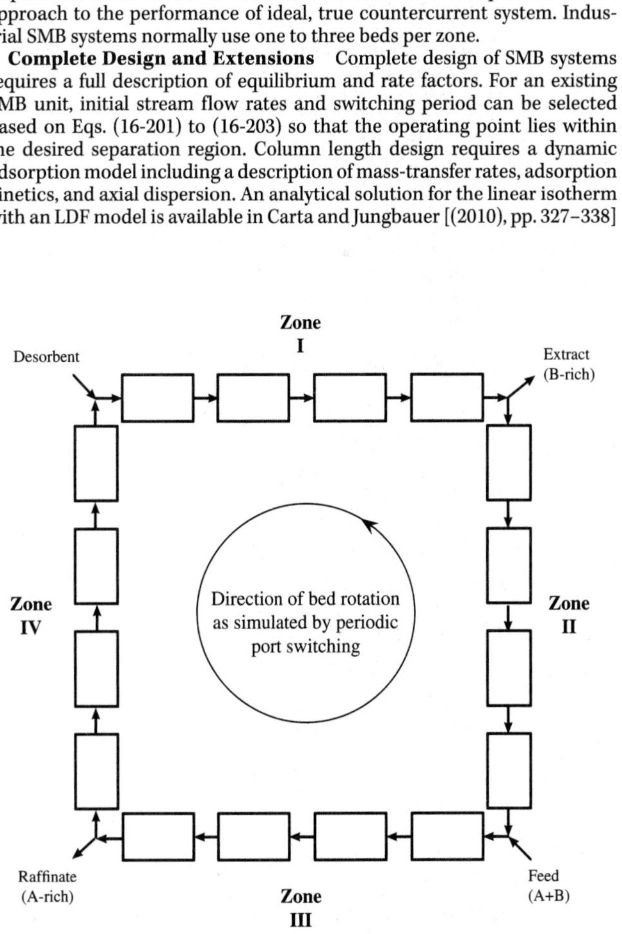

FIG. 16-40 General scheme of a simulated moving-bed (SMB) adsorption system. Bed rotation is simulated by periodic switching of ports in the direction of fluid flow. A is less strongly retained than B.

and Dunnebier et al., *Ind. Eng. Chem. Res.* **39:** 2290 (2000). Operation with a nonlinear isotherm is analyzed in a similar manner. In this case, the right triangle defining the complete separation region in Fig. 16-39 is distorted, acquiring one or more curved sides and further restricting the range of conditions leading to complete separation. Operating conditions can be selected on this basis, but a complete design typically requires a numerical solution. As is evident from the analysis above, only binary separations are achievable with a four-zone system. Multicomponent separations require multiple SMB units or integrated units comprising more than four zones. Useful references covering SMB design for linear and nonlinear isotherms are by Ruthven and Ching, *Chem. Eng. Sci.* **44:** 1011 (1989); Storti et al., *Chem. Eng. Sci.* **44:** 1329 (1989); Zhong and Guiochon, *Chem. Eng. Sci.* **51:** 4307 (1996); Mazzotti et al., *J. Chromatogr. A* **769:** 3 (1997); Mazzotti et al., *AIChE J.* **40:** 1825 (1994); and Minceva and Rodrigues, *Ind. Eng. Chem. Res.* **41:** 3454 (2002).

OTHER ADSORPTION CYCLES

Hybrid Recycle Systems Liquid chromatography has been used commercially to separate glucose from fructose and other sugar isomers, for recovery of nucleic acids, and for other uses. Sanmatsu Kogyo Co., Ltd. [Yoritomi, Kezuka, and Moriya, U.S. Patent number 4,267,054, (1981)] developed an improved chromatographic process that is simpler to build and operate than simulated moving-bed processes. Figure 16-41 [see Keller, Anderson, and Yon (1987) in General References] diagrams its use for a binary separation. It is a displacement-purge cycle where pure component cuts are recovered, while cuts that contain both components are recycled to the feed end of the column.

The UOP Cyclesorb^{SM} is another adsorptive separation process with semicontinuous recycle. It uses a series of chromatographic columns to separate fructose from glucose. A series of internal recycle streams of impure and dilute portions of the chromatograph are used to improve the efficiency [Gerhold, U.S. Patent numbers 4,402,832 (1983) and 4,478,721 (1984)]. A schematic diagram of a six-vessel UOP Cyclesorb process is shown in Fig. 16-42 [Gembicki, Oroskar, and Johnson (1991), p. 595 in General References]. The process has four external streams and four internal recycles: dilute raffinate and impure extract are like displacement steps; and impure raffinate and dilute extract are recycled from the bottom of an adsorber to its top. Feed and desorbent are fed to the top of each column in a predetermined sequence. The switching of the feed and desorbent are accomplished by the same rotary valve used for Sorbex switching (see hereafter). A chromatographic profile is established in each column that is moving from top to bottom, and all portions of an adsorber are performing a useful function at any time.

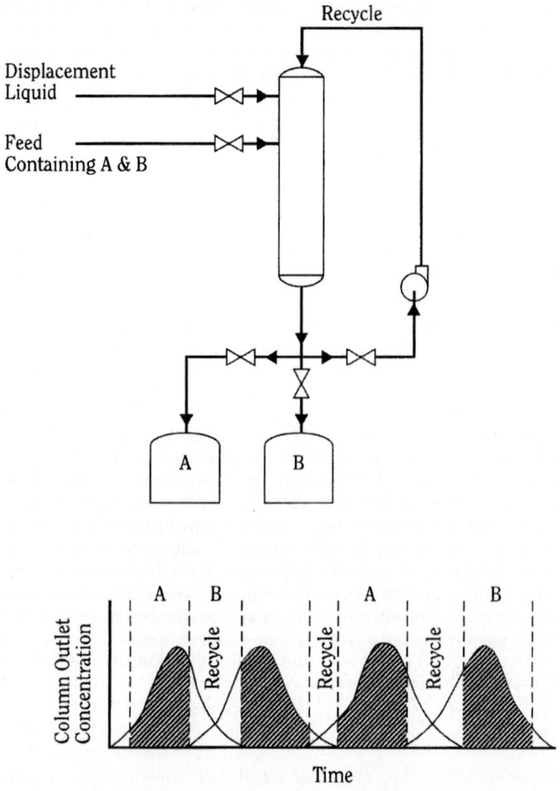

FIG. 16-41 Sanmatsu Kogyo chromatographic process. (*Reprinted with permission of Wiley. Reference: Keller, Anderson, and Yon, Chap. 12 in Rousseau,* Handbook of Separation Process Technology, *John Wiley & Sons, Inc., New York, 1987.*)

Steam Regeneration When steam is used for regeneration of activated carbon, it is desorbing by a combination of thermal swing and displacement purge (described earlier in this section). The exothermic heat released when the steam is adsorbed (condensed in pores) supplies the thermal energy much

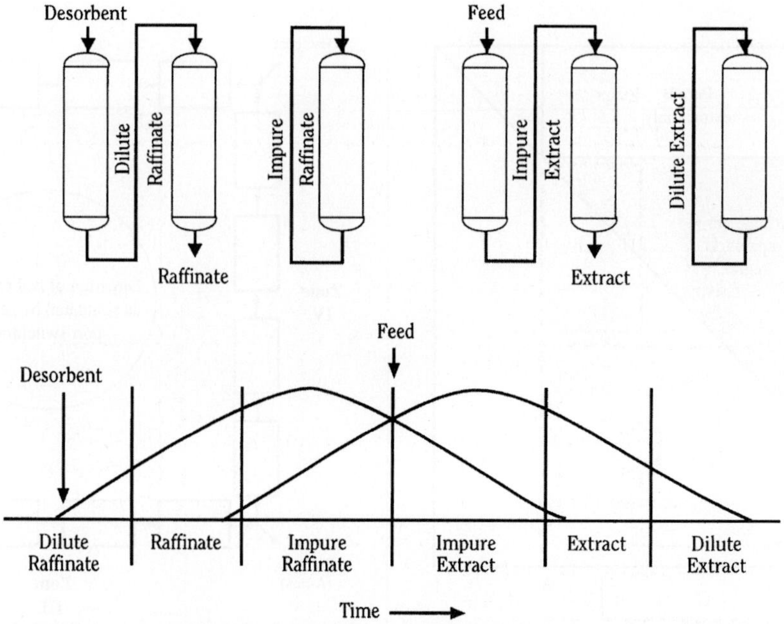

FIG. 16-42 UOP Cyclesorb process. (*Reprinted with permission of John Wiley & Sons, Inc. Reference: Gembicki, Oroskar, and Johnson, "Adsorption, Liquid Separation," in* Kirk-Othmer Encyclopedia of Chemical Technology, *4th ed., Wiley, New York, 1991.*)

more efficiently than is possible with heated purge gas. Slightly superheated steam at about 130°C is introduced into the bed countercurrent to adsorption; for adsorbates with high boiling points, the steam temperature must be higher. Adsorbates are desorbed and purged out of the bed with the steam. Steam and desorbates then go to a condenser for subsequent separation. The water phase can be further cleaned by air stripping, and the sorbate-laden air from that stripper can be recycled with the feed to the adsorption bed.

Steam regeneration is most commonly applied to activated carbon that has been used in the removal and/or recovery of solvents from gases. At volatile organic compound (VOC) concentration levels from 500 to 15,000 ppm, recovery of the VOC from the stream used for regeneration is economically justified. Below about 500 ppm, recovery is not economically justifiable, but environmental concerns often dictate adsorption followed by destruction. While activated carbon is also used to remove similar chemicals from water and wastewater, regeneration by steam is not usual (because the water-treatment carbon contains 1 to 5 kg of water per kg of adsorbent that must be removed by drying before regeneration or an excessive amount of superheated steam will be needed). In water treatment, there can also be significant amounts of nonvolatile compounds that do not desorb during steam regeneration and that residual will reduce the adsorption working capacity. There is a growing use of reticulated styrene-type polymeric resins for VOC removal from air [Beckett et al., *Environ. Technol.* **13:** 1129 (1992); Heinegaard, *Chem.-Ing.-Tech.* **60:** 907 (1988)]. LeVan and Schweiger [in Mersmann and Scholl, eds., *Fundamentals of Adsorption,* United Engineering Trustees, New York, 1991, pp. 487–496] tabulate reported steam utilizations (kg steam/kg adsorbate recovered) for a number of processes.

Energy Applications Desiccant cooling is a means for more efficiently providing air conditioning for enclosures such as supermarkets, ice rinks, hotels, and hospitals. Adsorbers are integrated with evaporative and electric vapor compression cooling equipment into an overall air handling system. Air conditioning is comprised of two cooling loads, latent heat for water removal and sensible heat for temperature reduction. The energy savings derive from shifting the latent heat load from expensive compression cooling (chilling) to cooling tower load. Early desiccant cooling used adsorption wheels (see hereafter) impregnated with the hygroscopic salt, LiCl. More recently, these wheels are being fabricated with zeolite and/or silica gel. They are then incorporated into a system such as the example shown in Fig. 16-43 [Collier et al., in Harrimam, *Desiccant Cooling and Dehumidification,* ASHRAE, Atlanta (1992)]. Process air stream 6, to be conditioned, passes through the adsorbent wheel, where it is dried. This is a nonisothermal process due to the release of the heat of adsorption and transfer of heat from a wheel that may be above ambient temperature. The dry but heated air (7) is cooled in a heat exchanger that can be a thermal wheel. This stream (8) is further cooled, and the humidity adjusted back up to a comfort range by direct contact evaporative cooling to provide supply air. Regeneration air stream 1, which can be ambient air or exhausted air, is evaporatively cooled to provide a heat sink for the hot, dry air. This warmed air (3) is heated to the desired temperature for regeneration of the adsorbent wheel and exhausted to the atmosphere. Many other combinations of drying and cooling are used to accomplish desiccant cooling [Belding, in *Proceedings of AFEAS Refrigeration and Air Conditioning Workshop,* Breckenridge, CO (June 23–25, 1993)].

Adsorption refrigeration technology (ART), often referred to as adsorptive heat pumps, is another developing application of adsorbents with some limited commercial success [Wang et al., *Adsorption Refrigeration Technology: Theory and Application,* Wiley, Singapore (2014)]. The single adsorbate–adsorbent working pair is central to their operation. Common pairs include methanol–activated carbon, ammonia–activated carbon, water–zeolite (e.g., NaX and high-silica NaY), water–silica gel, hydrogen–metal hydrides, ammonia–calcium chloride, and ammonia–strontium chloride. All of these adsorbate–adsorbent working pairs provide a means of transferring heat from a low temperature to a higher, more valuable level. Data have demonstrated that hydrothermally stable Na-mordenite and dealuminated NaY can be used with water in chemical heat pumps to upgrade 100°C heat sources by 50°C to 80°C using a 20°C heat sink [Fujiwara et al., *J. Chem. Eng. Japan* **23:** 738 (1990)]. Other work has shown that integration of two adsorber beds can achieve heating coefficients of performance of 1.56 for the system NaX/water, upgrading 150°C heat to 200°C with a 50°C sink [Douss et al., *Ind. Eng. Chem. Res.* **27:** 310 (1988)].

Gas storage for onboard vehicular fuel is important to providing alternatives to gasoline and diesel fuel. Natural gas is cleaner burning, and hydrogen would burn essentially pollution-free. Onboard storage of natural gas is typically as a high-pressure compressed gas. Adsorbed natural gas systems are a desirable solution because they could operate at lower pressures while maintaining the same capacities. The major problem currently impeding commercialization is the development of adsorbent materials with desirable isotherm capacities and shapes. Also, the exothermic nature of physical adsorption has a negative impact on charge and discharge in a gas storage cycle. Heat released during adsorption will increase the temperature of the adsorbent, thereby lowering the total amount of gas that can be stored. The vessel will cool during the discharge step, decreasing the amount of gas that can be delivered. Technological solutions are being developed and should appear in coming years [Chang and Talu, *Appl. Therm. Eng.* **16:** 359 (1996); Mota, *AIChE J.* **45:** 986 (1999)].

Energy Conservation Techniques The major use of energy in an adsorption cycle is associated with the regeneration step, whether it is thermal energy for TSA or compression energy for PSA. Since the regeneration energy per pound of adsorbent tends to be about constant, the first step in minimizing consumption is to maximize the operating loading. When the mass-transfer zone (MTZ) is a large portion of the adsorber relative to the

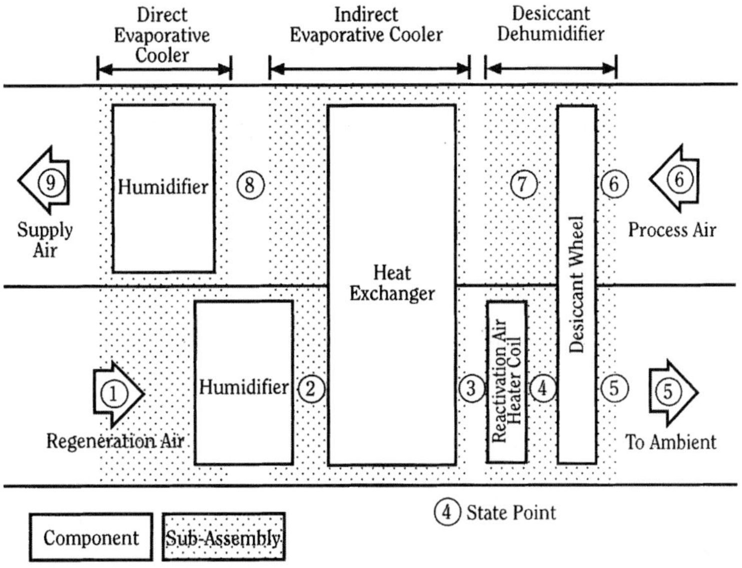

FIG. 16-43 Flow diagram of desiccant cooling cycle. [*Reprinted with permission of American Society of Heating, Refrigeration and Air Conditioning Engineers, Inc. (ASHRAE). Reference: Collier, Cohen, and Slosberg in Harrimam,* Desiccant Cooling and Dehumidification, ASHRAE, *Atlanta, 1992.*]

equilibrium section, the fraction of the bed being fully utilized is small. Most fixed-bed adsorption systems have two adsorbers so that one is on stream while the other is being regenerated. One means of improving adsorbent use is to use a lead/trim (or cascade, or merry-go-round) cycle. Two (or more) adsorbent beds in series treat the feed. The feed enters the lead bed first and then the trim bed. The trim bed is the one that has most recently been regenerated. The MTZ is allowed to proceed through the lead beds but not to break through the trim bed. In this way the lead bed can be almost totally used before being regenerated. When a lead bed is taken out of service, the trim bed is placed in the lead position, and a regenerated bed is placed in the trim position.

A thermal pulse cycle is a means of conserving thermal energy in heating-limited desorption. A process cycle that is heat-limited needs only a very small time (dwell) at temperature to achieve satisfactory desorption. If the entire bed is heated before the cooling is begun, every part of the bed will dwell at temperature for the entire time it takes the cooling front to traverse the bed. Thus, much of the heat in the bed at the start of cooling would be swept from the bed. Instead, cooling is begun before any heat front has exited the bed creating a thermal pulse that moves through the bed. The pulse expends its thermal energy for desorption so that only a small temperature peak remains at the end of regeneration and no excess heat has been wasted. If the heating step is stripping-limited, a thermal pulse is not applicable.

A series cool/heat cycle is another way in which the heat that is purged from the bed during cooling can be conserved. Sometimes the outlet fluid is passed to a heat sink where energy is stored to be reused to preheat heating fluid, or cross exchanged against the purge fluid to recover energy. However, there is also a process cycle that accomplishes the same effect. Three adsorbers are used, with one on adsorption, one on heating, and one on cooling. The regeneration fluid flows in series, first to cool the bed just heated and then to heat the bed to be desorbed. Thus, all of the energy swept from the adsorber during heating can be reused to reduce the heating requirement. Unlike thermal pulse, this cycle is applicable to both heat- and stripping-limited heating.

TABLE 16-15 Process Descriptors

Number	Statement
1	Feed is a vaporized liquid or a gas.
2	Feed is a liquid that can be fully vaporized at less than about 200°C.
3	Feed is a liquid that cannot be fully vaporized at 200°C.
4	Adsorbate concentration in the feed is less than about 3 wt%.
5	Adsorbate concentration in the feed is between about 3 and 10 wt%.
6	Adsorbate concentration in the feed is greater than about 10 wt%.
7	Adsorbate must be recovered in high purity (> than 90–99% rejection of nonadsorbed material).
8	Adsorbate can be desorbed by thermal regeneration.
9	Practical purge or displacement agents cannot be easily separated from the adsorbate.

Keller, Anderson, and Yon in Rousseau (ed.), *Handbook of Separation Process Technology*, John Wiley & Sons, Inc., New York, 1987; reprinted with permission.

Process Selection The preceding sections present many process cycles and their variations. It is important to have some guidelines for design engineers to narrow their choice of cycles to the most economical for a particular separation. Keller and coworkers [see Keller, Anderson, and Yon (1987) in General References] have presented a method for choosing appropriate adsorption processes. Their procedure considers the economics of capital, energy, labor, and other costs. Although these costs can vary from site to site, the procedure is robust enough to include most scenarios. In Table 16-15, nine statements are made about the character of the separation being considered. The numbers of the statements that are true (i.e., applicable) are used in the matrix in Table 16-16. A "no" for any true statement under a given process should remove that process from further consideration. Any process having all "yes" answers for true statements deserves strong consideration. Entries other than "yes" or "no" provide a means of prioritizing processes when more than one cycle is satisfactory.

EQUIPMENT

ADSORPTION

General Design Adsorbents can be used in adsorbers with fixed inventory, with intermittent solids flow, or with continuous-moving solids flow. The most common are fixed beds operating as batch units or as beds of adsorbent through which the feed fluid passes, with periodic interruption for regeneration. Total systems consist of pressure vessels or open tanks along with the associated piping, valves, controls, and auxiliary equipment needed to accomplish regeneration of the adsorbent. Gas treating equipment includes blowers or compressors with a multiplicity of paths to prevent dead-heading. Liquid treating equipment includes pumps with surge vessels as needed to assure continuous flow.

Adsorber Vessel The most frequently used method of fluid–solid contact for adsorption operations is in cylindrical, vertical vessels, with the adsorbent particles in a fixed and closely but randomly packed arrangement. The adsorber must be designed with consideration for pressure drop and must contain a means of supporting the adsorbent and a means of assuring that the incoming fluid is evenly distributed to the face of the bed. There are additional design considerations for adsorbers when the streams are liquid and for high-performance separation applications using very small particles (<0.05 mm) such as in HPLC.

For most large-scale processes, adsorbent particle size varies from 0.06 to 6 mm (0.0025 to 0.25 in), but the adsorbent packed in a fixed bed will have a fairly narrow particle size range. Pressure drop in adsorbers can be changed by changing the diameter to bed depth ratio and by changing the particle size (see Sec. 5). Adsorbent size also determines separation performance of adsorbent columns, increasing efficiency with decreasing particle size. In liquid-phase processing, the total cost of the adsorption step can sometimes be reduced by designing for overall pressure drops as large as 300 to 600 kPa (45 to 90 psi) because pumping is not the major utility cost.

TABLE 16-16 Process Selection Matrix

Statement number, Table 16-15	Gas- or vapor-phase processes					Liquid-phase processes		
	Temperature swing	Inert purge	Displacement purge	Pressure swing	Chromatography	Temperature swing*	Simulated moving bed	Chromatography
1	Yes	Yes	Yes	Yes	Yes	No	No	No
2	Not likely	Yes	Yes	Yes	Yes	Yes	Yes	Yes
3	No	No	No	No	No	Yes	Yes	Yes
4	Yes	Yes	Not likely	Not likely	Not likely	Yes	Not likely	Maybe
5	Yes	Yes	Yes	Yes	Yes	No	Yes	Yes
6	No	Yes	Yes	Yes	Yes	No	Yes	Yes
7	Yes	Yes	Yes	Maybe†	Yes	Yes	Yes	Yes
8	Yes	No	No	No	No	Yes‡	No	No
9	Maybe§	Not likely	Not likely	N/A	Not likely	Maybe§	Not likely	Not likely

*Includes powdered, fixed-bed, and moving-bed processes.
†Very high ratio of feed to desorption pressure (>10:1) will be required. Vacuum desorption will probably be necessary.
‡If adsorbate concentration in the feed is very low, it may be practical to discard the loaded adsorbent or reprocess off-site.
§If it is not necessary to recover the adsorbate, these processes are satisfactory.
Keller, Anderson, and Yon in Rousseau (ed.), *Handbook of Separation Process Technology*, John Wiley & Sons, Inc., New York, 1987; reprinted with permission.

In special, high-resolution applications (HPLC), pressure drops as high as 5000–25,000 kPa (800–4000 psi) are sometimes used, requiring special pumping and column hardware [Colin, in Ganetsos and Barker, *Preparative and Production Scale Chromatography,* Marcel Dekker, New York, 1993, pp. 11–45]. However, the cost of compressing gases is significant. Since blowers are limited to about 5 kPa (20 in wc) of lift, atmospheric gas applications are typically designed with adsorbent pressure drops of 1 to 4 kPa (4 to 16 in wc). To keep the compression ratio low, compressed gas adsorption pressure drops are 5 to 100 kPa (0.7 to 15 psi), depending on the pressure level.

Besides influencing how much pressure drop is allowable, the operating pressure determines other design features. When adsorption and/or regeneration is to be performed at pressures above atmospheric, the adsorber vessels are designed like process pressure vessels (see Fig. 16-44) [EPA (1973) in General References]. Their flow distributors can consume more gas momentum at higher pressure. On the other hand, for applications near atmospheric pressure, any pressure drop can be costly, and most design choices are made in the direction of minimizing head loss. Beds have large face areas and shallow depth. Many times, the choice is to fabricate a horizontal (horizontal axis) vessel where flows are radial rather than axial as in conventional vertical beds. Figure 16-45 [Leatherdale in Cheremisinoff and Ellerbusch (1978) in General References] depicts how a rectangular, shallow adsorber bed is oriented in a horizontal vessel. Flow distributors, especially for large units, are often elaborate in order to evenly divide the flow.

There are two types of support systems used for fixed beds of adsorbent. The first is a progressive series of grid and screen layers. In this system, each higher-layer screen has successively smaller openings, with the last small enough to prevent particles from passing through. Each lower layer has greater strength. A series of I-beams can be used to support a layer of subway grating that, in turn, supports several layers of screening. In other cases, special support grills such as Johnson™ screens may rest on the I-beams or on clips at the vessel wall and thus directly support the adsorbent. The topmost screen must retain the original-size particles and some reasonable size of broken particles. The second type of support is a graded system of particles such as ceramic balls or gravel that rests directly on the bottom of the adsorber. A typical system might consist of 100 mm (4 in) or 50 mm (2 in) diameter material, covered by succeeding layers of 25, 12, and 6 mm (1, ½, and ¼ in) of support material for a 3 mm (⅛ in) adsorbent. This type of multi-segmented flow distributor for PSA systems has been described in detail by Baksh et al., US 2005/0155492 A1 (2005). In water treatment, the support may actually start with filter blocks and have an upper layer of sand (see Fig. 16-46) [EPA (1973)].

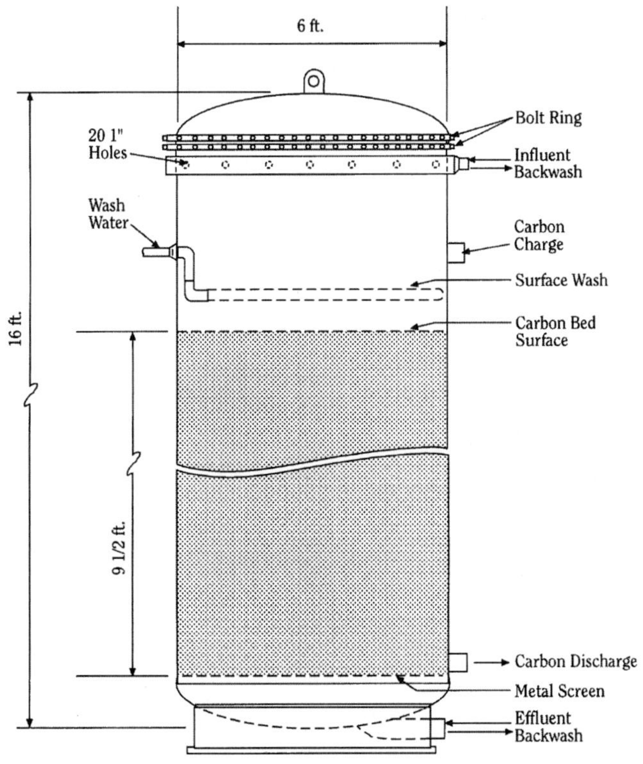

FIG. 16-44 Pressurized adsorber vessel. (*Reprinted with permission of EPA. Reference: EPA,* Process Design Manual for Carbon Adsorption, *U.S. Envir. Protect. Agency, Cincinnati, 1973*)

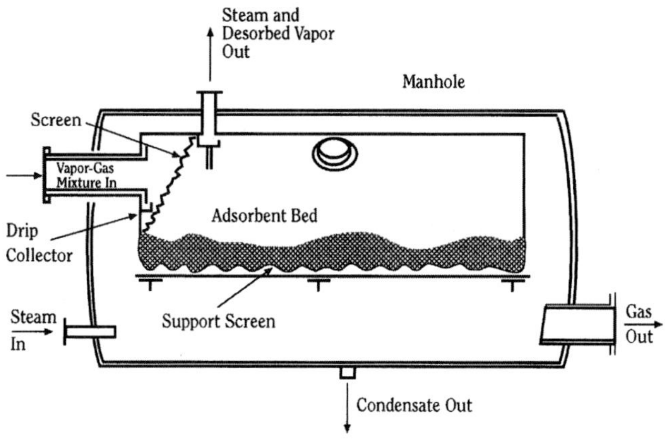

FIG. 16-45 Ambient pressure adsorber vessel. (*Reprinted with permission of Ann Arbor Science. Reference: Leatherdale in Cheremisinoff and Ellerbusch,* Carbon Adsorption Handbook, *Ann Arbor Science, Ann Arbor, 1978.*)

If flow is not evenly distributed throughout the bed of adsorbent, there will be less than maximum utilization of the adsorbent during adsorption and of the desorption fluid during regeneration. Incoming fluids from the nozzles are at a much higher velocity than the average through the bed and may have asymmetric momentum components due to the piping manifold. The simplest means of allowing flow to redistribute across the face of the bed is to employ ample plenum space above and below the fixed bed. In many situations this excess dead volume in the bed may be detrimental to the separation performance [Baksh et al., US 2005/0155492 A1 (2005)]. A much more cost-effective method is to install simple baffle plates with symmetrically placed inlet and outlet nozzles. The solid, or perforated, baffles are designed to break the momentum of the incoming fluid and redistribute it to prevent direct impingement on the adsorbent. When graded bed support is installed at the bottom, the baffles should be covered by screening to restrain the particles. An alternative to screened baffles is slotted metal or Johnson™ screen distributors. Shallow horizontal beds often have such a large flow area that multiple inlet and outlet nozzles are required. These nozzle headers must be carefully designed to assure balanced flow to each nozzle. In liquid systems, a single inlet may enter the vessel and branch into several pipes that are often perforated along their length (Fig. 16-44). Such "spiders" and "Christmas trees" often have holes that are not uniformly spaced and sized but are distributed to provide equal flow per bed area.

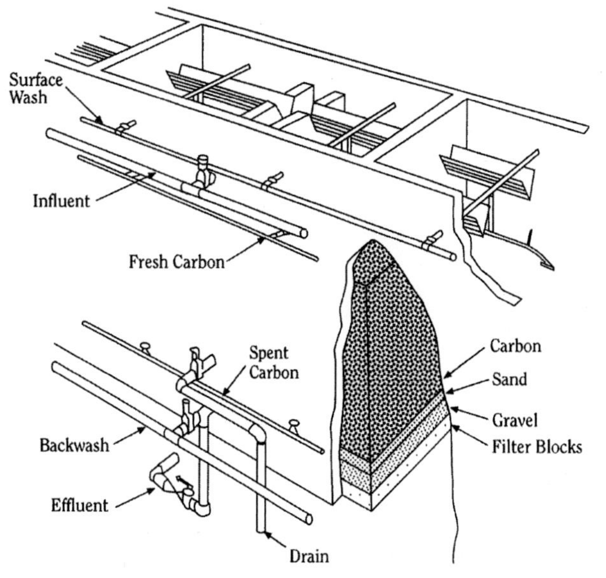

FIG. 16-46 Adsorber vessel with graded support system. (*Reprinted with permission of EPA. Reference: EPA,* Process Design Manual for Carbon Adsorption, *U.S. Envir. Protect. Agency, Cincinnati, 1973.*)

The gas flows through PSA vessels must be given special consideration. The feed flow is usually up through a PSA bed, so the bed support systems just described are all effective and used in practice. However, flow reversal is essential to any PSA process and happens every few seconds to minutes, and the gas necessarily expands significantly as the pressure in a PSA bed decreases during any of the pressure changing steps. As a result, the velocities leaving and/or entering a vessel from the top header are usually much greater than that associated with the feed step, making flow distruibution and bed retention at the top of a PSA vessel critical to prevent the adsorbent particles from moving. In fact, for upflow through a PSA vessel, the velocity should be less than 70 to 80 percent of the particle fluidization velocity, and for downflow it should be less than 180 percent of it [Ruthven (1984)]. This is significant because the ratio of the downflow to upflow velocity in a PSA bed can easily exceed 2.0. So, when baffle plates with Johnson™ screens are used to distribute the flow entering the top of the bed to go radially outward toward the vessel walls before going down into the bed, a bed retention system is usually installed on top of the bed to prevent movement of the adsorbent particles. A typical bed retention system in this case would consist of a four- to six-inch layer of ceramic stone or other dense media. Alternatively, a multisegmented flow distributor system similar to that described by Baksh et al. [US 2005/0155492 A1 (2005)] circumvents the use of baffle plates and screens, thereby minimizing the plenum space and also serving as the bed retention system.

Although allowable pressure loss with liquids is not a restricting factor, there are also special considerations for liquid treating systems. Activated carbon adsorbers used in water and wastewater treatment are designed and constructed using the same considerations used for turbidity removal by sand or multilayer filters. A typical carbon bed is shown in Fig. 16-46. Such contactors for liquids at ambient pressure are often nothing more than open tanks or concrete basins, with flow distribution simply an overflow weir. In liquid treating, the adsorbers must be designed with a means for liquid draining and filling occasionally or during every cycle when a gas is used for regeneration. Draining is by gravity, sometimes assisted by a 70–140 kPa (10–20 psig) pressure pad. Even with time to drain, there will be significant liquid holdup to recover. As much as 40 cc of liquid per 100 g of adsorbent is retained in the micro- and macropores and bridged between particles. When drain is cocurrent to adsorption, the drained liquid can be recovered as treated product. When drain is countercurrent, drained fluid must be returned to feed surge. Minimizing other holdup in dead volume is especially important for liquid separation processes such as chromatography because it adversely affects product recovery and regeneration efficiency. In filling an adsorber, there must be sufficient time for any gas trapped in the pores to escape. The fill step is preferably upflow to sweep the vapor out and to prevent gas pockets that could cause product contamination, bed lifting, or flow maldistribution. In liquid upflow, the buoyancy force of the liquid plus the pressure drop must not exceed the gravitational forces if bed lifting is to be prevented. Because there is very little increase in pressure drop beyond the lifting (or fluidization) velocity, some liquid systems are designed with bed expansion to limit pressure drop. Upflow-adsorption expanded beds are also preferred when the liquid contains suspended solids, so that the bed does not act as a filter and become plugged. Since increased expansion causes the adsorbent to become increasingly well mixed, with accompanying drop in removal efficiency, expansion is usually limited to about 10 percent. Higher velocities also tend to cause too much particle turbulence, abrasion, attrition, and erosion.

Regeneration Equipment Sometimes it is economically justified to remove the adsorbent from the adsorber when it is exhausted and have an outside contractor regenerate it rather than install on-site regeneration equipment. This is feasible only if the adsorbent can treat feed for weeks or months rather than only hours or days. In other cases, the process conditions during regeneration are so much more severe than those for adsorption that a single regenerator with materials of construction capable of handling the conditions is more cost-effective than to make all adsorbers of the expensive material. This is true for most water and wastewater treatment with thermally reactivated carbon. Otherwise, desorption is conducted in situ with any additional equipment connected to the adsorbers.

Figure 16-47 [*Engineering Data Book*, 10th ed., Gas Processors Suppliers Association, Tulsa, 1988, Sec. 20, p. 22] depicts the flow scheme for a typical two-bed TSA dryer system showing the auxiliary equipment associated with regeneration. Some of the dry product gas is externally heated and used countercurrently to heat and desorb water from the adsorber not currently drying feed. The wet, spent regeneration gas is cooled; the water is condensed out; and the gas is recycled to feed for recovery.

The thermal reactivation of spent activated carbon may require the same high temperatures and reaction conditions used for its manufacture. Although the exact conditions to be used for reactivation depend on the type of carbon and the adsorbates to be removed, the objective is to remove the adsorbed material without altering the carbon structure. This occurs in four stages: (*a*) drying, (*b*) desorption, (*c*) pyrolysis and carbonization, and (*d*) burnoff. Each of these steps is associated with a particular temperature range [see LeVan and Carta (2008)].

Cycle Control Valves are the heart of cycle control for cyclic adsorption systems. These on/off valves switch flows among beds so that external to the system it appears as if operation is continuous. In general, one valve is needed for each bed at each end for each step that is performed (e.g., for a two-bed system with an adsorption step plus heating and cooling step [carried out in the same direction and during the same step], only $2 \times 2 \times 2 = 8$ valves would be needed [see Fig. 16-47]). In some cycles such as pressure-swing systems, it may be possible to use valves for more than one function (e.g., repressurization with feed gas using the same manifold as adsorption feed). Without multiple use, the cycle in Fig. 16-48 would need $9 \times 2 \times 5 = 90$ valves instead of 55 to accommodate the five steps of feed (adsorption), cocurrent depressurization, countercurrent depressurization, light reflux, and light product repressurization (even without the equalization steps). For some applications, three- and four-way valves can replace two and four valves, respectively. The ultimate integration of switching valves is the UOP rotary valve discussed below. For long step times (eight hours or more), it is possible for the valves to be manually switched by operators. For most systems, it is advantageous that the opening and closing of the valves be controlled by automatic timers. The same logic controller can be responsible for maintaining flows and pressure, proceeding only on completion of events, and safety bypass or shutdown. Automatic control can provide for

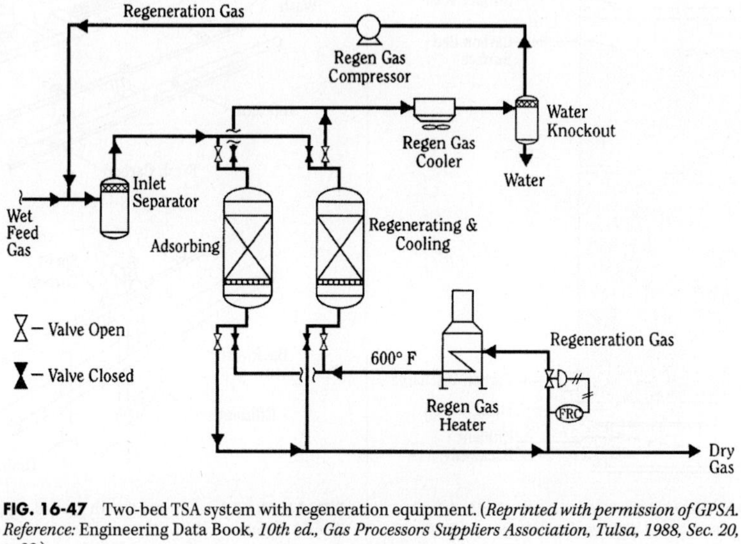

FIG. 16-47 Two-bed TSA system with regeneration equipment. (*Reprinted with permission of GPSA. Reference:* Engineering Data Book, *10th ed., Gas Processors Suppliers Association, Tulsa, 1988, Sec. 20, p. 22.*)

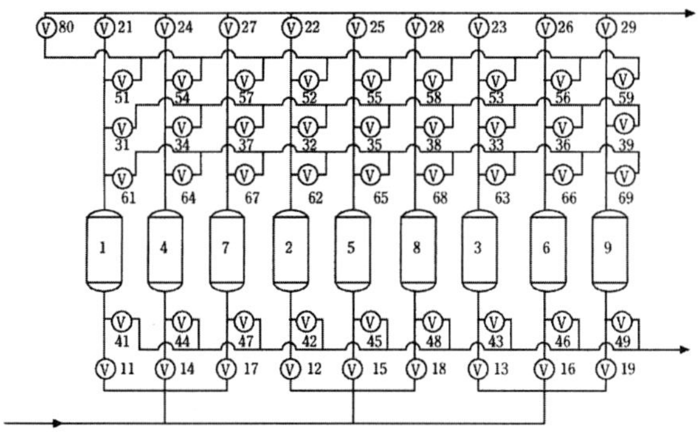

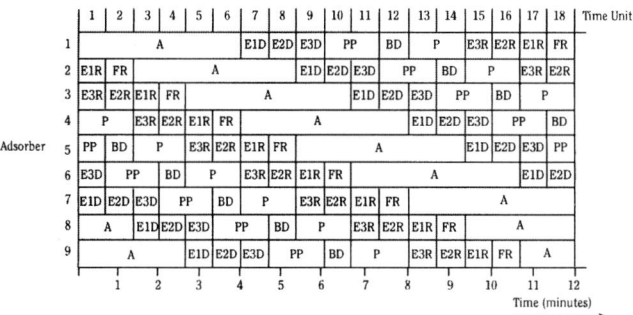

Adsorber	1	2	3	4	5	6	7	8	9	10	11	12	13	14	15	16	17	18
1	A						E1D	E2D	E3D	PP		BD	P		E3R	E2R	E1R	FR
2	E1R	FR	A						E1D	E2D	E3D	PP		BD	P		E3R	E2R
3	E3R	E2R	E1R	FR	A						E1D	E2D	E3D	PP		BD	P	
4	P		E3R	E2R	E1R	FR	A						E1D	E2D	E3D	PP		BD
5	PP	BD	P		E3R	E2R	E1R	FR	A						E1D	E2D	E3D	PP
6	E3D	PP		BD	P		E3R	E2R	E1R	FR	A						E1D	E2D
7	E1D	E2D	E3D	PP		BD	P		E3R	E2R	E1R	FR	A					
8	A		E1D	E2D	E3D	PP		BD	P		E3R	E2R	E1R	FR	A			
9	A				E1D	E2D	E3D	PP		BD	P		E3R	E2R	E1R	FR	A	

Time Unit (columns 1–18); Time (minutes): 1 2 3 4 5 6 7 8 9 10 11 12

FIG. 16-48 Column and valving flow scheme for the UOP nine-bed polybed PSA H_2 unit. (*Reference: Fuderer and Rudelstorfer, U.S. patent number 3,986,849, 1976.*)

a period of parallel flow paths to assure transitions. In some applications, process analyzers can interface with the controller to initiate bed switching when adsorbate is detected breaking through into the effluent.

Continuous Countercurrent Systems Most adsorption systems use fixed-bed adsorbers. However, if the fluid to be separated and that used for desorption can be countercurrently contacted by a moving bed of the adsorbent, there are significant efficiencies to be realized. Because the adsorbent leaves the adsorption section essentially in equilibrium with the feed composition, the inefficiency of the mass-transfer zone is eliminated. The adsorption section only needs to contain an MTZ length of adsorbent compared to an MTZ plus an equilibrium section in a fixed bed. Likewise, countercurrent regeneration is more efficient. Since the adsorbent is moved from an adsorption chamber to another chamber for regeneration, only the regeneration section is designed for the often more severe conditions. Countercurrent adsorption can take advantage of an exceptionally favorable equilibrium in water softening, and the regeneration step can be made favorable by the use of relatively concentrated elutent. Continuous units generally require more headroom but much less footprint. The foremost problems to be overcome in the design and operation of continuous countercurrent sorption operations are the mechanical complexity of equipment and the attrition of the sorbent.

Cross-Flow Systems There are at least three implementations of moving-bed adsorption that are cross-flow rather than fixed beds or countercurrent flow: (1) panel beds, (2) adsorbent wheels, and (3) rotating annular beds. *Cross-flow* means that the adsorbent is moving in a direction perpendicular to the fluid flow. All of these employ moving adsorbent—the first, a downflowing solid; and the others, a constrained solid. Panel beds of activated carbon have been applied to odor control [Lovett and Cunniff, *Chem. Eng. Progr.* **70**(5): 43 (1974)] and to the desulfurization of waste gas [Richter et al., *Verfahrenstechnik (Mainz)* **14:** 338 (1980)]. The spent solid falls from the bottom panel into a load-out bin, and fresh regenerated carbon is added to the top; gas flows across the panel.

The heart of an adsorbent wheel system is a rotating cylinder containing the adsorbent. Figure 16-49 illustrates two types: horizontal and vertical. In some adsorbent wheels, the adsorbent particles are placed in basket segments (a multitude of fixed beds) to form a horizontal wheel that rotates around a vertical axis. In other instances, the adsorbent is integral to the monolithic wheel or coated onto a metal, paper, or ceramic honeycomb substrate. These monolithic or honeycomb structures rotate around either a vertical or a horizontal axis. The gas to be treated usually flows through the wheel parallel to the axis of rotation, although some implementations use radial flow configurations. Most of the wheel is removing adsorbates. The remaining (smaller) portion of the wheel is undergoing thermal regeneration—usually countercurrently. The wheel constantly rotates to provide a continuous treated stream and a steady concentrated stream. Adsorbent wheels are most often used to treat ambient air because they have very low pressure drop. One application of wheels is the removal of VOC where the regeneration stream is usually sent to an incinerator for destruction of VOC. Another use is in desiccant cooling (see previously). They do suffer from low efficiency due to the short contact time, mechanical leakage at seals, and the tendency to allow the wheel to exceed breakthrough in order to get better adsorbent utilization. Some adsorbent wheels are operated in an intermittent manner such that the wheel periodically indexes to a new position; this is particularly true of radial flow wheels.

The rotating annular bed system for liquid chromatographic separation (two-dimensional chromatography) is analogous to the horizontal adsorbent wheel for gases [see LeVan and Carta (2008)]. Feed to be separated flows to a portion of the top face of an annular bed of sorbent. A displacement purge in the form of a solvent or carrier gas flows to the rest of the annulus. The less strongly adsorbed components travel downward through the sorbent at a higher rate. Thus, they will exit at the bottom of the annulus at a smaller angular distance from the feed sector. The more strongly adsorbed species will exit at a greater angle. Several potential applications are reviewed by Carta and Byers (*Chromatographic and Membrane Processes in Biotechnology,* NATO ASI Proceeding, Kluwer, 1991).

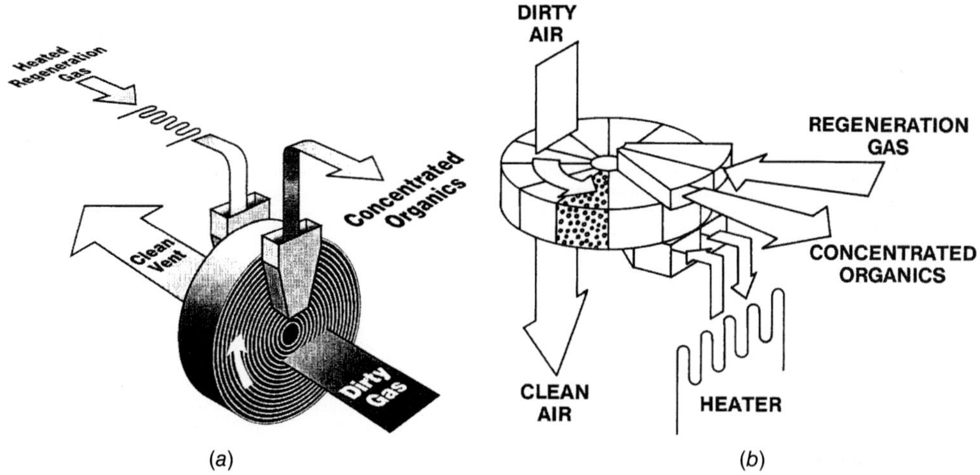

FIG. 16-49 Adsorbent wheels for gas separation: (*a*) horizontal with fixed beds; (*b*) vertical monolith. (*Reprinted with permission of UOP.*)

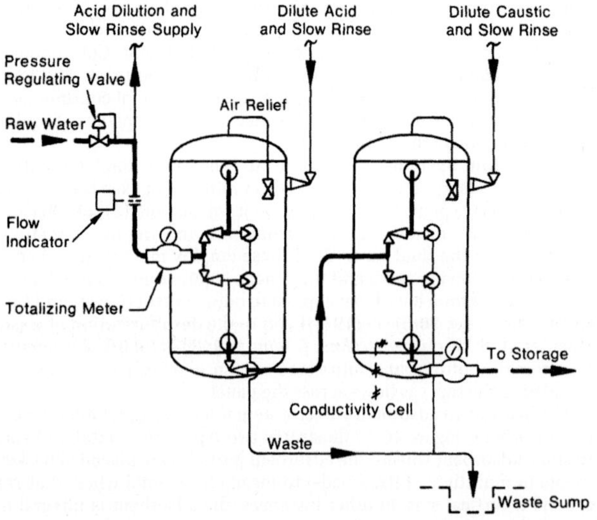

FIG. 16-50 Typical two-bed deionizing system. (*Infilco Degremont Inc.*)

ION EXCHANGE

A typical fixed-bed ion exchanger consists of a vertical cylindrical vessel of lined steel or stainless steel. Linings are usually of natural or synthetic rubber. Spargers are provided at the top and bottom, and often a separate distributor is used for the regenerant solution. The resin bed, consisting of a meter or more of ion-exchange particles, is supported by the screen of the bottom distributor. Externally, the unit is provided with a valve manifold to permit downflow loading, upflow backwash, regeneration, and rinsing. For deionization, a two-exchanger assembly comprising a cation and an anion exchanger is a common configuration (Fig. 16-50).

Column hardware designed to allow countercurrent, upflow regeneration of ion-exchange resins is available. An example is given in Fig. 16-51. During upflow of the regenerant, bed expansion is prevented by withdrawing the effluent by applying vacuum. A layer of drained particles is formed at the top of the bed, while the rest of the column functions in the usual way.

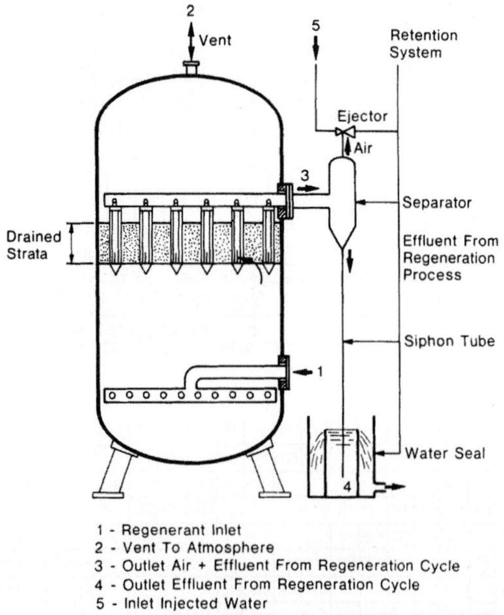

FIG. 16-51 Internals of an upflow regenerated unit. (*Infilco Degremont Inc.*)

Typical design data for fixed-bed ion-exchange columns are given in Table 16-17. These should be used for preliminary evaluation purposes only. Characteristic design calculations are presented and illustrated by Applebaum (*Demineralization by Ion Exchange*, Academic, New York, 1968). Large amounts of data are available in published literature and in bulletins of manufacturers of ion exchangers. In general, however, laboratory testing and pilot-plant work are advisable to determine usable exchanger capacities, regenerant quantities, exchanger life, and quality of product for any application not already thoroughly proven in other plants for similar conditions. Firms that manufacture ion exchangers and ion-exchange equipment will often cooperate in such tests.

TABLE 16-19 Design Data for Fixed-Bed Ion Exchanger*

Type of resin	Maximum and minimum flow, m/h [gal/(min·ft²)]	Minimum bed depth, m (in)	Maximum operating temperatures, °C (°F)	Usable capacity, g-equivalent/L†	Regenerant, g/L resin‡
Weak acid cation	20 max. (8) 3 min. (1)	0.6 (24)	120 (248)	0.5–2.0	110% theoretical (HCl or H_2SO_4)
Strong acid cation	30 max. (12) 3 min. (1)	0.6 (24)	120 (248)	0.8–1.5 0.5–1.0 0.7–1.4	80–250 NaCl 35–200 66° Bé. H_2SO_4 80–500 20° Bé. HCl
Weak and intermediate base anions	17 max. (7) 3 min. (1)	0.75 (30)	40 (104)	0.8–1.4	35–70 NaOH
Strong-base anions	17 max. (7) 3 min. (1)	0.75 (30)	50 (122)	0.35–0.7	70–140 NaOH
Mixed cation and strong-base anion (chemical-equivalent mixture)	40 max. (16)	1.2 (47)	50 (122)	0.2–0.35 (based on mixture)	Same as cation and anion individually

*These figures represent the usual ranges of design for water-treatment applications. For chemical-process applications, allowable flow rates are generally somewhat lower than the maximums shown, and bed depths are usually somewhat greater.

†To convert to capacity in terms of kilograins of $CaCO_3$ per cubic foot of resin, multiply by 21.8.

‡To convert to pounds of regenerant per cubic foot of resin, multiply by 0.0625.

Section 17

Gas–Solid Operations and Equipment

Ted M. Knowlton, Ph.D. *Technical Consultant and Fellow, Particulate Solid Research, Inc.; Member, American Institute of Chemical Engineers (Section Editor, Fluidized Bed Systems, Cyclone Separators)*

Shrikant Dhodapkar, Ph.D. *Fellow, The Dow Chemical Company; Fellow, American Institute of Chemical Engineers (Gas–Solids Separations)*

FLUIDIZED-BED SYSTEMS

GAS–SOLIDS SEPARATIONS

FLUIDIZED-BED SYSTEMS

Consider a bed of particles in a column that is supported by a distributor plate with small holes in it. If gas is passed through the plate so that the gas is evenly distributed across the column, the drag force on the particles produced by the gas flowing through the particles increases as the gas flow through the bed is increased. When the gas flow through the bed causes the drag forces on the particles to equal the weight of the particles in the bed, the particles are fully supported, and the bed is said to be fluidized. Further increases in gas flow through the bed cause bubbles to form in the bed, much as in a fluid, and early researchers noted that this resembled a fluid and called this a fluidized state.

When fluidized, the particles are suspended in the gas, and the fluidized mass (called a fluidized *bed*) has many properties of a liquid. Like a liquid, the fluidized particles seek their own level and assume the shape of the containing vessel. Large, heavy objects sink when added to the bed, and light particles float. If the bed is tilted, the surface of the bed will remain parallel to the surface of the earth.

Fluidized beds are used successfully in many processes, both catalytic and noncatalytic. Among the catalytic processes are fluid catalytic cracking and reforming, oxidation of naphthalene to phthalic anhydride, the production of polyethylene, and ammoxidation of propylene to acrylonitrile. Some of the noncatalytic uses of fluidized beds are in the roasting of sulfide ores, coking of petroleum residues, calcination of ores, combustion of coal, incineration of sewage sludge, and drying and classification.

Although it is possible to fluidize particles as small as about 1 μm and as large as 4 cm, the range of the average size of solid particles which are more commonly fluidized is about 30 μm to over 2 cm. Particle size affects the operation of a fluidized bed more than particle density or particle shape. Particles with an average particle size of about 40 to 125 μm fluidize smoothly because bubble sizes are relatively small in this size range. Larger particles (125 μm and larger) produce larger bubbles when fluidized. The larger bubbles result in a less homogeneous fluidized bed, which can manifest itself in large pressure fluctuations. If the bubble size in a bed approaches approximately one-half to two-thirds the diameter of the bed, the bed will slug. A slugging bed is characterized by large pressure fluctuations that can result in instability and severe vibrations in the system. Small particles (smaller than about 30 μm in diameter) have large interparticle forces (generally van der Waals, electrostatic, or capillary forces) that cause the particles to stick together, as flour particles do. These types of solids fluidize poorly because of the agglomerations caused by the cohesive forces. At velocities that would normally fluidize larger particles, channels, or spouts (commonly called "ratholes"), form in the bed of these small particles, resulting in severe gas bypassing. To fluidize these small particles, it is generally necessary to operate at very high gas velocities so that the shear forces are larger than the cohesive forces of the particles. Using a high-shear gas distributor that contains many high-velocity horizontal jets can also prevent cohesive agglomeration from occurring for some materials. Adding finer-sized particles to a coarse bed, or coarser-sized particles to a bed of cohesive material (i.e., increasing the particle size range of a material), usually results in better (smoother) fluidization.

Gas velocities in fluidized beds generally range from 0.1 to 3 m/s (0.33 to 9.9 ft/s). The gas velocities referred to in fluidized beds are superficial gas velocities—the volumetric flow through the bed divided by the bed area. More detailed discussions of fluidized beds can be found in Kunii and Levenspiel, *Fluidization Engineering,* 2d ed., Butterworth Heinemann, Boston, 1991; Pell, *Gas Fluidization,* Elsevier, New York, 1990; Geldart, ed., *Gas Fluidization Technology,* Wiley, New York, 1986; Yang, ed., *Handbook of Fluidization and Fluid Particle Systems,* Marcel Dekker, New York, 2003; and papers published in periodicals, transcripts of symposia, and the American Institute of Chemical Engineers symposium series.

GAS–SOLID SYSTEMS

Researchers in the fluidization field have long recognized that particles of different size behave differently in fluidized beds, and several have tried to define these differences. Some of these characterizations are described below.

Types of Solids Perhaps the most widely used categorization of particles is that of Geldart [*Powder Technol.* **7:** 285–292 (1973)]. Geldart categorized solids into four different groups (Geldart Groups A, B, C, and D) that exhibited different properties when fluidized with a gas. He classified the four groups in his famous plot, shown in Fig. 17-1. This plot defines the four groups as a function of average particle size d_{sv}, μm, and density difference $\rho_p - \rho_g$, kg/m³, where ρ_p = particle density, kg/m³; ρ_g = gas density, lb/ft³; and d_{sv} = surface volume diameter (also called the Sauter mean diameter) of the particles, μm. Generally, d_{sv} is the preferred average particle size for fluid-bed applications because it is based on the surface area of the particle. The drag force used to generate the pressure drop used to fluidize the bed is proportional to the surface area of the particles. Another widely used average particle size is the median particle size, $d_{p,50}$.

When the gas velocity through a bed of Geldart Group A, B, C, or D particles increases, the pressure drop through the bed also increases. The pressure drop increases until it equals the weight of the bed (W) divided by the cross-sectional area of the column (A). The gas velocity at which this occurs is called the minimum fluidizing velocity, U_{mf}. After the minimum fluidization velocity is achieved, increases in gas velocity for a bed of Geldart Group A particles (generally in the particle size range between 30 and 120 μm) will result in a uniform expansion of the particles without bubbling until at some higher gas velocity (called the minimum bubbling velocity, U_{mb}) gas bubbles start to form. For Geldart Group B particles (between 120 and about 1000 μm) and Geldart Group D particles (about 1000 μm and larger), bubbles start to form immediately after U_{mf} is achieved, so that U_{mf} and U_{mb} are essentially equal for these two Geldart groups. Geldart Group C particles (generally smaller than 30 μm) are termed cohesive particles and clump together in particle agglomerates because of interparticle forces (generally van der Waals, electrostatic, or capillary forces). When gas is passed through beds of cohesive solids, the gas tends to channel or "rathole" through the bed. Instead of fluidizing the particles, the gas opens channels that extend from the gas distributor to the surface of the bed. At higher gas velocities where the shear forces are great enough to overcome the interparticle forces, or with mechanical agitation or vibration, cohesive particles will fluidize, but generally with larger clumps or clusters of particles formed in the bed.

Two-Phase Theory of Fluidization The two-phase theory of fluidization assumes that all gas in excess of the minimum bubbling velocity passes through the bed as bubbles [Toomey and Johnstone, *Chem. Eng. Prog.* **48:** 220 (1952)]. In this view of the fluidized bed, the gas flowing through the emulsion phase in the bed is at the minimum bubbling velocity, while the gas flow above U_{mb} is in the bubble phase. This view of the bed is an approximation, but it is a helpful way of understanding what happens as the gas velocity is increased through a fluidized bed. As the gas velocity is increased above U_{mb}, more and larger bubbles are formed in the bed. As more bubbles are produced in the bed, the bed expands, and the bed density decreases.

For Geldart groups (A, B, C, and D) that are satisfactorily fluidized, as the gas velocity is increased, the fluidized-bed density is decreased and the turbulence, or agitation, of the bed is increased. In smaller-diameter beds, but especially with Geldart Group B and D powders, slugging will occur as the bubbles increase in size to a diameter greater than one-half to two-thirds of the bed diameter. Slugging does not occur in large, commercial beds because the bubbles will not grow to a size that is one-half to two-thirds the diameter of the commercial unit. Bubbles grow by vertical and lateral merging (called coalescence), and they increase in size as the gas velocity is increased (Whitehead, in *Fluidization,* ed. Davidson and Harrison, Academic, London and New York, 1971). As the gas velocity is increased further, the stable bubbles break down into smaller, unstable voids, which only exist for a few seconds before breaking up and reforming.

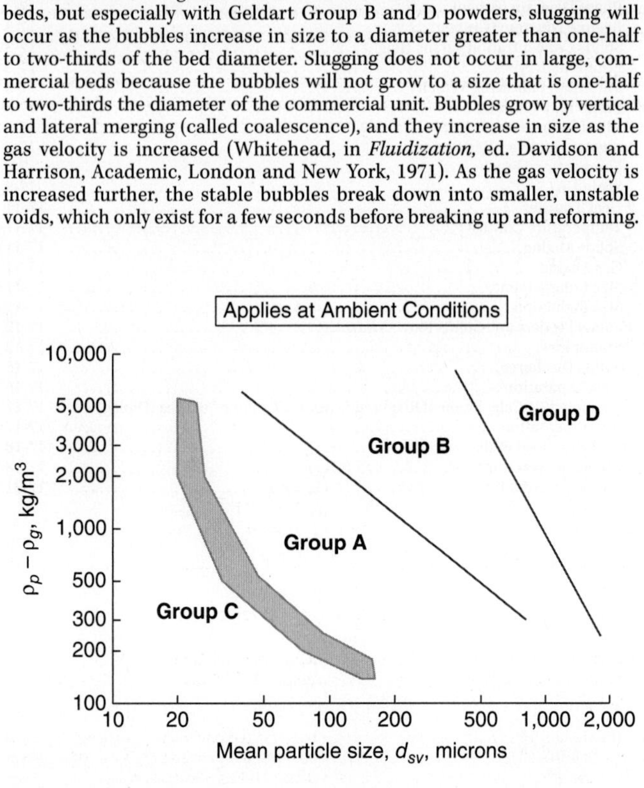

FIG. 17-1 Power-classification diagram for fluidization by air (ambient conditions). (*Courtesy of PSRI, Chicago Ill.*)

When the unstable voids characterize the gas phase in fluidized beds, the bed is not in the bubbling regime anymore, but is said to be in the turbulent regime. The turbulent regime is a much more vigorous regime than the bubbling regime. Because of the high rates of void breakup and recoalescence, the turbulent regime is characterized by higher heat- and mass-transfer rates than bubbling fluidized beds, and the pressure fluctuations in the bed are reduced relative to bubbling beds as the voids are smaller. As the gas velocity is increased above the turbulent fluidized regime, the turbulent bed gradually changes into the pneumatic conveying regime.

Phase Diagram (Upward Gas Flow) A phase diagram for upward gas flow in gas/particle systems is shown in Fig. 17-2. The diagram indicates how the pressure drop per unit length $(P_2 - P_1)$ across a length (L) of vertical line varies as a function of gas velocity with solids mass flux (G) as a parameter. Line OD in Fig. 17-2 is the pressure drop versus gas velocity curve for a packed bed, and line DE is the curve for a fluidized bed with no net solids flow through it. Line OK is the pressure drop versus velocity curve for a vertical line with no solids flow through it. Line ABC is the curve for the dilute-phase, pneumatic conveying region at a constant mass flux of G_1. Zenz (Zenz and Othmer, *Fluidization and Fluid Particle Systems*, Reinhold, New York, 1960) indicated that there was an instability between points C and E because with no solids flow, all the particles will be entrained from the bed. However, if solids are added to replace those entrained, curve ABC prevails.

Phase Diagram (Grace) Grace [*Can. J. Chem. Eng.* **64:** 353–363 (1986)] has correlated the various types of gas–solid systems in which the gas is flowing vertically upward in a status graph using the parameters of the Archimedes number, Ar, for the particle size and a nondimensional velocity, U^*, for the gas effects. This plot is shown in Fig. 17-3. By means of this plot, the fluidization regime for various operating systems can be roughly approximated. This plot is a good guide to estimate the fluidization regime for various particle sizes and operating conditions. However, it should not be substituted for more exact methods of determining the actual fluidization operating regime.

Regime Diagram (Grace) Grace [*Can. J. Chem. Eng.* **64:** 353–363 (1986)] approximated the appearance of the different regimes of fluidization as shown in the schematic drawing of Fig. 17-4. This drawing shows the fluidization regimes that occur as superficial gas velocity is increased from the low-velocity, packed-bed regime to the high-velocity, pneumatic conveying transport regime. As the gas velocity is increased, the bed transitions from the packed-bed regime (also called the fixed bed regime) into the fluidization regime. At the fluidization regime, the velocity reaches a value U_{mf} (the minimum fluidization velocity), where the drag forces on the particles equal the weight of the bed particles, and the particles become suspended in the gas stream. At this point, the particles are said to be fluidized. If the particles are Geldart Group A particles, then a "bubbleless" particulate fluidization regime is first formed called the particulate regime. As the gas

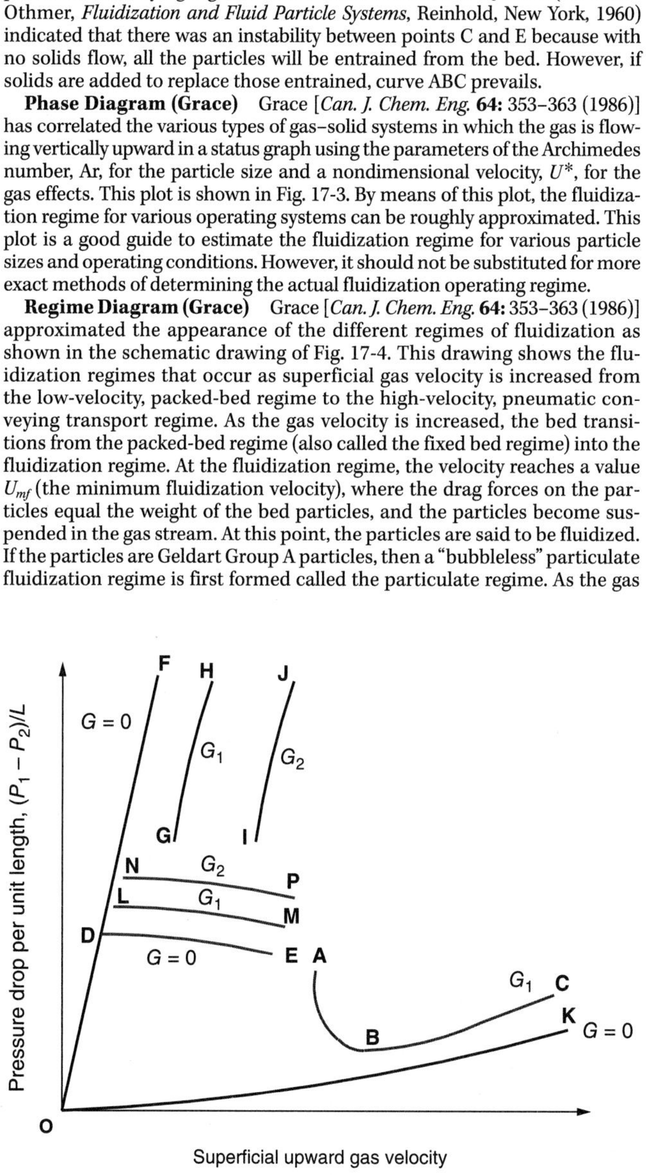

FIG. 17-2 Schematic phase diagram in the region of upward gas–solid flow. G = solids mass flux, lb/(s-ft²); L = length, ft; P_1 and P_2 = pressure, lb/ft². (*Courtesy of PSRI, Chicago, Ill.*)

Key:

OD = packed bed	DF = constrained packed bed (no solids flow)
DE = fluidized bed	GH, IJ = constrained packed bed (with solids flow)
OK = gas flow only	LM, NP = dense, fluidized solids flow
ABC = dilute-phase pneumatic conveying	

velocity is further increased, bubbles start to form in the bed. The velocity at which bubbles start to form is called the minimum bubbling velocity, U_{mb}. For Geldart Group B and D particles, the particulate fluidization regime does not form. Instead, the bed passes directly from the packed-bed regime to the bubbling fluidized-bed regime. As the gas velocity is increased above U_{mb}, the bubbles in the bed grow in size. In small laboratory beds, if the bubble size grows to a value equal to approximately one-half to two-thirds the diameter of the fluidization column, the bed will slug. The slugging fluidized bed is characterized by severe pressure fluctuations and limited solids mixing. It only occurs with small-diameter fluidization columns. Commercial fluidized beds are too large for bubbles to grow to the size where slugging will occur.

At high gas velocities in the bed, the stable bubbles break down into unstable voids that continuously disintegrate and reform. This type of bed is said to be operating in the turbulent fluidized-bed regime, and it is characterized by higher heat- and mass-transfer rates than in the bubbling bed. As the gas velocity is increased further, the bed transitions from the turbulent bed into the dilute-phase pneumatic transport regime. The dilute-phase, pneumatic transport regime is composed of two basic regions: the lower-velocity, fast fluidized-bed regime, and the higher-velocity pneumatic conveying regime. The pneumatic transport regime is a very important regime and is defined by the curve ABC for the constant solids mass flux, G_1, in Fig. 17-2. A more detailed drawing of the pneumatic transport regime is shown in Fig. 17-5. In this figure, it can be seen that as the gas velocity is decreased from point C, the pressure drop per unit length begins to decrease. This occurs because the total pressure drop in the transport regime is composed of two types of terms—a term composed of frictional pressure drops (gas/wall friction, solid/wall friction, and gas/solids friction) and a term required to support the solids in the vertical line (the static head of solids term). At high gas velocities the frictional terms dominate; and as the gas velocity is decreased from point C, the frictional terms begin to decrease in magnitude. As this occurs, the concentration of solids in the line starts to increase. At some gas velocity, the static head of solids term and the frictional pressure drop term are equal (the minimum point on the curve). As the gas velocity is decreased below the minimum point (point B), the static head of solids term begins to dominate as the concentration of solids in the line increases. The static head of solids pressure drop increases until it is no longer possible for the gas to fully support the solids in the line. The gas velocity at which the solids cannot be supported at solids flow rate G_1 for curve ABC is known as the choking velocity (U_{ch}) for solids mass flux, G_1. Curve RST is a pneumatic transport curve operating at a higher solids mass flux, G_2. It has a higher choking velocity than curve ABC. Because many Geldart Group A beds in the turbulent and all Geldart groups in the transport regime operate above the terminal velocity of some or all of the particles, a solids collection and return system is needed to maintain a stable fluidized bed with these regimes.

Solids Concentration versus Height As previously described, there are several fluidization regimes. In order of increasing gas velocity, these regimes are: the particulate fluidization regime (for Geldart Group A *only*), the bubbling (aggregative) fluidization regime, the turbulent fluidization regime, the fast-fluidization regime, and the pneumatic transport regime. Each of these regimes has a characteristic solids concentration profile as shown in Fig. 17-6. Only the bubbling fluidization regime can be said to have a distinct top to its bed. The other regimes have essentially continuously changing solids concentrations as a function of height.

Equipment Types Fluidized-bed systems take many forms. Figure 17-7 shows some of the more common concepts and configurations with approximate ranges of gas velocities indicated in the legend.

Minimum Fluidizing Velocity U_{mf}, the minimum fluidizing velocity, is often used in fluid-bed calculations. This parameter is often measured experimentally in small-scale equipment at ambient conditions by measuring the pressure drop across the bed as a function of superficial gas velocity through the bed. The most accurate measurements of U_{mf} are determined from a decreasing-velocity curve where the bed is first fluidized and then the gas velocity is decreased systematically. This type of curve avoids the problem of different packed-bed compactions experienced in an increasing-velocity curve. The correlation to predict U_{mf} by Wen and Yu [*A.I.Ch.E.J.* **12:** 610–612 (1966)], shown below, can then be used to back-calculate an effective particle size, d_{peff}. This gives a particle size that takes into account the effects of size distribution and particle shape, or sphericity. The correlation can then be used to estimate U_{mf} at process conditions using this effective particle size. If U_{mf} cannot be determined experimentally, then it can be calculated using the following Wen and Yu correlation:

$$\text{Re}_{mf} = (1135.7 + 0.0408\text{Ar})^{0.5} - 33.7 \quad (17\text{-}1)$$

where $\text{Re}_{mf} = d_{\text{peff}}\rho_g U_{mf}/\mu$ (17-2)

$$\text{Ar} = d_{sv}\rho_g(\rho_p - \rho_g)g/\mu^2 \quad (17\text{-}3)$$

$$d_{sv} = 1/\Sigma(x_i/d_{pi}) \quad (17\text{-}4)$$

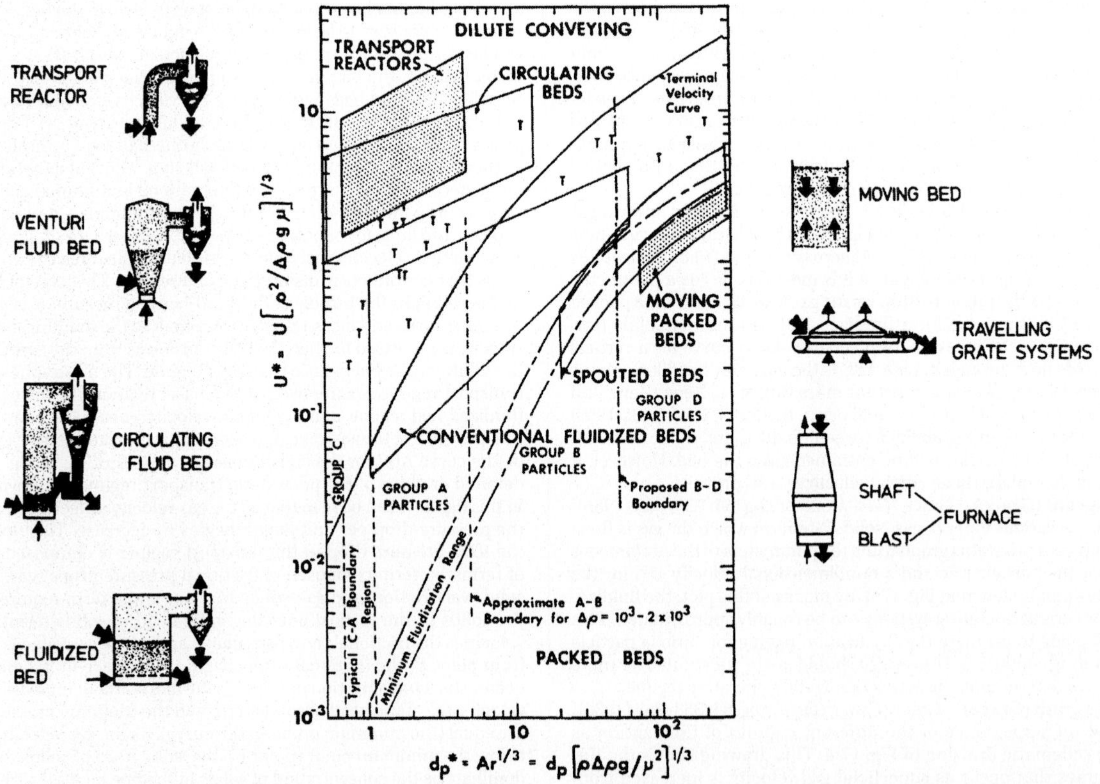

FIG. 17-3 Simplified fluidized-bed regime plot. [From *Grace*, Can J. Chem. Eng. **64:** 353–363 (1986); *sketches from Reh*, Ger. Chem. Eng. **1:** 319–329 (1978).]

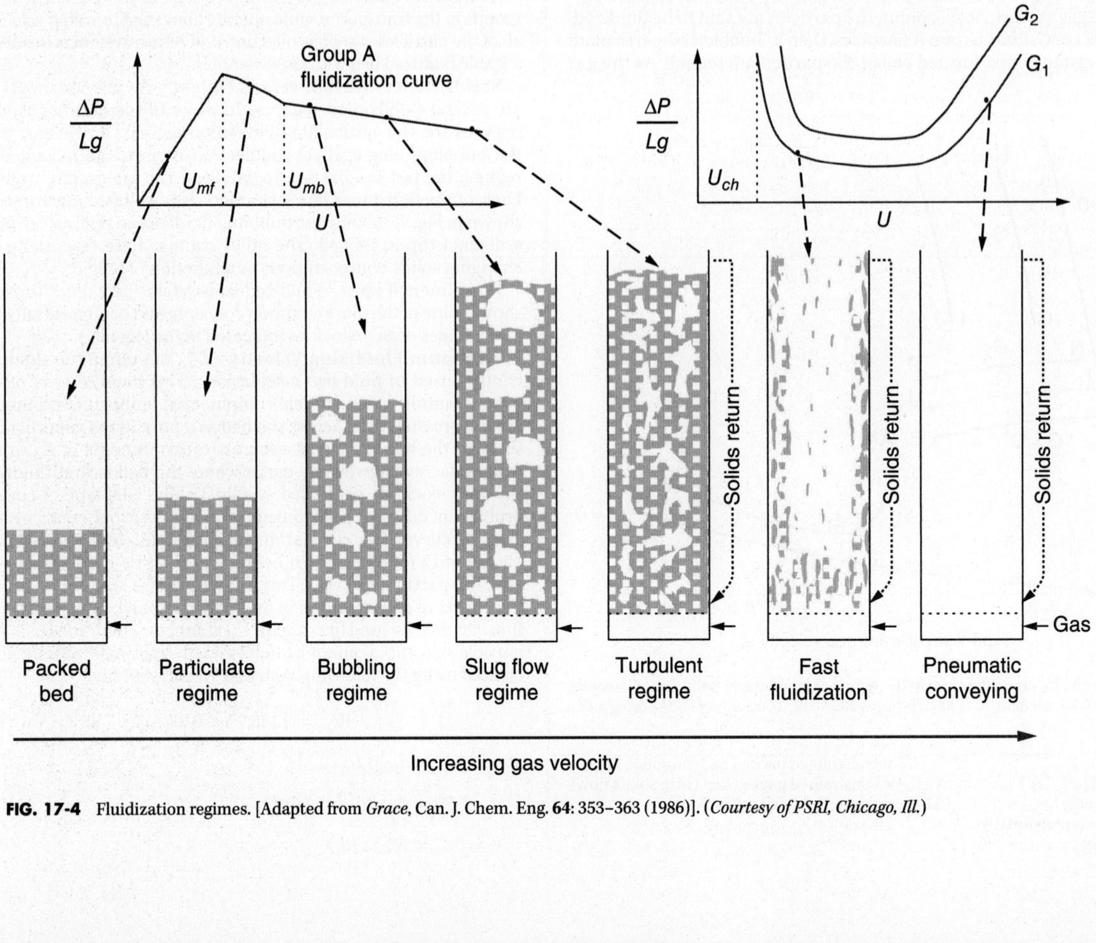

FIG. 17-4 Fluidization regimes. [Adapted from *Grace*, Can. J. Chem. Eng. **64:** 353–363 (1986)]. (*Courtesy of PSRI, Chicago, Ill.*)

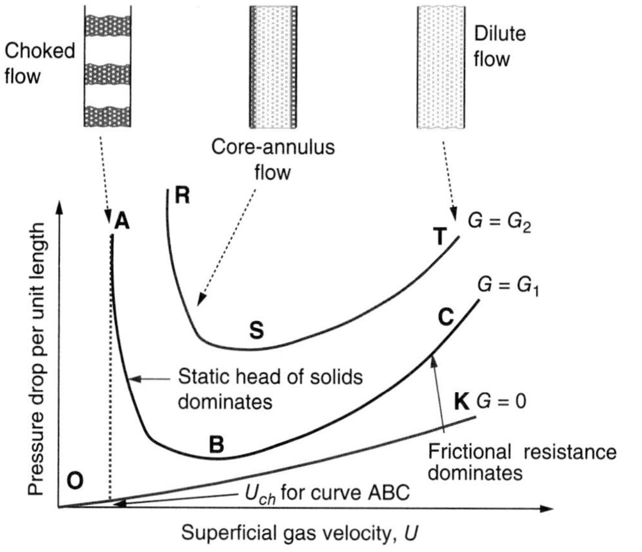

FIG. 17-5 Pneumatic (dilute-phase) transport regime. (*Courtesy of PSRI, Chicago, Ill.*)

For wide particle-size distributions of group B and D materials, U_{mf} does not apply. With these materials, the largest materials in the distribution may not be fluidized while most of the bed is fluidized. For materials such as this, the minimum velocity to completely fluidize the entire particle-size distribution is called the complete fluidization velocity, U_{cf}. This velocity can be estimated by applying the Wen and Yu correlation for U_{mf} to the largest particle in the mixture, not the average particle size.

The gas velocity required to maintain a completely homogeneous bed of solids in which coarse or heavy particles will not segregate from the fluidized portion is very different from the minimum fluidizing velocity. The bed may be completely fluidized, but segregation can still occur. See Nienow and Chiba, *Fluidization,* 2d ed., Wiley, 1985, pp. 357–382, for a discussion of segregation or mixing mechanisms as well as the means of predicting this flow; also see Baeyens and Geldart, *Gas Fluidization Technology,* Wiley, 1986, 97–122.

Particulate Fluidization Fluid beds of Geldart Group A powders that are operated at gas velocities above the minimum fluidizing velocity (U_{mf}) but below the minimum bubbling velocity (U_{mb}) are said to be particulately fluidized. As the gas velocity is increased above U_{mf}, the bed further expands. Decreasing ($\rho_p - \rho_g$), d_p and/or increasing μ_f increases the spread between U_{mf} and U_{mb}. Richardson and Zaki [*Trans. Inst. Chem. Eng.* **32:** 35 (1954)] showed that $U/U_i = \varepsilon^n$, where n is a function of system properties, ε = void fraction, U = superficial fluid velocity, and U_i = theoretical superficial velocity from the Richardson and Zaki plot when $\varepsilon = 1$.

Vibrofluidization It is possible to fluidize a bed mechanically by imposing vibration to throw the particles upward using a cyclical force. This enables the bed to operate with either no upward gas velocity or a vastly reduced gas flow. Entrainment can also be greatly reduced using this technique compared to unaided fluidization. This technique is used commercially in drying and other applications [Mujumdar and Erdesz, *Drying Tech.* **6:** 255–274 (1988)], and chemical reaction applications are also possible. See Sec. 12 for more on drying applications of vibrofluidization.

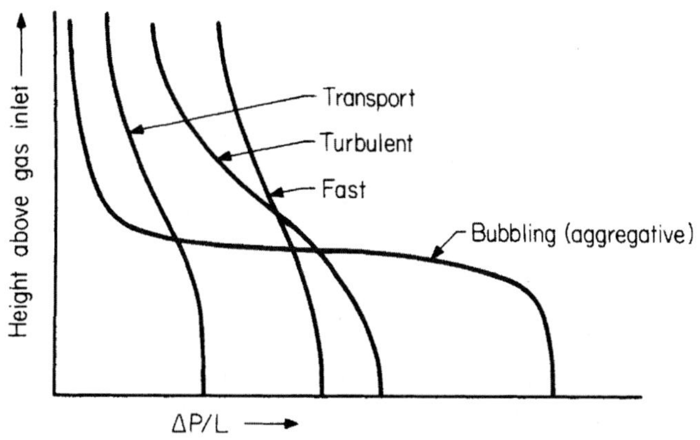

FIG. 17-6 $\Delta P/L$ (proportional to solids concentration) versus height above the gas distributor for regimes of fluidization.

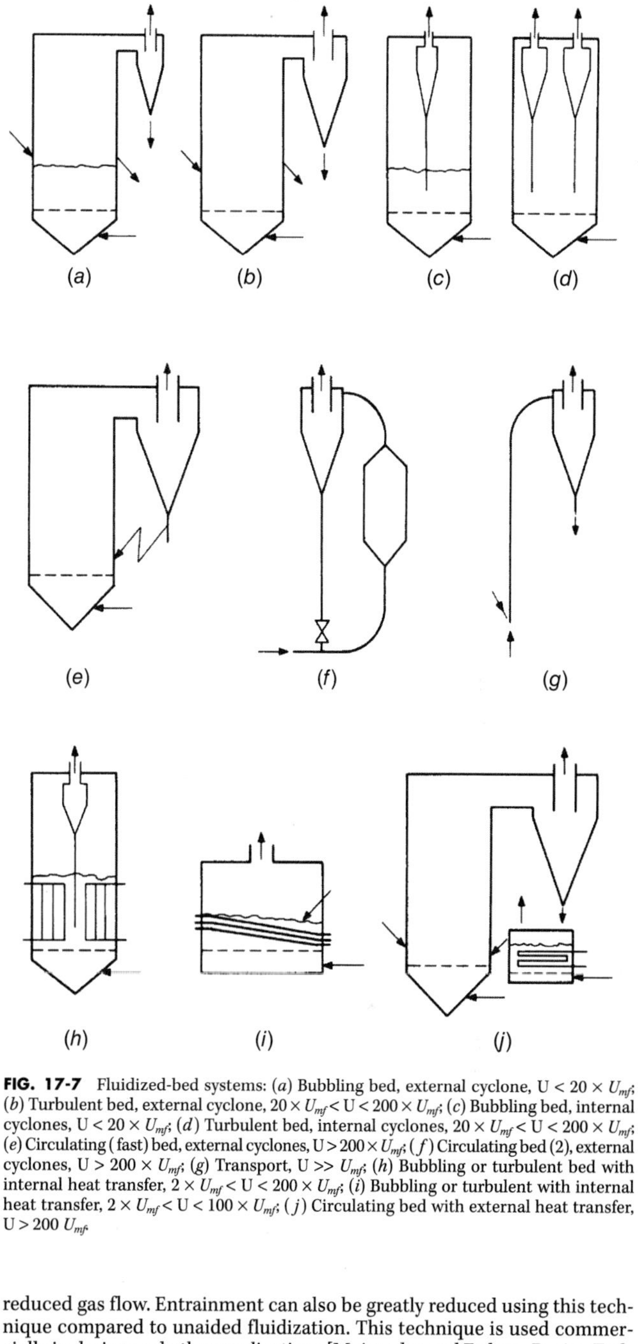

FIG. 17-7 Fluidized-bed systems: (*a*) Bubbling bed, external cyclone, $U < 20 \times U_{mf}$; (*b*) Turbulent bed, external cyclone, $20 \times U_{mf} < U < 200 \times U_{mf}$; (*c*) Bubbling bed, internal cyclones, $U < 20 \times U_{mf}$; (*d*) Turbulent bed, internal cyclones, $20 \times U_{mf} < U < 200 \times U_{mf}$; (*e*) Circulating (fast) bed, external cyclones, $U > 200 \times U_{mf}$; (*f*) Circulating bed (2), external cyclones, $U > 200 \times U_{mf}$; (*g*) Transport, $U >> U_{mf}$; (*h*) Bubbling or turbulent bed with internal heat transfer, $2 \times U_{mf} < U < 200 \times U_{mf}$; (*i*) Bubbling or turbulent with internal heat transfer, $2 \times U_{mf} < U < 100 \times U_{mf}$; (*j*) Circulating bed with external heat transfer, $U > 200\ U_{mf}$.

DESIGN OF FLUIDIZED-BED SYSTEMS

The use of the fluidization technique requires (in almost all cases) the use of a fluidized-bed system rather than an isolated piece of equipment. Figure 17-8 illustrates the arrangement of components of some systems.

The major parts of a fluidized-bed system are:

1. Fluidization vessel
   a. Fluidized-bed portion
   b. Disengaging space or freeboard
   c. Gas distributor
2. Solids feeder or flow control
3. Solids discharge
4. Dust separator for the exit gases
5. Instrumentation
6. Gas supply

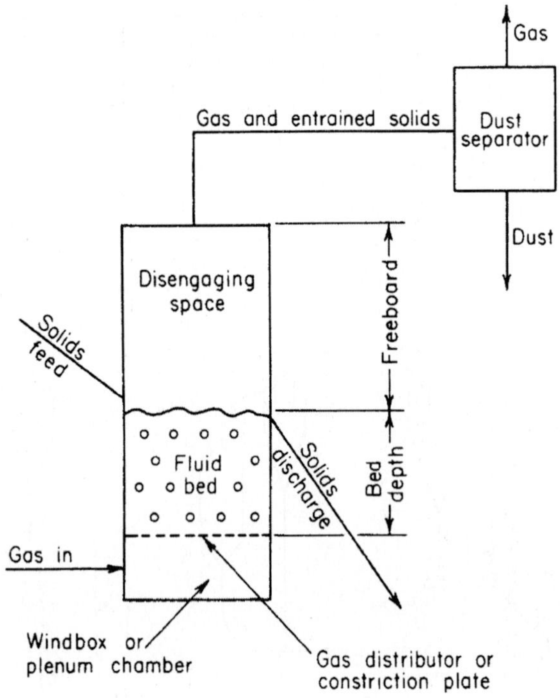

FIG. 17-8 Noncatalytic fluidized-bed system.

Fluidization Vessel The most common shape for a fluidized-bed vessel is a vertical cylinder. Just as for a vessel designed for boiling a liquid, space must be provided in the vessel for vertical expansion of the solids and for disengaging splashed and entrained material. The volume above a bubbling fluidized bed is called the disengaging space. The cross-sectional area of the vessel is determined by the volumetric flow of gas and the allowable or required fluidizing velocity of the gas at operating conditions. In some cases the lowest permissible velocity of gas is used, and in others the greatest permissible velocity is used. The maximum volumetric flow is often determined by the carryover or entrainment of solids, and this is related to the dimensions of the disengaging space (cross-sectional area and height).

Bed Bed height is determined by a number of factors, either individually or collectively, such as:

1. Gas-contact time
2. *L*/*D* ratio required to provide staging
3. Space required for internal heat exchangers
4. Solids-retention time

Fluidized-bed heights can range from less than 0.3 m (12 in) to more than 25 m (82 ft).

Although the reactor is usually a vertical cylinder, generally there is really no limitation on shape. The specific design features vary with operating conditions, available space, and use. The lack of moving parts results in simple, clean designs.

Most fluidized-bed units operate at elevated temperatures. For this use, refractory-lined steel is the most economical design. The refractory (typically an insulting refractory with a hard, attrition-resistant layer in contact with the particles) serves two main purposes: (1) it insulates the metal shell from the elevated temperatures, and (2) it protects the metal shell from abrasion by the motion of the bed particles, and particularly the splashing solids at the top of the bed that result from bursting bubbles. Depending on specific conditions, several different refractory linings are used [Van Dyck, Chem. Eng. Prog. (December): 46–51 1979)]. Generally, for the moderate temperatures encountered in catalytic cracking of petroleum, a reinforced-gunnite lining has been found to be satisfactory. This also permits the construction of larger units than would be permissible if self-supporting ceramic domes were to be used for the roof of the reactor.

When heavier refractories are required because of operating conditions, insulating brick is installed next to the shell, and firebrick is installed to protect the insulating brick. Industrial experience in many fields of application has demonstrated that such a lining will successfully withstand the abrasive conditions in the bed for many years without replacement. Most serious refractory wear occurs with coarse particles at high gas velocities and is usually most pronounced near the operating level of the bubbling fluidized bed.

Gas leakage behind the refractory has plagued a number of units. Care should be taken in the design and installation of the refractory to reduce the possibility of the formation of "chimneys" in the refractories. A small flow of solids and gas behind the refractory can quickly erode large passages in soft insulating brick, or even in dense refractory. Gas stops are often attached to the shell and project into the refractory lining. Care in the design and installation of openings in shell and lining is also required.

In many cases, cold spots on the reactor shell will result in condensation and high corrosion rates. Sufficient insulation is needed to keep the shell and appurtenances above the dew point of the reaction gases. Hot spots can occur where refractory cracks allow heat to permeate to the shell. These can sometimes be repaired by pumping castable refractory into the hot area from the outside.

The violent motion of a bubbling or turbulent fluidized bed requires an ample foundation and a sturdy supporting structure for the reactor. Even a relatively small differential movement of the reactor shell with the lining will materially shorten refractory life. The lining and shell must be designed as a unit. Structural steel should not be supported from a vessel that is subject to severe vibration.

Freeboard and Entrainment The freeboard, or disengaging height, is the distance between the top of the fluid bed and the gas-exit nozzle in bubbling- or turbulent-bed units. The distinction between bed and freeboard is difficult to determine in turbulent, fast, and transport units (see Fig. 17-6).

At least two actions can take place in the freeboard: (1) classification of solids and (2) reaction of solids and gases.

As a bubble reaches the upper surface of a fluidized bed, the bubble breaks through the thin upper envelope composed of solid particles and entrains some of these particles. The crater-shaped void formed by the erupting bubble is rapidly filled by flowing solids. When these solids meet at the center of the void, solids are geysered upward. The downward pull of gravity and the upward pull of the drag force of the upward-flowing gas act on the particles. The larger and denser particles return to the top of the bed, and the finer and lighter particles are carried upward. The distance above the bed at which the entrainment becomes constant with height is called the transport disengaging height (TDH). Cyclones and vessel gas outlets are usually located just above the TDH. Figure 17-9 graphically estimates how TDH changes as a function of superficial gas velocity in the freeboard and bed size.

The higher the concentration of an entrainable particle size in the bed, the greater its rate of entrainment. Finer particles have a greater rate of entrainment than coarse ones. These principles are embodied in the method of Geldart (*Gas Fluidization Tech.*, Wiley, 1986, pp. 123–153) via the equation, $E(i) = K^*(i)x(i)$, where $E(i)$ = entrainment flux for size *i*, kg/(m$^2\cdot$s); $K^*(i)$ = entrainment rate constant for particle size *i*; and $x(i)$ = weight fraction for particle size *i*. K^* is a function of operating conditions given by $K^*(i)/(P_f u) = 23.7 \exp[-5.4\, U_t(i)/U]$. The composition and the total entrainment are calculated by summing over all of the entrainable size fractions. A different way to calculate the entrainment rate is to use the method of Zenz, as reproduced by Pell (*Gas Fluidization*, Elsevier, Amsterdam, 1990, pp. 69–72).

In batch classification, the removal of fines (particles less than any arbitrary size) can be correlated by treating carryover as similar to a second-order reaction $K = (F/\theta)[1/x(x-F)]$, where K = rate constant, F = fines removed in time θ, and x = original concentration of fines.

Gas Distributor The gas distributor (also often called the grid of a fluidized bed) has a considerable effect on proper operation of the fluidized bed. For good fluidized-bed operation, it is absolutely necessary to have a properly designed gas distributor. Gas distributors can be used both when the gas is clean and when the gas contains a small loading of solids. The primary purpose of the gas distributor is to cause uniform gas distribution across the entire bed cross section. It should support the solids in the bed, operate for years without plugging or breaking, minimize sifting of solids back into the gas inlet to the distributor (called "weeping"), and minimize the attrition of the bed material. When the gas is clean, the gas distributor is often designed to prevent backflow of solids during normal operation, and in many cases it is designed to prevent backflow during shutdown. To provide good operation of the distributor, it has been found by experience that the grid should have a pressure drop equal to about one-third of the bed pressure drop. Because of pressure fluctuations in the bed caused by bubbles, the pressure fluctuations in the bed can be as much as about one-third of the bed pressure drop. If the grid pressure drop is not at least equal to the bed pressure fluctuations for upward-pointing grid nozzles, solids can be forced downward through these holes into the plenum below the grid. When the solids eventually flow back upward through the grid, excessive erosion of the grid holes can occur. Good gas distribution through the grid can be achieved a grid pressure drop of at least as low as one-tenth of the pressure drop across the bed. However, to prevent weeping of solids through the grid for upward-pointing nozzles, the grid pressure drop should be at least one-third of the bed pressure drop.

For gas distributors with downward-pointing nozzles, the grid pressure drop can be as low as one-tenth of the pressure drop across the bed to

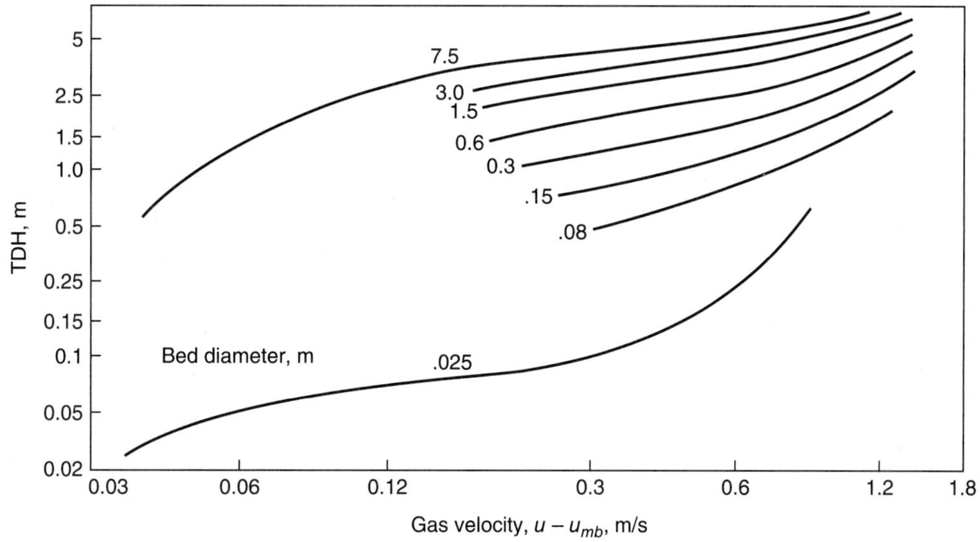

FIG. 17-9 Estimating transport disengaging height (TDH).

prevent solids backflow into the distributor. If the pressure drop across the bed is extremely low, gas maldistribution can result, with the bed being fluidized in one area and not fluidized in another. In units with shallow beds, such as dryers, or where gas distribution is less crucial, lower gas distributor pressure drops can be used.

When both solids and gas pass through the distributor, such as in some fluidized catalytic cracking (FCC) units, a number of different gas distributor designs have been used. Because the inlet gas contains solids, it is much more erosive than gas alone, and care has to be taken to minimize the erosion of the grid openings as the solids flow through them. Generally, this is done by decreasing the inlet gas/solids velocity through the holes so that erosion of the grid openings is low. Some examples of grids that have been used with both solids and gases in the inlet gas are concentric rings in the same plane, with the annuli open (Fig. 17-10*a*); concentric rings in the form of a cone (Fig. 17-10*b*); grids of T bars or other structural shapes (Fig. 17-10*c*); flat metal perforated plates supported or reinforced with structural members (Fig. 17-10*d*); and dished and perforated plates concave both upward and downward (Fig. 17-10*e* and *f*). The distributors shown in Fig. 17-10*d*, *e*, and *f* also can be used with no solids in the gas to the distributor. The curved distributors of Fig. 17-10*d* and *e* are often used because they minimize thermal expansion effects.

There are three basic types of clean-gas distributors: (1) a perforated plate distributor, (2) a bubble cap type of distributor, and (3) a sparger or pipe-grid type of gas distributor. The perforated plate distributor (Fig. 17-10*d*) is the simplest type of gas distributor and consists of a flat or curved plate containing a series of vertical holes. The gas flows upward into the bed from a chamber below the bed called a plenum. This type of distributor is easy and economical to construct. However, when the gas is shut off, the solids can sift downward into the plenum and may cause erosion of the holes when the bed is started up again. The bubble cap type of distributor is designed to prevent backflow of solids into the plenum chamber or inlet line of the gas distributor on start-up or shutdown. The cap or tuyere type of distributor generally consists of a vertical pipe containing several small horizontal holes or holes angled downward from 30° to 45° from the horizontal (Fig. 17-11*a* and *b*). It is difficult for the solids to flow back through such a configuration when the fluidizing gas is shut off.

The pipe distributor (often called a sparger) differs from the other two distributor types because it consists of pipes with distribution holes in them that are inserted into the bed (Fig. 17-12). This type of distributor will often have solids below it that are not fluidized. If this is not acceptable for a process, then this type of distributor cannot be used. However, the pipe distributor has certain advantages. It does not require a large plenum, the holes in the pipe can be positioned at any angle (although most often they are pointed in downward at about a 45° angle), and it can be used in cases when multiple gas injections are required in a process (Fig. 17-12*c*). The most common type of pipe distributor is the multiple-pipe (manifold sparger) grid shown in Fig. 17-12*a*. Less common, but also used is the "wagon wheel" type of sparger distributor shown in Fig. 17-12*(b)*.

To generate a sufficient pressure drop for good gas distribution, a high velocity through the grid openings may be required. It is best to limit this velocity to less than about 45 m/s (150 ft/s) to minimize attrition of the bed material. The maximum hole velocity allowable may be even lower for very soft materials that attrit easily. The pressure drop and the gas velocity through the hole in the gas distributor are related by the equation

$$\Delta P = \frac{u^2 \rho_g}{2c^2 g_c} \quad \text{for fps units} \tag{17-5}$$

$$\Delta P = \frac{u^2 \rho_g}{2c^2} \quad \text{for SI units} \tag{17-6}$$

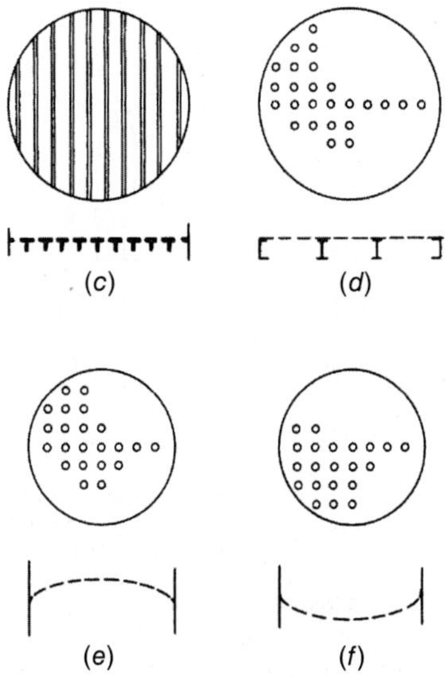

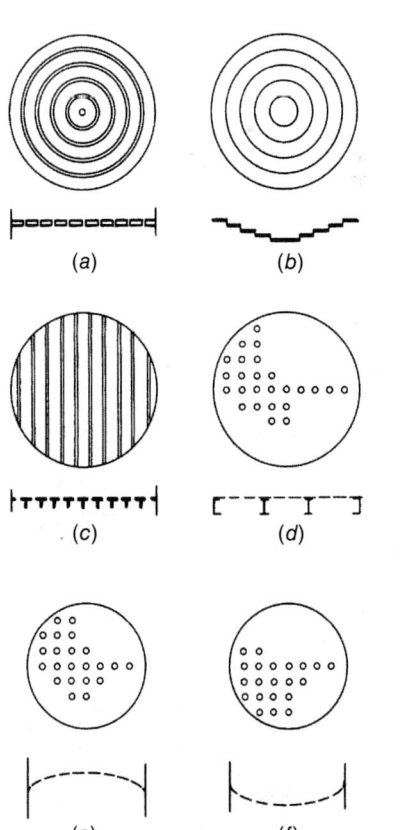

FIG. 17-10 Gas distributors for gases containing solids.

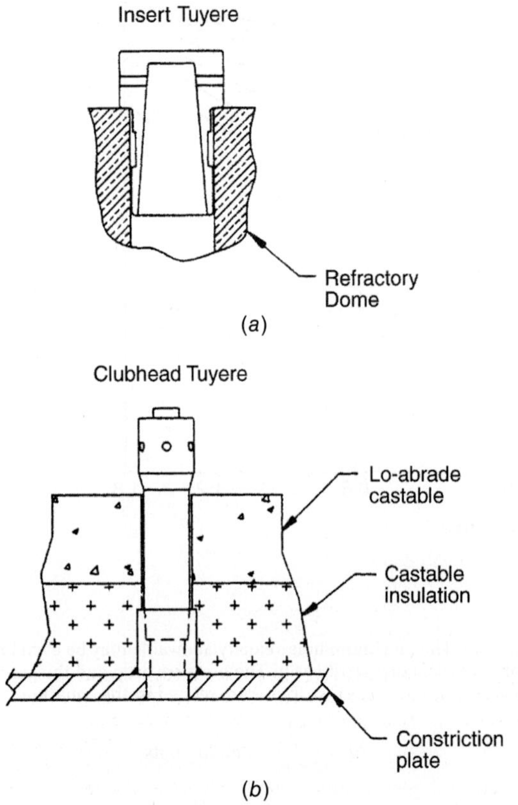

FIG. 17-11 Gas inlet nozzles designed to prevent backflow of solids: (*a*) Insert tuyere; (*b*) clubhead tuyere. (*Dorr-Oliver, Inc.*)

where u = superficial gas velocity in the grid hole at inlet conditions
ρ_g = gas density in hole at conditions in inlet to hole, kg/m³ or lb/ft³
ΔP = pressure drop in consistent units, kPa or lb/ft²
C = orifice constant, dimensionless (typically 0.8 for gas distributors)
g_c = gravitational conversion constant, ft-lb$_m$/(s²·lb$_f$)

Due to the pressure drop requirements across the gas distributor for good gas distribution, the velocity through the grid hole may be higher than desired in order to minimize or limit particle attrition. Therefore, it is common industrial practice to place a length of pipe (called a shroud) over the gas distributor hole such that the diameter of the pipe is larger than the diameter of the distributor hole. This technique effectively allows a smaller hole to give the required pressure drop, and the larger hole diameter of the shroud reduces the exit gas velocity into the bed so that particle attrition at the grid will be minimized. This technique is applied to both plate and pipe spargers.

Experience has shown that a concave-downward (Fig. 17-10*f*) gas distributor is a better arrangement than a concave-upward (Fig. 17-10*e*) gas distributor because it tends to increase the flow of gases in the outer portion of the bed. This counteracts the normal tendency of the gas to flow into the center of the bed after it exits the gas distributor. In addition, the concave-downward type of gas distributor tends to assist the general solids flow pattern in the bed, which is up in the center and down near the walls. The concave-upward gas distributor tends to have a slow-moving region at the bottom near the wall. If solids are large (or if they are slightly cohesive), they can build up in this region.

Structurally, distributors must withstand the differential pressure across the restriction during normal and abnormal flow conditions. In addition, during a shutdown, all or a portion of the bed will be supported by the distributor until sufficient backflow of the solids has occurred into the plenum to reduce the weight of solids above the distributor and to support some of this remaining weight by transmitting the force to the walls and bottom of the reactor. During start-up, a considerable upward thrust can be exerted against the distributor as the settled solids under the distributor are carried up into the normal reactor bed.

When the feed gas is devoid of or contains only small quantities of fine solids, more sophisticated designs of gas distributors can be used to realize economies in initial cost and maintenance. This is most pronounced when the inlet gas is cold and noncorrosive. When this is the case, the plenum chamber gas distributor and distributor supports can be fabricated of mild steel by using normal temperature design factors. The first commercial fluidized-bed ore roaster [Mathews, *Trans. Can. Inst. Min. Metall.* **L11:** 97 (1949)], supplied by the Dorr Co. (now Dorr-Oliver Inc.) in 1947 to Cochenour-Willans, Red Lake, Ontario, was designed with a mild steel constriction plate covered with castable refractory to insulate the plate from the calcine, and to provide cones in which refractory balls were placed to act as ball checks. The balls eroded unevenly, and the castable cracked. However, when the unit was shut down by closing the air control valve, the runback of solids was negligible because of bridging. If, however, the unit was shut down by de-energizing the centrifugal blower motor, the higher pressure in the reactor would relieve through the blower, and fluidizing gas plus solids would run back through the constriction plate. Figure 17-11 illustrates two designs of gas inlets that have been successfully used to prevent the flowback of solids. For best results, irrespective of the design, the gas flow should be stopped and the pressure relieved from the bottom upward through the bed. Some units have been built and successfully operated with simple slot-type distributors made of heat-resistant steel. This requires a heat-resistant plenum chamber but eliminates the often encountered problem of corrosion caused

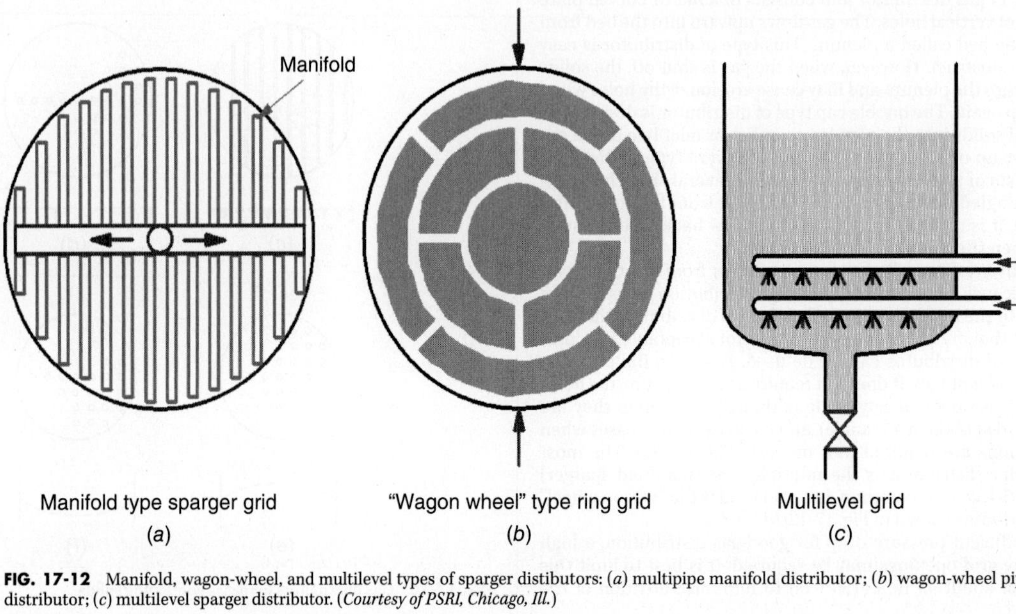

FIG. 17-12 Manifold, wagon-wheel, and multilevel types of sparger distributors: (*a*) multipipe manifold distributor; (*b*) wagon-wheel pipe distributor; (*c*) multilevel sparger distributor. (*Courtesy of PSRI, Chicago, Ill.*)

by condensation of acids and water vapor on the cold metal of the distributor. When the inlet gas is hot, such as in dryers or in the upper distributors of multibed units, ceramic arches or heat-resistant metal grates are generally used. Self-supporting ceramic domes have been in successful use for many years as gas distributors when temperatures range up to 1100°C. Some of these domes are fitted with alloy-steel orifices to regulate air distribution. However, the ceramic arch presents the same problem as the dished head positioned concave upward. Either the holes in the center must be smaller, so that the sum of the pressure drops through the distributor plus the bed is constant across the entire cross section, or the top of the arch must be flattened so that the bed depths in the center and outside are equal. This is especially important when shallow beds are used.

It is important to consider thermal effects in the design of the grid-to-shell seal. Bypassing of the grid at the seal point is a common problem caused by situations such as uneven expansion of metal and ceramic parts, a cold plenum and hot solids in contact with the grid plate at the same time, and start-up and shutdown scenarios. When the atmosphere in the bed is sufficiently benign, a sparger-type distributor may be used (Fig. 17-12). In some cases, it is impractical to use a plenum chamber under the constriction plate. This condition arises when a flammable or explosive mixture of gases is being introduced to the reactor. One solution is to pipe the gases to a multitude of individual gas inlets in the floor of the reactor. In this way it may be possible to maintain the gas velocities in the pipes above the flame velocity or to reduce the volume of gas in each pipe to the point at which an explosion can be safely contained. Another solution is to provide separate inlets for the different gases and to rely on the rapid axial mixing of the fluidized bed (Fig. 17-12*c*). The inlets should be fairly close to one another, as lateral gas mixing in fluidized beds is poor.

Much attention has been paid to the effect of gas distribution on bubble growth in the bed and the effect of this on catalyst utilization, space-time yield, etc., in catalytic systems. It would appear that the best gas distributor would be a porous membrane because of its even distribution. However, this type of distributor is seldom practical for commercial units because of structural limitations and the fact that it requires absolutely clean gas. Practically, the limitations on hole spacing in a gas distributor are dependent on the particle size of the solids, materials of construction, and type of distributor. If easily worked metals are used, then punching, drilling, and welding are not expensive operations, and they permit the use of a large number of holes. The use of tuyeres or bubble caps permits horizontal distribution of the gas so that a smaller number of gas inlet ports can still achieve good gas distribution. If a ceramic arch is used, generally only one hole per brick is permissible, and brick dimensions must be reasonable.

Scale-Up

Bubbling or Turbulent Beds Scale-up problems in fluidized beds usually occur when the reaction rate is very fast. In this case, bubbling fluidized-bed hydrodynamics limit the rate of the reaction of gas and solids because the mass transfer of gas from the bubble to the emulsion (dense) phase is slower than the reaction rate. Therefore, hydrodynamics can limit the reaction, and it is of interest to try to increase the mass transfer rate. For Geldart Group A solids, most reactions are carried out in the turbulent fluidized-bed regime because of the increased mass transfer rate in that regime relative to the bubbling regime. For reactions that are slower than the mass transfer of bubble gas to the emulsion phase, scale-up fluidized beds is more straightforward, and this can sometimes be carried out on an area basis. However, scale-up even with slow reactions is not simple, and care must be taken at each step of the scale-up.

In a typical scale-up, small-scale tests are first made to determine reaction kinetics and physical limitations, such as sintering, agglomeration, and solids-holdup time required. Scale-up is typically conducted in several steps, from laboratory to commercial size. The hydrodynamics of gas–solids flow and contacting is quite different in small-diameter high-L/D fluid beds as compared with large-diameter moderate-L/D beds. In small-diameter beds, bubbles will be small, and they cannot grow larger than the vessel diameter. If they grow to a size approximately 2/3 of the bed diameter, the fluidized bed is said to be in a slugging mode, with large pressure fluctuations across it. In larger, deeper units, bubbles can grow very large. This is especially so for Geldart Group B and D particle beds.

The size of a bubble in the bed as a function of bed height was given by Darton et al. [*Trans. Inst. Chem. Eng.* **55:** 274–280 (1977)] as:

$$d_b = \frac{0.54(U-U_{mb})^{0.4}\left(h+4\sqrt{A_t/N_0}\right)^{0.8}}{g^{0.2}} \tag{17-7}$$

where d_b = bubble diameter, m
h = height above the grid, m
A_t/N_0 = grid area per hole, m^2/hole
U = superficial gas velocity, m/s
U_{mf} = minimum bubbling velocity, m/s

Bubble growth in fluidized beds will be limited by the diameter of the containing vessel and bubble hydrodynamic stability. Bubbles in Geldart Group B and D systems can quickly grow to over 1 m in diameter if the gas velocity and the bed height are sufficient. Bubbles in Geldart Group A materials with a high percentage of fines (defined to be material less than 44 μm in size) will reach a maximum stable bubble size in a range of about 8 to 20 cm. Furthermore, solids and gas backmixing are much lower in deep (high-L/D) beds (whether they are slugging or bubbling) than in shallow (low-L/D) beds. Thus, the conversion or yield in large, unstaged reactors can sometimes be considerably lower than in small, high-L/D units. To overcome some of the problems of scale-up, staged fluidized-bed units are often used (Fig. 17-13). A brief history of fluidization, fluidized-bed scale-up, and modeling will illustrate some problems involved with scaling up fluidized beds.

Fluidized beds were first used commercially in Germany in the late 1920s to gasify coal. Scale-up problems either were insignificant or were not publicized. During World War II, fluidized catalytic cracking of oil to produce gasoline was successfully commercialized by scaling up from pilot-plant size (a few centimeters in diameter) to commercial size (several meters in diameter). It is fortunate that the ratio of crude oil to catalyst is determined by thermal balance and the required catalyst circulation rates, and that the crude oil feed point was in the dilute-phase riser, which gives less

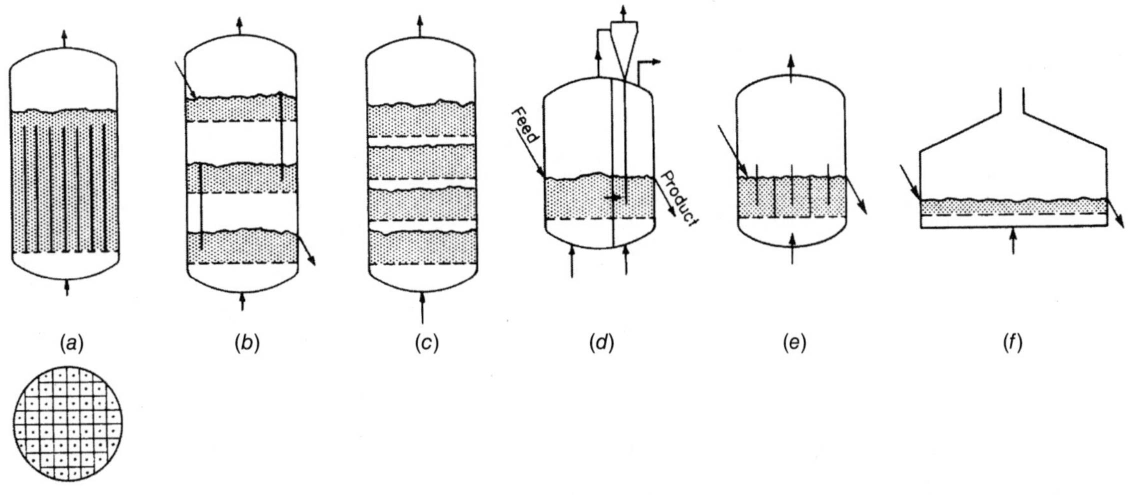

FIG. 17-13 Methods of staging or minimizing backmixing in fluidized beds: (*a*) vertical, cross-hatched baffles to give high L/D; (*b*) staged beds with standpipes between stages; (*c*) staged beds with no standpipes between stages; (*d*) divided bed with small opening between the beds; (*e*) vertical baffles to give over/under solids flow pattern; (*f*) very shallow bed to minimize horizontal mixing.

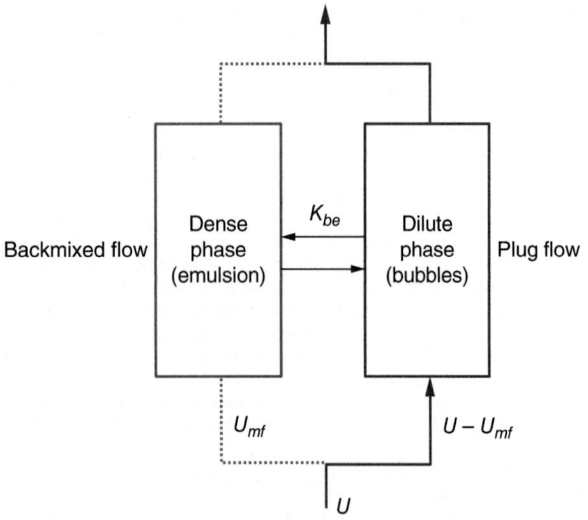

FIG. 17-14 Two-phase model according to May [*Chem. Eng. Prog.* **55**(12): 49–55 (1959)] and Van Deemter [*Chem. Eng. Sci.* **13:** 143–154 (1961)]. U = superficial gas velocity, U_{mf} = minimum fluidizing velocity, D_{ax} = axial dispersion coefficient, and K_{be} = mass transfer coefficient. (*Courtesy of PSRI, Chicago, Ill.*)

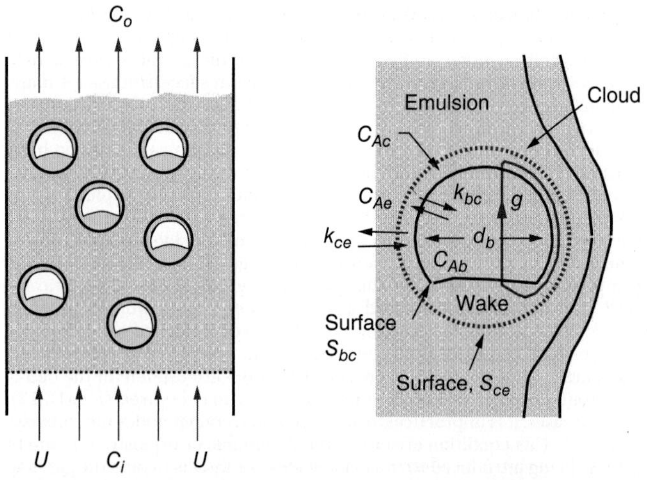

FIG. 17-15 Bubbling-bed model of Kunii and Levenspiel. d_b = effective bubble diameter, C_{ab} = concentration of species A in emulsion, q = volumetric gas flow rate into or out of bubble, k_{bc} = mass-transfer coefficient between the bubble and the cloud, k_{ce} = mass transfer coefficient between the cloud and the emulsion. (*From Kunii and Levenspiel,* Fluidization Engineering, *Wiley, New York, 1969.*) (*Courtesy of PSRI, Chicago, Ill.*)

backmixing than in a bubbling fluidized bed. The first experience of problems with scale-up of fluidized beds was associated with the production of gasoline from natural gas using the Fischer-Tropsch process. The results from a 0.10-m- (4-in-), 0.20-m- (8-in-), and 0.30-m- (12-in-) diameter pilot plants using a Group B iron catalyst were scaled to a 7-m-diameter commercial unit, where the yield was only about 50 percent of that achieved in the pilot units.

The problem was that the smaller reactors were operating in the slugging fluidized-bed mode (Fig. 17-4). In this mode, the slugs travel upward more slowly than in the bubbling fluidized-bed mode in the larger reactor, where the bubbles were much larger than the slugs and traveled much more rapidly through the bed. This resulted in a much shorter gas/solids contact time and the significantly lower conversion.

Immediately after this unfortunate experience, people were reluctant to use fluidized beds, and this slowed their development for some time. However, it was later shown that going to a smaller particle size (Geldart Group A material) and operating in the turbulent mode (for high mass transfer) solved most of the scale-up problems for the Fischer-Tropsch process.

Many bubbling fluidized-bed models have been developed; these basically are of two types, the two-phase model [May, *Chem. Eng. Prog.* **55**(12): 49–55 (1959); and Van Deemter, *Chem. Eng. Sci.* **13:** 143–154 (1961)] and the Kunii and Levenspiel bubble model (Kunii and Levenspiel, *Fluidization Engineering,* Wiley, New York, 1969). The two-phase model according to May and Van Deemter is shown in Fig. 17-14. In the two-phase model, all or most of the gas passes through the bed in plug flow in the bubble phase, which does not contain solids. The solids form a dense-suspension emulsion phase in which gas and solids mix according to an axial dispersion coefficient (D_{ax}). Cross-flow between the two phases is predicted by a mass-transfer coefficient, K_{be}.

Conversion of a gaseous reactant can be given by $C/C_0 = \exp[-Na \times Nr/(Na + Nr)]$, where C = the exit concentration, C_0 = the inlet concentration, Na = diffusional driving force, and Nr = reaction driving force. Conversion is determined by both reaction and diffusional terms. It is possible for reaction to dominate in a lab unit with small bubbles, and for diffusion to dominate in a plant-size unit. It is this change of limiting regimes that makes scale-up so difficult. Refinements of the basic model and predictions of mass-transfer and axial-dispersion coefficients are the subject of many papers [Van Deemter, *Proc. Symp. Fluidization,* Eindhoven (1967); de Groot, ibid.; Van Swaaij and Zuidweg, *Proc. 5th Eur. Symp. React. Eng.,* Amsterdam, B9–25 (1972); DeVries, Van Swaaij, Mantovani, and Heijkoop, ibid., B9–59 (1972); Werther, *Ger. Chem. Eng.* **1:** 243–251 (1978); and Pell, *Gas Fluidization,* Elsevier, Amsterdam, 75–81 (1990)].

The Kunii and Levenspiel (K-L) bubbling bed model (Kunii and Levenspiel, *Fluidization Engineering,* Wiley, New York, 1969; Fig. 17-15) assumes constant-sized bubbles (with an effective bubble size d_b) rising through the emulsion phase. Gas is transferred from the bubble void to the cloud and wake with a mass-transfer coefficient of k_{bc} and from the cloud to the emulsion phase with a mass-transfer coefficient k_{ce}. Experimental results have been fitted to theory by means of adjusting the effective bubble size. As mentioned previously, bubble size changes from the bottom to the top of the bed, and thus this model is not completely realistic, although it can be of considerable use in evaluating reactor performance. Several bubble models using bubbles of increasing size from the distributor to the top of the bed and gas interchange between the bubbles and the emulsion phase according to Kunii and Levenspiel have been proposed [Kato and Wen, *Chem. Eng. Sci.* **24:** 1351–1369 (1969); and Fryer and Potter, in *Fluidization Technology,* vol. I, ed. Keairns, Hemisphere, Washington, 1975, pp. 171–178].

There are several methods available to reduce scale-up loss. These are summarized in Fig. 17-16 for a process operating with Geldart Group A solids. The conversion efficiency of a fluid-bed reactor has been found to typically decrease as the size of the reactor increases. This decrease in reactor efficiency can be minimized by the use of a high gas velocity, fine solids, staging methods, and a high L/D. A high gas velocity maintains the reactor in the turbulent mode, where gas void breakup is rapid and frequent. A smaller particle size was found to lead to also promote turbulent fluidization. Maintaining a high L/D minimizes backmixing, as does the use of baffles in the reactor. By using these techniques, Mobil was able to scale up its methanol to gasoline technology with little difficulty [Krambeck, Avidan, Lee, and Lo, *A.I.Ch.E.J.* **33:** 1727–1734 (1987)].

Another way to conduct the scale-up of bubbling fluidized-bed hydrodynamics is to build a cold and/or hot scale model of the commercial design and conduct testing in the models. Typically, a pilot plant reactor (typically 6 inches to 24 inches in diameter) will be built to obtain scale-up

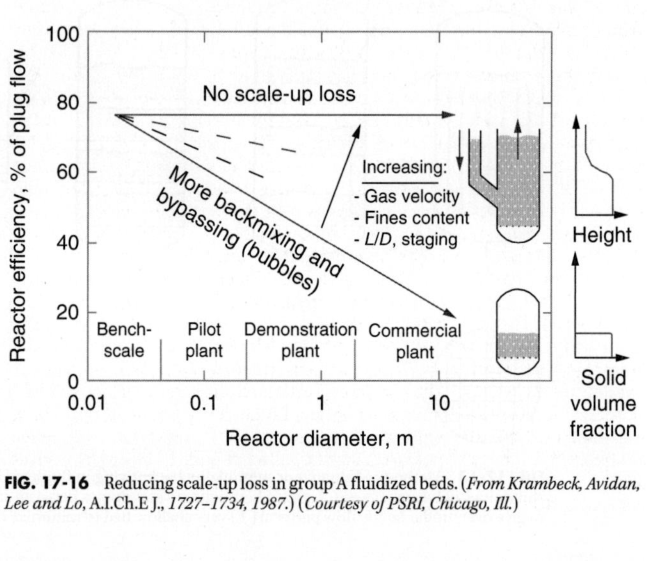

FIG. 17-16 Reducing scale-up loss in group A fluidized beds. (*From Krambeck, Avidan, Lee and Lo,* A.I.Ch.E J., *1727–1734, 1987.*) (*Courtesy of PSRI, Chicago, Ill.*)

information in parallel with information from a relatively large cold model. Scaling parameters have also been developed such that dimensionless groups are the same in a small cold-flow model as well as with the high-temperature, large-scale units. However, designing a cold-flow model with dimensionless groups the same as in the large, high-temperature unit almost invariably results in the material in the cold model being different in particle size, particle density, or both.

Circulating or Fast Fluidized Beds The circulating, or fast fluidized bed, is actually a misnomer in that it is not an extension of the turbulent bed, but is actually a part of the transport regime, as previously discussed. However, the fast fluidized bed operates in that part of the transport regime that is dominated by the static head of solids pressure drop term (the part of the regime where the solids concentration is the highest). The solids may constitute up to 10 percent of the volume of the system in this regime. There are no bubbles, mass-transfer rates are high, and there is little gas backmixing in the system. The high velocity in the system results in a high gas throughput, which minimizes reactor cost. Because there are no bubbles, scale-up is also less of a problem than with bubbling beds.

Many circulating systems (especially circulating fluidized-bed combustor systems) are characterized by an external cyclone return system that can have as large a footprint as the reactor itself. The axial solids density profile is relatively flat, as indicated in Fig. 17-6. There is a parabolic radial solids density profile in fast fluidized beds that is termed core annular flow. In the center of the reactor, the gas velocity and the solids velocity may be double the average. The solids in the center of the column (often termed a riser) are in dilute flow, traveling at their expected slip velocity $U_g - U_t$. Near the wall in the annulus, the solids are somewhat lower than their fluidized-bed density. The solids at the wall can flow either upward or downward. Whether they do so is determined primarily by the gas velocity used in the system. In circulating fluidized-bed combustor systems, the gas velocity in the rectangular riser is generally in the range of 4 to 6 m/s, and the solids flow downward at the wall. In fluid catalytic cracking, the velocity in the riser is typically in the range of 13 to 20 m/s, and the solids flow upward at the wall. Engineering methods for evaluating the hydrodynamics of the circulating bed are given by Kunii and Levenspiel (*Fluidization Engineering,* 2d ed., Butterworth, Oxford, UK, 1991, pp. 195–209), Werther (*Circulating Fluid Bed Technology IV,* Mobil Research and Development Corporation, Paulsboro Research Laboratory, 1994), and Avidan, Grace, and Knowlton, eds. (*Circulating Fluidized Beds,* Blackie Academic, New York, 1997).

Pneumatic Conveying Pneumatic conveying systems can generally be scaled up on the principles of dilute-phase transport. Mass and heat transfer can be predicted on both the slip velocity during acceleration and the slip velocity at full acceleration. The slip velocity increases as the solids concentration is increased.

Heat Transfer Heat-exchange surfaces have been used to provide the means of removing or adding heat to fluidized beds. Usually, these surfaces are provided in the form of vertical or horizontal tubes manifolded at the tops and bottom or in a trombone shape manifolded exterior to the vessel. Horizontal tubes are extremely common as heat-transfer tubes. In any such installation, adequate provision must be made for abrasion of the exchanger surface by the bed. The prediction of the heat-transfer coefficient for fluidized beds is covered in Secs. 5 and 11.

Normally, the heat-transfer rate is between 5 and 25 times that for the gas alone. Bed-to-surface heat-transfer coefficients vary according to the type of solids in the bed. Group A solids have bed-to-surface heat-transfer coefficients of approximately 300 J/(m^2·s·K) [150 Btu/(h·ft^2·°F)]. Group B solids have bed-to-surface heat-transfer coefficients of approximately 100 J/(m^2·s·K) [50 Btu/(h·ft^2·°F)], while Group D solids have bed-to-surface heat-transfer coefficients of about 60 J/(m^2·s·K) [30 Btu/(h·ft^2·°F)]. These heat-transfer coefficients are approximate only, and they can vary considerably depending on gas velocity, system temperature, and pressure.

The large area of the solids per cubic foot of bed, 5000 m^2/m^3 (15,000 ft^2/ft^3) for 60-μm particles of about 600 kg/m^3 (40 lb/ft^3) bulk density, and the vast difference in heat capacity of the solids relative to the gas, results in the rapid approach of gas and solids temperatures near the bottom of the bed. Equalization of gas and solids temperatures generally occurs within 4 to 15 cm (1.5 to 6 in) of the top of the distributor in a bubbling fluidized bed.

Bed thermal conductivities in the vertical direction have been measured in the laboratory in the range of 40 to 60 kJ/(m^2·s·K) [20,000 to 30,000 Btu/(h·ft^2·°F·ft)]. Horizontal thermal conductivities for 3-mm (0.12-in) particles in the range of 2 kJ/(m^2·s·K) [1000 Btu/(h·ft^2·°F·ft)] have been measured in large-scale experiments. Except for extreme L/D ratios or in some beds containing many horizontal baffles, the temperature in the fluidized bed is uniform—with the temperature at any point in the bed generally being within 5°C (10°F) or less of any other point. In fact, temperature difference is a good indicator of whether the bed is fluidized well. If two temperatures at any point in the bed differ by more than about 10°C (20°F), the bed is considered to have fluidization problems.

Temperature Control Because of the rapid equalization of temperatures in fluidized beds, temperature control can be accomplished in a number of ways:

1. *Adiabatic.* Control gas flow and/or solids feed rate so that the heat of reaction is removed as sensible heat in off-gases and solids or heat supplied by gases or solids.
2. *Solids circulation.* Remove or add heat by circulating solids.
3. *Gas circulation.* Recycle gas through heat exchangers to cool or heat.
4. *Liquid injection.* Add volatile liquid so that the latent heat of vaporization equals excess energy.
5. *Cooling or heating surfaces in bed.*

Solids Mixing Solids mixing in fluidized beds occurs because of bubbles. Solids are carried upward in the wake and the drift (or tail) of the bubble. When the bubble reaches the top of the bed, the solids are ejected upward and outward, and the solids are then circulated downward at the walls (Rowe and Patridge, "Particle Movement Caused by Bubbles in a Fluidized Bed," Third Congress of European Federation of Chemical Engineering, London, 1962). Thus, no mixing will occur at incipient fluidization, and solids mixing increases as the gas velocity through the bed is increased. Naturally, particles brought to the top of the bed must displace particles toward the bottom of the bed. Generally, solids upflow is upward in the center of the fluidized bed and downward at the wall.

At high ratios of fluidizing velocity to minimum fluidizing velocity, tremendous solids circulation from top to bottom of the bed assures rapid mixing of the solids. For all practical purposes, beds with L/D ratios of from 0.1 to 4 can be considered to be completely mixed continuous-reaction vessels insofar as the solids are concerned.

Batch mixing using fluidization has been successfully employed in many industries. In this case there is practically no limitation on vessel dimensions.

All the foregoing pertains to solids of approximately the same physical characteristics. There is evidence that solids of widely different characteristics will classify one from the other at certain gas flow rates [Geldart, Baeyens, Pope, and van de Wijer, *Powder Technol.* **30**(2): 195 (1981)]. Two fluidized beds, one on top of the other, may be formed, or a lower static bed with a fluidized bed above may result. The latter often occurs when agglomeration takes place because of either fusion of particles in the bed or poor dispersion of sticky feed solids. Increased gas velocity in the bed sometimes overcomes this problem. However, improved feeding techniques or a change in operating conditions may be required. Another solution is to remove agglomerates either continuously or periodically from the bottom of the bed.

Gas Mixing The mixing of gases as they pass vertically up through the bed has never been considered a problem. However, horizontal mixing is often inadequate, and it requires effective distributors if two gases are to be mixed in the fluidized bed.

In bubbling beds operated at velocities of a few multiples of U_{mf}, the gases will flow upward in both the emulsion and the bubble phases. At higher gas velocities, the downward velocity of the solids in the emulsion phase is sufficient to carry the contained gas downward. The gas velocity where the gas in the emulsion begins to flow downward depends primarily on the particle size of the material. The back mixing of gases increases as U/U_{mf} is increased until the circulating or fast regime is reached. In the fast fluidization regime gas back mixing decreases as the velocity is further increased.

Size Enlargement Under proper conditions, solid particles can be caused to increase in size in the bed. This can be advantageous or disadvantageous. Particle growth is usually associated with the melting or softening of some portion of the bed material (e.g., addition of soda ash to calcium carbonate feed in lime reburning, tars in fluidized-bed coking, or lead or zinc roasting causes agglomeration of dry particles in much the same way as binders act in rotary pelletizers). The motion of the particles, one against the other, in the bed results in spherical pellets. If the size of these particles is not controlled, rapid agglomeration and segregation of the large particles from the bed can occur. Control of agglomeration can be achieved by crushing a portion of the bed product and recycling it to form nuclei for new growth. Often, liquids or slurries are fed via a spray nozzle into the bed to cause particles to grow. In drying solutions or slurries of solutions, the location of the feed injection nozzle (spray nozzle) has a large effect on the size of the particles formed in the bed. Also of importance are the operating temperature, relative humidity of the off-gas, and gas velocity in the bed. Particle growth can occur as agglomeration (two or more particles sticking together) or by the particle growing in layers, often called *onion skinning.*

Size Reduction *Attrition* is the term used to describe particle reduction in the fluidized bed. Three major attrition mechanisms occur in the fluidized bed: particle abrasion, particle fragmentation (particle fracture), and particle thermal decrepitation. Particle abrasion occurs when the protruding edges on individual particles are broken off in the bed. These particle sizes are very small—usually on the order of 2 to 10 μm. Particle fragmentation occurs when particle interaction is severe enough to cause the particles to break up into smaller individual pieces, but much greater than the particles produced by abrasion.

Particle attrition occurs near the grid because of particles being accelerated by the gas jets and then impacting the particles in the bed. Particle attrition in cyclones and risers occurs because the particles hit the cyclone wall or the bend at the top of the riser, respectively.

Because of the random motion of the solids, some abrasion of the particle surface occurs in the fluidized bed itself. However, this abrasion is extremely small relative to the particle breakup caused by the high-velocity jets at the distributor or the high inlet velocities in cyclones, and is often neglected. Typically, particle abrasion has been determined in some catalytic processes to be about 0.25 to 1 percent of the solids per day. Whether attrition occurs by abrasion or by fragmentation depends on the strength of the particles. In many catalytic processes, nearly all of the attrition occurring is due to abrasion. In other processes, fragmentation is the dominant mechanism. In the area of high gas velocities at the distributor, greater rates of attrition will occur because of fracture of the particles by impact. As mentioned previously, particle fracture of the grid is reduced by adding shrouds to the gas distributor.

Generally, particle attrition is unwanted. However, at times, controlled attrition is desirable. For example, in fluidized-bed coking units where agglomeration due to wet particles is frequent, jets are used to attrit particles to control particle size [Dunlop, Griffin, and Moser, *J. Chem. Eng. Prog.* **54:** 39–43 (1958)].

Thermal decrepitation occurs often when crystals are rearranged because of transition from one form to another, or when new compounds are formed (e.g., calcination of limestone). Sometimes the stresses on particles in cases such as this are sufficient to reduce the particle to almost the basic crystal size. All these mechanisms will cause the completion of fractures that were started before the introduction of the solids into the fluidized bed.

Solids Feeders and Solids Flow Control Several designs of valves for solids flow control are used. These should be chosen with care to suit the specific conditions. Figure 17-17 shows (schematically) some of the devices used for solids flow control. The devices shown are a slide valve, also known as a knife-gate valve (Fig. 17-17*a*), a rotary valve (Fig. 17-17*b*), a table feeder (Fig. 17-17*c*), a screw feeder (Fig. 17-17*d*), a cone valve (Fig. 17-17*e*), and an L-valve (Fig. 17-17*f*). All of the feeders are mechanical feeders except for the L-valve, which is a nonmechanical valve. This type of valve uses only injected aeration gas to control the flow of solids through it.

Not shown in Fig. 17-17 is the flow-control arrangement used in the Exxon Research & Engineering Co. model IV catalytic-cracking units. This device consists of a U-bend. A variable portion of regenerating air is injected into the riser leg. Changes in air-injection rate change the fluid density in the riser part of the U-bend and thereby achieve control of the solids flow rate. Catalyst circulation rates of 1200 kg/s (70 tons/min) have been reported using these bends.

When the solid is one of the reactants, such as in ore roasting, the flow must be continuous and precise in order to maintain constant conditions in the reactor. Feeding of free-flowing granular solids into a fluidized bed is not difficult. Standard commercially available solids-weighing and -conveying equipment can be used to control the rate and deliver the solids to the feeder. Screw conveyors (Fig. 17-17*d*), dip pipes, seal legs, and injectors are used to introduce the solids into a reactor. Difficulties arise and special techniques must be used when the solids are not free-flowing, as is the case with most filter cakes. One solution to this problem was developed at Cochenour-Willans. After much difficulty in trying to feed a wet and sometimes frozen filter cake into the reactor by means of a screw feeder, experimental feeding of a water slurry of flotation concentrates was attempted. This trial was successful, and this method has been used in almost all cases in which the heat balance, particle size of solids, and other considerations have permitted. Gilfillan et al. (*J. Chem. Metall. Min. Soc. S. Afr.*, May 1954) present complete details on the use of this system for feeding.

When slurry feeding is impractical, recycling of solids product to mix it with the feed, both to dry and to achieve a better-handling material, has been used successfully. Also, the use of a rotary table feeder mounted on top of the reactor, discharging through a mechanical disintegrator, has been successful. The wet solids generally must be broken up into discrete particles of very fine agglomerates either by mechanical action before entering the bed, or by rapidly vaporizing water. If lumps of dry or semidry solids are fed, the agglomerates do not break up but tend to fuse together. Because the size of the agglomerate is many times the size of the largest individual particle, these agglomerates will segregate out in the bed, and in time the whole of the fluidized bed may be replaced with a static bed of agglomerates.

Standpipes In a fluid catalytic cracking (FCC) unit, hot Group A catalyst is added to aspirated crude oil feed in a riser to crack the feed oil into gasoline and other light and heavy hydrocarbons. The catalyst activity is reduced by this contact as carbon is deposited on the catalyst. The catalyst is then passed through a steam stripper to remove the gas product in the interstices of the catalyst and is transported to a regenerator. The carbon on the catalyst is burned off in the fluidized-bed regenerator, and then the regenerated, hot catalyst is transported back to the bottom of the riser to crack the feed oil. Large FCC units have to control solids flow rates from 10 to 80 tons/min. The units require makeup catalyst to be added to replace solids losses due to attrition and other losses. The amount of catalyst makeup is small, and need not be continuous. Therefore, the makeup catalyst is fed into the commercial unit from pressurized hoppers into one of the conveying lines. However, the primary solids flow control problem in this FCC unit is to maintain the correct temperature in the riser reactor by controlling the flow of hot regenerated catalyst around the test unit. This is done by using large, 1.2-m (4-ft)-diameter slide valves (also known as knife-gate valves) located in standpipes to control the flow rates of catalyst.

In the FCC process, the solids are transferred out of the fluidized-bed regenerator into the bottom of the riser via a standpipe. The purpose of a standpipe is to transfer solids from a low-pressure region to a high-pressure region via gravity. The point of removal of the solids from the regenerator bed is at a lower pressure than the point of feed introduction into the riser. Therefore, the transfer of solids from the regenerator bed to the bottom of the riser is accomplished with a standpipe. The standpipes in FCC units can be as large as 1.5 m (5 ft) in diameter, and as long as about 30 m (100 ft). They can be either vertical or angled (generally approximately 60° from the horizontal). The pressure is higher at the bottom of a standpipe due to the relative flow of gas counter to the solids flow. The gas in the standpipe may be flowing either downward relative to the pipe wall, but more slowly than the solids (the most common occurrence), or upward relative to the pipe wall. The standpipe may be fluidized, or the solids may be in moving packed-bed flow.

There are two basic types of standpipe configurations: the overflow standpipe and the underflow standpipe (Fig. 17-18). The overflow standpipe is so named because the solids "overflow" from the top of the fluidized bed into the standpipe, and there is no bed of solids above the standpipe. In the underflow standpipe, the solids are introduced into the standpipe from the underside, or bottom, of the bed or hopper, and a bed of solids is present above the standpipe. With this definition, a cyclone dipleg is classified as an overflow standpipe because there is no bed of solids above the entrance to the dipleg.

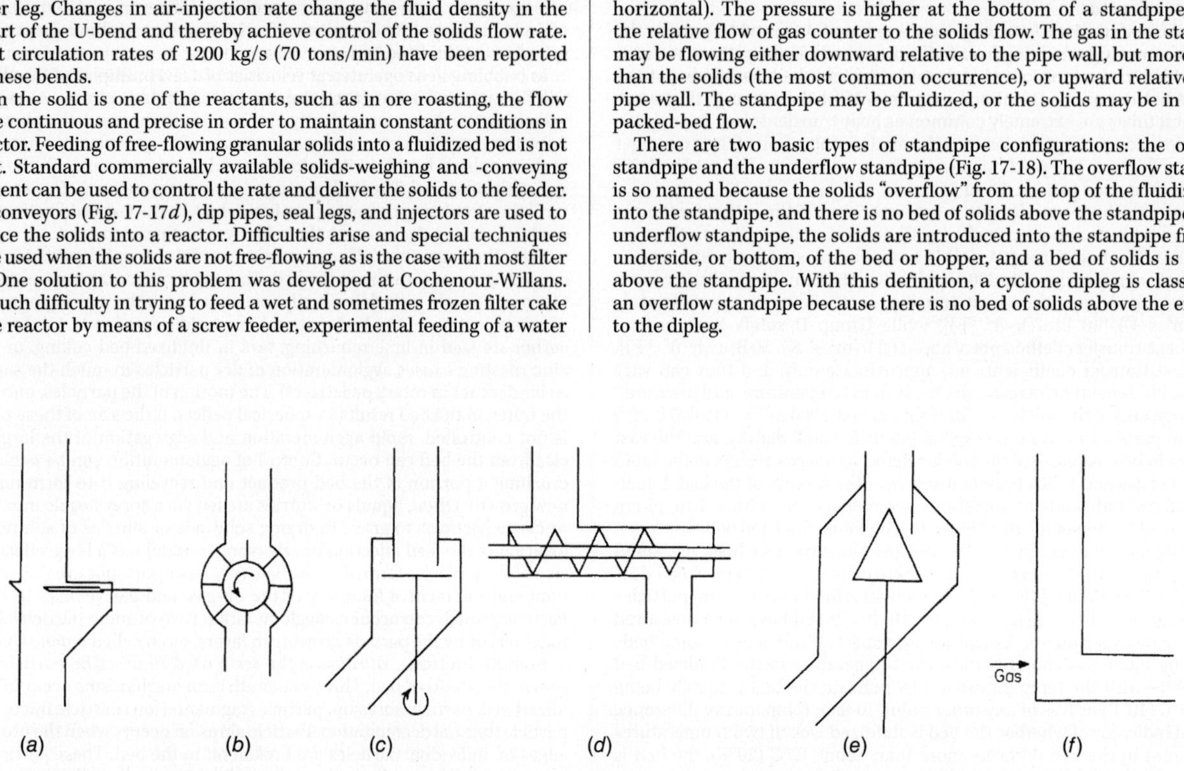

FIG. 17-17 Solids flow control devices: (*a*) slide valve, (*b*) rotary valve, (*c*) table feeder, (*d*) screw feeder, (*e*) cone valve, (*f*) L-valve.

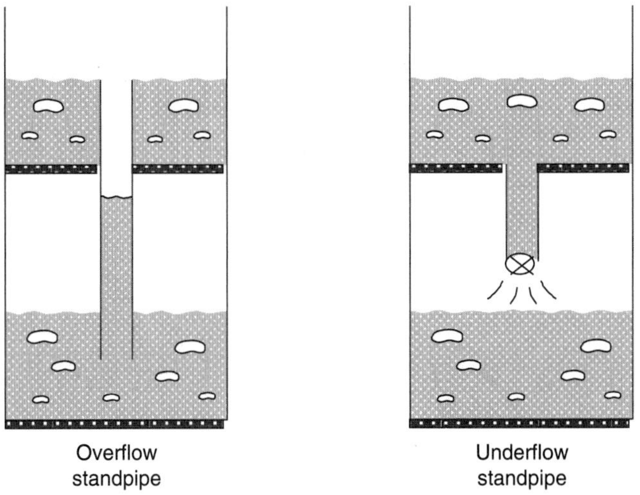

FIG. 17-18 Schematic depiction of overflow and underflow standpipes. (*Courtesy of PSRI, Chicago, Ill.*)

With the two types of standpipe configurations and the two typical standpipe flow regimes (fluidized and packed bed), there are four different types of standpipes:

1. An underflow packed-bed standpipe
2. An underflow fluidized-bed standpipe
3. An overflow fluidized-bed standpipe
4. An overflow packed-bed standpipe

All of these standpipes are used extensively in industry except for the overflow packed-bed standpipe. It is possible for this type of standpipe to operate, but it is much harder to operate and control than the others. Therefore, it is not used.

Fluidized standpipes can accommodate a much higher solids flow rate than moving packed-bed standpipes because the friction of the solids flow on the wall of the standpipe is much less in fluidized standpipes.

One of the most common standpipes in industry is the fluidized underflow standpipe (Fig. 17-19). With the fluidized underflow standpipe, aeration gas is added to the standpipe to maintain the solids in a fluidized state as they flow down the standpipe. As the solids flow down the fluidized underflow standpipe from a low pressure to a higher pressure, the gas in the standpipe is compressed, which causes the solids to move closer together. When the standpipe is operating at low pressures, the percentage change in gas density from the top of the standpipe to the bottom can be significant. If aeration is not added to the standpipe to prevent this, the solids can defluidize near the bottom of the standpipe. Defluidization of solids in this standpipe results in less pressure buildup in the standpipe and a reduction in the solids flow rate through it.

To keep the solids in a fluidized underflow standpipe in a fluidized state, aeration gas is added to the standpipe. Adding the correct amount of gas uniformly (every 1.5 to 2 m) in a commercial fluidized underflow standpipe will prevent defluidization at the bottom of the standpipe. If the material flowing in the standpipe is a Geldart Group A material, it is required that the aeration be added uniformly along the standpipe. If the aeration is added at only one location (e.g., at the bottom of the standpipe), a large bubble will form in the standpipe at the aeration point (Fig. 17-20*a*). If the bubble is large enough, it can restrict the flow of solids down the standpipe. The large bubble forms because it is difficult for the aeration gas to permeate the very fine solids moving through the standpipe. Therefore, it requires a significant area for the gas to dissipate through the very fine particles at the same rate that it is being added through the aeration tap. If the aeration gas is added at several locations, then the bubble size is significantly reduced, and standpipe operation is significantly improved (Fig. 17-20*b*). Aeration bubbles extend downward from the aeration point in the direction of flow of the solids (Fig. 17-20*b*). This occurs because the momentum of the solids is much greater than the buoyancy force of the bubble, and it elongates the aeration bubble in the direction of flow.

Typically, the pressure drop across the solids control valve in a commercial fluidized underflow standpipe should be designed for a minimum of approximately 2 psi (14 kPa) for good control. A maximum of no more than 10 to 12 psi (70 to 84 kPa) is recommended to prevent excessive erosion of the valve at high pressure drops [Zenz, *Powder Technol.* **2:** 105–113 (1986)].

For Geldart Group B solids, it is often unnecessary to add aeration at several locations along the standpipe to maintain the standpipe in fluidized flow. Adding aeration at the bottom of the standpipe operating with Group B solids is generally sufficient. This is because the gas can permeate through the larger Group B particles more easily than through the Group A particles (Group A particles have a significantly larger surface area and produce more drag for the same gas flow conditions).

The amount of aeration required to maintain solids in a fluidized state throughout a fluidized underflow standpipe was presented by Karri and Knowlton [*Circulating Fluidized Bed Technology* **IV:** 253 (1993)] to be:

$$Q = 2000\left[\frac{P_b}{P_t}\left(\frac{1}{\rho_{mf}} - \frac{1}{\rho_{sk}}\right) - \left(\frac{1}{\rho_t} - \frac{1}{\rho_{sk}}\right)\right] \tag{17-8}$$

where Q is the aeration required in actual cubic feet per ton of solids flowing in the standpipe, P_b is the pressure at the bottom of the standpipe in psia, P_t is the pressure at the top of the standpipe in psia, ρ_{mf} is the fluidized bed density at minimum fluidization in lb/ft^3, ρ_{sk} is the skeletal density of the particles in lb/ft^3, and ρ_t is the density at the top of the standpipe in lb/ft^3.

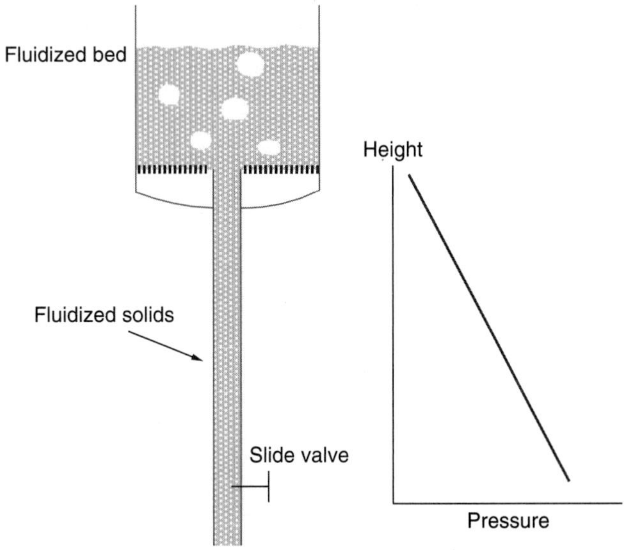

FIG. 17-19 Schematic depiction of a fluidized, underflow standpipe. (*Courtesy of PSRI, Chicago, Ill.*)

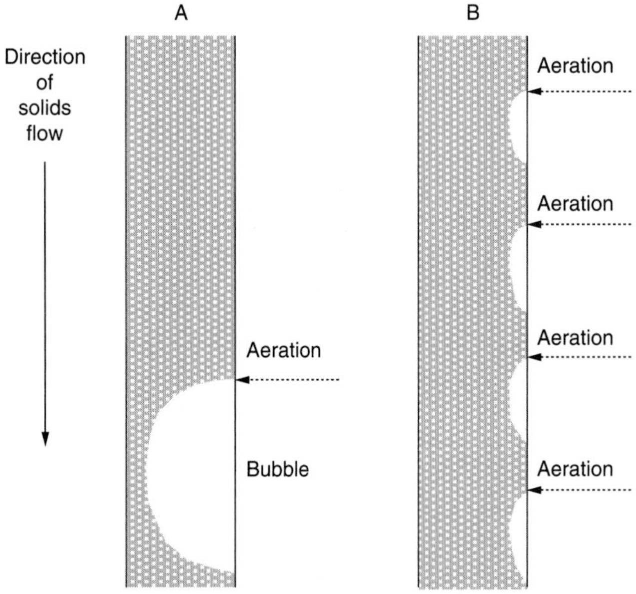

FIG. 17-20 The effect of adding aeration at a single point or multiple points in a Geldart group A underflow fluidized standpipe. (*Courtesy of PSRI, Chicago, Ill.*)

This is an estimate of the theoretical amount of aeration that should be added to the standpipe. In practice, it has been found that about 70 percent of the theoretical amount is a better estimate of the actual aeration required. In a commercial fluidized underflow standpipe, the amount of aeration theoretically required is added in equal increments via aeration taps located approximately 1.5 to 2 m apart. Care should be taken not to overaerate the standpipe. If this occurs, large bubbles are generated in the standpipe that hinder solids flowing down the standpipe. Thus, standpipes can be overaerated as well as underaerated.

As indicated above, it is detrimental to have bubbles in standpipes. For fluidized-bed underflow standpipes with the standpipe entrance in the fluidized bed, bubbles can be "sucked" down the standpipe at its entrance if nothing is done to prevent this from occurring. This is especially true when the bed consists of Geldart Group A solids. When solids flow from a fluidized bed into the top of an underflow fluidized-bed standpipe, the solids are accelerated from a low velocity near 0 ft/s in the bed to as much as 6 ft/s in the standpipe. This sudden increase in solids velocity can carry bubbles with the solids down into the standpipe and degrade standpipe operation. To prevent this, a cone (Fig. 17-21) is often added to the top of the standpipe to minimize the solids velocity at the standpipe entrance and minimize bubble "carryunder." Experience has shown that the diameter of the standpipe inlet cone should be four to six times the area of the standpipe [King, *Fluidization VII:* 15, (1991)]. Standpipe inlet cones typically have an included angle of from 25° to 35°.

In many fluidized beds, a sparger type of gas distributor is used to fluidize the bed. The sparger consists of a pipe with nozzles in it inserted into the bottom of the fluidized bed. Solids flow down through the distributor and into the standpipe. Another technique to prevent bubbles from entering the top of the standpipe can be used with sparger grids. Instead of having the standpipe entrance in the bed, the standpipe entrance is located below the sparger grid. As the solids flow between the sparger grid and the standpipe entrance, the bubbles dissipate and do not enter the standpipe. Generally, an aeration ring is added around the standpipe to ensure that the solids are fluidized as they enter the standpipe. With this configuration, a cone at the entrance of the standpipe is not required.

Underflow fluidized standpipes are operated in either a vertical configuration, a completely angled configuration, or a hybrid configuration in which both vertical and angled sections are present (Fig. 17-22). Angling a standpipe is a very convenient way to transfer solids between two points that are separated horizontally as well as vertically. However, it has been found by Karri and Knowlton [*Circulating Fluidized Bed Technology* **IV:** 253 (1993)] and Yaslik [*Circulating Fluidized Bed Technology* **IV:** 484 (1993)] that long, angled underflow fluidized standpipes do not perform as well as vertical standpipes.

Sauer et al. [*AIChE Symposium Series 234* **80:** 1 (1984)] and Karri and Knowlton [*Circulating Fluidized Bed Technology* **IV:** 253 (1993)] studied hybrid angled standpipe operation using transparent standpipes to allow visual observation of the flow. Both found that the gas and solids separated in the standpipe, with the gas bubbles flowing up along the upper portion of the standpipe while the solids flowed down along the bottom portion of the standpipe (Fig. 17-23). The pressure buildup in the hybrid standpipe was lower than that in the vertical standpipe, and Karri and Knowlton [*Circulating Fluidized Bed Technology* **IV:** 253 (1993)] reported that the maximum solids mass flux possible in a hybrid angled underflow fluidized standpipe was less than that attainable in a vertical underflow fluidized standpipe. The principal reason for this is that the rising bubbles in the angled section of the standpipe become relatively large at a low solids flow rate (and low aeration rate). At a certain solids mass flux, the bubbles become large enough to bridge across the vertical section at the top of the standpipe, hindering the solids flow. When this occurs, the maximum solids flow rate in the hybrid angled standpipe has been achieved.

Karri et al. [*Fluidization* **VII:** 1075 (1995)] showed that the solids flow rate through a hybrid angled standpipe can be increased if a bypass line is added between the top of the angled section of the standpipe and the freeboard of the bed above it. The bypass line allows the bubble gas from the angled section to bypass the vertical section of the pipe so that large bubbles are not formed there. Thus, the solids flow rate can be increased. Karri et al. [*Fluidization* **VII:** 1075 (1995)] reported that if the bypass was used, the solids flow rate could be increased to such a value that the solids velocity in the hybrid standpipe became greater than the bubble rise velocity, and the bubbles were carried down the standpipe with the solids. When the bubbles were being carried down the standpipe by the solids, the bypass line could then be closed, and the standpipe would operate without slugging in the vertical section.

Even though vertical standpipes can transfer solids more efficiently than hybrid angled standpipes, true angled standpipes (those containing no vertical section) are commonly operated satisfactorily in large FCC units with Geldart Group A catalyst. However, these standpipes are relatively short, and they are designed so that the mass flux through them is not too high so that they can be operated satisfactorily. Yaslik [*Circulating Fluidized Bed Technology* **IV:** 484 (1993)] found that a long, angled standpipe had a limited solids circulation rate relative to vertical standpipes. Thus, when operating a hybrid angled standpipe or a true angled standpipe it is essential to: (1) keep the solids mass flux through the standpipe below a value which will lead to slugging, and (2) keep the line as short as possible so that the large gas slugs will not have as great a length in which to form.

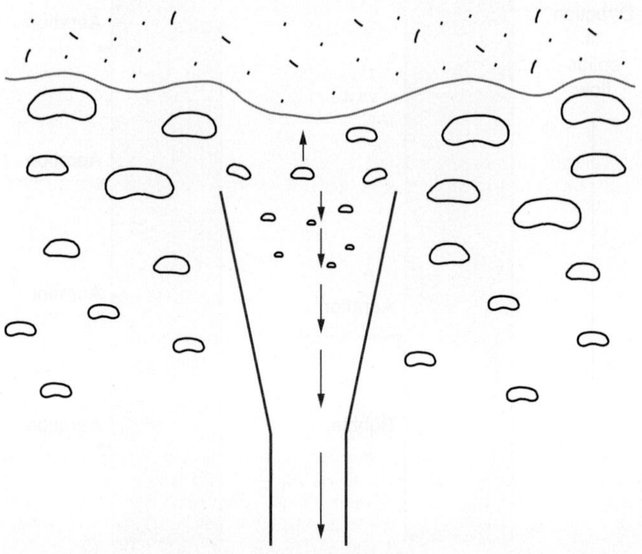

FIG. 17-21 Schematic depiction of an inlet cone for a fluidized underflow standpipe. (*Courtesy of PSRI, Chicago, Ill.*)

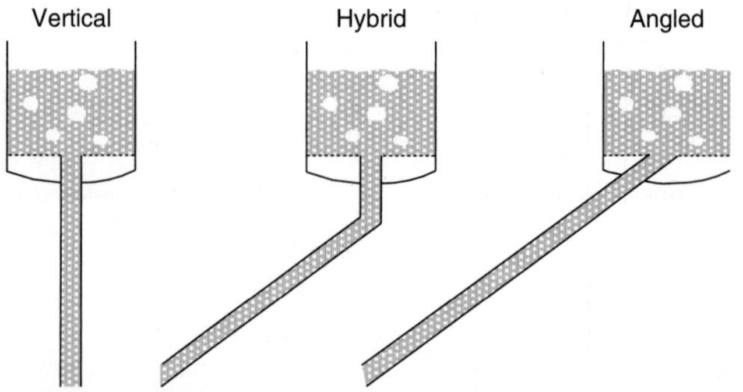

FIG. 17-22 Schematic depiction of vertical, angled, and hybrid standpipes. (*Courtesy of PSRI, Chicago, Ill.*)

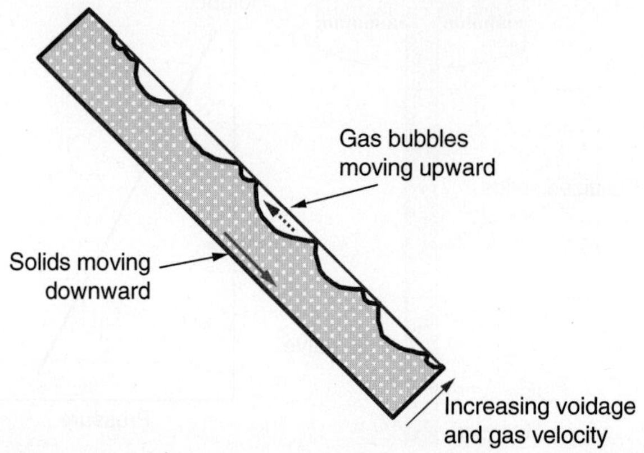

FIG. 17-23 Depiction of gas and solids flow in an angled standpipe. (*Courtesy of PSRI, Chicago, Ill.*)

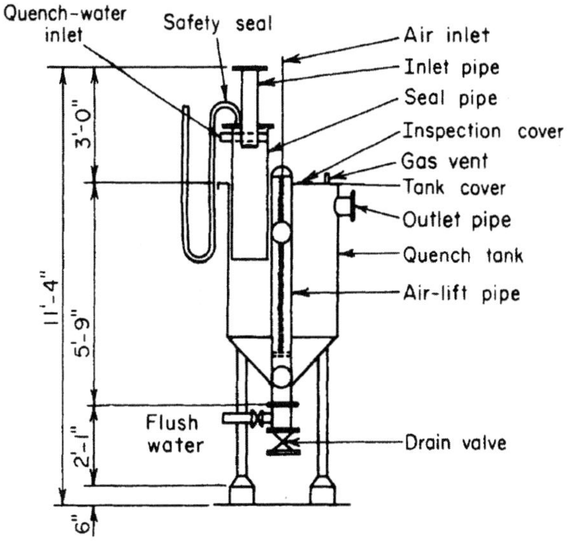

FIG. 17-24 Quench tank for overflow or cyclone solids discharge. [*Gilfillan et al., "The FluoSolids Reactor as a Source of Sulphur Dioxide,"* J. Chem. Metall. Min. Soc. S. Afr. (*May 1954*).]

Solids Discharge The type of fluidized-bed discharge mechanism used is dependent on the necessity of sealing the atmosphere inside the fluidized-bed reactor and the subsequent treatment of the solids. One of the simplest solids discharge types is an overflow weir. This can be used only when the escape of fluidizing gas does not present any hazards due to its nature or dust content, or when the leakage of gas into the fluidized-bed chamber from the atmosphere into which the bed is discharged is permitted. Solids will overflow from a fluidized bed through a port even though the pressure above the bed is maintained at a slightly lower pressure than the exterior pressure. When it is necessary to restrict the flow of gas through the opening, a simple flapper valve can sometimes be used. Overflow to a combination seal and quench tanks (Fig. 17-24) is used when it is permissible to wet the solids and when disposal or subsequent treatment of the solids in slurry form is desirable. The fluidized seal pot and a loop seal (sometimes called a FluoSeal) are simple and effective ways of sealing and purging gas from the solids when an overflow-type discharge standpipe is used (Fig. 17-25). The upleg of the loop seal must be fluidized in order for it to work properly. A more recent loop seal design has the downleg angled at about 45° to 60° from the horizontal (Fig. 17-26) so that it connects directly with the upleg. This angled design eliminates the horizontal flow section of the loop seal.

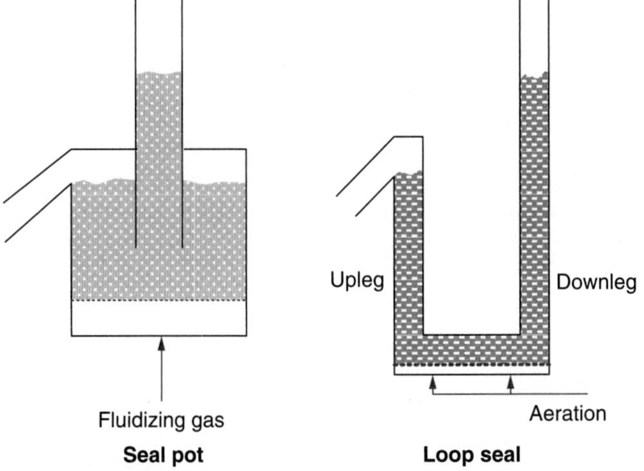

FIG. 17-25 Schematic drawing of a seal pot and a loop seal. (*Courtesy of PSRI, Chicago, Ill.*)

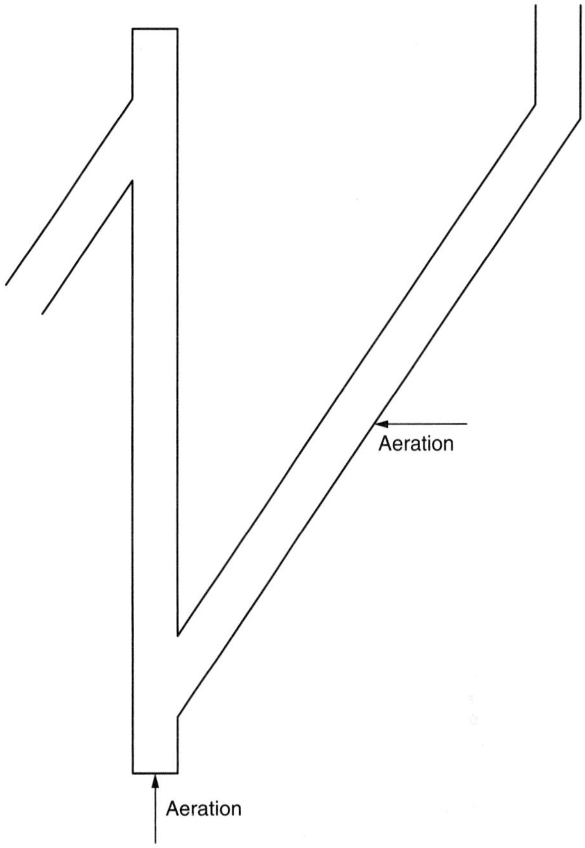

FIG. 17-26 Angled loop seal. (*Courtesy of PSRI, Chicago, Ill.*)

A star (rotary) valve is an effective sealing device for solids discharge. It functions well with a head of solids above it. Bottom-of-the-bed discharge is also acceptable via a slide valve with a head of solids.

Seal legs are often used in conjunction with both mechanical and nonmechanical solids-flow-control valves to equalize pressures and to strip trapped or adsorbed gases from the solids. The operation of a seal leg is shown schematically in Fig. 17-27. The solids flow by gravity from the fluidized bed into the seal leg or standpipe. Seal and/or stripping gas is introduced near the bottom of the leg. If designed properly, this gas flows both upward and downward. Pressures indicated in the illustration have no absolute value but are only relative. The seal legs can be designed for either fluidized or nonfluidized solids.

The L-valve is shown schematically in Fig. 17-28. The L-valve can increase or decrease the flow rate of solids through it by adding more or less aeration gas to it, respectively. In the control mode, the standpipe above the L-valve must be an underflow, moving (nonfluidized) bed [Knowlton, T. M., *Standpipes and Nonmechanical Valves*, chap. 21 in *Handbook of Fluidization and Fluid-Particle Systems,* ed. W. C. Yang, Marcel Dekker, New York, (2003), p. 575].

The L-valve can also act like a nonsolids flow control seal where it does not control the flow rate of solids through it. When operating in this mode, the standpipe above it should be an overflow, fluidized-bed standpipe [Knowlton, T. M., *Standpipes and Nonmechanical Valves,* chap. 21 in *Handbook of Fluidization and Fluid-Particle Systems,* ed. W. C. Yang, Marcel Dekker, New York, (2003), p. 575]. In this mode it is called an "automatic" L-valve. Thus, the L-valve can act both as a seal and as a solids-flow control valve. When used to control the solids flow rate, it is only practical to use the L-valve for solids that deaerate quickly (Geldart Group B and D solids). Generally, the average particle size of the solids must be greater than about 100 to 120 μm for the L-valve to work well. The height at which aeration is added to the L-valve in Fig. 17-28 is usually 1.5 pipe diameters above the centerline of the horizontal section of the L-valve. For L-valve design equations, see Yang and Knowlton [*Powder Tech.***77:** 49–54 (1993)] and [Knowlton, T. M., *Standpipes and Nonmechanical Valves,* chap. 21 in *Handbook of Fluidization and Fluid-Particle Systems,* ed. W. C. Yang, Marcel Dekker, New York, (2003), p. 575].

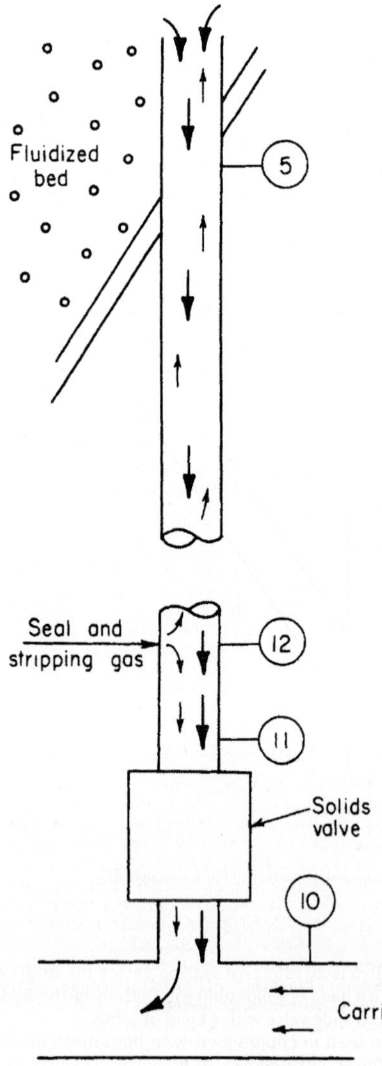

FIG. 17-27 Schematic drawing of a fluidized seal leg. Circled numbers refer to pressure taps.

In the sealing mode, the standpipe above the L-valve (a dipleg if the "automatic" L-valve is located below a cyclone) should be fluidized as indicated previously. Gas introduced below the normal solids level and above the discharge port will usually flow upward and downward—but this depends on the solids velocity in the standpipe. The relative flow in each direction is self-adjusting, depending on the differential pressure between the point of solids feed and discharge and the level of solids in the leg. The length and diameter of the horizontal section of the L-valve are selected

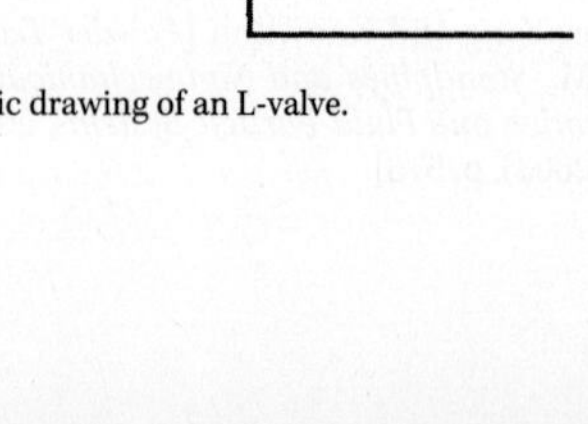

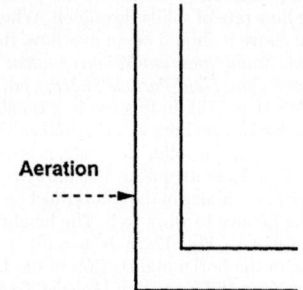

FIG. 17-28 Schematic drawing of an L-valve.

so that the undisturbed angle of repose of the solids will prevent discharge of the solids. As solids rate into the dipleg increases, the height, *H,* of solids in the standpipe increases. This increase in the solids level in the standpipe reduces the flow of gas in the upward direction and increases the flow of gas in the downward direction. When the flow of gas downward through the "automatic" L-valve increases, it causes more solids to flow around the automatic L-valve. Usually, the level of solids above the point of gas introduction will float.

Dust Separation It is usually necessary to recover the solids carried by the gas leaving the disengaging space or freeboard of the fluidized bed and return them to the bed. Generally, cyclones are used to collect these solids (see the subsection Gas–Solids Separation). However, in a few cases, filters are employed without the use of cyclones to reduce the loading of solids in the off-gas. For high-temperature usage, either porous ceramic or sintered metal filters have been employed. Multiple units must be provided so that one or more units can be blown back with clean gas while one or more are filtering.

Cyclones are arranged generally in any one of the arrangements shown in Fig. 17-29. The effect of cyclone arrangement on the height of the vessel and the overall height of the system is apparent. Details regarding cyclone design and collection efficiencies are to be found in another part of this section.

Discharging of the solids collected by a cyclone back into the fluidized bed requires some care. It is necessary to pressure seal the bottom of the cyclone so that the collection efficiency of the cyclone will not be impaired by the passage of appreciable quantities of gas up through the dipleg and into the bottom of the cyclone. This is usually done by (1) sealing the dipleg in the fluid bed, or (2) adding a trickle or flapper valve to the bottom of the dipleg if the dipleg is terminated in the freeboard of the fluidized bed.

Many processes start up their fluidized beds from an empty bed. Solids are added to the bed gradually as the unit temperature is gradually increased. There is usually no problem with primary cyclone diplegs during this period because the solids flux through the primary dipleg is very high, and it carries gas downward with the solids. However, the secondary cyclone diplegs have a very low flux through them, and the gas can travel back up the secondary dipleg, preventing solids from making the pressure seal in the secondary dipleg. To prevent too much gas from flowing up the secondary dipleg at start-up, secondary cyclone termination devices are used. The most common of these termination devices are shown in Fig. 17-30.

Because the solids fluxes through primary cyclone diplegs can be very high [up to 150 lb/s/ft^2 (750 kg/s/m^2)] a horizontal plate called a splash plate (somewhat larger in diameter than the dipleg) is often attached to the bottom of the primary cyclone dipleg to (1) help disperse the solids, and (2) shield internals from being eroded due to the momentum of the solids exiting the primary cyclone dipleg. Care must be taken to ensure that the horizontal plate (sometimes called a "dollar" plate) is located far enough away from the dipleg outlet that the solids discharge from the dipleg is not constricted.

In addition to an open dipleg (a dipleg with no termination device) immersed into a fluidized bed, various other devices have been used to seal cyclone dipleg returns, especially for second-stage cyclones. Several of these are shown in Fig. 17-30. One of the most often used is the trickle valve (Fig. 17-30*a*). Trickle valves work best when they are immersed in the fluidized bed. Dipleg operation is more stable when the trickle valve is immersed, and solids losses are generally reduced. However, trickle valves can discharge solids into the freeboard if necessary.

Another common dipleg termination device is the flapper valve (Fig. 17-30*b*). This device is similar to the trickle valve, but has its "flapper plate" located horizontally instead of vertically. A counterweight is used to ensure that the fluidized solids leg above the flapper can cause the flapper to open relative easily.

An "automatic" L-valve (Fig. 17-30*c*) can also be used as a dipleg termination. These devices are most commonly used at the bottom of external cyclone diplegs. It is necessary to add aeration gas to the automatic L-valve to ensure the best performance.

The loop seal is shown in Fig. 17-30*d*. This device is very common at the bottom of the external cyclone dipleg for circulating fluidized-bed combustors. This device works smoothly to transfer the solids from the cyclone back into the fluidized bed. The upleg of the loop seal must be fluidized in order for it to function properly. The downleg should be fluidized as well for practical, effective operation, but it is not absolutely necessary that it be fluidized.

A simple open-ended dipleg is shown in Fig. 17-30*d*. The open-ended dipleg can be used at the bottom of primary cyclone diplegs, but not general secondary cyclone diplegs. It is the simplest of all of the dipleg/termination devices. A splash plate is shown below the open-ended dipleg in the figure. The splash plate is used to disperse the solids into the bed more evenly, and to protect any bed internals that may be located immediately below the dipleg discharge.

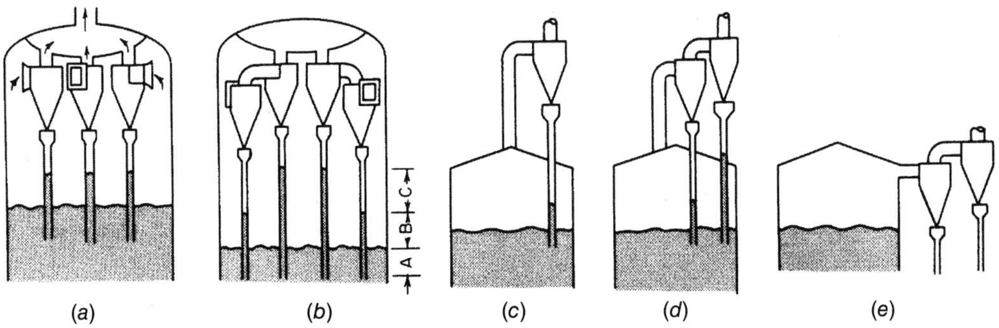

FIG. 17-29 Fluidized-bed cyclone arrangements. (*a*) Single-stage internal cyclone. (*b*) Two-stage internal cyclone. (*c*) Single-stage external cyclone; dust returned to bed. (*d*) Two-stage external cyclone; dust returned to bed. (*e*) Two-stage external cyclone; dust collected externally.

In any event, the diplegs must be large enough to accommodate momentarily high rates of solids flows (primary dipleg) and must provide solid seals to overcome cyclone pressure drops as well as to allow for differences in fluid density of solids in the bed and the solids in the diplegs. It has been reported that, in the case of catalytic processes operating with Geldart group A solids, the fluid density of the solids collected by the primary cyclone is essentially the same as that in the fluidized bed. This is so because in most of these processes, the bed fluidizing velocity is so high and the particles in the bed are so small that generally all are entrained. However, as a general rule, the fluidized density of solids collected by the second-stage cyclone is significantly less than the fluidized density of the bed. Each succeeding cyclone collects finer solids, and the smaller the solids, the lower the fluidized density.

The dipleg of both the primary and secondary cyclone must be long enough to seal the imposed pressure drop across them. A representative calculation showing the length of dipleg required for one specific installation is given in Example 17-1.

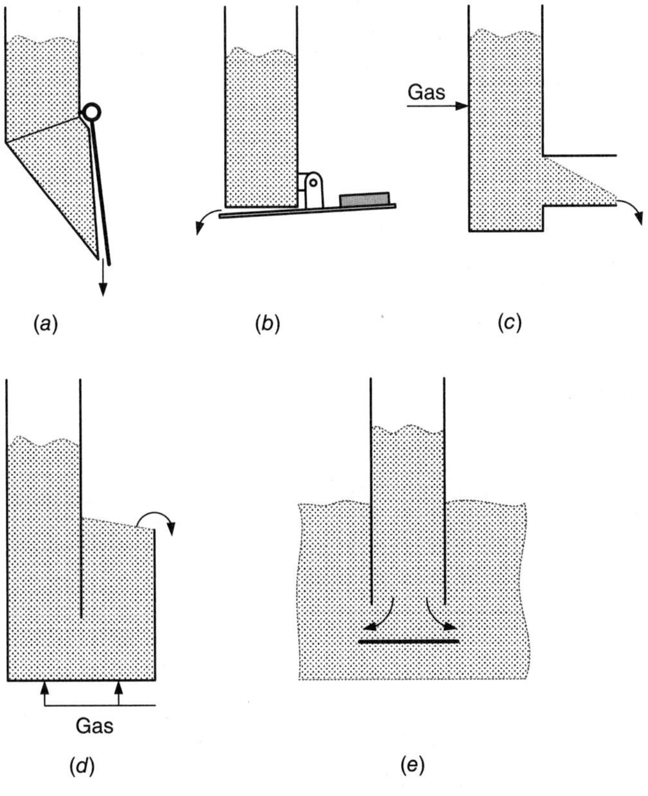

FIG. 17-30 Common cyclone dipleg terminations. (*a*) trickle valve, (*b*) flapper valve, (*c*) "automatic" L-valve, (*d*) loop seal, (*e*) open-ended dipleg with splash plate. (*Courtesy of PSRI, Chicago, Ill.*)

Example 17-1 Calculation of Required Length of Cyclone Seal Leg (Dipleg) The length of the required cyclone solids seal height in the seal leg (dipleg) for a single cyclone can be calculated as shown here.

Given:

Fluidized density of the fluidized bed = 1100 kg/m^3 (68.8 lb/ft^3)

Fluidized density of solids in the cyclone dipleg = 650 kg/m^3 (40.6 lb/ft^3)

Settled bed depth = 1.5 m (5 ft)

Fluidized bed depth = 2.4 m (8 ft)

Pressure drop through the cyclone = 1.4 kPa (0.2 lbf/in^2)

In order to ensure that the dipleg is sealed at start-up, the bottom of the seal leg (dipleg) is submerged 0.9 m (2.95 ft) in the fluidized bed.

From the information given, the total pressure differential to be balanced by the fluidized seal leg in the cyclone dipleg is:

$$(0.9 \times 1100 \times 9.81)/1000 + 1.4 = 11.1 \text{ kPa} \qquad \text{(in SI units)}$$

$$[(2.95 \times 68.8)/144 + 0.2 = 1.61 \text{ lb/in}^2] \qquad \text{(in fps units)}$$

To balance this differential pressure, the height of solids in the dipleg must be = $(11.1 \times 1000)/(650 \times 9.81) = 1.74$ m [$(1.61 \times 144)/40.6 = 5.7$ ft]; therefore, the bottom of the separator pot on the cyclone must be at least 1.74 + 1.5 or 3.24 m (5.7 + 5 or 10.7 ft) above the gas distributor. To take into account various contingencies, upsets, changes in size distribution, etc., this distance should be increased. Normally, the calculated solids height should be approximately 1.5 m (5 feet) below the bottom of a commercial cyclone. Therefore, the cyclone discharge level should be at least 3.24 + 1.5 = 4.74 m (10.7 + 5 = 15.7 ft) above the gas distributor.

Cyclones are less effective as the particle size entering them decreases, so secondary collection units are often required, such as filters, electrostatic precipitators, and scrubbers. When dry collection is not required, elimination of cyclones is possible using scrubbers if allowance is made for heavy solids loads in the scrubber (see the subsection Gas–Solids Separations; see also Sec. 14).

Instrumentation

Temperature Measurement Temperature measurement in fluidized beds is usually simple, and standard temperature-sensing elements (usually thermocouples) are adequate for continuous use. Because of the high abrasion wear on horizontal thermocouple protection tubes, vertical installations are often used. In highly corrosive atmospheres in which metallic protection tubes cannot be used, short, heavy ceramic tubes have been used successfully.

Pressure Measurement Although successful pressure measurement probes or taps have been fabricated by using porous materials, the most universally-accepted pressure tap consists of a purged tube projecting into the bed. Minimum internal diameters of the tube are 0.6 to 1.2 cm (0.25 to 0.5 in). A purge velocity of 2 to 4 m/s (6 to 12 ft/s) is usually required to prevent solids from plugging the signal lines. Bubbling fluidized-bed density is determined directly from $\Delta P/L$, the pressure drop inside the bed itself ($\Delta P/L$ in units of weight/area $\times$ $1/L$). The overall bed weight of a bubbling fluidized bed is obtained from a ΔP taken between a point just above the gas distributor and a point in the freeboard. This ΔP reading is equal to the weight of the bed (W) divided by the bed area (A). Multiplying the overall bed pressure drop by the area yields the weight of the bed. Nominal bed height of a bubbling fluidized bed is determined by dividing the ΔP across the entire bed by the $\Delta P/L$ over a section of the bed length. Splashing of the solids by bubbles bursting at the bed surface will eject solids well above the nominal bed height in most cases. The pressure drop signal from fluidized beds fluctuates due to bubble effects and the generally statistical nature of fluid-bed flow parameters. A fast Fourier transform of the pressure drop signals from the bed transforms the pressure perturbations to a frequency-versus-amplitude plot with a maximum at about 3 to 5 Hz. Changes in frequency and amplitude are associated with changes in the quality of the fluidization.

Experienced operators of fluidized beds can often predict what is happening in the bed from changes in the ΔP fluctuation signal. However, one of the best indications of whether a fluidized bed is operating properly is to monitor the temperatures inside the bed. If the difference between any two thermocouple readings is greater than 20°F (about 10°C), then one should be worried about the nature of the fluidization of the bed.

Flow Measurements Measurement of flow rates of clean gases entering the fluidized bed presents no problem. Flow measurement of gas streams containing solids is almost always avoided. The flow of solids in very large fluidized-bed processes is usually controlled, but not measured, except for the solids flows added to or taken from the system. This is because measuring solids flows is very difficult even in laboratory units, and it is not practical in larger units using present technology. Solids flows in the system are usually adjusted and determined on an inferential basis using variables such as temperature, pressure level, catalyst activity, gas analysis, and heat balance. In many roasting operations, the color of the calcine discharge material indicates whether the solids feed rate is too high or too low.

USES OF FLUIDIZED BEDS

There are many uses of fluidized beds. A number of applications have become commercial successes; others are in the pilot-plant stage, and others in bench-scale stage. Generally, the fluidized bed is used for gas–solids contacting. Uses of fluidized beds are listed below:

I. Chemical reactions
- *A.* Catalytic
- *B.* Noncatalytic
  1. Homogeneous
  2. Heterogeneous

II. Physical contacting
- *A.* Heat transfer
  1. To and from fluidized bed
  2. Between gases and solids
  3. Temperature control
  4. Between points in the bed
- *B.* Solids mixing
- *C.* Gas mixing
- *D.* Drying
  1. Solids
  2. Gases
- *E.* Size enlargement
- *F.* Size reduction
- *G.* Classification
  1. Removal of fines from solids
  2. Removal of fines from gas
- *H.* Adsorption-desorption
- *I.* Heat treatment
- *J.* Coating

Chemical Reactions

Catalytic Reactions The use of fluidized beds to optimize chemical catalytic reactions has provided the greatest impetus for the use of fluidized solids. Some of the details pertaining to this use are to be found in the preceding pages of this section. Reference should also be made to Sec. 21. Several of the catalytic process that use fluidized beds are described in the following text.

Fluidized Catalytic Cracking The fluidized catalytic cracking (FCC) process using Geldart Group A solids is the oldest and, still today, the most important commercial application of fluidized beds. The evolution of fluidized catalytic cracking since the early 1940s has resulted in several fluidized-bed process configurations. The high solids transfer rate between the fluidized-bed regenerator and the riser reactor in this process permits a balancing of the exothermic burning of carbon and tars in the regenerator and the endothermic cracking of petroleum in the reactor. Therefore, the temperature in both units can usually be controlled without resorting to auxiliary heat control. The high catalyst circulation rate also permits the maintenance of the catalyst at a constantly high activity. The early fluidized-bed regenerators were considered to be completely backmixed units. Newer processes now have staged regenerators to improve conversion (see Fig. 17-31). The use of the riser reactor operating in the fast fluid-bed mode also results in much lower gas and solids backmixing due to the more plug-flow nature of the riser. Staging and operating in the fast fluidized-bed mode facilitates cracking of the oil in the riser, and it allows for much more efficient operation in the regenerator.

The first fluid catalytic cracking unit was operated by Exxon and was called the Model I. It started operation in Baytown, Texas, in 1942. This was a low-pressure, 14- to 21-kPa (2- to 3-psig) unit operating in what is now called the turbulent fluidized-bed mode with a gas velocity of 1.2 to 1.8 m/s

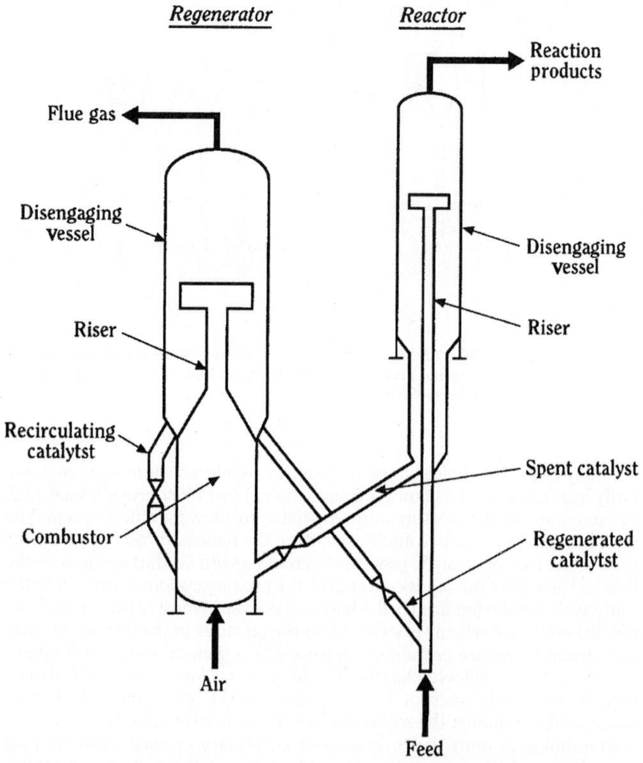

FIG. 17-31 UOP fluid catalytic cracking unit. (*Reprinted with permission of UOP.*)

(4 to 6 ft/s). Before the start-up of the Model I cracker, it was realized that by lowering the gas velocity in the bed, a dense, bubbling, or turbulent fluidized bed, with a bed density of 300 to 400 kg/m^3 (20 to 25 lb/ft^3), would be formed. The increased gas–solids contacting time in the denser bed allowed completion of the cracking reaction. System pressure was eventually increased to 140 to 210 kPa (20 to 30 psig) over the years.

In the 1970s, extremely active zeolite catalysts were developed so that the cracking reaction could be conducted in the transport riser itself and not in a dense, fluidized bed. Recently, heavier crude feedstocks have resulted in higher coke production in the cracker. The extra coke causes higher temperatures in the regenerator than are desired. This has resulted in the addition of catalyst cooling to the regeneration step, as shown in Fig. 17-32.

Many companies have participated in the development of the fluid catalytic cracker, including ExxonMobil Research & Engineering Co., UOP, Kellogg Brown and Root, Chevron, Gulf Research Development Co., and Shell Oil Company. Many of these companies provide designs and/or licenses to operate these units to others. For further historical and technical details, see Luckenbach et al., "Cracking, Catalytic," in *Encyclopedia of Chemical Processing and Design,* vol. 13, ed. McKetta, Marcel Dekker, New York, 1981, pp. 1–132.

Alkyl Chloride In this process, olefins are chlorinated to alkyl chlorides in a single fluidized bed. HCl reacts with O_2 over a copper chloride catalyst to form chlorine. The chlorine reacts with the olefin to form alkyl chloride. The process was developed by the Shell Development Co., and uses a recycle of catalyst fines in aqueous HCl to control the temperature [*Chem. Proc.* **16:** 42 (1953)].

Phthalic Anhydride To produce phthalic anhydride, naphthalene is oxidized by air to phthalic anhydride in a bubbling/turbulent fluidized reactor. Even though the naphthalene feed is in liquid form, the reaction is highly exothermic. Temperature control is achieved by removing heat by placing vertical tubes in the bed to raise steam [Graham and Way, *Chem. Eng. Prog.* **58:** 96 (January 1962)].

Acrylonitrile Acrylonitrile is produced by reacting propylene, ammonia, and oxygen (air) in a single fluidized bed using a complex catalyst. Known as the SOHIO process, this process was first operated commercially in 1960. In addition to acrylonitrile, significant quantities of HCN and acetonitrile are produced. This process is also exothermic, and temperature control is achieved by raising steam inside vertical tubes immersed in the bed [Veatch, *Hydrocarbon Process. Pet. Refiner* **41:** 18 (November 1962)].

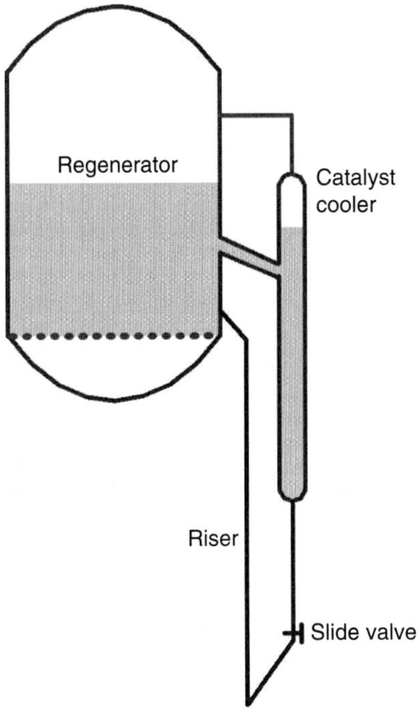

FIG. 17-32 Schematic drawing of a catalyst cooler attached to a regenerator in an FCC unit. (*Courtesy of PSRI, Chicago, Ill.*)

Fischer-Tropsch Synthesis One of the early attempts to scale up a bubbling bed reactor to produce gasoline from CO and H_2 was unsuccessful (see Scale-Up in the subsection Design of Fluidized-Bed Systems). However, Kellogg Co. later developed a successful Fischer-Tropsch synthesis reactor based on a dilute-phase transport-reactor concept. Kellogg, in its design, prevented excessive gas bypassing by using a transport (riser) reactor and maintained temperature control of the exothermic reaction by inserting heat exchangers in the transport line. This process has been very successful, and has been repeatedly improved upon at the South African Synthetic Oil Limited (SASOL) plant in the Republic of South Africa, where politics and economics favored the conversion of coal to gasoline and other hydrocarbons. Refer to Jewell and Johnson, U.S. Patent 2,543,974, Mar. 6, 1951 for more information. The process has been successfully modified to use a simpler, less expensive turbulent bed catalytic reactor system (Silverman, R. W., et al., in *Fluidization V*, ed. K. Østergaard and A. Sørensen, Engineering Foundation, New York, 1986, pp. 441–448).

Polyethylene The first commercial fluidized-bed polyethylene plant was constructed by Union Carbide in 1968. Modern units operate at a temperature of approximately 100°C and a pressure of about 2500 kPa (360 psig). The bed is fluidized with ethylene at about 0.5 to 0.7 m/s (1.65 to 2.3 ft/s) and operates in the turbulent fluidization regime. Small catalyst is added to the bed, and the ethylene polymerizes on the catalyst to form polyethylene particles of approximately 600- to 1000-μm average size, depending on the type of polyethylene product being produced. The excellent mixing provided by the fluidized bed is necessary to prevent hot spots, since the unit is operated just slightly below the melting point of the product. A model of the reactor (Fig. 17-33) that couples polyethylene reaction kinetics to the hydrodynamics in the fluidized reactor bed was given by Choi and Ray [*Chem. Eng. Sci.* **40:** 2261 (1985)].

Additional Catalytic Processes Nitrobenzene is hydrogenated to aniline (U.S. Patent 2,891,094). Melamine and isophthalonitrile are produced in catalytic fluidized-bed reactors. Badger developed a process to produce maleic anhydride by the partial oxidation of butane (Schaffel, Chen, and Graham, "Fluidized Bed Catalytic Oxidation of Butane to Maleic Anhydride," presented at Chemical Engineering World Congress, Montreal, Canada, 1981). Du Pont developed a circulating bed process for production of maleic anhydride (Contractor, *Circulating Fluidized Bed Tech. II*, Pergamon, Oxford, UK, 1988, pp. 467–474). Mobil developed a commercial process to convert methanol to gasoline (Grimmer et al., *Methane Conversion*, Elsevier, Amsterdam, 1988, pp. 273–291).

Noncatalytic Reactions

Homogeneous Reactions Homogeneous noncatalytic reactions are normally carried out in a fluidized bed to achieve good mixing of the gases and

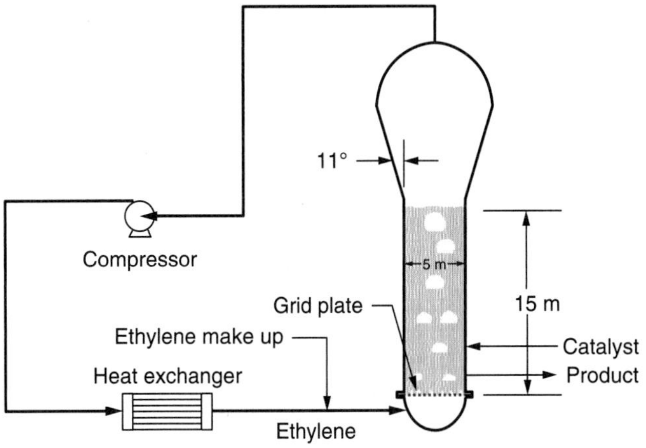

FIG. 17-33 High-pressure polyethylene reactor.

for good temperature control. The solids in the bed act as a heat sink (or source) and facilitate heat transfer from or to the gas or from or to heat-exchange surfaces. Reactions of this type include chlorination of hydrocarbons or oxidation of gaseous fuels.

Heterogeneous Reactions This category covers the greatest commercial use of fluidized beds other than fluid catalytic cracking. Roasting of ores in fluidized beds is very common. Roasting of sulfide, arsenical, and/or antimonial ores to facilitate the release of gold or silver values; the roasting of pyrite, pyrrhotite, or naturally occurring sulfur ores to provide SO_2 for sulfuric acid manufacture; and the roasting of copper, cobalt, and zinc sulfide ores to solubilize the metals are the major metallurgical uses. Figure 17-34 shows the basic items in the roasting process.

Thermally efficient calcination of lime, dolomite, and clay can be carried out in a multicompartment fluidized bed (Fig. 17-35). Fuels are burned in a fluidized bed of the product to produce the required heat. Bunker C oil, natural gas, and coal are used in commercial units as the fuel. Temperature control is accurate enough to permit production of lime of very high quality with close control of slaking characteristics. Also, half calcination of dolomite is an accepted practice in fluidized beds. The requirement of large crystal size for the limestone limits this application. Small crystals in the limestone result in low yields due to high dust losses from the fluidized bed.

Phosphate rock is calcined to remove carbonaceous material before being digested with sulfuric acid. Several different fluidized-bed processes have been commercialized for the direct reduction of hematite to high-iron,

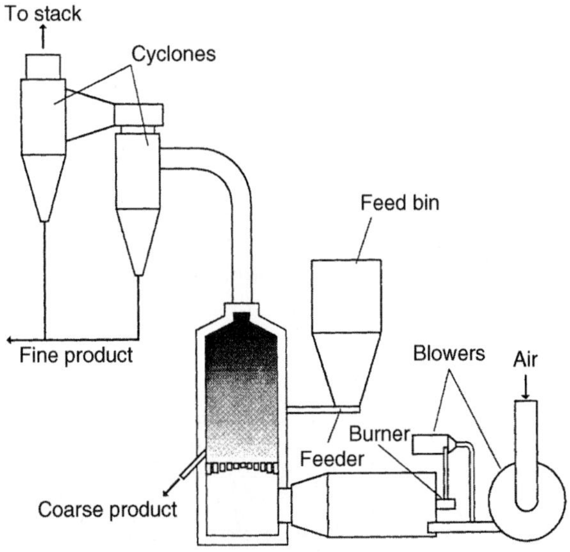

FIG. 17-34 Single-stage FluoSolids roaster or dryer. (*Dorr-Oliver, Inc.*)

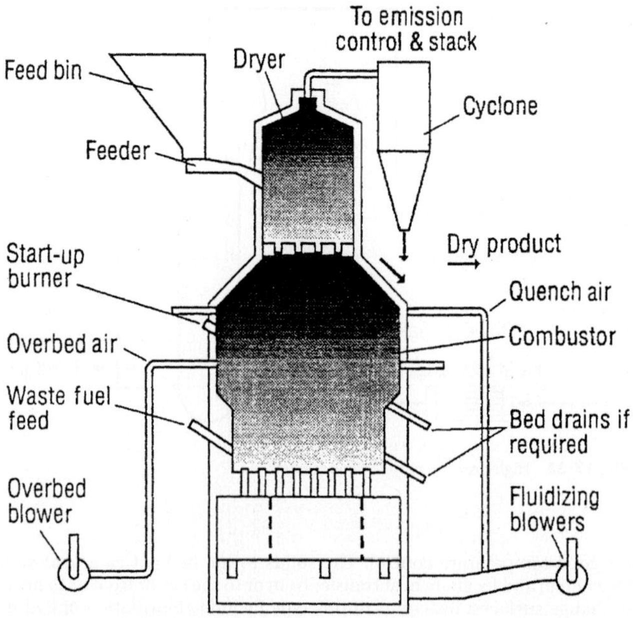

FIG. 17-35 FluoSolids multicompartment fluidized bed. (*Dorr-Oliver, Inc.*)

low-oxide products. Foundry sand is also calcined to remove organic binders and release fines. The calcination of $Al(OH)_3$ to Al_2O_3 in a circulating fluidized process produces a high-grade product. The process combines the use of circulating, bubbling, and transport beds to achieve high thermal efficiency (see Fig. 17-36).

An interesting feature of these high-temperature-calcination applications is the direct injection of heavy oil, natural gas, or fine coal into the fluidized bed. Combustion takes place at well below flame temperatures without atomization. Considerable care in the design of the fuel and air supply system is necessary to take full advantage of the fluidized bed, which serves to mix the air and fuel.

Coal can be burned in fluidized beds in an environmentally acceptable manner by adding limestone or dolomite to the bed to react with the SO_2 to form $CaSO_4$. Because of moderate combustion temperatures, about 800°C

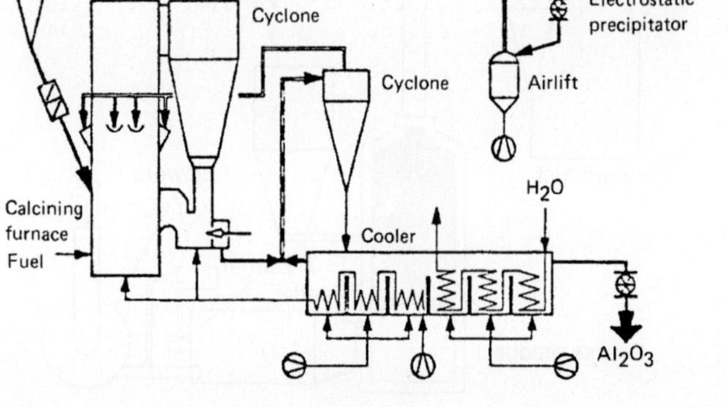

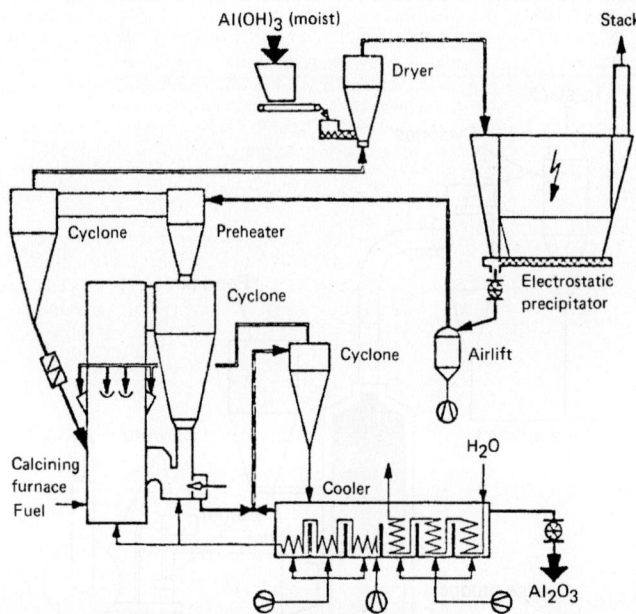

FIG. 17-36 Circulating fluid-bed calciner. (*Lurgi Corp.*)

to 900°C, NO_x formation, which results from the oxidation of nitrogen compounds contained in the coal, is kept at a low level. NO_x is increased by higher temperatures and higher excess oxygen contents. Two-stage air addition reduces NO_x. Several concepts of fluidized-bed combustion have been or are being developed. Atmospheric fluidized-bed combustion (AFBC), in which most of the heat-exchange tubes are located in the bed, is illustrated in Fig. 17-37. This type of unit is most commonly used for industrial applications up to about 50 t/h of steam generation. Larger units are generally of the circulating bed type, as shown in Fig. 17-37. Circulating fluidized-bed combustors have many advantages. The gas velocity is significantly higher than in bubbling or turbulent beds, which results in greater throughput. Since all the solids are recycled, fine limestone and coal can be fed to the combustor, which gives better limestone utilization and greater latitude in specifying coal sizing. Because of erosion due to high-velocity coarse solids, heat-transfer surface is usually not designed into the bottom of the combustion zone.

Pressurized fluidized-bed combustion (PFBC) is, as the name implies, operated at above atmospheric pressures. The beds and heat-transfer surfaces are stacked to conserve space and to reduce the size of the pressure vessel. This type of unit is usually conceived as a cogeneration unit. Steam raised in the boilers is employed to drive turbines or for other uses. The hot pressurized gases after cleaning are let down through an expander coupled to a compressor to supply the compressed combustion air and/or electric generator. Also see Sec. 24, Energy Resources, Conversion, and Utilization.

Incineration There are hundreds of units in operation that are used for the incineration of biological sludges. These units can be designed to operate autogenously with wet sludges containing as little as 6 MJ/kg (2600 Btu/lb) heating value (Fig. 17-38). Depending on the calorific value of the feed, heat can be recovered as steam either by means of waste heat boilers or by a combination of waste heat boilers and the heat-exchange surface in the fluid bed. Several units are used for sulfite papermill waste liquor disposal. Several units are used for oil refinery wastes, which sometimes include a mixture of liquid sludges, emulsions, and caustic waste [Flood and Kernel, *Chem. Proc.* (Sept. 8, 1973)]. Miscellaneous uses include the incineration of sawdust, carbon-black waste, pharmaceutical waste, grease from domestic sewage, spent coffee grounds, and domestic garbage.

Toxic or hazardous wastes can be disposed of in fluidized beds by either chemical capture or complete destruction. In the former case, bed material, such as limestone, will react with substances such as halides, sulfides, and metals to form stable compounds that can be landfilled. Contact times of up to 5 to 10 s at 1200 K (900°C) to 1300 K (1000°C) ensure complete destruction of most compounds.

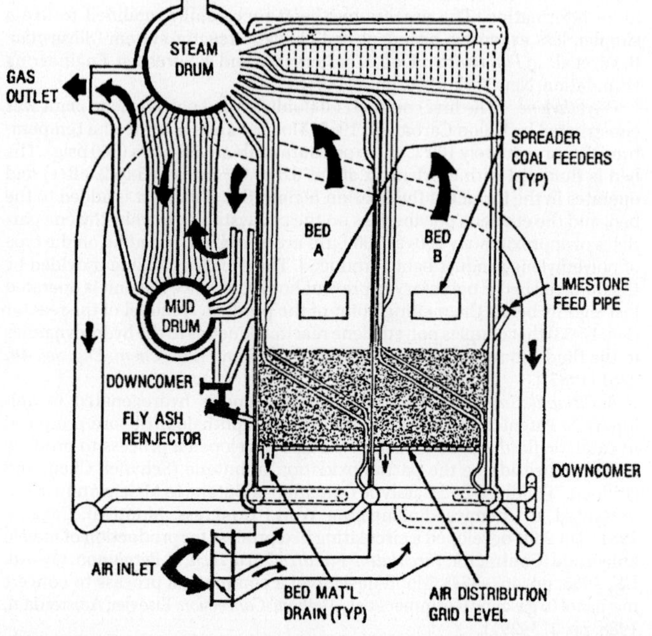

FIG. 17-37 Fluidized-bed steam generator at Georgetown University; 12.6-kg/s (100,000-lb/h) steam at 4.75-MPa (675-psig) pressure. (*From Georgetown Univ. Q. Tech. Prog. Rep. METC/DOE/10381/135, July–September 1980.*)

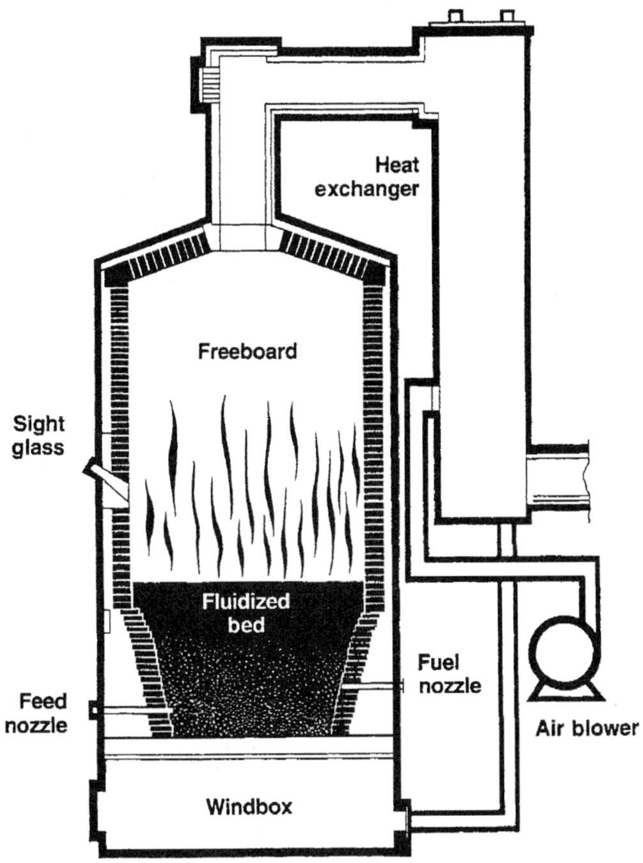

FIG. 17-38 Hot windbox incinerator/reactor with air preheating. (*Dorr-Oliver, Inc.*)

Physical Contacting

Drying Fluidized bed units for drying solids, particularly coal, cement, rock, and limestone, are in wide use. Economic considerations make these units particularly attractive when large tonnages of solids are to be handled. Fuel requirements are 3.3 to 4.2 MJ/kg (1500 to 1900 Btu/lb of water removed), and total power for blowers, feeders, etc., is about 0.08 kWh/kg of water removed. The maximum coal feed size is approximately 6 cm (2.4 in) × 0. One of the major advantages of this type of dryer is the close control of operating conditions so that a predetermined amount of free moisture may be left with the solids to prevent dusting of the product during subsequent material handling operations. The fluidized-bed dryer is also used as a classifier so that drying and classification operations are accomplished simultaneously. Wall and Ash [*Ind. Eng. Chem.* **41:** 1247 (1949)] state that in drying 4.8-mm (−4-mesh) dolomite with combustion gases at a superficial velocity of 1.2 m/s (4 ft/s), the following removals of fines were achieved:

Particle size	% removed
−65 + 100 mesh	60
−100 + 150 mesh	79
−150 + 200 mesh	85
−200 + 325 mesh	89
−325 mesh	89

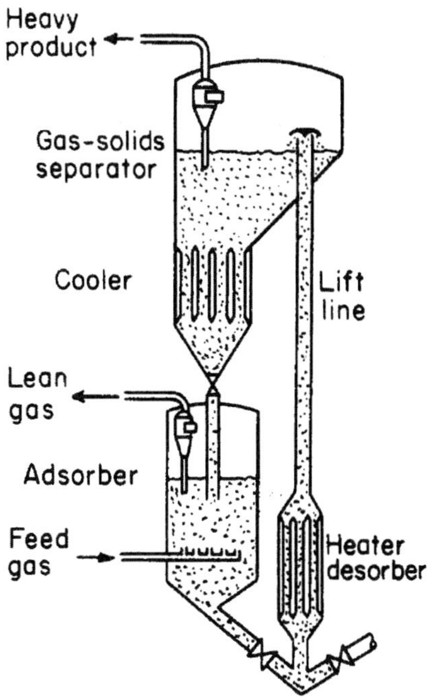

FIG. 17-39 Fluidized bed for gas fractionation. [*Sittig.* Chem. Eng. (*May 1953*).]

Classification The separation of fine particles from coarse particles can be accomplished by the use of a fluidized bed (see Drying). However, for economic reasons (e.g., initial cost, power requirements for compression of fluidizing gas), it is doubtful, except in special cases, that a fluidized-bed classifier would be built for this purpose alone.

It has been proposed that fluidized beds be used to remove fine solids from a gas stream. This is possible under special conditions.

Adsorption-Desorption An arrangement for gas fractionation is shown in Fig. 17-39.

The effects of adsorption and desorption on the performance of fluidized beds are discussed elsewhere. Adsorption of carbon disulfide vapors from airstreams is as great as 300 m^3/s (540,000 ft^3/min) in a 17-m-(53-ft-) diameter unit as reported by Avery and Tracey ("The Application of Fluidized Beds of Activated Carbon to Recover Solvent from Air or Gas Streams," Tripartate Chemical Engineering Conference, Montreal, Sept. 24, 1968).

Heat Treatment Heat treatment can be divided into two types: treatment of fluidizable solids and treatment of large, usually metallic objects in a fluid bed. The former is generally accomplished in multicompartment units to conserve heat. The heat treatment of large metallic objects is accomplished in long, narrow, heated beds. The objects are conveyed through the beds by an overhead conveyor system. Fluid beds are used because of the high heat-transfer rate and uniform temperature. See Reindl, "Fluid Bed Technology," *American Society for Metals,* Cincinnati, Sept. 23, 1981; Fennell, *Ind. Heat.* **48:** 9, 36 (September 1981).

Coating Fluidized beds of thermoplastic resins have been used to facilitate the coating of metallic parts. A properly prepared, heated metal part is dipped into the fluidized bed, which permits complete immersion in the dry solids. The heated metal fuses the thermoplastic, forming a continuous uniform coating.

GAS–SOLIDS SEPARATIONS

This subsection is concerned with the application of particle mechanics (see Sec. 6, Fluid and Particle Dynamics) to the design and application of dust-collection systems. It includes wet collectors, or scrubbers, for particle collection. Scrubbers designed for purposes of mass transfer are discussed in Secs. 14 and 18. Equipment for removing entrained liquid mist from gases is described in Sec. 18.

Nomenclature

Except where otherwise noted here or in the text, either consistent system of units (SI or U.S. customary) may be used. Only SI units may be used for electrical quantities, since no comparable electrical units exist in the U.S. customary system. When special units are used, they are noted at the point of use.

Symbols	Definition	SI units	U.S. customary system units	Special units
A_{in}	Area of cyclone inlet duct	m^2	ft^2	
A_{out}	Area of cyclone outlet pipe	m^2	ft^2	
B_c	Width of rectangular cyclone inlet duct	m	ft	
B_e	Spacing between wire and plate, or between rod and curtain, or between parallel plates in electrical precipitators	m	ft	
B_s	Width of gravity settling chamber	m	ft	
C^*	Dry scrubber pollutant gas equilibrium concentration over sorbent			
C_1	Dry scrubber pollutant gas inlet concentration			
C_2	Dry scrubber pollutant gas outlet concentration			
c_d	Dust concentration in gas stream	g/m^3	lb/ft^3	$grains/ft^3$
c_h	Specific heat of gas	$J/(kg \cdot K)$	$Btu/(lb_m \cdot °F)$	
c_{hb}	Specific heat of collecting body	$J/(kg \cdot K)$	$Btu/(lb_m \cdot °F)$	
c_{hp}	Specific heat of particle	$J/(kg \cdot K)$	$Btu/(lb_m \cdot °F)$	
D_b	Diameter or other representative dimension of collector body or device	m	ft	
D_{b1}, D_{b2}	Other characteristic dimensions of collector body or device	m	ft	
D_c	Cyclone diameter	m	ft	
D_d	Outside diameter of wire or discharge electrode of concentric-cylinder type of electrical precipitator	m	ft	
D_e	Diameter of cyclone gas exit duct	m	ft	
D_o	Volume/surface-mean-drop diameter			µm
D_p	Diameter of particle	m	ft	µm
d_{pth}	Cut diameter, diameter of particles of which 50 percent of those present are collected	m	ft	µm
d_P	Particle diameter of fraction number c'	m	ft	
D_t	Inside diameter of collecting tube of concentric-cylinder type of electrical precipitator	m	ft	
D_v	Diffusion coefficient for particle	m^2/s	ft^2/s	
DI	Decontamination index = $\log_{10}[1/(1-\eta)]$	Dimensionless	Dimensionless	
e	Natural (napierian) logarithmic base	2.718 . . .	2.718 . . .	
E	Potential difference	V		
E_c	Potential difference required for corona discharge to commence	V		
E_d	Voltage across dust layer	V		
E_s	Potential difference required for sparking to commence	V		
E_L	Cyclone collection efficiency at actual loading			
E_O	Cyclone collection efficiency at low loading			
F_E	Effective friction loss across wetted equipment in scrubber	kPa		in water
F_k	Packed-bed friction loss			
g	Acceleration due to gravity	m/s^2	ft/s^2	
g_c	Conversion factor		$32.17\ (lb_m/lb_f)(ft/s^2)$	
H_c	Height of rectangular cyclone inlet duct	m	ft	
H_s	Height of gravity settling chamber	m	ft	
I	Electrical current per unit of electrode length	A/m		
j	Corona current density at dust layer	A/m^2		
J_c	Cyclone solids discharge diameter, or dipleg diameter	m	ft	
$k\rho$	Density of gas relative to its density at 0°C, 1 atm	(–)	(–)	(–)
k_t	Thermal conductivity of gas	$W/(m \cdot K)$	$Btu/(s \cdot ft \cdot °F)$	
k_{tb}	Thermal conductivity of collecting body	$W/(m \cdot K)$	$Btu/(s \cdot ft \cdot °F)$	
k_{tp}	Thermal conductivity of particle	$W/(m \cdot K)$	$Btu/(s \cdot ft \cdot °F)$	
K	Empirical proportionality constant for cyclone pressure drop or friction loss	(–)	(–)	(–)
K_1	Resistance coefficient of "conditioned" filter fabric	kPa/(m/min)	in water/(ft/min)	
K_2	Resistance coefficient of dust cake on filter fabric	$\frac{kPa}{(m/min)(g/m^2)}$	$\frac{in\ water}{(ft/min)(lb_m/ft^2)}$	
K_a	Proportionality constant, for target efficiency of a single fiber in a bed of fibers	(–)	(–)	(–)
K_c	Resistance coefficient for "conditioned" filter fabric			$\frac{in\ water}{(ft/min)(cP^2)}$
K_d	Resistance coefficient for dust cake on filter fabric		$\frac{in\ water}{(ft/min)(gr/ft^2)(cP)}$	
K_e	Electrical-precipitator constant	s/m	s/ft	
K_F	Resistance coefficient for clean filter cloth			$\frac{in\ water}{(ft/min)(cP)}$
K_o	"Energy-distance" constant for electrical discharge in gases	m		
K_m	Stokes-Cunningham correction factor	(–)	(–)	(–)
L	Thickness of fibrous filter or of dust layer on surface filter	m	ft	
L_c	Barrel length	m	ft	
L_e	Length of collecting electrode in direction of gas flow	m	ft	
L_s	Length of gravity settling chamber in direction of gas flow	m	ft	(–)
ln	Natural logarithm (logarithm to the base e)	(–)	(–)	
M	Molecular weight	kg/mol	lb_m/mol	

Nomenclature *(Continued)*

Symbols	Definition	SI units	U.S. customary system units	Special units
n	Exponent	(–)	(–)	(–)
N_{Kn}	Knudsen number = λ_m/D_b	(–)	(–)	(–)
N_{Ma}	Mach number	(–)	(–)	(–)
N_o	Number of elementary electrical charges acquired by a particle	(–)	(–)	(–)
N_{Re}	Reynolds number = $(D_p \rho V_o/\mu)$ or $(D_p \rho u_t/\mu)$	(–)	(–)	(–)
N_{sc}	Interaction number = $18\,\mu/K_m \rho_p D_v$	(–)	(–)	(–)
N_{sd}	Diffusional separation number	(–)	(–)	(–)
N_{sec}	Electrostatic-attraction separation number	(–)	(–)	(–)
N_{sei}	Electrostatic-induction separation number	(–)	(–)	(–)
N_{sf}	Flow-line separation number	(–)	(–)	(–)
N_{sg}	Gravitational separation number	(–)	(–)	(–)
N_{si}	Inertial separation number	(–)	(–)	(–)
N_{st}	Thermal separation number	(–)	(–)	(–)
N_t	Number of transfer units = $\ln [1/(1-\eta)]$	(–)	(–)	(–)
N_s	Number of turns made by gas stream in a cyclone separator	(–)	(–)	(–)
Δp	Gas pressure drop	kPa	lb_f/ft^2	in water
Δp_i	Gas pressure drop in cyclone or filter			in water
p_F	Gauge pressure of water fed to scrubber	kPa		lbf/in^2
P_G	Gas-phase contacting power	$MJ/1000\ m^3$		$hp/(1000\ ft^3/min)$
P_L	Liquid-phase contacting power	$MJ/1000\ m^3$		$hp/(1000\ ft^3/min)$
P_M	Mechanical contacting power	$MJ/1000\ m^3$		$hp/(1000\ ft^3/min)$
P_T	Total contacting power	$MJ/1000\ m^3$		$hp/(1000\ ft^3/min)$
q	Gas flow rate	m^3/s	ft^3/s	
Q_G	Gas flow rate		ft^3/s	ft^3/min
Q_L	Liquid flow rate		ft^3/s	gal/min
Q_p	Electrical charge on particle	C		
r	Radius; distance from centerline of cyclone separator; distance from centerline of concentric-cylinder electrical precipitator	m	ft	
t_m	Time			min
S_c	Gas outlet tube distance below the bottom of the inlet	m	ft	
T	Absolute gas temperature	K	°R	
T_b	Absolute temperature of collecting body	K	°R	
u_s	Velocity of migration of particle toward collecting electrode	m/s	ft/s	
u_t	Terminal settling velocity of particle under action of gravity	m/s	ft/s	ft/s
v_m	Average cyclone inlet velocity, based on area A_c	m/s	ft/s	ft/s
v_p	Actual particle velocity	m/s	ft/s	
V_f	Filtration velocity (superficial gas velocity through filter)	m/min		ft/min
V_o	Gas velocity	m/s	ft/s	
V_s	Average gas velocity in gravity settling	m/s	ft/s	
V_{ct}	Tangential component of gas velocity in cyclone	m/s	ft/s	
w	Loading of collected dust on filter	g/m^2	lb_m/ft^2	gr/ft^2
Z_c	Cyclone cone length	m	ft	
	Greek symbols			
α	Empirical constant in equation of scrubber performance curve	$\left[\frac{MJ}{100\ m^3}\right]^{-\gamma}$		$\left[\frac{hp}{100\ ft^3/min}\right]^{-\gamma}$
γ	Empirical constant in equation of scrubber performance curve	(–)	(–)	(–)
δ	Dielectric constant	(–)	(–)	(–)
δ_g	Dielectric constant at 0°C, 1 atm	(–)	(–)	(–)
δ_o	Permittivity of free space	F/m		
δ_b	Dielectric constant of collecting body	(–)	(–)	(–)
δ_p	Dielectric constant of particle	(–)	(–)	(–)
Δ	Fractional free area (for screens, perforated plates, grids)	(–)	(–)	(–)
ε	Elementary electrical charge	1.60210×10^{-19} C		
ε_b	Characteristics potential gradient at collecting surface	V/m		
ε_v	Fraction voids in bed of solids	(–)	(–)	(–)
ζ	$=1+2\frac{\delta-1}{\delta+2}$ ranges from a value of 1 for materials with a dielectric constant of 1 to 3 for conductors	(–)	(–)	(–)
η	Collection efficiency, weight fraction of entering dispersoid collected	(–)	(–)	(–)
η_o	Target efficiency of an isolated collecting body, fraction of dispersoid in swept volume collected on body	(–)	(–)	(–)
η_t	Target efficiency of a single collecting body in an array of collecting bodies, fraction of dispersoid in swept volume collected on body	(–)	(–)	(–)
λ_i	Ionic mobility of gas	(m/s)/(V/m)		
λ_p	Particle mobility = u_e/E	(m/s)/(V/m)		
μ, μ_g	Gas viscosity	Pa · s	$lb_m/(s \cdot ft)$	cP
μ_L	Liquid viscosity			cP
ρ, ρ_g	Gas density	g/m^3	lb/ft^3	
ρ_d	Resistivity of dust layer	Ω · m		
ρ_L	Liquid density	kg/m^3	lb_m/ft^3	lb_m/ft^3

Nomenclature *(Continued)*

Symbols	Definition	SI units	U.S. customary system units	Special units
ρ_s, ρ_p	True (*not* bulk) density of solids or liquid drops	kg/m^3	lb$_m$/ft^3	lb$_m$/ft^3
ρ'	Density of gas relative to its density at 25°C, 1 atm	(–)	(–)	(–)
σ	Ion density	Number/m^3		
σ_{avg}	Average ion density	Number/m^3		
σ_L	Liquid surface tension			dyn/cm
ϕ_s	Particle shape factor = (surface of sphere)/(surface of particle of same volume)	(–)	(–)	(–)
	Script symbols			
$\mathcal{E}$	Potential gradient	V/m		
$\mathcal{E}_c$	Potential gradient required for corona discharge to commence	V/m		
$\mathcal{E}_i$	Average potential gradient in ionization stage	V/m		
$\mathcal{E}_o$	Electrical breakdown constant for gas	V/m		
$\mathcal{E}_p$	Average potential gradient in collection stage	V/m		
$\mathcal{E}_s$	Potential gradient required for sparking to commence	V/m		

General References: Burchsted, Kahn, and Fuller, *Nuclear Air Cleaning Handbook,* ERDA 76-21, Oak Ridge, Tenn., 1976; Cadle, *The Measurement of Airborne Particles,* Wiley, New York, 1975; Davies, *Aerosol Science,* Academic, New York, 1966; Davies, *Air Filtration,* Academic, New York, 1973; Dennis, *Handbook on Aerosols,* ERDA TID-26608, Oak Ridge, Tenn., 1976; Drinker and Hatch, *Industrial Dust,* 2d ed., McGraw-Hill, New York, 1954; Friedlander, *Smoke, Dust, and Haze,* Wiley, New York, 1977; Fuchs, *The Mechanics of Aerosols,* Pergamon, Oxford, 1964; Green and Lane, *Particulate Clouds: Dusts, Smokes, and Mists,* Van Nostrand, New York, 1964; Lapple, *Fluid and Particle Mechanics,* University of Delaware, Newark, 1951; Licht, *Air Pollution Control Engineering—Basic Calculations for Particle Collection,* Marcel Dekker, New York, 1980; Liu, *Fine Particles—Aerosol Generation, Measurement, Sampling, and Analysis,* Academic, New York, 1976; Lunde and Lapple, *Chem. Eng. Prog.* **53:** 385 (1957); Lundgren et al., *Aerosol Measurement,* University of Florida, Gainesville, 1979; Mercer, *Aerosol Technology in Hazard Evaluation,* Academic, New York, 1973; Nonhebel, *Processes for Air Pollution Control,* CRC Press, Cleveland, 1972; Shaw, *Fundamentals of Aerosol Science,* Wiley, New York, 1978; Stern, *Air Pollution: A Comprehensive Treatise,* vols. 3 and 4, Academic, New York, 1977; Strauss, *Industrial Gas Cleaning,* 2d ed., Pergamon, New York, 1975; Theodore and Buonicore, *Air Pollution Control Equipment: Selection, Design, Operation, and Maintenance,* Prentice-Hall, Englewood Cliffs, N.J., 1982; White, *Industrial Electrostatic Precipitation,* Addison-Wesley, Reading, Mass., 1963; White and Smith, *High-Efficiency Air Filtration,* Butterworth, Washington, 1964; ASME Research Committee on Industrial and Municipal Wastes, *Combustion Fundamentals for Waste Incineration,* American Society of Mechanical Engineers, New York, 1974; Buonicore and Davis, eds., *Air Pollution Engineering Manual,* Air & Waste Management Association, Van Nostrand Reinhold, New York, 1992; Burchsted, Fuller, and Kahn, *Nuclear Air Cleaning Handbook,* ORNL for the U.S. Energy Research and Development Administration, NTIS Report ERDA 76-21, 1976; Dennis, ed., *Handbook on Aerosols,* GCA for the U.S. Energy Research and Development Administration, NTIS Report TID-26608, 1976; Stern, *Air Pollution,* 3d ed., Academic Press, New York, 1977 (supplement 1986).

PURPOSE OF GAS-SOLIDS SEPARATION

Gas–solids separation is concerned with the removal or collection of solids dispersed in gases for purposes of:

1. Air pollution control, as in fly ash removal from power plant flue gases, dust control for dryer effluent air, diesel particulates from internal combustion engines
2. Equipment maintenance reduction, as in filtration of engine intake air or pyrites furnace gas treatment prior to its entry to a contact sulfuric acid plant
3. Safety, or health hazard elimination, as in collection of siliceous and metallic dusts around grinding and drilling equipment and in some metallurgical operations and flour dusts from milling or bagging operations; filtration of particulate matter from ambient air to improve indoor air quality, removal of combustible dust from air
4. Product quality improvement, as in air cleaning in the production of pharmaceutical products and photographic film, and manufacturing of microelectronics
5. Recovery of a valuable product in dry state, as in collection of dusts from dryers and smelters
6. Powdered-product collection, as in pneumatic conveying; the spray drying of milk, eggs, and soap; and manufacture of high-purity zinc oxide and carbon black; separation of catalyst in FCC reactors and separation of elutriated solids from fluidized-bed processes

PROPERTIES OF DISPERSED SOLIDS

An understanding of the fundamental properties and characteristics of solids dispersed in gas is essential to the design of industrial dust-control equipment. Figure 17-40 shows characteristics of dispersed solids and other particles together with the types of gas-cleaning equipment that are applicable to their control. Two types of solid dispersed in gases are shown: (1) dust, which is composed of particles larger than 1 μm in diameter; and (2) fume, which consists of particles generally smaller than 1 μm in diameter. Dusts usually result from mechanical disintegration of matter. They may be redispersed from the settled, or bulk, condition by an air blast. Fumes are submicrometer dispersed solids formed by processes such as combustion, sublimation, and condensation. Once collected, they cannot be redispersed from the settled condition to their original state of dispersion by air blasts or mechanical dispersion equipment.

The primary distinguishing characteristic of solids dispersed in gas is the particle size. The generally accepted unit of particle size is the micrometer, μm. (Prior to the adoption of the SI system, the same unit was known as the micron and was designated by μ.) The size of a spherical particle is unambiguously defined by its diameter, which can be considered the characteristic dimension. However, particles encountered in nature and industrial processes are rarely spherical or regular (e.g., cuboid, ellipsoid, cylindrical) in shape. The characteristic dimension of an irregular particle can be obtained by relating the relevant process response or geometric feature to an "equivalent" sphere (see Fig. 17-41). Therefore, each derived equivalent diameter represents a mechanism or characteristic relevant to the behavior of the particle in the process of interest (Trottier and Dhodapkar, *Chemical Engineering Progress,* July 2014). It should not simply be based on the dominant physical dimension of the particle. A summary of equivalent diameters is shown in Table 17-1.

From the standpoint of gas–solid separation, the most important size-related property of a dust particle is its dynamic behavior. Particles larger than 100 μm are readily collectible by simple inertial or gravitational methods. For particles under 100 μm, the range of principal difficulty in dust collection, the resistance to motion in a gas is viscous (see Sec. 6, Fluid and Particle Dynamics), and for such particles, the most useful size specification is commonly the Stokes settling diameter, which is the diameter of the spherical particle of the same density that has the same terminal velocity in viscous flow as the particle in question. It is yet more convenient in many circumstances to use the "aerodynamic diameter," which is the diameter of the particle of unit density (1000 kg/m^3 or 1 g/cm^3) that has the same terminal settling velocity. Use of the aerodynamic diameter permits direct comparisons of the dynamic behavior of particles that are actually of different sizes, shapes, and densities [Raabe, *J. Air Pollut. Control Assoc.* **26:** 856 (1976); Cooper and Alley, *Air Pollution Control—A Design Approach,* Waveland Press, Long Grove, Ill., 2011]. The Stokes diameter and the aerodynamic diameter are equal for particles of unit density (1000 kg/m^3 or 1 g/cm^3).

When the size of a particle approaches the same order of magnitude as the mean free path of the gas molecules, the settling velocity is greater than predicted by Stokes' law because of molecular slip. The mean free path for air at standard conditions is about 0.065 μm. Submicron particles in an air suspension can therefore slip by without impacting the molecules. The slip-flow correction is appreciable for particles smaller than 1 μm and is allowed for by the Cunningham correction for Stokes' law (Cooper in *Handbook of Powder Science & Technology,* ed. Fayed and Otten, Chapman and Hall, London, 1997; Cooper and Alley, *Air Pollution Control,* 4th ed., Waveland Press, Long Grove, Ill., 2011). The correction factors for particle sizes of 0.01, 0.1, 1.0, and 10.0 mm at 1 atm and 25°C are 22.5, 2.89, 1.17, and 1.02, respectively (Cooper and Alley, *Air Pollution Control,* 4th ed., Waveland Press, Long Grove, Ill., 2011). The effect of operating temperature and pressure must also be taken into account. The Cunningham correction is applied in calculations of the aerodynamic diameters of particles that are in the appropriate size range.

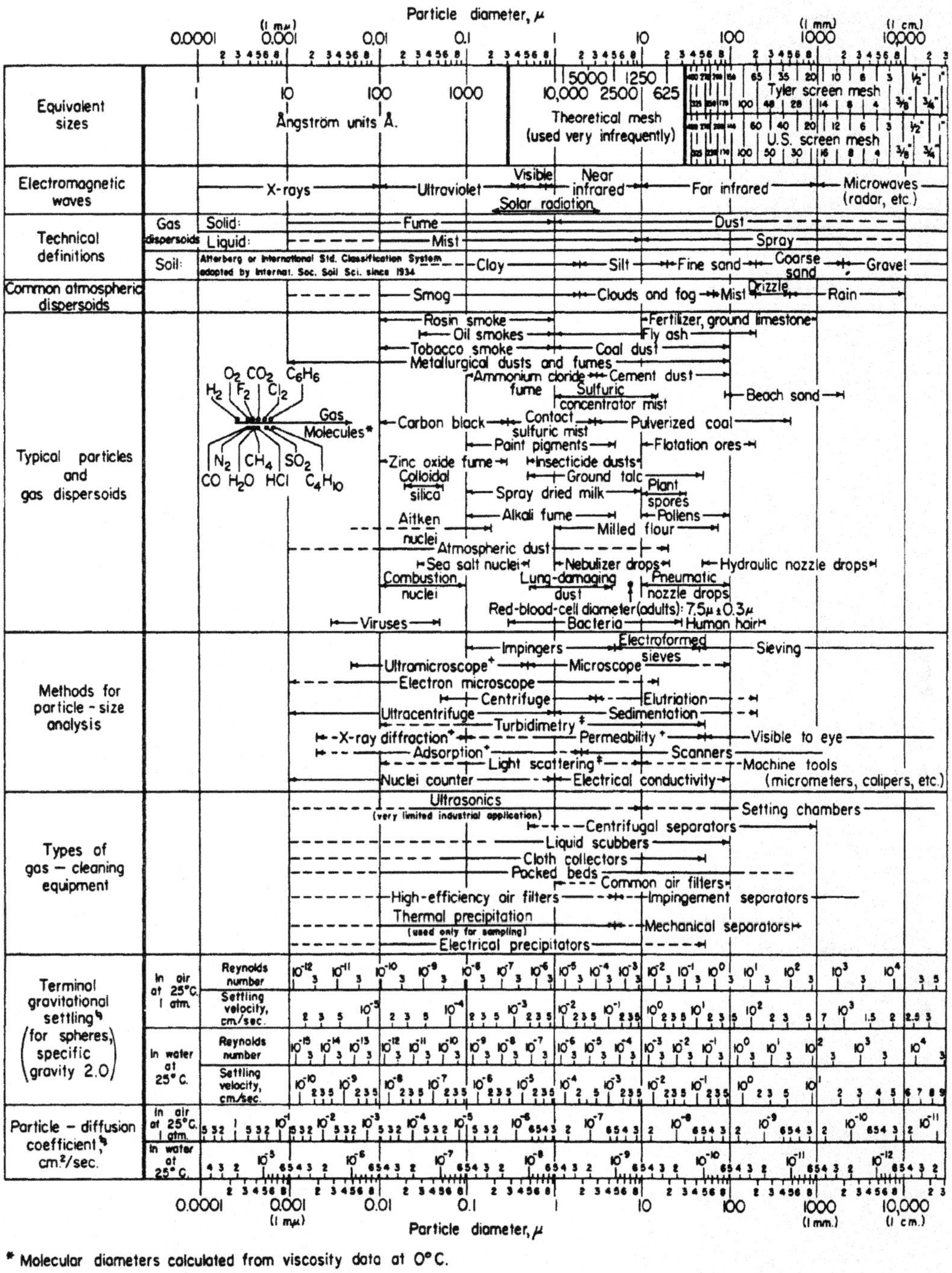

* Molecular diameters calculated from viscosity data at 0°C.

\+ Furnishes average particle diameter but no size distribution.

‡ Size distribution may be obtained by special calibration.

§ Stokes – Cunningham factor included in values given for air but not included for water.

FIG. 17-40 Characteristics of particles and particle dispersoids. (*Courtesy of the Stanford Research Institute; prepared by C. E. Lapple.*)

Although solid fume particles may range in size down to perhaps 0.001 μm, fine particles effectively smaller than about 0.1 μm are not of much significance in industrial dust and fume sources because their aggregate mass is only a very small fraction of the total mass emission. At the concentrations present in such sources (e.g., production of carbon black) the coagulation, or flocculation, rate of the ultrafine particles is extremely high, and the particles speedily grow to sizes of 0.1 μm or greater. The most difficult collection problems are thus concerned with particles in the range of about 0.1 to 2 μm, in which forces for deposition by inertia are small. For collection of particles under 0.1 μm, diffusional deposition becomes increasingly important as the particle size decreases.

In a gas stream carrying dust or fume, some degree of particle flocculation will exist, so both discrete particles and clusters of adhering particles will be present. The discrete particles composing the clusters may be only loosely attached to each other, as by van der Waals forces [Lapple, *Chem. Eng.* **75**(11): 149 (1968)]. Flocculation tends to increase with increases in particle concentration and may strongly influence collector performance.

PARTICLE MEASUREMENTS

Measurements of the concentrations and characteristics of dust dispersed in air or other gases may be necessary (1) to determine the need for control

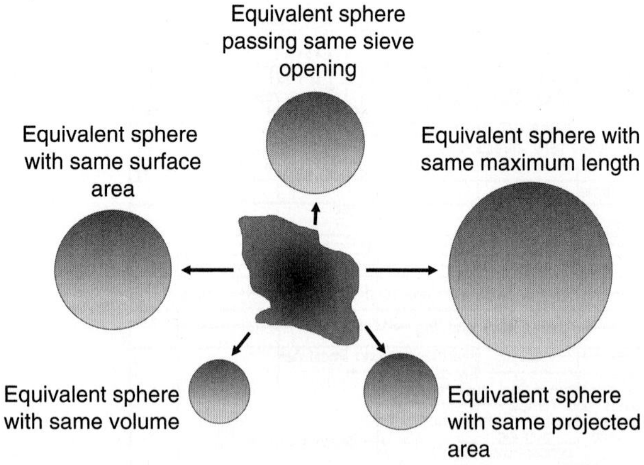

FIG. 17-41 Examples of equivalent sphere diameters for an irregular particle.

measures, (2) to establish compliance with legal requirements, (3) to select appropriate gas cleaning technology, (4) to obtain information for collector design, and (5) to determine collector performance.

Atmospheric-Pollution Measurements The dust-fall measurement is one of the common methods for obtaining a relative long-period evaluation of particulate air pollution. Stack-smoke densities are often graded visually by means of the Ringelmann chart. Plume opacity may be continuously monitored and recorded by a photoelectric device which measures the amount of light transmitted through a stack plume. Equipment for local atmospheric dust concentration measurements fall into five general types: (1) the impinger, (2) the hot-wire or thermal precipitator, (3) the electrostatic precipitator, (4) the filter, and (5) impactors and cyclones. The filter is the most widely used, in the form of either a continuous tape, or a number of filter disks arranged in an automatic sequencing device, or a single, short-term, high-volume sampler. Samplers such as these are commonly used to obtain mass emission and particle-size distribution. Impactors and small cyclones are commonly used as size-discriminating samplers and are usually followed by filters for the determination of the finest fraction of the dust (Lundgren et al., *Aerosol Measurement,* University of Florida, Gainesville, 1979; and Dennis, *Handbook on Aerosols,* U.S. ERDA TID-26608, Oak Ridge, Tenn., 1976; Willeke and Baron, *Aerosol Measurement,* Van Nostrand Reinhold, New York, 1993).

Process-Gas Sampling In sampling process gases either to determine dust concentration or to obtain a representative dust sample for composition, particle size, or density measurements, it is necessary to take special precautions to avoid inertial segregation of the particles. To prevent such classification, a traverse of the duct may be required, and at each point the sampling nozzle must face directly into the gas stream with the velocity in the mouth of the nozzle equal to the local gas velocity at that point. This is called isokinetic sampling. If the sampling velocity is too high, the dust sample will contain a lower concentration of dust than the mainstream, with a greater percentage of fine particles; if the sampling velocity is too low, the dust sample will contain a higher concentration of dust with a greater percentage of coarse particles. The sampling probe must be maintained parallel to the streamlines if concentration is being measured. The sampling point should be located such that classification or separation caused by bends, valves, or transitions does not affect the measurements. The measurement can be made upstream of these elements or at least 8 to 10 diameters downstream. The critical elements of a sampling system are nozzle design, dust extraction system, and sample collector. However, it is not always possible to achieve isokinetic and isoaxial sampling conditions in the field. Specific corrections must then be made to the measured data (Allen, *Particle Size Measurement,* 4th ed., Chapman and Hall, London, 1990). Additional details can be found in these references [Lapple, *Heat. Piping Air Cond.* **16:** 578 (1944); *Manual of Disposal of Refinery Wastes,* vol. V, American Petroleum Institute, New York, 1954; and Dennis, *Handbook on Aerosols,* U.S. ERDA TID-26608, Oak Ridge, Tenn., 1976; EPA Report, *Principles and Practices of Air Pollution Control—Student Manual,* APTI Course 452, U.S. EPA, July 2003; EPA Report, *Quality Assurance Handbook for Air Pollution Measurement System,* 454/B-13-003, May 2013; Allen, *Particle Size Measurement,* 4th ed., Chapman and Hall, London, 1990].

Particle-Size Analysis Methods for particle-size analysis are shown in Fig. 17-40, and various techniques for particle-size analysis are summarized in Table 17-2. More detailed information may be found in Lapple, *Chem. Eng.* **75**(11): 140 (1968); Lapple, "Particle-Size Analysis," in *Encyclopedia of Science and Technology,* 5th ed., McGraw-Hill, New York, 1982; Cadle, *The Measurement of Airborne Particles,* Wiley, New York, 1975; Lowell, *Introduction to Powder Surface Area,* 2d ed., Wiley, New York, 1993; Allen, *Particle Size Measurement,* 4th ed, Chapman and Hall, London, 1990; Allen, *Powder Sampling and Particle Size Determination,* Elsevier, 2003; and Baron et al., *Aerosol Measurement: Principles, Techniques, and Applications,* 3d ed., John Wiley & Sons, Inc., 2011.

TABLE 17-1 Commonly Used Equivalent Diameters

Basis	Equivalent diameter	Symbol*	Definition
Volume-based diameter	Volume diameter	d_v	Diameter of a sphere having the same volume as the particle
	Surface diameter	d_s	Diameter of a sphere having the same external surface area as the particle
	Surface volume diameter (Sauter diameter)	d_{sv}	Diameter of a sphere having the same external surface-area-to-volume ratio as the particle
	Sieve diameter	d_A	Diameter of a sphere equivalent to the size of the minimum square aperture through which the particle will pass
	Drag diameter	d_d	Diameter of a sphere that has the same resistance to motion in the same fluid (having same viscosity), moving at the same velocity as the particle
	Free-falling diameter	d_f	Diameter of a sphere having the same density and the same free-falling speed as the particle in the same fluid
	Stokes' diameter	d_{St}	Diameter of a free-falling sphere in laminar region
	Aerodynamic diameter	d_a	Stokes' diameter of a sphere of unit specific gravity (water)
Equivalent circle	Projected area diameter	d_{ps}	Diameter of a circle having the same area as the projected area of the particle at rest in a stable position
	Projected area diameter	d_{pr}	Diameter of a circle having the same area as the projected area of the particle at rest in a random position
	Perimeter diameter	d_c	Diameter of a circle having the same perimeter as the projected area (outline) of the particle, irrespective of the particle orientation
Statistically derived	Feret's diameter	d_F	Mean value of the distance between pairs of parallel tangents to the projected outline of the particle
	Martin's diameter	d_M	The mean chord length of the projected outline of the particle
	Unrolled diameter	d_R	The mean chord length through the center of gravity of the particle
	Maximum chord diameter	d_{CH}	The maximum length of a line limited by the particle contour
	Shear diameter	d_{SH}	Particle width obtained with an image shearing device

*It is a standard practice in the United States to use the letter *d* for particle size, whereas European literature, including ISO standards, uses the letter *x*.

Adapted from NIST Recommended Practice Guide [*Particle Size Characterization,* Special Publication 960-1, January 2001; Allen, *Powder Sampling and Particle Size,* Elsevier, Amsterdam, 2003].

TABLE 17-2 Summary of Particle Size Measurement Techniques

Principle/technique	Instrument type	Wet/dry/both	Measurement basis	Typical size range, mm	Typical sample size, g
Ensemble	Laser diffraction	Both	Mass	0.04–2000	<1–wet >5–dry
Ensemble	Quasi-elastic light scattering (dynamic light scattering, photon correlation spectroscopy)	Wet	Intensity	0.003–6	< 0.1
Ensemble	Ultrasonic spectroscopy	Wet	Mass	0.01–1000	0.1–20
Fractionation	Wet sieving	Wet	Mass	>5	0.5–500
Fractionation	Dry sieving	Dry	Mass	>45	0.5–500
Fractionation	Sedimentation X-ray gravitational	Wet	Mass	0.5–100	0.1–1
Fractionation	Centrifugal sedimentation X-ray	Wet	Mass	0.01–100	0.1–1
Fractionation	Centrifugal sedimentation Optical	Wet	Mass	0.01–100	0.1–1
Fractionation	Field-flow fractionation	Wet	Mass	0.003–3	< 0.1
Fractionation	Hydrodynamic chromatography (HDC) and capillary HDC	Wet	Mass	0.02–1	<0.1
Fractionation	Cascade impactor	Dry	Mass	0.1–20	< 1
Single particle counting	Dynamic image analysis	Both	Number	>5	<2 wet >2 dry
Single particle counting	Optical particle counter	Both	Number	0.2–2500	<2 wet >2 dry
Single particle counting	Electrozone counter	Wet	Number	0.4–1200	<2
Single particle counting	Aerodynamic particle sizer	Dry	Number	0.5–20	<0.5

Adapted from Trottier and Dhodapkar, *Chemical Engineering Progress,* July 2014.

With the exception of image analysis techniques, every particle sizing technique measures some type of equivalent diameter, which is then reported as the particle diameter. In the past 25 years, the particle-size measurement devices have become sophisticated and yet easier to use. Often, the users do not understand the underlying principles and take the results as absolute values. A direct comparison of particle-size distributions obtained from instruments based on different measurement techniques should be avoided.

Particle-size distribution may be presented on either a frequency or a cumulative basis; the various methods are discussed in the references just cited. The most common method presents a plot of particle size versus the cumulative weight percent of material larger or smaller than the indicated size. The distributions can be reported by number, surface, or volume. However, the conversion between number, surface, and volume distributions done by the instrument software does usually take the particle shape into account. For nonspherical particles, the measurement techniques must always be reported along with the particle size analysis.

As discussed earlier, the dynamic response of a particle in a fluid medium is best represented by the Stokes diameter, the aerodynamic diameter, or the Sauter mean diameter. For determination of the aerodynamic diameters of particles, the most commonly applicable methods for particle-size analysis are those based on inertia: aerosol centrifuges, cyclones, and inertial impactors (Lundgren et al., *Aerosol Measurement,* University of Florida, Gainesville, 1979; Liu, *Fine Particles—Aerosol Generation, Measurement, Sampling, and Analysis,* Academic, New York, 1976; and Baron et al., *Aerosol Measurement: Principles, Techniques, and Applications,* 3d ed., John Wiley & Sons, Inc., 2011), and time of flight aerodynamic particle sizers. Cascade impactors are most commonly used. It should be noted that the impactor measurements are subject to many errors [Rao and Whitby, *Am. Ind. Hyg. Assoc. J.* **38:** 174 (1977); Marple and Willeke, "Inertial Impactors," in *Aerosol Measurement,* Lundgren et al., University of Florida, Gainesville, 1979; and Fuchs, "Aerosol Impactors," in *Fundamentals of Aerosol Science,* Shaw, Wiley, New York, 1978]. Reentrainment due to particle bouncing and blowoff of deposited particles makes a dust appear finer than it actually is, as does the breakup of flocculated particles. The processing of cascade-impactor data also presents possibilities for substantial errors (Fuchs, *The Mechanics of Aerosols,* Pergamon, Oxford, 1964) and is laborious as well. Lawless (Rep. No. EPA-600/7-78-189, U.S. EPA, 1978) discusses problems in analyzing and fitting cascade-impactor data to obtain dust-collector efficiencies for discrete particle sizes.

The measured diameters of particles should as nearly as possible represent the effective particle size of a dust as it exists in the gas stream. The ideal, but also the most challenging, approach would be on line sampling and particle-size analysis. However, most particle sizing methods are suitable for off-line analysis of the collected samples. When the sample is redispersed for particle-size analysis, the agglomerates of fine particles, which might exist in the process stream, will be broken down to their constituent particles. The fines fraction from such analysis would report a higher value than actually observed in the process stream.

For dust-control work, it is recommended that a preliminary qualitative examination of the dust first be made without a detailed particle count. A visual estimate of particle-size distribution will often provide sufficient guidance for a preliminary assessment of requirements for collection equipment.

Sampling It is often assumed that the sample obtained and analyzed from the process is a representative sample. Care and expertise expended on analyzing a nonrepresentative sample will not yield useful data. To obtain a representative sample for particle-size analysis, one must follow good sampling practices (Allen, *Particle Size Measurement,* 4th ed., Chapman and Hall, 1990; Allen, *Powder Sampling and Particle Size Determination,* Elsevier, Amsterdam, 2003; NIST Practice Guide, *Particle Size Characterization,* Special Publication 960-1, 2001). The gross sample so obtained is usually much larger than the sample required for analysis (Table 17-2). Further reduction in the sample size from gross sample to laboratory sample, and finally the analytical sample, must be done without introducing bias or random variability. Various methods, such as cone and quartering, scoop sampling, table sampling, chute riffling, and spin riffling, are commonly practiced. Allen and Khan [*Chem. Engr.* 238 (1970)] have suggested that a spinning riffler is the most reproducible technique for sample subdivision. More details can be found in Allen (*Powder Sampling and Particle Size Determination,* Elsevier, Amsterdam, 2003).

MECHANISMS OF GAS-SOLIDS SEPARATION

The basic operations in dust collection by any device are (1) separation of the gas-borne particles from the gas stream by deposition on a collecting surface, (2) retention of the deposit on the surface, and (3) removal of the deposit from the surface for recovery or disposal. The separation step

TABLE 17-3 Summary of Mechanisms and Parameters in Aerosol Deposition

Deposition	Origin of force field	Deposition mechanism measureable in terms of		System parameters
		Basic parameter	Specific modifying parameters	
Flow-line interception*	Physical gradient*	$N_{sf} = \left(\frac{D_p}{D_b}\right)$		
Inertial deposition	Velocity gradient	$N_{st} = \left(\frac{K_m \rho s D_p^2 V_o}{18 \mu D_b}\right)$	$N_{sc} = \left(\frac{N_{sf}^2}{N_{st} N_{sd}}\right) = \left(\frac{18\mu}{K_m \rho_p D_\upsilon}\right)^{\dagger}$	Geometry: (D_{b1}/D_b), (D_{b2}/D_b), etc. ϵ_v Δ
Diffusional deposition	Concentration gradient	$N_{sd} = \left(\frac{D_\upsilon}{V_o D_b}\right)$		
Gravity settling	Elevation gradient	$N_{sg} = \left(\frac{u_t}{V_o}\right)$		Flow pattern: N_{Re}‡ N_{Ma} N_{Kn}
Electrostatic precipitation	Electric-field gradient§	$N_{sec} = \left(\frac{K_m Q_p \varepsilon_b}{\mu D_p V_o}\right)$	δ_p, δ_b¶	Surface accommodation
	a. Attraction			
	b. Induction	$N_{set} = \left(\frac{\delta_p - 1}{\delta_p + 2}\right)\left(\frac{K_m D_p^2 \delta_o \varepsilon_b 2}{\mu D_b V_o}\right)$		
Thermal precipitation	Temperature gradient	$N_{st} = \left(\frac{T - T_b}{T}\right)\left(\frac{\mu}{K_m \rho D_b V_o}\right)\left(\frac{k_t}{2k_t + k_{tp}}\right)$	(T_b/T), (T_p/T),† (N_{Pr}), (k_{tp}/k_t), (k_{tb}/k_t),¶ (c_{hp}/c_h), (c_{hb}/c_h)¶	

SOURCE: Lunde and Lapple, *Chem. Eng. Prog.*, **53,** 385 (1957).

*This has also commonly been termed "direct interception" and in conventional analysis would constitute a physical boundary condition imposed upon the particle path induced by action of other forces. By itself it reflects deposition that might result with a hypothetical particle having finite size but no mass or elasticity.

†This parameter is an alternative to N_{sf}, N_{si}, or N_{ad} and is useful as a measure of the interactive effect of one of these on the other two. It is comparable with the Schmidt number.

‡When applied to the inertial deposition mechanism, a convenient alternative is $(K_m \rho_s / 18\rho) = N_{si}/(N_{sf}^2 N_{\mathrm{Re}})$.

§In cases in which the body charge distribution is fixed and known, ϵ_b may be replaced with Q_{bs}/δ_o.

¶Not likely to be significant contributions.

requires (1) application of a force that produces a differential motion of a particle relative to the gas and (2) a gas retention time sufficient for the particle to migrate to the collecting surface. The principal mechanisms of aerosol deposition that are applied in dust collectors are (1) gravitational deposition, (2) flow-line interception, (3) inertial deposition, (4) diffusional deposition, and (5) electrostatic deposition. Thermal deposition is only a minor factor in practical dust-collection equipment because the thermophoretic force is small. Table 17-3 lists these six mechanisms and presents the characteristic parameters of their operation [Lunde and Lapple, *Chem. Eng. Prog.* **53:** 385 (1957)]. The actions of the inertial-deposition, flow-line-interception, and diffusional-deposition mechanisms are illustrated in Fig. 17-42 for the case of a collecting body immersed in a particle-laden gas stream.

Two other deposition mechanisms, in addition to the six listed, may be in operation under particular circumstances. Some dust particles may be collected on filters by sieving when the pore diameter is less than the particle diameter. Except in small membrane filters, the sieving mechanism is probably limited to surface-type filters, in which a layer of collected dust is itself the principal filter medium.

The other mechanism appears in scrubbers. When water vapor diffuses from a gas stream to a cold surface and condenses, there is a net hydrodynamic flow of the noncondensable gas directed toward the surface. This flow, termed the Stefan flow, carries aerosol particles to the condensing surface (Goldsmith and May, in Davies, *Aerosol Science*, Academic, New York, 1966) and can substantially improve the performance of a scrubber. However, there is a corresponding Stefan flow directed away from a surface at which water is evaporating, and this will tend to repel aerosol particles from the surface.

In addition to the deposition mechanisms themselves, methods for preliminary conditioning of aerosols may be used to increase the effectiveness of the deposition mechanisms subsequently applied. One such conditioning method consists of imposing on the gas high-intensity acoustic vibrations to cause collisions and flocculation of the aerosol particles, producing large particles that can be separated by simple inertial devices such as cyclones. This process, termed *sonic* (or *acoustic*) *agglomeration*, has attained only limited commercial acceptance.

Another conditioning method, adaptable to scrubber systems, consists of inducing condensation of water vapor on the aerosol particles as nuclei, increasing the size of the particles and making them more susceptible to collection by inertial deposition.

Most forms of dust-collection equipment use more than one of the collection mechanisms, and in some instances the controlling mechanism may change when the collector is operated over a wide range of conditions. Consequently, collectors are most conveniently classified by type rather than according to the underlying mechanisms that may be operating.

PERFORMANCE OF GAS-SOLIDS SEPARATORS

The performance of a gas–solids separation device is most commonly expressed as the collection efficiency η, the weight ratio of the dust collected to the dust entering the apparatus. However, the collection efficiency is usually related exponentially to the properties of the dust and gas and the operating conditions of most types of collectors and hence is an insensitive function of the collector operating conditions as its value approaches 1.0. Performance in the high-efficiency range is better expressed by the penetration $1 - \eta$, the weight ratio of the dust escaping to the dust entering. Particularly in reference to collection of radioactive aerosols, it is common to express performance in terms of the reciprocal of the penetration $1/(1 - \eta)$, which is termed the *decontamination index* (DI). The number of transfer units N_t, which is equal to $\ln [1/(1 - \eta)]$ in the case of dust collection, was first proposed for use by Lapple (Wright, Stasny, and Lapple, "High Velocity Air Filters," WADC Tech. Rep. 55-457, ASTIA No. AD-142075, October 1957) and is more commonly used than the DI. Because of the exponential form of the relationship between efficiency and process variables for most dust collectors, the use of N_t (or DI) is particularly suitable for correlating collector performance data.

In comparing alternative collectors for a given service, a figure of merit is desirable for ranking the different devices. Since power consumption is one of the most important characteristics of a collector, the ratio of N_t to power consumption is a useful criterion. Another is the ratio of N_t to capital investment.

Gas–solids separation devices are often installed in series to achieve the highest possible efficiency and performance reliability. A precleaner may be installed to reduce the solids loading in the primary collector to improve collection efficiency and reduce abrasive wear. A secondary collector downstream may be installed to mitigate chances of emissions during a process upset in the primary collector. The overall efficiency of the combined system can be calculated by multiplying the penetration value of each device that is connected in series (Fig. 17-43).

$$Pt_{\text{system}} = Pt_1, Pt_2, Pt_3 \tag{17-9}$$

$$\eta_{\text{system}} = 1 - Pt_{\text{system}} \tag{17-10}$$

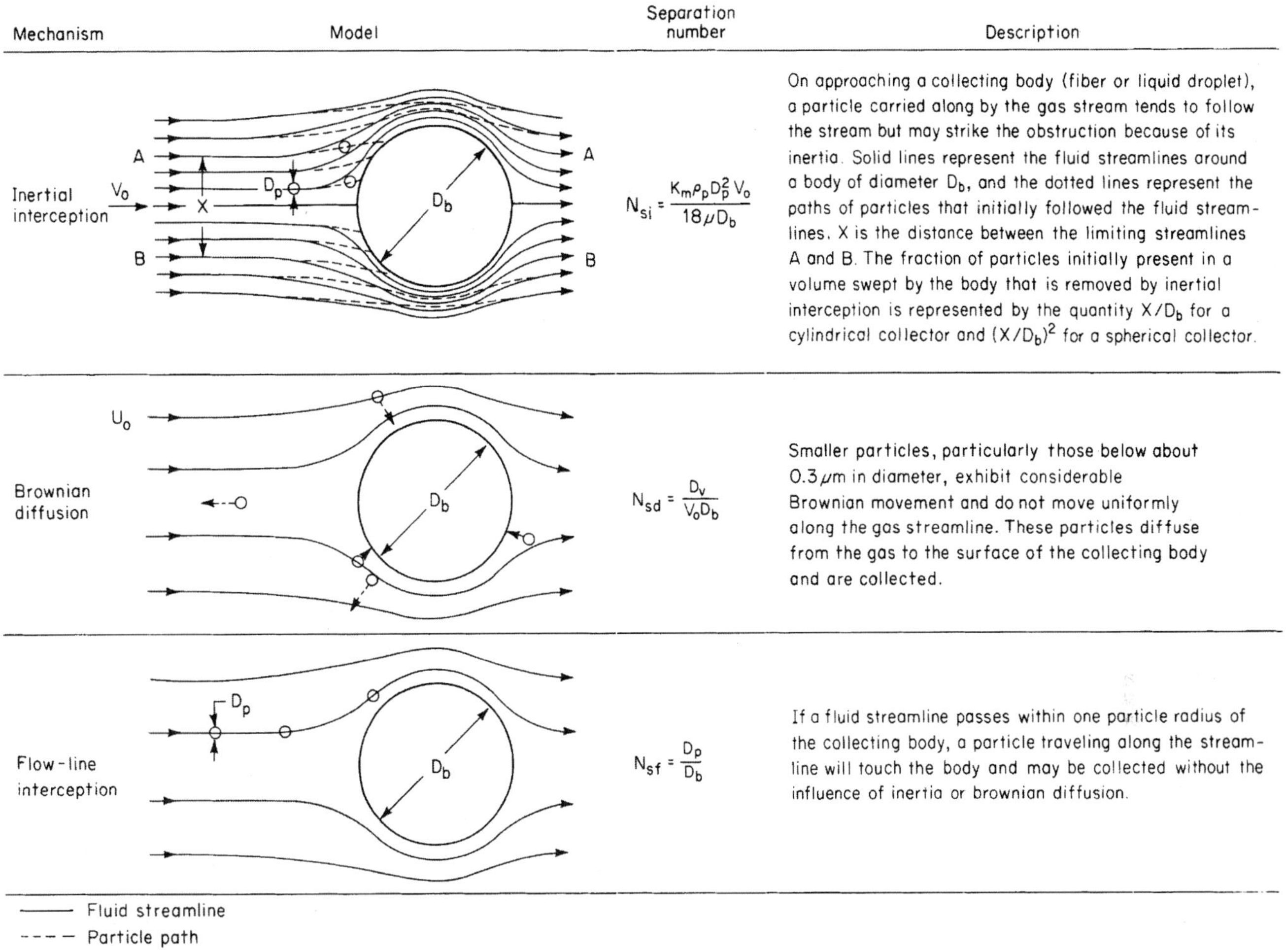
Mechanism
Model
Separation number
Description
Inertial interception
A
B
X
V_o
D_p
D_b
$N_{si} = \frac{K_m \rho_p D_p^2 V_o}{18 \mu D_b}$
On approaching a collecting body (fiber or liquid droplet), a particle carried along by the gas stream tends to follow the stream but may strike the obstruction because of its inertia. Solid lines represent the fluid streamlines around a body of diameter D_b, and the dotted lines represent the paths of particles that initially followed the fluid streamlines. X is the distance between the limiting streamlines A and B. The fraction of particles initially present in a volume swept by the body that is removed by inertial interception is represented by the quantity X/D_b for a cylindrical collector and $(X/D_b)^2$ for a spherical collector.
Brownian diffusion
U_o
D_b
$N_{sd} = \frac{D_v}{V_o D_b}$
Smaller particles, particularly those below about 0.3 μm in diameter, exhibit considerable Brownian movement and do not move uniformly along the gas streamline. These particles diffuse from the gas to the surface of the collecting body and are collected.
Flow-line interception
D_p
D_b
$N_{sf} = \frac{D_p}{D_b}$
If a fluid streamline passes within one particle radius of the collecting body, a particle traveling along the streamline will touch the body and may be collected without the influence of inertia or brownian diffusion.
Fluid streamline
Particle path

FIG. 17-42 Particle deposition on collector bodies.

Gravimetric Grade (or Fractional) Efficiency The performance of gas–solids separation devices is quantified by the overall collection efficiency or the fraction of the total mass of incoming dust that is separated from the gas stream. The overall collection efficiency depends on performance characteristics of the separator and the particle-size distribution of the incoming dust. It is possible for a coarse dust to be collected at 100 percent efficiency, whereas a fine dust may be separated with lower efficiency in the same unit operating at exactly the same conditions. Therefore, specification of an efficiency without qualifying the operating conditions and inlet dust characteristics is misleading.

The performance characteristics of a separator are represented by the gravimetric grade (or fractional) efficiency, or simply the grade efficiency. Grade efficiency can be conceptualized as a probabilistic function where every incoming particle has a certain chance of being removed from the gas stream. The flow path of various particles, even at identical operating conditions, will not be identical. Larger particles have a higher probability of being collected, while the smaller particles have a lower probability. The grade efficiency curve ($G(d_p)$) is typically S-shaped (Fig. 17-44*a*). The characteristic particle size for the abscissa should be relevant to the principle of separation. Stokes' diameter, Sauter mean diameter, or aerodynamic diameter are commonly used. The particle size corresponding to 50 percent efficiency is called the cut size (d_{p50}). It represents the size fraction which has equal probability of being collected or elutriated. A step-function at the cut size, which is often assumed but rarely realized, would represent the ideal shape

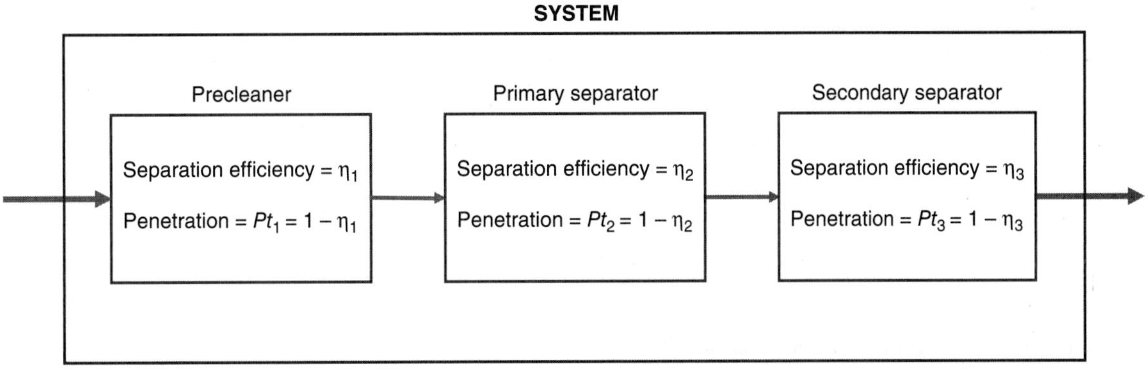

FIG. 17-43 Overall efficiency of separation devices connected in series.

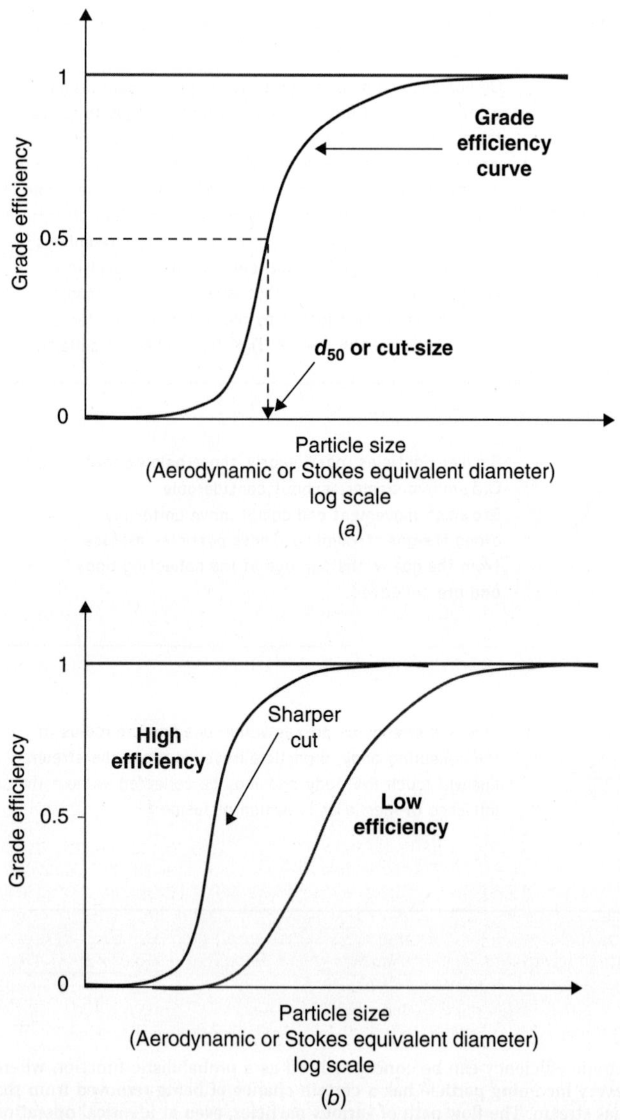

FIG. 17-44 (*a*) Grade efficiency curve and cut size. (*b*) Comparison of different grade efficiency curves.

of a grade or fractional efficiency curve. The departure from ideality can be attributed to turbulent dispersion, reentrainment, fines agglomeration, attrition, and other nonidealities within the separator.

Each curve is specific for a given design, size, and the operating conditions. Grade efficiency curves of two different devices can be plotted on the same basis for comparison (see Fig. 17-44*b*).

The overall gravimetric efficiency (or collection efficiency, η) is a product of grade efficiency and the weight fraction of incoming dust for various particle size fractions (Fig. 17-45).

$$\eta = \sum G(d_{pi}) w(d_{pi}) * 100 \tag{17-11}$$

where η = overall collection efficiency, %
$G(d_{pi})$ = grade (or fractional) efficiency at particle size d_{pi} (0 – 1)
$w(d_{pi})$ = weight fraction of particles corresponding to particle size d_{pi} (0 – 1)

The grade efficiency curve reaches an asymptotic limit on both ends, at smaller and larger particles. It is difficult to determine the particle size that has a 100 percent chance of separation (limit of separation). Therefore, 98 or 99 percent is usually chosen as the upper limit. On the lower end of the particle size, the curve can terminate at a finite value on the abscissa for separators with defined purge streams (e.g., uniflow cyclones, baffle separators) or for other inefficiencies in the operation (e.g., leakage in cyclones and bypassing in electrostatic precipitators).

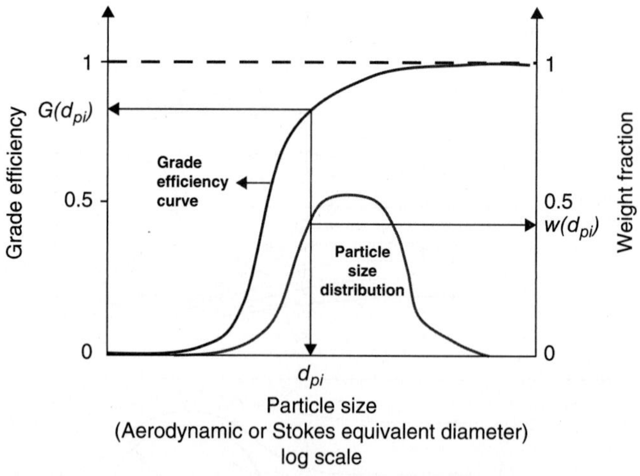

FIG. 17-45 Calculation of overall efficiency from grade efficiency and particle size distribution.

While this concept is very useful, and is equally applicable to solid–liquid separation and solid–solid separation, it is limited to unit operations where the mechanism of separation does not change with time or during operation. The performances of gravity settling chambers, inertial collectors, cyclones, electrostatic precipitators, and scrubbers are well represented by this concept, whereas filters (fabric, granular, media) are unsuitable due to changing mechanisms during operation.

For design and performance prediction, one must estimate (experimentally or theoretically) the grade efficiency curve of a device. In practice, experimental data are used to validate and adjust the parameters in a model. A generalized schematic of gas–solids separation is shown in Fig. 17-46.

The three streams are feed, coarse, and fines. Each stream is described by the solids rate and the particle-size distribution. Only two out of three streams must be defined, and the third one can be calculated from the mass balance. Unless there is a process advantage for measurements, Svarovsky (*Solid-Gas Separation, Handbook of Powder Technology*, vol. 3, Elsevier, Amsterdam, 1981) has shown that estimation based on fines and coarse streams yields the best results, or at least one of the streams should be the fines stream (Hoffman and Stein, *Gas Cyclones and Swirl Tubes—Principles, Design and Operation*, Springer, Berlin, 2002). The overall efficiency of separation can be estimated by Eq. (17-11).

For selection purposes, the entire efficiency curve need not be estimated. Approximations based on cut size and analytical cut diameter, which represent equipment performance and inlet dust characteristics respectively, can provide a useful guide.

Cut Size (d_{p50}): As discussed earlier, it is the particle size corresponding to 50 percent on the grade efficiency curve (Fig. 17-44*a*). Since a grade efficiency curve depends on operating conditions and equipment size/geometry, the family of curves can be condensed into a range of cut sizes for a given technology. These ranges are available from manufacturers and are well documented for commonly practiced devices. It is essential to understand the definition of particle size and its associated measurement method used to generate these grade efficiency curves.

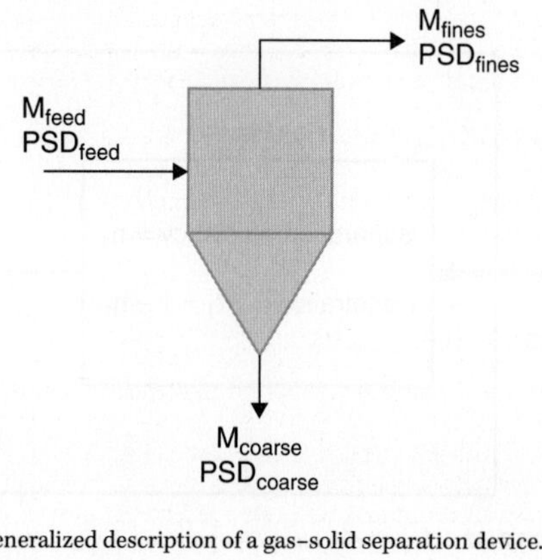

FIG. 17-46 Generalized description of a gas–solid separation device.

Analytical Cut Diameter ($d_{p\text{-}ac}$): This is the particle size on a cumulative weight percent oversize distribution curve that is *numerically* equal to the desired overall collection efficiency of the gas–solid separation device. Again, the selection of a characteristic particle size for the abscissa must be relevant to the principle of separation and the same as the grade efficiency curves under consideration.

The suitability of a technology for gas–solid separation is determined by its ability to achieve the desired collection or separation efficiency. For preliminary selection, the analytical cut diameter can be estimated from the particle-size distribution of the incoming dust. The analytical cut diameter is compared against typical ranges for various technologies (e.g., gravity settling, cyclones, scrubbers, electrostatic precipitators). The choice is acceptable if d_{50} is smaller than $d_{p\text{-}ac}$; otherwise the choice is unacceptable. There is an inherent assumption of the ideal shape of the curve. Nonetheless, it is a useful concept for preliminary process selection since it allows comparison of performances for a specific dust.

Particle Density In addition to the particle size, the density of particulate matter (PM) plays an important role in its ability to be entrained and remain in suspension. The particle density may be a function of particle size if the mechanism of particle formation for various size fractions is different. Very fine particles tend to agglomerate into a loose, porous structure, whereas coarse particles may be nonporous and dense. The density of nonporous solids can unambiguously be defined in the classical sense (mass per unit volume), and measured with pycnometers. On the other hand, porous particles will exhibit different dynamic behavior in fluids as compared to nonporous particles of the same size and composition. The equivalent or effective density, also known as the envelope density, is a measure that includes the pores and reflects the dynamic response of the particle in the fluid. True density or skeletal density is the density of the matter the particles are composed of. For gas–solids separation operations, we are interested in measuring the effective or envelope density of the particulate matter.

DESIGN OF GAS–SOLID SEPARATORS

For any separation device, the relative contribution of various collection mechanisms (Fig. 17-42) depends on the particle and gas characteristics, the geometry of the equipment, and the fluid-flow pattern. Although the general case is exceedingly complex, it is usually possible in specific instances to determine which mechanism or mechanisms may be controlling. Nevertheless, the difficulty of theoretical treatment of dust-collection phenomena has made necessary simplifying assumptions, with the introduction of corresponding uncertainties. Theoretical studies have been hampered by a lack of adequate experimental techniques for verification of predictions. Although theoretical treatment of collector performance has been greatly expanded in the past 30 years with the widespread application of computational fluid dynamics (CFD) coupled with the discrete element method (DEM), few of the resulting performance models have received adequate experimental confirmation because of experimental limitations. CFD is increasingly used to understand the gas and particle dynamics within the collector and for applying the learnings to optimize and troubleshoot systems.

The best-established models of collector performance are those for fibrous filters and fixed-bed granular filters, in which the structures and fluid-flow patterns are reasonably well defined. These devices are also adapted to small-scale testing under controlled laboratory conditions. Realistic modeling of full-scale electrostatic precipitators and scrubbers is incomparably more difficult. Confirmation of the models has been further limited by a lack of monodisperse aerosols that can be generated on a scale suitable for testing equipment of substantial sizes. When a polydisperse test dust is used, the particle-size distributions of the dust both entering and leaving a collector must be determined with extreme precision to avoid serious errors in the determination of the collection efficiency for a given particle size.

The design of industrial-scale collectors still rests essentially on empirical or semiempirical methods, although it is increasingly guided by concepts derived from theory. Existing theoretical models often embody constants that must be evaluated by experiment and may actually compensate for deficiencies in the models.

COMMON GAS–SOLID SEPARATORS

Gravity Settling Chambers The gravity settling chamber is probably the simplest and earliest type of dust-collection equipment, consisting of a chamber in which the gas velocity is reduced to enable dust to settle out by the action of gravity. Its simplicity lends it to almost any type of construction. Practically, however, its industrial utility is limited to removing particles larger than 325 mesh (43-μm diameter). For removing smaller particles, the required chamber size is generally excessive. As precleaners, gravity settling chambers can handle heavy dust loading and can process coarse abrasive particles at high temperatures and with low pressure drop.

Gravity collectors are generally built in the form of long, empty, horizontal, rectangular chambers with an inlet at one end and an outlet at the side or top of the other end. The dust-laden gas stream traverses the length of the settling chamber at low velocities (0.3 m/s typical, 3 m/s maximum), thereby allowing sufficient time for the coarse fraction to settle and to prevent reentrainment. The grade efficiency of gravity settling chambers can be calculated based on particle trajectory within the unit. Svarovsky [*Solid–Gas Separation, Handbook of Powder Technology*, vol. 3, Elsevier, Amsterdam, 1981] has outlined three calculation approaches, depending on the nature of flow:

- Laminar flow (Re < 2300)
- Turbulent flow with lateral mixing
- Turbulent flow with lateral and longitudinal mixing

Laminar Flow By assuming a low degree of turbulence relative to the settling velocity of the dust particle in question, the performance (grade efficiency) of a gravity settling chamber is given by

$$G(d_p)=\frac{B_sL_s}{q}U_t(d_p)=\frac{L_s}{V_sH_s}U_t(d_p)=\frac{\rho_s g d_p^2}{18\mu}\frac{L_s}{V_sH_s}B \qquad (17\text{-}12)$$

where V_s = average gas velocity. The dimensionless group B allows comparison of models for laminar and turbulent mixing. Expressing U_t in terms of particle size (equivalent spherical diameter), the smallest particle that can be completely separated out corresponds to $\eta = 1.0$ and, assuming Stokes' law, is given by

$$d_{p,\min}=\sqrt{\frac{18\mu H_sV_s}{gL_s(\rho_s-\rho)}} \qquad (17\text{-}13)$$

where ρ = gas density and ρ_s = particle density. For a given volumetric airflow rate, collection efficiency depends on the total plan cross section of the chamber and is independent of the height. The height needs to be made only large enough so that the gas velocity V_s in the chamber is not so high as to cause reentrainment of separated dust.

Turbulent Flow The footprint of a gravity settling chamber can become excessively large if laminar flow must be maintained. In practice, turbulent flow (Re > 2300) is often observed in these units. Turbulent mixing will reduce the collection efficiency, however, and a balance between economics (footprint) and efficiency must be sought. Simple mechanistic models proposed by Svarovsky [*Solid–Gas Separation, Handbook of Powder Technology*, vol. 3, Elsevier, Amsterdam, 1981] provide insight into the effect of turbulent mixing on performance. For an assumption of turbulent flow with complete lateral mixing and no longitudinal mixing, the grade efficiency is given by

$$G(d_p)=1-\exp\left(-\frac{U_t(d_p)L_s}{V_sH_s}\right)=1-\exp[-B]=1-\exp\left(-\frac{U_t(d_p)B_sL_s}{q}\right) \qquad (17\text{-}14)$$

For the case of lateral and longitudinal mixing in the settling chamber, the efficiency can be calculated by

$$G(d_p)=\frac{B}{1+B} \qquad (17\text{-}15)$$

As shown in Fig. 17-47, the performance is progressively diminished with increased mixing. These models do not account for reentrainment or recirculation flows within the unit. As the models suggest, reducing the settling height (H_s) will increase the collection efficiency. Horizontal plates arranged as shelves within the chamber will give a marked improvement in collection. This arrangement is known as the Howard dust chamber (Fume Arrester, U.S. Patent 896,111, 1908). The disadvantage of the unit is the difficulty of cleaning owing to the close shelf spacing and warpage at elevated temperatures.

The pressure drop through a settling chamber is small, consisting primarily of entrance and exit losses. Because low gas velocities are used, the chamber is not subject to abrasion and may, therefore, be used as a precleaner to remove very coarse particles and thus minimize abrasion on subsequent equipment.

Impingement Separators Impingement separators are a class of inertial separators in which particles are separated from the gas by inertial impingement on collecting bodies arrayed across the path of the gas stream. Fibrous-pad inertial impingement separators for the collection of wet particles are the main applications for this technology. With the growing need for very high-performance dust collectors, there is little application anymore for dry impingement collectors.

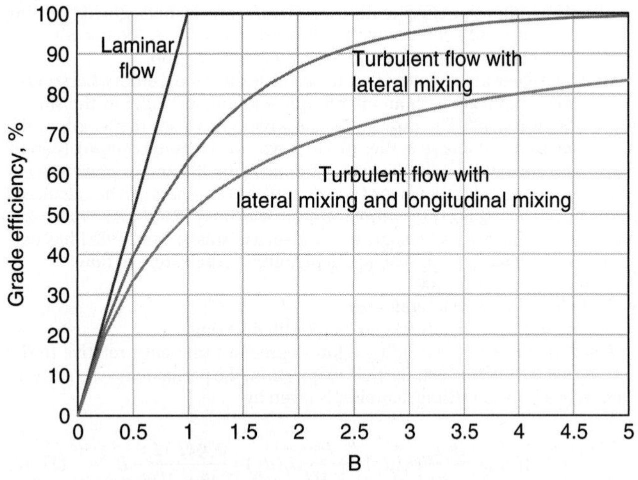

FIG. 17-47 Comparison of grade efficiency curves of settling chamber models.

Louver and Baffle Collectors With these devices (Fig. 17-48), particles are separated from a dust-laden stream when the direction of gas flow is abruptly changed. Due to inertia, the particles do not follow the gas streamlines but continue to be concentrated in the dust stream. A typical cut size of 10 to 20 μm can be achieved by these collectors. The pressure drop is slightly higher than with gravity settlers, but comparatively higher efficiency and lower space requirements make them a viable alternative. They are used as precleaners (e.g., for an engine intake) or concentrators (e.g., fly ash removal). The inside surface can be irrigated with water to reduce reentrainment and further increase the performance.

Cyclone Separators Cyclones have been widely used in process industries as primary separators or as precleaners to reduce dust loading on primary separators. The lack of moving parts, simplicity of construction, ability to operate at high temperature and pressure, separation of dust in dry state, and low capital cost make them an attractive option for gas cleaning. Successful applications of cyclones can be found in many industries involved with fluid–particle processing, including

- Petroleum (e.g., FCC cracking and fluid coking processes)
- Chemical (e.g., pneumatic conveying, drying, grinding)

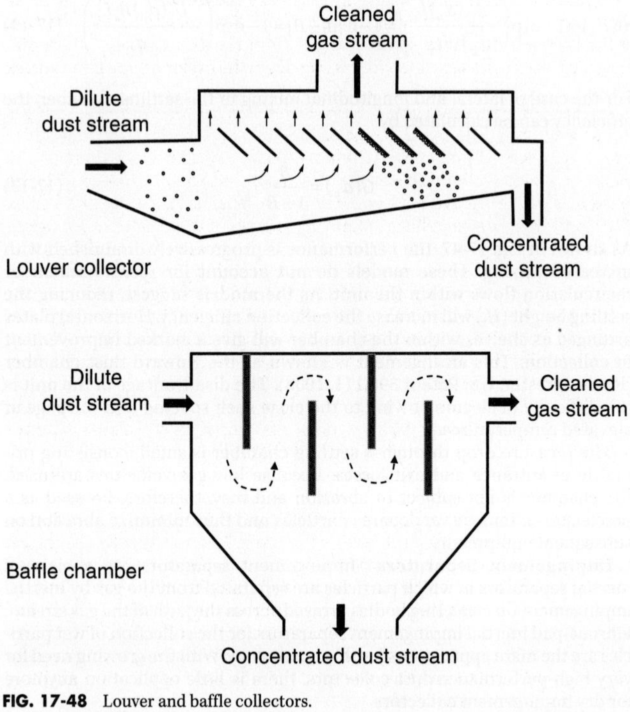

FIG. 17-48 Louver and baffle collectors.

- Agricultural (e.g., grain processing and conveying)
- Food (e.g., conveying, classification, grinding)
- Power (e.g., precleaners, emission control)
- Mineral (e.g., smelting, ore refining, dust control)

Within the range of their performance capabilities, cyclone collectors offer one of the least expensive means of dust collection from the standpoint of both investment and operation. Their major limitation is that their efficiency is low for the collection of particles smaller than 5 to 10 microns. However, third-stage separators (TSS) in fluid catalytic cracking (FCC) units can achieve very high efficiencies for particles as small as 2 μm. Typically, the TSS units collect everything larger than 5 μm, and approximately 90 percent of the particles in the loss stream will be smaller than 2 μm. If the loading is low to a TSS, (typically less than about 200 mg/Nm^3), the TSS in conjuction with a fourth-stage separator (FSS) can meet the emission requirements from a plant. If the loading is higher than about 200 mg/Nm^3, a hot gas filter is required after the TSS to meet the emission requirements. For emission control applications, especially for PM2.5 or PM1 (particulate matter with size less than 2.5 μm and 1 μm, respectively), cyclones must be followed by a secondary separator (e.g., fabric filter or electrostatic precipitator). However, highly efficient cyclones have been successfully designed for lower flow rates. Dyson's innovative (U.S. Patent 4593429, 1986) bagless vacuum cleaners based on cyclone technology have proven to be commercially successful. Small cyclones are also used for stack sampling and particle-size analysis of fines.

Although cyclones may be used to collect particles larger than 200 μm, gravity settling chambers or simple inertial separators (such as gas-reversal chambers) are usually satisfactory for this size of particle and are less subject to abrasion. In special cases in which the dust is highly agglomerated or where concentrations over 230 g/m^3 (100 gr/ft^3) are encountered, cyclones will remove dusts having small particle sizes. In certain instances, efficiencies as high as 98 percent have been attained with dusts having ultimate particle sizes of 0.1 to 2.0 μm because of the predominant effect of particle agglomeration due to large interparticle forces. Cyclones are used to remove both solids and liquids from gases, and they have been operated at temperatures as high as 1200°C and pressures as high as 50,700 kPa (500 atm).

Cyclones can be very small or very large. The smallest cyclones range from approximately 1 to 2 cm in diameter and the largest up to about 10 m in diameter. The number of cyclones used for a single fluidized bed can vary from 1 to up to 22 sets of first-stage and second-stage cyclones (44 cyclones total). Cyclones in process duty can be installed internally or externally to a reactor, horizontally or vertically, in series and in parallel, and in pressure/vacuum operation. Their adaptability to suit various applications has resulted in innumerable designs, but the underlying mechanisms of gas–solid separation are similar.

Mechanism of Separation In a conventional reverse-flow cyclone, the dust-laden gas usually enters a cylindrical or conical chamber tangentially at one or more entrances (usually rectangular in cross section) and leaves through a central opening (Fig. 17-49). The tangential entry of gas creates a swirling flow within the cyclone body that imparts a substantial centrifugal separating force on the particles. The force exerted in the particles relative to the gravitational force is proportional to $(U_i)^2/gr_o$, where g is the gravitational acceleration, U_i is the inlet velocity in the cyclone inlet, and r_o is the radius of the cyclone. For a cyclone with an inlet width that is 20 percent of the diameter of the cyclone (40 percent of the radius), the number of g's that the particles experience (assuming the particles are all at the centerline of the cyclone inlet) is

$$\frac{U_i^2}{g\;0.8r_o} = \frac{U_i^2}{9.81(0.4D_b)} \tag{17-16}$$

For a cyclone with a barrel diameter of 1 m and operating with an inlet gas velocity of 20 m/s, over 100 g's will be exerted on the entering particles. As a result, the particles move outward toward the cyclone wall, and aided by the outer vortex they migrate toward the dust outlet. The centrifugal force is opposed by the drag force that is directed inward. The cleaned gas stream leaves from the gas exit at the top of the cyclone.

Uniflow cyclones, also known as straight-though cyclones or swirl tubes, generate the swirling action with inlet vanes instead of a tangential inlet. The classified stream near the wall with higher dust concentration is discharged from the cyclone along with 1 to 3 percent of the incoming gas.

Flow Pattern In a reverse flow cyclone, the gas moves in a double vortex, with the gas initially spiraling downward at the outside after it enters the inlet, then flowing upward in the center of the cyclone before it exits. It should be noted that there is an inward flux of gas from outer vortex to inner vortex, which is not necessarily uniform in the axial direction. Due to the swirling nature of the flow, a high static pressure region is created near the wall in the entrance region. The resulting pressure gradient causes inward leakage, called "lip-leakage," along the roof and outer wall of the vortex finder

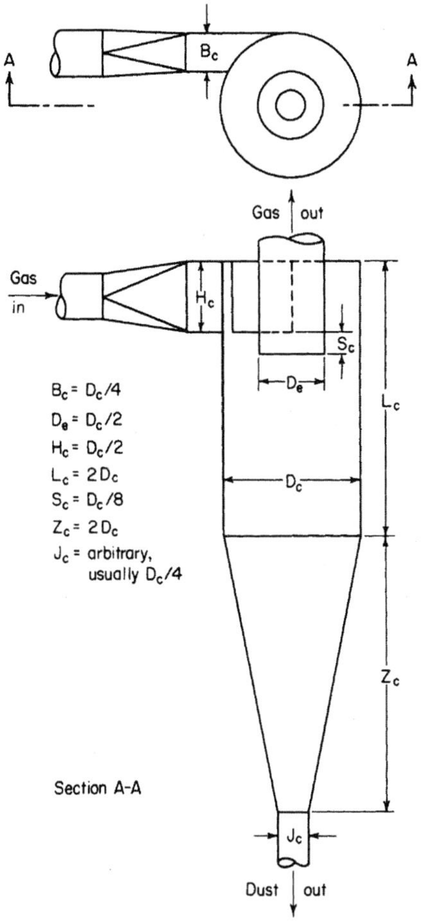

FIG. 17-49 Cyclone separator proportions.

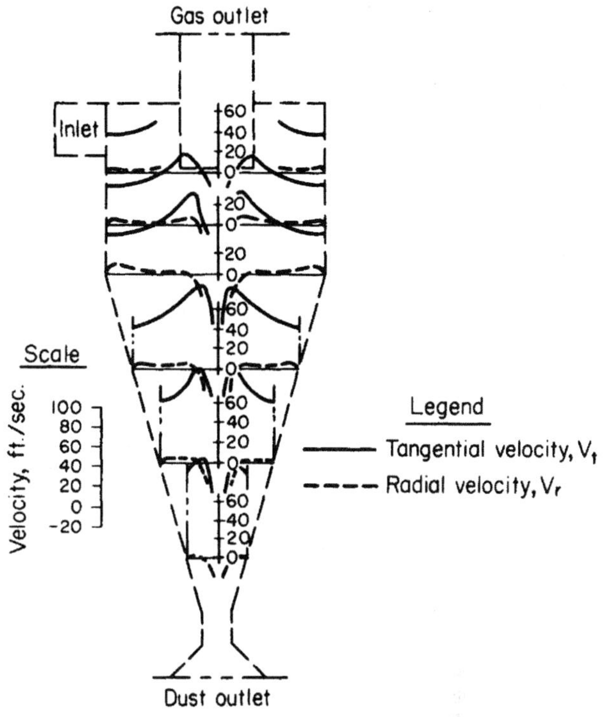

FIG. 17-50 Variation of tangential velocity and radial velocity at different points in a cyclone. [Ter Linden, *Inst. Mech. Eng. J.* **160:** 235 (1949).]

[Hoffman and Stein, *Gas Cyclones and Swirl Tubes—Principles, Design and Operation*, Springer, Berlin, 2002]. For practical purposes, it can be assumed that approximately 10 percent of incoming gas will bypass the cyclone due to this leakage [Muschelknautz, in *VDI Heat Atlas*, 2d ed. (English), Springer, Berlin, 2010]. The negative impact of this circulation pattern on collection efficiency can be significant for high-pressure cyclones [Heumann, *Chemical Engineering* (June): 118–123 (1991)] with a dished top (as compared to a flat top). This issue can be resolved by installing an internal false roof so the cyclone effectively operates with a flat roof.

When the gas enters the cyclone, its velocity undergoes a redistribution so that the tangential component of velocity increases with decreasing radius. The tangential velocity in a cyclone typically may reach a value approximately two to three times the average inlet gas velocity. Although the gas velocity approaches zero at the wall, the boundary layer is sufficiently thin that pitot-tube measurements show relatively high tangential velocities there, as shown in Fig. 17-50. The radial velocity, V_r, is directed toward the center throughout most of the cyclone, except at the center, where it is directed outward. Superimposed on the "double spiral," there may be a "double eddy" [Van Tongran, *Mech. Eng.* **57:** 753 (1935); and Wellmann, *Feuerungstechnik* **26:** 137 (1938)] similar to that encountered in pipe coils. Measurements on cyclones of the type shown in Fig. 17-49 indicate, however, that such double-eddy velocities are small compared with the tangential velocity [Shepherd and Lapple, *Ind. Eng. Chem.* **31:** 972 (1939); **32:** 1246 (1940)]. Recent analyses of flow patterns can be found in Hoffman et al., *Powder Technol.* **70:** 83 (1992); and Trefz and Muschelknautz, *Chem. Eng. Technol.* **16:** 153 (1993).

The preceding observations were made on smooth-wall cyclones. In practice, there are nonuniformities on the wall surface due to scaling, fouling, corrosion, weld seams, refractory, thermocouples, sampling probes, access ports, sight glasses, and sometimes due to hammer dents caused during unplugging operations [Hoffman and Stein, *Gas Cyclones and Swirl Tubes—Principles, Design and Operation*, Springer, Berlin, 2002]. Any such nonidealities can result in disruption of the boundary layer and cause reentrainment of collected solids, thereby reducing the collection efficiency. The effect of the roughness applies to low loading (i.e., second-stage) cyclones. High-loading cyclones (with loadings above about 1 kg of solids/kg of gas) are not as sensitive to wall roughness.

The inner vortex (often called the core of the vortex) rotates at a higher velocity than the outer vortex. In the absence of solids, the diameter of this inner vortex has been measured to be 0.8 to 0.85 times the diameter of the gas outlet tube. With axial inlet cyclones, the inner core vortex is aligned with the axis of the gas outlet tube. With tangential or volute cyclone inlets, however, the vortex is not exactly aligned with the axis. The asymmetric entry of the tangential or volute inlet causes the axis of the vortex to be slightly off center from the axis of the cyclone. This means that the bottom of the vortex is displaced some distance from the axis and can "pluck off" and reentrain dust from the solids sliding down the cyclone cone if the vortex gets too close to the wall of the cyclone cone.

At the bottom of the vortex, there is substantial turbulence as the gas flow reverses and flows up the middle of the cyclone into the gas outlet tube. As indicated previously, if this region is too close to the wall of the cone, substantial reentrainment of the separated solids can occur. If the cyclone is located above a solids collection hopper, even if the inner vortex does not touch the cone but extends beyond the dust outlet, it can entrain the collected dust from the hopper in a manner similar to a tornado. Just as important is an inner vortex that reverses within the body of the cylinder, which creates a secondary recirculation pattern below the vortex that can lead to reentrainment near the dust exit (Hoffman and Stein, *Gas Cyclones and Swirl Tubes—Principles, Design and Operation*, Springer, Berlin, 2002). Therefore, the relationship between the cyclone dimensions and vortex length must be taken into account during design.

If the cyclone is part of a recirculating system with a dipleg below the cyclone, then this reentrainment does not occur in the same manner because the solids are moving downward in the dipleg.

The vortex of a cyclone will precess (or wobble) about the center axis of the cyclone. This motion can bring the vortex into close proximity of the wall of the cone of the cyclone and "pluck" off and reentrain the collected solids flowing down along the wall of the cone. The vortex may also cause erosion of the cone if it touches the cone wall. Sometimes an inverted cone or a similar device called a vortex stabilizer is added to the bottom of the cyclone in the vicinity of the cone and dipleg to stabilize and "fix" the vortex. If it is placed correctly, the vortex will attach to the vortex stabilizer and the vortex movement will be stabilized, thus minimizing the efficiency loss due to plucking the solids off the wall and also minimizing the erosion of the cyclone cone.

Hugi and Reh [*Chem. Eng. Technol.* **21**(9): 716–719 (1998)] have reported that (at high solids loadings) enhanced cyclone efficiency occurs when the solids form a coherent, stable strand at the entrance to a cyclone. The formation of such a strand is dependent on several factors. They reported a higher cyclone efficiency for smaller ($d_{p50} < 40$ micron) solids than for larger solids (d_{p50}= 125 μm). This is not what theory would predict. However, they also found that the smaller particles formed coherent, stable strands more readily than the larger particles, which explained the reason for the apparent discrepancy.

Cyclone Efficiency The overall efficiency of any gas–solid separation device is given by the ratio of mass of solids collected to the mass of solids in the inlet stream. The overall collection efficiency of a cyclone is affected by the particle-size distribution of the feed stream even if all the operating conditions remain the same (see Fig. 17-45). The collection efficiency of each size fraction, known as the fractional or grade efficiency, must be determined when evaluating the performance of a cyclone. It is also important to specify the definition of the particle size along with its measurement technique. The aerodynamic particle size or a similar sedimentation-based method is recommended for cyclone applications. However, in many processes, the Sauter mean diameter is used to evaluate cyclones.

A reliable prediction of collection efficiency or emission rate is critical for manufacturers of pollution control equipment where performance warranties are required. Despite all the scientific progress made in the past 100 years, no generalized theory or calculation method from first principles is available that can calculate the performance (efficiency and pressure drop) with certifiable certainty. In practice, manufacturers conduct extensive experimental studies on a family of cyclone designs at various scales to generate the necessary scale-up information or fit parameters in the existing models for better predictive capability.

The cyclone efficiency can be estimated by theoretical models, scaling approach, or computational fluid dynamics (CFD). Each approach has its own merits and limitations. The theoretical models can be classified as (1) equilibrium orbit models [Barth, *Brennstoff-Warme-Kraft 8*, Heft 1 (1956); Muschelknautz, *Chemie-Ing.-Techn.* **44:** 63–71 (1972); Licht, *Air Pollution Control Engineering*, Marcel Dekker, New York (1980)] and (2) residence time or time-of-flight models [Rosin et al., *Zeit Ver. Deutscher Ing.* **76:** 433 (1932); Reitema, *De Ingenieur 71 jaargang No. 39*, ch 59–ch 65 (1959); Zenz, "Cyclone Design," in W. C. Yang, ed., *Fluidization Solids Handling and Processing Industrial Applications*, Noyes, Devon, UK (1999)]. Dietz [*AIChE Journal* **27:** 888–892 (1981)] and Mothes and Loffler [*Int. Chem. Eng.* **28:** 231–240 (1988)], among others, have proposed models that combine both approaches.

In this section, the approach outlined by Zenz and the so-called Stokesian scaling approach are presented.

Zenz Method The methods described here for calculation of pressure drop and efficiency were given by Zenz in *Manual on Disposal of Refinery Wastes—Atmospheric Emissions,* chap. 11 (1975), American Petroleum Institute Publ. 931 and improved by Particulate Solid Research Inc. (PSRI), Chicago. Cyclones work by using centrifugal force to increase the gravity field experienced by the solids. The solids then move to the wall under the influence of their effectively increased weight. Movement to the wall is improved as the path the solids traverse under centrifugal flow is increased. This path is equated with the number of spirals the solids make in the cyclone barrel. Figure 17-51 gives the number of spirals N_s as a function of the maximum velocity in the cyclone. The maximum velocity may be either the inlet or the outlet velocity, depending on the design. The equation for d_{pth}, the theoretical size of a particle collected by the cyclone at 50 percent collection efficiency, is

$$d_{pth} = \sqrt{\frac{9\mu_g B_c}{\pi N_s v_{max}(\rho_p - \rho_g)}} \tag{17-17}$$

This equation is a result of the residence time theory of particle collection. In this theory, the time that it takes for a particle to reach the wall is balanced by the time that a particle spends in the cyclone. The particle size that makes it to the wall by the time that it exits the cyclone is the particle size collected at 50 percent collection efficiency, d_{pth}.

When consistent units are used, the particle size calculated by Eq. (17-17) will be in either meters or feet. The equation contains the effects of cyclone size, gas velocity, gas viscosity, gas density, and particle density of the solids. In practice, a design curve such as that given in Fig. 17-52 uses d_{pth} as the size at which 50 percent of solids of a given size are collected by the cyclone. The material entering the cyclone is divided into fractional sizes, and the collection efficiency for each size is determined. The total low loading efficiency of collection is the sum of the product of the individual collection efficiencies of the cuts, E_{oi}, and the weight fraction of the cuts.

Equation (17-17) for d_{pth} applies for very dilute systems, usually on the order of 1 gr/ft^3, or 2.3 g/m^3 where 1 gr = (1/7000) lb. When denser flows of solids are present in the inlet gas, cyclone efficiency increases dramatically. This is thought to be due to the coarse particles carrying a large percentage of the finer particles along with them in their interstices as they flow to the wall of the cyclone. Other explanations are that the solids have a lower drag coefficient or tend to agglomerate in multiparticle environments, thus effectively becoming larger particles. At very high inlet solids loadings, it is believed the gas simply cannot hold that much solid material in suspension at high centrifugal forces, and the bulk of the solids simply "condenses" out of the gas stream.

The phenomenon of increasing efficiency with increasing loading is represented by Fig. 17-53. The initial efficiency of a cyclone operating at low loading (E_0) is found on the left, *y*-axis of the chart, and the parametric line is followed to the proper overall solids loading. The efficiency for that cut size is then read from the graph.

A single cyclone can sometimes give sufficient gas–solids separation for a particular process or application. However, solids collection efficiency can usually be enhanced by placing cyclones in series. Cyclones in series are typically necessary for most processes to minimize particulate emissions or to minimize the loss of expensive solid reactant or catalyst. Two cyclones in series are most common, but very often three cyclones in series are used. Some processes even have four stages of cyclones. Cyclones placed

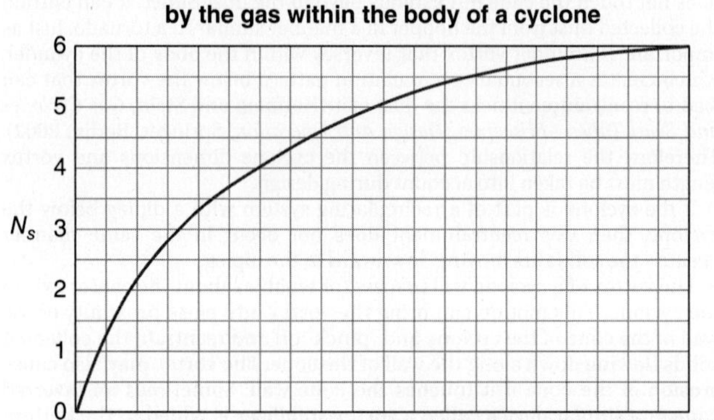

FIG. 17-51 N_s versus gas velocity, where the larger of either the inlet or outlet gas velocity is used.

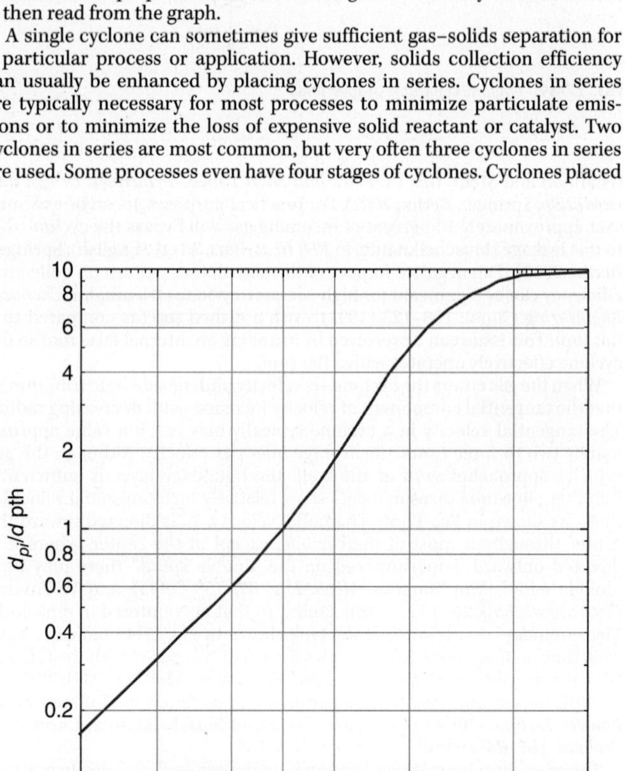

FIG. 17-52 Low loading ("single particle") cyclone collection efficiency curve. (*Courtesy of PSRI, Chicago.*)

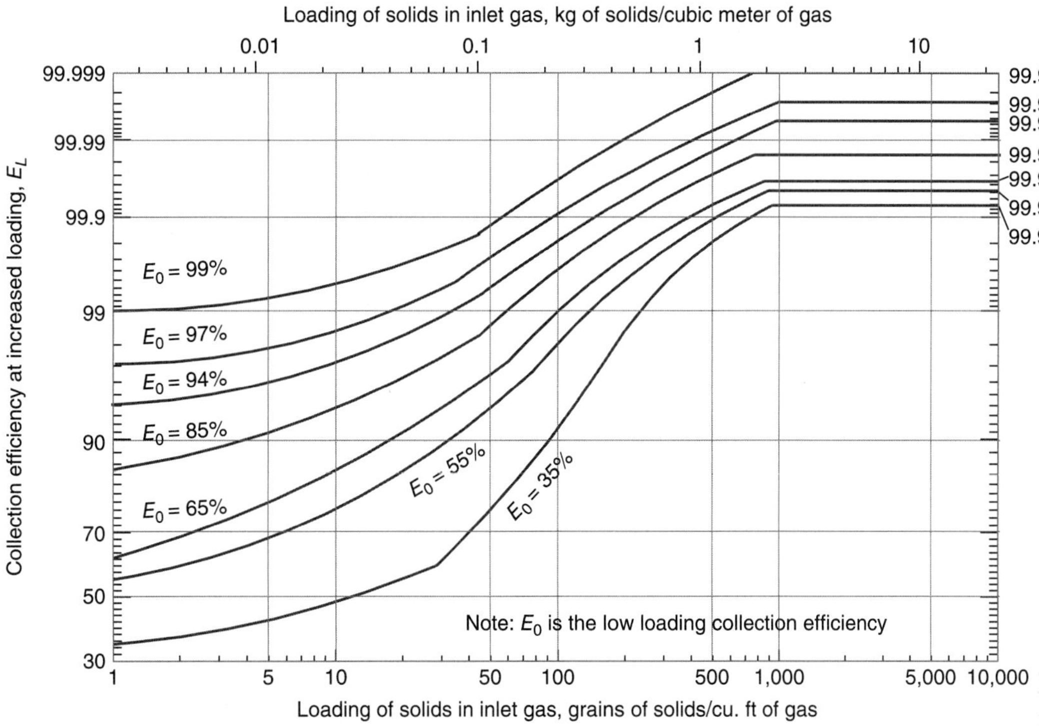

FIG. 17-53 Effect of inlet solids loading on cyclone collection efficiency. (*Courtesy of PSRI, Chicago.*)

in series can be very efficient. In fluidized catalytic cracking regenerators, two stages of cyclones can give efficiencies of up to and even greater than 99.999 percent.

Typically, first-stage cyclones will have an inlet gas velocity less than that of second-stage cyclones. The lower inlet velocity of first-stage cyclones results in lower particle attrition rates and lower wall erosion rates. After most of the solids are collected in the first stage, a higher velocity is generally used in second-stage cyclones to increase the centrifugal force on the solids and increase collection efficiency. Inlet erosion rates are generally low in the second stage because of the vastly reduced flux of solids into the second-stage cyclone. However, cone erosion rates in second-stage cyclones are much greater than in first-stage cyclones. Cone erosion rates can be most effectively reduced or eliminated by adding a vortex stabilizer to the cyclone cone.

Pressure Drop Cyclone pressure drop can be determined by summing five pressure drop components associated with a cyclone.

1. *Inlet contraction*

$$\Delta p = 0.5\rho_g\left(v_{\text{in}}^2 - v_{\text{vessel}}^2 + Kv_{\text{in}}^2\right) \qquad (17\text{-}18)$$

where K is taken from Table 17-4, and v_{in} and v_{vessel} are the velocities in the cyclone inlet duct and the velocity in the freeboard of the reactor vessel, respectively. The area ratios in Table 17-4 are either: (1) the area of the inlet duct to the area of the reactor freeboard, or (2) the cross-sectional area of the gas outlet tube to the cross-sectional area of the cyclone barrel. Using SI units gives the pressure drop in Pa. In U.S. conventional units, the factor of 32.2 for g must be included. This pressure loss is primarily associated with cyclones located in the freeboard of a fluidized bed. If the cyclone is located externally to a vessel and the high-pressure tap used to measure the cyclone pressure drop is in the inlet pipe before the cyclone, the measured pressure drop will generally not include this pressure loss, and this term should not be used to calculate total cyclone pressure drop. However, if the high-pressure tap to measure the cyclone pressure drop is located in the freeboard of the bed, this component will be included in the measured pressure drop, and it should be included in the calculation of the total cyclone pressure drop.

TABLE 17-4 *K* versus Area Ratio

Area ratio	*K*
0	0.50
0.1	0.47
0.2	0.43
0.3	0.395
0.4	0.35

2. *Particle acceleration*

$$\Delta P = Lv_{\text{in}}(v_{p\text{in}} - v_{p\text{vessel}}) \qquad (17\text{-}19)$$

For small particles, the particle velocity of the solids in the cyclone inlet is taken to be equal to the gas velocity, and L is the solids loading, kg/m^3.

3. *Barrel friction* The inlet diameter, d_{in}, is taken to be the hydraulic diameter, which is 4 × (inlet area)/inlet perimeter. Then

$$\Delta P = \frac{2f\rho_g v_{\text{in}}^2 \pi D_c N_s}{d_{\text{in}}} \qquad (17\text{-}20)$$

where the Reynolds number for determining the Fanning friction factor, f, is based on the cyclone inlet area. Values of f are typically between 0.003 and 0.008.

4. *Gas flow reversal*

$$\Delta P = \frac{\rho_g v_{\text{in}}^2}{2} \qquad (17\text{-}21)$$

5. *Exit contraction*

$$\Delta p = 0.5\rho_g\left(v_{\text{exit}}^2 - v_c^2 + Kv_{\text{exit}}^2\right) \qquad (17\text{-}22)$$

where K is again determined from Table 17-4 based on the area ratio of the area of the gas outlet tube to the area of the barrel of the cyclone. The total pressure drop is the sum of the five individual pressure drops.

However, the actual pressure drop observed turns out to be a function of the solids loading. The cyclone pressure drop is high when the inlet gas is free of solids and then *decreases* as the solids loading increases up to about 3 kg/m^3 (0.2 lb/ft^3). This is unusual because adding solids to most flowing gas streams results in an increase of the pressure drop. The cause of the initial decline is that the presence of solids decreases the tangential velocity of the gas [Yuu, *Chem. Eng. Sci.* **33:** 1573 (1978)]. Figure 17-54 gives the actual pressure drop based on the cyclone loading. When solids are absent, the observed pressure drop can be 2.5 times the calculated pressure drop with solids present.

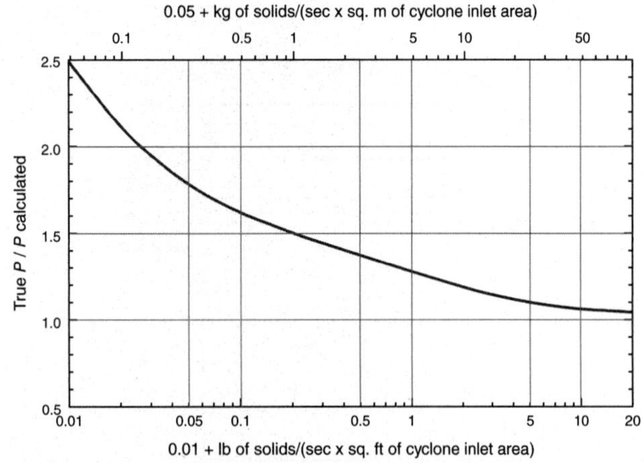

FIG. 17-54 Effect of cyclone inlet loading on pressure drop. (*Courtesy of PSRI, Chicago.*)

The sum of the pressure drops calculated above is assumed to be at a loading greater than 20 lb/s/ft^2 of cyclone inlet area (on the x-axis). If the loading is greater than this value, then the pressure drop calculation from adding the five different terms above is correct. However, if the loading is significantly lower (and the pressure drop is higher), then the calculated pressure drop must be multiplied by the value on the x-axis in Fig. 17-54 to give the corrected pressure drop. For example, if the inlet solids flux is approximately 1, then the calculated pressure drop should be multiplied by approximately 1.3.

Scaling Approach Theoretically, performance characteristics of a new large cyclone can be estimated using lab test data on a geometrically similar smaller cyclone of the same family. Dimensional analysis has been used to derive relationships for separation efficiency and pressure drop [Hoffman and Stein, *Gas Cyclones and Swirl Tubes—Principles, Design and Operation*, Springer, Berlin, 2002; Svarovsky, *Solid–Gas Separation, Handbook of Powder Technology*, vol. 3, Elsevier, Amsterdam, 1981]. From dimensional analysis, it can be shown that separation efficiency is a function of Reynolds number and Stokes number.

$$\eta(d_p) = f(\text{Re}, \text{Stk}) \tag{17-23}$$

For cut size (d_{p50}), the efficiency is 0.5, and Eq. (17-23) can be rewritten as

$$\text{Stk}(d_{p50}) = \frac{(\rho_p - \rho_g) d_{p50}^2 v_{\text{in}}}{18 \mu D_c} = f(\text{Re}) \tag{17-24}$$

Comparison of cyclones of various geometries [Overcamp and Scarlett, *Aerosol Science and Technology* **19:** 362–370 (1993)] suggests that the Stokes number [Stk(d_{p50})] is largely independent of Reynolds number > 2×10^4, and each design can be assigned a unique value of Stk(d_{p50}). It implies

$$[\text{Stk}(d_{p50})]_{\text{lab-cyclone}} = [\text{Stk}(d_{p50})]_{\text{field-cyclone}} \tag{17-25}$$

This analysis suggests that the cut size of a larger cyclone in a process can be estimated from a reasonably smaller scale lab unit. This relationship was derived for conventional cyclones with smooth walls and low dust concentrations (<1 g /m^3). The effect of loading can be estimated from Smolik's empirical equation (Svarovksy, *Solid-Gas Separation Handbook of Powder Technology*, vol. 3, Elsevier, Amsterdam, 1981):

$$\eta(c_1) = 1 - (1 - \eta(c_1)) \left(\frac{c_1}{c_2} \right)^{0.182} \tag{17-26}$$

The Euler number is used to relate pressure losses in a process unit to characteristic velocity.

Qualitative Trends The separation efficiency has been found to:

- Increase with an increase in inlet velocity
- Increase with an increase in particle specific gravity
- Increase with an increase in solids loading
- Increase with a decrease in gas temperature
- Increase with a decrease in cyclone diameter
- Increase with an decrease in gas outlet diameter (vortex finder)
- Increase with an increase in particle residence time in cyclone

Cyclone Design Factors Cyclones are sometimes designed to meet specified pressure drop limitations. For ordinary installations, operating at approximately atmospheric pressure, fan limitations generally dictate a maximum allowable pressure drop corresponding to a cyclone inlet velocity in the range of 8 to 30 m/s (25 to 100 ft/s). Consequently, cyclones are usually designed for an inlet gas velocity of 15 to 20 m/s (50 to 65 ft/s), although there are exceptions where the inlet gas velocity would be required to be lower and higher than indicated in this range.

Because of the relatively high gas velocities at the inlet of cyclones, particle attrition in fluidized-bed systems is generally dominated by the attrition produced in the cyclone. In some catalytic systems with very expensive catalysts, the economics of the process can be dependent on low attrition losses. In such cases, reducing the inlet velocity of the cyclone will significantly reduce the attrition losses in the process. To compensate for the reduction in inlet velocity, the exit gas velocity will generally be increased (by reducing the diameter of the outlet tube) in order to maintain high cyclone efficiencies. Reducing the outlet tube diameter increases the outlet gas velocity and increases the velocity in the vortex of the cyclone (and the centrifugal force in the vortex), increasing collection efficiency. However, as the vortex velocity is increased, its length is also increased. Therefore, care must be taken to ensure that the cyclone is long enough to contain the increased vortex length. If it is not, the vortex can extend far into the cone and can entrain solids flowing on the sides of the cone as it comes near them.

Cyclone Roughness Large weld beads and other factors can also reduce cyclone efficiency. If the solids flow along the wall of a cyclone encounters a large protuberance such as a weld bead, the weld bead acts as a type of "ski jump" and causes the solids to be deflected farther into the center of the cyclone, where they can be thrown into the vortex and carried out of the cyclone. In small pilot or research cyclones, this is especially common, because the distance between the wall of the cyclone and the vortex tube is very small. Because of their detrimental effect on cyclone efficiency, weld beads should be ground off to make the cyclone inner wall smooth.

In high-temperature processes, cyclones are often lined with refractory to both minimize heat loss and protect the metal surfaces from abrasion. These refractory surfaces are not as smooth as metal, but after a few days of operation, the refractory becomes smoother because of the abrasive action of the solids.

With very small laboratory or pilot cyclones, some solids (large polymer beads, spherical particles, etc.) can sometimes bounce off the cyclone wall immediately across from the cyclone inlet and be deflected into the vortex. Very large particles can be found in the gas outlet stream of the cyclone with these very small cyclones and with particles that bounce. To increase cyclone efficiency with these types of solids, the cyclone barrel diameter can be increased. This increases the distance between the cyclone vortex and the wall and prevents most of the solids from bouncing back into the vortex.

Theoretically, a primary design factor that can be used to control collection efficiency is the cyclone diameter. A smaller-diameter unit operating at a fixed pressure drop generally has a higher efficiency than a larger-diameter cyclone [Anderson, *Chem. Metall.* **40:** 525 (1933); Drijver, *Wärme* **60:** 333 (1937); and Whiton, *Power* **75:** 344 (1932); *Chem. Metall.* **39:** 150 (1932)]. In addition, smaller-diameter cyclones have a much smaller overall length. Small-diameter cyclones, however, will require multiple units in parallel to give the same capacity as a large cyclone. In such cases, the smaller cyclones generally discharge the dust into a common receiving hopper [Whiton, *Trans. Am. Soc. Mech. Eng.* **63:** 213 (1941)]. However, when cyclones discharge into a common hopper, there is a tendency of the gas to produce "cross-talk." This occurs when the gas exiting from one small cyclone passes up the exit of an adjoining cyclone, thus reducing efficiency. Various types of mechanical devices are generally added to the bottom of these small cyclones in parallel to reduce the cross-talk. The final cyclone design involves a compromise between collection efficiency and the complexity of the equipment. It is customary to design systems for a single cyclone for a given capacity, resorting to multiple parallel units only if the predicted collection efficiency is inadequate for a single unit or single units in series.

Reducing the gas outlet diameter should increase both collection efficiency and pressure drop. To exit the cyclone, gas must enter the cyclonic flow associated with the outlet tube. If the outlet diameter is reduced, the outlet vortex increases in length to compensate. Therefore, when the outlet area is less than the inlet area, the length of the cyclone must increase. Too short a cyclone is associated with erosion of the cone and reentrainment of solids into the exit flow. Table 17-5 gives the required increase in cyclone length as a function of outlet-to-inlet area. The cyclone length is measured centrally along an "extended vortex," a cylinder 10 cm larger in diameter than the inner diameter of the outlet tube, to prevent interference with the cone. If the cone interferes with this extended vortex, the cyclone must be lengthened.

As previously discussed, theoretically a smaller-diameter cyclone should be able to collect smaller particles because it can develop a higher centrifugal force. However, using smaller cyclones generally means that many have

TABLE 17-5 Required Cyclone Length as a Function of Area Ratio

A_{out}/A_{in}	Length below outlet tube/D_c
>1.0	2.0
0.8	2.2
0.6	2.6
0.4	3.2

to be used in parallel to accommodate large gas flows. The problem with parallel cyclones (as indicated above) is that it is difficult to get even distribution of solids into all the cyclones. If maldistribution occurs, this can cause inefficiencies that can negate the natural advantage of the smaller cyclones.

Cyclone diameters can be very large. Perhaps the largest cyclones are those used in circulating fluidized-bed combustors, where cyclone diameters approach or exceed 10 m. Large-diameter cyclones also result in very long cyclones, and so these large-diameter, long-length cyclones are really not feasible as internal cyclones in fluidized beds (they make the vessel too tall).

The minimum cone angle of the cyclone should be 60°. It is generally much greater (usually 70° to 75°), with steeper cone angles appropriate to materials that are more cohesive. The cyclone inlet is almost always rectangular (more efficient at getting material to the wall). In any case, the projection of the inlet flow path should never strike the wall of the outlet tube. This means that the inlet width of a cyclone should always be less than the distance between the wall and the outside diameter of the outlet tube. If a very heavy solids loading is anticipated, the barrel diameter should be increased slightly to minimize solids interference with the outlet gas tube.

Collection efficiency is normally increased by increasing the gas throughput [Drijver, *Wärme* **60:** 333 (1937)]. However, if the entering dust is agglomerated, high gas velocities may cause breakup of the agglomerated solids in the cyclone, so efficiency remains the same or actually decreases. Also, variations in design proportions that result in increased collection efficiency with dispersed dusts may be detrimental with agglomerated dusts. Kalen and Zenz [*Am. Inst. Chem. Eng. Symp. Ser.* **70**(137): 388 (1974)] report that collection efficiency increases with increasing gas inlet velocity up to a minimum tangential velocity at which dust is either reentrained or not deposited because of saltation. Koch and Licht [*Chem. Eng.* **84**(24): 80 (1977)] estimate that for typical cyclones the saltation velocity is consistent with cyclone inlet velocities in the range of 15 to 27 m/s (50 to 90 ft/s). Lapple (private communication) reports that in cyclone tests with talc dust, collection efficiency increased steadily as the inlet velocity was increased up to a maximum of 52 m/s (170 ft/s). With ilmenite dust, which was much more strongly flocculated, efficiency decreased over the same inlet velocity range. In later experiments with well-dispersed talc dust, collection efficiency continued to increase at inlet velocities up to the maximum used, 82 m/s (270 ft/s).

Another effect of increasing the cyclone inlet gas velocity is that friable materials may disintegrate (or attrit) as they hit the cyclone wall at high velocity. Thus, the increase in efficiency associated with increased velocity may be more than lost due to the generation of fine attrited material that the cyclone cannot collect. High inlet gas velocities in a primary cyclone can also lead to erosion of the cyclone where the solids in the inlet stream hit the wall.

Cyclones can be either placed in the freeboard above the fluidized bed or located outside of the fluidized-bed vessel. There are advantages and disadvantages to each type of placement, and the optimum type of placement depends on what is best for a particular process.

Internal cyclones have the advantages that they require no inlet piping (their inlets can be open to the freeboard), they require no high-pressure shell, and they normally have straight cyclone diplegs. Internal cyclones are generally smaller in diameter than external cyclones because their size is limited by the headspace available in the freeboard above the fluidized bed. These size limitations generally result in using several smaller cyclones in parallel instead of one large cyclone. However, it is difficult to aerate second-stage cyclone diplegs (generally an advantageous technique) when internal cyclones are used. Aerating secondary cyclone diplegs can improve the operation of the diplegs significantly.

The advantages of external cyclones are that (1) they can be much larger than internal cyclones, (2) they are more accessible than internal cyclones, and (3) their diplegs can be aerated more easily. The primary disadvantage of external cyclones is that they require a pressure shell.

Cyclone Inlets The design of the cyclone inlet can greatly affect cyclone performance. It is generally desired to have the width of the inlet, B_c, be narrow so that the entering solids will be as close as possible to the cyclone wall where they can be collected. However, narrow inlet widths require that the height of the inlet H be very long in order to give an inlet area required for the desired inlet gas velocities. Therefore, a balance between narrow inlet widths and the length of the inlet height has to be struck. Typically, low-loading cyclones (cyclones with inlet loadings less than approximately 2 to 5 kg/m^3) have height/width ratios, H/B_c, of between 2.5 and 3.0. For high-loading cyclones, this inlet aspect ratio can be increased to as high as 7 or so with the correct design. Such high inlet aspect ratios require that the cyclone barrel length increase.

A common cyclone inlet is a rectangular tangential inlet with a constant area along its length. This type of inlet is satisfactory for many cyclones, especially those operating at low solids loadings. However, a better type of inlet is one in which the inner wall of the inlet is angled toward the outer cyclone wall at the cyclone inlet. This induces solids momentum toward the outer wall of the cyclone. The bottom wall of this inlet is angled downward so that the area decrease along the inlet flow path is not too rapid and acceleration is controlled. In addition, the entire inlet can be angled slightly downward to give enhanced efficiencies. This type of inlet is superior to the constant-area tangential inlet, especially for higher solids loadings (greater than 2 to 5 kg/m^3).

Hugi and Reh [*Chem. Eng. Technol.* **21**(9): 716–719 (1998)] report that continuous acceleration of the solids throughout the inlet is desired for improved efficiency, and that the angled inlet previously described achieves this. If the momentum of the solids is sufficient and the solids are continuously accelerating along the length of the inlet, the stable, coherent strand important for high collection efficiencies is produced.

The best type of inlet for high solids loadings is the volute cyclone inlet. At high inlet loadings (above approximately 2 to 3 kg/m^3) in a *tangential* cyclone inlet, the gas–solids stream expands rapidly from its minimum width at the point of contact. This rapid expansion disturbs the laminar gas flow around the gas outlet tube and causes flow separation around the tube. At some loadings, the inlet stream can expand to such an extent that the solids can hit the gas outlet tube. Both effects result in lowered cyclone efficiency. However, when a volute inlet is used, the expanding solids stream is farther from the gas outlet tube and enters at an angle so that the solids do not induce as much flow separation or asymmetric flow around the gas outlet tube. Therefore, cyclone efficiency is not affected to as great a degree. If a tangential cyclone is used at high solids loadings, an extra distance between the gas outlet tube and the cyclone wall should be designed into the cyclone to prevent the solids from striking the gas outlet tube. At low solid loadings, the solids do not hit the gas outlet tube. Because tangential cyclone inlets are less expensive than volute inlets, the tangential cyclone is often typically used for low loadings, and the volute inlet cyclone is used for high loadings.

The nature of the gas solids flow in the inlet ducting to the cyclone can affect cyclone efficiency significantly. If the solids in the inlet salt out on the bottom and result in dune formation and the resulting unsteady or pulsing flow, cyclone efficiency is adversely affected. To minimize the possibility of this occurring, it is recommended that the inlet line to the cyclone operate above the saltation velocity (Gauthier et al., in *Circulating Fluidized Bed Technology III*, ed. Basu, Horio, and Hasatani, Pergamon Press, Oxford, 1990, pp. 639–644), which will prevent the solids from operating in the dune or pulsing flow regime. If this is not possible, then the inlet line can be angled downward (approximately 15° to 20°) to let gravity assist in the flow of the solids. Keeping the inlet line as short as possible can also minimize any pulsing of the solids flow.

The presence of bends before and in close proximity to the cyclone inlet can be beneficial or detrimental, depending on the orientation of the bend. A bend that turns in the opposite direction from the gas entering the cyclone will decrease the collection efficiency since it concentrates the solids away from the cyclone wall at the inlet.

Discharge Outlet A cyclone will operate equally well on the suction or pressure side of a fan. One of the greatest causes of poor cyclone performance is the leakage of air into the dust outlet of the cyclone. A slight air leak at this point can result in a tremendous drop in collection efficiency, particularly with fine dusts. For batch operation, an airtight hopper or receiver may be used. For continuous withdrawal of collected dust from a hopper, a rotary star valve, a double-lock valve, or a screw conveyor may be used. The size of the outlet at the bottom of the hopper or receiver should be large enough to prevent flow impediment due to upward air leakage.

For continuous solids flow operation, mechanical trickle and flapper valves at the end of cyclone diplegs can also be used for continuous withdrawal into fluidized beds or into the freeboard of fluidized beds. Open-ended diplegs simply immersed into a fluidized bed can be used in cases where start-up losses are not excessive, and they are the simplest type of discharge system for returning solids into a fluidized bed (see Solids Discharge in the subsection Fluidized-Bed Systems).

Solids Loading Range Cyclones can collect solids over a wide range of loadings. Traditionally, solids loadings have been reported as either kilograms of solids per cubic meter of gas (kg/m^3), or as kilograms of solids per kilogram of gas (kg_s/kg_g). However, loading based on mass is probably not the best way to report solids loadings for cyclones. This is so because the volume of solids processed by a cyclone at the same mass loading can vary greatly, depending on the density of the solids. For example, many polymers have a bulk density of approximately 400 kg/m^3, and iron ore has a bulk density of approximately 2400 kg/m^3. This difference is a factor of 6. Therefore, a cyclone operating with polymer would have to process six times the volume of solids that a cyclone operating with iron ore would process at the same mass loading. If the cyclone operating with the polymer were designed to operate at high loadings on a mass basis, it could plug. In addition, the diplegs below the cyclone operating with the polymer may experience operational problems because of the high volumetric loading.

At ambient conditions, cyclones have been operated at solids loadings as low as 0.002 kg/m^3 (0.00125 kg/kg) and as high as 64 kg/m^3 (50 kg_s/kg_g). This is a factor of 32,000. In general, cyclone efficiency increases with increasing solids loading. This is so because at higher loadings, very fine particles are trapped in the interstices of the larger particles, and this entrapment increases the collection efficiency of the small particles. Even though collection efficiencies are increased with increased loading, cyclone loss rates are also increased as loading is increased. This is so because the cyclone efficiency increase is almost always less than the efficiency due to the increase in the solids loading.

A heavy chain suspended from the gas outlet duct has been found to be beneficial to minimize dust buildup on the cyclone walls in certain circumstances. Such a chain should be suspended from a swivel so that it can rotate without twisting. Substantially all devices that have been reported to reduce pressure drop do so by reducing spiral velocities in the cyclone chamber and consequently result in reduced collection efficiency.

A cyclone will operate as well in a horizontal position as in a vertical position. However, departure from the normal vertical position results in an increasing tendency to plug the dust outlet line. If the dust outlet becomes plugged, collection efficiency will, of course, be low. If the cyclone exit duct must be reduced to tie in with proposed duct sizes, the transition should be made at least three pipe diameters downstream from the cyclone outlet and preferably after a bend. In the event that the transition must be made closer to the cyclone, a Greek cross can be installed in the transition piece to avoid excessive pressure drop.

Cyclone Length As described previously, the cyclone length should be great enough to contain the vortex below the gas outlet tube. It is generally advisable to have the cyclone somewhat longer than required so that modifications to the gas outlet tube can be made in the future if required. Either the barrel or the cone can be increased in length to contain the vortex. However, cyclone barrels can be made too long. If the barrel is too long, the rotating spiral of solids along the wall can lose its momentum. When this happens, the solids along the wall can be reentrained into the rotating gas in the barrel, and cyclone efficiency will be reduced.

Hoffman et al. [*AIChE J.* **47**(11): 2452–2460 (2001)] studied the effect of cyclone length on cyclone efficiency and showed that the efficiency of a cyclone increases with length. However, they also found that after a certain length, cyclone efficiency decreased. They reported that cyclone efficiency suddenly decreased after a certain cyclone length, which in their cyclone was at a length-to-diameter ratio of 5.65, although recent tests at PSRI with larger cyclones have shown no reduction in cyclone efficiency at length-to-diameter ratios of 7. Hoffman et al. also reported that cyclone pressure drop decreased with increasing cyclone length. This probably occurs for the same reason that cyclone pressure drop decreases with increasing cyclone loading. For long cyclones, the increased length of the cyclone wall results in a longer path for the gas to travel. This creates greater resistance to the flow of the gas in the cyclone (much as a longer pipe produces greater resistance to gas flow than a shorter pipe) that results in reducing the tangential velocity in the cyclone and, therefore, the cyclone pressure drop.

Commercial Equipment and Operation Simple cyclones are available in a wide variety of shapes ranging from long, slender units to short, large-diameter units. The body may be conical or cylindrical, and entrances may be involute or tangential and round or rectangular.

Some of the special types of commercial cyclones are shown in Fig. 17-55. In the multiclone, a spiral motion is imparted to the gas by annular vanes, and it is furnished in multiple units of 15.2- and 22.9-cm (6- and 9-in) diameter. Its largest field of application has been in the collection of fly ash from steam boilers.

In some cyclones, the roof is helically shaped. This is done to ensure that the rotating solids stream does not collide with the inlet solids stream. Axial cyclones make use of the annular inlet vanes as discussed for multiclones. The spiral, or volute, inlet adds solids more gradually to the cyclone. This type of inlet is a better inlet for high-loading cyclones because it minimizes any interaction of the solids with the inlet solids stream.

In addition to the conventional reverse-flow cyclones, some use is made of uniflow, or straight-through, cyclones, in which the gas and solids discharge at the same end (Fig. 17-56 in a vertical position and in Fig. 17-55

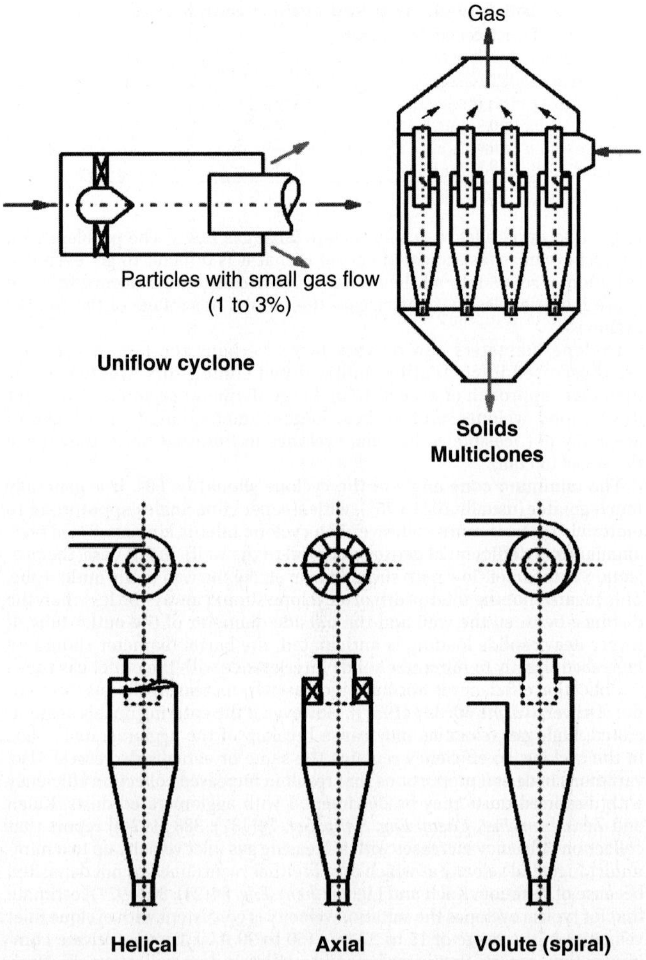

FIG. 17-55 Different types of cyclone designs. (*Courtesy of PSRI, Chicago.*)

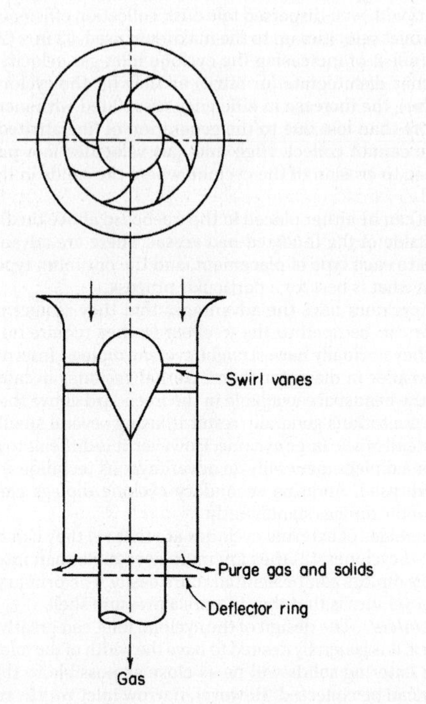

FIG. 17-56 Uniflow cyclone. [*Ter Linden,* Inst. Mech. Eng. J., ***160,*** *233* (*1949*).]

in a horizontal position). These devices generally act as concentrators; the concentrated dust, together with the inlet gas, is discharged at the periphery, while the clean gas passes out through the center tube. The straight-through cyclones are usually designed into multiple-tube units. They are also used to try to minimize the residence time of gas and solids to limit unwanted reactions.

Most processes operate at high temperatures and/or high pressures. Therefore, it is important to know how cyclones operate at these conditions. Efficient cyclones can collect very small particles sizes. Therefore, cyclone efficiency is proportional to $1/d_{pth}$.

$$E_0 \propto \frac{1}{d_{pth}} = \sqrt{\frac{\pi N_s U(\rho_p - \rho_g)}{9\mu B_c}} \quad (17\text{-}27)$$

The effects of temperature and pressure manifest themselves in how they affect the gas density and gas viscosity. From Eq. (17-27), it can be seen that cyclone efficiency is theoretically related to gas density and gas viscosity as

$$E_0 \propto \sqrt{\frac{\rho_p - \rho_g}{\mu}} \quad (17\text{-}28)$$

As pressure is increased, gas density will increase. However, the term $(\rho_p - \rho_g)$ does not change with increases in gas density because particle density is so much greater than the gas density (typically about 2000 kg/m^3 versus approximately 20 kg/m^3 at high pressure) that it dominates this term. Therefore, it is expected that gas density would have a small effect on cyclone efficiency. Conversely, cyclone efficiency would be expected to decrease with system temperature because gas viscosity increases with increasing pressure.

Knowlton and Bachovchin [*Coal Processing Technol.* **4:** 122–127 (1978)] studied the effect of pressure on cyclone performance and found little change in overall cyclone efficiency with pressure over a pressure range from 0 to 55 barg. However, fractional efficiency curves from the same study showed that cyclone efficiency decreased with pressure for particle sizes less than about 20 μm. For particle sizes greater than about 20 μm there was no effect of pressure on cyclone efficiency.

The effect of temperature on cyclone efficiency was studied by both Parker et al. [*J. Environ. Sci. and Technol.* **15**(4): 451 (1981)] and Patterson and Munz [*Canadian J. Chem. Eng.* **67:** 321 (1989)]. Both studies showed that cyclone efficiency decreased with increasing gas viscosity. Similar to the studies at high pressure, Patterson and Munz (1989) reported that only the collection efficiency of particles less than about 10 μm was reduced because of operation at high temperature.

Troubleshooting Key operational issues for cyclones are related to efficiency, pressure drop, and erosive wear. Table 17-6 provides basic guidance on troubleshooting.

Mechanical Centrifugal Separators A number of collectors in which the centrifugal field is supplied by a rotating member are commercially available. In the typical unit shown in Fig. 17-57, the exhauster or fan and dust collector are combined as a single unit. The blades are specially shaped to direct the separated dust into an annular slot leading to the collection hopper, while the cleaned gas continues to the scroll. The performance characteristics of the impeller blades are similar to a forward-curve blade fan with a medium pressure range. The concentrated dust suspension in the secondary air circuit enters a sealed collection hopper which serves as a gravity settling chamber. The returning air from the hopper is fed back into the separators. The hopper operates under a slight positive pressure; therefore, an airlock at the hopper outlet is required for continuous discharge of collected dust.

These separators offer a unitized approach for dust separation in exhaust systems for ceramics, chemicals, food products, rubber, plastics, woodworking, grains, and metalworking. They have a compact construction, low pressure drop, low power consumption, minimal maintenance, and can maintain constant airflow regardless of dust loading. For dust streams with large particles or for heavy loading, additional precleaners may be installed to improve efficiency and reduce wear.

Although no comparative data are available, the collection efficiency of units of this type is probably comparable with that of the single-unit, high-pressure-drop-cyclone installation. The clearances are smaller and the centrifugal fields higher than in a cyclone, but these advantages are probably compensated for by the shorter gas path and the greater degree of turbulence with its inherent reentrainment tendency. The chief advantage of these units lies in their compactness, which may be a prime consideration for large installations or plants requiring a large number of individual collectors. Caution should be exercised when trying to apply this type of unit to a dust that shows a marked tendency to build up on solid surfaces because of the high maintenance costs that may be encountered from plugging and rotor unbalancing.

Particulate Scrubbers Wet collectors, or scrubbers, form a class of devices in which a liquid (usually water) is used to assist or accomplish the separation of dusts, mist, or fumes from a gas stream. Such devices have been in use for well over 100 years, and innumerable designs have been or are offered commercially or constructed by users. Remarkably, most of the designs are based on the similar fundamental mechanisms of particle capture and removal, and associated energy losses. They collect particulates on droplets, liquid film, or liquid-coated surfaces.

The primary advantages of particle scrubbers (wet collectors) include:

- Both gaseous and particulate contaminants can be removed.
- Suitable for sticky, hygroscopic, combustible, and explosive dust particles.
- Able to process gaseous streams with high humidity or condensable components.
- Suitable for high-temperature gas streams that can be quenched and scrubbed.
- Smaller footprint as compared to dry collectors.
- Constant pressure drop during operation.
- Generally, less expensive than electrostatic precipitators and filter collectors unless a wastewater treatment facility is needed.

Some of the disadvantages are:

- A new process stream (slurry) is generated that must be treated for product recovery or waste treatment. This will adversely affect the process economics.
- Higher energy consumption (operating cost) for fine particle removal.
- Submicron particles are difficult to remove without significant energy expenditure.
- Operational problems, such as fouling, scaling, and corrosion, must be dealt with.
- The scrubbing liquid (water) requires freeze protection in cold climates.

For more detailed discussion, refer to *ASHRAE Handbook 2016*; Cooper, in *Handbook of Powder Science & Technology*, ed. Fayed and Otten, Chapman and Hall, London, 1997; Cooper and Alley, *Air Pollution Control—A Design Approach*, 4th ed., Waveland Press, Long Grove, Ill., 2011; Schifftner and Hesketh, *Wet Scrubbers*, 2d ed., Technomic Pub., Lancaster, Pa., 1996.

Separation Mechanism In particulate scrubbers, the liquid is dispersed into the gas as a spray, and the liquid droplets are the principal collectors for the dust particles. Depending on their design and operating conditions, particulate scrubbers can be adapted to collecting fine as well as coarse particles. Collection of particles by the drops follows the same principles illustrated in Fig. 17-42. Many investigations of the relative contributions of the various mechanisms have led to the conclusion that the predominant mechanism is inertial impaction. Flow-line interception is only a minor mechanism in the collection of the finer dust particles by liquid droplets of the sizes encountered in scrubbers. Diffusion is indicated to be a relatively minor mechanism for the particles larger than 0.1 μm that are of principal concern. Thermal deposition is negligible. Gravitational settling is ineffective because of the high gas velocities and short residence times used in scrubbers. Electrostatic deposition is unlikely to be important except in cases in which the dust particles or the water, or both, are being deliberately charged from an external power source to enhance collection. Deposition produced by Stefan flow can be significant when water vapor is condensing in a scrubber. A similar mechanism is in play when the particles are captured on a liquid film or liquid-coated surfaces in fluidized beds, mobile beds, fibrous beds, and packed-bed scrubbers.

Despite many claims or speculations that wetting of dust particles by the scrubbing liquid plays a major role in the collection process, there is no unequivocal evidence that this is the case. The issue is whether wetting is an important factor in the adherence of a particle to a collecting droplet upon impact. From the body of general experience, it can be inferred that wettable particles probably are not collected much, if any, more readily than nonwettable particles of the same size. However, the available experimental techniques have not been adequate to permit any direct test to resolve the question. Changing from a wettable to a nonwettable test aerosol or from one scrubbing liquid to another is virtually certain to introduce other (and possibly unknown) factors into the scrubbing process. The most informative experimental studies appear to be some by Weber [*Staub*, English trans. **28:** 37 (November 1968); **29:** 12 (July 1969)], who bombarded single drops of various liquids with dust particles at different velocities and studied the behavior at impact by means of high-speed photography. Dust particles hitting the drops were invariably retained by the latter, regardless of their wettability by the liquid used. The use of wetting agents in scrubbing water is equally controversial, and there has been no clear demonstration that it is beneficial. The hypothesis is that wettable particles are ensconced inside the droplet, whereas the nonwettable particles are attached to the surface.

A particulate scrubber may be considered to consist of two stages: (1) a contactor stage, in which a spray is generated and the dust-laden gas stream is brought into contact with it; and (2) an entrainment separation stage, in which the spray and deposited dust particles are separated from

TABLE 17-6 Troubleshooting Guide for Cyclone Separators

Symptom	Possible cause	Solution
Collection efficiency is lower than expected.	Design basis is wrong.	Verify that the specified design conditions and vendor performance predictions are correct. If higher ΔP can be provided by the system air mover, and the collection efficiency is close to the desired level, and the dipleg seals can accommodate the higher pressure drops, and erosion and attrition are not a concern, modify cyclone inlet and/or outlet to increase the velocity. Replace the cyclone with a better cyclone.
	Gas leakage into the cyclone.	Check and repair any leaks or holes. Check to make sure flange connections are properly gasketed and tight. In pneumatic conveying systems, check and repair feeder valves for proper operation and gas tightness. For a low-loaded cyclone, with a trickle valve at the end of the dipleg, the gas flow up the dipleg may be too high. In this case, immerse the dipleg end into the fluidized bed.
	Inlet or outlet ductwork is improperly designed.	Check and repair inlet and/or outlet ductwork if any flow disturbance is induced into the cyclone.
	There is an internal obstruction.	Ensure that any access doors are flush and smooth. Ensure that there are no instruments or probes sticking into the cyclone flow stream. If the cyclone is lined, check for and repair any major erosion that causes a sharp edge disturbance to the flow stream. If plugging is occurring, see below.
Plugging	Feeder valve is sized improperly for the particulate loading and density.	Resize and replace the feeder valve.
	Cyclone discharge diameter or dipleg is too small for the particulate loading and apparent density.	Redesign and replace lower sections.
	Dipleg plugs.	Add dipleg purges (although the introduction of purge gas itself can reduce collection efficiency, this is preferable to 0% collection resulting from a plug). Check and repair dipleg discharge valve.
	Particulate build up on surfaces.	If caused by condensation, insulate and/or heat trace. Consider non-stick coatings or polished surfaces. If caused by condensation, insulate and/or heat trace. Periodic vibration and/or air cannon. Consider non-stick coatings or polished surfaces. Replace with a cyclone with greater internal clearances, periodic vibration, and/or air cannon. Provide easy access for cleaning and maintenance. Replace with a cyclone with greater internal clearances. If buildup is on the backside of the outlet tube, consider using an eccentric gas outlet to minimize eddy formation. Provide easy access for cleaning and maintenance.
Erosion	Cyclone inlet velocity is too high.	Replace the cyclone or modify the inlet so that the inlet velocity is as low as possible. If buildup is on the backside of the outlet tube, consider using an eccentric gas outlet to minimize eddy formation. Reduce the gas flow rate, if possible. Replace the cyclone or modify the inlet so that the inlet velocity is as low as possible. Use a vortex stabilizer to reduce erosion of the cone in low-loading cyclones. Reduce the gas flow rate if possible.
	Particulate is abrasive.	Use the lowest possible inlet velocity. Use a vortex stabilizer to reduce erosion of the cone in low-loading cyclones.
	Particulate is abrasive.	Add an inside lining to the cyclone made out of an abrasion-resistant material. If a combination of corrosion is occurring with erosion, then the materials of construction must first be corrosion resistant since virtually all materials will abrade away rapidly when in an oxidized state. Use the lowest possible inlet gas velocity. Design the installation and cyclone itself so that worn parts can be replaced and/or repaired as economically as possible.
ΔP is too high	Design basis is wrong.	Verify that specified design conditions and vendor performance predictions are correct. Design the installation and cyclone itself so that worn parts can be replaced and/or repaired as economically as possible. If the high ΔP is not causing any real problem, leave it alone. The cyclone should be providing higher collection efficiency than specified. Verify that specified design conditions and vendor performance predictions are correct. Modify the air-moving portion of the system to accommodate higher ΔP. If the high ΔP is not causing any real problem, leave it alone. The cyclone should be providing higher collection efficiency than specified. Enlarge the cyclone inlet or outlet pipe to reduce velocity (note this will reduce the cyclone collection efficiency). Modify the air-moving portion of the system to accommodate higher ΔP. Replace the cyclone. Enlarge the cyclone inlet or outlet pipe to reduce velocity (note this will reduce the cyclone collection efficiency). Check ΔP instrumentation. Replace the cyclone.
	Excess air leaking into upstream ductwork.	Repair ductwork. Check ΔP instrumentation.
ΔP is too low.	Design basis is wrong. Excess air leaking into upstream ductwork.	Verify that specified design conditions and vendor performance predictions are correct. Repair ductwork.
	Design basis is wrong.	If the low ΔP is not causing any real problem, leave it alone. Verify that specified design conditions and vendor performance predictions are correct. If the cyclone efficiency is too low by a small margin, modify the inlet and/or outlet to increase velocity. If the low ΔP is not causing any real problem, leave it alone. Increase the gas flow rate to the cyclone. If the cyclone efficiency is too low by a small margin, modify the inlet and/or outlet to increase velocity.
	Leaks into the cyclone.	Repair. Increase the gas flow rate to the cyclone. Check ΔP instrumentation. Repair.
	Internal obstruction.	Inspect and clear or repair. Check ΔP instrumentation.
Particle attrition	Inlet gas velocity is too high. Internal obstruction.	Reduce the inlet gas velocity by increasing the inlet area. sInspect and clear or repair.
	Inlet gas velocity is too high.	Reduce the inlet gas velocity by increasing the inlet area.

Adapted from Dhodapkar and Heumann, *Chemical Engineering*, May 2011.

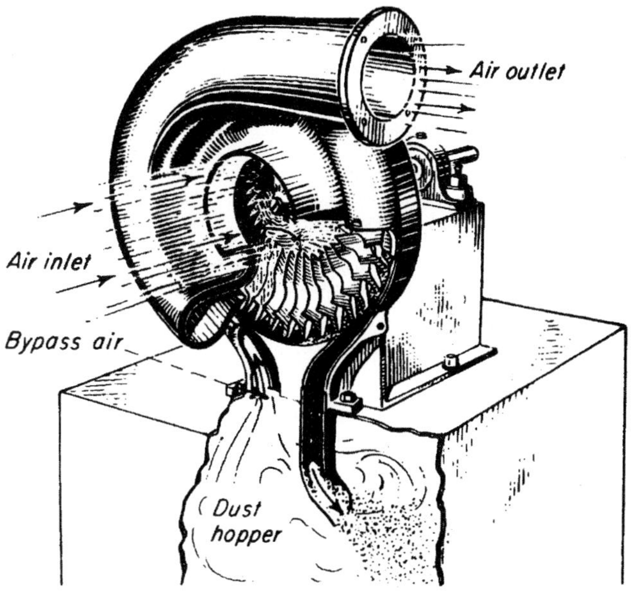

FIG. 17-57 Typical mechanical centrifugal separator. Type D Rotoclone (cutaway view). (*American Air Filter Co., Inc.*)

the cleaned gas. These two stages may be separate or physically combined. The contactor stage may be of any form intended to bring about effective contacting of the gas and spray. The spray may be generated by the flow of the gas itself in contact with the liquid, by spray nozzles (pressure-atomizing or pneumatic-atomizing), by a motor-driven mechanical spray generator, or by a motor-driven rotor through which both gas and liquid pass. The droplet size is typically between 50 and 500 microns.

Contactor Stage In this zone, the particulates in the dust-laden gas are brought into contact with the liquid (water or slurry) droplets. The velocity difference between the two phases facilitates the impact and capture of particles on the droplets, thereby increasing their effective size. Higher liquid-to-gas ratios increase the probability of impact and capture. There is an optimal size of the droplets as well. While fine droplets will have a higher surface-area-to-volume ratio for better capture, they will have a lower slip velocity, which reduces the probability of impaction. The transition from the dry to the wet zone must be carefully designed to avoid fouling, scaling, and build-up problems. Proper distribution of liquid is critical for low- and medium-pressure scrubber designs.

Entrainment Separation Stage The entrainment separator is a critical element of a scrubber, since the collection efficiency of the scrubber depends on essentially total removal of the spray from the gas stream. The sprays generated in scrubbers are generally large enough in droplet size that they can be readily removed by properly designed inertial separators, which are usually cyclones or impingement separators of various forms. If properly designed, these devices can remove virtually all droplets of the sizes produced in scrubbers.

Primary collection of the spray is seldom the critical limitation on separator performance, but reentrainment is a common problem. Reentrainment of the liquid can take place in poorly designed or overloaded separators. Although the most common cause of reentrainment is simply the use of excessive gas velocities, few data are available on the gas-handling capacities of separators. In the absence of good data, there is a frequent tendency to underdesign separators in an effort to reduce costs.

In dust scrubbers, it is essential that the entrainment separator not be of a form readily subject to blockage by solids deposits, and that it be readily cleared of deposits if they should occur. Cyclone separators are advantageous in this respect and are widely used with venturi contactors. However, they cannot readily be made integral with scrubbers of some other configurations, which can be more conveniently fitted with various forms of impingement separators.

Wet-film collectors logically form a separate subcategory of devices. They comprise inertial collectors in which a film of liquid flows over the interior surfaces, preventing reentrainment of dust particles and flushing away the deposited dust. Wetted-wall cyclones are an example [Stairmand, *Trans. Inst. Chem. Eng.* **29:** 356 (1951)]. Wet-film collectors have not been studied systematically, but they can probably be expected to perform much as do equivalent dry inertial collectors, except for the benefit of reduced reentrainment.

Scrubber Types and Performance The diversity of particulate scrubber designs is so great as to defy any detailed and consistent system of classification based on configuration or principle of operation. However, it is convenient to characterize scrubbers loosely according to prominent constructional features, even though the modes of operation of different devices in a group may vary widely.

The relationship of power consumption to collection efficiency is characteristic of all particulate scrubbers. Attaining increased efficiency requires increased power consumption, and the power consumption required to attain a given efficiency increases as the particle size of the dust decreases. Experience generally indicates that the power consumption required to provide a specific efficiency on a given dust does not vary widely even with markedly different devices. The extent to which this generalization holds true has not been fully explored, but the known extent is sufficient to suggest that the underlying collection mechanism may be essentially the same in all types of particulate scrubbers.

Since some relationship of power consumption to performance appears to be a universal characteristic of particulate scrubbers, it is useful to characterize such devices broadly according to the source from which the energy is supplied to the gas–liquid contacting process. The energy may be drawn from (1) the gas stream itself, (2) the liquid stream, or (3) a motor driving a rotor. For convenience, devices in these classes may be termed, respectively, (1) gas-atomized spray scrubbers, (2) spray, or preformed-spray, scrubbers, and (3) mechanically aided or disintegrator scrubbers. The gas-atomized spray scrubbers include venturi, orifice, impingement-plate, tray (plate), self-induced spray, packed-bed, and mobile-bed scrubbers. Examples of scrubbers with preformed spray are spray, cyclone, ejector venturi, and some baffle-type scrubbers.

Particulate scrubbers may also be classed broadly into low-energy, medium-energy, and high-energy scrubbers (see Table 17-7). The distinction between the classes is arbitrary, since the devices are not basically different and the same device may fall into either class, depending on the amount of power it consumes. However, some differences in configuration are sometimes necessary to adapt a device for high-energy service. Typically, high-energy scrubbers may be regarded as those using sufficient power to give substantial efficiencies on submicrometer particles.

Calvert (*Scrubber Handbook*, 1977) has organized scrubbers into generic categories according to the particle collection mechanisms: plate, packed bed, fiber bed, preformed spray, gas-atomized spray, centrifugal, baffle, impingement-and-entrainment, mechanically aided, moving bed, and combinations. A brief discussion of salient design features and range of operating variables for select scrubbers is included here (see Table 17-8). More details can be found in Calvert [*Chem. Eng.*, (August 29): 54–68 (1977); *Scrubber Selection Guide, Calvert Environmental Equipment* (1988)], Schifftner and Hesketh (*Wet Scrubbers*, 2d ed., Technomic Pub., Lancaster, Pa., 1996); Cooper and Alley (*Air Pollution Control*, 4th ed., Waveland Press, Long Grove, Ill., 2011); L. Theodore (*Air Pollution Control Equipment Calculations*, Wiley, New York, 2008).

Venturi Scrubbers The venturi scrubber is one of the most widely used types of particulate scrubbers. The designs have become generally standardized, and units are manufactured by many companies. Venturi scrubbers may be used as either high- or low-energy devices, but they are most commonly employed as high-energy units. The units originally studied and used were designed to the proportions of the classical venturis used for metering, but since it was discovered that these proportions have no special merits, simpler and more practical designs have been adopted. Most "venturi" contactors in current use are in fact not venturis but variable orifices of one form or another. Any of a wide range of devices can be used, including a simple pipe-line contactor. Although the venturi scrubber is not inherently more efficient at a given contacting power than other types of devices, its simplicity and flexibility favor its use. It is also useful as a gas absorber for relatively soluble gases, but because it is a cocurrent contactor, it is not well suited to absorption of gases having low solubilities.

Current designs for venturi scrubbers generally use the vertical downflow of gas through the venturi contactor and incorporate three features:

TABLE 17-7 Classification of Scrubbers Based on Energy Consumption and Pressure Drop

Class	Energy consumption	Range of pressure drop
Low energy	<2 kJ/m^3 (<1 W/cfm)	0.25–1.5 kPa (1–6 in WC)
Medium energy	2–6 kJ/m^3 (1–3 W/cfm)	1.5–4.5 kPa (6–18 in WC)
High energy	>6 kJ/m^3 (>3 W/cfm)	> 4.5 kPa (>18 in WC)

ASHRAE Handbook—HVAC Systems and Equipment, 2016, Chapter 30, "Industrial Gas Cleaning and Air Pollution Control," p. 30.15.

TABLE 17-8 Comparison of Performance Characteristics of Various Scrubbers

Scrubber type	Particle diameter,* μm	Typical collection efficiency, %	Comparative energy requirement†	Gas pressure loss, Pa	Liquid pressure loss, Pa
Gravity spray	10	70	5	25–250	140–690
Centrifugal	5	90	12–26	500–2000	140–690
Impingement	5	95	9–31	500–2000	140–690
Packed bed	5	90	4–34	125–2500	35–210
Dynamic (mechanically aided)	2	95	30–200	Provides pressure	35–210
Self-inducted spray	2	90	9–21	500–1500	None
Jet	2	90	15–30	Provides pressure	345–610
Venturi	0.1	95–99	30–300	2500–15000	70–210

*Minimum particle diameter for which the device is effective
†Compared to dry inertial collector–settling chamber
Adapted from ASHRAE Handbook—HVAC Systems and Equipment, Ch. 30, 2016.

(1) a "wet-approach" or "flooded-wall" entry section, to avoid dust buildup at a wet/dry junction; (2) an adjustable throat for the venturi (or orifice), to provide for adjustment of the pressure drop; and (3) a "flooded elbow" located below the venturi and ahead of the entrainment separator, to collect large droplets and coarse particles via the impingement mechanism and to reduce wear by abrasive particles. The venturi throat is sometimes fitted with a refractory lining to resist abrasion by dust particles. The entrainment separator is commonly, but not invariably, of the cyclone type. An example of the "standard form" of venturi scrubber is shown in Fig. 17-58.

The scrubbing liquid, which is injected directly at the entrance of the throat or distributed as a liquid film before the entrance, is atomized into droplets by the gas turbulence in the throat region. The wet-approach entry section has made practical the recirculation of slurries. In some designs, supplementary nozzles are used, but they are prone to plugging or wear unless clean liquid is used. Various forms of adjustable throats, which may be under manual or automatic control, permit maintaining a constant pressure drop and constant efficiency under conditions of varying gas flow.

The grade (or fractional) efficiency of venturi scrubbers is affected primarily by throat velocity and liquid-to-gas ratio, and to a much lesser extent by the mode of liquid distribution and the shape of the converging and diverging sections. Calvert [*Chem. Eng.*, (August 29): 54–68 (1977)] suggested that it is a good practice to use high liquid-to-gas ratios rather than high gas velocities to achieve a given cut diameter.

The orifice plate scrubbers can be considered a special case of venturi scrubbers where the throat length is zero. The orifice plate with multiple orifices is irrigated with scrubbing liquid, and the droplets are formed at the edge of an orifice due to shear induced by the gas flow. They can be designed for high- or medium-energy operation.

Venturi scrubbers are used to control particulate emissions from boilers and incinerators and from chemical, mineral products, wood, pulp and paper, rock products, asphalt, lead, aluminum, iron and steel, and grey iron manufacturing industries (*Air Pollution Control Technology Fact Sheet*, EPA-452/F-03-017).

Ejector-Venturi Scrubbers In the ejector-venturi scrubber (Fig. 17-59) the cocurrent water jet from a spray nozzle serves both to scrub the gas and to provide the draft for moving the gas. No fan is required, but the equivalent

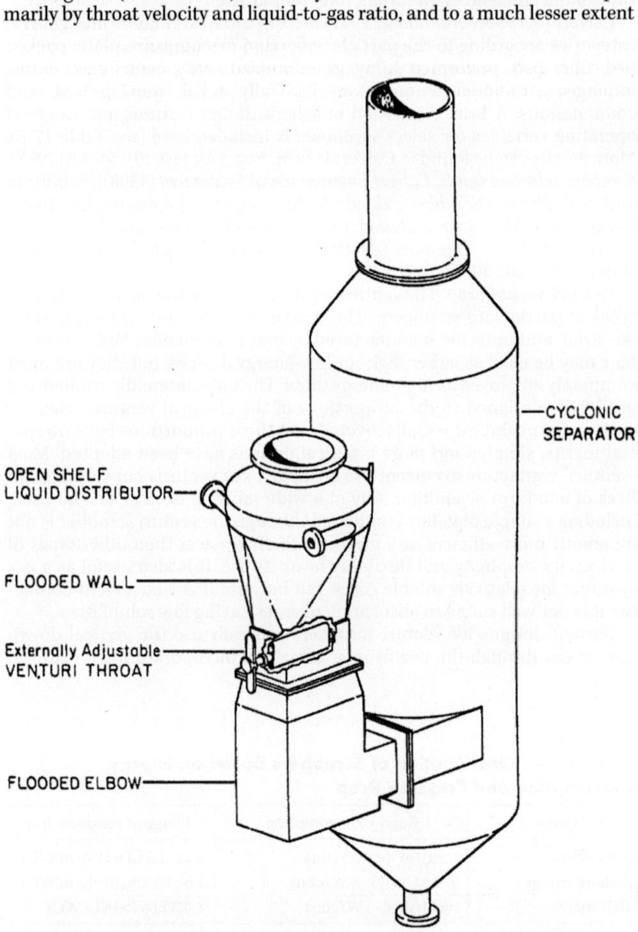

FIG. 17-58 Venturi scrubber. (*Nepture AirPol.*)

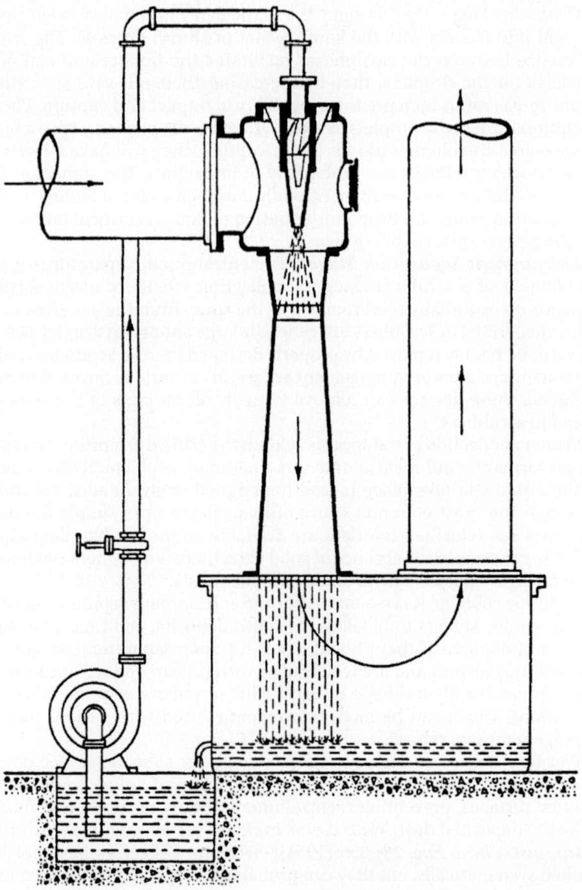

FIG. 17-59 Ejector-venturi scrubber. (*Schutte & Koerting Division, Amtek, Inc.*)

power must be supplied to the pump that delivers water to the ejector nozzle. The water must be supplied in sufficient volume and at high enough pressure to provide both adequate draft and enough contacting power for the required scrubbing operation. Considered as a gas pump, the ejector is not a very efficient device, but the dissipated energy that is not effective in pumping does serve in gas–liquid contacting. The energy equivalent to any gas pressure rise across the scrubber is not part of the contacting power (Semrau et al., EPA-600/2-77-234, 1974).

The ejector-venturi scrubber is widely used as a gas absorber, but the combinations of water pressure and flow rate that are sufficient to provide the required draft usually do not also yield enough contacting power to give high collection efficiencies on submicrometer particles. Other types of ejectors have been employed to provide higher contacting-power levels. In one, superheated water is discharged through the nozzle, and part flashes to steam, increasing the mechanical energy available for scrubbing [Gardenier, *J. Air Pollut. Control Assoc.* **24:** 954 (1974)]. Some units use two-fluid nozzles, with either compressed air or steam as the compressible fluid [Sparks, *J. Air Pollut. Control Assoc.* **24:** 958 (1974)]. Most of the energy for gas movement and for atomizing the liquid and scrubbing the gas is derived from the compressed air or steam. In some ejector-venturi scrubber installations, part of the draft is supplied by a fan [Williams and Fuller, *TAPPI* **60**(1): 108 (1977)].

Self-Induced Spray Scrubbers Self-induced spray scrubbers form a category of gas-atomized spray scrubbers in which a tube or a duct of some other shape forms the gas–liquid contacting zone. An inverted cone may be placed at the bottom to provide an annular opening that can be adjusted. The gas stream flowing at high velocity through the contactor atomizes the liquid in essentially the same manner as in a venturi scrubber. The dust-laden gas makes a turn through the liquid surface into the central tube or duct. However, the liquid is fed into the contactor and later recirculated from the entrainment separator section by gravity instead of being circulated by a pump as in venturi scrubbers. The scheme is well illustrated in Fig. 17-60*a*. The pressure drop, hence the efficiency, is controlled by the level of the liquid and the annular space between the inverted cone and the tube. A static pool of liquid must be continuously maintained during operation.

A great many such devices using contactor ducts of various shapes, as in Fig. 17-60*b*, are offered commercially. Oftentimes, they are also referred to as orifice scrubbers, gas-induced spray scrubbers, or entrainment scrubbers. Although self-induced spray scrubbers can be built as high-energy units and sometimes are, most such devices are designed for medium to low-energy service.

The principal advantage of self-induced spray scrubbers is the elimination of a pump for recirculation of the scrubbing liquid. However, the designs for high-energy service are somewhat more complex and less flexible than those for venturi scrubbers. The design and maintenance of the sludge removal system is a significant challenge. The industrial applications include food and pharmaceutical processing and packing, and the manufacture of chemicals, rubber, plastics, ceramics, and fertilizers.

Spray Scrubbers Spray scrubbers consist of empty chambers of some simple form in which the gas stream is contacted with liquid droplets generated by spray nozzles. A common form is a spray tower, in which the gas flows upward through a bank or successive banks of spray nozzles. Similar arrangements are sometimes used in spray chambers with horizontal gas flow. Such devices have very low gas pressure drops, and all but a small part of the contacting power is derived from the liquid stream. The required contacting power is obtained from an appropriate combination of liquid pressure and flow rate. All the energy may be supplied from the liquid, using a pressure nozzle, but some or all may also be provided by compressed air or steam in a two-fluid nozzle or by a motor driving a spray generator.

Most spray scrubbers are low-energy units. Collection of fine particles is possible but may require very high liquid-to-gas ratios, liquid feed pressures, or both. Removal efficiencies up to 90 percent for particles larger than 5 μm, 60 to 80 percent for particle size between 3 and 5 μm, and less than 50 percent for particles smaller than 3 μm are typically observed (*EPA Report*, EPA/452/B-02-001-Section 6 Particulate Matter Controls).

Plugging of small nozzles can be a persistent maintenance problem. Plugging on the gas side is rare, so these scrubbers are suitable for streams with high dust loadings. Entrainment separators are necessary to prevent carryover of spray into the exit gas, especially when the superficial velocity exceeds 1 m/s (3.3 ft/s).

Cyclone Scrubbers The vessels of cyclone scrubbers are all in the form of cyclones, which provide for compact integral entrainment separation. The gas–liquid contacting section normally uses some sort of spray generator to disperse liquid throughout the gas. These devices have higher relative velocities between the droplets and the particles compared to simple tray towers, and the efficient capture of droplets due to cyclonic action can improve collection efficiency in certain applications (Calvert, S., and R. Parker, EPA-600/8-78-005c, U.S. EPA, June 1978). Collection efficiencies as high as 95 percent for particles larger than 5 μm and 60 to 75 percent for submicron particles have been reported (EPA Report, EPA/452/B-02-001).

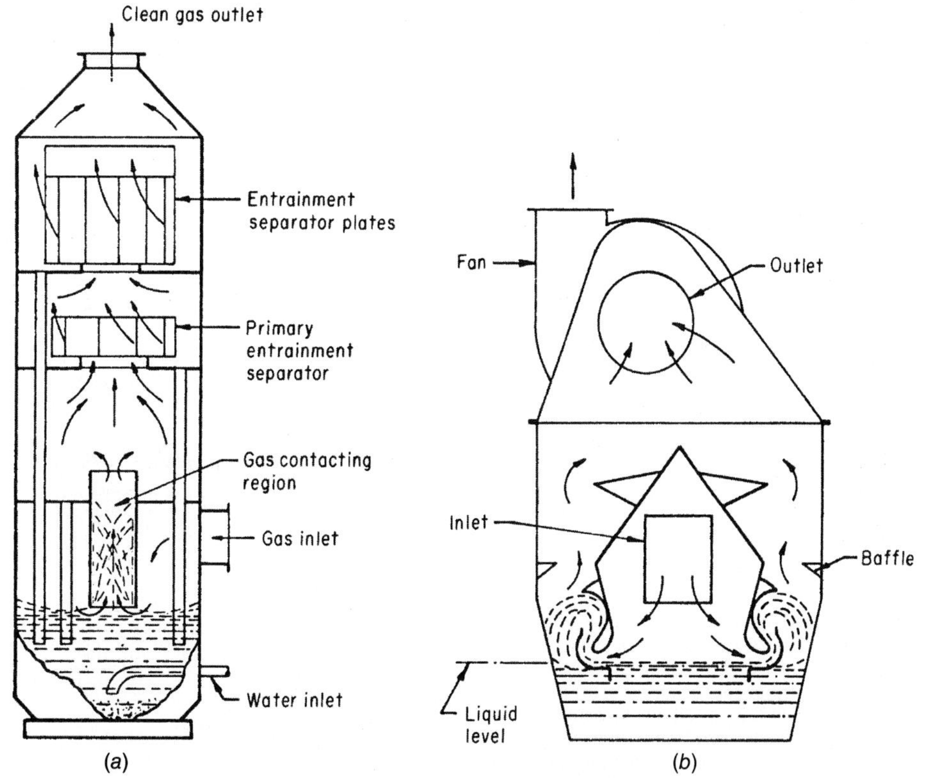

FIG. 17-60 Self-induced spray scrubbers. (*a*) Blaw-Knox Food & Chemical Equipment, Inc. (*b*) American Air Filter Co., Inc.

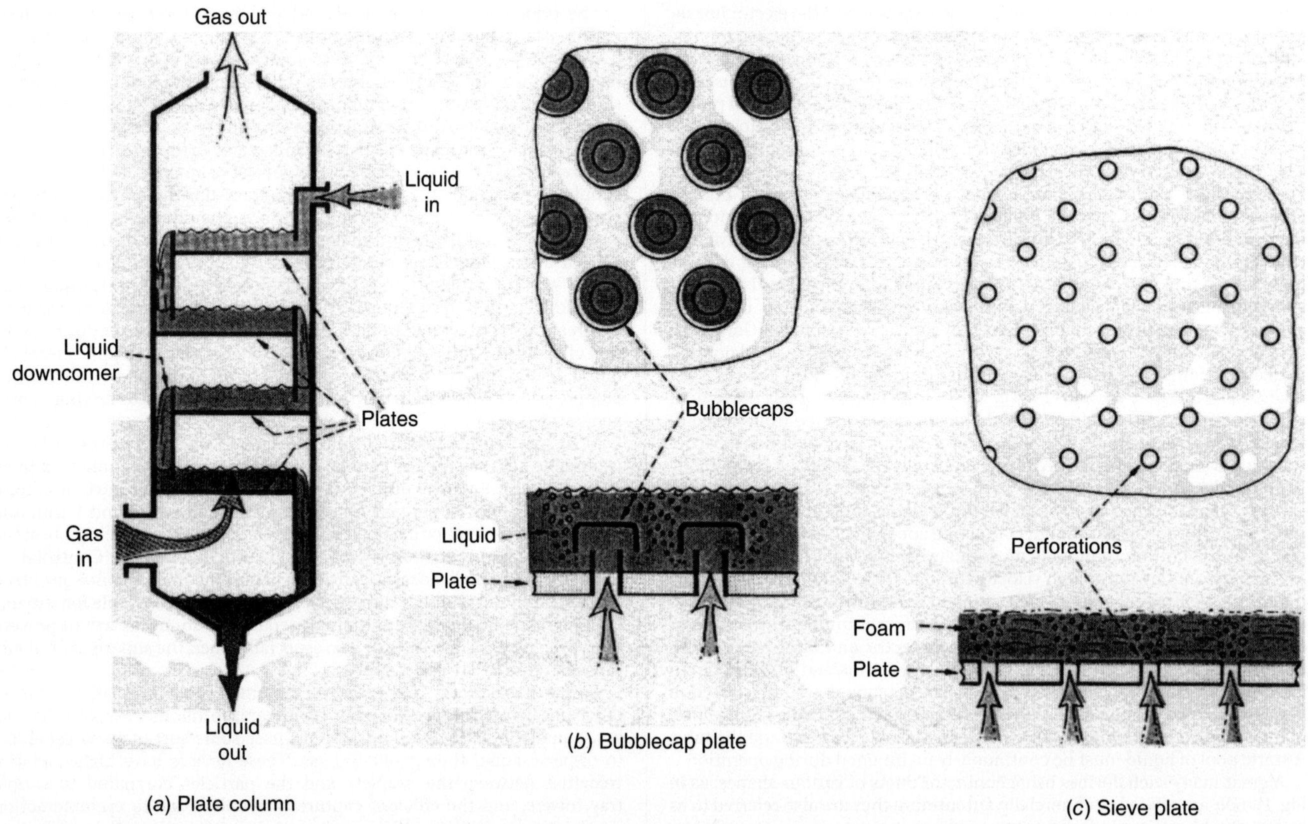

FIG. 17-61 Plate (tray) scrubber. (Calvert, S., *Chemical Engineering*, p. 54, August 29, 1977.)

Plate Scrubbers Plate (tray) scrubbers are countercurrent gas-atomized spray scrubbers using one or more plates for gas–liquid contacting. They are essentially the same as, if not identical to, the devices used for gas absorption, and are often employed in applications in which gases are to be absorbed simultaneously with the removal of dust. Except possibly in cases in which condensation effects are involved, countercurrent operation is not significantly beneficial in dust collection.

The plates may be any of several types, including sieve, bubble-cap, and valve trays (Fig. 17-61). For sieve trays, the liquid flow into the holes is prevented by the upward flow of dust-laden gas. The gas flow rate must be maintained in a certain range to prevent weeping on the low side and flooding on the high side. The bubble-caps provide an impingement surface that enhances the collection. The impingement surfaces are continuously washed by the liquid. Valve trays constitute multiple self-adjusting orifices that provide a nearly constant gas pressure drop over considerable ranges of variation in gas flow. The gas pressure drop that can be taken across a single plate is necessarily limited, so units designed for high contacting power must use multiple plates. The efficiency of plate scrubbers for particles larger than 5 mm can be as high as 97 percent (Cooper, C.D., and F.C. Alley, *Air Pollution Control: A Design Approach*, Waveland Press, Long Grove, Ill., 2011).

Plate towers are more subject to plugging and fouling than venturi-type scrubbers that have large passages for gas and liquid.

Packed-Bed Scrubbers Packed-bed scrubbers using gas-absorption-type tower packings may also be used for collecting soluble solids, but they are not irrigated intensively enough to avoid plugging by deposits of insoluble solids. Mobile-bed and fiber-bed scrubbers can be self-cleaning if they are adequately flushed.

Mobile-Bed Scrubbers Mobile-bed scrubbers (Fig. 17-62) are constructed with one or more beds of low-density spheres that are free to move between upper and lower retaining grids. The spheres are commonly 1.0 in (2.5 cm) or more in diameter and made from rubber or a plastic such as polypropylene. The plastic spheres may be solid or hollow. Gas and liquid flows are countercurrent, and the spherical packings are fluidized by the upward-flowing gas. The movement of the packings is intended to minimize fouling and plugging of the bed. Mobile-bed scrubbers were first developed for absorbing gases from gas streams that also carry solid or semisolid particles.

The spherical packings are too large to serve as effective targets for the deposition of fine dust particles. In dust-collection service, the packings actually serve as turbulence promoters, while the dust particles are collected primarily by the liquid droplets.

The gas pressure drop through the scrubber may be increased by increasing the gas velocity, the liquid-to-gas ratio, the depth of the bed, the density of the packings, and the number of beds in series. In an experimental study,

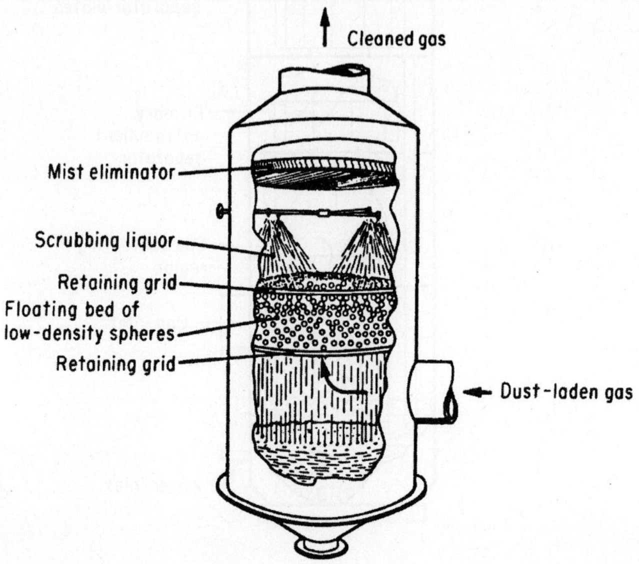

FIG. 17-62 Mobile-bed scrubber. (*Air Correction Division, UOP.*)

Yung et al. (EPA-600/7-79-071, 1979) determined that the collection efficiency of a mobile-bed scrubber was dependent only on the gas pressure drop and was not influenced independently by the gas velocity, the liquid-to-gas ratio, or the number of beds except as these factors affected the pressure drop. Yung et al. also reported that the mobile-bed scrubber was less efficient at a given pressure drop than scrubbers of the venturi type, but without offering comparable experimental supporting evidence.

Fiber-Bed Scrubbers Fibrous-bed structures are sometimes used as gas–liquid contactors, with cocurrent flow of the gas and liquid streams. In such contactors, both scrubbing (particle deposition on droplets) and filtration (particle deposition on fibers) may take place. If only mists are to be collected, small fibers may be used, but if solid particles are present, the use of fiber beds is limited by the tendency of the beds to plug. For dust-collection service, the fiber bed must be composed of coarse fibers and have a high void fraction, so as to minimize the tendency to plug. The fiber bed may be made from metal or plastic fibers in the form of knitted structures, multiple layers of screens, or random-packed fibers. However, the bed must have sufficient dimensional stability so that it will not be compacted during operation. Fiber-bed scrubbers are not suitable for applications where the dust loading of insoluble solids is high.

Lucas and Porter (U.S. Patent 3,370,401, 1967) developed a fiber-bed scrubber in which the gas and scrubbing liquid flow vertically upward through the fiber bed (Fig. 17-63). The beds tested were composed of knitted structures made from fibers with diameters ranging from 89 to 406 μm. Lucas and Porter reported that the fiber-bed scrubber gave substantially higher efficiencies than venturi-type scrubbers tested with the same dust at the same gas pressure drop. In similar experiments, Semrau et al. (Semrau and Lunn, "Performance of Particulate Scrubbers as Influenced by Gas–liquid Contactor Design and by Dust Flocculation," EPA-600/9-82-005c, 1982, p. 43) also found that a fiber-bed contactor made with random-packed steel wool fibers gave higher efficiencies than an orifice contactor. However, there were indications that the fiber bed would have little advantage in the collection of submicron particles, presumably because of the large fiber size feasible for dust-collection service.

Alliger (U.S. Patents 3,659,402, 1972, and 3,905,788, 1975) describes fiber-bed structures that are not random, but are rather built up from flat mesh sheets offset angularly from one layer to the next and then compressed and bonded. Such bonded beds of relatively coarse hydrophobic fibers both are remarkably flushable, to prevent fouling by insoluble solids, and have surprisingly high collection efficiency per unit pressure drop for submicron particles, approaching that of irrigated fine hydrophobic fiber filters such as those described by Fair (U.S. Patent 3,135,592, 1964) and Vosseller (U.S. Patent 3,250,059, 1966).

Electrically Augmented Scrubbers In some types of wet collectors, attempts are made to apply the electrostatic-deposition mechanism by charging the dust particles, the water droplets, or both. The objective is to combine in one scrubber both the high efficiency for collecting fine particles and the moderate power consumption characteristic of an electrical precipitator. Successful devices of this type have been essentially wet electrical precipitators and should properly be discussed in that category (see the subsection Electrical Precipitators). So far, there has been no clear demonstration of a device that combines the small size, compactness, and high efficiency of a high-energy scrubber with the relatively low power consumption of an electrical precipitator.

Mechanical Scrubbers Mechanical scrubbers or dynamic scrubbers comprise those devices in which a power-driven rotor produces the fine spray and the contacting of gas and liquid. The liquid is injected at the hub of the fan where it is drawn into the rotating blades. In the disintegrator concept, a partially submerged impeller is used to generate the droplets. As in other types of scrubbers, it is the droplets that are the principal collecting bodies for the dust particles. The rotor acts as a turbulence producer. An entrainment separator, such as a cyclonic separator, must be used to prevent carryover of spray. Among potential maintenance problems are unbalancing of the rotor by buildup of dust deposits and abrasion by coarse particles. A precleaner is often used to remove the coarse particles and to reduce the dust loading. These devices are not preferred for high-temperature applications.

The simplest commercial devices of this type are essentially fans upon which water is sprayed. The unit shown in Fig. 17-64 is adapted to light duty, and heavy dust loads are avoided to minimize buildup on the rotor.

Scrubber Performance Models A number of investigators have made theoretical studies of the performance of venturi scrubbers and have sought to produce performance models, based on first principles, that can be used to design a unit for a given duty without recourse to experimental data other than the particle size and size distribution of the dust. Among these workers are Calvert [*Am. Inst. Chem. Eng. J.* **16:** 392 (1970); *J. Air Pollut. Control Assoc.* **24:** 929 (1974)], Boll [*Ind. Eng. Chem. Fundam.* **12:** 40 (1973)], Goel and Hollands [*Atmos. Environ.* **11:** 837 (1977)], Yung et al. [*Environ. Sci. Technol.* **12:** 456 (1978)], Azzopardi et al. [*Filtr. Sep.* **21:** 196 (1984); *Trans. IChemE* **69B:** 237 (1991)], and Hesketh [*Journal of Air Pollution Control Association* **24:** 10 (1974)]. Comparatively few efforts have been made to model the performance of scrubbers of types other than the venturi, but a number of such models are summarized by Yung and Calvert (U.S. EPA-600/8-78-005b, 1978).

The various venturi-scrubber models embody a variety of assumptions and approximations. The solutions of the equations for particulate collection must in general be determined numerically, although Calvert et al. [*J. Air Pollut. Control Assoc.* **22:** 529 (1972)] obtained an explicit equation by making some simplifying assumptions and incorporating an empirical constant that must be evaluated experimentally; the constant may absorb some of the deficiencies in the model. Although other models avoid direct incorporation of empirical constants, the use of empirical relationships is necessary to obtain specific estimates of scrubber collection efficiency. One of the areas of greatest uncertainty is the estimation of droplet size.

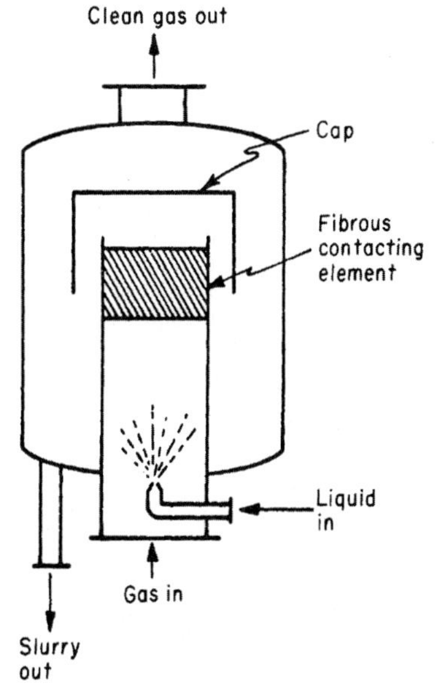

FIG. 17-63 Fibrous-bed scrubber. (*Lucas and Porter, U.S. Patent 3,370,401, 1967.*)

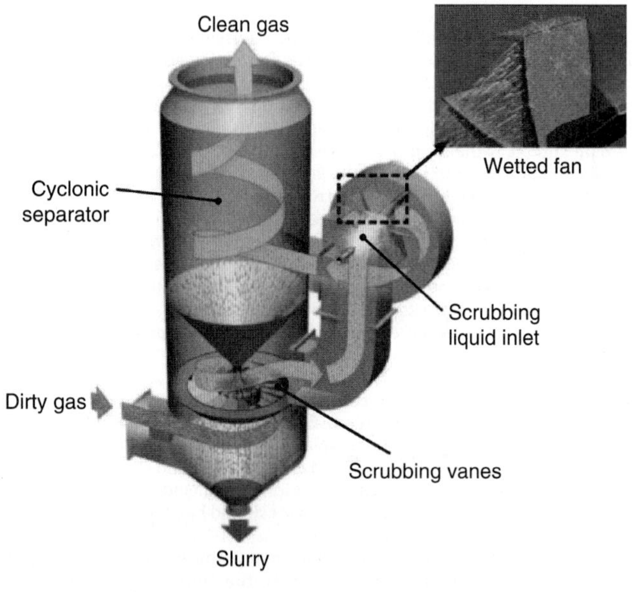

FIG. 17-64 Mechanical scrubber. (*Courtesy of Nederman Mikropul.*)

Most of the investigators have assumed the effective drop size of the spray to be the Sauter (surface-mean) diameter, and they have used the empirical equation of Nukiyama and Tanasawa [*Trans. Soc. Mech. Eng., Japan,* **5:** 63 (1939)] to estimate the Sauter diameter:

$$D_o = \frac{1920\sqrt{\sigma_L}}{V_o\sqrt{\rho_L/62.3}} + 75.4\left(\frac{\mu_L}{\sqrt{\sigma_L\rho_L/62.3}}\right)^{0.45}\left(\frac{1000Q_L}{Q_G}\right)^{1.5} \tag{17-29}$$

where D_o = drop diameter, μm; V_o = gas velocity, ft/s; σ_L = liquid surface tension, dyn/cm; ρ_L = liquid density, lb/ft^3; μ_L = liquid viscosity, cP; Q_L = liquid flow rate, ft^3/s; and Q_G = gas flow rate, ft^3/s.

The Nukiyama-Tanasawa equation, which is not dimensionally homogeneous, was derived from experiments with small, internal-mix pneumatic atomizing nozzles with a concentric feed of air and liquid (Lapple et al., "Atomization: A Survey and Critique of the Literature," Stanford Res. Inst. Tech. Rep. No. 6, AD 821-314, 1967; Lapple, "Atomization," in *McGraw-Hill Encyclopedia of Science and Technology,* 5th ed., vol. 1, McGraw-Hill, New York, 1982, p. 858). The effect of nozzle size on the drop size is undefined. Even within the range of parameters for which the relationship was derived, the drop sizes reported by various investigators have varied by twofold to threefold from those predicted by the equation [Boll, *Ind. Eng. Chem. Fundam.* **12:** 40 (1973)]. The Nukiyama-Tanasawa equation has, nevertheless, been applied to large venturi and orifice scrubbers with configurations radically different from those of the atomizing nozzles for which the equation was originally developed.

Primarily because of the lack of adequate experimental techniques (particularly, the production of appropriate monodisperse aerosols), there has been no comprehensive experimental test of any of the venturi-scrubber models over wide ranges of design and operating variables. The models for other types of scrubbers appear to be essentially untested.

The pressure drop across venturi scrubbers is generally expressed as:

$$\Delta P = KV_t^2\rho_g\left(\frac{Q_L}{Q_G}\right) \tag{17-30}$$

where V_t = throat velocity and K is design specific.

Contacting Power Correlation A scrubber design method that has achieved wide acceptance and use is based on correlation of the collection efficiency with the power dissipated in the gas–liquid contacting process, which is termed "contacting power." The method originated from an investigation by Lapple and Kamack [*Chem. Eng. Prog.* **51:** 110 (1955)] and has been extended and refined in a series of papers by Semrau and coworkers [*Ind. Eng. Chem.* **50:** 1615 (1958); *J. Air Pollut. Control Assoc.* **10:** 200 (1960); **13:** 587 (1963); U.S. EPA-650/2-74-108, 1974; U.S. EPA-600/2-77-234, 1977; *Chem. Eng.* **84**(20): 87 (1977); and "Performance of Particulate Scrubbers as Influenced by Gas–Liquid Contactor Design and by Dust Flocculation," EPA-600/9-82-005c, 1982, p. 43]. Other workers have made extensive independent studies of the correlation method [Walker and Hall, *J. Air Pollut. Control Assoc.* **18:** 319 (1968)], and numerous studies of narrower scope have been made. The major conclusion from these studies is that the collection efficiency of a scrubber on a given dust is essentially dependent only on the contacting power and is affected to only a minor degree by the size or geometry of the scrubber or by the way in which the contacting power is applied. This contacting-power rule is strictly empirical, and the full extent of its validity has still not been explored. It has been best verified for the class of gas-atomized spray scrubbers, in which the contacting power is derived from the gas stream and takes the form of gas pressure drop. Tests of the equivalence of contacting power supplied from the liquid stream in pressure spray nozzles have been far less extensive and are strongly indicative but not yet conclusive. Evidence for the equivalence of contacting power from mechanically driven devices is also indicative, but extremely limited in quantity.

Contacting power is defined as the power per unit of volumetric gas flow rate that is dissipated in gas–liquid contacting and is ultimately converted to heat. In the simplest case, in which all the energy is obtained from the gas stream in the form of pressure drop, the contacting power is equivalent to the friction loss across the wetted equipment, which is termed *effective friction loss,* F_E. The pressure drop may reflect kinetic-energy changes rather than energy dissipation, and pressure drops that result solely from kinetic-energy changes in the gas stream do not correlate with performance. Likewise, any friction losses taking place across equipment that is operating dry do not contribute to gas–liquid contacting and do not correlate with performance. The gross power input to a scrubber includes losses in motors, drive shafts, fans, and pumps that obviously should be unrelated to scrubber performance.

The effective friction loss, or "gas-phase contacting power," is easily determined by direct measurements. However, the "liquid-phase contacting power," supplied from the stream of scrubbing liquid, and the "mechanical contacting power," supplied by a mechanically driven rotor, are not directly measurable; the theoretical power inputs can be estimated, but the portions of these quantities effectively converted to contacting power can only be inferred from comparison with gas-phase contacting power. Such data as are available indicate that the contributions of contacting power from different sources are directly additive in their relation to scrubber performance.

Contacting power is variously expressed in units of MJ/1000 m^3 (SI), kWh/1000 m^3 (meter-kilogram-second system), and hp/(1000 ft^3/min) (U.S. customary). Relationships for conversion to SI units are

$$1.0 \text{ kWh}/1000 \text{ m}^3 = 3.60 \text{ MJ}/1000 \text{ m}^3$$

$$1.0 \text{ hp}/(1000 \text{ ft}^3/\text{min}) = 1.58 \text{ MJ}/1000 \text{ m}^3$$

The gas-phase contacting power, P_G, may be calculated from the effective friction loss by the following relationships:

SI units:

$$P_G = 1.0F_E \tag{17-31}$$

where F_E = kPa.

U.S. customary units:

$$P_G = 0.1575F_E \tag{17-32}$$

where F_E = in of water.

The power input from a liquid stream injected with a hydraulic spray nozzle may usually be taken as approximately equal to the product of the nozzle feed pressure, p_F, and the volumetric liquid rate. The liquid-phase contacting power, P_L, may then be calculated from the following formulas:

SI units:

$$P_L = 1.0p_F\,(Q_L/Q_G) \tag{17-33}$$

where p_F = kPa gauge, and Q_L and Q_G = m^3/s.

U.S. customary units:

$$P_L = 0.583p_F\,(Q_L/Q_G) \tag{17-34}$$

where p_F = lb$_f$/in^2 gauge, Q_L = gal/min, and Q_G = ft^3/min.

The correlation of efficiency data is based on the total contacting power, P_T, which is the sum of P_G, P_L, and any power P_M that may be supplied mechanically by a power-driven rotor.

In general, the liquid-to-gas ratio does not have an influence independent of contacting power on the collection efficiency of scrubbers of the venturi type. This is true at least of operation with liquid-to-gas ratios above some critical lower value. However, several investigations [Semrau and Lunn, "Performance of Particulate Scrubbers as Influenced by Gas–Liquid Contactor Design and by Dust Flocculation," EPA-600/9-82-005c, 1982, p. 43; and Muir et al., *Filtr. Sep.* **15:** 332 (1978)] have shown that at low liquid-to-gas ratios, relatively poor efficiencies may be obtained at a given contacting power. Such regions of operation are obviously to be avoided.

It has sometimes been asserted that multiple gas–liquid contactors in series will give higher efficiencies at a given contacting power than will a single contacting stage. However, there is little experimental evidence to support this contention. Lapple and Kamack [*Chem. Eng. Prog.* **51:** 110 (1955)] obtained slightly higher efficiencies with a venturi and an orifice in series than they did with a venturi alone. Muir and Mihisei [*Atmos. Environ.* **13:** 1187 (1979)] obtained somewhat higher efficiencies on two redispersed dusts when using two venturis in series rather than one. The improvement obtained with two-stage scrubbing was greatest with the coarser of the two dusts and was relatively small with the finer dust. Flocculation or deflocculation of the dusts may have been responsible for some of the behavior encountered. Semrau et al. (EPA-600/2-77-234, 1977) compared the performance of a four-stage, multiple-orifice contactor with that of a single-orifice contactor, using well-dispersed aerosols generated from ammonium fluorescein. The multiple-orifice contactor gave about the same efficiency as the single-orifice in the upper range of contacting power, but lower efficiencies in the lower range. The deviations in performance in this case were probably characteristic of the particular multiple-orifice contactor rather than of multistage contacting as such.

Most scrubbers actually incorporate more than one stage of gas–liquid contacting even though these may not be identical (e.g., the contactor and the entrainment separator). The preponderance of evidence indicates that multiple-stage contacting is not inherently either more or less efficient than single-stage contacting. However, two-stage contacting may have practical benefits in dealing with abrasive or flocculated dusts.

Some investigators have proposed, mostly on the basis of mathematical modeling, to optimize the design of scrubbers to obtain a given efficiency with a minimum power consumption [e.g., Goel and Hollands, *Atmos. Environ.* **11:** 837 (1977)]. In fact, no optimum in performance appears to exist; so long as increased entrainment is prevented, increased contacting power yields increased efficiency.

Scrubber Performance Curves The scrubber performance curve, which shows the relationship of scrubber efficiency to the contacting power, has been found to take the form

$$N_t = \alpha P_T^{\gamma} \tag{17-35}$$

where α and γ are empirical constants that depend primarily on the aerosol (dust or mist) collected. In a log-log plot of N_t versus P_T, γ is the slope of the performance curve and α is the intercept at $P_T = 1$. Figure 17-65 shows such a performance curve for the collection of coal fly ash by a pilot-plant venturi scrubber (Raben "Use of Scrubbers for Control of Emissions from Power Boilers," United States–U.S.S.R. Symposium on Control of Fine-Particulate Emissions from Industrial Sources, San Francisco, 1974). The scatter in the data reflects not merely experimental errors but actual variations in the particle-size characteristics of the dust. Because the characteristics of an industrial dust vary with time, the scrubber performance curve necessarily must represent an average material, and the scatter in the data is often greater than is shown in Fig. 17-65. For best definition, the curve should cover as wide a range of contacting power as possible. Obtaining the data thus requires pilot-plant equipment with the flexibility to operate over a wide range of conditions. Because scrubber performance is not greatly affected by the size of the unit, it is feasible to conduct the tests with a unit handling no more than 170 m^3/h (100 ft^3/min) of gas. Davis (*Air Pollution Engineering Manual*, 2d ed., Wiley, 2000) has summarized values of empirical constants α and γ for various scrubbing applications.

A clear interpretation of γ, the slope of the curve, is still lacking. Presumably, it should be related to the particle-size distribution of the dust. Because scrubbing preferentially removes the coarser particles, the fraction of the dust removed (or the increment of N_t) per unit of contacting power should decrease as the contacting power and efficiency increase, so that the value of γ should be less than unity. In fact, the value of γ has been less than unity for most dusts. Nevertheless, some data in the literature have displayed values of γ greater than unity when plotted on the transfer-unit basis, indicating that the residual fraction of dust became more readily collectible as contacting power and efficiency increased. More recent studies by Semrau et al. (EPA-650/2-74-108 and EPA-600/2-77-237) have revealed performance curves having two branches (typified by Fig. 17-66), the lower having a slope greater than unity and the upper a slope less than unity. This suggests that had the earlier tests been extended into higher contacting-power ranges, performance curves with flatter slopes might have appeared in those ranges.

Among the aerosols that gave performance curves with $\gamma > 1$, the only obvious common characteristic was that a large fraction of each was composed of submicron particles.

The manufacturers can provide performance curves for their scrubber designs based on lab and operational data (see Figs. 17-67 and 17-68). These are essentially grade efficiency curves, as discussed previously. The overall collection efficiency can be calculated by integrating it with the particle-size distribution of the incoming dust.

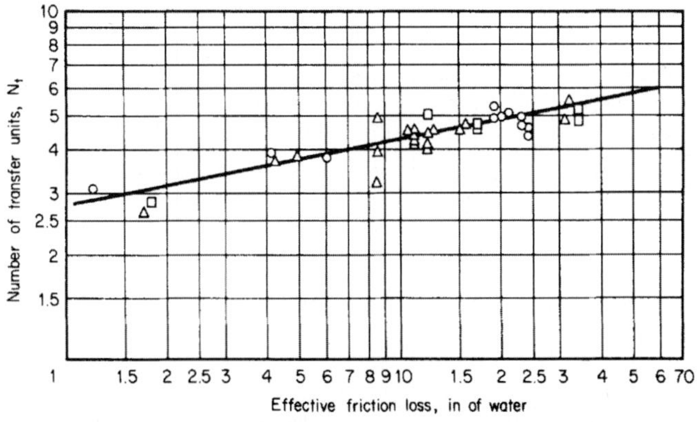

FIG. 17-65 Performance of a pilot-plant venturi scrubber on fly ash. Liquid-to-gas ratio, gal/1000 ft^3: ○ 10; △ 15; □ 20. (*Raben, United States–U.S.S.R. Symposium on Fine-Particulate Emissions from Industrial Sources, San Francisco, 1974.*)

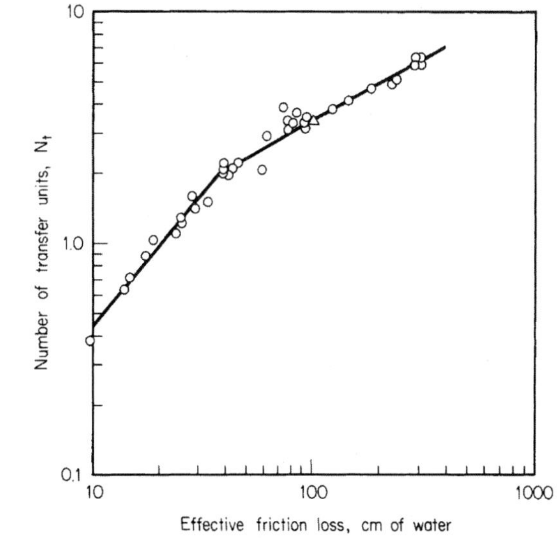

FIG. 17-66 Performance curve for orifice scrubber collecting ammonium fluorescin aerosol. (*Semrau et al. EPA 600/2-77-237, 1977.*)

Cut-Power Correlation Another design method, also based on scrubber power consumption, is the cut-power method of Calvert [*J. Air Pollut. Control Assoc.* **24:** 929 (1974); *Chem. Eng.* **84**(18): 54 (1977)]. Since most scrubbers collect particles by inertial impaction, the relationship between penetration $(1 - \eta)$ and the aerodynamic size can be modeled by

$$Pt_i = e(-Ad_a^B) = \frac{c_o}{c_i} \tag{17-36}$$

where Pt_i = particle penetration for ith fraction, d_a = aerodynamic particle size, c_o = outlet particle concentration, and c_i = inlet particle concentration (g/m^3). A and B are empirical constants. The overall efficiency can be calculated by integrating the penetration of individual size fractions with the size distribution. Calvert suggests that the value of B for gas-atomizing, packed-bed and plate-type scrubbers is 2.0, whereas it is 0.7 for centrifugal cyclonic scrubbers. A value of 2.0 can be assumed for all practical purposes. Calvert further proposed fitting the aerodynamic particle-size distribution data to a log-normal distribution, which can be functionally represented by the mass median diameter (d_{pg}) and geometric standard deviation (σ_g). Integrated values of penetration are plotted in Fig. 17-69 as a function of cut-ratio (d_{RC}/d_{pg}); where d_{RC} is the required cut diameter. The required cut diameter can be estimated based on the desired overall collection efficiency and inlet particle-size distribution.

The cut diameter (the particle diameter for which the collection efficiency is 50 percent) as a function of the gas pressure drop or of the power input per unit of volumetric gas flow rate for various scrubbers is shown in Fig. 17-70. The functional relationship is presented as a log-log plot of the cut diameter versus the pressure drop (or power input). In principle, the function could be constructed by experimentally determining scrubber performance curves for discrete particle sizes and then plotting the particle sizes against the corresponding pressure drops necessary to give efficiencies of 50 percent. In practice, Calvert and coworkers have in most cases constructed the cut-power functions for various scrubbers by modeling (Yung and Calvert, U.S. EPA-600/8-78-005b, 1978). They show a variety of curves, whereas empirical studies have indicated that different types of scrubbers generally have about the same performance at a given level of power consumption.

Condensation Scrubbing The collection efficiency of scrubbing can be increased by the simultaneous condensation of water vapor from the gas stream. Water-vapor condensation assists in particle removal by two entirely different mechanisms. One is the deposition of particles on cold-water droplets or other surfaces as the result of Stefan flow. The other is the condensation of water vapor on particles as nuclei, which enlarges the particles and makes them more readily collected by inertial deposition on droplets. Both mechanisms can operate simultaneously. However, for the buildup of particles by condensation to be effective, there must be adequate time for the particles to grow substantially before the principal gas–liquid contacting operation takes place. Hence, if particle buildup is to be sought, the scrubber should be preceded by an appropriate gas-conditioning section. On the other hand, particle collection by Stefan flow can be induced

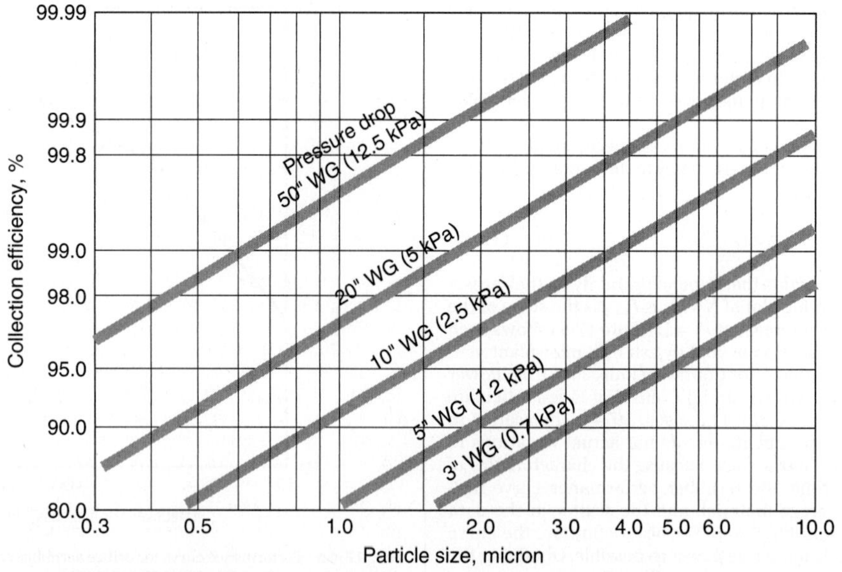

FIG. 17-67 Typical scrubber performance curve provided by manufacturer.

simply by scrubbing the hot, humid gas with sufficient cold water to bring the gas below its initial dew point. Any practical method of inducing condensation on the dust particles will incidentally afford opportunities for the operation of the Stefan-flow mechanism. The hot gas stream must, of course, have a high initial moisture content, since the magnitude of the effects obtained is related to the quantity of water vapor condensed.

Although there is a considerable body of literature on particle collection by condensation mechanisms, most of it is either theoretical or, if experimental, treats basic phenomena in simplified cases. Few studies have been made to determine what performance may be expected from condensation scrubbing under practical conditions in industrial applications. In a series of studies, Calvert and coworkers investigated several types of equipment for condensation scrubbing, generally emphasizing the use of the condensation center effect to build up the particles for collection by inertial deposition (Calvert and Parker, EPA-600/8-78-005c, 1978). From early estimates, they predicted that a condensation scrubber would require only about one-third or less of the power required by a conventional high-energy scrubber. A subsequent demonstration-plant scrubber system consisted of a direct-contact condensing tower fed with cold water followed by a venturi scrubber fed with recirculated water (Chmielewski and Calvert, EPA-600/7-81-148, 1981). The condensation and particle buildup took place in the cooling tower. In operation on humidified iron-foundry-cupola gas, this system still required about 65 percent as much power as for conventional high-energy scrubbing.

Semrau and coworkers [*Ind. Eng. Chem.* **50:** 1615 (1958); *J. Air Pollut. Control Assoc.* **13:** 587 (1963); EPA-650/2-74-108, 1974] investigated condensation scrubbing in pilot-plant studies in the field and, later, under laboratory conditions. Hot, humid gases were scrubbed directly with cold water under conditions that were favorable for the Stefan-flow mechanism but offered little or no opportunity for particle buildup. Some of the field studies indicated a contacting-power saving of as much as 50 percent for condensation scrubbing of Kraft-recovery-furnace fume. Laboratory tests on a predominantly submicrometer synthetic aerosol showed contacting-power savings of up to 40 percent with condensation scrubbing.

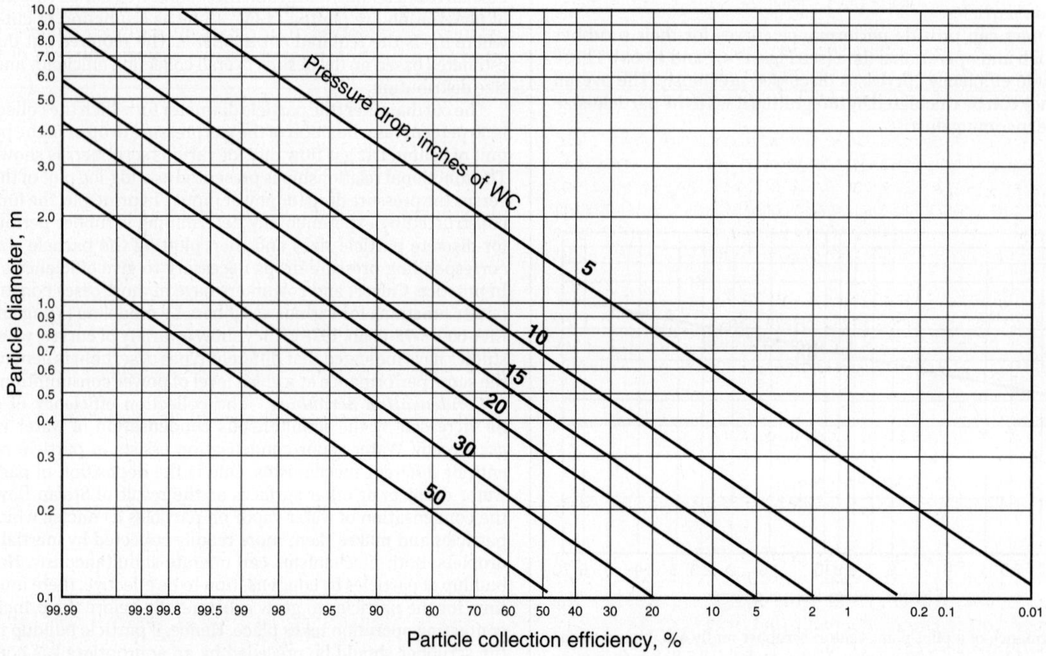

FIG. 17-68 Performance characteristics of a venturi scrubber.

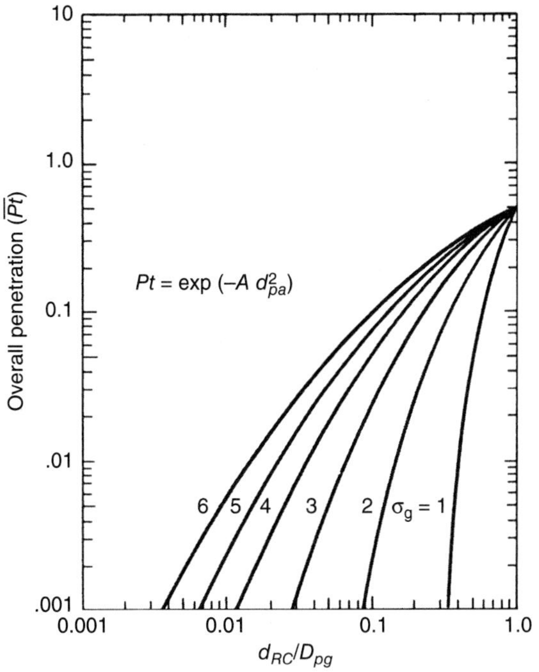

FIG. 17-69 Performance characteristic for scrubbers ($B = 2.0$). [*Calvert, Chem. Eng., 84 (18), 54 (1977).*]

In the scrubbing of hot gases with high water content, condensation reduces contacting power and affords a direct power saving through the reduction of the gas volume by cooling and water-vapor condensation, but it incurs other costs for power and equipment for heat transfer and water cooling. However, condensation scrubbing may offer a net economic advantage if recovery of low-level heat is practical. It should also be advantageous when a hot gas must not only be cleaned but cooled and dehumidified as well; examples are the cleaning of blast-furnace gas for use as fuel and of SO_2-bearing waste gases for feed to a sulfuric acid plant.

Fabric Filters Fabric filters, commonly termed "bag filters" or "baghouses," are collectors in which dry particles are removed from the gas stream by passing the dust-laden gas through a filter medium of some type (e.g., woven cloth, felt, paper or porous membrane). The filter media can be flexible (bags or envelops), semirigid (cartridges), or rigid (ceramic/metal cartridges).

Fabric filters are chosen as gas–solid separation equipment when used as:

1. Pollution control devices where health and safety are concerned
2. In-process applications to separate solids from a gas (e.g., pneumatic conveying systems)

The performance requirements for these two applications are very different. The inlet dust loading for pollution control devices is typically low (<80 g/m³), and the operation can be intermittent. In-process applications may require continuous operation where operational reliability and product recovery are of prime concern. Fabric filters can be used as standalone units or in conjunction with precleaners (e.g., cyclones) to reduce dust loading.

Because of their inherently high efficiency with dusts in all particle-size ranges, fabric filters have been used as primary gas cleaning equipment for the collection of fine dusts and fumes for over 100 years. Some of the key advantages over alternate technologies, such as wet scrubbers or electrostatic precipitators, are:

a. Infinite turndown ratio.
b. Almost complete collection for particles larger than 2 μm.
c. Solids are collected in a dry state.
d. Works with a wide range of particle sizes.
e. Collection efficiency is insensitive to particle properties or process upsets.

Limitations:

a. High-temperature applications: The greatest limitation on filter application has been imposed by the temperature limits of available fabric materials. The upper limit for natural fibers is about 90°C (200°F). The major new developments in filter technology that have been made since 1945 have followed the development of fabrics made from glass and synthetic fibers, which has extended the temperature limits to about 230°C to 260°C (450°F to 500°F). The capabilities of available fibers to resist high temperatures are still among the most severe limitations on the possible applications of fabric filters.

b. Presence of liquid droplets, moist and sticky particles will blind the pores of the media and result in an irrecoverable increase in the pressure drop. Fabric filters are not suitable for such applications. To prevent the condensation of vapors in the filter body, the temperature of the gas and all surfaces in contact with the filter media must be kept above the dew point.

c. The presence of very fine particles may blind the filter media due to penetration. Special fabrics, such as PTFE membranes, must be used. This will limit the operability window.

d. The life of filter media can be severely affected by the abrasive nature of dust, the presence of alkaline or acidic components in gas, or high operating temperatures. The long-term operability must be taken into consideration with the factors we have described.

Mechanism of Separation These devices are "surface" filters in that dust collects in a layer on the surface of the filter medium, and the dust layer itself becomes the effective filter medium. The pores in the medium (particularly in woven cloth) are usually many times the size of the dust particles, so collection efficiency is low until enough particles have been collected to build up a "precoat" in the fabric pores (Billings and Wilder, *Handbook of Fabric Filter Technology,* vol. I, EPA No. APTD-0690, NTIS No. PB-200648, 1979). During this initial period, particle deposition takes place mainly by inertial and flow-line interception, inertial deposition, diffusion, and electrostatic attrition. The agglomerates of fine particles can bridge over openings as large as 10 times the particle diameter, thereby creating a layer of dust upon which subsequent particles deposit. Some of the fine particles penetrate into the fabric during the initial phase (clean media) of cake deposition, and they continue to increase in amount over the life of the filter. Once the dust layer (or dust cake) has been fully established, sieving or depth filtration becomes the dominant mechanism. The pressure drop is largely incurred by gas flow through the dust cake, which acts as a compressible packed bed. The thickness of the dust cake continues to build until the cleaning cycle is initiated.

The process of dust removal on a fabric filter is inherently a batch process. The dust cake is attached to the filter media by adhesive forces, and it is held together with cohesive forces. These forces must be overcome to break the cake and dislodge it from the filter media. All cleaning mechanisms are functionally designed to overcome the adhesion and cohesion. The nature of the cake upon detachment and the extent of cleanliness of the filter after a cleaning cycle will depend on the forces generated during the cleaning cycle. Starting from a clean medium, the removal of particles after each cycle is not 100 percent—the cleaning leaves a small amount of fines adhered to the filter medium. The residual fines amount eventually reaches an equilibrium, and the filter is assumed to be conditioned with a stable pressure drop after the cleaning cycle. Since the filtration mechanism changes with time of operation, the concept of grade efficiency is not suitable for fabric filters. The overall collection efficiency is usually very high after the initial dust layer has been deposited. The lowest efficiencies during operation are observed during start-up, during and immediately after the cleaning cycle, and with problems with media integrity or leakage due to improper installation.

Classification of Fabric Filters (or Baghouses) Fabric filters can be broadly classified as static or fixed filters and regenerative or cleanable filters. The static filters are usually discarded after one-time use and are not expected to be cleanable. An inlet cartridge filter for a blower is a typical example. The majority of fabric filters in process industries are the cleanable type, where the dust burden on the filter is maintained at an equilibrium by periodic cleaning. Fabric filters or baghouses must be evaluated as a *unit operation* where the gas flow patterns, dust distribution, gas–solid separation at the filter interface, cake filtration, cake adhesion/cohesion, and reentrainment of dust are key factors influencing the performance. The performance metrics include pressure drop, collection efficiency, life of filter media, capital cost, and operating cost. A wide array of designs has evolved over the past 100 years as the range of applications has grown. The key differentiating features for various design categories are as follows:

Location of deposited dust: Interior or exterior surface of the filter bag (see Fig. 17-71)

Cleaning mechanisms: Shaker, reverse air, pulse (reverse air pulse, pulse jet) and acoustic

Direction of gas flow: Upflow (countercurrent to the separated dust), downflow, and cross-flow (see Fig. 17-71)

Mode of operation: Batch, intermittent, continuous

Filter media type: Bags/envelopes, cartridges (paper, felt, ceramic, sintered metal), rigid media made with sintered polymer

The most common approach is to categorize by the method of cleaning. Certain design features are feasible only in select combinations of other features, which allow the narrowing down of design choices.

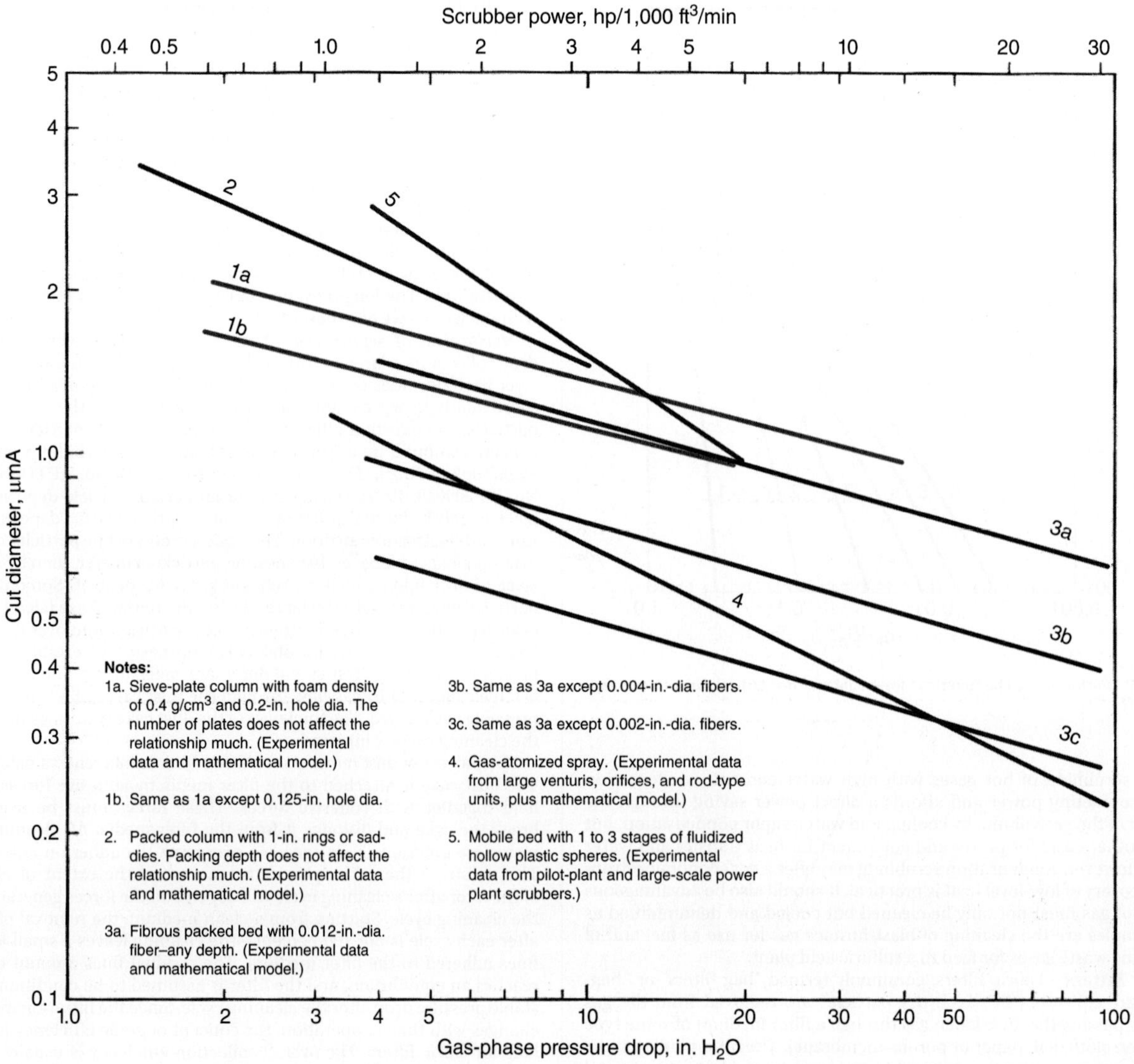
Scrubber power, hp/1,000 ft³/min
0.4
0.5
1.0
2
3
4
5
10
20
30
Cut diameter, μmA
5
4
3
2
1.0
0.5
0.4
0.3
0.2
0.1
2
5
1a
1b
3a
3b
4
3c
Notes:
1a. Sieve-plate column with foam density of 0.4 g/cm³ and 0.2-in. hole dia. The number of plates does not affect the relationship much. (Experimental data and mathematical model.)
1b. Same as 1a except 0.125-in. hole dia.
2. Packed column with 1-in. rings or saddies. Packing depth does not affect the relationship much. (Experimental data and mathematical model.)
3a. Fibrous packed bed with 0.012-in.-dia. fiber–any depth. (Experimental data and mathematical model.)
3b. Same as 3a except 0.004-in.-dia. fibers.
3c. Same as 3a except 0.002-in.-dia. fibers.
4. Gas-atomized spray. (Experimental data from large venturis, orifices, and rod-type units, plus mathematical model.)
5. Mobile bed with 1 to 3 stages of fluidized hollow plastic spheres. (Experimental data from pilot-plant and large-scale power plant scrubbers.)
1.0
2
3
4
5
10
20
30
40
50
100
Gas-phase pressure drop, in. H₂O

FIG. 17-70 Cut diameter as a function of power consumption or pressure drop. [*Calvert, Chem. Eng., 84 (18), 54 (1977).*]

Location of Deposited Dust/Mode of Filtration As shown in Fig. 17-71, the filtered dust can be deposited either on the inside surface or on the outside surface of the bag. Inside collection was prevalent in the earlier designs with shaker cleaning mechanisms. The pulse-jet collectors are exclusively used with outside collection, whereas reverse air collectors can be operated either way.

Cleaning Mechanisms As the separated dust deposits on the surface of a filter medium, it increases the resistance to gas flow (pressure drop), thereby increasing the energy consumption. All cleaning mechanisms rely on imparting sufficient force to break the adhesive bond between filter media and dust layer (cake), and the cohesive bonds between progressive layers of the deposited dust. The detached material falls under the force of gravity toward the discharge hopper, where it is removed. During each cleaning cycle, the entire layer of dust on the filter medium may not be detached. However, over the long term, an equilibrium dust burden, which is unique for a given set of conditions, is established.

There are three major designs for cleaning mechanisms in common practice: (1) shaker (includes oscillation, vibration, and tapping), (2) reverse air cleaning, and (3) reverse pulse cleaning (high-volume, low-pressure and low-volume, high-pressure). Shaking may be combined with reverse air cleaning or with acoustic energy cleaning to enhance its effectiveness.

The shaker mechanism, which was commercially introduced in late 1800s, was the earliest attempt at cleaning filter media. The most common design entails a shaking mechanism at the top where the bags are attached directly or to a common frame connected to a shaking mechanism (see Fig. 17-72). The oscillations can be horizontal, vertical, or a combination of the two. The amplitude (half-stroke) of the oscillations is usually between 10 mm (0.4 in) and 64 mm (2.5 in), and frequencies between 4 Hz and 8 Hz are common. Various modes of shaking have been summarized in an EPA report (*Fabric Filter Cleaning Studies*, p. 37, EPA-650/2-75-009, January 1975). Numerous patents related to shaking techniques have been issued for improving dust removal. The effective duration of cleaning depends on particle properties and filter media, and in the absence of specific experience, 200 to 300 cycles can be a good starting point.

The acceleration imparted to the media by the shaker mechanism causes the pores to open momentarily. This allows the fines to penetrate more deeply into the media and increases abrasive wear. Proper bag tension is critical to prevent the dampening of oscillatory motion. Excessive tension, on the other hand, will reduce the life of bags and increase emissions due to opening of the pores.

The filtration airflow must be stopped during a shaking cycle. If the filter is to be operated continuously, it must be constructed with multiple compartments so that the individual compartments can be sequentially taken off line for cleaning while the other compartments continue in operation. The section undergoing cleaning will not participate in filtration, and the design calculations must account for it. The relative magnitude of cleaning time and filtration time dictates the number of compartments a shaker-type filter must be designed for.

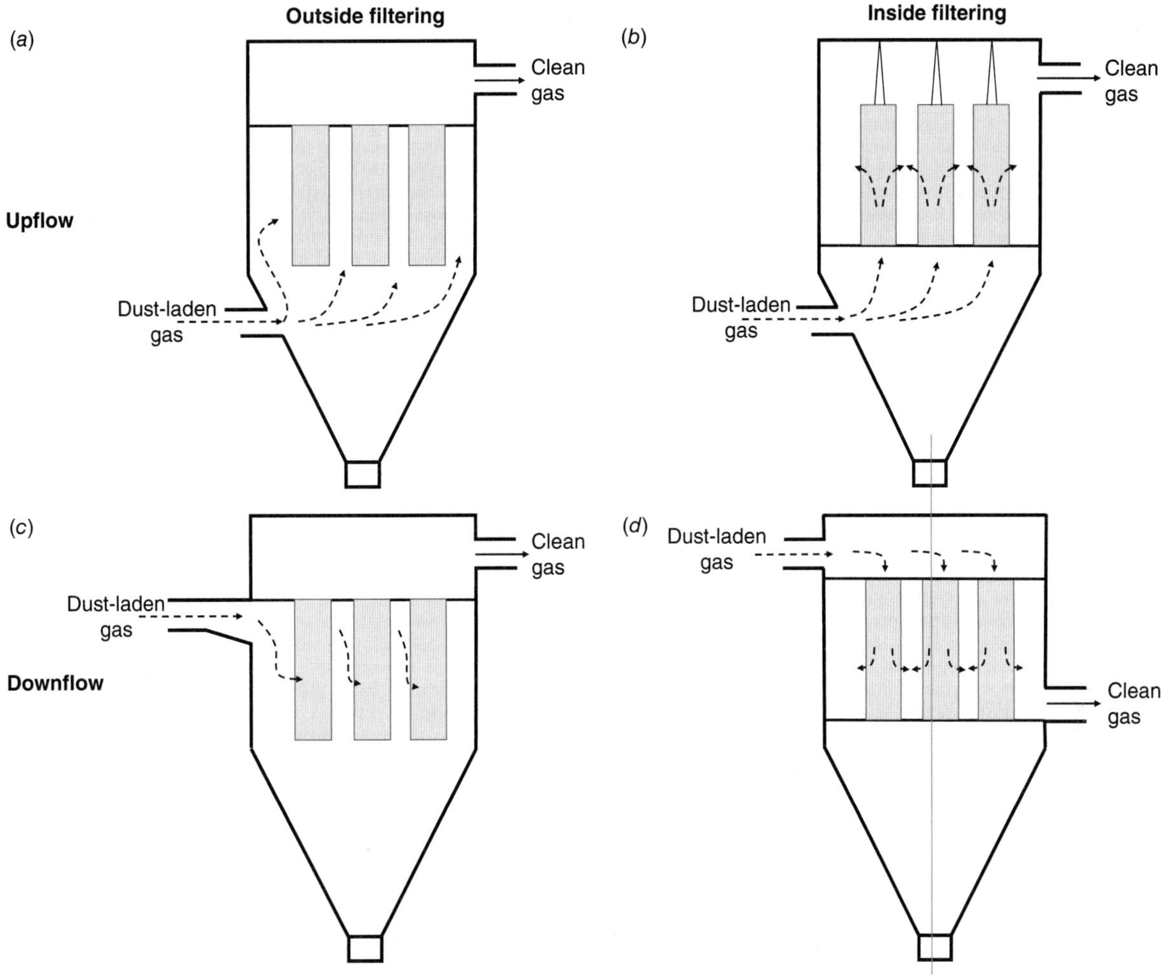

FIG. 17-71 Flow modes and directions of filtration in dust collectors.

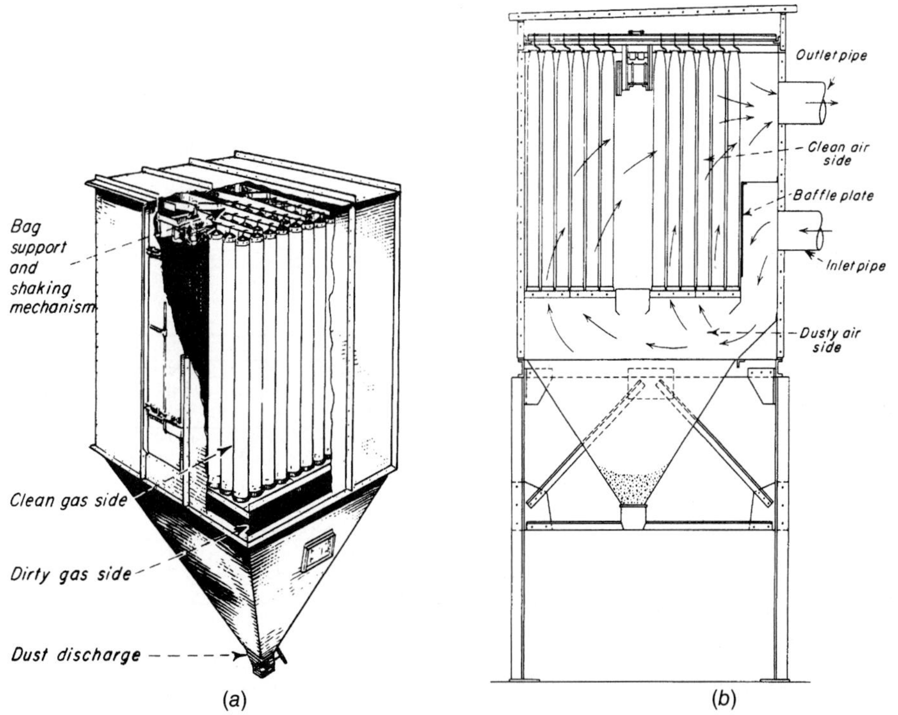

FIG. 17-72 Typical shaker-type fabric filters. (*a*) *Buell Norblo* (cutaway view). (*b*) *Wheelabrator-Frye, Inc.* (sectional view).

Shaker-cleaned filters are available as standard commercial units, although large baghouses for heavy-duty service are commonly custom-designed and fabricated. The oval or round bags used in the standard units are usually 12 to 20 cm (5 to 8 in) in diameter and 2.5 to 5 m (8 to 17 ft) long. The large, heavy-duty baghouses may use bags up to 30 cm (12 in) in diameter and 9 m (30 ft) long. The bags must be made of woven fabrics to withstand the flexing and stretching involved in shaking. The fabrics may be made from natural fibers (cotton or wool) or synthetic fibers. Fabrics of glass or mineral fibers are generally too fragile to be cleaned by shaking and are usually used in reverse-flow-cleaned filters. Felted fabrics operate at higher face velocities than woven fabrics.

Large units (other than custom units) are usually built up of standardized rectangular sections in parallel. Each section contains on the order of 1000 to 2000 ft^2 of cloth, and the sections are assembled in the field to form a single filter housing. In this manner, the filter can be partitioned so that one or more sections at a time can be cut out of service for shaking or general maintenance.

Ordinary shaker-cleaned filters may be shaken every one-quarter to 8 h, depending on the service. A fixed time cycle requires one to estimate the rate of pressure rise based on inlet loading. Excessive cleaning will result in premature failure of bags, whereas insufficient cleaning can result in higher energy consumption and blinding. On-demand cleaning based on measurement of pressure drop is the recommended approach.

It is essential that the gas flow through the filter be stopped when shaking in order to permit the dust to fall off. With very fine dust, it may even be necessary to equalize the pressure across the cloth [Mumford, Markson, and Ravese, *Trans. Am. Soc. Mech. Eng.* **62:** 271 (1940)]. In practice this can be accomplished without interrupting the operation by cutting one section out of service at a time, as shown in Fig. 17-73. In automatic filters this operation involves closing the dampers, shaking the filter units either pneumatically or mechanically, sometimes with the addition of a reverse flow of cleaned gas through the filter, and lastly reopening the dampers. For compressed-air-operated automatic filters, this entire operation may take only 2 to 10 s. For ordinary mechanical filters equipped for automatic control, the operation may take as long as 3 min.

Shaker-cleaned filters are generally operated at filtration velocities of 0.3 to 2.5 m/min (1 to 8 ft/min) and at pressure drops of 0.5 to 1.5 kPa (2 to 6 inches water). For very fine dusts or high dust concentrations, filtration velocities should not exceed 1 m/min (3.3 ft/min). For fine fumes and dusts in heavy-duty installations, filtration velocities of 0.3 to 0.6 m/min (1 to 2 ft/min) have long been accepted on the basis of operating experience. The concept of filtration velocity is discussed in greater depth later.

Where the dust is collected on the inside surface of the filter, the gas velocity at the inlet of the filter bag is the highest because the cross-sectional area of the bag is substantially smaller than the overall filter area. Excessive inlet velocity can lead to abrasive wear on the bag.

Cyclone precleaners are sometimes used to reduce the dust load on the filter or to remove large, hot cinders or other materials that might damage the bags. However, reducing the dust load on the filter by this means may not reduce the pressure drop, since the fines exiting the cyclone can create a dust layer with lower permeability and hence negate the advantage of reduction in the fabric dust loading.

Reverse-flow-cleaned filters are generally similar to the shaker-cleaned filters except for the elimination of the shaker. It was developed as a gentler option for cleaning fragile bags (e.g., glass-fiber fabric). After the flow of dirty gas has stopped, a fan is used to force clean gas through the bags from the clean-gas side. This flow of gas flexes the bags and dislodges the collected dust, which falls into the dust hopper. The reverse airflow also reduces the cohesion within the dust layer, which enhances removal efficiency.

Reverse air cleaning can be implemented for (1) inside-out (dust deposition inside the bag; see Fig. 17-71*b* or *d*) or (2) outside-in (dust deposition outside the bag, see Fig. 17-71*a* or *c*) operation. For the former mode, rings are usually sewn into the bags at intervals along the length to prevent complete collapse, which would obstruct the fall of the dislodged dust. In the latter case, the collector is designed with filter elements laid out radially in the tube sheet (see Fig. 17-74). As the rotating manifold aligns itself with a row of bags, the airflow through the bags is momentarily reversed for cleaning. The filter bag forms a sleeve that is drawn over a wire cage or perforated tube, which is usually cylindrical. The cage supports the fabric on the clean-gas side, and the dust is collected on the outside of the bag. During the filtration cycle, the bag clings to the wire cage (or support tube) which prevents it from collapsing. The shape of the bag becomes convex outward during the reverse flow cleaning cycle, thereby dislodging the accumulated dust.

Medium-sized collectors can be cleaned on-line continuously using this approach. However, large collectors must be compartmentalized into modules for off-line cleaning (see Fig. 17-73). As with shaker-cleaned filters, compartments of the baghouse are taken off line sequentially for bag cleaning. The use of butterfly valves or poppet valves is common for the isolation of various modules and automatic cleaning sequences (Heumann, *Industrial Air Pollution Control System*, McGraw-Hill, New York, 1997).

The principal applications of reverse-flow cleaning are in units using fiberglass fabric bags for dust collection at temperatures above 150°C (300°F). Collapsing and reinflation of the bags can be made sufficiently gentle to avoid putting excessive stresses on the fiberglass fabrics [Perkins and Imbalzano,

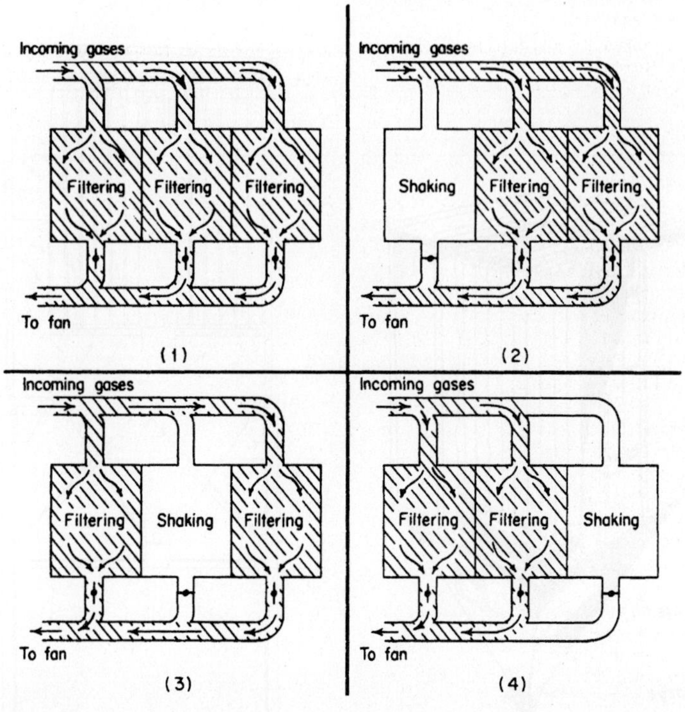

FIG. 17-73 Three-compartment bag filter at various stages in the cleaning cycle. (*Wheelabrator-Frye, Inc.*)

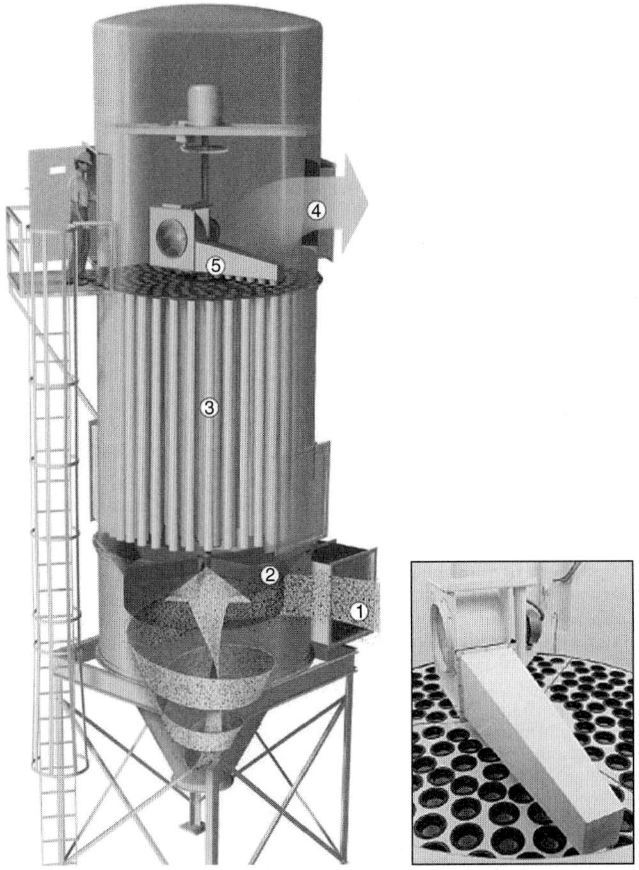

FIG. 17-74 Reverse air dust collector: (1) air inlet; (2) plenum; (3) filter bags; (4) rotating manifold; (5) air outlet. (*Courtesy of Nederland Mikropul.*)

"Factors Affecting the Bag Life Performance in Coal-Fired Boilers," 3d APCA Specialty Conference on the User and Fabric Filtration Equipment, Niagara Falls, N.Y., 1978; and Miller, *Power* **125**(8): 78 (1981)].

The gas for reverse-flow cleaning is commonly supplied in an amount necessary to give a superficial velocity through the bags that is the same or higher than the filtration velocities for a given application (EPA Report, EPA/452/B-02-001, December 1998). A separate blower provides sufficient cleaning air for one or two compartments. Typical pressures are between 14 and 28 kPa (2 to 4 psi). The operational filtration face velocities (air-to-cloth ratio) for reverse air filters are similar to shaker filters.

Reverse-pulse filters (often termed reverse-jet filters) were developed in the 1950s and have become popular due to their lower capital cost and ability to operate continuously (Fig. 17-75). They combine the two previously discussed mechanisms, flexing of filter media and reverse airflow through the dust cake, to disrupt adhesive and cohesive forces. Two distinct pulse cleaning methods have evolved in practice: low-pressure/high-volume and high-pressure/low-volume cleaning modes (Heumann, *Industrial Air Pollution Control System*, McGraw-Hill, New York, 1997). This cleaning mechanism is used for filters where the dust is collected on the outer surface of the bag or filter element.

The low-pressure/high-volume approach is similar to reverse air cleaning (see Fig. 17-74) for outside-in flow except that a reservoir is used to accumulate the air from the blower at typical pressures of 41 to 55 kPa (6 to 8 psi). A large diaphragm valve is used to discharge the air in a few milliseconds. The high-volume pressure pulse momentarily stops the gas flowing through the bank of filters and inflates the bags, causing the separated dust to fall into the hopper. The cleaning efficacy is better than conventional reverse air cleaning.

The high-pressure/low-volume cleaning approach is the most common approach practiced today. It requires a source of clean (oil-free) and dry compressed air (60 psig to 120 psig) to create the reverse pulse. A venturi nozzle is located in the clean-gas outlet from the bag. For cleaning, a jet of high-velocity air is directed through the venturi nozzle and into the bag, inducing a flow of cleaned gas to enter the bag and flow through the fabric to the dirty-gas side. The high-velocity jet is released in a sudden, short pulse

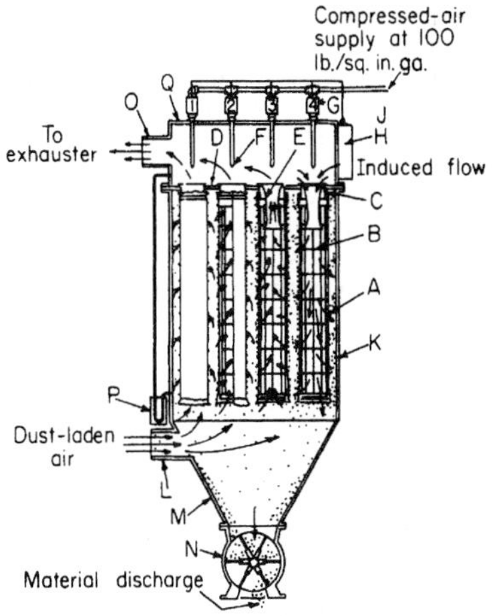

FIG. 17-75 Reverse-pulse fabric filter: (A) filter cylinders; (B) wire retainers; (C) collars; (D) tube sheet; (E) venturi nozzle; (F) nozzle or orifice; (G) solenoid valve; (H) timer; (J) air manifold; (K) collector housing; (L) inlet; (M) hopper; (N) air lock; (O) upper plenum; (P) manometer; (Q) top of filter housing. (*Mikropul Division, U.S. Filter Corp.*)

(typical duration 100 ms or less) from a compressed-air line by a solenoid valve. The propagating pulse of air expands the bag as it travels downward and dislodges the collected dust. The pulse intensity naturally decays as it travels downward, which reduces the cleaning efficiency toward the bottom of the bag. A perforated tube, instead of the cage, may be used to maintain the uniformity of pulse intensity where appropriate.

Rows of bags are cleaned in a timed sequence by programmed operation of the solenoid valves. The pressure of the pulse is sufficient to dislodge the dust without cessation of the gas flow through the filter unit.

Some manufacturers are using relatively low-pressure air (100 kPa, or 15 lbf/in^2, instead of 690 kPa, or 100 lbf/in^2) and are eliminating the venturi tubes for clean-gas induction. Others have eliminated the separate jet nozzles located at the individual bags and use a single jet to inject a pulse into the outlet-gas plenum.

Reverse-pulse filters are typically operated at higher filtration velocity, also called air-to-cloth ratio, than shaker or reverse-flow filters designed for the same duty (Table 17-9). Recommended values of filtration velocities

TABLE 17-9 Examples of Recommended Filtration and Can Velocity

Material	Filtration velocity shaker/reverse air cleaning, ft/min	Filtration velocity pulse cleaning, ft/min	Maximum recommended can velocity, ft/min
Activated carbon	2	6	175
Aluminum oxide	2	9	200
Baking powder	2.5	9	175
Bauxite	2.5	8	200
Cement	2.5	8	225
Feeds, grain	3.5	14	200
Fertilizer	3	8	200
Fly ash	2.5	7	200
Graphite	2	5	200
Iron oxide	2.5	7.5	200
Lime	2.5	9	200
Metallurgical fumes	1.5	6	150
Mica	2.7	9	175
Plastic pellets	3.5	10	225
Quartz	2.8	9	200
Sand	2.5	10	225
Sawdust, fines	4	10	150
Soda ash	3	9	200
Sugar	2	8	200
Talc	2.5	9	200
Titanium dioxide	2	7	175
Zinc oxide	2	7	200

Adapted from Croom, M., Filter Dust Collectors, McGraw Hill, 1995.

for various products have been tabulated in Croom, *Filter Dust Collectors,* McGraw-Hill, New York, 1995; Heumann, *Industrial Air Pollution Control System*, McGraw-Hill, New York, 1997; and EPA Report, EPA/452/B-02-001, December 1998. Filtration velocities may range from 1 to 4.5 m/min (3 to 15 ft/min), depending on the dust being collected, but for most dusts the commonly used range is about 1.2 to 2.5 m/min (4 to 8 ft/min). The frequency of cleaning is also dependent on the nature and concentration of the dust, with the intervals between pulses varying from about 2 to 15 min.

The cleaning action of the pulse is so effective that the dust layer may be completely removed from the surface of the fabric. Consequently, the fabric itself must serve as the principal filter medium for at least a substantial part of the filtration cycle. Woven fabrics are unsuitable for such service, and felts of various types must be used. The bulk of the dust is still removed from the surface layer, but the felt ensures that an adequate collection efficiency is maintained until the dust layer has formed. This cleaning mechanism is applicable to various types of filter elements, such as bags, envelopes, nonwoven cartridges, ceramic, and metal cartridges.

It has been a common practice to clean the bags on-line (i.e., without stopping the flow of dirty gas into the filter), and reverse-pulse bag filters have been built without division into multiple compartments. However, investigations [Leith et al., *J. Air Pollut. Control Assoc.* **27:** 636 (1977)] and experience have shown that, with on-line cleaning of reverse-pulse filters, a large fraction of the dust dislodged from the bag being cleaned may redeposit on neighboring bags rather than fall into the dust hopper. The gas flow through a bag (or row of bags) is substantially higher after it has been cleaned. When the adjacent row is cleaned, all the dislodged dust simply migrates to the previously cleaned row. To mitigate this problem, the reverse-pulse cleaning sequence is randomized so that adjacent rows are not sequentially cleaned.

Acoustic Cleaning Acoustic or sonic energy can be used to dislodge dust cake from filter media. It is used either as a supplement to the shaker to enhance cleaning efficiency or as a standalone method of cleaning. The acoustic energy is generated by a diaphragm using compressed air, which is then channeled through a horn. A typical frequency range is 60 Hz to 250 Hz, with lower frequencies being more effective but tending to diffuse more rapidly. Higher frequencies are directional and tend to be an irritant for operational personnel in the process area. The power of sound energy is expressed in decibels, which is a logarithmic scale. The output of sonic horns is generally between 120 and 150 decibels. It is also possible to clean the accumulation in dead zones or initiate flow in the hopper using sonic horns.

Direction of Gas Flow The relative locations of the dust-laden gas inlet and the clean air outlet will result in different flow regimes within the body of the collector, namely upflow, downflow, and cross-flow. Since the dust is always removed by gravity settling, the bulk airflow in the filter can be cocurrent or countercurrent to the separated dust. Downflow or cross-flow configurations are advantageous when collecting fine, light, or fluffy dust (e.g., fumes, wood dust, polymer dust) because the dust removed from the filter elements is swept by the flowing gas toward the collection hopper. For an upflow configuration, it is necessary to keep the approach velocity of dust-laden air, called the *can velocity*, below a critical value to prevent reentrainment of dust. The allowable can velocity is higher than the single-particle terminal velocity because it is assumed that the cake falls as chunks or large agglomerates. The guidance provided in Table 17-9 is based on operational experience.

The *can velocity* is based on the actual flow area, which is calculated by subtracting the total area occupied by the filter elements (bags/cartridges) from the overall cross section of the filter housing (ASHRAE Handbook—2016, Chapter 30).

$$V_c = \frac{q}{A_h - NA_f} \qquad (17\text{-}37)$$

where V_c = can velocity, m/s (ft/min)
q = gas flow being cleaned, m^3/s (ft^3/min)
A_h = cross-sectional area of housing, m^2 (ft^2)
N = number of filter bags (filter elements) in the collector
A_f = cross-sectional area, perpendicular to gas flow, of each filter bag (or filter element), m^2 (ft^2)

For the upflow configuration, the can velocity will be maximum in the plane below the bottom of the filter elements. The can velocity is not relevant for downflow configuration.

Optimal design of a collector must balance can velocity, filtration velocity, pressure drop, footprint, and capital cost.

Mode of Operation Certain process applications, such as collection of nuisance dust, allow intermittent stoppage for the filter to undergo a cleaning cycle. On the other hand, filters for pneumatic conveying systems in a production process must operate continuously without the possibility of a shutdown for cleaning. Shaker and reverse collectors operate in batch or intermittent mode, while pulse-jet collectors are most suited for operation in continuous mode.

Type of Fabric Filter Elements There are three major types of fabric filter elements in common practice: bags, envelopes, and cartridges. Bag, sock, or tube filters are the most common filter elements in use today. They can be sewn from woven or felted media with a seam along the length. The design of the open end is adapted for ease of installation (e.g., snap bands). Grounding wires are incorporated in the design for combustible dust applications.

Envelope-type elements are usually found in smaller unit collectors where multiple banks of elements can be closely packed horizontally or vertically to achieve high filtration area (see Fig. 17-76). The cartridge type collectors are limited to outside-in flow and reverse air or pulse-jet cleaning mechanisms. They have gained greater market share in the past 20 years due to their compactness and ability to function at high efficiencies. Cartridges

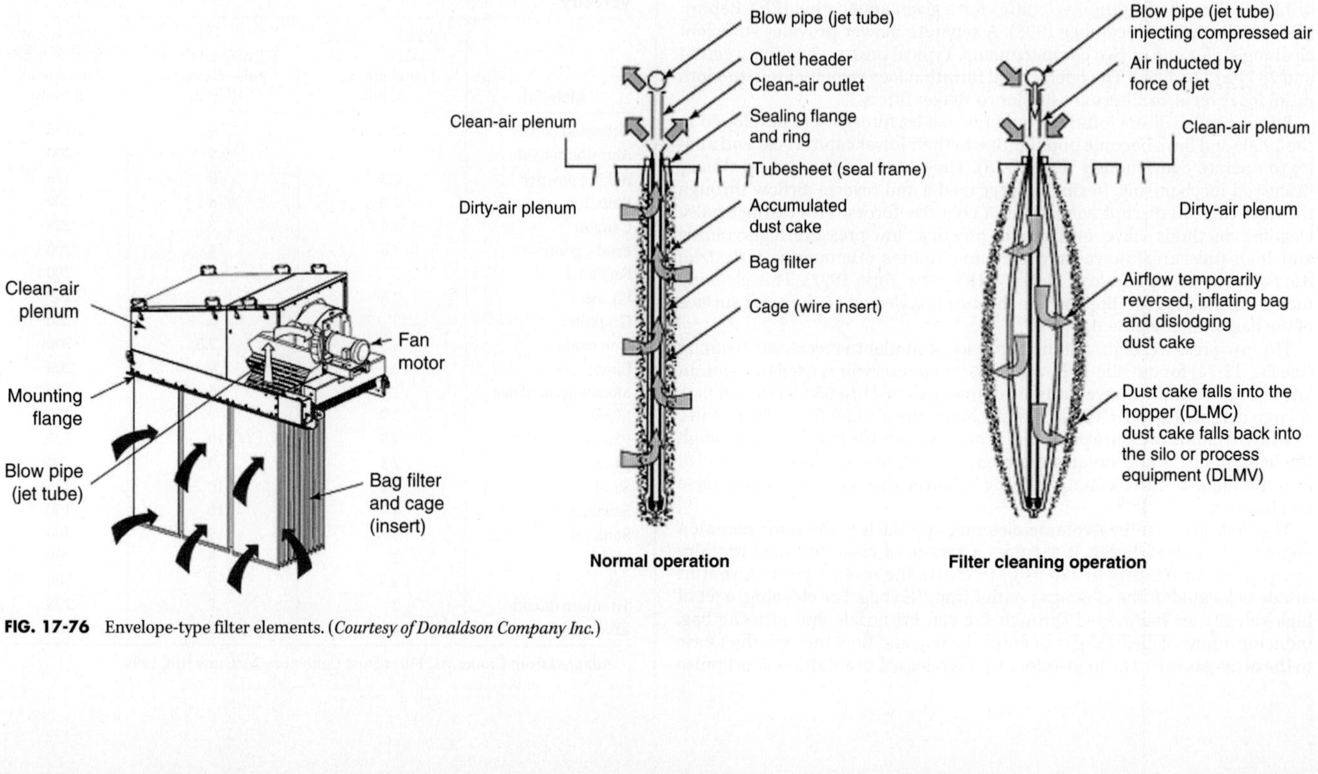

FIG. 17-76 Envelope-type filter elements. (*Courtesy of Donaldson Company Inc.*)

are typically manufactured from pleated paper or synthetic microfibers that are spunbonded (e.g., spunbonded polyester). Common construction is 150 to 350 mm (6–14 in) in diameter and 400 to 900 mm (14–36 in) long. An open pleated structure (4–8 pleats/in) is recommended for cohesive dust and sticky dusts, whereas higher pleat density (12–16 pleats/in) will be more economical for coarser particles. The pleat depth is between 25 and 75 mm (1–3 in). Deeper pleats are difficult to clean, and dust can get wedged deep into the pleat. The upper limits on operating temperatures (200°F–350°F) are due to the adhesives used for bonding the end caps, even if high-temperature filter media are used.

Cartridges are often recommended as a retrofit upgrade to existing baghouses where the filtration area (cloth area) can be easily doubled. The cartridges are also shorter than bags, which can reduce abrasive wear due to high gas inlet velocities. Unlike bags, the cartridges do not require cages for reinforcement, and they are designed for easier installation. The user is cautioned to check the can velocity in a filter housing with new operating conditions. Cartridges of two different lengths can be used as a practical remedy to this problem.

The high porosity of these media, in combination with small pore size, results in low pressure drop and high collection efficiency for submicron particulates. The paper cartridges can collect 2.5 μm particles at 99.999+ percent efficiency, while the nonwoven (spunbonded polyester) cartridges have an efficiency of 99.99 percent at 2.5 μm. However, limited choices of filter media, high-temperature limitations, and plugging issues with fine/sticky/cohesive powders have hindered broader applicability nonwoven cartridges.

Selection of Filter Media Inappropriate selection of fabric filter media can lead to high operating and maintenance costs due to premature failures and higher emissions. The selection must be based on (1) maximum operating temperature, (2) presence of acidic components in the gas, (3) presence of alkaline components in the gas, (4) presence of moisture or condensed droplets, (5) presence of sticky, hygroscopic, or submicron particles, and (6) abrasiveness of dust. The characteristics of common fabric filter media are summarized in Table 17-10.

The choice of woven versus felted media is largely dictated by the chosen method of cleaning. Woven media (cloth) is used with shaker and reverse air cleaning systems, whereas felt media is suitable for reverse air and pulse-jet systems. Woven cloth is unsuitable for pulse-jet cleaning because the flexing action due to the pressure pulse increases the pore size, which results in a spike in emissions, and it shortens the bag life due to abrasive wear.

Woven media are made from weaving yarn (filament, staple, or spun) into filter cloth of geometric weave patterns (e.g., plain weave, twill weave, satin weave). The multitude of combinations of raw materials, spinning methods, and weaving processes results in a broad range of filtration characteristics. For example, woven media made with spun or heavyweight staple yarn fabric are suitable for shaker systems, whereas those made with lighter-weight filament yarn fabric are used for reverse air cleaning systems.

The felts made from synthetic fibers are needle felts (i.e., felted on a needle loom) and are normally reinforced with a woven insert. In felted media, there is no clear open path for the particles to pass through, which results in greater efficiencies after the cleaning phase.

The selection of fabric media for a given application must be done in collaboration with the manufacturers, and by leveraging prior experience with similar applications. The key parameters specified for woven and non-woven (felt) media are:

- Fiber type (e.g., Polyester)
- Weight (oz./yd^2 or g/m^2)
- Construction (e.g., weave type, felt structure, spunbonding)
- Thickness
- Count
- Air permeability
- Mullen burst strength
- Tensile strength—warp direction and fill direction
- Maximum operating temperature
- Thermal stability (shrinkage)
- Surface finish (e.g., glazing, singeing, calendaring, membrane)

The cost of the filter bags represents a substantial part of the erected cost of a bag filter—typically 5 to 20 percent, depending on the bag material [Reigel and Bundy, *Power* **121**(1): 68 (1977)]. The cost of bag repair and replacement is the largest component of the cost of bag-filter maintenance. Consequently, the proper choice of filter fabric is critical to both the technical performance and the economics of operating a filter. With the advent of synthetic fibers, it has become possible to produce fabrics having a wide range of properties. However, demonstrating the acceptability of a fabric still depends on experience with prolonged operation under the actual or simulated conditions of the proposed application. The choice of a fabric material for a given service is necessarily a compromise, since no single material possesses all the properties that may be desired. An example of cost analysis for fabric filters can be found in Turner et al., EPA/452/B-02-001 (1998).

Specification of Filtration Velocity or Air-to-Cloth Ratio The specification of filtration face velocity (V_f), better known as the air-to-cloth (AC) ratio in commercial practice, is critical for sizing a filter collector. It is defined as

$$V_f = q/A_f \tag{17-38}$$

where A_f is the active filtration area (excluding the filter area that is compartmented for cleaning).

As is evident from Eq. (17-38), it has the units of velocity (ft/min or m/min). Omission of units while reporting this critical parameter can lead to major design errors. The values reported by U.S. manufacturers are often in ft/min units.

The face velocity (or air-to-cloth ratio) has a direct bearing on filter performance, specifically:

i. Pressure drop and the rate of pressure rise between cleaning cycles, hence cleaning frequency
ii. Media blinding and operational life of filter elements
iii. Emissions
iv. Capital cost, energy cost, and maintenance cost

A first-principles approach to calculate the filtration velocity for a given product, filter media, operating conditions, and cleaning mechanism does not exist because of the complexity of interactions inherent in the filtration process. However, tables of recommended values are available in various texts and references [Croom, *Filter Dust Collectors,* McGraw-Hill, New York, 1995; Turner et al., EPA/452/B-02-001 (1998); Heumann, *Industrial Air Pollution Control System*, McGraw-Hill, New York, 1997; *ASHRAE Handbook 2016*; manufacturers]. These values provide a good starting point. For a new application with no prior experience, pilot-scale testing is highly recommended.

Croom (*Filter Dust Collectors,* McGraw-Hill, New York, 1995, p. 24) has summarized a factored approach that is widely practiced in the industry.

$$V_{fe} = V_f \cdot A \cdot T \cdot P \cdot D$$

where V_{fe} = effective filtration velocity, m/min (ft/min)

TABLE 17-10 Temperature Limits and Characteristics of Filter Media (ASHRAE Handbook—2016)

Filter media	Maximum operating temperature, °F	Resistance to acid	Resistance of base or alkaline	Resistance to moisture and humidity	Resistance to oxidation	Flex abrasion	Approximate cost compared to cotton
Cotton	180	Poor	Very good	Poor	Fair	Good	X
Acrylic	270	Good	Fair	Very good	Very good	Good	2–3X
Glass Fiber	500	Fair	Fair	Excellent	Excellent	Poor	3–5X
Polyester	275	Good	Good	Fair	Very Good	Very good	2–3X
Polypropylene	180	Very good	Very good	Very good	Very good	Very good	1–2X
Polyamide (Nylon)	200	Poor	Very good	Good	Fair	Excellent	2–3X
Polyaramide (Nomex™)	400	Fair	Good	Good	Very good	Very good	4–8X
PTFE	500	Excellent	Excellent	Excellent	Excellent	Fair	10–30X
Wool	195	Fair	Poor	Poor	Poor	Good	X

TABLE 17-11 Correction Factors for Filtration Velocity or Air-to-Cloth Ratio

A = Application	0.8	Oily, moist, or agglomerating
	0.9	Product collection
	1.0	Nuisance dust collection
T = Temperature	0.8	Above 225°F (107°C)
	0.9	Between 110°F and 225°F (43°C–107°C)
	1.0	110°F (43°C) and lower
P = Particle size	0.8	Less than 3 μm
	0.9	For 3 μm to 9 μm
	1.0	For 10 μm to 50 μm
	1.1	For 51 μm to 100 μm
	1.2	For above 100 μm
D = Dust loading	0.8	Above 80 gr/ft^3
	0.9	For 50 to 80 gr/ft^3
	1.0	For 20 to 50 gr/ft^3
	1.1	For 10 to 20 gr/ft^3
	1.2	Less than 10 gr/ft^3

Appropriate values of the factors A (application), T (temperature), P (particle size), and D (dust loading) must be selected from Table 17-11.

Calculation of Pressure Drop Across Filter Media The filtration, or superficial face, velocities used in fabric filters are generally in the range of 0.3 to 4.6 m/min (1 to 15 ft/min), depending on the types of fabric, fabric supports, and cleaning methods used. In this range, gas pressure drops conform to Darcy's law for streamline flow in porous media, in which the pressure drop is directly proportional to the flow rate. The pressure drop across the fabric and the collected dust layer may be expressed (Billings and Wilder, *Handbook of Fabric Filter Technology*, vol. I, EPA No. APTD-0690, NTIS No. PB-200648, 1979) by

$$\Delta p = K_1V_f + K_2wV_f$$

$$\Delta p = K_1V_f + K_2wV_f = (K_1 + K_2w)V_f = SV_f \qquad (17\text{-}39)$$

where Δp = kPa, or in of water; V_f = superficial velocity through filter, m/min, or ft/min; w = dust loading on filter, g/m^2, or lbm/ft^2; and K_1 and K_2 are resistance coefficients for the "conditioned" fabric and the dust layer, respectively. The conditioned fabric is that fabric in which a relatively consistent dust load remains deposited in a certain depth following cycles of filtration and cleaning. K_1, expressed in units of kPa/(m/min) or in water/(ft/min), may be more than 10 times the value of the resistance coefficient for the original clean fabric. If the depth of the dust layer on the fabric is greater than about 0.2 cm (1/16 in), corresponding to a fabric dust loading on the order of 200 g/m^2 (0.04 lbm/ft^2), the pressure drop across the fabric (including the dust in the pores) is usually negligible relative to that across the dust layer. It should be noted that increasing the filter area for a given duty (gas flow rate and solids concentration) will reduce both the filtration velocity and dust loading on the filter, and, therefore, has a significant impact on the pressure drop. The rate of increase in pressure drop will also be lower, which reduces the cleaning frequency. Lower cleaning frequency results in longer filter life and lower overall emissions.

The combined term, S, is called the system drag, which accounts for pressure drop from the inlet to the outlet [Turner et al., EPA/452/B-02-001 (1998)]. The pressure drop across the fabric filter, and the system drag, is a function of time (see Fig. 17-77). The pressure drop characteristics of a conditioned filter medium is very different than an unused filter medium, and further, the actual dust loading on the filter and the specific resistance coefficient of the dust layer are difficult to estimate from first principles. From a practical perspective, it is best to conduct pilot-scale testing to obtain the drag values of a conditioned filter and the specific resistance coefficient of the dust layer. The performance data from an existing installation can also be used to achieve the same objective.

Equation (17-39) indicates that for filtration at a given velocity, the pressure drop is a linear function of the fabric dust loading, w. In some cases, particularly with smooth-surfaced fabrics, this is approximately the case, but in other instances the function displays an upward curvature with increases in w, indicating compression of the dust layer, the fabric, or both, and a consequent increase in K_2 [Snyder and Pring, *Ind. Eng. Chem.* **47:** 960 (1955); K. T. Semrau, unpublished data, SRI International, Menlo Park, Calif., 1952–1953]. Several investigations have shown K_2 to be increased by increases in the filtration velocity [Billings and Wilder, *Handbook of Fabric Filter Technology*, vol. I, EPA No. APTD-0690, NTIS No. PB-200648, 1979; Spaite and Walsh, *Am. Ind. Hyg. Assoc. J.* **24:** 357 (1968)]. However, the various investigators do not agree on the magnitude of the velocity effect. Billings and Wilder suggest assuming as an approximation that K_2 is directly proportional to the filtration velocity, but the actual relationship is probably dependent on the nature of the fabric and fabric surface, the characteristics of the dust, the dust loading on the fabric, and the pressure drop.

Clearly, the factors determining K_2 are far more complex than is indicated by a simple application of the Kozeny-Carman equation, and when possible, filter design should be based on experimental determinations made under conditions approximating those expected in the planned installation.

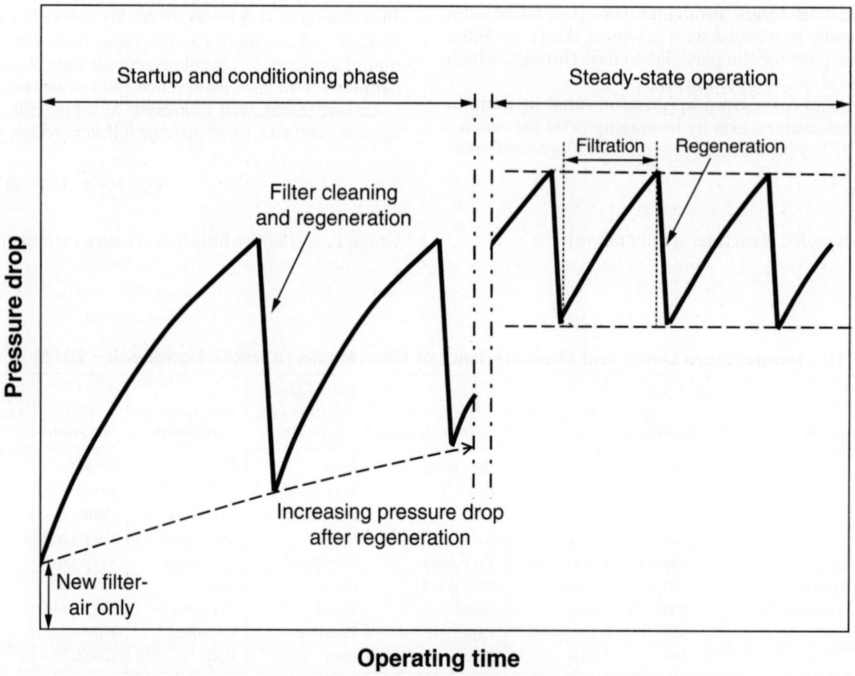

FIG. 17-77 Typical pressure drop characteristics of a filter collector during operation.

Collection Efficiency The inherent collection efficiency of fabric filters is usually so high that, for practical purposes, the precise level has not commonly been the subject of much concern. Furthermore, for collection of a given dust, the efficiency is usually fixed by the choices of filter fabric, filtration velocity, method of cleaning, and cleaning cycle, leaving few if any controllable variables by which efficiency can be further influenced. Inefficiency usually results from bags that are poorly installed, torn, or stretched from excessive dust loading and pressure drop. Of course, certain types of fabrics may simply be unsuited for filtration of a particular dust, but usually this will soon become obvious.

Few basic studies of the efficiency of bag filters have been made. Increased dust penetration immediately following cleaning has been readily observed while the dust layer is being reestablished. However, field and laboratory studies have indicated that during the rest of the filtration cycle, the effluent-dust concentration tends to remain constant regardless of the inlet concentration [Dennis, *J. Air Pollut. Control Assoc.* **24:** 1156 (1974)]. In addition, there has been little indication that the penetration is strongly related to dust-particle size, except possibly in the low-submicrometer range. These observations appear to be generally consistent with sieving being the principal collection mechanism.

Leith and First [*J. Air Pollut. Control Assoc.* **27:** 534 (1977); **27:** 754 (1977)] studied the collection efficiency of reverse-pulse filters and concluded that once the dust cake has been established, "straight-through" penetration by dust particles that pass through the filter without being stopped is negligible by comparison with penetration by dust that actually deposits initially and then seeps through the fabric to be reentrained into the exit airstream. They also noted that pinholes may form in the dust cake, particularly over pores between yarns in a woven fabric, and that particles may subsequently penetrate straight through at the pinholes. The formation of pinholes, or "cake puncture," had been observed earlier by Stephan et al. [*Am. Ind. Hyg. Assoc. J.* **21:** 1 (1960)], but without measurement of the associated loss of collection efficiency. When a supported flat filter medium with extremely fine pores (e.g., glass-fiber paper, membrane filter) was used, no cake puncture took place, even with very high pressure differentials across the cake. However, puncture did occur when a cotton-sateen filter fabric was used as the cake support. The formation of pinholes with certain combinations of dusts, fabrics, and filtration conditions was also observed by Koscianowski et al. (EPA-600/7-78-056, 1978). Evidently, puncture occurs when the local cake structure is not strong enough to maintain a bridge over the aperture represented by a large pore, and the portion of the cake covering the pore is blown through the fabric. This suggests that the formation of pinholes will be highly dependent on the strength of the surface forces between particles that produce flocculation of dusts. The seepage of a dust through a filter is probably also closely related to the strength of the surface forces.

Surface pores can be greatly reduced in size by coating what will become the dusty side of the filter fabric with a thin microporous membrane that is supported by the underlying fabric. That has the effect of decreasing the effective penetration, both by eliminating cake pinholes and by preventing the seepage of dust that is dragged through the fabric by successive cleanings. A variety of different membrane-forming polymers can be used in compatible service. The most versatile and effective surface filtration membranes are microfibrous Teflon, as already described by Brinckman and Maresca [*ASME Med. Waste Symp.* (1992)] in the section on dry scrubbing.

Design of Filter Collector The most common design objectives are collection efficiency (emissions), operating cost (pressure drop, maintenance, life of filter elements), and capital cost (installed cost). Listed here is a step-by-step approach for designing filter collectors.

Step 1: Define design objectives, performance expectations, and constraints (location, available space, utilities).

Step 2: Define process conditions (gas flow rate range, pressure, temperature, humidity, gas composition, corrosivity, chemical reactivity) and dust properties (particle-size distribution, abrasiveness, explosivity, stickiness, density, bulk density, shape).

Step 3: Select (or shortlist) suitable fabric media (material of construction).

Step 4: Based on prior industry experience and range of gas flow rate, select the *mode of filtration* (inside-out vs. outside-in collection).

- If inside-out flow is selected, the gas flow pattern will most likely be upflow, and shaker or reverse air cleaning mechanisms will be applicable.
- If outside-in flow is selected, the gas flow pattern can be upflow, downflow, or cross-flow based on particle size and abrasiveness. Downflow and cross-flow patterns are preferable for dusts in the micron range, but abrasive dust can cause higher erosive wear.

Step 5: Select suitable air-to-cloth ratio from Table 17-9 or prior experience. Use factors from Table 17-11.

Step 6: Calculate the cloth area (media surface area) required for filtration. For compartmentalized design, the total filter area must include the offline modules.

Step 7: Working with vendors and based on standard available filter element sizes, select appropriate size and number of elements.

Step 8: Estimate the pressure drop. Typical operating range of pressure drops is 3 to 20 inches of water.

Step 9: Based on recommended maximum can velocity, calculate the dimensions of filter housing. This defines the smallest footprint that is feasible.

Step 10: Finalize the design features, namely

a. Media replacement: Clean side vs. dirty side
b. Gas inlet design: tangential, diffusors
c. Design of dust hopper and airlock
d. Instrumentation and automation for cleaning
e. Tracing/insulation to avoid condensation: In filter operation, it is essential that the gas be kept above its dew point to avoid water-vapor condensation on the bags and the resulting plugging of the bag pores.
f. Explosion venting

Practical Operational Aspects While it is difficult to cover all combinations of filter designs, product characteristics, and failure modes, an attempt has been made to capture the most frequent problems, potential causes, and remedies in the troubleshooting guides (Tables 17-12, 17-13, and 17-14).

TABLE 17-12 Troubleshooting Guide: High Pressure-Drop Problems in Dust Collectors

Symptom	Possible causes	Solutions
Normal pressure drop with conditioned bags, which rapidly increases during a cycle.	High dust loading on filters	Install a precleaner before the dust collector. Use cyclonic inlet or velocity reduction in transition piece.
	Insufficient cloth area	Increase bag length, if possible. Consider pleated bags/cartridges.
Pressure drop with new clean filter media is low, but increases more than tenfold after conditioning with progressive increase in short term. Does not restore even after prolonged off-line cleaning.	Media is blinding due to fines penetration	Choose different filter media with smaller pore size. Consider PTFE membrane filtration. Decrease air-to-cloth ratio or face velocity by choosing longer bags. Moisture or condensation in the gas; operate above dew point. Oil droplet or mist in the air, especially cleaning air in pulse-jet; install filters.
Under steady-state operation, the pressure drop does not appreciably change after a cleaning cycle has been completed.	Instrument error or lines plugged	Clean lines, recalibrate ΔP sensor, verify with pressure gage, check for leaks and condensation in lines.
	Cleaning mechanism is not effective	Check operation of cleaning system for changes, esp. after media changes. Check all mechanical, electrical, and pneumatic subsystems. Increase duration of cleaning for shaker and reverse air; increase pressure for pulse-jet. Consider augmented cleaning, e.g., using sonic horns.
	High reentrainment after cleaning impedes dust removal from filters	Avoid cleaning adjacent filter elements in sequence. Find ways to reduce the can velocity (e.g., filter elements of different lengths). Modify airflow patterns below the filter elements to prevent reentrainment of dust. Cyclonic action or deflectors are useful. For compressible dust, allow dust cake to compact by reducing the cleaning frequency. Confirm reentrainment problem by executing the cleaning cycle off-line. The pressure should recover back to conditioned media.

TABLE 17-13 Troubleshooting Guide: High Emission Problems in Dust Collectors

Symptom	Possible causes	Solutions
Emissions are consistently high	Incorrect choice of filter media	Select media with higher efficiency (e.g., woven bags with tighter weave, thicker felt, finer fabric, membrane filtration)
	Filter media operated at higher than recommended face velocity	Increase filter area by choosing longer bags or converting bags to pleated elements
	Filter elements are damaged—physical integrity	Check filter quality (workmanship); check if the damage is due to shrinkage caused by high temperature—select alternate filter media; reduce intensity of cleaning
	Filter elements are damaged—abrasive wear	Check flow pattern within the collector; remove selected filter in the path of high velocity incoming jet; reduce inlet velocity; install baffles
	Filter elements are damaged—chemical degradation	Select compatible media
	Filter elements are incorrectly installed or have come loose during operation	Check for localized dust accumulation on tubesheet; improve filter design and clamping mechanism; check for high dust level in the hopper; check for damaged cages
	Damaged or leaky gaskets	Replace gaskets
	Missing filter elements	Confirm with inspection
	Leakage at tubesheet due to warping or cracks	Investigate root cause. Mechanical reinforcement and repairs needed
Emmissions high only during cleaning cycle	Excessive cleaning energy	Reduce cleaning intensity depending on the cleaning mechanism
	Excesive cleaning frequency	Reduce cleaning frequency if based on timer; use dP-based approach to trigger cleaning cycle; use pre-cleaner to reduce dust load; increase filter area to reduce dust loading on filter element
	Incorrect selection or specification of filter media	Select thicker media with lower porosity; use membrane filtration
	Filter element is damaged—phyiscal integrity	Confirm with inspection; replace

TABLE 17-14 Troubleshooting Guide: Filter Failure Problems in Dust Collectors

Symptom	Possible causes	Solutions
Premature failure of filter element resulting in high emissions	Excessive cleaning energy	Reduce cleaning intensity depending on the cleaning mechanism
	Excesive cleaning frequency	Reduce cleaning frequency if based on timer; use dP-based approach to trigger cleaning cycle; use pre-cleaner to reduce dust load; increase filter area to reduce dust loading on filter element
	Filter elements are damaged—physical integrity	Check filter quality (workmanship); check if the damage is due to shrinkage caused by high temperature—select alternate filter media; reduce intensity of cleaning
	Filter elements are damaged—abrasive wear	Check flow pattern within the collector; remove selected filter in the path of high velocity incoming jet; reduce inlet velocity; install baffles to protect bags or distribute the air uniformly
	Filter elements are damaged—chemical degradation	Select compatible media
	Exposure to high stresses on the filter elements	Check for localized dust accumulation on tubesheet; improve filter design and clamping mechanism; check for high dust level in the hopper; check for damaged cages
	Damaged or leaky gaskets	Replace gaskets
	Missing filter elements	Confirm with inspection
	Leakage at tubesheet due to warping or cracks	Investigate root cause; mechanical reinforcement and repairs needed
Media blinding resulting in high pressure drop	Condensation on filter media	Operate above dew point; increase reverse air or pulse-jet air temperature
	Moisture or oil on the clean air side	Oil or water in the compressed air–install filters
	Incorrect selection or specification of filter media	Select thicker media with lower porosity; use membrane filtration or other surface treatments
	Incorrect choice of filter media	Select media with higher efficiency (e.g., woven bags with tighter weave, thicker felt, finer fabric, membrane filtration); condition the filter with precoating dust
	Filter media operated at higher than recommended face velocity	Increase filter area by choosing longer bags or converting bags to pleated elements
	Dust leakage to the cleanside can get embedded into the media during cleaning cycles	Visual inspection of tubesheet to confirm; analyze the filter bag using microscopy; fix the leak

The three most common problems encountered in filter operation are related to pressure drop, collection efficiency, and the life of filter elements.

Granular-Bed Filters Granular-bed filters may be classified as "depth" filters, since dust particles deposit in depth within the bed of granules. The granules themselves present targets for the deposition of particles by inertia, diffusion, flow-line interception, gravity, and electrostatic attraction, depending on the dust and filter characteristics and the operating conditions. Other deposition mechanisms are minor at most. Although it is physically possible under some circumstances for a dust layer to form on the inlet face of the filter, the practical limits of gas pressure drop will normally have been reached long before a surface dust layer can be established. Typical granular-bed media includes sand, gravel, activated carbon, glass, and ceramics.

Granular-bed filters are synonymous with loose-surface (LS) filters, gravel-bed filters, moving-bed filters, sand filters, expandable-bed filters, panel-bed filters, and porous-bed filters in the literature. All these filters have the same working principle. They are suitable and competitive in high-temperature (>1000°F) and high-pressure gas streams with abrasive, sticky, and combustible dust; especially when combined with the removal of gaseous pollutants by adsorption. They have been used with mixed success in the cement, power, chemical, and nuclear industries. A competitive evaluation of granular-bed filters with media filtration (fabric or ceramic cartridge filters) and electrostatic precipitators should be based on collection efficiency, filtration capacity, and capital and maintenance costs. While the concept of granular-bed filtration is simple, the ancillary system required to reliably regenerate the media substantially increases the capital and operating costs.

Excellent reviews of theoretical and commercial developments in granular-bed filtration can be found in Tien (*Granular Filtration of Aerosols and Hydrosols*, Butterworth, Boston/London, 1989) and Tardos and Zenz (*Handbook of Powder Science and Technology*, 2d ed., chap. 17, ed. Fayed and Otten, Chapman and Hall, London, 1997).

Granular-bed filters may be divided into three classes:

1. *Fixed-bed, or packed-bed, filters.* These units are not cleaned when they become plugged with deposited dust particles but are broken up for disposal or simply abandoned. If they are constructed from fine granules (e.g., sand particles), they may be designed to give high collection efficiencies on fine dust particles. However, if such a filter is to have a reasonable operating life, it can be used only on a gas containing a low concentration of dust particles.

2. *Cleanable granular-bed filters.* In these devices, provisions are made to separate the collected dust from the granules either continuously or periodically, so that the units can operate continuously on gases containing moderate to high dust concentrations. The necessity for cleaning and recycling the granules generally restricts the practical lower granule size to about 3 to 10 mm. This in turn makes it difficult to attain high collection efficiencies on fine particles with granule beds of reasonable depth and gas pressure drop.

3. *Fluidized-bed filters.* Fluidized beds of granules have received considerable study on theoretical and experimental levels but have not been applied on a practical commercial scale.

Fixed Granular-Bed Filters Fixed-bed filters composed of granules have received considerable theoretical and experimental study [Thomas and Yoder, *AMA Arch. Ind. Health,* **13:** 545 (1956); **13:** 550 (1956); Knettig and Beeckmans, *Can. J. Chem. Eng.* **52:** 703 (1974); Schmidt et al., *J. Air Pollut. Control Assoc.* **28:** 143 (1978); Tardos et al., *J. Air Pollut. Control Assoc.* **28:** 354 (1978); and Gutfinger and Tardos, *Atmos. Environ.* **13:** 853 (1979)]. The theoretical approach is the same as that used in the treatment of deep-bed fibrous filters.

Fibers for filter applications can be produced with diameters smaller than it is practical to obtain with granules. Consequently, most concern with filtration of fine particles has been focused on fibrous-bed rather than granular-bed filters. However, for certain specialized applications, granular beds have shown some superior properties, such as greater dimensional stability. Granular-bed filters of special design (deep-bed sand filters) have been used since 1948 for removing radioactive particles from waste air and gas streams in atomic energy plants (Lapple, "Interim Report—200 Area Stack Contamination," U.S. AEC Rep. HDC-743, Oct. 11, 1948; Juvinall et al., "Sand-Bed Filtration of Aerosols: A Review of Published Information," U.S. AEC Rep. ANL-7683, 1970; and Burchsted et al., *Nuclear Air Cleaning Handbook,* U.S. ERDA 76-21, 1976). The filter characteristics needed included high collection efficiency on fine particles, large dust-holding capacity to give long operating life, and low maintenance requirements. The sand filters are as much as 2.7 m (9 ft) in depth and are constructed in graded layers with about a 2:1 variation in the granule size from one layer to the next. The airflow direction is upward, and the granules decrease in size in the direction of the airflow. The bottom layer is composed of rocks about 5 to 7.5 cm (2 to 3 in) in diameter, and granule sizes in successive layers decrease to 0.3 to 0.6 mm (50 to 30 mesh) in the finest layer. With superficial face velocities of about 1.5 m/min (5 ft/min), gas pressure drops of clean filters have ranged from 1.7 to 2.8 kPa (7 to 11 in water). Collection efficiencies of up to 99.98 percent with a polydisperse dioctyl phthalate aerosol of 0.7-μm mean diameter have been reported (Juvinall et al., "Sand-Bed Filtration of Aerosols: A Review of Published Information," U.S. AEC Rep. ANL-7683, 1970). Operating lives of five years or more have been attained.

Cleanable Granular-Bed Filters The principal objective in the development of cleanable granular-bed filters is to produce a device that can operate at temperatures above the range that can be tolerated with fabric filters. In some of the devices, the granules are circulated continuously through the unit, then are cleaned of the collected dust and returned to the filter bed. In others, the granular bed remains in place but is periodically taken out of service and cleaned by some means, such as backflushing with air. Zenz (*Handbook of Powder Science and Technology,* 2d ed., chap. 17, ed. Fayed and Otten, Chapman and Hall, London, 1997) has provided an insightful chronological review of commercial developments of this technology during the past century and has identified the key challenges for its successful application.

A number of moving-bed granular filters have used cross-flow designs. One form of cross-flow moving-granular-bed filter, produced by the Combustion Power Company (Fig. 17-78), is currently in commercial use in some applications. The granular filter medium consists of one-eighth- to one-quarter-inch (3- to 6-mm) pea gravel. Gas face velocities range from 30 to 46 m/min (100 to 150 ft/min), and reported gas pressure drops are in the range of 0.5 to 3 kPa (2 to 12 in water). The original form of the device [Reese, *TAPPI* **60**(3): 109 (1977)] did not incorporate electrical augmentation. Collection efficiencies for submicron particles were low, and the electrical augmentation was added to correct the deficiency (Parquet, "The Electroscrubber Filter: Applications and Particulate Collection Performance," EPA-600/9-82-005c, 1982, p. 363). The electrostatic grid immersed in the bed of granules is charged to a potential of 20,000 to 30,000 V, producing an electric field between the grid and the inlet and outlet louvers that enclose the bed. No ionizing electrode is used to charge particles in the incoming gas; reliance is placed on the existence of natural charges on the dust particles. Individual dust particles commonly carry either positive or negative charges even though the net charge on the dust as a whole is normally neutral. Depending on their charges, dust particles are attracted or repelled by the electrical field and are therefore caused to deposit on the rocks in the bed.

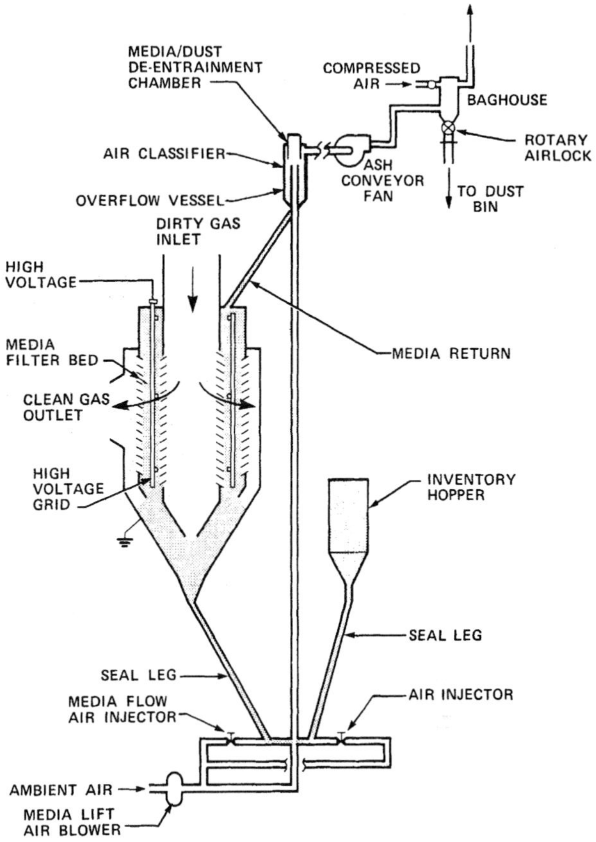

FIG. 17-78 Electrically augmented granular-bed filter. (*Combustion Power Company.*)

Self et al. ("Electrical Augmentation of Granular Bed Filters," EPA-600/9-80-039c, 1980, p. 309) demonstrated in theoretical studies and laboratory experiments that such an augmentation system should yield substantial increases in the collection efficiency for fine particles if the particles carry significant charges. Significant improvements in the performance of the combustion power units with electrical augmentation have been reported by the manufacturer (Parquet, "The Electroscrubber Filter: Applications and Particulate Collection Performance," EPA-600/9-82-005c, 1982, p. 363).

Another type of gravel-bed filter, developed by GFE in Germany, has had limited commercial application in the United States [Schueler, *Rock Prod.* **76**(7): 66 (1973); **77**(11): 39 (1974)]. After precleaning in a cyclone, the gas flows downward through a stationary horizontal filter bed of gravel. When the bed becomes loaded with dust, the gas flow is cut off, and the bed is backflushed with air while being stirred with a double-armed rake that is rotated by a gear motor. The backflush air also flows backward through the cyclone, which then acts as a dropout chamber. Multiple filter units are constructed in parallel so that individual units can be taken off-line for cleaning. The dust dislodged from the bed and carried by the backflush air is flocculated, and part is collected in the cyclone. The backflush air with the remaining suspended dust is cleaned in the other gravel-bed filter units that are operating on-line. Performance tests made on one installation for the U.S. Environmental Protection Agency (EPA-600/7-78-093, 1978) did not give clear results, but indicated that collection efficiencies were low on particles under 2 μm and that some of the dust in the backflush air was redispersed sufficiently to penetrate the operating filter units.

Air Filters The types of equipment previously described are intended primarily for the collection of process dusts, whereas air filters comprise a variety of filtration devices designed for the collection of particulate matter at low concentrations, usually atmospheric dust. The difference in the two categories of equipment is not in the principles of operation but in the adaptations required to deal with the different quantities of dust. Process-dust concentrations may run as high as several hundred grams per cubic meter (or grains per cubic foot), but they usually do not exceed 45 g/m^3 (20 gr/ft^3). Atmospheric-dust concentrations that may be expected in various types of locations are shown in Table 17-15 and are generally below 12 mg/m^3 (5 gr/1000 ft^3).

TABLE 17-15 Average Atmospheric-Dust Concentrations*

1 gr/1000 ft^3 = 2.3 mg/m^3

Location	Dust concentration, gr/1000 ft^3
Rural and suburban districts	0.02–0.2
Metropolitan districts	0.04–0.4
Industrial districts	0.1–2.0
Ordinary factories or workrooms	0.2–4.0
Excessive dusty factories or mines	4.0–400

**Heating Ventilating Air Conditioning Guide*, American Society of Heating, Refrigerating and Air-Conditioning Engineers, New York, 1960, p. 77.

The harmful consequences to human health of inhaling particulate matter has been well established by many epidemiological studies. The removal of particulate matter (PM) from urban, rural, and industrial sources to achieve acceptable indoor air quality (IAQ) can be achieved by air filtration. Particulate matter can be a complex mixture of airborne dry solids, wet solids, and liquid droplets. The particles less than 10 μm in size (PM_{10}) are considered to be respirable particulate matter. The harmful impact of particulates smaller than 2.5 μm ($PM_{2.5}$) on the respiratory system has been particularly noted. Further, the finer fraction ($PM_{2.5}$) will also contribute to atmospheric haze and visibility.

The World Health Organization (WHO) has proposed the following guidelines for air quality:

PM_{10}: 20 $\mu g/m^3$ – annual mean value; 50 $\mu g/m^3$ for a 24-hour mean value

$PM_{2.5}$: 10 $\mu g/m^3$ – annual mean value; 25 $\mu g/m^3$ for a 24-hour mean value

Various regional ambient air quality standards have been developed and enforced. For instance, guidelines from the U.S. EPA are summarized in Table 17-16.

The most frequent application of air filters is in cleaning atmospheric air for building ventilation, which usually requires only moderately high collection-efficiency levels. However, a variety of industrial operations require air of extreme cleanliness, sometimes for pressurizing enclosures such as clean rooms and sometimes for use in a process itself. The presence of particulate matter can adversely affect certain production processes (e.g., the manufacture and assembly of semiconductors, pharmaceutical manufacturing, photographic and optical equipment manufacturing, painting) and the reliability of machinery (e.g., internal combustion engines, turbines, and rotating equipment). High-efficiency air filters are sometimes used for emission control when particulate contaminants are low in concentration but present special hazards; cleaning of ventilation air and other gas streams exhausted from nuclear plant operations is an example.

The recognition of concentration and size of particulate matter has a direct bearing on the performance expectations for the air filtration technology. The recovery of usable product is rarely expected, and in most cases the filter elements are meant for single use.

Air Filter Types Various types of air filters are commonly used, namely

a. Panel filters
b. Viscous oil-coated panel filters
c. Filter or mesh pads
d. Bag or pocket filters
e. Pleated pockets or panel filters
f. Cartridge filters
g. Roll filters (dry and viscous oil-coated) for automatic renewal
h. High-efficiency filters (EPA—Efficiency Particulate Air filter, HEPA—High Efficiency Particulate Air filter, and ULPA—Ultra Low Penetration Air filter)

Panel filters are constructed in units of convenient size to facilitate installation, maintenance, and cleaning. Each unit consists of a cleanable or replaceable cell or filter pad in a substantial frame that may be bolted to the frames of similar units to form an airtight partition between the source of the dusty air and the destination of the cleaned air.

Panel filters may use either viscous or dry filter media. Viscous filters are so called because the filter medium is coated with a tacky liquid of high viscosity (e.g., mineral oil and adhesives) to retain the dust. The filter pad consists of an assembly of coarse fibers (now usually metal, glass, or polymer). Because the fibers are coarse and the media are highly porous, resistance to airflow is low, and high filtration velocities can be used. Media with decreasing porosity in the direction of flow are used to achieve depth filtration.

Dry filters are usually deeper than viscous filters. The dry filter media have finer fibers and have much smaller pores than the viscous media and need not rely on an oil coating to retain collected dust. Because of their greater resistance to airflow, dry filters must use lower filtration velocities to avoid excessive pressure drops. Hence, dry media must have larger surface areas and are usually pleated or arranged in the form of pockets (Fig. 17-79), generally sheets of cellulose pulp, cotton, felt, or spun glass. A PTFE membrane may be further added to increase collection efficiency. Recent introduction of nanofiber technology with a fiber diameter less than 0.5 μm to fabricate fibrous filter media has resulted in higher collection efficiencies and lower pressure drops compared to conventional media.

Bag or pocket filters are suitable for dust streams with high loading. They are usually fabricated as an array with the same frontal dimensions as the flat panel, but they provide much higher filter area and dust retention capacity. They can be designed for HEPA and ULPA performance duty.

Automatic filters are made with either viscous-coated or dry filter media. However, the cleaning or disposal of the loaded medium is essentially continuous and automatic. In most such devices, the air passes horizontally through a movable filter curtain. As the filter loads with dust, the curtain is continuously or intermittently advanced to expose clean media to the airflow and to clean or dispose of the loaded medium. Movement of the curtain can be provided by a hand crank or a motor drive. Movement of a motor-driven curtain can be actuated automatically by a differential-pressure switch connected across the filter.

Selection of Air Filters The selection of a suitable air filter for a given application is about matching the process conditions and performance requirements with the performance characteristics of the filter.

Process conditions: Airflow rate, temperature, dust loading, particle-size distribution of dust, hazards (combustibility, toxicity), presence of liquid/mist

Performance requirement: Minimum acceptable efficiency (or maximum penetration) for various size fractions (PM_{10}, $PM_{2.5}$ and submicron), measurement of efficiency (mass vs. number of particles)

The quantification of the performance characteristics of air filters has been an evolutionary process. Having a verifiable and a meaningful rating system that is indicative of filter performance in real-world conditions has been the main goal of various standards. There are four key standards (national and international) that are relevant to air filters (see Table 17-17):

a. ANSI/ASHRAE 52.2-2012 (Method of Testing General Ventilation Air Cleaning Devices for Removal Efficiency by Particle Size) proposed by American Society of Heating, Refrigeration and Air-conditioning Engineers (ASHRAE)

b. EN 779:2012 (Particulate air filters for general ventilation—Determination of the filtration performance) proposed by CEN (Comite Europeen des Normalisations) and EUROVENT (European Committee of Air Handling & Refrigerating Equipment Manufacturers)

TABLE 17-16 National Air Quality Standards from U.S. EPA

Final rule/decision	Primary (health based)/ secondary (welfare based)	Indicator	Averaging time	Level	Form
2012 78 FR 3085 Jan 15, 2013	Primary	$PM_{2.5}$	Annual	12.0 $\mu g/m^3$	Annual arithmetic mean, averaged over 3 years
2012 78 FR 3085 Jan 15, 2013	Secondary	$PM_{2.5}$	Annual	15.0 $\mu g/m^3$	Annual arithmetic mean, averaged over 3 years
2012 78 FR 3085 Jan 15, 2013	Primary and secondary	$PM_{2.5}$	24 hour	35 $\mu g/m^3$	98th percentile, averaged over 3 years
2012 78 FR 3085 Jan 15, 2013	Primary and secondary	PM_{10}	24 hour	150 $\mu g/m^3$	Not to be exceeded more than once per year on average over a 3-year period

FIG. 17-79 Typical air filters: (*a*) panel, (*b*) pleated, (*c*) bag or pocket, (*d*) V-filter. (*American Air Filter Co.*)

c. Eurocode 1822 from European Committee for Standardization (see Table 17-18): This standard is applied for very high efficiency filters for ventilation and air-conditioning, such as clean rooms.

- CSN EN 1822-1: 2009—High-efficiency air filters (EPA, HEPA, and ULPA)—Part 1: Classification, performance testing, marking
- CSN EN 1822-2: 2009—High-efficiency air filters (EPA, HEPA, and ULPA)—Part 2: Aerosol production, measuring equipment, particle counting statistics
- CSN EN 1822-3: 2009—High-efficiency air filters (EPA, HEPA, and ULPA)—Part 3: Testing flat sheet filter media
- CSN EN 1822-4: 2009—High-efficiency air filters (EPA, HEPA, and ULPA)—Part 4: Determining leakage of filter elements (scan method)
- CSN EN 1822-5: 2009—High-efficiency air filters (EPA, HEPA, and ULPA)—Part 5: Determining the efficiency of filter elements

d. ISO 16890 (2016) for air filters for general ventilation by international standards

- ISO16890-1, Air filter for general ventilation—Part 1: Technical specifications, requirements and efficiency classification systems based on particulate matter (PM)
- ISO16890-2, Air filter for general ventilation—Part 2: Measurement of fractional efficiency and airflow resistance
- ISO16890-3, Air filter for general ventilation—Part 3: Determination of the gravimetric efficiency and the airflow resistance versus the mass of test dust captured
- ISO16890-4, Air filter for general ventilation—Part 4: Conditioning method to determine the minimum fractional test efficiency

A detailed discussion and comparison of these standards is beyond the scope of this chapter. These standards are interrelated and have followed similar evolutionary paths by addressing the deficiencies and limitations in previous versions. The reader is advised to review all the standards along with the recent prior versions for completeness.

Air-Filtration Theory Current high-efficiency air- and gas-filtration methods and equipment have resulted largely from the development of filtration theory since about 1930 and particularly since the 1940s. Much of the theoretical advance was originally encouraged by the requirements of the military and atomic energy programs. The fibrous filter has served both as a practical device and as a model for theoretical and experimental investigation. Extensive reviews and new treatments of air-filtration theory and experience have been presented by Chen [*Chem. Rev.* **55:** 595 (1955)], Dorman ("Filtration," in *Aerosol Science,* ed. Davies, Academic, New York, 1966), Pich (*Theory of Aerosol Filtration by Fibrous and Membrane Filters,* in in *Aerosol Science,* ed. Davies, Academic, New York, 1966), Davies (*Air Filtration,* Academic, New York, 1973), Kirsch and Stechkina ("The Theory of Aerosol Filtration with Fibrous Filters," in *Fundamentals of Aerosol Science,* ed. Shaw, Wiley, New York, 1978), and Hinds (*Aerosol Technology: Properties, Behavior, and Measurement of Airborne Particles,* Wiley, New York, 1998).

The theoretical treatment of filtration starts with the processes of dust-particle deposition on collecting bodies, as outlined in Fig. 17-42 and Table 17-3. All the mechanisms shown in Table 17-3 may come into play, but inertial deposition, flow-line interception, and diffusional deposition are usually dominant. Electrostatic precipitation may become a major mechanism if the collecting body, the dust particle, or both, are charged. Gravitational settling is a minor influence for particles in the size range of usual interest. Thermal precipitation is nil in the absence of significant temperature gradients. Sieving is a possible mechanism only when the pores in the filter medium are smaller than or approximately equal to the particle size, and they will not be encountered in fibrous filters unless they are loaded sufficiently for a surface dust layer to form.

Filtration theory assumes that a dust particle that touches a collector body adheres to it. This assumption appears to be valid in most cases, but evidence of nonadherence, or particle bouncing, has appeared in some

TABLE 17-17 Comparison of U.S and European Standards

ASHRAE 52.2-2012 Standard

Composite Average Particle Size Efficiency, %

Minimum efficiency reporting value (MERV)	E1 Size range 1 0.3–1.0 μm	E2 Size range 2 1.0–3.0 μm	E3 Size range 3 3.0–10.0 μm	Average arrestance by ASHRAE 52.1, %	Minimum final pressure drop, Pa
1	—	—	E3 < 20	Avg. < 65	75
2	—	—	E3 < 20	70 > Avg. ≥ 65	75
3	—	—	E3 < 20	75 > Avg. ≥ 70	75
4	—	—	E3 < 20	Avg. ≥ 75	75
5	—	—	35 > E3 ≥ 20	—	150
6	—	—	50 > E3 ≥ 35	—	150
7	—	—	70 > E3 ≥ 50	—	150
8	—	E2 ≥ 20	E3 ≥ 70	—	150
9	—	E2 ≥ 35	E3 ≥ 75	—	250
10	—	65 > E2 ≥ 50	E3 ≥ 80	—	250
11	E1 ≥ 20	80 > E2 ≥ 65	E3 ≥ 85	—	250
12	E1 ≥ 35	E2 ≥ 80	E3 ≥ 90	—	250
13	E1 ≥ 50	E2 ≥ 85	E3 ≥ 90	—	350
14	85 > E1 ≥ 75	E2 ≥ 90	E3 ≥ 95	—	350
15	95 > E1 ≥ 85	E2 ≥ 90	E3 ≥ 95	—	350
16	E1 ≥ 95	E2 ≥ 95	E3 ≥ 95	—	350

European EN 779:2012 Standard

Group	Class	Average efficiency of 0.4 μm particles, %	Minimum efficiency of 0.4 μm particles, %	Average arrestance of synthetic dust, %	Final test pressure, Pa
Coarse	G1	—	—	65 > Avg. ≥ 50	250
Coarse	G2	—	—	80 > Avg. ≥ 65	250
Coarse	G3	—	—	90 > Avg. ≥ 80	250
Coarse	G4	—	—	Avg. ≥ 90	250
Medium	M5	60 > E ≥ 40	—		450
Medium	M6	80 > E ≥ 60	—		450
Fine	F7	90 > E ≥ 40	35		450
Fine	F8	95 > E ≥ 90	55		450
Fine	F9	E ≥ 95	70		450

instances. Wright et al. ("High Velocity Air Filters," WADC TR 55-457, ASTIA Doc. AD-142075, 1957) investigated the performance of fibrous filters at filtration velocities of 0.091 to 3.05 m/s (0.3 to 10 ft/s), using 0.3-μm and 1.4-μm supercooled liquid aerosols and a 1.2-μm solid aerosol. The collection efficiencies agreed well with theoretical predictions for the liquid aerosols, and apparently also for the solid aerosol at filtration velocities under

TABLE 17-18 Classification of High-Efficiency Filters per EN 1822:2009

Filter class description	Filter class code	Integral value of collection efficiency,* %	Local value of penetration,* %
EPA: efficient particulate air filter	E10	85	—
	E11	95	—
	E12	99.5	—
HEPA: high efficiency particulate air filter	H13	99.95	0.25
	H14	99.995	0.025
ULPA: ultra low penetration air filter	U15	99.9995	0.0025
	U16	99.99995	0.00025
	U17	99.999995	0.0001

*Measured for most penetrating particle size (MPPS).

0.3 m/s (1 ft/s). But at filtration velocities above 0.3 m/s, some of the solid particles failed to adhere. With a filter composed of 30-μm glass fibers and a filtration velocity of 9.1 m/s (30 ft/s), there were indications that 90 percent of the solid aerosol particles striking a fiber bounced off.

Bouncing may be regarded as a defect in the particle-deposition process. However, particles that have been deposited in filters may subsequently be blown off and reentrained into the airstream (Corn, "Adhesion of Particles," in *Aerosol Science*, ed. Davies, Academic, New York, 1966; and Davies, 1966).

The theories of filtration by a fibrous filter relate only to the initial efficiency of the clean filter in the "static" period of filtration before the deposition of any appreciable quantity of dust particles. The deposition of particles in a filter increases the number of targets available to intercept particles, so collection efficiency increases as the filter loads. At the same time, the filter undergoes clogging and the pressure drop increases. No theory is available for dealing with the "dynamic" period of filtration in which collection efficiency and pressure drop vary with the loading of collected dust. The theoretical treatment of this filtration period is incomparably more complex than that for the static period. Investigators have noted that both the increase in collection efficiency and the increase in pressure drop are exponential functions of the loading of collected dust or are at least roughly so (Davies, 1966). Some empirical relationships have been derived for correlating data in particular instances.

The dust particles collected by a fibrous filter do not deposit in uniform layers on fibers, but tend to deposit preferentially on previously deposited particles (Billings, "Effect of Particle Accumulation in Aerosol Filtration," Ph.D. dissertation, California Institute of Technology, Pasadena, 1966), forming chainlike agglomerates called *dendrites.* The growth of dendritic deposits on fibers has been studied experimentally [Billings, op. cit.; Bhutra and Payatakes, *J. Aerosol Sci.* **10:** 445 (1979)], and Payatakes and coworkers [Payatakes and Tien, *J. Aerosol Sci.***7:** 85 (1976); Payatakes, *Am. Inst. Chem. Eng. J.* **23:** 192 (1977); Payatakes and Gradon, *Chem. Eng. Sci.* **35:** 1083 (1980)] have tried to model the growth of dendrites and its influence on filter efficiency and pressure drop.

Electrical Precipitators When particles suspended in a gas are exposed to gas ions in an electrostatic field, they will become charged and migrate under the action of the field. The functional mechanisms of electrical precipitation may be listed as follows:

1. Gas ionization
2. Particle collection
   a. Production of electrostatic field to cause charging and migration of dust particles
   b. Gas retention to permit particle migration to a collection surface
   c. Prevention of reentrainment of collected particles
   d. Removal of collected particles from the equipment

There are two general classes of electrical precipitators: (1) single-stage, in which ionization and collection are combined; (2) two-stage, in which ionization is achieved in one portion of the equipment, followed by collection in another. Various types in each class differ essentially in the details by which each function is accomplished.

The underlying theory presented in the following paragraphs assumes that the dust concentration is small, since only very incomplete evaluations for conditions of high dust concentration have been made.

Field Strength Whereas the applied potential or voltage is the quantity commonly known, it is the field strength that determines behavior in an electrostatic field. When the current flow is low (i.e., before the onset of spark or corona discharge), these are related by the following equations for two common forms of electrodes:

Parallel plates:

$$\mathscr{E} = E/B_e \tag{17-40}$$

Concentric cylinders (wire-in-cylinder):

$$\mathscr{E} = \frac{E}{r\ln(D_t/D_d)} \tag{17-41}$$

The field strength is uniform between parallel plates, whereas it varies in the space between concentric cylinders, being highest at the surface of the central cylinder. After corona sets in, the current flow will become appreciable. The field strength near the center electrode will be less than given by Eq. (17-41), and that in the major portion of the clearance space will be greater and more uniform [see Eqs. (17-46) and (17-47)].

Potential and Ionization In order to obtain gas ionization, it is necessary to exceed, at least locally, the electrical breakdown strength of the gas. Corona is the name applied to such a local discharge that fails to propagate itself. Sparking is essentially an advanced stage of corona in which complete breakdown of the gas occurs along a given path. Since corona represents a local breakdown, it can occur only in a nonuniform electrical field (Whitehead, *Dielectric Phenomena—Electrical Discharge in Gases,* Van Nostrand, Princeton, N.J., 1927, p. 40). Consequently, for parallel plates, only sparking occurs at a field strength or potential difference given by the empirical expressions

$$\mathscr{E}_c = \mathscr{E}_o = \left[k_p\left(1+\sqrt{\frac{K_o}{k_p B_e}}\right)\right] \tag{17-42}$$

$$E_s = \mathscr{E}_o k_p B_e + K_o \mathscr{E}_o \tag{17-43}$$

For air in the range of k_pB_e from 0.1 to 2, $\mathscr{E}_o = 111.2$ and $K_o = 0.048$. Thornton [*Phil. Mag.* **28**(7): 666 (1939)] gives values for other gases. For concentric cylinders (Loeb, *Fundamental Processes of Electrical Discharge in Gases,* Wiley, New York, 1939; Peek, *Dielectric Phenomena in High-Voltage Engineering,* McGraw-Hill, New York, 1929; and Whitehead, *Dielectric Phenomena—Electrical Discharge in Gases,* Van Nostrand, Princeton, N.J., 1927, p. 40), corona sets in at the central wire when

$$\mathscr{E}_c = \mathscr{E}_o\, k_p\left(1+\sqrt{\frac{K_o}{k_p D_d}}\right) \tag{17-44}$$

$$E_c = \left(\frac{\xi_o k_p D_d}{2}\right)\left(1+\sqrt{\frac{K_o}{k_p D_d}}\right)\ln\left(\frac{D_t}{D_d}\right) \tag{17-45}$$

TABLE 17-19 Sparking Potentials* (Small Wire Concentric in Pipe)

Pipe diameter, in	Sparking potential,† volts: Peak	Sparking potential,† volts: Root mean square
4	59,000	45,000
6	76,000	58,000
9	90,000	69,000
12	100,000	77,000

*Data reported by Anderson in Perry, *Chemical Engineers' Handbook,* 2d ed., p. 1873, McGraw-Hill, New York, 1941.

†For gases at atmospheric pressure, 100°F, containing water vapor, air, CO_2, and mist, and negative-discharge-electrode polarity.

For air, approximate values are $\mathscr{E}_o = 110$, $K_o = 0.18$. Corona, however, will set in only if $(D_t/D_d) > 2.718$. If this ratio is less than 2.718, no corona occurs, and only sparking will result, following the laws given by Eqs. (17-44) and (17-45) (Peek, *Dielectric Phenomena in High-Voltage Engineering,* McGraw-Hill, New York, 1929).

In practice, precipitators are usually operated at the highest voltage practicable without sparking, since this increases both the particle charge and the electrical precipitating field. The sparking potential is generally higher with a negative charge on the discharge electrode and is less erratic in behavior than a positive corona discharge. It is the consensus, however, that ozone formation with a positive discharge is considerably less than with a negative discharge. For these reasons, negative discharge is generally used in industrial precipitators, and a positive discharge is used in air-conditioning applications. Table 17-19 shows some typical values for the sparking potential for the case of small wires in pipes of various sizes. The sparking potential varies approximately directly with the density of the gas but is very sensitive to the character of any material collected on the electrodes. Even small amounts of poorly conducting material on the electrodes may markedly lower the sparking voltage. For positive polarity of the discharge electrode, the sparking voltage will be very much lower. The sparking voltage is greatly affected by the temperature and humidity of the gas, as shown in Fig. 17-80.

Current Flow Corona discharge is accompanied by a relatively small flow of electric current, typically 0.1 to 0.5 mA/m^2 of collecting-electrode area (projected, rather than actual area). Sparking usually involves a considerably larger flow of current, which cannot be tolerated except for occasional periods of a fraction of a second duration, and then only when suitable electrical controls are provided to limit the current. However, when suitable controls are provided, precipitators have been operated continuously with a small amount of sparking to ensure that the voltage is in the correct range to ensure corona. Besides disruptive effects on the electrical equipment and electrodes, sparking will result in low collection efficiency because of reduction in applied voltage, redispersion of collected dust, and current channeling. Although an exact calculation can be made for the current flow

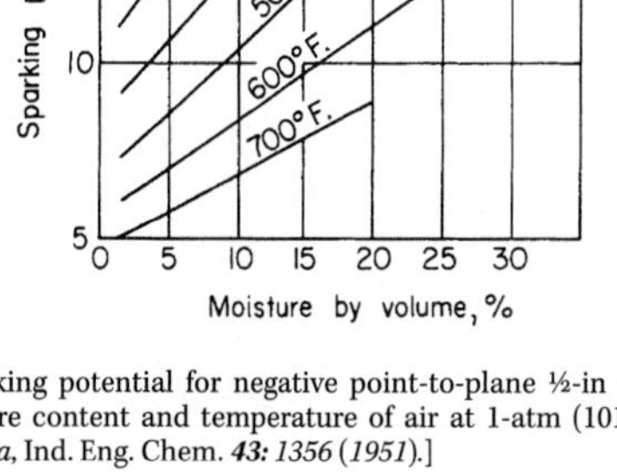

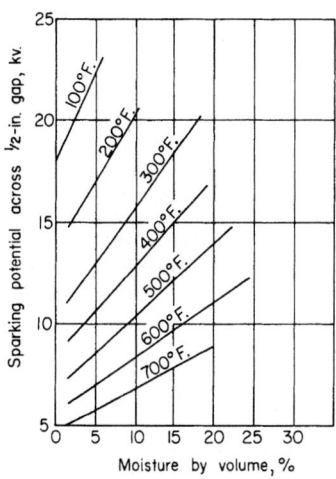

FIG. 17-80 Sparking potential for negative point-to-plane ½-in (13-cm) gap as a function of moisture content and temperature of air at 1-atm (101.3 kPa) pressure. [*Sproull and Nakada,* Ind. Eng. Chem. *43: 1356 (1951).*]

for a direct-current potential applied between concentric cylinders, the following simpler expression, based on the assumption of a constant space charge or ion density, gives a good approximation of corona current [Ladenburg, *Ann. Phys.* **4**(5): 863 (1930)]:

$$I = \frac{8\lambda_i E(E - E_c)}{D_t^2 \ln(D_t/D_d)} \tag{17-46}$$

and the average space charge is given by (Whitehead, *Dielectric Phenomena—Electrical Discharge in Gases,* Van Nostrand, Princeton, N.J., 1927, p. 40)

$$\sigma_{\text{avg}} = \frac{4(E - E_c)}{\pi D_t^2 \varepsilon} \tag{17-47}$$

In the space outside the immediate vicinity of corona discharge, the field strength is sensibly constant, and an average value is given by

$$\mathscr{E} = \sqrt{2I/\lambda_i} \tag{17-48}$$

which applies if the potential difference is above the critical potential required for corona discharge so that a significant current flows.

Ionic mobilities are given by Loeb (*International Critical Tables,* vol. 6, McGraw-Hill, New York, 1929, p. 107). For air at 0°C, 760 mmHg, λ_i = 624 (cm/s)/(statV/cm) for negative ions. Positive ions usually have a slightly lower mobility. Loeb (*Basic Processes of Gaseous Electronics,* University of California Press, Berkeley and Los Angeles, 1955, p. 53) gives a theoretical expression for ionic mobility of gases which is probably good to within ±50 percent:

$$\lambda_i = \frac{100.0}{k_\rho \sqrt{(\delta_g - 1)M}} \tag{17-49}$$

In general, ionic mobilities are inversely proportional to gas density. Ionic velocities in the usual electrostatic precipitator are on the order of 30.5 m/s (100 ft/s).

Electric Wind By virtue of the momentum transfer from gas ions moving in the electrical field to the surrounding gas molecules, a gas circulation, known as the "electric" or "ionic" wind, is set up between the electrodes. For conditions encountered in electrical precipitators, the velocity of this circulation is on the order of 0.6 m/s (2 ft/s). Also, as a result of this momentum transfer, the pressure at the collecting electrode is slightly higher than at the discharge electrode (Whitehead, *Dielectric Phenomena—Electrical Discharge in Gases,* Van Nostrand, Princeton, N.J., 1927, p. 167).

Charging of Particles [Deutsch, *Ann. Phys.* **68**(4): 335 (1922); **9**(5): 249 (1931); **10**(5): 847 (1931); Ladenburg, *Ann. Phys.* **4**(5): 863 (1930); and Mierdel, *Z. Tech. Phys.* **13:** 564 (1932).] Three forces act on a gas ion in the vicinity of a particle: attractive forces due to the field strength and to the ionic image, and repulsive forces due to the Coulomb effect. For spherical particles larger than 1 μm diameter, the ionic image effect is negligible, and charging will continue until the other two forces balance according to the equation

$$N_o = \left(\frac{\zeta \xi D_p^2}{4\varepsilon}\right)\left(\frac{\pi\sigma\varepsilon t \lambda_i}{1 + \pi\sigma\varepsilon\lambda_i t}\right) \tag{17-50}$$

The ultimate charge acquired by the particle is given by

$$N_o = \zeta \mathscr{E} D_p^2 / 4\varepsilon \tag{17-51}$$

and is very nearly attained in a fraction of a second. For particles smaller than 1 μm diameter, the initial charging will occur according to Eq. (17-50). However, owing to the ionic-image effect, the ultimate charge will be considerably greater because of penetration resulting from the kinetic energy of the gas ions. For charging times of the order encountered in electrical precipitation, the ultimate charge acquired by spherical particles smaller than about 1 μm diameter may be approximated (±30 percent) by the empirical expression

$$N_o = 3.4 \times 10^3 D_p T \tag{17-52}$$

Values of N_o for various sizes of particles are listed in Table 17-20 for 70°F, $\zeta = 2$, and $\mathscr{E}$ = 10 statV/cm.

Particle Mobility By equating the electrical force acting on a particle to the resistance due to air friction, as expressed by Stokes' law, the particle velocity or mobility may be expressed by

1. For particles larger than 1 μm diameter:

$$\lambda_p = \left(\frac{u_e}{E_p}\right) = \frac{\zeta D_p E_i K_m}{12\pi\mu} \tag{17-53}$$

TABLE 17-20 Charge and Motion of Spherical Particles in an Electric Field

For $\zeta = 2$, and $\varepsilon = \varepsilon_i = \varepsilon_p$ = 10 statV/cm

Particle diam., μ	Number of elementary electrical charges, N_0	Particle migration velocity,* u_e, ft/sec
0.1	10	0.27
.25	25	.15
.5	50	.12
1.0	105	.11
2.5	655	.26
5.0	2,620	.50
10.0	10,470	.98
25.0	65,500	2.40

NOTE: To convert feet per second to meters per second, multiply by 0.3048.

2. For particles smaller than 1 μm diameter:

$$\lambda_p = \left(\frac{u_e}{\xi_p}\right) = \frac{360 K_m \varepsilon T}{\mu} \tag{17-54}$$

For single-stage precipitators, $\mathscr{E}_i$ and $\mathscr{E}_p$ may be considered essentially equal. It is apparent from Eq. (17-54) that the mobility in an electric field will be almost the same for all particles smaller than about 1 μm diameter, and hence, in the absence of reentrainment, collection efficiency should be almost independent of particle size in this range. Very small particles will actually have a greater mobility because of the Stokes-Cunningham correction factor. Values of u_e are listed in Table 17-20 for 70°F, $\zeta = 2$, and $\mathscr{E} = \mathscr{E}_i = \mathscr{E}_p$ = 10 statV/cm.

Collection Efficiency Although actual particle mobilities may be considerably greater than would be calculated on the basis given in the preceding paragraph because of the action of the electric wind in single-stage precipitators, the latter acts in a compensating fashion, and the overall effect of the electric wind is probably to provide an equalization of particle concentration between the electrodes similar to the action of normal turbulence [Mierdel, *Z. Tech. Phys.* **13:** 564 (1932)]. On this basis, Deutsch (op. cit.) has derived the following equations for collection efficiency, the form of which had previously been suggested by Anderson on the basis of experimental data:

$$\eta = 1 - e^{-(u_e A_e/q)} = 1 - e^{-K_e u_e} \tag{17-55}$$

For the concentric-cylinder (or wire-in-cylinder) type of precipitator, $K_e = 4L_e/D_t V_e$; for rod-curtain or wire-plate types, $K_e = L_e/B_e V_e$. Strictly speaking, Eq. (17-55) applies only for a given particle size, and the overall efficiency must be obtained by an integration process for a specific dust distribution, as described in the subsection Cyclone Separators. However, over limited ranges of performance conditions, Eq. (17-55) has been found to give a good approximation of overall collection efficiency, with the term for particle migration velocity representing an empirical average value. Such values, calculated from overall collection-efficiency measurements, are given in Table 17-21 for specific installations.

For two-stage precipitators with close collecting-plate spacings (see Fig. 17-91), the gas flow is substantially streamlined, and no electric wind exists. Consequently, with reentrainment neglected, collection efficiency may be expressed as [Penny, *Electr. Eng,* **56:** 159 (1937)]

$$\eta = u_e L_e / V_e B_e \tag{17-56}$$

which holds for values of $\eta \le 1.0$. In practice, however, extraneous factors may cause the actual efficiency to approach a relationship of the type given by Eq. (17-55).

Application The theoretical considerations that have been expounded should be used only for order-of-magnitude estimates, since a number of extraneous factors may enter into actual performance. In actual installations, rectified alternating current is employed. Hence the electric field is not fixed but varies continuously, depending on the waveform of the rectifier, although Schmidt and Anderson [*Electr. Eng.* **57:** 332 (1938)] report that the waveform is not a critical factor. Allowances for high dust concentrations have not been fully studied, although Deutsch (op. cit.) has presented a theoretical approach. In addition, irregularities on the discharge electrode will result in local discharges. Such irregularities can readily result from dust incrustation on the discharge electrodes due to charging of particles with opposite polarity within the thin but appreciable flow or ionization layer surrounding this electrode. Very high dust loadings increase the potential difference required for corona and reduce the current due to the space

TABLE 17-21 Performance Data on Typical Single-Stage Electrical Precipitator Installations*

Type of precipitator	Type of dust	Gas volume, cu ft/min	Average gas velocity, ft/sec	Collecting electrode area, sq ft	Over-all collection efficiency, %	Average particle migration velocity, ft/sec
Rod curtain	Smelter fume	180,000	6	44,400	85	0.13
Tulip type	Gypsum from kiln	25,000	3.5	3,800	99.7	.64
Perforated plate	Fly ash	108,000	6	10,900	91	.40
Rod curtain	Cement	204,000	9.5	26,000	91	.31

*Research-Cottrell, Inc. To convert cubic feet per minute to cubic meters per second, multiply by 0.00047; to convert feet per second to meters per second, multiply by 0.3048; and to convert square feet to square meters, multiply by 0.0929.

charge of the particles. This tends to reduce the average particle charge and reduces collection efficiency. This can be compensated for by increasing the potential difference when high dust loadings are involved.

Several investigators have tried to modify the basic Deutsch equation so that it would more nearly describe precipitator performance. Cooperman ("A New Theory of Precipitator Efficiency," Pap. 69-4, APCA meeting, New York, 1969) introduced correction factors for diffusional forces arising from variations in particle concentration along the precipitator length and also perpendicular to the collecting surface. Robinson [*Atmos. Environ.* **1**(3): 193 (1967)] derived an equation for collection efficiency in which two erosion or reentrainment terms are introduced.

An analysis of precipitator performance based on theoretical considerations was undertaken by the Southern Research Institute for the National Air Pollution Control Administration (Nichols and Oglesby, "Electrostatic Precipitator Systems Analysis," AIChE annual meeting, 1970). A mathematical model was developed for calculating the particle charge, electric field, and collection efficiency based on the Deutsch-Anderson equation. The system diagram is shown in Fig. 17-81. This system-analysis method, using high-speed computers, makes it possible to analyze what takes place in each increment of precipitator length. Collection efficiency versus particle size is computed for each 1 ft (0.3 m) of gas travel, and the inlet particle-size distribution is modified accordingly. Computed overall efficiencies compare well with measured values on three precipitators. The model assumes that field charging is the only charging mechanism. The authors considered the addition of several refinements to the program: the influence of diffusion charging; reentrainment effects due to rapping and erosion; and loss of efficiency due to maldistribution of gas, dust resistivity, and gas-property effects. The modeling technique appeared promising, but much more work was needed before it could be used for design. The same authors prepared a general treatise (Oglesby and Nichols, *A Manual of Electrostatic Precipitator Technology,* parts I and II, Southern Research Institute, Birmingham, Ala., U.S. Government Publications PB196360, 196381, 1970).

High-Pressure, High-Temperature Electrostatic Precipitation In general, increased pressure increases precipitation efficiency, although a somewhat higher potential is required, because it reduces ion mobility and hence increases the potential required for corona and sparking. Increased temperature reduces collection efficiency because ion mobility is increased, lowering critical potentials, and because gas viscosity is increased, reducing migration velocities.

Precipitators have been operated at pressures up to 5.5 MPa (800 psig) and temperatures to 800°C.

The effect of increasing gas density on sparkover voltage has been investigated by Robinson [*J. Appl. Phys.* **40:** 5107 (1969); *Air Pollution Control,* part 1, Wiley-Interscience, New York, 1971, chap. 5]. Figure 17-82 shows the effect of gas density on corona-starting and sparkover voltages for positive and negative corona in a pipe precipitator. The sparkover voltages are experimental and are given by the solid points. Experimental corona-starting voltages are given by the hollow points. The solid lines are corona-starting voltage curves calculated from Eq. (17-57). This is an empirical relationship developed by Robinson.

$$\frac{E_c}{\rho'} = A \frac{B}{\sqrt{D_d \rho'/2}} \tag{17-57}$$

E_c is the corona-starting field, kV/cm. ρ' is the relative gas density, equal to the actual gas density divided by the density of air at 25°C, 1 atm. D_d is the

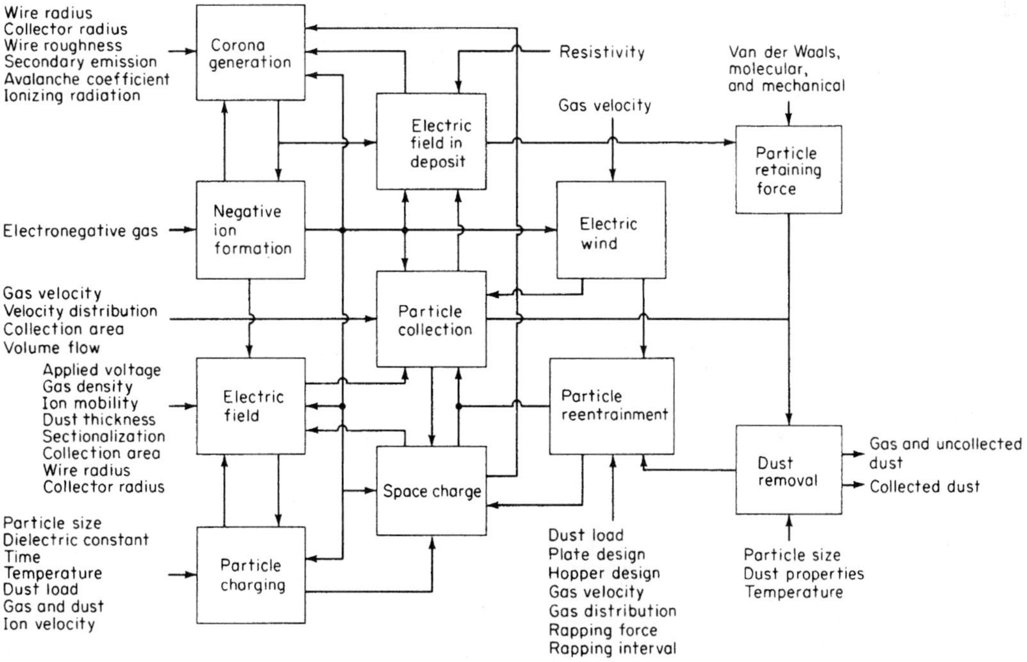

FIG. 17-81 Electrostatic precipitator-system model. (*Nichols and Oglesby. "Electrostatuc or 300°Precipitator Systems Analysis," AIChE Annual Meeting, 1970.*)

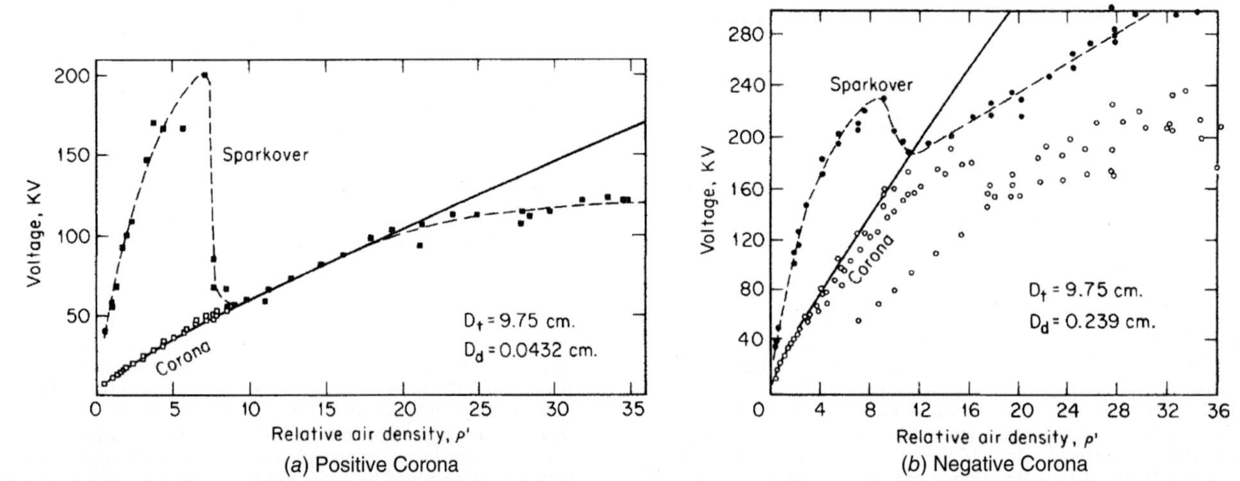

FIG. 17-82 Corona-starting and sparkover voltages for coaxial wire-pipe electrodes in air (25°C). D_t and D_d are the respective pipe and wire diameter. The voltage is unvarying direct current. (*Robinson,* Air Pollution Control, *part 1, Wiley-Interscience, New York, 1971, chap. 5.*)

diameter of the ionizing wire, cm. A and B are constants that are characteristics of the gas. In dry air, $A = 32.2$ kV/cm and $B = 8.46$ kV/cm$^{1/2}$. Agreement between experimental and calculated starting voltages is good for the case of positive corona, but in the case of negative corona the calculated line serves as an upper limit for the data. This lower-than-expected starting-voltage characteristic of negative corona is confirmed by Hall et al. [*Oil Gas J.* **66:** 109 (1968)] in a report of an electrostatic precipitator that removes lubricating-oil mist from natural gas at 5.5 MPa (800 psig) and 38°C (100°F). The use of electrostatic precipitators at elevated pressure is expected to increase because the method requires very low pressure drop [approximately 69 Pa (0.1 lbf/in^2)]. This results from the fact that the electric separation forces are applied directly to the particles themselves rather than to the entire mass of the gas, as in inertial separators. The use of electrostatic precipitators at temperatures up to 400°C is well developed for the powerhouse fly-ash application, but in the range of 600°C to 800°C they are still in the experimental phase. The U.S. Bureau of Mines has tested a pilot-scale tubular precipitator for fly ash. See Shale [*Air Pollut. Control Assoc. J.* **17:** 159 (1967)] and Shale and Fasching (*Operating Characteristics of a High-Temperature Electrostatic Precipitator,* U.S. Bur. Mines Rep. 7276, 1969). It operated over a temperature range of 27°C to 816°C (80°F to 1500°F) and a pressure range of 552 kPa (35 to 80 psig). Initial collection efficiencies ranged from 90 to 98 percent at 793°C (1460°F), 552 kPa (80 psig), but continuous operation was not achieved because of excessive thermal expansion of internal parts.

Resistivity Problems Optimum performance of electrostatic precipitators is achieved when the electrical resistivity of the collected dust is sufficiently high to result in electrostatic pinning of the particles to the collecting surface, but not so high that dielectric breakdown of the dust layer occurs as the corona current passes through it. The optimum resistivity range is generally considered to be from 10^8 to 10^{10} Ω·cm, measured at operating conditions. As the dust builds up on the collecting electrode, it impedes the flow of current, so a voltage drop is developed across the dust layer:

$$E_d = j\rho_d L_d \tag{17-58}$$

If E_d/L_d exceeds the dielectric strength of the dust layer, sparks occur in the deposit and form back-corona craters. Ions of both polarities are formed. Positive ions formed in the craters are attracted to the negatively charged particles in the gas stream, whose charge level is reduced so that collection efficiency decreases. Some of the positive ions neutralize part of the negative-space-charge cloud normally present near the wire, thereby increasing total current. Collection efficiency under these conditions will not correlate with total power input (Owens, E. I. du Pont de Nemours & Co. internal communication, 1971). Under normal conditions, collection efficiency is an exponential function of corona power (White, *Industrial Electrostatic Precipitation,* Addison-Wesley, Reading, Mass., 1963). With typical ion density in the range of 10^9/cm^3, overall voltage gradient would be about 4000 V/cm, and current about 1 μA/cm^2. Dielectric breakdown of the dust layer (at about 10,000 V/cm) would therefore be expected for dusts with resistivities above 10^{10} Ω·cm.

Problems due to high resistivity are of great concern in fly-ash precipitation because air-pollution regulations require that coals have low (<1 percent) sulfur content. Figure 17-83 shows that the resistivity of low-sulfur coal ash exceeds the threshold of 10^{10} Ω·cm at common operating temperatures. This has resulted in the installation of a number of precipitators that have failed to meet guaranteed performance. This has occurred to an alarming extent in the United States, but it has also been encountered in Australia, where the sulfur content is typically 0.3 to 0.6 percent. Maartmann (Pap. EN-34F, 2d International Clean Air Congress, Washington, 1970) reports the installation of a number of precipitators that performed below guarantees, so the Electricity Commission of New South Wales decided that each manufacturer wishing to bid on a new station must first make pilot tests to prove performance on the actual coal to be burned in that station. Problems of back corona and excessive sparking with low-sulfur coal usually require that the operating voltage be reduced. This reduces the migration velocity and

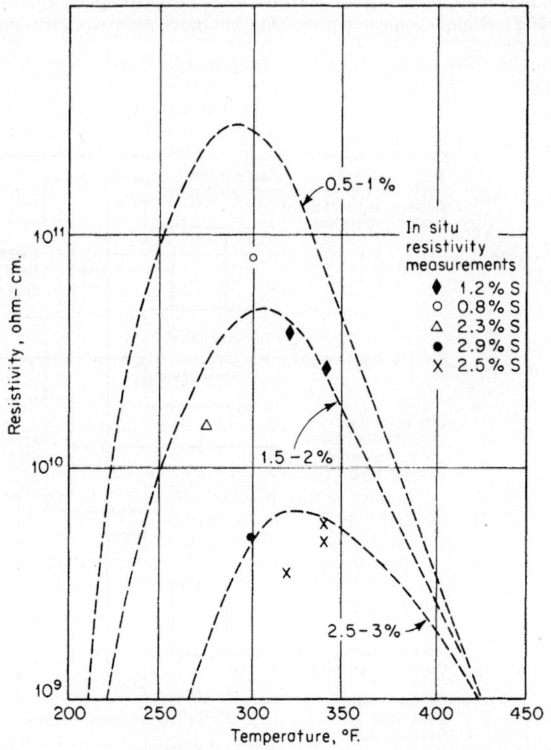

FIG. 17-83 Trends in resistivity of fly ash with variations in the flue-gas temperature and coal sulfur content. °C = (°F − 32) × 5/9. (*Oglesby and Nichols,* A Manual of Electrostatic Precipitator Technology, *part II, Southern Research Institute, Birmingham, Ala., 1979.*)

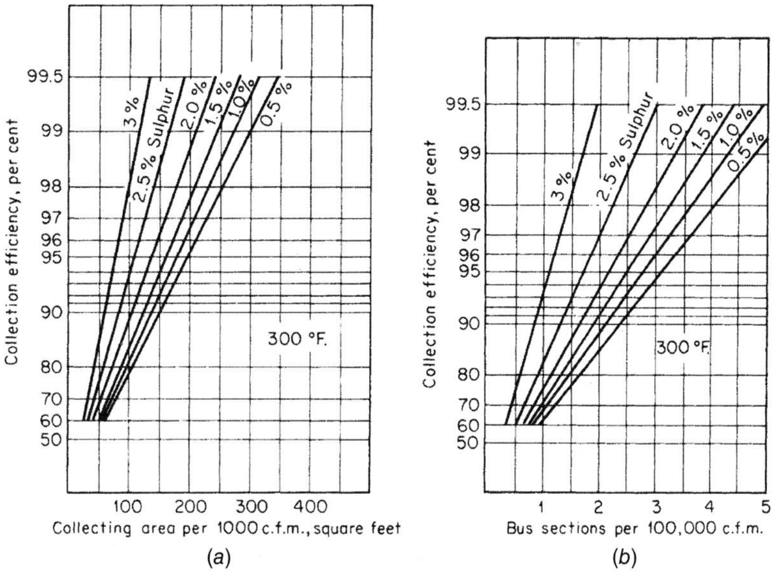

FIG. 17-84 Design curves for electrostatic precipitators for fly ash. Collection efficiency for various levels of percent sulfur in coal versus (*a*) specific collecting surface, and (*b*) bus sections per 100,000 ft^3/min (4.7 m^3/s). °C = (°F − 32) × 5/9. (*Ramsdell,* Design Criteria for Precipitators for Modern Central Station Power Plants. *American Power Conference, Chicago, Ill., 1968.*)

leads to larger precipitators. Ramsdell (*Design Criteria for Precipitators for Modern Central Station Power Plants,* American Power Conference, Chicago, 1968) developed the curves in Fig. 17-84. They show the results of extensive field tests by the Consolidated Edison Co. In another paper ("Anti-Pollution Program of Consolidated Edison Co. of New York," ASCE, May 13–17, 1968), Ramsdell traces the remarkable growth in the size of precipitators required for high efficiency on low-sulfur coals. The culmination of this work was the precipitator at boiler 30 at Ravenswood Station, New York. Resistivity problems were avoided by operating at high temperature [343°C (650°F)]. The mechanical (cyclone) collector was installed after the precipitator to clean up puffs due to rapping.

Maartmann (Pap. EN-34F, 2d International Clean Air Congress, Washington, 1970) agrees that sulfur content is important but feels that it should not be the sole criterion for the determination of collecting surface. He points to specific collecting-surface requirements as high as 500 $ft^2/(1000\ ft^3 \cdot min)$ for 95 percent collection efficiency with high-resistivity Australian ash.

Schmidt and Anderson [*Electr. Eng.* **57:** 332 (1938)] and Anderson [*Physics,* **3:** 23 (July 1932)] claim that resistivity of the collected dust may be a controlling factor that is very sensitive to moisture. They state that an increase in relative humidity of 5 percent may double the precipitation rate because of its effect on the conductivity of the collected dust layer.

Conditioning agents have been added to the flue gas to alter dust resistivity. Steam, sodium chloride, sulfur trioxide, and ammonia have all been successfully used. Research by Chittum and others [Schmidt, *Ind. Eng. Chem.* **41:** 2428 (1949)] led to a theory of conditioning by alteration of the moisture-adsorption properties of dust surfaces. Chittum proposed that an intermediate chemical-adsorption film, which was strongly bound to the particle and which in turn strongly adsorbed water, would be an effective conditioner. This explains how acid conditioners, such as SO_3, help resistivity problems associated with basic dusts, such as many types of fly ash, whereas ammonia is a good additive for acidic dusts such as alumina. Moisture alone can be used as a conditioning agent. This is shown in Fig. 17-80. Moisture is beneficial in two ways: It reduces the electrical resistivity of most dusts (an exception is powdered sulfur, which apparently does not absorb water), and it increases the voltage, which may safely be employed without sparking, as shown in Fig. 17-80.

Low resistivity can sometimes be a problem. If the resistivity is below $10^4\ \Omega \cdot cm$, the collected particles are so conductive that their charges leak to ground faster than they are replenished by the corona. The particles are no longer electrostatically pinned to the plate, and they may then be swept away and reentrained in the exit gas. The particles may even pick up positive charges from the collecting plate and then be repelled. Low-resistivity problems are common with dusts of high carbon content and may also occur in fly-ash precipitators that handle the ash from high-sulfur coal and operate at low gas temperatures. Low resistivity in this case results from excessive condensation of electrically conductive sulfuric acid.

Single-Stage Precipitators The single-stage type of unit, commonly known as a Cottrell precipitator, is most generally used for dust or mist collection from industrial-process gases. The corona discharge is maintained throughout the precipitator and, besides providing initial ionization, serves to prevent redispersion of precipitated dust and recharges neutralized or discharged particle ions. Cottrell precipitators may be divided into two main classes, the so-called plate type (Fig. 17-85), in which the collecting electrodes consist of parallel plates, screens, or rows of rods, chains, or wires; and the pipe type (Fig. 17-86), in which the collecting electrodes consist of a nest of parallel pipes that may be square, round, or of any other shape. The discharge or precipitating electrodes in each case are wires or rods, either round or edged, which are placed midway between the collecting electrodes or in the center of the pipes and may be either parallel or perpendicular to the gas flow in the case of plate precipitators. When the collecting electrodes are screens or rows of rods or wires, the gases are usually passed parallel to the plane of each, but they may also be passed through it. In pipe precipitators, the gas flow is generally vertical up through the pipe, although downflow is not unusual. The pipe-type precipitator is usually used for the removal of liquid particles and volatilized fumes [Cree, *Am. Gas J.* **162:** 27 (March 1945)], and the plate type is used mainly on dusts. In the pipe type, the discharge electrodes are usually suspended from an insulated support and kept taut by a weight at the bottom. Cree [*Am. Gas J.* **162:** 27 (March 1945)] discusses the application of electrical precipitators to tar removal in the gas industry.

Rapping Except when liquid dispersoids are being collected or, in the case of film precipitators, when a liquid is circulated over the collecting-electrode surface (Fig. 17-87), thus continuously removing the precipitated

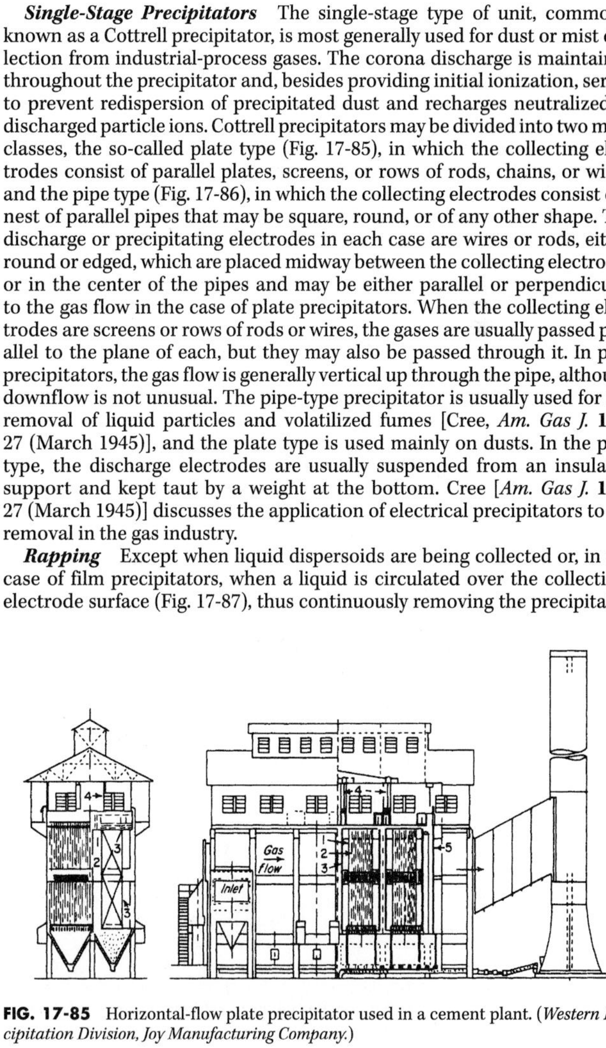

FIG. 17-85 Horizontal-flow plate precipitator used in a cement plant. (*Western Precipitation Division, Joy Manufacturing Company.*)

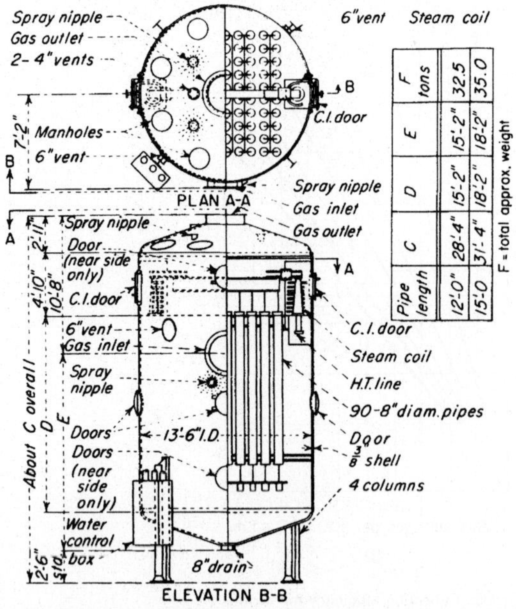

FIG. 17-86 Blast-furnace pipe precipitator. (*Research-Cottrell, Inc.*)

material, the collected dust is dislodged from the electrodes either periodically or continuously by mechanical rapping or scraping, which may be performed automatically or manually. Automatic rapping with either impact-type or vibrator-type rappers is common practice. White (*Industrial Electrostatic Precipitation*, Addison-Wesley, Reading, Mass., 1963) recommends fairly continuous rapping with magnetic-impulse rappers. Rapping with excessive force leads to dust reentrainment and possible mechanical failure of the plates, while insufficient rapping leads to excessive dust buildup with poor electrical operation and reduced collection efficiency. Intermittent rapping at intervals of an hour or more causes heavy puffs of reentrained dust. Sproull [*Air Pollut. Control Assoc. J.* **15:** 50 (1965)] reports the importance of electrode acceleration and shows that it varies with the type of dust, whether the electrode is rapped perpendicularly (normally) to the plate or parallel to it. Figure 17-88 shows the accelerations required for

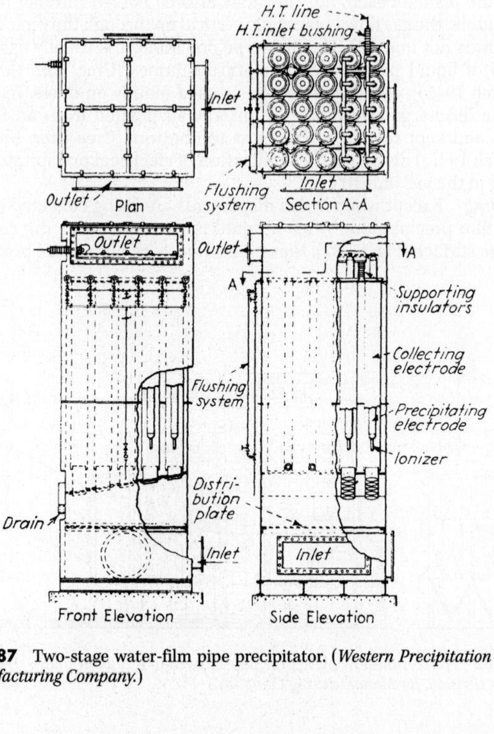

FIG. 17-87 Two-stage water-film pipe precipitator. (*Western Precipitation Division, Joy Manufacturing Company.*)

rapping normally to the plate. Difficult dusts may require as much as 100 G acceleration for 90 percent removal, and even higher accelerations are required when the vibrating force is applied in the plane of the plate.

Perforated-plate or rod-curtain precipitators are often rapped without shutting off the gas flow and with the electrodes energized. This procedure, however, results in a tendency for reentrainment of collected dust. Sectional or composite-plate collecting electrodes (sometimes known as hollow, pocket, or tulip electrodes) are used to minimize this tendency in the continuous removal of the precipitated material, provided that it is free-flowing. These are generally designed for vertical gas flow and comprise a collecting electrode containing a dead air space and provided with horizontal protruding slots that guide the dust into this space (see Fig. 17-89), although some types use horizontal flow.

The choice of size, shape, and type of electrode is based on economic considerations and is usually determined by the characteristics of the gas and suspended matter and by mechanical considerations such as flue arrangement, the available space, and previous experience with the electrodes on similar problems. The spacing between collecting electrodes in plate-type precipitators and the pipe diameter in pipe-type precipitators usually ranges from 15 to 38 cm (6 to 15 in). The smaller the spacing, the lower the necessary voltage and overall equipment size, but the greater the difficulties involved in maintaining proper alignment and resulting from disturbances due to collected material. Large spacings are usually associated with high dust concentration in order to minimize sparkover due to dust buildup. For very high dust concentrations, such as those encountered in fluid-catalyst plants, it is advantageous to use greater spacings in the first half of the precipitator than in the second half. Precipitators, especially of the plate type, are often built with groups of collecting electrodes in series in a common housing. Collecting electrodes are generally on the order of 0.9 to 1.8 m (3 to 6 ft) wide and 3 to 5.5 m (10 to 18 ft) high in plate-type precipitators and 1.8 to 4.6 m (6 to 15 ft) high in pipe types. It is essential for good collection efficiency that the gas be evenly distributed across the various electrode elements. Although this can be achieved by proper gas-inlet transitions and guide vanes, perforated plates or screens located on the upstream side of the electrodes are generally used for distribution. Perforated plates or screens located on the downstream side may be used in special cases.

Electrical precipitators are generally designed for collection efficiency in the range of 90 to 99.9 percent. It is essential, however, that the units be properly maintained in order to achieve the required collection efficiency. Electric power consumption is generally 0.2 to 0.6 kW/(1000 $ft^3 \cdot min$) of gas handled, and the pressure drop across the precipitator unit is usually less than 124 Pa (0.5 in water), ranging from 62 to 248 Pa (¼ to 1 in) and representing primarily distributor and entrance-exit losses. Applied potentials range from 30,000 to 100,000 V. Gas velocities and retention times are generally in the range of 0.9 to 3 m/s (3 to 10 ft/s) and 1 to 15 s, respectively. Velocities are kept low in conventional precipitators to avoid reentrainment of dust. There are, however, precipitator installations on carbon black in which the precipitator acts to flocculate the dust so that it may be subsequently collected in multiple small-diameter cyclone collectors. By not trying to collect the particles in the precipitator, higher velocities may be used with a correspondingly lower investment cost.

Power Supply Electrical precipitators are generally energized by rectified alternating current of commercial frequency. The voltage is stepped up to the required value by means of a transformer and then rectified. The rectifying equipment has undergone an evolution that began with the synchronous mechanical rectifier in 1904 and was followed by mercury-vapor rectifiers in the 1920s; the first solid-state selenium rectifiers were introduced about 1939. Silicon rectifiers are the latest and most widely used type, since they provide high efficiency and reliability. Automatic controls commonly are tied to voltage, current, spark rate, or some combination of these parameters. Modern precipitators use control circuits similar to those on Fig. 17-90. A high-voltage silicon rectifier is used together with a saturable reactor and means for limiting current and controlling voltage and/or spark rate. One popular method adjusts the voltage to give a specified sparking frequency (typically 50 to 150 sparks per minute per bus section). Half-wave rectification is sometimes used because of its lower equipment requirements and power consumption. It also has the advantage of longer decay periods for sparks to extinguish between current pulses.

Electrode insulators must also be designed for a particular service. The properties of the dust or mist and gas determine their design as well as the physical details of the installation. Conducting mists require special allowances such as oil seals, energized shielding cups, or air bleeds. With saturated gas, steam coils are often used to prevent condensation on the electrodes.

Typical applications in the chemical field include detarring of manufactured gas, removal of acid mist and impurities in contact sulfuric acid plants, recovery of phosphoric acid mists, removal of dusts in gases from roasters, sintering machines, calciners, cement and lime kilns, blast furnaces, carbon-black furnaces, regenerators on fluid-catalyst units, chemical-recovery

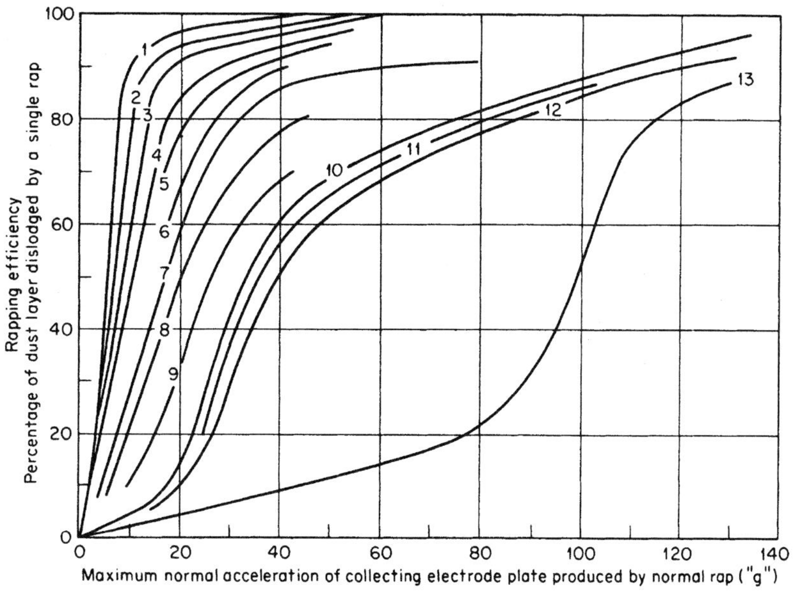

FIG. 17-88 Normal (perpendicular) rapping efficiency for various precipitated dust layers having about 0.03 g dust/cm^2 (0.2 g dust/in^2) as a function of maximum acceleration in multiples of *g*. Curve 1, fly ash, 200°F or 300°F, power off. Curve 2, fly ash, 70°F, power off; also 200°F or 300°F, power on. Curve 3, fly ash, 70°F, power on. Curve 4, cement-kiln feed, 300°F, power off. Curve 5, cement dust, 300°F, power off. Curve 6, same as 5, except power on. Curve 7, cement-kiln feed, 300°F, power on. Curve 8, cement dust, 200°F, power off. Curve 9, same as 8, except power on. Curve 10, cement-kiln feed, 200°F, power off. Curve 11, same as 10, except at 70°F. Curve 12, cement-kiln feed, 200°F, power on. Curve 13, cement-kiln feed, 70°F, power on. °C = (°F − 32) × 5/9. [*Sproull,* Air Pollut. Control Assoc. J. ***15:*** *50 (1965).*]

furnaces in soda and sulfate pulp mills, and gypsum kettles. Figure 17-89 shows a vertical-flow steel-plate-type precipitator similar to a type used for catalyst-dust collection in certain fluid-catalyst plants.

A development of interest to the chemical industry is the tubular precipitator of reinforced-plastic construction (Wanner, *Gas Cleaning Plant after TiO_2 Rotary Kilns,* technical bulletin, Lurgi Corp., Frankfurt, Germany, 1971). Tubes made of polyvinyl chloride plastic are reinforced on the outside with polyester-fiber glass. The use of modern economical materials of

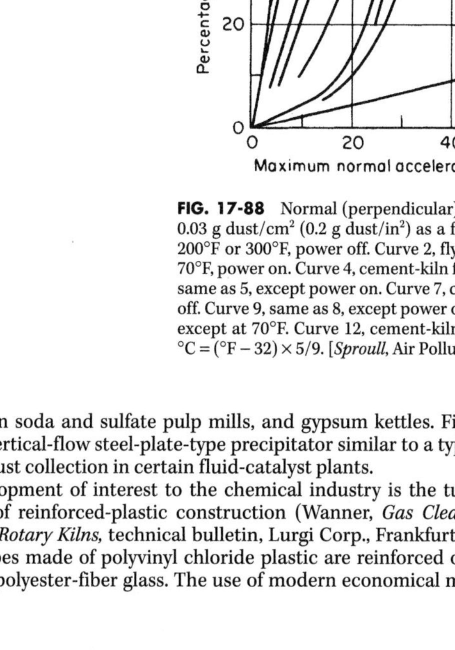

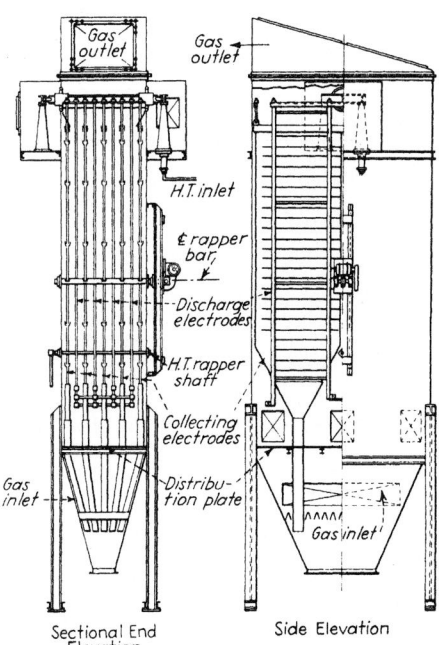

FIG. 17-89 Vertical-flow heavy-duty plate precipitator. (*Western Precipitation Division, Joy Manufacturing Company.*)

construction to replace high-maintenance materials such as lead has been long awaited for corrosive applications.

Electrical precipitators are probably the most versatile of all types of dust collectors. Very high collection efficiencies can be obtained regardless of the fineness of the dust, provided that the precipitators are given proper maintenance. The chief disadvantages are the high initial cost and, in some cases, high maintenance costs. Furthermore, caution must be exercised with dusts that are combustible in the carrier gas.

Two-Stage Precipitators In two-stage precipitators, corona discharge takes place in the first stage between two electrodes having a nonuniform field (see Fig. 17-91). This is generally obtained by a fine-wire discharge electrode and a large-diameter receiving electrode. In this stage the potential difference must be above that required for corona discharge. The second stage involves a relatively uniform electrostatic field in which charged particles are caused to migrate to a collecting surface. This stage usually consists of either alternately charged parallel plates or concentric cylinders with relatively close clearances compared with their diameters. The only voltage requirement in this stage is that no sparking occur, though higher voltages will result in increased collection efficiency. Since collection occurs in the absence of corona discharge, there is no way of recharging reentrained and discharged particles. Consequently, some means must be provided for avoiding reentrainment of particles from the collecting surface. It is also essential that there be sufficient time and mixing between the first and second stages to the secure distribution of gas ions across the gas stream and proper charging of the dust particles.

A unit is available in which electrostatic precipitation is combined with a dry-air filter of the type shown in Fig. 17-79. In another unit an electrostatic field is superimposed on an automatic filter. In this case, the ionizer wires are located on the leading face of the unit, and the collecting electrodes consist of alternate stationary and rotating parallel plates. Cleaning in this case is automatic and continuous.

Although intended primarily for air-conditioning applications, these units have been successfully applied to the collection of relatively nonconducting mists such as oil. However, other process applications have been limited largely to experimental installations. The large cost advantage of these units over the Cottrell precipitator lies in the smaller equipment size made possible by the close plate spacing, in the lower power consumption due to the two-stage operation, and primarily in the mass production of standardized units. In process applications, the close plate spacing is objectionable because of the relatively high dust concentrations involved. Special material or weight requirements for the structural members may eliminate the mass-production advantage except for individual wide applications.

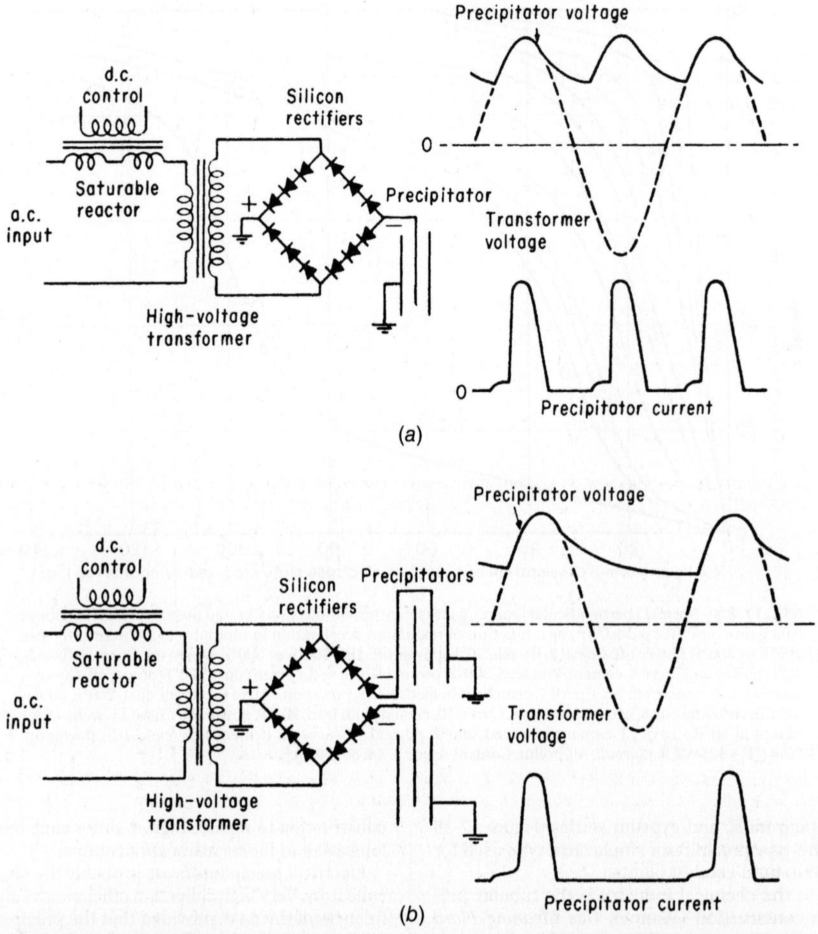

FIG. 17-90 Schematic circuits for silicon rectifier sets with saturable reactor control. (*a*) Full-wave silicon rectifier. (*b*) Half-wave silicon rectifier. (*White,* Industrial Electrostatic Precipitation. *Addison-Wesley. Reading, Mass., 1963.*)

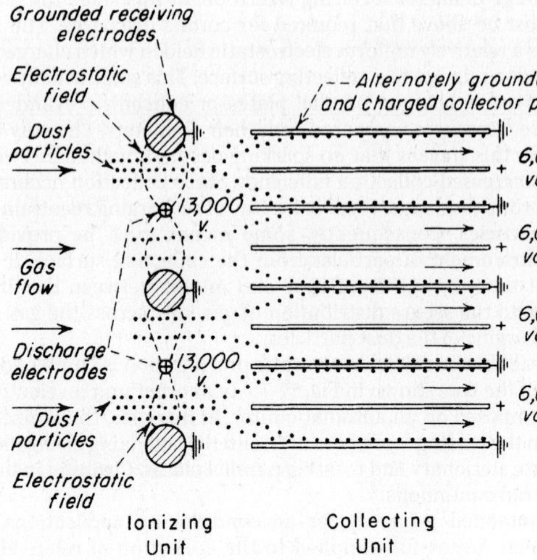

FIG. 17-91 Two-stage electrical precipitation principle.

Alternating-Current Precipitators High-voltage alternating current may be employed for electrical precipitation. Corona discharge will result in a net rectification, provided that no spark gaps are used in series with the precipitator. However, the equipment capacity for a given efficiency is considerably lower than for direct current. In addition, difficulties due to induced high-frequency currents may be encountered. The simplicity of an ac system, on the other hand, has permitted very satisfactory adaptation for laboratory and sampling purposes [Drinker, Thomson, and Fitchet, *J. Ind. Hyg.* **5:** 162 (September 1923)].

Some promising work with alternating current has been undertaken at the University of Karlsruhe. Lau [*Staub,* English ed. **29:** 10 (1969)] and coworkers found that ac precipitators operated at 50 Hz were more effective than dc precipitators for dusts with resistivities higher than 1011 Ω·cm. An insulating screen covering the collecting electrode permitted higher-voltage operation without sparkover.

Section 18

Liquid-Solid Operations and Equipment

Wayne J. Genck, Ph.D. *President, Genck International; consultant on crystallization and precipitation; Member, American Chemical Society, American Institute of Chemical Engineers, Association for Crystallization Technology, International Society of Pharmaceutical Engineers (ISPE) (Section Editor, Crystallization and Selection of Solid-Liquid Separators)*

Brooke Albin, M.S.E. *Chemical Engineer, MATRIC (Mid-Atlantic Technology, Research and Innovation Center), Charleston, WV; Member, American Institute of Chemical Engineers, American Filtration Society (Crystallization from the Melt)*

Frank A. Baczek, B.S. *Sr. Research Advisor, FLSmidth USA, Inc. (Gravity Sedimentation Operations)*

David S. Dickey, Ph.D. *Consultant, MixTech, Inc.; Fellow, American Institute of Chemical Engineers; Member, North American Mixing Forum (NAMF); Member, American Chemical Society; Member, American Society of Mechanical Engineers; Member, Institute of Food Technology (Mixing and Processing of Liquids and Solids & Mixing of Viscous Fluids, Pastes, and Doughs)*

Craig G. Gilbert, B.Sc. *Global Product Manager-Paste, FLSmidth USA, Inc.; Member, Society for Mining, Metallurgy, and Exploration; Mining and Metallurgical Society of America; Registered Professional Engineer (Gravity Sedimentation Operations)*

Taryn Herrera, B.S. *Process Engineer, Manager Separations Laboratory, FLSmidth USA, Inc. (Gravity Sedimentation Operations)*

Tim J. Laros, M.S. *Owner, Filtration Technologies, LLC, Park City, UT; Member, Society for Mining, Metallurgy, and Exploration (Filtration)*

Wenping Li, Ph.D. *R&D Director, Agrilectric Research Company; Member, American Filtration and Separations Society, American Institute of Chemical Engineers (Expression)*

Paul McCurdie, B.S. *Product Manager-Vacuum Filtration, FLSmidth USA, Inc. (Filtration)*

James K. McGillicuddy, B.S. *Product Specialist, Centrifuges, Andritz Separation Inc.; Member, American Institute of Chemical Engineers (Centrifuges)*

Terence P. McNulty, Ph.D. *President, T. P. McNulty and Associates, Inc.; consultants in mineral processing and extractive metallurgy; Member, National Academy of Engineering; Member, American Institute of Mining, Metallurgical, and Petroleum Engineers; Member, Society for Mining, Metallurgy, and Exploration; Member, The Metallurgical Society; Member Mining and Metallurgical Society of America (Leaching)*

Charles G. Moyers, Ph.D. *Senior Chemical Engineering Consultant, MATRIC (Mid-Atlantic Technology, Research and Innovation Center), Charleston, WV; Fellow, American Institute of Chemical Engineers (Crystallization from the Melt)*

Fred Schoenbrunn, B.S. *Director-Sedimentation Products, Member, Society of Metallurgical and Exploration Engineers of the American Institute of Minting, Metallurgical and Petroleum Engineers; Registered Professional Engineer (Gravity Sedimentation Operations)*

Todd W. Wisdom, M.S. *Director-Separations Technology, FLSmidth USA, Inc.; Member, American Institute of Chemical Engineers (Filtration)*

Wu Chen, Ph.D. *Principal Research Scientist, The Dow Chemical Company; Fellow, American Filtration and Separations Society (Expression)*

MIXING AND PROCESSING OF LIQUIDS AND SOLIDS

MIXING OF VISCOUS FLUIDS, PASTES, AND DOUGHS

CRYSTALLIZATION FROM SOLUTION

CRYSTALLIZATION FROM THE MELT

LEACHING

GRAVITY SEDIMENTATION OPERATIONS

FILTRATION

CENTRIFUGES

EXPRESSION

SELECTION OF A SOLIDS-LIQUID SEPARATOR

Nomenclature

Symbol	Definition	SI units	U.S. Customary units
c	Specific heat	J/(kg · k)	Btu/(lb · °F)
C	Off-bottom impeller clearance	m	in
C_o	Orifice coefficient	Dimensionless	Dimensionless
d_o	Orifice diameter	m	in
$d_{p,\max}$	Drop diameter	m	ft
d_t	Pipe diameter	m	in
d_t	Tube diameter	m	ft
D	Impeller diameter	m	ft
D_a	Impeller diameter	m	ft
D_j	Diameter of jacketed vessel	m	ft
D_T	Tank diameter	m	ft
g	Acceleration of gravity	9.8 m/s^2	32.2 ft/s^2
g_c	Dimensional constant	$g_c = 1$ when using SI units	32.2 (ft · lb)/(lbf · s^2)
h	Local individual coefficient of heat transfer, equals $dq/(dA)(\Delta T)$	J/(m^2 · s · K)	Btu/(h · ft^2 · °F)
H	Velocity head	m	ft
k	Thermal conductivity	J/(m · s · K)	(Btu · ft)/(h · ft^2 · °F)
k_La	Overall mass transfer coefficient	s^{-1}	s^{-1}
N	Agitator rotational speed	s^{-1}, (r/s)	s^{-1}, (r/s)
N_{Fr}	Froude number	Dimensionless	Dimensionless
N_{js}	Agitator speed for just suspension	s^{-1}	s^{-1}
N_{Re}	$D_a^2N\rho/\mu$ impeller Reynolds number	Dimensionless	Dimensionless
N_p	Power number = $(g_cP)/\rho N^3D_a^5$	Dimensionless	Dimensionless
N_Q	Impeller pumping coefficient = Q/ND_a^3	Dimensionless	Dimensionless
N_r	Impeller speed	s^{-1}	s^{-1}
N_t	Impeller speed	s^{-1}	s^{-1}
P	Power	(N · m/s)	ft · lb$_f$/s
Q	Impeller flow rate	m^3/s	ft^3/s
T	Tank diameter	m	ft
υ	Average fluid velocity	m/s	ft/s
υ'	Fluid velocity fluctuation	m/s	ft/s
V	Bulk average velocity	m/s	ft/s
Z	Liquid level in tank	m	ft
	Greek symbols		
γ	Rate of shear	s^{-1}	s^{-1}
Δp	Pressure drop across orifice		lbf/ft^2
μ	Viscosity of liquid at tank temperature	Pa · s	lb/(ft · s)
μ	Stirred liquid viscosity	Pa · s	lb/(ft · s)
μ_b	Viscosity of fluid at bulk temperature	Pa · s	lb/(ft · s)
μ_c	Viscosity, continuous phase	Pa · s	lb/(ft · s)
μ_D	Viscosity of dispersed phase	Pa · s	lb/(ft · s)
μ_f	Viscosity of liquid at mean film temperature	Pa · s	lb/(ft · s)
μ_{wt}	Viscosity at wall temperature	Pa · s	lb/(ft · s)
ρ	Stirred liquid density	g/m^3	lb/ft^3
ρ	Density of fluid	kg/m^3	lb/ft^3
ρ_{av}	Density of dispersed phase	kg/m^3	lb/ft^3
ρ_c	Density	kg/m^3	lb/ft^3
σ	Interfacial tension	N/m	lbf/ft
Φ_D	Average volume fraction of discontinuous phase	Dimensionless	Dimensionless

MIXING AND PROCESSING OF LIQUIDS AND SOLIDS

General References: Harnby, N., M. F. Edwards, and A. W. Neinow, eds., *Mixing in the Process Industries,* Butterworth, Stoneham, Mass., 1986; Kresta, S. M., A. W. Etchells III, D. S. Dickey, and V. A. Atiemo-Obeng, eds., *Advances in Industrial Mixing: A Companion to the Handbook of Industrial Mixing,* Wiley, Hoboken, N.J., 2015; Nagata, S., *Mixing: Principles and Applications,* Kodansha Ltd., Tokyo, Wiley, New York, 1975; Oldshue, J. Y., *Fluid Mixing Technology,* McGraw-Hill, New York, 1983; Paul, E. L., V. A. Atiemo-Obeng, and S. M. Kresta, eds., *Handbook of Industrial Mixing, Science and Practice,* Wiley, Hoboken, N.J., 2004; Tatterson, G. B., *Fluid Mixing and Gas Dispersion in Agitated Tanks,* McGraw-Hill, New York, 1991; Uhl, V. W., and J. B. Gray, eds., *Mixing, Theory and Practice,* vol. I, Academic Press, New York, 1966; vol. II, Academic Press, New York, 1987; vol. III, Academic Press, Orlando, Fla., 1986; Ulbrecht, J. J., and G. K. Paterson, eds., *Mixing of Liquids by Mechanical Agitation,* Gordon & Breach Science Publishers, New York, 1985.

Proceedings: *Fluid Mixing,* vol. I, Inst. Chem. Eng. Symp., Ser. No. 64 (Bradford, England), The Institute of Chemical Engineers, Rugby, England, 1984; *Mixing—Theory Related to Practice,* AIChE, Inst. Chem. Eng. Symp. Ser. No. 10 (London), AIChE and The Institute of Chemical Engineers, London, 1965; *Proc. First (1974), Second (1977), Third (1979), Fourth (1982), Fifth (1985), and Sixth (1988) European Conf. on Mixing,* ed. N. G. Coles, BHRA Fluid Eng., Cranfield, England; *Process Mixing, Chemical and Biochemical Applications,* G. B. Tatterson, and R. V. Calabrese, eds., AIChE Symp. Ser. No. 286 (1992).

FLUID MIXING TECHNOLOGY

Fluid mixers are found in almost every process industry. Processes may center around one key mixer essential to the process or numerous mixers each contributing to the total process. The size of mixers can range from small laboratory mixers to large industrial mixers capable of handling many thousands of liters of product and powered by motors exceeding 750 kW (1000 hp). In many cases the power of the drive is secondary to the amount of torque (power divided by speed) transmitted by the mixer to the fluid. The diversity of applications in nearly all categories makes a simple description of mixing and agitation impossible. About the only common characteristic of fluid mixing is the promotion and improvement of uniformity in composition, dispersion, suspension, temperature, and/or other fluid properties.

Fluid mixers perform a key role in process industries, such as chemicals, polymers, pharmaceuticals, mineral processing, corn wet milling, food, pulp and paper, water and wastewater treatment, and many others. The diversity and uniqueness of fluid mixing become evident when process categories are broken down into subgroups. For instance, the mixing requirements and equipment for chemical production might be quite different depending on whether the chemical is a mass quantity of a primary chemical or a small batch of a specialty chemical. If the reaction rate is fast, intense local mixing may affect not only the productivity of the process, but also the selectivity of the products. The control of a process may depend on secondary processes, such as heat removal from an exothermic reaction. All of which may depend in part on the agitation contributed by the fluid mixer.

As further examples of different types of fluid mixing, polymerization may be done in bulk, by suspension, or in an emulsion. Each type of process has different fluid mixing requirements that must be quantified and controlled within an acceptable range. Mineral processing typically involves large quantities of material that must be processed. In some cases the processes involve the suspension of solids for dissolution or chemical reaction, including the dispersion of air for the purpose of floatation. In other mineral processing, a liquid-liquid dispersion created by fluid mixing equipment uses an ion exchange to extract the valuable metal from the ore. Corn wet milling involves several processes linked together as different parts of the corn are cooked, separated, and converted into different products. Each product stream involves multiple mixers. Pulp processing involves several chemical and physical steps to extract the fibers that go into the paper product. The chemical steps involve the addition of chemicals to help break down pulp sources, adjust the pH, and dilute or control the moisture content of the pulp on its way to the paper machine. Other fluid mixers are found beneath the paper machine, where they add water and repulp the paper if it fails to meet acceptable quality standards. Waste water processing is a necessary part of environmental control, often involving extremely large quantities of liquid, with variable composition and flow rate, but of limited intrinsic value. The objective is to get the best possible results from the minimum amount of expended energy. In all of these cases, the fluid mixers are the key processing equipment needed to convert the raw materials into the desired product. Poor mixing is rarely an acceptable alternative.

One of the difficulties in fluid mixing is describing the desired result in a way that can be related to the effects of a mixer. Mixers move fluids, which may be various combinations of liquids, solids, gases, and other liquids. The problem is that the liquid motion is not the end product. The conversion of the liquid motion into a result that promotes the production of the desired product must follow some categorization and quantification to be a practical method for the design or evaluation of mixing equipment.

One method of categorizing mixing separates applications according to the phases of the material present. One category may include just liquids, but even liquids may represent two different categories, depending on whether liquids are miscible or immiscible. Miscible liquids may involve a blending process to create uniformity, while immiscible liquids may involve a dispersion process. The dispersion might need to be separable for mass transfer or stable to form an emulsion. Applications involving both solids and liquids may be suspensions for various applications, including suspension products, dissolving solids, or even crystal formation for separation. Sometimes the mixing problem is creating the solids suspension in the first place. The addition of a powder to a liquid is sometimes quite difficult and may focus on surface motion. Gas dispersion into a liquid may have applications in mass transfer processes or even final products containing dispersed bubbles, such as air dispersions in a viscous or solid oil to make shortening for cooking. The problems only get more complicated when the mixing applications involve combinations of multiple phases, since almost any combination makes the mixing more difficult.

Combinations of miscible liquids typically are the simplest mixing category, especially for low-viscosity fluids. However, even the mixing requirements for miscible liquids will involve some description of the motion. Certainly, no motion at any place in a tank is an insufficient condition for effective mixing. Once all of the liquid is moving, some applications will require intense local turbulence to promote a fast chemical reaction. In other cases any amount of motion able to promote general uniformity may be sufficient. However, even with a single liquid, the effects of viscosity may become a problem for uniform mixing. A rotating impeller is capable of directly moving only the liquid near the impeller. However, that liquid must have sufficient momentum to move surrounding liquid, especially reaching to the sides of the tank or the surface of the liquid. Even combining two miscible liquids can be difficult if the viscosities are different. A simple blending operation may take both more time and intensity if the liquids have different viscosities. However, adding a high-viscosity liquid to a low-viscosity liquid almost always goes faster than adding the low-viscosity liquid to the high-viscosity liquid.

In addition to categorizing mixing applications by the phases of the materials present, processes can be categorized by the predominant fluid-dynamic mechanism. A simple blending application may be controlled by the bulk motion of the liquid. That motion might be characterized by the time required to accomplish blending to some degree of uniformity. Blend time correlations will be discussed later in this section. Rapid mixing in the turbulent region near the impeller can be quantified in terms of the energy dissipation, especially as power per mass. Other dynamic mixing characteristics include dispersion, whether liquid-liquid dispersion or gas-liquid dispersion. Mixer applications involving solid-liquid processes are most likely to be limited by solids suspension requirements. However, fine particles may settle slowly enough to be treated as part of the moving fluid while the mixer is running and effectively change the limitations to fluid motion, when behaving as a viscous or nonnewtonian fluid. The dispersion that might apply to solid-liquid systems is most likely to involve the breakup of agglomerates as opposed to actual particle breakage. The shear effect produced by most mixers is hydraulic shear, as opposed to mechanical shear. Hydraulic shear rarely has sufficient force to break any but the most fragile solids or weak biological cells.

Other ways of categorizing mixing include continuous or batch processes, stirred-tank or in-line mixers, rapid or slow process requirements, chemical reactions or formulations, and many other more complicated or subtle requirements. Continuous processes are expected to keep the mixer contents near a steady-state condition, with ingredient materials constantly entering, a steady conversion or blending operation, and consistent product materials constantly leaving. Batch processes may go through many different conditions, involving different liquid levels as ingredients are added or products are removed. In addition to quantity changes, the physical properties may change. Often processes designed to make products with unique physical properties may experience increased viscosity during the process, as polymer blends or emulsions are formed. Stirred tanks may operate continuously or in batches. In-line mixers always involve continuous flow through them, but they may be either single pass or recirculated flow. Some processes require rapid mixing for chemical reactions, other processes are relatively slow and may require consistent mixing for long periods of time. Chemical reactions often cause physical properties to change as a result of the reaction. Mixing will not directly cause reactions to take place, but mixing will assist the combination of reactants, circulation of products, transport of heat, or promote uniformity. Many processes done by mixing do not involve any chemical reactions, or at least no more than pH adjustment. Such formulation processes involve the combination and blending of multiple ingredients, sometimes accompanied by dispersion or other physical mixing processes.

In all cases, mixing equipment design comes down to a combination of three basic considerations, described by: (1) the quantity of material to be processed, (2) the difficulty of the material being processed, and (3) the required mixing intensity. Each of those three factors are independent process considerations. The more material to be processed, the larger the mixer needs to be, even if the other factors are identical. Some materials are more difficult to mix—for example, a high-viscosity fluid will be more difficult to mix than a low-viscosity liquid. Similarly, rapidly settling particles will be more difficult to suspend or larger quantities of gas will be more difficult to disperse. The third factor involves how much fluid motion or mixing intensity is required for a successful process. Mixing intensity is difficult to quantify or describe on a measurable scale. Most methods for estimating mixing intensity are based on experience with similar processes or small-scale experimentation. All mixing characteristics are developed in some way from empirical studies. The empiricism is simply observation that has been sufficiently documented to be repeatable and useful.

Another problem with mixer design comes because the mixer is often purchased separately from the tank, at least in the United States. The problem is not just one of preference, but also one of practicality. Transporting a large tank is much more difficult and expensive than transporting a large mixer. Consequently, tanks are typically built and shipped regionally, but mixers can be built and shipped nationally and even internationally. Making sure that the mixer fits the tank and the mixer design in that tank provides adequate intensity for the application is part of a communication problem, requiring considerable coordination between the suppliers and users.

The final complexity of mixer design is strictly mechanical. Mechanical mixers are rotating machines with all of the design requirements for adequate strength and reliability. Most mixers have some sort of speed reduction, which must provide efficient power transmission and increased torque. The speed reduction may be accomplished by a gear reducer or belt drive or a combination of both. The output from the drive must be transferred to the impeller located some distance away in the fluid. The shaft between the drive and impeller must have adequate strength to handle torque and bending loads and be capable of operating without any destructive vibration caused by a natural frequency. Impeller blades and their attachment to the shaft must have adequate strength to operate successfully. The materials of construction must be compatible with the process fluids. The mechanical design of a mixer can be as important and sometimes as difficult as the process design.

Two important measures of power transmission that directly affect the process results are power and torque. Power input to the fluid relates to turbulent energy dissipation. Torque input is more closely related to momentum transfer to the fluid. Momentum can be interpreted as the ability of a mixer to create a velocity and move some quantity of fluid. All mixers supply power and momentum to mix fluid. The relative quantities of power and momentum will affect mixing performance, equipment design, and cost.

The available power for mixing is set by the motor size. The motor power needs to be greater than the power required to rotate the impeller(s) in the fluid. The only power applied to mixing is that which is transmitted to the fluid by the impellers. The impeller power may be considerably less than the available motor power for some mixers. For liquid blending, impeller power is a function of impeller type, impeller diameter, rotational speed, fluid density, and viscosity. Reasonably accurate estimates of impeller power can be calculated from known or measured power numbers, operating conditions, and fluid properties. The available power from the motor will have internal motor losses, friction losses through speed reducers, and possible losses through shaft seals, before it is available to the impellers and process. All of these losses must be a small percentage of the motor power for typical process mixing equipment because any power losses will create heat. High power losses are not only inefficient, but the heat may cause mechanical failures.

The amount of torque delivered to the fluid depends on both the power and the speed reduction. With the same power, the greater the speed reduction, the more torque can be transmitted to the fluid. Of course, for the torque to be applied to the fluid at a lower speed, the impeller diameter must be larger for the same amount of power. Higher torque also requires stronger speed reducers, mixer shafts, and impeller blades. Speed reducers can be seen as constant power and increased torque transmission devices.

In simplest terms common drive components should give nearly constant power transmission. Power requirements will be closely related to the operating cost of a mixer. Speed reducers will increase torque and require larger, heavier, and stronger mechanical components. Torque requirements will be closely related to the capital cost of a mixer.

INTRODUCTION TO FLUID MECHANICS

An understanding of fluid dynamics is essential to understanding and using mixing equipment. However, some of the applicable fluid dynamics are different from those of other engineering situations. For instance, the fluid dynamics applied to the design of aircraft and naval ships both involve linear motion through an initially stagnant fluid. The boundary conditions at the wing of an aircraft or hull of a ship can be critical to performance, but the flow past the surfaces is not as complicated as the flow around a mixing impeller. Other fluid dynamics applications, such as river flow or pipe flow, involve forms of constrained, but again essentially linear, flow. Physical models such as wind tunnels, flow channels, and pump loops have been used successfully for many years to model constrained flow processes. The physical models have provided data for the validation of computer models. The successful design and improvement of aircraft and ships have been widely demonstrated using both physical and computer models.

Many of the same modeling techniques have been applied to mixing applications, with increasing success as hardware and software improve, but the limitations are often greater than the successes. Mixer geometry is more complicated, with a rotating impeller and a stationary tank providing boundary conditions for both physical and computer models. The many combinations of geometry and operating conditions are only the beginning of the complexities. In process applications, the detailed requirements of local velocities or turbulence intensities are not always well understood with respect to the corresponding process performance. It is easier to observe solids suspension performance on the basis of off-bottom suspension in a transparent tank than it is to quantify the equivalent local velocities needed at the bottom of the tank. Most mixing requirements are a cumulative average of many individual fluid dynamic effects. A successful chemical reaction carried out in a stirred tank may be a function of bulk fluid motion needed for uniformity or heat transfer, local turbulence intensity needed to bring reactants together, and flow directions needed to move reactants from one location to another.

As with other fluid dynamics applications, dimensionless groups can be developed to generalize empirical results based on the interaction between fluid motion and physical properties. The dimensionless groups used in mixing applications are different from those used in other applications. The dimensionless groups in mixing are based on the same physical principles and often represent a ratio of forces or dimensions, which generalize results, especially for scale-independent behavior.

Impeller Reynolds Number As with other fluid dynamics applications, mixing has a Reynolds number, which can be interpreted as the ratio of inertial to viscous forces on the impeller and is defined as:

$$N_{Re} \equiv \frac{D^2 N \rho}{\mu} \qquad (18\text{-}1)$$

where D is the impeller diameter, N is the rotational speed (as in revolutions/time, not radians/time), ρ is the fluid density, and μ is the dynamic viscosity of the fluid. The impeller Reynolds number differs from a pipe Reynolds number because instead of the pipe diameter, the impeller diameter is the length dimension and instead of flow velocity, the product of impeller diameter times the rotational speed is used as a representation of impeller tip speed, not including a factor of pi. The impeller Reynolds number is a dimensionless group defined as it appears in Eq. (18-1). The magnitudes of the Reynolds number are also considerably different from pipe flow Reynolds numbers. For impeller Reynolds numbers, ranges are typically represented as:

Turbulent	$N_{Re} \geq 20{,}000$
Transition	$20{,}000 > N_{Re} > 10$
Laminar	$N_{Re} \leq 10$

These values may adequately represent the conditions in the impeller region and have been found to effectively correlate other dimensionless groups, such as the impeller power number. However, conditions away from the impeller, such as in the recirculation zone in the top third of the tank, may require a $N_{Re} \geq 300{,}000$ to be fully turbulent [Machado et al., *Chem. Eng. Sci.* **98:** 218–230 (2013).]

Another essential dimensionless group for mixing is the impeller power number, which is effectively a ratio of imposed forces to inertial forces:

$$N_P \equiv \frac{P}{\rho N^3 D^5} \qquad (18\text{-}2)$$

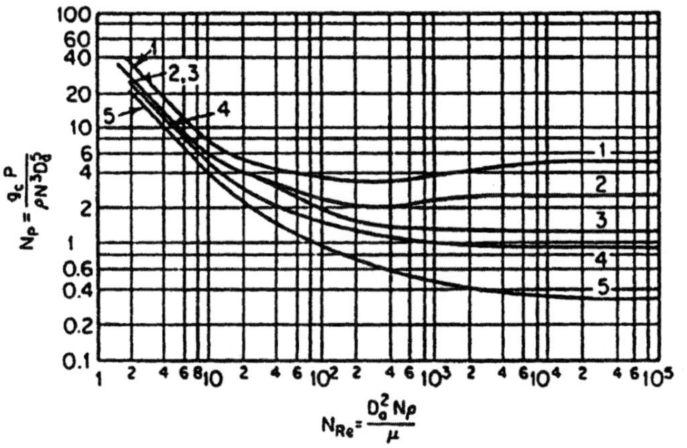

FIG. 18-1 Impeller power number correlations as a function of Reynolds number for different impellers: Curve 1, disk-style turbine, $W/D = 1/5$, like Fig. 18-30, with six blades, in a tank with four baffles each $T/12$ wide; Curve 2, straight-blade, $W/D = 1/8$, similar to Fig. 18-4, but with six blades, in a tank with four baffles each $T/12$ wide; Curve 3, 45° pitched-blade turbine similar to Fig. 18-5, but with six blades, in a tank with four baffles, each $T/12$ wide; Curve 4, marine-style propeller, like Fig. 18-7, with pitch $2.0D$, in a tank with four baffles, each $T/10$ wide, also the same propeller in angle offset position, like Fig. 18.13, in a tank with no baffles; Curve 5, marine-style propeller, like Fig. 18-7, with pitch $1.0D$ or hydrofoil impeller like Fig. 18-6, in tank with four baffles, each $T/10$ wide, also the same propeller or hydrofoil in an angle offset position, like Fig. 18-13, in a tank with no baffles; Where D = impeller diameter, T = tank diameter, g_c = gravitational force constant, N = impeller rotational speed, P = impeller power requirement, W = impeller blade height, μ = dynamic viscosity of stirred fluid, and ρ = density of stirred fluid. Any set of consistent units may be used, but N must be rotations (rather than radians) per unit time. In metric system, g_c is dimensionless and unity. [Curves 1, 2, and 3 from Bates, Fondy, and Corpstein, *Ind. Eng. Chem. Process Des. Dev.* **2:** 310 (1963) by permission of American Chemical Society, Curves 4 and 5 from Rushton, Costich, and Everett, *Chem Eng. Prog.* **46:** 395 & 467 (1950) by permission of American Institute of Chemical Engineers.]

To make the power number dimensionless, the numerator may require a gravitational force constant to convert units of force to mass.

Power numbers are shown for several different impeller types in Fig. 18-1.

The power number is defined relative to inertial forces. For baffled tanks, the impeller power number is a constant under turbulent conditions. The turbulent power number is often used as a descriptor for different types of impellers. Radial-flow impellers typically have higher power numbers than axial-flow impellers. Other impeller features have anticipated effects: wider blades increase the power number, more blades increase the power number, and so on. Power number is not the only characteristic of an impeller, but it is usually the most important characteristic and is often related to other features, such as pumping, blend time, and heat transfer.

Impeller Power Number and Power Draw In the turbulent range, $N_{Re} \geq 20{,}000$, the power number is a constant in a baffled tank, which means that power is proportional to the power number, density, speed cubed, and impeller diameter to the fifth power:

$$P = \rho N^3 D^5 \tag{18-3}$$

Therefore, the impeller power is a function of the impeller type as reflected in the power number. Turbulent power is directly proportional to liquid density, which is similar to a centrifugal pump for turbulent conditions.

The obvious missing effect on power is viscosity. For turbulent conditions, viscosity has no effect on power. However, as viscosity increases and Reynolds number decreases, viscosity gradually becomes a factor as indicated by the increasing power numbers in Fig. 18-1. In the viscous range at low Reynolds numbers, $N_{Re} \leq 10$, the power number becomes inversely proportional to the Reynolds number. At those conditions, impeller power becomes proportional to viscosity, but independent of density, and proportional to rotational speed squared and impeller diameter cubed.

Dimensionless Groups Another dimensionless group of potential interest in mixing is the Froude number, which is a ratio of inertial to gravitational forces and is expressed for impeller mixing as:

$$N_{Fr} \equiv \frac{N^2 D}{g} \tag{18-4}$$

where g is the acceleration of gravity (not the gravitational force constant). The significance of the Froude number to mixing applications should be limited, since the force of gravity is a constant. The most relevant effect of the Froude number is deformation of the liquid surface, whether it is the depth of a central vortex or the height of ripples on the surface. Most other correlations involving the Froude number are more likely fortuitous relationships involving a strong function of rotational speed, rather than a gravitational effect.

Many other dimensionless ratios can be formed from mixing equipment dimensions. Geometric similarity can be a useful method for constructing and evaluating model systems. All of the length variables can be used to form geometric ratios, the most common of which are the impeller-diameter-to-tank-diameter ratio or D/T, the liquid-level-to-tank-diameter ratio or H/T, and the impeller off-bottom-clearance ratio that can be expressed in terms of tank diameter, C/T, impeller diameter, C/D, or liquid level, C/H.

Other dimensionless groups can be formed for process variables, such as volumetric flow rates or fluid velocities. An impeller pumping number can be written in terms of a pumping capacity as:

$$N_Q \equiv \frac{Q}{N\,D^3} \tag{18-5}$$

A dimensionless velocity is often written as:

$$N_v \equiv \frac{v}{N\,D} \tag{18-6}$$

where v can be a fluid velocity measured at any location in the tank. The dimensionless velocity can be considered like a flow pattern for turbulent conditions, where all velocities are proportional to each other. Velocities near the impeller will be high, and velocities near the wall will be lower. For a given impeller type, axial or radial, the corresponding velocities will represent a typical recirculating pattern. Local velocities will be some proportion of the impeller tip speed, $v_{tip} = \pi ND$. Conservation of momentum and fluid dynamics should mean that doubling the tip speed, or rotational speed for a given impeller, should double local velocities elsewhere in a tank. Combinations of geometric similarity and dimensionless groups can be used to do scale-up from small-scale mixing tests. Other dimensionless and dimensional variables will be used to evaluate and design mixing equipment.

MIXING EQUIPMENT

One of the essential tasks in the design of mixing equipment is accurate and effective communication. Therefore, knowing and understanding some commonly used nomenclature and impeller descriptions are an important start for the communication process. The communication must also include an accurate description of the desired process and how mixing is expected to contribute the process. For instance, "good mixing" or "well mixed" are probably not adequate descriptions for defining mixing requirements. What is good mixing for solids suspension may be quite different from good mixing for blending, especially for viscous blending. The operating conditions with similar equipment or the preferred impeller type may be different depending on the actual process requirements.

Nomenclature for a Mixed Tank Some widely used nomenclature for a mixed tank is shown in Fig. 18-2.

The capital letters commonly used for primary mixer dimensions are D for impeller diameter, T for tank diameter, and H for liquid level, measured in meters for metric units. Liquid level is commonly measured from the

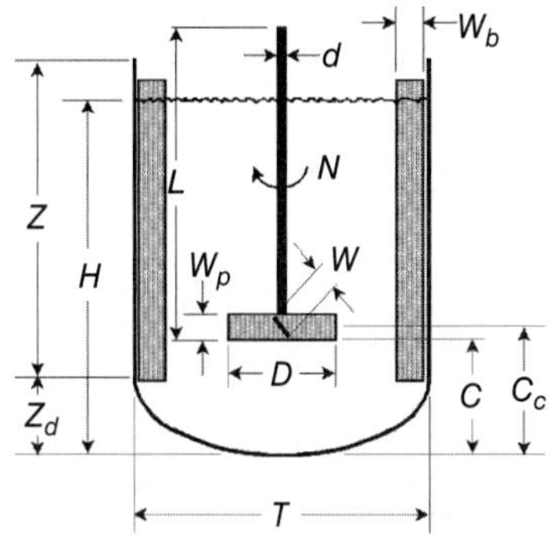

FIG. 18-2 Typical nomenclature for an agitated tank with a center-mounted mixer.

center of the bottom of a tank. Other units either metric or imperial can be used for lengths, provided appropriate factors are used in evaluating or using dimensionless groups. A capital letter N is used for rotational speed, as in revolutions per second (1/s). Other variables shown in Fig. 18-2 use forms of the capital letter Z to represent the vertical dimensions of the tank, such as straight side and head depth. The baffle width, W_b, may sometimes be represented by the capital letter B. The off-bottom clearance of an impeller can be measured to either the bottom of the impeller or the centerline of the impeller. Impeller blade widths can be either actual width, W, or projected blade width, W_P, for angled blades.

A subtle, but potentially important, distinction can be made for measuring the impeller diameter. Some impeller diameters are measured as the centerline distance between opposite blades or in the case of an odd number of blades, twice the radial distance from the center of the impeller to the center of the blade tips. In the case of angled or shaped blades, the center diameter and the maximum diameter of a blade may be different, leading to some confusion about the best measurement of the diameter. This question of the diameter becomes more complicated, even with rectangular blades, which are mounted at an angle. The leading and trailing edge tips are further from the center of the impeller than the center of the blade tip. A different measure that eliminates some confusion, but can be difficult to measure, is the swept diameter of the impeller. The swept diameter is the maximum diameter measured to any point on the tip of a blade. The measurement of an impeller diameter would not be such a problem if it were not that an impeller diameter enters the power calculation to the fifth power for turbulent conditions. Taking the dimensions of a standard pitched-blade turbine (PBT), the diameter based on the edge tips of the blade is about 1 percent greater than the diameter based on the center of the blade tips. This difference results in about a 5 percent difference in a power number, depending on the impeller diameter definition. Knowing what diameter measurement was used in a power number can be important when doing power calculations. Other factors, such as baffles, impeller-to-tank-diameter ratio, and off-bottom-clearance ratio, may have a greater effect on the impeller power number [Mack and Kroll, *Chem. Eng. Prog.* **44:** 189 (1948); Bates, Fondy, and Corpstein, *I&EC Proc Des. Develop.* **2:** 310 (1963); Chapple, Kresta, Wall, and Afacan, *Trans. I. Chem. E.* **80:** 364 (2002)].

Influence of Baffles The largest tank effect on both mixing intensity and impeller power in turbulent agitation is the effect of baffles. In low-viscosity applications, for $N_{Re} > 1000$, baffles are essential for good mixing flow patterns with center-mount mixers. Without baffles, the predominant flow pattern is solid-body rotation of the liquid, which gives only minimal radial and axial mixing. Impeller power without baffles may be as little as one-third of the power input of a fully baffled tank. "Standard" baffles in a cylindrical tank with a center-mounted mixer, as shown in Fig. 18-2, are four vertical plates, spaced at 90-degree intervals around the tank, one-twelfth the tank diameter in width, set a short distance, about one-sixth the baffle width, off the wall of the tank. In some cases three baffles are used instead of four with little loss of performance, or baffles one-tenth of the tank diameter are used with a minor increase in the required power. Without baffles, the flow pattern often looks like what is shown in Fig. 18-3.

In Fig. 18-3, the rotational flow does not mix the fluid, and the surface vortex may draw air into the liquid. The presence of a deep vortex, especially one that reaches the impeller, is a sign of poor mixing. Baffles restrict the naturally occurring rotational flow caused by the rotating impeller. Redirection of the rotational flow by the baffles creates vertical motion at the tank wall, which also results in radial mixing because of recirculating flow patterns, as discussed in the next subsection.

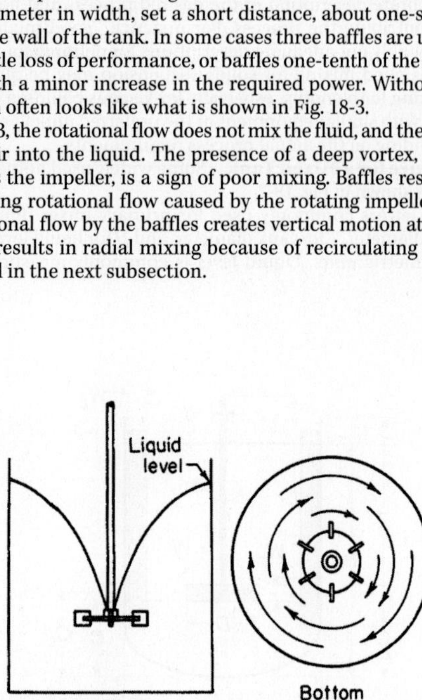

FIG. 18-3 Typical flow pattern for either axial- or radial-flow impellers in an unbaffled tank.

FIG. 18-4 Chemineer straight-blade turbine. (*Mixing Technologies Group of NOV.*)

Depending on the initial discharge direction from an impeller, impellers are typically categorized as either radial flow or axial flow. The straight-blade turbine shown in Fig. 18-4 drives flow outward toward the walls of a tank.

Radial-flow turbines have relatively high power numbers as shown in curves 1 and 2 in Fig. 18-1. The high power number and high power input may have advantages in applications where local energy dissipation is needed, such as for fast chemical reactions and liquid or gas dispersion. Radial-flow impellers are some of the older designs, which have been useful for many years.

By angling the blades, the pitched-blade turbine, Fig. 18-5, creates a more axial discharge.

However, because the discharge from the impeller begins to spread almost immediately, it is sometimes called a mixed-flow impeller, creating a mix of axial and radial flow. The pitched-blade turbine is almost always used to create a down-pumping flow pattern. Better circulation is achieved when the flow is directed at the solid bottom of the tank, rather than at the free surface of the liquid. Pitched-blade turbines have advantages over straight-blade turbines in liquid blending and solids suspension applications.

Further improvements in axial flow can be achieved with hydrofoil impellers, Fig. 18-6.

The term *hydrofoil* comes from the curved cross section of the blades, called camber in propeller design. The shape acts like the airfoil design of an aircraft wing by increasing the velocity across the top of the blade, gradually directing the flow downward, and increasing the axial discharge from the impeller. The three-blade, narrow-blade design shown in Fig. 18-6 is the most common hydrofoil design used in low- to moderate-viscosity liquid mixing. The combination of narrow blades and shallow pitch gives these hydrofoil impellers a low power number while efficiently creating axial flow.

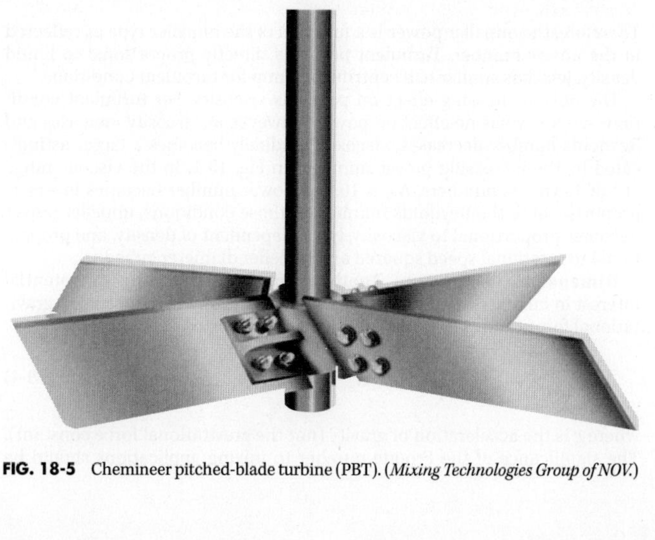
FIG. 18-5 Chemineer pitched-blade turbine (PBT). (*Mixing Technologies Group of NOV.*)

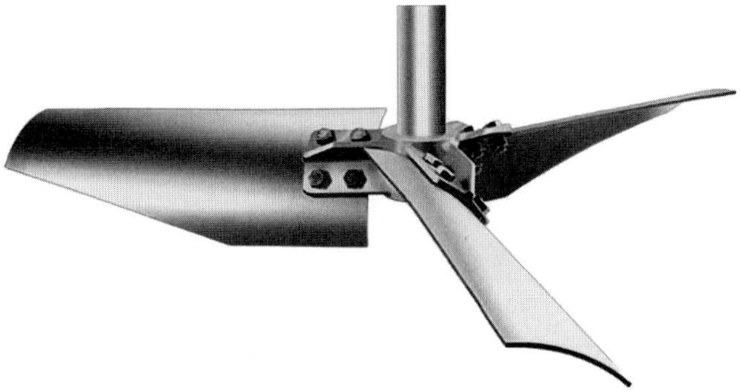

FIG. 18-6 Chemineer narrow-blade hydrofoil impeller. (*Mixing Technologies Group of NOV.*)

These impellers are often more efficient than pitched-blade turbines in liquid blending and solids suspension applications. The shallow angle and narrow blades have limitations in some mixing applications, such as moderately viscous fluids or gas dispersion. The higher power numbers with steeper angles and wider blades on hydrofoil impellers make smaller-diameter impellers possible for side-entry mixers.

The original basis for the hydrofoil designs was the marine-type mixing impeller, Fig. 18-7.

The marine propeller efficiently converts rotational motion into axial fluid flow when applied in a mixing application. The three-blade design is common. The propeller is usually a casting, so blade shape can be almost anything, with smooth curves, helical pitch (a steeper angle nearer the hub than at the tip), and variable cross section (thicker at the center and tapered at the leading and trailing edges). While a casting has advantages for shape, castings tend to be heavy and more expensive than fabrications. The hydrofoil designs have replaced most marine propellers in mixing applications. If nothing else, large hydrofoil impellers more than 3.0 m (120 in) in diameter can be fabricated and applied in mixing applications.

Glass-Lined Agitators Many reactors are glass-lined for an inert, low-adhesion surface. The glass lining of the vessel and glass coating of the impeller are fragile and have a potential to crack with rapid temperature changes or at sharp corners. The tradition impeller design for glass-lined reactors is the retreat-curve impeller (RCI), Fig. 18-8.

Improved glass and coating techniques have allowed greater flexibility for impeller design in recent years, but the retreat-curve impeller remains in wide use. Also related to the limitations of glass lining, the impeller is placed near the bottom of the vessel, and a single baffle is mounted from a nozzle in the vessel head. This configuration results in a circulating flow pattern near the bottom and mixed flow near the top [Dickey et al., *CEP* **11** (2004)].

High-Shear Devices Applications involving the dispersion of immiscible liquids to form emulsions, the dispersion of solids for dissolving, the dispersion of particle agglomerates, such as those found in pigments, and similar dispersion processes often require special impellers. Typical dispersion impellers operate at high rotational speeds, with high tip speeds and relatively low pumping rates. One type of open-style impeller used for dispersion is a sawtooth impeller similar to the one in Fig. 18-9.

The sawtooth impellers come in many different forms with smaller or larger teeth, some aligned around the circumference of a disk, others with teeth set at angles to act more like radial-flow turbines. Because of the high tip speeds and abrasive characteristics of many dispersions, especially powder agglomerates, sawtooth blades wear out and must be replaced from time to time. Another type of high-shear device, called a rotor-stator mixer, is often used for even more intense dispersion. The rotor-stator style dispersers/homogenizers have a rotating impeller inside a close-fitting, but nonrotating, housing. The rotor and stator may have various combinations of blades, slots, and holes through which rapid changes in velocity and direction result in velocity gradients (shear) to cause dispersion. Mixers may have adequate shear to disperse immiscible liquids and break particle agglomerates, but most do not actually grind solid particles. The shear is usually more hydraulic than mechanical.

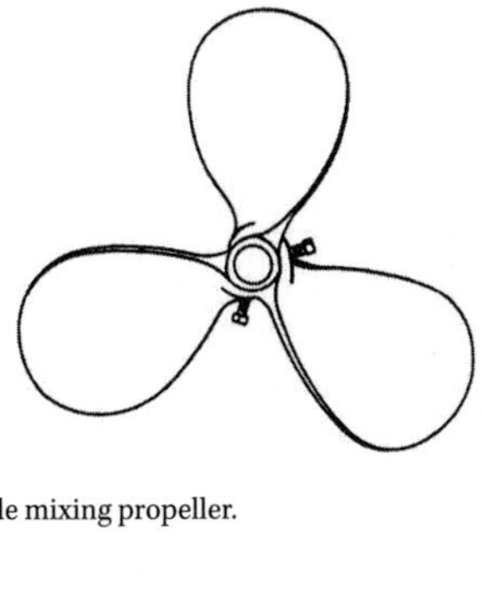

FIG. 18-7 Marine-style mixing propeller.

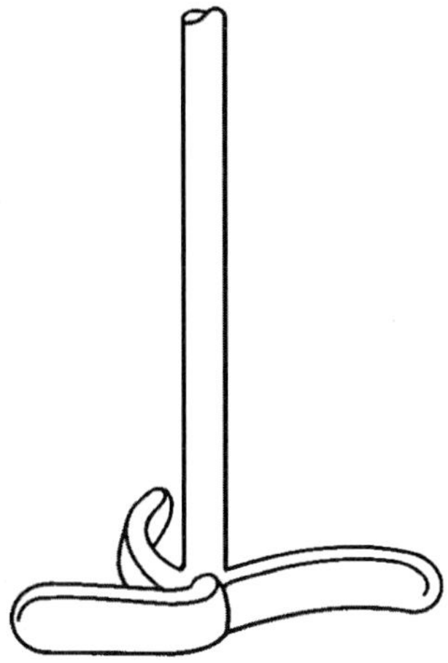

FIG. 18-8 Retreat-curve impeller (RCI). (*The Pfaudler Company.*)

For high-viscosity and high-concentration slurry applications, several types of close-clearance impellers are used for mixing. Close-clearance impellers are typically 90 to 95 percent of the tank diameter. The use of these impellers will be discussed in more detail in the next subsection on Mixing of Viscous Fluids, Pastes, and Doughs.

FLUID BEHAVIOR IN MIXING VESSELS

An essential part of mixer design is understanding what is needed for a process result and how the mixer will accomplish that result. One of the most important parts of understanding the mixer performance is knowing about the flow pattern and energy dissipation. Testing and understanding mixing is empirical, which means it is by observation. The observation can be visual in the laboratory using transparent vessels, indirect by instruments in pilot-plant or production equipment, or aided by computer modeling. In all cases, the more experience or sources available for evaluation, the better the mixing analysis will be.

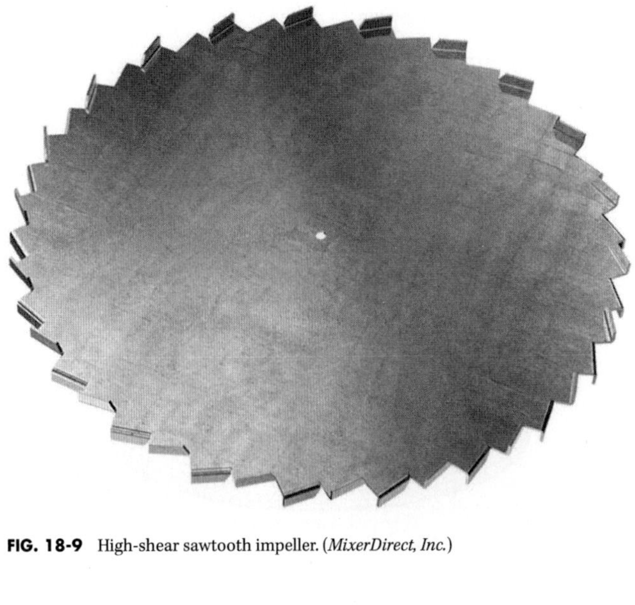

FIG. 18-9 High-shear sawtooth impeller. (*MixerDirect, Inc.*)

Design for a process application in a stirred tank usually starts with the tank dimensions and internals. Then mixer selections need to be made for the impeller type, an impeller diameter, a rotational speed, an off-bottom clearance, and other variables. A final step should include the mechanical design of the mixer and tank support. The quantity of material to be mixed will come from the tank dimensions, or the tank dimensions will come from the desired quantity of material to be mixed. The fluid properties, such as viscosity and density for blending, will establish how difficult the mixing will be. Finally, the intensity of the mixing will be established by the impeller type, size, and rotational speed. The last three mixer characteristics will establish the power and torque input to the fluid, which will in turn provide information about the flow pattern, fluid velocities, and energy dissipation.

Mixing Flow Patterns and CFD Mixing flow patterns are a good place to start understanding fluid behavior in mixing vessels. Through visual observation in transparent vessels, our eyes integrate fluid motion into general flow patterns, such as axial flow with a pitched-blade turbine as shown in the computational fluid dynamics (CFD) vector plot in Fig. 18-10.

The CFD vector plot is intended to show the turbulent flow pattern of a pitched-blade turbine (PBT) in a baffled tank with the liquid level equal to the tank diameter. High velocities in the impeller discharge are represented by longer arrows, and the flow pattern is indicated by the direction of the arrows. The impeller depicted has a diameter one-third of the tank diameter and is located about the same distance off the bottom of the tank. These conditions are typical, but not essential for good mixing. For a similar tank, the impeller diameter would normally be between 25 and 50 percent of the tank diameter, with optimal diameters between 30 and 40 percent of the tank diameter, but in the extreme diameters could be between 15 and 60 percent of the tank diameter. The off-bottom clearance might be between 25 and 33 percent of the liquid level, but less for mixing partially filled tanks. A second impeller about halfway between the bottom impeller and the liquid surface may give better general mixing or extend the successful operating range to higher viscosities. Multiple impellers are needed in tall tanks.

This vector plot is easy to understand, but it presents some questionable results. If the vectors show velocities and directions, then why are the vectors at the bottom center of the tank so small? Is there a dead spot? The vectors show a downward flow pattern from the impeller to the bottom of the tank, and then up the sides of the tank to recirculate back to the top of the impeller. Why are the velocities near the surface and in the center of the recirculation loops so small? Does this impeller provide good mixing throughout the tank? The general response to all of these questions is that the CFD vector plot is a time average of velocities, but turbulent mixing is anything but average. Velocity magnitudes fluctuate greatly, probably plus or minus 75 percent or more in most locations. Velocity directions can make similar changes. The net effect of similar velocity magnitudes in opposite directions results in a zero average velocity. Average velocities may also approach zero where flow directions make sharp turns. Small velocity vectors away from the impeller may represent a combined result of a wide range of velocity magnitudes and fluctuating directions. Actual mixing is impossible to adequately represent in a still picture. Whether visually watching mixing or doing computer model calculations, the amount of local information about velocity magnitude and direction can be beyond comprehension. To make a decision about whether a mixing pattern or rotational speed is sufficient, the thousands of local velocities represented in the CFD plot must be distilled into a few key values or into an integrated quantity that represents a successful range of operating conditions. In nearly all cases, some minimum level of mixing intensity is needed to be sure that all of the tank contents move. In other situations, too much mixing intensity can be a problem.

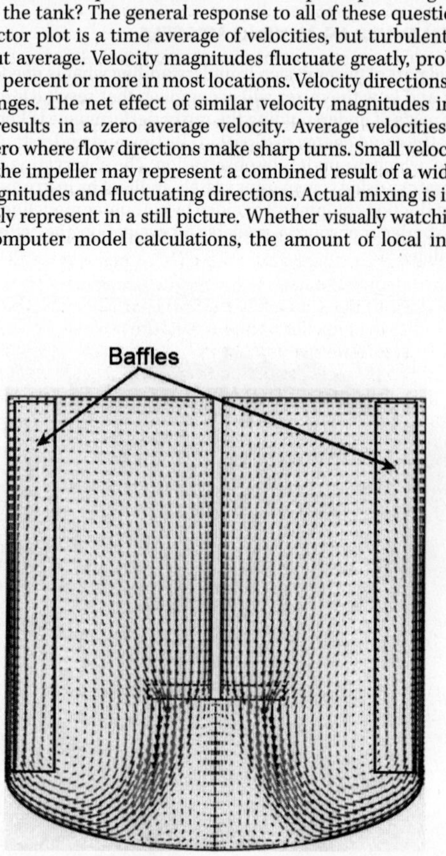

FIG. 18-10 Typical CFD flow pattern in a baffled tank with a pitched-blade turbine.

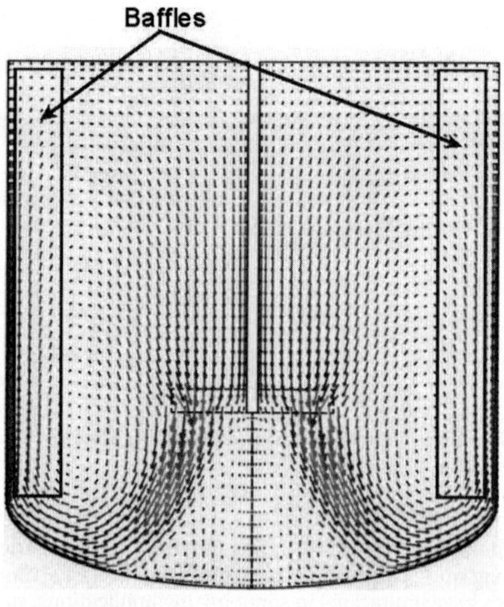

FIG. 18-11 Typical CFD flow pattern in a baffled tank with a hydrofoil impeller.

The CFD vector plot in Fig. 18-11 represents the flow pattern for a hydrofoil impeller.

Comparing the flow pattern for the hydrofoil impeller with the pitched-blade turbine shows that the hydrofoil discharge does not spread as much as the PBT, and more of the tank bottom appears to be swept by high velocities, although vector length does not represent identical velocity magnitudes in these hydrofoil and PBT vector plots. The narrow axial discharge of the hydrofoil impeller provides excellent solids suspension with less sensitivity to off-bottom clearance than the PBT. Other axial-flow impellers have flow patterns similar to the ones shown in Figs. 18-10 and 18-11, with variations for factors like different degrees of axial flow, impeller-to-tank-diameter ratio, and off-bottom clearance.

The radial-flow pattern for a straight-blade turbine is shown in Fig. 18-12.

The radial-flow pattern shows velocities extending outward from the blade tips toward the tank walls. At the tank wall, part of the flow goes upward and part goes downward. The two directions of flow create two

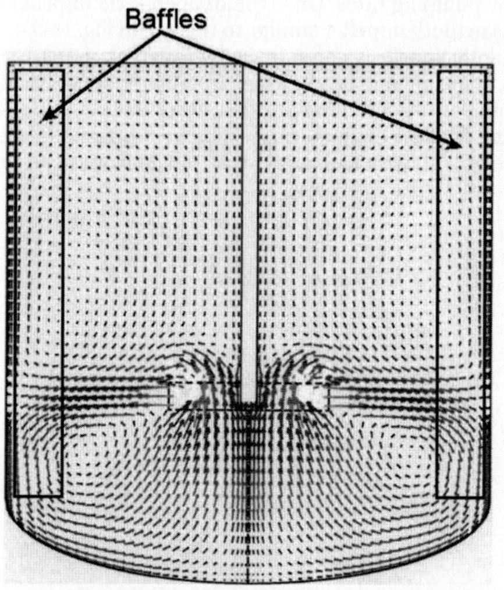

FIG. 18-12 Typical CFD flow pattern in a baffled tank with a straight-blade turbine.

recirculating loops, one in the bottom portion of the tank and one in the top portion. The loop in the bottom of the tank appears to be stronger, primarily because the loop is smaller and tighter. The presence of two loops can create a staging effect in the tank. If a quantity of material is added to the liquid surface it will mix more quickly into the upper portion of the tank, followed by a modest delay as the material is exchanged in the impeller region, and then blend into the loop in the lower portion of the tank. This staging effect may be advantageous for liquid or gas dispersion, but disadvantageous for blending and solids suspension. If the radial-flow impeller is placed near the bottom of the tank, the discharge outward to the tank wall has only one path upward, which creates a circulation loop similar to the axial-flow impeller patterns.

Various other ways of measuring and representing flow velocities are available for the investigation of mixing patterns. Laser Doppler anemometry (LDA) uses crossed laser beams to measure velocity in a small region of a tank. For LDA to be used, the vessel and the fluid must be transparent, and obvious distortion caused by looking through a curved tank wall must be corrected to make effective measurements.

Another way to represent computationally the more complicated flow patterns in real mixing is through the simulation of tracer particle paths. These tracer paths show some of the random and variable direction of flow simulated in a dynamic model of stirred tank mixing.

Unbaffled Tanks Not all cylindrical tanks have baffles, and with an angled and/or an off-center mount they may not need baffles for moderate mixing requirements. Remember that an unbaffled tank with a center-mounted mixer in a low-viscosity liquid creates solid body rotation and poor mixing, as shown in Fig. 18-3. The presence of a strong surface vortex is indicative of this poorly mixed condition. To counteract the rotational flow, an off-center, angle-mounted mixer with a hydrofoil impeller or marine propeller can use the discharge flow from the impeller to counteract the natural rotational flow, Fig. 18-13.

The angle mounting provides flow that sweeps across the bottom of the tank. The off-center mounting uses the axial discharge from the impeller to counteract the inherent rotational flow. The resulting flow pattern is as close as possible to the axial-flow pattern in a baffled tank, Fig. 18-11. This type of mounting works well with small mixers [less than 2 kW (3 hp)] in small tanks (less than 5000 L). Mixing larger and taller tanks is not practical with angle-mounted mixers. Mounting a mixer at an angle with a long shaft may cause the shaft to bend or may place excessive loads on the mixer mount.

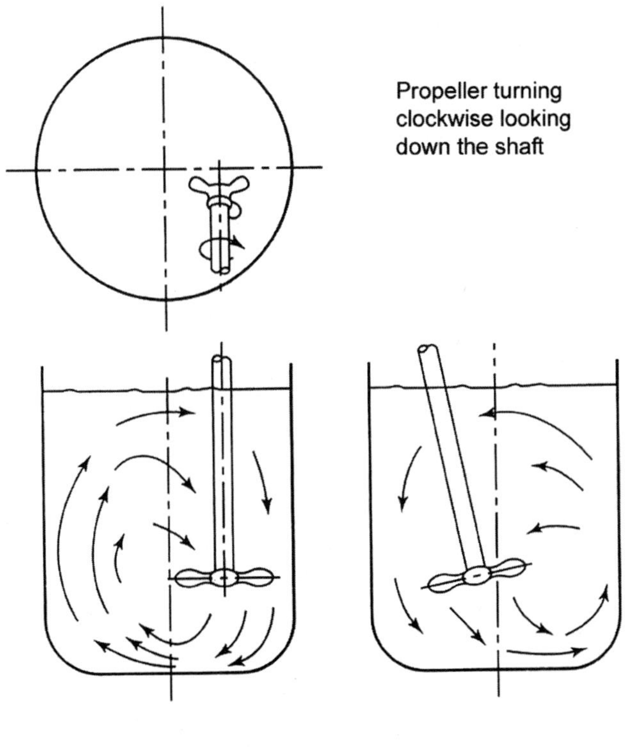

FIG. 18-13 Typical flow pattern with a propeller or hydrofoil in an angled off-center position without baffles.

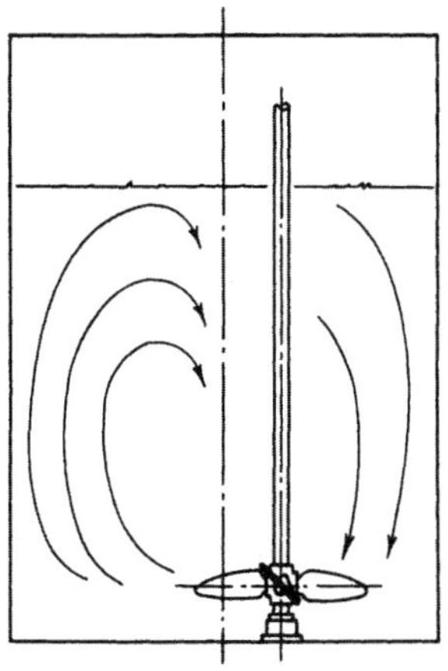

FIG. 18-14 Typical flow pattern with propeller or hydrofoil in vertical off-center position without baffles.

An alternative mounting found in some liquid storage applications uses off-center, vertical mounting. All of these mountings can provide moderate axial, radial, and vertical mixing in tanks without baffles. For intense mixing, baffles are necessary.

Another variation for use in unbaffled vessels is a vertical off-center mount, Fig. 18-14.

These mixers are often found in pulp stock chests. The stock chest may be a concrete chamber, lined with corrosion resistant material, such as brick backed by rubber. The chest can be square or rectangular with a flat bottom. To reduce or eliminate dead spots in the corners, a concrete fillet is built to mimic the otherwise dead spot that could collect pulp. If the pulp is not moving, it will eventually rot and flake dark spots into the white paper coming off the machine.

Many other types of impellers and tank configurations are used for mixing. Draft tubes are sometimes used to create a more controlled circulation pattern in a tank. The draft tube is an open-ended cylinder perhaps half to two-thirds of the tank diameter, with an impeller placed in or below the tube to create a vertical circulation pattern, Fig. 18-15.

The draft tube often has baffles inside it to restrict rotational flow and create strong axial flow. Some draft tube impellers look like sophisticated hydrofoil impellers to take advantage of the restricted flow pattern. Draft tube mixers are often used in crystallization applications.

Simplified Descriptions of Mixing Some of the simplified methods representing fluid mixing intensity are power per volume, tip speed, torque per volume, turnover rate, and bulk fluid velocity. Each of these methods tries to take quantifiable impeller inputs, distribute them in the tank, and estimate how effective the mixing will be. The impeller type will have a power number and a pumping number, which will be primary factors in the estimation of power, torque, and pumping. Turnover rate and bulk fluid velocity depend on both the impeller pumping and the tank variables, like volume and impeller-to-tank-diameter ratio. The impeller size and rotational speed will establish the tip speed, or peripheral velocity, of the impeller.

The impeller diameter relative to the tank diameter, tank baffles, and discharge direction of the impeller will all influence the recirculation pattern. The recirculation influences the effectiveness of the other mixing measures. None of these fluid motion variables assure adequate or effective mixing for all applications. Some applications, such as blending, may be more influenced by liquid motion and circulation. Solids suspension applications may

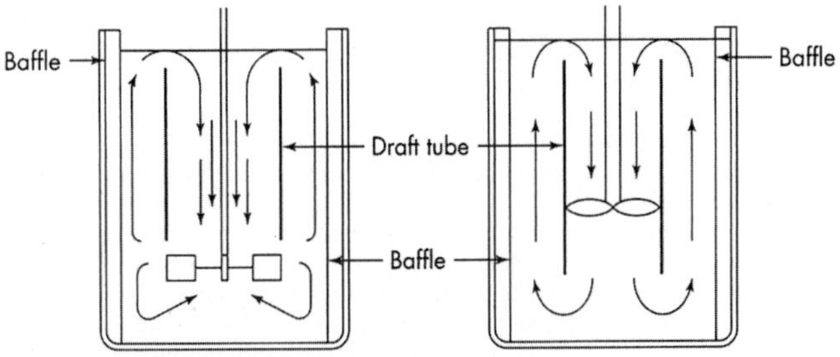

FIG. 18-15 Different arrangements for draft tube agitation.

depend more on local velocities near the bottom of the vessel or vertical velocities in the upper part of the tank. Chemical reactions and other processes may depend on the local turbulence in the liquid.

Power per Volume Power per volume, or more scientifically power per mass, would seem to be an effective measure of the energy dissipation in the tank. However, power will not be evenly distributed throughout the tank. More of the power will be dissipated near the impeller than at other locations in the tank, especially near the surface. If carrying out a chemical reaction is the process objective, one or more of the reactants may need to be introduced near the impeller. Power per volume gives a relative measure for power input in different-sized vessels or quantities of fluid. The volume used to compute power per volume is typically either the total volume of the fluid or the swept volume for the impeller rotation. Swept volume accounts for both the impeller diameter and the blade width.

Impeller Tip Speed Impeller tip speed will have an obvious effect on fluid velocities near the impeller. Tip speed will be a function of both impeller diameter and rotational speed. Larger impellers and higher rotational speeds will result in higher tip speeds. The tip speed not only establishes the fluid velocity near the end of the blade, but the tip speed will also influence the relative velocity between the impeller blade and the surrounding fluid. Tip vortices are shed from each blade tip in turbulent flow. The velocity gradient between the rotational flow in the vortex and the surrounding fluid can be a major contributor to fluid shear. Velocity gradients can be related to fluid shear and may have an effect on dispersion processes. However, with a constant flow pattern in turbulent conditions, local velocities at other locations in the tank should be some fraction of the impeller tip speed. Increasing the rotational speed of a given impeller in a specific tank will also increase local fluid velocities, which may also improve mixing performance for blending and solids suspension.

Torque per Volume Torque per volume is not a direct measure of mixing behavior, but it can be indirectly related to momentum transfer. Momentum transfer from the impeller to fluid does relate to the fluid motion, and conservation of momentum is a basic fact of fluid motion. On a per volume, or more accurately per mass basis, the mixing input is related to the total quantity of fluid present in a tank. The difference between a volume and mass basis is usually indistinguishable because the forces exerted on the fluid are proportional to the density, as is the fluid momentum. The unique feature of torque as a measure of mixing intensity is that the amount of fluid motion generated by torque per volume is less dependent on the impeller-diameter-to-tank-diameter ratio and more dependent on impeller type than some other measures of mixing intensity. Different impeller types require different levels of torque per volume to achieve similar amounts of fluid motion.

Turnover Rate Turnover rate is pumping rate, a volumetric flow, divided by the tank volume to get a measure of the time required for impeller pumping to move a volume equivalent to the contents of the tank. The turnover rate may approximate some fraction of blend time, since multiple circulations of fluid would be required for uniform blending. The measure is not accurate because it fails to account for a flow pattern, which can be important in accomplishing a blend. However, the measure does take the pumping rate and relate it to fluid volume, much like other per-volume measures.

Bulk Fluid Velocity Bulk fluid velocity is an artificial measure of mixing intensity, which uses an impeller-to-tank-diameter-ratio influenced pumping rate and averages it over the cross-sectional area of the tank. The result of a pumping rate divided by an area does have the units of a velocity. Velocity is in fact an observed measure of mixing intensity. Fluid velocity, whether observed on the surface of a production vessel or through the side of a transparent laboratory vessel, is often equated visually to mixing intensity. Higher velocities look like more intense mixing and similar velocities in different-sized vessels may appear to have similar mixing intensities. The relationship between velocity and mixing intensity is a reason for using tip speed as a scale-up criterion for liquid mixing, primarily in geometrically similar vessels.

While power per mass may have some direct relevance to energy dissipation, power per volume, tip speed, torque per volume, turnover rate, and bulk fluid velocities are all indirect measures of mixing intensity. However, these measures can be more quantifiable and accessible than more direct measures of mixing performance, like reaction rate, blend time, or solids suspension. In any case, the real measures of mixing success depend on the process result, which is often a complicated combination of mixer input and fluid dynamic effects.

Macro Mixing Macro mixing or bulk motion is an essential and often primary mechanism for mixing. Without transport from one location in a tank, such as on the surface, to another location, such as the region near the impeller, uniform mixing can never be achieved. A rotating impeller always creates rotational bulk motion, but radial and axial-flow impellers require baffles in low-viscosity fluids to achieve radial and vertical fluid motion. Axial flow is perhaps the most important aspect of macro mixing. Axial flow brings surface additions to the impeller region for recirculation and dispersion. Axial flow takes bottom velocity for solids suspension and moves settling solids from the bottom into the upper portion of the tank. Without effective macro mixing, stagnant or slowly moving portions of a fluid batch will not be mixed to uniformity.

Micro Mixing Micro mixing describes the smallest turbulent eddies before they degenerate into molecular motion, which is simply heat. All power transferred from a mixer to the fluid results in heat, regardless of the impeller type, tank size, fluid flow mechanism, or other mechanism descriptor. All mixers are 100 percent effective in converting power applied by the impeller to heat in the fluid. In the case of viscous fluids, where power requirements for mixing may be high and heat transfer may be low, fluid temperatures may rise measurably even over relatively short periods of time. Estimates for the temperature rise in a batch of fluid can be made from the applied impeller power and the heat capacity of the fluid batch. Micro mixing may be a critical mechanism for bringing reactants for a chemical reaction together.

Meso Mixing Meso mixing describes the mixing mechanisms between macro and micro mixing, but it has many important effects. Perhaps the best description of meso mixing involves flow structures, such as tip vortices shed by impeller blades. Tip vortices are well defined and observable forms of meso mixing. A trailing vortex leaving the tip of an axial-flow impeller blade creates a rotating velocity that moves through the surrounding fluid in a helical path. The higher velocity in the vortex passes through regions of lower velocities in the surrounding fluid. The difference between the vortex velocity and the surrounding fluid accounts for the velocity gradient that can be called fluid shear. This fluid shear may create dispersions, such as immiscible liquid-droplet dispersion, gas-bubble dispersion, or solids-agglomerate dispersion. Such mechanisms may be essential in some process mixing requirements or secondary in others. The relative importance of macro, micro, and meso mixing effects depends on the specific mixing requirements for a process.

DESIGN OF AGITATION EQUIPMENT

Perhaps the single biggest problem in the design of agitation equipment is the diversity of mixing applications for which agitation equipment can be used. In addition to the extreme range of applications, a large diversity of equipment sizes and shapes can be used to provide the needed agitation. To further complicate the design process, different fluid properties, materials of construction, and ultimately equipment cost must all be considered in design. Agitation equipment design is different from equipment rating. In design, the starting point is effectively a clean sheet of paper onto which a series of decisions or restrictions develops the equipment configuration. In rating, the starting point is existing equipment for which mixing performance needs to be evaluated, used, improved, or modified. A *well-designed* agitator should take advantage of as many optimum characteristics as possible. A properly *rated* agitator may involve some compromises in the process performance because of equipment limitations.

Tank Dimensions The design of most agitation equipment begins with selection of the tank or vessel. The most widely used and studied tank design is a vertical cylinder. To serve as a fluid container, at least the bottom of the cylinder must be enclosed, and it typically has either a flat or a dished bottom, although sloped, conical, and hemispherical heads may be used. If the vessel is expected to contain a vapor or pressure, then a top head is also necessary. In the case of a pressure requirement, the head must be dished or shaped in some way to effectively transmit forces to the walls of the cylinder. Other vessels, such as square or rectangular chests and horizontal cylinders, sometimes are used for agitation applications.

Most applications and studies involve a fluid batch size where the liquid level is approximately equal to the diameter of the tank. A cylindrical tank with the liquid level equal to the tank diameter is called a square batch, which has nothing to do with the shape of the tank cross section. Most of the studied, reported, and correlated relationships for either design or rating of agitation equipment are based on the square batch geometry. In the real world, large tanks are often tall tanks because of transportation limitations on the tank's diameter. In batch processes, liquid levels may change at different points in the process, increasing as the batch is being created or decreasing as the product is being emptied. In agitation equipment design, a square batch will usually provide the best results or at least the most information about potential results.

Impact of Baffles For applications involving low-viscosity fluids, especially for turbulent conditions, baffles are needed to effectively control rotational flow with vertical, center-mounted agitation equipment. Standard baffles are typically four vertical plates, one-twelfth of the tank diameter in width, extending up the entire straight side of the vessel. A similar effect is created by three baffles, one-tenth the tank diameter in width. As critical as

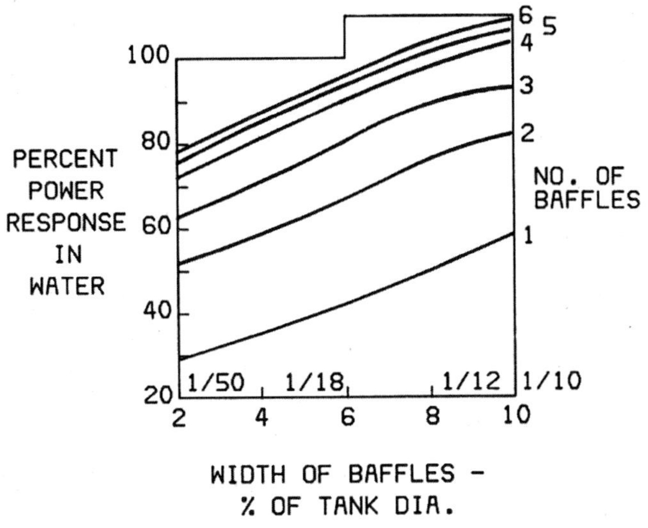

FIG. 18-16 Baffle number and width effects on power.

impeller diameter can be for defining mixing intensity, baffle dimensions are much less important. Baffle width, length, and number can change considerably without a loss of essential function.

Figure 18-16 shows how the number of baffles and the baffle width affect impeller power in turbulent conditions. In this graph, 100 percent power is identified with standard baffles, which are four baffles each with a width of one-twelfth the tank diameter. More baffles or wider baffles will increase the impeller power requirement slightly. Narrower baffles down to 1/50 of the tank diameter are shown to decrease the power gradually. Changing the number of baffles has a relatively small effect between four, five, and six baffles, but the effect becomes more pronounced with three, two, and one baffle. The power requirement without any baffles is effectively represented by the curve for one baffle at the narrowest width, which is less than 30 percent of the power with standard baffles.

While decreased power with fewer baffles may reduce the motor load, the process results are also reduced.

Figure 18-17 shows the effect of no baffles, one baffle, and four baffles on some simple solids suspension with a pitched-blade turbine set one-quarter the liquid level off the bottom. Without any baffles, the primarily rotational flow creates a vortex on the surface and leaves a pile of unsuspended solids at the center of the bottom. The small quantity of solids near the wall of the tank is rotating around the tank with little vertical or recirculating motion. The presence of just one baffle almost eliminates the vortex on the surface and the pile of solids on the bottom. Four baffles completely suspend the solids and drive them well up into the upper part of the tank. The increased power for multiple baffles and the development of vertical motion effectively made the solids suspension successful. A strong vortex on the surface and rotational flow is usually a sign of poor mixing. The addition of baffles can drastically improve mixing by creating vertical motion and using power to create turbulence. Other information about the effect of baffles can be found in the following references: Myers, K. J., M. F. Reeder, and J. B. Fasano, *CEP* (February 2002), pp. 42-47; Fořt, I., A. Gračková, and V. Koza, *Coll. Czech. Chem. Commun.* **37:** 2371–2385 (1972).

Mounting of Equipment Mounting of agitation equipment has many options. The most common is top mounting, which can be either on an open tank or a closed tank. Open tank mounting can either be a vertical center mount, typically on a beam bridge support, or an offset angle mount with a clamp or external support. Closed tank mounting is usually on a nozzle, which is gusseted for support. Seals for a closed tank can be a lip seal, for dust exclusion, or a stuffing box or mechanical seal, for pressure containment. Mechanical seals can be single or double seals. Double seals with pressurized seal fluid between them can provide positive leak protection. Other mounting options include side or bottom mounting, both of which require some type of seal that can be similar to the top-mounted seal options. An additional option, primarily for a bottom entering mixer, is a magnetic drive. The magnetic drive has an external motor with rotating magnet, and on the inside of the tank is a magnetic impeller, which can link through a stationary can arrangement. The magnetic drive has no rotating seal penetration, and it provides a positive seal to prevent leakage or contamination.

Identifying Process Requirements Identifying process requirements can be the most difficult step in the design of agitation equipment, primarily because understanding the connection between fluid motion and process results can involve multiple factors. A simple blending problem, such as uniform storage of multiple batches of the same product, requires that all of the fluid must move and that the movement must involve transport from different regions to all other regions. Fluid motion may not be sufficient if the motion does not penetrate stratified layers or adequately move material from the bottom to the top of the tank. The presence of solids may make the problem more difficult, as will be discussed in a later subsection. Agitation problems become more difficult when formulation processes involve multiple ingredients with different properties, changing liquid levels, and various addition steps. In such cases, identifying the critical or limiting step is important, but because the agitation intensity required for one step may exceed the allowable intensity for another step, multiple impellers or variable speed may be required for successful process results. Multiple impellers may include different types of impellers, even multiple drives operating at different speeds and at different locations in the tank. In nearly all agitation applications, some minimum level of mixing intensity is required. Above that level, hopefully the mixing intensity is sufficient for good results and can be maintained over a range of process conditions. At high mixing intensity, adverse process conditions may develop, even as simply as air being drawn into a liquid batch. Understanding agitation requirements means identifying the minimum and maximum conditions for process success. Operating below the minimum or above maximum conditions may be accompanied by a rapid decline in process performance. Identifying and avoiding those rapid changes can mean the difference between process success and failure.

Process material properties directly influence the performance of agitation equipment with respect to process success. One of the most obvious material properties that must be considered in agitation is fluid viscosity. Even simple newtonian viscosities are temperature dependent, which may influence blending or heat transfer results. Viscosity is not necessarily an easy property to measure or describe, because high viscosities often include other variable properties, such as an apparent viscosity that is shear or time dependent. Nonnewtonian fluid behavior includes time-independent properties of shear thinning or shear thickening. The shear rate affecting fluid viscosity is proportional to the rotational speed of the mixer. Such shear-dependent behavior may be accompanied by a yield stress, often exhibited by a "gel" characteristic. Time-dependent fluid behavior may also be shear

FIG. 18-17 Effects of baffles on solids suspension with a pitched-blade turbine.

thinning or thickening, usually resulting in a hysteresis effect, which exhibits different viscosities at the same operating conditions, but depends on previous operations. Some materials also exhibit elastic return, like bread dough. Each of these viscosity characteristics will influence the selection of the most appropriate agitation equipment.

Fluid density is also an important property in agitation because impeller power is directly proportional to density in turbulent conditions. Liquid density is a relatively easy property to measure until it involves different phases, such as solids dispersed or suspended in liquid. In the case of power requirements for a solids suspension problem, fluid density includes the suspended solids at the conditions in the region of the impeller. Even though liquid density is usually a constant for agitator design, slurry density may depend on the concentration and uniformity of the suspension. Dispersed gas has an even greater effect on impeller power than just a bulk density. Gas bubbles formed in low-pressure regions behind impeller blades will alter local drag, resulting in much greater power reduction than predicted by density alone. In all multiphase agitation applications, the material properties are only part of the problem. The interaction between the phases, whether considered as dispersions or suspensions, influences the process results and the agitation requirements. Factors such as particle density relative to liquid density and particle shape are factors in solids suspension. Interfacial tension in immiscible liquid or gas dispersion is an important effect. Viscosity differences between liquid phases can be even more important than surface tension forces.

Materials of construction are also important aspects of agitator design. Materials of construction are usually chosen on the basis of the fluid chemistry or the application requirements. Because metals and metal alloys are commonly used in both tanks and mixer components, corrosion and erosion resistance can be extremely important. Corrosion resistance may be determined by previous experience or in special cases by testing. Erosion problems are usually limited to applications that involve suspended particles. Cavitation is rarely a problem in agitator design because the tip speeds are not very high, and the liquid head above the impeller is sufficiently large to prevent formation of vapor bubbles.

Equipment cost is always a consideration, but with mixing equipment the cost should not be the primary concern. In most cases, the agitation equipment is used to combine and convert raw materials into the product. The importance of selecting the right equipment can make the difference between a successful process or an expensive failure. The value of a little more than the minimum agitation intensity can mean a more rapid start-up, more reliable operation, and even additional capacity.

LIQUID BLENDING

Uniform liquid blending is typically a minimum requirement for all types of agitation, even in multiphase processes. Few processes will cause miscible liquids to separate to an appreciable degree once they are mixed. Most multiphase processes involving mass transfer improve with uniform concentration in the continuous liquid phase.

Fluid Motion Fluid motion is the direct result of rotating a mixer in a quantity of fluid. The consequences of that fluid motion will hopefully provide the desired process results. Understanding the flow patterns created by a mixer is an essential first step in deciding what mixer design will accomplish specific process results. Most of the understanding of flow patterns in a stirred tank comes from experience, through observation, modeling, and process evaluation. Simply observing mixing patterns in a transparent pilot-scale vessel, using some suspended solids for flow followers, can provide insight into both the basic patterns and the complicated motion that provides effective blending uniformity.

One measure of solids suspension is called complete off-bottom suspension. This condition is observed when none of the particles remain on the bottom for more than one second. More about other degrees of solids suspension will be discussed in a following subsection. A photo of off-bottom suspension created by a pitched-blade turbine is shown in Fig. 18-18.

In this photo, a transparent baffled tank is filled with water, and a modest quantity of plastic beads are added to show both solids suspension and flow patterns. The diameter of the pitched-blade turbine is approximately one-third of the tank diameter. The tank has a dished bottom, and the impeller is located about one-fourth of the liquid level from the center of the dished bottom. The four-blade, pitched-blade turbine is operated at a speed sufficient to achieve off-bottom suspension. The picture shows that the solids are swept cleanly from the bottom of the tank and driven approximately two-thirds of the way to the liquid surface. A pitched-blade turbine is often called a mixed-flow, axial-flow impeller because the discharge flow is not strictly axial and has a significant radial component that spreads the flow across the tank cross section.

For comparison, the flow pattern and suspension capability of a hydrofoil impeller are shown in Fig. 18-19.

FIG. 18-18 Off-bottom suspension with a pitched-blade turbine in a baffled tank.

The hydrofoil impeller easily achieves off-bottom suspension and drives the suspended particles further into the upper part of the tank than the pitched-blade turbine. The hydrofoil impeller has more axial flow with less of a radial component than the pitched-blade turbine. The resulting flow pattern does a more effective job of lifting settled particles from the bottom and creates a higher vertical recirculation loop in the upper part of the tank.

FIG. 18-19 Solids suspension with a hydrofoil impeller at the same power, speed, and torque as off-bottom suspension with a pitched-blade turbine in Fig. 18-18.

The method for comparing the hydrofoil impeller to the pitched-blade turbine is intended to be industrially significant, even if a bit unconventional for academic design of experiments. The hydrofoil impeller is operated at the same speed as the pitched-blade turbine, with the same power input, which also means the same torque. With the same power, speed, and torque, the mixer drive could be identical for the two impellers. However, because the hydrofoil impeller has a lower power number ($N_P = 0.3$) than the pitched-blade turbine ($N_P = 1.3$), the hydrofoil impeller must have a larger diameter (34 percent larger) than the pitched-blade turbine. While the impeller diameter and the impeller-to-tank diameter change in the comparison, the comparison is made as if the mixer motor, drive, and essential parts of the mixer are unchanged; only the impeller is replaced and sized properly for the mixer. This type of comparison seems more practical than keeping the impeller diameter constant and trying to explain a comparison where the speed, power, or torque must be adjusted to obtain a similar level of solids suspension. While this comparison is based on solids suspension, similar comparisons can be done for other process results, such as blend time, heat transfer, or gas dispersion, with different relative results for different impellers.

A third impeller comparison with a straight-blade, radial-flow turbine is provided in Fig. 18-20.

The straight-blade turbine is running at the same speed, power, and torque as the pitched-blade turbine in Fig. 18-18. Because the straight-blade turbine has a higher power number ($N_P = 3.96$) than the pitched-blade turbine, the straight-blade turbine is smaller (80 percent of the pitched-blade diameter). However, the poorer solids suspension, with a pile sitting in the bottom center of the tank, is not just a function of the potential mixing capability of the straight-blade turbine, but rather a function of the flow pattern. The radial-flow pattern goes outward toward the tank wall and then both upward and downward. The downward pattern does not sweep the settled solids off the bottom as the axial impellers did, but rather tries to draw the solids upward off the bottom. The upward flow under the straight-blade turbine is much less effective than the downward flow from the axial impellers. The radial-flow impellers can be more effective when placed lower in the tank, as will be shown in the solids suspension subsection. Radial-flow impellers can be more effective for liquid-liquid dispersion and gas dispersion than the axial impellers. Different impeller types have different functions and are used successfully in different applications.

Blend Time Blend time for miscible liquids can be an effective measure of process performance for single-phase liquid applications. A typical "blend time" is considered to be the time required to blend to some degree of uniformity, for example, 95 percent, the surface addition of a small quantity of miscible liquid with similar density and viscosity to an agitated batch of liquid. The two most common measurement techniques are either a color change observation in a transparent tank or a concentration measuring probe located at a slowly mixed location in the tank. Blend time measurements have been made for several impeller types, liquid levels, and fluids of different viscosities. The product of rotational speed times blend time forms a dimensionless blend time, which can be correlated with geometry and fluid property variables. In effect, the rotational speed of the impeller becomes the clock for blending, and the uniformity is a function of the number of revolutions of the impeller. For turbulent conditions, this dimensionless blend time is a constant for turbine impellers rotating in a baffled tank.

FIG. 18-20 Solids suspension with a straight-blade turbine at the same power, speed, and torque as off-bottom suspension with a pitched-blade turbine in Fig. 18-18.

The simplest visual observation method for blend time involves just the addition of a quantity of dye to the agitated batch of liquid. This method, while simple and quick, has limitations. The most obvious limitation, even in a clear liquid, is that the dye will obstruct a view of the last area of clear liquid. Not knowing when the last location of incomplete mixing disappeared gives only an approximation to the total mixing time. However, the addition of a dye will give a quick indication of blend time or blending problems, even in an opaque liquid.

The better method for visual observation of blend time is with a color change indication, going from color to clear. With this method, the final location of complete mixing is the place where the color is last to disappear. One of the simplest color change methods is using a pH indicator, such as phenolphthalein. Phenolphthalein goes from a pink color to clear around a pH of 7.0. So the addition of a quantity of acid to a caustic solution with an indicator will cause the color indicator to disappear at the final point of mixing. To give a strong color change without uncertainty, the change is often done at 50 percent uniformity, resulting from a sufficient acid addition to change the blended pH from 8.0 to 6.0. Then, by assuming that the blending process involved an exponential decay from unmixed to mixed percentages of uniformity, we can estimate other degrees of uniformity. The time for 95 percent uniformity can be estimated from the 50 percent color change by the following formula:

$$\Theta_{95} = \Theta_{50}\frac{\ln(1-0.95)}{\ln(1-0.50)} = 4.32\Theta_{50} \qquad (18\text{-}7)$$

Achieving 95 percent uniformity will take 4.32 times as long as the observed 50 percent uniformity blend time. Other degrees of uniformity can be estimated by adjusting the fractions in Eq. (18-7). The two obvious limitations of observing blending color changes are (1) that the experiments need to be conducted in a transparent tank and (2) the liquid must also be transparent.

To conduct blend time experiments in a metal tank, some type of measurement probe is more practical. Studies have been done using ionic solutions and temperature changes. Of course, the response time of the measurement device needs to be considered depending on the tank size and anticipated blend time. For turbulent conditions ($N_{Re} > 6400$) dimensionless blend times measured by a conductivity probe were found to correlate with the following expression:

$$N\Theta_{95} = \frac{5.20}{N_P^{1/3}}\left(\frac{T}{D}\right)^2 \qquad (18\text{-}8)$$

by Grenville (Grenville, R. K., Ph.D. dissertation, Cranfield Institute of Technology, 1992). The effect of impeller type is interpreted in the power number (N_P) for several turbine-style impellers. This correlation applies to newtonian liquids in a tank with the liquid level equal to the tank diameter.

In the transition regime, $530 < N_{Re} < 6400$, the dimensionless blend time becomes a function of both power number and Reynolds number.

$$N\Theta_{95} = \frac{33{,}490}{N_P^{2/3}\, N_{Re}}\left(\frac{T}{D}\right)^2 \qquad (18\text{-}9)$$

The transition from the turbulent blend time correlation, Eq. (18-8), to the transition blend time correlation, Eq. (18-9), occurs at a transitional Reynolds number:

$$N_{Re,TT} = \frac{6370}{N_P^{1/3}} \qquad (18\text{-}10)$$

Other variables, such as liquid level, location of addition, rate and quantity of addition, and property differences all may affect the blend time, but a few general correlations exist. The blend time estimates for liquid additions should be used as a guide to understanding a blending operation and rarely are sufficient for accurate process estimates.

Heat Transfer In general, the fluid mechanics of the film on the mixer side of the heat transfer surface is a function of what happens at that surface rather than the fluid mechanics going on around the impeller. The impeller largely provides flow across and adjacent to the heat-transfer surface, and that is the major consideration of the heat-transfer result. Many of the correlations are in terms of traditional dimensionless groups in heat transfer, while the impeller performance is often expressed as the impeller Reynolds number.

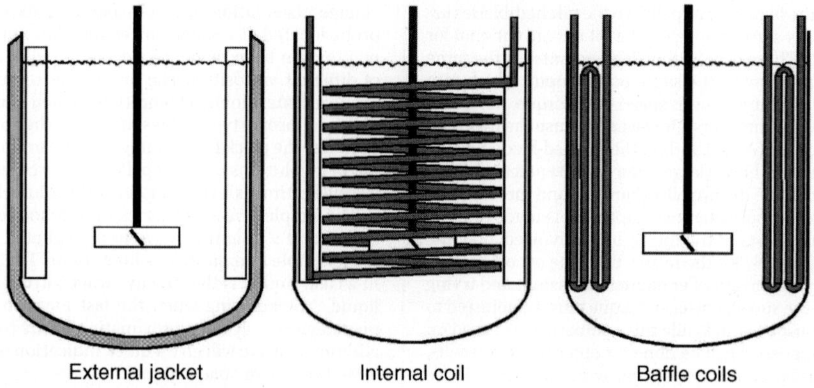

FIG. 18-21 Typical vessel heat transfer surfaces.

The hydrofoil impellers (shown in Fig. 18-6) usually give more flow for a given power level than the traditional axial- or radial-flow turbines. More flow and greater temperature uniformity are advantages for heat transfer. The heat-transfer surface generates some turbulence to provide the film coefficient. Different types of heat transfer surfaces are used for agitated tanks (Fig. 18-21). Local turbulence is true to a limited degree in jacketed tanks. Internal helical coils may restrict recirculation flow, so a better option for an internal heat transfer surface is to add coils as baffles. Heat transfer baffles provide both additional surface area and flow direction control.

HEAT TRANSFER

Jackets and Coils of Agitated Vessels Most of the correlations for heat transfer from the agitated liquid contents of vessels to jacketed walls are in a dimensionless form, with the Nusselt number written as a function of impeller Reynolds number, the Prandtl number, and a bulk-to-wall viscosity ratio:

$$\frac{hT}{k} = a\left(\frac{D^2N\rho}{\mu}\right)^b \left(\frac{C_p\mu}{k}\right)^{1/3} \left(\frac{\mu_b}{\mu_w}\right)^m \tag{18-11}$$

The film coefficient h is for the inside wall of the vessel; T is the inside diameter of the vessel. The Reynolds number for mixing involves D, the impeller diameter, and N, the rotational speed of the agitator. Recommended values of the constants a, b, and m are given in Table 18-1.

A wide variety of configurations exist for coils in agitated vessels. Correlations of data for heat transfer to helical coils have been of two forms, of which the following are representative:

$$\frac{hT}{k} = 0.87\left(\frac{D^2N\rho}{\mu}\right)^{0.62} \left(\frac{C_p\mu}{k}\right)^{1/3} \left(\frac{\mu_b}{\mu_w}\right)^{0.14} \tag{18-12}$$

Where the agitator is a paddle, the Reynolds number range is 300 to 400,000 [Chilton, Drew, and Jebens, *Ind. Eng. Chem.* **36:** 510 (1944)], and

$$\frac{hd_t}{k} = 0.17\left(\frac{D^2N\rho}{\mu}\right)^{0.67} \left(\frac{C_p\mu}{k}\right)^{0.37} \left(\frac{D}{T}\right)^{0.1} \left(\frac{d_t}{T}\right)^{0.5} \tag{18-13}$$

where the agitator is a disc flat-blade turbine, and the Reynolds number range is 400 to 200,000 [Oldshue and Gretton, *Chem. Eng. Prog.* **50:** 615 (1954)]. The term d_t is the outside diameter of the coil tube.

TABLE 18-1 Values of Constants for Use in Eq. (18-11)

Agitator	a	b	m	Range of Reynolds numbers
Paddle*	0.36	2/3	0.21	300 < Re < 300,000
Pitched-blade turbine†	0.53	2/3	0.24	80 < Re < 200
Disc, flat-blade turbine‡	0.54	2/3	0.14	40 < Re < 300,000
Propeller§	0.54	2/3	0.14	Re = 2000 (one point)
Anchor†	1.0	1/2	0.18	10 < Re < 300
Anchor†	0.36	2/3	0.18	300 < Re < 40,000
Helical ribbon¶	0.633	1/2	0.18	8 < Re < 100,000

*Chilton, Drew, and Jebens, *Ind. Eng. Chem.* **36:** 510 (1944), with constant m modified by Uhl.
†Uhl, *Chem. Eng. Progr.* Symp. Ser. 17, **51:** 93 (1955).
‡Brooks and Su, *Chem. Eng. Progr.* **55**(10): 54 (1959).
§Brown et al., *Trans. Inst. Chem. Engrs.* (London), **25:** 181 (1947).
¶Gluz and Pavlushenko, *J. Appl. Chem.* U.S.S.R., **39:** 2323 (1966).

The most comprehensive correlation for heat transfer to vertical baffle-type coils is for a disc flat-blade turbine over the Reynolds number range 1000 to 2,000,000:

$$\frac{hd_t}{k} = 0.09\left(\frac{D^2N\rho}{\mu}\right)^{0.65} \left(\frac{C_p\mu}{k}\right)^{0.33} \left(\frac{D}{T}\right)^{0.33} \left(\frac{2}{n_b}\right)^{0.2} \left(\frac{\mu_b}{\mu_w}\right)^{0.4} \tag{18-14}$$

where n_b is the number of baffle-type coils and μ_w is the fluid viscosity at the mean film temperature [Dunlop and Rushton, *Chem. Eng. Prog. Symp. Ser.* 5, **49:** 137 (1953)].

Chapman and Holland (*Liquid Mixing and Processing in Stirred Tanks*, Reinhold, New York, 1966) review heat transfer to low-viscosity fluids in agitated vessels. Uhl ("Mechanically Aided Heat Transfer," in *Mixing: Theory and Practice*, vol. I, ed. Uhl and Gray, Academic, New York, 1966, chap. V.) surveys heat transfer to low- and high-viscosity agitated fluid systems. This review includes scraped-wall units and heat transfer on the jacket and coil side for agitated vessels.

A more recent survey and summary of agitated heat transfer film coefficient correlations with other impeller types and broader Reynolds number ranges can be found in Dream [Dream, R. F., *Chem. Eng.* (January 1999), pp. 90–96]. That reference also provides a correlation for the jacked-side film coefficient with turbulent flow (Re > 10,000).

$$\frac{h_jd_e}{k} = 0.027\left(\frac{d_eV\rho}{\mu}\right)^{0.8} \left(\frac{C_p\mu}{k}\right)^{0.33} \left(\frac{\mu_b}{\mu_w}\right)^{0.14} \left(1+3.5\frac{d_e}{d_c}\right) \tag{18-15}$$

The film coefficient h_j is the jacket side film coefficient for the outside of the vessel wall. For a spiral baffle jacket, the equivalent heat transfer diameter, d_e, for the rectangular cross section is equal to four times the width of the annular space, w, and d_c is the mean or centerline diameter of the jacket. The flow velocity, V, is calculated for the actual cross section in the jacket and the spiral baffle pitch, even though the leakage around the spiral baffles can amount to 35 to 50 percent of the total flow through the jacket. The same correlation can be applied to a half-pipe coil, where d_c is the mean diameter of the coil. This correlation probably gives a conservative estimate of the jacket side coefficient in a dimple jacket because of the turbulence created in the intersecting flow passages.

SOLID-LIQUID PROCESSING

Solid-liquid processing is done in a number of commercial processes, most of which use some type of rotating-impeller mixing equipment. The mixing equipment is only capable of moving fluid, which is a combination of a liquid and dispersed particles. The effects of the mixer on the dispersed particles will depend on the properties of the liquid, density and viscosity, and the properties of the particles, size, density, shape, and concentration. The dispersed particles will often settle rapidly enough that achieving or maintaining a suspension may be the primary purpose of the mixing equipment. In applications requiring particle suspension, the processes may also involve mass transfer or particle transport. Slowly settling particles and even floating particles are found in some situations. Particles may settle slowly in the liquid because of small particle size, a minor density difference, or the viscosity of the liquid. Floating particles or those difficult to add into the liquid can be lower density or nonwetting particles.

Some processes involving solid-liquid systems include suspension and dispersion of solids to make a slurry. Although the slurry is rarely the final product, a well-dispersed slurry may involve mass transfer for dissolution or leaching. Crystallization goes in the opposite direction of a dissolution, as particles are created or enlarged out of a liquid solution. Solid catalyzed reactions typically involve mass transfer going in both directions between the solid and liquid, as do adsorption, desorption, and ion exchange processes. Suspension polymerization involves bulk polymerization of dispersed monomer droplets to form solid polymer particles, requiring solids suspension. Storage applications may also involve solids suspension for either the purpose of uniformity in batch processes or transport or solids to a following step in the process. In some situations the agglomeration or deagglomeration of particle aggregates may be an objective of a solid-liquid process. With the exception of highly loaded slurries or suspensions in a viscous liquid, most slurries behave as low-viscosity fluids and require baffles for effective solids suspension.

Particle Suspension and Dispersion Particle suspension and dispersion are a necessary feature of most other solid-liquid process objectives. Most of the literature and research in the mixing of particle-solid systems focus on either off-bottom suspension or degree of uniformity. The conditions at which particles are moved or lifted from the bottom of a vessel is an essential element of all particle suspension processes. Once particles are lifted off the bottom of the vessel, then the degree of uniformity of the suspension becomes a factor in the process. The three most commonly used descriptions for degree of solids suspension are on-bottom motion, off-bottom suspension, and uniform suspension (Fig. 18-22).

On-Bottom Motion On-bottom motion occurs when only a portion of the solids are suspended. All of the solids remaining on the bottom are in motion. The motion of the solids is typically seen as a sliding motion, with clusters of particulate solids moving together. The essential increment for on-bottom motion is the elimination of permanently settled groups of solids. The bottom locations where settled solids are last mobilized depend on the shape of the bottom. In a vessel with a flat or sloped bottom, the place where the bottom joins the sidewall of the tank is almost always the last point where solids begin to move. In vessels with dished bottoms or shallow conical bottoms, the center of the bottom is usually the last point for suspension. Deep conical bottoms can be extremely difficult to get the solids in motion or off the bottom. On-bottom motion is an acceptable degree of suspension in some typically large-volume applications, like mineral processing or wastewater treatment, where a limited accumulation of solids does not pose a critical process problem. In such applications, the accumulation of solids tends to be self-limiting, once an initial accumulation fills the point of least effective suspension and forms a gradual transition from one direction of flow to another.

Off-Bottom Suspension Off-bottom suspension is the most studied and well-defined degree of solids suspension. The condition of off-bottom or complete suspension occurs when none of the solids rests on the bottom of the vessel for more than one second. The primary difference between off-bottom suspension and on-bottom motion is that with suspension, all of the particles are lifted off the bottom frequently. The important effect that off-bottom suspension has on solid-liquid applications is that all surfaces of the particles are continuously or frequently exposed to the liquid. This liquid exposure is essential for good mass transfer, as for dissolving particles.

Uniform Suspension Uniform suspension is a bit of a misnomer, since settling particles are almost never completely uniform at the free surface of a liquid. The degree of suspension associated with uniform suspension is effectively as uniform as the suspension will get, both vertically and radially. Depending on the settling characteristics of the suspended particles, a little or a lot of additional power may be required when going from off-bottom suspension to uniform suspension. Rapidly settling particles may require several times as much power as required for off-bottom suspension. The additional power input may promote some liquid phase reactions associated with the process, but the increase in mass transfer between the particles and liquid is not likely to add enough benefit to justify the cost of the increased mixing intensity.

SOLIDS SUSPENSION BY MIXERS

Determining the degree of solids suspension may be used to evaluate the capabilities of existing equipment or the design of new equipment. In either case, the tank and mixer geometry are crucial and interrelated factors for solids suspension. Although many studies have focused on solids suspension, most of those studies have actually been directed at a rather limited range of equipment and solids often found to be the most effective. Most of the solids suspension studies have involved axial-flow impellers, either pitched-blade turbines (Fig. 18-5) or hydrofoil impellers (Fig. 18-6) [Grenville, R. K., A. T. C. Mak, and D. A. R. Brown, *Chem. Eng. Res. Des.* **100:** 282–291 (2015)]. Those studies have been done in baffled, cylindrical tanks with the liquid level equal to the tank diameter. The impeller diameters are most often between 33 and 45 percent of the tank diameter. Typically, the liquid suspending the particles is water, and the particles are sandlike with a relatively narrow particle size distribution. While these conditions are quite representative of some processes that require solids suspension, other factors such as particle-size distributions and density distributions are not well studied. Differences in impeller or tank geometry can have a significant effect on the capabilities of the equipment. In general, any change in impeller or tank geometry will have an observable effect on the degree of solids suspension.

Just-Suspended Speed Just-suspended speed is the mixer speed at which off-bottom suspension occurs. The definition and beginning for most technical work on solids suspension comes from a study by Zwietering [Zwietering, T. N., *Chem. Eng. Sci.* **8:** 244–253 (1958)]. In this paper, the definition of off-bottom suspension is, "When no deposits remained on the bottom for more than 1 sec, the suspension was considered complete." The transition from on-bottom motion to off-bottom suspension has been found to be sufficiently identifiable that other studies have used the method and found similar results [Armenante, P. M., E. U. Nagamine, and J. Susanto, *Can. J. Chem. Eng.* **76:** 413–419 (1998); Ayranci, I., T. Ng, A. W. Etchells, and S. M. Kresta, *Chem. Eng. Res. Des.* in press (2015)]. Zwietering developed a correlation for a dimensionless constant, *S*, which can be expressed with dimensionless variables as follows:

$$S = N_{\mathrm{Re}^{0.1}} N_{\mathrm{Fr}^{0.45}} \left(\frac{D}{d_p}\right)^{0.2} X^{0.13} \tag{18-16}$$

where X is the ratio of solids mass to liquid mass in the suspension, multiplied by 100, for a percent mass ratio of solids in liquid. The constant S contains the effects of all the geometry variables associated with impeller type, relative size, and location.

The Zwietering correlation is often written in a dimensional form to obtain the just-suspended speed:

$$N_{js} = S \frac{\nu^{0.1} d_p X^{0.13} \left(\dfrac{g \Delta\rho}{\rho_L}\right)^{0.45}}{D^{0.85}} \tag{18-17}$$

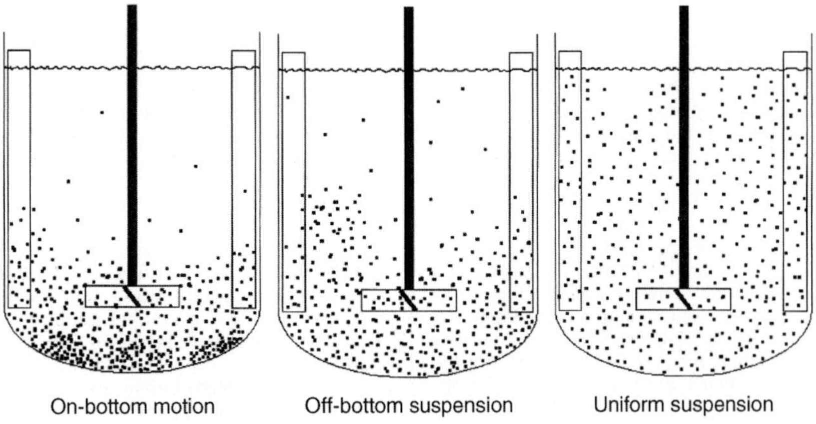

FIG. 18-22 Common descriptions for degrees of solids suspension.

The kinematic viscosity, ν, appears in this expression only because the Reynolds number was assumed to be an appropriate dimensionless group for data correlation. The viscosity was not varied by Zwietering, and it has been found by other investigators to have only a minor data-scattering effect with liquid viscosities less than about 200 cP. The particle size, d_p, has a stronger effect on the just-suspended speed than the particle density, which is the opposite of the effect that particle size and density have on terminal steeling velocity. The stronger effect of particle size is probably because the limiting mechanism for solids suspension is the turbulent velocity effect of lifting particles from the bottom, rather than the upward velocity flow for keeping the particles suspended [Ayranci, I., et al., *Chem. Eng. Sci.* **79:** 163–176 (2012)].

The three most important mixer geometric effects on solids suspension in a baffled tank are impeller type, impeller-diameter-to-tank-diameter ratio, and off-bottom clearance. For each combination of these three variables, a different *S* parameter is needed to use the Zwietering correlation, Eq. (18-17), to estimate a just-suspended speed, N_{js}. A study by Ayranci and Kresta [Ayranci, I., and S. M. Kresta, *Chem. Eng. Res. Des.* **89**(10): 1961–1971 (2011)] identified *S* values for a number of combinations of impeller type, *D/T*, and *C/T*; see Table 18-2.

That same reference discusses several other forms of the correlation for N_{js}, along with adjustments for solids loading, particle size distribution, and other effects.

Mixer Geometry Impeller type, size, and location are obvious geometry factors affecting the solids suspension capabilities of a mixer. Based on the mixer and tank geometry, the rotational speed can increase or reduce the degree of suspension. The mixer speed has a direct effect on the power and torque required by the mixer to achieve a necessary degree of suspension. A factor like off-bottom location of an impeller may have a relatively minor effect on blend time, heat transfer, or other liquid mixing requirement. Off-bottom location of the main or lower impeller will have a significant effect on solids suspension. Within a practical range, the closer an axial-flow impeller is placed to the bottom of the tank, the less power and torque are required for off-bottom suspension, although vertical uniformity may be reduced at low impeller clearances. Impeller location is important because the primary mechanism for suspension is lifting the particles off the bottom.

Off-bottom distance is a critical design variable for off-bottom suspension [Armenante, P. M., and E. U. Nagamine, *Chem. Eng. Sci.* **53**(9): 1757–1775 (1998)]. In an earlier comparison of a radial-flow straight-blade turbine (Fig. 18-12) with an axial-flow pitched-blade turbine (Fig. 18-10) at equal power, torque, and speed, the straight-blade turbine failed to achieve off-bottom suspension. The solids suspension problem for the straight-blade turbine can be solved by placing the impeller close to the bottom of the vessel as shown in Fig. 18-23.

The discharge flow from the impeller must sweep across the bottom of the vessel with sufficient turbulence and flow to lift the suspended particles off the bottom.

The effect of off-bottom clearance with other impeller types also emphasizes how important the flow pattern can be in solids suspension results. The off-bottom suspension demonstrated by a pitched-blade turbine in Fig. 18-10 was with an off-bottom clearance of $C/T = 1/4$, which works well for many solids suspension applications. Clearances less than $C/T = 1/4$ may not be as effective for blending liquids or suspending solids in the upper part of the tank. The pitched-blade turbine works well at close clearance and fairly well at $C/T = 1/3$, but it fails to suspend solids at $C/T = 1/2$, as shown in Fig. 18-24.

The discharge flow from the pitched-blade turbine spreads enough at large clearance that it does not sweep the bottom of the tank, and solids suspension is lost. However, with the more axial flow from a hydrofoil impeller, the discharge spread is less, and the hydrofoil impeller still suspends solids at $C/T = 1/2$, as shown in Fig. 18-25.

Caution must be exercised when using any correlation for mixing performance, especially for solids suspension, to avoid using equipment parameters outside the range of values covered by the correlation.

To further emphasize that a strong flow pattern across the bottom of the tank is needed to lift and suspend solids, see the comparison of the typical down-pumping, pitched-blade turbine compared with the up-pumping turbine in Fig. 18-26.

Impeller location can be even more important when considering multiple impellers. Two impellers may be needed in situations where the liquid level is greater than the tank diameter, but also advantageous when the liquid level is equal to the tank diameter and greater vertical uniformity is needed. The lower impeller does most of the work of lifting the particles off the bottom, while the upper impeller helps distribute the particles more evenly in the upper part of the tank [Montante, G., D. Pinelli, and F. Magelli, *Chem. Eng. Sci.* **58:** 5363–5372 (2003)].

TABLE 18-2 Zwietering *S* Values for Various Impellers and Geometries in Flat-Bottom Vessels

Impeller	*D/T*	*C/T*	*S*	$Pjs/P_{HE\text{-}3}$
A310	0.417	0.250	6.90	1.017
(*NP* = 0.3)	0.500	0.250	7.10	1.712
	0.520	0.170	6.39	1.373
	0.520	0.250	7.03	1.831
	0.520	0.330	7.71	2.424
45/PBT	0.333	0.167	4.87	0.881
(*NP* = 1.3)	0.333	0.250	5.58	1.339
	0.333	0.333	6.39	2.000
	0.500	0.167	2.72	0.424
	0.500	0.250	2.77	0.441
	0.500	0.333	3.40	0.814
	0.714	0.125	4.50	4.542
	0.714	0.250	5.40	7.847
HE-3	0.350	0.170	7.07	0.831
(*NP* = 0.35)	0.350	0.250	7.39	0.949
	0.350	0.330	8.17	1.271
	0.390	0.170	6.60	0.881
	0.390	0.250	6.88	1.000
	0.390	0.330	7.82	1.458
	0.440	0.170	6.49	1.119
	0.440	0.250	6.64	1.203
	0.440	0.330	7.23	1.542
	0.470	0.170	6.26	1.186
	0.470	0.250	6.25	1.169
	0.470	0.330	6.81	1.525
	0.520	0.170	6.89	2.017
	0.520	0.250	6.88	2.000
	0.520	0.330	7.72	2.831
PBT Down 6 Blade	0.380	0.170	4.24	1.051
(*NP* = 1.7)	0.380	0.250	3.99	0.881
	0.380	0.330	4.78	1.508
	0.520	0.170	5.39	4.695
	0.520	0.250	5.72	5.610
	0.520	0.330	6.52	8.305
PBT Up 6 Blade	0.520	0.170	5.14	4.068
(*NP* = 1.7)	0.520	0.250	5.19	4.186
	0.520	0.330	5.30	4.458
RT (Rushton)	0.250	0.143	8.70	8.288
(*NP* = 5.5)	0.250	0.167	9.20	10.949
	0.250	0.200	9.90	13.644
	0.330	0.170	5.42	4.610
	0.330	0.250	6.96	9.729
	0.330	0.330	8.37	16.949
	0.333	0.143	5.80	4.966
	0.333	0.167	6.10	6.593
	0.333	0.200	6.60	8.525
	0.500	0.143	3.20	2.305
	0.500	0.167	3.40	3.203
	0.500	0.170	4.34	6.542
	0.500	0.200	3.60	3.881
	0.500	0.250	4.44	7.000
	0.500	0.330	4.69	8.254
Intermig Single	0.600	0.170	6.78	4.763
(*NP* = 0.61)	0.600	0.250	6.85	4.898
	0.600	0.330	7.55	6.559
Intermig Double	0.600	0.170	7.44	6.271
(*NP* = 0.74)	0.600	0.250	8.30	8.729
	0.600	0.330	8.72	10.119

From Ayranci and Kresta 2011.

Not all solids suspension is done in baffled cylindrical tanks. Tanks with square or rectangular cross sections are found in several applications. The corners of a square tank may provide some baffling effect for an axial-flow mixer, but the bottom corners are the most likely places for solids to accumulate. Some modified design considerations are necessary for square tanks [Mitchell, E. T., K. J. Myers, E. Janz, and J. B. Fasano, *Can. J. Chem. Eng.* **86:** 110–116 (2008)].

The importance of geometry effects on solids suspension cannot be understated, as demonstrated by the effect of baffle off-bottom clearance on solids suspension in a flat-bottom tank [Myers, K. J., and J. B. Fasano, *Can. J. Chem. Eng.* **70:** 596–599 (1992)]. The most successful baffle clearance was one-half of the baffle width off the bottom of the tank.

While correlated values of N_{js} provide a numerical measure for agitation intensity required for solids suspension, the tangible effects can be better understood by observation of the solids suspension (Fig. 18-27).

A small pitched-blade turbine, $D/T = 0.2$, provides off-bottom suspension in a baffled tank with a dished bottom. At equal power, a larger impeller, $D/T = 0.4$, keeps solids off the bottom and also drives the suspension further up in the tank. However, at equal power, the larger impeller operates at a lower speed, which means higher torque and a bigger drive, typically with a higher cost. In the extreme of a large impeller, $D/T = 0.6$, the discharge flow from the pitched-blade turbine no longer sweeps across the bottom of the tank, and it fails at any practical speed to get off-bottom suspension.

Close clearance 1/4 Clearance

FIG. 18-23 Straight-blade turbine at close clearance gives better solids suspension.

Close clearance 1/3 Clearance 1/2 Clearance

FIG. 18-24 Off-bottom clearance affects solids suspension with a pitched-blade turbine.

Close clearance 1/3 Clearance 1/2 Clearance

FIG. 18-25 Off-bottom clearance has little effect on solids suspension with a hydrofoil impeller.

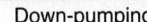

Down-pumping Up-pumping

FIG. 18-26 Up-pumping impeller does not suspend solids.

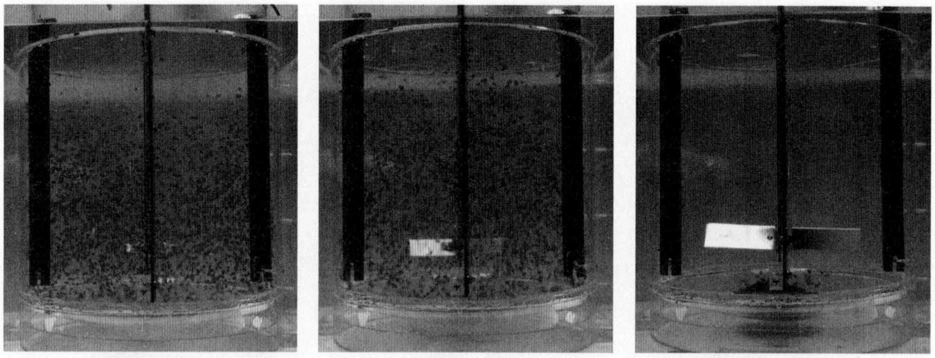

FIG. 18-27 Solids suspension with a pitched-blade turbine at equal power but with different diameter impellers in the same tank.

Cloud Height Cloud height provides a visual description of solids suspension in the upper part of an agitated vessel. Moderate concentrations of similar size and density solids can be suspended in what appears to be a cloud of particles. The height of the top of the cloud can be measured as the cloud height [Bittorf, K. J., and S. M. Kresta, *Chem. Res. Des.* **81**(5): 568–577 (2003); Hicks, M. T., K. J. Myers, and A. Bakker, *Chem. Eng. Commun.* **160:** 137–155 (1997)]. The cloud height is potentially a measure of intermediate solids suspension between off-bottom suspension and uniform suspension. However, the cloud height is primarily a visual observation associated with transparent tanks used for pilot-scale studies. If a cloud height is clearly defined, it may even act as an interfacial barrier between the moving suspension and the relatively clear upper layer. This suspension/clear interface may delay vertical blending of liquid additions.

Properties of Solids All of the properties of particulate solids have some effect on their suspension. Some properties are more important than others, and some effects are a result of combined factors. The concentration of solids has an effect on both the difficulty to suspend and the properties of the fluid suspension. Slurries of suspended solids are more difficult to handle than the liquid component alone for a number of reasons [Merrow, E. W., *Chemical Innovation* **30:** 35–41 (2000)]. Several studies have looked at different particle properties and their effect on solids suspension [Myers, K., J. Fasano, and R. Corpstein, *Can. J. Chem. Eng.* **72:** 745–748 (1994); Myers, K., E. E. Janz, and J. Fassano, *Can. J. Chem. Eng.* **91:** 1508–1512 (2013); Shamlou, P. A., *I. Chem. E. Symp. Ser.* **121:** 367–413 (1990); Ditl, P., and B. Nauman, *AIChE J.* **38**(6): 959–965 (1992)].

Particle size has a greater effect on solids suspension than particle density, partly because of the greater range of particle sizes. Particles can easily range from the submicron size to millimeter size, representing four orders of magnitude. Density differences are almost always less than a factor of 5 and at most a factor of about 20. However, particle size may not be an easy dimension to establish because particle shape also enters the definition. Many particles can be irregular shapes, and some suspensions contain combinations of differently shaped particles. On a simple scale, most particles can be approximated by a sphere (diameter), a rod (diameter and length), a plate (height, width, and length) or irregular shapes, like agglomerates or fibers. Within the sphere, rod, and plate categories, an approximation to an equivalent length dimension is usually measured by the smallest of the main dimensions. Rods or plates falling through a liquid will tend to align with the narrowest face in the lead because of drag minimization. In general, spherical particles of equal mass tend to be the most difficult to suspend, which makes an equivalent spherical diameter a good starting point for particle diameter in solids suspension estimates.

The concentration of solids has a relatively minor effect on solids suspension, as demonstrated by the small exponent, 0.13, on the mass ratio in the Zwietering expression for N_{js}, Eq. (18-17). Doubling the solids concentration only increases the N_{js} by about 9 percent, which increases the power in turbulent conditions by 31 percent [Choudhury, N. H., W. R. Penney, K. Myers, and J. B. Fasano, *AIChE Symp. Ser.* **305**(91): 131–138 (1995)].

Solids Suspension Scale-Up Because most of the understanding of solids suspension comes from empirical observation of pilot-scale test results, scale-up is also an empirical process. The mechanisms by which suspension is initiated and carried out are a combination of factors involving both local turbulence and an effective flow pattern across the bottom of the tank. That combination of factors does not lead to simple hydrodynamic mechanisms. While geometric similarity is often used for mixing scale-up, it is especially important for solids suspension because of the many geometry factors affecting degree of suspension. The one aspect of geometric similarity that does not apply to solids suspension is size of the suspension particles. All solids suspension evaluations treat the liquid and suspended solids as "the fluid" being agitated. Even as equipment becomes larger in scale-up, the particle size and concentration are kept the same, so that the fluid properties do not change.

With geometric similarity scale-up, all of the linear dimensions of the large-scale mixer are effectively set by the dimensional ratios of the small-scale mixer. With scale down, the dimensions of the small-scale test should be set by the geometric ratios of the large-scale mixer being evaluated. The only remaining variable in scale change with geometric similarity is the rotational speed. The adjustment to the speed should be in some proportion with respect to the speed in the other scale. The scale ratio can be calculated for any of the length dimensions because all of the ratios will be the same with geometric similarity. The scale ratio between test sizes is usually raised to an exponent to hold some mixing characteristic constant as represented in the following equation:

$$N_2 = N_1\left(\frac{T_1}{T_2}\right)^n = N_1\left(\frac{D_1}{D_2}\right)^n \tag{18-18}$$

The exponent n on the scale ratio decides which operating variable is held constant as the scale changes from size 1 to size 2, by either scale-up or scale-down. An exponent of one, $n = 1$, will keep impeller tip speed constant and all other velocities in the flow pattern the same. An exponent of two-thirds, $n = 2/3$, will keep power per volume constant between scales for turbulent conditions. Constant power per volume is also constant power per mass with the same fluid density. The smaller exponent on the scale ratio will make a smaller-speed change between scales. Therefore, a power per volume scale-up will result in a higher large-scale speed than equal tip speed. The higher large-scale speed also represents a higher power and torque in the large scale, which is a more conservative scale-up criterion than equal tip speed. The opposite power and torque comparisons are true for scale-down.

The off-bottom suspension speed from Zwietering's correlation, Eq. (18-17), shows the impeller diameter with an exponent of 0.85. With geometric similarity, the length ratios are all the same, so the scale ratio exponent should be 0.85. However, Zwietering recommends using equal power per volume, which is an exponent of 2/3. Other studies have tried to correlate scale change results using power per volume as a parameter, which by default makes scale-up by equal power per volume. The article by Corpstein et al. presents the scale-ratio exponent as a function of particle settling rate, as shown in Fig. 18-28 [Corpstein, R. R., J. B. Fasano, and K. J. Myers, "The High-Efficiency Road to Liquid-Solid Agitation, *Chem. Eng.* (October 1994), pp. 138–144].

The variable exponent offers approximately a power per volume effect for particles in the settling range between 0.05 and 0.10 m/s, which is a typical range for test work and many industrial applications. The variable exponent also shows that at low settling rates, when the particles tend to follow the flow of the liquid, the scale-change exponent results in equal velocity, $n = 1$. Exponents smaller than 2/3 result in requirements for more than equal power per volume, approaching equal Froude number, $n = 1/2$.

Solids Incorporation Getting solids into a liquid can be the limiting performance criterion for solids suspension, especially with low-density solids that might be easily suspended. Solids that are less dense than the liquid will result in floating solids, making incorporation a difficult and continuous

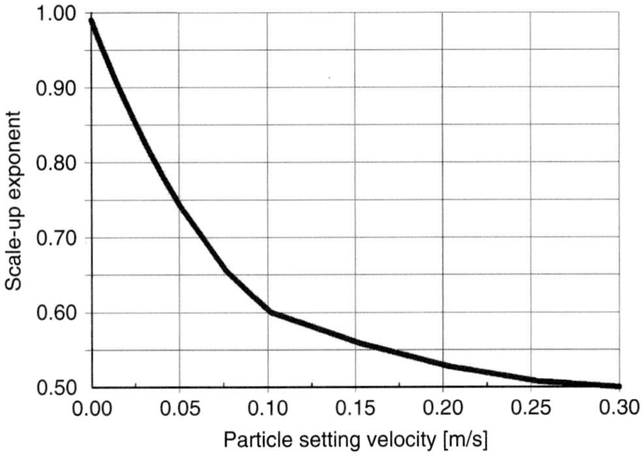

FIG. 18-28 Scale-ratio exponent may change with solids settling rate.

operating problem. Solids with a density only slightly greater than the liquid may be difficult to add and incorporate because of surface tension, nonwetting solid properties, or agglomerates containing air bubbles.

Studies involving floating solids have found different methods to help incorporate solids, including reduced number of baffles, baffles cut off below the surface, and up-pumping impellers [Edwards, M. F., and D. I. Ellis, *Fluid Mixing II, I. Chem. E. Symp. Ser.* **89:** 1–13 (1984); Khazam, O., and S. M. Kresta, *Can. J. Chem. Eng.* **86**(4): 622–634 (2008); Khazam, O., and S. M. Kresta, *Chem. Eng. Res. Des.* **87**(3): 280–290 (2009); Özcan-Taşkin, N. G., and D. Wei, *Chem. Eng. Sci.* **58:** 2011–2022 (2003); Özcan-Taşkin, N. G., *Chem. Eng. Sci.* **61:** 2871–2879 (2006)]. The basic idea is to create a sufficiently active surface so that the floating solids can break through the surface tension, then be drawn into the flow pattern and down through the impeller region for dispersion. The surface motion is a strong function of the amount of liquid above the impeller closest to the surface. Less impeller coverage results in more vigorous surface motion and better solids incorporation. Problems may develop when a large quantity of solids is added, causing the liquid level to increase. The result is less surface motion after most of the solids have been added, which might also be the conditions causing the greatest difficulty in adding solids.

Crystallization Crystallization is a process by which one chemical can be brought out of a solution and made into a solid. The process is a practical method for obtaining pure commercial substances in a form that is more suitable for handling. The primary objectives are crystal yield and purity, with secondary objectives of crystal size and shape. The primary steps in crystallization are nucleation and growth. Nucleation has several physical steps, but it is essentially the spontaneous formation of new crystals from a supersaturated solution. Growth is the process by which small crystals become larger crystals. The objective of most crystallization processes is the formation of uniform size, large crystals for appearance, filtering, consistent behavior, and minimal caking. For uniformity, crystal growth is desired over nucleation. Strong single crystals are sought over aggregates of crystals, which are likely to be fragile and break, forming small pieces. To achieve uniform growth, the circulation pattern created by the mixer should be as uniform and consistent as possible. A uniform circulation pattern is often enhanced by the use of a draft tube, Fig. 18-15. Other mixer features include smooth surfaces, minimized mechanical energy, hot feed below the surface, and a dense slurry to encourage growth and minimize nucleation.

GAS-LIQUID SYSTEMS

Gas-Liquid Dispersion Gas-liquid dispersion involves the physical dispersion of gas bubbles by the impeller and the effect of gas flow on the impeller power. Many gas-liquid systems also involve the simultaneous suspension of solids. The solids may be microorganisms in fermentation processes for the production of pharmaceuticals or chemicals. The solids may also be catalysts used to convert chemicals in the liquid.

Gas-liquid-solid applications, such as industrial fermentation, are often done in large vessels. The large size makes tall vessels easier to build and ship, so the vessels are often two or three times as tall as the diameter. A typical large gas-dispersion vessel is shown in Fig. 18-29.

The gas, most often air, enters through a sparge ring near the bottom of the vessel and underneath the bottom impeller. In the 1960s and before, most gas-liquid operations were conducted using multiple flat-blade, disk-style turbines like the one in Fig. 18-30.

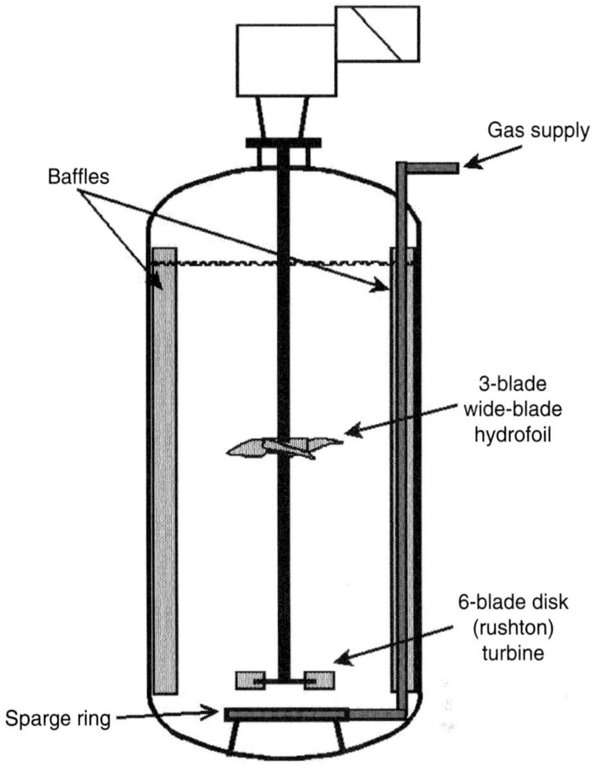

FIG. 18-29 Typical gas dispersion arrangement with bottom radial-flow, disk-style turbine, and upper wide-blade hydrofoil impeller.

More recently, the lower flat-blade turbines have been replaced by curved or cupped blades, like the impeller shown in Fig. 18-31, to reduce the tendency of gas bubbles to streamline the back of the flat blade.

This design change gives the impeller greater gas-handling capacity and reduces the change in power caused by the dispersed gas compared with power at zero gas rate. This impeller usually gives similar mass transfer rates at the same power levels as the flat-blade design and higher power at the same gas rate.

FIG. 18-30 Chemineer radial-flow disk-style (Rushton) turbine. (*Mixing Technologies Group of NOV.*)

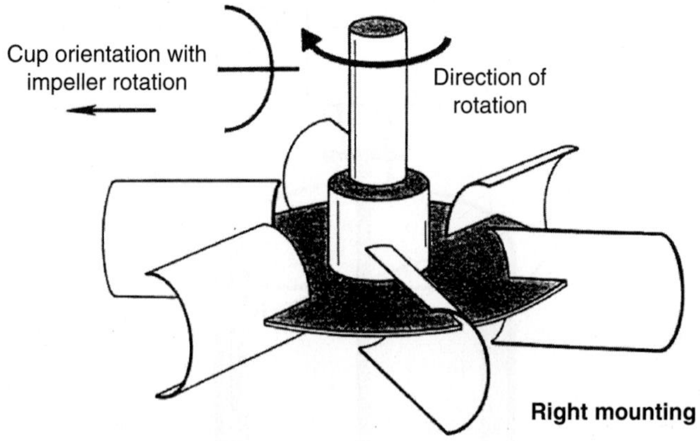

FIG. 18-31 Radial-flow cupped-blade (Smith) turbine.

Because of the high power number for the radial-flow impellers, a large amount of power would be required for blending uniformity in the upper part of the tank. In order to improve the blending and solid-suspension characteristics, hydrofoil impellers (typified by the A315, Fig. 18-32) have been used as upper impellers in tall tanks.

The wide-blade hydrofoil impellers provide both gas dispersion and blending uniformity. These impellers typically have a very high solidity ratio, on the order of 0.85 or more, and produce a strong axial downward flow at a low gas rate.

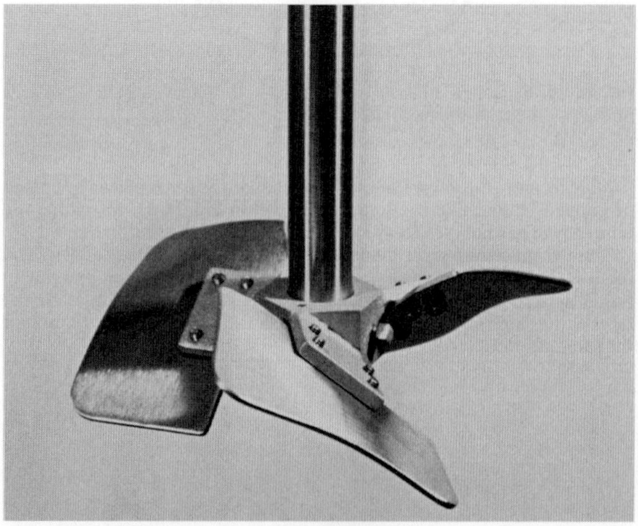

FIG. 18-32 Wide-blade hydrofoil impeller (A315) designed for gas dispersion and mass transfer.

As the gas rate increases, the flow pattern becomes more radial due to the upward flow of the gas counteracting the downward flow of the impeller. Some of the upper impellers are designed to pump upward in support of the rising gas flow. The up or down flow depends on the application and gas rate. Radial impellers are used for initial gas dispersion near the bottom of the tank, and axial impellers are used at as many upper locations as necessary to control the entire batch.

Effective design for simultaneous gas dispersion and solids suspension is difficult. The gas dispersion and radial-flow impeller are less effective at suspending solids than the axial-flow impellers in liquid-solid-only systems. However, the intense agitation necessary to disperse gas and promote mass transfer usually can overcome the solid suspension difficulties.

Gas-Liquid Mass Transfer Gas-liquid mass transfer normally is correlated by means of an overall mass-transfer coefficient, k_La, which is a function of power input and superficial gas velocity. The superficial gas velocity is the volume of gas at the local temperature and pressure divided by the cross-sectional area of the vessel. In order to obtain a mass transfer driving force, an assumption must be made about the partial pressure in equilibrium with the concentration of gas in the liquid. Many times this must be assumed, but if Fig. 18-33 is obtained in the pilot plant and the same assumption principle is used in evaluating the mixer in the full-scale tank, the error from the assumption is limited.

In the plant-size unit, Fig. 18-33 must be translated into a mass-transfer-rate curve for the particular tank volume and operating conditions selected. Every time a new physical condition is selected, a different curve is obtained. Typical exponents on the effect of power and gas rate on k_La tend to be around 0.5 for each variable, ±0.1.

Viscosity markedly changes the process. Usually, increasing the viscosity lowers the mass-transfer coefficient. For the common application of waste treating and for some of the published data on biological slurries, some data for k_La may be found in the literature. For a completely new gas or liquid in a liquid slurry system, data must be obtained by an experiment.

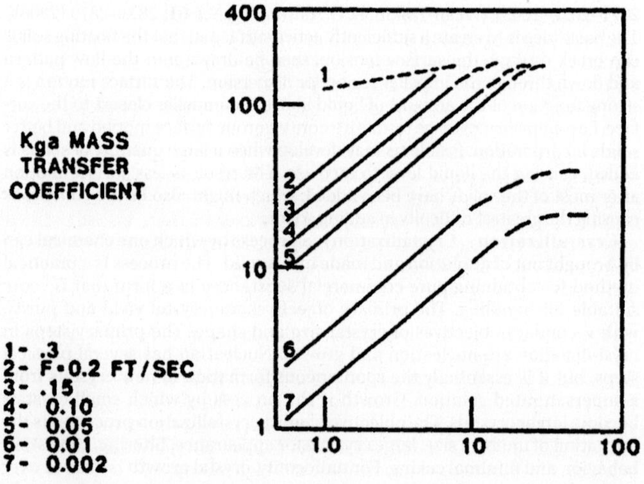

FIG. 18-33 Typical curves for mass transfer coefficient, k_La, as a function of mixer power and superficial gas velocity.

MIXING OF VISCOUS FLUIDS, PASTES, AND DOUGHS

General References: Harnby, N., M. F. Edwards, and A. W. Nienow, eds., *Mixing in the Process Industries,* 2d ed., Butterworth-Heinemann, Boston, 1992; Kresta, S. M., A. W. Etchells III, D. S. Dickey, and V. A. Atiemo-Obeng, eds., *Advances in Industrial Mixing: A Companion to the Handbook of Industrial Mixing,* Wiley, Hoboken, N.J., 2015; Oldshue, J. Y., *Fluid Mixing Technology,* McGraw-Hill, New York, 1983; Ottino, J. M., *The Kinematics of Mixing: Stretching, Chaos, and Transport,* Cambridge University Press, New York, 1999; Paul, E. L., V. A. Atiemo-Obeng, and S. M. Kresta, eds., *Handbook of Industrial Mixing, Science and Practice,* Wiley, Hoboken, N.J., 2004; Tatterson, G. B., *Fluid Mixing and Gas Dispersion in Agitated Tanks,* McGraw-Hill, New York, 1991; Zlokarnik, M., *Stirring, Theory and Practice,* Wiley-VCH, New York, 2001.

INTRODUCTION

Even the definition of mixing for viscous fluids, pastes, and doughs is complicated. While mixing can be defined simply as increasing or maintaining uniformity, the devices that cause mixing to take place may also accomplish deagglomeration, dispersion, extrusion, heat transfer, or other process objectives. Fluids with viscosities greater than 10 Pa · s (10,000 cP) can be considered *viscous.* However, nonnewtonian fluid properties are often as important in establishing mixing requirements. Viscous fluids can be polymer melts, polymer solutions, and a variety of other high-molecular-weight or low-temperature materials. Many polymeric fluids are shear thinning. Pastes are typically formed when particulate materials are wetted by a fluid to the extent that particle-particle interactions create flow characteristics similar to those of viscous fluids. The particle-particle interactions may cause shear-thickening effects. Doughs have the added characteristic of elasticity. Viscous materials often exhibit a combination nonnewtonian characteristics, and other characteristics such as a yield stress.

One common connection between viscous fluids, pastes, and doughs is the types of equipment used to mix or process them. While often designed for a specific process objective or a certain fluid characteristic, most types of

The contribution of the late Dr. J. Y. Oldshue, who authored part of this and many editions, is acknowledged.

viscous mixing equipment have some common characteristics. The nature of all viscous materials is their resistance to flow. This resistance is usually overcome by a mixer that will eventually contact or directly influence all the material in a container, particularly material near the walls or in corners. Small clearances between rotating and stationary parts of a mixer create regions of high local shear. Intermeshing blades or stators prevent material from rotating as a solid mass. Such equipment provides greater control of fluid motion than equipment used for low-viscosity fluids, but typically at greater cost and complexity.

The one failure common to all mixing equipment is any region of stagnant material. With a shear-thinning material, the relative motion between a rotating mixer blade and adjacent fluid will reduce the local viscosity. However, away from the mixer blade, shear will decrease and the viscosity will increase, leading to the possibility of stagnation. With a shear-thickening material, high shear near a mixer blade will result in high viscosity, which may reduce either local relative motion or the surrounding bulk motion. Yield stress requires some minimum shear stress to accomplish any motion at all. Viscoelastic characteristics cause motion normal to the applied stresses. Thus all major nonnewtonian characteristics reduce effective mixing and increase the possibility of local stagnation.

Blade shape and mixing action can have significant impacts on the mixing process. A scraping action is often necessary to promote heat transfer or prevent adhesion to equipment surfaces. A smearing action can improve dispersion. A combination of actions is necessary to accomplish the random or complicated pattern necessary for complete mixing. No one mixing effect or equipment design is ideal for all applications.

Because of high viscosity, the mixing Reynolds number ($N_{Re} = D^2N\rho/\mu$, where D is impeller diameter, N is rotational speed, ρ is density, and μ is viscosity) may be less than 100. At such viscous conditions, mixing occurs because of laminar shearing and stretching. Turbulence is not a factor, and complicated motion is a direct result of the mixer action. The relative motion between moving parts of the mixer and the walls of the container or other mixer parts creates both shear and bulk motion. The shear effectively creates thinner layers of nonuniform material, which diminishes striations or breaks agglomerates to increase homogeneity. Bulk motion redistributes the effects of the stretching processes throughout the container.

Often as important as or more important than the primary viscosity is the relative viscosity of fluids being mixed. When a high-viscosity material is added to a low-viscosity material, the shear created by the low-viscosity material may not be sufficient to stretch and interact with the high-viscosity material. When a low-viscosity material is added to a high-viscosity material, the low-viscosity material may act as a lubricant, thus allowing slippage between the high-viscosity material and the mixer surfaces. Viscosity differences can be orders of magnitude different. Density differences are smaller and typically less of a problem in viscous mixing.

Besides mixing fluids, pastes, and doughs, the same equipment may be used to create those materials. Viscous fluids such as polymers can be created by reaction from low-viscosity monomers in the same equipment described for viscous mixing. Pastes may be created by either the addition of powders to liquids or the removal of liquids from slurries, again using the same type of equipment as for bulk mixing. Doughs are usually created by the addition of a powder to liquid and the subsequent hydration of the powder. The addition process itself becomes a mixer application, which may fall somewhere between low-viscosity and high-viscosity mixing, but often including both types of mixing.

BATCH MIXERS

Anchor Mixers Anchor mixers are the simplest and one of the more common types of high-viscosity mixers (Fig. 18-34). The diameter of the anchor D is typically 90 to 95 percent of the tank diameter T. The result is a small clearance C between the rotating impeller and the tank wall. Within this gap, the fluid is sheared by the relative motion between the rotating blade and the stationary tank wall. The shear near the wall typically reduces the buildup of stagnant material and promotes heat transfer. To reduce buildups further, flexible or spring-loaded scrapers, typically made of polymeric material, can be mounted on the rotating blades to move material physically away from the wall.

The benefits of an anchor mixer are limited by the fact that the vertical blades provide very little vertical fluid motion between the top and bottom of the tank. Ingredient additions at the surface of the fluid may make many rotations before gradually being spread and circulated to the bottom of the tank. To promote top-to-bottom fluid motion, angled blades on the anchor or helical ribbon blades, described in the next subsection, make better mixers for uniform blending. Significant viscosity differences between fluids may extend mixing times to unacceptable limits with the basic anchor.

Anchor mixers may be used in combination with other types of mixers, such as turbine mixers, high-shear mixers, or rotor-stator mixers, which were described in the previous subsection. Such mixers can be placed on a vertical shaft midway between the anchor shaft and blade. A secondary mixer can promote top-to-bottom motion and also limit bulk rotation of the fluid. A stationary baffle is sometimes placed between the anchor shaft and rotating blade to limit fluid rotation and enhance shear.

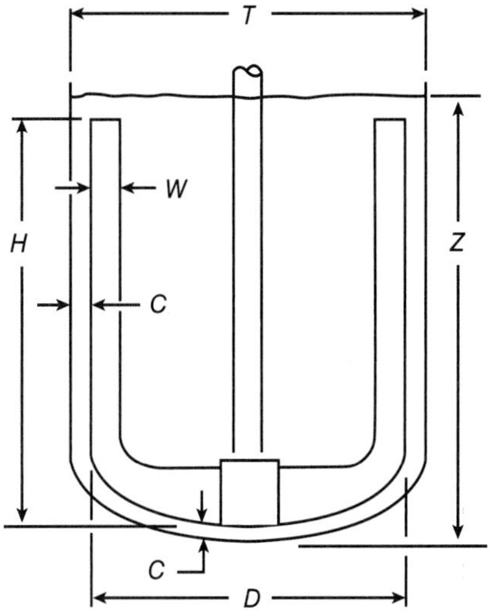

FIG. 18-34 Anchor impeller with nomenclature.

A dimensionless group called the power number is commonly used to predict the power required to rotate a mixing impeller. The *power number* is defined as $P/(\rho N^3 D^5)$, where P is power, ρ is fluid density, N is rotational speed, and D is impeller diameter. To be dimensionless, the units of the variables must be coherent, such as SI metric; otherwise appropriate conversion factors must be used. The conversion factor for common engineering units gives the following expression for power number:

$$N_p = \frac{1.524 \times 10^{13} P}{\text{sp gr}\, N^3 D^5} \tag{18-19}$$

where P is power in horsepower, sp gr is fluid-specific gravity based on water, N is rotational speed in rpm, and D is impeller diameter in inches. The power number is an empirically measured value that describes geometrically similar impellers. Power number is a function of Reynolds number, which accounts for the effects of fluid properties. Impeller Reynolds number, as defined earlier, is another dimensionless group. A conversion factor is needed for common engineering units:

$$N_{Re} = \frac{10.4 D^2 \text{sp gr}}{\mu} \tag{18-20}$$

where D is the impeller diameter in inches, N is rotational speed in rpm, sp gr is specific gravity based on water, and μ is viscosity in centipoise.

Power can be calculated by rearranging the definition of power number; see the following example. A value for the appropriate power number must be obtained from empirically derived data for geometrically similar impellers. Power number correlations for anchor impellers are shown in Fig. 18-35. The typical anchor impellers have two vertical arms with a blade width W equal to one-tenth of the impeller diameter D, and the arm height H equal to the impeller diameter D. Correlations are shown for typical impellers 95 and 90 percent of the tank diameter. The clearance C is one-half of the difference between the impeller diameter and the tank diameter, or 2.5 and 5.0 percent of the tank diameter for the respective correlations. An additional correlation is shown for an anchor with three vertical arms and a diameter equal to 95 percent of the tank diameter. The correlation for a three-arm impeller that anchors 90 percent of the tank diameter is the same as that for the typical anchor that is 95 percent of the tank diameter.

The power number and corresponding power of an anchor impeller are proportional to the height of the vertical arm. Thus, an anchor with a height H equal to 75 percent of the impeller diameter would have a power number equal to 75 percent of the typical values shown in Fig. 18-35. Similarly, a partially filled tank with a liquid level Z that covers only 75 percent of the vertical arm will also have a power number that is 75 percent of the typical correlation value. The addition of scrapers will increase the power

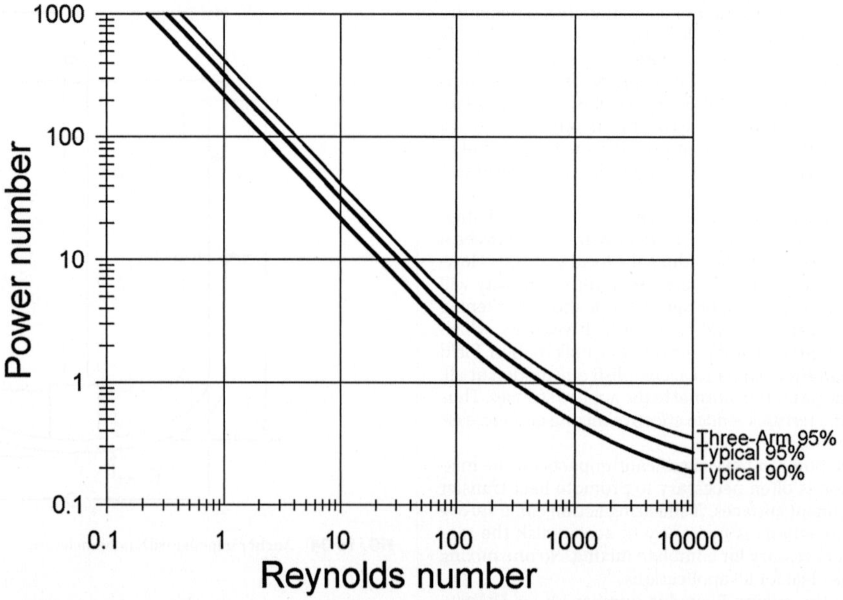

FIG. 18-35 Power numbers for anchor impellers: typical two-arm impeller anchors 95 percent of tank diameter *T* and 90 percent of *T*; three-arm impeller anchors 95 percent of *T*; and three-arm impeller anchors 90 percent of *T*, similar to two-arm impeller that anchors 95 percent of *T*.

requirement for an anchor impeller, but the effect depends on the clearance at the wall, the design of the scrapers, processed material, and many other factors. Correlations are not practical or available.

Unfortunately, the power number only provides a relationship between impeller size, rotational speed, and fluid properties. The power number does not tell whether a mixer will work for an application. Successful operating characteristics for an anchor mixer usually depend on experience with a similar process or experimentation in a pilot plant. Scale-up of pilot-plant experience is most often done for a geometrically similar impeller and equal tip (peripheral) speed.

Helical Ribbon Mixers Helical ribbon mixers (Fig. 18-36), or simply helix mixers, have major advantages over the anchor mixer because they force strong top-to-bottom motion even with viscous materials. These impellers are some of the most versatile mixing impellers but also some of the most expensive. Besides having a formed helical shape, the blades must be rolled the hard way, with the thick dimension normal to the direction of the circular rolled shape. Helical ribbon mixers will work with most viscous fluids up to the limits of a flowable material, as high as 4,000,000 cP or more, depending on nonnewtonian characteristics. While not cost-effective for low-viscosity materials, they will adequately mix, and even suspend solids, in low-viscosity liquids. These characteristics make helical ribbon mixers effective for batch processes, such as polymerization or other processes beginning with low-viscosity materials and changing to high-viscosity products. Helical ribbon mixers will even work with heavy pastes and flowable powders. Usually the helix pumps down at the tank wall with fluids and up at the wall with pastes or powders.

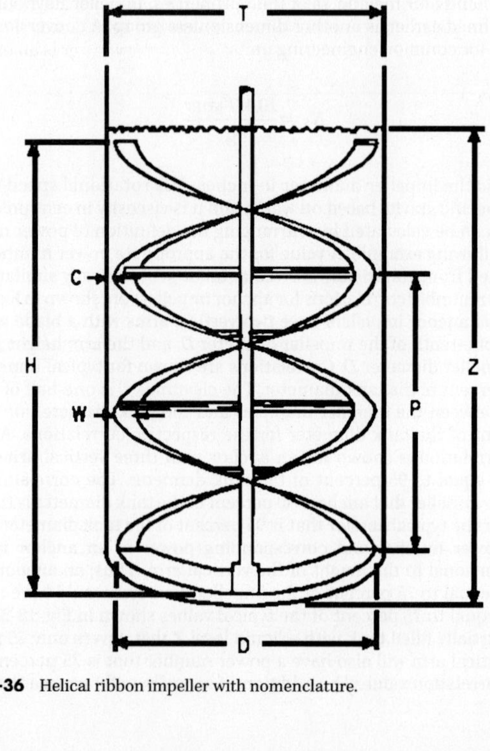

FIG. 18-36 Helical ribbon impeller with nomenclature.

The helical ribbon power numbers are a function of Reynolds number similar to the correlations for anchor impellers. Figure 18-37 shows correlations for some typical helical ribbon power numbers. The upper curve is for a double-flight helix with the blade width *W* equal to one-tenth the impeller diameter *D*, the pitch *P* equal to the impeller diameter, and the impeller diameter at 95 percent of the tank diameter *T*. The height *H* for this typical helix is equal to the impeller diameter and pitch, not 15 times the pitch, as shown in Fig. 18-37. A second curve shows the power number correlation for a helical ribbon impeller that is 90 percent of the tank diameter. The curve marked "Single 90%" is for a single-flight helix, 90 percent of the tank diameter. Each ribbon beginning at the bottom of the impeller and spiraling around the axis of the impeller is called a flight. Single-flight helixes are theoretically more efficient, but a partially filled tank can cause imbalanced forces on the impeller. The correlation for a 95 percent diameter single-flight helix is the same as the correlation for the double-flight 90 percent diameter helix.

Example 18-1 Calculate the Power for a Helix Impeller Calculate the power required to rotate a double-flight helix impeller that is 57 in in diameter, 57 in high, with a 57-in pitch operating at 30 rpm in a 60-in-diameter tank. The tank is filled 85 percent full with a 100,000-cP fluid, having a 1.05 specific gravity.

$$N_{\mathrm{Re}} = \frac{10.4 D^2 N \,\mathrm{sp\,gr}}{\mu} = \frac{10.4(57)^2(30)(1.05)}{100{,}000} = 10.6$$

Referring to Fig. 18-37, the power number N_P for the full-height helix impeller is 27.5 at $N_{\mathrm{Re}} = 10.6$. At 85 percent full, the power number is $0.85 \times 27.5 = 23.4$. Power can be calculated by rearranging Eq. (18-19).

$$P = \frac{N_p \,\mathrm{sp\,gr}\, N^3 D^5}{1.524 \times 10^{13}} = \frac{23.4(1.05)(30)^3(57)^5}{1.524 \times 10^{13}} = 26.2 \text{ hp}$$

Helical ribbon mixers can also be formed to fit in conical bottom tanks. While not as effective at mixing as in a cylindrical tank, the conical bottom mixer can force material to the bottom discharge. By more effectively discharging, a higher yield of the product can be obtained.

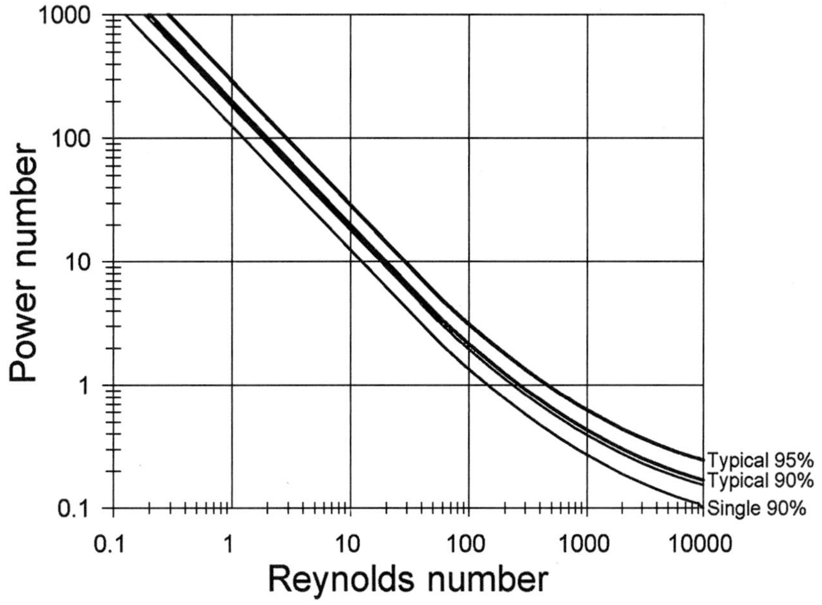

FIG. 18-37 Power numbers for helical ribbon impeller: typical double-flight helixes 95 percent of tank diameter T and 90 percent of T; single-flight helix 90 percent of T; single-flight 95 percent of T similar to double-flight 90 percent of T.

Planetary Mixers A variation on the single-anchor mixer is essentially a double-anchor mixer with the impellers moving in a planetary pattern. Each anchor impeller rotates on its own axis, while the pair of intermeshing anchors also rotates on the central axis of the tank. The intermeshing pattern of the two impellers gives a kneading action, with blades alternately wiping each other. The rotation around the central axis also creates a scraping action at the tank wall and across the bottom. With successive rotations of the impellers, all the tank contents can be contacted directly. A typical planetary mixer is shown in Fig. 18-38.

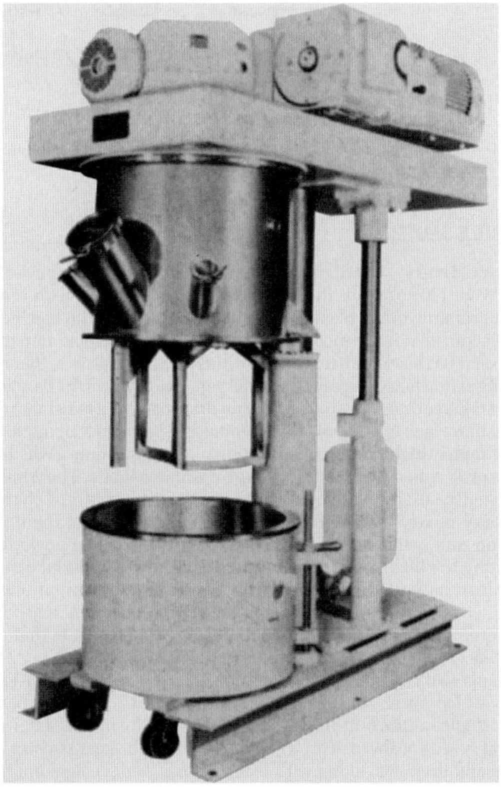

FIG. 18-38 Planetary mixer. (*Charles Ross & Son Company.*)

The intimate mixing provided by the planetary motion means that the materials need not actively flow from one location in the tank to another. The rotating blades cut through the material, creating local shear and stretching. Even thick pastes and viscoelastic and high-viscosity fluids can be mixed with planetary mixers. The disadvantage of poor top-to-bottom motion still exists with conventional planetary mixers. However, some new designs offer blades with a twisted shape to increase vertical motion.

To provide added flexibility and reduce batch-to-batch turnaround or cross-contamination, a change-can feature is often available with planetary and other multishaft mixers. The container (can) in a change-can mixer is a separate part that can be rapidly exchanged between batches. Batch ingredients can even be put in the can before it is placed under the mixing head. Once the mixing or processing is accomplished, the container can be removed from under the mixer and taken to another location for packaging and cleaning. After one container is removed from the mixer and the blades of the impeller are cleaned, another batch can begin processing. Because the cans are relatively inexpensive compared with the cost of the mixer head, a change-can mixer can be better utilized, and processing costs can be reduced.

Double- and Triple-Shaft Mixers The planetary mixer is an example of a double-shaft mixer. However, many different combinations of mixing actions can be achieved with multishaft mixers. One variation on planetary motion involves replacing one anchor-style impeller with a high-shear impeller. The high-shear mixer can be used to incorporate powdered material effectively or create a stable emulsion leading to a final batch of viscous paste or fluid.

Many types of multishaft mixers do not require planetary motion. Instead the mixers rely on an anchor-style impeller to move and shear material near the tank wall, while another mixer provides a different type of mixing. The second or third mixer shafts may have a pitched-blade turbine, hydrofoil impeller, high-shear blade, rotor-stator mixer, or other type of mixer. The combination of multiple impeller types adds to the flexibility of the total mixer. Many batch processes involve different types of mixing over a range of viscosities. Some mixer types provide the top-to-bottom motion that is missing from the anchor impeller alone.

Double-Arm Kneading Mixers A double-arm kneader consists of two counter-rotating blades in a rectangular trough with the bottom formed like two overlapping or adjacent half-cylinders (Fig. 18-39). The blades are driven by gearing at one or both ends. The older-style kneaders emptied through a door or valve at the bottom. Those mixers are still used where complete discharge or thorough cleaning between batches is not essential. More commonly, double-arm kneaders are tilted for discharge. The tilting mechanism may be manual, mechanical, or hydraulic, depending on the size of the mixer and weight of the material.

A variety of blade shapes have evolved for different applications. The mixing action is a combination of bulk movement, shearing, stretching, folding, dividing, and recombining. The material being mixed is also squeezed and stretched against the blades, bottom, and sidewalls of the mixer. Clearances

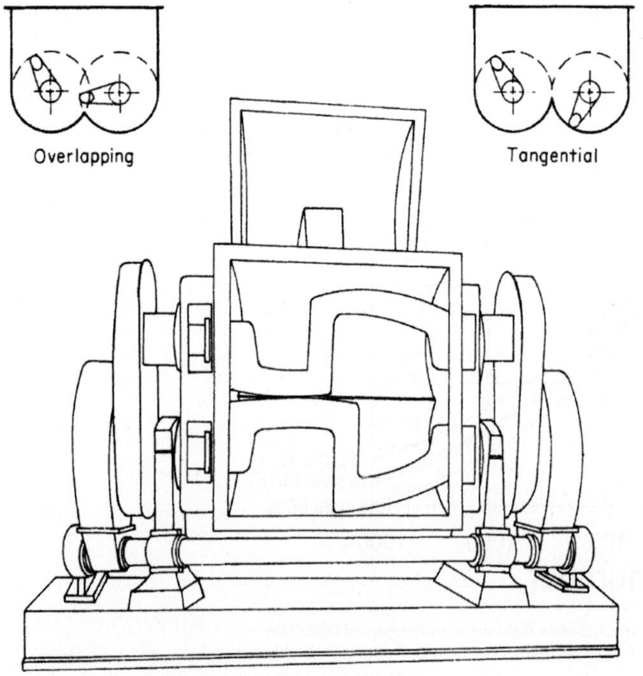

FIG. 18-39 Double-arm kneader. (*APV Baker Invensys.*)

may be as close as 1 mm (0.04 in). Rotation is usually such that the material is drawn down in the center between the blades and up at the sidewalls of the trough. Most of the blades are pitched to cause end-to-end motion.

The blades can be tangential or overlapping. Tangential blades can run at different speeds with the advantages of faster mixing caused by changes in the relative position of the blades, greater heat-transfer surface area per unit volume, and less tendency for the material to ride above the blades. Overlapping blades can reduce the buildup of material sticking to the blades.

Because the materials most commonly mixed in kneaders are very viscous, often elastic or rubbery materials, a large amount of energy must be applied to the mixer blades. All that energy is converted to heat within the material. Often the material begins as a semisolid mass, with liquid or powder additives, and the blending process both combines the materials and heats them to create uniform bulk properties.

The blade design most commonly used is the sigma blade (Fig. 18-40*a*). The sigma-blade mixer can start and operate with either liquids or solids, or a combination of both. Modifications to the blade faces have been introduced to increase particular effects, such as shredding or wiping. The sigma blades can handle elastic materials and readily discharge materials that do not stick to the blades. The sigma blades are easy to clean, even with sticky materials.

The dispersion blade in Fig. 18-40*b* was developed to provide higher compressive shear than the standard sigma blade. The blade shape forces material against the trough surface. The compressive action is especially good for dispersing fine particles in a viscous material. Rubbery materials have a tendency to ride the blades, and a dispersion blade is often used to keep the material in the mixing zone.

Multiwiping overlapping (MWOL) blades (Fig. 18-40*c*) are commonly used for mixtures that start tough and rubberlike. The blade shape initially cuts the material into small pieces before plasticating it.

The single-curve blade (Fig. 18-40*d*) was developed for incorporating fiber reinforcement into plastics. In this application, the individual fibers, such as glass, must be wetted with the polymer without undue fiber breakage.

Many other designs have been developed for specific applications. The double-naben blade (Fig. 18-40*e*) is good for mixtures that "ride," meaning they form a lump that bridges across the sigma blade.

Screw-Discharge Batch Mixers A variant of the sigma-blade mixer has an extrusion-discharge screw located at the center of the trough, just below the rotating blades. During the mixing cycle, the screw moves the material within the reach of the mixing blades, thus accelerating the mixing process. At discharge time, the screw extrudes the finished material through a die opening in the end of the machine. The discharge screw is driven independently of the mixer blades.

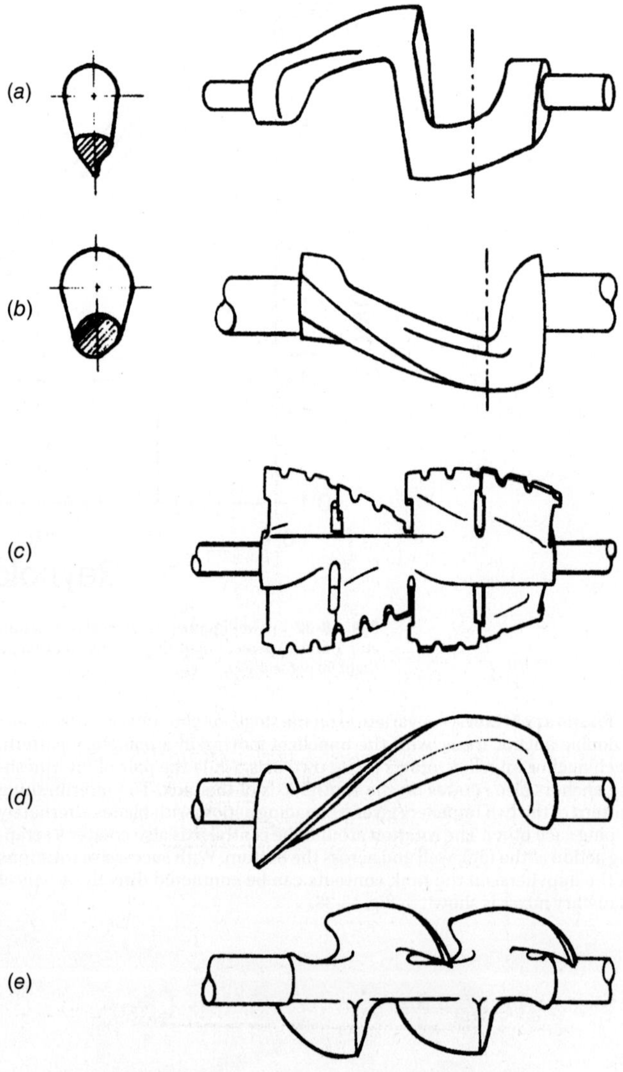

FIG. 18-40 Agitator blades for double-arm kneader: (*a*) sigma; (*b*) dispersion; (*c*) multiwiping; (*d*) single-curve; (*e*) double-naben. (*APV Baker Invensys.*)

INTENSIVE MIXERS

Banbury Mixers The dominant high-intensity mixer, with power input up to 6000 kW/m^3 (30 hp/gal), is the Banbury mixer made by Farrel Co. (Fig. 18-41). It is used primarily in the plastics and rubber industries. The batch charge of material is forced into the mixing chamber by an air-operated ram at the top of the mixer. The clearance between the rotors and the walls is extremely small. The mixing action takes place in that small gap. The rotors of the Banbury mixer operate at different speeds, so one rotor can drag material against the rear of the other and thus clean ingredients from behind and between the rotors.

The extremely high power consumption of these machines, which operate at speeds of 40 rpm or less, requires large-diameter shafts. The combination of heavy shafts, stubby blades, close clearances, and a confined charge limits the Banbury mixer to small batches relative to the size of the mixer. The production rate is increased as much as possible by using powerful drives and rotating the blades at the highest speed that the material can tolerate without degradation. The heat added by the high-power input often limits operating conditions because of temperature limits on the material being mixed. Equipment is available from laboratory size to a mixer that can handle a 450-kg (1000-lb) charge and applying 2240 kW (3000 hp).

High-Intensity Mixers Mixers such as the one shown in Fig. 18-42 combine a high-shear zone with a fluidized vortex for mixing pastes and powders. Blades at the bottom of the vessel scoop the material upward with peripheral speeds of about 40 m/s (130 ft/s). The high shear stresses between the blade and the bowl, along with blade impact, reduce agglomerates and create an intimate dispersion of powders and liquids. Because the energy input is high, 200 kW/m^3 (8 hp/ft^3), even powdery material can heat rapidly.

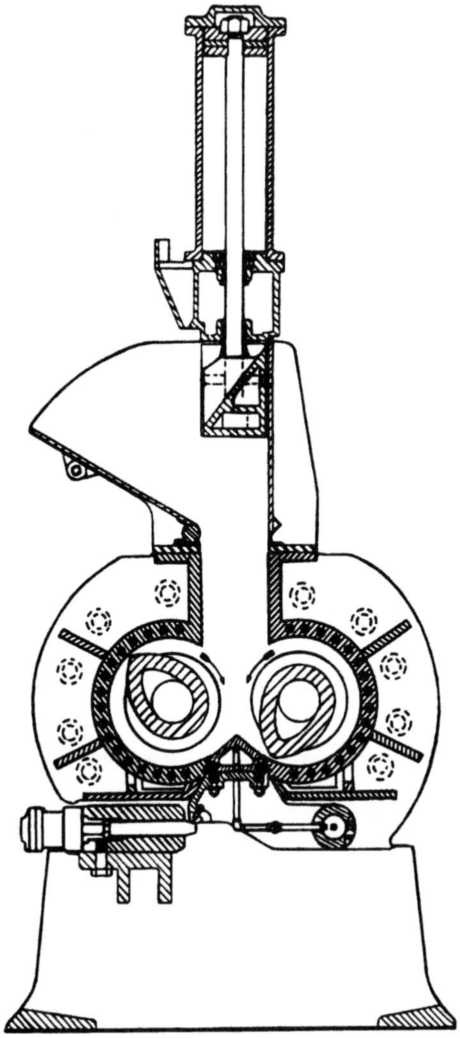

FIG. 18-41 Banbury mixer. (*Farrel Co.*)

These mixers are particularly suited for the rapid mixing of powders and granules with liquids, for dissolving resins or solids in liquids, or for removing volatiles from pastes under a vacuum. Scale-up is usually based on the constant peripheral speed of the impeller.

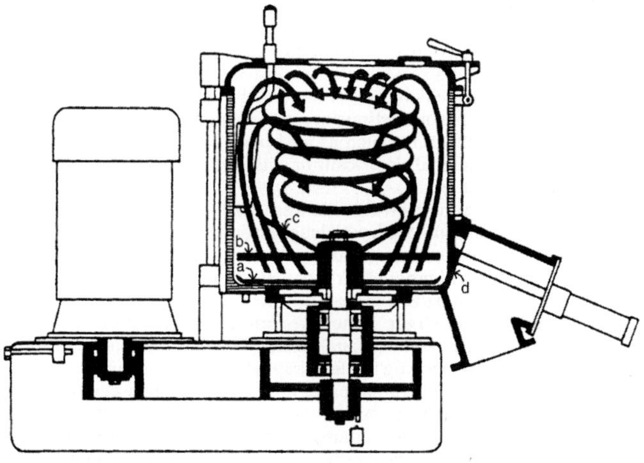

FIG. 18-42 High-intensity mixer: (*a*) bottom scraper; (*b*) fluidizing tool; (*c*) horn tool; (*d*) flush-mounted discharge valve. (*Henschel Mixers America, Inc.*)

Roll Mills Roll mills can provide extremely high localized shear while retaining extended surface area for temperature control. A typical roll mill has two parallel rolls mounted in a heavy frame with provisions for accurately regulating the pressure and distance between the rolls. Since one pass between the rolls does only a little blending, the mills are usually used as a series of mixers. Only a small amount of material is in the high-shear zone at a time, thus allowing time and exposure for cooling.

To increase the shearing action, the rolls are usually operated at different speeds. The material passing between the rolls can be returned to the feed by the rotation of the rolls. If the rolls are at different temperatures, the material will usually stick to the hotter roll and return to the feed point as a thick layer.

At the end of a period of batch mixing, heavy materials may be discharged by simply dropping from between the rolls. Thin, lighter mixes may be removed by a scraper bar pressing against the descending surface of one of the rolls. Roll mixers are used primarily for preparing color pastes for inks, paints, and coatings. A few applications in heavy-duty blending of rubber stocks use corrugated rolls for masticating the material.

Miscellaneous Batch Mixers Many mixers used for solids blending (Sec. 21 of eighth edition) are suitable for liquid-solids blending. Some solids processing applications involve the addition of liquids, and the same blenders may transition from dry powders to cohesive pastes.

Ribbon blenders typically have multiple helical ribbons with opposing pitches operating in a horizontal trough with a half-cylinder bottom. These mixers can be used for wetting or coating a powder. The final product may have a paste consistency, but it must remain at least partially flowable for removal from the blender.

Plowshare mixers have plow-shaped blades mounted at the ends of arms on a horizontal rotating shaft in a cylindrical chamber. The shaft rotates at a sufficient speed to toss the material into the free space in the vessel. The angled surfaces of the plow-shaped blades provide additional intermixing and blending in the bed of solids. High-speed (3600-rpm) chopper blades mounted in the lower side of the mixing chamber can disperse fine particles or break agglomerates. Mixers are available in sizes from 0.03- to 30-m^3 (1.0- to 1000-ft^3) working capacity. Plowshare mixers can be used for either batch or continuous processing. Paddle mixers are a variation of horizontal mixers where the plow blades are replaced with flat angled paddles.

Conical mixers are also known as Nauta mixers (Fig. 18-43). Material placed in the conical bin is lifted by the rotation of the helical screw, which in turn is rotated around the wall of the cone. The lifting actions of the screw combined with motion around the cone provide bulk mixing for flowable dry powders, paste materials, and even viscous fluids. The specific energy input is relatively small, and the large volume of the mixers can even provide storage capacity. The mixers may have multiple screws, tapered screws, and high-speed dispersers for different applications. At constant speed, both the mixing time and power scale up with the square root of volume. Sizes from 0.1 to 20 m^3 (3.3 to 700 ft^3) are available.

Pan mullers are the modern industrial equivalent of the traditional mortar and pestle. Typical mullers have two broad wheels (M1 and M2) on an axle (Fig. 18-44). The mixer rotates about the approximate midpoint of the axle, so that the wheels both rotate and skid over the bottom of the mixing chamber (A). Plow blades (P1 and P2), which rotate with the mixer, push material from the center (T) and walls (C) of the mixing chamber into the path of the rollers. The mixing action combines both crushing and shearing to break lumps or agglomerates and evenly distribute moisture.

Mullers can be used if the paste is not too fluid or sticky. The main application of muller mixers is now in the foundry industry to mix small amounts of moisture and binder with sand for both core and molding sand. Muller mixers also handle such diverse materials as clay, storage-battery paste, welding-rod coatings, and chocolate coatings. Standard muller mixers range in capacity from 0.01 to 1.7 m^3 (0.4 to 60 ft^3), with power requirements from 0.2 to 56 kW (⅓ to 75 hp).

A continuous muller design employs two intersecting and communicating chambers, each with its own mullers and plows. At the point of intersection of the two chambers, the outside plows give an approximately equal exchange of material from one chamber to the other. Material builds in the first chamber until the feed rate and the discharge rate of the material are equal. The quantity of material in the muller is regulated by adjusting the outlet gate.

CONTINUOUS MIXERS

Some batch mixers previously described can be modified for continuous processing. Product uniformity may be limited because of broad residence time distributions. If ingredients can be accurately metered, which can be a problem with powdered or viscous materials, several continuous mixers are available. Continuous mixers often consist of a closely fitting agitator element rotating within a stationary housing.

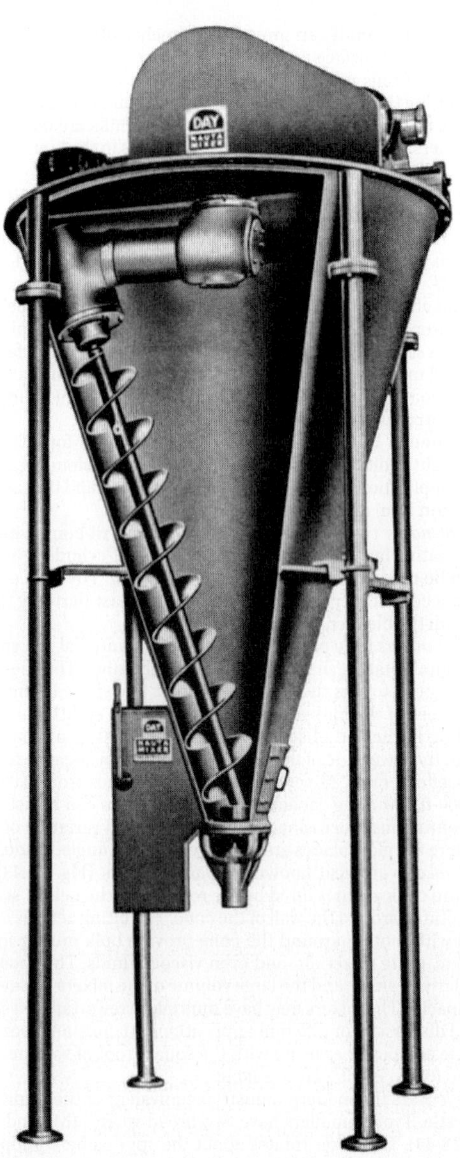

FIG. 18-43 Day Nauta conical mixer. (*Littleford Day, Inc.*)

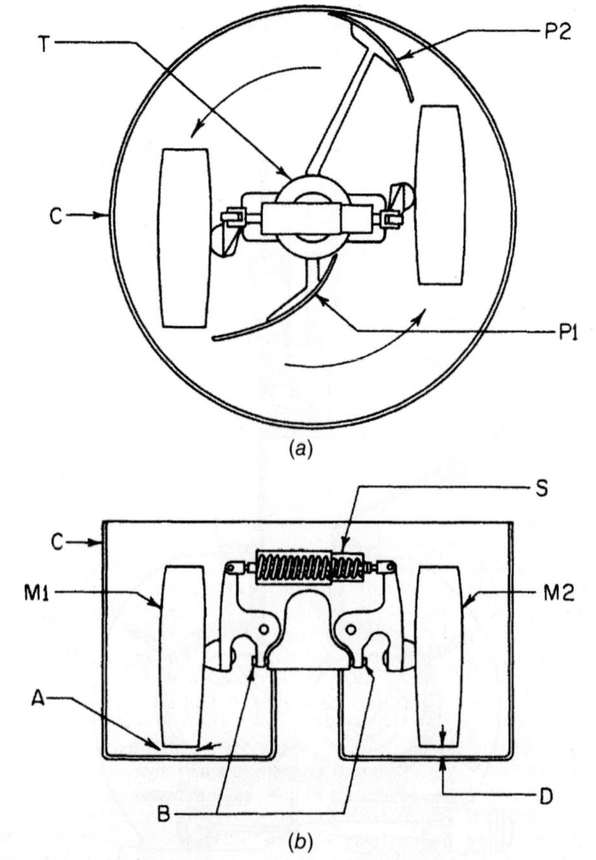

FIG. 18-44 Pan muller: (*a*) plan view; (*b*) sectional elevation. [*Bullock*, Chem. Eng. Prog. ***51:** 243 (1955); by permission.*]

Single-Screw Extruders The use of extruders, like the one shown in Fig. 18-45, is widespread in the plastic industries. The quality and utility of the product often depend on the uniformity of additives, stabilizers, fillers, and other ingredients. A typical extruder combines the process functions of melting the base resin, mixing in additives, and developing the pressure required for shaping the product into pellets, sheet, or profiles. Dry ingredients, sometimes premixed in a batch blender, are fed into the feed throat where the channel depth is deepest. As the root diameter of the screw is increased, the plastic is melted by a combination of friction and heat transfer from the barrel. Shear forces can be very high, especially in the melting zone. The mixing is primarily a laminar shear action.

Single-screw extruders can be built with a long length-to-diameter ratio to permit sufficient space and residence time for a sequence of process operations. Capacity is determined by diameter, length, and power. Most extruders are in the 25- to 200-mm-diameter range. Larger units have been made for specific applications, such as polyethylene homogenization. Mixing enhancers (Fig. 18-46) are used to provide both elongation and shearing action to enhance dispersive (axial) and distributive (radial) mixing.

The maximum power (P in kilowatts) supplied for single-screw extruders varies with the screw diameter (D in millimeters) approximately as

$$P = 5.3 \times 10^{-3} D^{2.25} \tag{18-21}$$

The energy required for most polymer mixing applications is from 0.15 to 0.30 kWh/kg (230 to 460 Btu/lb).

Twin-Screw Extruders Two screws in a figure-eight barrel have the advantage of interaction between the screws plus action between the screws and the barrel. Twin-screw extruders are used to melt continuously, mix, and homogenize different polymers and additives. Twin-screw extruders can also be used to provide the intimate mixing needed to carry out chemical reactions in high-viscosity materials. The screws can be either tangential or intermeshing, with the latter either co-rotating or counter-rotating. Tangential designs allow variability in the channel depth and permit longer lengths.

The most common twin-screw extruder is the counterrotating intermeshing type. The counterrotating intermeshing screws provide a dispersive milling action between the screws and can generate pressure efficiently. The two keyed or splined shafts are fitted with pairs of slip-on kneading or conveying elements, as shown in Fig. 18-47. Each pair of kneading paddles causes an alternating compression and expansion effect that massages the contents and provides a combination of shearing and elongational mixing actions. The arrays of elements can be varied to provide a wide range of mixing effects. The barrel sections are also segmented to allow for optimum positioning of features such as feed ports, vents, and barrel valves. The barrels may be heated electrically or with oil or steam and cooled with air or water.

Counterrotating twin-screw extruders are available in diameters ranging from 15 to 300 mm (0.5 to 12 in), with length-to-diameter ratios up to 50 and throughput capacities to 7 kg/s (55,000 lb/h). Screw speeds can be as high as 8 r/s (500 rpm) in small production extruders. Residence times for melting are usually less than 120 s (2 min).

Farrel Continuous Mixer The Farrel mixer consists of rotors similar in cross section to the Banbury batch mixer. The first section of the rotor acts a screw conveyor, moving the feed ingredients into the mixing section. The mixing action is a combination of intensive shear between the rotor and the chamber wall, kneading between the rotors, and a rolling action within the material itself. The amount and quality of mixing are controlled by the adjustment of speed, feed rate, and discharge orifice opening. Mixers are available with chamber volumes up to 4.2 ft^3. With speeds up to 200 rpm, the power range is from 7.5 to 300 hp.

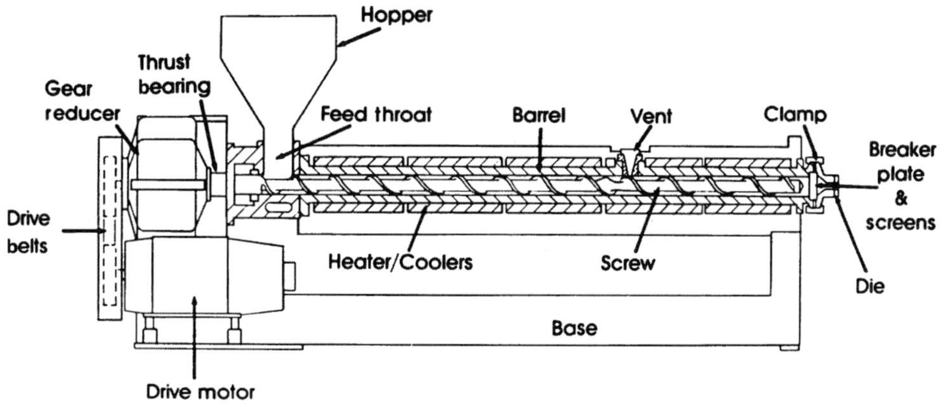

FIG. 18-45 Single-screw extruder. (*Davis Standard.*)

Miscellaneous Continuous Mixers Because of the diversity of material properties and process applications involving viscous fluids, pastes, and doughs, the types of mixers are almost as diverse.

Trough-and-screw mixers usually consist of a single rotor or twin rotors that continually turn the feed material over as it progresses toward the discharge end of the mixer. Some mixers have been designed with extensive heat-transfer surface area. The continuous-screw, *Holo-Flite processor* (Fig. 18-48), is used primarily for heat transfer, since the hollow screws provide extended surfaces without creating much shear. Two or four screws may be used.

Another type of trough-and-screw mixer is the *AP Conti paste mixer,* shown in Fig. 18-49. These self-cleaning mixers are particularly appropriate when the product being handled goes through a sticky stage, which could plug the mixer or foul the heat-transfer surfaces.

Pug mills have one or two shafts fitted with short, heavy paddles, mounted in a cylinder or trough holding the material to be processed. In the two-shaft mills, the shafts are parallel and may be either horizontal or vertical. The paddles may or may not intermesh. Clearances are wide, so considerable mass mixing takes place. Unmixed or partially mixed ingredients are fed at one end of the machine, which is usually totally enclosed. Liquid may be added to the material entering the mixer. The paddles push the material forward as they cut through it. The action of the paddles carries the material toward the discharge end of the mixer. The product may discharge through one or two open ports or through extrusion nozzles. The nozzles create roughly shaped continuous strips of material. Automatic cutters may be used to make blocks or pellets from the strips. Pug mills are most often used for mixing mineral or clay products.

Motionless mixers are an alternative to rotating impeller mixers. Motionless or static mixers use stationary-shaped elements inside pipes or conduits to divide, divert, twist, and recombine flowing material. The dividing, stretching, and recombining processes lead to thinner and thinner striations in viscous materials to achieve uniformity.

The twisted-element mixers, such as the Kenics static mixer (Fig. 18-50), create 2^n layers in n divisions. Each element twists the flow, moving material from the center to the wall and from the wall to the center. The twisting also stretches striations having different properties and reorients the material before the next division. Each successive element twists the divided material in the opposite direction. The more viscous the material, the more mixing elements are required for uniformity.

Other motionless designs, such as the *Sulzer static mixer* (Fig. 18-51), accomplish mixing by making multiple divisions at each element transition. The flowing material follows a wavy path to stretch and distort the striations. The number of divisions and distorted paths causes more rapid mixing, but at the expense of a greater pressure drop per unit length of the mixer.

The power required to accomplish mixing in a motionless mixer is provided by the pump used to force the fluid through the mixer. The pressure drop through a motionless mixer is usually expressed as a multiplier K of the open pipe loss or as a valve coefficient C_V. The value of the multiplier is strongly dependent on the detail geometry of the mixer, but is usually available through information from the supplier. Fluid properties are taken into account by the value of the Reynolds number for the open pipe. Motionless mixers are usually sized to match the diameter of the connecting pipe. Pumping adjustments are made when necessary to handle the increased pressure drop.

Because motionless mixers continuously interchange fluid between the walls and the center of the conduit, they also provide good heat transfer, especially with the twisted-element style of mixers. Sometimes, high-viscosity heat exchange is best accomplished with a static mixer.

Distributive (radial) mixing is usually excellent; dispersive (axial) mixing is often poor. The result can be a good plug-flow mixer or reactor, with corresponding benefits and limitations.

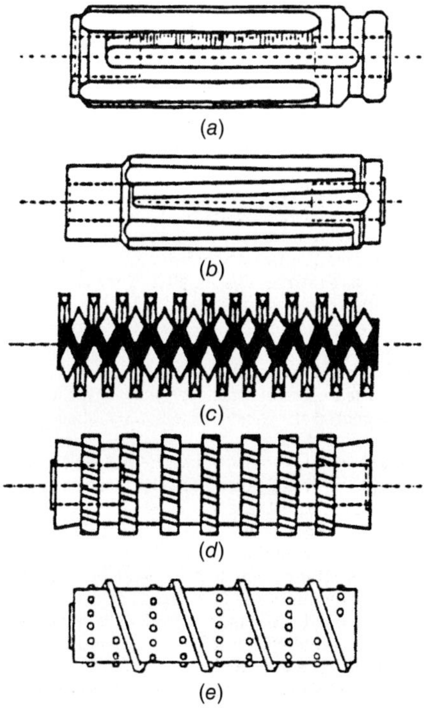

FIG. 18-46 Mixing enhancers for single-screw extruders: (*a*) Maddock, straight; (*b*) Maddock, tapered; (*c*) pineapple; (*d*) gear; (*e*) pin.

PROCESS DESIGN CONSIDERATIONS

Scale-Up of Batch Mixers While a desirable objective of scale-up might be equal blending uniformity in equal time, practicality dictates that times for blending are longer with larger batches. Scale-up of many processes and applications can be successfully done by holding constant the peripheral speed of the rotating element in the mixer. Equal peripheral speed, often called *equal tip speed,* essentially means that the maximum velocity in the mixer remains constant.

Perhaps one of the most difficult concepts to grasp about viscous mixing is that, unlike in turbulent mixing, greater mixer speed does not always translate to better mixing results. If a rotating mixer blade cuts through a viscous fluid or heavy paste too quickly, the stretching process that reduces

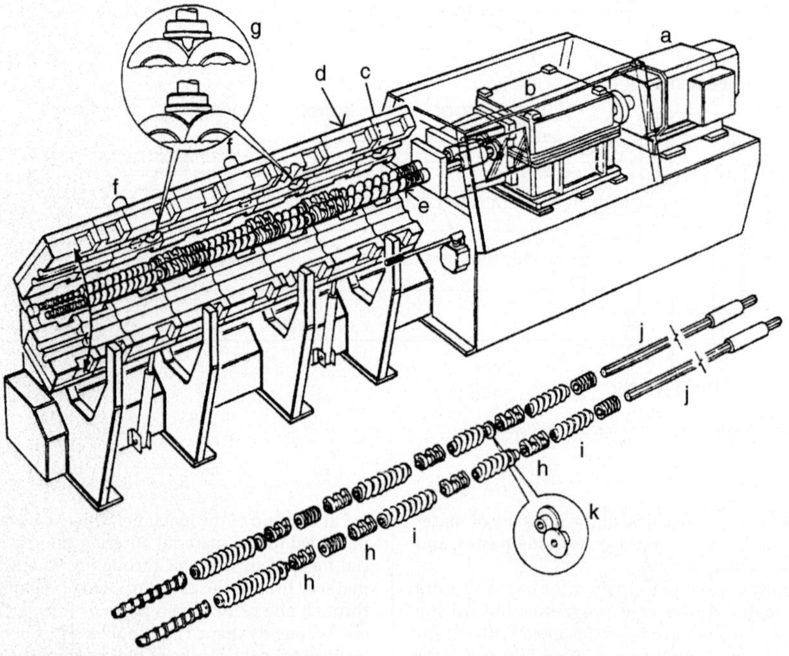

FIG. 18-47 Intermeshing corotating twin-screw extruder: (*a*) drive motor; (*b*) gearbox; (*c*) feed port; (*d*) barrel; (*e*) assembled rotors; (*f*) vent; (*g*) barrel valve; (*h*) kneading paddles; (*i*) conveying screws; (*j*) splined shafts; (*k*) blister rings. (*APV Chemical Machinery, Inc.*)

striation thickness does not take place throughout the material. At high rotational speeds, rapid shearing between a blade tip and the wall or housing may take place, but flow to create bulk motion may not have time to occur. Thus, slower speeds may actually give better mixing results.

With geometric similarity, equal tip speed means that velocity gradients are reduced, and blend times become longer. However, power per volume is also reduced, and viscous heating problems are likely to be more controllable. With any geometric scale-up, the surface-to-volume ratio is reduced, which means that any internal heating, whether by viscous dissipation or chemical reaction, becomes more difficult to remove through the surface of the vessel.

In many applications, the blend time is closely related to the actual number of revolutions made by the mixing device. Thus if mixing were successfully accomplished in 5 min at 60 rpm in a small mixer, the same uniformity could be achieved in 10 min at 30 rpm in a larger mixer. Other factors, such as the rate of heating, could limit scale-up and mixing times.

The physical properties of a paste are difficult to define because a combination of yield stress, shear dependence, time dependence, and even elasticity may be present. Further, many process applications involve the formation or modification of the physical properties. To relate accurately specific material properties to mixing characteristics or power requirements can be extremely difficult. Actual observation and measurement in small-scale equipment or comparison with similar existing processes may be the only practical way of predicting successful operating conditions. Power measurements in small-scale equipment are often essential to predict large-scale conditions and may form the basis for operating production equipment.

Scale-Up of Continuous Mixers While geometric similarity may be practical for most batch mixers, changes in the length-to-diameter ratio or other geometry may be necessary with continuous mixers. The most common problem is heat generation by friction and heat removal by surface transfer.

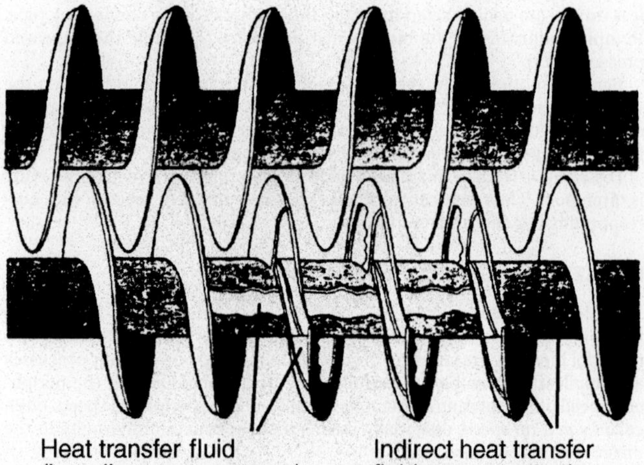

FIG. 18-48 Holo-Flite processor. (*Metso Minerals.*)

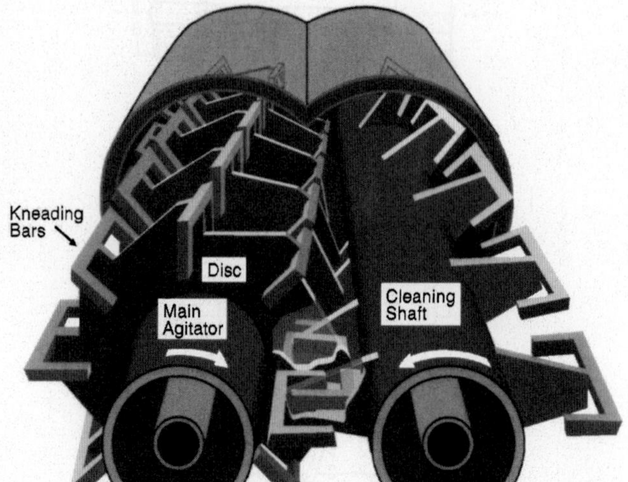

FIG. 18-49 AP Conti paste mixer. (*LIST, Inc.*)

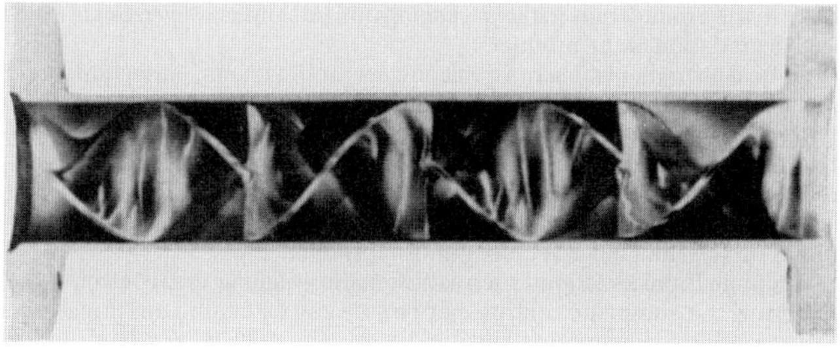

FIG. 18-50 Kenics static mixer. (*Chemineer, Inc.*)

In single-screw extruders (see Fig. 18-45), channel depth in the flights cannot be increased in proportion to the screw diameter because the distribution of heat generated by friction at the barrel wall requires more time as the channel depth becomes greater. With constant retention time, therefore, a nonhomogeneous product would be discharged from a geometrically similar large-scale extruder.

As the result of the departure from geometric similarity, the throughput rate of single-screw extruders scales up with the diameter to 2.0 to 2.5 power, instead of the diameter cubed, at constant length-to-diameter and screw speed. The throughput rates of twin-screw extruders (Fig. 18-47) and the Farrel continuous mixer are scaled up with the diameter to about the 2.6 power.

The extent of axial dispersion through a continuous mixer can be characterized either by an axial diffusion coefficient or by analogy to a number of well-mixed stages in series. Retention time can control the performance of a mixing system. As the number of apparent stages increases, the greater is the assurance that all the material will have the required residence time. Under conditions requiring uniform retention time, the feed streams must enter at the correct ratio on a time scale much shorter than the average residence time of the mixer. Otherwise, variations in the feed will appear as changes in the product. Different types of continuous mixers have different degrees of axial dispersion. Thus, appropriate feed conditions must be considered. Single-screw extruders have an equivalent number of stages equal to approximately one-half the length-to-diameter ratio.

HEATING AND COOLING MIXERS

Heat Transfer Pastes and viscous fluids are often heated or cooled by heat transfer through the walls of the mixing container or hollow mixing arms. A uniform temperature throughout the bulk material is almost as important for good heat transfer as a large heat-transfer-surface-to-mixer-volume ratio. Bulk temperature uniformity will maximize the temperature-difference driving force for heat transfer. Surface area is a direct factor in overall heat transfer. Effective motion near the surface promotes convection over conduction for better heat transfer. Most mixers for pastes or viscous fluids have some sort of scraper or close-clearance device to move stagnant material away from heat-transfer surfaces.

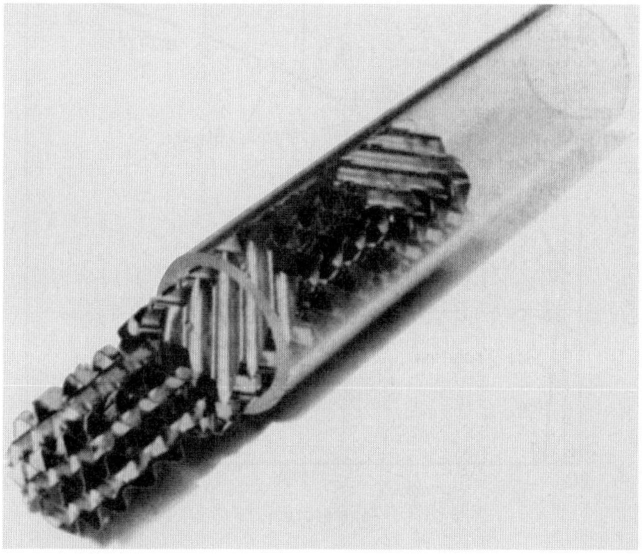

FIG. 18-51 Sulzer static mixer. (*Sulzer Chemtech.*)

Typical overall heat-transfer coefficients are between 20 and 200 J/($m^2 \cdot s \cdot K$) [4 to 35 Btu/($h \cdot ft^2 \cdot °F$)].

Heating Methods Steam heating is widely used because it is economical, safe, and easily controlled. The mixer shell must be designed to withstand both the positive pressure of steam and a vacuum caused when the steam condenses. Transfer liquid heating, using water, oil, special organic liquids, or molten salts, permits good temperature control and provides insurance against overheating the process material. Jackets for transfer liquids are usually baffled to provide good circulation. Higher temperatures can be achieved without the heavy vessel construction required by steam pressures.

Electric heating requires that the elements be electrically insulated from the vessel while still providing good thermal contact. The heaters must be designed for uniform heating to avoid creating hot spots. Temperature control can be precise and maintenance costs low, but utility costs can be very high for large mixers. Electrical heating may be excluded when flammable vapors or dusts are present.

Friction or viscous heating develops rapidly in some mixers, such as a Banbury mixer. The first temperature rise may be beneficial in softening the materials and accelerating chemical reactions. Because energy inputs can be high, higher temperatures detrimental to the products may develop rapidly. So cooling may be required during other portions of a process.

Cooling Methods Air cooling with air blown over external surfaces or external fins may be sufficient for some mixers. Evaporation of excess water or solvent under a vacuum or ambient pressure provides good cooling. A small amount of evaporation produces a large amount of cooling. However, removing too much solvent may damage the product. Some mixers are cooled by circulating water or refrigerants through jackets or hollow agitators. With viscous fluids, lower temperatures near the cooled surfaces increase viscosity and make heat transfer more difficult.

EQUIPMENT SELECTION

The most common and sometimes the only available approach is by analogy. Many companies manufacture similar products, either of their own or those of competitors. With similar products, both good and bad features of existing or typical mixing equipment need to be considered carefully. Some types of mixing equipment are commonly used throughout certain industries. Sometimes existing equipment can be adapted to a new process. Otherwise, new equipment will be needed.

If new equipment is needed, laboratory or pilot-plant studies are recommended. Often unique product features involve unusual or special fluid properties, which makes the prediction of mixer performance almost impossible. The objective is to find potentially suitable equipment and test available mixers. Most equipment vendors have equipment to rent or a demonstration laboratory to test their mixers.

The following list provides some characteristics of a new process that must be considered:

1. List all materials in the process and describe their characteristics.
   a. Method of delivery to the mixer: bags, drums, tote sacks, bulk, pipeline, etc.
   b. Storage and/or weighing requirements at the mixer
   c. Physical form of the material
   d. Specific gravity and bulk characteristics
   e. Particle size or size range
   f. Viscosity
   g. Melting, boiling, or degradation point
   h. Corrosive properties
   i. Abrasive characteristics
   j. Toxicity
   k. Fire or explosion hazards
   l. Irritant characteristics, to skin, eyes, or lungs
   m. Sensitivity of materials when exposed to air, moisture, or heat
2. List pertinent information related to production.
   a. Quantity to be produced per batch
   b. Formulation and order of addition
   c. Analysis required
   d. Cleaning requirements between batches or products
   e. Preceding and/or following process steps
   f. Any changes in physical state during process
   g. Any chemical reactions—exothermic or endothermic
   h. Temperature requirements
   i. Physical form of final product
   j. Removal of product from mixer—pumping or gravity flow through piping, chute, or dumping
3. Describe the controlling features of the finished product.
   a. Degree of uniformity: solution, aggregates, particle size, etc.
   b. Stability of emulsion or dispersion

c. Ultimate color requirements
d. Uniformity of active ingredients, as in a pharmaceutical product
e. Degree of moisture content control

Preparation and Addition of Materials To ensure product quality and productivity, ingredient preparation is important. Order of addition, method and rate of addition, and even preprocessing must be considered.

Some finely powdered materials, such as carbon black or silica, contain a lot of air. If possible, such materials should be compacted, wetted, or agglomerated before being added to the mixture. Air bubbles can be extremely difficult to remove from viscous materials. Holding the product under a vacuum may help release some air or trapped gases. The presence of air in the product may make packaging difficult and may even cause eventual degradation of the product.

Critical ingredients, such as vulcanizers, antioxidants, surfactants, and active agents, are often present in small proportions. If these materials form lumps or aggregates, milling or screening of the materials may be necessary to ensure a uniform product. If small ingredients are soluble in liquid ingredients, adding them as a solution may improve blending. Master batching small quantities of an ingredient into part of a major ingredient often simplifies mixing and makes a more uniform product.

Additional considerations, such as automatic weighing, feed control, liquid metering, and automatic control, may be essential for continuous processes.

CRYSTALLIZATION FROM SOLUTION

General References: *AIChE Testing Procedures: Crystallizers,* American Institute of Chemical Engineers, New York, 1970; *AIChE Testing Procedures: Evaporators,* American Institute of Chemical Engineers, New York, 1961; Bennett, *Chem. Eng. Prog.* **58**(9): 76 (1962); Buckley, *Crystal Growth,* Wiley, New York, 1951; Campbell and Smith, *Phase Rule,* Dover, New York, 1951; Larson, M. A., ed., "Crystallization from Solution: Factors Influencing Size Distribution," *Chem. Eng. Prog. Symp. Ser.* **67**(110), 1971; De Jong and Jancic, eds., *Industrial Crystallization,* North-Holland Publishing Company, Amsterdam, 1979; Genck, *Chem. Eng. Prog.* **99**(6): 36 (2003) and *Chem. Eng. Prog.* **100**(10): 26 (2004); Jancic and Grootscholten, *Industrial Crystallization,* D. Reidel Publishing, Boston, 1984; Jones, *Crystallization Process Systems,* Butterworth-Heinemann, Boston, 2002; Mersmann, ed., *Crystallization Technology Handbook,* Marcel Dekker, New York, 1995; Mullin, ed., *Industrial Crystallization,* 4th ed., Butterworth-Heinemann, Boston, 2001; Newman and Bennett, *Chem. Eng. Prog.* **55**(3): 65 (1959); Palermo and Larson, eds., "Crystallization from Solutions and Melts," *Chem. Eng. Prog. Symp. Ser.* **65**(95), 1969; Randolph, ed., "Design, Control and Analysis of Crystallization Processes," *Am. Inst. Chem. Eng. Symp. Ser.* **76**(193), 1980; Randolph and Larson, *Theory of Particulate Processes,* 2d ed., Academic Press, New York, 1988; Seidell, *Solubilities of Inorganic and Metal Organic Compounds,* American Chemical Society, Washington, D.C., 1965.

Crystallization is important as an industrial process because of the number of materials that are and can be marketed in the form of crystals. Its wide use is due to the highly purified and favorable form of a chemical solid that can be obtained from relatively impure solutions in a single processing step. In terms of energy requirements, crystallization requires much less energy for separation than do distillation and other commonly used methods of purification. In addition, it can be performed at relatively low temperatures and on a scale that varies from a few grams up to thousands of tons per day.

Crystallization may be carried out from a vapor, from a melt, or from a solution. Most of the industrial applications of the operation involve crystallization from solutions. Nevertheless, crystal solidification of metals is basically a crystallization process, and much theory has been developed in relation to metal crystallization. This topic is highly specialized and is outside the scope of this subsection, which is limited to crystallization from solution.

PRINCIPLES OF CRYSTALLIZATION

Crystals A crystal may be defined as a solid composed of atoms or molecules arranged in an orderly, repetitive array. The interatomic distances in a crystal of any definite material are constant and are characteristic of that material. Because the pattern or arrangement of the atoms or molecules is repeated in all directions, there are definite restrictions on the kinds of symmetry that crystals can possess.

There are five main types of crystals, and these types have been arranged into seven crystallographic systems based on the crystal interfacial angles and the relative length of its axes. The treatment of the description and arrangement of the atomic structure of crystals is the science of *crystallography*. The material in this discussion will be limited to a treatment of the growth and production of crystals as a unit operation.

Solubility and Phase Diagrams Equilibrium relations for crystallization systems are expressed in the form of solubility data, which are plotted as phase diagrams or solubility curves. Solubility data are ordinarily given as parts by weight of anhydrous material per 100 parts by weight of total solvent. In some cases, these data are reported as parts by weight of anhydrous material per 100 parts of solution. If water of crystallization is present in the crystals, this is indicated as a separate phase. The concentration is normally plotted as a function of temperature and has no general shape or slope. It can also be reported as a function of pressure, but for most materials the change in solubility with change in pressure is very small. If there are two components in solution, it is common to plot the concentration of these two components on the *x* and *y* axes and represent the solubility by isotherms. When three or more components are present, there are various techniques for depicting the solubility and phase relations in both three-dimensional and two-dimensional models. For a description of these techniques, refer to Campbell and Smith (*Phase Rule,* Dover, New York, 1951). Shown in Fig. 18-52 is a phase diagram for magnesium sulfate in water. The line *p–a* represents the freezing points of ice (water) from solutions of magnesium sulfate. Point *a* is the eutectic, and the line *a–b–c–d–q* is the solubility curve of the various hydrates. Line *a–b* is the solubility curve for $MgSO_4 \cdot 12H_2O$, *b–c* is the solubility curve for $MgSO_4 \cdot 7H_2O$, *c–d* is the solubility curve for $MgSO_4 \cdot 6H_2O$, and *d–q* is the portion of the solubility curve for $MgSO_4 \cdot H_2O$.

As shown in Fig. 18-53, the mutual solubility of two salts can be plotted on the *x* and *y* axes with temperatures as isotherm lines. In the example shown, all the solution compositions corresponding to 100°C with solid-phase sodium chloride present are shown on the line *DE.* All the solution compositions at equilibrium with solid-phase KCl at 100°C are shown by the line *EF.* If both solid-phase KCl and NaCl are present, the solution composition at equilibrium can only be represented by point *E,* which is the invariant point (at constant pressure). Connecting all the invariant points results in the mixed-salt line. The locus of this line is an important consideration in making phase separations.

There are many solubility data in the literature; the standard reference is by Seidell (*Solubilities of Inorganic and Metal Organic Compounds,* American Chemical Society, Washington, D.C., 1965). Valuable as they are, they nevertheless must be used with caution because the solubility of compounds is often influenced by pH or by the presence of other soluble impurities, which usually tend to depress the solubility of the major constituents. While exact values for any system are often best determined by actual

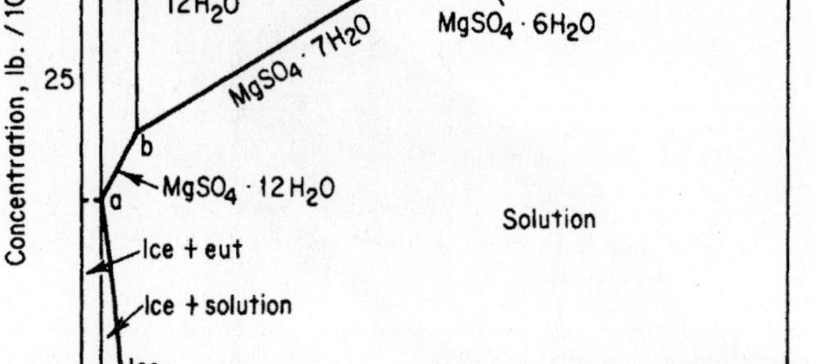

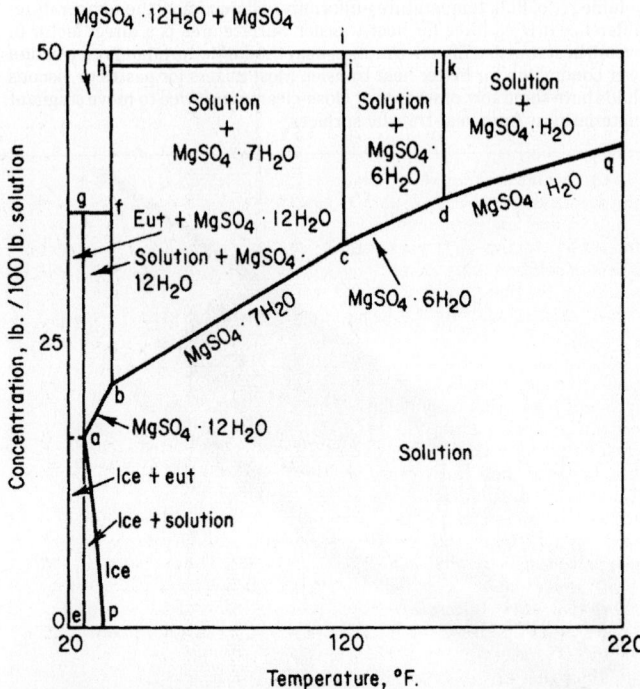

FIG. 18-52 Phase diagram. $MgSO_4 \cdot H_2O$. To convert pounds to kilograms, divide by 2.2; K = (°F + 459.7)/1.8.

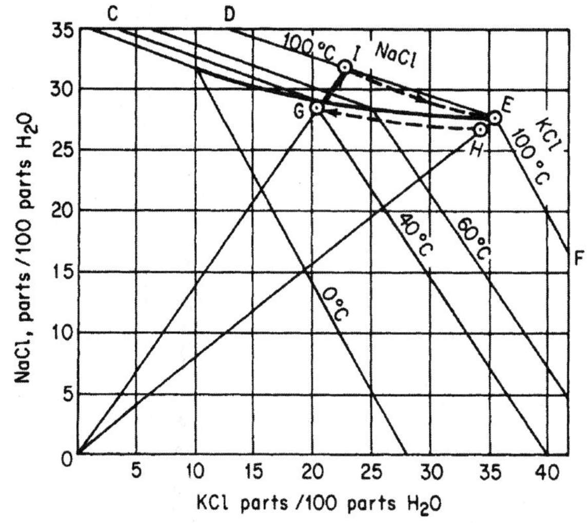

FIG. 18-53 Phase diagram, KCl – NaCl – H_2O. K = °C + 273.2.

composition measurements, the difficulty of reproducing these solubility diagrams should not be underestimated. To obtain data that are readily reproducible, elaborate pains must be taken to be sure the system sampled is at equilibrium, and often this means holding a sample at constant temperature for a period of from 1 to 100 h. While the published curves may not be exact for actual solutions of interest, they generally will be indicative of the shape of the solubility curve and will show the presence of hydrates or double salts.

Heat Effects in a Crystallization Process The heat effects in a crystallization process can be computed by two methods: (1) a heat balance can be made in which individual heat effects such as sensible heats, latent heats, and the heat of crystallization can be combined into an equation for total heat effects; or (2) an enthalpy balance can be made in which the total enthalpy of all leaving streams minus the total enthalpy of all entering streams is equal to the heat absorbed from external sources by the process. In using the heat-balance method, it is necessary to make a corresponding mass balance, since the heat effects are related to the quantities of solids produced through the heat of crystallization. The advantage of the enthalpy-concentration-diagram method is that both heat and mass effects are taken into account simultaneously. This method has limited use because of the difficulty in obtaining enthalpy-concentration data. This information has been published for only a few systems.

With compounds whose solubility increases with increasing temperature, there is an absorption of heat when the compound dissolves. In compounds with decreasing solubility as the temperature increases, there is an evolution of heat when solution occurs. When there is no change in solubility with temperature, there is no heat effect. The solubility curve will be continuous as long as the solid substance of a given phase is in contact with the solution, and any sudden change in the slope of the curve will be accompanied by a change in the heat of solution and a change in the solid phase. Heats of solution are generally reported as the change in enthalpy associated with the dissolution of a large quantity of solute in an excess of pure solvent. Tables showing the heats of solution for various compounds are given in Sec. 2.

At equilibrium, the heat of crystallization is equal and opposite in sign to the heat of solution. Using the heat of solution at infinite dilution as equal but opposite in sign to the heat of crystallization is equivalent to neglecting the heat of dilution. With many materials, the heat of dilution is small in comparison with the heat of solution, and the approximation is justified; however, there are exceptions. Relatively large heat effects are usually found in the crystallization of hydrated salts. In such cases, the total heat released by this effect may be a substantial portion of the total heat effects in a cooling-type crystallizer. In evaporative-type crystallizers, the heat of crystallization is usually negligible when compared with the heat of vaporizing the solvent.

Yield of a Crystallization Process In most cases, the process of crystallization is slow, and the final mother liquor is in contact with a large enough crystal surface that the concentration of the mother liquor is substantially that of a saturated solution at the final temperature in the process. In such cases, it is normal to calculate the yield from the initial solution composition and the solubility of the material at the final temperature. If evaporative crystallization is involved, the solvent removed must be taken into account in determining the final yield. If the crystals removed from solution are hydrated, account must be taken of the water of crystallization in the crystals, since this water is not available for retaining the solute in solution. The yield is also influenced in most plants by the removal of some mother liquor, with the crystals being separated from the process. Typically, with a product separated on a centrifuge or filter, the adhering mother liquor would be in the range of 2 to 10 percent of the weight of the crystals.

The actual yield may be obtained from algebraic calculations or trial-and-error calculations when the heat effects in the process and any resultant evaporation are used to correct the initial assumptions on calculated yield. When calculations are made by hand, it is generally preferable to use the trial-and-error system since it permits easy adjustments for relatively small deviations found in practice, such as the addition of wash water, or instrument and purge water additions. The following calculations are typical of an evaporative crystallizer precipitating a hydrated salt. If SI units are desired, kilograms = pounds × 0.454; K = (°F + 459.7)/1.8.

Example 18-2 Yield from a Crystallization Process A 10,000-lb batch of a 32.5 percent $MgSO_4$ solution at 120°F is cooled without appreciable evaporation to 70°F. What weight of $MgSO_4 \cdot 7H_2O$ crystals will be formed (if it is assumed that the mother liquor leaving is saturated)?

From the solubility diagram in Fig. 18-52, at 70°F the concentration of solids is 26.3 lb $MgSO_4$ per 100-lb solution.

The mole weight of $MgSO_4$ is 120.38.

The mole weight of $MgSO_4 \cdot 7H_2O$ is 246.49.

For calculations involving hydrated salts, it is convenient to make the calculations based on the hydrated solute and the "free water."

$$0.325 \text{ weight fraction} \times \frac{246.94}{120.38} = 0.6655 \text{ MgSO}_4 \cdot 7\text{H}_2\text{O in the feed solution}$$

$$0.263 \times \frac{246.94}{120.38} = 0.5385 \text{ MgSO}_4 \cdot 7\text{H}_2\text{O in the mother liquor}$$

Since the free water remains constant (except when there is evaporation), the final amount of soluble $MgSO_4 \cdot 7H_2O$ is calculated by the ratio of $\dfrac{0.538 \text{ lb MgSO}_4 \cdot 7\text{H}_2\text{O}}{(1-0.538) \text{ lb free water}}$

	Total	$MgSO_4 \cdot 7H_2O$	Free water	$MgSO_4 \cdot 7H_2O$ / Free water
Feed	10,000	6655	3345	1.989
Mother liquor	7249	3904*	3345	1.167
Yield	2751	2751		

*3345 × (0.538/0.462) = 3904

A formula method for calculation is sometimes used where

$$P = R\frac{100W_0 - S(H_0 - E)}{100 - S(R-1)}$$

where P = weight of crystals in final magma, lb
R = mole weight of hydrate/mole weight of anhydrous = 2.04759
S = solubility at mother-liquor temperature (anhydrous basis) in lb per 100 lb solvent. [0.263/(1 – 0.263)] × 100 = 35.68521
W_0 = weight of anhydrous solute in the original batch. 10,000(0.325) = 3250 lb
H_0 = total weight of solvent at the beginning of the batch. 10,000 – 3250 = 6750 lb
E = evaporation = 0

$$P = 2.04\,\frac{(100)(3250) - 35.7(6750)}{100 - 35.7(2.04-1)} = 2751 \text{ lb}$$

Note that taking the difference between large numbers in this method can increase the chance for error.

Fractional Crystallization When two or more solutes are dissolved in a solvent, it is often possible to (1) separate these into the pure components or (2) separate one and leave the other in the solution. Whether or not this can be done depends on the solubility and phase relations of the system under consideration. Normally alternative 2 is successful only when one of the components has a much more rapid change in solubility with temperature than does the other. A typical example that is practiced on a large scale is the separation of KCl and NaCl from water solution. A phase diagram for this system is shown in Fig. 18-53. In this case, the solubility of NaCl is plotted on the y axis in parts per 100 parts of water, and the solubility of KCl is plotted on the x axis. The isotherms show a marked decrease in solubility for each component as the amount of the other is increased. This is typical for most inorganic salts. As explained earlier, the mixed-salt line is CE, and to make a separation of the solutes into the pure components it is necessary to be on one side of this line or the other. Normally a 95 to 98 percent approach to this line is possible. When evaporation occurs during a cooling or concentration process, this can be represented by movement away from the origin on a straight line through the origin. Dilution by water is represented by movement in the opposite direction.

A typical separation might be represented as follows: Starting at E with a saturated brine at 100°C, a small amount of water is added to dissolve any traces of solid phase present and to make sure the solids precipitated initially are KCl. Evaporative cooling along line HG results in the precipitation of KCl. During this evaporative cooling, part of the water evaporated must be added back to the solution to prevent the coprecipitation of NaCl. The final composition at G can be calculated by the NaCl/KCl/H_2O ratios and the known amount of NaCl in the incoming solution at E. The solution at point G may be concentrated by evaporation at 100°C. During this process, the solution will increase in concentration with respect to both components until point I is reached. Then NaCl will precipitate, and the solution will become more concentrated in KCl, as indicated by the line IE, until the original point E is reached. If concentration is carried beyond point E, a mixture of KCl and NaCl will precipitate.

Example 18-3 Yield from Evaporative Cooling Starting with 1000 lb of water in a solution at H on the solubility diagram in Fig. 18-53, calculate the yield on evaporative cooling and concentrate the solution back to point H so the cycle can be repeated, indicating the amount of NaCl precipitated and the evaporation and dilution required at the different steps in the process.

In solving problems of this type, it is convenient to list the material balance and the solubility ratios. The various points on the material balance are calculated by multiplying the quantity of the component that does not precipitate from solution during the transition from one point to another (normally the NaCl in cooling or the KCl in the evaporative step) by the solubility ratio at the next step, illustrated as follows:

Basis. 1000 lb of water at the initial conditions.

Solution component	KCl	NaCl	Water	Solubility ratios KCl	Solubility ratios NaCl	Solubility ratios Water
H	343	270	1000	34.3	27.0	100
$G(a)$	194	270	950	20.4	28.4	100
KCl yield	149					
Net evaporation			50			
$I(b)$	194	270	860	22.6	31.4	100
$E(c)$	194	153	554	35.0	27.5	100
NaCl yield		117				
Evaporation			306			
Dilution			11			
H'	194	153	565	34.3	27.0	−100

The calculations for these steps are:

a. 270 lb NaCl (100 lb water/28.4 lb NaCl) = 950 lb water
950 lb water (20.4 lb KCl/100 lb water) = 194 lb KCl
b. 270 lb NaCl (100 lb water/31.4 lb NaCl) = 860 lb water
860 lb water (22.6 lb KCl/100 lb water) = 194 lb KCl
c. 194 lb KCl (100 lb water/35.0 lb KCl) = 554 lb water
554 lb water (27.5 lb NaCl/100 lb water) = 153 lb NaCl

Note that during the cooling step, the maximum amount of evaporation permitted by the material balance is 50 lb for the step shown. In an evaporative-cooling step, however, the actual evaporation that results from adiabatic cooling is more than this. Therefore, water must be added back to prevent the NaCl concentration from rising too high; otherwise, coprecipitation of NaCl will occur.

Inasmuch as only mass ratios are involved in these calculations, kilograms or any other unit of mass may be substituted for pounds without affecting the validity of the example.

Although the figures given are for a step-by-step process, it is obvious that the same techniques will apply to a continuous system if the fresh feed containing KCl and NaCl is added at an appropriate part of the cycle, such as between steps G and I for the case of dilute feed solutions.

Another method of fractional crystallization, in which advantage is taken of different crystallization rates, is sometimes used. Thus, a solution saturated with borax and potassium chloride will, in the absence of borax seed crystals, precipitate only potassium chloride on rapid cooling. The borax remains behind as a supersaturated solution, and the potassium chloride crystals can be removed before the slower borax crystallization starts.

Crystal Formation There are two steps involved in the preparation of crystal matter from a solution. The crystals must first form and then grow. The formation of a new solid phase either on an inert particle in the solution or in the solution itself is called *nucleation.* The increase in size of this nucleus with a layer-by-layer addition of solute is called *growth.* The growth process involves two steps, diffusion of the solute to the crystal interface followed by incorporation of the same into the lattice. One of these will control, depending on factors such as the degree of agitation and temperature. Nucleation can be classified as primary or secondary. The former usually occurs at high supersaturation and does not involve product crystals. Secondary nucleation involves nuclei generation from product crystals by contact with the agitator, with the crystallizer internals and with one another. Each system has a metastable zone where growth is encouraged in the presence of supersaturation. Secondary nucleation can occur within the zone. Both nucleation and crystal growth have supersaturation as a common driving force. Unless a solution is supersaturated, crystals can neither form nor grow. Supersaturation refers to the quantity of solute present in solution compared with the quantity that would be present if the solution were kept for a very long period of time with solid phase in contact with the solution. The latter value is the equilibrium solubility at the temperature and pressure under consideration. The supersaturation coefficient can be expressed as

$$S = \frac{\text{parts solute/100 parts solvent}}{\text{parts solute at equilibrium/100 parts solvent}} \geq 1.0 \qquad (18\text{-}22)$$

Solutions vary greatly in their ability to sustain measurable amounts of supersaturation. With some materials, such as sucrose, it is possible to develop a supersaturation coefficient of 1.4 to 2.0 with little danger of nucleation. With some common inorganic solutions, such as sodium chloride in water, the amount of supersaturation that can be generated stably is so small that it is difficult or impossible to measure.

Certain qualitative facts in connection with supersaturation, growth, and the yield in a crystallization process are readily apparent. If the concentration of the initial solution and the final mother liquor are fixed, the total weight of the crystalline crop is also fixed if equilibrium is obtained. The particle-size distribution of this weight, however, will depend on the relationship between the two processes of nucleation and growth. Considering a given quantity of solution cooled through a fixed range, if there is considerable nucleation initially during the cooling process, the yield will consist of many small crystals. If only a few nuclei form at the start of the crystallization (or seeds are added) and the resulting yield occurs uniformly on these nuclei or seeds without significant secondary nucleation, a crop of large uniform crystals will result. Obviously, many intermediate cases of varying nucleation rates and growth rates can also occur, depending on the nature of the materials being handled, the rate of cooling, agitation, and other factors.

When a process is continuous, nucleation often occurs in the presence of a seeded solution by the combined effects of mechanical stimulus and nucleation caused by supersaturation (heterogeneous nucleation). If such a system is completely and uniformly mixed (i.e., the product stream represents the typical magma circulated within the system) and if the system is operating at steady state, the particle-size distribution has definite limits that can be predicted mathematically with a high degree of accuracy, as will be shown later in this section.

Geometry of Crystal Growth Geometrically, a crystal is a solid bounded by planes. The shape and size of such a solid are functions of the interfacial angles and of the linear dimension of the faces. As the result of the constancy of its interfacial angles, each face of a growing or dissolving crystal, as it moves away from or toward the center of the crystal, is always parallel to its original position. This concept is known as the "principle of the parallel displacement of faces." The rate at which a face moves in a direction perpendicular to its original position is called the translation velocity of that face or the rate of growth of that face.

From the industrial point of view, the term *crystal habit* or *crystal morphology* refers to the relative sizes of the faces of a crystal. The crystal habit is determined by the internal structure and external influences on the crystal such as the growth rate, solvent used, and impurities present during the crystallization growth period. The crystal habit of commercial products is of very great importance. Long, needlelike crystals tend to be easily broken during centrifugation and drying. Flat, platelike crystals are very difficult to wash during filtration or centrifugation and result in relatively low filtration rates. Complex or twinned crystals tend to be more easily broken in transport than chunky, compact crystal habits. Rounded or spherical crystals (caused generally by attrition during growth and handling) tend to give considerably less difficulty with caking than do cubical or other compact shapes.

Internal structure (unit cell) can be different in crystals that are chemically identical. This is called *polymorphism.* Polymorphs can vary substantially in physical and chemical properties such as bioavailability and solubility. They can be identified by analytical techniques such as X-ray diffraction, infrared, Raman spectro, and microscopic techniques. For the same internal structure, very small amounts of foreign substances will often completely change the crystal habit. The selective adsorption of dyes by different faces of a crystal or the change from an alkaline to an acidic environment will often produce pronounced changes in the crystal habit. The presence of other soluble anions and cations often has a similar influence. In the crystallization of ammonium

TABLE 18-3 Some Impurities Known to Be Habit Modifiers

Material crystallized	Additive(s)	Effect	Concentration	References
$Ba(NO_2)^2$	Mg, Te^{+4}	Helps growth	—	1
$CaSO_4 \cdot 2H_2O$	Citric, succinic, tartaric acids	Helps growth	Low	
	Sodium citrate	Forms prisms	—	5
$CuSO_4 \cdot 5H_2O$	H_2SO_4	Chunky crystals	0.3%	5
KCl	$K_4Fe(CN)_6$	Inhibits growth, dendrites	1000 ppm	4
	Pb, Bi, Sn^{+2}, Ti, Zr, Th, Cd, Fe, Hg, Mg	Helps growth	Low	1
$KClO_4$	Congo red (dye)	Modifies the 102 face	50 ppm	6
K_2CrO_4	Acid magenta (dye)	Modifies the 010 face	50 ppm	6
KH_2PO_4	$Na_2B_4O_7$	Aids growth	—	1
KNO_2	Fe	Helps growth	Low	1
KNO_3	Acid magenta (dye)	Tabular crystals		7
	Pb, Th, Bi	Helps growth	Low	1
K_2SO_4	Acid magenta (dye)	Forms plates	2000 ppm	6
	Cl, Mn, Mg, Bi, Cu, Al, Fe	Helps growth	Low	1
	Cl_3	Reduces growth rate	1000 ppm	4
	$(NH_4)_3Ce(NO_3)_6$	Reduces growth rate	1000 ppm	4
$LiCl \cdot H_2O$	$Cr \cdot Mn^{+2}$, Sn^{+2}, Co, Ni, Fe^{+3}	Helps growth	Low	1
$MgSO_4 \cdot 7H_2O$	Borax	Aids growth	5%	1
$Na_2B_4O_7 \cdot 10H_2O$	Sodium oleate	Reduces growth & nuc.	5 ppm	
	Casein, gelatin	Promotes flat crystals	—	2, 5
	NaOH, Na_2CO_3	Promotes chunky crystals	—	
$Na_2CO_3 \cdot H_2O$	SO_4^{2-}	Reduces L/D ratio	0.1–1.0%	Canadian Patent 812,685
	Ca^{+2} and Mg^{+2}	Increase bulk density	400 ppm	U.S. Patent 3,459,497
$NaCO_3 \cdot NaHCO_3 \cdot 2H_2O$	D-40 detergent	Aids growth	20 ppm	U.S. Patent 3,233,983
NaCl	$Na_4Fe(CN)_6$, CdBr	Forms dendrites	100 ppm	4
	Pb, Mn^{+2}, Bi, Sn^{+2}, Ti, Fe, Hg	Helps growth	Low	1
	Urea, formamide	Forms octahedra	Low	2
	Tetraalkyl ammon. salts	Helps growth & hardness	1–100 ppm	U.S. Patent 3,095,281
	Polyethylene-oxy compounds	Helps growth & hardness	—	U.S. Patent 3,000,708
$NaClO_3$	Na_2SO_4, $NaClO_4$	Tetrahedrons	—	3
$NaNO_3$	Acid green (dye)	Flattened rhombahedra		7
Na_2SO_4	NH_4SO_4 @ pH 6.5	Large single crystals	Low	
	$CdCl_2$	Inhibits growth	1000 ppm	4
	Alkyl aryl sulfonates	Aids growth	—	2
	Calgon	Aids growth	100 ppm	
NH_4Cl	Mn, Fe, Cu, Co, Ni, Cr	Aid growth	Low	1
	Urea	Forms octahedra		5
NH_4ClO_4	Azurine (dye)	Modifies the 102 face	22 ppm	6
NF_4F	Ca	Helps growth	Low	1
$(NH_4)NO_3$	Acid magenta (dye)	Forms 010 face plates	1%	6
$(NH_4)_2HPO_4$	H_2SO_4	Reduces L/D ratio	7%	
$NH_4H_2PO_4$	Fe^{+3}, Cr, Al, Sn	Helps growth	Traces	1
$(NH_4)_2SO_4$	Cr^{+3}, Fe^{+3}, Al^{+3}	Promotes needles	50 ppm	
	H_2SO_4	Promotes needles	2–6%	U.S. Patent 2,092,073
	Oxalic acid, citric acid	Promotes chunky crystals	1000 ppm	U.S. Patent 2,228,742
	H_3PO_4, SO_2	Promotes chunky crystals	1000 ppm	
$ZnSO_4 \cdot 7H_2O$	Borax	Aids growth	—	1
Adipic acid	Surfactant-SDBS	Aids growth	50–100 ppm	2
Fructose	Glucose, difructose	Affects growth		8
L-asparagine	L-glutamic acid	Affects growth		8
Naphthalene	Cyclohexane (solvent)	Forms needles	—	2
	Methanol (solvent)	Forms plates		
Pentaerythritol	Sucrose	Aids growth	—	1
	Acetone (solvent)	Forms plates	—	2
Sodium glutamate	Lysine, CaO	Affects growth		8
Sucrose	Raffinose, KCl, NaBr	Modify growth rate		
Urea	Biuret	Reduces L/D & aids growth	2–7%	
	NH_4Cl	Reduces L/D & aids growth	5–10%	

1. Gillman, *The Art and Science of Growing Crystals*, Wiley, New York, 1963.
2. Mullin, *Crystallization*, Butterworth, London, 1961.
3. Buckley, *Crystal Growth*, Wiley, New York, 1961.
4. Phoenix, L., *British Chemical Engineering*, vol. II, no. 1 (Jan. 1966), pp. 34–38.
5. Garrett, D. E., *British Chemical Engineering*, vol. I, no. 12 (Dec. 1959), pp. 673–677.
6. Buckley, *Crystal Growth*, (Faraday Soc.) Butterworths, 1949, p. 249.
7. Butchart and Whetstone, *Crystal Growth*, (Faraday Soc.) Butterworths, 1949, p. 259.
8. Nyvlt, J., *Industrial Crystallization*, Verlag Chemie Publishers, New York, 1978, pp. 26–31.

sulfate, the reduction in soluble iron to below 50 ppm of ferric ion is sufficient to cause significant change in the habit of an ammonium sulfate crystal from a long, narrow form to a relatively chunky and compact form. Additional information is available in the patent literature, and Table 18-3 lists some of the better-known additives and their influences.

Since the relative sizes of the individual faces of a crystal vary between wide limits, it follows that different faces must have different translational velocities. A geometric law of crystal growth known as the *overlapping principle* is based on those velocity differences: in growing a crystal, only those faces having the lowest translational velocities survive, and in dissolving a crystal, only those faces having the highest translational velocities survive.

For example, consider the cross sections of a growing crystal as in Fig. 18-54. The polygons shown in the figure represent varying stages in the growth of the crystal. The faces marked A are slow-growing faces (low translational velocities), and the faces marked B are fast-growing (high translational velocities). It is apparent from Fig. 18-54 that the faster B faces tend to disappear as they are overlapped by the slower A faces.

It has been suggested that crystal habit or crystal morphology is related to the internal structure based on energy considerations and speculated that it should be possible to predict the growth shape of crystals from the slice energy of different flat faces. One can predict the calculated attachment energy for various crystal species. Recently, computer programs have been developed that predict crystal morphology from attachment energies. These techniques are particularly useful in dealing with organic or molecular crystals, and rapid progress in this area is being made by companies such as Molecular Simulations of Cambridge, England.

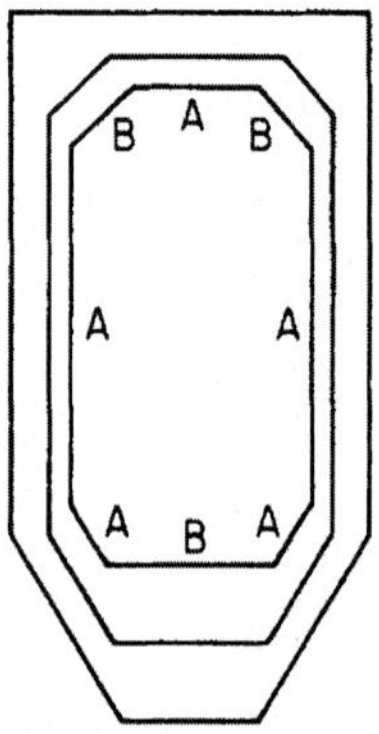

FIG. 18-54 Overlapping principle.

Purity of the Product If a crystal is produced in a region of the phase diagram where a single-crystal composition precipitates, the crystal itself will normally be pure, provided that it is grown at relatively low rates and constant conditions. With many products, these purities approach a value of about 99.5 to 99.8 percent. The difference between this and a purity of 100 percent is generally the result of small pockets of mother liquor called inclusions trapped within the crystal. Although often large enough to be seen with an ordinary microscope, these inclusions can be submicroscopic and represent dislocations within the structure of the crystal. They can be caused by either attrition or breakage during the growth process or by slip planes within the crystal structure caused by interference between screw-type dislocations and the remainder of the crystal faces. To increase the purity of the crystal beyond the point where such inclusions are normally expected (about 0.1 to 0.5 percent by volume), it is generally necessary to reduce the impurities in the mother liquor itself to an acceptably low level so that the mother liquor contained within these pockets will not contain sufficient impurities to cause an impure product to be formed. It is normally necessary to recrystallize material from a solution that is relatively pure to surmount this type of purity problem.

In addition to the impurities within the crystal structure itself, there is normally an adhering mother-liquid film left on the surface of the crystal after separation in a centrifuge or on a filter. Typically, a centrifuge may leave about 2 to 10 percent of the weight of the crystals as adhering mother liquor on the surface. This varies greatly with the size and shape or habit of the crystals. Large, uniform crystals from low-viscosity mother liquors will retain small quantities of mother liquor, while nonuniform or small crystals crystallized from viscous solutions will retain a considerably larger proportion. Comparable statements apply to the filtration of crystals, although normally the amounts of mother liquor adhering to the crystals are considerably larger. It is common practice when crystallizing materials from solutions that contain appreciable quantities of impurities to wash the crystals on the centrifuge or filter with either fresh solvent or feed solution. In principle, such washing can reduce the impurities quite substantially. It is also possible in many cases to reslurry the crystals in fresh solvent and recentrifuge the product in an effort to obtain a longer residence time during the washing operation and better mixing of the wash liquors with the crystals. Mother liquor inclusions and residual moisture after drying can present caking problems.

Coefficient of Variation One of the problems confronting any user or designer of crystallization equipment is the expected particle-size distribution of the solids leaving the system and how this distribution may be adequately described. Most crystalline-product distributions plotted on arithmetic-probability paper will exhibit a straight line for a considerable portion of the plotted distribution. In this type of plot, the particle diameter should be plotted as the ordinate and the cumulative percent on the log-probability scale as the abscissa.

It is common practice to use a parameter characterizing crystal-size distribution called the coefficient of variation. This is defined as follows:

$$CV = 100\frac{PD_{16\%} - PD_{84\%}}{2PD_{50\%}} \tag{18-23}$$

where CV = coefficient of variation, as a percentage
PD = particle diameter from intercept on ordinate axis at percent indicated

In order to be consistent with normal usage, the particle-size distribution when this parameter is used should be a straight line between approximately 10 percent cumulative weight and 90 percent cumulative weight. By giving the coefficient of variation and the mean particle diameter, a description of the particle-size distribution is obtained that is normally satisfactory for most industrial purposes. If the product is removed from a mixed-suspension crystallizer, this coefficient of variation should have a value of approximately 50 percent (Randolph and Larson, *Theory of Particulate Processes,* 2d ed., Academic Press, New York, 1988, chap. 2).

CRYSTAL NUCLEATION AND GROWTH

Rate of Growth Crystal growth is a layer-by-layer process, and since growth can occur only at the face of the crystal, material must be transported to that face from the bulk of the solution. Diffusional resistance to the movement of molecules (or ions) to the growing crystal face, as well as the resistance to integration of those molecules into the face, must be considered. As discussed earlier, different faces can have different rates of growth, and these can be selectively altered by the addition or elimination of impurities.

If L is a characteristic dimension of a crystal of selected material and shape, the rate of growth of a crystal face that is perpendicular to L is, by definition,

$$G \equiv \lim_{\Delta L \to 0} \frac{\Delta L}{\Delta t} = \frac{dL}{dt} \tag{18-24}$$

where G is the growth rate over time interval t. It is customary to measure G in the practical units of millimeters per hour. It should be noted that growth rates so measured are actually twice the facial growth rate.

The delta L *law.* It has been shown by McCabe [*Ind. Eng. Chem.* **21**(30): 112 (1929)] that all geometrically similar crystals of the same material suspended in the same solution grow at the same rate if growth rate is defined as in Eq. (18-24). The rate is independent of crystal size, provided that all crystals in the suspension are treated alike. This generalization is known as the delta L law. Although there are some well-known exceptions, they usually occur when the crystals are very large or when movement of the crystals in the solution is so rapid that substantial changes occur in diffusion-limited growth of the faces.

The delta L law does not apply when similar crystals are given preferential treatment based on size. It also fails when surface defects or dislocations significantly alter the growth rate of a crystal face. Nevertheless, it is a reasonably accurate generalization for a surprising number of industrial cases. When it is, it is important because it simplifies the mathematical treatment in modeling real crystallizers and is useful in predicting crystal-size distribution in many types of industrial crystallization equipment.

Important exceptions to McCabe's growth-rate model have been noted by Bramson, by Randolph, and by Abegg. These are discussed by Canning and Randolph, *Am. Inst. Chem. Eng. J.* **13:** 5 (1967).

Nucleation The mechanism of crystal nucleation from solution has been studied by many scientists, and their work suggests that—in commercial crystallization equipment, at least—the nucleation rate is the sum of contributions by (1) primary nucleation and (2) nucleation due to contact between crystals and (*a*) other crystals, (*b*) the walls of the container, and (*c*) the impeller. If B^0 is the net number of new crystals formed in a unit volume of solution per unit of time,

$$B^0 = B_{ss} + B_{cc} + B_{ci} \tag{18-25}$$

where B_{ci} is the rate of nucleation due to crystal-impeller contacts, B_{cc} is that due to crystal-crystal contacts, and B_{ss} is the primary nucleation rate due to the supersaturation driving force. The mechanism of the last-named is not precisely known, although it is obvious that molecules forming a nucleus not only have to coagulate, resisting the tendency to redissolve, but also must become oriented into a fixed lattice. The number of atoms or molecules required to form a stable crystal nucleus has been variously estimated at from 80 to 100 (with ice), and the probability that a stable nucleus will result depends on many factors, such as activation energies and supersaturation. In commercial crystallization equipment, in which supersaturation is low and agitation is used to keep the growing crystals suspended, the predominant mechanism is contact nucleation or, in extreme cases, attrition.

In order to treat crystallization systems both dynamically and continuously, a mathematical model has been developed that can correlate the nucleation rate to the level of supersaturation or the growth rate. Because the growth rate is more easily determined and because nucleation is sharply nonlinear in the regions normally encountered in industrial crystallization, it has been common to assume

$$B^0 = ks^b \tag{18-26}$$

where s, the supersaturation, is defined as $(C - C_s)$, C being the concentration of the solute and C_s its saturation concentration; and the exponent b and dimensional coefficient k are values characteristic of the material.

While Eq. (18-26) has been popular among those attempting correlations between nucleation rate and supersaturation, it has become common to use a derived relationship between nucleation rate and growth rate by assuming that

$$G = k's^g \tag{18-27}$$

whence, in consideration of Eq. (18-26),

$$B^0 = k''G^i \tag{18-28}$$

where the dimensional coefficient k' and exponent g are characteristic of the material and the conditions of crystallization and $k'' = k/(k')^i$ with $i = b/g$, a measure of the relative dependence of B^0 and G on supersaturation. Feeling that a model in which nucleation depends only on supersaturation or growth rate is simplistically deficient, some have proposed that contact nucleation rate is also a power function of slurry density and that

$$B^0 = k_n G^i M_T^j \tag{18-29}$$

where M_T is the density of the crystal slurry in g/L.

Although Eqs. (18-28) and (18-29) have been adopted by many as a matter of convenience, they are oversimplifications of the very complex relationship that is suggested by Eq. (18-25); Eq. (18-29) implicitly and quite arbitrarily combines the effects of homogeneous nucleation and those due to contact nucleation. They should be used only with caution.

In work pioneered by Clontz and McCabe [*Chem. Eng. Prog. Symp. Ser.* **67**(110): 6 (1971)] and subsequently extended by others, contact nucleation rate was found to be proportional to the input of energy of contact and frequency of contact and a function of contact area and supersaturation. This observation is important to the scaling up of crystallizers. At the laboratory or bench scale, particle contact frequency with the agitator is high, while in commercial equipment the contact energy input is higher at the impeller, but the contact frequency is less. Scale-up modeling of a crystallizer, therefore, must include its mechanical characteristics as well as the physiochemical driving force.

Nucleation and Growth From the preceding, it is clear that no analysis of a crystallizing system can be truly meaningful unless the simultaneous effects of nucleation rate, growth rate, heat balance, and material balance are considered. The most comprehensive treatment of this subject is by Randolph and Larson (1988), who developed a mathematical model for continuous crystallizers of the mixed-suspension or circulating-magma type [*Am. Inst. Chem. Eng. J.* **8:** 639 (1962)] and subsequently examined variations of this model that include most of the aberrations found in commercial equipment. Randolph and Larson showed that when the total number of crystals in a given volume of suspension from a crystallizer is plotted as a function of the characteristic length as in Fig. 18-55, the slope of the line is usefully identified as the crystal population density, *n:*

$$n = \lim_{\Delta L \to 0} \frac{\Delta N}{\Delta L} = \frac{dN}{dL} \tag{18-30}$$

where N = total number of crystals up to size L per unit volume of magma. The population density thus defined is useful because it characterizes the nucleation-growth performance of a particular crystallization process or crystallizer.

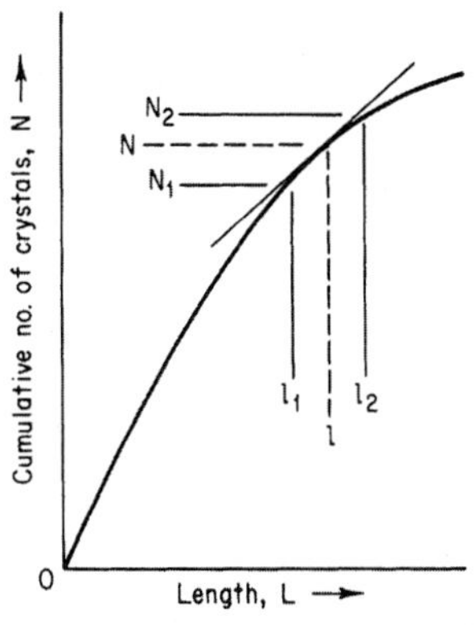

FIG. 18-55 Determination of the population density of crystals.

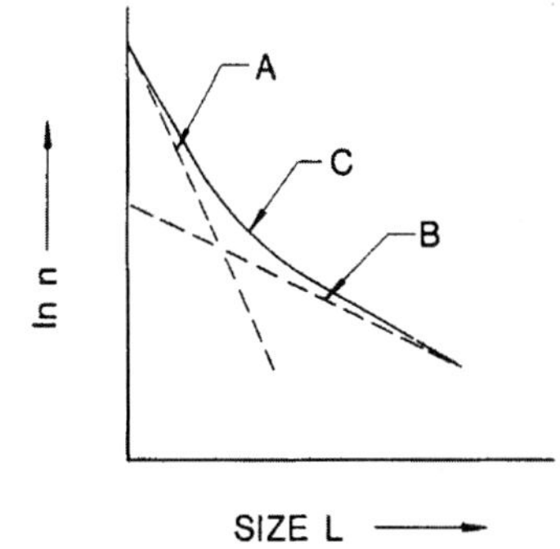

FIG. 18-56 Population density of crystals resulting from Bujacian behavior.

The data for a plot like Fig. 18-56 are easily obtained from a screen analysis of the total crystal content of a known volume (e.g., a liter) of magma. The analysis is made with a closely spaced set of testing sieves (or intervals for a particle counter), the cumulative number of particles smaller than each sieve in the nest being plotted against the aperture dimension of that sieve. The fraction retained on each sieve is weighed, and the mass is converted to the equivalent number of particles by dividing by the calculated mass of a particle whose dimension is the arithmetic mean of the mesh sizes of the sieve on which it is retained and the sieve immediately above it.

In industrial practice, the size-distribution curve usually is not actually constructed. Instead, a mean value of the population density for any sieve fraction of interest (in essence, the population density of the particle of average dimension in that fraction) is determined directly as $\Delta N/\Delta L$, ΔN being the number of particles retained on the sieve and ΔL being the difference between the mesh sizes of the retaining sieve and its immediate predecessor. It is common to employ the units of $(\text{mm} \cdot \text{L})^{-1}$ for *n*.

For a steady-state crystallizer receiving solids-free feed and containing a well-mixed suspension of crystals experiencing negligible breakage, a material-balance statement yields negligible agglomeration and breakage to a particle balance (the Randolph-Larson general-population balance); in turn, it simplifies to

$$\frac{dn}{dL} + \frac{n}{Gt} = 0 \tag{18-31}$$

if the delta L law applies (i.e., G is independent of L) and the draw-down (or retention) time is assumed to be invariant and calculated as $t = V/Q$. Integrated between the limits n^0, the population density of nuclei (for which L is assumed to be zero), and n, that of any chosen crystal size L, Eq. (18-31) becomes

$$\int_{n^0}^{n} \frac{dn}{n} = -\int_0^L \frac{dL}{Gt} \tag{18-32}$$

$$\ln n = \frac{-L}{Gt} + \ln n^0 \tag{18-33a}$$

or

$$n = n^0 e^{-L/Gt} \tag{18-33b}$$

It can be shown that

$$B^0 = n^0 G \tag{18-33c}$$

A plot of ln n versus L is a straight line whose intercept is ln n^0 and whose slope is $-1/Gt$. (For plots on base-10 log paper, the appropriate slope correction must be made.) Thus, from a given product sample of known slurry density and retention time, it is possible to obtain the nucleation rate and growth rate for the conditions tested if the sample satisfies the assumptions of the derivation and yields a straight line. A number of derived relations that describe the nucleation rate, size distribution, and average properties are summarized in Table 18-4.

If a straight line does *not* result (Fig. 18-56), at least part of the explanation may be violation of the delta L law [Canning and Randolph, *Am. Inst. Chem. Eng. J.* **13:** 5 (1967)]. The best current theory about what causes size-dependent growth suggests what has been called growth dispersion or "Bujacian behavior." In the same environment, different crystals of the same size can grow at different rates owing to differences in dislocations or other surface effects. The graphs of "slow" growers (Fig. 18-56, curve *A*) and

TABLE 18-4 Common Equations for Population-Balance Calculations

Name	Symbol	Units	Systems without fines removal	Systems with fines removal		References
				Fines stream	Product stream	
Drawdown time (retention time)	t	h	$t = V/Q$	$t_F = V_{\text{liquid}}/Q_F$	$t = V/Q$	
Growth rate	G	mm/h	$G = dL/dt$	$G = dL/dt$	$G = dL/dt$	
Volume coefficient	K_υ	1/no. (crystals)	$K_\upsilon = \frac{\text{volume of one crystal}}{L^3}$	$K_\upsilon = \frac{\text{volume of one crystal}}{L^3}$	$K_\upsilon = \frac{\text{volume of one crystal}}{L^3}$	
Population density	n	No. (crystals)/mm	$n = dN/dL$	$n = dN/dL$	$n = dN/dL$	1
Nuclei population density	n^o	No. (crystals)/mm	$n^o = K_M M^j G^{i-1}$			2
Population density	n	No. (crystals)/mm	$n = n^o e^{-L/Gt}$	$n_F = n^o e^{-L/Gt_F}$	$n = n^{o-L_e/Gt_F} e^{-L/Gt}$	1, 3
Nucleation rate	B_0	No. (crystals)/h	$B_0 = Gn^o = K_M M^j G^i$	$B_0 = G_n^o$		4
Dimensionless length	x	None	$x = \frac{L}{Gt}$	$x_F = \frac{L}{Gt_F}, L_0 \to L_f$	$x = \frac{L}{Gt}, L_f \to L$	1
Mass/unit volume (slurry density)	M_T	g/L	$M_t = K_\upsilon \rho \int_0^\infty nL^3\,dL$ $M_t = K_\upsilon \rho 6\,n^o (Gt)^4$	$M_{T_F} = K_\upsilon \rho \int_0^{L_f} n^o e^{-L/Gt_F} L^3\,dL$	$M_T = K_\upsilon \rho \int_{Lf}^\infty n^o e^{-L/Gt_F} e^{-L/Gt} L^3\,dL$	1
$\frac{\text{Cumulative mass to } x}{\text{Total mass}}$	W_x	None	$W_x = 1 - e^{-x}\left(\frac{x^3}{6} + \frac{x^2}{2} + x + 1\right)$	$W_F = \frac{e^{-x}(x^3 + 3x^2 + 6x + 6) - 6}{e^{-x}c(x_c^3 + 3x_c^2 + 6x_c + 6) - 6}$	$W = \frac{6K_\upsilon \rho n^o e^{-Le/Gt_F} GT)^4 \left[1 - e^{-x}\left(\frac{x^3}{6} + \frac{x^2}{2} + x + 1\right)\right]}{\text{Slurry density } M,\ g/L}$ when $L_c \approx 0$, compared with L_a	5
Dominant particle	L_d	mm	$L_d = 3Gt$			
Average particle, weight	L_a	mm	$L_a = 3.67Gt$			6
Total number of crystals	N_T	No./L	$N_T = \int_0^\infty n\,dl$	$N_F = \int_0^{L_f} n_F\,dL$	$N_T = \int_{Lf}^\infty n\,dL$	1, 3

1. Randolph and Larson, *Am. Inst. Chem. Eng. J.* **8:** 639 (1962).
2. Timm and Larson, *Am. Inst. Chem. Eng. J.* **14:** 452 (1968).
3. Larson, private communication.
4. Larson, Timm, and Wolff, *Am. Inst. Chem. Eng. J.* **14:** 448 (1968).
5. Larson and Randolph, *Chem. Eng. Prog. Symp. Ser.* **65**(95): 1 (1969).
6. Schoen, *Ind. Eng. Chem.* **53:** 607 (1961).

"fast" growers (curve *B*) sum to a resultant line (curve *C*), concave upward, that is described by Eq. (18-34) (Randolph, in deJong and Jancic, eds., *Industrial Crystallization,* North-Holland Publishing Company, Amsterdam, 1979, p. 254):

$$n = \sum \frac{B^0 i}{G_i} e^{(-L/G_i t)} \tag{18-34}$$

Equation (18-31) contains no information about the crystallizer's influence on the nucleation rate. If the crystallizer is of a mixed-suspension, mixed-product-removal (MSMPR) type, satisfying the criteria for Eq. (18-31), and if the model of Clontz and McCabe is valid, the contribution to the nucleation rate by the circulating pump can be calculated [Bennett, Fiedelman, and Randolph, *Chem. Eng. Prog.* **69**(7): 86 (1973)]:

$$B_e = K_e \left(\frac{I^2}{P}\right) \rho G \int_0^\infty nL^4\, dL \tag{18-35}$$

where I = tip speed of the propeller or impeller, m/s
ρ = crystal density, g/cm^3
P = volume of crystallizer/circulation rate (turnover), m^3/(m^3/s) = s

Since the integral term is the fourth moment of the distribution (m_4), Eq. (18-35) becomes

$$B_e = K_e \rho G \left(\frac{I^2}{P}\right) m_4 \tag{18-36}$$

Equation (18-36) is the general expression for impeller-induced nucleation. In a fixed-geometry system in which only the speed of the circulating pump is changed and in which the flow is roughly proportional to the pump speed, Eq. (18-36) may be satisfactorily replaced with

$$B_e = K_e'' \rho G (S_R)^3 m_4 \tag{18-37}$$

where S_R = rotation rate of impeller, r/min. If the maximum crystal-impeller impact stress is a nonlinear function of the kinetic energy, shown to be the case in at least some systems, Eq. (18-37) no longer applies.

In the specific case of an MSMPR exponential distribution, the fourth moment of the distribution may be calculated as

$$m_4 = 4! n^0 (Gt)^5 \tag{18-38}$$

Substitution of this expression into Eq. (18-36) gives

$$B_e = kn^0 G (S_R)^3 L_D^5 \tag{18–39}$$

where $L_D = 3Gt$, the dominant crystal (mode) size.

Equation (18-39) displays the competing factors that stabilize secondary nucleation in an operating crystallizer when nucleation is due mostly to impeller/crystal contact. Any increase in particle size produces a fifth-power increase in nucleation rate, tending to counteract the direction of the change and thereby stabilizing the crystal-size distribution. From dimensional argument alone, the size produced in a mixed crystallizer for a (fixed) nucleation rate varies as $(B^0)^{1/3}$. Thus, this fifth-order response of contact nucleation does not wildly upset the crystal size distribution but instead acts as a stabilizing feedback effect.

Nucleation due to crystal-to-crystal contact is greater for equal striking energies than crystal-to-metal contact. However, the viscous drag of the liquid on particle sizes normally encountered limits the velocity of impact to extremely low values. The assumption that only the largest crystal sizes contribute significantly to the nucleation rate by crystal-to-crystal contact permits a simple computation of the rate:

$$B_c = K_c \rho G m_j^2 \tag{18-40}$$

where m_j = the fourth, fifth, sixth, or higher moments of the distribution.

A number of different crystallizing systems have been investigated by using the Randolph-Larson technique, and some of the published growth rates and nucleation rates are included in Table 18-5. Although the usefulness of these data is limited to the conditions tested, the table gives a range of values that may be expected, and it permits resolution of the information gained from a simple screen analysis into the fundamental factors of growth rate and nucleation rate. Experiments may then be conducted to determine the independent effects of operation and equipment design on these parameters.

Although this procedure requires laborious calculations because of the number of samples normally needed, these computations and the determination of the best straight-line fit to the data are readily programmed for digital computers.

Example 18-4 Population Density, Growth, and Nucleation Rate Calculate the population density, growth, and nucleation rates for a crystal sample of urea for which there is the following information. These data are from Bennett and Van Buren [*Chem. Eng. Prog. Symp. Ser.* **65**(95): 44 (1969)].

Slurry density = 450 g/L
Crystal density = 1.335 g/cm^3
Drawdown time t = 3.38 h
Shape factor k_v = 1.00

Product size:

−14 mesh, +20 mesh	4.4 percent
−20 mesh, +28 mesh	14.4 percent
−28 mesh, +35 mesh	24.2 percent
−35 mesh, +48 mesh	31.6 percent
−48 mesh, +65 mesh	15.5 percent
−65 mesh, +100 mesh	7.4 percent
−100 mesh	2.5 percent

n = number of particles per liter of volume
14 mesh = 1.168 mm, 20 mesh = 0.833 mm, average opening 1.00 mm
Size span = 0.335 mm = ΔL

$$n_{20} = \frac{(450\text{ g/L})(0.044)}{(1.335/1000)\text{g/mm}^3(1.00^3\text{ mm}^3/\text{particle})(0.335\text{ mm})(1.0)}$$

$$n_{20} = 44{,}270$$

$$\ln n_{20} = 10{,}698$$

Repeating for each screen increment:

Screen size	Weight, %	k_v	ln n	L, average diameter, mm
100	7.4	1.0	18.099	0.178
65	15.5	1.0	17.452	0.251
48	31.6	1.0	16.778	0.356
35	24.2	1.0	15.131	0.503
28	14.4	1.0	13.224	0.711
20	4.4	1.0	10.698	1.000

Plotting ln n versus L as shown in Fig. 18-57, a straight line having an intercept at zero length of 19.781 and a slope of −9.127 results. As mentioned in discussing Eq. (18-24), the growth rate can then be found.

$$\text{Slope} = -1/Gt \text{ or } -9.127 = -1/[G(3.38)]$$

or

$$G = 0.0324\text{ mm/h}$$

and

$$B_0 = Gn^0 = (0.0324)(e^{19.781}) = 12.65 \times 10^6 \frac{n^0}{\text{L}\cdot\text{h}}$$

and

$$L_a = 3.67(0.0324)(3.38) = 0.40\text{ mm}$$

An additional check can be made of the accuracy of the data by the relation

$$M_T = 6k_v \rho n^0 (Gt)^4 = 450\text{ g/L}$$

$$M_T = (6)(1.0)\,\frac{1.335\text{ g/cm}^3}{1000\text{ mm}^3/\text{cm}^3}\,e^{19.78}[(0.0324)(3.38)]^4$$

$$M_T = 455\text{ g/L} \approx 450\text{ g/L}$$

Had only the growth rate been known, the size distribution of the solids could have been calculated from the equation

$$W_f = 1 - e^{-x}\left(\frac{x^3}{6} + \frac{x^2}{2} + x + 1\right)$$

where W_f is the weight fraction *up to* size L and $x = L/Gt$.

$$x = \frac{L}{(0.0324)(3.38)} = \frac{L}{0.1095}$$

Screen size	L, mm	x	W_f*	Cumulative % retained 100 $(1-W_f)$	Measured cumulative % retained
20	0.833	7.70	0.944	5.6	4.4
28	0.589	5.38	0.784	21.6	18.8
35	0.417	3.80	0.526	47.4	43.0
48	0.295	2.70	0.286	71.4	74.6
65	0.208	1.90	0.125	87.5	90.1
100	0.147	1.34	0.048	95.2	97.5

*Values of W_f as a function of x may be obtained from a table of Wick's functions.

TABLE 18-5 Growth Rates and Kinetic Equations for Some Industrial Crystallized Products

Material crystallized	G, m/s × 10^8	Range t, h	Range M_T, g/L	Temp., °C	Scale*	Kinetic equation for B_0 no./(L·s)	References†
$(NH_4)_2SO_4$	1.67	3.83	150	70	P	$B_0 = 6.62 \times 10^{-25} G^{0.82} p^{-0.92} m_2^{2.05}$	Bennett and Wolf, *AIChE,* SFC, 1979.
$(NH_4)_2SO_4$	0.20	0.25	38	18	B	$B_0 = 2.94(10^{10})G^{1.03}$	Larsen and Mullen, *J. Crystal Growth* **20:** 183 (1973).
$(NH_4)_2SO_4$	—	0.20	—	34	B	$B_0 = 6.14(10^{-11})S_R^{7.84} M_T^{0.98} G^{1.22}$	Youngquist and Randolph, *AIChE J.* **18:** 421 (1972).
$MgSO_4 \cdot 7H_2O$	3.0–7.0	—	—	25	B	$B_0 = 9.65(10^{12})M_T^{0.67} G^{1.24}$	Sikdar and Randolph, *AIChE J.* **22:** 110 (1976).
$MgSO_4 \cdot 7H_2O$	—	—	Low	29	B	$B_0 = f(N, L^4, N^{4.2}, S^{2.5})$	Ness and White, *AIChE Symposium Series 153,* vol. 72, p. 64.
KCl	2–12	—	200	32	P	$B_0 = 7.12(10^{39})M_T^{0.14} G^{4.99}$	Randolph et al., *AIChE J.* **23:** 500 (1977).
KCl	3.3	1–2	100	37	B	$B_0 = 5.16(10^{22})M_T^{0.91} G^{2.77}$	Randolph et al., *Ind. Eng. Chem. Proc. Design Dev.* **20:** 496 (1981).
KCl	0.3–0.45	—	50–147	25–68	B	$B_0 = 5 \times 10^{-3} G^{2.78}(M_T TIP^2)^{1.2}$	Qian et al., *AIChE J.* **33**(10): 1690 (1987).
KCr_2O_7	1.2–9.1	0.25–1	14–42	—	B	$B_0 = 7.33(10^4)M_T^{0.6} G^{0.5}$	Desari et al., *AIChE J.* **20:** 43 (1974).
KCr_2O_7	2.6–10	0.15–0.5	20–100	26–40	B	$B_0 = 1.59(10^{-3})S_R^3 M_T G^{0.48}$	Janse, Ph.D. thesis, Delft Technical University, 1977.
KNO_3	8.13	0.25–0.050	10–40	20	B	$B_0 = 3.85(10^{16})M_T^{0.5} G^{2.06}$	Juraszek and Larson, *AIChE J.* **23:** 460 (1977).
K_2SO_4	—	0.03–0.17	1–7	30	B	$B_0 = 2.62(10^3)S_R^{2.5} M_T^{0.5} G^{0.54}$	Randolph and Sikdar, *Ind. Eng. Chem. Fund.* **15:** 64 (1976).
K_2SO_4	2–6	0.25–1	2–20	10–50	B	$B_0 = 4.09(10^6)\exp\left(\frac{10900}{RT}\right) M_T G^{0.5}$	Jones, Budz, and Mullin, *AIChE J.* **33:** 12 (1986).
K_2SO_4	0.8–1.6	—	—	—	B	$\frac{G}{G_0} = 1 + 2L^{2/3}$ (L in μm)	White, Bendig, and Larson, *AIChE Mtg.,* Washington, D.C., Dec. 1974.
NaCl	4–13	0.2–1	25–200	50	B	$B_0 = 1.92(10^{10})S_R^2 M_T G^2$	Asselbergs, Ph.D. thesis, Delft Technical University, 1978.
NaCl	—	0.6	35–70	55	P	$B_0 = 8 \times 10^{10} N^2 G^2 M_T$	Grootscholten et al., *Chem. Eng. Design* **62:** 179 (1984).
NaCl	0.5	1–2.5	70–190	72	P	$B_0 = 1.47(10^2)\left(\frac{I^2}{P}\right) m_4^{0.84} G^{0.98}$	Bennett et al., *Chem. Eng. Prog.* **69**(7): 86 (1973).
Citric acid	1.1–3.7	—	—	16–24	B	$B_0 = 1.09(10^{10})m_4^{0.084} G^{0.84}$	Sikdar and Randolph, *AIChE J.* **22:** 110 (1976).
Fructose	0.1–0.25	—	—	50	B	—	Shiau and Berglund, *AIChE J.* **33:** 6 (1987).
Sucrose	—	—	—	80	B	$B_0 = 5 \times 10^6 N^{0.7} M_T^{0.3} G^{0.4}$	Berglund and deJong, *Separations Technology* **1:** 38 (1990).
Sugar	2.5–5	0.375	50	45	B	$B_0 = 4.38(10^6)M_T^{1.01}(\Delta C - 0.5)^{1.42}$	Hart et al., *AIChE Symposium Series 193,* vol. 76, 1980.
Urea	0.4–4.2	2.5–6.8	350–510	55	P	$B_0 = 5.48(10^{-1})M_T^{-3.87} G^{1.66}$	Bennett and Van Buren, *Chem. Eng. Prog. Symposium Series* **95**(7): 65 (1973).
Urea	—	—	—	3–16	B	$B_0 = 1.49(10^{-31})S_R^{2.3} M_T^{1.07} G^{-3.54}$	Lodaya et al., *Ind. Eng. Chem. Proc. Design Dev.* **16:** 294 (1977).

*B = bench scale; P = pilot plant.
†Additional data on many components are in Garside and Shah, *Ind. Eng. Chem. Proc. Design Dev.,* **19,** 509 (1980).

Note that the calculated distribution shows some deviation from the measured values because of the small departure of the actual sample from the theoretical coefficient of variation (i.e., 47.5 versus 50 percent).

The critical value of i, which is defined in Eq. (18-28) as the ratio of b/g or the relative dependence of nucleation and growth on supersaturation, can be determined by a few extra experiments. This is done by varying the residence time of the crystals (changing feed rate) while keeping everything else constant. The B^0 and G values are determined at each residence time, and a plot of ln B^0 versus ln G should yield a straight line of slope i. High values of i indicate a propensity to nucleate versus grow and dictate the need to ensure low values of supersaturation.

Had sufficient data indicating a change in n^0 for various values of M_T at constant G been available, a plot of ln n^0 versus ln M_T at corresponding G's would permit determination of the power j.

Crystallizers with Fines Removal In Example 18-4, the product was from a forced-circulation crystallizer of the MSMPR type. In many cases, the product produced by such machines is too small for commercial use; therefore, a separation baffle is added within the crystallizer to permit the removal of unwanted fine crystalline material from the magma, thereby controlling the population density in the machine so as to produce a coarser crystal product. When this is done, the product sample plots on a graph of ln n versus L as shown in line P, Fig. 18-58. The line of steepest slope, line F, represents the particle-size distribution of the fine material, and samples that show this distribution can be taken from the liquid leaving the fines-separation baffle. The product crystals have a slope of lower value, and typically there should be little material present smaller than L_f, the size that the baffle is designed to separate. However, this is not to imply that there are no fines in the product stream. The effective nucleation rate for the product material is the intersection of the extension of line P to zero size.

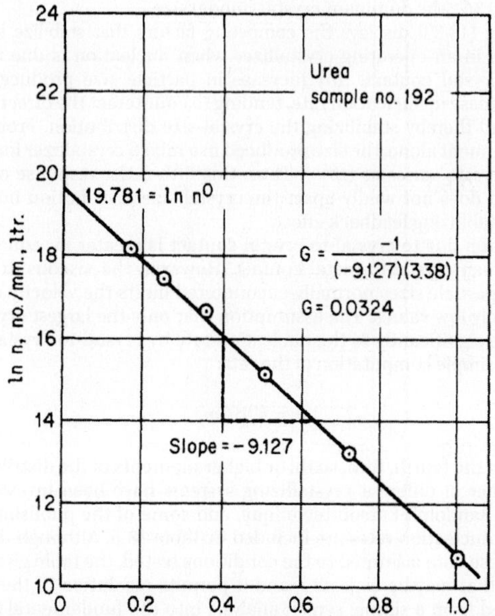

FIG. 18-57 Population density plot for Example 18-4.

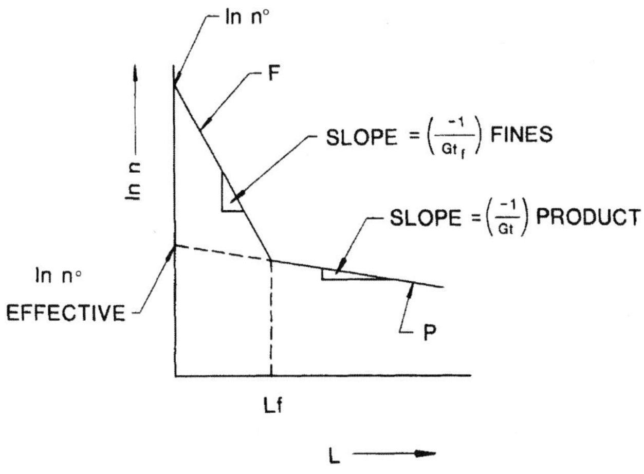

FIG. 18-58 Plot of log N against L for a crystallizer with fines removal.

As long as the largest particle separated by the fines-destruction baffle is small compared with the mean particle size of the product, the seed for the product may be thought of as the particle-size distribution corresponding to the fine material that ranges in length from zero to L_f, the largest size separated by the baffle.

The product discharged from the crystallizer is characterized by the integral of the distribution from size L_f to infinity:

$$M_T = k_v \rho \int_{L_f}^{\infty} n^0(-L_f/Gt_f)\exp(L/GT)L^3 dL \tag{18-41}$$

The integrated form of this equation is shown in Table 18-4.

For a given set of assumptions, it is possible to calculate the characteristic curves for the product from the crystallizer when it is operated at various levels of fines removal as characterized by L_f. This has been done for an ammonium sulfate crystallizer in Fig. 18-59. Also shown in that figure is the actual size distribution obtained. In calculating theoretical size distributions in accordance with the Eq. (18-41), it is assumed that the growth rate is a constant, whereas in fact larger values of L_f will interact with the system driving force to raise the growth rate and the nucleation rate. Nevertheless, Fig. 18-59 illustrates clearly the empirical result of the operation of such equipment, demonstrating that the most significant variable in changing the particle-size distribution of the product is the size removed by the baffle. Conversely, changes in retention time for a given particle-removal size L_f make a relatively small change in the product-size distribution. Jancic and Grootscholten (*Industrial Crystallization,* D. Reidel Publishing, Boston, 1984, p. 318) have found that the size enlargement is dependent on the fines size, the relative kinetic order *i*, and the rate of flow to the fines circuit versus product flow.

It is implicit that increasing the value of L_f will raise the supersaturation and growth rate to levels at which mass nucleation can occur, thereby leading to periodic upsets of the system or cycling [Randolph, Beer, and Keener, *Am. Inst. Chem. Eng. J.* **19:** 1140 (1973)]. That this could actually happen was demonstrated experimentally by Randolph, Beckman, and Kraljevich [*Am. Inst. Chem. Eng. J.* **23:** 500 (1977)], and that it could be controlled dynamically by regulating the fines-destruction system was shown by Beckman and Randolph [ibid., (1977)]. Dynamic control of a crystallizer with a fines-destruction baffle and fine-particle-detection equipment using a light-scattering (laser) particle-size-measurement instrument is described in U.S. Patents 4,263,010 and 5,124,265.

CRYSTALLIZATION EQUIPMENT

Whether a vessel is called an evaporator or a crystallizer depends primarily on the criteria used in arriving at its sizing. In an evaporator of the salting-out type, sizing is done on the basis of vapor release. In a crystallizer, sizing

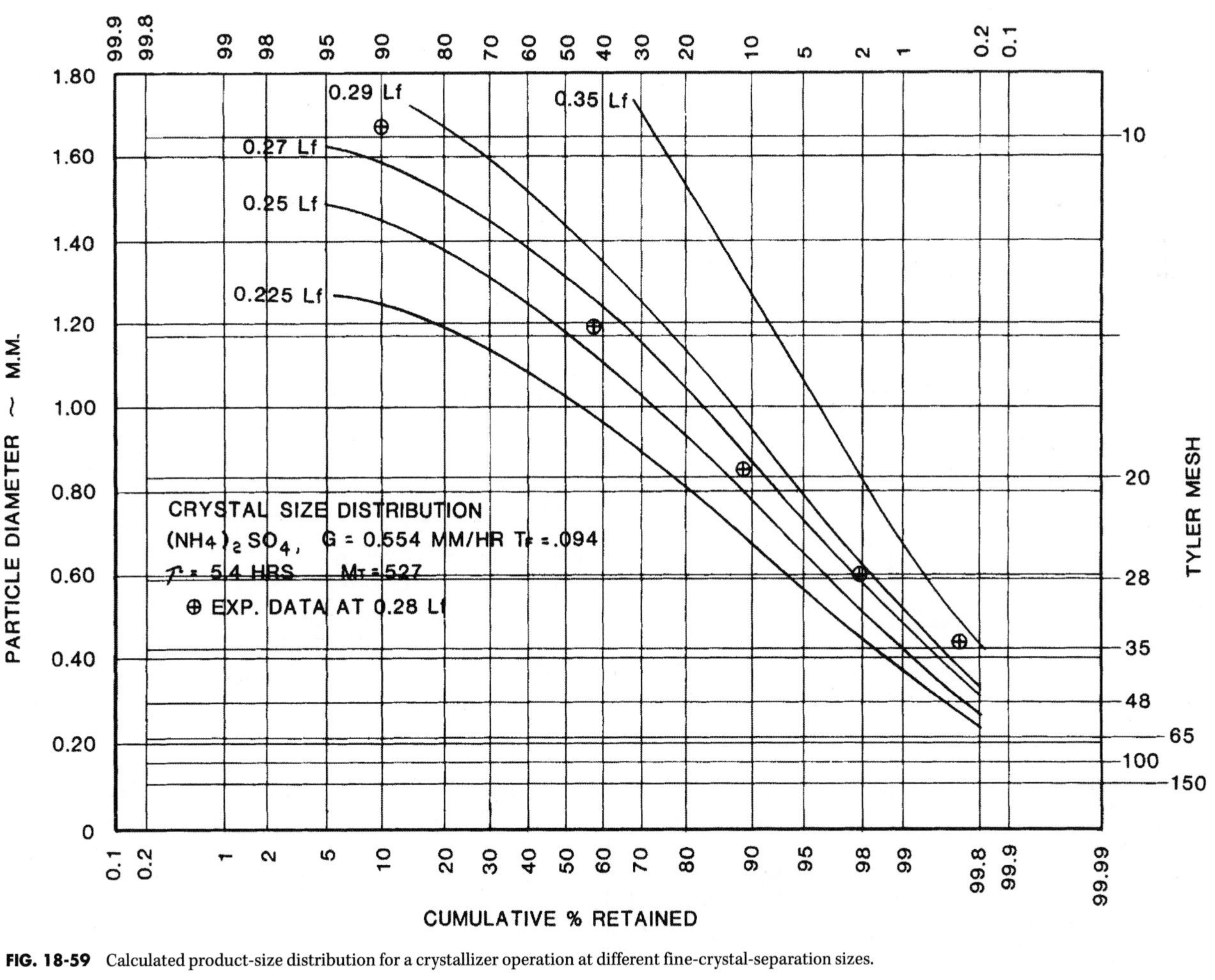

FIG. 18-59 Calculated product-size distribution for a crystallizer operation at different fine-crystal-separation sizes.

is normally done on the basis of the volume required for crystallization or for special features required to obtain the proper product size. In external appearance, the vessels could be identical. Evaporators are discussed in Sec. 11. Genck [*Chem. Eng. Prog.* **100**(10): 26 (2004)] provides a detailed discussion of guidelines for crystallizer selection and operation.

In the discussion that follows, crystallization equipment has been classified according to the means of suspending the growing product. This technique reduces the number of major classifications and segregates those to which Eq. (18-31) applies.

Mixed-Suspension, Mixed-Product-Removal Crystallizers This type of equipment, sometimes called the circulating-magma crystallizer, is by far the most important in use today. In most commercial equipment of this type, the uniformity of suspension of product solids within the crystallizer body is sufficient for the theory [Eqs. (18-31) to (18-33*c*)] to apply. Although a number of different varieties and features are included within this classification, the equipment operating with the highest capacity is the kind in which the vaporization of a solvent, usually water, occurs.

Although surface-cooled types of MSMPR crystallizers are available, most users prefer crystallizers that use vaporization of solvents or of refrigerants. The primary reason for this preference is that heat transferred through the critical supersaturating step is through a boiling-liquid-gas surface, avoiding the troublesome solid deposits that can form on a metal heat-transfer surface. In this case, very low LMTDs are required to stay within the metastable zone to promote growth and reduce scaling. The result is multipass, large-surface-area heat exchangers.

A *forced-circulation (FC) evaporator-crystallizer* is shown in Fig. 18-60. Slurry leaving the body is pumped through a circulating pipe and through a tube-and-shell heat exchanger, where its temperature increases by about 2°C to 6°C (3°F to 10°F). Since this heating is done without vaporization, materials of normal solubility should produce no deposition on the tubes. The heated slurry, returned to the body by a recirculation line, mixes with the body slurry and raises its temperature locally near the point of entry, which causes boiling at the liquid surface. During the consequent cooling and vaporization to achieve equilibrium between liquid and vapor, the supersaturation that is created causes growth on the swirling body of suspended crystals until they again leave via the circulating pipe. Severe vortexing must be eliminated to ensure that the supersaturation is relieved. The quantity and the velocity of the recirculation, the size of the body, and the type and speed of the circulating pump are critical design items if predictable results are to be achieved. A further discussion of the parameters affecting this type of equipment is presented by Bennett, Newman, and Van Buren [*Chem. Eng. Prog.* **55**(3): 65 (1959); *Chem. Eng. Prog. Symp. Ser.* **65**(95): 34, 44 (1969)].

If the crystallizer is not of the evaporative type but relies only on *adiabatic evaporative cooling* to achieve the yield, the heating element is omitted. The feed is admitted into the circulating line after withdrawal of the slurry, at a point sufficiently below the free-liquid surface to prevent flashing during the mixing process.

FC units typically range from 2 to 20 ft in diameter. They are especially useful for high evaporation loads. For example, a unit used to evaporate water at 380 mm Hg can typically be designed to handle 250 to 300 lb/(h · ft^2). Other than allowing one to adjust the residence time or slurry density, the FC affords little opportunity to change the size distribution.

In a *draft-tube-baffle (dtb) evaporator-crystallizer,* because mechanical circulation greatly influences the level of nucleation within the crystallizer, a number of designs have been developed that use circulators located within the body of the crystallizer, thereby reducing the head against which the circulator must pump. This technique reduces the power input and circulator tip speed and therefore the rate of nucleation. A typical example is the draft-tube-baffle (DTB) evaporator-crystallizer shown in Fig. 18-61. The suspension of product crystals is maintained by a large, slow-moving propeller surrounded by a draft tube within the body. The propeller directs the slurry to the liquid surface so as to prevent solids from short-circuiting the zone of the most intense supersaturation. Slurry that has been cooled is returned to the bottom of the vessel and recirculated through the propeller. At the propeller, heated solution is mixed with the recirculating slurry.

The design of Fig. 18-61 contains a fines-destruction feature comprising the settling zone surrounding the crystallizer body, the circulating pump, and the heating element. The heating element supplies sufficient heat to meet the evaporation requirements and to raise the temperature of the solution removed from the settler so as to destroy any small crystalline particles withdrawn. Coarse crystals are separated from the fines in the settling zone

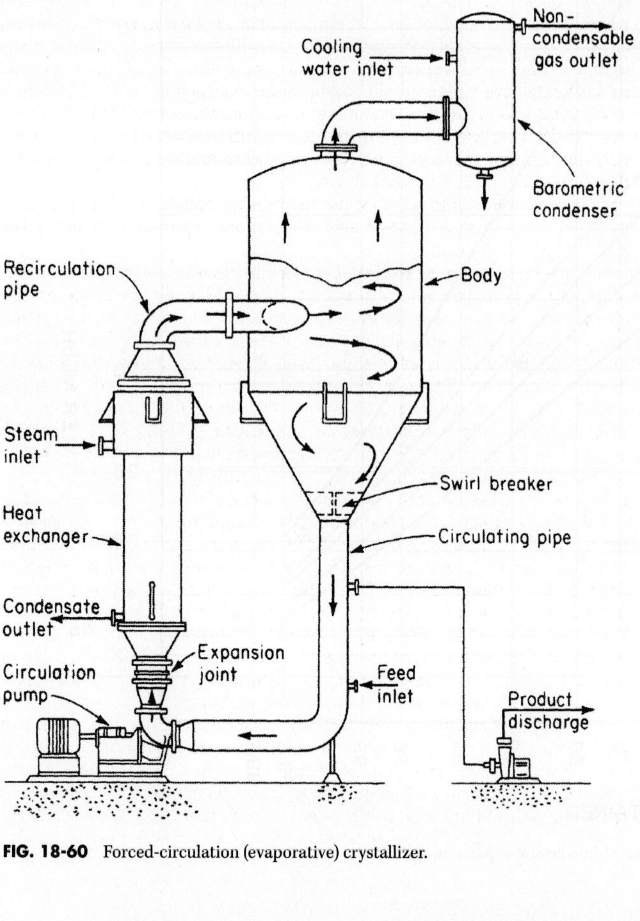

FIG. 18-60 Forced-circulation (evaporative) crystallizer.

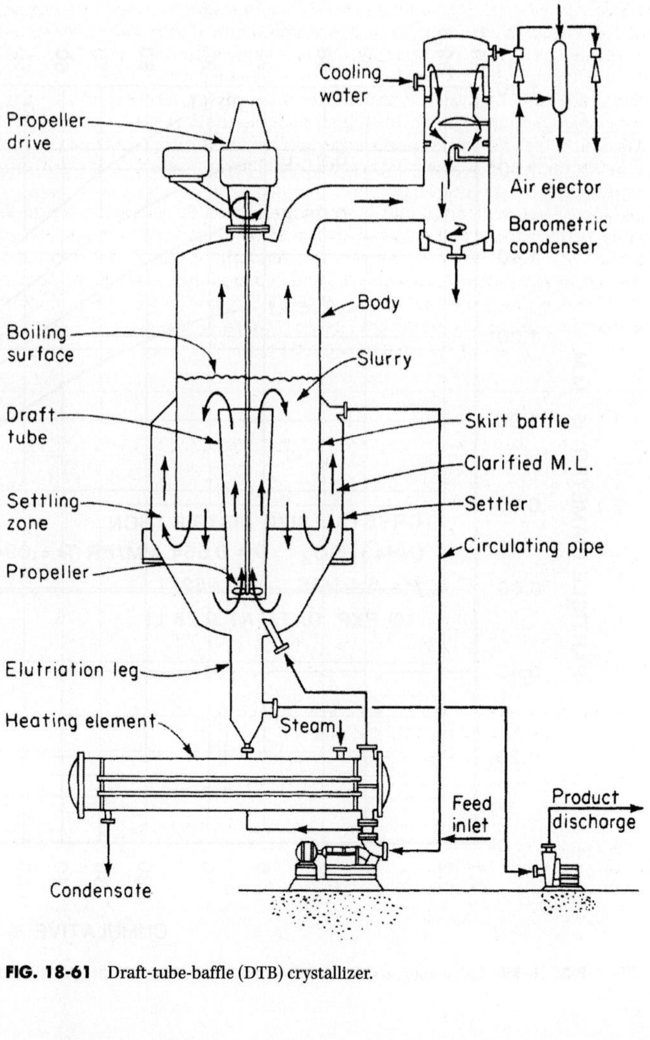

FIG. 18-61 Draft-tube-baffle (DTB) crystallizer.

by gravitational sedimentation, and therefore this fines-destruction feature is applicable only to systems in which there is a substantial density difference between crystals and mother liquor.

This type of equipment can also be used for applications in which the only heat removed is that required for adiabatic cooling of the incoming feed solution. When this is done and the fines-destruction feature is to be employed, a stream of liquid must be withdrawn from the settling zone of the crystallizer, and the fine crystals must be separated or destroyed by some means other than heat addition—for example, either dilution or thickening and physical separation.

In some crystallization applications, it is desirable to increase the solids content of the slurry within the body above the natural make, which is that developed by equilibrium cooling of the incoming feed solution to the final temperature. This can be done by withdrawing a stream of mother liquor from the baffle zone, thereby thickening the slurry within the growing zone of the crystallizer. This mother liquor is also available for removal of fine crystals for size control of the product.

A *draft-tube (DT) crystallizer* may be used in systems in which fines destruction is not needed or wanted. In such cases, the baffle is omitted, and the internal circulator is sized to have the minimum nucleating influence on the suspension.

In DTB and DT crystallizers, the circulation rate achieved is generally much greater than that available in a similar forced-circulation crystallizer. The equipment therefore finds application when it is necessary to circulate large quantities of slurry to minimize supersaturation levels within the equipment. In general, this approach is required to obtain long operating cycles with material capable of growing on the walls of the crystallizer. The draft-tube and draft-tube-baffle designs are commonly used in the production of granular materials such as ammonium sulfate, potassium chloride, photographic hypo, and other inorganic and organic crystals for which product in the range 8 to 30 mesh is required.

A *surface-cooled crystallizer* can be used for some materials, such as sodium chlorate, where it is possible to use a forced-circulation tube-and-shell exchanger in direct combination with a draft-tube-crystallizer body, as shown in Fig. 18-62. Careful attention must be paid to the temperature difference between the cooling medium and the slurry circulated through the exchanger tubes. In addition, the path and rate of slurry flow within the crystallizer body must be such that the volume contained in the body is "active." That is to say, crystals must be so suspended within the body by the turbulence that they are effective in relieving supersaturation created by the reduction in temperature of the slurry as it passes through the exchanger. Obviously, the circulating pump is part of the crystallizing system, and careful attention must be paid to its type and its operating parameters to avoid undue nucleating influences.

The use of the internal baffle permits operation of the crystallizer at a slurry consistency other than that naturally obtained by the cooling of the feed from the initial temperature to the final mother-liquor temperature. The baffle also permits fines removal and destruction.

With most inorganic materials, this type of equipment produces crystals in the range 30 to 100 mesh. The design is based on the allowable rates of heat exchange and the retention required to grow the product crystals.

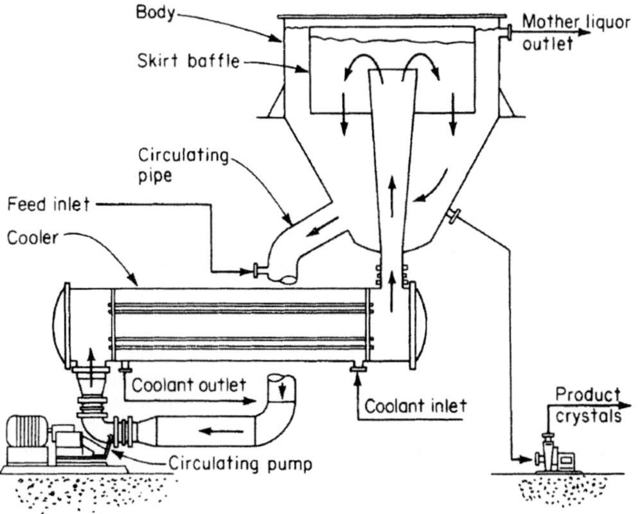

FIG. 18-62 Forced-circulation baffle surface-cooled crystallizer.

A *direct-contact-refrigeration crystallizer* can be used for some applications, such as the freezing of ice from seawater, where it is necessary to go to such low temperatures that cooling by the use of refrigerants is the only economical solution. In such systems, it is sometimes impractical to use surface-cooled equipment because the allowable temperature difference is so small (under 3°C) that the heat-exchanger surface becomes excessive or because the viscosity is so high that the mechanical energy put in by the circulation system requires a heat-removal rate greater than can be obtained at reasonable temperature differences. In such systems, it is convenient to admix the refrigerant with the slurry being cooled in the crystallizer so that the heat of vaporization of the refrigerant cools the slurry by direct contact. The successful application of such systems requires that the refrigerant be relatively immiscible with the mother liquor and be capable of separation, compression, condensation, and subsequent recycle into the crystallizing system. The operating pressures and temperatures chosen have a large bearing on power consumption.

This technique has been very successful in reducing the problems associated with buildup of solids on a cooling surface. The use of direct-contact refrigeration also reduces overall process-energy requirements, since in a refrigeration process involving two fluids, a greater temperature difference is required on an overall basis when the refrigerant must first cool some intermediate solution, such as calcium chloride brine, and that solution in turn cools the mother liquor in the crystallizer.

Equipment of this type has been successfully operated at temperatures as low as −59°C (−75°F).

Reaction-Type Crystallizers In chemical reactions in which the end product is a solid-phase material, such as a crystal or an amorphous solid, a reaction crystallizer may be used. By mixing the reactants in a large circulated stream of mother liquor containing suspended solids of the equilibrium phase, it is possible to minimize the driving force created during their reaction and to remove the heat of reaction through the vaporization of a solvent, normally water. Depending on the final particle size required, it is possible to incorporate a fines-destruction baffle to take advantage of the control over particle size afforded by this technique. In the case of ammonium sulfate crystallization from ammonia gas and concentrated sulfuric acid, it is necessary to vaporize water to remove the heat of reaction, and this water so removed can be reinjected after condensation into the fines-destruction stream to afford a very large amount of dissolving capability.

Other examples of this technique are where a solid material is to be decomposed by mixing it with a mother liquor of a different composition. Carnallite ore ($KCl \cdot MgCl_2 \cdot 4H_2O$) can be added to a mother liquor into which water is also added so that decomposition of the ore into potassium chloride (KCl) crystals and magnesium chloride–rich mother liquor takes place. Circulated slurry in the draft tube suspends the product crystals as well as the incoming ore particles until the ore can decompose into potassium chloride crystals and mother liquor. By taking advantage of the fact that water must be added to the process, the fines-bearing mother liquor can be removed behind the baffle, and then water can be added so that the finest particles are dissolved before being returned to the crystallizer body.

Other examples of this technique involve neutralization reactions such as the neutralization of sulfuric acid with calcium chloride to result in the precipitation of gypsum.

Mixed-Suspension, Classified-Product-Removal Crystallizers Many of the crystallizers just described can be designed for classified-product discharge. Classification of the product is normally done by means of an elutriation leg suspended beneath the crystallizing body as shown in Fig. 18-61. The introduction of clarified mother liquor to the lower portion of the leg fluidizes the particles prior to discharge and selectively returns the finest crystals to the body for further growth. A relatively wide distribution of material is usually produced unless the elutriation leg is extremely long. Inlet conditions at the leg are critical if good classifying action or washing action is to be achieved.

If an elutriation leg or other product-classifying device is added to a crystallizer of the MSMPR type, the plot of the population density versus L is changed in the region of largest sizes. Also, the incorporation of an elutriation leg destabilizes the crystal-size distribution and under some conditions can lead to cycling. To reduce cycling, fines destruction is usually coupled with classified product removal. The theoretical treatment of both the crystallizer model and the cycling relations is discussed by Randolph, Beer, and Keener [*Am. Inst. Chem. Eng. J.* **19:** 1140 (1973)]. Although such a feature can be included on many types of classified-suspension or mixed-suspension crystallizers, it is most common to use this feature with the forced-circulation evaporative crystallizer and the DTB crystallizer.

Classified-Suspension Crystallizer This equipment is also known as the *growth* or *Oslo crystallizer* and is characterized by the production of supersaturation in a circulating stream of liquor. Supersaturation is developed in one part of the system by evaporative cooling or by cooling in a heat exchanger, and it is relieved by passing the liquor through a fluidized bed of crystals. The fluidized bed may be contained in a simple tank or in a more

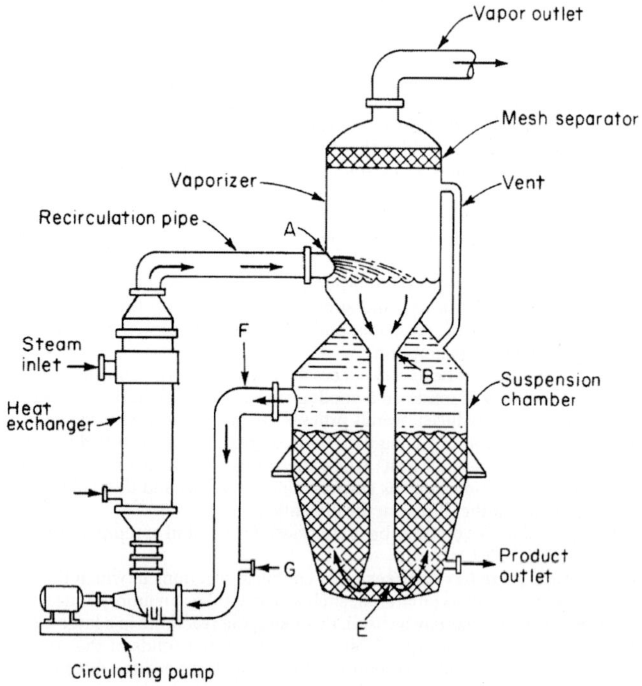

FIG. 18-63 OSLO evaporative crystallizer.

sophisticated vessel arranged for a pronounced classification of the crystal sizes. Ideally this equipment operates within the metastable supersaturation field described by Miers and Isaac, *J. Chem. Soc.* **89:** 413–454 (1906).

In the *evaporative crystallizer* of Fig. 18-63, solution leaving the vaporization chamber at *B* is supersaturated slightly within the metastable zone so that new nuclei will not form. The liquor contacting the bed at *E* relieves its supersaturation on the growing crystals and leaves through the circulating pipe *F.* In a cooling-type crystallization, hot feed is introduced at *G,* and the mixed liquor flashes when it reaches the vaporization chamber at *A.* If further evaporation is required to produce the driving force, a heat exchanger is installed between the circulating pump and the vaporization changer to supply the heat for the required rate of vaporization.

The transfer of supersaturated liquor from the vaporizer (point *B,* Fig. 18-63) can cause salt buildup in the piping and reduction of the operating cycle in equipment of this type. The rate of buildup can be reduced by circulating a thin suspension of solids through the vaporizing chamber; however, the presence of such small seed crystals tends to rob the supersaturation developed in the vaporizer, thereby lowering the efficiency of the recirculation system.

The decrease in temperature due to flashing is typically less than 4°F to 6°F, and the increase in solute concentration in the circulating liquor is often around 1 to 3 g/L solvent. Care must be taken to ensure that the liquid velocities in the tapered cross section of the lower body allow classification of the solids. One must know the settling rates and morphologies of the crystals for proper design and operation. An unclassified operation will perform as an FC unit.

The Oslo crystallizer is best suited for use with compounds with high settling velocities, such as those greater than 20 to 40 mm/s. If the crystals have high settling rates, the larger particles will settle out quickly. Crystals with low settling velocities require large cross-sectional areas, which implies large crystallizers and low crystal production rates.

The suggested productivities for concentration driving forces depend on the settling velocities of the crystals. For a given ΔC, the higher the settling velocity, the higher the allowable crystallizer productivity. For example, for a change in concentration of 2 g/L, the recommended productivity increases from 125 to 250 kg/(h · m^3) as the settling rate increases from 20 to 30 mm/s.

The Oslo/Krystal can also function as a surface-cooled unit.

Scraped-Surface Crystallizer A number of crystallizer designs using direct heat exchange between the slurry and a jacket or double wall containing a cooling medium have been developed. The heat-transfer surface is scraped or agitated in such a way that the deposits cannot build up. The scraped-surface crystallizer provides an effective and inexpensive method of producing slurry in equipment that does not require expensive installation or supporting structures. At times these units are employed to provide auxiliary cooling capacity for existing units.

A *double-pipe scraped-surface crystallizer* consists of a double-pipe heat exchanger with an internal agitator fitted with spring-loaded scrapers that wipe the wall of the inner pipe. The cooling liquid passes between the pipes, this annulus being dimensioned to permit reasonable shell-side velocities. The scrapers prevent the buildup of solids and maintain a good film coefficient of heat transfer. Crystal growth is in the bulk of the liquid. The equipment can be operated in a continuous or in a recirculating batch manner.

Such units are generally built in lengths to above 12 m (40 ft). They can be arranged in parallel or in series to give the necessary liquid velocities for various capacities. Heat-transfer coefficients have been reported in the range of 170 to 850 W/(m^2 · K) [30 to 150 Btu/(h · ft^2 · °F)] at temperature differentials of 17°C (30°F) and higher [Garrett and Rosenbaum, *Chem. Eng.* **65**(16): 127 (1958)]. Equipment of this type is marketed as the Votator and the Armstrong crystallizer.

Batch Crystallization Batch crystallization has been practiced longer than any other form of crystallization in both atmospheric tanks, which are either static or agitated, and in vacuum or pressure vessels. It is widely practiced in the pharmaceutical and fine chemical industry or in those applications where the capacity is very small. This supersaturation can be generated by a number of modes, including antisolvent addition, cooling, evaporation, pH adjustment, and chemical reaction.

A typical batch process involves charging the crystallizer with concentrated or near-saturated solution, producing supersaturation by means of a cooling temperature profile or evaporation profile, and seeding the batch in the metastable zone or by allowing spontaneous nucleation to occur. The final mother-liquor temperature and concentration are achieved by a time-dependent profile, and the batch is then held for ripening followed by transferring the same to downstream processing such as centrifuging, filtration, and drying.

Control of a batch crystallizer is critical to achieve the desired size distribution. It is necessary to have some way to determine when the initial solution is supersaturated so that seed of the appropriate size, quantity, and habit may be introduced into the batch. After seeding, it is necessary to limit the cooling or evaporation in the batch to that which permits the generated supersaturation to be relieved on the seed crystals. This means that the first cooling or evaporation following seeding must be at a very slow rate, which is increased nonlinearly in order to achieve the optimum batch cycle and product properties. Often such controls are operated by cycle timers or computers in order to achieve the required conditions. Shown in Fig. 18-64 is a typical batch crystallizer comprising a jacketed closed tank with top-mounted agitator and feed connections. The tank is equipped with a short distillation column and surface condenser so that volatile materials may be retained in the tank and solvent recycled to maintain the batch integrity. Provisions are included so that the vessel may be heated with steam added to the shell or cooling solution circulated through the jacket to control the temperature. Tanks of this type are intended to be operated with a wide variety of chemicals under both cooling and solvent evaporation conditions.

A detailed discussion of crystallization practice is provided by Genck in the following articles: Genck, *Chem. Eng.* **104**(11): 94 (1997); Genck, *Chem. Eng.* **107**(8): 90 (2000), and Genck, *Chem. Eng. Progress* **99**(6): 36 (2003).

Recompression Evaporation-Crystallization In all types of crystallization equipment wherein water or some other solvent is vaporized to produce supersaturation or cooling, attention should be given to the use of mechanical vapor recompression, which by its nature permits the substitution of electrical energy for evaporation and solvent removal rather than requiring the direct use of heat energy in the form of steam or electricity. A typical recompression crystallizer flow sheet is shown in Fig. 18-65, which shows a single-stage evaporative crystallizer operating at approximately atmospheric pressure. The amount of heat energy needed to remove 1 kg of water to produce the equivalent in crystal product is approximately 550 kilocalories. If the water evaporated is compressed by a mechanical compressor of high efficiency to a pressure where it can be condensed in the heat exchanger of the crystallizer, it can thereby supply the energy needed to sustain the process. Then the equivalent power for this compression is about 44 kilocalories (Bennett, *Chem. Eng. Progress,* 1978, pp. 67–70).

Although this technique is limited economically to those large-scale cases where the materials handled have a relatively low boiling point elevation and in those cases where a significant amount of heat is required to produce the evaporation for the crystallization step, it nevertheless offers an attractive technique for reducing the use of heat energy and substituting mechanical energy or electrical energy in those cases where there is a cost advantage for doing so. This technique finds many applications in the crystallization of sodium sulfate, sodium carbonate monohydrate, and sodium chloride. Shown in Fig. 18-66 is the amount of vapor compressed per kilowatt-hour for water vapor at 100°C and various ΔTs. The amount of water vapor compressed per horsepower decreases rapidly with increasing ΔT and, therefore, normal design considerations dictate that the recompression evaporators have a relatively large amount of heat-transfer surface to minimize the power cost. Often this technique is used only with the initial

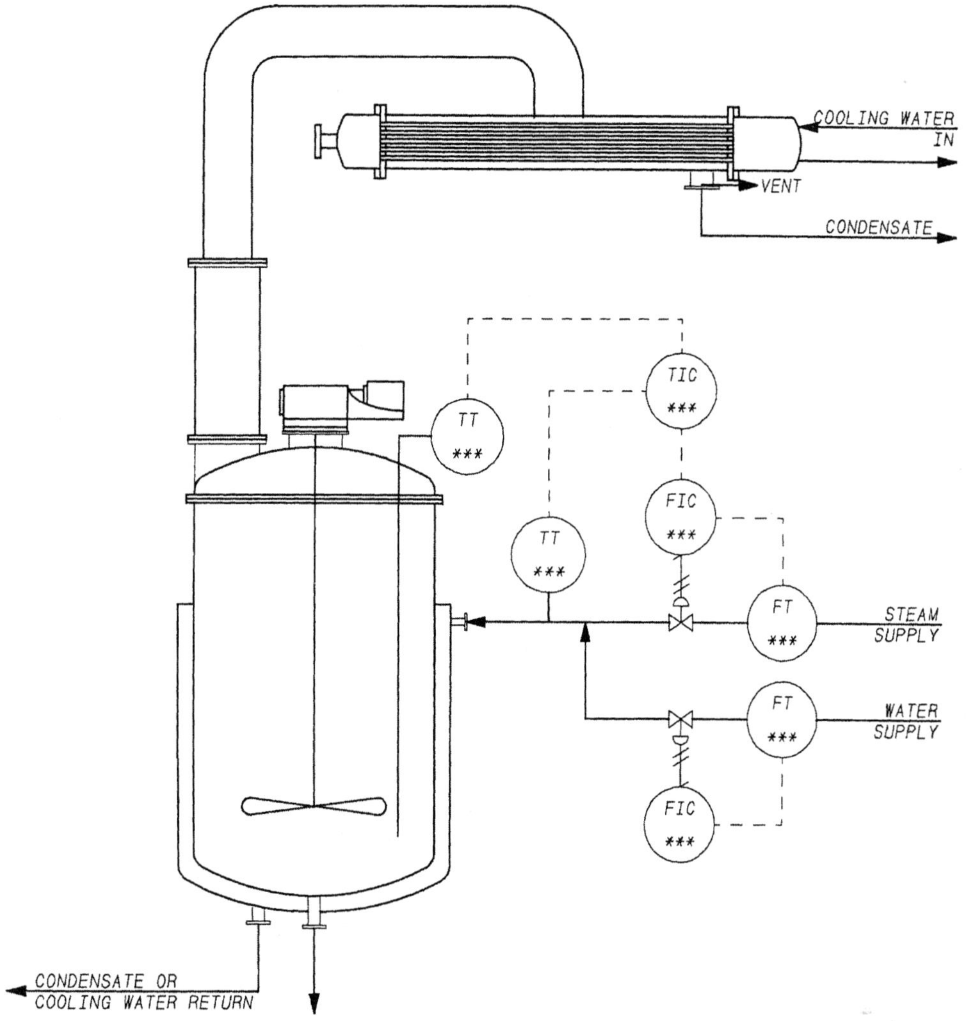

FIG. 18-64 Typical agitated batch crystallizer.

stages of evaporation, where concentration of the solids is relatively low and, therefore, the boiling-point elevation is negligible. In order to maintain adequate tube velocity for heat transfer and suspension of crystals, the increased surface requires a large internal recirculation within the crystallizer body, which consequently lowers the supersaturation in the fluid pumped through the tubes. One benefit of this design is that with materials of flat or inverted solubility, the use of recompression complements the need to maintain low ΔTs to prevent fouling of the heat-transfer surface.

INFORMATION REQUIRED TO SPECIFY A CRYSTALLIZER

The following information regarding the product, properties of the feed solution, and required materials of construction must be available before a crystallizer application can be properly evaluated and the appropriate equipment options identified. Is the crystalline material being produced a hydrated or an anhydrous material? What is the solubility of the compound in water or in other solvents under consideration, and how does this change with temperature? Are there other compounds in solution that coprecipitate with the product being crystallized, or do these remain in solution, increasing in concentration until some change in product phase occurs? What will be the influence of impurities in the solution on the crystal habit, growth, and nucleation rates? What are the physical properties of the solution and its tendency to foam? What is the heat of crystallization of the product crystal? What is the production rate, and what is the basis on which this production rate is computed? What is the tendency of the material to grow on the walls of the crystallizer? What materials of construction can be used in contact with the solution at various temperatures? What utilities will be available at the crystallizer location, and what are the costs associated with the use of these utilities? Is the final product to be blended or mixed with other crystalline materials or solids? What size of product and what shape of product are required to meet these requirements? How can the crystalline material be separated from the mother liquor and dried? Are there temperature requirements or wash requirements to be met? How can these solids or mixtures of solids be handled and stored without undue breakage and caking? Is polymorphism an issue?

Another basic consideration is whether crystallization is best carried out on a batch basis or on a continuous basis. The present tendency in most processing plants is to use continuous equipment whenever possible. Continuous equipment permits the adjustment of the operating variables to a relatively fine degree in order to achieve the best results in terms of energy usage and product characteristics. It allows the use of a smaller labor force and results in a continuous utility demand, which minimizes the size of boilers, cooling towers, and power-generation facilities. It also minimizes the capital investment required in the crystallizer and in the feed-storage and product-liquor-storage facilities.

Materials that have a tendency to grow readily on the walls of the crystallizer require periodic washout, and therefore an otherwise continuous operation would be interrupted once or even twice a week for the removal of these deposits. The impact that this contingency may have on the processing-equipment train ahead of the crystallizer must be considered.

A batch operation usually has economic application only on small scale, or when multiple products are produced in common facilities.

CRYSTALLIZER OPERATION

Crystal growth is a layer-by-layer process, and the retention time required in most commercial equipment to produce crystals of the size normally desired is often on the order of 2 to 6 h. Growth rates are usually limited to

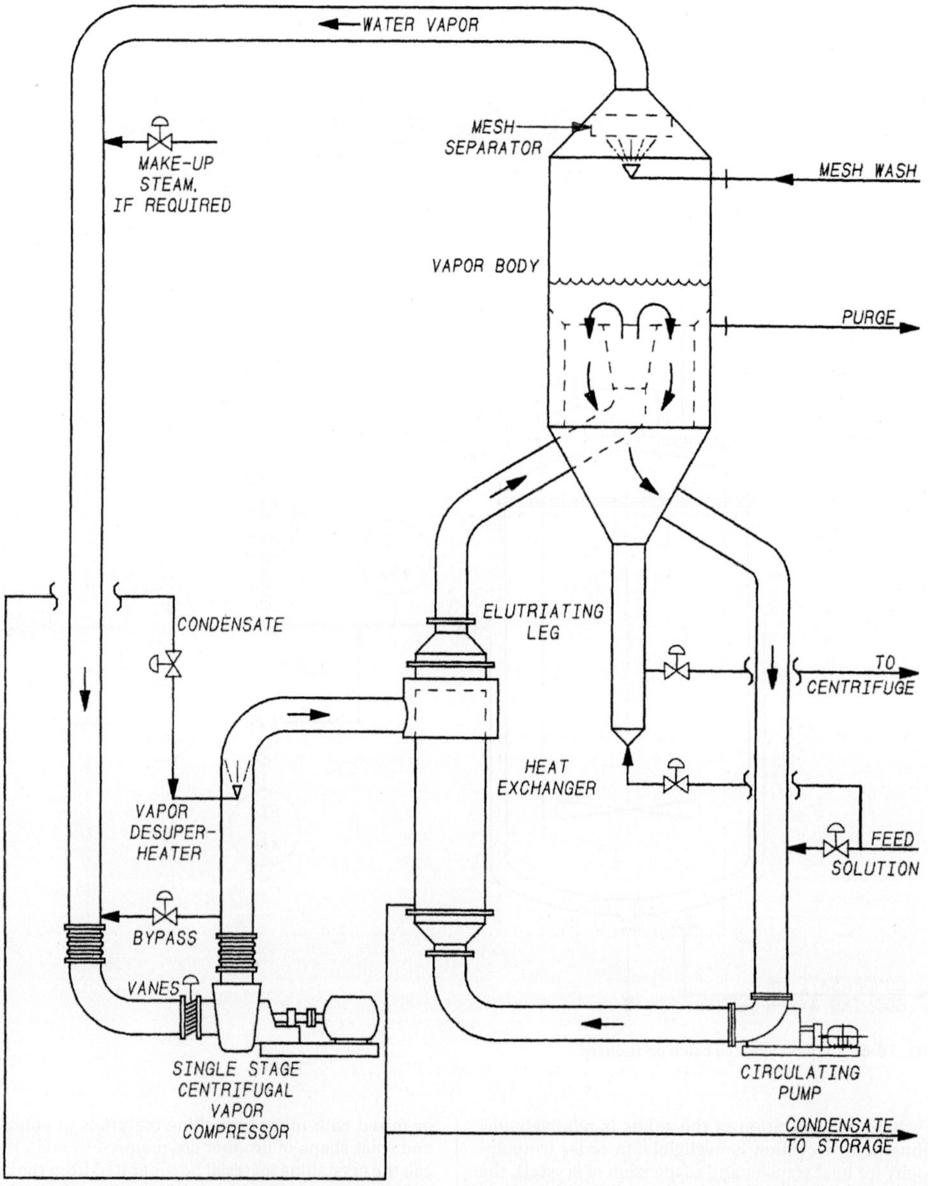

FIG. 18-65 Single-stage recompression evaporator.

less than 1 to 2 μm/min. On the other hand, nucleation in a supersaturated solution can be generated in a fraction of a second. The influence of any upsets in operating conditions, in terms of the excess nuclei produced, is very short-term in comparison with the total growth period of the product removed from the crystallizer. A worst-case scenario for batch or continuous operation occurs when the explosion of nuclei is so severe that it is impossible to grow an acceptable crystal size distribution, requiring redesolution or washout of the system. In a practical sense, this means that steadiness of operation is much more important in crystallization equipment than it is in many other types of process equipment.

It is to be expected that six to nine retention periods will pass before the effects of an upset will be damped out. Thus, the recovery period may last from 12 to 54 h.

The rate of nuclei formation required to sustain a given product size decreases exponentially with increasing size of the product. Although when crystals in the range of 100 to 50 mesh are produced, the system may react quickly, the system response when generating large crystals in the 14-mesh size range is quite slow. This is because a single pound of 150-mesh seed crystals is enough to provide the total number of particles in a ton of 14-mesh product crystals. In any system producing relatively large crystals, nucleation must be carefully controlled with respect to all internal and external sources. Particular attention must be paid to preventing seed crystals from entering with the incoming feed stream or being returned to the crystallizer with recycle streams of mother liquor coming back from the filter or centrifuge.

Experience has shown that in any given body operating at a given production rate, control of the magma (slurry) density is important to the control of crystal size. Although in some systems a change in slurry density does not result in a change in the rate nucleation, the more general case is that an increase in the magma density increases the product size through a reduction in nucleation and increased retention time of the crystals in the growing bed. The reduction in supersaturation at longer retention times, together with the increased surface area at higher percent solids, appears to be responsible for the larger product.

A reduction in the magma density will generally increase nucleation and decrease the particle size. This technique has the disadvantage that crystal formation on the equipment surfaces increases because lower slurry densities create higher levels of supersaturation within the equipment, particularly at the critical boiling surface in a vaporization-type crystallizer.

High levels of supersaturation at the liquid surface or at the tube walls in a surface-cooled crystallizer are the dominant cause of wall salting. Although some types of crystallizers can operate for several months continuously when crystallizing KCl or $(NH_4)_2SO_4$, most machines have much shorter operating cycles. Second only to control of particle size, the extension of operating cycles is the most difficult operating problem to be solved in most installations.

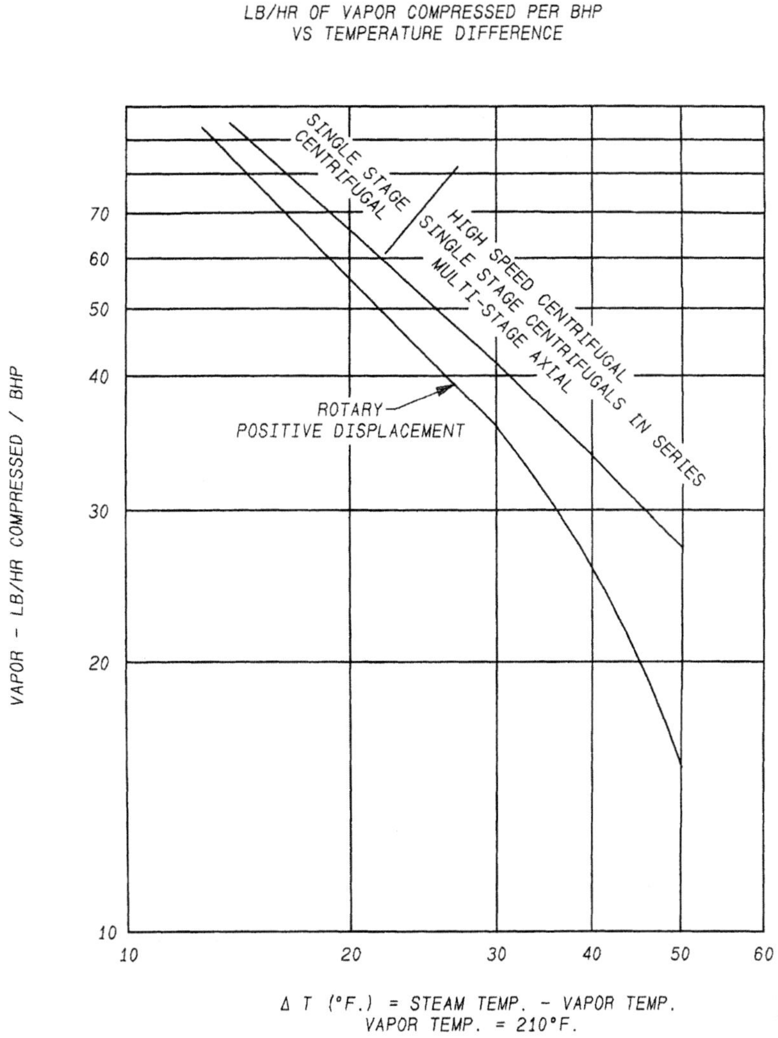

FIG. 18-66 Recompression evaporator horsepower as a function of overall ΔT.

In the forced-circulation-type crystallizer (Fig. 18-60), primary control over particle size is exercised by the designer in selecting the circulating system and volume of the body. From the operating standpoint there is little that can be done to an existing unit other than to supply external seed, classify the discharge crystals, or control the slurry density. Nevertheless, machines of this type are often carefully controlled by these techniques and produce a predictable and desirable product-size distribution.

When crystals cannot be grown sufficiently large in forced-circulation equipment to meet product-size requirements, it is common to use one of the designs that allow some influence to be exercised over the population density of the finer crystals. In the DTB design (Fig. 18-61), this is done by regulating the flow in the circulating pipe in order to withdraw a portion of the fines in the body in the amount of about 0.05 to 0.5 percent by settled volume. The exact quantity of solids depends on the size of the product crystals and on the capacity of the fines-dissolving system. If the machine is not operating stably, this quantity of solids will appear and then disappear, indicating changes in the nucleation rate within the circuit. At steady-state operation, the quantity of solids overflowing will remain relatively constant, with some solids appearing at all times. Should the slurry density of product crystals circulated within the machine rise to a value higher than about 50 percent settled volume, large quantities of product crystals will appear in the overflow system, disabling the fines-destruction equipment. Too high a circulating rate through the fines trap will produce this same result. Too low a flow through the fines circuit will remove an insufficient number of particles and will result in a smaller product-size crystal. To operate effectively, a crystallizer of the type using fines-destruction techniques requires more sophisticated control than does the simpler forced-circulation equipment.

The classifying crystallizer (Fig. 18-63) requires approximately the same control of the fines-removal stream and, in addition, requires control of the fluidizing flow circulated by the main pump. This flow must be adjusted to achieve the proper degree of fluidization in the suspension chamber, and this quantity of flow varies as the crystal size varies between start-up operation and normal operation. As with the draft-tube-baffle machine, a considerably higher degree of skill is needed to operate this equipment than the forced-circulation type.

While most of the industrial designs in use today are built to reduce the problems due to excess nucleation, it is true that in some crystallizing systems a deficiency of seed crystals is produced and the product crystals are larger than are wanted or needed. In such systems, nucleation can be increased by increasing the mechanical stimulus created by the circulating devices or by seeding through the addition of fine crystals from some external source.

CRYSTALLIZER COSTS

Because crystallizers can come with such a wide variety of attachments, capacities, materials of construction, and designs, it is very difficult to present an accurate picture of the costs for any except certain specific types of equipment, crystallizing specific compounds. This is illustrated in Fig. 18-67, which shows the prices of equipment for crystallizing two different compounds at various production rates, one of the compounds being produced in two alternative crystallizer modes. Installed cost (including cost of equipment and accessories, foundations and supporting steel, utility piping, process piping and pumps, electrical switchgear, instrumentation, and labor, but excluding cost of a building) will be approximately twice these price figures.

Most crystallization equipment is custom designed, and costs for a particular application may vary greatly from those illustrated in Fig. 18-67. Realistic estimation of installation costs also requires reference to local labor rates, site-specific factors, and other case specifics.

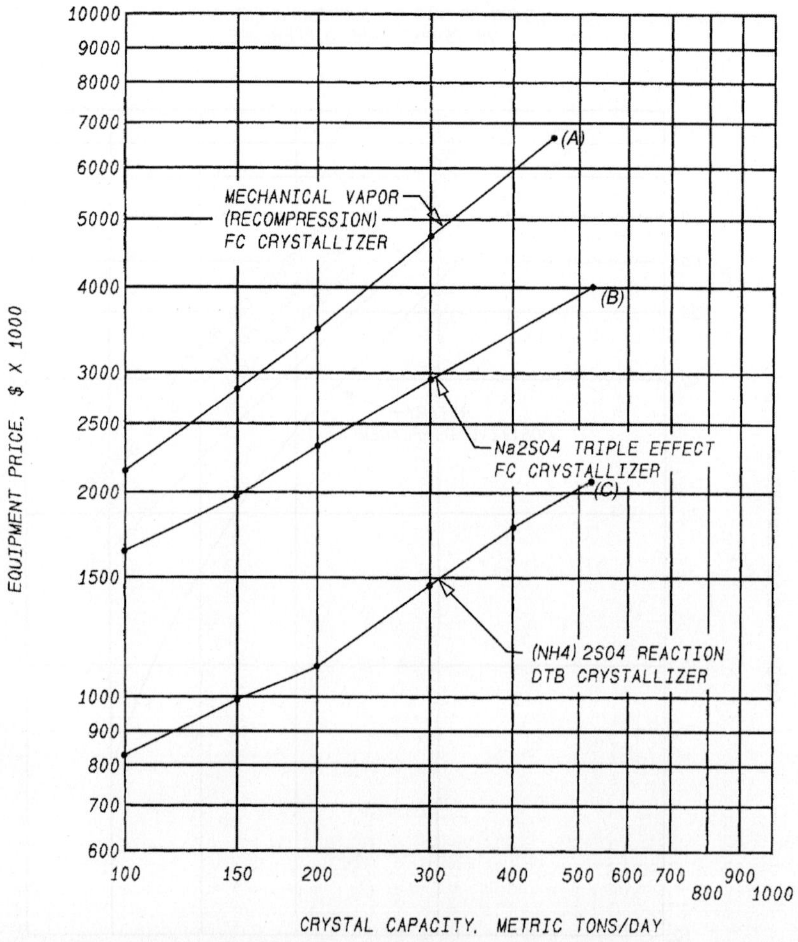

FIG. 18-67 Equipment prices, FOB point of fabrication, for typical crystallizer systems. Prices are for crystallizer plus accessories including vacuum equipment (2016). (*A*), (*B*) Na_2SO_4 production from Glauber's salt. Melting tank included. (*C*) Reaction of $NH_3 + H_2SO_4$ to make $(NH_4)_2SO_4$.

CRYSTALLIZATION FROM THE MELT

General References: Arkenbout, *Melt Crystallization Technology*, Technomic Publishing Company, Basel, 1995; Mullin, *Crystallization*, 4th ed., Butterworth-Heinemann, Oxford, UK, 2001; Myerson, *Handbook of Industrial Crystallization*, 2d ed., Butterworth-Heinemann, Oxford, UK, 2001; Pfann, *Zone Melting*, 2d ed., Wiley, New York, 1966; Saxer and Papp, "The MWB Crystallization Process," *Chem. Eng. Prog.* **4:** 64–66 (1980); Sloan and McGhie, *Techniques of Melt Crystallization*, Wiley, New York, 1988; Van't Land, *Industrial Crystallization of Melts*, Marcel Dekker, Enschede, The Netherlands, 2005; Zief and Wilcox, *Fractional Solidification*, Marcel Dekker, New York, 1967.

INTRODUCTION

Purification of a chemical species by solidification from a liquid mixture can be termed either *solution crystallization* or crystallization from the melt. The distinction between these two operations is somewhat subtle. The term *melt crystallization* has been defined as the separation of components of a binary mixture without the addition of solvent, but this definition is somewhat restrictive. In *solution crystallization*, a diluent solvent is added to the mixture; the solution is then directly or indirectly cooled, or solvent is evaporated to effect crystallization. The solid phase is formed and maintained somewhat below its pure component freezing point temperature. In melt crystallization, no diluent solvent is added to the reaction mixture, and the solid phase is formed by cooling of the melt. Product is often maintained near its pure-component freezing point in the refining section of the apparatus.

Many techniques are available for carrying out crystallization from the melt. An abbreviated list includes partial freezing and solids recovery in crystallizer-centrifuge systems, partial melting (e.g., sweating), zone melting, and falling-film column crystallization. A description of all these methods is not within the scope of this discussion. Zief and Wilcox (*Fractional Solidification*, Marcel Dekker, New York, 1967) and Myerson (*Handbook of Industrial Crystallization*, 2d ed., Butterworth-Heinemann, Oxford, UK, 2001) describe many of these processes. Two of the more common methods—zone melting and melt crystallization from the bulk—are discussed here to illustrate the techniques used for practicing crystallization from the melt.

High or ultrahigh product purity is obtained with many of the melt purification processes. Table 18-6 compares the product quality and product form that are produced from several of these operations. Zone refining can produce very pure material when operated in a batch mode; however, other melt crystallization techniques also provide high purity and become attractive if continuous high-capacity processing is desired. A comparison of the features of melt crystallization and distillation is shown in Table 18-7.

TABLE 18-6 Comparison of Processes Involving Crystallization from the Melt

Processes	Approximate upper melting point, °C	Materials tested	Minimum purity level obtained, ppm, weight	Product form
Progressive freezing	1500	All types	1	Ingot
Zone melting				
Batch	3500	All types	0.01	Ingot
Continuous	500	SiI_4	100	Melt
Melt crystallization				
Continuous	300	Organic	10	Melt
Cyclic	300	Organic	10	Melt

Abbreviated from Zief and Wilcox, *Fractional Solidification*, Marcel Dekker, New York, 1967, p. 7.

TABLE 18-7 Comparison of Features of Melt Crystallization and Distillation

Distillation	Melt crystallization
Phase equilibria	
Both liquid and vapor phases are totally miscible.	Liquid phases are totally miscible; solid phases are not.
Conventional vapor/liquid equilibrium.	Eutectic system.
Neither phase is pure.	Solid phase is pure, except at eutectic point.
Separation factors are moderate and decrease as purity increases.	Partition coefficients are very high (theoretically, they can be infinite).
Ultrahigh purity is difficult to achieve.	Ultrahigh purity is easy to achieve.
No theoretical limit on recovery.	Recovery is limited by eutectic composition.
Mass-transfer kinetics	
High mass-transfer rates in both vapor and liquid phases.	Only moderate mass-transfer rate in liquid phase, zero in solid.
Close approach to equilibrium.	Slow approach to equilibrium; achieved in brief contact time. Included impurities cannot diffuse out of solid.
Adiabatic contact assures phase equilibrium.	Solid phase must be remelted and refrozen to allow phase equilibrium.
Phase separability	
Phase densities differ by a factor of 100–10,000:1.	Phase densities differ by only about 10%.
Viscosity in both phases is low.	Liquid phase viscosity moderate, solid phase rigid.
Phase separation is rapid and complete.	Phase separation is slow; surface-tension effects prevent completion.
Countercurrent contacting is quick and efficient.	Countercurrent contacting is slow and imperfect.

Wynn, *Chem. Eng. Prog.*, **88**, 55 (1992). Reprinted with permission of the American Institute of Chemical Engineers.

A brief discussion of solid-liquid phase equilibrium is presented before we discuss specific crystallization methods. Figures 18-68 and 18-69 illustrate the phase diagrams for binary solid solution and eutectic systems, respectively. In the case of binary solid solution systems, illustrated in Fig. 18-68, the liquid and solid phases contain equilibrium quantities of both components in a manner similar to vapor-liquid phase behavior. This type of behavior causes separation difficulties since multiple stages are required. In principle, however, high purity and yields of both components can be achieved since no eutectic is present.

If the impurity or minor component is completely or partially soluble in [illegible] an interfacial distribution coefficient k, defined by Eq. (18-42):

$$k = C_s/C_\ell \tag{18-42}$$

C_s is the concentration of impurity or minor component in the solid phase, and C_ℓ is the impurity concentration in the liquid phase. The distribution coefficient generally varies with composition. The value of k is greater than 1 when the solute raises the melting point and less than 1 when the melting point is

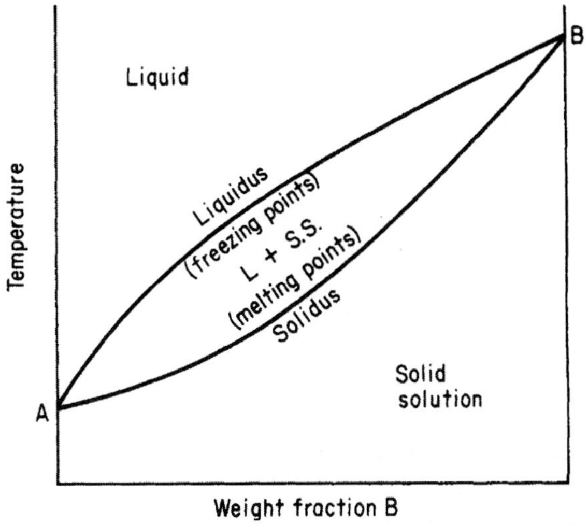

FIG. 18-68 Phase diagram for components exhibiting complete solid solution. (*Zief and Wilcox,* Fractional Solidification, *vol. 1, Marcel Dekker, New York, 1967, p. 31.*)

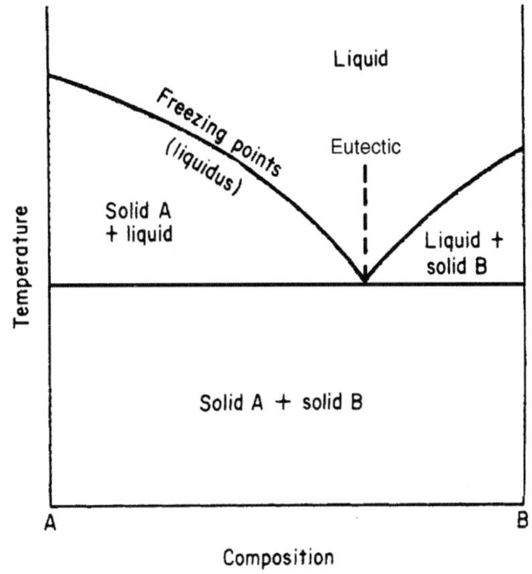

FIG. 18-69 Simple eutectic-phase diagram at constant pressure. (*Zief and Wilcox,* Fractional Solidification, *vol. 1, Marcel Dekker, New York, 1967, p. 24.*)

depressed. In the regions near pure A or B the liquidus and solidus lines become linear; that is, the distribution coefficient becomes constant. This is the basis for the common assumption of constant k in many mathematical treatments of fractional solidification in which ultrapure materials are obtained.

In the case of a simple eutectic system shown in Fig. 18-69, a pure solid phase is obtained by cooling if the composition of the feed mixture is not at the eutectic composition. If liquid composition is eutectic, then separate crystals of both species will form. In practice it is difficult to attain perfect separation of one component by crystallization of a eutectic mixture. The solid phase will always contain trace amounts of impurity because of incomplete solid-liquid separation, slight solubility of the impurity in the solid phase, or volumetric inclusions. It is difficult to generalize on which of these mechanisms is the major cause of contamination because of analytical difficulties in the ultrahigh purity range.

The distribution coefficient concept is commonly applied to f[illegible] phase diagram. If the quantity of impurity entrapped in the solid phase for whatever reason is proportional to that contained in the melt, then the assumption of a constant k is valid. It should be noted that the theoretical yield of a component exhibiting binary eutectic behavior is fixed by the feed composition and the position of the eutectic. Also, in contrast to the case of a solid solution, only one component can be obtained in a pure form.

There are many types of phase diagrams in addition to the two cases presented here; these are summarized in detail by Zief and Wilcox (1967, p. 21). Solid-liquid phase equilibria must be determined experimentally for most binary and multicomponent systems. Predictive methods are based mostly on ideal phase behavior and have limited accuracy near eutectics.

PROGRESSIVE FREEZING

Progressive freezing, sometimes called *normal* freezing, is the slow, directional solidification of a melt. Basically, this involves slow solidification at the bottom or sides of a vessel or tube by indirect cooling. The impurity is rejected into the liquid phase by the advancing solid interface. This technique can be used to concentrate an impurity or, by repeated solidifications and liquid rejections, to produce a very pure ingot. Figure 18-70 illustrates a progressive freezing apparatus. The solidification rate and interface position are controlled by the rate of movement of the tube and the temperature of the cooling medium. There are many variations of the apparatus; for example, the residual liquid portion can be agitated and the directional freezing can be carried out vertically, as shown in Fig. 18-70, or horizontally (see Richman et al. in Zief and Wilcox 1967, p. 259). In general, there is a solute redistribution when a mixture of two or more components is directionally frozen.

Component Separation by Progressive Freezing When the distribution coefficient is less than 1, the first solid that crystallizes contains less solute than the liquid from which it was formed. As the fraction that is frozen increases, the concentration of the impurity in the remaining liquid is increased, and hence the concentration of impurity in the solid phase increases (for $k < 1$). The concentration gradient is reversed for $k > 1$. Consequently, in the absence of diffusion in the solid phase, a concentration gradient is established in the frozen ingot.

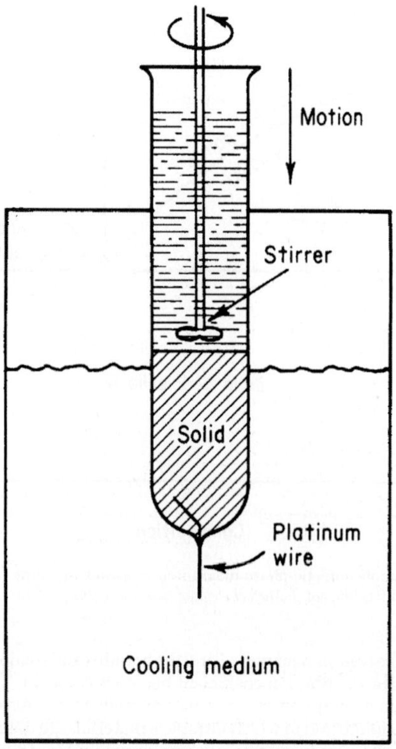

FIG. 18-70 Progressive freezing apparatus.

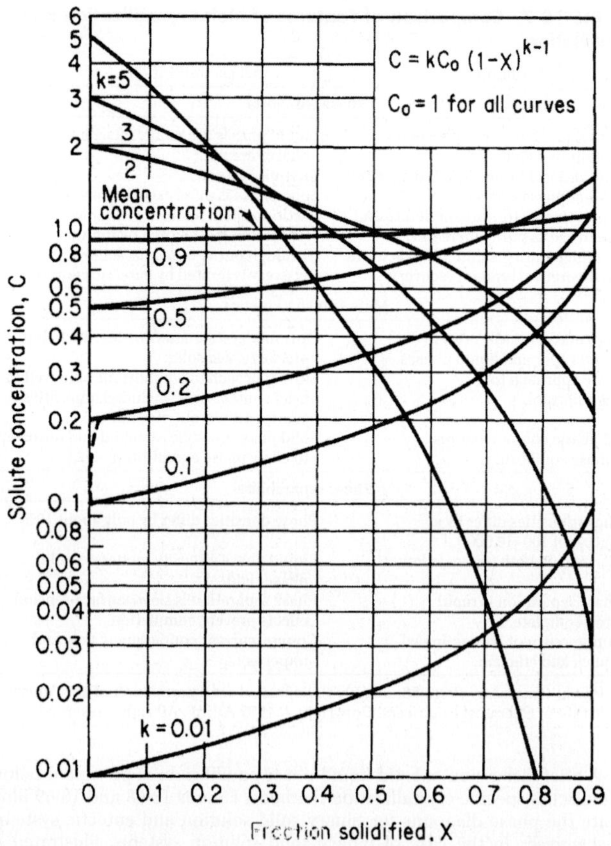

FIG. 18-71 Curves for progressive freezing, showing solute concentration C in the solid versus fraction solidified X. (*Pfann, Zone Melting, 2d ed. Wiley, New York, 1966, p. 12.*)

One extreme of progressive freezing is equilibrium freezing. In this case, the freezing rate must be slow enough to permit diffusion in the solid phase to eliminate the concentration gradient. When this occurs, there is no separation if the entire tube is solidified. Separation can be achieved, however, by terminating the freezing before all the liquid has been solidified. Equilibrium freezing is rarely achieved in practice because the diffusion rates in the solid phase are usually negligible (Pfann, *Zone Melting,* 2d ed., Wiley, New York, 1966, p. 10).

If the bulk liquid phase is well mixed and no diffusion occurs in the solid phase, a simple expression relating the solid phase composition to the fraction frozen can be obtained for the case in which the distribution coefficient is independent of composition and fraction frozen [Pfann, *Trans. Am. Inst. Mech. Eng.* **194:** 747 (1952)].

$$C_s = kC_0(1-X)^{k-1} \tag{18-43}$$

C_0 is the solution concentration of the initial charge, and X is the fraction frozen. Figure 18-71 illustrates the solute redistribution predicted by Eq. (18-43) for various values of the distribution coefficient.

There have been many modifications of this idealized model to account for variables such as the freezing rate and the degree of mixing in the liquid phase. For example, Burton et al. [*J. Chem. Phys.* **21:** 1987 (1953)] reasoned that the solid rejects solute faster than it can diffuse into the bulk liquid. They proposed that the effect of the freezing rate and stirring could be explained by the diffusion of solute through a stagnant film next to the solid interface. Their theory resulted in an expression for an effective distribution coefficient k_{eff} that could be used in Eq. (18-43) instead of k

$$k_{\text{eff}} = \frac{1}{1+(1/k-1)e^{-f\delta/D}} \tag{18-44}$$

where f = crystal growth rate, cm/s; δ = stagnant film thickness, cm; and D = diffusivity, cm²/s. No further attempt is made here to summarize the various refinements of Eq. (18-43). Zief and Wilcox (1967, p. 69) have summarized several of these models.

Pertinent Variables in Progressive Freezing The dominant variables that affect solute redistribution are the degree of mixing in the liquid phase and the rate of solidification. It is important to attain sufficient mixing to facilitate diffusion of the solute away from the solid-liquid interface to the bulk liquid. The film thickness δ decreases as the level of agitation increases. Cases have been reported in which essentially no separation occurred when the liquid was not stirred. The freezing rate, which is controlled largely by the lowering rate of the tube (see Fig. 18-70), has a pronounced effect on the separation achieved. The separation is diminished as the freezing rate is increased. Also, fluctuations in the freezing rate caused by mechanical vibrations and variations in the temperature of the cooling medium can decrease the separation.

Applications Progressive freezing has been applied to both solid solution and eutectic systems. As Fig. 18-71 illustrates, large separation factors can be attained when the distribution coefficient is favorable. Relatively pure materials can be obtained by removing the desired portion of the ingot. Also, in some cases progressive freezing provides a convenient method of concentrating the impurities; for example, in the case of $k < 1$, the last portion of the liquid that is frozen is enriched in the distributing solute.

Progressive freezing has been applied on the commercial scale. For example, aluminum has been purified by continuous progressive freezing [Dewey, *J. Metals* **17:** 940 (1965)]. The Proabd refiner described by Molinari (Zief and Wilcox 1967, p. 393) is also a commercial example of progressive freezing. In this apparatus, the mixture is directionally solidified on cooling tubes. Purification is achieved because the impure fraction melts first; this process is called sweating. This technique has been applied to the purification of naphthalene and *p*-dichlorobenzene.

ZONE MELTING

Zone melting also relies on the distribution of solute between the liquid and solid phases to effect a separation. In this case, however, one or more liquid zones are passed through the ingot. This extremely versatile technique, which was invented by W. G. Pfann, has been used to purify hundreds of materials. Zone melting in its simplest form is illustrated in Fig. 18-72. A molten zone can be passed through an ingot from one end to the other by either a moving heater or by slowly drawing the material to be purified through a stationary heating zone.

Progressive freezing can be viewed as a special case of zone melting. If the zone length were equal to the ingot length and if only one pass were used, the operation would become progressive freezing. In general, however, when the zone length is only a fraction of the ingot length, zone melting has the advantage that a portion of the ingot does not have to be discarded after each solidification. The last portion of the ingot that is frozen in progressive freezing must be discarded before a second freezing.

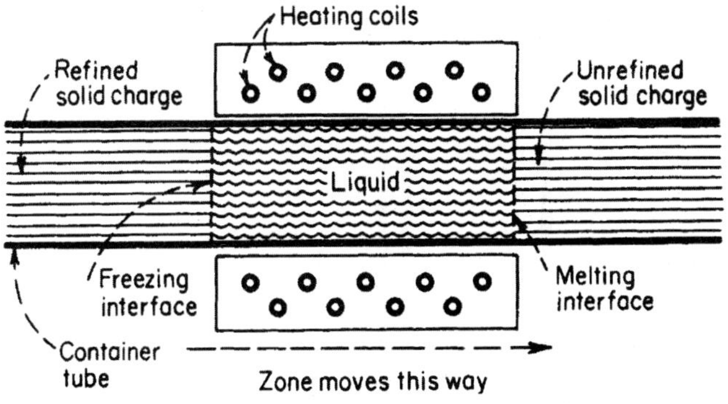

FIG. 18-72 Diagram of zone refining.

Component Separation by Zone Melting The degree of solute redistribution achieved by zone melting is determined by the zone length l, ingot length L, number of passes n, the degree of mixing in the liquid zone, and the distribution coefficient of the materials being purified. The distribution of solute after one pass can be obtained by material balance considerations. This is a two-domain problem; that is, in the major portion of the ingot of length $L - l$, zone melting occurs in the conventional sense. The trailing end of the ingot of length l undergoes progressive freezing. For the case of a constant distribution coefficient, perfect mixing in the liquid phase, and negligible diffusion in the solid phase, the solute distribution for a single pass is given by Eq. (18-45) [Pfann, *Trans. Am. Inst. Mech. Eng.* **194:** 747 (1952)].

$$C_s = C_0[1-(1-k)e^{-kx/\ell}] \qquad (18\text{-}45)$$

The position of the zone x is measured from the leading edge of the ingot. The distribution for multiple passes can also be calculated from a material balance, but in this case the leading edge of the zone encounters solid corresponding to the composition at the point in question for the previous pass. The multiple pass distribution has been numerically calculated (Pfann, *Zone Melting,* 2d ed., Wiley, New York, 1966, p. 285) for many combinations of k, L/l, and n. Typical solute composition profiles are shown in Fig. 18-73

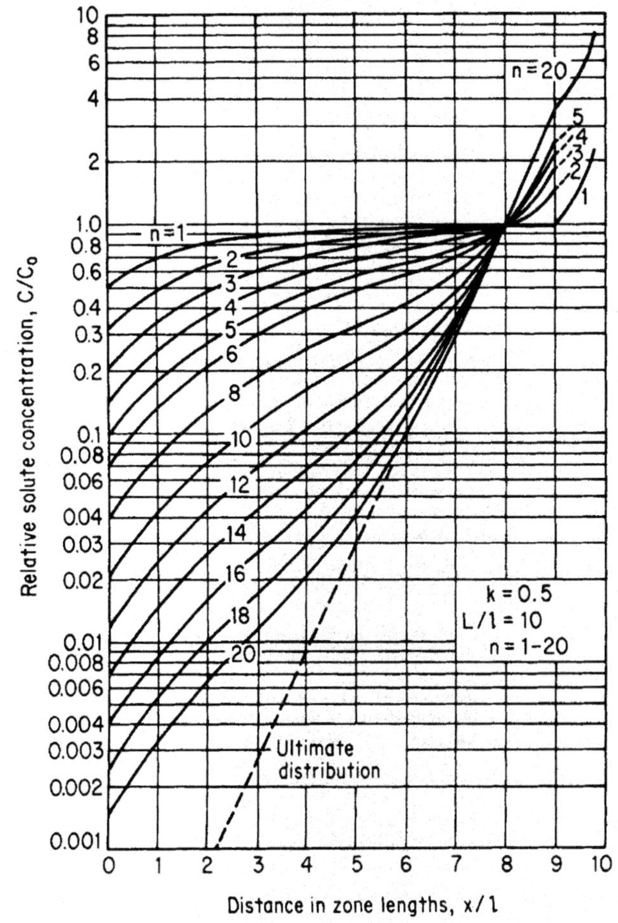

FIG. 18-73 Relative solute concentration C/C_0 (logarithmic scale) versus distance in zone lengths x/l from beginning of charge, for various numbers of passes n. L denotes

The ultimate distribution after an infinite number of passes is also shown in Fig. 18-73 and can be calculated for $x < (L - l)$ from the following equation (Pfann 1966, p. 42):

$$C_s = Ae^{BX} \qquad (18\text{-}46)$$

where A and B can be determined from the following relations:

$$k = B\ell/(e^{B\ell} - 1) \qquad (18\text{-}47)$$

$$A = C_0BL/(e^{BL} - 1) \qquad (18\text{-}48)$$

The ultimate distribution represents the maximum separation that can be attained without cropping the ingot. Equation (18-46) is approximate because it does not include the effect of progressive freezing in the last zone length.

As in progressive freezing, many refinements of these models have been developed. Corrections for partial liquid mixing and a variable distribution coefficient have been summarized in detail (Zief and Wilcox 1967, p. 47).

Pertinent Variables in Zone Melting The dominant variables in zone melting are the number of passes, ingot length–zone length ratio, freezing rate, and degree of mixing in the liquid phase. Figure 18-73 illustrates the increased solute redistribution that occurs as the number of passes increases. Ingot length–zone length ratios of 4 to 10 are commonly used (Zief and Wilcox 1967, p. 624). An exception is encountered when one pass is used. In this case the zone length should be equal to the ingot length; progressive freezing provides the maximum separation when only one pass is used.

The freezing rate and degree of mixing have effects in solute redistribution similar to those discussed for progressive freezing. Zone travel rates of 1 cm/h for organic systems, 2.5 cm/h for metals, and 20 cm/h for semiconductors are common. In addition to the zone travel rate, the heating conditions affect the freezing rate. A detailed summary of heating and cooling methods for zone melting has been outlined by Zief and Wilcox (1967, p. 192). Direct mixing of the liquid region is more difficult for zone melting than for progressive freezing. Mechanical stirring complicates the apparatus and increases the probability of contamination from an outside source. Some mixing occurs because of natural convection. Methods have been developed to stir the zone magnetically by using the interaction of a current and a magnetic field (Pfann 1966, p. 104) for cases in which the charge material is a reasonably good conductor.

Applications Zone melting has been used to purify hundreds of inorganic and organic materials. Many classes of inorganic compounds, including semiconductors, intermetallic compounds, ionic salts, and oxides, have been purified by zone melting. Organic materials of many types have been zone melted. Zief and Wilcox (1967, p. 624) have compiled tables that give operating conditions and references for both inorganic and organic materials with melting points ranging from −115°C to over 3000°C.

Some materials are so reactive that they cannot be zone melted to a high degree of purity in a container. Floating zone techniques in which the molten zone is held in place by its own surface tension have been developed by Keck et al. [*Phys. Rev.* **89:** 1297 (1953)].

A continuous zone melting apparatus has been described by Pfann (1966, p. 171). This technique offers the advantage of a close approach to the ultimate distribution, which is usually impractical for batch operation.

Performance data have been reported by Kennedy et al. (*The Purification of Inorganic and Organic Materials,* Marcel Dekker, New York, 1969, p. 261) for continuous-zone refining of benzoic acid.

MELT CRYSTALLIZATION (PRODUCTION OF REFINED LIQUID)

Conducting crystallization inside a vertical or horizontal column with a countercurrent flow of crystals and liquid can produce a higher product purity than conventional crystallization or distillation. Traditional melt crystallization produces a purified product that is a liquid melt. This is in contrast to progressive freezing and zone melting, which typically produce a refined solid product. The working concept is to form a crystal phase from the bulk liquid, either internally or externally, and then transport the solids through a countercurrent stream of enriched reflux liquid obtained from melted product. The problem in practicing this technology is the difficulty of controlling solid-phase movement. Unlike distillation, which exploits the significant specific gravity differences between liquid and vapor phases, melt crystallization involves the contacting of liquid and solid phases that have nearly identical physical properties. Phase densities are often very close, and gravitational settling of the solid phase may be slow and ineffective. The challenge of designing equipment to accomplish crystallization in a column has resulted in myriad configurations to achieve reliable solid-phase movement, high product yield and purity, and efficient heat addition and removal.

In general, all vertical and horizontal column configurations except the falling-film column have proven unreliable for a number of reasons, primarily solids conveying difficulties and unwanted solids buildup on cool surfaces, which lead to plugging issues. Successful implementation of the falling-film concept has proven to be a reliable and scalable approach.

FALLING-FILM CRYSTALLIZATION

Falling-film crystallization uses progressive freezing principles to purify melts and solutions. The technique established to practice the process is inherently cyclic. Figure 18-74 depicts the basic working concept. First a crystalline layer is formed by subcooling a liquid film on a vertical surface inside a tube. This coating is then grown by extracting heat from a falling film of melt (or solution) through a heat transfer surface. Impure liquid is then drained from the crystal layer, and the product is reclaimed by melting. The process is staged to provide both high product recovery and purity. Variants of this technique have been perfected and are used commercially for many types of organic materials. Both static and falling-film techniques have been described by Wynn [*Chem. Eng. Progr.* (1992)]. Mathematical models for both static and dynamic operations have been presented by Gilbert [*AIChE J.* **37:** 1205 (1991)].

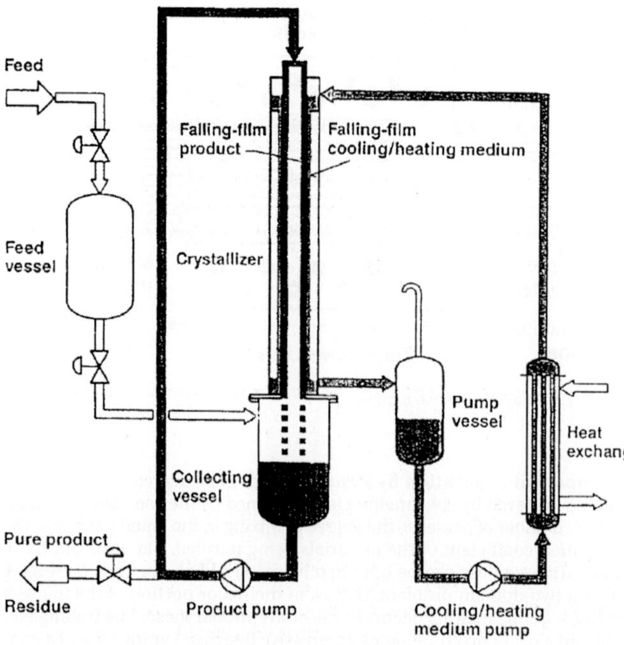

FIG. 18-74 Dynamic crystallization system. (*Sulzer Chemtech.*)

Figure 18-75 demonstrates the mathematical derivation of a falling-film freezing model. The terms depicted in the illustration can be used to obtain the interfacial partition coefficient, *k*, previously defined in Eq. (18-42).

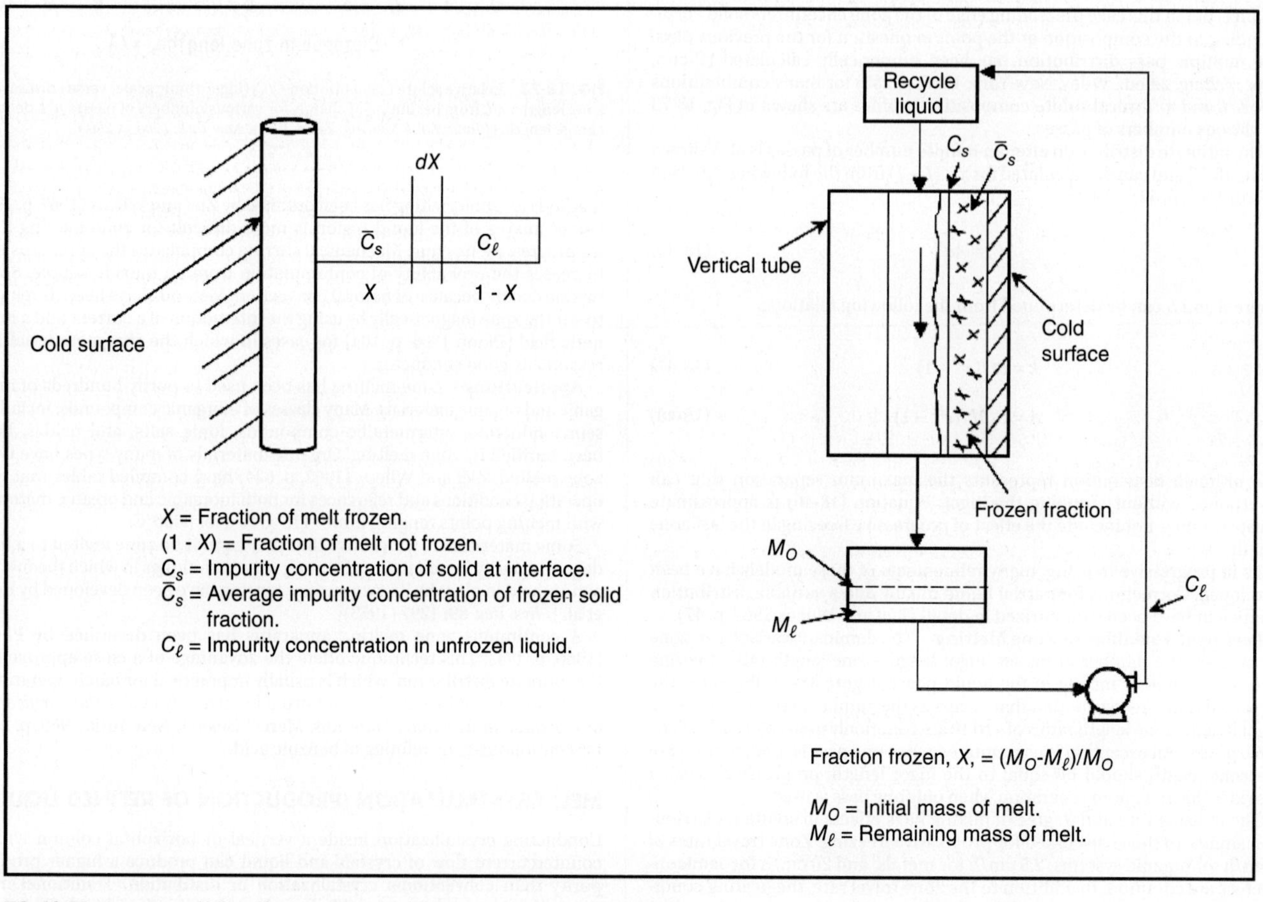

FIG. 18-75 Derivation of falling-film freezing model. (*Original drawing by C. G. Moyers.*)

TABLE 18-8 Melt Crystallization Equipment Patents

Patent no.	Date	Title	Inventor(s)	Assignee
US 3,621,664	1971	Fractional crystallization process	K. Saxer	Buchs Metallwerk Ag
US 3,796,060	1974	Process for purification by crystallization	C. G. Moyers	Union Carbide Corp.
US 2,613,136	1952	Fractional crystallizer	D. L. McKay	Phillips Petroleum Co.
US 3,174,832	1965	Pulsating fractional crystallizer	B. B. Bohrer	Sun Oil Co.
US 2,617,273	1952	Continuous crystallization apparatus and process	R. A. Findlay	Phillips Petroleum Co.
US 3,261,170	1966	Multi-stage crystallization process and apparatus	W. C. McCarthy, G. H. Dale	Phillips Petroleum Co.
US 3,501,275	1970	Apparatus for continuously separating mixtures of substances	R. Sailer, H. Philipp, C. Berther	Inventa A.G.

The interfacial coefficient k can be determined from experimental data using the following relation:

$$\frac{\bar{C}_S}{C_{\ell f}} = \frac{(1-X)^{(1-k)} - (1-X)}{X} \qquad (18\text{-}49)$$

where $\bar{C}_s$ is the average concentration of impurity in the frozen solid, and $C_{\ell f}$ is the final impurity concentration in the unfrozen liquid. The fraction frozen, X, is calculated from the overall mass balance as indicated in Fig. 18-75. Coefficient k is either backed out of Eq. (18-49) or can be solved for directly using Eq. (18-50):

$$k = 1 - \frac{\ln\left[\frac{\bar{C}_S}{C_{\ell f}} X + (1-X)\right]}{\ln(1-X)} \qquad (18\text{-}50)$$

Example 18-5 Calculation of k Using the following experimental data, an example calculation has been included to demonstrate the use of Eq. (18-49). To compute the value of k, substitute the experimental data into Eq. (18-49):

Example experimental data

$C_{\ell f} = 0.1258$ g impurity/g liquid after freezing

$\bar{C}_s = 0.0130$ g impurity/g frozen solid

$X = 0.6312$ g frozen/g total feed

$$\frac{\bar{C}_S}{C_{\ell f}} = \frac{0.0130}{0.1258} = \frac{(1-0.6312)^{(1-k)} - (1-0.6312)}{0.6312}$$

Then after simplifying and rearranging the terms as follows:

$$0.06523 = (0.3688)^{(1-k)} - (0.3688)$$
$$0.4340 = (0.3688)^{(1-k)}$$

one can solve for *k = 0.163.*

A sampling of patent references related to melt crystallization equipment and processes can be found in Tables 18-8 and 18-9, respectively.

Principles of Operation Figure 18-76 describes a typical three-stage falling-film crystallization process for purification of MCA (monochloroacetic acid). Crystallizer E-8 consists of a number of vertical tubular elements working in parallel enclosed in a shell. Normal tube length is 12 m, with a 50- to 75-mm tube inside diameter, although commercial demands extend both tube diameter and length. Feed enters stage 2 of the sequential operation, is added to the kettle (T-5), and is then circulated to the top of the crystallizer and distributed as a falling film inside the tubes. Nucleation is induced at the inside walls, and a crystal layer starts to grow. The temperature of the coolant is progressively lowered to compensate for reduced heat transfer and lower melt freezing point until the thickness inside the tube is between 5 and 20 mm, depending on the product. Kettle liquid is evacuated to the first-stage holding tank (T-3) for eventual recrystallization at a lower temperature to maximize product yield and to strip product from the final liquid residue. Semirefined product frozen to the inside of the tube during stage 2 is first heated above its melting point and slightly melted (sweated). This semipurified melted material (sweat) is removed from the crystallizer kettle, stored in a stage tank (T-4), and then added to the next batch of fresh feed. The remaining material inside the crystallizer is then melted, mixed with product sweat from stage 3, recrystallized, and sweated to upgrade the purity even further (stage 3).

Commercial Equipment and Applications Sulzer Chemtech has successfully developed and applied a variety of melt crystallization systems. A partial list of products successfully processed in a falling-film-type crystallizer is presented in Table 18-10.

An overview of Sulzer Chemtech's melt crystallization technology includes:

- Dynamic: Inside falling film or outside falling film (which uses a heat pump to control freezing and melting). No addition of solvent is required.
- Static: Small capacity or high contaminant concentration in feed. Vertical plates are immersed in a stagnant melt. Impure liquid is drained, and the frozen layer is purified by sweating.
- Hybrid: Dynamic plus static. High recovery and purity.

Sulzer Chemtech, GEA Messo PT, and TNO/SoliQz B.V. (Netherlands) have developed a type of non layer based melt process that is continuous rather than batch as in the Sulzer batch sequential layer-based process. The suspension-based processes grow crystals suspended in bulk liquid and wash and remove crystals in a hydraulic (TNO/SoliQz B.V., GEA) or mechanical wash column (GEA and Sulzer). The wash column employs filtration to remove impure liquid while simultaneously compressing and washing the solids. The purified solids are melted and recovered in liquid form. There are several designs available, which differ in the mode of crystal transport.

Sulzer Chemtech suspension crystallization plants consist of two loops. The crystallization loop is comprised of a crystallizer and a growth vessel. After crystals are formed in the crystallization loop, they are melted and recovered in a mechanical piston-type wash column.

GEA Messo PT offers various wash column designs, including screw-type and piston-type mechanical wash columns and a hydraulic-type wash column. Crystal suspension enters a wash column assembly, and separation occurs as the solids are compressed and contacted with product melt in a filter/scraper assembly.

TNO-Netherlands has used a Thijssen hydraulic wash column equipped with filter tubes. Performance is continuous and similar to the mechanical GEA and Sulzer wash columns.

TABLE 18-9 Melt Crystallization Process Patents

Patent no.	Date	Title	Inventor(s)	Assignee
US 6,310,218	2001	Melt crystallziation purification of lactides	W. O'Brien, G. Sloan	E.I. du Pont de Nemours and Company
US 5,675,022	1997	Recovery of dioxanone by melt crystallization	C. G. Moyers, M. P. Farr	Union Carbide Chemicals & Plastics Technology Corp.
US 4,321,334	1982	Melt crystallization of butene-1 polymers	A. Chatterjee	Shell Oil Company
US 5,430,194	1995	Process for improving enantiomeric purity of aldehydes	B. A. Barner, J. R. Briggs, J. J. Kurland, C. G. Moyers	Union Carbide Chemicals & Plastics Technology Corp.

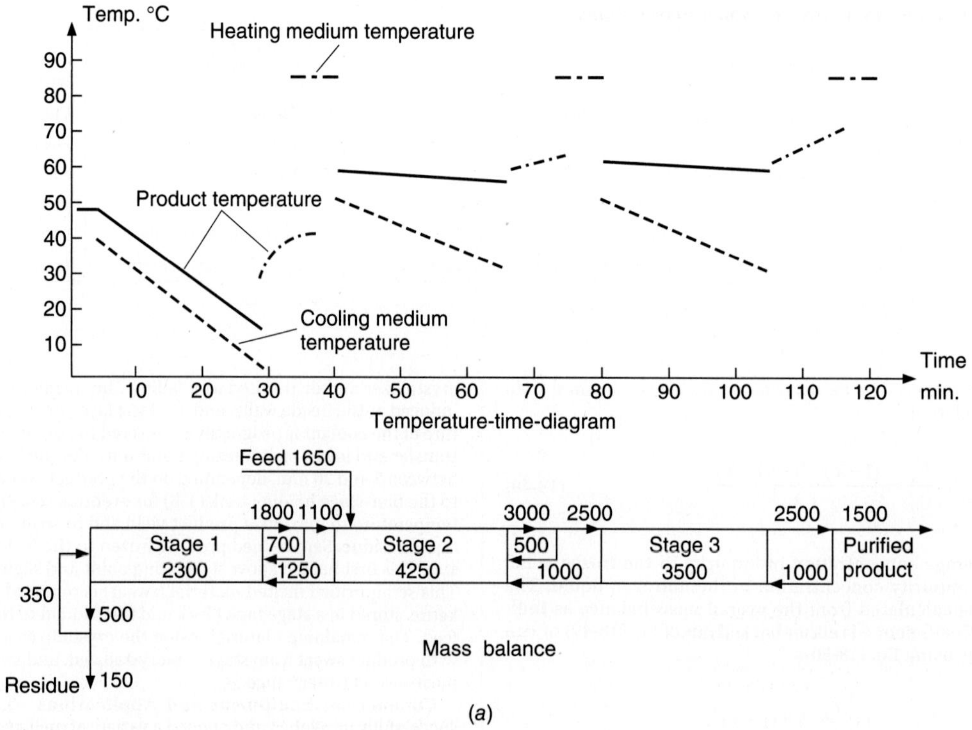

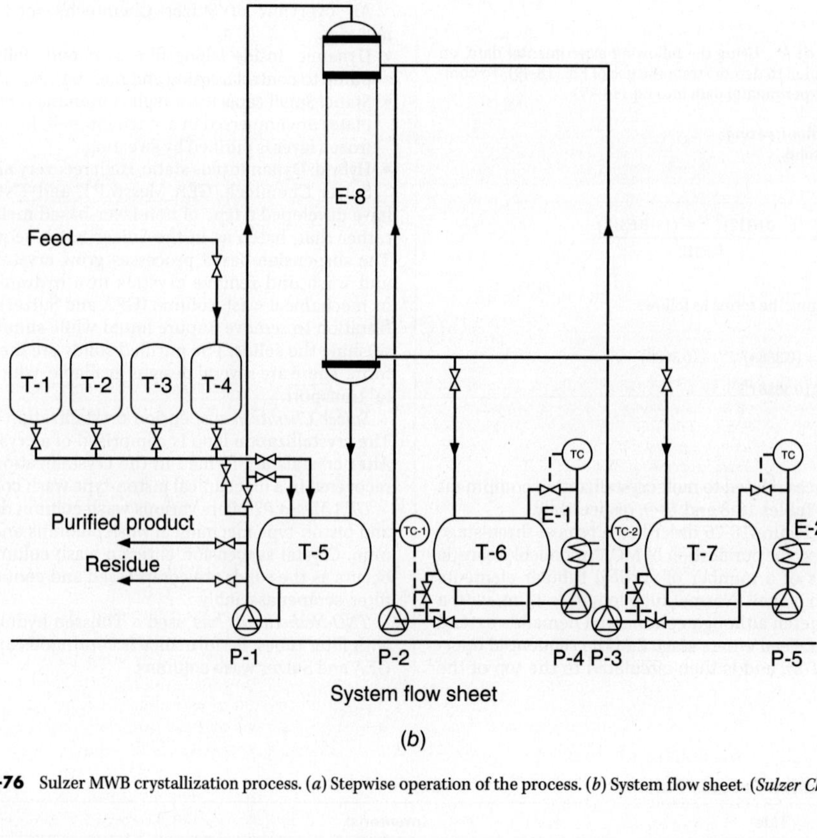

FIG. 18-76 Sulzer MWB crystallization process. (*a*) Stepwise operation of the process. (*b*) System flow sheet. (*Sulzer Chemtech.*)

TABLE 18-10 Fractional Crystallization Reference List

Product	Main characteristics	Capacity, tons/year	Purity	Type of plant	Country	Client
Acrylic acid	Very low aldehyde content, no undesired polymerization in the plant	Undisclosed	99.95%	Falling film	Undisclosed	Undisclosed
		Undisclosed	99.9%	Falling film	Undisclosed	Undisclosed
Benzoic acid	Pharmaceutical grade, odor- and color free	4500	99.97%	Falling film	Italy	Chimica del Friuli
Bisphenol A	Polycarbonate grade, no solvent required	150,000	Undisclosed	Falling film	USA	General Electric
Carbonic acid		1200	Undisclosed	Falling film	Germany	Undisclosed
Fatty acid	Separation of tallow fatty acid into saturated and unsaturated fractions	20,000	Stearic acid: Iodine no. 2 Oleic acid: Cloud pt 5°C	Falling film	Japan	Undisclosed
Fine chemicals		<1000	Undisclosed	Falling film	GUS	Undisclosed
		<1000	Undisclosed	Static	Switzerland	Undisclosed
		<1000	Undisclosed	Falling film	Switzerland	Undisclosed
		<1000	Undisclosed	Falling film	Switzerland	Undisclosed
		<1000	Undisclosed	Falling film	USA	Undisclosed
		<1000	Undisclosed	Falling film	Germany	Undisclosed
		<1000	Undisclosed	Falling film	Japan	Undisclosed
Hydrazine	Satellite grade	3	>99.9%	Falling film	Germany	ESA
Monochloro acetic acid (MCA)	Low DCA content	6000	>99.2%	Falling film	USA	Undisclosed
Multipurpose	Separation or purification of two or more chemicals, alternatively	1000	Various grades	Falling film	Belgium	UCB
		1000	Undisclosed	Falling film	Belgium	Reibelco
Naphthalene	Color free and color stable with low thionaphthene content	60,000	99.5%	Falling film	Germany	Rütgers-Werke
		20,000	99.5%	Falling film/static	P.R. China	Anshan
		10,000	99.8%	Falling film/static	P.R. China	Jining
		12,000	Various grades	Falling film	The Netherlands	Cindu Chemicals
p-Dichlorobenzene	No solvent washing required	40,000	99.95%	Falling film	USA	Standard Chlorine
		5000	99.98%	Falling film	Japan	Toa Gosei
		4000	99.8%	Falling film/distillation	Brazil	Nitroclor
		3000	99.95%	Falling film	P.R. China	Fuyang
		3000	>97%	Static	P.R. China	Shandong
p-Nitrochlorobenzene		18,000	99.3%	Falling film/distillation	P.R. China	Jilin Chemical
		10,000	99.5%	Static	India	Mardia
Toluene diisocyanate (TDI)	Separation of TDI 80 into TDI 100 & TDI 65	20,000	Undisclosed	Falling film	Undisclosed	Undisclosed
Trioxan		<1000	99.97%	Falling film	Undisclosed	Undisclosed

LEACHING

General References: Coulson and Richardson, "Leaching," chap. 10 in *Chemical Engineering,* 5th ed., vol. 2, Butterworth-Heinemann, Oxford, UK, 2002, pp. 502–541; Harriott, *Chemical Reactor Design,* Marcel Dekker, New York, 2003, pp. 89–99; McCabe, Smith, and Harriott, *Unit Operations of Chemical Engineering,* 7th ed., McGraw-Hill, New York, 2005; Mular, Halbe, and Barratt, *Mineral Processing Plant Design, Practice, and Control,* vols. 1 and 2, Society for Mining, Metallurgy, and Exploration, Englewood, Colo., 2002; Prabhudesai, sec. 5.1 in Schweitzer, *Handbook of Separation Techniques for Chemical Engineers,* 3d ed., McGraw-Hill, New York, 1996; Rickles, *Chem. Eng.* **72**(6): 157 (1965); *SME Mineral Processing and Extractive Metallurgy Handbook,* Society for Mining, Metallurgy, and Exploration, Englewood, Colo., 2016; Section on Reactors in annual issues of *Chemical Engineering Buyers' Guide;* Wakeman, "Extraction (Liquid-Solid)" in *Kirk-Othmer Encyclopedia of Chemical Technology,* 4th ed., vol. 10, Wiley, New York, 1993, p. 186; Wilkes, *Fluid Mechanics for Chemical Engineers,* Prentice-Hall, Englewood Cliffs, N.J., 1999; Young, et al., eds., *Hydrometallurgy 2008: Proceedings of the Sixth International Symposium,* Society for Mining, Metallurgy, and Exploration, Englewood, Colo., 2008.

DEFINITION

Leaching is the removal of a soluble fraction, in the form of a solution, from an insoluble, usually permeable, solid phase with which it is associated. Leaching generally involves selective dissolution with or without diffusion. In the case of simple washing, it requires only displacement (with some mixing) of one interstitial liquid by another with which it is miscible. The soluble constituent may be solid or liquid, and it may be incorporated within, chemically combined with, adsorbed upon, or bound mechanically in the pore structure of the insoluble material. Sometimes, the insoluble phase may be massive and porous, but usually it is particulate; the particles may be openly porous, cellular with selectively permeable cell walls, or surface-activated.

By convention, the removal of a surface-adsorbed solute is treated as a special case of adsorption, rather than leaching. The washing of filter cakes is also excluded, as in the case of leaching of soil or sewage sludges.

Due to its great breadth of application and its importance to some ancient processes, leaching is known by many names, including *digestion, extraction, solid-liquid extraction, lixiviation, percolation, infusion, washing,* and *decantation-settling.* If the stream of solids being leached is densified by settling, it is often called *underflow.* Oil seed processors may refer to the solids as *marc.* The liquid stream containing the leached solute is called *overflow* (following densification), *extract, solution, pregnant leach solution (PLS), lixiviate, leachate,* or *miscella.*

Mechanism Leaching may simply result from the solubility of a substance in a liquid, or it may be enabled by a chemical reaction. The rate of transport of solvent into the mass to be leached, or of the soluble fraction into the solvent, or of extracted solution out of the insoluble material, or of some combination of these rates, may influence overall leaching kinetics, as may an interfacial resistance or a chemical reaction rate.

Since the overflow and underflow streams are not immiscible phases but streams based on the same solvent, the concept of equilibrium for leaching is not the one applied in other mass-transfer separations. If the solute is not adsorbed on the inert solid, true equilibrium is reached only when all of the solute is dissolved and distributed uniformly throughout the solvent in both underflow and overflow, or when the solvent is uniformly saturated with the solute. The practical interpretation of leaching equilibrium is the state in which the overflow and underflow liquids are of the same composition; on a y-x diagram, the equilibrium line will be a straight line through the origin with a slope of unity. It is customary to calculate the number of ideal (equilibrium) stages required for a given leaching application and to adjust the number by applying a stage efficiency factor, although local efficiencies, if known, can be applied stage by stage.

Usually, however, it is not feasible to establish a stage or overall efficiency or a leaching rate index (e.g., overall coefficient) without testing small-scale models of pertinent apparatus. In fact, the results of such tests may have to be scaled up empirically, without explicit evaluation of rate or quasi-equilibrium indices.

Methods of Operation Leaching systems are distinguished (1) by operating cycle (batch, continuous, or multibatch intermittent); (2) by direction of streams (cocurrent, countercurrent, or hybrid flow); (3) by staging (single-stage, multistage, or differential-stage); or (4) by method of contacting (sprayed percolation, immersed percolation, or solids dispersion). In general, descriptors from all four categories must be assigned to characterize a leaching system completely.

Whatever the mechanism and the method of operation, the effectiveness of the leaching process will be enhanced by increased surface per unit volume of the solids to be leached and by decreased diffusion path lengths within the solids, both of which are favored by decreased particle size. Excessively fine solids, on the other hand, cause a slow percolation rate during quiescent leaching, difficult solid-liquid separation, and possible poor quality of the solid product due to inadequate washing. The basis for an optimum particle size distribution is established by laboratory quantification of these characteristics.

LEACHING EQUIPMENT

There are two primary categories of contacting methods according to which leaching equipment is classified: (1) leaching may be accomplished by percolation and (2) the particulate solids may be dispersed into a liquid phase and then separated from it. Each may be operated in a batch or continuous manner. Materials that disintegrate during leaching are treated in the second class of equipment. An important exception to this classification is *in-situ* leaching, as discussed below.

Percolation *Heap leaching,* as shown in Fig. 18-77 (see Keane et al., *SME Mineral Processing and Extractive Metallurgy Handbook*, sec. 9.3, Society for Mining, Metallurgy and Exploration, Englewood, Colo., 2016), is very widely applied to the ores of copper and precious metals, but percolation is also conducted on a smaller scale in batch tanks or vats. In the heap leaching of low-grade oxidized gold ores, for instance, a dilute alkaline solution of sodium cyanide is distributed over a heap of ore that typically has been crushed finer than one inch and the fines agglomerated by the addition of portland cement at conveyor transfer points. Heap leaching of very low-grade gold ores and oxide copper ores is conducted on run-of-mine (uncrushed) material. Heap leaching is the least expensive form of leaching, but recovery of dissolved values is slow. In virtually all cases, an impervious polymeric membrane, typically 40- to 80-mil (0.04–0.08-in) high-density polyethylene (HDPE), is installed over a compacted clay liner before the heap is constructed.

In-situ leaching (see McNulty, et al., *SME Mineral Processing and Extractive Metallurgy Handbook*, sec. 9.2, Society for Mining, Metallurgy and Exploration, Englewood, Colo., 2016) is a special case of solution mining and depends on the existing porosity and permeability of a subsurface deposit containing minerals or compounds that are to be dissolved and extracted. Holes ("wells") are drilled into the deposit and are lined with plastic or stainless steel pipe that is perforated at depths corresponding to mineralized strata. The leaching solution is pumped down the injection wells and flows through the deposit or "formation," and the "pregnant leach solution" (PLS) is extracted from production wells, treated for solute recovery, reconstituted, and reinjected. *In-situ* leaching (ISL) is used for the extraction of halite (NaCl), potash (KCL), compounds of lithium, boron, and magnesium, sulfur, copper, and uranium, as well as for the removal of toxic or hazardous constituents from contaminated soil or groundwater.

In 2016, 48 percent of the world's uranium was produced by ISL technology in Kazakhstan, Uzbekistan, the United States, Australia, China, and Russia. (See World Nuclear Association 2017, *World Uranium Mining Production.* www.world-nuclear.org/information-library/nuclear-fuel-cycle/mining-of-uranium/world-uranium-mining-production.aspx.) From a solution management standpoint, regulatory agencies require that the deposit be confined within an aquifer and below the natural water table (Fig. 18-78*a*). This allows the wellfield operator to maintain gradual inward movement of aquifer water by producing a few percent more solution than is being injected "hydraulic control"). A well completion method similar to that used in hydrocarbon recovery is employed (Fig. 18-78*b*). Monitoring wells situated outside the perimeter of the wellfield are sampled continuously and analyzed for potential contaminants. In Kazakhstan, the solvent is dilute sulfuric acid, and air lift pumping is preferred due to its low cost and easy maintenance. In the United States, submerged reversible turbine pumps are used, and the solvent is amended site water. Additives to the injected water include sufficient dissolved oxygen to oxidize insoluble tetravalent uranium minerals to the soluble hexavalent form of uranium and sufficient carbon dioxide to complex the dissolved uranium. The PLS treatment plant uses resin ion exchange, precipitation, thickening, centrifuging, and drying to produce "yellowcake." A solution bleed stream controls impurity levels and is either evaporated in a lined pond or injected into a deep disposal well.

A *batch percolator tank* is a large circular or rectangular tank with a false bottom. The solids to be leached are dumped into the tank to a uniform depth. They may be sprayed with solvent until their solute content is reduced to an economic minimum and are then excavated. Alternatively, solvent up-flow may be preferred if fine particles impair percolation and cannot be agglomerated. Countercurrent flow of the solvent through a series of tanks is common, with fresh solvent entering the tank containing most nearly exhausted material. So-called *vat leaching* was practiced in oxide copper ore processing prior to 1980; the vats were typically 37 by 42 by 6 m (120 ft wide by 135 ft long by 20 ft deep) and treated about 10,900 Mg (12,000 U.S. tons per day) of ore on a 6- to 10-day cycle. Sometimes, in other applications, tanks are operated under pressure to contain volatile solvents or to increase the percolation rate. A series of pressure tanks operating with countercurrent solvent flow is sometimes called a *diffusion battery.*

Continuous Percolators Coarse solids are also leached by percolation in moving-bed equipment, including single-deck and multideck rake classifiers, baskets, bucket-elevator contactors, and horizontal-belt conveyors.

The *Bollman-type extractor* is a bucket-elevator unit designed to handle about 2000 to 20,000 kg/h (50 to 500 U.S. tons per day) of flaky solids (e.g., soybeans). Buckets with perforated bottoms are held on an endless moving belt. Dry flakes, fed into the descending buckets, are sprayed with partially enriched solvent ("half miscella") pumped from the bottom of the column of ascending buckets. As the buckets rise on the other side of the unit, the solids are sprayed with a countercurrent stream of pure solvent. Exhausted flakes are dumped from the buckets at the top of the unit into a

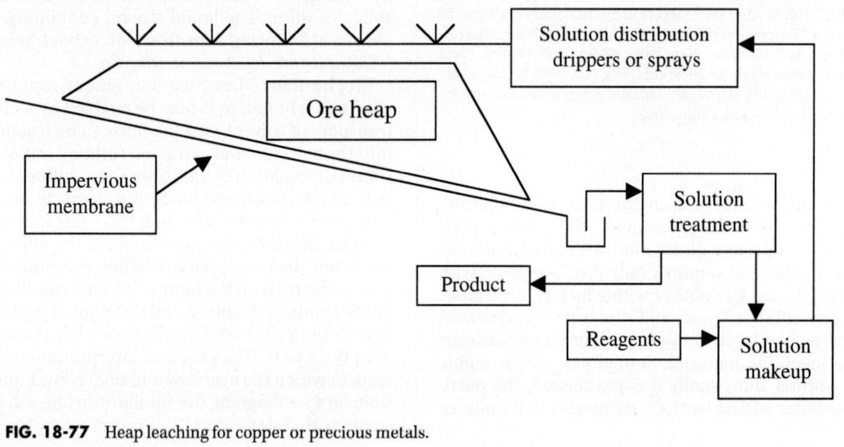

FIG. 18-77 Heap leaching for copper or precious metals.

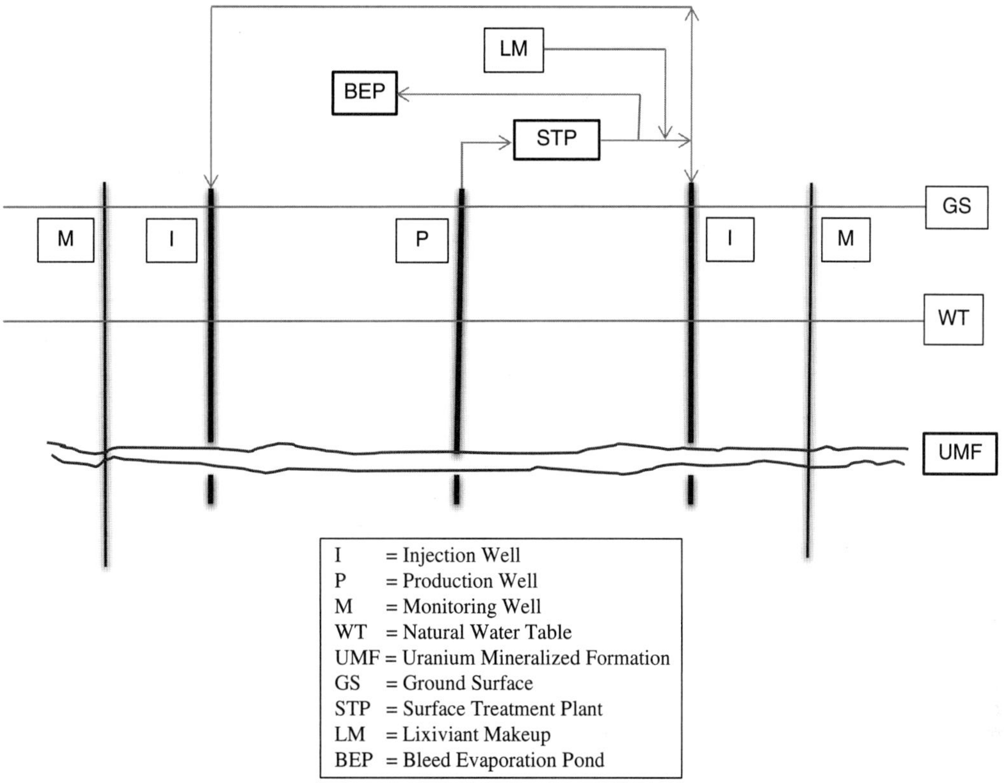

(*a*)

FIG. 18-78*a* Uranium in-situ leaching.

paddle conveyor, while enriched solvent, the "full miscella," is pumped from the bottom of the casing. It has been largely displaced in the oil extraction industry by horizontal basket, pan, or endless-belt percolators.

In the *horizontal-basket design*, illustrated by the *Rotocel extractor* (Fig. 18-79), walled compartments in the form of annular sectors with liquid-permeable floors revolve about a central axis. The compartments successively pass a feed point, a number of solvent sprays, a drainage section, and a discharge station (where the floor opens to discharge the extracted solids). The discharge station is circumferentially contiguous to the feed point. Countercurrent extraction is achieved by feeding fresh solvent only to the last compartment before dumping occurs and by washing the solids in each preceding compartment with the effluent from the succeeding one. The Rotocel is simple and inexpensive, and it requires little headroom. This type of equipment is made by a number of manufacturers. Horizontal table and tilting-pan vacuum filters, of which it is the gravity counterpart, are used as extractors for leaching processes that involve difficult solution-residue separation. Detailed descriptions of the Bollman-type and Rotocel extractors are presented on pp. 765 and 766 of McCabe, Smith, and Harriott, *Unit Operations of Chemical Engineering*, 7th ed., McGraw-Hill, New York, 2005.

The *endless-belt percolator* [Wakeman, "Extraction (Liquid-Solid)" in *Kirk-Othmer Encyclopedia of Chemical Technology*, 4th ed., vol. 10, Wiley, New York, 1993, p. 186] is similar in principle, but the successive feed, solvent spray, drainage, and dumping stations are linearly rather than circularly disposed. Horizontal-belt vacuum filters, which resemble endless-belt extractors, are sometimes used for leaching.

The *Kennedy extractor*, also requiring little headroom, operates substantially as a percolator that moves the bed of solids through the solvent rather than the conventional opposite. Because the solids are subjected to mechanical action that is somewhat more intense than in other types of continuous percolators, the Kennedy extractor is now seldom used for fragile materials such as flaked oil seeds.

Dispersed-Solids Leaching Equipment for batch leaching of fine solids in a liquid suspension is now confined mainly to batch tanks with rotating impellers. For a detailed discussion of all aspects of the suspension of solid particles in fluids, refer to Solid-Liquid Processing and Solids Suspension.

Batch stirred tanks agitated by coaxial impellers (turbines, paddles, or propellers) are commonly used for batch dissolution of solids in liquids and may be used for leaching fine solids. Since the controlling rate in mass transfer is the rate of transfer of material into or from the interior of the solid particles, rather than the rate of transfer to or from the surface of particles, the main function of the agitator is to supply unexhausted solvent to the particles while they reside in the tank long enough for the diffusion process to be completed. The agitator does this most efficiently if it just gently circulates the solids across the tank bottom or barely suspends them above the bottom. However, if the slurry contains particles having significantly different settling velocities, it is usually necessary to introduce sufficient mixing power to ensure full suspension of all particles. Failure to do so will result in an accumulation of the larger or denser particles unless provision is made to drain the settled material often.

The leached solids must be separated from the extract by settling and decantation or by external filters, centrifuges, or thickeners, all of which are treated elsewhere in Sec. 18. The difficulty of solids-extract separation and the fact that a batch stirred tank provides only a single equilibrium stage are its major disadvantages.

Continuous stirred tanks, often called continuous stirred-tank reactors (CSTRs), can be operated singly or in series. Figure 18-80 illustrates three tanks in series, each with a mechanical agitator. Nearly all stirred tanks are equipped with vertical baffles to prevent swirling and ineffective energy use by the agitator. Advancing of slurry from one stage to the next may be by overflow if successive stages are lower, or interstage pumps may be used. As with batch stirred tanks, insufficient agitation intensity (see subsection Fluid Mixing Technology, above) may allow larger or denser particles to accumulate. Resolution may require more efficient upstream comminution and classification or provisions for draining the accumulated solids. A succession of continuously operated tanks or reactors or extractors is called a train or battery or circuit.

Autoclaves, as shown schematically in Fig. 18-81, are closed, usually multi-compartmented vessels often designed for operation at pressures in excess of 600 psig (40 bar) and temperatures of 600°F or higher. Internal baffles allow the isolation of stages to minimize short-circuiting, and each stage

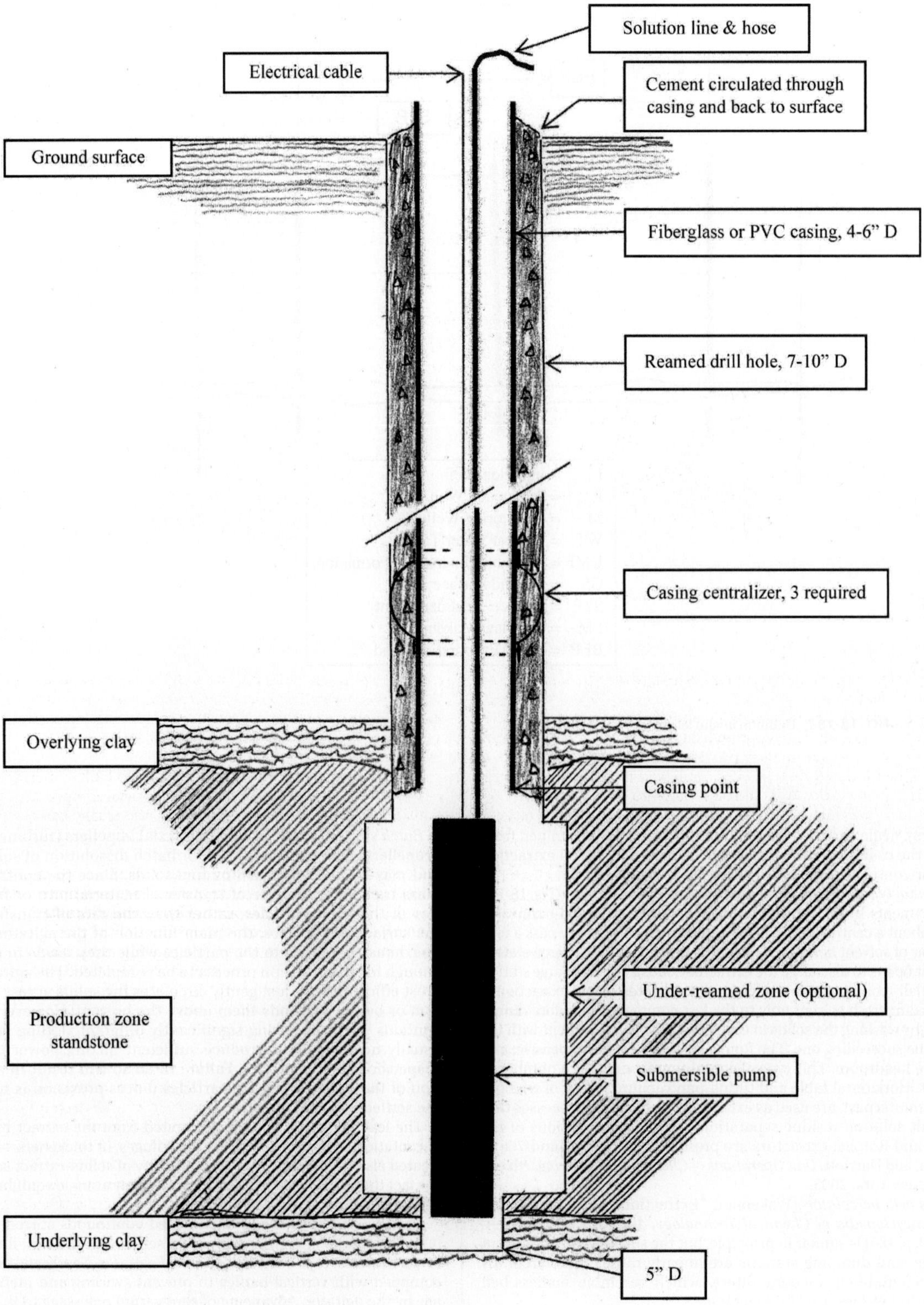

(*b*)

FIG. 18-78*b* In-situ well completion method.

is agitated. The purpose of some autoclaves is simply to effect aqueous oxidation, for example, of organic wastes or sulfide minerals. In the latter case, an example is oxidation of pyrite, followed by cyanide leaching of precious metals under ambient conditions. Other autoclaves are designed to effect leaching, as in the case of sulfuric acid leaching of nickel and cobalt from lateritic ores. The feed slurry is delivered to the autoclave by one or more positive displacement pumps, usually of the piston diaphragm type. If oxidation is required, oxygen may be used instead of air to reduce operating pressure or to improve kinetics. Flash cooling of the autoclave product is usually accomplished in one or more pressure reduction ("flash letdown")

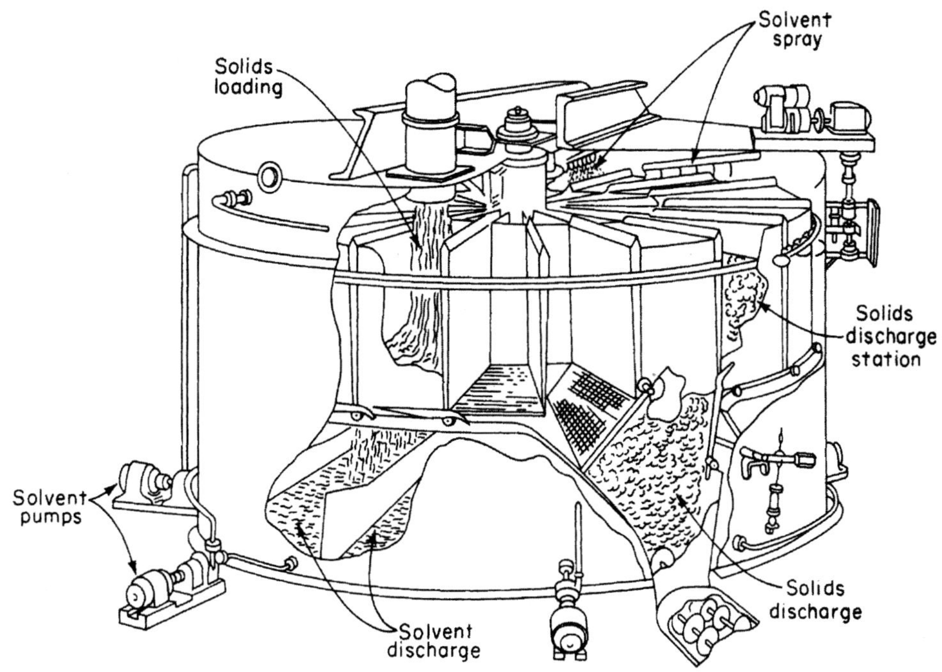

FIG. 18-79 Rotocel extractor. [Rickles, Chem. Eng. **72**(6): 164 (1965). Used with permission of McGraw-Hill, Inc.]

vessels with replaceable abrasion-resistant nozzles and targets; the resulting steam preheats the feed slurry.

The design of a production autoclave is almost invariably based on a comprehensive pilot-scale testing program using a small, continuous, multi (four to six)-compartment autoclave. See Fig. 18-82, courtesy of Hazen Research, Inc. When leaching is to be accomplished in the autoclave, the feed slurry pH is adjusted prior to its introduction into the autoclave.

Continuous autoclaves have been used for decades to produce refined alumina (Al_2O_3) by aqueous caustic leaching of mined bauxite. Other applications include the following:

- Oxidation of refractory pyritic gold ores prior to aqueous cyanide leaching in continuous stirred tanks at ambient conditions
- Simultaneous oxidation and leaching of concentrates of sulfide minerals of nickel, cobalt, zinc, and Platinum Group Metals
- Leaching of oxidized (lateritic) ores of nickel and cobalt with either ammonium carbonate or sulfuric acid
- Alkaline leaching of uranium/vanadium ores with high carbonate content that renders sulfuric acid uneconomical
- Limited commercial applications to copper sulfide concentrates, zinc oxide ores and residues, spent automotive and hydrodesulphurization catalysts, and hydrochloric acid leaching of ilmenite ($FeO \cdot TiO_2$) to yield synthetic rutile (TiO_2)

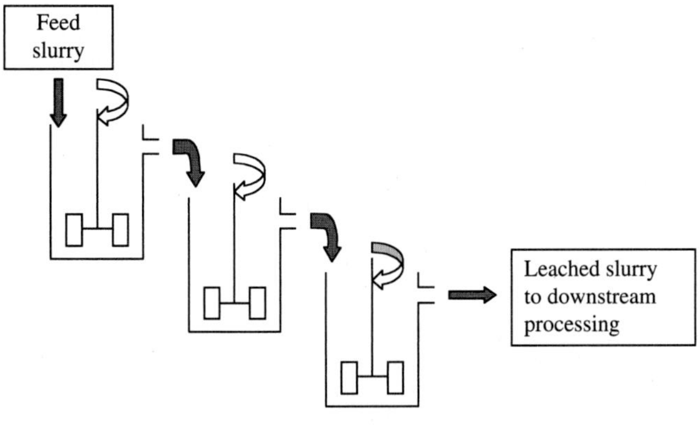

FIG. 18-80 Stirred tanks, three in series, with gravity overflow.

Autoclave oxidation and hydrometallurgical processing of the sulfide ores and concentrates of nickel, cobalt, copper, and zinc must compete economically with processes that use roasting or smelting. Hydrometallurgical plants generally consume more energy and usually are less effective in recovering by-product gold and silver.

Autoclaves are usually custom designed and built to meet a specific client's requirements, and they are expensive. However, operating problems and downtime for maintenance and repair are usually confined to peripheral equipment like feed pumps and vapor letdown apparatus, not to the autoclave itself.

Continuous Dispersed-Solids Leaching A *vertical-plate extractor*, exemplified by the *Bonotto extractor*, consists of a column divided into cylindrical compartments by equally spaced horizontal plates. Each plate has a radial opening staggered 180° from the openings of the plates immediately above and below it, and each is wiped by a rotating radial blade. Alternatively, the plates may be mounted on a coaxial shaft and rotated past stationary blades. The solids are fed to the top plate, fall to each lower plate in succession, and are discharged by a screw conveyor and compactor. Like the Bollman extractor, the Bonotto extractor has been virtually displaced by horizontal belt or tray percolators for the extraction of oil seeds.

Gravity Sedimentation Tanks Operated as thickeners, these tanks can serve as continuous contacting and separating devices in which fine solids may be leached continuously. A series of such units, properly connected, can provide true continuous countercurrent washing of fine solids. If needed, a mixing tank may be associated with each thickener to improve the contact between the solids and liquid being fed to that stage. Gravity sedimentation thickeners are described under the subsection Gravity Sedimentation Operations. Of all continuous leaching equipment, gravity thickeners require the most area, and they are limited to relatively fine solids.

Screw-Conveyor Extractors The *Hildebrandt total-immersion extractor*, the *De Danske Sukkerfabriker*, the *BMA diffusion tower*, and the *tray classifier* are devices that once were used extensively in the food processing industry—for example, for sugar beets and flaked oil seeds—but they have generally been supplanted by gentler machines mentioned earlier. Discussions and illustrations of these obsolescent machines can be found in the eighth edition of *Perry's Handbook*.

SELECTION OR DESIGN OF A LEACHING PROCESS

At all stages of the design of a leaching plant, the core issue is process engineering of the extraction unit or train. The essential aspects pertaining to the leaching operation are the selection of process and operating conditions and the sizing of the extraction equipment.

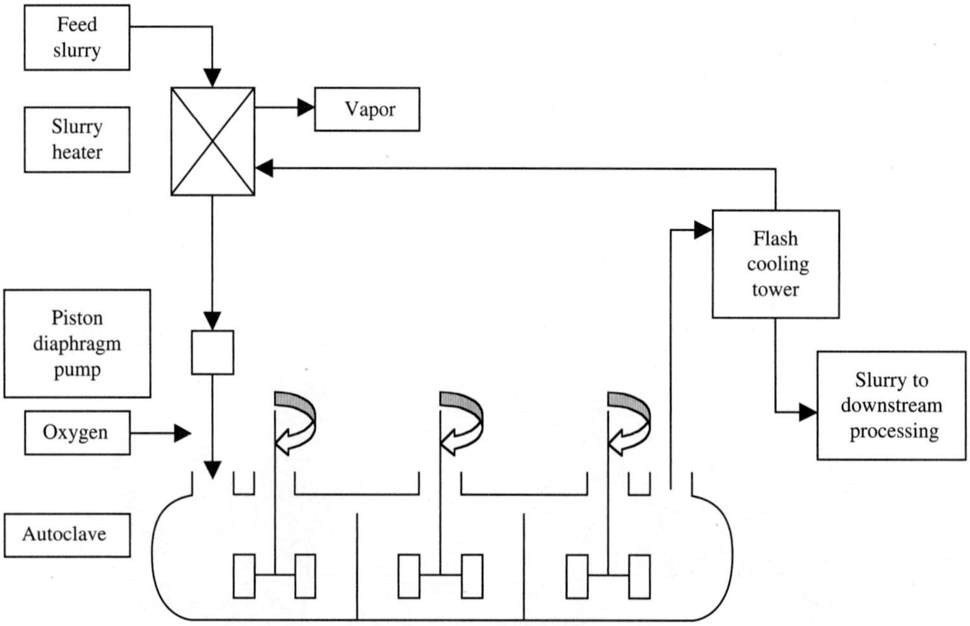

FIG. 18-81 Three-compartment autoclave.

Process and Operating Conditions The major parameters that must be fixed or identified are the solvent to be used, the temperature, the terminal stream compositions and quantities, leaching cycle (batch or continuous), optimum retention time, contacting method, and specific extractor choice.

The *choice of solvent* must offer an economical and practical balance of a number of desirable characteristics: high saturation limit and selectivity for the solute to be extracted, ability to produce extracted material of quality unimpaired by the solvent, chemical stability under process conditions, low viscosity, low vapor pressure, low toxicity and flammability, appropriate density and surface tension, ease and economy of recovery from the extract stream, consideration of wastewater treatment regulations and technologies, and price. The specific needs of each application determine the interactions and relative significance of these characteristics.

Temperature of the extraction should be chosen for the best balance of solubility, solvent-vapor pressure, solute diffusivity, solvent selectivity, effect of viscosity on solid/liquid separation, and sensitivity of the product

FIG. 18-82 Four-compartment autoclave.

to thermal degradation. In some cases, the temperature sensitivity of construction materials to corrosion or erosion attack may be significant.

Terminal stream composition and quantity parameters are linked to the production capacity of the leaching plant (rate of extract production or rate of raw material purification by extraction). When options exist, the degree of solute removal and the concentration of the extract stream are the variables that maximize process economy while satisfying other constraints. It is imperative that a material balance be developed that allows estimation of the effects of circulating loads on stream flow rates and compositions.

Continuous or intermittent operation is largely a matter of the size and nature of the process of which the extraction is a part. Selection of a percolation or solids-dispersion technique depends principally on the amenability of the extraction process to effective and sufficiently rapid percolation. Fine solids or a wide range of solids density may rule out percolation.

The *type of reactor* that is most compatible (or least incompatible) with the chosen combination of the preceding parameters is often not a clear choice, but this is a common challenge facing a process engineer. The ultimate criteria are simplicity, ruggedness, reliability, and profitability.

Extractor Sizing Calculations The design capacity (throughput rate) and retention time determine the total volume needed in the leaching circuit. The required active volume of the tanks can then be calculated if the number of stages is known. In practice, this involves determining the number of ideal stages needed and applying appropriate stage efficiencies. The methods of calculation resemble those for other mass-transfer operations (see Secs. 13, 14, and 15), involving equilibrium data and contact conditions and based on material balances. They are discussed briefly in the following paragraphs with specific application to countercurrent contacting.

In the case of continuous stirred-tank reactors, short-circuiting is minimized by maximizing the number of stages (reactors) in a train. Practically, three stages may be insufficient, and six stages may be overly conservative. If the reactor unit volume is based on a batch laboratory vessel, it is common to apply a safety factor of 25 to 33 percent. Parallel trains are often used to allow continued operation of one train during maintenance of another. Multiple trains are also common in very large leaching plants where a single train would require tanks and agitators larger than commercially proven.

For equipment costs and a discussion of cost estimation and process economics, see Sec. 9. When possible, budgetary purchased prices should be based on a material balance and obtained from a vendor. Installed cost will typically be on the order of three times the purchased price of a piece of equipment, and the range of that multiple is roughly 2 to 5.

Software Packages Widespread use is made of software developed for modeling and simulation of all types of unit operations, including leaching.

Proprietary programs especially useful to mineral processors can be found on the *JKSimMet* and *MetSim* websites. These software developers have been in business since the 1980s, and their products are regularly updated. Equipment suppliers including Metso and FLSmidth also offer modeling and design services. Monthly issues of *Chemical Engineering Progress* (*CEP*) usually contain a page entitled "Software" that announces new packages for various applications, and the same publication may contain an annual summary of programs useful to chemical and mineral processing engineers.

Composition Diagrams A leaching system typically consists of three components: inert, insoluble solids; a single nonadsorbed liquid or solid solute; and a single solvent. *It is a special type of ternary system because of the total mutual "insolubility" of two of the phases and the simple nature of equilibrium.

The composition of a typical system is satisfactorily represented by diagrams. Those diagrams most often used are a right-triangular plot of mass fraction of solvent against mass fraction of solute (Fig. 18-83*a*) and a plot suggestive of a Ponchon-Savarit diagram, with inerts taking the place of enthalpy (Fig. 18-83*b*). A third diagram, less frequently used, is a modified McCabe-Thiele plot in which the overflow solution (free of inerts) and the underflow solution (traveling out of a stage with the inerts) are treated as pseudo phases, and the mass fraction of solute in overflow, y, is plotted against the mass fraction of solute in underflow, x. (An additional representation, the equilateral-triangular diagram often used for liquid-liquid ternary systems, is seldom used because the field of leaching data is confined to a small portion of the triangle.)

With reference to Fig. 18-83 (both graphs), EF represents the locus of overflow compositions for the case in which the overflow stream contains no inert solids. $E'F'$ represents the overflow streams containing some inert solids, either by entrainment or by partial solubility in the overflow solution. Lines GF, GL, and GM represent the loci of underflow compositions for the three different conditions indicated on the diagram. In Fig. 18-83*a*, the constant underflow line GM is parallel to EF, the hypotenuse of the triangle, whereas GF passes through the right-hand vertex representing 100 percent solute. In Fig. 18-83*b*, underflow line GM is parallel to the abscissa, and GF passes through the point on the abscissa representing the composition of the clear solution adhering to the inert solids.

Compositions of overflow and underflow streams leaving the same stage are represented by the intersection of the composition lines for those streams with a tie line (AC, AC', BD, BD'). Equilibrium tie lines (AC, BD) pass through the origin (representing 100 percent inerts) in Fig. 18-83*a*, and they are vertical (representing the same inert-free solution composition in both streams) in Fig. 18-83*b*. For nonequilibrium conditions with or without adsorption or for equilibrium conditions with selective adsorption, the tie lines are displaced, such as AC' and BD'. Point C' is to the right of C if the solute concentration in the overflow solution is less than that in the underflow solution adhering to the solids. Unequal concentrations in the two solutions indicate insufficient contact time and/or preferential adsorption of one of the components on the inert solids. Tie lines such as AC' may be considered "practical tie lines" (i.e., they represent actual, rather than ideal, stages) if data on underflow and overflow composition have been obtained experimentally under conditions simulating actual operation, particularly with respect to contact time, agitation intensity, percent solids in the slurry, and particle size distribution of the solids.

The illustrative construction lines of Fig. 18-83 have been made with the assumption of constant underflow composition. In the more realistic case of variable underflow composition, the points C, C', D, D' would lie along line GL. Like the practical tie lines, GL is a plot of experimental data.

Algebraic Computation The *algebraic computation* method starts with the calculation of the quantities and compositions of all the terminal streams, assigning a convenient value, like 100, to one of the streams. A material balance and the stream compositions are then computed for a terminal ideal stage at either end of an extraction train (i.e., at point A or point B in Fig. 18-83), using equilibrium and solution-retention data. Calculations are repeated for each successive ideal stage from one end of the train to the other until an ideal stage corresponding to the desired conditions is obtained. Essentially any solid-liquid extraction problem can be solved by this method.

*The solubility of a noninert solid, adsorption of solute on that solid, and complexity of solvent and extracted material can be taken into account if necessary, but their consideration is beyond the scope of this handbook.

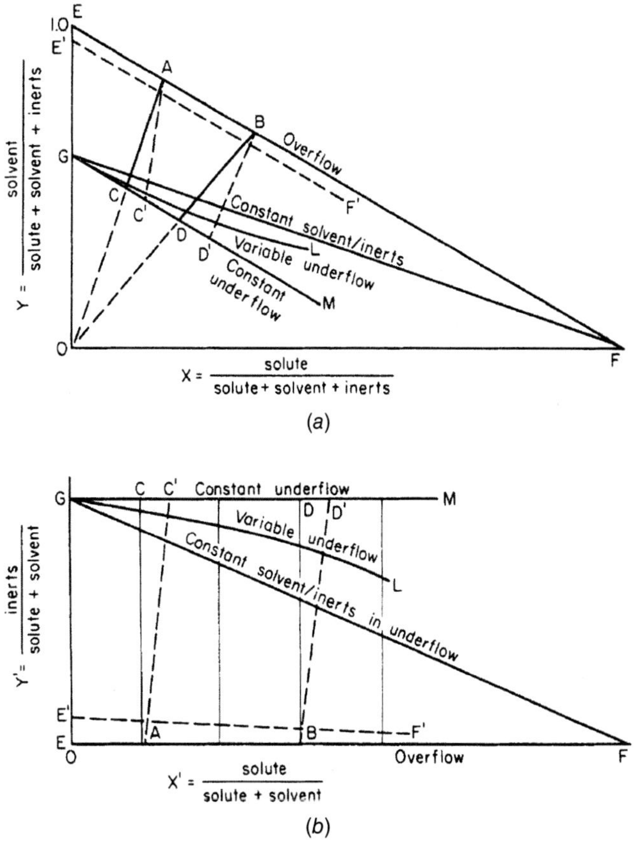

FIG. 18-83 Composition diagrams for leaching calculations: (*a*) right-triangular diagram; (*b*) modified Ponchon-Savarit diagram.

For certain simplified cases, it is possible to calculate directly the number of stages required to attain a desired product composition for a given set of feed conditions. For example, if equilibrium is attained in all stages, and if the underflow mass flow rate is constant, both the equilibrium and operating lines on a modified McCabe-Thiele diagram are straight. It then is possible to calculate directly, using Eq. (18-51), the number of ideal stages, N, required to accommodate any rational set of terminal flows and compositions. (Chapter 20 in McCabe, Smith, and Harriott, *Unit Operations of Chemical Engineering*, 7th ed., McGraw-Hill, New York, 2005, offers details and includes sample problems.)

$$N = \frac{\log[(y_b - x_b)/(y_a - x_a)]}{\log[(y_b - y_a)/(x_b - x_a)]} \qquad (18\text{-}51)$$

Even when each stage is at equilibrium and underflow compositions are constant, Eq. (18-51) normally is not valid for the first stage because the unleached solids entering that stage usually are not premixed with solution to produce the underflow mass that will leave the stage. This is easily rectified by calculating the exit streams for the first stage and using those values in Eq. (18-51) to calculate the number of stages required after stage 1.

Graphical Method This method of calculation is simply a diagrammatic representation of all the possible compositions in a leaching system, including equilibrium values, on which material balances across ideal (or, in some cases, nonideal) stages can be evaluated in the graphical equivalent of the stage-by-stage algebraic computation. It usually is simpler than manual calculation of the algebraic solution, and it permits visualization of the process variables and their effects on the operation. Any of the four types of composition diagrams described previously can be used, but modified Ponchon-Savarit or right-triangular plots (Fig. 18-83) are most convenient for leaching calculations.

GRAVITY SEDIMENTATION OPERATIONS

General References: Albertson, *Fluid/Particle Sep. J.* **7:** IS (1994); Jewell, Fourie, and Lord, *Paste and Thickened Tailings—A Guide*, Australian Centre for Geomechanics, 2002, pp. 49–79; Mular, Halbe, and Barratt, *Mineral Processing Plant Design, Practice, and Control*, vol. 2, Society for Mining, Metallurgy, and Exploration, Englewood, Colo., 2002, pp. 1295–1312 and 2164–2173; Pearse, *Gravity Thickening Theories: A Review*, LR 261(MP), Warren Spring Laboratory, Hertfordshire, England, 1977; Sankey and Payne, *Chemical Reagents in the Mineral Processing Industry*, Society for Mining, Metallurgy, and Exploration, Englewood, Colo., 1985, p. 245; Schweitzer, *Handbook of Separation Techniques for Chemical Engineers*, 2d ed., McGraw-Hill, New York, 1988, pp. 4-121 to 4-147; Talmage and Fitch, *Ind. Eng. Chem.* **47:** 38 (1955); Wilhelm and Naide, *Min. Eng.* (Littleton, Colo.), 1710 (1981).

Sedimentation is the partial separation and concentration of suspended solid particles from a liquid by gravity settling. This field may be divided into the functional operations of thickening and clarification. The primary purpose of thickening is to increase the concentration of suspended solids in a feed stream, while that of clarification is to remove a relatively small quantity of suspended particles and produce a clear effluent. These two functions are similar and occur simultaneously, and the terminology merely makes a distinction between the primary process results desired. Generally, thickener mechanisms are designed for the heavier-duty requirements imposed by a large quantity of relatively concentrated pulp, while clarifiers usually will include features that ensure essentially complete suspended-solids removal, such as greater depth, special provision for coagulation and flocculation of the feed, and greater overflow-weir length. Sedimentation equipment is designed to meet performance specifications for many differing applications. This section will present information on the approach to design, equipment features, and options, and will give some guidelines.

CLASSIFICATION OF SETTLEABLE SOLIDS AND THE NATURE OF SEDIMENTATION

The types of sedimentation encountered in process technology will be affected by the obvious factors of particle size, liquid viscosity, solid and solution densities, and the particle characteristics within the slurry. These properties, as well as the process requirements, determine both the type of equipment that will achieve the desired performance and the design of the equipment.

Figure 18-84 illustrates the relationship between solids concentration, interparticle cohesiveness, and the type of sedimentation that may exist. "Totally discrete" particles include many mineral particles (usually greater in diameter than 20 μm), salt crystals, and similar substances that have little tendency to cohere. "Flocculent" particles generally will include those smaller than 20 μm (unless present in a dispersed state owing to surface charges), metal hydroxides, many chemical precipitates, and most organic substances other than true colloids.

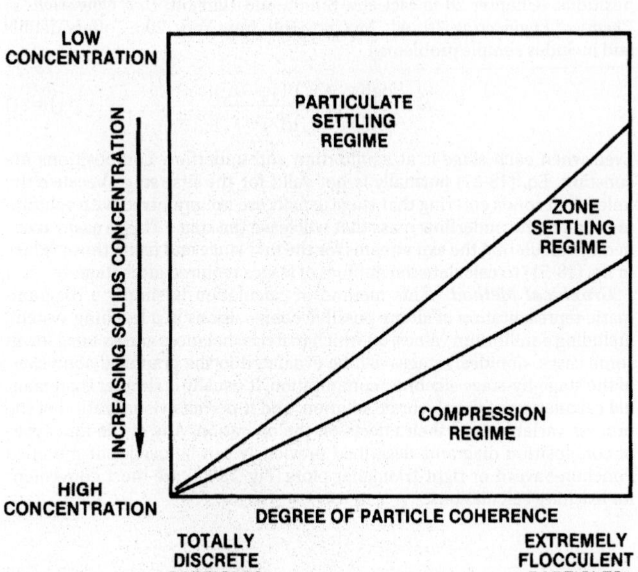

FIG. 18-84 Combined effect of particle coherence and solids concentration on the settling characteristics of a suspension.

At low solids concentrations, the type of sedimentation encountered is called particulate settling. Regardless of their nature, particles are sufficiently far apart to settle freely. Faster-settling particles may collide with slower-settling ones and, if they do not cohere, continue downward at their own specific rate. Those that do cohere will form floccules of a larger diameter that will settle at a rate greater than that of the individual particles.

There is a gradual transition from particulate settling into the zone-settling regime, where the particles are constrained to settle as a mass. The principal characteristic of this zone is that the settling rate of the mass will be a function of its solids concentration (for any particular condition of flocculation, particle density, etc.).

The solids concentration will ultimately reach a level at which particle descent is restrained not only by hydrodynamic forces but also partially by mechanical support from the particles below; therefore, the weight of particles in mutual contact can influence the rate of sedimentation of those at lower levels. This compression, as it is termed, will result in further solids concentration because of compaction of the individual floccules and partial filling of the interfloc voids by the deformed floccules. Accordingly, the rate of sedimentation in the compression regime is a function of both the solids concentration and the depth of pulp in this particular zone. As indicated in Fig. 18-84, granular, nonflocculent particles may reach their ultimate solids concentration without passing through this regime.

As an illustration, coarse-size (45 μm) aluminum oxide trihydrate particles produced in the Bayer process would be located near the extreme left of Fig. 18-84. These solids settle in a particulate manner, passing through a zone-settling regime only briefly, and reach an ultimate solids concentration without any significant compressive effects. At this point, the solids concentration may be as much as 80 wt%. The same compound, but of a gelatinous nature when precipitated in water treatment as aluminum hydroxide, would be on the extreme right-hand side of the figure. This flocculent material enters into a zone-settling regime at a low concentration (relative to the ultimate concentration it can reach) and gradually thickens. With sufficient pulp depth present, and aided by gentle raking, the compression-zone effect will occur and the sludge will attain its maximum solids concentration, around 10 wt%.

A feed stream to be clarified or thickened can exist at any state represented within this diagram. As it becomes concentrated owing to sedimentation, it may pass through all the regimes, and the settling rate in any one may be the size-determining factor for the required equipment.

Sedimentation Testing Data from full-scale sedimentation equipment, operating in the application under consideration, are always a first choice for designing and sizing new equipment. The characteristics of the feed stream for the new application (e.g., solids and process liquor characteristics, particle size, viscosities, pH, use of flocculants) must be identical to the existing application. It is also necessary to know how close to "capacity" the existing equipment is operating. If the application under question deviates sufficiently from reference installations, bench- or pilot-scale testing is recommended to size and design a new sedimentation unit.

To properly design and size sedimentation equipment, several pieces of information are required. Some information is unique to the job site (application, feed rate, etc.), while other data are supplied from similar references or from test work. Site-specific information from the plant site includes:

- Equipment application objectives
- Feed flow rate and solids concentration range
- Feed stream characteristics
- Site-specific requirements: seismic zone, weather-related specifications, local mechanical design codes, and the user's preferred design specifications
- Local operating practices

Testing must be structured to produce all or some of the following information:

- Feed stream characteristics
- Feed chemical treatment (type, solution concentration, dose, etc.): coagulants and flocculants (organic or inorganic); acid/base for treatment and pH adjustment
- Coagulation and flocculation (mixing time and energy requirements, optimum solids concentration)
- Expected sedimentation objectives: underflow slurry density; overflow solids concentration (suspended solids and/or turbidity); chemical treatment for soluble components (i.e., hardness, metals, anions, pH, etc.)
- Required vessel area and depth
- Underflow slurry rheology (for raking mechanism design, drive torque specification and underflow pump selection)

There are three basic approaches to testing for sedimentation equipment:

- *Batch bench-scale settling tests* A relatively small amount of sample is tested in a controlled environment using laboratory equipment under static conditions.

- *Semicontinuous bench-scale tests* Laboratory pumps are used to feed slurry and treatment chemicals into settling cylinders from which overflow liquor and underflow slurry are collected.
- *Continuous piloting* A small-diameter thickener or clarifier of the same design as the full-scale equipment being considered is tested at the industrial site.

TESTING COMMON TO CLARIFIERS AND THICKENERS

Feed Characterization Sample characterization is necessary. Without these data included in the basis of design, the sizing and predicted performance cannot be validated for the specified feed stream. Characterization requires the following measurements as a minimum:

- General chemical makeup of the solids and liquor phases
- Feed solids concentration
- Particle size distribution—include coarse (+100 μm) and fine (−20 μm) particle diameters
- Solids specific gravity
- Liquid specific gravity
- Liquid-phase dissolved materials concentration
- Temperature
- pH

Coagulant and/or Flocculant Selection Coagulants and flocculants are widely used to enhance the settling rate, which reduces thickener and clarifier size and improves overflow clarity or underflow density. The terms *coagulation* and *flocculation* each describe separate functions in the particle agglomeration process.

Coagulation is a preconditioning step that may be required to destabilize the solids suspension to allow complete flocculation to occur in clarification applications. Flocculation is the bridging and binding of destabilized solids into larger particles. As particle size increases, settling rate generally increases. Descriptions of the science of flocculation can be found in many texts and in the literature available from flocculant vendors.

Both coagulation and flocculation are typically considered in designing clarifiers, whereas flocculation is normally the only feed treatment in designing thickeners.

Coagulants may be either organic, such as polyelectrolytes, or inorganic, such as alum. Coagulants can be used alone or in conjunction with flocculants to improve the performance of the flocculant or reduce the quantity of the flocculant required.

There are two primary types of flocculants:

- *Natural flocculants* Starch, guar, and other natural materials have historically been used for flocculation, but they have been replaced by more effective synthetic polymers.
- *Synthetic polymeric flocculants* There are many synthetic polymers developed for specific applications. Because of the variety, a screening program is necessary, and the choice can be narrowed by the following:
  - Prior experience with flocculants on the feed stream under evaluation.
  - Test one each of the major types of flocculant charge: anionic, nonionic, and cationic.
  - Test one each of the synthetic polymer molecular weight and charge density.

The purpose of the screening test is to select a coagulant or flocculant whose generic type will most likely be effective in plant operation, and to develop a basis for the required dose (g/tonne solids or mg/L liquor). Although a thickener or clarifier may be started up on the flocculant selected in the testing, it is very common to conduct further tests on the full-scale machine to further optimize dosage or flocculant type. The flocculant manufacturer can be a source of great assistance in both the testing and the full-scale optimization of flocculant use.

Coagulant or flocculant solutions should be prepared according to the manufacturer's instructions and used within the shelf life recommended. The solution concentration recommended for testing is typically more dilute than the "neat" concentration so that the viscosity will be lower to make dispersion more rapid during testing.

In the screen tests, each coagulant or flocculant is added to beaker samples of representative slurry or liquor in a dropwise fashion, while the sample is mixed with a spatula, stirrer, or three-to-six-jar stirrer apparatus. The amount of coagulant or flocculant required to initiate floc particle formation is recorded, along with notes as to the size of the floc, capture of fines, resultant liquor clarity, and stability of the floc structure. The dosage is typically noted in g/t solids if the sample is primarily solids (thickener design), or in mg/L liquor if the sample is for clarification and the solids concentration is low.

TESTING SPECIFIC TO CLARIFICATION

Detention Test The test uses a 1- to 4-L beaker or similar vessel. The sample is placed in the container, coagulated and/or flocculated by suitable means, if required, and allowed to settle. Small samples for suspended-solids analysis are withdrawn from a point approximately midway between the liquid surface and settled solids interface, taken with sufficient care that settled solids are not resuspended. Sampling times may be at consecutively longer intervals, such as 5, 10, 20, 40, and 80 min.

The suspended-solids concentration is plotted on log-log paper as a function of the sample (detention) time. A straight line will usually result, and the required static detention time t to achieve a certain suspended-solids concentration C in the overflow of an ideal basin can be taken directly from the graph. If the plot is a straight line, the data are described by the equation

$$C = Kt^m \tag{18-52}$$

where the coefficient K and exponent m are characteristic of the particular suspension.

Should the suspension contain a fraction of solids that are "unsettleable," the data are more easily represented by using the second-order procedure. This depends on the data being reasonably represented by the equation

$$Kt = \frac{1}{C - C_\infty} - \frac{1}{C_0 - C_\infty} \tag{18-53}$$

where C_∞ is the unsettleable solids concentration and C_0 is the concentration of suspended solids in the feed sample. The residual solids concentration remaining in suspension after a sufficiently long detention time (C_∞) must be determined first, and the data then plotted on linear paper as the reciprocal concentration function $1/(C - C_\infty)$ versus time.

Bulk Settling Test After the detention test is completed, a bulk settling test determines the maximum overflow rate. This is done by carrying out a settling test in which the solids are first concentrated to a level at which zone settling just begins. This is usually marked by a very diffuse interface during initial settling. Its rate of descent is measured with a graduated cylinder of suitable size, preferably at least 1 L, and the initial straight-line portion of the settling curve is used for specifying a bulk-settling rate. The design overflow rate generally should not exceed half of the bulk settling rate. From the two clarifier tests, detention time, and bulk settling rate, the more conservative results will govern the size of the clarifier.

Solids Contact Clarification In many instances, the rate of clarification is enhanced by increasing the solids concentration in the flocculation zone of the clarifier. This is done in a full-scale operation by internally or externally recycling previously settled solids into the flocculation zone where they are mixed with fresh, coagulated feed. The higher population of solids improves the flocculation efficiency and clarification rate.

To conduct these tests, a sample of feed is first treated at the chemical dosages and mixing intensity determined in the screening tests. The solids are allowed to settle, and the supernatant is carefully decanted. The settled solids are then transferred to a new fresh sample, and tests are conducted again, using the same chemical dosages and mixing intensity. Recycle testing continues with subsequent tests until the suspended solids in the sample have concentrations of 1, 2, 3, and 5 g/L. Bulk settling rate, suspended solids, and other effluent parameters are measured with each test until an optimal treatment scenario is found.

In some suspensions, very fine colloidal solids are present and are very difficult to coagulate, and it is typically necessary to adjust for coagulation mixing intensity and time to obtain coagulated solids that are more amenable to flocculation.

Detention Efficiency Conversion from the ideal basin sized by detention-time procedures to an actual clarifier requires the inclusion of an efficiency factor to account for the effects of turbulence and nonuniform flow. Efficiencies vary greatly, being dependent not only on the relative dimensions of the clarifier and the means of feeding but also on the characteristics of the particles. The curve shown in Fig. 18-85 can be used to scale up laboratory data in sizing circular clarifiers. The static detention time determined from a test to produce a specific effluent solids concentration is divided by the efficiency (expressed as a fraction) to determine the nominal detention time, which represents the volume of the clarifier above the settled pulp interface divided by the overflow rate. Different diameter–depth combinations are considered by using the corresponding efficiency factor. In some cases, area may be determined by factors other than the bulk-settling rate, such as practical tank-depth limitations.

TESTING SPECIFIC TO THICKENING

Optimization of Flocculation Conditions After a flocculant type is selected, the next step is to conduct a range of tests, using the selected flocculant, to gather data on the effects of feed slurry solids concentrations on flocculant dosage and settling rate. There are a range of solids concentrations for which flocculation effectiveness is maximized, resulting in

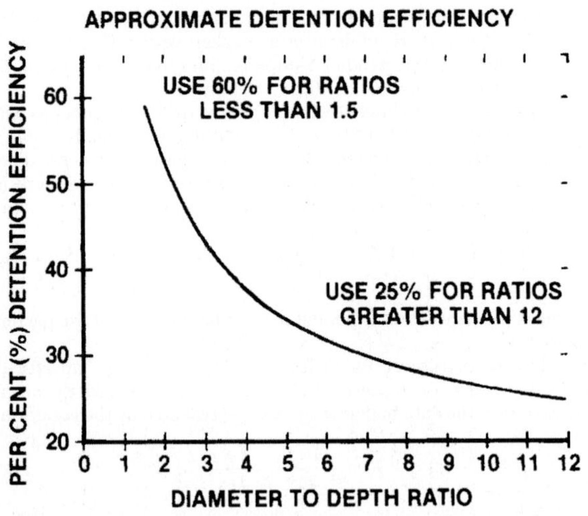

FIG. 18-85 Efficiency curve for scale-up of batch clarification data to determine nominal detention time in a continuous clarifier.

improved settling characteristics. Operating within this feed solids range results in smaller equipment sizes, higher underflow slurry densities, better overflow liquor clarity, and lower flocculant dosages.

The tests are conducted using a series of samples prepared at solids concentrations decreasing incrementally in concentration from the expected thickener feed concentration. Typically, the samples are prepared in 250- to 500-mL graduated cylinders that give some distance to measure the settling rate more accurately. For some very fine solids samples (e.g., alumina red mud, clays, leached nickel laterites), it is recommended to also check a sample diluted to 2 to 3 wt% solids. Begin adding the flocculant solution dropwise; make notes on the dosage at which flocculation begins and the settling velocity. Continue adding flocculant incrementally and noting the floc structure, fines capture, liquor clarity, and settling velocity. Once the settling velocity remains constant for a few tests, sample testing can be stopped. From the tests, the plot shown in Fig. 18-86 can be drawn and the results used to set conditions for the larger and final tests for sizing the thickening equipment. The test procedure for the design tests should be structured to span the optimum solids concentration and two points slightly higher and lower.

Determination of Thickener Basin Area Many procedures to determine thickener unit area (m^2/tpd) have been developed. A good historical overview of the various approaches can be found in Pearse [*Gravity Thickening Theories: A Review*, LR 261(MP), Warren Spring Laboratory, Hertfordshire, England, 1977]. The current methods most commonly used by designers and suppliers of thickening equipment are presented here.

If polymeric flocculants are used, a batch cylinder test approach based on the Kynch theory is most common. In this method, the test is carried out at the optimum feed solids and flocculant dose (as determined in tests described earlier) and continued until underflow concentration is achieved in the cylinder. The flocculant solution should be added to the slurry under conditions that promote rapid dispersion and uniform, complete mixing with a minimum of shear using an apparatus consisting of a syringe, a tube, and an inverted rubber stopper. To determine the unit area, Talmage and

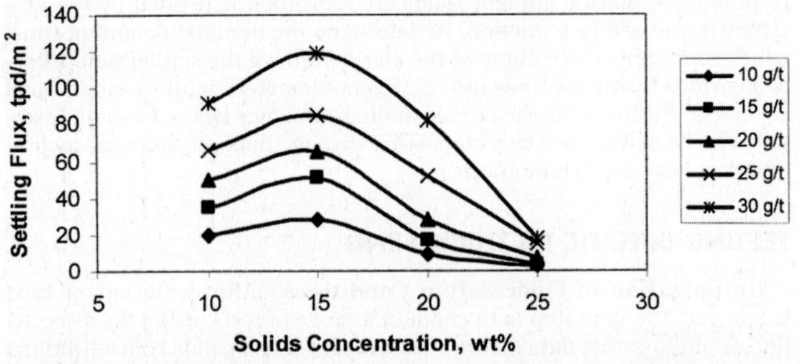

FIG. 18-86 Data showing that slurry solids concentration affects flocculation efficiency, thus improving solids settling flux.

Fitch [*Ind. Eng. Chem.* **47:** 38 (1955)] proposed an equation derived from a relationship equivalent to that shown in Eq. (18-54):

$$\text{Unit area} = \frac{t_u}{C_0 H_0} \tag{18-54}$$

where t_u is the time, days; C_0 is the initial solids concentration in the feed, t/m^3; and H_0 is the initial height of the slurry in the test cylinder, m. The term t_u is taken from the intersection of a tangent to the curve at the critical point and a horizontal line representing the depth of pulp at underflow concentration. There are various means for selecting this critical point, all of them empirical, and the unit area value determined cannot be considered precise. The review by Pearse (1977) presents many of the different procedures used in applying this approach to laboratory settling test data.

Two other approaches avoid using the critical point by computing the area requirements from the settling conditions existing at the underflow concentration. The Wilhelm and Naide procedure [*Min. Eng.* (Littleton, Colo.), 1710 (1981)] applies zone-settling theory to the entire thickening regime. Tangents drawn to the settling curve are used to calculate the settling velocity at all concentrations obtained in the test. This permits the construction of a plot (Fig. 18-87) showing unit area as a function of underflow concentration.

A second, "direct" approach that yields a similar result, since it also takes compression into account, uses the value of settling time t_x taken from the settling curve at a particular underflow concentration. This value is used to solve the Talmage and Fitch equation (18-54) for unit area.

In applying either of these two procedures, it is necessary to run the test in a vessel having an average bed depth close to that expected in a full-scale thickener. This requires a very large sample, and it is best to carry out the test in a cylinder having a volume of 4 L or more. The calculated unit area value from this test can be extrapolated to full-scale depth by carrying out similar tests at different average bed depths to determine the effect on unit area. Alternatively, an empirical relationship can be used that is effective in applying a depth correction to laboratory cylinder data over normal operating ranges. The unit area calculated by either the Wilhelm and Naide approach or the direct method is multiplied by a factor equal to $(h/H)^n$, where h is the average depth of the pulp in the cylinder, H is the expected full-scale compression zone depth, usually taken as 1 m, and n is the exponent calculated from Fig. 18-88. For conservative design purposes, the minimum value of this factor that should be used is 0.25.

It is essential to use a slow-speed (0.1 r/min) picket rake in all cylinder tests to prevent particle bridging and promote solids consolidation to reach the underflow density that is obtainable in a full-scale thickener.

Continuously operated, small-scale or pilot-plant thickeners, ranging from 75 mm diameter by 400 mm depth to several meters in diameter, are also effectively used for sizing full-scale equipment. This approach requires a significantly greater volume of sample, such as would be available in an operating installation or a pilot plant. Continuous units and batch cylinders will produce equivalent results if proper procedures are followed with either system.

Thickener-Basin Depth The pulp depth required in the thickener will be greatly affected by the role that compression plays in determining the rate of sedimentation. If the zone-settling conditions define the area needed, then depth of pulp will be unimportant and can be largely ignored, as the

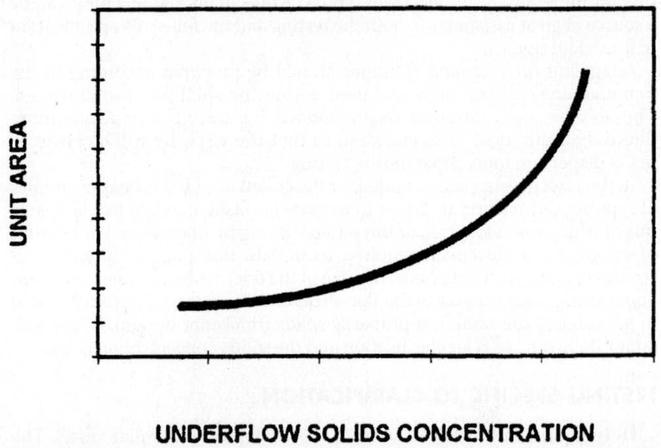

FIG. 18-87 Characteristic relationship between thickener unit area and underflow solids concentration (fixed flocculant dosage and pulp depth).

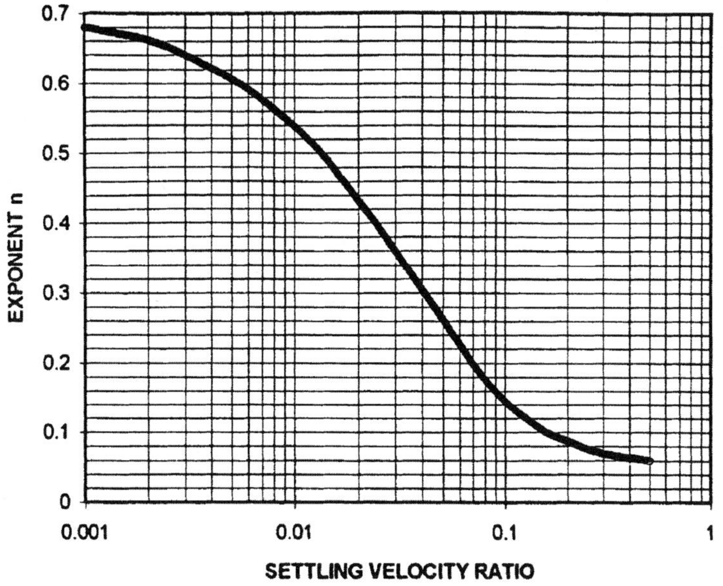

FIG. 18-88 Depth correction factor to be applied to unit areas determined by Wilhelm-Naide and "direct" methods. Velocity ratio calculated using tangents to settling curve at a particular settled solids concentration and at start of test.

"normal" depth found in the thickener will be sufficient. On the other hand, with the compression zone controlling, depth of pulp will be significant, and it is essential to measure the sedimentation rate under these conditions. This is the case for the new deep-bed, high-density/paste thickeners.

To determine the compression-zone requirement in a thickener, a test should be run in a deep cylinder in which the average settling pulp depth approximates the depth anticipated in the full-scale basin. The average density of the pulp in compression is calculated and used in Eq. (18-55) to determine the required compression-zone volume:

$$V = \frac{\theta_c(\rho_s - \rho_l)}{\rho_s(\rho_{sl} - \rho_l)} \tag{18-55}$$

where V is the volume, m^3, required per ton of solids per day; θ_c is the compression time, days, required in the test to reach underflow concentration; and ρ_s, ρ_l, ρ_{sl} are the densities of the solids, liquid, and slurry (average), respectively, t/m^3. This value divided by the average depth of the pulp during the period represents the unit area defined by compression requirements. If it exceeds the value determined from the zone-settling tests, it is the quantity to be used.

The side depth of the thickener is determined as the sum of the depths needed for the compression zone and for the clear zone. Normally, 1.5 to 2 m of clear liquid depth above the expected pulp level in a thickener will be sufficient for stable, effective operation. When the location of the pulp level cannot be predicted in advance or it is expected to be relatively low, a thickener sidewall depth of 2 to 3 m is usually safe. Greater depth may be used in order to provide better clarity, although in most thickener applications the improvement obtained by this means will be marginal.

Scale-Up Factors Factors used in thickening will vary, but typically, a 1.2 to 1.3 multiplier applied to the unit area calculated from laboratory data is sufficient if proper testing procedures have been followed and the samples are representative.

Torque Requirements Sufficient torque must be available in the raking mechanism of a full-scale thickener to allow it to move through the slurry and assist solids movement to the underflow outlet. Granular, particulate solids that settle rapidly and reach a terminal solids concentration without going through any apparent compression or zone-settling region require a maximum raking capability because they must be moved to the outlet solely by the mechanism. At the other end of the spectrum, extremely fine materials, such as clays and precipitates, require a minimum of raking, for most of the solids may reach the underflow outlet hydrodynamically. The rakes prevent a gradual buildup of some solids on the bottom, and the gentle stirring action from the rake arm often aids the thickening process. As the underflow concentration approaches its ultimate limit, the consistency or yield stress will increase greatly, resulting in a higher raking requirement and an increase in torque.

Test methods to specify torque from small-scale tests are of questionable value since it is difficult to duplicate full-scale conditions. Manufacturers of sedimentation equipment select torque ratings from experience with similar substances and will recommend a torque capability on this basis. In addition, rheological properties of the thickened slurry (e.g., yield stress) can be measured on test samples and used as a guide to the magnitude of operating torque that can be anticipated. The rheological tests are usually conducted using a vane-type viscometer. Figure 18-89 shows an example of thickener underflow yield stress behavior as the solids are concentrated to higher slurry densities in the various types of thickeners.

Definitions of *operating torque* vary with the manufacturer, and the user should ask the supplier to specify the *B-10 life* for bearings and to reference appropriate mechanical standards for continuous operation of the selected gear set at specific torque levels. This will provide guidelines for plant operators and help avoid premature failure of the mechanism. Abnormal conditions above the normal operating torque are inevitable, and a thickener should be provided with sufficient torque capability for short-term operation at higher levels in order to ensure continuous performance.

Generic underflow slurry characteristics that affect the selection of drive torque are presented in Table 18-11. Duty classification gives the user a feeling as to the level of raking duty the drive will be required to provide to manage the movement of thickened solids to a discharge point in the thickener.

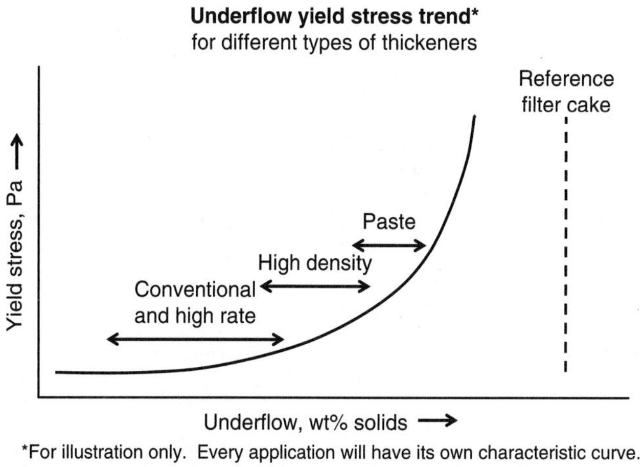

FIG. 18-89 An illustration of typical slurry yield stress increase as solids concentration increases, which affects raking torque and underflow pump selection and design.

TABLE 18-11 Factors Affecting Design Torque in a Thickener

Duty classification	K factor, N/m	Underflow, wt% dry solids	% Total solids <74 micron	% Total solids >210 micron	Specific gravity of solids	Examples
Light	15–60	<5	100	0	1.0–1.25	River water clarification, brine clarification
Standard	70–130	5–30	85–100	0–5	1.25–3.0	Lime softening, brine softening
Heavy	145–290	30–50	50–85	5–15	3.0–4.0	Copper tailings, iron tailings, zinc concentrate, phosphate tailings
Extra heavy	>290	>50	<50	>15	>4.0	Iron ore concentrate, titanium ilmenite
Paste	>2150	Max flowable	No upper limit on fines	As much as supported by rheology	>2.5	Lead/zinc tailings, red mud, kimberlite, platinum tailings, trona tailings, coal refuse

NOTE: Torque, Nm= $K * D^2$, where D = thickener diameter in meters. Nm/1.356 = lb · ft

Underflow Pump Requirements Many suspensions will thicken to a concentration higher than that which can be handled by conventional slurry pumps. The unit area to produce the maximum concentration that can be pumped is the usual design basis, and thickening tests should be performed with this in mind. Since most thickened slurries will be non-newtonian, determination of the ultimate pumpable concentration can be approximated by measuring the yield stress and rheological behavior of the thickened slurry and then consulting with pump experts and suppliers who have databases for a wide range of materials.

If the testing is for the design of a paste thickener, the shear-thinning characteristics at the maximum underflow concentration should be studied. After measuring the unsheared yield stress, the sample should be sheared with high-intensity mixing for a few minutes and the yield stress measured and noted. If shear thinning is then beneficial for removal of the underflow from the thickener, the thickener and pump suppliers should be consulted about adding a shear-thinning system to the thickener.

To obtain a sample from the test cylinder for rheology tests, the supernatant should be decanted following a test, and the total settled solids transferred to a container where they can be gently mixed until homogeneous. Samples for rheology tests are then taken from this batch.

THICKENERS

The primary function of a continuous thickener is to concentrate suspended solids by gravity settling so that a steady-state material balance is achieved: solids being withdrawn continuously in the underflow at the rate they are supplied in the feed. An inventory of pulp is maintained in order to achieve the desired underflow concentration. This volume will vary somewhat as operating conditions change; on occasion, this inventory can be used to store solids when feed and underflow rates are reduced or temporarily suspended.

A thickener has several basic components: a tank to contain the settling slurry and clarified process liquor, feed piping and a feedwell to allow the feed stream to enter the tank, a rake mechanism to assist in moving the concentrated solids to the withdrawal points, an underflow solids-withdrawal system, and an overflow launder. The basic design of a bridge-supported thickener mechanism is illustrated in Fig. 18-90.

Thickener Types Thickeners are classified as *conventional, high-rate, ultrahigh-rate,* or *high-density/paste* based on the particular design purpose and the method in which flocculants are applied and mixed. These designations can be confusing in that they imply sharp distinctions between each type, which is *not* the case.

In *high-rate thickeners*, the greater capacity is due solely to the effective use of flocculant to maximize throughput. In most applications, there is a threshold dosage and feed solids concentration at which a noticeable increase in capacity begins to occur, as shown in Fig. 18-91. This effect will continue up to a limit, at which point the capacity will be at a maximum unless a lower underflow solids concentration is acceptable, as illustrated in Fig. 18-87. Flocculant is usually added to a thickener in either the feed line or the feedwell, and there are a number of proprietary feedwell designs that are used in high-rate thickeners to optimize flocculation by introducing feed dilution liquor from the thickener overflow. The other components of these units are not materially different from those of a conventional thickener.

Ultrahigh-rate thickeners use a tall, deep tank with a steep bottom cone, and they may be used with or without a raking mechanism. It combines the functions of a thickener (to provide a dense underflow) and a clarifier (to provide a clear overflow or supernatant) but is considerably taller, and generally one-half to one-third the diameter of a conventional or high-rate thickener. Figure 18-92 illustrates the internals of these types of units, showing the use of dewatering cones, whose function is similar to that of the lamellar inclined plates of the tilted-plate thickeners.

High-density/paste thickeners can be designed to produce underflows having very high apparent viscosity and exhibiting a yield stress (Fig. 18-90), permitting the disposal of waste slurries at a concentration that avoids segregation of fines and coarse particles or the formation of a substantial free-liquid pond on the surface of the waste deposit. This practice is applied in *dry-stacking* systems and underground *paste-fill* operations for the disposal of mine tailings and similar materials. The thickener mechanism requires a special rake design and provides a torque capability much higher than normal for a particular diameter thickener (Fig. 18-93), together with a deeper sidewall and steeper floor slope. Underflow slurries will be at a higher concentration than for conventional or high-rate thickeners, typically being about 15 percent lower than vacuum filter cake from the same material. Shear thinning and special pumping requirements are necessary if the slurry is to be transported a significant distance, with line pressure drop typically in the range of 3 to 4 kPa/m of pipeline.

Design Features There are four mechanical classes of thickeners, each differentiated by its drive mechanism: (1) bridge-supported, (2) center-column supported, (3) peripheral-traction drives, and (4) without drives. The diameter of the tank will range from 2 to 150 m (6.5 to 492 ft), and the drive support structure is often determined by the diameter required. These classes are described in detail in the subsection Components and Accessories for Sedimentation Units, which covers components common to both thickeners and clarifiers.

Operation When operated correctly, thickeners require minimum attention and, if the feed characteristics do not change radically, they can be expected to maintain design performance consistently. In this regard, it is usually desirable to monitor feed and underflow rates and solids concentrations, flocculant dosage rate, and pulp interface level with dependable instrumentation systems. Process variations are then easily handled by changing the principal operating controls—underflow rate and flocculant dose—to maintain stability.

Starting up a thickener is usually the most difficult part of the operation, and there is more potential for mechanical damage to the mechanism at this stage than at any other time. In general, two conditions require special attention at this point: underflow pumping and mechanism torque. If possible, the underflow pump should be in operation as soon as feed enters the system, recirculating underflow slurry at a reduced rate, until the bed builds, or advancing it to the next process step (or disposal) if the feed contains a considerable quantity of coarse solids, such as more than 20 percent + 75 μm particles. At this stage of the operation, coarse solids separate from the pulp and produce a difficult raking and pumping situation. Torque can rise rapidly if this material accumulates faster than it is removed. If the torque reaches a point where the automatic control system raises the rakes, it is usually preferable to reduce or cut off the feed completely until the torque drops and the rakes are returned to the lowest position. As the fine fraction of the feed slurry begins to thicken and accumulate in the basin, providing both buoyancy and fluidity, torque will drop, and normal feeding can be continued. This applies whether the thickener tank is empty at start-up or filled with liquid. The latter approach contributes to coarse-solids raking problems but at the same time provides conditions more suited to good flocculation, with the result that the thickener will reach stable operation much sooner.

As the solids inventory in the thickener reaches an equilibrium level with underflow slurry at the desired concentration, the torque will reach a normal operating range. Special note should be made of the torque reading at this time. Torque levels that increase without any change in operating conditions can almost always be attributed to island formation, and corrective action can be taken early, before serious problems develop. *Island* is the name given to a mass of semi-solidified solids that have accumulated on or in front of the rakes, often as a result of excessive flocculant use. This mass will usually continue to grow, eventually producing a torque spike that can shut down the thickener, and often resulting in lower underflow densities than would otherwise be achievable.

An island is easily detected, usually by the higher-than-normal, gradually increasing torque reading. Probing the rake arms near the thickener center with a rigid rod will confirm this condition—the mass is easily distinguished by its cohesive, claylike consistency. At an early stage, the island is readily removed by raising the rakes until the torque drops to a minimum value. The rakes are then lowered gradually, a few centimeters at a time, so as to

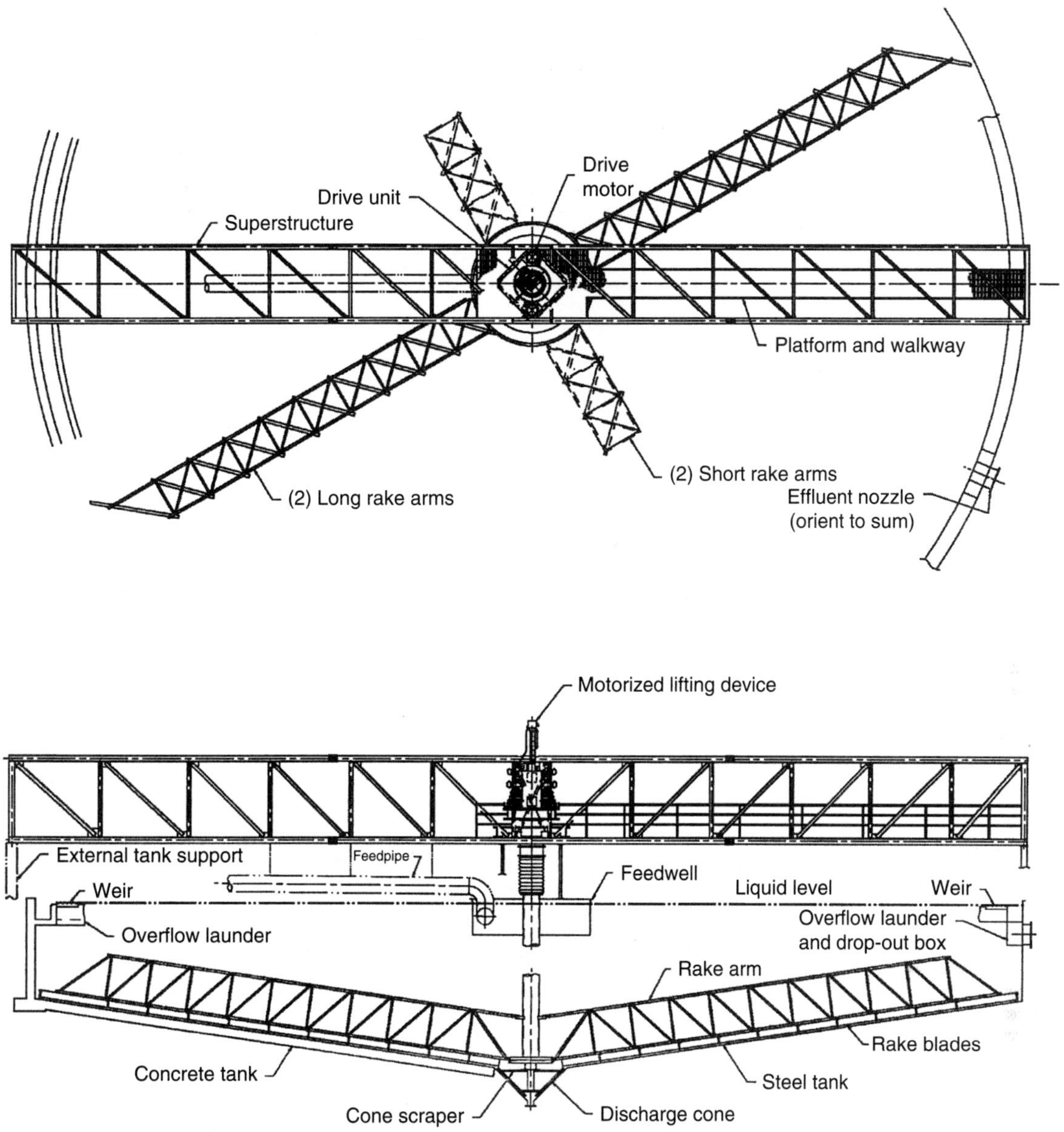

FIG. 18-90 Unit thickener bridge-mounted mechanism. (*FLSmidth.*)

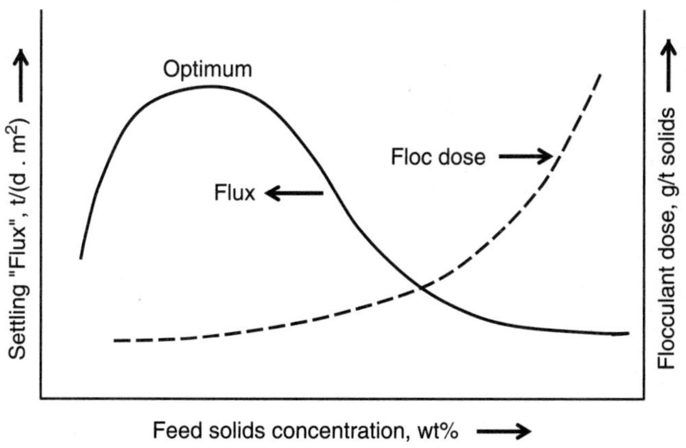

FIG. 18-91 Settling flux curve.

shave off the mass of solids and discharge this gelled material through the underflow. This operation can take several hours, and if island formation is a frequent occurrence, the procedure should be carried out on a regular basis, typically once a day, preferably with an automatic system to control the entire operation.

Stable thickener performance can be maintained by carefully monitoring operating conditions, particularly the pulp interface level and the underflow rate and concentration. As process changes occur, the pulp level can vary; regulation of the underflow pumping rate will keep the level within the desired range. If the underflow varies in concentration, this can be corrected by adjusting the flocculant dosage. Response will not be immediate, and care should be taken to make only small step changes at any one time. Procedures for use of automatic control are described in the section on instrumentation.

CLARIFIERS

Continuous clarifiers' primary purpose is to produce a relatively clear overflow; they are generally employed with dilute suspensions, principally industrial process streams and domestic municipal wastes. They are basically identical to thickeners in design and layout except that they use a mechanism of lighter construction and a drive head with a lower torque capability. These differences are permitted because the thickened pulp is

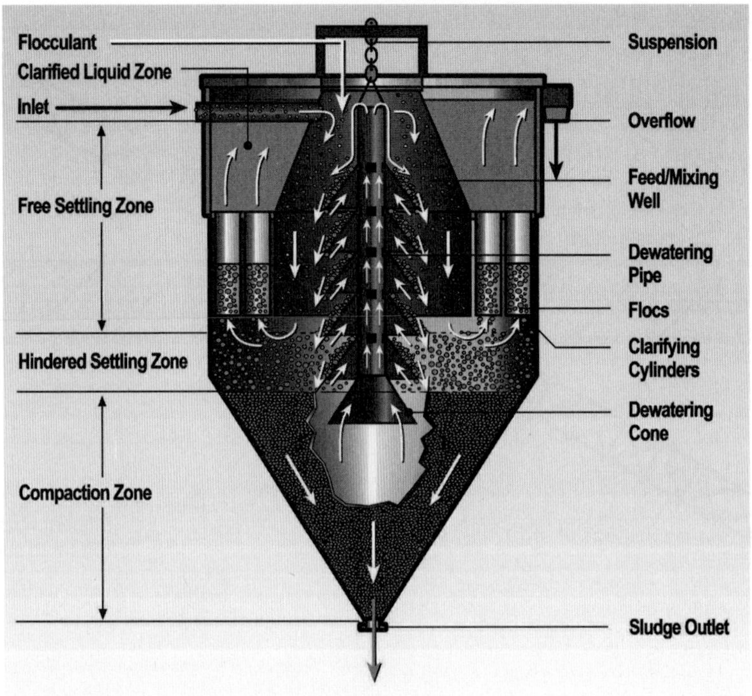

FIG. 18-92 Ultrahigh-rate thickener. (*FLSmidth.*)

smaller in volume and appreciably lower in suspended solids concentration, owing in part to the large percentage of relatively fine solids. The installed cost of a clarifier, therefore, is approximately 5 to 10 percent less than that of a thickener of equal tank size, as given in Fig. 18-98.

Rectangular Clarifiers Rectangular clarifiers are used primarily in municipal water and waste treatment plants, as well as in certain industrial waste plants. The raking mechanism employed in many designs consists of a chain-type drag, and suction systems are used for light-duty applications. The drag moves the deposited pulp to a sludge hopper located on one end by means of scrapers fixed to endless chains. During their return to the sludge raking position, the flights may travel near the water level and act as skimming devices for removing surface scum. Rectangular clarifiers are available in widths of 2 to 10 m (6 to 33 ft). The length is generally three to five times the width. The larger widths have multiple raking mechanisms, each with a separate drive.

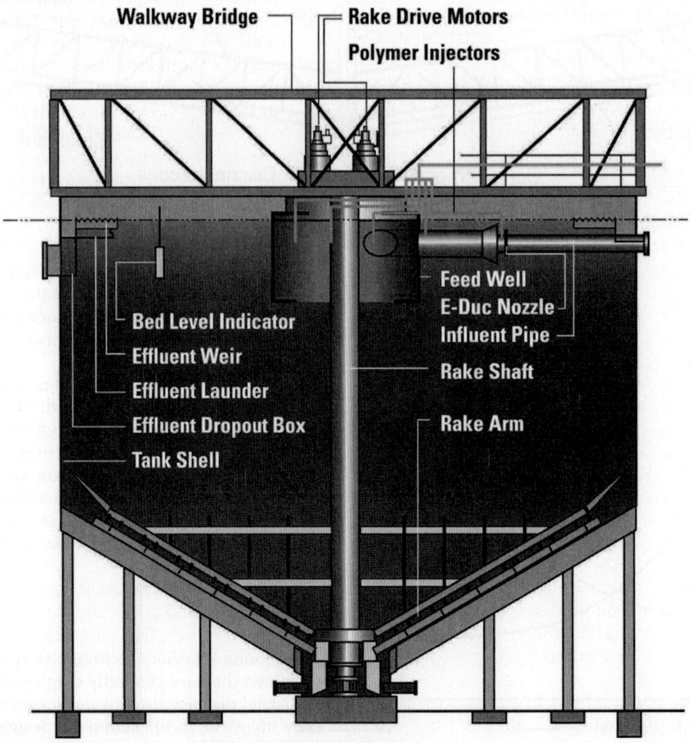

FIG. 18-93 Deep Cone™ paste thickener. (*FLSmidth.*)

This type of clarifier is used in applications such as preliminary oil-water separations in refineries and clarification of waste streams in steel mills. When multiple units are employed, common walls are possible, reducing construction costs and saving floor space. Overflow clarities are generally not as good as with circular clarifiers, due primarily to reduced overflow weir length for equivalent areas.

Circular Clarifiers Circular units are available in the same basic types as thickeners: bridge, center-column, and peripheral-traction. Because of economic considerations, the bridge-supported type is generally limited to tanks less than 40 m in diameter.

A circular clarifier is often equipped with a surface-skimming device, which includes a rotating skimmer, scum baffle, and scum-box assembly. In sewage and organic-waste applications, squeegees are normally provided for the rake-arm blades because it is desirable that the bottom be scraped clean to preclude the accumulation of organic solids with their resultant septicity and floating decomposing material.

Center-drive mechanisms are also installed in square tanks. This mechanism differs from the standard circular mechanism in that a hinged corner blade is provided to sweep the corners that lie outside the path of the main mechanism.

Clarifier-Thickener Clarifiers can serve as thickeners, achieving additional densification in a center, deep sludge sump that provides adequate retention time and pulp depth to compact the solids to a higher density. Drive mechanisms on this type of clarifier have higher torque capability than would be supplied on a standard clarifier.

Industrial Waste Secondary Clarifiers Many plants that formerly discharged organic wastes to the sewer have installed their own treatment facilities to reduce municipal treatment plant charges. For organic wastes, the waste-activated sludge process is a preferred approach, using an aeration basin for the bio-oxidation step and a secondary clarifier to produce a clear effluent and to concentrate the biomass for recycling to the basin. To produce an acceptable effluent and achieve sufficient concentration of the low-density solids that make up the biomass, certain design criteria must be followed. Typical design parameters include the following:

Feed pipe velocity: ≤ 1.2 m/s
Energy-dissipating feed entry velocity (tangential): ≤ 0.5 m/s
Downward velocity from feedwell: ≤ 0.5–0.75 (peak) m/min
Feedwell depth: entry port depth +1–3 m
Tank depth: typically 3–5 m
Radial velocity below feedwell: ≤ 90 percent of downward velocity

Overflow rate can range between 0.68 and 2.0 m/h, depending on the application. Consult an equipment supplier and a manual of practices for recommended overflow rates for specific applications.

Inclined-Plate Clarifiers Lamella or inclined-plate separators contain a multiplicity of plates inclined at 45° to 60° from the horizontal. Various feed designs are used so that the influent passes into each inclined channel. The geometry of the plates results in the solids having to settle only a short distance in each channel before sliding down the base to the collection zone beneath the plates. The clarified liquid passes in the opposite direction beneath the ceiling of each channel to the overflow connection.

The area that is theoretically available for separation is equal to the sum of the projected areas or all channels on the horizontal plane. Figure 18-94 shows the horizontally projected area A_S, of a single channel in a clarifier of unit width.

For a settling length L and width W, inclined at angle α, the horizontally projected area A_S can be calculated as

$$A_S = LW \cos \alpha \tag{18-56}$$

Multiply A_S by the total number of plates in the clarifier to calculate the total clarification area available. Plate angle α must be larger than the angle of repose of the sludge so that it will slide down the plate; the most common range is 55° to 60°. Plate spacing must be large enough to accommodate the opposite flows of liquid and sludge, reducing interference and preventing plugging, and to provide enough residence time for the solids to settle to the bottom plate. Usual X values are 50 to 75 mm (2 to 3 in).

Operating capacities range from 0.5 to 3.0 $m^3/h \cdot m^2$ (0.2 to 1.2 gpm/ft^2) projected horizontal area.

The principal advantage of the inclined-plate clarifier is the increased solids capacity per unit of plane area. Major disadvantages are an underflow solids concentration that is generally lower than in other gravity clarifiers and difficulty of cleaning when scaling or fouling occur. The design does not permit feed flocculation within the unit, and external mixers and tankage are required.

Ultrahigh-Rate (Rakeless) These thickeners use internal cones to achieve the inclined-plate effect. The design allows internal flocculation. The tank is tall, with a 60° bottom cone, providing sludge compression height and volume, resulting in a higher-density underflow.

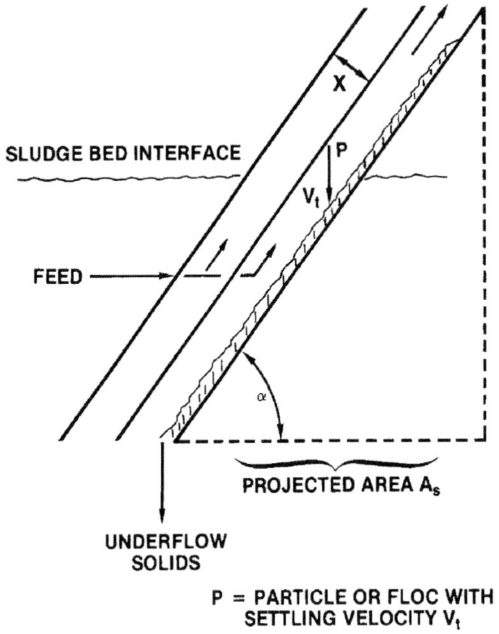

FIG. 18-94 Basic concept of the inclined-plate type of clarifier.

Solids-Contact Clarifiers These are used to allow chemical treatment and sedimentation in a single tank. Those employing mechanically assisted mixing in the reaction zone are the most efficient. They generally permit the highest overflow rate at a minimum chemical dosage while producing the best effluent quality. The unit illustrated in Fig. 18-95*a* consists of a combination dual drive that has a low-speed rake mechanism and a high-rate, low-shear turbine located in the top portion of the center well for internal solids recirculation. The influent, dosed with chemicals, is contacted with previously settled solids in a recirculation draft tube within the reaction well by means of the pumping action of the turbine. Owing to the higher concentration of solids being recirculated, chemical reactions are more rapid, and flocculation is improved. Outside of the reaction well the flocculated particles settle to the bottom and are raked to the center to be used again in the recirculation process. When particles are too heavy to be circulated up through the draft tube (as in the case of metallurgical pulps), a modified design (see Fig. 18-95*b*) using external recirculation of a portion of the thickened underflow is chosen. These units employ a special mixing impeller in a reaction well with a controlled outlet.

Solids-contact clarifiers are advantageous for clarifying turbid waters or slurries that require coagulation and flocculation for the removal of bacteria, suspended solids, or color. Applications include lime softening of water, clarifying industrial-process streams and industrial wastewaters; tertiary treatment for removal of phosphates, BOD_5, and turbidity; and, silica removal from produce water, cooling tower makeup, and geothermal brines.

COMPONENTS AND ACCESSORIES FOR SEDIMENTATION UNITS

Sedimentation systems consist of a collection of components, each of which can be supplied in a number of variations. The variations selected are based on the application, its sedimentation characteristics, and the desired performance. The basic components are the same for thickeners or clarifiers: tank, drive-support structure, drive unit and lifting device, rake structure, feedwell, overflow arrangement, underflow arrangement, instrumentation, and flocculation facilities.

Tanks Tanks or basins are constructed of materials such as steel, concrete, wood, compacted earth, plastic sheeting, and soil cement. The selection of the materials of construction is based on cost, availability, topography, water table, ground conditions, climate, operating temperature, and chemical-corrosion resistance. Typically, industrial tanks up to 45 m (150 ft) in diameter are made of steel. Concrete is generally used in municipal and large industrial applications. Extremely large units employing earthen basins with impermeable liners have proved to be economical.

The rakeless ultrahigh-rate thickeners use elevated tanks up to 12 m in diameter. Advantages are no drive, high throughput rate, and the small footprint. Disadvantages are the height of the elevated tank.

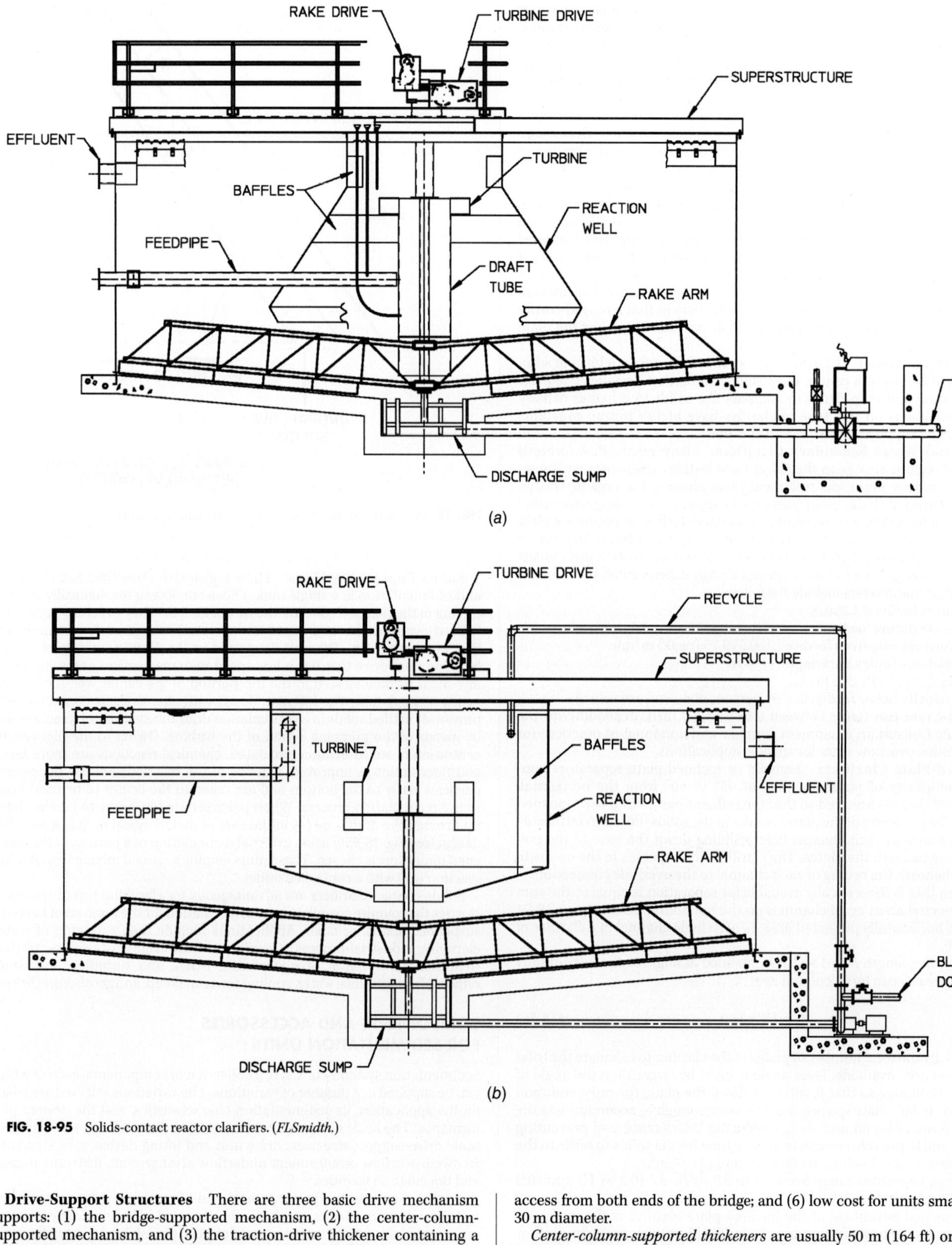

FIG. 18-95 Solids-contact reactor clarifiers. (*FLSmidth.*)

Drive-Support Structures There are three basic drive mechanism supports: (1) the bridge-supported mechanism, (2) the center-column-supported mechanism, and (3) the traction-drive thickener containing a center-column-supported mechanism with the driving arm attached to a motorized carriage at the tank periphery.

Bridge-supported thickeners (Fig. 18-90) are common in diameters up to 30 m, the maximum being about 45 m (150 ft). They offer advantages over a center-column-supported design: (1) ability to transfer loads to the tank periphery; (2) ability to give a denser and more consistent underflow concentration with the single draw-off point; (3) a less complicated lifting device; (4) fewer structural members subject to mud accumulation; (5) maintenance access from both ends of the bridge; and (6) low cost for units smaller than 30 m diameter.

Center-column-supported thickeners are usually 50 m (164 ft) or more in diameter. The mechanism is supported by a stationary steel or concrete center column, and the raking arms are attached to a driving cage that rotates around the center column.

Traction thickeners are most adaptable to tanks larger than 60 m (200 ft) in diameter. Maintenance is generally less difficult than with other types of thickeners. The drive may be supported on the concrete wall (the wall would be a structural member) or supported outside the wall on the ground (a standard tank wall could be used). Disadvantages of the traction thickener are that (1) no practical lifting device can be used, (2) operation may be difficult in climates where snow and ice can accumulate on the traction drive rail, causing loss of friction (rail sanders and heat cables can be applied to attempt to minimize friction loss), and (3) the driving-torque effort must be transmitted from the tank periphery to the center, where the heaviest raking conditions occur.

Drive Assemblies The drive assembly is the key component of a sedimentation unit. The drive assembly provides (1) the force to move the rakes through the thickened pulp and to move settled solids to the point of discharge, (2) the support for the mechanism that permits it to rotate, (3) adequate reserve capacity to withstand upsets and temporary overloads, and (4) a reliable control that protects the mechanism from damage when a major overload occurs.

Drives usually have steel or iron main spur gears mounted on bearings, alloy-steel pinions, or a planetary gear. Direct-drive hydraulic systems are also used. The gearing components are preferably enclosed for maximum service life. The drive typically includes a torque-measuring system with torque indicated on the mechanism and often transmitted to a remote indicator. If the torque becomes excessive, it can automatically activate safeguards against structural damage such as sounding an alarm, raising the rakes, and stopping the drive.

Rake-lifting mechanisms should be provided when abnormal thickener operation is probable. Abnormal thickener operation or excessive torque may result from insufficient underflow pumping, surges in the solids feed rate, excessive amounts of large particles, sloughing of solids accumulated between the rakes and the bottom of the tank or on structural members of the rake mechanism, or miscellaneous obstructions falling into the thickener. The lifting mechanism may be set to raise the rakes automatically when a specific torque level (e.g., 40 percent of design) is encountered, continuing to lift until the torque returns to normal or until the maximum lift height is reached. Generally, corrective action must be taken to eliminate the cause of the upset. Once the torque returns to normal, the rake mechanism is lowered slowly to plow gradually through the excess accumulated solids until these are removed from the tank.

Motorized rake-lifting devices are designed to allow for a vertical lift of the rake mechanism of up to 90 cm (3 ft).

The cable arm design uses cables attached to a truss above or near the liquid surface to move the rake arms, which are hinged to the drive structure, allowing the rakes to lift when excessive torque is encountered. An advantage of this design is the relatively small surface area of the raking mechanism, which reduces the solids accumulation and downtime in applications in which scaling can occur. The disadvantages of the hinged-arm self-lifting design are that there is very little lift at the center, where the overload usually occurs, and the difficulty of returning the rakes to the lowered position in settlers containing solids that compact firmly.

The *rake mechanism* assists in moving the settled solids to the point of discharge. It also aids in thickening the pulp by disrupting bridged floccules, permitting trapped fluid to escape and allowing the floccules to become more consolidated. Rake mechanisms are designed for specific applications, usually having two long rake arms with an option for two short rake arms for bridge-supported and center-column-supported units. Traction units usually have one long arm, two short arms, and one intermediate arm.

Figure 18-96 illustrates types of rake-arm designs. The conventional design is typically used in bridge-supported units, while the dual-slope design is used for units of larger diameter.

Rake blades can have attached spikes or serrated bottoms to cut into solids that have a tendency to compact. Lifting devices typically are used with these applications.

Rake-speed requirements depend on the types of solids being thickened. Peripheral speed ranges used are: for slow-settling solids, 3 to 8 m/min (10 to 25 ft/min); for fast-settling solids, 8 to 12 m/min (25 to 40 ft/min); for coarse solids or crystalline materials, 12 to 30 m/min (40 to 100 ft/min).

Feedwell The feedwell is designed to allow the feed to enter the thickener with minimum turbulence and uniform distribution while dissipating most of its kinetic energy. Feed slurry enters the feedwell, usually located in the center of the thickener, through a pipe or launder suspended from the bridge. To avoid excess velocity, an open launder normally has a slope no greater than 1 to 2 percent. Pulp should enter the launder at a velocity that prevents sanding at the inlet. With nonsanding pulps, the feed may also enter upward through the center column from a pipeline installed beneath the tank.

The standard feedwell for a thickener is designed for a maximum vertical outlet velocity of about 1.5 m/min (5 ft/min). High turbidity caused by short-circuiting the feed to the overflow can be reduced by increasing the depth of the feedwell. When overflow clarity is important or the solids specific gravity is close to the liquid specific gravity, deep feedwells of large diameter are used, and they are designed to reduce the velocity of the entering feed slurry.

Shallow feedwells may be used when overflow clarity is not important, the overflow rate is low, or solids density is appreciably greater than that of water.

When flocculants are used, the optimum feed solids concentration for flocculation may be considerably less than the normal concentration, and significant savings in reagent cost will be made possible by dilution of the feed prior to flocculation. This can be achieved by recycling overflow or,

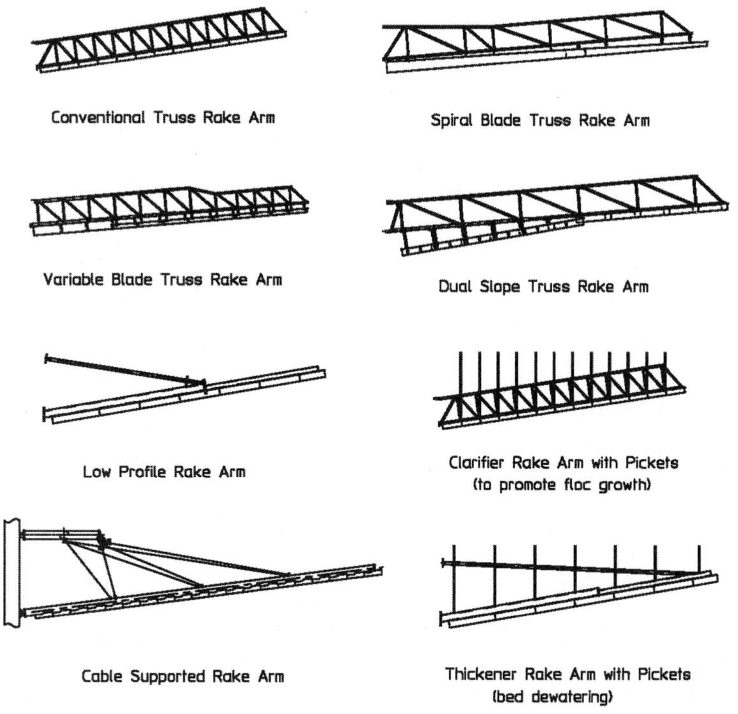

FIG. 18-96 Rake-mechanism designs for specific applications and duties. (*FLSmidth.*)

FIG. 18-97 E-Duc® Feed dilution system installed on a 122-m-diameter thickener. (*FLSmidth.*)

more efficiently, by feedwell designs that provide for feed dilution internal to the tank. One design for achieving the dilution prior to flocculant addition is pictured in Fig. 18-97. This design uses the energy available in the incoming feed stream to achieve the dilution by momentum transfer and requires no additional energy expenditure to dilute the slurry by as much as three to four times.

Overflow Arrangements Clarified effluent typically is removed in a peripheral launder located inside or outside the tank. The effluent enters the launder by overflowing a V-notch or level flat weir, or through submerged orifices in the bottom of the launder. Uneven overflow rates caused by wind blowing across the liquid surface in large thickeners can be better controlled when submerged orifices or V-notch weirs are used. Radial launders are used when uniform upward liquid flow is desired in order to improve clarifier detention efficiency. This arrangement provides an additional benefit in reducing the effect of wind, which can seriously impair clarity in applications that employ basins of large diameter.

The hydraulic capacity of a launder must be sufficient to prevent flooding, which can cause short-circuiting of the feed and deterioration of overflow clarity. Standards are occasionally imposed on weir overflow rates for clarifiers used in municipal applications; typical rates are 3.5 to 15 $m^3/(h \cdot m)$ [7000 to 30,000 gal/(day · ft)], and they are highly dependent on clarifier sidewater depth. Industrial clarifiers may have higher overflow rates, depending on the application and the desired overflow clarity. Launders can be arranged in a variety of configurations to achieve the desired overflow rate. Several alternatives to improve clarity include an annular launder inside the tank (the liquid overflows both sides), radial launders connected to the peripheral launder (providing the very long weir that may be needed when abnormally high overflow rates are encountered and overflow clarity is important), and Stamford baffles, which are located below the launder to direct flow currents back toward the center of the clarifier.

In many thickener applications, on the other hand, complete peripheral launders are not required, and no difference in either overflow clarity or underflow concentration will result through the use of launders extending over only a fraction (e.g., one-fifth) of the perimeter. For design purposes, a weir-loading rate in the range of 7.5 to 30.0 $m^3/(h \cdot m)$ [10 to 40 gpm/ft] can be used, the higher values being employed with well-flocculated, rapidly settling slurries. The overflow launder required may occupy only a single section of the perimeter rather than consisting of multiple, shorter segments spaced uniformly around the tank.

Underflow Arrangements Concentrated solids are removed from the thickener by the use of centrifugal or positive displacement pumps or, particularly with large-volume flows, by gravity discharge through a flow control valve or orifice suitable for slurry applications. Due to the risk to the thickener operation of a plugged underflow pipe, it is recommended that duplicate underflow pipes and pumps be installed in all thickening applications. Provision to recycle underflow slurry back to the feedwell or position near the tank wall and floor intersection (tank knuckle) is also useful, particularly if solids are to be stored temporarily in the thickener as capacity and torque allow. With the advent of high-density/paste thickening, it has also become common to use pumps under the thickener to recirculate the high-yield-stress thickened underflow back into the discharge cone, cylinder, or other external mixing vessel in order to reduce the yield stress before it is pumped to the next stage. These pumps can also be used to recirculate the underflow back through the mud bed, reinjecting at the tank knuckle, while the thickener is not in operation and the mud bed has not been pumped out.

There are three basic underflow arrangements: (1) the underflow pump adjacent to the thickener sidewall with buried piping from the discharge cone, (2) the underflow pump under the thickeners at the discharge cone or adjacent to the sidewall with the piping from the discharge cone in a tunnel, and (3) the underflow pump located in the center of the thickener on the bridge, or using piping up through the center column.

The *pump adjacent to thickener with buried piping* arrangement of buried piping from the discharge cone is the least expensive system but the most susceptible to plugging. It is used only when the solids do not compact to an unpumpable slurry and can be easily backflushed if plugging occurs. Typically, two or more underflow pipes are installed from the discharge cone to the underflow pump so that solids removal can continue if one of the lines plugs. Valves should be installed to permit flushing with water and compressed air in both directions to remove blockages.

A *tunnel* may be constructed under the thickener to provide access to the discharge cone when underflow slurries are difficult to pump and have characteristics that cause plugging. The underflow pump may be installed underneath the thickener or at the perimeter. In many cases, thickeners are installed on legs or piers, making tunneling for access to the center unnecessary. A tunnel or an elevated thickener is more expensive than the other underflow arrangements, but there are certain operational and maintenance advantages. The hazards of working in a tunnel (flooding and interrupted ventilation, for example) and related safety regulations must be considered.

A *center-column pumping* arrangement may be used instead of a tunnel. Several designs are available. One is a bridge-mounted pump with a suction line through a wet or dry center column. The pump selection may be limiting, requiring special attention to priming, net positive suction head, and the maximum density that the pump can handle. Another design has the underflow pump located in a room under the thickener mechanism and connected to openings in the column. Access is through the drive gear at the top of the column.

INSTRUMENTATION

The following types of devices are commonly applied to measure the various operational parameters of thickeners and clarifiers. They have been used in conjunction with automatic valves and variable-speed pumps to achieve automatic operation as well as to simply provide local or remote indications.

Torque Rake torque is an indication of the force necessary to rotate the rakes. Higher rake torque is an indication of higher underflow density or viscosity, deeper mud bed, higher fraction of coarse material, island formation, or heavy scale buildup on the rake arms.

Rake torque measurement is usually provided by the thickener manufacturer as part of the rake drive mechanism. Typical methods involve load cells, motor power measurement, hydraulic pressure, or mechanical displacement against a spring. Torque-measuring devices are designed to produce a signal that may be used for alarming or control. At a minimum, they are used to prevent mechanical damage to the rake structure by stopping the drive at high torque levels. In practice, the torque signal is used extensively to help prevent bogging of the rakes and subsequent shutdown.

Rake Height Rake lifting devices are used to minimize the torque on the arms by lifting them out of heavy bed solids and to enable the rake to continue running during upset conditions. Rake drives should be prevented from running for extended periods at torques above 50 to 60 percent to prevent accelerated wear on the drive. Lifting the rakes a small distance is usually effective in reducing the torque. Because of this, in using "torque indication" in a control strategy, one must also consider the rake height to effectively control the thickener. The two most common rake height indicators are the ultrasonic and the potentiometer type with a reeling cable. Lifting of the rakes allows a short period of time to make corrections before one is forced to shut down the thickener.

Bed Level There are several general types of bed level detection instruments, and each has advantages and disadvantages, which are discussed next. There is no standard bed level sensor that is recommended for all applications.

- *Ultrasonic* bed level sensors work by sending a pulse down from just under the overflow surface, which bounces off the bed and back to the receiver. Elapsed time is used to calculate the distance. Advantages are noninterfering location, measurement over a large span, and relatively low cost. The downside is that they do not work on all applications. If the overflow is cloudy, it can interfere with the transmission or cause too much reflection to give a reliable signal. Scaling affects accuracy and can cause drifting or loss of signal. Applications in which froth can form on the liquid surface have proved to be particularly troublesome.
- *Nuclear* bed level sensors work by sensing either the background radiation level or attenuation between a source and a detector, depending on whether the solids have a natural background radiation level. The sensor is comprised of a long rod that extends down into the bed with radiation detectors spaced along the length. If mineral ore changes from not having radiation to having it, there will be problems. The advantages are that it is relatively reliable when properly applied. The downside is that it measures over a limited range, may interfere with rakes (a hinged version that will swing out of the way when the rakes pass by is available), and is relatively expensive.
- *Float and rod* types work with a ball with a hollow sleeve that slides up and down on a rod that extends down into the bed. The ball weight can be adjusted to float on top of the bed of solids. These are subject to fouling and sticking, and they can be installed and measured only in the area above the rakes; however, they are relatively inexpensive.
- *Reeling devices* work by dropping a sensor down on a cable and sensing the bed level by optical, conductivity, or point ultrasonic sensors. In theory they are nonfouling and retract as the rakes approach, but in practice, they have often become entangled with the rakes. The price is midrange to high end, depending on the design. Freezing wind and cold temperatures can lead to icing problems.
- *Vibrating or tuning fork sensors* are designed to sense a difference in the vibrating frequency in different masses of solids.
- *Bubble tube or differential pressure* is an old, but tried and true, method of bed level detection. There may be some plugging or fouling of the tube over time.
- *External density* through sample ports: slurry samples are taken from nozzles on the side of the tank and pass through a density meter to determine the presence of solids. This system can be set up with automated valves to measure several different sample points. This system requires external piping and disposal of the sample stream. Line pluggage is often a problem.

Bed Pressure Because thickeners maintain a constant liquid level, the pressure at the bottom of the thickener is an indication of the overall specific gravity in the tank. If the liquor specific gravity is constant, the overall specific gravity is an indication of the amount of solids in the tank and can be converted to a rough solids inventory. This can be a very effective tool for thickener control because it is relatively low cost and highly reliable. Because of relative height-to-diameter ratios, it is considered less useful for large-diameter thickeners, but many newer large thickeners are equipped with them.

Differential pressure sensors are used to measure the bed pressure, leaving one leg open to the atmosphere to compensate for barometric pressure variations. Care must be taken in the installation to minimize plugging with solids. This is normally done by tilting the tank nozzle on which the DP cell is mounted downward from the sensor, so the solids tend to settle away from the sensor surface. A shutoff valve and washer flush tap are recommended to allow easy maintenance.

Flow Rate Flow rates of feed and underflow lines are useful, particularly when combined with density measurements to generate solids mass flow rates. Since flocculant is usually dosed on a solids mass basis, knowing the feed mass flow rate is very useful for flocculant control, providing a fast response system. Flow rate measurement is an absolute necessity for the newer generation of ultrahigh-rate and high density/paste thickeners. The streams being measured are usually slurries, and the flow rate is usually measured by either magnetic flowmeters or Doppler-type flowmeters. As long as these instruments are properly installed in suitable full straight-pipe sections, avoiding air if possible, they are accurate and reliable. If the feed stream is in an open launder, flow measurement is more difficult, but it can be accomplished using ultrasonic devices.

Density Nuclear gauges are the norm for density measurement. Nuclear density instruments require handling permits in most countries, and there are now some types that use very low-level sources that do not require nuclear licensing. Density gauges should be recalibrated regularly because they are subject to drifting. Small flow applications may be able to use a Coriolis meter to measure both mass flow and percent solids with one instrument.

Settling Rate The settling rate of the flocculated solids in the feedwell is a good indication of the degree of flocculation, and it can be used to maintain consistent flocculation over widely varying feed conditions. A settleometer is a device that automatically pulls a sample from the feedwell and measures the settling rate. The flocculant can then be adjusted to maintain a constant settling rate.

Overflow Turbidity Overflow turbidity can be used as feedback to control flocculant or coagulant. There is generally some significant lag time between the actual flocculation process and when the clarified liquor reaches the overflow discharge point where the sensor is typically positioned. These sensors and meters are generally used as alarms or for trim only.

CONTROL STRATEGIES

Thickeners Thickener control philosophies are usually based on the idea that the underflow density obtained is the most important performance criterion. The overflow clarity is also a consideration, but this is generally not as critical. Additional factors that must be considered are optimization of flocculant usage and protection of the raking mechanism.

Automated control schemes employ one or more sets of controls, which will fit into three categories: (1) control loops that are used to regulate the addition of flocculant, (2) control loops to regulate the withdrawal of underflow, and (3) rake drive controls. Often, the feed to a thickener is not controlled, and most control systems have been designed with some flexibility to deal with changes in feed characteristics.

Flocculant addition rate can be regulated in proportion to the thickener volumetric feed rate or solids mass flow in a feed-forward mode, or in a feedback mode on either rake torque, underflow density, settling solids (sludge) bed level, or solids settling rate.

Underflow is usually withdrawn continuously on the basis of bed mass, bed level, rake torque, or underflow solids concentration in a feedback mode. Some installations incorporate two or more of these parameters in their underflow withdrawal control philosophy. For example, the continuous withdrawal may be based on underflow solids density with an override to increase the withdrawal rate if either the rake torque or the bed level reaches a preset value. In some cases, underflow withdrawal has been regulated in a feed-forward mode on the basis of thickener feed solids mass flow rate. Any automated underflow pumping scheme should incorporate a lower limit on volumetric flow rate as a safeguard against line pluggage.

It is also important to consider the level of the sludge bed in the thickener. Although this can be allowed to increase or decrease within moderate limits, it must be controlled enough to prevent solids from overflowing the thickener or from falling so low that the underflow density becomes too dilute. The settling slurry within the sludge bed is normally free flowing and will disperse to a consistent level across the thickener diameter.

Rake drive controls protect the drive mechanism from damage and usually incorporate an alarm to indicate high torque with an interlock to shut down the drive at a higher torque level. Rake lift mechanisms are normally set up to raise the rakes above a torque setpoint to help limit the torque level. Lowering of the rakes can be either automatic at lower torque levels or done manually.

A complete automated control scheme incorporates controls from each of the three categories. It is important to consider the interaction of the various controls, especially of the flocculant addition and underflow withdrawal control loops, when designing a system. The lag and dead times of any feedback loops as well as the actual response of the system to changes in manipulated variables must be considered. For example, in some applications, it is possible that excessive flocculant addition may produce an increase in the rake torque (due to island formation or viscosity increase) without a corresponding increase in underflow density. Additionally, sludge bed level sensors generally require periodic cleaning to produce a reliable signal. In many cases, it has not been possible to effectively maintain the sludge bed level sensors, requiring a change in the thickener control logic after start-up. Some manufacturers offer complete thickener control packages.

Clarifiers Control philosophies for clarifiers are based on the premise that the overflow is the most important performance criterion. Underflow density or suspended solids content is a consideration, as is the optimal use of flocculation and pH control reagents. Automated controls are of three basic types: (1) control loops that optimize coagulant, flocculant, and pH control reagent additions; (2) those that regulate underflow removal; and (3) rake drive controls. Equalization of the feed is provided in some installations, but the clarifier feed is usually not a controlled variable with respect to the clarifier operation.

Automated controls for flocculating reagents can use a feed-forward mode based on feed turbidity and feed volumetric rate, or a feedback mode incorporating a streaming current detector on the flocculated feed. Attempts to control coagulant addition on the basis of overflow turbidity generally have been less successful. Control for pH has been accomplished by feed-forward modes on the feed pH and by feedback modes on the basis of clarifier feedwell, reaction well, or external reaction tank pH. Control loops based on the measurement of reaction well pH are useful for control in applications in which flocculated solids are internally recirculated within the clarifier reaction well.

Automated sludge withdrawal controls are usually based on the sludge bed level or pressure. These can operate in on-off or continuous modes and can use either single-point or continuous sludge level indication sensors. In many applications, automated control of underflow withdrawal does not provide an advantage, since so few settled solids are produced that it is only necessary to remove sludge for a short interval once a day or less often. In applications in which the underflow is recirculated internally within the feedwell, it is necessary to maintain enough sludge inventory to properly feed the recirculation turbine. This can be handled in an automated system with a single-point *low sludge* bed level sensor in conjunction with a low-level alarm or pump shutoff solenoid. Some applications require continuous external recirculation of the underflow direct to the feedwell or external reaction tanks, and an automated control loop can be used to maintain recirculation based on flow measurement, with a manually adjusted setpoint.

Control philosophies applied to continuous countercurrent decantation (CCD) thickeners are similar to those used for thickeners in other applications, but they emphasize keeping the CCD circuit in balance. It is important to prevent any one of the thickeners from pumping out too fast, otherwise an upstream unit could be starved of wash liquor while at the same time too much underflow could be placed in a downstream unit, disrupting the operation of both units as well as reducing the circuit washing efficiency. Several control configurations have been attempted, and the more successful schemes have linked the solids mass flow rate of underflow pumping to that of the upstream unit or to the CCD circuit solids mass feed rate. Wide variations in the solids feed rate to a CCD circuit will require some means of dampening these fluctuations if design wash efficiency is to be maintained.

CONTINUOUS COUNTERCURRENT DECANTATION

The system of separation or washing of solid-phase material from an associated solution by repeated stages of dilution and gravity sedimentation is adapted for many industrial-processing applications through an operation known as continuous countercurrent decantation (CCD). The flow of solids proceeds in a direction countercurrent to the flow of solution diluent (water, usually), with each stage composed of a mixing step followed by settling of the solids from the suspension. The number of stages ranges from two to as many as 10, depending on the degree of separation required, the amount of wash fluid added (which influences the final solute concentration in the first-stage overflow), and the underflow solids concentration attainable. Applications include processes in which the solution is the valuable component (as in alumina extraction), or in which purified solids are sought (magnesium hydroxide from seawater), or both (as often encountered in the chemical-processing industry and in base-metal hydrometallurgy).

The factors that may make CCD a preferred choice over other separation systems include: rapidly settling solids, assisted by flocculation; a relatively high ratio of solids concentration between underflow and feed; moderately high wash ratios allowable (2 to 4 times the volume of liquor in the thickened underflows); a large quantity of solids to be processed; and the presence of fine-size solids that are difficult to concentrate by other means. A technical feasibility and economic study is useful in making the best choice.

Flow-Sheet Design Thickener-sizing tests, as described earlier, will determine unit areas, flocculant dosages, and underflow densities for the various stages. For most cases, unit areas will not vary significantly throughout the circuit; similarly, underflow concentrations should be relatively constant. In practice, the same unit area is generally used for all thickeners in the circuit to simplify construction. Serious consideration should be given to the design underflow density since operating at the higher, manageable densities will offer the benefits of improved wash efficiency. Many CCD installations, alumina red mud washing in particular, have installed paste thickeners and reduced the number of stages or lowered the required volume of wash water.

Equipment The equipment selected for CCD circuits may consist of multiple-compartment washing-tray thickeners or a train of individual unit thickeners. The washing-tray thickener consists of a vertical array of coaxial trays connected in series, contained in a single tank. The advantages of this design are smaller floor-area requirements, less pumping equipment and piping, and reduced heat losses in circuits operating at elevated temperatures. However, operation is generally more difficult, and user preference has shifted toward high-rate, ultrahigh-rate and high-density/paste thickeners.

Underflow Pumping Diaphragm pumps with open discharge are employed in some low-volume cases, primarily because underflow densities are readily controlled with these units. Disadvantages include the generally higher maintenance and initial costs than for other types and their inability to transfer the slurry any great distance. Large flows often are best handled with variable-speed, rubber-lined centrifugal pumps, using automatic control to maintain the underflow rate and density.

Overflow Pumps These can be omitted if the thickeners are located at increasing elevations from first to last so that overflows are transferred by gravity or if the mixture of underflow and overflow proceeding to the next CCD unit is to be pumped. Overflow pumps are necessary, however, when maximum flexibility and control are sought.

Interstage Mixing Efficiencies Mixing or stage efficiencies rarely achieve the ideal 100 percent, in which solute concentrations in overflow and underflow liquor from each thickener are identical. Part of the deficiency is due to insufficient blending of the two streams, and attaining equilibrium will be hampered further by heavily flocculated solids. In systems in which flocculants are used, interstage efficiencies often will drop gradually from first to last thickener, and typical values will range from 98 percent to as low as 70 percent. In some cases, operators will add the flocculant to an overflow solution that is to be blended with the corresponding underflow. While this is very effective for good flocculation, it can result in reflocculation of the solids before the entrained liquor has had a chance to blend completely with the overflow liquor. The preferable procedure is to recycle a portion of the overflow back to the feed line of the same thickener, adding the reagent to this liquor.

The usual method of interstage mixing consists of a relatively simple arrangement in which the flows from preceding and succeeding stages are added to a feed box at the thickener periphery. A nominal detention time in this mixing tank of 30 to 60 s and sufficient energy input to avoid solids settling will ensure interstage efficiencies greater than 95 percent.

The performance of a CCD circuit can be estimated through the use of the following equations, which assume 100 percent stage efficiency:

$$R = \frac{O/U[O/U]^N - 1]}{[(O/U)^{N+1} -]} \qquad (18\text{-}57)$$

$$R = 1 - \left(\frac{U}{O'}\right)^N \qquad (18\text{-}58)$$

For O/U and $U/O' \neq 1$, R is the fraction of dissolved value in the feed that is recovered in the overflow liquor from the first thickener, O and U are the overflow and underflow liquor volumes per unit weight of underflow solids, and N is the number of stages. Equation (18-58) applies to a system in which the circuit receives dry solids with which the second-stage thickener overflow is mixed to extract the soluble component. In this instance, O' refers to the overflow volume from the thickeners following the first stage.

For more precise values, computer programs can be used to calculate soluble recovery as well as solution compositions for conditions that are typical of a CCD circuit, with varying underflow concentrations, stage efficiencies, and solution densities in each of the stages. The calculation sequence is easily performed by using material-balance equations around each thickener.

DESIGN SIZING CRITERIA

Table 18-12 has the typical design sizing criteria and operating conditions for a number of applications. It is presented for purposes of illustration or preliminary estimate. Actual thickening performance is dependent on particle-size distribution, specific gravity, sludge bed compaction characteristics, and other factors. Final design should be based on bench scale or pilot tests.

THICKENER COSTS

Equipment Costs vary widely for a given diameter because of the many types of construction. Figure 18-98 shows the approximate installed costs of thickeners up to 100 m (330 ft) in diameter. These costs are to be used

TABLE 18-12 Typical Thickener and Clarifier Design Criteria and Operating Conditions

	Percent solids		Unit area,	Overflow rate,
	Feed	Underflow	$m^2/(t/d)$*[†]	$m^3/(m^2 \cdot h)$*
Alumina Bayer process				
Red mud, primary	3–4	10–25	2–5	
Red mud, washers	6–8	15–35	1–4	
Hydrate, fine or seed	1–10	20–50	1.2–3	0.07–0.12
Brine purification	0.2–2.0	8–15		0.5–1.2
Coal, refuse	0.5–6	20–40	0.5–1	0.7–1.7
Coal, heavy-media (magnetic)	20–30	60–70	0.05–0.1	
Cyanide, leached-ore	16–33	40–60	0.3–1.3	
Flue dust, blast-furnace	0.2–2.0	40–60		1.5–3.7
Flue dust, BOF	0.2–2.0	30–70		1–3.7
Flue-gas desulfurization sludge	3–12	20–45	0.3–3[†]	
High-density paste thickeners				
Red mud, washers[‡]	3–4	45–50	0.05–0.08	
Coal, refuse[‡]	6–8	50–54	0.08–0.1	
Cyanide, leached ore[‡]	10–15	65–70	0.05–0.08	
Copper tailings[‡]	10–20	50–75	0.07–0.15	
Tailings (magnetic)[‡]	10–20	60–75	0.07–0.1	
Tailings (nonmagnetic)[‡]	10–20	60–70	0.07–0.1	
Magnesium hydroxide from brine	8–10	25–40	5–10	
Magnesium hydroxide from seawater	1–4	15–20	3–10	0.5–0.8
Metallurgical				
Copper concentrates	14–50	40–75	0.2–2	
Copper tailings	10–30	45–65	0.4–1	
Iron ore				
Concentrate (magnetic)	20–35	50–70	0.01–0.08	
Concentrate (nonmagnetic), coarse: 40–65% −325	25–40	60–75	0.02–0.1	
Concentrate (nonmagnetic), fine: 65–100% −325	15–30	60–70	0.15–0.4	
Tailings (magnetic)	2–5	45–60	0.6–1.5	1.2–2.4
Tailings (nonmagnetic)	2–10	45–50	0.8–3	0.7–1.2
Lead concentrates	20–25	60–80	0.5–1	
Molybdenum concentrates	10–35	50–60	0.2–0.4	
Nickel, $(NH_4)_2CO_3$ leach residue	15–25	45–60	0.3–0.5	
Nickel, acid leach residue	20	60	0.8	
Zinc concentrates	10–20	50–60	0.3–0.7	
Zinc leach residue	5–10	25–40	0.8–1.5	
Municipal waste				
Primary clarifier	0.02–0.05	0.5–1.5		1–1.7
Thickening				
Primary sludge	1–3	5–10	8	
Waste-activated sludge	0.2–1.5	2–3	33	
Anaerobically digested sludge	4–8	6–12	10	
Phosphate slimes	1–3	5–15	1.2–18	
Pickle liquor and rinse water	1–8	9–18	3.5–5	
Plating waste	2–5	5–30		1.2
Potash slimes	1–5	6–25	4–12	
Potato-processing waste	0.3–0.5	5–6		1
Pulp and paper				
Green-liquor clarifier	0.2	5		0.8
White-liquor clarifier	8	35–45	0.8–1.6	
Kraft waste	0.01–0.05	2–5		0.8–1.2
Deinking waste	0.01–0.05	4–7		1–1.2
Paper-mill waste	0.01–0.05	2–8		1.2–2.2
Sugarcane defecation			0.5[§]	
Sugar-beet carbonation	2–5	15–20	0.03–0.07[‡]	
Uranium				
Acid-leached ore	10–30	25–65	0.02–1	
Alkaline-leached ore	20	60	1	
Uranium precipitate	1–2	10–25	5–12.5	
Water treatment				
Clarification (after 30-min flocculation)				1–1.3
Softening lime-soda (high-rate, solids-contact clarifiers)				3.7
Softening lime-sludge	5–10	20–45	0.6–2.5	

*$m^2/(t/d) \times 9.76 = ft^2/(\text{short ton/day})$; $m^1/(m^2 \cdot h) \times 0.41 = gal/(ft^2 \cdot min)$; 1 t = 1 metric ton.
[†]High-rate thickeners using required flocculant dosages operate at 10 to 50 percent of these unit areas.
[‡]Typical design using high Density/Paste thickeners. Feed per cent solids are that diluted for flocculation.
[§]Basis: 1 t of cane or beets.

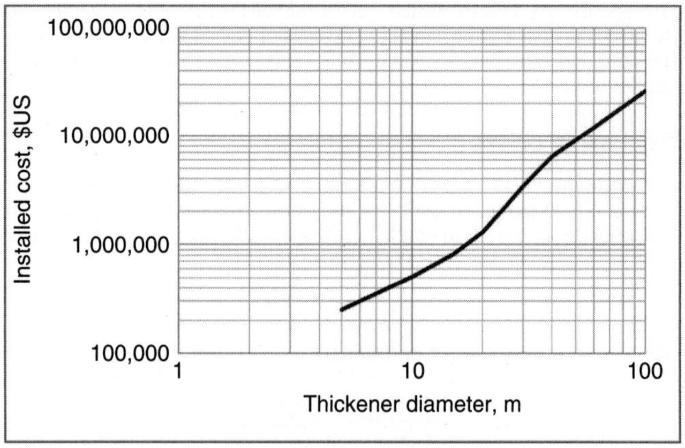

FIG. 18-98 Approximate installed cost of high-rate thickeners (2016 US $).

only as a guide, and they are based on elevated steel tanks up to 40 m diameter and on-ground tanks above 40 m diameter. They include the erection of mechanism and tank plus normal uncomplicated site preparation, excavation, reinforcing bar placement, backfill, and surveying. The price does not include any electrical work, pumps, piping, or external instrumentation. Special design modifications, which are not in the price, could include high-density or deep-cone thickener designs, insulation or drive enclosures required because of climatic conditions, and mechanism designs required because of scale buildup tendencies.

Operating Costs The power cost for a continuous thickener is an almost insignificant item. For example, a unit thickener 60 m (200 ft) in diameter with a torque rating of 1.0M N·m (0.74M lb·ft) will normally require 12 kW (16 hp). The low power consumption is due to the very slow rotative speeds. Normally, a mechanism will be designed for a peripheral speed of about 10 m/min (0.5 ft/s), which corresponds to only 3 r/h for a 60-m (200-ft) unit. This low speed also means very low maintenance costs. Operating labor cost is low because little attention is normally required after initial operation has balanced the feed and underflow. If chemicals are required for flocculation, the chemical cost often dwarfs all other operating costs.

FILTRATION

General References: Cheremisinoff and Azbel, *Liquid Filtration*, Ann Arbor Science, Woburn, Mass., 1983; Moir, *Chem. Eng.* **89**(15): 46 (1982); Orr, ed., *Filtration: Principles and Practice*, part I, Marcel Dekker, New York, 1977; part II, 1979; Purchas, ed., *Solid/Liquid Separation Equipment Scale-Up*, Uplands Press, Croydon, England, 1977; Schweitzer, ed., *Handbook of Separation Techniques for Chemical Engineers*, part 4, McGraw-Hill, New York, 1979; Shoemaker, ed., "What the Filter Man Needs to Know about Filtration," *Am. Inst. Chem. Eng. Symp. Ser.* **73**(171), (1977); Talcott et al., in *Kirk-Othmer Encyclopedia of Chemical Technology*, 3d ed., vol. 10, Wiley, New York, 1980, p. 284; Tiller et al., *Chem. Eng.* **81**(9): 116–136 (1974); also published as McGraw-Hill Repr. R203.

DEFINITIONS AND CLASSIFICATION

Filtration is the separation of a fluid-solids mixture involving passage of most of the fluid through a porous barrier that retains most of the solid particulates contained in the mixture. This subsection deals only with the filtration of solids from liquids; gas filtration is treated in Sec. 17. *Filtration* is the term for the unit operation. A filter is a piece of unit-operations equipment by which filtration is performed. The *filter medium* or *septum* is the barrier that lets the liquid pass while retaining most of the solids; it may be a screen, cloth, paper, or bed of solids. The liquid that passes through the filter medium is called the *filtrate*.

Filtration and filters can be classified several ways:

1. *By driving force.* The filtrate is induced to flow through the filter medium by hydrostatic head (gravity), pressure applied upstream of the filter medium, vacuum or reduced pressure applied downstream of the filter medium, or centrifugal force across the medium. Centrifugal filtration is closely related to centrifugal sedimentation, and both are discussed later under the subsection Centrifuges.

2. *By filtration mechanism.* Two models are generally considered and are the basis for the application of theory to the filtration process: when solids are stopped at the surface of a filter medium and pile upon one another to form a cake of increasing thickness, the separation is called *cake filtration*; when solids are trapped within the pores or body of the medium, it is termed *depth*, *filter-medium*, or *clarifying filtration*.

3. *By objective.* The process goal of filtration may be dry solids (the cake is the product of value), clarified liquid (the filtrate is the product of value), or both. Good solids recovery is best obtained by cake filtration, while clarification of the liquid is accomplished by either depth or cake filtration.

4. *By operating cycle.* Filtration may be intermittent (batch) or continuous. Batch filters may be operated with constant-pressure driving force, at constant rate, or in cycles that are variable with respect to both pressure and rate. Batch cycle can vary greatly, depending on filter area and solids loading.

5. *By nature of the solids.* Cake filtration may involve an accumulation of solids that is compressible or substantially incompressible, corresponding roughly in filter-medium filtration to particles that are deformable and to those that are rigid. The particle or particle-aggregate size may be of the same order of magnitude as the minimum pore size of most filter media (1 to 10 μm and greater), or may be smaller (1 μm down to the dimension of bacteria and even large molecules). Most filtrations involve solids of the former size range; those of the latter range can be filtered, if at all, only by filter-medium-type filtration or by ultrafiltration unless they are converted to the former range by aggregation prior to filtration.

These methods of classification are not mutually exclusive. Thus filters usually are divided first into the two groups of cake and clarifying equipment, then into groups of machines using the same kind of driving force, then further into batch and continuous classes. This is the scheme of classification underlying the discussion of filters of this subsection. Within it, the other aspects of operating cycle, the nature of the solids, and additional factors (e.g., types and classification of filter media) will be treated explicitly or implicitly.

FILTRATION THEORY

While research has developed a significant and detailed filtration theory, it is still too difficult to define a given liquid-solid system, and it is both faster and more accurate to determine filter requirements by performing small-scale tests. Filtration theory does, however, show how the test data can best be correlated, and extrapolated when necessary, for use in scale-up calculations.

In cake or surface filtration, there are two primary areas of consideration: continuous filtration, in which the resistance of the filter cake (deposited process solids) is very large with respect to that of the filter media and filtrate drainage, and batch pressure filtration, in which the resistance of the filter cake is not very large with respect to that of the filter media and filtrate drainage. Batch pressure filters are generally fitted with heavy, tight filter cloths, and sometimes a layer of precoat, and these represent a significant resistance that must be taken into account. Continuous filters, except for precoat filters, use relatively open cloths that offer little resistance compared to that of the filter cake.

Simplified theory for both batch and continuous filtration is based on the time-honored Hagen-Poiseuille equation:

$$\frac{1}{A}\frac{dV}{d\Theta}=\frac{p}{\mu(\alpha wV/A+r)} \qquad (18\text{-}59)$$

where V is the volume of filtrate collected, Θ is the filtration time, A is the filter area, P is the total pressure across the system, w is the weight of cake solids/unit volume of filtrate, μ is the filtrate viscosity, α is the cake-specific resistance, and r is the resistance of the filter cloth plus the drainage system.

CONTINUOUS FILTRATION

Since testing and scale-up are different for batch and continuous filtration, discussion in this section will be limited to continuous filtration.

It is both convenient and reasonable in continuous filtration, except for precoat filters, to assume that the resistance of the filter cloth plus filtrate drainage is negligible compared to the resistance of the filter cake and to assume that both pressure drop and specific cake resistance remain constant throughout the filter cycle. Equation (18-59), integrated under these conditions, may then be manipulated to give the following relationships:

$$W=\sqrt{\frac{2wp\Theta_f}{\mu(\alpha wV/A+r)}} \qquad (18\text{-}60)$$

$$V_f=\sqrt{\frac{2p\Theta_f}{\mu\alpha w}} \qquad (18\text{-}61)$$

$$\Theta_w = \frac{WV_w\mu\alpha}{p_w} \tag{18-62}$$

$$\Theta_W \propto NW^2 \tag{18-63}$$

$$\frac{\Theta_w}{\Theta_f} = 2\frac{V_w}{V_f} \tag{18-64}$$

where W is the weight of dry filter cake solids/unit area, V_f is the volume of cake formation filtrate/unit area, V_w is the volume of cake wash filtrate/unit area, Θ_f is the cake formation time, Θ_w is the cake wash time, and N is the wash ratio, the volume of cake wash/volume of liquid in the discharged cake.

As long as the suspended solids concentration in the feed remains constant, these equations lead to the following convenient correlations:

$$\log W \text{ vs. } \log \Theta_f \tag{18-65}$$

$$\log V_f \text{ vs. } \log \Theta_f \tag{18-66}$$

$$\Theta_w \text{ vs. } WV_w \tag{18-67}$$

$$\Theta_w \text{ vs. } NW^2 \tag{18-68}$$

$$\Theta_{w/f} \text{ vs. } V_w/V_f \tag{18-69}$$

There are two other useful empirical correlations as follows:

$$W \text{ vs. cake thickness} \tag{18-70}$$

$$\log R \text{ vs. } N \tag{18-71}$$

where R is percent remaining—the percent of solute in the unwashed cake that remains after washing.

FACTORS INFLUENCING FILTER SELECTION AND OPERATION

(Purchas, ed., *Solid/Liquid Separation Equipment Scale-Up,* Uplands Press, Croydon, England, 1977.)

Vacuum or Pressure The vast majority of all continuous filters use vacuum to provide the driving force for filtration. However, if the feed slurry contains a highly volatile liquid phase, or if it is hot, saturated, and/or near the atmospheric pressure boiling point, the use of pressure for the driving force may be required. Pressure filtration might also be used where the required cake moisture content is lower than that obtainable with vacuum.

The objective of most continuous filters is to produce a dry or handleable cake. Most vacuum filters easily discharge a "dry" consolidated cake as they are usually operated in an open or semi-open environment. However, whenever the filter must operate under pressure or within a vapor-tight enclosure, either because of the need for a greater driving force or because of the vapor pressure of the liquid phase, a dry-cake discharge becomes difficult. The problem of removing a dry cake from a pressurized enclosure generally requires a lock hopper system to maintain pressure in the enclosure.

Cake Discharge For any filter application to be practical, it must be possible to produce a cake thick enough to discharge. Table 18-13 gives the minimum acceptable cake thickness required for discharge for various types of filters and discharge mechanisms

TABLE 18-13 Minimum Cake Thickness for Discharge

Filter type	Minimum design thickness, mm	Minimum design thickness, in
Drum		
Belt	3–5	⅛–3/16
Roll discharge	1	1/32
Std. scraper	6	¼
Coil	3–5	⅛–3/16
String discharge	6	¼
Precoat	0–3 max.	0–⅛ max.
Horizontal belt	3–5	⅛–3/16
Horizontal table	20	¾
Tilting pan	20–25	¾–1
Disc	10–13	⅜–½

Feed Slurry Temperature Temperature can be both an aid and a limitation. As temperature of the feed slurry is increased, the viscosity of the liquid phase is decreased, causing an increase in filtration rate and a decrease in cake moisture content. The limit to the benefits of increased temperature occurs when the vapor pressure of the liquid phase starts to materially reduce the allowable vacuum. If the liquid phase is permitted to flash within the filter internals, various undesired results may ensue: disruption in cake formation adjacent to the medium, scale deposit on the filter internals, a sharp rise in pressure drop within the filter drainage passages due to increased vapor flow, or decreased vacuum pump capacity. In most cases, the vacuum system should be designed so that the liquid phase does not boil.

In some special cases, steam filtration can be used to gain the advantages of temperature without having to heat the feed slurry. Where applicable, dry steam is passed through the deliquored cake to raise the temperature of the residual moisture, reduce its viscosity, and lower its content. The final drying or cooling period that follows steam filtration uses the residual heat left in the cake to evaporate some additional moisture.

Cake Thickness Control Sometimes the rate of cake formation with bottom feed–type filters is rapid enough to create a cake too thick for subsequent operations. Cake thickness may be controlled by adjusting the bridge blocks in the filter valve to decrease the effective submergence, by reducing the slurry level in the vat, and by reducing the vacuum level in the cake formation portion of the filter valve. If these measures are inadequate, it may be necessary to use a top-loading filter.

Cake thickness must often be restricted when cake washing is required or the final cake moisture content is critical. Where the time required for cake washing is the rate-controlling step in the filter cycle, maximum filtration rate will be obtained when using the minimum cake thickness that gives good cake discharge. Where minimum cake moisture content is the controlling factor, there is usually some leeway with respect to cake thickness, although the minimum required for cake discharge is controlling in some cases. Since a relatively constant quantity of moisture is transferred from the medium to the filter cake when the vacuum is released prior to cake discharge, very thin cakes will sometimes be wetter than thicker cakes.

Filter Cycle Each filter cycle is composed of cake formation plus one or more of the following operations: deliquoring (dewatering or drying), washing, thermal drying, steam drying, and cake discharge. The number of these operations required by a given filtration operation depends on the process flowsheet. It is neither possible nor necessary to consider all of these operations at once. The basic testing program is designed to look at each operation individually. The requirements for each of the steps are then fit into a single filter cycle.

All filters using a rotary filter valve have their areas divided into a number of sections, sectors, or segments (see Fig. 18-112). When a drainage port passes from one portion of the filter valve to another, the change at the filter medium does not occur instantaneously, nor does it occur at some precise location on the filter surface. The change is relatively gradual and occurs over an area, as the drainage port at the filter valve first closes by passing onto a stationary bridge block and then opens as it passes off that bridge block on the other side.

On a horizontal belt filter, the equivalent sections extend across the filter in narrow strips. Therefore, section changes occur rapidly and may be considered as happening at a particular point along the length of the filter.

Representative Samples for Sizing and Design The results that are obtained in any bench-scale testing program can be only as good as the sample that is tested. It is absolutely essential that the sample used be representative of the slurry in the full-scale plant and that it be tested under the conditions that prevail in the process. If there is to be some significant time between taking or producing the sample and commencing the test program, due consideration must be given to what effect this time lapse may have on the characteristics of the slurry. If the slurry is at a temperature different from ambient, the subsequent heating and/or cooling could change the particle size distribution. Even sample age itself may exert a significant influence on particle size. If there is likely to be an effect, the bench-scale testing program should be carried out at the plant or laboratory site on fresh material.

Whenever a sample is to be held for some time or shipped to a distant laboratory for testing, some type of characterizing filtration test should be run on the fresh sample and then duplicated at the time of the test program. A comparison of the results of the two tests will indicate how much of a change there has been in the sample. If the change is too great, it will be necessary to make arrangements to work on a fresh sample. Any shipped sample must be protected from freezing, as freezing can substantially change the filtration characteristics of a slurry, *particularly hydrated* materials.

The slurry should always be defined as completely as possible by noting suspended solids concentration, particle size distribution, viscosity, density of solids and liquid, temperature, chemical composition, and so on.

Feed Solids Concentration Feed slurries that are so dilute that they settle rapidly usually yield reduced solids filtration rates and produce stratified cakes with higher moisture contents than would normally be obtained with a homogeneous cake. It is well known that an increase in feed solids concentration is generally an effective means of increasing solids filtration rate, assisting in forming a homogeneous suspension and thereby minimizing cake moisture content. Equipment required to concentrate a slurry sample and the tests needed to predict how far a slurry will thicken were discussed previously in this section.

Pretreatment Chemicals Even though the suspended solids concentration of the slurry to be tested may be correct, it is often necessary to modify the slurry in order to provide an acceptable filtration rate, washing rate, or final cake moisture content. The most common treatment, and one that may provide improvement in all three of these categories, is the addition of flocculating agents, either inorganic chemicals or natural or synthetic polymers. The main task at this point is to determine which is the most effective chemical and the quantity of chemical that should be used.

There are a number of commercially available surfactants that can be employed as an aid in filter cake moisture reduction. These reagents can be added to the filter feed slurry or to the filter cake wash water, if washing is used. Since these reagents have a dispersing effect, flocculation may be required subsequently. Typical moisture reductions of 2 to 4 percentage points are obtained at reagent dosages of 200 to 500 g/t of solids.

Cloth Blinding Continuous filters, except for precoats, use filter medium to effect the separation of the solid and filtrate phases. Since the medium is in contact with the process solids, there is a probability of medium blinding. The term *blinding* refers to blockage of the fabric itself, either by the wedging of process solids or by solids precipitated in and around the yarn.

The filter medium chosen should be as open as possible yet still able to maintain the required filtrate clarity. Those fabrics that will produce a clear filtrate and yet do not have rapid blinding tendencies are often lightweight (woven from thin filaments or yarn) and will not wear as long as some of the heavier, more open fabrics (woven from heavy filaments or yarn). Whenever the filter follows a gravity thickening or clarification step, it is advisable to return the filtrate to the thickener or clarifier so that the filtrate clarity requirements may be relaxed in favor of using a heavier, more open cloth with reduced blinding tendencies. Excessively dirty filtrates should be avoided because the solids may be abrasive and detrimental to the internals of the filter or perhaps may cut the fabric yarn.

It should be noted at this point that an *absolutely* clear filtrate can rarely be obtained on a cloth-covered continuous filter. The passages through the medium are invariably larger than some of the solids in the slurry, and there will be some amount of solids passing through the medium. Once the pores of the fabric have been bridged, the solids themselves form the septum for the remaining particles, and the filtrate becomes clear. It is this bridging action of the solids that permits the use of a relatively open filter medium, while at the same time maintaining a reasonably clear filtrate.

Filters with media in the form of an endless belt have greatly reduced the concern about blinding. Most synthetic fabrics can be successfully cleaned of process solids by washing the medium after cake discharge, and the rate of blinding due to chemical precipitation also can be drastically reduced. Current practice suggests that the belt-type filter with continuous-medium washing be the first choice unless experience has shown that medium blinding is not a factor or if the belt-type system cannot be successfully applied.

Sealing of the belt along the edges of the filter drum is never perfect, and some leakage should be expected. If good clarity is essential, it may be preferable to use a drum filter with the cloth caulked in place and design the system to contend with the effects of blinding.

The one exception to the points noted above is the continuous precoat filter. Here the purpose of the filter medium is to act as a support for the sacrificial bed of precoat material. Thus, the medium should be tight enough to retain the precoat solids and prevent bleeding of the precoat solids through the filter medium during operation, yet open enough to permit easy cleaning at the end of each cycle. Lightweight felt media work well in these respects.

Use of Steam or Hot Air The cycle might include steam filtration or thermal drying using hot air. While effective use is made of both steam and hot air, the applications are rather limited. As a general rule, steam application will reduce cake moisture 2 to 4 percentage points. Hot-air drying can produce a bone-dry cake, but generally it is practical only if the air rate is high, greater than about 1800 $m^3/m^2 \cdot h$ (98 cfm/ft²).

BENCH-SCALE TEST PROCEDURES

Information and guidelines on testing procedures, apparatus, data correlation, scale-up, and design can be found in: Purchas, Derek B., *Solid/Liquid Separation Equipment Scale-Up*, chap. 11, "Continuous Vacuum and Pressure Filtration" (D.A. Dahlstrom, C.E. Silverblatt), Uplands Press, Croydon, England, 1977, and in *Perry's Chemical Engineer's Handbook*, 8th ed., McGraw-Hill, New York, 2008.

BATCH FILTRATION

Since most batch-type filters operate under pressure rather than vacuum, the following discussion will apply primarily to pressure filtration and the various types of pressure filters.

To use Eq. (18-53), one must know the pattern of the filtration process, that is, the variation of the flow rate and pressure with time. Generally the pumping mechanism determines the filtration flow characteristics and serves as a basis for the following three categories [Tiller and Crump, *Chem. Eng. Prog.* **73**(10): 65 (1977)]:

1. *Constant-pressure filtration.* The actuating mechanism is compressed gas maintained at a constant pressure.
2. *Constant-rate filtration.* Positive-displacement pumps of various types are employed.
3. *Variable-pressure, variable-rate filtration.* The use of a centrifugal pump results in this pattern: the discharge rate decreases with increasing back pressure.

Flow rate and pressure behavior for the three types of filtration are shown in Fig. 18-99. Depending on the characteristics of the centrifugal pump, widely differing curves may be encountered, as suggested by the figure.

Constant-Pressure Filtration For constant-pressure filtration, Eq. (18-59) can be integrated to give the following relationships between total time and filtrate measurements:

$$\frac{\theta}{V/A} = \frac{\mu\alpha}{2P}\frac{V}{A} + \frac{\mu r}{P} \tag{18-72}$$

$$\frac{\theta}{V/A} = \frac{\mu\alpha w}{2P}\frac{V}{A} + \frac{\mu r}{P} \tag{18-73}$$

$$\frac{\theta}{V/A} = \frac{\mu\alpha\rho c}{2P(1-mc)}\frac{W}{A} + \frac{\mu r}{P} \tag{18-74}$$

For a given constant-pressure filtration, these may be simplified to

$$\frac{\theta}{V/A} = K_p\frac{W}{A} + C = K_p'\frac{V}{A} + C \tag{18-75}$$

where K_p, K_p', and C are constants for the conditions employed. It should be noted that K_p, K_p', and C depend on filtering pressure not only in the obvious explicit way but also in the implicit sense that α, m, and r are generally dependent on P.

Constant-Rate Filtration For substantially incompressible cakes, Eq. (18-53) may be integrated for a constant rate of slurry feed to the filter to give the following equations, in which filter-medium resistance is treated as the equivalent constant-pressure component to be deducted from the

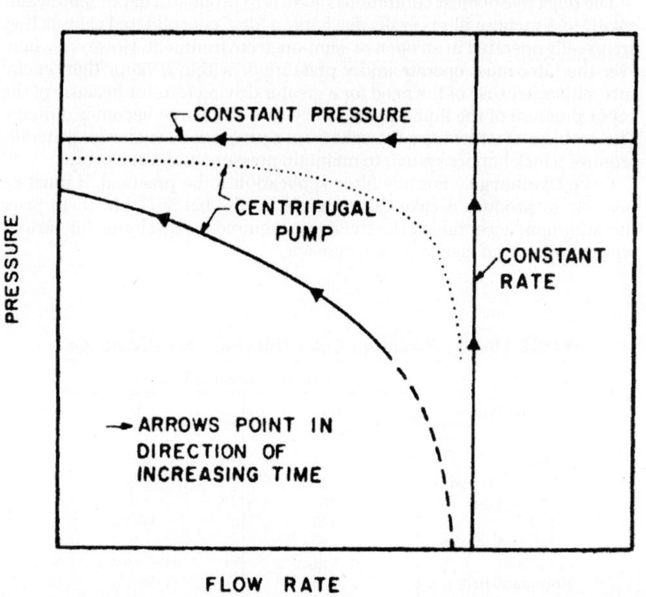

FIG. 18-99 Typical filtration cycles. [*Tiller and Crump, Chem. Eng. Prog. 73(10): 72(1977), by permission.*]

rising total pressure drop to give the variable pressure through the filter cake [Ruth, *Ind. Eng. Chem.* **27:** 717 (1935)]:

$$\frac{\theta}{V/A} = \frac{1}{\text{rate per unit area}} = \frac{\mu\alpha}{P-P_1}\frac{W}{A} \qquad (18\text{-}76)$$

which may also be written

$$\frac{\theta}{V/A} = \frac{\mu\alpha w}{P-P_1}\frac{V}{A} = \frac{\mu\alpha\rho c}{(P-P_1)(1-mc)}\frac{V}{A} \qquad (18\text{-}77)$$

In these equations, P_1 is the pressure drop through the filter medium.

$$P_1 = \mu r(V/A\theta)$$

For a given constant-rate run, the equations may be simplified to

$$V/A = P/K_r + C' \qquad (18\text{-}78)$$

where K_r and C' are constants for the given conditions.

Variable-Pressure, Variable-Rate Filtration The pattern of this category complicates the use of the basic rate equation. The method of Tiller and Crump [*Chem. Eng. Prog.* **73**(10): 65 (1977)] can be used to integrate the equation when the characteristic curve of the feed pump is available.

In the filtration of small amounts of fine particles from liquid by means of bulky filter media (such as pierced felt), it has been found that the preceding equations based on the resistance of a cake of solids do not hold, since no cake is formed. For these cases, in which filtration takes place on the surface or within the interstices of a medium, analogous equations have been developed [Hermans and Bredée, *J. Soc. Chem. Ind.* **55T:** 1 (1936)]. These are usefully summarized, for both constant-pressure and constant-rate conditions, by Grace [*Am. Inst. Chem. Eng. J.* **2:** 323 (1956)]. These equations often apply to the clarification of such materials as sugar solutions, viscose and other spinning solutions, and film-casting dopes.

FILTER MEDIA

All filters require a filter medium to retain solids, whether the filter is for cake filtration or for filter-medium or depth filtration. Specification of a medium is based on retention of some minimum particle size at good removal efficiency and on acceptable life of the medium in the environment of the filter. The selection of the type of filter medium is often the most important decision in success of the operation. For cake filtration, medium selection involves an optimization of the following factors:

1. Ability to bridge solids across its pores quickly after the feed is started (i.e., minimum propensity to bleed)
2. Low rate of entrapment of solids within its interstices (i.e., minimum propensity to blind)
3. Minimum resistance to filtrate flow (i.e., high production rate)
4. Resistance to chemical attack
5. Sufficient strength to support the filtering pressure
6. Acceptable resistance to mechanical wear
7. Ability to discharge cake easily and cleanly
8. Ability to conform mechanically to the kind of filter with which it will be used

Filter-medium selection embraces many types of construction: fabrics of woven fibers, felts, and nonwoven fibers, porous or sintered solids, polymer membranes, or particulate solids in the form of a permeable bed. Media of all types are available in a wide choice of materials.

Fabrics of Woven Fibers For cake filtration, these fabrics are the most common type of medium. A wide variety of materials are available and listed in Table 18-14, with ratings for chemical and temperature resistance. In addition to the material of the fibers, a number of construction

TABLE 18-14 Characteristics of Filter-Fabric Materials*

Generic name and description	Breaking tenacity, g/denier	Abrasion resistance	Resistance to acids	Resistance to alkalies	Resistance to oxidizing agents	Resistance to solvents	Specific gravity	Maximum operating temperature, °F†
Acetate—cellulose acetate. When not less than 92% of the hydroxyl groups are acetylated, "triacetate" may be used as a generic description.	1.2–1.5	G	F	P	G	G	1.33	210
Acrylic—any long-chain synthetic polymer composed of at least 85% by weight of acrylonitrile units.	2.0–4.8	G	G	F	G	E	1.18	300
Glass—fiber-forming substance is glass.	3.0–7.2	P	E	P	E	E	2.54	600
Metallic—composed of metal, metal-coated plastic, plastic-coated metal, or a core completely covered by metal.	—	G						
Modacrylic—fiber-forming substance is any long-chain synthetic polymer composed of less than 85% but at least 35% by weight of acrylonitrile units.	2.5–3.0	G	G	G	G	G	1.30	180
Nylon—any long-chain synthetic polyamide having recurring amide groups as an integral part of the polymer chain.	3.8–9.2	E	F–P	G	F–P	G	1.14	225
Polyester—any long-chain synthetic polymer composed of at least 85% by weight of an ester of a dihydric alcohol and terephthalic acid (p—HOOC—C_6H_4—COOH).	2.2–7.8	E–G	G	G–F	G	G	1.38	300
Polyethylene—long-chain synthetic polymer composed of at least 85% weight of ethylene.	1.0–7.0	G	G	G	F	G	0.92	165‡
Polypropylene—long-chain synthetic polymer composed of at least 85% by weight of propylene.	4.8–8.5	G	E	E	G	G	0.90	250§
Cotton—natural fibers.	3.3–6.4	G	P	F	G	E–G	1.55	210
Fluorocarbon—long-chain synthetic polymer composed of tetrafluoroethylene units.	1.0–2.0	F	E	E	E	G	2.30	550¶

*Adapted from Mais, *Chem. Eng.*, **78**(4), 51 (1971). Symbols have the following meaning: E = excellent, G = good, F = fair, P = poor.

†°C = (°F − 32)/1.8; K = (°F + 459.7)/1.8.

‡Low-density polymer. Up to 230°F for high-density.

§Heat-set fabric; otherwise lower.

¶Requires ventilation because of release of toxic gases above 400°F.

characteristics describe the filter cloth: (1) weave, (2) style number, (3) weight, (4) count, (5) ply, and (6) yarn number. Of the many types of weaves available, only four are extensively used as filter media: plain (square) weave, twill, chain weave, and satin.

All these weaves may be made from any textile fiber, natural or synthetic. They may be woven from spun staple yarns, multifilament continuous yarns, or monofilament yarns. The performance of the filter cloth depends on the weave and the type of yarn.

A recently developed medium known as a double weave incorporates different yarns in warp and fill in order to combine the specific advantages of each type. An example of this is Style 99FS, made by Madison Filtration, in which multifilament warp yarns provide good cake release properties and spun staple fill yarns contribute to greater retentivity.

Metal Fabrics or Screens These are available in several types of weave in nickel, copper, brass, bronze, aluminum, steel, stainless steel, Monel, and other alloys. In the plain weave, 400 mesh is the closest wire spacing available, thus limiting use to coarse crystalline slurries, pulps, and the like. The "Dutch weaves" employing relatively large, widely spaced, straight warp wires and relatively small crimped filling wires can be woven much more closely, providing a good medium for filtering fine crystals and pulps. This type of weave tends to plug readily when soft or amorphous particles are filtered and makes the use of filter aid desirable. Good corrosion and high temperature resistance of properly selected metals makes filtrations with metal media desirable for long-life applications. This is attractive for handling toxic materials in closed filters to which minimum exposure by maintenance personnel is desirable.

Pressed Felts and Nonwoven Media These materials are used to filter gelatinous particles from paints, spinning solutions, and other viscous liquids. Filtration occurs by deposition of the particles in and on the fibers throughout the mat.

Nonwoven media consist of web or sheet structures that are composed primarily of fibers or filaments bonded together by thermal, chemical, or mechanical (such as needle-punching) means. Needled felts are the most commonly used nonwoven fabric for liquid filtration. Additional strength is often provided by including a scrim of woven fabric encapsulated within the nonwoven material. The surface of the medium can be calendered to improve particle retention and assist in filter cake release. Weights range from 270 to 2700 gm/m^2 (8 to 80 oz/yd^2). Because of their good retentivity, high strength, moderate cost, and resistance to blinding, nonwoven media have found wide acceptance in filter press use, particularly in mineral concentrate filtration applications. They are used often on horizontal belt filters where their dimensional stability reduces or eliminates wrinkling and biasing problems often encountered with woven belts.

Filter Papers These papers come in a wide range of permeability, thickness, and strength. As a class of material, they have low strength, however, and require a perforated backup plate for support.

Rigid Porous Media These are available in sheets or plates and tubes. Materials used include sintered stainless steel and other metals, graphite, aluminum oxide, silica, porcelain, and some plastics—a gamut that allows a wide range of chemical and temperature resistance. Most applications are for clarification.

Polymer Membranes These are used in filtration applications for fine-particle separations such as microfiltration and ultrafiltration (clarification involving the removal of 1-μm and smaller particles). The membranes are made from a variety of materials, the commonest being cellulose acetates and polyamides. Membrane filtration, discussed in Sec. 22, has been well covered by Porter (in Schweitzer, *Handbook of Separation Techniques for Chemical Engineers,* 3d ed., McGraw-Hill, New York, 1996, sec. 2.1).

Media made from woven or nonwoven fabrics coated with a polymeric film, such as Primapor, and Primapor II made by Clear Edge Filtration, Gore-Tex, made by W. L. Gore and Associates, and Tetratex, made by Donaldson Company, combine the high retentivity characteristics of a membrane with the strength and durability of a thick filter cloth. These media are used on both continuous and batch filters where excellent filtrate clarity is required.

Granular Beds of Particulate Solids Beds of solids like sand or coal are used as filter media to clarify water or chemical solutions containing small quantities of suspended particles. Filter-grade grains of desired particle size can be purchased. Often beds will be constructed of layers of different materials and different particle sizes.

Various types of filter media and the materials of which they are constructed are surveyed extensively by Purchas (*Industrial Filtration of Liquids,* CRC Press, Cleveland, 1967, chap. 3), and characterizing measurements (e.g., pore size, permeability) are reviewed in detail by Rushton and Griffiths (in Orr, ed., *Filtration: Principles and Practice,* part I, Marcel Dekker, New York, 1977, chap. 3). Briefer summaries of classification of media and of practical criteria for the selection of a filter medium are presented by Shoemaker ["What the Filter Man Needs to Know about Filtration," *Am. Inst. Chem. Eng. Symp. Ser.* **73**(171): 26 (1977)] and Purchas [*Filtr. Sep.* **17:** 253, 372 (1980)].

FILTER AIDS

The use of filter aids is often applied to filtrations in which problems of slow filtration rate, rapid medium blinding, or unsatisfactory filtrate clarity arise. Filter aids are granular or fibrous solids capable of forming a highly permeable filter cake in which very fine solids or slimy, deformable flocs may be trapped. The application of filter aids may allow the use of a much more permeable filter medium than the clarification would require to produce filtrate of the same quality by depth filtration.

Filter aids should have low bulk density to minimize settling and to aid good distribution on a filter-medium surface during application. They should also be porous and capable of forming a porous cake to minimize flow resistance, and they must be chemically inert to the filtrate. These characteristics are all found in the two most popular commercial filter aids: diatomaceous earth (diatomite) and expanded perlite, particles of "puffed" lava that are principally aluminum alkali silicate. Cellulosic fibers (ground wood pulp) are sometimes used when siliceous materials cannot be used but are much more compressible. The use of other less effective aids (e.g., carbon and gypsum) may be justified in special cases. Sometimes a combination of carbon and diatomaceous earth permits adsorption in addition to filter-aid performance. Various other materials, such as salt, fine sand, starch, and precipitated calcium carbonate, are employed in specific industries where they represent either waste material or inexpensive alternatives to conventional filter aids.

Diatomaceous Earth Filter aids of diatomaceous earth have a dry bulk density of 128 to 320 kg/m^3 (8 to 20 lb/ft^3), contain particles mostly smaller than 50 μm, and produce a cake with porosity in the range of 0.9 (volume of voids/total filter-cake volume). The high porosity (compared with a porosity of 0.38 for randomly packed uniform spheres and 0.2 to 0.3 for a typical filter cake) is indicative of its filter-aid ability. Different methods of processing the crude diatomite result in a series of filter aids having a wide range of permeability.

Perlite Perlite filter aids are somewhat lower in bulk density (48 to 96 kg/m^3, or 3 to 6 lb/ft^3) than diatomaceous silica and contain a higher fraction of particles in the 50- to 150-μm range. Perlite is also available in a number of grades of differing permeability and cost, the grades being roughly comparable to those of diatomaceous earth. Diatomaceous earth will withstand slightly more extreme pH levels than perlite, and it is said to be somewhat less compressible.

Filter aids are used in two ways: (1) as a precoat and (2) mixed with the slurry as a "body feed." Precoat filtration, employing a thin layer of about 0.5 to 1.0 kg/m^2 (0.1 to 0.2 lb/ft^2) deposited on the filter medium prior to beginning feed to the filter, is in wide use to protect the filter medium from fouling by trapping solids before they reach the medium. It also provides a finer matrix to trap fine solids and assure filtrate clarity. Body-feed application is the continuous addition of filter aid to the filter feed to increase the porosity of the cake. The amount of addition must be determined by trial, but in general, the quantity added should at least equal the amount of solids to be removed. For solids loadings greater than 1000 ppm this may become a significant cost factor. An acceptable alternative might be to use a rotary vacuum precoat filter [Smith, *Chem. Eng.* **83**(4): 84 (1976)]. Further details of filter-aid filtration are set forth by Cain (in Schweitzer 1996, sec. 4.2) and Hutto [*Am. Inst. Chem. Eng. Symp. Ser.* **73**(171): 50 (1977)]. Figure 18-100 shows a flow sheet indicating

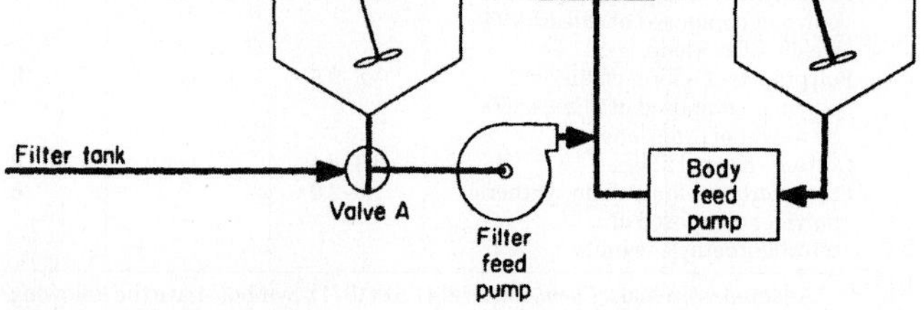

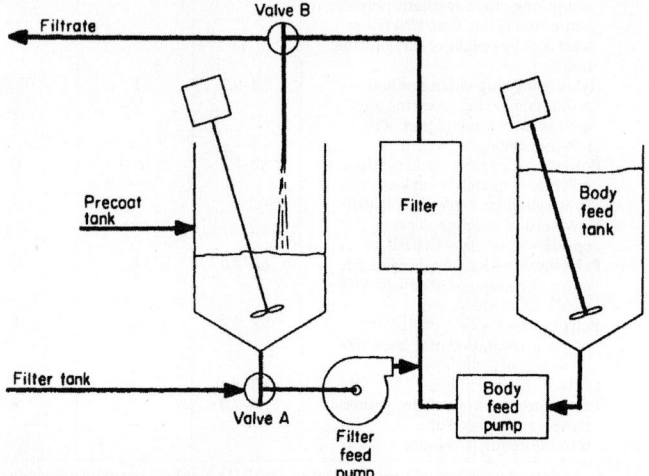

FIG. 18-100 Filter-aid filtration system for precoat or body feed. (*Schweitzer,* Handbook of Separation Techniques for Chemical Engineers, *p. 4-12. Copyright 1979 by McGraw-Hill, Inc. and used with permission.*)

arrangements for both precoat and body-feed applications. Most filter aid is used on a one-time basis, although some techniques have been demonstrated to reuse precoat filter aid on vertical-tube pressure filters.

FILTRATION EQUIPMENT

Cake Filters Filters that accumulate appreciable visible quantities of solids on the surface of a filter medium are called cake filters. The slurry feed may have a solids concentration from about 1 percent to greater than 40 percent. The filter medium on which the cake forms is relatively open to minimize flow resistance; once the cake forms, it becomes the effective filter medium. The initial filtrate will contain fine solids until the cake is formed. This situation may be made tolerable by recycling the filtrate until acceptable clarity is obtained or by using a downstream polishing filter (clarifying type).

Cake filters are used when the desired product of the operation is the solids, the filtrate, or both. When the filtrate is the product, the degree of removal from the cake by washing or blowing with air or gas becomes an economic optimization. When the cake is the desired product, the incentive is to obtain the desired degree of cake purity by washing, blowing, and sometimes mechanical expression of residual liquid.

Implicit in cake filtration is the removal and handling of solids, since the cake is usually relatively dry and compacted. Cakes can be sticky and difficult to handle; therefore, the ability of a filter to discharge the cake cleanly is an important equipment selection criterion.

In the operational sense, some filters are batch devices, whereas others are continuous. This difference provides the principal basis for classifying cake filters in the discussion that follows. The driving force by which the filter functions—hydrostatic head ("gravity"), pressure imposed by a pump or a gas blanket, or atmospheric pressure ("vacuum")—will be used as a secondary criterion.

Batch Cake Filters A *nutsche filter* is one of the simplest batch filters. It is a tank with a false bottom, perforated or porous, which may either support a filter medium or act as the filter medium. The slurry is fed into the filter vessel, and separation occurs by gravity flow, gas pressure, vacuum, or a combination of these forces.

The filter is often used in laboratory, pilot-plant, or small-plant operation. For large-scale processing, the excessive floor area per unit of filtration area and the difficulty of cake removal are strong deterrents.

Thorough displacement washing is possible in a nutsche if the wash is added before the cake begins to be exposed to air displacement of filtrate. If washing needs to be more effective, an agitator can be provided in the nutsche vessel to reslurry the cake to allow adequate diffusion of solute from the solids.

The *horizontal plate filter*, a horizontal multiple-plate pressure filter, consists of a number of horizontal circular drainage plates and guides placed in a stack in a cylindrical shell (Fig. 18-101). In normal practice the filtering pressure is limited to 345 kPa (50 psig), although special filters have been designed for shell pressures of 2.1 MPa (300 psig) or higher.

The *filter press*, one of the most commonly used filters in the early years of the chemical industry, is still widely used. Often referred to generically as the plate-and-frame filter, it has probably over 100 design variations.

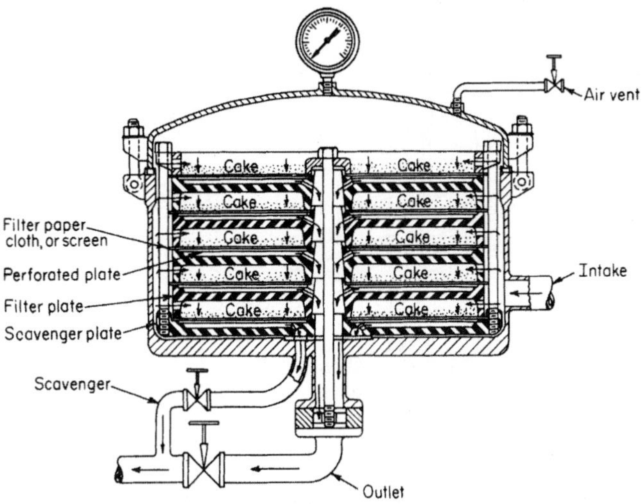

FIG. 18-101 Elevation section of a Sparkler horizontal plate filter. (*Sparkler Filters, Inc.*)

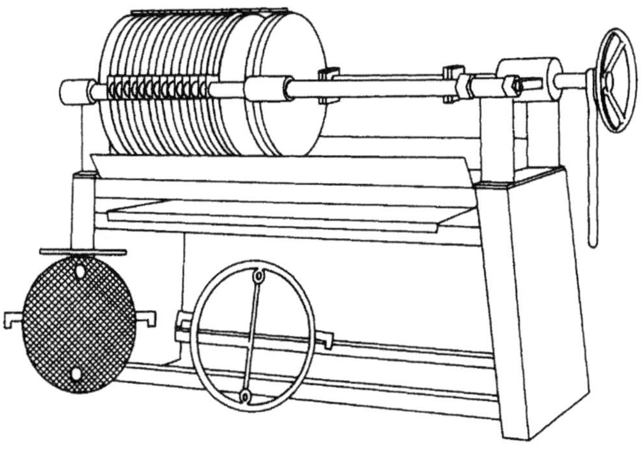

FIG. 18-102 Circular-plate fabricated-metal filter press. (*Star Filter, Hilliard Corp.*)

Two basic popular designs are the plate-and-frame and the recessed-plate press. Both are available in a wide range of materials, including metals, coated metals, and plastics.

A *plate-and-frame press* is an alternate assembly of plates covered on both sides with a filter medium, usually a cloth, and hollow frames that provide space for cake accumulation during filtration. The frames have feed and wash manifold ports, while the plates have filtrate drainage ports. The plates and frames are usually rectangular; circles and other shapes are also used (Fig. 18-102).

Two wash techniques are used in plate-and-frame filter presses, illustrated in Fig. 18-103. In simple washing, the wash liquor follows the same path as the filtrate. If the cake is not extremely uniform and highly permeable, this type of washing is ineffective in a well-filled press. A better technique is thorough washing, in which the wash is introduced to the faces of alternate plates (with their discharge channels valved off). The wash passes through the entire cake and exits through the faces of the other plates. This improved technique requires a special design and the assembly of the plates in proper order. Thorough washing should be used only when the frames are well filled, since an incomplete fill of cake will allow cake collapse during the wash entry. The remainder of the wash flow will bypass through cracks or channels opened in the cake.

Filter presses are made in plate sizes from 10 by 10 cm (4 by 4 in) to 2 by 4 m (79 by 158 in). Frame thickness ranges from 0.3 to 20 cm (0.125 to 8 in). Operating pressures up to 689 kPa (100 psig) are common, with some presses designed for 6.9 MPa (1000 psig). Some metal units have cored plates for steam or refrigerant. The maximum pressure for plastic frames is 410 to 480 kPa (60 to 70 psig).

The filter press has the advantage of simplicity, low capital cost, flexibility, and ability to operate at high pressure in either a cake-filter or a clarifying-filter application. Floor-space and headroom needs per unit of filter area are small, and capacity can be adjusted by adding or removing plates and frames. Filter presses are easily cleaned, and the filter medium is easily replaced. With proper operation, a denser, drier cake compared with that of most other filters is obtained.

There are several serious disadvantages, including imperfect washing due to variable cake density, relatively short filter cloth life due to the mechanical wear of emptying and cleaning the press (often involving scraping the cloth), and high labor requirements. Presses often drip or leak and thereby create housekeeping problems, but the biggest problem arises from the requirement to open the filter for cake discharge. The operator is thus exposed routinely to the contents of the filter, and this is becoming an increasingly severe disadvantage as more and more materials once believed safe are given restricted exposure limits.

A recessed-plate filter press is similar to the plate-and-frame press in appearance but consists only of plates (Fig. 18-104). Both faces of each plate are hollowed to form a chamber for cake accumulation between adjacent plates. This design has the advantage of about 50 percent fewer joints than a plate-and-frame press, making a tight closure more certain. Figure 18-105 shows some of the features of one type of recessed-plate filter that has a gasket to further minimize leaks. Air can be introduced behind the cloth on both sides of each plate to assist in cake removal.

Some interesting variations of standard designs include the ability to roll the filter to change from a bottom to a top inlet or outlet and the ability to add blank dividers to convert a press to a multistage press for further clarification of the filtrate or to do two separate filtrations simultaneously in the

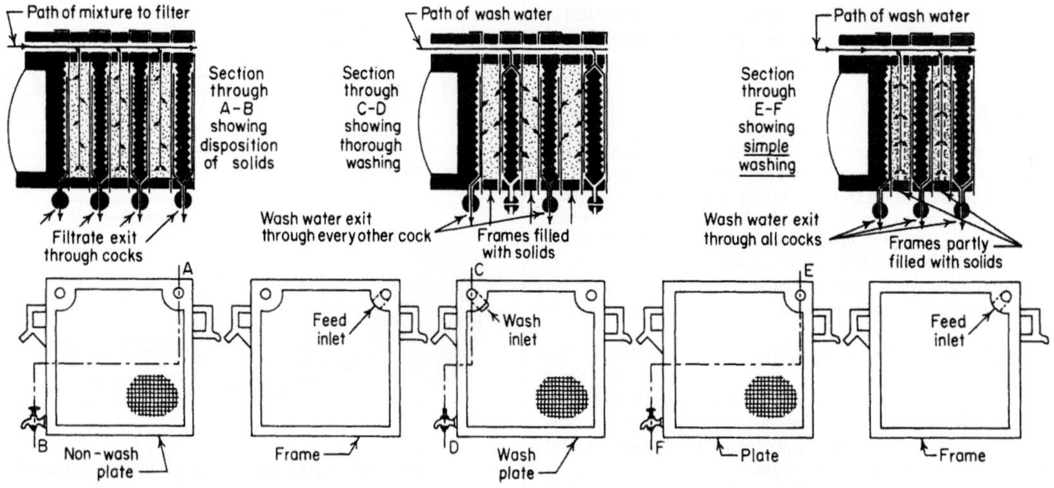

FIG. 18-103 Filling and washing flow patterns in a filter press. (*D. R. Sperry & Co.*)

same press. Some designs have elastomer membranes between plates that can be expanded when filtration is finished to squeeze out additional moisture. Some designs feature automated opening and cake-discharge operations to reduce labor requirements. Examples of this type of pressure filter include Pneumapress, Larox, Vertipress, and Oberlin.

Internal cake tube filters or *liquid bag filters,* such as those manufactured by Industrial Filter and Pump Mfg. Co. and many others, use one or more perforated tubes supported by a tube sheet or by the lip of the pressure vessel. A cylindrical filter bag, sealed at one end, is inserted into the perforated tube. The open end of the filter bag generally has a flange or special seal ring to prevent leakage.

Slurry under pressure is admitted to the chamber between the head of the shell and the tube sheet, whence it enters and fills the tubes. Filtration occurs as the filtrate passes radially outward through the filter medium and the wall of each tube into the shell and out the filtrate discharge line, depositing cake on the medium. The filtration cycle is ended when the tubes have filled with cake or when the media have become plugged. The cake can be washed (if it has not been allowed to fill the tubes completely) and air-blown. The filter has a removable head to provide easy access to the tube sheet and mouth of the tubes; thus "sausages" of cake can be removed by taking out the filter bags or each tube and bag assembly together. The tubes themselves are easily removed for inspection and cleaning.

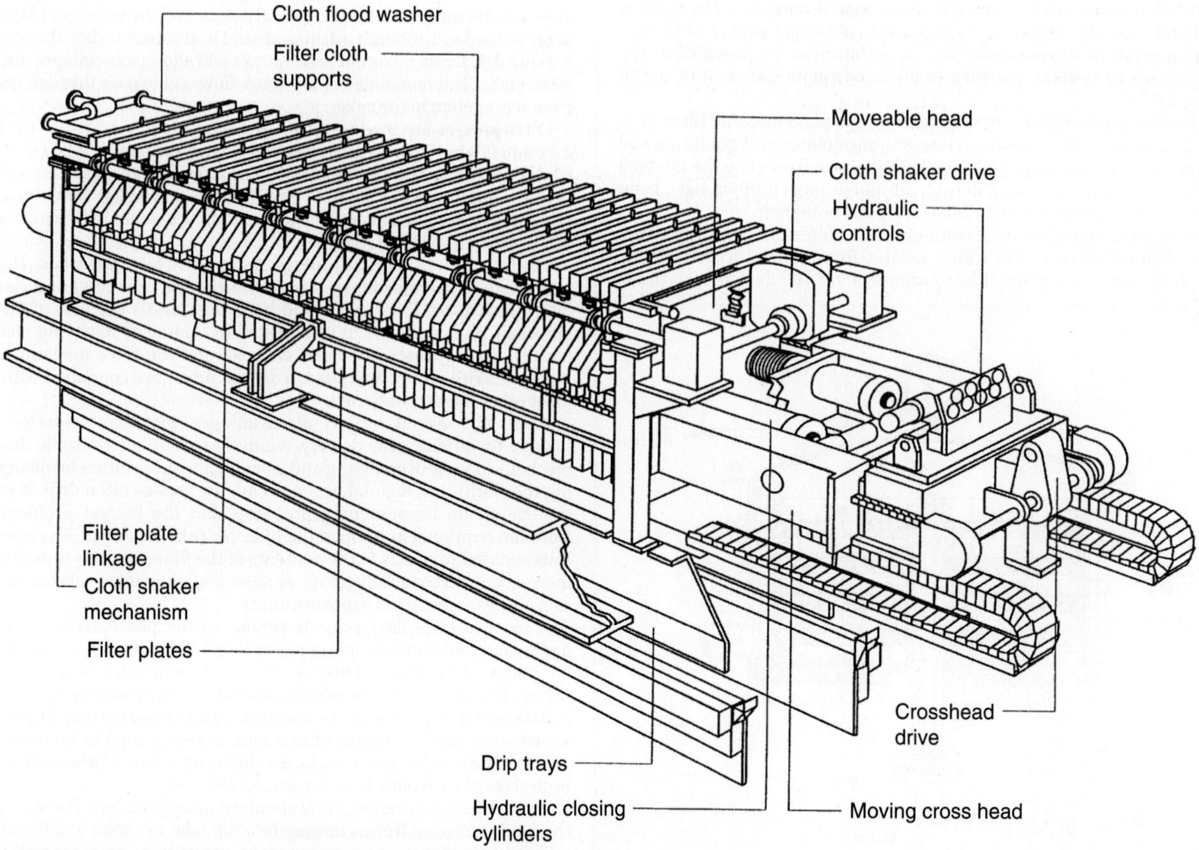

FIG. 18-104 Automated recessed-plate filter press used in mineral applications. (*FLSmidth.*)

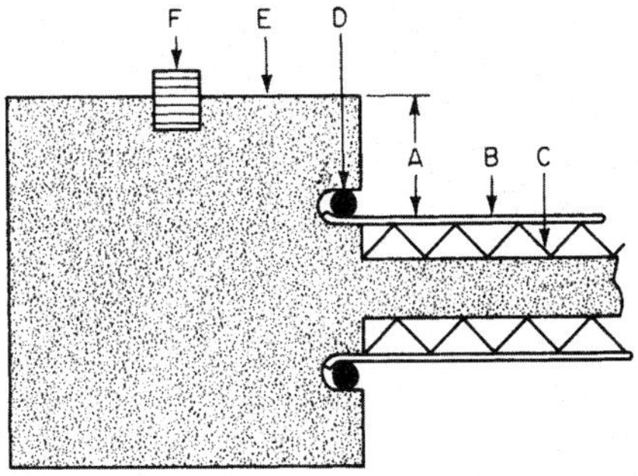

FIG. 18-105 Section detail of a caulked-gasketed-recessed filter plate: (*a*) cake recess; (*b*) filter cloth; (*c*) drainage surface of plate; (*d*) caulking strip; (*e*) plate joint; (*f*) sealing gasket. (*FLSmidth.*)

The advantages of the tubular filter are that it uses an easily replaced filter medium, its filtration cycle can be interrupted and the shell can be emptied of prefilt at any time without loss of the cake, the cake is readily recoverable in dry form, and the inside of the filter is conveniently accessible. There is also no unfiltered heel. Disadvantages are the need for and the labor requirements of emptying by hand and replacing the filter media and the tendency for heavy solids to settle out in the header chamber. Applications are as scavenger filters to remove fines not removed by other equipment in a previous filtration stage, to handle the runoff from other filters, and in semiworks and small-plant operations in which the filter's size, versatility, and cleanliness are required.

External-cake tubular filter designs are available with vertical tubes supported by a filtrate-chamber tube sheet in a vertical cylindrical vessel (Fig. 18-106). The tubes may be made of wire cloth; porous ceramic, carbon, plastic, or metal; or closely wound wire. The tubes may have a filter cloth on the outside. Often a filter-aid precoat will be applied to the tubes. The prefilt slurry is fed near the bottom of the vertical vessel. The filtrate passes from the outside to the inside of the tubes and into a filtrate chamber at the top or the bottom of the vessel. The solids form a cake on the outside of the tubes with the filter area actually increasing as the cake builds up, partially compensating for the increased flow resistance of the thicker cake. The filtration cycle continues until the differential pressure reaches a specified level, or until about 25 mm (1 in) of cake thickness is obtained.

Cake-discharge methods are the chief distinguishing feature among the various designs. That of the Industrial Filter & Pump Hydra-Shoc, for example, removes cake from the tubes by filtrate backflushing assisted by the "shocking" action of a compressed-gas pocket formed in the filtrate chamber at the top of the vertical vessel. Closing the filtrate outlet valve while continuing to feed the filter causes compression of the gas volume trapped in the dome of the

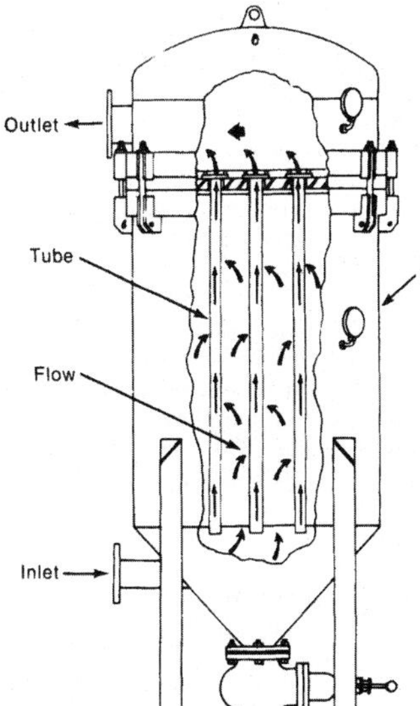

FIG. 18-106 Top-outlet tubular filter. (*Industrial Filter & Pump Mfg. Co.*)

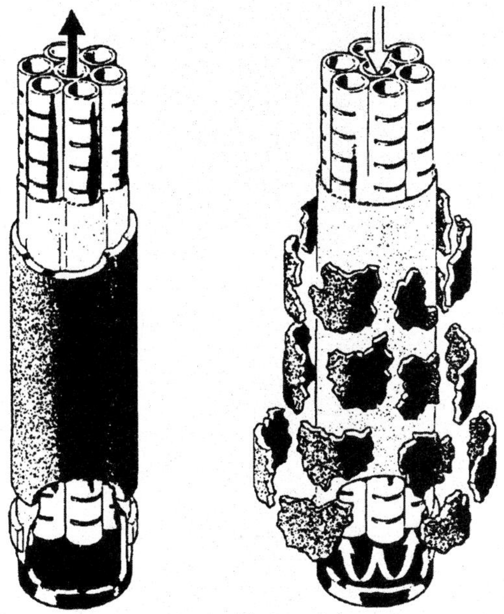

FIG. 18-107 Cake formation and discharge with the Fundabac filter element. (*Dr. Müller AG, Switzerland.*)

vessel until, at the desired gas pressure, quick-acting valves stop the feed and open a bottom drain. The compressed gas rapidly expands, forcing a rush of filtrate back across the filter medium and dislodging the cake, which drains out the bottom with the flush liquid; this technique may be used only when wet-cake discharge is permitted.

Dry-cake discharge can be achieved with a Fundabac candle-type filter manufactured by DrM, Dr. Müller, AG, of Switzerland. This filter uses a candle made up of six small-diameter tubes around a central filtrate delivery tube. This design allows the filter cloth to be flexed outward upon blowback, easily achieving an effective dry-cake discharge (Fig. 18-107).

Pressure leaf filters, sometimes called tank filters, consist of flat filtering elements (leaves) supported in a pressure shell. The leaves are circular, arc-sided, or rectangular, and they have filtering surfaces on both faces. The shell is a cylindrical or conical tank. Its axis may be horizontal or vertical, and the filter type is described by its shell axis orientation.

A filter leaf consists of a heavy screen or grooved plate over which a filter medium of woven fabric or fine wire cloth may be fitted. Textile fabrics are more commonly used for chemical service and are usually applied as bags that may be sewed, zippered, stapled, or snapped. Wire-screen cloth is often used for filter-aid filtrations, particularly if a precoat is applied. It may be attached by welding, riveting, bolting, or caulking or by the clamped engagement of two 180° bends in the wire cloth under tension, as in Multi Metal's Rim-Lok leaf. The filter medium, regardless of material, should be as taut as possible to minimize sagging when it is loaded with a cake; excessive sag can cause cake cracking or dropping. Leaves may be supported at top, bottom, or center and may discharge filtrate from any of these locations. Figure 18-108 shows the elevation section of a precoated bottom-support wire leaf.

Pressure leaf filters are operated batchwise. The shell is locked, and the prefilt slurry is admitted from a pressure source. The slurry enters in such a way as to minimize settling of the suspended solids. The shell is filled, and filtration occurs on the leaf surfaces, the filtrate discharging through an individual delivery line or into an internal manifold, as the filter design dictates. Filtration is allowed to proceed only until a cake of the desired thickness has formed, since to overfill will cause cake consolidation with consequent difficulty in washing and discharge. The decision of when to end the filtering cycle is largely a matter of experience, guided roughly by the rate in a constant-pressure filter or pressure drop in a constant-rate filter. This judgment may be supplanted by the use of a detector that "feels" the thickness of cake on a representative leaf.

If the cake is to be washed, the slurry heel can be blown from the filter and wash liquor can be introduced to refill the shell. If the cake tends to crack during air blowing, it may be necessary to displace the slurry heel with wash gradually so as never to allow the cake to dry. Upon the completion of filtration and washing, the cake is discharged by one of several methods, depending on the shell and leaf configuration.

In horizontal pressure leaf filters, the leaves may be rectangular and run parallel to the axis, and they are of varying sizes since they form chords of

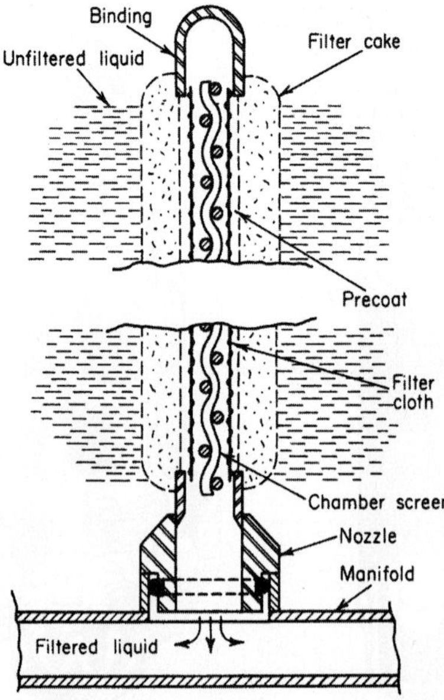

FIG. 18-108 Section of precoated wire filter leaf.

the shell; or they may be circular or square elements parallel to the head of the shell, and all of the same dimension. The leaves may be supported in the shell from an independent rack, individually from the shell, or from a filtrate manifold. Horizontal filters are particularly suited to dry-cake discharge.

Most of the currently available commercial horizontal pressure filters have leaves that are parallel to the shell head. Cake discharge may be wet or dry and can be achieved by sluicing with liquid sprays, vibration of the leaves, or leaf rotation against a knife, wire, or brush. If a wet-cake discharge is allowable, the filters will probably be sluiced with high-pressure liquid. If the filter has a top or bottom filtrate manifold, the leaves are usually in a fixed position, and the spray header is rotated to contact all filter surfaces. If the filtrate header is center-mounted, the leaves are generally rotated at about 3 r/min and the spray header is fixed. Some units may be wet-cake-discharged by mechanical vibration of the leaves with the filter filled with liquid. Dry-cake discharge normally will be accomplished by vibration if leaves are top- or bottom-manifolded and by rotation of the leaves against a cutting knife, wire, or brush if they are center-manifolded.

In many designs the filter is opened for cake discharge, and the leaf assembly is separated from the shell by moving one or the other on rails (Fig. 18-109). For processes involving toxic or flammable materials, a closed filter system can be maintained by sloping the bottom of the horizontal cylinder to the drain nozzle for wet discharge or by using a screw conveyor in the bottom of the shell for dry discharge.

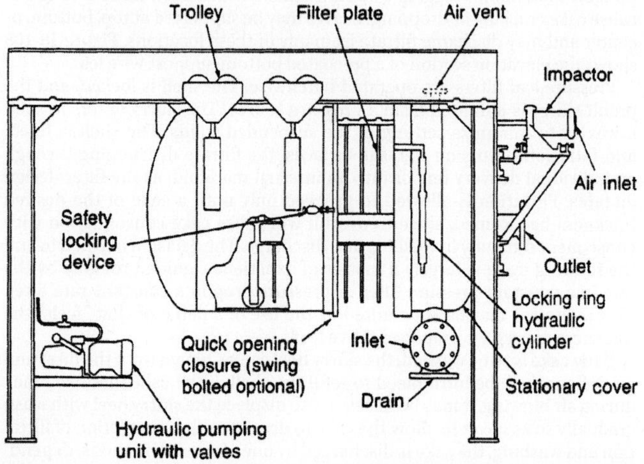

FIG. 18-109 Horizontal-tank pressure leaf filter designed for dry cake discharge. (*Sparkler Filter, Inc.*)

Vertical pressure leaf filters have vertical, parallel, rectangular leaves mounted in an upright cylindrical pressure tank. The leaves are usually different widths to allow them to conform to the curvature of the tank. The leaves often rest on a filtrate manifold, the connection being sealed by an O ring, so that they can be lifted individually from the top of the filter for inspection and repair. A scavenger leaf is often installed in the bottom of the shell to allow virtually complete filtration of the slurry heel at the end of a cycle.

Vertical filters are not convenient for the removal of dry cake, although they can be used in this service if they have a bottom that can be retracted to permit the cake to fall into a bin or hopper. They are adapted rather to wet-solids discharge, a process that may be assisted by leaf vibration, air or steam sparging of a filter full of water, sluicing from fixed, oscillating, or traveling nozzles, and blowback. They are made by many companies, and they enjoy their widest use for filter-aid precoat filtration.

The advantages and uses of pressure leaf filters are their considerable flexibility (up to the permissible maximum, cakes of various thickness can be formed successfully), low labor charges, particularly when the cake may be sluiced off or the dry cake discharged cleanly by blowback, the basic simplicity of many of the designs, and their adaptability to quite effective displacement washing. Pressure leaf filters provide complete process isolation as the vessel is the barrier to contact with the process and the filter cannot function without this barrier in place. These filters offer very good temperature control by jacketing the vessel. Their disadvantages are the requirement of exceptionally intelligent and watchful supervision to avoid cake consolidation or dropping, their inability to form as dry a cake as a filter press, their tendency to classify vertically during filtration and to form misshapen nonuniform cakes unless the leaves rotate, and the restriction of most models to 610 kPa (75 psig) or less.

Pressure leaf filters are used in similar applications as filter presses and are used much more extensively than filter presses for filter-aid filtrations. They should be seriously considered whenever uniformity of production permits long-time operation under essentially constant filtration conditions, when thorough washing with a minimum of liquor is desired, or when vapors or fumes make enclosed construction desirable. Under such conditions, if the filter medium does not require frequent changing, they may show a considerable advantage in cycle and labor economy over a filter press. Pressure leaf filters are available with filtering areas of 930 cm^2 (1 ft^2) (laboratory size) up to about 440 m^2 (4734 ft^2) for vertical filters and 158 m^2 (1700 ft^2) for horizontal ones. Leaf spacing ranges from 5 to 15 cm (2 to 6 in) but are seldom less than 7.5 cm (3 in) since 1.3 to 2.5 cm (0.5 to 1 in) should be left open between surfaces.

In a *centrifugal-discharge filter*, horizontal top-surface filter plates may be mounted on a hollow motor-connected shaft that serves both as a filtrate-discharge manifold and as a drive shaft to permit centrifugal removal of the cake. An example is the Funda filter (marketed in the United States by Steri Technologies), illustrated schematically in Fig. 18-110. The filtering surface may be a textile fabric or a wire screen, and the use of a precoat is optional. The Funda filter is driven from the top, leaving the bottom unobstructed for inlet and drainage lines; a somewhat similar machine that employs a bottom drive, providing a lower center of mass and ground-level access to the drive system, is the Pall Sietz *Schenk filter* available from Pall Corporation.

During filtration, the vessel that coaxially contains the assembly of filter plates is filled with prefilt under pressure, the filtrate passes through the plates and out the hollow shaft, and cake is formed on the top surfaces of the plates. After filtration, the vessel is drained, or the heel may be filtered by recirculation through a cascade ring at the top of the filter. The cake may be washed—or it may be extracted, steamed, air-blown, or dried by hot gas. It is discharged, wet or dry, by rotation of the shaft at sufficiently high speed to sling away the solids. If flushing is permitted, the discharge is assisted by a backwash of appropriate liquid.

The operating advantages of the centrifugal-discharge filter are those of a horizontal-plate filter and, further, its ability to discharge cake without being opened. It is characterized by low labor, easy adaptability to automatic control, and amenability to the processing of hazardous, noxious, or sterile materials. Its disadvantages are its complexity and maintenance (stuffing boxes, high-speed drive) and its cost. The Funda filter is made in sizes that cover the filtering area range of 1 to 50 m^3 (11 to 537 ft^2). The largest Schenk filter provides 100 m^2 (1075 ft^2) of area.

Continuous Cake Filters Continuous cake filters are applicable when cake formation is fairly rapid.

Rotary Drum Filters The rotary drum filter is the most widely used of the continuous filters. There are many design variations, including operation as either a pressure filter or a vacuum filter. The major difference between designs is in the technique for cake discharge, to be discussed later. All the alternatives are characterized by a horizontal-axis drum covered on the cylindrical portion by filter medium over a grid support structure to allow drainage to manifolds. Basic materials of construction may be metals or plastics. Sizes (in terms of filter areas) range from 0.37 to 186 m^2 (4 to 2000 ft^2).

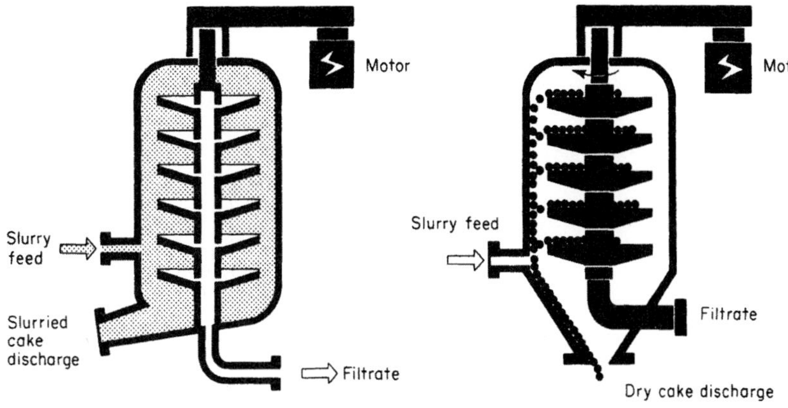

FIG. 18-110 Schematic of a centrifugal-discharge filter. (*Steri Technologies.*)

All drum filters (except the single-compartment filter) use a rotary-valve arrangement in the drum-axis support trunnion to facilitate the removal of filtrate and wash liquid and to allow the introduction of air or gas for cake blowback if needed. The valve controls the relative duration of each cycle as well as providing "dead" portions of the cycle through the use of bridge blocks. A typical valve design is shown in Fig. 18-111. Internal piping manifolds connect the valve with various sections of the drum.

Most drum filters are fed by operating the drum with about 35 percent of its circumference submerged in a slurry trough, although submergence can be set for any desired amount between zero and almost total. Some units contain an oscillating rake agitator in the trough to aid solids suspension. Others use propellers, paddles, or no agitator.

Slurries of free-filtering solids that are difficult to suspend are sometimes filtered on a top-feed drum filter or filter-dryer. An alternative for slurries of extremely coarse, dense solids is the internal drum filter. In the chemical-process industry, both top-feed and internal drums (which are described briefly by Emmett in Schweitzer 1996, p. 4-41) have largely been displaced by the horizontal vacuum belt or pan filters.

Most drum filters operate at a rotation speed in the range of 0.1 to 10 r/min. Variable-speed drives are provided to allow adjustment for changing cake-formation and drainage rates.

Drum filters are commonly classified according to the feed arrangement and the cake-discharge technique. The characteristics of the slurry and the filter cake usually dictate the cake-discharge method.

The *scraper-discharge filter* medium is usually caulked into grooves in the drum grid, with cake removal facilitated by a scraper blade just prior to the resubmergence of the drum (Fig. 18-112). The scraper serves mainly as a deflector to direct the cake, dislodged by an air blowback, into the discharge chute, since actual contact with the medium would cause rapid wear. In some cases the filter medium is held by circumferentially wound wires spaced 50 mm (2 in) apart, and a flexible scraper blade may rest lightly against the wire winding. A taut wire in place of the scraper blade may be used in some applications in which physical dislodging of sticky, cohesive cakes is needed.

For a given slurry, the maximum filtration rate is determined by the minimum cake thickness that can be removed—the thinner the cake, the less the flow resistance and the higher the rate. The minimum thickness is about 6 mm (0.25 in) for relatively rigid or cohesive cakes. Solids that form friable cakes composed of less cohesive materials will usually require a cake thickness of 13 mm (0.5 in) or more. Filter cakes composed of fine solids, which often produce cakes that crack or adhere to the medium, usually need a thickness of at least 10 mm (0.38 in).

A *string-discharge filter*, a system of endless strings or wires spaced about 13 mm (0.5 in) apart pass around the filter drum but are separated tangentially from the drum at the point of cake discharge, lifting the cake off as they leave contact with the drum. The strings return to the drum surface guided by two rollers, the cake separating from the strings as they pass over the rollers. Success depends on the ability of the cake to be removed with the strings and must be determined experimentally. Applications are mainly in the starch and pharmaceutical industries. *Removable-medium filters* exist in some drum filters to provide for the filter medium to be removed and reapplied as the drum rotates. This feature permits the complete discharge of thin or sticky cake and provides the regenerative washing of the medium to reduce blinding.

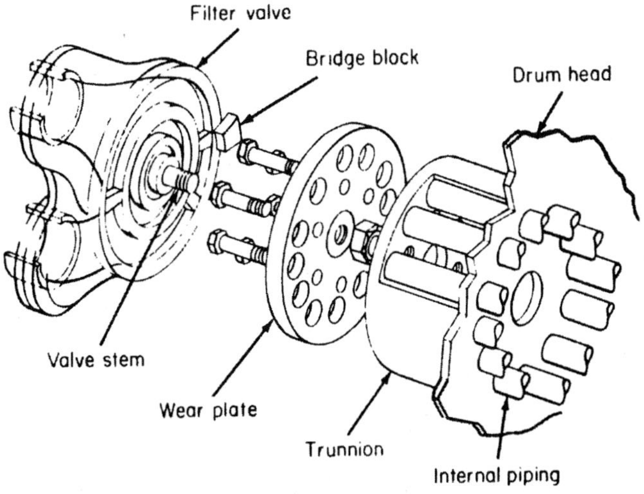

FIG. 18-111 Component arrangement of a continuous-filter valve. (*FLSmidth.*)

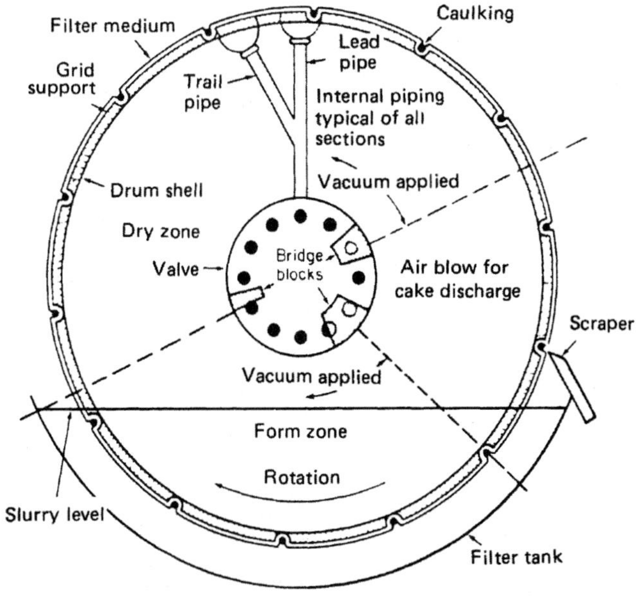

FIG. 18-112 Schematic of a rotary-drum vacuum filter with scraper discharge, showing operating zones. (*Schweitzer,* Handbook of Separation Techniques for Chemical Engineers, *p. 4-38. Copyright 1979 by McGraw-Hill, Inc. and used with permission.*)

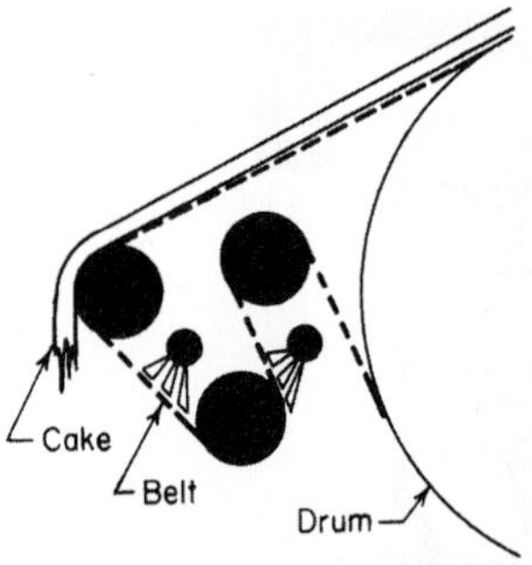

FIG. 18-113 Cake discharge and medium washing on an EIMCO belt filter. (*FLSmidth.*)

A *belt-discharge filter* is a drum filter carrying a fabric that is removed, passed over rollers, washed, and returned to the drum. Figure 18-113 shows the path of the medium while it is off the drum. A special aligning device keeps the medium wrinkle-free and in proper line during its travel. Thin cakes of difficult solids that may be slightly soluble are good applications. When acceptable, a sluice discharge makes cakes as thin as 1.5 to 2 mm (about 1/16 in) feasible.

The Coilfilter (Komline-Sanderson Engineering Corp.) is a drum filter with a medium consisting of one or two layers of stainless-steel helically coiled springs, about 10 mm (0.4 in) in diameter, placed in a corduroy pattern around the drum. The springs follow the drum during filtration with cake forming the coils. They are separated from the drum to discharge the cake and undergo washing; if two layers are used, the coils of each layer are further separated from those of the other, passing over different sets of rolls. The use of stainless steel in spring form provides a relatively permanent medium that is readily cleaned by washing and flexing. Filtrate clarity is poorer than with most other media, and a relatively large vacuum pump is needed to handle greater air leakage than is characteristic of fabric media. Material forming a slimy, matlike or fibrous cake (e.g., raw sewage, pulp and paper waste) is the typical application.

In *roll-discharge filters*, a roll in close proximity to the drum at the point of cake discharge rotates in the opposite direction at a peripheral speed equal to or slightly faster than that of the drum (Fig. 18-114). If the cake on the drum is adequately tacky and cohesive for this discharge technique, it adheres to cake on the smaller roll and separates from the drum. A blade, taut wire, or wire "combs" remove the material from the discharge roll. This design is especially good for thin, sticky cakes. If necessary, a slight air blow may be provided to help release the cake from the drum. Typical cake thickness is 1 to 3 mm (0.04 to 0.12 in).

A *single-compartment drum filter,* such as the Bird-Young filter, (Andritz Group) differs from most drum filters in that the drum is not compartmented, and there is no internal piping or rotary valve. The entire inside of the drum is subjected to vacuum, with its surface perforated to pass the filtrate. Cake is discharged by an air blowback applied through a "shoe" that covers a narrow discharge zone on the inside surface of the drum to interrupt the vacuum, as illustrated in Fig. 18-115. The internal drum surface must be machined to provide close clearance of the shoe to avoid leakage. The filter is designed for high filtration rates with thin cakes. Rotation speeds to 40 r/min are possible, with cakes typically 3 to 6 mm (0.12 to 0.24 in) thick. Filter sizes range from 930 cm^2 to 19 m^2 (1 to 207 ft^2) with 93 percent of the area active. The slurry is fed into a conical feed tank designed to prevent solids from settling without the use of mechanical agitators. The proper liquid level is maintained by overflow, and submergence ranges from 5 to 70 percent of the drum circumference.

Wash sprays may be applied to the cake, with collection troughs or pans inserted inside the drum to keep the wash separate from the filtrate. Filtrate is removed from the lower section of the drum by a pipe passing through the trunnions.

The major advantages of the Bird-Young filter are its ability to handle thin cakes and operate at high speeds, its washing effectiveness, and its low internal resistance to air and filtrate flow. An additional advantage is the possibility of construction as a pressure filter with up to 1.14-MPa (150-psig) operating pressure to handle volatile liquids. The chief disadvantages are its high cost and the limited flexibility imposed by not having an adjustable rotary valve. The best applications are on free-draining, nonblinding materials such as paper pulp or crystallized salts.

Continuous pressure filters consist of conventional drum or disc filters totally enclosed in pressure vessels and are generally referred to as hyperbaric filters. Filtration takes place with the vessel pressurized up to 6 bar and the filtrate discharging either at atmospheric pressure or into a receiver maintained at a suitable backpressure. Cake discharge is facilitated through a dual valve and lock-hopper arrangement in order to maintain vessel pressure. Alternatively, the discharged filter cake can be reslurried within the filter or in an adjoining pressure vessel and removed through a control valve.

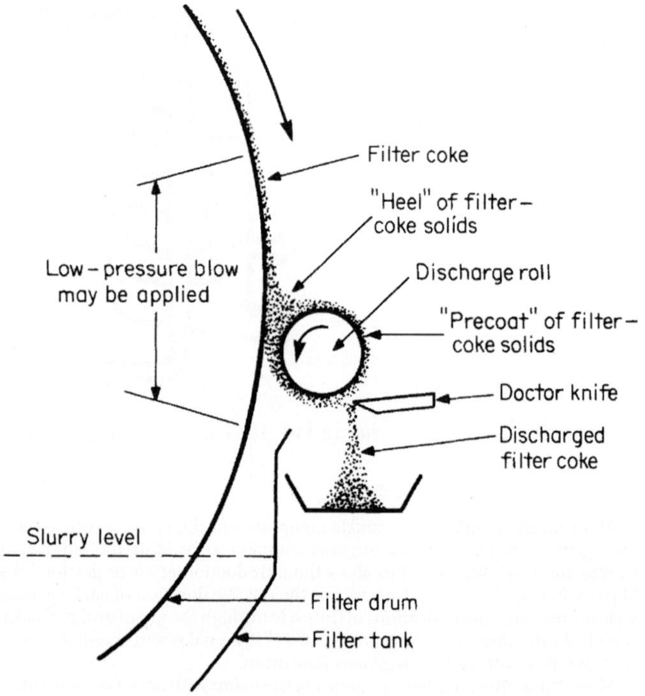

FIG. 18-114 Operating principles of a roll-discharge mechanism. (*Schweitzer,* Handbook of Separation Techniques for Chemical Engineers, *p. 4-40. Copyright 1979 by McGraw-Hill, Inc. and used with permission.*)

One variation in design, the LAROX CC filter offered by Outotec, employs "gasless" ceramic media instead of traditional filter fabrics, relying partly on capillary action to achieve low moistures. This results in a significant drop in power consumption by greatly reducing the compressed air requirements.

Other types of continuous pressure filters include the belt press filter commonly used for dewatering sludges and the BHS FEST rotary pressure filter.

Continuous precoat filters may be operated as either pressure or vacuum filters, although vacuum operation prevails. The filters are not totally continuous but have an extremely long batch cycle (1 to 10 days). Applications are for continuous clarification of liquids from slurries containing 50 to 5000 ppm of solids when only very thin, unacceptable cakes would form on other filters and where "perfect" clarity is required.

Construction is similar to that of other drum filters, except that vacuum is applied to the entire rotation. Before feeding slurry, a precoat layer of filter aid or other suitable solids, 75 to 125 mm (3 to 5 in) thick, is applied. The feed slurry is introduced and trapped in the outer surface of the precoat, where it is removed by a progressively advancing doctor knife that trims a thin layer of solids plus precoat (Fig. 18-116). When the precoat has been cut to a predefined minimum thickness, the filter is taken out of service, washed, and freshly precoated.

Disc Filters A disc filter is a vacuum filter consisting of a number of vertical discs attached at intervals on a continuously rotating horizontal hollow central shaft (Fig. 18-117). Each disc consists of 10 to 30 sectors of metal, plastic, or wood, ribbed on both sides to support a filter cloth and provide drainage via an outlet nipple into the central shaft. Each sector may be replaced individually. The filter medium is usually cloth, and for some heavy-duty applications, stainless-steel screens may be used.

The discs are typically 30 to 50 percent submerged in a troughlike vessel containing the slurry. Another horizontal shaft running beneath the discs may contain agitator paddles to maintain suspension of the solids, as in the FLSmidth Eimco *Agidisc® filter.* In some designs, feed is distributed through nozzles below each disc. Vacuum is supplied to the sectors as they rotate into the liquid to allow cake formation. Vacuum is maintained as the sectors emerge from the liquid and are exposed to air. Wash may be applied with sprays, but most applications are for dewatering only. As the sectors rotate to the discharge point, the vacuum is cut off, and a slight air blast is used to loosen the cake. This allows scraper blades to direct the cake into discharge chutes positioned between the discs. Vacuum and air blowback is controlled by an automatic valve as in rotary-drum filters.

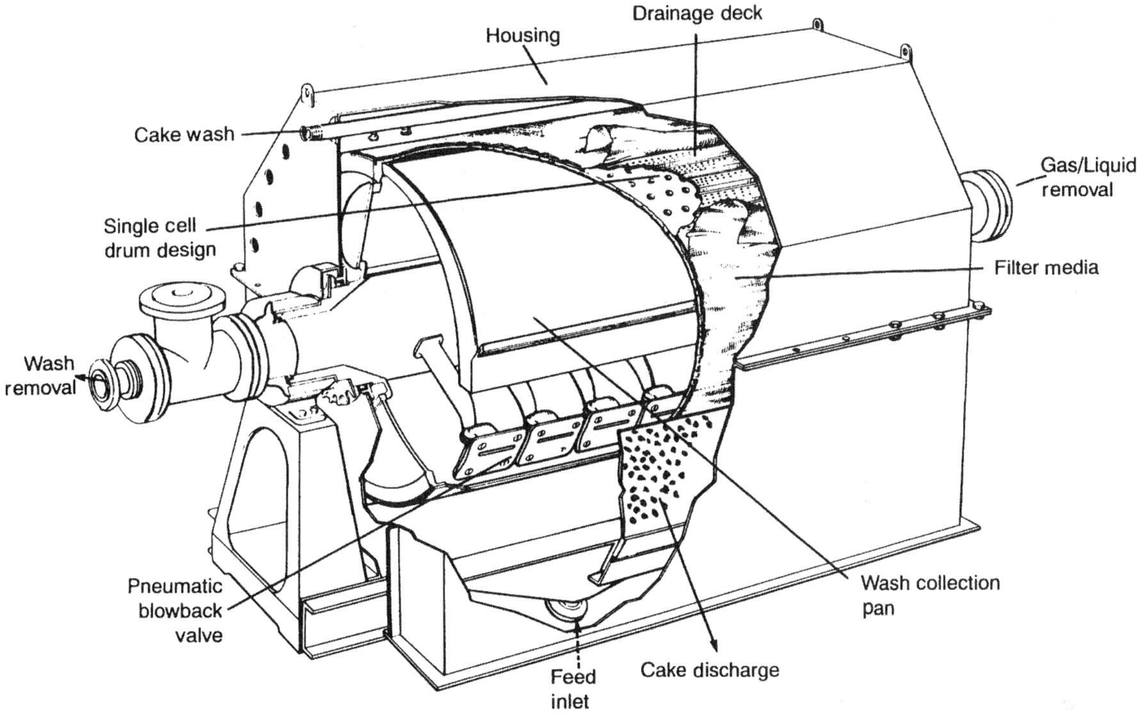

FIG. 18-115 Cutaway of the single-compartment drum filter. (*Andritz Group.*)

Of all continuous filters, the vacuum disc is the lowest in cost per unit area of filter when mild steel or similar materials of construction may be used. It provides a large filtering area with minimum floor space, and it is used mostly in high-tonnage dewatering applications in sizes up to about 300 m^2 (3300 ft^2) of filter area.

The main disadvantages are the inadaptability to have effective wash and the difficulty of totally enclosing the filter for hazardous-material operations.

Horizontal Vacuum Filters These filters are generally classified into two broad classes: rotary circular and belt-type units. Regardless of geometry, they have similar advantages and limitations. They provide flexibility of choice of cake thickness, washing time, and drying cycle. They effectively handle heavy, dense solids, allow flooding of the cake with wash liquor, and are easily designed for true countercurrent leaching or washing. The disadvantages are that they are more expensive to build than drum or disc filters, they use a large amount of floor space per filter area, and they are difficult to enclose for hazardous applications.

Horizontal-table, scroll-discharge, and pan filters are all basically revolving annular tables with the top surface a filter medium (Fig. 18-118). The table is divided into sectors, each of which is a separate compartment. Vacuum is applied through a drainage chamber beneath the table that leads to a large rotary valve. Slurry is fed at one point, and cake is removed by a horizontal scroll conveyor that elevates the cake over the rim of the filter. A clearance of about 10 mm (0.4 in) is maintained between the scroll and the filter medium to prevent damage to the medium. Residual cake, called a heel cake, on the medium may be loosened by an air blow from below, with high-velocity liquid sprays from above, or with a combination of the two. The heel cake is unique to the horizontal pan filter and, besides protecting the filter media, it is the wettest part of the cake and is NOT discharged by the scroll; thus, the discharged cake is typically drier than that from a similarly sized vacuum filter of another type. Unit sizes range from about 0.9 to 9 m (3 to 30 ft) in diameter, with about 80 percent of the surface available for filtration.

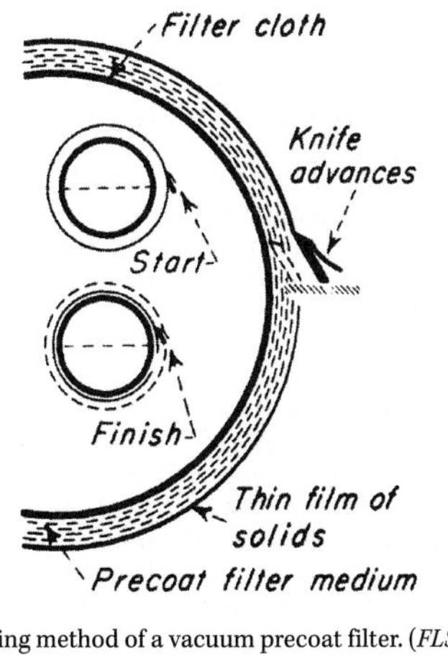

FIG. 18-116 Operating method of a vacuum precoat filter. (*FLSmidth.*)

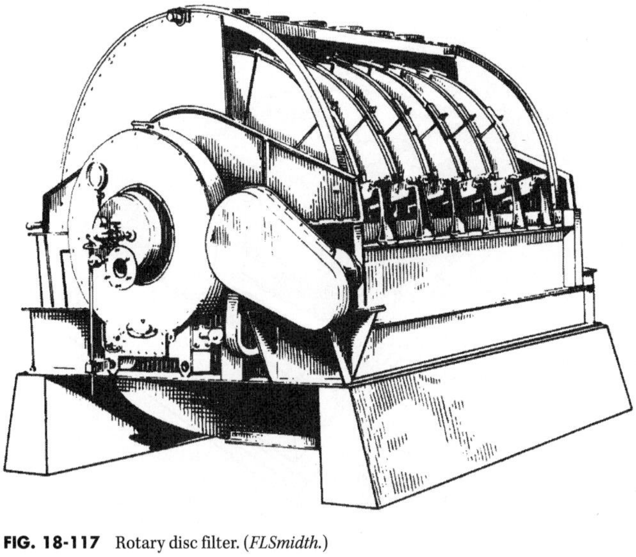

FIG. 18-117 Rotary disc filter. (*FLSmidth.*)

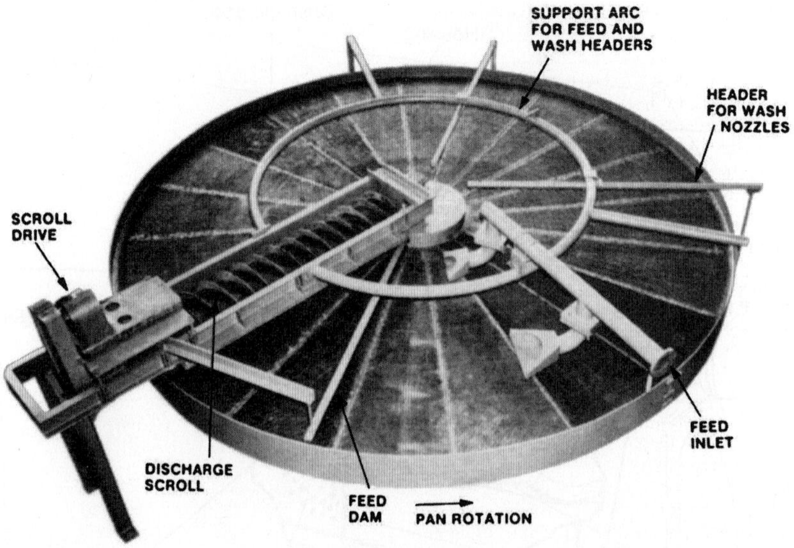

FIG. 18-118 Continuous horizontal vacuum table filter. (*FLSmidth.*)

A *tilting-pan filter* is a modification of the table or pan filter in which each of the sectors is an individual pan pivoted on a radial axis to allow its inversion for cake discharge, usually assisted by an air blast. Filter-cake thicknesses of 50 to 100 mm (2 to 4 in) are common. Most applications involve free-draining inorganic-salt dewatering. In addition to the advantages and disadvantages common to all horizontal continuous filters, tilting-pan filters have the relative advantages of complete wash containment per sector, good cake discharge, filter-medium washing, and feasibility of construction in very large sizes, up to about 25 m (80 ft) in diameter, with about 75 percent of the area usable. Relative disadvantages are high capital cost (especially in smaller sizes) and mechanical complexity leading to higher maintenance costs.

A *horizontal-belt filter* filter consists of a slotted or perforated elastomer drainage belt driven as a conveyor belt carrying a filter fabric belt (Fig. 18-119). Both belts are supported by and pass across a support deck. A vacuum pan, aligned with the slots in the elastomer belt, forms a continuous vacuum surface that may include multiple zones for cake formation,

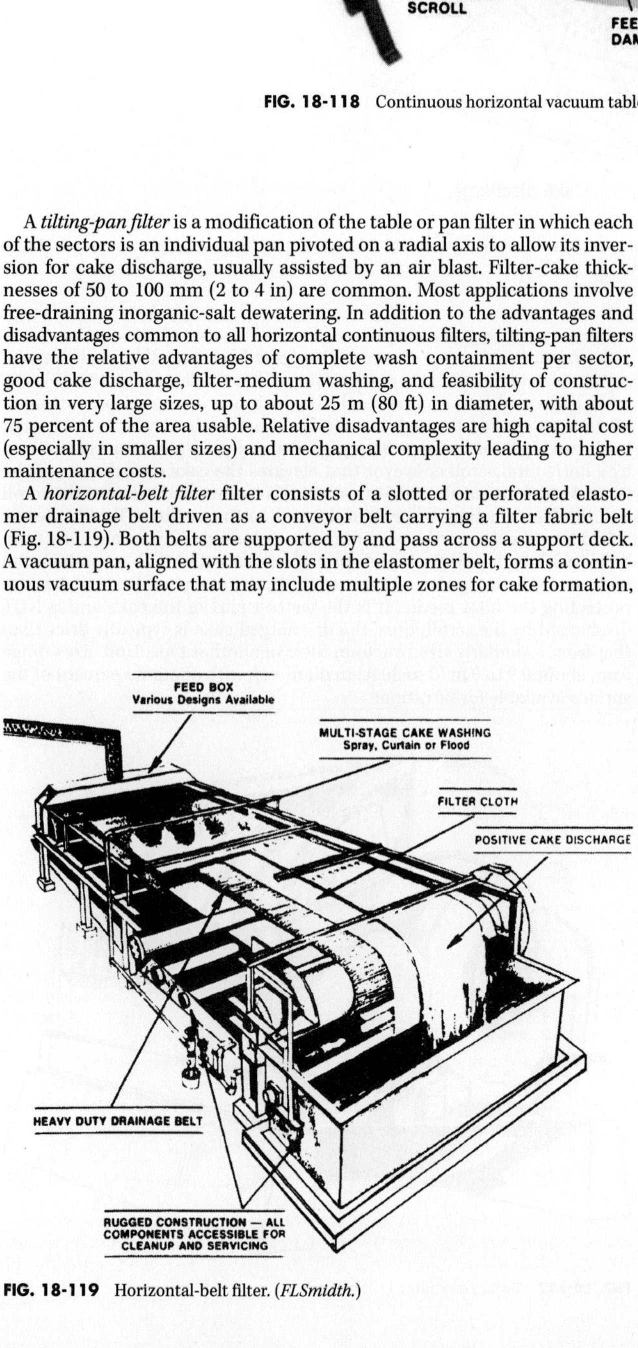

FIG. 18-119 Horizontal-belt filter. (*FLSmidth.*)

washing and final dewatering. Several manufacturers provide horizontal-belt filters, the major differences among which lie in the construction of the drainage belt, the method of retaining the slurry/cake on the belt, and the method of maintaining the alignment of the filter medium. The filters are rated according to the available active filtration area. *Indexing horizontal-belt filters* do away with the elastomer drainage belt of the original design in favor of large drainage pans directly beneath the filter medium. Either the pans or the filter medium is indexed to provide a pseudo-continuous filtration operation. The applied vacuum is cycled with the indexing operation to minimize wear to the sliding surfaces, and the indexing filter must be de-rated for the indexing cycle. The indexing horizontal-belt filter avoids the problem of process compatibility with the elastomer drainage belt. The major differences among the indexing machines of several manufacturers lie in the method of indexing and the method of cycling the applied vacuum.

The method of feeding, washing, dewatering, and discharging is essentially the same with all horizontal-belt filters. Slurry is fed at one end by overflow weirs or a fantail chute; wash liquor, if required, is applied by sprays or weirs at one or more locations as the formed cake moves along the filter. Wiping dams and separations in the drainage pan(s) provide controlled wash application. The cake is discharged as the filter-medium belt passes over the end pulley after separation from the drainage surface. Separating the filter medium from the drainage surface allows thorough spray cleaning of the filter-medium belt. The duration of the filtration cycle is controlled by belt speed that may be as high as 1 m/s (3.3 ft/s) and is typically variable. The minimum possible cake thickness, at a given solids loading, which can be effectively discharged limits the belt speed from a process point of view. The maximum cake thickness is dependent on the method used to retain the slurry during cake formation and can be 100 to 150 mm (4 to 6 in) with fast-draining materials.

Some of the advantages of horizontal-belt filters are the precise control of the filtration cycle, including the capability for countercurrent washing of the cake, effective cake discharge, and thorough cleaning of the filter medium belt. The horizontal-belt filter's primary disadvantage is that at least half of the filtration medium is always idle during the return loop. This contributes to a significantly higher capital cost that can be two to four times that of a drum or disc filter with equal area. Horizontal-belt filters with active filtration area ranging from 0.18 m^2 to 330 m^2 (2 ft^2 to 3550 ft^2) on a single machine have been installed.

Filter Thickeners Thickeners are devices that remove a portion of the liquid from a slurry to increase the concentration of solids in suspension. The most common method of thickening is by gravity sedimentation. However, occasions may arise in which a filter may be called upon for thickening service. The filter press with special plates containing flow channels that keep velocity high enough to prevent cake buildup, cycled tube or candle filters with the cake discharge into the filter tank, and continuous leaf filters that use rotating elements adjacent to the filtering surfaces to limit filter cake buildup can be used as thickeners. Examples of these filters include the Shriver Thickener, the Industrial Hydra-Shoc Filter employing Back Pulse Technology, the DrM, the Dr. Müller AG, and the Contibac Thickener. Crossflow membrane filters may also be used as filter thickeners.

Clarifying Filters Clarifying filters are used to separate small quantities of solids from a liquid. When the solids are finely divided enough to be observed only as a haze, the filter that removes them is sometimes called a polishing filter. The prefiltered slurry generally contains no more than 0.10 percent solids, the size of which may vary widely (0.01 to 100 μm). The filter usually produces no visible cake, sometimes because the amount of solids removed is so small, sometimes because the particles are removed by being entrapped within rather than upon the filter medium. Clarifying filters are primarily used in beverage and water polishing, pharmaceutical filtration, fuel- and lubricating-oil clarification, electroplating-solution conditioning, and dry-cleaning-solvent recovery. Clarifying filters may be classified as disc and plate presses, cartridge clarifiers, precoat pressure filters, deep-bed filters, and miscellaneous types. Membrane filters constitute a special class of plate presses and cartridge filters.

Disc Filters and Plate Presses Filters employing cakes of cotton fibers (filter-masse) or sheets of paper or other media are used widely for the polishing of beverages, plating solutions, and other low-viscosity liquids that contain small quantities of suspended matter. The term *disc filter* is applied to assemblies of discs made of synthetic or cellulose fibers and sealed into a pressure case. The discs may be preassembled into a self-supporting unit (Fig. 18-120), or each disc may rest on an individual screen or plate against which it is sealed as the filter is closed (Fig. 18-121). The liquid flows through the discs, and into a central or peripheral discharge manifold. Flow rates are on the order of 122 L/(min · m^2) [3 gal/(min · ft^2)], and the operating pressure normally does not exceed 345 kPa (50 psig) (usually it is less). Disc filters are almost always operated as pressure filters. Individual units deliver up to 378 L/min (6000 gal/h) of low-viscosity liquid.

Disc-and-plate assemblies somewhat resemble horizontal-plate pressure filters, which, in fact, may be used for polishing. In one design (Sparkler VR filter), both sides of each plate are used as filtering surfaces, having paper or other media clamped against them.

Pulp filters employ one or more packs of filter-masse (cellulose fibers compressed to a compact cylinder) stacked into a pressure case. The packs are sometimes supported in individual trays that provide drainage channels and sometimes rest on one another with a loose spacer plate between each two packs and with a drainage screen buried in the center of each pack. The liquid being clarified flows under a pressure of 345 kPa (50 psig) or less through the pulp packs and into a drainage manifold. Flow rates are somewhat less than for disc filters, on the order of 20 L/(min · m^2) [0.5 gal/(min · ft^2)]. Pulp filters are used chiefly to polish beverages. The filter-masse may be washed in special washers and re-formed into new cakes.

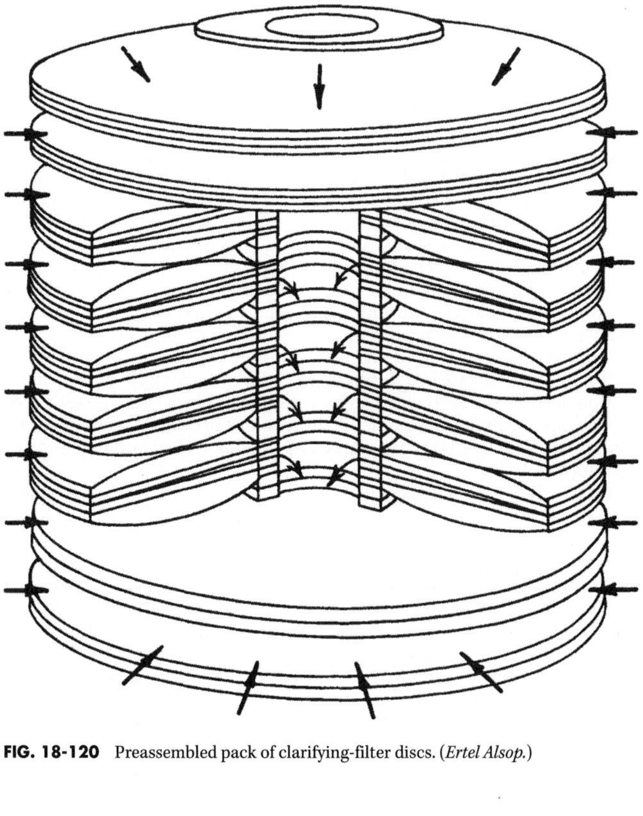

FIG. 18-120 Preassembled pack of clarifying-filter discs. (*Ertel Alsop.*)

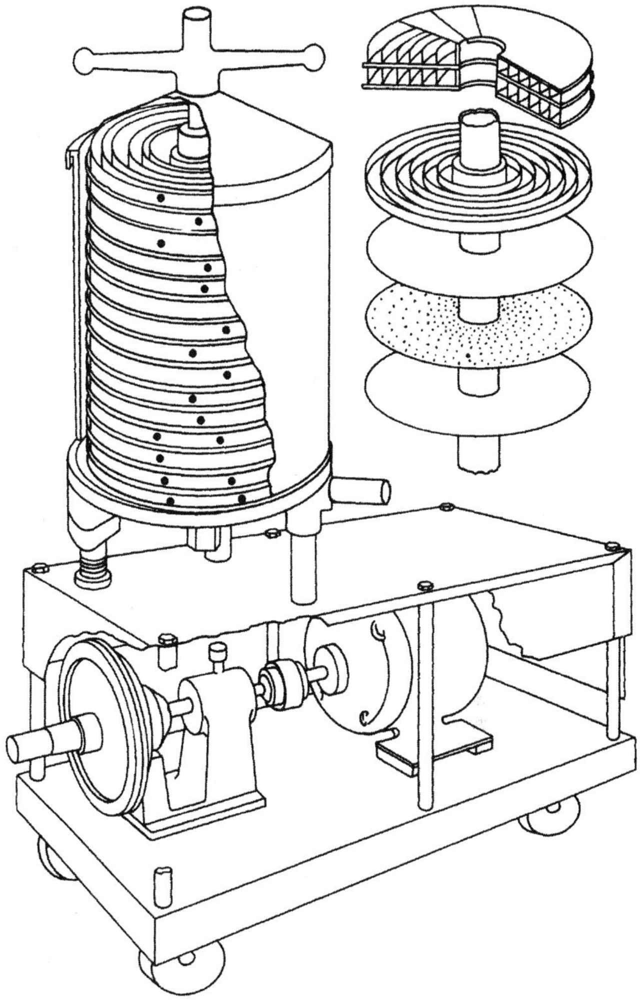

FIG. 18-121 Disc-and-plate clarifying-filter assembly. (*Ertel Alsop.*)

Plate presses, sometimes called sheet filters, are assemblies of plates, sheets of filter media, and sometimes screens or frames. They are essentially modified filter presses with practically no cake-holding capacity. A press may consist of many plates or of a single filter sheet between two plates, the plates may be rectangular or circular, and the sheets may lie in a horizontal or vertical plane. The operation is similar to that of a filter press, and the flow rates are about the same as for disc filters. The operating pressure usually does not exceed 138 kPa (20 psig). The presses are used most often for low-viscosity liquids.

Disc, pulp, and sheet filters accomplish extreme clarification. Not infrequently, their mission is complete removal of particles above a stipulated cut size, which may be much less than 1 μm. They operate over a particle-size range of four to five orders of magnitude, contrasting with two orders of magnitude for most other filters. Therefore, they involve a variety of kinds and grades of filter media, often in successive stages. One unique medium, the Zeta Plus filter medium from 3M Purification, consists of a composite of cellulose and inorganic filter aids that have a positive charge and provide an electrokinetic attraction to hold colloids (usually negatively charged). These media therefore provide both mechanical straining and electrokinetic adsorption.

Cartridge Clarifiers *Cartridge clarifiers* are units that use one or more replaceable or renewable cartridges containing the active filter element. The unit is usually placed in-line, and clarification occurs while the liquid is in transit.

Mechanical or edge filters consist of stacks of discs separated to precise intervals by spacer plates, or a wire wound on a cage in grooves of a precise pitch, or a combination of the two. The liquid to be filtered flows radially between the discs, wires, or layers of paper, and particles larger than the spacing are screened out. Edge filters can remove particles down to 0.001 in (25 μm). They have small solids-retaining capacity and hence must be cleaned often to avoid plugging, and continuous cleaning is provided in some filters.

Micronic clarifiers include the greatest number of cartridge clarifiers of the micronic class, with elements of fiber, resin-impregnated filter paper, porous stone, or porous stainless steel of controlled porosity. The elements may be chosen to remove particles larger than a fraction of a micrometer, although many are made to pass 10-μm solids and smaller. By proper choice of multiple-cylinder cartridges or multiple cartridges in parallel, any desired flow rate can be obtained at a reasonable pressure drop, often less than 138 kPa (20 psig).

When the pressure rises to the permissible maximum, the cartridge must be opened and the element replaced. Micronic elements of the fiber type cannot be cleaned, and they are priced so that they can be discarded or the filter medium replaced economically. Stone elements usually must be cleaned, a process best accomplished by the manufacturer of the porous ceramic or in accordance with the manufacturer's directions. The user can clean stainless-steel elements by chemical treatment.

Cartridge filters are flexible: cartridges of different ratings and materials of construction can be interchanged, permitting modifications for shifting conditions. They have the disadvantage of very limited solids-handling capability, and concentrations in the feed are limited to about 0.01 percent solids. The biggest limitation is the need to open the filter to replace cartridges. Some manufacturers—for example, the Hydraulic Research Division of Textron Inc. and the Fluid Dynamics Division of Brunswick Corp.—have designed cartridges of bonded metal fibers that can be backflushed or chemically cleaned without opening the unit. These filters, which can operate at temperatures to 482°C (900°F) and at pressures of 33 MPa (325 atm) or greater, are particularly useful for filtering polymers.

Granular Media Filters *Granular media filters*, of which there are many types, are used for clarification, operating either as gravity or pressure filters. Gravity filters rely on a difference in elevation between inlet and outlet to provide the driving force needed to force the liquid through the granular media. Pressure filters use closed vessels operating at relatively low pressure differentials, on the order of 50 to 70 kPa (7 to 10 psig), which may function in either an upflow or a downflow mode.

The media may be a single material, such as sand, but more often will consist of two or even three layers of different materials, such as anthracite coal in the top layer and sand in the lower one. Solids are captured throughout the bed depth, rather than on the surface, and the gradient in void size provides substantially more solids-holding capacity. The anthracite layer, typically employing 1-mm grain size, serves as a roughing filter and also provides a flocculating action that helps the finer sand, about 0.5-mm particle size, to serve as an effective polishing zone. Media depths vary, but 0.7 to 1.0 m is typical of a dual media installation. Deeper beds of up to 2.5 m (8 ft) are used in some cases involving special applications where greater solids-holding capacity is desired.

Filtrate is collected in the underdrain system, which may be as simple as a network of perforated pipes covered by graded gravel or a complex structure with slotted nozzles or conduits that will retain the finest sand media while maintaining high flow rates. This latter design allows the use of both air and liquid for the backwashing and cleaning operations.

Backwashing usually is carried out when a limiting pressure drop is reached and before the bed becomes nearly filled with solids, which would lead to a deterioration in filtrate clarity. Cleaning the media is greatly aided by the use of an air scour that helps break loose the trapped solids and provides efficient removal of this material in the subsequent backflushing step. The filtration action tends to agglomerate the filtered solids and, as a result, these generally will settle out readily from the backwash fluid. If the filter is handling a clarifier overflow, it is usually possible to discharge the backwash liquid into the clarifier without risk of these solids returning to the filter. Filter media consumption is low, with normal replacement usually being less than 5 percent per year.

These filters are best applied on relatively dilute suspensions, <150 mg/L suspended solids, allowing operation at relatively high rates, 7.5 to 15.0 $m^3/m^2/h$ (3 to 6 gpm/ft^2). Solids capture will range from 90 to 98 percent in a well-designed system. Typical operating cycles range from 8 to 24 h of filtration (and up to 48 h in municipal water treatment), followed by a backwash interval of 15 to 30 min. Applications are principally in municipal and wastewater treatment, but granular media filters also have been employed in industrial uses such as pulp and paper plant inlet water treatment; removal of oil, grease, and scale from steelmaking process wastewater; and clarification of electrolyte in copper electrowinning operations.

Veolia Water Technologies, Maxi-Flo Filter. The Maxi-Flo Filter is an example of the upflow closed-vessel design. Filtration rates to 0.0081 $m^3/(m^2 \cdot s)$ [12 gal/($ft^2 \cdot$ min)] and filter cross section areas up to 10.5 m^2 (113 ft^2) are possible. Deep-bed filtration has been reviewed by Tien and Payatakes [*Am. Inst. Chem. Eng. J.* **25:** 737 (1970)] and by Oulman and Baumann [*Am. Inst. Chem. Eng. Symp. Ser.* **73**(171): 76 (1977)].

Dyna Sand Filter. A filter that avoids batch backwashing for cleaning, the Dyna Sand Filter is available from Parkson Corporation. The bed is continuously cleaned and regenerated by recycling solids internally through an air-lift pipe and a sand washer. Thus a constant pressure drop is maintained across the bed, and the need for parallel filters to allow continued on-stream operation, as with conventional designs, is avoided.

Miscellaneous clarifiers of various types such as cartridge, magnetic, and bag filters are widely used in polishing operations, generally to remove trace amounts of suspended solids remaining from prior unit operations. A thorough discussion of cartridge and felt strainer bag filters is available in Schweitzer 1996, sec. 4.6 (Nickolaus) and sec. 4.7 (Wrotnowski).

SELECTION OF FILTRATION EQUIPMENT

If a process developer who must provide the mechanical separation of solids from a liquid has cleared the first decision hurdle by determining that filtration is the way to get the job done (see the final subsection of Sec. 18, Selection of a Solids-Liquid Separator)—or that it must remain in the running until some of the details of equipment choice have been settled—choosing the right filter and the right filtration conditions may still be difficult. Much as in the broader determination of which unit operation to employ, the selection of filtration equipment involves the balancing of process specifications and objectives against capabilities and characteristics of the various equipment choices (including filter media) available. The important process-related factors are slurry character, production throughput, process conditions, performance requirements, and permissible materials of construction. The important equipment-related factors are type of cycle (batch or continuous), driving force, production rates of the largest and smallest units, separation sharpness, washing capability, dependability, feasible materials of construction, and cost. The estimated cost must account for installed cost, equipment life, operating labor, maintenance, replacement filter media, and costs associated with product-yield loss (if any). In between the process and equipment factors are considerations of slurry preconditioning and use of filter aids.

Slurry characteristics determine whether a clarifying or a cake filter is appropriate; and if the latter, they determine the rate of formation and nature of the cake. They affect the choice of driving force and cycle as well as specific design of machine.

There are no absolute selection techniques available to come up with the "best" choice since there are so many factors involved, many of them difficult to make quantitative and, not uncommonly, some contradictory in their demands. However, there are some published general suggestions to guide the thinking of the engineer who faces the task of selecting filtration equipment. Figure 18-122 is a decision tree designed by Tiller [*Chem. Eng.* **81**(9): 118 (1974)] to show the steps to be followed in solving a filtration problem. It is erected on the premise that the rate of cake formation is the most important guide to equipment selection. A filter-selection process proposed by Purchas (*Solid/Liquid Separation Equipment Scale-Up,* Uplands Press, Croydon, England, 1977, pp. 10–14) employs additional criteria and is based on a combination of process specifications and the results of simple tests. The filter application is coded by the use of Figs. 18-123, 18-124, and 18-125, and the resulting codes are matched against Table 18-15 to identify possible filters. Information needed for Fig. 18-124 can be obtained by observing the settling of a slurry sample (Purchas suggests 1 L) in a graduated cylinder. The filter-cake growth rate (Fig. 18-134) is determined by small-scale leaf or funnel tests as described earlier.

Almost all types of continuous filters can be adapted for cake washing. The effectiveness of washing is a function of the number of wash displacements applied, and this, in turn, is influenced by the ratio of wash time to cake-formation time. Countercurrent washing, particularly with three or more stages, is usually limited to horizontal filters, although a two-stage countercurrent wash sometimes can be applied on a drum filter handling freely filtering material, such as crystallized salts. Cake washing on batch filters is commonly done, although, generally, a greater number of wash displacements may be required in order to achieve the same degree of washing obtainable on a continuous filter.

Continuous filters are most attractive when the process application is a steady-state continuous one, but the rate at which cake forms and

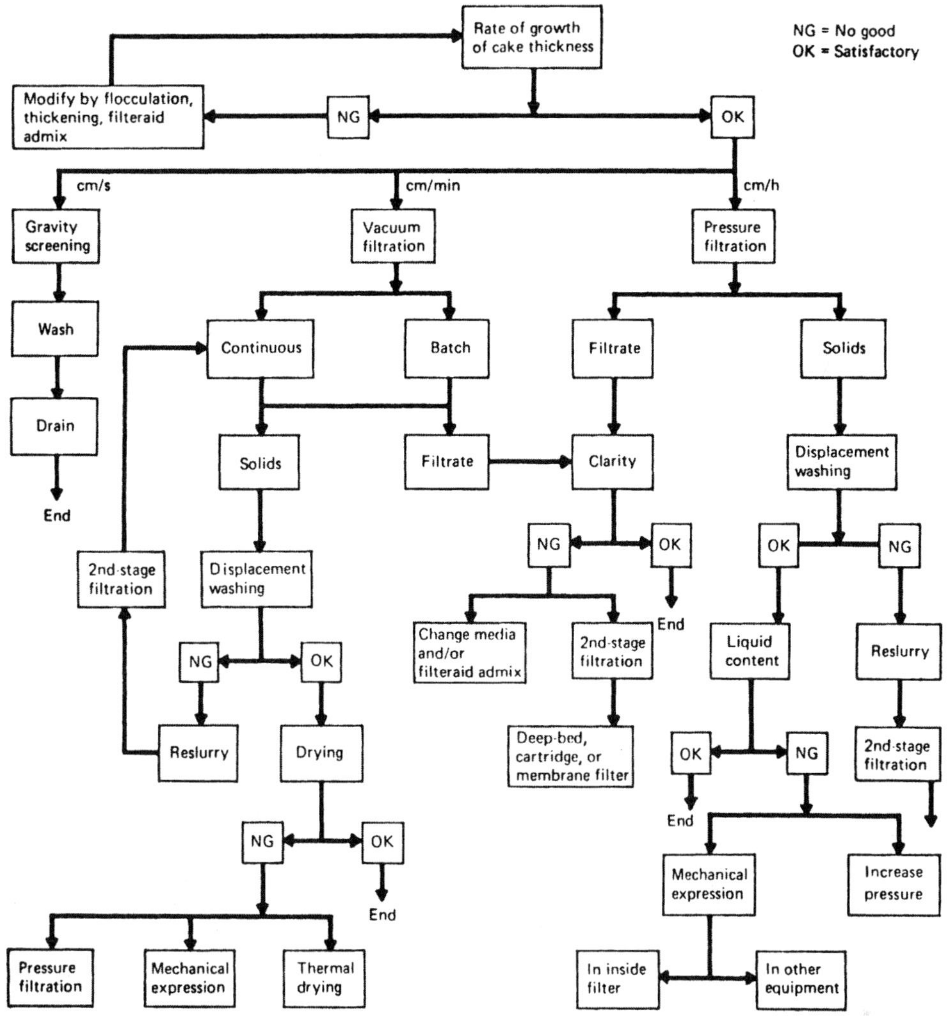

FIG. 18-122 Decision pattern for solving a filtration problem. [*Tiller,* Chem. Eng. ***81***(*9*)*: 118* (*1974*)*, by permission.*]

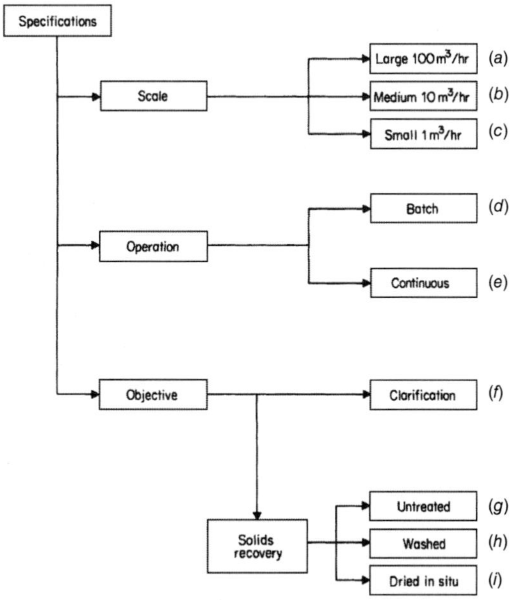

FIG. 18-123 Coding the problem specification. (*Purchas, Solid/Liquid Separation Equipment Scale-Up, Uplands Press, Croydon, England, 1977, p. 10, by permission.*)

the magnitude of production rate are sometimes overriding factors. A rotary vacuum filter, for example, is a dubious choice if a 3-mm (0.12-in) cake will not form under normal vacuum in less than 5 min and if less than 1.4 m^3/h (50 ft^3/h) of wet cake is produced. Upper production-rate limits to the practicality of batch units are harder to establish, but any operation above 5.7 m^3/h (200 ft^3/h) of wet cake should be considered for continuous filtration if it is at all feasible. Again, however, other factors such as the desire for flexibility or the need for high pressure may dictate batch equipment.

For estimating filtration rate (therefore, operating pressure and size of the filter), washing characteristics, and other important features, small-scale tests such as the leaf or pressure bomb tests described earlier are usually essential. In the conduct and interpretation of such tests, and for advice on labor requirements, maintenance schedule, and selection of accessory equipment, the assistance of a dependable equipment vendor is advisable.

FILTER COSTS

One of the factors affecting the selection of a filter system is total cost of carrying out the separation with the selected machine. An important component of this cost item is the installed cost of the filter, which starts with the purchase price.

Costs per unit of filtration area, based on 2015 prices, are presented in Fig. 18-126 for some of the primary types of industrial filters. They have a claimed accuracy of ±10 percent, and they should be used confidently only with study-level cost estimations (±25 percent) at best. The cost of delivery to the plant can be approximated as 7 to 10 percent of the FCA points of manufacture price.

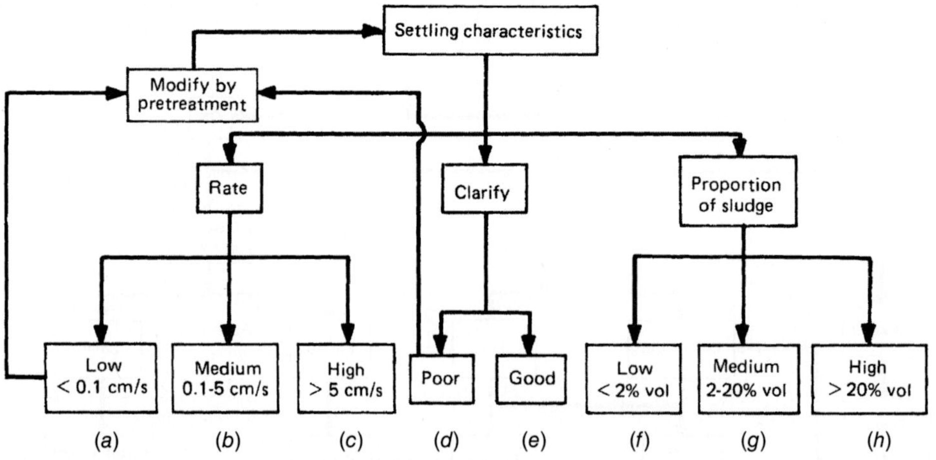

FIG. 18-124 Coding the settling characteristics of a slurry. (*Purchas, Solid/Liquid Separation Equipment Scale-Up, Uplands Press, Croydon, England, 1977, p. 11, by permission.*)

The cost of the filter station includes only the cost of the filter, packed, FCA points of manufacture. Accessories typically add 20 to 30 percent to the price of the filter, depending on the application and the filter type. Examples of accessories are feed pumps and storage facilities, precoat tanks, vacuum systems (often a major cost factor for a vacuum filter station), and compressed-air systems. Costs to deliver the accessories to the plant are 7 to 10 percent of the accessory costs. Large items that must be shipped break-bulk will incur higher shipping costs. Installation costs must be estimated with reference to local labor costs and site-specific considerations.

The absence of data for some types of filters, in particular the filter press, is due to the complex variety of individual features and materials of construction. For information about filters not represented in Fig. 18-126, and for firmer estimates for those types presented, vendors should be consulted. In all cases of serious interest, consultation should take place early in the evaluation procedure so that it can yield timely advice on testing, selection, and price.

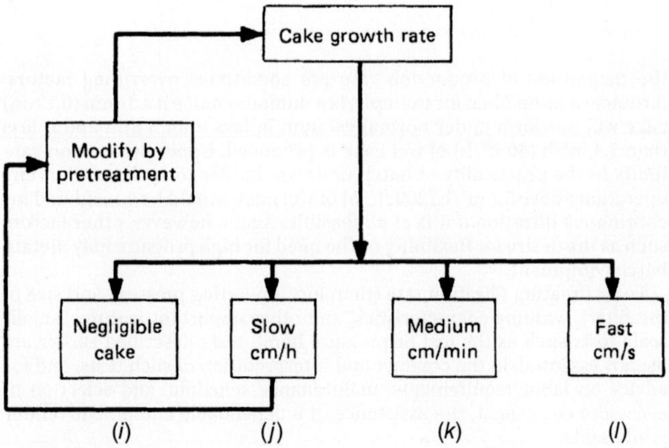

FIG. 18-125 Coding the filtration characteristics of a slurry. (*Purchas, Solid/Liquid Separation Equipment Scale-Up, Uplands Press, Croydon, England, 1977, p. 12, by permission.*)

TABLE 18-15 Classification of Filters According to Duty and Slurry-Separation Characteristics*

Type of equipment	Suitable for duty specification†	Required slurry-separation characteristics‡: Slurry-settling characteristics	Slurry-filtering characteristics
Deep-bed filters	a or b e f	A D F	I
Cartridges	b or c d f	A or B D or E F	
Batch filters			
Pressure vessel with vertical elements	a, b, or c d f, g, h, or i	A or B D or E F or G	I or J
Pressure vessel with horizontal elements	b or c d g or h	A or B D or E F or G	J or K
Filter presses	a, b, or c d f, g, h, or i	A or B D or E F, G, or H	I or J
Variable-volume filters	a, b, or c d or e g (or h)	A or B D or E G or H	J or K
Continuous filters			
Bottom-fed drum or belt drum	a, b, or c e f, g, h, or i	A or B D or E F, G, or H	I, J, K, or L
Top-fed drum	a, b, or c e g, i (or h)	C E G or H	L
Disc	a, b, or c e g, i (or h)	A or B) D or E G or H	J or K
Horizontal belt, pan, or table	a, b, or c d or e g or h	A, B, or C D or E F, G, or H	J, K, or L

Adapted from Purchas, *Solid/Liquid Separation Equipment Scale-Up*, Uplands Press, Croydon, England, 1977, p. 13, by permission.
†Symbols are identified in Fig. 18-123.
‡Symbols are identified in Figs. 18-124 and 18-125.

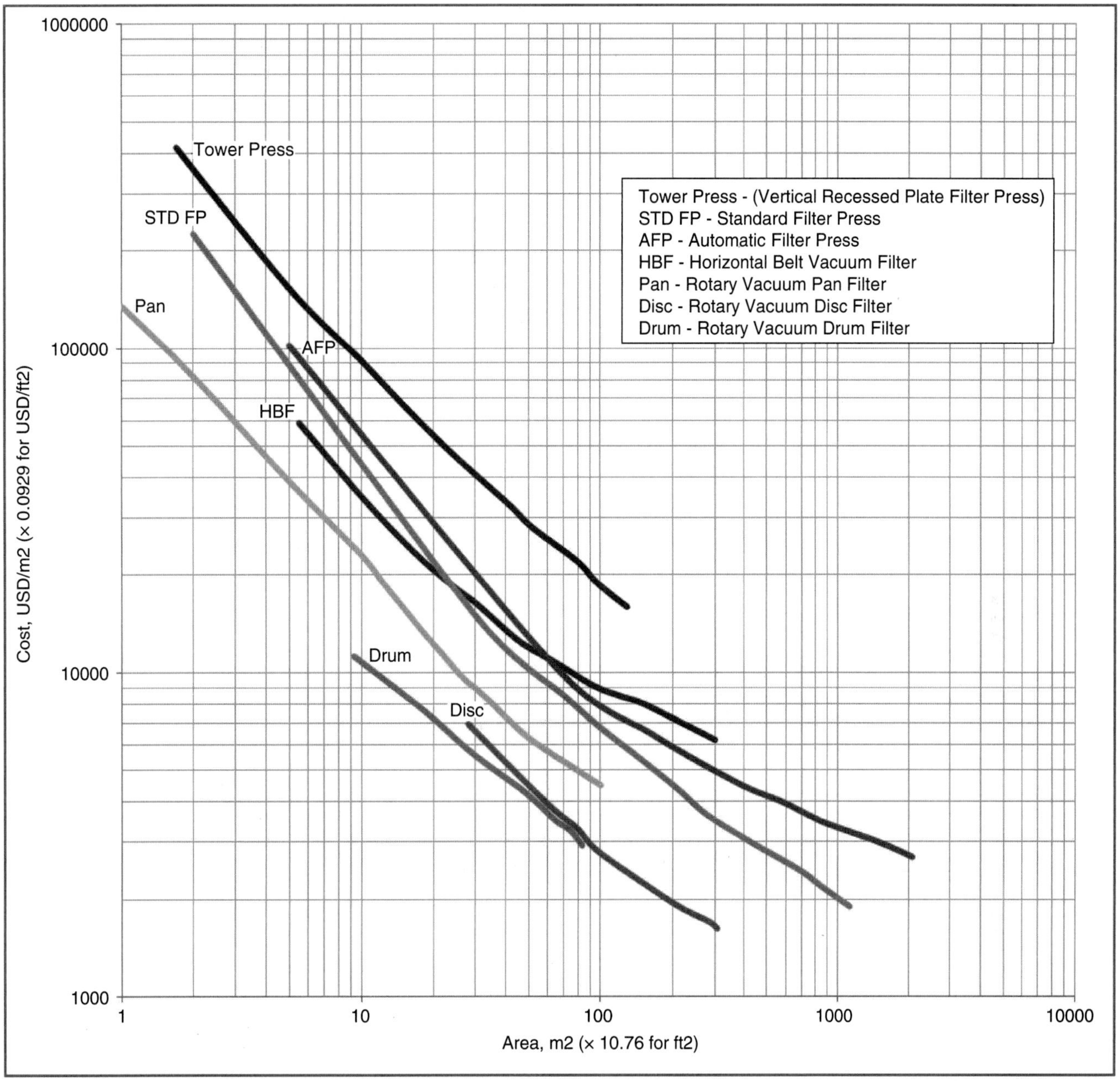

FIG. 18-126 Prices of some types of primary industrial filters, FCA points of manufacture. (*FLSmidth.*)

CENTRIFUGES

General References: Ambler, in McKetta, *Encyclopedia of Chemical Processing and Design,* vol. 7, Marcel Dekker, New York, 1978; also sec. 4 in Schweitzer, *Handbook of Separation Techniques for Chemical Engineers,* McGraw-Hill, New York, 1979; Ambler and Keith, in Perry and Weissberger, *Separation and Purification Techniques of Chemistry,* 3d ed., vol. 12, Wiley, New York, 1978; Flood, Porter, and Rennie, *Chem. Eng.* **73**(13): 190 (1966); Greenspan, *J. of Fluid Mech.* **127**(9): 91 (1983); Gerl, Stadager, and Stahl, *Chemical Eng. Progress* **91**: 48–54 (May 1995); Hultsch and Wilkesmann, chap. 12 in *Solid/Liquid Separation Equipment Scale-Up,* ed. Purchas, Uplands Press, Croydon, England, 2d ed., 1986; Karolis, A., *The Technology of Solid-Bowl Scroll Centrifuges;* McGillicuddy, *Chem. Process. Mag.* **59**(12): 54–59 (Dec. 1996); Leung, *Chem. Eng.* (1990); Leung, *Fluid-Particle Sep. J.* **5**(1): 44 (1992); Leung, *10th Pittsburgh Coal Conf. Proceed.* (1993); Leung and Shapiro, *Filtration and Separation Journal,* Sept. and Oct. 1996; Leung and Shapiro, U.S. Patents 5,520,605 (May 28, 1996), 5,380,266 (Jan. 10, 1995), and 5,401,423 (March 28, 1995); Mayer and Stahl, *Aufbereitungs-Technik* **11**: 619 (1988); Moyers, *Chem. Eng.* **73**(13): 182 (1966); Records, chap. 6 in *Solid/Liquid Separation Equipment Scale-Up,* ed. Purchas, Uplands Press, Croydon, England, 2d ed., 1986; Smith, *Ind. Eng. Chem.* **43**: 439 (1961); Sullivan and Erikson, ibid., p. 434; Svarovsky, ed., chap. 7 in *Solid-Liquid Separation,* 3d ed., Butterworths, London, 1990; Tiller, *AIChE J.* **33**(1): (1987); Zeitsch, chap. 14 in *Solid-Liquid Separation,* 3d ed., ed. Svarovsky, Butterworths, London, 1990.

Nomenclature

Symbol	Definition	SI units	U.S. Customary units
a	Acceleration	m/s^2	ft/s^2
B_o	Bond number	Dimensionless	Dimensionless
b	Basket axial length	m	ft
C_f	Frictional coefficient	Dimensionless	Dimensionless
D	Bowl/basket diameter	m	in
d	Particle diameter	m	ft
Ek	Ekman number	Dimensionless	Dimensionless
F	Cumulative fraction	Dimensionless	Dimensionless
G	Centrifugal gravity	m/s^2	ft/s^2
g	Earth gravity	m/s^2	ft/s^2
h	Cake height	m	ft
K	Cake permeability	m^2	ft^2

Nomenclature *(Continued)*

Symbol	Definition	SI units	U.S. Customary units
L	Length	m	ft
M	Mass	kg	lb
m	Bulk mass rate	kg/s	lb/h
N_c	Capillary number	Dimensionless	Dimensionless
P	Power	kw	hp
Q	Flow rate	L/s	gpm
R_m	Filter media resistance	m^{-1}	ft^{-1}
Ro	Rossby number	Dimensionless	Dimensionless
r	Radius	m	ft
Rec	Solids recovery	Dimensionless	Dimensionless
S	Liquid saturation in cake (= volume of liquid/ volume cake void)	Dimensionless	Dimensionless
sg	Specific gravity	Dimensionless	Dimensionless
t	Time	s	s
t_d	Time	Dimensionless	Dimensionless
u	Velocity	m/s	ft/s
$V\theta$	Circumferential velocity	m/s	ft/s
V_c	Bulk cake volume	m^3	ft^3
V	Velocity	m/s	ft/s
W	Weight fraction of solids	Dimensionless	Dimensionless
Y	Yield	Dimensionless	Dimensionless
Z	Capture efficiency	Dimensionless	Dimensionless
	Greek symbols		
ε	Cake void volume fraction	Dimensionless	Dimensionless
ε_s	Cake solids volume fraction	Dimensionless	Dimensionless
μ	Liquid viscosity	Pa·s	P
ρ_s	Solid density	kg/m^3	lb/ft^3
ρ_L	Liquid density	kg/m^3	lb/ft^3
σ	Surface tension	N/m	lbf/ft
σ_h	Hoop stress	Pa	psi
θ	Angle	Radian	degree
Σ	Scale-up factor (equivalent sedimentation area)	m^2	ft^2
ξ	Time	Dimensionless	Dimensionless
ϕ	Feed solids volume fraction	Dimensionless	Dimensionless
τ_y	Yield stress	Pa	psi
Δ	Differential speed	1/s	r/min
Ω	Angular speed	1/s	r/min
	Subscripts		
b	Bowl or basket		
c	Cake		
e	Centrate		
f	Feed		
f	Filtrate		
acc	Acceleration		
p	Pool		
con	Conveyance		
t	Tangential		
L	Liquid		
s	Solid		

INTRODUCTION

Centrifuges for the separation of solids from liquids are of two general types: (1) sedimentation centrifuges, which require a difference in density between the two or three phases present (solid-liquid or liquid-liquid or liquid-liquid-solid or solid-liquid-solid) and (2) filtering centrifuges (for solid-liquid separation), in which the solid phase is retained by the filter medium through which the liquid phase is free to pass. The following discussion is focused on solid-liquid separation for both types of centrifuges; however, a dispersed liquid phase in another continuous liquid phase as used in sedimenting centrifuges exhibits similar behavior to that of solid in liquid, and therefore the results developed are generally applicable. The use of centrifuges covers a broad range of applications, from separation of fine calcium carbonate particles of less than 10 μm to coarse coal of 0.013 m (½ in).

GENERAL PRINCIPLES

Centripetal and Centrifugal Acceleration A centripetal body force is required to sustain a body of mass moving along a curve trajectory. The force acts perpendicular to the direction of motion and is directed radially inward. The centripetal acceleration, which follows the same direction as the force, is given by the kinematic relationship:

$$a = \frac{V_\theta^2}{r} \tag{18-79}$$

where V_θ is the tangential velocity at a given point on the trajectory and r is the radius of curvature at that point. This analysis holds for the motion of a body in an inertial reference frame, such as a stationary laboratory. It is most desirable to consider the process in a centrifuge, and the dynamics associated with such, in a noninertial reference frame such as in a frame rotating at the same angular speed as the centrifuge. Here, additional forces and accelerations arise, some of which are absent in the inertial frame. Analogous to centripetal acceleration, an observer in the rotating frame experiences a centrifugal acceleration directed radially outward from the axis of rotation with magnitude:

$$a = \Omega^2 r \tag{18-80}$$

where Ω is the angular speed of the rotating frame and r is the radius from the axis of rotation.

Solid-Body Rotation When a body of fluid rotates in a solid-body mode, the tangential or circumferential velocity is linearly proportional to radius:

$$V_\theta = \Omega r \tag{18-81}$$

as with a system of particles in a rigid body. Under this condition, the magnitude of the centripetal acceleration, Eq. (18-79), equals that of the centrifugal acceleration, Eq. (18-80), despite the fact that these accelerations are considered in two different reference frames. Hereafter, the rotating frame attached to the centrifuge is adopted. Therefore, centrifugal acceleration is exclusively used.

***G*-Level** Centrifugal acceleration G is measured in multiples of earth gravity g:

$$\frac{G}{g} = \frac{\Omega^2 r}{g} \tag{18-82}$$

With the speed of the centrifuge Ω in r/min and D the diameter of the bowl,

$$\frac{G}{g} = 0.000559\Omega^2 D, D(m) \tag{18-83}$$

With D in inches, the constant in Eq. (18-83) is 0.0000142. G can be as low as 100g for slow-speed, large basket units to as much as 10,000g for high-speed, small decanter centrifuges and 15,000g for disc centrifuges. Because G is usually very much greater than g, the effect due to earth's gravity is negligible. In analytical ultracentrifuges used to process small samples, G can be as much as 500,000g to effectively separate two phases with a very small density difference.

Coriolis Acceleration The Coriolis acceleration arises in a rotating frame, which has no parallel in an inertial frame. When a body moves at a linear velocity u in a rotating frame with angular speed Ω, it experiences a Coriolis acceleration with magnitude:

$$a = 2\Omega u \tag{18-84}$$

The Coriolis vector lies in the same plane as the velocity vector and is perpendicular to the rotation vector. If the rotation of the reference frame is counterclockwise, then the Coriolis acceleration is directed 90° clockwise from the velocity vector, and vice versa when the frame rotates clockwise. The Coriolis acceleration distorts the trajectory of the body as it moves rectilinearly in the rotating frame.

Effect of Fluid Viscosity and Inertia The dynamic effect of viscosity on a rotating liquid slurry as found in a sedimenting centrifuge is confined in very thin fluid layers, known as Ekman layers. These layers are adjacent to rotating surfaces that are perpendicular to the axis of rotation, such as bowl heads, flanges, and conveyor blades. The thickness of the Ekman layer δ is of the order

$$\delta = \sqrt{\frac{\mu/\rho}{\Omega}} \tag{18-85}$$

where μ/ρ is the kinematic viscosity of the liquid. For example, with water at room temperature, μ/ρ is 1×10^{-6} m^2/s and for a surface rotating at $\Omega = 3000$ r/min, δ is 0.05 mm! These layers are very thin; nevertheless, they are responsible for the transfer of angular momentum between the rotating surfaces to the fluid during acceleration and deceleration. They worked together with the larger-scale inviscid bulk flow transferring momentum in a rather complicated way. This is demonstrated by the teacup example in which the cup's contents are brought to speed when it is stirred and brought to a halt after undergoing solid-body rotation. The viscous effect is characterized by the dimensionless Ekman number:

$$\text{Ek} = \frac{\mu/\rho}{\Omega L^2} \tag{18-86}$$

where L is a characteristics length. The Ekman number measures the scale of the viscous effect versus that of the bulk flow.

The effect of fluid inertia manifests during abrupt changes in velocity of the fluid mass. It is quantified by the Rossby number:

$$\text{Ro} = \frac{u}{\Omega L} \tag{18-87}$$

Typically, Ro is small to the order of 1, with the high end of the range showing possible effects due to inertia, whereas Ek is usually very small, 10^{-6} or smaller. Therefore, the viscous effect is confined to thin boundary layers with thickness $\text{Ek}^{1/2}L$.

Sedimenting and Filtering Centrifuges Under centrifugal force, the solid phase assumed to be denser than the liquid phase settles out to the bowl wall—sedimentation. Concurrently, the lighter, more buoyant liquid phase is displaced toward the smaller diameter—flotation. This is illustrated in Fig. 18-127*a*. Some centrifuges run with an air core, that is, with free surface, whereas others run with slurry filled to the center hub or even to the axis in which pressure can be sustained.

In a sedimenting centrifuge, the separation can be in the form of *clarification*, wherein solids are separated from the liquid phase and the clarity of the liquid phase is of prime concern. For biological sludge, polymers are used to agglomerate fine solids to facilitate clarification. Separation can also be in the form of *classification* and degritting, in which separation is effected by means of particle size and density. Typically, the finer solids (such as kaolin) of smaller size or density in the feed slurry are separated in the centrate stream as product (for example 90 percent of particles less than 1 μm), whereas the larger or denser solids are captured in cake as reject. Furthermore, separation can be in the form of *thickening*, where solids settle under centrifugal force to form a stream with concentrated solids. In *dewatering* or *deliquoring*, the objective is to produce dry cake with high solids consistency by centrifugation.

In a filtering centrifuge, separating solids from liquid does not require a density difference between the two phases. Should a density difference exist between the two phases, sedimentation is usually at a much more rapid rate than filtration. In both cases, the solid and liquid phases move toward the bowl under centrifugal force. The solids are retained by the filter medium, while the liquid flows through the cake solids and the filter. This is illustrated in Fig. 18-127*b*.

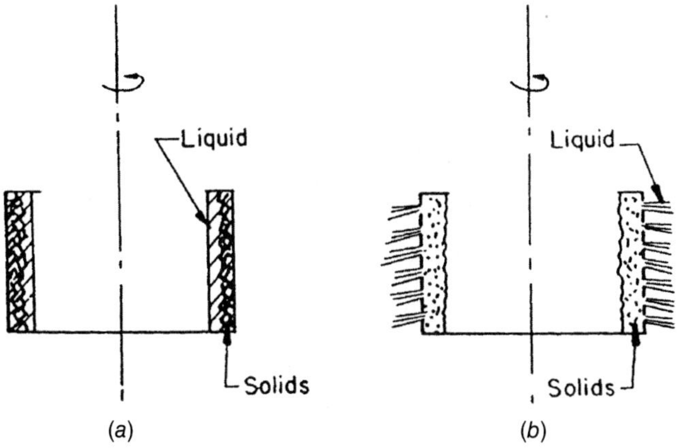

FIG. 18-127 Principles of centrifugal separation and filtration: (*a*) sedimentation in rotating imperforate bowl; (*b*) filtration in rotating perforate basket.

Performance Criteria Separation of a given solid-liquid slurry is usually measured by the purity of the separated liquid phase in the centrate (or liquid effluent) in sedimenting mode or the filtrate in filtering mode, and the separated solids in the cake. In addition, there are other important considerations. Generally, a selected subset of the following criteria are used, depending on the objectives of the process:

Cake dryness or moisture content
Total solids recovery
Polymer dosage
Size recovery and yield
Volumetric and solids throughput
Solid purity and wash ratio
Power consumption

Cake dryness by dewatering is required because the cake usually needs to be as dry as possible. Cake dryness is commonly measured by the solids fraction by weight W or by volume ε_s. The moisture content is measured by the complement of W or ε_s. The volume fraction of the pores and void in the wet cake is measured by the cake porosity $\varepsilon = (1 - \varepsilon_s)$, whereas the volume fraction of the liquid in the pores of the cake is measured by the saturation S. For well-defined solids in the cake with solid density (bone dry) ρ_s and liquid density ρ_L, and given that the cake volume V_c and the mass of solids in the cake w_s are known, the cake porosity is determined by

$$\varepsilon = 1 - \frac{w_s}{\rho_s V_c} \tag{18-88}$$

For undersaturated cake with $S < 1$, saturation can be inferred from the weight fraction of solids and the porosity of the cake, together with the solid and liquid densities:

$$S = \left(\frac{1-W}{W}\right)\left(\frac{1-\varepsilon}{\varepsilon}\right)\frac{\rho_s}{\rho_L} \tag{18-89}$$

When the cake is saturated $S = 1$, the cake porosity can be determined from Eq. (18-90) as

$$\varepsilon = \left(1 + \frac{\rho_L}{\rho_s}\frac{W}{1-W}\right)^{-1} \tag{18-90}$$

Cake dewatering by compression and rearrangement of the solids in the cake matrix reduce ε, yet the cake is still saturated with $S = 1$. (Assuming cake solids are ideal spheres of uniform size, the maximum packing, in rhombohedral arrangement, is such that $\varepsilon_s = 74$ percent or $\varepsilon = 26$ percent.) Drainage of liquid within the cake by centrifugation further reduces S to be less than 1. There is a lower limit on S that is determined by the cake height, dewatering time, centrifugal force as compared to the capillary and surface forces, and the surface roughness and porosity of the particles.

Total solids recovery is important in clarification as the clarity of the effluent is measured indirectly by the total solids recovered in the cake as

$$\text{Rec} = \frac{m_c W_c}{m_f W_f} \tag{18-91}$$

where subscripts c and f denote, respectively, the cake and the feed. m is the bulk mass flow rate in kg/s (lb/h).

Under steady state, the mass balance on both solids and liquid yield, respectively:

$$m_f W_f = m_c W_c + m_e W_e \tag{18-92}$$

$$m_f = m_c + m_e \tag{18-93}$$

From the preceding, it follows that

$$\text{Rec} = \frac{1-(W_e/W_f)}{1-(W_e/W_c)} \tag{18-94}$$

where subscript e represents liquid centrate. Stringent requirements on centrate quality or capture of valuable solid product often require the recovery to exceed 90 percent and, in some cases, 99+ percent. In such cases, the centrate solids are typically measured in ppm.

Polymer dosage by cationic and anionic polymers has been commonly used to coagulate and flocculate fine particles in the slurry. This is especially pertinent to biological materials such as those found in wastewater treatment. In the latter, cationic polymers are often used to neutralize the negative-charge ions left on the surface of the colloidal particles. Polymer dosage is measured by kg of dry polymer/1000 kg of dry solids cake (lbm of dry polymer/ton of dry solids cake). With liquid polymers, the equivalent (active) dry solid polymer is used to calculate the dosage. There is a minimum polymer dosage to agglomerate and capture the fines in the cake. Overdose can be undesirable for recovery and cake dryness. The range of optimal dosage is dictated by the types of solids in the slurry, slurry physical properties such as pH and ionic strength, and the operating condition and characteristics of the centrifuge. It is known that flocculated particles or flocs obtained from certain polymers may be more sensitive to shear than others, especially during feed acceleration in the centrifuges. A more gentle feed accelerator is beneficial for this type of polymer. Also, polymers can be introduced to the feed at various locations, either within or outside the centrifuge.

For *size recovery and yield*, centrifuges have been applied to classify polydispersed fine particles. The size distribution of the particles is quantified by the cumulative weight fraction F less than a given particle size d for both the feed and the centrate streams. It is measured by a particle size counter that operates based on principles such as sedimentation or optical scattering.

In kaolin classification, the product is typically measured with a certain percentage less than a given size (example 90 percent or 95 percent less than 1 or 2 μm). Each combination of percent and size cut represents a condition by which the centrifuge would have to tune to yield the product specification.

The yield Y is defined as the fraction of feed particles of a given size below which they report to the centrate product. Thus,

$$Y = \frac{m_e W_e F_e}{m_f W_f F_f} \tag{18-95}$$

From material balance, the particle size distribution of the feed and centrate, as well as the total solids recovery, determine the yield,

$$Y(d) = \frac{F_e(d)}{F_f(d)}\,(1 - \text{Rec}) \tag{18-96}$$

The complement is the cumulative capture efficiency Z (= 1 − Y), which is defined as the feed particles of a given size and smaller that are captured in the cake, which in most dewatering applications is the product stream.

Volumetric and solids throughput to a centrifuge are dictated by one or several governing factors; the most common ones are the centrate solids, cake dryness, and capacity (torque and power) of the drive/gear unit. The solids throughput is also governed by other factors, such as solids conveyance and discharge mechanisms for continuous and batch centrifuges. The settling rate, which may be significantly reduced by increasing feed solids concentration, also becomes crucial to solids throughput, especially if it has to meet a certain specification on centrate quality.

Regarding *solids purity and wash ratio*, cake washing in a centrifuge is used to remove dissolved impurities on the solids particle surface. It is most effective in filtering centrifuges—typically a single wash step with a wash ratio of 0.05 to 0.3 kg wash/kg solids in continuous centrifuges, although higher ratios and/or multiple cocurrent or countercurrent wash steps can be achieved with derated capacities to provide sufficient residence time. Batch filtering centrifuges are unlimited in the wash quantity that can be applied. Solid soluble impurities generally cannot be washed out in situ due to insufficient contact time, in which case repulp washing may be more effective. Repulp is often used with sedimenting centrifuges where wash is required.

Power consumption is required to overcome windage and bearing (and seal) friction, to accelerate the feed stream from zero speed to full tangential speed in order to establish the required *G*-force for separation, and to convey and discharge cake. The power to overcome windage and bearing friction is usually established through tests for a given centrifuge geometry at different rotation speeds. It is proportional to the mass of the centrifuge, to the first power of the speed for the bearing friction, and to the second power of speed for windage. It is also related to the bearing diameter. The seal friction is usually small.

The horsepower for feed acceleration is given by

$$P_{\text{acc}} = 5.984(10^{-10})\,\text{sg}\,Q(\Omega r_p)^2 \tag{18-97}$$

where sg is the specific gravity of the feed slurry, Q the volumetric flow rate of feed in gpm(l/s), and Ω the speed in r/min; r_p in meters corresponds to the radius of the pool surface for sedimenting centrifuge, or to the radius of the cake surface for filtering centrifuge. *Note:* To convert horsepower to kilowatts, multiply by 0.746.

The horsepower for cake conveyance for scroll centrifuge is

$$P_{\text{con}} = 1.587\,(10^{-5})\,T\Delta \tag{18-98}$$

where Δ is the differential speed in r/min (s^{-1}) between the scroll conveyor and the bowl, and T is the conveyance torque in in · lb_f (N · m). For centrifuge where cake is discharged differently, the conveyance power is simply

$$P_{\text{con}} = MGC_f V \tag{18-99}$$

where M is the mass of the cake, G the centrifugal acceleration, C_f the coefficient of friction, and V the cake velocity. Comparing Eqs. (18-98) and (18-99), the conveyance torque is inversely related to the differential speed and directly proportional to the G acceleration, cake velocity, and cake mass.

Stress in the Centrifuge Rotor The stress in the centrifuge rotor is quite complex. Analytical methods, such as the finite element method, are used to analyze the mechanical integrity of a given rotor design. Without getting into an involved analysis, some useful knowledge can be gained from a simple analysis of the hoop stress of a rotating bowl under load. At equilibrium, the tensile hoop stress σ_h of the cylindrical bowl wall with thickness t is balanced by the centrifugal body force due to the mass of the bowl wall with density ρ_m and its contents (cake or slurry or liquid) with equivalent density ρ_L. Consider a circular wall segment with radius r, unit subtended angle, and unit axial length. A force balance requires

$$\sigma_h = \rho_m V_t^2 \left[1 + 0.5\frac{\rho_L}{\rho_m}\left(1 - \frac{r_s^2}{r_b^2}\right)\frac{r_b}{t}\right] \tag{18-100}$$

$V_t = \Omega r_b$ is the tip speed of the bowl. The term in the bracket is typically of order 1. Typically the maximum allowed σ_h is designed to be no more than 60 percent of the yield stress of the bowl material, which for steel is about $2.07(10^8)$ Nm^{-2}(30,000 lb_f/in^2). Given that the ρ_m of stainless steel is 7867 kg/m^3 (0.284 lbm/in^3), and there is no liquid load, then $(V_t)_{max} = \{\sigma_h/\rho_m\}^{1/2}$ = 126 m/s (412 ft/s). With additional liquid load, ρ_L = 1000 kg/m^3 (0.0361 lbm/in^3), r_b/t = 10, and further assuming the worst case with liquid filling to the axis, the term in the curly bracket is 1.636. Using Eq. (18-100), $(V_t)_{max} = \{\sigma_h/\rho_m/1.636\}^{1/2}$ = 98 m/s (322 ft/s). Indeed, almost all centrifuges are designed with top rim speed about 91 m/s (300 ft/s). With special construction materials for the rotor, such as duplex ferritic/austenitic stainless steel, with higher yield stress, the maximum rim speed under full load can be over 122 m/s (400 ft/s).

***G*-Force versus Throughput** The *G*-acceleration can be expressed as

$$G = \Omega^2 r_b = \frac{(V_t)^2_{\max}}{r_b} \tag{18-101}$$

Figure 18-128 shows the range of diameter of commercial centrifuges and the range of maximum G developed in each type. It demonstrates an inverse relationship between G and r_b at $V_t = (V_t)_{max}$, which is constant for a given material. Figure 18-129 shows a log-log plot of G versus Ω for various bowl diameters, Eq. (18-101). Also, the limiting conditions as delineated by $G = \Omega^2 R = \Omega (V_t)_{max}$ with various $(V_t)_{max}$ are superimposed on these curves. These two sets of curves dictate the operable speed and G for a given diameter and a given construction material for the bowl. The throughput capacity of a machine, depending on the process need, is roughly proportional to the nth power of the bowl radius,

$$Q = C_1 r_b^n \tag{18-102}$$

where n is normally between 2 and 3, depending on clarification, classification, thickening, or dewatering. Thus,

$$G = c_2 (V_t)^2_{\max} Q^{1/n} \tag{18-103}$$

where c_1 and c_2 are constants. It follows that large centrifuges can deliver high flow rates, but separation is at a lower *G*-force; on the other hand, smaller centrifuges can deliver lower flow rates, but separation is at a higher *G*-force. Also, using higher-strength material for construction of the rotating assembly permits higher maximum tip speed, thus allowing higher *G*-forces for separation at a given feed rate.

Centrifuge bowls are made of almost every machinable alloy of reasonably high strength. Preference is given to those alloys having 1 percent elongation to minimize the risk of cracking at stress-concentration points. Typically the list includes (in increasing cost) rubber-lined carbon steel, SS316L, SS317LMN, duplex SS (SAF2205), Alloy 904L, AL6 XN, Inconel, Hastelloy C22, Hastelloy B, nickel, titanium and zirconium. Most coatings (such as Halar and PTFE) cause problems with stressed components such as baskets, but they can be used for static components such as housings. Vertical-basket centrifuges are often made of carbon steel or stainless steel coated with rubber, neoprene, Penton, or Kynar. Casings and feed, rinse, and

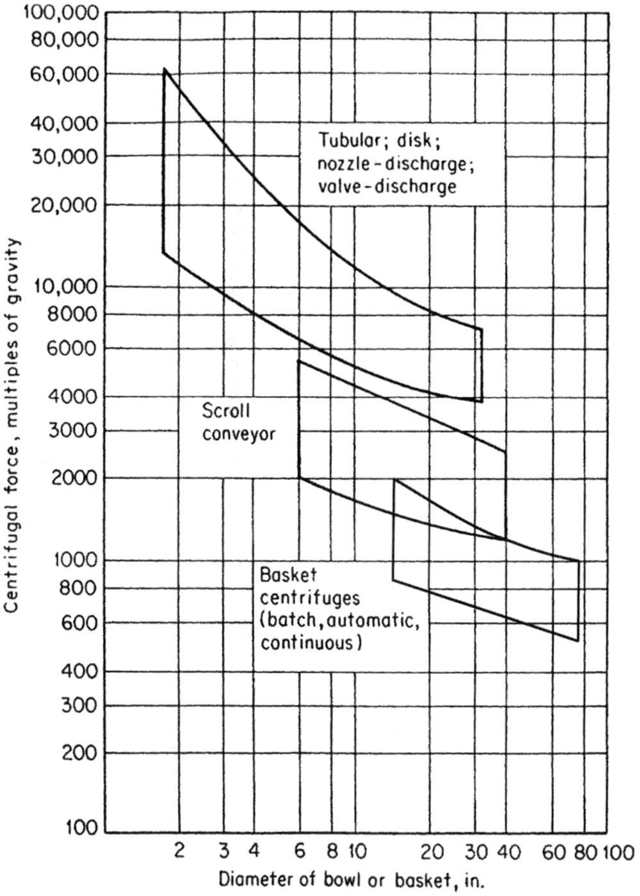

FIG. 18-128 Variation of centrifugal force with diameter in industrial centrifuges.

discharge lines that are stationary and lightly stressed may be made of any suitable rigid corrosion-resistant material. Wear-resistant materials—tungsten and ceramic carbide, hard-facing, and others—are often used to protect the bare metal surfaces in high-wear areas such as the blade tips of the decanter centrifuge.

Critical Speeds In the design of any high-speed rotating machinery, attention must be paid to the phenomenon of critical speed. This is the speed at which the frequency of rotation matches the natural frequency of

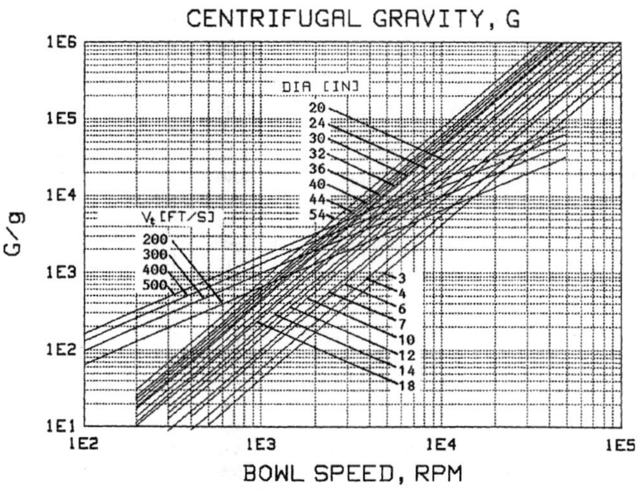

FIG. 18-129 Variation of centrifugal force with r/min.

the rotating part. At this speed, any vibration induced by slight unbalance in the rotor is strongly reinforced, resulting in large deflections, high stresses, and even failure of the equipment. Speeds corresponding to harmonics of the natural frequency are also critical speeds, but they give relatively small deflections and are much less troublesome than the fundamental frequency. The critical speed of simple shapes may be calculated from the moment of inertia; with complex elements such as a loaded centrifuge bowl, it is best found by tests.

Nearly all centrifuges operate at speeds well above the primary critical speed and therefore must pass through this speed during acceleration and deceleration. To permit them to do so safely, there must be some degree of damping in their mounting. A machine's natural frequency is purposely lowered to below the operating speed range by increasing the inertial mass of the machine frame or by adding a supplemental inertia block. Most modern designs use isolation systems to reduce dynamic forces transmitted to the structure. Structural design (especially for large centrifuges) should always include a dynamic analysis to ensure sufficient stiffness and to ensure that structural resonance does not occur in the operating speed range.

SEDIMENTATION CENTRIFUGES

When a spherical particle of diameter d settles in a viscous liquid under earth gravity g, the terminal velocity V_s is determined by the weight of the particle-balancing buoyancy and the viscous drag on the particle in accordance to Stokes' law. In a rotating flow, Stokes' law is modified by the "centrifugal gravity" $G = \Omega^2 r$, thus

$$V_s = \frac{1}{18\mu}\Omega^2 r(\rho_s - \rho_L)d^2 \tag{18-104}$$

In order to have good separation or high settling velocity, a combination of the following conditions is generally sufficient:

1. High centrifuge speed
2. Large particle size
3. Large density difference between solid and liquid
4. Large separation radius
5. Low liquor viscosity

Among the five parameters, the settling velocity is very sensitive to change in speed and particle size. It varies as the square of both parameters. The maximum achievable rotational speed of a centrifuge is normally dictated by the stresses exerted by the processing medium on the bowl and the stresses of the bowl on periphery equipment, most notably the drive system, which consists of a gear unit or hydraulic pump. If the particles in the feed slurry are too small to be separated in the existing G-field, coagulation and flocculation by polymers are effective ways to create larger agglomerated particles for settling. Unlike separation under a constant gravitational field, the settling velocity under a centrifugal field increases linearly with the radius. The greater the radius at which the separation takes place in a given centrifuge at a given rotational speed, the better the separation. Sedimentation of particles is favorable in a less viscous liquid. Some processes are run under elevated temperature where liquid viscosity drops to a fraction of its original value at room temperature.

Laboratory Tests In *spin-tube tests*, the objective is to check the settleability of solids in a slurry under centrifugation. The clarity of the supernatant liquid and the solids concentration in the sediment can also be evaluated. A small and equal amount of feed is introduced into two diametrically opposite test tubes (typically plastic tubes in a stainless steel holder) with a volume of 15 to 50 mL. The samples are centrifuged at a given G and for a period of time t. The supernatant liquid is decanted off from the spin tubes from which the clarity (in the form of turbidity or any measurable solids, dissolved and suspended) is measured. The integrity—more precisely, the yield stress—of the cake can be determined approximately by the amount of penetration of a rod into the cake under its weight and accounting for the buoyancy effect due to the wet cake. It is further assumed that the rod does not lean on the sides of the tube. The yield stress τ_y of the centrifuged cake can be determined from:

$$\tau_y = \frac{1}{2}\left(\frac{\rho_r L}{h} - \rho_L\right) g r_d \tag{18-105}$$

where r_d and L are, respectively, the radius and length of the solid circular rod; ρ_r is the density of the rod; and h is the penetration of the rod into the cake. By using rods of various sizes and densities, yield stress, which is indicative of cake handling and integrity, can be measured for a wide range of conditions.

The solids recovery in the cake can be inferred from measurements using Eq. (18-94). It is shown as a function of G-seconds for different feed solids concentration in Fig. 18-130; see also the transient centrifugation theory discussed below.

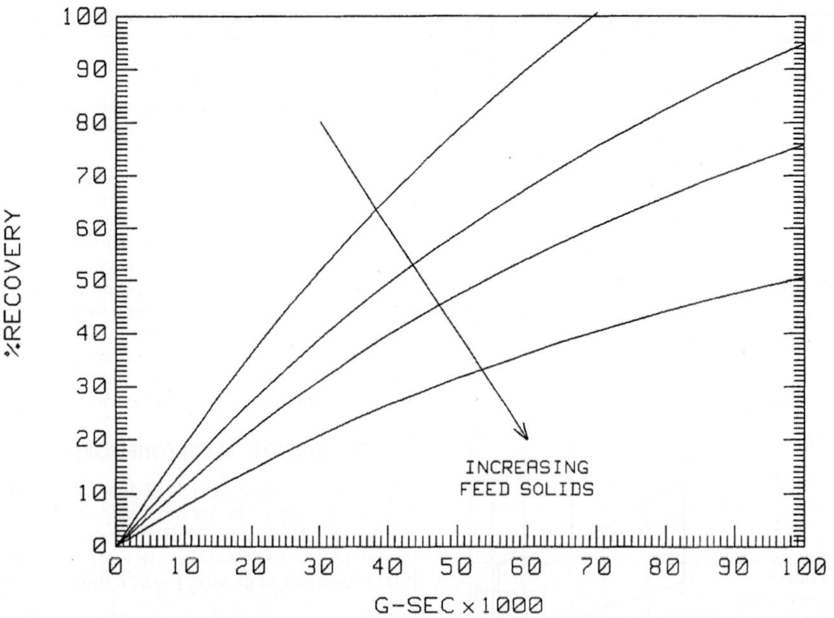

FIG. 18-130 Recovery as a function of *G*-seconds for centrifugal sedimentation.

In *imperforate bowl tests*, the amount of supernatant liquid from spin tubes is usually too small to warrant accurate gravimetric analysis. A fixed amount of slurry is introduced at a controlled rate into a rotating imperforate bowl to simulate a continuous sedimentation centrifuge. The liquid is collected as it overflows the ring weir. The test is stopped when the solids in the bowl build up to a thickness that affects centrate quality. The solids concentration of the centrate is determined similarly to that of the spin tube.

Transient Centrifugation Theory As in gravitational sedimentation, three layers exist during batch settling of a slurry in a centrifuge: a clarified liquid layer closest to the axis, a middle feed slurry layer with suspended solids, and a cake layer adjacent to the bowl wall with concentrated solids. Unlike with constant gravity *g*, the centrifugal gravity *G* increases linearly with radius. It is highest near the bowl and is zero at the axis of rotation. Also, the cylindrical surface area through which the particle has to settle increases linearly with radius. Both of these effects give rise to some rather unexpected results.

Consider the simple initial condition $t = 0$ where the solid concentration ϕ_{so} is constant across the entire slurry domain $r_L \le r \le r_b$, where r_L and r_b are, respectively, the radii of the slurry surface and the bowl. At a later time $t > 0$, three layers coexist: the top clarified layer, a middle slurry layer, and a bottom sediment layer. The air–liquid interface remains stationary at radius r_L, while the liquid–slurry interface with radius r_s expands radially outward, with t with r_s given by:

$$\frac{r_s}{r_L} = \sqrt{\frac{\phi_{so}}{\phi_s}} \tag{18-106}$$

Equation (18-106) can be derived from conservation of angular momentum as applied to the liquid–slurry interface.

Interestingly, the solid concentration in the slurry layer ϕ_s does not remain constant with time as in gravitational sedimentation. Instead, ϕ_s decreases with time uniformly in the entire slurry layer in accordance with:

$$\frac{\phi_s}{\phi_{s\max}} = 1 - \left[1 - \left(\frac{\phi_{so}}{\phi_{s\max}}\right)\right] e^{2\xi} \tag{18-107}$$

where ξ is a dimensionless time variable:

$$\xi = \left(\frac{V_{go}t}{r_b}\right)\left(\frac{G}{g}\right) \tag{18-108}$$

In Eqs. (18-107) and (18-108), under hindered settling and 1*g*, the solids flux $\phi_s V_s$ is assumed to be a linear function of ϕ_s decreasing at a rate of V_{go}. Also, the solids flux is taken to be zero at the "maximum" solids concentration ϕ_{smax}. As $G/g >> 1$, this solids flux behavior based on 1*g* is assumed to be ratioed by G/g.

Concurrent with the liquid–slurry interface moving radially outward, the cake layer builds up with the cake–slurry interface moving radially inward, with radial position given by:

$$\frac{r_c}{r_b} = \sqrt{\frac{(\varepsilon_s - \phi_{so})}{(\varepsilon_s - \phi_s)}} \tag{18-109}$$

where ε_s is a constant cake solids concentration. Sedimentation stops when the growing cake–slurry interface meets the decreasing slurry–liquid interface with $r_c = r_s$. This point is reached at $\phi_s = \phi_s^*$ and $t = t^*$ when

$$\frac{1}{\phi_s^*} = \frac{1}{\varepsilon_s} + \left(\frac{r_b}{r_L}\right)^2 \left(\frac{1}{\phi_{so}} - \frac{1}{\varepsilon_s}\right) \tag{18-110}$$

$$t^* = \frac{1}{2}\left(\frac{gr_b}{GV_{go}}\right) \ln\left(\frac{\phi_{s\max} - \phi_s^*}{\phi_{s\max} - \phi_{so}}\right) \tag{18-111}$$

Example 18-6 ***Calcium Carbonate-Water Slurry***

$G/g = 2667$
$V_{go} = 1.31 \times 10^{-6}$ m/s (5.16×10^{-5} in/s)
$\phi_{smax} = 0.26$ (with $\phi_s V_g = 0$)
$\phi_{so} = 0.13$
$r_L = 0.0508$ m (2 in)
$r_b = 0.1016$ m (4 in)
$\xi = 2667\,(1.31 \times 10^{-6})\,(1/0.1016) = 0.0344$
$\varepsilon_s = 0.52$

t(s)	ξ	ϕ_s/ϕ_{smax}	r_s (m)	r_c (m)
0.0	0	0.50	0.051	0.102
1.0	0.034	0.46	0.053	0.100
5.0	0.173	0.29	0.067	0.095
7.6	0.261	0.16	0.091	0.091

There are six types of industrial sedimenting centrifuges:

- Tubular-bowl centrifuges
- Multichamber centrifuges
- Skimmer pipe/knife-discharge centrifuges
- Disc centrifuges
- Decanter centrifuges
- Screen bowl centrifuges

The first three types, including the manual-discharge disc, are batch-feed centrifuges, whereas the latter three, including the intermittent and nozzle-discharge discs, are continuous centrifuges.

Tubular-Bowl Centrifuges The tubular-bowl centrifuge is widely employed for purifying used lubricating and other industrial oils and in the food, biochemical, and pharmaceutical industries. Industrial models

TABLE 18-16 Specifications and Performance Characteristics of Typical Sedimenting Centrifuges

Type	Bowl diameter	Speed, r/min	Maximum centrifugal force × gravity	Throughput		Typical motor size, hp
				Liquid, gal/min	Solids, tons/h	
Tubular	1.75	50,000*	62,400	0.05–0.25		*
	4.125	15,000	13,200	0.1–10		2
	5	15,000	15,900	0.2–20		3
Disc	7	12,000	14,300	0.1–10		⅓
	13	7500	10,400	5–50		6
	24	4000	5500	20–200		7½
Nozzle discharge	10	10,000	14,200	10–40	0.1–1	20
	16	6250	8900	25–150	0.4–4	40
	27	4200	6750	40–400	1–11	125
	30	3300	4600	40–400	1–11	125
Scroll conveyor	6	8000	5500	To 20	0.03–0.25	5
	14	4000	3180	To 75	0.5–1.5	20
	18	3500	3130	To 100	1–3	50
	24	3000	3070	To 250	2.5–12	125
	30	2700	3105	To 350	3–15	200
	36	2250	2590	To 600	10–25	300
	44	1600	1600	To 700	10–25	400
	54	1000	770	To 750	20–60	250
Knife discharge	20	1800	920	†	1.0‡	20
	36	1200	740	†	4.1‡	30
	68	900	780	†	20.5‡	40

*Turbine drive, 100 lb/h (45 kg/h) of steam at 40 lbf/in^2 gauge (372 KPa) or equivalent compressed air.
†Widely variable.
‡Maximum volume of solids that the bowl can contain, ft^3.
NOTE: To convert inches to millimeters, multiply by 25.4; to convert revolutions per minute to radians per second, multiply by 0.105; to convert gallons per minute to liters per second, multiply by 0.063; to convert tons per hour to kilograms per second, multiply by 0.253; and to convert horsepower to kilowatts, multiply by 0.746.

have bowls 102 to 127 mm (4 to 5 in) in diameter and 762 mm (30 in) long (Table 18-16). They are capable of delivering up to 18,000*g*. The smallest size, 44 mm × 229 mm (1.75 in × 9 in bowl), is a laboratory model capable of developing up to 65,000*g*. It is also used for separating difficult-to-separate biological solids with very small density difference, such as cells and virus.

The bowl is suspended from an upper bearing and drive (electric or turbine motor) assembly through a flexible-drive spindle with a loose guide in a controlled damping assembly at the bottom. The unit finds its axis of rotation if it becomes slightly unbalanced due to process load.

The feed slurry is introduced into the lower portion of the bowl through a small orifice. Immediately downstream of the orifice is a distributor and a baffle assembly that distribute and accelerate the feed to circumferential speed. The centrate discharges from the top end of the bowl by overflowing a ring weir. Solids that have sedimented against the bowl wall are removed manually from the centrifuge when the buildup of solids inside the bowl is sufficient to affect the centrate clarity.

The liquid-handling capacity of the tubular-bowl centrifuge varies with use. The low end shown in Table 18-16 corresponds to stripping small bacteria from a culture medium. The high end corresponds to purifying transformer oil and restoring its dielectric value. The solids-handling capacity of this centrifuge is limited to 4.5 kg (10 lb) or less. Typically, the feed stream solids should be less than 1 percent in practice.

Multichamber Centrifuges While the tubular bowl has a high aspect ratio (length-to-diameter ratio) of 5 to 7, the multichamber centrifuges have aspect ratios of 1 or less. The bowl driven from below consists of a series of short tubular sections of increasing diameter nested to form a continuous tubular passage of stepwise increases in diameter for the liquid flow. The feed is introduced at the center tube and gradually finds its way to tubes with larger diameters. The larger and denser particles settle out in the smaller-diameter tubes, while the smaller and lighter particles settle out in the larger-diameter tubes. Classification of particles can be conveniently carried out. Clarification may be significantly improved by spacing tubes, especially the outer tubes, more closely together to reduce the settling distance, a concept that is fully exploited by the disc-centrifuge design. This also serves to maintain a constant velocity of flow between adjacent tubes. As much as six chambers can be accommodated. The maximum solids-holding capacity is 0.064 m^3 (17 gal). The most common use is for clarifying fruit juices, wort, and beer. For these services it is equipped with a centripetal pump at effluent discharge to minimize foaming and contact with air.

Knife-Discharge Centrifugal Clarifiers Knife-discharge centrifuges with solid instead of perforated bowls are used as sedimenting centrifuges. The liquid flow is usually continuous until the settled solids start to interfere with the effluent liquid. The feed enters the hub end and is accelerated to speed before its introduction to the separation pool. The solids settle out to the bowl wall, and the clarified liquid overflows the ring weir or discharges through a skimmer pipe. In some designs, internal baffles in the bowl are required to stop wave action, primarily along the axial direction. When enough thick solid layer has built up inside the bowl, the supernatant liquid is skimmed off by moving the opening of the skimmer pipe radially inward. After the liquid is sucked out, the solids are knifed out as with centrifugal filters. However, unlike centrifugal filters, the cake is always fully saturated with liquid, S = 100 percent. These centrifuges are used for coarse, fast-settled solids. When clarification needs to be more effective, the operation may be totally batchwise, with prolonged spinning of each batch. If the solids content in the feed is low, several batches may be successively charged and the resulting supernatant liquor skimmed off before unloading of the accumulated solids.

Commercial centrifuges of this type have bowl diameters ranging from 0.3 to 2.4 m (12 to 96 in). The large sizes are used on heavy-duty applications such as coal dewatering and are limited by stress considerations to operate at 300*g*. The intermediate sizes for chemical process service develop up to 1000*g* (see Table 18-16).

Disc Stack Centrifuges One of the most common types of commercially used centrifuges is a vertically mounted disc machine, one type of which is shown in Fig. 18-131. Feed is introduced proximate to the axis of the bowl, accelerated to speed typically by a radial vane assembly, and flows through a stack of closely spaced conical discs in the form of truncated cones. Generally 50 to 150 discs are used. They are spaced 0.4 to 3 mm (0.015 to 0.125 in) apart to reduce the distance for solid/liquid separation. The angle made by conical discs with the horizontal is typically between 40° and 55° to facilitate solids conveyance. Under centrifugal force, the solids settle against the underside of the disc surface and move down to the large end of the conical disc and subsequently to the bowl wall. Concurrently, the clarified liquid phase moves up the conical channel. Each disc carries several holes spaced uniformly around the circumference. When the disc stack is assembled, the holes provide a continuous upward passage for the lighter clarified liquid released from each conical channel. The liquid collects at the top of the disc stack and discharges through overflow ports. To recover the kinetic energy and avoid foaming due to discharging of a high-velocity jet against a stationary casing, the rotating liquid is diverted to a stationary impeller from which the kinetic energy of the stream is converted to hydrostatic pressure. Unlike most centrifuges operating with a slurry pool in contact with a free surface, disc centrifuges with a rotary seal arrangement can operate under high pressure. The settled solids at the bowl wall are discharged in different forms, depending on the type of disc centrifuge.

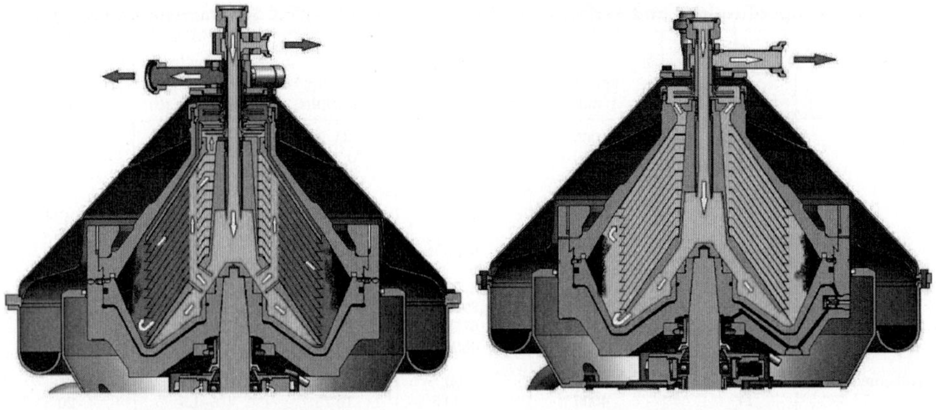

FIG. 18-131 Disc stack centrifuge. (*Andritz Separation.*)

For *manual discharge disc stack centrifuges,* in the simplest design shown in Fig. 18-132*a*, the accumulated solids must be removed manually on a periodic basis, similar to that for the tubular-bowl centrifuge. This requires stopping and disassembling the bowl and removing the disc stack. Although the individual discs rarely require cleaning, manual removal of solids is economical only when the fraction of solids in the feed is very small.

Self-cleaning disc centrifuges are more commonly known as clarifiers (two-phase) and separators (three-phase). These centrifuges, which also contain a conical disc stack inside the bowl, automatically discharge accumulated solids on a timed cycle while the bowl is at full speed. Feed is introduced into the bowl through a nonrotating feed pipe and into a distributor, which evenly distributes the slurry to the appropriate disc stack channels. Slurry is forced up through the disc stack, where solids accumulate on the underside of the discs and slide down the discs, where they are forced to the sludge holding area just inside the maximum diameter of the double cone-shaped bowl, as shown in Fig. 18-131. When the solids chamber is full, the bottom of the bowl, which is held closed to the top portion hydraulically, drops by evacuating the hydraulic operating fluid. The solids are discharged at full speed in a very short time into an outer housing, where they are diverted out of the machine. The liquid or liquids (in a three-phase separator) are normally discharged via stationary impellers under pressure. These types of centrifuges are commonly used in the clarification of beverages and the purification of mineral and edible oils.

In *disc nozzle centrifuges,* solids are discharged continuously, along with a portion of the liquid phase, through nozzles spaced around the periphery of the bowl, which are tapered radially outward, providing a space for solids storage (see Fig. 18-132*b*). The angle of repose of the sedimented solids determines the slope of the bowl walls for satisfactory operation. Clarification efficiency is seriously impaired if the buildup of solids between nozzles reaches into the disc stack. The nozzle diameter should be at least twice the diameter of the largest particle to be processed, and prescreening of extraneous solids is recommended. Typically, nozzle diameters range from 0.6 to 3 mm (0.25 to 0.125 in). Large disc centrifuges may have as many as 24 nozzles spaced out at the bowl.

For clarification of a single liquid phase with controlled concentration of the discharged slurry, a centrifuge that provides recirculation is used (see Fig. 18-132*b*). A fraction of the sludge discharged out of the machine is returned to the bowl to the area adjacent to the nozzles through lines external to the machine as well as built-in annular passages at the periphery of the bowl. This has the effect of preloading the nozzles with sludge that has already been separated, and it reduces the net flow of liquid with the newly sedimented solids from the feed. Increased concentration can also be obtained by recycling a portion of the sludge to the feed, but this increases solids loading at the disc stack, with a corresponding sacrifice in the effluent clarity for a given feed rate.

With proper rotary seals, the pressure in the machine can be contained up to 1.1 MPa (150 psig) or higher. Also, operating temperature can be as high as 315°C (about 600°F). The rotating parts are made of stainless steel, with the high-wear nozzles made of tungsten carbide. The bowls may be underdriven or suspended, and they range from several centimeters to over 1 m (3.3 ft) in outer diameter. The largest size capable of clarifying up to 1920 L/m (500 gpm) requires 112 kW (150 hp). These types of centrifuges are commonly used in applications such as corn wet milling (starch separation, gluten thickening), classifying kaolin clay particles, washing terephthalic acid crystals, and dewaxing lube oils.

Decanter Centrifuges The decanter centrifuge (also known as the solid-bowl or scroll centrifuge) consists of a solid exterior bowl with an internal screw or scroll conveyor (see Fig. 18-133). Both the bowl and the conveyor rotate at a high speed, yet there is a difference in speed between the two,

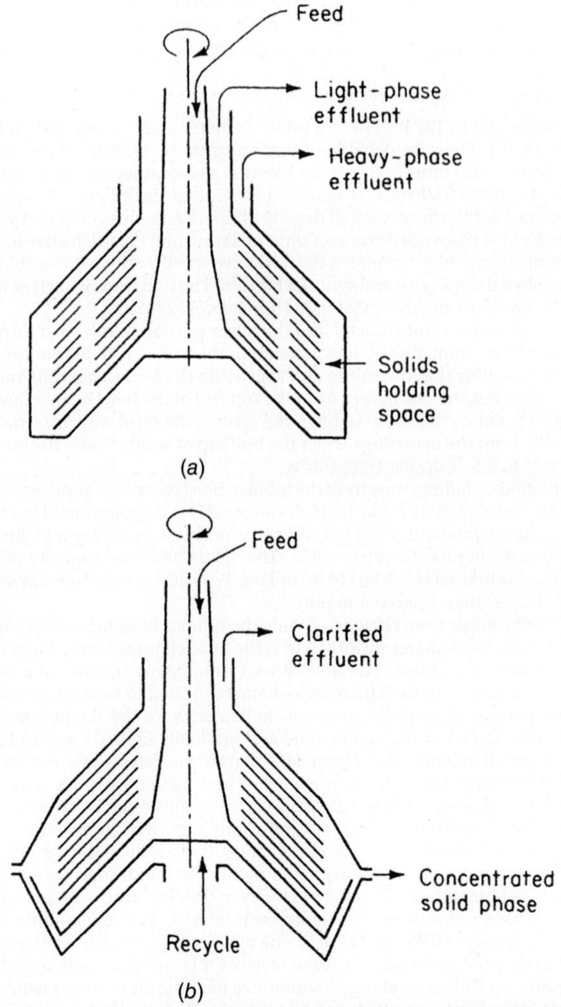

FIG. 18-132 Disc stack centrifuge bowls: (*a*) separator, solid wall; (*b*) recycle clarifier, nozzle discharge.

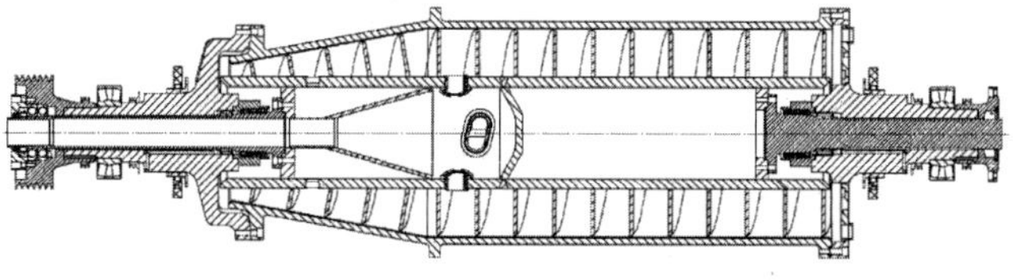

FIG. 18-133 Standard two-phase decanter centrifuge. (*Andritz Separation.*)

which is responsible for conveying the sediment along the machine from the cylinder to the conical discharge end. The rotating assembly is commonly mounted horizontally, with bearings on each end. Some centrifuges are vertically mounted with the weight of the rotating assembly supported by a single bearing at the bottom or with the entire machine suspended from the top. With the former configuration, the weight of the rotating assembly provides a good sealing surface at the bearing for high-pressure applications. The bowl may be conical or, more commonly, a combination of conical and cylindrical sections (see Fig. 18-133).

Slurry is fed through a stationary pipe into the feed zone located near the center of the scroll. The product is then accelerated circumferentially and passes through distribution ports into the bowl. The bowl has a cylindrical/conical shape and rotates at a preset speed optimal for the application. The slurry rotates with the bowl at the operating speed and forms a concentric layer at the bowl wall. In the separation pool or pond, under centrifugal gravity the solids, which are heavier than the liquid, settle toward the bowl wall, while the clarified liquid moves radially toward the pool surface. Subsequently, the liquid flows along the helical channel (or channels, if the screw conveyor has multiple leads) formed by adjacent blades of the conveyor to the liquid bowl head, from which it discharges over the weirs. The annular pool/pond height can be changed by adjusting the radial position of the weir openings, which take the form of circular holes or crescent-shaped slots, or by adjusting a stationary impeller, which will discharge the liquid under pressure.

The solids contained in the slurry are deposited against the bowl wall by centrifugal force. The length of the cylindrical bowl section and the cone angle are selected to meet the specific requirements of an application. The scroll conveyor rotates at a slightly different speed from the bowl, and it conveys the deposited solids toward the conical end of the bowl, also known as the beach. The half cone angle ranges between 5° and 20°. The cake is submerged in the pool when it is in the cylinder and at the beginning of the beach. In this region, liquid buoyancy helps to reduce the effective weight of the cake under centrifugal gravity, resulting in lower conveyance torque. Farther up the beach, the cake emerges above the pool and moves along the "dry beach," where buoyancy force is absent, resulting in more difficult conveyance and higher torque. But it is also in this section that the cake is dewatered, with expressed liquid returned to the pool. The centrifugal force helps to dewater, yet at the same time it hinders the transport of the cake in the dry beach. Therefore, a balance in cake conveyance and cake dewatering is the key in setting the pool and the *G*-force for a given application. Also, clarification is important in dictating this decision.

The cylindrical section provides clarification under high centrifugal gravity. In some cases, the pool should be shallow to maximize the *G*-force for separation. In other cases, when the cake layer is too thick inside the cylinder, the settled solids—especially the finer particles at the cake surface—entrain into the fast-moving liquid stream above, which eventually ends up in the centrate. A slightly deeper pool becomes beneficial in these cases because there is a thicker buffer liquid layer to ensure settling of resuspended solids. This can be at the expense of cake dryness due to reduction of the dry beach. Consequently, there is again a compromise between centrate clarity and cake dryness. Another reason for the tradeoff of centrate clarity with cake dryness is that, in losing fine solids to the centrate (i.e., classification), the cake with larger particles, having less surface-to-volume ratio, can dewater more effectively, resulting in drier cake. It is best to determine the optimal pool for a given application through tests.

The speed with which the cake transports is controlled by the differential speed. High differential speed facilitates high solids throughput where the cake thickness is kept to a minimum so as not to impair centrate quality due to the entrainment of fine solids. Also, high differential speed improves cake dewatering due to a reduction in the drainage path with smaller cake height; however, this is offset by the fact that higher differential speed also reduces cake residence time, especially in the dry beach. The opposite holds for low differential speed. Therefore, an optimal differential speed is needed to balance centrate clarity and cake dryness. The desirable differential speed is usually maintained by using a two-stage planetary gearbox, the housing of which rotates with the bowl speed, with a fixed first-stage pinion shaft. In some applications, the pinion is driven by an electrical backdrive (dc or ac) or hydraulic backdrive, or braked by an eddy-current device at a fixed rotation speed. The differential speed is then the difference in speed between the bowl and the pinion divided by the gear ratio. This also applies to the case when the pinion arm is held stationary, in which the pinion speed is zero. The torque at the spline of the conveyor, conveyance torque, is equal to the product of the pinion torque and the gear ratio. A higher gear ratio gives a lower differential speed, and vice versa; a lower gear ratio gives a higher differential for higher solids capacity. The torque at the pinion shaft has been used to control the feed rate or to signal an overload condition by the shearing of a safety pin. Under this condition, both the bowl and the conveyor are bound to rotate at the same speed (zero differential) with no conveyance torque and no load at the pinion.

Soft solids, most of which are biological waste such as sewage, are difficult to convey up the beach. Annular baffles or dams are commonly used to provide a pool-level difference wherein the pool is deeper upstream of the baffle toward the clarifier and lower downstream of the baffle toward the beach. The pool-level difference across the baffle, together with the differential speed, provides the driving force to convey the compressible sludge up the beach. This has been used effectively in thickening of waste-activated sludge and in some cases of fine clay with dilatant characteristics.

High solids decanter centrifuges have been used to dewater mixed raw sewage sludge (with a volume ratio of primary to waste-activated sludge such as 50 percent to 50 percent or 40 percent to 60 percent), aerobically digested sludge, and anaerobically digested sludge. Cake solids as dry as 28 percent to 35 percent by weight are obtained for raw mixed sludge and 20 percent to 28 percent for the digested sludges, with the aerobic sludge at the lower end of the range. The typical characteristics of high-solids applications are: low differential speed (0.5 to 3 r/min), high conveyance torque, high polymer dosage (10 to 30 lb dry polymer/ton dry solids, depending on the feed sewage), and slightly lower volumetric throughput rate. An electrical (dc or ac with variable-frequency drive) or hydraulic backdrive on the conveyor with high torque capacity is essential to operate these conditions at steady state.

The horizontal decanter centrifuge is operated below its critical speed. The bowl is mounted between fixed bearings anchored to a rigid frame. The gearbox is cantilevered outboard of one of these bearings, and the feed pipe enters the rotating assembly through the other end. The frame is isolated from the support structure by spring-type or rubber vibration isolators. In the vertical configuration, the bowl and the gearbox are suspended from the drive head, which is connected to the frame and casing through vibration isolators. A clearance bushing at the bottom limits the excursion of the bowl during start-up and shutdown but does not provide the radial constraint of a bearing under normal operating conditions.

Decanter centrifuges with mechanical shaft-to-casing seals are available for pressure containment up to 1.1 MPa (150 psig), similar to the nozzle-disc centrifuge. They can be built to operate at temperatures from −87°C to +260°C (−125°F to +500°F).

When abrasive solids are processed, the points of wear are protected with replaceable inserts or tiles made from silica carbide, tungsten carbide, ceramic, or other abrasive-resistant materials. These high-wear areas include the feed zone, including feed ports; the conveyor blade tip, especially the pressure or pushing face; the conical beach; and the solids discharge ports. The transport of solids is encouraged in some applications by longitudinal strips or grooves at the inner diameter of the bowl, especially at the beach, to enhance the frictional characteristics between the sediment and the bowl surface, and by polished conveyor faces to reduce frictional drag. For fluidlike sediment cake, by using the strips in the beach, a much tighter gap between the conveyor blade tip and the bowl surface is possible with a cake heel layer trapped by the strips. This reduces leakage of the fluid sediment flowing through an otherwise larger gap opening to the pool. Gypsum coating on the bowl wall at the beach section has been used to achieve the same objective.

Various bowl configurations with a wide range of aspect ratios (length-to-diameter ratios) from less than 1 to 4 are available for specific applications, depending on whether the major objective is maximum clarification, classification, or solids dryness. Generally, the movement of liquid and solids is in countercurrent directions, but in the cocurrent design, the movement of liquid is in the same direction as that of the solids. In this design, the feed is introduced at the large end of the machine, and the centrate is taken by a skimmer at the beach–cylinder junction. The settled solids transverse the entire machine and discharge at the beach exit. Compound angle beaches are used in specific applications, such as washing and drying of polystyrene beads. The pool level is located at the intersection of the two angles at the beach, the steeper angle being under the pool and the shallower angle above the pool (i.e., dry beach), allowing a longer dewatering time. The wash is applied at the pool side of the beach–angle intersection and functions as a continuously replenished annulus of wash liquid through which the solids are conveyed. The size of decanter centrifuge ranges from 6-in diameter to 54 in. The larger the machine, the slower the speed, and the less *G*-force.

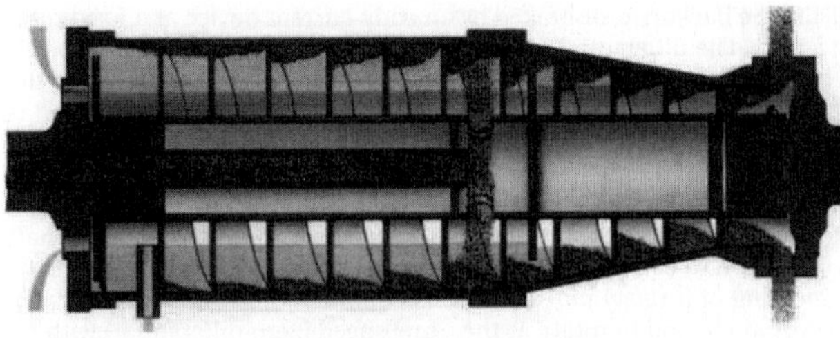

FIG. 18-134 Three-phase decanter centrifuge. (*Andritz Separation.*)

However, the larger machine provides a much higher throughput capacity, which cannot be accommodated with smaller machines (see Table 18-16). Decanter centrifuges are used in many industries and applications where large amounts of solids need to be separated from liquids continually. These industries include, but are not limited to, food, beverage (including dairy), chemical, pharmaceutical, oil, edible oil, industrial, and municipal wastewater.

Three-Phase Decanter Centrifuges Three-phase decanter centrifuges are similar in principle to decanter centrifuges (see Fig. 18-134), but they separate feed into three phases, two immiscible liquids and one sedimenting/suspended solids phase. The sedimenting solids that collect on the bowl wall are conveyed out of the centrifuge and discharged as in a standard decanter centrifuge. The two liquid phases are discharged either by gravity over two sets of adjustable weir plates or rings or through a dual discharge system where the heavy liquid phase (typically water) is discharged by a stationary impeller under pressure and the light liquid phase (typically fat or oil) is discharged by gravity over a ring dam. The benefit of the dual discharge system is that the liquid interface zone (and ultimately the pool/pond height) can be adjusted while the machine is operating at full speed. These types of centrifuges are commonly used in the fish, animal by-products, oil sludge, and edible oil (e.g., olive and palm) industries.

Specialty Decanter Centrifuges Decanter technology has evolved over the past 20 years to include machines that can perform separations that standard decanter centrifuges normally cannot. These specialty two- and three-phase decanters use the same basic premise of solids discharge via an internal scroll, but with specific machine geometries that allow for specialty separations. These specialty decanters include the Sedicanter (shown in Fig. 18-135) and the CENSOR sorting decanter (shown in Fig. 18-136). The Sedicanter, which has a double-cone cocurrent bowl design and specialized scroll geometry, is capable of achieving higher rotational speeds (up to 7750 r/min and 10,000*G*) and can, therefore, increase clarification efficiency and effectively discharge fine, pasty solids where a normal decanter is inefficient and ineffective. The Sedicanter is commonly used in certain biotechnology, vitamin, soy, and yeast separations.

The CENSOR sorting decanter has a scroll with reversing pitch on one side, which skims a floating solids layer off the top of the carrier liquid, usually an aqueous brine of intermediate density. Sinking solids are scrolled out as in normal decanter centrifuges, and liquid is discharged under pressure by a stationary but variable impeller. The CENSOR sorting decanter is used in the carpet and plastics recycling industries.

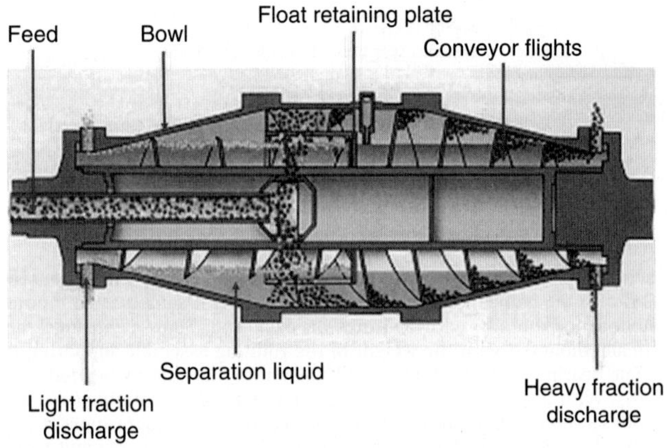

FIG. 18-136 CENSOR sorting decanter centrifuge. (*Andritz Separation.*)

Screen Bowl Decanter Centrifuges The screen bowl centrifuge consists of a solid-bowl decanter to which, at the smaller conical end, a cylindrical screen has been added (see Fig. 18-137). The scroll spans the entire bowl, conforming to the profile of the bowl. It combines a sedimenting centrifuge with a filtering centrifuge. Therefore, the solids that are processed are typically larger than 23 to 44 μm.

As in a decanter, an accelerated feed is introduced to the separation pool. The denser solids settle toward the bowl wall and the effluent escapes through the ports at the large end of the machine. The sediment is scrolled toward the beach, typically with a steeper angle than in the decanter centrifuge. As the solids are conveyed to the screen section, the liquid in the sediment cake further drains through the screen, resulting in drier cake. Washing of the sediment in the first half of the screen section and occasionally in the conical beach area is very effective in removing impurities, with the second half of the screen section reserved for dewatering of mother liquor and wash liquid.

The screen is typically constructed of a wedge-bar with an aperture between adjacent bars, which opens up to a larger radius. This prevents solids from blinding the screen and also reduces conveyance torque. For abrasive materials such as coal, the screens are made of wear-resistant materials such as tungsten carbide.

A variation of the screen bowl decanter incorporates a short cylindrical screen followed by a conical screen diverging toward the discharge end (Fig. 18-138). The conveying scroll ends at the beginning of the conical section, so in theory in that section the solids are transported by displacement. In practice, the main benefit is a reduction in conveying torque and increased solids throughput. This design has demonstrated over 100 MTPH capacity in potash applications. Increased capacity has been shown for other applications, as well as improved reliability due to fewer torque-related shutdowns.

Continuous Centrifugal Sedimentation Theory The Stokes settling velocity of a spherical particle under a centrifugal field is given by Eq. (18-104). Useful relationships have been established on continuous sedimentation by studying the kinematics of settling of a spherical particle of diameter d in an

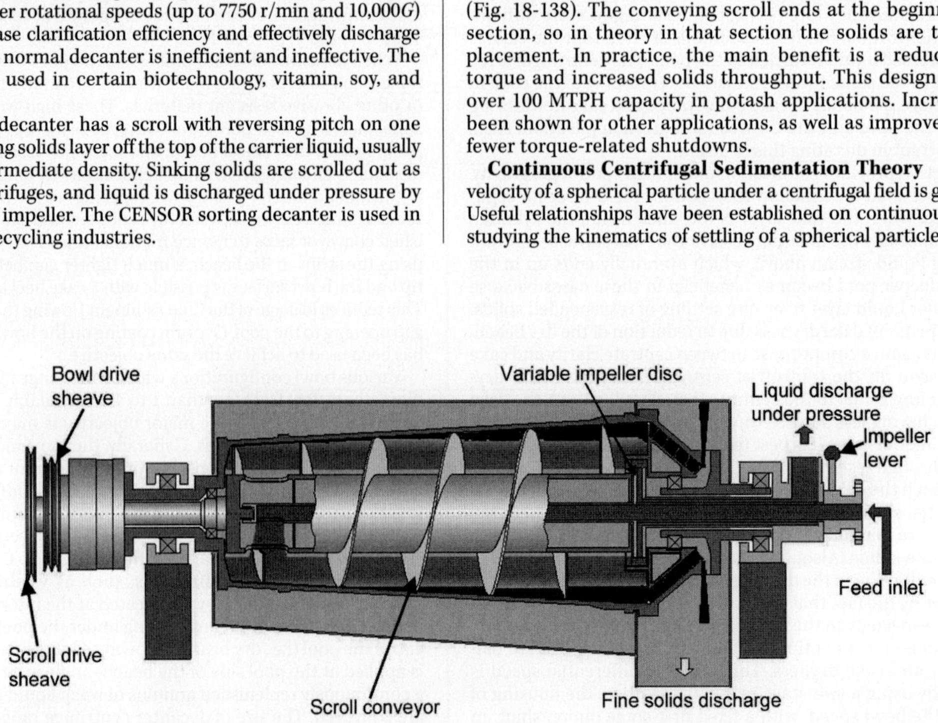

FIG. 18-135 Sedicanter centrifuge. (*Flottweg Separation Technologies.*)

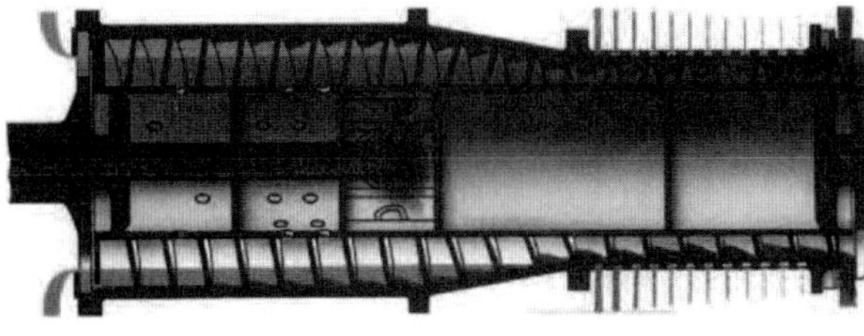

FIG. 18-137 Cylindrical screen bowl centrifuge. (*Andritz Separation.*)

annular rotating pool. Equating the time rate of change in a radial position to the settling velocity, and the rate of change in an axial position to bulk-flow velocity, thus gives

$$\frac{dr}{dt} = crd^2 \qquad (18\text{-}112)$$

$$\frac{dx}{dt} = \frac{Q}{\pi(r_b^2 - r_p)} \qquad (18\text{-}113)$$

where $c = (\rho_s - \rho_L)\,\Omega^2/18\mu$, x is distance along the axis of the bowl, Q is the volumetric feed rate, and r_b and r_p are, respectively, the bowl and pool surface radii. For a particle located at one end of the bowl at radius r with $r_p < r < r_b$, after transversing the full bowl length, it settles out and is captured by the bowl wall. Solving the preceding equations with these boundary conditions, the limiting trajectory is:

$$\frac{r}{r_b} = \exp\left[\frac{-\pi cL\,(r_b^2 - r_p^2)d^2}{Q}\right] \qquad (18\text{-}114)$$

If the same size particle d is located at an initial starting radius less than r given by Eq. (18-114), it is assumed to escape from being captured by the bowl, whereas it would have been captured if it had been at an initial radius greater than r. Assuming that the number of particles with size d is uniformly distributed across the annular pool, the recovery Rec_d (known also as *grade efficiency*) is the differential of the cumulative recovery $Z = 1 - Y$, with Y given in Eq. (18-94) for particles with size d, as the ratio of the two annular areas:

$$\text{Rec}_d = \frac{r_b^2 - r^2}{r_b^2 - r_p^2} \qquad (18\text{-}115)$$

Combining Eqs. (18-113) and (18-114), the maximum Q to the centrifuge, so as to meet a given recovery Rec_d of particles with diameter d, is

$$\frac{Q_d}{2V_{gd}} = \left(\frac{\pi\Omega^2 L}{g}\right)\left(\frac{r_p^2 - r_b^2}{\ln\{1 - \text{Rec}_d[1 - (r_p/r_b)^2]\}}\right) = \sum\nolimits_{\text{Rec}_d} \qquad (18\text{-}116)$$

Note in Eq. (18-116) that V_{gd} is the settling rate under 1*g*, and it is a function of the particle size and density and fluid properties. The ratio $Q_d/2V_{gd}$ is then related only to the operating speed and geometry of the centrifuge, as well as to the size recovery. It measures the *required* surface area for settling under centrifugal gravity to meet a specified Rec_d. When the size recovery Rec_d is set at 50 percent, the general result, Eq. (18-116), reduces to the special case, which is the well-known Ambler's sigma factor, which for a straight rotating bowl (applicable to bottle centrifuge, decanter centrifuge, and others) is:

$$\sum = \left[\frac{\pi\Omega^2 L}{g}\right]\left[\frac{r_b^2 - r_p^2}{\ln[2r_b^2/(r_b^2 + r_p^2)]}\right] \qquad (18\text{-}117)$$

It can be simplified to:

$$\sum = \pi\Omega^2 L\frac{(3r_b^2 + r_p^2)}{2g} \qquad (18\text{-}118)$$

For a disc centrifuge, a similar derivation results in

$$\sum = \frac{2\pi\Omega^2(N-1)(r_2^3 - r_1^3)}{3g\tan\theta} \qquad (18\text{-}119)$$

where N is the number of discs in the stack, r_1 and r_2 are the outer and inner radii of the disc stack, and θ is the conical half-angle.

Typical Σ factors for the three types of sedimenting centrifuges are given in Table 18-17. In scale-up from laboratory tests, sedimentation performance should be the same if the value of Q/Σ is the same for the two machines. This is a widely used criterion for the comparison of centrifuges of similar geometry and liquid-flow patterns developing approximately the same *G;* however, it should be used with caution when comparing centrifuges of different configurations. In general, the shortcomings of the theory are due to the oversimplified assumptions being made, such as (1) there is an idealized plug-flow pattern; (2) sedimentation abides by Stokes' law as extended to many *g*'s; (3) feed solids are uniformly distributed across the surface of the bowl head at one end of the clarifier, and capture implies that the particles' trajectory intersect the bowl wall; (4) the feed reaches full tangential speed as it is introduced to the pool; and (5) the recovery of given-size particles is at 50 percent. This criterion also does not account for (6) possible entrainment of already settled particles in the liquid stream and (7) entrance and exit effects.

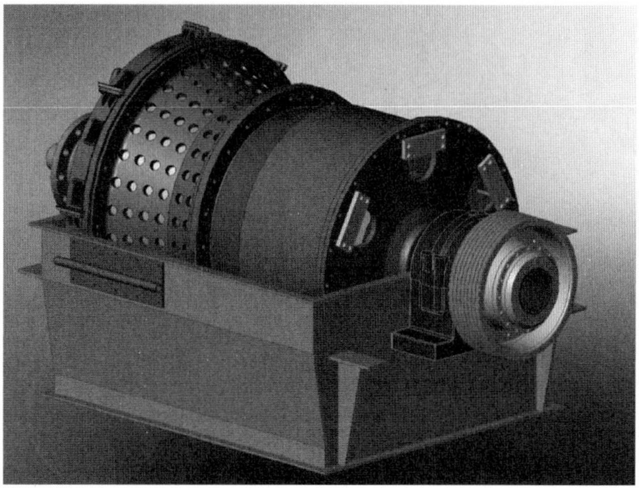

FIG. 18-138 Conical screen bowl centrifuge. (*Andritz Separation.*)

Experience in using the Σ concept has demonstrated that the calculated Σ factor should be modified by an efficiency factor to account for some of the aforementioned effects that are absent in the theory, and this factor depends on the type of centrifuge. It is nearly 100 percent for simple spin-tube bottle centrifuges, 80 percent for tubular centrifuges, and less than 55 percent for disc centrifuges. The efficiency varies widely for decanter centrifuges, depending on cake conveyability and other factors.

FILTERING CENTRIFUGES

Filtering centrifuges are broadly categorized as continuous operating and batch operating. Both continuous and batch filtering centrifuges use some type of filtration media fitted against the basket (bowl) wall. As the solid-liquid mixture is introduced, the liquid filters through the solids, through the filter media, and typically through perforations in the basket shell (except with rotational siphon designs, discussed later). Filtering centrifuges are primarily chosen over sedimenting types where high cake purity through cake washing is a requirement, or where minimal residual cake moisture is desired. Typical solids retention times range from 5 to 45 s for continuous operating filtering centrifuges and 5 to 180 min for batch operating filtering centrifuges.

Usually, the solids phase is of a higher specific gravity than the liquid; but unlike with sedimenting centrifuges, this is not an absolute requirement. In the nontypical case where the opposite is true, then filtration, be it centrifugal, vacuum, or pressure, is the only option for solid-liquid separation. However, for batch filtering centrifuges with the solids phase lighter than the liquid, care must be taken that liquid is not allowed to build in the basket during the feed step, or else buoyancy forces may float the settled solids, resulting in uneven filtration and high vibration. This is not of concern for continuous filtering centrifuges since the filtration rate must inherently exceed the liquid feed rate for stable operation. Refer to Table 18-18 for typical operating ranges of filtering centrifuges.

BATCH FILTERING CENTRIFUGES

Although continuous centrifuges are often preferred for reasons of lowest capital cost, high unit capacity, and ease of integration into continuous upstream and downstream processes, batch filtering centrifuges with cyclic

TABLE 18-17 Scale-up Factors for Sedimenting Centrifuges

Type of centrifuge	Inside diameter, in	Disc diameter, in/number of discs	Speed, r/min	Σ value, units of 10^4 ft²	Recommended scale-up factors*
Tubular	1.75	—	23,000	0.32	1†
Tubular	4.125	—	15,000	2.7	21
Tubular	4.90	—	15,000	4.2	33
Disc	—	4.1/33	10,000	1.1	1
Disc	—	9.5/107	6500	21.5	15
Disc	—	12.4/98	6250	42.5	30
Disc	—	13.7/132	4650	39.3	25
Disc	—	19.5/144	4240	105	73
Scroll conveyor	6	—	6000	0.27	1
Scroll conveyor	14	—	4000	1.34	5
Scroll conveyor	14‡	—	4000	3.0	10
Scroll conveyor	18	—	3450	3.7	12.0
Scroll conveyor	20	—	3350	4.0	13.3
Scroll conveyor	25	—	3000	6.1	22
Scroll conveyor	25	—	2700	8.6	31

*These scale-up factors are relative capacities of centrifuges of the same type but different sizes when performing at the same level of separation achievement (e.g., same degree of clarification). These factors must not be used to compare the capacities of different types of centrifuges.

†Approaches 2.5 at rates below mL/min.

‡Long bowl configuration.

NOTE: To convert inches to millimeters, multiply by 25.4; to convert revolutions per minute to radians per second, multiply by 0.105; and to convert 10^4 square feet to square meters, multiply by 929.

operation will always have a role in the CPI for reasons of highest possible product purity, lowest possible cake moisture, and highest product recovery. The slurry's physical properties, such as particle size distribution or liquid viscosity, may require long retention times and may drive the selection from continuous to batch. Generally, when the value of the product is high, batch operating centrifuges are preferred.

A development over the last 10 to 20 years is the introduction of true cGMP designs suitable for high-purity fine chemicals and pharmaceutical applications. Requirements in this sector usually favor batch operation due to demands for batch identity; typical properties of the slurry dictate batch operation, high wash requirements, cleanability and inspectability, avoidance of cross-contamination for multipurpose applications, and elimination of operator exposure to the process.

Most modern batch filtering centrifuges are provided with ac variable-frequency drives (VFDs) for speed variation, often with power regenerative braking. Even in the case of peeler centrifuges capable of constant-speed operation throughout the cycle, in most cases they are still fitted with VFDs to meet starting requirements (accelerating a high inertial load), operating flexibility, and regenerative braking in hazardous areas. However, there are still units available that use either a two-speed motor (generally considered obsolete) or hydraulic drive.

The main subcategories of batch filtering centrifuges are vertical basket centrifuges (both top unloading and bottom unloading of various configurations), horizontal peeler centrifuges, and horizontal inverting filter centrifuges.

Vertical Basket Centrifuge—Operating Method and Mechanical Design The vertical basket centrifuge is equipped with a cylindrical basket rotating about the vertical axis. The basket shell is normally perforated and lined with filtration media consisting of filter cloth, and backing screens to provide liquid drainage paths to the basket holes. Securing the filter cloth may be by hook and loop attachment or with snap-in retainers.

Feed slurry is introduced into the basket through either single or multiple feed pipes, or by other means such as rotating feed cones to help distribute the solids on the basket wall. In most cases, feed slurry is introduced at an intermediate speed, although in some applications (FGD gypsum, for one), feeding is done at full speed. There are several methods available to control the feed and cake level such as mechanical, paddle-type feelers, capacitance probes, ultrasonic sensors, feed totalizer, or load cells.

The solids distribution profile may tend to be parabolic, with thicker cakes near the bottom of the basket, tapering down toward the top, since the *G*-field is perpendicular to the force of gravity. This is especially true with fast-sedimenting solids that will settle toward the basket bottom before the slurry is fully accelerated by the basket. The coarser solids can settle toward the basket bottom, while the finer solids deposit preferentially toward the top. This can result in uneven filtration resistance in the cake, affecting the wash pattern and efficiency of the wash. In cases where this is a concern, a rotating feed cone may be better for even distribution, or a horizontal peeler centrifuge may be better suited to the application.

During and after feeding, filtrate passes through the cake, filter media, and out through the basket shell and is collected in a housing surrounding the basket and discharged through a tangential nozzle. The solids build up on the basket wall during the feed step until the desired loading is achieved.

After a spin time to filter the mother liquor through the solids, wash liquor is commonly applied in either a single step or various combinations, typically via a wash pipe with nozzles. The cake is spun for a time at high speed, then the machine ramps down to discharge. Solids removal can be accomplished by one of several methods.

Top Unloading Vertical Basket Centrifuges This is one of the oldest types of centrifuges, dating back to about 1900 or even earlier. With this design, the perforated basket is fitted with either a filter cloth or a filter bag. The basket has a solid bottom. After the dry spin portion of the cycle is completed, the machine is stopped. Solids removal is either by manually digging out the cake or by removing (lifting) the filter bag from the top of the unit. Except for small pilot-scale units and some specialty applications, this design is no longer commonly marketed or desired due to labor intensiveness, incompatibility with solvent wet or toxic products (operator exposure), and overall inefficiency of operation.

TABLE 18-18 Operating Range of Filtering Centrifuges

Type of centrifuge	*G/g*	Minimum feed solids concentration by wt.	Minimum mean particle size, μm	Minimum V_{fo}, m/s
Vibratory	30–120	50	300	5×10^{-4}
Tumbler	50–300	50	200	2×10^{-4}
Screen scroll	500–2000	35	75	1×10^{-5}
Pusher	300–800	40	120	5×10^{-5}
Screen bowl	500–2000	20	45	2×10^{-6}
Peeler	300–2000	5	10	2×10^{-7}
Vertical	200–1000	5	5	1×10^{-7}
Inverting filter	300–1000	5	2	5×10^{-8}

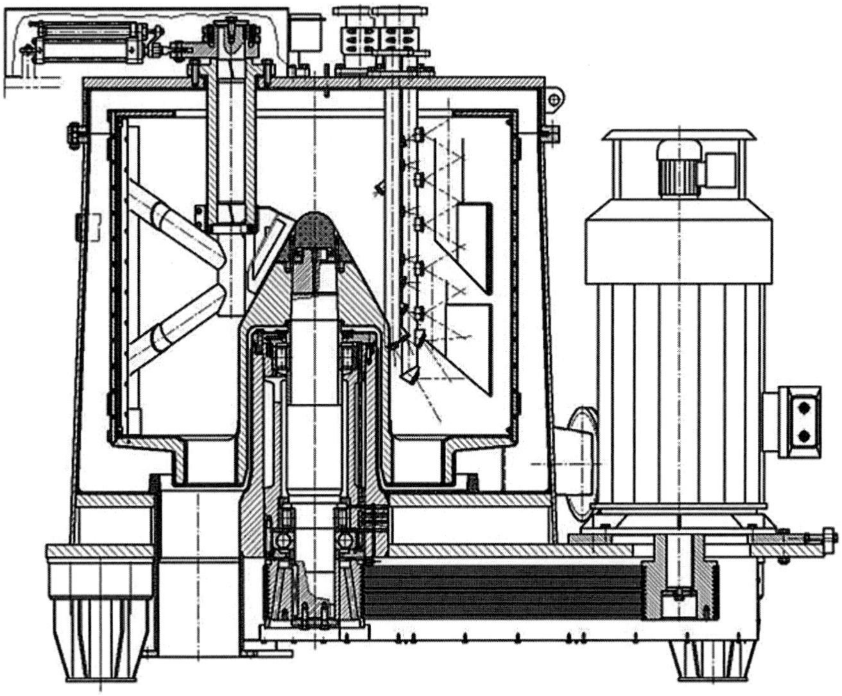

FIG. 18-139 Typical bottom unloading vertical basket centrifuge. (*Andritz KMPT GmbH.*)

Bottom Unloading Vertical Basket Centrifuges Most common for modern machines is the bottom discharge design, which incorporates either a swiveling scraper mechanism that cuts the cake in a single motion or a two-motion, oscillating scraper for finer or stickier cakes or pharmaceuticals. In every case, solids discharge must be at low speed, which necessitates ramping the machine up and down every cycle. After discharge, a thin cake layer or heel remains on the filter cloth. See Fig. 18-139.

Heel removal can be automated by dissolving the heel, flushing the heel out the solids discharge chute with subsequent downstream diverting away from solids handling equipment, or pneumatically removing the heel with blowoff nozzles, discharging it out the solids discharge. Pneumatic heel removal can be accomplished either from within the basket (often incorporated with the knife) or from the outside of the basket.

There are different cover arrangements to access the interior, such as hinged or pivoting manway, half or full opening covers. Filter cloth maintenance or component adjustments usually require entering the unit in any case, except for small sizes. See Fig. 18-140.

Isolation of the load imbalances from the structure has historically been by link, or three-column suspension. This system is relatively inefficient and transmits substantial dynamic forces to the building foundation, limiting operating speeds and performance. In response to the common shortcomings of this design, some manufacturers redesigned the suspension/isolation

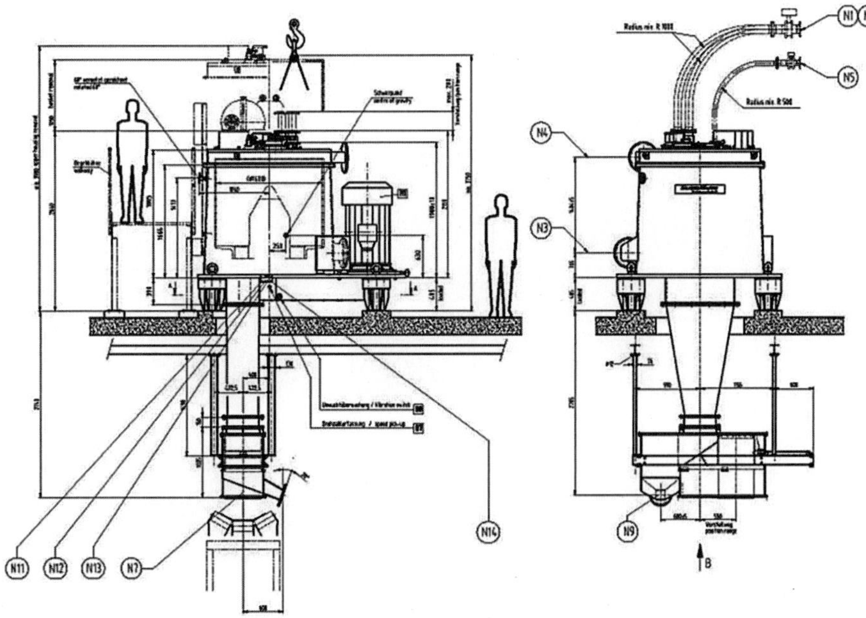

FIG. 18-140 Typical vertical basket centrifuge installation. (*Andritz KMPT GmbH.*)

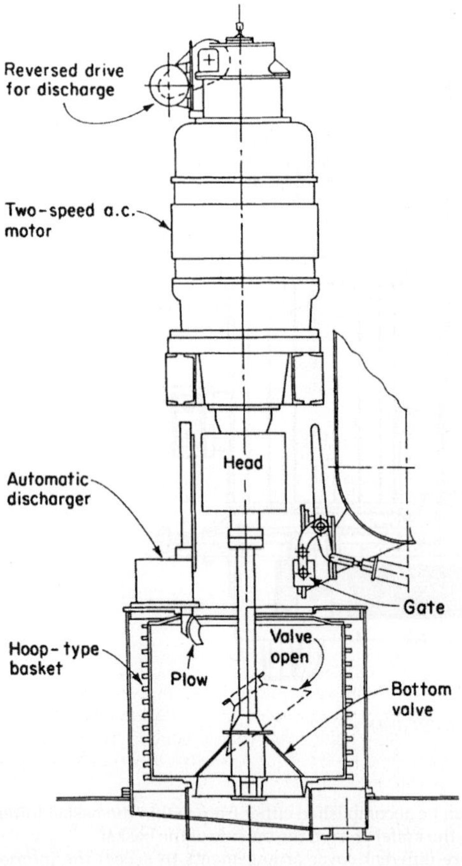

FIG. 18-141 Top-suspended vertical centrifuge. (*Western States Machine Co.*)

system to a massive inertial baseplate and housing supported on tuned coil springs and dampers at each corner. This greatly improved the dynamic force attenuation and stability of the machine. This system has enabled the design of very large machines (1600-mm-diameter by 1250-mm-high baskets) for processing materials with high solids density, such as wallboard-grade gypsum from FGD systems with a unit capacity in excess of 10 Mtons/h.

Top-Suspended Vertical Centrifuges A special type of top-suspended centrifuge is widely used in sugar processing and is shown in Fig. 18-141. Conventionally, the drive is suspended from a horizontal bar supported at both ends from two A-frames. The drive head, which is connected to the motor or a driven pulley through a flexible coupling, carries the thrust and radial bearings that support the basket, shaft, and load. These units can be equipped with large 112-kW (150-hp) drive motors, and on white sugar they can process about 350 kg/cycle with 24 cycles/h.

Horizontal Peeler Centrifuge—Operating Method and Mechanical Design The chemical design peeler centrifuge was developed in the 1920s and has many applications (Fig. 18-142). Like the vertical basket, it has a rotating filtration basket, except that it rotates about the horizontal axis. Early machines supported the basket from both ends, but virtually all modern machines are cantilever-supported for purposes of accessibility to the basket and internal components. Some machines are equipped with a fully opening door that swings away with internal components, while other designs incorporate a fully opening housing that also provides access to the basket exterior. This unit has an extremely rugged construction compared to vertical baskets, required due to the full-speed feeding and discharge capability of the peeler centrifuge. It is often provided with high-power ac VFD drives for accelerating the feed slurry at full speed and for optimum operating flexibility. Gastight construction to 400-mm (16-in) water column is usually standard, and higher pressure ratings can easily be accomplished.

By reorienting the axis to horizontal, many advantages become possible, such as superior wash capability with the more uniform solids distribution compared to the vertical basket, with the potential for uneven, parabolic cake profile resulting in uneven wash penetration.

The peeler centrifuge costs more than a vertical machine with a comparable basket size, although often a smaller peeler centrifuge can outperform a larger vertical basket. In addition, the higher capital cost is offset by many process and mechanical advantages, such as these:

- A fully opening door contains the feed, wash, feed control, and solids discharge components. Easily swung open, it then provides complete access to the basket interior and all internal components mounted on the door. Filter cloth exchange does not require vessel entry.
- Isolation of dynamic forces is far superior in horizontal machines compared to vertical.
- The peeler centrifuge will distribute the solids more evenly because it is not feeding perpendicular to gravity as is the vertical basket. This provides smoother operation, better wash effect, and an ability to handle faster-draining materials.
- The ability to discharge at high speed eliminates or minimizes dead cycle time required for acceleration and braking for higher capacity, lower power consumption, and lower wear and tear. The cycle time savings is particularly beneficial with short-cycle (fast-filtering) requirements.
- The peeler centrifuge can provide higher centrifugal forces than can vertical baskets, for increased performance and flexibility.
- Peeler centrifuges are available in larger sizes than vertical centrifuges—up to 2100-mm (83-in) diameter.
- The peeler centrifuge is also capable of automatic heel removal by several methods: dissolving the heel, reslurrying and discharging the heel wet and diverting downstream, or dry heel removal pneumatically.

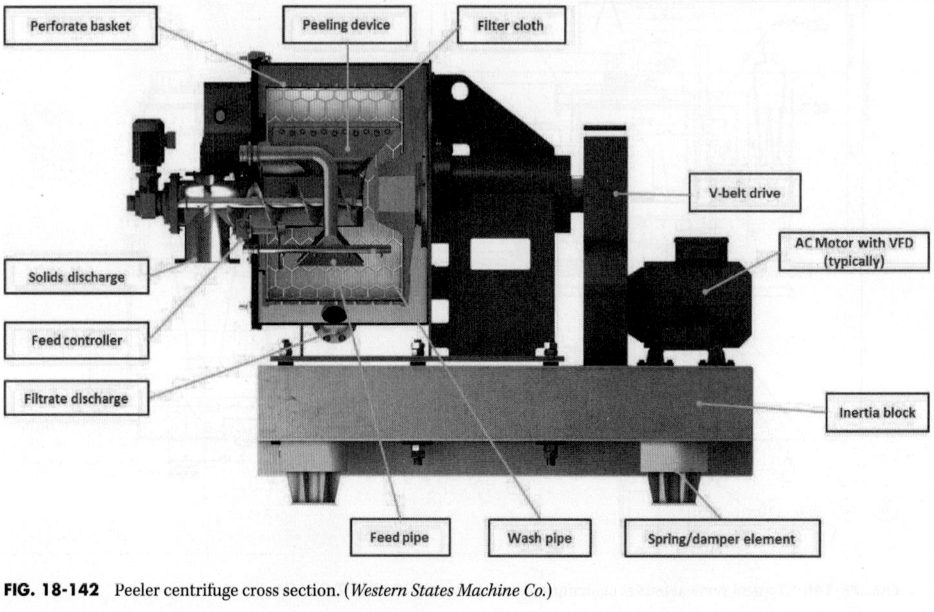

FIG. 18-142 Peeler centrifuge cross section. (*Western States Machine Co.*)

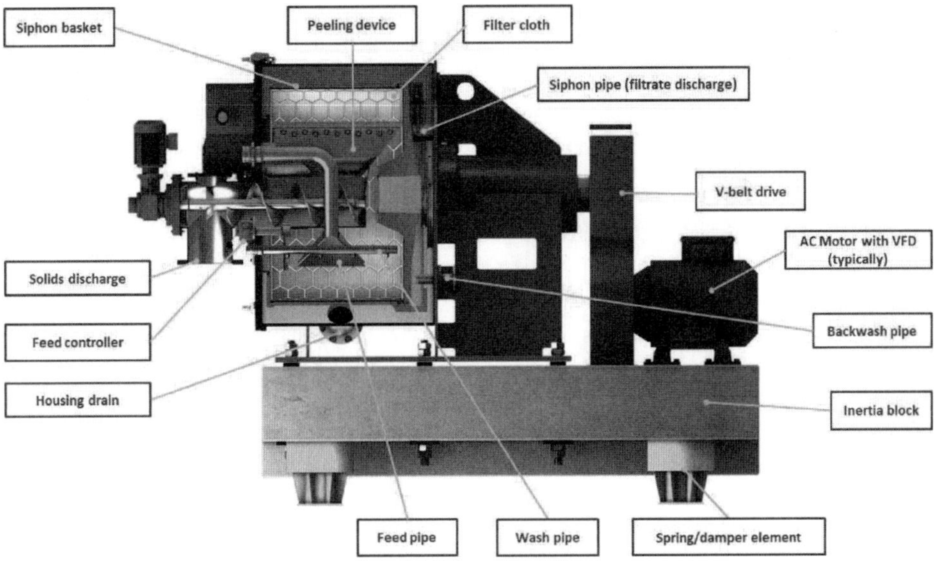

FIG. 18-143 Siphon peeler centrifuge cross section. (*Andritz KMPT GmbH.*)

Pneumatic heel removal can be accomplished either from within the basket or from outside the basket.

Siphon Peeler Centrifuge The siphon peeler centrifuge (Fig. 18-143) was developed and patented by Krauss-Maffei in the 1970s. Instead of using only centrifugal pressure as the driving force, as do all perforated units both vertical and horizontal, the rotational siphon centrifuge provides an increased pressure gradient by reducing the pressure behind the filter media and thereby increasing the driving force for filtration.

As in the perforated basket design, the liquid filters through the cake and filter media, but instead of discharging through perforations in the basket shell, the basket wall is solid, and the liquid flows axially to the basket rear and into a separate chamber. At this point, the filtrate is skimmed out with a radially adjustable skimmer.

In perforated baskets, the driving force for filtration is approximately the hydrostatic pressure established by the liquid column. The driving force diminishes as the liquid column height decreases, often causing a wet layer near the base of the cake due to capillary pressure balancing the centrifugal pressure. In siphon baskets, in addition to the centrifugal pressure, by skimming at a radius greater than the filter cloth, a rotational siphon is established. Due to the gravitational field in which it is working, a height difference Δh of only 20 to 30 mm is sufficient to lower the pressure behind the cloth to the vapor pressure of the liquid. This additional vacuum remains in place until all the interstitial liquid is drawn through the cake and will overcome the cake capillary pressure, thus preventing this wet layer. Once the supernatant and interstitial liquid drain from the cake, the siphon chamber behind the cloth drains, and filtration characteristics are like those of a perforated basket. See Fig. 18-144.

To reestablish the siphon for the next cycle, a priming step precedes feeding where the siphon skimmer is pivoted inward near the rim of the siphon chamber, and liquid is introduced into the siphon chamber that backflows up through the heel cake, displacing gas from the chamber. Feeding then begins with the heel submerged. After a time delay, the siphon swivels downward to the working position (δh of 20 to 30 mm), where it remains for the remainder of the cycle. Besides increased driving force for filtration, other benefits of the rotational siphon include the following:

- Accurate control of the filtration rate is useful during feed and wash. For fast-filtering products, filtration rates can be throttled, ensuring even solids distribution.
- Backwashing the residual heel after each cycle rejuvenates the heel to maintain good permeability. Heel life is often extended.
- Feeding into a liquid bath helps lay down a more porous heel layer since the larger particles sediment faster than finer particles.

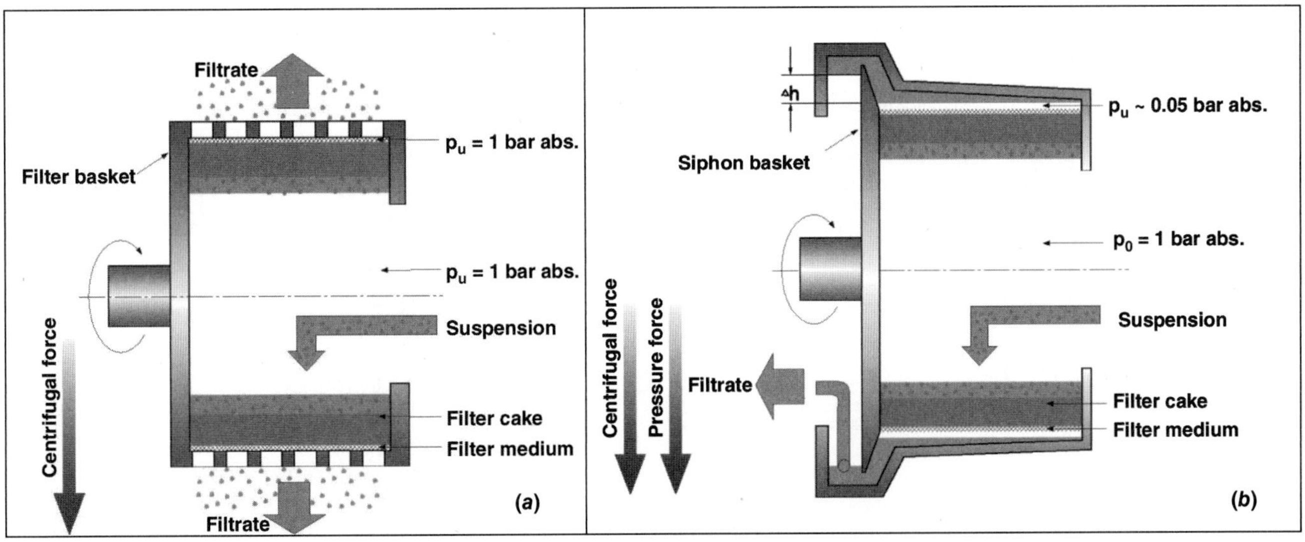

FIG. 18-144 Schematic representation of (*a*) perforate versus (*b*) siphon centrifuge. (*Andritz KMPT GmbH.*)

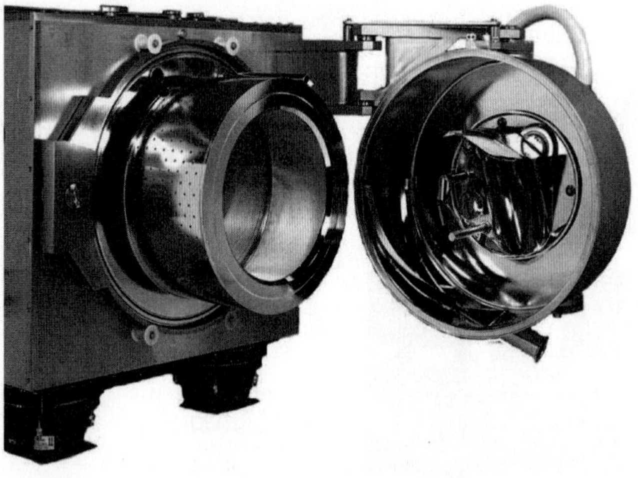

FIG. 18-145 Pharma peeler centrifuge. (*Andritz KMPT GmbH.*)

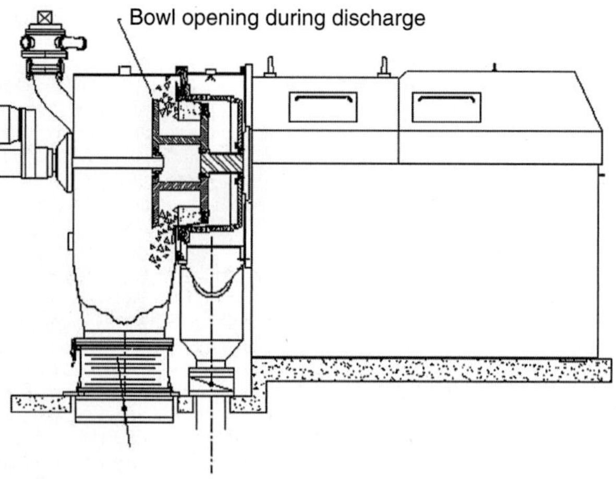

FIG. 18-147 Inverting filter centrifuge. (*Heinkel USA.*)

- Separate discharge of filtrate from splash/overflow provides better product yield.
- Deep siphon chambers with cake backwashing capability have been successfully used to completely submerge the cake and indefinitely increase wash contact time.

Pressurized Siphon Peeler Centrifuge Theoretically, the same principle can further increase driving force with overpressure in the process housing; for example, 3-bar overpressure would produce up to 4-bar pressure gradient across the cake. To date this has not been used in practice due to the complexity of the installation.

Pharma Peeler Centrifuge For applications requiring hygienic operation, a special type of peeler centrifuge was developed in the 1990s (Fig. 18-145). The primary applications for this type of machine are in fine chemicals and pharmaceuticals, often in multipurpose use where cross-contamination must be avoided. It provides for ease of cleanability and inspectability with automatic CIP/SIP, access to every wetted surface, pressure-tight construction suitable for steam sterilization, automatic heel removal, and separation of mechanical components from the process end, making it suitable for through-the-wall clean-room installation. Operation is contained, thereby eliminating operator exposure.

Inverting Filter Centrifuge The inverting filter centrifuge was introduced in the late 1970s to provide a means of ensuring that all the filter cake is discharged from the filter medium. By turning the cloth inside out to achieve solids discharge, the problems of operator exposure and variable product quality associated with manual cake removal or residual-heel blinding were largely eliminated. By the late 1980s, this style of centrifuge had found widespread application in pharmaceutical and agricultural chemical production. See Figs. 18-146 to 18-148.

The inverting filter design comprises a horizontal axis shaft with a two-part bowl attached. The perforated cylindrical bowl remains in a fixed axial position throughout the operation, while a bowl insert can move along the horizontal axis. The filter cloth is attached at one end to the axially fixed bowl and at the other to the axially movable bowl insert. Therefore, moving the bowl insert causes the filter cloth to turn inside out. In the filtering position, the bowl insert sits inside the bowl, with the filter cloth covering perforations. As with other filtering centrifuges, the cake builds up on the cloth during filling. It is washed using a true positive-displacement, plug-flow wash, and cake dewatering can be achieved simply by spinning (often at the maximum speed) for a time.

As this cake-discharging mechanism involves little or no risk of cloth blinding, inverting filter centrifuges usually operate at optimum conditions with relatively thin cakes and frequent discharges. (Cake thicknesses are typically 1 to 3 in, and cycle times are typically 8 to 14 min.) This style of operation is particularly effective with compressible materials where the filtration rate drops off dramatically with increasing cake thickness. By operating with thin cakes and short cycle times, the average filtration flux throughout the batch operation is maximized for these difficult applications.

Inverting filter centrifuges come in bowl diameters ranging from 300 to 1300 mm and achieve *g*-forces of 3000 − 900 × gravity.

CONTINUOUS FILTERING CENTRIFUGES

Where processing conditions and objectives allow, continuous filtering centrifuges offer the combination of high processing capacities and good wash capabilities. Inherently they are less flexible than batch filtering centrifuges; they are primarily constrained by much shorter retention time, and in some

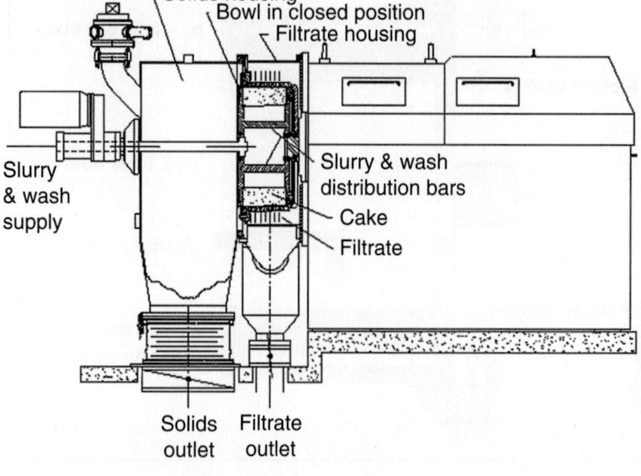

FIG. 18-146 Inverting filter centrifuge. (*Heinkel USA.*)

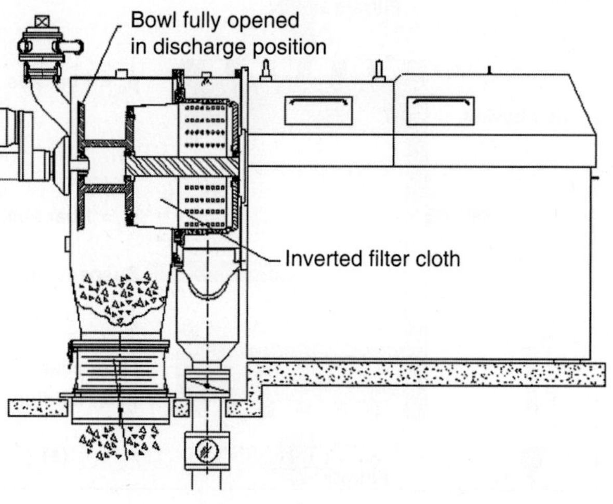

FIG. 18-148 Inverting filter centrifuge. (*Heinkel USA.*)

cases liquid handling capacity requires upstream preconcentration of the slurry. Fines loss to the filtrate is also greater with continuous designs than with batch.

Conical-Screen Centrifuges When a conical screen in the form of a frustum is rotated about its axis, the component of the centrifugal force normal to the screen surface impels the liquid to filter through the cake and the screen, whereas the component of the centrifugal force parallel to the screen in the longitudinal direction conveys the cake to the screen at a larger diameter. The sliding of the solids on the cone is favored by smooth perforated plates or wedge-wire sections with slots parallel to the axis of rotation, rather than woven wire mesh.

Wide-Angle Conical Screen Centrifuges If the half-angle of the cone screen is greater than the angle of repose of the solids, the solids will slide across it with a velocity that depends on frictional properties of the cake but not on feed rate. The frictional property of the cake depends on the solid property, such as shape and size, as well as on moisture content. If the half-angle of the cone greatly exceeds the angle of repose, the cake slides across the screen at a high velocity, thereby reducing the retention time for dewatering. The angle selected is therefore highly critical with respect to performance on a specific application. Wide-angle and compound-angle centrifuges are used to dewater coarse coal and rubber crumb and to dewater and wash crude sugar and vegetable fibers such as those from corn and potatoes.

Shallow-Angle Conical Screen Centrifuges By selecting a half-angle for the conical screen that is less than the angle of repose of the cake and providing supplementary means for the controlled conveyance of the cake across the conical screen from the small to large diameter, longer retention time is available for cake dewatering. Three methods are in common use for cake conveyance:

1. *Vibrational conveyance.* This is called the vibratory centrifuge. A relatively high-frequency force is superimposed on the rotating assembly. This can be either in-line with the axis of rotation or torsional, around the driveshaft. In either case, the cake under inertial force from the vibration is partly "fluidized" and propelled down the screen under a somewhat steady pace toward the large end, where it is discharged.

2. *Oscillating or "tumbling" conveyance.* This is commonly known as the tumbling centrifuge. The driveshaft is supported at its lower end on a pivot point. A supplementary power source causes the shaft and the rotating bracket it carries to gyrate about the pivot at a controlled amplitude and at a frequency lower than the rate of rotation of the basket. The inertia force generated also provides partial fluidization of the bed of solids in the basket, causing the cake to move toward the large end, as in the vibrational conveyance.

3. *Scroll conveyance.* Another type of continuous filtering centrifuge is the scroll screen centrifuge, as shown in Fig. 18-149. The scroll screen centrifuges are also sometimes called worm screen centrifuges. The design consists of a fixed-angle rotating basket and a concentric screw conveyor to control the transport and discharge of solids. Common applications include crystal, fiber, and mineral separations.

Scroll screen centrifuges are typically used for continuous feeds of slurries of at least 10 percent solids by volume, of materials with an average size of 100 μm or greater. This design offers some tolerance to process variation and typically removes the bulk of surface moisture.

The scroll and the screen are rotating in the same direction with a small differential speed of typically less than 100 r/min. The feed is deposited into the acceleration cone of the scroll, then passes through the feed openings of the scroll. The solids are retained on the screen; as the liquid migrates through, the cake passes the screen media and the basket. The discharge housing collects the liquid, and the solids are conveyed to the large diameter of the rotating basket and are continuously discharged.

An internal product wash is also available in the scroll screen centrifuges. Wash liquid is added in a chamber midway along the basket, and the wash liquid migrates through the cake prior to final drying and discharge.

The rotating basket is used to retain the screen media. Wedge-wire as well as sheet metal screens are available, but they are typically limited to a minimum opening size of 70 μm or larger. Common basket designs include 10°, 15°, and 20°.

The scroll acts as a screw conveyor and discharges the solids. The typical solids retention time in the centrifuge ranges from 0.5 to 6 s. A close tolerance, 0.3 to 1 mm, is common between the scroll and screen; therefore, little material remains on the screen. This minimizes the potential for imbalances.

Pusher Centrifuges—Operating Method and Mechanical Design Pusher centrifuges (Fig. 18-150) are continuous filtering centrifuges used for dewatering and washing free-draining bulk crystalline, polymer, or fibrous materials. Where suited, they provide the best washing characteristics of any continuous centrifuge due to control of retention time, uniform cake bed, and essentially plug flow of solids through the unit. For a typical application such as salt, they range in capacity from about 1 ton/h for small (250-mm-diameter) units up to about 120 tons/h or more in the largest units (1250-mm-diameter). They are generally applied where the mean particle size is at least 150 μm. Typical solids retention time is between 10 and 30 s. Normally the machine is fed by a feed pipe, but it can also be used as post-treatment of a prior dewatering step such as a vacuum filter. In this case it is fed by a feed screw. Due to the gentle handling of the product, pusher centrifuges are better suited for fragile crystals than are other types of continuous filtering centrifuges.

Generally there are three limitations to capacity in pusher centrifuges: (1) solids volumetric throughput, (2) liquid filtration capacity, and (3) retention time needed to achieve desired objectives for cake purity and residual cake moisture. In most cases (2) dictates; therefore, to optimize capacity and performance, preconcentrating the feed slurry as high as possible is desired. Some designs have a short conical section at the feed end for prethickening within the unit, but generally it is preferable to thicken ahead of the centrifuge with gravity settlers, hydrocyclones, or inclined screens.

As depicted schematically in Fig. 18-150, the rotating assembly consists of a belt-driven outer rotor that rotates at constant speed. The outer shaft (hollow shaft) is fixed to the main or outer basket. Within the hollow shaft is the pusher shaft, which is keyed together with the hollow shaft but

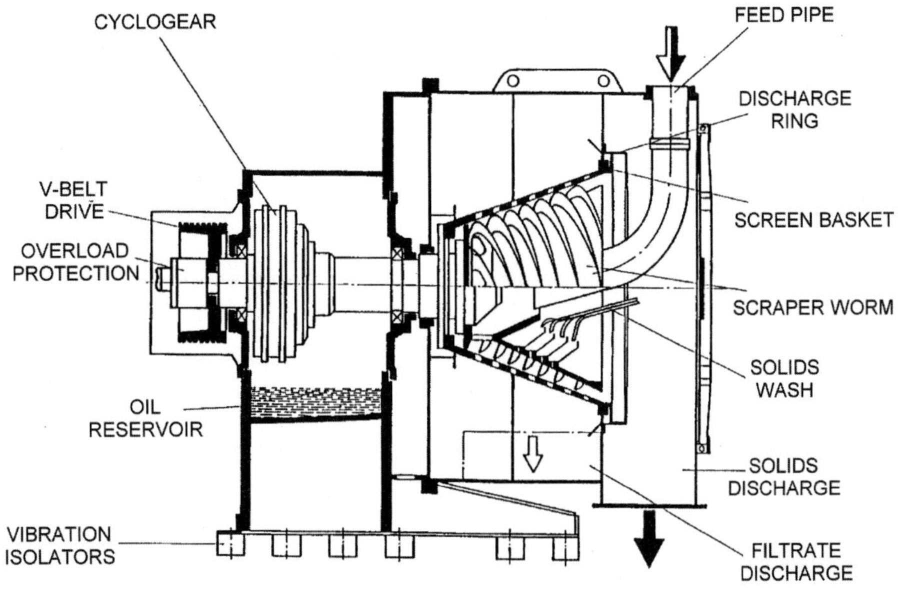

FIG. 18-149 Scroll screen centrifuge. (*TEMA Systems, Inc.*)

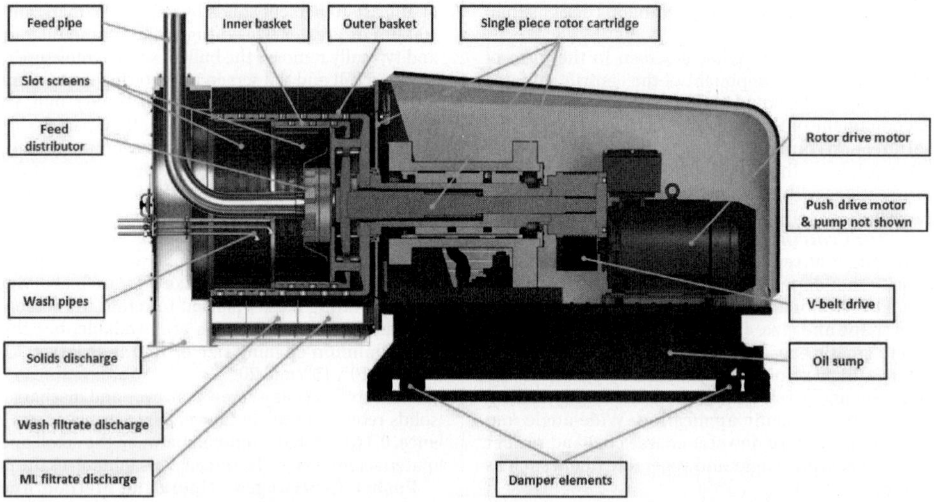

FIG. 18-150 Pusher centrifuge cross section. (*Andritz KMPT GmbH.*)

also oscillates. The reciprocal motion is provided by a mechanical gearbox for smaller units (400-mm-diameter and less) or hydraulically in larger units. The depicted schematic is of a two-stage design in which the pusher shaft is fixed to the inner basket and the pusher plate is attached to the outer basket by posts. The stroke length is between 30 and 85 mm, depending on machine size, and stroke frequency is usually between 45 and 90 strokes per minute.

The feed slurry enters through a stationary central pipe into a feed accelerator/distributor, then is introduced onto the (in this case) oscillating inner basket just in front of the pusher plate. In the feed zone, most of the liquid is drained, forming a cake sufficiently stiff to transfer the push force through the bed of solids and transport the cake without shearing. This is why it requires fast-draining materials and is liquid-limited, since it must form a cake within the period of one stroke.

With designs that use a simple feed cone or plate for feed distribution, most of the slurry acceleration takes place on the screen surface, with lower effective slurry speed and driving force for filtration. More advanced designs use an impeller-type feed accelerator that largely preaccelerates the slurry prior to its introduction on the screen for higher capacity and lower screen wear.

With each stroke of the pusher, the material in the feed zone is pushed up to a certain height, primarily depending on the friction coefficient between the solids and the screen and the screen deck length and secondarily depending on *G*-force and loading. Once the cake in the feed zone is compressed and has formed a ring with this height, it transmits the push force to the stationary bed of cake in the basket, which begins to move the cake bed forward until the forward end of the stroke. This cake height is often referred to as the natural cake height. The schematic in Fig. 18-151 shows what is taking place in the feed zone.

The distance the cake ring moves forward divided by the stroke length is defined as the *push efficiency.* The push efficiency varies with solids volumetric loading, resulting in a self-compensating control of varying rates. Depending on the cake properties, primarily compressibility, up to about 90 percent push efficiency is achievable. In some cases, volumetric throughput can be further increased beyond the volumetric push capacity at the natural cake height, in which case the push efficiency remains almost constant, and the cake height increases with increasing load, commonly referred to as the forced cake height. This realm of operation is usually only possible with multistage designs.

As the cake bed is transported through the basket, it passes through the various process steps shown in Fig. 18-152 with product moisture gradient as shown in Fig. 18-153.

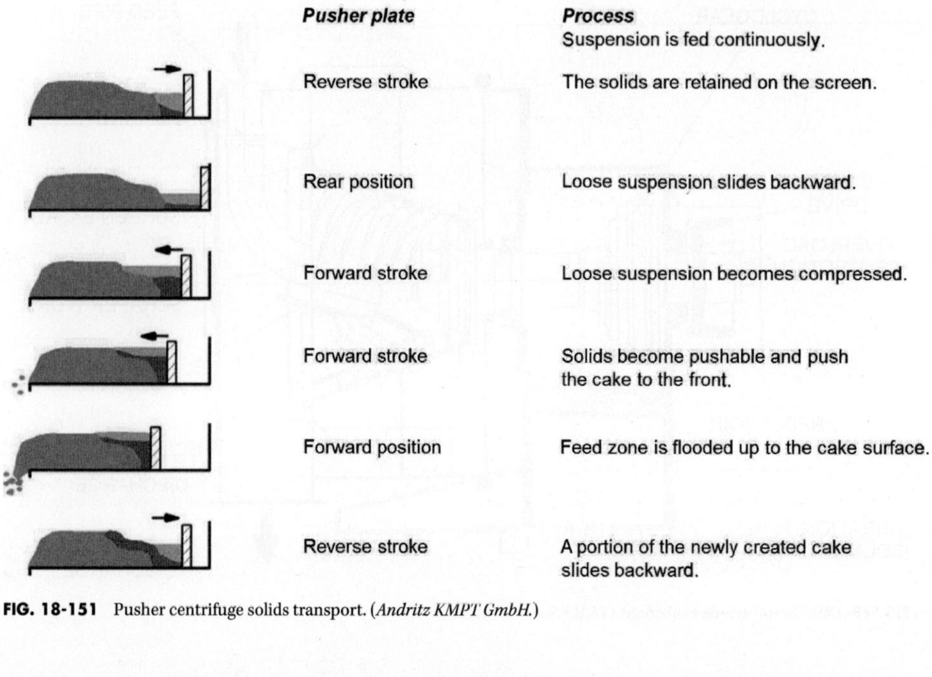
Pusher plate
Process
Suspension is fed continuously.
Reverse stroke
The solids are retained on the screen.
Rear position
Loose suspension slides backward.
Forward stroke
Loose suspension becomes compressed.
Forward stroke
Solids become pushable and push the cake to the front.
Forward position
Feed zone is flooded up to the cake surface.
Reverse stroke
A portion of the newly created cake slides backward.

FIG. 18-151 Pusher centrifuge solids transport. (*Andritz KMPT GmbH.*)

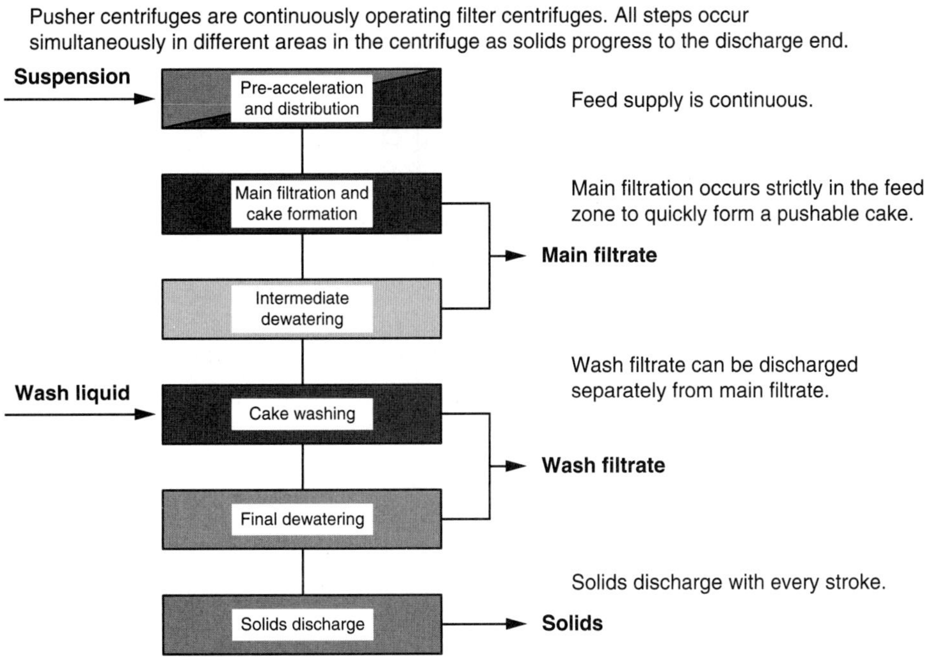

FIG. 18-152 Pusher centrifuge process steps. (*Andritz KMPT GmbH.*)

Usually, cake wash ratios of about 0.1 to 0.3 kg wash/kg solids are possible within the normal residence time of the wash zone. This usually can displace at least 95 percent of the mother liquor and impurities. In some cases, higher wash ratios or even multistage countercurrent washes are used, in which case sufficient residence time via throughput reduction must be considered.

Single-Stage versus Multistage Pusher centrifuges can be single-stage configuration with a single long basket and screen, two-stage (as shown schematically in Fig. 18-153), three-, or four-stage designs. Cake height and push force are primarily influenced by screen deck length and cake friction coefficient.

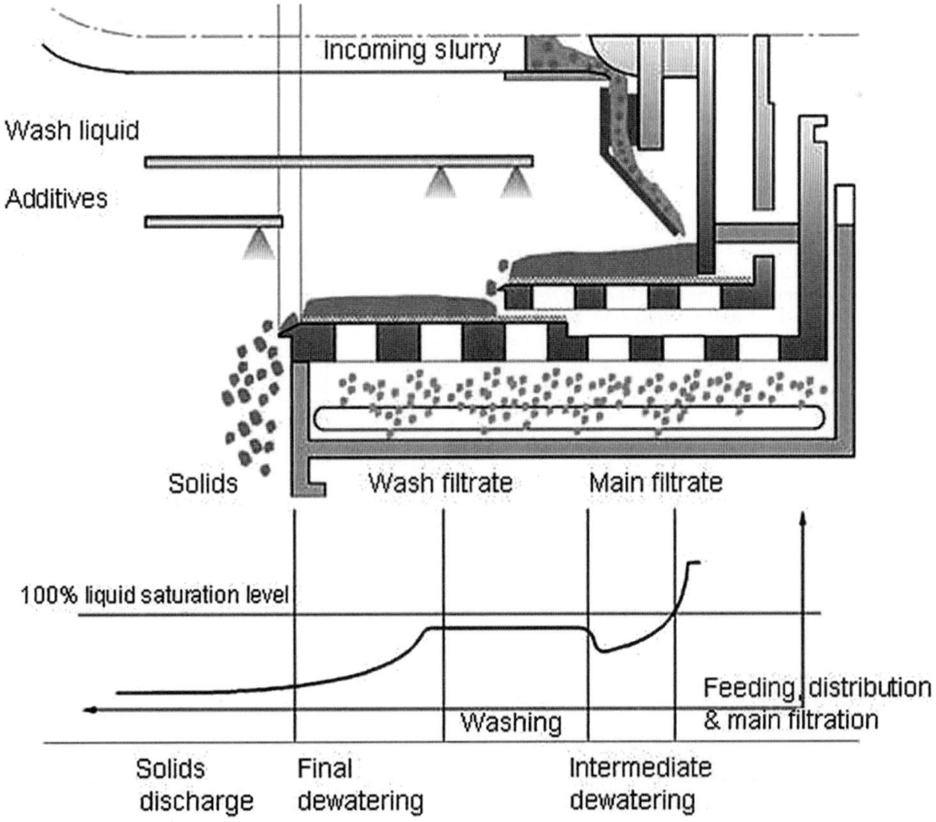

FIG. 18-153 Pusher centrifuge product moisture gradient. (*Andritz KMPT GmbH.*)

Single-Stage Where single-stage units are appropriate (ammonium sulfate is one example due to very large crystal size and good cake shear strength), the solids volumetric capacity can be maximized. However, because the push force requirement increases with screen length, cake shear or buckling can be the result with unstable operation. Because the average cake thickness in the feed zone is higher, filtration capacity may be slightly less than with multistage units. Fines losses can be slightly less with single-stage units since a smaller proportion of the cake bed is in contact with the slotted screen, and there is no reorientation of crystals between stages. These units are often limited to low-speed operation for stability.

Two-Stage Most pusher centrifuges sold today are two-stage. They provide greater flexibility than single-stage types; they offer greater filtration capacity, a lower tendency for cake shear, and higher speed. When it is possible to operate with a forced cake, capacities can approach those of single-stage designs. Wash typically is applied on the latter portion of the first stage and through the transition onto the second stage. During this transition, the crystals are reoriented and the capillaries opened, which can enhance the wash effect. With even-stage units, the feed acceleration system is not oscillating relative to the feed pipe. Some advanced designs of feed acceleration systems incorporating impellers benefit from this constant relationship.

Three- and Four-Stage These designs are generally reserved for the largest sizes that have long baskets that need to be subdivided into reasonable-length stages as well as for very special applications with very high friction coefficients, low internal cake shear strength, or fairly high compressibility. For example, in processing high-rubber ABS, four-stage units have been used, but the deck lengths are so short, with the corresponding thin cakes and short retention time, that capacity and performance are severely reduced. Other types of machines (such as peeler centrifuges or cylindrical/conical pushers) can be better suited.

Cylindrical/Conical A variation of single- and two-stage designs uses a cylindrical section or stage at the feed end followed by a conical section or stage sloping outward to the discharge end. The benefit of this design is that the axial component of force in the conical end assists with solids transport. Care must be taken that the cone angle not exceed the sliding friction angle of the cake, or else the cake will short-circuit the zone, resulting in poor performance and high vibration. Fabrication costs of the baskets are higher than those of cylindrical designs, and slotted screen construction is complicated, with high replacement costs.

Theory of Centrifugal Filtration The behavior of solid-liquid mixtures in a filtering centrifuge are more difficult to predict theoretically than slurry behavior in pressure and gravity filtration. The area of flow and the driving force are both proportional to the radius, and the specific resistance and porosity may also change markedly within the cake. Filtering centrifuges are nearly always selected by scale-up from lab tests on materials to be processed, such as using bucket centrifuges where a wide range of test conditions (cake thickness, time, and *G*-force) can be controlled. Although tests with the bucket centrifuge provide some quantitative data to scale-up, the results include the wall effect from buckets, which are not representative of actual cylindrical basket geometry; bucket centrifuges are not useful in quantifying filtration rates. A modified version of the buckets or even a cylindrical perforated basket can be used. In the latter, there is less control of cake depth and circumferential uniformity. The desired quantities to measure are filtration rate, washing rate, spinning time, and residual moisture. Also, with filtering centrifuges such as the screen-bowl centrifuge, screen-scroll centrifuge, and to some extent in multistage pushers, the cake is constantly disturbed by the scroll conveyor or conveyance mechanism; liquid saturation due to capillary rise as measured in bucket tests is absent. While bucket centrifuge tests are very useful for first-look feasibility, it is always recommended to follow with pilot-scale testing of the actual equipment type being considered.

Bulk Filtration Rate, Q For a basket with axial length *b*, when the centrifuge cake is submerged in a pool of liquid, as in the case of a fast-sedimenting, solids-forming cake, the rate of filtration becomes limiting and is given by:

$$Q = \frac{\pi b \rho K \Omega^2 (r_b^2 - r_p^2)}{\mu \left(\ln\left[\frac{r_b}{r_c}\right] + \frac{KR_m}{r_b} \right)} \tag{18-120a}$$

where μ and ρ are, respectively, the viscosity and density of the liquid; Ω is the angular speed; K is the average permeability of the cake and is related to the specific resistance α by the relationship $\alpha K \rho_s = 1$, with ρ_s being the solids density; and r_p, r_c, and r_b are, respectively, the radii of the liquid pool surface, the cake surface, and the filter medium adjacent to the perforated bowl. Here, the pressure drop across the filter medium, which also includes that from the cake heel, is $\Delta p_m = \mu R_m(Q/A)$, with R_m being the combined resistance. The permeability K has a unit m^2, α m/kg, and R_m m^{-1}. The driving force is due to the hydrostatic pressure difference across the bowl wall and the pool surface—that is, the numerator of Eq. (18-120*a*), and the resistance is due to the cake layer and the filter medium—that is, the denominator of Eq. (18-120*a*). Figure 18-154 shows the pressure distribution in the cake and the liquid layer above. The pressure (gauge) rises from zero to a maximum at the cake surface; thereafter, it drops monotonically within the cake in overcoming resistance to flow. There is a further pressure drop across the filter medium, the magnitude dependent on the combined resistance of the medium and the heel at a given flow rate. This scenario holds, in general, for incompressible as well as for compressible cake. For the latter, the pressure distribution also depends on the compressibility of the cake.

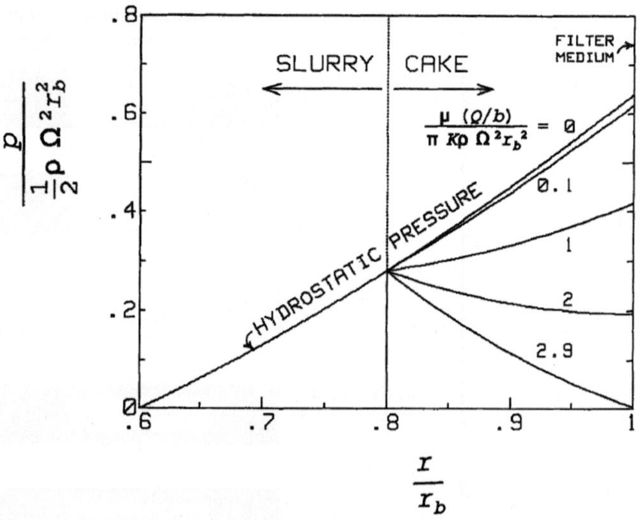

FIG. 18-154 Pressure distribution in a basket centrifuge under bulk filtration.

For incompressible cake, the pressure distribution and the rate depend on the resistance of the filter medium and the permeability of the cake. Figure 18-154 shows several possible pressure profiles in the cake with increasing filtration rates through the cake. It is assumed that $r_c/r_b = 0.8$ and $r_p/r_b = 0.6$. The pressure at $r = r_b$ corresponds to pressure drop across the filter medium Δp_m, with the ambient pressure taken to be zero. The filtration rate and the pressure distribution depend on the medium resistance and that of the cake. High medium resistance or blinding of the medium results in a greater penalty on filtration rate.

In most filtering centrifuges, especially the continuous-feed ones, the liquid pool above the cake surface should be at a minimum to prevent liquid from running over the cake. Setting $r_p = r_c$ in Eq. (18-119*a*), the dimensionless filtration flux is plotted in Fig. 18-155 against r_c/r_b for different ratios of

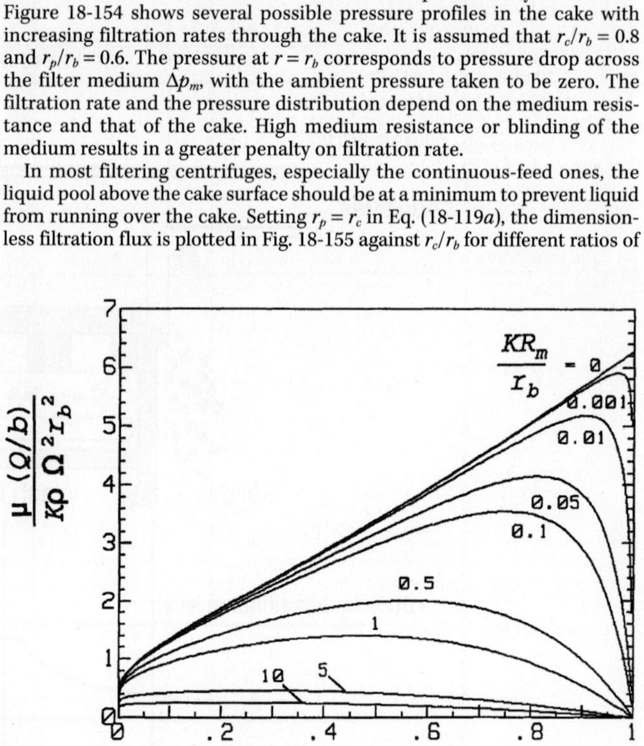

FIG. 18-155 Centrifugal filtration rate as a function of both cake and medium resistance.

filter-medium resistance to cake resistance, KR_m/r_b. For negligible medium resistance, the flux is a monotonic decreasing function with increasing cake thickness, that is, smaller r_c. With finite medium resistance, the flux curve for a range of different cake thicknesses has a maximum. This is because for thin cake the driving liquid head is small and the medium resistance plays a dominating role, resulting in lower flux. For very thick cake, despite the increased driving liquid head, the resistance of the cake becomes dominant; therefore, the flux decreases again. The medium resistance to cake resistance should be small, with $KR_m/r_b < 5$ percent. However, the cake thickness, which is directly proportional to the throughput, should not be too small, despite the fact that the machine may have to operate at somewhat less than the maximum flux condition.

It is known that the specific resistance for centrifuge cake, especially for compressible cake, is greater than that of the pressure or vacuum filter. Therefore, the specific resistance has to be measured from centrifuge tests for different cake thicknesses so as to scale up accurately for centrifuge performance. It cannot be extrapolated from pressure and vacuum filtration data. For cake thickness that is much smaller than the basket radius, Eq. (18-120a) can be approximated by

$$V_f = V_{fo}\left(\frac{h}{h_c}\right) \tag{18-120b}$$

where $h = r_b - r_p$ is the liquid depth, $h_c = r_b - r_c$ is the cake thickness, and $V_{fo} = (\rho GK/\mu)$ is a characteristic filtration velocity. Table 18-18 shows some common filtering centrifuges and the application with respect to the G-level, minimum feed-solids concentration, minimum mean particle size, and typical filtration velocity. The vibratory and tumbler centrifuges have the largest filtration rate of 5×10^{-4} m/s (0.02 in/s) for processing 200-μm or larger particles, followed by screen-scroll, pusher, and screen bowl centrifuges, whereas batch filtering centrifuges, vertical and peeler, have the lowest filtration rate of 1×10^{-7} m/s (4×10^{-6} in/s) for processing 5-μm particles with increased cycle time. For faster filtering batch centrifuge applications, siphon peelers are preferred for better control of filtration and solids distribution.

Film Drainage and Residual Moisture Content Desaturation of the liquid cake ($S < 1$) begins as the bulk filtration ends, at which point the liquid level starts to recede below the cake surface. Liquids are trapped in (1) cake pores between particles that can be drained with time (free liquid), (2) particle contact points (pendular liquid), (3) fine pores that form continuous capillaries (capillary rise), and (4) particle pores, or they are bound by particles (bound liquid). Numbers 1 to 3 can be removed by centrifugation, and thus each of these components depends on G to a different extent. Only desaturation of the free liquid, and to a much lesser entent the liquid at contact points, is a function of time. The wet cake starts from a state of being fully saturated, $S = 1$, to a point where $S < 1$, depending on the dewatering time. At a very large amount of time, it approaches an equilibrium point S_∞, which is a function of G, capillary force, and the amount of bound liquid trapped inside or externally attached to the particles.

The following equations, which have been tested in centrifugal dewatering of granular solids, prove useful:

Total saturation:

$$S_{\text{total}} = S_\infty + S_T(t) \tag{18-121}$$

Equilibrium component:

$$S_\infty = S_c + (1 - S_c)(S_p + S_z) \tag{18-122}$$

Transient component:

$$S_T(t) = (1 - S_c)(1 - S_p - S_z)S_t(t) \tag{18-123}$$

The details of the mathematical model of these four components are given below.

Drainage of free liquid in thin film:

$$S_t(t) = \left(\frac{4}{3}\right)\left(\frac{1}{t_d^n}\right), \quad t_d > 0 \tag{18-124}$$

where for smooth-surface particles, $n = 0.5$, and for particles with rough surfaces, n can be as low as 0.25.

Bound liquid saturation:

$$S_p = \text{function(particle characteristics)} \tag{18-125}$$

Pendular saturation:

$$S_z = 0.075, \quad N_c \le 5 \tag{18-126a}$$

$$S_z = \frac{5}{(40 + 6N_c)}, \quad 5 \le N_c \le 10 \tag{18-126b}$$

$$S = \frac{0.5}{N_c}, \quad N_c \ge 10 \tag{18-126c}$$

Often when $N_c < 10$, S_p and S_z are combined for convenience; the sum is typically 0.075 for smooth particles and can be as high as 0.35 for rough-surface particles. This has to be determined from tests.

Saturation due to capillary rise:

$$S_c = \frac{4}{B_o} \tag{18-127}$$

where the dimensionless time t_d, capillary number N_c, and Bond number B_o are, respectively:

$$t_d = \frac{\rho G d_h^2 t}{\mu H} \tag{18-128}$$

$$N_c = \frac{\rho G d_h^2}{\sigma \cos\theta} \tag{18-129}$$

$$B_o = \frac{\rho G H d_h}{\sigma \cos\theta} \tag{18-130}$$

where ρ and μ are, respectively, the density and viscosity of the liquid; θ is the wetting angle of the liquid on the solid particles; σ is the interfacial tension; H is the cake height; d is the mean particle size; and t is the dewatering time. The hydraulic diameter of the particles can be approximated by either $d_h = 0.667\,\varepsilon d/(1 - \varepsilon)$ or $d_h = 7.2(1 - \varepsilon)K^{1/2}/\varepsilon^{3/2}$, where ε is the cake porosity.

Example 18-7 Calculation of Moisture Weight Fraction

Given: $\rho = 1000$ kg/m^3, $\rho_s = 1200$ kg/m^3, $\mu = 0.004$ N·s/m^2, $\sigma \cos\theta = 0.068$ N/m, $H = 0.0254$ m, $d = 0.0001$ m, $\varepsilon = 0.4$, $G/g = 2000$, $t = 2$ s, $S_p = 0.03$.

Calculate: $d_h = 4.4 \times 10^{-5}$ m, $t_d = 748$, $N_c = 0.56$, $B_o = 322$, $S_t = 0.048$, $S_z = 0.075$, $S_c = 0.012$, $S_\infty = 0.116$, $S_T = 0.043$, $S_{\text{total}} = 0.158$, $W = 0.919$.

Note that W is the solids fraction by weight and is determined indirectly from Eq. (18-89). The moisture weight fraction is 0.081.

The transient component depends not only on G, cake height, and cake properties, but also on dewatering time, which ties to solids throughput for a continuous centrifuge and cycle time for batch centrifuge. If the throughput is too high or the dewatering cycle is too short, the liquid saturation can be high and becomes limiting. Given that time is not the limiting factor, dewatering of the liquid lens at particle contact points requires a much higher G-force. The residual saturation depends on the G-force to the capillary force, as measured by N_c, the maximum of which is about 7.5 percent, which is quite significant. If the cake is not disturbed (scrolled and tumbled) during conveyance and dewatering, liquid can be further trapped in fine capillaries due to liquid rise, the amount of which is a function of B_o, which weighs the G-force to the capillary force. This amount of liquid saturation is usually smaller than the capillary force associated with liquid-lens (also known as pendular) saturation. Lastly, liquid can be trapped by chemical force at the particle surface or by physical capillary or interfacial force in the pores within the particles. Because the required desaturating force is extremely high, this portion of moisture cannot be removed by mechanical centrifugation. Fortunately, for most applications it is a small percentage, if it exists.

SELECTION OF CENTRIFUGES

Table 18-19 summarizes the several types of commercial centrifuges, their manner of liquid and solids discharge, their unloading speed, and their relative volumetric capacity. When either the liquid or the solids discharge is not continuous, the operation is said to be cyclic. Cyclic or batch centrifuges are often used in continuous processes by providing appropriate upstream and downstream surge capacity.

Sedimentation Centrifuges These centrifuges often are selected on the basis of tests on tubular, disc, or helical-conveyor centrifuges of small size.

TABLE 18-19 Characteristics of Commercial Centrifuges

Method of separation	Rotor type	Centrifuge type	Manner of liquid discharge	Manner of solids discharge or removal	Centrifuge speed for solids discharge	Capacity*
Sedimentation	Batch	Ultracentrifuge				1 mL
		Laboratory, clinical	Batch	Batch manual	Zero	To 6 L
	Tubular	Supercentrifuge	Continuous†	Batch manual	Zero	To 1200 gal/h
		Multipass clarifier	Continuous†	Batch manual	Zero	To 3000 gal/h
	Disc	Solid wall	Continuous†	Batch manual	Zero	To 30,000 gal/h
		Light-phase skimmer	Continuous	Continuous for light-phase solids	Full	To 1200 gal/h
		Peripheral nozzles	Continuous	Continuous	Full	To 24,000 gal/h
		Peripheral valves	Continuous	Intermittent	Full	To 3000 gal/h
		Peripheral annulus	Continuous	Intermittent	Full	To 12,000 gal/h
	Solid bowl (scroll conveyor)	Constant-speed horizontal	Continuous†	Cyclic	Full (usually)	To 60 ft^3
		Variable-speed vertical	Continuous†	Cyclic	Zero or reduced	To 16 ft^3
		Continuous decanter	Continuous	Continuous screw conveyor	Full	To 54,000 gal/h To 100 tons/h solids
Sedimentation and filtration		Screen-bowl decanter	Continuous	Continuous	Full	To 60,000 gal/h To 125 tons/h solids
Filtration	Conical screen	Wide-angle screen	Continuous	Continuous	Full	To 40 tons/h solids
		Differential conveyor	Continuous	Continuous	Full	To 80 tons/h solids
		Vibrating, oscillating, and tumbling screens	Continuous	Essentially continuous	Full	To 250 tons/h solids
		Reciprocating pusher	Continuous	Essentially continuous	Full	Limited data
	Cylindrical screen	Reciprocating pusher, single and multistage	Continuous	Essentially continuous	Full	To 100 tons/h solids
		Horizontal	Cyclic	Intermittent, automatic	Full (usually)	To 25 tons/h solids
		Vertical, underdriven	Cyclic	Intermittent, automatic, or manual	Zero or reduced	To 10 tons/h solids
		Vertical, suspended	Cyclic	Intermittent, automatic, or manual	Zero or reduced	To 10 tons/h solids

*To convert gallons per hour to liters per second, multiply by 0.00105; to convert tons per hour to kilograms per second, multiply by 0.253; and to convert cubic feet to cubic meters, multiply by 0.0283.

†Feed and liquid discharge interrupted while solids are unloaded.

The centrifuge should be of a configuration similar to that of the commercial centrifuge it is proposed to be used for. The results in terms of capacity for a given performance (effluent clarity and solids concentration) may be scaled up by using the sigma concept of Eqs. (18-118) to (18-120). Spin-tube tests may be used for information on systems that contain well-dispersed solids. Such tests are totally unreliable on systems that contain a dispersed phase that agglomerates or flocculates during the time of centrifugation.

Filtering Centrifuges These filters often can be selected on the basis of batch tests on a laboratory unit, preferably one at least 12 in (305 mm) in diameter. A bucket centrifuge test would be helpful to study the effect of *G*, cake height, and dewatering time, but not filtration rates. It is always recommended to follow bucket tests with pilot-scale testing of the actual equipment type being considered. Caution is needed in correcting for capillary saturation, which may be absent in large continuous centrifuges with scrolling conveyances.

Unless operating data on similar material are available from other sources, continuous centrifuges should be selected and sized only after tests on a centrifuge of similar configuration.

It is often forgotten that test results are valid only to the extent that the slurry and the test conditions duplicate what will exist in the operating plant. This may involve testing on a small scale (or even on a large one) with a slipstream from an existing unit, but the dependability of the data is often worth the extra effort involved. Most centrifuge manufacturers provide testing services and demonstration facilities in their own plants and maintain a supply of equipment for field-testing in the customer's plant, such as with a pilot centrifuge module with associated peripheral equipment. Larger-scale pilot equipment provides better scale-up accuracy, as in evaluations of the effect of cake thickness in batch filtering centrifuges.

COSTS

Neither the investment cost nor the operating cost of a centrifuge can be directly correlated with any single characteristic of a given type of centrifuge. The costs depend on how the features of the centrifuge apply to the physical and chemical nature of the materials being separated, the degree and difficulty of separation, the flexibility and capability of the centrifuge and its auxiliary equipment, the environment in which the centrifuge is located, and many other nontechnical factors, including market competition.

The useful parameter for value analysis is the installed cost of the number of centrifuges required to produce the demanded separative effect (end product) at the specified capacity of the plant. The possible benefits of adjustments in the upstream and downstream components of the plant and the process should be carefully examined in order to minimize the total plant costs; the systems approach should be used.

Purchase Price Due to rapidly changing conditions for material cost and availability, equipment manufacturing locations, consolidation of equipment suppliers, currency exchange, volatility, and other factors, CAPEX and OPEX must be evaluated on a case-by-case basis for the project at hand. Prices will vary upward with the use of more exotic materials of construction, the need for explosion-proof electrical gear, the type of enclosure required for vapor containment, and the degree of portability, and this holds for all types of centrifuges.

Installation Costs Installation costs of centrifuges vary over an extremely wide range, depending on the type of centrifuge, on the area and kind of structure in which it is installed, and on the details of installation. Some centrifuges, such as portable tubular and disc oil purifiers, are shipped as package units and require no foundation and a minimum of connecting piping and electrical wiring. Others, such as large batch automatic and continuous scroll-type centrifuges, may require substantial foundations and even building reinforcement, extensive interconnecting piping with required flexibility, auxiliary feed and discharge tanks and pumps and other facilities, and elaborate electrical and process-control equipment. Minimum installation costs, covering a simple foundation and minimum piping and wiring, are about 5 to 10 percent of purchase price for tubular and disc centrifuges, and up 10 to 25 percent for bottom drive, batch automatic, and continuous-scroll centrifuges. If the cost of all auxiliaries—such as special foundations, tanks, pumps, conveyors, electrical and control equipment—is included, the installation cost may well range from one to two times the purchase price of the centrifuge itself.

Maintenance Costs Because of the care with which centrifuges are designed and built, their maintenance costs are in line with those of other slower-speed separation equipment, averaging in the range of 1 to 4 percent for batch machines, 3 to 8 percent for pusher centrifuges, and 5 to 10 percent for decanters and disc centrifuges per year of the purchase price for centrifuges in light to moderate duty. For centrifuges in severe service and on highly corrosive fluids, the maintenance cost may be several times these values. Maintenance costs are likely to vary from year to year, with lower costs for general maintenance and periodic large expenses for major overhaul. Centrifuges are subject to erosion from abrasive solids such as sand, minerals, and grits. When these solids are present in the feed, the centrifuge components are subject to wear. Parts such as feed and solids discharge ports, unloader knives, and helical scroll blade tips should be protected with replaceable wear-resistant materials. Excessive out-of-balance forces strongly contribute to maintenance requirements and should be avoided.

Operating Labor Centrifuges run the gamut from completely manual control to fully automated operation. For the former, one operator can run several centrifuges, depending on their type and the application. Fully automatic centrifuges usually require little direct operator attention. In most production environments, PLC- or DCS-based automatic controls are the norm.

EXPRESSION

GENERAL REFERENCES: Chen, W., F. J. Parma and W. Schabel, "Testing Methods for Belt Press Biosludge Dewatering," *Filtration J.* **5**(1): 29–32 (2005); Shirato, M., T. Murase, and T. Aragaki, "Slurry Deliquoring by Expression," in *Progress in Filtration & Separation*, vol. 4, 181–288, Elsevier, Amsterdam, 1986; Tiller, F. M., and W. F. Leu, "Basic Data Fitting in Filtration," *J. Chinese Inst. Chem. Engr.* **11**: 61–70 (1980); Tiller, F. M., and W. Li, "Determination of the Critical Pressure Drop for Filtration of Supercompactible Cakes," *Water Sci. and Technol.* **44**(10): 171–176 (2001); Tiller, F. M., and W. Li, "Dangers of Lab-Plant Scale Up for Solid/Liquid Separation Systems," *Chem. Eng. Commun.* **190**(1): 128–150 (2003); Tiller, F. M., and C. S. Yeh, The Role of Porosity in Filtration XI: Filtration Followed by Expression," *AIChE J.* **33**: 1241–1256 (1987).

FUNDAMENTALS OF EXPRESSION

Definition Expression refers to the deliquoring of a particulate bed by squeezing. It is normally the last stage of a solid-liquid separation process before thermal drying. It has been widely applied in a variety of fields, such as in food industries to increase product yield, in wastewater treatment plants to reduce transportation and disposal cost by decreasing sewage sludge moisture content, and in chemical processes to eliminate liquid content in the solid product prior to drying. The energy required to express liquid from solid-liquid mixtures is much less than that of any thermal method. The driving force for expression can be hydraulic pressure, mechanical contact, gas blowing, gravity, or centrifugal drainage. The commonality is that physical contacts among particles must be present to transmit the squeezing force. In this section, the term *expression* refers to mechanical compression of a solid-liquid mixture by mechanical devices like diaphragms, rolls, pistons, or screws on the surface of cakes.

Filtration and Expression of Compactible Filter Cakes

Filtration A filter cake can be incompactible, moderately compactible, highly compactible, or supercompactible (Tiller and Li 2003). Porosity is not uniformly distributed in a compactible cake, and a layer of skin cake with high resistance to liquid flow is developed next to the filter medium, as shown in Fig. 18-156 (Tiller and Li 2001) for the case of an activated sludge. The skin deters frictional forces needed to consolidate the cake and increase solid content in the large portion of the cake away from the skin layer. As a result, as illustrated by Fig. 18-157, increasing filtration pressure on highly compactible filter cakes only increases the pressure drop in the skin layer and cannot attain substantial deliquoring of the whole cake (e.g., flocculated latex), while increasing filtration pressure does help to make a drier cake on a less compactible material (e.g., Kaolin Flat D).

Expression In mechanical expression, the pressure is applied directly on filter cakes rather than relying on flow frictions generated by hydraulic pressure drop to deliquor the cake. The effects of stress distribution in a compactible filter cake by these two different mechanisms are shown in Fig. 18-158. The stress distribution of a mechanical expression is more uniform across the cake than that of a pressure filtration, leading to a more uniform cake. Expression is therefore a better choice for deliquoring compactible filter cakes.

Fundamental Theory A theoretical model was developed by Shirato (1986) based on Terzaghi's and Voigt's consolidation model in soil mechanics. Shirato's theory includes a filtration stage followed by a consolidation.

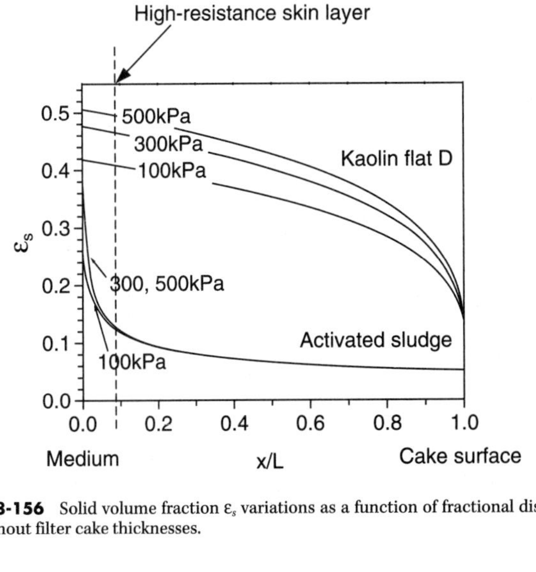

FIG. 18-156 Solid volume fraction ε_s variations as a function of fractional distance throughout filter cake thicknesses.

Average consolidation ratio U_c is given as a function of consolidation time θ_c and other characteristic parameters of an expression process, including true solids density, liquid density, liquid viscosity, specific cake resistance (or permeability) versus pressure, porosity versus pressure, and frictional stress on solids throughout cake thickness versus applied pressure. The relationships of specific cake resistance and porosity versus pressure, and local frictional stress on solids throughout cake thickness during the primary consolidation stage, are given by empirical constitutive equations (Tiller and Leu 1980), and can be determined by a compression-permeability cell test (Tiller and Leu 1980), as shown in Fig. 18-159.

Factors Affecting Expression Operations Based on the fundamental theory, variables affecting expression include characteristics of suspending particles, properties of liquid, properties of filter cake, and expression operation conditions, as summarized in Fig. 18-160. Expression efficiency is determined by the properties of the filter cake, which very much depend on characteristics of the suspending particles, properties of liquid, and operation conditions. Interrelationships of the preceding parameters are described by empirical equations covering restrictive ranges.

EXPRESSION EQUIPMENT

This type of equipment uses mechanical expression rather than pump pressure for cake compression. Drier cakes and faster cycles rate can be achieved compared to pressure filters. Low- to high-pressure (up to 2000 psi) units are available for expression. They can be divided into two categories: batch expression equipment, which allows higher compression pressure and has lower slurry-handling capacity, and continuous expression equipment, which uses lower compression pressure but offers higher slurry-handling capacities.

Batch Expression Equipment In batch expression equipment, the cake is initially formed by pressure filtration just as in a pressure filter. After the filtration stage, a squeezing device such as a diaphragm is inflated with gas or liquid to compress the cake. Batch expression equipment allows longer compression time and higher compression pressure. The cake can be very dry.

Diaphragm Presses Diaphragm presses, also called membrane presses, are derived from filter presses, which were described in the pressure filtration section. In a diaphragm press, a diaphragm (Fig. 18-161*a*) is attached to the filter plate. The operation of a diaphragm press is the same as that of a filter press during the filtration step. At the end of filtration, the diaphragm is inflated (Fig. 18-161*b*) to squeeze the filter cake. After the squeezing, the diaphragm is deflated and the filter chamber opened to discharge the cake.

The diaphragm can be made of polypropylene or rubber, but polypropylene is most often used. Both air and water can be used as the inflating medium for the diaphragm. Because the inflating medium needs to be brought into the filter plates by hoses, a dangerous condition can exist if a hose is broken with air under pressure. Therefore, hydraulic fluid (mostly water) is generally used to inflate the diaphragm. Air is only used occasionally in small pilot units.

As in filter presses, one disadvantage of the diaphragm press is the manual operation for filter cake discharge. Automatic cake discharge devices are available from most filter manufacturers. However, the reliability of automatic cake discharge needs to be verified by actual field operation.

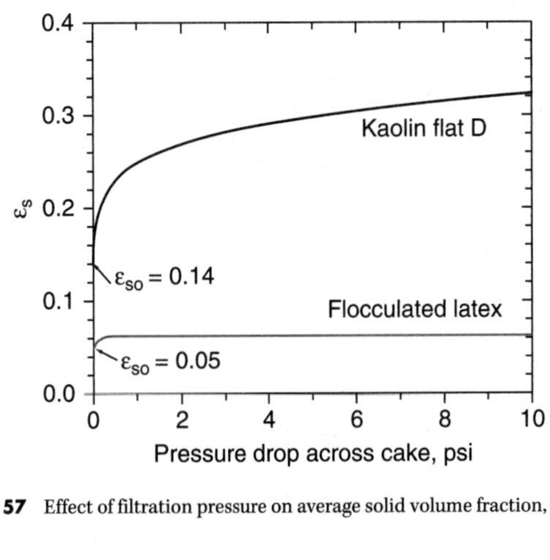

FIG. 18-157 Effect of filtration pressure on average solid volume fraction, ε_s.

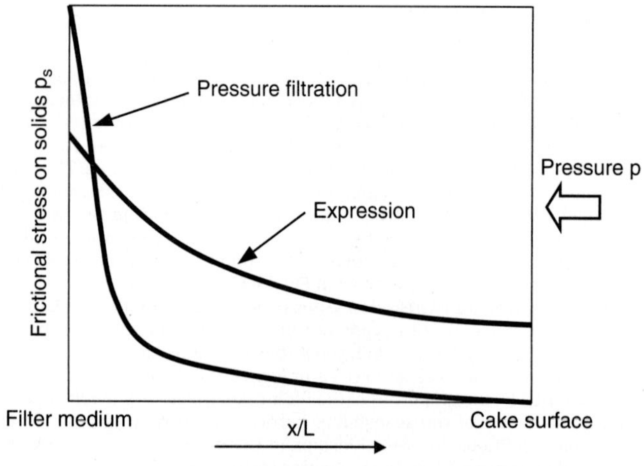

FIG. 18-158 Comparisons of frictional stress distributions in expression and pressure filtration.

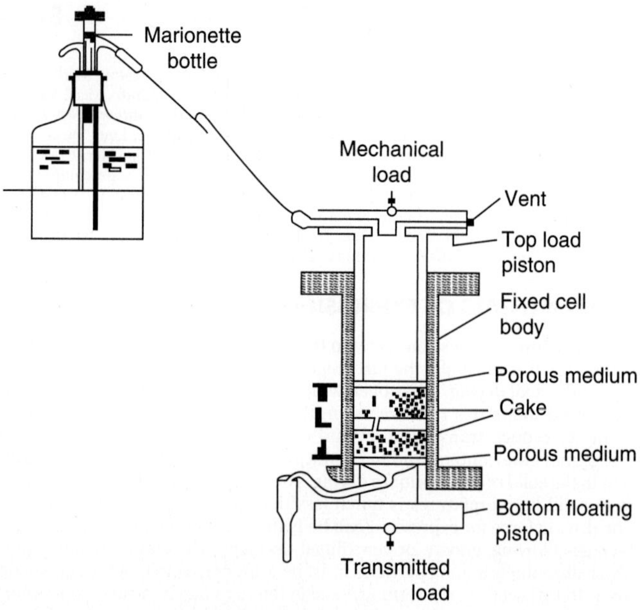

FIG. 18-159 Compression-permeability (C-P) cell.

Normally, automatic cake discharge has a better chance of success in diaphragm presses than in filter presses because the cakes are normally drier in diaphragm presses.

The cake deliquoring is primarily done during the expression step, so the cake formation period is normally carried out under low pressure, and a high-pressure slurry pump is not necessary; it helps to reduce floc damage during pumping. The normal expression pressure used in a diaphragm press is 110 or 220 psi; in some designs, pressures up to 800 psi are used.

Diaphragm presses are superior to filter presses in deliquoring compactible cakes (such as biological sludge, pulps, or highly flocculated materials). Because a diaphragm press costs more than a regular filter press, the use of a diaphragm press may not be advantageous if solids are not very compactible. There are laboratory and pilot tests available to determine the need for a diaphragm press.

The best way to evaluate diaphragm presses for an application is to run tests with a small pilot unit. Although smaller test units are available, pilot units with 1-ft^2 filter plate area are more common and are recommended.

A laboratory pressure filter (Fig. 18-162) equipped with a piston can provide a simple feasibility test. In this device, the suspension is poured into the filter cylinder, and the first stage of the test is just like a pressure filtration test. After the filtration, compressed air or water is used to push the piston down to squeeze the filter cake. The filtration rate, final cake thickness, and dryness are recorded for evaluation and comparison with the same test without the compression by the piston.

Horizontal Diaphragm Presses This is similar to the diaphragm press except the filter plates lie horizontally (while in a diaphragm press, the filter plates are operated vertically). The press can be a single-chamber unit, or multiple chambers can be stacked to achieve greater filtration area.

In each filter plate, the filter medium is attached to a moving belt (Fig. 18-163). The slurry is fed into the filter chamber, and the operation starts as a pressure filtration. After filtration, the diaphragm is inflated to squeeze the cake. After expression, the filter chamber opens, and the belt moves the cake out of the filter chamber for discharging. The filter chamber is then closed and ready for the next filtration cycle. Permanent filter belts or disposable media can be used as filter media. The disposable media are especially useful when handling particles that have a high tendency to foul the filter media. With the moving belt, the press operation is fully automatic and is another advantage of this equipment.

The testing for evaluating the horizontal diaphragm press is the same as that described earlier for the vertical diaphragm presses. To ensure automatic operation, the cake solids should not stick to the medium or seal of the filter chamber, and they need to be carefully evaluated during testing.

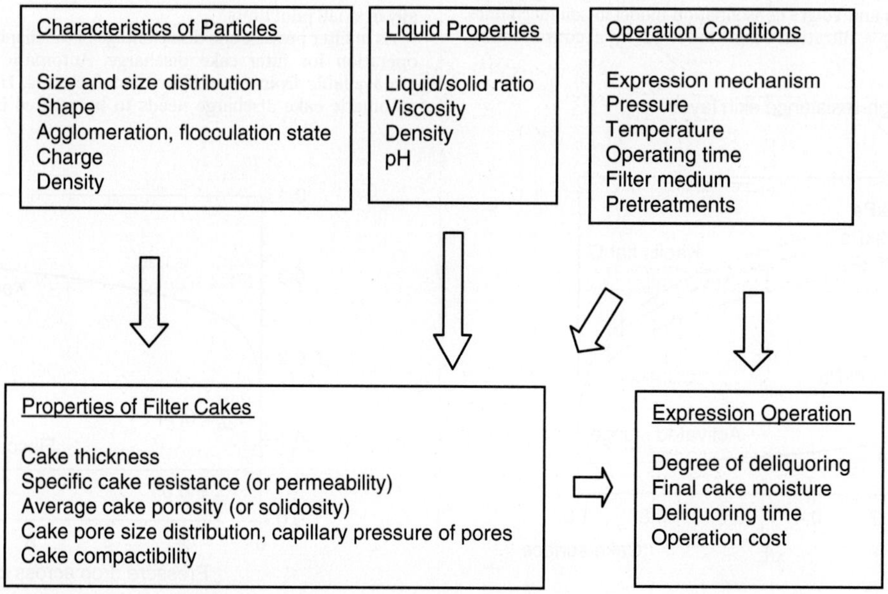

FIG. 18-160 Variables affecting expression.

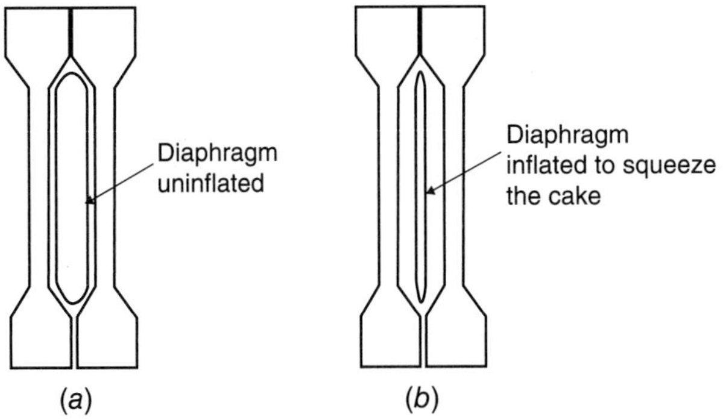

FIG. 18-161 Diaphragm press plate.

Tower Presses The tower press is similar to the stacked horizontal diaphragm presses, but only one filter belt is used (Fig. 18-164). The operation is also fully automatic. The primary applications are in chemical, mineral, and pharmaceutical industries. The testing method is the same as for the diaphragm presses.

Tubular Presses A tubular press is a high-pressure, variable chamber, batch, diaphragm-type filter press. Designed to operate at pressures up to 1500 psi, the tube press is suitable for dewatering fine particle slurries, generally <10 μm. Typically the tube press operates at cycle times between 6 and 10 minutes and produces very dry, friable cakes.

The tube press (Fig. 18-165) is composed of an outer cylindrical casing and an inner candle supporting the filter medium. The chamber between

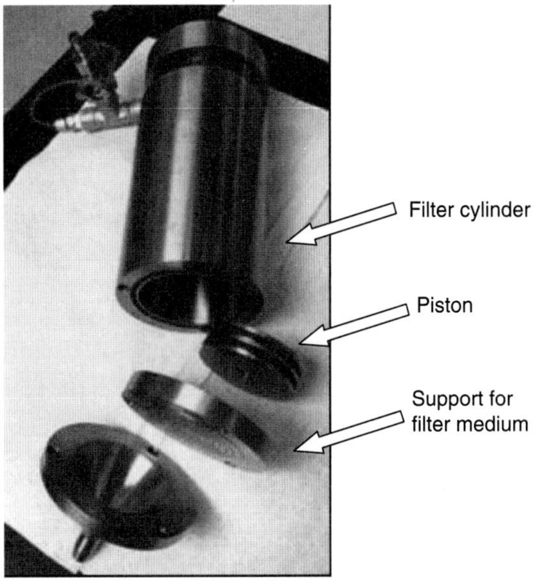

FIG. 18-162 Laboratory pressure filter with a piston to compress the cake.

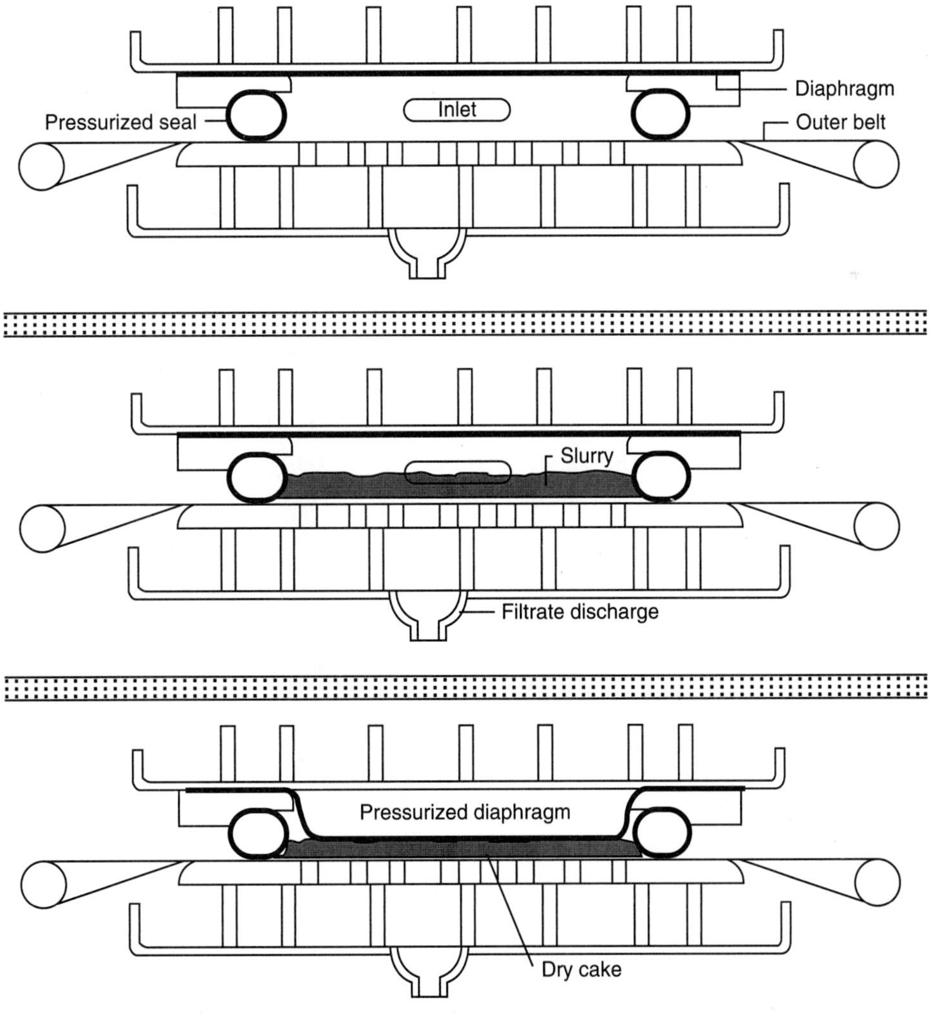

FIG. 18-163 A horizontal diaphragm press. (*Courtesy of Filtra-Systems.*)

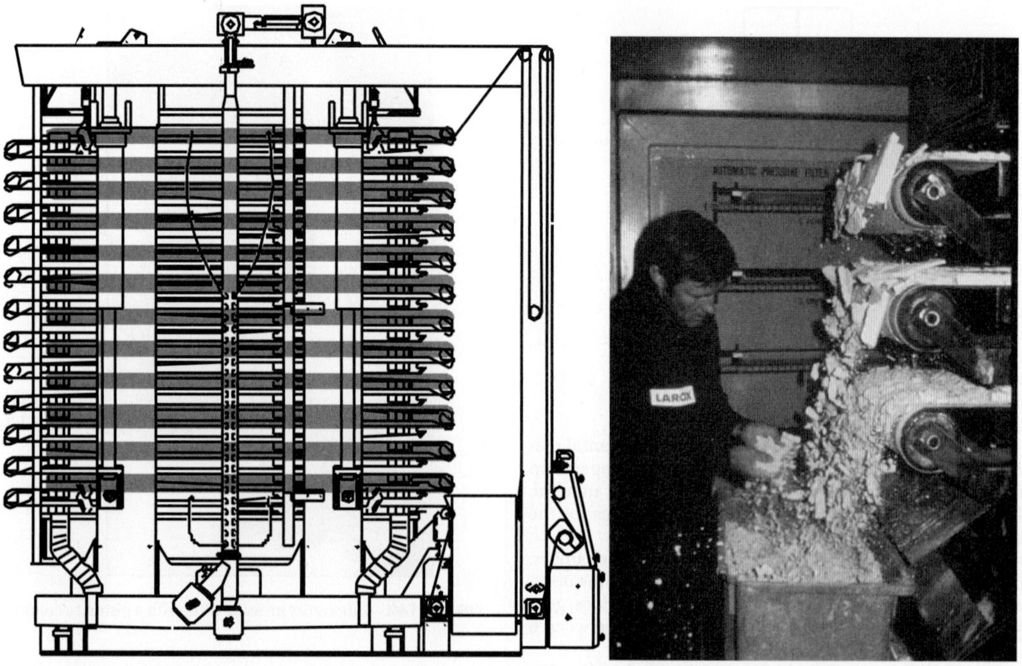

FIG. 18-164 Tower press. (*Courtesy of Outotec.*)

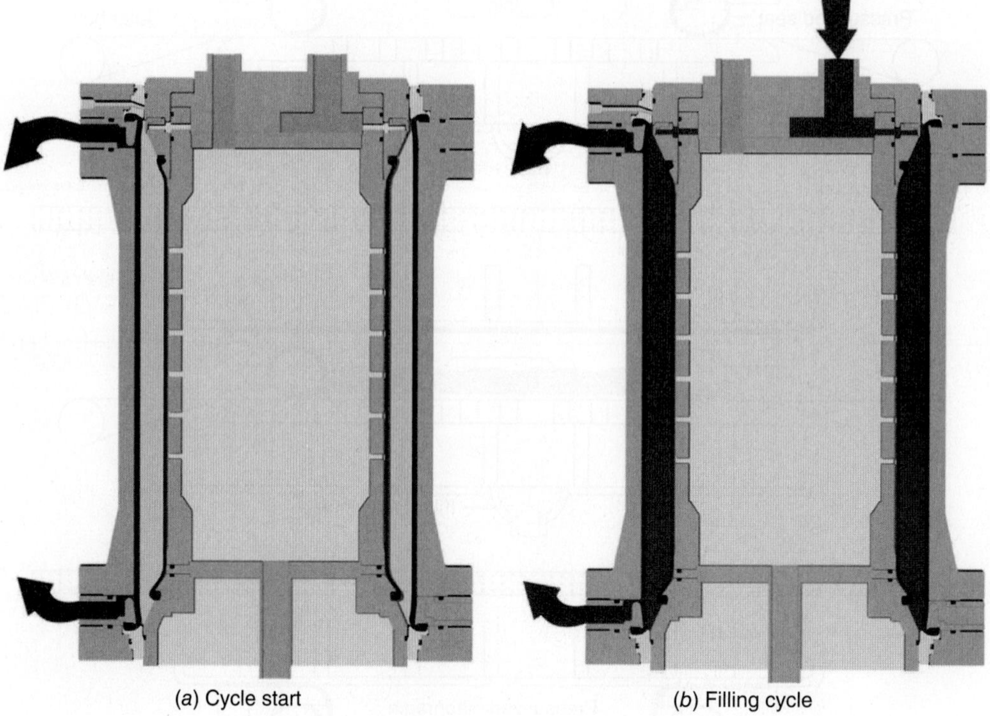

FIG. 18-165 A tubular press and its operation cycle. (*Courtesy of Metso Minerals.*)

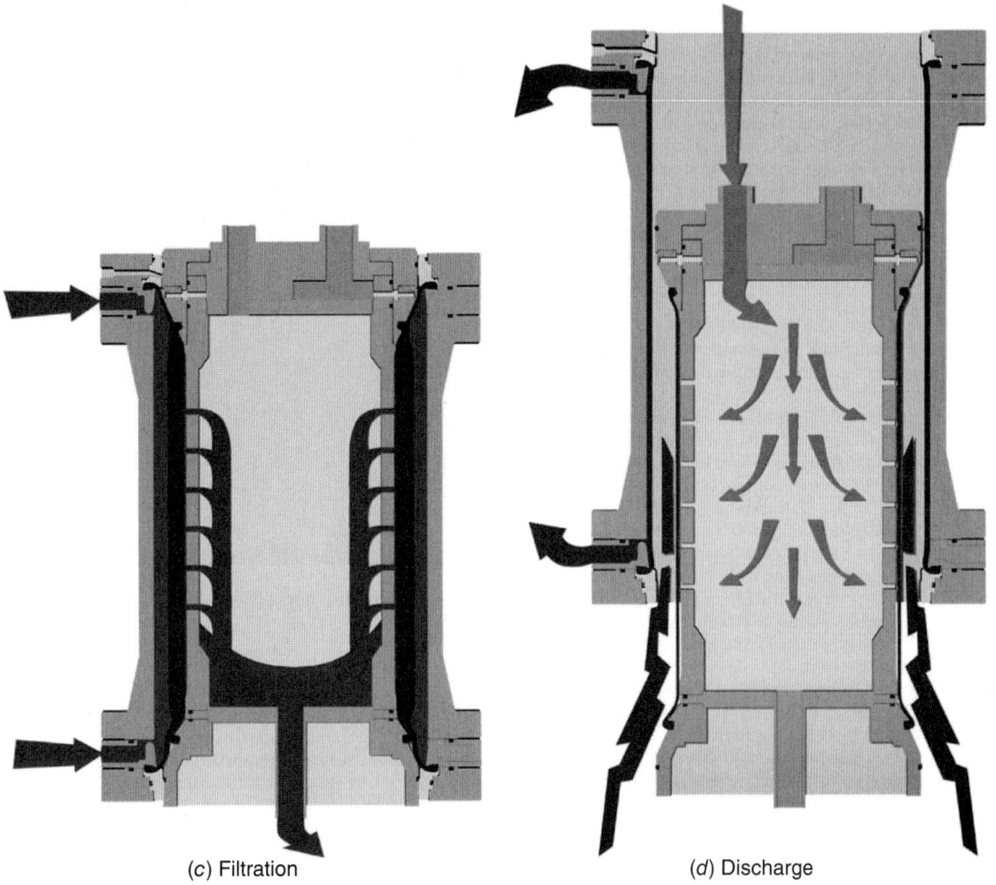

FIG. 18-165 (*Continued*)

the outer casing and the candle is divided by a diaphragm (bladder). Slurry is pumped into the chamber, then water is pumped into the outer side of the bladder to inflate the bladder and compress the cake. The final filter cake is discharged by retracting the bladder under vacuum and opening the tube press. Both air purge and liquid wash are possible options to improve the cake dryness and cake purity.

The same laboratory testing equipment as in the diaphragm press can be used but with a higher pressure. A commercially available piston press can also be used.

Continuous Expression Equipment Continuous expression equipment has the advantage of large capacity and automatic operation. Compared to batch expression equipment, lower pressure is used to squeeze the cake in the continuous expression equipment. As a result, the cakes are not as dry as those from the batch expression devices.

Belt Filter Presses Belt presses (Fig. 18-166) have two filter belts that move around rollers of different sizes to dewater the slurry. A typical belt press may have one or more of the following stages: a preconditioning zone, a gravity drainage zone, a linear compression zone (low pressure), and a roller compression zone (high pressure).

The slurry is fed into the belt press at the preconditioning zone (a tank or pipe), where coagulant and flocculant are added to condition the slurry. The slurry then goes to a horizontal section, where the slurry is thickened

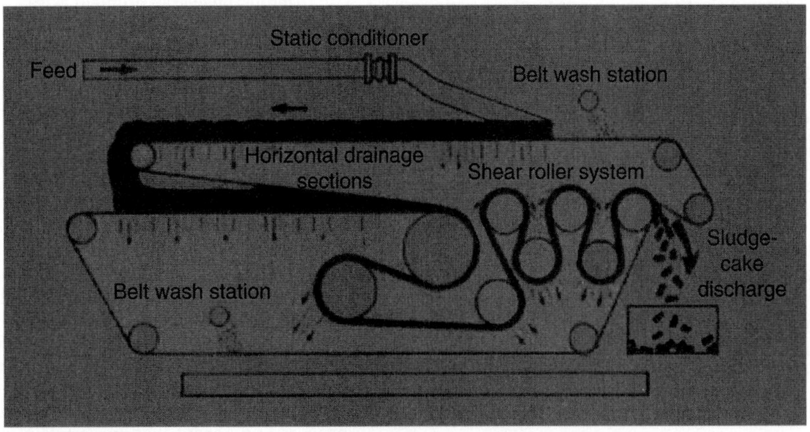

FIG. 18-166 A belt filter press. (*Courtesy of Alfa Laval.*)

by gravity drainage. At the end of the gravity drainage section, the thickened slurry (or wet cake) drops into a wedge section, where the wet cake starts to be squeezed by both belts under pressure. At the end of the wedge section, both belts come together with the cake sandwiched in between and moved through a series of rollers. The final dewatering is accomplished by moving the cake through these rollers in the order of decreasing roller diameters. As the roller diameter gets smaller, the pressure exerted on the cake gets higher. After the final roller, the two belts are separated to release the cake. Each belt goes through a few washing nozzles to clean off any remaining solids on the belt.

It is important to condition the slurry by coagulation and/or flocculation before it is fed into the belt press. Insufficiently flocculated slurries will not dewater properly, and the cake might be squeezed out through the belts or from the side (both sides of a belt press are open). Good conditioned flocs look like cottage cheese. The key challenges in operating a belt press are in the slurry conditioning and in the optimization of flocculant dosage. Flocculant consumption can add significant operational costs if proper control strategy is not used.

The pressure applied on the cake in a belt press is low compared to that in other expression filters. The applied pressures are commonly expressed in pli (pound per linear inch), which is not straightforward in translating to a commonly recognized pressure unit. As a rough comparison, the pressures used in belt presses are around 10 to 20 psi. This pressure can be controlled by the belt and roller tension but are seldom adjusted by operators in the field.

Belt presses have the advantage of large capacity and automatic operation. The initial capital cost is also low. They were originally developed in the pulp and paper industry. Any slurry with fibers will do well in a belt press, and high-fiber material can be added to the slurry as a filter aid for belt press operation. Today, in addition to pulp dewatering, the belt press is widely used in wastewater sludge dewatering.

Due to the relatively low pressure used, the final cakes are not very dry. The dryness of biological sludge cakes from a belt press ranges from 10 to 20 wt%. As fiber content goes up, the cake can be as dry as 40 wt%.

Testing for applications in belt presses is most commonly done by flocculation in beakers and visual observation of the size and strength of the formed flocs. The conditioned slurry can be poured into a filter for a gravity drainage test. These tests can be useful for an experienced person. However, it is not possible to simulate the final cake dryness with these methods. The most effective testing is done with a commercially available apparatus called the Crown press (Fig. 18-167). This device can simulate the roller actions in the belt press and can provide very accurate cake dryness predictions (Chen et al. 2005).

Screw Presses A screw press is a screw conveyor turning inside a perforated or slotted cylinder. The slurry is fed into the feed tank at one end and moves through the screw, and the cake is discharged from the other end. The screw has a smaller diameter at the feed end, and the diameter gradually increases and the screw pitch is shortened toward the discharge end. This design allows gradually decreasing space for slurry/cake and also increasing squeezing pressure on the cake. As the cake moves toward the outlet, the water is squeezed out through the perforated cylinder.

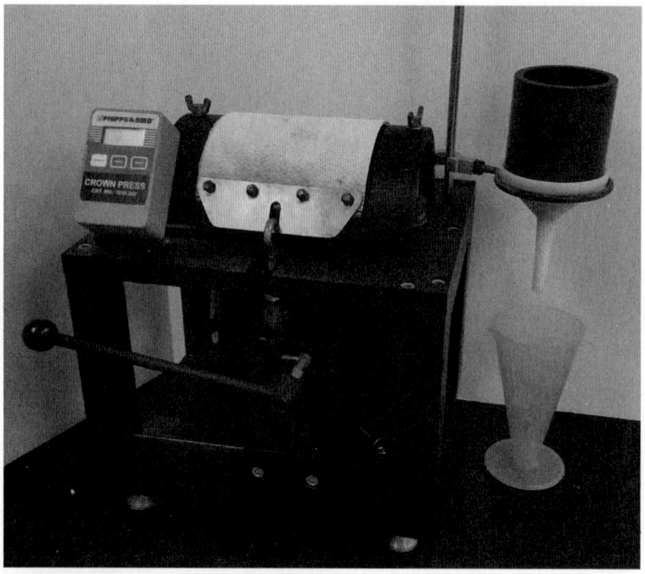

FIG. 18-167 The Crown press.

Screw presses also have the advantage of continuous and automatic operation. Screw presses are primarily used in the pulp and paper, citrus, and dairy industries. Applications also exist in many other industries such as dewatering of synthetic rubbers and wastewater sludge. Three pressure (high, medium, and low) ranges are used. High-pressure screw presses are used for vegetable and animal oil; the capacities are relatively smaller. Medium-pressure units are used to dewater deformable particles (such as plastic pellet and synthetic rubber) and paper pulp. Wastewater sludge applications normally use the low-pressure range.

SELECTION OF A SOLIDS-LIQUID SEPARATOR

A good solids-liquid separator performs well in service, both initially and over time. It operates reliably day after day, with enough flexibility to accommodate to normal fluctuations in process conditions, and it does not need frequent maintenance and repair. Selection of such a separator begins with a preliminary listing of a number of possible devices, which may solve the problem at hand, and it usually ends with the purchase and installation of one or more commercially available machines of a specific type, size, and material of construction. Rarely is it worthwhile to develop a new kind of separator to fill a particular need.

In selecting a solids-liquid separator, it is important to keep in mind the capabilities and limitations of commercially available devices. Among the multiplicity of types on the market, many are designed for fairly specific applications, and unthinking attempts to apply them to other situations are likely to meet with failure. The danger is the more insidious because failure often is not of the clean no-go type; rather it is likely to be in the character of underproduction, subspecification product, or excessively costly operation—the kinds of limping failure that may be slowly detected and difficult to analyze for cause. In addition, it should be recognized that the performance of mechanical separators—more, perhaps, than most chemical-processing equipment—strongly depends on preceding steps in the process. A relatively minor upstream process change, one that might be inadvertent, can change the optimal separator choice.

PRELIMINARY DEFINITION AND SELECTION

The steps in solving a solids-liquid separation problem, in general, are:

1. Define the overall problem, with expert assistance if necessary.
2. Establish process conditions.
3. Identify appropriate separator types; make preliminary selections.
4. Develop a test program.
5. Take representative samples.
6. Make simple tests.
7. Modify process conditions if necessary.
8. Consult with equipment manufacturers.
9. Make a final selection; obtain quotations.

Problem Definition Intelligent selection of a separator requires a careful and complete statement of the nature of the separation problem. Focusing narrowly on the specific problem, however, is not sufficient, especially if the separation is to be one of the steps in a new process. Instead, the problem must be defined as broadly as possible, beginning with the chemical reactor or other source of material to be separated and ending with the separated materials in their desired final form. In this way the influence of preceding and subsequent process steps on the separation step will be illuminated. Sometimes, of course, the new separator is proposed to replace an existing unit; the new separator must then fit into the current process and accept feed materials of more or less fixed characteristics.

At other times the separator is only one item in a train of new equipment, all parts of which must work in harmony if the separator is to be effective.

Assistance in problem definition and in developing a test program should be sought from persons experienced in the field. If your organization has a consultant in separations of this kind, by all means make use of the expertise available. If not, it may be wise to employ an outside consultant, whose special knowledge and guidance can save time, money, and headaches. It is important to do this early; after the separation equipment has been installed, there is little a consultant can do to remedy the sometimes disastrous effects of a poor selection. Often it is best to work with established equipment manufacturers throughout the selection process, unless the problem is unusually sensitive or confidential. Their experience with problems similar to yours may be most helpful and may avoid many false starts.

Preliminary Selections Assembling background information permits a tentative selection of promising equipment and rules out clearly unsuitable types. If the material to be processed is a slurry or pumpable suspension of solids in a liquid, several methods of mechanical separation may be suitable, and these are classified into settling and filtration methods as shown in Fig. 18-168. If the material is a wet solid, removal of liquid by various methods of expression should be considered.

Settling does not give a complete separation: one product is a concentrated suspension and the other is a liquid that may contain fine particles of suspended solids. However, settling is often the best way to process very large volumes of a dilute suspension and remove most of the liquid. The concentrated suspension can then be filtered with smaller equipment than would be needed to filter the original dilute suspension, and the cloudy liquid can be clarified if necessary. Settlers can also be used for classifying particles by size or density, which is usually not possible with filtration.

Screens may sometimes be used to separate suspensions of coarse particles, but they are not widely applicable. For separating fine solids from liquids, cake filtration or the newer systems of crossflow filtration should be considered. Crossflow filtration includes ultrafiltration, where the solids are macromolecules or very fine solids ($D_p \leq 0.1$ μm), and microfiltration, where the particle size generally ranges from 0.1 to 5 μm. In microfiltration, a suspension is passed at high velocity of 1 to 3 m/s (3 to 10 ft/s) and moderate pressure (10 to 30 lb_f/in^2 gauge) parallel to a semipermeable membrane in sheet or tubular form. Organic membranes are made of various polymers, including cellulose acetate, polysulfone, and polyamide; they are usually asymmetric, with a thin selective skin supported on a thicker layer that has larger pores. Inorganic membranes of sintered metal or porous alumina are also available in various shapes, with a range of average pore sizes and permeabilities. Most membranes have a wide distribution of pore sizes and do not give complete rejection unless the average pore size is much smaller than the average particle size in the suspension.

In microfiltration, particles too large to enter the pores of the membrane accumulate at the membrane surface as the liquid passes through. They form a layer of increasing thickness that may have appreciable hydraulic resistance and cause a gradual decrease in permeate flow. A decline in liquid flow may also result from small particles becoming embedded in the membrane or plugging some of the pore mouths. The particle layer may reach a steady-state thickness because of shear-induced migration of particles back into the mainstream, or the liquid flux may continue to decline, requiring frequent backwashing or other cleaning procedures. Because of the high velocities, the change in solids concentration per pass is small, and the suspension is either recycled to the feed tank or sent through several units in series to achieve the desired concentration. The products are a clear liquid and a concentrated suspension similar to those produced in a settling device, but the microfiltration equipment is much smaller for the same production rate.

SAMPLES AND TESTS

Once the initial choice of promising separator types is made, representative liquid-solid samples should be obtained for preliminary tests. At this point, a detailed test program should be developed, preferably with the advice of a specialist.

Establishing Process Conditions Step 2 is taken by defining the problem in detail. Properties of the materials to be separated, the quantities of feed and products required, the range of operating variables, and any restrictions on materials of construction must be accurately fixed, or reasonable assumptions must be made. Accurate data on the concentration of solids, the average particle size or size distribution, the solids and liquid densities, and the suspension viscosity should be obtained *before* selection is made, not after an installed separator fails to perform. The required quantity of the liquid and solid may also influence separator selection. If the solid is the valuable product and crystal size and appearance are important, separators that minimize particle breakage and permit nearly complete removal of fluid may be required. If the liquid is the more valuable product, can minor amounts of solid be tolerated, or must the liquid be sparkling clear? In some cases, partial or incomplete separation is acceptable and can be achieved simply by settling or by crossflow filtration. Where clarity of the liquid is a key requirement, the liquid may have to be passed through a cartridge-type clarifying filter after most of the solid has been removed by the primary separator.

Table 18-20 lists the pertinent background information that should be assembled. It is typical of data requested by manufacturers when they are asked to recommend and quote on a solid-liquid separator. The more accurately and thoroughly these questions can be answered, the better the final choice is likely to be.

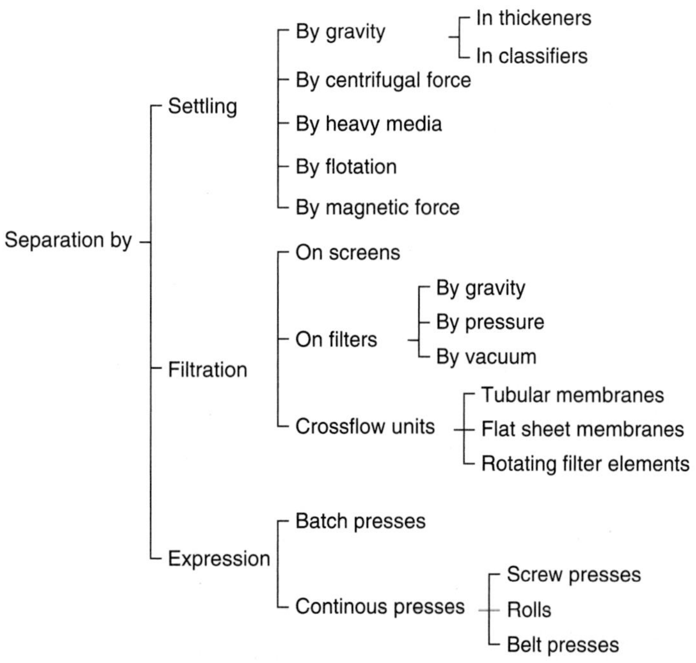

FIG. 18-168 Main paths to solids-liquid separation.

Representative Samples For meaningful results, tests must be run on representative samples. In liquid-solids systems, good samples are hard to get. Often a liquid-solid mixture from a chemical process varies significantly from hour to hour, from batch to batch, or from week to week. A well-thought-out sampling program over a prolonged period, with samples spaced randomly and sufficiently far apart, under the most widely varying process conditions possible, should be formulated. Samples should be taken from all shifts in a continuous process and from many successive batches in a batch process. The influence of variations in raw materials on the separating characteristics should be investigated, as should the effect of reactor or crystallizer temperature, intensity of agitation, or other process variables.

Once samples are taken, they must be preserved unchanged until tested. Unfortunately, cooling or heating the samples or the addition of preservatives may markedly change the ease with which solids may be separated from the liquid. Sometimes they make the separation easier, sometimes harder; in either case, tests made on deteriorated samples give a false picture of the capabilities of separation equipment. Even shipping of the samples can have a significant effect. Often it is so difficult to preserve liquid-solids samples without deterioration that accurate results can be obtained only by incorporating a test separation unit directly in the process stream.

Simple Tests It is usually profitable, however, to make simple preliminary tests, recognizing that the results may require confirmation through subsequent large-scale studies.

Preliminary gravity settling tests are made in a large graduated cylinder in which a well-stirred sample of slurry is allowed to settle, the height of the interface between clear supernatant liquid and concentrated slurry being recorded as a function of settling time. Centrifugal settling tests are normally made in a bottle centrifuge in which the slurry sample is spun at various speeds for various periods of time, and the volume and consistency of the settled solids are noted. In gravity settling tests in particular, it is important to evaluate the effects of flocculating agents on settling rates.

Preliminary filtration tests may be made with a Büchner funnel or a small filter leaf, covered with canvas or other appropriate medium and connected to a vacuum system. Usually the suspension is poured carefully into the vacuum-connected funnel, whereas the leaf is immersed in a sample of the slurry and vacuum is applied to pull filtrate into a collecting flask. The time required to form each of several cakes in the range of 3 to 25 mm (⅛ to 1 in) thick under a given vacuum is noted, as is the volume of the collected filtrate. Properly conducted tests with a Büchner or a vacuum leaf closely simulate the action of rotary vacuum filters of the top- and bottom-feed variety, respectively, and they may give the experienced observer enough information for complete specification of a plant-size filter. Alternatively, they may point to pressure-filter tests or, indeed, to a search for an alternative to filtration. Centrifugal filter tests are made in a perforated basket centrifugal filter 254 or 305 mm (10 or 12 in) in diameter lined with a suitable filter medium. Slurry is poured into the rotating basket until an appropriately thick cake—say, 25 mm (1 in)—is formed. Filtrate is recycled to the basket at such a rate that a thin layer of liquid is just visible on the surface of the cake. The discharge rate of the liquor under these conditions is the draining rate. The test is repeated with cakes of other thicknesses to establish the productive capacity of the centrifugal filter.

Batch tests of microfiltration may be carried out in small pressurized cells with a porous membrane at the bottom and a magnetic stirrer to provide high shear at the membrane surface. These tests may quickly show what type of membrane, if any, gives satisfactory separation, but scaling up to large production units is difficult. Small modules with hollow-fiber, tubular, or spiral-wound membranes are available from equipment vendors, so tests can be made with continuous flow at pressures and velocities likely to be used for large-scale operation. The permeate flux should be measured as a function of time for different slurry concentrations, pressure drops, and solution velocities or Reynolds numbers. Often a limiting flux will be reached as the pressure drop is increased, but operation at a lower pressure drop is often desirable since the flux decline may not be as great and the average permeation rate over a batch cycle may be greater.

More detailed descriptions of small-scale sedimentation and filtration tests are presented in other parts of this section. Interpretation of the results and their conversion into preliminary estimates of such quantities as thickener size, centrifuge capacity, filter area, sludge density, cake dryness, and wash requirements also are discussed. Both the tests and the data treatment must be in experienced hands if error is to be avoided.

Modification of Process Conditions Relatively small changes in process conditions often markedly affect the performance of specific solids-liquid separators, making possible their application when initial test results indicated otherwise or vice versa. Flocculating agents are an example; many gravity settling operations are economically feasible only when flocculants are added to the process stream. Changes in precipitation or crystallization steps may greatly enhance or diminish filtration rates and hence filter capacity. Changes in the temperature of the process stream, the solute content, or the chemical nature of the suspending liquid also influence solids-settling rates. Occasionally it is desirable to add a heavy, finely divided solid to form a pseudo-liquid suspending medium in which the particles of the desired solid will rise to the surface. Attachment of air bubbles to solid particles in a flotation cell, using a suitable flotation agent, is another way of changing the relative densities of liquid and solid.

Consulting the Manufacturer Early in the selection campaign—certainly no later than the time at which the preliminary tests are completed—manufacturers of the more promising separators should be asked for assistance. Additional tests may be made at a manufacturer's test center; again a

TABLE 18-20 Data for Selecting a Solids-Liquid Separator*

1. Process
   a. Describe the process briefly. Make up a flowsheet showing places where liquid-solid separators are needed.
   b. What are the objections to the present process?
   c. Briefly, what results are expected of the separator?
   d. Is the process batch or continuous?
   e. Number the following objectives in order of importance in your problem: (*a*) separation of two different solids __; (*b*) removal of solids to recover valuable liquor as overflow __ ; (*c*) removal of solids to recover the solids as thickened underflow __ or as "dry" cake __; (*d*) washing of solids __; (*e*) classification of solids __; (*f*) clarification or "polishing" of liquid __; (*g*) concentration of solids __.
   f. List the available power and current characteristics.

2. Feed
   a. Quantity of feed:
   Continuous process: __ gal/min; __ h/day; __ lb/h of dry solids.
   Batch process: volume of batch: __; total batch cycle: __ h.
   b. Feed properties: temp. __ ; pH __ ; viscosity __ .
   c. What maximum feed temperature is allowable?
   d. Chemical analysis and specific gravity of carrying liquid.
   e. Chemical analysis and specific gravity of solids.
   f. Percentage of solids in feed slurry.
   g. Screen analysis of solids: wet __ dry __
   h. Chemical analysis and concentration of solubles in feed.
   i. Impurities: form and probable effect on separation.
   j. Is there a volatile component in the feed? __ Should the separator be vapor-tight? __ Must it be under pressure? __ If so, how much? __

3. Filtration and settling rates
   a. Filtration rate on Büchner funnel: __ gal/(min)(ft^2) of filter area under a vacuum of __ in Hg. Time required to form a cake __ in thick: __ s.
   b. At what rate do the solids settle by gravity?
   c. What percentage of the total feed volume do the settled solids occupy after settling is complete? After how long?

4. Feed preparation
   a. If the feed tends to foam, can antifoaming agents be used? If so, what type?
   b. Can flocculating agents be used? If so, what agents?
   c. Can a filter aid be used?
   d. What are the process steps immediately preceding the separation? Can they be modified to make the separation easier?
   e. Could another carrying liquid be used?

5. Washing
   a. Is washing necessary?
   b. What are the chemical analysis and specific gravity of wash liquid?
   c. Purpose of wash liquid: to displace residual mother liquor or to dissolve soluble material from the solids?
   d. Temperature of wash liquid.
   e. Quantity of wash allowable, in lb/lb of solids.

6. Separated solids
   a. What percentage of solids is desired in the cake or thickened underflow?
   b. Is particle breakage important?
   c. Amount of residual solubles allowable in solids.
   d. What further processing will have to be carried out on the solids?

7. Separated liquids
   a Clarity of liquor: what percentage of solids is permissible?
   b. Must the filtrate and spent wash liquid be kept separate?
   c. What further processing will be carried out on the filtrate and/or spent wash?

8. Materials of construction
   a. What metals look most promising?
   b. What metals must not be used?
   c. What gasket and packing materials are suitable?

*U.S. customary engineering units have been retained in this data form. The following SI or modified-SI units might be used instead: centimeters = inches × 2.54; kilograms per kilogram = pounds per pound × 1.0; kilograms per hour = pounds per hour × 0.454; liters per minute = gallons per minute × 3.785; liters per second · square meter = gallons per minute · square foot × 0.679; and pascals = inches mercury × 3377.

major problem is to obtain and preserve representative samples. As much process information as tolerable should be shared with the manufacturers to make full use of their experience with their particular equipment. Full-scale plant tests, although expensive, may well be justified before a final selection is made. Such tests demonstrate operation on truly representative feed, show up long-term operating problems, and give valuable operating experience.

In summary, separator selection calls for clear problem definition, in broad terms; thorough cataloging of process information; and preliminary and tentative equipment selection, followed by refinement of the initial selections through tests on an increasingly larger scale. Reliability, flexibility of operation, and ease of maintenance should be weighed heavily in the final economic evaluation; rarely is purchase price, by itself, a governing factor in determining the suitability of a liquid-solids separator.

Section 19

Reactors

Carmo J. Pereira, Ph.D., M.B.A. *DuPont Fellow, E. I. du Pont de Nemours and Company; Fellow, American Institute of Chemical Engineers (Section Editor)*

Tiberiu M. Leib, Ph.D. *Principal Consultant, The Chemours Company (Retired); Fellow, American Institute of Chemical Engineers (Section Editor)*

The editors would like to thank Stanley M. Walas, Ph.D., Professor Emeritus (deceased), Department of Chemical and Petroleum Engineering, University of Kansas (Fellow, American Institute of Chemical Engineers), for editing this section in the sixth and seventh editions; Dennie T. Mah, M.S.Ch.E., Senior Consultant (retired), E. I. du Pont de Nemours and Company (Senior Member, American Institute of Chemical Engineers; Member, Industrial Electrolysis and Electrochemical Engineering; Member, The Electrochemical Society), for his contributions to the Electrochemical Reactors subsection in the eighth edition, that was reviewed and carried over to the current edition; John Villadsen, Ph.D., Senior Professor, Department of Chemical Engineering, Technical University of Denmark, for his contributions to the Bioreactors subsection in the eighth edition, that was reviewed and carried over to the current edition; and John R. Richards, Ph.D., Research Fellow (retired), E. I. du Pont de Nemours and Company (Fellow, American Institute of Chemical Engineers) for reviewing the Polymerization Reactors subsection.

SOME CASE STUDIES

Nomenclature and Units

In this section, the concentration is represented by C. Mass balance accounting in terms of the number of moles and the fractional conversion is discussed in Sec. 7 and can be very useful. The rate of reaction is r; the flow rate in moles for species A is N_a; the volumetric flow rate is V'; reactor volume is V_r. Several equations are presented without specification of units. The use of any consistent unit set is appropriate.

Following is a listing of typical nomenclature expressed in SI and U.S. Customary System units. Specific definitions and units are stated at the place of application in this section.

Symbol	Definition	SI units	U.S. Customary System units
a	Surface area per volume	1/m	1/ft
A_k	Heat-transfer area	m^2	ft^2
C	Concentration of substance	kg · mol/m^3	lb · mol/ft^3
C_0	Initial mean concentration	kg · mol/m^3	lb · mol/ft^3
c_p	Heat capacity at constant pressure	kJ/(kg · K)	Btu/(lbm · °F)
CSTR	Continuous stirred tank reactor		
d	Diameter	m	ft
D	Diameter, diffusivity		
D_{eff}	Effective diffusion coefficient	m^2/s	ft^2/s
D_e	Effective dispersion coefficient	m^2/s	ft^2/s
D_a	Damkohler number		
E	Activation energy	kJ/(kg · mol)	Btu/(lb · mol)
$E(t)$	Residence time distribution		
$E(t_r)$	Normalized residence time distribution		
f_a	Fraction of A remaining unconverted, C_a/C_{a0} or n_a/n_{a0}		
g	Gravitational acceleration	m/s^2	ft/s^2
$F(t)$	Age function of tracer		
Fr	Froude number		
h	Heat-transfer coefficient	kJ/(s · m^2 · °C)	Btu/(h · ft^2 · °F)
H	Height of tank	m	ft
He	Henry constant	Pa · m^3/(kg · mol)	psi · ft^3/lb · mol)
ΔH_r	Heat of reaction	kJ/(kg · mol)	Btu/(lb · mol)
k	Specific rate constant		
k_m	Mass-transfer coefficient	m/s	ft/s
L	Length of path in reactor	m	ft
m	Magnitude of impulse	kg · mol	lb · mol
n	Number of stages in a CSTR battery, reaction order, parameter of Erlang or gamma distribution		
N	Speed of agitator	rpm	rpm
N_{Eo}	Eotvos number		
Nu	Nusselt number		
Ha	Hatta number		
p_a	Partial pressure of substance A	Pa	psi
P	Total pressure, power		
Pe	Peclet number for dispersion		
PFR	Plug flow reactor		
q	Heat flux, reaction order, impeller-induced flow		
Q	Volumetric flow rate, heat addition or removal		
r	Rate of reaction per unit volume, radius		
R	Radius	m	ft
Re	Reynolds number		
Sc	Schmidt number		
Sh	Sherwood number		
St	Stanton number		
t	Time	s	s
$\bar{t}$	Mean residence time	s	s
t_r	Reduced time, $t/\bar{t}$		
T	Temperature	°C	°F
TFR	Tubular flow reactor		
u	Linear velocity	m/s	ft/s
$u(t)$	Unit step input		
U	Overall heat-transfer coefficient	kJ/(s · m^2 · °C)	Btu/(h · ft^2 · °F)
υ	Volumetric flow rate during semibatch operation	m^3/s	ft^3/s
υ_{ij}	Stoichiometric coefficients		
V'	Volumetric flow rate	m^3/s	ft^3/s
V_r	Volume of reactor	m^3	ft^3
w	Catalyst loading		
x	Conversion		
X	Lockhart-Martinelli parameter		
Y	Yield coefficient		
z	Axial position in reactor	m	ft
	Greek letters		
α	Fraction of feed that bypasses reactor		
β	Fraction of reactor volume that is stagnant, Prater number		
$\delta(\tau)$	Unit impulse input, Dirac delta function		
δ_L	Distance or film thickness	m	ft
ε	Void fraction in a packed bed, particle porosity, void fraction		
η	Effectiveness factor of porous catalyst		
λ	Thermal conductivity	kJ/(s · m · °C)	Btu/(h · ft · °F)
$\Lambda(t)$	Intensity function		
μ	Viscosity	Pa · s	lbm/(ft · s)
ν	Kinematic viscosity, μ/ρ	m^2/s	ft^2/s
ρ	Density	kg/m^3	lbm/ft^3
σ	Surface tension	N/m	lbf/ft
$\sigma^2(\tau)$	Variance		
$\sigma^2(t_r)$	Normalized variance		
ξ	Fractional conversion		
τ	Tortuosity		
ϕ	Thiele modulus		
ϕ_σ	Shape factor		
$\phi\bar{\varepsilon}$	Local rate of energy dissipation		
	Subscripts		
a	Agitator, axial, species A		
b	Bed, species B		
c	Critical value, catalyst, coolant, continuous phase		
cir	Circulation		
d	Dispersed phase		
f	Fluid, feed		
G	Gas phase		
i	Interface		
L	Liquid phase		
ma	Macro		
me	Meso		
mi	Micro		
p	Pellet, product		
r	Reaction, reduced		
R	Reactor		
s	Surface, substrate		
t	Tank		
u	Step function		
W	Wall		
x	Biomass		
0	Inlet, initial		
δ	Delta function		
	Superscripts		
0	Initial condition		

General References: The General References listed in Sec. 7 are applicable for Sec. 19. References to specific topics are made throughout this section.

INTRODUCTION

A chemical reactor is a controlled volume in which a chemical reaction can occur in a safe and controllable manner. A reactor typically is a piece of equipment; however, it can also be a product (such as a coating or a protective film). One or more reactants may react together at a desired set of operating conditions, such as temperature, pressure, and composition. There may be a need for appropriate mixing, control of flow distribution and residence time, contacting between the reactants (sometimes in the presence of a catalyst or biocatalyst), removal (or addition) of heat, and integration of the reactor with the rest of the downstream process. Depending on the nature of the rate-limiting step(s), a reactor may serve primarily as a holding tank, a heat exchanger, or a mass-transfer device. Chemical reactions generate desired products and also by-products that have to be separated and disposed. A successful commercial unit is an economic balance of all these factors. A variety of reactor types are used in the chemical, biochemical, petrochemical, and pharmaceutical industries. Some of these reactors are listed in Table 19-1. They include gas, liquid, or multiphase batch reactors, stirred tank reactors, and tubular rectors.

There are a number of textbooks on chemical reaction engineering. Davis and Davis (*Fundamentals of Chemical Reaction Engineering*, McGraw-Hill, New York, 2003) provide a lucid discussion of kinetics and principles. A more comprehensive treatment together with access to CD-ROM and web resources is in the text by Fogler (*Elements of Chemical Reaction Engineering*, 3d ed., Prentice-Hall, New Jersey, 1999). A chemistry-oriented perspective is provided by Schmidt (*The Engineering of Chemical Reactions*, Oxford University Press, 1999). The book by Froment and Bischoff (*Chemical Reactor Analysis and Design*, Wiley, New York, 1990) provides a thorough discussion of reactor analysis and design. A practical manual on reactor design and scale-up is by Harriott (*Chemical Reactor Design*, Marcel Dekker, New York, 2003). Levenspiel (*Chemical Reaction Engineering*, 3d ed., Wiley, New York, 1999) was among the first to present a phenomenological discussion of fundamentals. The mathematical underpinnings of reactor modeling are covered by Bird et al. (*Transport Phenomena*, 2d ed., Wiley, New York, 2002). This section contains a number of illustrations and sketches from books by Walas (*Chemical Process Equipment Selection and Design*, Butterworth, Mass., 1990) and Ullmann [*Encyclopedia of Chemical Technology* (in German), vol. 3, Verlag Chemie, Weinheim, Ger., 1973, pp. 321–518].

Mathematical models may be used to design reactors and analyze their performance. Detailed models have mainly been developed for large-scale commercial processes. A number of software tools are now available. This chapter will discuss some of the reactors used commercially together with how mathematical models may be used. For additional details, a number of books on reactor analysis cited in this section are available. The discussion will indicate that logical choices aimed at maximizing reaction rate and selectivity for a given set of kinetics can lead to rational reactor selection. While there has been progress in recent years, reactor design and modeling are largely an art.

REACTOR CONCEPTS

A primary purpose of a reactor is to provide desirable conditions for reaction. The *reaction rate per unit volume* of reactor is important in analyzing or sizing a reactor. For a given production rate, it determines the reactor volume required to effect the desired transformation. The residence time in a reactor is inversely related to the term *space velocity* (defined as volumetric feed rate/reactor volume). The fraction of reactants converted to products and by-products is the *conversion*. The fraction of desired product in the material converted on a molar basis is referred to as *selectivity*. The product of conversion and the fractional selectivity provides a measure of the fraction of reactants converted to product, known as *yield*. The product yield provides a direct measure of the level of (atom) utilization of the raw materials and may be an important component of operating cost. A measure of reactor utilization called *space time yield* (*STY*) is the ratio of product generation rate to reactor volume. When a catalyst is used, the reactor has to make product without major process interruptions. The catalyst may be homogeneous or heterogeneous, and the latter can be a living biological cell. A key aspect of catalyst performance is the *durability* of the active site. Since a chemical or biochemical process has a number of unit operations around the reactor, it is often beneficial to minimize the variability of reactant and product flows. This typically means that the reactor is operated at a *steady state*. Interactions between kinetics, fluid flow, transport resistances, and heat effects sometimes result in *multiple steady states* and *transient* (*dynamic*) *behavior*. Reactor dynamics can also result in *runaway behavior*, where reactor temperature continues to increase until the reactants are depleted, or *wrong-way behavior*, where reducing inlet temperature (or reactant flow rate) can result in temperature increases farther downstream and a possible runaway. Since such behavior can result in large perturbations in the process and possibly safety issues, a reactor *control strategy* has to be implemented. The need to operate safely under all conditions calls for a thorough analysis to ensure that the reactor is *inherently safe* and that all possible unsafe outcomes have been considered and addressed. Since various solvents may be used in chemical processes and reactors generate both products and by-products, solvent and by-product emissions can cause *emission* and *environmental footprint* issues that must be considered.

Reactor design is often discussed in terms of independent and dependent variables. Independent variables are choices such as reactor type and internals, catalyst type, inlet temperature, pressure, and fresh feed composition. Dependent variables result from independent variable selection. They may be constrained or unconstrained. Constrained dependent variables often include pressure drop (limited due to compressor cost), feed composition (dictated by the composition of the recycle streams), temperature rise (or decline), and local and effluent composition. The reactor design problem is often aimed at optimizing independent variables (within constraints) to maximize an objective function (such as conversion and selectivity).

Since the reactor feed may contain inert species (e.g., nitrogen and solvents) and since there may be unconverted feed and by-products in the reactor effluent, a number of unit operations (such as distillation and filtration) may be required to produce the desired product(s). In practice, the flow of mass and energy through the process is captured by a process flow sheet. The flow sheet may require recycle (of unconverted feed, solvents, etc.) and purging that may affect reaction chemistry. Reactor design and operation influence the process and vice versa.

REACTOR TYPES

Reactors may be classified according to the mode of operation, the end-use application, the number of phases present, whether (or not) a catalyst is used, and whether some other function (e.g., heat transfer, separations) is conducted in addition to the reaction.

Classification by Mode of Operation

Batch Reactors A "batch" of reactants is introduced into the reactor operated at the desired conditions until the target conversion is reached. Batch reactors are typically tanks in which stirring of the reactants is achieved using internal impellers, or a pump-around loop where a fraction of the reactants is removed and externally recirculated back to the reactor. Temperature is regulated via internal cooling surfaces (such as coils or tubes), jackets, reflux condensers, or a pump-around loop that passes through a heat exchanger. Batch processes are suited to small production rates, to long reaction times, to achieve desired selectivity, and for flexibility in campaigning different products.

Continuous Reactors Reactants are added and products removed continuously at a constant mass flow rate. Large daily production rates are mostly conducted in continuous equipment.

A *continuous stirred tank reactor* (CSTR) is a vessel to which reactants are added and products removed while the contents within the vessel are vigorously stirred using internal agitation or by internally (or externally) recycling

TABLE 19-1 Residence Times and/or Space Velocities in Industrial Chemical Reactors*

Product (raw materials)	Type	Reactor phase	Catalyst	*T*, °C	*P*, atm	Residence time or space velocity	Source and page†
Acetaldehyde (ethylene, air)	FB	L	Cu and Pd chlorides	50–100	8	6–40 min	[2] 1, [7] 3
Acetic anhydride (acetic acid)	TO	L	Triethylphosphate	700–800	0.3	0.25–5 s	[2]
Acetone (*i*-propanol)	MT	LG	Ni	300	1	2.5 h	[1] 1 314
Acrolein (formaldehyde, acetaldehyde)	FL	G	MnO, silica gel	280–320	1	0.6 s	[1] 1 384, [7] 33
Acrylonitrile (air, propylene, ammonia)	FL	G	Bi phosphomolybdate	400	1	4.3 s	[3] 684, [2] 47
Adipic acid (nitration of cyclohexanol)	TO	L	Co naphthenate	125–160	4–20	2 h	[2] 51, [7] 49
Adiponitrile (adipic acid)	FB	G	H_3BO_3	370–410	1	3.5–5 s	[1] 2 152
			H_3PO_4			350–500 GHSV	[7] 52
Alkylate (*i*-C_4, butenes)	CST	L	H_2SO_4	5–10	2–3	5–40 min	[4] 223
Alkylate (*i*-C_4, butenes)	CST	L	HF	25–38	8–11	5–25 min	[4] 223
Allyl chloride (propylene, Cl_2)	TO	G	NA	500	3	0.3–1.5 s	[1] 2 416, [7] 67
Ammonia (H_2, N_2)	FB	G	Fe	450	150	28 s	[6] 61
						7800 GHSV	
Ammonia (H_2, N_2)	FB	G	Fe	450	225	33 s	[6] 61
						10,000 GHSV	
Ammonia oxidation	Flame	G	Pt gauze	900	8	0.0026 s	[6] 115
Aniline (nitrobenzene, H_2)	B	L	$FeCl_2$ in H_2O	95–100	1	8 h	[1] 3 289
Aniline (nitrobenzene, H_2)	FB	G	Cu on silica	250–300	1	0.5–100 s	[7] 82
Aspirin (salicylic acid, acetic anhydride)	B	L	None	90	1	>1 h	[7] 89
Benzene (toluene)	TU	G	None	740	38	48 s	[6] 36
						815 GHSV	[9] 109
Benzene (toluene)	TU	G	None	650	35	128 s	[1] 4 183, [7] 98
Benzoic acid (toluene, air)	SCST	LG	None	125–175	9–13	0.2–2 h	[7] 101
Butadiene (butane)	FB	G	Cr_2O_3, Al_2O_3	750	1	0.1–1 s	[7] 118
Butadiene (1-butene)	FB	G	None	600	0.25	0.001 s	[3] 572
						34,000 GHSV	
Butadiene sulfone (butadiene, SO_2)	CST	L	*t*-Butyl catechol	34	12	0.2 LHSV	[1] 5 192
i-Butane (*n*-butane)	FB	L	$AlCl_3$ on bauxite	40–120	18–36	0.5–1 LHSV	[4] 239, [7] 683
i-Butane (*n*-butane)	FB	L	Ni	370–500	20–50	1–6 WHSV	[4] 239
Butanols (propylene hydroformylation)	FB	L	PH_3-modified Co carbonyls	150–200	1000	100 g L·h	[1] 5 373
Butanols (propylene hydroformylation)	FB	L	Fe pentacarbonyl	110	10	1 h	[7] 125
Calcium stearate	B	L	None	180	5	1–2 h	[7] 135
Caprolactam (cyclohexane oxime)	CST	L	Polyphosphoric acid	80–110	1	0.25–2 h	[1] 6 73, [7] 139
Carbon disulfide (methane, sulfur)	Furn.	G	None	500–700	1	1.0 s	[1] 6 322, [7] 144
Carbon monoxide oxidation (shift)	TU	G	Cu-Zn or Fe_2O_3	390–220	26	4.5 s	[6] 44
						7000 GHSV	
Portland cement	Kiln	S		1400–1700	1	10 h	[11]
Chloral (Cl_2, acetaldehyde)	CST	LG	None	20–90	1	140 h	[7] 158
Chlorobenzenes (benzene, Cl_2)	SCST	LG	Fe	40	1	24 h	[1] 8 122
Coking, delayed (heater)	TU	LG	None	490–500	15–4	250 s	[1] 10 8
Coking, delayed (drum, 100 ft max height)	B	LG	None	500–440	4	0.3–0.5 ft/s vapor	[1] 10 8
Cracking, fluid catalytic	Riser	G	Zeolite	520–540	2–3	2–4 s	[14] 353
Cracking, hydro (gas oils)	FB	LG	Ni, SiO_2, Al_2O_3	350–420	100–150	1–2 LHSV	[11]
Cracking (visbreaking residual oils)	TU	LG	None	470–495	10–30	450 s, 8 LHSV	[11]
Cumene (benzene, propylene)	FB	G	H_3PO_4	260	35	23 LHSV	[11]
Cumene hydroperoxide (cumene, air)	CST	L	Metal porphyrins	95–120	2–15	1–3 h	[7] 191
Cyclohexane (benzene, H_2)	FB	G	Ni on Al_2O_3	150–250	25–55	0.75–2 LHSV	[7] 201
Cyclohexanol (cyclohexane, air)	SCST	LG	None	185–200	48	2–10 min	[7] 203
Cyclohexanone (cyclohexanol)	CST	L	N.A.	107	1	0.75 h	[8] (1963)
Cyclohexanone (cyclohexanol)	MT	G	Cu on pumice	250–350	1	4–12 s	[8] (1963)
Cyclopentadiene (dicyclopentadiene)	TJ	G	None	220–300	1–2	0.1–0.5 LHSV	[7] 212
DDT (chloral, chlorobenzene)	B	L	Oleum	0–15	1	8 h	[7] 233
Dextrose (starch)	CST	L	H_2SO_4	165	1	20 min	[8] (1951)
Dextrose (starch)	CST	L	Enzyme	60	1	100 min	[7] 217
Dibutylphthalate (phthalic anhydride, butanol)	B	L	H_2SO_4	150–200	1	1–3 h	[7] 227
Diethylketone (ethylene, CO)	TO	L	Co oleate	150–300	200–500	0.1–10 h	[7] 243
Dimethylsulfide (methanol, CS_2)	FB	G	Al_2O_3	375–535	5	150 GHSV	[7] 266
Diphenyl (benzene)	MT	G	None	730	2	0.6 s	[7] 275
						3.3 LHSV	[8] (1938)
Dodecylbenzene (benzene, propylene tetramer)	CST	L	$AlCl_3$	15–20	1	1–30 min	[7] 283
Ethanol (ethylene, H_2O)	FB	G	H_3PO_4	300	82	1800 GHSV	[2] 356, [7] 297
Ethyl acetate (ethanol, acetic acid)	TU, CST	L	H_2SO_4	100	1	0.5–0.8 LHSV	[10] 45, 52, 58
Ethyl chloride (ethylene, HCl)	TO	G	$ZnCl_2$	150–250	6–20	2 s	[7] 305
Ethylene (ethane)	TU	G	None	860	2	1.03 s	[3] 411
						1880 GHSV	[6] 13
Ethylene (naphtha)	TU	G	None	550–750	2–7	0.5–3 s	[7] 254
Ethylene, propylene chlorohydrins (Cl_2, H_2O)	CST	LG	None	30–40	3–10	0.5–5 min	[7] 310, 580
Ethylene glycol (ethylene oxide, H_2O)	TO	LG	1% H_2SO_4	50–70	1	30 min	[2] 398
Ethylene glycol (ethylene oxide, H_2O)	TO	LG	None	195	13	1 h	[2] 398
Ethylene oxide (ethylene, air)	FL	G	Ag	270–290	1	1 s	[2] 409, [7] 322
Ethyl ether (ethanol)	FB	G	WO_3	120–375	2–100	30 min	[7] 326
Fatty alcohols (coconut oil)	B	L	Na, solvent	142	1	2 h	[8] (1953)
Formaldehyde (methanol, air)	FB	G	Ag gauze	450–600	1	0.01 s	[2] 423
Glycerol (allyl alcohol, H_2O_2)	CST	L	H_2WO_4	40–60	1	3 h	[7] 347

TABLE 19-1 Residence Times and/or Space Velocities in Industrial Chemical Reactors* *(Continued)*

Product (raw materials)	Type	Reactor phase	Catalyst	T, °C	P, atm	Residence time or space velocity	Source and page[†]
Hydrogen (methane, steam)	MT	G	Ni	790	13	5.4 s 3000 GHSV	[6] 133
Hydrodesulfurization of naphtha	TO	LG	Co-MO	315–500	20–70	1.5–8 LHSV 125 WHSV	[4] 285 [6] 179 [9] 201
Hydrogenation of cottonseed oil	SCST	LG	Ni	130	5	6 h	[6] 161
Isoprene (*i*-butene, formaldehyde)	FB	G	HCl, silica gel	250–350	1	1 h	[7] 389
Maleic anhydride (butenes, air)	FL	G	V_2O_5	300–450	2–10	0.1–5 s	[7] 406
Melamine (urea)	B	L	None	340–400	40–150	5–60 min	[7] 410
Methanol (CO, H_2)	FB	G	ZnO, Cr_2O_3	350–400	340	5000 GHSV	[7] 421
Methanol (CO, H_2)	FB	G	ZnO, Cr_2O_3	350–400	254	28,000 GHSV	[3] 562
o-Methyl benzoic acid (xylene, air)	CST	L	None	160	14	0.32 h 3.1 LHSV	[3] 732
Methyl chloride (methanol, Cl_2)	FB	G	Al_2O_3 gel	340–350	1	275 GHSV	[2] 533
Methyl ethyl ketone (2-butanol)	FB	G	ZnO	425–475	2–4	0.5–10 min	[7] 437
Methyl ethyl ketone (2-butanol)	FB	G	Brass spheres	450	5	2.1 s 13 LHSV	[10] 284
Nitrobenzene (benzene, HNO_3)	CST	L	H_2SO_4	45–95	1	3–40 min	[7] 468
Nitromethane (methane, HNO_3)	TO	G	None	450–700	5–40	0.07–0.35 s	[7] 474
Nylon-6 (caprolactam)	TU	L	Na	260	1	12 h	[7] 480
Phenol (cumene hydroperoxide)	CST	L	SO_2	45–65	2–3	15 min	[7] 520
Phenol (chlorobenzene, steam)	FB	G	Cu, Ca phosphate	430–450	1–2	2 WHSV	[7] 522
Phosgene (CO, Cl_2)	MT	G	Activated carbon	50	5–10	16 s 900 GHSV	[11]
Phthalic anhydride (*o*-xylene, air)	MT	G	V_2O_5	350	1	1.5 s	[3] 482, 539, [7] 529
Phthalic anhydride (naphthalene, air)	FL	G	V_2O_5	350	1	5 s	[9] 136, [10] 335
Polycarbonate resin (bisphenol-A, phosgene)	B	L	Benzyltriethylammonium chloride	30–40	1	0.25–4 h	[7] 452
Polyethylene	TU	L	Organic peroxides	180–200	1000–1700	0.5–50 min	[7] 547
Polyethylene	TU	L	Cr_2O_3, Al_2O_3, SiO_2	70–200	20–50	0.1–1000 s	[7] 549
Polypropylene	TO	L	R_2AlCl, $TiCl_4$	15–65	10–20	15–100 min	[7] 559
Polyvinyl chloride	B	L	Organic peroxides	60	10	5.3–10 h	[6] 139
i-Propanol (propylene, H_2O)	TO	L	H_2SO_4	70–110	2–14	0.5–4 h	[7] 393
Propionitrile (propylene, NH_3)	TU	G	CoO	350–425	70–200	0.3–2 LHSV	[7] 578
Reforming of naphtha (H_2/hydrocarbon = 6)	FB	G	Pt	490	30–35	3 LHSV 8000 GHSV	[6] 99
Starch (corn, H_2O)	B	L	SO_2	25–60	1	18–72 h	[7] 607
Styrene (ethylbenzene)	MT	G	Metal oxides	600–650	1	0.2 s 7500 GHSV	[5] 424
Sulfur dioxide oxidation	FB	G	V_2O_5	475	1	2.4 s 700 GHSV	[6] 86
t-Butyl methacrylate (methacrylic acid, *i*-butene)	CST	L	H_2SO_4	25	3	0.3 LHSV	[1] 5 328
Thiophene (butane, S)	TU	G	None	600–700	1	0.01–1 s	[7] 652
Toluene diisocyanate (toluene diamine, phosgene)	B	LG	None	200–210	1	7 h	[7] 657
Toluene diamine (dinitrotoluene, H_2)	B	LG	Pd	80	6	10 h	[7] 656
Tricresyl phosphate (cresyl, $POCl_3$)	TO	L	$MgCl_2$	150–300	1	0.5–2.5 h	[2] 850, [7] 673
Vinyl chloride (ethylene, Cl_2)	FL	G	None	450–550	2–10	0.5–5 s	[7] 699
Aldehydes (diisobutene, CO)	CST	LG	Co Carbonyl	150	200	1.7 h	[12] 173
Allyl alcohol (propylene oxide)	FB	G	Li phosphate	250	1	1.0 LHSV	[15] 23
Automobile exhaust	FB	G	Pt-Pd: 1–2 g/unit	400–600+	1		
Gasoline (methanol)	FB	G	Zeolite	400	20	2 WHSV	[13] 3 383
Hydrogen cyanide (NH_3, CH_4)	FB	G	Pt-Rh	1150	1	0.005 s	[15] 211
Isoprene, polymer	B	L	Al(*i*-Bu)$_3$·$TiCl_4$	20–50	1–5	1.5–4 h	[15] 82
NO_x pollutant (with NH_3)	FB	G	V_2O_5·TiO_2	300–400	1–10		[14] 332
Automobile emission control	M	G	Pt/Rh/Pd/Al_2O_3	350–500	1	20,000 GHSV	[16] 69
Nitrogen oxide emission control	M	G	V_2O_5-WO_3/TiO_2	300–400	1	4–10,000 GHSV	[16] 306
Carbon monoxide and hydrocarbon emission control	M	G	Pt-Pd/Al_2O_3	500–600	1	80–120,000 GHSV	[16] 334
Ozone control from aircraft cabins	M	G	Pd/Al_2O_3	130–170	1	~10^6 GHSV	[16] 263
Vinyl acetate (ethylene + CO)	MT	LG	Cu-Pd	130	30	1 h L, 10 s G	[12] 140

*Abbreviations: reactors: batch (B), continuous stirred tank (CST), fixed bed of catalyst (FB), fluidized bed of catalyst (FL), furnace (Furn.), monolith (M), multitubular (MT), semicontinuous stirred tank (SCST), tower (TO), tubular (TU). Phases: liquid (L), gas (G), both (LG). Space velocities (hourly): gas (GHSV), liquid (LHSV), weight (WHSV). Not available, NA. To convert atm to kPa, multiply by 101.3.

[†]1. J. J. McKetta, ed., *Encyclopedia of Chemical Processing and Design,* Marcel Dekker, 1976 to date (referenced by volume).
2. W. L. Faith, D. B. Keyes, and R. L. Clark, *Industrial Chemicals,* revised by F. A. Lowenstein and M. K. Moran, John Wiley & Sons, 1975.
3. G. F. Froment and K. B. Bischoff, *Chemical Reactor Analysis and Design,* John Wiley & Sons, 1979.
4. R. J. Hengstebeck, *Petroleum Processing,* McGraw-Hill, New York, 1959.
5. V. G. Jenson and G. V. Jeffreys, *Mathematical Methods in Chemical Engineering,* 2d ed., Academic Press, 1977.
6. H. F. Rase, *Chemical Reactor Design for Process Plants,* Vol. 2: Case Studies, John Wiley & Sons, 1977.
7. M. Sittig, *Organic Chemical Process Encyclopedia,* Noyes, 1969 (patent literature exclusively).
8. Student Contest Problems, published annually by AIChE, New York (referenced by year).
9. M. O. Tarhan, *Catalytic Reactor Design,* McGraw-Hill, 1983.
10. K. R. Westerterp, W. P. M. van Swaaij, and A. A. C. M. Beenackers, *Chemical Reactor Design and Operation,* John Wiley & Sons, 1984.
11. Personal communication (Walas, 1985).
12. B. C. Gates, J. R. Katzer, and G. C. A. Schuit, *Chemistry of Catalytic Processes,* McGraw-Hill, 1979.
13. B. E. Leach, ed., *Applied Industrial Catalysts,* 3 vols., Academic Press, 1983.
14. C. N. Satterfield, *Heterogeneous Catalysis in Industrial Practice,* McGraw-Hill, 1991.
15. C. L. Thomas, *Catalytic Processes and Proven Catalysts,* Academic Press, 1970.
16. Heck, Farrauto, and Gulati, *Catalytic Air Pollution Control: Commercial Technology,* Wiley-Interscience, 2002.

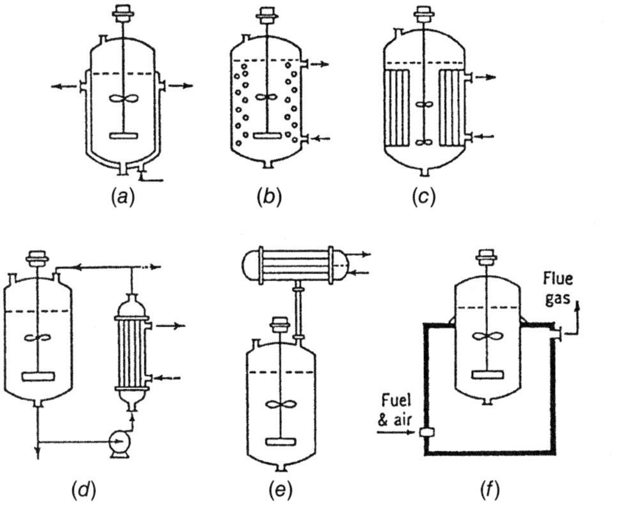

FIG. 19-1 Stirred tank reactors with heat transfer. (*a*) Jacket. (*b*) Internal coils. (*c*) Internal tubes. (*d*) External heat exchanger. (*e*) External reflux condensor. (*f*) Fired heater. (*Fig. 11.1 in Walas,* Reaction Kinetics for Chemical Engineers, *McGraw-Hill, New York, 1959.*)

the contents. CSTRs may be employed in series or in parallel. An approach to employing CSTRs in series is to have a large cylindrical tank with partitions: feed enters the first compartment and over (or under) flows to the next compartment, and so on. The composition is maintained as uniform as possible in each individual compartment; however, a stepped concentration gradient exists from one CSTR to the next. When the reactants have limited solubility (miscibility) and a density difference, the vertical staged reactor with countercurrent operation may be used. Alternatively, each CSTR in a series or parallel configuration can be an independent vessel. Examples of stirred tank reactors with heat transfer are shown in Fig. 19-1.

A *tubular flow reactor* (TFR) is a tube (or pipe) through which reactants flow and are converted to product. The TFR may have a varying diameter along the flow path. In such a reactor, there is a continuous gradient (in contrast to the stepped gradient characteristic of a CSTR-in-series battery) of concentration in the direction of flow. Several tubular reactors in series or in parallel may also be used. Both horizontal and vertical orientations are common. When heat transfer is needed, individual tubes are jacketed or a shell-and-tube construction is used. The reaction side may be filled with solid catalyst or internals such as static mixers (to improve interphase contact in heterogeneous reactions or to improve heat transfer by turbulence). Tubes that have 3- to 4-in diameter and are several miles long may be used in polymerization service. Large-diameter vessels, with packing (or trays) used to regulate the residence time in the reactor, may also be used. Some of the configurations in use are axial flow, radial flow, multi-shell with built-in heat exchangers, and so on.

A reaction battery of CSTRs in series, although both mechanically and operationally more complex and expensive than a tubular reactor, provides flexibility. Relatively slow reactions are best conducted in a stirred tank reactor battery. A tubular reactor is used when heat transfer is needed, where high pressures and/or high (or low) temperatures occur, and when relatively short reaction times suffice.

Semibatch Reactors Some of the reactants are loaded into the reactor, and the rest of the reactants are fed gradually. For instance, one reactant is loaded into the reactor, and the other reactant is fed continuously. Once the reactor is full, it may be operated in a batch mode to complete the reaction. Semibatch reactors are especially favored when there are large heat effects and heat-transfer capability is limited. Exothermic reactions may be slowed down and endothermic reactions controlled by limiting reactant concentration. In bioreactors, the reactant concentration may be limited to minimize toxicity. Other situations that may call for semibatch reactors include the control of undesirable by-products or when one of the reactants is a gas of limited solubility that is fed continuously at the dissolution rate.

Classification by End Use *Chemical reactors* are typically used for the synthesis of chemical intermediates for a variety of specialty (e.g., agricultural, pharmaceutical) or commodity (e.g., raw materials for polymers) applications. *Polymerization reactors* convert raw materials to polymers having a specific molecular weight distribution and functionality. The difference between polymerization and chemical reactors is artificially based on the size of the molecule produced. *Bioreactors* use (often genetically manipulated) organisms to catalyze transformations either aerobically (in the presence of air) or anaerobically (without air present). *Electrochemical reactors* use electricity to drive desired reactions. Examples include synthesis of Na metal from NaCl and Al from bauxite ore. A variety of reactor types are employed for specialty materials synthesis applications (e.g., electronic, defense, and other).

Classification by Phase Despite the generic classification by operating mode, reactors are designed to accommodate the reactant phases and provide optimal conditions for reaction. Reactants may be fluid(s) or solid(s), and as such, several reactor types have been developed. *Single-phase* reactors have one moving phase and are typically gas- (or plasma-) or liquid-phase reactors. *Two-phase* reactors have two moving phases and may be gas–liquid, liquid–liquid, gas–solid, or liquid–solid reactors. *Multiphase* reactors typically have two or more moving phases present. The most common type of multiphase reactor is a gas–liquid–solid reactor; however, liquid–liquid–solid and gas–liquid–liquid reactors are also used. The classification by phases will be used to develop the contents of this section.

In addition, a reactor may perform a function other than reaction alone. *Multifunctional* reactors may provide both reaction and mass transfer (e.g., reactive distillation, reactive crystallization, and reactive membranes), or reaction and heat transfer. This coupling of functions within the reactor inevitably leads to additional operating constraints on one or the other function. Multifunctional reactors are often discussed in the context of process intensification. The primary driver for multifunctional reactors is functional synergy and equipment cost savings.

REACTOR MODELING

As discussed in Sec. 7, chemical kinetics may be mathematically described by rate equations. Reactor performance is also amenable to quantitative analysis. The quantitative analysis of reaction systems is dealt with in the field of chemical reaction engineering.

The level of mathematical detail that can be included in the analysis depends on the level of understanding of the physical and chemical processes that occur in a reactor. As a practical matter, engineering data needed to build a detailed model for some new chemistry typically are unavailable early in the design phase. Reactor designers may use similarity principles (e.g., dimensionless groups), rules of thumb, trend analysis, design of experiments (DOE), and principal-component analysis (PCA) to scale up laboratory reactors. For hazardous systems in which compositional measurements are difficult, surrogate indicators such as pressure or temperature may be used. As more knowledge becomes available, however, a greater level of detail may be included in a mathematical model. A detailed reactor model may contain information on vessel configuration, stoichiometric relationships, kinetic rate equations, correlations for thermodynamic and transport properties, contacting efficiency, residence time distribution, and so on.

Models may be used for analyzing data, estimating performance, reactor scale-up, simulating start-up and shutdown behavior, troubleshooting, and control. The level of detail in a model depends on the need, and this is often a balance between value and cost. Very elaborate models are justifiable and have been developed for certain widely practiced and large-scale processes, or for processes where operating conditions are especially critical.

Modeling Considerations A useful reactor model allows the user to predict performance or to explore uncertainties not easily or cost-effectively investigated through experimentation. Uncertainties that may be explored through modeling may include scale-up options, explosion hazards, runaway reactions, environmental emissions, reactor internals design, and so on. As such, the model must contain an optimal level of detail (principle of *optimal sloppiness*) required to meet the desired objective(s). For example, if mixing is critical to performance, the model must include flow equations that reflect the role of mixing. If heat effects are small, an isothermal model may be used.

A key aspect of modeling is to derive the appropriate momentum, mass, or energy conservation equations for the reactor. These balances may be used in lumped systems or derived over a differential volume within the reactor and then integrated over the reactor volume. Species mass conservation equations have the following general form (e.g., for species *A*):

$$\begin{bmatrix}\text{Amount of } A\\ \text{introduced}\\ \text{per unit time}\end{bmatrix} - \begin{bmatrix}\text{Amount of } A\\ \text{leaving per unit}\\ \text{time}\end{bmatrix} - \begin{bmatrix}\text{Amount of } A\\ \text{converted per}\\ \text{unit time}\end{bmatrix} = \begin{bmatrix}\text{Amount of } A\\ \text{accumulated}\\ \text{per unit time}\end{bmatrix} \tag{19-1}$$

The general form for the energy balance equation is

$$\begin{bmatrix}\text{Amount of}\\ \text{energy added}\\ \text{per unit time}\end{bmatrix} - \begin{bmatrix}\text{Amount of}\\ \text{energy removed}\\ \text{per unit time}\end{bmatrix} + \begin{bmatrix}\text{Energy}\\ \text{generated per}\\ \text{unit time}\end{bmatrix} = \begin{bmatrix}\text{Accumulation}\\ \text{of energy}\\ \text{per unit time}\end{bmatrix} \tag{19-2}$$

The model defines each of these terms. Solving the set of equations provides outputs that can be validated against experimental observations and then used for predictive purposes. Mathematical models for ideal reactors that are generally useful in estimating reactor performance will be presented. Additional information on these reactors is available also in Sec. 7.

Batch Reactor Since there is no addition or removal of reactants, the mass and energy conservation equations for a single reacting species in a batch reactor with a constant reactor volume are

$$V_r r(C,T) + V_r \frac{dC}{dt} = 0 \tag{19-3}$$

$$-q\,A_k - V_r[-\Delta H_r(T)]\,r(C,T) + V_r \rho c_p \frac{dT}{dt} = 0 \tag{19-4}$$

where qA_k is the addition (or removal) of heat from the reactor and r is the specific rate of reaction per reaction volume (a function of concentration and temperature). The heat of reaction ΔH_r is negative for exothermic reactions and positive for endothermic reactions. Mean values of physical properties are used in Eqs. (19-3) and (19-4). For an isothermal first-order reaction $r(C,T) = kC$, the mass conservation is the only relevant equation, and the solution is

$$C = C_0 e^{-kt} \tag{19-5}$$

Typically, batch reactors may have complex kinetics, mixing, and heat-transfer issues. In such cases, detailed momentum, mass, and energy balance equations will be required.

Semibatch Reactor For a reaction $A+B \rightarrow P$, feed of reactant A is added for a fixed time to a batch initially containing only reactant B, and the reaction proceeds as the feed is added. The reactor equations governing the feed addition portion of the process are

$$-V'C_{a0} + V_r r(C,T) + \frac{d(V_r C_a)}{dt} = -V'C_{a0} + V_r r(C,T) + C_a \frac{dV_r}{dt} + V_r \frac{dC_a}{dt} = 0 \tag{19-6}$$

$$-\rho_0 c_p V' T_0 - qA_k - V_r[-\Delta H_r(T)] r(C,T) + \rho c_p V_r \frac{dT}{dt} = 0 \tag{19-7}$$

The material balance for species B is the same as Eq. (19-6) minus the feed term. For a constant reactant flow rate $V' = \upsilon_0$,

$$\frac{dV_r}{dt} = \upsilon_0 \tag{19-8}$$

Given initial conditions $V_r = V_r^0$ and $C_a = 0$ at $t = 0$,

$$V_r(t) = V_r^0 + \upsilon_0 t \tag{19-9}$$

For an isothermal first-order reaction in A and zero-order reaction in B (e.g., large excess B), substitution of this relationship in Eq. (19-6) yields the concentration of A versus time in the reactor

$$C_a = \frac{C_{a0}\upsilon_0}{k(V_r^0 + \upsilon_0 t)}(1 - e^{-kt}) \tag{19-10}$$

The concentration of B versus time can be calculated from the reaction stoichiometry, the concentration of A, and the initial reactor concentration of B. After feed addition is completed, the reactor may be operated in a batch mode. In this case, Eqs. (19-3) and (19-4) may be used for both species with the concentration at the end of feed addition serving as the initial concentration for the batch phase.

Continuous Stirred Tank Reactor In a CSTR, reactants are fed into and removed from an ideally mixed tank. As a result, the concentration within the tank is uniform and identical to the concentration of the effluent. The mass and energy conservation equations for an ideal constant-volume or constant-density CSTR with constant volumetric feed rate V' may be written as

$$V'C - V_0' C_0 + V_r r(C,T) + V_r \frac{dC}{dt} = 0 \tag{19-11}$$

$$-Q(T) + V_0' \rho_0 c_p (T - T_0) - V_r r(C,T)[-\Delta H_r(T)] + V_r \rho c_p \frac{dT}{dt} = 0 \tag{19-12}$$

where $Q(T)$ represents any addition or removal of heat from the reactor, and mean values of physical properties are used. For example, if heat is transferred through the reactor wall, $Q(T) = A_k U(T_c - T)$, where A_k is the heat-transfer area, U is the overall heat-transfer coefficient, and T_c is the temperature of the heat-transfer fluid.

The preceding *ordinary differential equations* (ODEs), Eqs. (19-11) and (19-12), can be solved with an initial condition. For an isothermal first-order reaction and an initial condition, $C(0) = 0$, the linear ODE may be solved analytically. At steady state, the accumulation term is zero, and the solution for the effluent concentration becomes

$$\frac{C}{C_0} = \frac{1}{1 + kV_r/V'} = \frac{1}{1 + k\bar{t}} \tag{19-13}$$

Since the contents of an ideal CSTR are perfectly mixed, the concentration in the reactor is uniform and the same as the concentration in the reactor effluent. In practice, CSTRs may not be ideally mixed. In such cases, the reactor may be modeled as having a fraction of the feed α in bypass and a fraction β of the reactor volume stagnant. The material balance is

$$C = \alpha C_0 + (1-\alpha) C_1 \tag{19-14}$$

$$(1-\alpha)V'C_0 = (1-\alpha)V'C_1 + (1-\beta)kV_r C_1^n \tag{19-15}$$

where C_1 is the concentration leaving the active zone of the tank. Elimination of C_1 will relate the input and overall output concentrations. For a first-order reaction,

$$\frac{C_0}{C} = \frac{1 + \dfrac{k\bar{t}\,(1-\beta)}{(1-\alpha)}}{1 - \alpha + \alpha\dfrac{k\bar{t}\,(1-\beta)}{(1-\alpha)}} \tag{19-16}$$

The two parameters α and β may be expected to depend on reactor internals and the amount of agitation.

Plug Flow Reactor A plug flow reactor (PFR) is an idealized tubular reactor in which each reactant molecule enters and travels through the reactor as a "plug," that is, each molecule enters the reactor at the same velocity and has exactly the same residence time. As a result, the concentration of reactants (and products) is determined by the distance downstream of the inlet, and it is uniform perpendicular to flow. The mass and energy balance for a differential volume between position V_r and $V_r + dV_r$ from the inlet may be written as *partial differential equations* (PDEs), which for a constant-density system:

$$\frac{\partial(V'C)}{\partial V_r} + r(C,T) + \frac{\partial C}{\partial t} = 0 \tag{19-17}$$

$$\frac{-Q(T)}{V_r} + V'\rho c_p \frac{\partial T}{\partial V_r} - [-\Delta H_r(T)] r(C,T) + \rho c_p \frac{\partial T}{\partial t} = 0 \tag{19-18}$$

where $Q(T)$ represents any addition of heat to (or removal from) the reactor wall and mean values of physical properties are used. The preceding PDEs can be solved with an initial condition, such as $C(z,0) = C_{t=0}(z)$ with $z = V_r/A_c$ being the distance from inlet, and a boundary condition, such as $C(0,t) = C_0(t)$, which is the concentration at the inlet. At steady state, the accumulation term is zero, and the solution for an isothermal first-order reaction is the same as that for a batch reactor, Eq. (19-5):

$$C = C_0 \exp\left(-k\frac{V_r}{V'}\right) = C_0 e^{-k\bar{t}} \tag{19-19}$$

A tubular reactor will likely deviate from plug flow in most practical cases, for example, due to backmixing in the direction of flow or reactor internals. A way of simulating axial backmixing is to represent the reactor volume as a series of n stirred tanks in series. The steady-state solution for a single ideal CSTR may be extended to find the effluent concentration after two ideal CSTRs and then to n ideal stages as

$$\frac{C_n}{C_0} = \frac{1}{(1 + kV_r/V')^n} = \frac{1}{(1 + k\bar{t})^n} \tag{19-20}$$

In this case, V_r is the volume of each individual reactor in the battery. In modeling a reactor, n is empirically determined based on the extent of reactor backmixing obtained from tracer studies or other experimental data. In general, the number of stages n required to approach an ideal PFR depends on the rate of reaction (e.g., the magnitude of the specific rate constant k for the first-order reaction above).

An alternate way of generating backmixing is to recycle a fraction of the product from a PFR back to the inlet. This reactor, known as a recycle reactor, has been described in Sec. 7. As the recycle ratio (i.e., recycle flow to product flow) is increased, the effective backmixing is increased, and the recycle reactor approaches an ideal CSTR.

Tubular Reactor with Dispersion An alternative approach to describe deviation from ideal plug flow due to backmixing is to include a term that allows for axial dispersion D_e in the plug flow reactor equations. The reactor mass balance equation now becomes

$$V'\frac{\partial C}{\partial V_r} - D_e\frac{\partial^2 C}{\partial V_r^2} + r(C,T) + \frac{\partial C}{\partial t} = 0 \qquad (19\text{-}21)$$

The model is referred to as a *dispersion model*, and the value of the dispersion coefficient D_e is determined empirically based on correlations or experimental data. In a case where Eq. (19-21) is converted to dimensionless variables, the inverse of the coefficient of the second derivative is referred to as the Peclet number ($\mathrm{Pe} = uL/D_e$), where L is the reactor length, $u = V'/A_c$ is the linear velocity, A_c being the reactor cross section area. For plug flow, $D_e = 0$ ($\mathrm{Pe} \Rightarrow \infty$) while for a CSTR, $D_e = \infty$ ($\mathrm{Pe} = 0$). To solve Eq. (19-21), one initial condition and two boundary conditions are needed. The "closed-ends" boundary conditions are $uC_0 = (uC - D_e\partial C/\partial z)_{z=0}$ and $(\partial C/\partial z)_{z=L} = 0$, where $z = V_r/A_c$ is the distance from the reactor inlet (e.g., see Wen and Fan, *Models for Flow Systems in Chemical Reactors*, Marcel Dekker, New York, 1975). Figure 19-2 shows the performance of a tubular reactor with dispersion compared to that of a plug flow reactor and a CSTR for first- and second-order reactions.

Ideal chemical reactors typically may be modeled using a combination of ideal CSTR, PFR, and dispersion model equations. In the case of a single phase, the approach is relatively straightforward. In the case of two-phase flow, a bubble column (fluidized-bed) reactor may be modeled as containing an ideal CSTR liquid (emulsion) phase and a plug flow (with dispersion) gas phase containing bubbles. Given inlet gas conditions, the concentration in the liquid (emulsion) may be calculated using mass transfer from the bubbles to the liquid (emulsion) along with reaction in the liquid (emulsion) phase along the length of the reactor. In flooded gas–liquid reactors where the gas and liquid are countercurrent to each other, a plug flow (with dispersion) model may be used for both phases. A solution technique called the shooting method is required since the concentration of each phase is known at its feed end only and its concentration at the other end has to be initially estimated to start the integration. Using mass-transfer correlations and reaction kinetics together with a plug flow (with dispersion) model, the unknown effluent concentrations are calculated and the initial estimates are updated with calculated ones. This is an iterative process that continues until the concentrations from consecutive iterations are the same within the required convergence tolerance. Often more complex zone models are needed to describe multiphase reactors, with multiple CSTRs with backmixing for each phase.

Reactor Selection Ideal CSTR and PFR models are extreme cases of complete axial dispersion ($D_e = \infty$) and no axial dispersion ($D_e = 0$), respectively. As discussed earlier, staged ideal CSTRs may be used to represent intermediate axial dispersion. Alternatively, within the context of a PFR, the dispersion (or a PFR with recycle) model may be used to represent increased dispersion. Real reactors inevitably have a level of dispersion in between that for a PFR or an ideal CSTR. The level of dispersion may depend on fluid properties (e.g., whether the fluid is newtonian), fluid flow (e.g., the level of mixing), transport properties (e.g., the diffusivity of reactants in the fluid), and reactor geometry. The effect of dispersion in a real reactor is discussed within the context of an ideal CSTR and PFR model in Fig. 19-2.

Figure 19-2*a* shows the effect of dispersion on the reactor volume required to achieve a certain exit concentration (or conversion). As the Pe number increases (i.e., dispersion decreases), the reactor begins to approach plug flow, and the reactor volume required to achieve a certain conversion approaches the volume for a PFR. At lower Pe numbers, reactor performance approaches that of an ideal CSTR, and the reactor volume required to achieve a certain concentration is much higher than that of a PFR. This behavior can be observed in Fig. 19-2*b* that shows the effect of exit concentration on reaction rate. At a given rate, an ideal CSTR has the highest exit concentration (lowest conversion) and a PFR has the lowest exit concentration (highest conversion). As Fig. 19-2*c* shows, since the concentration in an ideal CSTR is the same as the exit concentration, there is a sharp drop in concentration from the inlet to the bulk concentration. In contrast, the concentration in the reactor drops continuously from the inlet to the outlet for a PFR. At intermediate values of Pe, the "closed-ends" boundary condition $uC_0 = (uC - D_e\partial C/\partial z)_{z=0}$ and $(\partial C/\partial z)_{z=L} = 0$ in the dispersion model causes a drop in concentration to levels lower than for an ideal CSTR.

As discussed in Fig. 19-2, for a given conversion, the reactor residence time (or reactor volume required) for a positive order reaction with dispersion will be greater than that of a PFR. This need for a longer residence time is illustrated for a first-order isothermal reaction in a PFR versus an ideal CSTR using Eqs. (19-13) and (19-19).

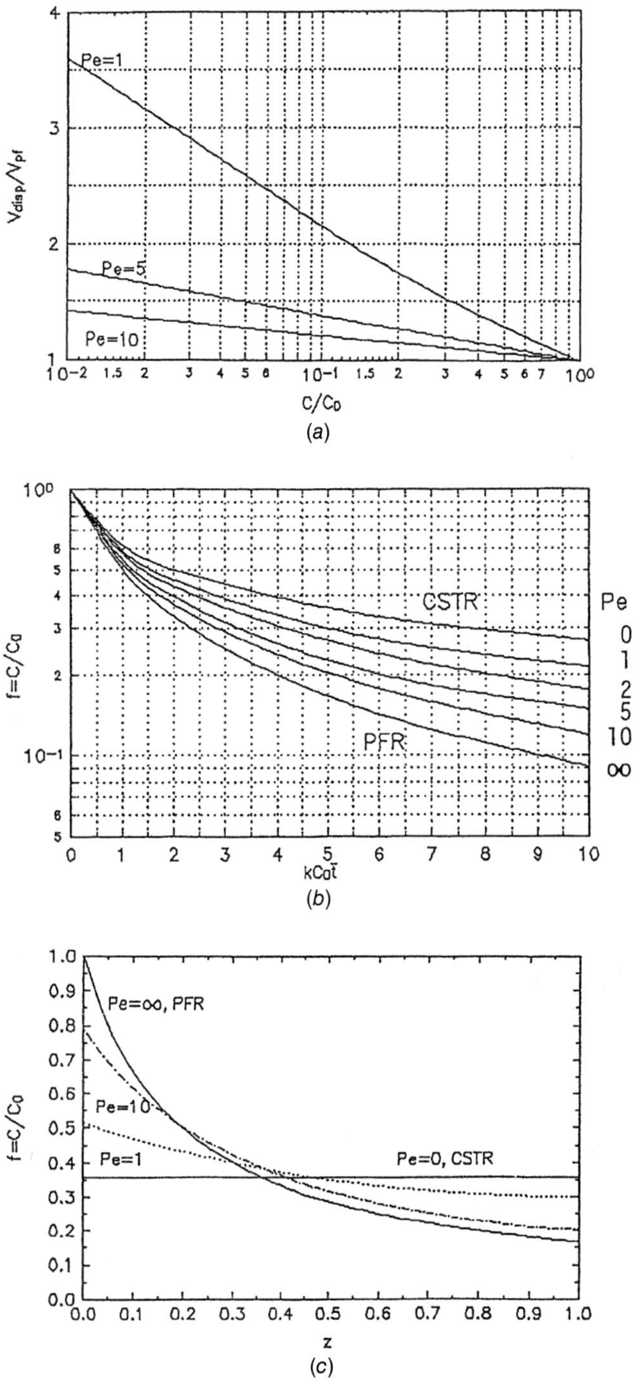

FIG. 19-2 Chemical conversion by the dispersion model. (*a*) Volume relative to plug flow against residual concentration ratio for a first-order reaction. (*b*) Residual concentration ratio against kC_0t for a second-order reaction. (*c*) Concentration profile at the inlet of a closed-ends vessel with dispersion for a second-order reaction with $kC_0t = 5$.

$$\frac{\bar{t}_{\text{ideal CSTR}}}{\bar{t}_{\text{PFR}}} = \frac{C/C_0 - 1}{(C/C_0)\ln(C/C_0)} \qquad (19\text{-}22)$$

Equation (19-22) indicates that, for a nominal 90 percent conversion, an ideal CSTR will need nearly four times the residence time (or volume) of a PFR. This result is also worth bearing in mind when batch reactor experiments are used to model a reactor in the field as a battery of ideal

CSTRs in series. The performance of a completely mixed batch reactor and a steady-state PFR having the same residence time are the same [Eqs. (19-5) and (19-19)]. At a given residence time, if a batch reactor provides a nominal 90 percent conversion for a first-order reaction, a single ideal CSTR will only provide a conversion of 70 percent. The preceding discussion addresses conversion. Product selectivity in complex reaction networks may be profoundly affected by dispersion. This aspect has been addressed from the standpoint of parallel and consecutive reaction networks in Sec. 7.

Reactors may contain one or more fluid phases. The level of dispersion in each phase may be represented mathematically by using some of the preceding thinking.

In industrial practice, the laboratory equipment used in chemical synthesis can influence reaction selection. As issues relating to kinetics, mass transfer, heat transfer, and thermodynamics are addressed, reactor design evolves to commercially viable equipment. Often, more than one type of reactor may be suitable for a given reaction. For example, in the partial oxidation of butane to maleic anhydride over a vanadium pyrophosphate catalyst, heat-transfer considerations dictate reactor selection and choices may include fluidized beds or multitubular reactors. Both types of reactors have been commercialized. Often, experience with a particular type of reactor within the organization can play an important part in selection.

There are several books on reactor analysis and modeling, including those by Froment and Bischoff (*Chemical Reactor Analysis and Design*, Wiley, New York, 1990), Fogler (*Elements of Chemical Reaction Engineering*, Prentice-Hall International Series, Upper Saddle River, N.J., 2005), Levenspeil (*Chemical Reaction Engineering*, Wiley, New York, 1999), and Walas (*Modeling with Differential Equations in Chemical Engineering*, Butterworth-Heineman, Oxford, UK, 1991).

Chemical Kinetics Reactor models include chemical kinetics in the mass and energy conservation equations. The two basic laws of kinetics are the *law of mass action* for the rate of a reaction and the *Arrhenius equation* for its dependence on temperature. Both of these strictly apply to elementary reactions. More often, laboratory data are used to develop mathematical relationships that describe reaction rates that are then used. These relationships require analysis of the laboratory reactor data, as discussed in Sec. 7. Reactor models will require that kinetic rate information be expressed on a unit reactor volume basis. Two-phase or multiphase reactors will require a level of detail (e.g., heat and mass transport between phases) to capture the relevant physical and chemical processes that affect rate.

Pressure Drop, Mass and Heat Transfer Pressure drop is important in reactor design. The size of the compressor is dictated by pressure drop across the reactor, especially in the case of gas recycle. Compressor costs can be significant and can influence the aspect ratio of a packed or trickle bed reactor. Pressure drop correlations often may depend on the geometry, the scale, the fluids used in data generation, and the particle size and shape for packed beds. Prior to using literature correlations, it often is advisable to validate the correlation with measurements on a similar system at a relevant scale.

Depending on the type of reactor, appropriate mass-transfer correlations may have to be used to connect intrinsic chemical kinetics to the reaction rate per unit reactor volume. A number of these correlations have already been discussed in Sec. 5, Heat and Mass Transfer. The determination of intrinsic kinetics has already been discussed in Sec. 7. In the absence of a correlation validated for a specific use, the analogy between momentum, heat, and mass transfer may often be invoked.

The local reactor temperature affects the rates of reaction, equilibrium conversion, and catalyst deactivation. As such, the local temperature has to be controlled to maximize reaction rate and to minimize deactivation. In the case of an exothermic (endothermic) reaction, higher (lower) local temperatures can cause suboptimal local concentrations. Heat will have to be removed (added) to maintain more uniform temperature conditions. The mode of heat removal (addition) will depend on the application and on the required heat-transfer rate.

Examples of stirred tank reactors with heat transfer are shown in Fig. 19-1. If the heat of reaction is not significant, an adiabatic reactor may be used. For modest heat addition (removal), a jacketed stirred tank is adequate (Fig. 19-1*a*). As the heat exchange requirements increase, internal coils or internal tubes that contain a heat-transfer fluid may be required (Fig. 19-1*b* and *c*). In special cases, where the peak temperature has to be tightly controlled (e.g., in bioreactors) or where fouling may be an issue, the liquid may be withdrawn, circulated through an external heat exchanger, and returned to the reactor (Fig. 19-1*d*). In some cases, the vapor above the liquid may be passed through an external reflux condenser and returned to the reactor (Fig. 19-1*e*). In highly endothermic reactors, the entire reactor may be placed inside a fired heater (Fig. 19-1*f*), or the reactor shell may be heated to high temperatures by using induction heat.

Several of the heat-transfer options for packed beds are illustrated in Fig. 19-3. Again, if heat requirements are modest, an adiabatic reactor is adequate (Fig. 19-3*a*). If pressure drop through the reactor is an issue, a radial flow reactor may be used (Fig. 19-3*b*). There are few examples of radial flow reactors in industry. Potential problems include gas maldistribution, especially in the case of catalyst attrition or settling. A common way of dealing with more exothermic (endothermic) reactions is to split the reactor into several beds and then provide interbed heat exchange (Fig. 19-3*c*). For highly exothermic (endothermic) reactors, a shell-and-tube multitubular reactor concept may be used (Fig. 19-3*d*). The reactor now begins to look more like a heat exchanger. If multiple beds are needed, rather than using interbed heat exchangers, cold feed may be injected (also called cold shot) in between beds (Fig. 19-3*e*). In some cases, the heat exchanger may be outside the reactor (Fig. 19-3*f*). The concept of a reactor as a heat exchanger may be extended to an autothermal multitubular reactor in which heat integration is achieved when an exothermic reaction is coupled with an endothermic reaction. For example, the reactants are preheated on the endothermic shell side with an exothermic reaction occurring in the tubes (Fig. 19-3*g*). Such reactors can have control issues. A common approach is to have multiple adiabatic reactors with cooling in between reactors (Fig. 19-3*h*). If the reaction is endothermic, heat may be added by passing the effluents from each reactor through tubes placed inside a common process heater (as is the case for a petroleum reforming reactor shown in Fig. 19-3*i*). For highly endothermic reactions, a fuel–air mixture or raw combustion gases may be introduced into the reactor. In an extreme situation, the entire reactor may be housed within a furnace (as in the case of steam reforming for hydrogen synthesis or ethane cracking for ethylene production).

When the reaction is exothermic, the conversion is limited by thermodynamic equilibrium. In such cases, packed beds in series with interstage cooling may be used as well. The performance enhancement associated with this approach is shown for two cases in Table 19-2. Such units can take advantage of initial high rates at high temperatures and higher equilibrium conversions at lower temperatures. For SO_2 oxidation, the typical conversion attained in the fourth bed is 97.5 percent, compared with an adiabatic single-bed value of 74.8 percent. With the three-bed ammonia reactor, final ammonia concentration is 18.0 percent, compared with the one-stage adiabatic value of 15.4 percent.

Since reactors come in a variety of configurations, use a variety of operating modes, and may handle mixed phases, design provisions for temperature control may draw on a large body of heat-transfer theory and data. These extensive topics are treated in other sections of this Handbook and in other references. Some of the high points pertinent to reactors are covered by Rase (*Chemical Reactor Design for Process Plants,* Wiley, New York, 1977). Two encyclopedic references, *Heat Exchanger Design Handbook* (5 vols., Begell House, 1983–1998) and Cheremisinoff, ed. (*Handbook of Heat and Mass Transfer,* 4 vols., Gulf, Houston, 1986–1990), have several articles addressed specifically to reactors.

Reactor Dynamics Continuous reactors are designed to operate at or near a steady state by controlling the operating conditions. In addition, process control systems are designed for safety and to minimize fluctuations from the target conditions. Batch and semibatch reactors are designed to operate under predefined protocols based on the best understanding of the process. However, the potential for large and unexpected deviations from steady state as a result of process variable fluctuations is significant due to the complexity and nonlinearity of reaction kinetics and of the relevant mass and heat-transfer processes. For a set of operating conditions (pressure, temperature, composition, and phases present), more than one steady state can exist. Which steady state is actually reached depends on the initial condition. Not all steady states are stable states, and only those that are stable can be reached without special control schemes. More complex behavior such as self-sustained oscillations and chaotic behavior has also been observed with reacting systems. Further, during start-up, shutdown, and abrupt changes in process conditions, the reactor dynamics may result in conditions that exceed reactor design limits (e.g., of temperature, pressure, and materials of construction) and can result in a temperature runaway, reactor blowout, and even an explosion (or detonation). Parametric sensitivity deals with the analysis of reactor dynamics in response to abrupt changes.

Steady-State Multiplicity and Stability A simple example of *steady-state multiplicity* is due to the interaction between kinetics and heat transport in an adiabatic CSTR. For a first-order reaction at steady state, Eq. (19-13) gives

$$r(C,T) = kC = \frac{kC_f}{1+k\bar{t}} = \frac{C_f \exp(a+b/T)}{1+\bar{t}\exp(a+b/T)} \tag{19-23}$$

where C_f is the feed concentration and a and b are constants related to Arrhenius rate expression. The energy balance equation at steady state is given by

$$Q_G(T) = -\Delta H_r V_r r(C,T) = V'\rho C_p (T-T_f) = Q_H(T) \tag{19-24}$$

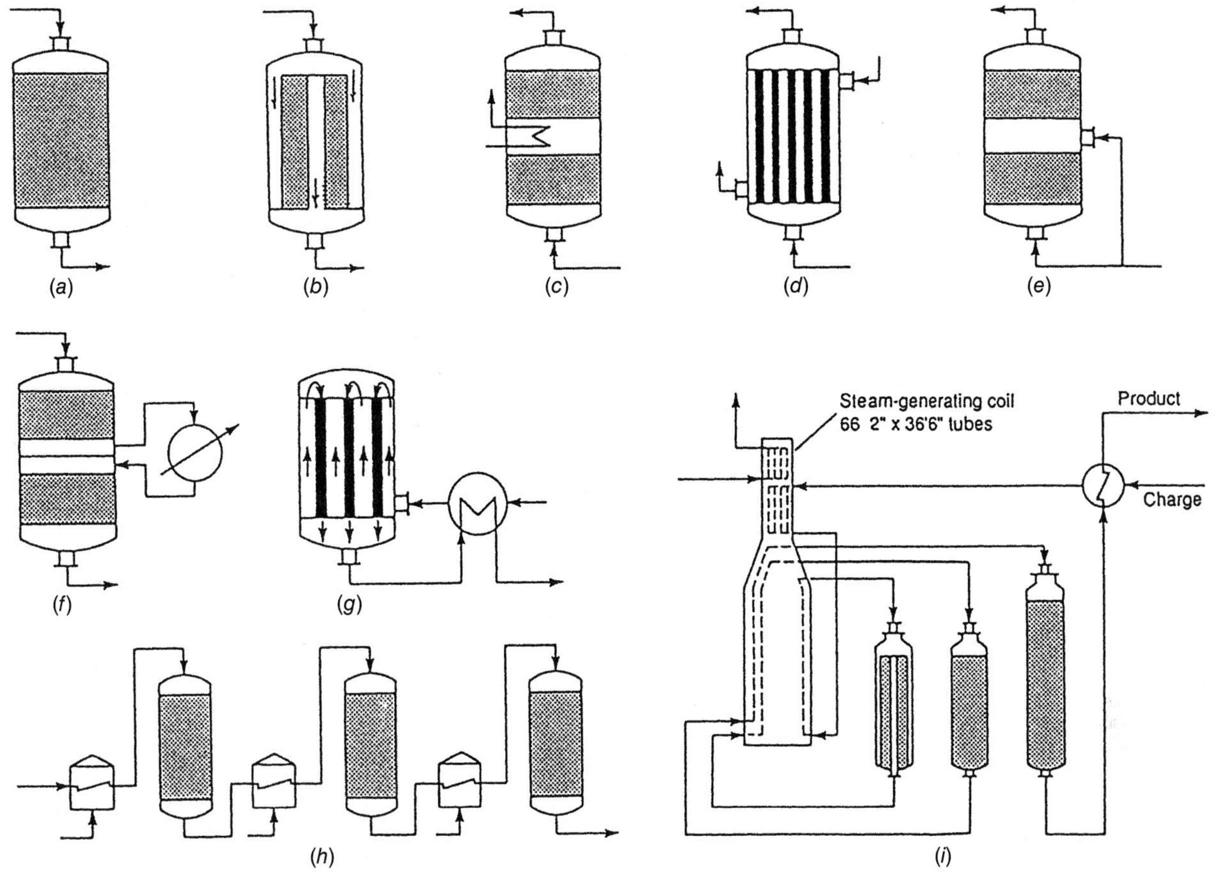

FIG. 19-3 Fixed-bed reactors with heat exchange. (*a*) Adiabatic downflow. (*b*) Adiabatic radial flow, low ΔP. (*c*) Built-in interbed exchanger. (*d*) Shell and tube. (*e*) Interbed cold-shot injection. (*f*) External interbed exchanger. (*g*) Autothermal shell, outside influent/effluent heat exchanger. (*h*) Multibed adiabatic reactors with interstage heaters. (*i*) Platinum catalyst, fixed-bed reformer for 5000 BPSD charge rates. Reactors 1 and 2 are 5.5 by 9.5 ft, and reactor 3 is 6.5 by 12.0 ft; temperatures 502°C ⇒ 433°C, 502°C ⇒ 471°C, 502°C ⇒ 496°C. To convert feet to meters, multiply by 0.3048; BPSD to m³/h, multiply by 0.00662.

where Q_G is the *heat generation by reaction*, Q_H is the *heat removal by flow*, T is the reactor temperature at steady state, and T_f is the feed temperature. Plotting the heat generation and heat removal terms versus temperature gives the result shown in Fig. 19-4. As shown, as many as three steady states are possible at the intersection of Q_G and Q_H.

Another example of multiplicity is shown in Fig. 19-15 for an adiabatic catalyst pellet, indicating that three effectiveness factor values can be obtained for a given Thiele modulus for a range of Prater numbers and Thiele modulus values, leading to three potential steady states. Multiple steady states can occur in different reactor types, including isothermal systems with complex nonlinear kinetics and systems with interphase transfer, the main requirement being the existence of a feedback mechanism—hence, a homogeneous PFR (without backmixing) will not exhibit multiplicity due to the lack of a feedback mechanism. Depending on the various physical and chemical interactions in a reactor, oscillatory and chaotic behavior can also occur.

TABLE 19-2 Multibed Reactors, Adiabatic Temperature Rises, and Approaches to Equilibrium*

Oxidation of SO_2 at atmospheric pressure in a four-bed reactor. Feed 6.26% SO_2, 8.3% O_2, 5.74% CO_2, and 79.7% N_2.

°C		Conversion, %	
In	Out	Plant	Equilibrium
463.9	592.8	68.7	74.8
455.0	495.0	91.8	93.4
458.9	465.0	96.0	96.1
435.0	437.2	97.5	97.7

Ammonia synthesis in a three bed reactor at 225 atm. Feed 22% N_2, 66% H_2, 12% inerts.

°C		Ammonia, %	
In	Out	Calculated	Equilibrium
399	518.9	13.0	15.4
427	488.9	16.0	19.0
427	470.0	18.0	21.7

*To convert atm to kPa multiply by 101.3.

SOURCE: Plant data and calculated design values from Rase, *Chemical Reactor Design for Process Plants*, Wiley, 1977, Case study 106 Table CS-6.2 and Case study 107 Table CS-7.1.

There is a voluminous literature on steady-state multiplicity, oscillations (and chaos), and derivation of bifurcation points that define the conditions that lead to the onset of these phenomena. For example, see Morbidelli et al. ("Reactor Steady-State Multiplicity and Stability," in *Chemical Reaction and Reactor Engineering*, ed. Carberry and Varma, Marcel Dekker, New York, 1987), Luss ("Steady State Multiplicity and Uniqueness Criteria for Chemically Reacting Systems," in *Dynamics and Modeling of Reactive Systems*, ed. Stewart et al., Academic Press, New York, 1980), Schmitz [*Adv. Chem. Ser.* **148:** 156, ACS (1975)], and Razon and Schmitz [*Chem. Eng. Sci.* **42** (1987)]. However, many of these criteria for specific reaction and reactor systems have not been validated experimentally.

Linearized or asymptotic stability analysis examines the stability of a steady state to small perturbations from that state. For example, when heat generation is greater than heat removal (as at points $A-$ and $B+$ in Fig. 19-4), the temperature will rise until the next stable steady-state temperature is reached (for $A-$ it is A, for $B+$ it is C). In contrast, when heat generation is less than heat removal (as at points $A+$ and $B-$ in Fig. 19-4*a*), the temperature will fall to the next-lower stable steady-state temperature (for $A+$ and $B-$ it is A). A similar analysis can be done around steady-state C, and the result indicates that A and C are stable steady states since small perturbations from the vicinity of these return the system to the corresponding stable points. Point B is an unstable steady state, since a small perturbation moves the system away to either A or C, depending on the direction of the perturbation. Similarly, at conditions where a unique steady state exists, this steady state is always stable for the adiabatic CSTR. Hence, for the adiabatic CSTR considered in Fig. 19-4, the slope condition $dQ_H/dT > dQ_G/dT$ is a necessary and sufficient condition for asymptotic stability of a steady state. In general

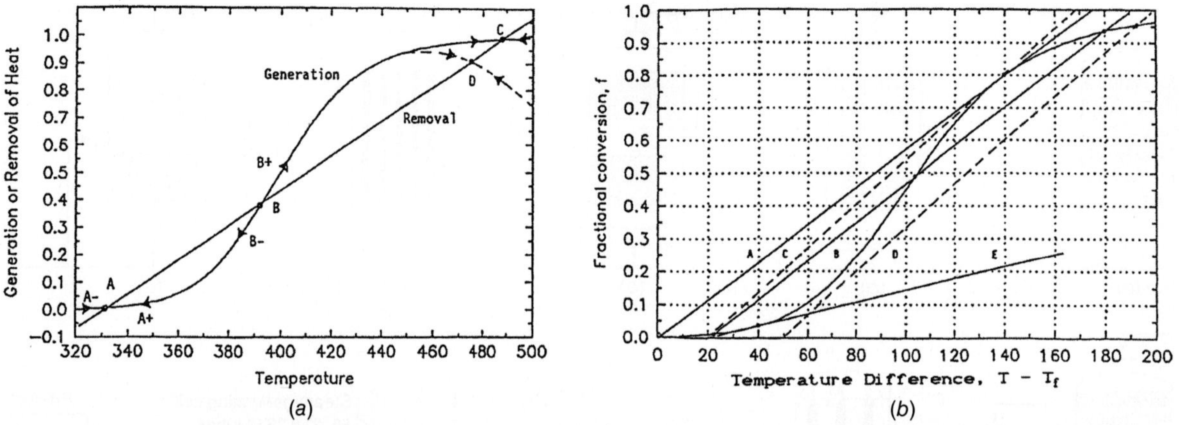

FIG. 19-4 Multiple steady states of CSTRs, stable and unstable, adiabatic. (*a*) First-order reaction, A and C stable, B unstable; the dashed line is for a reversible reaction. (*b*) One, two, or three steady states depending on the combination (C_f, T_f).

(e.g., for an externally cooled CSTR), however, the slope condition is a necessary but not a sufficient condition for stability; that is, violation of this condition leads to asymptotic instability, but its satisfaction does not ensure asymptotic stability. For example, in selected reactor systems, even a unique steady state can become unstable, leading to oscillatory or chaotic behavior.

Local asymptotic stability criteria may be obtained by first solving the steady-state equations to obtain steady states and then linearizing the transient mass and energy balance equations in terms of deviations of variables around each steady state. The determinant (or slope) and trace conditions derived from the matrix A in the set of equations obtained are necessary and sufficient for asymptotic stability:

$$\frac{d}{dt}\begin{pmatrix} x \\ y \end{pmatrix} = A\begin{pmatrix} x \\ y \end{pmatrix} \quad x = C - C_{ss} \quad y = T - T_{ss} \tag{19-25}$$

$$\Delta = \det(A) > 0 \quad \sigma = \text{trace}(A) < 0$$

where x and y are the deviation variables around the steady state (C_{ss}, T_{ss}). The approach may be extended to systems with multiple concentrations and complex nonlinear kinetics. For additional references on asymptotic stability analysis, see Denn (*Process Modeling*, Longman, Essex, UK, 1986) and Morbidelli et al. ("Reactor Steady-State Multiplicity and Stability," in *Chemical Reaction and Reactor Engineering*, ed. Carberry and Varma, Marcel Dekker, New York, 1987).

Parametric Sensitivity and Dynamics The global stability and sensitivity to abrupt changes in parameters cannot be determined from an asymptotic analysis. For instance, for the simple CSTR, a key question is whether the temperature can run away from a lower stable steady state to a higher one. The critical temperature difference ΔT_c is useful in designing for globally stable operation for the case of a single exothermic reaction:

$$T - T_j < \Delta T_c = \frac{RT^2}{E} \tag{19-26}$$

where T is the reactor temperature, T_j is the cooling jacket temperature, E is the activation energy, and R is the universal gas constant. Similarly, for a jacketed PFR, a conservative criterion for stability is $T_{max} - T_j < \Delta T_c$, where T_{max} is the temperature of the hot spot. These criteria require knowledge of the activation energy, which may not always be available.

Another example of sensitivity to abrupt changes is the *wrong-way effect*, exhibited, for instance, in packed-bed reactors, where an abrupt reduction in feed rate or in feed temperature results in a dramatic increase in reactor peak temperature for exothermic reactions. The reactor may eventually return to the original steady state, or if a higher-temperature steady state exists, the reactor may establish a temperature profile corresponding to the new high steady state. Such a dynamic excursion can result in an increase of undesirable by-products concentration, catalyst deactivation, permanent reactor damage, and safety issues; see work by Luss and coworkers ["Wrong-Way Behavior of Packed-Bed Reactors: I. The Pseudo-Homogeneous Model," *AIChE J.* **27:** 234–246 (1981)]. For more complex systems, the transient model equations are solved numerically. A more detailed discussion of parametric sensitivity is provided by Varma et al. (*Parametric Sensitivity in Chemical Systems*, Cambridge University Press, Cambridge, UK, 1999).

Reactor Models As discussed earlier, reactor models try to strike a balance between the level of detail included and the usefulness of the model. Too many details in the model may require a larger number of adjustable model parameters, increase computational requirements, and limit how widely the model may be used. Too few details, on the other hand, increase ease of implementation but may compromise the predictive or design capabilities of the model. Figure 19-5 is a schematic of the inherent tradeoff between ease of implementation and the insight that may be obtained from the model.

Increases in computational power are allowing a more cost-effective inclusion of a greater number of details. Computational fluid dynamics (CFD) models provide detailed flow information by solving various approximations of the Navier-Stokes equations for mass, momentum, and heat balances. The user will, however, need to be familiar with the basic elements of the respective software and may need a license. A typical numerical solution of the governing transport equations is obtained within the Eulerian framework, using a large number of computational cells (or finite volumes) that represent reactor geometry. Current capabilities in commercial CFD software can be used to resolve the flow, concentration, and temperature patterns in a single phase with sufficient detail and reasonable accuracy for all length and time scales. The ability to visualize flow, concentration, and temperature inside a reactor is useful in understanding performance and in designing reactor internals.

Addition of transport properties and more than one phase (as is the case with solid catalysts) within a CFD framework complicates the problem in that the other phase(s) also may have to be included in the calculations. This may require additional transport equations to address a range of complexities associated with the dynamics and physics of each phase, the interaction between and within phases, sub-grid-scale heterogeneities (such as size distributions within each phase), and coupling with kinetics at the molecular level. For example, one needs the bubble size distribution in a bubble column reactor to correctly model interfacial area and local mass-transfer coefficients, which can further affect the chemical kinetics. Although phenomenological models describing such physical effects have greatly improved over the years, this area still lacks reliable multiphase turbulence closures, or experimentally validated intraphase and interphase transport models. Mathematical modeling in industrial practice will continue to involve compromises between computational complexity, experimental data needs, ability to validate the model, cost, and the time frame in which the work may be useful to the organization.

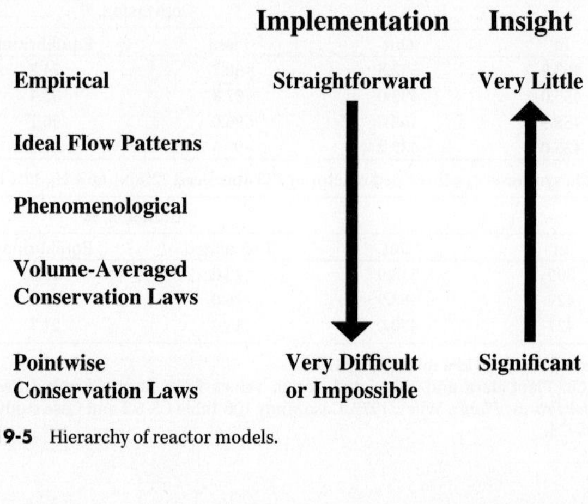

FIG. 19-5 Hierarchy of reactor models.

RESIDENCE TIME DISTRIBUTION AND MIXING

The time spent by reactants and intermediates at reaction conditions determines conversion (and perhaps selectivity). It is therefore often important to understand the residence time distribution (RTD) of reaction species in the reactor. This RTD could be considerably different from what is expected from ideal reactor behavior. Reasons for the deviation could be channeling of fluid, recycling of fluid, or creation of stagnant regions in the reactor, as illustrated in Fig. 19-6.

This section introduces how tracers are used to establish the RTD in a reactor and to contrast against RTDs of ideal reactors. The section ends with a discussion of how reactor performance may be connected to RTD information.

TRACERS

Tracers are typically nonreactive substances used in small concentration that can be easily detected. The tracer is typically injected at the inlet of the reactor along with the feed or by using a carrier fluid, according to some definite time sequence. The inlet and outlet concentrations of the tracer are recorded as a function of time. These data are converted to a residence time distribution of feed in the reactor vessel. Tracer studies may be used to detect and define regions of nonideal behavior, develop phenomenological zone models, calculate reactor performance (conversion, selectivity), and synthesize optimal reactor configurations for a given process. The RTD does not represent the mixing behavior in a vessel uniquely. Several arrangements of reactors or internals within a vessel may provide the same tracer response. For example, any series arrangement of the same number of CSTR and plug flow reactor elements will provide the same RTD. This lack of uniqueness may limit direct application of tracer studies to first-order reactions with constant specific rates. For other reactions, the tracer curve may determine the upper and lower limits of reactor performance. When this range is not too broad, or when the purpose of the tracer test is to diagnose maldistribution or bypassing in the reactor, the result can be useful. Tracer data also may be taken at several representative positions in the vessel in order to develop a better understanding for the flow behavior.

Inputs Although some arbitrary variation of input concentration with time may be employed, five mathematically simple tracer input signals meet most needs. These are *impulse, step, square pulse* (started at time a and kept constant for an interval), *ramp* (increased at a constant rate for a period of interest), and *sinusoidal.* Sinusoidal inputs are difficult to generate experimentally.

Types of Responses The key relationships associated with tracers are provided in Table 19-3. Effluent concentrations resulting from impulse and step inputs are designated C_δ and C_u, respectively. The mean concentration resulting from an impulse of magnitude m into a vessel of volume V_r is $C^0 = m/V_r$. The mean residence time is the ratio of the vessel volume to the volumetric flow rate:

$$\bar{t} = \frac{V_r}{V'} \quad \text{or} \quad \bar{t} = \frac{\int_0^\infty tC_\delta\, dt}{\int_0^\infty C_\delta\, dt} \tag{19-27}$$

The reduced time is $t_r = t/\bar{t}$. Residence time distributions are used in two forms: normalized, $E(t_r) = C_\delta / C^0$; or plain, $E(t) = C_\delta / \int_0^\infty C_\delta\, dt$. The area under either RTD is unity: $\int_0^\infty E(t_r)\, dt_r = \int_0^\infty E(t)\, dt = 1$, and the relation between them is $E(t_r) = \bar{t}E(t)$. The area between the ordinates at t_1 and t_2 is the fraction of the total effluent that has spent the period between these times in the vessel. The age function is defined in terms of these steps as

$$F(t) = \frac{C_u}{C_f} = \int_0^t E(t)\, dt \tag{19-28}$$

Reactor Tracer Responses

Continuous Stirred Tank Reactor (CSTR) With a step input of magnitude C_f, the unsteady material balance of tracer

$$V_r \frac{dC}{dt} + V'C = V'C_f \tag{19-29}$$

can be integrated to yield

$$\frac{C}{C_f} = F(t_r) = 1 - e^{-t_r} \tag{19-30}$$

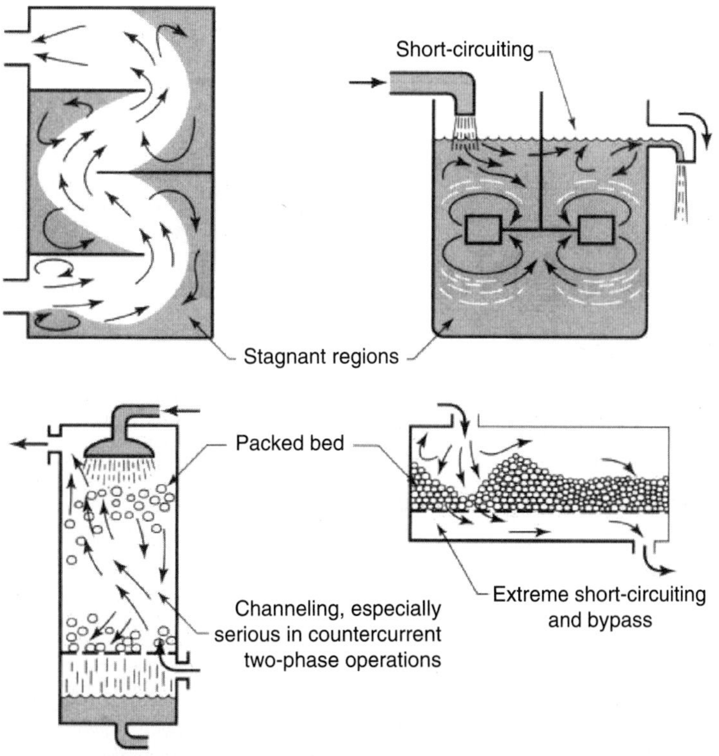

FIG. 19-6 Some examples of nonideal flow in reactors. (*Fig. 11.1 in Levenspiel,* Chemical Reaction Engineering, *John Wiley & Sons, 1999.*)

TABLE 19-3 Tracer Response Functions

Mean residence time:

$$\bar{t}=\frac{\int_0^\infty tC_\delta\,dt}{\int_0^\infty C_\delta\,dt}=\frac{\int_0^{C_{u\infty}} t\,dC_u}{C_{u\infty}}$$

Initial mean concentration with impulse input,

$$C^0=\frac{m}{V_r}=\left(\frac{V'}{V_r}\right)\int_0^\infty C_\delta\,dt=\frac{\int_0^\infty C_\delta\,dt}{\bar{t}}$$

Reduced time:

$$t_r=\frac{t}{\bar{t}}$$

Residence time distribution:

$$E(t)=\frac{C_\delta}{\int_0^\infty C_\delta\,dt}=\frac{E(t_r)}{\bar{t}}=\frac{dF(t)}{dt}$$

Residence time distribution, normalized,

$$E(t_r)=\frac{\text{impulse output}}{\text{initial mean concentration}}=\frac{C_\delta}{C^0}=\frac{\bar{t}C_\delta}{\int_0^\infty C_\delta dt}=\bar{t}E(t)=\frac{dF(t)}{dt}$$

Age:

$$F(t)=\frac{\text{step output}}{\text{step input}}=\frac{C_u}{C_f}=\frac{\int_0^t C_\delta\,dt}{\int_0^\infty C_\delta\,dt}=F(t_r)$$

Internal age:

$$I(t)=1-F(t)$$

Intensity:

$$\Lambda(t)=\frac{E(t)}{1-F(t)}=\frac{E(t)}{I(t)}$$

Variance:

$$\sigma^2(t)=\int_0^\infty (t-\bar{t})^2E(t)dt=-\bar{t}^2+\frac{\int_0^\infty t^2C_\delta\,dt}{\int_0^\infty C_\delta\,dt}$$

Variance, normalized:

$$\sigma^2(t_r)=\frac{\sigma^2(t)}{\bar{t}^2}=-1+\frac{\int_0^\infty t^2C_\delta\,dt}{\int_0^\infty C_\delta\,dt}=\int_0^1(t_r-1)^2\,dF(t_r)$$

Skewness, third moment:

$$\gamma^3(t_r)=\int_0^\infty (t_r-1)^3E(t_r)dt_r$$

With an impulse input of magnitude m or an initial mean concentration $C^0=m/V_r$, the material balance is

$$\frac{dC}{dt_r}+C=0 \quad \text{with} \quad C=C^0, t=0 \tag{19-31}$$

And integration gives

$$\frac{C}{C^0}=E(t_r)=e^{-t_r} \tag{19-32}$$

These results show that

$$E(t_r)=\frac{dF(t_r)}{dt_r} \tag{19-33}$$

Multistage CSTR Since tubular reactor performance can be simulated by a series of CSTRs, multistage CSTR tracer models are useful in analyzing data from empty tubular and packed-bed reactors. The solution for a tracer through n CSTRs in series is found by mathematical induction from the solution of one stage, two stages, and so on.

$$E(t_r)=\frac{C_n}{C^0}=\frac{n^n}{(n-1)!}t_r^{n-1}e^{-nt_r} \tag{19-34}$$

The solution for a step response can be obtained by integration

$$F(t_r)=\int_0^{t_r}E(t_r)\,dt_r=1-e^{-nt_r}\sum_{j=0}^{n-1}\frac{(nt_r)^j}{j!} \tag{19-35}$$

where $E(t_r)$ and $F(t_r)$ for various values of n are shown in Fig. 19-7.

The theoretical RTD responses in Fig. 19-7*a* are similar in shape to the experimental responses from pilot and commercial reactors shown in Fig. 19-8. The value of n in Fig. 19-8 represents the number of CSTRs in series that provide a similar RTD to that observed commercially. Although not shown in the figure, a commercial reactor having a similar space velocity as a pilot reactor and a longer length typically has a higher n value than a pilot reactor due to greater linear velocity.

The variance of the RTD of a series of CSTRs, σ^2, is the inverse of n.

$$\sigma^2=\int_0^\infty (t_r-1)^2E(t_r)\,dt_r=\frac{1}{n} \tag{19-36}$$

Plug Flow Reactor The tracer material balance over a differential reactor volume dV_r is

$$\frac{\partial C}{\partial t}+V'\frac{\partial C}{\partial V_r}=0 \tag{19-37}$$

With step input $u(t)$, the initial and boundary conditions are

$$C(0,t)=C_f u(t) \quad \text{and} \quad C(V_r,0)=0 \tag{19-38}$$

The solution is

$$\frac{C}{C_f}=F(t)=u(t-\bar{t})=\begin{cases}0 & \text{when } t\le\bar{t}\\ 1 & \text{when } t>\bar{t}\end{cases} \tag{19-39}$$

As discussed earlier, the response to an impulse input is the derivative of $F(t)$.

$$\frac{C}{C_\delta}=E(t)=\delta(t_r-1) \tag{19-40}$$

The effluent RTD is an impulse that is delayed from the input impulse by $t_r=1$, or $t=\bar{t}$.

Tubular Reactor with Dispersion As discussed earlier, a multistage CSTR model can be used to simulate the RTD in pilot and commercial reactors. The *dispersion model*, similar to Fick's molecular diffusion law with an empirical dispersion coefficient D_e replacing the diffusion coefficient, may also be used.

$$\frac{\partial C}{\partial t}+V'\frac{\partial C}{\partial V_r}-D_e\frac{\partial^2 C}{\partial V_r^2}=0 \tag{19-41}$$

Equation (19-41) is often converted to dimensionless variables and solved. The solution of this partial differential equation is recorded in the literature [Otake and Kunigata, *Kagaku Kogaku* **22:** 144 (1958)]. The plots of $E(t_r)$ versus t_r are bell-shaped, similar to the response for a series of n CSTRs model (Fig. 19-7). A relation between $\sigma^2(t_r)$, n, and Pe (for the closed-ends condition) is

$$\sigma^2(t_r)=\frac{2[\text{Pe}-1+\exp(-\text{Pe})]}{\text{Pe}^2}=\frac{1}{n} \tag{19-42}$$

Examples of values of Pe are provided in Fig. 19-8. When Pe is large, $n\Rightarrow \text{Pe}/2$, and the dispersion model reduces to the PFR model. For small values of Pe, the preceding equation breaks down since the lower limit on n does not reduce to $n=1$ as it should for a single CSTR. To better represent dispersion behavior, a series of CSTRs with backmixing between each CSTR may be used; for example, see Froment and Bischoff (*Chemical Reactor Analysis and Design*, Wiley, New York, 1990). A model analogous to the dispersion model may be used when there are velocity profiles across the reactor cross section (e.g., for laminar flow). In this case, a relationship such as in Eq. (19-42) will contain terms associated with the radial position in the reactor.

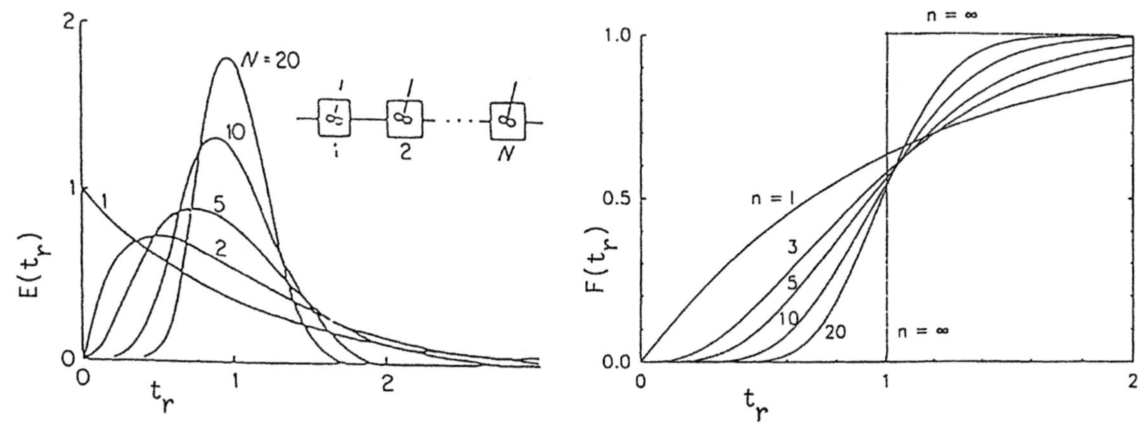

(a) (b)

FIG. 19-7 Tracer responses to n-stage continuous stirred tanks in series: (a) Impulse inputs. (b) Step input.

Understanding Reactor Flow Patterns As discussed previously, an RTD obtained using a nonreactive tracer may not uniquely represent the flow behavior within a reactor. For diagnostic and simulation purposes, however, tracer results may be explained by combining the expected tracer responses of ideal reactors combined in series, in parallel, or both, to provide an RTD that matches the observed reactor response. The most commonly used ideal models for matching an actual RTD are PFR and CSTR models. Figure 19-9 illustrates the responses of CSTRs and PFRs to impulse or step inputs of tracers.

Since the tracer equations are linear differential equations, Laplace transform $L\{f(t)\}=\int_0^\infty f(t)e^{-st}dt$ may be used to relate tracer inputs to responses. The concept of a transfer function facilitates the combination of linear elements.

$$\bar{C}_{\text{output}}(s)=(\text{transfer function})\,\bar{C}_{\text{input}}(s)=G(s)\,\bar{C}_{\text{input}}(s) \qquad (19\text{-}43)$$

Some common Laplace transfer functions are listed in Table 19-4.

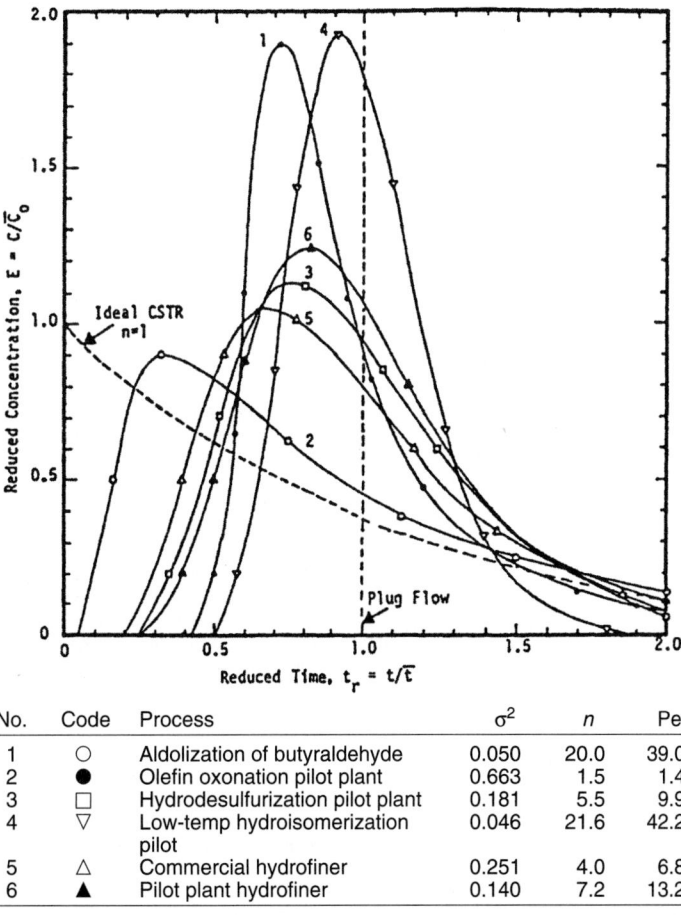

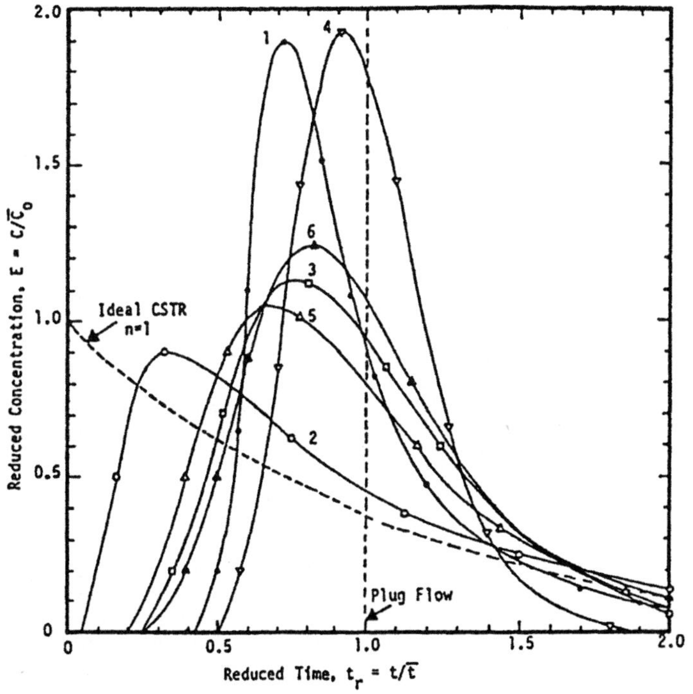

No.	Code	Process	σ^2	n	Pe
1	○	Aldolization of butyraldehyde	0.050	20.0	39.0
2	●	Olefin oxonation pilot plant	0.663	1.5	1.4
3	□	Hydrodesulfurization pilot plant	0.181	5.5	9.9
4	▽	Low-temp hydroisomerization pilot	0.046	21.6	42.2
5	△	Commercial hydrofiner	0.251	4.0	6.8
6	▲	Pilot plant hydrofiner	0.140	7.2	13.2

FIG. 19-8 Residence time distributions of pilot and commercial reactors. σ^2 = Variance of the residence time distribution, n = number of stirred tanks with the same variance, Pe = Peclet number. (*Fig 17.1 in Walas,* Chemical Process Equipment, *Butterworth, Stoneham, Mass., 1990.*)

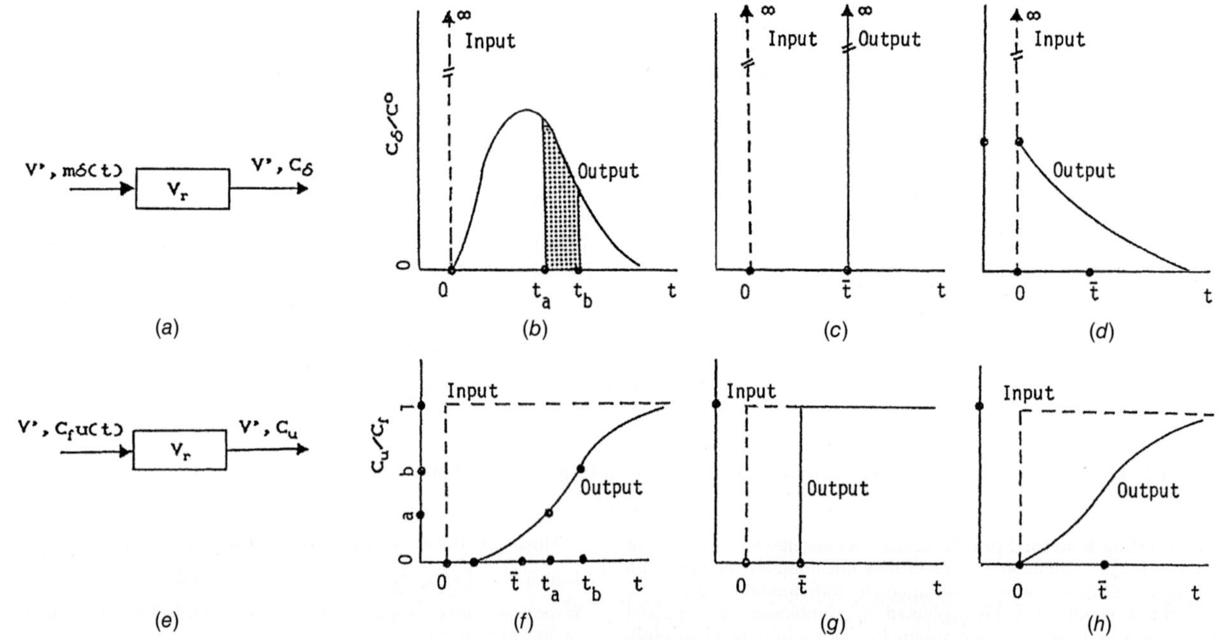

FIG. 19-9 Tracer inputs and responses for PFR and CSTR. (*a*) Experiment with impulse input of tracer. (*b*) Generic behavior; area between ordinates at t_a and t_b equals the fraction of the tracer with residence time in that range. (*c*) Plug flow behavior. (*d*) Completely mixed vessel. (*e*) Experiment with step input of tracer. (*f*) Generic behavior; fraction with ages between t_a and t_b equals the difference between the ordinates, $b - a$. (*g*) Plug flow behavior. (*h*) Completely mixed behavior.

The Laplace transform may be inverted to provide a tracer response in the time domain. In many cases, the overall transfer function cannot be analytically inverted. Even in this case, moments of the RTD may be derived from the overall transfer function. For instance, if G_0' and G_0'' are the limits of the first and second derivatives of the transfer function $G(s)$ as $s \Rightarrow 0$, the mean residence time and variance are

$$\bar{t} = G_0' \quad \text{and} \quad \sigma^2(t) = G_0'' - (G_0')^2 \tag{19-44}$$

In addition to understanding the flow distribution, tracer experiments may be conducted to predict or explain reactor performance based on a particular RTD. To do this, a mathematical expression for the RTD is needed. A PFR model, or a dispersion model with a small value of the dispersion coefficient, may be used to simulate an empty tubular reactor. In some cases, to fit the measured RTD, the model may have to be modified by taking account of bypass zones, stagnant zones, or other features associated with the geometry and operation of the reactor. Sometimes the vessel can be visualized as a zone of complete mixing in the vicinity of impellers followed by plug flow zones elsewhere, such as CSTRs followed by PFRs. Packed beds usually deviate substantially from plug flow. The dispersion model and some combination of PFRs and CSTRs or multiple CSTRs in series may approximate their behavior. Fluidized beds in small sizes approximate CSTR behavior, but large ones exhibit bypassing, stagnancy, nonhomogeneous regions, and several varieties of contact between particles and fluid. The additional parameters required to simulate such mixing behavior can increase the mathematical complexity of the model.

The characteristic bell shape of many RTDs can be fit to well-known statistical distributions. Hahn and Shapiro (*Statistical Models in Engineering*, Wiley, New York, 1967) discuss many of the standard distributions and conditions for their use. The most useful distributions are the gamma (or Erlang) and the Gaussian together with its Gram-Charlier extension.

TABLE 19-4 Some Common Laplace Transform Functions

Element	Transfer function G(s)
Ideal CSTR	$\dfrac{1}{1+\bar{t}s}$
PFR	$\exp(-\bar{t}s)$
n-stage CSTR (Erlang)	$\dfrac{1}{(1+\bar{t}s)^n}$
Erlang with time delay	$\dfrac{\exp(-\bar{t}_1 s)}{(1+\bar{t}_2 s)^n}$

These distributions are represented by only a few parameters that can be used to determine, for instance, the mean and the variance.

Qualitative inspection of the tracer response can go a long way toward identifying flow distribution problems. Additional references on tracers are Wen and Fan (*Models for Flow Systems in Chemical Reactors*, Marcel Dekker, New York, 1975) and Levenspiel (*Chemical Reaction Engineering*, 3d ed., Wiley, New York, 1999).

CONNECTING RTD TO CONVERSION

When the flow pattern is known, the conversion for a given reaction mechanism may be evaluated from the appropriate material and energy balances. When only the RTD is known (or can be calculated from tracer response data), however, different networks of reactor elements can match the observed RTD. In reality, reactor performance for a given reactor network will be unique. The conversion obtained by matching the RTD is, however, unique only for linear kinetics. For nonlinear kinetics, two additional factors have to be accounted for to fully describe the contacting or flow pattern: the degree of segregation of the fluid and the earliness of mixing of the reactants.

Segregated Flow The degree of segregation relates to the tendency of fluid particles to move together as aggregates or clumps (e.g., bubbles in gas–liquid reactors, particle clumps in fluidized beds, polymer striations in high-viscosity polymerization reactors) rather than each molecule behaving independently (e.g., homogeneous gas, low-viscosity liquid). A system with no aggregates may be called a microfluid, and the system with aggregates a macrofluid (e.g., see Levenspiel, *Chemical Reaction Engineering*, 3d ed., Wiley, New York, 1999). In an ideal plug flow or in an ideal batch reactor, the segregated particles in each clump spend an equal time in the reactor, and therefore the behavior is no different from that of a microfluid that has individual molecules acting independently. The reactor performance is therefore unaffected by the degree of segregation, and the PFR or ideal batch model equations may be used to estimate performance. This is not the case for a CSTR where the performance equation for a microfluid is the same as that of an ideal CSTR, while that of a CSTR with segregated flow is not.

In segregated flow, the molecules travel as distinct groups. All molecules that enter the vessel together leave together. The groups are small enough that the RTD of the whole system is represented by a smooth curve. Each group of molecules reacts independently of any other group, that is, as a batch reactor. For a batch reactor with a power law kinetics,

$$\left(\frac{C}{C_0}\right)_{\text{batch}} = \begin{cases} e^{-kt} = e^{-k\bar{t}\theta_r} & \text{1st order} \\ \left[\dfrac{1}{1+(q-1)kC^{q-1}\bar{t}t_r}\right]^{\frac{1}{q-1}} & q \text{ order} \end{cases} \tag{19-45}$$

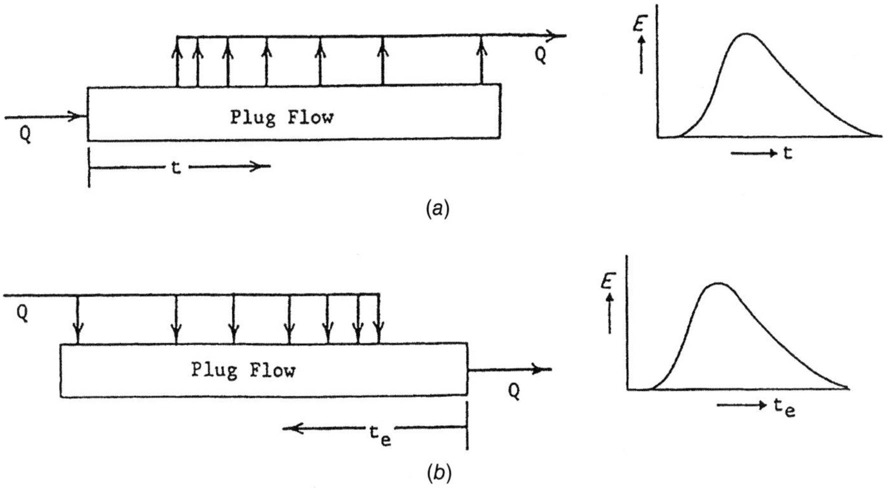

FIG. 19-10 Two limiting flow patterns with the same RTD. (*a*) Segregated flow. (*b*) Maximum mixedness flow.

For other rate equations, a numerical solution may be needed. The mean conversion of all the groups is the sum of the products of the individual conversions and their volume fractions of the total flow. Since the groups are small, the sum may be replaced by an integral. Thus,

$$\left(\frac{C}{C_0}\right)_{\text{segregated}} = \int_0^\infty \left(\frac{C}{C_0}\right)_{\text{batch}} E(t)\,dt = \int_0^\infty \left(\frac{C}{C_0}\right)_{\text{batch}} E(t_r)\,dt_r \qquad (19\text{-}46)$$

When a conversion and an RTD are known, a value of k may be estimated by regression so the segregated integral is equal to the known value. If a series of conversions are known at several residence times, the order of the reaction that matches the data may be estimated by regression. One has to realize, however, that the RTD may change with residence time due to flow pattern changes. Alternatively, for known intrinsic kinetics, a combination of ideal reactors that reasonably match both RTD and performance may be considered.

Early versus Late Mixing—Maximum Mixedness The concept of early versus late mixing may be illustrated using a plug flow reactor and an ideal CSTR in series. In the case of early mixing, the ideal CSTR precedes the plug flow reactor. In the case of late mixing, the plug flow reactor precedes the CSTR. Both early mixing and late mixing cases have the same RTD.

The concept of maximum mixedness and completely segregated flow is illustrated in Fig. 19-10. In maximum mixedness (or earliest possible mixing), the feed is intimately mixed with elements of fluid of different ages, for instance, using multiple side *inlets* at various points along a plug flow reactor. The amount and location of the inlet flows are designed to match the RTD. This means that each portion of fresh material is mixed with all the material that has the same life expectation, regardless of the actual residence time in the vessel up to the time of mixing. The life expectation under plug flow conditions is related to the distance remaining to be traveled before leaving the vessel. Segregated flow is represented as a plug flow reactor with multiple side *outlets* and, as indicated, has the same RTD as the maximum mixedness case. In segregated flow the mixing occurs only after each side stream leaves the reactor and the location and amount of side streams is designed to match the RTD.

These two mixing extremes—as late as possible and as early as possible—both having the same RTD, correspond to extremes of reactor performance.

The mathematical model for maximum mixedness has been provided by Zwietering [*Chem. Eng. Sci.* **11:** 1 (1959)].

$$\frac{dC}{dt} = r_c - \frac{E(t)}{1-F(t)}(C_0 - C) \qquad (19\text{-}47)$$

where r_c is the chemical reaction rate; for example, for an order q, $r_c = kC^q$. The preceding differential equation in dimensionless variable form (where $f = C/C_0$ and $t_r = t/\bar{t}$) becomes

$$\frac{df}{dt_r} = k\bar{t}\,C_0^{q-1} f^q - \frac{E(t_r)}{1-F(t_r)}(1-f) \qquad (19\text{-}48)$$

with boundary condition

$$\frac{df}{dt_r} = 0 \quad \text{for } t_r \Rightarrow \infty \qquad (19\text{-}49)$$

which makes

$$k\bar{t}\,C_0^{q-1} f_\infty^q - \frac{E(\infty)}{1-F(\infty)}(1-f_\infty) = 0 \qquad (19\text{-}50)$$

The conversion achieved in the vessel is obtained by the solution of the differential equation at the exit of the vessel where the life expectation is $t = 0$. The starting point for the integration is (f_∞, t_∞). When integrating numerically, however, the RTD becomes essentially 0 by the time t_r approaches 3 or 4. Accordingly, the integration interval is from $(f_\infty, t_r \le 3 \text{ or } 4)$ to $(f_{\text{effluent}}, t_r = 0)$ with f_∞ obtained from Eq. (19-50).

The conversion is a maximum in *segregated flow* and a minimum under *maximum mixedness* conditions, for a given RTD and reaction orders >1. A few comparisons are made in Fig. 19-11. In some ranges of the parameters n or r_c, the differences in reactor volume for a given conversion, when segregated or maximum mixedness flow is assumed, are substantial. If only the RTD is known, these two extremes bracket reactor performance. As a general trend, for reaction orders >1, conversion increases as maximum mixedness < late mixing of microfluids < segregated flow (and the opposite is the case for orders <1). Increased deviation from ideal plug flow increases the effect of segregation on conversion. At low conversion, the conversion is insensitive to the RTD and to the extent of segregation.

Novosad and Thyn [*Coll. Czech. Chem. Comm.* **31:** 3710–3720 (1966)] solved the maximum mixedness and segregated flow equations (fit with the Erlang model) numerically. There are few experimental confirmations of these mixing extremes. One study with a 50-gal stirred tank reactor found segregation at low agitation and was able to correlate complete mixing and maximum mixedness in terms of the power input and recirculation within the vessel [Worrell and Eagleton, *Can. J. Chem. Eng.* (December): 254–258 (1964)].

REACTION AND MIXING TIMES

Reactants may be premixed or fed directly into the reactor. To the extent that the kinetics are limiting (i.e., reaction rate is slow), the rate of mixing plays a minor role in determining conversion or selectivity. If the mixing rate of reactants is comparable to or slower than the reaction rate, however, mixing can have a significant impact.

The characteristic chemical reaction time t_r or characteristic time scale of the chemistry may be calculated from the reaction rate expression. For a single irreversible reaction,

$$t_r = \frac{C_0}{r(C_0, T_0)} \qquad (19\text{-}51)$$

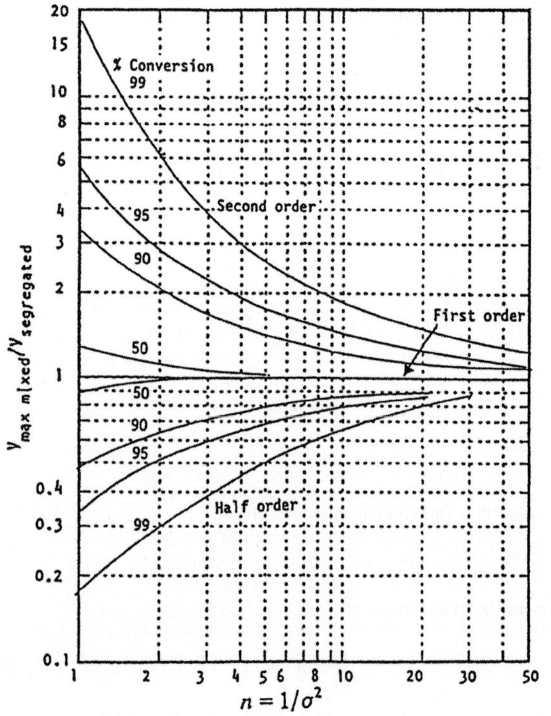

FIG. 19-11 Ratio of reactor volume for maximum mixedness and segregated flow models as a function of the variance (or n), for several reaction orders.

where C_0 is a characteristic concentration of the limiting reactant and T_0 is a characteristic temperature. For instance, characteristic values can be feed, initial, or reactor average values. For a first-order reaction, $t_r = 1/k$, where k (s^{-1}) is the rate constant.

Mixing may occur on several scales: on the reactor scale (macro), on the scale of dispersion from a feed nozzle or pipe (meso), and on a molecular level (micro). Examples of reactions where mixing is important include fast consecutive-parallel reactions, where reactant concentrations at the boundaries between zones rich in one or the other reactant being mixed can determine selectivity.

Much of the literature around mixing times has been developed around the mixing of two liquids in agitated stirred tanks. The macromixing time t_{ma} can be defined as the time for the concentration to settle within, say, ±2 percent of its final value (98 percent homogeneity). With a standard turbine in a baffled tank and Re $(= N D_a^2 \rho/\mu) > 5000$,

$$t_{ma} \cong \frac{4}{N}\left(\frac{D_t}{D_a}\right)^2\left(\frac{H}{D_t}\right) \tag{19-52}$$

where N is the stirrer speed, D_t is the tank diameter, D_a is the agitator diameter, and H is the height of the tank. In a case of a tank with an aspect ratio of unity and $D_a/D_t = \frac{1}{3}$, $Nt_{ma} \cong 36$. For a stirrer speed of 120 rpm, the macromixing time is 18 s.

The circulation time t_{cir} is the time to circulate the reactor contents once:

$$t_{cir} = \frac{V_r}{q} \tag{19-53}$$

where q is the flow induced by the impeller. The induced flow is about two times the direct discharge from the turbine, creating uncertainty in estimating q; t_{cir} is roughly one-fourth of the macromixing time for 98 percent homogeneity.

The micromixing time t_{mi} is the time required for equilibration of the smallest eddies by molecular diffusion, engulfment, and stretching. For liquid–liquid mixing, stretching and engulfment are limiting factors, and t_{mi} depends on the kinematic viscosity (μ/ρ) and the local rate of energy dissipation $\phi\bar{\varepsilon}$:

$$t_{mi} = 17\left(\frac{\mu/\rho}{\phi\bar{\varepsilon}}\right)^{1/2} \tag{19-54}$$

For a kinematic viscosity of 10^{-6} m^2/s and an energy dissipation of 1.0 W/kg, $t_{mi} = 0.017$ s. The local energy dissipation will vary greatly with position in the tank, with its greatest value near the tip of the impeller. Injection of reactant at the point of greatest turbulence minimizes t_{mi}.

The mesomixing time t_{me} is the time for "significant mixing" of an incoming jet of feed liquid with the surrounding fluid. A formula for estimating t_{me} is the time for turbulent diffusion to transport liquid over a distance equal to the feed pipe diameter d_0.

$$t_{me} \cong \frac{5.3\, d_0^2}{(\phi\bar{\varepsilon})^{1/3} D_a^{4/3}} \tag{19-55}$$

If the diameter of the pipe is proportional to the agitator diameter, t_{me} increases as $d_0^{2/3}$. Since t_{me} depends on the local energy dissipation, it is sensitive to location. Typically, t_{me} ($> t_{mi}$) is a fraction of a second or so.

A parameter used to diagnose mixing issues for reactive systems is the *Damköhler number* Da, which is the ratio of the mixing time to the reaction time, Da = t_{mixing}/t_r. Small Da numbers (Da << 1) indicate relatively rapid mixing compared to the reaction, so mixing is less important. In contrast, large Da numbers (Da >> 1) indicate a need to consider mixing issues. A more complete discussion of the topic is provided in Baldyga and Bourne (*Turbulent Mixing and Chemical Reactions*, Wiley, New York, 1998), and in Harriott (*Chemical Reactor Design*, Marcel Dekker, New York, 2003).

SINGLE-PHASE REACTORS

Section 7 presents the theory of reaction kinetics that deals with homogeneous reactions in batch and continuous equipment. Single-phase reactors typically contain a liquid or a gas with (or without) a homogeneous catalyst that is processed in a reactor at conditions required to complete the desired chemical transformation.

LIQUID PHASE

Batch reactions of single or miscible liquids are often done in stirred or pump-around tanks. Agitation is needed to mix multiple feeds and to enhance heat exchange with cooling (or heating) during the process. Topics that require special importance on an industrial scale are the quality of mixing in tanks and the residence time distribution in vessels where plug flow may be the goal. A special case is that of laminar and related flow distributions characteristic of nonnewtonian fluids, which often occurs in polymerization and bioreactors. The information about agitation and heat transfer in tanks is described in Secs. 11 and 18.

Homogeneous Catalysis A catalyst is a substance, usually used in small amounts relative to the reactants, that increases the rate of a reaction without being consumed in the process. Liquid-phase reactions are often conducted in the presence of homogeneous catalysts. Typically, homogeneous catalysts are ions or metal coordination complexes or enzymes in aqueous solution. The specific action of a particular metal complex can be altered by varying the ligands (or coordination number) of the complex or the oxidation state of the central metal atom. Some examples of homogeneous catalysts in industrial practice include hydrolysis of esters by hydronium (H_3O^+) or hydroxyl (OH^-) ions, hydroformylation of olefins using Rh or Co carbonyls, decomposition of hydrogen peroxide by ferrous ions, decomposition of nitramides catalyzed by acetate ion, inversion of sucrose by HCl, halogenation of acetone by H^+ and OH^-, and hydration of isobutene by acids. A characteristic of homogeneous catalysis is that, compared to solid catalysis, the reactions proceed under relatively mild conditions. A key issue associated with homogeneous catalysis is the difficulty of separating product and catalyst.

In stirred tanks, the power input to agitate the tank will depend on the physical properties of the liquid. In tubular reactors, the axial dispersion in empty tubes may be estimated [e.g., Wen in *Residence Time Distribution Theory in Chemical Engineering*, ed. Petho and Noble, Verlag Chemie, Weinheim, Ger., 1982] as

$$\frac{1}{\text{Pe}} = \frac{1}{(\text{Re})(\text{Sc})} + \frac{(\text{Re})(\text{Sc})}{192} \qquad 1 \le \text{Re} \le 2000 \text{ and } 0.2 \le \text{Sc} \le 1000 \tag{19-56}$$

$$\frac{1}{Pe} = \frac{3 \times 10^7}{(Re)^{2.1}} + \frac{1.35}{(Re)^{0.125}} \qquad Re \geq 2000$$

In a general case, the velocity may also be a function of radius. One such case is that of laminar flow, which is characterized by a parabolic velocity profile. The velocity at the wall is zero, while that at the centerline is twice the average velocity. In such cases, a momentum balance equation is solved along with the equations for heat and mass transfer, and each equation contains terms for the radial contribution. Laminar flow can be avoided by mixing over the cross section. For this purpose, *in-line* static mixers can be provided. For very viscous materials and pastes, screws of the type used for pumping and extrusion are used as reactors. When the temperature of the reactants changes during the course of the reaction (due to either the heat of reaction or the work required to keep the contents well mixed), material and energy balance equations have to be solved simultaneously.

Examples

- Crude oil is heated to temperatures at which it thermally cracks into gasoline and distillate products and lower-molecular-weight gases. This liquid cracking process is referred to as visbreaking. A schematic of the process and the effect of operating variables on performance is shown in Fig. 19-12.
- The Wacker process for the oxidation of ethylene to acetaldehyde with $PdCl_2/CuCl_2$ at 100°C (212°F) with 95 percent yield and 95 to 99 percent conversion per pass.
- The OXO process for higher alcohols: $CO + H_2 + C_3H_6 \Rightarrow$ *n*-butanal $\Rightarrow$ higher molecular weight alcohols. The catalyst is a rhodium triphenylphosphine coordination compound at 100°C (212°F), 30 atm (441 psi).
- Acetic acid from methanol by the Monsanto process, $CH_3OH + CO \Rightarrow CH_3COOH$, rhodium iodide catalyst, 3 atm (44 psi), 150°C (302°F), 99 percent selectivity.

See a review of industrial processes that employ homogeneous catalysts by Jennings, ed., *Selected Developments in Catalysis,* Blackwell Scientific, Oxford, UK, 1985.

GAS PHASE

There are few examples of industrial processes with pure gas-phase reactions. The most common and oldest example is combustion. Although termed *homogeneous,* most gas-phase reactions take place in contact with solids, either the vessel wall or particles as heat carriers. With inert solids, the only complication is with heat transfer. Several of these reactions are listed in Table 19-1. Whenever possible, liquefaction of gas-phase systems is considered in order to take advantage of the higher rates of liquid reactions, to use liquid homogeneous catalysts, or to keep equipment size down.

The specific type of equipment used for gas-phase reactions depends on the conditions required for undertaking the reaction. Examples of noncatalytic gas-phase reactions are shown in Fig. 19-13. In general, mixing of feed gases and temperature control are major process requirements. Gases are usually mixed by injecting one of the streams into the rest of the gases using a high-speed nozzle, as in the flame reactor (Fig. 19-13*d*).

Examples

- In the cracking of light hydrocarbons and naphtha to olefins, heat is supplied from combustion gases through tubes in fired heaters at 800°C (1472°F) and sufficiently above atmospheric pressure to overcome pressure drop. Superheated steam is injected to bring the temperature up quickly and retard coke deposition. The reaction time is 0.5 to 3.0 s, followed by rapid quenching. The total tube length of an industrial furnace may be more than 1000 m. Some other important gas-phase cracking processes include conversion of toluene to benzene, diphenyl to benzene, dicyclopentadiene to cyclopentadiene, and 1-butene to butadiene. Figure 19-13*a* shows a cracking furnace.
- The Wulf process for acetylene by pyrolysis of natural gas uses a heated brick checkerwork on a 4-min cycle of heating and reacting. Heat is transferred by direct contact with solids that have been preheated by combustion gases. The process is a cycle of alternate heating and reacting periods. The temperature play is 15°C (27°F), peak temperature is 1200°C (2192°F), and residence time is 0.1 s, of which 0.03 s is near the peak (Faith, Keyes, and Clark, *Industrial Chemicals,* vol. 27, Wiley, New York, 1975).
- The Wisconsin process for the fixation of nitrogen from air operates at 2200°C (3992°F), followed by extremely rapid quenching to freeze the small equilibrium content of nitrogen oxide that is made [Ermenc, *Chem. Eng. Prog.* **52:** 149 (1956)]. A pebble heater recirculates refractory pebbles continuously through heating and reaction zones. Such moving-bed units have been proposed for cracking to olefins but have been obsolesced like most moving-bed reactors.
- Acetylene may be produced from light hydrocarbons and naphthas by injecting inert combustion gases directly into the reacting stream in a flame reactor. Figure 19-13*a* and *d* shows two such devices; Fig. 19-13*e* shows a temperature profile (with reaction times in milliseconds).
- Oxidative pyrolysis of light hydrocarbons to acetylene is conducted in a special burner, at 0.001- to 0.01-s reaction time, peak at 1400°C (2552°F), followed by rapid quenching with oil or water. A portion of a combustible reactant is burned by adding a small amount of air or oxygen to generate the reaction temperatures needed.
- Chlorination reactions of methane and other hydrocarbons typically result in a mixture of products whose relative amounts can be controlled by varying the Cl/hydrocarbon ratio and recycling unwanted derivatives. For example, one can recycle the mono and di derivatives when only the tri and tetra derivatives are of value or keep the chlorine ratio low when emphasizing the lower derivatives. Temperatures are normally kept in the range of 230°C to 400°C (446°F to 752°F) to limit carbon formation but may be raised to 500°C (932°F) when favoring CCl_4. Exothermic processes use cooling through heat-transfer surfaces or cold shots. Shell-and-tube reactors with small-diameter tubes, towers with internal recirculation of gases, or multiple stages with intercooling may be used for these reactions.

SUPERCRITICAL CONDITIONS

At near-critical or supercritical conditions, a heterogeneous reaction mixture (e.g., of water, organic compounds, and oxygen) becomes homogeneous and has some liquid and gaseous properties. The rate of reaction may be considerably accelerated because of (1) the higher gas-phase diffusivity, (2) increase of concentration due to liquidlike density, (3) enhanced solubility, and (4) increase of the specific rate of reaction by pressure. The mole fraction solubility of naphthalene in ethylene at 35°C (95°F) goes from 0.004 at 20 atm (294 psi) to 0.02 at 100 atm (1470 psi) and 0.05 at 300 atm (4410 psi). High destructive efficiencies (above 99.99 percent) of complex organic pollutant compounds in water can be achieved with residence times of under 5 min at near-critical conditions. The critical properties of water are 374°C (705°F) and 218 atm (3205 psi).

We are not aware of industrial implementation of supercritical conditions in reactors. Two areas of potential interest are wastewater treatment (for instance, removal of phenol or organic compounds) and reduction of coke on refining catalysts by keeping heavy oil decomposition products in solution. A pertinent reference is by Kohnstam ("The Kinetic Effects of Pressure," in *Progress in Reaction Kinetics,* Pergamon, Oxford, UK, 1970). More recent reviews of research progress are by Bruno and Ely, eds., *Supercritical Fluid Technology,* CRC Press, Boca Raton, Fla., 1991; Kiran and Brennecke, eds., *Supercritical Engineering Science,* ACS, Washington, D.C., 1992.

POLYMERIZATION REACTORS

Polymerization reactors contain one or more phases. There are examples using solvents in which the reactants and products are in the liquid phase, the reactants are fed as a liquid (gas) but the products are solid, or the reactants are a slurry and the products are soluble. Phase transformations can occur, and polymers that form from the liquid phase may remain dissolved in the remaining monomer or solvent, or they may precipitate. Sometimes beads are formed and remain in suspension; sometimes emulsions form. In some processes, solid polymers precipitate from a gas phase into a fluidized bed containing product solids. Polymers are thought of as organic materials; however, inorganic polymers may be also synthesized (e.g., using crystallization and precipitation). Examples of inorganic polymers are zeolites.

The structure of the polymer determines its physical properties, e.g., crystallinity, refractive index, tensile strength, glass transition temperature (at which the specific volume changes slope), and processability. The average molecular weight can cover a wide range between 10^4 and 10^7. Given the change in molecular weight, the viscosity can change dramatically as conversion increases. For example, in styrene polymerization, the viscosity increases by a factor of 10^6 as conversion increases from 0 to 60 percent. Initiators of chain polymerization reactions have concentration as low as 10^{-8} g · mol/L, so they are highly sensitive to small concentrations of poisons and impurities. The reaction time can also vary. Reaction times for butadiene-styrene rubbers are 8 to 12 h; polyethylene molecules continue to grow for 30 min, whereas ethyl acrylate in 20 percent emulsion reacts in less than 1 min, so monomer must be added gradually to keep the temperature within limits. In some cases, the adiabatic temperature rise may be very high. For example, in the polymerization of ethylene, a high adiabatic temperature rise may lead to reactor safety issues by initiating runaway ethylene decomposition reactions. The reactor operating conditions have to be controlled such that the possibility of *ethylene decomposition* is eliminated.

Since it is impractical to fractionate the products and reformulate them into desirable ranges of molecular weights, immediate attainment of desired properties must be achieved through the correct choice of reactor type and operating conditions, notably of distributions of residence time and temperature. Reactor selection may be made on rational grounds, for historical reasons, or to obtain a proprietary position.

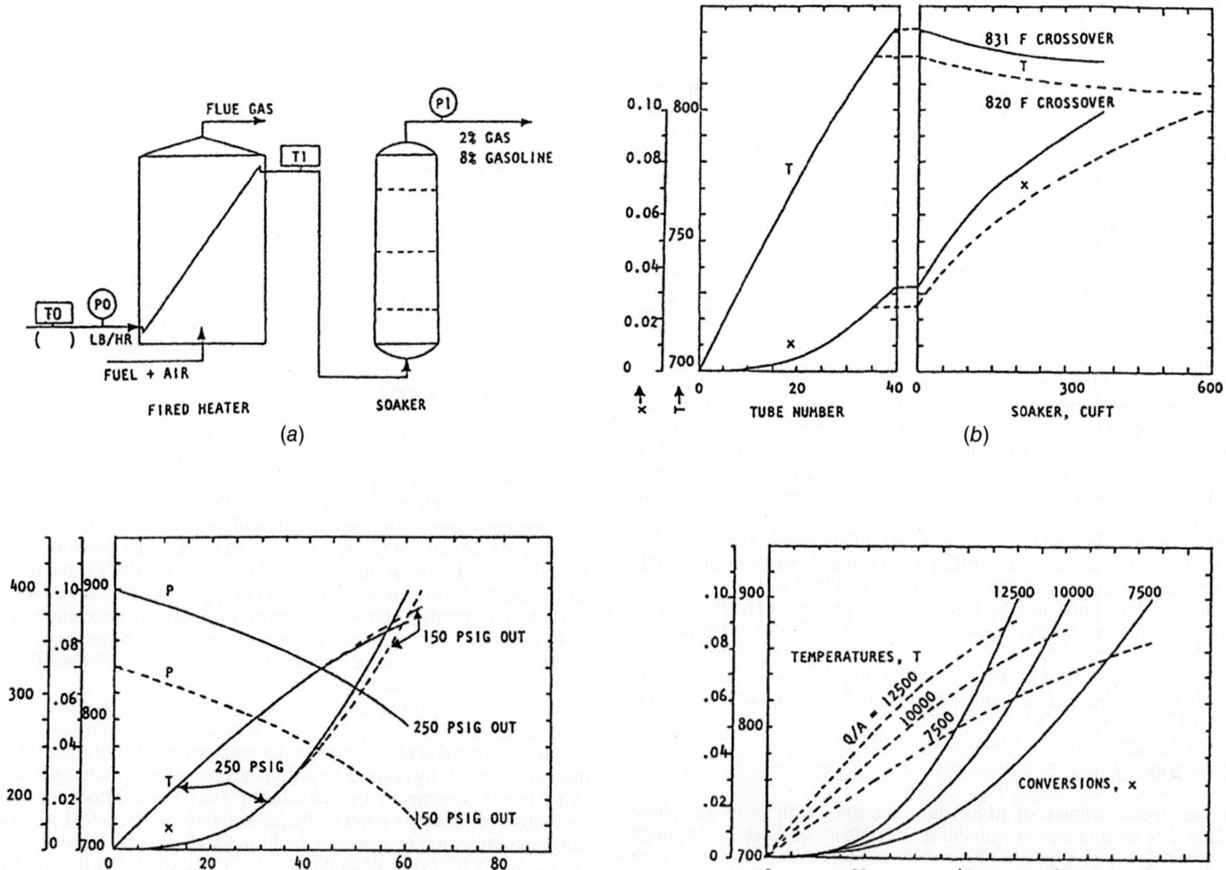

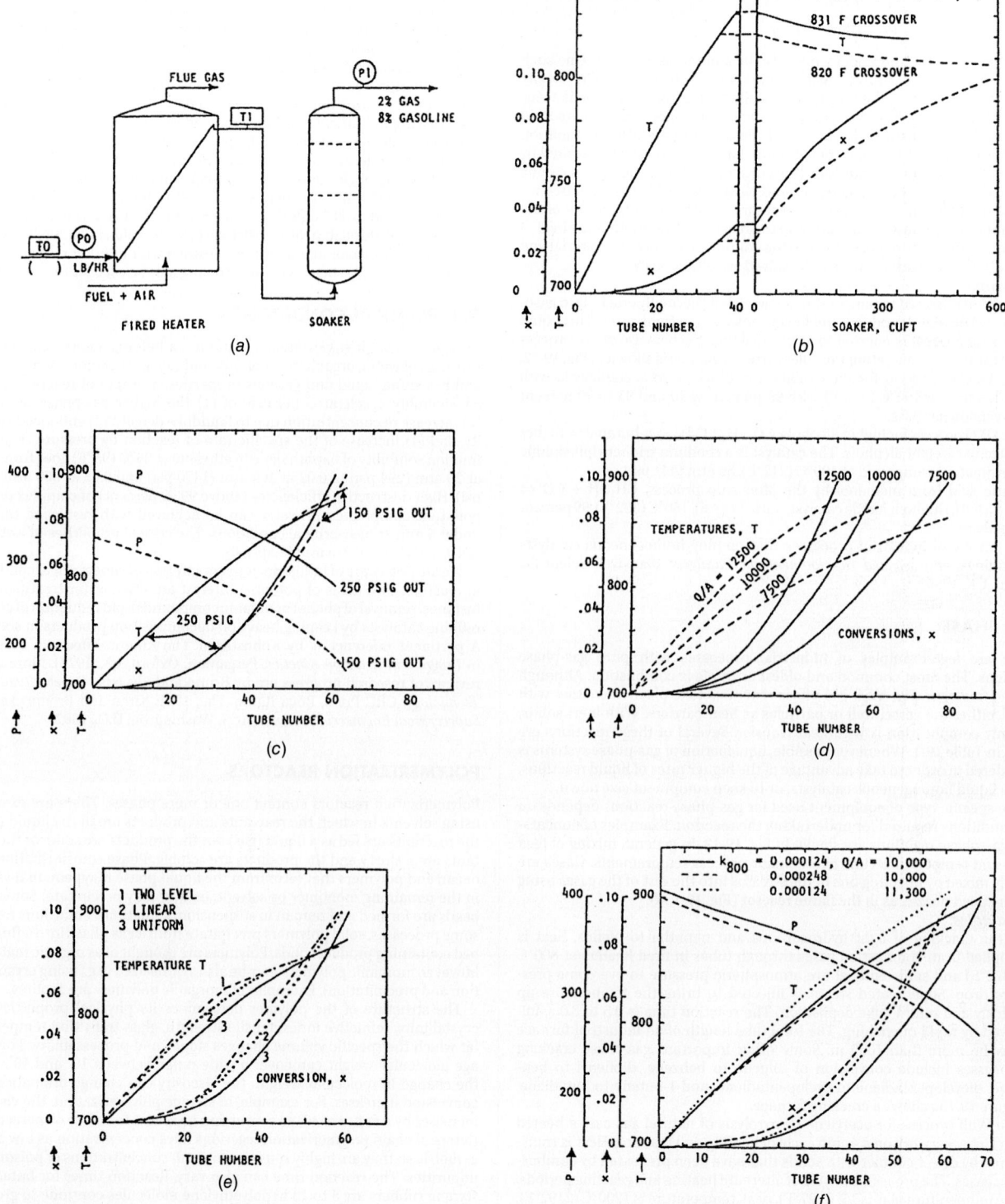

FIG. 19-12 (*a*) Visbreaking flow sketch, feed 160,000 lbm/h, $k_{800} = 0.000248$/s, tubes 5.05-in ID by 40 ft. (*b*) $Q/A = 10{,}000$ Btu/(ft$^2\cdot$h), $P_{out} = 250$ psig. (*c*) $Q/A = 10{,}000$ Btu/(ft$^2\cdot$h), $P_{out} = 150$ or 250 psig. (*d*) Three different heat fluxes, $P_{out} = 250$ psig. (*e*) Variation of heat flux, average 10,000 Btu/(ft$^2\cdot$h), $P_{out} = 250$ psig. (*f*) Halving the specific rate. T in °F. To convert psi to kPa, multiply by 6.895; ft to m, multiply by 0.3048; in to cm, multiply by 2.54.

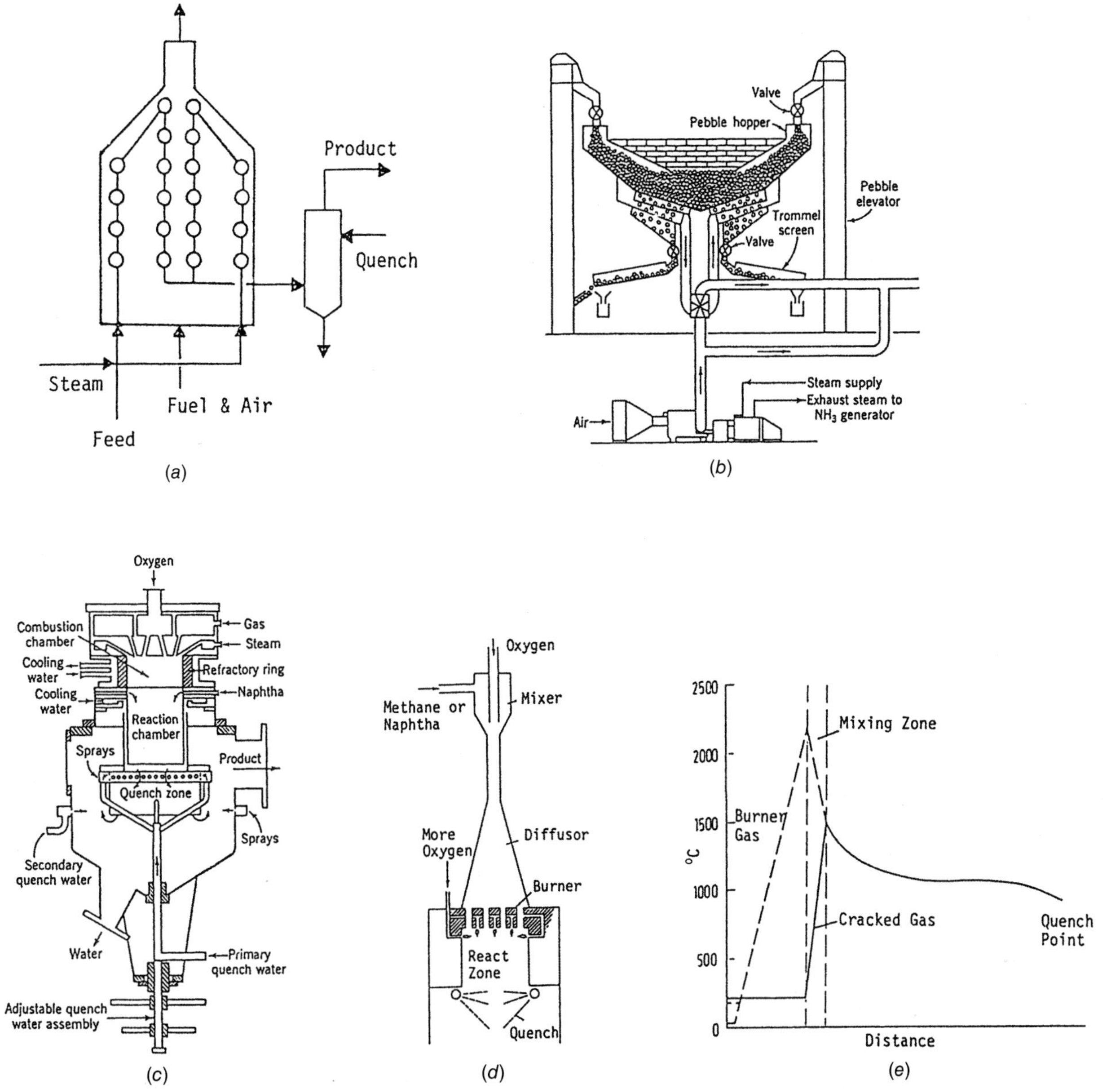

FIG. 19-13 Noncatalytic gas-phase reactions. (*a*) Steam cracking of light hydrocarbons in a tubular fired heater. (*b*) Pebble heater for the fixation of nitrogen from air. (*c*) Flame reactor for the production of acetylene from hydrocarbon gases or naphthas. [*Patton, Grubb, and Stephenson,* Pet. Ref. ***37****(11)*: *180* (*1958*).] (*d*) Flame reactor for acetylene from light hydrocarbons (*BASF*). (*e*) Temperature profiles in a flame reactor for acetylene (Ullmann Encyclopadie der Technischen Chemie, *vol. 3, Verlag Chemie, Weinheim, Ger., 1973, p. 335*).

Each reactor is designed based on the need for mass transfer, heat transfer, and reaction. Stirred batch (autoclave) and continuous tubular reactors are widely used because of their flexibility. In stirred tanks, ideal mixing is typically not achieved, wide variations in temperatures may result, and stagnant zones and bypassing may exist. Devices that counteract these unfavorable characteristics include inserts that cause radial mixing, scraping impellers, screw feeders, hollow-shaft impellers (with coolant flow through them), recirculation using internal and external draft tubes, and so on. The high viscosity of bulk and melt polymerization reactions is avoided with solution, bead, or emulsion polymerization, and more favorable RTDs are obtained. In tubular reactors, such as those for low-density polyethylene production, there are strong temperature gradients in the radial direction, and cooling may become an issue. These reactors are operated in a single phase, often with multiple catalyst injection points, and the reactor can be several miles in length. Examples of polymerization reactors are illustrated in Fig. 19-14. As previously indicated, some of these polymerization reactors are multiphase and these are handled in accordance with the subsequent discussion on multiphase reactors.

A number of terms unique to polymerization are discussed in Sec. 7. A general reference on polymerization is Rodriguez (*Principles of Polymer Systems,* McGraw-Hill, New York, 1989), and a reference guide on polymerization reactors is available by Gerrens [*German Chem. Eng.* **4:** 1–13 (1981); *ChemTech* pp. 380–383, 434–443] and Meyer and Keurentjes (*Handbook of Polymer Reaction Engineering,* Wiley VCH, New York, 2005).

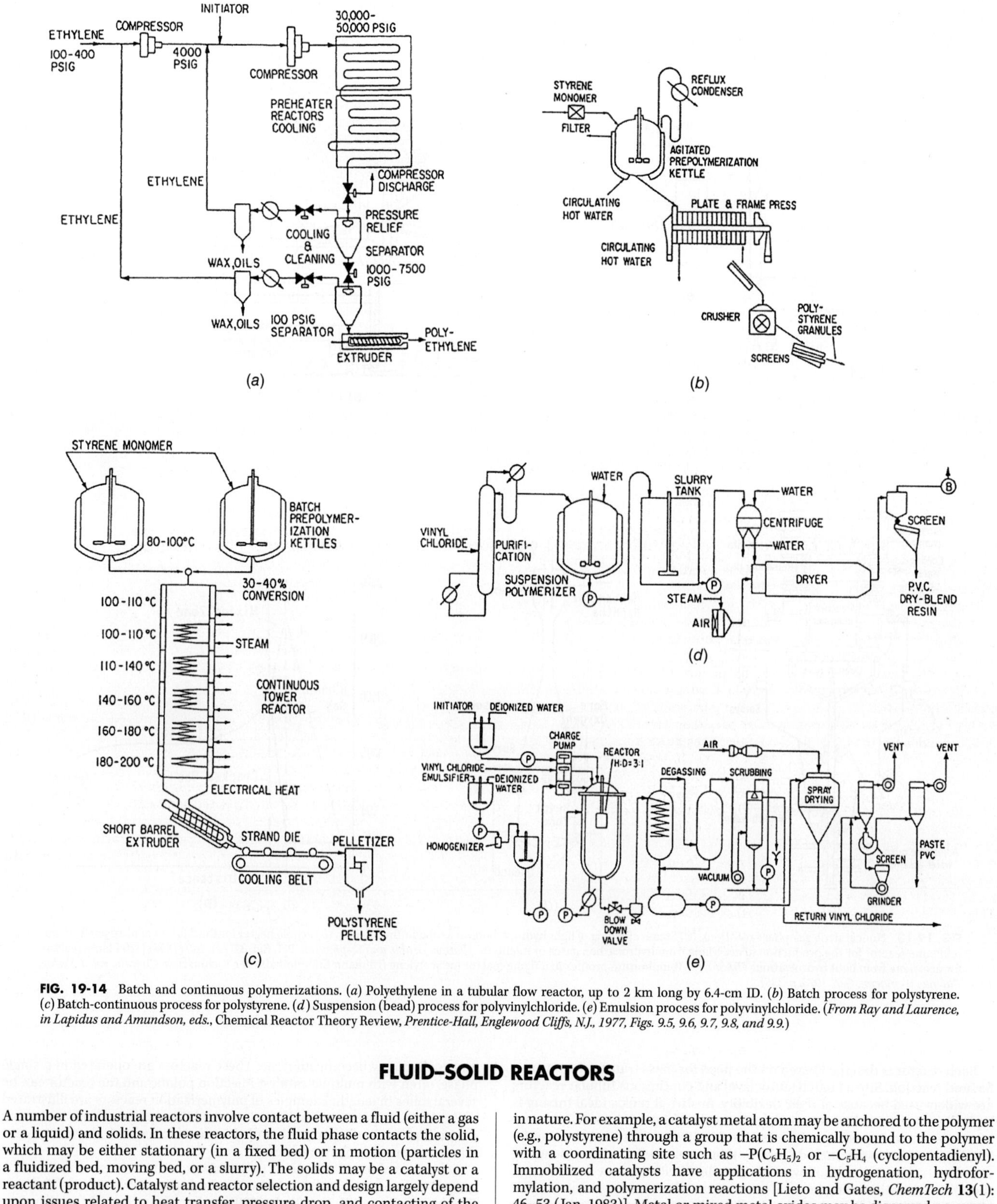

FIG. 19-14 Batch and continuous polymerizations. (*a*) Polyethylene in a tubular flow reactor, up to 2 km long by 6.4-cm ID. (*b*) Batch process for polystyrene. (*c*) Batch-continuous process for polystyrene. (*d*) Suspension (bead) process for polyvinylchloride. (*e*) Emulsion process for polyvinylchloride. (*From Ray and Laurence, in Lapidus and Amundson, eds.,* Chemical Reactor Theory Review, *Prentice-Hall, Englewood Cliffs, N.J., 1977, Figs. 9.5, 9.6, 9.7, 9.8, and 9.9.*)

FLUID-SOLID REACTORS

A number of industrial reactors involve contact between a fluid (either a gas or a liquid) and solids. In these reactors, the fluid phase contacts the solid, which may be either stationary (in a fixed bed) or in motion (particles in a fluidized bed, moving bed, or a slurry). The solids may be a catalyst or a reactant (product). Catalyst and reactor selection and design largely depend upon issues related to heat transfer, pressure drop, and contacting of the phases. In many cases, continuous regeneration or periodic replacement of deteriorated or deactivated catalyst may be needed.

HETEROGENEOUS CATALYSTS

Solid catalysts may have a homogeneous catalyst (or enzyme) or catalytic ingredients dispersed on a support. The support may be organic or inorganic in nature. For example, a catalyst metal atom may be anchored to the polymer (e.g., polystyrene) through a group that is chemically bound to the polymer with a coordinating site such as $-P(C_6H_5)_2$ or $-C_5H_4$ (cyclopentadienyl). Immobilized catalysts have applications in hydrogenation, hydroformylation, and polymerization reactions [Lieto and Gates, *ChemTech* **13**(1): 46–53 (Jan. 1983)]. Metal or mixed metal oxides may be dispersed on amorphous materials (such as carbon, silica, or alumina) or exchanged into the cages of a zeolite. Expensive catalytic metal ingredients, such as Pt or Pd, may be <1 percent of catalyst weight. Catalysts may be shaped as monoliths, shaped pellets, spheres, or powders. Some exceptions are bulk catalysts, such as platinum gauzes for the oxidation of ammonia to nitric oxides (in nitric acid production) and synthesis of hydrogen cyanide, which are in the form of several layers of fine-mesh catalyst gauze.

The catalyst support may either be inert or play a role in catalysis. Supports typically have a high internal surface area. Special shapes (e.g., trilobed particles) are often used to maximize the geometric surface area of the catalyst per reactor volume (and thereby increase the reaction rate per unit volume for diffusion-limited reactions) or to minimize pressure drop. Smaller particles may be used instead of shaped catalysts; however, the pressure drop increases and compressor costs become an issue. For fixed beds, the catalyst size range is 1 to 5 mm (0.04 to 0.197 in). In reactors where pressure drop is not an issue, such as fluidized and transport reactors, particle diameters can average less than 0.1 mm (0.0039 in). Smaller particles improve fluidization; however, they are entrained and have to be recovered. Very small particles may become cohesive and difficult to fluidize. In slurry beds the diameters can be from about 1.0 mm (0.039 in) down to 10 μm or less.

The support has an internal pore structure (i.e., pore volume and pore size distribution) that facilitates the transport of reactants (products) into (out of) the particle. Low pore volume and small pores limit the accessibility of the internal surface because of increased (intraparticle) diffusion resistance. Diffusion of products outward also is decreased, and this may cause product degradation or catalyst fouling within the catalyst particle. As discussed in Sec. 7, the effectiveness factor η is the ratio of the actual reaction rate to the rate in the absence of any intraparticle diffusion limitations. When the rate of reaction greatly exceeds the rate of diffusion, the effectiveness factor is low, and the internal volume of the catalyst pellet is not used effectively for catalysis. In such cases, expensive catalytic metals are best placed as a shell around the pellet. The rate of diffusion may be increased by optimizing the pore structure to provide larger pores (or macropores) that transport the reactants (products) into (out of) the pellet and smaller pores (micropores) that provide the internal surface area needed for effective catalyst dispersion. Micropores typically have volume-averaged diameters of 50 to 200 Å with macropore diameters of 1000 to 5000 Å. The pore volume and the pore size distribution within a porous support determine its surface area. The surface area of supports can range from 0.06 m^2/mL (18,300 ft^2/ft^3) to 600 m^2/mL (1.83×10^8 ft^2/ft^3) and above. Higher-pore-volume catalysts have higher diffusion rates at the expense of reduced crush strength and increased particle attrition.

The effective diffusion coefficient D_{eff} determines the rate of diffusion of reactants into the volume of the catalyst. The coefficient is determined by the nature of the diffusing species and the pore structure of the catalyst. It has been found to be directly proportional to the product of diffusivity and porosity, ε, and inversely proportional to the tortuosity, τ (which is empirically determined). In large pores of >1000 Å, where molecules collide with one another and the interaction with the pore walls is minimal, molecular (or bulk) diffusion is important. For pore diameters in the range of 50 to 200 Å, collision with the pore walls becomes more important, and this regime is called the Knudsen diffusion regime. In an extreme case where the size of the molecule is comparable to the size of the pore, the size and configuration of the pores themselves affect diffusivity. This happens when the diffusing molecule is very large (as in transporting large organometallitic molecules through catalyst pores in heavy oil hydrotreating) or the pore is very small (as in diffusion in zeolites), or both (e.g., see Sec. 7 for diffusion regimes). ε ranges from 0.1 to 0.5, and τ ranges from 1 to 7. In the absence of other information, a τ value of 3 to 4 may be used; however, it is best measured for the catalyst of interest. Expressions for estimating the effective diffusion coefficient are available in textbooks such as Satterfield (*Heterogeneous Catalysis in Practice,* McGraw-Hill, New York, 1991).

The theoretical basis for the effectiveness factor, η, in a porous catalyst has been discussed in Sec. 7. For example, for an isothermal first-order reaction

$$r_c = k\eta C_i \tag{19-57}$$

where C_i is the pellet external surface concentration (equal to the bulk concentration in absence of external mass-transfer limitation) of the reactant. As discussed previously, η is a function of the ratio of the rate of reaction to diffusion, also called the Thiele modulus ϕ. As the rate constant increases, η decreases and eventually reaches an asymptotic value (which depends on ϕ). Under these conditions, the effective rate constant $k\eta$ varies as $k^{1/2}$. The role of diffusion and reaction in porous catalysts, however, is more complicated in a case where heat effects are present. In addition to the mass conservation equation around the pellet, an energy balance equation is required. Two additional dimensionless parameters are needed for estimating an effectiveness factor:

$$\beta = -\frac{\Delta H_r D_{\text{eff}} C_0}{\lambda T_s} \quad \text{and} \quad \gamma = \frac{E}{RT_s} \tag{19-58}$$

where ΔH_r is the heat of reaction, λ is the thermal conductivity of the catalyst, E is the activation energy, and R is the universal gas constant. The dimensionless parameter β, known as the Prater number, is the ratio of the heat generation to heat conduction within the pellet and is a measure of the intraparticle temperature increase; γ is the dimensionless activation energy for the reaction. For an exothermic reaction, the temperature inside the catalyst pellet is greater than or equal to the surface temperature. The maximum steady-state temperature inside the pellet is $T_s(1+\beta)$. Figure 19-15 is one of several cases examined by Weisz and Hicks for a first-order reaction in an adiabatic catalyst pellet [*Chem. Eng. Sci.* **17:** 263 (1962)]. Although this predicts some very large values of η in some ranges of the parameters, these values are often not realized in commercial reactors (see Table 19-5). The modified Lewis number, defined as $Lw' = \lambda_s/\rho_s C_{ps} D_{\text{eff}}$, can determine the transient temperature inside the pellet, which can be much larger than the steady-state temperature.

The concept of an effectiveness factor is useful in estimating the reaction rate per catalyst pellet volume (or mass). It is, however, mainly useful in cases where the reaction rate is a simple function of reactant concentration. When there are complex reaction pathways, the concept of effectiveness factor is no longer easily applicable, and species and energy balance equations inside the particle may have to be solved to obtain the reaction rates per unit volume of catalyst. Dumesic et al. (*The Microkinetics of Heterogeneous Catalysis,* American Chemical Society, Washington, D.C., 1993) use microkinetic analysis to elucidate reaction pathways of several commercial catalysts.

Another complication is the fact that Fig. 19-15 was developed for the constant-concentration boundary condition, $C|_{r=R} = C_0$. In a more general case, external mass-transfer limitations will need to be included,

$$k_m a(C_0 - C_i) = r_c(C_i) = k\eta C_i \tag{19-59}$$

where k_m is the external mass-transfer coefficient obtained from literature correlations and a is the external surface area per unit pellet volume. Equation (19-59) will have to be solved for C_i, the concentration of the reactant on the external surface of the catalyst, so that the rate per pellet can be obtained. The reaction rate per unit reactor volume then becomes $r_c(1-\varepsilon_b)$, where ε_b is the bed void fraction.

A further complication is that catalyst activity declines with time. Catalysts may deactivate chemically (via poisons and masking agents), thermally (via support sintering), or mechanically (through attrition). Commercial catalyst life can range from a second to several years. For example, in refinery fluid catalytic cracking, the catalyst may lose most of its activity in less than 10 s, and a transport bed reactor coupled with a fluidized-bed regenerator is used to regenerate catalyst. In contrast, a refinery hydroprocessing catalyst deactivates very slowly, and a fixed-bed reactor may be used without catalyst replacement for one or more years. The deactivation rate expression may often be inferred from aging experiments undertaken in a pilot-plant under conditions of constant temperature or conversion. Since accelerated-aging experiments are often difficult (especially when the concentration of reactant or products affects the deactivation rate), the volume of catalyst required to provide the guaranteed performance between regeneration cycles requires good basic data and experience. The literature describes approaches aimed at managing deactivation. In the case of platinum reforming with fixed beds, a large recycle of hydrogen prevents coke deposition, while a high temperature compensates for the retarding effect of hydrogen on this essentially dehydrogenating process. Fluidized beds are largely isothermal and can be designed for continuous regeneration; however, they are more difficult to operate, require provisions for dust recovery, suffer from backmixing, and are more expensive. Catalyst deactivation mechanisms and kinetics are discussed in detail in Sec. 7.

A catalyst for a particular chemical transformation is selected using knowledge of similar chemistry and some level of empirical experimentation. Solid catalysts are widely used due to lower cost and ease of separation from the reaction medium. Their drawbacks include a possible lack of specificity and deactivation that can require reactor shutdown for catalyst regeneration or replacement.

There are a number of useful books on catalysis. Information on catalysts and processes is presented by Thomas (*Catalytic Processes and Proven Catalysts,* Academic Press, New York, 1970), Pines (*Chemistry of Catalytic Conversions of Hydrocarbons,* Academic Press, New York, 1981), Gates et al. (*Chemistry of Catalytic Processes,* McGraw-Hill, New York, 1979), Matar et al. (*Catalysis in Petrochemical Processes,* Kluwer Academic Publishers, Dordrecht, Nl., 1989), and Satterfield (*Heterogeneous Catalysis in IndustrialPractice,* McGraw-Hill, New York, 1991). The books by Thomas (*Catalytic Processes and Proven Catalysts,* Academic Press, New York, 1970), Butt and Petersen (*Activation, Deactivation and Poisoning of Catalyst,* Academic Press, New York, 1988), and Delmon and Froment (*Catalyst Deactivation,* Elsevier, Amsterdam, 1980) provide several examples of catalyst deactivation. Catalyst design is

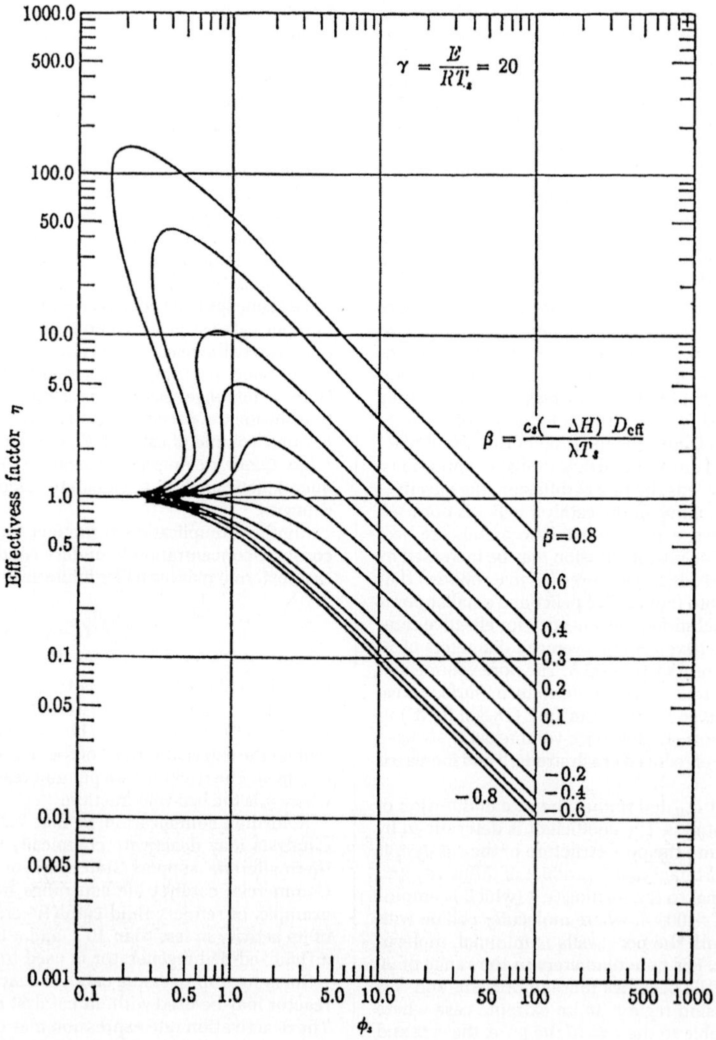

FIG. 19-15 Effectiveness factors versus Thiele modulus for a first-order reaction in spheres under adiabatic conditions. [*Weisz and Hicks,* Chem. Eng. Sci. **17:** 265 (1962).]

discussed by Trimm (*Design of Industrial Catalysts,* Elsevier, Amsterdam, 1980), Hegedus et al. (*Catalyst Design Progress and Perspectives,* Wiley, New York, 1987), and Becker and Pereira (*Catalyst Design,* Marcel Dekker, New York, 1993). A thorough review of catalytic reactions and catalysts arranged according to the periodic table is in a series by Roiter, ed. [*Handbook of Catalytic Properties of Substances* (in Russian), Academy of Sciences, Ukrainian SSR, 1968]. Stiles (*Catalyst Manufacture,* Dekker, New York, 1983) discusses catalyst manufacture.

CATALYTIC REACTORS

Due to the considerations we have noted, reactor selection will depend on the type of catalyst chosen and its activity, selectivity, and deactivation behavior. Some reactors with solid catalysts are represented in Figs. 19-16, 19-17, and 19-18.

Wire Gauzes Wire screens are used for very fast catalytic reactions or reactions that require a bulk noble metal surface for reaction. The nature

TABLE 19-5 Parameters of Some Exothermic Catalytic Reactions

Reaction	β	γ	γβ	Lw′	ϕ
NH_3 synthesis	0.000061	29.4	0.0018	0.00026	1.2
Synthesis of higher alcohols from CO and H_2	0.00085	28.4	0.024	0.00020	—
Oxidation of CH_3OH to CH_2O	0.0109	16.0	0.175	0.0015	1.1
Synthesis of vinyl chloride from acetylene and HCl	0.25	6.5	1.65	0.1	0.27
Hydrogenation of ethylene	0.066	23–27	2.7–1	0.11	0.2–2.8
Oxidation of H_2	0.10	6.75–7.52	0.21–2.3	0.036	0.8–2.0
Oxidation of ethylene to ethylenoxide	0.13	13.4	1.76	0.065	0.08
Dissociation of N_2O	0.64	22.0	1.0–2.0	—	1–5
Hydrogenation of benzene	0.12	14–16	1.7–2.0	0.006	0.05–1.9
Oxidation of SO_2	0.012	14.8	0.175	0.0415	0.9

SOURCE: After Hlavacek, Kubicek, and Marek, *J. Catal.,* **15,** 17, 31 (1969).

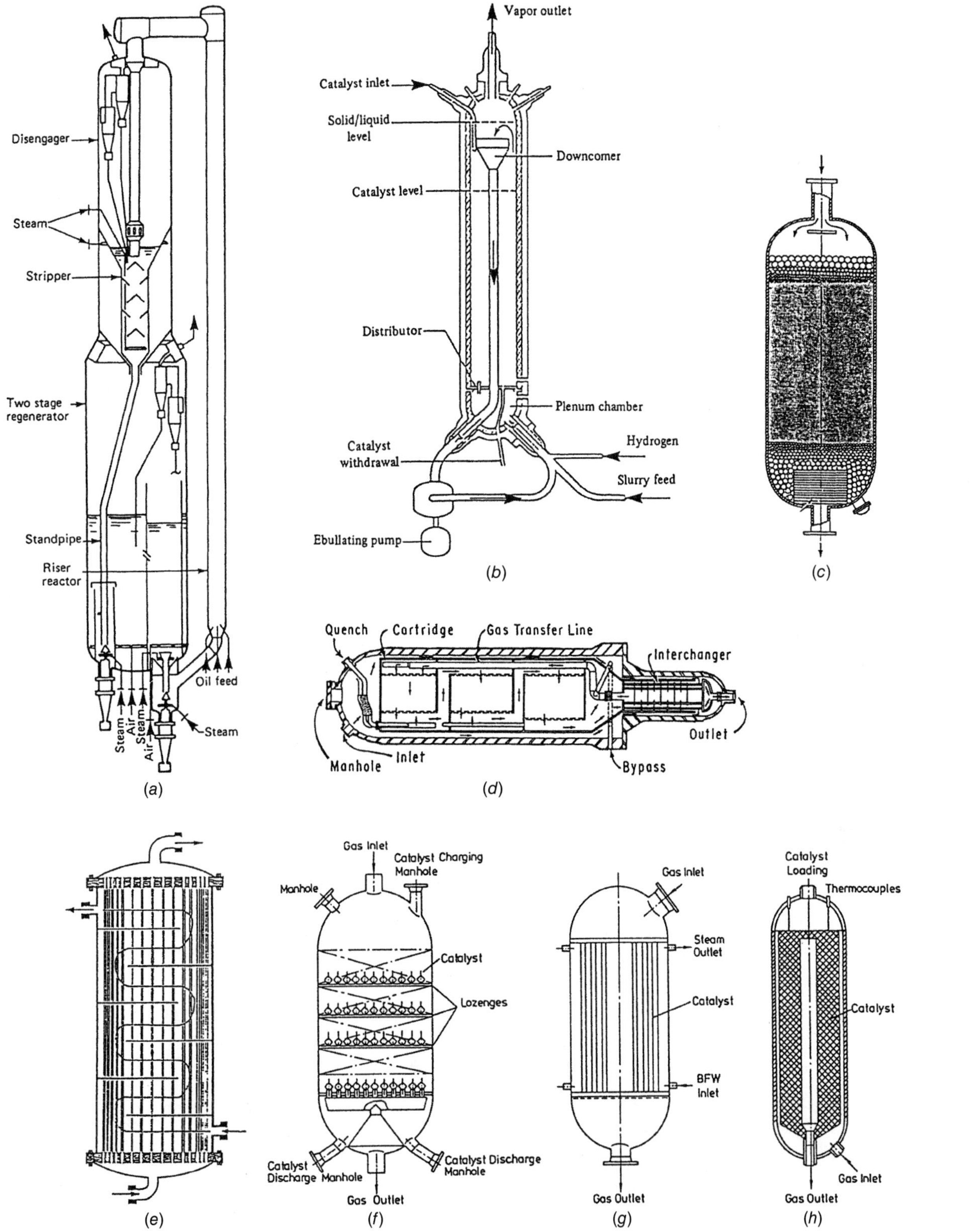

FIG. 19-16 Reactors with solid catalysts. (*a*) Fluid catalytic cracking riser-regenerator with fluidized zeolite catalyst, 540°C. (*b*) Ebullating fluidized bed for conversion of heavy stocks to gas and light oils. (*c*) Fixed-bed unit with support and hold-down zones of larger spheres. (*d*) Horizontal ammonia synthesizer, 26 m long without the exchanger (*M W Kellogg Co.*). (*e*) Shell-and-tube vessel for hydrogenation of crotonaldehyde has 4000 packed tubes, 30-mm ID, 10.7 m long [*after Berty, in Leach, ed.*, Applied Industrial Catalysis, *vol. 1, Academic Press, 1983, p. 51*]. (*f*), (*g*), (*h*) Methanol synthesizers, 50 to 100 atm, 230°C to 300°C, Cu catalyst; ICI quench type, Lurgi tubular, Haldor Topsoe radial flow (*Marschner and Moeller, in Leach, loc. cit.*). To convert atm to kPa, multipy by 101.3.

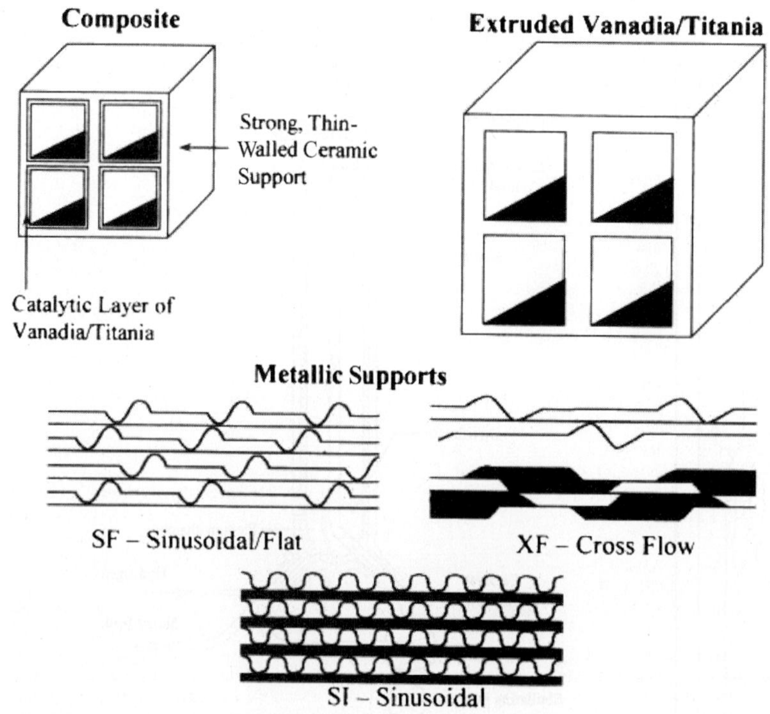

FIG. 19-17 Types of monolith catalysts. (*Fig. 12.9 in Heck, Farrauto, and Gulati,* Catalytic Air Pollution Control: Commercial Technology, *Wiley-Interscience, New York, 2002.*)

and morphology of the gauze or the finely divided catalyst are important in reactor design. Reaction temperatures are typically high, and the residence times are on the order of milliseconds.

Since noble metals are expensive, the catalyst cost is typically high. The physical properties of the gauze pack are important to determine performance, selectivity, and catalyst replacement strategy. The gauze is typically mounted over the top of a heat exchanger tube sheet or over porous ceramic bricks that are laid over the tube sheet. The gauze pack may be covered with a ceramic blanket to minimize radiation losses. From a modeling standpoint, the external surface area per gauze volume and the external mass-transfer coefficient for each component are important parameters, and the reaction rate per unit volume of catalyst may be limited by the rate of external mass transfer. The reaction rate can then be included into a corresponding PFR or dispersion model to obtain estimates of conversion and selectivity.

Examples

- In ammonia oxidation, a 10 percent NH_3 concentration in air is oxidized by flow through a fine-gauze catalyst made of 2 to 10 percent Rh in Pt, 10 to 30 layers, 0.075–mm-diameter (0.0030-in) wire. Contact time is 0.0003 s at 750°C (1382°F) and 7 atm (103 psi). The product is rapidly quenched to prevent further undesirable reactions.
- In hydrogen cyanide synthesis using the Andrussow process, air, methane, and ammonia are fed over 15 to 50 layers of noble metal gauze at 1050°C to 1150°C at near atmospheric pressure.

Monolith Catalysts For fast reactions that may require a slightly higher residence time than gauzes or that do not benefit from the bulk noble metal gauze structure, monoliths may be used. Most often, the monolith catalyst is an extruded ceramic honeycomb structure that has discrete channels that traverse its length. The catalytic ingredients may be dispersed on a high surface area support and coated on an inert honeycomb. In some cases, the catalyst paste itself may be extruded into a monolith catalyst. Monoliths may also be made of metallic supports. Stainless steel plates (or wire mesh) with ridges may be coated with catalysts and stacked one against the other in a reactor. Corrugated stainless steel layers may alternate in between flat sheets to form the structure. A variant is a stainless steel sheet that is corrugated in a herringbone pattern, coated with catalyst and then rolled

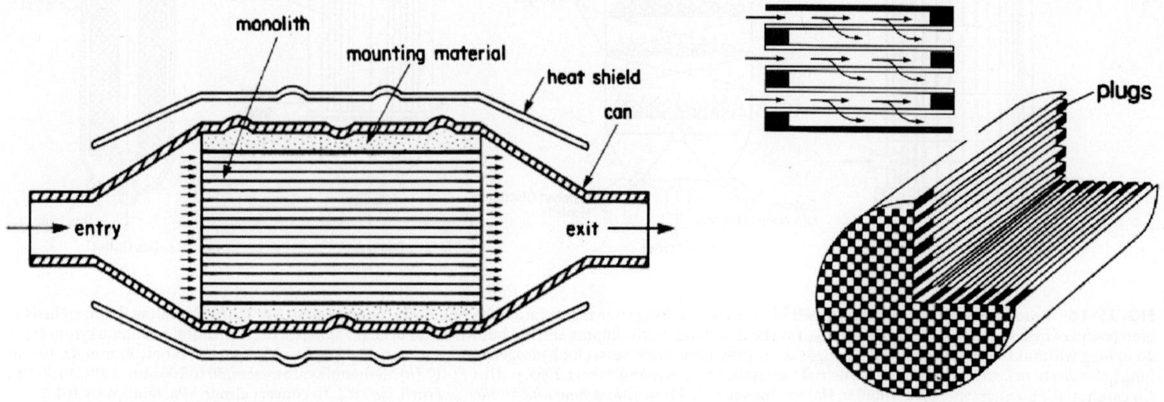

FIG. 19-18 Monolith catalysts. (*a*) Schematic of an automobile catalytic converter for the three-way removal of CO, hydrocarbons, and NO_x. (*b*) Schematic of a diesel trap. (*Figs. 7.10 and 9.6 in Heck, Farrauto, and Gulati,* Catalytic Air Pollution Control: Commercial Technology, *Wiley-Interscience, New York, 2002.*)

(or folded back and forth onto itself) into a reactor module. Examples of cross sections of the types of monoliths used in industry are shown in Fig. 19-17.

The thickness of monolith walls is adjusted according to the materials of construction (ceramic honeycombs have thicker walls to provide mechanical strength). The size of the channels is selected according to the application. For example, for particulate-laden gases, a larger-channel-size ceramic monolith and a higher linear velocity allow the particles to pass through the catalyst without plugging the channel. In contrast, for feed that does not contain particles, smaller-channel monoliths may be used. The cell density of the monolith may vary between 9 and 600 cells per square inch.

A monolith catalyst has a much higher void fraction (between 65 and 91 percent) than does a packed bed (which is between 36 and 45 percent). In the case of small channels, monoliths have a high geometric surface area per unit volume and may be preferred for mass-transfer-limited reactions. The higher void fraction provides the monolith catalyst with a pressure drop advantage compared to fixed beds.

A schematic of a monolith catalyst is shown in Fig. 19-18*a*. In cases where low pressure drop is important, such as for CO oxidation in cogeneration power plant exhausts, monolith catalyst panels may be stacked to form a thin (3- to 4-in-thick) wall. The other dimensions of the wall can be on the order of 35 × 40 ft. CO conversion is over 90 percent, with a pressure drop across the catalyst of 1.5 in of water. Alternatively, the monolith may be used as a catalyst and filter, as is the case for a diesel particulate filter. In this case, monolith channels are blocked, and the exhaust gases from a diesel truck are forced through the walls (Fig. 19-18*b*). The filter is a critical component in a continuous regenerable trap. NO in the exhaust gases is oxidized into NO_2 that reacts with the soot trapped in the walls of the filter to regenerate it *in situ*.

Modeling considerations for monoliths are similar to those of gauze catalysts; however, since the flow and temperature in each channel may be assumed to be identical to those in the next channel, the solution for a single channel may reflect the performance of the reactor. For an application in which the reaction rate is mass transfer limited, the reactant concentration at the wall of the catalyst is much lower than in the bulk and may be neglected. In such a case, the fractional conversion ξ is

$$\xi = 1 - e^{-k_m at} = 1 - \exp\left(-\frac{\text{Sh}\, aL}{\text{Sc Re}}\right) \tag{19-60}$$

where Sh $(= k_m d_{ch}/D)$ is the Sherwood number, Sc $(= \mu/\rho D)$ is the Schmidt number, and Re $(= u d_{ch}\rho/\mu)$ is the channel Reynolds number; a is the geometric surface area per unit volume of monolith. A number of correlations for Sh are available for various types of monoliths. For example, in the case of extruded ceramic monoliths, a correlation for estimating the external mass-transfer coefficient is provided by Uberoi and Pereira [*Ind. Eng. Chem. Res.* **35:** 113–116 (1996)]:

$$\text{Sh} = 2.696\left(1 + 0.139\,\text{Sc Re}\frac{d}{L}\right)^{0.81} \tag{19-61}$$

Since typical monolith catalysts have a thin coating of catalytic ingredients on the channel walls, they can be susceptible to poisoning. The various mechanisms for catalyst poisoning have been discussed in Sec. 7. The nature and shape of a monolith light-off curve for a facile hydrocarbon oxidation often indicate the poisoning mechanism, as shown in Fig. 19-19. The figure shows the light-off curve for a fresh catalyst. A reduction in the number of active sites (due to either poisoning or sintering of the catalytic metal) results in movement of the curve to the right. In contrast, when the pores within the catalyst become plugged with reactants or products (such as coke), the light-off curve shifts to the right and downward. In the case of deactivation due to masking, the active sites are covered with masking agents that may also plug the pores (such as in the case of silica deposition), resulting in more severe deactivation. Understanding the root cause of deactivation may allow for the design of improved catalysts, contaminant guard beds, catalyst regeneration procedures, and catalyst replacement protocols.

A good reference on monolith applications is by Heck, Farrauto, and Gulati (*Catalytic Air Pollution Control: Commercial Technology*, Wiley-Interscience, New York, 2002).

Examples

- For the control of carbon monoxide, hydrocarbon, and nitrogen oxide emissions from automobiles, oval-shaped extruded cordierite or metal monolith catalysts are wrapped in ceramic wool and placed inside a stainless steel casing (Fig. 19-18*a*). The catalytic metals are Pt-Rh or Pd-Rh, or combinations. Cell sizes typically range between 400 and 600 cells per square inch. The catalysts achieve over 90 percent reduction in all three pollutants.
- Monolith catalysts are used for the control of carbon monoxide and hydrocarbon (known as volatile organic compounds or VOCs) emissions from chemical plants and cogeneration facilities. In this case, square bricks are stacked on top of one another in a wall perpendicular to the flow of exhaust gases at the appropriate temperature location within the heat recovery boiler. The size of the brick can vary from 6 in (ceramic) to 21 in (metal). Pt and Pd catalysts are used at operating temperatures between 600°F and 1200°F. Cell sizes typically range between 100 and 400 cells per square inch. Typical pressure drop requirements for monoliths are less than 2 in of water.
- *Selective catalytic reduction* (SCR) catalysts are used for controlling nitrogen oxide emissions from power plants. The reducing agent is ammonia, and the active ingredients are $V_2O_5/WO_3/TiO_2$. Operating temperatures are 300°C to 450°C. Cell sizes vary between 9 and 50 cells per square inch. The paper by Beeckman and Hegedus [*Ind. Chem. Eng. Res.* **30:** 969 (1991)]) is a good reaction engineering reference on SCR catalysts.

Fixed Beds A fixed-bed reactor typically is a cylindrical vessel that is uniformly packed with catalyst pellets. Nonuniform packing of catalyst may cause channeling that could lead to poor heat transfer, poor conversion, and catalyst deactivation due to hot spots. The bed is loaded by pouring and manually packing the catalyst or by sock loading. As discussed earlier, catalysts may be regular or shaped porous supports, uniformly impregnated with the catalytic ingredient or containing a thin external shell of catalyst. Catalyst pellet sizes usually are in the range of 0.1 to 1.0 cm (0.039 to 0.39 in).

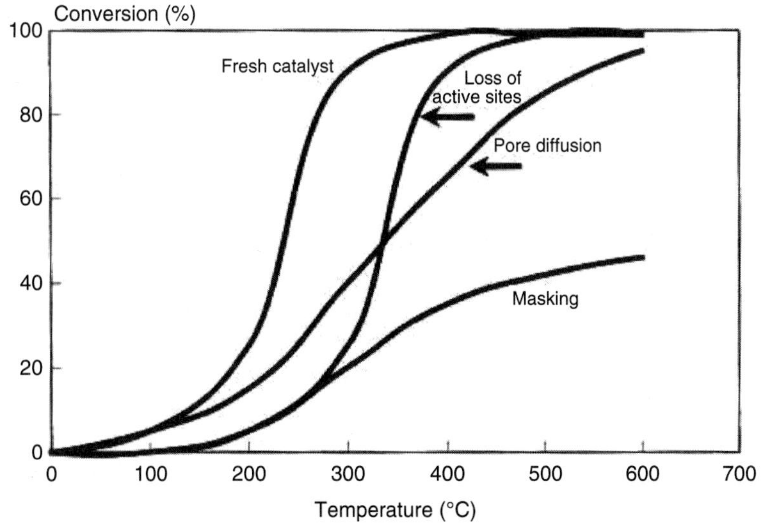

FIG. 19-19 Relative changes in conversion versus temperature behavior for various deactivation models. (*Fig. 5.4 in Heck, Farrauto, and Gulati,* Catalytic Air Pollution Control: Commercial Technology, *Wiley-Interscience, New York, 2002.*)

Packed-bed reactors are easy to design and operate. The reactor typically contains a manhole for vessel entry and openings at the top and bottom for loading and unloading catalyst. A metal support grid is placed near the bottom, and screens are placed over the grid to support the catalyst and prevent the particles from passing through. Typically, inert ceramic balls are placed above and below the catalyst bed to distribute the feed uniformly and to prevent the catalyst from passing through. One has to guard the bed from sudden pressure surges because they can disturb the packing and cause maldistribution and bypassing of feed.

As discussed earlier, heat management is an important issue in the design of fixed-bed reactors. A series of adiabatic fixed beds with interbed cooling (heating) may be used. For very highly exothermic (endothermic) reactions, a multitubular reactor with catalyst packed inside the tubes and cooling (heating) fluids on the shell side may be used. The tube diameter is typically greater than eight times the diameter of the pellets (to minimize flow channeling), and the length is limited by allowable pressure drop. The heat transfer required per volume of catalyst may impose an upper limit on diameter as well. Multitubular reactors require special procedures for catalyst loading that charge the same amount of catalyst to each tube at a definite rate to ensure uniform loading, which in turn ensures uniform flow distribution from the common header. After filling, each tube is checked for pressure drop. In addition to the high surface area for heat transfer/volume, the advantage of a multitubular fixed-bed reactor is its easy scalability. A bench-scale unit can be a full-size single tube, a pilot plant can be several dozen tubes, and a large-scale commercial reactor can have thousands of tubes. Disadvantages include high cost and a limit on maximum size (tube length and diameter, and number of tubes).

As discussed in Sec. 7, the intrinsic reaction rate and the reaction rate per unit volume of reactor are obtained based on laboratory experiments. The kinetics are incorporated into the corresponding reactor model to estimate the required volume to achieve the desired conversion for the required throughput. The acceptable pressure drop across the reactor often can determine the reactor aspect ratio. The pressure drop may be estimated by using the Ergun equation

$$\frac{\Delta P}{L} = \frac{150\,u_s\mu}{(\varphi_s d_p)^2}\frac{(1-\varepsilon_b)^2}{\varepsilon_b^3} + \frac{1.75\rho u_s^2}{\varphi_s d_p}\left(\frac{1-\varepsilon_b}{\varepsilon_b^3}\right) \tag{19-62}$$

where u_s is the superficial velocity, ε_b is the bed void fraction, φ_s is the shape factor, and d_p is the particle diameter. Correlations that provide estimates for the heat-transfer and mass-transport properties are available in the literature. For example, if the dispersion model is used to simulate concentration and temperature profiles along the reactor, the axial dispersion coefficient may be estimated from Wen and Fan, *Models For Flow Systems and Chemical Reactors*, Marcel Dekker, New York, 1975:

$$\frac{1}{\text{Pe}} = \frac{0.3}{\text{Re}\,\text{Sc}} + \frac{0.5}{1+3.8(\text{Re}\,\text{Sc})^{-1}} \qquad 0.008 \le \text{Re} \le 400 \quad \text{and} \quad 0.28 \le \text{Sc} \le 2.2 \tag{19-63}$$

where $\text{Pe} = d_p u_s/(\varepsilon_b D_e)$, $\text{Re} = d_p \rho u_s/\mu$, u_s is the superficial velocity, and d_p is the particle diameter.

Mathematical Models Catalytic packed-bed reactors are used for exothermic (e.g., hydrogenations, Fischer-Tropsch synthesis, oxidations) and endothermic (e.g., steam reforming) reactions. The two primary modes of heat management are (1) adiabatic operation, usually in a single or series of packed zones, the latter with interstage cooling or heating, and (2) multitubular reactors with cooling (e.g., shell-and-tube heat exchange with a coolant) or heating (e.g., locating the tubes in a furnace with heat supplied by combustion of a fuel). Other more complex schemes can include heat exchange between the feed and the effluent, or reverse flow operation, and these are discussed in the multifunctional reactors section.

The mechanism for heat transfer includes the following steps: (1) conduction in the catalyst particle; (2) convection from the particle to the gas phase; (3) conduction at contact points between particles; (4) convection between the gas and vessel wall; (5) radiation heat transfer between the particles, the gas, and the vessel wall; (6) conduction in the wall; and (7) convection to the coolant. There are a number of ways, through reactor models, that these steps are correlated to provide design and analysis estimates (and criteria) for preventing runaway in exothermic reactors.

The temperature profile depends on the relative rates of heat generation by reaction and heat transfer. The temperature rise (or drop) in a reactor affects catalyst life, product selectivity, and equilibrium conversion, and excessive heat release can lead to reaction runaway. Hence, reactor design and analysis requires a good understanding of the coupling of reaction and heat transfer. Mathematical models for fixed-bed reactors can vary in their level of detail, depending on the end use. For more details, see Froment and Bischoff (*Chemical Reactor Analysis and Design*, Wiley, New York, 1990) and Harriott (*Chemical Reactor Design*, Marcel Dekker, New York, 2003).

Homogeneous one-dimensional model This is the simplest description of a packed bed, with an overall heat-transfer coefficient U. The particle and gas temperatures are identical, and only axial variation in composition and temperature is considered, giving the following mass and energy balance equations for any species C_i:

$$\begin{aligned} &\frac{d(uC_i)}{dz} = \sum_j \upsilon_{ij} r_j \qquad C_i = C_{i0} \quad \text{at } z=0 \\ &u\rho c_p \frac{dT}{dz} = \sum_j \upsilon_{ij}(-\Delta H_r)_j r_j - \frac{4U}{d_R}(T-T_c) \\ &T = T_0 \quad \text{at } z=0 \end{aligned} \tag{19-64}$$

Equation (19-64) is similar to the steady-state version of generic PFR Eqs. (19-17) and (19-18). The overall heat-transfer coefficient U is based on the bed-side heat-transfer area A_R and includes three terms: heat transfer on the bed side b, thermal conduction in the vessel wall, and heat transfer on the coolant side c:

$$\begin{aligned} &\frac{1}{U} = \frac{1}{h_b} + \frac{d_R}{k_w}\frac{A_R}{A_m} + \frac{1}{h_c}\frac{A_R}{A_c} \\ &A_m = \text{log mean}\,(A_R, A_c) = \frac{A_c - A_R}{\ln(A_c/A_r)} \end{aligned} \tag{19-65}$$

Here, h_i are the heat-transfer coefficients on the bed side and the coolant side, k_w is the wall thermal conductivity, and A_i are the heat-transfer areas. The coolant side heat-transfer coefficient can be obtained from general heat-transfer correlations that depend on the coolant side geometry (see any heat-transfer text and Secs. 5 and 11). For the process-side heat-transfer coefficient, there is a large body of literature with a variety of correlations. There is no clear advantage of one correlation over another because these depend on the particle and fluid properties, the temperature range, and other factors; for example, see the correlation of Leva, *Chem. Eng.* **56:** 115 (1949):

$$\begin{aligned} &\text{Nu}_R = \begin{cases} 0.813\,\text{Re}_p^{0.9}\,e^{-6dp/d_R} & \text{for heating} \\ 3.5\,\text{Re}_p^{0.7}\,e^{-4.6dp/d_R} & \text{for cooling} \end{cases} \\ &\text{Re}_p = \frac{\rho u d_p}{\mu} \qquad \text{Nu}_R = \frac{h_b d_R}{\lambda_f} \end{aligned} \tag{19-66}$$

The one-dimensional homogeneous model is useful for preliminary estimates and when lab (or pilot-plant) data for the same diameter tube are available. This simple model does not provide information on the effect of the tube diameter on the effective radial temperature gradients.

Fixed-bed reactors may exhibit axial dispersion. If axial dispersion is important for reactor simulation, analysis, or design, a variant of the one-dimensional homogeneous model that contains an axial dispersion term may be used. Approximate criteria to determine if mass and heat axial dispersion have to be considered are available (see, e.g., Froment and Bischoff, *Chemical Reactor Analysis and Design*, Wiley, New York, 1990).

Homogeneous two-dimensional model This model accounts for radial variation of composition and temperature in the bed that may be present for large heats of reaction. Corresponding material and energy balances are:

$$\begin{aligned} &\frac{1}{r}\frac{\partial}{\partial r}\left(rD_{er}\frac{\partial C_i}{\partial r}\right) - \frac{\partial(uC_i)}{\partial z} + \sum_j v_{ij} r_j = 0 \\ &\frac{1}{r}\frac{\partial}{\partial r}\left(rk_{er}\frac{\partial T}{\partial r}\right) - u\rho c_p \frac{dT}{dz} - \sum_j v_{ij}(-\Delta H_r)_j r_j = 0 \\ &C_i = C_{i0} \quad T = T_0 \quad \text{at } z=0 \\ &\frac{\partial C_i}{\partial r} = 0 \quad \frac{\partial T}{\partial r} = 0 \quad \text{at } r=0 \\ &\frac{\partial C_i}{\partial r} = 0 \quad \frac{\partial T}{\partial r} = -\frac{h_w}{k_{er}}(T_R - T_w) \qquad \text{at } r=R \end{aligned} \tag{19-67}$$

The effective radial diffusivity D_{er} is normally different from the axial diffusivity. It is often safe to neglect the radial variation of species concentration due to the relatively fast radial mixing. The effective conductivity k_{er} has to be determined from heat-transfer experiments, preferably with the actual bed and fluids. This coefficient can be either a constant (averaged radially) or a function of the radial position with a higher value for the core and a lower value near the wall due to different velocity and void fraction near the wall. There are a number of correlations for k_{er} and h_w, and there is significant variability in their utility; for example, see the correlation of De Wash and Froment, *Chem. Eng. Sci.* **27:** 567 (1972):

$$k_{er} = k_{er0} + \frac{0.0105\,\mathrm{Re}}{1+46(dp/d_R)^2}$$
$$h_w = h_{w0} + 0.0481\,\mathrm{Re}\frac{d_R}{d_p} \tag{19-68}$$

The static contribution k_{er0} incorporates heat transfer by conduction and radiation in the fluid present in the pores, conduction through particles, at the particle contact points and through stagnant fluid zones in the particles, and radiation from particle to particle. Figure 19-20 compares various literature correlations for the effective thermal conductivity and wall heat-transfer coefficient in fixed beds [Yagi and Kunii, *AIChE J.* **3:** 373 (1957)].

The two-dimensional model can be used to develop an equivalent one-dimensional model with a bed-side heat-transfer coefficient defined as [see, e.g., Froment, *Chem. Eng. Sci.* **7:** 29 (1962)]

$$\frac{1}{h_b} = \frac{1}{h_w} + \frac{R}{4k_{er}} \tag{19-69}$$

The objective is to have the radially averaged temperature profile of the 2D model match the temperature profile of the 1D model.

Heterogeneous one-dimensional model The heterogeneous model allows resolution of composition and temperature differences between the catalyst particle and the fluid.

$$\frac{d(uC_i)}{dz} = k_G a\,(C_i - C_{is})$$
$$k_G a\,(C_i - C_{is}) = \sum_j v_{ij} r_j$$

$$u\rho c_p \frac{dT}{dz} = ha(T_s - T) - \frac{4U}{d_R}(T - T_C) \tag{19-70}$$
$$ha(T_s - T) = \sum_j v_{ij}(-\Delta H_r)_j r_j$$
$$C_i = C_{i0} \quad T = T_0 \quad \text{at } z = 0$$

Mears developed a criterion that provides conditions for limiting interphase temperature gradients, which for a single exothermic reaction is [Mears, *J. Catal.* **20:** 127 (1971) and *I&EC Proc. Des. Dev.* **10:** 541 1971)]

$$\frac{d_p(-\Delta H_r)r}{2hT} < 0.15\frac{RT}{E} \tag{19-71}$$

An extension of the one-dimensional heterogeneous model is to consider intraparticle diffusion and temperature gradients, for which the lumped equations for the solid are replaced by second-order diffusion/conduction differential equations. Effectiveness factors can be used as applicable and are discussed in previous parts of this section and in Sec. 7 (see also Froment and Bischoff, *Chemical Reactor Analysis and Design*, Wiley, New York, 1990).

Typically the interphase temperature gradients are substantially smaller than the radial and axial temperature gradients, being on the order of 1°C to 3°C, and they can often be neglected.

Heterogeneous two-dimensional model Two-dimensional heterogeneous models have been developed—for example, see De Wash and Froment, *Chem. Eng. Sci.* **27:** 567 (1972). Figure 19-21 compares the various models. The results indicate that the homogeneous and heterogeneous models predict similar temperature profiles; however, the heterogeneous model contains additional information on interparticle concentration and temperature gradients that may be useful in catalyst or reactor design. The 2D models predict substantially higher peak temperatures than the corresponding 1D models. The pseudo-homogeneous 2D model may contain valuable information on radial temperature profiles, especially in the case of highly exothermic reactions. The heterogeneous 2D model also contains additional radial interparticle mass and heat-transfer information. The heterogeneous 2D model with no heat transfer through the solid shows a very steep temperature rise. This case illustrates the notion that a reasonably complicated model may indeed provide unrealistic results if inappropriate assumptions are made.

Examples

- Oxidation of SO_2 in large adiabatic packed-bed reactors (Fig. 19-22*a*). The catalyst is nominally 1/2 in V_2O_5/SiO_2 pellets that may be promoted with Cs. The inlet temperature is between 390°C and 450°C. The temperature increase across a bed is between 20°C and 210°C. The oxidation is thermodynamically limited, with lower temperatures favoring higher SO_3 conversion. After each bed, the exhaust is cooled (using a heat exchanger) and returned to the next bed. The advantage of interstage cooling is shown in Table 19-2.
- Phosgene synthesis from CO and Cl_2 in a multitubular reactor (Fig. 19-22*b*). The activated carbon catalyst is packed inside the tubes with cooling water on the shell side. Reaction by-products include CCl_4. The temperature profile in a tube (shown in the figure) is characterized by a hot spot. The position of the hot spot moves toward the exit of the reactor as the catalyst deactivates.

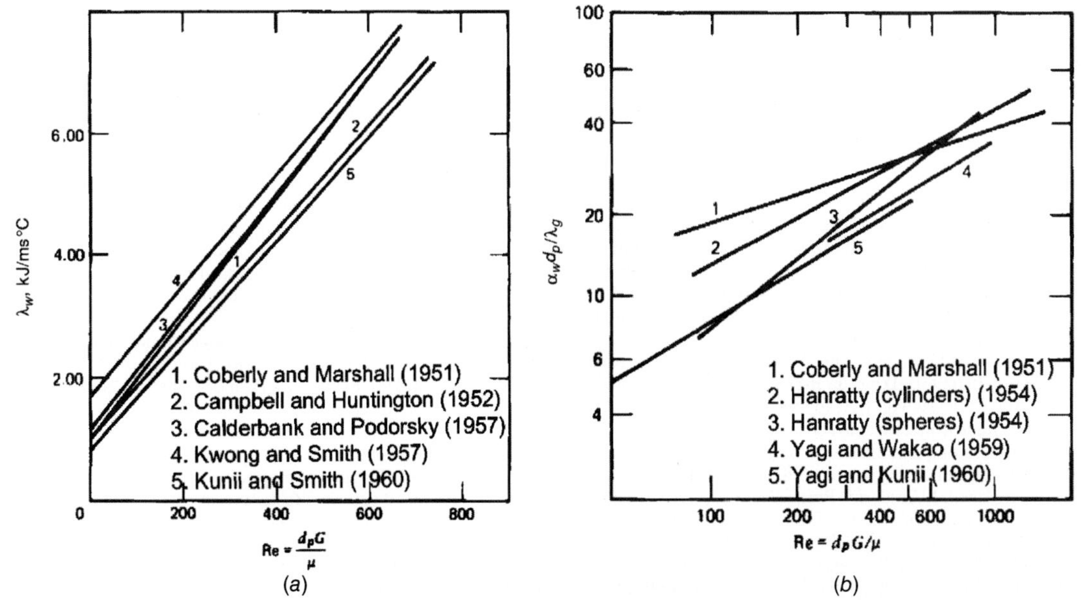

FIG. 19-20 Thermal conductivity and wall heat transfer in fixed beds. (*a*) Effective thermal conductivity. (*b*) Nusselt number for wall heat transfer. (*Figs. 11.7.1-2 and 11.7.1-3 in Froment and Bischoff,* Chemical Reactor Analysis and Design, *Wiley, New York, 1990.*)

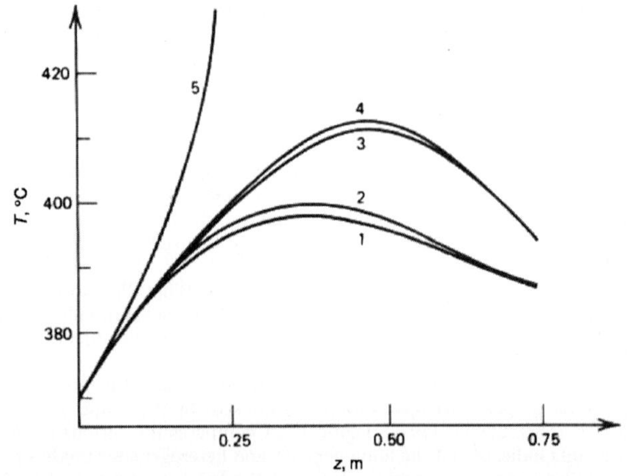

FIG. 19-21 Comparison of model predictions for radial mean temperature as a function of bed length. (1) Basic pseudohomogeneous one-dimensional model. (2) Heterogeneous model with interfacial gradients. (3) Pseudohomogeneous two-dimensional model. (4) Two-dimensional heterogeneous model with appropriate boundary conditions. (5) Two-dimensional heterogeneous model with no heat transfer through solid. (*Fig. 11.10-1 in Froment and Bischoff,* Chemical Reactor Analysis and Design, *Wiley, New York, 1990.*)

- Production of cumene from benzene and propylene using a phosphoric acid on quartz catalyst (Fig. 19-22*c*). There are four reactor beds with interbed cooling with cold feed. The reactor operates at 260°C.
- Vertical ammonia synthesizer at 300 atm with five cold shots and an internal exchanger (Fig. 19-22*d*). The nitrogen and hydrogen feeds are reacted over an Al_2O_3-promoted spongy iron catalyst. The concentration of ammonia is also shown in the figure.
- Vertical methanol synthesizer at 300 atm (Fig. 19-22*e*). A Cr_2O_3-ZnO catalyst is used with six cold shots totaling 10 to 20 percent of the fresh feed, and the temperature profile is shown in the figure.
- Methanol is oxidized to formaldehyde in a thin layer of finely divided silver or a multilayer screen, with a contact time of 0.01 s at 450°C to 600°C (842°F to 1112°F). A shallow bed of silver catalyst is also used in the DuPont process for in situ production of methyl isocyanate by reacting monomethylformamide and oxygen at 500°C.
- The Sohio process for vapor-phase oxidation of propylene to acrylic acid uses two beds of bismuth molybdate catalyst at 20 to 30 atm (294 to 441 psi) and 290°C to 400°C (554°F to 752°F).
- Oxidation of ethylene to ethylene oxide is done in two stages with supported silver catalyst, the first stage to 30 percent conversion, the second to 76 percent, with a total of 1.0 s contact time.
- Steam reforming reactors have supported nickel on alpha-alumina catalyst packed in tubes with the heat (for this endothermic reaction) supplied from a furnace on the shell side. The feed is natural gas (or naphtha) and water vapor heated to over 800°C (1056°F).
- Vinyl acetate reactors have a supported Pd/Au/alumina catalyst packed in ~25-mm (0.082-ft) ID tubes and heat (for this exothermic reaction) is removed by generating steam on the shell side. The feed contains ethylene, oxygen, and acetic acid in the vapor phase at 150°C to 175°C (302°F to 347°F).
- Maleic anhydride is made by oxidation of butane with air above 350°C (662°F) with V-Mo catalyst in a multitubular reactor with 2-cm diameter tubes. The heat-transfer medium is a molten salt eutectic mixture at 375°C (707°F). Even with small tubes, the heat transfer is so limited that a peak temperature rise of 100°C (212°F) above the shell side is developed and moves along the tubes.
- Butanol is made by the hydrogenation of crotonaldehyde in a reactor with 4000 tubes, 28 mm (0.029-ft) ID by 10.7 m (35.1 ft) long [Berty, in Leach, ed., *Applied Industrial Catalysis,* vol. 1, Academic Press, New York, 1983, p. 51].
- Vinyl chloride is made from ethylene and chlorine with Cu and K chlorides. The Stauffer process employs three multitubular reactors in series with 25-mm (0.082-ft) ID tubes [Naworski and Velez, in Leach, ed., *Applied Industrial Catalysis,* vol. 1, Academic Press, New York, 1983, p. 251].

Moving Beds In a moving-bed reactor, the catalyst, in the form of large granules, circulates by gravity and gas lift between reaction and regeneration zones (Fig. 19-23). The first successful operation was the Houdry cracker that replaced a plant with fixed beds that operated on a 10-min cycle between reaction and regeneration. Handling of large (hot) solids is difficult. The Houdry process was soon made obsolete by fluid catalytic cracking units (FCCUs). The only currently publicized moving-bed process is a UOP platinum reformer (Fig. 19-23*c*) that regenerates a controlled quantity of catalyst on a continuous basis.

Fluidized Beds Fluidized beds are reactors in which small particles (with average size typically below 0.1 mm) are fluidized by the reactant gases or liquids. When the linear velocity exceeds the minimum required for fluidization, a dense fluidized bed is obtained. As the superficial velocity increases, the bed expands and becomes increasingly dilute. At a high enough linear velocity, the particles entrain from the bed and have to be separated from the exhaust gases and recycled. The various fluidization flow regimes are described in Sec. 17.

Advantages of fluidized beds are temperature uniformity, good heat transfer, and the ability to continuously remove catalyst for regeneration. Disadvantages are solids backmixing, catalyst attrition, and complex recovery of fines. Baffles have been used often to reduce backmixing.

Fluidized beds contain a gas distributor or grid at the bottom of the reactor used to feed the reacting gas that is also the fluidizing medium. A sketch of a fluidized bed is shown in Fig. 19-23*d*. These reactors employ a wide range of particle sizes and densities. Geldardt (*Gas Fluidization Technology,* Wiley, New York, 1986) developed a widely accepted particle classification system based on fluidization characteristics. Type A powders are employed in many refinery and chemical processes, such as catalytic cracking, acrylonitrile synthesis, and maleic anhydride synthesis. Type B powders are also used, for instance, in fluidized-bed combustion. The properties of different powders are summarized in the subsection Fluidized-Bed Systems in Sec. 17. Good distributor design and the presence of a substantial fraction of fines (mainly for processes employing group A powders) are essential for good fluidization, to eliminate maldistribution, and for good performance. Internals for heat transfer (e.g., cooling tubes) and other baffling for improved performance provide design challenges as their effect is not yet well understood (in spite of the voluminous literature).

The *particle size distribution* and the linear velocity are important in reactor design. The *minimum fluidization* velocity is the velocity at the onset of fluidization, while the *terminal* velocity is the velocity above which a particle can become entrained from the bed. The nature of the particles and the linear velocity determine fluid bed properties such as *gas holdup, equilibrium bubble size* (for bubbling systems), *mass transfer from gas to solid, entrainment rate* of particles from the bed, and the flow regime *transition velocities.* The height beyond which the concentration of entrained particles does not vary significantly is called the *transport disengagement height.* Knowledge of this height is required for the design and location of *cyclones* for solids containment by their separation from the effluent gas. In addition to the velocity and the nature of the particles, the layout of the equipment can determine the particle *attrition rate.*

The two-phase theory of fluidization has been extensively used to describe fluidization (e.g., see Kunii and Levenspiel, *Fluidization Engineering,* 2d ed., Wiley, New York, 1990). For type A powders, the fluidized bed is assumed to contain a bubble (or dilute) and an emulsion (or dense) phase. The bubble phase may be modeled by a plug flow (or dispersion) model, and the emulsion phase is assumed to be well mixed and may be modeled as a CSTR. Correlations for the size of the bubbles and the heat and mass transport from the bubbles to the emulsion phase are available in Sec. 17 and in textbooks on the subject. Davidson and Harrison (*Fluidization,* 2d ed., Academic Press, New York, 1985), Geldart (*Gas Fluidization Technology,* Wiley, New York, 1986), Kunii and Levenspiel (*Fluidization Engineering,* Wiley, New York, 1969), and Zenz (*Fluidization and Fluid-Particle Systems,* Pemm-Corp Publications, 1989) are good reference books.

Examples

- The original fluidized bed reactor was the Winkler coal gasifier (patented 1922), followed in 1940 by the Esso cracker that has now been replaced by riser reactors with zeolite catalysts.
- Combination of fluidized bed and transport reactor is used for the Sasol Fischer-Tropsch process (Fig. 19-23*a*). The regenerator is a conventional fluidized bed and the reactor is a transport bed (see Transport Reactors subsection below).
- Esso-type combined riser/conventional fluidized bed is used for cracking petroleum oils (Fig. 19-23*b*). The regenerator is a conventional fluidized bed and the reactor is a riser or transport bed (see Transport Reactors subsection below).
- Acrylonitrile is made in a fixed fluidized bed by reacting propylene, ammonia, and oxygen at 400°C to 510°C (752°F to 950°F) over a Bi-Mo oxide catalyst. The good temperature control with embedded heat exchangers permits a catalyst life of several years.
- Vinyl chloride is produced by chlorination of ethylene at 200°C to 300°C (392°F to 572°F), 2 to 10 atm (29.4 to 147 psi), with a supported cupric chloride catalyst in a fluidized bed.

Slurry Reactors Slurry reactors are one example of liquid fluidized beds. In some cases (e.g., for hydrogenation), a limited amount of hydrogen may be dissolved in the liquid feed. The solid material is maintained in a fluidized state by agitation or by internal or external recycle of the liquid.

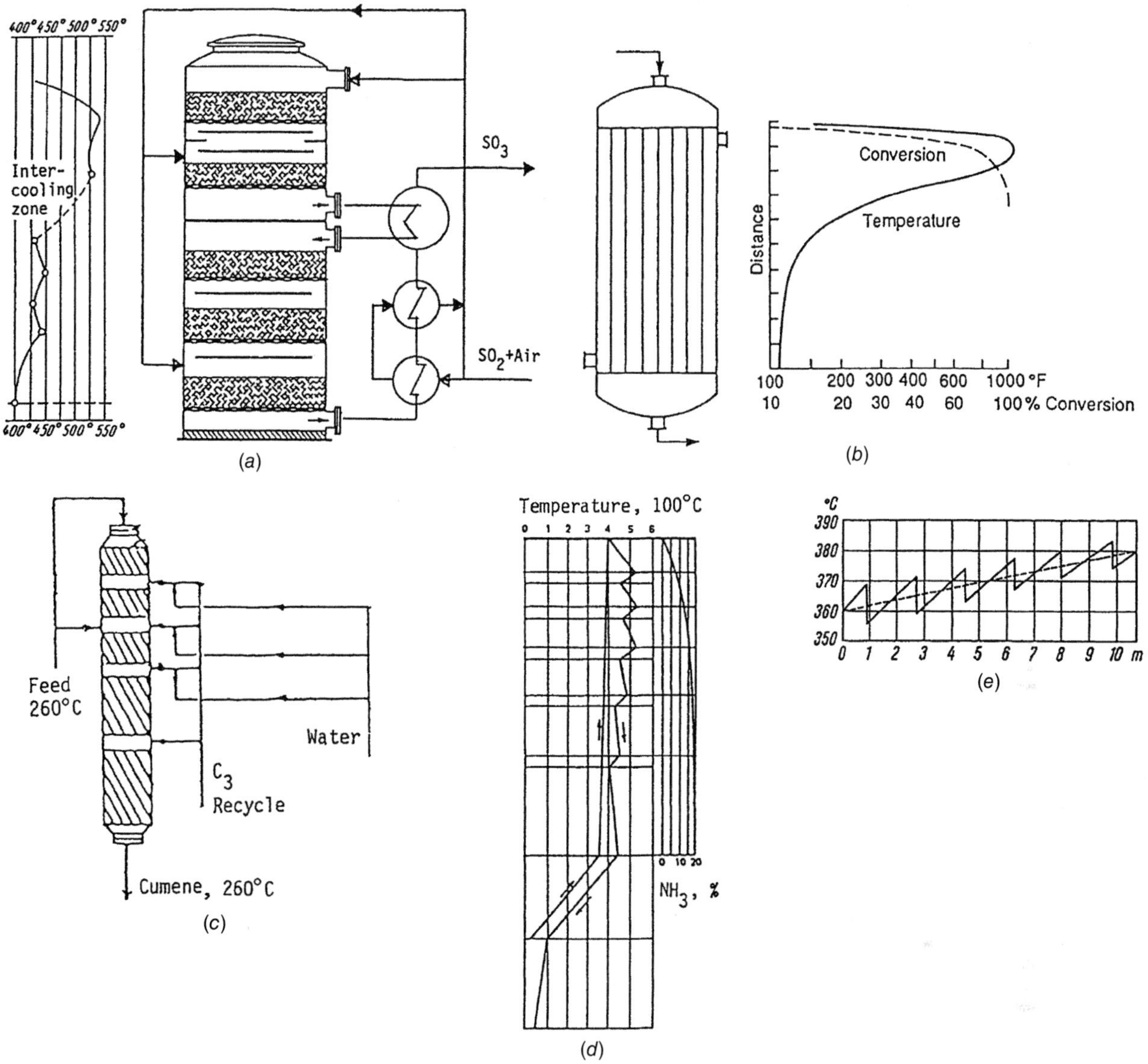

FIG. 19-22 Temperature and composition profiles. (*a*) Oxidation of SO_2 with intercooling and two cold shots. (*b*) Phosgene from CO and Cl_2, activated carbon in 2-in tubes, water-cooled. (*c*) Cumene from benzene and propylene, phosphoric acid on quartz with four quench zones, 260°C. (*d*) Vertical ammonia synthesizer at 300 atm, with five cold shots and an internal exchanger. (*e*) Vertical methanol synthesizer at 300 atm, Cr_2O_3-ZnO catalyst, with six cold shots totaling 10 to 20 percent of the fresh feed. To convert psi to kPa, multiply by 6.895; atm to kPa, multiply by 101.3.

Most industrial processes with slurry reactors also use a gas in reactions such as chlorination, hydrogenation, and oxidation. This is discussed in the subsection Multiphase Reactors.

Transport Reactors The superficial velocity of the gas exceeds the terminal velocity of the solid particles, and the particles are transported along with the gas. Usually, there is some "slip" between the gas and the solids, and the solid velocity is slightly lower than the gas velocity. Transport reactors are typically used when the required residence time is small and the fluid reactant (or the solid reactant) can be substantially converted (consumed). They may also be used when the catalyst is substantially deactivated during its time in the reactor and has to be regenerated.

Advantages of transport reactors include low gas and solid backmixing (compared to fluidized beds) and, as in fluidized beds and slurry reactors, the ability to continuously remove deactivated catalyst (and add fresh catalyst), thereby maintaining catalyst activity. The fluid and catalyst are separated downstream by using settlers, cyclones, or filters.

Transport reactors are typically cylindrical pipes. The reactants may be injected at a tee or by using injection pipes at the bottom of the reactor. The size of the pipe may be increased along the reaction path to accommodate volumetric changes that may occur during reaction. Both solid and gas phases may be modeled using a PFR model with exchange between the gas and solid phases. A core-annular concept is often used to describe transport or riser reactors, with most of the particles rising at the center and some flowing back down along the walls.

Examples

- A transport reactor is used in the Sasol Synthol Fischer-Tropsch process. The catalyst is promoted iron. It circulates through the 1.0-m (3.28-ft) ID riser at 72,600 kg/h (160,000 lbm/h) at 340°C (644°F) and 23 atm (338 psi) and has a life of about 50 days. Figure 19-23*a* shows an in-line heat exchanger in the Sasol unit.
- The FCCU riser cracks crude oil into gasoline and distillate range products in a transport bed reactor using a zeolite-Y catalyst. The riser residence time is 4 to 10 s. The riser top temperature is between 950°F and 1050°F. The ratio of catalyst to crude oil is between 4 and 8 on a weight basis. During its stay in the riser, the catalyst is deactivated by coke, which is burned in the regenerator. The heat generated by burning the coke heats the catalyst and is used to vaporize the crude oil feed. A schematic of the FCCU is shown in Fig. 19-23*b*.

Multifunctional Reactors Reaction may be coupled with other unit operations to reduce capital and/or operating costs, increase selectivity, and improve safety. Examples are reaction with distillation, and reaction with heat transfer. Concepts that combine reaction with membrane separation, extraction, and crystallization are also being explored. In each case, while possibly reducing cost, the need to accommodate both reaction and the additional operation constrains process flexibility by reducing the operating envelope.

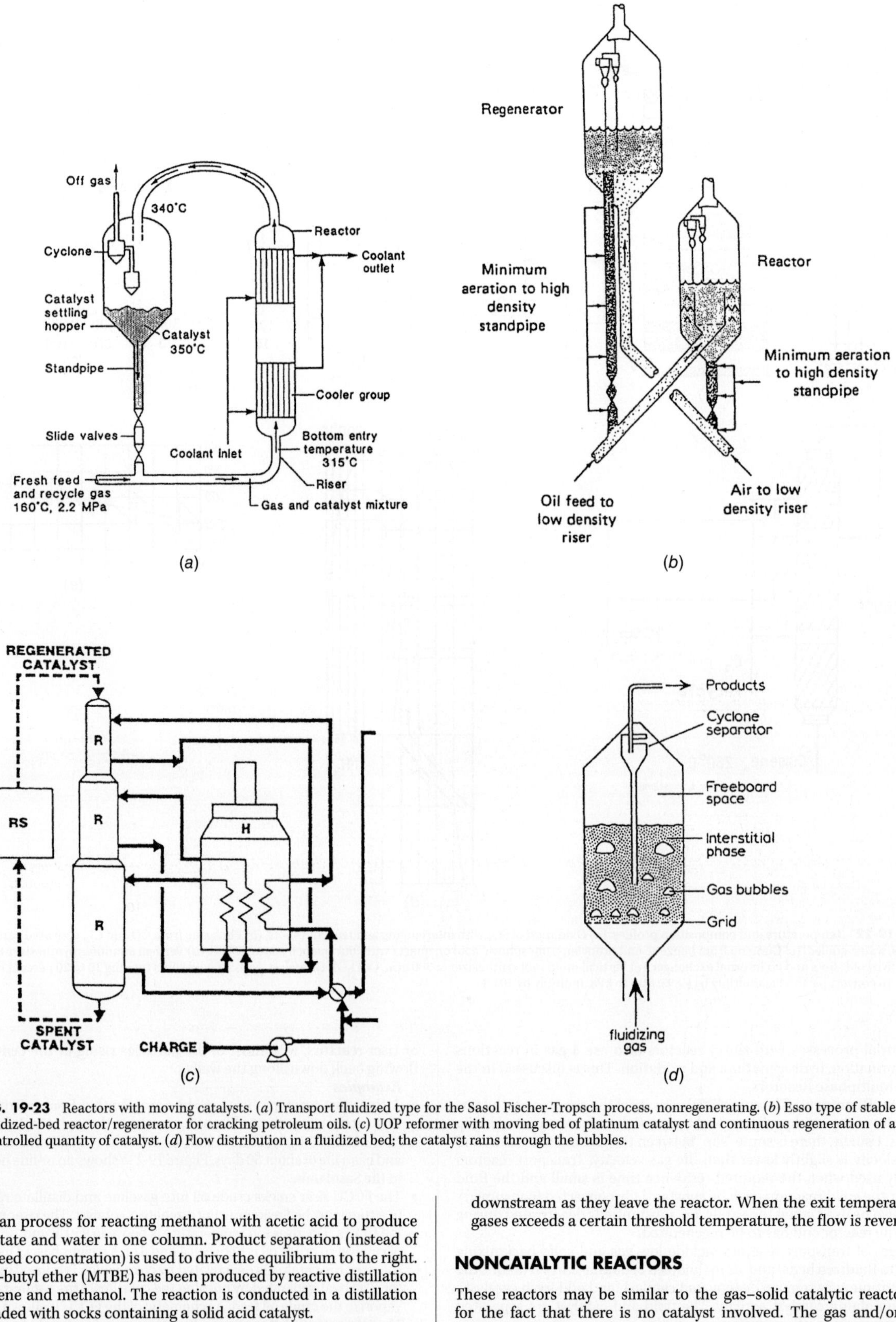

FIG. 19-23 Reactors with moving catalysts. (*a*) Transport fluidized type for the Sasol Fischer-Tropsch process, nonregenerating. (*b*) Esso type of stable fluidized-bed reactor/regenerator for cracking petroleum oils. (*c*) UOP reformer with moving bed of platinum catalyst and continuous regeneration of a controlled quantity of catalyst. (*d*) Flow distribution in a fluidized bed; the catalyst rains through the bubbles.

Examples

- The Eastman process for reacting methanol with acetic acid to produce methyl acetate and water in one column. Product separation (instead of increased feed concentration) is used to drive the equilibrium to the right.
- Methyl *tert*-butyl ether (MTBE) has been produced by reactive distillation of isobutylene and methanol. The reaction is conducted in a distillation column loaded with socks containing a solid acid catalyst.
- VOC emissions from printing and chemical plants are oxidized in reverse flow reactors that couple reaction with regenerative heat transfer. The concept here is to maintain a catalyst zone in the center of a packed bed with inert heat-transfer packing on either side. Feed is heated to the desired temperature as it travels through the hot inert bed to the catalyst zone. After the catalyst, the outlet gases lose heat to the cooler packing downstream as they leave the reactor. When the exit temperature of the gases exceeds a certain threshold temperature, the flow is reversed.

NONCATALYTIC REACTORS

These reactors may be similar to the gas-solid catalytic reactors, except for the fact that there is no catalyst involved. The gas and/or the solid may be reactants and/or products. Section 7 provides greater discussion on reaction types and corresponding kinetics for a range of gas-solid reactions. The oldest examples of gas-solid noncatalytic reactors are kilns. A solid is heated with hot combustion gases (that may contain a reactant) to form a desired product. Some of the equipment in use is presented in Fig. 19-24. Temperatures are usually high, so the equipment is refractory-lined.

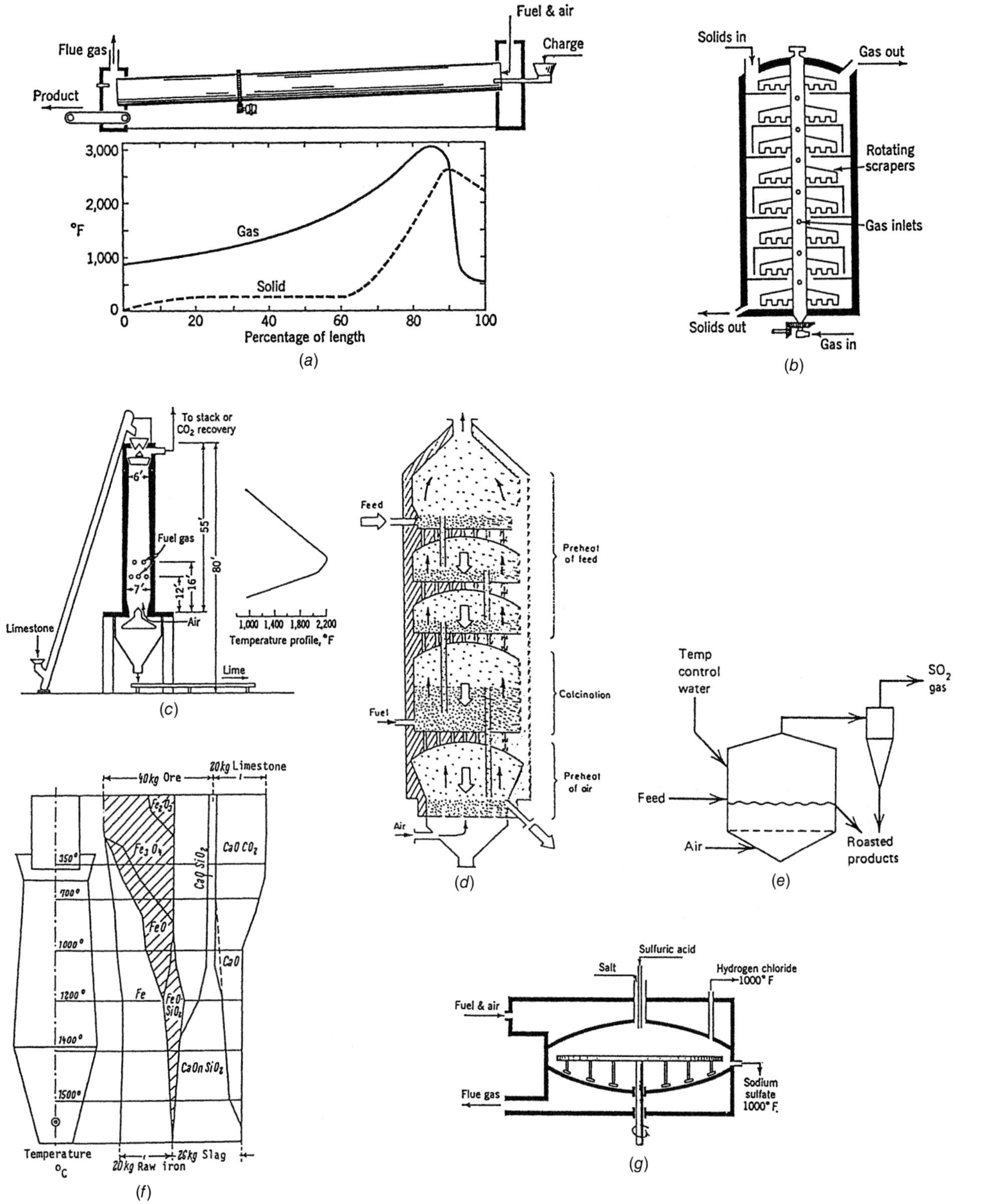

FIG. 19-24 Reactors for solids. (*a*) Temperature profiles in a rotary cement kiln. (*b*) A multiple-hearth reactor. (*c*) Vertical kiln for lime burning, 55 ton/d. (*d*) Five-stage fluidized-bed lime burner, 4 by 14 m, 100 ton/d. (*e*) A fluidized bed for roasting iron sulfides. (*f*) Conditions in a vertical moving bed (blast furnace) for reduction of iron oxides. (*g*) A mechanical salt cake furnace. To convert ton/d to kg/h, multiply by 907.

The solid is in granular form, at most a few millimeters or centimeters in diameter. Historically, much of the equipment was developed for the treatment of ores and the recovery of metals.

Rotary Kilns A rotary kiln is a long, narrow cylinder inclined 2° to 5° to the horizontal and rotated at 0.25 to 5 rpm. It is used for the decomposition of individual solids, for reactions between finely divided solids, and for reactions of solids with gases or even with liquids. The length/diameter ratio ranges from 10 to 35, depending on the reaction time needed. The solid is in granular form and may have solid fuel mixed in. The granules are scooped into the vapor space and are heated as they cascade downward. Holdup of solids is 8 to 15 percent of the cross section. For most free-falling materials, the solids pattern approaches plug flow axially and complete mixing laterally. Rotary kilns can tolerate some softening and partial fusion of the solid. For example, CaF_2 is reacted with SO_3 in a rotary kiln to make hydrofluoric acid. The morphology of the CaF_2 solids can change considerably as they travel downward through the kiln. Approximate ranges of space velocities in rotary kilns are shown in Table 19-6.

Vertical Kilns Vertical kilns are used primarily where no fusion or softening occurs, as in the burning of limestone or dolomite, although rotary kilns may also be used for these operations. A cross section of a continuous 50,000-kg/d (110,000-lbm/d) lime kiln is shown in Fig. 19-24*c*. The diameter range of these kilns is 2.4 to 4.5 m (7.9 to 14.8 ft), and height is 15 to 24 m (49 to 79 ft). Peak temperatures in lime calcination are 1200°C (2192°F), although decomposition proceeds freely at 1000°C (1832°F). Fuel supply may be coke mixed and fed with the limestone or other fuel. Space velocity of the kiln is 14 to 485 kg CaO/(m^3·h) [0.87 to 99 lbm/(ft^3·h)]. Factors that influence kiln size include its vintage, the method of firing, and the lump size, which is in the range of 10 to 25 cm (3.9 to 9.8 in).

The blast furnace (Fig. 19-24*f*) is a vertical kiln in which fusion takes place in the lower section. This is a vertical moving-bed device; iron oxides and coal are charged at the top and flow counter-currently to combustion and reducing gases. Units of 1080 to 4500 m^3 (38,000 to 159,000 ft^3) may produce up to 9×10^6 kg (20×10^6 lbm) of molten iron per day. Figure 19-24*f* identifies the temperature and composition profiles. Reduction is with CO and H_2 that are made from coal, air, and water within the reactor. Another type of vertical kiln is a hearth furnace shown in Fig. 19-24*b*. These high-temperature reactors convert minerals for easier separation from gangue or for easier recovery of metal.

Fluidized Beds Fluidized beds are used for the combustion of solid fuels, and some 30 installations are listed in the *Encyclopedia of Chemical Technology* (vol. 10, Wiley, New York, 1980, p. 550). Another application is a five-stage fluidized-bed calciner sketched in Fig. 19-24*d*. Such a unit, 4 m (13 ft) in diameter and 14 m (46 ft) high, has a production of 91,000 kg CaO/d (200,000 lbm/d). The roasting of iron sulfide in fluidized beds at 650°C to 1100°C (1202°F to 2012°F) is analogous. The pellets have a 10-mm (0.39-in) diameter. There are numerous plants, but they are threatened with obsolescence because cheaper sources of sulfur are available for making sulfuric acid.

There are a number of references on gas–solid noncatalytic reactions, including Brown, Dollimore, and Galwey ["Reactions in the Solid State," in Bamford and Tipper (eds.), *Comprehensive Chemical Kinetics,* vol. 22, Elsevier, Amsterdam, 1980], Galwey (*Chemistry of Solids,* Chapman and Hall, London, 1967), Sohn and Wadsworth (eds.) (*Rate Processes of Extractive Metallurgy,* Plenum Press, New York, 1979), Szekely, Evans, and Sohn (*Gas–Solid Reactions,* Academic Press, New York, 1976), and Ullmann (*Enzyklopaedie der technischen Chemie,* "Uncatalyzed Reactions with Solids," vol. 3, 4th ed., Verlag Chemie, Weinheim, Ger., 1973, pp. 395–464).

In recent years, a new gas–solid application, called chemical vapor deposition (CVD), has been introduced commercially. In CVD, gases are reacted to deposit solid particles on surfaces. The particles can include organic, inorganic, metallic, nonmetallic, semiconductor, or insulation materials. The surfaces are used in the production of electronic and optical devices and coatings. Due to the need for surface uniformity, CVD processes often require special reactor technology and clean room environments. Komiyama et al. have reviewed the chemical reaction engineering issues in CVD [see *Chem. Eng. Sci.* **54:** 1941–1957 (1999)]. Reactor types are described in the handbook by Pierson (*Handbook of Chemical Vapor Deposition: Principles, Technology and Applications,* 2nd edition, Noyes Publications, Westwood, N.J., 1999) and in *Principles of Chemical Vapor Deposition,* Dobkin and Zuraw (eds.), Springer Science-Business Media, Berling, 2003. The processes are classified based on the energy input type (e.g., thermal, laser, photo and plasma CVD) or the operation mode (e.g., as closed, or batch, and open, or continuous).

TABLE 19-6 Approximate Ranges of Space Velocities in Rotary Kilns

Process	Space velocity, m tons/(m^{3}d)
Cement, dry process	0.4–1.1
Cement, wet process	0.4–0.8
Cement, with heat exchange	0.6–1.9
Lime burning	0.5–0.9
Dolomite burning	0.4–0.6
Pyrite roasting	0.2–0.35
Clay calcination	0.5–0.8
Magnetic roasting	1.5–2.0
Ignition of inorganic pigments	0.15–2.0
Barium sulfide preparation	0.35–0.8

Examples

- Cement kilns are up to 6 m (17 ft) in diameter and 200 m (656 ft) long. Inclination is 3° to 4°, and rotation is 1.2 to 2.0 rpm. Typical temperature profiles are shown in Fig. 19-24*a*. Near the flame, the temperature is 1800°C to 2000°C (3272°F to 3632°F). The temperature of the solid reaches 1350°C to 1500°C (2462°F to 2732°F), which is necessary for clinker formation. In one smaller kiln, a length of 23 m (75 ft) was allowed for drying, 34 m (112 ft) for preheating, 19 m (62 ft) for calcining, and 15 m (49 ft) for clinkering. Total residence time is 40 min to 5 h, depending on the type of kiln. The time near the clinkering temperature of 1500°C (2732°F) is 10 to 20 min. Subsequent cooling is as rapid as possible. A kiln 6 m (20 ft) in diameter by 200 m (656 ft) can produce 2.7×10^6 kg/d (6×10^6 lbm/d) of cement. For production rates less than 270,000 kg/d (600,000 lbm/d), shaft kilns are used. These are vertical cylinders 2 to 3 m (6.5 to 10 ft) by 8 to 10 m (26 to 33 ft) high, fed with pellets and finely ground coal.
- Chlorination of ores ($MeO + Cl_2 + C \Rightarrow MeCl_2 + CO$, where Me is Ti, Mg, Be, U, and Zr, whose chlorides are water-soluble). For titanium, carbon is roasted with ore, and chlorine is sparged through the bed ($TiO_2 + C + 2Cl_2 \Rightarrow TiCl_4 + CO_2$). The chlorine can be supplied indirectly, as in $Cu_2S + 2NaCl + O_2 \Rightarrow 2CuCl + Na_2SO_4$.
- Oxidation of sulfide ores ($MeS + 1.5O_2 \Rightarrow MeO + SO_2$, where Me is Fe, Mo, Pb, Cu, or Ni). Iron sulfide (pyrite) is burned with air for recovery of sulfur and to make the iron oxide from which the metal is more easily recovered. Sulfides of other metals also are roasted. A multiple-hearth furnace, as shown in Fig. 19-24*b*, is used. In some designs, the plates rotate; in others, the scraper arms rotate or oscillate and discharge the material to lower plates. Material charged at the top drops to successively lower plates, while reactant and combustion gases flow upward. A reactor with 9 trays 5 m (16 ft) in diameter and 12 m (39 ft) high can roast about 600 kg/h (1300 lbm/h) of pyrite. A major portion of the reaction is found to occur in the vapor space between trays. A unit in which most of the trays are replaced by empty space is called a *flash roaster;* its mode of operation is like that of a spray dryer. Molybdenum sulfide is roasted at the rate of 5500 kg/d (12,000 lbm/d) in a unit with 9 stages, 5-m (16-ft) diameter, at 630 ± 15°C (1166 ± 27°F), and the sulfur is reduced from 35.7 percent to 0.006 percent. A Dorr-Oliver fluidized-bed roaster is 5.5 m (18 ft) in diameter, 7.6 m (25 ft) high, with a bed height of 1.2 to 1.5 m (3.9 to 4.9 ft). It operates at 650°C to 700°C (1200°F to 1300°F) and has a capacity of 154,000 to 200,000 kg/d (340,000 to 440,000 lbm/d) (Kunii and Levenspiel, *Fluidization Engineering,* Butterworth, Stoneham, Mass., 1991). Two modes of operation can be used for a fluidized-bed unit like that shown in Fig. 19-24*e*. In one mode, a stable fluidized-bed level is maintained. The superficial gas velocity of 0.48 m/s (1.6 ft/s) is low. A reactor is 4.8 m (16 ft) in diameter, 1.5 m (4.9 ft) bed depth, and 3 m (9.8 ft) freeboard. The capacity is 82,000 kg/d (180,000 lbm/d) pyrrhotite of 200 mesh. It operates at 875°C (1600°F), and 53 percent of the solids are entrained. In the other mode, the superficial gas velocity of 1.1 m/s (3.6 ft/s) is higher and results in 100 percent entrainment. This reactor operates in the *transport fluidization regime,* and such a design can process 545,000 kg/d (1.2×106 lbm/d) of 200 mesh material at 780°C (1436°F).
- Sodium sulfate. A single-hearth furnace like that shown in Fig. 19-24*g* is used. Sodium chloride and sulfuric acid are charged continuously to the center of the pan, and the rotating scrapers gradually work the reacting mass to the periphery, where the sodium sulfate is discharged at 540°C (1000°). Pans are 3.3 to 5.5 m (11 to 18 ft) in diameter and can handle 5500 to 9000 kg/d (12,000 to 20,000 lbm/d) of salt. Rotary kilns also are used for this purpose. Such a unit 1.5 m (4.9 ft) in diameter by 6.7 m (22 ft) has a capacity of 22,000 kg/d (48,000 lbm/d) of salt cake. A pan furnace also is used, for instance, in the Leblanc soda ash process and for making sodium sulfide from sodium sulfate and coal.
- Magnetic roasting. In this process, ores containing Fe_2O_3 are reduced with CO to Fe_3O_4, which is magnetically separable from gangue. Rotary kilns are used, with temperatures of 700°C to 800°C (1292°F to 1472°F). Higher temperatures form FeO. The CO may be produced by incomplete combustion of a fuel. A unit for 2.3×10^6 kg/d (5×10^6 lbm/d) has a power consumption of 0.0033 to 0.0044 kWh/kg (3 to 4 kWh/ton) and a heat requirement of 180,000 to 250,000 kcal/ton (714,000 to 991,000 Btu/ton). The magnetic concentrate can be agglomerated for further treatment by pelletizing or sintering.
- Other examples include *calcination reactions* ($MeCO_3 \Rightarrow MeO + CO_2$, where Me is Ca, Mg, and Ba), *sulfating reactions* ($CuS + 2O_2 \Rightarrow CuSO_4$, of which the sulfate is water-soluble), and *reduction reactions* ($MeO + H_2 \Rightarrow Me + H_2O$, $MeO + CO \Rightarrow Me + CO_2$, where Me is Fe, W, Mo, Ge, and Zn).
- The deposition of polycrystalline silicon in microelectronic circuit fabrication ($SiH_4 \Rightarrow Si + 2H_2$) or the deposition of hard TiC films on machine tool surfaces ($TiCl_4 + CH_4 \Rightarrow TiC + 4HCl$).
- In reactive etching, a patterned film is selectively etched by reacting it with a gas such as chlorine ($Si + 2Cl_2 \Rightarrow SiCl_4$).

FLUID-FLUID REACTORS

Industrial fluid–fluid reactors may be broadly divided into gas–liquid and liquid–liquid reactors. Gas–liquid reactors typically may be used for the manufacture of pure products (such as sulfuric acid, nitric acid, nitrates, phosphates, adipic acid, and other chemicals) where all the gas and liquid react. They are also used in processes where gas-phase reactants are sparged into the reactor and the reaction takes place in the liquid phase (such as hydrogenation, halogenation, oxidation, nitration, alkylation, fermentation, oxidation of sludges, production of proteins, biochemical oxidations, and so on). Gas purification (in which relatively small amounts of impurities such as CO_2, CO, COS, SO_2, H_2S, NO, and others are removed from reactants) is also an important class of gas–liquid reactions. Liquid–liquid reactors are used for the synthesis of chemicals (or fuels). One of the liquids may serve as the catalyst, or the liquids may react with one another across the interface. In the latter case, the product may be soluble in one of the liquids or precipitate out as a solid. Often a phase transfer homogeneous catalyst is employed.

GAS-LIQUID REACTORS

Since the reaction rate per unit reactor volume depends on the transfer of molecules from the gas to the liquid, the mass-transfer coefficient is important. As discussed in Sec. 7, the mass-transfer coefficient in a nonreacting system depends on the physical properties of the gas and liquid and the prevailing hydrodynamics. Here D_G and D_L are diffusivities of the absorbing species in the gas and liquid phases, respectively; $p_i = f(C_{Li})$ or $p_i = \mathrm{He}C_{Li}$, is the equilibrium relation at the gas–liquid interface; a = interfacial area/unit volume of liquid; and using the film theory, δ_G, δ_L are film thicknesses on the gas and liquid sides, respectively. The steady rates of solute transfer are

$$r = k_G a (p_G - p_i) = k_L a (C_{Li} - C_L) \qquad (19\text{-}72)$$

where $k_G = D_G/\delta_G$ and $k_L = D_L/\delta_L$ are the mass-transfer coefficients of the individual films. Overall coefficients are defined by

$$r = K_G a (p_G - p_L) = K_L a (C_G - C_L) \qquad (19\text{-}73)$$

These equations are valid when the reaction is slow enough so that no appreciable reaction occurs in the liquid film. Upon introducing the equilibrium relation at the interface, the relation between the various mass-transfer coefficients is

$$\frac{1}{K_G a} = \frac{\mathrm{He}}{K_L a} = \frac{1}{k_G a} + \frac{\mathrm{He}}{k_L a} \qquad (19\text{-}74)$$

When the solubility is low, the Henry constant He is large, and $k_L \Rightarrow K_L$; when the solubility is high, He is small, and $k_G \Rightarrow K_G$. The reaction rate in the liquid phase determines the relative importance of the mass-transfer coefficient. Reaction-diffusion regimes are detailed in Sec. 7. Thus in the slow reaction regime, for very slow reactions, reaction rate in the liquid phase determines the overall rate (kinetic regime). In contrast, still in the slow reaction regime, when the reaction is faster than the interphase mass-transfer rate, interphase mass transfer is rate-determining. In the fast reaction regime, the reaction also occurs in the film along with diffusion, thus enhancing the mass transfer. The relative role of mass transfer (across the gas–liquid interface) versus kinetics is important in gas–liquid reactor selection and design. Three modes of contacting gas with liquid are possible: (1) The gas is dispersed as bubbles in the liquid; (2) the liquid is dispersed as droplets in the gas; and (3) the liquid and gas are brought together as thin films over a packing or wall. Considerations that influence reactor selection include the reaction rate, the magnitude and distribution of the residence times of the phases, the power requirements, the scale of the operation, the opportunity for heat transfer, and so on.

As we have indicated, for purely physical absorption, the mass-transfer coefficients depend on the hydrodynamics, the physical properties of the phases, and the diffusivity of the reactants (products). The literature contains measured values of mass-transfer coefficients and correlations (see discussion that follows on agitated tanks and bubble columns). Tables 19-7 and 19-8 present experimental information on apparent mass-transfer coefficients for the absorption of select gases. On this basis, a tower for absorption of SO_2 with NaOH is smaller than that with pure water by a factor of roughly 0.317/7.0 = 0.045. Table 19-9 lists the main factors that are needed for mathematical representation of $K_G a$ in a typical case of the absorption

TABLE 19-7 Typical Values of $K_G a$ for Absorption in Towers Packed with 1.5-in Intalox Saddles at 25% Completion of Reaction*

Absorbed gas	Absorbent	$K_G a$, lb mol/(h·ft³·atm)
Cl_2	H_2O·NaOH	20.0
HCl	H_2O	16.0
NH_3	H_2O	13.0
H_2S	H_2O·MEA	8.0
SO_2	H_2O·NaOH	7.0
H_2S	H_2O·DEA	5.0
CO_2	H_2O·KOH	3.10
CO_2	H_2O·MEA	2.50
CO_2	H_2O·NaOH	2.25
H_2S	H_2O	0.400
SO_2	H_2O	0.317
Cl_2	H_2O	0.138
CO_2	H_2O	0.072
O_2	H_2O	0.0072

*To convert in to cm, multiply by 2.54; lb mol/(h·ft³·atm) to kg mol/(h·m³·kPa), multiply by 0.1581.
SOURCE: From Eckert et al., *Ind. Eng. Chem.*, **59,** 41 (1967).

TABLE 19-8 Selected Absorption Coefficients for CO_2 in Various Solvents in Towers Packed with Raschig Rings*

Solvent	$K_G a$, lb mol/(h·ft³·atm)
Water	0.05
1-*N* sodium carbonate, 20% Na as bicarbonate	0.03
3-*N* diethanolamine, 50% converted to carbonate	0.4
2-*N* sodium hydroxide, 15% Na as carbonate	2.3
2-*N* potassium hydroxide, 15% K as carbonate	3.8
Hypothetical perfect solvent having no liquid-phase resistance and having infinite chemical reactivity	24.0

*Basis: L = 2500 lb/(h·ft²); G = 300 lb/(h·ft²); T = 77°F; pressure, 1.0 atm. To convert lb mol/(h·ft³·atm) to kg mol/(h·m³·kPa), multiply by 0.1581.
SOURCE: From Sherwood, Pigford, and Wilke, *Mass Transfer,* McGraw-Hill, 1975, p. 305.

TABLE 19-9 Correlation of $K_G a$ for Absorption of CO_2 by Aqueous Solutions of Monoethanolamine in Packed Towers*

$$K_G a = F\left(\frac{L}{\mu}\right)^{2/3}[1 + 5.7(C_e - C)Me^{0.0067T - 3.4p}]$$

where $K_G a$ = overall gas-film coefficient, lb mol/(h·ft³·atm)
μ = viscosity, centipoises
C = concentration of CO_2 in the solution, mol/mol monoethanolamine
M = amine concentration of solution (molarity, g mol/L)
T = temperature, °F
p = partial pressure, atm
L = liquid-flow rate, lb/(h·ft²)
C_e = equilibrium concentration of CO_2 in solution, mol/mol monoethanolamine
F = factor to correct for size and type of packing

Packing	F	Basis for calculation of F
5- to 6-mm glass rings	7.1×10^{-3}	Shneerson and Leibush data, 1-in column, atmospheric pressure
⅜-in ceramic rings	3.0×10^{-3}	Unpublished data for 4-in column, atmospheric pressure
¾- by 2-in polyethylene Tellerettes	3.0×10^{-3}	Teller and Ford data, 8-in column, atmospheric pressure
1-in steel rings		
1-in ceramic saddles	2.1×10^{-3}	
1½- and 2-in ceramic rings	$0.4\text{–}0.6 \times 10^{-3}$	Gregory and Scharmann and unpublished data for two commercial plants, pressures 30 to 300 psig

*To convert in to cm, multiply by 2.54.
SOURCE: From Kohl and Riesenfeld, *Gas Purification*, Gulf, 1985.

of CO_2 by aqueous monoethanolamine. Other than Henry's law, $p = HeC_L$, which holds for some fairly dilute solutions, there is no general simple form of equilibrium relation. A typically complex equation is that for CO_2 in contact with sodium carbonate solutions [Harte, Baker, and Purcell, *Ind. Eng. Chem.* **25:** 528 (1933)], which is

$$p_{CO_2} = \frac{137 f^2 N^{1.29}}{S(1-f)(365-T)^2} \tag{19-75}$$

where f = fraction of total base present as bicarbonate
N = normality, 0.5 to 2.0
S = solubility of CO_2 in water at 1 atm, g · mol/L
T = temperature, 65°F to 150°F

The mass-transfer coefficient with a reactive solvent can be represented by multiplying the purely physical mass-transfer coefficient by an *enhancement factor E* that depends on a parameter called the *Hatta number* (analogous to the *Thiele* modulus in porous catalyst particles).

$$\text{Ha}^2 = \frac{\text{reaction rate in film}}{\text{diffusional transport rate through film}} \tag{19-76}$$

For example, for the reaction $A(g) + bB(l) \rightarrow P$ with liquid reactant B in excess and the reaction first order in A and B,

$$\text{Ha}^2 = \frac{kC_{aLi}C_{bL}\delta_L}{D_{aL}(C_{aLi}-0)/\delta_L} = \frac{kC_{bL}\delta_L^2}{D_{aL}} = \frac{kC_{bL}D_{aL}}{k_L^2} \tag{19-77}$$

When Ha >> 1 (fast reaction regime), all the reaction occurs in the film and the amount of interfacial area is controlling, necessitating equipment that generates a large interfacial area. When Ha << 1 (kinetic or slow reaction regime), no reaction occurs in the film, and the bulk liquid volume is controlling. As guidance, the following criteria may be used for reactor selection:

Ha < 0.3	Reaction needs large bulk liquid volume.
0.3 < Ha < 3.0	Reaction needs large interfacial area and large bulk liquid volume.
Ha > 3.0	Reaction needs large interfacial area.

Of the parameters making up the Hatta number, liquid diffusivity and mass-transfer coefficient data and measurement methods are well reviewed in the literature.

As discussed in Sec. 7, the factor E represents an enhancement of the rate of mass transfer of A caused by the reaction compared with physical absorption, that is, k_L, *the physical mass transfer coefficient*, is replaced by Ek_L. The theoretical variation of E with Hatta number for a first- and second-order reaction in a liquid film is shown in Fig. 19-25. The uppermost line on the upper right represents the pseudo first-order reaction, for which E = Ha coth (Ha). Three regions are identified with different requirements of liquid holdup ε and interfacial area a, and for which particular kinds of contacting equipment may be best:

Region I, Ha > 2. Reaction is fast and occurs mainly in the liquid film, so $C_{aL} \Rightarrow 0$. The rate of reaction $r_a = k_L a E C_{aLi}$ will be large when a is large, but liquid holdup is not important. Packed towers or stirred tanks will be suitable.

Region II, 0.02 < Ha < 2. Most of the reaction occurs in the bulk of the liquid. Both interfacial area and holdup of liquid should be high. Stirred tanks or bubble columns will be suitable.

Region III, Ha < 0.02. Reaction is slow and occurs in the bulk liquid. Interfacial area and liquid holdup should be high, especially the latter. Bubble columns will be suitable.

The preceding analysis and Fig. 19-25 provide a theoretical foundation similar to the Thiele-modulus effectiveness factor relationship for fluid–solid systems. However, there are no generalized closed-form expressions of E for the more general case of a complex reaction network, and its value has to be determined by solving the complete diffusion-reaction equations for known intrinsic mechanism and kinetics, or else estimated experimentally. Note that the enhancement factor in Fig. 19-25 is defined so that the observed mass transfer equals EC_{Li}, with E ranging from zero to infinity; a different definition of the enhancement factor is often used where the observed mass transfer equals $E'(C_{Li} - C_L)$, and in this case E' ranges from one to infinity.

Some of this theoretical thinking may be used in reactor analysis and design. Unfortunately, some of the parameter values required to undertake a rigorous analysis often are not available. Illustrations of gas–liquid reactors are shown in Fig. 19-26. As discussed in Sec. 7, the intrinsic rate constant k for a liquid-phase reaction without the complications of diffusional resistances may be estimated from properly designed laboratory experiments. Gas and liquid phase holdups may be estimated from correlations or measured. The interfacial area per unit reactor volume a may be estimated from correlations or measurements that use techniques of transmission or reflection of light, though these are limited to small diameter columns. The combined volumetric mass-transfer coefficient on the liquid side, $k_L a$, can be also directly measured in reactive or nonreactive systems (see, for example, Charpentier, *Advances in Chemical Engineering*, vol. 11, Academic Press, New York, 1981, pp. 2–135). Mass-transfer coefficients, interfacial areas, and liquid holdup typical for various gas–liquid reactors are provided in Tables 19-10 and 19-11.

There are many examples of commercial gas–liquid reactions in the literature. These include common operations such as the absorption of ammonia to make fertilizers and of carbon dioxide to make soda ash. Other examples are the recovery of phosphine from off-gases of phosphorous

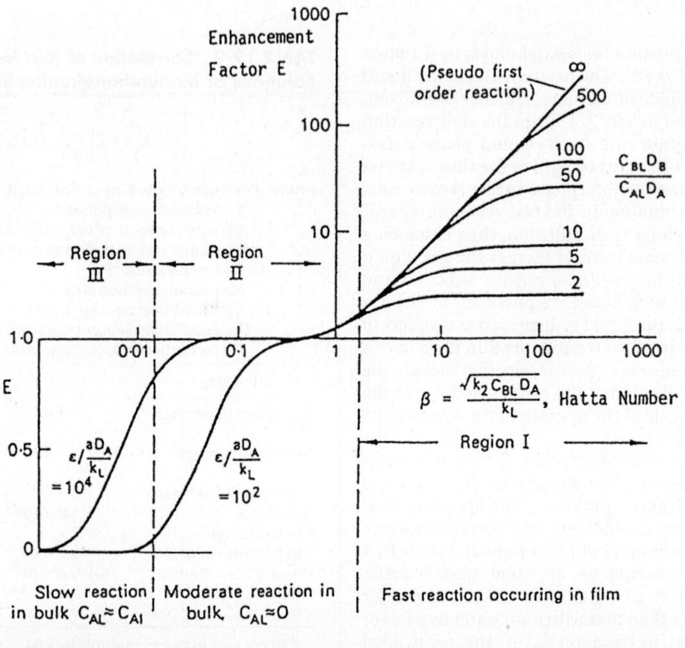

FIG. 19-25 Enhancement factor E and Hatta number of first- and second-order gas–liquid reactions. (*Coulson and Richardson,* Chemical Engineering, *vol. 3, Pergamon, Oxford, UK, 1971, p. 80.*)

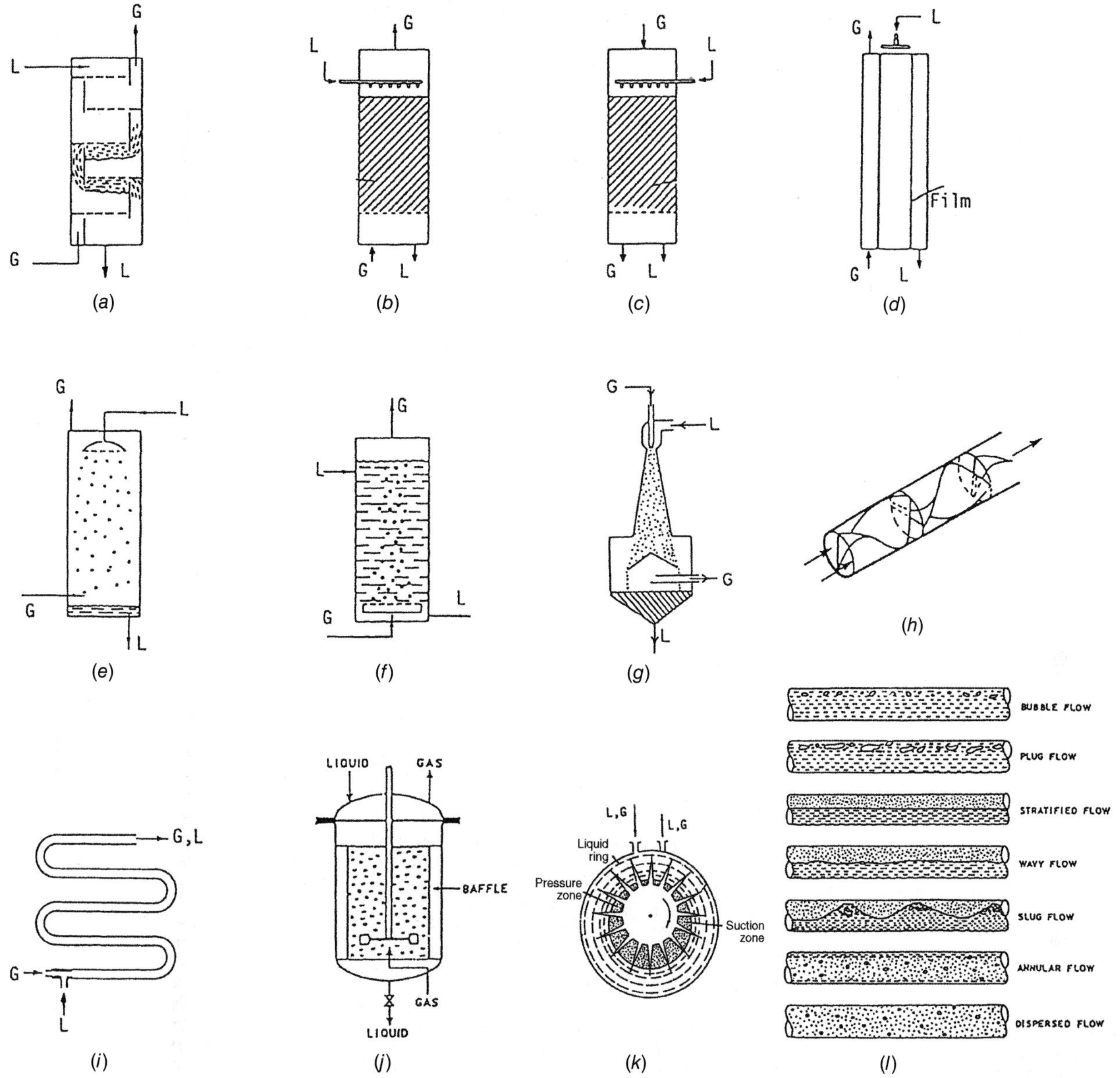

FIG. 19-26 Types of industrial gas-liquid reactors. (*a*) Tray tower. (*b*) Packed, countercurrent. (*c*) Packed, co-current. (*d*) Falling liquid film. (*e*) Spray tower. (*f*) Bubble tower. (*g*) Venturi mixer. (*h*) Static in-line mixer. (*i*) Tubular flow. (*j*) Stirred tank. (*k*) Centrifugal pump. (*l*) Two-phase flow in horizontal tubes.

plants; the recovery of HF; oxidation, halogenation, and hydrogenation of various organics; the hydration of olefins to alcohols; the oxo reaction for higher aldehydes and alcohols; ozonolysis of oleic acid; the absorption of carbon monoxide to make sodium formate; the alkylation of acetic acid with isobutylene to make *tert*-butyl acetate, the absorption of olefins to make various products; and HCl and HBr plus higher alcohols to make alkyl halides. By far the greatest number of applications is for the removal or recovery of mostly small concentrations of acidic and other components from air, hydrocarbons, and hydrogen. Two lists of gas-liquid reactions of industrial importance have been compiled. The literature survey by Danckwerts (*Gas-Liquid Reactions,* McGraw-Hill, New York, 1970) cites 40 different systems. A supplementary list by Doraiswamy and Sharma (*Heterogeneous Reactions: Fluid-Fluid-Solid Reactions,* Wiley, New York, 1984) cites another 50 cases and indicates the most suitable kind of reactor to be used for each. A number of devices have been in use for estimating mass-transfer coefficients, and correlations are available. This topic is reviewed in books, for example, by Danckwerts (*Gas-Liquid Reactions,* McGraw-Hill, New York, 1970) and Charpentier [in Ginetto and Silveston (eds.), *Multiphase Chemical Reactor Theory, Design, Scaleup,* Hemisphere, Washington, D.C., 1986]. One of the issues associated with designing commercial reactors is to properly understand whether data obtained on the laboratory scale are applicable or whether larger-scale data are needed to reduce the scale-up risk.

LIQUID-LIQUID REACTORS

Much of the thinking on gas-liquid reactors is also applicable to liquid-liquid reactors. The liquids are usually not miscible, and the transport of reactants can determine the specific reaction rate. Liquid-liquid reactors require dispersion of one of the liquid phases to provide sufficient interfacial area for mass transfer. This can be achieved by the use of static mixers, jets, or mechanical means.

In a stirred tank, either liquid can be made continuous by charging that liquid first, starting the agitator, and introducing the liquid to be dispersed. For other reactor types, the choice of which phase is continuous and which is dispersed will depend on the physicochemical properties of the phases and operating conditions (such as temperature, pressure, and flow rates). Equipment suitable for reactions between liquids is represented in Fig. 19-27. Almost invariably, one of the phases is aqueous and the other organic, with reactants distributed between phases. Such reactions can be carried out in any kind of equipment that is suitable for physical extraction,

TABLE 19-10 Mass-Transfer Coefficients, Interfacial Areas, and Liquid Holdup in Gas-Liquid Reactions

Type of reactor	ε_L, %	k_G, gm mol/(cm²/s·atm) × 10⁴	k_L, cm/s × 10²	a, cm²/cm³ reactor	k_La, s⁻¹ × 10²
Packed columns					
Countercurrent	2–25	0.03–2	0.4–2	0.1–3.5	0.04–7
Cocurrent	2–95	0.1–3	0.4–6	0.1–17	0.04–102
Plate columns					
Bubble cap	10–95	0.5–2	1–5	1–4	1–20
Sieve plates	10–95	0.5–6	1–20	1–2	1–40
Bubble columns	60–98	0.5–2	1–4	0.5–6	0.5–24
Packed bubble columns	60–98	0.5–2	1–4	0.5–3	0.5–12
Tube reactors					
Horizontal and coiled	5–95	0.5–4	1–10	0.5–7	0.5–70
Vertical	5–95	0.5–8	2–5	1–20	2–100
Spray columns	2–20	0.5–2	0.7–1.5	0.1–1	0.07–1.5
Mechanically agitated bubble reactors	20–95	—	0.3–4	1–20	0.3–80
Submerged and plunging jet	94–99	—	0.15–0.5	0.2–1.2	0.03–0.6
Hydrocyclone	70–93	—	10–30	0.2–0.5	2–15
Ejector reactor	—	—	—	1–20	—
Venturi	5–30	2–10	5–10	1.6–25	8–25

SOURCE: From Charpentier in *Advances in Chemical Engineering*, vol. 11, Academic Press, 1981, pp. 2–135, Table XVIII.

including mixer-settlers and towers of various kinds: empty or packed, still or agitated, either phase dispersed, provided that adequate heat transfer can be incorporated. Mechanically agitated tanks are favored because the interfacial area can be made large, as much as 100 times that of spray towers, for instance. Power requirements for liquid–liquid mixing are about 5 hp/1000 gal. Agitator tip speed of turbine-type impellers is 4.6 to 6.1 m/s (15 to 20 ft/s). Table 19-12 provides data for common types of liquid–liquid contactors. As shown, the given range of k_La is more than 100/1 even for the same equipment. It is provided merely for guidance, and correlations need to be validated with data at some reasonable scale.

Efficiencies of several kinds of small-scale extractors are shown in Fig. 19-28. Larger-diameter equipment may have less than one-half these efficiencies. Spray columns are inefficient and are used only when other kinds of equipment may become clogged. Packed columns as liquid–liquid reactors are operated at 20 percent of flooding. Their *height equivalent to theoretical stage* (HETS) range is from 0.6 to 1.2 m (1.99 to 3.94 ft). Sieve trays minimize backmixing and provide repeated coalescence and redispersion. Mixer-settlers provide approximately one theoretical stage, but several stages can be incorporated in a single shell, although with some loss of operating flexibility. The HETS of a rotating disk contactor (RDC) is 1 to 2 m (3.2 to 6.4 ft). More elaborate staged extractors bring this down to 0.35 to 1.0 m (1.1 to 3.3 ft).

When liquid–liquid contactors are used as reactors, values of their mass-transfer coefficients may be enhanced by reaction, analogously to those of gas–liquid processes. Reactions can occur in either or both phases or near the interface. Nitration of aromatics with HNO_3-H_2SO_4 occurs in the aqueous phase [Albright and Hanson (eds.), *Industrial and Laboratory Nitrations,* ACS Symposium Series 22, American Chemical Society, Washington, D.C., (1975)]. An industrial example of reaction in both phases is the oximation of cyclohexanone, a step in the manufacture of caprolactam for nylon (Rod, *Proc. 4th Int./6th European Symp. Chemical Reactions,* Heidelberg, Pergamon, Oxford, UK, 1976, p. 275). The dioxane forms from isobutene in a hydrocarbon phase and aqueous formaldehyde is preponderantly in the aqueous phase; however, the rate equation is first order in formaldehyde and is also proportional to the concentration of isobutene in the organic phase [Hellin et al., *Genie. Chim.* **91:** 101 (1964)]. Doraiswamy and Sharma (*Heterogeneous Reactions,* Wiley, New York, 1984) have compiled a list of 26 classes of reactions. The reactions include examples such as making soap with alkali, nitration of aromatics to make explosives, and alkylation of C4s with sulfuric acid to make gasoline alkylate.

REACTOR TYPES

The discussion is centered around gas–liquid reactors. If the dissolved gas content exceeds the amount needed for the reaction, the liquid may be first saturated with gas and then sent through a stirred tank or tubular reactor as a single phase. If the residence times for the liquid and gas are comparable, both gas and liquid may be pumped in and out of the reactor together. If the gas has limited solubility, it is bubbled through the reactor, and the residence time for gas is much smaller. Figure 19-29 provides examples of gas–liquid reactors for specific processes.

Stirred Tanks Stirred tanks are common gas–liquid reactors. Reaction requirements dictate whether the gas and liquid are in a batch or continuous mode. For a liquid-phase reaction with a long time constant, a batch mode may be used. The reactor is filled with liquid, and gas is continuously fed into the reactor to maintain pressure. If by-product gases form, these gases may need to be purged continuously. If gas solubility is limiting, a higher-purity gas may be continuously fed (and, if required, recycled). As the required liquid residence time decreases, product may be continuously removed as well. A hybrid reactor type is the semibatch reactor. Gas and liquid are continuously fed to the reactor until the reactor is full of liquid. The reactor then operates as a batch reactor.

Stirred tanks are preferred when high gas–liquid interfacial area is needed. Disadvantages include maintenance of the motor and seals, potential for contamination in biological and food applications, and higher cost.

A basic stirred tank design is shown in Fig. 19-30. Height/diameter ratio is $H/D = 1$ to 3. Heat transfer may be provided through a jacket or internal coils. Baffles prevent movement of the mass as a whole. A draft tube can enhance vertical circulation. The vapor space is about 20 percent of the total volume. A hollow shaft and impeller increase gas circulation by entraining the gas from the vapor space into the liquid. A splasher can be attached to the shaft at the liquid surface to improve the entrainment of gas. A variety of impellers may be used. The pitched propeller moves the liquid axially, the flat blade moves it radially, and inclined blades move it both axially and

TABLE 19-11 Order-of-Magnitude Data of Equipment for Contacting Gases and Liquids

Device	k_La, s⁻¹	V, m³	k_LaV, m³/s (duty)	a, m⁻¹	ε_L	Liquid mixing	Gas mixing	Power per unit volume, kW/m³
Baffled agitated tank	0.02–0.2	0.002–100	10⁻⁴–20	~200	0.9	~Backmixed	Intermediate	0.5–10
Bubble column	0.05–0.01	0.002–300	10⁻⁵–3	~20	0.95	~Plug	Plug	0.01–1
Packed tower	0.005–0.02	0.005–300	10⁻⁵–6	~200	0.05	Plug	~Plug	0.01–0.2
Plate tower	0.01–0.05	0.005–300	10⁻⁵–15	~150	0.15	Intermediate	~Plug	0.01–0.2
Static mixer (bubble flow)	0.1–2	Up to 10	1–20	~1000	0.5	~Plug	Plug	10–500

SOURCE: From J. C. Middlcton, in Harnby, Edwards, and Nienow, *Mixing in the Process Industries*, Butterworth, 1985.

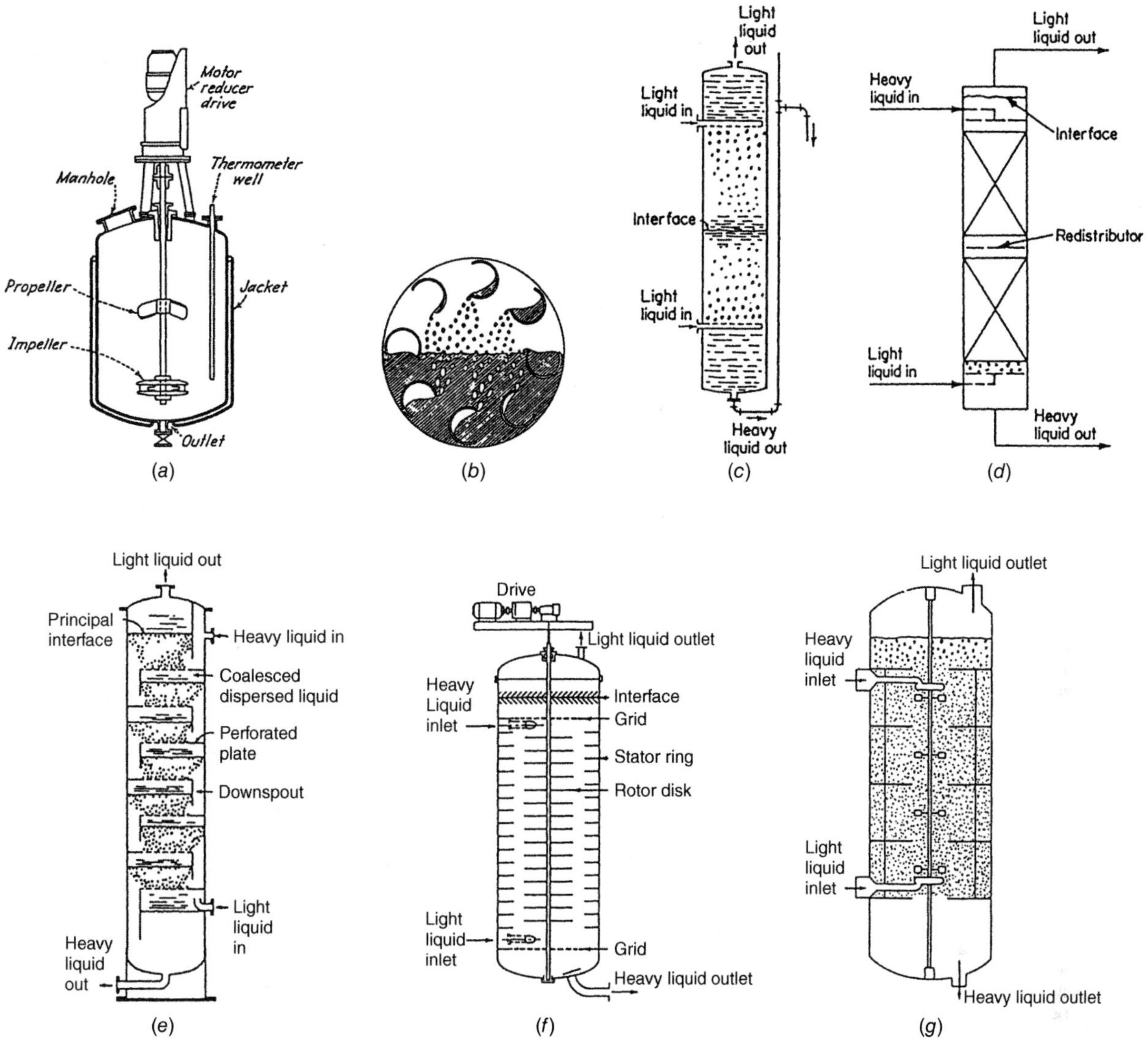

FIG. 19-27 Equipment for liquid–liquid reactions. (*a*) Batch stirred sulfonator. (*b*) Raining bucket (*RTL S A, London*). (*c*) Spray tower with both phases dispersed. (*d*) Two-section packed tower with light phase dispersed. (*e*) Sieve tray tower with light phase dispersed. (*f*) Rotating disk contactor (RDC) (*Escher B V, Holland*). (*g*) Oldshue-Rushton extractor (*Mixing Equipment Co.*).

TABLE 19-12 Continuous-Phase Mass-Transfer Coefficients and Interfacial Areas in Liquid–Liquid Contactors*

Type of equipment	Dispersed phase	Continuous phase	ε_D	τ_D	$k_L \times 10^2$, cm/s	a, cm²/cm³	$k_L a \times 10^2$, s⁻¹
Spray columns	P	M	0.05–0.1	Limited	0.1–1	1–10	0.1–10
Packed columns	P	P	0.05–0.1	Limited	0.3–1	1–10	0.3–10
Mechanically agitated Contactors	PM	M	0.05–0.4	Can be varied over a wide range	0.3–1	1–800	0.3–800
Air-agitated liquid/ liquid contactors	PM	M	0.05–0.3	Can be varied over a wide range	0.1–0.3	10–100	1.0–30
Two-phase cocurrent (horizontal) contactors	P	P	0.05–0.2	Limited	0.1–1.0	1–25	0.1–25

*P = plug flow, M = mixed flow, ε_D = fractional dispersed phase holdup, τ_D = residence time of the dispersed phase.

SOURCE: From Doraiswamy and Sharma, *Heterogeneous Reactions,* Wiley, 1984, Table 14.2.

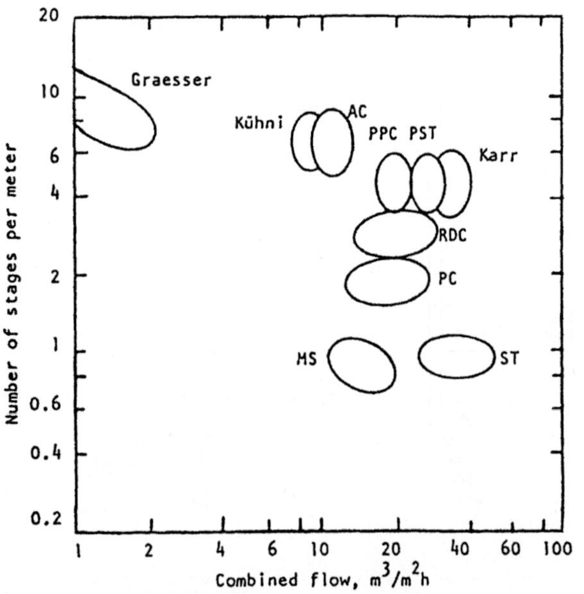

FIG. 19-28 Efficiency and capacity range of small-diameter extractors, 50- to 150-mm diameter. Acetone extracted from water with toluene as the disperse phase, $V_d/V_c = 1.5$. Code: AC = agitated cell; PPC = pulsed packed column; PST = pulsed sieve tray; RDC = rotating disk contactor; PC = packed column; MS = mixer-settler; ST = sieve tray. [*Stichlmair,* Chem. Ing. Tech. ***52****(3): 253–255 (1980).*]

radially. The anchor propeller and some other designs are suited to viscous liquids. For gas dispersion, the six-bladed turbine is preferred. When the ratio of liquid height to diameter is $H/D \le 1$, a single impeller suffices, and in the range $1 \le H/D \le 1.8$ two are needed.

Gases may be dispersed in liquids by spargers or nozzles. However, more intensive dispersion and redispersion are obtained by mechanical agitation. The gas is typically injected at the point of greatest turbulence near the injector tip. Agitation also provides the heat transfer and, if needed, keeps catalyst particles (in a three-phase or slurry reactor) in suspension. Power inputs of 0.6 to 2.0 kW/m^3 (3.05 to 10.15 hp/1000 gal) are suitable. Bubble sizes depend on agitation as well as on the physical properties of the liquid. They tend to be greater than a minimum size regardless of power input due to coalescence. Pure liquids are of a coalescing type; solutions with electrolytes are noncoalescing. Agitated bubble size in air/water is about 0.5 mm (0.020 in), and holdup fractions are about 0.10 for coalescing and 0.25 for noncoalescing liquids; however, more elaborate correlations are available and required for reactor sizing. The reactor may be modeled as two ideal reactors, one for each phase, with mass transfer between the phases. More elaborate models use series of CSTRs for each phase to describe departure from ideal flow patterns. CFD models have also been used, but these models are limited mostly to a description of hydrodynamics, and they are used to a lesser extent for reactions. For example, if the gas has limited solubility and is sparged through a liquid, the gas may be modeled as a PFR and the liquid as a CSTR. Mass-transfer coefficients vary—for example, as the 0.7 exponent on the power input per unit volume (with the dimensions of the vessel and impeller and the superficial gas velocity as additional factors). A survey of such correlations is made by van't Riet [*Ind. Eng. Chem. Proc. Des. Dev.* **18:** 357 (1979)]. Also, Charpentier [in *Multiphase Chemical Reactors,* ed. Gianetto and Silveston, Hemisphere, Washington, D.C., 1986, pp. 104–151] discusses hydrodynamic parameters for stirred tank (and other) reactors, and typical values are shown in Tables 19-10 and 19-11.

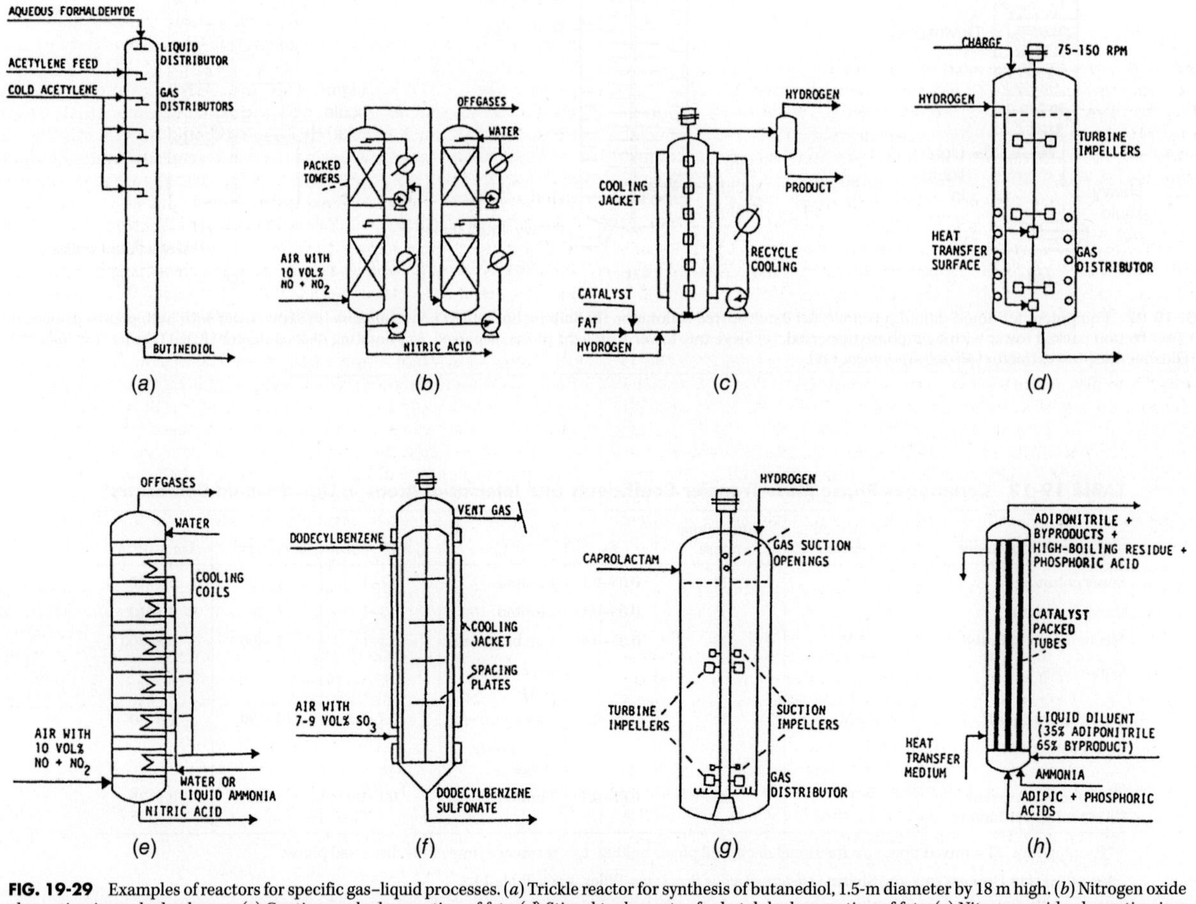

FIG. 19-29 Examples of reactors for specific gas–liquid processes. (*a*) Trickle reactor for synthesis of butanediol, 1.5-m diameter by 18 m high. (*b*) Nitrogen oxide absorption in packed columns. (*c*) Continuous hydrogenation of fats. (*d*) Stirred tank reactor for batch hydrogenation of fats. (*e*) Nitrogen oxide absorption in a plate column. (*f*) A thin-film reactor for making dodecylbenzene sulfonate with SO_3. (*g*) Stirred tank reactor for the hydrogenation of caprolactam. (*h*) Tubular reactor for making adiponitrile from adipic acid in the presence of phosphoric acid.

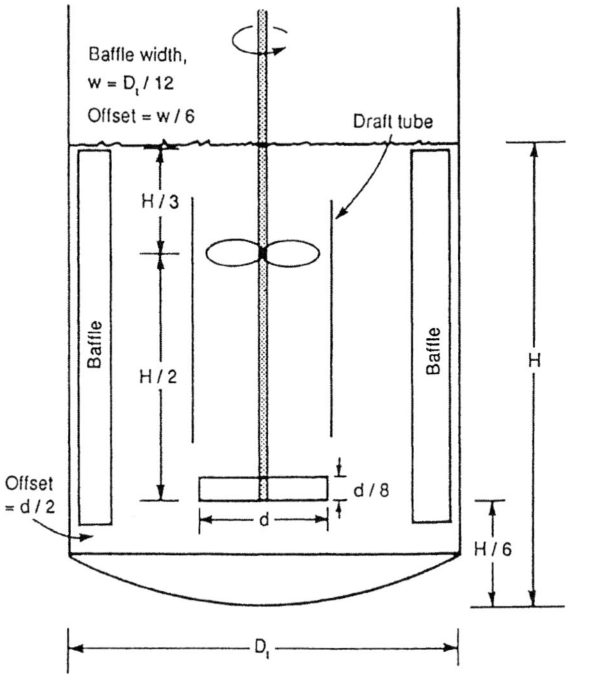

FIG. 19-30 A basic stirred tank design, not to scale, showing a lower radial impeller and an upper axial impeller housed in a draft tube. Four equally spaced baffles are standard. H = height of liquid level, D_t = tank diameter, d = impeller diameter. For radial impellers, $0.3 \le d/D_t \le 0.6$.

Examples

- *Production of penicillin.* An agitated stirred tank is used for the large-scale aerobic fermentation of penicillin by the growth of a specific mold. Commercial vessel sizes are 40,000 to 200,000 L (1400 to 7000 ft^3). The operation is semibatch in that the lactose or glucose nutrient and air are charged at controlled rates to a pre-charged batch of liquid nutrients and cell mass. Reaction time is 5 to 6 d. The broth is limited to 7 to 8 percent sugars, which is all the mold will tolerate. Solubility of oxygen is limited, and air must be supplied over a long period as it is used up. The air is essential to the growth. Dissolved oxygen must be kept at a high level for the organism to survive. Air also serves to agitate the mixture and to sweep out the CO_2 and any noxious by-products that are formed. Air supply is in the range of 0.5 to 1.5 volumes/(volume of liquid)(min). For organisms grown on glucose, the oxygen requirement is 0.4 g/g dry weight; on methanol it is 1.2 g/g. The pH is controlled at about 6.5 and the temperature at 24°C (75°F). The heat of reaction requires cooling water at the rate of 10 to 40 L/(1000 L holdup)(h). Vessels under about 500 L (17.6 ft^3) are provided with jackets, larger ones with coils. For a 55,000-L vessel, 50 to 70 m^2 of heat-transfer area may be taken as average. Mechanical agitation is needed to break up the gas bubbles, but it must avoid rupturing the cells. A disk turbine with radial action is most suitable. It can tolerate a superficial gas velocity up to 120 m/h (394 ft/h) without flooding [whereas a propeller is limited to about 20 m/h (66 ft/h)]. When flooding occurs, the impeller is working in a gas phase and cannot assist the transfer of gas to the liquid phase. Power input by agitation and air sparger is 1 to 4 W/L [97 to 387 Btu/($ft^3 \cdot h$)] of liquid.
- *Refinery alkylation.* C3-C4 olefins are reacted with isobutane in the presence of concentrated acid to form higher-molecular-weight hydrocarbons that may be blended into the gasoline pool. Commercial alkylation processes are catalyzed by either sulfuric or hydrofluoric acid. For both processes, alkylate product quality and acid consumption are affected by temperature, isobutane/olefin ratio, space velocity, and acid concentration. DuPont Stratco's contactor reactor is a horizontal pressure vessel containing an inner circulation tube, a tube bundle to remove the heat of reaction, and a mixing impeller. The hydrocarbon feed and sulfuric acid enter on the suction side of the impeller inside the circulation tube, producing an emulsion. The reaction emulsion is partially separated in a settler, and the acid emulsion is recycled to the contactor's shell side. The hydrocarbon effluent is directed to the contactor's tube bundle, where flash vaporization removes the heat of reaction. Contactor arrangements are also used when the alkylation reaction is conducted using hydrofluoric acid.

Bubble Columns Nozzles or spargers disperse the gas. The mixing and mass transfer are due to rising bubbles, not mechanical agitation. Bubble action provides agitation about equivalent to that of mechanical stirrers (and similar mass- and heat-transfer coefficients) at the same power input per volume. The reaction medium may be a liquid (or a slurry containing a heterogeneous catalyst). To improve the operation, redispersion of gas in liquid or an approach to plug flow may be achieved by using static mixers (such as perforated plates) at regular intervals. Because of their large volume fraction of liquid, bubble column reactors are suited to slow reactions where the rate of reaction is limiting. Major advantages are an absence of moving parts, the ability to handle solid particles without erosion or plugging, good heat transfer at the wall or coils, high interfacial area, and high mass-transfer coefficients. A disadvantage is backmixing in the liquid phase and some backmixing in the gas phase. The static head of the liquid will increase gas pressure drop, and this may be undesirable. Generally, the bubble column height can be greater than for tray or packed towers.

From a mechanical standpoint, a bubble column reactor is a vertical cylindrical vessel with nozzles or a sparger grid at the bottom. The sparger grid is an array of parallel pipes connected to a manifold or several radial arms in a spider pattern or concentric circles, all with downward-facing holes every few inches or so. The holes are sized to give exit velocities of 100 to 300 ft/s, and the gas enters the liquid as jets that break up into bubbles after a short distance. The height/diameter ratio of the vessel is at least 1.5, and it may be as large as 20. Depending on the heat-transfer requirements, coils or a jacket may be needed.

The liquid may be in batch mode or it may enter from the top or bottom. The simplest mathematical model may assume that the liquid is well mixed and the gas is in plug flow. Liquid backmixing may have a detrimental effect on selectivity. In the oxidation of liquid *n*-butane, for instance, the ratio of methyl ethyl ketone to acetic acid is much higher in plug flow than in backmixed reactors. Similarly, in the air oxidation of isobutane to *tert*-butyl hydroperoxide, where *tert*-butanol also is obtained, plug flow is more desirable. Backmixing in the liquid may be reduced with packing or perforated plates. Packed bubble columns operate with flooded packing, in contrast with normal packed columns that usually operate below 70 percent of the flooding point. With packing, liquid backmixing is reduced, and interfacial area is increased 15 to 80 percent, but the true mass-transfer coefficient remains the same. At relatively high superficial gas velocities [10 to 15 cm/s (0.33 to 0.49 ft/s)] and for taller columns, backmixing is reduced, so the vessel performs as a CSTR battery. Radial baffles (also called disk-and-doughnut baffles) are also helpful. A rule of thumb is that the hole should be about 0.7 times the vessel diameter, and the spacing should be 0.8 times the diameter.

The literature may provide guidance on several parameters: bubble diameter and bubble rise velocity, gas holdup, interfacial area, mass-transfer coefficient k_L, axial liquid-phase and gas-phase dispersion coefficients, and heat-transfer coefficient to the wall. The key design variable is the superficial velocity of the gas that affects the gas holdup, the interfacial area, the mass-transfer coefficient, and backmixing. Each of these has been described in some detail by Deckwer (*Bubble Column Reactors*, Wiley, New York, 1992). The effect of vessel diameter on these parameters is not well understood beyond $D \ge 0.15$ to 0.3 m (0.49 to 1 ft), the range for most of the existing literature correlations. From a qualitative standpoint, increasing the superficial gas velocity increases the holdup of gas, the interfacial area, and the overall mass-transfer coefficient. The ratio of height to diameter is not very important in the range of 4 to 10. Decreasing viscosity and decreasing surface tension increase the interfacial area. Electrolyte solutions have smaller bubbles, higher gas holdup, and higher interfacial area. Sparger design is unimportant for superficial gas velocities > 5 to 10 cm/s (0.16 to 0.32 ft/s) and tall columns. Liquid entrainment considerations (discussed in Sec. 14) provide an upper bound on gas superficial velocity; however, gas conversion falls off at higher superficial velocities, so values under 10 cm/s (0.32 ft/s) are often desirable. Some examples of bubble column reactor types are illustrated in Fig. 19-31. Figure 19-31*a* is a conventional bubble column with no internals. Figure 19-31*b* is a tray bubble column. The trays are used to redistribute the gas into the liquid and to induce staging to approximate plug flow. Figure 19-31*c* is a packed bubble column, with the packing being either an inert or a catalyst. Bubble columns are further discussed in the multiphase reactor section.

An excellent reference is Deckwer (*Bubble Column Reactors,* Wiley, New York, 1992). Two complementary reviews of this subject are by Shah et al. [*AIChE J.* **28:** 353–379 (1982)] and Deckwer [in de Lasa, ed., *Chemical Reactor Design and Technology,* Martinus Nijhoff, 1985, pp. 411–461]. Useful comments are made by Doraiswamy and Sharma (*Heterogeneous Reactions,* Wiley, New York, 1984).

Examples

- A number of reactions in the production of pharmaceuticals or crop protection chemicals are conducted in bubble columns. Oxygen and chlorine may be the reactant gases.

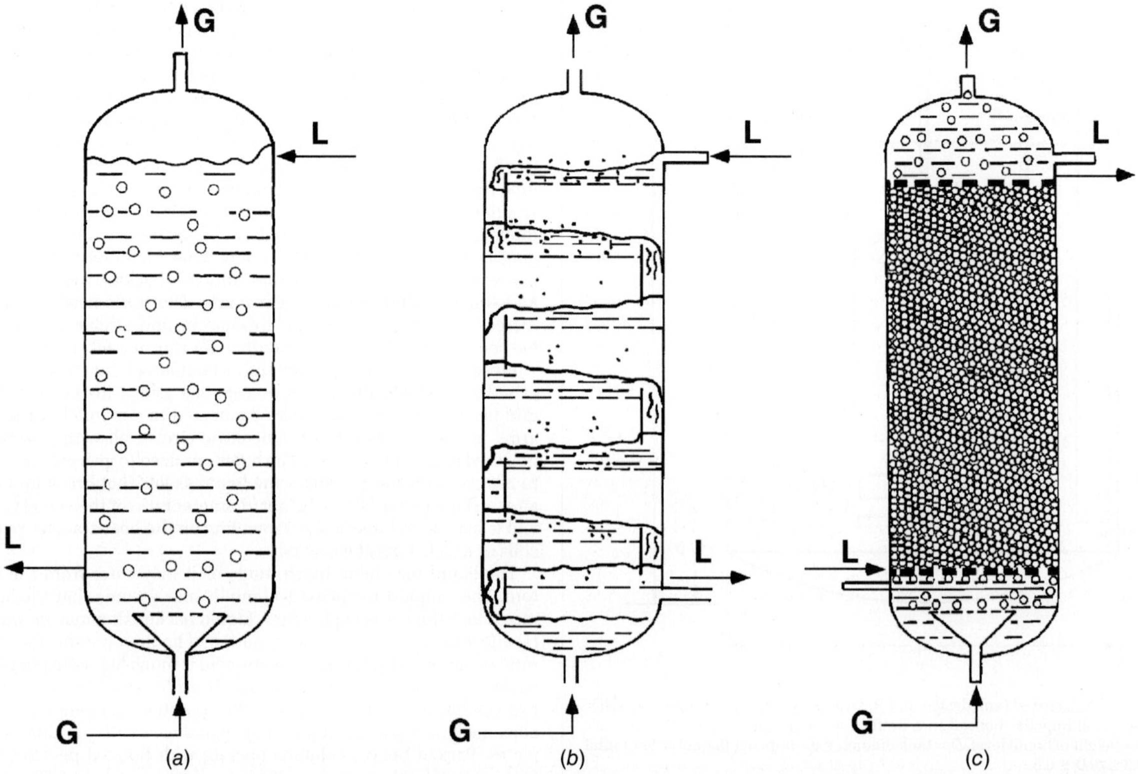

FIG. 19-31 Some examples of bubble column reactor types. (*a*) Conventional bubble column with no internals. (*b*) Tray bubble column. (*c*) Packed bubble column with the packing being either an inert or a catalyst. [*From Mills, Ramachandran, and Chaudhari, "Multiphase Reaction Engineering for Fine Chemicals and Pharmaceuticals,"* Reviews in Chemical Engineering, ***8****(1-2), 1992, Figs. 2, 3, and 4.*]

- Hydrogenation reactions may be carried out in bubble column reactors. Often a slurry catalyst may be used, which makes it a multiphase reactor.
- Aerobic fermentations are carried out in bubble columns when scale advantage is required, and the cells can be considered a third phase, making these multiphase reactors.

Tubular Reactors In a tubular or pipeline reactor, gas and liquid flow concurrently. A variety of flow patterns, ranging from a small quantity of bubbles in the liquid to small quantities of droplets in the gas, are possible, depending on the flow rate of the two streams. Figure 19-26*l* shows the patterns in horizontal flow; those in vertical flow are a little different.

Two-phase tubular reactors offer opportunities for temperature control, accommodate wide ranges of T and P, and approach plug flow, and the high velocities prevent settling of slurries or accumulations on the walls. Mixing of the phases may be improved by helical in-line static mixing inserts. Idealized models use a PFR for both gas and liquid phases.

Depending on the gas and liquid residence times required, the reactor may be operated horizontally or vertically with either downflow or upflow. Weikard (in Ullmann, *Enzyklopaedie,* 4th ed., vol. 3, Verlag Chemie, Weinheim, Ger., 1973, p. 381) discusses possible reasons for operating an upflow concurrent flow tubular reactor for the production of adipic acid nitrile (from adipic acid and ammonia). The reactor has a liquid holdup of 20 to 30 percent and a residence time of 1.0 s for gas and 3 to 5 min for liquid.

1. The process has a large Hatta number; that is, the rate of reaction is much greater than the rate of diffusion, so a large interfacial area is desirable for carrying out the reaction.
2. With typical excess ammonia, the gas/liquid ratio is about 3500 m^3/m^3. At this high ratio there is danger of fouling the surface with tarry reaction products. The ratio is brought down to a more satisfactory value of 1000 to 1500 by recycle of some of the effluent.
3. High selectivity of the nitrile is favored by short contact time.
4. The reaction is highly endothermic, so heat input must be at a high rate.

Points 2 and 4 are the main ones governing the choice of reactor type. The high gas/liquid ratio restricts the choice to types *d, e, i,* and *k* in Fig. 19-26. Due to the high rate of heat transfer needed, the choice is a falling film or tubular reactor.

A horizontal tubular loop reactor is used for the bioconversion of methane to produce biomass used, for example, as fish meal. This is a large-diameter pipe operated at high liquid circulation velocity with the O_2/CH_4 feed injected at several locations along the reactor. Cooling of the exothermic aerobic fermentation is accomplished by external heat exchangers. Static mixers are used to maintain gas dispersion in the liquid.

Packed, Tray, and Spray Towers Packed and tray towers have been discussed in the subsection Mass Transfer in Sec. 5. Typically, the gas and liquid are countercurrent to each other, with the liquid flowing downward. Each phase may be modeled using a PFR or dispersion (series of stirred tanks) model. The model is solved numerically.

Spray columns are used with slurries or when the reaction product is a solid. The coefficient k_L in spray columns is about the same as in packed columns, but the spray interfacial area is much lower. Considerable backmixing of the gas also takes place, which makes the spray volumetrically inefficient. An entrainment control device (e.g., mist eliminator) usually is needed at the outlet. In the treatment of phosphate rock with sulfuric acid, off-gases contain HF and SiF_4. In a spray column with water, solid particles of fluorosilic acid are formed but do not harm the spray operation.

In venturi scrubbers, the gas is the motive fluid. This equipment is of simple design and can handle slurries and large volumes of gas, but the gas pressure drop may be high. When the reaction is slow, further holdup in a spray chamber is necessary.

In liquid ejectors or aspirators, the liquid is the motive fluid, so the gas pressure drop is low. Flow of slurries in the nozzle may be erosive. Otherwise, the design is as simple as that of the venturi. Kohl and Riesenfeld (*Gas Purification,* Gulf, Houston, 1985, pp. 268–288) describe the application of liquid dispersion reactors to the absorption of fluorine gases.

Examples

- Process effluent gas emissions of CO_2 and H_2S are controlled in packed or tray towers. Aqueous solutions of monoethanolamine (MEA), diethanolamine (DEA), and K_2CO_3 are the principal reactive solvents for the removal of acidic constituents from gas streams [Danckwerts and Sharma, *The Chemical Engineer* **202:** CE244 (1966)]. These solvents are all regenerable. Absorption proceeds at a lower temperature or higher pressure, and regeneration is done in a subsequent vessel at higher temperature or lower pressure, usually with some assistance from stripping steam. The CO_2 can be recovered to make dry ice. H_2S is treated for recovery of the sulfur. Vessel diameters and allowable gas and liquid flow rates are established by the same correlations as for physical absorption. The calculation of tower heights uses vapor–liquid equilibrium data and enhanced mass-transfer coefficients for the particular system. Such calculations are complex enough to warrant the use of the professional methods of tower design that are available from a number of service companies. Partly because of their low cost, aqueous solutions of sodium or potassium carbonate also are used for CO_2 and H_2S removal. Potassium bicarbonate has the higher solubility, so the potassium salt is preferred. In view of the many competitive amine and carbonate plants that are in operation, the economics of alternative options have to be reviewed rather carefully. Additives are often used to affect equilibria and enhance absorption coefficients. Sodium arsenite is the major additive in use; however, sodium hypochlorite and small amounts of amines also are effective. *Sterically hindered amines* as promoters are claimed by Say et al. [*Chem. Eng. Prog.* **80**(10): 72–77 (1984)] to result in 50 percent more capacity than ordinary amine promoters of carbonate solutions. Kohl and Riesenfeld (*Gas Purification,* Gulf, Houston, 1985) cite operating data for carbonate plants. Pilot-plant tests are reported on 0.10 and 0.15 m (4 and 6 in) columns packed to depths of 9.14 m (30 ft) of Raschig rings by Benson et al. [*Chem. Eng. Prog.* **50:** 356 (1954)].

- SO_2 emissions from sulfuric acid plants are controlled in spray towers. Effluent gases contain less than 0.5 percent SO_2. The SO_2 emissions have to be controlled (or recovered as elemental sulfur by, for example, the Claus process). An approach is to absorb the SO_2 in a lime (or limestone) slurry (promoted by small amounts of carboxylic acids, such as adipic acid). Flow is in parallel downward. The product calcium salt is sent to a landfill or sold as a by-product. Limestone is pulverized to 80 to 90 percent through 200 mesh. Slurry concentrations of 5 to 40 percent have been used in pilot plants. Rotary wheel atomizers require 0.8 to 1.0 kWh/1000 L. The lateral throw of a spray wheel requires a large diameter to prevent accumulation on the wall; the ratio of length to diameter of 0.5 to 1.0 is in use in such cases. The downward throw of spray nozzles permits smaller diameters but greater depths; L/D ratios of 4 to 5 or more are used. Spray vessel diameters of 15 m (50 ft) or more are known. Liquid/gas ratios are 0.2 to 0.3 gal/MSCF. Flue gas enters at 149°C (300°F) at a velocity of 2.44 m/s (8 ft/s). Utilization of 80 percent of the solid reagent may be approached. Residence times are 10 to 12 s. At the outlet the particles are made just dry enough to keep from sticking to the wall, and the gas is within 11°C to 28°C (20°F to 50°F) of saturation. The fine powder is recovered with fabric filters. In one test facility, a gas with 4000 ppm SO_2 had 95 percent removal with lime and 75 percent removal with limestone.
- A study on the hydrolysis of fats with water was conducted at 230°C to 260°C (446°F to 500°F) and 41 to 48 atm (600 to 705 psi) in a continuous commercial spray tower. A small amount of water dissolved in the fat and reacted to form an acid and glycerine. Most of the glycerine migrated to the water phase. The tower was operated at about 18 percent of flooding, at which condition the HETS was found to be about 9 m (30 ft) compared with an expected 6 m (20 ft) for purely physical extraction [Jeffreys, Jenson, and Miles, *Trans. Inst. Chem. Eng.* **39:** 389–396 (1961)].
- There are instances where an extractive solvent is used to force completion of a reversible homogeneous reaction by removing the reaction product. In the production of KNO_3 from KCl and HNO_3, for instance, the HCl can be removed continuously from the aqueous phase by contact with amyl alcohol, thus forcing completion [Baniel and Blumberg, *Chim. Ind.* **4:** 27 (1957)].

SOLIDS REACTORS

Reactions of solids are typically feasible only at elevated temperatures. High temperatures are achieved by direct contact with combustion gases. Often, the product of reaction is a gas. The gas has to diffuse away from the reactant, sometimes through a solid product. Thermal and mass-transfer resistances are major factors in the performance of solids reactors. There are a number of commercial processes that use solids reactors. Reactor analysis and design appear to rely on empirical models that are used to fit the kinetics of solids decomposition. Most of the information on commercial reactors is proprietary.

General references on solids reactions include Brown, Dollimore, and Galwey ("Reactions in the Solid State," in *Comprehensive Chemical Kinetics,* vol. 22, ed. Bamford and Tipper, Elsevier, Amsterdam, 1980), Galwey (*Chemistry of Solids,* Chapman and Hall, London, 1967), Sohn and Wadsworth, eds. (*Rate Processes of Extractive Metallurgy,* Plenum Press, New York, 1979), Szekely, Evans, and Sohn (*Gas–Solid Reactions,* Academic Press, New York, 1976), and Ullmann, ed. (*Enzyklopaedie der technischen Chemie,* "Uncatalyzed Reactions with Solids," vol. 3, 4th ed., Verlag Chemie, Weinheim, Ger., 1973, pp. 395–464).

THERMAL DECOMPOSITION

Thermal decompositions may be exothermic or endothermic. Solids that decompose on heating without melting often form gaseous products. When the product is a gas, the reaction rate can be affected by diffusion, so particle size can be important. Aging of solids can result in crystallization of the surface. Annealing reduces strains and slows the decomposition rate. The decomposition of some fine powders follows a first-order rate law. Otherwise, empirical rate equations are available (e.g., in Galwey, *Chemistry of Solids,* Chapman and Hall, London, 1967).

A few organic compounds decompose before melting. These decomposition processes are highly exothermic and may cause explosions. Decomposition kinetics may be autocatalytic. The temperature range for decomposition is 100°C to 200°C (212°F to 392°F). The decomposition of oxalic acid (m.p. 189°C) obeyed a zero-order law at 130°C to 170°C (266°F to 338°F). The decomposition of malonic acid has been measured for both the solid and the supercooled liquid.

Exothermic decompositions are nearly always irreversible. When several gaseous products are formed, the reverse reaction would require that these products all combine together, which is unlikely. Commercial interest in such materials lies more in their energy storage properties than as a source of desirable products. These are often nitrogen-containing compounds such as azides, diazo compounds, and nitramines. Ammonium nitrate, an important explosive, decomposes into nitrous oxide and water. In the solid phase, decomposition begins at about 150°C (302°F) but becomes extensive only above its melting point (170°C) (338°F). The reaction is first order, with activation energy of about 40 kcal/(g·mol) [72,000 Btu/(lb·mol)]. Traces of moisture and Cl^- lower the decomposition temperature. Many investigations have reported on the decomposition of azides of barium, calcium, strontium, lead, copper, and silver in the range of 100°C to 200°C (212°F to 392°F). Activation energies were found to be 30 to 50 kcal/(g·mol [54,000 to 90,000 Btu/(lb·mol)] or so. Some difficulties with data reproducibility were encountered with these hazardous materials. Lead styphnate (styphnic acid contains nitrogen) monohydrate was found to detonate at 229°C (444°F). The course of decomposition could be followed at 228°C and below. Sodium azide is a propellant in most motor vehicle SRS systems (airbags). Silver oxalate decomposes smoothly and completely in the range of 100°C to 160°C (212°F to 320°F). Ammonium chromates and some other solids exhibit aging effects. Material that has been stored for months or years follows a different decomposition rate than a fresh material. Examples of such materials are available in the review by Brown et al. ("Reactions in the Solid State," in *Comprehensive Chemical Kinetics,* vol. 22, ed. Bamford and Tipper, Elsevier, Amsterdam, 1980).

Endothermic decompositions are generally reversible. Hydroxides (which give off water) and carbonates (which give off CO_2) have been the most investigated compounds. Activation energies are nearly the same as reaction enthalpies. As the reaction proceeds, the rate of reaction may be limited by diffusion of the water through the product layer. Since a particular compound may have several hydrates, the level of dehydration will depend on the partial pressure of water vapor in the gas. For example, $FeCl_2$ combines with 4, 5, 7, or 12 molecules of water with melting points ranging from about 75°C to 40°C (167°F to 104°F). The dehydration of $CuSO_4$ pentahydrate at 53°C to 63°C (127°F to 145°F) and of the trihydrate at 70°C to 86°C (158°F to 187°F) obeys the Avrami-Erofeyev equation [$-\ln(1-x) = kt^n$, $n = 3.5, 4$]. The rate of water loss from $Mg(OH)_2$ at lower temperatures is sensitive to the partial pressure of water. Its decomposition above 297°C (567°F) yields appreciable amounts of hydrogen and is not reversible. Carbonates decompose at relatively high temperatures—for example, 660°C to 740°C (1220°F to 1364°F) for $CaCO_3$. When deep beds are used, the rate of heat transfer or the rate of CO_2 removal controls the decomposition rate. Some ammonium salts decompose reversibly and release ammonia, such as $(NH_4)_2SO_4 \Leftrightarrow NH_4HSO_4 + NH_3$ at 250°C (482°F). Further heating can release SO_3 irreversibly. The decomposition of silver oxide was one of the earliest solid reactions studied. It is smoothly reversible below 200°C (392°F). The reaction is sensitive to the presence of metallic silver at the start (indicating autocatalysis) and to the presence of silver carbonate, which was accidentally present in some investigations.

SOLID–SOLID REACTORS

In solid–solid reactions, ions or molecules in solids diffuse to the interface prior to reaction. This diffusion takes place through the normal crystal lattices of reactants and products as well as in channels and fissures of imperfect crystals. Solid diffusion is slow compared to liquids, even at the elevated temperatures at which these reactions have to be conducted. Solid–solid reactions are conducted in powder metallurgy. Typical particle sizes are 0.1 to 1000 μm, and pressures are 138 to 827 MPa (20,000 to 60,000 psi). Reactions of solids occur in ceramic, metallurgical, and other industries. Even though cement manufacture has been discussed in the gas–solid reactor section, solid–solid reactions take place as well. Large contact areas between solid phases are essential. These may be obtained by forming and mixing fine powders and compressing them. Reaction times are 2 to 3 h at 1200°C to 1500°C (2192°F to 2732°F), even with 200-mesh particles.

The literature reports several examples of laboratory solid–solid reactions. The mechanism of zinc ferrite formation ($ZnO + Fe_2O_3 \Rightarrow ZnFe_2O_4$) has been studied up to temperatures of 1200°C (2192°F). At lower temperatures, ZnO is the mobile phase that migrates and coats the Fe_2O_3 particles. Similarly, MgO is the mobile phase in the $MgO + Fe_2O_3 \Rightarrow MgFe_2O_4$ reaction. Smaller particles (<1 μm) obey the power law $x = k \ln t$, but larger ones have a more complex behavior. In the reaction $2AgI + HgI_2 \Rightarrow Ag_2HgI_4$, nearly equivalent amounts of the ions Ag^+ and Hg_2^+ were found to migrate in opposite directions and arrive at their respective interfaces after 66 days at 65°C (149°F).

Several reactions that yield gaseous products have attracted attention because their progress is easily followed. Examples include $MnO_3 + 2MoO_3 \Rightarrow 2MnMoO_4 + 0.5O_2$ (where MoO_3 was identified as the mobile phase) and $Ca_3(PO_4)_2 + 5C \Rightarrow 3CaO + P_2 + 5CO$. For the reaction $KClO_4 + 2C \Rightarrow KCl + CO_2$, fine powders were compressed to 69 MPa (10,000 psi) and reacted at 350°C (662°F), well below the 500°C (932°F) melting point. The reaction $CuCr_2O_4 + CuO \Rightarrow Cu_2Cr_2O_4 + 0.5O_2$ eventually becomes diffusion controlled and is described by the relationship $[1 - (1 - x)^{1/3}]^2 = k \ln t$. In the reaction $CsCl + NaI \Rightarrow CsI + NaCl$, two solid products are formed. The rate-controlling step is the diffusion of iodide ion in CsCl.

Carbothermic reactions are solid–solid reactions with carbon that apparently take place through intermediate CO and CO_2. The reduction of iron oxides has the mechanism $Fe_xO_y + yCO \Rightarrow xFe + yCO_2$, $CO_2 + C \Rightarrow 2CO$. The reduction of hematite by graphite at 907°C to 1007°C in the presence of lithium oxide catalyst was correlated by the equation $1 - (1 - x)^{1/3} = kt$. The reaction of solid ilmenite ore and carbon has the mechanism $FeTiO_3 + CO \Rightarrow Fe + TiO_2 + CO_2$, $CO_2 + C \Rightarrow 2CO$. A similar case is the preparation of metal carbides from metal and carbon, $C + 2H_2 \Rightarrow CH_4$, $Me + CH_4 \Rightarrow MeC + 2H_2$.

Self-Propagating High-Temperature Synthesis (SHS) Conventional methods of synthesizing materials via solid reactions involve multiple grinding, heating, and cooling of suitable precursor compounds. Reactions need extended time periods mainly because interdiffusion in solids is slow, even at high temperatures. By contrast, in SHS, highly reactive metal particles ignite in contact with boron, carbon, nitrogen, and silica to form boride, carbide, nitride, and silicide ceramics. Since the reactions are extremely exothermic, the reaction fronts propagate rapidly through the precursor powders. Usually, the ultimate particle size can be controlled by the particle size of the precursors. In recent years, several commercial and semicommercial facilities have been built (in Russia, the United States, Spain, and Japan) to synthesize TiC powders, nitrided ferroalloys, silicon nitride (β-phase) and titanium hydride powders, high-temperature insulators, lithium niobate, boron nitride, and others (see Weimer, *Carbide, Nitride and Boride Materials Synthesis and Processing*, Chapman & Hall, London, 1997).

MULTIPHASE REACTORS

Multiphase reactors include, for instance, gas-liquid-solid and gas-liquid-liquid reactions. In many important cases, reactions between gases and liquids occur in the presence of a porous solid catalyst. The reaction typically occurs at a catalytic site on the solid surface. The kinetics and transport steps include dissolution of gas into the liquid, transport of dissolved gas to the catalyst particle surface, and diffusion and reaction in the catalyst particle. Say the concentration of dissolved gas A in equilibrium with the gas-phase concentration of A is C_{aLi}. Neglecting the gas-phase resistance, and assuming that no catalyst particles are located in the liquid film (so reaction cannot occur there), the series of rates involved are from the liquid side of the gas–liquid interface to the bulk liquid where the concentration is C_{aL}, and from the bulk liquid to the surface of catalyst where the concentration is C_{as} and where the reaction rate is ηwkC_{as}^n. At steady state,

$$r_a = k_L a(C_{aLi} - C_{aL}) = k_s a_s(C_{aL} - C_{as}) = \eta wkC_{as}^n \qquad (19\text{-}78)$$

where w is the catalyst loading (mass of catalyst per slurry volume). For a first-order reaction, $n = 1$, the catalyst effectiveness η is independent of C_{as}, so that after elimination of C_{aL} and C_{as} the explicit solution for the observed specific rate is

$$r_{a,\text{observed}} = C_{aLi}\left(\frac{1}{k_L a} + \frac{1}{k_s a_s} + \frac{1}{\eta wk}\right)^{-1} \qquad (19\text{-}79)$$

More complex chemical rate equations will require numerical solution. Ramachandran and Chaudhari (*Three-Phase Chemical Reactors*, Gordon and Breach, Philadelphia, 1983) apply such rate equations to the sizing of plug flow, CSTR, and dispersion reactors. They list 75 reactions and identify reactor types, catalysts, temperature, and pressure for processes such as hydrogenation of fatty oils, hydrodesulfurization, Fischer-Tropsch synthesis, and miscellaneous hydrogenations and oxidations. A list of 74 gas–liquid–solid reactions with literature references has been compiled by Shah (*Gas–Liquid–Solid Reactions*, McGraw-Hill, New York, 1979), classified into groups where the solid is a reactant, a catalyst, or an inert. Other references include de Lasa (*Chemical Reactor Design and Technology*, Martinus Nijhoff, 1986), Gianetto and Silveston, eds. (*Multiphase Chemical Reactors*, Hemisphere, Washington, D.C., 1986), Ramachandran et al., eds. (*Multiphase Chemical Reactors*, vol. 2, Sijthoff & Noordhoff, Leiden, NL, 1981), and Satterfield ["Trickle Bed Reactors," *AIChE J.* **21**: 209–228 (1975)]. Some contrasting characteristics of the main kinds of three-phase reactors are summarized in Table 19-13.

BIOREACTORS

Bioreactors use live cells or enzymes to perform biochemical transformations of feedstocks to desired products. Bioreactor operation is restricted to conditions at which these biological systems can function. Most plant and animal cells live at moderate temperatures and do not tolerate extremes of pH. The vast majority of microorganisms also prefer mild conditions, but some thrive at temperatures above the boiling point of water or at pH values far from neutral. Some can endure concentrations of chemicals that most

TABLE 19-13 Characteristics of Gas–Liquid–Solid Reactors

Property	Trickle bed	Flooded	Stirred tank	Entrained solids	Fluidized bed
Gas holdup	0.25–0.45	Small	0.2–0.3		
Liquid holdup	0.05–0.25	High	0.7–0.8		
Solid holdup	0.5–0.7		0.01–0.10		0.5–0.7
Liquid distribution	Good only at high liquid rate		Good	Good	Good
RTD, liquid phase	Narrow	Narrower than for entrained solids reactor	Wide	Wide	Narrow
RTD, gas phase	Nearly plug flow		Backmixed	Backmixed	Narrow
Interfacial area	20–50% of geometrical	Like trickle bed reactor	100–1500 m^2/m^3	100–400 m^2/m^3	Less than for entrained solids reactor
MTC, gas/liquid	High		Intermediate		
MTC, liquid/solid	High		High		
Radial heat transfer	Slow		Fast	Fast	Fast
Pressure drop	High with small d_p		Hydrostatic head		

RTD = residence time distribution; MTC = mass-transfer coefficient.

other cells find highly toxic. Commercial operations depend on having the correct organisms or enzymes and preventing death (or deactivation) or the entry of foreign organisms that could harm the process.

The pH, temperature, redox potential, and nutrient medium may favor certain organisms and discourage the growth of others. In mixed culture systems, especially those for biological waste treatment, there is an ever-shifting interplay between microbial populations and their environments that influences performance and control. Although open systems may be suitable for hardy organisms or for processes in which the conditions select the appropriate culture, many bioprocesses are closed, and elaborate precautions including sterilization and cleaning are taken to prevent contamination. The optimization of the complicated biochemical activities of isolated strains, of aggregated cells, of mixed populations, and of cell-free enzymes or components presents engineering challenges. Performance of a bioprocess can suffer from changes in any of the many biochemical steps functioning in concert, and genetic controls are subject to mutation. Offspring of specialized mutants, especially bioengineered ones that yield high concentrations of product, tend to revert during propagation to less productive strains—a phenomenon called *rundown.*

Developments such as immobilized enzymes and cells are being exploited for increased productivity, and genetic manipulations through recombinant DNA techniques are leading to practical processes for molecules that could previously be found only in trace quantities in plants or animals.

Bioreactors may have either two phases (liquid–solid, e.g., in anaerobic processes) or in most cases three phases (gas–liquid–solid, e.g., aerobic processes and anaerobic processes that evolve CO_2). The solid phase typically contains cells that serve as the biocatalyst. The solid can be either the free biocatalyst (e.g., bacteria, fungi, algae), also called the biotic phase (with density close to water), or an immobilized version, in which case the cells are immobilized on a solid structure (e.g., porous particles). The liquid is primarily water with dissolved feed (usually a sugar together with mineral salts and trace elements) and products (referred to as metabolites). In aerobic bioreactors, the gas phase is primarily air, with the product gas containing product CO_2 (produced by the organism) and evaporated water. Bioreactors are mainly mechanically agitated tanks, bubble columns, and air lift reactors. For low-biomass concentrations (e.g., less than 60 g/L), bioreactor design is similar to that of a gas–liquid reactor. For some specialized applications, such as in some wastewater treatment processes, packed beds or slurry reactors with immobilized biocatalyst are used. Figure 19-32 shows some typical bioreactors.

While bioreactors do not differ fundamentally from other two- and three-phase reactors, there are more stringent requirements regarding control of temperature, pH, contamination (presence and growth of other microorganisms or phage), and toxicity (which may result from high feed and product concentrations). In aerobic processes, since O_2 is required for respiration, it must be properly distributed and managed. Whereas bacteria and yeast cells are very robust, cultivations of filamentous fungi and especially animal cell cultures and plant cell cultures are quite shear-sensitive. To maintain a robust culture of animal and plant cells, very gentle stirring either by a mechanical stirrer or by gas sparging is usually necessary. Unlike chemical catalysis, one of the (main) bioreaction products is biomass (new cells), leading to autocatalytic behavior; i.e., the rate of production of new cells per liquid volume is proportional to the cell concentration. Section 7 presents more details on the kinetics of bioreactions.

Bioreactors mainly operate in batch or semibatch mode, which allows better control of the key variables. However, an increasing number of bioprocesses are operated in continuous mode, typically processes for treating wastewater, but also large-scale processes such as lactic acid production, conversion of natural gas to biomass (single-cell protein production), and production of human insulin using genetically engineered yeast. Continuous operation requires good process control, especially of the sterility of the feed, but also that the biocatalyst be robust and its traits (especially for bioengineered strains) persist over many generations.

Several special terms are used to describe traditional reaction engineering concepts. Examples include *yield coefficients* for the generally fermentation environment-dependent stoichiometric coefficients, *metabolic network* for reaction network, *substrate* for feed, *metabolite* for secreted or in-cell accumulated bioreaction products, *biomass* for cells, *broth* for the fermenter medium, *aeration rate* for the rate of air addition, *vvm* for volumetric airflow rate per broth volume, OUR for O_2 uptake rate per broth volume, and CER for CO_2 evolution rate per broth volume. For continuous fermentation, *dilution rate* stands for feed or effluent rate (equal at steady state) per broth volume, and *washout* for a condition where the feed rate exceeds the cell growth rate, resulting in washout of cells from the reactor. Section 7 discusses a simple model of a CSTR reactor (called a chemostat) using empirical kinetics.

The mass conservation equations for a batch reactor are as follows:

$$\text{Cells:}\quad V_r \frac{dC_x}{dt} = (r_x - r_d)V_r \tag{19-80}$$

$$\text{Substrate:}\quad V_r \frac{dC_s}{dt} = -Y_{xs}\, r_x V_r - r_{sm} V_r \tag{19-81}$$

$$\text{Product:}\quad V_r \frac{dC_p}{dt} = Y_{xp} r_x V_r \tag{19-82}$$

Several of these terms have been discussed in Sec. 7: r_x and r_d are the specific rates (per broth volume) for cell growth and death, respectively; r_{sm} is the specific rate of substrate consumed for cell maintenance, and Y_{xi} are the stoichiometric yield coefficient of species i relative to biomass x. The maintenance term in Eq. (19-81) can also result in an increased production of product p from the extra substrate consumption for maintenance [an additional term is then required in Eq. (19-82)]. In many cases, a semibatch reactor is used, where the reactants are added with initial ingredients, cells and sugar concentration, and a certain feed profile or recipe is used—this is also called *fed batch* operation mode.

Further modeling details are available in the books by Villadsen, Nielsen, and Liden (*Bioreaction Engineering Principles*, 3d ed., Springer, New York, 2011) and Fogler (*Elements of Chemical Reaction Engineering*, 3d ed., Prentice-Hall, Upper Saddle River, N.J., 1999). Bioreactors and bioreaction engineering are discussed in detail by Bailey and Ollis (*Biochemical Engineering Fundamentals*, 2d ed., McGraw-Hill, New York, 1986), Clark (*Biochemical Engineering*, Marcel Dekker, New York, 1997), and Schugerl and Bellgardt (*Bioreaction Engineering, Modeling and Control*, Springer, Berlin, 2000).

ELECTROCHEMICAL REACTORS

Electrochemical reactors are used for electrolysis (conversion of electric energy to chemicals, e.g., chlor-alkali), power generation (conversion of chemicals to electric energy, e.g., batteries or fuel cells), electrochemical machining and forming, or chemical separations (electrodialysis). An electrochemical cell contains at least two electronically conducting electrode phases and one ionic conducting electrolyte phase. The electrolyte phase separates the two electrode phases. The electrode phases are also connected to each other through an electronically conducting pathway typically external of the electrochemical cell; but in the case of corrosion, the electrode phases may be localized regions on the same piece of metal, the bulk metal allowing electron flow between the regions. Thus a series electric circuit is completed beginning at one electrode through the electrolyte to the second electrode and then out of the reactor through the external circuit back into the starting electrode.

An electrochemical cell reaction involves the transfer of electrons across an electrode/electrolyte interface. There are two types of electrochemical cell reactions. In one reaction, the electron transfer is from an electrode to a chemical species within the electrolyte, resulting in a reduction process, and in this case the electrode is defined as the cathode. The second electrochemical reaction involves the electron transfer from a chemical species within the electrolyte to an electrode, resulting in an oxidation process; in this case the electrode is defined as the anode. Each of these cathode (reduction) or anode (oxidation) electrochemical reactions is considered a half-cell reaction. Since an electrochemical cell requires a complete series electric circuit, the overall electrochemical cell reaction is the stoichiometric sum of the electrochemical half-cell reactions, and all electrochemical cell reactions are close-coupled to maintain the conservation of electric charge. Electrochemical cell reactions are considered heterogeneous reactions since they occur at the interface of the electrode surface and electrolyte. Sometimes the electrochemical product species is employed, in turn, as a reducing or oxidizing species, either in the bulk electrolyte or in a separate external process vessel. Subsequently, the spent reducing or oxidizing species is regenerated within the electrochemical reactor. This augmentation is known as a mediated (or indirect) electrochemical process.

An electrochemical reactor is a controlled volume containing the electrolyte and two electrodes. The electrode phases may be a solid, such as carbon or metal, or a liquid, such as mercury. The geometry of the electrodes is optimized to maximize energy efficiency and/or cell life, and it usually consists of parallel plates or concentric cylinders. The electrolyte may be a liquid (such as concentrated brine in the production of caustic or a molten salt in the production of aluminum) or a solid (such as a proton-conducting Nafion® membrane in fuel cells). As the electric current passes through the electrolyte, a voltage drop occurs that represents an energy loss; therefore, the gap or spacing between the electrodes is usually minimized. The electrodes may also be separated by a membrane, a diaphragm, or a separator so as to prevent the unwanted mixing of chemical species, ensure process safety, and maintain product purity and yield. One or both of the electrodes may evolve a gas (e.g., chlorine); or alternatively, one or both of the electrodes may be fed with a gas (e.g., hydrogen or oxygen) to reduce cell voltage or use gaseous feedstocks. Examples of electrochemical reactors are shown in Fig. 19-33.

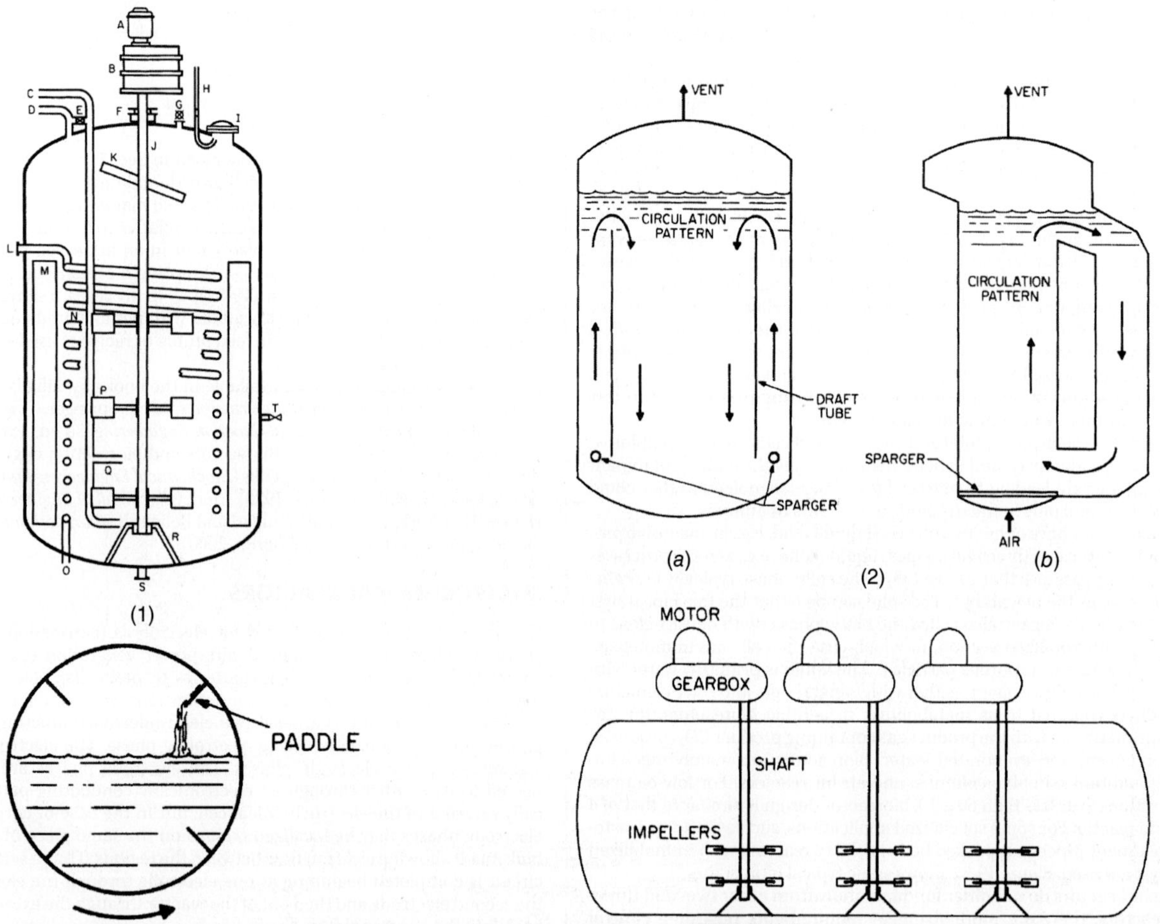

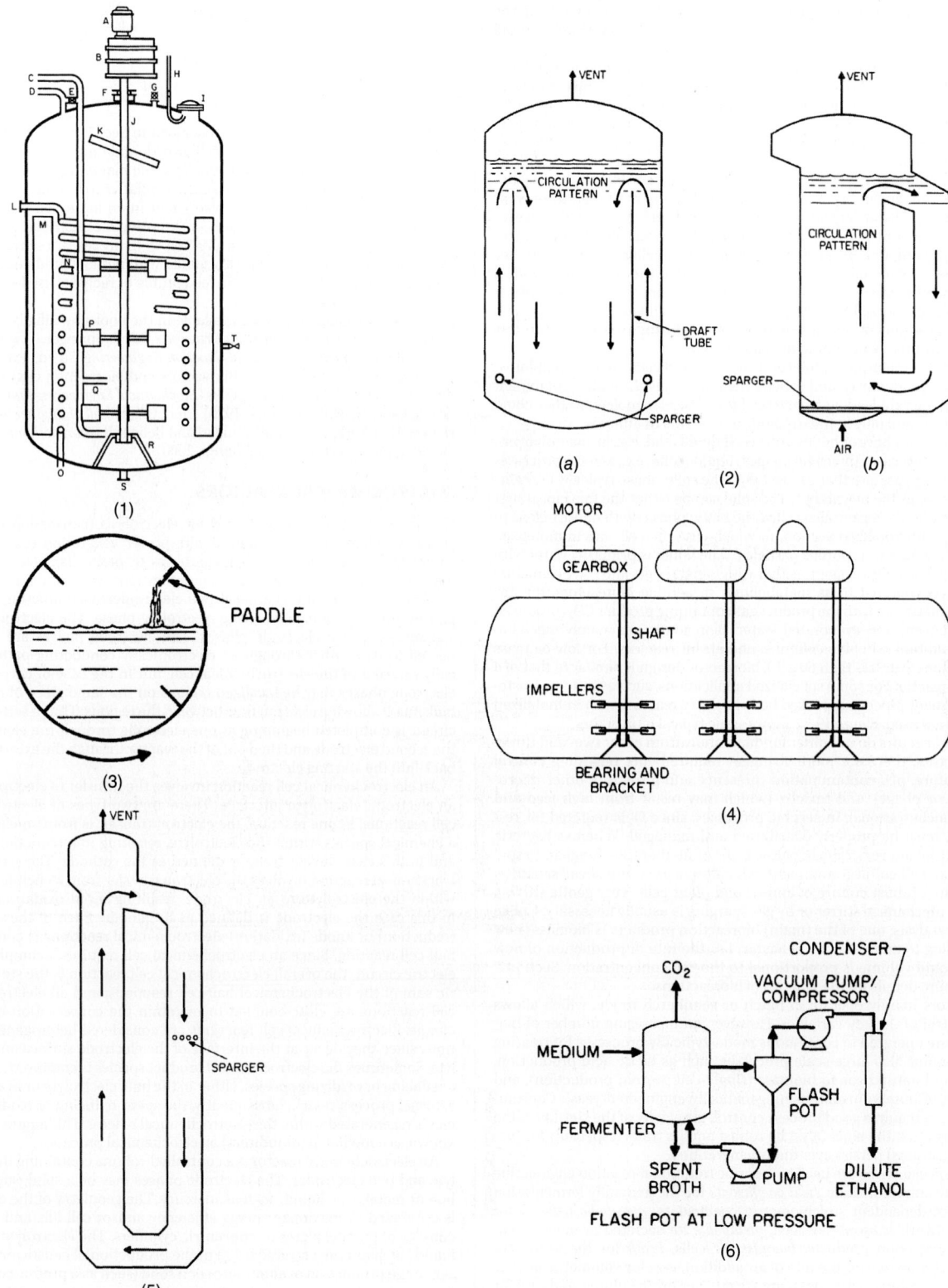

FIG. 19-32 Some examples of fermenters. (1) Conventional batch fermenter. (2) Air lift fermenters: (*a*) Concentric cylinder or bubble column with draft tube; (*b*) external recycle. (3) Rotating fermenter. (4) Horizontal fermenter. (5) Deep-shaft fermenter. (6) Flash-pot fermenter.

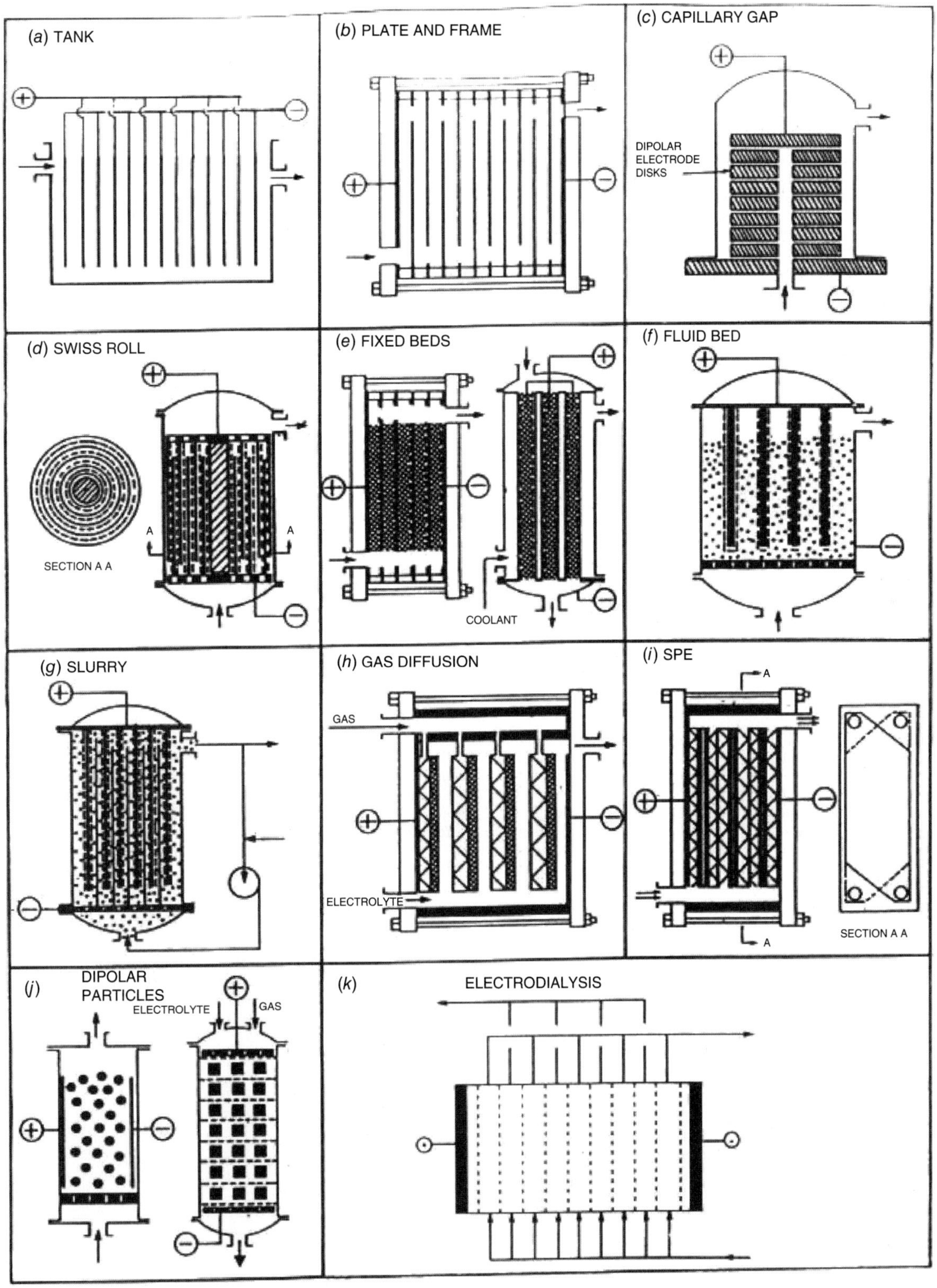

FIG. 19-33 Electrochemical reactor configurations. [*From Oloman,* Electrochemical Processing for the Pulp and Paper Industry, *The Electrochemical Consultancy, 1999, p. 79, Fig. 2.10; printed in Great Britain by Alresford Press Ltd. Referring to "Tutorial Lectures in Electrochemical Engineering and Technology" (D. Chin and R. Alkire, eds.),* AIChE Symposium Series 229, *vol. 79, 1983; reproduced with permission.*]

The size of an electrochemical reactor may be determined by evaluating the capital costs and the operating costs (on a dollar per unit mass basis) as a function of the operating current density (production rate per unit electrode area basis). Typically, the capital costs decrease with increasing current density, and the operating costs increase with current density. Thus, there is a minimum point in the total costs, and this can serve as a basis for the sizing of the electrochemical reactor. Given an optimal current density, the electrochemical reactor design is refined to minimize voltage losses and maximize current efficiency. This is done by taking into consideration the component availability (e.g., membrane widths), the management of the excess heat, the minimization of pressure drops (due to liquid and gas traffic within the electrochemical reactor), and the maintenance costs (associated with reactor rebuilding). The largest, most cost-effective reactor size is then replicated to meet production capacity needs. An electrochemical reactor usually has a shorter operating life than the rest of the plant facility, requiring the periodic rebuilding of the reactors.

In electrochemical engineering, several terms share similar definitions to those in traditional reaction engineering. These include fractional conversion, yield, selectivity, space velocity, and space time yield. Several terms are unique to electrochemical reaction engineering, such as *cell voltage* (the electric potential difference between the two electrodes within the electrochemical cell) and *cell overpotentials* (voltage losses within the electrochemical cell). Voltage losses include (1) *ohmic overpotential* (associated with the passage of electric current in the bulk of the electrolyte phase and the bulk electrode phases, and the electrical conductors between the electrochemical cell and the power supply or electrical load); (2) *activation overpotential* (associated with the limiting rates at which some steps in the electrode reactions can proceed); and (3) *concentration overpotential* (generated from the local depletion of reactants and the accumulation of products at the electrode/electrolyte interface relative to the bulk electrolyte phase due to mass transport limitations). The *current density* is the current per unit surface area of the electrode. Typically, the geometric or projected area is used because the true electrode area is difficult to estimate due to surface roughness or porosity. It is related to the production rate of the electrolytic cell through the Faraday constant. The *current efficiency* is the ratio of the theoretical electric charge (coulombs) required for the amount of product obtained to the total amount of electric charge passed through the electrochemical cell. Many of these and other terms are discussed in Sec. 7, in Pletcher and Walsh (*Industrial Electrochemistry*, 2d ed., Chapman and Hall, London, 1984) and in Gritzner and Kreysa ["Nomenclature, Symbols and Definitions in Electrochemical Engineering," *Pure & Appl. Chem.* **65**(5): 1009–1020 (1993)].

A discussion of electrochemical reactors is available in books by Prentice (*Electrochemical Engineering Principles*, Prentice-Hall, Englewood Cliffs, N.J., 1991), Hine (*Electrode Processes and Electrochemical Engineering*, Plenum Press, New York, 1985), Oloman (*Electrochemical Processing for the Pulp and Paper Industry*, The Electrochemical Consultancy, 1996), and Goodridge and Scott (*Electrochemical Process Engineering: A Guide to the Design of Electrolytic Plant*, Plenum, New York, 1995). See the review of Mah, Taylor, Inman, Botte, Reimer, and Orazem ["Electrochemical Manufacturing in the 21st Century," *ECS Interface* **23**(3): 47–67 (2014)]. See also Electrochemical Reactions in Sec. 7.

REACTOR TYPES

Multiphase reactors are typically mechanically agitated vessels, bubble columns, trickle beds, flooded fixed beds, gas–liquid–solid fluidized beds, and entrained solids reactors. *Agitated reactors* keep solid catalysts in suspension mechanically; the overflow may be a clear liquid or slurry, and the gas disengages from the vessel. *Bubble column reactors* keep the solids in suspension as a result of agitation caused by the sparging gas. In *trickle bed reactors*, both gas and liquid phases flow down through a packed bed of catalyst. The reactor is gas continuous, with liquid "trickling" as a film over the solid catalyst. In *flooded reactors or packed bubble columns,* the gas and liquid flow upward through a fixed bed. The reactor is liquid continuous. In gas–liquid–solid fluidized beds, as the superficial velocity is increased, the solids first become suspended (as a dense fluidized bed) and may eventually be entrained and the effluent separated into its phases in downstream equipment. When the average residence time of solids approaches that of the liquid, the reactor becomes an *entrained solids reactor.*

Agitated Slurry Reactors The gas reactant and solid catalyst are dispersed in a continuous liquid phase by mechanical agitation using stirrers. Most issues associated with gas–liquid–solid stirred tanks are analogous to the gas–liquid systems. In addition to providing good gas–liquid contacting, the agitation has to be sufficient to maintain the solid phase suspended. Catalytic reactions in stirred gas–liquid–solid reactors are used in a large number of applications, including hydrogenations, oxidations, halogenations, and fermentations.

The benefits of using a mechanically agitated tank include nearly isothermal operation, excellent heat transfer, good mass transfer, and the use of high-activity powder catalyst with minimal intraparticle diffusion limitations. The reactors may be operated in a batch, semibatch, or continuous mode; and catalyst deactivation may be managed by on-line catalyst makeup. Scale-up is relatively straightforward through geometric similarity and by providing the agitator power and volume required to produce the same volumetric mass-transfer coefficient and adequate mixing at larger scales, while also accounting for the respective flow regime at different scales. The hydrodynamics are decoupled from the gas flow rate. Some downsides of stirred gas–liquid–solid reactors include difficulty with catalyst/liquid product separation and lower volumetric productivity than fixed beds (due to lower catalyst loading per reactor volume). In addition, the reactor size may be limited due to high power consumption (due to horsepower limitations on the agitator motor)—typically the limit is at around 50 m^3, though volumes as large as 200 m^3 have been used for specific applications. Sealing of the agitator system can also be challenging for large reactors (magnetic coupling is used for small to midrange units). These result in increased capital and operating costs.

Solid particles are in the range of 0.01 to 1.0 mm (0.0020 to 0.039 in), the minimum size limited by filterability. Small particle diameters are used to provide as large an interface as possible to minimize the liquid–solid mass-transfer resistance and intraparticle diffusion limitations. Solids concentrations up to 30 percent by volume may be handled; however, lower concentrations may be used as well. For example, in the hydrogenation of oils with Ni catalyst, the solids content is about 0.5 percent. In the manufacture of hydroxylamine phosphate with Pd-C, the solids content is 0.05 percent.

The hydrodynamic parameters that are required for stirred tank design and analysis include phase holdups (gas, liquid, and solid); volumetric gas–liquid mass-transfer coefficient; liquid–solid mass-transfer coefficient; liquid, gas, and solid mixing; and heat-transfer coefficients. The hydrodynamics are driven primarily by the stirrer power input and the stirrer geometry and type, and not by the gas flow, though the gas flow affects agitation efficiency. Key agitator characteristics include the required power input and the pumping flow rate of the stirrer.

The reactant gas either is sparged below the stirrer or is induced from the vapor space by a *gas-inducing* agitator that has a hollow shaft with suction orifices on the shaft and discharge orifices on the impeller. Impellers vary with applications. For low-viscosity applications, flat-bladed Rushton turbines are widely used, and they provide radial mixing and gas dispersion. Pitched-blade turbines may also be used to induce axial flow. Often multiple impellers are provided on one shaft, sometimes with a mix of flat-blade and pitched-blade type agitators. Additional information may be obtained from Sec. 18 and from Baldyga and Bourne (*Turbulent Mixing and Chemical Reactions*, Wiley, New York, 1998).

As the stirrer speed is increased, different flow regimes are observed, depending on the stirrer type and geometry and the nature of the gas–liquid system considered. For example, for a Rushton turbine with a low-viscosity liquid, three primary flow regimes are observed (Fig. 19-34). Regime I (Fig. 19-34*a*) has single bubbles that rise, and the gas is not dispersed uniformly. Regime II (Fig. 19-34*b*) has the gas dispersed radially as the bubbles ascend. Regime III (Fig. 19-34*c*) has the gas recirculated to the stirrer in an increasingly complex pattern [see, e.g., Baldi, "Hydrodynamics and Gas–Liquid Mass Transfer in Stirred Slurry Reactors," in *Multiphase Chemical Reactors*, ed. Gianetto and Silveston, Hemisphere, Washington, D.C., 1986].

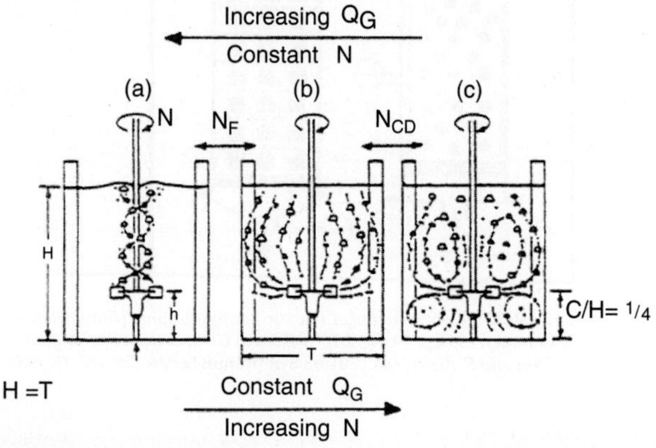

FIG. 19-34 Gas circulation as a function of stirrer speed. (*From Nienow et al.,* 5th European Conference on Mixing, *Wurzburg, 1985; published by BHRA, The Fluid Engineering Centre, Cranfield, England, Fig. 1.*)

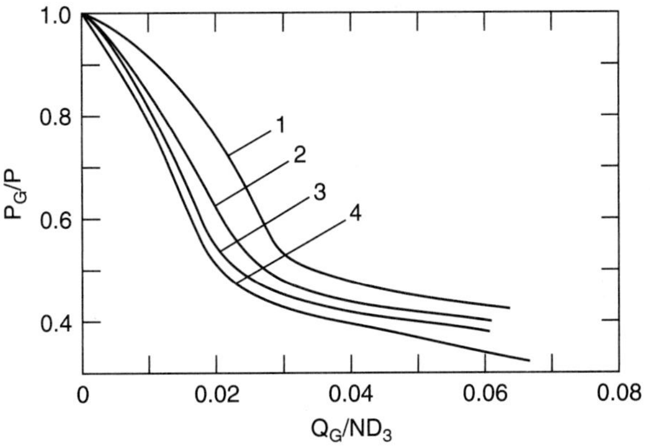

FIG. 19-35 Effect of aeration number and stirrer speed on gassed power—N increases in order of $N_1 < N_2 < N_3 < N_4$. [*Adapted from Baldi, "Hydrodynamics and Mass Transfer in Stirred-Slurry Reactors," in* Multiphase Chemical Reactors, ed. *Gianetto and Silveston, Hemisphere Publishing Corp., 1986, Fig. 14.8.*]

For gas-liquid systems, the power dissipated by the stirrer at the same stirrer speed N is lower than the corresponding power input for liquid systems due to reduced drag on the impeller. The power consumed by the gassed system P_G is estimated from correlations using the power consumed by the ungassed system P_0 and the aeration number N_a:

$$P_0 = N_p \rho_L N^3 D_I^5 \tag{19-83}$$

$$N_a = \frac{Q_G}{ND_I^3} \tag{19-84}$$

The power number N_p depends on impeller geometry/type; the ratio of the gassed to ungassed power is a decreasing function of the aeration rate, as shown in Fig. 19-35.

For instance, Hughmark [*Ind. Eng. Chem. Proc. Des. Dev.* **19:** 638 (1980)] developed a correlation for the power number of Rushton turbines that correlates a large database:

$$\frac{P_G}{P_0} = 0.1 N_a^{-0.25}\left(\frac{D_I^3}{V_L}\right)^{-0.25}\left(\frac{N^2 D_I^4}{gH_I V_L^{2/3}}\right)^{-0.2} \tag{19-85}$$

Increasing the solids content increases the power, as indicated, for example, by Wiedman et al. [*Chem. Eng. Comm.* **6:** 245 (1980)].

With solids present, a minimum agitator speed is required to suspend all the solids; see the correlation of Baldi et al. [*Chem. Eng. Sci.* **33:** 21 (1978)]:

$$N_m = \frac{\beta_2 \mu_L^{0.17}\left[g(\rho_p - \rho_L)\right]^{0.42} d_p^{0.14} w^{0.125}}{\rho_L^{0.58} D_I^{0.89}} \tag{19-86}$$

where w is the catalyst loading in weight percent and parameter β_2 depends on reactor/impeller diameter ratio, for example, from Nienow [*Chem. Eng. J.* **9:** 153 (1975)], $\beta_2 = 2(d_R/D_I)^{1.33}$.

Gas holdup and volumetric gas-liquid mass-transfer coefficients are correlated with the gassed power input/volume and with the aeration rate (actual gas superficial velocity); see the correlation of van't Riet [*Ind. Eng. Chem. Proc. Des. Dev.* **18:** 357 (1979)] for the volumetric mass-transfer coefficient of coalescing and noncoalescing systems:

$$k_L a = \begin{cases} 2.6\times 10^{-2}\left(\dfrac{P_G}{V_L}\right)^{0.4} u_G^{0.5} & \text{for coalescing nonviscous liquids} \\ 2.6\times 10^{-3}\left(\dfrac{P_G}{V_L}\right)^{0.7} u_G^{0.2} & \text{for noncoalescing nonviscous liquids} \end{cases} \tag{19-87}$$

For the gas holdup, a similar correlation was developed by Loiseau et al. [*AIChE J.* **23:** 931 (1977)]:

$$\varepsilon_G = \begin{cases} 0.011\sigma^{-0.36}\mu_L^{-0.056}\left(\dfrac{P'_G}{V_L}\right)^{0.27} u_G^{0.36} & \text{for nonfoaming systems} \\ 0.0051\left(\dfrac{P'_G}{V_L}\right)^{0.57} u_G^{0.24} & \text{for foaming system} \end{cases} \tag{19-88}$$

$$\frac{P'_G}{V_L} = \frac{P_G}{V_L} + 0.03\frac{Q_G\rho_G u_0^2}{V_L} + \frac{Q_G\rho_G RT}{V_L MW_G}\ln\frac{P_1}{P_2}$$

The last two terms of the power/volume equation include the power/volume ratio from the isothermal expansion of the gas through the gas distributor holes having a velocity u_0 and the power/volume ratio to transfer the gas across the hydrostatic liquid head.

Increasing the solids loading leads to a decrease in gas holdup and gas-liquid volumetric mass-transfer coefficient at the same power/volume ratio [e.g., Inga and Morsi, *Can. J. Chem. Eng.* **75:** 872 (1997)].

Liquid-solid mass transfer is typically not limiting because of the small particle size that results in a large particle surface area/volume of reactor, unless the concentration of the particles is very low, or larger particles are used. In the latter case, intraparticle mass-transfer limitations would also occur. Ramachandran and Chaudhari (*Three-Phase Catalytic Reactors,* Gordon and Breach, Philadelphia, 1983) present several correlations for liquid-solid mass transfer, typically as a Sherwood number versus particle Reynolds and Schmidt numbers, e.g., the correlation of Levins and Glastonbury [*Trans. Inst. Chem. Engrs.* **50:** 132 (1972)]:

$$\text{Sh} = \frac{k_s d_p}{D} = 2 + 0.44\,\text{Re}_p^{0.5}\,\text{Sc}^{0.38} \quad \text{Re}_p = \frac{\rho_L u_c d_p}{\mu_L} \quad \text{Sc} = \frac{\nu_L}{D} \tag{19-89}$$

Here u_c is a characteristic velocity, and the velocity terms composing it are estimated from additional correlations.

There is good discussion of heat transfer in agitated gas-liquid-solid slurry reactors; see, for example, van't Riet and Tramper for correlations (*Basic Bioreactor Design*, Marcel Dekker, New York, 1991).

Additional information on mechanically agitated gas-liquid-solid reactors can be obtained in van't Riet and Tramper (*Basic Bioreactor Design*, Marcel Dekker, New York, 1991), Ramachandran and Chaudhari (*Three-Phase Catalytic Reactors*, Gordon and Breach, Philadelphia, 1983), and Gianetto and Silveston (*Multiphase Chemical Reactors*, Hemisphere, Washington, D.C., 1986).

Examples

- Liquid benzene is chlorinated in the presence of metallic iron turnings or Raschig rings at 40°C to 60°C (104°F to 140°F).
- Carbon tetrachloride is made from CS_2 by bubbling chlorine into it in the presence of iron powder at 30°C (86°F).
- Substances that have been hydrogenated in slurry reactors include nitrobenzene with Pd-C, butynediol with Pd-$CaCO_3$, chlorobenzene with Pt-C, toluene with Raney® Ni, and acetone with Raney® Ni.
- Some oxidations in slurry reactors include cumene with metal oxides, cyclohexene with metal oxides, phenol with CuO, and *n*-propanol with Pt.
- Aerobic fermentations.

For many hydrogenations, semibatch operations often are preferred to continuous ones because of the variety of feedstocks or product specifications, or long reaction times, or small production rates. A batch hydrogenator is shown in Fig. 19-36.

The vegetable oil hydrogenator, which is to scale, uses three impellers. The best position for inlet of gas is at a point of maximum turbulence near the impeller, or at the bottom of the draft tube. A sparger is desirable; however, an open pipe is often used. A two-speed motor is desirable to prevent overloading. Since the gassed power requirement is significantly less than ungassed, the lower speed is used when the gas supply is cut off but agitation is to continue. In tanks of 5.7 to 18.9 m^3 (1500 to 5000 gal), rotation speeds are from 50 to 200 rpm, and power requirements are 2 to 75 hp; both depend on superficial velocities of gas and liquid [Hicks and Gates, *Chem. Eng.* (July): 141–148 (1976)]. As a rough guide, power requirements and impeller tip speeds are shown in Table 19-14.

Edible oils are mixtures of unsaturated compounds with molecular weights in the vicinity of 300. The progress of the hydrogenation reaction is expressed in terms of iodine value (IV), which is a measure of unsaturation. The IV is obtained by a standard procedure in which the iodine adds to the unsaturated double bond in the oil. IV is the ratio of the amount of iodine absorbed per 100 g of oil.

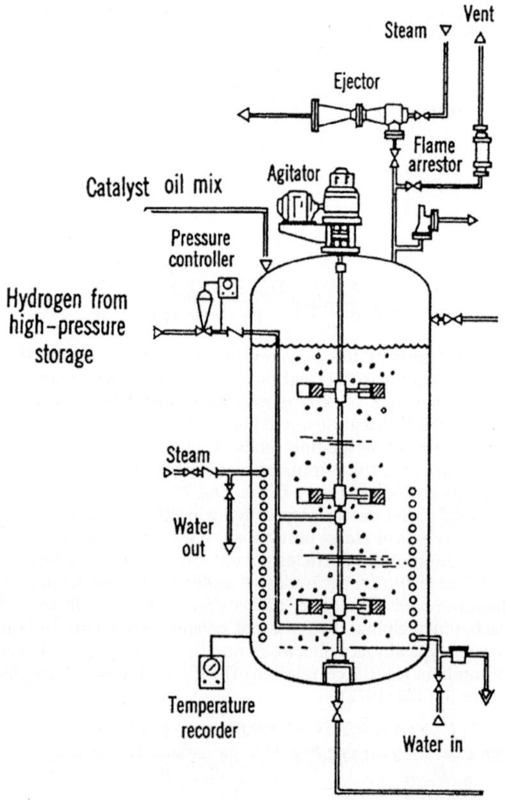

FIG. 19-36 Stirred tank hydrogenator for edible oils. (*Votator Division, Chemetron Corporation.*)

TABLE 19-14 Power Requirements and Impeller Tip Speed Guidelines

Operation	hp/1000 gal*	Tip speed, ft/s
Homogeneous reaction	0.5–1.5	7.5–10
With heat transfer	1.5–5	10–15
Liquid-liquid mixing	5	15–20
Gas-liquid mixing	5–10	15–20

*1 hp/1000 gal = 0.197 kW/m³.

To start a hydrogenation process, the oil and catalyst are charged, then the vessel is evacuated for safety, and hydrogen is continuously added and maintained at some fixed pressure, usually in the range of 1 to 10 atm (14.7 to 147 psi). Internal circulation of hydrogen is provided by axial and radial impellers or with a hollow impeller that throws the gas out centrifugally and sucks gas in from the vapor space through the hollow shaft. Some plants have external gas circulators. Reaction times are 1 to 4 h. For edible oils, the temperature is kept at about 180°C (356°F). Since the reaction is exothermic and because space for heat-transfer coils in the vessel is limited, the process is organized to give a maximum IV drop of about 2.0/min. The rate of reaction drops off rapidly as the reaction proceeds, so a process may take several hours. The endpoint of a hydrogenation is a specified IV of the product. Hardness or refractive index also can be measured to follow reaction progress.

Saturation of the oil with hydrogen is maintained by agitation. The rate of reaction depends on agitation and catalyst concentration. Beyond a certain agitation rate, resistance to mass transfer is eliminated, and the rate becomes independent of pressure. The effect of catalyst concentration also reaches limiting values. The effects of pressure and temperature on the rate are indicated by Fig. 19-37.

A supported nickel catalyst (containing 20 to 25 weight percent Ni on a porous silica particle) is typically used. The pores allow access of the reactants to the extended pore surface, which is in the range of 200 to 600 m^2/g (977×10^3 to 2931×10^3 ft^2/lbm), of which 20 to 30 percent is catalytically active. The concentration of catalyst in the slurry can vary over a wide range but is usually under 0.1 percent Ni. After the reaction is complete, the catalyst can be easily separated from the product. Catalysts are subject to degradation and poisoning, particularly by sulfur compounds. Accordingly, 10 to 20 percent of the recovered catalyst is replaced by fresh catalyst before reuse. Other catalysts are applied in special cases. Expensive palladium has about 100 times the activity of nickel and is effective at lower temperatures. A case study of the hydrogenation of cottonseed oil was made by Rase (*Chemical Reactor Design for Process Plants,* vol. 2, Wiley, New York, 1977, pp. 161–178).

Slurry Bubble Column Reactors As in the case of gas–liquid slurry agitated reactors, bubble column reactors may also be used when solids are present. Most issues associated with multiphase bubble columns are analogous to the gas–liquid bubble columns. In addition, the gas flow has to be sufficient to maintain the solid phase suspended. In the case of a bubble column fermenter, the sparged air is partly used to grow biomass that serves as the catalyst in the system. Many bubble columns operate in semibatch mode, with gas sparged continuously and liquid and catalyst in batch mode.

The benefits of using slurry bubble columns include nearly isothermal operation, excellent heat transfer, good mass transfer, and the use of high-activity powder catalyst with minimal intraparticle diffusion limitations. The reactors may be operated in semibatch (continuous gas) or continuous mode, and they require less power input than mechanically agitated reactors. Catalyst deactivation may be managed by on-line catalyst makeup. The reactor (essentially an empty shell with a sparger grid at the bottom) is easy to design, and the capital investment can be low. Some downsides of slurry bubble column reactors include catalyst/liquid product separation difficulty and lower volumetric productivity than fixed beds (due to lower catalyst loading per reactor volume), and catalyst distribution can be skewed with higher

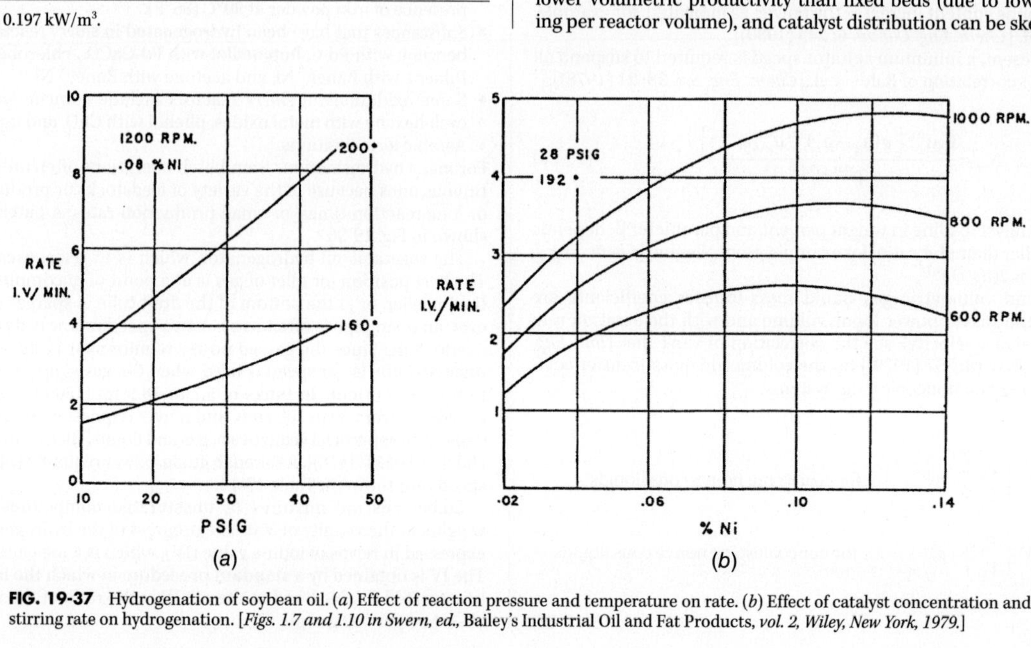

FIG. 19-37 Hydrogenation of soybean oil. (*a*) Effect of reaction pressure and temperature on rate. (*b*) Effect of catalyst concentration and stirring rate on hydrogenation. [*Figs. 1.7 and 1.10 in Swern, ed.,* Bailey's Industrial Oil and Fat Products, *vol. 2, Wiley, New York, 1979.*]

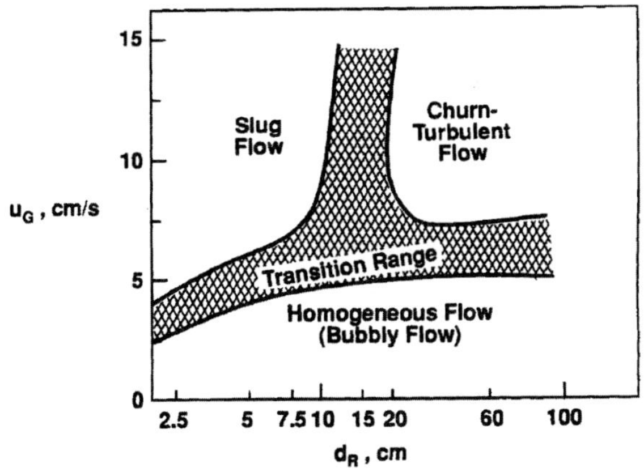

FIG. 19-38 Flow regime map for gas–liquid bubble columns. [*Fig. 16 in Deckwer et al.,* Ind. Eng. Chem. Process Des. Dev. ***19:*** *699–708 (1980).*]

concentrations at the bottom than at the top of the reactor. Also, accounting for the effect of internals (e.g., heat exchange tubes) and of increased diameter on the hydrodynamics is not well understood. Hence, gradual scale-up is often required over multiple intermediate scales before commercialization. Cold-flow models can also be useful in determining hydrodynamics in the absence of reaction.

As is the case for reactors with two or more mobile phases, a variety of flow regimes exist, depending primarily on the gas superficial velocity (the driver for bubble column hydrodynamics) and column diameter. A qualitative flow regime map is shown in Fig. 19-38.

In the homogeneous flow regime at low gas superficial velocity, bubbles are relatively small, and they rise at a constant rate (about 20 to 25 cm/s). As the flow rate is increased, bubbles become larger and more irregular in shape, they frequently coalesce and break up, and the transition to churn turbulent regime is obtained. In small-diameter columns, the larger bubbles may bridge the column, creating slugs—hence the slug flow regime. The large transition zones in Fig. 19-38 are indicative of the lack of accurate knowledge of the dependence of the transition region on conditions (temperature, pressure) and physical properties of the gas and liquid.

Hydrodynamic parameters that are required for bubble column design and analysis include phase holdups (gas, liquid, and solid for slurry bubble columns); volumetric gas–liquid mass-transfer coefficient; liquid–solid mass-transfer coefficient; liquid, gas, and solid axial and radial mixing; and heat-transfer coefficients. These parameters depend strongly on the prevailing flow regime.

Correlations for gas holdup and the volumetric gas–liquid mass-transfer coefficient can have the general form

$$\varepsilon_G = \alpha u_G^{\beta} \qquad k_L a = \gamma u_G^{\delta} \tag{19-90}$$

where u_G is the superficial gas velocity, ε_G is the gas holdup (fraction of gas volume), k_L is the liquid-side gas–liquid mass-transfer coefficient, and a is the interfacial area per volume of either the liquid or the expanded liquid (liquid + gas). The exponents are β, $\delta \sim 1$ for the homogeneous bubbly flow regime and β, $\delta < 1$ for heterogeneous turbulent flow regime. The correlations depend on the gas–liquid–solid system properties. Gas–liquid systems can be classified as coalescing, leading to increased bubble size with distance from the sparger, and noncoalescing, leading to larger gas holdup and volumetric mass-transfer coefficients for the latter. There is a voluminous literature for these parameters, and there is substantial variability in estimated values—one should be careful to validate the parameters with data applicable to the real system considered. For instance, for gas holdup see the correlation of Yoshida and Akita [*AIChE J.* **11:** 9 (1965)]

$$\frac{\varepsilon_G}{(1-\varepsilon_G)^4} = \alpha\left(\frac{\rho_L g d_R^2}{\sigma}\right)^{1/8}\left(\frac{\rho_L^2 g d_R^3}{\mu_L^2}\right)^{1/12}\frac{u_G}{\sqrt{g d_R}}$$

$$\alpha = \begin{cases} 0.2 & \text{for pure liquids and nonelectrolytes} \\ 0.25 & \text{for salt solutions} \end{cases} \tag{19-91}$$

and for volumetric gas–liquid mass-transfer coefficient, see the correlation of Akita and Yoshida [*I&EC Proc. Des. Dev.* **12:** 76 (1973)]:

$$k_L a = 0.6\frac{D}{d_R^2}\left(\frac{\mu_L}{\rho_L D}\right)^{0.5}\left(\frac{\rho_L g d_R^2}{\sigma}\right)^{0.62}\left(\frac{\rho_L^2 g d_R^3}{\mu_L^2}\right)^{0.31}\varepsilon_G^{1.1} \tag{19-92}$$

More recent correlations for gas holdup and mass transfer include the effect of pressure and bimodal bubble size distribution (small and large bubbles), in a manner analogous to the treatment of dilute and dense phases in fluidized beds [see, e.g., Letzel et al., *Chem. Eng. Sci.* **54**(13): 2237 (1999)].

Increasing the catalyst loading decreases the gas holdup and the volumetric gas–liquid mass-transfer coefficient [see, e.g., Maretto and Krishna, *Catalysis Today* **52:** 279 (1999)].

Axial mixing in the liquid, induced by the upflow of the gas bubbles, can be substantial in commercial-scale bubble columns, especially in the churn turbulent regime. Due to typically small particle size, the axial dispersion of the solid catalyst in slurry bubble columns is expected to follow closely that of the liquid; exceptions are high-density particles. The liquid axial mixing can be represented by an *axial dispersion coefficient,* which typically has the form

$$D_{aL} = \alpha u_G^{\beta} d_R^{\gamma} \tag{19-93}$$

Based on theoretical considerations (Kolmogoroff's theory of isotropic turbulence), $\beta = 1/3$ and $\gamma = 4/3$. For example, Deckwer et al. [*Chem. Eng. Sci.* **29:** 2177 (1973)] developed the following correlation:

$$D_{aL} = 2.7 u_G^{0.3} d_R^{1.4} \tag{19-94}$$

It is expected that the strong dependence on reactor diameter only extends up to a maximum diameter beyond which there is no effect of diameter; however, there is disagreement among experts as to what that maximum diameter may be. There are a large number of correlations for liquid axial dispersion with widely different predictions, and care must be exerted to validate the predictions with data at some significant scale, even if only in a cold-flow mockup.

The gas axial mixing is due to the bubble size distribution, resulting in a distribution of bubble rise velocities, which varies along the column due to bubble breakup and coalescence. There are a variety of correlations in the literature, with varying results and reliability, such as the correlation of Mangartz and Pilhofer [*Verfahrenstechn.* **14:** 40 (1980)].

$$D_{aG} = 5\times 10^{-4}\left(\frac{u_G}{\varepsilon_G}\right)^3 d_R^{1.5} \tag{19-95}$$

This equation is dimensional, and cm/s for u_G, cm for d_R, and cm^2/s for D_{aG} should be used. The radial mixing can be represented by radial dispersion coefficients for the gas and the liquid. For instance, the liquid radial dispersion coefficient is estimated at less than one-tenth of the axial one.

Correlations for the heat-transfer coefficient have the general form

$$\text{St} = f(\text{Re Fr Pr}^2)$$

$$\text{Re Fr} = \frac{u_G^3 \rho_L}{\mu_L g} \qquad \text{Pr} = \frac{c_{pL}\mu_L}{\lambda_L} \qquad \text{St} = \frac{h_w}{u_G \rho_L c_{pL}} \tag{19-96}$$

For instance, see the correlation of Deckwer et al. [*Chem. Eng. Sci.* **35**(6): 1341–1346 (1980)].

$$\text{St} = 0.1(\text{Re Fr Pr}^2)^{-1/4} \tag{19-97}$$

Additional information on the hydrodynamics of bubble columns and slurry bubble columns can be obtained from Deckwer (*Bubble Column Reactors,* Wiley, New York, 1992), Nigam and Schumpe (*Three-Phase Sparged Reactors,* Gordon and Breach, Philadelphia, 1996), Ramachandran and Chaudhari (*Three-Phase Catalytic Reactors,* Gordon and Breach, Philadelphia, 1983), and Gianetto and Silveston (*Multiphase Chemical Reactors,* Hemisphere, Washington, D.C., 1986). Computational fluid mechanics approaches have also been used recently to estimate mixing and mass-transfer parameters [e.g., see Gupta et al., *Chem. Eng. Sci.* **56**(3): 1117–1125 (2001)].

Examples There are a number of examples, including Fischer-Tropsch synthesis in the presence of Fe or Co solid catalysts, methanol synthesis in the presence of Cu/Zn solid catalyst, and hydrocracking in the presence of zeolite catalyst. Fermentation reactions are conducted in bubble column reactors when there is a benefit for increased scale and for reduced cost. The oxygen is sparged from the bottom, and the liquid reactants are added in a semibatch mode. The absence of reactor internals is an advantage because it prevents contamination. Heat transfer has to be managed through a cooling jacket. If heat removal is an issue, cooling coils may be installed, or the broth can be circulated through an external heat exchanger.

Fluidized Gas–Liquid–Solid Reactors In a gas–liquid–solid fluidized-bed reactor, gas and liquid enter at the bottom of the reactor and leave at the top; the gas disengages from the top of the expanded slurry; and the liquid is removed, for instance, through internal filters located at the top of the expanded slurry or through other design concepts, ensuring catalyst containment in the reactor. Liquid is continuous, gas is dispersed. Particles are larger than in bubble columns, with a typical range of 0.2 to 1.0 mm (0.008 to 0.04 in). Bed expansion can be small. Bed temperatures are uniform within 2°C (3.6°F) in medium-size beds, and heat transfer to embedded surfaces and/or the wall is excellent. Catalyst may be bled off and replenished continuously, or reactivated continuously. Figure 19-39*a* illustrates a conventional gas–liquid–solid fluidized-bed reactor.

Figure 19-39*b* shows an *ebullating bed* reactor for the hydroprocessing of heavy crude oil. A stable fluidized bed is maintained by recirculation of the mixed fluid through the bed and a draft tube. Reactor temperatures may range from 350°C to 600°C (662°F to 1112°F) and 200 atm (2940 psi). An external pump sometimes is used instead of the built-in impeller shown. Such units were developed for the liquefaction of coal.

A biological treatment process (Dorr-Oliver Hy-Flo) employs a vertical column filled with fluidized inert solids such as sand on which bacterial growth takes place while waste liquid and air are charged from the bottom. A large interfacial area for reaction is provided, about 33 cm^2/cm^3 (84 in^2/in^3). BOD removal of 85 to 90 percent is claimed in 15 min compared with 6 to 8 h in conventional units.

In entrained beds, the three-phase mixture flows through the vessel and is separated downstream. These reactors are used in preference to fluidized beds where catalyst particles are very fine or subject to disintegration or if the catalyst deactivates rapidly in the process.

Trickle Bed Reactors Reactant gas and liquid flow cocurrently downward through a packed bed of solid catalyst particles. The most common use of trickle bed reactors is for hydrogenation reactions. The solubility of feed hydrogen in the liquid even at the higher pressure is insufficient to provide the stoichiometric needs of the reaction, and a gas flow exceeding the hydrogen need is fed into the reactor. High hydrogen partial pressures can prevent catalyst deactivation due to undesirable reactions, such as coking. Cooling (or heating) is typically done between stages, either with heat transfer to a coolant outside the reactor or through direct cooling with a cold reactant gas or liquid.

Advantages of a trickle bed are ease of installation, low liquid holdup (and therefore less undesirable homogeneous reactions), minimal catalyst handling issues, low catalyst attrition, and catalyst life of one to four years. The liquid and gas flow in trickle beds approaches plug flow (leading to higher conversion than slurry reactors for the same reactor volume). The downsides of trickle beds include flow maldistribution (bypassing), sensitivity to packing uniformity and prewetting (leading to hot spots), incomplete contacting and wetting, intraparticle diffusion resistance, the potential for fouling and bed plugging due to particulate matter in the feed, and high pressure drop. A significant fraction of the flow is gas that has to be compressed and recycled (i.e., increased compressor costs).

A schematic of a trickle bed reactor is shown in Fig. 19-40. The reactor is a high-pressure vessel equipped with a drain and a manhole for vessel entry. Typical vessel diameters may range from 3 to 30 ft with height from 6 to 100 ft. The liquid enters the reactor and is distributed across the cross section by a distributor plate. The liquid feed flows downward due to gravity, helped along by the drag of the gas, at such a low rate that it is distributed over the catalyst as a thin film. The gas enters at the top and is distributed along with the liquid. In the simplest arrangement, the liquid distributor is a perforated plate with about 10 openings/dm^2 (10 openings/15.5 in^2), and the gas enters through several risers about 15 cm (5.9 in) high. More elaborate distributor caps also are used. Uniform distribution of liquid across the reactor is critical to reactor performance. The aspect ratio of the reactor can vary between 1 and 10, depending on the pressure drop that can be accommodated by the compressor. It is not uncommon to redistribute the liquid using a redistribution grid every 8 to 15 ft.

The catalyst is often loaded on screens supported by a stainless steel grid near the bottom of the reactor. Often, large inert ceramic balls are loaded at the very bottom, with slightly smaller ceramic balls above the first layer, and then the catalyst. Smaller inert ceramic balls can also be loaded above the catalyst bed and topped off with the larger balls. The layer of inert balls can be 6 in to 2 ft in depth. The balls restrict the movement of the bed and distribute the liquid across the catalyst.

As is the case when two or more mobile phases are present, cocurrent gas–liquid downflow through packed beds produces a variety of flow regimes, depending on the gas and liquid flow rates and the physical properties of the gas and the liquid. In Fig. 19-41, a flow regime map for trickle beds of foaming and nonfoaming systems is presented. Here L and G are the liquid and gas *fluxes* (mass flow rate per total flow cross-sectional area). In the low interaction or trickle flow regime, gas is the continuous phase, and the liquid is flowing as rivulets. Increasing the liquid and gas flow results in high interaction or pulse flow, with the liquid and gas alternatively bridging the bed voids. At high liquid flow and low gas flow, the liquid becomes the continuous phase and the gas is the dispersed phase, called dispersed bubble flow. Finally at high gas flow and low liquid flow, the spray flow regime exists, with liquid being the dispersed phase.

The literature contains a number of references to other flow regime maps; however, there is no clear advantage to using one map versus another. Wall

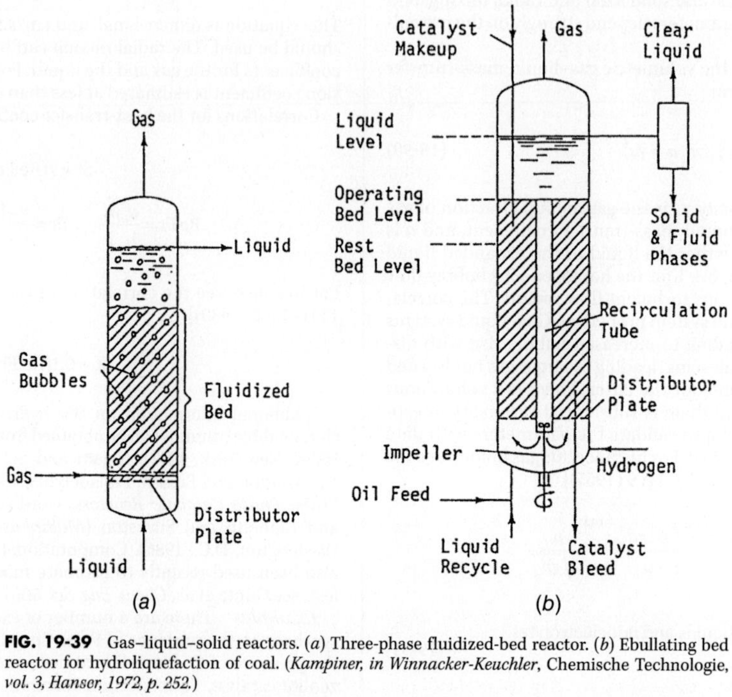

FIG. 19-39 Gas–liquid–solid reactors. (*a*) Three-phase fluidized-bed reactor. (*b*) Ebullating bed reactor for hydroliquefaction of coal. (*Kampiner, in Winnacker-Keuchler,* Chemische Technologie, *vol. 3, Hanser, 1972, p. 252.*)

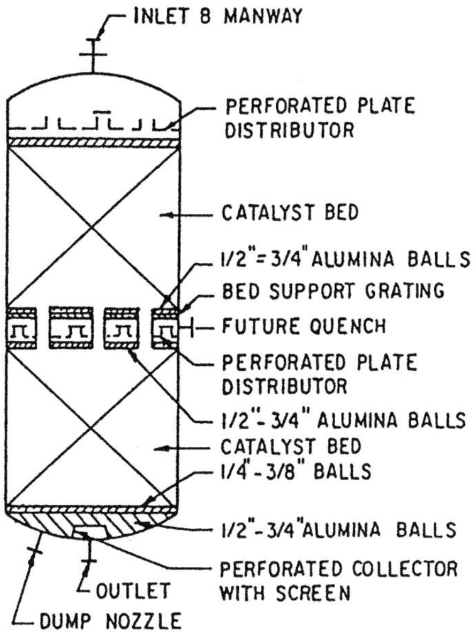

FIG. 19-40 Trickle bed reactor for hydrotreating 20,000 bbl/d of light catalytic cracker oil at 370°C and 27 atm. To convert atm to kPa, multiply by 101.3. (*Fig. 16.6 in Gianetto and Silveston, eds.,* Multiphase Chemical Reactors, *Hemisphere, Washington, D.C., 1986*)

effects can have a major effect on the hydrodynamics of trickle bed reactors. Most of the data reported in the literature are for small laboratory units of 2-in diameter and under, and there is no evidence that pulse flow occurs in large-diameter commercial reactors in the high interaction regime.

Hydrodynamic parameters that are required for trickle bed design and analysis include *bed void fraction, phase holdups* (gas, liquid, and solid), *wetting efficiency* (fraction of catalyst wetted by liquid), *volumetric gas–liquid mass-transfer coefficient, liquid–solid mass-transfer coefficient* (for the wetted part of the catalyst particle surface), *gas–solid mass-transfer coefficient* (for the unwetted part of the catalyst particle surface), *liquid and gas axial mixing, pressure drop,* and heat-transfer coefficients. These parameters vary with the flow regime (i.e., for the low and high interaction regimes).

There are a number of pressure drop correlations for two-phase flow in packed beds originating from the Lockhart-Martinelli correlation for two-phase flow in pipes. These correlate the two-phase pressure drop to the single-phase pressure drops of the gas and the liquid obtained from the Ergun equation. See, for instance, the Larkins correlation [Larkins, White, and Jeffrey, *AIChE J.* **7:** 231 (1967)]

$$\log \frac{\Delta P_{GL}}{\Delta P_L + \Delta P_G} = \frac{0.416}{0.666 + (\log X)^2} \qquad \text{where } X = \sqrt{\frac{\Delta P_L}{\Delta P_G}} \quad 0.05 \le X \le 30 \tag{19-98}$$

Since some of the published pressure drop correlations can differ by an order of magnitude, it is best to verify the relationship with actual data before designing a reactor. Other approaches to two-phase pressure drop include the relative permeability method of Saez and Carbonell [*AIChE J.* **31**(1): 52–62 (1985)].

The bed void volume available for flow and for gas and liquid holdup is determined by the particle size distribution and shape, the particle porosity, and the packing effectiveness. The total voidage and the total liquid holdup can be divided into external and internal terms corresponding to interparticle (bed) and intraparticle (porosity) voidage. The external liquid holdup is further subdivided into *static holdup* ε_{Ls} (holdup remaining after bed draining due to surface tension forces) and *dynamic holdup* ε_{Ld}. Additional expressions for the liquid holdup are the pore fillup F_i and the liquid saturation S_L:

$$\begin{aligned} &\varepsilon_t = \varepsilon_B + \varepsilon_p(1 - \varepsilon_B) && \text{total voidage} \\ &\varepsilon_L = \varepsilon_{Le} + \varepsilon_{Li} && \text{total liquid holdup} \\ &\varepsilon_{Le} = \varepsilon_{Ld} + \varepsilon_{Ls} && \text{external liquid holdup} \\ &\varepsilon_{Li} = F_i\varepsilon_p(1 - \varepsilon_B) && \text{internal liquid holdup} \\ &S_L = \frac{\varepsilon_L}{\varepsilon_B} && \text{liquid saturation} \end{aligned} \tag{19-99}$$

The static holdup can be correlated with the *Eotvos number* N_{Eo} as it results from a balance of surface tension and gravity forces on the liquid held up in the pores in the absence of flow:

$$N_{Eo} = \frac{\text{gravity force}}{\text{surface tension force}} = \frac{\rho_L g d_p^2}{\sigma_L} \tag{19-100}$$

For instance, Fig. 19-42 illustrates the dependence of the static holdup on the Eotvos number for porous and nonporous packings.

A variety of correlations have been developed for the total and the dynamic liquid holdup. For instance, the total liquid holdup has been correlated with the Lockhardt-Martinelli parameter X for spherical and cylindrical particles [Midou, Favier, and Charpentier, *J. Chem. Eng. Japan* **9:** 350 (1976)]

$$\frac{\varepsilon_L}{\varepsilon_b} = \frac{0.66\, X^{0.81}}{1 + 0.66\, X^{0.81}} \tag{19-101}$$

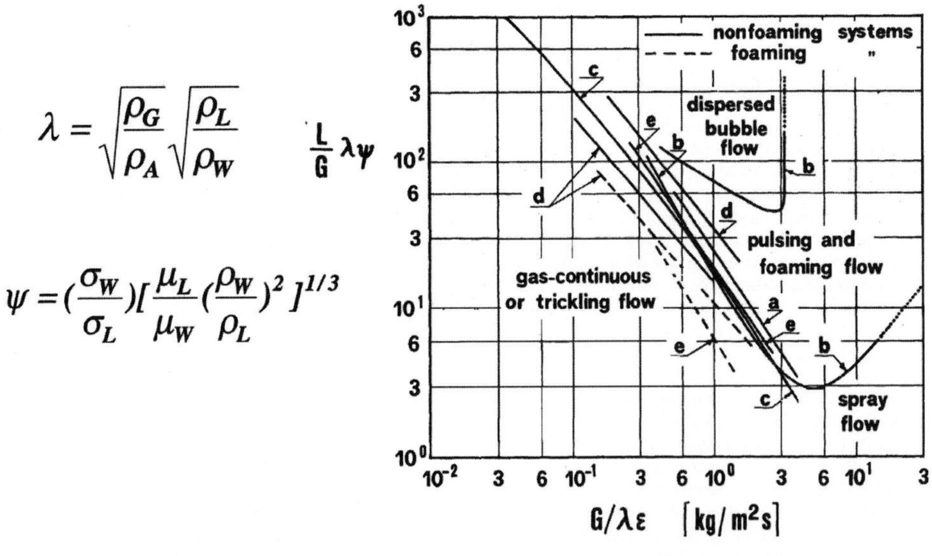

FIG. 19-41 Trickle bed flow regime map. [*Gianetto et al.,* AIChE J. *24(6): 1087–1104 (1978); reproduced with permission.*]

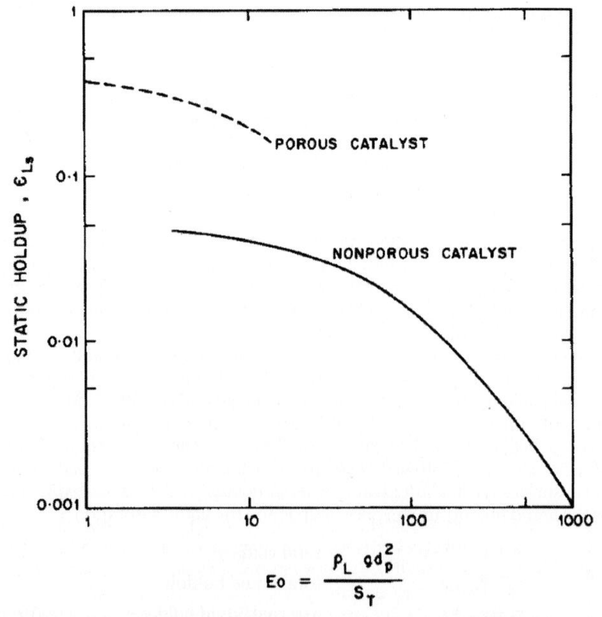

FIG. 19-42 The static liquid holdup for porous and nonporous solids. (*Fig. 7.7 in Ramachandran and Chaudhari,* Three-Phase Catalytic Reactors, *Gordon and Breach, Philadelphia, 1983.*)

Correlations for the dynamic liquid holdup have also been developed as a function of various dimensionless numbers, including the liquid and gas Reynolds number, and the two-phase pressure drop (see, e.g., Ramachandran and Chaudhari, *Three-Phase Catalytic Reactors*, Gordon and Breach, Philadelphia, 1983; and Hofmann, "Hydrodynamics and Hydrodynamic Models of Fixed Bed Reactors," in *Multiphase Chemical Reactors*, ed. Gianetto and Silveston, Hemisphere, Washington, D.C., 1986).

The various volumetric mass-transfer coefficients are defined in a manner similar to that discussed for gas–liquid and fluid–solid mass transfer in previous sections. There are a large number of correlations obtained from different gas–liquid–solid systems. For more details, see Shah (*Gas-Liquid-Solid Reactor Design*, McGraw-Hill, New York, 1979), Ramachandran and Chaudhari (*Three-Phase Catalytic Reactors*, Gordon and Breach, Philadelphia, 1983), and Shah and Sharma ("Gas-Liquid-Solid Reactors," in *Chemical Reaction and Reactor Engineering*, ed. Carberry and Varma, Marcel Dekker, New York, 1987).

Axial mixing of the liquid is an important factor in the design of trickle bed reactors, and criteria were proposed to establish conditions that limit axial mixing. Mears [*Chem. Eng. Sci.* **26:** 1361 (1971)] developed a criterion that, when satisfied, ensures that the conversion will be within 5 percent of that predicted by plug flow:

$$\mathrm{Pe} = \frac{u_L L}{D} > 20n \ln \frac{1}{1-x} \tag{19-102}$$

where n is the order of the reaction with respect to the limiting reactant and x is the fractional conversion of that reactant. Correlations for axial dispersion can be found in Ramachandran and Chaudhari, *Three-Phase Catalytic Reactors*, Gordon and Breach, Philadelphia, 1983.

Incomplete wetting can be also a critical factor in reactor design and analysis leading usually to lower performance due to incomplete utilization of the catalyst bed. In a few select cases, the opposite may be the case; for example, when a volatile reactant reacts faster from the gas in contact with the dry catalyst than from the liquid in contact with the wetted catalyst because it is not limited by the gas–liquid mass-transfer resistance and because of higher gas diffusivity. Correlations for the fraction of catalyst surface wetted are available, although they are not very reliable and they are strongly system-dependent (e.g., Shah, *Gas-Liquid-Solid Reactor Design*, McGraw-Hill, New York, 1979).

Due to the complex hydrodynamics and the dependence of the hydrodynamic parameters on the flow regime, trickle beds are notoriously difficult to scale up. Laboratory units (used for kinetics and process development) and commercial units typically are operated at the same *liquid hourly space velocity* (LHSV). Since the LHSV represents the ratio of the superficial liquid velocity to the reactor length, the superficial velocity in a laboratory reactor will be lower than in a commercial reactor by the ratio of reactor lengths, which is often well over an order of magnitude. This means that heat and mass transport parameters may be considerably different in laboratory reactors operated at the target LHSV. This also shifts the flow regime from trickle flow (low interaction) in the lab and small pilot plants to the high-interaction regime in large-scale commercial reactors.

Wall effects in lab units of 50-mm (1.97-in) diameter can be important, while these are negligible for commercial reactors of 1 m or more diameter. Wall effects in the lab can be reduced by using reactor/particle diameter ratios greater than 8. If that is not possible, inert fines such as sand are added to the catalyst bed to reduce wall effects. Also, in large-diameter beds, uniform liquid distribution is difficult, even with a large number of distributor nozzles, and unless the flow is redistributed, the nonuniformity can persist along the bed, leading to potential hot spots that can cause by-products and fast catalyst deactivation. In trickle beds that are not pre-wetted, a *hysteresis* phenomenon related to wetting occurs, where the behavior with increasing flow of the liquid phase is not retraced with decreasing liquid flow. This can often be avoided by prewetting the reactor before start-up.

In practice, the thickness of liquid films in trickle beds has been estimated to vary between 0.01 and 0.2 mm (0.004 and 0.008 in). The dynamic liquid holdup fraction is 0.03 to 0.25, and the static fraction is 0.01 to 0.05. The high end of the static fraction includes the liquid that partially fills the pores of the catalyst. The effective gas–liquid interface is 20 to 50 percent of the geometric surface of the particles, but it can approach 100 percent at high liquid loading. This results in an increase of reaction rate as the amount of wetted surface increases (i.e., when the gas–solid reaction rate is negligible).

Examples Hydrodesulfurization of petroleum oils was the first large-scale application of trickle bed reactors commercialized in 1955. In this application, organosulfur species contained in refinery feeds are removed in the presence of hydrogen and a catalyst, and they are released as hydrogen sulfide. Conditions depend on the quality and boiling range of the oil. The reactor pressure is optimized to increase the solubility of the hydrogen and minimize catalyst deactivation due to coking. Over the life of the catalyst, the temperature is increased to maintain a constant conversion. Temperatures are in the range of 345°C to 425°C (653°F to 797°F) with pressures of 34 to 102 atm (500 to 1500 psi). A large commercial reactor may have 20 to 25 m (66 to 82 ft) of total depth of catalyst, and it may be up to 3 m (9.8 ft) in diameter or above in several beds of 3- to 6-m (9.8- to 19.7-ft) depth. Bed depth is often limited by pressure drop, the catalyst crush strength, and the maximum adiabatic temperature increase for stable operation. The need to limit pressure drop is driven by the capital and operating costs associated with the hydrogen recycle compressor. Catalyst granules are 1.5 to 3.0 mm (0.06 to 0.12 in), sometimes a little bigger. Catalysts are 10 to 20 percent Co and Mo (or Ni and W) on alumina. The adiabatic temperature rise in each bed usually is limited to 30°C (86°F) by the injection of cold hydrogen between beds. Since the liquid trickles over the catalyst, the wetting efficiency of the catalyst is important in determining the volumetric reaction rate. As expected, wetting efficiency increases with increasing liquid rate. Catalyst effectiveness of particles 3 to 5 mm (0.12 to 0.20 in) in diameter has been found to be about 40 to 60 percent.

Packed Bubble Columns (Cocurrent Upflow) These reactors are also called flooded-bed reactors. In contrast to trickle beds, both gas and liquid flow cocurrently upward. A screen is needed at the top to retain the catalyst particles. Such a unit has been used for the hydrogenation of nitro and double-bond compounds and nitriles [Ovcinnikov et al., *Brit. Chem. Eng.* **13:** 1367 (1968)]. High gas rates can cause movement and attrition of the particles. Accordingly, such equipment is restricted to low gas flow rates, for instance, where a hydrogen atmosphere is necessary but the consumption of hydrogen is slight. The liquid is the continuous phase, and the gas, the dispersed phase. Benefits of cocurrent upflow versus trickle (cocurrent downflow) include high wetting efficiency (resulting in good liquid–solid contacting), good liquid distribution, and better heat and mass transfer. Disadvantages include higher pressure drop and liquid backmixing, the latter resulting in more undesirable homogeneous reactions.

A number of flow regime maps are available for packed bubble columns [see, e.g., Fukushima and Kusaka, *J. Chem. Eng. Japan* **12:** 296 (1979)]. Correlations for the various hydrodynamic parameters can be found in Shah (*Gas-Liquid-Solid Reactor Design*, McGraw-Hill, New York, 1979), Ramachandran and Chaudhari (*Three-Phase Catalytic Reactors*, Gordon and Breach, Philadelphia, 1983), and Shah and Sharma ("Gas-Liquid-Solid Reactors," in *Chemical Reaction and Reactor Engineering*, ed. Carberry and Varma, Marcel Dekker, New York, 1987).

Countercurrent Flow Packed Beds The gas flows up countercurrent with the downflow liquid. This mode of operation is not as widely used for catalytic reactions since operation is limited by flooding at high gas velocity: at flooding conditions, increasing the liquid flow does not result in an increase of the liquid holdup.

For more details, see Shah (*Gas-Liquid-Solid Reactor Design*, McGraw-Hill, New York, 1979) and Hofmann (*Hydrodynamics and Hydrodynamic Models of Fixed Bed Reactors*, in *Multiphase Chemical Reactors*, ed. Gianetto and Silveston, Hemisphere, Washington, D.C., 1986).

SOME CASE STUDIES

The literature contains case studies that may be useful for the analysis or design of new reactors. Several of these are listed for reference.

Rase (*Case Studies and Design Data,* vol. 2 of *Chemical Reactor Design for Process Plants,* Wiley, New York, 1977):
- Styrene polymerization
- Cracking of ethane to ethylene
- Quench cooling in the ethylene process
- Toluene dealkylation
- Shift conversion
- Ammonia synthesis
- Sulfur dioxide oxidation
- Catalytic reforming
- Ammonia oxidation
- Phthalic anhydride production
- Steam reforming
- Vinyl chloride polymerization
- Batch hydrogenation of cottonseed oil
- Hydrodesulfurization

Rase (*Fixed Bed Reactor Design and Diagnostics,* Butterworth, Stoneham, Mass., 1990) has several case studies and a general computer program for reactor design:
- Methane-steam reaction
- Hydrogenation of benzene to cyclohexane
- Dehydrogenation of ethylbenzene to styrene

Tarhan (*Catalytic Reactor Design,* McGraw-Hill, New York, 1983) has computer programs and results for these cases:
- Toluene hydrodealkylation to benzene and methane
- Phthalic anhydride by air oxidation of naphthalene
- Trickle bed reactor for hydrodesulfurization

Ramage et al. (*Advances in Chemical Engineering,* vol. 13, Academic Press, New York, 1987, pp. 193–266):
- Mobil's kinetic reforming model

Dente and Ranzi [in *Pyrolysis Theory and Industrial Practice,* ed. Albright et al., Academic Press, New York, 1983, pp. 133–175]:
- Mathematical modeling of hydrocarbon pyrolysis reactions

Shah and Sharma [in *Chemical Reaction and Reaction Engineering Handbook,* ed. Carberry and Varma, Marcel Dekker, New York, 1987, pp. 713–721]:
- Hydroxylamine phosphate manufacture in a slurry reactor

Trambouze et al. (*Chemical Reactors, Design/Engineering/Operation,* Editions Technip, Paris, 1988) summarizes 20 years of experience of the Institut Francais du Petrole (IFP) on reactor scale-up and design for the refinery and petrochemical industry.

Tominaga and Tamaki, eds. (*Chemical Reaction and Reactor Design,* Wiley, New York, 1998) has several reactor design case studies:
- Naphtha cracking (Yagi)
- Tubular steam reforming (Rostrup-Nielsen and Christiansen)
- Epoxy resin production (Soma and Hosomo)
- Hydrotreating reactor design (Bridge and Blue)
- Fluid catalytic cracking (Takatsuka and Minami)
- Wet flue gas desulphurization (Yanagioka and Sugiya)

Salmi et al., *Chemical Reaction Engineering and Reactor Technology,* CRC Press, Boca Raton, Fla., 2011, has several novel reactor case studies and also discusses new paradigms in reaction engineering:
- Catalytic three-phase hydrogenation of citral in the monolith reactor
- Hydrogenation of triglycerides
- Delignification of wood

Exploration for an acceptable or optimum design for a new reactor may require consideration of several feed and product specifications, reactor types, catalysts, operating conditions, and economic evaluations. Modifications to an existing process likewise may need to consider many cases. Commercial software may be used to facilitate the examination of options. A typical package can handle a number of reactions in various ideal reactors under isothermal, adiabatic, or heat-transfer conditions in one or two phases. Outputs can provide profiles of composition, pressure, and temperature as well as vessel size.

Thermodynamic software packages may be used to find equilibrium compositions at prescribed temperatures and pressures. Such calculations require a knowledge of feed components and products and their thermodynamic properties, and they are based on Gibbs free energy minimization techniques. Examples of thermodynamic packages may be found in Smith and Missen (*Chemical Reaction Equilibrium Analysis Theory and Algorithms,* Wiley, New York, 1982) and in Walas (*Phase Equilibria in Chemical Engineering,* Butterworth, Stoneham, Mass., 1985).

For some widely practiced processes, especially in the petroleum industry, computer models are available from a number of vendors or, by license, from proprietary sources. Such processes include fluid catalytic cracking, hydrotreating, hydrocracking, alkylation with HF or H_2SO_4, reforming with Pt or Pt-Re catalysts, tubular steam cracking of hydrocarbon fractions, noncatalytic pyrolysis to ethylene, and ammonia synthesis. Catalyst vendors may sometimes also provide simple process models. The reader is advised to peruse some of the process simulation packages listed for sale in the *CEP Software Directory* (e.g., AIChE, 1994) that gets periodically updated with new offerings.

Section 20

Bioreactions and Bioprocessing

Gregory Frank, Ph.D. *Principal Engineer, Amgen Inc.; Fellow, American Institute of Chemical Engineers; Member, Society of Biological Engineering; North American Mixing Forum; Pharmaceutical Discovery, Development, and Manufacturing Forum (Section Coeditor, Bioreactions and Bioprocessing; Bioengineering Overview and Concepts, Bioreactors, and Upstream Processes)*

Jeffrey Chalmers, Ph.D. *Professor of Chemical and Biomolecular Engineering, The Ohio State University; Member, American Institute of Chemical Engineers; American Chemical Society; Fellow, American Institute for Medical and Biological Engineering (Section Coeditor, Bioreactions and Bioprocessing; Bioengineering Overview and Concepts; Bioreactors and Upstream Processes)*

Roger G. Harrison, Ph.D. *Professor of Chemical, Biological, and Materials Engineering and Professor of Biomedical Engineering, University of Oklahoma; Member, American Institute of Chemical Engineers, American Chemical Society, American Society for Engineering Education, Oklahoma Higher Education Hall of Fame; Fellow, American Institute for Medical and Biological Engineering (Downstream Processing: Primary Recovery and Purification)*

Paul W. Todd, Ph.D. *Chief Scientist Emeritus, Techshot, Inc.; Member, American Institute of Chemical Engineers (Downstream Processing: Primary Recovery and Purification)*

Scott R. Rudge, Ph.D. *Chief Operating Officer and Chairman, RMC Pharmaceutical Solutions, Inc.; Adjunct Professor, Chemical and Biological Engineering, University of Colorado; Vice President, Margaux Biologics, Scientific Advisory Board, Sundhin Biopharma (Downstream Processing: Primary Recovery and Purification); Member, American Chemical Society, International Society of Pharmaceutical Engineers, American Association for the Advancement of Science, Parenteral Drug Association (Downstream Processing: Primary Recovery and Purification)*

Demetri P. Petrides, Ph.D. *President, Intelligen, Inc.; Member, American Institute of Chemical Engineers, American Chemical Society (Downstream Processing: Primary Recovery and Purification)*

Clint Pepper, Ph.D. *Director, Lonza; Member, American Institute of Chemical Engineers (Product Attribute Control)*

Brandon Downey, B.A.Sc. *Principal Engineer, R&D, Lonza; Member, American Institute of Chemical Engineers (Product Attribute Control)*

Jeffrey Breit, Ph.D. *Principal Scientist, Capsugel; Member, American Association of Pharmaceutical Scientists (Product Attribute Control)*

Michael J. Betenbaugh, Ph.D. *Professor of Chemical and Biomolecular Engineering, Johns Hopkins University; Member, American Institute of Chemical Engineers (Emerging Biopharmaceutical and Bioprocessing Technologies and Trends)*

Nathan Calzadilla, M.S.E. *Research Program Assistant, Johns Hopkins Medicine, Chemical and Biomolecular Engineering, Johns Hopkins University; Member, American Institute of Chemical Engineers (Emerging Biopharmaceutical and Bioprocessing Technologies and Trends)*

BIOENGINEERING OVERVIEW AND CONCEPTS

BIOREACTORS AND UPSTREAM PROCESSES

DOWNSTREAM PROCESSING: SEPARATION AND PURIFICATION

PRODUCT QUALITY ATTRIBUTE CONTROL

EMERGING BIOPHARMACEUTICAL AND BIOPROCESS TECHNOLOGIES AND TRENDS

Nomenclature and Units

A	Constant in equations for blend time [Eq. (20-9*b*)] and linear growth rate [Eq. (20-75)]
A	Cross-sectional area of column, m^2
a	Specific area of bubbles, m^{-1}
a_p	Radius of particle, m
AE	Analysis equipment
A_H	Heat exchange area, m^2
AR	Aspect ratio, H/D_T
ANOVA	Analysis of variation
ATF	Alternating tangential flow
B	Volumetric breakage rate of particles in the size range L to ΔL, number s^{-1} L m^{-1}
B	Magnetic flux density, T
c	Concentration of solute in liquid phase, M or kg L^{-1}
c_k	Concentration in the adsorbent pore, M or kg L^{-1}
C_L	Concentration in the liquid phase, M or kg L^{-1}
c^*	Equilibrium concentration of solute in the liquid phase, M or kg L^{-1}
c	Controlled variable
CEX	Cation exchange chromatography
CFD	Computational fluid dynamic
CHO	Chinese hamster ovary
CIP	Clean in place
CPQA	Critical product quality attribute
CPA	Critical process attribute
CPP	Critical process parameter
CTR	Carbon dioxide transfer rate, mol m^{-1} s^{-1}
dCO_2	Dissolved CO_2 concentration, percent
D	Diameter of tube or column, m
D_h	Equivalent diameter of channel, m
d_i	Impeller diameter, m
d_m	Diameter of pore in membrane, m
d_p	Stationary phase particle diameter, m
$\mathcal{D}$	Diffusion coefficient of solute, m^2 s^{-1}
$\mathcal{D}_s$	Shear-induced diffusion coefficient for a suspended particle, m^2 s^{-1}
D_I	Impeller diameter, m
D_L	Axial dispersion coefficient, m^2 s^{-1}
D_{LK}	Liquid axial dispersion coefficient in zone k, m^2 s^{-1}
D_{SK}	Absorbed particle axial dispersion coefficient in zone k, m^2 s^{-1}
$\mathcal{D}_p$	Intraparticle diffusion coefficient, m^2 s^{-1}
DO	Dissolved O_2 concentration, percent
DOE	Design of experiments
D_T	Bioreactor diameter, m
E	Applied electric field strength (voltage per length) given by $E = I/k_eA$, V m^{-1}
EDR	Energy dissipation rate, W m^{-3}
F_m	Magnetic force, N
Fabs	Fragment antigen-binding antibodies
FIC	Flow indicator controller
g	Slope of the elution gradient, M L^{-1}, or g L^{-2}
g	Gravitational acceleration, 9.8066 m s^{-2}
g	Linear control model
G	Multiple of gravitational acceleration $\left(=\dfrac{\omega^2 R}{g}\right)$
G	Particle linear growth rate, m s^{-1}
G	Normalized slope of the gradient, M or kg L^{-1}
h	Half height of rectangular flow channel, m
H	Bioreactor liquid level, m; or Henry's law constant, atm L mol^{-1}
H	Magnetizing field, A m^{-1}
HCP	Host cell protein
H_B	Height of impeller off the vessel bottom, m
H_G	Height of aerated liquid, m
HPLC	High-performance liquid chromatography
I	Ionic strength, M
J_m	Transmembrane fluid flux, L m^{-2} h^{-1}
k	Mass-transfer coefficient ($=\mathcal{D}/\delta$), m s^{-1}
k	Constant in displacement model for particle breakage, m^{1-n} s^{-1} [Eq. (20-69)]
k_B	Boltzmann constant, 1.3807×10^{-23} J K^{-1}
ke	Solution electrical conductivity, S/m
k_{fk}	Film mass-transfer coefficient in zone k, s^{-1}
kg	Gas film mass-transfer coefficient, mol s^{-1} m^{-2} atm^{-1}

k_L	Liquid film mass-transfer coefficient, m s^{-1}
k_La	Specific O_2 mass-transfer coefficient, s^{-1} or h^{-1}
k_La_{CO2}	Specific CO_2 mass-transfer coefficient, s^{-1} or h^{-1}
k_P	Mass-transfer coefficient, s^{-1}
$K_{\text{van't}}$	Constant used in van't Riet correlation
K	Constant in Eq. (20-76) for volumetric breakage rate, s^2 m^{-2} kg^{-1}
K_i	Salting-in constant, M^{-1}
K_s	Salting-out constant, M^{-1}
K'	Constant in the Juckes equation, M^{-1}, or L kg^{-1} [Eq. (20-46)]
K'_s	Constant in the Cohn equation, M^{-1} [Eq. (20-52)]
K_1	Constant in equation for volumetric breakage rate, L^{-1} s^{-1} [Eq. (20-76)]
K_A	Rate constant for particle growth by diffusion, L s^{-1} [Eq. (20-59)]
K_a	Overall mass-transfer coefficient, s^{-1}
l_e	Mean eddy length, or Kolmogoroff length, m [Eq. (20-56)]
L	Length of flow channel over membrane, m
L	Length of the column, m
L	Particle diameter, m
L_e	Equilibrium particle diameter, m
$\bar{L}_m$	Mass-averaged particle diameter, m
L_{mol}	Molecular diameter, m
L_o	Critical particle diameter where particles start to grow by colliding and sticking together, m
LCMS	Liquid chromatography mass spectroscopy
M_o	Molecular mass of solute, Da
M_T	Total mass concentration of particles, kg L^{-1}
mAb	Monoclonal antibody
MPC	Model predictive control
n	Number of required cell doublings
n	Exponent in Eq. (20-40) for hindered settling
n	Population density distribution function, number L^{-1} cm^{-1} [Eq. (20-72)]
n	Order of displacement model for particle breakage [Eq. (20-69)]
n_o	Population density distribution function at $L = L_o$, number L^{-1} m^{-1}
N	Particle number concentration, number L^{-1}
N	Agitator speed, s^{-1} or rpm
N	Specific rate of mass transfer, mol m^{-3} s^{-1}
N_i	Impeller rotation rate, revolutions s^{-1}
N_b	Blend number, dimensionless
N_Q	Impeller flow number, dimensionless
NIR	Near-infrared
OD	Oxygen demand, mol m^{-3} s^{-1}
OTR	Oxygen transfer rate, mol m^{-3} s^{-1}
OUR	Oxygen uptake rate, mol m^{-3} s^{-1}
p	Pressure, Pa, atm or mm Hg
p_x	Partial pressure of species x, Pa, atm or mm Hg
P	Power, W
P_g	Gassed power number, dimensionless
P_o	Ungassed power number, dimensionless
P_T	Total pressure, Pa, atm or mm Hg
P/V	Power input per volume of liquid, W L^{-1}
PAT	Process analytical technology
PQA	Product quality attribute
PQAC	Product quality attribute control
PLS	Partial least squares
q_{O2}	Specific cellular oxygen demand, mol s^{-1} $cell^{-1}$
q	Concentration of a solute in stationary phase averaged over an adsorbent particle, M or kg L^{-1}
q_{sat}	Concentration of solute on stationary phase at equilibrium with feed concentration c_0, averaged over an adsorbent particle, M or kg L^{-1}
$q_{i,n}$	Separand concentration in adsorbent phase of column n averaged over an adsorbent particle, M or kg L^{-1}
$\bar{q}_k$	Average adsorbent phase concentration, M or kg L^{-1}
Q	Impeller pumping capacity, m^3 s^{-1}
Q	Volumetric flow rate, L s^{-1}
$\bar{Q}$	Flow rate/(column volume), s^{-1}
Q_{sg}	Sparge gas flow rate, m^3 s^{-1}
Q_H	Heat evolution rate, W m^{-3}
QbD	Quality by design
R	Distance from center of rotation, cm
R	Gas constant, 0.082058 L atm $gmol^{-1}$ K^{-1} or 8.3145×10^6 Pa m^3 kg-mol^{-1} K^{-1}

Nomenclature and Units *(Continued)*

R_k	Radius of adsorbent in zone k, m
Re	Reynolds number
r_s	Fraction of solute rejected by membrane
r_r, r_s	Radius of polymer rod, radius of protein solute, m
RQ	Respiratory quotient, dimensionless
RTRT	Real time release testing
s	Sedimentation coefficient, s [Eq. (20-41)]
s	system variable
S	Protein solubility, M or kg L^{-1}
S_o	Solubility of dipolar ion at zero ionic strength, M or kg L^{-1}
Sc	Schmidt number $\left(=\frac{\mu}{\rho \mathcal{D}}\right)$
Sh	Sherwood number $\left(=\frac{kD_h}{\mathcal{D}}\right)$
S_p	Solubility of dipolar ion, M or kg L^{-1}
SIP	Steam in place
SS	Stainless-steel
SU	Single-use
SUB	Single-use bioreactor
t	Time, s
t_b	Time at breakthrough, s
t^*	Ideal adsorption time, s
T	Temperature, K
t_c	Circulation time, s
t_D	Cell doubling time, h
t_m	Mixing time, s
t_{st}	Total seed train elapsed time from vial to production bioreactor, h
T	Bioreactor diameter, m
T	Temperature, °C or K
TFF	Tangential flow filtration
u	Dipole moment, C m
u_k	Superficial velocity, m s^{-1}
u	Velocity of fluid through interstices of bed, m s^{-1}
u_b	Bulk fluid velocity, m s^{-1}
U	Heat-transfer coefficient, W m^{-2} K^{-1}
U_T	Impeller tip speed, m s^{-1}
U_o	Electro-osmotic mobility of chamber walls due to zeta (surface) potential, m^2 s^{-1} V^{-1}
U_{el}	The electrophoretic mobility of charged particle, m^2 s^{-1} V^{-1}
UPLC	Ultrahigh-performance liquid chromatography
UV	Ultraviolet
v	Sedimentation velocity, m s^{-1}
v	Mobile phase superficial velocity, m s^{-1}
v_c	Sedimentation velocity in concentrated suspension, m s^{-1}
V	Volume of retentate in cross-flow filtration, L
V	Precipitation reactor volume, L
V	Volume of fixed-bed column, L
V_0	Column void volume, L
$\bar{V}$	Partial specific volume of polymer, L m^{-1}
v_{sg}	Superficial gas velocity, m s^{-1}
vvm	Specific volumetric flow rate, min^{-1}
V	Volume of cell culture medium, m^3
VCD	Viable cell density
WCB	Working cell bank
y	Mol fraction
X_o	Number of cells used to initiate seed train
X_{inoc}	Number of cells required to inoculate production bioreactor
X_{max}	Maximum viable cell density, cells m^{-3}
X_{viab}	Viable cell density, cells m^{-3}
z_i	Charge of component i
Z	Axial distance from column entrance, m
Greek letters	
α, β	Exponents
α	Particle collision effectiveness factor
β	Constant in Cohn equation [Eq. (20-52)]
β'	Constant in Juckes' equation [Eq. (20-46)]
$\gamma, \bar{\gamma}$	Shear rate, average shear rate, s^{-1}
γ_w	Shear rate at membrane surface, s^{-1}
δ	Diffusion distance, m
δ	Boundary layer thickness, m
ΔC	Concentration driving force, mol m^{-3}
Δp	Pressure drop, atm
$\Delta \pi$	Osmotic pressure, atm
ΔT	Temperature differential, K
ε	Dielectric constant of solvent
ε	Column void fraction
ε_B	Bed voidage in column
ε_{BK}	Local bed voidage in zone k
ε_{crit}	Critical specific energy dissipation rate, W kg^{-1} or W m^{-3}
ε_{max}	Maximum specific energy dissipation rate, W kg^{-1} or W m^{-3}
ε_{mean}	Mean specific energy dissipation rate, W kg^{-1} or W m^{-3}
ε_p	Adsorbent porosity
θ_m	Mixing time, s
λ_K	Kolmogoroff turbulence scale, μm
μ	Viscosity of fluid, atm s
μ_o	Magnetic permeability free space, T m A^{-1}
ρ	Liquid density, kg m^{-3}
ρ_o	Density of medium in sedimentation, kg m^{-3}
ρ	Density of particle or molecule, kg m^{-3}
ν	Kinematic viscosity, m^2 s^{-1}
σ, τ	Exponents
σ	Membrane reflection coefficient for a solute
τ	Mean residence time, s
τ_p	Time constant of the oxygen probe, s
ϕ	Particle volume fraction in bulk suspension
ϕ	Volume fraction of particles
ϕ_1	Volume fraction of submicrometer-size "primary particles"
ω	Angular velocity, rad s^{-1}
Subscripts	
CO_2	Carbon dioxide
crit	Critical value
in	At the entrance
g	When air is sparged; or in the gas phase
k	Column number
i	Proteins and salt
max	Maximum
out	At the exit
O_2	Oxygen
s	Surface
Superscript	
*	At equilibrium

General References: New developments in bioprocessing are found in journals and professional society publications such as *Biotech. Bioeng., Biotech. Progress, J. Membrane Science, AIChE J.,* and professional society conferences (e.g., AIChE, SBE, ISPE, or ACS). Engineering standards are available from the American Society of Mechanical Engineers in *Bioprocessing Equipment Standards.* Additional references include Alberts, *Essential Cell Biology*, 4th ed., Garland Science, New York, 2014; Paul, *Handbook of Industrial Mixing*, Wiley, Hoboken, N.J., 2004; Shuler, Kargi, and DeLisa, *Bioprocess Engineering*, 3d ed., Prentice Hall, Upper Saddle River, N.J., 2017; Lutz, *Ultrafiltration for Bioprocessing*, Woodhead Publishing, Cambridge, UK, 2015; Flickinger, *Downstream Industrial Biotechnology*, Wiley, Hoboken, N.J., 2013; Flickinger, *Upstream Industrial Biotechnology*, vols. 1 & 2, Wiley, Hoboken, N.J., 2013; Olsson, *Advances in Biochemical Engineering/Biotechnology Biofuels*, vol. 108, Springer, Medford, Mass., 2007; Grunwald, *Industrial Biocatalysis*, CRC Press, Boca Raton, Fla., 2015; Pandey, *Industrial Biorefineries & White Biotechnology*, Elsevier, Boca Raton, Fla., 2015; Lutz, Section 20, *Membrane Separation Processes, Perry's Chemical Engineers' Handbook*, 8th ed., McGraw Hill, New York, 2007; Ahuja (ed.), *Handbook of Bioseparations*, Academic Press, Cambridge, Mass., 2000; Belter, Cussler, and Hsu, *Bioseparations*, Wiley, Hoboken, N.J., 1988; Forciniti, *Industrial Bioseparations: Principles and Practice*, Wiley, Hoboken, N.J., 2008; García, Bonen, Ramírez-Vick, Sadaka, and Vappu, *Bioseparation Process Science*, Blackwell Science, Hoboken, N.J., 1999; Ghosh, *Principles of Bioseparations Science and Engineering*, World Scientific, Singapore, 2006. Gottschalk (ed.), *Process Scale Purification of Antibodies*, Wiley, Hoboken, N.J., 2009; Harrison, Todd, Rudge, and Petrides, *Bioseparations Science and Engineering*, 2d ed., Oxford University Press, Oxford, UK, 2015; Harrison (ed.), *Protein Purification Process Engineering*, Marcel Dekker, New York, 1993; Ladisch, *Bioseparations Engineering*, Wiley, Hoboken, N.J., 2001; Shukla, Etzel, and Gadam (eds.), *Process Scale Bioseparations for the Biopharmaceutical Industry*, CRC Press, Boca Raton, Fla., 2007; Sivasankar, *Bioseparations: Principles and Techniques*, Prentice Hall, India, 2005; Subramanian, *Bioseparations and Bioprocessing: A Handbook*, vols. 1 & 2, Wiley-VCH, Weinheim, Germany, 2007. Zborowski and Chalmers (eds.), *Magnetic Cell Separation*, Elsevier B. V., Amsterdam, Netherlands, 2008.

BIOENGINEERING OVERVIEW AND CONCEPTS

BIOPROCESSING INTRODUCTION AND OVERVIEW

Bio-based reactions and processes have been around for as long as humankind has been cultivating crops and processing food (i.e., fermented foods); however, it has only been since the second half of the 20th century that large-scale, commercial processes for nonfood products have grown to significant scales (i.e., ethanol for fuel). Bioprocesses utilize living organisms and biological materials derived from them, such as enzymes, to produce medicines, vaccines, food, beverages, fuels, chemicals, polymers, and other unique materials. The complex physicochemical properties and sensitivity to process conditions of cells and enzymes used to carryout bioreactions require specific technologies and controls that differentiates bioprocessing from more traditional chemical processing. This complexity and sensitivity carries over to many of the protein products made by living cells as well.

As with most new, complex, and/or expensive technology, the initial commercial applications focused on low-volume, high-value products, such as human biopharmaceuticals. An early example is the treatment of Gaucher's disease, a genetic disorder caused by an enzyme deficiency. The first treatment was glucocerebrosidase, an enzyme purified from human tissue and then enzymatically modified. Such treatments, known as biopharmaceuticals or biotherapeutics, are effective because they have complex structures and modes of interaction with disease targets that cannot be matched by small molecule pharmaceuticals. This complexity is also what makes them difficult to develop and produce, resulting in treatments that can cost upward of hundreds of thousands of dollars per year. Subsequent to the first treatment for Gaucher's disease, recombinant engineered cells were developed to produce non-human-derived versions of the original biopharmaceutical. This eliminated safety concerns related to harvesting enzymes from human tissue, and it also resulted in a more robust, predictable supply chain to ensure the therapeutic was available to patients.

As both fundamental and applied biological and process sciences have advanced, in some cases in very large leaps forward, the commercial production of lower-value, higher-volume products has emerged and is continuing to emerge. In an attempt to categorize this evolution, the production of biotherapeutics has been called the *first wave* of modern biotechnology; the genetic engineering of agriculture, primarily for food, has been called the *second wave*; and the production of products for non-health-related use is the *third wave.* This third wave has also been called *industrial biotechnology.* The first wave was motivated by human need and had to achieve stringent product quality and safety requirements, while this third wave, or industrial biotechnology, is almost exclusively cost-driven. Biopharma, which arguably fits in this first wave, as of 2015 has a global revenue of \$163 billion/yr, which accounts for 20 percent of total pharma revenue; this percentage continues to grow. As of 2015 it has been estimated that agricultural biotechnology's global revenue is \$28 billion/yr, and accounts for 10 percent of agricultural crops globally. In comparison, biotechnology's third wave, industrial biotechnology, has been estimated to have global revenues of \$140 billion/yr. This includes traditional products such as ethanol, bulk antibiotics for animals, amino acids, and industrial enzymes. In this contribution we are ignoring plant biotechnology, such as genetically modified crops, since in most cases these plants are grown and processed as typical nonengineered crops, while this section is primarily focused on those aspects of engineering and processing unique to biotechnology.

An example of the dramatic advances in biotechnology, in this case in terms of productivity improvements and corresponding decrease in price per unit amount, is seen in penicillin production over a span of 40 yr (Fig. 20-1). Figure 20-2 superimposes increased penicillin titer over time along with human biopharmaceutical titers produced from Chinese hamster ovary (CHO) cells. What is not shown in this figure is a corresponding increase in the safety profile of these biopharmaceuticals as better cell lines were developed and more was understood regarding immunogenicity and its relationship to specific biopharmaceutical protein quality attributes.

Biopharmaceuticals and biofuels represent two ends of the bioprocessing spectra with respect to value, quantity produced, regulatory oversight, and production process technology. The quantity of each biopharmaceutical produced in a year is measured in tens or hundreds of kilograms per year. In contrast to these high-value products, over 14 billion gal of ethanol is produced per year in the United States as a biofuel. Table 20-1 classifies commercial bioproducts from low value to high value in terms of unit price.

While both enzymes and cells are used in various commercial manufacturing processes, this section focuses on production processes using cells. Fundamental to commercial production using cells is the large-scale cultivation of cells in a bioreactor or fermenter, followed by separation, isolation, and purification of the product. In contrast to many other chemical processes, these cell-based processes are conducted in aqueous environments at moderate temperatures, and the final cellular concentration and corresponding product are at a very low to low concentration relative to a traditional organically synthesized product. Further, all but the smallest biotechnology-derived products, such as organic acids and amino acids, are often sensitive to temperature, pH, reducing environments, and other environmental factors. These can affect the structure, efficacy, and even safety of the protein product. This sensitivity presents substantial challenges in the separation and purification process compared to traditional chemical processing. Further, purification processes and final product purity requirements are dependent on its final use. The most extreme example is a final product that is injected into humans as a biotherapeutic.

Despite the wide range of biotechnology-derived products, a generic flow sheet can be used to describe the various steps in the bioreaction, separation, and purification process. Figure 20-3 is one example of such a flow sheet. A number of salient points can be made. First, a major branch point in this flow diagram, after the cells have been cultivated in the bioreactor, is whether the product (1) is secreted by the cell into the suspending fluid (supernatant), (2) is contained within the cell with subsequent need to rupture the cell to remove the product, or (3) is the cell itself. This decision point has a significant impact on the overall process, especially the process economics and overall recovery, and can determine the overall financial feasibility. Clearly, the choice of the cell type used to make the product has a major impact on the process. Whether the final product produced by the cell is in its final soluble form, or an insoluble form needing significant postproduction processing, is another major branch point. Last, the final use of the product will determine not only the number and types of purification steps, but also their level of sophistication and expense. A classic example is the comparison of the production processes used to create the same product (i.e., an antibiotic or vaccine), but with the first producing an animal use product and the second a human use product.

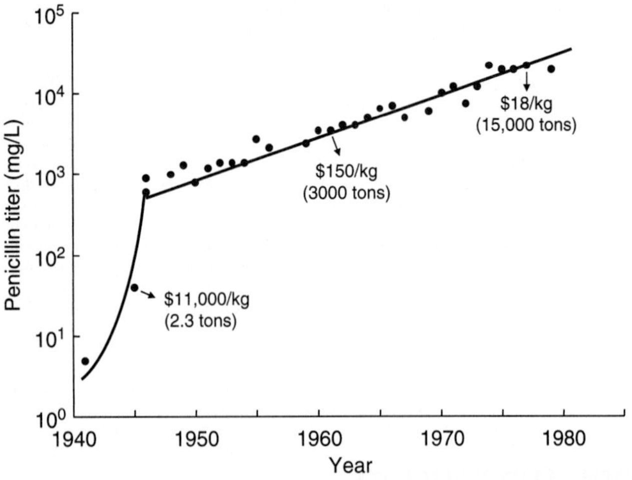

FIG. 20-1 Penicillin titer improvements over 40 years and corresponding cost reduction. (Reproduced from Seth, G., Hossler, P., Yee, J., and Hu, W-S., "Engineering Cells for Cell Culture Bioprocessing—Physiological Fundamentals," *Adv. Biochem. Eng./Biotechnol.*, **101:** 121 (2006) with permission of Springer. Copyright Springer-Verlag, Berlin-Heidelberg, 2006.)

BIOPROCESS PRODUCTS

While the actual number of commercial biotechnology products number is in the hundreds if not thousands, they can generally be broken into major categories based on cost, markets served, and processing characteristics, as shown in Table 20-1. An alternative way to approach classifying biotechnology products and/or processes is to consider the chemically synthesized alternative. Obviously, in the case of complex biopharmaceuticals, a synthesized alternative does not exist. However, as biotechnology develops and expands into markets in which chemically synthesized products already exist, direct economic competition exists between a typically petrochemical-based approach and a biotechnological approach. Since one of the primary costs of chemically synthesized compounds is feedstock, a careful look at these costs is in order. Most polymers and manufactured chemicals are based on five categories of hydrocarbon feedstocks: methane, ethane, propane, butane, and aromatics. The abundance of light carbon hydrocarbons (three carbons and smaller) and existing depreciated petroleum processing infrastructure create significant barriers for biotechnology to replace their use without significant economic and political incentives, such as exist for ethanol. An exception is Brazil's sugar cane–based ethanol industry. Recent analysis of algal-based biofuel production indicates that it is not currently cost competitive, with significant costs related to feedstocks (Davis, *Process Design and Economics for the Conversion of Algal Biomass to Biofuels,* National Renewable Energy Laboratory, Golden, C.O., 2014, pp.73–76).

Four-carbon-based feedstocks, which include butadiene, isobutylene, butene-2, and butene-1, represent the transition from natural gas and natural gas liquids to crude oil sources. These four-carbon feedstocks are typically obtained from cracking gas, oil, and naphtha or directly separating it from crude. As an example of a biotechnology-based process exploiting an opportunity created by the increasing price of a hydrocarbon feedstock, Genomatica and DuPont Tate & Lyle have successfully demonstrated biological production of 1,4-butanediol on a commercial scale. Interestingly, this three- and four-carbon price barrier is not rigid; DuPont Tate & Lyle Bio Products company has been commercially producing 1,3 propanediol at a capacity of 140 million lb/yr using a biological-based process since 2011.

With an increasing number of carbon atoms per hydrocarbon molecule, further alternative biotechnological synthesis opportunities arise. The aromatic hydrocarbons, benzene, toluene, and xylene, BTX, are actively being targeted for bio-based alternative synthesis. For example, Gevo Inc. in partnership with Coca-Cola is developing a process to produce para-xylene, a feedstock for polyethylene terephthalate bottles.

Examples of other C4 or larger feedstock chemicals whose biotechnological synthesis is either currently commercial or being developed for commercial production include these: polyethylene, manufactured by Brashkem in Trifo, Brazil; isoprene, under development by Brashkem and Amyris; and three key nylon intermediates—hexamethylenediamine, caprolactone, and adipic acid—under development by Genomatica. Table 20-2, summarizing a set of opinion and review articles published in 2016, attempts to compile available public information with respect to the big-picture view of the current state of biotechnology's third wave.

Biopharmaceuticals Over the past 20 years biopharmaceuticals or, as they are also called, biotherapeutics have grown from a minor to a significant proportion of new therapeutics becoming available. This is due to their ability to treat previously unmet medical needs, higher success rates in clinical trials, and advances in biotechnology and bioprocessing.

Biopharmaceuticals are classified here as a biologically derived treatment for diseases and biologically based in vivo and in vitro diagnostics. They range from naturally occurring (native) proteins, the earliest biopharmaceuticals, such as erythropoietin to treat anemia, to engineered monoclonal

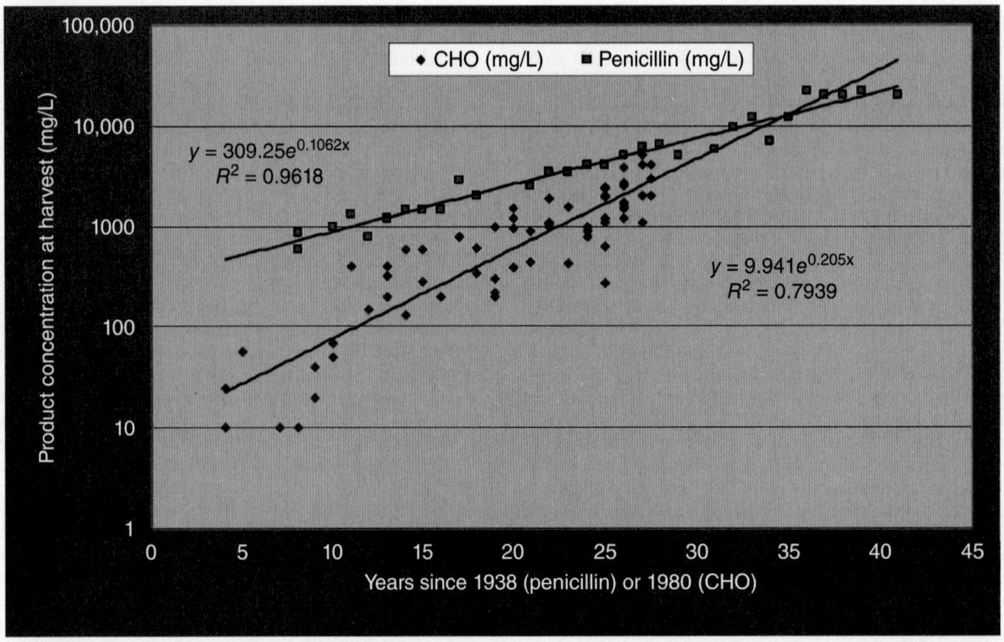

FIG. 20-2 Productivity gains of human biopharmaceuticals produced by CHO cells and penicillin produced by mold host cells over time. Data shown are representative of companies having prior experience producing these therapeutics. (Reproduced with permission of Matt Croughan, all rights retained.)

TABLE 20-1 Biologically Derived Products Range from Low-Cost Commodities to High-Value Biopharmaceuticals

Market segment	Examples	Rel. value	Product characteristics and market valuation ($ U.S.)	Process characteristics
Bulk chemicals	Ethanol, acetone, butanol, organic acids (acetic, butyric, propionic, citric), glycerol	$	Low cost, commodity	Low-cost feedstock often requires pretreatment; typically uses conventional purification unit operations.
Fine chemicals	Enzymes, additives (detergents), enantiomerically pure amino acids, biologically derived active pharmaceutical ingredients (APIs)	$–$$	Low- to moderate-cost enzymes and chemical compounds Global Enzymes: $3–4 billion[a, b] APIs: $16 billion[c]	Fermentation process is followed by recovery of secreted compounds, or cell disruption to recover and purify enzymes.
Biofuels	Ethanol, biodiesel	$	Commercial feasibility tied to government subsidies, fuel mandates, and global oil market U.S. ethanol market: $40 billion[d]	Feedstock pretreatment is required and is a critical cost factor; product recovery often utilizes traditional chemical engineering unit operations such as distillation.
Food	Dairy products (yogurt, cheese), baker's yeast, beverages (beer), food additives, vitamins, protein	$–$$	Low cost, commodity	Bioprocesses in food processing applications are highly diversified.
Biopharmaceuticals	Native proteins (insulin, EPO), engineered proteins, antibodies and antibody variants, antibody-drug conjugates, vaccines, cell therapeutics	$$–$$$$$	Highly regulated process and product quality; specialized licensed facilities Global $200 billion in 2013 Reaching $500 billion in 2020[e]	Complex, expensive medium, smaller scales, sophisticated process monitoring and controls, and downstream processing.

SOURCES: [a]Global Industry Analysts, Inc., *Industrial Enzymes: Global Strategic Business Report,* Global Industry Analysts, Inc., San Jose, C.A. 2012, pp. II-6, II-18; [b]Business Communications Co., *Research about Enzymes in Industrial Applications*, BCC Research, Wellesley, Mass. 2011; [c]Carlson, R., "Estimating the Biotech Sector's Contribution to the US Economy," *Nature Biotechnology* **34:** 247–255 (2016); [d]Bunge, J., and Newman, J., *Wall Street Journal,* Jan. 2, 2015; [e]Research and Markets, Report Code BT001, Industry Experts, Chino, C.A., 2013.

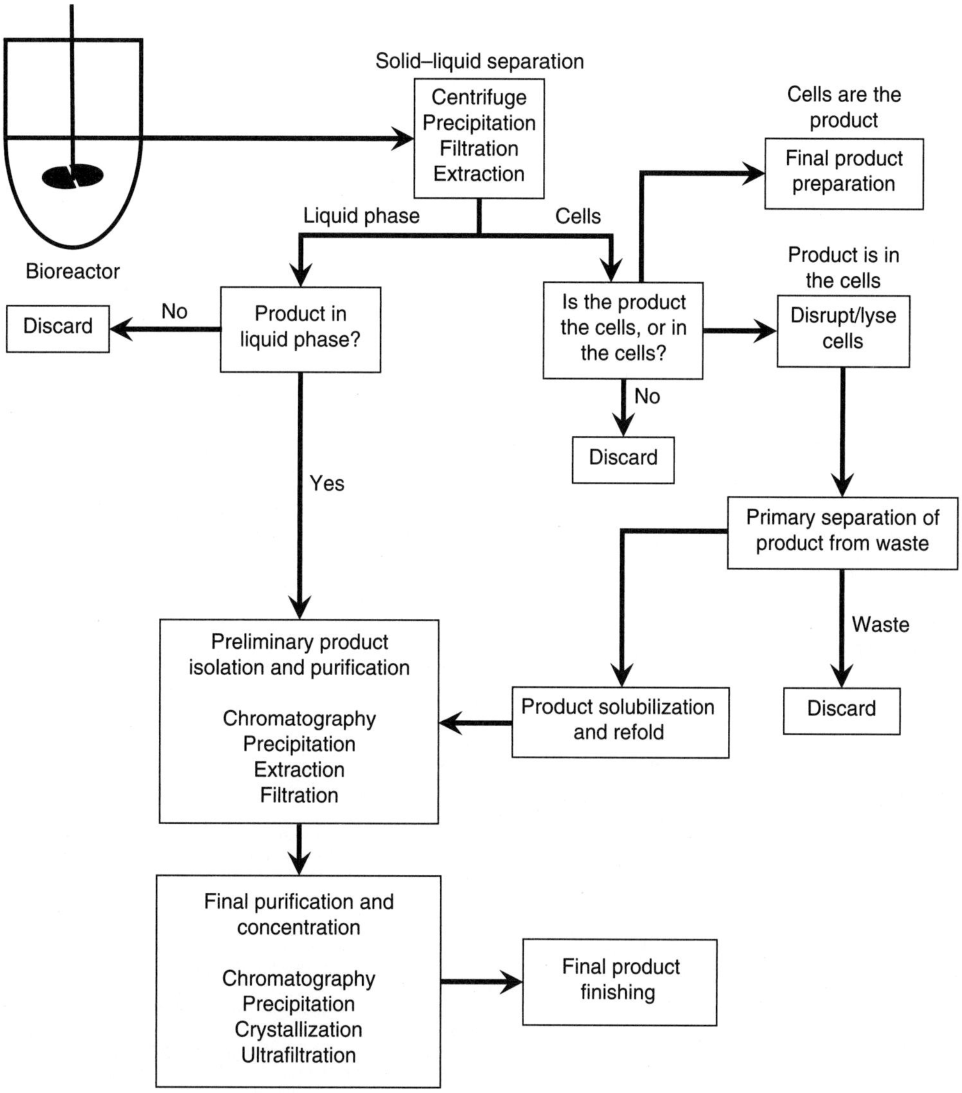

FIG. 20-3 General bioprocessing scheme from bioreactor to final product. Branch points at each step illustrate processing choices that are dependent on the product and process step objectives.

TABLE 20-2 Summary of Industrial Biotechnology Projects*

From raw biomass to product				From processed biomass to product			
Type of biomass	No. of projects	Type of product	No. of projects	Biomass-derived feedstock	No. of projects	Categories of products from process biomass	No. of projects
Unspecified	59	Ethanol	31	Sugars	25	Polymers, rubber	10
Wood	20	Dicarboxylic acid	14	Lignin	9	Dicarboxylic acids	8
Corn plant parts	12	Polymers, rubber	13	Ethanol	5	Carboxylic acids	7
Sugar cane	11	Jet fuel	12	Levulinic acid	3	Diols	3
Oils	10	Unspecified fuel	11	Glycerin	3	BTX	3
Cellulosic biomass	10	Alcohols	9	Black liquor	3	Butadiene	2
Algae	10	Diesel	9	Succinic acid	3	Ethanol	1
Municipal solid waste	10	Carboxylic acid	5	Biogas	2	Jet fuel	1
Guayle plant	5	Oils	5	Other	7	Alcohol	1
Agricultural residues	4	Diols	4			Diesel	1
Other	24	Other	62			Gasoline	1
						Other	21

*Of these 235 total projects, a little less than 50 percent are at commercial scale.
SOURCE: Cooker, B,, *Chem. Eng. Progress* **112:** 31 (June 2016).

antibodies (mAbs), glycoproteins, and other hybrid constructs such as antibody-drug conjugates, which is a combination of an antibody with a small molecule drug. Other examples include native proteins that have been engineered to improve therapeutic effectiveness (erythropoiesis and pegfilgrastim), hormones (insulin, human growth hormones), lymphokines (interferons, interleukins, granulocyte-macrophage colony-stimulating factor), therapeutic enzymes (factor VIII, beta-glucocerebrosidase, iduronidase), inhibitors (alpha-galactosidase), antibody variants (antibody fragments (FAbs) and related constructs (bifunctional T-cell engagers), as well as vaccines (hepatitis B and human papillomavirus vaccines). There are also whole-cell therapies, such as immunotherapies for cancer treatment using whole living T-cells as the therapeutic agent, and regenerative medicine using bone marrow concentrates, or autologous stem cells.

Wild type (i.e., naturally occurring) recombinant viruses or *viruslike particles* (VLPs) for therapeutic applications can be produced in mammalian, plant, fungal, and bacterial cell cultures. Virtually all the original human vaccines were commercially produced viruses. Recent vaccines, however, can simply be virus protein fragments produced through genetic engineering. Recombinant viruses have begun to find applications as therapies themselves, either as vectors to introduce genetic material or to infect a targeted cell, such as a cancer cell, to elicit an immune response to attack the infected cell. A potentially emerging new type of biopharmaceutical is the use of exosomes to deliver therapeutic cargos to cells.

Enzymes Enzymes are catalytic proteins, many of which have been engineered to improve performance and stability to produce a given product. In the industrial biotechnology segment, they are used in a wide range of processes, including food and beverage, dairy, fuel, bioethanol, leather, textiles, detergents, and others (Grunwald, *Industrial Biocatalysis*, Pan Stanford, Singapore, 2015, pp. 9–11). Enzymes are used to catalyze a range of reactions, including some used for therapeutic protein production. Oxidative enzymes, which catalyze oxidation reduction reactions, have been used for the conversion and production of alcohols and have potential applications to pharmaceutical intermediates, fine and specialty chemicals, and materials. Cyanobacteria enereductases can be used to produce enantiomerically pure pharmaceuticals and fine chemicals. Phospholipases have been used for removal of phospholipids, and lipases and have been evaluated as an alternative to synthetic catalysts in the transesterification processes in biodiesel processing.

HOST CELL TYPES AND CHARACTERISTICS

Central to bioprocessing are the cellular organisms themselves that produce the product, convert raw materials into more useful products, or are themselves the product. Cells used in bioprocessing are, for the most part, either naturally occurring as single-cell organisms or have been adapted from multicellular organisms to function as single cells in suspension cell bioreactors. In some cases, such as stem cells, the cells are adherent, meaning they require a surface upon which to attach and grow, such as a microcarrier. Bioprocessing sometimes uses multicellular organisms such as algae, or even whole plants or transgenic animals that have been genetically engineered to produce a recombinant product. While interesting, the use of whole plants or animals for commercial biopharmaceutical production is not common. Producing the right product using living cells outside of whole plants or animals requires systems to provide strict control of environmental conditions, including optimal temperature, pH, oxygen, nutrients, and other processing conditions described later in this section.

Cellular organisms are grouped into two broad classes, those having chromosomes surrounded by a membrane forming a nucleus and those without a nucleus. Cells with a membrane-enclosed nucleus are known as *eukaryotic*. Simpler organisms lacking a nucleus are termed *prokaryotic*.

Prokaryote and Eukaryote Host Cells Bacteria are a very simple host cell class that does not contain a nucleus; they are classified as prokaryotes. Archaea is a second type of prokaryote that is generally found in anaerobic conditions, but has not been used in biopharmaceutical production. Bacteria are important both for production of a wide variety of products and as a potential contaminant to other bioprocesses. Eukaryotic organisms are the second major class of cellular organisms and are characterized by having a nucleus containing their genetic material (DNA). There are a wide range of eukaryotic cell types, including mammalian, plant, fungal, and others. They are the workhorses of modern biopharmaceutical production. Table 20-3 lists general characteristics of each host cell class, although eukaryotes

TABLE 20-3 Characteristics of Prokaryotic and Eukaryotic Cells

Prokaryotic cells	Eukaryotic cells
No nucleus	Has a nucleus
Small, typically a few micrometers	Larger, typically 10 to 30 μm
Has cell wall (tough protective coat)	Mammalian cells have flexible cellular membrane, a few have cell wall
High metabolic rate; population doubling time as little as 20 min	Lower metabolic rate; doubling time typically 24 h, although some (e.g., yeasts) have higher metabolism
Most often single-cell organisms; some grow as multicellular or filamentous structures	Multicellular anchorage-dependent cells, but can be adapted to single- cell, anchorage-independent suspension cultures
Naturally occurring as immortal cells	Naturally limited cellular life span; can be genetically modified as immortalized cells
Aerobic or anaerobic growth	Aerobic growth
Growth medium—A prokaryote can be found to utilize almost any carbon-based substrate ranging from wood to petroleum. Medium and feedstock pretreatment is sometimes required.	Mammalian cells can have complex medium requirements. Trend in biotherapeutics is to use chemically defined medium to reduce risk of animal derived component related contaminations such as viruses.
Examples *Escherichia coli*—biopharmaceuticals, e.g., insulin *Clostridium Acetobutylicum*—acetone and butanol *Lactobacillus*—food production	Examples Mammalian (CHO)—majority of mAb biopharmaceuticals Plant cells—tobacco Fungus including yeast *S. cerevisiae, Aspergillus niger*—food, pharmaceuticals, fuels *Pichia pastoris*—biopharmaceuticals, e.g., insulin Mold *penicillium sp.*—antibiotics Algae *Chlamydomonas reinhardtii*—fuels and methane

have diverse characteristics making it difficult to catalog all their characteristics. For example, a few have cell walls, while most do not. The choice of what type of cell, or host organism, is used for bioprocessing has significant implications for the product that can be made; bioreactor design; growth medium cost and complexity; bioreactor susceptibility to viral, phage, and bacterial contaminations; recovery and purification process design; and product quality. Of particular importance are the biotherapeutic product quality attributes that can be incorporated into the protein being produced. Eukaryotes have the cellular machinery to make post-translational modifications to the recombinant protein being produced, whereas prokaryotic cells lack this capability. Post-translation modifications are essential for all but the simplest recombinant protein biotherapeutic.

Host Cell Characteristics Although there are millions of cellular species, they have many commonalities, the most fundamental being that all life utilizes the same genetic code system. While humankind has exploited biological systems for personal and commercial purposes for nearly as long as civilization has existed, the rate of growth of the use of these systems has greatly accelerated with the realization that DNA is the fundamental message encoding system in all life and the subsequent technological ability (genetic or recombinant engineering) to introduce specific DNA coding sequences into a researcher's choice of organism. The initial applications of genetic engineering targeted the low-hanging fruit, the making of very high-value, relatively simple therapeutic protein products that were either nonexistent or in very short supply. As the technology developed, the products have become increasingly complex, and in some cases they have become the cells themselves. As genetic engineering continued to progress, especially the commercial applications, it became increasing obvious that fully functioning and noninhibitory products made through the genetic engineering of a host organism are not always as simple as just moving the genetic code from a human to a bacterium, such as *Escherichia coli.*

There are some basic features of cells that have implications for bioprocessing. Every cell contains cytoplasm, a colloidal system of large biochemicals in a complex solution of smaller organic molecules and inorganic salts. The cytoplasm is bounded by a semielastic, selectively permeable cell membrane that controls the transport of molecules into and out of the cell. In some cases, these transport mechanisms facilitate secretion of the desired product out of the cell, while some cells lack the capability or only marginally secrete the desired product. Prokaryotic cells are typically protected by rigid cell walls external to the cell membranes. These cell walls can be very robust and require additional processing to recover products that are not secreted, but are retained inside the cells as dense protein masses called *inclusion bodies*. In contrast to the simple prokaryotic cells, more-complex cells derived from mammals, such as the highly utilized Chinese hamster ovary (CHO) cell lines, can produce more complex proteins such as monoclonal antibodies (mAbs) that require their more sophisticated cellular machinery. CHO cells typically secrete these recombinant products into the suspending fluid, which simplifies their subsequent separation and purification.

Table 20-4 presents common host cell types used to produce commercial products along with their differences, advantages, and disadvantages.

TABLE 20-4 Host Cell Characteristics and Implications for Biopharmaceutical Production

	Host organism		
Characteristic	Bacterial	Yeasts lower eukaryotic	Mammalian higher eukaryotic
Host Traits			
Examples	*Escherichia coli*	*Saccharomyces cerevisiae* *Pichia pastoris*	Chinese Hamster Ovary (CHO) cell lines
Cell size	Small, several micrometers Tough cell wall	On the order of 10 μm	Typically 10 to 30 μm
Genetics	Readily sequenced, single DNA molecule	More complex, about 10-fold more base pairs than bacteria Some *E. coli* well characterized, while others not as complete	Very complex, 1000-fold more base pairs than bacteria Very flexible biotherapeutic platform CHO cell genome well characterized
Metabolism	Very high; population doubling time as little as 20 min	High, but only about 25 percent of *E. coli*	Low, population doubling time up to about 24 h
Expression levels	Excellent	Very good	Fair to good
Glycosylation	None	Limited capability may result in suboptimal serum half-life, potency, and/or immunogenic response	Excellent post-translational modifications Complex glycosylation reduces immunogenicity risk, resulting in safer therapeutics with better potency
Product Traits			
Types	Industrial chemicals Nonglycosylated protein therapeutics	Fuels and chemicals Limited therapeutics	Complex biotherapeutics
Secretion	Small molecules secreted, Proteins generally retained as dense inclusion bodies	Small molecules and proteins secreted	Proteins secreted
Confirmation and folding	Simple structures Protein refolding required after inclusion bodies recovered from lysed cells	Good, better than *E. coli*	Excellent
Reactor Design*			
Media	Simple, low cost Utilizes wide range of carbon-based substrate	Simple, low cost Methylotrophic yeast can use methanol as growth media	Complex and costly Chemically defined media used to eliminate animal-derived component-related contamination risks
Cell densities	Very high	Very high	Relatively low to moderate
Heat and mass transfer	High cooling and oxygen demands	High cooling and oxygen demands	Relatively low demands
Run duration	Generally short	Generally short	Runs range from a week to months for perfusion; Seed train can take 30 days to produce sufficient cells to initiate production
Other	Post-production cell disruption and protein refolding required to recover active therapeutic complicates downstream processing	*P. pastoris* can use methanol as inducer and substrate, but may trigger hazardous location design requirements	Requires strict bacterial and viral contamination control; External devices needed to obtain high cell densities; Continuous perfusion process becoming more common to boost productivity Sophisticated monitoring and control strategies to control critical product quality attributes may be used

*The term *bioreactor* is commonly used to describe mammalian cell culture reactors. *Fermenter* is used to describe reactors used with host cells having high metabolic activity such as bacterial and yeast cells.

The goal is to select the host that is the most efficient and robust to produce the required product that meets safety and other product requirements. Different hosts can produce the same product. Insulin, for example, has been produced by *E. coli* bacteria, as well as yeasts *S. cerevisiae* and *P. pastoris*. In addition to factors such as product secretion or retention as an inclusion body, companies may choose one over the other based on their in-house experience and intellectual property portfolio.

Host Cell Types Cell types of greatest value to bioprocessing include mammalian cells, yeasts, bacteria, algae, and molds. While progress continues in the development of various types of hosts, in practice over the last couple of decades, the actual number of hosts used commercially has actually remained steady. It is unlikely that a new host system will appear and replace currently used systems. When it comes to human biopharmaceuticals, significant inertia exists and needs to be overcome to introduce a new host, not the least of which is safety-related regulatory issues. An understanding of how a host cell can affect the biotherapeutic product and its potential to impact product safety and efficacy is built up over years of practical experience. Adoption of a new host for biotherapeutic production would have to have significant advantages to justify the time and expense needed to satisfy regulatory requirements. The most probable impact of a new host will be in the area of bulk chemicals or biofuels production in which significant economic pressure exists.

Bacterial Bacteria are microscopic single-cell organisms ranging in size from 5 to 20 μm, although some can be smaller or larger. They have a cell wall that imparts a characteristic shape which can be round, ovoid, rod, or spiral. The shape can vary depending on culture conditions and is termed *pleomorphism*. Certain species may arrange themselves into clusters, chains, or discrete packets.

Bacterial cytoplasm may contain storage granules consisting of carbohydrates, lipids, or other excess proteins and are commonly called *inclusion bodies*. The cytoplasm also can contain plasmids that are pieces of genetic material existing outside the primary genome. Plasmids can be used as a vehicle to introduce foreign DNA (genes) into the bacteria to impart new synthetic capabilities to an otherwise "wild" type of bacterial strain. The genome of *E. coli* has been well sequenced. This, combined with its rapid doubling time of 20 min and ease of growth, has led to *E. coli* being widely used to produce foreign proteins.

While human genes are routinely inserted into bacteria to produce human proteins, bacteria lack metabolic pathways capable of making post-translation modifications, such as glycosylation, that are required for many human therapeutic proteins to function correctly and safely. This lack of post-translational modification capability, when combined with the typical precipitation of the targeted protein into inclusion bodies, making downstream processing more challenging, has limited the use of *E. coli.* Even so, the first recombinant protein approved by the FDA was insulin produced in *E. coli.* Insulin has a relatively simple structure which did not require post-translational modifications. This combined with the ability to insert plasmids, relatively simple fermentation process, and high productivity made *E. coli* an attractive host cell.

Beyond therapeutic proteins produced by *E. coli*, engineered strains of *Corynebacterium glutamicum* bacterium have been used to produce organic acids, alcohols, diamines, polyhydroxyalkanoates, and proteins. *Clostridium acetobutylicum* produces acetone and butanol, and *Lactobacillus* has wide application in food production.

Adverse Attributes of Bacteria In addition to their use to produce various products, bacteria have the potential to contaminate mammalian cell cultures and utilities that support biopharmaceutical production. These contaminations can result in product loss, triggering intensive and costly root cause determination of the contamination source and remediation. Under adverse conditions some microorganisms produce endospores that remain in stasis until favorable conditions are restored and they germinate. Endospores are a very stable form of bacteria that may survive in dry, nutrient-starved conditions. If unwanted bacteria contaminate a cell culture, the possible presence of endospores has to be considered in developing a decontamination strategy.

Some bacterial species produce a gelatinous material that surrounds them as a capsule in certain circumstances. This provides them with a means of attachment and some protection from other organisms. If many cells share the same gelatinous covering, it is called a *slime layer*, or *biofilm*. If they contaminate a purified water system, such as those used in the production of biopharmaceuticals, biofilms can adversely impact production and present a significant decontamination challenge.

At 0.3 μm, *Mycoplasma* strains are one of the smallest bacteria. They are important to bioprocessing since they can infect mammalian cell cultures, reducing cell viability and productivity. These infections can be hard to detect in cell cultures. Screening cells propagated from the original cell clone and subsequently used to create a "cell bank" are required to prevent *Mycoplasma* contamination and propagation throughout development and production facilities.

Mammalian Cells Mammalian cells are used to produce the majority of biotherapeutics and many viral vaccines. Processes utilizing mammalian cells have much in common with cell cultures utilizing other microorganisms such as bacteria and yeasts. There are, however, some critical differences that translate into specific bioreactor and recovery design requirements depending on the microorganism being cultured. These cells are less robust than bacteria due to the lack of a cell wall and may need to be operated within narrower pH and temperature ranges, although their susceptibility to mechanical hydrodynamic forces due to agitation has generally been overstated. Mammalian cells are characterized by their large number of chromosomes, leading to more complex genetic modification strategies. Further, they have significantly lower growth rates compared to bacteria and require a more complex growth medium. Early use of mammalian cell cultures required attachment-dependent bioreactors, commonly known as roller bottles. However, as the field progressed and the commercial need to produce complex, post-translationally modified human therapeutics became apparent, advances in mammalian cell culture development have led to strains of animal cells that produce very high secreted product titers in suspension culture, which greatly simplifies production. Researchers were also able to produce "humanized" mAbs, reducing immunological reaction risks compared to murine-derived antibodies. These two advances greatly contributed to the current commercial success of human therapeutic antibodies.

Fungi and Yeast Among the most useful fungal microorganisms are yeasts. They are typically unicellular organisms surrounded by a cell wall and have a distinct nucleus. Under certain circumstances they may produce spores. Yeasts are probably the most important, and the longest used, organism for commercial purposes, given the prevalence of fermented alcoholic beverages and bread. While significant inertia exists in genetically modifying yeast for alcoholic beverages, significant effort to develop optimum yeast for bread production has been, and continues to be, conducted. Beyond food and beverages, genetically engineered yeasts continue to be used to produce many biopharmaceuticals such as hepatitis vaccines and insulin. Application of yeasts has been limited by their inability to produce glycan structures required to produce safe, effective human biotherapeutics such as antibodies.

Virus Viruses are comprised mostly of genetic material, either DNA or RNA, and surrounded by a protein sheath. Lacking metabolic machinery, viruses exist only as intracellular, highly host-specific parasites. While technically not a cell line, when used in combination with the appropriate cell line, viruses can be used to produce biotherapeutics. However, specific and unique challenges are present in their bioprocessing.

Initial commercial vaccine production consisted of the production of the viruses responsible for a disease for which the vaccine was targeted to prevent, such as polio, with post-production processing steps to inactivate or reduce the virus's infection capabilities. While many modern vaccines have focused on only producing immunogenic proteins associated with the virulent virus, the advent of potential gene therapy and several cell therapy applications of fully functioning viruses has renewed interest in virus production processes.

Adverse Attributes of Viruses In terms of negative impact on commercial biopharmaceutical processes, viruses that attack bacteria are called bacteriophages and must be managed to prevent contaminating production facilities utilizing bacterial host cells. Once a facility is contaminated, completely remediating a facility contaminated by a phage is challenging, expensive, and disruptive to production. Viruses may be either virulent or temperate (lysogenic). Virulent bacteriophages divert cellular resources to the manufacture of phage particles. New phage particles are released to the medium as the host cell dies and lyses. Temperate ones have no immediate effect upon the host cell; they become attached to the bacterial chromosome and may be carried through many generations before being triggered to virulence by some environmental event.

With respect to animal cell culture, rodent viruses, such as the hantavirus carried by the deer mouse, can be a significant challenge since a number of cell lines are from the rodent family. Of equal concern are cell lines derived from primates and humans; the potential two-way transfer of viruses from humans to cell culture, and from cell culture to humans, exists. Virus contaminations have impacted facilities using mammalian cell culture to produce human biotherapeutics. Impaired cell growth in commercial bioreactors used to produce a biotherapeutic to treat Gaucher disease was determined to be the result of a viral contamination. This contamination was linked to a raw material used in the medium formulation, and it led to the complete shutdown of the facility until it could be sanitized. This was the second time one of the company's facilities was shut down due to a viral contamination (DePalma, *Genetic Eng. News* **30:** 2010).

Specific processing steps are needed to eliminate viral contaminants from the biotherapeutic as well as to ensure that host cells and facilities are not contaminated. Viruses are significantly smaller than the other microorganisms described here, and typically they range in size from as small as 10 nm up to 200 nm. They are classified as enveloped, having an outer lipid bilayer, or nonenveloped. Both size and presence or absence of an envelope are key considerations in designing processes to eliminate viral contaminants (World Health Organization, WHO Tech. Report No. 924, 2004).

Algae Algae are a diverse group of photosynthetic organisms that range from microscopic to giant kelp, reaching lengths of up to 20 m. Algae has the potential for producing food additives such as omega-3 fatty acids, biodiesel, biopolymers, animal feed, and fertilizers (Ogden, *Chem. Eng. Prog.*, Nov. 2014, pp. 63–66). Green algae *Dunaliella salina* has been used to produce beta-carotene, and *Halomonas elongata* produces ectoine, which has varied uses in biotechnology, cosmetics, and biomedical applications (Grunwald, *Industrial Biocatalysis*, Pan Stanford, Singapore, 2015, pp. 1007–1064). Algae can grow in environments ranging from low to high salinity. This trait allows algae to grow in a wide range of aquatic environments including salt water, fresh water, or wastewater. Outdoor ponds are suitable for growing algae because large surfaces and high illumination are required. One strain, Dunaliella, that grows in high-salinity conditions has been used to produce glycerol. The glycerol accumulates within the cells to counter high external osmotic pressure.

Molds The most famous mold, *Penicillium chrysogenum*, was isolated and used to produce penicillin. Through strain development the productivity was substantially increased, leading to modern production processes (Fig. 20-1). Keys to becoming an effective antibiotic included developing techniques to grow *P. chrysogenum* as a deep-culture fermentation instead of an adherent film culture, and downstream processing using centrifugal liquid-liquid extraction to retain efficacy and stability. Other important pharmaceuticals derived from molds include lovastatin (a cholesterol-reducing statin) produced by *Aspergillus terreus*.

Plants Plants can be genetically modified to express commercially important proteins, including biopharmaceuticals. Uses of transgenic plants and insect cells have been proposed as alternatives to mammalian cell culture to produce mAbs in order to reduce impurities, lower production costs, and reduce risk of viral contamination. They can be grown as whole plants, such as tobacco, rice, and soybean, or as plant cells in suspension culture bioreactors. The latter has advantages of being contained, reducing contamination risk to food crops, and secreting glycosylated proteins into culture fluid, reducing recovery and purification costs, making them potentially viable expression systems for biotherapeutics. The tobacco plant *Nicotiana benthamiana* was used in the early development stages of a potential Ebola virus treatment ZMapp. To reduce the cycle time needed to produce the antibody, thereby speeding development, production was switched to a mammalian host cell line. A second example is paclitaxel, a treatment for breast, ovarian, and other cancers. Initially, paclitaxel was extracted from the bark of yew trees, and later it was produced using a semisynthetic route beginning with an extract from the needles of yew trees. Given the limited amounts of naturally occurring yew trees, Phyton Biotech developed a fermentation process using plant cells to produce the initial plant-derived feedstock.

Insect Insect cell lines such as *Spodoptera frugiperda* and *Trichoplusiani* can be used to produce recombinant proteins. These processes are unique in that actual protein production is the result of infecting host cells with a genetically engineered virus, called a baculovirus. The actual production process is similar to a suspended mammalian cell culture, except that instead of genetically engineering cells to produce the protein, that step is replaced by the addition and subsequent infection of the cells with the modified baculovirus. Like plant cells, insect cells have the advantage of being free of infectious human viruses and have the capability to impart post-translational modifications not possible with bacterial host cells [Drumond, *Biotech. Advan.* **30:** 1140–57 (2011)]. While not widely adopted for commercial mAb production, insect cells have been used to rapidly produce large amounts of complex proteins for subsequent development studies.

HOST CELL APPLICATIONS AND PROCESS DESIGN IMPLICATIONS

The type of host cell utilized for biopharmaceutical production has significant implications both for the design and operation of the bioreactor or fermenter used for cell culture and for the subsequent product recovery and purification unit operations. These host-driven differences are discussed at a high level here and in greater detail in subsequent sections.

Reactors used for the cultivation of yeasts and bacteria are commonly called *fermenters* (reactors specifically designed for microbial and yeast organisms). They range in size from tens of liters to very large systems of more than 100,000 gal for industrial biotechnology applications. Due to the typically large scales, high cell densities and high metabolic rates of these organisms, heat removal is a major design consideration. Fermenters also have to achieve high mass-transfer rates to satisfy cellular oxygen demand and carbon dioxide removal. This is accomplished through the design of gas sparging systems, agitation, and ancillary equipment such as air compressors for gas supply. Combined, these requirements result in fermenter vessels with high height/diameter ratios capable of maintaining a sterile environment, high power mixing systems, and internal heat-transfer coils or other means to maximize the heat-transfer surface area. Fermenters used to produce biopharmaceuticals tend to avoid internal coils to facilitate stringent cleaning requirements.

Recovery of bacterial products formed as inclusion bodies retained within the cells requires some type of separation equipment, such as centrifugation, to remove the cells from the spent medium. This is followed by cell breakage to recover the inclusion bodies. The inclusion bodies are then solubilized, followed by a subsequent protein refolding step to attain the active biopharmaceutical. Additional purification steps are still required to produce the biopharmaceutical. In some cases, the refolding step will require the largest equipment used in the production process due to the low product concentrations at that stage. Secreted yeast products, however, employ a simpler recovery process without the need for cell breakage and protein refolding steps to recover the active biopharmaceutical. Cells are simply removed from the product containing spent medium, which is then further processed to recover and purify the product.

In contrast to microbial and yeast cells, mammalian cells have lower metabolic rates, resulting in bioreactors (reactors designed specifically for mammalian and other lower-metabolic-rate cell cultures) with less demanding heat and mass-transfer design requirements, simplified ancillary equipment for temperature control and gas supply, and no need for additional product recovery unit operations for cell breakage and product refolding. The trade-off is that a series of smaller-scale upstream cell culture process steps, called a *seed train*, is required to generate enough cell mass to inoculate the production bioreactor. (While seed trains exist for both bacterial and animal cells, there are typically more steps in the seed train for an animal cell system.) Establishing and maintaining strict sterile conditions is also required. Any bacterial contamination will quickly overwhelm the mammalian cells, resulting in a lost production run. The requirement to exclude even the smallest microbial contamination results in a more complex bioreactor design and operations supported by complex supporting utilities suppling high-purity water, clean steam, and high-purity process gases.

While traditional fermentation and mammalian cell culture processes often use some variant of the traditional fed batch process to increase product titer and cell densities, thereby increasing process productivity, perfusion processes using external devices such as membranes, are increasingly being considered and adopted, especially for mammalian cell cultures. While appealing, perfusion processes require continually introducing and removing a significant amount of fresh culture medium. This results in more complex equipment design and operation and higher medium costs. As with any move from batch to perfusion and perfusion-enabled continuous processing, the potential productivity and economic advantages must overcome the increased complexity and expenses. These processing options are discussed in further detail below.

BIOPROCESSING OVERVIEW

Biopharmaceutical Protein Structure Biopharmaceuticals are differentiated by their large molecular weight, ranging from 10,000 to more than 300,000 Da, and complex structures compared to small molecule therapeutics usually produced by synthetic organic chemical processes. Proteins are examples of biopharmaceuticals that have complex spatial structures (Fig. 20-4). The basic chemical composition, typically described as the primary structure, is a linear sequence of amino acid building blocks. The secondary structure results from the interaction of the amino acids in close proximity to one another. Examples of such structures include alpha helices (a coiled or spiral confirmation) and beta sheets (connected, pleated amino acid strands). Tertiary structure refers to the three-dimensional conformation of the protein. Quaternary structures refer to how multiple protein subunits interact with one another.

These structures are a result of 20 different amino acids that make up the biotherapeutic protein, each having different properties. Amino acids can be grouped based on certain characteristics. These include charged, polar, nonpolar, hydrophobic, and other physicochemical properties such as specific bonding characteristics (Fig. 20-5). When combined into a protein, they impart both the sequence and the higher-order structure (folding) of the protein. When in its proper form, the

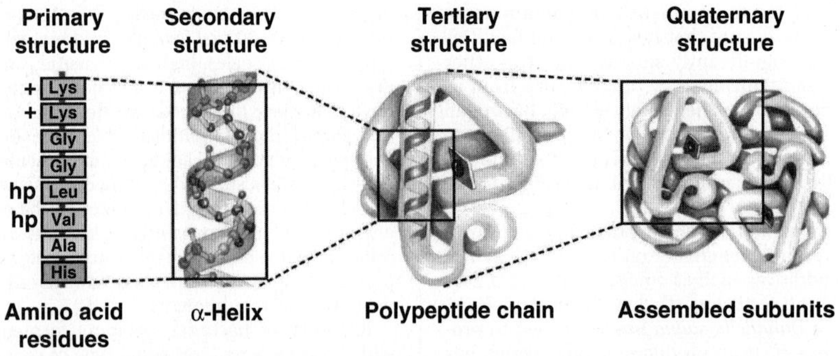

FIG. 20-4 Illustration of protein structural hierarchy from primary amino acids to quaternary protein-protein interactions typical of human biopharmaceuticals. (Adapted from Shapiro, FDA presentation titled "Quality Considerations for Biosimilars," Aug. 8, 2012.)

protein drug has its desired therapeutic effect. When this form is disrupted, through environmental conditions, mechanical processes, or the host cell forming protein inclusion bodies, the protein can lose its effectiveness, or even have undesired side effects such as triggering an unwanted immunological response.

The complex physicochemical properties of the amino acids, such as size, charge, polarity, and hydrophobicity, can be exploited when designing downstream purification processes. For example, the use of pH to change charge state and salt concentrations to manipulate electrostatic interactions are key considerations in designing chromatographic separations. Manipulating the protein environment to affect a purification step can also, however, affect protein stability. Therefore, each process step must take into account the conditions needed to achieve the desired process objective, but also manage unwanted impacts to the protein being processed.

Biopharmaceutical Regulation and Attributes Biopharmaceuticals are subject to regulatory approvals and oversight worldwide to ensure safety and efficacy. The regulatory framework is more complex, and the required structural, chemical, and biophysical characterization more demanding, compared to purely chemically synthesized pharmaceuticals. This is due to the complex nature of these protein products. Regulatory objectives include ensuring that the product and process used to produce it are adequately understood and characterized, and that it adequately addresses the risk of contamination by viruses, bacteria, nonproduct host cell proteins, and other contaminants to ensure a safe, consistent therapeutic.

Achieving these requirements together with the needed protein structure and functionality begins with selecting a well-defined host cell line which is then engineered to produce the desired product. Choice of host cell to produce mAb biopharmaceuticals is a major factor in minimizing immunogenicity while producing a glycosylation structure (sugar residues added to the mAb protein after it is formed) to achieve optimum pharmacokinetics. Host cells such as yeast and bacteria lack the cellular machinery to make the necessary complex post-translational modifications needed to generate glycan structures, which has led to the widespread use of mammalian cells to produce biopharmaceuticals.

Regulatory agencies require a protein therapeutic to be produced from a single clone that is reproduced from a single ancestral cell, and not from a pool of engineered cells derived from the same cloning step that may have different ancestral cells. Such pools will exhibit genetic variations from one cell clone to another contained within the pool. From a processing perspective, certain clones will have attributes that make them more suited for large-scale manufacturing. Clones are typically put through a formal manufacturability assessment to ensure the best clone is selected from both a product attribute and manufacturing perspective. The specific cell clone derived from the host cell line, bioreactor conditions, process, and process control strategies can all affect critical product quality attributes such as post-translational glycosylation. Even small changes to a product such as a mAb can significantly impact the effectiveness and safety of these complex molecules.

Producing a protein therapeutic having the desired pharmacokinetics (solubility, biological activity, stability, and achieving and sustaining adequate levels in the blood) often requires post-translation modifications such as glycosylation of the protein inside the cell after it has been produced. It is an important consideration in human therapeutics as it affects blood serum half-life, solubility, stability, and immunogenicity. Obtaining the right glycan structure has been the target of significant cell line and process development efforts over many years.

Due to their ability to perform post-translational modifications, mammalian cells of various types have been used to produce therapeutic proteins such as monoclonal antibodies, with muromonab-CD3 being the first approved for therapeutic use in 1986. Cell lines to produce monoclonal antibody therapeutics and variants such as bi-functional mAbs, fragment antigen-binding (Fab) antibodies, and others have undergone an evolution from mouse (or murine) antibodies, chimeric antibodies (combination of murine and human) to humanized, and finally fully human antibodies to address potential immunogenic responses while achieving target pharmacokinetic properties [Ryu, *Biotech. Bioproc. Eng.* **17:** 900–911 (2012)]. These advantages have made mammalian cell culture producing fully human monoclonal antibodies the dominant modality for biotherapeutic production.

Bioreactor conditions can also affect product-related impurities that are generated only to be removed later by the purification process. Early in the process development phase, the cell line clones and bioreactor conditions are screened for product-related impurity generation to ensure that the therapeutic production process is suitable for large-scale commercial production. This manufacturability assessment often utilizes high-throughput screening tools where possible.

As the initial wave of innovator biotherapeutics began to lose patent protection, other companies were able to seek regulatory approval for their own versions of the innovator therapeutic, known as *biosimilar biotherapeutics.* Having a firm understanding of how the critical product quality attributes can be monitored and possibly controlled is central to producing an approvable biosimilar therapeutic. In the case of a biosimilar, achieving a close match of the originator's mAb glycosylation diversity is even more important in order to match the original biopharmaceutical's critical product attributes.

BIOPROCESSING STRATEGIES—EXAMPLES

Biopharmaceuticals Development of a biotherapeutic begins with selection of modality, such as monoclonal antibody, choice of host cell to produce the product, and development of the bioreactor and purification processes. Each preceding step has implications for the others. The type of product influences the host cell choice, which in turn has implications for bioreactor design. The product type and host cell choice then determine the downstream recovery and purification steps. Experience shows that for a given class of therapeutic, the process may be very similar from one therapeutic molecule to the next. For example, the production of monoclonal antibodies generally utilizes CHO cell culture, followed by a cell removal recovery step, an initial chromatographic purification capture step, viral inactivation, additional chromatographic purification, viral filtration, final ultrafiltration-diafiltration, and drug substance formulation (Fig. 20-6). This is termed a *platform process* and has the advantage of applying lessons learned to improve costs, speed to clinic, and product quality to subsequent mAb biopharmaceutical development projects.

By contrast the recovery and purification of protein therapeutics produced as inclusion bodies within bacterial cells is more complex. This is especially true at the initial stages in which the protein inclusion bodies are released from the cells, the nonnative protein is then recovered, and it undergoes refold steps required to produce an active therapeutic protein, as shown in Fig. 20-7.

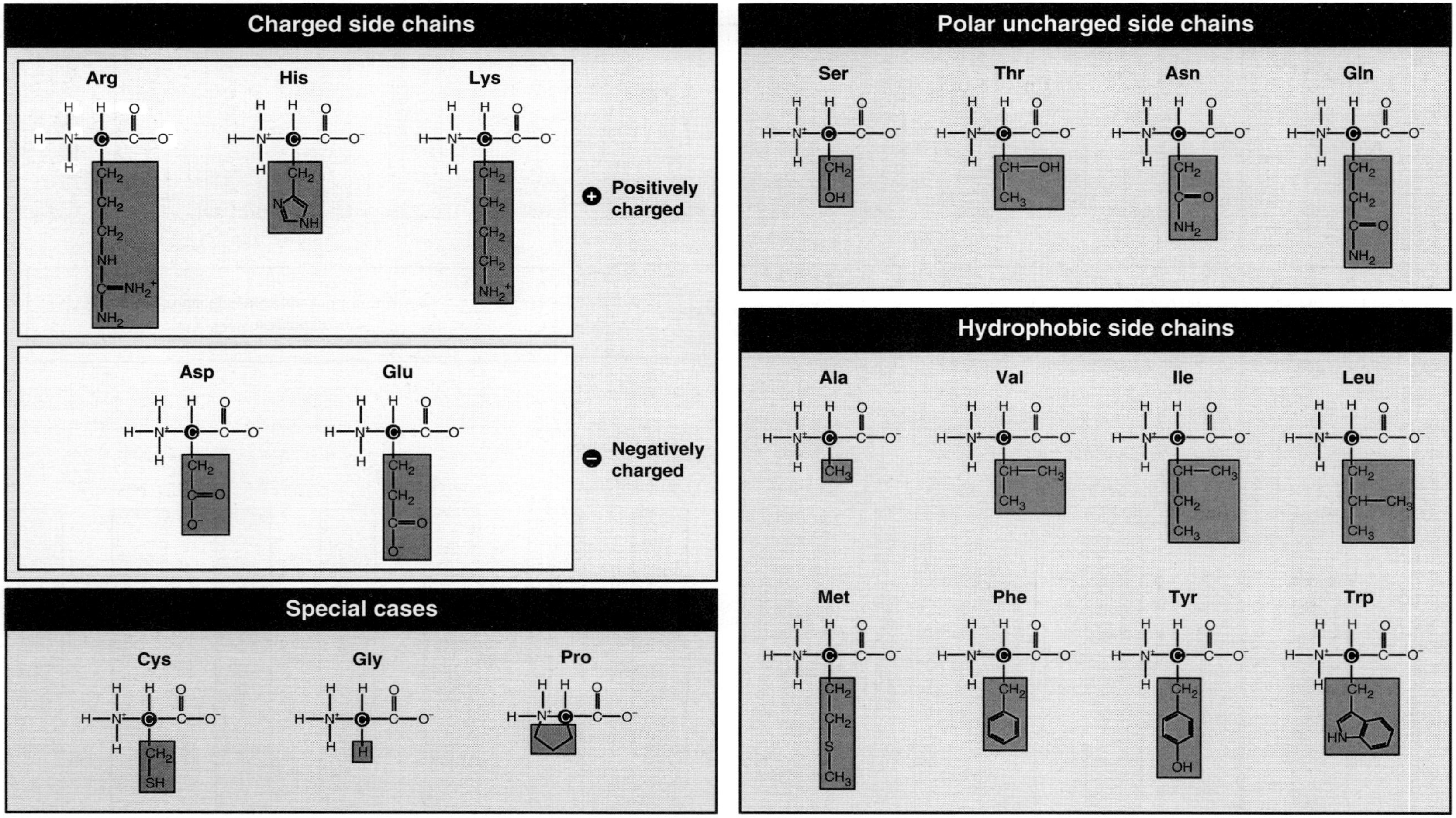

FIG. 20-5 Amino acid structures and general chemical properties. (Amino acid diagram reproduced from *Making Biologic Medicines for Patients: The Principles of Biopharmaceutical Manufacturing.* https://www.edx.org/course/making-biologic-medicines-patients-mitx-10-03x-0. Used with permission. Copyright © 2015 MIT.)

Industrial Biotechnology Industrial biotechnology processes are more diverse than biopharmaceutical processes. Process design depends on the feedstock, feedstock pretreatment requirements, organism or enzyme used in the process, and physicochemical properties of the product being produced. It is likely, however, that many of the unit operations employed will include ones familiar to most chemical engineers, especially in the production of small molecule products from biomass. These include centrifugation, liquid-liquid extraction, chemical transformations, leaching, and distillation.

Figure 20-8 shows examples of industrial biotechnology processes using biomass feedstocks. Conversion of biomass into usable products entails

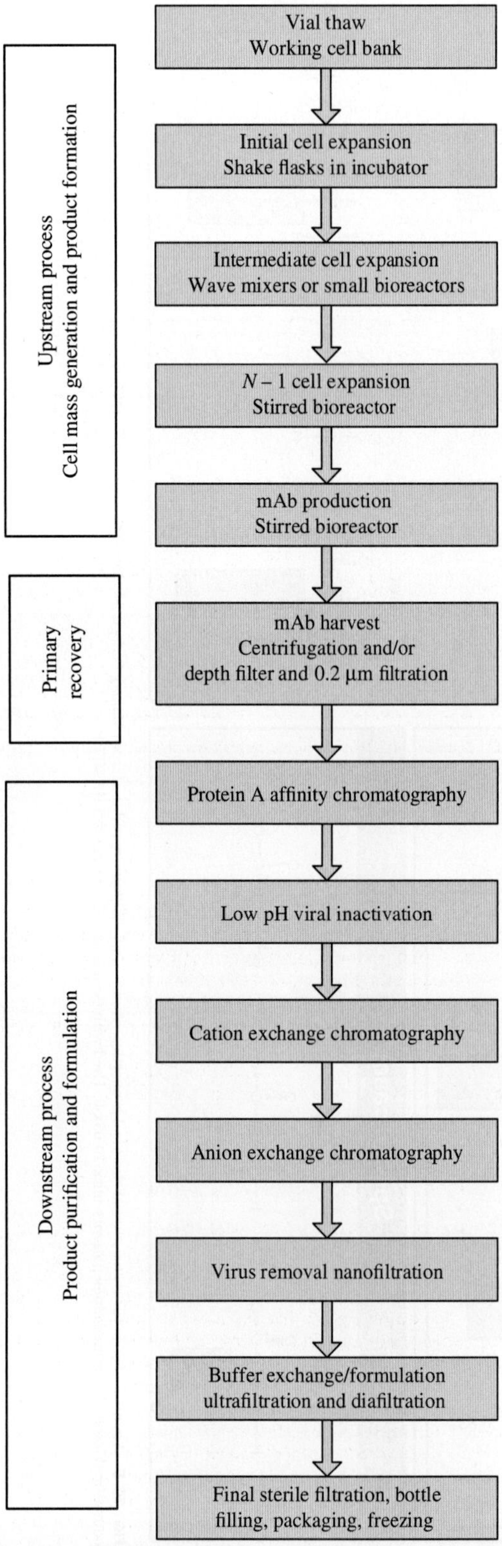

FIG. 20-6 Generic mAb manufacturing flowsheet. Protein product is secreted from the CHO host cells in the production bioreactor.

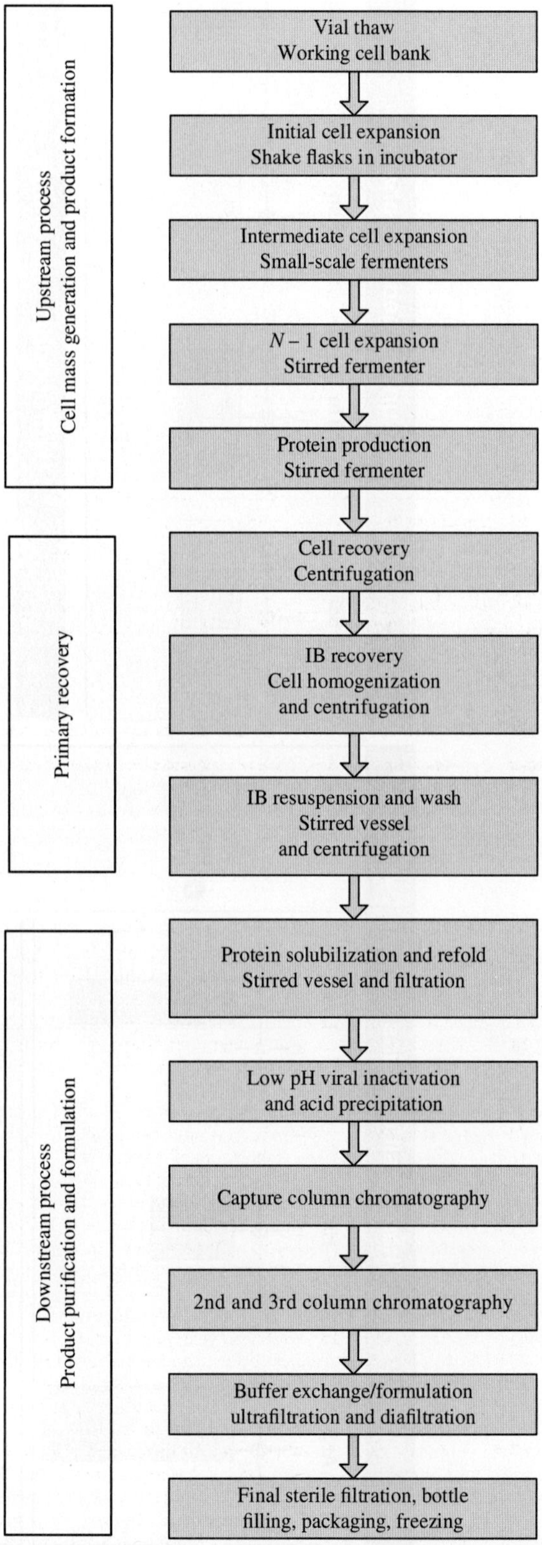

FIG. 20-7 Generic microbial biopharmaceutical manufacturing flowsheet. Product formed in the fermenter is retained within the host cells as inclusion bodies until the cells are homogenized, releasing the protein product.

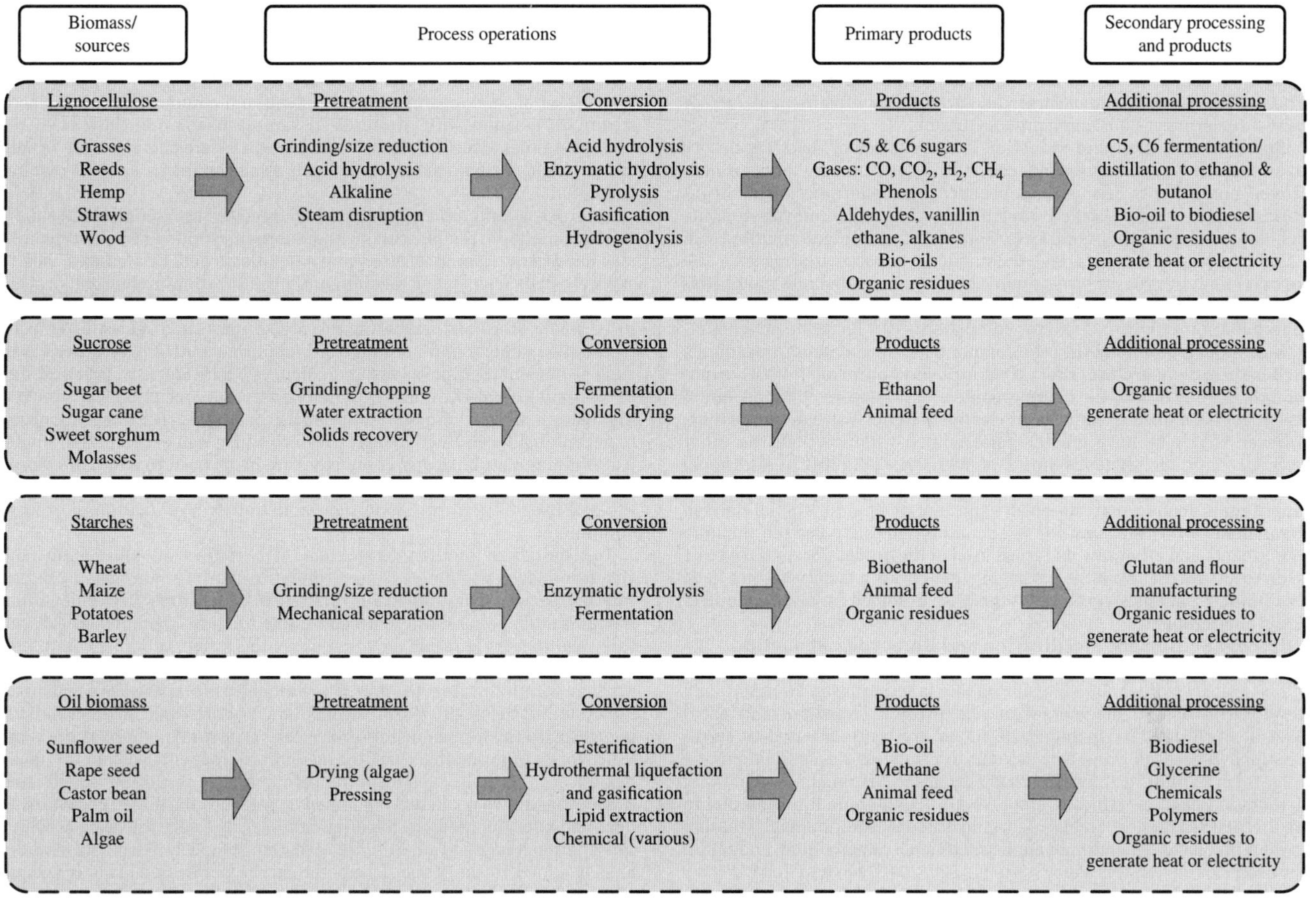

FIG. 20-8 Overview of biomass conversion processes starting with raw material pretreatment required prior to conversion followed by primary product formation and products that can be produced through additional secondary processing.

feedstock selection, feedstock pretreatment, and subsequent processes. Feedstock can be either energy crops cultivated for purpose, such as starch crops and short rotation forestry, or biomass residues and agriculture by-products such as straw, bark, wood chips, used cooking oils, and bioprocessing waste streams. Feedstock conversion or pretreatment can be biochemical processes (fermentation, enzymatic), thermochemical (gasification, pyrolysis), chemical (acid hydrolysis, esterification), and mechanical (fractionation, pressing, size reduction).

Facilities designed to convert biomass into a number of products have been referred to as *biorefineries*. Biorefineries have been defined as the sustainable processing of biomass into a spectrum of marketable products and energy. Moving beyond the R&D, pilot, or small-scale demonstration stage to commercial implementation depends on economic conditions and cost competitiveness relative to petrochemicals. Coproduction of biofuels such as ethanol, chemicals, and animal feed may be necessary to achieve cost competitiveness (Lynd, *Strategic Biorefinery Analysis*, National Renewable Energy Laboratory, Golden, C.O., 2002, p. 15).

Enzyme Production Most industrially important enzymes will be produced in fermenters utilizing bacterial, yeast, or fungi as extracellular proteins. To be viable commercial products, the fermenters and medium have to be relatively inexpensive, yet the control of both the process and the medium ingredients needs to be sufficient to induce high productivity and optimal cell densities. A common approach is to obtain inexpensive medium sources such as starches, whey, or other agricultural or food production by-products, such as animal slaughterhouse by-products. These may be used with or without some sort of pretreatment depending on the application. Enzymes used as human diagnostics or in pharmaceutical production to perform biotransformations require higher purity and quality control standards, requiring better-defined, more complex medium and purification schemes (Dordick, *Biocatalysts for Industry*, Plenum Press, New York, 1991, pp. 6–13).

Whole-Cell Biocatalysts Whole-cell biocatalysis is an intermediate between enzyme-catalyzed reactions and product synthesis by medium conversion via the complex metabolic pathways of actively growing host cells. Whole-cell biocatalysis often entails conversion or functionalization of a substrate component that has been added to the medium by either growing or resting cells. These cell-mediated transformations usually involve only one or a few reaction steps; however, these steps are either too complex or too expensive to be carried out through synthetic organic chemistry routes. Engineered cells have been used in this way to produce pharmaceutics such as statins and specialty chemicals.

CONTAMINATION RISK MITIGATION IN BIOPHARMACEUTICAL MANUFACTURING

Contamination Reduction and Control Contamination prevention, inactivation, and removal through various methodologies are essential elements of both upstream and downstream bioprocessing. Microbial and viral contaminants can be present in a production facility, process equipment, a compromised utility system, raw materials, raw material packaging, or even a contaminated host cell line (e.g., mycoplasma). The impact of a contamination ranges from a lost bioreactor run, a final purified protein therapeutic product rendered unusable, to the shutdown of an entire facility.

Facilities producing biopharmaceuticals will have contamination control plans to protect the product, meet regulatory agency requirements, and reduce business risk as a consequence of contamination by an adventitious agent. A well-designed, controlled process will ensure sterility of the bioreactor and include a downstream process designed and operated to achieve reproducibly safe bioburden levels, a measure of microbial contaminants, through bioburden reduction steps. Bioburden can be reduced by physically removing microbial contamination by filtering liquid and gas streams, by inactivation through exposure to heat or irradiation, or by chemical means such as sodium hydroxide sanitization. In addition to bioburden reduction, microbial contamination is excluded from the process by establishing a sterile boundary in the case of a bioreactor, and by operating as a closed process which essentially ensures the surrounding environment cannot contact the product-containing process stream.

Viral contamination and reduction is the subject of strict regulatory requirements when mammalian cell cultures or human- and animal-derived components are used to produce a biopharmaceutical. Regulations require that validated viral inactivation and removal steps be included as part of the bioprocess. Sources of viral contamination can include incoming raw materials, the manufacturing environment, or the host cell line. In-process virus removal technologies include nanofiltration and chemical inactivation such as low pH or detergent solutions. Viral contamination risk from medium components can be reduced by eliminating the use of animal-derived components, and medium treatments such as high-temperature, short time exposure, ultraviolet, UV-C, exposure, or nanofiltration. Chromatographic steps (capture columns, anion exchange) have been used for viral clearance in downstream processing; however, they may have greater performance variance than the other methods [Cipriano, *Methods Mol. Biol.* **899:** 277–292 (2012)]. Additional measures to mitigate viral contamination risks include the raw material supply chain management to ensure that no, or limited, animal-derived components which may harbor viral contaminants are used in medium components and other raw materials used in the process. Steps have to be taken to ensure that even the packaging that enters a facility has not been contaminated at any stage, or are otherwise excluded from the manufacturing environment.

Process Equipment Sterilization and Bioburden Reduction If a mammalian cell culture is contaminated by bacteria, the mammalian cells will be quickly overwhelmed, resulting in a failed process. To maintain a sterile environment, the bioreactor and incoming medium and gases must be sterilized. Sterilizing-grade filters are routinely used to sterilize supplied gases (air and oxygen) and to provide a sterile boundary by using them to filter the bioreactor exhaust gases. Mammalian cell culture medium sterilization typically employs sterilizing-grade filtration due to the thermal sensitivity of many of its constituents. Additional information on membranes and membrane applications such as gas, liquid, and nanofiltration can be found in Perry's 8th ed., Sec. 20 (Green and Perry, *Perry's Chemical Engineering Handbook*, 8th ed., McGraw-Hill, New York, 2008, pp. 20-36–20-71).

Stainless-steel bioreactors are specially designed and validated to be sterilized using saturated clean stream, which is more lethal to contaminants than superheated steam. These bioreactors are also designed to facilitate cleaning by being highly polished and devoid of any crevices or other areas that could harbor soilants or contaminating organisms that could resist the sterilization cycle. Prior to use, but after the bioreactor has been cleaned with all components needed to establish a sterile boundary attached, clean steam is introduced into the empty stainless-steel bioreactor in a process called *steam in place* (SIP) sterilization. The sterilization process continues at a specific temperature and time that have been determined to result in a high probability that microbial contaminants have been inactivated. Methods have been developed to predict the right combination of time and temperature needed to inactivate various contaminants (Clark, *Bioprocessing Engineering*, Woodhead Publishing, Cambridge, UK, 2013, pp. 189–208). The advent of single-use bioreactor systems and components has replaced steam sterilization with presterilized gamma irradiation. Just as in the case of stainless-steel bioreactor sterilization where the SIP cycle is validated, single-use bioreactor sterilization using gamma irradiation requires validation.

Bioprocess unit operations used to recover the biopharmaceutical product made in the bioreactor and subsequent purification steps may also use SIP or other means to reduce the bioburden. Centrifuges used to separate cells from spent medium and ultrafiltration membrane process skids for buffer exchanges or product concentration have been designed to be steam-sterilized, excluding the membranes themselves, when used to produce clinical and commercial biopharmaceuticals. In other cases, such as chromatography skids and columns where SIP is impractical due to equipment and chromatographic resin limitations, chemical sanitization using sodium hydroxide solutions has been used to reduce bioburden to acceptable levels. Increasingly presterilized, gamma-irradiated, single-use equipment, tubing, and connectors are being used to configure closed downstream process operations to limit or even eliminate the need for the end-user to perform bioburden reduction or sterilization of process equipment.

The design of modern fermenters with respect to cleanability and steam-in-place capabilities is very similar to that of mammalian cell culture bioreactors. One important difference stems from the fermentation medium often being more thermally stable than the mammalian cell culture medium. As a result, in some cases the fermentation medium and fermenters can be sterilized together as part of the fermenter sterilization process, within certain size limits. The fermenter is prepared as usual for an SIP cycle, but before SIP starts, the medium is added. Next steam is directly injected into the fermenter to sterilize the vessel and medium at the same time. This is only possible when the medium components will not thermally degrade, or thermally sensitive components can be aseptically added later. A potential drawback to this combined approach is the added cycle time to heat and cool the medium-filled fermenter. This limits the scale at which this approach is practical. In those cases, continuously sterilizing medium as it is added to the sterile fermenter can be performed. This is accomplished by exposing medium to high temperatures for short durations using in-line heat exchangers. This is similar to the approach that has been used to treat complex mammalian cell culture medium to deactivate potential viral contaminants.

Biopharmaceutical processes are supported by clean utility systems providing gases (O_2, CO_2, N_2, air), purified water, and clean steam generation systems. Design and operation of such systems require very strict attention to detail and ongoing maintenance to mitigate contamination risk. Design best practices and standards for bioprocessing equipment, including bioreactors, fermenters, and utilities, are available as part of the ASME standards series [ASME BPE-2016, American Society of Mechanical Engineers, New York, 2016].

BIOREACTORS AND UPSTREAM PROCESSES

Bioreactors play a variety of roles in biopharmaceutical production. Most importantly they produce the products, be they therapeutic proteins, vaccines, or whole-cell therapies. In most cases, the actual commercial production is achieved in a series of successively larger bioreactors until the final product production bioreactor is reached. Further, bioreactors are used to generate a "bank" of frozen cells which are later thawed to initiate a production run. Process characterization studies are carried out in scale-down reactors to satisfy regulatory requirements. Laboratory and pilot plant scale-down bioreactors are typically used to evaluate proposed changes to a commercial production process, to evaluate unusual processing events and excursions (nonconformances), and to assess the potential effect and controllability of critical process parameters and their relation to product quality attributes. Process optimization, cell clone screening, and production of small quantities of a potential therapeutic molecule for use in research are all carried out using various scales and types of bioreactors. In this section we focus primarily on *bioreactors*, the term commonly used to describe vessels used for mammalian cell production; but we point out where significant differences exist for *fermenters*, the term commonly used to describe vessels used for bacterial or yeast cell bioreactions. Similar principles can also be applied to stirred-tank bioreactors and fermenters utilized for other purposes, products, and host cells.

CELL LINE DEVELOPMENT, CELL BANKING, AND PROCESS REQUIREMENTS

Cell Line Development Biopharmaceutical production requires engineered cells to make nonnative proteins. This entails genetically modifying a host cell line to create multiple clones which can produce the desired product. Development of a cell culture process begins at this clone development and screening stage. Cell culture processes can be optimized through cell line engineering to improve productivity, yield, product purity, and achievement of target quality attributes. These cellular engineering processes are covered in detail in Hauser (*Animal Cell Culture*, Springer, New York, 2015, pp. 1–28). Cell line development includes selecting a host, genetically modifying the cells yielding a polyclonal pool of cells, and screening individual clones to select the specific one having the best combination of characteristics that will be used to develop the process and product.

Early-stage development of a bioengineered product uses a clone screening process to assess each clone for product expression levels and stability, as well as assessing the protein product for required biological activity, safety profiles, product variants, contaminants, and other critical quality attributes. Typically, this screening process begins with microplates, and it progresses to micro/mini stirred bioreactors or shake flasks. It is often automated to the degree possible to efficiently and rapidly screen large numbers of clones. This initial work will include manufacturability assessments designed to identify the clone that best achieves a combination of product quality, potency, therapeutic target binding, yield, titer, and low levels of product-related impurities and aggregates that may prove difficult to remove later. Additional assessments are performed to characterize clone robustness, including being able to reproducibly initiate a production seed train from a thawed vial, stable genetic profile over many generations, and consistently producing the target protein.

To more fully develop manufacturability assessment information on cell culture performance and acceptable process parameter ranges, potential clones are grown under representative production conditions and the results are assessed against qualities facilitating commercial-scale production. This stage of the manufacturability assessment encompasses high yields, cell densities, viability, protein stability, aggregate formation (or lack thereof), ease of product recovery and purification, and process economics. Stirred bioreactors, on the scale of 1 to 5 L, are typically used at this stage, frequently with many in parallel to develop statistically relevant data. These studies further evaluate medium formulation and optimization such as feeding strategies, amino acids, and other trace elements. The result is an optimized process with acceptable ranges for bioreactor conditions (temperature, pH, dissolved oxygen, and carbon dioxide concentrations) based on product quality and manufacturability.

Cell Banking Once a cell line has been engineered and the desired clone selected, the chosen clone will be propagated to generate the master cell line, which is then further propagated and subaliquoted into hundreds of small vials and frozen. These vials form the *master cell bank* (MCB) and will be used in the manufacture of the biopharmaceutical over its lifetime. Each MCB is used to generate a *working cell bank* (WCB), with each WCB containing hundreds of vials. Each time a production campaign is initiated, an inoculum vial from this WCB is used to initiate the production seed train. Generation of these cell banks can use a combination of vessels including shake flasks, rocking bags (i.e., a Wave), and/or small-scale stirred bioreactors. In many cases single-use systems will be used to reduce cleaning and sterilization requirements and to minimize cross-contamination risk. Recent efforts to streamline the cell culture production scaled-up train have sought to increase the initial cell numbers beyond what a typical vial provides. This has included using membranes to increase the cell density when propagating the cell bank, or by banking cells in larger-format, single-use bags. These techniques allow several production process steps to be eliminated and are described more fully in the Seed Train Optimization subsection.

Process Requirements Figure 20-9 depicts the design, processing, and control elements to be accounted for in developing and executing a typical mAb upstream process. It also generally applies to many other protein-based biopharmaceutical modalities with small changes. A small subset of process parameters can impact critical product quality attributes, while others may affect only important process attributes (e.g., titer or yield, but not quality); still others may pose little risk to the process or product. Determining which parameters are critical and the ranges that mitigate the highest risks requires the use of proven scale-down bioreactors coupled with *design of experiment* (DOE), and other experimental designs. An acceptable operating space for critical and important parameters is defined and incorporated into design and operating requirements that ensure bioreactor conditions are maintained within acceptable ranges. This is termed the *design space*, and it provides a high degree of assurance that the process efficiently yields a product meeting quality requirements.

SEED TRAIN AND PRODUCTION BIOREACTORS

Bioreactors have two primary functions in biopharmaceutical production. The first is to generate biomass as part of a seed train, and it utilizes a variety of types and scales of bioreactors. The second function is the actual product formation as the final stage of the seed train, primarily in production-scale stirred vessels. Producing biomass consists of several sequential stages starting with thawing a WCB vial, which is used to inoculate the first stage of the seed train. Flasks placed on orbital mixers housed in incubators designed to closely control temperature and CO_2 levels are often used for

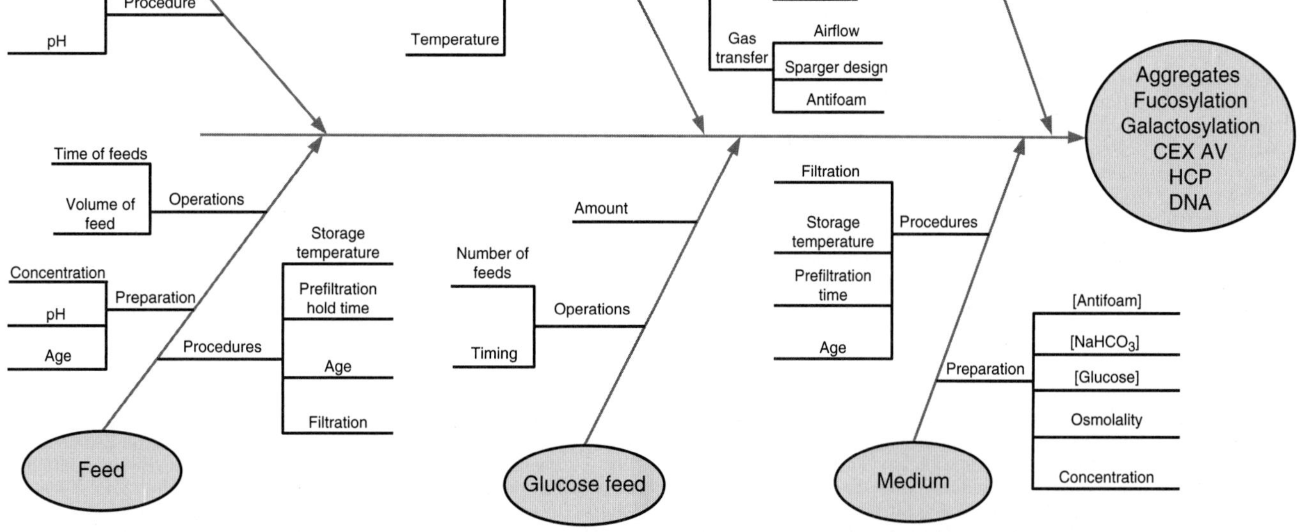

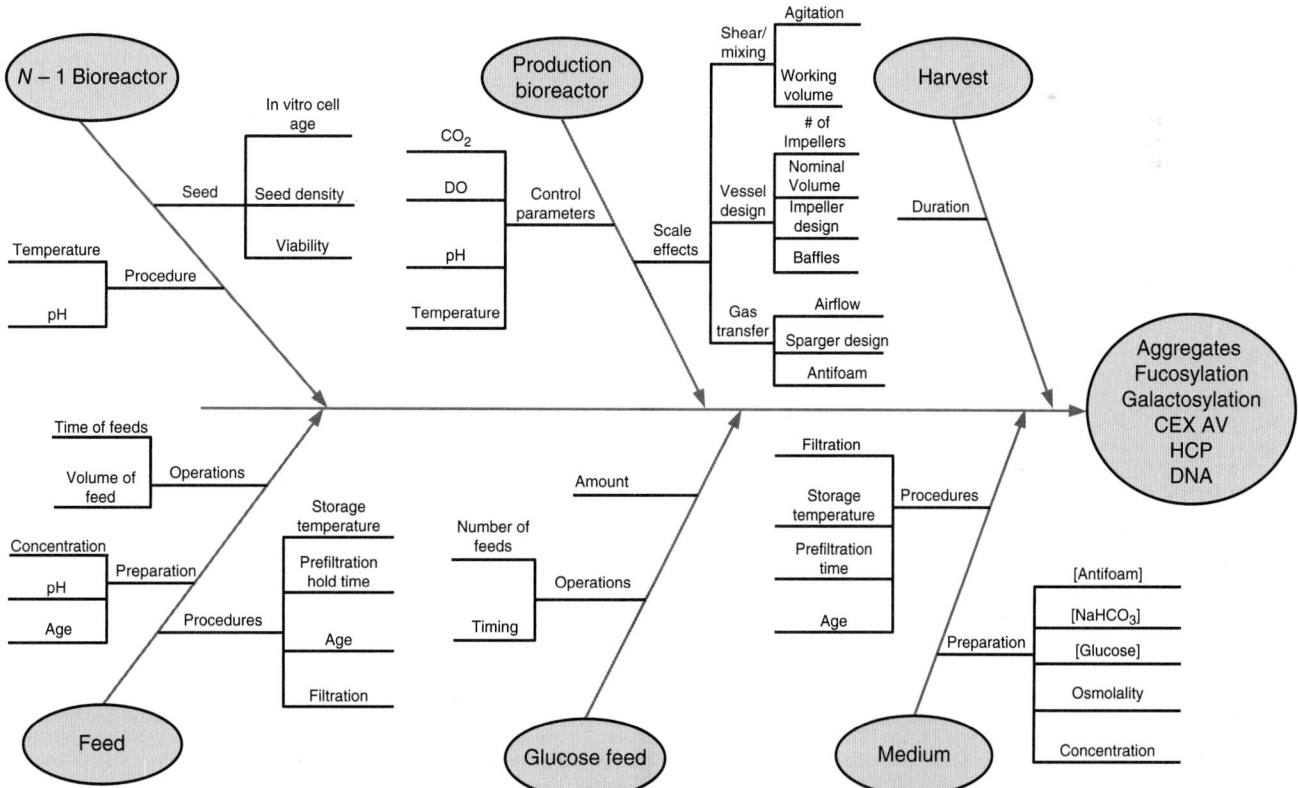

FIG. 20-9 Overview of design, processing, and control elements important in the development and execution of a typical mAb process. [CMC Biotech W.G., A-Mab: "A Case Study," *Bioproc. Dev.*, Ver. 2.1, p.74, in Public Domain at ISPE, www.ispe.org (2009).]

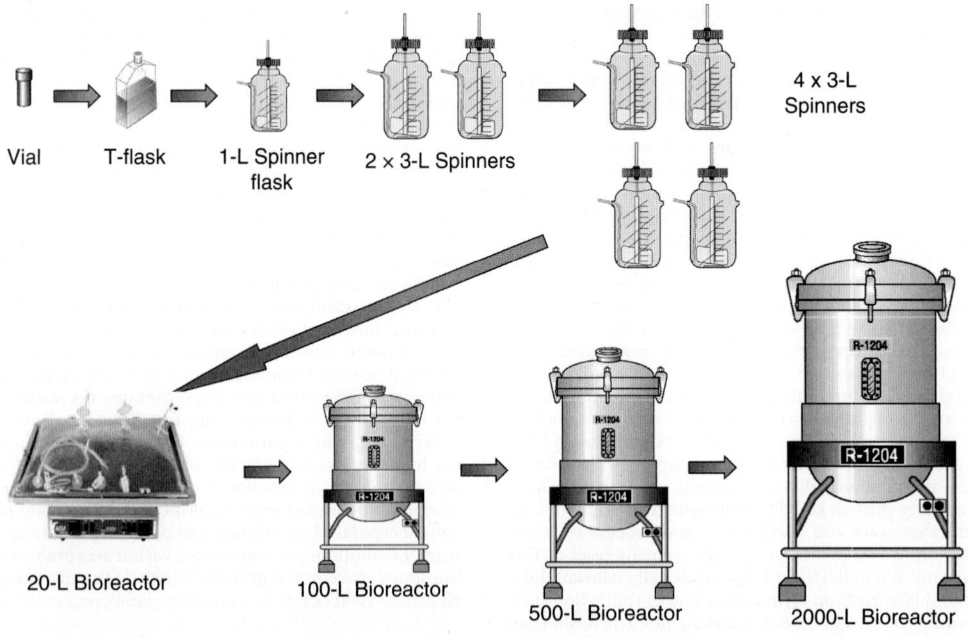

FIG. 20-10 Mammalian cell culture seed train for a typical 2000-L mAb production process. Open operations from vial thaw through spinner stages would be performed in a biosafety cabinet. Postspinner operations would be performed using closed processing in either stainless or single-use bioreactors to maintain sterility.

this first stage. After a growth period these flasks are then used to inoculate the next stage, further increasing cell mass. These subsequent stages consist of either increasing numbers of larger flasks or small bioreactors. This process continues until the cell mass reaches a target cell density and volume sufficient to inoculate the production bioreactor, designated as the N stage, as shown in Fig. 20-10. At this stage the objective changes from biomass generation to both biomass and product formation. The number of stages and volumes used in the seed train is driven by plant capacity utilization, the scale of the production bioreactor, and economics. Microbial fermentation processes require fewer seed train stages, each having larger volumetric increments compared to mammalian cell seed train due to their higher sustained metabolic rates that rapidly produce high cell numbers.

While most mammalian production runs use batch or fed-batch processes to optimize biomass production, recent efforts have also focused on using tangential flow membrane perfusion, discussed further later, to increase seed train cell density at the $N - 1$ stage (step immediately preceding production bioreactor) [Yang et al., *Biotech. Prog.* **30:** 616–625 (2014)]. This results in a higher initial cell density to inoculate the production bioreactor, reducing the less productive cell mass generation period. Because the production bioreactor is on the critical path for upstream production, shifting cell mass generation to the comparatively shorter duration $N - 1$ stage from the longer-duration production bioreactor effectively debottlenecks batch and fed-batch manufacturing.

Cell culture production medium includes glucose, amino acids, trace elements, and other components to produce the target product. The use of less defined medium components such as soy hydrolysates or animal-derived components such as fetal bovine serum is being phased out in biopharmaceutical production in favor of better-defined medium components. While seed trains and production bioreactors use different medium formulations, chemically defined medium is almost always used for both operations in the biopharmaceutical industry. This eliminates contamination risk from animal-derived products that may harbor potential viral or other undesirable species, leading to more consistent process performance and product quality by eliminating medium variability. Cell culture medium commonly consists of a chemically defined basal medium formulation containing amino acids, vitamins, glucose, and other components required to support the cell culture with only recombinant proteins such as growth factors or human insulin being used. It also contains Kolliphor (formerly Pluronic F68) to protect cells from damage at air-liquid interfaces. Engineered CHO cells are also modified to be resistant to the cytotoxic action of methotrexate. Adding methotrexate to the cell expansion medium up to, but not including, the production bioreactor stage ensures that only the desired engineered clone propagates by elimination of nonclonal variants. Antifoam is added to the bioreactor as separate aliquots as needed to avoid foaming that could cause exhaust filter fouling.

The seed train can affect process performance if it does not produce sufficient cell mass, but rarely impacts product quality. At the product production bioreactor stage, medium composition, bioreactor temperature, pH, and dissolved gases (O_2 and CO_2) can all impact the process and product quality to varying degrees [Huong, *J. Biotech.* **162:** 210–223 (2012)]. The extent to which these process parameters can be allowed to vary without impacting the process depends on the parameter. Defining target ranges required to reliably produce quality biopharmaceutical products is the subject of intense effort during the development phase. This is discussed further in subsequent subsections.

BIOREACTOR DESIGN, MIXING, AERATION, AND SCALING CONCEPTS

Bioreactor Overview Mammalian cell bioreactors as well as bacterial and yeast cell fermenters used for biopharmaceutical production are predominantly stirred vessels, and they share many of the same functional design requirements. The major differences relate to the high cell densities and metabolic rates of the fermentation process. This results in greater demands on heat removal, gas sparging, mixing power requirements, and the ancillary equipment supplying utilities.

Figures 20-11, 20-12, and 20-13 show key features of fermenters and bioreactors. Figure 20-11 is more typical of legacy and industrial biotechnology fermenter installations. Figures 20-12 and 20-13 are more representative of biopharmaceutical production bioreactors and fermenters. Section 19 includes descriptions of other types of bioreactors and fermenters that may be encountered. Due to the large aspect ratio (H/D_T) and high agitation levels needed to satisfy mass-transfer and mixing requirements, fermenters often have multiple Rushton impellers. Mammalian bioreactors typically have a single impeller, although larger scales may require two. Commercial-scale fermenters are commonly equipped with internal cooling coils to provide the required heat-transfer surface area needed to control the temperature of bacteria and yeast fermentations, although the trend is to move away from coils in the biopharmaceutical industry over cleaning concerns. A probe belt (not shown) accommodating instrumentation and sampling ports is typically installed in the lower portion of the vessel. Maintaining sterility for biopharmaceutical production requires inlet and exhaust gas filters. The exhaust stream typically has a condenser to limit water loss followed by a heater to prevent wetting out the exhaust filters. Smaller seed and mammalian cell bioreactors may simply use exhaust gas heating to eliminate filter wetting without the need for a condenser to limit water loss.

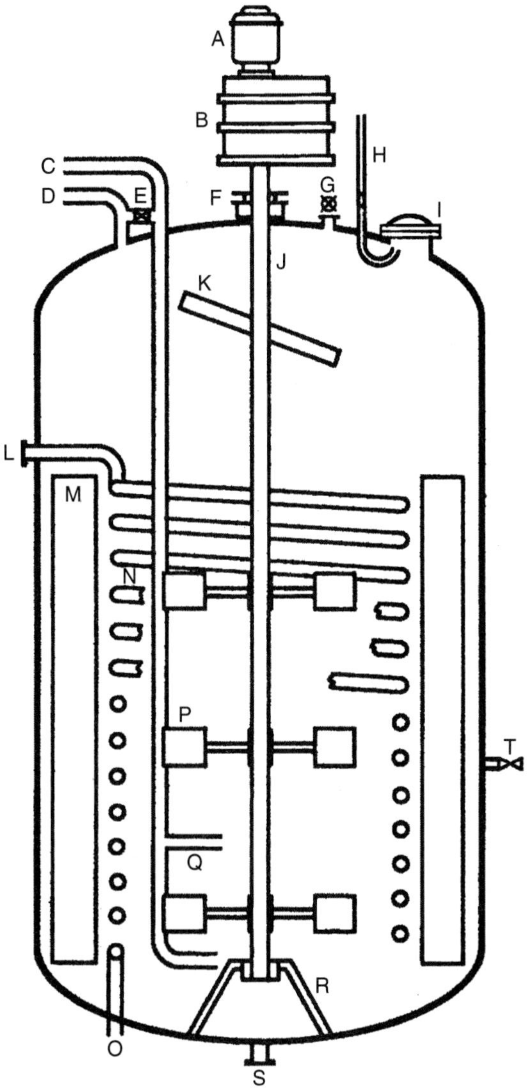

FIG. 20-11 Conventional batch fermenter with internal coils. A, agitator motor; B, speed reduction gears; C, air inlet; D, air outlet; E, air bypass valve; F, shaft seal; G, sight glass with light; H, sight glass clean-off line; I, manway with sight glass; J, agitator shaft; K, paddle to break foam; L, cooling-water outlet; M, baffle; N, cooling coils; O, cooling-water inlet; P, mixer; Q, sparger; R, shaft bearing and bracket; S, outlet (steam seal not shown); T, sample valve (steam seal not shown). Not shown are ladder rungs for maintenance, antifoam probe, antifoam system, and sensors (pH, dissolved oxygen, temperature, etc.). Note that (1) coils may be between baffles and tank wall, (2) coils may connect to top to minimize openings below liquid level, and (3) bottom-entering mixers are feasible.

Bioreactor Design Bioreactor design encompasses the vessel, mixing system, gas sparging system design, and other elements that, when combined, support the intended bioreaction and achieves the required level of sterility and cleaning. Primary design parameters are those that have a direct impact on bioreactor performance, while secondary parameters have a lesser impact. A summary of primary and secondary design elements for a bioreactor capable of supporting a high-density mammalian cell culture process is shown in Table 20-5. These design requirements are based on published literature, bioreactor engineering industry best practices, and extensive prior experience with cell culture operations at multiple scales. Key design objectives in bioreactor design include achieving homogeneity while meeting aerobic culture metabolic demands. One of the challenges in bioreactor design is that low oxygen solubility in the growth medium can result in rapid O_2 depletion in production bioreactors unless sufficient mass transfer is provided for. This presents significant design constraints in order to achieve the proper combination of agitation for mixing and mass transfer, with a gas sparging system for oxygen delivery and carbon dioxide removal, which together can maintain targeted dissolved gas levels and overall homogeneity.

In designing scaled-down or scaled-up production bioreactors, maintaining geometrically similar designs facilitates process scaling. In this approach, key dimensions and aspect ratios are proportionally held constant, or nearly constant. These include vessel, baffles (if installed), and impeller placement and dimensions. Sensor and sampling design and placement are also important considerations along with the analytical instruments and methods used to process the samples and data.

The *aspect ratio* (AR) is the liquid height to vessel diameter ratio H/D_T. Maintaining constant AR from bench to pilot to commercial scale is commonly viewed as leading to a more scalable design, although it may not be the most cost-effective design approach as scale increases. The aspect ratio typically used for bioreactors up to 15,000 L is 1.0 to 1.6. At larger scales it is more economical to have higher aspect ratios (2 to 3), especially for very large-scale fermenters. The overall vessel height also needs to include sufficient freeboard (height above liquid level) to accommodate some foaming without impacting the exhaust system. When very large fermenters are designed, transportation and constructability considerations can result in vessels with high AR that are not geometrically similar to smaller-scale fermenters used at pilot or small production scales.

Addition and monitoring port design and location are chosen to provide accessibility for operators, while avoiding artifacts such as nonrepresentative measurements. For pH and nutrients, addition of concentrated feeds or base must be mixed rapidly to prevent localized regions of high concentrations. Rapid dispersion is obtained by ensuring that the addition ports are located in a well-mixed region, preferably subsurface and within the impeller mixing zone. In practice, this has rarely been done in bioreactors due to concerns about additional complexity or sterility.

Agitation using impellers, and to some extent gas sparing, provides the fluid energy needed to generate the mixing required for homogeneous, or nearly homogeneous, nutrient, product, and cell distribution throughout the bioreactor. Given the significantly higher gas mass-transfer demands in microbial culture compared to animal cell culture, the fraction of the mixing energy coming from gas sparging is significantly higher in fermentation compared to animal cell bioreactors. With respect to mechanical mixing, design considerations include the ratio of impeller diameter to vessel diameter, impeller dimensions and type, placement and orientation, the ratio of impeller diameter to distance from bottom, impeller revolutions per minute (rpm), and the corresponding impeller power number. Multiple impellers are used at larger scales, requiring the designer to take into account impeller interactions. Related to impeller efficiency, baffles (typically 3 to 4 with a width of $D_T/12$) greatly assist in the transmission of the mixing power from the impeller to the bulk liquid by way of minimizing solid body rotation of the liquid. Some single-use bioreactor designs that do not include baffles can be more prone to this effect. The result may be significantly reduced mixing effectiveness unless compensated for by other means such as an offset impeller.

Mixing Concepts When scaling or transferring processes between bioreactors, key considerations include mixing, mixing times, and the related gas mass transfer. Mixing is often characterized in terms of power per unit volume, blend times, Reynolds number, mass-transfer rates, and hydrodynamic shear. Factors to consider in selecting impellers to satisfy mixing requirements include flow characteristics which affect gas dispersion, bulk liquid motion and mixing, and power characteristics. Impellers are often characterized by their power number P_o, an indication of how much power is transmitted to the medium, and the impeller flow number N_Q, the ratio of impeller pumping capacity to the product of impeller rotational speed N_i and diameter D_i, cubed.

In the absence of gas sparging, the power input by the impeller depends on its power number P_o, as defined by

$$P_o = P/(\rho N_i^3 D_i^5) \tag{20-1}$$

where P is the power input into the liquid being mixed. This relationship is used to determine P_o values by directly measuring power transmitted to a mixed vessel. In turbulent sparged gas-liquid mixing systems, the impeller power number can decrease with increasing gas flow. When P_o, the ungassed power number, is determined while sparging with gas, it is referred to as the *gassed power number* P_g. The difference between P_o and P_g is dependent on the impeller type and the gas flow rate. In most bioreactors the difference will be small due to relatively low gas flow; however, in fermenters the difference needs to be taken into account.

When operating in the turbulent regime, the impeller P_o is essentially constant for all Re numbers and independent of scale, provided geometric similarity is maintained. Once P_o is known, the total power draw P and power per unit volume P/V can be calculated.

$$P = P_o \rho N_i^3 D_i^5 \tag{20-2}$$

$$P/V = (P_o \rho N_i^3 D_i^5)/V \tag{20-3}$$

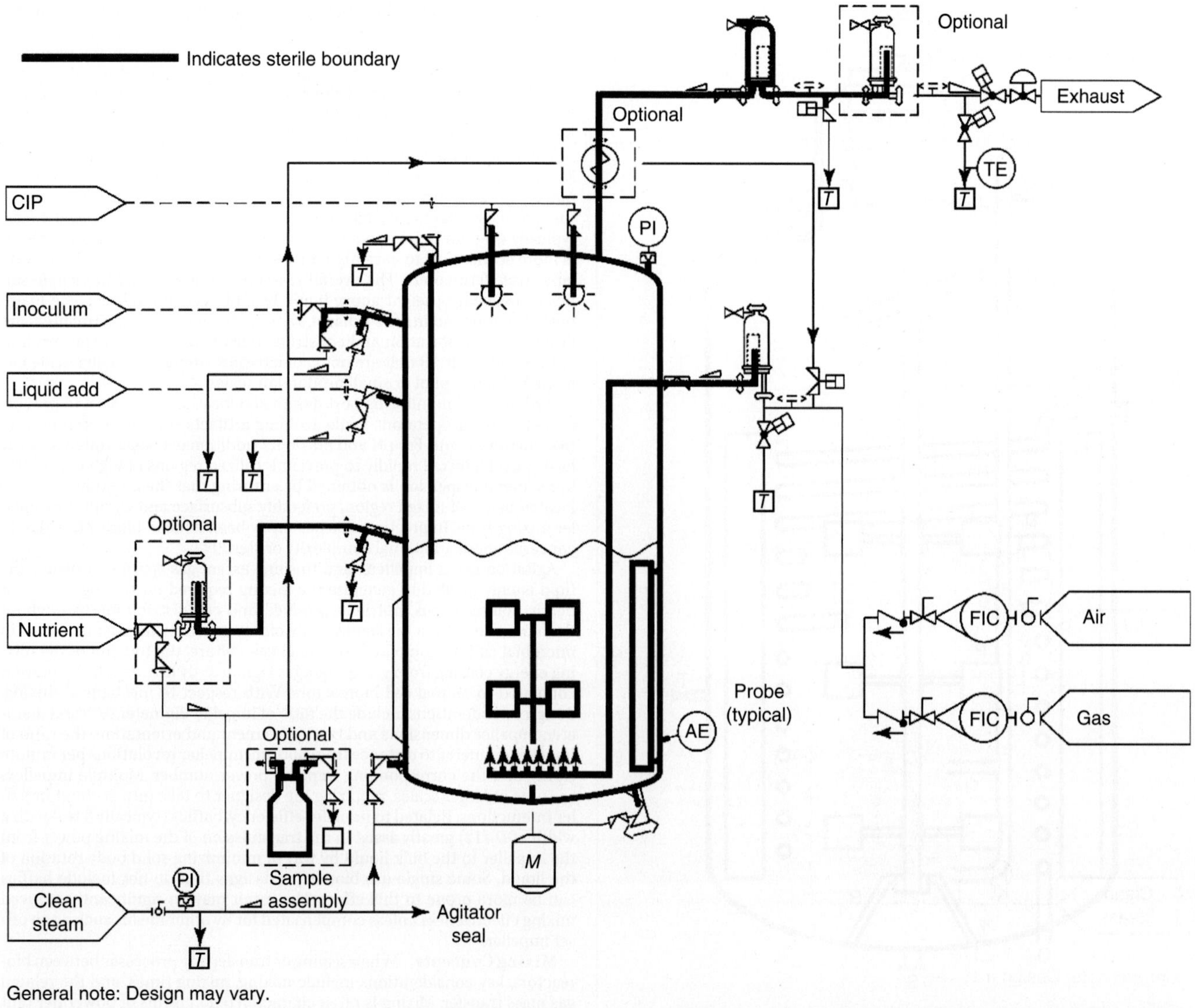

FIG. 20-12 Fermenter designed for biopharmaceutical production. Design features include clean steam and steam traps (*T*) for vessel sterilization, clean in place (CIP) solution addition through top-mounted spray balls, mass flow controllers (FIC) to supply air, oxygen, and carbon dioxide. Sterilizing-grade filters are shown on gas and nutrient supplies. Exhaust line includes optional heat exchanger and up to two filters. Liquid addition lines incorporate clean steam block and bleed arrangement to accommodate connections for sterile additions. Vessel is jacketed to provide heating and cooling in lieu of coils to facilitate cleaning. Single-use fermenters would share many of the same functional elements, except there would be no clean steam or CIP-related elements. Gas and liquid connections would be made using sterile tubing welders and/or sterile single-use connectors. (Reprinted from ASME BPE-2014, by permission of The American Society of Mechanical Engineers. All rights reserved.)

The mean turbulent *energy dissipation rate* (EDR), denoted by ε_{mean}, is simply P/V divided by fluid density [Eq. (20-4)]. In low-cell-density mammalian cultures P/V and ε_{mean} can be used interchangeably.

$$\varepsilon_{mean} = P/(\rho V) \qquad (20\text{-}4)$$

Microbial fermentations with their high metabolism and cell densities require high gas flow rates to provide sufficient oxygen and, to a lesser extent, to remove CO_2. The side effects of this increased gas sparging include higher gas holdup, which in turn reduces the power imparted to the broth due to the impact on P_g. In extreme cases the impeller becomes flooded. This is where a majority of the fluid surrounding the impeller is gas, thereby significantly reducing mass transfer and mixing efficiency. As shown in Eq. (20-1), the power input into a sparged bioreactor depends on the agitator type, speed, and size. However, the resistance that the agitator experiences is not necessarily constant; increased gas flow rate as well as lowering of broth viscosity can effectively reduce the power number, with the magnitude of the loss depending on impeller type. For Rushton impellers at high aeration rates seen in fermenters, P_g falls significantly, up to 50 percent. High solidity hydrofoil impellers (impellers designed for high viscosity or high solids suspension) such as the A320 shown in Fig. 20-14 suffer less power loss, especially when operated in the upflow mode [Galindo, *Chem. Eng. Tech.* **16:** 102–108 (1993)]. At the relatively low sparge rates and cell densities seen in mammalian cell cultures, the difference between gassed and ungassed power may not be significant. However, as mammalian cell densities are pushed higher, this effect becomes more significant and the gassed (P_g) versus ungassed (P_o) values may need to be accounted for. Characterization studies and correlations to predict relative values of P_o/P_g for various impellers have been reported (Middleton, *Handbook of Industrial Mixing*, Wiley, Hoboken, N.J., 2004, pp. 585–638).

An aerated fermenter using Rushton impellers may be designed to operate at specific sparge rates to optimize equipment and operating costs. Operation at a lower sparge rate, however, or if sparging is interrupted, can result in a corresponding significant increase in agitator power draw. System designs should take into account such operational deviations and failure modes by installing a system capable of operating at power levels greater than the targeted sparged condition.

Occasionally bioreactors using bottom-mount agitators, such as a Novaseptic impeller, that is designed to entrain liquid flow down through the impeller (down pumping) for solids mixing, may lead to adverse impacts

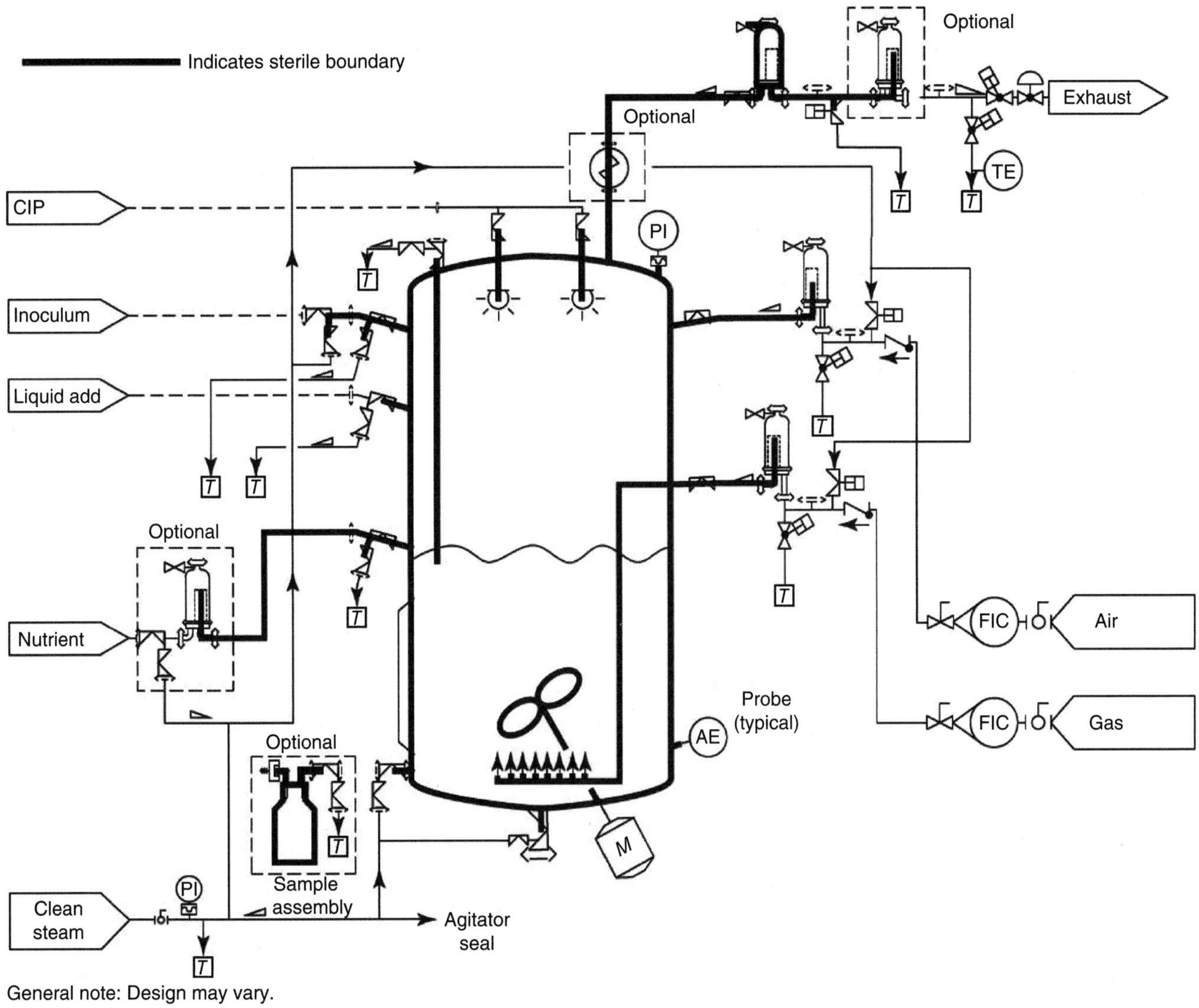

FIG. 20-13 Bioreactor designed for biopharmaceutical production. Functional design elements are nearly the same as for the biopharmaceutical fermenter. Differences are mainly in the scale of the equipment resulting from the lower metabolic rates of mammalian cells. Bioreactors require less power, hence smaller motors, and smaller gas supply elements (FIC, filters, heat exchanger, if used). Heating and cooling loads will be lower as well. The agitator is shown as an axial-flow type, which is typical of bioreactors. An angled impeller is required to minimize solid body rotation in this case since the bioreactor is not equipped with baffles. It is more typical to have baffles in stainless-steel bioreactors, although they are not common in single-use bioreactors. (Reprinted from ASME BPE-2014, by permission of The American Society of Mechanical Engineers. All rights reserved.)

if installed in a bioreactor. As impeller speed increases, this specific design can result in sparge gas bubbles being recycled back through the impeller, breaking them down into finer bubbles. Although mass transfer increases, this design has been observed to generate large amounts of undesirable stable foam when used in a bioreactor at high agitation rates.

Most bioreactor scaling approaches used in the biopharmaceutical industry have some reliance on P/V, which in turn depends on accurate determination of the impeller power number P_o. The P_o values are often provided by the vessel or impeller fabricator for stainless-steel bioreactors, or the equipment vendor in the case of single-use bioreactors. Values may also be estimated based on similar impeller installations or from literature values, in which case the accuracy will be uncertain; or they can be measured by the end-user. The power imparted to the bioreactor, and hence power number, is a function of the impeller design, but also depends on the impeller installation and vessel design details that affect liquid flow and turbulence (e.g., baffles; impeller type, placement, and orientation; height off the bottom). Because of these dependencies, relying on P_o values measured in a system that does not reflect the one being scaled has an element of uncertainty. An example of this was encountered in the case of two geometrically similar bioreactors equipped with the same type of bottom-mounted impellers, but different installation details. One had the impeller mounted close to the vessel bottom, while the second one had an eightfold greater impeller clearance. This resulted in the measured P_o values being 40 percent greater for the impeller mounted closer to the vessel bottom.

Power number P_o can be derived by measuring agitator electric power draw or directly measuring impeller shaft torque. The accuracy of electric power measurements is better in systems having large power requirements, such as fermenters, where power loss due to friction in the agitator drive and seals is low compared to power required for mixing. In the case of animal cell cultures, friction losses may be significant relative to the mixer power draw, so P_o measurements using this approach may not be as accurate, even when compensating for empty vessel friction losses. In such situations, torque measurements may provide better results, although it is not always possible to directly measure torque. For some magnetically coupled agitators, often used in single-use bioreactors (SUBs), electric power measurements can have excessive error, and it is not possible to attach a torque sensor. By temporarily replacing the magnetic drive with a mechanically coupled motor for testing purposes, it is possible to more accurately measure P_o values using either the power draw or the torque measurement technique.

In the case of magnetic mixers and other situations where accurate P_o values cannot be directly measured, it is also possible to use CFD modeling to estimate P_o values [Ascanio, *Chem. Eng. Res. and Design* **82:** 1282–1290 (2004)]. The accurate predictive capability of CFD has been debated by

TABLE 20-5 Design, Processing, and Control Elements Key in the Development and Execution of a Typical mAb Process

	Bioreactor performance indicators													
Design parameters	Hydrodynamic shear	Power per volume	Mixing time	Culture heterogeneity	Gas transfer (k_La)	Superficial gas velocity	Gas holdup volume	CO_2 stripping time	Bioburden control	pH	DO	pCO_2	Temperature	Osmolarity
Aspect ratio	P	S	P	P	P	S					P		P	
Baffles			P	P	S									
Impeller design/size	P	P	P	P	P	S								
Number of impellers	P	P	P	P	P	S								
Agitation rate	P	P	P	P	P	S	S	S		S	S	S	S	S
Gas composition, flow rates, control				P	P	P	P	P			P	P		
Sparger design and location	S	P		P	P	P	P	P		S	P	P		
Location of addition ports/tubes				P						S				
Feed addition rates				P						S			S	S
Vessel pressure									P	S	S	S		
Probe locations				S						P	P	S	P	S
DO control loop						S				S	P	S		S
pH control loop						S				P	S	S		S
Temperature control loop										S	S	S	P	
Foam control						P	P	S						

P = primary design consideration expected to impact bioreactor capability. Impact assessment based on prior knowledge, engineering fundamentals, and/or modeling studies (e.g., computational fluid dynamics).

S = secondary design indirectly impacts bioreactor capability, based on prior knowledge and engineering standard design.

SOURCE: CMC Biotech W.G., A-Mab: A Case Study in Bioproc. Dev. Ver 2.1, p. 74, in public domain at ISPE (www.ispe.org) (2009).

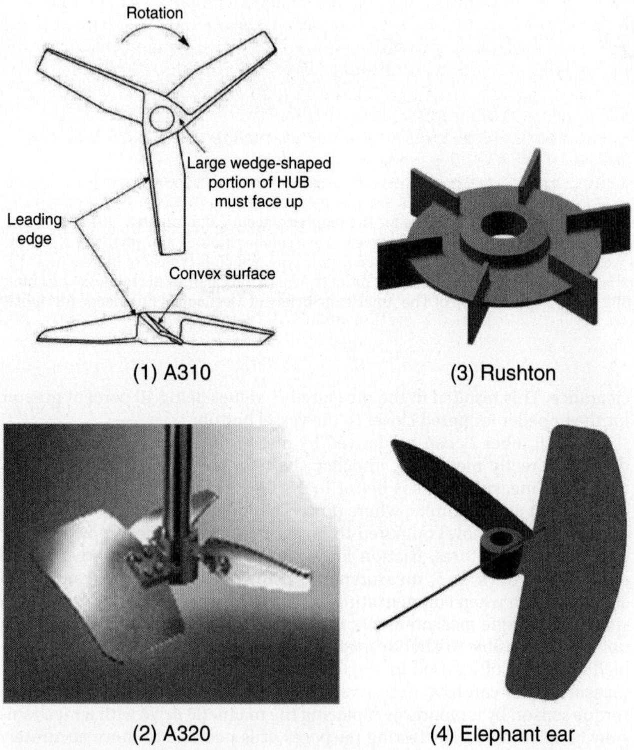

FIG. 20-14 Impellers 1, 2, and 4 are axial-flow impellers. Impeller 3 is a radial flow Rushton impeller more commonly used in fermenters. Rushton impellers are sometimes used in larger-scale bioreactors having multiple impellers to distribute sparge gases. In those cases, they are combined with axial-flow impellers to promote top-to-bottom mixing. (Images 1 and 2 courtesy of SPX Corp. Lightnin Mixers; images 3 and 4 courtesy of ABEC, Inc.)

many in industry, however. CFD results using most common software require the right choice of turbulence model, mesh size and distribution, and whether power input is determined from summing turbulent dissipation rates or from the torque acting on the impeller. Modeling of large-scale bioreactors using mesh-dependent CFD numerical approaches historically has been challenged to provide mesh-independent results due to the computing requirements, especially modeling large-scale bioreactors [Bartels, *Comput. Fluids* **31:** 69–97 (2002)]. An alternative CFD modeling approach that has shown improved predictive capability for stirred vessel uses *large-eddy-simulations* [Murthy, *Chem. Eng. Sci.* **63:** 5468–5495 (2008); Feng, *Chem. Eng. Sci.* **69:** 30–44 (2012)]. Recently this approach has been shown to predict P_o for unique geometries within 5 to 15 percent using the Lattice-Boltzman LES modeling approach, although it was also shown that if the model impeller geometry was varied, the predicted value could differ from measured values by up to 25 percent (Thomas and Frank, paper 487f, AIChE Annual Meeting, 2016). It can be expected that as computing power continues to increase in combination with opportunities to both compare and optimize CFD bioreactor models against experimental results, the real and perceived accuracy of CFD will continue to grow.

Given the importance of determining an accurate measurement of the power number to bioreactor scaling and agitation system design, DECHEMA's Expert Group Single Use Technologies has developed a standard method to measure P_o based on torque measurement (Meusel, *Recommendations for Process Engineering Characterization of Single-Use Bioreactors and Mixing Systems*, DECHEMA, Frankfurt, D.E., 2016, pp. 22–31). This method was recommended since it can be applied to a variety of equipment including stirred vessels, orbital shakers, and oscillating equipment, and it is similar to methods used to characterize stainless-steel bioreactors. A complete guide to mixing equipment testing is available (AIChE, *Equipment Testing Procedure—Mixing Equipment,* 3d ed., Wiley, Hoboken, N.J., 2001).

Impellers The agitator-sparger combination is one of the most important bioreactor design considerations due to low oxygen solubility and high oxygen transfer rates required to meet metabolic demand. Gas dispersion by the agitator depends on the size, speed, design, and relative placement of the impeller and sparger. The impeller design must also generate sufficient flow of the bulk liquid combined with adequate mixing intensity to homogenize the bioreactor contents. It is well established that different impeller designs have a significant effect on the degree of blending.

Impellers are generally categorized as radial flow, high P_o (i.e., Rushton), or axial flow, relatively lower P_o (hydrofoil, pitched blade). In this current discussion, the terms *high-shear impeller* and *low-shear impeller* are avoided since they can be misleading as applied to impellers. Although the specific impeller design can vary (Fig. 20-14), the general choice for mammalian cell cultures is a wide-blade, high-flow, low-power-number impeller that provides good axial flow, liquid blending, and sparged gas dispersion. This choice is largely based on historical concerns that mammalian cells are "shear-sensitive," hence the need for low-shear impellers. Typical examples are the informally named *elephant ear* or hydrofoil impellers. Conventional high AR bioreactor designs using multiple impellers often have hydrofoil upper impellers with a Rushton impeller at the bottom near the sparger, based on the historical belief that a high-shear impeller is needed to generate high gas mass transfer. However, other types may be more economical with the same or better performance, and especially for generating good axial mixing [Nienow, *Animal Cell Culture*, Springer, New York, 2015, p. 160).

Fermenters commonly use Rushton turbines because they have a relatively high power number ($P_o \sim 5$) compared to a similarly sized hydrofoil or pitched blade impellers. This is so because the higher the power number, the higher the amount of energy transferred to the fluid at a given rpm. Specification of this impeller is based on the idea that fermenters need so-called high-shear impellers to achieve high rates of mixing and mass transfer. The trade-off is that Rushton impellers have lower axial flow compared to pitched blade impellers. As a consequence, when multiple Rushton impellers are used, the rate of fluid exchange (mixing) between each impeller zone will be poorer than if an axial-flow, pitched blade impeller were used (Fig. 20-15). There are industrial examples where different "zones" associated with each Rushton impeller are formed due to the mixing patterns, and these individual zones have limited interchange between them, potentially leading to inhomogeneity at large scales.

It has been shown that a Rushton turbine operating at a power input P and speed N_i can be replaced by a larger-diameter hydrofoil impeller having a lower P_o and a higher flow number N_Q, and it will achieve better blending while providing the same mass-transfer rates. Further, such an impeller

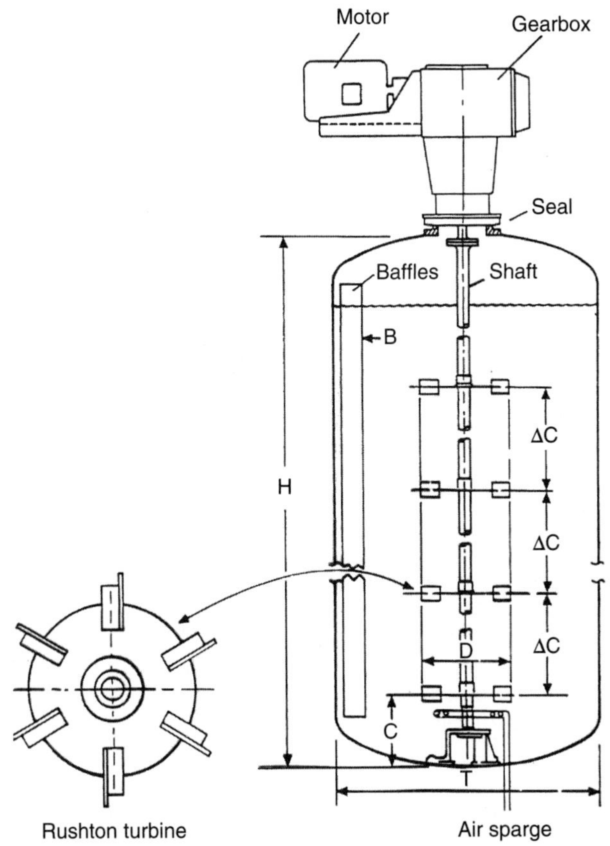

FIG. 20-15 Schematic representation of multiple Rushton impellers in high-aspect-ratio fermenter ($D_I/T = 1/3$; $B/T = 1/10$; $C/T = 1/4$; $H/T \sim 3$; 4 baffles). Gas dispersion and mixing intensity are high near each impeller, but vertical flow between each impeller zone is lower than for axial impellers, which may lead to heterogeneous conditions in very large fermenters. [Reproduced from Nienow, "Hydrodynamics of Stirred Bioreactors," *Appl. Mech. Rev.* **51:** 5 (1998) with permission by ASME. Copyright American Society of Mechanical Engineers, 1998.]

design will suffer less power loss under high sparge rate conditions. This consistency simplifies system design and predictability. These advantages have led to hydrofoil impellers being adopted for some fermenters in the biopharmaceutical industry [Buckland, *Biotechnol. Bioeng.* **31:** 737–742 (1988)]. Further, when used in an up-pumping mode under high sparge conditions common to fermenters, the susceptibility of high-solidity hydrofoils to instabilities and associated mechanical wear has been shown to be reduced while still achieving good gas dispersion [Hari-Prajitno et al., *Can. J. Chem. E.* **76:** 1056–1068 (1998)]. Nonetheless, because of the successful experience companies have had using Rushton impellers, they continue to be widely used for fermenters.

Specifying the correct placement and size of an impeller is critical to achieving the required mixing performance. These dimensions are typically specified relative to the vessel geometry. Impeller diameter to vessel diameter ratios can range from 0.3 to 0.5 depending on the impeller and characteristics of the fluid being mixed. Mixing at larger scales is ensured by using multiple vertically spaced impellers, especially if the aspect ratio increases with scale. When multiple impellers are installed in a bioreactor, they are typically spaced one D_I apart. Consideration also needs to be given to the submergence of the top impeller. If it is too close to the liquid surface, splashing or vortexing may result. A deep vortex can cause air entrainment and foaming in protein solutions. The minimum submergence requirement depends on the impeller type, pumping direction, and baffles/baffle width. It will also be affected by whether the impeller is offset. For example, a submergence greater than $D_I/2$ may be sufficient for down-pumping axial-flow impellers in a vessel equipped with conventional baffles (width = $D_T/12$). However, a minimum of D_I submergence is required for narrow baffles. In the absence of baffles, as in the case of offset impellers in nonbaffled mixers common in single-use bioreactors, submergence will likely be at least D_I, but will depend on a number of factors such as impeller type, operating speed, and agitator shaft angle. A unique example is bottom-mounted mixers that have been designed specifically to entrain solids which may float on the liquid surface. Vortexing may occur at higher rpm even when they are installed at the bottom of the vessel. As part of bioreactor design, changing medium levels from the beginning of the run to the end, as is the case in fed-batch processes, must also be accounted for. In most cases, the submergence will be greater than one impeller diameter to account for changing liquid levels.

When multiple impellers are used, the proper clearance between them and height off the bottom of the vessel must be specified. Table 20-6 shows recommended impeller spacing for liquid blending and gas dispersion applications in stirred vessels. The former may be representative of mammalian cell cultures operated at low sparge rates, while gas dispersions could better describe fermentations. Provided they are separated by about ~0.6H or more, the total power input is approximately the sum of the individual power input from each impeller.

Detailed design guidance is provided in the literature (Paul, *Handbook of Industrial Mixing*, Wiley, Hoboken, N.J., 2004).

Aeration and Mass Transfer The overall oxygen demand of cells in any bioreactor operated at a significant production scale must be met by supplying oxygen from a sparged gas. The sparge gas can be air, oxygen, or oxygen-enriched air. Design and operation of the bioreactor must be capable of achieving an oxygen mass-transfer rate equal to cell culture oxygen demand at the maximum *viable cell density* (VCD). The mass-transfer rate will be a function of the mass-transfer coefficient between the liquid and gas phases and the oxygen driving force. The mass-transfer coefficient for a sparged bioreactor will be a function of agitation and sparge gas velocity for a given sparging system and impeller.

Oxygen is sparingly soluble in water-based medium but is required for cellular growth and product formation for aerobic cultures. Without a constant oxygen supply from supplied gases, its concentration would fall below the target range in a matter of minutes. Below about 10 percent saturated dissolved oxygen (DO) in an air sparged system, mammalian cell metabolism can begin to overproduce lactate and ammonia, impacting cell growth and product formation.

TABLE 20-6 Recommended Impeller Clearance and Spacing

Mixing	Max. liquid height H/D_T	No. of impellers	Impeller height above vessel bottom: Bottom	Upper
Liquid blending	1.4	1	$H/3$	—
	2.1	2	$D_T/3$	$2H/3$
Gas dispersion	1.0	1	$D_T/6$	—
	1.8	2	$D_T/6$	$2H/3$

Adapted from Hemrajani, *Handbook of Industrial Mixing*, Wiley, Hoboken, N.J., 2004, p. 372, with permission.

In mammalian cell cultures for every 1 mol of O_2 taken up by a cell, approximately 1 mol of CO_2 is produced. This ratio is the *respiratory quotient* RQ. Because CO_2 solubility in the medium is much greater than that of oxygen [27 times greater in water at 25°C (Lide, *CRC Handbook of Chemistry and Physics*, 75th ed., CRC Press, Boca Raton, Fla., 1994, p. 6-4)] relatively larger quantities of CO_2 are dissolved in the medium (dCO_2) that needs to be stripped out to avoid excessive accumulation. Effective CO_2 stripping is important because high dCO_2 levels can impact productivity, cell growth, viability, or product quality attributes such as glycosylation. This dCO_2 threshold has been identified to be between 120 and 150 mm Hg for mammalian cell cultures. Low dCO_2 levels (< 40 to 50 mm Hg) have also been linked to poor culture performance [Zhu et al., *Biotech. Prog.* **21:** 70–77 (2005); Mostafa and Gu, *Biotech. Prog.* **19:** 45–51 (2003); Schmelzer and Miller, *Biotech. Prog.* **18:** 346–353 (2002)].

Since CO_2 sparging is commonly used in concert with sodium bicarbonate buffers as part of the pH control scheme, excessive dCO_2 can lead to higher osmolality as base is added to offset the acidifying effect of high dCO_2 to maintain the pH set point. High osmolality (>400 to 450 mOsm kg^{-1}) together with high dCO_2 (>140 to 160 mm Hg) can result in a significant drop in viable cell count (Godoy-Silva, *Encyclopedia of Industrial Biotechnology: Bioprocess, Bioseparation, and Cell Technology*, Wiley, Canada, 2010, p. 7).

Mass-Transfer Coefficient Mass transfer to supply O_2 and remove CO_2 is characterized by the respective mass-transfer coefficients k_La (h^{-1}). To allow meaningful comparison and scaling between bioreactors, a standard k_La measurement protocol should be applied to minimize measurement variability. Mass transfer will be affected by process parameters such as the test medium composition, temperature, impeller speed, sparge gas velocity, and working volume. Equipment design factors affecting mass-transfer coefficients include vessel geometry, impeller number and spacing, impeller type and pumping direction, baffles, and sparger design, including sparger position, and the number and size of the holes. Once determined, mass-transfer data can be used to model and optimize sparging and mixing strategies to satisfy mass-transfer requirements between different scales, geometries, designs, and operating conditions.

For gas-liquid mass transfer in a bioreactor, the liquid-phase mass transfer will be rate-limiting, and the overall mass-transfer coefficient is simply equal to the liquid-phase mass-transfer coefficient k_La. For practical purposes, bioreactor mass transfer at the liquid surface can be ignored for most production bioreactors greater than ~300 to 500 L when operated at typical sparge rates and impeller speeds. In this case the mass-transfer rate is given by

$$\frac{dC_L}{dt} = k_L a \times (C^* - C_L) = k_L a \times \Delta C_L \qquad (20\text{-}5)$$

Referring to Fig. 20-16, the oxygen mass-transfer driving force ΔC_L is the difference between the oxygen concentration in the liquid film surrounding the bubbles (C^*_{O2}) that is in equilibrium with the oxygen partial pressure in the bubble (p_{O2}), and that in the bulk liquid (C_{O2}). Usually C_{O2} is usually expressed as a percent of saturation (% DO) with respect to air as measured by a dissolved oxygen probe. It is also common to refer to dissolved gas concentrations in terms of their equilibrium partial pressures in units of mm Hg. For example, sparging with air at 1 bar (760 mm Hg), the percent oxygen

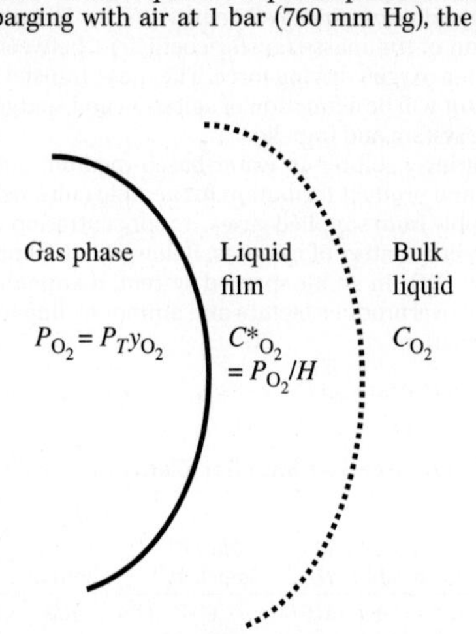

FIG. 20-16 Stagnant film theory applied to sparged gas bubbles. A stagnant liquid film is assumed to surround the gas bubble. The dissolved oxygen concentration C^*_{O2} in that film is assumed to be in equilibrium with the oxygen partial pressure within the bubble. The oxygen gas mass-transfer driving force is the difference between C^*_{O2} and the bulk liquid dissolved oxygen concentration C_{O2}.

concentration is 21 percent, so p_{O2} = 159.6 mm Hg. Therefore, at a reading of 30 percent DO, C_{O2} can be expressed as either 47.9 mm Hg or 30 percent DO. The actual liquid-phase oxygen concentration in mmol L^{-1} can be estimated using Henry's law coefficients, $C_{O2} = p_{O2}/H$ [Hu, *Cell Culture Bioprocess Engineering*, 2012, p. 217].

The most frequently used methods to measure k_La are referred to as the cell-free gassing out method and the dynamic method which is performed with cells present. In the past the sodium sulfite method was used to estimate k_La, but it has been found to overestimate k_La. In addition, it utilizes copper catalysts, which is undesirable in a reactor used to produce human therapeutics (Clark, *Bioprocess Engineering: An Introductory Engineering and Life Science Approach*, Woodhead Pub. Ltd., Cambridge, UK, 2013, p. 164).

In the gassing out method, cell-free test medium is first added to the bioreactor up to the volume of interest. Oxygen is then stripped from the medium using nitrogen, followed by sparging under the conditions of gas flow, composition, and agitation to be characterized. The dissolved oxygen concentration is measured between 20 and 80 percent dissolved oxygen over time. This is repeated for as many conditions as required to fully characterize the mass transfer. Equation (20-6) is used to calculate k_La for each condition. The accuracy of this method depends, in part, on the probe response lag time being negligible compared to the rate of change of the dissolved oxygen concentration. This is often the case for mammalian cell bioreactors.

$$\ln(1 - C_L/C^*) = -k_L a \times t \qquad (20\text{-}6)$$

A similar dynamic method can be applied to an operational bioreactor. At a given point in time during a run, the sparge gas is turned off. From the rate of dissolved oxygen decline dC_L/dt, and knowing the viable cell density X_{viab}, an estimate of cellular respiration quotient q_{O2} can be made during this part of the test using Eq. (20-7) since k_La = 0. Once the oxygen concentration reaches a lower level, but still within a range that will not impact cellular metabolism, sparging is resumed. The resulting measurement of C_L over time is used to determine dC_L/dt by fitting a polynomial expression to the data. From Eq. (20-7), knowing X, q_{O2}, and $C^* - C_L$ over time, k_La can be determined by using numerical methods. Alternative methods can be found in the literature [Shuler, *Bioprocess Engineering: Basic Concepts*, 3rd edition, Prentice Hall, Upper Saddle River, N.J., 2017].

$$\frac{dC_L}{dt} = k_L a \times (C^* - C_L) - q_{o2} \times X_{viab} \qquad (20\text{-}7)$$

This method has the advantage of measuring mass transfer under actual operating conditions, but is not often used due to concerns about impacting the cell culture, especially in high-cell-density cultures where oxygen depletion is rapid.

DECHEMA has developed a recommended procedure for the determination of mass-transfer coefficients using the gassing out method (Meusel, *Recommendations for Process Engineering Characterization of Single-Use Bioreactors and Mixing Systems*, DECHEMA, Frankfurt, 2016, pp. 45–56). Although the method includes a recommended test medium, it is prudent to develop a practical test medium composition that mimics the medium used in the bioreactor and fermenter of interest. This is discussed further later. DECHEMA also notes the importance of characterizing the potential impact of DO probe response times on k_La measurements. In the case of mammalian cell cultures (k_La up to 20 h^{-1}) the probe response time will not affect measurements; however, for fermentations (k_La = 300 to 500 h^{-1}) probe response time may result in significant measurement errors. In cases where the sensor data are recorded in an automated data historian for later analysis, the potential impact of data filtering or compression should be considered.

Carbon Dioxide Stripping In principle, methods used to measure oxygen mass-transfer coefficients, such as the cell-free gassing out method, should be applicable to CO_2 stripping mass-transfer coefficient measurements. When sodium bicarbonate buffers are used, the chemical equilibrium with H_2CO_3 and its dissociation species as a function of pH need to be accounted for. This approach was demonstrated using bioreactors ranging from 80- to 2000-L working volumes and included development of a model test medium [Matsunaga, *J. Biosci. Bioeng.* **107:** 419–424 (2009)].

An alternative is to estimate the CO_2 mass-transfer coefficient from the oxygen mass-transfer coefficient. Based on molecular diffusivities, the ratio of the O_2 mass-transfer coefficient k_La to the CO_2 stripping mass-transfer coefficient (k_La_{CO2}) is about 0.9 [Sperandio, *Biotech. Bioeng.* **53:** 243–252 (1997)]. The risk of this approach to estimate CO_2 stripping is that under the actual operating conditions the stripping gas may become saturated with CO_2, and its removal is no longer governed by mass transfer.

At 1 bar, CO_2 solubility in water is 33 mmol L^{-1} compared to the solubility of O_2 (in air), in water of 0.2 mmol L^{-1}. As an air bubble travels up through a bioreactor, the O_2 concentration within the bubble will gradually fall, whereas the CO_2 concentration increases more quickly due to the higher driving force. If the bubble residence time is long enough and the sparge rate is insufficient, then the gas bubbles can become saturated with CO_2 even before they rise to the liquid surface. The following examples illustrate this.

A CO_2 stripping study was performed using a YSI CO_2 probe to directly measure dCO_2 in solution [Xing, *Biotech. Bioeng.* **103:** 733–746 (2009)]. This study demonstrated that CO_2 stripping was unaffected by headspace airflow at the 5000-L scale. Stripping was a strong function of sparge rates at all P/V levels tested. However, dCO_2 was independent of P/V levels below 24 W m^{-3}. Since commercial-scale bioreactors could only operate at $P/V < 24$ W m^{-3}, CO_2 stripping was effectively governed by sparge rates. This suggests that CO_2 stripping gases may be approaching saturation levels at commercial scales.

Using pH to measure dCO_2 levels, stripping rates and $k_L a_{CO2}$ values were determined for a 330-L bioreactor operating at a sparge rate of 0.09 vvm [Sieblist, *Biotech. J.* **6:** 1547–1556 (2011)]. From these experiments combined with a numerical model, it was demonstrated that the bubble residence time was sufficient for them to become saturated with CO_2. Based on their study the authors state that CO_2 stripping rates are expected to fall with increasing scales due to saturation effects resulting from increased bubble residence time and decreased driving force as the bubbles rise through the bioreactor. This implies that in this case dCO_2 stripping at larger scales is expected to be, for the most part, a function of sparge gas flow rates. This may be why companies have reported that using open-pipe spargers to generate larger bubbles and operating them at higher aeration rates can lead to better CO_2 stripping at large scales. Meeting metabolic oxygen demands, however, favors generating smaller bubbles that provide higher mass-transfer area. These competing requirements represent an optimization challenge for design and scale-up. As a result, many bioreactors are equipped with two sparger types. One is an open-pipe or drilled tube sparger that produces large bubbles (hence larger volume and lower mass-transfer area) to provide dCO_2 stripping capability. This is combined with a second sparger optimized to produce smaller, higher-surface-area bubbles to ensure sufficient oxygen mass-transfer capacity.

Under conditions where bubbles become saturated with CO_2, $k_L a_{CO2}$ measurements are inaccurate and the apparent $k_L a_{CO2}$ will be lower than the actual value (Nienow, *Animal Cell Culture,* Springer, New York, 2015, pp. 151–152). Because the gas phase within the bubble becomes saturated, $k_L a_{CO2}$ does not increase with increased agitation speed, although oxygen $k_L a$ measured under the same conditions does increase (Fig. 20-17). Increasing sparge gas flow rates, however, does increase dCO_2 stripping capacity. This results in the apparent $k_L a_{CO2}$ increasing with increased sparge rate, but not with agitator speed. Under these conditions, which can be expected in large stirred vessels (300 to 10,000 L or more), stripping is primarily a function of the gas throughput and not dependent on bubble size or impeller configuration according to the authors [Sieblist, *Biotech. J.* **6:** 1547–1556 (2011)]. In practice, bioreactor designs need to be individually assessed to determine if dCO_2 stripping is governed by gas throughput or mass transfer for the anticipated operating ranges.

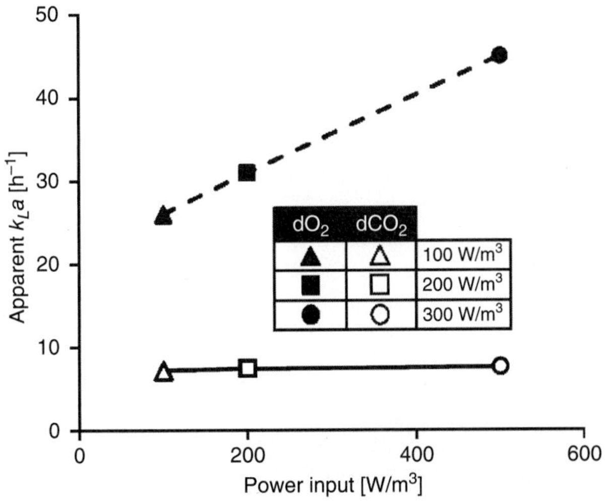

FIG. 20-17 The O_2 $k_L a$ (dashed line) and apparent CO_2 $k_L a$ (solid line) as a function of P/V at a constant sparge rate. Sparge gas has become saturated with CO_2 so its apparent mass-transfer coefficient is constant even as agitation is increased [Sieblist, *Biotech. J.* **6:** 1551 (2011)]. (Reproduced with permission of John Wiley & Sons, © 2011 Wiley-VCH, Verlag GmbH & Co. KGaA, Weinheim, Germany.)

Mass-Transfer Coefficient Correlations The van't Riet equation [Eq. (20-8)] is an empirical expression often used to correlate the sparge gas-liquid mass-transfer coefficient to P/V and the sparge rate expressed as superficial velocity v_{sg}. The coefficients α and β are determined by measuring $k_L a$ at various conditions, using the methods described above. For geometrically equivalent bioreactors whose designs result in the same α and β coefficients, operating at the same sparge gas flow rate per reactor volume (vvm), the superficial flow rates are equal. It follows that the mass-transfer rates will be the same when both scales are operated at the same P/V.

$$k_L a = K_{\text{van't}} \left(\frac{P}{V}\right)^{\alpha} (v_{sg})^{\beta} \tag{20-8}$$

The van't Riet correlation coefficients for several bioreactors and fermenters are shown in Table 20-7. Although the coefficients for the mammalian cell bioreactors listed here are similar to one another, this is not typically the case and the coefficients should be determined for the particular medium and bioreactor of interest. The coefficient $K_{\text{van't}}$ is affected by medium composition. Both α and β are dependent on the specifics of the vessel geometry, sparger and sparger-impeller interactions, and the operating ranges tested to develop the correlation.

In general, $k_L a$ can be difficult to determine experimentally due to the time and expense required, especially at large scale. It is impractical to use actual medium in $k_L a$ studies due to cost and test complexity, so development of a standard test medium is needed. Since the numerical value of $K_{\text{van't}}$ is sensitive to composition small-scale tests, comparing the test medium to the actual medium should be part of the test method development. The same test medium should then be used to characterize bioreactors being compared. Bubble coalescence properties of the test medium will affect the measured mass-transfer coefficient, as shown in Table 20-7. Two general medium types have been used for $k_L a$ measurements. The coalescence medium does not contain any surfactants and has very low ionic (salt) concentrations. Gas bubbles readily coalesce in this medium. Noncoalescing medium will contain ionic species such as salts or phosphate-buffered saline (PBS) that will inhibit bubble coalescence. The latter type is almost always more representative of cell culture medium. In some studies, addition of antifoam has been reported to reduce $k_L a$ by a factor of up to 2, although in other studies the effect of varying antifoam concentration was not as significant above a threshold value. Ionic strength can have significant impact on mass transfer, with salts increasing it up to fourfold. Therefore, in designing a $k_L a$ test medium it is best to add antifoam and ionic species at concentrations representative of the medium to be used.

The van't Riet equation does not always adequately correlate $k_L a$ data. This has been observed for some sintered (highly porous) sparger designs, as well as in the case of some magnetically coupled bottom-mount agitators. In the latter case the agitator had significant impact on the sparge gas bubbles and hence mass transfer. In these cases with their complex nonlinear interactions, a higher-order equation generated by a design of experiments (DOE) surface response model can be used to more accurately correlate mass-transfer dependency on power and gas flow.

Adding to any errors inherent in the mass-transfer rate measurement method itself, such as instrument tolerances, there are other possible sources of error that are difficult to quantify. These include how closely the test medium represents the actual bioreactor conditions, or whether the vendor has provided dependable power numbers P_o. Designs and scaling correlations based on these factors will therefore also have a degree of uncertainty. To account for this, the designer should also incorporate prior experience with similar systems and ensure that flexibility is provided for when specifying operating ranges, and equipment start-up plans should include engineering runs (i.e., shakedown runs) when possible to tune the mixing and mass transfer to produce the required process outcome.

TABLE 20-7 van't Riet Coefficients for Bioreactors and Fermenters for $k_L a$ [=] h^{-1}, P/V [=] W m^{-3}, and v_{sg} [=] m s^{-1}

Parameter	BR1*	BR2*	BR3†	Ferm1‡	Ferm2‡
K	495	482	270	9.2	93.6
α	0.39	0.25	0.47	0.7	0.4
β	0.77	0.79	0.8	0.2	0.5

*2 m^3 bioreactors; medium: salt, pluronic F68, antifoam; drilled hole sparger; axial flow impeller.

†3- to 4.4-m^3 medium in a 5-m^3 bioreactor; medium: NaCl, sodium bicarb., pluronic F68 A100 impeller; pipe sparger (millimeter-scale bubbles) [Xing, *Biotech. Bioeng.* **103:** 733–746 (2009)].

‡Ferm1: noncoalescing salt solutions, Ferm2: coalescing aqueous salt solutions [van't Riet, *Ind. Eng. Chem. Proc. Des. Dev.* **18:** 357–364 (1979)].

Sparger Design A key design requirement is that the sparger design, in combination with adequate mixing, meet cellular oxygen uptake rates while providing sufficient CO_2 stripping to avoid impacting product quality or critical process performance parameters. The role of a sparger is to produce and disperse an initial bubble size distribution of the sparged gases entering the bioreactor. Spargers are almost always located below the agitator which further distributes sparged gases. This can also lead to cases where the impeller shears the bubbles, affecting bubble size distribution and mass transfer. The degree to which an impeller interacts with the sparged gases varies depending on vessel design and operating conditions. An example is a fermenter operated at high P/V levels, where k_La has a greater dependency on P/V than on v_{sg}, indicating a strong impeller effect. Ferm1 in Table 20-7 is an example where the impeller has a dominant effect on mass transfer. In these cases, simpler sparger designs may suffice. Bioreactors operating at lower power input compared to fermenters have a greater reliance on sparger design and placement to generate gas mass transfer.

Design considerations include sparger type, such as sintered, drilled hole, or open-pipe, as shown in Fig. 20-18, material of construction, and placement relative to impeller. Placement is especially important if it serves double duty as a clean-in-place (CIP) spray nozzle to clean the underside of an impeller as part of the post-use CIP cycle. This is often the case for stainless-steel vessels used for biopharmaceutical production, as shown in Fig. 20-19.

Drilled hole sparger design choices include hole size, number, distance between holes, and gas velocity through the sparger holes (Table 20-8). Open-pipe spargers or multiple tube drilled hole spargers are used more frequently in fermenters which are more dependent on high-power mixing to generate the high mass-transfer rates needed to achieve required oxygen transfer rates (OTRs). Drilled hole tube and disk spargers are frequently used in bioreactors. Sintered spargers have been used in small pilot- or bench-scale bioreactors in order to meet required OTR demands, but are avoided where possible at larger scales.

In mammalian cell culture, open-pipe and drilled hole spargers produce larger bubbles and consequently lower k_La. Sintered spargers produce very small bubbles yielding a much higher k_La. Successful DO and dCO_2 control strategies have employed both types of sparger elements, but the sintered elements have a number of drawbacks including increased foaming tendency, susceptibility to cracking, difficulty cleaning, and reports of negative impacts on cell culture viability.

Bioreactor sparger design has the potential to contribute to cell death in a number of ways. Sintered elements generate very small bubbles. When bubbles rupture at the liquid surface, they release a significant amount of energy, and the energy content increases rapidly as the bubble size decreases (i.e., from 10 kW m^{-3} to 10^5 kW m^{-3} as the bubble diameter goes from 6.5 to 1.7 mm). Cell damage or death may occur when the local energy exceeds 10^4 kW m^{-3} in sparged bioreactors [Ma N., *Biotech. Bioeng.* **80:** 428–437 (2002)], even when a cell protectant such as Pluronic F68 or Kolliphore is used. There are also reports that cell damage can occur if the exit velocity from sparger holes is >30 m s^{-1} [Jobses, *Biotech. Bioeng.* **37:** 484–490 (1991); Zhu, *Biotech. Bioeng.* **101:** 751–760 (2008); Liu, *Biotech. Progr.* **30:** 48–58 (2014)]. A general guideline with respect to drilled hole sparger designs is to avoid extremely small bubble sizes while ensuring that the sparged gas exit velocity does not exceed 30 m s^{-1}.

Hole spacing for drilled spargers is important to both distribute gas flow and minimize bubble coalescence as the bubbles form at the sparger. As the sparge gas flow rate increases, gas velocity through the sparger holes increases until a point is reached where bubble coalescence and chaotic bubble formation can occur. In these cases, after a steady increase as the gas flow rate increases, the mass-transfer coefficient trend versus sparge rate begins to plateau. At still higher sparge rates, a point is reached where further increase results in increased mass transfer (Fig. 20-20). Little design guidance has been published on hole spacing criteria, although spacing of ~1 to 3 cm is common. The bubble size produced from a sparge hole is dependent on the gas flow rate and velocity, material of construction, and physicochemical properties of the medium [Tilton, *Chem. Eng.* **89:** 61–68 (1982)]. Bubbles produced by a drilled-hole sparger are generally larger than the hole size. After bubbles are formed at the sparger surface, fluid forces generated by the impeller may cause bubble breakup and coalescence, resulting in a stable, but different size distribution than that initially generated at the sparger. Because of these interactions that are difficult to characterize, drilled-hole sparger designs tend to be empirical based on prior experience and in-house characterization of design alternatives.

DO control set points are typically set in the range of 30 to 50 percent. This ensures sufficient driving force while DO remains above the critical value throughout the bioreactor even if inhomogeneity exists. Achieving a combination of P/V and sparging that simultaneously meets the *oxygen uptake rate* (OUR) and dCO_2 stripping requirements requires optimizing the design against sometimes conflicting requirements. Increasing OTR to satisfy OUR demand can be accomplished with spargers having small hole or pore sizes that generate smaller bubbles, and by operating at higher P/V. While increasing OTR, the total gas flow Qsg will be lower. Stripping can be

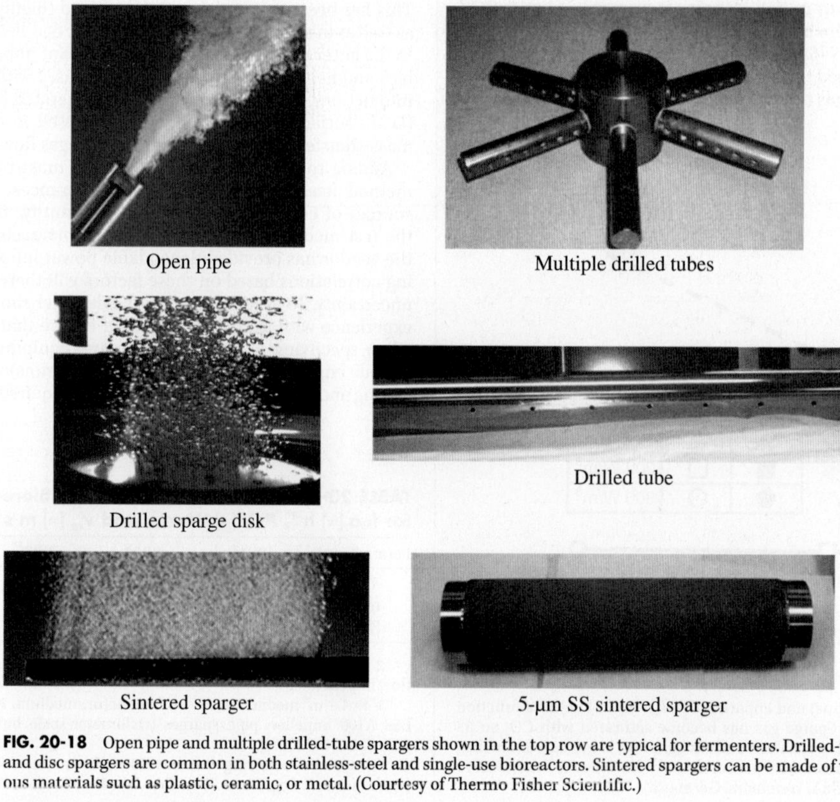

FIG. 20-18 Open pipe and multiple drilled-tube spargers shown in the top row are typical for fermenters. Drilled-tube and disc spargers are common in both stainless-steel and single-use bioreactors. Sintered spargers can be made of various materials such as plastic, ceramic, or metal. (Courtesy of Thermo Fisher Scientific.)

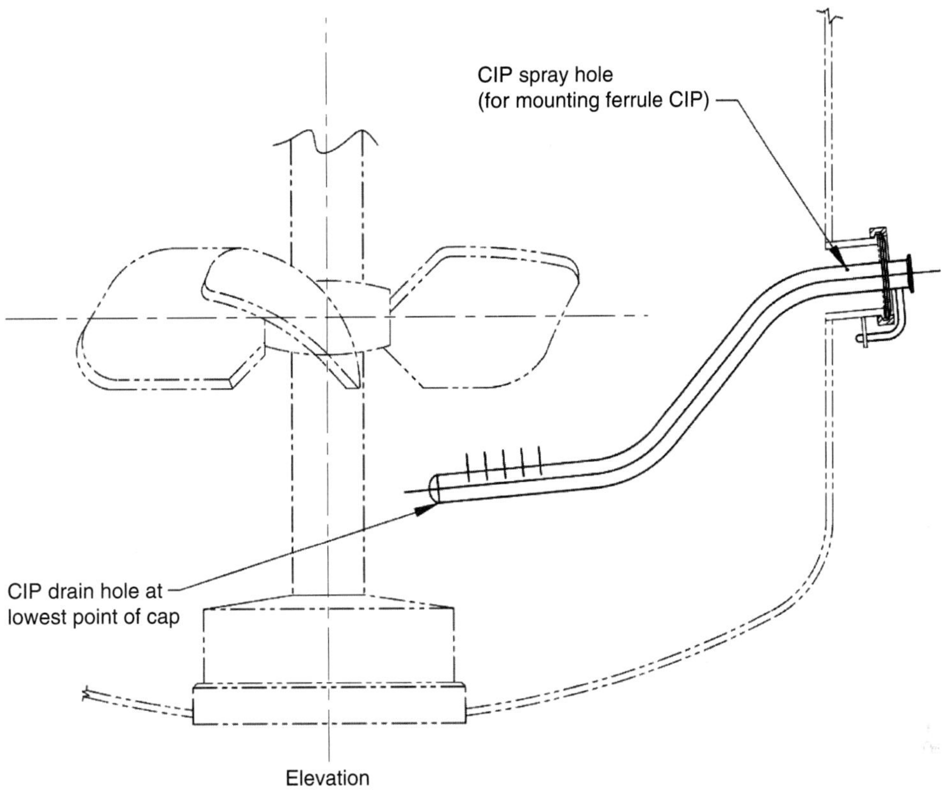

FIG. 20-19 Sparger is located beneath the impeller to achieve good gas dispersion and also to clean the underside of the impeller. Note additional holes to clean the mounting ferrule port and to drain the CIP solution. (Reprinted from ASME BPE-2014, by permission of The American Society of Mechanical Engineers. All rights reserved.)

impacted as the bubbles are quickly saturated with CO_2, leading to higher dissolved CO_2. In this case the design is essentially too efficient at providing oxygen, thereby reducing the CO_2 stripping capacity. Low CO_2 stripping efficiency has been reported when high-efficiency sintered spargers have been used [Marks, *Cytotech.* **42:** 21–33 (2003)]. CO_2 stripping can be enhanced by generating larger bubbles which can take significantly longer to become saturated compared to the small bubbles produced by sintered spargers. Larger sparge gas bubbles result in higher gas flows in order to maintain adequate k_La values to satisfy the O_2 demand, consequently providing greater CO_2 stripping capacity. These design trade-offs are illustrated in Fig. 20-21.

Options to increase OTR without increasing sparge rates may be needed for high-cell- density bioreactors, or if foaming is an issue due to high superficial velocity. Enriching air with oxygen raises p_{O2}, increasing the driving force. In stainless-steel vessels, increasing exhaust backpressure increases the sparge gas pressure, and hence the oxygen partial pressure. Each of these methods for increasing OTR reduces the relative CO_2 stripping capacity. At the other extreme, increased sparger hole size increases bubble size, favoring CO_2 stripping, although OTR can be lower unless compensated for. Models to determine stripping rates, OTR, and dissolved CO_2 concentrations over time for sparged fed-batch bioreactors have been used to predict overall sparger performance [Xing, *Biotech. Bioeng.* **103:** 733–746 (2009)].

TABLE 20-8 Sparger Types and Characteristics

Sparger type	Sintered	Membrane	Open pipe	Drilled hole
Avg. bubble diameter (mm)	<< 1	< 1 to > 3	> 2	> 1
Size distribution	Wide bell	Narrow to wide bell	Wide bell	Uniform to wide bell
Predistribution	Low	Low	Low	Low to high
Efficiency	High	Low to high	Low	Low to mid
Flow rate	Low	Low to high	Mid to high	Mid to high
Foaming potential	High	Low	Low to mid	Low to mid

Adapted from Jones and Brau, "Improving Gas Liquid Mass Transfer Scale-Up," presented at AIChE 2013 Annual Meeting, courtesy of Thermo Fisher Scientific.

The general relationship between sparger hole size and bubble size in a bioreactor depicted in Fig. 20-21, i.e., larger holes resulting in larger bubbles throughout the bioreactor, holds for mammalian cell culture in many cases, except for certain impeller designs. Bottom-mounted agitators located just above the sparge elements can cause significant bubble breakup at higher rpm by shearing the bubbles as they are formed at the sparger. The relationship between bubble size and drilled tube hole size is also less likely to be maintained in fermenters. Operating at high *P/V*, bubble breakup

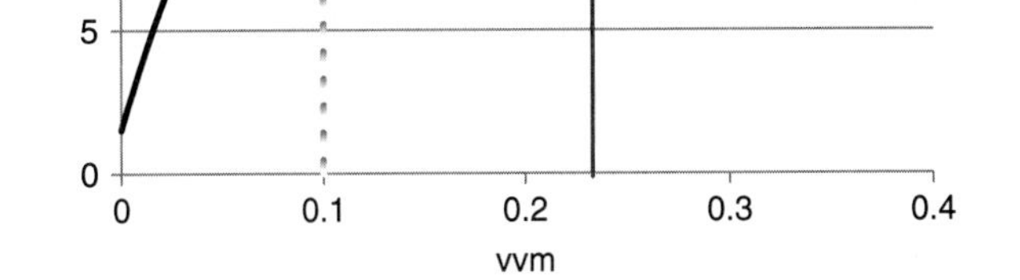

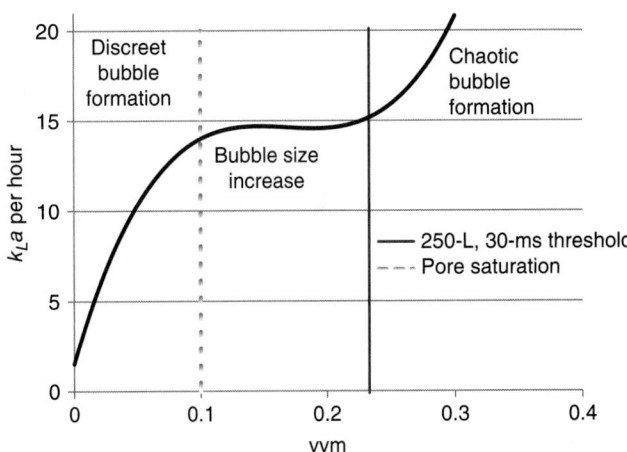

FIG. 20-20 Mass-transfer coefficient as a function of sparge gas flow rate expressed as bioreactor volumes per minute (vvm). The 6.7-mm drilled-hole disk sparger with 500 × 0.203-mm holes (pores) mounted in a 250-L scale single-use bioreactor operated at 139 rpm. Dashed line shows where sparged bubble coalescence begins. Solid line indicates where the sparge gas linear velocity exiting the sparge disc holes reaches 30 m s^{-1}. (Adapted from Jones and Brau, "Improving Gas Liquid Mass Transfer Scale-Up," presented at 2013 AIChE Annual Meeting. Courtesy of Thermo Fisher Scientific.)

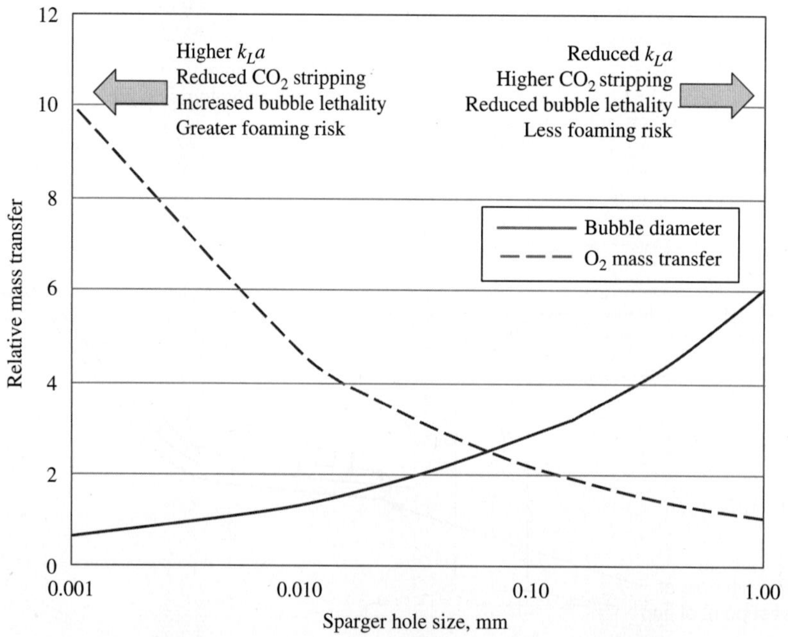

FIG. 20-21 General trends resulting from changes to sparger hole size on O_2 mass transfer, CO_2 stripping, foaming tendency, and potential cell death as sparge gas bubble size increases from micrometer to millimeter scale. This figure is illustrative of likely trends. Actual results will be dependent on a variety of factors such as detailed sparger design, gas velocity through the sparge holes, medium surface tension and ionic strength, agitation, and others.

can dominate the sparge gas bubble size. In a recent study using Rushton turbines operating at 100 to 500 W m^{-3}, the same k_La values were achieved regardless of whether an open jet sparger or a drilled tube ring sparger was used. From this the authors determined that sparger design was not the determining factor in generating mass-transfer area; rather the mean bubble size was determined by the agitator [Sieblist, *Biotech. J.* **6:** 1547–1556 (2011)].

As mammalian cell culture densities have increased, multiple sparge elements have been incorporated into bioreactors. One sparger will be optimized to meet cell culture OUR demand, striking a balance between high mass transfer while managing foaming tendency and minimizing impact on cell viability. A second sparger optimized for CO_2 stripping, usually meaning larger bubbles, can be added to maintain target CO_2 levels. This combination of two spargers—one that produces smaller bubbles and/or uses oxygen enriched air to increase OTR and a second one using only air and producing larger bubbles for CO_2 stripping has been successfully applied in both single-use and traditional stainless-steel reactors. The spargers can either be two drilled tube spargers, or a combination of a drilled tube and sintered spargers, although the latter is more frequently used at small scales. In some cases, CO_2 reduction has been accomplished through changes to the pH chemistry that reduces or eliminates bicarbonate [Goudar, *Biotech. Bioeng.* **96:** 1107–1117 (2007)].

The effectiveness of open-pipe and drilled-hole spargers for CO_2 stripping and O_2 supply was compared at various sparge rates in a stirred-tank bioreactor (Jones, AIChE Annual Meeting, Nov. 3, 2013). The drilled-hole sparger achieved better O_2 and CO_2 mass transfer, with k_La values up to 4 times higher than the open-pipe sparger (Table 20-9). Mass-transfer rates as a function of sparge gas flow also increased more for the drilled-hole sparger. These results demonstrate that, in contrast to fermenters using Rushton operating at high *P/V* where agitation is dominant, the type of sparger used in mammalian cell culture can have an equally large effect on mass transfer. In the latter case, bubble formation at the sparger to create mass-transfer area is more effective than using an open-pipe sparger and relying on the impeller to break up larger bubbles emitting from the open pipe into smaller ones. This design approach significantly decouples mass-transfer area generation by the sparger from impeller mixing and gas distribution. This provides more degrees of freedom in the design and operation of the bioreactor.

TABLE 20-9 Mass Transfer (k_La, h^{-1}) versus Sparger Type

Spare rate, vvm	Open pipe	Drilled hole
	Carbon dioxide stripping	
0.025	~1.4	~3.3
0.050	~1.9	~5.9
	Oxygen transfer	
0.025	~3.0	~8.0
0.050	~3.5	~13.0

Adapted from Jones and Brau, "Improving Gas Liquid Mass Transfer Scale-Up," presented at AIChE 2013 Annual Meeting, courtesy of Thermo Fisher Scientific.

Blending Bulk liquid mixing is another critical factor in scale-up. Insufficient blending, or homogenization, can lead to pH, temperature, and nutrient gradients in large-scale bioreactors [Langheinrich, *Biotechnol. Bioeng.* **66:** 17; 179 (1999); Bylund, *Bioproc. Eng.* **18:** 171–180 (1998)]. Blending to eliminate such gradients is characterized by both the quality of mixing, such as how close the concentration is throughout the vessel compared to its final concentration, and the time it takes to achieve it. Mixing time θ_m can be defined as the time it takes for a vessel to reach 95 percent of its final concentration following rapid addition of a tracer, such as a base or salt solution. The tracer is usually added at several locations in a series of tests to determine the effect of addition location on measured mixing time. Tracer studies, literature correlations, and CFD modeling are all used to generate mixing time estimates. A further discussion of the use of residence time distribution to characterize bioreactors can be found in [Bailey and Ollis, *Biochemical Engineering Fundamentals*, 2nd ed., McGraw-Hill, New York, 1986, pp. 553–560].

Due to variations in bioreactor designs, and especially for single-use bioreactors with nontraditional designs, θ_m measurements are preferred over literature correlations. DECHEMA has developed a recommended procedure for measuring mixing times using two methods, decolorization and tracers monitored by sensors, that can be applied to any single-use or stainless-steel vessel (Meusel, *Recommendations for Process Engineering Characterization of Single-Use Bioreactors and Mixing Systems*, DECHEMA, Frankfurt, pp. 32–44, 2016). Decolorization methods utilize redox reactions to effect a change in color as the redox reagents are mixed throughout the vessel contents. This method can use an in situ probe immersed in the liquid, or a noncontact probe using an accessible section of the transparent film used in single-use systems. The second method, conductivity sensors, uses a salt tracer whose concentration is tracked with a sensor following salt solution addition to the vessel. To provide accurate mixing times, the sensor must have sufficiently high response times and transmitter sampling rates. In cases where the data are recorded in an automated data historian for later analysis, the potential impact of data filtering or compression should be considered.

In the absence of experimental data, Eqs. (20-9*a*) and (20-9*b*) can be used to estimate θ_m for bioreactors operating in the turbulent regime, Re > ~10^4. Correlation Eq. (20-9*a*) was developed for bioreactors having an aspect ratio of 1:1 and a single impeller, independent of whether it is an axial or radial flow type (Nienow, *Animal Cell Culture*, Springer, New York, 2015, pp. 155–156). Equation (20-9*b*), where the constant *A* is defined for each specific bioreactor geometry, is a more general form of Eq. (20-9*a*) that was suggested for bioreactors having an AR > 1, geometric similarity, and multiple impellers operated at similar *P/V* [Nienow, *Chem. Eng. Sci.* **52:** 2557–2565 (1997)]. It was later shown that the form of Eq. (20-9*b*) can be more generally applied to large-scale mammalian cell cultures with a single impeller and AR = 1.3 [Nienow, *Cytotechnol.* **50:** 9–33 (2006)]. The term $(H/D_T)^{2.43}$ originally developed for multiple-impeller systems was found to apply to liquid level in single-impeller stirred vessels.

$$\theta_m = 5.9(P/V)^{-1/3}(D_I/D_T)^{-1/3}D_T^{2/3} \tag{20-9a}$$

$$\theta_m = A(P/V)^{-1/3}(D_I/D_T)^{-1/3}(H/D_T)^{2.43}D_T^{2/3} \tag{20-9b}$$

It has also been shown that Eq. (20-9*b*) can be used to correlate mixing times for single-use bioreactors up to 2000 L with *A* = 3.5 (Meusel, *Recommendations for Process Engineering Characterization of Single-Use Bioreactors and Mixing Systems*, DECHEMA, Frankfurt, DE, 2016, p. 17). Recent CFD studies have shown that Eq. (20-9*b*) can correlate SUB mixing times in Sartorius SUBs equipped with multiple impellers ranging in scale from 2 up to 50 L with the value of the constant *A* between 6.5 and 8.7 (Minn, *Computational Fluid Dynamics Technology and Applications, CFD for Characterizing Standard and Single-Use Stirred Cell Culture Bioreactors*, INTECH, Rijeka, Croatia, 2011, pp. 97–122).

For multiple radial impellers, or a combination of axial and radial impellers, Eq. (20-10) has been used to correlate mixing times where the proportionality depends on the type of impeller and other factors:

$$\theta_m \propto (P/V)^{-1/3} D_T^{2/3} \tag{20-10}$$

For fermenters with AR = 3 using three radial impellers spaced D_T apart with the upper impeller submerged ~$2/3D_T$, Eq. (20-11) has been used to correlate θ_m.

$$\theta_m = 3.3\,N^{-1}(P_g)^{-1/3}(H/D_I)^{2.43} \tag{20-11}$$

By substituting axial flow impellers for the upper radial flow impellers, the constant 3.3 decreases up to 50 percent due to the improved overall mixing pattern (Nienow, *Animal Cell Culture*, Springer, 2015, p. 156). As the name implies, axial flow impellers create flow that runs in the direction of the shaft (axial), while a radial flow impeller creates flow in the radial direction (perpendicular to the shaft).

For a single-use bioreactor using a bottom-mounted impeller, mixing time was correlated by Eq. (20-12) where N_b is a constant determined for a given bioreactor geometry (S. Westerhout, personal communication, 2010). This clearly has a different form than seen for traditional bioreactors, an indication of how dependent θ_m can be on specific bioreactor designs.

$$\theta_m = 2.11N_b\,N_Q^{-1}\,VN^{-1}\,D_I^{-3} \tag{20-12}$$

In general, unless they have been developed for specific bioreactors, θ_m correlations are at best estimates of mixing times for a specific bioreactor, or a means to estimate how θ_m may change with scale for similar bioreactors. For example, Eq. (20-11) was found to underpredict θ_m by 20 to 70 percent when scaling from 20 to 5000 L [Xing, *Biotech. Bioeng.* **103:** 733–746 (2009)]. In that case experimental data for the specific bioreactor design were correlated by

$$\theta_m = 13{,}546\;V^{2.398}\,\mathrm{Re}^{-0.7455}\,D_I^{-1} \tag{20-13}$$

with the Re numbers using the impeller diameter and impeller rate. Significant variations in mixing times may occur within very large-scale bioreactors, especially farthest from the impeller, and close to the liquid surface in high-AR bioreactors. This has important implications when base to control pH is added near the liquid surface. A better location is close to the impeller where dispersion and mixing are better. In practice this is not commonly done due to design challenges or sterility concerns. Aside from creating inhomogeneity, adding base in a poorly mixed zone can result in pH control loop instabilities, if not properly accounted for. This requires controlling the rate of base addition and tuning the control loop to accommodate lag times due to the longer mixing times. Finally, addition of high-concentration base in a poorly mixed zone can actually result in cell death [Ozturk, *Cytotechnol.* **22:** 3–16 (1996)].

Temperature Control Cell growth rate depends on precise temperature control with mammalian cell cultures often operating between 32°C and 40°C with a tolerance of ±0.5°C or less. Metabolic activity correlates with the oxygen uptake rate and can be used to estimate the metabolic heat release Q_H (W m^{-3}) and cooling load per Eqs. (20-14) and (20-15) (Nienow, *Animal Cell Culture*, Springer, 2015, p. 155)

$$Q_H \approx 4.6 \times 10^5\ \mathrm{OUR} \tag{20-14}$$

$$Q_H V = UA_H\,\Delta T \tag{20-15}$$

where *V* (m^3) is the medium volume in the bioreactor, ΔT (°C) is the difference between the coolant and broth temperature, and A_H (m^2) is the heat-transfer area. For jacketed stainless-steel vessels the overall heat-transfer coefficient *U* is usually ~2000 to 3000 W m^{-2} °C^{-1}. Impeller speed generally has little effect on *U*, although larger D_I/D_T impellers with lower P_o provide higher *U* values than smaller D_I/D_T impellers with higher P_o at the same *P/V*.

For mammalian cell bioreactors, heat transfer is generally not an issue. Typical design requirement for a 2-m^3 bioreactor is 300 M Btu h^{-1}. Aerobic fermentation differs from mammalian cell culture in several ways that dramatically affect the design. To support high metabolic activity, microbial fermenter agitation power requirements range from 1.6 to 5.7 kW m^{-3}. This requires mass-transfer rates of 250 to 400 h^{-1} with air sparge rates on the order of 0.5 vvm to meet high OUR requirements. As seen from Eq. (20-14), this also results in high heat release rates. Heat transfer to maintain temperature in microbial fermenters is often a problem since cooling requirements scale with volume, while jacketed vessel cooling surface area scales with diameter. Internal cooling coils have been used at larger scales to increase the heat-transfer surface area, although in more viscous fermentations they can interfere with mixing patterns, and in biopharmaceutical applications cleaning is an issue. An alternate approach is to incorporate the heat-transfer surface area into the vessel baffles, thereby facilitating cleaning. This design has been implemented in microbial fermenters producing biopharmaceuticals up to 10 m^3 achieving *U* values in excess of 200 Btu h^{-1} ft^{-2} °F^{-1} delivering up to 2 TBtu h^{-1} cooling capability (Knight, *Heat Transfer Baffle System and Uses Thereof*, United States Patent 8658419, 2014).

BIOREACTOR SCALING CONCEPTS

Product life cycle management depends on being able to predictably scale bioreactors ranging from small-scale bioreactors used to support cost-effective process and product development, up to the larger scales required for economical commercial manufacturing. Scale-up and scale-down integrate cell culture performance requirements, such as maximum cellular O_2 demand and dCO_2 stripping, with bioreactor capabilities, including k_La, OTR, and blend times. In scaling or transferring processes, existing or proposed design capabilities are evaluated against process requirements to identify design gaps. Equally important is identifying any unique cell culture sensitivities, and screening the design against these potential sensitives. To be successful across scales requires consistent process operations, including procedures, materials, monitoring, and controls. The scaled process will match both process performance (cell viability, viable cell density, and metabolic parameters such as lactate profiles) and product quality attributes as assessed by well-defined analytical methods. Product quality attributes important in biopharmaceutical production may include protein aggregation, charge heterogeneity, and glycosylation.

When applied to biopharmaceuticals, a scale-down model needs to be scientifically sound and may require evidence that establishes scalability if it is going to support regulatory objectives. For example, the FDA requires establishing *limits of in-vitro cell age* (LIVCA) for cell culture processes. Regulatory guidelines allow these studies to be done at commercial or pilot scale, but there must be evidence that the latter is representative of commercial scale. Scale-down bioreactors are typically used for process characterization studies, determination of critical quality attributes, evaluating changes to licensed products, or nonconformance investigations that may be subject to regulatory scrutiny.

Fluid Dynamics and Mixing Dependencies There are two types of scaling factors: scale-independent and scale-dependent. Examples of the former include bioreactor temperature, medium composition, cell density, and DO. These intensive properties can be maintained constant across scales. Scale-dependent parameters include bioreactor height and width, working volume, gas flows, impeller diameter, and impeller rotation rate. Mixing-related parameters most essential for scale-up and design includes an agitation system that provides sufficient power per volume to achieve adequate mass-transfer rates at reasonable gas sparge rates, and an impeller design that provides good sparge gas dispersion and effective bulk fluid mixing. It turns out that the various mixing parameters that are considered when scaling bioreactors often have different functional relationships to

TABLE 20-10 Effect of Different Scale-Up Criteria Using a Linear Scale-Up Factor of 10 Maintaining Geometric Similarity with Re > 10^4

	Scale-up criteria					
Large-scale/ small-scale value	Equal P/V	Equal N	Equal U_T	Equal Re	Equal k_La and vvm	Equal k_La and v_{sg}
$P \propto N^3 D_I^5$	1000	10^5	100	0.1	829	1000
$P/V \propto N^3 D_I^2$	1	100	0.1	10^{-4}	0.8	1
N or t_m^{-1}	0.22	1	0.1	0.01	0.3	0.22
$U_T \propto ND_I$	2.2	10	1	0.1	2.7	2.2
$\text{Re} \propto ND_I^2$	22	100	10	1	27.2	22
$Q \propto ND_I^3$	220	1000	100	10	272	220
$\text{Fr} \propto N^2 D_I$	0.48	10	0.1	10^{-3}	0.5	0.48
$t_c \propto N^{-1}$	4.55	1	10	100	9.4	4.55
k_La at equal vvm	1.59	39.8	0.32	2.5×10^{-5}	1	—
k_La at equal v_{sg}	1	25.1	0.20	1.6×10^{-3}	—	1

Adapted from Nienow, *Handbook of Industrial Mixing*, Wiley, Hoboken, N.J., 2004, p. 1088 with permission.

scale-dependent factors, as shown in Table 20-10. Because of these interdependencies, scaling bioreactor mixing, including gas-liquid contacting, is more complex when compared to other design aspects, such as scaling heat transfer for temperature control.

Scaling mechanically mixed bioreactors inevitably requires compromises since not all mixing and mass-transfer-related parameters can remain constant across scales. Depending on which parameter is fixed across scales, others will vary. This is illustrated in Table 20-10. Holding P/V constant results in equal k_La, provided the sparge gas superficial velocity is also maintained. If scaling is based on constant Reynolds number, however, k_La will be significantly different unless other changes (sparger design, sparge gas composition) are made to compensate, making this an overly complex strategy. Because of these interdependencies, the bioreactor designer must have a knowledge of the primary process requirements and the mixing parameters needed to achieve them, and which are secondary and can be allowed to vary.

The fluid dynamic implications of scaling based on various mixing parameters are covered in detail in Nienow, *Handbook of Industrial Mixing*, Wiley, Hoboken, N.J., 2004, pp. 1076–1080. For bioreactor and fermenter design purposes, the liquid viscosity will be relatively close to that of water, and the Reynolds number will be >10^4, unless one is designing a fermenter for fungal, algal, or other highly viscous processes. At reasonable mixing speeds, the assumption of turbulent fluid conditions can be applied across stirred-tank pilot and commercial scales since Re increases on scale-up.

Scaling based on equal impeller tip speeds or blend times is impractical for all but the smallest scales. Scaling based on equal impeller tip speeds results in unreasonably high impeller speeds at small scale when attempting to match large-scale tip speeds. Blend time increases going from bench to commercial scales since the impeller speeds needed to match small-scale blending would be unreasonably high. Effects of increased blend times with scale, such as spatial heterogeneity, are minimized by operating at sufficiently high P/V in a well-designed bioreactor. Achieving the shortest practical blend times requires the use of baffles or an offset impeller to promote mixing. Axial flow impellers and/or adding liquids close to the impeller region has been used to mitigate negative impacts to the cell culture. For example, it has been reported in one case that adding concentrated base to the liquid surface, where mixing is poor, led to locally high concentrations resulting in cell lysis [Ozturk, *Cytotechnol.* **22:** 3–16 (1996)]. Generally, however, this has not been an issue in most bioreactor applications.

Scaling Principles and Objectives A scaling methodology begins by recognizing the most important cell culture performance and product quality parameters to be achieved for a given process. This understanding is factored into a scaling strategy designed to achieve these outcomes, while avoiding conditions that can impact process performance. This strategy can be based on a combination of bioreactor characterization and modeling across scales, literature correlations, computational fluid dynamics, successful prior experience in bioreactor scaling, and scale-down bioreactor experiments. Scale-down models are useful for identifying key process requirements, unusual process sensitivities, and operating ranges removed from the edge of failure. Process characterization using such scale-down models has become an essential part of the development and commercialization of biopharmaceutical processes for these reasons.

Various scaling approaches have been suggested for animal cell culture bioreactors. Some of these are directly related to process requirements, such as achieving constant k_La or OTR across scales to meet metabolic oxygen demand. In this approach, achieving targeted DO and dCO_2 in a scaled system can be directly correlated to agitation, gassing (air, supplemental oxygen, stripping gas), and sparger design. Other strategies such as scaling based on constant P/V, i.e., mixing intensity, are less obviously tied to specific process outcomes, although there is a relationship to blend time, mass transfer, and other important factors.

In practice, it is generally observed that a scaling strategy will be founded on successful, practical in-house experience scaling to commercial manufacturing and knowledge of relevant manufacturing operational experience and limitations that together define the design space. This design space includes maximum impeller speed, process and bioreactor characterization at relevant scales, a gassing strategy (including the ability to predict required gas flow rates), and performance of sensitivity analysis of potential cell culture responses to environmental conditions as the design evolves. While relying on previous successful scaling experience is often a good starting point, the designer will need to be aware of how the cell culture processes, process understanding, and operations have evolved that may require modifications to the scaling approach.

Constant P/V Scaling Historically mammalian cell cultures have been operated at relatively low P/V levels compared to microbial fermentation. Mammalian cell cultures typically operate at 10 to 50 W m^{-3} [CMC Biotech W.G., A-Mab: A Case Study in Bioproc. Dev. Ver 2.1, p. 74, in Public Domain at ISPE (www.ispe.org) (2009)] although reported values range from 2 to 100 W m^{-3}. In contrast, aerobic fermentations operate from 1000 to 6000 W m^{-3}. These differences are driven by cell culture metabolic rates and cell densities, and the associated higher mass-transfer requirements.

Scaling bioreactors based on constant P/V may be reasonable if there is a proven track record that the operating conditions can support metabolic requirements at larger scales while resulting in a reasonable bioreactor design. If, on the other hand, P/V targets have evolved from experience with low-cell-density, low-metabolic-rate cell cultures, constant P/V scaling could result in larger-scale bioreactors being underpowered if the cell culture has significantly changed. As mammalian cell densities continue to increase, now up to 1×10^7 cells/mL and higher in some cases, low mixing intensities suitable for cell densities an order of magnitude less may not be able to support the increased metabolic demands without resorting to unreasonably high sparge gas flow rates, or using sintered frit spargers that may themselves impact cell growth.

Basing scaling on constant P/V defines only part of the scaling considerations. Elements such as the impeller and sparging systems will also need to be appropriately scaled. Arriving at the final bioreactor design is often an iterative process. If there is a strong basis and successful experience operating within a certain P/V range, this range can serve as a good starting point. As the design evolves, design space limitations may be identified, or significant cell culture changes may have evolved over time that now need to be supported, which can cause the design to move away from a strictly constant P/V scaling approach. This may be the case when scaling between existing bioreactors having different designs and design limitations.

Models have been developed to aid in the design process (Fig. 20-22). By incorporating bioreactor design and operational capabilities, bioprocess characteristics, pH control, agitation, and gassing control strategies, these models are able to predict process performance for various operating and design conditions. Some models have been developed utilizing virtual control loops which, when combined with cell culture performance characteristics and predictive correlations of bioreactor operation, are capable of generating a dynamic process simulation from inoculation to harvest. In addition, models can be used to estimate shear-related characteristics such as local energy dissipation rates and Kolmogorov eddy length scales. These capabilities can be useful to evaluate potential hydrodynamic sensitivities, as discussed further later.

P/V Scaling and Hydrodynamic Shear Mammalian cell culture bioreactors have generally been operated at low P/V levels due to concerns that the cells are easily damaged by the hydrodynamic stresses present in a bioreactor. In mammalian cell suspension cultures, the risk of cell damage due to hydrodynamic stress is less likely than generally thought. Nonetheless it is common to scale bioreactors based on constant, relatively low P/V based on the reasoning that it reduces the risk of damaging cells while providing sufficient mass transfer and homogeneity. Scaling based on lower, overly conservative P/V values due to concerns about hydrodynamic shear may result in spatial inhomogeneity at larger scales, negatively impacting performance. A more holistic approach is to understand and optimize process performance and bioreactor design, taking into account which cell cultures may be susceptible to hydrodynamic stresses, such as cells grown on microcarriers and some stem cells, and which are less susceptible, such as CHO cells which are more resilient than often realized.

Purely hydrodynamic shear damage to mammalian cells grown in suspension culture has been shown to occur only at much higher P/V values than those typically used in commercial biopharmaceutical cell culture processes (Table 20-11). From these results it is expected that hydrodynamic turbulent shear damage will be minimal for most mammalian suspension

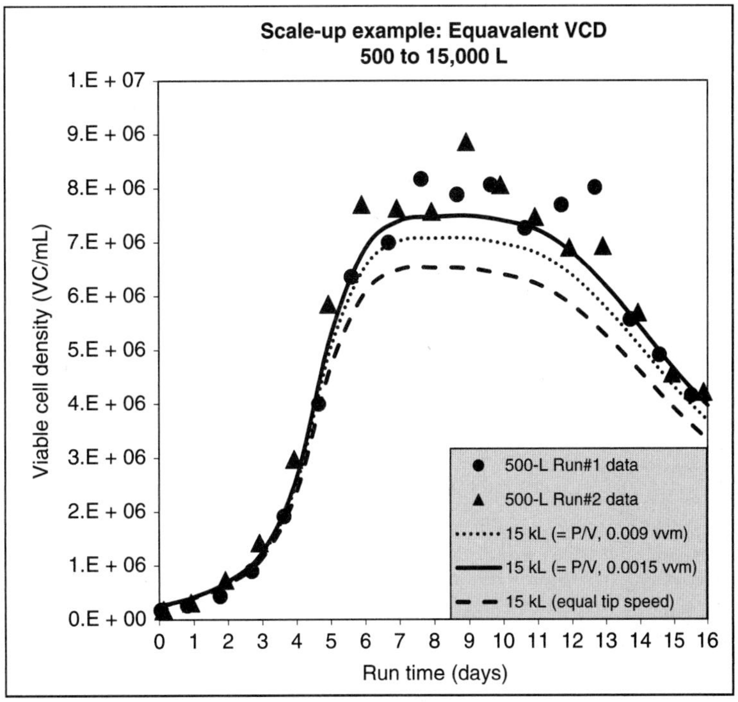

Simulation results

Parameter	500 L	15 kL = P/V, 0.009 vvm	15 kL = P/V, 0.0015 vvm	15 kL = tip speed
RPM	105	45	45	30
Tip Speed (m/s)	1.90	2.87	2.87	1.92
P/V (W/m^3)	76	75	75	22
Shear Rate (1/s)	14	22	22	14
TAS (1/s)	3.6	1.5	1.5	1.0
Blend Time (s)	19	31	31	75
k_La (1/h)	3.7	2.9	2.5	1.2
Air Sparge (vvm)	0.009	0.009	0.0015	0.009
Air Sparge (slpm)	2.5	111.4	18.6	111.4
O_2 Sparge (slpm)	1.2	71.5	98.8	65.7
% O_2 (sparge gas)	46.2	51.9	87.5	50.3
Max. CO_2 (mm Hg)	36	42	65	48

FIG. 20-22 Scale-up from 500- to 15,000-L bioreactor using either constant P/V or constant tip speed. Simulation results show scaling based on equivalent P/V and constant air sparge (vvm) achieves better performance than scaling based on tip speed. Performance at the 15,000-L scale was further improved by reducing the air sparge (0.0015 vvm) while increasing the O_2 sparge rate and concentration, thereby improving driving force and OTR. (Data courtesy of Ted Deloggio, BioProcess Consultants.)

cell cultures used to produce biopharmaceuticals. However, concerns over cell shear damage even at relatively low P/V levels continue to appear in the literature [Chalmers, *Cur. Opinion in Chem. Eng.* **10:** 94–102 (2015)]. When scaling to very large bioreactors, one should recognize that being overly conservative in agitation rates due to unfounded hydrodynamic shear concerns could result in greater temperature and composition heterogeneity. It can also limit mass-transfer rates, mixing, and gas dispersion, since these are related to impeller speed and power input, with the possible result being impact on cell growth rate and metabolism (Nienow, *Handbook of Industrial Mixing*, Wiley, Hoboken, N.J., 2004, pp. 1073, 1088). Hydrodynamic shear sensitivity, including sublethal effects, may need to be carefully evaluated in specific cases such as cell cultures using microcarriers or certain stem cell lines.

Constant Mass-Transfer Coefficient and OTR Scaling Because of the low oxygen solubility in water, failure to constantly satisfy mammalian cell culture OUR requirements will quickly result in DO levels falling below the critical oxygen concentration. Cellular oxygen uptake rate (OUR) is a function of viable cell density X_{viab}, as shown in Eq. (20-16),

$$\text{OUR} = q_{O2} \times X_{viab} \tag{20-16}$$

where q_{O2} is the specific oxygen uptake rate per cell. Table 20-12 shows typical q_{O2} values for various host cells.

For mammalian cell cultures the maximum oxygen demand occurs at the maximum viable cell density X_{max}. For $X_{max} = 50 \times 10^6$ cells/mL and $q_{O2} = 2 \times 10^{-16}$ mol O_2/cell·s, the DO level would fall below 10 percent in less than 15 min if the O_2 supply were lost (personal observations). At that point lack of oxygen begins to impact cellular metabolism and growth, making achievement of adequate mass transfer a critical requirement when developing a scaling approach.

TABLE 20-11 Effects of Hydrodynamic Conditions on Cells

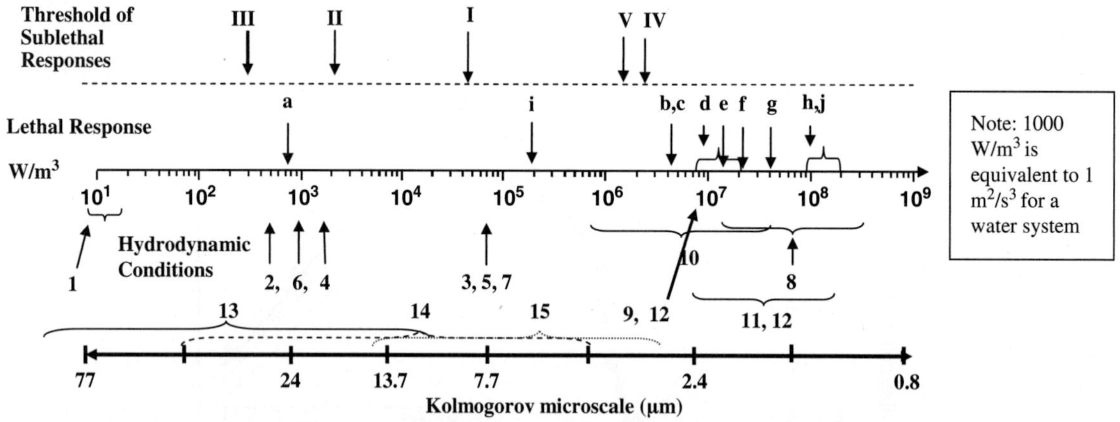

Threshold of sublethal physiological responses

Symbol	Cell	Mode of growth	Mode of test	Response	Reference
I	CHO	Suspended	10 days' repetitive exposure	Recombinant protein glycosylation profile change	[51]
II	LnCap	Attached	Time exposure to well-defined shear stress	Membrane integrity failure, change in receptor number	[60]
III	CHO	Suspended	2-L Applicon bioreactor	Recombinant protein rate of production and glycosylation effect	[40]
IV	CHO	Suspended	3-L bioreactor with external loop containing a nozzle	Threshold above which inhibitory effects begin	[41]
V	Sp2/0	Suspended	3-L bioreactor with external loop containing a nozzle	Threshold above which inhibitory effects begin	[41]

Lethal responses (necrosis including LDH release)

Symbol	Cell	Mode of growth	References
a	CHO-K1	Anchorage for growth and test	[61]
b	PER.C6	Suspended (naïve or adenovirus infected)	[32]
c	CHO (GS)	Suspended	[51]
d	Hybridoma	Suspended	[37, 38]
e	MCF-7	Suspended	[32]
f	Mouse myeloma	Suspended	[36]
g	Hela S3, mouse L929	Suspended	[62]
h	CHO-K1, Hybridoma HB-24	Suspended	[32]
i	CHO-K1, apoptosis	Anchorage for growth, suspended during test	[63]
j	CHO-K1	Suspended (wild type and bcl2 transfected)	[63]

Hydrodynamic conditions

Symbol	Process	Description	References
1	Agitation	Volume average energy dissipation rate in typical animal cell culture bioreactors	[64, 9]
2	Agitation	Volume average energy dissipation rate in a 10-L mixing vessel (Rushton Turbine impeller, 700 rpm)	[65]
3	Agitation	Maximum local energy dissipation rate in a 10-L mixing vessel (Rushton Turbine impeller, 700 rpm)	[65]
4	Agitation	Volume average energy dissipation rate in a 22,000-L fermenter (Rushton Turbine impeller, 140 rpm)	[66]
5	Agitation	Maximum local energy dissipation rate in a 22,000-L fermenter (Rushton Turbine impeller, 140 rpm)	[66]
6	Agitation	Maximum local energy dissipation rate in a spinner vessel	[67]
7	Bubble rupture	Pure water (bubble diameter: 6.32 mm)	[27]
8	Bubble rupture	Pure water (bubble diameter: 1.7 mm)	[27, 28]
9	Membrane filtration	CHO suspension pumped through Millipore membrane and capillary tubes	[68]
10	FACS	CHO cell damage sorted through a FACS	[69]
11	centrifugation	Bowl and disk centrifuge	[70, 71, 72]
12	Capillary	Scale-down of industrial continuous centrifuge	[73]
13	Pump	PuraLev 200; 1500 rpm, 3.4 L/min, ΔP ~35 mm Hg	[35]
14	Pump	PuraLev 200; 3500 rpm, 3.4 L/m, ΔP ~300 mm Hg	[35]
15	Pump	raLev 200; 5000 rpm, 3.4 L/min, ΔP ~650 mm Hg	[35]

Reproduced from Chalmers, *Cur. Opinion in Chem. Eng.*, 10, pp. 94–102 (2015) with permission.

TABLE 20-12 Specific Oxygen Uptake Rates (q_{O2}) for Cells in Culture*

Organism	q_{O2}, mmol O_2/(g-dw h) Mammalian: mol O_2/(cell·s)
Bacteria	
E. coli	10–12
Azotobacter sp.	30–90
Streptomyces sp.	2–4
Yeast	
S. cerevisiae	8
Molds	
Penicillium sp.	3–4
Aspergillus niger	ca. 3
Plant cells	
Acer pseudoplatanus (sycamore)	0.3
Saccharum (sugar cane)	1–3
Mammalian cells	
Sf-9	$0.25–4.5 \times 10^{-16}$
Hybridoma (various)	$0.4–6.9 \times 10^{-16}$
HeLa	1.1×10^{-16}
Diploid embryo WI-38	0.04×10^{-16}
CHO cells (various)	$0.5–2.2 \times 10^{-16}$

*Data from Flickinger, *Upstream Industrial Biotechnology—Aeration, Mixing, and Hydrodynamics,* Wiley, Hoboken, N.J., 2013, p. 79; and Shuler, *Bioprocess Engineering: Basic Concepts,* 2d ed., Prentice-Hall, Upper Saddle River, N.J., 2002, p. 293.

To avoid oxygen becoming the rate-limiting reagent, bioreactor design must be capable of supporting the maximum cellular oxygen uptake rate (OUR_{max}). This establishes the minimum OTR design requirement since $OTR_{max} = OUR_{max}$. Achieving this requires optimizing agitation and sparging, including sparger design and gassing strategy. The latter includes sparge gas composition, flow rates, and how they are allowed to vary over time. Various approaches to achieve OTR_{max} are possible, such as increasing the driving force (ΔC_L) by providing supplemental oxygen, changing the mass-transfer rate by adjusting P/V and/or sparge rates, or even modifying sparger designs. Several scaling approaches have been suggested to address this complexity.

Scaling by constant k_La has been proposed since it takes into account the interaction of sparging rate and P/V. This approach, however, does not factor in the O_2 driving force, which can be different across scales depending on sparge gas capabilities. Constant k_La scaling also breaks down if sparger or agitator design limitations preclude achieving constant mass-transfer coefficients across scales. This can be the case when scaling between existing bioreactors, or in developing a scaled-down model where design limits may constrain maximum achievable k_La.

A more generalized approach is scaling based on equal OTR, in that it specifically takes into account the driving force, in addition to the mass-transfer dependency on P/V and sparge flow rate. Adopting this approach ensures that cellular metabolic oxygen demand is satisfied. Even with this approach, the design problem can start with the premise of constant P/V, which can then be modified as the design assesses how to satisfy OTR demands. In the special case where geometric similarity is maintained across scales, maintaining constant P/V and v_{sg} results in equal k_La across scales. By keeping ΔC_L constant, OTR will also be constant on scaling. This shows that, in this case, maintaining constant v_{sg} with scale is sufficient to meet metabolic demands, provided the design does not result in saturating sparge gas with CO_2. This approach, however, does not provide a safety factor to account for uncertainties in bioreactor characterization data, or cell culture parameter changes, unless a safety factor is incorporated into the design to support higher OTR on scaling to account for these uncertainties.

Because the quantity of cells scales linearly with bioreactor volume, the total O_2 demand and CO_2 evolution will also scale with volume. Performing a gas mass balance on the bioreactor shows that to ensure sufficient CO_2 stripping and O_2 supply, a constant volumetric gas flow rate per unit bioreactor volume (vvm = Q_{sg}/V) should be maintained across scales if maximum cell densities are to be achieved (Nienow, *Animal Cell Culture*, Springer, 2015, p. 153). Based on this mass balance approach, scaling at less than constant vvm presents a risk of insufficient O_2 supply and could also result in higher dCO_2 levels. However, as shown in Table 20-13, by maintaining constant vvm and P/V with geometric similarity across scales, v_{sg} and OTR actually increase with scale. If the bioreactor aspect ratio also increases with scale, this effect is further amplified. This analysis suggests that scaling with constant vvm results in an overdesigned bioreactor at larger scales. The reason for this apparent contradiction is that the gas-phase mass balance approach does not factor in the effect of v_{sg} on k_La. The designer needs to be aware, however, that in cases where CO_2 is effectively saturating the stripping gases (i.e., small bubbles), CO_2 removal will be governed by the mass balance, not by mass transfer.

TABLE 20-13 Effect of Constant vvm Scaling Normalized to 500-L Scale Bioreactor*

Volume, L	500	2000	15,000
H/D_T	1.2	1.2	1.2
v_{sg} or D_T	1.0	1.6	3.1
k_La or OTR	1.0	1.4	2.5
H/D_T	1.2	2.5	2.5
v_{sg} or D_T	1.0	2.6	5.1
k_La or OTR	1.0	2.1	3.7

*Scaling based on typical high density mammalian cell culture bioreactor operated at 20 W m^{-3} with air sparging.

A conservative approach to this design problem is to simply size the bioreactor systems to accommodate both constant vvm and P/V scaling. Although the bioreactor will be overdesigned, it provides a safety factor against uncertainties in P_o and k_La measurement accuracy that may form the basis for scaling, as well as providing flexibility to accommodate future improvements such as increased cell density. It is up to the bioreactor designer to weigh the costs versus benefits related to potential component oversizing (mass flow controllers, exhaust systems) against uncertainties and risks in the design basis and future needs.

In practice, even if the design can accommodate constant vvm on scaling, the actual gas flow rates will be determined by the DO control loop. This control loop will act by adjusting operating parameters as needed to maintain the DO at the set point. Depending on the control scheme, it could reduce overall sparge rates (potentially increasing CO_2 levels), reduce O_2 enrichment of sparged air, or decrease impeller speed. The latter may result in poorer mixing if the process design was overly conservative with respect to P/V. Of these choices, reducing O_2 enrichment, if possible, to achieve DO set point retains CO_2 stripping capacity and mixing. Incorporating a second CO_2 stripping sparger can provide additional flexibility in scaling, especially when the primary sparger utilizes O_2 enriched air, or 100 percent O_2.

Scale-Up and Scale-Down Heterogeneity and Sensitivity Ideally the same microscopic environment for cells would be achieved across scales; however, this is unlikely to be the case. Heterogeneity exists near the liquid surface where mixing is often less, where bubbles are formed at the sparger, at the liquid surface where they burst, and in trailing vortices associated with the impeller. Energy is not dissipated uniformly throughout bioreactors. The maximum energy dissipation ε_{max} occurs in the impeller discharge region, and specifically at the impeller tip and trailing vortex. As P/V increases, energy dissipation rates in the impeller region can become much higher than the volume average ε_{mean}, with as much as 70 percent of the energy being dissipated in the impeller discharge region of a pitched blade turbine (Godoy-Silva, *Encyclopedia of Industrial Biotechnology*, Wiley, Hoboken, N.J., 2010, pp. 802–803). The ratio of $\varepsilon_{max}/\varepsilon_{mean}$ can be as high as 50 to 70, or even higher for some impellers used in bioreactors and fermenters (Patterson, *Handbook of Industrial Mixing*, Wiley, Hoboken, N.J., 2004, p. 781). If local ε values exceed a certain level, defined here as the critical EDR level (ε_{crit}), cells may be damaged. For CHO cells, ε_{crit} values are generally quite high, as shown in Table 20-11.

Energy Dissipation Because much of the energy dissipation is localized in the impeller region, the bioreactor volume where ε_{crit} values may exist is expected to be a relatively small fraction of the total bioreactor volume. Increasing P/V increases not only ε values, but also possibly the volume fraction of the bioreactor where ε exceeds ε_{crit}. If the volume of this region where cell damage may occur increases enough, it can result in noticeable impacts to the overall bioreactor process performance. Cell damage, if it occurs, will be a function of both the frequency and residence time of cells being exposed to a region where $\varepsilon > \varepsilon_{crit}$. If, however, this high-EDR region remains isolated around the small region of the impeller tip, then any cell damage, if it happens at all, may be negligible and undetectable. Ideally we could characterize how ε varies throughout the bioreactor and quantify the regions where $\varepsilon > \varepsilon_{crit}$. Then this could be used to predict whether significant cell culture impacts would be higher or lower at different scales, or the effect of different impeller geometry and vessel designs. To do this, both ε_{crit} and characterization of the extent of the ε_{crit} region must be known.

Various approaches to determine cell impacts related to EDR have been suggested. One approach uses a device specifically designed and fully characterized so the energy dissipation rates to which cells are exposed are known. This may be used to assess and quantify hydrodynamic shear impacts on cells [Ma, *Biotech. Bioeng.* **80:** 428–437 (2002)].

Another approach is to design scale-down bioreactor experiments operated at various conditions spanning the range of EDR levels expected at larger scales. EDR levels resulting in cell damage may also be inferred from previous production and pilot-scale cell culture experience where cell impacts were observed. Translating these results into specific values of $\varepsilon_{max}/\varepsilon_{mean}$ or ε_{crit} that result in cell damage depends on having a well-characterized bioreactor, including a detailed understanding of the energy distribution. This is generally only available through CFD modeling, or by extrapolating EDR levels from literature values, as discussed further later.

One approach to quantify the bioreactor volume where high ε levels exist by using CFD modeling has been suggested as a way to compare bioreactors. In this method, the critical volume is defined as that region where $\varepsilon > 0.5\varepsilon_{max}$, with ε_{max} defined as the ε value at the impeller tip. Although there is a degree of arbitrariness to this approach, it does not require knowledge of ε_{crit}. As a first-pass estimate to compare different bioreactors and designs, this method can provide insight as to whether significant EDR differences exist that warrant further consideration.

CFD models can be used to map ε levels throughout a bioreactor. By using these results, the percentage of the bioreactor volume where ε exceeds a specific level, such as ε_{crit}, can be estimated, as shown in Fig. 20-23, and differences related to bioreactor design can be assessed (Fig. 20-24). The

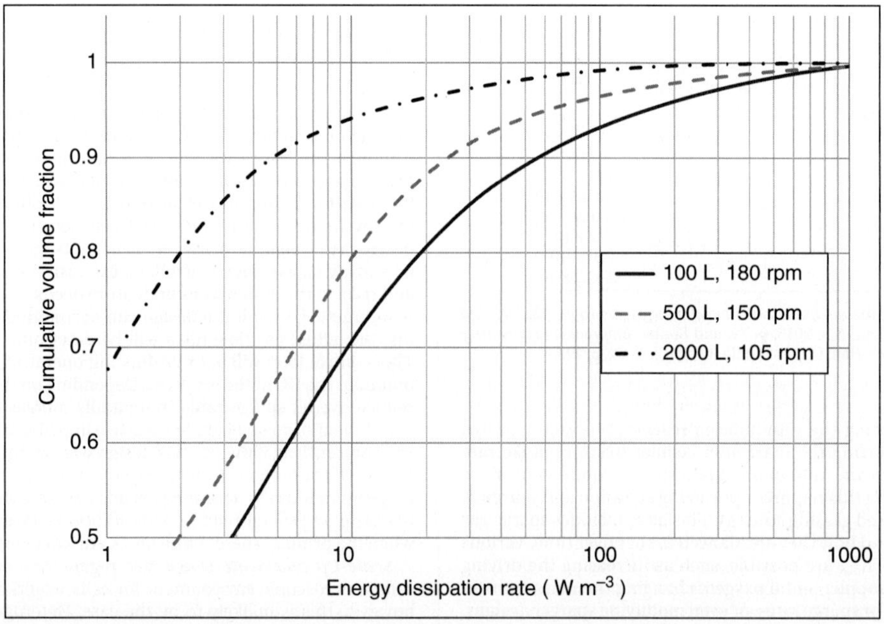

FIG. 20-23 Computational fluid dynamic modeling of energy dissipation rate levels for 100-, 500-, and 2000-L scale bioreactors operating at various agitator speeds. The graph shows the volume fraction of the bioreactor at or above specific energy dissipation levels. For example, the percentage of the bioreactor volume exceeding 100 W m^{-3} is 7 percent at the 100-L scale, 3.5 percent at the 500-L scale, and 0.8 percent at the 2000-L scale. (CFD modeling provided by J. Thomas, M-Star Simulations, LLC, and Research Professor at Johns Hopkins University.)

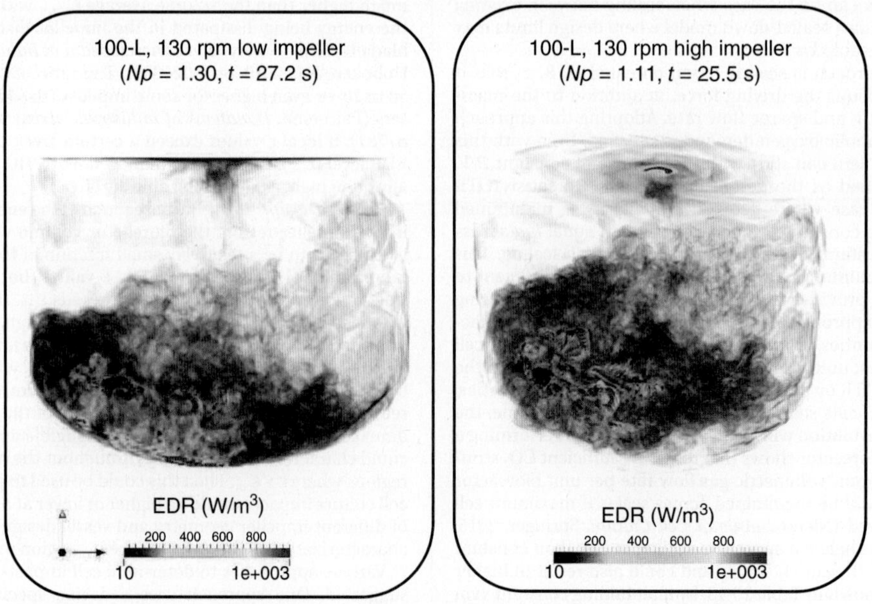

FIG. 20-24 CFD modeling results show energy dissipation rate distribution in the impeller region of a 100-L bioreactor as a function of impeller position off the bottom. The darker color in the impeller region represents highest energy dissipation rates. Also shown are the model-derived impeller power numbers *Np* for each case. When the impeller is located in close proximity to the vessel bottom (~1/12th impeller diameter, left-hand figure), the power number is nearly 20 percent higher compared to when the impeller is located approximately ½ impeller diameter above the bottom of the vessel (right-hand figure). The overall energy dissipation ε_{mean} and ε_{max} are also greater at a given rpm when the impeller is located in close proximity to the vessel bottom (data not shown). (CFD modeling provided by John Thomas, M-Star Simulations, LLC, and Research Professor at Johns Hopkins University.)

CFD model can then be used to evaluate the effects of various design choices, such as repositioning an agitator, or to understand other scale and equipment-related differences including impeller types, geometries, and positioning in the bioreactor. Using this approach, CFD studies have shown that even though the $\varepsilon_{max}/\varepsilon_{mean}$ ratio was expected to be the same between a small and a 20-fold larger production scale bioreactor due to design similarities, the percentage of the small-scale bioreactor experiencing high-ε conditions, defined here as >100 W m^{-3}, was nearly 10 times greater than that at the large scale (Fig. 20-23). This CFD study also showed that both ε distribution and impeller *Np* were affected by the impeller height off the bottom of the bioreactor (Fig. 20-24). The impact on N_o resulting from geometric differences across scales predicted by the CFD model was confirmed through direct measurements of the impeller power input.

A further refinement using CFD predicts the exposure of cells to different ε environments by combining the effects of ε levels, residence time, and frequency of cellular exposure to those environments. Due to the higher circulation rate at small scales and proportionately larger high-ε environments, cells may have greater exposure rates to high-ε conditions compared to larger scales. There continues to be debate regarding the applicability and predictive capabilities of bioreactor CFD models; however, it can be expected that as computing power continues to increase in combination with opportunities to both compare and optimize CFD bioreactor models against experimental results, the real and perceived accuracy of CFD will continue to grow. The application and potential limitations of CFD modeling to stirred vessels are discussed in greater detail in the Mixing Concepts subsection above.

Beyond CHO cell cultures there are cases where cells may be expected to be more sensitive to hydrodynamic shear, such as some stem cells or adherent cells grown on microcarriers. To determine whether EDR levels expected at large scale may negatively impact a cell culture process, a bench-scale bioreactor can be used to evaluate potential process sensitivities related to ε_{max} exposure [Nienow, J., *Biotech.* **171:** 82–84 (2014)]. The approach is based on the $\varepsilon_{max}/\varepsilon_{mean}$ ratio and its spatial distribution being scale-independent when operated at the same ε_{mean} values, and for $\varepsilon_{max}/\varepsilon_{mean}$ to be similar across many impeller types when the D_T/D_I ratios are the same. In addition, it was shown that the total time cells are exposed to areas of high and low ε is also scale-independent. On this basis, the recommended rule of thumb is to operate the bench scale-down model at an impeller speed ~2 times higher than needed to match the commercial scale ε_{mean} values. This results in ε_{max} at the small scale being about 10-fold higher than that expected at large scale, which should be sufficient to identify potential cell damage upon scaling while providing a reasonable safety margin to account for uncertainties in this approach.

Kolmogorov Length Scale The Kolmogorov length scale is frequently used to estimate the size of turbulent, microscale eddies in the impeller region, thereby characterizing the potential to cause hydrodynamic shear damage to cells The minimum turbulent eddy size λ_K is based on ε_{max} [Eq. (20-17)]. In lieu of actual measured values of ε_{max}, its value is typically estimated as some multiplier of the overall *P/V* or through CFD modeling. as described in the EDR discussion earlier. As a result, there is a degree of uncertainty regarding the actual microscale eddy length.

$$\lambda_K = (\varepsilon_{max} / \nu^3)^{-0.25} \tag{20-17}$$

The commonly accepted idea is that cell (or microcarrier) damage does not occur if λ_K is greater than the cell diameter (Nienow, *Encyclopedia of Industrial Biotechnology*, vol. 5, Wiley, Hoboken, N.J., 2010, pp. 2959–2971). If the turbulence eddy length scale is similar in size to the cells or microcarriers being cultured, then there is a risk of hydrodynamic shear damage. In a mammalian cell bioreactor operating at *P/V* values of 20 to 50 W m^{-3}, λ_K is approximately 90 to 70 μm based on ε_{mean}, but $\lambda_K \sim 40$ to 30 μm based on estimated ε_{max}. Since the characteristic turbulent eddy length λ_K is larger than the cell diameter (~20 μm), theoretically they will not be subjected to high levels of turbulent shear, but rather will be swept along with the bulk fluid motion. Note that λ_K is not based on a physical measurement. It is based on the theory that turbulent energy is dissipated isotropically. Unless measured ε_{max} values are available [Zhou and Kresta, *AIChE J.* **42:** 2476–2490 (1996)], the calculated value depends on the volume in which the energy is assumed to dissipate. It has been suggested that ε_{max} can be estimated by assuming that power input is primarily dissipated in the impeller swept volume. Alternatively, it can be assumed for $D_I/D_T = 0.33$ to 0.4 that $\varepsilon_{max}/\varepsilon_{mean}$ ~30. This ratio is largely independent of impeller type [Nienow, *Biochem. Eng. J.* **108:** 24–29 (2016)]. As a scaling parameter, λ_K can be used to compare bioreactors, but as an indication of the absolute eddy length scale, it should be regarded as an approximation.

Bioreactor Scaling Case Studies There are a variety of strategies that can be employed to scale bioreactors, although not all will have the same degree of success as measured by consistent cell density, productivity, process, and product quality. The choice will be driven by experience, achievable operational ranges, cell sensitives, and other factors such as constraints related to existing infrastructure. Table 20-14 and the accompanying text describe several case studies that illustrate the complexities involved and the concepts described earlier in this section.

The first case in Table 20-14 is the most straightforward and is a common approach in that it starts by maintaining constant *P/V*. It also assumes geometric similarity, although it is less common to be able to scale between geometrically similar bioreactors when scaling involves existing bioreactors. Scaling between geometrically consistent bioreactors at constant *P/V* and v_{sg} achieves consistent k_La and OTR (provided driving force remains constant) while exposing the cells to similar energy dissipation rates. The risk with this approach is that by scaling with constant v_{sg} the design may not be robust due to uncertainties in bioreactor characterization (P_o or k_La). In addition, the design may lack flexibility to accommodate advances in cell culture. In practice, as a rule of thumb, the design sparge rate should be between v_{sg} and vvm, although it is up to the designer to determine the best balance between design capabilities and cost related to adopting a conservative design. If the sparge gas is likely to be saturated with CO_2, then at a minimum vvm scaling could be needed to achieve consistent dCO_2 levels, unless other design changes are made.

The second case shows the complexity of transferring processes between dissimilar bioreactors, even at the same scale. Two scaling approaches were evaluated for the SUB, although to maximize mass transfer both options used a small nominal sparger pore size (2 mm), along with higher gas flow rates. The first condition tested used higher sparging rates and operated at a lower *P/V*, but the overall performance was poor. The second condition tested used equal *P/V* between the two bioreactors, but the gassing strategy was modified to achieve required OTR and dCO_2 levels. Because the maximum achievable k_La in the SUB was still only 35 percent of the SS bioreactor k_La, supplemental oxygen combined with the small sparger nominal pore size (2 μm) was needed. In addition, a separate CO_2 stripping sparger was added to attain target dCO_2 levels. The VCD trended lower in the SUB, however, and could have been a result of the small nominal pore size used in the sparger design, although other factors may also have contributed.

The third case study demonstrates a pragmatic approach to creating a scale-down model when the bioreactors are not fully characterized. Since the large-scale operating conditions were already established, the goal was to match process and product quality performance by adjusting parameters at the small scale. Engineering runs, which are runs performed to demonstrate equipment functionality before committing to GMP manufacturing, began by operating the scale-down bioreactor at the same *P/V* as the large scale, while adjusting sparge rates to achieve required OTR to maintain target DO. This resulted in overstripping dCO_2. Then *P/V* was increased, which resulted in lower O_2 sparge rates. This also had the effect of reducing dCO_2 stripping. Agitation was adjusted until the dCO_2 profiles matched the large-scale target range. It cannot be stated unequivocally that the scale-down bioreactor sparge gases were saturated with CO_2, but the fact that scaling required constant vvm and that despite an increase in mass transfer dCO_2 increased suggests that may have been the case. This approach yielded a scale-down model matching all primary process and product quality parameters.

A number of the case studies center on developing representative scale-down models of commercial production bioreactors to support biopharmaceutical development. Case study four shows how several scaling strategies were assessed before settling on a constant OTR strategy in order to match process and product quality performance. To match the larger scale, the scale-down model was operated at what is considered to be a very high-power input for a mammalian cell culture bioreactors (290 W m^{-3}). This was required to overcome limited mass-transfer capabilities resulting from the use of a drilled-hole sparger at low agitation speeds. The drilled hole sparger was implemented after it was found that using a sintered sparger to increase k_La led to significant impacts on cell viability. Case study five also used a constant OTR scaling approach, after a constant *P/V* scaling approach was not able to achieve required performance. Agitation rates were varied during engineering runs until large-scale performance was matched, similar to the approach used in the third case study.

Case study six is instructive as it illustrates the challenges related to identifying the commercial manufacturing site late in the process development cycle. A bioreactor process was developed at small scale, but it was later determined it could not be properly scaled due to manufacturing operating limits. The low *P/V* combined with low sparge rates used at small scale to avoid cell damage led to an operating window that could not achieve the required mass-transfer rates at large scale without major modifications to the bioreactor. This would not only delay implementation but also incur additional costs, making it economically unfeasible. The bench scale bioreactor model was used to reoptimize the process to operate within the manufacturing scale capabilities. Factorial design of experiments using

TABLE 20-14 Mammalian Cell Culture Production Bioreactor Scale-Up and Scale-Down Case Studies

Case study	Method and constraints	Description
1. Best case scaling	Geometric similarity Consistent bioreactor design and operation across scales	Bioreactor designs use same internals (sparger, impeller), automation control loops (e.g., DO and pH), operations, and aligned instrumentation and analytical procedures at both scales. Geometric similarity with constant P/V and v_{sg} achieves same k_La and OTR. As a safety factor to account for bioreactor characterization uncertainty, and for future needs, bioreactors can be designed to support constant vvm across scales. This case will be the exception most often. The advent of well-designed SUB bioreactor families makes this case more realistic in new installations, but the use of SUBs may limit the maximum working volume. Geometric scaling is favored from a scaling perspective, but increasing the aspect ratio with scale has other advantages, such as conserving floor space.
2. Process transfer from 500-L SS to 500-L SU bioreactor	Constant OTR scaling Maximum SUB k_La limited to 35% of SS	500-L SS: Rushton impeller (P_o=10), 4 baffles; 20-mm sparger, 500-L SUB: Pitched blade impeller (P_o ~ 1.2) offset 15° 2-mm O_2 sparger and a separate CO_2 stripping sparger Two conditions of impeller speeds and sparging were evaluated for the SUB: Condition 1*: rpm = 200%, P/V = 25%, vvm = 560%, k_La = 35%, tip speed = 35% Condition 2*: rpm = 300%, P/V = 100%, vvm = 200%, k_La = 35%, tip speed = 90% Condition 2 was implemented since process performance and OTR were maintained, and mixing was better at higher P/V compared to condition 1. Process parameters including viability and lactate were consistent between the SUB and SS reactors, although VCD was ~15% lower in the SUB. That may be due to the sintered sparger.
3. Scale-down from 2000-L to 500-L SUB	The k_La data were unavailable Agitation adjusted to achieve consistent dissolved gas profiles	500-L scale engineering run used to match CO_2 profiles by adjusting agitation speed while maintaining OTR. Starting conditions used equal P/V, but 500-L agitation was adjusted until CO_2 profiles matched across scales. Final conditions†: P/V = 162%; stripping sparger vvm = 100%, O_2 sparger vvm = 100%, tip speed = 100%, λ_K = 80% (~3 times > cell diameter). Process and product quality consistency was achieved across scales.
4. 5-L scale-down model of 2000 L	Qualify 5 L as a scale-down model to characterize scale-up design space	Evaluated four scaling strategies. Constant mixing time was not feasible. P/V scaling was not successful due to low k_La. The k_La matching required sintered sparger, resulting in excessive cell damage. OTR matching strategy using same drilled-hole sparger design was successful. Final 5-L scale P/V was 4 times greater† (290 W m^{-3}).
5. 200-L scale-down model of 2000 L	Qualify 200-L SUB as pilot-scale bioreactor to be used for development and confirmation runs	Both scales used an initial air sparge with a maximum flow rate cap. Sparge gas was enriched with O_2 once the airflow rate cap was reached. Agitation speed was manually adjusted during initial pilot-scale run until full-scale O_2 sparge flow rate profiles were matched on a vvm basis. Adjusting agitation at the 200-L scale resulted in equal OTR when using equal vvm sparge rates. The strategy of adjusting agitation in lieu of further enriching air sparging with O_2 was adopted to match OTR while ensuring adequate CO_2 stripping.
6. Scaling bench process to 100-L and 250-L scales	Reduced sparge rates by increasing P/V to offset bubble-induced cell damage. Process was modified so it could operate within larger-scale constraints.	An optimized process was developed at bench scale using low P/V and low sparge rates to avoid cell damage. Later it was determined the process was not feasible in manufacturing due to inability to achieve required k_La. The 5-L DOE studies reassessed cell culture sensitivity to increased P/V and sparge rates. A scalable set of operating parameters including two-fold higher P/V and higher sparge rates was identified that allowed the large-scale bioreactors to achieve required k_La while minimizing impact to cell viability. An additional transfer complication led to low titers at large scale. Inaccurate viable cell counts led to inconsistent bioreactor inoculation cell densities, feeds, and temperature shifts. An alternative cell-counting method that was less sensitive to cell clumping errors improved titer consistency.

*Values expressed as percentage of stainless-steel bioreactor.
†Values expressed as percentage of large-scale bioreactor.

the scale-down model identified critical process parameters and ranges, as well as which conditions would negatively impact cell growth and product formation. When compared against the known large-scale capabilities, a set of scalable process operating conditions was identified. An additional lesson learned was that poor instrumentation initially caused significant commercial scale process variability until the root cause was identified and corrected.

These case studies demonstrate the variety of approaches that have been used for process scale-up, scale-down, and process transfers between bioreactors. Taken together, a number of themes emerge. Bioreactor scaling often focuses on mass-transfer-related factors, even though initial designs often begin by assuming constant P/V. Achieving consistent k_La is not always achievable across bioreactors. In those cases, scaling focused on achieving an OTR that could support the process. Although sintered spargers could be used to increase k_La, it can also result in decreased viability and is often not preferred in clinical and commercial manufacturing. In driving toward enhancing k_La or OTR, it should not be lost that improved oxygen mass-transfer rates may result in sparge gases being saturated with CO_2 due to lower overall sparge gas flow rates. Independently controlled dCO_2 stripping spargers are fairly common in order to provide an additional degree of freedom to overcome such sparging limitations. In several case studies it was seen that operating scale-down bioreactors at significantly higher P/V levels proved to be more representative, which runs counter to a common view that constant P/V is important to avoid hydrodynamic shear impacting cell culture performance. That is not to say, however, that assessing potential sensitivity to hydrodynamic shear is unimportant. For this reason and others, having adequate scale-down models, employing DOE studies, and numerical bioreactor models are important scaling tools. Although not explicitly discussed in Table 20-14, facility fit issues, such as ability to handle feed volumes and other ancillary equipment, need to be considered in process scaling and transfers. Finally, it was shown that successful scaling and process transfers depend on having consistent instrumentation and operations to achieve consistency across scales.

In comparison to bioreactors, to scale fermenters is a simpler process engineering exercise, although more demanding from a mechanical perspective due to the high metabolic demands. Issues such as hydrodynamic shear and sparger-related cell damage are largely nonexistent. The challenge of satisfying metabolic oxygen demand can be met by increasing power input. In addition, the power input by the agitator is such that it dominates gas dispersion and bubble formation/breakup to the extent that, in comparison to bioreactors, sparger design details often have less effect on k_La. This dominance of the impeller in generating mass transfer also means that constant P/V scaling is more likely to achieve consistent results across scales, provided mixing and dCO_2 stripping are sufficient.

Still, to scale fermentations, one has to consider specific host cells that have their own requirements and limitations. Mycelial fermentation scaling is often based on tip speed due to concern about high velocities at the impeller tip damaging these filamentous organisms (Nienow, *Handbook of Industrial Mixing,* Wiley, Hoboken, N.J., 2004, p. 1086). Some mycelial fermentations exhibit early sporulation, mycelium breakup, and low yields if hydrodynamic shear at the impeller is excessive. In these fermentations, tip speed has been limited to 2.5 to 5 m s^{-1}. In other cases, the opposite is true, and the design has to provide sufficient tip speed to minimize pellet formation that would otherwise limit the O_2 supply to some host microbial cell systems.

STRATEGIES TO OPTIMIZE PRODUCTION

Cell Culture Optimization Cell culture performance has undergone dramatic improvements in productivity, achieving target *product quality attributes* (PQAs) and robustness. These advances have been achieved through cell line engineering to increase productivity while sustaining or improving PQAs, medium development, optimized cell culture conditions, and improvements to seed train and production bioreactor technology and operation. One advantage of the latter is that these improvements can often be broadly applied, as compared to the former strategies which may have limited applicability to specific cell lines.

Due to the number of cell variants and conditions that need to be screened, performing cell line development and process optimization manually is no longer economically feasible. The advent of small-scale, automated high-throughput development tools has made cell line and medium optimization more efficient, thereby reducing resource requirements and development time. These range from flow cytometry, to nonstirred microwell plates, to robotically operated, parallel stirred microscale, presterilized single-use bioreactor systems such as ambr™ which span a broad range of scales. An evaluation of the ambr™ system has been performed [Moses, *Adv. Biosci. Biotechnol.* **3:** 918–927 (2012)] which concluded that these systems can provide consistent performance across scales in the ambr™ family, although the 250-mL bioreactor was not part of that study. Techniques and technology used to optimize cell lines have been thoroughly reviewed [Kuystermans, *Animal Cell Culture,* Springer, 2015, pp. 327–372].

Seed Train Optimization Mammalian cell culture seed train operations and productivity have been improved in some cases by application of gamma-irradiated, presterilized single-use cryobags in lieu of autoclaved glass vials for the preservation of *working cell banks* (WCBs) [Heidemann, *Biotech. Prog.* **26:** 1154–1163 (2010)]. Each cell line using this technology should be assessed for potential sensitivity to DMSO cryo-protectant exposure times prior to the freezing step. Cryobags can hold up to 100 mL compared to typical 1-mL WCB vials, thereby increasing the number of cells used to initiate the seed train process. This shortens the number of seed train steps by eliminating the initial lower volume steps that are required when using a 1-mL vial to initiate the seed train. The gain in productivity resulting from starting with increased cell numbers when using a cryobag versus a traditional vial can be determined by comparing the times required to execute the seed train process from the beginning of the seed train to inoculation of the production bioreactor, i.e., seed train elapsed time t_{st}. This time can be calculated from Eq. (20-18) given the initial number of cells contained in either the vial or cryobag X_0, the cell doubling time t_D (typically 24 h), and the required number of cells needed to inoculate the production bioreactor X_{innoc} (Chuck, Erythropoietins, Erythropoietic Factors, and Erythropoiesis, Birkhauser Verlag, 2009, p. 94)

$$t_{st} = n \times t_D = \ln(X_{\text{innoc}}/\,X_0)/\ln 2 \qquad (20\text{-}18)$$

where n is the number of required cell doublings. Using the example of a 10,000-L bioreactor with an initial working volume of 7500 L, a required initial production reactor inoculation cell density of 1×10^6 cells/mL, and a cell density of 1×10^7 cells/mL in the vial or cryobag, $X_{\text{innoc}} = 7.5 \times 10^{12}$ cells. This results in $t_{st} = 19.5$ days for the standard 1-mL vial-initiated seed train. By using a cryobag to initiate the seed train, the elapsed time is reduced to $t_{st} = 12.8$ days, a 35 percent improvement.

In addition to reducing the seed train duration, cryobags may simplify seed train operations. Cryobags have sufficient cell numbers that the seed train can be initiated by transferring the cells directly into a standard small-scale bioreactor. This eliminates the need to use the early-stage spinner or shake flasks which require manual aseptic operations in a biological safety cabinet. An additional advantage is that the transfer from the cryobag can be carried out as a closed process in a standard upstream production environment. This is in contrast to spinner and shake flask seed train steps which need incubators to provide the required cell growth environment, and biological safety cabinets to conduct related cell transfer operations. Due to regulatory requirements, these cabinets need to be housed in specially constructed clean rooms designed to adhere to stringent requirements. This results in significant initial and ongoing facility, equipment, and operational costs. Elimination of these open operations through adoption of closed processing reduces costs as well as ongoing facility environmental monitoring and quality oversight. Replacing manual aseptic operations with closed processing also reduces the likelihood of seed train contamination.

Further seed train process optimization leading to overall facility productivity gains is possible by the application of a perfusion process to select seed train steps. A perfusion process continually adds fresh medium and removes waste product while retaining cells. Perfusion is described in greater detail below. It was shown that using a perfusion process for the bioreactor generating the WCB, the cell density being aliquoted into vials or cryobags could be significantly increased, further streamlining the seed train [Heidemann, *Cytotech.* **38:** 99–108 (2002)].

Although promising, these advances in generating high-cell-mass cryobags has been slow to be widely adopted in biopharmaceutical production. At least one European facility is using this approach, including the use of single-use, closed processing (Ribault, 2013 *AIChE Annual Meeting Conf. Proceedings,* presentation 548a, 2013). The slow adoption may be related to the existing infrastructure and procedures in established biotech companies, and the need to maintain two GMP cell banking systems in order to adapt to higher-productivity options while maintaining legacy products.

Perfusion processing to increase cell density has also been applied to the seed train step used to inoculate the production bioreactor. This step is commonly designated as the $N - 1$ bioreactor, with the production reactor being the N bioreactor. Using *alternating tangential flow* (ATF) membrane units to retain the cells (Fig. 20-27), the perfused $N - 1$ stage was able to achieve a cell density > 40×10^6 cells mL^{-1} [Yang, *Biotech. Prog.* **30:** 616–625 (2014)] using a process optimized to limit total perfused medium volume. It was reported that the cells remained in the exponential growth phase throughout the $N - 1$ process and the production reactor duration was reduced by 30 percent while achieving product quality comparable to the low-seed process. Using a high seed density to inoculate the production reactor reduces the initial unproductive period of time where cell mass is being generated and protein productivity is low. Since production bioreactors are the upstream bottleneck for mammalian cell culture manufacturing, reducing the duration of this step leads to an improvement in volumetric and facility productivity, provided a new bottleneck is not created in downstream harvest and purification operations. Adoption of this technology may be limited for existing facilities and processes due to the added complexity and space. This includes ATF sterilization, space required for additional medium storage and the ATF unit, and modifications to the $N - 1$ seed bioreactor if its design restricts the gas flows needed to support the higher cell densities.

Production Operating Modes Mammalian cell culture has predominantly used batch and fed-batch bioreactors. More recently perfusion processes that operate as high-density cell cultures and/or longer-duration perfusion-enabled continuous upstream processes are being considered or implemented to boost productivity. In a batch process, all the medium and the nutrients are added to the bioreactor at the beginning of a run, which is then inoculated with cells from the $N - 1$ stage. As shown in Fig. 20-25, the cell density begins to increase, consuming nutrients until a maximum viable cell density is attained, followed by a drop in productivity and cell viability as the nutrients are further consumed and waste products accumulate. The run is terminated and the product is recovered through harvest operations when the bioreactor productivity becomes uneconomical, or if unwanted product characteristics begin to emerge. This mode of operation, while simple, has a number of drawbacks. Since all the nutrients, such as the glucose needed for cell growth, cell maintenance, and product formation, are added at the start of a run, their concentration may be limited by either solubility limits or if higher component concentrations negatively affect cell growth or product formation. These limitations result in shorter run durations and low product concentrations. In addition, it has become clear that high and variable glucose concentrations can affect glycosylated protein product quality attributes. At the end of each run, the bioreactor must be cleaned and resterilized, resulting in facility downtime and loss of productivity. Even with these limitations, batch processes have been used to successfully produce many biopharmaceuticals.

To overcome batch process limitations, fed-batch processes have been developed. In this mode batch medium is added at the start of the run. This initial batch medium volume will be less than in the batch process to allow for subsequent additions. It also has lower nutrient concentrations, including glucose, compared to the batch process. Fresh medium is periodically added to the cell culture as bolus feeds throughout the run. These feeds can be tailored to optimize production throughout the run, and will often have more concentrated glucose and less salt to support cellular metabolism while minimizing osmolality. It has also become more common to initiate a shift to a lower temperature partway through the run, usually after the maximum growth phase. This sustains protein production while limiting growth, reducing the need for additional feeds. Fed-batch culture with a temperature shift enables the culture to be run for longer durations without exhausting essential nutrients compared to batch processes. This achieves higher product concentrations by maintaining higher viable cell densities through the course of the run, resulting in higher facility productivity. Periodic feeding also provides the flexibility to have different medium formulations for each bolus feed to further optimize the process. This can be used to control maximum nutrient concentration, such as glucose and other components, which may improve product characteristics and consistency. Due to the fed-batch processing advantages, it is more common for biopharmaceutical production than batch processing.

Feeding strategies for fed-batch processes can range from very simple time-based bolus additions based on historical data, to bolus, or continuous feed strategies based on culture conditions such as periodic VCD measurements, or direct measurements of nutrient concentrations and process parameters. This can be done through manual or automated sampling combined with off-line or real-time on-line measurements. The feeds themselves can be manually performed or can be fully automated control loops using multivariate, model predictive control schemes. These more complex and expensive strategies are normally reserved for cases where tight control is required to achieve consistent product quality objectives such as glycosylation, amino acid concentration targets to ensure correct incorporation

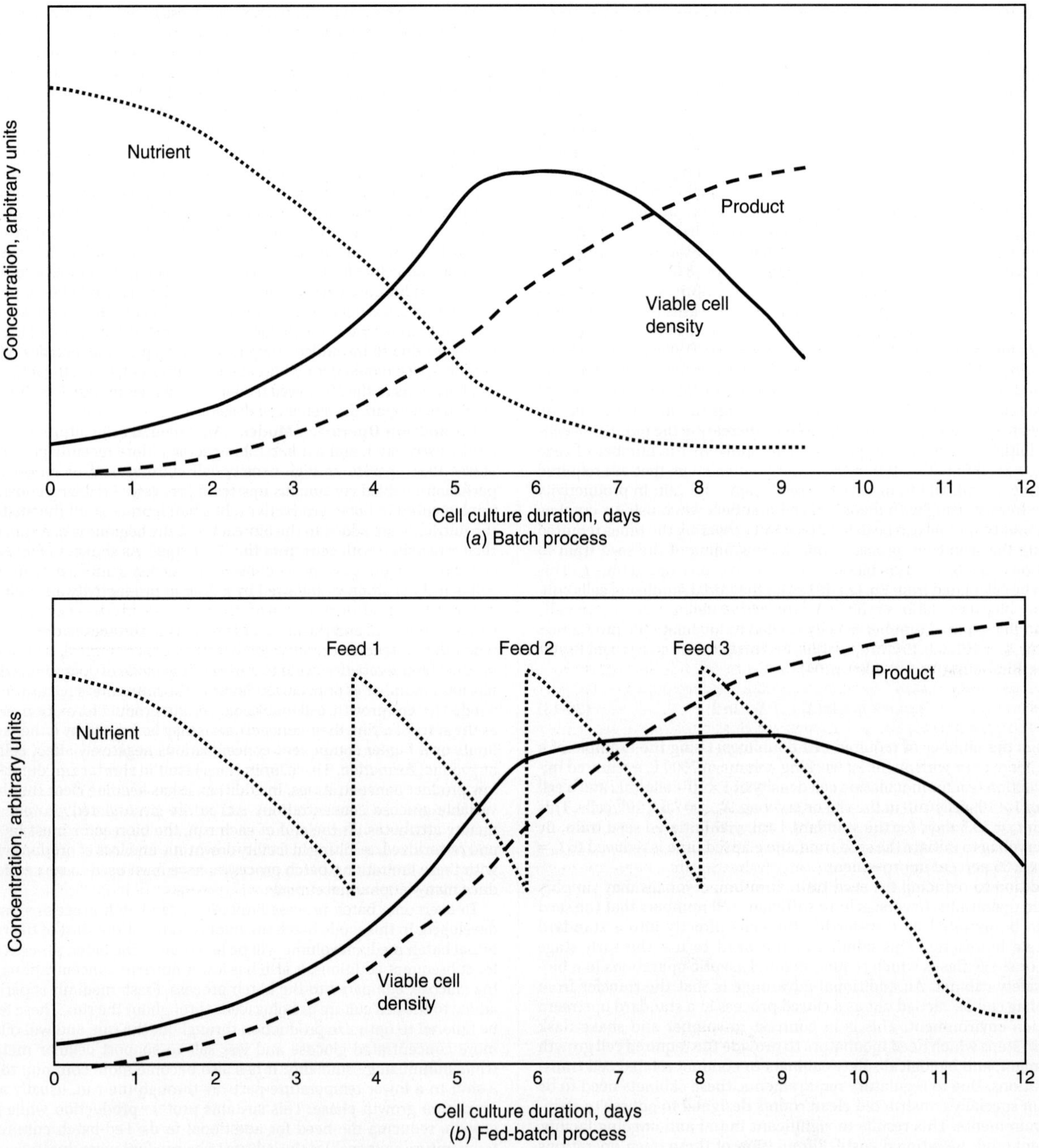

FIG. 20-25 Typical (*a*) batch and (*b*) fed-batch mammalian cell culture process trends.

into the protein, or to control deamination after the protein has been secreted from the cells.

Perfusion Culture and Process Intensification Perfusion bioreactors operate by continually adding fresh medium to the bioreactor while medium is constantly being withdrawn, maintaining a constant level in the bioreactor. Medium removed from the bioreactor can contain product, cellular debris, unused medium components, and cells unless some means to retain them in the bioreactor is put into place. Perfusion alleviates some of the shortcomings of batch and fed-batch processes since it provides a consistent environment without waste products accumulating. Since product can be continually removed from the bioreactor, perfusion has been used to produce unstable biotherapeutics and enzymes in increasing numbers (Fig. 20-26). One of the earliest applications of perfusion processing was for factor VIII production, an unstable protein [Kim, *J. Biosci. Bioeng.* **121:** 561–565 (2016)]. The advantages of perfusion processing to achieve higher productivities have led to it being applied to stable products as previous technical challenges have been overcome.

To achieve a commercially viable perfusion process, cells need to be either retained in the bioreactor or recycled from the effluent stream. Operating as a chemostat, where cells are not retained in the bioreactor by some means, the perfusion rate would be limited by the maximum cell growth rate; otherwise all the cells will eventually be washed out. This limitation is overcome by adding a device to retain the cells, or to concentrate the cells and recycle them back into the bioreactor. The cell retention device is critical to implementing a perfusion process since it allows the perfusion rate to be adjusted to optimize cell culture productivity, medium cost, and product quality.

Biopharmaceutical companies were slow to widely adopt perfusion processes as a means to achieve high cell density, high-productivity cultures due to shortcomings in available suspended cell retention options. Cell recycling from spent medium has been accomplished using inclined plates that employ gravity for cell separation, hydrocyclones, or more recently acoustic resonance settlers that induce cell agglomeration, leading to faster cell settling and more efficient cell recycle. These devices have been limited

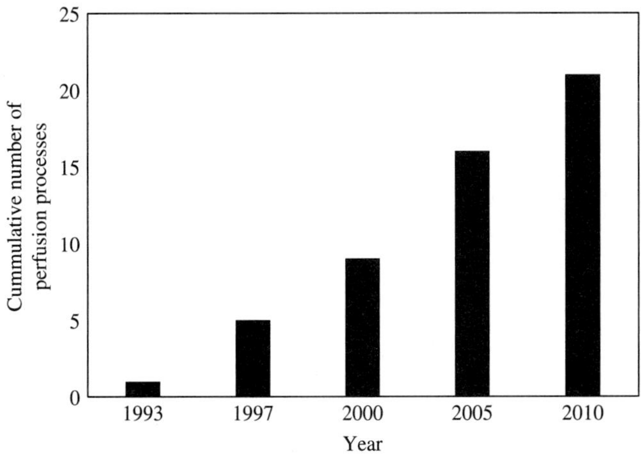

FIG. 20-26 Cumulative number of biopharmaceuticals produced using continuous perfusion bioreactors. [Data from Le, *Chemical Eng. Prog.* **111:** 34 (2015).]

by the maximum feasible cell settling rates, recycle capacity, and limited scalability. Some of these devices, however, are more efficient for separating cells grown on microcarriers due to their higher densities, or may see application at small bioreactor scales.

With any cell retention device, the time spent outside of the controlled bioreactor environment needs to be minimized to avoid oxygen depletion and metabolic shifts. This is especially important as these devices significantly increase cell densities above the bioreactor cell densities. Devices such as spin filters have been evaluated as in situ cell retention devices to eliminate this concern, but they suffer from fouling, complexity, and unpredictable scaling.

Tangential flow microfiltration has been used for continuous enzyme production processes for some time. Application to perfusion bioreactors has been limited due to membrane fouling, cell damage related to recirculation through the membranes, and damage from the pump used to recirculate the cell culture fluid [Stressmann, *Biotech. Prog.* **24:** 890–897 (2008)]. Alternating tangential flow membrane filtration has proved to be more successful for perfusion processes than recirculating TFF [Karst, *Biochem. Eng. J.* **110:** 17–26 (2016)]. ATF operates by repeatedly drawing fluid out of and back into the bioreactor through a hollow fiber membrane module. This is done using a flexible diaphragm located below the membrane module, acting as a pump, as shown in Fig. 20-27. This action creates fluid flow inside the hollow fibers (the lumen) tangential to the membrane surface; at the same time fluid is drawn across the membrane as the permeate stream. A pump connected to the permeate side creates the driving force to draw fluid through the membrane. This pump also acts to control the permeate flow rate to achieve the desired bioreactor perfusion rate. ATF's success for cell culture perfusion has been attributed to two mechanisms. One is the tangential flow through the lumen to limit the buildup of a fouling layer on the inner membrane surface. The second is a self-cleaning mechanism where the flow going through the membrane pores is periodically reversed. This can occur when fluid is being drawn into the ATF unit from the bioreactor. Although few data have appeared in the literature to specifically demonstrate this mechanism, it is thought to reduce fouling by limiting the buildup of cells, cell debris, medium components such as antifoam on the membrane surface, or adsorption onto membrane surfaces including pores [Kelly, *Biotech. Prog.* **30:** 1291–1300 (2014)]. While reasonable for larger membrane pore sizes, it can be expected that this mechanism will be less effective at smaller nominal pore sizes due to increased resistance to backflow across the membrane (Craig Robinson, personal communication, 2014). Other fouling mechanisms include adsorption of medium components or DNA onto the membrane [Kelly, *Biotech. Prog.* **30:** 1291–1300 (2014)].

ATF membranes are available in nominal pore sizes ranging from small pore sizes (ultrafiltration) to more open, larger pore membranes (microfiltration) depending on the process objectives (Fig. 20-28). If the product is to be continually removed from the bioreactor, as is the case for unstable proteins or when operating as a continuous production process, then more open microfiltration membranes are used. These are typically 0.1- to 1-µm nominal pore sizes to retain cells, yet allowing protein product to pass through the membranes. If the product is to be retained in the bioreactor for later harvest, small-pore-size ultrafiltration membranes are required. In either case, ATF membrane studies such as flux excursion and permeability studies performed over the course of an ATF process are required to identify the best membrane choice and potential fouling mechanisms. The basic principles for characterizing membrane process performance are described in van Reis [*Biotech. Bioeng.* **55:** 736–746 (1997)] and in Sec. 20 of *Perry's Chemical Engineer's Handbook*, 8th edition.

Following inoculation of the perfusion bioreactor, there is an initial growth period where the cell density increases over time until the target cell density is reached. At that point a cell bleed is initiated to remove excess cells to maintain the target VCD as the cells continue to grow. This can be accomplished by either periodic off-line cell counts (typically done by manually withdrawing a certain culture volume), or through application of in situ instrumentation such as capacitance probes to determine cell density integrated with an automated cell bleed control loop. Temperature shifts

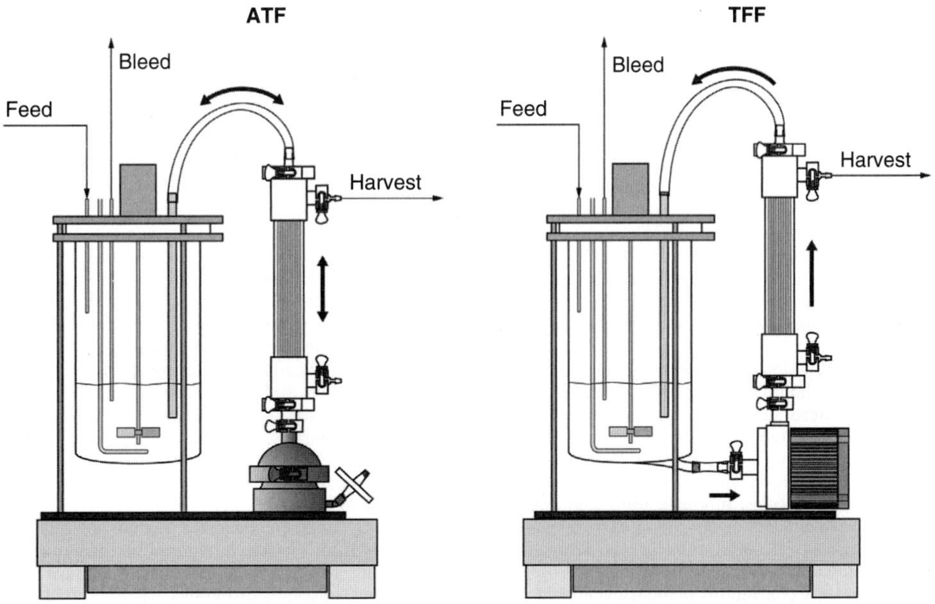

FIG. 20-27 An alternating tangential flow hollow fiber device for bioreactor perfusion and cell retention, along with a recirculating tangential flow filter perfusion setup. [Reprinted from Karst, D., Serra, E., Villiger, T., Soos, M., and Morbidelli, M., "Characterization and Comparison of ATF and TFF in Stirred Bioreactors," *Biochem. Eng. J.* **110:** 19 (2016). Reproduced with permission of Elsevier, © 2016 Elsevier B.V. All rights reserved.]

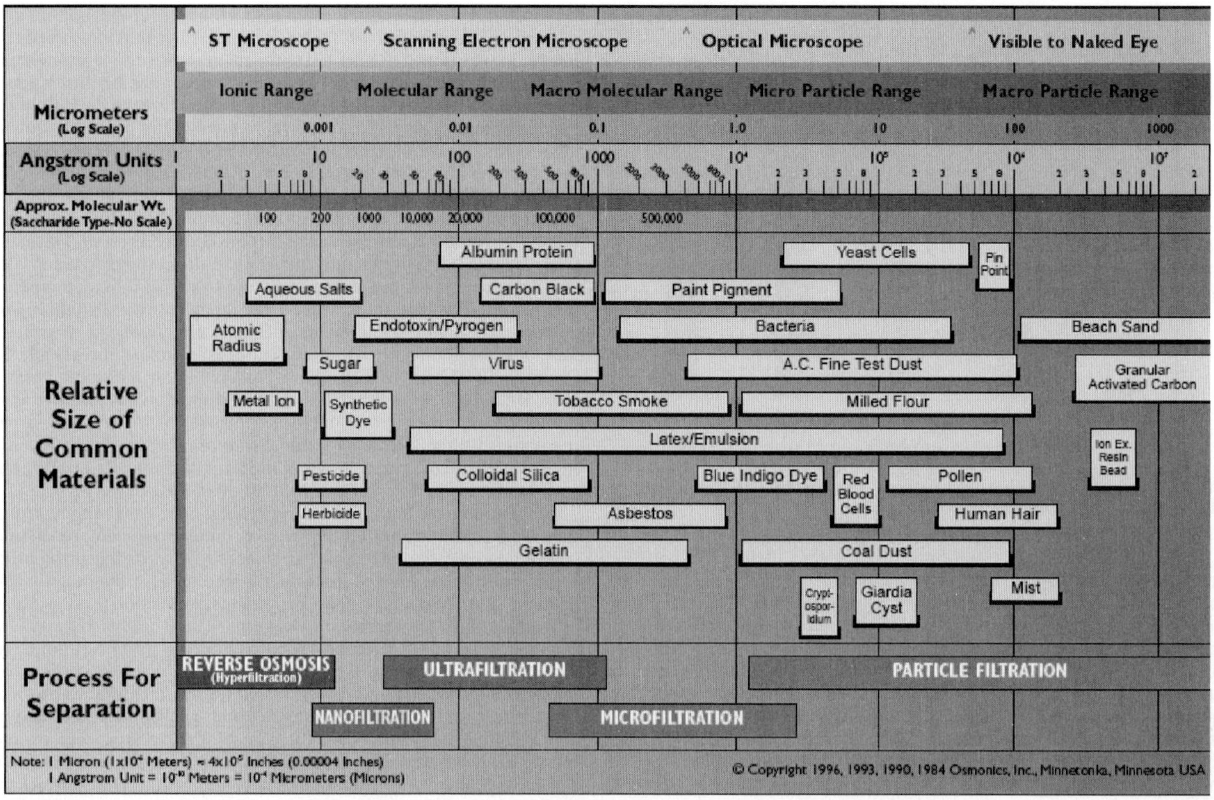

FIG. 20-28 Common membrane pore and particle sizes. (Reprinted by permission of Osmonics, Minnetonka, Minn. Copyright 1996.)

can also be used to maximize protein production while managing growth rate and reducing perfusion rates.

Cell retention coupled with constant optimal protein production conditions leads to higher productivity as a direct result of higher cell density, higher viability, and more consistent product quality throughout the course of a run, in contrast to batch and fed-batch operations with their constantly changing conditions. Applying these techniques cell densities of up to 80 to 100 × 10^{-6} cells mL^{-1} for 2 to 3 months have been reported [Vernardis, *Metabolic Eng.* **19:** 1–9 (2013)]. Extending the process duration reduces the time and resources needed to turn a bioreactor around between production runs, boosting facility productivity. Extended processing brings additional challenges, however, including the need to assess potential cellular genetic drift over the extended process times and the additional time required to develop and demonstrate a perfusion process, which is often on the critical path to moving a new molecular entity into clinical studies. Other potential limitations involve the concern that high-cell-density cultures (up to100 × 10^{-6} cells mL^{-1} with productivities up to 3.5 g L^{-1} day^{-1}) can result in higher cell culture viscosities. This could result in the loss of ATF cross-flow, effectively limiting the maximum cell density [Clincke, *Biotech. Prog.* **29:** 768–777 (2013)]. An additional design consideration is to be able to change an ATF unit during a run in the event of a failure. The bioreactor and its connection to the ATF need to be set up to allow the filtration unit to be changed without compromising sterility.

Process intensification and optimization using perfusion or other means lead to higher volumetric productivity, allowing demand to be met by smaller bioreactors. This reduction in bioreactor footprint will be partially offset by the need for additional perfusion medium storage requirements. Smaller bioreactors have other advantages, such as minimizing bioreactor scaling factors, facilitating scale-up from process development scale bioreactors, and process transfers to commercial scales. Intensification also enables production demand to be met with smaller, lower-capital-cost, single-use bioreactors, rather than larger 10,000- to 15,000-L stainless bioreactors commonly used for batch and fed-batch biopharmaceutical production.

Adoption of small, high-cell-density continuous biopharmaceutical production technology will be gradual. Most companies have large stainless-steel bioreactors already installed, validated, and in many cases fully depreciated. It is expected from an economic perspective that this existing infrastructure will continue to be used to produce many current and future products. High-cell-density perfusion bioreactors are more likely to be considered as an option as new facilities come on-line. From a regulatory perspective, the FDA has recognized that biopharmaceutical production can benefit from the flexibility and cost benefits that may be realized by adopting continuous, high-productivity cell culture manufacturing. The FDA has been working with the biopharmaceutical industry to develop guidance and a regulatory framework needed for widespread adoption of this technology [Woodcock, "Modernizing Pharm. Manuf.—Continuous Manuf. as a Key Enabler," *International Symp. on Cont. Manuf. of Pharm.*, May 20–21, United States, http://iscmp.mit.edu/sites/default/files/documents/ISCMP2014_Keynote_Slides.pdf (2014)]. Until downstream harvest and purification can be adapted to become more continuous processes, the full benefit of continuous upstream processing may not be realized.

SINGLE USE BIOREACTORS

The biopharmaceutical industry is rapidly shifting from traditional materials of construction, such as glass and stainless steel, to alternatives constructed from engineered, single-use plastic films and components. These come presterilized and ready for use, after which they are disposed of or recycled. Small-scale flasks and bioreactors used in seed trains and process development were the first single-use (SU) applications. As single-use technology developed, larger and more complex vessels could be fabricated, expanding single-use applications to medium formulation mixers, medium storage, and production bioreactors, commonly called *single-use bioreactors* (SUBs). The maximum scale is limited by physical characteristics of the plastics, ability to set up increasingly larger SU vessels, and other factors. Where the manufacturing scale is a fit with single-use alternatives, adoption has been driven by speed, simplicity, flexibility, and cost considerations.

A SUB is constructed from a flexible polymer bag, often comprised of a multilayer film to achieve required performance, as shown in Table 20-15. These films have to be engineered to achieve the right physicochemical properties. These include biological compatibility, tensile strength, puncture resistance, flexibility, transportability, gas and water vapor permeability, pH stability, leachables and extractables, and product or medium adsorption. International standards have been developed for most of these properties by recognized organizations including United States Pharmacopoeia (USP),

TABLE 20-15 Single-Use Bioreactor Film* Construction and Characteristics

Layer†	Thickness, mils	Characteristics
Outermost layer—polyester	0.8	Flexible, provides strength. Permeable to gases and water vapor. Requires additional layers if application requires long-term storage of aqueous solutions.
Tie layer	0.9	Adhesive layer joining materials that do not adhere.
Core layer—ethylene vinyl alcohol (EVOH)	1.0	Gas barrier, moderate water vapor properties.
Tie layer	0.9	Adhesive layer joining materials that do not adhere.
Innermost layer—polyethylene (PE)	10.4	Cleaner material with lower extractable/leachable levels compared to EVA. Inert to a wide range of chemicals. Good water barrier properties.

*Film consists of multiple layers, each serving a different purpose.
†Film layers are listed starting from outermost layer to the innermost layer that contacts the cell culture fluid. Layers shown are for Thermo Scientific CX5-14 film.

International Organization for Standardization (ISO), ASTM International, and European Pharmacopoeia (EP).

The shift to SUBs brings with it new concerns related to polymer chemistry and interactions with cell culture fluids. It was found that in some cases medium contacting films during storage, or during use in SUBs, would take up compounds from the polymer film. Such compounds are referred to as *leachables*. A group of biopharmaceutical companies jointly developed a standard method that allowed direct comparison of the relative impact, if any, of SU films on cell growth [Horvath, *Biopharm. Int.* **23:** 34–41 (2013)]. This test showed that in some cases, but not all, medium contacting films could impact cell growth across a number of cell lines. This ultimately led some suppliers to utilize similar methods to screen different film formulations for possible cell cytotoxic effects [Jurkiewicz, *Biotech. Prog.* **30:** 1171–1178 (2014)].

The simplest SUB is the Wave bioreactor that consists of a two-dimensional cell culture bag placed on a platform that rocks back and forth (Fig. 20-10). This motion generates mixing and gas mass transfer sufficient to support typical seed train cell densities by varying the frequency and angle of rocking. Temperature is controlled using small electrical heaters. Attached tubing accommodates gas flow, small additions, harvest, and sampling.

To construct larger, three-dimensional storage vessels and bioreactors, film sections are cut and welded together to form what will become a fully closed, single-use vessel. All functional components needed in a bioreactor, such as connection ports, tubing for gas supply and exhaust, agitation, medium additions, harvest, probe and instrument ports for monitoring and sampling, are added by welding them to the film, much as done when adding ports to a stainless vessel. The open end of the attached single-use tubing has to be closed prior to sterilization by either seal-welding or using single-use aseptic connectors. The entire assembly is then carefully packaged both to eliminate transportation damage and for ease of deployment at the end-user's site. Once packaged, the SUB is sterilized using a validated gamma irradiation cycle. End users should expect vendors to validate the SUB bag fabrication methods and to perform periodic tests to ensure the fabrication process continues to meet specifications. Integrity test units have been developed to provide end users with a final check of the installed assembly. These can detect significant breaches of the sterile boundary before the cell culture medium is added.

Small single-use SUBs such as the Wave are self-supporting, but 50-L SUBs and greater use an open-top stainless-steel shell to support the flexible SUB liner. The shell is typically jacketed to provide heat transfer and has openings to accommodate ports for instrumentation, sampling, and lines for feeds and gas supply exhaust. SUBs can accommodate a range of instrumentation, from single-use optical oxygen sensors to standard pH probes that can be separately autoclaved and aseptically inserted into the SUB. As monitoring technology develops, it is relatively easy to accommodate changes into a redesigned bag. Most SUB designs use an offset agitator to promote mixing in the absence of baffles. SUB capabilities are being extended through the development of higher-power mixers and internal baffles leading to SUBs capable of supporting fermentation processes.

The maximum SUB size has been limited by polymer film strength, ease of handling and setup, as well as an economic breakpoint beyond which traditional stainless-steel bioreactors become more cost-effective, especially for installations in existing facilities. Even with this limitation, operated as high-cell-density perfusion bioreactors, smaller SUB production rates rival those of many larger traditional stainless-steel bioreactors. A 2000-L SUB operated at high cell densities (80×10^6 cell-mL^{-1}) with cell-specific product productivity of 25 pg cell^{-1} day^{-1} can produce more than 1400 kg/yr. This is as much as a standard 15,000-L fed-batch bioreactor operating at 10×10^6 cells mL^{-1}. Using smaller SUBs in lieu of larger traditional stainless vessels results in a smaller facility footprint yet achieves the same productivity. As a result, facility construction will have shorter timelines and lower costs. Using SUBs lowers initial bioreactor capital cost, and associated cleaning and steam sterilization infrastructure is eliminated. This in turn simplifies start-up and validation requirements, further reducing costs.

FDA guidance has recognized that single-use technologies are available that can facilitate conformance with cGMP requirements and streamline product development. These include disposable equipment, components, and presterilized closed processing assemblies (Frank, PD2M Plenary paper 39d, AIChE Annual Meeting, 2013). Single-use technology not only reduces requirements for cleaning and sterilization but can also reduce facility environmental controls and monitoring requirements. Connections from SUBs to other SU equipment such as medium storage vessels and gas supplies can be made in the open, using aseptic welders or single-use sterile connectors. This enables biopharmaceuticals to be made in less complex facilities where the need for stringent air filtration and environmental monitoring is reduced compared to facilities using stainless-steel reactors which are periodically opened to the manufacturing room environment.

Personalized medicines include cell-based immunotherapies where a patient's own T-cells are genetically modified to specifically target their cancer. Individual patient centered therapeutics such as these require significantly different approaches and technologies. These include how starting materials are sourced (harvesting a patient's T-cells), tracked, produced (genetic modifications), stored (cryopreservation), shipped, and administered back to the patient. Processing this new immunotherapy modality is reliant on single-use technologies to a large degree because they provide presterilized closed processing while largely eliminating cross-contamination risk, resulting in a simplified facility that has the flexibility and agility to adapt to evolving production demands and technology advances. SU technologies used in immunotherapy processing include small scale bioreactors, centrifugal cell separation, final cell product storage, and transportation back to the patient [Levine, *Mol. Ther. Methods. Clin. Dev.* **4:** 92–100 (2017); Wang, *Mol. Ther. Oncolytics* **3:** 1–7 (2016)].

Adoption of SUBs and related single-use technology has been gradual when companies have an installed base of stainless-steel bioreactors, but has become common in new facilities. Along with the benefits accrued by using SUBs, practices need to be put into place to provide oversight of the single-use components and equipment supply chain, similar to what has been done for other raw materials. This includes a change management system in partnership with the vendor. As the single-use industry matures, it remains to be seen to what degree suppliers focus on providing defined product offerings, or retain flexibility to customize and prototype SUBs to meet specific bioreactor requirements as cell lines and processes continue to advance.

MONITORING AND CONTROLLING BIOREACTORS

Process Parameter Monitoring and Control Bioreactor instrumentation is used to measure process parameters affecting process performance and metabolic activity. Table 20-16 summarizes many of the common bioreactor monitoring instruments. In addition to in situ instruments, they can be categorized as at-line near real-time measurements, or as off-line instrumentation. Upstream control strategies rely on instrument data to maintain process parameters within target ranges. This ensures that the bioreactor produces quality product, and that it consistently achieves process requirements and economic targets including titer, yield, and productivity.

Instruments are used to continuously monitor process parameters including pH, DO, osmolality, temperature, and agitation speed. These measurements are integrated into automated control loops that maintain

TABLE 20-16 Typical Process Parameters and Quality Attributes for Animal Cell Cultures. Additional Parameters May Be Important for Other Organisms or Applications

Parameter	Typical monitoring device	Typical monitoring method
	Physicochemical parameters	
Temperature	Resistance thermometer (PT 100 probe)	In situ
pH value	Electrochemical Ingold pH glass electrode/optical sensor	In situ
Dissolved oxygen (DO)	Amperometric Clark electrode/optical sensor	In situ
Pressure	Membrane sensor	In situ
Stirrer speed	Revolution counter	In situ (on stirrer motor)
Foam	Conductivity probe	In situ
Fluid addition/weight	Balance/load cell	In situ (on vessel)
Gas addition	Mass flow meters	In situ (gas lines)
DCO_2	DCO_2 probe (pH sensor in saturated bicarbonate)	In situ
Redox potential	Redox electrode	In situ
Osmolarity	Freezing point depression meter	Offline
	Physiological/biological quality attributes	
Biomass	Light scattering/absorbance probes	In situ
Cell volume	Packed cell volume (PCV) tube	Offline
Cell number and viability	Dye-based and dye-free methods	Offline, online, or in situ
Substrates and metabolites	Enzymatic, spectroscopic, and other methods	Offline or in situ
Lactate dehydroge- nase activity	Enzymatic/photometric methods	Offline
Product concentration	HPLC, photometric, spectroscopic, and other methods	Offline or in situ
Product quality	Various methods, e.g., SEC, IEC, LCMS, HPLC	Offline
Apoptosis	Flow cytometry	Offline
Metabolic state	Mass spectrometry (MS), nuclear magnetic resonance spectroscopy (NMR)	Offline
Transcriptomic state	Microarrays, RNA-seq (next-generation sequencers)	Offline

Reproduced from Schwamb, *Animal Cell Culture, Monitoring of Cell Culture*, Springer, 2015, p. 189, with permission of Springer.

the cell culture environment within well-defined ranges to achieve consistent cell growth, viability, dissolved gases, and product quality. Except for agitation speed, these are in situ, real-time measurements utilizing probes that directly monitor the cell culture fluid. Advantages include no lag time between sampling and sample analysis, and no measurement impact due to sample handling artifacts that can occur if samples need to be removed from the bioreactor for sample preparation prior to analysis using off-line instrumentation.

An example of a control loop utilizing in situ measurement is dissolved oxygen monitoring and control. This control loop automatically maintains the process within an acceptable DO range that has been identified through bench or micro-bioreactor scale-down studies using DOE and other statistical methods. It includes an in situ sensor that continuously monitors DO. This sensor is integrated with a transmitter that converts the sensor signal into a process value, which is then passed onto a supervisory control system. This automated controller compares the real-time process value to the target DO setpoint. It then adjusts sparge gases using air and oxygen mass flow controllers, which are the control elements. The automated controller is usually configured to adjust gas flow rates or sparge gas composition as needed to maintain the target DO setpoint. There are examples where agitation is also varied to adjust mass-transfer rates, although in most cases agitation remains constant over the course of a process.

Not all measurements can be integrated into automated control loops, however. Off-line measurements involve removing a sample from the bioreactor using aseptic techniques and taking it to instrumentation that may be located in another room for analysis. Glucose, VCD, viability, dissolved CO_2, amino acids, and other medium components are measured using off-line analysis, though on-line sensors are in development for most of these. When off-line measurements are outside acceptable target control ranges, adjustments have to be made by manually changing automation parameters, or by adjusting feed rates or composition to bring the process back into an acceptable state.

When considering cell culture monitoring options, the effects of cell density and metabolic rates on the measurement results must be taken into account. Samples removed from a bioreactor will change over time, especially with high cell density cultures. Time delays due to sampling, sample preparation, and finally running the sample measurement can significantly impact some measurements. Sample handling and analysis needs to be rapid and consistent. Off-line analysis is also limited by the periodic nature of such measurements. Off-line analysis is typically performed once or twice per day to manage resources and to minimize bioreactor contamination risk.

A developing approach is to integrate traditional off-line instrumentation with the bioreactor. This is termed at-line instrumentation and is able to perform process measurements in near real-time that in situ instruments are incapable of. At-line instruments can be fully automated measurement systems, although some still require operator manipulation. These instruments require samples to be periodically removed from the bioreactor and presented to instrumentation for analysis. Automated sampling systems have been designed to periodically take samples from the bioreactor and direct them to various at-line instruments. If needed, these systems can be designed to automatically remove cells from a sample, so a cell-free sample is provided to the instrumentation. A well-designed system will minimize time delays between sampling and measurement to avoid significant changes in dissolved gases, pH, and other parameters.

Unlike off-line instrumentation, the near real-time data that at-line analysis provides can be integrated into automated control loops. Because of the sophistication of the at-line instrumentation, these control loops can go beyond measuring and controlling simple process parameters, such as DO and pH, and can directly measure cell culture parameters such as glycosylation patterns that have been linked to critical product quality attributes. Once these attributes can be measured, complex control schemes coupling multivariate analysis with model predictive control can be implemented. Examples include modulation of antibody galactosylation through controlled feeding of uridine, manganese chloride, and galactose [Gramer, *Biotech. Bioeng.* **108:** 1591–1602 (2011)]. Other studies [Amand, *Biotech. Bioeng.* **111:** 1957–1970 (2014)] have shown that mAb glycan distribution can be predictably and reliably controlled by controlling various medium components. The ability to manipulate glycan structure is especially important for biosimilar development where it is necessary to closely match the innovator glycan structure. As sampling and monitoring technology continues to develop, the choice to implement complex active product quality attribute control (i.e., glycosylation), or to instead rely on consistently operating a process to reproducibly yield target product quality, will need to be considered on a case-by-case basis. The choice will likely depend on the maturity of the sampling and monitoring technology, the ability to predictably manipulate and control critical product quality attributes, and how reliably and reproducibly a process operated in a well-characterized design space yields the target product quality.

As very high-cell-density processes continue to be developed and requirements to closely control product quality attributes' real-time increases, the potential to deviate from target control tolerances by relying on off-line analysis or manual process adjustments increases. This is leading to the development of more advanced in situ instruments such as Raman spectroscopy that have the ability to perform multiple in situ measurements currently performed off-line. This enables implementation of real-time control loops while eliminating reliance on manual sampling, analysis, and, in many cases, manual data entry and calculations to determine bolus feed schedules and other set point adjustments. Achieving this automated measurement and control leads to a more robust bioprocess, and it is consistent with regulatory agency guidance for biopharmaceutical production. In situ Raman is relatively expensive, but by multiplexing a Raman instrument to monitor multiple bioreactors at a time, the cost per control element can be reduced.

Bioprocess Instrumentation Requirements In addition to typical instrumentation requirements, such as sensitivity, accuracy, stability, and ease of use, instrumentation used in biopharmaceutical production must be robust and suitable for a production environment. If it requires contact with the medium, it cannot compromise sterility. In situ probes used in stainless-steel bioreactors will need to be cleanable and withstand steam sterilization. Single-use sensors, such as optical DO and pH probes, need to be compatible with gamma irradiation that is used to sterilize single-use systems. Reusable probes, such as pH and conductivity, can be separately autoclaved and aseptically inserted into a SUB, using a single-use adapter such as Pall Corporation's KleenPac.

In a regulated application, the instrument has to be capable of being validated and calibrated. It should be stable for long durations to minimize calibration requirements, reducing calibration costs, record keeping, and process risk. If an instrument used to measure critical quality or process parameters is found to be out of tolerance, then any material produced relying on that instrument going back to the time of the last known in-tolerance calibration will need to be assessed for potential impact.

Biopharmaceutical process data generated and stored in instruments, control systems, and data historians are subject to additional regulatory requirements. These systems may be subject to CFR 21 Part 11 compliance in the United States and similar regulations in other countries. Included in these regulations are data integrity assessments to provide evidence that the electronic data generated and stored on a system are the original unaltered records. There are also security risk assessments to demonstrate that properly managed password protection is in place, and unauthorized personnel cannot access or alter systems covered under CFR 21 Part 11.

Additional information on cell culture monitoring and control within the context of regulatory agency expectations and industry best practices is provided below in the Product Quality Attribute Control subsection. More detailed instrumentation details can be found in the literature (Schamb, *Animal Cell Culture,* Springer, 2015, pp. 185–221).

DOWNSTREAM PROCESSING: SEPARATION AND PURIFICATION*

INTRODUCTION

Purification of bioproducts typically involves a long sequence of steps, and each step requires the use of one or more *unit operations,* such as filtration, precipitation, and chromatography. Each bioproduct has unique properties that differentiate it from other bioproducts. For bioseparation purposes, important properties include thermal stability, solubility, diffusivity, charge, isoelectric pH, molecular mass, hydrodynamic radius, hydrophobicity, reaction rate constants, and separation thermodynamics. A considerable amount of downstream process planning is based on the lability of most bioproducts. Temperature, pH, and concentration must be kept within specific limits to maintain product bioactivity.

Removal of solids (or recovery), isolation of product, purification, and polishing constitute a sequence of processing stages applied to nearly every product as it is prepared. These four stages make use of various unit operations. A *unit operation* is a single process and the equipment used therefor. These operations are applied in varying combinations to the four stages, approximately as indicated in Table 20-17.

The development and optimization of a bioseparation process involve the integration of the various unit operations in the most efficient way possible. The strategy for how to design a process to meet the desired objectives is presented along with a brief discussion of each unit operation. This is followed by a discussion of measures of process/product quality and the types of bioproducts that are on route to market. After this introduction section, more detailed treatments of key unit operations used in bioseparations are given that are not discussed elsewhere in this handbook.

Bioseparation Process Development The development of a process for the recovery and purification of a biological product is a creative activity that draws on the experience and imagination of the engineer. Attempts have been made to capture that experience on the computer in the form of expert systems [Petrides, *Comput. Chem. Eng.* **18:** S621 (1994)] and automate to some extent the process synthesis tasks. Experienced engineers rely heavily on certain *rules of thumb,* also known as *heuristics,* for putting together the skeleton of a recovery and purification process [Nfor et al., *J. Chem. Technol. Biotechnol.* **83:** 124 (2008)]. A few such heuristics follow:

1. Remove the most plentiful impurities first.
2. Remove the easiest-to-remove impurities first.
3. Make the most difficult and expensive separations last.
4. Select processes that make use of the greatest differences in the properties of the product and its impurities.
5. Select and sequence processes that exploit different separation driving forces.
6. Sequence unit operations such that the product of one can serve as the feedstock for the next without dilution or intermediate treatment.

Approaches to process development (also known as *process synthesis*) must be made with future scale-up in mind. A biotechnology product production process is a set of sequential unit operations that produces a final product able to meet defined levels of quality, yield, and cost. The following criteria should therefore be used in evaluating and developing a bioseparation process:

- Product purity
- Cost of production as related to yield
- Scalability
- Reproducibility and ease of implementation
- Robustness with respect to process stream variables

Examples of the integration of unit operations for the efficient synthesis of bioseparation processes are presented later in the subsection Bioprocess Design and Economics.

TABLE 20-17 Objectives and Typical Unit Operations of the Four Stages in Bioseparations

Stage	Objective(s)	Typical unit operations
Separation of insolubles	Remove or collect cells, cell debris, or other particulates. Reduce volume (depends on unit operation).	Filtration, sedimentation, extraction, adsorption
Isolation of product	Remove materials having properties widely different from those desired in product. Reduce volume (depends on unit operation).	Extraction, adsorption, ultrafiltration, precipitation, affinity methods
Purification	Remove remaining impurities, which typically are similar to the desired product in chemical functionality and physical properties.	Chromatography, affinity methods, crystallization, fractional precipitation
Polishing	Remove liquids. Convert the product to crystalline or other solid or concentrated form (not always possible).	Drying, crystallization

Reproduced from *Bioseparations Science and Engineering,* 2d. ed., by Roger G. Harrison, Paul W. Todd, Scott R. Rudge, and Demetri P. Petrides (2015). By Permission of Oxford University Press.

*Portions of this section have been reproduced from *Bioseparations Science and Engineering,* 2d ed., by Roger G. Harrison, Paul W. Todd, Scott R. Rudge, and Demetri P. Petrides (2015). By Permission of Oxford University Press.

Figure 20-29 provides a generalized structure for creating an initial block diagram representation of a product recovery process. For each product category (intracellular or extracellular), several branches exist in the main pathway. Selection among the branches and alternative unit operations is based on the properties of the product, the properties of the impurities, and the properties of the producing microorganisms, cells, or tissues. Bioprocess synthesis thus consists of sequencing steps according to the six heuristics above and the structure of Fig. 20-29. The majority of bioprocesses, especially those employed in the production of high-value and low-volume products, operate in batch mode. Conversely, continuous bioseparation processes are utilized in the production of commodity biochemicals, such as organic acids and biofuels.

Primary Recovery Stage Primary recovery encompasses the first steps of downstream processing where some purification and broth volume reduction occurs. Primary recovery includes both the solids separation stage and parts of the product isolation stages shown in Table 20-17. According to Fig. 20-29, the selection of the first step depends on whether the product is intracellular (remains inside the microorganism or eukaryotic cell after its expression) or extracellular (is secreted into the solution). Almost all low-molecular-weight bioproducts are extracellular, as are many that have a high molecular weight. Recovery and purification are easier for these bioproducts than for intracellular products because of the lower amount of impurities present. Most recombinant eukaryotic proteins produced by prokaryotic microorganisms are intracellular products. These

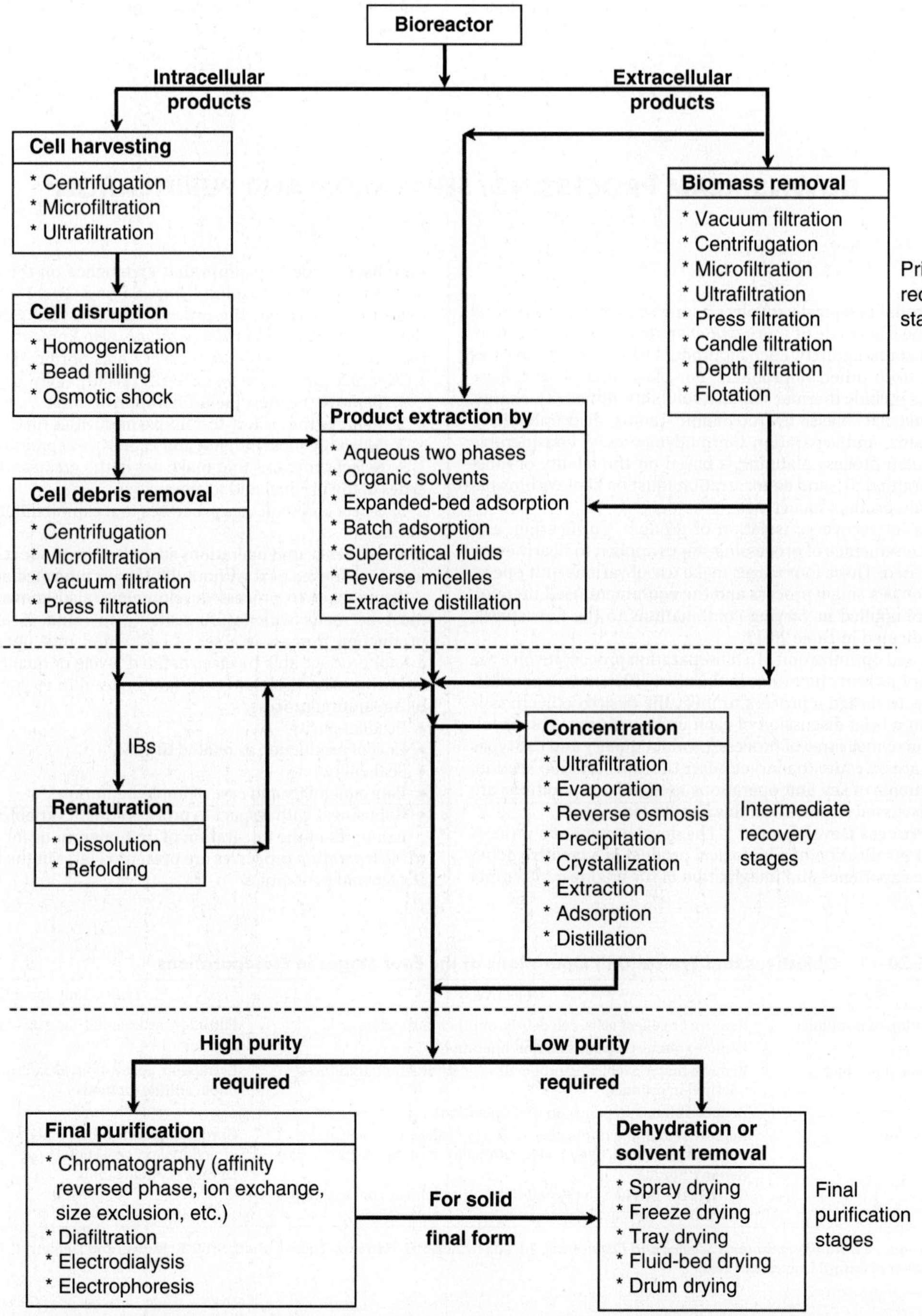

FIG. 20-29 Generalized block diagram of downstream processing. [Petrides et al., in Shuler (ed.), *Chemical Engineering Problems in Biotechnology,* American Institute of Chemical Engineers, 1989, p. 351.]

proteins accumulate inside the host cell in either native or denatured form; the denatured intracellular products often form insoluble *inclusion bodies* (IBs). A brief review of the most common primary recovery steps follows, and various rationales for unit operation selection are included. The human insulin process analyzed in this section provides additional information on the recovery and purification of intracellular products.

The first purification step for intracellular products is cell harvesting. Removal of the extracellular liquid is in agreement with the first general heuristic: *Remove the most plentiful impurities first.*

As seen in Fig. 20-29, centrifugation and membrane filtration (either microfiltration or ultrafiltration) are the primary techniques used for large-scale cell harvesting. Centrifugation has advantages for large, sturdy, and dense microorganisms (diameter >2 μm and density >1.03 g cm^{-3}), and it avoids the fouling that can occur with membrane filtration. For instance, centrifugation is very efficient for harvesting yeast. Centrifugation can lead to cell disruption (releasing additional impurities for extracellular products) with fragile cells, such as mammalian cells in suspension. Filtration has advantages for harvesting small, fragile, and light cells. Filtration operates with much lower shear rates compared to centrifugation. Another advantage of filtration is its superior product recovery. Cell loss and/or supernatant loss during centrifugation can be as high as 50 percent and is rarely less than 5 percent. However, with membrane filtration, essentially all cells are recovered unless there is cell disruption (lysis) or unless the membranes fail.

The second step for intracellular products is usually cell disruption, which serves to break open the host cells and release the intracellular product. Disruption of bacteria and yeast is carried out either by high-pressure homogenizers or bead mills [Kula and Horst, *Biotechnol. Prog.* **3:** 31 (1987)]. For large capacities (several cubic meters per hour) only high-pressure homogenizers are practical. Osmotic shock is often used for release of periplasmic products that accumulate between the cell membrane and the cell wall.

Prior to disruption, the concentrate is often diluted (to 5 to 10 percent by weight) with a "lysis buffer" to create conditions that minimize product denaturation upon release from the cell and maximize product solubility. For hard-to-disrupt microorganisms, multiple homogenization passes at 500 to 1000 bar are required. Multiple passes are also required if the product forms inclusion bodies. This allows the IBs to be released and breaks the cell debris into very small particles, facilitating the separation of IBs from cell debris further downstream. Some soluble product degradation often occurs during cell disruption as a result of high shear at interfaces and oxidation.

The cell debris generated by cell disruption is usually removed by centrifugation or microfiltration. Other options include rotary vacuum filtration, press filtration, depth filtration, extraction, and expanded-bed adsorption (EBA).

When the product is soluble, it is recovered during cell debris removal either in the light phase of a centrifuge or in the permeate stream of a filter. Centrifuges are only able to efficiently separate fairly large particles of cell debris (>0.5-μm Stokes diameter). Therefore, when a centrifuge is used for cell debris removal, a polishing filtration step must follow the centrifugation in order to remove small debris particles that might otherwise cause severe problems in processes downstream, such as chromatography. Filters of various types (e.g., depth, press, candle, rotary vacuum, membrane microfilters) can be used for polishing. Alternatively, these filters can be used for cell debris removal with no preceding centrifugation step. It is very difficult to predict a priori which filter will perform best for a specific product, so lab and pilot-scale testing are typically used to make this decision. In addition, when microfilters are used for cell debris removal, some degree of diafiltration is required to achieve an acceptable product recovery yield.

When the product is insoluble and forms inclusion bodies, it must first be separated from the cell debris particles, then dissolved and refolded (see insulin example later in this section for additional information on the subject). Fortunately, inclusion bodies usually have a large diameter (0.3 to 1.0 μm) and high density (1.3 to 1.5 g cm^{-3}) [Taylor et al., *Bio/Technol.* **4:** 553 (1986)] and can be separated from cell debris with a disk-stack centrifuge. The inclusion bodies are recovered in the heavy phase of the centrifuge, while most cell debris particles remain in the light phase. The heavy phase is usually resuspended and recentrifuged two or three times to reach a high degree of inclusion body purity. Resuspension in a solution of a low concentration of a chaotropic agent such as a detergent is often practiced to facilitate the removal of other contaminants without solubilizing the inclusion bodies. The pH and the ionic strength of the solution are adjusted to reduce the hydrophobicity of the cell debris particles and to enhance their removal in the light phase. Final product purity exceeding 70 percent is quite common.

Separation of soluble product from cell debris can be carried out by extraction and/or adsorption. Organic solvents are commonly used as extractants for low-molecular-weight products, such as various antibiotics. Aqueous two-phase systems have found applications for recovery of proteins. The criteria for extractant selection are as follows: (1) the partition coefficient of the product should be higher than the partition coefficient of the contaminants, (2) the extractant should not degrade the product, and (3) the extractant should not be expensive and should be easy to recover or dispose of.

Alternatively, product separation from debris and simultaneous concentration can be achieved by adsorptive techniques [Palmer, *Process Biochem.* **12:** 24 (1977)]. Adsorbents of various types (e.g., ion exchange, reversed phase, affinity) can be used. This type of purification requires the disrupted cells and product to be mixed in a stirred tank with an adsorbent. A washing step, where most of the cell debris particles and contaminants are washed out, follows product adsorption. Expanded-bed adsorption is an alternative technology for separating proteins from cell debris particles [Chang and Chase, *Biotechnol. Bioeng.* **49:** 512 (1996); Jin, *Pharm. Eng.*, Jan./Feb., p. 66 (2015)]. The feed is pumped upward through an expanded bed. Target proteins are bound to the adsorbent while cell debris and other contaminants pass through. A washing step removes all weakly retained material. An elution step follows that releases and further purifies the product.

In agreement with the second generic heuristic (*Remove the easiest-to-remove impurities first*), biomass removal is usually the first step of downstream processing of extracellular products. This step can be accomplished by using one (or more) of the following unit operations: rotary vacuum filtration, disk-stack or decanter centrifugation, press filtration, depth filtration, microfiltration, ultrafiltration, and flotation. Since each unit operation has advantages and disadvantages for different products and microorganisms, the selection of the best unit operation(s) for a given system can be difficult.

Rotary vacuum filtration, especially with precoat, is the classical method for removal of mycelial organisms [Dlouhy and Dahlstrom, *Chem. Eng. Prog.* **64:** 116 (1968)]. Rotary vacuum filters can operate continuously for long periods. In addition, the filtrate flux in these units is usually higher than 200 L m^{-2} h^{-1} and may reach 1000 L m^{-2} h^{-1}. The most important disadvantage of this type of unit is the problem with disposal of the mixture of filter aid and biomass. Filter aid is added in equal or higher amounts than biomass. Stringent environmental laws have made it costly to dispose of such solid materials. Therefore, if the disposal cost of filter aid is relatively high where a new plant is going to be built, alternative unit operations should be considered for biomass separation. However, if the disposal cost of filter aid is relatively low, a rotary vacuum filter is a good choice. The citric acid process, which is described later in this section, offers an example where rotary vacuum filtration is used for biomass removal.

Disk-stack and decanter centrifuges are frequently used at large scale [Brunner and Hemfort, in Mizrahi and Alan (eds.), *Advances in Biotechnological Processes*, vol. 8, Liss Inc., New York, 1988, p. 1]. Disk-stack centrifuges operate at higher rotational speeds and remove smaller and lighter microorganisms than decanters. However, with the use of flocculating agents, the decanter centrifuge performance improves, and choosing between the two types becomes more difficult. It appears that the only criterion being applied when disk-stack is chosen instead of decanter is the ability to remove small, light microorganisms. Centrifugation does not require filter aid, which is a significant advantage over rotary vacuum filtration. In general, the centrifuge paste contains 40 to 60 percent v/v extracellular liquid. To recover the product dissolved in that liquid, the paste is usually washed and recentrifuged.

With membrane filters (microfilters and ultrafilters), the extracellular product passes through the membrane while biomass and other particulate components remain in the concentrate. Concentration is usually followed by diafiltration to prevent product degradation and/or to improve the performance of the subsequent step. Membrane filters are used for biomass removal mainly in recovery of low-molecular-weight products, such as antibiotics from mycelia. For high-molecular-weight products, gel layer formation often limits application to cases in which the amount of solids is rather small (e.g., cell culture).

Depth filtration is commonly used in the recovery of antibodies from cell culture. New depth filtration media contains a layer of fiber and a layer of filter aid contained within another filter. These filters have shown a combination of high solids load and excellent retention of small particles at micrometer and submicrometer scales. Depth filtration can process solids-containing fluids at 1700 to 25,000 L m^{-2} min^{-1} [http://www.pall.com/main/biopharmaceuticals/product.page?id=28210].

Intermediate Recovery Stages The primary recovery stages just described are followed by the intermediate stages, where the product is concentrated and further purified. Intermediate recovery has similarities to the product isolation stages shown in Table 20-17. If the product is soluble, product capture and concentration is usually the first step. If the product is denatured and insoluble, it is first dissolved and refolded and then captured, concentrated, and purified.

After primary separation, the product is usually in a dilute solution. Volume reduction by concentration is in agreement with heuristics 1 and 2.

Common concentration options include ultrafiltration, reverse osmosis, evaporation, adsorption, precipitation, extraction, and distillation.

Ultrafiltration is used extensively for protein solution concentration. The molecular weight cutoff of the membrane is selected to retain the product while allowing undesirable impurities (mainly low-molecular-weight solutes) to pass through the membrane. The low operating temperature and the purification achieved along with concentration are some of the advantages of ultrafiltration over evaporation. The typical operating transmembrane pressure is 2 to 5 bar, and the average flux is 20 to 50 L m^{-2} h^{-1}.

Membranes with smaller pore sizes are used for reverse-osmosis filters. The process of reverse osmosis may be used when concentrating medium- to low-molecular-weight products (e.g., antibiotics, certain amino acids).

Agitated film evaporators can operate at relatively low temperatures (40 to 50°C) under vacuum. These units compete in the market with ultrafiltration and reverse osmosis for concentrating both low- and high-molecular-weight compounds. Unlike ultrafiltration, however, evaporation lacks the capability to provide purification during concentration. Advantages include the ability to concentrate to a higher final solids concentration and the ability to handle large throughputs [Freese, in Vogel and Todaro (eds.), *Fermentation and Biochemical Engineering Handbook*, Noyes, Westwood, N.J., 1993, p. 227].

Precipitation is often used for concentration and purification. Blood protein fractionation and citric acid production (see later subsection Bioprocess Design and Economics) constitute typical applications. Addition of salts, solvents, and polymers and changes in pH, ionic strength, and temperature are commonly used to selectively precipitate compounds of interest [Chan et al., *Biotechnol. Bioeng.* **28:** 387 (1986)]. Precipitation often follows an extraction carried out in a polymer/salt (e.g., PEG and potassium phosphate) aqueous two-phase system. When the product is recovered in the polymer-rich phase, precipitation is accomplished by adding more polymer. It is important for economic reasons to recover and recycle the precipitating materials. Precipitation is also used to remove contaminants (i.e., nucleic acids) by adding manganese sulfate ($MnSO_4$) and streptomycin sulfate.

The process of distillation is used for concentrating and purifying low-molecular-weight and volatile compounds, such as acetone, ethanol, butanol, acetic acid, etc.

Pervaporation is a membrane-based process that has found applications in biofuels for the dehydration of ethanol and other alcohols. One component from a liquid mixture selectively permeates a membrane, driven by a gradient in partial vapor pressure and leaving the membrane as a vapor (Bruschke, in Nunes and Peinemann, *Membrane Technology in the Chemical Industry*, Wiley-VCH, Weinheim, Germany, 2006, p. 127). Dehydration of ethanol-water azeotropic solutions (around 90 percent ethanol) is facilitated by the use of hydrophilic membranes. Hydrophobic membranes are used for removal/recovery of small amounts of organics from aqueous solutions.

Eukaryotic proteins produced by prokaryotic microorganisms often form insoluble inclusion bodies in the host cell, and renaturation of the protein is required when this occurs. Inclusion bodies first need to be dissolved and usually reduced (to break disulfide bonds that have formed incorrectly in the inclusion body) and then oxidized to form the proper disulfide bonds. Inclusion bodies can be dissolved rapidly by using solutions of strong chaotropes, such as 6 *M* guanidine hydrochloride or urea. Detergents and surfactants can also be used. Common reducing agents are 0.5 *M* 2-mercaptoethanol or 50 *mM* dithiothreitol [Fish et al., *Biochemistry* **24:** 1510 (1985)]. The dissolved protein is then allowed to refold to its native conformation. This can be done with air or oxygen in the presence of copper ion as a catalyst, and by removing the chaotropic agents through diafiltration, dilution, or chromatography, with final protein concentrations in the range of 10 to 100 mg L^{-1}. Dilution is sometimes necessary for minimizing intermolecular interactions, especially when there are multiple cysteines in the protein. Intermolecular interactions that occur during product refolding can lead to product inactivation, dimerization, crosslinking, and poor yield. Regulation of the redox state of the refold through addition of small amounts of thiols such as reduced glutathione (1 to 5 *mM*) and oxidized glutathione (0.01 to 0.5 *mM*) and incubation at 35 to 40°C for 5 to 10 h completes the refolding process. Thus, choosing an upstream process that forms IBs entails consideration of the large volumes, hence large waste streams, that are produced. More information on IB solubilization and protein refolding can be found in the insulin example (see later subsection Bioprocess Design and Economics).

Final Purification Stages The final purification steps are dependent on the required final product purity. Pharmaceutical products require high purity, while commodity products require lower purity. For products of relatively low purity, such as industrial enzymes, the final purification step is dehydration or more generally a solvent removal step. For high-purity products, the final purification stages usually involve a combination of chromatographic and filtration steps [Bjurstrom, *Chem. Eng.* **92:** 126 (1985)]. If the final product is required in solid form, a dehydration or solvent removal step follows.

Chromatography is typically done later in a process in agreement with the third generic heuristic (*Make the most difficult and expensive separations last*). With the preceding separation steps, a large fraction of contaminants is removed, thereby reducing the volume of material that needs to be treated further. A sequence of chromatographic steps is usually required to achieve the desired final product purity, and the fourth and fifth generic heuristics are good guides for selecting and sequencing such steps [Wheelwright, *Bio/Technol.* **5:** 789 (1987)]. For instance, according to the fifth heuristic, an ion exchange step should not be followed by another step of the same type. Instead, it should be followed by a reversed-phase, affinity, or any other chromatography type that takes advantage of a different separation driving force. The insulin and monoclonal antibody examples presented later in this section provide additional information on selection, sequencing, and operation of chromatographic separation units.

Membrane adsorption units combine the high flux of membrane filters with the selective binding of chromatographic resins [Zhou and Tressel, *BioProcess Int.* **3:** 32 (2005)]. As a result, they have advantages compared to traditional column chromatography when operated in the flow-through mode for removing small amounts of specific contaminants. The membrane retains certain impurities (e.g., DNA molecule fragments) while the product molecules pass through the membrane. Such membrane systems are typically used as the last step of biopharmaceutical protein purification processes. More information about membrane chromatography is given in the subsection Liquid Chromatography and Adsorption.

Simulated moving-bed (SMB) chromatography is the method of choice for handling large volumes of material [Holzer et al., *BioProcess Int.* **6:** 74 (2008)]. It has found applications in the purification of amino acids, high-fructose corn syrup, cheese whey proteins, lactic acid, succinic acid, etc. In SMB systems, multiple columns operate out of phase using a complex system of valves. The feed stream usually passes through two columns, resulting in increased yield and resolution. One column is always out of use for cleaning/regeneration. SMB systems can handle feed streams of continuous flow, which is the preferred method of operation for the production of high-volume biochemicals. More information about SMB can be found in the subsection Simulated Moving-Bed Chromatography.

Membrane filtration can be used between chromatographic steps to exchange buffers, concentrate the dilute product solutions, and control bioburden. Dead-end filtration can also be used to remove contaminating particles in chromatography feed streams.

Crystallization and fractional precipitation can sometimes result in significant purification. Because these processes are cheaper to operate than chromatography, they should always be considered. The crystalline form of a bioproduct is especially advantageous, since the purity can be quite high and crystals can usually be stored for long periods. The citric acid process that is analyzed later in this section is a good example of a product that is recovered and purified using precipitation and crystallization.

Dehydration or solvent removal is achieved with dryers. Spray, fluidized-bed, and tray dryers are used when products can withstand temperatures of 50 to 100°C. Freeze dryers are used for products that degrade at high temperatures. Freeze dryers require high capital expenditures and should be avoided, if possible.

Pairing of Unit Operations in Process Synthesis It is often advantageous to consider how two different unit operations can be paired to improve process efficiency, which follows the sixth heuristic (*Sequence unit operations such that the product of one can serve as the feedstock for the next without dilution or intermediate treatment*). The following gives some examples of operations that are logical to pair:

- In the pairing of extraction and precipitation, the bioproduct is extracted with a solvent and then precipitated. To increase the yield, it is often desirable to concentrate the extract before the precipitation. The major hurdle to overcome for this pairing is to find a solvent that will work with both extraction and precipitation.
- The pairing of precipitation and hydrophobic interaction chromatography is usually accomplished for protein purification by using ammonium sulfate to precipitate impurities, leaving the desired bioproduct in the mother liquor. The ammonium sulfate is added to a concentration just below that needed to precipitate the bioproduct. After removal of precipitated impurities, the mother liquor can be applied directly to a hydrophobic interaction chromatography column, which was equilibrated to the concentration of ammonium sulfate in the mother liquor prior to the loading. The bioproduct adsorbs to the column under these conditions. The column is eluted with a decreasing gradient of ammonium sulfate, and the desired bioproduct is recovered in a fraction from the elution.
- When the bioproduct is contained in the filtrate after filtration, it can often be extracted with an immiscible solvent. For the extraction of small molecules such as antibiotics with organic solvents, the pH must usually be

adjusted to obtain the bioproduct in either its free base or free acid form so it will partition into the organic phase. For the aqueous two-phase extraction of proteins, two polymers or a salt and a polymer must be added. If the additions to the filtrate can be made in-line, the filtration and extraction steps can be carried out simultaneously, reducing the processing time.

Process and Product Quality In the development of bioseparation processes, important measures of product quality due to processing are purity, fold purification, specific activity, and yield. Purity is defined as follows:

$$\text{Purity} = \frac{\text{amount of product}}{\text{amount of product} + \text{amount of total impurities}} \tag{20-19}$$

Fold purification is the ratio of the purity at any stage in the process to the purity at the start of the purification process. However, this factor is usually expressed in terms of the impurity itself, and is given in logs; for example,

$$\text{Fold DNA removal} = -\log\left(\frac{\text{DNA concentration}_{\text{stage of interest}}}{\text{DNA concentration}_{\text{beginning}}}\right) \tag{20-20}$$

Another measure of purity is

$$\text{Specific activity} = \frac{\text{units of biological activity}}{\text{mass}} \tag{20-21}$$

where units of biological activity are assayed by means of a biological test, such as moles of substrate converted per second per liter or fraction of bacterial cells killed. For proteins, the mass in Eq. (20-21) is usually total protein; on this basis, the specific activity reaches a constant value when the protein is pure.

Yield is given by

$$\text{Yield} = \frac{\text{amount of product produced}}{\text{amount of product in feed}} \tag{20-22}$$

Purity is a strictly quantitative measure and not always an expression of the quality of the product. A therapeutic protein can be 99.99 percent pure but still unacceptable if any pyrogen (a substance that produces a fever) is present. However, if the product is not a therapeutic protein but an industrial enzyme, then practically any impurities that do not inhibit the activity of the product or endanger the user are allowed. Figure 20-30 indicates this principle by comparing the level and range of purity acceptable for three different protein products. In short, two kinds of measurements of purity are required: activity, composition, and structure on the product itself; and host cell materials, degraded product, and excipients (additives) on the impurities.

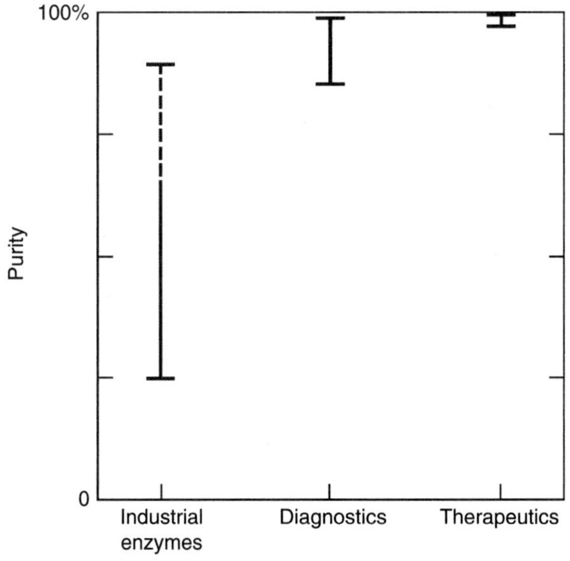

FIG. 20-30 Different levels of purity are required for different products. [Reproduced from *Bioseparations Science and Engineering*, 2d ed., by Roger G. Harrison, Paul W. Todd, Scott R. Rudge, and Demetri P. Petrides (2015). By Permission of Oxford University Press.]

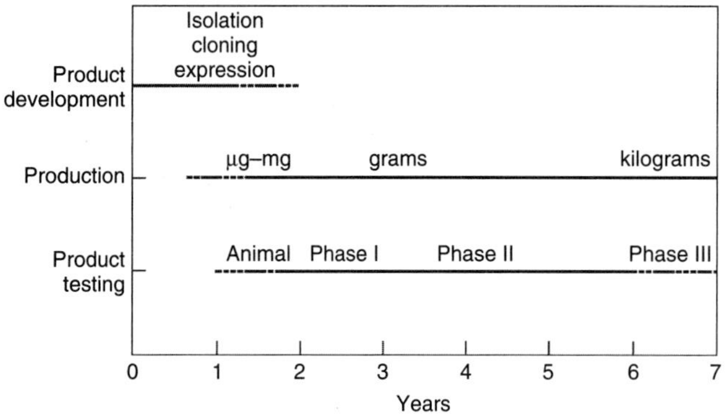

FIG. 20-31 Pathway to market beginning with the discovery of a biotechnology product. [Reproduced from *Bioseparations Science and Engineering*, 2d ed., by Roger G. Harrison, Paul W. Todd, Scott R. Rudge, and Demetri P. Petrides (2015). By Permission of Oxford University Press.]

The Route to Market The route for a biotechnology product to be on the market is complex and can be lengthy. Figure 20-31 charts the pathway to market for a biotechnology product. Note that several years are required for the completion of the route from genetic discovery to marketing.

Besides identifying the market for the product and determining the optimal process for the expected production rate, the personnel involved in this effort must make sure that all the regulatory requirements are met. This is clearly an interdisciplinary task, involving not only engineers and bioscientists but also marketing and regulatory personnel. Often applications personnel, such as medical doctors and veterinarians, must also be involved.

As mentioned earlier, modern biotechnology produces the full range of substances for a wide range of applications; some well-known examples help emphasize this point:

Aspartame, a dipeptide nonnutritive sweetener

Paclitaxel, a triterpene from plants used in cancer treatment

Erythropoietin, a peptide hormone that stimulates red cell production in bone marrow

Oligonucleotides, which can be made to function according to shape and according to catalytic activity

Bacillus thuringiensis (Bt), whole bacterial cells that, when dried, can be sprayed on crops to prevent insect damage

Bioethanol, which can be produced by the anaerobic fermentation of starch

Note the categories: food, medicine, agriproducts, and biofuels. While most of the excitement in biotechnology and bioprocessing today is in the area of biopharmaceuticals produced by genetically engineered organisms, a much wider variety of products exist, usually with lower value per mass unit, including biopesticides, novel food products, polymer components, hydrogel polymers, industrial enzymes, commodity enzymes (laundry products), and fine chemicals. To know what product purity, cost, and FDA requirements must be met, the exact application of the proposed bioproduct must be identified early in process development, or even beforehand.

FILTRATION

Filtration is an operation that has found an important place in the processing of biotechnology products. In general, filtration is used to separate particulate or solute components in a fluid suspension or solution according to size by flowing under a pressure differential through a porous medium. There are two broad categories of filtration, which differ according to the direction of the fluid feed in relation to the filter medium. In *conventional* or *dead-end filtration,* the fluid flows perpendicular to the medium, which generally results in a cake of solids depositing on the filter medium. In *cross-flow filtration* (which is also called *tangential flow filtration*), the fluid flows parallel to the medium to minimize buildup of solids on the medium. Conventional and cross-flow filtration are illustrated schematically in Fig. 20-32.

Conventional filtration (Fig. 20-32*a*) is typically used when a product has been secreted from cells, and the cells must be removed to obtain the product that is dissolved in the liquid. Antibiotics and steroids are often processed by using conventional filtration to remove the cells. Conventional filtration is also commonly used for sterile filtration in biopharmaceutical production. Cross-flow filtration (Fig. 20-32*b*) has been used in a wide variety of applications, including the separation of cells from a product that has been secreted, the concentration of cells, the removal of cell debris

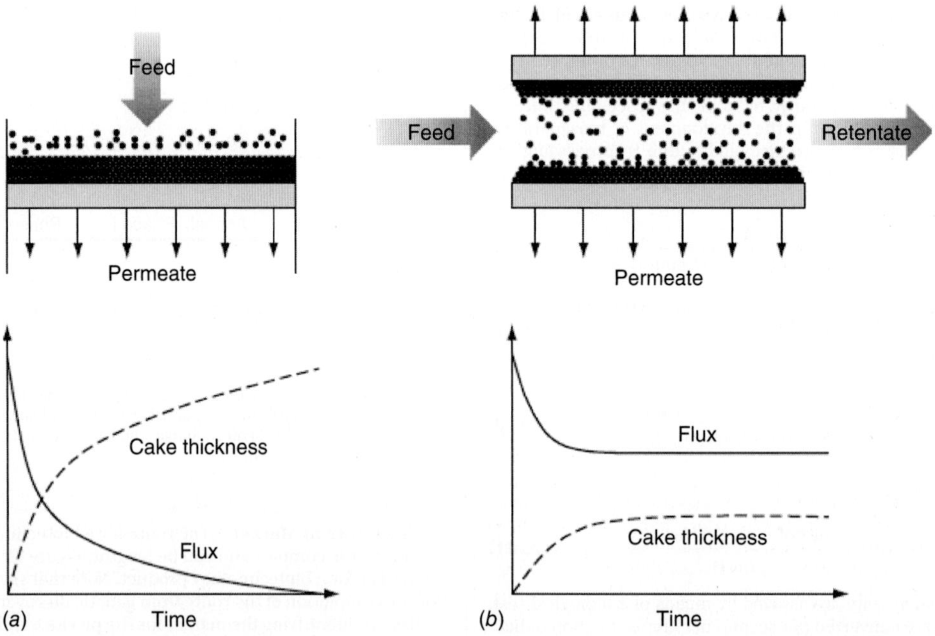

FIG. 20-32 Schematic diagrams for (*a*) dead-end or conventional filtration and (*b*) cross-flow filtration. For dead-end filtration the thickness of the solids buildup increases, and the permeate flux decreases with time, ultimately reaching zero. In cross-flow filtration, the feed can contain either a soluble or a solid solute, which becomes concentrated at the membrane surface; the permeate flux reaches a constant value at steady state. [Reproduced from *Bioseparations Science and Engineering*, 2d ed., by Roger G. Harrison, Paul W. Todd, Scott R. Rudge, and Demetri P. Petrides (2015). By Permission of Oxford University Press.]

from cells that have been lysed, the concentration of protein solutions, the exchange or removal of a salt or salts in a protein solution, and the removal of viruses from protein solutions.

Filtration often occurs in the early stages of bioproduct purification, in keeping with the process design heuristic *Remove the most plentiful impurities first* (see Introduction). At the start of purification, the desired bioproduct is usually present in a large volume of aqueous solution, and it is desirable to reduce the volume as soon as possible to reduce the scale and thus the cost of subsequent processing operations. Filtration, along with sedimentation and extraction, is an effective means of accomplishing volume reduction.

The theory, equipment, and scale-up of conventional filtration are described in the subsection Filtration in Sec. 18. Here, additional topics that relate specifically to the filtration operations for bioproducts are discussed: (1) the principles, filter media, equipment, and scale-up of cross-flow filtration; and (2) the design of sterile filters, especially important in the pharmaceutical industry where products often need to be in sterile form.

Cross-Flow Filtration As illustrated in Fig. 20-32*b*, the fluid in cross-flow filtration flows parallel to the membrane surface, resulting in constant permeate flux at steady state. The analysis of cross-flow filtration can be divided into two categories depending on whether the component being filtered is soluble or insoluble—see the discussion of filtration principles for each category later. When dissolved species such as proteins are being filtered, ultrafiltration membranes are generally used. The ultrafiltration membrane is selected so that the species of interest will not pass through these membranes. The retained species is carried to the surface of the membrane by the convective flow of fluid, and the concentration of the species builds up next to the membrane surface. The concentration of the species can be so high that it precipitates on the membrane surface, further impeding the flow of fluid through the membrane. The resulting layer of solids on the membrane surface has been called a *gel layer*. In addition, even without precipitation, the increased osmolarity near the membrane surface creates a solvent flux in the opposite direction of the flux due to the transmembrane pressure gradient.

When suspended micrometer-size particles are present, these particles are carried to the membrane surface and may form a thin cake layer at the surface, but do not significantly impact the osmotic pressure. In this situation, microfiltration membranes are generally utilized. These membranes let dissolved components pass through but retain particles above a certain size. There are instances, such as in the cross-flow filtration of cells that have been ruptured, where the layer at the membrane surface contains both suspended particles and precipitated solutes. More complete information about ultrafiltration and microfiltration membranes is given later in this section.

Dissolved Species In cross-flow filtration, a solution under pressure flows across the surface of a membrane. As a result of this pressure, fluid is forced through the membrane. This flow toward the surface of the membrane causes species dissolved in the solution also to be carried toward the membrane's surface. For a solute that is rejected by the membrane, there will be a concentration gradient of this solute across a stagnant boundary layer next to the surface of the membrane, as indicated in Fig. 20-33. The elevation of the solute concentration at the membrane surface (c_w) compared to that in the bulk solution (c_b) is known as *concentration polarization.*

At steady state, the rate of convective mass transfer of solute toward the membrane surface must be equal to the rate of mass transfer of solute by

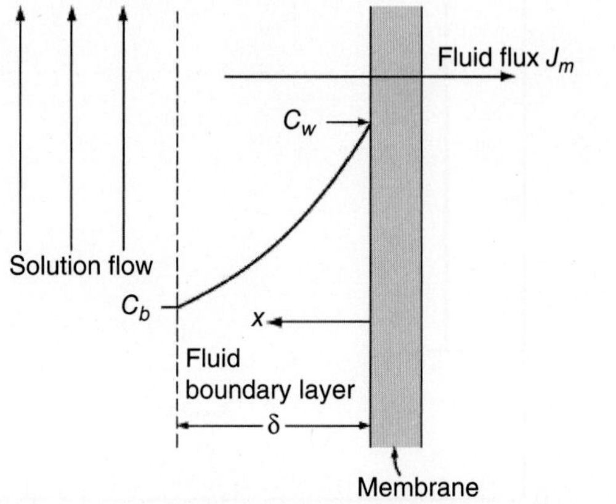

FIG. 20-33 Schematic representation of the boundary layer in cross-flow filtration with a dissolved solute in the feed. A solution flows parallel to the membrane surface, and fluid flows through the membrane under the influence of pressure. A fluid boundary layer forms next to the membrane surface, creating a gradient in the concentration *c* of solute. [Reproduced from *Bioseparations Science and Engineering*, 2d ed., by Roger G. Harrison, Paul W. Todd, Scott R. Rudge, and Demetri P. Petrides (2015). By Permission of Oxford University Press.]

diffusion away from the membrane surface, which is described as [Michaels, *Chem. Eng. Prog.* **64:** 31 (1968)]

$$J_m c = -\mathcal{D}\frac{dc}{dx} \tag{20-23}$$

where J_m is the transmembrane fluid flux, c is the concentration of the solute, and $\mathcal{D}$ is the diffusion coefficient of the solute. For a boundary layer thickness of δ, the solution of Eq. (20-23) is

$$J_m = \left(\frac{\mathcal{D}}{\delta}\right)\ln\left(\frac{c_w}{c_b}\right) \tag{20-24}$$

which can be also written as

$$\frac{c_w}{c_b} = \exp\left(\frac{J_m\delta}{\mathcal{D}}\right) \tag{20-25}$$

The term $\mathcal{D}/\delta$ can also be defined as a mass transfer coefficient k. The ratio c_w/c_b is sometimes called the *polarization modulus* and indicates the extent of concentration polarization. From Eq. (20-25), it can be seen that the polarization modulus is particularly sensitive to changes in J_m, δ, and $\mathcal{D}$ because of the exponential functionality involved. For high-molecular-weight solutes (small $\mathcal{D}$) and membranes with high solvent permeability (high J_m), concentration polarization can become severe, with $c_w/c_b > 10$. At high concentration polarization levels, the solubility of the solute can be exceeded, resulting in the precipitation of the solute and the formation of a solids or gel layer on the membrane surface.

Experimental data have been obtained to support this simple model of concentration polarization. Figure 20-34 shows data for the ultrafiltration of the protein casein and dextran with a molecular weight of 110,000. For both species, a plot of ultrafiltration flux versus log of bulk concentration gives a straight line for bulk concentration varying by an order of magnitude. Estimates of the concentration of solute at the membrane surface can be obtained from the intercept of the straight line in Fig. 20-34 at zero flux. These estimates of c_w appear to be physically reasonable.

Correlations have been developed for the mass-transfer coefficient k. For laminar flow, boundary layer theory has been applied to yield analytical solutions (known as the Lévêque or Graetz solutions) for k [Blatt et al., in Flinn (ed.), *Membrane Science and Technology*, Plenum Press, New York, 1970, p. 47]:

$$k = 0.816\left(\gamma_w \frac{\mathcal{D}^2}{L}\right)^{1/3} \tag{20-26}$$

where γ_w is the fluid shear rate at the membrane surface and L is the length of the flow channel over the membrane. The constant 0.816 is applicable for the gel-polarized condition of concentration polarization where the solute concentration at the wall is constant. For a rectangular slit of height $2h$ and bulk stream velocity u_b,

$$\gamma_w = \frac{3u_b}{h} \tag{20-27}$$

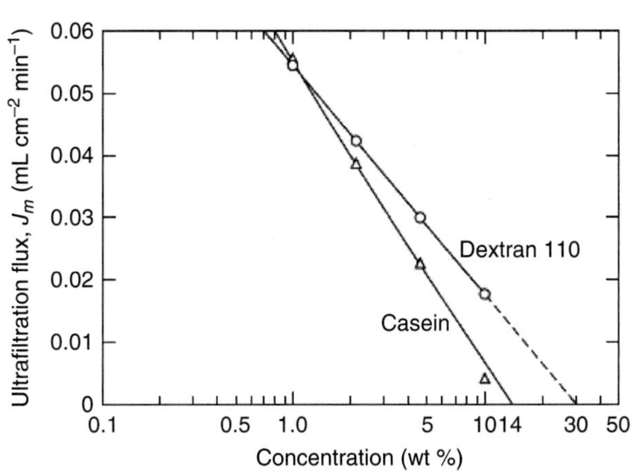

FIG. 20-34 Decline in ultrafiltration flux with increasing concentration for solutions of casein and Dextran (MW 110,000). Data were obtained in thin-channel recirculating flow cells. [Data from Blatt et al., in Flinn (ed.), *Membrane Science and Technology*, Plenum Press, 1970, p. 63.]

and the equation for a circular tube of diameter D is

$$\gamma_w = \frac{8u_b}{D} \tag{20-28}$$

For turbulent flow, empirical correlations have been developed for the mass-transfer coefficient. These correlations are based on dimensional analysis of the equations of change for forced convection mass transfer in a closed channel, which gives (Bird et al., *Transport Phenomena*, Wiley, Hoboken, N.J., 2002, p. 681):

$$\text{Sh} = \frac{kD_h}{\mathcal{D}} = f(\text{Re}, \text{Sc}, L/D_h) \tag{20-29}$$

where $\text{Re} = \frac{D_h u_b \rho}{\mu}$ = Reynolds number

$\text{Sc} = \frac{\mu}{\rho\mathcal{D}}$ = Schmidt number

D_h = equivalent diameter of the channel

$= 4\left(\frac{\text{cross-sectional area}}{\text{wetted perimeter}}\right)$

u_b = bulk stream velocity

ρ, μ = density and viscosity of fluid, respectively

A typical correlation that has been developed for this Sherwood number [Calderbank and Moo-Young, *Chem. Eng. Sci.* **16:** 39 (1961)] is

$$\text{Sh} = 0.082\,\text{Re}^{0.69}\text{Sc}^{0.33} \tag{20-30}$$

Note that the L/D_h term in the generalized dimensionless analysis solution is left out of the empirical correlation in Eq. (20-30).

Example 20-1 Concentration Polarization in Ultrafiltration Equipment is available for ultrafiltration of a protein solution at constant volume to remove low-molecular-weight species (achieved by the addition of water or buffer to the feed in an operation called diafiltration—see Example 20-3). The flow channels for this system are tubes 0.1 cm in diameter and 100 cm long. The protein has a diffusion coefficient of 9×10^{-7} cm^2/s. The solution has a viscosity of 1.2 cP and a density of 1.1 g/cm^3. The system is capable of operating at a bulk stream velocity of 300 cm/s. At this velocity, determine the polarization modulus for a transmembrane flux of 45 L m^{-2} h^{-1}.

We see from Eq. (20-25) that we can determine c_w/c_b if we know $J_m\delta/\mathcal{D} = J_m/k$, where J_m is as given in the problem statement. We can determine k from either Eq. (20-26) or Eqs. (20-29) and (20-30), depending on whether the flow is laminar or turbulent. We first need to calculate the Reynolds number to characterize the flow regime:

$$\text{Re} = \frac{D_h u_b \rho}{\mu} = \frac{0.1\,\text{cm} \times \frac{300\,\text{cm}}{\text{s}} \times 1.1\frac{\text{g}}{\text{cm}^3}}{\frac{0.012\ \text{g}}{\text{cm}\cdot\text{s}}} = 2750$$

In the calculation of the Reynolds number, we see that the equivalent diameter of the channel D_h is the same as the diameter of the tubes. The flow is turbulent, since the Reynolds number is greater than 2100. For turbulent flow, we need to know Sc, the Schmidt number:

$$\text{Sc} = \frac{\mu}{\rho\mathcal{D}} = \frac{\frac{0.012\ \text{g}}{\text{cm}\cdot\text{s}}}{1.1\frac{\text{g}}{\text{cm}^3} \times 9 \times 10^{-7}\frac{\text{cm}^2}{\text{s}}} = 1.21 \times 10^4$$

From Eqs. (20-29) and (20-30),

$$k = \frac{\mathcal{D}\text{Sh}}{D_h}$$

$$= \frac{\mathcal{D}}{D_h}(0.082)\text{Re}^{0.69}\,\text{Sc}^{0.33}$$

$$= \left(\frac{9 \times 10^{-7}\frac{\text{cm}^2}{\text{s}}}{0.1\ \text{cm}}\right)(0.082)(2750)^{0.69}(12{,}100)^{0.33} = 3.88 \times 10^{-3}\,\text{cm/s}$$

From Eq. (20-25), the polarization modulus can be estimated:

$$\frac{c_w}{c_b} = \exp\left(\frac{J_m\delta}{\mathcal{D}}\right) = \exp\left(\frac{J_m}{k}\right)$$

$$= \exp\left(\frac{45\frac{\text{L}}{\text{m}^2\,\text{h}} \times \frac{10^3\,\text{cm}^3}{\text{L}} \times \frac{1\ \text{m}^2}{10^4\,\text{cm}^2} \times \frac{1\,\text{h}}{3600\ \text{s}}}{3.88 \times 10^{-3}\frac{\text{cm}}{\text{s}}}\right) = 1.38$$

Thus, the concentration polarization is not severe.

Concentration polarization at the surface of the membrane can become great enough that it creates a significant resistance to fluid flow. This effect on the transmembrane fluid flux J_m can be modeled by using Darcy's law with the flow resistance made up of the sum of the membrane resistance R_m and the resistance R_p of the polarized boundary layer and any gel layer next to the surface of the membrane. For membranes, a correction needs to be made for the osmotic pressure $\Delta\pi$ of the solute that is being filtered, which results in the following equation:

$$J_m = \frac{\Delta p - \sigma\Delta\pi}{\mu_o(R_m + R_p)} \tag{20-31}$$

where Δp is the pressure difference between the bulk fluid and the permeate, μ_0 is the viscosity of the permeate, and σ is the reflection coefficient for the solute. A reflection coefficient of 1.0 indicates no passage of solute through the membrane, while a coefficient of 0 indicates free passage of the solute with the solvent through the membrane. For an ideal dilute solution,

$$\Delta\pi = RT\,c_w \tag{20-32}$$

where R is the ideal gas constant, T is the absolute temperature, and c_w is the solute concentration at the surface of the membrane.

Data for J_m versus Δp for the ultrafiltration of serum albumin in a stirred filtration cell are shown in Fig. 20-35 as a function of both cell stirrer speed and protein concentration. At high values of Δp for 3.9 percent and 6.5 percent protein concentrations, the ultrafiltration flux J_m is constant with increasing Δp. Constant flux as Δp increases can be interpreted as a condition in which the solute concentration at the membrane surface c_w has reached a solubility limit; this behavior is supported by Eq. (20-24), where J_m becomes constant when c_w reaches a constant.

Suspended Particles In the cross-flow filtration of mixtures with suspended particles, the fluid flux toward the membrane carries particles to the membrane surface, where they are rejected and form a cake layer that is analogous to the gel layer in the ultrafiltration of dissolved species. The cake layer initially grows with time, thus reducing the permeate flux and constricting the channel. At steady-state conditions, the layer reaches a constant thickness, which is relatively thin if the shear exerted by the fluid flowing tangentially to the membrane surface is high enough.

The theory for the cross-flow filtration of dissolved species has been found to hold only for very small suspended particles up to approximately 1 μm in size [Belfort et al., *J. Membrane Sci.* **96:** 1 (1994)]. Beyond this size, experimental membrane flux values are often 1 to 2 orders of magnitude higher than those obtained from the Lévêque or Graetz theory given by Eq. (20-26) [Porter, *Ind. Eng. Chem. Prod. Res. Dev.* **11:** 234 (1972)]. Two different theories have been developed to more accurately predict membrane fluxes for particles larger than 1 μm, depending on the size of the particles: a shear-induced diffusion theory for small particles and an inertial lift theory for larger particles.

In the shear-induced diffusion theory, the concentration polarization model can be applied to the cross-flow filtration of suspended particles if the Brownian diffusivity is replaced by a shear-induced hydrodynamic diffusivity [Zydney and Colton, *Chem. Eng. Commun.* **47:** 1 (1986)]. In shear-induced hydrodynamic diffusion, the particles are randomly displaced from the streamlines in a shear flow and collide with other particles. In their model, Zydney and Colton employed an approximate relationship for the shear-induced diffusion $\mathcal{D}_s$ of spherical particles measured by Eckstein et al. [*J. Fluid Mech.* **79:** 191 (1977)] for $0.2 < \varphi < 0.45$

$$\mathcal{D}_s = 0.3\gamma_w a_p^2 \tag{20-33}$$

where φ is the particle volume fraction in the bulk suspension, a_p is the particle radius, and γ_w is the fluid shear rate at the membrane surface. Substituting for the diffusivity in the equation for the laminar flow mass-transfer coefficient in the concentration polarization model [Eq. (20-26)], we obtain

$$k = 0.366\gamma_w\left(\frac{a_p^4}{L}\right)^{1/3} \tag{20-34}$$

where L is the tube or channel length. Comparing this equation for k with the one for dissolved species in laminar flow, we see that the mass-transfer coefficient is much more strongly dependent on the shear rate at the wall. Shear-induced diffusion has been found to dominate for particles up to 30 to 40 μm, while inertial lift dominates for larger particles [Belfort et al., *J. Membrane Sci.* **96:** 1 (1994)].

Inertial lift arises when the Reynolds number based on the particle size is not negligible. Inertial lift produces a velocity that carries particles away from the membrane surface. For fast laminar flow of dilute suspensions with thin fouling layers, the steady-state transmembrane flux predicted by the inertial lift theory is [Belfort et al., *J. Membrane Sci.* **96:** 1 (1994)]:

$$J_m = \frac{0.036\rho_o a_p^3 \gamma_w^2}{\mu_o} \tag{20-35}$$

where ρ_0 and μ_0 are the density and viscosity, respectively, of the permeate. Thus, for transport by inertial lift, the transmembrane flux is strongly dependent on the particle size and shear rate at the membrane surface and not dependent at all on the length of the filter or on the concentration of particles in the bulk suspension.

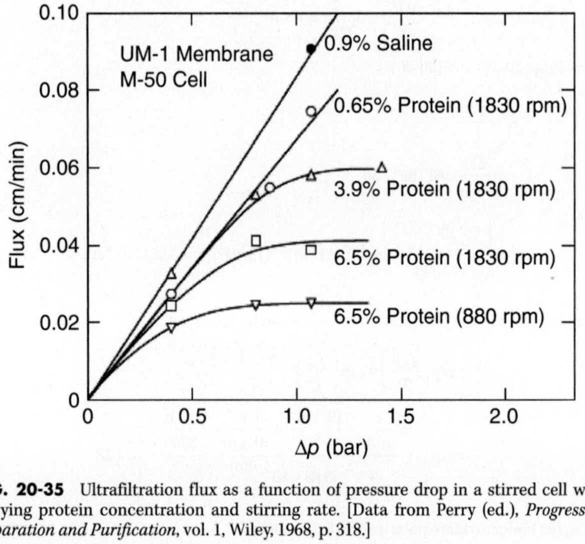

FIG. 20-35 Ultrafiltration flux as a function of pressure drop in a stirred cell with varying protein concentration and stirring rate. [Data from Perry (ed.), *Progress in Separation and Purification*, vol. 1, Wiley, 1968, p. 318.]

Example 20-2 Comparison of Mass-Transfer Coefficient Calculated by Boundary Layer Theory versus by Shear-Induced Diffusion Theory Cells that are 5 μm in diameter are being concentrated by cross-flow filtration using hollow fibers that are 1.1 mm in diameter (inside) and 20 cm long. The velocity of the fluid in the hollow fibers is 100 cm/s. The fluid can be assumed to have the same density and viscosity as water. The temperature is 4°C. Compare the mass-transfer coefficient k calculated by using boundary layer theory with that using shear-induced diffusion theory.

Boundary layer theory We first calculate the Reynolds number in order to determine the flow regime. At 4°C for water, $\rho = 1.00$ g/cm³ and $\mu = 1.56$ cP (from Tables 2-92 and 2-139, respectively). Therefore,

$$\text{Re} = \frac{D u_b \rho}{\mu} = \frac{0.11\,\text{cm} \times \dfrac{100\ \text{cm}}{\text{s}} \times 1.00\dfrac{\text{g}}{\text{cm}^3}}{\dfrac{0.0156\,\text{g}}{\text{cm}\cdot\text{s}}} = 705$$

The upper limit of the Reynolds number for laminar flow in tubes is about 2100. Therefore, the flow is laminar in our case. For laminar flow in tubes, we can calculate k using Eqs. (20-26) and (20-28). Equation (20-26) requires the diffusion coefficient for the cells, which can be estimated from the Stokes-Einstein equation for spheres, Eq. (20-62):

$$\mathcal{D} = \frac{k_B T}{6\pi\mu a_p} = \frac{\left(1.38\times10^{-23}\dfrac{\text{J}}{\text{K}}\right)(277\ \text{K})\left(\dfrac{10^7\,\text{g cm}^2\ \text{s}^{-3}}{\text{J s}^{-1}}\right)}{6\pi\left(0.0156\dfrac{\text{g}}{\text{cm}\cdot\text{s}}\right)(2.5\times10^{-4}\ \text{cm})} = 5.20\times10^{-10}\text{cm}^2\ \text{s}^{-1}$$

From Eqs. (20-28) and (20-26), respectively,

$$\gamma_w = \frac{8u_b}{D} = \frac{8\times100\dfrac{\text{cm}}{\text{s}}}{0.11\,\text{cm}} = 7273\,\text{s}^{-1}$$

$$k = 0.816\left(\gamma_w\frac{\mathcal{D}^2}{L}\right)^{1/3}$$

$$= 0.816\left(\frac{7273\dfrac{1}{\text{s}}\times\left(5.20\times10^{-10}\dfrac{\text{cm}^2}{\text{s}}\right)^2}{20\ \text{cm}}\right)^{1/3} = 3.77\times10^{-6}\ \text{cm s}^{-1}$$

Shear-induced diffusion For laminar flow with shear-induced diffusion, Eq. (20-34) applies:

$$k = 0.366\gamma_w\left(\frac{a_p^4}{L}\right)^{1/3}$$

$$= 0.366\left(7273\frac{1}{\text{s}}\right)\left[\frac{(2.5\times10^{-4})^4\,\text{cm}^4}{20\,\text{cm}}\right]^{1/3} = 1.54\times10^{-2}\ \text{cm}\cdot\text{s}^{-1}$$

Therefore, k is more than 4000 times greater when calculated using shear-induced diffusion theory than when calculated using Fickian diffusion theory for 5-μm cells.

Membrane Fouling Fouling of membranes results from physical and/or chemical interactions between the membrane and various components present in the process stream. Fouling leads to a decline in the permeate flux and a change in the membrane selectivity. Because of the high affinity of proteins for solid surfaces, proteins are the leading contributor to fouling in ultrafiltration (UF) and microfiltration (MF). The flux decrease caused by proteins can be attributed to one or more of the following processes [Belfort et al., *J. Membrane Sci.* **96:** 1 (1994)]:

- Protein adsorption, which involves the interaction of the proteins and the membrane and occurs with no convective flow through the membrane
- Protein deposition, which is the addition of more protein that is associated with the membrane, over and above the protein that would be adsorbed in a nonflowing system
- Protein accumulation on the membrane surface exposed to the flowing process stream as a result of concentration polarization, which can lead to formation of a gel layer

This last effect is governed by the mass-transfer coefficient [see Eqs. (20-26) and (20-30)] and thus can be minimized by increasing the fluid shear rate at the filtering surface of the membrane. These three processes can lead to pores in membranes that are constricted and even completely blocked, as illustrated in Fig. 20-36.

Protein adsorption studies have been performed for both MF and UF membranes to determine the extent to which the pores are constricted by adsorbed protein. To eliminate concentration polarization effects, these studies were done in the absence of flow. The work with MF membranes and a variety of proteins showed that there was approximately monolayer adsorption throughout the internal pores [Belfort et al., *J. Membrane Sci.* **96:** 1 (1994)]. Protein adsorption was reduced on the more hydrophilic membranes such as the hydrophilic PVDF type, with only a fraction of monolayer adsorption. A study of the adsorption of bovine serum albumin (molecular weight 69,000 Da) on UF membranes indicated the following extents of protein coverage within the ultrathin skin portion of the membrane: monolayer adsorption for the 300,000-Da MWCO membrane, only about half a monolayer for the 100,000-Da MWCO membrane, and no measurable protein adsorption for the 50,000-Da MWCO membrane [Robertson and Zydney, *J. Colloid Interface Sci.* **134:** 563 (1990)]. Thus, protein fouling within the membrane can be completely eliminated by making the pore size small enough.

To account for fouling caused by irreversible adsorption and deposition of proteins and other biological molecules to the membrane surface, it is useful to write Darcy's law [Eq. (20-31)] for the transmembrane flux in terms of the clean membrane resistance R_m, the resistance due to concentration polarization and a gel layer on the membrane surface R_p, and the resistance caused by irreversible fouling R_{if} [Rajabzadeh et al., *J. Membrane Sci.* **361:** 191 (2010)]:

$$J_m = \frac{\Delta p - \sigma\Delta\pi}{\mu_o(R_m + R_p + R_{if})} \tag{20-36}$$

In this model, irreversible fouling is considered the fouling that can only be removed by cleaning agents such as acids, bases, enzymes, surfactants, and disinfectants; and the resistance caused by concentration polarization and any gel layer is that which can be removed by physical means, such as circulation of feed through the system at zero transmembrane pressure and relatively high shear rate, or by backflushing using negative transmembrane pressure to remove foulants from the membrane surface [Zeman and Zydney, *Microfiltration and Ultrafiltration*, Dekker, New York, 1996, p. 452]. A comprehensive review of methods to control fouling in microfiltration and ultrafiltration membranes has been written by Hilal et al. [*Sep. Sci. Tech.* **40:** 1957 (2005)].

Filter Media and Equipment In the development of a filtration process, much effort is often devoted to the evaluation of the filter media and equipment that are available. The filter media and equipment selected can have a large impact on the process economics, in terms of both the capital outlay required and the operating expenses.

Filter media for cross-flow filtration are generally referred to as *membranes*. There are two general categories of membranes: ultrafiltration membranes and microporous membranes. The separation between ultrafiltration (UF) and microfiltration (MF) is based on the pore size of the membrane, with membranes having pores 0.1 μm and larger considered to be microporous. The pore sizes of UF membranes are in the range of 0.001 to 0.1 μm [Zeman and Zydney, *Microfiltration and Ultrafiltration*, Dekker, New York, 1996, p. 13], although these membranes are usually classified by their *molecular weight cutoff* (MWCO), which is the molecular weight of a globular solute at which the solute is rejected by the membrane. Typically, a 90 percent rejection level is used for establishing the MWCO. Ultrafiltration membranes can be obtained down to an MWCO level of 1000 daltons (Da) and up to as high as 1,000,000 Da. Ultrafiltration membranes are commonly used to concentrate and for changing the buffer for biomolecules such as proteins, while microfiltration is typically used to concentrate and wash cell suspensions. There is a class of membranes known as *nanofilters* that are used to remove virus particles based on their size. These membranes are typically operated in cross-flow mode due to the high resistance of the membrane. Virus filters are typically operated at about 100 kPa and achieve fluxes of about 60 L m^{-2} h^{-1}. Finally, there is also a category of membranes called *reverse-osmosis* (RO) or hyperfiltration membranes that pass only water and a very low flux of solutes. A very good RO membrane rejects 99.7 percent or more of sodium chloride (Henry et al., in *Perry's Chemical Engineers' Handbook*, 7th ed., McGraw-Hill, New York, 1997, pp. 22–49).

Three basic structures are commonly used for membranes: homogeneous, asymmetric, and composite. These three types are illustrated in Fig. 20-37. The homogeneous structure has no significant variation in pore diameter from the feed side to the filtrate side. In the asymmetric structure, there is a thin layer on the feed side of the filter that has very small pores. Below this layer is a much thicker layer that has much larger pores and serves as structural support for the membrane. The composite membrane is similar to the asymmetric membrane in having a thin layer containing very small pores next to the filtering surface; however, the thin and thick layers of this membrane are made of two different types of material.

Backflushing composite filters is generally not recommended. The major element in pressure drop across the filter is due to the thin layer with very small pores, and so significantly more mechanical force is applied to the area of the thin layer, compared to the support layer. When operated in the intended direction, the thin layer is forced toward the support layer. But when operated in the opposite direction, the thin layer is forced away from the support layer, which can lead to delamination.

Filtration membranes are made from a wide variety of polymers and inorganic materials. The polymers that are used include cellulose acetate,

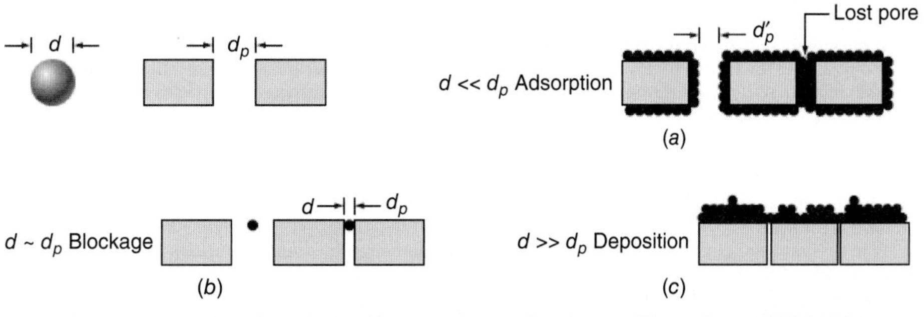

FIG. 20-36 Mechanisms of membrane fouling. (*a*) Pore narrowing and constriction. (*b*) Pore plugging. (*c*) Solute deposition and formation of a gel or cake layer. Dimensions: *d*, protein or particle diameter; d_p, clean pore diameter; d'_p, effective pore diameter in the presence of adsorbed proteins or particles. [Reproduced from *Bioseparations Science and Engineering*, 2d ed., by Roger G. Harrison, Paul W. Todd, Scott R. Rudge, and Demetri P. Petrides (2015) By Permission of Oxford University Press.]

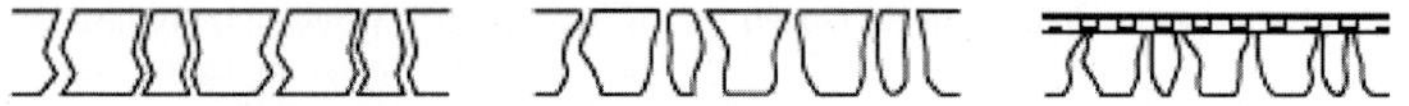

FIG. 20-37 Three commonly used membrane structures: (*a*) homogeneous: uniform pore profile through filter; (*b*) asymmetric: finer filtering surface faces feed suspension; and (*c*) composite: two types of materials used. [Reproduced from *Bioseparations Science and Engineering*, 2d ed., by Roger G. Harrison, Paul W. Todd, Scott R. Rudge, and Demetri P. Petrides (2015). By Permission of Oxford University Press.]

polyamide, polyether, polycarbonate, polyester, polypropylene, polyethylene, regenerated cellulose, poly(vinyl chloride), poly(vinylidene fluoride) (PVDF), poly(tetrafluoroethylene) (PTFE), acrylonitrile copolymers, and polyethersulfones. The inorganic materials used include ceramics, zirconium oxide, borosilicate glass, stainless steel, and silver.

Cross-flow filtration membranes are available in a variety of configurations. The membranes are housed in a physical unit called a *module.* The membrane module must satisfy a number of mechanical, hydrodynamic, and economic requirements, the most important of which are the following [Zeman and Zydney, *Microfiltration and Ultrafiltration*, Dekker, New York, 1996, p. 327]:

Mechanical: It must obtain effective (physical) separation of the feed and permeate streams; provide the necessary physical support for the membrane (including the ability of the module to endure the required pressure drops and any backflushing).

Hydrodynamic: It must minimize pressure drops through the module (to reduce pumping costs); optimize solute mass transfer (reduce concentration polarization); minimize particulate plugging or fouling; avoid dead spots (for sanitary design); allow for or promote turbulence (for sanitary design and to reduce boundary layer thickness).

Economic: It must maximize the membrane packing density (ratio of membrane area to module volume); minimize manufacturing costs; permit easy access for cleaning and/or membrane replacement; provide sufficient chemical resistance and operational lifetime; incorporate modularity of design for easy scale-up, staging, or cascading.

Several of these criteria are in mutual opposition: for example, modules with high membrane packing density tend to be highly susceptible to plugging with particulates. Therefore, the choice of a particular module involves balancing these criteria to arrive at the most economic system for each particular application.

There are five types of module configuration for cross-flow filtration (Fig. 20-38): hollow fiber, tubular, flat plate, spiral wound, and rotating. The tubular module has the same general configuration as the hollow fiber module, but the tubes have much larger diameters than the fibers. Some key characteristics of these modules are compared in Table 20-18.

Hollow fiber modules consist of an array of narrow-bore, self-supporting fibers that generally have an asymmetric membrane structure. The dense skin layer is usually on the lumen side of the fiber, but it can be placed on the outside. The feed flows through the lumen of the fibers when the dense skin layer is on the lumen side, and the flow is typically laminar, although hollow fiber systems by some manufacturers can be operated in turbulent flow. Since the hollow fibers are self-supporting, they can be cleaned by backflushing, that is, by reversing the direction of the permeate flow. One disadvantage of hollow fiber modules is that the entire module usually needs to be replaced upon the rupture of even a single fiber. Also, to avoid plugging of the small-diameter fibers, feed streams generally need to be prefiltered.

Except for some inorganic membranes, tubular membranes are not self-supporting and are usually cast in place within a porous support tube made of fiberglass, ceramic, plastic, or stainless steel. The feed flows through the inside of the tubes, while the permeate flows radially outward across the membrane and support tube. Inorganic membranes are usually constructed in a honeycomb monolith unit in which the membranes are arranged in a parallel array. Tubular systems are typically operated in the turbulent flow regime. Because of the large diameter of the tubes in these systems, the pumping costs are relatively high compared to other module configurations operated in turbulent flow. In addition, large-diameter tubes decrease the filter surface area to module volume ratio, leading to bigger equipment footprints. The chief advantages of tubular modules are that they are highly resistant to plugging by particulates, and they can be cleaned by backflushing. The diameter of the tubes can be selected to avoid prefiltration of the feed.

Most flat plate systems are in a rectangular configuration. The flow channels can be open or can have separator screens to improve the mass-transfer characteristics. Units with open channels are usually operated in laminar flow, while those with separator screens are often operated with flow in the turbulent regime. One disadvantage of these modules is that backflushing is often impractical because the membranes are effectively supported only on one side.

Spiral-wound modules are constructed from flat sheet membranes separated by spacer screens. The feed solution is fed into one end of the module and flows through the separator screens along the surface of the membranes. The retentate is then collected on the other end of the module. The permeate spirals radially inward, eventually to be collected through a central tube. The flow in these systems tends to be turbulent. The main disadvantage of spiral-wound modules is that they are susceptible to fouling by particulates because of the narrow and irregular flow through the spacer screens.

One effective rotating-type module is the rotating cylinder device originally proposed by Hallstrom and Lopez-Leiva [*Desalination* **24:** 273 (1978)]. The feed flows into a thin annular region between two concentric cylinders (Fig. 20-38*b*). Either cylinder or both can be porous and can have a membrane bound to the surface. With the membrane on the inner cylinder, permeate is collected in the central chamber. When the membrane is on the outer cylinder, permeate is collected in a separate annular permeate region adjacent to the outer porous cylinder. The inner cylinder is rotated at high speed (typically > 3000 rpm) to induce the formation of Taylor vortices, which effectively mix the fluid near the surface of the membrane and thereby increase the mass-transfer coefficient for the solutes/particles in the feed. The vortices also help reduce the thickness of the cake or gel layer at the membrane surface. One advantage of this system is that the rate of mass transfer is determined almost entirely by the rate of rotation of the inner cylinder, which effectively decouples the mass-transfer characteristics from the feed flow rate. As a result, the rotating devices can be operated at very low flow rates and with minimal pumping costs. Since, however, the energy requirements for rotating the device are usually very high, capital costs are high, and scale-up is very difficult, these systems have been limited to small-scale operation.

Scale-Up and Design Experimental testing is generally needed in the scale-up and design of cross-flow filtration systems. Cross-flow filtration modules are available from manufacturers for carrying out laboratory or pilot plant tests. The size of a plant unit can be determined by a direct scale-up of the filtration area based on the feed or permeate flow rate. For this scale-up, however, it is important that the following variables be kept constant [Datar and Rosen, in Rehm and Reed (eds.), *Biotechnology*, vol. 3, *Bioprocessing*, VCH, Weinheim, Germany, 1993, p. 486]:

- Inlet and outlet pressures
- Cross-flow (or tangential) velocity
- Flow channel sizes (height and width)
- Feed stream properties—test slurries should be representative of the actual process streams
- Membrane type and configuration—test data from one design cannot directly be used to design another geometry

Maintaining the inlet and outlet pressures, the cross-flow velocity, and the flow channel size means that the length of the flow path is constant as well. It is also important to make an assessment of the rate of fouling of the membranes. Tests should be performed for several cycles of operation to estimate the fouling rate and to determine how fouling can be kept to a minimum.

There are four basic modes of operation of cross-flow filtration (Fig. 20-39): batch concentration, diafiltration, purification, and complete recycle. In the batch concentration mode, the retained stream containing the product suspended particles or dissolved macromolecules is reduced in volume. In diafiltration, the volume of the retained stream is kept constant by the continuous addition of water or buffer, which results in the removal of low-molecular-weight solutes into the permeate. Diafiltration is commonly used when salt removal or exchange is desired. In the purification mode, a low-molecular-weight product passes into the permeate and is thus separated from higher-molecular-weight impurities; or the product can be retained and impurities removed in the permeate. In the complete recycle mode, both the retained stream and the permeate are returned to the feed tank. Systems may be operated in complete recycle at start-up to reach steady state, saturate the membranes, test for leaks and blockage, and adjust the feed rate.

Diafiltration is similar to dialysis in that salts are exchanged or removed in both modes. Dialysis is a laboratory method that involves placing the

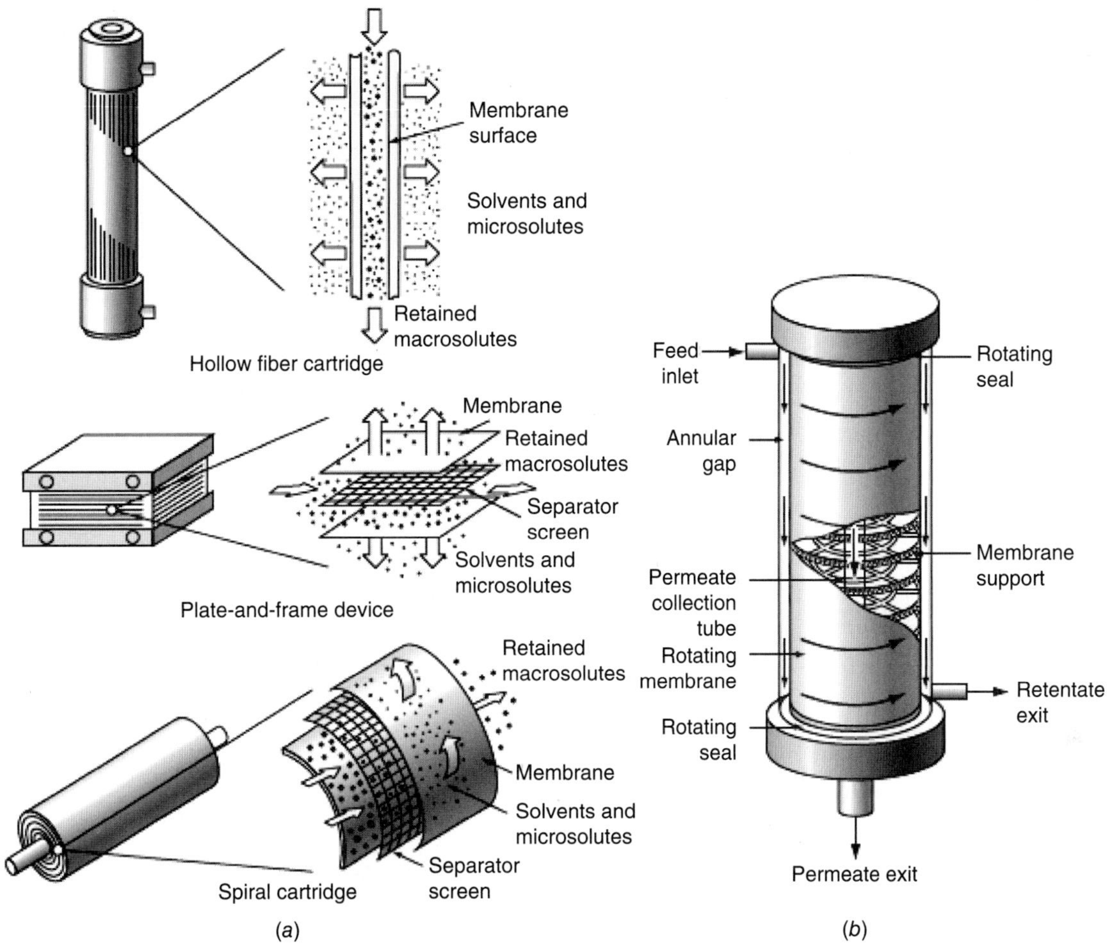

FIG. 20-38 Schematic representations of filter modules. (*a*) Hollow fiber, plate, and spiral-wound membrane modules. (*b*) A rotating cylinder module. [Reproduced from *Bioseparations Science and Engineering*, 2d ed., by Roger G. Harrison, Paul W. Todd, Scott R. Rudge, and Demetri P. Petrides (2015). By Permission of Oxford University Press.]

feed in a bag made with an ultrafiltration membrane. The bag is placed in a large volume of the new solvent desired, and the new solvent is mixed until the system comes to equilibrium or near equilibrium. Complete equilibrium takes 3 to 5 h, while 90 percent equilibrium can be reached in 2 to 3 h (Scopes, *Protein Purification*, Springer-Verlag, 1982, p. 16). For a complete change of solvent, it is necessary to do the dialysis at least twice. It is frequently the job of the engineer to scale up a dialysis done in the laboratory to a diafiltration in the pilot plant or plant. For scaling and planning, it is customary to perform material balances on solids mass and liquid volume around any component of the operation.

Example 20-3 Diafiltration Mode in Cross-Flow Filtration It is desired to use a cross-flow filtration system to desalt 1000 L of a protein solution containing NaCl. The system is capable of operating at a transmembrane flux of 30 L m^{-2} h^{-1}. A membrane is used that will allow complete passage of the salt but no passage of the protein. To remove 99.99 percent of the salt, determine the time required and the volume of water required using a cross-flow filtration unit with a membrane area of 100 m^2.

We operate the system in the diafiltration mode, which means that the volume of the solution being desalted is maintained constant. A material balance on the salt in the retained volume gives

$$V\frac{dc_s}{dt} = -Qc_s(1-r_s)$$

where V = volume of solution being desalted
c_s = concentration of salt in volume V
r_s = fraction of salt rejected by the membrane
Q = filtration rate (volume/time)

For this case there is no rejection of salt by the membrane, so that $r_s = 0$. Integrating the differential equation with the initial time equal to 0 gives

$$\ln\frac{c_s}{c_{so}} = -\frac{Qt}{V}$$

TABLE 20-18 Comparison of Key Characteristics of Cross-Flow Membrane Modules*

Module type	Channel spacing (cm)	Packing density (m^2/m^3)	Energy costs	Particulate plugging	Ease of cleaning
Hollow fiber	0.02–0.25	1200	Low	High	Fair
Tubular	1.0–2.5	60	High	Low	Excellent
Flat plate	0.03–0.25	300	Moderate	Moderate	Good
Spiral wound	0.03–0.1	600	Low	Very high	Poor to fair
Rotating	0.05–0.1	10	Very high	Moderate	Fair

*Zeman and Zydney, *Microfiltration and Ultrafiltration*, Dekker, New York, 1996. p. 331.
Reproduced from *Bioseparations Science and Engineering*, 2d ed., by Roger G. Harrison, Paul W. Todd, Scott R. Rudge, and Demetri P. Petrides (2015). By Permission of Oxford University Press.

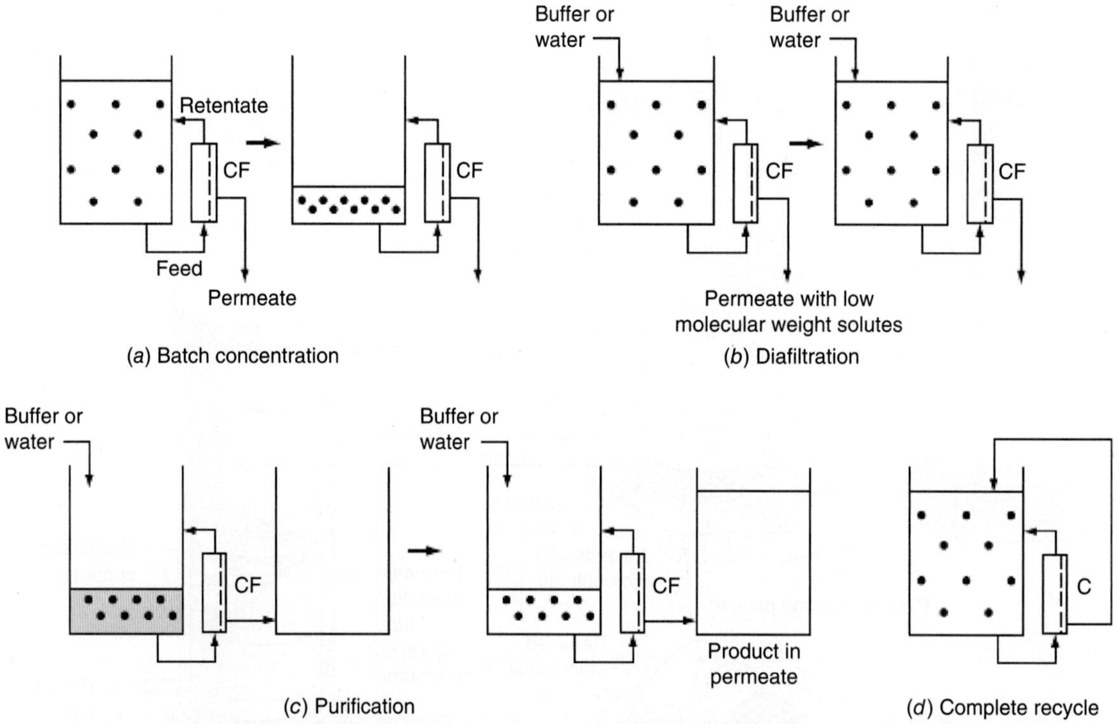

FIG. 20-39 The four basic modes of operation of cross-flow filtration (CF = cross-flow filter): before the arrow, start of the operation; after the arrow, end of the operation. [Reproduced from *Bioseparations Science and Engineering*, 2d ed., by Roger G. Harrison, Paul W. Todd, Scott R. Rudge, and Demetri P. Petrides (2015). By Permission of Oxford University Press.]

where c_{so} = initial salt concentration. Solving for t, we obtain

$$t = -\frac{V \ln \frac{c_s}{c_{so}}}{Q} = -\frac{(1000\ \text{L})(\ln 0.0001)}{\left(30 \frac{\text{L}}{\text{m}^2\,\text{h}}\right)(100\ \text{m}^2)} = 3.07\ \text{h}$$

$$\text{Volume of water required} = (3.07\ \text{h})\left(30 \frac{\text{L}}{\text{m}^2\text{h}}\right)(100\ \text{m}^2) = 9210\ \text{L}$$

Thus, a relatively large volume of wastewater will be generated by this process.

In designing a diafiltration process, a decision must be made about the concentration of retained product at which to operate. As this concentration is increased, the filtration flux will decrease according to Eq. (20-24), and the total volume of filtrate will decrease for the removal of a given percentage of a low-molecular-weight solute. This leads to an optimum concentration to minimize the time required, which can be determined mathematically if the relation between filtrate flux and concentration in the bulk fluid (c_b) is known [Tutunjian, in Moo-Young (ed.), *Comprehensive Biotechnology*, vol. 2, *Principles of Biotechnology*, Pergamon, 1985, p. 411].

The basic components in the design of a cross-flow filtration system are shown in Fig. 20-40. A pump flows the feed through the filtration module to give a permeate and a retentate or retained stream. The pump needs to be sized to provide the desired flow velocity and pressure. The transmembrane pressure is controlled by a back-pressure valve on the retentate stream exiting from the filtration module. Thus the transmembrane pressure drop is estimated by

$$\Delta p_{TM} = \frac{1}{2}(p_i + p_o) - p_p \tag{20-37}$$

where p_i and p_o are the retentate pressures in and out of the module, respectively, and p_p is the pressure of the outlet permeate. In designing a cross-flow filtration system, it is important to minimize the occurrence of gas-liquid interfaces, since bioproduct denaturation, especially of proteins, can occur at these interfaces in the presence of mechanical shear and turbulent flow [Virkar et al., *Biotechnol. Bioeng.* **23:** 425 (1981)].

Cross-flow filtration systems may be designed to operate in the batch or the continuous mode (Fig. 20-41). In a batch system, feed is pumped through the filtration module and then back to the feed tank. In a variation of this mode (called *semibatch*) for diafiltration, fluid is continuously added to the feed tank to keep the feed volume constant (Fig. 20-39*b*).

In the continuous mode of operation, also sometimes called the *feed-and-bleed* mode, or *retentate bleed* mode, feed is added to a recirculation loop by the feed pump, and concentrate exiting in the retained stream is withdrawn from the system so that the concentration factor (i.e., concentration in the retentate divided by the concentration in the feed) is at the desired value. When steady state has been achieved, the concentrate will be at its maximum concentration, which means that the filtration flux will be at a minimum throughout the run [Tutunjian, in Moo-Young (ed.), *Comprehensive Biotechnology*, vol. 2, *Principles of Biotechnology*, Pergamon, 1985, p. 411]. It is generally more economical to use a multistage system in a continuous process (Fig. 20-42). As more stages are added, the average filtration flux approaches that for a batch system, and thus the total filtration area decreases as more stages are added. This is illustrated in Table 20-19, where batch ultrafiltration operation is compared with continuous operation using one, two, three, and five stages.

Because of the economic advantages of continuous operation and reduced tankage, this scheme is preferable to batch operation for most large-scale ultrafiltration operations (Henry et al., in *Perry's Chemical Engineers'*

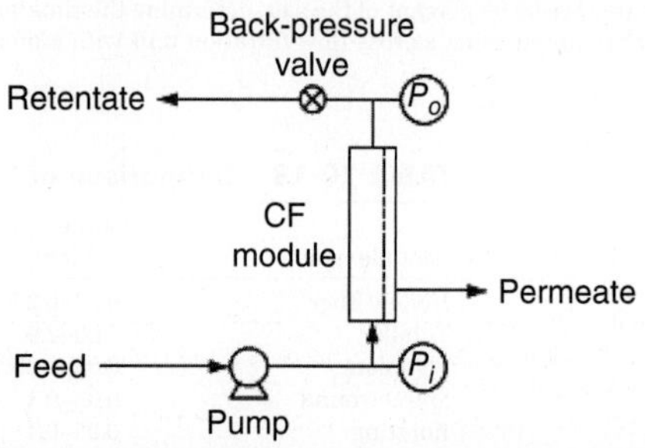

FIG. 20-40 Basic components of a cross-flow filtration system. [Reproduced from *Bioseparations Science and Engineering*, 2d ed., by Roger G. Harrison, Paul W. Todd, Scott R. Rudge, and Demetri P. Petrides (2015). By Permission of Oxford University Press.]

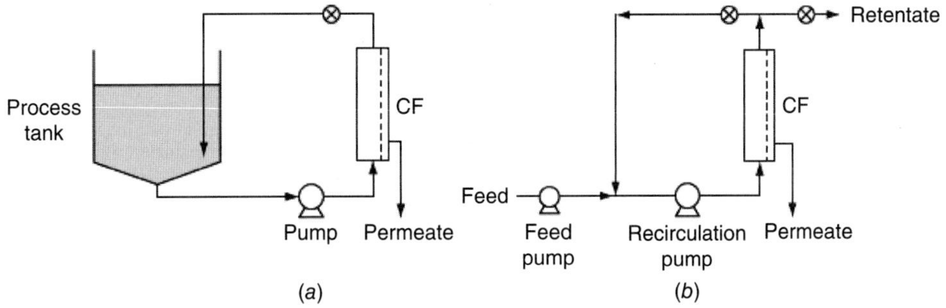

FIG. 20-41 Comparison of (*a*) batch and (*b*) single-stage continuous (feed-and-bleed) cross-flow filtration systems. [Reproduced from *Bioseparations Science and Engineering,* 2d ed., by Roger G. Harrison, Paul W. Todd, Scott R. Rudge, and Demetri P. Petrides (2015). By Permission of Oxford University Press.]

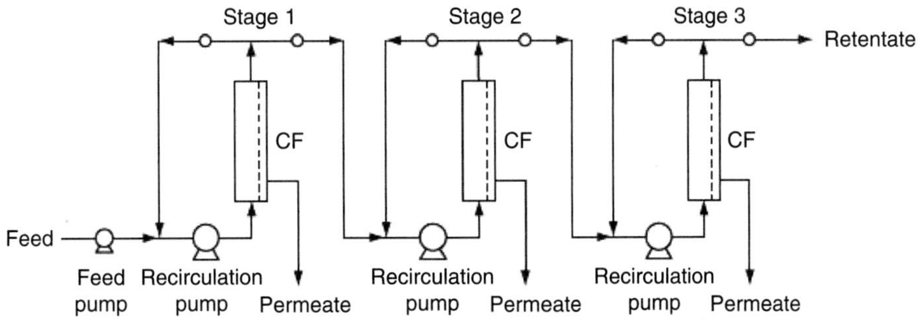

FIG. 20-42 Multistage cross- flow filtration system using the retentate bleed mode. [Reproduced from *Bioseparations Science and Engineering,* 2d ed., by Roger G. Harrison, Paul W. Todd, Scott R. Rudge, and Demetri P. Petrides (2015). By Permission of Oxford University Press.]

Handbook, 7th ed., McGraw-Hill, New York, 1997, pp. 22–56). Another advantage of continuous operation is that it permits the minimization of the residence time of the product in the cross-flow filtration unit, which is important for products that are sensitive to heat or shear. Dairy and food proteins are generally processed continuously, while most pharmaceutical and biological products are processed using batch operation [Tutunjian, in Moo-Young (ed.), *Comprehensive Biotechnology,* vol. 2, *Principles of Biotechnology,* Pergamon, 1985, p. 411].

Sterile Filter Design Sterile filtration is an important application of conventional filters in the pharmaceutical industry, where many products are sold in sterile form. It is sometimes desirable that other bioseparation processing steps besides the final one be carried out under sterile conditions, which requires the filtration of both liquids and air.

Sterile Liquid Filtration In typical sterile liquid filtration applications, there is no appreciable filter load, no cake buildup, and no contribution to filtration resistance from the filtrate that is reliable or reproducible. Further, there is no opportunity to add filter aids to form a cake, since this is usually the last step in a clean process.

Sizing of the filter is almost always based on the maximum allowable load in the solution to be filtered as well as the desired sterility assurance limit. The sterility assurance limit is the calculated probability of a single unit of product containing a single microorganism. This is expected to be a maximum of 10^{-4} for aseptic processes, and one usually designs in an extra order of magnitude at least. One determines the maximum load on the filter by knowing the bioburden or viral load specification of the purified pool from the preceding step, or by knowing the acceptable microbial or viral levels in the raw materials that make up that pool. The retentive requirement for the filter is then divided by the retention capability (per area) of the filter to be used. The resulting filter area is the minimum required to meet the sterility goal of fewer than one unit dose contaminated with one organism per million doses manufactured. The next step is to ensure that the filter is large enough to allow reasonable processing times. Typically, one does not want to filter for more than one shift (8 h). At 24 h, questions will be raised about "grow-through" (where retained microbes will have colonized the filter and "grown through" to the other side). The filtration time for the filter size is determined by finding the clean water flow rate for the filter, either through laboratory experiment or through the vendor's published data, and adjusting for the viscosity of the solution to be filtered relative to water (a correction for the effect of temperature on viscosity may be necessary). Once the flow rate is known, the time to filter a given size batch can be determined. The area of the filter can be increased to reduce the filtration time, but cannot be decreased without jeopardizing the sterility assurance.

For the ultimate filtration step, in which the product is rendered sterile, a depth filter can be added upstream of the membrane filter. This protects the sterile filter from fouling during use. This is a critical consideration in high-concentration process streams. It is very risky to replace a sterile filter once in place, with all the connections downstream sterilized. Depth filters with high "dirt loads" can keep particulate matter from reaching the membrane surface, keeping it clean for the removal of microbes. Two sterile filters are sometimes used in series, for added sterility assurance. This gives very high probability that dosage units will not be contaminated, but the arrangement sometimes becomes difficult to manage in other respects. For instance, multiple sterility assurance tests must be performed, a requirement that adds manipulations and increases chances for breaches in sterility.

TABLE 20-19 Comparison of Batch and Continuous Ultrafiltration Systems Using a Model of Flux as a Function of Bulk Concentration*

System†	Flux (L m^{-2} h^{-1})	Total area (m^2)
Batch	33.1 (average)	136
Continuous		
One-stage	8.1	555
Two-stage	31.1	243
	8.1	
Three-stage	38.7	194
	23.4	
	8.1	
Five-stage	44.7	165
	35.6	
	26.4	
	17.3	
	8.1	

*Tutunjian, in M. Moo-Young (ed.), *Comprehensive Biotechnology,* vol. 2, *Principles of Biotechnology,* Pergamon, Oxford, UK, 1985, p. 411.

†System design for 10× concentration factor and feed rate of 5000 L/h.

Flux from $J_m = 20 \ln (30/c_b)$.

Reproduced from *Bioseparations Science and Engineering,* 2d ed., by Roger G. Harrison, Paul W. Todd, Scott R. Rudge, and Demetri P. Petrides (2015). By Permission of Oxford University Press.

Filters used for the purpose of sterilizing pharmaceutical products must be tested to ensure that they have been properly installed and have no defects that could allow sterility to be compromised. The two most common tests for a sterile filter for an aqueous product are the *bubble point* and the *pressure hold* or *diffusion* test. In both tests, a wetted, sterilized filter is pressurized with air or nitrogen. In the bubble point test, this pressure is increased until the lowest pressure is found at which a gas passes freely through the filter. It is assumed that this is the pressure required to overcome the surface tension and viscosity of water in the largest membrane pore (where the forces are the weakest). As the pore size of the membrane decreases, the bubble point increases. This test will also detect a sliced or unseated O-ring or other seal failure. The pressure hold test pressurizes a wetted filter to a pressure lower than the bubble point and measures either the total flow of gas through the membrane (by diffusion) or the rate of pressure decay. Both should be low, and both should meet the manufacturer's specifications for an intact, properly installed sterile filter.

Compatibility and extractable tests need to be performed on each filter type. Compatibility means that the filter does not adsorb an active ingredient or any other type of active excipient, such as a preservative. Such adsorption to the filter medium can change the strength (or biological potency) of the solution after the filtration step. Extractables and leachable chemicals, such as monomers from the membrane or support medium, filter coatings or mold release agents, and storage solutions, such as ethanol, must also be tested for in pilot studies with the product. These chemicals would be considered impurities in the final product and should be removed or prevented from leaching. Compatibility and extractables should be tested in a sample of the product solution. When this is not possible, because product or excipients would interfere with the analysis of the impurities, a placebo or similar solution should be used.

Air Sterile Filtration Air filters are also sized based on the anticipated flow rate. Airflows are specified with one or more of the following criteria: pressure differential between areas, air changes per hour, or linear flow rate. Pressure differentials are usually measured between rooms, or between rooms and corridors. These are specified when it is important for air to flow from a "cleaner" area to a "dirtier" area when doors, windows, or pass-throughs are opened between them. Pressure differentials are typically 50 to 150 Pa [= 0.5 to 1.5 cm of water, or 0.4 to 1.1 mm Hg (torr)]. Air changes per hour are often specified for rooms based on hazardous materials or clean operations the room is designed for. *Air changeover* is the airflow rate divided by the volume of the room. Air changeovers are different from *makeup rate*, which is the fraction of fresh air injected in the room's air handler.

Clean rooms often have very high changeover rates but low makeup rates. Rooms handling solvents or hazardous materials have high makeup rates (a 100 percent makeup rate is called *once-through air*) but low air cleanliness requirements. The changeover rate varies with the hazard of the material or the cleanliness requirement for the room. Linear flow rate is usually specified in hoods and smaller spaces. Linear flow rates of 5 to 20 ft/s (= 2.4 to 6.1 m/s) are found in laminar flow hoods, in aseptic areas where pharmaceuticals are sterile-processed, and in clean rooms where microelectronics are manufactured. Linear flow rates are established to prevent airborne microbes and particulate matter from settling in or on the materials being processed.

The air quality of a room or area is typically classified according to the number of airborne particles that can be measured in 1 m^3. Increasingly, rooms are being classified according to the International Organization for Standardization (ISO), which has published a series of documents numbered 14644 (volumes 1 to 5). These standards classify clean room space as 1 (cleanest) through 9 (dirtiest), with each classification allowing approximately one log more particles in a cubic meter of air than the previous. For biotechnology and pharmaceuticals, the most commonly used grades are ISO Class 5 (similar to the military standard Class 100) to ISO Class 8 (similar to the military standard class 100,000). ISO Class 5 is used for aseptic processing where products may be briefly exposed to the environment, and ISO Class 8 is used for areas containing primarily closed operations. High-efficiency particulate air (HEPA) filters are routinely tested for integrity by spraying a polyalphaolefin (PAO) aerosol on the feed side of the filters. The typical specification for a PAO test is a 10^3 reduction in aerosol particles.

Vent filters filter air on tanks and drainable equipment. Vent filters typically follow the same specification as the liquid filters used in the process. Vent filters are also sized based on the maximum airflow rate anticipated. The maximum flow rate is often encountered when the tanks are being filled or cleaned. Vent filters must allow flow in both directions. Integrity testing of vent filters is an evolving technology. These filters are typically hydrophobic, so aqueous-based bubble points and pressure holds are not appropriate. The PAO tests will not work because vent filters are typically membrane filters (which are the most compact), while HEPA filters are depth filters. Testing bubble points in solvents is feasible, and often performed, but is considered a destructive form of measurement.

SEDIMENTATION

Sedimentation is the movement of particles or macromolecules in an inertial field. Its applications in separation technology are widespread. Extremes of applications range from the settling due to gravity of tons of solid waste and bacteria in wastewater treatment plants to the high speed centrifugation of a few microliters of blood to determine packed blood cell volume in the clinical laboratory. Accelerations range from 1*g* in flocculation tanks to 100,000*g* in ultracentrifuges for measuring the sedimentation rates of macromolecules. In bioprocessing, the most frequent applications of sedimentation include the clarification of broths and lysates, the collection of cells and inclusion bodies, and the separation of fluids having different densities.

Unit operations in sedimentation include settling tanks and tubular centrifuges for batch processing, continuous centrifuges such as disk centrifuges, and less frequently used unit operations such as field-flow fractionators and inclined settlers. Bench-scale centrifuges that accommodate small samples can be found in most research laboratories and are frequently applied to the processing of bench-scale cell cultures and enzyme preparations. Certain high-speed ultracentrifuges are used as analytical tools for the estimation of molecular weights and diffusion coefficients.

The common types of production centrifuges are shown in Fig. 20-43, and a comparison of the advantages and disadvantages of the different centrifuge designs is given in Table 20-20 [Bell et al., in Fiechter (ed.), *Advances in Biochemical Engineering/Biotechnology*, vol. 26, Springer-Verlag, Berlin, 1983, p. 1]. The principles of sedimentation and more details about equipment for centrifugation are discussed in the Centrifuges subsection. Here, additional information on this topic that is specific for a bioproduct separation process is given, including the range of particle sizes and densities that are encountered, the concept of the sedimentation coefficient and its application, and the concept of equivalent time, which has been found to be useful for scale-up.

Fluid Dynamics During the sedimentation of a particle in the presence of a centrifugal field, the velocity of the particle reaches a constant value when all the forces are balanced. For a centrifugal field of acceleration $\omega^2 R$, where ω is the centrifuge's angular velocity and R is the distance of the particle from the center of rotation, at constant particle velocity v for a particle of radius a_p and density ρ in a fluid of density ρ_0 and viscosity μ, an analysis using the equation of motion gives (Harrison et al., *Bioseparations Science and Engineering*, Oxford University Press, Oxford, UK, 2015, p. 187)

$$v = \frac{2a_p^2(\rho - \rho_o)\omega^2 R}{9\mu} \tag{20-38}$$

Creeping flow conditions are usually satisfied in sedimentation. The Reynolds number for sedimenting spherical particles is

$$\mathrm{Re} = \frac{2a_p v\rho}{9\mu} \tag{20-39}$$

Creeping flow occurs at Reynolds numbers less than about 0.1 (Bird et al., *Transport Phenomena*, Wiley, Hoboken, N.J., 2002, p. 59).

Table 20-21 shows the sedimentation velocities for important bioparticles and biomolecules calculated from Eq. (20-38) with ρ_0 = 1.0 g/cm^3, μ = 0.01 g cm^{-1} s^{-1} (poise), and representative values of ρ and a_p. The sedimentation velocity and Reynolds number results shown in Table 20-21 for yeast cells and bacterial cells at gravitation acceleration can be multiplied by a centrifuge's centrifugal acceleration to give the corresponding values for operation in the centrifuge; for example, at a dimensionless acceleration of 10,000, the Reynolds number for yeast cells is 0.07, which means that the flow is still creeping.

It is clear from Table 20-21 that gravitational sedimentation is too slow to be practical for bacteria, and conventional centrifugation is too slow for protein macromolecules. In the case of true particles, flocculation is often used to increase the Stokes radius a_p, while ultracentrifugation is used in macromolecular separations (Harrison et al., *Bioseparations Science and Engineering*, Oxford University Press, Oxford, UK, 2015, pp. 203, 205).

When particle density and solvent density are equal, the sedimentation velocity v is zero, and the process is called *isopycnic* or *equilibrium* sedimentation. This fact is exploited in the determination of molecular densities and in the separation of living cells. A density gradient or a density shelf is employed in such cases. Densities of representative cells, organelles, and biomolecules measured by this method are given in Table 20-22.

An example of a density shelf used for the preparation of cells is the preparation of lymphocytes by sedimentation. The goal of this separation

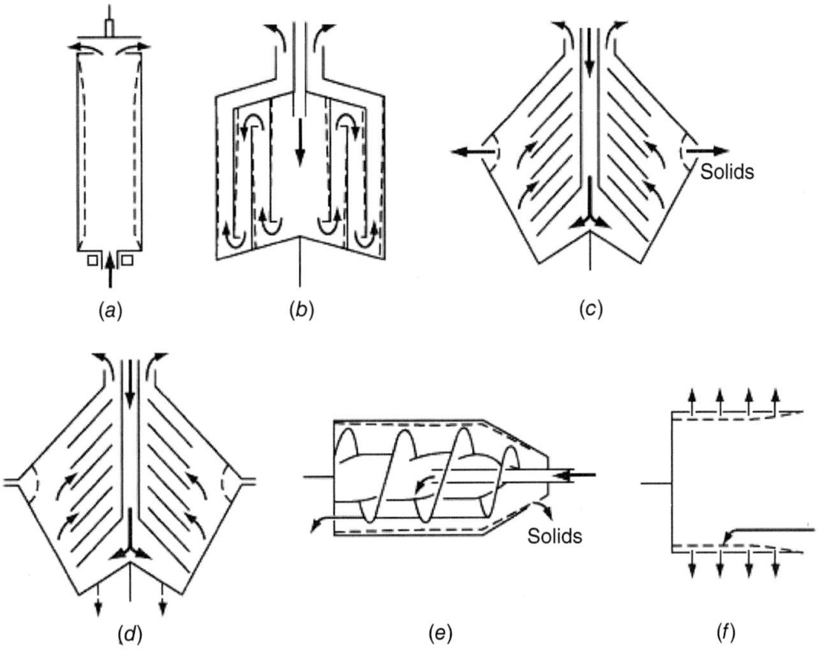

FIG. 20-43 Common types of production centrifuges: (*a*) tubular bowl; (*b*) multichamber; (*c*) disk, nozzle; (*d*) disk, intermittent discharge; (*e*) scroll; and (*f*) basket. Arrows indicate the path of the liquid phase; dashed lines show where the solids accumulate. [Reproduced from *Bioseparations Science and Engineering*, 2d ed., by Roger G. Harrison, Paul W. Todd, Scott R. Rudge, and Demetri P. Petrides (2015). By Permission of Oxford University Press.]

TABLE 20-20 Comparison of Production Centrifuges*

System	Advantages	Disadvantages
Tubular bowl	High centrifugal force Good dewatering Easy to clean Simple dismantling of bowl	Limited solids capacity Foaming unless special skimming or centripetal pump used Recovery of solids difficult
Chamber bowl	Clarification efficiency remains constant until sludge space full Large solids-holding capacity Good dewatering Bowl cooling possible	No solids discharge Cleaning more difficult than with tubular bowl Solids recovery difficult
Disk centrifuge	Solids discharge possible Liquid discharge under pressure eliminates foaming Bowl cooling possible	Poor dewatering Difficult to clean
Scroll or decanter centrifuge	Continuous solids discharge High feed solids concentration	Low centrifugal force Turbulence created by scroll
Basket centrifuge	Solids can be washed well Good dewatering Large solids holding capacity	Not suitable for soft biological solids No solids discharge Recovery of solids difficult

*Bell et al., in Fiechter (ed.), *Advances in Biochemical Engineering/Biotechnology*, vol. 26, Springer-Verlag, Berlin, 1983, p. 1

Reproduced from *Bioseparations Science and Engineering*, 2d ed., by Roger G. Harrison, Paul W. Todd, Scott R. Rudge, and Demetri P. Petrides (2015). By Permission of Oxford University Press.

is to remove erythrocytes from a leukocyte population on the basis of a density shelf. By combining Ficoll, a high-molecular-weight polymer, and Hypaque, a heavily iodinated benzoic acid derivative, in appropriate proportions in aqueous buffers, it is possible to achieve a density around 1.07 g/cm^3 in isotonic solutions. At this density, most white blood cell subpopulations will float and nearly all red blood cells will sediment, which is in agreement with what would be expected for these cells based on their densities, shown in Table 20-22.

When the concentration of sedimenting particles increases, the sedimentation velocity has been found to decrease, a phenomenon known as *hindered settling.* This effect has been quantified by the following expression for particles of any shape [Richardson and Zaki, *Trans. Inst. Chem. Eng.* **32:** 36 (1954)]:

$$v_c = v(1-\varphi)^n \tag{20-40}$$

where v_c is the sedimentation velocity of particles in a concentrated suspension, v is the velocity of individual particles [Eq. (20-38)], φ is the volume fraction of the particles, and n is a function of only the shape of the particle and the Reynolds number. For spherical particles with Re < 0.2 (usually satisfied during sedimentation), the exponent n has been found to be 4.65. Equation (20-40) may also be applied to particles of any size in a polydisperse system, using the volume fraction for all the particles in the calculation [Richardson and Shabi, *Trans. Inst. Chem. Eng.* **38:** 33 (1960)].

The magnitude of the hindered settling effect for spherical particles as a function of the particle volume fraction φ can be seen in Table 20-23. Note

TABLE 20-21 Calculated Settling Velocities and Reynolds Number for Example Bioproducts (Assumes ρ_0 = 1.0 g/cm^3 and μ = 1.0 cP)

Bioparticle or biomolecule	Sedimentation radius a_p (μm)	Density ρ (g/cm^3)	Dimensionless acceleration ($G=\omega^2R/g$)	Sedimentation velocity v (cm/h)	Reynolds number Re
Yeast cell	2.5	1.1	1	0.5	7×10^{-6}
Bacteria cell	0.5	1.1	1	0.02	6×10^{-8}
Protein	0.005	1.3	10^4	0.06	2×10^{-9}

Reproduced from *Bioseparations Science and Engineering*, 2d ed., by Roger G. Harrison, Paul W. Todd, Scott R. Rudge, and Demetri P. Petrides (2015). By Permission of Oxford University Press.

TABLE 20-22 Measured Values of Density of Representative Cells, Organelles, and Biomolecules

Cell, organelle, or biomolecule	Density ρ (g/cm³)	Ref.
Escherichia coli	1.09*	1
Bacillus subtilis	1.12	2
Arthrobacter sp.	1.17	3
Saccharomyces pombe	1.09	1
Saccharomyces cerevisiae	1.11*	1
Amoeba proteus	1.02	1
Murine B cells	1.06*	1
Chinese hamster ovary (CHO) cells	1.06	1
Red blood cells	1.10*	4
White blood cells	1.02*	5
Peroxisomes	1.26*	6
Mitochondria	1.20*	6
Plasma membranes	1.15*	6
Proteins	1.30*	6
Ribosomes	1.57*	6
DNA	1.68*	6
RNA	2.00*	6

*Average value.

References

1. Kubitschek, *Crit. Rev. Microbiol.* **14:** 73 (1987).
2. Hart and Edwards, *Arch. Microbiol.* **147:** 68 (1987).
3. Illmer, *FEMS Microbiol. Lett.* **196:** 85 (2001).
4. Ponder, *J. Biol. Chem.* **144:** 333 (1942).
5. Barnkob et al., 15th International Conference on Miniaturized Systems for Chemistry and Life Sciences, Oct. 2–6, Seattle, Wash., 2011.
6. Sheeler, *Centrifugation in Biology and Medicine*, Wiley-Interscience, New York, 1981.

that hindered settling can be significant for particle concentrations of a few percent or greater.

Sedimentation Coefficient When a body force is applied, velocity through a viscous medium is usually proportional to the accelerating field (examples are electric, magnetic, and inertial). In the case of sedimentation, the resulting constant, a property of both the particle and the medium, is the *sedimentation coefficient*, which is defined as

$$s \equiv \frac{v}{\omega^2 R} \qquad (20\text{-}41)$$

where v is the steady state velocity of the particle, R is the distance of the particle from the center of rotation, and ω is the angular velocity. Comparing this equation with Eq. (20-38), we see that

$$s = \frac{2a_p^{\,2}(\rho - \rho_o)}{9\mu} \qquad (20\text{-}42)$$

which defines s in terms of only properties of the particle and the medium. This coefficient is usually expressed at 20°C and under conditions (viscosity and density) of pure water as

$$s_{20,w} \text{ (units of S)}$$

The sedimentation coefficient is often expressed in svedberg units, where 10^{-13} seconds = 1 svedberg unit (S), named after the inventor of the ultracentrifuge, Theodor Svedberg.

TABLE 20-23 Effect of Particle Volume Fraction φ on the Particle Sedimentation Velocity for Spherical Particles

φ	$\frac{v_c}{v}$
0.01	0.95
0.05	0.79
0.10	0.61
0.20	0.35

Example 20-4 Application of the Sedimentation Coefficient In 1974, D. E. Koppel determined the sedimentation coefficient $s_{20,w}$ for the smaller ribosomes from *Escherichia coli* to be 70 S [Koppel, *Biochemistry* **13:** 2712 (1974)]. Estimate how long it would take to completely clarify a suspension of these ribosomes in a high-speed centrifuge operating at 10,000 rpm with a tube containing the ribosome suspension in which the maximum distance of travel of particles radially outward is 1 cm and the initial distance from the center of rotation to the particles nearest the center of rotation is 4 cm.

We can write Eq. (20-41) as

$$s = \frac{dR}{dt}\frac{1}{\omega^2 R}$$

or

$$\omega^2 s\,dt = \frac{dR}{R}$$

We integrate this equation with the initial condition at $t = 0$, $R = R_0$ (distance from center of rotation to the particles nearest the center of rotation) to give

$$\omega^2 st = \ln\frac{R}{R_o}$$

To determine the maximum time required, we evaluate R at the maximum travel of the cells measured from the center of rotation (5 cm):

$$t = \frac{\ln\left(\frac{R}{R_o}\right)}{\omega^2 s} = \frac{\ln(5/4)\times\frac{1\text{ h}}{3600\text{ s}}}{\left(10{,}000\frac{\text{rev}}{\text{min}}\times\frac{2\pi\text{ rad}}{\text{rev}}\times\frac{1\text{ min}}{60\text{ s}}\right)^2(70\times10^{-13}\text{ s})} = 8.1\text{ h}$$

This should not be an unreasonable amount of time to centrifuge the ribosomes. However, since the time varies inversely with the square of the rotation speed, the time can be reduced to 2 h by doubling the speed.

Equivalent Time To assess the approximate properties of a particle type to be separated, it is sometimes convenient to calculate an "equivalent time." To do this, we first define a dimensionless acceleration G, the ratio of the centrifugal to gravitational acceleration for a particular centrifuge:

$$G \equiv \frac{\omega^2 R}{g} \qquad (20\text{-}43)$$

where R is usually defined as the radius of the centrifuge bowl. Thus, this dimensionless unit is measured in "g's"—multiples of the earth's gravitational acceleration. A rough approximation of the difficulty of a given separation by centrifugation is the product of the dimensionless acceleration and the time required for the separation. This product is called the *equivalent time* for the separation and is written as

$$\text{Equivalent time} = \frac{\omega^2 R}{g}t \qquad (20\text{-}44)$$

Typical values of equivalent time are as follows: 0.3×10^6 s for eukaryotic cells, 9×10^6 s for protein precipitates, 18×10^6 s for bacteria, and 1100×10^6 s for ribosomes (Belter et al., *Bioseparations*, Wiley, Hoboken, N.J., 1988, p. 63).

The equivalent time for the centrifugation of cells or biological particles of unknown sedimentation properties may be estimated in a laboratory centrifuge. Samples are centrifuged for various times until a constant volume of packed cells is reached. The equivalent time Gt is calculated as the product of the G for the particular centrifuge and the time required to reach constant packed cell volume. A centrifuge that has commonly been used for this determination is the Gyro-Tester (Alfa Laval, Inc.).

One approach to scale-up of a centrifugal operation is to assume constant equivalent time:

$$(Gt)_1 = (Gt)_2 \qquad (20\text{-}45)$$

where the subscripts refer to centrifuges 1 and 2, respectively.

Example 20-5 Scale-Up Based on Equivalent Time If bacterial cell debris has $Gt = 54 \times 106$ s (Belter et al., Bioseparations, Wiley, Hoboken, N.J., 1988, p. 63), how large must the centrifuge bowl be and what centrifuge speed is needed to effect a full sedimentation in 2 h?

From Eq. (20-44) for Gt, we can estimate the centrifuge speed ω if we know the centrifuge bowl size and the time of centrifuging. It is reasonable to have a centrifuge that is 10 cm in diameter. Solving Eq. (20-44) for ω by using these values gives

$$\omega = \left(\frac{Gtg}{Rt}\right)^{1/2} = \left(\frac{54\times10^6\text{ s}\times 9.81\frac{\text{m}}{\text{s}^2}}{0.05\text{ m}\times 2(3600)\text{ s}}\right)^{1/2}$$

$$= 1213\frac{\text{rad}}{\text{s}}\times\frac{1\text{ rev}}{2\pi\text{ rad}}\times\frac{60\text{ s}}{\text{min}} = 11{,}590\text{ rpm}$$

This speed can be achieved in a production tubular bowl centrifuge.

Another commonly used method to scale up centrifugation is the *sigma analysis* method, as discussed under the subsection Sedimentation Centrifuges in Sec. 18. This method uses the velocity of particles at 1*g* and a sigma factor that represents the geometry and speed of the centrifuge.

PRECIPITATION

Precipitation, which is the process of coming out of solution as a solid, is an important method in the purification of proteins that usually comes early in the purification process. Precipitation is frequently used in the commercial separation of proteins. The primary advantages of precipitation are that it is relatively inexpensive, can be carried out with simple equipment, can be done continuously, and leads to a form of bioproduct that is often stable in long-term storage. Since precipitation is quite tolerant of various impurities, including nucleic acids and lipids, it is used early in many bioseparation processes.

The goal of precipitation is often concentration to reduce volume, although significant purification can sometimes be achieved. For example, all the protein in a stream might be precipitated and redissolved in a smaller volume, or a fractional precipitation might be carried out to precipitate the protein of interest and leave many of the contaminating proteins in the mother liquor.

In this section, first the focus is upon protein solubility, which is the basis of separations by precipitation. Then we discuss the basic concepts of particle formation and breakage and the distribution of precipitate particle sizes. The specific methods that can be used to precipitate proteins are treated next. Finally, the methodology to use for the design of precipitation systems is discussed.

Protein Solubility The most important factors affecting the solubility of proteins are structure and size, protein charge, and the solvent. Explanations follow for each of these factors.

Structure and Size In the native state, a protein molecule in an aqueous environment assumes a structure that minimizes the contact of the hydrophobic amino acid residues with the water solvent molecules and maximizes the contact of the polar and charged residues with the water. The major forces acting to stabilize a protein in its native state are hydrogen bonding, van der Waals interactions, and solvophobic interactions. In aqueous solution, these forces tend to push the hydrophobic residues into the interior of the protein and the polar and charged residues to the protein's surface. For example, one study of 36 globular proteins has shown that 95 percent of the ionizable groups are solvent accessible [Rashin and Honig, *J. Mol. Biol.* **173:** 515 (1984)]. In other studies of 69 proteins, the average solvent-(water-) accessible atomic surface was found to be 57 percent nonpolar, 25 percent polar, and 19 percent charged [Miller et al., *J. Mol. Biol.* **196:** 641 (1987); Janin et al., *J. Mol. Biol.* **204:** 155 (1988)]. Thus, in spite of the forces operating to force hydrophobic residues to the protein's interior, the surface of proteins usually contains a significant fraction of nonpolar atoms.

The forces acting on a protein lead to the achievement of a minimum Gibbs free energy. For a protein in its native configuration, the net Gibbs free energy is on the order of only 10 to 20 kcal/mol. This is a relatively small net free energy, which means that the native structure is only marginally stable and can be destabilized by relatively small environmental changes [Privalov, *Annu. Rev. Biophys. Biophys. Chem.* **18:** 47 (1989)].

Water molecules bind to the surface of the protein molecule because of association of charged and polar groups and immobilization by nonpolar groups. For example, a study of the hydration of human serum albumin found two layers of water around the protein [Van Oss and Good, *J. Protein Chem.* **7:** 179 (1988)]. In the layer next to the protein, the water molecules are almost totally oriented, with the hydrogen atoms adjacent to and facing the albumin surface, while the oxygen atoms face away from the protein surface. In the second layer of water molecules, most of the water molecules (70 percent) are nonoriented. These hydration layers are thought to promote solubility of the protein by maintaining a distance between the surfaces of protein molecules.

The size of a protein becomes important with respect to solubility when the protein is excluded from part of the solvent. This can happen when nonionic polymers that are added to the solution result in steric exclusion of protein molecules from the volume of solution occupied by the polymer. Juckes [*Biochim. Biophys. Acta* **229:** 535 (1971)] developed a model for this phenomenon based on the protein molecule being in the form of a solid sphere and the polymer molecule in the form of a rod, which gave the following equation for *S*, the solubility of the protein:

$$\ln S = \beta' - K'c_p \tag{20-46}$$

where

$$K' = \frac{\overline{V}}{2.303}\left(\frac{r_s + r_r}{r_r}\right)^3 \tag{20-47}$$

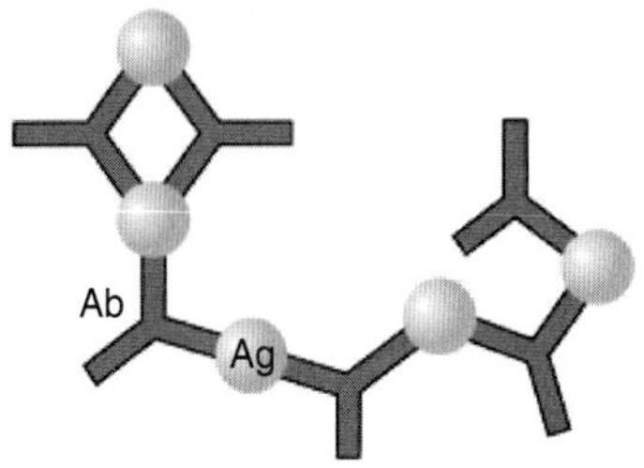

FIG. 20-44 Schematic representation of antibody–antigen (Ab–Ag) interaction. [Reproduced from *Bioseparations Science and Engineering*, 2d ed., by Roger G. Harrison, Paul W. Todd, Scott R. Rudge, and Demetri P. Petrides (2015). By Permission of Oxford University Press.]

Here, r_s and r_r are the radius of the protein solute and polymer rod, respectively, $\overline{V}$ is the partial specific volume of the polymer, c_p is the polymer concentration, and β' is a constant. Based on this model, we can expect the lowest protein solubility for large proteins.

Molecular size is the predominant factor in a type of precipitation known as *affinity precipitation.* When affinity groups or antibodies to a specific biomolecule (antigen) are added to a solution, the antibody-antigen interaction can form large multimolecular complexes, as shown in Fig. 20-44. Such complexes are usually insoluble and cause selective precipitation of the antigen. The average size of the complex agglomerates is maximized when there is a 1:1 stoichiometric ratio of antibody and antigen. If either is present in great excess, only bimolecular complexes will be formed; and there may be no precipitation, or low recovery, even if a precipitate is formed.

Charge The net charge of a protein has a direct bearing upon the protein's solubility. The solubility of a protein increases as its net charge increases, a result of greater interaction with dipolar water molecules. A repulsive reaction between protein molecules of like charge further increases solubility.

A simple way to vary the charge on a protein is by changing the pH of the solution. The pH of the solution in which a protein has zero net charge is called the *isoelectric pH* or *isoelectric point.* The isoelectric pH is known as the pI of the protein. The solubility of a protein is, in general, at its minimum at the isoelectric point. A typical example is shown in Fig. 20-45. Nonuniform charge distribution, however, results in a dipole moment on the molecule, which leads to an increase in solubility and a move in the minimum solubility away from the isoelectric point. The effect of the dipole moment is discussed further in the following subsection.

The net charge of a protein is determined by the following factors: the total number of ionizable residues, the accessibility of the ionizable residues to the solvent, the dissociation constants (or pK_a values) of the ionizable groups, and the pH of the solution [Rothstein, in Harrison (ed.), *Protein Purification Process Engineering*, Dekker, New York, 1994, p. 115]. Besides the chemical makeup of the ionizable groups, factors that can influence the pK_a values are the chemical nature of the neighboring groups (e.g., inductive effects), the temperature, the chemical nature of the solvent as partially reflected by its dielectric constant, and the ionic strength of the solvent.

Solvent The solvent affects the solubility of proteins primarily through two parameters, hydrophobicity and ionic strength. The first parameter has been well studied through observations of single-phase solutions of water and primary alcohols. Although these solutions can cause protein denaturation at room temperature, denaturation can be avoided at sufficiently low temperatures. Studies of primary alcohols have shown that denaturing efficiency is as follows:

methanol < ethanol < propanol < butanol

This led to the conclusion that the alcohols with longer alkyl chains are binding more effectively to nonpolar groups on the protein, weakening intraprotein hydrophobic interactions and thus leading to denaturation [Bull and Breese, *Biopolymers* **17:** 2121 (1978)]. It is thought that when the temperature is low, the primary alcohols compete for the water of hydration on the protein and cause the protein molecules to approach more closely, so that van der Waals interactions lead to aggregation.

The ionic strength of the solvent can have both solubilizing and precipitating effects. The solubilizing effects are referred to as *salting in*, while the precipitating actions are called *salting out.* The salting-in effect has been observed for several proteins, including the class of proteins named *euglobulins.* These proteins are insoluble in the absence of salt at their isoelectric points, but become soluble when salt is added. In contrast, members of another class of proteins, the *albumins,* are very soluble in water as well as in high concentrations of salt. It is believed that the solubilities of the euglobulins and the albumins differ because the euglobulins have a

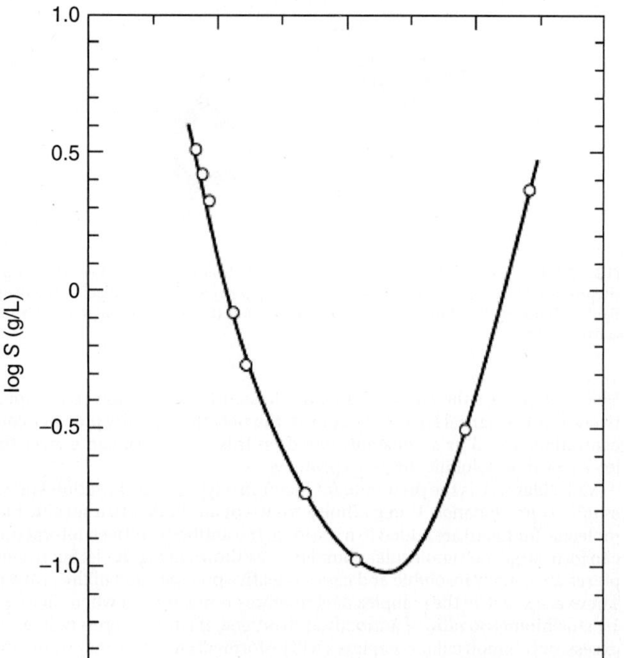

FIG. 20-45 The solubility S of insulin in 0.1 N NaCl as a function of pH. The charge Z is the average protonic charge per 12,000 g of insulin at the pH values indicated. Insulin's pI corresponds to a zero net charge where its solubility is near minimum. (Data from Tanford, *Physical Chemistry of Macromolecules*, Wiley, New York, 1961.)

much higher dipole moment than the albumins [Oncley, in Cohn and Edsall (eds.), *Proteins, Amino Acids, and Peptides*, Reinhold, New York, 1943, p. 543].

A theoretical treatment of the interactions between ions and dipoles developed in 1943 by Kirkwood [Cohn and Edsall (eds.), *Proteins, Amino Acids, and Peptides*, Reinhold, New York, 1943, p. 276] accounts for salting-in effects by considering the solute size, solute shape, solute dipole moment, solvent dielectric constant, solution ionic strength, and temperature. One of Kirkwood's models was for a spherical dipolar ion with a point dipole moment u located at the center of the sphere. The equation derived to describe the interactions is as follows:

$$\ln\left(\frac{S_p}{S_o}\right) = K_i I - K_s I \tag{20-48}$$

where S_p is the solubility of the dipolar ion at ionic strength I, S_0 is the solubility of the dipolar ion in the absence of salt, K_i is the salting-in constant, and K_s is the salting-out constant. Ionic strength is defined by

$$I = \frac{1}{2}\sum_i c_i z_i^2 \tag{20-49}$$

where c_i is the molar concentration of any ion and z_i is its charge. The salting-in and salting-out constants can be related to other variables as follows:

$$K_i \propto \left(\frac{u}{\varepsilon T}\right)^2 \tag{20-50}$$

$$K_s \propto \frac{V_e}{\varepsilon T} \tag{20-51}$$

where ε is the dielectric constant of the solvent, T is absolute temperature, and V_e is the excluded volume of the dipolar ion. Equations (20-48) and (20-50) confirm the observed strong relationship between the solubility of dipolar proteins and the size of their dipole moment u, as well as the greater salting-in effect observed with proteins with high dipole moments. Also it can be seen that the salting-in term increases more than the salting-out term as the dielectric constant decreases. The dielectric constant decreases as the polarity of the solvent decreases. Therefore, the salting-in effect tends to predominate in relatively nonpolar solvents, while the salting-out effect is more dominant in aqueous solvents.

At high ionic strength, the salting-out effect becomes predominant and can be described empirically by the Cohn equation [Cohn, *Physiol. Rev.* **5:** 349 (1925)]:

$$\ln S = \beta - K_s' I \tag{20-52}$$

where S is the solubility of the protein and K_s' is a salting-out constant characteristic of the specific protein and salt that is independent of temperature and pH above the isoelectric point. The constant β, the hypothetical solubility of the protein at zero ionic strength, depends on only temperature and pH for a given protein and is a minimum at the isoelectric point. It is interesting that the Cohn equation is identical in form to the equation that describes the precipitation of proteins by the addition of nonionic polymers, Eq. (20-46). In addition, the Kirkwood equation for the solubility of dipolar ions [Eq. (20-48)] can be arranged to give

$$\ln S_p = \ln S_o - (K_s - K_i)I \tag{20-53}$$

which is also identical in form to the Cohn equation, with

$$\beta = \ln S_o \tag{20-54}$$

$$K_s' = K_s - K_i \tag{20-55}$$

Both salting in and salting out are illustrated in Fig. 20-46 for hemoglobin with ammonium sulfate or sodium sulfate being added. From zero ionic strength, the solubility of the protein increases to a maximum as salt is added and then continuously decreases as even more salt is added.

Example 20-6 Salting Out of a Protein with Ammonium Sulfate Data were obtained on the precipitation of a protein by the addition of ammonium sulfate. The initial concentration of the protein was 15 g/L. At ammonium sulfate concentrations of 0.5 and 1.0 M, the concentrations of the protein remaining in the mother liquor at equilibrium were 13.5 and 5.0 g/L, respectively. From this information, estimate the ammonium sulfate concentration to give 95 percent recovery of the protein as precipitate.

We can use the Cohn equation [Eq. (20-52)], to solve this problem if we can determine the constants in the equation. Since ionic strength is directly proportional to concentration c for a given salt [Eq. (20-49)], we can rewrite the Cohn equation as

$$\ln S = \beta - K_s'' c$$

Substituting the experimental data into this equation gives

$$\ln(13.5) = \beta - 0.5K_s''$$

$$\ln(5.0) = \beta - 1.0K_s''$$

Solving these equations for the constants yields

$$\beta = 3.60$$

$$K_s'' = 1.99 \text{ M}^{-1}$$

For 95 percent recovery, the protein solubility in the mother liquor at equilibrium is 5 percent of the initial protein concentration. At this solubility, from the Cohn equation

$$c = \frac{\beta - \ln S}{K_s''} = \frac{3.60 - \ln(0.05 \times 15)}{1.99} = 1.95 \text{ M}$$

Precipitate Formation Phenomena By studying the phenomena of precipitate formation, we can maximize control over the characteristics of the final protein precipitate. Important characteristics of protein precipitates are the particle size distribution, density, and mechanical strength. Normally, it is desired to avoid having a large fraction of particles in the small size range. For proteins, particle sizes near or below 1 μm are considered to be small. Protein precipitates that consist largely of particles with small particle sizes can be difficult to filter or centrifuge. Low particle densities also can lead to filtration or centrifugation problems and can give excessive bulk volumes of the final dried precipitate. Particles with low mechanical strength can give problems with excessive attrition when the dry particles are moved. Low strength can also be interpreted as gel formation, which leads to major problems in filtration and centrifugation.

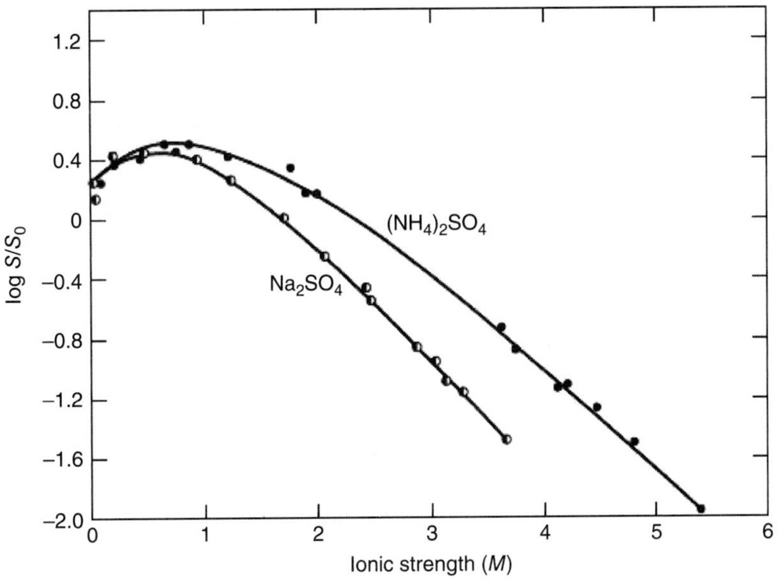

FIG. 20-46 The effect of $(NH_4)_2SO_4$ and Na_2SO_4 on the solubility of hemoglobin: S_0 is the solubility in pure water, and S is the solubility in the salt solution. [Data from Tanford, *Physical Chemistry of Macromolecules*, Wiley, 1961, p. 244.]

Precipitates form by a series of steps that occur in sequence, which are the following: (1) initial mixing, (2) nucleation, (3) growth governed by diffusion, and (4) growth governed by fluid motion. There is often some overlap between these steps. The final size of the precipitate particles during step 4 is subject to the limits imposed by particle breakage during mixing. The completion of the growth by fluid motion step can be followed by an "aging" step, where the particles are mixed until reaching a stable size.

Initial Mixing Initial mixing is the mixing required to achieve homogeneity after the addition of a component to cause precipitation. It is important to bring precipitant and product molecules into collision as soon as possible. This is a problem in micromixing, which is also important in fermentation. This subject was studied by the Russian statistician Kolmogoroff in the form of the homogeneous isotropic turbulence model, which assumes that mixing between randomly dispersed eddies is instantaneous and the mixing within eddies is diffusion-limited. It is therefore important to know the mean length of eddies, also known as the *Kolmogoroff length*, here designated l_e. It can be calculated from [Bell et al., in Fiechter (ed.), *Advances in Biochemical Engineering/Biotechnology*, vol. 26, Springer-Verlag, 1983, p. 1].

$$l_e = \left(\frac{\rho \nu^3}{P/V} \right)^{1/4} \qquad (20\text{-}56)$$

where ρ is the liquid density, ν is the liquid kinematic viscosity, and P/V is the agitator power input per unit volume of liquid.

It is necessary to mix until all molecules have diffused across all eddies. This time can be estimated from the Einstein diffusion relationship

$$t = \frac{\delta^2}{2\mathcal{D}} \qquad (20\text{-}57)$$

where δ is the diffusion distance and $\mathcal{D}$ is the diffusion coefficient for the molecule being mixed. For spherical eddies of diameter l_e, this becomes

$$t = \frac{l_e^2}{8\mathcal{D}} \qquad (20\text{-}58)$$

Thus, precipitation is initiated in a well-stirred tank for a period of time determined on the basis of isotropic turbulence.

Nucleation Nucleation is the generation of particles of ultramicroscopic size. For particles of a given solute to form, the solution must be *supersaturated* with respect to the solute. In a supersaturated solution, the concentration of the solute in solution is greater than the normal equilibrium solubility of the solute. The difference between the actual concentration in solution and the equilibrium solubility is called the *degree of supersaturation*, or just *supersaturation*. The rate of nucleation increases exponentially up to the maximum level of supersaturation, or supersaturation limit, which is illustrated in Fig. 20-47. Note that the rate of nucleation increases to a very high value at the supersaturation limit.

High supersaturations generally have negative consequences in carrying out precipitations. When the supersaturation is high, the precipitate tends to be in the form of a colloid, a gel, or a highly solvated precipitate. To obtain precipitate particles having desirable characteristics, the supersaturation should be kept relatively low.

Growth Governed by Diffusion The growth of precipitate particles is limited by diffusion immediately after nucleation and until the particles grow to a limiting particle size defined by the fluid motion, which generally ranges from 0.1 to 10 μm for high and low shear fields, respectively [Ives, in Ives (ed.), *Scientific Basis of Flocculation*, Sijthoff and Noordhoff, Alphen aan den Rijn, 1978, p. 37]. In a dispersion of particles of uniform size that are growing as dissolved solute diffuses to the particles, the initial rate of decrease of particle number concentration N can be described by a second-order rate equation that was first derived by Smoluchowski [*Z. Phys. Chem.* **92:** 129 (1917)]:

$$-\frac{dN}{dt} = K_A N^2 \qquad (20\text{-}59)$$

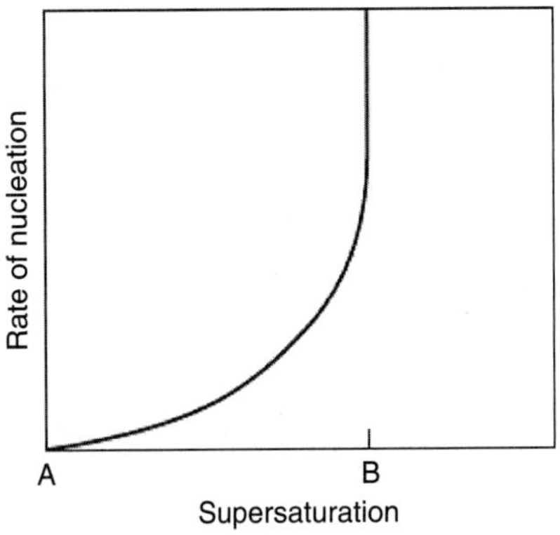

FIG. 20-47 Nucleation rate as a function of degree of supersaturation. The normal equilibrium solubility is at A, and the supersaturation limit is at B. [Reproduced from *Bioseparations Science and Engineering*, 2d ed., by Roger G. Harrison, Paul W. Todd, Scott R. Rudge, and Demetri P. Petrides (2015). By Permission of Oxford University Press.]

Here N is the number of mono-sized particles at any given time t. The constant K_A is determined by the diffusivity $\mathcal{D}$ and diameter L_{mol} of the molecules that are adding to the particles as follows:

$$K_A = 8\pi \mathcal{D} L_{mol} \tag{20-60}$$

Integrating Eq. (20-59) gives

$$N = \frac{1}{K_A t + 1/N_o} \tag{20-61}$$

For convenience, N_o is taken as the initial number concentration of dissolved solute molecules. The Stokes-Einstein equation can be used to estimate the diameter of globular proteins, which can be modeled as spheres

$$L_{mol} = \frac{k_B T}{3\pi \mathcal{D} \mu} \tag{20-62}$$

where k_B is the Boltzmann constant, T is the absolute temperature, and μ is the liquid viscosity.

Equation (20-61) can be rewritten as

$$\frac{N}{N_o} = \frac{1}{1 + N_o K_A t} \tag{20-63}$$

With M as the molecular weight of particles at time t and M_o as the molecular weight of the solute,

$$\frac{M}{M_o} = \frac{N_o}{N} \tag{20-64}$$

so that

$$M = M_o(1 + K_A N_o t) \tag{20-65}$$

This equation has been verified experimentally by measuring the molecular weight of precipitating α-casein. The data plotted in Fig. 20-48 indicate good agreement with Eq. (20-65) after an initial lag time.

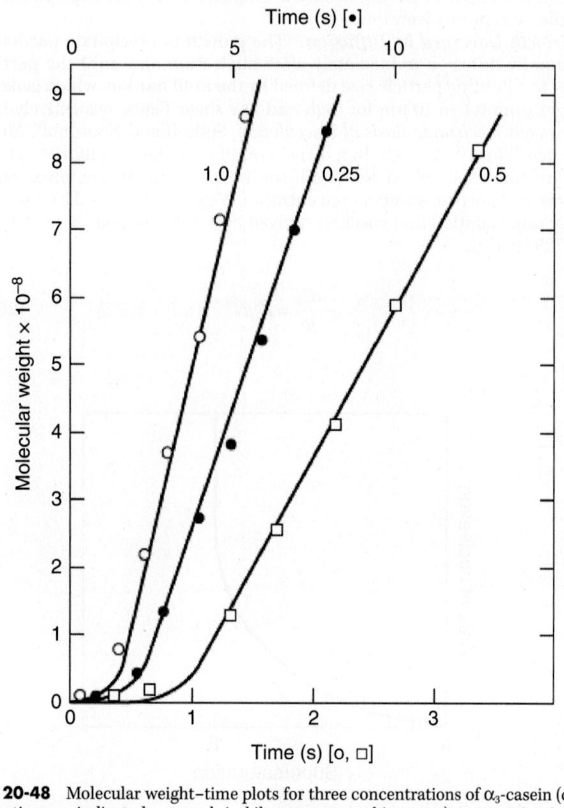

FIG. 20-48 Molecular weight–time plots for three concentrations of α_3-casein (concentrations as indicated on graph in kilograms per cubic meter), aggregating in the presence of 0.008 M $CaCl_2$. Molecular weight was determined from light-scattering and turbidity measurements. (Data from Bell et al., *Adv. Biochem. Eng.* **26:** 20, Springer-Verlag, 1983, p. 20.)

Growth Governed by Fluid Motion Growth of particles is governed by fluid motion after the particles have reached a critical size, typically 1 μm in diameter [Bell et al., in Fiechter (ed.), *Advances in Biochemical Engineering/Biotechnology*, vol. 26, Springer-Verlag, 1983, p. 1]. In this growth regime, particles tend to grow by colliding and then sticking together. This is a flocculation process, which is enhanced when electrostatic repulsion between particles is reduced in comparison to the attractive van der Waals force. This can be accomplished by raising the ionic strength and lowering the temperature, to reduce the thickness of the electrical double layer, or Debye length, around particles.

For particles of uniform size in a suspension, the initial rate of decrease of particle number concentration N due to collisions can be described by a second-order rate equation

$$-\frac{dN}{dt} = \frac{2}{3}\alpha L^3 \gamma N^2 \tag{20-66}$$

which was also derived by Smoluchowski [*Z. Phys. Chem.* **92:** 129 (1917)]. Here α is the collision effectiveness factor (fraction of collisions that result in permanent aggregates), L is the diameter of the particles, and γ is the shear rate (velocity gradient). Assuming that the volume fraction of the particles ($\varphi = \pi L^3 N/6$) is constant during particle growth governed by fluid motion and integrating yields

$$\frac{N}{N_o} = \exp(-4\alpha\phi\gamma t/\pi) \tag{20-67}$$

where N_0 is now the particle number concentration at the time [t = 0 in Eq. (20-67)] at which particle growth starts to be governed by fluid motion. For turbulent flow, the average shear rate $\bar{\gamma}$ can be estimated by the following equation, developed by Camp and Stein [*J. Boston Soc. Civ. Engrs.* **30:** 219 (1943)]:

$$\bar{\gamma} = \left(\frac{P/V}{\mu}\right)^{1/2} \tag{20-68}$$

where P/V is power dissipated per unit volume and μ is the viscosity of the liquid, respectively. Information about the determination of the collision effectiveness factor α has been given by Bell [Bell et al., in Fiechter (ed.), *Advances in Biochemical Engineering/Biotechnology*, vol. 26, Springer-Verlag, 1983, p. 1].

Precipitate Breakage When precipitate particles grow large enough by colliding and sticking together, they become susceptible to breakage during collisions. The rate of precipitate breakage has been shown to depend on the shear rate and particle concentration. In a study of soy protein precipitate particles, for example, particle breakup dominated at sizes greater than 16 μm, and breakup became negligible at low particle volume fractions (<0.0002) [Brown and Glatz, *Chem. Eng. Sci.* **42:** 1831 (1987)].

A model that has successfully described the breakup of protein precipitates is the displacement model, which depicts the rate of aggregate size change as a function of displacement from an equilibrium aggregate diameter L_e [Twineham et al., *Chem. Eng. Sci.* **39:** 509 (1984)]:

$$\frac{dL}{dt} = k(L_e - L)^n \tag{20-69}$$

where the rate constant k would be expected to depend on the volume fraction of particles φ and the shear rate γ. This model with $n = 1$ (first-order) fits data well for the mean diameter of soy protein particles at constant shear and various particle volume fractions (Fig. 20-49). The equilibrium diameter L_e has been shown to depend on the shear rate. For isoelectric soy protein precipitate in laminar Couette shear flow,

$$\bar{L}_{e,v} \propto \gamma^{-0.14} \qquad 2000\ \text{s}^{-1} \le \gamma \le 80{,}000\ \text{s}^{-1} \tag{20-70}$$

and for casein precipitated by salting out in a continuous stirred-tank reactor [Twineham et al., *Chem. Eng. Sci.* **39:** 509 (1984)],

$$\bar{L}_{e,v} \propto \bar{\gamma}^{-0.21} \qquad 12\ \text{s}^{-1} \le \bar{\gamma} \le 154\ \text{s}^{-1} \tag{20-71}$$

where $\bar{L}_{e,v}$ is the equilibrium particle size at the volume mean of the particle size distribution. It is remarkable that both correlations are relatively similar, given the varation in protein precipitation type, protein type, and shear rate range. Particle breakage is also discussed later in the analysis of the particle size distribution in a continuous flow stirred-tank reactor.

Precipitate Aging As indicated in Fig. 20-49, protein precipitate particles reach a stable size after a certain length of time in a shear field. The time period for reaching this stable size is called the aging time. The strength of

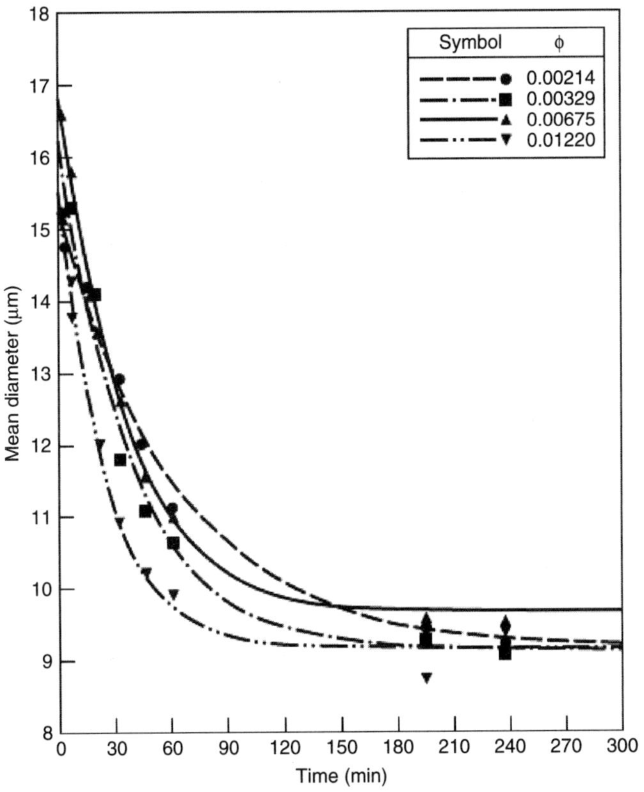

FIG. 20-49 Volume mean aggregate diameter as a function of time for soy precipitate particles exposed to shear rate of 1340 s^{-1} at different particle volume fractions φ. Lines are drawn for the displacement model. Points are experimental data. [Data from Brown and Glatz, *Chem. Eng. Sci.* **42:** 1831 (1987).]

protein particles has been correlated with the product of the mean shear rate and the aging time, $\bar{\gamma}t$, which is known as the *Camp number*. As indicated in Fig. 20-50 for soy protein particles, the mean particle size becomes approximately constant after reaching a Camp number of 105. Aging of precipitates helps the particles withstand processing in pumps and centrifuge feed zones without further size reduction.

Particle Size Distribution in a Continuous Flow Stirred-Tank Reactor The equations describing precipitate formation and breakage in the preceding subsection are convenient to use when the size of the precipitate is changing with time. In addition to being applicable to the batch reactor, these equations can be used to describe precipitate growth in the tubular reactor (see the Design of Precipitation Systems subsection). However, for one important type of reactor—the *continuous flow stirred-tank reactor* (CSTR)—operation is at steady state. Because precipitated protein leaves the CSTR at the same time as the feed protein solution enters, the product stream leaving the CSTR will have a *range of particle sizes*, which does not vary with time for operation at steady state. The distribution of particle sizes can be characterized by performing a population balance for precipitate particles in the CSTR (Randolph and Larson, *Theory of Particulate Processes*, Academic Press, Cambridge, Mass., 1971, p. 49). Before writing the population balance, it is first necessary to define the population density distribution function $n(L)$ and the linear growth rate G:

$$n(L) = \frac{dN}{dL} \tag{20-72}$$

$$G = \frac{dL}{dt} \tag{20-73}$$

Thus, if we know $n(L)$, we can take the integral of $n(L)dL$ to find the total number of particles bracketed by two given particle diameters, a minimum and a maximum, for example.

Making a balance on the number ΔN of precipitate particles within a given size range ΔL and taking the limit as $\Delta L \to 0$, we obtain for constant reactor volume V

$$\frac{d(nG)}{dL} + \frac{n}{\tau} + B = 0 \tag{20-74}$$

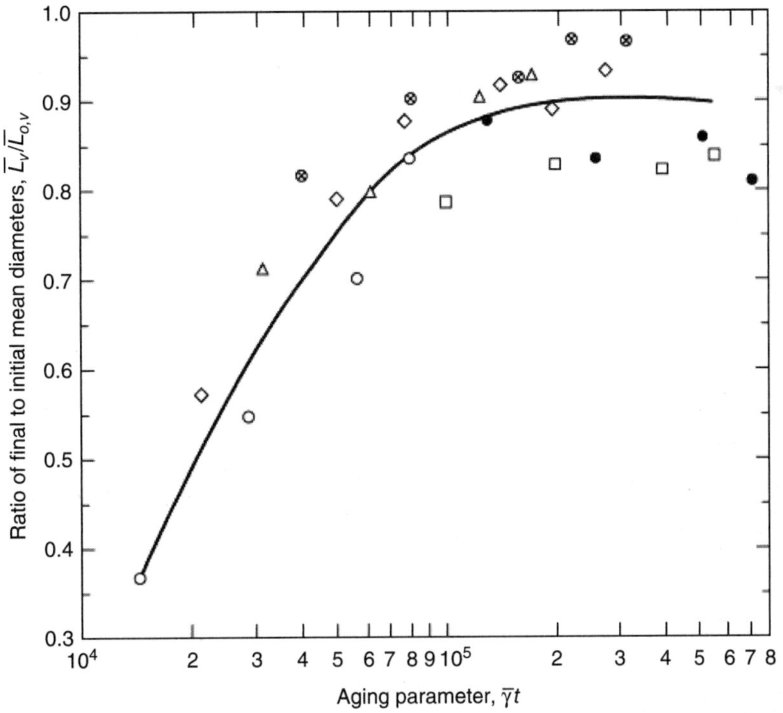

FIG. 20-50 The effect of aging on the change in the volume mean diameter of soy protein precipitate particles exposed to capillary shear. Average capillary rate of shear, 1.7×10^4 s^{-1}; average time of exposure in capillary, 0.065 s; protein concentration, 30 kg m^{-3}; initial mean diameter $\overline{L}_{0,v}$ (μm) prior to exposure to capillary shear: ○, 53.5; Δ, 23.4; ◊, 19.5; ⊗, 15.2; □, 10.2; •, 8.8. [Data from Bell and Dunnill, *Biotechnol. Bioeng.* **24:** 1271 (1982).]

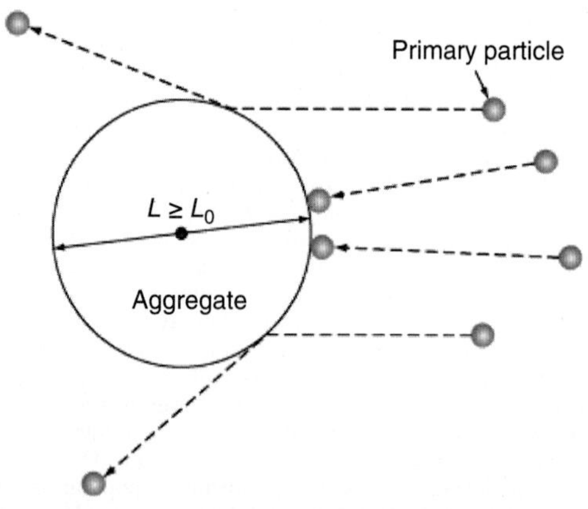

FIG. 20-51 Schematic drawing of the growth of a precipitate aggregate by collisions of the aggregate with small primary particles: L_0 is the critical diameter above which particles grow by colliding with and sticking to small primary particles. [Reproduced from *Bioseparations Science and Engineering*, 2d ed., by Roger G. Harrison, Paul W. Todd, Scott R. Rudge, and Demetri P. Petrides (2015). By Permission of Oxford University Press.]

where τ is the mean residence time (= V/Q) and B is the volumetric breakage or death rate of particles in the size range L to ΔL, such that the net rate of disappearance of particles due to breakage per volume is $B\,\Delta L$.

The growth of protein precipitate particles in a CSTR has been successfully described by Petenate and Glatz [*Biotechnol. Bioeng.* **25:** 3059 (1983)], who used the following expression for the linear growth rate:

$$G = \frac{dL}{dt} = \left(\frac{A}{4\pi}\right)\gamma\phi_1 L = K_o L \tag{20-75}$$

where A is a constant and ϕ_1 is the volume fraction of submicrometer-sized "primary" particles. In this model, aggregates above a critical diameter L_0 grow because small primary particles collide with aggregates and stick (see Fig. 20-51). Primary particles typically have diameters of about 0.2 μm for protein precipitation in a CSTR [Petenate and Glatz, *Biotechnol. Bioeng.* **25:** 3059 (1983)].

The larger a precipitate aggregate becomes, the more susceptible it is to being broken up by the local shear stresses encountered in turbulent flow. For the conditions under which local shear stresses are dominant, the following equation for the volumetric breakage or death rate B has been successfully used by Petenate and Glatz to describe the breakup of protein aggregates:

$$B = K\mu\gamma^2 nL^3 = K_1 nL^3 \tag{20-76}$$

where K is a constant and μ is the viscosity.

A useful parameter to know for the precipitation in a CSTR is the mass-averaged particle size. For spherical particles, the total mass concentration M_T in terms of the mass-averaged diameter $\bar{L}_m$ is

$$M_T = \rho_p \pi\left(\frac{\bar{L}_m^3}{6}\right)N \tag{20-77}$$

where ρ_p is the density of individual particles and N is the total particle number concentration. From $n(L)$, we can determine M_T and N for spherical particles equal to or greater than L_o in size as follows:

$$M_T\Big|_{L\ge L_o} = \frac{\rho_p\pi}{6}\int_{L_o}^{\infty} L^3 n(L)\,dL \tag{20-78}$$

$$N\Big|_{L\ge L_o} = \int_{L_o}^{\infty} n(L)\,dL \tag{20-79}$$

Combining Eqs. (20-77), (20-78), and (20-79) gives the mass-averaged particle diameter for $L \ge L_0$:

$$\bar{L}_m\Big|_{L\ge L_o} = \left[\frac{\int_{L_o}^{\infty} L^3 n(L)\,dL}{\int_{L_o}^{\infty} n(L)\,dL}\right]^{1/3} \tag{20-80}$$

Example 20-7 Dependence of Population Density on Particle Size and Residence Time in a CSTR For the precipitation of protein in a CSTR, determine the dependence of the population density distribution n on the particle diameter L and residence time τ. Consider both particle growth by aggregation and particle breakup.

Use the expressions for the linear growth rate and particle breakup [Eqs. (20-75) and (20-76)] in the population balance equation [Eq. (20-74)]. This gives

$$\frac{d(K_o Ln)}{dL} + \frac{n}{\tau} + K_1 nL^3 = 0$$

Differentiating and rearranging, we have

$$\frac{dn}{n} = -dL\left[\frac{\left(1+\frac{1}{K_o\tau}\right)}{L} + \frac{K_1 L^2}{K_o}\right]$$

With lower limits of integration of n_0 (the population density function at $L = L_0$) and critical diameter L_0, we obtain

$$\frac{n}{n_o} = \left(\frac{L_o}{L}\right)^{\left(1+\frac{1}{K_o\tau}\right)} \exp\left[\frac{-K_1}{3K_o}(L^3 - L_o^3)\right]$$

Methods of Precipitation Methods that have been developed to precipitate proteins are based on a knowledge of the solubility of proteins. Based on the previous discussion of protein solubility, the most obvious methods that emerge are pH adjustment to the isoelectric point of the protein (called *isoelectric precipitation*), addition of organic solvents, salting out, and addition of nonionic polymers.

Isoelectric precipitation is based on the fact that the solubility of a given protein is generally at a minimum at the isoelectric point (pI) of the protein, which is the pH at zero charge (see Fig. 20-45). This is a convenient method to use when fractionating a protein mixture. For this situation the pH should be adjusted above the highest pI or below the lowest pI of all the proteins present. The pH is then changed to the nearest pI, where precipitate is allowed to form and is then removed. This procedure is repeated for as many proteins as one desires to precipitate. Local extremes of pH should be avoided during pH adjustment to minimize protein denaturation (Harrison et al., *Bioseparations Science and Engineering*, Oxford University Press, Oxford, UK, 2015, p. 32). There are two advantages of isoelectric precipitation when acids are added to cause precipitation: mineral acids are cheap, and several acids (e.g., phosphoric, hydrochloric, sulfuric) are acceptable in protein food products. This method, however, will not work for all proteins; for example, gelatin, which is a very hydrophilic protein, does not precipitate at its isoelectric point in solvents having low ionic strength [Bell et al., in Fiechter (ed.), *Advances in Biochemical Engineering/Biotechnology*, vol. 26, Springer-Verlag, 1983, p. 1].

Several organic solvents have been used to precipitate proteins, including alcohols, acetone, and ether. Alcohols, however, have been the most widely used in industry. One of the most important processes utilizing alcohol to precipitate proteins is the Cohn process to purify therapeutic proteins from human plasma [Strong, in Kirk and Othmer (eds.), *Encyclopedia of Chemical Technology*, Interactive Encyclopedia System, 1948, p. 566]. This process uses ethanol at temperatures below 0°C to minimize denaturation by the organic solvent. The variables that are manipulated in the Cohn process are pH, ionic strength, and ethanol concentration. Ionic strength is kept low, which leads to a salting-in effect (see the Protein Solubility subsection). This salting-in effect is enhanced when ethanol is added. Cohn's method has been used to obtain albumin, plasminogen, prothrombin, isoagglutinins, and γ-globulin starting with blood plasma.

In the salting out of proteins, salt is dissolved in the solution containing the proteins. The protein solubility decreases as the salt ionic strength rises according to the Cohn equation [Eq. (20-52)]. The most important consideration in salting out is the type of salt used. Salts with multiply charged anions such as sulfate, phosphate, and citrate are the most effective; for the cation, monovalent ions should be used (Scopes, *Protein Purification*, Springer-Verlag, 1982, p. 47). Following the Hofmeister or lyotropic series, the salting-out ability of the common multiply charged anions is citrate^{2-} > phosphate^{3-} > sulfate^{2-}; for the common monovalent cations, the order is $NH_4^+ > K^+ > Na^+$.

The salt that has the most desirable properties in general for precipitating proteins is ammonium sulfate. The solubility of this salt is very high (approximately 4 *M* in pure water) and varies very little in the range of 0 to 30°C. The density of a saturated solution of ammonium sulfate is 1.235 g cm^{-3}, which is enough below the density of protein aggregates (approximately 1.29 g cm^{-3}) to allow centrifugation. Another advantage of using ammonium sulfate is that protein precipitates are often very stable for years in 2 to 3 *M* salt; in fact, many commercial enzymes are normally sold in ammonium sulfate solution at high molarity. Furthermore, proteolysis and bacterial action are prevented in concentrated ammonium sulfate solutions. The only disadvantage of ammonium sulfate is that it cannot be used above pH 8 because of the buffering action of ammonia. Sodium citrate is very soluble and is a good alternative to ammonium sulfate when the precipitation must be performed above pH 8 (Scopes, *Protein Purification*, Springer-Verlag, 1982, p. 48).

Several nonionic polymers have been used to precipitate proteins, including dextran, poly(vinyl pyrrolidone), poly(propylene glycol), and poly(ethylene glycol) (PEG). Of these polymers, by far the most extensively studied is PEG. Several guidelines have been developed for the use of PEG as a protein precipitant. Solutions of PEG up to 20 percent w/v can be used without viscosity becoming a problem. PEGs with molecular weights above 4000 have been found to be the most effective (Scopes, *Protein Purification*, Springer-Verlag, 1982, p. 59). Protein destabilization in PEG solutions does not occur until the temperature is significantly higher than room temperature (>40°C) [Rothstein, in Harrison (ed.), *Protein Purification Process Engineering*, Dekker, New York, 1994, p. 115].

Design of Precipitation Systems Once a precipitating agent has been chosen for the protein of interest, it may be necessary to design a large-scale precipitation process. The safest procedure is to base the design on a laboratory or pilot plant system that has given acceptable results. Important considerations in obtaining the best possible plant design are the following: the type of precipitation reactor, processing conditions (flow rates, concentrations, etc.), and assumptions used to scale up to the plant scale. There are three basic types of precipitation reactor: the batch reactor, the continuous stirred-tank reactor (CSTR), and the tubular reactor. These types are discussed and compared.

The batch reactor is the simplest of the three types and is often the one that is tried first at small scale. Batch precipitation is carried out by slowly adding the precipitating agent to a protein solution that is being mixed. Addition of the precipitating agent continues until the desired level of supersaturation is reached with respect to the protein being precipitated. At this point nucleation begins, and precipitation proceeds through the steps of particle growth and aggregation. Mixing continues until the precipitation is complete. The mixing in a batch reactor is generally turbulent. Protein particles precipitated in a batch reactor tend to be more compact and regular in shape than those precipitated in a tubular reactor, apparently because of the different shear profiles existing in the two reactors and the length of time the particles are exposed to this shear [Bell and Dunnill, *Biotechnol. Bioeng.* **24:** 2319 (1982)]. The shear field in a tubular reactor is essentially homogeneous; by contrast, in the batch reactor the precipitate particles are exposed to a very wide range of shears and to much longer times of exposure than in the tubular reactor, resulting in improved precipitate mechanical stability.

In a tubular reactor, precipitation takes place in volume elements that approach plug flow as they move through the tube. Thus, the distance–particle size distribution history of the particles in a volume element moving through a tubular reactor is comparable to the time–particle size distribution history of a stationary volume element in a batch reactor. In the tubular reactor, the feed protein solution and the precipitating agent are contacted in a zone of efficient mixing at the reactor inlet. Good mixing can be accomplished, for example, by flowing the protein solution through a convergent nozzle to a biplanar grid and then introducing the precipitating agent just downstream of the biplanar grid [Virkar et al., *Biotechnol. Bioeng.* **24:** 871 (1982)]. The flow pattern in the reactor can be turbulent, a property that can be promoted by wire meshes at intervals along the reactor. In comparison to either the batch reactor or the CSTR, the tubular reactor has the advantages of short fluid residence times, an absence of moving mechanical parts, uniformity of flow conditions throughout the reactor, a simple and inexpensive design, and a relatively small holdup of fluid. For particles that grow relatively slowly, however, the length of the tubular reactor can be excessive.

In the CSTR, fresh protein feed contacts a mixed slurry containing precipitate aggregates. The mixing conditions in a CSTR are similar to those in a batch reactor. Upon entering the CSTR, fresh protein feed nucleates, the nucleate particles grow by diffusion, and the submicrometer-size "primary" particles collide with and adhere to growing aggregates. The degree of supersaturation can be more easily controlled than in the batch or tubular reactor, which means that the formation of precipitates with undesirable properties is less likely.

A few general statements can be made regarding the processing conditions in precipitation systems. Flows are normally turbulent; flow must be high enough to avoid inadequate mixing and high supersaturation but low enough to avoid excessive particle breakage, leading to particles that are smaller than desirable. For both the batch and tubular reactors, the flow regime can be changed from turbulent to laminar during the particle growth phase to avoid excessive particle breakage. The rate of addition of

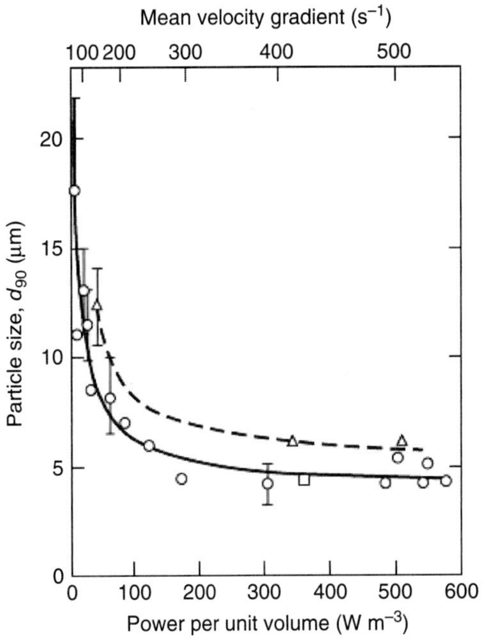

FIG. 20-52 The relationship between aggregate size and power per unit volume for the preparation of isoelectric soy protein precipitate in batch-stirred tanks: d_{90} is the particle size for which all particles that are larger account for 90 wt% of the total. Protein concentration is 30 kg m^{-3}. Vessel volumes: △, 0.27 L; ○, 0.67 L; □, 200 L. [Data from Bell et al., *Adv. Biochem. Eng.* **26:** 40, Springer-Verlag, 1983, p. 40.]

precipitant is especially important. This rate should be kept low enough to avoid high supersaturations that lead to colloidal, highly solvated precipitates. The concentration of the precipitant being added is also important, with lower concentrations leading to lower supersaturations [Rothstein, in Harrison (ed.), *Protein Purification Process Engineering*, Dekker, New York, 1994, p. 115].

The key parameter for scale-up of precipitation is mixing. One recommended approach to scale up mixing is to first consider using geometric similarity and constant power per unit volume (P/V) (Oldshue, *Fluid Mixing Technology*, McGraw-Hill, New York, 1983, p. 198). For geometric similarity, all important dimensions are similar and have a common constant ratio. As seen in Fig. 20-52, this method gave reasonable agreement for the d_{90} particle size (the particle size for which all particles that are larger account for 90 wt% of the total; d_{90} is also referred to as *D10* in describing particle size distributions) as a function of P/V for the batch isoelectric precipitation of soy protein for vessels ranging in size from 0.27 to 200 L. If the precipitate is susceptible to shear breakage, however, the assumption of constant P/V for scale-up may not be satisfactory. The impeller tip speed, which determines the maximum shear rate, rises when P/V is held constant upon scale-up of the reactor volume, as seen in Table 20-24. These results assume turbulent flow, where the power number is constant, so that

$$P \propto N_i^3 d_i^5 \tag{20-81}$$

Here, N_i is the impeller rotation rate, and d_i is the impeller diameter. It is also assumed that the maximum shear rate is proportional to the impeller tip velocity ($= \pi N_i d_i$) and that the volume is proportional to d_i^3. From Table 20-24 we see that even for a volume scale-up of 10,000 for constant P/V, the tip speed increases by less than a factor of 3. A volumetric scale-up factor of 10,000 is typical in scaling up from the lab to the plant.

TABLE 20-24 Scale-Up of Turbulent Agitation, Assuming Constant *P/V*

Volume scale-up factor	$(\text{Tip speed})_{\text{large}}$ / $(\text{Tip speed})_{\text{small}}$
10	1.3
100	1.7
1000	2.2
10,000	2.8

Reproduced from *Bioseparations Science and Engineering*, 2d ed., by Roger G. Harrison, Paul W. Todd, Scott R. Rudge, and Demetri P. Petrides (2015). By Permission of Oxford University Press.

LIQUID CHROMATOGRAPHY AND ADSORPTION

As discussed in the introduction, adsorption is typically used for the removal of insolubles such as cells or cell debris as well as for the initial isolation of the product, while chromatography is often used for the final purification steps. Different types of chromatography are used to achieve the high purity required, as is the case for bioproducts such as monoclonal antibodies (see the subsection Bioprocess Design and Economics for a discussion of the purification of monoclonal antibodies). Detailed information about the theory and equipment for liquid chromatography and adsorption can be found in Sec. 16. In the current section, chromatography and adsorption topics are discussed that have particular relevance to bioproducts: (1) adsorbents that are commonly used to purify biomolecules; (2) the length of unused bed (LUB) method for scaling up adsorption of biomolecules that obey the Langmuir isotherm, which has often been used to correlate equilibrium adsorption data for proteins; (3) agitated-bed adsorption for processing cell suspensions in which it is desired to selectively remove a bioproduct that has been secreted by the cells; (4) membrane chromatography, which has applications for macromolecules such as proteins and DNA; and (5) the scale-up of the chromatography of proteins.

Adsorbents for Purification of Bioproducts Many adsorbent resins have been developed for adsorptive and chromatographic separation of bioproducts. There are two basic resin materials: polymer and silica. Many types of ligand chemistries can be conjugated to either resin material. Typically, however, silica resins most commonly have hydrophobic coatings and are used for reversed-phase chromatography. Polymer resins are more often used in aqueous applications and are conjugated with ion exchange, hydrophobic interaction, or affinity-type ligands. More detailed information about polymer and silica adsorbents can be found in the subsection Adsorbents and Ion Exchangers in Sec. 16.

Surface area and particle size are very important properties of adsorbent resins. The resin provides the surface area for the adsorption, which is generally 100 to 1500 m^2/g. The surface area on the outer surface of a 10-μm-diameter solid sphere is 1.7 m^2/g, so it follows that most of the surface area is in the internal porosity of the particle. Since this surface area is accessed by molecular diffusion (one class of resins, called *perfusion resins*, allows convection through some macropores), the path length for this diffusion is important. This path length is the radius of the resin particle. Therefore, both the diameter of the particle and its internal surface area are important for the resin performance. Resins have large ranges of surface areas and particle sizes, and they do not necessarily co-vary.

The various adsorbents used for adsorption and chromatography of biomolecules are discussed below based on the type of operation that is carried out.

Ion Exchange Adsorption and Chromatography A comprehensive discussion of ion exchange adsorbents is given in the subsection Adsorbents and Ion Exchangers in Sec. 16.

Reversed-Phase Chromatography Reversed-phase chromatography employs a hydrophobic phase bonded to the surface of the resin. Typically, reversed-phase resins are silica based. Reversed phase is so named because the partitioning of solutes between the mobile phase and stationary phase is opposite to that observed with bare silica. In other words, hydrophobic solutes bind in higher proportions in reversed phase, while hydrophilic solutes bind in higher proportion in "normal phase." Solutes are typically introduced into reversed-phase columns in water, or with minimal amounts of organic solvent, so that most solutes partition to the stationary phase. The organic content of the mobile phase is slowly increased, typically as a percent of acetonitrile, methanol, or isopropanol, thereby decreasing the polarity of the mobile phase.

Hydrophobic phases that are bonded to silica are typically aliphatic octyl (C_8), octyldecyl (C_{18}), phenyl (C_6 aromatic), and methyl (C_1). The different chain lengths and densities (called *coverage* in the commercial literature) of the different bonded phases obviously lead to more or less hydrophobicity. The entire surface of the silica cannot be fully covered with a monolayer of the desired phase, however, because of steric effects. Bare silica remains exposed, and this bare silica can participate in the separation by interacting with hydrophilic molecules, or hydrophilic domains of large molecules, thereby altering the binding. The strategy employed for covering this exposed silica surface depends almost as much on the specifics of the separation achieved as it does on the chain length of the bonded phase. Polymerized phases represent an attempt to cover the surface by polymerizing the alkyl chains together at their point of attachment to the silica. End-capped resins use a short chain length group, such as methyl or ethyl, to cover the unreacted surface sites. Resins that do not utilize either method are left so intentionally to take advantage of the "mixed-mode" separation that may result. The separation characteristics of these resins are difficult to reproduce precisely, leading to considerable variation in separation between different manufacturers' resins, and within one manufacturer's resin from one lot of material to another.

Ions do not partition well in hydrophobic phases. It is common, therefore, to choose a counterion for reversed-phase chromatography. For biological mixtures, the counterion is nearly always a strong acid anion, such as trifluoroacetate, acetate, or chloride. The counterion has a strong effect on the separation, by partitioning along with the co-ion of interest. The counterion can be used to make solutes more or less hydrophobic and will not affect all solutes equally.

Hydrophobic Interaction Chromatography Hydrophobic interaction chromatography is often used for protein separations. It employs derivatized polymer resins, with phenyl, butyl, or octyl ligand groups, typically. Proteins adhere to the hydrophobic surface under high-salt conditions and redissolve into the mobile phase as the salt concentration is reduced. Hydrophobic interaction chromatography differs from reversed-phase in that the mobile phase is kept aqueous (polar), and the salt concentration is used to effect the partitioning to the surface. The mechanism of partition is related to precipitation as opposed to two-phase partitioning (reversed-phase) or ionic interaction (ion exchange). This is due to the use of salts that strip proteins of their solvation water. As the concentration of "lyophilic" salts increases, the probability increases that the protein will aggregate, or nucleate on the surface of the resin [Roettger et al., *Biotechnol. Prog.* **5:** 79 (1989)]. Hydrophobic interaction chromatography is sensitive to pH, salt used, buffer type, and temperature. Each of these must be carefully controlled to achieve reproducible separation but also to represent an opportunity for increased selectivity.

Affinity Chromatography and Adsorption Affinity chromatography and adsorption take advantage of biological interactions to effect binding of specific solutes. Antibodies, antigens, or dyes are conjugated to polymer resins for the purpose of binding specific solutes from a mixture. For instance, an antibody could be raised against a target protein. The antibody would be conjugated to a resin, usually via cyanogen bromide activation. The antibody then captures the solute out of the mixture, and the impurities flow through the column. The solute can be recovered by changing the pH, increasing the salt concentration, or adding a displacer, that is, a molecule that also has some affinity for the antibody, or other binding agent (affinity ligand) on the stationary phase.

Other examples of specific interactions that can be used to isolate proteins are enzyme–ligand, enzyme–cofactor, and receptor–agonist (antagonist). In each case, one member of any of these pairs may be immobilized to a resin to isolate the desired partner. Affinity chromatography is often coupled with cloning techniques to synthesize the target molecule with an "epitope" or recognition sequence that can be captured by the affinity ligand. Affinity chromatography is used from small-scale research (e.g., high-throughput screening) to large-scale purification.

An example of a type of affinity chromatography that has had a major impact on the purification of monoclonal antibodies is the use of immobilized protein A in process scale purification [Vunnum, in U. Gottschalk (ed.), *Process Scale Purification of Antibodies*, Wiley, Hoboken, N.J., 2009, p. 79]. Protein A is often used as the first chromatography step in the purification, which gives a large purification factor and results in simplifying the purification process. Protein A has affinity for the constant region of Ig-G antibodies, so it is generally applicable to the capture and purification of most monoclonal antibodies. It is a polypeptide with a molecular mass of 54 kDa that is found in the cell wall of *Staphylococcus aureus.* Recombinant protein A that lacks the membrane-binding region is used in affinity purification.

One of the drawbacks of the protein A and other affinity methods is that low levels of the affinity ligand or its fragments can co-elute with the target molecule. In particular, when an affinity step is used as the first step in the purification process, it is exposed to significant levels of extracellular proteases that liberate fragments of the ligand into the column effluent. However, trace amounts of affinity ligands and others such as high-molecular-weight aggregates of the antibody and host cell protein can be removed in one or two subsequent chromatography steps. More information about the process scale purification of monoclonal antibodies using protein A is found in the Bioprocess Design and Economics subsection.

Immobilized Metal Affinity Chromatography Some proteins have high affinities for specific metals. This affinity may be either structural, as in the case of metalloproteins, which require metal centers for their biological activity, or based on the content of specific amino acid residues, such as histidine, tryptophan, and cysteine, which have increased affinity for transition metals such as nickel and copper. Techniques have been developed to immobilize metal ions with spacer arms onto polymer resins. These resins are referred to as IMAC resins, and they are used to purify proteins that have one of the two characteristics mentioned above. Genetic engineering has also been used to enable IMAC to be performed, with cloned target biomolecules being fused to polyhistidine tags [Arnold, *Bio/Technology* **9:** 152 (1991)].

Size Exclusion Chromatography Sometimes referred to as *gel filtration*, size exclusion chromatography (SEC) separates solutes on the basis of their size. There is no derivatization of the polymer gel, and there is no binding between the solutes and the resin. Molecules larger than the largest pores in the gel (the exclusion limit) cannot enter the gel and are eluted first. Smaller molecules enter the gel to varying extents, depending on their size and shape, and thus are retarded on their passage through the bed. In general, SEC resins are hydrophilic polymer gels with a broad distribution of pore sizes. The pore size is dependent on the degree of polymerization of the gel. Size exclusion chromatography is a useful technique, especially for changing buffers or for removing small molecules from protein solutions. Because of the lack of binding between the solute and the resin, however, the capacity of this technique is low. Size exclusion effects may be in action in all the above-described techniques, since all resins are macroporous.

Agitated-Bed Adsorption Agitated-bed adsorption processes have been developed to allow removal of a product secreted by the cells without first having to remove the cells. In this type of process, cell culture broth is passed through a series of agitated columns containing an adsorbent, as shown in Fig. 20-53. Each column has screens at the inlet and outlet that are designed to retain the adsorbent within the column but allow the broth to pass through. When the concentration of the product in the effluent of the last column in the series reaches a certain value, the flow is stopped, and the lead column is taken out of the train. Periodic countercurrent operation is obtained by advancing each of the remaining columns in the train, placing a regenerated column of adsorbent in the last position, and restarting the feed flow. The lead column taken out of the train is washed with the adsorbent agitated to remove the broth solids, and the product is eluted from the adsorbent, usually in the fixed-bed mode.

This process has advantages over filtration, in that there is no filter aid to dispose of and it is not necessary to wash a filter cake containing the cells, so losses of the product are often less. The equipment for this process is less expensive and easier to maintain than that used for centrifugation. The disadvantage is that expensive solid adsorbents are more easily fouled by the dirtier feed stream and so require harsher or more expensive regeneration procedures and more frequent replacement compared to adsorbents utilized with streams with fewer impurities. Also, resin attrition can be an issue.

A useful mathematical model for this process has been developed [Belter et al., *Biotechnol. Bioeng.* **15:** 533 (1973)]. The continuity equation for the nth column in the train can be written for separand i as

Rate of separand in − rate of separand out = rate of accumulation of separand

$$Qc_{i,n-1} - Qc_{i,n} = V_L \frac{dc_{i,n}}{dt} + V_R \frac{dq_{i,n}}{dt} \tag{20-82}$$

where Q = volumetric flow rate
$c_{i,n-1}$, $c_{i,n}$ = separand concentration in feed to and effluent from column n, respectively
V_L = liquid volume in column
V_R = volume of adsorbent in column
$q_{i,n}$ = separand concentration in adsorbent phase of column n averaged over an adsorbent particle
t = time

The rate of mass transfer of separand to the adsorbent phase is described by a linear driving force expression

$$\frac{dq_{i,n}}{dt} = K_a(c_{i,n} - c^*_{i,n}) \tag{20-83}$$

where $c^*_{i,n}$ is the separand concentration in the bulk liquid when it is at equilibrium with $q_{i,n}$, and K_a is an overall mass-transfer coefficient that can be correlated to experimental data as

$$K_a = Ae^{-B(q_i/q_{i,\text{sat}})} + De^{-E(q_i/q_{i,\text{sat}})} \tag{20-84}$$

Here A, B, D, and E are constants, and $q_{i,\text{sat}}$ is the adsorbent phase concentration which is in equilibrium with the separand concentration c_{i0} in the feed to the train of mixed columns. In the use of this model for the recovery of an antibiotic, equilibrium was modeled by the Freundlich isotherm written in the form

$$c^*_{i,n} = bq^a_{i,n} \tag{20-85}$$

where a and b are constants.

Equations (20-82) to (20-85) constitute a set of mathematical relationships that govern the performance of each column in the train. These simultaneous equations can be solved by the Runge-Kutta numerical method to predict the effluent and adsorbent concentrations as a function of time. Excellent agreement between predicted and experimental adsorption data for the recovery of the antibiotic novobiocin in a three-stage train has been obtained using this method (Belter et al., *Bioseparations*, Wiley, Hoboken, N.J., 1988, p. 63).

The scale-up of a series of agitated columns containing adsorbent (Fig. 20-53), operated in the periodic countercurrent mode, follows directly from the mathematical model presented for this process [Eqs. (20-82) to (20-85)] (Belter et al., *Bioseparations*, Wiley, Hoboken, N.J., 1988, p. 63). Besides the increases in the flow rate and volume that occur upon scale-up, the mixing patterns may also change (Harrison et al., *Bioseparations Science and Engineering*, Oxford University Press, Oxford, UK, 2015, p. 305).

Length of Unused Bed (LUB) Method for Scale-Up of Fixed-Bed Adsorption This method allows scale-up of fixed-bed adsorption based on data from laboratory columns, keeping the particle size and superficial velocity constant (Thomas and Crittenden, Adsorption Technology and Design, Butterworth-Heinemann, Waltham, Mass., 1998, p. 165). In discussing the LUB method of scale-up, it is necessary to define the break-point time t_b and the ideal adsorption time t^* on a breakthrough curve, which are indicated in Fig. 20-54. The break-point time is usually taken at the relative concentration c_i/c_{i0} = 0.05 or 0.10, where c_{i0} is the feed concentration (McCabe et al., *Unit Operations of Chemical Engineering*, McGraw-Hill, New York, 1993, p. 819). Since only the fluid last exiting the column has this concentration, the average fraction of the solute removed from the start of feeding to the break-point time is usually 0.99 or higher. The ideal adsorption time is the time for breakthrough that would occur if the solute were in perfect equilibrium with the bed of adsorbent, which would give a vertical breakthrough curve. For a symmetrical breakthrough curve, the ideal adsorption time is the time at which c_i/c_{i0} = 0.5. At the ideal adsorption time for a bed initially free of the solute to be adsorbed, based on a unit area of bed cross section,

$$vc_{i0}t^* = L\rho_b q_{i,\text{sat}} \tag{20-86}$$

where L = bed length
$q_{i,\text{sat}}$ = average adsorbent phase concentration of solute i in equilibrium with feed concentration c_{i0}, based on the adsorbent weight (weight of solute i per weight of adsorbent)
v = superficial, or linear, velocity (flow rate divided by the cross-sectional area)
c_{i0} = concentration of solute i in feed
ρ_b = bulk density of adsorbent

The ideal adsorption time is therefore given by

$$t^* = \frac{L\rho_b q_{i,\text{sat}}}{vc_{i0}} \tag{20-87}$$

The amount of the solute adsorbed at the break point can be determined by integrating the breakthrough curve up to time t_b, as indicated in Fig. 20-55. The width of the breakthrough curve defines the width of the *mass-transfer zone* in the bed.

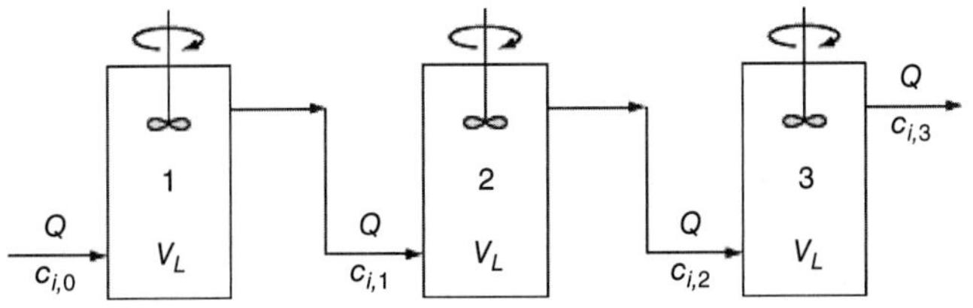

FIG. 20-53 Train of agitated-bed adsorption columns for the processing of cell culture broth. [Reproduced from *Bioseparations Science and Engineering*, 2d ed., by Roger G. Harrison, Paul W. Todd, Scott R. Rudge, and Demetri P. Petrides (2015). By Permission of Oxford University Press.]

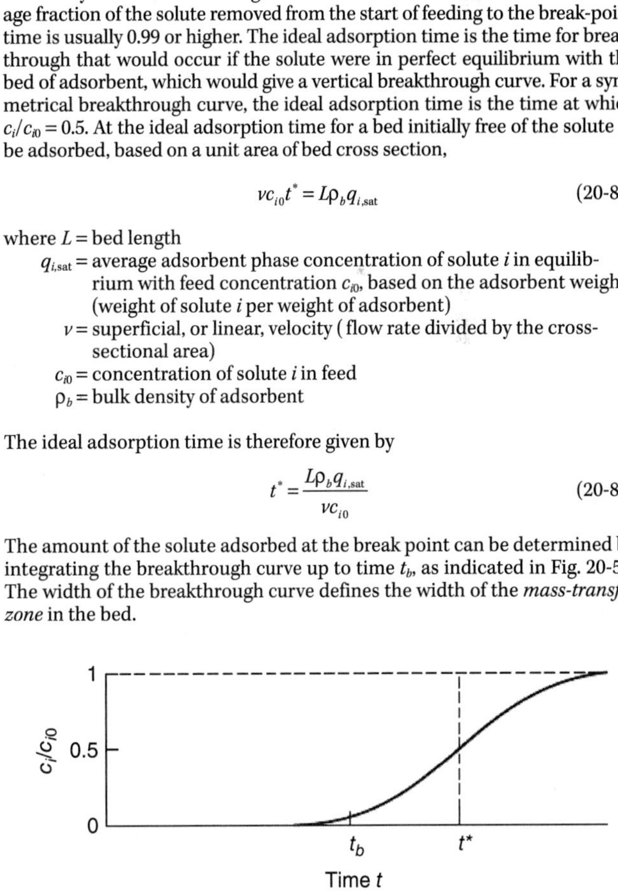

FIG. 20-54 Breakthrough curve for a fixed-bed adsorber, showing the break-point time t_b, chosen at 5 percent of the solute feed concentration, and the ideal adsorption time t^*. The ratio of mobile phase concentration of solute to solute concentration in feed to the adsorber is c_i/c_{i0}. [Reproduced from *Bioseparations Science and Engineering*, 2d ed., by Roger G. Harrison, Paul W. Todd, Scott R. Rudge, and Demetri P. Petrides (2015). By Permission of Oxford University Press.]

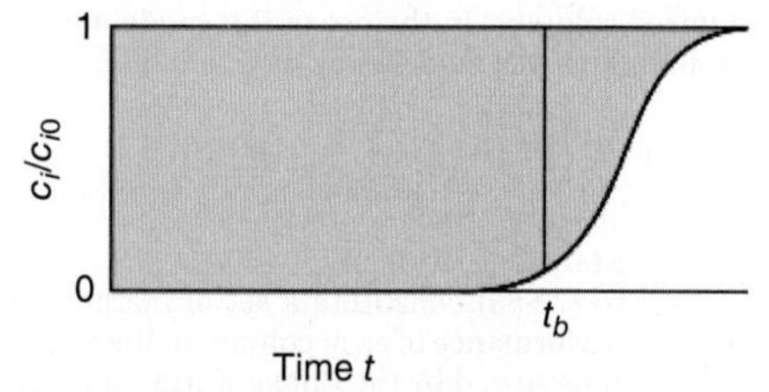

FIG. 20-55 Integration of the breakthrough curve for a fixed-bed adsorber. The area of integration to the left of a vertical line at time t is proportional to the amount of solute adsorbed up to that time. The ratio of mobile phase concentration of solute to solute concentration in feed to the adsorber is c_i/c_{i0}. [Reproduced from *Bioseparations Science and Engineering*, 2d ed., by Roger G. Harrison, Paul W. Todd, Scott R. Rudge, and Demetri P. Petrides (2015). By Permission of Oxford University Press.]

For adsorption where the equilibrium isotherm is favorable, which is true for the Langmuir isotherm that can often be used for protein adsorption, the concentration profile in the mass-transfer zone takes on a characteristic shape that does not change as the zone propagates through the bed [Ruthven, in Kroschwitz (ed.), *Kirk-Othmer Encyclopedia of Chemical Technology*, vol. 1, 4th ed., Wiley-Interscience, Hoboken, N.J., 1991, p. 493]. At the break-point time, the adsorbent between the inlet of the bed and the beginning of the mass-transfer zone is completely saturated (in equilibrium with the solute in the feed). The adsorbent in the mass-transfer zone goes from being completely saturated to being almost free of solute, and the adsorbent could be assumed to be on the average about half-saturated. This would be equivalent to half the adsorbent in the mass-transfer zone being saturated and the other half being unused. The scale-up principle is that the length of the unused bed in the mass-transfer zone does not change as the bed length is changed [Collins, *AIChE Symp.* **74**(63): 31 (1967)].

The length of unused bed (LUB) can be determined directly from the breakthrough curve obtained experimentally. If $q_{i,b}$ is the average adsorbent phase concentration of solute i at the break-point time, then the fraction of bed capacity utilized at the break-point time is $q_{i,b}/q_{i,\text{sat}}$. Therefore, the unused fraction of the bed is $1 - q_{i,b}/q_{i,\text{sat}}$. From Eq. (20-86),

$$\frac{q_{i,b}}{q_{i,\text{sat}}} = \frac{t_b}{t^*} \tag{20-88}$$

so that LUB can be written as

$$\text{LUB} = \left(1 - \frac{q_{i,t_b}}{q_{i,\text{sat}}}\right)L = \left(1 - \frac{t_b}{t^*}\right)L \tag{20-89}$$

where t_b and t^* are stoichiometric times determined by integration of the breakthrough curve:

$$t^* = \int_0^\infty \left(1 - \frac{c_i}{c_{i0}}\right) dt \tag{20-90}$$

$$t_b = \int_0^{t_b} \left(1 - \frac{c_i}{c_{i0}}\right) dt \tag{20-91}$$

In scale-up calculations, the length of column required can easily be found by adding the LUB to the length calculated by assuming local equilibrium, with a shock wave concentration front [Ruthven, in Kroschwitz (ed.), *Kirk-Othmer Encyclopedia of Chemical Technology*, vol. 1, 4th ed., Wiley-Interscience, Hoboken, N.J., 1991, p. 493].

Example 20-8 Scale-Up of the Fixed-Bed Adsorption of a Pharmaceutical Product The breakthrough data given in the table below were obtained for the adsorption of a pharmaceutical product in a laboratory column (5-cm diameter × 15-cm high) at a feed flow rate of 400 mL/h and feed concentration of 0.75 U/L, where U is units of biological activity of the pharmaceutical product. It is desired to scale up the process to operate in a column 30 cm high. What break-point time can be expected in the 30-cm-high column?

t(h)	c_i (U/L)
20.5	0.01
26.7	0.20
32.0	0.39
36.0	0.53

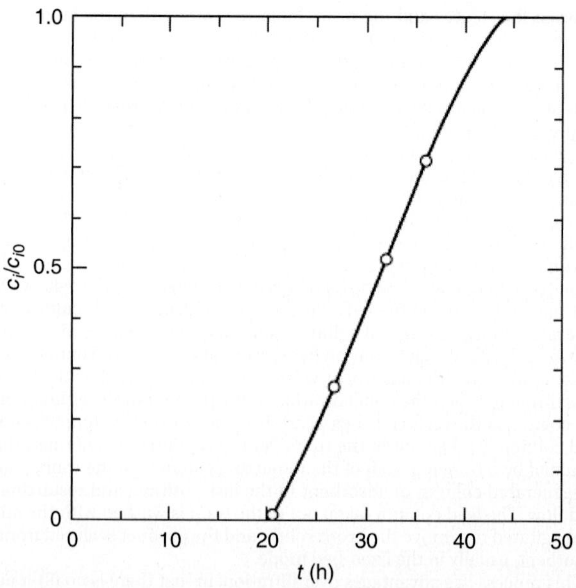

FIG. 20-56 Breakthrough curve for the fixed-bed adsorption of a pharmaceutical product in Example 20-8. [Reproduced from *Bioseparations Science and Engineering*, 2d ed., by Roger G. Harrison, Paul W. Todd, Scott R. Rudge, and Demetri P. Petrides (2015). By Permission of Oxford University Press.]

We use the LUB method, which involves graphical integrations of the breakthrough curve to determine the amount of solute adsorbed at various times. The breakthrough curve with concentration in dimensionless form is therefore plotted, with the curve extended to $c_i/c_{i0} = 1.0$, assuming that the curve is symmetric about $c_i/c_{i0} = 0.5$ (Fig. 20-56). By doing graphical integrations of this curve, we find the following:

Total solute adsorbed at saturation ($c_i/c_{i0} = 1.00$)

$$= Qc_{i0}\int_0^{44.0}\left(1 - \frac{c_i}{c_{i0}}\right)dt = Qc_{i0}(31.5\text{ h})$$

Total solute adsorbed at break point ($c_i/c_{i0} = 0.05$)

$$= Qc_{i0}\int_0^{21.5}\left(1 - \frac{c_i}{c_{i0}}\right)dt = Qc_{i0}(21.2\text{ h})$$

where Q is the volumetric flow rate. Thus, t^* and t_b are 31.5 and 21.2 h, respectively, for this column. From Eq. (20-89)

$$\frac{\text{LUB}}{L} = 1 - \frac{Qc_{i0}(21.2\text{ h})}{Qc_{i0}(31.5\text{ h})} = 1 - 0.673 = 0.327$$

$$\text{LUB} = 0.327(15\text{ cm}) = 4.90\text{ cm}$$

The adsorbent loading at saturation per total bed volume is

$$\rho_b q_{i,\text{sat}} = \frac{400\dfrac{\text{cm}^3}{\text{h}} \times 0.00075\dfrac{\text{U}}{\text{cm}^3} \times 31.5\text{ h}}{\dfrac{3.14 \times 5^2\text{ cm}^2 \times 15\text{ cm}}{4}} = 0.0321\frac{\text{U}}{\text{cm}^3}$$

For the column 30 cm high, the LUB does not change, so that

$$\frac{\text{LUB}}{L} = \frac{4.90}{30} = 0.163$$

The superficial velocity stays the same and is

$$v = \frac{400\dfrac{\text{cm}^3}{\text{h}}}{\dfrac{3.14 \times 5^2\text{ cm}^2}{4}} = 20.4\frac{\text{cm}}{\text{h}}$$

From Eq. (20-87),

$$t^* = \frac{L\rho_b q_{i\text{sat}}}{vc_{i0}} = \frac{30\text{ cm} \times 32.1\dfrac{\text{U}}{\text{L}}}{20.4\dfrac{\text{cm}}{\text{h}} \times 0.75\dfrac{\text{U}}{\text{L}}} = 62.9\text{ h}$$

Therefore, from Eq. (20-89),

$$t_b = t^*\left(1-\frac{\text{LUB}}{L}\right) = (62.9\text{ h})(1-0.163) = 52.6\text{ h}$$

Membrane Chromatography The use of filters and membranes to adsorb substances from a process stream has a long history. Indeed, the classic elementary school experiment, in which India ink is separated into a rainbow of colors as it ascends a wetted strip of filter paper, is a form of membrane chromatography, in which the pigment chemicals have differential binding affinities for the cellulose fibers that form the filter material. However, the use of membranes in chromatography received renewed interest with the development of microporous adsorptive membranes, in which the surfaces of the pores are chemically derivatized with an interactive ligand. Membrane chromatography was first performed in 1988 by [Brandt et al., *Bio/Technology* **6:** 779 (1988)] using hollow fiber membranes functionalized with either the protein gelatin for isolating fibronectin from human plasma or protein A for purifying an IgG antibody. Compared to the equivalent chromatography column packed with 100-μm adsorbent particles, the membrane chromatography column volume and fluid residence time were less by more than 3 orders of magnitude.

The major difference between membrane and packed-bed chromatography is that in membrane chromatography the transport of solutes to their binding sites takes place predominantly by convection through the pores (rather than by diffusion in the stagnant fluid inside the adsorbent particles for packed-bed chromatography), thereby reducing both process time and liquid volume. Much higher flow rates can be used in membrane chromatography than in packed beds, and the binding efficiency is relatively independent of the feed flow rate over a wide range [Ghosh, *J. Chromatogr. A* **952:** 13 (2002)]. Although they have a shorter flow path, membrane chromatography systems have generally been found to achieve resolutions comparable to those with packed beds of adsorbent [Yamamoto and Sano, *J. Chromatogr.* **597:** 173 (1992); Gerstner et al., *J. Chromatogr.* **596:** 173 (1992)]. However, membrane chromatography systems typically are used for adsorption of contaminants in a process stream, such as DNA or endotoxin, or low levels of host cell proteins, rather than for the binding and elution of a product molecule and its closely related variants, such as those caused by oxidation or amino acid misincorporation. For example, in an antibody purification process, a final step may be the reduction of DNA, endotoxin, and certain host cell proteins. In this case, the monoclonal antibody need not bind to the chromatography media (resin or membrane); therefore, a pH may be chosen where the antibody does not bind, while the contaminants do. This is called the *flow-through mode,* or *frontal chromatography.*

Another application where membrane chromatography is particularly suitable is in the purification of large proteins (molecular weight > 250,000) [Ghosh, *J. Chromatogr. A* **952:** 13 (2002)]. Proteins this large diffuse slowly into the pores of chromatography adsorbent media, but this is not a limitation in membrane chromatography since diffusion paths are very short.

In membrane chromatography, there is dispersion in the system due to the flow found in open pores, and the diffusion of molecules in the radial direction across the parabolic flow profile and also in the axial direction. There is some dispersion due to the kinetics of adsorption and desorption, which are a feature of conventional chromatography as well. Finally, there is dispersion due to mixing volumes upstream and downstream of the membrane for flow distribution.

Membranes for chromatography have been fabricated and are commercially available as thin sheets, flat disks, hollow fibers, and monolithic rods, disks, and tubes [Avramescu et al., in A. Pabby et al. (eds.), *Handbook of Membrane Separations*, CRC Press, Boca Raton, Fla., 2009, p. 25]. Hollow fibers are usually bundled within a shell using potting material. For large-scale production, thin membrane sheets are usually stacked or spiral-wound. Various configurations of membrane systems are illustrated in Fig. 20-57. The volumetric binding capacities are lower than in conventional chromatography, and the depths are not typically high.

The case typically made for membrane chromatography is that the disposable nature of the filters and the large volumes of water required to operate a fixed column make membrane chromatography a viable alternative to column chromatography for polishing applications for removal of contaminants such as viruses, DNA, and endotoxins [Fraud et al., *BioPharm. Int.* **23:** 44 (2010)]. However, in analyses such as these, the chromatography column is sized at 100 times the size of the membrane; in other words, the chromatography column will be used at 1 percent of its dynamic binding capacity, where the membrane will be used at 100 percent dynamic binding capacity. Furthermore, the dynamic binding capacity of the chromatography resin typically exceeds that of the membrane configuration [70 mg/mL for Sepharose Q-HP (GE Life Sciences Technical Literature, available at www.gelifesciences.com) versus 50 mg/mL for the Sartobind Q (Sartorius technical literature, available at www.sartorius.com), for example]. The actual ligand density in the two formats is approximately 50 *mM* for the Sartobind Q versus 140 *mM* for the Sepharose Q-HP. This demonstrates two related points. First, the ligand density achievable in membrane chromatography is much lower (three times in this comparison) because the surface area for adsorption is lower than in resin chromatography. Second, because this internal surface area in resin chromatography is less accessible than for membrane chromatography, the proportion of dynamic binding capacity available to a large molecule such as a protein is much smaller for resin than for membranes.

Another perceived limitation of using membranes as chromatography media is the cleanliness of the feed stream required. For the specific application of stripping extremely low-concentration contaminants from a concentrated antibody solution as the last step (with the cleanest feed stream) in the process, a high solids load could potentially foul the membrane surface and blind pores. Ultimately, only the largest pores will be left open, diffusion path lengths will increase, and flow will deteriorate. This issue is easily overcome by adding a prefilter upstream of the membrane, just as is frequently done for conventional chromatography. The contaminant load on the membrane itself is addressed by sizing the membrane for the anticipated load, again, just as one would with conventional chromatography. For example, in the polishing of antibody solutions using anion exchange, many contaminants, including viruses, nucleic acids, and endotoxins, tend to be negatively charged at neutral pH, while antibodies are positively charged at this pH and flow through without binding [Zhou et al., *BioPharm Int.* **20:** 26 (2007)].

Scale-up of membrane chromatography is typically achieved by increasing the diameter of the filter, conserving the interstitial velocity from scale to scale. However, it is possible to scale up by maintaining the ratio of membrane thickness to interstitial velocity constant [Etzel and Riordan, in Shukla et al. (eds.), *Process Scale Bioseparations for the*

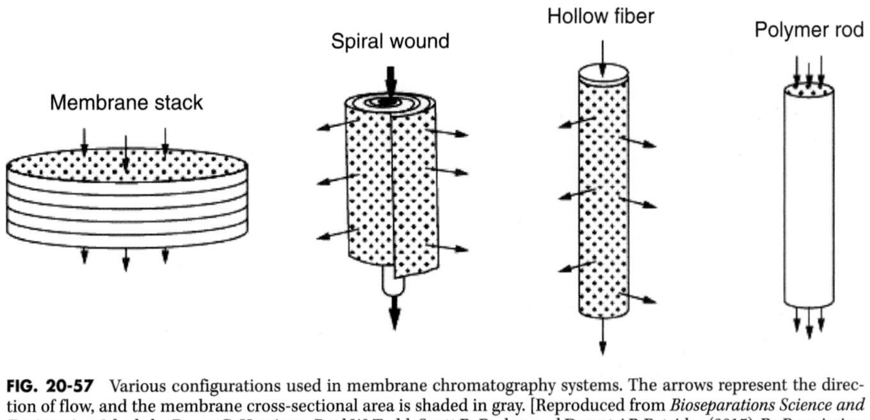

FIG. 20-57 Various configurations used in membrane chromatography systems. The arrows represent the direction of flow, and the membrane cross-sectional area is shaded in gray. [Reproduced from *Bioseparations Science and Engineering*, 2d ed., by Roger G. Harrison, Paul W. Todd, Scott R. Rudge, and Demetri P. Petrides (2015). By Permission of Oxford University Press.]

Biopharmaceutical Industry, Taylor and Francis, 2007, p. 278], and when this is done, the longer or thicker membranes would have to be able to perform with higher pressure drops and higher velocities than their scaled-down counterparts.

Example 20-9 Comparison of Time for Diffusion Mass Transfer in Conventional Chromatography and Membrane Chromatography Estimate the time for diffusion mass transfer in conventional chromatography and in membrane chromatography. Assume a typical particle diameter for an adsorbent used in preparative chromatography of 100 μm, pore diffusion coefficient $\mathcal{D}_p$ of 3×10^{-7} cm^2 s^{-1} of a representative protein bovine hemoglobin in size exclusion media at 20°C [Athalye, *J. Chromatog.* **589:** 71 (1992)], superficial velocity of 600 cm/h in the pores in membrane chromatography possible in membrane chromatography with no compromise in efficiency [Zhou et al., *BioPharm Int.* **20:** 26 (2007)], void fraction of 0.6 for the membranes, and a membrane pore size of 1 μm. Assume the fluid has the properties of water.

The time required for diffusion can be calculated using Einstein's diffusion equation

$$t = \frac{\delta^2}{2\mathcal{D}_p}$$

For conventional chromatography, the diffusion length is 50 μm (one-half the particle diameter). The type of convective flow regime in the pores in membrane chromatography can be determined by calculating the Reynolds number in a cylindrical pore:

$$\text{Re} = \frac{d_m\left(\frac{v}{\varepsilon}\right)\rho}{\mu} = \frac{1\ \mu\text{m} \times \dfrac{600\frac{\text{cm}}{\text{h}}}{0.6} \times 1.0\dfrac{\text{g}}{\text{cm}^3} \times \dfrac{1\text{ cm}}{10^4\ \mu\text{m}} \times \dfrac{1\text{ h}}{3600\text{ s}}}{0.01\dfrac{\text{g}}{\text{cm}\cdot\text{s}}} = 0.003$$

where d_m is the diameter of a pore in the membrane, v is the fluid superficial velocity, ε is the void fraction, and ρ and μ are the density and viscosity of the fluid, respectively. Thus, this is laminar, creeping flow since Re is less than 0.1, and the velocity in the pore has a parabolic profile. Since the protein is being adsorbed on the surface of the pore, there will be a radial concentration gradient from the surface to the center of the pore, and the maximum diffusion length will therefore be 0.5 μm (one-half the pore diameter).

Therefore, for the conventional chromatography the diffusion time from the Einstein diffusion equation is

$$t = \frac{(50\ \mu\text{m})^2 \times \left(\dfrac{1\text{ cm}}{10^4\ \mu\text{m}}\right)^2}{2 \times 3.0 \times 10^{-7}\dfrac{\text{cm}^2}{\text{s}}} = 42\text{ s}$$

For membrane chromatography, assuming the diffusion coefficient in the pores of the membrane is the same as in pores of the conventional chromatography adsorbent,

$$t = \frac{(0.5\ \mu\text{m})^2 \times \left(\dfrac{1\text{ cm}}{10^4\ \mu\text{m}}\right)^2}{2 \times 3.0 \times 10^{-7}\dfrac{\text{cm}^2}{\text{s}}} = 0.004\text{ s}$$

These calculations show that the diffusion time is four orders of magnitude less for membrane chromatography compared to conventional chromatography. A similar result was obtained by Klein (*Affinity Membranes*, Wiley, Hoboken, N.J., 1991, p. 135) for the diffusion time of IgG antibody in membrane chromatography with hollow fibers compared to a fast-flow gel matrix, both using immobilized protein A.

In summary, membrane chromatography is a technology that is growing in importance in bioseparation science. There is a place for low-capacity, high-throughput adsorption media for the stripping of low-level contaminants from a process stream. The membranes are available with virtually all the same chemistries as conventional chromatography. They can be made inexpensively and fit well with disposable technologies currently in favor in the industry.

Scale-Up of the Chromatography of Proteins Chromatography scale-up algorithms typically account for changes in bed height and diameter, linear and volumetric flow rate, and particle size. A general approach to scale-up is based on keeping the resolution constant. Yamamoto et al. [*J. Chromatog.* **409:** 101 (1987); *Ion-Exchange Chromatography of Proteins*, Dekker, 1988, p. 263] have developed the following proportionality for resolution R_s of proteins in linear gradient elution ion exchange chromatography and hydrophobic interaction chromatography:

$$R_s \propto \left[\frac{\mathcal{D}_m L}{g(V - V_0)ud_p^2}\right]^{1/2} \tag{20-92}$$

where $\mathcal{D}_m$ = diffusion coefficient of protein in solution
L = column length
g = slope of gradient (change in concentration of gradient per volume of gradient)
V = column volume
V_0 = column void volume
u = interstitial fluid velocity = superficial velocity v/column void fraction ε
d_p = particle diameter

This equation has been found to be valid over a wide range of experimental conditions (Yamamoto et al., ibid.). To remove the volume terms from the expression for resolution, we make the definitions

$$\bar{Q} = \frac{Q}{V} = \frac{\varepsilon u A_c}{V} \tag{20-93}$$

$$G = Vg \tag{20-94}$$

where Q is the inlet flow rate, ε is the column void fraction, and A is the column cross-sectional area. These substitutions lead to

$$R_s \propto \left(\frac{\mathcal{D}\varepsilon}{G(1-\varepsilon)\bar{Q}d_p^2}\right)^{1/2} \tag{20-95}$$

Thus, for scale-up with constant resolution from scale 1 to scale 2 for the same product and the same column void fraction, the scale-up equation is

$$G_1\bar{Q}_1 d_{p1}^2 = G_2\bar{Q}_2 d_{p2}^2 \tag{20-96}$$

Thus, as the particle size increases on scale-up, the flow rate relative to the column volume must decrease and/or the gradient slope must decrease to maintain constant resolution, which seems correct intuitively.

In practice, the gradient and the stationary phase size and chemistry are not changed upon scale-up. This is so because it is easy to develop lab-scale processes that use the same resin and same gradient that can be used at the commercial process scale. Therefore, in practice, only the ratio between column volume and flow rate needs to be addressed:

$$\frac{Q_1}{V_1} = \frac{Q_2}{V_2} \tag{20-97}$$

Scale-up from a well-designed process development or preparatory column is reasonably straightforward. When the bed height can be maintained on scale-up, the mobile phase linear velocity remains the same, and the column is simply scaled by diameter. Scaling from a column 1 cm in diameter to a column having a diameter of 10 cm would constitute a 100-fold increase in scale. This is the most conservative way to scale up a column, but is not a necessary constraint.

Example 20-10 Scale-Up of Protein Chromatography A column 20 cm long, with an internal diameter of 5 cm, gives sufficient purification to merit scale-up. The column produces 3.2 g of purified protein per cycle, and a cycle takes 6 h, from equilibration through regeneration. To design a throughput of 10 g/h, what are the new column's dimensions if the superficial velocity is held constant and it is assumed that the gradient and particle size are not changed on scale-up?

Equation (20-97) is applicable since the gradient and particle size remain the same. According to this equation, the residence time is the same at small and large scales, and therefore the cycle time is not changed. Thus, the scaled-up column must produce 6 h/cycle ×10 g/h = 60 g/cycle. Since the flow rate is proportional to the throughput of protein,

$$\frac{Q_2}{Q_1} = \frac{60\text{ g/cycle}}{3.2\text{ g/cycle}} = 18.75$$

From Eq. (20-97),

$$\frac{Q_1}{A_1 L_1} = \frac{Q_2}{A_2 L_2}$$

which leads to

$$\frac{v_1}{L_1} = \frac{v_2}{L_2}$$

Since the superficial velocity is constant,

$$L_1 = L_2$$

Therefore,

$$\frac{Q_2}{Q_1} = \frac{V_2}{V_1} = \frac{\pi\left(\dfrac{D_2}{2}\right)^2 L_2}{\pi\left(\dfrac{D_1}{2}\right)^2 L_1} = \left(\frac{D_2}{D_1}\right)^2 = 18.75$$

where D_1 and D_2 are the column diameters for columns 2 and 1, respectively. Since $D_1 = 5.0$ cm, we obtain

$$D_2 = (18.75)^{0.5} D_1 = 21.6 \text{ cm}$$

It is not always necessary to scale according to bed diameter. The flow rate may be normalized against the volume of the empty column, to give units of time^{-1}. This normalized flow rate is held constant from one scale to another, as shown in Eq. (20-97). Considerable research and practice indicate that this technique is equally effective and less restrictive compared to holding linear velocity constant. In this case, column bed height may be increased or decreased, depending on the requirements of pressure drop and mechanical seals. A shallower bed gives a lower linear velocity and a wider diameter. A deeper bed gives a higher linear velocity (and higher pressure drop) and a proportionally narrower bed diameter. Therefore, bed height may need to be scaled on the basis of pressure drop constraints.

BIOPROCESS DESIGN AND ECONOMICS

Given a product and a desired annual production rate (process throughput), bioprocess design endeavors to answer the following and other related questions: What are the required amounts of raw materials and utilities needed for a single batch? What is the total amount of resources consumed per year? What is the required size of process equipment and supporting utilities? Can the product be produced in an existing facility, or is a new plant required? What is the total capital investment? What is the manufacturing cost? What is the optimum batch size? How long does a single batch take? How much product can be generated per year? Which process steps or resources constitute scheduling and throughput bottlenecks? What changes can increase throughput? What is the environmental impact of the process (i.e., amount and type of waste materials)? Which design is the "best" among several plausible alternatives?

Definitions and Background Process design is the conceptual work done prior to building, expanding, or retrofitting a process plant. It consists of two main activities, process synthesis and process analysis. *Process synthesis* is the selection and arrangement of a set of unit operations (process steps) capable of producing the desired product at an acceptable cost and quality. *Process analysis* is the evaluation and comparison of different process synthesis solutions. In general, a synthesis step is usually followed by an analysis step, and the results of analysis determine the subsequent synthesis step.

Process design and project economic evaluation require integration of knowledge from many different scientific and engineering disciplines and are carried out at various levels of detail. Table 20-25 presents a common classification of design and cost estimates and typical engineering costs for a $50 million capital investment for a new plant.

Figure 20-58 presents the need for design estimates of various types during the life cycle of product development and commercialization. The trapezoidal shape of the diagram represents the drastic reduction in product candidates as we move from feasibility studies to commercialization. In fact, the chances of commercialization at the research stage for a new product are only about 1 to 3 percent, at the development stage they are about 10 to 25 percent, and at the pilot plant stage they are about 40 to 60 percent (Douglas, *Conceptual Design of Chemical Processes*, McGraw-Hill, New York, 1988, p. 4).

Order-of-magnitude estimates are usually practiced by experienced engineers who have worked on similar projects in the past. They take minutes or hours to complete, but the error in the estimate can be as high as 50 percent. Table 20-26 presents a sample of data typically used by experienced

TABLE 20-25 Types of Design Estimates, Their Cost and Accuracy for a $50 Million Project*

Level	Type of estimate	Error (%)	Cost ($1000)
1	Order-of-magnitude estimate (ratio estimate) based on similar previous projects	≤ 50	
2	Project planning estimate (budget estimation) based on knowledge of major equipment items	≤ 30	30–100
3	Preliminary engineering (scope estimate) based on sufficient data to permit the estimate to be budgeted	≤ 25	250–750
4	Detailed engineering (capital approval stage) based on almost complete process data	≤ 15	1250–2000
5	Procurement and construction (contractor's estimate) based on complete engineering drawings, specifications, and site surveys	≤ 10	3500–7000

*Douglas, *Conceptual Design of Chemical Processes*, McGraw-Hill, New York, 1988, p. 7.

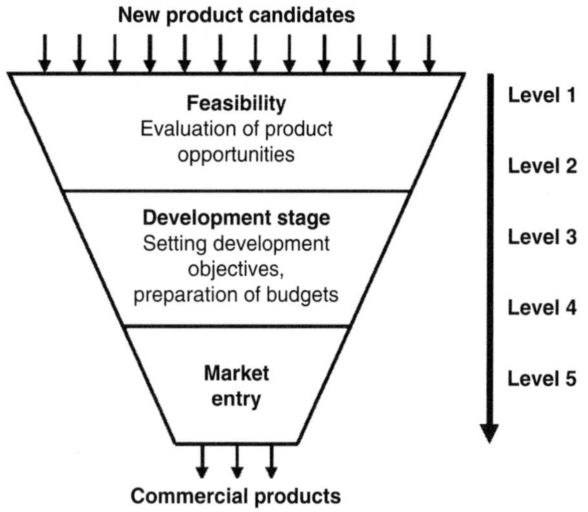

FIG. 20-58 Types of design estimates during the life cycle of a product [Frohlich, in *Biopharmaceutical Process Economics and Optimization* (conference proceedings), Sept. 30–Oct. 1, 1999, International Business Communications, Westborough, Mass].

engineers for order-of-magnitude estimates. The table lists the capital investment for five large-scale facilities built to manufacture therapeutic monoclonal antibodies using cell culture (by growing mammalian cells in stirred-tank bioreactors). Column 2 displays the number of the production bioreactors, the working volume of each, and the total working volume. For instance, a Genentech facility includes six production bioreactors, each having a working volume of 15 m^3. Column 4 displays the total capital investment, and column 5 displays the ratio of the total capital investment to the total production bioreactor volume. The ratio ranges from 3.3 to 6.2 with an average value of $4.6 million per cubic meter of bioreactor volume. Based on the data of Table 20-26, an engineer may conclude with some confidence that the capital investment for a new 100 m^3 (total production bioreactor volume) cell culture facility would be in the range of $330 million to $620 million and most likely around $460 million. Please note, however, that advances in technology (e.g., cell lines that generate higher product titers and the use of single-use systems) and other factors may render such data obsolete and reduce the accuracy of order-of-magnitude estimates. As a result, cost estimates are progressively refined as new product candidates move through the development life cycle shown in Fig. 20-58.

Most engineers employed by operating companies usually perform level 2 and 3 studies. Such studies take weeks or months to complete using appropriate computer aids. The main objective of such studies is to evaluate alternatives and pinpoint the most cost-sensitive areas—the economic "hot spots"—of a complex process. The results of such analyses are used to plan future research and development and to generate project budgets.

Level 4 and 5 studies are usually performed by the engineering and construction companies hired to build new plants for promising new products that are at an advanced stage of development. Such estimates are beyond the scope of this section. Instead, the focus of the material in the rest of this section will be on level 1, 2, and 3 studies. Also note that opportunities for creative process design work are usually limited to preliminary studies. By the time detailed engineering work has been initiated, a process is more

TABLE 20-26 Capital Investments for Cell Culture Facilities

	Bioreactor capacity (m^3)	Completion year	Investment ($ millions)	$ million per m^3
Boehringer Ingelheim (Germany)	6 × 15 = 90	2003	296	3.3
Lonza Biologics (Portsmouth, N.H.)	3 × 20 = 60	2004	207	3.4
Genentech (Oceanside, Calif.)	6 × 15 = 90	2005	450	5.0
Bristol Myers Squibb (Devens, Mass.)	6 × 20 = 120	2009	750	6.2
Roche Pharmaceuticals (Switzerland)	6 × 12.5 = 75	2009	375	5.0

than 80 percent fixed. Furthermore, the vast majority of important decisions for capital expenditures and product commercialization are based on results of preliminary process design and cost analysis. This explains why it is so important for a new engineer to master the skills of preliminary process design and cost estimation.

Environmental impact assessment is an activity closely related to process design and cost estimation. Biochemical plants generate a wide range of liquid, solid, and gaseous waste streams that require treatment prior to discharge. The cost associated with waste treatment and disposal has skyrocketed in recent years due to increasingly strict environmental regulations. This cost can be reduced through minimization of waste generation at the source. However, generation of waste from a chemical or biochemical process is dependent on the process design and the manner in which the process is operated. Thus, reducing waste in an industrial process requires intimate knowledge of the process technology. In contrast, waste treatment is essentially an add-on at the end of the process. In addition, minimization of waste generation must be considered by process engineers at the early stages of process development. Once a process has undergone significant development, it is difficult and costly to make major changes. Furthermore, regulatory constraints that are unique to the pharmaceutical industry restrict process modifications after clinical efficacy of the drug has been established. These are only some of the reasons why process synthesis and analysis must be initiated at the early stages of product development.

Synthesis of Bioseparation Processes Rules of thumb, or heuristics, were given in the introduction for use in developing flowsheets for the recovery and purification of biological products, and examples were given of the application of these rules of thumb to the placement of various bioseparation unit operations.

Process Analysis The flowsheets created during process synthesis must be analyzed and compared on the basis of capital investment, manufacturing cost, environmental impact, and other criteria to decide which ideas to consider further. Methodologies for estimating capital investment and manufacturing cost are presented later in the Process Economics subsection. In both cases, estimation is based on the results of material and energy balances and equipment sizing. These calculations are typically done using spreadsheets or process simulators. These tools allow the process design team to characterize a processing scenario, and then quickly and accurately redo the entire series of calculations for a different set of assumptions and other input data.

Spreadsheets Spreadsheet applications, such as Microsoft Excel, have become as easy to use as word processors and graphics packages. In its simplest form, a spreadsheet is an electronic piece of paper with empty boxes, known as *cells*. The user can enter data in those *cells*, perform calculations, and generate results. Results from spreadsheets can be easily plotted in a variety of graphs.

Process Simulators and Their Benefits Process simulators are software applications that enable the user to readily represent and analyze integrated processes. They have been in use in the petrochemical industries since the early 1960s. Established simulators for those industries include Aspen Plus and Aspen HYSYS from Aspen Technology, Inc. (Burlington, Mass.), ChemCAD from Chemstations, Inc. (Houston, Tex.), and PRO/II from SimSci-Esscor, Inc. (Lake Forest, Calif.).

The simulators mentioned above have been designed to model primarily continuous processes and their transient behavior. Most biological products, however, are produced in batch and semicontinuous mode (Korovessi and Linningerr, *Batch Processes*, Taylor and Francis, Oxford, UK, 2006; Heinzle et al., *Development of Sustainable Bioprocesses*, Wiley, Hoboken, N.J., 2006). Such processes are best modeled with batch process simulators that account for time dependency and sequencing of events. The first simulator designed specifically for batch processes was called Batches (from Batch Process Technologies in West Lafayette, Ind.). It was commercialized in the mid-1980s. All its operation models are dynamic, and simulation always involves integration of differential equations over a period of time. In the mid-1990s, Aspen Technology (Burlington, Mass.) introduced Batch Plus (now called Aspen Batch Process Developer), a recipe-driven simulator that targeted batch pharmaceutical processes. Around the same time, Intelligen, Inc. (Scotch Plains, N.J.) introduced SuperPro Designer [Petrides et al., *Pharm. Eng.* **22:** 56 (2002); Toumi et al., *Pharm. Eng.* **20**, March/April 2010]. SuperPro Designer is a flowsheet-driven simulator that handles material and energy balances, equipment sizing and costing, economic evaluation, environmental impact assessment, process scheduling, and debottlenecking of batch and continuous processes.

Discrete-event simulators have also found applications in the bioprocessing industries. Established tools of this type include ProModel from ProModel Corporation (Orem, Utah), Arena and Witness from Rockwell Automation, Inc. (Milwaukee, Wis.), Extend from Imagine That, Inc. (San Jose, Calif.), and FlexSim from FlexSim Software Products, Inc. (Orem, Utah). The focus of models developed with such tools is usually on the minute-by-minute time dependency of events and the animation of the process. Material balances, equipment sizing, and cost analysis tasks are usually out of the scope of such models.

The benefits from the use of process simulators depend on the type of product, stage of development, and size of the investment. For commodity biological products, such as biofuels, minimization of capital and operating costs is the primary benefit. For high-value biopharmaceuticals, systematic process development that shortens the time to commercialization is the primary motivation. Figure 20-59 shows a pictorial representation of the benefits from the use of such tools at the various stages of the commercialization process.

When product and process ideas are first conceived, process modeling tools are used for project screening, selection, and strategic planning based on preliminary economic analyses. During this phase, the company's process development groups are looking into the various options available for synthesizing, purifying, characterizing, and formulating the final product. The process undergoes constant changes during development. Typically, a large number of scientists and engineers are involved in the improvement and optimization of individual processing steps. The use of process simulators at this stage can introduce a common language of communication and facilitate team interaction. A computer model of the entire process can provide a common reference and evaluation framework to facilitate process development. The impact of process changes can be readily evaluated and documented in a systematic way. Once a reliable model is available, it can be used to pinpoint the cost-sensitive areas of a complex process. These are usually steps of high capital and operating cost or low yield and production throughput. The findings from such analyses can focus further lab and pilot plant studies to optimize those portions of the process. The ability to experiment on the computer with alternative process setups and operating conditions reduces the costly and time-consuming laboratory and pilot plant effort. A simulator can also evaluate the environmental impact of a process. For instance, material balances calculated for the projected large-scale manufacturing reveal environmental hot-spots. These are usually process steps that utilize organic solvents and other regulated materials with high disposal costs. Environmental issues not addressed during process development may lead to serious drawbacks during manufacturing.

With process development near completion at the pilot plant level, simulation tools are used to systematically design and optimize the process for commercial production. Availability of a good computer model can greatly facilitate the transfer of a new process from the pilot plant to the large-scale facility. If a new facility needs to be built, process simulators can size process equipment and supporting utilities and estimate the required capital

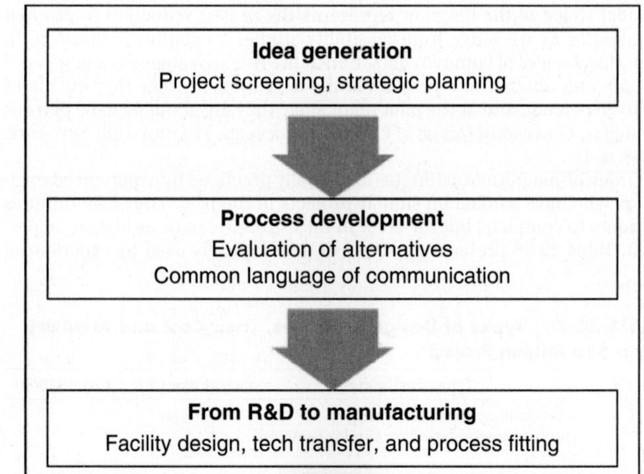

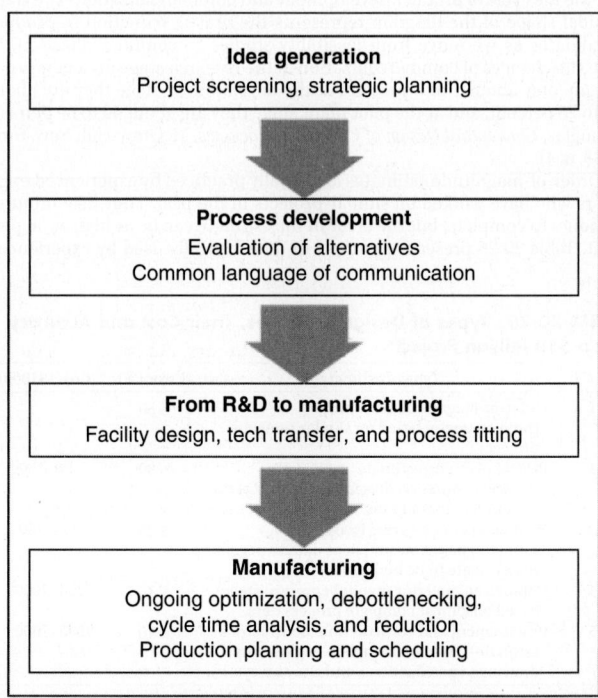

FIG. 20-59 Benefits of using process simulators. [Reproduced from *Bioseparations Science and Engineering*, 2d ed., by Roger G. Harrison, Paul W. Todd, Scott R. Rudge, and Demetri P. Petrides (2015). By Permission of Oxford University Press.]

investment. In transferring production to existing manufacturing sites (technology transfer), process simulators can be used to evaluate the various sites from a capacity and cost point of view and to select the most appropriate one.

In large-scale manufacturing, simulation tools are mainly used for ongoing process optimization and debottlenecking studies. Furthermore, tools that are equipped with batch process scheduling capabilities can be used to generate production schedules on an ongoing basis in a way that does not violate constraints related to the limited availability of equipment, labor resources, utilities, inventories of materials, etc.

Using a Biochemical Process Simulator The minimum requirements for a biochemical process simulator are the ability to handle batch as well as continuous processes and the ability to model the unit operations that are specific to bioprocessing. Because SuperPro Designer (from Intelligen, Inc.) has the ability to satisfy these requirements, we will use it to illustrate the role of such tools in bioprocess design. A functional evaluation version of SuperPro Designer and additional information on bioprocess simulation can be obtained at the website www.intelligen.com. Tutorial videos on the use of SuperPro Designer can be viewed at www.intelligen.com/videos.

To model an integrated process using a simulator, the user starts by developing a flowsheet that represents the overall process. For instance, the flowsheet of a hypothetical process can be drawn on the main window of SuperPro Designer. The flowsheet is developed by putting together the required unit operations (sometimes referred to as *unit procedures,* as explained later in this section) and joining them with material flow streams. Next, the user initializes the flowsheet by registering (selecting from the component database) the various materials used in the process and specifying operating conditions and performance parameters for the various operations.

Most biochemical processes operate in batch or semicontinuous mode. This is in contrast to continuous operation, which is typical in the petrochemical and other industries that handle large throughputs. In continuous operations, a piece of equipment performs the same action all the time, which is consistent with the notion of unit operations. In batch processing, however, a piece of equipment goes through a cycle of operations. For instance, a typical chromatography cycle includes *equilibration, loading, washing, elution,* and *regeneration.* In SuperPro Designer, the set of operations that comprise a processing step is called a *unit procedure* (versus a *unit operation*). Each unit procedure contains individual tasks (e.g., equilibration, loading) called *operations.* A unit procedure is represented on the screen with a single equipment icon. In essence, a unit procedure is the recipe of a processing step that describes the sequence of actions required to complete that step. The significance of the unit procedure is that it enables the user to describe and model the various activities of batch processing steps in detail.

For every operation within a unit procedure, SuperPro includes a mathematical model that performs material and energy balance calculations. Based on the material balances, SuperPro performs equipment-sizing calculations. If multiple operations within a unit procedure dictate different sizes for a certain piece of equipment, the software reconciles the different demands and selects an equipment size that is appropriate for all operations. In other words, the equipment is sized to ensure that it will not be overfilled during any operation but is no larger than necessary (to minimize capital costs). In addition, the software checks to ensure that the vessel contents will not fall below a user-specified minimum volume (e.g., a minimum impeller volume) for applicable operations.

Before any simulation calculations can be done, the user must initialize the various operations by specifying operating conditions and performance parameters through appropriate dialog windows. After initialization of the operations, the simulator performs material and energy balances for the entire process and estimates the required sizes of equipment. Optionally, the simulator may be used to carry out cost analysis and economic evaluation calculations. The fundamentals of process economics are described in the next subsection.

Other tasks that can be handled by process simulators include process scheduling, environmental impact assessment, debottlenecking, and throughput analysis. Issues of process scheduling and environmental impact assessment are addressed in the Illustrative Examples subsection. Throughput analysis and debottlenecking is the analysis of the capacity and time utilization of equipment and resources (e.g., utilities, labor, raw materials). The objective is to identify opportunities for increasing throughput with the minimum possible capital investment (see Illustrative Examples for additional information on the subject).

Having developed a good model using a process simulator or a spreadsheet, the user may conduct virtual experiments with alternative process setups and operating conditions. This may potentially reduce costly and time-consuming laboratory and pilot plant effort. One must be aware, however, that the GIGO (garbage in, garbage out) principle applies to all computer models. More specifically, if some assumptions and input data are incorrect, the outcome of the simulation will not be reliable. Consequently, validation of the model is necessary. In its simplest form, a review of the results by an experienced engineer can play the role of validation.

Process Economics The preliminary economic evaluation of a project for manufacturing a biological product usually involves estimation of capital investment, estimation of operating costs, and analysis of profitability. For biopharmaceuticals, another figure worth considering is the average cost of new drug development, which is in the range of \$500 million to \$1 billion [DiMasi et al., *J. Health Econ.* **22:** 151 (2003); Gilbert et al., *The Business and Medicine Report* **21:** Nov. 2003]. Much of this figure represents research and development (R&D) spending for all unsuccessful products. In other words, the actual average development cost per successful drug may be \$100 to \$200 million, but because more than 90 percent of new projects never reach commercialization, the average overall R&D cost skyrockets. This order-of-magnitude cost increase reinforces the need for effective process design tools and methodologies that assist engineers and scientists in efficiently evaluating and eliminating nonpromising project ideas at the very early stages of product and process development.

Capital Cost Estimation The capital investment for a new plant includes three main items: *direct fixed capital* (DFC), working capital, and start-up and validation cost. The DFC for small to medium biotechnology facilities is usually in the range of \$50 to \$200 million, whereas for large facilities it is in the range of \$250 to \$750 million. For preliminary design purposes, the various items of DFC are estimated based on the total equipment *purchase cost* (PC) using several multipliers sometimes called *Lang factors.* Table 20-27 provides ranges and average values for the multipliers and a skeleton for the calculations. Detailed definitions of the various cost items and additional information can be found in traditional process design textbooks and the technical literature [Turton et al., *Analysis, Synthesis, and Design of Chemical Processes,* 5th ed., Pearson, Upper Saddle River, N.J., 2018; Douglas, *Conceptual Design of Chemical Processes,* McGraw-Hill, New York, 1988; Peters and Timmerhaus, *Plant Design and Economics for Chemical Engineers,* 5th ed., McGraw-Hill, New York, 2002; Ulrich et al., *A Guide to Chemical Process Design and Economics, Wiley,* Hoboken, N.J., 1984; Valle-Riestra et al., *Project Evaluation in the Chemical Process Industries,* McGraw-Hill, New York, 1983; Garrett, *Chemical Engineering Economics,* Van Nostrand Reinhold, New York, 1989; Seider et al., *Process Design Principles—Synthesis, Analysis, and Evaluation,* Wiley, Hoboken, N.J., 1999; Towler and Sinnott, *Chemical Engineering Design; Principles, Practice and Economics of Plant and Process Design,* Butterworth-Heinemann (Elsevier), Oxford, U.K., 2008].

Notice the wide range of multiplier values in Table 20-27 for estimating the cost of buildings. Plants for commodity biochemicals, such as ethanol and citric acid, fall on the low end of the range. Conversely, biopharmaceutical facilities with their expensive heating, ventilation, and air conditioning (HVAC) requirements fall on the high end. The average value of 0.45 corresponds to relatively large plants that produce medium- to high-value products (e.g., industrial enzymes).

For more accurate estimation of building costs, it is necessary to estimate the process area required based on the footprint of the equipment and the space required around the equipment for safe and efficient operation and

TABLE 20-27 Fixed Capital Cost Estimation

Cost item	Average multiplier	Range of multiplier values
Total plant direct cost (TPDC)		
Equipment purchase cost (PC)		
Installation	0.50 × PC	0.2–1.5
Process piping	0.40 × PC	0.3–0.6
Instrumentation	0.35 × PC	0.2–0.6
Insulation	0.03 × PC	0.01–0.05
Electrical	0.15 × PC	0.1–0.2
Buildings	0.45 × PC	0.1–3.0
Yard improvement	0.15 × PC	0.05–0.2
Auxiliary facilities	0.50 × PC	0.2–1.0
Total plant indirect cost (TPIC)		
Engineering	0.25 × TPDC	0.2–0.3
Construction	0.35 × TPDC	0.3–0.4
Total plant cost (TPC)	TPDC + TPIC	
Contractor's fee	0.05 × TPC	0.03–0.08
Contingency	0.10 × TPC	0.07–0.15
Direct fixed capital (DFC)	TPC + contractor's fee and contingency	

Reproduced from *Bioseparations Science and Engineering,* 2d ed., by Roger G. Harrison, Paul W. Todd, Scott R. Rudge, and Demetri P. Petrides (2015). By Permission of Oxford University Press.

TABLE 20-28 Building Cost Estimation (Year 2012 Prices)*

Space function	Unit cost ($/m^2)	Air circulation rates (volume changes/h)
Process area[†]		
ISO Grade 8	3000–3750	10–20
ISO Grade 7	3750–5200	30–70
ISO Grade 6	6700–9000	70–160
ISO Grade 5	9000–12,000	200–600
Mechanical room (utilities)	450–900	
Laboratory	1500–3000	
Office	750–900	

*Frolich, in *Biopharmaceutical Process Economics and Optimization* (conference proceedings), International Business Communications, Sept. 30–Oct. 1, 1999.

[†]The ISO Grade refers to ISO 14644, "Cleanrooms and associated controlled environments." An earlier system uses a "class number" that refers to the maximum number of particles 0.5 μm or larger per cubic foot. Classes 100,000, 10,000, 1000, and 100 are similar to ISO Grades 8, 7, 6, and 5, respectively.

Reproduced from *Bioseparations Science and Engineering,* 2d ed., by Roger G. Harrison, Paul W. Todd, Scott R. Rudge, and Demetri P. Petrides (2015). By Permission of Oxford University Press.

maintenance. Then the building cost is estimated by multiplying the area of the various sections (e.g., process, laboratory, office) of a plant by an appropriate unit cost provided in Table 20-28. This table, which was developed by DPS Biometrics (Framingham, Mass.), also provides information on air circulation rates for the various process areas, which determine the sizing and power requirements of HVAC systems.

Table 20-27 indicates a wide range in the equipment installation cost multipliers. Using multipliers that are specific to individual equipment items leads to the most accurate estimates. In general, equipment delivered mounted on skids has a lower installation cost.

For preliminary cost estimates, Table 20-27 clearly shows that the fixed capital investment of a plant is a multiple (usually 3 to 10 times) of its equipment purchase cost. The low end of the range applies to large-scale facilities that produce biofuels and commodity biochemicals. The high end applies to biopharmaceutical facilities.

The equipment purchase cost can be estimated from vendor quotations, published data, company data compiled from earlier projects, and the use of process simulators that are equipped with appropriate costing capabilities. Vendor quotations are time-consuming to obtain and are, therefore, usually avoided for preliminary cost estimates. Instead, engineers tend to rely on the other three sources. Figures 20-60 to 20-63 provide equipment cost data for disk-stack centrifuges, membrane filtration systems, chromatography columns, and vertical agitated tanks that meet the specifications of the biopharmaceutical industry. The cost of the membrane filtration systems includes the cost of the skid, tank, pumps, and automation hardware and software. The tanks are appropriate for buffer preparation. They include a low-power agitator, but no heating/cooling jacket. The data represent average values from several vendors.

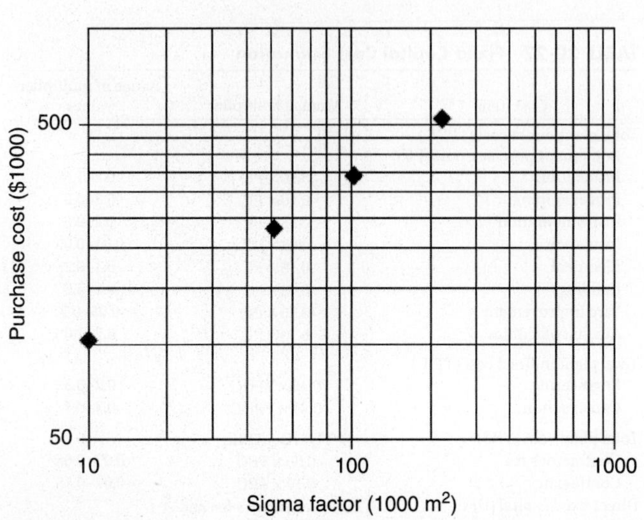

FIG. 20-60 Purchase cost of disk-stack centrifuges versus Σ factor (2012 prices). [Reproduced from *Bioseparations Science and Engineering,* 2d ed., by Roger G. Harrison, Paul W. Todd, Scott R. Rudge, and Demetri P. Petrides (2015). By Permission of Oxford University Press.]

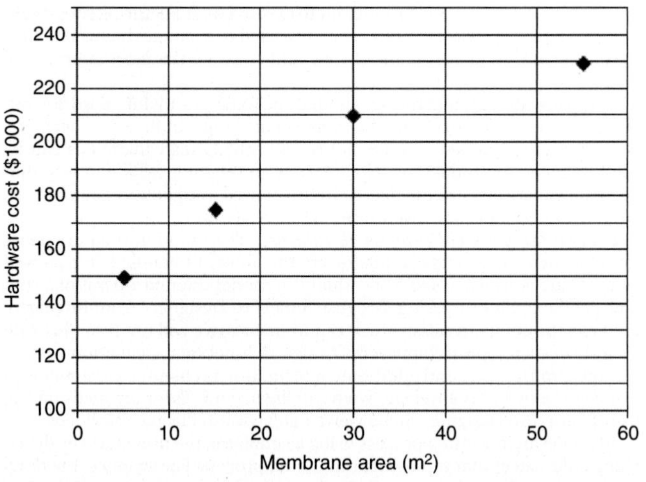

FIG. 20-61 Purchase cost of membrane filtration systems (2012 prices). [Reproduced from *Bioseparations Science and Engineering,* 2d ed., by Roger G. Harrison, Paul W. Todd, Scott R. Rudge, and Demetri P. Petrides (2015). By Permission of Oxford University Press.]

Note that equipment purchase cost is a strong function of industrial application and plant location. The data of Figs. 20-60 to 20-63 are applicable to biopharmaceutical facilities in developed countries. The cost of membrane filtration systems used in the food, biofuel, and water purification industries is more than an order of magnitude lower compared to the biopharmaceutical industry. The much larger equipment scale and the less stringent equipment specifications relative to biopharmaceuticals are responsible for the large difference in cost. The same trend applies to the cost of chromatography columns, storage tanks, reactor vessels, and most other equipment items. A good source of cost data for equipment used in the biofuel and biomaterial industries is available from the U.S. Department of Energy [DOE/NETL-2002/1169 (2002). Process Equipment Cost Estimation. Available at http://www.osti.gov/bridge/purl.cover.jsp?purl=/797810-Hmz80B/native/]. Additional sources for bioprocessing equipment cost data are available in the literature [Kalk and Langlykke, in Demain and Solomon (eds.), *Manual of Industrial Microbiology and Biotechnology,* American Society for Microbiology, Washington D.C., 1986, p. 363; Reisman, *Economic Analysis of Fermentation Processes,* CRC Press, Boca Raton, FL, 1988, p. 55].

Often, cost data for one or two discrete equipment sizes are available, but the cost for a different-size piece of equipment must be estimated. In such cases, the *scaling law* (expressed by the following equation) can be used:

$$\text{Cost}_2 = \text{Cost}_1\left(\frac{\text{Size}_2}{\text{Size}_1}\right)^a \tag{20-98}$$

The mathematical form of the scaling law explains why cost-versus-size data graphed on logarithmic coordinates tend to fall on a straight line. The value of the exponent a in Eq. (20-98) ranges between 0.5 and 1.0, with an average value for vessels of around 0.6 (this explains why the scaling law is also known as the "0.6 rule"). According to this rule, when the size of a vessel doubles, its cost will increase by a factor of $(2/1)^{0.6}$, or approximately 52 percent. This result is often referred to as the *economy of scale.* In using the scaling law, it is important to make sure that the piece of equipment whose cost is being estimated has a size that does not exceed the maximum available size for that type of equipment.

The prices of equipment change with time owing to inflation and other market conditions. That change in price is captured by the Chemical Engineering Plant Cost Index (CE Index) that is published monthly by *Chemical Engineering* magazine. The index I is used to update equipment cost data according to the following equation:

$$\text{Cost}_2 = \text{Cost}_1\left(\frac{I_2}{I_1}\right) \tag{20-99}$$

Another factor that affects equipment purchase cost is the material of construction. For instance, a tank made of stainless steel costs approximately 2.5 to 3 times as much as a carbon-steel tank of the same size. A tank made of titanium costs around 15 times the cost of a carbon-steel tank of the same size. Other factors that affect equipment cost include the

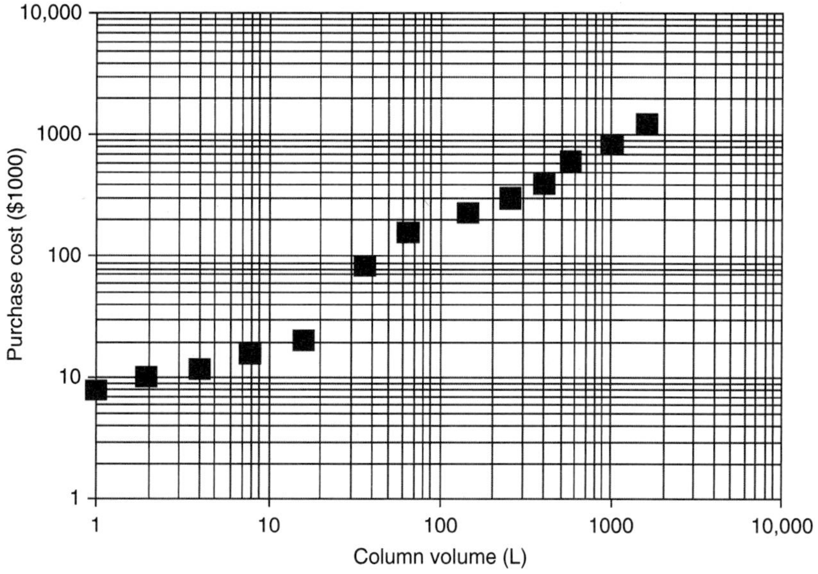

FIG. 20-62 Purchase costs of chromatography columns made of acrylic tube and stainless-steel bed supports (2012 prices). [Reproduced from *Bioseparations Science and Engineering*, 2d ed., by Roger G. Harrison, Paul W. Todd, Scott R. Rudge, and Demetri P. Petrides (2015). By Permission of Oxford University Press.]

finishing of the metal surface and the instrumentation that is provided with the equipment.

Working capital accounts for cash that must be available for investments in ongoing expenses and consumable materials. These expenses may include raw materials for 1 to 2 months, labor for 2 to 3 months, utilities for 1 month, waste treatment/disposal for 1 month, and other miscellaneous expenses. The required amount of working capital for a process is usually 10 to 20 percent of the DFC.

Start-up and validation costs can also represent a significant capital investment for a biopharmaceutical plant. A value of 20 to 30 percent of DFC is quite common.

Operating Cost Estimation The operating cost to run a biochemical plant is the sum of all ongoing expenses including raw materials, labor, consumables, utilities, waste disposal, and facility overhead. Dividing the annual operating cost by the annual production rate yields the unit production cost (e.g., in dollars per kilogram). The unit cost and selling prices of bioproducts are inversely proportional to market size (Harrison et al., *Bioseparations Science and Engineering*, Oxford University Press, Oxford, UK, 2015, p. 2). Low-molecular-weight commodity biochemicals and biofuels that are produced in large quantities cost around $1 to $5/kg to make. Citric acid, whose production is analyzed later in this section, is a product of this type. Specialty biochemicals that are used as food supplements (e.g., vitamins) and flavoring agents have a manufacturing cost of $5 to $100/kg. The manufacturing cost of therapeutic proteins produced in large quantities is in the range of $1/g to $1000/g. Human serum albumin (HSA) which is extracted from blood plasma and has an annual production volume of more than 500 metric tons lies close to the low end. The manufacturing cost of therapeutic proteins with annual production volume ranging from a few hundreds of kilograms to a few metric tons is in the range of $50 to $1000/g. The insulin and monoclonal antibody processes analyzed later in this section

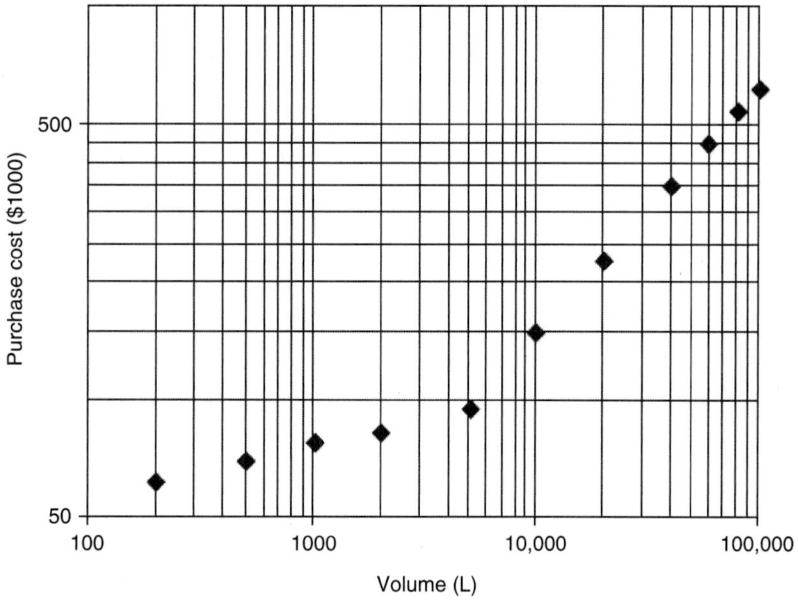

FIG. 20-63 Purchase cost of agitated tanks made of stainless steel (2012 prices). [Reproduced from *Bioseparations Science and Engineering*, 2d ed., by Roger G. Harrison, Paul W. Todd, Scott R. Rudge, and Demetri P. Petrides (2015). By Permission of Oxford University Press.]

TABLE 20-29 Operating Cost Items and Ranges

Cost item	Type of cost	Range of values (% of total)
Raw materials	Direct	10–80
Labor	Direct	10–50
Consumables	Direct	1–50
Lab/QC/QA	Direct	1–50
Waste disposal	Direct	1–20
Utilities	Direct	1–30
Facility overhead	Indirect	10–70
Miscellaneous	Indirect	0–20

Reproduced from *Bioseparations Science and Engineering,* 2d ed., by Roger G. Harrison, Paul W. Todd, Scott R. Rudge, and Demetri P. Petrides (2015). By Permission of Oxford University Press.

represent products of this type. The manufacturing cost of interferons, erythropoietin (EPO), and other therapeutic proteins with very low annual production volume (from hundreds of grams to a few kilograms) is more than $10,000/g [Jagschies, *BioPharm Int.* **21:** 72 (2008)].

Table 20-29 displays the various types of operating cost, their direct or indirect nature, and ranges for their values relative to the total operating cost. Sometimes cost items are categorized as either fixed or variable. Fixed costs are incurred regardless of the volume of product output. The clearest case of a fixed cost is depreciation, which is part of the equipment-dependent cost. The clearest case of a variable cost would be the cost of raw materials. Most other costs have a fixed component and a variable component. It is obvious from the wide range of values in Table 20-29 that industry averages cannot predict the operating cost of a process; a certain level of detailed calculations is required.

The raw materials cost includes the cost of all fermentation media, recovery chemicals, and cleaning materials. For commodity biochemicals, such as ethanol, the cost of fermentation media is the main component. For high-value products, the solutions used for product recovery and equipment cleaning can be a major part of the raw materials cost. Table 20-30 provides a list of commonly used raw materials in the biochemical industries. Note that the price of a raw material can vary widely depending on its required purity. This can be clearly seen in the case of water. *Water for injection* (WFI), for instance, costs 100 to 500 times as much as city water. Raw materials for biopharmaceuticals are typically required to be USP/NF grade. Prices for a wide range of chemicals are available online at www.spectrumchemicals.com.

TABLE 20-30 Common Bioprocessing Raw Materials (Year 2012 Prices)

Raw material	Comments	Price ($/kg)
C Source		
Glucose	Solution 70% w/v	0.30–0.40
Corn syrup	95% Dextrose equivalent	0.40–0.50
Molasses	50% Fermentable sugars	0.12–0.20
Soybean oil	Refined	1.10–1.30
Corn oil	Refined	1.30–1.40
Ethanol	USP tax-free	0.80–0.90
Methanol	Gulf Coast	0.40–0.45
n-Alkanes		0.75–0.90
N Source		
Ammonia	Anhydrous, fertilizer grade	0.30–0.60
Soybean flour	44% protein	0.45–0.50
Cottonseed flour	62% protein	0.50–0.60
Casein	13.5% w/w total N	10.00–12.00
Ammonium sulfate	Technical	0.17–0.25
Ammonium nitrate	Fertilizer grade 33.5% N, bulk	0.20–0.30
Urea	46% N, agricultural grade	0.55–0.65
Yeast	Brewers, debittered	1.25–1.40
Whey	Dried, 4.5% w/w N	1.25–1.40
Salts		
KH_2PO_4	USP, granular	1.65–1.85
K_2SO_4	Granular, purified	2.80–3.00
Na_2HPO_4		1.40–1.80
$MgSO_4 \cdot 7H_2O$		0.45–0.55
$ZnSO_4 \cdot 7H_2O$	Agricultural grade, powder	0.65–0.75
Other		
Process water		0.0001–0.001
RO water		0.005–0.01
Water for injection		0.02–0.5
H_3PO_4 (85% w/w)	Food grade	3.5 – 4.5
NaOH		0.2–0.5
HCl (37% w/w)		0.7–0.8
H_2SO_4 (98% w/w)		0.15–0.25

Reproduced from *Bioseparations Science and Engineering,* 2d ed. by Roger G. Harrison, Paul W. Todd, Scott R. Rudge, and Demetri P. Petrides (2015). By Permission of Oxford University Press.

Labor is estimated based on the total number of operators, which in turn is calculated by summing the operator requirements of the various operations as a function of time. As will become clear in the examples discussed later, the labor requirement in a batch manufacturing facility varies with time. In a single-product facility, the number of operators in each shift must be based on maximum demand during that shift. In multiproduct facilities, each product line can employ a certain number of dedicated operators and rely on floating operators during periods of peak demand. In general, smaller facilities tend to utilize a larger number of operators per processing step because these plants are less automated. For instance, a small biotech company may utilize two or three operators to set up a fermenter, whereas in a large, highly automated fermentation facility, a single operator may handle the setup of six different fermenters remotely from the control room. In general, a typical biotech company that deals with high-value products will allocate at least one operator to each processing step (centrifugation, membrane filtration, chromatography, etc.) during its operation. The setup of a step may require multiple operators for a short period. The annual cost of an operator (including salary and benefits) varies widely around the globe. It is in the range of $4000 to $10,000 in developing nations and can exceed $50,000 in developed countries (Pollak et al., *Contract Pharma,* Jan./Feb. 2012).

Consumables are items that may be used up, fouled, or otherwise damaged during processing, such as membranes, chromatography resins, and activated carbon. These items must be replaced periodically. As the examples later in this section will illustrate, the high unit cost of chromatography resins and their frequent replacement can make them a major component of the manufacturing cost. The unit cost of typical ion exchange and hydrophobic interaction chromatography resins used for the purification of proteins is in the range of $500 to $2000 per liter of resin. The unit cost of protein A affinity resins that are commonly used for the purification of monoclonal antibodies is in the range of $5000 to $15,000 per liter of resin. The replacement frequency of such resins is in the range of 50 to 200 cycles of usage (the high-end resins have a longer useful life). In contrast, the unit cost of polymeric chromatography resins used for the purification of small biomolecules (e.g., amino acids) is substantially lower (under $100 per liter of resin), and their life is longer (1000 to 2000 h of operation). Likewise, the unit cost of silica-based resins used for water demineralization is around $0.5 per liter, and their life is in the range of 2000 to 6000 h of operation (the life strongly depends on the composition of the treated materials).

Regarding membrane filtration operations, the unit cost of MF/UF membranes used in the biopharmaceutical industry (in the form of hollow fiber cartridges or cassettes) is in the range of $300 to $800/m². Such membranes typically handle 10 to 50 filtration cycles before disposal. The unit cost of related membranes used in industrial biotechnology (e.g., for production of industrial enzymes) is considerably lower (under $200/m²), and the expected life is more than 2000 h of operation. The cost of membranes used for large-scale water purification is under $50/m², and their useful life is at least 6000 h of operation. In general, ceramic membranes cost more than polymeric ones, but they last longer.

The cost of disposable bags or containers, also known as single-use systems, is part of the consumables cost as well. Disposable bags have become popular in biopharmaceutical manufacturing because they eliminate the need for cleaning and sterilization in place (Papavasileiou et al., *BioPharm Int. Suppl.,* Nov. 2008, p. 16). Other advantages of single-use systems include increased processing flexibility and shorter validation, start-up, and commercialization times. Table 20-31 provides information on disposable bags used for the preparation and storage of buffer solutions and fermentation

TABLE 20-31 Disposable Bags for Preparation and Storage of Solutions (Year 2012 Prices)

Volume (L)	Bags for storage ($)	Bags for mixing ($)
50	310	600
100	340	690
200	360	820
500	460	930
1000	650	1180

Reproduced from *Bioseparations Science and Engineering,* 2d ed., by Roger G. Harrison, Paul W. Todd, Scott R. Rudge, and Demetri P. Petrides (2015). By Permission of Oxford University Press.

media. Bags with mixing capability are required for solution preparation. Similar bags are used for inoculum preparation in rocking and stirred-tank bioreactors. Bags for stirred-tank bioreactors are available with working volumes of up to 2000 L. A number of biopharmaceuticals are produced exclusively in single-use systems. Note that large disposables bags (larger than 50 L) utilize appropriate supporting skids.

Laboratory, QC, and QA activities include off-line analysis, quality control (QC), and quality assurance (QA) costs. Chemical and biochemical analysis and physical property characterization, from raw materials to final product, are a vital part of biochemical operations. The laboratory/QC/QA cost is usually 10 to 20 percent of the operating labor cost. However, for certain biopharmaceuticals that require a large number of very expensive assays, this cost can be as high as the operating labor. For such cases, it is important to account for the number and frequency of the various assays in detail, since changes in lot size that can reduce the frequency of analysis can have a major impact on profit margins.

The treatment of wastewater and the disposal of solid and hazardous materials are other important operating costs. The amount and composition of the various waste streams are derived from the material balances. Multiplying the amount of each waste stream by the appropriate unit cost yields the cost of treatment and disposal. Treatment of low biological oxygen demand (BOD) wastewater (< 1000 mg/L) by a municipal wastewater treatment facility usually costs \$0.2 to \$0.5/m^3. This is not a major expense for most biotech facilities that deal with high-value products. However, disposal of contaminated solvents (typically generated by chromatography steps) and other regulated compounds can become a major expense because the unit disposal cost can be more than \$1/kg. Waste disposal may also become a problem if an unwanted by-product is generated as part of the recovery chemistry of a process (see the citric acid example). Disposal of single-use systems via incineration costs \$100 to \$200 per metric ton of material.

Utilities costs include the cost of heating and cooling agents as well as electricity. The amounts are calculated as part of the material and energy balances. Aerobic fermenters are major consumers of electricity, but downstream processing equipment generally does not consume much electricity. In terms of unit cost, electricity costs \$0.05 to \$0.15/kWh. The cost of heat removal using cooling water is in the range of \$0.002 to \$0.01 per 1000 kcal of heat removed. The cost of cooling using chilled water and refrigerants is in the range of \$0.05 to \$0.1 per 1000 kcal of heat removed. The cost of producing steam for use as a heating medium is around \$5 to \$15/1000 kg depending on pressure (low, medium, high), type of fuel used for its generation, and scale of production. The cost of clean steam (generated utilizing highly purified water) is around \$50 to \$100/1000 kg (depending on the scale of production and level of water purity). Clean steam is used in biopharmaceutical facilities for sterilizing equipment as part of equipment cleaning (e.g., "steam-in-place" or SIP operations). Note that manufacturers often classify purified water used for buffer preparation and equipment cleaning as a utility and not as a raw material, thus increasing the cost contribution of utilities. The insulin example, presented later in this section, describes a methodology for the systematic sizing of systems that supply purified water.

Facility overhead costs account for the depreciation of the fixed capital investment, maintenance costs for equipment, insurance, local (property) taxes, and possibly other overhead-type expenses. For preliminary cost estimates, the entire fixed capital investment is usually depreciated linearly over a 10-year period. In the real world, the U.S. government allows corporations to depreciate equipment in 5 to 7 years and buildings in 25 to 30 years. The value of land cannot be depreciated. The annual maintenance cost can be estimated as a percentage of the equipment's purchase cost (usually 10 percent) or as a percentage of the overall fixed capital investment (usually 3 to 5 percent). Insurance rates depend to a considerable extent on the maintenance of a safe plant in good repair condition. A value for insurance in the range of 0.5 to 1 percent of DFC is appropriate for most bioprocessing facilities. The processing of flammable, explosive, or highly toxic materials usually results in higher insurance rates. The local (property) tax is usually 2 to 5 percent of DFC. The factory expense represents overhead cost incurred by the operation of non-process-oriented facilities and organizations, such as accounting, payroll, fire protection, security, and cafeteria. A value of 5 to 10 percent of DFC is appropriate for these costs.

Included in miscellaneous costs are ongoing R&D, process validation, and other overhead-type expenses that can be ignored in preliminary cost estimates. Other general expenses of a corporation include royalties, advertising, and selling. If any part of the process or any equipment used in the process is covered by a patent not assigned to the corporation undertaking the new project, permission to use the technology covered by the patent must be negotiated, and some form of royalty or license fee is usually required. Advertising and selling covers expenses associated with the activities of the marketing and sales departments.

Profitability Analysis Estimates of capital investment, operating cost, and revenues of a project provide the information needed to assess its profitability and attractiveness from an investment point of view. There are various measures for assessing profitability. The simplest ones include gross margin, return on investment (ROI), and payback time, and they are calculated by using the following equations:

$$\text{Gross margin} = \frac{\text{Gross profit}}{\text{Revenues}} \tag{20-100}$$

$$\text{Return on investment (ROI)} = \frac{\text{Net profit per year}}{\text{Total investment}} \times 100\,\% \tag{20-101}$$

$$\text{Payback time (yr)} = \frac{\text{Total investment}}{\text{Net profit per year}} \tag{20-102}$$

where gross profit is equal to annual revenues minus the annual operating cost, and net profit is equal to gross profit minus income taxes plus depreciation. All variables are averaged over the lifetime of a project.

Other measures that are more involved, such as the net present value (NPV) and internal rate of return (IRR), consider the cash flows of a project over its evaluation life and the time value of money. Detailed definitions for NPV and IRR can be found in the literature [Peters and Timmerhaus, *Plant Design and Economics for Chemical Engineers*, 4th ed., McGraw-Hill, New York, 1991, p. 327; Towler and Sinnott, *Chemical Engineering Design; Principles, Practice and Economics of Plant and Process Design*, Butterworth-Heinemann (Elsevier), Oxford, U.K., 2008, p. 366]. The examples presented next demonstrate how these measures facilitate the decision-making process.

Illustrative Examples In this section, SuperPro Designer is used to illustrate the analysis and evaluation of the production of three biological products. The first example analyzes the production of citric acid, a commodity organic acid heavily used in the beverage industry. The second deals with the bacterial production of recombinant human insulin, the first commercial product of modern biotechnology. The third example focuses on the production of monoclonal antibodies (mAbs) from mammalian cells cultured in stirred-tank bioreactors. The generation of the flowsheets for the production of all three products was based on information available in the patent and technical literature combined with engineering judgment and experience with other biological products. These examples are used to draw general conclusions on the manufacturing cost of biological products. The computer files for these examples are available as part of the evaluation version of SuperPro Designer at the website www.intelligen.com. The flowsheets and charts of the examples are available in color format in the readme files of the corresponding SuperPro Designer examples. Additional examples and pertinent publications are available at www.intelligen.com/literature.

Citric Acid Production A number of organic acids are produced via fermentation. Of these, citric acid is produced in the largest amount [> 1,800,000 metric tons (t) per year]. Citric acid is marketed as citric acid monohydrate or as anhydrous citric acid. The majority of citric acid (> 60 percent) is used in the food and beverage industries to preserve and enhance flavor. In the chemical industries (which represent 25 to 30 percent of total utilization), the uses of citric acid include the treatment of textiles, softening of water, and manufacturing of paper. In the pharmaceutical industry (10 percent of total utilization), citrate is used as a buffer and formulation excipient, iron citrate is used as a source of iron, and citric acid is used as a preservative for stored blood, tablets, and ointments, and in cosmetic preparations (Crueger and Crueger, *Biotechnology—A Textbook of Industrial Microbiology*, 2d ed., Sinauer, Sunderland, Mass., 1989, p. 134). Citric acid is increasingly being used in the detergent industry as a replacement for polyphosphates.

Citric acid was first recovered in 1869 in England from calcium citrate, which was obtained from lemon juice. Its production by filamentous fungi has been known since 1893. The first production via surface culture fermentation was initiated in 1923. Production using stirred-tank fermenters began in the 1930s, and presently this is the preferred method for large-scale manufacturing. The plant considered in this example produces around 18,000 t/yr of crystal citric acid, which represents approximately 1 percent of the current world demand.

The entire flowsheet is shown in Fig. 20-64. Molasses, the carbon source of fermentation, is diluted with water from about 50 percent fermentable sugars content to 20 percent in a blending tank (V-101). Suspended particulate material is then removed by filtration (PFF-101). Metal ions, particularly iron, are subsequently removed by an ion exchange chromatography column (C-101), and the purified raw material solution is then heat-sterilized (ST-101). Nutrients (i.e., sources of ammonium, phosphorus, magnesium, potassium copper, and zinc) are dissolved in water (V-104) and heat-sterilized (ST-101). (It is necessary to sterilize the first three nutrients separately to avoid a precipitate.) The fermentation cycle is 7 days, and the production is handled by seven fermenters that operate in staggered mode.

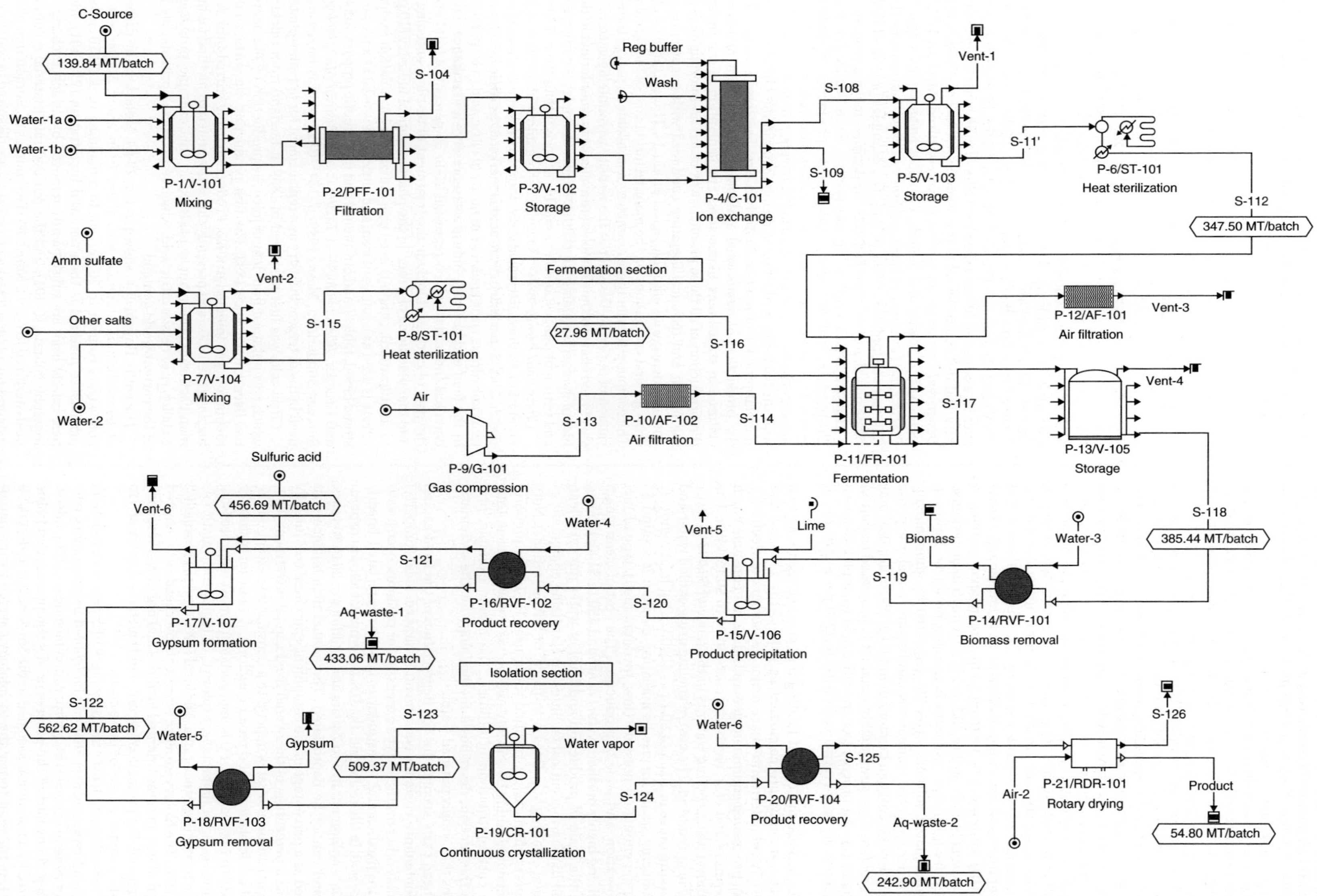

FIG. 20-64 Citric acid production flowsheet. [Reproduced from *Bioseparations Science and Engineering*, 2d ed., by Roger G. Harrison, Paul W. Todd, Scott R. Rudge, and Demetri P. Petrides (2015). By Permission of Oxford University Press.]

Since the plant operates around the clock, one fermentation batch is initiated daily and another one is completed daily. Each fermenter has a vessel volume of 350 m^3 and generates broth of around 315 m^3. A three-step seed fermenter train (not shown in the flowsheet) supplies inoculum to each production fermenter (FR-101). A pure culture of the mold *Aspergillus niger* is used to inoculate the smallest seed fermenter. When optimum growth of mycelium is reached, the contents of the seed fermenter are transferred to the next stage fermenter, which is approximately 10 times larger. Similarly, this larger seed fermenter inoculates the production fermenter with about 10 percent volume of actively growing mycelium broth. Air is supplied by a compressor (G-101) at a rate that gradually increases from 0.15 vvm (volume of air per volume of liquid per minute) to 1.0 vvm. Cooling water removes the heat produced by the exothermic process (2990 kcal/kg of citric acid formed) and maintains the temperature at 28°C. The fermented broth is discharged into the holding tank (V-105), which acts as a buffer tank between the batch upstream section and the continuous downstream section.

Purification starts with the removal of biomass by a rotary vacuum filter (RVF-101). The clarified fermentation liquor flows to an agitated reactor vessel (V-106) where approximately 1 part of hydrated lime, $Ca(OH)_2$, for every 2 parts of liquor is slowly added to precipitate calcium citrate. The lime solution must be very low in magnesium content to minimize losses due to generation of relatively soluble magnesium citrate. Calcium citrate is separated by a second rotary vacuum filter (RVF-102), and the citrate-free filtrate (Aq-Waste-1) is sent to a wastewater collection tank. The calcium citrate cake is sent to another agitated reactor vessel (V-107), where it is acidified with dilute sulfuric acid to form a precipitate of calcium sulfate (gypsum). A third filter (RVF-103) removes the precipitated gypsum and yields an impure citric acid solution in the filtrate. Careful control of the pH and temperature of each precipitation step is important for maximizing the yield of citric acid. The resulting solution is concentrated and crystallized using a continuous evaporator/crystallizer (CR-101). The crystals formed are separated by rotary vacuum filtration (RVF-104) and dried in a rotary dryer (RDR-101). If the final product is required in high purity, then treatment with activated carbon may precede crystallization to remove colorants. In addition, ion exchange is sometimes used to remove metal ions and other ionic species.

An equipment occupancy chart for consecutive batches can be developed using SuperPro Designer (Harrison et al., *Bioseparations Science and Engineering*, Oxford University Press, Oxford, UK, 2015, p. 475). The process batch time is approximately 200 h (or 8.3 days). This is the time elapsed from the preparation of raw materials to the final product of a single batch (excluding the time required for inoculum preparation). The duration of each fermentation batch is 160 h (6.7 days). The availability of seven production fermenters operating in staggered mode (out of phase) enables the plant to initiate a new batch every 24 h. The upstream portion of the process (i.e., raw material preparation and fermentation) operates in batch mode. The downstream section (product recovery and purification) operates continuously.

A summary of the overall material balances (expressed as input-output component balance) is given in Table 20-32. "CA crystal" stands for crystalline citric acid and represents the final product. Glucose represents the fermentable carbohydrates in molasses (50 percent w/w). Note the large amounts of $Ca(OH)_2$ and sulfuric acid consumed and gypsum (calcium sulfate) generated. The quantities of these compounds depend on the chemistry of the purification process and cannot be reduced without changing the recovery technology. Since this gypsum is contaminated with biomass, it has little or no commercial value. A disposal cost of \$50/MT (metric ton) was assumed in this example. The large amount of wastewater is also worth noting.

A list of major equipment items along with their purchase costs (generated by SuperPro Designer) is shown in Table 20-33. The total equipment cost for a plant of this capacity is around \$10.4 million. Note that approximately 30 percent of the equipment cost is associated with the seven production fermenters. The fermenters are made of stainless steel to minimize leaching of heavy metals that affect product formation. The final item, cost of unlisted equipment, accounts for the cost of the seed fermenters, pumps, and other secondary equipment that is not considered explicitly. Table 20-34 displays the various items of the direct fixed capital (DFC) investment. The total DFC for a plant of this capacity is around \$43.6 million or approximately 4.2 times the total equipment cost.

A summary of the operating costs is shown in Table 20-35. The raw materials cost is the most important, accounting for 40.9 percent of the overall operating cost. This is quite common for commodity biochemicals. Molasses is the most expensive raw material, accounting for 67 percent of the raw materials cost. The purification chemicals, sulfuric acid and calcium hydroxide, account for 18.7 percent and 9.3 percent of the overall raw

TABLE 20-32 Overall Material Balances for Citric Acid (CA) Production (kg/yr)

Component	In	Out	Out − In
Ammonium sulfate	278,000	26,000	−252,000
Biomass	0	2,014,000	2,014,000
CA crystal	0	18,250,000	18,250,000
$Ca(OH)_2$	11,717,000	558,000	−11,159,000
Calcium citrate	0	623,000	623,000
CO_2	0	3,861,000	3,861,000
Citric acid	0	657,000	657,000
Glucose	22,934,000	275,000	−22,659,000
Gypsum	0	19,972,000	19,972,000
Impurities	459,000	459,000	0
Nutrients	1,894,000	383,000	−1,511,000
Oxygen	65,171,000	57,994,000	−7,177,000
NaOH	185,000	185,000	0
Sulfuric acid	14,979,000	576,000	−14,403,000
Water	308,879,000	320,663,000	11,784,000
Total	426,496,000	426,496,000	0

TABLE 20-33 Equipment Specification and Purchase Costs for Citric Acid Production (Year 2012 Prices)

Quantity	Name	Description	Unit cost (\$)	Cost (\$)
1	AF-101	Air filter Rated throughput = 7.7 m^3/s	100,000	100,000
1	AF-102	Air filter Rated throughput = 3.0 m^3/s	37,000	37,000
1	C-101	Chromatography column Column volume = 4.66 m^3	150,000	150,000
1	CR-101	Crystallizer Vessel volume = 130 m^3	542,000	542,000
7	FR-101	Fermenter Vessel volume = 350 m^3	436,000	3,052,000
1	G-101	Centrifugal compressor Compressor power = 1,430 kW	1,560,000	1,560,000
1	PFF-101	Plate & frame filter Filter area = 335 m^2	145,000	145,000
1	RDR-101	Rotary dryer Drying area = 85 m^2	475,000	475,000
1	RVF-101	Rotary vacuum filter Filter area = 46 m^2	154,000	154,000
1	RVF-102	Rotary vacuum filter Filter area = 83 m^2	214,000	214,000
1	RVF-103	Rotary vacuum filter Filter area = 71 m^2	195,000	195,000
1	RVF-104	Rotary vacuum filter Filter area = 35 m^2	137,000	137,000
1	ST-101	Sterilizer Rated throughput = 34 m^3/h	308,000	308,000
1	V-101	Blending tank Vessel volume = 300 m^3	503,000	503,000
1	V-102	Blending tank Vessel volume = 300 m^3	503,000	503,000
1	V-103	Blending tank Vessel volume = 300 m^3	503,000	503,000
1	V-104	Blending tank Vessel volume = 35 m^3	139,000	139,000
2	V-105	Flat bottom tank Vessel volume = 350 m^3	198,000	396,000
1	V-106	Neutralizer Vessel volume = 42 m^3	126,000	126,000
1	V-107	Neutralizer Vessel volume = 15 m^3	94,000	94,000
		Unlisted Equipment		1,037,000
		Total		**10,370,000**

TABLE 20-34 Fixed Capital Estimate Summary for Citric Acid Production (Year 2012 Prices, $)

Total plant direct cost (TPDC)		
Equipment purchase cost	10,370,000	
Installation	3,726,000	
Process piping	3,111,000	
Instrumentation	2,074,000	
Insulation	311,000	
Electrical	1,037,000	
Buildings	2,074,000	
Yard improvement	1,556,000	
Auxiliary facilities	1,037,000	
TPDC		25,295,000
Total plant indirect cost (TPIC)		
Engineering	5,059,000	
Construction	7,589,000	
TPIC		12,648,000
Total plant cost (TPC = TPDC + TPIC)		37,943,000
Contractor's fee	1,897,000	
Contingency	3,794,000	
Direct fixed capital (DFC) TPC + contractor's fee and contingency		43,634,000

Reproduced from *Bioseparations Science and Engineering*, 2d ed., by Roger G. Harrison, Paul W. Todd, Scott R. Rudge, and Demetri P. Petrides (2015). By Permission of Oxford University Press.

materials cost, respectively. The following prices were assumed: \$0.15/kg of molasses, \$0.013/kg of 10 percent w/w H_2SO_4 solution, \$0.08/kg of $Ca(OH)_2$, and \$0.1/m^3 of process water. The facility-dependent cost is the second most important, accounting for 28.7 percent of the overall cost. Depreciation of the fixed capital investment and maintenance of the facility are the main contributors to this cost. Utilities are the third largest expense, accounting for 14.8 percent of the overall cost. Electricity and cooling water utilized by the fermenters are the main contributors to this cost. Labor lies in the fourth position, and the environmental cost (waste treatment/disposal) is fifth. Disposal unit costs of \$1/m^3 and \$50/MT (metric ton) were assumed for liquid and solid (gypsum) waste streams, respectively. The disposal of gypsum accounts for 85 percent of the overall environmental cost. The overall unit production cost is approximately \$1.4/kg, which is roughly equal to the early 2012 selling price of citric acid [Saleem, http://www.prlog.org/11293569-china-citric-acid-price-trend-outlook-2011.html (2011)]. This can be explained by noting the excess citric acid production capacity around the world (which keeps profit margins low), and the fact that most operating citric acid plants are rather old and partially depreciated. If depreciation is ignored, the facility-dependent cost is reduced by more than 80 percent, and the overall unit cost drops to around \$1/kg.

Based on the preliminary evaluation of this project idea, one should not recommend investing in new citric acid production capacity unless there is a combination of favorable conditions. Obviously, availability of inexpensive equipment (e.g., by acquiring an existing facility) and raw materials (e.g., by locating the plant near a source of low-cost molasses) are the most important factors. Development or adoption of a superior technology may also change the attractiveness of citric acid production. Such a technology is actually available; it utilizes extraction for citric acid recovery [Roberts, in McKetta and Cunningham (eds.), *Encyclopedia of Chemical Processing and Design*, vol. 8, Dekker, New York, 1979, p. 324]. Recovery by extraction

TABLE 20-35 Operating Cost Summary for Citric Acid Production (Year 2012 Prices)

Cost item	Citric acid crystals (\$/kg)	Annual cost (\$/yr)	Proportion of total (%)
Raw materials	0.57	10,310,000	40.92
Facility-dependent	0.40	7,223,000	28.67
Labor	0.12	2,102,000	8.34
Consumables	0.00	15,000	0.06
Lab/QC/QA	0.01	210,000	0.83
Waste treatment and disposal	0.09	1,611,000	6.39
Utilities	0.21	3,724,000	14.78
Total	1.40	22,195,000	100.00

Reproduced from *Bioseparations Science and Engineering*, 2d ed., by Roger G. Harrison, Paul W. Todd, Scott R. Rudge, and Demetri P. Petrides (2015). By Permission of Oxford University Press.

eliminates the consumption of $Ca(OH)_2$ and H_2SO_4 and the generation of the unwanted $CaSO_4$. Butanol has been used as an extractant, as has tributyl phosphate. Ion pair extraction by means of secondary or tertiary amines dissolved in a water-immiscible solvent (e.g., octyl alcohol) provides an alternative route. Developments in electrodialysis membranes could lead to recovering citric acid directly from the fermentation broth by this technique (Blanch and Clark, *Biochemical Engineering*, Dekker, New York, 1997, p. 611).

Human Insulin Production Insulin facilitates the metabolism of carbohydrates and is essential for the supply of energy to the cells of the body. Impaired insulin production leads to the disease diabetes mellitus, which is a significant cause of death and health problems in industrialized and developing countries [Barfoed, *Chem. Eng. Prog.* **83:** 49 (1987); World Health Organization, fact sheet no. 310, 2011].

Human insulin is a polypeptide consisting of 51 amino acids arranged in two chains: chain A with 21 amino acids and chain B consisting of 30 amino acids. The A and B chains are connected by two disulfide bonds. Human insulin has a molecular weight of 5808 and an isoelectric point of 5.4. Human insulin can be produced by four different methods:

- Extraction from human pancreas
- Chemical synthesis via individual amino acids
- Conversion of pork insulin, or "semisynthesis"
- Fermentation of genetically engineered microorganisms

Extraction from the human pancreas cannot be practiced because the availability of raw material is so limited. Total synthesis, while technically feasible, is not economically viable because the yield is very low. Production based on pork insulin, also known as *semisynthesis*, transforms the porcine insulin molecule into an exact replica of the human insulin molecule by substituting a single amino acid, threonine, for alanine in the G-30 position. This technology has been developed and implemented by Novo Nordisk A/S (Denmark). However, this option is also quite expensive because it requires the collection and processing of large amounts of porcine pancreases. In addition, the supply is limited by the availability of porcine pancreas.

At least three alternative technologies have been developed for producing human insulin based on fermentation and utilizing recombinant DNA technology [Ladisch and Kohlmann, *Biotechnol. Prog.* **8:** 469 (1992)]. The first successful technique of *biosynthetic human insulin* (BHI) production based on recombinant DNA technology was the two-chain method. This technique was developed by Genentech, Inc. (South San Francisco) and scaled up by Eli Lilly and Company (Indianapolis, Ind.). Each insulin chain is produced as a β-galactosidase fusion protein in *Escherichia coli,* forming inclusion bodies. The two peptide chains are recovered from the inclusion bodies (IBs), purified, and combined to yield human insulin. Later, the β-galactosidase operon was replaced with the tryptophan (Trp) operon, resulting in a substantial yield increase.

The so-called intracellular method of making proinsulin eliminates the need for the separate fermentation and purification trains required by the two-chain method. Intact proinsulin is produced instead. The proinsulin route has been commercialized by Eli Lilly [Kehoe, in Sikdar et al., (eds.), *Frontiers in Bioprocessing*, CRC Press, Boca Raton, Fla., 1989, p. 45]. Figure 20-65 shows the key transformation steps. The *E. coli* cells overproduce Trp-LE′-Met-proinsulin (Trp-LE′-Met is a 121-amino acid peptide signal sequence; proinsulin, with 82 amino acids, is a precursor to insulin) in the form of inclusion bodies, which are recovered and solubilized. Proinsulin is released by cleaving the methionine linker using cyanogen bromide (CNBr). The proinsulin chain is subjected to a folding process to allow intermolecular disulfide bonds to form; and the C peptide, which connects the A and B chains in proinsulin, is then cleaved with enzymes to yield human insulin. A number of chromatography and membrane filtration steps are required to purify the product.

A second (extracellular) method of producing proinsulin was developed by Novo Nordisk A/S. It is based on yeast cells that secrete insulin as a single-chain insulin precursor [Barfoed, *Chem. Eng. Prog.* **83:** 49 (1987); World Health Organization, fact sheet no. 310, 2011]. Secretion simplifies product isolation and purification. The precursor contains the correct disulfide bridges and is therefore identical to those of insulin. It is converted to human insulin by transpeptidation in organic solvent in the presence of a threonine ester and trypsin followed by deesterification. Another advantage of the secreted proinsulin technology is that by employing a continuous bioreactor–cell separator loop, it is possible to reuse the cells.

In this example, we analyze a process based on the intracellular proinsulin method. The economics of this process were studied previously by Datar and Rosen, in Asenjo (ed.), *Separation Processes in Biotechnology*, Dekker, New York, 1990, p. 741, and by Petrides et al., *Biotechnol. Bioeng.* **5:** 529 (1995).

The annual world demand for insulin and insulin analogs was over 50,000 kg in 2010 and was growing at an annual rate of around 20 percent [Nair, *BioPharm Int.* **24:** 11 (2011)]. The plant analyzed in this example has a capacity of around 1800 kg of purified *biosynthetic human insulin* (BHI) per year. This is a relatively large plant for producing polypeptide-based

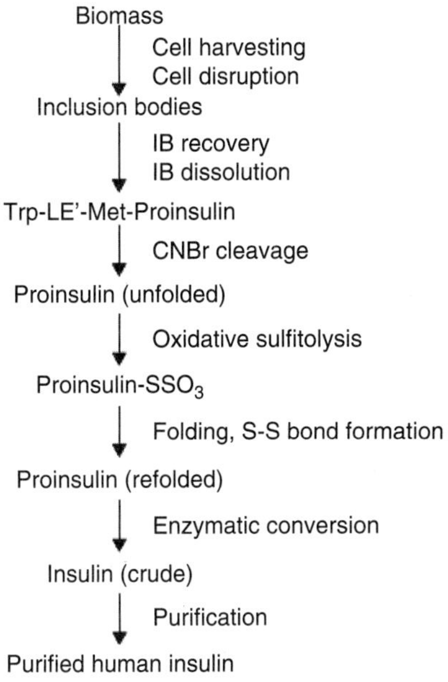

FIG. 20-65 Human insulin from proinsulin fusion protein. [Reproduced from *Bioseparations Science and Engineering*, 2d ed., by Roger G. Harrison, Paul W. Todd, Scott R. Rudge, and Demetri P. Petrides (2015). By Permission of Oxford University Press.]

biopharmaceuticals. The plant operates around the clock, 330 days a year. A new batch is initiated every 48 h, resulting in 160 batches per year. The fermentation broth volume per batch is approximately 37.5 m^3.

The entire flowsheet for the production of BHI is shown in Fig. 20-66. It is divided into four sections: fermentation, primary recovery, reactions, and final purification. *Note:* A "section" in SuperPro is simply a set of unit procedures (processing steps). If you open the file "insulin.spf" in SuperPro, you will see that all the unit procedures within a given section have their own distinctive color (blue, green, purple, and black for fermentation, primary recovery, reactions, and final purification, respectively).

Fermentation media are prepared in a stainless-steel tank (BT-101) and sterilized in a continuous heat sterilizer (ST-101). The axial compressor (G-101) and the absolute filter (AF-101) provide sterile air and ammonia to the fermenter at an average rate of 0.5 vvm. A two-step seed fermenter train (not shown in the flowsheet) is used to inoculate the 50-m^3 production fermenter (FR-101) with transformed *E. coli* cells. These cells are used to produce the Trp-LE′-Met-proinsulin precursor of insulin, which is retained in the cellular biomass. The fermentation time in the production fermenter is about 18 h, and the fermentation temperature is 37°C. The final concentration of *E. coli* in the production fermenter is about 30 g/L (dry cell weight). The Trp operon is turned on when the *E. coli* fermentation runs out of tryptophan. The chimeric protein Trp-LE′-Met-proinsulin accumulates intracellularly as insoluble aggregates (inclusion bodies), and this decreases the rate at which the protein is degraded by proteolytic enzymes. In the base case, it was assumed that the inclusion bodies (IBs) constitute 20 percent of total dry cell mass. At the end of fermentation, the broth is cooled down to 10°C to minimize cell lysis. After completion of each processing step in the fermentation section (and subsequent sections), the equipment is washed to prepare for the next batch of product.

After the end of fermentation, the broth is transferred into a surge tank (VT-101), which isolates the upstream section from the downstream section of the plant. Three disk-stack centrifuges (DS-101) operating in parallel are used for cell harvesting. Note that a single unit procedure icon in the SuperPro model may represent multiple equipment items operating in parallel. During centrifugation, the broth is concentrated from 37,000 L to 9157 L, and most of the extracellular impurities are removed. The cell recovery yield is 98 percent. The cell sludge is diluted with an equal volume of buffer solution (buffer composition: 96.4 percent w/w water for injection (WFI), 0.7 percent EDTA, and 2.9 percent Tris-base) in a blending tank (BT-102). The buffer facilitates the separation of the cell debris particles from inclusion bodies. Next, a high-pressure homogenizer (HG-101) is used to break the cells and release the inclusion bodies. The exit temperature is maintained at around 10°C. The same centrifuges as before (DS-101) are then used for inclusion body recovery (P-12). The reuse of these centrifuges can be seen by noting that procedures P-9 and P-12 have the same equipment name, DS-101. The IBs are recovered in the heavy phase (with a yield of 98 percent), while most of the cell debris particles remain in the light phase. This is possible because the density (1.3 g/cm^3) and size (diameter about 1 μm) of the IBs are significantly greater than those of the cell debris particles. The IB sludge, which contains approximately 20 percent solids w/w, is washed with WFI containing 0.66 percent w/w Triton-X100 detergent (the volume of solution is twice the volume of inclusion body sludge) and recentrifuged (P-14) using the same centrifuges as before (DS-101). The detergent solution facilitates purification (dissociation of debris and soluble proteins from inclusion bodies). The exit temperature is maintained at 10°C. The slurry volume at the end of the primary recovery section is around 1440 L.

The inclusion body suspension is transferred to a glass-lined reaction tank (V-101) and is mixed with urea and 2-mercaptoethanol to final concentrations of 300 g/L (5 *M*) and 40 g/L, respectively. Urea is a chaotropic agent that dissolves the denatured protein in the inclusion bodies, and 2-mercaptoethanol is a reductant that reduces disulfide bonds. A reaction time of 8 h is required to reach a solubilization yield of 95 percent. The inclusion bodies are composed of 80 percent w/w Trp-LE′-Met-proinsulin, with the remainder being other (contaminant) proteins. At the end of the solubilization reaction, a diafiltration unit (DF-101) is used to replace urea and 2-mercaptoethanol with WFI and to concentrate the solution. This operation is performed in 6 h with a recovery yield of 98 percent. All remaining fine particles (biomass, debris, and inclusion bodies) are removed by means of a polishing dead-end filter (DE-101). This polishing filter protects the chromatographic units that are used further downstream. The solution volume at this point is around 2200 L.

The fusion protein is cleaved with CNBr (cyanogen bromide) into the signal sequence Trp-LE′-Met, which contains 121 amino acids, and the denatured proinsulin (82 amino acids) in a glass-lined reactor (V-102). The reaction is carried out in a 70 percent formic acid solution containing 30-fold molar excess CNBr (stoichiometrically, 1 mol of CNBr is required per mole of Trp-LE′-Met-proinsulin). The reaction takes 12 h at 20°C and reaches a yield of 95 percent. The mass of the released proinsulin is approximately 30 percent of the mass of Trp-LE′-Met-proinsulin. A small amount of cyanide gas is formed as a by-product of the cleavage reaction. Detailed information on CNBr cleavage is available in the patent literature (Di Marchi, U.S. Patent 4,451,396, 1984). The formic acid, unreacted CNBr, and generated cyanide gas are removed by applying vacuum and raising the temperature to around 35°C (the boiling point of CNBr). This operation is carried out in a rotary vacuum evaporator (CSP-101) and takes 1 h. Since cyanide gas is toxic, all air exhausted from the vessels is scrubbed with a solution of hypochlorite, which is prepared and maintained in situ (Bobbit and Manetta, U.S. Patent 4,923,967, 1990).

Sulfitolysis of the denatured proinsulin takes place in a glass-lined reactor (V-103) under alkaline conditions (pH 9–11). This operation is designed to unfold proinsulin, break any disulfide bonds, and add SO_3 moieties to all sulfur residues on the cysteines. The product of interest is human proinsulin($S—SO_3—)_6$ (protein–S–sulfonate). The sulfitolysis step is necessary for two reasons: (1) The proinsulin probably is not folded in the correct configuration when expressed in *E. coli* as part of a fusion protein, and (2) the cyanogen bromide treatment tends to break existing disulfide bonds. The final sulfitolysis mixture contains 50 percent w/w guanidine HCl (6 *M*), 0.35 percent ammonium bicarbonate (NH_4HCO_3), 3 percent Na_2SO_3, and 1.5 percent $Na_2S_4O_6$ (Di Marchi, U.S. Patent 4,451,396, 1984). A reaction time of 12 h is required to reach a yield of 95 percent. The presence of the denaturing reagent (guanidine HCl) prevents refolding and cross-folding of the same protein molecule onto itself or two separate protein molecules onto each other. Urea may also be used as a denaturing reagent. Upon completion of the sulfitolysis reaction, the sulfitolysis solution is exchanged with WFI to a final guanidine HCl concentration of 20 percent w/w. This procedure, P-21, utilizes the DF-101 diafilter that also handles buffer exchange after IB solubilization. The human proinsulin($S—SO_3—)_6$ is then chromatographically purified by means of three ion exchange columns (C-101) operating in parallel and each running four cycles per batch. Each column has a diameter of 140 cm and a bed height of 25 cm. A cation exchange resin is used (SP Sepharose Fast Flow from GE Healthcare Biosciences) operating at pH 4.0. The eluant solution contains 69.5 percent w/w WFI, 29 percent urea, and 1.5 percent NaCl. Urea, a denaturing agent, is used to prevent incorrect refolding and cross-folding of proinsulin($S—SO_3—)_6$. The following operating assumptions were made: (1) The column is equilibrated for 30 min prior to loading. (2) The total resin binding capacity is 20 mg/mL. (3) The eluant volume is equal to 5 *column volumes* (CVs). (4) The total volume of the solutions for column wash, regeneration, and storage is 15 CVs. (5) The protein of interest is recovered in 1.5 CVs of eluant buffer with a recovery yield of 90 percent.

A refolding operation catalyzes the removal of the SO_3 moiety and then allows disulfide bond formation and correct refolding of the proinsulin to its native form. It takes place in a reaction tank (V-104). This process step

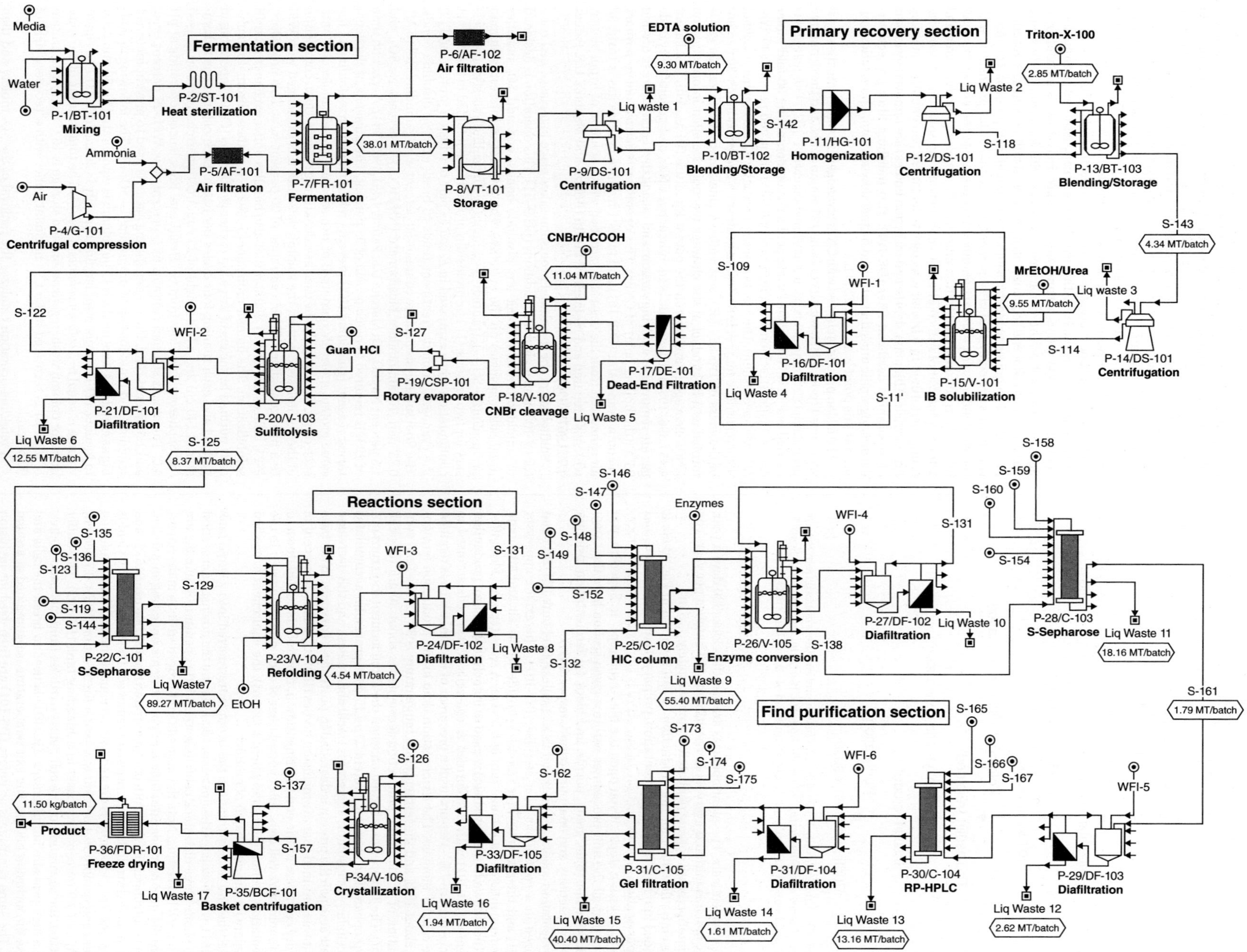

FIG. 20-66 Insulin production flowsheet. [Reproduced from *Bioseparations Science and Engineering*, 2d ed., by Roger G. Harrison, Paul W. Todd, Scott R. Rudge, and Demetri P. Petrides (2015). By Permission of Oxford University Press.]

involves treatment with mercaptoethanol (MrEtOH), a reductant that facilitates the disulfide interchange reaction. It is added at a ratio of 1.5 mol of mercaptoethanol to 1 mol of SO_3. Dilution to a proinsulin(S—SO_3—)$_6$ concentration of less than 1 g/L is required to prevent cross-folding of proinsulin molecules. The reaction is carried out at 8°C for 12 h and reaches a yield of 85 percent. After completion of the refolding step, the refolding reagents are replaced with WFI, and the protein solution is concentrated using a diafiltration unit (DF-102), which has a product recovery yield of 95 percent (5 percent of the protein denatures). The volume of the solution at this point is around 4500 L. Next, the human proinsulin is chromatographically purified in a *hydrophobic interaction chromatography* (HIC) column (C-102). The following operating assumptions were made: (1) The column is equilibrated for 30 min prior to loading. (2) The total resin binding capacity is 20 mg/mL. (3) The eluant volume is equal to 6 CVs. (4) The total volume of the solutions for column wash, regeneration, and storage is 15 CVs. (5) The protein of interest is recovered in 1 CV of eluant buffer with a recovery yield of 90 percent. (6) The material of a batch is handled in three cycles.

The removal of the C-peptide from human proinsulin is carried out enzymatically (using trypsin and carboxypeptidase B) in a reaction vessel (V-105). Trypsin cleaves at the carboxy-terminus of internal lysine and arginine residues, and carboxypeptidase B removes terminal amino acids. The amount of trypsin used is rate-limiting and allows intact human insulin to be formed. Carboxypeptidase is added to a final concentration of 4 mg/L, while trypsin is added to a final concentration of 1 mg/L. The reaction takes place at 30°C for 4 h and reaches a conversion yield of 95 percent. The volume of the solution at this point is around 4300 L.

A purification sequence based on multimodal chromatography, which exploits differences in molecular charge, size, and hydrophobicity, is used to isolate biosynthetic human insulin. A description of all the purification steps follows.

The enzymatic conversion solution is exchanged with WFI and concentrated by a factor of 4 in a diafilter (DF-102). An ion exchange column (C-103) is used to purify the insulin solution. The following operating assumptions were made: (1) The column is equilibrated for 30 min prior to loading. (2) The total resin binding capacity is 20 mg/mL. (3) The eluant volume is equal to 8 CVs, and the eluant is an 11.5 percent w/w solution of NaCl in WFI. (4) The total volume of the solutions for column wash, regeneration, and storage is 14 CVs. (5) The protein of interest is recovered in 1.5 CV of eluant buffer with a recovery yield of 95 percent. (6) The material from each batch is handled in four cycles. The solution volume at this point is around 1780 L.

Next, the ion exchange eluant solution is exchanged with WFI in a diafilter (DF-103) and is concentrated by a factor of 2.0. A recovery yield of 98 percent was assumed for this step (2 percent denatures).

The purification of the insulin solution proceeds with a *reversed-phase high-performance liquid chromatography* (RP-HPLC) step (C-104). Detailed information on the use of RP-HPLC for insulin purification is available in the literature. Analytical studies with a variety of reversed-phase systems have shown that an acidic mobile phase can provide excellent resolution of insulin from structurally similar insulinlike components. Minor modifications in the insulin molecule, resulting in monodesamido formation at the 21st amino acid of the A chain, or derivatization of amines via carbamylation or formylation results in insulin derivatives having significantly increased retention. Derivatives of this nature are typical of the kind of insulinlike components that are found in the charge stream going into the reversed-phase purification. The use of an acidic mobile phase results in the elution of all the derivatives after the insulin peak, while the use of mildly alkaline pH results in derivatives eluted on either side of the parent insulin peak. An ideal pH for insulin purification is in the region of 3.0 to 4.0, since this pH range is far enough below the isoelectric pH of 5.4 to provide for good insulin solubility. An eluant buffer with an acetic acid concentration of 0.25 *M* meets these operational criteria because it is compatible with the chromatography and provides good insulin solubility. A 90 percent insulin yield was assumed in the RP-HPLC step with the following operating conditions: (1) the column is equilibrated for 30 min prior to loading; (2) the total resin binding capacity is 15 mg/mL; (3) the column height is 25 cm; (4) the eluant volume is 6 CVs and its composition is 25 percent w/w acetonitrile, 1.5 percent w/w acetic acid, and 73.5 percent w/w WFI; (5) the total volume of the solutions for column wash, equilibration, regeneration, and storage is 6 CVs; and (6) the protein of interest is recovered in 1 CV of eluant buffer with a recovery yield of 90 percent.

The RP-HPLC buffer is exchanged with WFI and concentrated by a factor of 2.0 in a diafilter (DF-104) that has a product recovery yield of 98 percent (2 percent denatures). Purification is completed by a gel filtration chromatography column (C-105). The following operating assumptions were made: (1) The column is equilibrated for 30 min prior to loading. (2) The sample volume is equal to 5 percent of the column volume. (3) The eluant volume is equal to 4 CVs. (4) The total volume of the solutions for column wash, depyrogenation, stripping, and storage is 6 CVs. (5) The protein of interest is recovered in 0.5 CV of eluant buffer with a recovery yield of 90 percent. The mobile phase is a solution of acetic acid.

Next, a diafilter (DF-105) is used to concentrate the purified insulin solution by a factor of 10. The solution volume at this point is around 180 L, which contains approximately 12.8 kg of insulin. This material is pumped into a jacketed and agitated tank (V-106). Ammonium acetate and zinc chloride are added to the protein solution until each reaches a final concentration of 0.02 *M* [Datar and Rosen, in Asenjo (ed.), *Separation Processes in Biotechnology*, Dekker, New York, 1990, p. 741]. The pH is then adjusted to between 5.4 and 6.2. The crystallization is carried out at 5°C for 12 h. Insulin crystallizes with zinc with the following stoichiometry: insulin$_6$-Zn_2. Step recovery on insulin is around 90 percent.

The crystals are recovered with a basket centrifuge (BCF-101) with a yield of 95 percent. Finally, the crystals are freeze-dried (FDR-101). The purity of the crystallized end product is between 99.5 and 99.9 percent as measured by analytical high-performance liquid chromatography (HPLC). Approximately 11.5 kg of product is recovered per batch. The overall recovery yield is around 32 percent.

Table 20-36 displays the material requirements in kilograms per year, per batch, and per kilogram of main product (MP = purified insulin crystals). The solutions of H_3PO_4 (5 percent w/w) and NaOH (0.5 *M*) are used for equipment cleaning. WFI is used for preparing all the buffers utilized in product purification as well as all the cleaning solutions. Note the large amounts of formic acid, urea, guanidine hydrochloride, acetic acid, and acetonitrile required per kilogram of final product. All these materials end up in waste streams.

In the base case, this waste is treated and discarded. However, opportunities may exist for recycling some chemicals for in-process use and recovering others for off-site use. For instance, formic acid (HCOOH), acetonitrile, and urea are good candidates for recycling and recovery. Formic acid is used in large quantities (11 tons per batch) in the CNBr cleavage step (V-102), and it is removed by means of a rotary vacuum evaporator (CSP-101), along with small quantities of CNBr, H_2O, and urea. The recovered formic acid can be readily purified by distillation and recycled in the process. Around 2.2 metric tons per batch of urea is used for the dissolution of inclusion bodies (V-101), and 17 metric tons per batch is used in the first chromatography step (C-101) to purify proinsulin(S—SO_3)$_6$ before its refolding. Approximately 90 percent of the urea appears in just two waste streams (Liq Waste 4 and 8). It is unlikely that these urea-containing streams can be purified

TABLE 20-36 Raw Material Requirements for Human Insulin Production: 1 Batch = 11.5 kg Main Product (MP = Purified Insulin Crystals)

	Requirement		
Raw material	kg/year	kg/batch	kg/kg MP
Glucose	782,200	4,889	425.1
Salts	71,400	446	38.8
Water	9,715,000	60,719	5,279.9
H_3PO_4 (5% w/w)	3,979,000	24,869	2,162.5
NaOH (0.5 *M*)	3,842,000	24,013	2,088.0
WFI	51,890,000	324,313	28,201.1
Ammonia	81,600	510	44.3
EDTA	10,420	65	5.7
TRIS base	43,160	270	23.5
Triton-X-100	3,035	19	1.6
MrEtOH	98,660	617	53.6
Urea	3,054,000	19,088	1,659.8
CNBr	15,270	95	8.3
Formic acid	1,752,000	10,950	952.2
Guanidine-HCl	805,600	5,035	437.8
$Na_2O_6S_4$	24,160	151	13.1
NH_4HCO_3	5,551	35	3.0
Sodium sulfite	48,320	302	26.3
Sodium chloride	775,500	4,847	421.5
Acetic acid	975,700	6,098	530.3
Sodium hydroxide	137,200	858	74.6
Enzymes	3	0	0.0
Acetonitrile	764,700	4,779	415.6
Ammonium acetate acetate	181	1	0.1
Zinc chloride	320	2	0.2
Total	78,874,980	492,943	42,866.8

Reproduced from *Bioseparations Science and Engineering*, 2d ed., by Roger G. Harrison, Paul W. Todd, Scott R. Rudge, and Demetri P. Petrides (2015). By Permission of Oxford University Press.

economically for in-process recycling. However, these solutions can be concentrated, neutralized, and shipped off site for further processing and utilization as a nitrogen fertilizer.

Approximately 4.8 metric tons (MT) per batch of acetonitrile is used in the reversed-phase HPLC column (C-104), and most of it ends up in the waste stream of the column (Liq Waste 13) along with 6.8 MT of water, 1.85 MT of acetic acid, and small amounts of NaCl and other impurities. It is unlikely that acetonitrile can be recovered economically to meet the high-purity specifications for a step so close to the end of the purification train. However, there may be a market for off-site use.

The scheduling and equipment utilization for consecutive batches can be determined using SuperPro Designer (Harrison et al., *Bioseparations Science and Engineering,* Oxford University Press, Oxford, UK, 2015, p. 489). The batch time is approximately 11 days. This is the time required to go from the preparation of raw materials to final product for a single batch (excluding inoculum preparation). However, since most of the equipment items are utilized for much shorter periods within a batch, a new batch is initiated every 48 h. The equipment with the least idle time between consecutive batches is the *time (or scheduling) bottleneck* (V-104 in this case) that determines the maximum number of batches per year. Its occupancy time (approximately 45 h) is the minimum possible time between consecutive batches. The production line operates around the clock and processes 160 batches per year.

Process scheduling is closely related to the determination of the annual capacity of a batch process. The last part of this example discusses how changes in scheduling and installation of additional equipment can be used to increase process throughput and reduce manufacturing cost.

Another characteristic of batch processing is the variable demand for resources (e.g., labor, utilities, and raw materials) as a function of time. For example, SuperPro Designer can be used for tracking the use of WFI and for sizing the equipment for WFI needed to produce insulin by this process (Harrison et al., *Bioseparations Science and Engineering,* Oxford University Press, Oxford, UK, 2015, p. 490).

The results of the economic evaluation are shown in Table 20-37. The detailed tables for these calculations are available as part of the evaluation version of SuperPro Designer. For a plant of this capacity, the total capital investment is $178 million. The unit production cost is $61 per gram of purified insulin crystals. Assuming a selling price of $100/g, the project yields an after-tax internal rate of return (IRR) of 35 percent and a net present value (NPV) of $250 million (assuming a discount interest rate of 7 percent). Based on these results, this project represents a very attractive investment. However, if amortization of up-front R&D costs is considered in the economic evaluation, the numbers change drastically. For instance, a modest amount of $100 million for up-front R&D cost amortized over a period of 10 yr reduces the IRR to 16.8 percent and the NPV to $153 million.

The operating cost is broken down in Fig. 20-67. The cost of consumables is the most important, accounting for 31.9 percent of the overall manufacturing cost. Consumables represent the expense of periodically replacing the resins of the chromatography columns and the membranes of the membrane filters. The cost of raw materials is the second most important, accounting for 27.3 percent of the overall cost. The facility-dependent cost is third, accounting for 24.1 percent of the overall cost. This cost item accounts for the depreciation and maintenance of the facility and other overhead expenses. Labor and Lab/QC/QA account for 9.1 percent. The treatment and disposal of waste materials accounts for 7.3 percent of the total cost. As mentioned in the material balance section, recycling and reuse of some of the waste materials may reduce this cost.

The percentage of the operating cost associated with each flowsheet section is displayed in Fig. 20-68. Only 7.3 percent of the overall cost is

TABLE 20-37 Key Economic Evaluation Results for Human Insulin Production

Direct fixed capital	$145 million
Total capital investment	$178 million
Plant throughput	1803 kg/yr
Manufacturing cost	$110 million/yr
Unit production cost	$61/g
Selling price	$100/g
Revenues	$180 million/yr
Gross profit	$70 million/yr
Taxes (40%)	$28 million/yr
IRR (after taxes)	35%
NPV (for 7% discount interest rate)	$250 million

Reproduced from *Bioseparations Science and Engineering,* 2d ed., by Roger G. Harrison, Paul W. Todd, Scott R. Rudge, and Demetri P. Petrides (2015). By Permission of Oxford University Press.

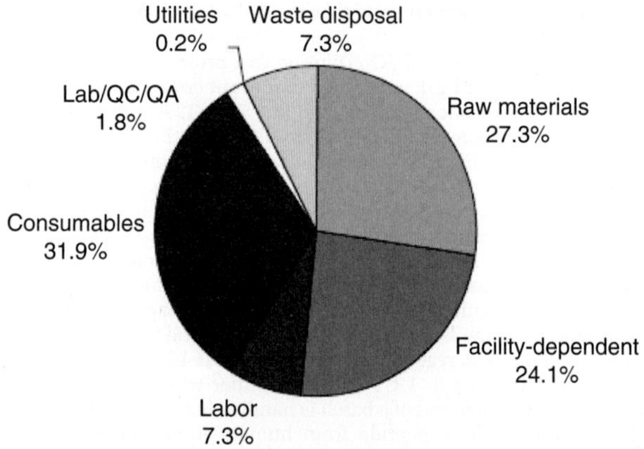

FIG. 20-67 Breakdown of manufacturing cost for human insulin production. [Reproduced from *Bioseparations Science and Engineering,* 2d ed., by Roger G. Harrison, Paul W. Todd, Scott R. Rudge, and Demetri P. Petrides (2015). By Permission of Oxford University Press.]

associated with fermentation. The other 92.7 percent is associated with the recovery and purification sections. Most of the cost is associated with the reactions section because of the large amounts of expensive chemicals and consumables required for purification.

Finally, for each raw material used in the process, Table 20-38 displays the price, annual cost, and contribution to the overall raw materials cost. WFI, urea, and H_3PO_4 (5 percent w/w) are the top three contributors to the raw materials cost. The H_3PO_4 and NaOH solutions are used for equipment cleaning.

Other assumptions for the economic evaluation include the following: (1) A new manufacturing facility will be built and dedicated to production of 1800 kg/yr of purified insulin. (2) The entire direct fixed capital is depreciated linearly over a period of 10 yr. (3) The project life-time is 15 yr. (4) The unit cost of membranes is $800/m^2 and they are replaced every 50 cycles. (5) The average unit cost of chromatography resins is $1500/L. (6) The waste disposal cost is $5/m^3 for low BOD streams and $150/m^3 for streams containing significant amounts of solvents and other regulated chemicals.

In the base case, a new batch is initiated every 48 h. Most of the equipment items, however, are utilized for less than 24 h per batch. If the market conditions are favorable, this provides the opportunity for increasing plant throughput without major capital expenditures. A realistic improvement is to initiate a batch every 24 h. This will require a new fermenter of the same size as the original fermenter, whose operation will be staggered relative to the existing unit so that one fermenter is ready for harvesting every day. Such a production change will also require additional equipment of the following types: (1) disk-stack centrifuges to reduce the occupancy of DS-101 to less than 24 h, (2) two reaction vessels to reduce the occupancy of V-103 and V-104, and (3) membrane filters to reduce the occupancy of DF-104 and DF-105.

The additional capital investment for such a change is around $55 million. This additional investment will allow the plant's capacity to be doubled, and

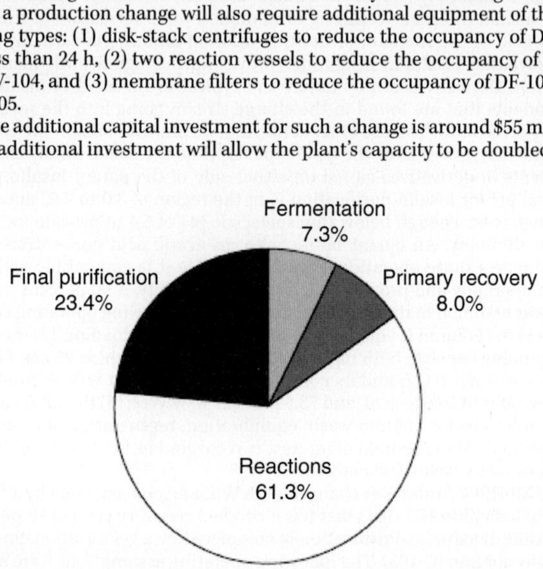

FIG. 20-68 Cost distribution per flowsheet section for human insulin production. [Reproduced from *Bioseparations Science and Engineering,* 2d ed., by Roger G. Harrison, Paul W. Todd, Scott R. Rudge, and Demetri P. Petrides (2015). By Permission of Oxford University Press.]

TABLE 20-38 Cost of Raw Materials for Human Insulin Production (Year 2012 Prices)

Bulk material	Unit cost ($/kg)	Annual cost ($)	Percent
Glucose	0.60	469,300	1.56
Salts	1.00	71,400	0.24
Water	0.05	485,800	1.61
H_3PO_4 (5% w/w)	1.00	3,979,000	13.19
NaOH (0.5 *M*)	0.50	1,921,200	6.37
WFI	0.10	5,188,500	17.20
Ammonia	0.70	57,100	0.19
EDTA	18.50	192,700	0.64
TRIS base	6.00	259,000	0.86
Triton-X-100	1.50	4,600	0.02
MrEtOH	3.00	296,000	0.98
Urea	1.52	4,642,200	15.38
CNBr	11.00	168,000	0.56
Formic acid	1.60	2,802,400	9.29
Guanidine-HCl	2.15	1,732,000	5.74
$Na_2O_6S_4$	0.60	14,500	0.05
NH_4HCO_3	1.00	5,600	0.02
Sodium sulfite	0.40	19,300	0.06
Sodium chloride	1.23	954,000	3.16
Acetic acid	2.50	2,439,400	8.08
Sodium hydroxide	3.50	480,300	1.59
Enzymes	500,000	1,691,100	5.60
Acetonitrile	3.00	2,294,200	7.60
Ammonium acetate	15.00	2,700	0.01
Zinc chloride	12.00	3,800	0.01
Total		30,174,100	100.00

Reproduced from *Bioseparations Science and Engineering,* 2d ed., by Roger G. Harrison, Paul W. Todd, Scott R. Rudge, and Demetri P. Petrides (2015). By Permission of Oxford University Press.

the new unit production cost will be around $55/g. The reduction in the unit production cost is rather small because the majority of the manufacturing cost is associated with consumables and raw materials that scale approximately linearly with production.

Therapeutic Monoclonal Antibody Production Monoclonal antibodies are large protein molecules that consist of two main regions, the Fab (fragment antigen binding) region and the Fc (fragment crystallizable) region. Monoclonal antibodies (mAbs) are the fastest-growing segment in the biopharmaceutical industry. Currently mAbs are used to treat various types of cancer, rheumatoid arthritis, psoriasis, severe asthma, macular degeneration, multiple sclerosis, and other diseases. More than 20 mAbs and Fc fusion proteins are approved for sale in the United States and Europe, and approximately 200 mAbs are in clinical trials for a wide variety of indications [Walsh, *Trends Biotechnol.* **23:** 553 (2005); Pavlou and Belsey, *Eur. J. Pharm. Biopharm.* **59:** 389 (2005)]. The market size for mAbs in 2010 was in excess of $35 billion [Rader, *Contract Pharma*, July/Aug. 2012].

The high-dose demand for several mAbs translates into annual production requirements for purified product in the metric ton range. This example illustrates the analysis of a large-scale mAb process. Again, the modeling and calculations are performed with SuperPro Designer.

The flowsheet of the overall process is displayed in Fig. 20-69. The generation of the flowsheet was based on information available in the patent and technical literature combined with the authors' engineering judgment and experience with such processes [Kelley, *Biotechnol. Prog.* **23:** 995 (2007)]. The process in this example produces 1544 kg/yr of purified mAb. The flow diagram of Fig. 20-69 is a simplified representation of the actual process because it lacks all the buffer preparation and holding activities. Such processes require 20 to 30 buffer solutions for product purification. These solutions are prepared in mixing tanks and then stored in holding tanks located close to the purification train. The tanks required for buffer preparation and holding add to the capital investment of the facility, while the required labor adds to the manufacturing cost. The model files for this example that are part of the evaluation version of SuperPro Designer (available at www.intelligen.com) include the tanks for buffer preparation and holding. In addition, the capital and operating costs associated with buffer preparation activities were considered in the cost analysis results presented in this example. Additional information on this example in the form of two tutorial videos is available at www.intelligen.com/videos.

The upstream part is split into two sections: the Inoculum Preparation section and the Bioreaction section. The inoculum is initially prepared in 225-mL T-flasks (TFR-101). Next, the material from the T-flasks is moved to 2.2-L roller bottles (RBR-101; shake flasks are frequently used instead of roller bottles for suspension cultures), then to 100-L and subsequently to 200-L rocking bioreactors that utilize disposable bags (BBS-101 and BBS-102). Sterilized media are fed at the appropriate amount in all four initial steps (3.6, 11.4, 43.6, 175.4 kg/batch, respectively). The broth is then moved to the first stirred seed bioreactor (DSBR-101), which utilizes 1000-L disposable bags. The second seed bioreactor (SBR-102) is a 5000-L stainless-steel vessel. For the two seed bioreactors, the media powder is dissolved in WFI in two prep tanks (MP-101 and MP-102) and then sterilized/fed to the reactors through 0.2-μm dead-end filters (DE-101 and DE-102). In the cell culture section, serum-free low-protein media powder is dissolved in WFI in a stainless-steel tank (MP-103). The solution is sterilized using a 0.2-μm dead-end polishing filter (DE-103). A 20,000-L stainless stirred-tank bioreactor (BR-101) is used to grow the cells, which produce the therapeutic monoclonal antibody (mAb). The production bioreactor operates in fed-batch mode. High media concentrations are inhibitory to the cells, so one-half of the media is added at the start of the process and the rest is fed at a variable rate during fermentation. The concentration of dry media powder in the initial feed solution is 17 g/L. The cell culture time is 12 days. The volume of broth generated per bioreactor batch is approximately 15,000 L, which contains roughly 30.5 kg of product (the product titer is approximately 2 g/L).

Between the downstream unit procedures there are 0.2-μm dead-end filters to ensure sterility. The generated biomass and other suspended compounds are removed using a disk-stack centrifuge (DS-101). During this step, roughly 5 percent of mAb is lost in the solids waste stream. The bulk of the contaminant proteins are removed using a protein A affinity chromatography column (C-101) which processes a batch of material in four cycles. The following operating assumptions were made for each chromatography cycle: (1) Resin binding capacity is 15 g of product per liter of resin. (2) The eluant or elution buffer is a 0.6 percent w/w solution of acetic acid, and its volume is equal to 5 column volumes (CVs). (3) The product is recovered in 2 CVs of eluant with a recovery yield of 90 percent. (4) The total volume of the solution required for column equilibration, wash and regeneration is 14 CVs. The entire procedure takes approximately 27 h and requires a resin volume of 502 liters. The protein solution is then concentrated fivefold and diafiltered with two volumes of buffer (in P-21/DF-101). This step takes approximately 8.4 h and requires a membrane of 21 m². The product yield is 97 percent. The concentrated protein solution is then chemically treated for 1.5 h with Polysorbate 80 to inactivate viruses (in P-22/V-111). The ion exchange (IEX) chromatography step (P-24/C-102) that follows processes one batch of material in three cycles. The following operating assumptions were made for each cycle: (1) The resin's binding capacity is 40 g of product per liter of resin. (2) A gradient elution step is used with a sodium chloride concentration ranging from 0.0 to 0.1 *M* and a volume of 5 CVs. (3) The product is recovered in 2 CVs of eluent buffer with a mAb yield of 90 percent. (4) The total volume of the solutions required for column equilibration, wash, regeneration, and rinse is 16 CVs. The step takes approximately 22.3 h and requires a resin volume of 210 L. Ammonium sulfate is then added to the IEX eluate (in P-25/V-109) to a concentration of 0.75 *M*. This increases the ionic strength of the eluate in preparation for the hydrophobic interaction chromatography (HIC, P-26/C-103) step that follows. Like the IEX step which preceded it, the HIC step processes one batch of material in three cycles. The following operating assumptions were made for each cycle of the HIC step: (1) The resin binding capacity is 40 g of product per liter of resin. (2) The eluant is a sodium chloride (4 percent w/w) sodium dihydrogen phosphate (0.3 percent w/w) solution, and its volume is equal to 5 CVs. (3) The product is recovered in 2 CVs of eluant buffer with a recovery yield of 90 percent. (4) The total volume of the solutions required for column equilibration, wash, and regeneration is 12 CVs. The step takes approximately 22 h and requires a resin volume of 190 L. A viral filtration step (DE-105) follows. It is a dead-end type of filter with a pore size of 0.02 μm. Finally, the HIC elution buffer is exchanged for the *product bulk storage* (PBS) buffer and concentrated 1.5-fold (in DF-102). This step takes approximately 20 h and requires a membrane area of 10 m². The approximately 800 L of final protein solution is stored in twenty 50-L disposable storage bags (DCS-101). The overall yield of the downstream operations is 63.2 percent, and 19.3 kg of mAb is produced per batch.

The equipment occupancy chart of the process for consecutive batches can be drawn using SuperPro Designer (Harrison et al., *Bioseparations Science and Engineering,* Oxford University Press, Oxford, UK, 2015, p. 498). The schedule represents a plant that has a single production train. The batch time is approximately 50 days. This is the time required from the start of inoculum preparation to the final product purification of a single batch. A new batch is initiated every 2 weeks (14 days). The production bioreactor (BR-101) is the time (scheduling) bottleneck. On an annual basis, the plant processes 20 batches and produces approximately 386 kg of purified mAb. The analysis revealed that under these conditions the downstream train is underutilized, and the cycle time of the process—the time between

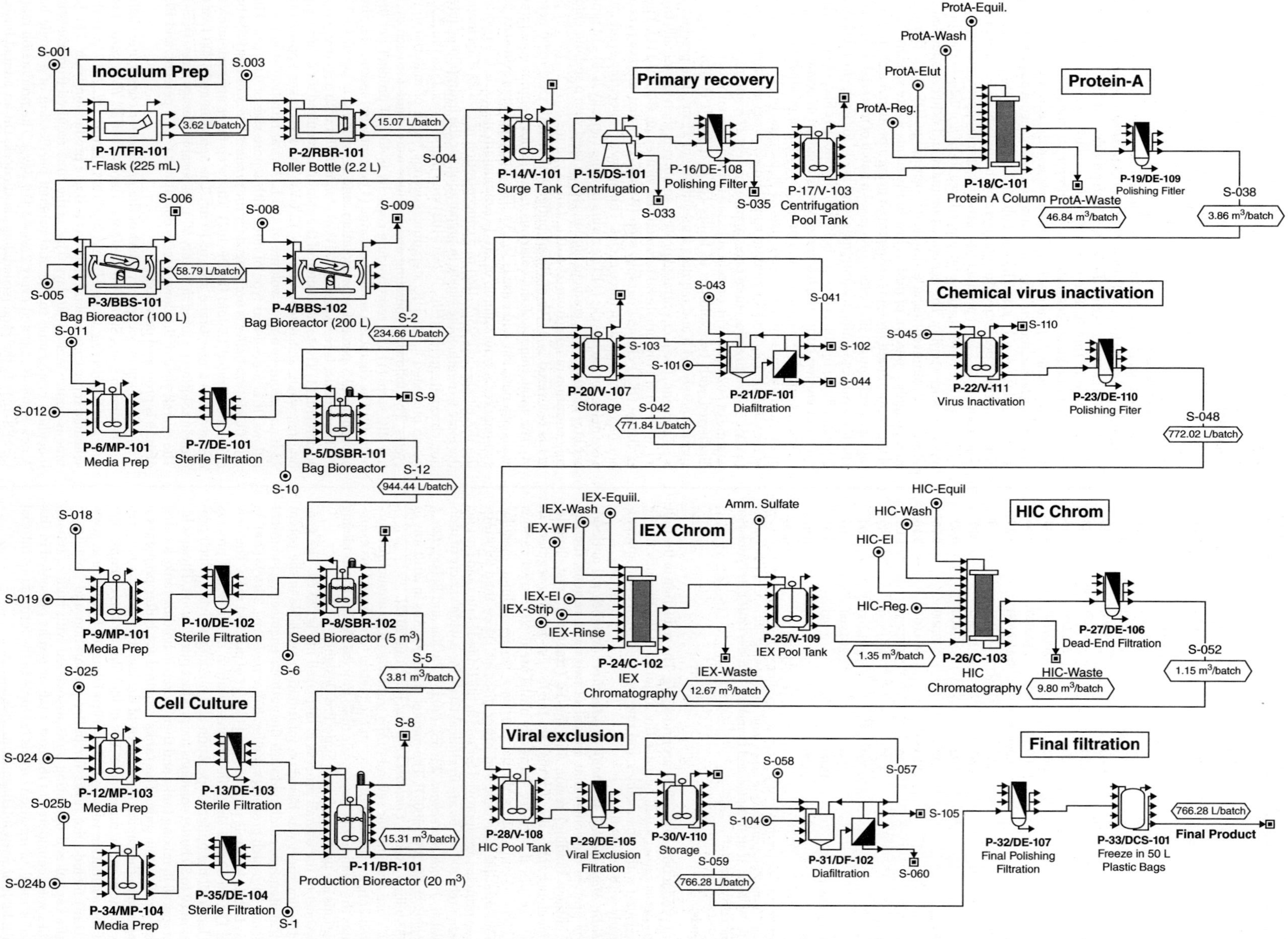

FIG. 20-69 Monoclonal antibody production flowsheet (L = liters). [Reproduced from *Bioseparations Science and Engineering,* 2d ed., by Roger G. Harrison, Paul W. Todd, Scott R. Rudge, and Demetri P. Petrides (2015). By Permission of Oxford University Press.]

TABLE 20-39 Raw Material Requirements for mAb Production: 1 Batch = 19.3 kg Main Product (MP= Purified mAb)

Material	kg/yr	kg/batch	kg/kg MP
Inoc. media solution	18,630	232.88	12.066
WFI	8,430,000	105,375.00	5,459.845
Serum-free media	37,340	466.75	24.184
H_3PO_4 (5% w/w)	2,277,000	28,462.50	1,474.741
NaOH (0.5 *M*)	2,054,000	25,675.00	1,330.311
NaOH (0.1 *M*)	7,815,000	97,687.50	5,061.528
Amm. sulfate	12,160	152.00	7.876
Polysorbate 80	6	0.08	0.004
Protein A equil. buffer	1,967,000	24,587.50	1,273.964
Protein A elution buffer	800,200	10,002.50	518.264
Protein A reg. buffer	480,400	6,005.00	311.140
NaCl (1 *M*)	184,200	2,302.50	119.301
IEX elution buffer	16,130	201.63	10.447
IEX equil. buffer	664,900	8,311.25	430.635
HIC elution buffer	239,200	2,990.00	154.922
HIC equil. buffer	449,600	5,620.00	291.192
Concentrated PBS	14,370	179.63	9.307
EtOH (10% w/w)	363,000	4,537.50	235.104
Total	**25,823,136**	**322,789.20**	**16,724.83**

Reproduced from *Bioseparations Science and Engineering*, 2d ed., by Roger G. Harrison, Paul W. Todd, Scott R. Rudge, and Demetri P. Petrides (2015). By Permission of Oxford University Press.

consecutive batches—is relatively long. The cycle time of the process can be reduced and the plant throughput increased by installing multiple bioreactor trains that operate in staggered mode (out of phase) and feed the same purification train. For a case of four bioreactor trains feeding the same purification train, the new cycle time is 3.5 days, which is one-fourth of the original. Under these conditions, the plant processes 80 batches per year and produces 1544 kg/yr of mAb. Some biopharmaceutical companies have installed more than four bioreactor trains per purification train in order to achieve cycle times as low as 2 days.

The material requirements of the process are summarized in Table 20-39. Note the large amount of WFI utilized per batch. The majority of WFI is consumed for cleaning and buffer preparation.

Key economic evaluation results generated using the built-in cost functions of SuperPro Designer are displayed in Table 20-40. The total capital investment (for the case with the four bioreactor trains) is around $477 million. The total annual operating cost is $130 million, resulting in a unit production cost of around $84/g (1544 kg of purified mAb is produced annually). Assuming a selling price of $200/g, the project yields an after-tax internal rate of return (IRR) of 24.3 percent and a net present value (NPV) of $560 million (assuming a discount interest rate of 7 percent).

A breakdown of the operating cost contributors is presented in Table 20-41. The facility-dependent cost is the most important item, accounting for 46.7 percent of the manufacturing cost or $39.2/g of final product. This is common for high-value products that are produced in small quantities in expensive facilities. Depreciation of the fixed capital investment and maintenance of the facility are the main contributors to this cost. Consumables are the second most important operating cost, accounting for 18.2 percent of the total or $15.3/g of final product. Consumables include

TABLE 20-40 Key Economic Evaluation Results for mAb Production

Direct fixed capital	$365 million
Total capital investment	$477 million
Plant throughput	1544 kg of mAb/yr
Manufacturing cost	$130 million/yr
Unit production cost	$84/g of mAb
Selling price	$200/g of mAb
Revenues	$309 million/yr
Gross profit	$179 million/yr
IRR (after taxes)	24.3%
NPV (for 7% discount interest rate)	$560 million

Reproduced from *Bioseparations Science and Engineering*, 2d ed., by Roger G. Harrison, Paul W. Todd, Scott R. Rudge, and Demetri P. Petrides (2015). By Permission of Oxford University Press.

TABLE 20-41 Breakdown of the Manufacturing Cost for mAb Production

Cost item	$ million/yr	$/g	%
Raw materials	16.67	10.8	12.86
Facility-dependent	60.54	39.2	46.71
Labor	18.89	12.2	14.58
Consumables	23.58	15.3	18.19
Lab/QC/QA	4.45	2.9	3.43
Miscellaneous	5.47	3.5	4.23
Total	**129.60**	**83.9**	**100.00**

Reproduced from *Bioseparations Science and Engineering*, 2d ed., by Roger G. Harrison, Paul W. Todd, Scott R. Rudge, and Demetri P. Petrides (2015). By Permission of Oxford University Press.

chromatography resins, membrane filters, and disposable bags that need to be replaced on a regular basis. Labor and raw materials costs come third and fourth, accounting for 14.6 percent and 12.9 percent of the total cost, respectively. The miscellaneous cost item (4.2 percent of total) accounts for heating/cooling utilities, electricity, and environmental costs. The cost of WFI, commonly classified as a utility cost in industry, is accounted for in the cost of raw materials in this example. In terms of cost distribution per section, 62 percent of the cost is associated with the upstream section and 38 percent with the downstream.

The economic evaluation relies on the following key assumptions: (1) A new manufacturing facility will be built and dedicated to production of 1544 kg/yr of mAb. (2) The entire direct fixed capital is depreciated linearly over a period of 12 yr. (3) The project lifetime is 16 yr. (4) The unit cost of WFI is $0.15/L. (5) The cost of the serum-free media (in powder form) is $300/kg. (6) All the chemicals used are high-purity grade. (7) The unit cost of membranes is $400/m^2. (8) The unit cost of chromatography resins is $6000/L, $1200/L, and $2050/L for columns C-101, C-102, and C-103, respectively. (9) The chromatography resins are replaced every 60, 50, and 50 cycles for columns C-101, C-102, and C-103, respectively.

After a model of the entire process has been developed on the computer, tools such as SuperPro Designer can be used to ask and readily answer "what if?" questions and to carry out sensitivity analysis with respect to key design variables. In this example, the impact of product titer (varied from 1 to 10 g/L) and bioreactor size (10,000 and 20,000 L) on unit production cost is evaluated. Figure 20-70 displays the results of the analysis. All points correspond to four production bioreactors feeding a single purification train. For low product titers, the bioreactor volume has a considerable effect on the unit production cost. For instance, for a product titer of 1 g/L, going from 10,000 to 20,000 L of production bioreactor volume reduces the unit cost from $230/g to $162/g. On the other hand, for high product titers (e.g., around 5 g/L), the impact of bioreactor scale is not as important. This can be explained by the fact that at high product titers, the majority of the manufacturing cost is associated with the purification train. It is therefore wise to shift R&D efforts from cell culture to product purification as the product titer in the bioreactor increases. A key assumption underlying the sensitivity analysis is that the composition and cost of the cell culture media are independent of product titer.

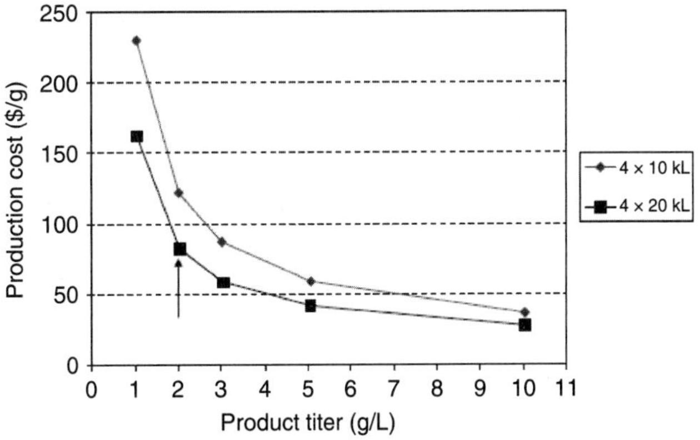

FIG. 20-70 The mAb production cost as a function of product titer and production bioreactor volume (L = liters). The arrow indicates the product titer for this example. [Reproduced from *Bioseparations Science and Engineering*, 2d ed., by Roger G. Harrison, Paul W. Todd, Scott R. Rudge, and Demetri P. Petrides (2015). By Permission of Oxford University Press.]

PRODUCT QUALITY ATTRIBUTE CONTROL

INTRODUCTION

This section focuses on the implementation of *product quality attribute control* (PQAC) for biopharmaceutical product manufacturing. PQAC of these products is mainly concerned with controlling aspects of the product that may impact the function of the product, such as patient outcomes. The majority of the discussion is focused on the upstream portion of the manufacturing process. Process perturbations and environmental conditions at this stage, where the macromolecules are actually created in the host cells, can have the greatest impact on *critical product quality attributes* (CPQAs). Application of PQAC in the downstream purification process, where unwanted species such as aggregates or charge variants are cleared from the product, is also discussed.

Compared to chemical and small molecule pharmaceutical manufacturing, PQAC in the biopharmaceutical industry has lagged by several decades. Application of PQAC to biopharmaceutical processing is challenging due to the complexity of the product and its mode of action, the cellular processes creating the product, unique issues with representative sampling and analysis, and the rigorous regulatory process they are subject to. Other bio-based processes, such as the manufacture of biofuel, can use many of the tools discussed here, but often enjoy a simplified PQAC approach due to the relative simplicity of the molecules produced and the governing regulatory requirements.

Biopharmaceutical processes are routinely controlled for temperature, pressure, pH, dissolved gases, and other *critical process parameters* (CPPs). Unlike in other industries, the linkage between the controlled process variables and product quality is not always clear. Without direct, near real-time CPQA measurement, implementation of PQAC is very difficult. As a result, upstream biopharmaceutical processes have often been operated as a black box. This is an open-loop or recipe-based control scheme with respect to CPQA. The commercial process is simply operated within a set of process parameter targets established during development with the expectation that, by operating within this design space, the process will consistently achieve required product quality targets. Biopharmaceutical manufacturing processes operate this way due to process and product complexities, coupled with limited analytics and bioreactor sampling capabilities.

In the case of chemical or synthetic organic pharmaceutical manufacturing, target species PQA and undesired by-products are, for the most part, well characterized, and their mode of action is understood. Sample collection and analysis methods have been well established, so that when they are combined with process optimization and control strategies, a robust PQAC program can be implemented. The biopharmaceutical industry is beginning to overcome technical hurdles, such as reliable automated sampling and integrated analytical analysis, that have historically hampered PQAC implementation.

Defining CPQA for a biopharmaceutical is significantly more challenging compared to defined chemical entities. Biopharmaceutical quality attributes include parameters that impact safety and efficacy in the patient that may only be known by their effects in clinical studies or through limited in vitro bioassays. Biopharmaceuticals are large, complex molecules whose structures are related to CPQA, but may also vary during the course of an upstream bioprocess. If biopharmaceutical lots having a distribution of product attributes with unique biological responses are used in clinical development, then the commercial process may be required to match that CPQA distribution, further challenging a deterministic PQAC approach. In the case of biosimilars, target PQA are defined by matching the innovator's molecule. These targets are determined by off-line analysis of attributes such as glycosylation, deamidation, and aggregates. The control of these parameters to match the innovator molecule is often a priority in the development and production of a biosimilar molecule and can benefit from a PQAC approach.

On-line sensors exist for temperature, pH, and dissolved gases, and in situ cell mass and medium composition can be determined with dielectric and Raman spectroscopy, respectively. However, there are no near-term on-line solutions to directly measure product quality attributes. Therefore, a key element of any PQAC scheme is a means to automatically collect and analyze representative bioreactor samples. Figure 20-71 shows an example PQAC process that employs a combination of online and at-line data collection approaches.

Biopharmaceutical PQAC schemes continue to become more sophisticated and refined. Many companies have committed to it in their development programs, and several regulatory agencies have voiced their support for PQAC strategies. With increased understanding of the PQA themselves, improved consistency in sample collection and analysis, and increased regulatory support, companies will continue to establish PQAC programs.

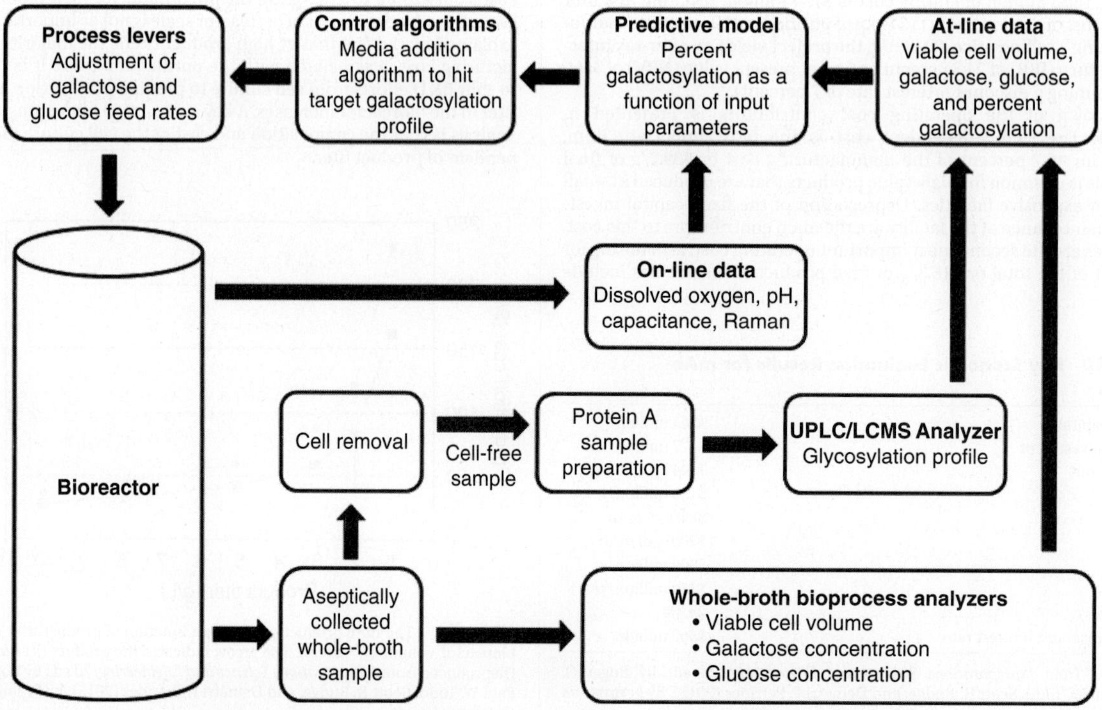

FIG. 20-71 PQAC application in which a target mAb galactosylation level is maintained by controlling galactose feed concentration.

REGULATORY GUIDANCE AND INDUSTRY TRENDS

PQAC concepts are driving significant changes in regulatory guidance and approaches to consistently achieve CPQA. Historically biopharmaceutical manufacturing has been performed as a series of batch unit operations. Product quality and release testing were performed only after process completion. This approach led to the development of process validation, where a process is developed and locked in. Consistency was then demonstrated over a relatively small number of runs, using the locked process. Once validated, manufacturers strove to minimize changes in an effort to maintain consistent process results. As a result, biopharmaceuticals have often been produced using a validated process that uses the same operational parameters from run to run on the basis that the manufactured product retains the same quality attributes. Having operated this way for decades, this has been an accepted biopharmaceutical regulatory compliance approach, although advances in technology and regulatory guidance are leading to changes.

The FDA and other regulatory agencies have recognized the importance of shifting toward a PQAC approach. As a result, several guidance documents addressing process analytical technology (PAT) and quality by design (QbD) have been published. These include *Guidance for Industry, PAT—A Framework for Innovative Pharmaceutical Development, Manufacturing and Quality Assurance,* FDA, 2004; ICH Q8, *Guidance for Industry, Q8 Pharmaceutical Development,* Rev. 2, FDA, Nov. 2009; ICH Q9, *Guidance for Industry, Q9 Quality Risk Management,* FDA, 2006; and ICH Q10, *Guidance for Industry,* Q10 *Pharmaceutical Quality System,* FDA, 2009. Together these guidance documents outline a system in which a design space where the process can be successfully operated is characterized, real-time bioreactor monitoring using online and at-line technologies is implemented, and the knowledge gained is used to actively control product quality. This fuller understanding may lead to a PQAC scheme and possibly a *real-time release testing* (RTRT) program, although regulatory agencies will continue to require rigorous validation of these processes.

QUALITY BY DESIGN (QbD)

The FDA's Office of Biotechnology Products states that the basic concept of QbD is to "provide guidance on pharmaceutical development to facilitate design of products and processes that maximizes the product's efficacy and safety profile while enhancing product manufacturability." Furthermore, it is stated that "knowledge gained during the pharmaceutical development program is critical for enhanced understanding of product quality and provides a basis for risk management and increased regulatory flexibility." QbD is a working methodology, focused on systematically evaluating, understanding, and refining manufacturing processes with the goal of delivering an optimized product with an understanding of the relationship of PCQAs and CPPs. This can lead to the establishment of a design space and appropriate control strategy.

QbD is a proactive scientific approach to "build" quality into a product rather than "test" quality into a product. Its objective is to identify process parameters that impact product attributes (desired and undesired) so that controls can be implemented so that the desired product is manufactured. The objectives include providing increased understanding of the process operating space, including linking critical process parameters to the required target product quality parameters. This knowledge allows greater flexibility in accommodating process changes, while also establishing control ranges for a PQAC approach. QbD strategies can include better understanding of manufacturing process inputs, including raw material testing, process parameters and PAT integration, and outputs, including product identification, efficacy, performance, and safety testing.

Protein therapeutics are inherently diverse, and variants may exist within each bioreactor production lot. A formulated drug product may be comprised of proteins having somewhat different attributes associated with both the active molecule and other components present in the final drug formulation. This variability can be due to process variations in upstream and downstream processing unit operations, or it may be the result of analytical variability in particular bioassays. Although CPQAs may be linked to manufacturing CPPs, the CPQAs themselves, including efficacy, immunogenicity, and other safety-related attributes, are uniquely associated with the protein product. Since CPP variations impact protein CQPAs, biomanufacturing processes are ideal opportunities to apply QbD principles.

Attributes most commonly affected in the bioreactor include glycosylation (sugar residues), charge variants (oxidation, deamidation), and protein molecular weight (reduction, aggregation). Batch and fed-batch bioreactor processes are dynamic wherein cell state is changing as a function of cell age, feeding strategy, medium composition, metabolite buildup, and biopharmaceutical production. Since these may be CPPs impacting CPQAs, implementing a QbD strategy incorporating process and analytical monitoring and at the bioreactor can lead to better understanding of the process parameters that correlate with CPQAs.

On-line, in-line, and automated sampling and analysis tools have been developed to facilitate development of a data-driven QbD strategy focused on process operations and PQA. Time course data of metabolites, glycosylation patterns, and other PQA can be generated using these tools. Examples include Raman, near IR, and dielectric spectroscopy (capacitance) probes installed in bioreactor probe ports. These data along with chromatography and mass spectroscopy data analysis can be used with metabolite profiling technologies to predict cell culture productivity endpoints. Small-scale bioreactor design of experiments (DOE) studies are also performed to gain a better understanding of factors that affect PQA. These data streams can inform predictive modeling approaches, which can lead to process control strategies focused on real-time PQA optimization in bioreactors.

Downstream protein purification unit operations may also benefit from a cohesive QbD strategy since they can also impact product attributes. Downstream purification often consists of a series of distinct unit operations. These downstream technologies may include host cell removal (depth filtration), primary purification (protein A bind and elute), viral load reduction through inactivation (low pH) and removal (nanofiltration), and polishing steps (ion exchange resins or membrane adsorbers). Additionally, a *tangential flow filtration* (TFF) step is typically included to perform a buffer exchange and then concentrate the protein to achieve the required antibody drug substance formulation. PQAs affected by these steps may include contaminants such as charge variants, host cell proteins, reduced variants, protein aggregation, viral load reduction, and product purity.

In contrast to upstream operations, downstream operations are of shorter duration, and the relationship between CPP and process step objectives, such as aggregate removal, is more readily characterized and measured compared to cell culture processes, making them ideal candidates for a QbD approach. Downstream processing is also more likely to be a standardized sequence of unit operations, often using a defined platform process. Upstream processes require customized approaches from product to product since cell lines, medium formulations, variable biomolecular constructs, and variable CPQA are driven by the therapeutic application. By using platform downstream chromatography processes, for example, the CPPs that dictate the quality outcomes may be readily defined and characterized by a set of bench-scale experiments. Process parameters to be evaluated may include pH, conductivity, temperature, buffer gradients, stability, peak collection criteria, resin lifetime, and process run times. Applying PAT tools during the development phase enables rapid generation of robust data sets as part of DOE studies, leading to well-characterized process design space and a QbD-enabled commercial process.

PQAC Implementation The biotech industry is moving toward incorporating PQAC into fed-batch, perfusion, and fully continuous processes. Accomplishing this requires knowledge of CPQAs and their relation to CPPs, a means to rapidly analyze CPQAs and process behavior and direct the results into models and control systems. For example, companies are developing automated mAb glycan profile control beginning at bench and pilot plant scales. As PQAC is demonstrated at those scales, it will progress into clinical manufacturing to support biopharmaceutical product development and regulatory approval. The time required for these programs to reach the approval stage and begin commercial manufacturing may mean it will be many years before biopharmaceuticals are manufactured using PQAC. Given the faster path to commercialization for veterinary products, coupled with cost control drivers, the first commercial biomolecule manufactured using a PQAC scheme may be a veterinary product.

Understanding of biopharmaceutical CPQAs and their impact on molecule function is rapidly increasing. For example, Ghaderi studied which mAb glycosylation patterns to target and which ones to avoid [Ghaderi, Zhang, Hurtado-Ziola, and Varki, *Biotech. and Genetic Eng. Rev.* **28:** 147–176 (2012)]. Once CPQAs are identified, supporting PQAC technology development can advance. A key requirement of PQAC is the ability to measure CPQAs and process behavior in near real time. Upstream and downstream sample collection and preparation systems are maturing, making it possible to routinely and automatically collect whole-broth samples and analyze them for viable cell density and volume, medium and metabolite concentrations (glucose, lactate etc.), and dissolved gases. Samples can also be prepared using automated cell removal, which is required to perform analytical protein A chromatography preprocessing, followed by HPLC/UPLC or LCMS analysis to measure PQA such as glycan profiles.

To complete a PQAC process, CPQA and other process analytical results can be fed into predictive models that drive control schemes to maintain CPQA at a desired level. Significant improvements in such predictive model development have been reported [Van Impe, Vercammen, and Derlinden, *Procedia Food Science* **1:** 965–971 (2001)].

BIOPROCESS ANALYSIS TECHNOLOGIES—GENERAL

Consistently delivering quality biopharmaceuticals typically requires one or more PAT tools to monitor and control processes in real time. The increasingly common use of bioprocess automation to achieve tighter control of product quality is driving the development and adoption of PAT for process monitoring and integration with feedback control of both CPP and CPQA.

Significant progress has been made in the development of analytical methods for measuring CPQA. These analytical methods, however, are often off-line analyses performed days or even weeks after sample collection. This analytical time delay makes PQAC difficult to implement because control adjustments to a bioprocess in any PQAC scheme cannot be implemented before the measurement of the CPQA is completed. The challenge to be addressed to allow implementation of any PQAC scheme therefore is to integrate and complete sample collection and analysis within a short enough time frame to make a PQAC scheme reasonable. Once CPQA results can be rapidly and automatically generated, the data can be fed into process models and control algorithms, which may lead to development of automated CPP and attribute control of biopharmaceutical manufacturing production.

What follows is a brief overview of currently adopted bioprocess PAT technologies. Additional details can be found in Al-Rubeai, *Animal Cell Culture*, Springer, New York, 2015, pp. 185–222.

Online versus At-Line Technologies Bioreactor sterility requirements have traditionally favored online PAT for upstream PAC, which does not require bioreactor sampling, over at-line PAT which requires sampling. This is so because automated PQAC schemes relying on sophisticated at-line technology require consistent and reliable sample collection, which presents several hurdles for any automated sampling system. First, bioreactor sterility must be maintained during sampling. In the case of mammalian cell culture, the smallest microbial contamination can result in product loss worth several million dollars. Second, once a sample is removed from the bioreactor, the system must consistently route viscous, high-solids-content samples to multiple instruments for weeks, or in the case of perfusion cell culture for months, consistently and without clogging. There are now automated aseptic sampling technologies commercially available that have demonstrated the ability to collect representative samples while maintaining bioreactor sterility, allowing traditionally off-line analytical techniques to be applied in a pseudo-online, or at-line fashion (Pepper, *Modernizing Biopharma Sampling*, Contract Pharma, Sept. 2014). Autosampling systems can collect whole-broth samples directly from the bioreactor and route those samples to multiple analytical destinations. Whole-broth or cell-retained samples are routed to bioanalyzers to measure cell density, viability, protein titer, metabolites, salts, and dissolved gases. Samples analyzed by chromatographic or mass spectrometer methods must be cell-free. Autosampling systems provide a mechanism for removing the cells, resulting in a cell-free sample. Prior to analysis, these cell-free samples can be processed in a liquid handler employing methods such as protein A processing, solid-phase extraction, dilution, digestion, or filtration. Cell-free, processed samples, analyzed with the full capability of HPLC/UPLC and LCMS types of systems, can generate the full spectrum of PQA data such as glycosylation profiles, amino acid concentrations, and levels of deamination and aggregation (Fig. 20-72).

Automated downstream PAC requires higher sampling frequencies compared to bioreactors to enable real-time monitoring and control, driving adoption of rapid at-line online PAT. Automated sampling technology is being applied to downstream processes to integrate traditionally off-line analysis, such as UPLC systems. Contamination control in automated downstream sampling systems is generally simpler than in upstream applications, since the processing buffers are not growth-promoting, and bioburden reduction can be achieved through chemical sanitization, or presterilized closed processing solutions. Additionally, the streams being sampled are relatively low-viscosity and free of suspended solids, making sample preparation less challenging.

For both upstream and downstream processes, the selection of which PAT to apply depends on which PQA must be controlled, knowledge of how these PQAs are related to process conditions, and which PAT measurements are needed to monitor relevant process conditions.

BIOPROCESS ANALYSIS TECHNOLOGIES—UPSTREAM TECHNOLOGIES

In cell culture processes, small variations in process conditions or medium formulations can cause wide variations in the quality and quantity of the product produced. For example, Le analyzed over 200 commercial cell culture runs with nominally identical process conditions [Le, Kabbur, Pollastrini, Sun, Mills, Johnson, and Hu, *J. of Biotech.* **162:** 210–223 (2012)]. They found that minor variations in the seed culture resulted in large run-to-run variations in cellular metabolism related to lactate production and consumption in the production bioreactor (Fig. 20-73). Conditions that led to the accumulation of lactate resulted in lower product titers, whereas conditions that favored cellular metabolism switching from lactate production to consumption yielded the highest product titers. Using multivariate analysis, it was shown that lactate could be used as a process performance indicator. If lactate levels could also be tied to CPP, it could form the basis of a real-time control strategy to maximize production. In this example, PAT to monitor lactate levels and control bioreactor conditions based on these measurements could be implemented to reduce process variability.

Established PAT Online technologies that provide continuous measurements include pH, dissolved oxygen, temperature, feed, and gas flow rates. These are established, well-understood technologies, and they may come in disposable or single-use compatible formats. These parameters are typically independently controlled to a target set point. Modulation of these parameters often affects the cell state globally and is therefore less desirable as a means of independent, targeted product attribute controls compared to controls that specifically target quality attributes.

Off-line measurements, which require manual bioreactor sampling, are needed to monitor other process parameters such as extracellular metabolite and medium concentrations (glucose, lactate, glutamine, glutamate, ammonium), electrolyte concentrations (sodium, potassium, calcium), osmolality, cell density, carbon dioxide, and product concentration. Cell culture medium composition (amino acids, trace metals) measurements are less frequently employed, although automated sampling is beginning to make them more commonplace.

Raman and NIR Spectroscopy Raman and *near-infrared* (NIR) spectroscopic technologies have been used to continuously monitor and control bioreactor processes, although not as often as other instruments. These technologies measure bioreactor broth composition via a probe inserted into the bioreactor which introduces a light source. The incident light from the probe interacts with the bioreactor contents within the measurement volume. The light is absorbed or scattered in a specific manner according to the specific mixture of chemical species present in the measurement volume of the probe. This light interaction can be detected and used to generate a spectrum consisting of various peaks quantifying the light interaction as a function of wavelength. This spectrum is a convolution of the response of all components that interact with the incident light. Multivariate analysis (MVA) or other models can be developed to interrogate the spectra to identify and quantify the various species that responded to the light source.

These spectroscopic technologies can measure many different analytes simultaneously and are available in single-use compatible formats. Due to the complexity of the bioreactor contents, interpreting the spectrum requires analysis and modeling approaches such as partial least squares (PLS). MVA and PLS model development requires generating spectra from known solutions having representative concentrations of all species present during production. Some of the data are used to generate the model, with the remainder used to validate it. Controlling mAb glycation levels using Raman spectroscopy integrated with a PLS model was demonstrated [Berry, *Biotech. Prog.* **32:** 224–234 (2016)].

Predictive modeling approaches have also been used to generate empirical "fingerprints" of the bioreactor contents from the raw spectral data and

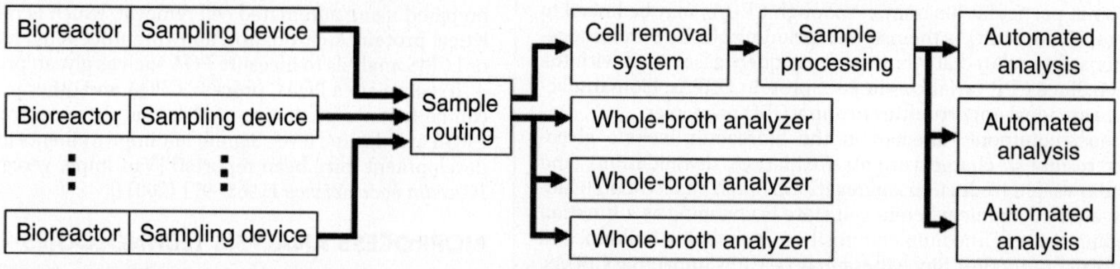

FIG. 20-72 An automated sampling and analysis system capable of collecting samples from multiple bioreactors and sending those samples to multiple analytical destinations.

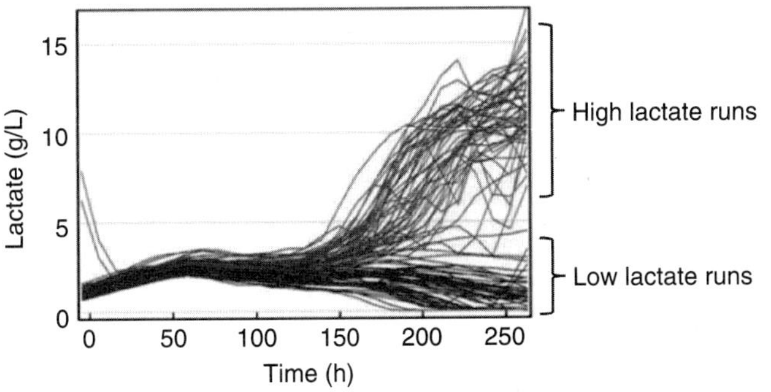

FIG. 20-73 Example of cell culture variability. Lactate concentrations are shown for nominally identical commercial-scale cell culture productions runs. Runs where lactate continued to increase over time had lower product titers, whereas runs where lactate levels stayed low or decreased had the highest titers. [Reprinted from Le, K., Pollastrini, S., Mills, J., and Hu, W., "Multivariate Analysis of Cell Culture Bioprocess Data—Lactate Consumption as a Process Indicator," *J. Biotech.* **162:** 214 (2012). Reproduced with permission of Elsevier. © 2012 Elsevier B.V. All rights reserved.]

to correlate them with specific culture states. More information on the generation and use of spectral fingerprints can be found in Lourenço, *Anal. Bioanal. Chem.* **404:** 1211–1237 (2012). This empirical bioreactor state information (a state estimator) can be used in a *model predictive control* (MPC) strategy. For both PLS and MPC, significant amounts of data are required to build, validate, and successfully apply these methods to bioreactors.

Dielectric Spectroscopy *Dielectric* (or capacitance) *spectroscopy* (DS) technology is often used to measure viable cell mass, which can be correlated to viable cell density for a range of host cells including CHO, microbial, yeast, and insect cells. DS relies on the fact that the cellular membrane of live cells is impermeable to ions. When exposed to an alternating electric field, the membranes become polarized, resulting in a capacitance related to the dielectric properties. Capacitance is proportional to the biomass, dependent on cell size, cell type, suspended versus adherent cells, temperature, and physiological cell state. Cell debris, gas bubbles, and most medium components are not detected by DS. In most cases nonviable cells are also not detected by DS, although there are cases where cells which are classified as nonviable do produce a capacitance signal (typically depending on the specific mode of cell death resulting in the nonviable cells). A method for detecting and correcting for the phenomenon has been published [Downey, *Biotech. Prog.* **30:** 479–487 (2014)].

The measurement frequency used depends on the host cell and is determined by comparing medium with and without cells present. Frequency-scanning DS instruments are available that may obviate the need for this step. To determine biomass or cell density from capacitance, the data are correlated to cell density determined by other instruments such as off-line image analysis cell counters. Changing bioreactor conditions that affect cell size, hence cell volume, will affect the instrument accuracy, unless accounted for in the correlations.

DS advantages include directly measuring viable cell mass, being a simple continuous online technology, and available in single-use compatible formats. Although measurement accuracy may be impacted in the declining culture phase as viability drops, and initial functional calibration for the specific application is required, the simplicity of this PAT tool has led to its wide application. Potential DS applications can include measuring specific cellular phenomena such as cell size, growth rates, physiological cell state, adherent cell density, and ion polarization [Justice, *Biotech. Adv.* **29:** 391–401 (2011)]. These advanced measurements will likely require data-driven, predictive modeling approaches similar to spectroscopic techniques.

Soft Sensors In addition to considering individual technologies for measuring specific aspects of a process, multiple measurements can be combined to predict the value of a process variable of interest. The process of using easily measured secondary variables to infer a primary measurement is known as a *soft sensor.* Robust application requires a data set to be generated, which can be used to train a predictive model, and a second set of data for verification. Soft sensor approaches are typically employed when direct in-process measurement of the parameter of interest is impractical, or where multiple physical measurements have to be analyzed together by application of a multivariate method to predict specific outcomes, such as product quality and titer [Gunther, *Comput. Chem. Eng.* **33:** 88–96 (2009)].

Extracellular Metabolites While many extracellular metabolites may be measured using online spectroscopic technologies, development of robust autosampling systems combined with automated analytical instruments has enabled many off-line metabolite assays to be used at-line. These data can be used for near real-time bioreactor monitoring and control in conjunction with cell density and other measurements. This provides insight into cell metabolic states over the course of a bioreactor process, which is known to be a key determiner of product quality attributes [Zhao, *Eng. Life Sci.* **15:** 459–468 (2015)]. Which specific extracellular metabolites are measured depends on the level of understanding of their effect on cellular metabolism and product attributes. Commonly measured metabolites include major inputs and outputs of central metabolism. Some examples include ammonia, lactate, glucose, galactose, amino acids, and nucleosides. Currently available analyzers include various commercial bioanalyzers, amino acid analyzers, as well as LC/UPLC instruments. These have all been implemented at-line using automated sampling systems. Some of these techniques, such as LC-based analysis, require automated cell removal as part of the sampling system prior to analysis.

Product Attributes A major goal of bioreactor PAT has been to enable consistent biopharmaceutical production meeting CPA targets. Direct measurement of these attributes and incorporation into a PAC scheme has benefited from the same robust automation platforms that enable at-line extracellular metabolite measurements. Many product attribute measurements have been deployed as at-line instruments to constantly monitor bioreactor CPA. Many product attribute measurements require cell removal and initial product purification, typically using preparative pro-A chromatography for mAbs and mAb variants, prior to analysis. These methods use sophisticated automated sampling and liquid handling systems to preprocess and direct samples to a wide range of analytical instruments. Product attribute-related measurements include product concentration, glycosylation patterns, high- and low-molecular-weight species, aggregates, and charge variants. The specific product attributes being measured depend greatly on understanding the impact of the product attributes on the safety and efficacy of the product.

BIOPROCESS ANALYSIS TECHNOLOGIES—DOWNSTREAM TECHNOLOGIES

Since downstream processes involves a series of unit operations, the application of PAT depends on the step being monitored. Typical downstream unit operations include solids-liquid separation/clarification, protein separation/purification, viral inactivation/clearance, product concentration/formulation, and sterile filtration. Each one has a specific role in meeting requirements for purity, impurities removal, and preparing the product stream for the next step in the production process.

Online PAT Online PAT tools routinely employed in downstream processes include pH, conductivity, UV, turbidity, flow rates, pressure, and temperature. Most of these can be a CPP for chromatography steps. For example, pressure may be a CPP for viral filtration, and pH for viral inactivation. PAT is used to monitor and document process performance and is integrated with process controllers to modulate process parameters. Chromatographic protein purification steps often use UV absorbance, often at 280 nm and/or 300 nm, to measure relative protein concentrations and as an indicator for product peak collection start and stop times, which can affect purity and yield. Conductivity and pH are routinely used to monitor buffer composition during chromatographic purification and are important parameters that can affect protein stability, binding affinity, and the separation of product from contaminants. Turbidity can be used to measure suspended solids in clarified liquid streams as part of bioreactor harvest operations. It has been used to monitor the separation efficiency of spent cell removal from the product stream by centrifuges and to monitor potential depth filtration breakthrough.

At-Line PAT Application of new technologies may lead to PAT for process related impurities (host cell proteins, DNA, antifoam, leached protein A) and product-related attributes (size, aggregates, acid/base variants, high/low molecular-weight species, and product concentrations). Automated sampling technologies are enabling offline measurements to be performed at-line. One example is controlling product loading onto chromatography columns using at-line analytical chromatography instruments to monitor product concentration in the feed stream. Additionally, impurities such as aggregate levels can be monitored and controlled to optimize step yield while meeting product specifications. Because downstream samples are cleaner, automated systems can be simplified. However, the need for rapid analysis is increased due to short step cycle times.

PRODUCT ATTRIBUTE CONTROL

Variations in raw materials, process parameters, and cell state can significantly affect bioprocess outcomes. The complexity of raw materials makes these variations very difficult or expensive to detect, and they may only manifest themselves when used in the process. This is driving an increasing emphasis on *product attribute control* (PAC) in the face of raw material variability.

Open-Loop Attribute Control Bioreactor PAC can be implemented using open-loop or closed-loop control, although some process parameters, such as temperature, will use closed-loop control independent of any PAC scheme. Open-loop control is the most commonly used PAC method, primarily due to an inadequate linkage between CPQA and CPP. In open-loop control, process parameters, cell lines, and medium are developed to ensure the process is robust against disturbances, and is intended to be consistently operated within an established design space. This process "recipe" is established using laboratory scale DOE that can vary from product to product, even with similar host cell lines. Because open-loop control has been the de facto standard for PAC, much of the required technology is commercially available, making it a simple, quick, low-effort way to implement a recipe-controlled process. This approach is limited by an inability to respond to unforeseen, or more extreme, process disturbances beyond what the process was designed and tested for. In extreme cases, this may lead to process failures. To increase process robustness against such disturbances, additional raw material vigilance and development may be required to minimize process variances, sometimes even after the process has reached commercial scale.

Closed-Loop Attribute Control Increasingly biopharmaceutical manufacturers are looking toward closed-loop PQAC as both a development and a commercial strategy to meet increasing regulatory expectations, and to compete in the biosimilar market segment. The goal of closed-loop control is a process that is more robust and has less product variability. Closed-loop control actively controls the process trajectory toward a desired CPQA endpoint by manipulating CPP. This requires an understanding of which and to what degree CPP affects CPQA. An additional complexity is that the degree to which a controlled parameter affects a CPQA can vary during the course of a batch or fed-batch process.

To be a candidate for closed-loop PAC, an attribute must be able to be measured in process; one or more controllable process parameters must exist that can positively and/or negatively manipulate the attribute (ideally independently of others), and detailed knowledge of how the manipulated parameters affect the attribute must be obtained. There is some debate as to whether closed-loop PAC is more suited as a development tool used to arrive at a recipe-driven commercial process where PAC can predictably achieve the target range, or if it can be used to routinely control commercial processes. It is likely that there will be a range of situations, from those where PAC is dependent on a few well-characterized variables so closed-loop control can be robustly and efficiently implemented, to more complex cases where either it is not feasible or the process variability is low and the impact on critical PQA low, so active control does not provide a clear benefit or advantage. It can be expected that as monitoring and development tools advance, so will the implementation of closed-loop PQAC for routine biopharmaceutical production.

Implementation of Product Attribute Control for Bio-Based Processes In many bioreactions the effect resulting from adjustments to a process parameter on a product attribute is both complex and dependent on the cellular metabolic state due to the highly interactive cellular processes. For example, Yang [*Biotech. Prog.* **16:** 751–759 (2000)] found that the state of ammonia accumulation (a major metabolic by-product of amino acid metabolism) in CHO cultures also had a significant effect on terminal sialylation of glycans, an important mAb PQA. Because of this complexity and state dependence, control schemes in cell culture processes often involve some form of *model-predictive control* (MPC) to account for multivariate interactions when anticipating the effects of manipulating control parameters. A simplified MPC scheme is shown in Fig. 20-74.

Although MPC is used in many applications (see Sec. 8, Process Control and Instrumentation), there are some unique considerations for biologics. Any feedback control scheme needs in-process measurements of the controlled variable. For multivariate, state-dependent bioreactor systems, a number of in-process measurements are needed to quantify the state of the system, or to use as model inputs, to enable closed-loop MPC. Process measurements must be taken often enough to allow the controller to adjust to dynamic changes in the process outputs. For bioreactor processes, the required sampling frequency is dependent on the system dynamics including host cell metabolic rates and the rate of change of the manipulated and controlled variables. In many cases, manual sampling may be insufficient, and at-line PAT is needed.

One or more control handles that can be manipulated to achieve desired changes in the product attributes of interest must also be identified. Unlike processes in more established industries, in cell culture processes the

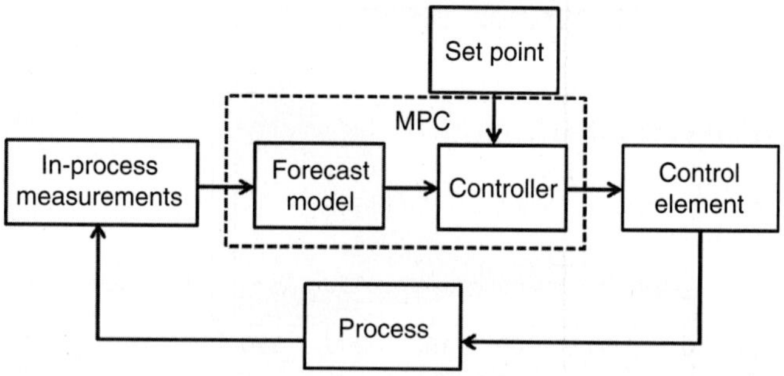

FIG. 20-74 Schematic representation of a model-predictive controller. The controller uses in-process measurements of the controlled variables and the culture state as inputs to a predictive model. The predictive model forecasts are used to determine the appropriate control element manipulations needed to achieve the desired process outcome.

control handles to achieve a given change in a product attribute are often not fully characterized. In addition to the fundamental metabolic-based studies used to identify control handles for many important biopharmaceutical PQA, empirically based methods can also be employed to identify the relationship of process inputs to a given output. Application of machine learning methods in conjunction with metabolic studies may also be used to identify suitable control handles from cell culture data. An example application of this type of methodology, where bioreactor data were mined to determine candidate CPP for titer, is presented here in Charaniya, *J. Biotechnol.* **147:** 186–197 (2010).

A process for identifying potential control handles and development of control schemes has been described for IgG1 mAb glycan distribution [St. Amand, *Biotech. Bioeng.* **111:** 1957–1970 (2014)]. In that example a DOE was executed followed by an *analysis of variation statistical analysis* (ANOVA) to determine main and interaction effects of manganese, galactose, and ammonia levels on glycan distribution. The results were used to perform a controllability analysis in which controllable modes—those that have a significant impact on glycan distributions—were identified and ranked according to their potential to effect changes in glycan distributions. This formed the basis of a controllability analysis which identified both input modes (i.e., suitable control parameters) and output modes, which together identify what biological processes are controllable to achieve a desired outcome. The full description of this analysis is provided in the supplemental information provided by Amad.

Last, MPC requires a model to predict the change of a specific output in response to a set of measured process inputs. Data-driven predictive models can use experimental data to derive these relationships in the form of either fundamental or empirical relationships. Models used for MPC must incorporate *causal and dynamic* predictions of how a given manipulated variable will affect the desired output variable. Experimental techniques commonly employed in the field of systems identification can be applied to generate such relationships. See the following reference for more information on applying system identification techniques to processes: Isermann, *Identification of Dynamic Systems: An Introduction with Applications,* 1st ed., Springer-Verlag, Berlin, 2011. Downey described the practical application of a simple system identification experiment to develop a model predictive controller of galactosylation of a mAb produced in CHO cell culture process [Downey, *Biotech. Prog.* **33:** 1647-1661 (2017)]. In this example, the causal, dynamic relationship between galactose feeds and galactosylation of a mAb were determined by subjecting a culture grown in a pseudo-steady-state culture to step increases in galactose concentration, and recording the resulting dynamic response of mAb galactosylation over time.

Generation of MPC models typically requires large amounts of experimental data that span the expected parameter ranges in order to provide reliable, robust control. The relationship between inputs and outputs in cell culture processes can be highly nonlinear, multivariate systems and therefore difficult to construct an accurate model from limited data. For MPC, however, a sufficiently accurate model may be obtained by approximating a system using linear methods. Zupke described the application of MPC to control high-mannose species generated in a cell culture process [Zupke, *Biotech. Prog.* **31:** 1433–1441 (2015)]. The functional, causal relationship between mannose feeds and high-mannose residues on a mAb were determined using small-scale batch experiments. A model predictive controller was constructed from this fundamental relationship using a receding-horizon method.

EMERGING BIOPHARMACEUTICAL AND BIOPROCESS TECHNOLOGIES AND TRENDS

HIGH CELL DENSITY CULTURES (HCDCs)

Introduction High cell density cultures are used to achieve higher bioreactor efficiencies. A high cell density culture is characterized by having at least 10 times the normal cell density of a common batch reactor. For yeasts, archaea, and bacteria, this means biomass concentrations greater than 100 g/L. Applying high-density culturing to mammalian cells can result in cell concentrations in excess of 10^7 cells/mL to 10^8 cells/mL. This approach is more economical since less growth medium and volume are required, enhancing cell productivity per volume, which improves reactor utilization and efficiency. These high-density cultures are also amenable to continuous cultivation, resulting in reduced size and capital costs. However, high-density cell reactors are difficult to design and challenging to maintain; therefore, specific strategies and issues associated with the high cell density approach are discussed.

Microbial Reactors A variety of reactor types are used for high-density microbial systems including stirred-tank reactors, dialysis membrane reactors, cyclone reactors, gas-lift reactors, and shaken ceramic flasks. Currently, the stirred-tank reactor operating under fed-batch conditions is the most widely used approach for large-scale high-density cell production. Stirred-tank reactors are commonly used, but for HCDC, internal or external cell retention methods are needed so that substrate can be refreshed. Internal retention has advantages because pumping cells out of the reactor causes stress, and filters or membranes can be used to hold back the cells. The dialysis membrane reactor also uses internal cell retention and is advantageous because it allows for continuous removal of toxic and inhibitory products without added stress on the cells. Dialysis fermentation can be achieved using two-vessels linked with a dialysis device or using one vessel with two chambers. The disadvantage to dialysis is the potential loss of valuable nutrients while filtering inhibitory ones.

An important consideration for growing high-density cultures is the need to maintain sufficient levels of nutrients to satisfy the increased number of cells while not inhibiting cells with too much substrate. Fed-batch and dialysis fermentation techniques are able to overcome these restrictions. Continuous or intermittent feeding is recommended to meet the demands of these cultures. For fed-batch reactors, mathematical models and indirect and direct feedback control systems are used to optimize the feed rate and nutrient concentrations. Direct control may not be sufficient because of the lack of sensors for fast and reliable measurement and the complexity of the cellular dynamics. Bioactivity patterns such as changes in pH or dissolved oxygen concentration are instead used as indirect measures to overcome these limitations. Noninvasive optical measurement methods are also attractive for industrial use. There are multiple and varied challenges encountered during HCDC of microbial cultures including solubility of solid and gaseous substrates in aqueous media. For example, low oxygen solubility and inhibition of substrates with respect to growth can be prevented by controlling the supply of the carbon or oxygen source and through metabolic engineering. Other issues include instability, volatility, and degradation of products; accumulation of products or metabolic by-products to a growth-inhibiting level; high evolution rates of CO_2, heat generation, and viscosity of the medium. For example, acetate production can become inhibitory in fed-batch, microbial cultures. One approach is to remove the inhibitory acetate through integration of fermentation and separation processes. Another physical limitation for HCDC is the reduced mixing efficiency that occurs with increasing fermenter size. To address issues involving substrates, a controlled feeding strategy can be applied in concert with control of pH and other process parameters. Limiting essential nutrients may reduce growth but also limit production of inhibitor by-products such as acetate. The fed-batch approach is now widely used to obtain high cell density with dialysis fermentation as an alternative. Nevertheless, due to the vast array of microorganisms, a number of unknowns remain for elucidating the optimal conditions and bioprocesses for HCDC. Further studies are required to evaluate microbial physiologies in order to develop optimal processes. The development of online sensor systems and monitoring and control algorithms will ensure further enhancement of HCDC processes. Additionally, intracellular regulatory mechanisms remain to be investigated and overcome, using metabolic engineering and synthetic biology. As a result of these advances, highly productive industrial HCDC bioprocesses will continue to be in demand and have a significant impact on the biotechnology industry.

Mammalian Bioprocessing High-yielding mammalian processes can lower process operating costs and increase biomanufacturing efficiencies. For mammalian cell cultures, high-density culture is employed to maximize the economical generation of biological products. Thus, high cell density cultures are ideal for yielding high volumetric production rates; but significant challenges exist.

A number of growth considerations need to be taken into account for high cell density mammalian cultures. Oxygenation, CO_2 removal, and media design are critical for maximizing cell count. Adjustment of substrate feed rate (or perfusion rate if a perfusion system is used) is important to meet the metabolic demand of increasing cell density for fed-batch and continuous cultures. Also, controlled nutrient feeding can contribute to consistent product qualities. In addition to growth considerations, there are a number of limitations to high cell density cultures. Mass-transfer and mixing demands become more prevalent for high cell density cultures. Increasing oxygen demand at high cell density requires higher oxygen delivery rates to the bioreactor. In addition, a significant amount of dissolved CO_2 accumulates at high cell density. Base addition, often used to combat high dissolved CO_2 (which can lower pH), can lead to cultures with high osmolarity, resulting in the loss of cell growth and even cell death. Nonuniform mixing can promote heterogeneity in concentrations, and cell aggregation can result in segregation of cells within different mixing zones. Physical space is another limitation to production in high cell density culture processes. Theoretically, the maximum cell density for mammalian cell cultures is approximately 10^9 cells/mL. Typical values on the order of 10^7 or 10^8 cells/mL occupy about 1 to 10 percent of the total volume area achieved.

While currently most high-density animal cultures are obtained through feed batch cultures, cell retention is one particular approach in reactor design to increase cell density. Most membrane systems can be challenging with respect to handling very high cell density cultures. Cell immobilization and encapsulation work are feasible for higher densities (on the order of 10^8 cells mL^{-1}), but the intensive labor and high preparatory cost make this approach impractical for industrial production. Many cell lines are adapted to grow in suspension as single-cell or small aggregates while certain attachment systems still require the presence of solid surfaces.

Media supplementation or fortification is essential for optimizing the number of viable cells attainable in HCDC, and the media exchange rate directly affects production levels and yield. In many cases, complete media exchange can occur periodically to maintain a cell density of 10^7 cells/mL while media exchange rates should be increased in order to maintain a density of 10^8 cells/mL. However, high perfusion rates can pose challenges to the efficiency of cell retention systems. Additionally, minimizing media perfusion rates is important to minimize processing costs. The optimal solution is often to develop special concentrated or fortified medium for fed-batch and perfusion reactors.

Process control is also required for consistent production. In high-density perfusion reactors, cell densities generally increase from 0.5×10^5 cells mL^{-1} to 10^7 to 10^8 cells/mL. Initially during cell accumulation, media can be provided in order to maximize cell growth. During the stationary or production phase, cell densities vary less and should be controlled at set points through media feeding and environmental control. Optimal conditions for high cell density also require proper pH, temperature, and dissolved oxygen control. Additionally, nutrient and metabolic by-product concentrations can be optimized through media exchange as needed to optimize reactor performance.

Another mixing issue is dissolved CO_2 accumulation. The CO_2 produced can be incorporated as bicarbonate, and CO_2 can be stripped from the medium and transferred to the gas phase by increased sparging and mass transfer. However, CO_2 stripping may proceed at a lower rate than oxygen delivery, creating a gradual increase in dissolved CO_2. See the Carbon Dioxide Stripping subsection for more detailed discussion of CO_2 control in bioreactors.

However, base addition can engender other issues including high osmolarity and additional mixing issues. When sufficient time is not allowed for thorough mixing and homogenization of the initial base, the continual addition of base can cause locally elevated pH values, which can then cause cell lysis. The resulting cell lysis can lead to viscous regions with high pH. Since bulk pH is not affected, incorporating additional base amplifies the issues and leads to formation of "snowball" structures of lysed cells. To resolve this, placement of a base addition line close to the impeller system can provide sufficient mixing for homogenization. Thus, correctly tuning agitation, base feed location and proper adjustment of the base pump feed rate can yield robust pH control.

Lack of adequate reactor performance and control can lead to heterogeneities in the bioreactor. Heterogeneities are generally a result of poor mixing as reactor sizes are scaled up and cell densities increase. These inconsistencies can result in not only nutrient and pH gradients, but also oxygen-starved dead zones which facilitate cell lysis and alter product quality and yield. Poor reactor performance can also result in cell aggregates and clumping at high densities. Aggregation can cause segregation of cells, where larger aggregates migrate to lower agitation zones, leading to lower productivity and reduced process efficiency. Heterogeneity in the cellular environment can lower cell viability, density, and productivity which will reduce the overall performance. Proper mixing, media feeding and exchanges, and control in suspension cell cultures will provide more consistent quality and product from mammalian HCDC in biotechnology.

INTEGRATED CELL CULTURE AND PRODUCT RECOVERY

The biotechnology industry has leveraged human ingenuity to manipulate and modify organisms to produce pharmaceuticals, antibodies, biofuels, metabolites, and other high-value biochemicals. Since the early 1990s, the industry has witnessed significant growth in production, commercialization, and scale. However, this growth is limited by costly and separate upstream cell culture/fermentation and downstream product recovery steps. Product compounds are typically produced by batch, fed-batch, or continuous fermentation, followed by filtration or sedimentation to separate cells from spent medium. The media are then subjected to a series of costly separation steps for secreted products. Unfortunately, this distinct separation phase can be inefficient and costly due to energy requirements, supplies, material transport, and equipment maintenance [Schügerl, Karl, *Biotechnology Advances* **18:** 581–599 (2000)].

In recent years, methodologies have been implemented to eliminate the product recovery bottleneck by integrating the separation and fermentation processes. Efforts to enhance production efficiency include techniques that incorporate the cell culture and product recovery steps into a single or a few steps in order to rapidly develop higher-quality products at lower cost and with improved bioprocess performance. Some of these methods include pervaporation, perstraction, extraction, adsorption, gas stripping, precipitation, and crystallization (Table 20-42). These methods of in situ product recovery are also desirable since product compounds such as ethanol, butanol, and others can be toxic or inhibit growth at the elevated concentrations present in a fermentation culture. In situ recovery allows for increased substrate consumption, higher cell densities, elevated product yields, and more efficient and economical integrated operations.

Pervaporation One method for bypassing the product recovery bottleneck is to include a pervaporation process (Fig. 20-75). During pervaporation a pressure differential is applied across a highly selective membrane which separates volatiles from the biologics recycle stream. Pervaporation is usually carried out with the feed stream set at atmospheric pressure while the permeate stream is subject to vacuum. The difference in partial pressure between the components in the feed stream causes the species with a higher vapor pressure to evaporate, thereby causing a concentration gradient within the membrane, which becomes the driving force for mass transfer. Permeate from the pervaporation process is condensed and subjected to a cold trap for further separation and improved process selectivity.

Perstraction Perstraction is similar to pervaporation in that a selectively permeable membrane separates the fermentation stream from the separation stream. However, perstraction membranes are hollow tubes that allow the flow of two liquid streams on separates sides with a solute material transferred between them. This process works by circulating fermentation broth through the membranes and back to the vessel, allowing only the component of interest to seep through into the harvesting stream. This harvesting stream will include a solvent, often organic, that solubilizes one of the metabolites of interest from the culture media to be purified. The target metabolite mixes and dissolves into solvent with a high solubility for the component of interest and is subsequently purified. One advantage of perstraction is that toxic products from a fermentation can be eliminated; however, the presence of an in-stream membrane separation can be costly and subject to clogging and fouling.

Extraction *Extraction* refers to the separation technique of extracting a compound from a mixture. In situ liquid-liquid extraction can also be used to recover volatile components from a fermentation process. This method

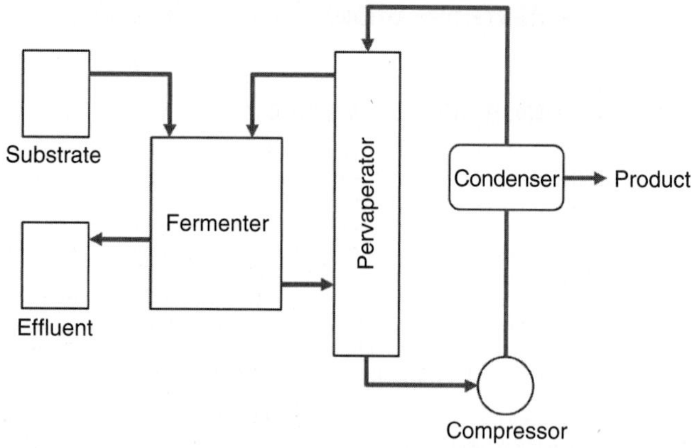

FIG. 20-75 Schematic of integrated fermentation-pervaporation process.

is often used to separate two or more immiscible liquids such as an organic liquid-phase solution from an aqueous liquid-phase fermentation solution, which are in direct contact unlike perstraction which separates the phases using a membrane. To perform this extraction, however, the extraction solvent must be nontoxic to the culture and have a high partition coefficient for the component of interest. Economic studies show that ethanol extraction from a continuous fermentation is more cost-effective than conventional separation methods for ethanol recovery. Other applications involving extraction include separations of antibiotics such as cephalosporins or other beta-lactam precursors. Extraction from culture, however, does not always remove the accumulated organic acids or metabolites efficiently. Furthermore, toxicity of the solvent, especially when organics are used, can be an issue, which can be reduced by either immobilizing the cells in culture or using protective agents such as propylene glycol.

To minimize the toxicity of an organic, two-phase aqueous extraction can be implemented in which both immiscible phases are water-based solutions. The two phases are distinct based on the presence of two distinct dissolved polymers, such as polyethylene glycol and dextran in the two aqueous phases, or include a polymer in one phase and a salt solution in the other. These aqueous two-phase extraction systems can be especially advantageous for recovering proteins or other sensitive biologics since the compounds will not denature in either of the two liquid phases. Ionic liquids, in which salts are present in a liquid state, represent another potential extraction solvent that eliminates organics, although their capabilities in integrated recovery systems are not demonstrated.

Gas Stripping and Adsorption Gas stripping is a separation process whereby a vapor stream is used to physically remove one or more components from a liquid stream. Adsorption is the adhesion of a substance known as an *adsorbate* to a surface known as an *adsorbent*. Gas stripping and adsorption have been applied to remove volatile metabolites from the vapor phase of fermentation processes. Recovery by stripping the fermentation culture can be achieved by sparging with an inert gas (Fig. 20-76). An inert gas bubbling throughout the broth forces volatile components into the gas phase.

The gas-phase mixture is then sent through either a condenser or an adsorption process, which removes target compounds and recycles the inert stream. Studies have been conducted to test for the recovery of ethanol and butanol produced by fermentation. Adsorbent media such as activated carbon, polystyrene, charcoal, and divinylbenzene-styrene beads have also been tested. In situ adsorption may be used in an effort to remove ammonia and other inhibitory side products as well as to improve the production of antibiotics by fermentation. However, in situ adsorption can be adversely affected by clogging due to proteins and cells. Once again, immobilizing

TABLE 20-42 Benefits and Limitations of the Proposed Processes

	Stripping	Adsorption	Extraction	Pervaporation	Perstraction
Capacity	Moderate	Low	High	Moderate	Low
Selectivity	Low	Low	High	Moderate	High
Fouling	Low	High	Moderate	Low	Low
Operational simplicity	High	Low	Low	High	High

When one is designing a process, the demand, selectivity, and process maintenance should be taken into account in order to maximize overall process effectiveness. From Groot, W. J., van der Lans, R. G. J. M., and Luyben, K. Ch. A. M., "Technologies for Butanol Recovery Integrated with Fermentations," *Process Biochemistry*, **27:** 64 (1991).

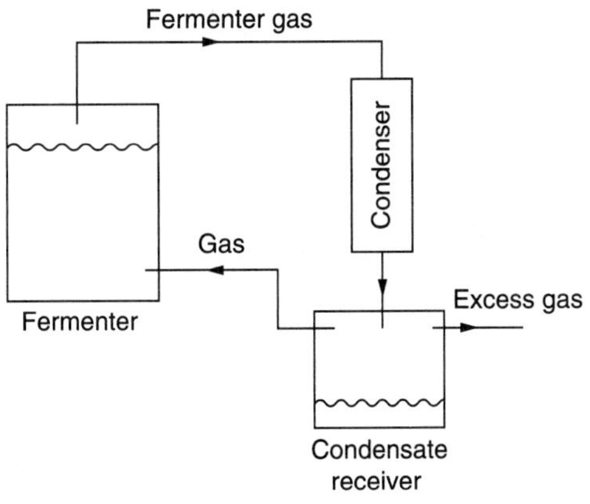

FIG. 20-76 Recovery by gas stripping and condensation.

cells in a matrix inside the reactor is a technique that can be used to reduce adsorbent clogging.

Precipitation with a Base Precipitation with an appropriate base is a simple and gentle method for recovering citric or lactic acid produced during fermentation. Usually a weak base such as calcium carbonate is added to the broth; in solution the calcium carbonate reacts with organic acids and makes calcium salts which precipitate out of solution due to a low aqueous solubility. The precipitate is then removed from the broth, and the organic acids are liberated by one of the following methods: treatment with a strong acid, solvent extraction, or electrodialysis.

Electrodialysis *Electrodialysis* (ED) is conducted by applying an electrical potential across a semipermeable membrane allowing for ions to flow to their respective poles. ED is an effective tool which takes advantage of the ionic character of target metabolites in order to separate and purify them. Studies have shown that this process can be as effective as precipitation without leading to significant waste products. The downside to electrodialysis is that cell-free broth may be required to preserve the bipolar ED membranes; therefore, a cell separation method consisting of immobilization or filtration must be employed before ED is performed.

Crystallization Crystallization is a phase separation and purification technique wherein crystals are precipitated from a liquid solution. Purification by crystallization is appropriate primarily with cell-free reaction mixtures; thus a process may be established with immobilized cultures. In situ crystallization applied to purification and separation of antibiotics can be performed where the antibiotic crystallizes in a fermenter. For instance, this phenomenon can be seen when producing tetracycline with *S. aureofaciens*.

Ion Exchange Ion exchange is a method used to exchange ions between two electrolytes or between an electrolytic solution and a substance. This method can be applied to the separation of compounds from a fermentation culture. Specifically, the ionic character of metabolites is exploited when subjecting a fermentation broth to a recovery process involving ion exchange. This process requires use of a specialized powerful ionic exchange resin, which will remove organic acids or other metabolites from a broth and can be used in situ. In situ studies on the incorporation of ion exchange revealed increased process productivity. However, because ion exchange resins can become clogged with suspended cells, these fermentations are also often carried out after immobilizing cells.

ADVANCED BIOREACTOR CONTROL TECHNIQUES

For industry, control of bioreactors can help solve many of the problems associated with inconsistent product quality and variable product output. By maintaining bioreactor operation at optimal operating conditions throughout the fermentation life cycle, much of the uncertainty that comes from bioreactor performance variability can be minimized. Eliminating uncertainty and maintaining product quality and consistency through optimal control of biomanufacturing operations are two of the major objectives of the highly regulated biopharmaceutical industry. The abilities to satisfy regulatory requirements, meet production goals, and speed up process development times are all vital to the success of an industrial biotechnology operation. For this reason, advanced process control methodologies will become increasingly important in biotechnology. Some of the advanced process control methods that can be used in bioreactors are summarized in Table 20-43. Advanced control of bioreactors is aimed at optimizing processes and controlling levels of three main parameters: biomass, substrate, and product output. To achieve these goals, researchers use a variety of both traditional control techniques and more advanced control techniques, such as model-predictive control, adaptive control, and habituating control.

Model-Predictive Control As shown in Fig. 20-77*a*, MPC functions by tuning control parameters based on a process model. Through the use of the model, the experimenter determines the optimal operating conditions and optimal controller settings for the system. The controller can then be tuned to desired settings to maintain these operating conditions. The model can be either an ad hoc form based on an off-line mathematical model or an adaptively determined linear dynamic on-line model. In the off-line model, theoretical laws and correlations are used to create a model. The use of theoretical predictions to create a static model is easier as no real-time data must be collected and the model only needs to be derived once. However, the downside is that this type of model can be inconsistent with the actual process when predictions do not match experimental output. Using an online model helps to eliminate this problem. For these models, the model is based on collected process data, and it can be altered based on the actual conditions and behaviors of the bioreactor system. The easiest models to create are linear models, and these are generated in both static and dynamic varieties. Linearization techniques are used for creating linear models for control. Local linearization is used when the perturbations around a steady-state operating condition are small. For example, for a single-variable system with x as the system variable, the change with time around the steady state x_s can be linearized with approximation as

$$g(x-x_s) \approx g(x_s) + \left.\frac{\partial g}{\partial x}\right]_{x_s} (x-x_s) \tag{20-103}$$

Similarly, in nonlinear systems with multiple variables such as a single-input, single-output system, with s as the system variable, c as the controlled variable, and a model given below, the change with time around the steady-state values (s_s, c_s) can be linearized with approximation as

$$\frac{ds}{dt} = g(s,c) \approx g(s_s,c_s) + \left.\frac{\partial g}{\partial s}\right]_{s_s} (s-s_s) + \left.\frac{\partial g}{\partial c}\right]_{c_s} (c-c_s) \tag{20-104}$$

TABLE 20-43 Comparison of Advanced Bioreactor Control Techniques

Method	Salient features	Pros	Cons
Model-predictive control	• Predicts corrective control action based on a mathematical model	• Provides systematic changes based on operating conditions • Offers effective control for multivariable systems	• Computational complexity • Often prone to providing ambiguous solution
Adaptive control	• Controls substrate concentration as well as production rate	• Gives real-time information on state of biomass • Eliminates errors quickly with fewer fluctuations	• Presents higher complexity than conventional PID controllers • Requires advanced expertise for running and maintenance
Habituating control	• Provides manipulation of more inputs than the controlled outputs	• Minimizes the input cost of affecting control in nonlinear bioreactor models • Provides efficient performance in closed loops by utilizing all available inputs	• Offers limited systematic mechanism for creating additional controlled outputs • Requires complex control efforts for introduction of outputs

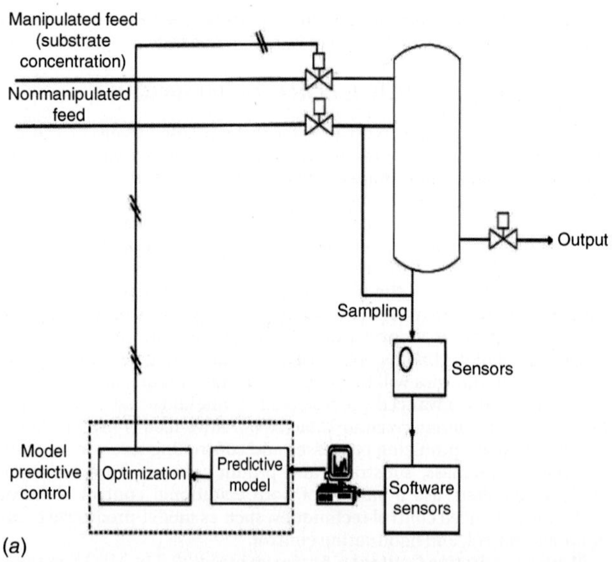

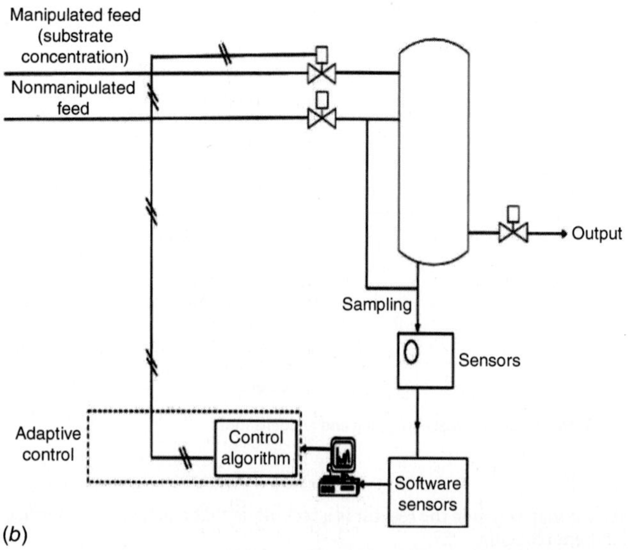

FIG. 20-77 Control systems: (*a*) model predictive control, (*b*) adaptive control.

Local linearization can often be applied to provide model flexibility. Moreover, custom linearization techniques are also used for special nonlinearities. Different factors are considered before choosing a linearization technique. These factors include, but are not limited to, the specific nature of the control problem, the time and cost incurred, and the quality of process model at hand. Once identified, these simplified process models are easier to work with and for many processes provide an adequate level of control. However, this does not necessarily provide a very robust control scheme. Alternative approaches can use mathematical techniques to transform nonlinear systems into more simplified linear systems that enhance controller design and application. The application of a nonlinear model is more challenging but may increase the validity of the controller. Optimizing control is typically applied in disturbance situations and in changing operating conditions. For example, model predictive control can be applied to optimize conditions such as biomass and substrate concentration or metabolite and protein production rates. The presence of many disturbances will cause the control to break down by disrupting the steady-state optimum on which control relies. One way to increase control robustness is to combine linearization techniques with adaptive control. The adaptive control, as described in the next subsection, helps to minimize some of the effects of the disturbances to allow for tighter control.

Adaptive Control As shown in Fig. 20-77*b*, adaptive control provides process control for systems with large perturbations from a steady-state operating condition by adjusting the control parameters to maintain performance objectives. This control technique can be applied to any dynamic process exhibiting parameter variations as well as to those processes which have variable parameters due to constantly changing physical conditions, such as fouling in heat exchangers. It operates as a self-tuning controller that works with or without the necessity of a process model. This control scheme relies upon measurements of certain indicator process variables. Because this method takes into account real-time data and tunes the controller based on these data, adaptive control schemes are often quite robust in maintaining functionality and accuracy even with high levels of uncertainty. This makes adaptive control a good option for processes with constantly changing parameters, and it can also be helped by the presence of a model, even though the control provided may not be the tightest. There are three main types of adaptive controllers: scheduled adaptive control, *model-reference adaptive control* (MRAC), and self-tuning adaptive control. Scheduled adaptive control works on a set of predetermined process control data and uses preprogrammed measures to achieve the control. However, MRAC exercises simulation of a predetermined process reference model with current process data to calculate the difference between actual process output and the model simulation output. The goal of MRAC is to reduce the difference between actual process output and the model simulation output by changing process parameters. By contrast, to achieve control, the self-tuning adaptive control allows a dynamic model-building process based on current process data as opposed to using a fixed predetermined model. This makes self-tuning control the most flexible adaptive control, but also most prone to errors as a poor model-building process could potentially lead to a highly unstable system. Adaptive control is effective when used both in start-up and disturbance situations. Thus, adaptive control can operate to bring a reactor from start-up to stability within normal operating conditions, or this control method can operate to alter the stability conditions of a reactor from outside to inside of normal and optimized operating conditions. Adaptive control is commonly used to control either the substrate concentration in the bioreactor or the production rate of biomass or product. These control schemes can also provide valuable indirect information about the state of the biomass within the reactor, thus helping to monitor growth and rate of generation of a desired product.

Habituation Control This approach is particularly useful for continuous processes; it is a more complex method in which multiple manipulated inputs are used to control a single output. A major advantage of this control is that the manipulation of several inputs in conjunction with minimizing deviations of inputs from steady-state values allows for tighter control of the output. For example, one of the commonly controlled parameters in a bioreactor operation is biomass concentration, which in turn is dependent on the substrate concentration. Substrate concentration in the bioreactor can be tightly controlled by using both the dilution rate and the feed substrate concentration as manipulated variables. Controlling the substrate concentration also enables control of other growth parameters such as the formation rates of the desired products. Habituating control offers the opportunity to control substrate concentration or biomass by a concerted and coordinated manipulation of both the substrate feed concentration and the dilution rate, which increases the rate of substrate into and out of the bioreactor. Furthermore, proper tuning of two parameters rather than a single one can facilitate tight regulation over the control objective and ultimately improved process performance and product consistency [Carrondo et al., *Biotechnology J.* **7:** 1522–1529 (2012)].

ADVANCED SEPARATION APPROACHES*

Electrophoresis Electrophoresis is a separation technique often applied to the analysis of biological or other polymeric samples. This technique is among the most powerful for estimating purity because of its simplicity, speed, and high resolution, and also because there is only a small probability that any of the components being analyzed will be lost during the process of analysis. Electrophoresis has frequent application to analysis of proteins and DNA fragment mixtures. The high resolution of electrophoresis has made it an important tool in the advancement of biotechnology. Variations of this methodology are being used for DNA sequencing and fingerprinting; for isolating active biological factors associated with diseases such as cystic fibrosis, sickle cell anemia, myelomas, and leukemia; and for establishing immunological reactions between individual compounds. Electrophoresis is an effective analytical tool because the electric field does not affect a molecule's structure, and it is highly sensitive to small differences in molecular charge, size, and sometimes shape. The electrophoresis systems with the highest capacity are free-flow electrophoresis, density gradient electrophoresis, recycling free-flow isoelectric focusing, and rotating isoelectric focusing. The principles of operation of these are discussed next.

*Portions of the Advanced Separation Approaches section have been reproduced from *Bioseparations Science and Engineering*, 2d ed., by Roger G. Harrison, Paul W. Todd, Scott R. Rudge, and Demetri P. Petrides (2015). By Permission of Oxford University Press.

Free-Flow Electrophoresis In free-flow electrophoresis, a sample to be separated is continuously injected into a thin rectangular channel through which a carrier fluid flows, with an electric field that is perpendicular to the flow (Fig. 20-78*a*). This results in differential deflection of charged solutes and particles, leading to separation by differences in electrophoretic mobility. Application of the electric field is initiated before sample loading and is achieved via two electrode compartments separated from the flow chamber by semipermeable membranes at the two vertical edges of the thin chamber. Each electrode compartment contains a single metal wire electrode and an electrolyte solution consisting of a nontoxic buffer and dissolved KCl or NaCl at an ionic strength higher than that of the carrier buffer. To maximize field strength (hence separand migration velocity) and minimize current density (hence heating and natural convection), the carrier buffer must have the lowest ionic strength possible that is compatible with separand solubility and stability (separands could be living cells). An upward-flowing buffer stream becomes warmer as it rises and is more stable against natural convection than is a downward-flowing stream. A thermostated water jacket can be in contact with the front and back planes, which are usually made of glass or polycarbonate.

In free-flow electrophoresis, sample solution is injected into one of the in-flowing channels or directly into a sample port in the chamber. There are limits on the concentration of sample based on solution density. A sample that is too dense will not rise with upward flow, and a sample that is too light will rise to the top of the chamber before significant separation can occur; vice versa for downward flow. Ideally the carrier buffer and sample solution should have the same density and very similar conductivities. Collection of separands occurs at the top (or receiving end) of the chamber at collection ports ("outlet fractions" in the diagram), which are connected by fine tubing to a collection vessel. The process is continuous, so carrier buffer is flowing out of outlets continuously during a separation. Separands exit according to their mobility. Separation may be quantified by identifying and measuring the concentration of each separand. The result is a series of non-Gaussian peaks, and yield may be low since separand distributed in the tails of the parabolic profiles might not be collected.

In the example shown in Fig. 20-78*a*, the flow is from bottom to top, but downward and horizontal configurations exist. Carrier fluid exits in thin outgoing channels at the opposite end of the chamber. The vertical flow is laminar, resulting in a parabolic velocity distribution in the *yz* plane using the axes specified by the diagram. Thus

$$v_z(y) = v_{\max}\left(1-\frac{y^2}{b^2}\right) \tag{20-105}$$

where *y* is the distance from the center of the chamber in the *y* direction and *b* is the chamber half-thickness. This velocity profile means that separands near the front and back chamber walls move vertically the most slowly, spend more time in the electric field, and are therefore carried farther horizontally in the field than their identical counterparts flowing faster at the center of the chamber. The time required for a volume element to reach the outlet is

$$t(y) = \frac{L}{v_z(y)} \tag{20-106}$$

where *L* is the distance from the inlet to the outlet.

The electric field also causes laminar electro-osmotic flow at the chamber walls in the *x* direction (toward the negative electrode) with the result

$$v_x(y) = \frac{1}{2}U_o E\left(\frac{3y^2}{b^2}-1\right) \tag{20-107}$$

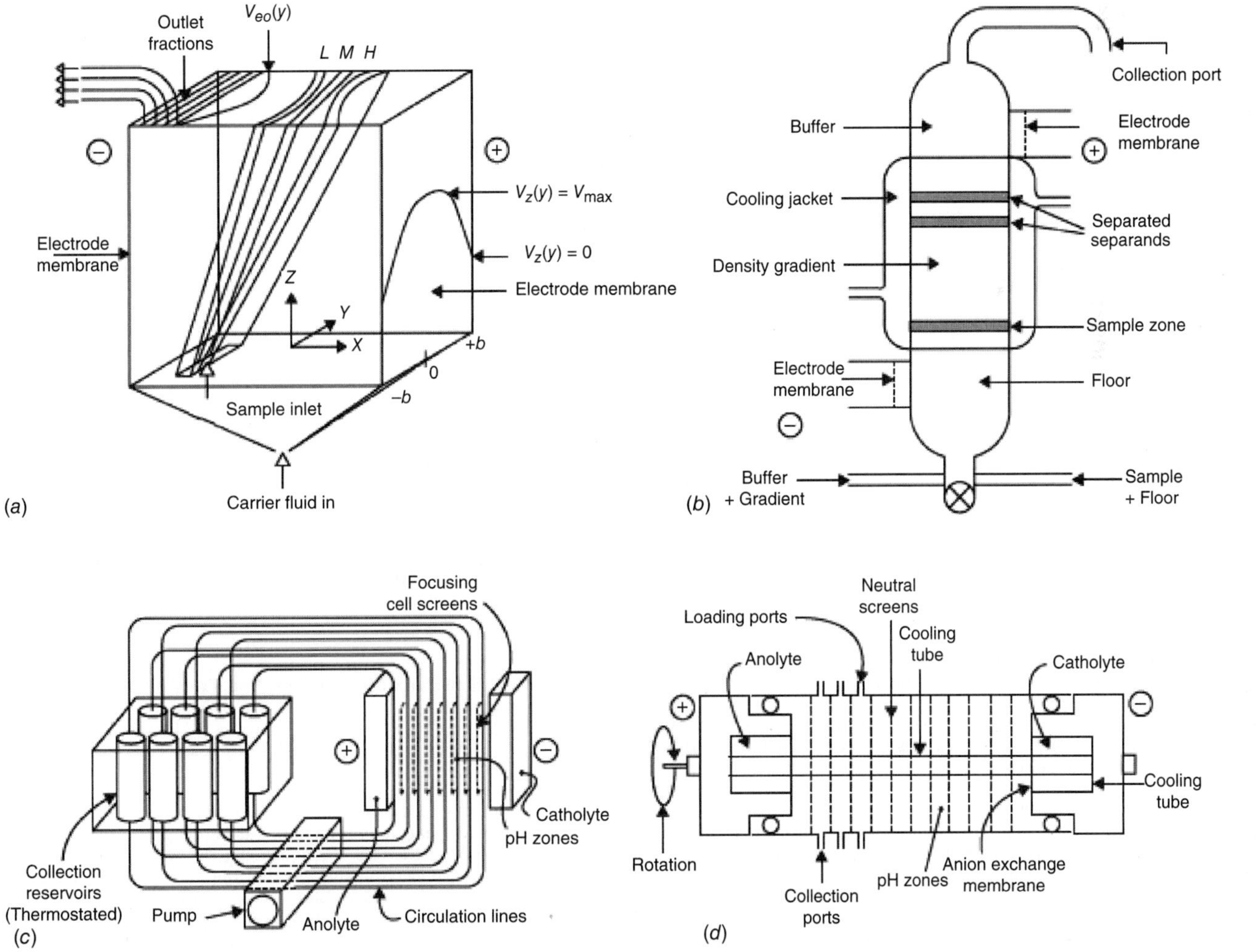

FIG. 20-78 Schematic drawings of four methods of preparative electrokinetic separation. (*a*) free-flow electrophoresis (FFE), (*b*) density-gradient electrophoresis (DGE), (*c*) recycling free-flow isoelectric focusing (RIEF), and (*d*) rotating isoelectric focusing (Rotofor). In (*a*), separand concentration trajectories are denoted for low- (L), medium- (M), and high- (H) mobility separands. [Reproduced from *Bioseparations Science and Engineering*, 2d ed., by Roger G. Harrison, Paul W. Todd, Scott R. Rudge, and Demetri P. Petrides (2015). By Permission of Oxford University Press.]

where U_o = electro-osmotic mobility of the chamber walls due to their zeta (surface) potential, E = applied electric field strength (voltage per length) given by $E = I/k_eA$, and k_e = solution electrical conductivity in siemens per meter.

Separand molecules near the wall will migrate more slowly toward their complementary electrode than will those at the center of the chamber. This relationship between laminar flow (y-z profile) and electro-osmotic flow (y-x profile) leads to an exact balancing of the two migrations for certain separands and gives a flat profile of the separand concentration across the thin chamber, as shown by the letter M on the diagram in Fig. 20-78*a*. And L and H correspond to lower and higher mobility separands, respectively. Separands are resolved on the basis of distance migrated X in the x direction according to the following equation:

$$X = \left[U_{el} t(y) + \frac{1}{2} U_o \left(\frac{3y^2}{b^2} - 1 \right) t(y) \right] E \tag{20-108}$$

Density Gradient Electrophoresis (DGE) DGE takes advantage of differences in the electrophoretic mobility and buoyancy of particles. In this type of electrophoresis, separands move through a density gradient in a static liquid (Fig. 20-78*b*). Separation of particulate separands is slowed due to increased upward buoyant force. In density gradient electrophoresis, carrier buffer is loaded (typically) from the bottom of a cylindrical column by a density gradient maker with low density followed by high density. Sample is loaded in a solution having higher density than the bottom of the density gradient, and it is followed by a very dense fluid that forms a "floor" under the fluids above it in the column. The electric field is applied after loading the sample. Applying an electric field to a column of suitable geometry results in negligible electro-osmotic flow relative to the electrophoretic velocity of the separands. However, Joule heating of the fluid is inevitable and is counteracted by thermostating the column with a water jacket. The density gradient counteracts the thermal convection due to Joule heating. The separands should be allowed to migrate upward to ensure adequate separation. For collection of separands, the contents of the column are either drained slowly out the bottom or pumped out the top of the column, and fractions are collected into individual containers as in preparative chromatography. The resolution of DGE is based on the number and size of fractions collected. Purity can be very high depending on fraction volume.

The forces acting on a particle in density gradient electrophoresis are the electric field, the gravitational field, and the buoyant force. Balancing these forces leads to a particle velocity for particles migrating upward toward the anode (positively charged electrode) as

$$v = U_{el} E - \frac{2}{9} \frac{a_p^2 g(\rho - \rho_o)}{\mu} \tag{20-109}$$

where U_{el} is the electrophoretic mobility of the charged particle, E is the field strength or voltage gradient (voltage V per length L between the electrodes), and the second term is the sedimentation velocity for spheres of density ρ and radius a_p in a fluid of density ρ_o and viscosity μ under the influence of gravity. This velocity will not be constant as a function of the distance migrated in the z direction (upward) because ρ_o changes with z.

Recycling Free-Flow and Rotating Isoelectric Focusing In recycling isoelectric focusing (RIEF, Fig. 20-78*c*), electrophoretic separation takes place as separands, in a solution containing ampholytes, flow through compartments in a focusing cell across which an electric field is applied. Separands and ampholytes pass repeatedly from compartment to compartment and back to a reservoir associated with each compartment by electrophoresis until they become neutral at their isoelectric pH in one of the compartments. At this point, electrophoretic migration ceases, and steady state has been reached. Then each of the reservoirs is drained for sample collection. The number of fractions equals the number of reservoirs. It is necessary to adjust flows to be the same in and out of all the reservoirs; flow balancing is very important.

The electric field is applied across the compartmented cell only, and the initial current is typically several milliamperes. While the electric current and pumps are on, fluid is circulated out of the reservoirs through the compartments of the electrophoresis cell and back to the reservoirs. Eventually the fluid in each compartment is at a different pH, resulting in a pH gradient across the separation cell. Each separand migrates to the compartment in the cell that is at its isoelectric pH, where $U_{el} = 0$, and therefore is found in the corresponding reservoir having that pH. At the end of the focusing period, the fluid contained in each reservoir is drained into a separate container, its pH is measured, and the contents are assayed for separands. The relatively small number of fractions will typically result in "single-fraction peaks" in the plot of concentration versus pH. A broad range of pHs, for example 2 to 10, will result in low resolution and purity; each reservoir could contain several separands. A narrow range of pHs, for example, 4.2 to 4.8, will result in high resolution and purity, but the pH range must be selected to include the isoelectric pH of the separand of interest.

The mixing effects of natural convection due to temperature and concentration gradients can be counteracted by rotation of the separation vessel at a carefully chosen velocity, resulting in a constantly changing gravity vector, hence rotating isoelectric focusing (Fig. 20-78*d*). The focusing chamber is filled with ampholyte solution with the sample mixed in using several of the loading ports. After filling without bubbles, the filling ports are sealed, and rotation is begun. The ampholytes and sample molecules will focus simultaneously. Cooling water is flowed through the central cooling tube, rotation is initiated, then power is switched on. After focusing, the solution in each compartment is rapidly drained through the collection ports without mixing.

MAGNETIC BIOSEPARATIONS

Magnetic capture is a powerful, highly selective method for collecting separands present in low abundance in a highly mixed suspension or solution. Once a separand has reacted with magnetic beads by extensive mixing, resulting in affinity adsorption, the beads are captured and held by a magnetic force so that all other components can be rinsed away, much as in a centrifugation wash step. The general principle is not unlike that of affinity chromatography, in which ligand molecules are affixed, usually by covalent reactions, to polymer microspheres (or nanoparticles) that contain iron oxide (from 3 to 80 percent by mass). Examples of ligand molecules include nucleic acids, glycoconjugates, avidin, and most often antibodies. Magnetic reagents have been manufactured for several purposes, and their sizes now range from some 20 nm up to 10 μm in diameter. While most are spherical, some may be irregular in shape, especially if the iron oxide content is high.

Physical Principles of Magnetic Separations It is desirable that much of the iron oxide be in the form of magnetite, Fe_3O_4, which provides unpaired electrons whose orbits can be oriented in a magnetic field, resulting in magnetization. This condition is referred to as paramagnetism, in which the material is magnetic if, and only if, it is in a magnetic field. This is important because materials that retain their magnetism after the magnetic field is removed are likely to aggregate and become difficult to handle. Paramagnetic materials do not exhibit remanent magnetization as ferromagnetic materials do, such as iron, cobalt, and permanent-magnet ceramics.

When a magnetizing field H (measured in amperes per meter, A m^{-1}) returns to zero, ferromagnets are still magnetic, but paramagnetic materials are not. The magnetization M is proportional to H so that

$$M = \chi H \tag{20-110}$$

where the constant of proportionality is the volume magnetic susceptibility χ. Magnetic flux density B is usually expressed in SI units of tesla (T, 1 T = 1 N A^{-1} m^{-1}) and is related to the applied field by the magnetic permeability, which in free space is $\mu_o = 4\pi \times 10^{-7}$ T m A^{-1}, a universal constant that can be expressed in the more convenient units of newtons per squared ampere (N A^{-2}):

$$B = \mu_0 H \tag{20-111}$$

In the presence of a large magnet having magnetizing field H (A m^{-1}), a small particle of volume V and susceptibility $\Delta\chi$ above that of the surrounding fluid will experience magnetization $M = \Delta\chi H$ (A m^{-1}) and a force F_m (in newtons) that is proportional to the gradient of the flux density produced by the external magnet ∇B (T m^{-1}) at the instantaneous location of the particle, given by Moeser et al., *AIChE J.* **50:** 2835 (2004),

$$F_m = V\Delta\chi H \cdot \nabla B \tag{20-112}$$

but since $B = \mu_o H$, the force can be expressed in terms of the strength of the permanent magnet:

$$F_m = V\Delta\chi \frac{B}{\mu_o} \cdot \nabla B \tag{20-113}$$

The features of this relationship are used to understand the equipment required to perform magnetic separations using magnetic reagents. The inducing field H plays the role of magnetizing the paramagnetic reagent by orienting unpaired electron spins so that the reagent will be attracted toward the magnetizing source by the gradient ∇B (see Fig. 20-79*a*).

Equipment and Reagents The volume of available commercial magnetic particles is a deliberate variable that may vary by more than 10^6-fold (4.0-μm versus 40-nm diameter, for example). The quality of magnetic material incorporated into the bead determines $\Delta\chi$, which is the relative susceptibility of the average bead material, that is, its susceptibility

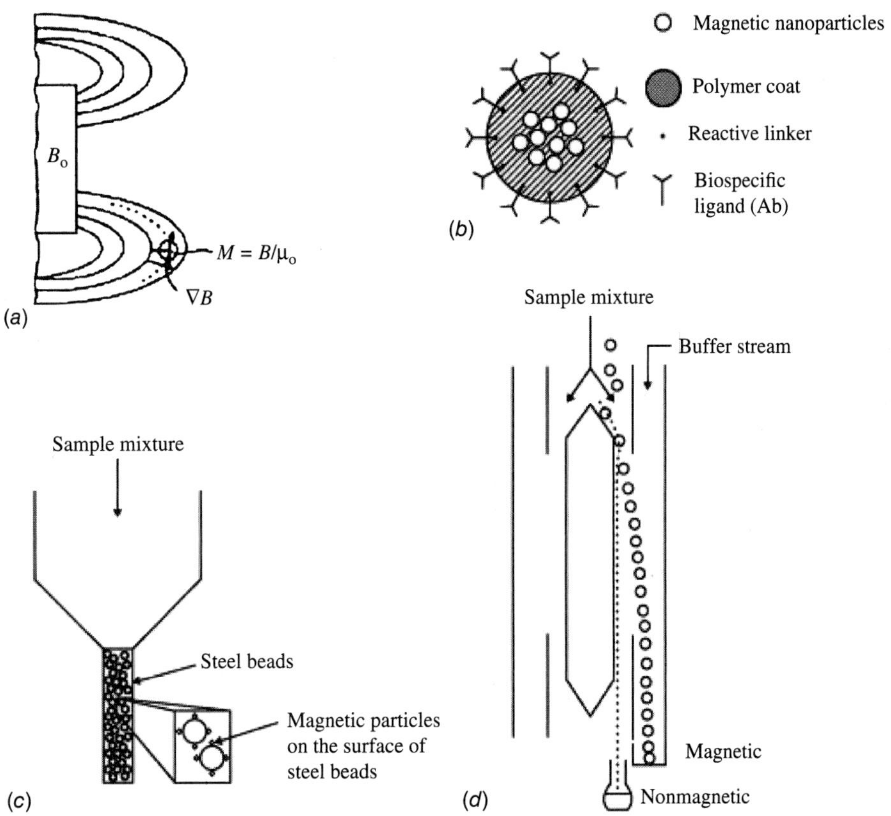

FIG. 20-79 Magnetic separations. (*a*) The role of magnetic flux density *B* and gradient ∇B in applying ponderomotive (motion-causing) force to a magnetically susceptible particle (represented as a circle with an arrow passing through it). The arrow through the particle represents the polarization of the magnetization *M*. (*b*) The general configuration of a magnetic reagent nanoparticle or microparticle showing susceptible core, nonsusceptible material coat, linker reactive molecule, and external ligand for reacting with separands with high specificity. (*c*) Operation of a column-based (equilibrium) separator showing packed column of steel beads or susceptible metal wool to which magnetic reagents adsorb and from which they are desorbed for batch collection by removing the magnetic field. (*d*) Operation of a flow-based (rate) cylindrical separator having a core and inlet and outlet splitters showing removal of magnetic reagents from an inner to an outer parallel annular flow for continuous collection of separands. [Reproduced from *Bioseparations Science and Engineering*, 2d ed., by Roger G. Harrison, Paul W. Todd, Scott R. Rudge, and Demetri P. Petrides (2015). By Permission of Oxford University Press.]

minus that of the surrounding fluid. Parenthetically, a solution containing a high concentration of paramagnetic solutes (ions) could diminish or even reverse this value. There are four features of magnetic reagents to consider besides bead diameter including the manner in which beads are manufactured, the magnetically susceptible material, the structural bead material (nonsusceptible), the conjugation site on the bead material, and the biospecific ligand (see Fig. 20-79*b*). Table 20-44 is a list of each of these choices, and the listed components can be combined in all possible permutations.

Magnetic separation devices fall into three broad categories: tube and magnet equilibrium separators, high-gradient batch column separators, and flowing continuous separators. Different methods of magnetic field gradient formation are utilized by the different magnets used in the different devices: (1) The simpler "tube and magnet" types expose the suspension of magnetic reagent (beads, ferrofluid) to one pole of a dipole magnet, which forms flux lines as shown in Fig. 20-79*a*. (2) When steel spheres or wires are placed in a column served by dipole magnets, flux lines enter the spheres essentially perpendicular to the sphere surface, creating intensely converging field lines and therefore a high flux gradient leading to the term *high-gradient magnetic separation* (HGMS) as in Fig. 20-79*c*. (3) Three forms of Halbach magnet assemblies have been used and are available for larger-volume separations. A Halbach array is a special arrangement of multiple permanent magnets that augments the magnetic field on one side of the array while canceling the field to near zero on the other side. A Halbach quadrupole assembly forms a central bore within which the channel of the quadrupole magnetic separator (Fig. 20-79*d*) is held.

TABLE 20-44 Examples of Susceptible Materials, Nonsusceptible Materials, Reactive Linkers, and Biospecific Ligands Used in Magnetic Reagents

Susceptible materials	Nonsusceptible materials	Reactive linkers	Biospecific ligands
Magnetite (Fe_3O_4)	Polystyrene	Carboxyl	Biotin
Maghemite (γ-Fe_2O_3)	Silica	Amino	Biotin (or streptavidin)
	Alginate	Epoxy	Antibodies
	Agarose	Tosyl	Glycoconjugates
			Complementary nucleic acids

Reproduced from *Bioseparations Science and Engineering*, 2d ed., by Roger G. Harrison, Paul W. Todd, Scott R. Rudge, and Demetri P. Petrides (2015). By Permission of Oxford University Press.

Magnetic Separation Applications Applications include bench-scale protein and nucleic acid purification, protein and nucleic acid diagnostics ("bead-based assays"), and magnetic cell sorting. Bench-scale purification is typically achieved using biospecific affinity ligands in much the same way as in affinity chromatography. Magnetic purification can also be used as a concentration step. A typical procedure consists of coupling the biospecific ligand to the surface of magnetic beads. Unreacted antibody (or other ligand) is washed away by magnetically aggregating the paramagnetic particles. Magnetic separation has unique advantages in cell purification; it may be used to remove, select, or enrich specific subsets in cell mixtures for application to regenerative medicine and rare-cell analysis. In the case of selected cells to be retained, chemical and enzymatic methods are available for the removal of magnetic reagents from cells after separation.

There are many ready-made magnetic reagents precoated with biospecific ligands available commercially. If desired reagents are unavailable,

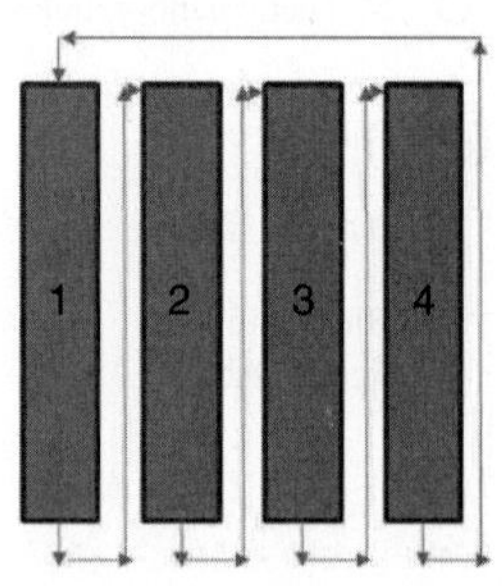

FIG. 20-80 Simulated moving-bed column illustration. [Li et al. *Continuous Processing in Pharmaceutical Manufacturing* **1:** 1–27 (2014).]

they can be synthesized by the user. Vendors of magnetic beads provide chemically activated surfaces and instructions for the conjugation of user-supplied biospecific ligands. It is easy to remove unreacted biospecific ligand and (if used) linker compounds by holding the beads in place with a magnet while the reactant solution is decanted and the beads are rinsed with final buffer before resuspending. To purify a separand, beads are added to the mixed sample solution and held—often at 4°C to preserve specificity—with gentle agitation while separand molecules or cells collide with the bead surfaces and react. Adequate agitation is required for this reaction, since magnetic beads can be considered "non-Brownian" and all reagents must be transported to them by diffusion and mixing,

SIMULATED MOVING-BED CHROMATOGRAPHY

Typically, downstream separation and purification of proteins relies on different physical properties, such as size, charge, hydrophobicity, and specific binding capability. *Simulated moving-bed* (SMB) chromatography is a semicontinuous or continuous alternative to conventional batch elution chromatography. Traditional chromatography is a time-dependent process: it involves collecting the components at different times from the same outlet. SMB, on the other hand, is space-dependent and accumulates the separate components simultaneously from different locations.

In moving-bed chromatography, the feed is provided and products are recovered simultaneously and continuously, but there are practical challenges to implementing this approach. Alternatively, in a simulated moving bed, the feed inlet, the eluent, or desorbent inlet and two output product positions are moved continuously, giving the impression of a moving bed (Figs. 20-80 and 20-81). When a mixture passes over a solid column, the components in a mixture react differently with the stationary phase and therefore move at different rates. They are also removed differently based on the individual components' interaction with the stationary phase relative to their interactions with an eluent, which is responsible for eluting the individual products from the mixture. In SMB, the column is set up so that the two products come out at different outlets due to their differing interactions with the stationary solid and eluent or desorbent liquid phase. Assuming the mixture only has two components, their velocities can be determined experimentally or predicted theoretically. The component with stronger interactions with the column will have net movement in one direction, and the component with weaker interactions will have net movement in the opposite direction. Assuming the input is located at the middle of the column and the column is long enough, the separated components can be collected at opposite ends. To keep the movement near continuous, each bed is split into subunits (Fig. 20-80). For small-scale separations, 4 to 8 subunits are used; for large-scale, up to 24. Instead of moving the bed, the desired and undesired product exit stages are continuously adjusted to create the simulation of a moving bed (Fig. 20-81).

Advantages of SMB relative to traditional elution chromatography include the fact that, because the process is only space-dependent, it can be run continuously. SMB is applicable at any scale; the only adjustments involve sizes of columns and equipment. Although SMB is usually conducted in the liquid phase, it can also be carried out in the gas phase, allowing for more efficient transfer. SMB also has lower solvent consumption and more efficient use of the solid adsorbent bed compared to traditional chromatography, as it can be run on overload. SMB also requires fewer theoretical plates to achieve the same purity as with standard elution chromatography. When traditional separation techniques are too harsh, SMB can be used instead of distillation, etc. The disadvantage to SMB, however, is that it is primarily used to split a mixture into two; so if there are more than two components to be separated, several units must be used [Juza et al., *Trends in Biotechnol.* **18:** 108–118 (2000)].

Theory A moving bed can be formulated to include salt gradient in ion exchange mode in which two initial sections have a higher salt concentration, and the next two sections have a lower salt concentration, thus creating a salt gradient [Li et al., *Continuous Processing in Pharmaceutical Manufacturing* **1:** 1–27 (2014)]. In ion exchange chromatography used for protein purification, the gradient allows for desorption of adsorbed proteins in the first two sections, and a higher adsorption of proteins in the two farther downstream sections. The gradient can be created by introducing lower salt concentration at the feed and higher salt concentration at the desorbent portion. Equation (20-114) represents the mass balance over volume of the bed k for proteins and salt. Equation (20-115) represents the mass balance in the particles for proteins and salt. These balances are described according to the linear driving force approximation.

Mass balance over a volume element of the bed k for proteins and salt is

$$\frac{\partial C_{ik}}{\partial t} = D_{Lk}\frac{\partial^2 C_{ik}}{\partial Z^2} - \frac{u_k}{\varepsilon_B}\frac{\partial C_{ik}}{\partial Z} - \frac{1-\varepsilon_B}{\varepsilon_B}k_{Pik}(q_{ik}^* - q_{ik}) \tag{20-114}$$

Mass balance in the particles for proteins and salt is

$$\frac{\partial q_{ik}}{\partial t} = k_{Pik}(q_{ik}^* - q_{ik}) \tag{20-115}$$

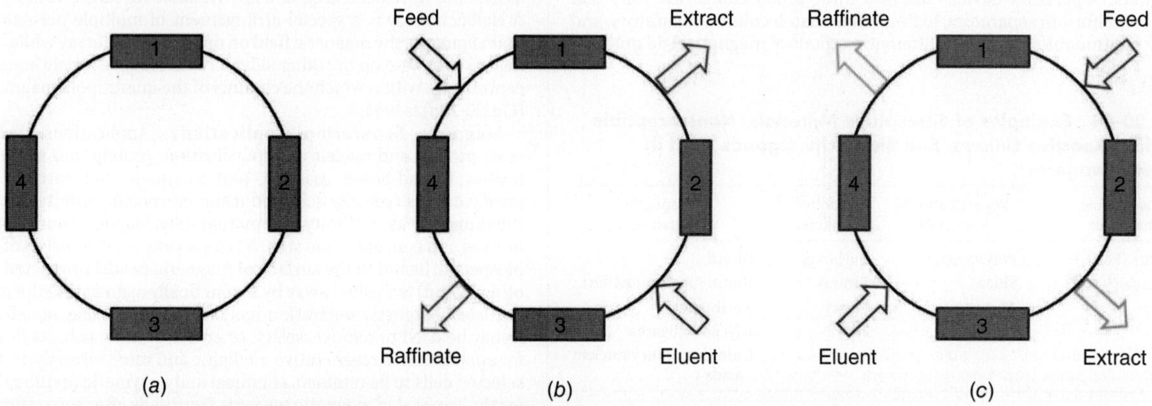

FIG. 20-81 Simulated moving-bed operation: (*a*) continuous rapid circulation of flow in one direction; (*b*) and (*c*) are examples of configurations that create the simulated movement by changing the feed valves. [Li et al. *Continuous Processing in Pharmaceutical Manufacturing* **1:** 1–27 (2014).]

EXPANDED-BED CHROMATOGRAPHY

Proteins are involved in a number of industries such as pharmaceuticals, food processing, textiles, leather goods, detergents, and even paper. Effective protein separation and purification is essential for economic production of quality product. A three-step separation and purification method is typically employed. These three steps include capture, intermediate separation, and final purification [Li et al., *Continuous Processing in Pharmaceutical Manufacturing* **1:** 1–27 (2014)].

The expanded bed utilizes particles in a fluidized environment, whereas classic column chromatography uses a solid phase in a packed bed. Expanded-bed chromatography utilizes a particle size gradient which allows for local zones of separation (Figs. 20-82 and 20-83). Following capture, the target proteins can be eluted through an elution solution separately from contaminants and impurities in the original entities. Strides are being made by using expanded-bed adsorption for primary recovery of proteins from a variety of sources including, but not limited to, *E. coli* homogenate, yeast fermentation, mammalian cell culture, milk, and animal tissue extracts [Hjorth, *Trends Biotechnol.* **15:** 230–235 (1997)].

Adsorbents When an adsorbent is chosen, there are a number of trade-offs to consider. The current trend to use dense solid-core materials as adsorbents has the benefits of allowing much higher feed velocities while not being sensitive to ionic strength. Low-density adsorbents have been shown to have a higher adsorption capacity in pilot-scale experiments, but that capacity has been shown to decrease in the presence of high ionic strength. An effective adsorbent will have a high-density matrix and a salt-tolerant ligand. The use of a high-density matrix minimizes the consumption of dilution buffer. The use of a salt-tolerant ligand is preferred because there is a lack of sensitivity to ionic strength and salt concentration does not decrease through dilution [Asghari and Jahanshahi, *J. Chromatogr.* A, **1257:** 89–97 (2012)]. A number of adsorbents have been utilized for research and commercial purposes, as shown in Table 20-45.

Modeling of the Expanded Bed Modeling an expanded bed is more complex than modeling a fixed bed. A three-zone model takes into account the variations in adsorption behavior throughout the bed. The variations are due to different adsorbent particle sizes, bed voidage, and liquid axial dispersion coefficients at the bottom, middle, and top zones of the columns (zones 1 to 3 in Fig. 20-83) and are described by Eqs. (20-116) through (20-118).

$$D_{Lk}\frac{\partial^2 C_k}{\partial Z^2} - \frac{u_k}{\varepsilon_{Bk}}\frac{\partial C_k}{\partial Z} - \frac{\partial C_k}{\partial t} - \frac{1-\varepsilon_{Bk}}{\varepsilon_{Bk}}\frac{3}{R_k}k_{fk}[C_k-(c_k)_{r=R_k}]=0 \qquad (20\text{-}116)$$

$$(1-\varepsilon_{Bk})\frac{\partial \bar{q}_k}{\partial t} = D_{Sk}\frac{\partial^2 \bar{q}_k}{\partial Z^2} + (1-\varepsilon_{Bk})\frac{3}{R_k}k_{fk}[C_k-(c_k)_{r=R_k}] \qquad (20\text{-}117)$$

$$\varepsilon_p\frac{\partial c_k}{\partial t}+\frac{\partial q_k}{\partial t} = \varepsilon_p D_p\left(\frac{\partial^2 c_k}{\partial r^2}+\frac{2}{r}\frac{\partial c_k}{\partial r}\right) \qquad (20\text{-}118)$$

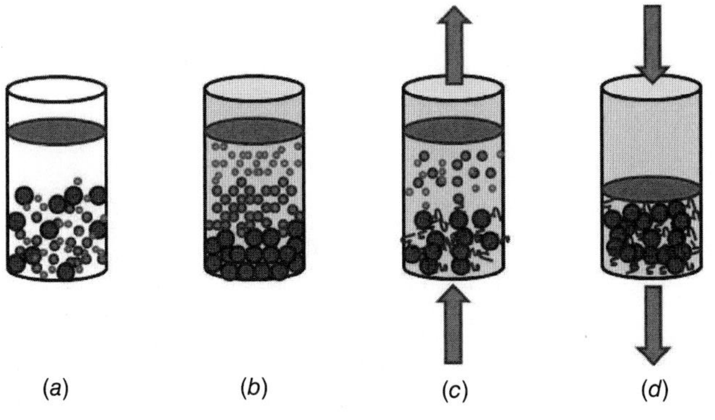

FIG. 20-82 Expanded-bed chromatography. (*a*) Adsorbent is embedded; (*b*) the column is fluidized and a concentration gradient is formed; (*c*) the feed is injected, and particulates and debris move past the bed and out of the column, while the product interacts with the adsorbent; (*d*) the column is repacked and the flow is reversed to collect the product. [Li et al. *Continuous Processing in Pharmaceutical Manufacturing* **1:** 1–27 (2014).]

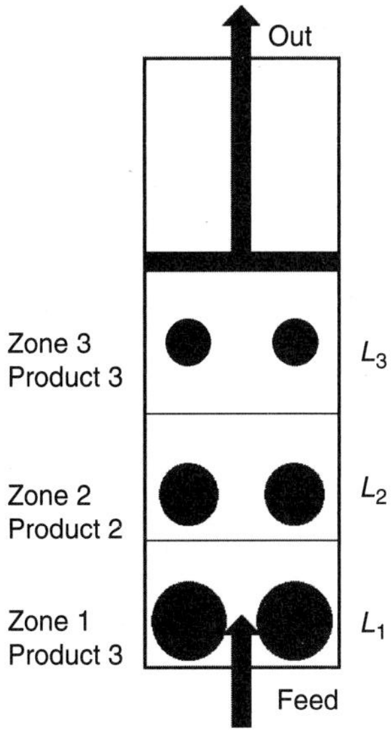

FIG. 20-83 Three-zone model for protein adsorption kinetics in expanded beds. Zone 1 is characterized by its location at the bottom of the column, large adsorbent particle size, small bed voidage, and significant liquid axial diffusion. Zone 2 is characterized by its location at the middle of the column, smaller adsorbent particle size, a larger bed voidage, and a weaker liquid axial dispersion than in zone 1. Zone 3 is characterized by its location at the top of the column, the smallest adsorbent particle size, high bed voidage, and smallest liquid axial dispersion. [Li et al. *Continuous Processing in Pharmaceutical Manufacturing* **1:** 1–27 (2014).]

Equations (20-116) through (20-118) correspond to the material balance for the bulk liquid phase in each zone, the mass balance for the adsorbent phase in each zone, and the pore diffusion for the adsorbent in each zone, respectively.

TABLE 20-45 Recent Commercial and Research Adsorbent Use in Expanded-Bed Chromatography*

Adsorbent	Core components	Year developed	Use
β-Cyclodextrin polymer	Tungsten carbide	2010	Research
Agarose	Tungsten carbide	2010	Research
6% Cross-linked agarose	Crystalline quartz	2010	Commercial
6% Cross-linked agarose	Crystalline quartz	2010	Commercial
Cellulose	Tungsten carbide	2011	Research
6% Cross-linked agarose	Crystalline quartz	2011	Commercial
Hydrogel-filled	Zirconium oxide	2011	Commercial
Agarose	Nickel (nanoporous)	2012	Research
Agarose	Zinc (nanoporous)	2012	Research
6% Cross-linked agarose	Crystalline quartz	2012	Commercial
6% Cross-linked agarose cover with polyvinyl pyrrolidone	Crystalline quartz	2012	Commercial
3% Agarose	Tungsten carbide	2013	Research
Polyacrylamide-based Cryogel	Titanium oxide	2013	Research
6% Cross-linked agarose	Tungsten carbide	2013	Commercial
Agarose	Tungsten carbide	2013	Commercial
Hydrogel-filled	Zirconium oxide	2013	Commercial
6% Cross-linked agarose	Crystalline quartz	2013	Commercial
Agarose	Tungsten carbide	2013	Commercial
Agarose	Tungsten carbide	2013	Commercial
Agarose	Tungsten carbide	2013	Commercial

*Adapted from Li, P., Gomes, P. F., Loureiro, J. M., and Rodrigues, A. E., "Proteins Separation and Purification by Expanded Bed Adsorption and Simulated Moving Bed Technology," *Continuous Processing in Pharmaceutical Manufacturing* **1:** 1–27 (2014).

Reproduced from *Bioseparations Science and Engineering,* 2d ed., by Roger G. Harrison, Paul W. Todd, Scott R. Rudge, and Demetri P. Petrides (2015). By Permission of Oxford University Press.

Section 21

Solids Processing and Particle Technology

Karl V. Jacob, B.S. *Fellow, The Dow Chemical Company; Lecturer, University of Michigan; Fellow, American Institute of Chemical Engineers (Section Editor, Particle Characterization, Pneumatic Conveying, Screening)*

Greg Mehos, Ph.D., P.E. *Senior Project Engineer, Jenike & Johanson, Inc. (Bulk Solids Flow and Hopper Design)*

John W. Carson, Ph.D. *President, Jenike & Johanson, Inc., Founding member and past chair of ASTM Subcommittee D18.24, "Characterization and Handling of Powders and Bulk Solids" (Bulk Solids Flow and Hopper Design)*

Yi Fan, Ph.D. *Associate Research Scientist, The Dow Chemical Company (Solids Mixing)*

Ben J. Freireich, Ph.D. *Technical Director, Particulate Solid Research, Inc. (Solids Mixing, Size Enlargement)*

James F. Koch, M.S. *Senior Process Engineering Specialist, The Dow Chemical Company (Size Reduction, Screening)*

Shrikant V. Dhodapkar, Ph.D. *Fellow, The Dow Chemical Company (Feeding, Metering, and Dosing); Fellow, American Institute of Chemical Engineers*

Pradeep Jain, M.S. *Senior Fellow, The Dow Chemical Company (Feeding, Metering, and Dosing)*

PARTICLE CHARACTERIZATION

BULK SOLIDS FLOW AND HOPPER DESIGN

The authors gratefully acknowledge the contributions of Wolfgang Witt, Ralf Weinekötter, Douglas Sphar, Erik Gommeran, Richard Snow, Terry Allen, Jim Litster, and Bryan Ennis, coauthors of the 8th Edition, Sec. 21.

SOLIDS MIXING

PNEUMATIC CONVEYING OF SOLIDS

SCREENING

SIZE REDUCTION

MODELING AND SIMULATION OF GRINDING PROCESSES

CRUSHING AND GRINDING EQUIPMENT: DRY GRINDING—IMPACT AND ROLLER MILLS

CRUSHING AND GRINDING EQUIPMENT: FLUID-ENERGY OR JET MILLS

CRUSHING AND GRINDING EQUIPMENT: WET/DRY GRINDING—MEDIA MILLS

SIZE ENLARGEMENT

FEEDING, METERING, AND DOSING

Nomenclature

Note that many symbols have multiple definitions. Also some symbols have only one set of units associated with them. This means that they are associated with an empirical equation.

Symbol	Definition	SI units	U.S. customary units
A	Pipe area	m^2	ft^2
A, B, C	Constants for Eq. (21-23)	—	—
A_0	Cross-sectional area of outlet	m^2	ft^2
a	Distance from the scatter to the detector	m	ft
a	Acceleration	m/s^2	ft/s^2
$a_{k,n}$	Constants for batch grinding eq.	dimensionless	dimensionless
B	Outlet size	m	ft
B	Bend factor, see Table 21-7	dimensionless	dimensionless
B	Nucleation rate	number/m^3/s	number/ft^3/s
B_{min}	Minimum outlet size	m	ft
$\Delta B_{k,u}$	Breakage function, Eq. (21-126)	dimensionless	dimensionless
C, n	Constants, Eq. (21-119)		
C_i	Impact crushing resistance		ft·lb/in
C_w	Concentration of water in air	kg/m^3	lb/ft^3
C_{PF}	Area concentration	1/cm^2	1/in^2
c	Average slug velocity	m/s	ft/s
c_D	Constant defined by Eq. (21-111)	dimensionless	dimensionless
D	Silo cylinder diameter	m	ft
D	Pipe diameter	m	ft
D	Rotor diameter of rotating mill	m	ft
D_F	Critical rathole diameter	m	ft
D_m	Media diameter	cm	in
d	Wire size	µm	in
d	Roll gap	cm	in
d_p	Particle diameter	m	ft
d_p	Geometric mean particle diameter	µm	in
d_p^*	Matsumoto critical diameter	m	ft
d_{50}	Cut size	µm	in
$d_{3,2}$	Sauter mean diameter	µm	in
E	Work done during grinding	J	ft·lb
E_i	Work index, Eq. (21-123)		kWh/ton
e	Elementary charge	C	C
e	Coefficient of restitution	dimensionless	dimensionless
f_c	Unconfined yield stress	Pa	lb/ft^2
ff	Flow factor	dimensionless	dimensionless
g	Gravitational constant	m/s^2	ft/s^2
G	Growth rate	m^3/s	ft^3/s
G_c	Solids flux at choking	kg/(m^2·s)	lb/(ft^2·s)
$G(x_i)$	Grade efficiency of ith fraction	dimensionless	dimensionless
$G(\phi_t)$	Rathole function, Fig. 21-64	dimensionless	dimensionless
H	Silo cylinder height	m	ft
H	Solids drop height	m	ft

Symbol	Definition	SI units	U.S. customary units
H	Absolute humidity	kg water/kg air	lb water/lb air
$H(\theta')$	Hopper function, Fig. 21-51	dimensionless	dimensionless
h	Height between hopper apex and cylinder-hopper transition	m	ft
h	Liquid layer thickness	µm	in
h_a	Asperity height	µm	in
I_0	Illuminating intensity	W/m^2	W/ft^2
$I(\theta)$	Intensity of scattered light	W/m^2	W/ft^2
K	Permeability	m/s	ft/s
K, a, b, c, d	Stegmaier constants (Table 21-8)	dimensionless	dimensionless
K_0	Permeability at reference density	m/s	ft/s
K	Wave number	dimensionless	dimensionless
k	Janssen coefficient	dimensionless	dimensionless
k	Stress concentration factor	dimensionless	dimensionless
k_B	Boltzmann constant	J/K	J/K
L	Length	m	ft
L	Roll crusher roll length	cm	in
l	Crack length	µm	in
l_p	Material layer thickness on screen	m	ft
M	Sample mass	kg	lb
$M_{k,r}$	kth moment of $q_r(x)$ distribution	µm	in
M_c, M_s	Mass flow rate of coarse & feed	kg/h	lb/h
m	Geometric coefficient (0 or 1)	dimensionless	dimensionless
$\dot{m}_s$	Solids discharge rate	kg/s	lb/s
N	Total number of particles in system	number	number
N_c	Critical rotation speed	rpm	rpm
n	As defined in Eq. (21-38)	dimensionless	dimensionless
n_a	Amount of absorbed gas, BET	mol/g	mol/lb
P	Pressure	Pa	lb/in^2
P	Settled weight	g	lb
P	Probability	dimensionless	dimensionless
p	Average value of particle property	—	—
p	Number of elementary charges	—	—
P_a, P_b, P_c	Probabilities of screen passage	dimensionless	dimensionless
P_t	Total pressure	Pa	lb/in^2
P_w^{sat}	Saturation pressure of pure water	Pa	lb/in^2
ΔP	Pressure drop	Pa	psia
Q	Roll crusher capacity	cm^3/min	
$Q_i(x)$	Cumulative distribution i = 0 number, i = 1 length i = 2 area, i = 3 volume	dimensionless	dimensionless
q	Modulus of the scattering vector	1/m	1/ft

Nomenclature (*Continued*)

Symbol	Definition	SI units	U.S. customary units
$q_i(x)$	Distribution density i = number, i = 1 length i = 2 area, i = 3 volume	dimensionless	dimensionless
q, q_F, q_R	Mill flows: total, feed, recycle	kg/h	lb/h
R	Radius	m	ft
R	Recycle ratio	dimensionless	dimensionless
R_{eff}	Effective pore radius	μm	in
R_H	Hydraulic radius	m	ft
r	Radial distance from hopper vertex	m	ft
RH	Relative humidity	dimensionless	dimensionless
S	Distance along chute	m	ft
SF	Stress factor		
SI	Stress intensity		
S_u	Grinding rate function for size u	1/s	1/s
S_V	Volume specific surface	1/μm	1/in
s	Peripheral roll speed	cm/min	ft/min
s_{max}	Maximum pore saturation	dimensionless	dimensionless
$s(\theta')$	Stress function, Eq. (21-61)	dimensionless	dimensionless
T	Temperature	°C or K	°F or K
t	Time	s	sec
U_g	Superficial gas velocity	m/s	ft/s
U_p	Particle velocity	m/s	ft/s
U_t	Terminal velocity	m/s	ft/s
u	Collision viscosity	m/s	ft/s
u_s	Superficial gas velocity	m/s	ft/s
V	Hopper volume	m^3	ft^3
V	Stream velocity	m/s	ft/s
V_0	Initial solids velocity on a chute	m/s	ft/s
V_1	Chute impact velocity	m/s	ft/s
V_2	Chute velocity after impact	m/s	ft/s
V_c	Choking velocity	m/s	ft/min
V_d	Droplet volume	cm^3	in^3
v_s	Saltation velocity	m/s	ft/min
V_t	Rotating mill tip speed	m/s	ft/s
W	Weight undersize	g	lb
W_s	Conveying system mass flow rate	kg/s	lb/s
w	Screen aperture size	μm	in
w	Solid-to-liquid mass ratio	dimensionless	dimensionless
w_k	Weight fraction on kth screen	dimensionless	dimensionless
X	Solid moisture content	kg water/kg solid	lb water/lb solid
X	Particle size, Eq. (21-119)	μm	in
X_F	Feed particle size	μm	in
X_P	Product particle size	μm	in
x or x_i	Particle size	μm	in
x_{St}	Stokes diameter	μm	in
$Z(x)$	Electron mobility of particle size x	C	C
z	Depth of solids	m	ft
	Greek letters		
α	Chute angle	degrees	degrees
α	Empirical constant, Eq. (21-51)	kg/m^3	lb/ft^3
β	Compressibility	degrees	degrees
β	Coalescence kernel	number/s	number/s
Γ	Decay rate	1/s	1/s
γ	Liquid surface tension	dyn/cm	
δ	Effective angle of friction	degrees	degrees
δ	Rizk constant, Eq. (21-85)	dimensionless	dimensionless
δ	Angle of disc granulator	degrees	degrees
ε	Voidage	dimensionless	dimensionless
ε_c	Voidage at choking	dimensionless	dimensionless
ε_{eff}	Effective porosity	dimensionless	dimensionless
ε_{min}	Minimum granule porosity	dimensionless	dimensionless
ε_{tap}	Tapped effective porosity	dimensionless	dimensionless
η	Hydrodynamic viscosity of liquid	Pa·s	P
η_k	Classifier efficiency	dimensionless	dimensionless
θ	Angle of stress rotation	degrees	degrees
θ	Solids impact angle	degrees	degrees
θ	Three phase contact angle	degrees	degrees
θ'	Hopper angle from vertical	degrees	degrees
θ_{end}	Pyramidal hopper end angle	degrees	degrees
θ_{side}	Pyramidal hopper side angle	degrees	degrees
θ_v	Hopper valley angle	degrees	degrees
κ	Rizk constant, Eq. (21-85)	dimensionless	dimensionless
λ_F	Fluid friction factor	dimensionless	dimensionless
λ_Z	Solids friction factor	dimensionless	dimensionless
μ	Solids loading ratio/phase ratio	dimensionless	dimensionless
μ	Liquid viscosity	Pa · s	
μs	Solids loading ratio at saltation	dimensionless	dimensionless
ρ	Slurry density	kg/m^3	lb/ft^3
ρ_b	Bulk density	kg/m^3	lb/ft^3
ρ_{ba}	Bulk density at zero pressure gradient, see Eq. (21-70)	kg/m^3	lb/ft^3
ρ_{bO}	Bulk density at outlet	kg/m^3	lb/ft^3
ρ_{b0}	Reference bulk density	kg/m^3	lb/ft^3
ρ_{bmin}	Minimum bulk density	kg/m^3	lb/ft^3
ρ_{bmax}	Maximum bulk density	kg/m^3	lb/ft^3
ρ_f	Fluid density	kg/m^3	lb/ft^3
ρ_g	Gas density	kg/m^3	lb/ft^3
ρ_m	Media density	kg/m^3	lb/ft^3
ρ_p	Particle density	kg/m^3	lb/ft^3
ρ_s	Particle density	kg/m^3	lb/ft^3
σ	Standard deviation of sample		
σ	Normal stress	Pa	lb$_f$/ft^2
σ	Impact pressure	Pa	lb$_f$/ft^2
σ_{avg}	Average stress	Pa	lb$_f$/ft^2
σ_h	Horizontal stress	Pa	lb$_f$/ft^2
σ_i	Local stress at tip of a crack	Pa	lb$_f$/ft^2
σ_N	Gross applied stress	Pa	lb$_f$/ft^2
σ_{ss}	Steady-state normal stress	Pa	lb$_f$/ft^2
σ_v	Vertical stress	Pa	lb$_f$/ft^2
σ_{vht}	Mean vertical stress	Pa	lb$_f$/ft^2
σ_w	Wall stress	Pa	lb$_f$/ft^2
σ_x, σ_y	Normal stresses	Pa	lb$_f$/ft^2
σ_Y	Granule yield strength	Pa	lb$_f$/ft^2
σ_0	Reference stress	Pa	lb$_f$/ft^2
σ_1	Major consolidation stress	Pa	lb$_f$/ft^2
σ_2	Minor consolidation stress	Pa	lb$_f$/ft^2
$\bar{\sigma}$	Abutment stress	Pa	lb$_f$/ft^2
τ	Shear stress	Pa	lb$_f$/ft^2
τ_{ss}	Steady-state shear stress	Pa	lb$_f$/ft^2
τ_{wall}	Wall stress	Pa	lb$_f$/ft^2
τ_{xy}, τ_{yx}	Shear stresses	Pa	lb$_f$/ft^2
ϕ	Kinematic angle of internal friction	degrees	degrees
ϕ	Sphericity	dimensionless	dimensionless
ϕ'	Wall friction angle	degrees	degrees
ω	Mill rotational speed	1/s	1/s
Ca	Capillary number	dimensionless	dimensionless
Fr$_p$	Particle Froude number	dimensionless	dimensionless
Fr$_s$	Solids Froude number	dimensionless	dimensionless
Kn	Knudsen number	dimensionless	dimensionless
L	Loschmidt number	1/mol	1/mol
Re$_p$	Particle Reynolds number	dimensionless	dimensionless
St$_{def}$	Deformation Stokes number	dimensionless	
St$_v$	Viscous Stokes number	dimensionless	dimensionless
St$_v^*$	Critical viscous Stokes number	dimensionless	dimensionless

PARTICLE CHARACTERIZATION

General References: Terence Allen, *Particle Size Measurement*, vol. 1: *Powder Sampling and Particle Size Measurement* and vol. 2: *Surface Area and Pore Size Determination*, 5th ed., Springer Netherlands, 1997. P. M. Gy, *Sampling of Particulate Materials: Theory and Practice*, Elsevier, 1979. Bart and Sun, "Particle Size Analysis Review," *Anal. Chem.* **57:** 151R (1985). Miller and Lines, "Critical Reviews in Analytical Chemistry," **20**(2): 75–116 (1988). Herdan, *Small Particles Statistics*, Butterworths, London, 1960. Orr and DalleValle, *Fine Particle Measurement*, 2d ed., Macmillan, New York, 1960. Kaye, *Direct Characterization of Fine Particles*, Wiley, New York, 1981. Van de Hulst, *Light Scattering by Small Particles*, Wiley, New York, 1957. K. Leschonski, "Representation and Evaluation of Particle Size Analysis Data," *Part. Part. Syst. Charact.* **1:** 89–95 (1984). Karl Sommer, *Sampling of Powders and Bulk Materials*, Springer, 1986. H. Merkus, *Particle Size Measurements: Fundamentals, Practice, Quality*, Springer Netherlands, 2009. M. Alderliesten, "Mean Particle Diameters, Part I: Evaluation of Definition Systems," *Part. Part. Syst. Charact.* **7:** 233–241 (1990); "Part II: Standardization of Nomenclature," *Part. Part. Syst. Charact.* **8:** 237–241 (1991); "Part III: An Empirical Evaluation of Integration and Summation Methods for Estimating Mean Particle Diameters from Histogram Data," *Part. Part. Syst. Charact.* **19:** 373–386 (2002); "Part IV: Empirical Selection of the Proper Type of Mean Particle Diameter Describing a Product or Material Property," *Part. Part. Syst. Charact.* **21:** 179–196 (2004); "Part V: Theoretical Derivation of the Proper Type of Mean Particle Diameter Describing a Product or Process Property," *Part. Part. Syst. Charact.* **22:** 233–245 (2005). ISO 9276, Representation of Results of Particle Size Analysis. H. C. van de Hulst, *Light Scattering by Small Particles*, Structure of Matter Series, Dover, 1981. Craig F. Bohren and Donald R. Huffman, *Absorption and Scattering of Light by Small Particles*, Wiley-Interscience, 2007. Bruce J. Berne and Robert Pecora, *Dynamic Light Scattering: With Applications to Chemistry, Biology, and Physics*, unabridged edition, Dover, 2000. J. R. Allegra and S. A. Hawley, "Attenuation of Sound in Suspensions and Emulsions: Theory and Experiment," *J. Acoust. Soc. America* **51:** 1545–1564 (1972).

PARTICLE SIZE

Specification for Particulates The behavior of dispersed matter is generally described by a large number of parameters, e.g., the powder's bulk density, flowability, and degree of aggregation or agglomeration. Each parameter might be important for a specific application. In solids processes such as comminution, classification, agglomeration, mixing, crystallization, or polymerization, or in related material handling steps, particle size plays an important role. Often it is the dominant quality factor for the suitability of a specific product in the desired application.

Particle Size As particles are extended three-dimensional objects, only a perfect spherical particle allows for a simple definition of the particle size x—as the diameter of the sphere. In practice, spherical particles are very rare. So usually *equivalent diameters* are used, representing the diameter of a sphere that behaves as the real (nonspherical) particle in a specific sizing experiment. Unfortunately, the measured size now depends on both the method used for assessing particle size and the subsequent data analysis. It is unreasonable to expect identical results for the particle size even for instruments using the same measurement method.

In most applications more than one particle is observed. As each individual may have its own particle size, methods for data reduction have been introduced. These include the particle size distribution, a variety of model distributions, and moments (or averages) of the distribution. Also note that these methods can be extended to other particle attributes. Examples include pore size, porosity, surface area, color, and electrostatic charge distributions, to name but a few.

Particle Size Distribution A *particle size distribution* (PSD) can be displayed as a table or a diagram. In the simplest case, one can divide the range of measured particle sizes into size intervals and sort the particles into the corresponding size class, as displayed in Table 21-1 (shown for the case of volume fractions).

Typically the fractions $\Delta Q_{r,i}$ in the different size classes i are summed and normalized to 100 percent, resulting in the *cumulative distribution* $Q(x)$, also known as the percentage undersize. For a given particle size x, the Q value represents the percentage of the particles finer than x.

If the quantity measure is "number," $Q_0(x)$ is called a *cumulative number distribution*. If it is length, area, volume, or mass, then the corresponding length $[Q_1(x)]$, area $[Q_2(x)]$, volume, or mass distributions are formed $[Q_3(x)]$; mass and volume are related by the specific density ρ. The index r in this notation represents the quantity measure (ISO 9276-1:1998, *Representation of Results—Part 1 Graphical Representation*). The choice of the quantity measured is of decisive importance for the appearance of the PSD, which changes significantly when the dimension r is changed. As, e.g., one 100-μm particle has the same volume as 1000 10-μm particles or 1 million 1-μm particles, a number distribution is always dominated by and biased to the fine fractions of the sample while a volume distribution is dominated by and biased to the coarse fractions.

The normalization of the fraction $\Delta Q_{r,i}$ to the size of the corresponding interval leads to the *distribution density* $\bar{q}_{r,i}$, or

$$\bar{q}_{r,i} = \frac{\Delta Q_{r,i}}{\Delta x_i} \quad \text{and} \quad \sum_{i=1}^{n} \Delta Q_{r,i} = \sum_{i=1}^{n} \bar{q}_{r,i} \Delta x_i = 1 = 100\% \tag{21-1}$$

TABLE 21-1 Tabular Presentation of Particle-Size Data

1	2	3	4	5	6	7
i	x_i, μm	$\Delta Q_{3,i}$	Δx_i, μm	$\bar{q}_{3,i} = \Delta Q_{3,i}/\Delta x_i$ 1/μm	$Q_{3,i}$	$\bar{q}^*_{3,i}$
0	0.063				0.0000	
1	0.090	0.0010	0.027	0.0370	0.0010	0.0028
2	0.125	0.0009	0.035	0.0257	0.0019	0.0027
3	0.180	0.0016	0.055	0.0291	0.0035	0.0044
4	0.250	0.0025	0.070	0.0357	0.0060	0.0076
5	0.355	0.0050	0.105	0.0476	0.0110	0.0143
6	0.500	0.0110	0.145	0.0759	0.0220	0.0321
7	0.710	0.0180	0.210	0.0857	0.0400	0.0513
8	1.000	0.0370	0.290	0.1276	0.0770	0.1080
9	1.400	0.0610	0.400	0.1525	0.1380	0.1813
10	2.000	0.1020	0.600	0.1700	0.2400	0.2860
11	2.800	0.1600	0.800	0.2000	0.4000	0.4755
12	4.000	0.2100	1.200	0.1750	0.6100	0.5888
13	5.600	0.2400	1.600	0.1500	0.8500	0.7133
14	8.000	0.1250	2.400	0.0521	0.9750	0.3505
15	11.20	0.0240	3.200	0.0075	0.9990	0.0713
16	16.000	0.0010	4.800	0.0002	1.0000	0.0028

If $Q_r(x)$ is differentiable, the distribution density function $q_r(x)$ can be calculated as the first derivative of $Q_r(x)$, or

$$q_r(x) = \frac{dQ_r(x)}{dx} \quad \text{or} \quad Q_r(x_i) = \int_{x_{\min}}^{x_i} q_r(x)\,dx \tag{21-2}$$

It is helpful in the graphical representation to identify the distribution type, as shown for the cumulative volume distribution $Q_3(x)$ and volume distribution density $q_3(x)$ in Fig. 21-1. If $q_r(x)$ displays one maximum only, the distribution is called a *monomodal size distribution*. If the sample is composed of two or more different-size regimes, then $q_r(x)$ shows two or more maxima and is called a *bimodal* or *multimodal size distribution*.

PSDs are often plotted on a logarithmic abscissa (Fig. 21-2). While the $Q_r(x)$ values remain the same, care has to be taken for the transformation of the distribution density $q_r(x)$, as the corresponding areas under the distribution density curve must remain constant (in particular the total area remains 1, or 100 percent) independent of the transformation of the abscissa. So the transformation has to be performed by

$$\bar{q}^*_r(\ln x_{i-1}, \ln x_i) = \frac{\Delta Q_{r,i}}{\ln(x_i/x_{i-1})} \tag{21-3}$$

This equation also holds if the natural logarithm is replaced by the logarithm to base 10.

Example 21-1 From Table 21-1 one can calculate, e.g.,

$$\bar{q}_{3,11} = \frac{\Delta Q_{3,11}}{\Delta x_{11}} = \frac{0.16}{0.8\ \mu\text{m}} = 0.2\ \mu\text{m}^{-1}$$

$$\bar{q}^*_{3,11} = \bar{q}^*_3(\ln x_{10}, \ln x_{11}) = \frac{\Delta Q_{3,11}}{\ln(x_{11}/x_{10})} = \frac{0.16}{\ln(2.8\ \mu\text{m}/2.0\ \mu\text{m})}$$

$$= \frac{0.16}{\ln 1.4} = 0.4755$$

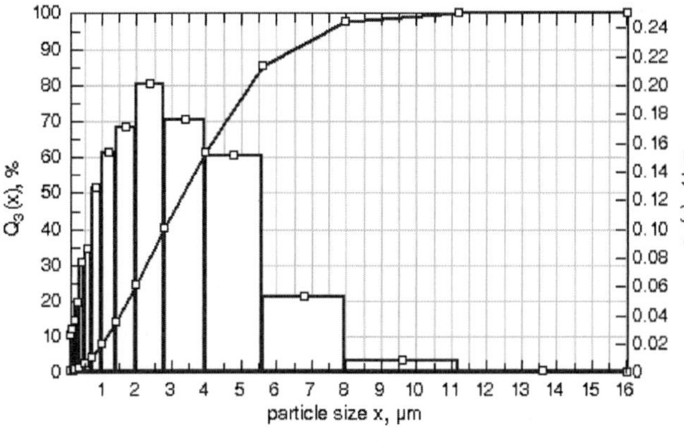

FIG. 21-1 Histogram $\bar{q}_3(x)$ and $Q_3(x)$ plotted with linear abscissa.

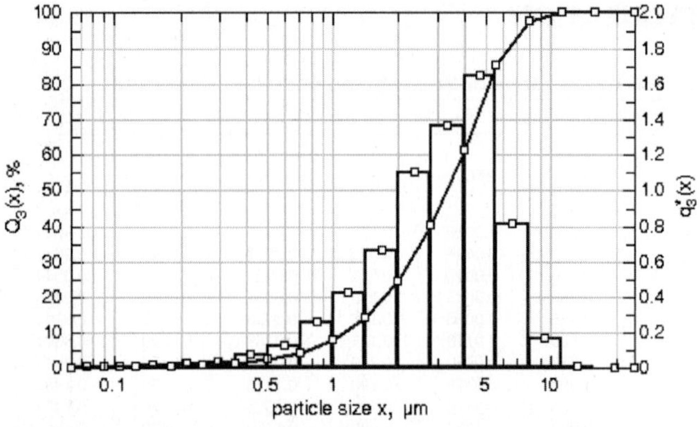

FIG. 21-2 Histogram $\bar{q}_3^*(x)$ and $Q_3(x)$ plotted with a logarithmic abscissa.

Model Distribution While a PSD with n intervals is represented by $2n + 1$ numbers, further data reduction can be performed by fitting the size distribution to a specific mathematical model. The logarithmic normal distribution or the logarithmic normal probability function is one common model distribution used for the distribution density, and it is given by

$$q_r^*(z) = \frac{1}{\sqrt{2\pi}} e^{-0.5 z^2} \quad \text{with} \quad z = \frac{1}{s} \ln\left(\frac{x}{x_{50,r}}\right) \tag{21-4}$$

The PSD can then be expressed by two parameters, namely, the mean size $x_{50,r}$ and, e.g., by the dimensionless standard deviation s (ISO 9276-5:2005, *Methods of Calculations Relating to Particle Size Analysis Using Logarithmic Normal Probability Distribution*). The data reduction can be performed by plotting $Q_r(x)$ on logarithmic probability graph paper or using the fitting methods described in ISO 9276-3:2008, *Adjustment of an Experimental Curve to a Reference Model.* This method is mainly used for the analysis of powders obtained by grinding and crushing and has the advantage that the transformation between PSDs of different dimensions is simple. The transformation is also log-normal with the same slope s.

Other model distributions used are the *normal distribution* (Laplace-Gauss), for powders obtained by precipitation, condensation, or natural products (e.g., pollens); the *Gates-Gaudin-Schuhmann distribution* (bilogarithmic), for analysis of the extreme values of fine particle distributions (Schuhmann, *Am. Inst. Min. Metall. Pet. Eng.*, Tech. Paper 1189 Min. Tech., 1940); or the *Rosin-Rammler-Sperling-Bennet distribution* for the analysis of the extreme values of coarse particle distributions, e.g., in monitoring grinding operations [Rosin and Rammler, *J. Inst. Fuel* **7:** 29–36 (1933); Bennett, ibid., **10:** 22–29 (1936)].

Moments Moments represent a PSD by a single value. With the help of moments, the average particle sizes, volume specific surfaces, and other mean values of the PSD can be calculated. The general definition of a moment is given by (ISO 9276-2:2014, *Calculation of Average Particle Sizes/Diameters and Moments from Particle Size Distributions*)

$$M_{k,r} = \int_{x_{\min}}^{x_{\max}} x^k q_r(x)\,dx \tag{21-5}$$

where $M_{k,r}$ is the kth moment of a $q_r(x)$ distribution density and k is the power of x.

Average Particle Sizes A PSD has many average particle sizes. The general equation is given by

$$\bar{x}_{k,r} = \sqrt[k]{M_{k,r}} \tag{21-6}$$

Two typically employed average particle sizes are the *arithmetic average particle size* $\bar{x}_{k,0} = M_{k,0}$ [e.g., for a number distribution ($r = 0$) obtained by counting methods], and the **weighted average particle size** $\bar{x}_{1,r} = M_{1,r}$ [e.g., for a volume distribution ($r = 3$) obtained by sieve analysis], where $\bar{x}_{1,r}$ represents the center of gravity on the abscissa of the $q_r(x)$ distribution.

Specific Surface The *specific surface area* can be calculated from size distribution data. For spherical particles this can simply be calculated by using moments. The volume specific surface is given by

$$S_V = \frac{6}{\bar{x}_{1,2}} \quad \text{or} \quad S_V = \frac{6}{M_{1,2}} = \frac{M_{2,0}}{M_{3,0}} = 6M_{1,3} \tag{21-7}$$

where $\bar{x}_{1,2}$ is the weighted average diameter of the area distribution, also known as *Sauter mean diameter.* It represents a particle having the same ratio of surface area to volume as the distribution, and it is also referred to as a *surface-volume average diameter.* The Sauter mean is an important average diameter used in solids handling and other processing applications where aspects of two-phase flow become important, as it appropriately weights the contributions of the fine fractions to surface area. For nonspherical particles, a shape factor has to be considered.

Example 21-2 The Sauter mean diameter and the volume-weighted particle size and distribution given in Table 21-1 can be calculated by using ISO 9276-2:2014, *Representation of Results of Particle Size Analysis—Part 2: Calculation of Average Particle Sizes/Diameters and Moments from Particle Size Distributions* via Table 21-2.

The Sauter mean diameter is

$$\bar{x}_{1,2} = M_{1,2} = \frac{M_{3,0}}{M_{2,0}} = \frac{1}{M_{-1,3}} \quad \text{with} \quad M_{1,3} = \sum_{i=1}^{n} \Delta Q_{3,i} \frac{\ln(x_i/x_{i-1})}{x_i - x_{i-1}}$$

which yields

$$\bar{x}_{1,2} = \frac{1}{0.473736} = 2.110882$$

The volume-weighted average particle size is

$$\bar{x}_{1,3} = M_{1,3} = \frac{1}{2}\sum_{i=1}^{n} \Delta Q_{3,i}(x_i + x_{i-1})$$

which yields

$$\bar{x}_{1,3} = \frac{1}{2}(7.280590) = 3.640295$$

PARTICLE SHAPE

For many applications not only the particle size but also the shape are of importance; e.g., toner powders should be spherical while polishing powders should have sharp edges. Traditionally in microscopic methods of size analysis, direct measurements are made on enlarged images of the particles by using a calibrated scale. While such measurements are always encouraged to gather a direct sense of the particle shape and size, care should be taken in terms of drawing general conclusions from limited particle images. Furthermore, with the strong progress in computing power, instruments have become available that acquire the projected area of many particles in short times, with a significant reduction in data manipulation times. Standardization of shape parameters is given in ISO 9276-6:2008 *Descriptive and Qualitative Representation of Particle Shape and Morphology.*

Equivalent Projection Area of a Circle Equivalent projection area of a circle is widely used for the evaluation of particle sizes from the projection area A of a nonspherical particle.

$$x_{EQPC} = 2\sqrt{A/\pi} \tag{21-8}$$

TABLE 21-2 Table for Calculation of Sauter Mean Diameter and Volume Weighted Particle Size

I	x_i, µm	$\Delta Q_{3,i}$	$\ln(x_i/x_{i-1})$	$\ln(x_i/x_{i-1})/(x_i - x_{i-1})$	$\Delta Q^*_{3,i} \ln(x_i/x_{i-1})/(x_i - x_{i-1})$	$\Delta Q^*_{3,i}(x_i + x_{i-1})$, µm
0	0.0630					
1	0.0900	0.0010	0.3567	13.2102	0.013210	0.000153
2	0.1250	0.0009	0.3285	9.3858	0.008447	0.000194
3	0.1800	0.0016	0.3646	6.6299	0.010608	0.000488
4	0.2500	0.0025	0.3285	4.6929	0.011732	0.001075
5	0.3550	0.0050	0.3507	3.3396	0.016698	0.003025
6	0.5000	0.0110	0.3425	2.3620	0.025982	0.009405
7	0.7100	0.0180	0.3507	1.6698	0.030056	0.021780
8	1.0000	0.0370	0.3425	1.1810	0.043697	0.063270
9	1.4000	0.0610	0.3365	0.8412	0.051312	0.146400
10	2.0000	0.1020	0.3567	0.5945	0.060635	0.346800
11	2.8000	0.1600	0.3365	0.4206	0.067294	0.768000
12	4.0000	0.2100	0.3567	0.2972	0.062418	1.428000
13	5.6000	0.2400	0.3365	0.2103	0.050471	2.304000
14	8.0000	0.1250	0.3567	0.1486	0.018577	1.700000
15	11.2000	0.0240	0.3365	0.1051	0.002524	0.460800
16	16.0000	0.0010	0.3567	0.0743	0.000074	0.027200
					Σ0.473736	7.280590

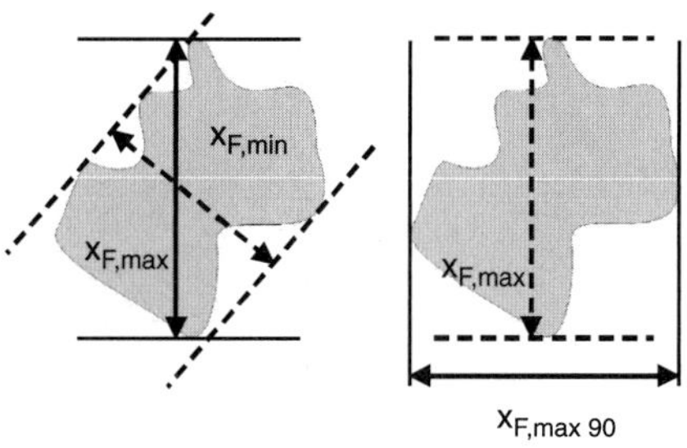

FIG. 21-3 Definition of Feret diameters.

Feret's Diameter Feret's diameter is determined from the projected area of the particles by using a slide gauge (Fig. 21-3). In general, it is defined as the distance between two parallel tangents of the particle at an arbitrary angle. In practice, the *minimum* $x_{F,\min}$ and *maximum Feret* diameters $x_{F,\max}$, the *mean Feret diameter* $\bar{x}_F$, and the Feret diameters obtained at 90° to the direction of the minimum and maximum Feret diameters $x_{F,\max 90}$ are used. The minimum Feret diameter is often used as the diameter equivalent to a sieve analysis.

Other diameters used in the literature include *Martin's diameter* or the edges of an *enclosing rectangle.* Martin's diameter is a line, parallel to a fixed direction, which divides the particle profile into two equal areas.

These diameters offer an extension over volume equivalent diameters to account for shape deviations from spherical. As with any other quality measure of size, many particles must be measured to determine distributions of these particle size diameters. With the advent of high-speed image processing, particle size and shape can be determined quickly. For shape characterization, these devices are able to generate galleries of particle shapes which can be very helpful in solving process and product problems. Particles can be sorted, for example, by fractal dimension, fiber length, or sphericity. The engineering challenge is to connect these shapes to product characteristics or plant processing issues.

Sphericity, Aspect Ratio, and Convexity Parameters describing the shape of the particles include the following:

The *sphericity* ψ_S $(0 < \psi_S \leq 1)$ is defined by the ratio of the perimeter of a circle with diameter x_{EQPC} to the perimeter of the corresponding projection area A. And $\psi_S = 1$ represents a sphere.

The *aspect ratio* ψ_A $(0 < \psi_A \leq 1)$ is defined by the ratio of the minimum to the maximum Feret diameter $\psi_A = x_{\text{Feret min}}/x_{\text{Feret max}}$. It gives an indication of the *elongation* of the particle. Some literature also used $1/\psi_A$ as the definition of sphericity.

The *convexity* ψ_C $(0 < \psi_C \leq 1)$ is defined by the ratio of the projection area A to the convex hull area $A + B$ of the particle, as displayed in Fig. 21-4.

In *Fourier techniques* the shape characteristic is transformed to a signature waveform. Beddow and coworkers (Beddow, *Particulate Science and Technology*, Chemical Publishing, New York, 1980) take the particle centroid as a reference point. A vector is then rotated about this centroid with the tip of the vector touching the periphery. A plot of the magnitude of the vector versus its angular position is a wave-type function. This waveform is then subjected to Fourier analysis. The lower-frequency harmonics constituting the complex wave correspond to the gross external morphology, whereas the higher frequencies correspond to the texture of the fine particle.

Fractal Dimension This was introduced into fine particle science by Kaye and coworkers (Kaye, *Direct Characterization of Fine Particles*, Wiley, New York, 1981), who show that the noneuclidean logic of Mandelbrot can be applied to describe the ruggedness of a particle profile. A combination of fractal dimension and geometric shape factors such as the aspect ratio can be used to describe a population of fine particles of various shapes, and these can be related to the functional properties of the particle.

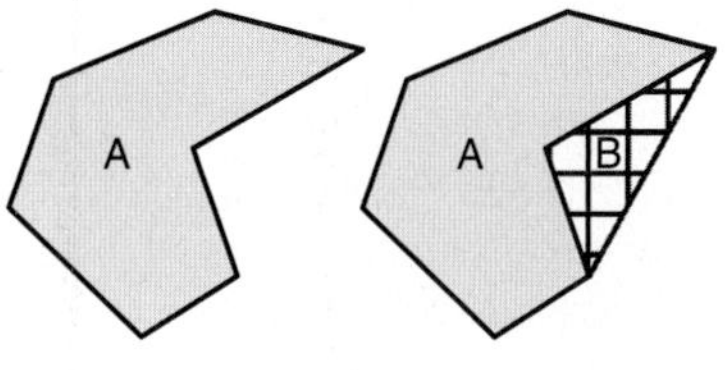

FIG. 21-4 Definition of the convex hull area $A + B$ for the projection area A of a particle.

SAMPLING AND SAMPLE SPLITTING

As most of the sizing methods are limited to small sample sizes, an important prerequisite to accurate particle size analysis is proper powder sampling and sample splitting (ISO 14488:2007, *Particulate Materials—Sampling and Sample Splitting for the Determination of Particulate Properties*).

When one is determining particle size (or any other particle attribute such as chemical composition or surface area), it is important to recognize that the error associated in making such a measurement can be described by its variance, or

$$\sigma^2_{\text{observed}} = \sigma^2_{\text{actual}} + \sigma^2_{\text{measurement}} \tag{21-9}$$

$$\sigma^2_{\text{measurement}} = \sigma^2_{\text{sampling}} + \sigma^2_{\text{analysis}} \tag{21-10}$$

That is, the *observed* variance in the particle size measurement is due to both the actual physical variance in size and the variance in the measurement. More importantly, the variance in measurement has two contributing factors: variance due to sampling, which would include systematic errors in the taking, splitting, and preparation of the sample; and variance due to the actual sample analysis, which would include not only the physical measurement at hand, but also how the sample is *presented* to the measuring zone, which can be greatly affected by instrument design and sample dispersion (discussed later). Successful characterization of the sample (in this discussion, taken to be measurement of particle size) requires that the errors in measurement be much less than actual physical variations in the sample itself, especially if knowledge of sample deviations is important. In this regard, great negligence is unfortunately often exhibited in sampling efforts. Furthermore, measured deviations in particle size or other properties are often incorrectly attributed to and reflect upon the measuring device, where in fact they are caused by inattention to proper sampling and sample splitting. Worse still, such deviations caused by poor sampling may be taken as true sample deviations, causing undue and frequent process corrections.

Powders may be classified as nonsegregating (cohesive) or segregating (free-flowing). Representative samples can be more easily taken from *cohesive powders,* provided that they have been properly mixed. For wet samples a sticky paste should be created and mixed from which the partial sample is taken.

In the case of *segregating powders,* four key rules should be followed, although some apply or can be equally employed for cohesive materials as well. These rules are especially important for in-line and on-line sampling, discussed below. Allen (Allen, *Particle Size Measurement, Volume 1: Powder Sampling and Particle Size Measurement* and *Particle Size Measurement, Volume 2: Surface Area and Pore Size Determination*, 5th ed., Springer, Netherlands, 1997) suggests the following:

1. The particles should be sampled while in motion. Transfer points are often convenient and relevant for this. Sampling a stagnant bed of segregating material by, e.g., thieves disrupts the state of the mixture and may be biased to coarse or fines.

2. The whole stream of powder should be taken in many short time intervals in preference to part of the stream being taken over the whole time, i.e., a complete slice of the particle stream. Furthermore, any mechanical collection point should not be allowed to overfill, since this will make the sample bias toward fines, and coarse material rolls off formed heaps.

3. The entire sample should be analyzed, splitting down to a smaller sample if necessary. In many cases, segregation of the sample will not affect the measurement, provided the entire sample is analyzed. There are, however, exceptions in that certain techniques may only analyze one surface of the final sample. In the case of chemical analysis, an example would be near infrared spectroscopy operated in reflectance mode as opposed to transmission. Such a technique may still be prone to segregation during the final analysis.

4. A minimum sample size exists for a given size distribution, generally determined by the sample containing a minimum number of coarse particles representative of the customer application. While many applications involving fine pharmaceuticals may only require milligrams to establish a representative sample, other cases such as detergents and coffee might require kilograms. Details are given in the standard ISO14488:2007, *Particulate Materials—Sampling and Sample Splitting for the Determination of Particulate Properties.*

In this regard, one should keep in mind that the sample size may also reflect variation in the degree of mixing in the bed, as opposed to true size differences. (See also the subsection Solids Mixing: Measuring the Degree of Mixing.) In fact, larger samples in this case help minimize the impact of segregation on measurements.

The estimated maximum sampling error on a 60:40 blend of free-flowing sand using different sampling techniques is given in Table 21-3.

TABLE 21-3 Reliability of Selected Sampling Method

Method	Estimated maximum sampling error, %
Cone and quartering	22.7
Scoop sampling	17.1
Table sampling	7.0
Chute splitting	3.4
Spinning riffling	0.42

The *spinning riffler* (Fig. 21-5) generates the most representative samples. In this device a ring of containers rotates under the powder feed. If the powder flows a long time with respect to the period of rotation, each container will be made up of many small fractions from all parts of the bulk. Many different configurations are commercially available. Devices with small numbers of containers (say, 8) can be cascaded *n* times to get higher splitting ratios $1{:}8^n$. This usually creates smaller sampling errors than does using splitters with more containers. A splitter simply divides the sample into two halves, generally pouring the sample into a set of intermeshed chutes. Figure 21-6 illustrates commercial rifflers and splitters.

For reference materials sampling errors of less than 0.1 percent are achievable (S. Röthele and W. Witt, *Standards in Laser Diffraction*, PARTEC, 5th European Symposium Particle Characterization, Nürnberg, 1992, pp. 625–642).

DISPERSION

Many sizing methods are sensitive to the agglomeration state of the sample. In some cases, this includes primary particles, possibly with some percentage of such particles held together as weak agglomerates by interparticle cohesive forces. In other cases, strong aggregates of the primary particles may also exist. Generally, the size of either the primary particles or the aggregates is the matter of greatest interest. In some cases, however, it may also be desirable to determine the level of agglomerates in a sample, requiring that the intensity of dispersion be controlled and variable. Often the agglomerates have to be dispersed smoothly without comminution of aggregates or primary particles. This can be done either in gas (dry) or in liquid (wet) by using a suitable dispersion device which is stand-alone or integrated in the particle-sizing instrument. If possible, dry particles should be measured in gas and wet particles in suspension.

Wet Dispersion Wet dispersion separates agglomerates down to the primary particles by a suitable liquid. Dispersing agents and optional cavitation forces induced by ultrasound are often used. Care must be taken that the particles not be soluble in the liquid, or that they not flocculate. Microscopy and zeta potential measurements may be of utility in specifying the proper dispersing agents and conditions for dispersion.

Dry Dispersion Dry dispersion uses mechanical forces for the dispersion. While a simple fall-shaft with impact plates may be sufficient for the dispersion of coarse particles, say, >300 μm, much higher forces have to be applied to fine particles.

In Fig. 21-7 the agglomerates are sucked in by the vacuum generated through expansion of compressed gas applied at an injector. They arrive at low speed in the dispersing line, where they are strongly accelerated. This creates three effects for the dispersion, as displayed in Fig. 21-8.

With suitable parameter settings, agglomerates can be smoothly dispersed down to 0.1 μm [K. Leschonski, S. Röthele, and U. Menzel, Entwicklung und Einsatz einer trockenen Dosier-Dispergiereinheit zur Messung von Partikelgrößenverteilungen in Gas-Feststoff-Freistrahlen aus Laser-Beugungsspektren; *Part. Charact.* **1:** 161–166 (1984)] without comminution of the primary particles.

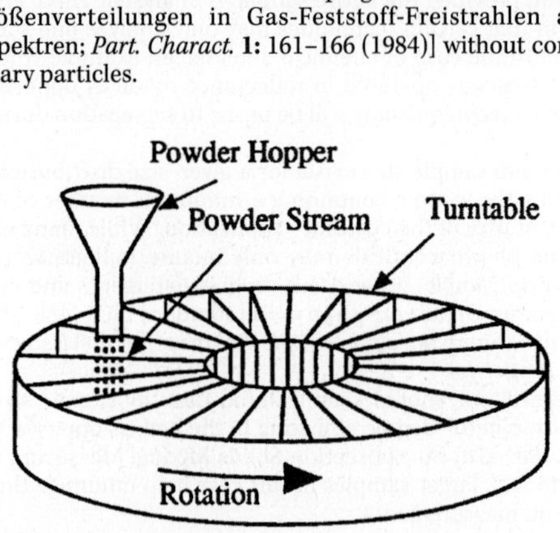

FIG. 21-5 Spinning riffler sampling device.

FIG. 21-6 Examples of commercial splitting devices. Spinning riffler and standard splitters. (*Courtesy of Retsch Corporation.*)

PARTICLE SIZE MEASUREMENT

There are many techniques available to measure the particle size distribution of powders or droplets. The wide size range, from nanometers to millimeters, of particulate products, however, cannot be analyzed by using only a single measurement principle. Added to this are the usual constraints of capital costs versus running costs, speed of operation, degree of skill required, and, most important, the end-use requirement.

If the particle size distribution of a powder composed of hard, smooth spheres is measured by any of the techniques, the measured values should be identical. However, many different size distributions can be defined for any powder made up of nonspherical particles. For example, if a rod-shaped particle is placed on a sieve, then its diameter, not its length, determines the size of aperture through which it will pass. If, however, the particle is allowed to settle in a viscous fluid, then the calculated diameter of a sphere of the same substance that would have the same falling speed in the same fluid (i.e., the Stokes diameter) is taken as the appropriate size parameter of the particle. Since the Stokes diameter for the rod-shaped particle will obviously differ from the rod diameter, this difference represents added information concerning particle shape. The ratio of the diameters measured by two different techniques is called the *shape factor.*

Historically methods primarily using mechanical, aerodynamic, or hydrodynamic properties for discrimination and particle sizing have been used, but today methods based on the interaction of the particles with electromagnetic waves (mainly light), ultrasound, or electric fields dominate.

Laser Diffraction Methods Over the past 30 years *laser diffraction* has developed into a leading principle for particle size analysis of all kinds of aerosols, suspensions, emulsions, and sprays in laboratory and process environments.

The scattering of unpolarized laser light by a single spherical particle can be mathematically described by

$$I(\theta) = \frac{I_0}{2k^2a^2}\{[S_1(\theta)]^2 + [S_2(\theta)]^2\} \tag{21-11}$$

where $I(\theta)$ is the total scattered intensity as function of angle θ with respect to the forward direction; I_0 is the illuminating intensity; k is the wave number $2\pi/\lambda$; a is the distance from the scatterer to the detector; and $S_1(\theta)$ and $S_2(\theta)$ are dimensionless, complex functions describing the change and amplitude in the perpendicular and parallel polarized light. Different algorithms have been developed to calculate $I(\theta)$. The *Lorenz-Mie theory* is based

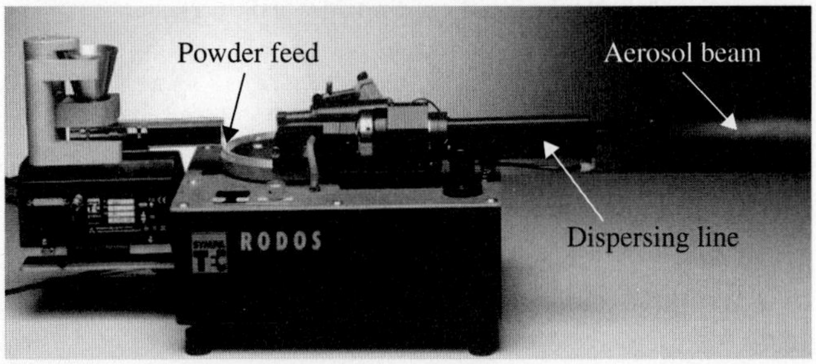

FIG. 21-7 Dry disperser RODOS with vibratory feeder VIBRI creating a fully dispersed aerosol beam from dry powder. (*Courtesy of Sympatec GmbH.*)

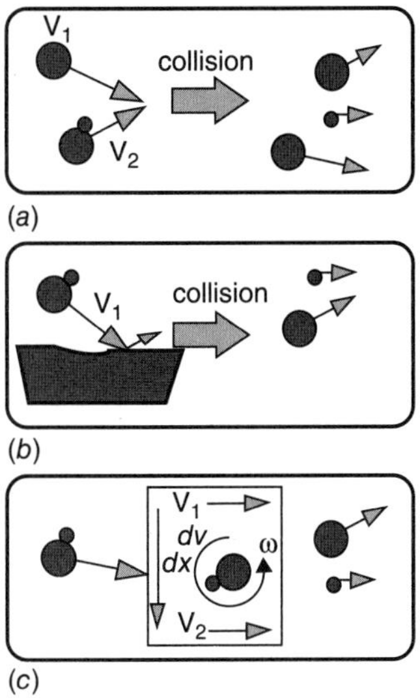

FIG. 21-8 Interactions combined for dry dispersion of agglomerates. (*a*) Particle-to-particle collisions. (*b*) Particle-to-wall collisions. (*c*) Centrifugal forces due to strong velocity gradients.

on the assumption of spherical, isotropic, and homogenous particles and that all particles can be described by a common complex refractive index $m = n - i\kappa$. Index m has to be precisely known for the evaluation, which is difficult in practice, especially for the imaginary part κ, and inapplicable for mixtures with components having different refractive indices.

The *Fraunhofer theory* considers only scattering at the contour of the particle and the near forward direction. No preknowledge of the refractive index is required, and $I(\theta)$ simplifies to

$$I(\theta)=\frac{I_0}{2k^2a^2}\alpha^4\left[\frac{J_1(\alpha\sin\theta)}{\alpha\sin\theta}\right]^2 \tag{21-12}$$

with J_1 as the Bessel function of the first kind and the dimensionless size parameter $\alpha = \pi x/\lambda$. This theory does not predict polarization or account for light transmission through the particle.

For a single spherical particle, the diffraction pattern shows a typical ring structure. The distance r_0 of the first minimum to the center depends on the particle size, as shown in Fig. 21-9*a*. In the particle sizing instrument, the acquisition of the intensity distribution of the diffracted light is usually performed with the help of a multielement photodetector.

Diffraction patterns of static nonspherical particles are displayed in Fig. 21-10. As all diffraction patterns are symmetric to 180°, semicircular detector elements integrate over 180° and make the detected intensity independent of the orientation of the particle.

Simultaneous diffraction on more than one particle results in a superposition of the diffraction patterns of the individual particles, provided that particles are moving and diffraction between the particles is averaged out. This simplifies the evaluation, providing a parameter-free and model-independent mathematical algorithm for the inversion process (M. Heuer and K. Leschonski, "Results Obtained with a New Instrument for the Measurement of Particle Size Distributions from Diffraction Patterns," *Part. Part. Syst. Charact.* **2:** 7–13, 1985).

Today the method is standardized (ISO 13320-1:2009, *Particle Size Analysis—Laser Diffraction Methods—Part 1: General Principles*), and many companies offer instruments, usually with the choice of Fraunhofer and/or Mie theory for the evaluation of the PSD. The size ranges of the instruments have been expanded by combining low-angle laser light scattering with 90° or back scattering, the use of different wavelengths, polarization ratio, and white light scattering, etc. It is now ranging from below 0.1 μm to about 1 cm. Laser diffraction is currently the fastest method for particle sizing at highest reproducibility. In combination with dry dispersion it can handle large amounts of sample, which makes this method well suited for process applications.

Instruments of this type are available, e.g., from Malvern Ltd. (Mastersizer), Sympatec GmbH (HELOS, MYTOS), Horiba (LA, LS series), Beckmann Coulter (LS 13320), or Micromeritics (Saturn).

Image Analysis Methods The extreme progress in image capturing and exceptional increase of the computational power within the last few years have revolutionized microscopic methods and made image analysis methods very popular for the characterization of particles, especially since,

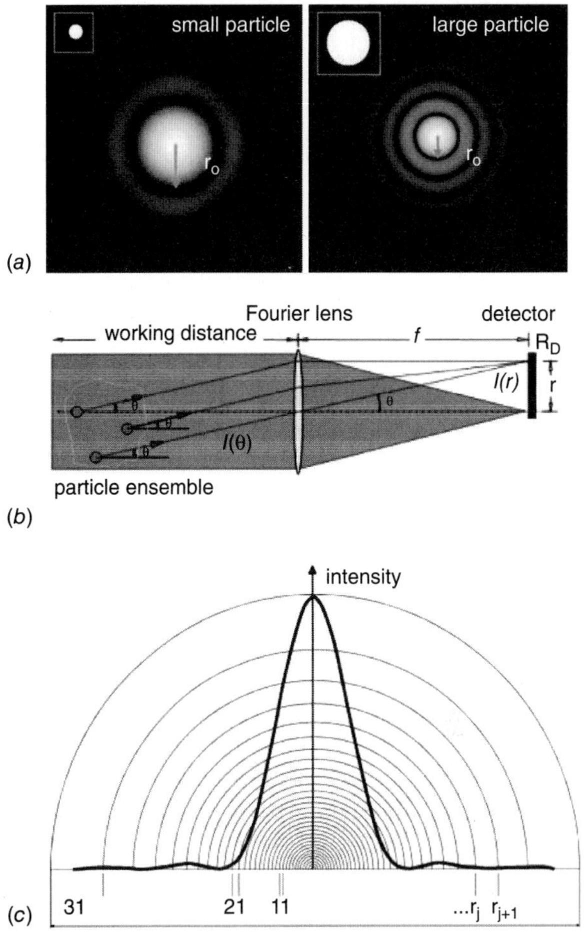

FIG. 21-9 (*a*) Diffraction patterns of laser light in forward direction for two different particle sizes. (*b*) The angular distribution $I(\theta)$ is converted by a Fourier lens to a spatial distribution $I(r)$ at the location of the photodetector. (*c*) Intensity distribution of a small particle detected by a semicircular photodetector.

in addition to size, relevant shape information becomes available by the method. Currently, mainly instruments creating a 2D image of the 3D particles are used. Two methods have to be distinguished.

Static image analysis is characterized by nonmoving particles, e.g., on a microscope slide (Fig. 21-11). The depth of sharpness is well defined, resulting in a high resolution for small particles. The method is well established and standardized (ISO 13322-1:2014, *Particle Size Analysis—Image Analysis Methods, Part 1: Static Image Analysis Methods*), but can handle only small amounts of data. The particles are oriented by the base; overlapping particles have to be separated by time-consuming software algorithms, and the tiny sample size creates a massive sampling problem, resulting in very low statistical relevance of the data. Commercial systems reduce these effects by using large or even stepping microscopic slides and the deposition of the particles via a dispersing chamber. As all microscopic techniques can be used, the size range is only defined by the microscope used.

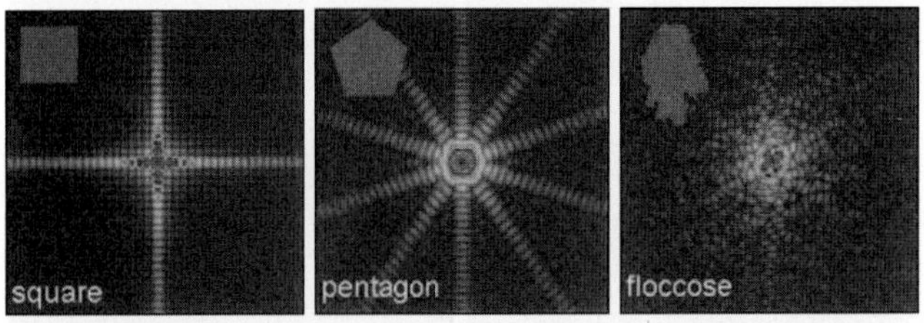

FIG. 21-10 Calculated diffraction patterns of laser light in forward direction for nonspherical particles: square, pentagon, and floccose. All diffraction patterns show a symmetry to 180°.

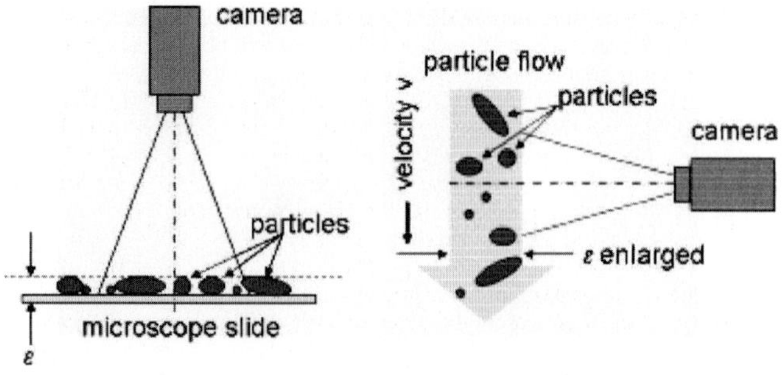

FIG. 21-11 Setup of static (*left*) and dynamic (*right*) image analysis for particle characterization.

Dynamic image analysis images a flow of moving particles. This allows for a larger sample size. The particles show arbitrary orientation, and the number of overlapping particles is reduced. Several companies offer systems which operate in either reflection or transmission, with wet dispersion or free fall, with matrix or line-scan cameras. The free-fall systems are limited to well-flowing bulk materials. Systems with wet dispersion only allow for smallest samples sizes and slow particles. As visible light is used for imaging, the size range is limited to about 1 μm at the fine end. This type of instruments has been standardized (ISO 13322-2:2006, *Particle Size Analysis—Image Analysis Methods, Part 2: Dynamic Methods*).

Common to all available instruments are small particle numbers, which result in poor statistics. Thus recent developments have yielded a combination of powerful dry and wet dispersion with high-speed image capturing. Particle numbers up to 10^7 can now be acquired in a few minutes. Size and shape analysis is available at low statistical errors [W. Witt, U. Köhler, and J. List, "Direct Imaging of Very Fast Particles Opens the Application of the Powerful (Dry) Dispersion for Size and Shape Characterization," PARTEC 2004, Nürnberg].

Dynamic Light Scattering Methods *Dynamic light scattering* (DLS) is now used on a routine basis for the analysis of particle sizes in the submicrometer range. It provides an estimation of the average size and its distribution within a measuring time of a few minutes.

Submicrometer particles suspended in a liquid are in constant *brownian motion* as a result of the impacts from the molecules of the suspending fluid, as suggested by W. Ramsay in 1876 and confirmed by A. Einstein and M. Smoluchowski in 1905/06.

In the Stokes-Einstein theory of brownian motion, the particle motion at very low concentrations depends on the viscosity of the suspending liquid, the temperature, and the size of the particle. If viscosity and temperature are known, the particle size can be evaluated from a measurement of the particle motion. At low concentrations, this is the *hydrodynamic diameter.*

DLS probes this motion optically. The particles are illuminated by a coherent light source, typically a laser, creating a diffraction pattern, showing in Fig. 21-12 as a fine structure from the diffraction between the particles, i.e., its near-order. As the particles are moving from impacts of the thermal movement of the molecules of the medium, the particle positions change with the time *t*.

The change of the position of the particles affects the phases and thus the fine structure of the diffraction pattern. So the intensity in a certain point of the diffraction pattern fluctuates with time. The fluctuations can be analyzed in the time domain by a correlation function analysis or in the frequency domain by frequency analysis. Both methods are linked by Fourier transformation.

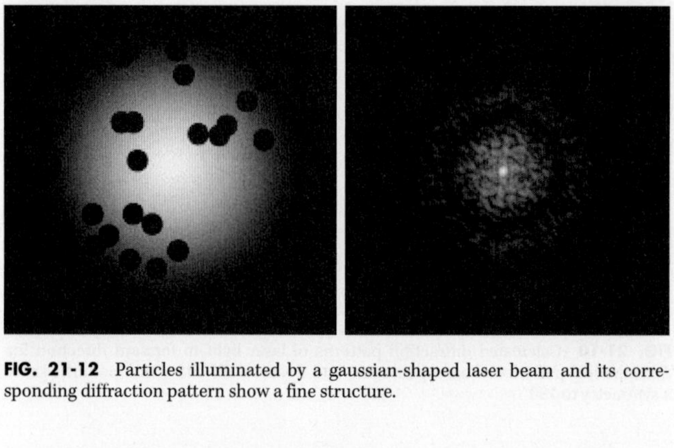

FIG. 21-12 Particles illuminated by a gaussian-shaped laser beam and its corresponding diffraction pattern show a fine structure.

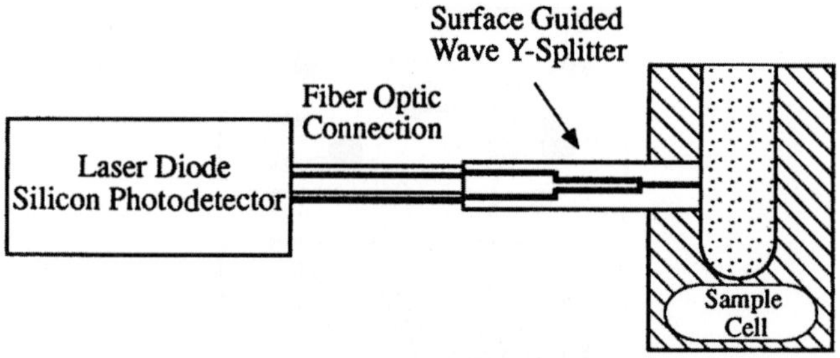

FIG. 21-13 Diagram of Leeds and Northrup Ultrafine Particle Size Analyzer (UPA), using fiber optics in a backscatter setup.

The measured decay rates Γ are related to the translational diffusion coefficients *D* of spherical particles by

$$\Gamma = Dq^2 \quad \text{with} \quad q = \frac{4\pi}{\lambda_0}\sin\frac{\theta}{2} \quad \text{and} \quad D = \frac{k_B T}{2\pi\eta x} \tag{21-13}$$

where q is the modulus of the scattering vector, k_B is the Boltzmann constant, T is the absolute temperature, and η is the hydrodynamic viscosity of the dispersing liquid. The particle size x is then calculated by the Stokes-Einstein equation from D at fixed temperature T and η known.

DLS covers a broad range of diluted and concentrated suspension. As the theory is only valid for light being scattered once, any contribution of multiple scattered light leads to erroneous PCS results and misinterpretations. So different measures have been taken to minimize the influence of multiple scattering.

The well-established *photon correlation spectroscopy* (PCS) uses highly diluted suspensions to avoid multiple scattering. The low concentration of particles makes this method sensitive to impurities in the liquid. So usually very pure liquids and a clean-room environment have to be used for the preparation and operation (ISO 13321:1996, *Particle Size Analysis—Photon Correlation Spectroscopy*).

Another technique (Fig. 21-13) utilizes an optical system which minimizes the optical path into and out of the sample, including the use of backscatter optics, a moving-cell assembly, or setups with the maximum incident beam intensity located at the interface of the suspension to the optical window (Trainer, Freud, and Weiss, Pittsburgh Conference, Analytical and Applied Spectroscopy, *Symp. Particle Size Analysis,* March 1990; ISO 22412:2008, *Particle Size Analysis—Dynamic Light Scattering*).

Photon cross-correlation spectroscopy (PCCS) uses a novel three-dimensional cross-correlation technique which completely suppresses the multiple scattered fractions in a special scattering geometry. In this setup two lasers *A* and *B* are focused to the same sample volume, creating two sets of scattering patterns, as shown in Fig. 21-14. Two intensities are measured at different positions but with identical scattering vectors.

$$\vec{\vec{q}} = \vec{k}_A - \vec{k}_1 = \vec{k}_B - \vec{k}_2 \tag{21-14}$$

Subsequent cross-correlation of these two signals eliminates any contribution of multiple scattering. So highly concentrated, opaque suspensions can be measured as long as scattered light is observed. High count rates result in short measuring times. High particle concentrations reduce the sensitivity of this method to impurities, so standard liquids and laboratory environments can be used, which simplifies the application [W. Witt, L. Aberle, and H. Geers, "Measurement of Particle Size and Stability of Nanoparticles in Opaque Suspensions and Emulsions with Photon Cross Correlation Spectroscopy," *Particulate Systems Analysis,* Harrogate (UK), 2003].

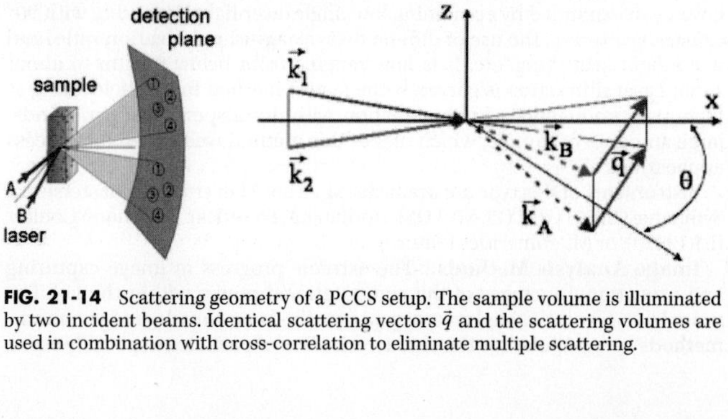

FIG. 21-14 Scattering geometry of a PCCS setup. The sample volume is illuminated by two incident beams. Identical scattering vectors $\vec{q}$ and the scattering volumes are used in combination with cross-correlation to eliminate multiple scattering.

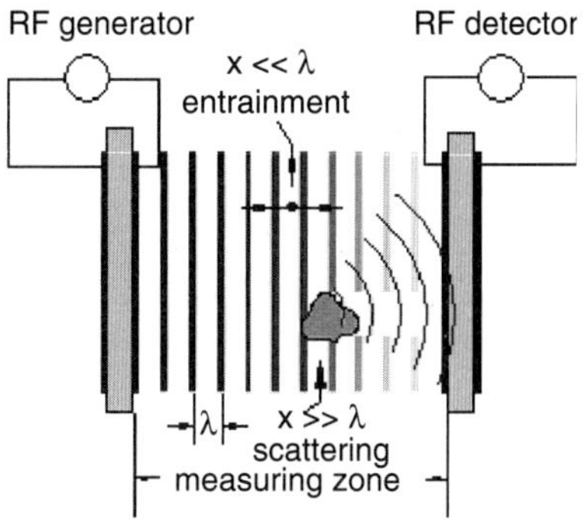

FIG. 21-15 Setup of an ultrasonic attenuation system for particle size analysis.

Acoustic Methods *Ultrasonic attenuation spectroscopy* is a method well suited to measuring the PSD of colloids, dispersions, slurries, and emulsions (Fig. 21-15). The basic concept is to measure the frequency-dependent attenuation or velocity of the ultrasound as it passes through the sample. The attenuation includes contributions from the scattering or absorption of the particles in the measuring zone and depends on the size distribution and the concentration of the dispersed material (ISO 20998-1:2006, *Particle Characterization by Acoustic Methods, Part 1: Ultrasonic Attenuation Spectroscopy*).

In a typical setup (see Fig. 21-15) an electric high-frequency generator is connected to a piezoelectric ultrasonic transducer. The generated ultrasonic waves are coupled into the suspension and interact with the suspended particles. After passing the measuring zone, the ultrasonic plane waves are received by an ultrasonic detector and converted to an electric signal, which is amplified and measured. The attenuation of the ultrasonic waves is calculated from the ratio of the signal amplitudes on the generator and detector sides.

PSD and concentration can be calculated from the attenuation spectrum by using either complicated theoretical calculations requiring a large number of parameters or an empirical approach employing a reference method for calibration. Following U. Riebel (Die Grundlagen der Partikelgrößenanalyse mittels Ultraschallspektrometrie, PhD thesis, University of Karlsruhe), the ultrasonic extinction of a suspension of monodisperse particles with diameter x can be described by Lambert-Beer's law. The extinction $-\ln(I/I_0)$ at a given frequency f is linearly dependent on the thickness of the suspension layer Δl, the projection area concentration C_{PF}, and the related extinction cross section K. In a polydisperse system the extinctions of single particles overlay:

$$-\ln\left(\frac{I}{I_0}\right)_{fi} \cong \Delta l \cdot C_{PF} \cdot \sum_j K(f_i, x_j) \cdot q_2(x_j)\Delta x \qquad (21\text{-}15)$$

When the extinction is measured at different frequencies f_i, this equation becomes a linear equation system, which can be solved for C_{PF} and $q_2(x)$. The key for the calculation of the particle size distribution is the knowledge of the related extinction cross section K as a function of the dimensionless size parameter $\sigma = 2\pi x/\lambda$. For spherical particles K can be evaluated directly from the acoustic scattering theory. A more general approach is an empirical method using measurements on reference instruments as input.

This disadvantage is compensated by the ability to measure a wide size range from below 10 μm to above 3 mm and the fact that PSDs can be measured at very high concentrations (0.5 to >50 percent of volume) without dilution. This eliminates the risk of affecting the dispersion state and makes this method ideal for in-line monitoring of, e.g., crystallizers (A. Pankewitz and H. Geers, LABO, "In-line Crystal Size Distribution Analysis in Industrial Crystallization Processes by Ultrasonic Extinction," May 2000).

Current instruments use different techniques for the attenuation measurement: with *static* or *variable* width of the measuring zone, measurement in *transmission* or *reflection*, with *continuous* or *swept* frequency generation, with *frequency burst* or *single-pulse* excitation.

For process environment, probes are commercially available with a frequency range of 100 kHz to 200 MHz and a dynamic range of >150 dB, covering 1 to 70 percent of volume concentration, 0 to 120°C, 0 to 40 bar, pH 1 to 14, and hazardous areas as an option.

Vendors of this technology include Sympatec GmbH (OPUS), Malvern Instruments Ltd. (Ultrasizer), Dispersion Technology Inc. (DT series), and Colloidal Dynamics Pty Ltd. (AcoustoSizer).

Single-Particle Light Interaction Methods Individual particles have been measured with light for many years. The measurement of the particle size is established by (1) the determination of the *scattered light* of the particle, (2) the measurement of the amount of *light extinction* caused by the particle presence, (3) the measurement of the *residence time* during motion through a defined distance, or (4) *particle velocity*.

Many commercial instruments are available, which vary in optical design, light source type, and means, and how the particles are presented to the light.

Instruments using *light scattering* cover a size range of particles of 50 nm to about 10 μm (liquid-borne) or 20 μm (gas-borne), while instruments using *light extinction* mainly address liquid-borne particles from 1 μm to the millimeter size range. The size range capability of any single instrument is typically 50:1. International standards are as follows: ISO 13323-1:2000, *Determination of Particle Size Distribution—Single-Particle Light Interaction Methods, Part 1: Light Interaction Considerations*; ISO 21501-2:2007, *Determination of Particle Size Distribution—Single Particle Light-Interaction Methods, Part 2: Light-Scattering Liquid-Borne Particle Counter*; ISO 21501-3:2007, *Part 3: Light-Extinction Liquid-Borne Particle Counter*; ISO 21501-4:2007, *Part 4: Light-Scattering Airborne Particle Counter for Clean Spaces*.

Instruments using the *residence time*, such as the aerodynamic particle sizers, or the *particle velocity*, as used by the phase Doppler particle analyzers, measure the particle size primarily based on the aerodynamic diameter.

Small-Angle X-Ray Scattering Method Small-angle X-ray scattering can be used in a size range of about 1 to 300 nm. Its advantage is that the scattering mainly results from the differences in the electron density between the particles and their surroundings. As internal crystallites of external agglomerates are not visible, the measured size always represents the size of the primary particles and the requirement for dispersion is strongly reduced [Z. Jinyuan, L. Chulan, and C. Yan, "Stability of the Dividing Distribution Function Method for Particle Size Distribution Analysis in Small Angle X-Ray Scattering," *J. Iron & Steel Res. Inst.* **3**(1): 1996; ISO 13762:2001, *Particle Size Analysis—Small Angle X-ray Scattering Method*].

Focused-Beam Techniques These techniques are based on a focused light beam, typically a laser, with the focal point spinning on a circle parallel to the surface of a glass window. When the focal point passes a particle, the reflected and/or scattered light of the particle is detected. The focal point moves along the particle on circular segments, as displayed in Fig. 21-16. Sophisticated threshold algorithms are used to determine the start point and endpoint of the chord, i.e., the edges of the particle. The chord length is calculated from the time interval and the track speed of the focal point. Focused-beam techniques measure a chord length distribution, which corresponds to the size and shape information of the particles typically in a complicated way (J. Worlische, T. Hocker, and M. Mazzoti, "Restoration of PSD from Chord Length Distribution Data Using the Method of Projections onto Convex Sets," *Part. Part. Syst. Char.* **22:** 81 ff.). So often the chord length distribution is directly used as the fingerprint information of the size, shape, and population status.

Instruments of this type are commercially available as robust finger probes with small probe diameters. They are used in on-line and preferably in in-line applications, monitoring the chord length distribution of suspensions and emulsions. Special flow conditions are used to reduce the sampling errors. Versions with fixed focal distance [Focused Beam Reflectance Measurement (FBRM®)] and variable focal distance (3D ORM technology) are available. The latter improves this technique for high concentrations and widens the dynamic range, as the focal point moves horizontally and vertically with respect to the surface of the window. For instruments refer, e.g., to Mettler-Toledo International Inc. (Lasentec FBRM probes) and Messtechnik Schwartz GmbH (PAT).

Electrical Sensing Zone Methods In the electric sensing zone method (Fig. 21-17), a well-diluted and well-dispersed suspension in an electrolyte is caused to flow through a small aperture [Kubitschek, *Research* **13:** 129 (1960)]. The changes in the resistivity between two electrodes on

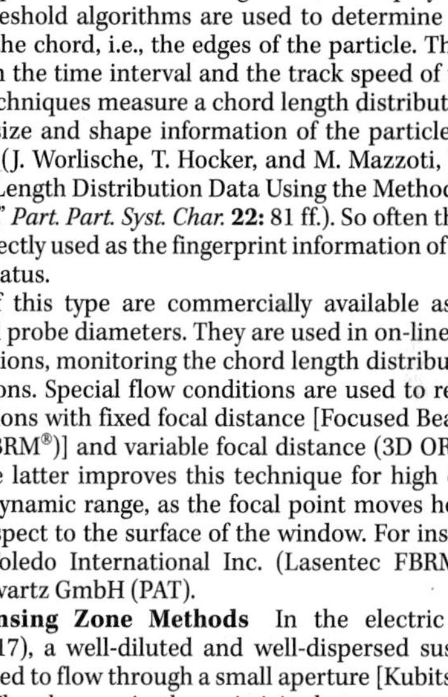

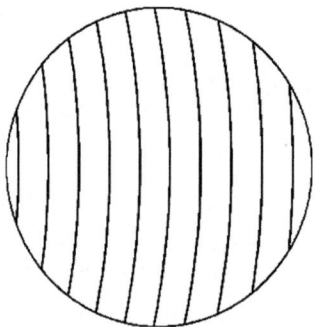

FIG. 21-16 Different chords measured on a constantly moving single spherical particle by focused-beam techniques.

FIG. 21-17 Multisizer™ 3 COULTER COUNTER® from Beckman Coulter, Inc., uses the electrical sensing zone method.

either side of the aperture, as the particles pass through, are related to the volumes of the particles. The pulses are fed to a pulse-height analyzer where they are counted and scaled. The method is limited by the resolution of the pulse-height analyzer of about 16,000:1 (corresponding to a volume diameter range of about 25:1) and the need to suspend the particles in an electrolyte (ISO 13319:2007, *Determination of Particle Size Distributions—Electrical Sensing Zone Method*).

Gravitational Sedimentation Methods In gravitational sedimentation methods, the particle size is determined from the settling velocity and the undersize fraction by changes of concentration in a settling suspension. The equation relating particle size to settling velocity is known as *Stokes' law* (ISO 13317-1:2001, *Part 1: General Principles and Guidelines*):

$$x_{St} = \sqrt{\frac{18\eta u}{(\rho_s - \rho_f)g}} \tag{21-16}$$

where x_{St} is the Stokes diameter, η is viscosity, u is the particle settling velocity under gravity, ρ_s is the particle density, ρ_f is the liquid density, and g is the gravitational acceleration.

The *Stokes diameter* is defined as the diameter of a sphere having the same density and the same velocity as the particle settling in a liquid of the same density and viscosity under laminar flow conditions. Corrections for the deviation from Stokes' law may be necessary at the coarse end of the size range. Sedimentation methods are limited to sizes above 1 μm due to the onset of thermal diffusion (brownian motion) at smaller sizes.

An experimental problem is to obtain adequate dispersion of the particles prior to a sedimentation analysis. For powders that are difficult to disperse, the addition of dispersing agents is necessary, along with ultrasonic probing. It is essential to examine a sample of the dispersion under a microscope to ensure that the sample is fully dispersed. (See the subsection Wet Dispersion.)

Equations to calculate size distributions from sedimentation data are based on the assumption that the particles sink freely in the suspension. To ensure that particle-particle interaction can be neglected, a volume concentration below 0.2 percent is recommended.

There are various procedures available to determine the changing solid concentration of a sedimenting suspension:

In the *pipette method,* concentration changes are monitored by extracting samples from a sedimenting suspension at known depths and predetermined times. The method is best known as *Andreasen modification* [Andreasen, *Kolloid-Z.* **39:** 253 (1929)], shown in Fig. 21-18. Two 10-mL samples are withdrawn from a fully dispersed, agitated suspension at zero time to corroborate the 100 percent concentration given by the known weight of powder and volume of liquid making up the suspension. The suspension is then allowed to settle in a temperature-controlled environment, and 10-mL samples are taken at time intervals in geometric 2:1 time progression starting at 1 min (that is, 1, 2, 4, 8, 16, 32, 64 min). The amount of powder in the extracted samples is determined by drying, cooling in a desiccator, and weighing. Stokes diameters are determined from the predetermined times and the depth, with corrections for the changes in depth due to the extractions. The cumulative mass undersize distribution comprises a plot of the normalized concentration versus the Stokes diameter. A reproducibility of

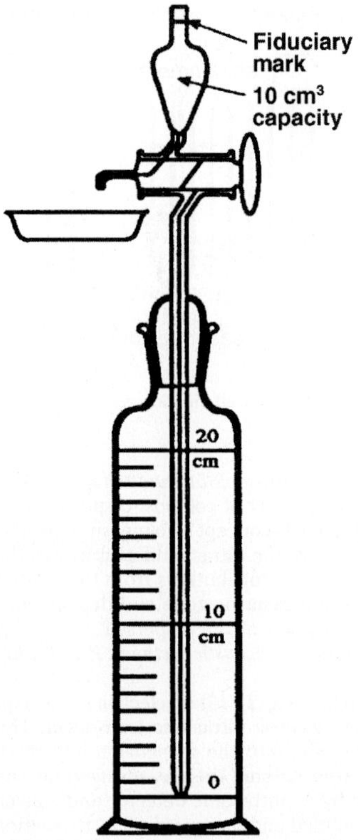

FIG. 21-18 Equipment used in the pipette method of size analysis.

±2 percent is possible by using this apparatus. The technique is versatile in that it is possible to analyze most powders dispersible in liquids; its disadvantages are that it is a labor-intensive procedure, and a high level of skill is needed (ISO 13317-2:2001, *Part 2: Fixed Pipette Method*).

The *hydrometer method* is simpler in that the density of the suspension, which is related to the concentration, is read directly from the stem of the hydrometer while the depth is determined by the distance of the hydrometer bulb from the surface (ASTM Spec. Pub. 234, 1959). The method has a low resolution but is widely used in soil science studies.

In *gravitational photo sedimentation methods,* the change of the concentration with time and depth of sedimentation is monitored by using a light point or line beam. These methods give a continuous record of changing optical density with time and depth and have the added advantage that the beam can be scanned to the surface to reduce the measurement time. A correction needs to be applied to compensate for a deviation from the laws of geometric optics (owing to diffraction effects the particles cut off more light than geometric optics predicts). The normalized measurement is a $Q_2(x)$ distribution (ISO 13317-4:2014, *Part 4: Photo Gravitational Method*).

In *gravitational X-ray sedimentation methods,* the change of the concentration with time and depth of sedimentation is monitored by using an X-ray beam. These methods give a continuous record of changing X-ray density with time and depth and have the added advantage that the beam can be scanned to the surface to reduce the measurement time. The methods are limited to materials having a high atomic mass (i.e., X-ray-opaque material) and give a $Q_3(x)$ distribution directly (ISO 13317-3:2001, *Part 3: X-ray Gravitational Technique*). See Fig. 21-19.

Sedimentation Balance Methods In sedimentation balances the weight of sediment is measured as it accumulates on a balance pan suspended in an initial homogeneous suspension. The technique is slow due to the time required for the smallest particle to settle out over a given height. The relationship between settled weight P, weight undersize W, and time t is given by

$$P = W - \frac{dP}{d\ln t} \tag{21-17}$$

Centrifugal Sedimentation Methods These methods extend sedimentation methods well into the submicrometer range. Alterations of the particle concentration may be determined space- and time-resolved during centrifugation (T. Detloff and D. Lerche, "Determination of Particle

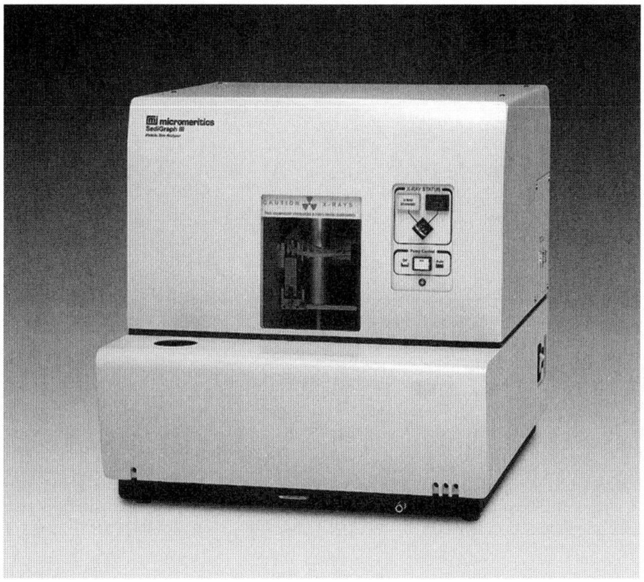

FIG. 21-19 The Sedigraph III 5120 Particle Size Analysis System determines particle size from velocity measurements by applying Stokes' law under the known conditions of liquid density and viscosity and particle density. Settling velocity is determined at each relative mass measurement from knowledge of the distance the X-ray beam is from the top of the sample cell and the time at which the mass measurement was taken. It uses a narrow, horizontally collimated beam of X-rays to measure directly the relative mass concentration of particles in the liquid medium.

Size Distributions Based on Space and Time Resolved Extinction Profiles in Centrifugal Field," *Proceedings of Fifth World Congress on Particle Technology, Session Particle Measurement,* Orlando, Fla., April 23–27, 2006). Sizes are calculated from a modified version of the Stokes equation:

$$x_{St} = \sqrt{\frac{18\eta u}{(\rho_s - \rho_f)\omega^2}} \tag{21-18}$$

where ω is the radial velocity of the centrifuge. The concentration calculations are complicated due to radial dilution effects (i.e., particles do not travel in parallel paths as in gravitational sedimentation but move away from each other as they settle radially outward). Particle velocities are given by

$$u = \frac{\ln(r/s)}{t} \tag{21-19}$$

where both the measurement radius r and the surface radius s can be varying. The former varies if the system is a scanning system, and the latter if the surface varies due to the extraction of samples.

Concentration undersize D_m is determined by

$$D_m = \int_{x_{\min}}^{x} \exp(-2ktz^2) q_3(x)\,dz \tag{21-20}$$

with

$$k = \frac{\rho_s - \rho_f}{18\eta}\omega^2 \tag{21-21}$$

where $q_3(x) = dQ_3(x)/dx$ is the volume or mass density distribution and z is the integration variable.

The solution of the integral for measuring the concentration at constant position over time is only approximately possible. A common way uses Kamack's equation [Kamack, *Br. J. Appl. Phys.* **5:** 1962–1968 (1972)] as recommended by ISO 13318-1:2001 (*Part 1: Determination of Particle Size by Centrifugal Liquid Sedimentation Methods*).

An analytical solution is provided by measuring the concentration to at least one time at different sedimentation heights:

$$Q_3(x) = \int_{1}^{D_m} \left(\frac{r_i}{s}\right)^2 dD_m \tag{21-22}$$

where r_i is the measurement position and s the surface radius; $Q_3(x)$ is the cumulative mass or volume concentration; and $(r_i/s_i)^2$ is the radial dilution correction factor.

The *disc centrifuge,* developed by Slater and Cohen and modified by Allen and Svarovsky [Allen and Svarovsky, *Dechema Monogram,* Nuremberg, nos. 1589–1625, pp. 279–292 (1975)], is essentially a centrifugal pipette device. Size distributions are measured from the solids concentration of a series of samples withdrawn through a central drainage pillar at various time intervals.

In the *centrifugal disc photodensitometer,* concentration changes are monitored by a light point or line beam. In one high-resolution mode of operation, the suspension under test is injected into clear liquid in the spinning disc through an entry port, and a layer of suspension is formed over the free surface of liquid (the line start technique). The analysis can be carried out using a homogeneous suspension. Very low concentrations are used, but the light-scattering properties of small particles make it difficult to interpret the measured data.

Several *centrifugal cuvette photocentrifuges* are commercially available. These instruments use the same theory as the photocentrifuges but are limited in operation to the homogeneous mode of operation (ISO 13318-1:2001, *Determination of Particle Size Distribution by Centrifugal Liquid Sedimentation Methods—Part 1: General Principles and Guidelines*; ISO 13318-2:2007 *Part 2: Photocentrifuge Method*).

The *X-ray disc centrifuge* is a centrifuge version of the gravitational instrument and extends the measuring technique well into the submicrometer range (ISO 13318-3:2004, *Part 3: Centrifugal X-ray Method*).

Sieving Methods Sieving is probably the most frequently used and abused method of analysis because the equipment, analytical procedure, and basic concepts are deceptively simple. In sieving, the particles are presented to equal-size apertures that constitute a series of go/no go gauges. Sieve analysis implies three major difficulties: (1) with woven-wire sieves, the weaving process produces three-dimensional apertures with considerable tolerances, particularly for fine-woven mesh; (2) the mesh is easily damaged in use; (3) the particles must be efficiently presented to the sieve apertures to prevent blinding.

Sieves are often referred to their mesh size, which is a number of wires per linear unit. Electroformed sieves with square or round apertures and tolerances of ±2 μm are also available (ISO 3310, *Test Sieves—Technical Requirements and Testing,* 2016: *Part 1: Test Sieves of Metal Wire Cloth*; 2013; *Part 2: Test Sieves of Perforated Metal Plate*; 1990; *Part 3: Test Sieves of Electroformed Sheets*).

For coarse separation, dry sieving is used, but other procedures are necessary for finer and more cohesive powders. The most aggressive agitation is performed with Pascal Inclyno and Tyler Ro-tap sieves, which combine gyratory and jolting movement, although a simple vibratory agitation may be suitable in many cases. With Air-Jet sieves, a rotating jet below the sieving surface cleans the apertures and helps the passage of fines through the apertures. The sonic sifter combines two actions, a vertical oscillating column of air and a repetitive mechanical pulse. Wet sieving is frequently used with cohesive powders.

Elutriation Methods and Classification In gravity elutriation the particles are classified in a column by a rising fluid flow. In centrifugal elutriation the fluid moves inward against the centrifugal force. A cyclone is a centrifugal elutriator, although it is not usually so regarded. The cyclosizer is a series of inverted cyclones with added apex chambers through which water flows. Suspension is fed into the largest cyclone, and particles are separated into different size ranges.

Differential Electrical Mobility Analysis (DMA) Differential electrical mobility analysis uses an electric field for the classification and analysis of charged aerosol particles ranging from about 1 nm to about 1 μm in a gas phase. It mainly consists of four parts: (1) A *preseparator* limits the upper size to a known cutoff size. (2) A *particle charge conditioner* charges the aerosol particles to a known electric charge (a function of particle size). A bipolar diffusion particle charger is commonly used. The gas is ionized either by radiation from a radioactive source (e.g., ^{85}Kr) or by ions emitted from a corona electrode. Gas ions of either polarity diffuse to the aerosol particles until charge equilibrium is reached. (3) A *differential electrical mobility spectrometer* (DEMS) discriminates particles with different electrical mobility by particle migration perpendicular to a laminar sheath flow. The voltage between the inner cylinder and the outer cylinder (GND) is varied to adjust the discrimination level. (4) An *aerosol particle detector* uses, e.g., a continuous-flow condensation particle counter (CPC) or an aerosol electrometer (AE).

A typical setup of the DEMS is shown in Fig. 21-20. It shows the flow rates of the sheath flow F_1, the polydisperse aerosol sample F_2, the monodisperse (classified) aerosol exiting the DEMS F_3, and the excess air F_4.

The electrical mobility Z depends on the particle size x and the number of elementary charges e:

$$Z(x) = \frac{p \cdot e}{3\pi\eta x}[1 + \mathrm{Kn}(A + Be^{C/\mathrm{Kn}})] \tag{21-23}$$

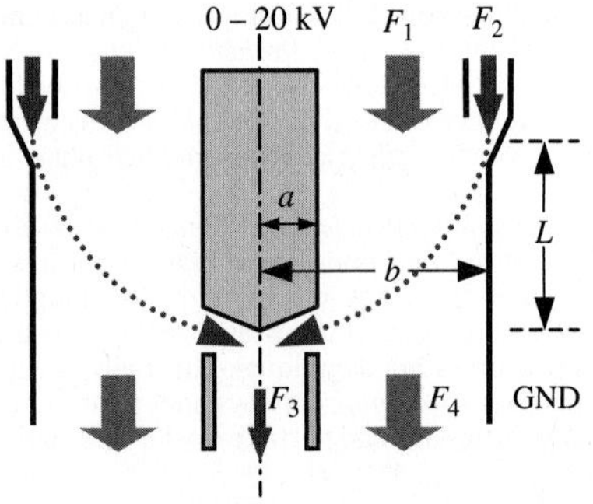

FIG. 21-20 Schematic of a differential electrical mobility analyzer.

with the number of elementary charges p, the Knudsen number Kn of $2l/x$, the mean path l of the gas molecule, the dynamic fluid viscosity η, and numeric constants A, B, C determined empirically.

Commercial instruments are available for a variety of applications in aerosol instrumentation, production of materials from aerosols, contamination control, etc. (ISO/CD 15900:2009, *Determination of Particles Size Distribution—Differential Electrical Mobility Analysis for Aerosol Particles*).

Surface Area Determination The surface-to-volume ratio is an important powder property since it governs the rate at which a powder interacts with its surroundings, e.g., in chemical reactions. The surface area may be determined from size distribution data or measured directly by flow through a powder bed or the adsorption of gas molecules on the powder surface. Other methods such as gas diffusion, dye adsorption from solution, and heats of adsorption have also been used. The most commonly used methods are as follows:

In *mercury porosimetry*, the pores are filled with mercury under pressure (ISO 15901-1:2005, *Pore Size Distribution and Porosity of Solid Materials—Evaluation by Mercury Porosimetry and Gas Adsorption—Part 1: Mercury Porosimetry*). This method is suitable for many materials with pores in the diameter range of about 3 nm to 400 μm (especially within 0.1 to 100 μm).

In *gas adsorption* for micro-, meso- and macropores, the pores are characterized by adsorbing gas, such as nitrogen at liquid-nitrogen temperature. This method is used for pores in the ranges of approximately <2 nm (*micropores*), 2 to 50 nm (*mesopores*), and >50 nm (*macropores*) (ISO 15901-2:2006, *Pore Size Distribution and Porosity of Solid Materials—Evaluation by Mercury Porosimetry and Gas Adsorption, Part 2: Analysis of Meso-pores and Macro-pores by Gas Adsorption;* ISO 15901-3:2007, *Part 3: Analysis of Micro-pores by Gas Adsorption*). An isotherm is generated of the amount of gas adsorbed versus gas pressure, and the amount of gas required to form a monolayer is determined.

Many theories of gas adsorption have been advanced. For mesopores the measurements are usually interpreted by using the BET theory [Brunauer, Emmet, and Teller, *J. Am. Chem. Soc.* **60:** 309 (1938)]. Here the amount of absorbed n_a is plotted versus the relative pressure p/p_0. The monolayer capacity n_m is calculated by the BET equation:

$$\frac{p/p_0}{n_a(1-p/p_0)} = \frac{1}{n_m C} + \frac{C-1}{n_m C}\cdot\frac{p}{p_0} \tag{21-24}$$

The specific surface per unit mass of the sample is then calculated by assessing a value a_m for the average area occupied by each molecule in the complete monolayer (say, $a_m = 0.162\ \text{nm}^2$ for N_2 at 77 K) and the Loschmidt number L:

$$a_s = n_m \cdot a_m \cdot L \tag{21-25}$$

PARTICLE SIZE ANALYSIS IN THE PROCESS ENVIRONMENT

The growing trend toward automation in industry has resulted in the development of particle sizing equipment suitable for continuous work under process conditions—even in hazardous areas (Fig. 21-21). The acquisition of particle size information in real time is a prerequisite for feedback control of the process.

Today the field of particle sizing in process environment is subdivided into three branches of applications.

At-Line At-line is the fully automated analysis in a laboratory. The sample is still taken manually or by stand-alone devices. The sample is transported to the laboratory, e.g., by pneumatic delivery. Several hundred samples can be measured per day, allowing for precise quality control of slow processes. At-line laser diffraction is widely used for quality control in the cement industry. See Fig. 21-22.

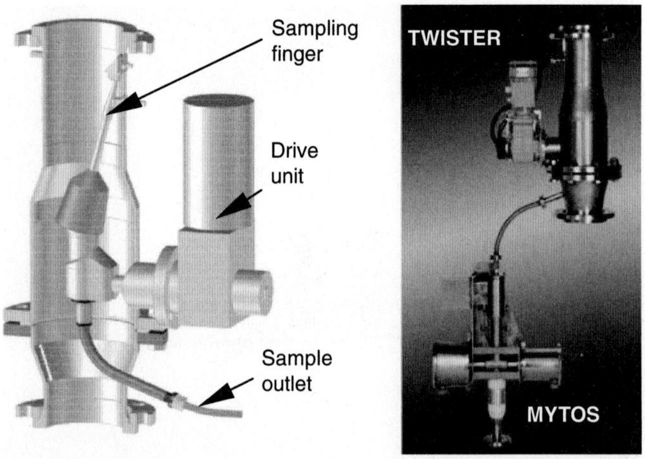

FIG. 21-21 A typical on-line application with a representative sampler (TWISTER) in a pipe of 150 mm, which scans the cross section on a spiral line, and dry disperser with particle-sizing instrument (MYTOS) based on laser diffraction. (*Courtesy of Sympatec GmbH.*)

On-Line On-line places the measuring device in the process environment close to, but not in, the production line. The fully automated system includes the sampling, but the sample is transported to the measuring device. Mainly laser diffraction, ultrasonic extinction, and dynamic light scattering are used. See Fig. 21-23.

In-Line In-line implements sampling, sample preparation, and measurement directly in the process, keeping the sample inside the production line. This is the preferred domain of laser diffraction (mainly dry), image analysis, focused-beam techniques, and ultrasonic extinction devices (wet). See Fig. 21-24.

VERIFICATION

The use of *reference materials* is recommended to verify the correct function of the particle sizing equipment. A simple electrical, mechanical, or optical test is generally not sufficient, as all functions of the measuring process, such as dosing, transportation, and dispersion, are only tested with sample material applied to the instrument.

Reference Materials Many vendors supply certified standard reference materials which address either a single instrument or a group of instruments. As these materials are expensive, it is often advisable to perform only the primary tests with these materials and perform secondary tests with a stable and well-split material supplied by the user. For best relevance, the size range and distribution type of this material should be similar to those of the desired application. It is essential that the total operational procedure be adequately described in full detail (S. Röthele and W. Witt, "Standards in Laser Diffraction," 5th European Symposium Particle Char., Nuremberg, March 24–26, 1992).

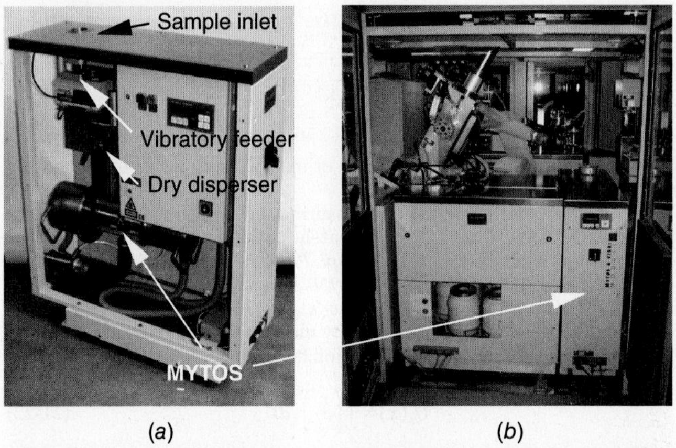

FIG. 21-22 (*a*) At-line particle sizing MYTOS module (*courtesy of Sympatec GmbH*) based on laser diffraction, with integrated dosing and dry dispersion stage. (*b*) Module integrated into a Polysius Polab® AMT for lab automation in the cement industry.

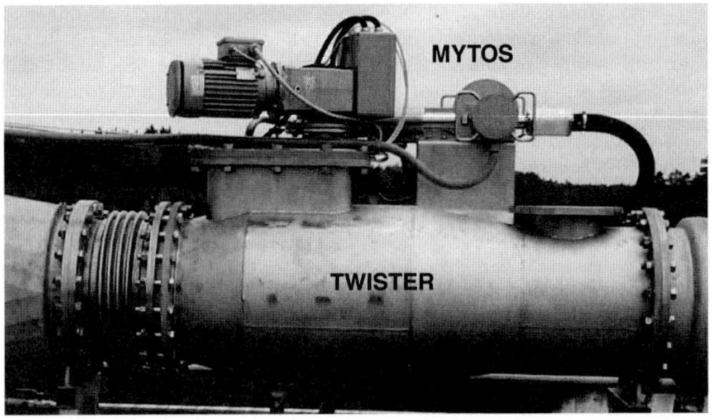

FIG. 21-23 Typical on-line outdoor application with a representative sampler TWISTER 440, which scans the cross section on a spiral line in a pipe of 440 mm, and a hookup dry disperser with laser diffraction particle sizer MYTOS. (*Courtesy of Sympatec GmbH.*)

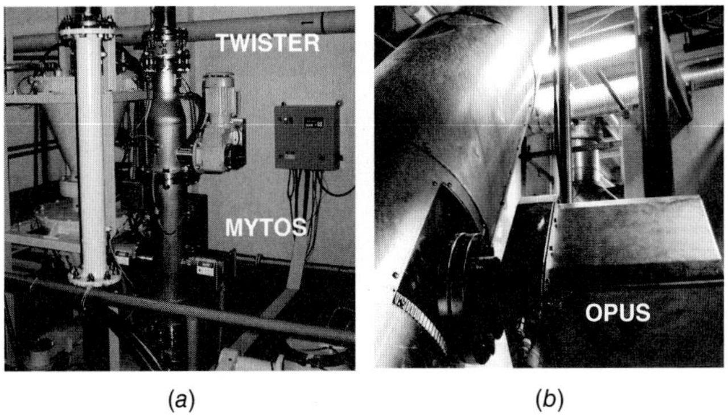

FIG. 21-24 (*a*) Typical in-line laser diffraction system with a representative sampler (TWISTER and MYTOS), all integrated in a pipe of 100 mm. (*b*) In-line application of an ultrasonic extinction (OPUS) probe monitoring a crystallization process in a large vessel. (*Both by courtesy of Sympatec GmbH.*)

BULK SOLIDS FLOW AND HOPPER DESIGN

General References: Jenike, *Storage and Flow of Solids,* Bulletin 123, University of Utah Engineering Station, 1964 (revised 1976); Schulze, *Powders and Bulk Solids—Behavior, Characterization, Storage, and Flow,* Springer, New York, 2007; Mehos, "Designing and Operating Gravity Dryers," *Chem. Eng.* **116**(5): 34 (May 2009); Maynard, "Ten Steps to an Effective Bin Design," *Chemical Engineering Progress,* pp. 25–32, (Nov. 2013); Carson, Pittenger, and Marinelli, "Characterize Bulk Solids to Ensure Smooth Flow," *Chem. Eng.* pp. 54–59, April 2016.

FUNDAMENTALS

Many industrial processes involve the transfer and feeding of bulk solids, and the ability of such materials to flow in a controlled manner during these operations is critical to product quality. Hoppers, bin, and silos are used to store granular raw materials, intermediates, and final products. With modifications, they can also be used as process vessels, such as purge columns, heaters and coolers, and moving-bed reactors. Transfer chutes are inclined or vertical assemblies in which bulk solids flow by gravity from one location to another. Unlike hoppers, bins, or silos, chutes are not filled with the bulk material.

Bulk solids have unique properties that cause them to flow differently from liquids. Bulk solids are frictional and in general are compressible. Liquids are frictionless and are nearly incompressible. Bulk solid flow properties are strongly dependent on the consolidation stresses applied and minimally if at all dependent on strain rate, whereas fluid flow properties are strongly dependent on strain rate and minimally dependent on absolute pressure. In addition, bulk solids are anisotropic whereas liquids are isotropic.

Definitions When discussing the storage and handling of granular materials, the following definitions are commonly accepted:

Bulk solid—a material consisting of discrete solid particles, handled in bulk form.

Hopper, bin, or silo—storage vessels for bulk solids. The terms are often used interchangeably. Silos usually refer to tall vessels that store several tons of material. Hoppers and bins frequently refer to smaller vessels. The converging section of a storage vessel is often called the *hopper section.* Examples of hopper, bin, and silo geometries are given in Fig. 21-25. For simplicity, the term *bin* will be used henceforth as a descriptor for a storage vessel of any size.

Chute—equipment used to transfer bulk material by gravity between other pieces of equipment. The cross-section must be only partially full of material; otherwise it acts as a sloping bin or hopper.

Cylinder—vertical part of a bin. The cylinder may be round or rectangular and has a constant cross section.

Expanded flow—flow pattern inside a bin, where all the bulk material is in motion in the bottom portion of the vessel when withdrawn, but flow only occurs in a flow channel in the top portion of the vessel centered over the outlet.

Feeder—device for modulating the withdrawal rate of bulk material, e.g., rotary valves, screw feeders, and belt feeders. Often, a valve or gate is used to stop and start flow, but such devices in general should not be used to control the discharge rate of bulk solids.

Flow channel—the space in a bin in which the bulk solid is actually flowing at any point in time during withdrawal.

Funnel flow—flow pattern inside a bin, where the bulk material only moves in a flow channel above the outlet when withdrawn.

Hopper section—the converging part of a storage vessel that has sloped walls and a variable cross section.

Mass flow—flow pattern inside a bin where all material is in motion when any material is withdrawn.

Pressure—force per unit area applied to an object in a direction perpendicular to the surface; same as *compressive stress.*

Process vessel—a bin that has been modified to allow heating, cooling, drying, reacting, or other processes to take place. Frequently they are equipped with heat exchangers or distributors for gas injection.

Stress—force per unit area, a tensor quantity.

Flow Problems Many storage vessels are fabricated from architectural or fabrication viewpoints (e.g., hopper walls sloped 30° from vertical for ease of fabrication or 45° to minimize headroom requirements and simplify design calculations). However, designing equipment without regard to the bulk material being handled often leads to flow problems. Common solids flow problems include the following:

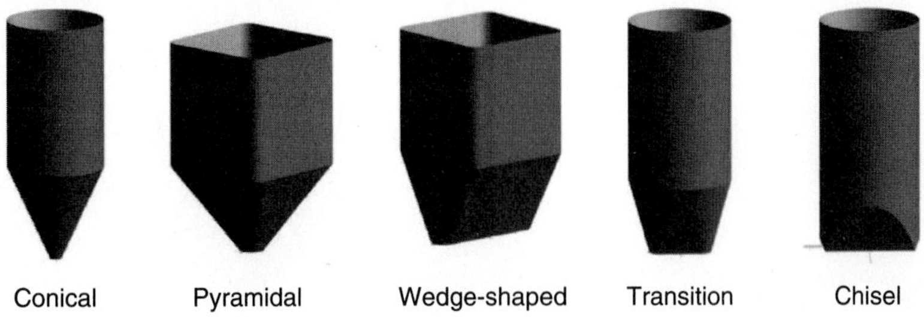

FIG. 21-25 Common bin geometries.

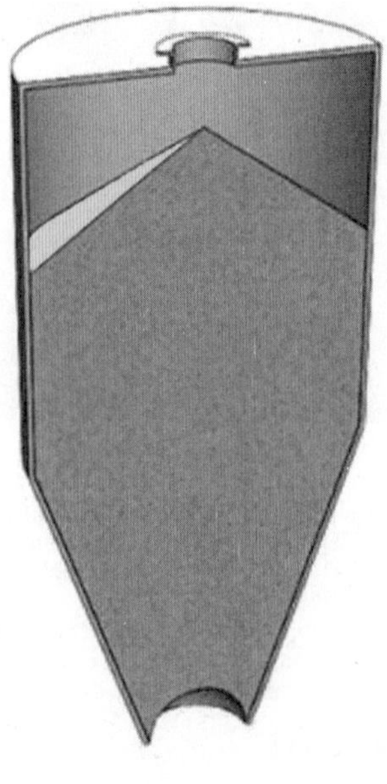

FIG. 21-26 Cohesive arch.

No flow. If a stable dome, bridge, or arch forms over the outlet of a bin, the bulk solid will not flow when the feeder is started or the gate is opened. If a stable rathole forms in a vessel in which flow only occurs in a narrow channel above the outlet, material will stop flowing when the flow channel empties. Obstructions to flow are illustrated in Figs. 21-26 and 21-27.

Erratic flow. Erratic flow occurs when both arching and ratholing occur. If a rathole collapses due to external vibration, the bulk solid may arch as it impacts the outlet. After the arch fails due to vibration or operator intervention, the flow channel will empty, leaving a rathole momentarily stopping flow until it eventually collapses, reforming a cohesive arch.

Flooding. If a stable rathole develops and fresh material is added or if a rathole collapses and falls into the channel, the material may become aerated or fluidized. Since most feeders are designed to handle solids and not fluids, the fluidized material may flood, that is, discharge uncontrollably in a fluidized state from the bin, and the feeder will not be able to control the rate of discharge.

Limited discharge rate. Because a fine powder dilates as it flows toward the outlet, vacuum will naturally develop inside the hopper above the outlet. As a consequence, air will flow counter to the solids, disrupting flow. Increasing the speed of the feeder will no longer increase the discharge rate of powder as the discharge rate has become limited.

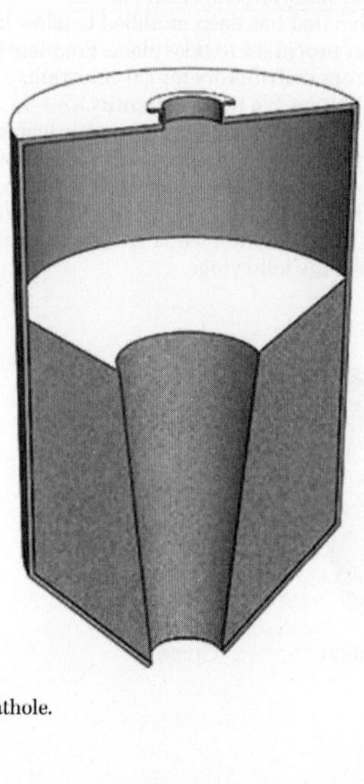

FIG. 21-27 Stable rathole.

Caking. Some materials will readily flow from a bin, portable container, or bag if handled continuously or even after time at rest. Other materials, however, will exhibit flow problems if allowed to remain at rest for a period of time. Given enough time at rest, some bulk solids will gain additional cohesive strength, and obstructions to flow (e.g., arches and ratholes) or hard lumps may become exceptionally difficult to eliminate.

Segregation. Some materials, when transferred into a bin or pile, will segregate; that is, particles of different size, shape, density, etc. will separate. Segregation can occur by a number of different mechanisms, depending on the physical characteristics of the particles and the method of handling.

Flow Patterns Three primary flow patterns can occur in a bin: *mass flow, funnel flow,* and *expanded flow.* Mass flow and funnel flow are illustrated in Fig. 21-28.

In *funnel flow,* an active flow channel forms above the outlet, with stagnant material remaining at the periphery. This occurs when the walls of the hopper section of the storage vessel are not steep enough or have low enough friction to allow flow along them. The size of the resultant flow channel is approximately the largest dimension of the outlet. It is equal to the diameter of a round outlet or the diagonal of a slotted outlet. In the case of conical funnel flow bins, the fraction of the vessel volume that is active can be dramatically small. If the bulk material is cohesive, a stable rathole will form, thereby reducing the effective capacity of the bin to a small fraction of its intended capacity.

A funnel flow bin typically exhibits a first-in, last-out flow sequence. Therefore, materials that readily cake or degrade over time should not be handled in funnel flow hoppers. Funnel flow can cause erratic flow and induce high loads (depending on vessel size) on the structure and downstream equipment due to collapsing ratholes and eccentric flow channels. If the bulk solid is cohesive, ratholes may become stable, and the vessel will not empty.

Funnel flow bins are best suited for bulk solids that are free-flowing and do not degrade or gain strength over time. They should not be used if segregation is a concern. Funnel flow bins require less headroom and in general are less expensive to build than mass flow bins since they can have shallower walls.

In *mass flow,* the entire bed of solids is in motion when material is discharged from the outlet, including material along the walls. Mass flow bins typically have steep and/or low-friction walls. Provided that the outlet is large enough to prevent arching, all material will be discharged from the hopper, since ratholes cannot form.

Mass flow bins are characterized by a first-in, first-out flow sequence and therefore are suitable for handling materials that degrade with time or are prone to caking. The steep hopper walls provide a more uniform flow than funnel flow bins, making mass flow hoppers suitable for process vessels. Discharge rates are predictable and steady, since the bulk density of the material at the outlet is nearly independent of the head of the material inside the vessel. Segregation by the dusting or sifting mechanism is minimized, as fine and course particles separated during filling are remixed at the outlet during discharge. A disadvantage of a mass flow bin is that it requires more headroom due to its steep hopper section. This is especially the case for conical mass flow bins.

Expanded flow is characterized by mass flow in the lowermost section of a bin and funnel flow in the upper section. An expanded flow bin is illustrated in Fig. 21-29.

The outlet of the funnel flow section must be large enough to prevent a stable rathole from developing. Because the bottom section is designed for mass flow, discharge rates are uniform and predictable. Expanded flow hoppers are frequently used when large bin diameters are required.

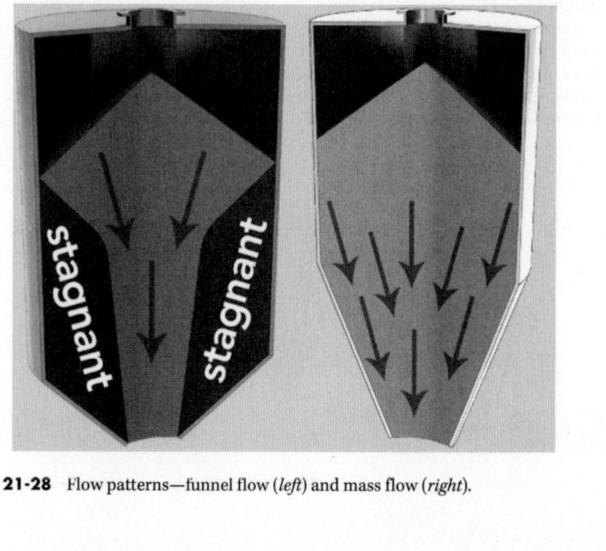

FIG. 21-28 Flow patterns—funnel flow (*left*) and mass flow (*right*).

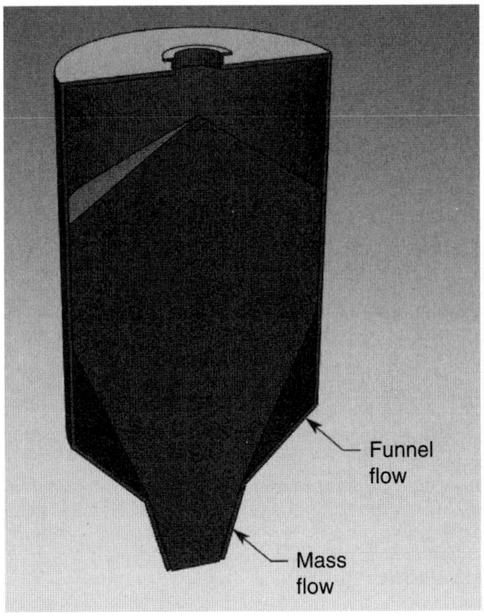

FIG. 21-29 Expanded flow pattern.

ANALYSIS OF STRESS

Bulk solids are anisotropic; their stresses vary with direction. Although a bulk solid consists of individual particles, it is convenient to describe a bulk material as a continuum. A bulk solid element is sketched in Fig. 21-30. The stresses acting normal (i.e., perpendicular to) the element in the x and y directions are denoted σ_x and σ_y, respectively. The shear stresses acting in the x and y directions are denoted τ_{yx} and τ_{xy}, respectively.

Given a state of stress, the magnitude of the normal and shear stresses acting on the bulk material will depend on the coordinate system used to describe the direction of these stresses. A new set of axes, denoted by x' and y', rotated an angle θ from the original axes, is shown in Fig. 21-31.

The stresses in terms of the new coordinate system are given by the following stress transformation equations:

$$\sigma_{x'} = \frac{\sigma_x + \sigma_y}{2} + \frac{\sigma_x - \sigma_y}{2}\cos 2\theta + \tau_{xy}\sin 2\theta \tag{21-26}$$

$$\sigma_{y'} = \frac{\sigma_x + \sigma_y}{2} - \frac{\sigma_x - \sigma_y}{2}\cos 2\theta - \tau_{xy}\sin 2\theta \tag{21-27}$$

$$\tau_{x'y'} = -\frac{\sigma_x - \sigma_y}{2}\sin 2\theta + \tau_{xy}\cos 2\theta \tag{21-28}$$

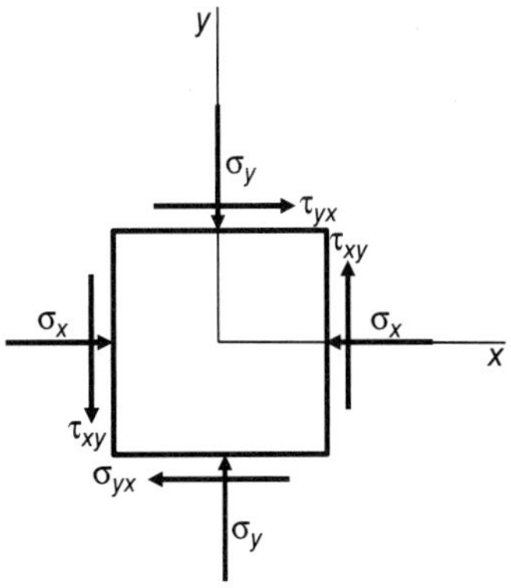

FIG. 21-30 Stress on an element of bulk solid.

When one is analyzing solids, the maximum and minimum stresses frequently must be identified. These stresses are called *principal stresses* and can be readily found using Mohr's circle, which can be derived from the stress transformation equations. Mohr's circle is illustrated in Fig. 21-32 and is described by

$$(\sigma_x - \sigma_{avg})^2 + \tau_{xy}^2 = R^2 \tag{21-29}$$

where

$$\sigma_{avg} = \frac{\sigma_x + \sigma_y}{2} \tag{21-30}$$

and

$$R^2 = \left(\frac{\sigma_x - \sigma_y}{2}\right)^2 + \tau_{xy}^2 \tag{21-31}$$

Note that Mohr's circle is centered at σ_{avg} and the two points (σ_x, τ_{xy}) and $(\sigma_y, -\tau_{xy})$ lie on opposite sides of the circle. To determine the stresses with respect to the rotated or transformed axes, the line connecting the two points (σ_x, τ_{xy}) and $(\sigma_y, -\tau_{xy})$ is rotated 2θ.

The maximum and minimum values of the normal stresses, i.e., the major and minor principal stresses, respectively, can be determined from the two intersection points of Mohr's circle and the horizontal axis. The major principal stress σ_1 and minor principal stress σ_2 can therefore be calculated from

$$\sigma_1 = \sigma_{avg} + R \tag{21-32}$$

$$\sigma_2 = \sigma_{avg} - R \tag{21-33}$$

SOLIDS-INDUCED LOADS

The geometry of the bin and the solids flow properties, which determine the solids flow pattern, prescribe the pressure profiles that develop within the bulk solids during initial fill and discharge. Solids-induced load analyses are used to determine vessel wall thicknesses and reinforcements for

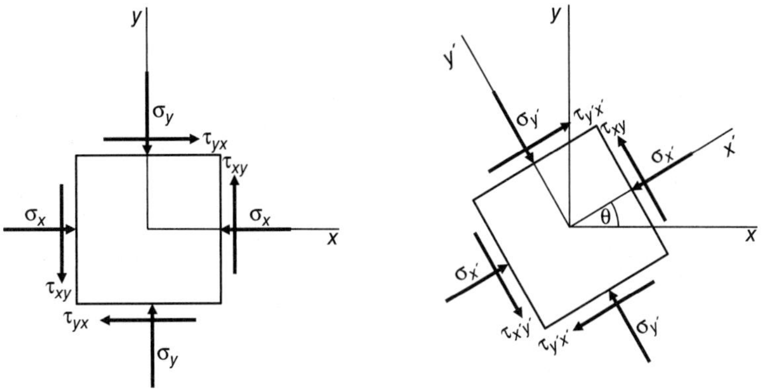

FIG. 21-31 Stresses on a rotated element.

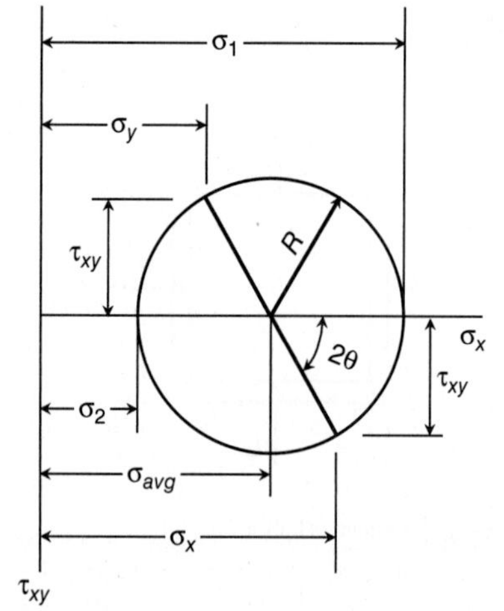

FIG. 21-32 Stress formation using Mohr's circle analysis.

structural integrity and the load on feeders, which is needed to determine power requirements.

A typical bin consists of a vertical (cylinder) section followed by a converging (hopper) section. Solids stresses are illustrated in Fig. 21-33.

In the cylindrical section, the vertical and wall stresses increase with depth, tending asymptotically toward a maximum. The wall stresses are smaller than the vertical stresses by a factor equal to k. At the centerline of the bin, the major principal stresses act vertically downward, and the minor principal stresses act horizontally.

When a previously empty bin is initially filled with a bulk solid, the major principal stresses in the converging section act vertically downward at the centerline. This stress state after initial fill is called the *active stress state.* A discontinuity exists in the wall stress profile. Both the wall stresses and vertical stresses decrease toward the hopper outlet.

When material is discharged from the bin, the stress conditions in the hopper section change if mass flow develops. In order to flow, the bulk solid is compressed laterally and expands vertically. As a result, the major principal stresses act horizontally at the centerline. This state of stress is called the *passive state.* A peak stress, called the *switch,* occurs at the hopper-cylinder interface.

Cylinder Stresses The stresses on the solids in the straight-walled section of a bin were originally calculated by Janssen [*Z. Ver. Dt. Ing.* **39:** 1045 (1895)]. An equilibrium force balance on a volume element of bulk solids yields the following differential equation:

$$\frac{d\sigma_v}{dz} + \frac{k\tan\phi'}{R_H}\sigma_v = \rho_b g \tag{21-34}$$

where σ_v is the vertical stress, z is the depth of the solids bed, R_H is the hydraulic radius, k is the Janssen coefficient, which is equal to the ratio of the stress acting perpendicular to the walls to the vertical stress, ρ_b is the bulk density, g is acceleration due to gravity, and ϕ' is the wall friction angle, which is equal to the inverse tangent of the wall friction coefficient. If a constant bulk density is selected, Eq. (21-34) can be integrated to give the Janssen equation:

$$\sigma_v = \frac{\rho_b g R_H}{k\tan\phi'}\left[1 - \exp\left(-\frac{k\tan\phi'}{R_H}z\right)\right] \tag{21-35}$$

The horizontal stress σ_h is given by

$$\sigma_h = \frac{\rho_b g R_H}{\tan\phi'}\left[1 - \exp\left(-\frac{k\tan\phi'}{R_H}z\right)\right] \tag{21-36}$$

The value of k is typically in the range of 0.3 to 0.6. Note that unlike for liquids, the maximum solids stress is proportional to the hydraulic radius of the cylinder and is independent of its height.

Hopper Section Walker [*Chem. Eng. Sci.* **21:** 11, 975 (1966)] and Walters [*Chem. Eng. Sci.* **28:** 1, 13 (1973)] analyzed the stresses in the hopper section by performing an equilibrium force balance on an elemental volume with converging sides. A differential equation results, which is generalized by Jenike [Loeffler, F. J., and C. R. Proctor (eds.), *Unit and Bulk Materials Handling, Effect of Solids Flow Properties and Hopper Configuration on Silo Loads,* ASME, 1980, pp. 97–106] as

$$\frac{d\sigma_v}{dz} - n\frac{\sigma_v}{z} = -g\rho_b \tag{21-37}$$

$$n = (m+1)\left[k\left(1 + \frac{\tan\phi'}{\tan\theta}\right) - 1\right] \tag{21-38}$$

where θ is the hopper angle (from vertical) and m is equal to 1 for a conical hopper and equal to 0 for a straight-walled hopper having a slotted outlet.

Integration yields the following [British Standard BS EN 1991-4:2006]:

$$\sigma_v = \frac{\rho_b g h}{n-1}\left[\frac{x}{h} - \left(\frac{x}{h}\right)^n\right] + \sigma_{vht}\left(\frac{x}{h}\right)^n \tag{21-39}$$

where x is the vertical coordinate upward from the hopper apex, h is the vertical height between the hopper apex and the cylinder-hopper transition, and σ_{vht} is the mean vertical stress on the solid at the transition after filling [as determined from Eq. (21-34)].

The wall stress σ_w is calculated from

$$\sigma_w = k\sigma_v \tag{21-40}$$

The frictional traction τ_w is determined from

$$\tau_w = k\sigma_w \tag{21-41}$$

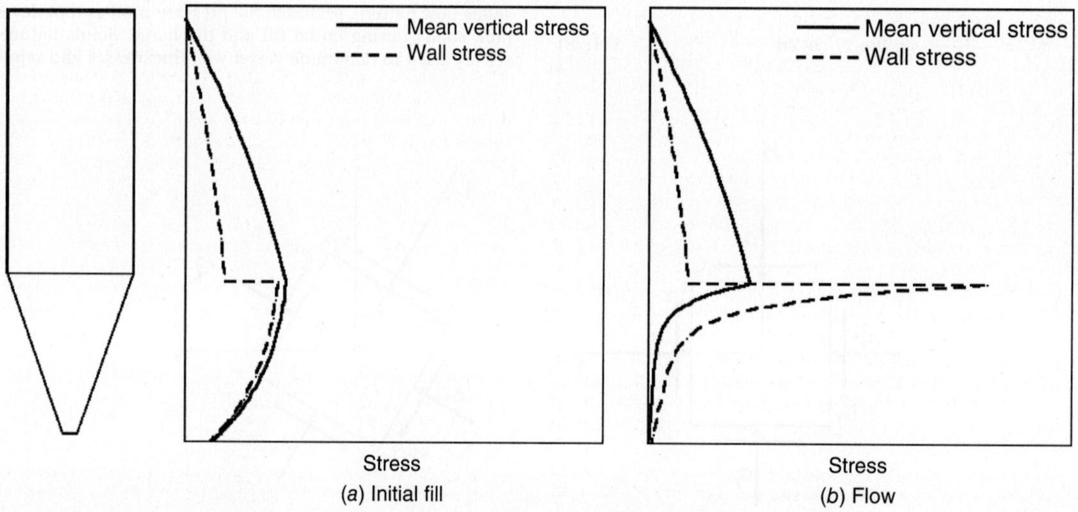

FIG. 21-33 Representative stress profiles in a mass flow bin; wall stress acts normal to the cylinder or hopper walls.

The value of the stress ratio k depends on the flow properties of the bulk material handled and the slope of the hopper walls. Methods described by Enstad [*Chem. Eng. Sci.* **40:** 10, 1273 (1975)] can be used to calculate the stress ratio.

Funnel Flow Hoppers In funnel flow hoppers, the wall friction is not fully mobilized. An effective wall friction angle is therefore determined per EN 1991-4:2006.

Eccentric Loads Although more common with funnel flow, eccentric discharge, in which the flow channel formation is not concentric with the vertical section of the bin, can occur in mass flow as well. Eccentric discharge can occur when a single outlet is not centered or when multiple outlets do not discharge at the same rate. Eccentric flow can also occur if gates or valves are partially opened or if feeder interfaces are not properly designed.

Asymmetric pressures develop on the walls when eccentric discharge takes place. A procedure for calculating discharge loads during eccentric flow is described in BS EN 1991-4:2006.

Note that the formulas for calculating normal pressures and shear tractions above do not include inherent factors of safety. A prudent designer must recognize that factors of safety are always necessary to account for unexpected loading conditions or deficiencies in the ability of the structure to withstand these loads. The designer is ultimately responsible for choosing the appropriate safety factor based on risks associated with structural failure and compliance with applicable design codes. Qualified engineers should review results of load analyses.

BULK SOLIDS FLOW PROPERTIES TESTING

Designing systems for bulk solids can be challenging since they have a wide range of characteristics, e.g., cohesive or free-flowing; fine or coarse; fluffy or dense; adhesive to surfaces or surface-repellant; easily aerated or nearly impermeable; and highly compressible or nearly incompressible. Defining a particle size distribution, density, or permeability is relatively straightforward. The best metric for cohesion or adhesion is not as obvious. Various characteristics or a combination of them are needed to define a bulk material's ease of flow or "flowability." However, even this is not enough, since flowability also depends on the design of the vessel from which the material is flowing—or is attempting to do so.

Several methods exist for measuring the *relative* flowability of bulk materials. The simplest is to determine the bulk solid's angle of repose by pouring it onto a horizontal surface and measuring the surcharge angle of the pile that is formed. A material that forms a steeper pile is believed to be less flowable than one that is shallow. However, as stated by Jenike [*Storage and Flow of Solids*, Bulletin 123, University of Utah, 1964 (revised, 1980)], "The angle of repose is not a measure of the flowability of solids. In fact, it is useful only in the determination of the contour of a pile, and its popularity among engineers and investigators is due not to its usefulness but to the ease with which it is measured."

A compressibility test is another relative measure of flowability. A sample of bulk solid is vibrated or compacted ("packed") inside a rigid container, and its change in bulk density is measured. Two common methods to analyze the results are the Hausner ratio and Carr compressibility. The former is the ratio of the "tapped" density to the aerated or loose bulk density. The latter is determined by dividing the difference between the packed and freely settled bulk density by the packed bulk density. A low Hausner ratio or Carr compressibility supposedly indicates that the material is easy to handle. However, these indices are of limited use, since at best they can be only loosely correlated to the flow behavior of similar bulk solids. In addition, these methods are deficient as the stress applied to the sample of bulk solid is unknown, the tests do not replicate the degree of consolidation that takes place when a material is stored in a vessel, and the gain in the material's strength during rest cannot be determined.

Solids rheometers of various designs are sometimes used to quantify the relative flowability of bulk solids. The material is placed in a cell equipped with an impeller, and the torque or energy required to rotate the agitator is measured. In some instruments, the vertical force on the agitator can also be directly measured. Flowability is deemed to correlate with the torque or the power drawn by the agitator. Unfortunately, the stresses acting in the shear zone during testing are unknown, and therefore the results cannot be applied to actual process conditions. In addition, both fluidization and agglomeration can occur inside the test cell, confounding the results. High torque or energy consumption may be the result of high friction between the bulk material and the walls of the cell, rather than an indication of the material's cohesive strength. Test methods based on stirred vessels therefore have questionable utility [Schulze, *Powders and Bulk Solids: Behavior, Characterization, Storage and Flow*, Springer, 2007].

There are five fundamental bulk solid flow properties that provide useful and reliable information from which one can design a bin for reliable flow: cohesive strength, internal friction, bulk density, wall friction, and permeability.

In contrast to fluids, materials having the same composition frequently have dramatically different fundamental flow properties. This is so because flow properties are often dependent on the material's particle size, shape, and particle size distribution. In addition, temperature, moisture content (or the relative humidity of the material's interstitial air), purity, surface energy, and morphology all can influence the flow behavior of a bulk solid. The flow properties of many materials change when they are stored at rest.

There is no substitute for measuring the flow properties of the actual materials that are to be handled, and testing should be performed over a range of temperatures, moisture contents, relative humidity levels, times at rest, and stress levels at which the bulk solid will be stored and handled. Using flow property data from the literature or assuming that the properties are the same as those of other materials whose properties are known is extremely risky. For example, "coal" is an extremely nonhomogeneous material; yet some textbooks and design codes provide values (a limited range or sometimes a single value) of important design parameters such as bulk density and wall friction. Such tabular data are, at a minimum, misleading and potentially worse than having no data at all.

Cohesive Strength and Internal Friction A bulk solid gains cohesive strength when consolidated in a bin. The size of the outlet of the vessel that will prevent arching or the formation of a stable rathole depends greatly on the bulk material's cohesive strength.

Figure 21-34 is a schematic of a uniaxial compressive strength tester. In a uniaxial test, a sample is placed in a cell with nearly frictionless walls and is then consolidated by applying a stress equal to σ_1. Next, the load and cell are removed. The compacted specimen is then loaded with increasing compressive stress until it breaks apart, i.e., fails. The failure stress is called the material's *cohesive strength* or the *unconfined yield strength* f_C.

Uniaxial compressive strength test results are often highly variable. Improvements have been made to uniaxial strength testers to reduce their variability; however, uniaxial compression tests usually do not provide a bulk material's true unconfined yield strength, and therefore the cohesive strength of a bulk solid is best measured by direct shear cell testing. Translational (Jenike), annular (ring), and rotational testers are frequently used. They are described in ASTM standards D-1628 (translational), D-6773 (annular), D-6682, and D7891 (rotational). Schematics of the testers are given in Fig. 21-35.

The direct translational shear tester was originally developed by Andrew Jenike [*Storage and Flow of Solids*, Bulletin 123, University of Utah, 1964 (revised, 1980)]. This tester is particularly robust in that its cell can be placed in extreme environments, allowing a material's cohesive strength to be measured over a full spectrum of process conditions. Its disadvantage is that significant operator training and experience are usually required to be able to obtain reproducible results.

Modern annular and rotational shear testers are computer-controlled and are thus more straightforward to operate and less prone to operator variability than manually controlled shear testers.

In a rotational shear cell, shear deformation of the specimen varies with radius in the cell: at the perimeter it is at its maximum, while at the center it is zero. This can result in data that differ from results obtained using a Jenike (translational) or annular shear cell. Rotational and annular shear cells permit infinite travel, so they are better suited than a Jenike shear cell for testing bulk solids that require large shear strain to reach steady state. The Jenike shear cell tester can be relatively easily modified to operate at high or low temperatures, whereas this is more difficult, if not impossible, with the other two types of testers.

Cohesive strength is measured by shear cell testing as described in ASTM D-1628, D-6682, D-6773, or D7891. A sample of bulk material is placed in a cell and then presheared, i.e., consolidated by exerting a normal stress and then shearing it until the measured shear stress is steady [as illustrated by the point (σ_{ss}, τ_{ss}) shown in Fig. 21-36. Next the shear step is conducted, in which the vertical compacting load is replaced with a smaller load, and the

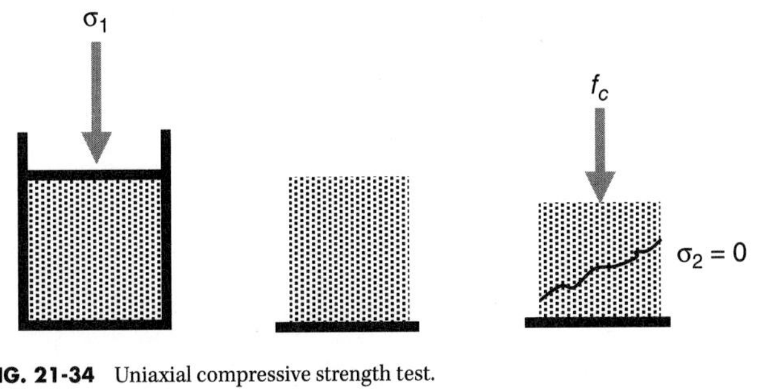

FIG. 21-34 Uniaxial compressive strength test.

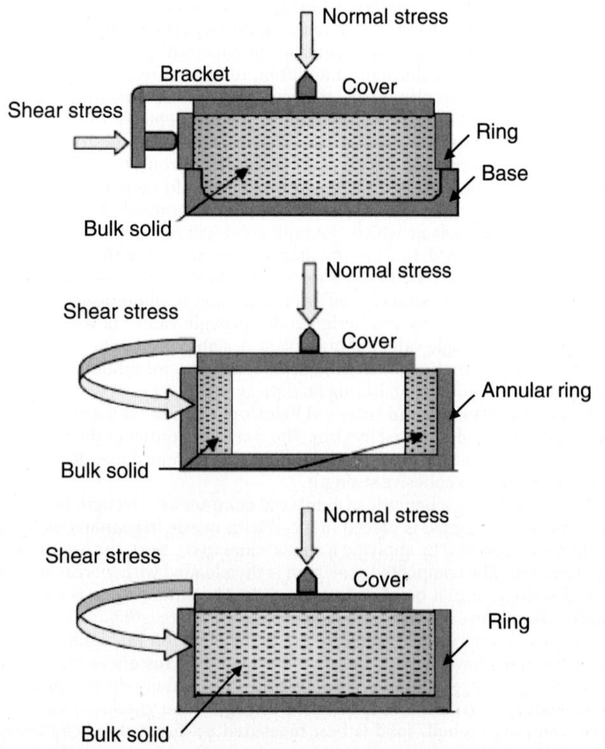

FIG. 21-35 Shear cell testers—Jenike direct (*top*), annular (*middle*), and rotational (*bottom*).

sample is again sheared until it fails. These preshear and shear steps are repeated at the same consolidation level for a number of reduced normal stresses, and the yield locus is then determined by plotting the failure shear stress versus normal stress (see Fig. 21-36).

Ideally, all measurements of the preshear shear stress τ_{ss} should be identical. However, because of unavoidable variability during testing, there is inevitably scatter in the τ_{ss} values. Prorating is used to account for the variability of the data [Jenike, *Storage and Flow of Solids*, Bulletin 123, University of Utah, 1964 (revised, 1980)].

To determine the major consolidation stress σ_1 and the unconfined yield shear strength f_C from the yield locus, a line is drawn through the shear step data. Mohr's semicircle is then drawn through the steady-state result (σ_{ss}, τ_{ss}) tangent to the yield locus line (see Fig. 21-37).

The larger point of intersection of the semicircle with the horizontal axis gives the value of the major consolidating stress σ_1. The unconfined yield strength f_C is determined by drawing Mohr's semicircle tangent to the yield locus and passing through the origin. The point of intersection of this circle and the horizontal axis is the unconfined yield strength, which can be considered the cohesive strength of the bulk solid. Note that all points on the yield locus should lie to the right of the point of tangency to the smaller Mohr's circle.

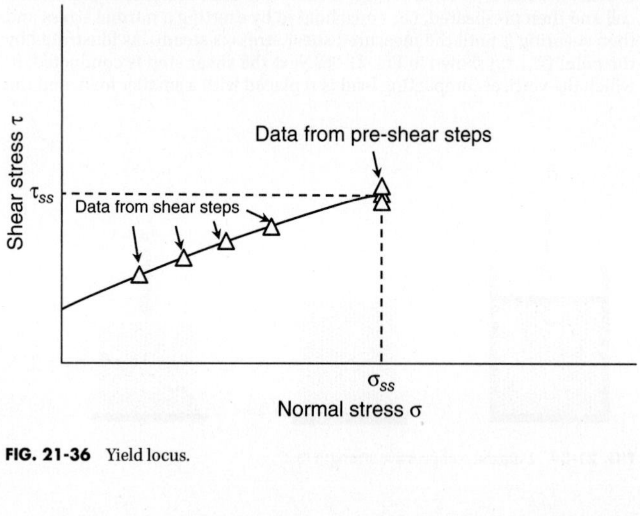

FIG. 21-36 Yield locus.

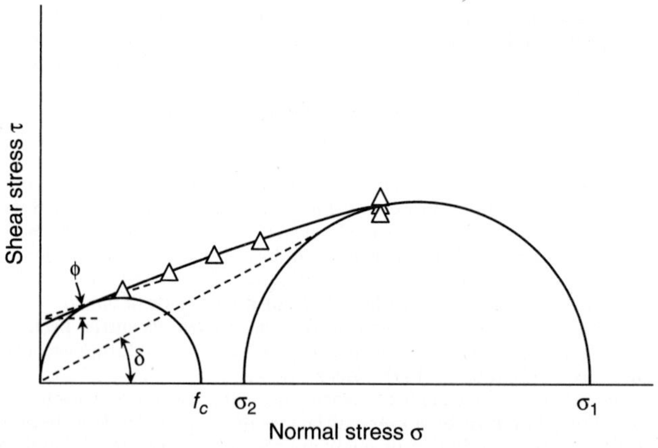

FIG. 21-37 Linear yield locus.

Also determined are the effective angle of friction and kinematic angle of internal friction (δ and ϕ, respectively). The effective angle of friction is found by constructing a line through the origin and tangent to the larger Mohr's semicircle. The kinematic angle of internal friction is the angle formed between a line that is horizontal and one drawn tangent to the smaller Mohr's circle at its intersection with the yield locus (see Fig. 21-37).

The yield locus generally is slightly concave downward, but with many particulate solids a straight line is a sufficient approximation. If the yield locus is approximated as a straight line for all particulate solids, then subsequent calculations are much simpler, but, in some cases, somewhat conservative results may be obtained, that is, a higher f_C value will be determined than when using a fitted curve. The shear data that make up the yield locus (i.e., all data points without the steady-state or preshear data) are regressed to give the following linear relation:

$$\tau = c + \sigma \tan \phi \tag{21-42}$$

where τ is the shearing stress and σ is the normal stress. Equation (21-42) is the Coulomb equation. The slope of the line is equal to the tangent of the kinematic angle of internal friction ϕ, and the intercept is equal to c, which is called the material's *cohesion*.

The unconfined yield strength and major consolidation stress are then calculated from

$$f_C = \frac{2c(1+\sin\phi)}{\cos\phi} \tag{21-43}$$

and

$$\sigma_1 = \left(\frac{A - \sqrt{A^2 \sin^2 \phi - \tau_{SS}^2 \cos^2 \phi}}{\cos^2 \phi} \right)(1+\sin\phi) - \frac{c}{\tan\phi} \tag{21-44}$$

respectively, where

$$A = \sigma_{SS} + \frac{c}{\tan\phi} \tag{21-45}$$

The minor consolidation stress σ_2 can be calculated from

$$\sigma_2 = \sigma_{SS} - \frac{\tau_{SS}^2}{\sigma_1 - \sigma_{SS}} \tag{21-46}$$

Finally, the effective angle of friction δ is calculated from

$$\delta = \sin^{-1}\left(\frac{\sigma_1 - \sigma_2}{\sigma_1 + \sigma_2} \right) \tag{21-47}$$

A straight line through the origin of the σ-τ diagram, tangent to the larger Mohr's circle, is the *effective yield locus* (EYL). The larger Mohr's circle can be constructed by drawing a circle having a radius R that passes through σ_1 on the horizontal axis. The radius R is given by

$$R = \frac{\sigma_1 - \sigma_2}{2} \tag{21-48}$$

Construction of the effective yield locus is illustrated in Fig. 21-38.

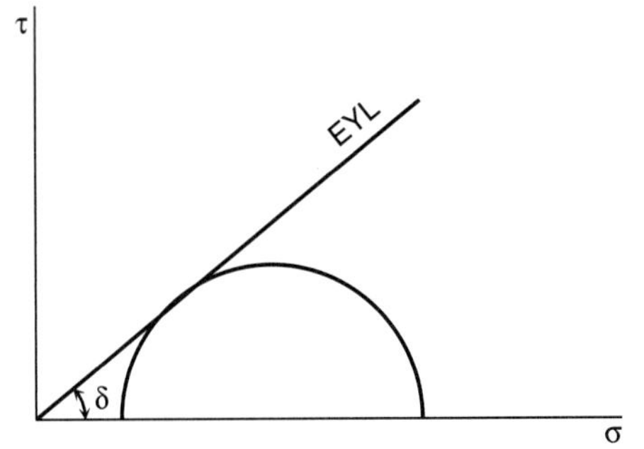

FIG. 21-38 Construction of the effective yield locus.

Plotting values of f_C versus the major consolidation stress σ_1 gives the *flow function* of the bulk solid. The flow function describes the relationship between a bulk material's unconfined yield strength and its state of consolidation. Construction of the flow function from a number of yield locus measurements is illustrated in Fig. 21-39.

Some bulk materials gain cohesive strength if stored at rest. Unless a bin is expected to be operated continuously, the time unconfined yield strength of the bulk material should be measured. To conduct a time test, a sample of bulk material is placed inside a cell and presheared using a normal stress σ_{ss} used during instantaneous testing. After preshear, the sample is then kept consolidated at that state of stress, typically by applying a vertically acting load equal to the major consolidation stress σ_1 associated with the corresponding instantaneous test. After the time of interest has passed (e.g., 2 to 3 days if the bulk material is to be stored at rest over a weekend), the vertical compacting load is replaced with a smaller load, and the shear step is conducted, in which the shearing force again is applied until the sample fails.

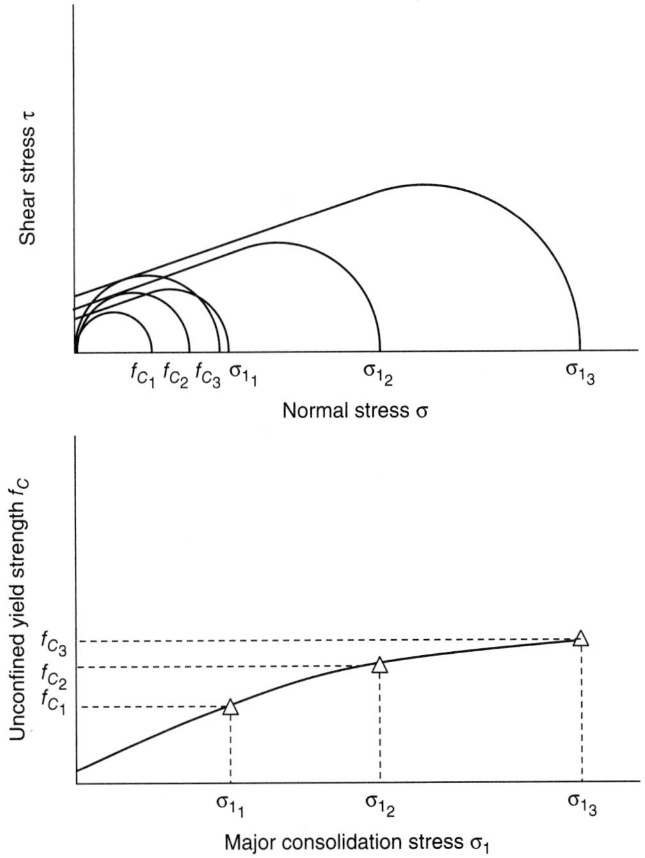

FIG. 21-39 Construction of flow function from yield loci.

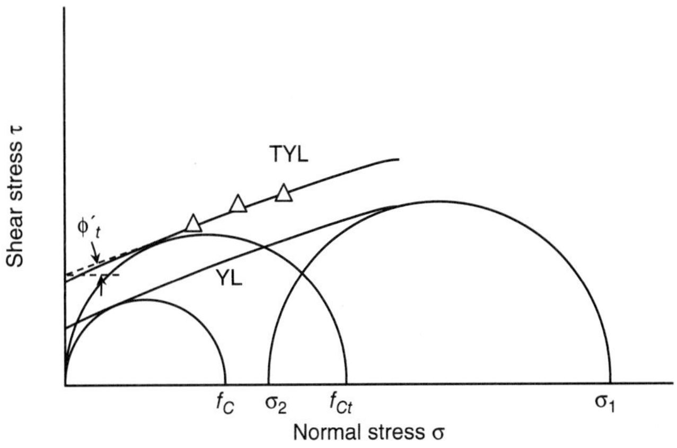

FIG. 21-40 Construction of the time yield locus.

The preshear, time consolidation, and shear steps are repeated at the same normal stress σ_{ss} for a number of normal stresses, and the *time yield locus* (TYL) is then determined by plotting the failure shear stress versus normal stress. An example of a time yield locus is given in Fig. 21-40.

To calculate the time unconfined yield strength, Mohr's circle is drawn through the origin and tangent to the time yield locus. The point of intersection with the horizontal axis is the material's time unconfined yield strength f_{Ct}. This value, along with the value of the major consolidation stress for instantaneous flow σ_1, becomes one point on the *time* flow function.

The *time angle* of internal friction ϕ_t is the angle formed between a horizontal line and a line drawn tangent to the smaller Mohr's circle at its intersection with the time yield locus (see Fig. 21-40).

By measuring time yield loci using other normal stress levels and corresponding major consolidation stresses, the time flow function can be determined by plotting the time unconfined yield strength f_{Ct} versus major consolidation stress σ_1. If a bulk material gains strength when stored at rest in a bin over time, its time flow function will lie above its instantaneous flow function, as illustrated in Fig. 21-41.

Because of the obvious time-consuming nature of a time consolidation test, extra test cells should be used so that the sample after preshear can be kept consolidated outside the shear tester.

Bulk Density A method to measure the bulk density of a material as a function of compressive stress (i.e., pressure) is given in ASTM D6683. A sample is placed in a cylinder of known volume, and its mass is recorded. A lid with a known weight is placed on the specimen, and the displacement is noted, allowing an updated volume to be calculated. The consolidation pressure is equal to the weight placed on the sample, divided by the cross-sectional area of the cylinder. The bulk density is equal to the mass of sample, divided by the volume. Increasing loads are placed on the lid, and the displacement is recorded for each load. From the data, the bulk density as a function of consolidation pressure is determined. A typical bulk density curve is shown in Fig. 21-42.

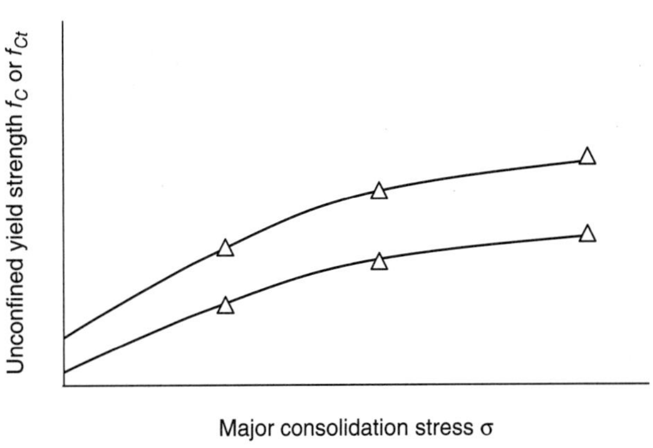

FIG. 21-41 Instantaneous and time flow functions.

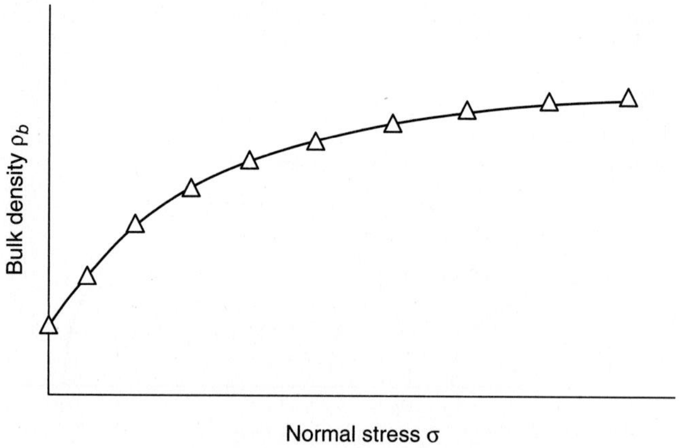

FIG. 21-42 Typical bulk density—consolidation stress relationship.

The relationship between bulk density and consolidation stress is nonlinear. The bulk density increases with increasing consolidation pressure, varying rapidly at low stress and less so at high stress. Data can be fit to a number of equations that describe the relationship between bulk density and consolidation pressure. Jenike and Johanson [*Powder Technol.* **5:** 133 (1971/72)] assumed a power law relationship:

$$\rho_b = \rho_{b0}\left(\frac{\sigma}{\sigma_0}\right)^{\beta} \tag{21-49}$$

where σ is the consolidation pressure, σ_0 is an arbitrarily chosen reference consolidation level, ρ_{b0} is the bulk density at that consolidation, and β is called the *compressibility*. A limitation of the model is that at zero stress, Eq. (21-49) gives a bulk density equal to zero. Hence, at a low consolidation pressure, the bulk density is limited to no less than $\rho_{b\min}$, the loose-fill bulk density.

Ideally, a model that describes the bulk density–consolidation pressure relationship should give a bulk density of $\rho_{\min}$ at zero consolidation. Some relationships that meet this criterion are [Gu et al., *Powder Technol.* **72:** 39 (1992)]

$$\rho_b = \rho_{b\min}(1+\alpha\sigma)^{\beta} \tag{21-50}$$

$$\rho_b = \rho_{b\min} + \alpha\sigma^{\beta} \tag{21-51}$$

$$\rho_b = \rho_{b\min} \qquad \sigma = 0$$

$$\rho_b = \rho_{b\min} + \frac{(\rho_{b0} - \rho_{b\min})\sigma}{\rho_{b0}} \qquad 0 < \sigma < \sigma_0 \tag{21-52}$$

$$\rho_b = \alpha\sigma^{\beta} \qquad \sigma \geq \sigma_0$$

$$\rho_b = \rho_{b\max} - (\rho_{b\max} - \rho_{b\min})\,e^{-\alpha\sigma} \tag{21-53}$$

$$\rho_b = \rho_{b\min}\rho_{b\max}\frac{1+\alpha\sigma}{\rho_{b\max} + \rho_{b\min}\alpha\sigma} \tag{21-54}$$

where α and β are empirical constants and $\rho_{b\max}$ is the material's maximum bulk density.

Wall Friction The flow pattern inside a bin depends on the friction between the bulk solid and the wall material. Therefore, measuring wall friction is one of the necessary steps to design reliable bulk solids storage vessels.

To measure the friction between a bulk solid and a wall material, the method described in ASTM D-6128 is usually followed. The test is conducted using a direct translation shear tester. A sample of bulk solid is placed inside a retaining ring on a flat coupon of wall material (see Fig. 21-43), and a load is then applied normal (i.e., perpendicular) to the bulk solid. The ring and bulk solid in the ring are forced to slide along the stationary wall material, and the resulting steady shear force is measured as a function of the applied normal load. The normal load is then reduced, and the test is continued until a new steady shear load is measured. The test is repeated for various normal loads.

In most industrial applications, the wall surface upon which a bulk solid is sliding is below the bulk solid. Due to the segregating nature of some bulk solids, differences in wall friction angles may be observed between testers

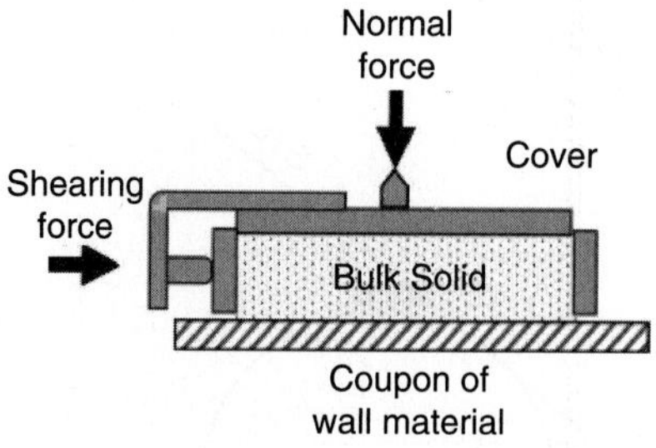

FIG. 21-43 Wall friction test equipment.

where the wall friction coupon is placed above the test cell and those, such as the Jenike shear cell, where the wall friction coupon is located below. The direction of grain orientation of a wall surface can have a significant effect on the wall friction angle. This effect can be captured in the Jenike shear cell, but not in rotational or annular shear cells.

After a number of steady shear load values have been recorded, the instantaneous *wall yield locus* (WYL) is constructed by plotting shear stress versus normal stress. The angle of wall friction ϕ' is the angle that is formed when a line is drawn from the origin to a point on the wall yield locus. A typical wall yield locus is shown in Fig. 21-44.

The wall yield locus is typically concave downward. In addition, the wall yield locus does not necessarily intersect the origin, as many bulk materials adhere to a wall surface in the absence of a normal stress. As a consequence, ϕ' is usually higher at lower applied stresses. This is important in the design of hoppers, since for mass flow, the stresses at the hopper outlet are low and the angle of wall friction is therefore usually higher near the outlet. Only when the yield locus is a straight line that passes through the origin is ϕ' constant.

To measure the static friction between a wall surface and a bulk solid after storage at rest, wall friction time tests are performed. A sample is sheared under a normal load until a steady shear load is observed. The normal load is then reduced by 10 to 20 percent, and shearing is continued until steady state is again reached. This step is analogous to the procedure of obtaining a point on the instantaneous wall yield locus.

The shear is then reduced to zero and the sample is stored in the cell for the required period of time. After the storage time, the sample is again sheared, and the maximum shear load is reported.

The pair of normal stress and maximum shear stress values provide one point on the *time wall yield locus* (TWYL). Repeating the test at different normal loads completes the time wall yield locus. The time angle of wall friction is the angle obtained by drawing a line from the time wall yield locus to the origin (see Fig. 21-45).

Changes in both the bulk solid's properties (for example, due to changes in moisture content, temperature, etc.) and the wall surface (for example, due to surface finish, corrosion, or abrasion) affect the wall friction angle. It is important that one not assume that a "smooth" surface, i.e., one with a low *Ra* (average surface roughness) value, will necessarily be low in wall friction angle. Sometimes a surface with small asperities (which result in a larger *Ra* value) will be less frictional than a smoother surface.

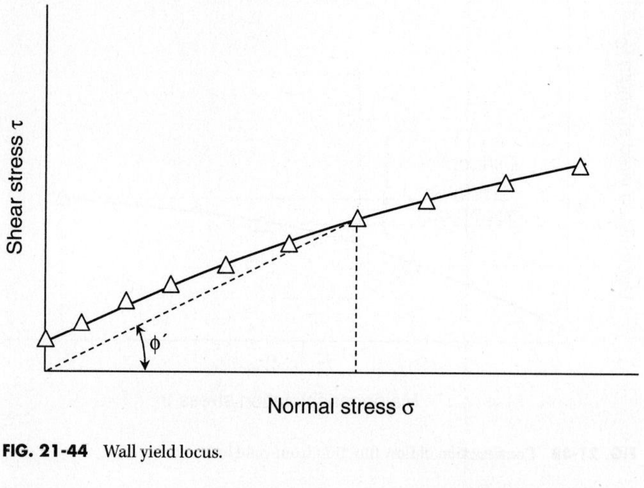

FIG. 21-44 Wall yield locus.

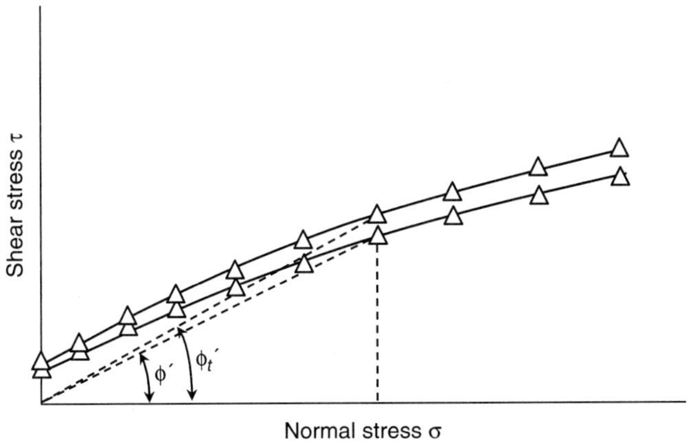

FIG. 21-45 Time wall yield locus.

Permeability The maximum achievable steady-state discharge rate of a bulk material from a bin depends on its permeability. The pressure drop that arises when a gas passes through a bed of bulk solid is described by Darcy's law if the gas flow is laminar:

$$u_s = -\frac{K}{\rho_b g}\frac{dP}{dz} \tag{21-55}$$

where u_s is the superficial gas velocity, K is the bulk material's permeability, and dP/dz is the gas pressure gradient. Permeability is determined by measuring the pressure drop and gas flow rate, using the apparatus shown in Fig. 21-46.

The permeability of a bulk solid is highly dependent on its bulk density. The relationship between permeability and bulk density can be determined by filling the tester with compacted material. Often the relationship between permeability and bulk density is described by a power-law relationship, e.g.,

$$K = K_0\left(\frac{\rho_b}{\rho_{b0}}\right)^{-\alpha} \tag{21-56}$$

where α is a constant, ρ_{b0} is a reference bulk density, and K_0 is the permeability of the bulk solid measured at the reference bulk density.

Abrasive Wear Abrasive wear of equipment is a problem that should not be overlooked. It is important that solids contact surfaces be durable, thick enough to provide reasonable wear life, and, if necessary, replaceable to minimize maintenance.

Prediction of the wear rate can be accomplished in a two-step process. The first step is to determine the bulk solids velocity and compressive stress acting on the wear surface. The second step is to determine the rate of wear of the wear surface as a function of bulk solids stress, using a suitable wear testing apparatus.

A suitable wear tester is one that continuously introduces a fresh supply of a bulk solid sample to a wear surface. The tester should continuously control and measure the relative velocity between the bulk solid and wear surface as well as the solids stress exerted on the wall material. Any changes in the wall friction angle can be determined by running a wall friction test before and after a certain number of cycles of the wear test.

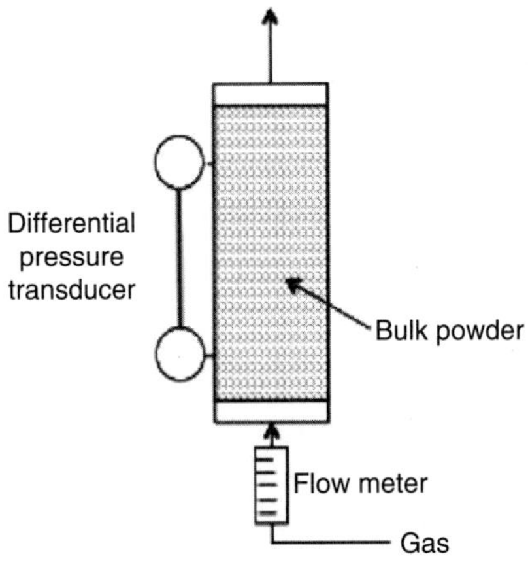

FIG. 21-46 Permeability tester.

A description of an abrasive wear tester is given in U.S. Patent No. 4,446,717. A sample of bulk solid is conveyed to a circular disc fabricated from the wear material, and a supporting frame applies and measures torque, thrust, and the number of revolutions. Tests are conducted over a range of expected solids stresses. Wear is measured as the weight loss of the sample disc after a certain number of revolutions of the disc.

It is convenient to define a *wear ratio* WR as the dimensionless ratio of unit thickness of material lost per unit distance travel of the bulk solid sliding relative to the wear surface.

The solids velocity at the hopper wall can be calculated using a method described by Johanson and Royal [*Bulk Solids Handling* **2:** 3, 517 (Sept. 1992)] and is inversely proportional to the cross-sectional area of the hopper. The wear ratio is a function of the applied solids stress. Methods to calculate solids-induced loads were described previously. The wear rate is equal to the product of the wear ratio WR and the solids velocity at the wall. Because the solids velocity and stress vary inside the hopper, the degree of abrasive wear will vary with location.

BIN DESIGN

When developing the *functional** design of a new bin, there are a number of factors that determine what type of bin is required. These factors include the cohesiveness of the bulk solid, headroom or footprint constraints, segregation concerns, the likelihood of degradation over time (e.g., caking, spoilage, spontaneous combustion), the location of the outlet relative to bin's centerline, and discharge rate requirements.

In general, for a given volume, mass flow hoppers and silos are taller than those designed for funnel flow. If there are headroom restrictions, designing a mass flow bin with the desired capacity may be challenging. Whenever this is the case, an engineer should confirm that the constraints are necessary or consider whether a funnel flow or expanded flow bin will suffice.

Mass Flow Hopper Angle Flow problems can often be prevented by ensuring that a mass flow pattern will develop in the vessel. The first step in achieving mass flow is to ensure that the converging walls are steep enough and have friction low enough to allow the bulk material to slide along them. This is accomplished by first testing the material to measure wall friction and then calculating the minimum hopper angle that will allow mass flow.

By assuming a radial stress field, Jenike (*Gravity Flow of Bulk Solids*, Bulletin 108, University of Utah, 1961) was able to calculate stresses in the region of the hopper outlet as a function of the effective angle of friction δ, hopper angle (from vertical) θ', and wall friction angle ϕ'. Jenike determined that when the boundary conditions are not compatible with the radial stress equations, mass flow cannot occur in a hopper, and a funnel flow pattern results.

Design charts originally developed by Jenike in 1961 provide allowable hopper angles for mass flow, given values of wall friction angle and the effective angle of friction. These charts are summarized in Figs. 21-47 and 21-48 for conical (or pyramidal) and planar hoppers (e.g., wedge-shaped hoppers and transition hoppers), respectively. The outlet of a wedge-shaped or transition hopper must be at least 2 times as long as it is wide for Fig. 21-48 to apply if it has vertical end walls and 3 times as long if its end walls are converging.

Values of the allowable hopper angle for mass flow θ' (measured from vertical) are on the abscissa, and values of the wall friction angle ϕ' are on the ordinate. Any combination of ϕ' and θ' that falls within the mass flow region of the chart will provide mass flow.

An analytical description of the theoretical boundary between mass flow and funnel flow regions for conical hoppers is given in Enstadt [*Chem. Eng. Sci.* **30:** 1273 (1975)]; an explicit equation that gives recommended mass flow angles for hoppers with slotted outlets is provided by Arnold et al. (*Bulk Solids: Storage, Flow, and Handling*, TUNRA Publications, 1980).

Hoppers with round or square outlets should not be designed at the theoretical mass flow hopper angle value. Otherwise, a small change in the bulk material's flow properties may cause the flow pattern inside the hopper to change from mass flow to funnel flow, with its associated risk of flow problems. A 3° to 5° margin of safety with respect to the mass flow hopper angle given in Fig. 21-47 is therefore recommended.

Sloping walls required for mass flow in wedge-shaped hoppers can be 10° to 12° less steep than those required to ensure mass flow in conical or

*A functional design includes hopper shape, hopper angles, outlet details, overall configuration, insert size and location (if one is needed), interior surface specifications for flow, liner material and installation considerations, details to avoid flow impediments, and considerations for proper gas flow where important.

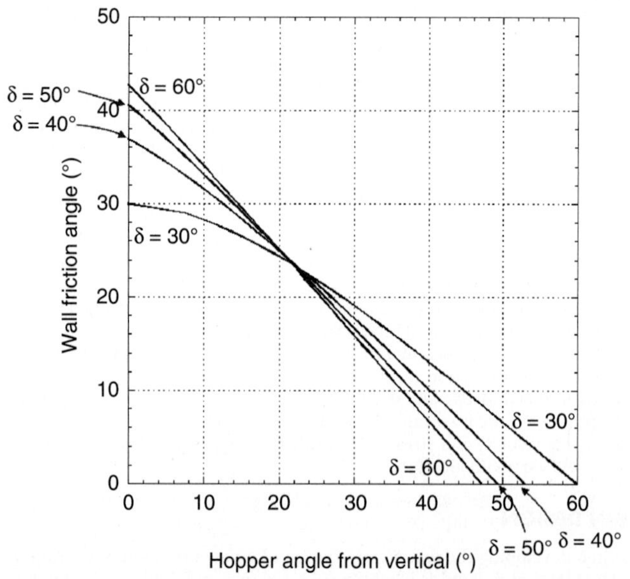

FIG. 21-47 Theoretical mass flow hopper angles for hoppers with round or square outlets. *Note: a minimum safety factor of 3 to 5° should be used.*

pyramidal hoppers. In fact, hoppers with angles less steep than those given in Fig. 21-48 may still allow flow along the walls. Planar flow hoppers are therefore highly suitable for materials that have high values of wall friction angle.

As illustrated in Fig. 21-49, transition hoppers have both straight sides (side walls) and round sides (end walls). The appropriate chart or equation must be used in specifying the angles of the end walls (Fig. 21-47) and side walls (Fig. 21-48) when designing a transition hopper for mass flow.

Additional care must be taken when designing a pyramidal hopper for mass flow. The angles that are formed at the intersections of the sloping walls of pyramidal hoppers are significantly less steep than those of the hopper walls themselves. The valley angle from vertical θ_v can be calculated from

$$\theta_v = \tan^{-1}\sqrt{\tan^2\theta_{\text{side}} + \tan^2\theta_{\text{end}}} \tag{21-57}$$

where θ_{side} and θ_{end} are the side and end wall angles from vertical, respectively. Side, end, and valley angles are defined in Fig. 21-50.

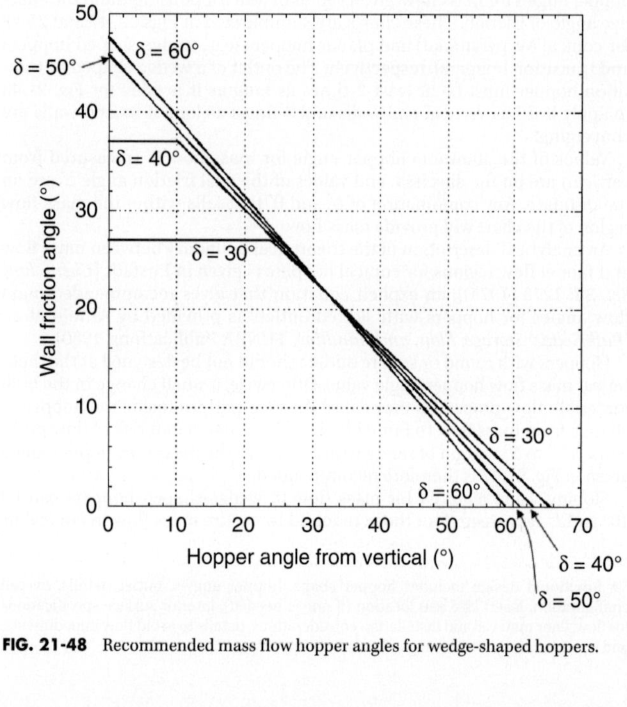

FIG. 21-48 Recommended mass flow hopper angles for wedge-shaped hoppers.

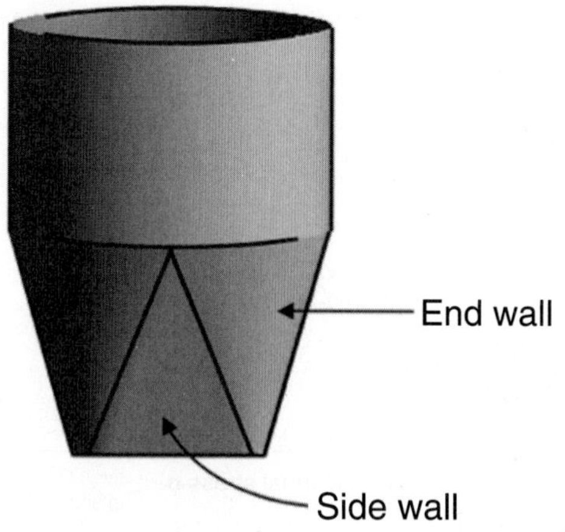

FIG. 21-49 Side and end walls of transition hopper.

Mass Flow Outlet Dimensions to Prevent Arching The outlet of the hopper must be large enough to prevent stable obstructions to flow from developing. The required outlet size depends on the unconfined yield strength, the effective angle of friction, the bulk density of the bulk solid, and the flow pattern.

For an obstruction to flow to develop, the cohesive strength that the bulk solid gains as a result of its consolidation in the hopper must be able to support the weight of the obstruction. Jenike's flow–no flow postulate is as follows [Jenike, *Storage and Flow of Solids*, Bulletin 123, University of Utah, 1964 (revised, 1980)]: "Gravity flow of a solid in a channel will take place provided the yield strength which the solid develops as a result of the action of the consolidating pressure is insufficient to support an obstruction to flow."

In a mass flow bin, as an element of bulk material flows downward, it becomes consolidated under a major consolidation stress σ_1 and develops an unconfined yield strength f_c. The consolidating stress follows the Janssen equation in the vertical section of the bin, sharply changes at the cylinder-hopper junction, and then decreases toward the outlet.

Jenike (*Gravity Flow of Bulk Solids*, Bulletin 108, University of Utah, 1961) calculated the stress on the abutment of a cohesive arch over the outlet $\bar{\sigma}$ as

$$\bar{\sigma} = \frac{\rho_b g B}{H(\theta')} \tag{21-58}$$

where B is the diameter of the outlet of a conical hopper or the width of the slotted outlet of a planar hopper and the function $H(\theta')$ given by Jenike is shown in Fig. 21-51; $H(\theta')$ can be calculated from [Arnold and McLean, *Powder Technol.* **13:** 255 (1976)]

$$H(\theta') = \frac{130° + \theta'}{65°} \tag{21-59}$$

for round outlets and

$$H(\theta') = \frac{200° + \theta'}{200°} \tag{21-60}$$

for slotted outlets.

The stress and strength profiles inside a bin are shown in Fig. 21-52. Note that there is a critical outlet size where the stress on the abutments of a

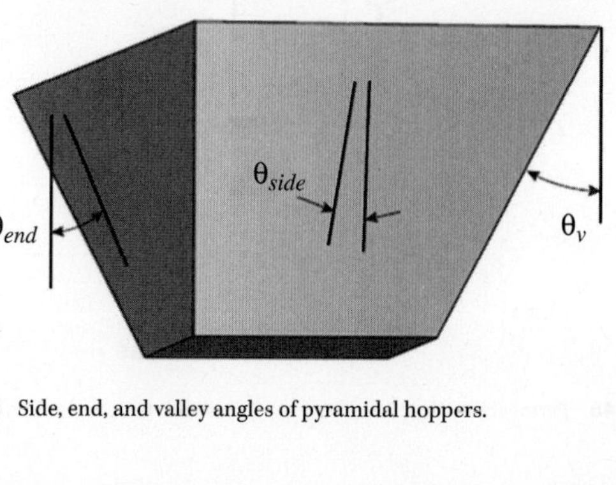

FIG. 21-50 Side, end, and valley angles of pyramidal hoppers.

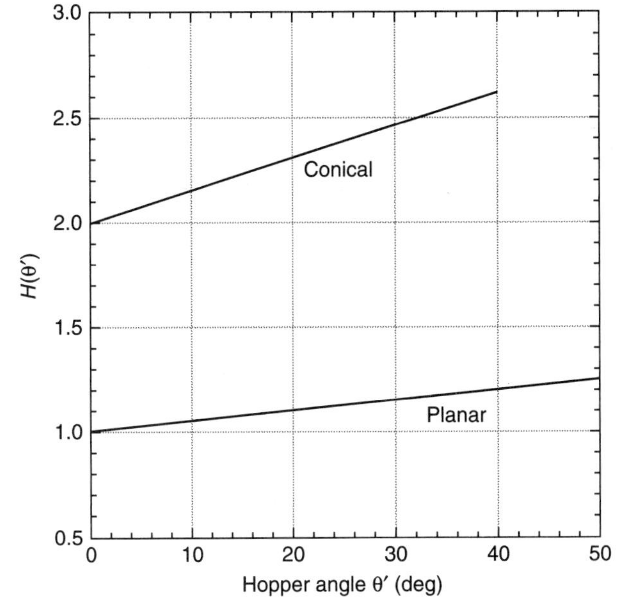

FIG. 21-51 Function $H(\theta')$.

cohesive arch is equal to the unconfined yield strength of the bulk solid. This outlet dimension represents the minimum outlet size that will prevent a stable cohesive arch from developing.

Jenike postulated that near the hopper outlet the stress distribution of the bulk solid could be described by a radial stress field, i.e., the stress distribution could be approximated by a straight line through the hopper vertex. The average stress was modeled as

$$\sigma_{avg} = r\rho_b g s(\theta') \qquad (21\text{-}61)$$

where r is the radial coordinate with the origin located at the vertex of the hopper, σ_{avg} is the average stress, and $s(\theta')$ is called the *stress function*. Jenike (1961) developed solutions to the stress function and presented them in chart form.

The major consolidation stress is related to the average stress by

$$\sigma_1 = \sigma_{avg}(1+\sin\delta) \qquad (21\text{-}62)$$

At the hopper outlet,

$$\sigma_1 = \frac{B\rho_b g s(\theta')(1+\sin\delta)}{2\sin\theta'} \qquad (21\text{-}63)$$

Jenike (1961) defined the ratio of the major consolidation stress to the arch support stress as the flow factor *ff*, that is,

$$ff = \frac{\sigma_1}{\bar{\sigma}} \qquad (21\text{-}64)$$

Hence, the flow factor is given by

$$ff = \frac{H(\theta')s(\theta')(1+\sin\delta)}{2\sin\theta'} \qquad (21\text{-}65)$$

The flow factor is a function of the hopper angle θ', angle of wall friction ϕ', and the effective angle of friction δ. The latter depends on the major consolidation stress σ_1 at the hopper outlet. The angle of wall friction depends on the stress normal to the hopper wall σ', which is *not* equal to σ_1.

Charts that provide flow factors for conical and planar flow hoppers based on Jenike's solutions to the stress function [Jenike, *Storage and Flow of Solids*, Bulletin 123, University of Utah, 1964 (revised, 1980)] are given in Figs. 21-53 through 21-60. Explicit expressions for the flow factor from an analytical form of the stress function were derived by Arnold and McLean [*Powder Technol.* **13:** 255 (1976); *Powder Technol.* **72:** 121 (1992)].

Superimposing the material's flow function and flow factor on the same graph allows the unconfined yield strength and arch stress to be compared. The flow factor is constructed by drawing a line having a slope equal to $1/ff$ through the origin.

The relationship between the effective angle of friction δ and the major consolidation stress σ_1 is provided by the effective yield locus. In a converging hopper, the stresses in the bulk solid are represented by Mohr's circle that is tangent to the material's effective yield locus. The intersections of Mohr's circle and the horizontal axis gives the principal stresses. In mass flow, the material is also slipping along the hopper wall, and therefore the wall stress σ' is represented by the wall yield locus. The shear and normal stresses at the wall are therefore located at the upper intersection of the wall yield locus and Mohr's circle. The relationship between $\sigma_1, \delta, \sigma'$, and ϕ' is illustrated in Fig. 21-61.

To determine the size of the outlet required to prevent arching, the flow function and flow factor are compared. The flow factor is dependent on the material's effective angle of friction δ and its angle of wall friction ϕ' as well as the hopper angle and geometry. The angle of wall friction is a function of the stress normal to the hopper wall σ'. Hence, unless the angle of wall friction and effective angle of friction are constant, calculation of the critical outlet diameter or width is iterative. The procedure is as follows

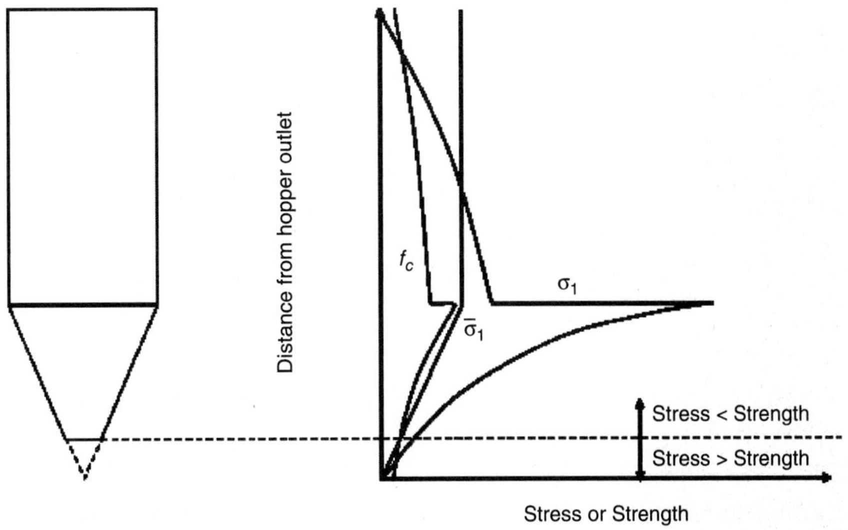

FIG. 21-52 Stress and strength profiles of mass flow hopper.

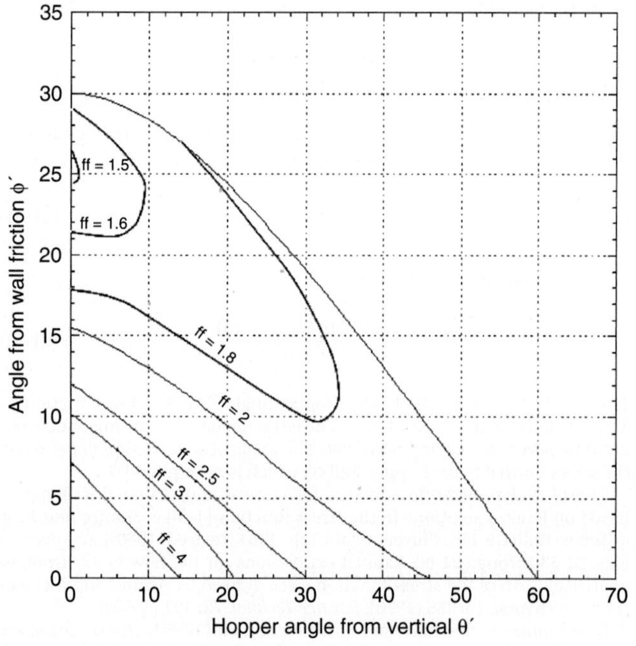

FIG. 21-53 Flow factors for conical hoppers, δ = 30°.

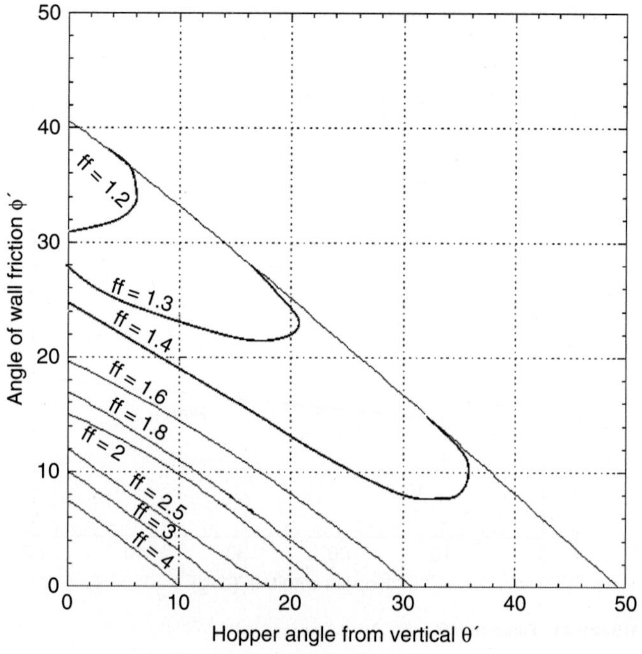

FIG. 21-55 Flow factors for conical hoppers, δ = 50°.

[Jenike, *Storage and Flow of Solids*, Bulletin 123, University of Utah, 1964 (revised, 1980)]:

1. The effective angle of friction δ, wall friction angle φ′, and bulk density ρ_b are estimated.
2. The hopper angle is selected, one that ensures mass flow, by using the appropriate charts (Fig. 21-47 or Fig. 21-48). Note that if a conical hopper is to be specified, a safety factor of at least 3° should be used with respect to the theoretical mass flow boundary.
3. The flow factor *ff* is determined from the appropriate chart given by Figs. 21-53 through 21-60.
4. The flow factor and flow function are plotted together. As shown in Fig. 21-62, there are three possibilities:

a. There is no intersection, and the flow function lies below the flow factor. A dimension *B* that is the minimum that prevents cohesive arching cannot be determined. Instead, *B* is selected based on other considerations such as discharge rate requirements, choice of feeder, or prevention of particle interlocking. The major consolidation stress σ_1 is determined from Eq. (21-66):

$$\sigma_1 = ff\,\frac{\rho_b g B}{H(\theta')} \tag{21-66}$$

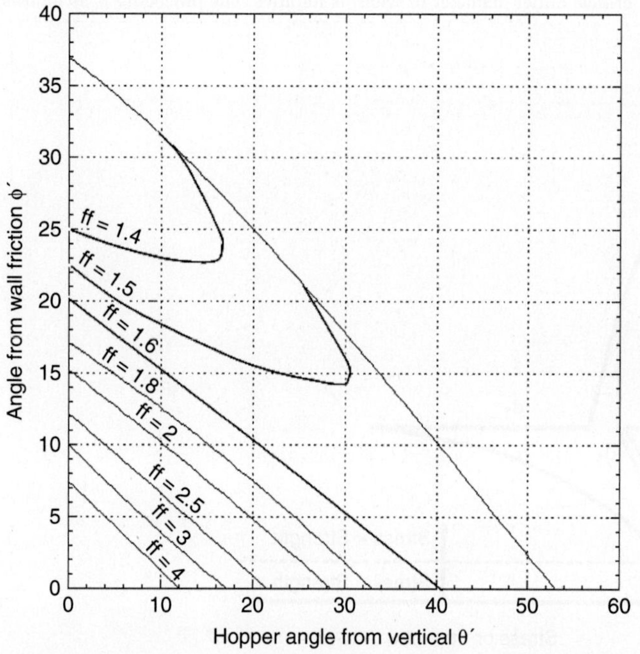

FIG. 21-54 Flow factors for conical hoppers, δ = 40°.

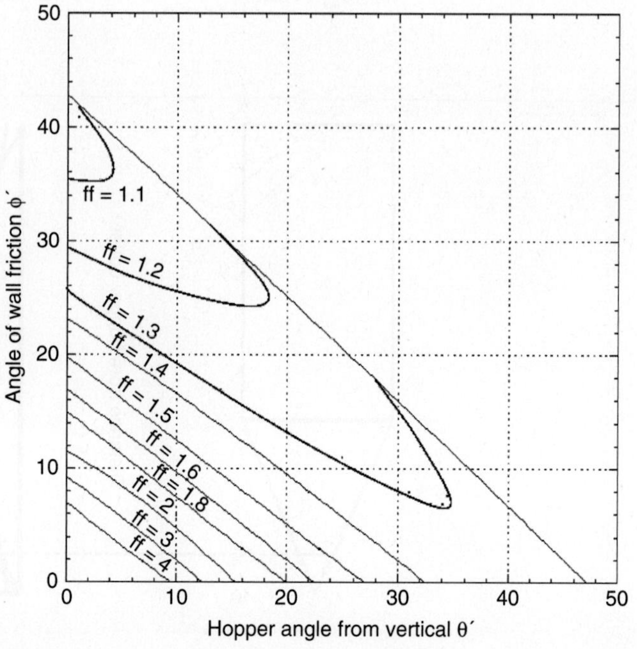

FIG. 21-56 Flow factors for conical hoppers, δ = 60°.

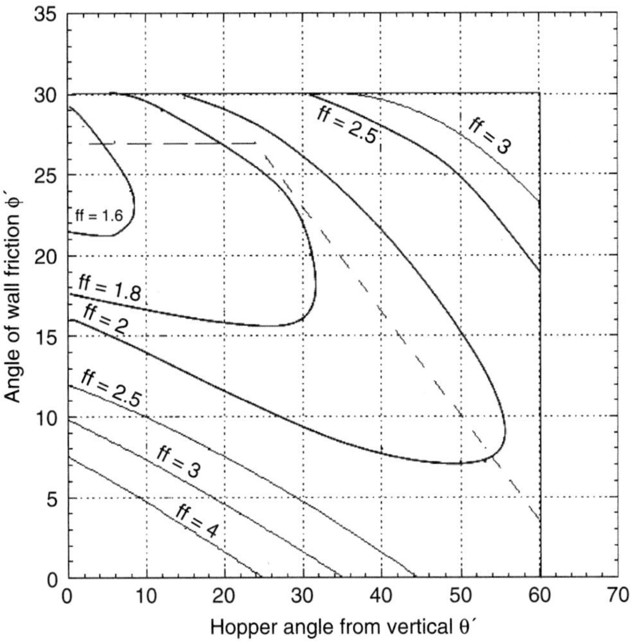

FIG. 21-57 Flow factors for planar flow hoppers with slotted outlets, $\delta = 30°$.

b. The flow factor and flow function intersect. The minimum outlet dimension $B_{\min}$ is calculated using Eq. (21-67):

$$B_{\min} = \frac{H(\theta')\sigma_{\text{crit}}}{\rho_b g} \tag{21-67}$$

Larger outlet diameters or widths of course can be used, and they are generally selected by considering standard feeder sizes or discharge rate requirements.

c. There is no intersection and the flow function lies above the flow factor. Gravity flow will no longer be possible in a hopper with converging walls. Consideration should be given to changing the flow properties of the material, such as increasing its particle size, reducing its moisture content, or using a flow aid.

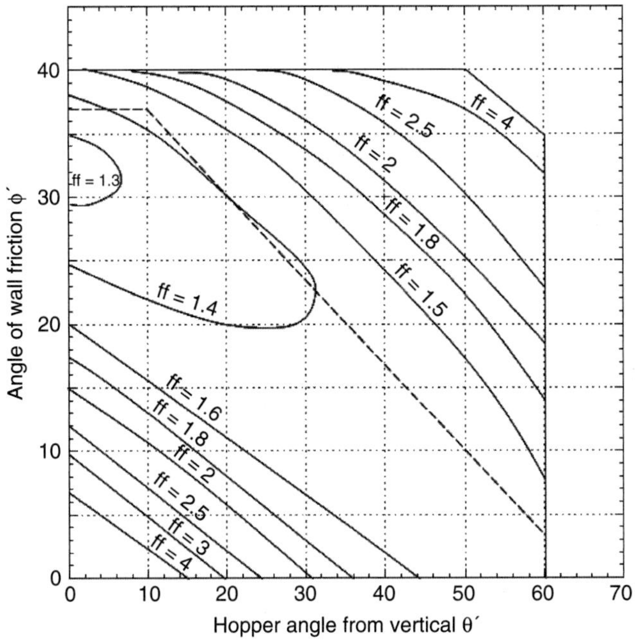

FIG. 21-58 Flow factors for planar flow hoppers with slotted outlets, $\delta = 40°$.

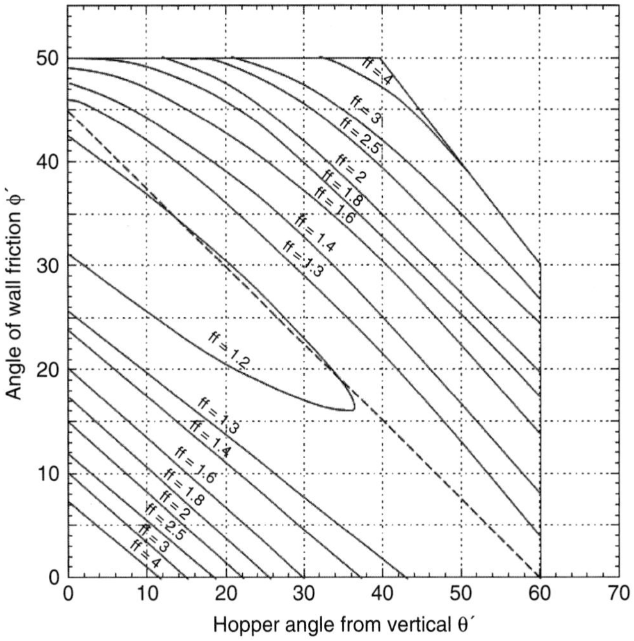

FIG. 21-59 Flow factors for planar flow hoppers with slotted outlets, $\delta = 50°$.

5. The value of ϕ' at the outlet is checked. The effective angle of friction is determined from a plot of δ versus σ_1, and the effective yield locus is drawn by drawing a straight line through the origin at an angle equal to δ. Mohr's circle is drawn through σ_1 that is tangent to the effective yield locus. The value of ϕ' is found from the intersection of Mohr's circle and the wall yield locus, as shown in Fig. 21-61.

6. The recommended hopper angle θ' is updated based on the new value of ϕ'. The steps are repeated until convergence is reached.

To prevent mechanical interlocking, the following rules of thumbs are used: for a conical hopper, the outlet diameter should be at least 6 to 8 times the size of the largest particle that will be handled; for hoppers with slotted outlets, the outlet width should be at least 3 to 4 times the largest particle size.

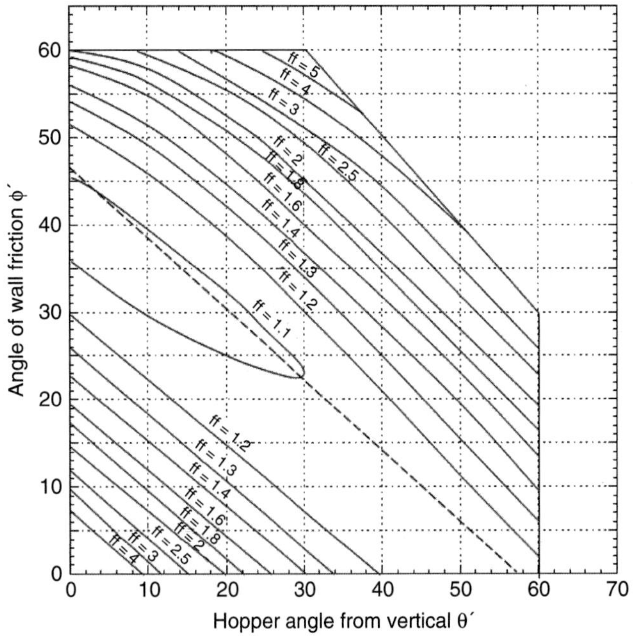

FIG. 21-60 Flow factors for planar flow hoppers with slotted outlets, $\delta = 60°$.

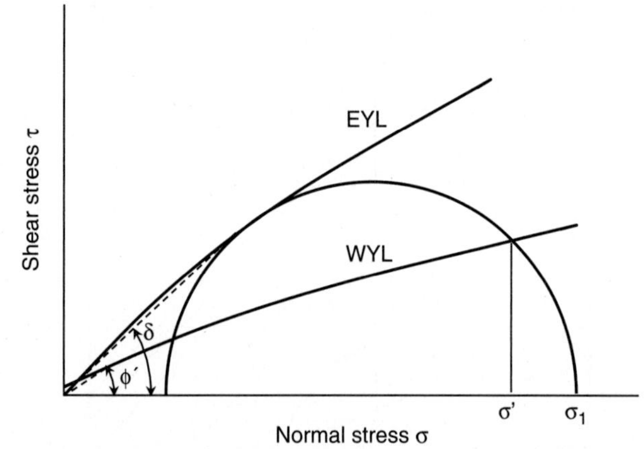

FIG. 21-61 Construction of effective yield locus and wall yield locus.

If a bulk solid is to be stored at rest in a bin, the flow function and wall yield locus must be based on time tests. The intersection of the time flow function and flow factor is used to determine the critical stress and hence the minimum outlet size.

Mass Flow Discharge Rates While an outlet diameter greater than the minimum will prevent cohesive arching, it may not necessarily be large enough to allow the desired discharge rate. From a force balance together with continuity of the solids, the following equation is derived for the discharge rate of coarse powders from mass flow hoppers:

$$\dot{m}_s = \rho_{bO} A_O \sqrt{\frac{Bg}{2(m+1)\tan\theta'}} \tag{21-68}$$

where $\dot{m}_s$ is the solids discharge rate, A_O is the cross-sectional area of the outlet, ρ_{bO} is the bulk density of the material at the hopper outlet, m is equal to 0 for conical hoppers and equal to 1 for hoppers with straight walls and slotted outlets, and B is the diameter of the outlet of a conical hopper or the width of a slotted outlet beneath a planar flow hopper.

The maximum flow rate of a fine powder can be several orders of magnitude lower than that of coarser materials. Two-phase flow effects are significant due to the movement of interstitial gas as the powder compresses and expands during flow. Figure 21-63 illustrates solids and gas pressure profiles in bins for coarse (high-permeability) and fine (low-permeability) powders.

In the cylindrical portion of a bin, the stress level increases with depth, causing the bulk density of the material to increase and the void fraction to decrease, squeezing out a portion of the interstitial gas. This gas leaves the top free surface of the bulk material. In the converging section of the vessel, the consolidated material expands as it moves toward the outlet, reducing the material's bulk density and increasing its void fraction. This expansion results in a reduction in interstitial gas pressure to below atmospheric (i.e., vacuum), causing gas counterflow through the outlet if the gas pressure below the outlet is atmospheric. At a critical solids discharge rate, the solids stress drops to zero, and efforts to exceed this limiting discharge rate will result in erratic flow.

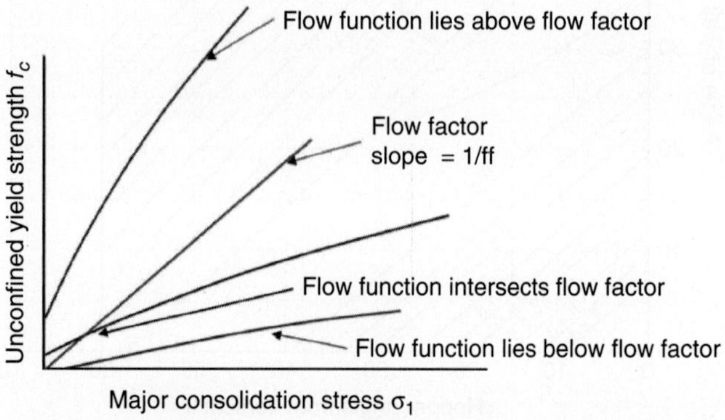

FIG. 21-62 Plot showing both flow factor and flow function.

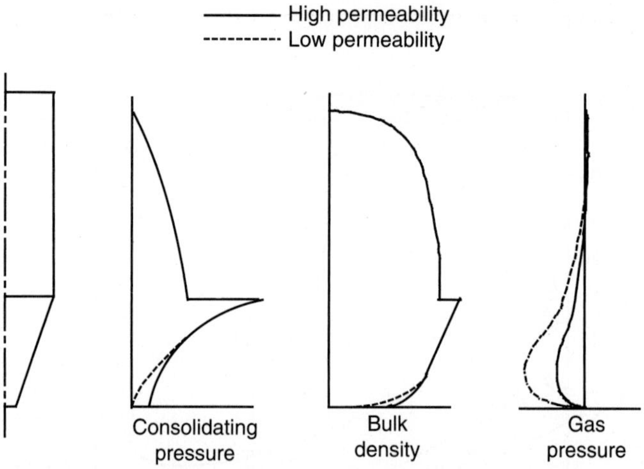

FIG. 21-63 Consolidating pressure, bulk density, and gas pressure profiles for coarse (high permeability) and fine (low permeability) powders.

For fine powders, the discharge rate from a mass flow hopper is given by

$$\dot{m}_s = \rho_{bO} A_O \sqrt{\frac{Bg}{2(m+1)\tan\theta'}\left(1+\frac{1}{\rho_b g}\frac{dP}{dz}\bigg|_O\right)} \tag{21-69}$$

Because the gas pressure gradient at the outlet $(dP/dz)|_O$ is often less than zero for fine powders, Eq. (21-69) shows they can have discharge rates dramatically lower than those of coarse powders.

The gas pressure gradient is related to the material's permeability and the rate of air counterflow by Darcy's law [Eq. (21-55)]. Gu et al. [*Powder Technol.* **72:** 39 (1992)] derived a relationship between the air and solids flow rates that when combined with Darcy's law gives

$$\frac{dP}{dz}\bigg|_O = \frac{\dot{m}_s \rho_{bO} g}{K_O A_O}\left(\frac{1}{\rho_{ba}} - \frac{1}{\rho_{bO}}\right) \tag{21-70}$$

where K_O is the permeability of the powder at the hopper outlet, ρ_{bO} is its bulk density at the outlet, and ρ_{ba} is the bulk density at a location inside the hopper where the gas pressure gradient is equal to zero. Calculating this value is challenging, and therefore an estimate that is proportional to the stress given by the Janssen equation is frequently used. Note that the calculated limiting mass flow discharge rate can be very sensitive to the choice of ρ_{ba}.

The stress balance used in the analysis does not account for the cohesiveness of the powder, and hence the expressions are only valid for outlet sizes that are significantly larger than the critical arching dimensions. Jenike & Johanson, Inc. (Tyngsboro, Mass., United States) uses a proprietary computer model based on unconfined yield strength, permeability, and compressibility data to calculate discharge rates from mass flow bins.

To increase the flow rate of fine powders, injection of a small amount of air above the hopper outlet is often effective, as it will eliminate the opposing air pressure gradient if injected at the correct rate and at the proper location.

Funnel Flow Outlet Size to Prevent Arching and Ratholing For funnel flow hoppers, the outlet must be large enough to prevent both a cohesive arch and stable rathole from developing. The critical rathole diameter is calculated by first determining the major consolidating pressure σ_1 on the bulk solid. The consolidating load can be estimated by the Janssen equation

$$\sigma_1 = \frac{\rho_b g R_H}{k\tan\phi'}\left[1-\exp\left(\frac{-k(\tan\phi')h}{R_H}\right)\right] \tag{21-71}$$

where R_H is the hydraulic radius of the vertical section of the hopper, h is its height, and k is the Janssen coefficient.

Jenike (*Gravity Flow of Bulk Solids*, Bulletin 108, University of Utah, 1961) calculated the stress on a rathole as

$$\bar{\sigma}_1 = \frac{\rho_b g D}{G(\phi_t)} \tag{21-72}$$

where D is the diameter of a round outlet or the diagonal of a slotted outlet and $G(\phi_t)$ is a function given by Jenike, which is plotted in Fig. 21-64.

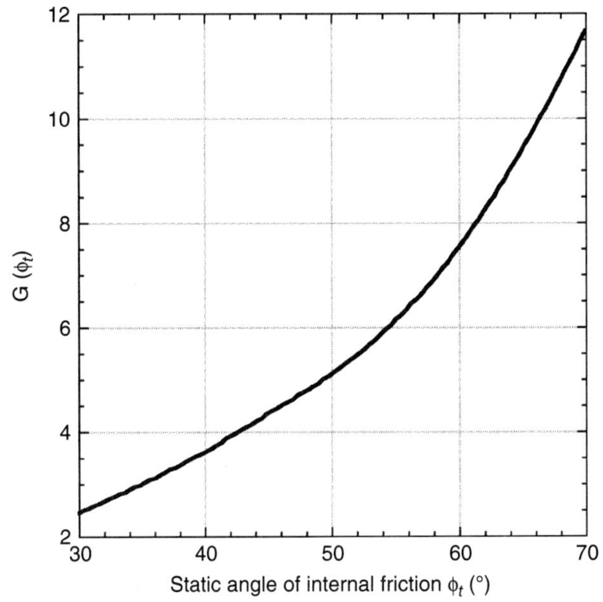

FIG. 21-64 Function $G(\phi_t)$.

Using Jenike's flow–no flow postulate, the rathole will collapse provided that the flow channel stress is greater than the cohesive strength of the bulk solid that makes up the rathole. The critical rathole diameter D_F can therefore be calculated as

$$D_F = \frac{G(\phi_t) f_C}{\rho_b g} \qquad (21\text{-}73)$$

where f_C is the unconfined yield strength of the bulk solid at the consolidation pressure given by the Janssen equation.

A conical funnel flow hopper with an outlet diameter smaller than D_F or a planar funnel flow hopper with an outlet whose diagonal is less than D_F will not empty completely. This is illustrated in Fig. 21-65. Because the major consolidation stress is higher in the lower part of the bin, the unconfined yield strength of the bulk solid will be correspondingly higher. As material discharges in a funnel flow pattern, ratholes that form in the upper part of the vessel may continually collapse, provided that the stress on the stagnant material is greater than its cohesive strength. However, if the size of the outlet is smaller than the critical rathole diameter, a level will be reached where the ratholes will no longer fail.

If a hopper with a square or round outlet is designed with an opening large enough to prevent development of a stable rathole, cohesive arching will not occur. When funnel flow hoppers with elongated outlets are designed, prevention of arching must also be considered, i.e., the width of the slotted outlet must be large enough to prevent a cohesive arch from developing. The same procedure that is used to determine the minimum outlet width to prevent arching in a planar flow mass flow hopper is followed, except that a flow factor of 1.7 is used.

Expanded Flow Hopper Dimensions An expanded flow hopper is essentially a funnel flow hopper above a mass flow hopper. The upper diameter of the mass flow section must be larger than the critical rathole diameter D_F, while its outlet size must be larger than the critical arching dimension. An example of an expanded flow hopper is shown in Fig. 21-66.

The main advantage of expanded flow compared to complete mass flow is that the overall bin height is less while at the same time a stable rathole cannot develop.

Capacity The mass of bulk material that must be stored is generally based on production rate, frequency of operation, required inventory, and, in some cases, residence time. Once the required mass has been determined, the necessary volumetric capacity can be determined. Because bulk solids are generally compressible, a suitable average bulk density should be used for converting mass to volume. To be conservative, a low bulk density value should be chosen to ensure that the volume of the bin is sufficient. Also note that the working capacity of a bin will be less than the actual volume since a pile will form when the bin is filled.

A reasonable height-to-diameter ratio (H/D) of the cylinder section should be used, with ratios between about 1.5 to 4 usually being the most economical. Height may be limited because of building constraints, zoning considerations, or constraints imposed by other structures or equipment.

The volume V and height H of some common hopper designs are given in Fig. 21-67.

The location and the number of inlets to a bin are not as important as the number and location of its outlets. In general, the use of a single outlet is preferable. If multiple outlets are required, the best option is to use a single, centered outlet beneath the hopper section and then split the stream beneath it.

If multiple outlets are to be located below a circular cylinder, each outlet should be located equidistant from the bin centerline. Examples of proper and improper multiple outlet designs are shown in Fig. 21-68. If a bin is center-filled, the outlet arrangement on the right is preferable to that on the left because similar material discharges from each outlet provided that all outlets are fully open and material is discharging at approximately the same rate from each one.

Useful Guidelines The quality of construction of a bin—particularly its interior surface—is critical to its ability to function as desired. If horizontal welds, incorrectly lapped liner plates, or poorly constructed mating flanges exist in a bin designed for mass flow, it may discharge in a funnel flow pattern. The lower of two mating flanges should be oversized to prevent any protrusions into the flowing solid. All flanges should be attached to the outside of the hopper, with the hopper wall material being the surface in contact with the flowing solids.

A feeder, slide gate, or both may be used below the hopper outlet. If a gate is used below a mass flow hopper, the gate must be either fully open or fully closed. A partially opened gate creates a flow obstruction and will convert what would otherwise be a mass flow design into funnel flow. A gate should never be used to modulate flow in a mass flow bin.

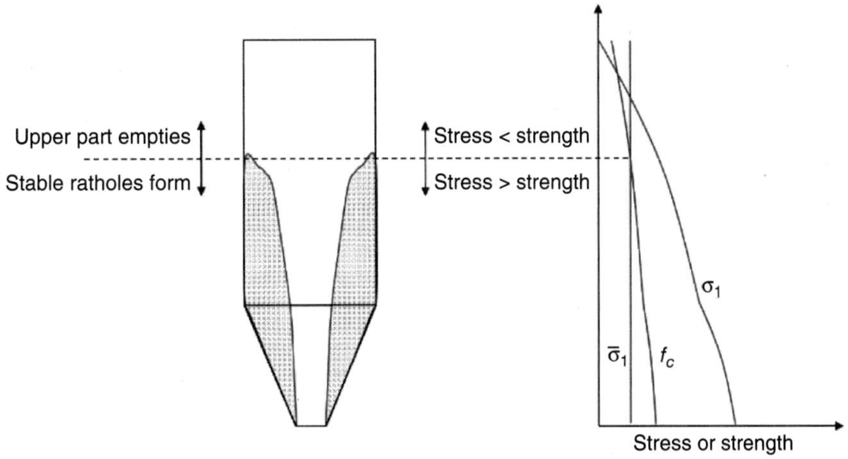

FIG. 21-65 Formation of a stable rathole in a funnel flow hopper.

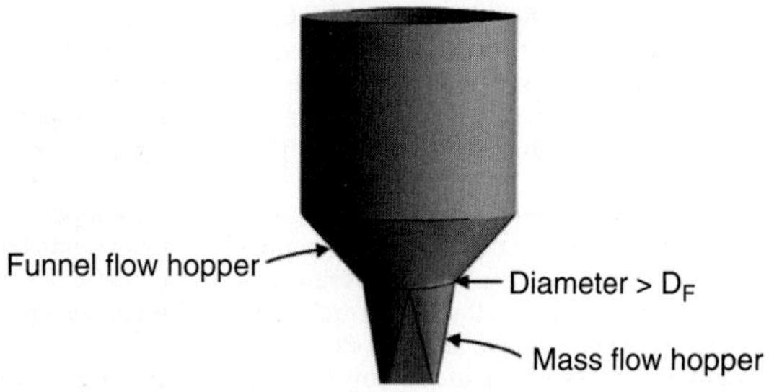

FIG. 21-66 Expanded flow hopper.

When using a feeder under a slotted outlet, its capacity must increase along the outlet length in the discharge direction. This is generally accomplished using a specially designed screw or belt. When a screw feeder is used, the screw should be comprised of a section having a tapered shaft and a constant shaft diameter, increasing pitch section. A properly designed interface between a slotted outlet of a hopper and a belt will progressively discharge more material onto the belt along its length.

A rotary valve is often used as a feeder under a conical hopper outlet to control flow. Common problems experienced with such use of a rotary valve are arching of material over the valve opening, erratic discharge, and preferential flow from one side of the hopper into one side of the valve. Most rotary valves have a transitioning throat area, which is significantly smaller than the nominal size of the valve and may be too small to prevent arching. In addition, this transition section may have a rougher than expected surface,

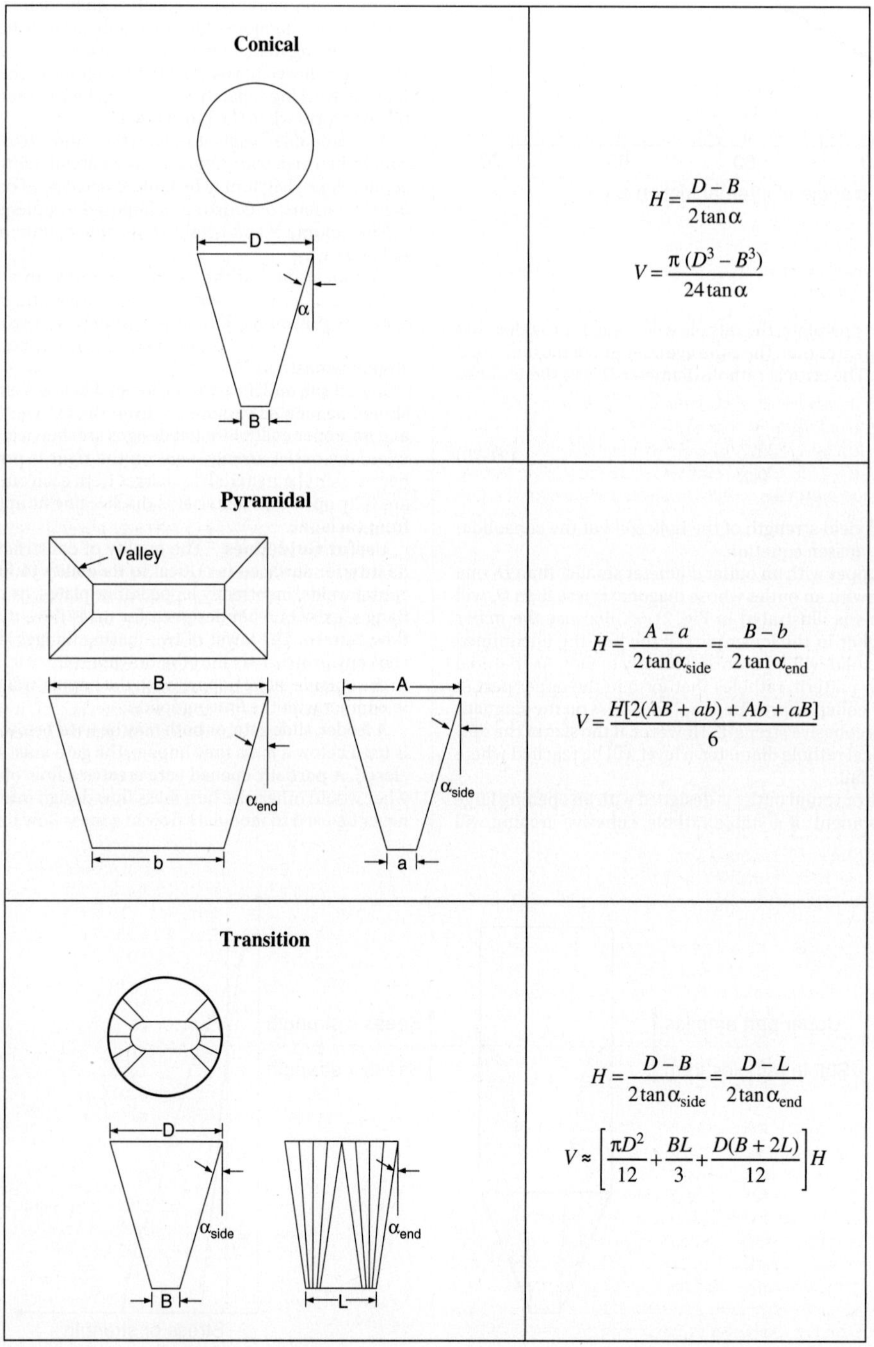

FIG. 21-67 Hopper capacities.

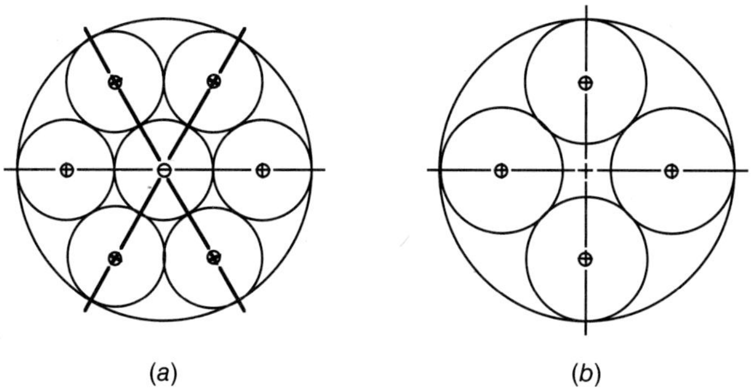

FIG. 21-68 Multiple outlets below a circular cylinder; (*a*) poor design; (*b*) recommended design.

preventing flow along it. Preferential flow can be overcome by having a vertical section with a length greater than its diameter.

Gas leakage through the rotary valve often restricts the material discharge rate, thereby lowering the efficiency of the valve, and it can even cause arching and erratic discharge. Proper venting of rotary valves is usually important, but it is critical if a fine powder is being handled.

Handling an abrasive bulk solid in a mass flow bin may result in significant abrasive wear of the wall material. Generally, the hopper surface becomes smoother with wear; however, occasionally the wall becomes rougher, which may upset mass flow. Wear is a function of the bulk solid flow properties, wall surface, solids pressure, and velocity. The life of a given wall material can be estimated by conducting wear tests in which a screw continuously feeds a fresh sample of bulk solid to a rotating coupon of wall material.

The material of construction of most bins is metal, with reinforced concrete the most common alternative. When because of its size the bin must be erected in the field, reinforced concrete should be considered. A double layer of reinforcing steel (rebar) is likely required, especially if eccentric loads are possible. Reinforced concrete is also preferred over metal if abrasion or corrosion is a concern.

FLOW AIDS

Flow aids are mechanical or pneumatic devices or chemical additives used to induce bulk solids to flow more easily. Vibrators and air cannons are two examples of common mechanical and pneumatic flow aids. Common chemical additives include silicates, stearates, and phosphates.

Vibrators impart forces to the bulk solid through the walls of the storage vessel, most frequently on the hopper walls. Some vibrators produce low-frequency, high-amplitude forces, while others deliver high-frequency, low-amplitude forces. Their effect on flow obstructions can vary. In some cases, they may be an effective means of restoring flow when a bin becomes plugged. In many cases, however, their effect is minimal or can even exacerbate flow problems. Applying sufficient but not excessive force where it is required to collapse an arch or rathole is difficult, particularly in the case of a rathole where the force must usually be transmitted through a significant mass of material to reach it.

The force required to overcome a cohesive arch depends on the bulk solid's cohesive strength and the size of the outlet. If the hopper outlet is slightly undersized, that is, its diameter or width is only marginally smaller than its critical arching dimension after storage at rest, a vibrator may be able to provide enough energy to restart flow. Because ratholes are inherently stable, the outlet diameter required to prevent a rathole from forming can be several times the outlet size of a bin; thus vibrators in general cannot be used to overcome ratholing.

A steep flow function is evidence of a bulk solid that is pressure-sensitive; i.e., its strength increases substantially when additional stresses are applied. Vibrating such a bulk solid often makes flow problems more severe.

A vibrating discharger or bin activator employs an inverted cone or dish that moves in a gyratory, horizontal, or vertical motion. The bulk solid then flows around the cone or dish into a conical section below it, which essentially operates as a chute. Vibratory dischargers can be effective in overcoming flow problems in some bins, provided that they are used appropriately. When used at the outlet of a funnel flow bin, the flow channel above it will be approximately the size of the top diameter of the discharger. If this diameter is smaller than the material's critical rathole diameter, a stable rathole will form, and the discharger will be ineffective in overcoming it.

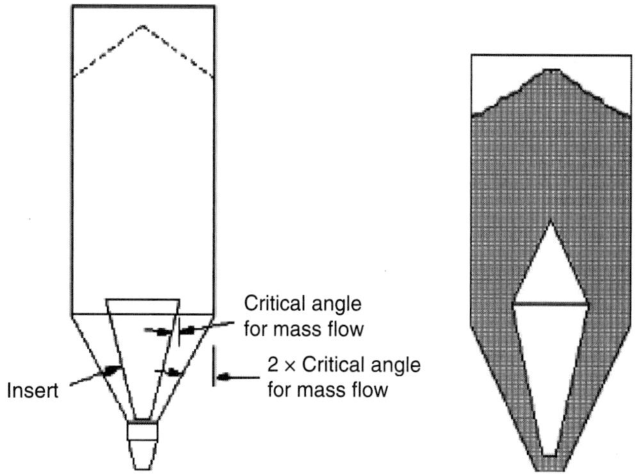

FIG. 21-69 Inserts: cone-in-cone (*left*) and bullet (*right*).

Air (or nitrogen) cannons operate differently in that they rely on a pressure wave to provide the stress required to break an arch. Cannons work by releasing a volume of high-pressure gas into the bin. The required size, number, and location of the cannons depend on the cohesive strength of the bulk material and the dimensions of the bin.

Air cannons are best used for reinitiating flow after a cohesive arch develops when the material is stored at rest. If flow problems occur without storage at rest, air cannons are unlikely to be a viable solution, as they must be fired repeatedly to maintain flow. Air cannons are usually not effective in preventing flow problems in funnel flow bins since ratholes are inherently stable.

Chemical flow aids are often used to prevent arching or the formation of a stable rathole. Parting agents such as silicates, stearates, and phosphates are effective as they increase the distance between adjacent particles, thereby reducing the magnitude of their cohesive forces. Note that while a flow aid may be effective in reducing a bulk solid's cohesive strength, the additive may increase wall friction, potentially resulting in flow problems associated with funnel flow.

Inserts A disadvantage of mass flow bins is that relatively steep hopper sections are generally required, so in some cases bins may be too tall for the available space. When properly designed, an insert can be used to allow mass flow in a bin with shallow hopper walls that, without modifications, would discharge in a funnel flow pattern.

Cone-in-cone and bullet designs are shown in Fig. 21-69. The cone-in-cone insert, or Binsert®, is designed to allow mass flow through the inner cone and also through the annular space between the inner and outer cones. The angle of the inner cone is equal to or steeper than the hopper angle recommended for mass flow in a conical hopper, and the angle of the outer cone is equal to double that of the inner cone. The outlet diameter of the inner cone must be greater than the critical arching diameter. For cohesive materials that would otherwise arch over the outlet of the inner cone of an insert, an inverted cone or "bullet" can be placed above the inner cone.

Note that the location of the inner cone and, when applicable, the size of the bullet greatly influence the effectiveness of an insert. An insert that is too high or too low relative to its optimum size and position will expand the flow channel very little, if at all. Also note that the loads acting on inserts are usually very high, and therefore the supports required to resist these loads may be large and can impede flow.

Air Assist and Fluidization Air pads and air nozzles are sometimes used to inject low-pressure air into a bin. While they are generally ineffective in correcting problems caused by arching or ratholing, they may be useful in increasing the discharge rate of fine powders by reducing the adverse pressure gradient that develops above a hopper outlet and the resulting counterflow of air.

A better technique to increase the discharge rate of fine powders is to use an air permeation system, which consists of a sloping shelf or insert through which a relatively low flow rate of air is introduced. The air reduces or eliminates the vacuum that naturally develops when a bulk solid dilates in the hopper section, increasing its void fraction. To be effective, the air should be distributed uniformly, its flow rate should be low enough to prevent fluidization, and the permeation system should not impede solids flow in the hopper.

Binsert is a registered trademark of Jenike & Johanson, Inc.

Air-assist dischargers are designed to reduce the wall friction angle to nearly zero, thereby allowing powders to flow along hopper walls. The hopper section either is lined with air panels or is fabricated using a permeable membrane through which a small amount of air is injected. Jenike [*Storage and Flow of Solids*, Bulletin 123, University of Utah, 1964 (revised, 1980)] recommends conical hopper angles between 40° and 50° from vertical, as steeper hoppers may require large outlet diameters to prevent arching. Shallow hopper angles can be used provided that a fully open, unrestricted on/off valve is used; or if enough gas is added to completely fluidize the bulk material, a fluidized discharger is used.

A fluidized discharger can be used when the bulk material is fluidizable and a low bulk density of the discharged material is acceptable. Fluidized dischargers can generally be used for Geldart group A, B, and C materials [*Powder Technol.* 7(5): 285 (1973)], although group C materials may require mechanical agitation. Discharge from a bin equipped with a fluidized discharger is typically controlled through use of a rotary valve. Note that if the entire contents of the bin are fluidized, then potential structural issues might develop due to the hydrostatic stress field.

STANDPIPES

A standpipe is frequently used to seal a bin outlet against gas pressures. It is composed of a vertical cylinder mounted between the bin and feeder. Its major advantages over other methods such as sealing screws is that it has no moving parts and it is inexpensive to fabricate and operate.

A standpipe provides a seal against gas pressure by providing a sufficient length to lower the pressure gradient such that the flow of gas through the bed of solids at the hopper outlet is reduced to an acceptable rate. Standpipe design is discussed by Carson and Marinelli [*Power Engr.* **85:** 11 (1981)].

CHUTES

Chutes are used to direct the flow of bulk solids. They need to be properly designed to avoid problems such as plugging, excessive wear, dust generation, and particle attrition.

A chute must be sufficiently steep and low enough in friction to permit sliding and clean off. Referring to Fig. 21-70, the velocity of a stream of particles (assuming no bouncing) after impacting a chute V_2 relative to its velocity before impact V_1 is

$$V_2 = V_1(\cos\theta - \sin\theta \tan\phi') \qquad (21\text{-}74)$$

where θ is the impact angle and ϕ' is the wall friction angle.

If the particles fall freely when they are dropped onto the chute, their velocity before impact V_1 is given by

$$V_1 = \sqrt{2gH} \qquad (21\text{-}75)$$

where H is the drop height.

If the sum of ϕ' and θ equals 90°, V_2 will be reduced to zero, and the bulk material will not slide on the chute surface unless its angle of inclination is greater than a minimum value. To determine this minimum value required to overcome adhesion at impact, chute tests [*Bulk Solids Handling* **12**(3): 447 (1992)] can be performed. A sample of the bulk material is loaded onto a wall coupon, and a load representing the impact pressure is briefly applied. The impact pressure σ is given by

$$\sigma \approx \rho_b g V_1^2 \sin^2\theta \qquad (21\text{-}76)$$

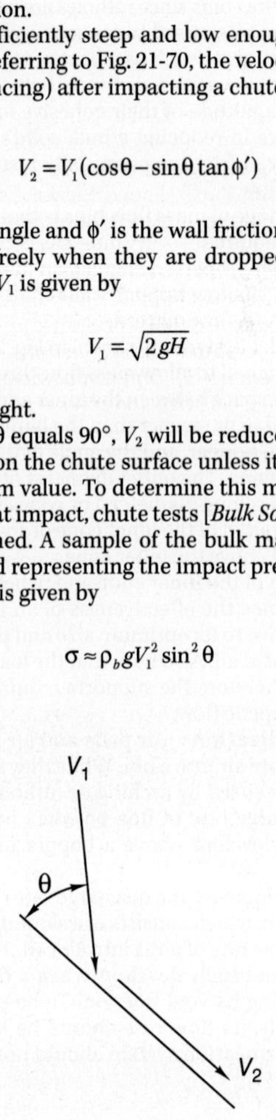

FIG. 21-70 Velocity of a particle after impact on a chute.

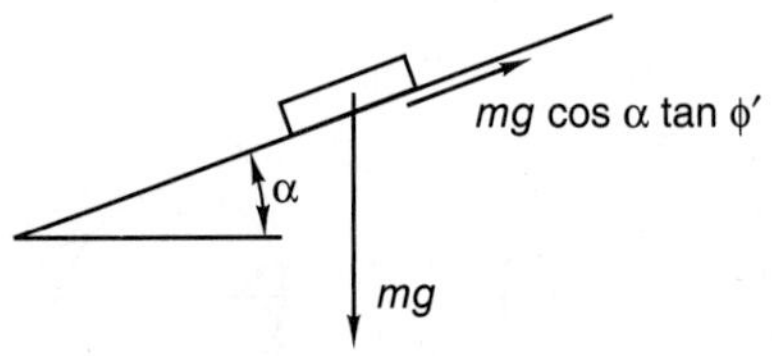

FIG. 21-71 Element of bulk solid sliding on a straight chute.

The coupon is inclined about a pivot point until it just starts to slide. Usually a safety factor of 5° is applied to this minimum value to ensure clean off.

While sliding on a straight surface, the particles will accelerate or decelerate, depending on the relative values of the chute angle α measured from horizontal and the wall friction angle ϕ' (see Fig. 21-71):

$$a = g(\sin\alpha - \cos\alpha \tan\phi') \qquad (21\text{-}77)$$

where a is the acceleration.

Assuming that the chute cross section does not decrease along a distance S on the chute surface, the stream velocity V is given by

$$V = \sqrt{V_0^2 + 2aS} \qquad (21\text{-}78)$$

where V_0 is its velocity at the starting point.

When the velocity of the stream changes as it passes through a chute, its cross-sectional area will change. To prevent flow stoppages, the chute should be sized such that it is no more than about one-third full at the point of minimum velocity.

While chutes can be fabricated and installed in rectangular sections, having curved surfaces upon which the material slides is advantageous. Chutes composed of cylindrical pipes or rounded surfaces control the stream well, as they can be used to center the load, allowing the momentum of the material to keep the chute clean. The path that the bulk material will flow depends on its frictional properties and flow rate. *Discrete element method* (DEM) models should be used to design chutes with complicated geometries.

Free-fall height and sudden changes in the direction of material flow should be minimized to reduce solids impact pressures, which can result in high abrasive wear, attrition, generation of dust, and potential adhesion of solids to the wall. Since impact pressure is proportional to $\sin\theta$ and V_1^2, reducing the impact angle θ and drop height H will reduce wear, and the momentum of the flowing material will keep the chute surface cleaned off.

Dust is created when air is entrained into the flowing material. To avoid dusting, the chute should be designed such that the material remains in contact with the chute surface, the material stream is concentrated, and the velocity through the chute is kept nearly constant. If the material is to land on a belt conveyor at the exit of the chute, the velocity of the stream should be in the direction of and equal to or greater than the velocity of the belt.

Attrition of friable particles is most likely to occur at impact points where the impact pressures are high. Therefore, attrition can be minimized by minimizing the impact angle θ, ensuring that the flowing stream is concentrated and remains in contact with the chute surface, and maintaining a constant stream velocity.

SEGREGATION

Some materials, when transferred into a bin, will segregate; that is, particles of different size, shape, density, etc. will separate. Segregation can occur by a number of different mechanisms, depending on the physical characteristics of the particles and the method of handling. The three most common mechanisms are *fluidization* (air entrainment), *dusting* (particle entrainment), and *sifting*. These segregation methods are illustrated in Fig. 21-72. They are discussed in detail in the Solids Mixing subsection.

Fluidization, or air entrainment, can cause vertical segregation, i.e., horizontal layers of fines and coarse material. Fine powders generally have a lower permeability than coarse materials and therefore retain air longer. Thus, when a bin is being filled, the coarse particles are driven into the bed while the fine particles remain fluidized near the surface. Air entrainment often develops in materials that contain a significant percentage of particles below 100 μm. Fluidization segregation is also likely to occur when a bin is filled or discharged at high rates or if gas counterflow is present. Segregation by the fluidization segregation mechanism is illustrated in Fig. 21-73.

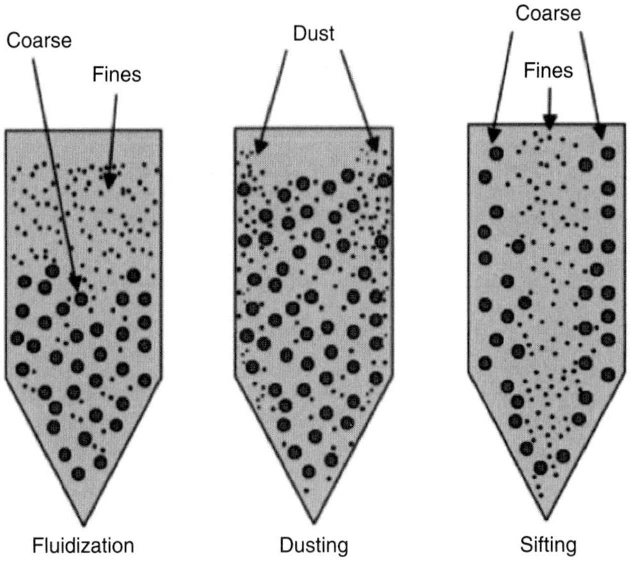

FIG. 21-72 Segregation by fluidization, dusting, and sifting.

Dusting, or particle entrainment, involves airborne particles, differences in settling velocities between particles, and air currents to cause movement of suspended particles. Dusting can occur when powder is dropped and impacts onto a pile surface, causing the release of finer particles into the air. Particles can also be reentrained in air if large pockets of air bubble up through a stationary bed of material from below. These particles will tend to remain suspended in the air and will be carried by air currents to the least active portion of the receiving vessel's area, generally the lowest part of the pile surface that is farthest away from the impact point. Generally, powders that are susceptible to this mechanism contain a portion of finer particles below 50 μm that do not readily adhere to larger particles. Dusting is illustrated in Fig. 21-74.

Sifting occurs when smaller particles move through a matrix of larger ones. Four conditions must exist for sifting to occur:

- A difference in particle size between the individual components, typically a minimum ratio of 2:1 or greater, but has been observed at size ratios as small as 1.2 to 1.3
- A sufficiently large mean particle size, typically one greater than approximately 500 μm
- Free-flowing material
- Interparticle motion

All four of these conditions must exist for sifting segregation to occur. If any one of these conditions does not exist, the mix will not segregate by this mechanism.

Sifting segregation is illustrated in Fig. 21-75, which is a photograph of a typical pile that forms when a vessel is filled. Because coarser particles tend to be more mobile, they roll downward toward the periphery of the pile. Fines percolate through the bed as they fall from the center and accumulate in the middle. The result is side-to-side separation of particles by size.

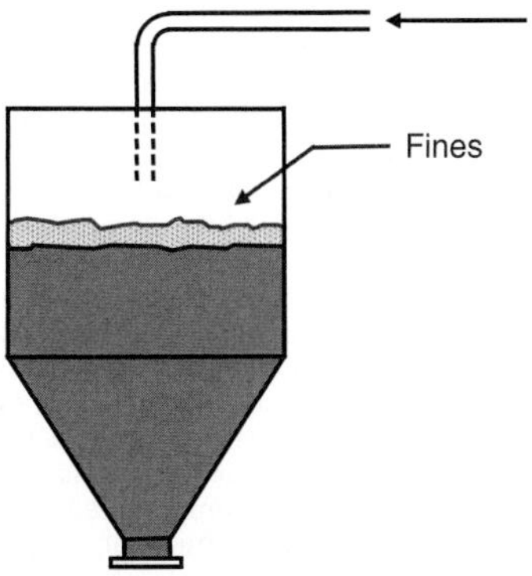

FIG. 21-73 Segregation due to fluidization of fine, light particles.

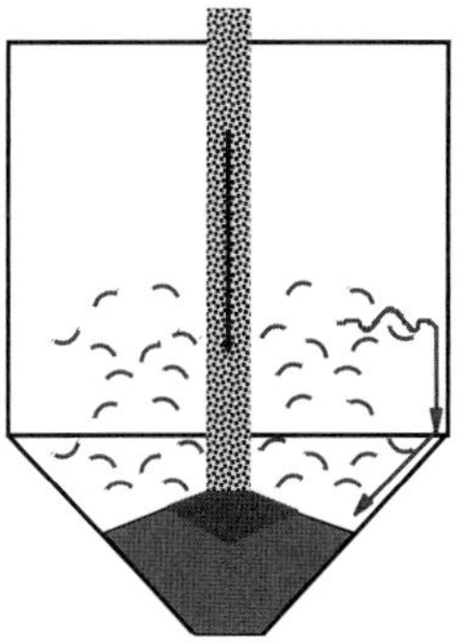

FIG. 21-74 Particle entrainment during filling of a bin.

CAKING

Caking is frequently moisture-induced. When the moisture content of a bulk material reaches a critical value, moisture will condense primarily at the contact points between adjacent particles, causing liquid bridges. If local drying occurs due to temperature swings during storage or transit, solid bridges may form when soluble components in the liquid precipitate. Water is also a plasticizer for many materials, and its presence can cause particles to deform and increase interparticle contact area. Elevated temperature and impurities also frequently increase the likelihood of a material to cake.

Caking occurs when the magnitude of interparticle forces increases significantly over time. These cohesive forces are primarily van der Waals forces, polar interactions, and forces associated with plastic creep or liquid bridges (when moisture is present). Van der Waals forces include all intermolecular forces that act between electrically neutral molecules. Polar interactions occur when adjacent particles contain regions that are permanently electron-rich or electron-poor. Van der Waals forces and polar interactions increase as the distance between particles decreases. Although these forces are proportional to particle size, the likelihood of caking generally decreases with increasing particle size since the number of inter-particle contacts is inversely proportional to the square of the particle diameter.

With some bulk materials, plastic creep, which is the tendency of a material to deform when under consolidation, may occur. Plastic creep can be severe if impurities that behave as plasticizers are present or if the bulk solid is subjected to high temperatures for long periods, especially when above its glass transition temperature T_g. *Differential scanning calorimetry* (DSC), *thermal mechanical analysis* (TMA), and *inverse gas chromatography* (IGC) are frequently used to measure T_g. IGC is preferable over the other methods if moisture is known to act as a plasticizer since it can be conducted at a constant relative humidity.

Liquid bridging occurs when moisture accumulates at the contact points between adjacent particles. The likelihood of liquid bridging can often be inferred from a powder's moisture sorption isotherm, which relates relative humidity and equilibrium moisture content. An example of an isotherm

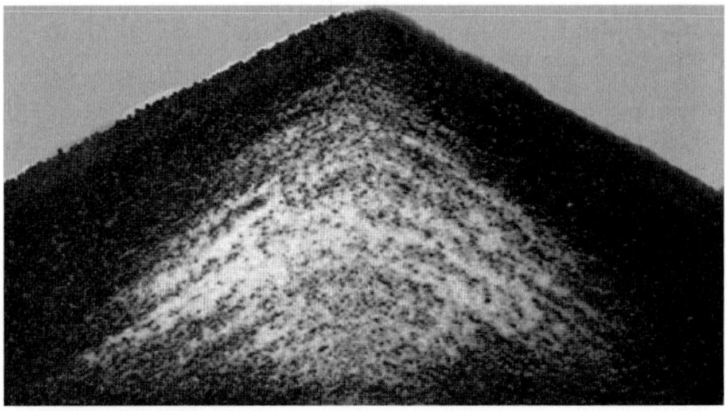

FIG. 21-75 Sifting segregation after formation of a pile.

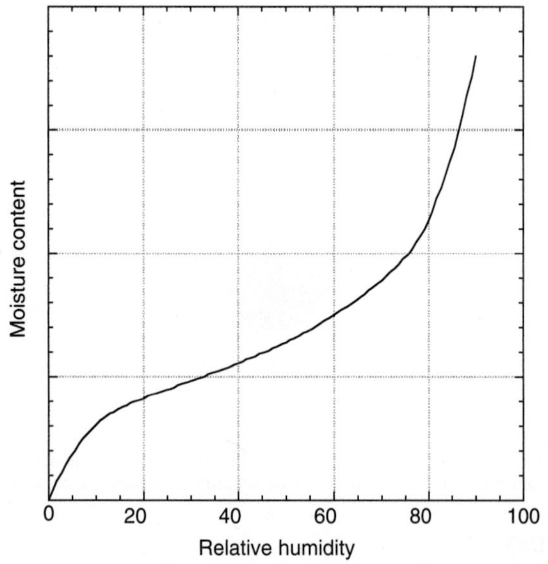

FIG. 21-76 Example isotherm.

that has the sigmoidal shape characteristic of many bulk materials prone to caking is shown in Fig. 21-76.

The moisture isotherm is initially linear as water molecules are adsorbed until a monolayer is formed. The effect of moisture on caking is generally negligible in this region. As relative humidity increases, multilayer adsorption takes place as a consequence of hydrogen bonding. In this region, the slope of the isotherm is initially shallow but steepens with increasing relative humidity. As moisture uptake increases, the particles become surrounded by moisture. If the solids are water-soluble, the layer of moisture can be viscous, and the bulk material may become cohesive.

The third region occurs at high relative humidity, where the equilibrium moisture content increases dramatically. In this region, most of the incremental condensation takes place at the contact points between particles. This phenomenon, which is known as *capillary condensation*, is accompanied by liquid bridging, and it results in strong forces between particles. This leads to caking over time. If soluble matter exists in the liquid bridges, and if the liquid evaporates, then strong, solid bridges may form.

To quantify caking, the temperature, relative humidity, and consolidation pressures used during the test must simulate those expected when the bulk material is stored. A material's unconfined yield strength is best measured using a shear cell tester or uniaxial compression tester. A significant change between the unconfined yield strength measured for continuous handling and after storage at rest indicates that caking has occurred.

Any moisture limits to avoid caking that are based on equilibrium moisture content should account for the possibility of moisture migration. This occurs when a temperature gradient exists during packaging, storage, or transit of powders. The mechanism of caking due to moisture migration is as follows:

- The relative humidity of the interstitial air at the warm boundary decreases.
- As a consequence, moisture desorbs from the warmer solids, as the solids and interstitial air are no longer in equilibrium.
- The *absolute* humidity of the interstitial air increases.
- The driving force in the gas phase leads to moisture migration toward the interior, which has a lower absolute humidity.
- The relative humidity of the cooler interstitial air increases.
- Moisture adsorbs onto solids in the interior in an effort to reestablish equilibrium.

Moisture migration is illustrated in Fig. 21-77.

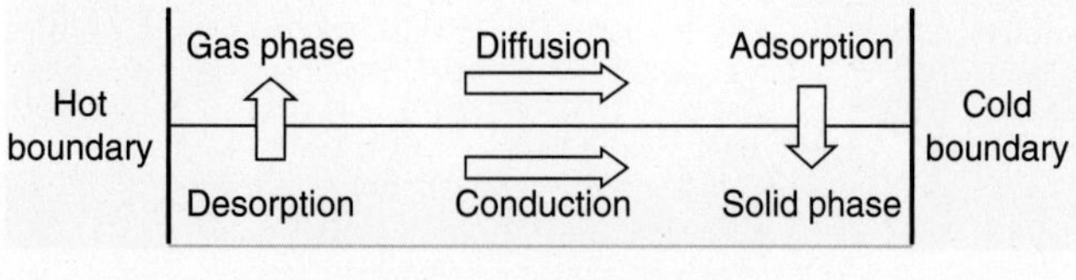

FIG. 21-77 Schematic describing moisture migration.

An analysis can be performed to determine the moisture distribution in a bulk solid that will result if a temperature gradient (e.g., if product pack-out temperatures exceed storage temperatures or if storage temperatures vary) is imposed. One assumes that the temperature gradient remains constant. Since this is not true, the analysis results in a conservative view of possible conditions that can exist if temperature differences were to remain for an extended time.

The analysis is as follows. If a bulk solid is exposed to a warm surface (temperature = T_H) on one side and a cool surface (temperature = T_C) on the other, the temperature profile at steady state would be given by

$$T = T_C + (T_H - T_C)z \tag{21-79}$$

where z is the ratio of the distance from the cold surface to the width between the hot and cold surfaces.

At steady state, the concentration of water in the interstitial air C_w is constant. The vapor phase moisture concentration is the product of the absolute humidity H and the dry air density ρ_a:

$$C_w = \rho_a H \tag{21-80}$$

The relative humidity RH is related to absolute humidity by

$$\frac{\text{RH}}{100} = \frac{29HP_t}{(18+29H)P_w^{\text{sat}}} \tag{21-81}$$

where P_t and P_w^{sat} are the total pressure and saturation pressure of pure water, respectively. Due to the temperature gradient, the relative humidity of the interstitial air will vary. As a result, the amount of condensed moisture that is in equilibrium with the interstitial air will also vary. The relationship between the solid's equilibrium moisture content X and the relative humidity of the interstitial air is given by the material's isotherm. Since the amount of moisture in the gas phase is negligible compared to that in the solid phase, the total amount of moisture in the solid after migration can be assumed to be equal to the initial solid moisture content X_0

$$\bar{X} = \int_z X(z)dz = X_0 \tag{21-82}$$

A specification for a bulk material's moisture content that, if exceeded, causes caking (as determined from unconfined yield strength measurements) can be determined by finding the value of C_w that satisfies Eqs. (21-80) to (21-82) and the material's moisture isotherm.

PROCESS VESSELS

Moving-bed process vessels are hoppers, bins, or silos modified to allow processing of a bulk solid. Examples of processes include heating, cooling, conditioning, drying, and conducting a chemical reaction. In some moving-bed processors, a gas is injected and passes countercurrent or perpendicular to the flow of solids. Others are equipped with plate-and-frame or tubular heat exchangers. Fluidization of the solids is not required, nor is it desired. Rather, the bulk material flows downward as a moving bed toward the processor's outlet.

To ensure trouble-free operation of moving-bed processors, the following design criteria must be met:

1. The outlet of the processor must be large enough to prevent solids flow stoppages and to allow the desired production rate.
2. The solids velocity should be as close to uniform across the processor's cross section as possible and practical. At a minimum, there must not be any regions of stagnant material inside the processor.
3. If required, the gas must be introduced in such a manner that it does not disrupt the flow of solids.
4. The processor must provide the required residence time.
5. If applicable, a driving force for heat or mass transfer must exist throughout the cylindrical section of the processor.

Counterflow, cross-flow, and radial-flow designs exist (see Fig. 21-78). In a counterflow system, the gas introduction system must be designed such that there are no regions with high gas velocities, as localized fluidization will cause channeling, bypassing of the solids, and flow instabilities. Designs that involve nozzles or perforated plates therefore should be avoided. Gas is most uniform when it is injected into the moving solids bed via an annulus and a set of crossbeams located near the intersection of the cylinder and hopper sections of the processor. Two examples of well-designed gas distributors are shown in Fig. 21-79 [Mehos, *Chem. Engr.* **116**(5): 34 (2009)].

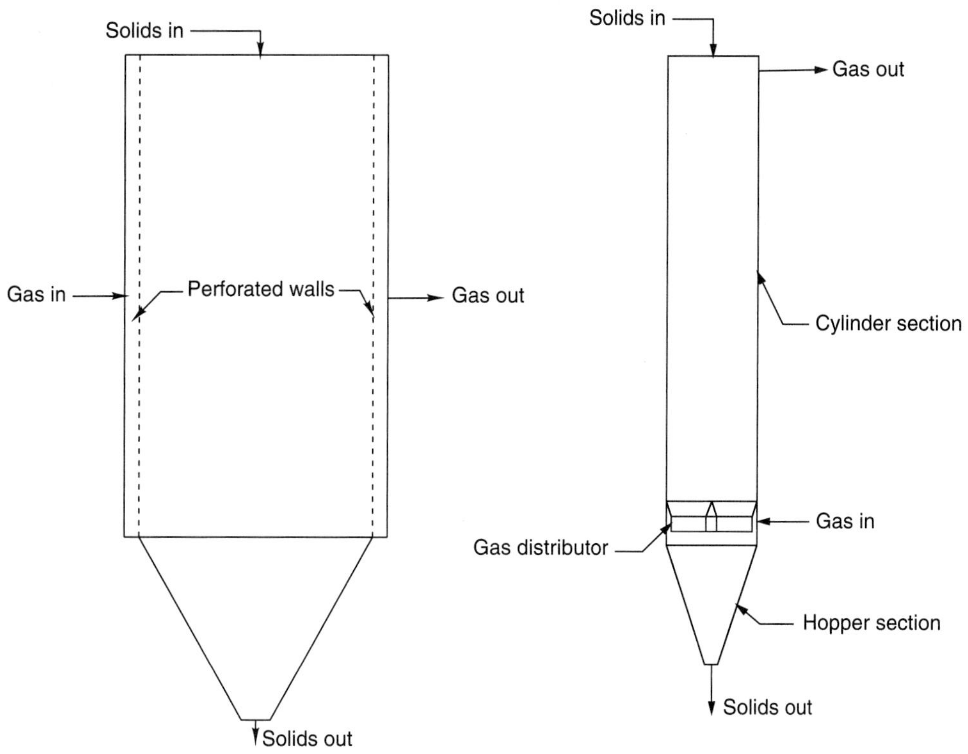

FIG. 21-78 Process vessels with gas injection—countercurrent (*left*) and cross-flow (*right*).

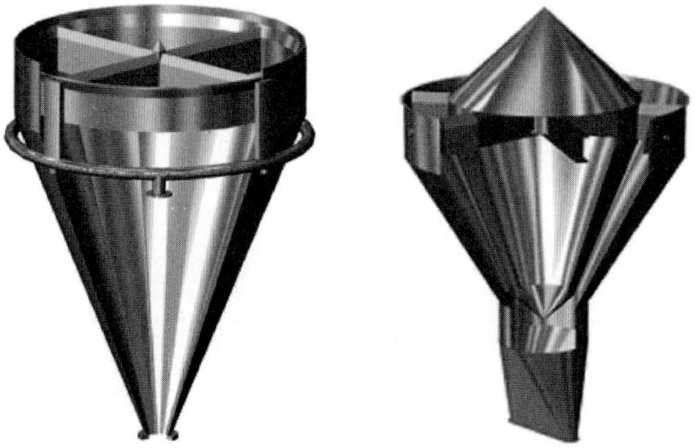

FIG. 21-79 Gas distributors with crossbeams.

In counterflow systems, the gas velocity must be low enough to prevent fluidization in the cylindrical section of the processor. A rule of thumb is that the vessel's cross-sectional area at the level of gas introduction should be such that the superficial gas velocity is no more than approximately one-third the bulk material's minimum fluidization velocity.

The gas flow rate must be low enough to prevent excessive entrainment of fines at the top of the column but high enough to ensure that a driving force for mass or heat transfer exists throughout the column if required.

Cross-flow and radial designs are preferred if the required gas flow rate is high because a low pressure drop will be realized. Cross-flow and radial-flow processors are fabricated with permeable walls through which the gas enters and exits the moving bed of solids.

For cross-flow and radial-flow designs, the gas velocity must be low enough to prevent pinning or cavity formation. Pinning occurs when the frictional force between the bulk solid and the permeable wall through which the gas exits is great enough to prevent the particles from flowing downward and along the wall. A cavity can develop if the pressure gradient that develops when gas is injected into the bed causes a gap to form between the bulk solid and the wall from which the gas is introduced. If a cavity forms, gas will flow preferably upward rather than across the bed.

SOLIDS MIXING

General References: L. T. Fan, Y.-M. Chen, and F. S. Lai, "Recent Developments in Solids Mixing," *Powder Technology* **61:** 255–287 (1990); N. Harnby, M. F. Edwards, and A. W. Nienow (eds.), *Mixing in the Process Industries*, 2d ed., Butterworth-Heinemann, 1992; B. Kaye, *Powder Mixing*, Springer, Netherlands, 1997; R. Weinekötter and H. Gericke, in B. Scarlett (ed.), *Mixing of Solids, Particle Technology Series*, Kluwer Academic Publishers, Dordrecht, 2000; J. M. Ottino and D. V. Khakhar, "Mixing and Segregation of Granular Materials," *Annul Review of Fluid Mechanics* **22**(1): 207–254 (2000); E. L. Paul, V. A. Atiemo-Obeng, S. M. Kresta (eds.), *Handbook of Industrial Mixing: Science and Practice*, Wiley, Hoboken, N.J., United States, 2003; A. Kudrolli, "Size Separation in Vibrated Granular Matter," *Reports on Progress in Physics* **67**(3): 209–247 (2004); J. Bridgwater, "Mixing of Powders and Granular Materials by Mechanical Means—A Perspective," *Particuology* **10**(4): 397–427 (2012).

PRINCIPLES OF SOLIDS MIXING

Industrial Relevance of Solids Mixing The mixing of powders, particles, flakes, and granules has substantial economic importance in a broad range of industries, including, e.g., the mixing of human and animal foodstuff, pharmaceutical products, detergents, chemicals, and plastics. As in most cases the mixing process adds significant value to the product, the process can be regarded as a key unit operation to the overall process stream.

The most important use of mixing is the production of a homogeneous blend of several ingredients in order to eliminate variations in concentration.

If a material consists of an ingredient or compound exhibiting fluctuations caused by an upstream process, or raw material, the term *homogenization* is used for the elimination of these fluctuations. By mixing, a new product or intermediate is created for which the quality and price are very often dependent on the efficiency of the mixing process. The efficiency is determined both by the materials to be mixed (e.g., particle size and particle size distribution, density, surface roughness) and by the process and equipment used for performing the mixing. The design and operation of the mixing unit itself have a strong influence on the quality produced, but upstream material handling process steps such as feeding, sifting, weighing, and transport determine also both the quality and the capacity of the mixing process. For example, the filling and discharge flow patterns in storage bins can aid mixing or cause demixing depending on their design. Downstream processing may also destroy the product quality due to segregation (demixing). Therefore, the mixture quality of the end product depends on the whole flowsheet, not just the mixer. A mixing process can fail, broadly, in two ways:

1. *The quality of the mix is poor.* In cases where the mixing produces the end product, this will be noticed immediately at the product's quality inspection. Frequently, however, mixing is only one in a series of further processing stages. In this case, the effects of unsatisfactory blending are less apparent, and might possibly be overlooked to the detriment of final product quality.
2. *Overmixing.* A batch may be overmixed if mixing proceeds longer than is necessary for content uniformity. Likewise, a continuous mixer may have residence time in excess of what is necessary for content uniformity. Excess residence time means either the throughput is lower than required or the vessel is larger than required. Both represent wasted resources. Note, for materials that degrade, attrite, or agglomerate, overmixing may also result in demixing due to segregation.

Mixing Mechanisms: Dispersive and Convective Mixing It is useful to consider the mixing process as two separate mechanisms: *dispersion* and *convection* (see Fig. 21-80). Dispersive mixing is a local effect, and it occurs when neighboring particles randomly change places with each other (e.g., micromixing). Therefore, dispersion is kinematically similar to thermal diffusion in liquids and gases. However, unlike with diffusion, in particulate systems there is no relative random motion unless energy is added by some external means (e.g., vibration or shearing). Dispersion then occurs when both a concentration gradient and agitation are supplied. Convection corresponds to relative movement of large groups of material (i.e., macromixing). For good convective mixing, the material should be continuously divided and then combined again after portions have changed places (Fig. 21-80). This forced convection can be achieved by bulk movement of material in a tumbling, fluidized, or static mixer; or by rotation or vibration of internals with agitated mixers. These two mechanisms work in conjunction as follows. Convection takes lumps of material of like components, separates them, and combines them with lumps of material of other components. Along with continuous agitation, dispersion then smears the composition gradients at the boundaries of these lumps, allowing for finer scales of mixing.

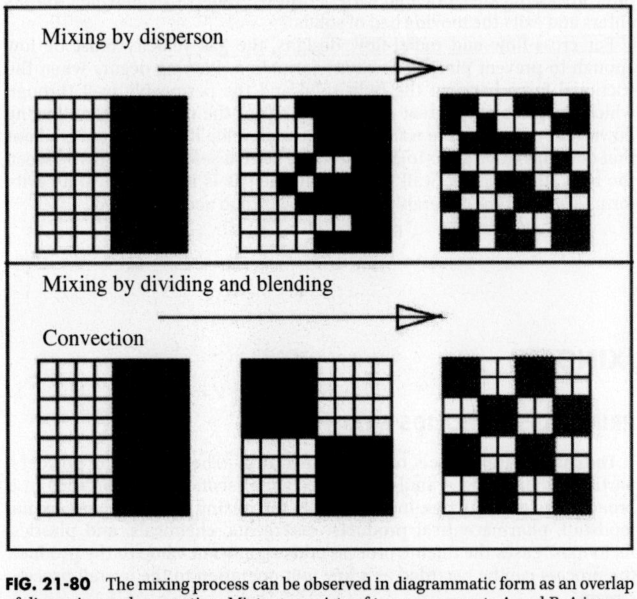

FIG. 21-80 The mixing process can be observed in diagrammatic form as an overlap of dispersion and convection. Mixture consists of two components A and B; A is symbolized by the white block and B by the hatched block. Dispersion results in a random arrangement of the particles; convection results in a regular pattern.

Segregation in Solids and Demixing If a solids mixture consists of particles of different size, shape, density, surface roughness, cohesion, or other material properties, then the mixture has propensity to segregate under external excitations. Segregation competes with both dispersive and convective mixing, resulting in a decrease of the mixing quality. Particle property differences along with process flow characteristics determine the degree and trend of segregation. Among all driving factors for segregation, the particle size difference is the most dominant factor for segregation (J. C. Williams, *Mixing, Theory and Practice,* vol. 3, V. W. Uhl and J. B. Gray (eds.), Academic Press, Orlando, Fla., 1986). In practice, there is always polydispersity of particle size, even in a single ingredient. Hence, nearly all industrial powders can be considered as solid mixtures of different size particles, and size segregation is one of the most ubiquitous problems in industrial solids handling. Figure 21-81 illustrates the four most common mechanisms for size segregation, and each is discussed below.

Percolation Segregation *Percolation segregation* or *sifting segregation* is by far the most important segregation effect, which occurs when finer particles percolate through the gaps between the larger ones under gravity (Fig. 21-81*a*). These gaps act as a sieve, so this effect is also called *kinetic sieving.* As a rule of thumb, a small sphere with diameter ~1/6.5 of larger spheres will just fit through the cracks, even in the larger sphere's densest, hexagonally close-packed, state. Beyond this limit, particles may "freely sift." For less ideal mixtures, sifting is still possible with motion promoting microstructural change. If a solids mixture is moved, gaps open up between the grains, allowing finer particles to selectively pass through the particle bed (see Fig. 21-81), resulting in segregation. Furthermore, percolation occurs even where there is a small difference in the size of the particles (250- and 300-μm particles) [J. C. Williams, *Fuel Soc. J.,* University of Sheffield, **14:** 29 (1963)]. The most significant economical example is segregation during

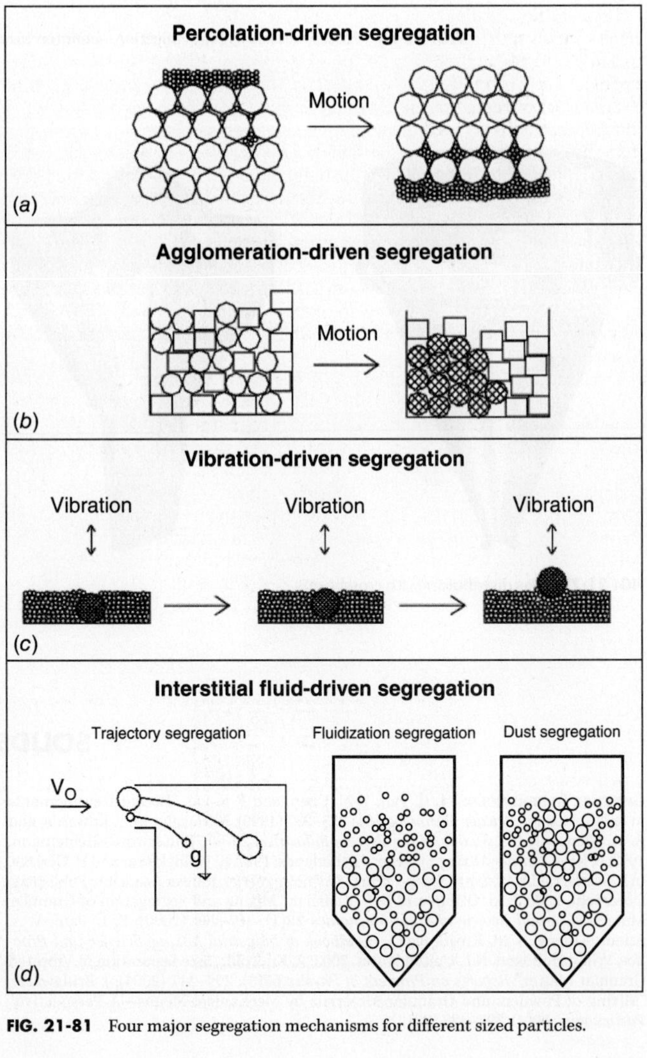

FIG. 21-81 Four major segregation mechanisms for different sized particles.

heap formation when filling solids into bunkers or silos. A mobile layer with rolling particles forms on the surface of such a pile, which restricts larger particles from passing into the core of the pile, so they remain near the free surface, slide or roll downward the heap, and eventually deposit near the sidewalls of the container. As a result, if filling a container at its center, segregation can result in a pattern in which fine particles concentrate in the middle region below the feed point and coarse particles concentrate near the sidewalls of the container. Thus filling a silo and then emptying it from a central discharge point are particularly critical. Remixing of such segregated heaps to certain extent can be achieved through *mass flow* discharge; i.e., the silo's contents move downward in blocks, slipping at the walls, rather than emptying from the central core (*funnel flow*).

Agglomeration Segregation This can occur if interparticle cohesion within a given species of particles is stronger than interparticle adhesion between species (Fig. 21-81*b*). In this case alike particles tend to agglomerate rather than disperse, forcing segregation and limiting mixing. Conversely, an ordered mixture may occur if adhesion is stronger than cohesion. In either case, more information on mechanisms of agglomeration can be found in the Agglomeration subsection.

Vibration-Induced Segregation This can occur if a solids mixture is vibrated, where the coarser particles often float up against the gravity force and accumulate near the top surface. This mechanism is often called the *Brazil nut effect* and is illustrated in Fig. 21-81*c* for the case of a large particle in a mix of finer material. During vibration, smaller particles flow into the vacant space created underneath large particles, preventing the large particles from reclaiming their original position. If the large particles have a higher density than the fines, they will compact the fines, further reducing their mobility and the ability of the large particles to sink. However, when the large particle is less dense than the fines, it may sink, called the *reverse Brazil nut effect*, but its underlying mechanism is still not well understood [Shinbrot and Muzzio, *Physical Review Letters* **81:** 4365–4368 (1998)]. When excessive vibration is applied, the powder bed can be fluidized and buoyancy will dominate segregation.

Interstitial Fluid-Driven Segregation This encompasses several effects which share the common factor of a fluid contributing to the segregation processes, including trajectory segregation, fluidization segregation, and dust segregation (Fig. 21-81*d*). Trajectory segregation often occurs in cyclones or conveying termination into a silo where particles are following the different trajectories due to different particle sizes. That is, fine particles follow fluid streamlines stronger than larger particles. Fluidization segregation occurs due to different size particles experiencing different drag and gravity forces, which may lead to segregation via elutriation, entrainment, or buoyancy. Dust segregation occurs when extremely fine particles move with air currents and separate from coarse particles.

Other Segregation Mechanisms These mechanisms include bouncing induced segregation, angle-of-repose induced segregation, and density segregation. They can also impact industrial processes, but occur less frequently and have fewer effects when compared to size segregation described above.

Predicting segregation quantitatively is difficult, because there are often multiple mechanisms affecting the final particle spatial distribution. For example, one or multiple types of the above-mentioned segregation can interplay with dispersive and convective mixing simultaneously in one process. Segregation, dispersion, and convection all depend on both particle properties and specific flow conditions in mixing, packaging, or transport processes, but there are no first principles theories to calculate them. Recently, significant progress has been made in quantifying how the percolation segregation interacts with dispersive and convective mixing by using a phenomenological continuum model [Fan et al., *J. Fluid Mech.* **741:** 252–279 (2014)].

To mitigate the segregation effect on mixing, a few methods can be used. Williams [J. C. Williams, *Fuel Soc. J.*, University of Sheffield, **14:** 29 (1963)] suggested adding a small quantity of water to form capillary bridges between the particles to reduce their mobility and thus stabilize the condition of the mixture. The bridges increase cohesion of particles, so the tendency to segregate decreases. However, increasing cohesion too much may reduce powder flowability and also reduce the dispersive mixing. Another method to limit segregation lies in changing the segregation direction or time scale. For example, using baffles in rotating drum mixers or using zigzag chutes can alter the flow direction to change the segregation direction [Shi et al., *Physical Review Letters* **99:** 148001 (2007)]. When filling a vessel, increasing the feed rate can reduce the segregation time scale relative to the dispersive and convective time scales so that a better mixing state can be achieved. However, in general, having ingredients of a uniform grain size (e.g., through granulation) and minimizing the motion of solids mixtures after blending are essential to avoid the segregation effect.

MIXTURE QUALITY: THE STATISTICAL DEFINITION OF HOMOGENEITY

To judge the quality of a solids mixture, the status of mixing has to be quantified. Thus a degree of mixing has to be defined. Here one has to specify what property characterizes the mixture (e.g., composition, particle size, temperature). Note there are circumstances in which a good mixture requires uniformity of several properties simultaneously, but for this discussion we assume only a single property of interest x. The mixture quality is traditionally checked by taking a number N of samples and analyzing the property of interest in each sample x_i. Note that one must adhere to proper sampling practices as described in the Particle Characterization subsection when extracting the samples. The arithmetic average value of the property of interest $p = \Sigma_i x_i/N$ will be near to the dosed value of that property to the batch. In fact, if the entire batch is analyzed, the average value of the property will be identical to the dosed value (barring any analytical uncertainty). However, regardless of whether the whole batch or a subset is analyzed, the variance of the property of interest over the samples, $\sigma^2 = \Sigma_i (x_i - p)^2/N$, is dependent on the quality of the mixture.

In general, higher-quality mixtures have lower variance. In some cases the sample *relative standard deviation* (RSD) = σ/p is reported as a measure of mixture quality. Regardless, the expected magnitude of the variance is strongly dependent on the size of the sample. In the illustration shown in Fig. 21-82, the numeric values represent x_i from a batch split into $N = 36$ samples. The variance for this small sample size is $\sigma^2 = 0.25$. If larger samples had been used instead, and only 9 samples were taken, the variance would have been $\sigma^2 \approx 0.04$, for the same batch. If even larger samples were collected so that only 4 samples were taken, the variance would have been only $\sigma^2 \approx 0.02$. Clearly the expected variance, therefore "mixed-ness," is dependent on the size of the sample collected. The size of the sample should then be tied directly to the scale at which the product must be mixed. Danckwerts termed this the *scale of scrutiny* [P. V. Danckwerts, *Appl. Sci. Res.* **279**(3): 279–296 (1952)]. Depending on the application, the scale of scrutiny can range from a single pharmaceutical tablet to an entire railcar or beyond. It is up to the engineer's understanding of the process, product, and application to determine the scale of scrutiny.

When computing the variance of a set of samples, it is useful to compare to some idealized scenarios. Three idealized mixtures of useful consideration are perfectly ordered, perfectly segregated, and ideally random, as shown in Fig. 21-83. In a perfectly ordered scenario (Fig. 21-83*a*), each component of the mixture is proportionally represented down to the individual particle scale. Theoretically such a mixture could spontaneously

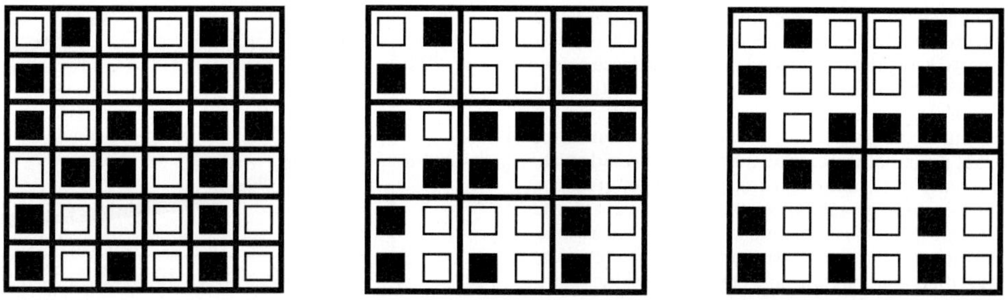

FIG. 21-82 Mixtures sampled based on increasing scale of scrutiny. From left to right the scale of scrutiny is 1x, 4x, 9x based on 36, 9, or 4 samples of the same mixture.

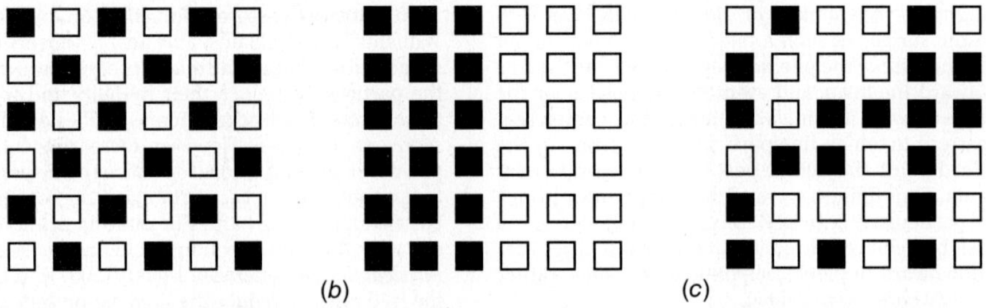

FIG. 21-83 Three idealized mixtures from left to right are perfectly ordered, perfectly segregated, and ideally random.

arise if intercomponent adhesion significantly outweighs cohesion between similar components. In such a case, the mixture variance will be equal to zero ($\sigma^2_{ordered} = 0$) for all scales of scrutiny above the smallest ordered pair of individual particles.

A perfectly segregated mixture (Fig. 21-83*b*) is quite the opposite. In this scenario all particles of one component are entirely unmixed from those of another. This may typically be the starting state for a batch mixture when multiple components are added sequentially. In such a case, the mixture variance can be calculated as $\sigma^2_{segregated} = P(1 - P)$ for all scales of scrutiny smaller than the entire vessel.

An ideally random mixture (Fig. 21-83*c*) occurs when a sample contains relative quantities of components in a completely random sense. For well-agitated, free-flowing yet nonsegregating systems, this configuration is optimal (as the name suggests). For the idealized case of a two-component mixture where each sample is composed of the same number of particles n, the variance of an ideally random mixture is $\sigma^2_{random} = P(1 - P)/n$. In practice most samples are not taken as fixed in number, but rather by bulk volume or mass. Furthermore, particles within a mixture typically vary in size such that a fixed mass may not contain a fixed number of particles. This conundrum may be solved by use of Stange's equation [K. Stange, *Chemie-Ing-Tech.* **26:** 331–337 (1954); R. Saunders, *The Chemical Engineering Journal* **39**(2): 129–132 (1988)]

$$\sigma^2_{random} = \frac{P(1-P)}{M}\left[Pm_1\left(1+c_1^2\right)+(1-P)m_2\left(1+c_2^2\right)\right] \tag{21-83}$$

where M is the mass of the sample, m_1 is the number average mass of a single particle of component 1, c_1 is the relative standard deviation of particle size (by number) of all particles of component 1, while m_2 and c_2 are the respective quantities for all particles that are not component 1.

It is useful to compare the measured value σ to the idealized values $\sigma_{ordered}$, σ_{random}, and $\sigma_{segregated}$. Mixtures where $\sigma_{ordered} < \sigma < \sigma_{random}$ can be called *pseudo-random.* In this case one can infer that some amount of ordering is occurring. For $\sigma_{random} < \sigma < \sigma_{segregated}$ the mixture is partially segregated; perhaps mixing is incomplete or spontaneous segregation does not allow complete mixing. Finally if $\sigma > \sigma_{segregated}$, a measurement or arithmetic error has occurred, as this is physically impossible. These bounds on σ suggest the Lacey mixing index $M = (\sigma^2_{segregated} - \sigma^2)/(\sigma^2_{segregated} - \sigma^2_{random})$. In general an increase of M corresponds to improved mixing. A segregated mixture results in a value of $M = 0$, while an ideally random mixture corresponds to $M = 1$.

A batch mixing process with components loaded in series begins with $M = 0$. If no segregation occurs, after mixing for some time M will increase toward 1. The engineer must decide a value of M for which the mixture quality is sufficient. Equivalently, the engineer must decide a value of σ for which the mixture quality is within acceptable limits for the given scale of scrutiny. The time at which the mixture reaches this quantity is deemed the *mixing time.* It must be determined experimentally for a given mixing system and mixture. The mixing time for a continuous system can be defined similarly by replacing the batch time with the continuous mixer residence time. Here it is worthwhile to note that overmixing may occur. In some systems, agglomeration or particle degradation due to mixing can force segregation. In these circumstances the mixing quality may actually pass through a maximum. That is, more mixing time does not always mean better mixing. To assess overmixing, examine the product for particle agglomeration or degradation in light of what potential segregation mechanisms could exist.

In almost all cases, the samples used to measure and then compute the mixture variance will not encompass the whole batch. In this case the sample variance $s^2 = \Sigma_i(x_i - p)^2/(N - 1)$ corresponds to a statistical estimator of the true batch variance σ^2. Under the assumption that the property of interest is normally distributed among the collected samples, the confidence limits of σ^2 can be computed using the normal methods of statistics. Because we are typically questioning if mixing is good enough, it is sufficient to only compute the upper bound confidence limit of σ^2. This value can be computed with the relation $\sigma^2_{upper} = s^2(N - 1)/F^{-1}_{N-1}(\alpha)$, where $F^{-1}k(\alpha)$ corresponds to the inverse chi-squared distribution with k degrees of freedom and a left-tail probability α. Figure 21-84 shows the upper confidence bound for a 95 percent confidence interval ($\alpha = 0.05$). For example, from the figure a sample size of $N = 10$ could have actual variance up to 1.7 times the measured sample variance. Likewise, for $N = 600$ one could have actual variance up to 1.1 times the measured sample variance.

When one is interpreting the sample variance, it is important to note that the measured variance is the summation of contributions from the mixture variance, variance due to the sampling method itself (see Particle Characterization subsection), and variance due to the analytical method used to determine the component concentration,

$$\sigma^2_{measured} = \sigma^2_{mixture} + \sigma^2_{sampling} + \sigma^2_{analytical} \tag{21-84}$$

The uncertainty from sampling and the analytical method must be determined separately based on the respective methods being used.

EQUIPMENT FOR MIXING OF SOLIDS

A wide variety of equipment is commercially available to suit a multiplicity of mixing tasks. In this overview, mixers and methods for mixing solids are divided into four groups: (1) bunker and silo mixers, (2) rotating tumbling mixers, (3) agitated mixers, and (4) other mixers or mixing methods.

Bunker and Silo Mixers These are sealed vessels (Fig. 21-85), which may serve to homogenize large quantities of solids. They are operated batchwise, continuously, or with partial recirculation of the mixture. Their sealed construction also enables material to be conditioned (e.g., humidified, granulated, dried, fluidized, or rendered inert) as well as mixed. Silo mixers can be divided into two groups: gravity silo mixers and pneumatic mixers.

In *gravity silo mixers*, an individual layer in the silo mixes with other layers for better homogenization by utilizing desired flow patterns from the specific design of the mixers. A variety of technologies have been developed to achieve homogenization in gravity silo mixers by using inserts, modified silo structure, multiple blending pipes, or mechanical activation to change flow patterns in the silo. Figure 21-85*a* shows the Binsert® hopper-in-hopper axisymmetric silo mixer. Placing an inner hopper modifies the

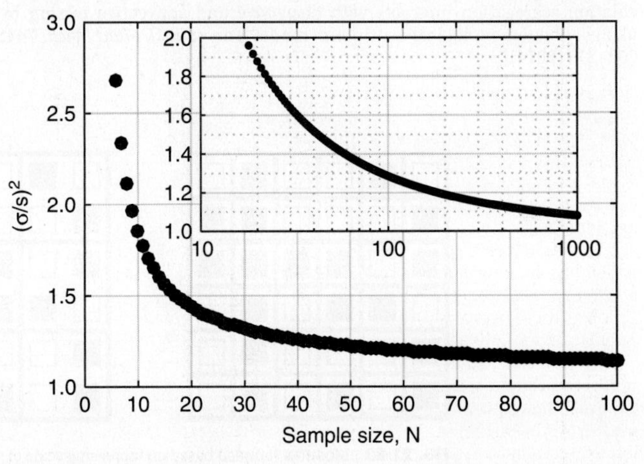

FIG. 21-84 Confidence interval for sample variance based on N samples.

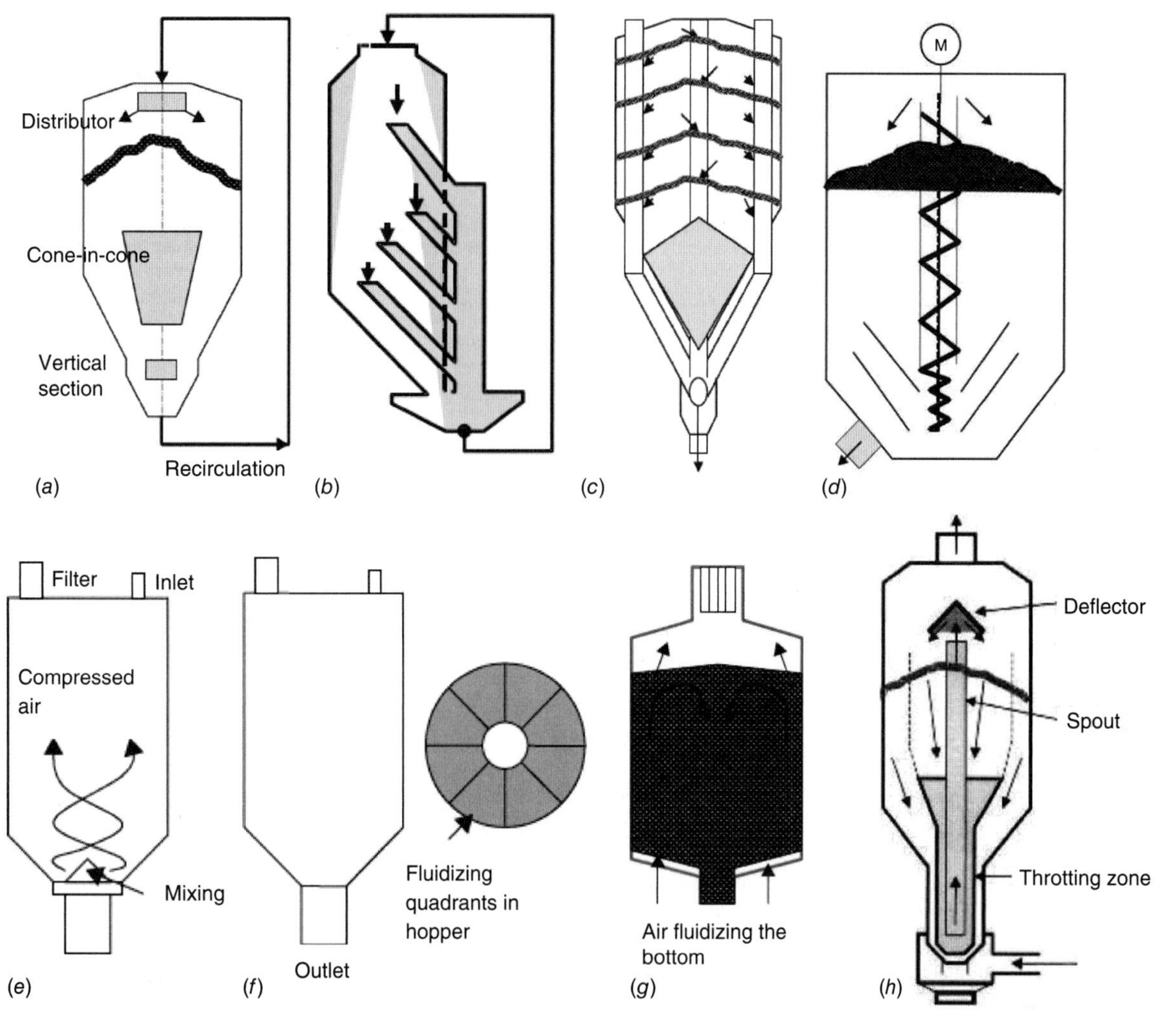

FIG. 21-85 Bunker and silo mixers. First row shows gravity silo mixers: (*a*) Binsert® blender; (*b*) mixing silo blender; (*c*) Phillips blender; and (*d*) mechanical blender. Second row shows pneumatic mixers: (*e*) air jet mixer; (*f*) air merge mixer; (*g*) fluidized bed mixers; and (*h*) pneumatic blender. (E. L. Paul, V. A. Atiemo-Obeng, S. M. Kresta (eds.), *Handbook of Industrial Mixing: Science and Practice*, Wiley, Hoboken, N.J., USA, 2003, with permission.)

radial velocity profiles in the silo to obtain a higher radial velocity gradient to promote mixing between layers. The aspect ratio of height to diameter of the cylindrical section is critical to the mixing efficiency in such mixers. Johanson [J. R. Johanson, *Chem. Eng. Prog.* **66**(6): 50–55 (1970)] suggested a ratio of 1.5 for hopper half angle below 35° and 0.5 for hopper half angle above 35°. Figure 21-85*b* shows a gravity silo mixer with modified internal structure [W. Muller, *Germ. Chem. Eng.* **5**: 263–277 (1982)]. Bulk solids flow through a system of tubes at various heights and radial locations and mix together afterward. Another type of silo mixer using the same mechanism (e.g., Zeppelin Centro mixer) uses a central takeoff tube into which the solids travel through openings arranged at various heights up this pipe. Figure 21-85*c* shows a Phillips mixer using multiple vertical pipes to homogenize bulk solids. Streams from the central feed point and the individual pipes mix at the outlets in the conical section of the silo. The mixing efficiency at one pass generally depends on the number of blending pipes, but installation of a large number of pipes requires additional consideration of structural support. Figure 21-85*d* shows a mechanical silo mixer, which has a screw shaft in the center. The rotating screw shaft generates a velocity profile across the silo diameter to promote mixing. It also lifts solids at the bottom of the silo to the top surface to generate an internal circulation for enhancing convective mixing.

If the mixing quality in gravity silo mixers within one pass does not meet requirements, the withdrawn material can be fed back into the bunker until the required homogeneity is achieved. The material drawn off in most cases is carried to the top of the bunker often by pneumatic conveying using an external circulation system. In addition, mass flow is necessary to allow reliable powder flow, avoid segregation during silo discharge, and allow full use of the bin contents during circulation. Gravity mixers are designed for free-flowing powders and are offered in capacities ranging between 5 and 200 m^3.

Pneumatic Mixers These mixers can be used for fluidizable bulk solids. Powders rise due to gas drag forces. As the gas velocity increases, it causes formation of bubbles in the powder bed. The minimum fluidization velocity to fluidize a powder bed is a function of both particle and gas properties. Detailed information can be found in Sec. 17. Powder mixing occurs during the formation, rise, and burst of the bubbles. The bubbles carry powders from the bottom to the surface of the powder bed and cause powder recirculation. Therefore, the mixing quality is closely related to the size, shape, and dynamics of bubbles, which depend on the powder properties and gas superficial velocity (see Sec. 17). Pneumatic mixers can be divided into different categories by how gas is injected. In *air jet mixers* (Fig. 21-85*e*), air is blown in through jets arranged around the circumference of a mixing head placed in the bottom of the vessel, which generates a swirling turbulent motion. The largest mixers have a capacity of 100 m^3. Alternatively, in *air merge mixers* (Fig. 21-85*f*), the vessel is divided into several segments, and each segment is fluidized independently. If gas flowing through a powder bed reaches a critical speed (minimum fluidization velocity), powders become fully fluidized in *fluidized-bed mixers* (Fig. 21-85*g*). Because of increased powder mobility, fluidized-bed mixers possess excellent mixing properties for solids in both a vertical and radial axis. Circulating fluidized beds are often used in reaction processes, which are combined with elevated heat transfer and material circulation to enhance the reaction processes. Figure 21-85*h* shows a *pneumatic mixer* with a central conveying tube to inject gas after powders are filled [W. Krambrock, *Powder Technology*

15: 199–206 (1976)]. A cone deflector on top of the tube is used to keep powders from falling into the tube during filling and also spread powders when they are carried through the tube during the mixing process. This mechanism generates both axial and vertical convective motion to achieve good mixing.

After blending, when one is discharging the mixtures from the mixer, a velocity smaller than the minimum fluidization velocity, called the *deaeration velocity*, is needed to better empty the mixer. It is also important to note that special care must be taken for the mechanical aspects of the pneumatic mixers. Under fluidized conditions, the solids exert hydrostatic stress on the silo, whereas under static or bulk flow conditions the Janssen effect vastly limits the radial stress. The largest fluidized-bed mixers as used in cement making reach a capacity of 10^4 m^3. The specific energy lies between 1 and 2 kWh/ton and air consumption rises sharply in the case of particle sizes above 500 μm.

Tumbling Mixers Tumbling mixers achieve solids mixing through the motion when particles roll down a sloping surface (called a *flowing layer*) as the mixer shell rotates on its own axis or eccentrically. Tumbling mixers are mostly suitable for free-flowing, nonsegregating bulk solids. For the simplest tumbling mixer, a *drum mixer* rotating along its horizontal axis (Fig. 21-86*a*), most mixing occurs in the radial direction driven by both convective mixing and dispersive mixing through random particle collisions in the flowing layer. Mixing in the axial direction is fairly slow, because no convective motion occurs and the mixing is only dispersive in this direction. For solids mixtures prone to segregation, drum mixers are incapable of achieving good mixing, because percolation-driven segregation often dominates convective and dispersive mixing in the flowing layer. To enhance mixing in the axial direction and minimize potential segregation in the radial direction, internal baffles with proper size and orientation are mounted on the shell to promote axial convective mixing and modify the segregation time scale. Axial mixing can also be improved in *asymmetrically rotating drum mixers* in which the cylinder is tilted obliquely to the main axis. Other tumbling mixers include the *V-blender, double-cone blender*, and *bin blender* (Fig. 21-86*b, c, d*). Compared to the rotating drum mixer, these mixers add convective mixing in the rotating axis direction due to the nonconstant cross-sectional area in this direction. Within each rotation, particle streamlines converge and diverge alternatingly, resulting in better mixing in the axial direction. Segregation can also occur in these mixers, so it is generally not recommended to blend segregating solid mixtures in tumbling mixers.

A proper flow pattern in tumbling mixers is critical for mixing efficiency and quality. Cascading flow is the ideal flow for tumbling mixers. In this flow regime, particles flow through a thin flowing layer near the free surface after being lifted by the mixer shell, and the rest of the particles have solid body rotation. The flow regimes in rotating tumbler mixers can be characterized by the *Froude number* $Fr = r\omega^2/g$, where r denotes the mixer radius, g denotes the gravitational acceleration, and ω denotes the angular velocity of the mixer. The Froude number represents the ratio of the centrifugal forces to the gravitational forces and depends on the rotation speed of the mixer. The cascading flow generally occurs when Fr ranges from 10^{-4} to 10^{-1} [J. Mellmann, *Powder Technology* **118**(3): 251–270 (2011)]. When $Fr < 10^{-4}$, the entire bulk solids slip on the wall and no mixing can occur. When $Fr > 10^{-1}$, particles start to leave the powder bed after being lifted. The centrifugal effect occurs at $Fr > 1$, where all particles are thrown to the shell, forming a ring of particles rotating with the mixer. In this flow condition ($Fr > 1$), good mixing quality cannot be achieved. The capacity of tumbling mixers is up to ~10 m^3 and they are typically used in batch processes.

Agitated Mixers Agitated mixers employ mechanical means to create mixing actions by rotating impellers (e.g., plows, paddles, or ribbons) along its shaft while the mixer shell remains stationary. The shear from the rotating parts can provide much stronger agitation compared to the shear due to gravity in tumbling mixers. This shear generates strong convective mixing and dispersive mixing in both axial and radial directions, so agitated mixers can handle a variety of bulk solids from free-flowing to cohesive or pastes. These mixers can also be used for agglomeration by injecting a binder. Agitated mixers include quite a few different styles as discussed below.

Paddle and Plow Mixers These mixers typically have a trough with an impeller, which is a single or twin shaft mounted with plows or paddles (Fig. 21-87*a, b*). The rotating plows or paddles lift bulk solids to generate convective motion, so these mixers are suitable for mixing free-flowing to cohesive powders, but not very cohesive powders. Plows of the plow mixers (Fig. 21-87*a*) have an inclined angle to the rotating shaft, which can move bulk solids in the axial direction to enhance axial mixing. The rotation speed of the impellers strongly impacts the mixing intensity of plow mixers. At lower rotation speed, similar to the tumbler mixers, bulk solids flow in the cascading regime, where particles roll down along the sloped surface. When the rotation speed increases, particles near the plow tips may fly away from the powder bed. Depending on the powder characteristics, aeration can take place at plow tips, and bulk solids may be locally fluidized, which enables particles to move freely and improve mixing. However, at very high rotation speed, most bulk solids are lifted toward the shell and slide off the wall without much rolling motion, so the mixing degree will not be promoted. Therefore, a proper rotation speed for operating plow mixers is critical for the mixing efficiency. Paddle mixers (Fig. 21-87*b*) are similar to the plow mixers except that paddles can rotate individually along their shaft to produce additional lateral and axial mixing. Paddle mixers generally operate at higher rotation speeds than plow mixers, which can cause segregation and some degree of attrition. It is recommended to determine the optimal rotation speed and blending time through test trials for plow and paddle mixers. Both plow and paddle types of mixers can have double counter-rotating shafts with two sets of horizontal impellers, where paddles or plows of one shaft overlap those on the other shaft. Both mixers can be operated in batch or continuous processes, and typical mixing time is up to 5 min.

Ribbon Mixers These mixers (Fig. 21-87*c*) use two sets of counter-rotating ribbon blades to transport bulk solids along the horizontal shaft in both directions to achieve axial mixing and displace bulk solids in the lateral direction mostly by centrifugal force. The outer ribbons transport material toward the center, and the inner ribbons push materials toward the end of the mixer. Compared to the plow and paddle mixers, the shearing generated from the ribbons and amount of bulk solids transported by the ribbons is smaller, so a longer mixing time is needed for ribbon mixers in the range from 15 to 30 min. Ribbon mixers can be used for agglomeration by spraying liquid and for solid-liquid mixing. A high-speed chopper can also be fitted to the mixers to reduce agglomerates. The clearances between the outer edge of ribbons and the trough are rarely less than 3 to 6 mm, but particle attrition may occur depending on the size and fragility of the particles. There are also vertical ribbon mixers. The shaft of such mixers is vertical, and outer ribbons move bulk solids upward near the shell; inner ribbons transport bulk solids downward in the center region of the mixers. The ribbon mixers can be used in both batch and continuous processes.

Fluidizing Paddle Mixers The fluidizing paddle mixers (e.g., Forberg mixers), as shown in Fig. 21-87*d*, are batch mixers consisting of paddles mounted on twin shafts in a twin trough. The counter-rotating paddles agitate bulk solids fed from the top of the mixers to fluidize them rapidly. The paddles also induce convective motion of particles to enhance mixing. After mixing, the mixture is discharged through a set of twin doors at the bottom of the mixer to minimize segregation. The tip speed of the paddles is about 1 m/s, and the mixing time is as quick as 1 minute [H. Forberg, *Powder Handling Process* **4**(3): 318–320 (1992)]. However, the mixing time depends on the cohesion of the powders. Higher cohesion needs longer mixing time, but can limit segregation during discharge. Test trials are recommended to determine the mixing time, tip speed, filling point, and discharge method. The Forberg mixers are much more efficient than the plow mixers in both mixing quality and time, because fluidization enables much stronger convective mixing. The capacity of Forberg mixers is up to 50 m^3. Forberg mixers have been adapted to continuous processes as well.

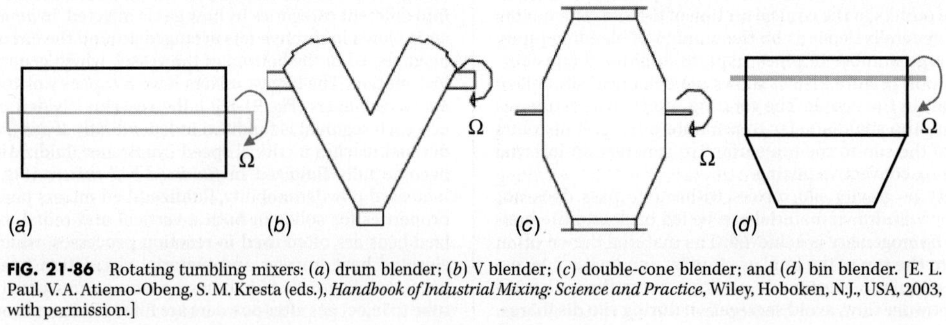

FIG. 21-86 Rotating tumbling mixers: (*a*) drum blender; (*b*) V blender; (*c*) double-cone blender; and (*d*) bin blender. [E. L. Paul, V. A. Atiemo-Obeng, S. M. Kresta (eds.), *Handbook of Industrial Mixing: Science and Practice*, Wiley, Hoboken, N.J., USA, 2003, with permission.]

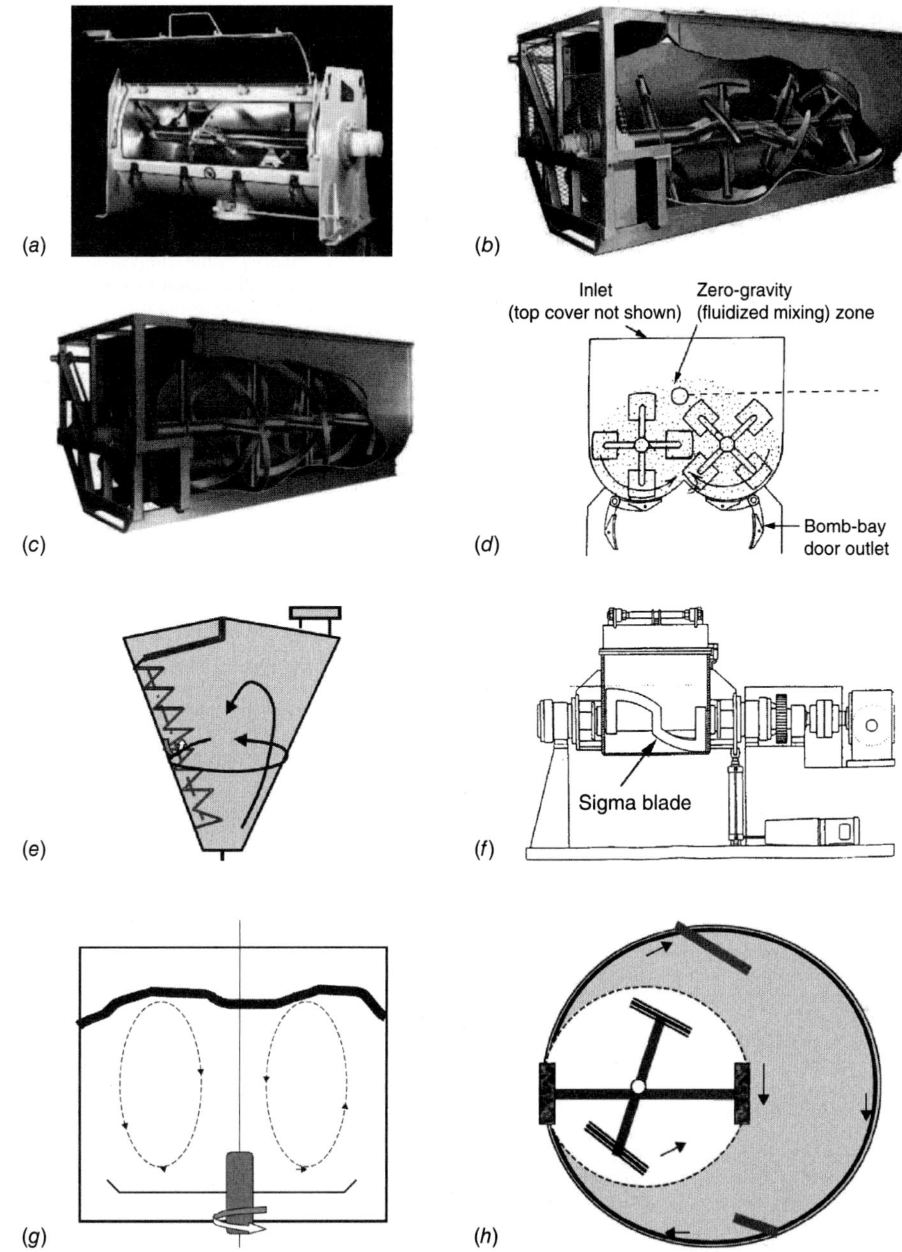

FIG. 21-87 Agitated mixers. (*a*) plow mixer; (*b*) paddle mixer; (*c*) ribbon mixer; (*d*) fluidizing paddle mixers; (*e*) screw mixer; (*f*) sigma-blade mixer; (*g*) Henschel mixer; and (*h*) Muller mixer. [E. L. Paul, V. A. Atiemo-Obeng, S. M. Kresta (eds.), *Handbook of Industrial Mixing: Science and Practice*, Wiley, Hoboken, N.J., USA, 2003, with permission.]

Screw Mixers The screw mixers (Fig. 21-87*e*) have a hopper-shaped vessel with a screw placed along the wall. During operation, the screw orbits around the hopper and also rotates along its own axis. This motion lifts bulk solids along the wall upward and spreads them on the surface, after which bulk solids flow downward in the center of the vessel. This circulating flow generates convective mixing for the bulk solids. The rotation of the screw also generates shearing to cause local dispersive mixing. A second short screw, called the *satellite screw*, can be used to further enhance mixing. The screw mixers can also have two screws in two hopper vessels joined along the wall. The screw mixers are suitable for mixing free-flowing, cohesive powders or pastes. The mixing time is a minimum of 10 min. Because of the small clearance between the screw and wall (3 to 6 mm), particle attrition may occur. The solids at the container wall are continuously replaced by the action of the screw so that bulk solids can be indirectly heated or cooled through the container's outer wall for drying applications. Spray nozzles can be used to add liquid for granulation. The screw mixers have capacity ranging from 25 L to 60 m^3.

Sigma-Blade Mixers or Z-Blade Mixers As shown in Fig. 21-87*f*, these mixers consist of twin troughs with a Z-shaped blade agitator in each trough. The blades can overlap and counter-rotate at the same speed or different speeds to fold and shear materials for mixing. The mixing time of Sigma-blade mixers ranges from 10 to 30 min. A spray bar can be placed above the blades for liquid addition. These mixers are normally used for producing dough or thick viscous pastes.

Impaction Mixers The impaction mixers (e.g., the *Henschel mixer* shown in Fig. 21-87*g*) resemble a typical kitchen food processor, where blades at the bottom of the mixer rotate at a high speed (2000 to 3000 rpm) to generate circulating flow patterns in the vertical plane to mix bulk solids through convective mixing. This mixer needs significant energy compared with other similar types of mixers due to its high rotation speed.

TABLE 21-4 Comparison of Agitated Mixers*

Factor	Ribbon/paddle	Plow	Fluidizing paddle	Sigma-blade
Material consistency	Powders/ granules	Powders/ granules	Powders/ granules	Pasty, sticky, gritty slurries up to 2×10^6 cP
Allowable fill level or batch size (% of total mixer capacity)	40–85	30–70	40–140†	40–65
Liquid addition configuration	Spray bar above ribbons	Spray nozzles at mixer top	Spray bar above paddles	Spray bar above blade
Delumping agitator configuration	High-speed chopper blades at sides	High-speed chopper blades at sides	Pin mills above paddles	None
Mixing cycle length (min)	15–20	< 5	< 1	10–30
Final moisture homogeneity (% of complete homogeneity)	90–95 or better	95–98 or better	98–99 or better	99 or better
Rotating or stationary vessel	Stationary	Stationary	Stationary	Stationary
Degree of particle shear	Some	High	Slight	Very high

*Reproduced from E. L. Paul, V. A. Atiemo-Obeng, S. M. Kresta (eds.), *Handbook of Industrial Mixing: Science and Practice*, John Wiley & Sons, Inc., Hoboken, N.J., USA, 2003.

†Percent fill more than 100% of the total capacity for another agitating batch mixer of equal volume.

High-Shear Mixers The high-shear mixers, such as the *Muller mixer* shown Fig. 21-87*h*, include a set of rollers and pans. While rotating, the rollers grind materials to very fine powders, which are mixed by the shearing from the pans. Usually, tumbling mixers are used to provide a reasonable mix before it is fed into the high-shear mixers.

Table 21-4 shows comparisons between different parameters for several common agitated mixers used in industry.

Other Mixing Methods The *mixed stockpiles* method achieves mixing by following a defined scheme for building up and emptying large stockpiles. A long stockpile is built up by a movable conveyor belt or other corresponding device traveling lengthwise. During loading the belt continuously travels up and down the whole length of the pile. As the pile forms, strata develop in order of the material's delivery. If the material is systematically removed *crosswise* to these layers, each portion removed from the stockpile will contain material from all the strata and therefore from the times it was supplied. This mixing method is mostly useful when the *scale of scrutiny* is larger than the thickness of the strata. The *mixing by feeding method* is metering different ingredients and bringing these streams of solids together locally. The quality of the metering determines the mix's homogeneity. Therefore, metered feeder units should ideally be used, preferably operated gravimetrically with appropriate feedback control of weight loss. There is limited axial mixing (transverse or back mixing) in this method. According to the requirements of the case in question, mixing is also required oblique to the direction of travel. If this oblique mixing is not sufficient, static mixers can be used for free-flowing powders or granules. The energy input into the mixer is very low, but such systems need sufficient height to achieve mix quality.

BLENDING TECHNOLOGY SELECTION

Here we offer advice for selecting the mixers described in the previous subsection. Before selection of a specific device can be made, it is necessary to broadly select whether a batch or continuous technology will be used. Table 21-5 offers a pointwise comparison between batch and continuous mixing.

In addition to the comparisons offered in the table, it is important to consider the batch versus continuous nature of the mixer's preceding and following unit operations. If both are continuous (batch), a continuous (batch) device may be favorable.

The cycle time of a batch mixer must be computed by dividing the batch size by the sum of the fill time, mixing time, discharge time, and any idle time. The fill and discharge times are dictated by the mixer but also by the device and whole system layout. The mixing time is dictated by the solids to be mixed and their relative concentrations. This time must be determined by testing or experience. In a batch mixing operation, the component concentrations are specified simply by their mass fill ratios.

In a continuous mixing process, the ingredients are continuously fed into the mixer, then mixed and prepared for the next processing stage. The operations of feeding, mixing, and discharging occur simultaneously within the mixer. In continuous mixing, the weighing and filling of the batch mixer are replaced by the component's controlled continuous feeding. The blending time is then equal to the material's residence time in the vessel. The residence time is equal to the mass of material in the mixer divided by the mass throughput of the mixer. Mixture quality is uncertain during start-up and shutdown of continuous mixers, making continuous mixers less flexible and less preferable in low production rate (<100 kg/h) situations.

The flowchart given in Fig. 21-88 provides a decision process for selecting mixer technology. The selection process starts at the upper left of the figure. For systems with low-proportion (<0.5 percent) minor components, it is advised to preblend the minor component. Note in some cases it is advised to blend materials by agglomeration or milling. These are not explicitly mixing unit operations, but must be used to mix when the segregation potential is high. In cases where these operations would be desirable but must be avoided, maintaining a uniform blend will be difficult.

TABLE 21-5 Comparison Between Batch and Continuous Mixing Processes

	Batch	Continuous
Number of ingredients	Unlimited	2 to 10; more should be premixed
Frequency of recipe change	Several per hour	Must remain constant for several hours
Frequency of cleaning or idle time	Several times a day	Once a day or less
Production rate or throughput	Any rate, limited by footprint	More than 100 kg/h
Footprint	Large, also requires intermediate storage	Generally lower footprint, even at higher throughput
Requirements of equipment	Simple feeding but high demands on mixer	Accurate feeding required, but low mixer demand
Safety	Quantity of material contained in the mixer relatively large	Smaller quantities of material are actually contained in mixer
Automation	Variable degree of automation	Contained in processing
Material type	Cohesive or free-flowing	Difficult with poorly feeding cohesive solids
Feeding	Straightforward; simply discharge contents into the feeder	Requires precision metering and monitoring
Maintenance and cleaning	Straightforward	More difficult
Material identity	Straightforward if batch integrity is maintained	Difficult, if not impossible
Application to minor components	Not good: minor fines can coat the interior	Mixing intensity
Segregation risk	High if material has segregation propensity; intermediate storage and transportation are required	Low if located immediately adjacent to the next unit operation
Flexibility	Generally versatile	Low flexibility
Robustness	Relatively high	Low due to necessity of several other pieces of equipment (e.g., feeders and monitoring equipment)

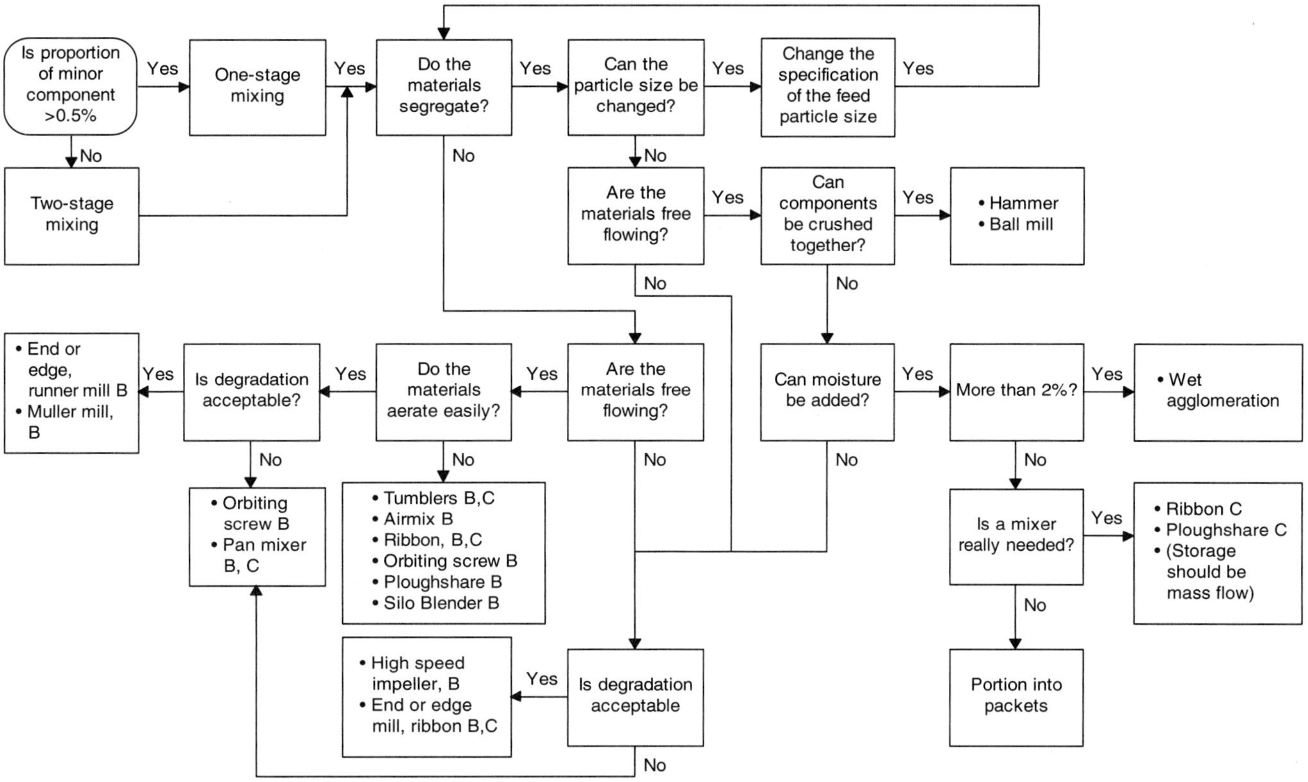

FIG. 21-88 Flowchart for mixer selection.

PNEUMATIC CONVEYING OF SOLIDS

General References: Klinzing, Rizk, Marcus, and Leung, *Pneumatic Conveying of Solids,* 3d ed., Springer, 2010; Dhodapkar and Jacob, "Fluid-Solids Flow in Ducts," in Crowe (ed.), *Multiphase Flow Handbook,* 2d ed., Taylor and Francis, CRC Press, 2016; Mills, *Pneumatic Conveying Design Guide,* Elsevier, 2016.

INTRODUCTION

When solids are introduced into a flowing gas stream in a pipe or tube at the proper conditions, the solids can be successfully moved from one point to another in a process plant, loading a ship, or to a mine, for example. On account of the advantages of pneumatic conveying over other bulk solids conveying technologies, it finds wide use in nearly every facet of modern production processes. The systems are usually highly reliable such that they can be operated round the clock for many months without equipment failures. The fact that the solids and gas are contained within the conveying pipe minimizes the contact of the conveyed material with the environment, and vice versa. Hence, while it is most common to convey solids with air, systems using other gases such as nitrogen or hydrogen or conditioned air are also routinely used. Typical conveying distances range from 5 to 10 m to 500 m. For short distances less than ~15 m, pneumatic conveying is considered alongside mechanical conveyors such as vibratory or screw conveyors. For longer distances greater than ~300 m, pneumatic conveying competes with belt conveyors and slurry systems depending on a variety of considerations such as attrition, particle size distribution, etc. Pneumatic conveying is used extensively across the world to move bulk solid materials from producers to customers. Trucks and railcars interface with dedicated conveying systems that move bulk solids with little effort into production facilities or intermediate storage areas. Materials such as cement, plastic pellets, grains, and foodstuffs are commonly transported in this manner.

Despite the numerous advantages of pneumatic conveying systems, they do have some distinct disadvantages. As discussed in detail later in this section, the gas expansion (and concurrent gas velocity increase) can limit system length due to pressure drop, velocity, or attrition considerations. For vacuum conveying systems, they are limited by the amount of vacuum that can be achieved by conventional air movers. For example, if a system operates at 1 bar when the solids are introduced into the system and the pressure at the destination point is 0.5 bar, then the gas velocity will be doubled at the destination point. For many materials this may not be a problem; however, attrition in conveying systems has been estimated to be proportional to velocity to the third to fifth power. Consequently, the doubling of velocity in a conveying system can result in significant degradation to either the bulk solids or the pipe in which they are conveyed.

THE GAS/SOLID SYSTEM

If solids are introduced into a pipe with sufficient gas velocity such that the solids are homogeneously suspended across the pipe cross section, the resulting condition is observed during dilute-phase conveying flows. This is depicted as condition 1 in Fig. 21-89*a*.

This situation is somewhat ideal as it is common to have inhomogeneities (such as in condition 2) during actual conveying flows possibly due to deviation of the velocity required for perfect homogeneous suspension or "roping" of the solids around an elbow, for example. As the gas velocity is further decreased, solids will begin to drop out of suspension and lay down on the bottom of horizontal sections of the conveying system. This condition is called *saltation.* These "salted out" solids may simply slide along the bottom of the pipe or may congregate as small dunes, as depicted in condition 3. Continued reduction of the gas velocity results in additional dune formation, as illustrated in conditions 4 and 5. Eventually, solids will begin to occupy the entire pipe cross section, as depicted in conditions 6 and 7, forming slugs of bulk solid. At this point, the resulting condition in the pipe depends on the particle size, permeability, and air retention characteristics of the conveyed solid. Fine solids, typically less than 150 μm, will have sufficient air retention such that they will continue to move through the system as a fluidized slug. In contrast, coarse bulk solids such as plastic pellets have sufficient air permeability that they will be conveyed easily in "dense phase" through the pipeline. For particles which deaerate easily and do not have sufficient air permeability (typical examples would be sand and granulated sugar), it is highly likely that the conveying system will plug or block upon reaching the conditions shown in condition 6 or 7. For these bulk solids, conveying systems rely on the use of air injection along the length of the pipe to control plug length. Flow conditions shown in conditions 1 and 2 are typical of dilute

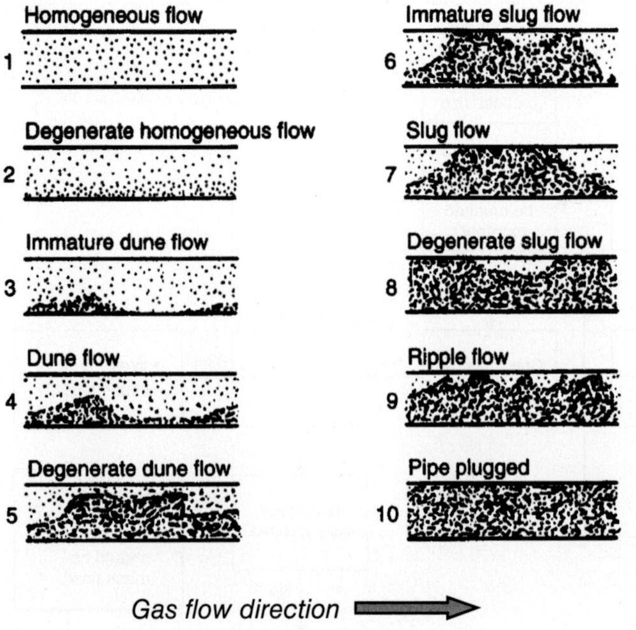

FIG. 21-89a Pattern of solids flow in pneumatic conveying. [*From Wen, U.S. Dept. of Interior, Bureau of Mines, PA, IC 8314 (1959) with permission.*]

phase while the conditions shown in conditions 3 through 9 are representative of dense-phase flow. Most bulk solids will not exhibit all the dense-phase flow regimes. For vertical flows, coarse solids such as seeds, grains, or plastic pellets will form uniform axisymmetric slugs. For fine, fluidizable solids, the entire pipe will be filled with bulk solid in a fluidized state.

Zenz or State Diagram Zenz (*Fluidization and Fluid Particle Systems,* Reinhold, 1960) proposed plotting pressure gradient versus velocity for a conveying system. This is shown in Fig. 21-89*b*.

This plot is frequently referred to as the Zenz or state diagram. Starting at the right-hand side (high velocity) of the plot, the system is operating in dilute phase. As the gas velocity decreases, so does the pressure gradient until the pressure minimum point is reached. At this point, particles begin to salt out on the bottom of the pipeline. The locus of pressure minimum points is referred to as the *pressure minimum curve* and serves as the basis for saltation velocity correlations. Additional reduction of gas velocity now results in dune formation along with a significant increase in pressure gradient. Visual observation of the conveying line often reveals considerable line vibration and shaking during the onset of dense-phase conveying. Figure 21-90 shows a Zenz diagram for three different mass flow rates of solids. In this case, the dashed line represents the "air only" pressure gradient. Three lines of constant mass flow rate are plotted where $M_3 > M_2 > M_1$.

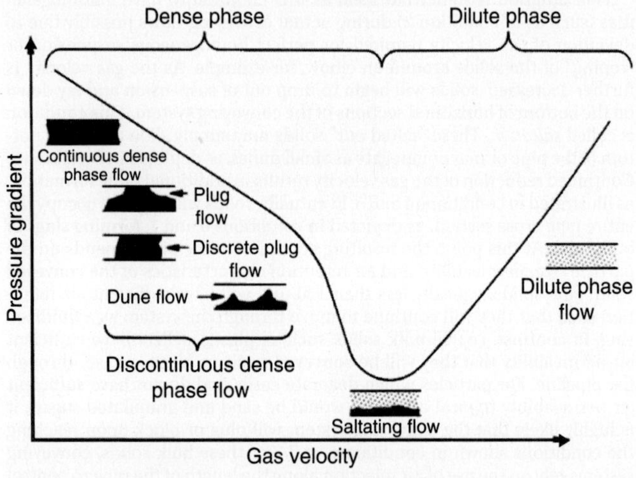

FIG. 21-89b Zenz diagram of pressure gradient versus velocity.

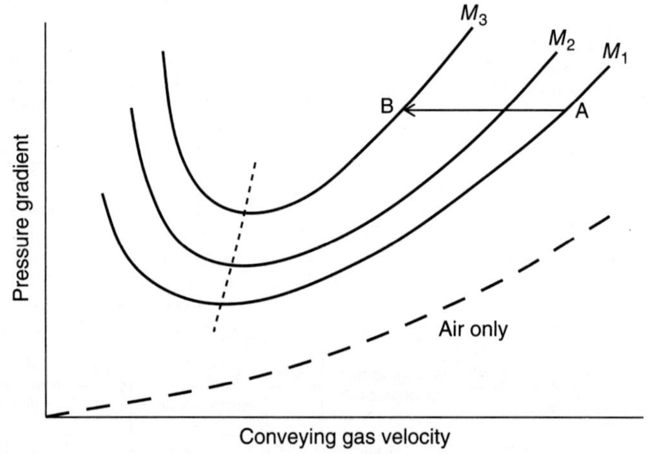

FIG. 21-90 Example of conveying system optimization.

The Zenz diagram can be useful in troubleshooting and optimization of conveying systems. One of the most common applications of the state diagram is the uprating of the mass flow of a dilute-phase conveying system. Consider a system operating at point A in Fig. 21-90 at mass flow rate M_1. As is the case for many systems, the maximum permissible pressure drop is constrained by the air mover. However, if the system gas velocity is *decreased* from point A to point B, the mass flow rate increases from M_1 to M_3. This counterintuitive result is explained by recognizing that the total system pressure drop is given by

$$\Delta P_{\text{total}} = \Delta P_{\text{air}} + \Delta P_{\text{solids}} \tag{21-85}$$

By reducing the gas velocity, additional solids can be conveyed without increasing overall pressure drop. The velocity can be decreased until the pressure minimum curve is reached. An experimental Zenz diagram for round, polystyrene Styropor resin is shown in Fig. 21-91 (d_p = 2.385 mm, pipe diameter = 52.6 mm, particle density = 1050 kg/m³).

Additionally, it is common for a well-operating dilute-phase system to progressively (or suddenly) begin to either experience high pressure drop or, in worst cases, block completely. A careful examination of Fig. 21-91 shows that as gas velocity is decreased below the pressure minimum curve, there is a significant increase in pressure gradient such that the system is at severe risk of blockage. The decrease in gas velocity can commonly be the result of gas leakage in a dilute-phase conveying system through rotary valves or diverter valves.

SALTATION VELOCITY

A key parameter for the design or analysis of a conveying system is the saltation velocity. As indicated previously, this velocity is the locus of points connecting the minima of the mass flow rate curves on the state diagram.

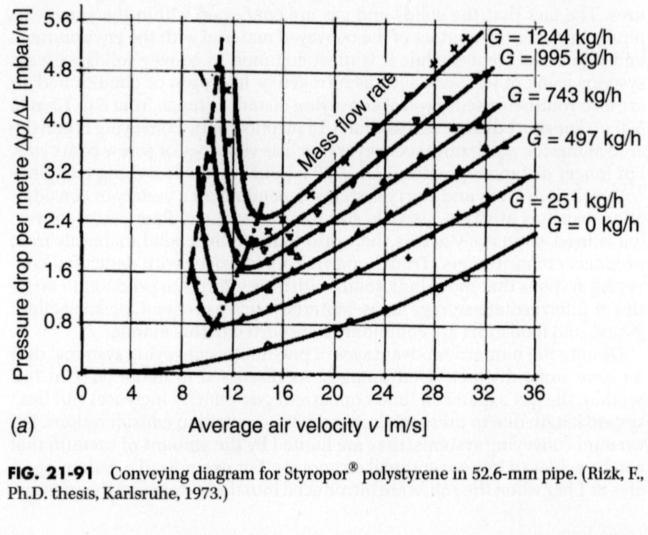

FIG. 21-91 Conveying diagram for Styropor® polystyrene in 52.6-mm pipe. (Rizk, F., Ph.D. thesis, Karlsruhe, 1973.)

There is an important physical distinction between *saltation velocity* and *pickup velocity*, even though the two terms tend to be used interchangeably in practice. Saltation represents the velocity at which particles just begin to lay down on a horizontal section of pipe. Pickup velocity is the point at which particles will be picked up from a stationary layer. Due to variations in particle size distribution and the visual determination of the exact saltation/pickup points, the velocities are rarely uniquely valued. In practice, the conveying system designer must consider the saltation of the largest particles of the size distribution to ensure successful conveyance of the bulk solids.

Numerous correlations have been proposed for the saltation velocity and have been reviewed by Leung and Jones (*Proceedings of Pneumotransport 4,* BHRA Fluid Engineering, 1978) and Gomes and Mesquita [*Braz. J. Chem. Eng.* **31**(1): 35–46 (2014)], Cabrejos and Klinzing [*Powder Technology* **79:** 173–186 (1994)] and Kalman and Rabinovich [*Powder Technology* **160:** 1003–1113 (2005)]. For large particle systems greater than 200 μm, the equation of Rizk (*Proceedings of Pneumotransport 3,* BHRA Fluid Engineering, 1973; and Ph.D. Thesis, University of Karlsruhe, 1973) finds broad applicability. Rizk's equation for saltation velocity is given by

$$\mu_s = \frac{1}{10^\delta}\left[\frac{v_s}{\sqrt{gD}}\right]^\kappa \tag{21-86}$$

where $\delta = 1.44d_p + 1.96$ and $\kappa = 1.1d_p + 2.5$ for d_p in millimeters, μ_s is the loading at saltation, D is the pipe diameter, and v_s is the saltation velocity. A very important distinction is that the equation as written is not explicit in v_s. The loading (mass flow rate of solids/mass flow rate of gas) at saltation is given by

$$\mu_s = \frac{W_s}{\rho_g A v_s} \tag{21-87}$$

where W_s is the mass flow rate of solids, A is the pipe cross-sectional area, and ρ_g is the density of the gas. Substituting the above expression into Eq. (21-86) and rearranging terms yields an explicit expression for the saltation velocity

$$v_s = \left[\frac{W_s\sqrt{gD}^{\,\kappa}\,10^\delta}{\rho_g A}\right]^{\frac{1}{\kappa+1}} \tag{21-88}$$

From Eq. (21-88), it can be deduced that saltation velocity increases both as the mass flow rate of solids increases and as the pipe diameter increases.

For finer particle sizes, Matsumoto et al. [*J. of Chem. Eng. of Japan* **10**(4): 273–279 (1977)] showed that the saltation velocity increases with decreasing particle size below a critical particle diameter d_p^*. This reflects the increased interparticle forces such as van der Waals forces, electrostatics, and capillary bonding forces at fine sizes. Matsumoto's saltation formulation first finds the critical particle diameter d_p^*:

$$\frac{d_p^*}{D} = 1.39\left(\frac{\rho_s}{\rho_g}\right)^{-0.74} \tag{21-89}$$

For $d_p \geq d_p^*$,

$$\mu_s = 0.373\left(\frac{\rho_s}{\rho_g}\right)^{1.06}\left(\frac{\mathrm{Fr}_p}{10}\right)^{-3.7}\left(\frac{\mathrm{Fr}_s}{10}\right)^{3.61} \tag{21-90}$$

where

$$\mathrm{Fr}_p = \frac{u_t}{\sqrt{gd_p}} \quad \text{and} \quad \mathrm{Fr}_s = \frac{v_s}{\sqrt{gD}}$$

For $d_p < d_p^*$,

$$\mu_s = 5560\left(\frac{d_p}{D}\right)^{1.43}\left(\frac{\mathrm{Fr}_s}{10}\right)^4 \tag{21-91}$$

Figure 21-92 shows a comparison of the saltation velocity for the Matsumoto and Rizk equations as a function of particle size.

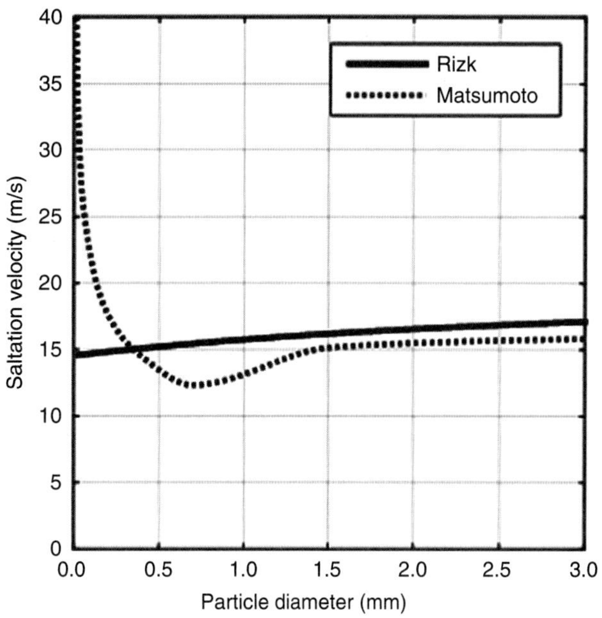

FIG. 21-92 Comparison of Rizk and Matsumoto correlations as a function of particle size.

Example 21-3 Calculate the saltation velocity for 10-μm and 2500-μm particles, using both correlations for a pressure conveying system at the pickup point. The absolute pressure at the pickup point is 150 kPa. The conveying gas is air at 22°C. The solids are high-density polyethylene with a density of 945 kg/m³. The solids conveying rate is 7500 kg/h through a 100-mm inner-diameter (ID) tube.

$$W_s = 7500\ \text{kg/h} = 2.08\ \text{kg/s} \qquad A = (\pi/4)D^2 = (\pi/4)(0.1\ \text{m})^2 = 0.0079\ \text{m}^2$$

$$\rho_g = \frac{P(Mw)}{RT} = \frac{(150\ \text{kPa})\left(29\ \dfrac{\text{kg}}{\text{kg}\cdot\text{mol}}\right)}{\left(8.314\ \dfrac{\text{kPa}\cdot\text{m}^3}{\text{kg}\cdot\text{mol}\cdot\text{K}}\right)(22+273.15)\ \text{K}} = 1.92\ \text{kg/m}^3$$

Rizk Eq. (21-88):

For 10-μm particles, $\delta = 1.44d_p + 1.96 = 1.44(0.01\ \text{mm}) + 1.96 = 1.9744$; $\kappa = 1.1d_p + 2.5 = 2.51$

For 2500-μm particles, $\delta = 1.44d_p + 1.96 = 1.44(2.5\ \text{mm}) + 1.96 = 5.56$; $\kappa = 1.1d_p + 2.5 = 5.25$

$$v_s = \left[\frac{W_s\sqrt{gD}^{\,\kappa}\,10^\delta}{\rho_g A}\right]^{\frac{1}{\kappa+1}} = \left[\frac{\left(2.08\ \dfrac{\text{kg}}{\text{sec}}\right)\left(\sqrt{\dfrac{9.81\ \text{m}}{\text{sec}^2}(0.1\ \text{m})}\right)^{2.511} 10^{1.9744}}{\left(1.92\ \dfrac{\text{kg}}{\text{m}^3}\right)(0.0079\ \text{m}^2)}\right]^{\frac{1}{2.511+1}} = 14.7\ \text{m/s}$$

Similarly for the 2500-μm particles, $v_s = 16.9$ m/s.

Matsumoto analysis, Eq. (21-89), needs to be calculated first.

$$\frac{d_p^*}{D} = 1.39\left(\frac{\rho_s}{\rho_g}\right)^{-0.74} = 1.39\left(\frac{945\ \text{kg/m}^3}{1.92\ \text{kg/m}^3}\right)^{-0.74} = 0.0142$$

$$d_p^* = 0.0142(0.1\ \text{m}) = 0.00142\ \text{m} = 1.42\ \text{mm}$$

Since 2.5 mm >1.42 mm, Eq. (21-90) is used.

$$\mu_s = 0.373\left(\frac{\rho_s}{\rho_g}\right)^{1.06}\left(\frac{\mathrm{Fr}_p}{10}\right)^{-3.7}\left(\frac{\mathrm{Fr}_s}{10}\right)^{3.61}$$

$$\mathrm{Fr}_p = \frac{u_t}{\sqrt{gd_p}} \quad \text{and} \quad \mathrm{Fr}_s = \frac{v_s}{\sqrt{gD}}$$

The terminal velocity for 2500-μm particles is 5.90 m/s.

$$\mathrm{Fr}_p = \frac{u_t}{\sqrt{gd_p}} = \frac{5.90\ \text{m/s}}{\sqrt{\left(9.81\ \dfrac{\text{m}}{\text{s}^2}\right)(0.0025\ \text{m})}} = 37.67$$

$$\mathrm{Fr}_s = \frac{v_s}{\sqrt{gD}} = \frac{v_s}{\sqrt{\left(9.81\ \dfrac{\text{m}}{\text{s}^2}\right)(0.1\ \text{m})}} = 1.01v_s$$

Substituting into Eq. (21-90) and solving for v_s give

$$\frac{W_s}{\rho_g A v_s} = \frac{2.08\ \text{kg/s}}{(1.92\ \text{kg/m}^3)(0.0079\ \text{m}^2)v_s} = 0.373\left(\frac{945}{1.92}\right)^{1.06}\left(\frac{37.67}{10}\right)^{-3.7}\left(\frac{1.01v_s}{10}\right)^{3.61}$$

Solving for v_s yields $v_s = 15.1$ m/s which is slightly lower than the 16.9 m/s obtained using the Rizk equation.

For 10 μm < 1420 μm, Eq. (21-91) is used.

$$\mu_s = 5560\left(\frac{d_p}{D}\right)^{1.43}\left(\frac{\mathrm{Fr}_s}{10}\right)^4$$

Substituting for μ_s and Fr_s yields

$$\frac{W_s}{\rho_g A v_s} = 5560\left(\frac{d_p}{D}\right)^{1.43}\left(\frac{v_s/\sqrt{gD}}{10}\right)^4$$

Substituting values and solving for v_s lead to

$$\frac{2.08\ \text{kg/s}}{(1.92\ \text{kg/m}^3)(0.0079\ \text{m}^2)v_s} = 5560\left(\frac{0.00001\ \text{m}}{0.1\ \text{m}}\right)^{1.43}\left(\frac{v_s\Big/\sqrt{\left(9.81\dfrac{\text{m}}{\text{s}^2}\right)(0.1\ \text{m})}}{10}\right)^4$$

so $v_s = 42.18$ m/s. This is considerably larger than the value obtained by using the Rizk equation.

CHOKING VELOCITY

For vertical lines, the limiting flow rate of solids occurs when solids can no longer be carried upward; this phenomenon is called *choking*. Chong and Leung [*Powder Technol.* **47**(1): 43–50 (1986)] reviewed choking correlations and recommended the equation of Yousfi and Gau [*Chem. Eng. Sci.* **29**(9): 1939–1953 (1974)] for fine materials (Geldart groups A and B, see Sec. 17 for a description) and Yang [Yang, *Powder Technol.* **35:** 143–150 (1983)] for larger group D materials. More recently, Klinzing et al. (*Pneumatic Conveying of Solids*, 3d ed., Springer, 2010) have commented that most of the choking work has been done in small bore (<75-mm) pipe and that most correlations give prediction in comparison with experimental data to +/−50 percent. They recommend the work of Leung et al. [*Ind. Eng. Chem. Proc. D.* **10:** 183–189 (1971)], Punwani et al. (*Proceedings of International Powder and Bulk Solids Handling and Processing Conference*, Chicago, 1979) and Yang. Here the work of Yousfi and Gau and the work of Yang are presented. Using experimental results for glass, polystyrene, and catalyst ranging from 20 to 290 μm, Yousfi and Gau developed an empirical relationship for the choking velocity for fine materials

$$v_c = 32\sqrt{gd_p}\,\mathrm{Re}_p^{-0.06}\,\mu^{0.28} \tag{21-92}$$

where $\mathrm{Re}_p = \rho_f U_t d_p/\mu_f$ and $\mu = W_s/W_g$ or the system phase ratio. Note that $W_g = \rho_g A v_c$ so Eq. (21-92) is not explicit in v_c. Using the data of Yousfi and others, Yang proposed the following system of equations for choking which Cheong and Leung have found to work well for larger particles.

$$\frac{2gD(\varepsilon_c^{-4.7}-1)}{(v_c-U_t)^2} = 6.81\times 10^5\left(\frac{\rho_f}{\rho_s}\right)^{2.2} \tag{21-93}$$

$$G_c = \frac{W_s}{A} = (v_c - U_t)\rho_s(1-\varepsilon_c) \tag{21-94}$$

This set of equations calculates both the voidage at choking ε_c and the choking velocity v_c. If the mass flow rate of system W_s is known, then the solution of the equations is fairly straightforward.

Example 21-4 Compare the choking velocity with the saltation velocities obtained in the previous example. For the 10-μm particles, it is prudent to use Eq. (21-92) by Yousfi and Gau. The terminal velocity can be calculated by using methods in Sec. 17. The terminal velocity for the 10-μm particles is 0.0028 m/s.

$$\mathrm{Re}_p = \frac{\rho_f U_t d_p}{\mu_f} = \frac{\left(1.92\dfrac{\text{kg}}{\text{m}^3}\right)(0.0028\ \text{m/s})(0.00001\ \text{m})}{1.84\times 10^{-5}\dfrac{\text{kg}}{\text{m}\cdot\text{s}}} = 0.0029$$

$$v_c = 32\sqrt{gd_p}\,\mathrm{Re}_p^{-0.06}\mu^{0.28}$$

$$= 32\sqrt{9.81\ \text{m/s}^2(0.00001\ \text{m})}\times(0.0029)^{-0.06}\left[\frac{2.08\ \text{kg/s}}{(1.92\ \text{kg/m}^3)(0.0079\ \text{m}^2)v_c}\right]^{0.28}$$

Solving the above for v_c gives a choking velocity of 1.57 m/s which is significantly smaller than the saltation velocity obtained from either correlation.

For the 2500-μm particles, the terminal velocity is 5.90 m/s. For Eq. (21-94), we have

$$G_c = \frac{W_s}{A} = \frac{2.08\ \text{kg/s}}{0.0079\ \text{m}^2} = (v_c - 5.90\ \text{m/s})\left(945\ \frac{\text{kg}}{\text{m}^3}\right)(1-\varepsilon_c)$$

For Eq. (21-93), we have

$$\frac{2\left(9.81\dfrac{\text{m}}{\text{s}^2}\right)(0.10\ \text{m})(\varepsilon_c^{-4.7}-1)}{(v_c-5.90)^2} = 6.81\times 10^5\left(\frac{1.92\ \text{kg/m}^3}{945\ \text{kg/m}^3}\right)^{2.2}$$

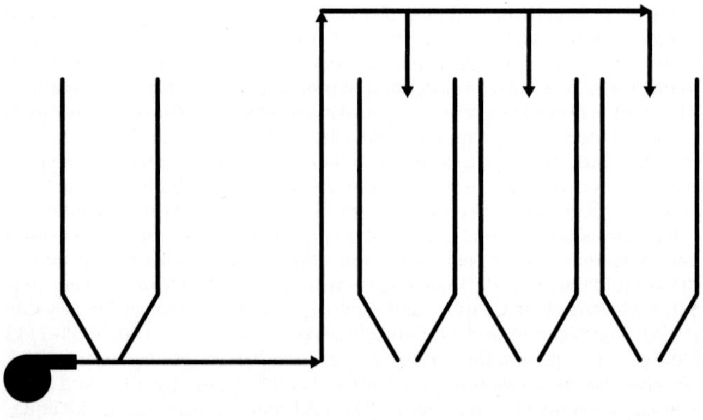

FIG. 21-93 Push or pressure conveying system.

The two equations can be solved simultaneously for ε_c and v_c, giving values of 0.84 and 7.65 m/s, respectively. This number is about 50 percent of the value obtained for the saltation velocity. Hence, usually saltation is the governing concept when designing or optimizing dilute-phase conveying systems.

TAXONOMY OF PNEUMATIC CONVEYING SYSTEMS

Pneumatic conveying systems can be classified generally into two broad divisions based on the flow type and system pressure. From the Zenz diagram, systems are either dense-phase or dilute-phase. Those systems operating above atmospheric pressure are known as *pressure systems* and are often called *push* systems as material is pushed from the pickup point to the system destination(s). Conversely, systems operating below atmospheric pressure are called *vacuum* or *pull* systems. Figures 21-93 and 21-94 show push and pull systems. The characteristics of each of the four combinations are examined here.

Dilute phase/pressure system This is one of the most common conveying systems used by industry. Conveying pressures are typically less than 1 bar gauge due to the widespread use of rotary lobe "Roots" style blowers. With the application of stepped lines, conveying distances of 300 m can be achieved. Loadings or phase ratios usually range from 1 to 10 kg/kg. Solids are in suspension when conveyed in this regime. As shown in Fig. 21-93, this system is preferred when there is a single source and multiple destinations because only one air mover is needed at the beginning of the system. It is possible to use a pressure system to pick up solids from multiple silos in a row, but one must ensure that air leakage is minimized from the conveying line into the silos.

Dilute phase/vacuum system This conveying is also very commonly used especially for short (typically 30 m or less) systems. These systems will typically operate at end of line pressures of no smaller than 0.5 bar absolute due to the considerable gas expansion along the length of the conveying line. For example, for a system starting at ambient pressure with 0.66-bar pressure drop, the gas velocity along the length of the line will triple. An added constraint is that rotary lobe blowers operating in vacuum configuration will only develop 0.5-bar pressure drop. Nevertheless, vacuum conveying systems find widespread use for batch loading operations, for example. A typical system is shown in Fig. 21-94. These systems operate by vacuum conveying a small batch of material into a destination hopper usually to a preset time or weight; the materials are then discharged from the destination hopper, and the cycle is repeated.

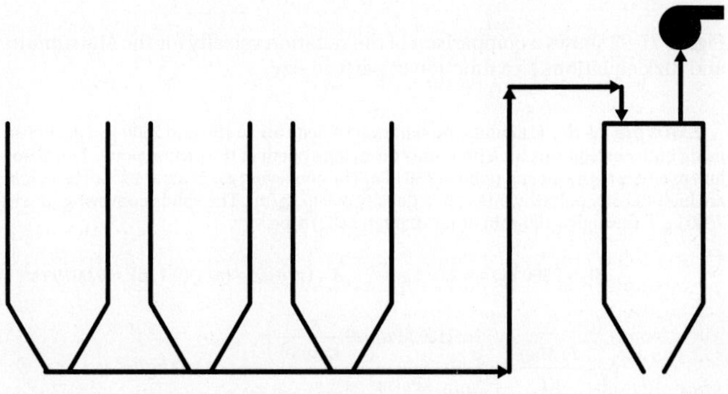

FIG. 21-94 Pull or vacuum-type conveying system.

Dense-phase/pressure system Numerous researchers and practitioners have shown that attrition in pneumatic conveying systems has been found to be proportional to the conveying velocity to the third to fifth power; hence velocity control in pneumatic conveying systems for attrition-prone materials is of utmost importance. Consequently, dense-phase or low-velocity conveying plays an important role in pneumatic conveying of bulk solids for the minimization of product degradation. Pressures in dense-phase systems range from 1 to 10 bar. For bulk solids with high air retention properties, phase ratios as high as 600 have been observed for bulk solids such as cement. As discussed in the earlier subsection on gas solid systems, the actual flow mode observed during dense-phase conveying is highly dependent on the air retention and permeability properties of the bulk solid. In general, these systems can be broken down into three broad classifications: (1) systems with fine materials which have sufficient air retention and permeability as to permit the solid-gas suspension to completely fill the line; (2) systems handling granular materials where active air management along the pipeline length is required to prevent pipeline blockage; and (3) systems with very permeable solids such as plastic pellets which naturally form slugs and require only proper pressure and velocity for successful dense-phase conveyance.

Dense-phase/Vacuum System While such systems are possible, they are not frequently used in practice due to the customary 0.5-bar pressure drop limitation. These systems are used for the transfer of fragile solids (for example, beaded carbon blacks) over short distances.

PNEUMATIC CONVEYING SYSTEM COMPONENTS

Air Movers Similar to a pump in the movement of liquids, air movers are a primary component of any pneumatic conveying system. Three distinct types of devices find use in conveying systems with each having distinct advantages and disadvantages with respect to cost and performance.

Fans Fans are the simplest devices and typically the least expensive. A typical centrifugal fan is shown in Fig. 21-95*a*. While units capable of delivering 0.3 bar are available, most units in practice range from 25 to 75 mbar, so this limits their utility in conveying systems. One unique advantage of fans is the ability of the bulk solid to pass through the fan itself (however, exercise caution when conveying combustible dusts). For light materials such as paper trim scrap, ground celluloses, films, etc., it is not uncommon for a fan to pull solids from a source, have them pass through the fan, and then be pressure-conveyed to a destination silo. The primary disadvantage for use of fans in pneumatic conveying systems is their comparatively poor pressure versus volume curve, as shown schematically in Fig. 21-96 (dotted line). From the dotted curve, it can be observed that for a small change in system pressure drop, the volume delivered can markedly increase or decrease.

Blowers Blowers are the most common air or prime mover for conveying systems. Invented in the mid-19th century by the brothers Philander and Francis Marion Roots, these positive displacement devices consist of two intermeshing lobes, as shown in Fig. 21-95*b*, and are often known as *Roots blowers.* While the two-lobe configuration is relatively common, three-lobe blowers are used to reduce pulsations and noise. In the pressure configuration, typical maximum pressures range from 0.5 to 1.2 bar, and in vacuum, typically 0.5-bar suction can be achieved. Since the process is essentially adiabatic compression, it is important to recognize that the temperature of the compressed gas can increase significantly (up to 80°C depending on discharge pressure), which may impact product degradation. A significant advantage of lobe-style blowers is their relatively flat performance curve, meaning that the volume delivered typically only decreases by ~15 percent over the expected range of operating pressures. Equipment suppliers provide volume delivered, adiabatic temperature rise, and brake horsepower as a function of operating pressure and blower speed. These blowers are often installed on trucks handling bulk solids and are connected to the engine power takeoff.

Compressors Compressors are routinely used for dense-phase systems where pressure drops in excess of 1 bar are not uncommon. They are the most expensive of the air movers, but do deliver nearly constant volume over a wide range of pressures.

Conveying Pipe and Connections Conventional pipe in a variety of different metals and plastics is used in conveying systems. While Schedule 40 pipe is sometimes used, lighter-gauge pipe (Schedule 10, for example) is more commonly used, given the amount of weight of material in the pipe at any given moment is relatively small. In addition to common ANSI pipe, tube is commonly used in smaller pipe sizes (100 mm). In the United States, for example, it is specified by its outside diameter (2, 2.5, 3, 3.5, and 4 in) with various wall thicknesses. While pipe and tubing suffice in most services, some materials demand special pipe alterations and treatment. For abrasive materials like sand, glass batch, and cement, hardened pipe and ceramic linings find common use. In other cases such as the conveying of plastics, pipe can be treated by grooving or shot-peening to prevent the

(*a*)

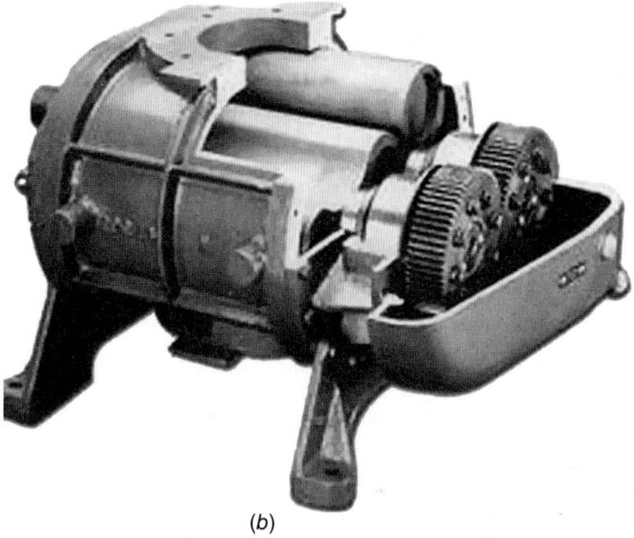

(*b*)

FIG. 21-95 (*a*) Typical centrifugal fan (https://www.indiamart.com/proddetail/industrial-centrifugal-fan-14497466212.html). (*b*) Roots blower (https://www.indiamart.com/proddetail/twin-lobe-industrial-root-blower-10456324097.html).

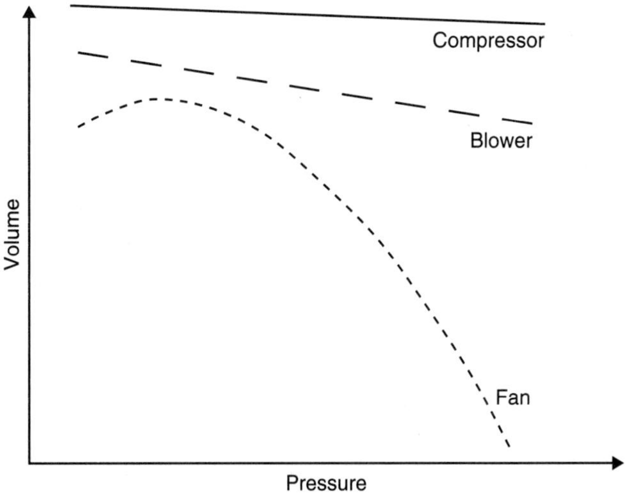

FIG. 21-96 Pressure versus volume characteristics for various air movers.

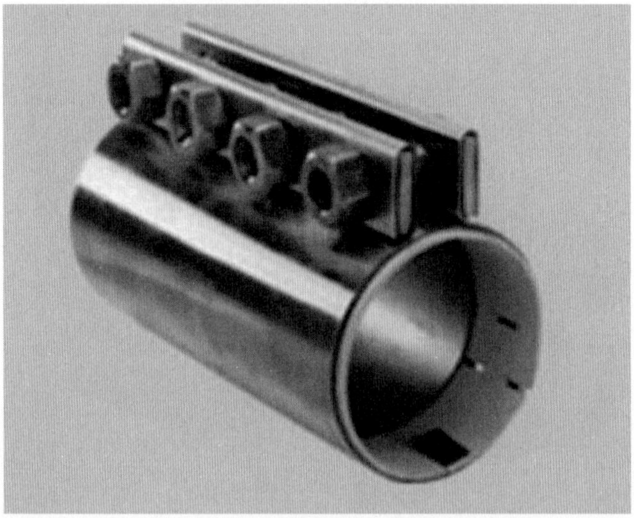

FIG 21-97 Typical compression coupling (http://www.morriscoupling.com/prod1.html).

formation of deposits such as streamers, angel hair, or snake skins. Flanges and compression couplings are the most common types of connectors used to join conveying pipe. For flanged pipe systems, it is not unusual to have self-aligning flanges to avoid impact of the solids with the lips/edges created with misaligned piping. The repeated impact of the bulk solids can result in attrition of the solids or deposition of the solids at the point of impact. A typical compression coupling is shown in Fig. 21-97.

Feeders for Pneumatic Conveying Systems Numerous devices exist to feed both dense- and dilute-phase conveying systems. Dhodapkar and Jacob ["Fluid-Solids Flow in Ducts," in Crowe (ed.), *Multiphase Flow Handbook*, 2d ed., Taylor and Francis, CRC Press, 2016] have summarized these in Table 21-6.

Despite the significant number of options available to the conveying system designer, three devices find widespread use in conveying system. The simple suction wand is commonly used for manual unloading of bulk solids from boxes using vacuum conveying. While not sophisticated in design, suction wands are usually designed with a "pipe in pipe" arrangement where the outer pipe permits the flow of air to the suction pipe and the inner pipe carries the bulk solid/air mixture. This is particularly critical for bulk solids with low permeability.

The rotary valve is the workhorse for both dilute- and, more recently, dense-phase conveying. A typical valve is shown in Fig. 21-98.

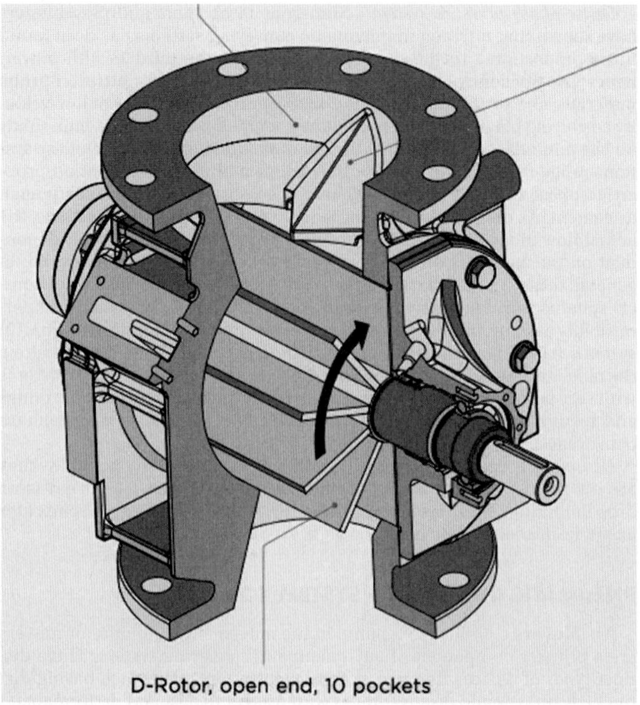

FIG. 21-98 Rotary valve. (*Courtesy of Coperion.*)

TABLE 21-6 Pneumatic Conveying System Feeders

Feeder	Suitable for system types	Operating pressure	Suitable materials	Unsuitable materials	Conveying mode	Comments
Suction nozzle	Vacuum	Up to 0.5 bar vacuum	All free flowing	Caking, large chunks, sticky, cohesive	Continuous dilute phase	Feed rate control is difficult
Rotary feeder	Vacuum and pressure	0.5 bar vacuum −3.5 bar pressure (6 bar maximum)	Wide range of materials	Abrasive, caking, sticky, very cohesive	Continuous dilute and dense phases	Most common feeder. Wide variety of design features
Eductor	Pressure	0.25 bar	Free flowing	Abrasive, friable, highly cohesive	Continuous dilute phase	Special design for abrasive materials, high-pressure motive gas may be required
Blow tank-bottom discharge	Pressure	0–10 bar	Wide range of materials, abrasive materials	Very cohesive sticky or compressible materials	Batch dilute and dense phases	Poorly flowing materials will require air injection in the cone, difficult to control conveying rate
Blow tank-fluidization type, top discharge	Pressure	0–10 bar	Fine fluidizable materials (Geldart class A/C, A)	Geldart class B, B/D, D or material with large chunks	Batch dilute and dense phases	
Double flapper valve	Pressure	0–0.5 bar	Abrasive and free-flowing material	Cohesive	Continuous dilute phase	Pulsing feed
Screw feeder	Pressure and vacuum	0.5 bar vacuum to 1 bar pressure	Wide range of materials	Sticky, compressible	Continuous dilute phase	Improper design will result in feed rate control (what do we mean by this?). Rate sensitive to material properties.
Fuller–Kinyon pump	Pressure	2 bar	Fine powders	Coarse materials	Continuous dilute and dense phases	Typically used in cement industry
Slide gate	Vacuum	Up to full vacuum	Coarse materials, preferably free-flowing	Cohesive, sticky materials	Continuous dilute phase	Simple, discharge rate depends on material properties
Other specialized feeders: diaphragm pump	Pressure	0–0.5 bar	Materials with high air retention or low deaeration, preferably low bulk density	Sticky, cohesive, low air retention	Continuous dilute and dense phases	

The rotary valve consists of a 6-, 8-, or 10-pocket rotor which fits closely inside the valve body. Typical rotor/body clearances are 75 to 150 μm. Rotors are available in a variety of manifestations: open-ended for free-flowing products; closed-ended rotors to prevent powder leakage out of the valve seal; replaceable flexible tip rotors for wear applications; and shallow pocket rotors for materials which are difficult to release from the pocket. Typical rotor speeds range from 10 to 50 rpm. Slower than 10 rpm results in a pulsatile conveying flow, while if the speed is greater than 50 rpm, it can be difficult for the solids to flow into the rotor pockets. The overall valve capacity is given by

$$\dot{V} = \dot{v}(\text{rpm})(\rho_b)(\text{fill efficiency}) \qquad (21\text{-}95)$$

where $\dot{v}$ is the volumetric displacement per revolution. Fill efficiency can be typically 70 to 90 percent, but in cases where the valve is not properly vented and is metering fine solids, the fill efficiency can drop to well below 50 percent. Typical performance curves for different sizes of a commercial valve are shown in Fig. 21-99. Additional detail on rotary valves can be found in the Feeding, Metering, and Dosing subsection.

Another key design factor is rotary valve leakage. In pressure conveying systems, gas will leak past the valve in two distinct ways. Even without rotation, gas will leak past the slits formed by the rotor tip and the valve body. In addition, as the valve rotates, gas will leak as a result of the gas-filled pockets rotating back to the low-pressure side. This gas volume loss must be accounted for during the design process or during process troubleshooting. Typical gas leakage curves are shown in Fig. 21-100.

Many suppliers will supply leakage curves for new valves, and it is important to recognize that these valves will increase during the life of the valve due to wear. Rotor to valve body clearance must be periodically checked with feeler gauges. Wear can be caused by abrasive solids, temperature changes, or deposition of solids on either the rotor or body. In the last several years, high-pressure rotary valves have been introduced and are now routinely used in dense-phase conveying applications.

Blow pots are likely the most common feed device used for dense-phase conveying. The most common arrangement consists of a conical hopper with block valves on the inlet and outlet. It is common to have some aeration in the cone in order to fluidize the solids and facilitate their exit out of the blow pot. Since the blow pot usually works in batch fashion (fill/pressurize/blow solids/de-pressurize), the actual "continuous" flow rate is lower than the instantaneous flow rate during the blow cycle. On account of their simple design, blow pots are generally very reliable with the exception of the inlet and outlet valves, which can see wear as a result of the number of cycles they see.

DESIGN AND ANALYSIS OF DILUTE-PHASE CONVEYING SYSTEMS

In many ways, analysis of dilute-phase systems is a relatively simple extension of incompressible flow through piping systems. However, there are four important differences between incompressible flow and dilute-phase conveying flow which make the solids flow situation more complicated.

1. The conveying gas expands continuously along the pipeline length. Hence, analysis of the system must be done in a piecewise fashion along the length of the pipe.
2. Pressure drop due to acceleration of the solids must be considered.
3. As opposed to the equivalent length concept for incompressible flow, each elbow is calculated separately, which serves to account for the deceleration and reacceleration of the solids as they transit the elbow.
4. In addition to the pressure drop associated with the fluid, the pressure drop due to the solids friction needs to be calculated.

The overall pressure drop equation for a dilute-phase conveying system is given by Eq. (21-96).

$$\Delta P_{\text{total}} = \Delta P_{\text{gas}} + \Delta P_{\text{acc}} + \Delta P_{\text{lift}} + \Delta P_{\text{bends}} + \Delta P_{\text{solids}} + \Delta P_{\text{solid-gas separation}} \qquad (21\text{-}96)$$

Each term will now be discussed in turn.

Gas Pressure Drop Gas-only pressure drop can be calculated by using the customary methods found in chemical engineering fluid mechanics textbooks or the corresponding section here. It is important to recognize that the gas pressure drop includes both the pressure drop across the conveying line and any additional pressure drop associated with the piping to and/or from the blower. As opposed to the methods for incompressible flow using a single calculation to calculate the pressure drop across the system, it is extremely important to segment the entire conveying line and calculate the pressure drop individually across each segment. This must be done because of the change in both the gas velocity and the gas density along the length of the piping system.

Acceleration Pressure Drop Both the gas and solids must be accelerated, and there is a pressure drop associated with this process. Klinzing et al. (*Pneumatic Conveying of Solids*, 3d ed., Springer, 2010) show by integrating a modified momentum balance that the acceleration of the solids can be represented by

$$\Delta P_{\text{acc-solids}} = \frac{U_p W_s}{A} \qquad (21\text{-}97)$$

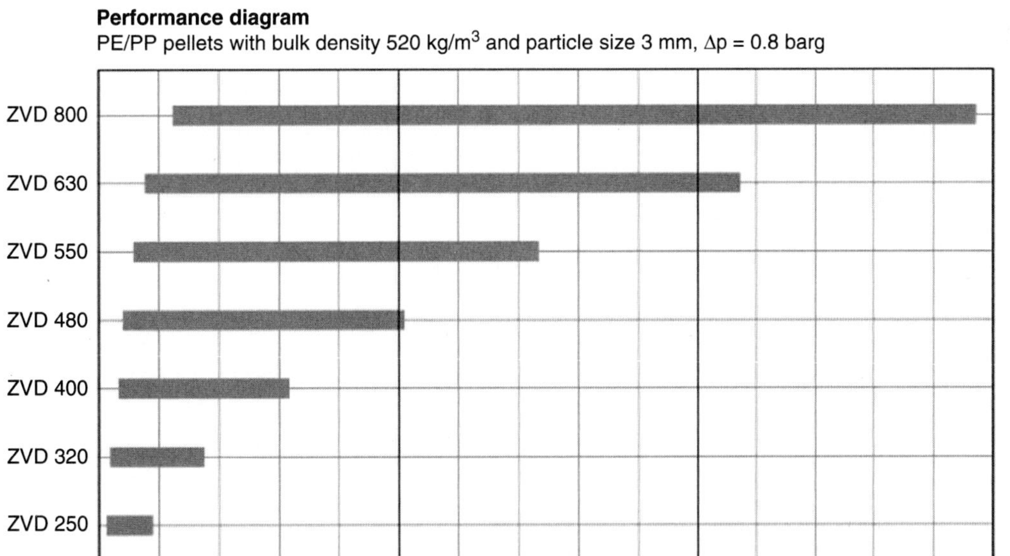

FIG. 21-99 Rotary valve performance chart. (*Courtesy of Coperion.*)

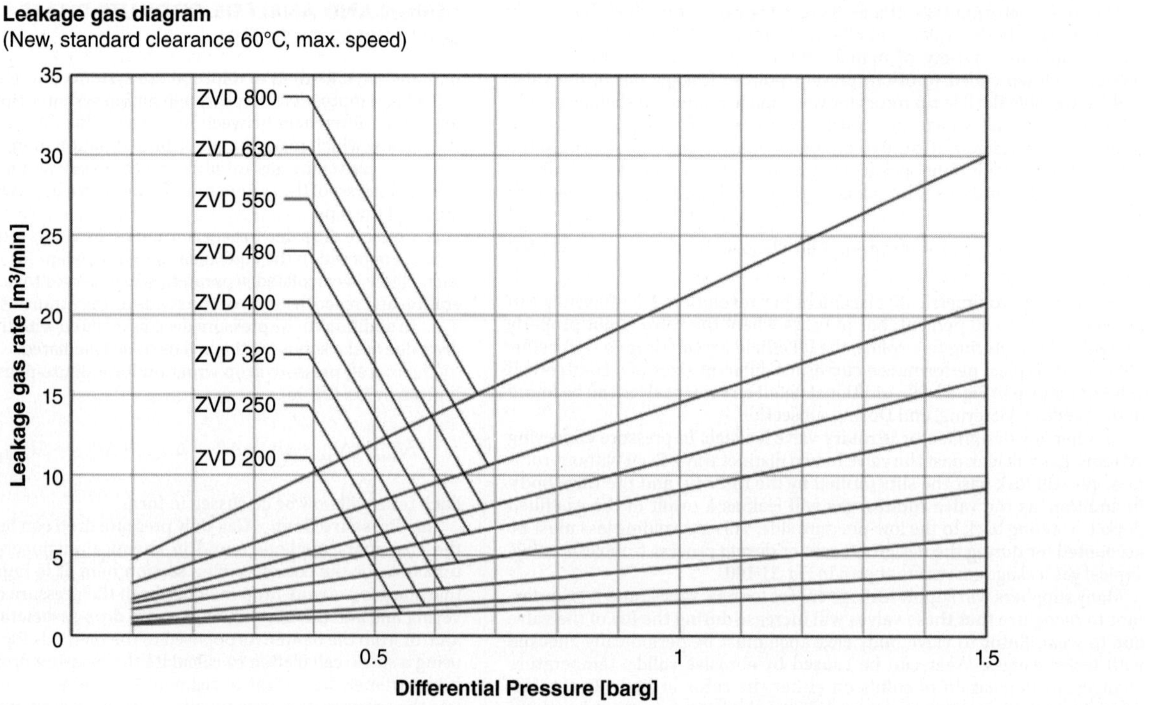

FIG. 21-100 Rotary valve leakage curves. (*Courtesy of Coperion.*)

The gas acceleration is

$$\Delta P_{\text{acc-gas}} = \frac{\rho_f U_g^2}{2} \tag{21-98}$$

Combining Eqs. (21-97) and (21-98), we find

$$\Delta P_{\text{acc}} = \Delta P_{\text{acc-solids}} + \Delta P_{\text{acc-gas}} = \frac{U_p W_s}{A} + \frac{\rho_f U_g^2}{2}$$

$$\Delta P_{\text{acc}} = \frac{\rho_f U_g^2}{2}\left[1 + \frac{2W_s}{A\rho_f U_g}\left(\frac{U_p}{U_g}\right)\right]$$

The term $W_s/A\rho_f U_g$ is the phase ratio or loading μ. So the final form for the acceleration pressure drop is as follows:

$$\Delta P_{\text{acc}} = \frac{\rho_f U_g^2}{2}\left[1 + \mu\left(\frac{U_p}{U_g}\right)\right] \tag{21-99}$$

For many solids, the particle-to-gas velocity ratio is given by either the Hinkle or IGT correlations.

$$\frac{U_p}{U_g} = 1 - 0.044 d_p^{0.3} \rho_s^{0.5} \quad \text{(Hinkle)} \tag{21-100}$$

$$\frac{U_p}{U_g} = 1 - 0.68 d_p^{0.92} \rho_s^{0.5} D^{-0.54} \rho_f^{-0.2} \quad \text{(IGT)} \tag{21-101}$$

(Hinkle, Ph.D. thesis, Georgia Institute of Technology, Atlanta, 1953; IGT, Dept. of Energy Contract 2286-32, October 1978). Typically, for large-scale industrial systems, the solid velocity will approach 70 to 80 percent of the gas velocity for hard particle systems. However, Vasquez et al. ("Effect of Elbow Geometry on Pressure Drop in Upstream Straight Sections in Dilute Phase Pneumatic Conveying," *RELPOWFLO IV Proceedings*, Tromso, Norway, June 2008, pp. 568–573) have shown that the velocity ratio for soft plastic materials can be as low as 25 to 30 percent due to repeated impact of the particles with the wall.

Pressure Drop Due to Lift There is a pressure loss associated with lifting the solids and gas from one elevation to another. As would be expected, the pressure drop associated with lifting the solid is significantly larger than the gas term. This pressure drop is given by the following relation:

$$\Delta P_{\text{lift}} = \Delta P_{\text{lift-gas}} + \Delta P_{\text{lift-solids}} = \varepsilon \rho_f \Delta Hg + (1-\varepsilon)\rho_s \Delta Hg \tag{21-102}$$

The gas-only term can be included here or as part of the ΔP_{gas} term. Since ε ranges from 0.97 to near 1, omitting ε in the ΔP_{gas} calculation results in very little error in the overall calculation. Using a mass balance around a control volume, the voidage is then

$$\varepsilon = 1 - \frac{W_s}{\rho_p A U_p} \tag{21-103}$$

One of the particle velocity correlations must be utilized to calculate U_p.

Bend Pressure Drop As solids enter a bend, they tend to form a rope of material which slides along the outer wall of the bend. This action slows the solids, and they are subsequently reaccelerated upon leaving the elbow, which results in additional pressure drop for several pipe diameters downstream of the elbow. This is shown in Fig. 21-101.

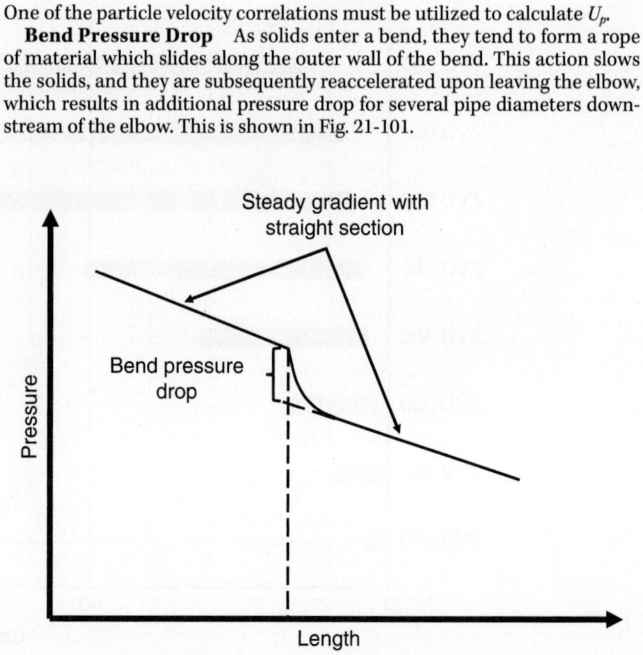

FIG. 21-101 Development of pressure drop around a bend.

TABLE 21-7 Bend Factors for Eq. (21-104)

R/D = bend radius/pipe diameter	B
2	1.50
4	0.75
≥6	0.50

If it is desired to measure the pressure drop across an elbow, it is important to measure 5 to 10 pipe diameters downstream of the elbow. It is important also to recognize that the roping phenomenon can be a source of pipe wear and is usually evidenced by scoring on the inside of the pipe. Pneumatic conveying elbows are specified by the elbow radius to pipe diameter ratio R/D. As compared to conventional pipe elbows used for liquids and gases which have R/D values of 1 to 2, conveying elbows typically have R/D value ranging from 4 to 12. Bradley [*Powder Handling and Processing* **2**(4): 315–321 (1990)] has shown that these ratios result in the lower pressure drops compared to short radius elbows. Several commercially available elbows are available (for example, the Gamma Bend® by Coperion and the Smart Elbow® by Hammertek) which promote the formation of a bed of solids in order to prevent pipe erosion.

Klinzing et al. (Klinzing, *Pneumatic Conveying of Solids*, 3d ed., Springer, 2010) have reviewed several different approaches to calculating the bend pressure drop. Chambers and Marcus ("Pneumatic Conveying Calculations," *Proc. 2nd Int. Conf. on Bulk Materials Storage, Handling and Transportation*, Wollongong, Australia, 1986, pp. 49–52) provide a simple, but effective, method to calculate the bend pressure drop.

$$\Delta P_{\text{bend}} = B(1+\mu)\frac{\rho_f U_g^2}{2} \tag{21-104}$$

where B is the bend factor in Eq. (21-104) and depends on the bend R/D factor as given in Table 21-7.

Solids Friction Pressure Drop The form of the equation for the pressure drop due to solids friction is similar in form to that of single-phase fluid flow except that the equation is multiplied by the solids loading μ.

$$\Delta P_{\text{solids}} = \mu\lambda_Z \frac{\rho_f U_g^2 L}{2D} \tag{21-105}$$

To calculate the pressure drop, it is imperative to know the solids friction factor λ_Z. There are two approaches here.

First, in the case where data either from an experimental system or from an operating system are available, then each of the other components in Eq. (21-96) can be calculated and then λ_Z is fit to the data. Typically the "air only" pressure drop will be measured along with the system pressure drop at several different solids throughputs. In the case of a pilot or laboratory system, it is very helpful to have specific test sections (both horizontal and vertical) where both the horizontal and vertical components of pressure drop can be measured. The experimental data can be used to calculate a single value of λ_Z, or it can be fit to one of the equations given below.

Second, if there is no access to experimental data, then λ_Z must be calculated based on one of a number of correlations. Over the last 70 years, numerous correlations have been proposed for the solids friction factor. Stegmaier [*fordern und heben* **28:** 363–366 (1978)] proposed the following equation for λ_Z for fine solids which permits the inclusion of particle properties into the solids friction equation:

$$\lambda_Z = K\mu^a \text{Fr}^b \text{Fr}_s^c \left(\frac{D}{d_p}\right)^d \quad \text{where} \quad \text{Fr} = \frac{U_g^2}{gD} \quad \text{and} \quad \text{Fr}_s = \frac{U_t^2}{gd_p} \tag{21-106}$$

Caution must be exercised to not confuse these Froude numbers with the Froude numbers used in the Matsumoto saltation analysis as these are squares of the former. Shortly thereafter, Weber [*Bulk Solids Handling* **2**(2): 231–233 (1982)] examined a number of conveying data sets for coarse solids and regressed the values for the constants K, a, b, c, and d. These are given in Table 21-8 for particle diameters greater than 500 μm (Weber) and for diameters less than 500 μm as given by Stegmaier.

For vertical conveying, the solids friction factor was determined by Weber to be

$$\lambda_Z = \frac{U_g}{1200U_p} + \frac{2U_g}{U_p \text{Fr}} \tag{21-107}$$

TABLE 21-8 Constants for Solids Friction Factors

Material particle size	K	a	b	c	d
>500 μm (Weber)	0.082	−0.3	−0.86	0.25	0.1
<500 μm (Stegmaier)	2.1	−0.3	−1	0.25	0.1

Pressure Drop Due to Solid Gas Separation In most cases, the terminus point in many pneumatic conveying systems may be some piece of solid-gas separation equipment. To determine the required blower performance, it is critical to include the pressure drop in conveying analysis calculations. Methods to estimate the pressure drop across cyclones and dust collectors can be found in Sec. 17.

Calculation Strategy for Dilute-Phase Conveying Since the gas velocity, solids velocity, and gas density change along the length of the conveying line, the calculations are usually broken into smaller segments. For example, a 6-m-long horizontal run might be broken in 6-m-long segments. It is easiest to begin the calculations from a known boundary condition, and usually that is the point of atmospheric pressure. Hence, calculations for vacuum systems will begin at the pickup point while for pressure systems, the calculations commence from the destination end and work their way back to the blower and the beginning of the system. While calculation by hand is possible, these calculations lend themselves well to either spreadsheet or programming tools such as MATLAB. Hence, there is less need to start the calculations from the known boundary condition since even simple manual iteration will allow the calculations to converge quickly.

Example 21-5 Using the values from the worked example in the saltation section for conveying 5000 kg/h of HDPE through 0.100-m ID pipe by vacuum for the case of 2500-μm particles for the configuration shown in Fig. 21-102, calculate both the pressure drop and the conveying gas flow rate.

In the saltation calculation, it was assumed that the pressure was 150 kPa. Since this is a vacuum conveying system, the pressure at the point where the solids are introduced into the line is atmospheric pressure, so P = 101.325 kPa. We can use Rizk's equation to calculate the saltation velocity since it works well for large particle systems. Following the approach above, the saltation velocity is as follows for a gas density of 1.20 kg/m³. Assume the pipe roughness ratio e/D is 0.0002.

Rizk equation:

For 2500-μm particles, $\delta = 1.44d_p + 1.96 = 1.44(2.5\text{ mm}) + 1.96 = 5.56$; $\kappa = 1.1d_p + 2.5 = 5.25$.

$$v_s = \left[\frac{W_s\sqrt{gD}^{\kappa}10^{\delta}}{\rho_g A}\right]^{\frac{1}{\kappa+1}} = \left\{\frac{\left(1.39\,\frac{\text{kg}}{\text{s}}\right)\left[\sqrt{\frac{9.81\text{ m}}{s^2}(0.1\text{ m})}\right]^{5.25} 10^{5.56}}{\left(1.20\,\frac{\text{kg}}{\text{m}^3}\right)(0.0079\text{ m}^2)}\right\}^{\frac{1}{5.25+1}} = 17.1\text{ m/s}$$

Since the cross-sectional pipe area is 0.0079 m², the inlet gas flow rate is

$$\dot{V} = v_s A = \left(17.1\,\frac{\text{m}}{\text{s}}\right)(0.0079\text{ m}^2) = 0.1351\,\frac{\text{m}^3}{\text{s}} = 486\,\frac{\text{m}^3}{\text{h}}$$

$$\dot{m} = \left(486\,\frac{\text{m}^3}{\text{h}}\right)\left(1.2\,\frac{\text{kg}}{\text{m}^3}\right) = 584\,\frac{\text{kg}}{\text{h}}$$

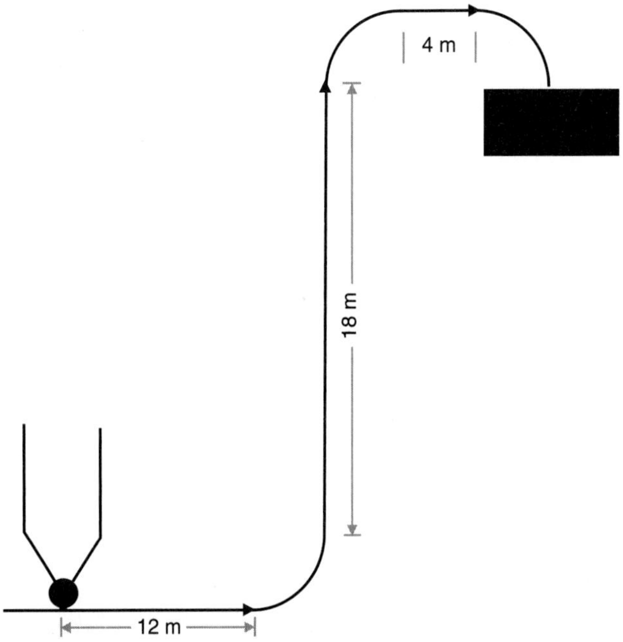

FIG. 21-102 Example 21-5 conveying system.

First, we have the 12-m-long horizontal section. Breaking this into two 6-m-long subsections, we now need to calculate each component in Eq. (21-96). Gas-only pressure drop can be calculated using the standard formulation for single-phase flow.

$$\Delta P_{\text{gas}} = \frac{\lambda_f \rho_f U_g^2 L}{2D}$$

where λ_f is the friction factor. It can be conveniently calculated by the equation of Swamee and Jain [*J. Hydraulic Eng.*, ASCE, **102:** 657–664 (1976)]

$$\lambda_f = \frac{1.325}{\left[\ln\left(\dfrac{e}{3.7D} + \dfrac{5.74}{\text{Re}^{0.9}}\right)\right]^2}$$

$$\text{Re} = \frac{\rho_f D U_g}{\mu} = \frac{\left(1.20\,\dfrac{\text{kg}}{\text{m}^3}\right)(0.100\text{ m})\left(17.1\,\dfrac{\text{m}}{\text{s}}\right)}{0.0000184\,\dfrac{\text{kg}}{\text{m}\cdot\text{s}}} = 111521$$

$$\lambda_f = \frac{1.325}{\left[\ln\left(\dfrac{0.0002}{3.7} + \dfrac{5.74}{118{,}695^{0.9}}\right)\right]^2} = 0.0187$$

Therefore, the gas pressure drop across this section is

$$\Delta P_{\text{gas}} = \frac{\lambda_f \rho_f U_g^2 L}{2D} = \frac{(0.0187)\left(\dfrac{1.2\text{ kg}}{\text{m}^3}\right)\left(17.1\,\dfrac{\text{m}}{\text{s}}\right)^2(6\text{ m})}{2(0.100\text{ m})} = 196\text{ Pa}$$

The acceleration pressure drop requires the calculation of the particle to gas velocity ratio.

$$\Delta P_{\text{acc}} = \frac{\rho_f U_g^2}{2}\left[1 + 2\mu\left(\frac{U_p}{U_g}\right)\right] \quad \text{where } \frac{U_p}{U_g} = 1 - 0.044 d_p^{0.3}\rho_s^{0.5}$$

$$\frac{U_p}{U_g} = 1 - 0.044 d_p^{0.3}\rho_s^{0.5} = 1 - 0.044(0.0025\text{ m})^{0.3}\left(945\,\frac{\text{kg}}{\text{m}^3}\right)^{0.5} = 0.7758$$

$$\Delta P_{\text{acc}} = \frac{\left(1.2\,\dfrac{\text{kg}}{\text{m}^3}\right)\left(17.1\,\dfrac{\text{m}}{\text{s}}\right)^2}{2}\left[1 + \frac{2\left(5000\,\dfrac{\text{kg}}{\text{h}}\right)}{584\,\dfrac{\text{kg}}{\text{h}}}0.7758\right] = 2532\text{ Pa}$$

Since this is a horizontal section, there are no bend or lift components to the pressure drop. The pressure drop due to solids friction is calculated as follows. Since the particles are greater than 500 μm, the Weber exponents are used in the λ_Z equation.

$$\Delta_Z = 0.082\mu^{-0.3}\,\text{Fr}^{-0.86}\,\text{Fr}_s^{0.25}\left(\frac{D}{d_p}\right)^{0.1} \quad \text{where Fr} = \frac{U_g^2}{gD} \text{ and Fr}_s = \frac{U_t^2}{g d_p}$$

At these conditions, U_t is 7.13 m/s.

$$\text{Fr} = \frac{\left(17.1\,\dfrac{\text{m}}{\text{s}}\right)^2}{\left(9.81\,\dfrac{\text{m}}{\text{s}^2}\right)(0.100\text{ m})} = 298 \quad \text{and} \quad \text{Fr}_s = \frac{\left(7.13\,\dfrac{\text{m}}{\text{s}}\right)^2}{\left(9.81\,\dfrac{\text{m}}{\text{s}^2}\right)(0.0025\text{ m})} = 2073$$

$$\lambda_Z = 0.082\left(\frac{5000\,\dfrac{\text{kg}}{\text{h}}}{584\,\dfrac{\text{kg}}{\text{h}}}\right)^{-0.3} 298^{-0.86}\,2073^{0.25}\left(\frac{0.100\text{ m}}{0.0025\text{ m}}\right)^{0.1} = 0.0031$$

So the solids friction pressure drop is

$$\Delta P_{\text{solids}} = \mu\lambda_Z\frac{\rho_f U_g^2 L}{2D} = \left(\frac{5000\,\dfrac{\text{kg}}{\text{h}}}{584\,\dfrac{\text{kg}}{\text{h}}}\right)(0.0031)\frac{\left(\dfrac{1.2\text{ kg}}{\text{m}^3}\right)\left(17.1\,\dfrac{\text{m}}{\text{s}}\right)^2(6\text{ m})}{2\,(0.100\text{ m})} = 282\text{ Pa}$$

So the total pressure drop across this first element is

$$\Delta P_{\text{total}} = \Delta P_{\text{gas}} + \Delta P_{\text{acc}} + \Delta P_{\text{solids}} = 196\text{ Pa} + 2532\text{ Pa} + 282\text{ Pa} = 3010\text{ Pa}$$

The pressure at the end of the first element is 101,325 Pa − 3010 Pa = 98,315 Pa.

The new gas density is

$$\rho_f = \frac{P(MW)}{RT} = \frac{(98{,}315\text{ Pa})\left(29\,\dfrac{\text{kg}}{\text{kg}\cdot\text{mol}}\right)}{8314\,\dfrac{\text{Pa}\cdot\text{m}^3}{\text{kg}\cdot\text{mol}\cdot\text{K}}(22+273.15)\text{ K}} = 1.162\,\frac{\text{kg}}{\text{m}^3}$$

The new gas velocity is

$$U_{g1} = U_{g0}\frac{P_0}{P_1} = \left(17.1\,\frac{\text{m}}{\text{s}}\right)\frac{101{,}325\text{ Pa}}{98{,}315\text{ Pa}} = 17.62\text{ m/s}$$

These new values for the gas density and gas velocity are used for the next 6-m-long horizontal section. The resulting values for the air and solids pressure drop are 201 Pa and 277 Pa, respectively, for a total element pressure drop of 478 Pa. At the end of this section, the gas velocity and gas density are 17.71 m/s and 1.610 kg/m³. Assuming a long radius bend ($B = 0.5$), the bend pressure drop is

$$\Delta P_{\text{bend}} = B(1+\mu)\frac{\rho_f U_g^2}{2} = 0.5\left(1 + \frac{5000\,\dfrac{\text{kg}}{\text{h}}}{584\,\dfrac{\text{kg}}{\text{h}}}\right)\frac{\left(1.610\,\dfrac{\text{kg}}{\text{m}^3}\right)\left(17.71\,\dfrac{\text{m}}{\text{s}}\right)^2}{2} = 873\text{ Pa}$$

The gas velocity and gas density at the end of the elbow are 17.87 m/s and 1.150 kg/m³, respectively.

The 18-m vertical section is now split into three 6-m sections. The gas pressure drop follows the methods outlined above and is 202 Pa. The solids friction coefficient is calculated using Weber's formulation for λ_Z in the vertical direction. The inverse of U_p/U_g can be used in both terms of the equation.

$$\lambda_Z = \frac{U_g}{1200U_p} + \frac{2U_g}{U_p\text{Fr}} = \frac{1/0.7758}{1200} + \frac{2(1/0.7758)}{\dfrac{(17.87\text{ m/s})^2}{\left(9.81\,\dfrac{\text{m}}{\text{s}^2}\right)(0.100\text{ m})}} = 0.009$$

The solids friction pressure drop is 851 Pa. For this element, the lift must be calculated.

$$\varepsilon = 1 - \frac{W_s}{\rho_p A U_p} = 1 - \frac{1.39\,\dfrac{\text{kg}}{\text{s}}}{\left(945\,\dfrac{\text{kg}}{\text{m}^3}\right)(0.0079\text{ m}^2)(0.7758)\left(17.87\,\dfrac{\text{m}}{\text{s}}\right)} = 0.9865$$

$$\Delta P_{\text{lift}} = \varepsilon\rho_f\,\Delta Hg + (1-\varepsilon)\rho_s\,\Delta Hg = (6\text{ m})\left(9.81\,\frac{\text{m}}{\text{s}^2}\right)$$
$$\times\left[\left(1.150\,\frac{\text{kg}}{\text{m}^3}\right)(0.9865) + (1-0.9865)\left(945\,\frac{\text{kg}}{\text{m}^3}\right)\right] = 817\text{ Pa}$$

So the total pressure drop across this element is 205 Pa + 851 Pa + 817 Pa = 1873 Pa. The resulting gas density and gas velocity at the end of this element are 1.128 kg/m³ and 18.22 m/s, respectively. The remaining vertical sections and the 2-m-long radius elbows and single 4-m horizontal run can be calculated using the methods outlined above; the results for all elements are found in Table 21-9.

The overall system pressure drop is 12.2 kPa.

DENSE-PHASE CONVEYING

As discussed in the Flow Modes subsection, the flow mode observed in dense phase depends on the particle properties. Fine, air-retentive materials will convey in full-line or full-bore mode where the entire pipe is full of solids and active slug management through injection of air at the beginning of the line is required. In contrast, coarse solids will convey in well-formed plugs in horizontal orientation with a stationary layer forming in the lower cross section of the pipe while a moving layer of solids passes over the stationary layer, simultaneously depositing and picking up solids from the nonmoving layer. Note that not all bulk solids will be conveyable in dense phase. For example, some fibrous materials such as ground cellulose will tend to knit together, making plug control very difficult.

TABLE 21-9 Results from Example 21-5

Element type	Length, m	Pressure at beginning of element, kPa	Gas density, kg/m³	Gas velocity, m/s	Pressure drop, Pa
Horizontal	6	101.3	1.2	17.1	3010
Horizontal	6	98.3	1.162	17.6	478
Bend		97.8	1.161	17.7	873
Vertical	6	97.0	1.150	17.9	1873
Vertical	6	95.1	1.128	18.2	1877
Vertical	6	93.2	1.106	18.6	1882
Bend		91.3	1.083	19.0	935
Horizontal	4	90.4	1.072	19.2	320
Bend		90.1	1.067	19.2	948

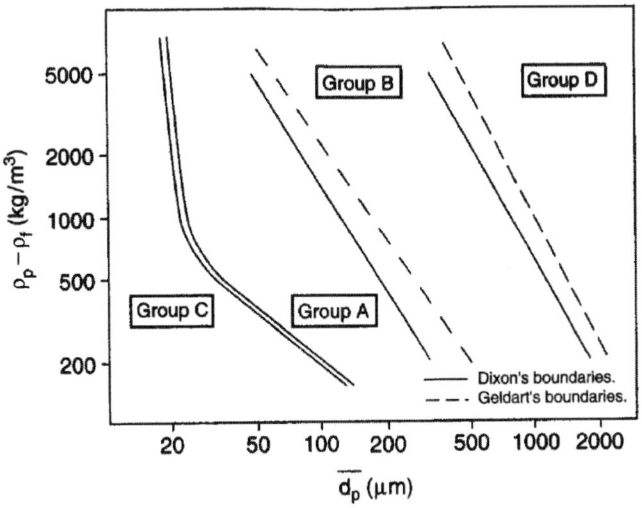

FIG. 21-103 Geldart's classification of aeration behavior with Dixon and Geldart boundaries. (*From Mason, Ph.D. thesis, Thomes Polytechnic, London, 1991, with permission.*)

Prediction of Dense-Phase Conveying Mode Prediction of dense-phase conveying mode has been the subject of considerable research. Early efforts focused on the use of single-particle properties such as particle size and particle density, as depicted on the Geldart diagram, Fig. 21-103 [Geldart, *Powder Technology* **7:** 285–292 (1973)] and later on a modified diagram proposed by Dixon (*Proceedings of the International Conference on Pneumatic Conveying*, London, 1979).

This work suggested that Geldart groups A and C (high gas retention materials) would exhibit fluidized dense phase, while group D would exhibit strong axisymmetric slugs. Group B would exhibit weak asymmetric slugs, making these materials unsuitable for dense phase without specially designed systems. In addition to the single-particle properties, several researchers have proposed the use of derived properties which mimic the behavior of the bulk solid in dense phase. Mainwaring and Reed [*Bulk Solids Handling* 7(3): 1987, 415–425 (1987)], Jones (Ph.D. Thesis, "The Influence of Bulk Particulate Properties on Pneumatic Conveying Performance," Thames Polytechnic, 1988), and Sanchez et al. [*Powder Technology* **138:** 93–117 (2003)] have proposed using permeability of the bulk solid and various forms of the deaeration factor as metrics to determine dense-phase conveying mode, as shown in Figs. 21-104 and 21-105. The lines of demarcation between the proposed flow modes should be used for guidance only, as numerous other properties such as particle cohesion, shape, and electrostatics will influence the actual observed flow mode.

The permeability is a relatively easy laboratory measurement (as shown in Fig. 21-46) and is obtained by passing gas through a bed of solids and measuring the pressure drop as a function of the gas velocity. In contrast, the measurement of deaeration of bulk solids is comparatively difficult, as it requires the test bulk solid to be fluidized by gas or mechanical means and then the gas is shut off simultaneously at both the inlet and outlet of the fluid bed. The subsequent settlement of the bed is used to derive the deaeration constant. For many bulk solids such as Geldart groups B and D, deaeration is very fast (<1 s); hence, obtaining exact values is difficult without the use of high-speed photography. For group C solids, settlement will occur over a long time (even days in some cases). Hence, Chambers et al. (*Proceedings of the 6th International Conference on Bulk Materials Storage, Handling and Transportation*, 1998, pp. 309–319) have suggested that only deaeration times between 1 and 10 s are of interest. Consequently, in their comprehensive review article on the subject, Jones and Williams [*Particuology* **6:** 289–300 (2008)] have proposed plotting the loose poured bulk density versus permeability as a predictor of dense phase flow mode as shown in Fig. 21-106.

The boundary between fluidized dense phase and dilute phase is given by $(P_f)(\rho_{lpb})^{0.75} \sim 300$ while the boundary between dilute phase and plug flow is $P_f \sim 20 \times 10^{-6}$ m$^3 \cdot$ s/kg, where P_f is the permeability. In their review article, Jones and Williams suggest that all methods have some merit with some techniques more suited to one Geldart group than another. This is summarized in Table 21-10.

Secondary Air Injection As indicated in the previous subsection, there is a middle region in most of the flow mode representations that suggests some bulk solids are suitable only for dilute phase. These materials form weak asymmetric plugs and dunes in horizontal conveying sections which may aggregate into fairly impermeable plugs, resulting in line blockage unless some active form of plug management is used to break up and control the plug length. Three primary methods are used to introduce gas into the pipeline to facilitate plug length control along the length of the line.

1. A secondary, but smaller, pipeline is run adjacent to the conveying line. At given intervals (typically 2 to 6 m), a valve (also known as a *booster*) is placed between the clean air line and the conveying line, and a constant flow of gas is bled into the conveying line. Determination of the spacing and the setting of the valve flow is made during pilot-scale conveying tests.

2. In somewhat similar fashion as for method 1, pressure sensors measure the pressure differential between the clean air line and the conveying pipe. These measurements are connected to control valves which meter

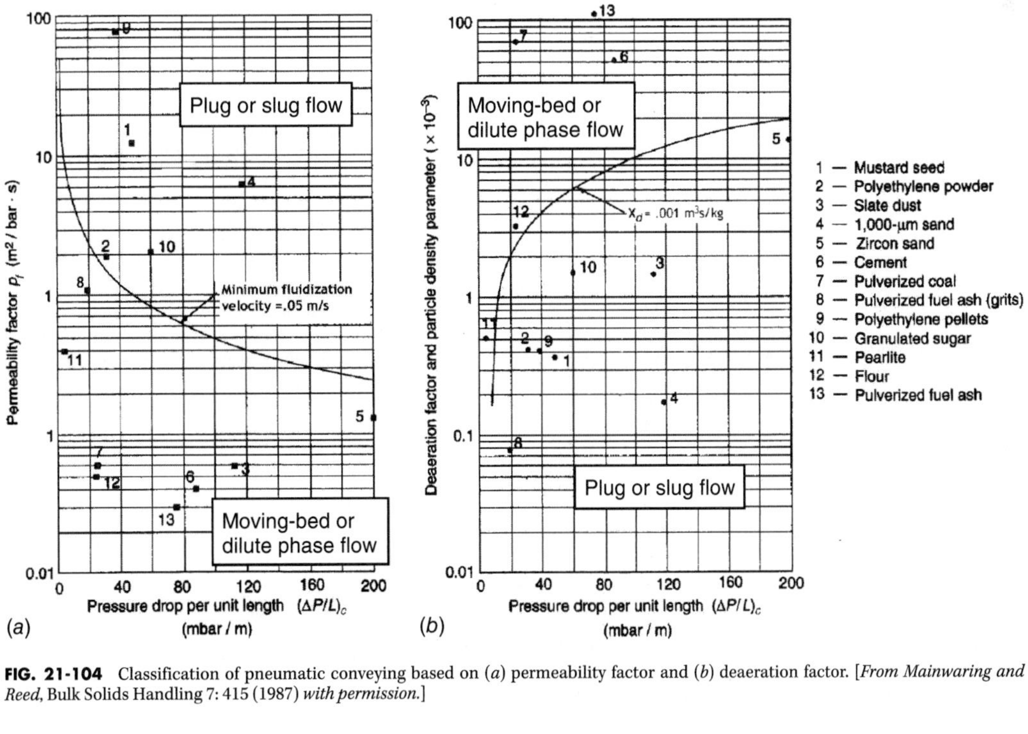

FIG. 21-104 Classification of pneumatic conveying based on (*a*) permeability factor and (*b*) deaeration factor. [*From Mainwaring and Reed,* Bulk Solids Handling 7: 415 (1987) *with permission.*]

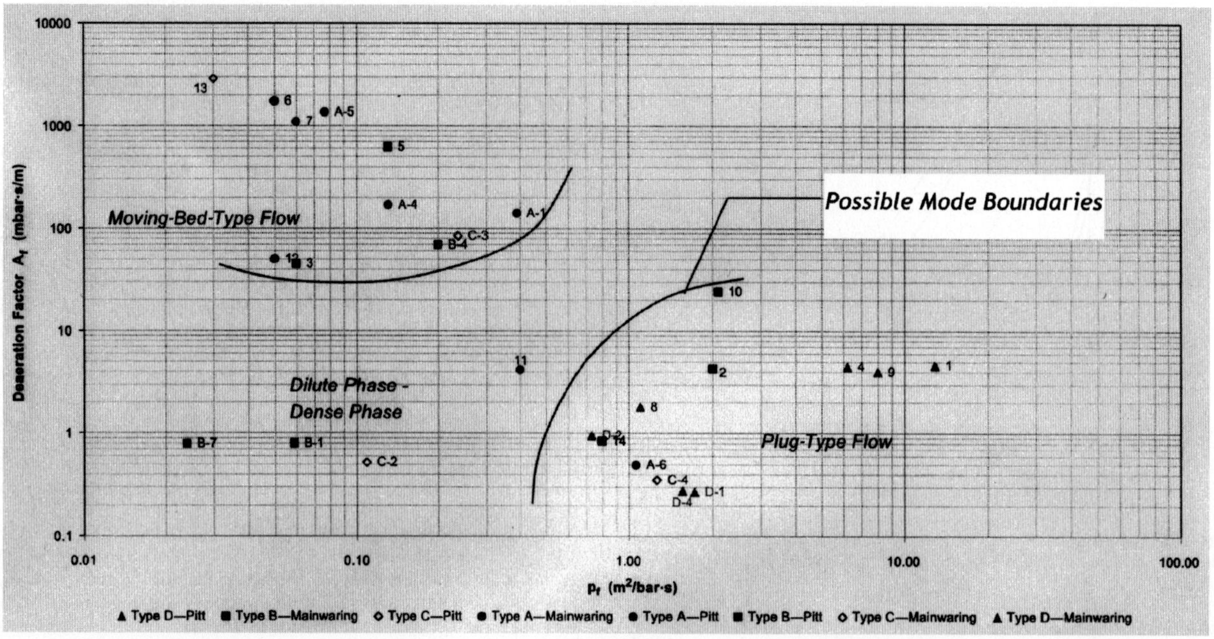

FIG. 21-105 Classification of pneumatic conveying based on combined permeability and deaeration factors, based on Jones and Miller. [*Sanchez et al.,* Powder Technology 138: 93 (2003), *with permission.*]

the additional booster gas into the conveying line. Clearly, an advantage with this method is that the amount of gas injected is typically less than the set booster method, but this comes at the expense of a more complex control system.

3. The third method is the insertion of a small pipe within the conveying pipe. This small pipe has slits or slots placed periodically along its length which permit the flow of gas to the point where material has defluidized and breaks the plug.

DENSE-PHASE SYSTEM DESIGN

While the methods for dilute-phase conveying design are relatively well established either for either a priori calculations or the use of existing data from a pilot or production plant, the state of the art for dense-phase system design is considerably less well defined. On account of the complexities of dense-phase flow such as conveying mode and use of secondary air injection, it is difficult to predict full system pressure drops. Typical design methods today consist of first performing a pilot-scale test and then use scaling methods such as those described in Mills (*Pneumatic Conveying Design Guide,* Elsevier, 2016) to adjust for system length, diameter, and throughput.

Nevertheless, it is still valuable to examine some of the work in the area of single-plug conveying as it provides insights into the controlling variables associated with dense-phase design. Building on his work and that of other collaborators at TU-Karlsruhe on fine, air-retentive materials (e.g., fumed silica, polymer powders, cement) in the 1970s [Weber, *Aufbereitungstechnik* **7:** 603 (1966); Weber, *Aufbereitungstechnik* **10:** 401 (1969), Muschelknautz and Krambrock, *Chem. Ing. Tech.* **41:** 1164 (1969)], Weber used a force balance around a single slug to predict the pressure drop across a single slug of solids, as given in Eq. (21-108)

$$P_i = P_f e^{\frac{\beta \mu g L}{RT(c/v_0)}} \tag{21-108}$$

where P_i and P_f are the upstream and downstream pressure, respectively, of the slug, β is an experimentally determined material coefficient, μ is the solids loading, L is the plug length, and c/v_0 is the ratio of the average slug

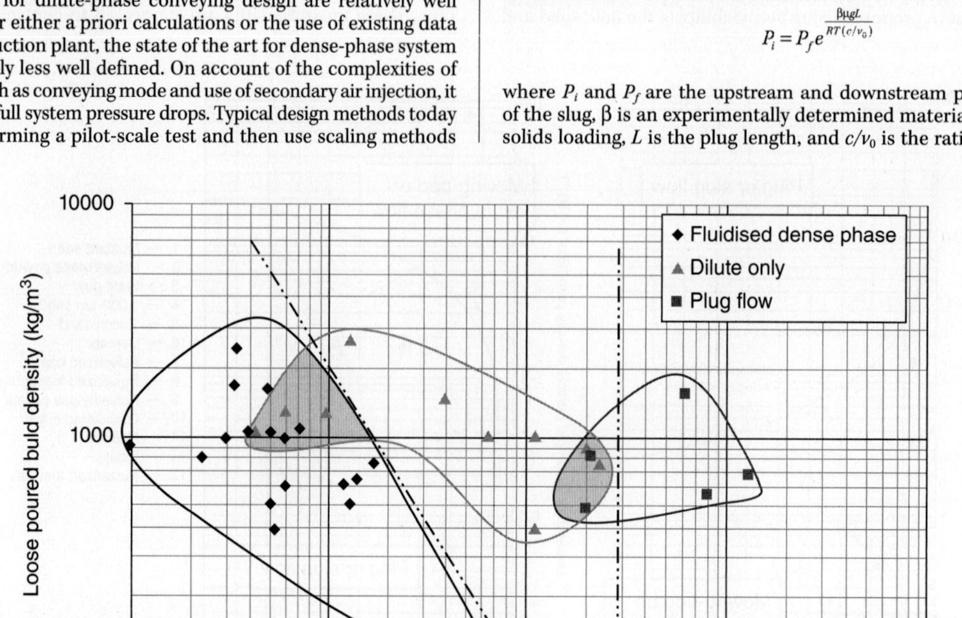

FIG. 21-106 Dense phase mode criteria after Jones and Williams. (*With permission.*)

TABLE 21-10 Dense-Phase Conveying Mode Models and Boundary Criteria

			Strong predictive regions			
Diagram	Boundaries		Fluidised dense phase	Dilute only	Plug flow	Transition regions
Geldart	A/C–B	$(\rho_p-\rho_g)\,d_p=225\times10^{-3}$	A and C powder groups	No prediction	D-type powders with $(\rho_s-\rho_a)<2000\ \text{kg/m}^3$	B-type powders, D-type powders with $(\rho_s-\rho_a)>2000\ \text{kg/m}^3$
	B–D	$(\rho_p-\rho_g)\,d_p^2=10^{-3}$				
Modified Geldart	A/C–B	$\rho_{\text{blp}}d_p=121\times10^{-3}$	A and C powder groups	B powders with $\rho_{\text{blp}}>1000\ \text{kg/m}^3$	D-type powders	B-type powders with $\rho_{\text{blp}}<1000\ \text{kg/m}^3$
	B–D	$\rho_{\text{blp}}d_p^2=539\times10^{-6}$				
Molerus	A/C–B	$(\rho_p-\rho_g)\,d_p^3=2.36\times10^{-9}$	A and C powder groups	No prediction	D-type powders with $(\rho_s-\rho_a)<2000\ \text{kg/m}^3$	B-type powders, D-type powders with $(\rho_s-\rho_a)>2000\ \text{kg/m}^3$
	B–D	$(\rho_p-\rho_g)\,d_p=1.56$				
Modified Molerus	A/C–B	$\rho_{\text{blp}}d_p^3=1.27\times10^{-9}$	A and C powder groups	B powders and $\rho_{\text{blp}}>1000\ \text{kg/m}^3$	D-type powders	B-type powders with $\rho_{\text{blp}}<1000\ \text{kg/m}^3$
	B–D	$\rho_{\text{blp}}d_p=0.841$				
Dixon (Ø50–100 mm pipes)	No slugging-asymmetric slugs	$\dfrac{d_p^{1.14}(\rho_s-\rho_a)^{0.714}}{D^{0.5}}=136\times10^3$	No slugging behaviour	No prediction	Axisymmetric slugging and $D>100$ mm i.d. $(\rho_s-\rho_a)<2000\ \text{kg/m}^3$	Slightly less accurate than the Geldart and Molerus models
	Asymmetric-axisymmetric slugs	$\dfrac{d_p(\rho_s-\rho_a)}{D^{0.5}}=\dfrac{1.64\times10^{-3}}{d_p}+2.68D^{0.5}-\dfrac{\rho_a d_p}{D}$				
Modified Dixon (Ø50–100 mm pipes)	No slugging-asymmetric slugs	$\dfrac{d_p^{1.14}\rho_{\text{blp}}^{0.714}}{D^{0.5}}=87.4\times10^3$	No slugging behaviour	Asymmetric slugging and $\rho_{\text{blp}}>1000\ \text{kg/m}^3$	Axisymmetric slugging and $D>100$ mm i.d.	Slightly less accurate than the Geldart and Molerus models
	Asymmetric-axisymmetric slugs	$\dfrac{d_p\rho_{\text{blp}}}{D^{0.5}}=\dfrac{0.885\times10^{-3}}{d_p}+1.44D^{0.5}$				
Pan	PC1–PC2/3	$d_p\rho_{\text{blp}}=0.1206$	PC1	PC3	PC2 and $d_p>1000\ \mu\text{m}$	PC2 and $d_p<1000\ \mu\text{m}$
	PC2–PC3	$\rho_{\text{blp}}=1000$				

velocity to the average gas velocity (Weber, *Strömungsfördertechnik*, Krausskopf-Verlag, 1974). Weber states that the material coefficient and the velocity ratio should be determined by experiments. Equation (21-108) shows that pressure drop will increase exponentially with both slug length and material loading, which makes intuitive sense. Hence the need to exact some measure of plug control for systems in fluidized dense-phase conveying.

For coarse materials such as two different types of plastic pellets, wheat and barley (particle diameters of 3.12, 3.76, 3.47, and 3.91 mm), Mi and Wypych [*Powder Technol.* **81:** 125–137 (1994)] performed experiments in a 105-mm-diameter steel pipeline of 52- and 96-m lengths. They developed a model based on a Janssen-type force balance (see Fig. 21-107) on an axisymmetric slug which recognized the fact that the force distribution on the slug is indeed asymmetric. For a horizontal conveying section of area A and length L with a mass flow rate of solids W_s and a slug velocity of U_s, the pressure drop is given by

$$\Delta P=(1+1.084\lambda\text{Fr}^{0.5}+0.542\text{Fr}^{-0.5})\frac{2g\mu_w W_s L}{AU_s} \qquad (21\text{-}109)$$

where $\text{Fr}=U_s^2/gD$, μ_w is the wall friction coefficient, and λ is the lateral stress coefficient. Mi and Wypych provide methods for calculating the lateral stress coefficient based on wall friction and internal friction measurements

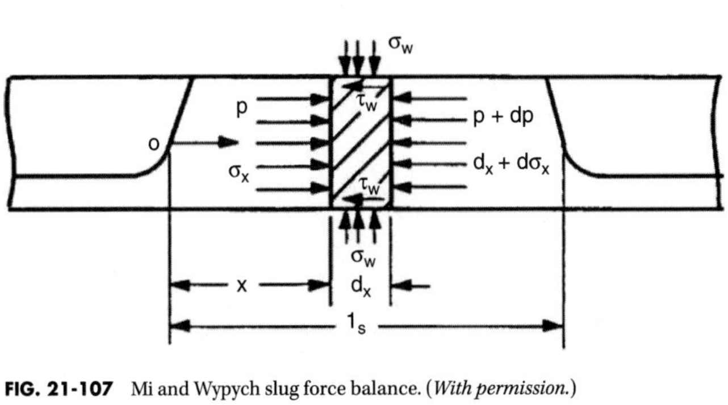

FIG. 21-107 Mi and Wypych slug force balance. (*With permission.*)

and calculations. However, the average values for λ for white plastic pellets, black plastic pellets, wheat, and barley are 0.756, 0.806, 0.572, and 0.655, respectively. And U_s can be measured directly in a test loop, or it can be calculated by the method proposed by Mi and Wypych.

$$U_s=k(U_g-U_{g,\min}) \qquad (21\text{-}110)$$

$$U_{g,\min}=\frac{\rho_s g(\tan\phi_w)\varepsilon^3 d_p^2}{180(1-\varepsilon)\mu} \qquad (21\text{-}111)$$

$$k=C_d\frac{\varepsilon d_p}{D}\left(\frac{\tan\phi_w}{\tan\phi}\right)^{0.33} \qquad (21\text{-}112)$$

In Eq. (21-112), Mi and Wypych have shown for plastic pellets that $c_d=105$.

CONVEYING SYSTEM TROUBLESHOOTING

Numerous issues can affect the performance of a pneumatic conveying system. Some of the more common conveying problems are reduced conveying rate, increased system pressure drop, system wear, and plugging of the conveying system. Conveying problems can be divided into three general categories: (1) conveying gas loss and management, (2) changes in conveyed solids, and (3) alterations to the conveying system.

The most common problem observed in conveying systems is the loss of system performance, usually manifest as a reduction in conveying rate, sometimes to the point of system plugging. For dilute-phase systems, it is important to establish the volumetric gas flow through the system and check these data against blower or fan curves. The gas flow rate can be measured by a variety of handheld devices (Pitot tube, hot wire anemometer, or vane anemometer). Pressure can obtained by handheld manometers or intrinsically safe pressure-recording devices. It can be very helpful to make measurements when the system is not conveying solids so that a comparison can be made to basic fluid mechanics calculations. In addition, measurements should be made at the end of a pressure system (suspected leaks out of the conveying system) or at the beginning of vacuum conveying system (suspected leaks into the conveying system). There are several common potential causes for system gas loss if a discrepancy between the air mover performance curve and field measurements is established: (1) rotary valve or diverter valve leakage; (2) holes in piping due to erosive wear; (3) misaligned piping, causing leakage at flanges or couplings; and (4) failure of relief devices to reset properly. In the latter three cases, proper plant maintenance can solve the problems. In the case of the rotary valve, the rotor tip clearance relative to the valve body needs to be measured with a feeler gauge and compared with valve specifications. Diverter valves are likely to not close properly due to solids binding in the valve or mechanical issues with the closure mechanism.

Another common area for problems in conveying systems is the feeder–conveying system interface. Three issues are often observed at this point in the system. First, for fine materials, if the rotary valve has excessive clearance, allowing the backflow of gas, then the solids are fluidized and their flow will be retarded. This can also occur if the pockets full of conveying gas are not vented. Solids feeding can also be reduced if the valve is not rotating in the correct direction. The flow of conveying gas and the incoming solids must not cross so that solids are not entrained back into the empty pockets of the valve. The last issue is the situation where the rotary valve is significantly oversized as compared to the required solids flow rate. The valve will consequently rotate slowly and dump large slugs of solids into the conveying pipe. If the conveying gas is unable to move the large slug down the pipeline, the system might plug.

Changes in the conveyed solids can drastically affect conveying system performance. If the particle size distribution shifts to smaller sizes, then an increase in material cohesion alters the flow mode or friction with the pipe wall. In contrast, shift toward the coarse will alter the largest particle in the pipeline with the potential for saltation. Moisture can have a very significant impact on the conveying characteristics of bulk solids. With increased moisture content, deposition of the solid on the wall can result in progressive reduction in pipe cross section with potential for plugging of the conveying pipe. In all cases, a holistic approach needs to be taken to solve these problems because the problem may not rest with the conveying system but rather with upstream unit operations.

Wear of the conveying system is an ever-present condition in most conveying systems. Even though it is somewhat counterintuitive, even comparatively "soft" materials such as polymer pellets can wear through aluminum elbows over time. In the more severe cases with improperly designed systems, elbows have been reported to wear through in as little as 20 min. For abrasive materials, either hardened metals or ceramic linings have proved successful in preventing pipe erosion.

The overall approach to troubleshooting can be summarized as follows. Additional information can be found in Dhodapkar and Jacob [Fluid-Solids Flow in Ducts," in Crowe (ed.), *Multiphase Flow Handbook*, 2d ed., Taylor and Francis, CRC Press, 2016], Mills (*Pneumatic Conveying Design Guide*, Elsevier, 2016), and Klinzing et al. (*Pneumatic Conveying of Solids*, 3d ed., Springer, 2010).

1. Compare the system to the "as designed" drawings and calculations.
2. Measure pressure and airflow and compare with blower curves (do for both air only and conveying solids cases). Consider sources of air loss.
3. Perform calculations as outlined in this subsection to understand expected values.
4. Check the mechanical integrity of the system (blower, valves, pipe joints, blower filters and silencers, etc.)

SCREENING

GENERAL REFERENCES: Schmidt, "Screening," in *Ullmann's Encyclopedia of Industrial Chemistry*, Wiley-VCH Verlag, **32:** 225–250 (2012); Dhodapkar, Bates, and Klinzing, "Dry Screening: Sorting Out the Basic Concepts," *Chem. Eng.*, Sept. 2007; Höffl, *Zerkleinerungs-und Klassiermaschinen*, Springer-Verlag, 1986; Schmidt, Körber, and Coppers, *Sieben und Siebmaschinen: Grundlagen und Anwendungen*, Wiley-VCH Verlag, 2003.

INTRODUCTION

Screening is a very common unit operation in many plants across a broad array of process industries including food, pharmaceuticals, mineral processing, fine chemicals, and polymers. Material flow rates can range from grams per hour for high-valued-added products to very high tonnages per hour for ore or coal processing. While it should always be the aim to produce the correct particle size or size distribution during the particle formation process, this is rarely achieved in practice. Screening is commonly used in concert with processes such as size reduction, agglomeration, pelletization, and drying where it is important to recycle or remove part of the size distribution. There are three typical screening problems: (1) remove unwanted oversize or coarse material; (2) remove the fine portion of a size distribution; and (3) classify a wide size distribution to produce a product (or products) with a narrower size distribution. These three cases are depicted in Fig. 21-108.

In the first task, the mass flow rate of the coarse stream is typically 10 percent or less of the feed stream. This process is frequently called *scalping*. The purpose is to remove unwanted oversize, such as agglomerates from a pelletizing process, oversize from a milling process, or even tramp metal parts from a process stream. Compared to the other two tasks, this process is relatively straightforward.

The second task is similar to the first task, but the screening takes place on the fine part of the particle size distribution. It is an important industrial problem because frequently the fines might need to be removed for industrial hygiene or product performance concerns. The task of removing the fines is more difficult than scalping the coarse fraction because of the preponderance of coarse particles on the screen deck, which tend to hinder the fines from reaching a screen aperture. With conventional screening equipment, the practical lower limit for fines removal is 325 or 400 mesh (45 or 38 μm). The development of the Kroosher® technology, which converts the monoharmonic oscillations of a vibratory screener into polyharmonic oscillations, has pushed the practical lower limit to ~20 μm.

The third task aims to take a specific size range of particles from the distribution. In Fig. 21-109, the hatched region might represent the desired fraction of the size distribution. The goal is to not lose any of the desired

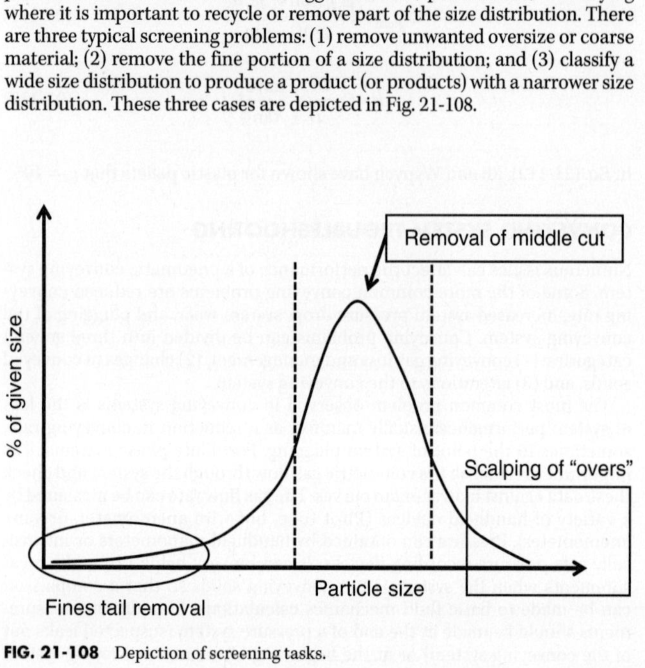

FIG. 21-108 Depiction of screening tasks.

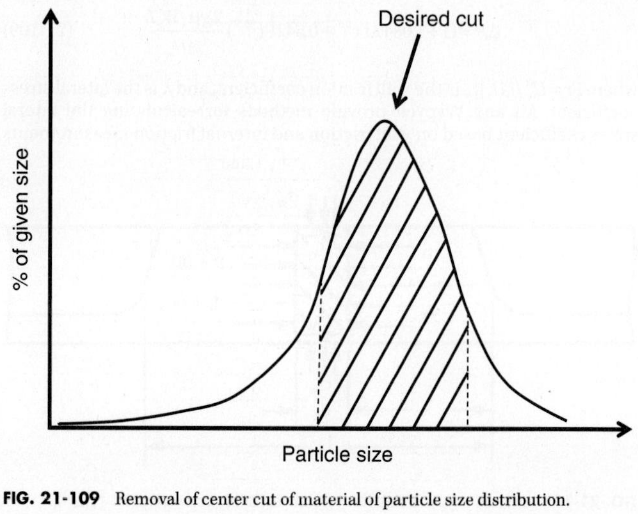

FIG. 21-109 Removal of center cut of material of particle size distribution.

fraction to either the fines or coarse fractions. Similarly, it is detrimental to allow either fine or coarse material to be misclassified in the product cut.

WIRE MESH/SCREEN CLOTH

Wire mesh or screen cloth is the essential element to any screening process. Screen cloth is available in a wide variety of sizes, weaves, and materials. It is usually specified by its mesh or mesh size. Traditionally, mesh means the number of apertures per inch. However, the actual aperture size will vary because of the different wire sizes used for different screen grades. In the United States, laboratory sieves conform to ASTM Standard E11-01. Table 21-11 gives the aperture size versus mesh size for this standard as well as the ISO standard. The ratio between consecutive sizes is given by $\sqrt[4]{2}$ from a base value of 1 mm. It is common in laboratory sieve analysis to use a $\sqrt{2}$ series; hence every other sieve in the E11-01 series is related by this ratio. This results in each consecutive sieve having an aperture size that has twice

TABLE 21-11 ASTM E11 and ISO 565/3310-1 Comparison Chart

Larger Sizes		
ASTM E11 Standard (mm)	Alternate (in)	ISO 565/3310-1 Size (mm)
125.0	5	125.0
106.0	4.24	106.0
100.0	4	100.0
90.0	3½	90.0
75.0	3	75.0
63.0	2½	63.0
—	—	56.0
53.0	2.12	53.0
50.0	2	50.0
45.0	1¾	45.0
—	—	40.0
37.5	1½	37.5
—	—	35.5
31.5	1¼	31.5
—	—	28.0
26.5	1.06	26.5
25.0	1	25.0
22.4	⅞	22.4
—	—	20.0
19.0	¾	19.0
18.0	—	18.0
16.0	⅝	16.0
—	—	14.0
13.2	0.53	13.2
12.5	½	12.5
11.2	7/16	11.2
—	—	10.0
9.5	⅜	9.5
—	—	9.0
8.0	5/16	8.0
—	—	7.1
6.7	0.265	6.7
6.3	¼	6.3

Finer Sizes		
ASTM E11 Standard (μm)	Alternate (mesh)	ISO 565/3310-1 Size (μm)
5600	3½	5600
—	—	5000
4750	4	4750
—	—	4500
4000	5	4000
3550	—	3550
3350	6	3350
3150	—	3150
2800	7	2800
—	—	2500
2360	8	2360
2000	10	2000
—	—	1800
1700	12	1700
—	—	1600
1400	14	1400
—	—	1250
1180	16	1180
—	—	1120
1000	18	1000
—	—	900
850	20	850
—	—	800
710	25	710
—	—	630
600	30	600
—	—	560
500	35	500
—	—	450
425	40	425
—	—	400
355	45	355
—	—	315
300	50	300
—	—	280
250	60	250
—	—	224
212	70	212
—	—	200
180	80	180
—	—	160
150	100	150
—	—	140
125	120	125
—	—	112
106	140	106
—	—	100
90	170	90
—	—	80
75	200	75
—	—	71
63	230	63
—	—	56
53	270	53
—	—	50
45	325	45
—	—	40
38	400	38
—	—	36
32	450	32
25	500	25
20	635	20

the open area as the previous sieve. Laboratory sieves have thick wires in order to withstand the rigors of day-to-day sieve analysis.

Commercial Screen Cloth Commercial screen cloth is manufactured in three different grades: market grade, mill grade, and *tensile bolting cloth* (TBC). Market grade has the thickest wire and smallest aperture while TBC has the thinnest wire and the largest aperture for a given mesh size. Table 21-12 shows the respective wire sizes and aperture size for the three commercial screen cloth grades. It is absolutely necessary to understand the grade of screen when specifying screen cloth. For example, a 14-mesh market grade screen has an aperture size of 1295 μm. By comparison, 14-mesh TBC has a 1575-μm aperture size which is over 20 percent larger. Market and mill grade screens are often used in high-wear applications with abrasive materials. TBC screens offer the possibility of considerably more open area and apertures per screen deck for the same aperture size. This fact can be used to advantage in situations where it is desired to increase screen capacity without replacing the screening machine itself. For example, take a screener operating with a

TABLE 21-12 Aperture and Wire Diameters for Commercial Screen Cloth

Mesh size	Wire diameter (inches)	Mesh opening (inches)	Open area (%)	(microns)	Mesh size	Wire diameter (inches)	Mesh opening (inches)	Open area (%)	(microns)
Mill 2	0.0540	0.4460	79.6	11328	TBC 48	0.0055	0.0153	54.2	389
MKT 2	0.0630	0.4370	76.4	11100	MKT 40	0.0100	0.0150	36.0	381
Mill 3	0.0410	0.2923	76.7	7424	TBC 50	0.0055	0.0145	52.6	368
MKT 3	0.0540	0.2790	70.1	7087	TBC 54	0.0040	0.0145	61.5	368
Mill 4	0.0350	0.2150	74.0	5461	Mill 45	0.0080	0.0142	40.8	361
MKT 4	0.0470	0.2030	65.9	5156	TBC 56	0.0040	0.0138	60.2	351
Mill 5	0.0320	0.1680	70.6	4267	TBC 52	0.0055	0.0137	51.0	348
MKT 5	0.0410	0.1590	63.2	4039	TBC 58	0.0040	0.0132	59.0	335
Mill 6	0.0280	0.1387	69.6	3523	TBC 54	0.0055	0.0130	49.4	330
MKT 6	0.0350	0.1320	62.7	3353	TBC 58	0.0045	0.0127	54.6	323
Mill 7	0.0280	0.1149	64.8	2918	TBC 60	0.0040	0.0127	57.8	323
MKT 7	0.0350	0.1080	57.2	2743	Mill 50	0.0075	0.0125	39.1	318
Mill 8	0.0250	0.1000	64.0	2540	TBC 60	0.0045	0.0122	53.3	310
MKT 8	0.0280	0.0970	60.2	2464	TBC 62	0.0040	0.0121	56.5	307
Mill 9	0.0230	0.0881	62.7	2238	TBC 62	0.0045	0.0116	51.7	295
Mill 10	0.0200	0.0800	64.0	2032	Mill 55	0.0070	0.0112	37.9	284
MKT 10	0.0250	0.0750	56.3	1905	TBC 66	0.0040	0.0112	54.2	284
MKT 11	0.0180	0.0730	64.5	1854	TBC 64	0.0045	0.0111	50.7	282
Mill 11	0.0200	0.0709	61.0	1801	MKT 50	0.0090	0.0110	30.3	279
Mill 12	0.0180	0.0653	60.8	1659	TBC 66	0.0045	0.0106	49.4	269
TBC 14	0.0090	0.0620	76.4	1575	TBC 70	0.0037	0.0106	54.9	269
MKT 12	0.0230	0.0600	51.8	1524	TBC 70	0.0040	0.0103	51.8	262
Mill 14	0.0170	0.0544	57.2	1382	Mill 60	0.0065	0.0102	37.5	259
TBC 16	0.0090	0.0535	73.3	1359	TBC 72	0.0037	0.0102	53.8	259
MKT 14	0.0200	0.0510	51.0	1295	TBC 72	0.0040	0.0099	50.7	251
TBC 18	0.0090	0.0466	70.2	1184	TBC 74	0.0037	0.0098	52.7	249
Mill 16	0.0160	0.0465	55.4	1181	TBC 74	0.0040	0.0095	49.6	241
MKT 16	0.0180	0.0445	50.7	1130	TBC 76	0.0037	0.0095	51.7	241
TBC 20	0.0090	0.0410	67.2	1041	MKT 60	0.0075	0.0092	30.5	234
Mill 18	0.0150	0.0406	53.4	1031	TBC 76	0.0040	0.0092	48.4	234
MKT 18	0.0170	0.0386	48.3	980	TBC 78	0.0037	0.0091	50.6	231
TBC 22	0.0075	0.0380	69.7	965	TBC 78	0.0040	0.0088	47.3	224
Mill 20	0.0140	0.0360	51.8	914	TBC 80	0.0037	0.0088	49.6	224
TBC 24	0.0075	0.0342	67.2	869	TBC 80	0.0040	0.0085	46.2	216
MKT 20	0.0160	0.0340	46.2	864	TBC 84	0.0035	0.0084	49.8	213
Mill 22	0.0135	0.0320	49.6	813	TBC 84	0.0040	0.0079	44.1	201
TBC 26	0.0075	0.0310	64.8	787	TBC 88	0.0035	0.0079	47.9	201
Mill 24	0.0130	0.0287	47.4	729	TBC 90	0.0035	0.0076	47.8	193
TBC 28	0.0075	0.0282	62.4	716	TBC 88	0.0040	0.0074	42.0	188
MKT 24	0.0140	0.0277	44.2	704	TBC 90	0.0040	0.0071	41.0	180
Mill 26	0.0110	0.0275	51.1	699	TBC 94	0.0035	0.0071	45.0	180
TBC 30	0.0065	0.0268	64.8	681	MKT 80	0.0055	0.0070	31.4	178
Mill 28	0.0100	0.0257	51.8	653	TBC 94	0.0040	0.0066	38.9	168
TBC 32	0.0065	0.0248	62.7	630	TBC 105	0.0030	0.0065	46.9	165
Mill 30	0.0095	0.0238	51.0	605	TBC 120	0.0026	0.0058	47.3	147
TBC 34	0.0065	0.0229	60.7	582	MKT 100	0.0045	0.0055	30.3	140
Mill 32	0.0090	0.0223	50.9	566	TBC 135	0.0023	0.0051	47.4	130
TBC 36	0.0065	0.0213	58.7	541	TBC 145	0.0022	0.0047	46.4	119
Mill 34	0.0090	0.0204	48.1	518	MKT 120	0.0037	0.0046	30.5	117
MKT 30	0.0130	0.0203	37.1	516	TBC 165	0.0019	0.0042	47.1	107
TBC 38	0.0065	0.0198	56.7	503	MKT 150	0.0026	0.0041	37.9	104
Mill 36	0.0090	0.0188	45.8	478	MKT 160	0.0024	0.0038	37.6	97
TBC 40	0.0065	0.0185	54.8	470	TBC 200	0.0016	0.0034	46.2	86
TBC 42	0.0055	0.0183	59.1	465	MKT 200	0.0021	0.0029	33.6	74
TBC 43	0.0050	0.0183	61.6	465	TBC 230	0.0014	0.0029	46.0	74
Mill 38	0.0085	0.0178	45.8	452	MKT 250	0.0016	0.0024	36.0	61
TBC 44	0.0055	0.0172	57.4	437	MKT 270	0.0016	0.0021	32.2	53
TBC 46	0.0045	0.0172	62.9	437	MKT 325	0.0014*	0.0017	30.5	43
Mill 40	0.0085	0.0165	43.6	419	MKT 400	0.0015*	0.0015	36.0	38
TBC 48	0.0045	0.0163	61.5	414	MKT 500	0.0010*	0.0010	25.0	25
TBC 46	0.0055	0.0162	55.8	411					
TBC50	0.0045	0.0155	60.1	394					

**Twilled*
MKT—Market Grade
Mill—Mill Grade
TBC—Tensile Bolting Cloth

All screen numbers based on data provided by the Southwestern Wire Cloth brochure
Micron numbers are calculated and may or may not match up with other data sets
Micron number obtained by the following calculation:
Microns = Mesh Opening Inches × 25400

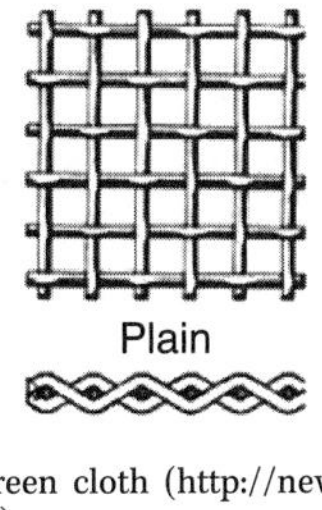

FIG. 21-110 Plain weave screen cloth (http://new-delhi.all.biz/plain-weave-wire-mesh-g235678#.WO97XGnysnQ).

20-mesh market grade (864-µm) screen which has 46.2 percent open area. If the screen can be replaced with a 24-mesh TBC (869 µm), the number of apertures increases by a factor of $(24^2/20^2) = 1.44$ and a corresponding 67.2 percent open area. Of course, there is the trade-off in screen wear since the wire diameter for a 24-mesh TBC (0.191 mm) is less than half the wire diameter for the 20-mesh market grade (0.406 mm).

Screen Cloth Types There are various screen cloth weaves, punched plates, and special screen decks available. Selection of the best option is based on prior experience, pilot-scale testing, screener supplier recommendations, or combination of all the above. The most common type of woven screen cloth uses a plain weave, as shown in Fig. 21-110. This gives a square opening for the aperture. It is also possible to get screen cloth with rectangular or slotted apertures. In addition, various weave patterns, such as twill or dutch, are available. Many screen cloth companies also offer custom fabrication of screen cloth, although this can be fairly expensive. Note that any screen cloth that is available can usually be fabricated into a screen for use in your screener. It is also common to use a punched plate as a screen. The opening size, shape, pattern, and thickness of the plate can be varied. They offer the benefit of extremely long service life. A number of specialty screening surfaces are also available, such as wedge wire; see Fig. 21-111. These can work very well with spherical particles since they tend to be resistant to screen blinding.

Screen Cloth Materials of Construction A variety of materials of construction are used. Carbon steel is commonly used in mining operations because of its low cost and the need for heavy-duty screens owing to issues with screen wear with the abrasive materials. Stainless steel is used in food, chemical, and pharmaceutical processing because of its better resistance to chemical attack and cleanliness concerns. If metal contamination in the product due to a broken screen is important, 400 series grades of stainless steel can be used for the screen cloth. This allows any broken wires from the screen to be removed with a magnet since some grades in the 400 series of stainless steel are magnetic. A number of polymers are also used to make screen cloth, such as nylon and polyester. These can be very effective in centrifugal sifters since the constant flexing of the screen surface will tend to fatigue metal screen cloth and lead to screen breakage.

FUNDAMENTALS OF SCREENING

While the process of passing a particle smaller than the mesh size would appear to be relatively simple, in practice this is quite frequently not so. Numerous issues such as blinding (blockage of apertures), misclassification of solids, breakage of screen wire, and device wear all plague screener operation. However, understanding a few concepts can aid in both the design and operation of screening equipment. When a stream of bulk solids enters a screener, the solids are spread across the screen deck as a result of the vibratory or gyratory motion of the device. Schmidt ("Screening," in *Ullmann's Encyclopedia of Industrial Chemistry,* Wiley-VCH Verlag, **32:** 225–250) examined the probability of a particle to pass through a screen by examining several individual mechanisms. For a particle to pass through the screen, the particle must first pass through the bed of solids to be presented to an aperture on the screen deck. If the layer of solids on the screen is of sufficient depth, then particles that are smaller than the screen apertures may never have the opportunity to pass through the deck and hence be misclassified to the coarse stream. If the aperture size is w and the thickness of the particle layer is given by l_p, then the probability of a particle passing through the screen is given by

$$P_a = w/l_p \tag{21-113}$$

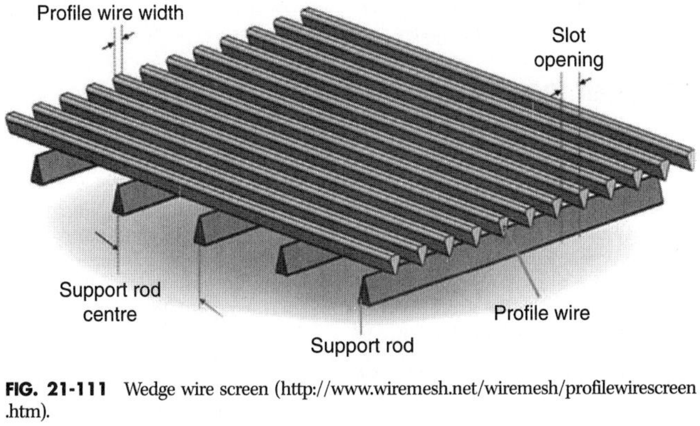

FIG. 21-111 Wedge wire screen (http://www.wiremesh.net/wiremesh/profilewirescreen.htm).

This supports the idea that shortly after the bulk solids are presented to the screen deck, it is imperative to spread them out across the screen cloth to aid in their passage.

As particles are presented to the apertures, they must be oriented correctly in order to pass through the hole. Long-aspect-ratio particles, such as needle-shaped crystals and some biomass materials such as straw, may need to be repeatedly presented to the apertures before they pass through. In addition, as the particle size approaches the aperture size, the probability of passage decreases significantly. This is the concept of *near mesh screening,* and the probability of passage is given by

$$P_b = \left(\frac{1 - \dfrac{x}{w}}{1 + \dfrac{d}{w}} \right)^2 \tag{21-114}$$

as described by Schmidt ("Screening," in *Ullmann's Encyclopedia of Industrial Chemistry,* Wiley-VCH Verlag, **32:** 225–250). Equation (21-114) recognizes two important physical phenomena for screening: (1) as the particle size x approaches the aperture size w, the probability for passage tends toward zero, as shown in the numerator; (2) as the wire size d increases, the probability for passage will also decrease.

Example 21-6 Compare the probability of passage of a 225-µm particle through a 76-mesh TBC and a 60-mesh market grade screen. The 76-mesh TBC has a 234-µm aperture and 101-µm wire. Applying Eq. (21-114) gives

$$P_b = \left(\frac{1 - \dfrac{225}{234}}{1 + \dfrac{101}{234}} \right)^2 = 0.00072$$

For 60-mesh market grade, 234-µm aperture and 165-µm wire,

$$P_b = \left(\frac{1 - \dfrac{225}{234}}{1 + \dfrac{161}{234}} \right)^2 = 0.00052$$

In this particular case, the larger wire size significantly reduces the probability of passage. Since P_b represents the probability of passage through the screen, $1/P_b$ represents the number of "events" that must occur before the particle passes through the screen. In the case of a vibrating screener operating at either 50 or 60 Hz, the particle will have 50 or 60 opportunities per minute to pass through the screen. For our example above, the times for passage are 23 and 32 min, respectively, which likely means the 225-µm particles will be mostly retained in the coarse stream.

Another important concept is the percentage of coarse in the screener feed. These particles act as an impediment to the passage of fines through the screen. Consequently, as the percentage of fines increases, so too does the probability increase for their passage through the screen, as given here

$$P_c = (\%F) \tag{21-115}$$

The three probabilities P_a, P_b, and P_c can now be combined to give the overall probability of passage of a given particle size through the screen

$$P_o = P_a P_b P_c \tag{21-116}$$

Example 21-7 Calculate the probability of passage of a 400-µm particle through a 26-mesh TBC screen where the percentage of fines is 83 percent and the layer thickness on the screen is 12 mm. For a 26-mesh TBC screen, the aperture size is 787 µm and the wire size is 190.5 µm.

Calculating each probability gives

$$P_a = \frac{787}{12\times1000} = 0.0656$$

$$P_b = \left(\frac{1-\frac{400}{787}}{1+\frac{191}{787}}\right)^2 = 0.1566$$

$$P_c = 0.83$$

$$P_o = (0.0656)(0.1566)(0.83) = 0.0085$$

If a screener is operating at 60 Hz, then the expected time for passage is (1/0.0085)/60 ~ 2 s. It is very important to recognize that the approach outlined above serves more as a theoretical framework than as an exact calculation method. In practice, numerous factors such as particle size distribution, particle shape, bulk solids moisture, screen blinding, screen wear, etc., will impact actual times for particle passage.

CLASSIFICATION EFFICIENCY

For any particulate separation process, it is important to be able to describe the separation or classification efficiency. This applies to all manner of separation processes: cyclones/hydrocyclones, screeners, magnetic separators, etc. In practice, it is often claimed that a certain separation process is 70 percent efficient. However, such claims do not describe the actual operation of the screener as a function of machine and process conditions. A generalized separator is shown in Fig. 21-112 with a feed stream S and two product streams of fine and coarse, F and C, respectively. Note that the diagram can represent any general separation: overflow versus underflow for a hydrocyclone, red particles versus blue particles; magnetic versus nonmagnetic particles, etc.

Each stream has a corresponding particle size distribution associated with it. The mass fraction of the ith size class can be described as x_{si}, x_{fi}, and x_{ci}, respectively. Figure 21-113 shows a plot of separation efficiency versus particle size. In the case of an ideal separation, there is no misclassification of either the fine or the coarse fractions, as shown by the dashed line in Fig. 21-113. This is rarely achieved in practice. The dotted line shows a more typical separation curve for a classification process.

This curve is most commonly called the *grade efficiency curve.* It can also be called the *Tromp curve.* The grade efficiency for the ith size $G(x_i)$ is described mathematically by

$$G(x_i) = \frac{M_{Ci}}{M_{Si}} \tag{21-117}$$

where M_{Ci} and M_{Si} are the mass in the ith fraction for the coarse and feed fractions. The particle size at which the grade efficiency is 50 percent is called the *cut size* d_{50}, and it physically represents the point at which there is equal probability for particles of diameter d_{50} to be processed to either the coarse or the fines stream. Region A in Fig. 21-113 represents the situation where coarse material has been misclassified to the fines stream. Conversely, region B represents fines in the coarse stream. For screening equipment, the cut size will often be close to the screen size. Hence, if the cut size is significantly greater than or less than the screen size, it is very likely that the screener is not performing as designed. Another important performance metric associated with the grade efficiency curve is the sharpness of cut. There are numerous definitions for the sharpness of cut, most of which seek to describe the slope of the grade efficiency curve. Common metrics include ratios such as d_{90}/d_{10}, d_{80}/d_{20}, and d_{75}/d_{25}, as well as the slope of the grade efficiency curve at the cut size. Note that the sharpness of cut ratios is not invariant with respect to particle size. For example, a d_{80}/d_{20}

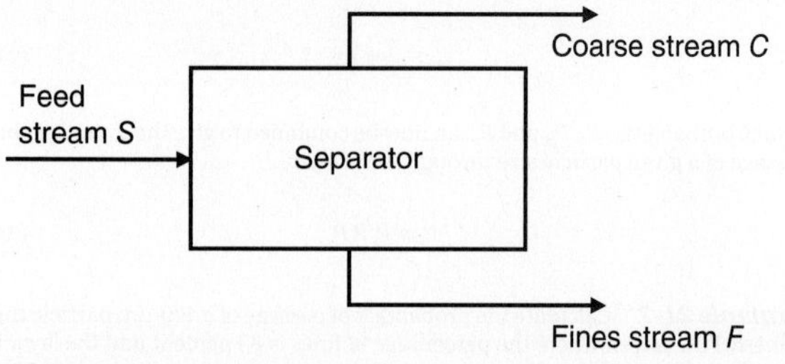

FIG. 21-112 Typical separator.

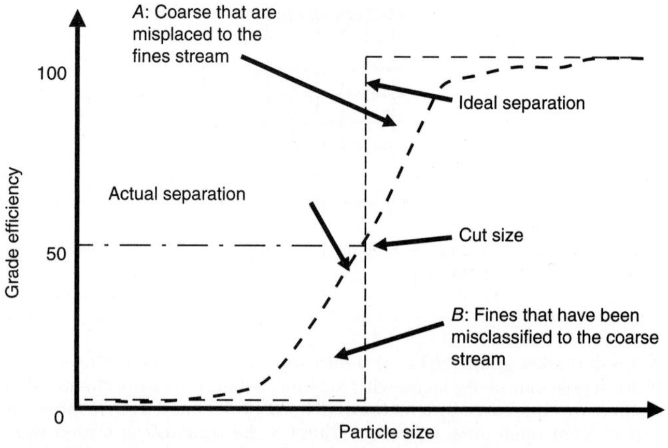

FIG. 21-113 Separation efficiency versus particle size.

ratio of 2 centered around a d_{50} of 60 μm will have a smaller difference in absolute particle size ($d_{80} - d_{20}$) as compared to a d_{50} of 600 μm. Consequently, comparison of sharpness of cut measures is typically only useful for a given unit operation and operating condition and is not useful in comparing different operations.

Grade Efficiency Measurement To calculate the grade efficiency curve for an operating screener, four of the six variables (mass flow for feed, coarse, and fines streams and their associated particle size distributions) must be known. The golden rules of sampling should be used to take the samples, and the samples should be appropriately riffled and analyzed. In some cases, the differences in the mass flow of the feed and that of the coarse or fines stream can be very small (for example, $M_c \sim 0.99M_s$). With the variance in the particle size distribution measurements, it is not uncommon for the mass balance to not close and grade efficiencies can exceed 100 percent. A worked example of this phenomenon can be found in Dhodapkar and Jacob, "Gas-Solid Separation," chap. 6, in Schweitzer (ed.), *Handbook of Separation Techniques for Chemical Engineers,* 3d ed., Wiley, 1997.

Example 21-8 Samples around a screener for the fines and coarse particle size distributions are given in Table 21-13. The mass flow rate of feed and fines streams are 2150 and 370 kg/h, respectively. Calculate the grade efficiency curve, cut size, and sharpness of cut d_{80}/d_{20}. Table 21-13 shows that the total amount of coarse can be calculated by closing the mass balance for this system. The amounts of each size class for the fines and coarse are then calculated and summed to give the amount for each size in the feed. The grade efficiency is calculated using Eq. (21-117). The resulting graph of the grade efficiency is given in Fig. 21-114. The d_{20}, d_{50} (cut size), and d_{80} can be read directly from the graph of the grade efficiency, and they are approximately 70, 95, and 130 μm, respectively. Hence the cut size is 95 μm, and the d_{80}/d_{20} is 130/70, or 1.86.

The grade efficiency curve can be very useful in determining the quality of the screener operation. Figure 21-115 shows the performance of a 48-in tumbling screener screening polymer resin operating at four different feed rates. The screen size is 813 μm. At the lowest feed rate of 2478 lb/h, the cut size is essentially the screen size, and the sharpness of cut is the best of the four feed rates. As the feed rate increases, the cut size progressively decreases from 813 to 700 μm for the 5797 lb/h feed rate. It is clear that as the feed rate increases beyond ~3800 lb/h, the screener is unable to process the fines through the screen and hence the fines are misclassified to the coarse stream, as is evidenced by both the decrease in cut size and a decrease in the sharpness of cut d_{80}/d_{20}. This points out the common trade-off of production capacity versus product quality. In many cases it is not necessary to have a nearly ideal separation. A separation that yields a product that meets the required end-use performance at the highest capacity will typically be the most economical, unless a narrower size distribution gives enhanced product performance and a premium can be obtained for that product.

SCREENER TROUBLESHOOTING

A number of common issues can arise when operating screening equipment. The operating or maintenance manual for your specific screener should be the first reference looked at when troubleshooting a problem. The manual will give specific guidance for the type of screener installed in your plant that cannot be covered in a general handbook. In addition, generation of the grade efficiency curve can be very helpful in troubleshooting screeners, as it helps to point to issues such as improper cut size and misclassification of material. The following are, however, some of the more common issues seen and reasons why they may occur.

Screen Blinding *Screen blinding* is the blockage of screen apertures with the product being screened. This is most often due to particles that are slightly larger than the screen aperture getting wedged in the screen

TABLE 21-13 Example 21-9: Grade Efficiency Calculation

Mass flow feed	kg/h	2150
Mass flow fines	kg/h	370
Mass flow coarse	kg/h	1780

Size (US mesh)	Micrometers	% Coarse	% Fines	Amt. coarse	Amt. fines	Amt. feed	$G(x)$
325	44	0.1%	30.0	1.8	111.0	112.8	0.016
200	75	3.0%	29.0	53.4	107.3	160.7	0.332
140	106	4.5%	13.0	80.1	48.1	128.2	0.625
100	150	11.7%	10.0	208.3	37.0	245.3	0.849
70	212	19.7%	8.0	350.7	29.6	380.3	0.922
50	300	27.0%	5.0	480.6	18.5	499.1	0.963
40	425	17.0%	2.8	302.6	10.4	313.0	0.967
30	600	10.0%	1.5	178.0	5.6	183.6	0.970
20	850	7.0%	0.7	124.6	2.6	127.2	0.980
Total		100.0%	100.0	1780	370	2150	

apertures. This can cause product quality issues because more fine material remains in the product or capacity issues because not enough product gets through the screen cloth.

The first thing to check is the condition of the screen cleaning system, which is most often *screener balls*. These elastomeric spheres are placed on a perforated deck located below the screen surface. They bounce on the deck and impact the bottom of the screen, which dislodges particles stuck in the screen apertures. The size and number of balls on each deck should be checked against the supplier specifications. These balls will wear with time and need to be changed periodically. In addition, the screener supplier offers recommendations on the number of balls that should be used. The authors are aware of an instance in which the screener balls were sufficiently worn to pass through the perforated deck and were later found in the product. When the screener was inspected, nearly all the balls were missing. Screen blinding can also be caused by buildup of small particles on the screen cloth.

In addition to checking the screen cleaning system, variation in the moisture content of the material fed to the screener should be verified. If the moisture content has increased, the material may exhibit greater adhesion to the screen cloth. Changes in the screener motion may also lead to changes in the performance of a screener. This can lead to both quality and capacity issues depending on the situation and how much the motion may have changed. Many screeners have adjustments that can be readily made to the amplitude and/or direction of the motion imparted to the screening surface. If the adjustment mechanism loosens, the screener motion can change during operation. Many screener manufacturers have stickers that are placed on the side of the screener to enable the user to visually check the amplitude and direction of the screener motion. Wear of screener parts, such as springs, can also change the motion of the screener.

The manufacturer's recommendations should be followed on the replacement schedule for these parts. Product quality issues with too much oversize in the product are typically due to a hole in the screen, but worn gaskets or improperly installed screens can also lead to bypassing of the oversize into the product stream. These same failures may also result in a larger than normal yield loss when fines are removed from the product. A visual inspection of the screen can usually be done to easily detect any holes. Many manufacturers install inspection ports for this purpose. If frequent screen breakage occurs, a screen cloth with larger wire diameter should be considered.

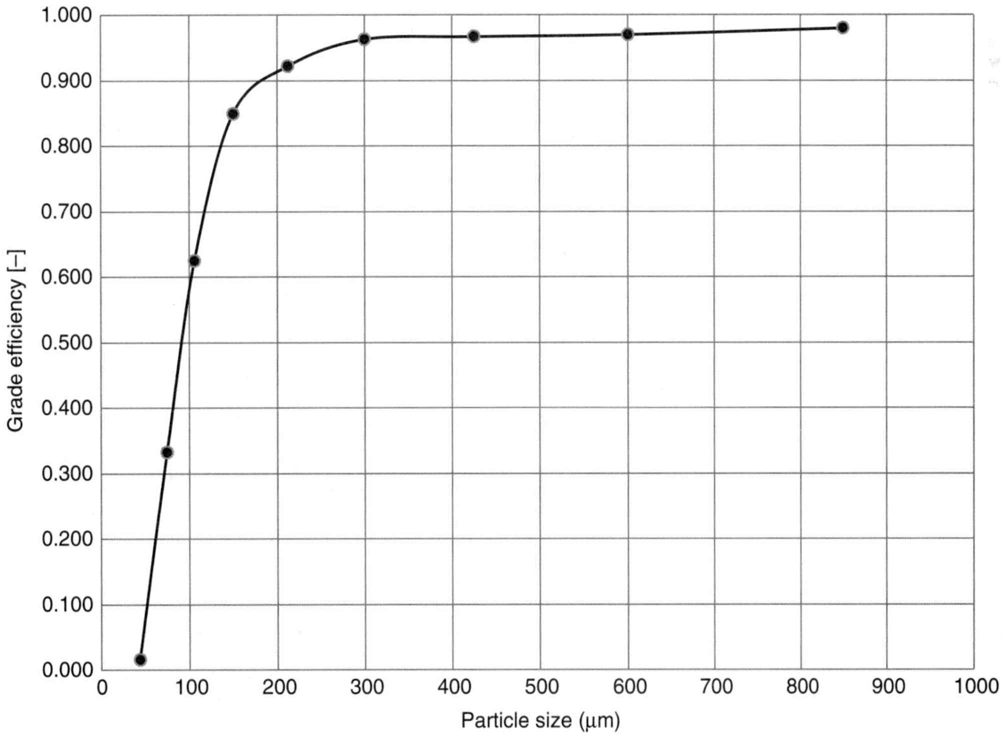

FIG. 21-114 Example 21-9: grade efficiency curve.

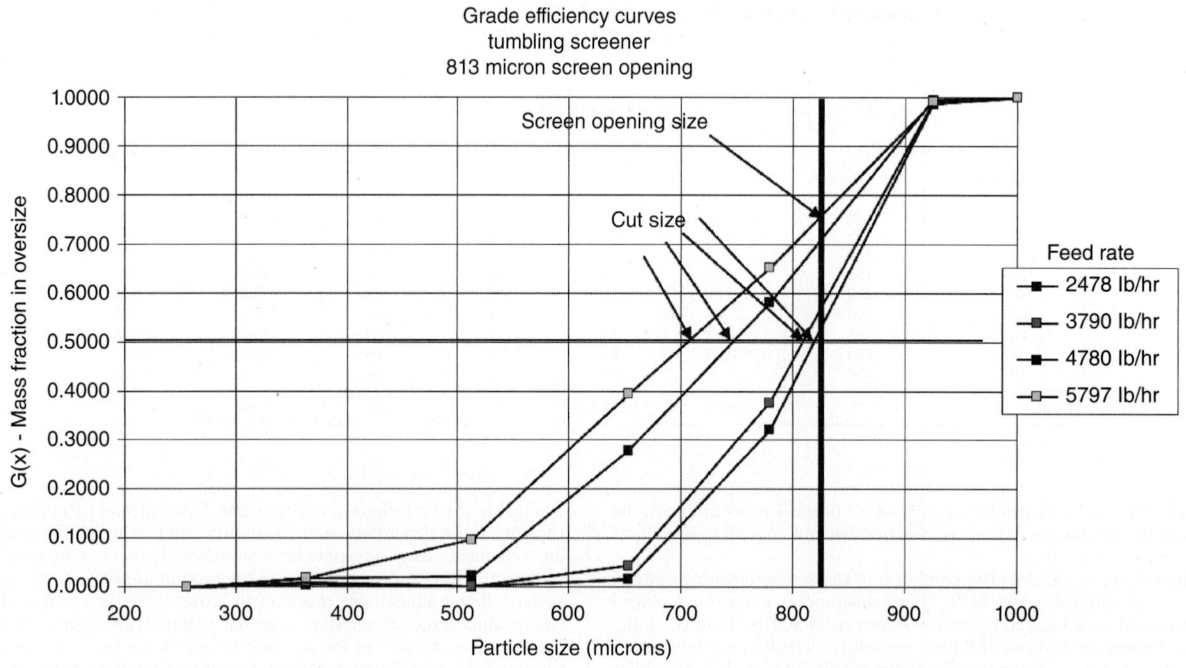

FIG. 21-115 Tumbling screener grade efficiency curves.

SCREENER PERFOMANCE

The two primary performance metrics are the quality of the separation and the screener throughput. As Fig. 21-115 shows, these two factors are intimately related as it is possible to overload the screen and allow fine material to be misclassified to the coarse stream. A priori prediction of screener throughput and the resulting grade efficiency curve is very difficult on account of myriad physical effects that impact screener performance. Höffl (*Zerkleinerungs-und Klassiermaschinen*, Springer-Verlag, 1986) proposed the general performance equation as

$$\dot{V} = \dot{V}_a A K_1 K_2 \cdots K_n \tag{21-118}$$

where $\dot{V}$ is the total throughput, $\dot{V}_a$ is the specific throughput for a screener, A is the deck or screener area, and the values of K represent correction factors based on the screening task at hand. The specific throughput needs to be referenced to a specific screening task such as dry sand of a given particle size distribution. Höffl proposed the following general correction factors: amount of oversize; number of fines less than one-half the aperture size; single- or double-deck screening; type of solid (ground solids, sand, coal); moisture; and dry or wet screening. This method provides an overall methodology for performance prediction; however, it is imperative to perform pilot-scale tests to determine the actual scaling parameters for a given screening task. Table 21-14 offers general guidance on how screener performance changes with respect to process changes.

SCREENER TYPES

A wide variety of different types of screeners are available commercially. These are typically classified according to the motion of the screen surface. Each type of screener has benefits and downsides based on the material properties and type of separation that needs to be performed. Therefore, a thorough evaluation and test program should be done prior to selecting a screener for a given application. A brief description of commercially available units and their type of motion, as shown in Fig. 21-116, follows.

Vibrating Screens Vibrating screeners are the most widely used screeners and are especially good where large capacity and high efficiency are desired. Vibrating screeners are horizontal (good for operations with low headroom) or inclined, as shown in Figs. 21-117 and 21-118, and can be mechanically or electromagnetically vibrated. Inclined screens normally have rectangular openings to compensate for the foreshortening of the aperture; oblong particles may, therefore, filter through the inclined screen that would normally leave with the oversize particles on a horizontal screen. High-speed vibrating screens can make size separations for some particles smaller than 75 μm, which is the usual limit for separation with the lower-speed screens, although at much reduced efficiencies. Several screening surfaces are used if multiple separations are required.

Reciprocating Screens Reciprocating screens, as shown in Fig. 21-119, are inclined slightly and operate with a shaking motion in the plane of the main frame. The material travels along the screen due to the forward motion of the unit with the finer flat springs. Reciprocating screens require minimal headroom and low power, but have low capacities compared to vibrating screens. Reciprocating screens are good for accurate sizing of large lumps.

Sifters Sifter screens operate with a rotary motion in the plane of the screening surface. There are three types of motion for sifters: circular motion, gyratory motion, and circular vibratory motion. Screens of each type appear in Figs. 21-120, 21-121, and 21-122*a* and *b*. The circular motion sifter has a circular flow pattern over the screen surface, the gyratory sifter varies from a circular flow pattern at the feed end to a reciprocal pattern at the discharge end, and the circular vibrating sifter has a spiral flow pattern over the screen surface with a peripheral discharge. The circular and gyratory motion sifters are used for fine separations, and the circular vibrator is used for wet or dry screening operations—fine separations, scalping, and dewatering. Multiple decks are frequently used for both single and multiple separations.

Sifters can be used for fines removal and are more accurate than vibrating screens; however, the capacity of sifters is normally much less. Increased sifter capacity is possible in a compact space by combining several units, as shown in Fig. 21-123. An example of this multiseparation system is the Super Sifter, which is suspended from the ceiling and ranges in size from two sections side by side of 12 sieves each up to six sections of 30 sieves each.

Tumbling Screens This type of screener, shown in Fig. 21-124, has a unique motion. It is best described as a *wobble*. This three-dimensional motion can provide very good separations of fine products. Separations using

TABLE 21-14 Screener Performance versus Process Variables

By increasing...	Capacity	Efficiency
Near size	↓	↓
Feed rate	↑	↓
Angle of inclination	↑	↓
Depth of bed	↑	↓
Dryness	↑	↑
Opening	↑	↑

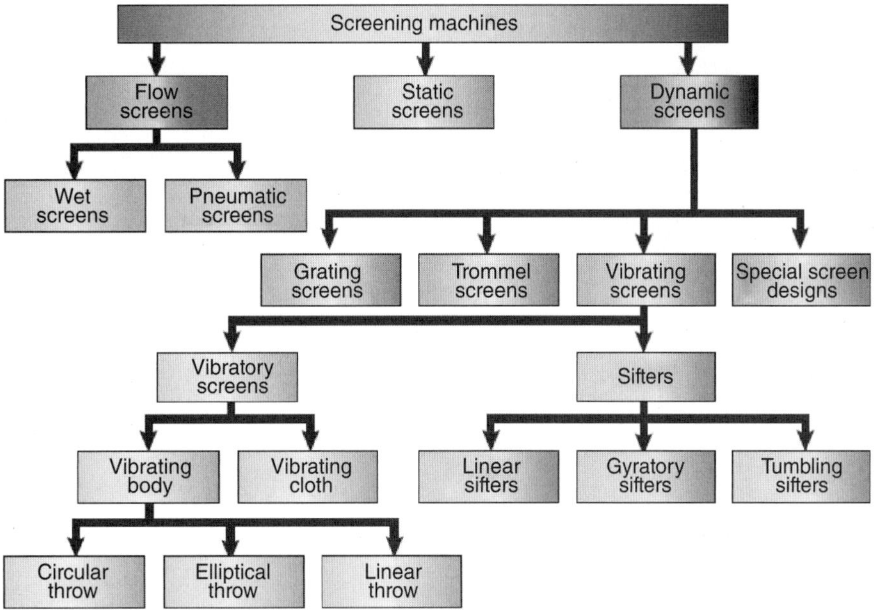

FIG. 21-116 Screener types sorted by screening motion. [*Reprinted by special permission from* Chemical Engineering (*September 2007*). *Copyright 2007, by Access Intelligence, New York, NY 10005.*]

FIG. 21-117 Inclined vibrating screener (http://www.conn-weld.com/incline-vibrating-screens.html).

FIG. 21-118 Horizontal vibrating screener (http://www.aggregatequip.com/en/products_xx.asp?id=66&products_BigClass=16).

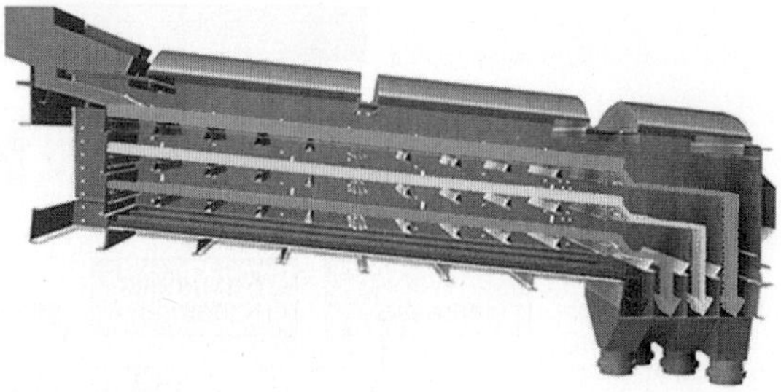

FIG. 21-119 Reciprocating screener (http://www.sssdynamics.com/equipment/screening/texas-shaker-screeners).

FIG. 21-120 Circular motion sifter (https://www.911metallurgist.com/equipment/vibratory-sifter-screen-separator/).

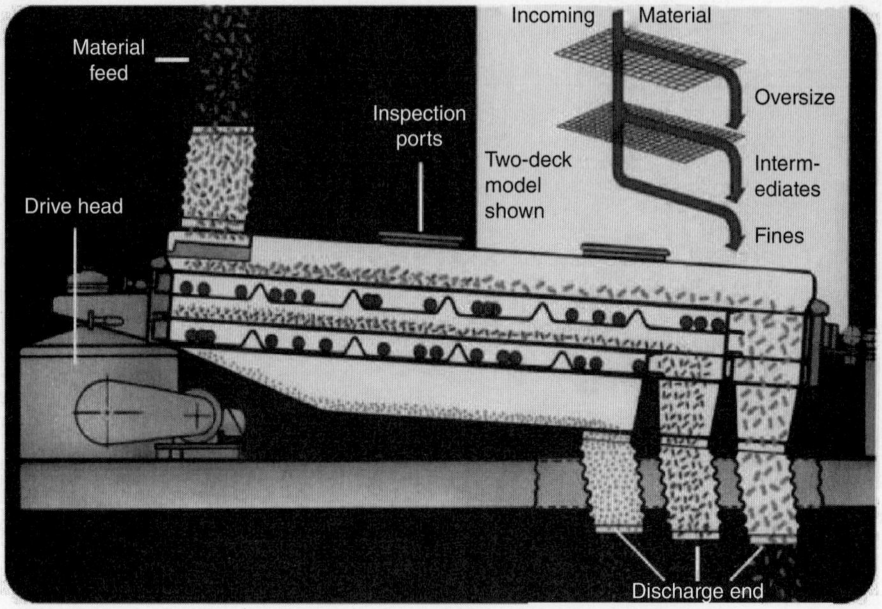

FIG. 21-121 Gyratory sifter (http://www.arcon-minerals.cz/en-cs/applications/screening).

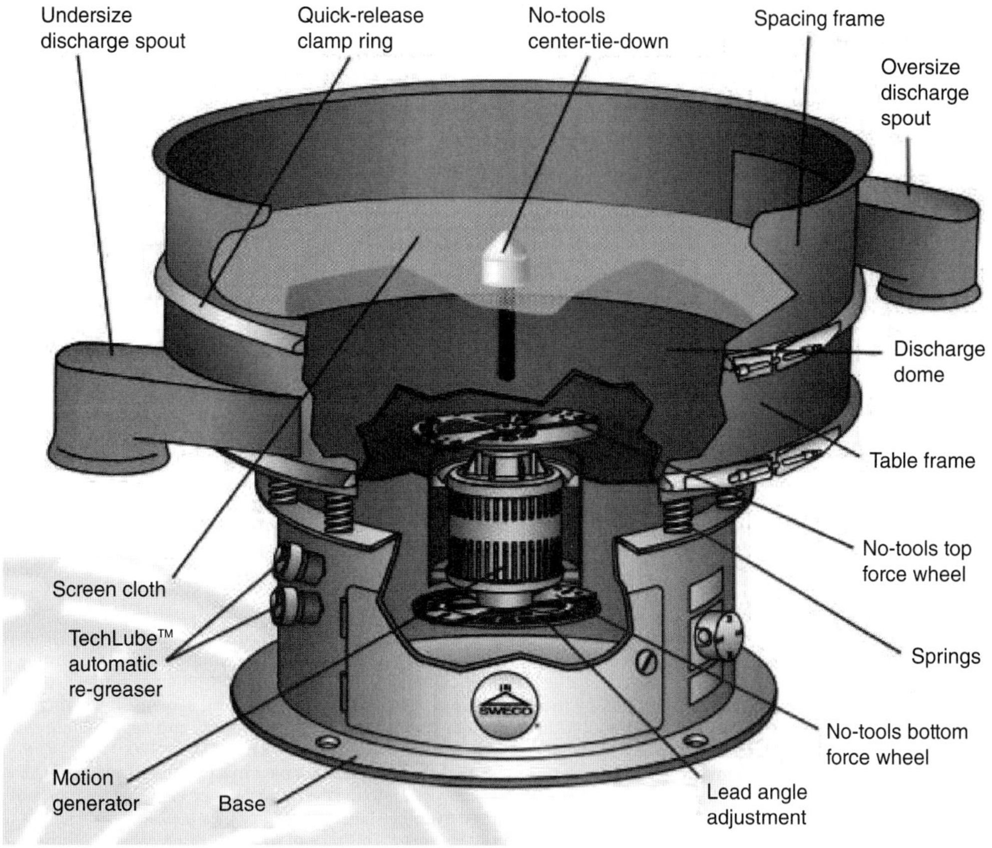

FIG. 21-122a Circular/vibratory sifter internals. (*Courtesy of Sweco.*)

400- and 500-mesh screens have been claimed by the manufacturer. This type of screener can have driven rotating brushes and air wands under the screen surface to facilitate online screen cleaning. The use of the rotating brushes enables the use of this type of screener for spherical products, which are very difficult to screen due to severe screen blinding issues. The downside to this type of screener is the initial cost and the mechanical complexity, but it can be a good option for narrow particle size distributions of fine products and for screening of spherical products.

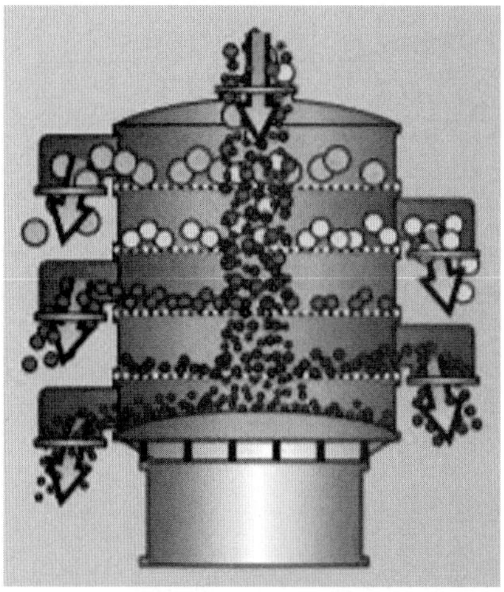

FIG. 21-122b Circular/vibratory sifter particle motion (http://www.sweco.com/tldr.aspx?fid=0C1C32AE6029096785257126004C2788). (*Courtesy of Sweco.*)

FIG. 21-123 Multideck sifter (http://www.buhlergroup.com/global/en/downloads/Sugar_MU16004_en.pdf).

FIG. 21-124 Tumbling screener.

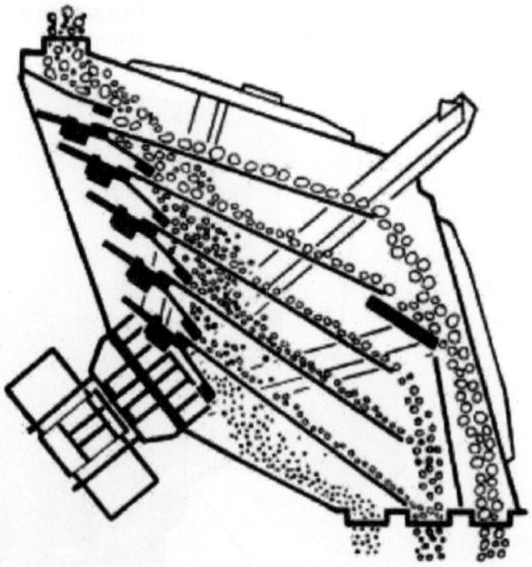

FIG. 21-125 Mogensen sizer.

Probability Screens This type of screener is based on the probability of a free-falling particle passing through the openings in the screen cloth. The screen cloth is typically declined, and the particles will bounce along the surface of the screen cloth if they don't pass through an opening. Since the screen cloth is declined, the opening in the screen cloth "appears" smaller to a vertically falling particle than the opening actually is. This allows the use of coarser screen cloth than would typically be used for a given separation. The downside is that this makes it more difficult to perform a sharp separation. A unit of this type typically has a number of screen decks. A common manufacturer of this type of screener is Mogensen. Figure 21-125 shows a schematic of a Mogensen Sizer.

Revolving Screens A revolving screen consists of a cylindrical screen surface of perforated steel plate or screen cloth that is mounted on rollers, is inclined slightly downward from the feed end, and is rotated around its longitudinal axis at speeds of about 20 rpm. These screens are simple and compact, with no vibration problems, but have low capacities and low efficiencies compared to vibrating screens. Revolving screens are used for rough size separations, dewatering, and agglomerate removal. An example of a revolving screen is shown in Fig. 21-126.

Paddle-Driven Screens In a paddle-driven screen, dry or moist material is conveyed into the cylindrical sifting chamber by a screw flight and is distributed over the screening surface by a paddle rotor. There is no airflow through the chamber. This type of screen is often called a *centrifugal sifter* or a *cyclone sifter* and is shown in Fig. 21-127. Paddle-driven screens are used to remove fines or lumps from powder or granular material. Because of the rotating paddle, it can also be effective at breaking up soft lumps or loose agglomerates.

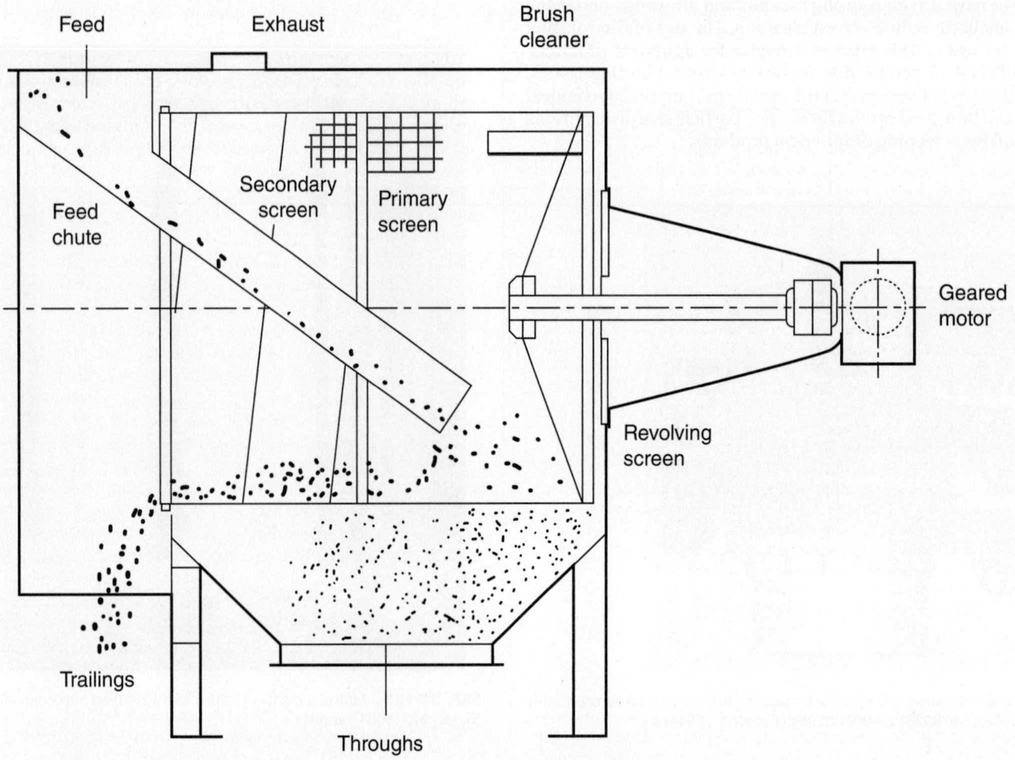

FIG. 21-126 Revolving screener (http://www.satake.com.au/pdf/Drum_Sieve.pdf).

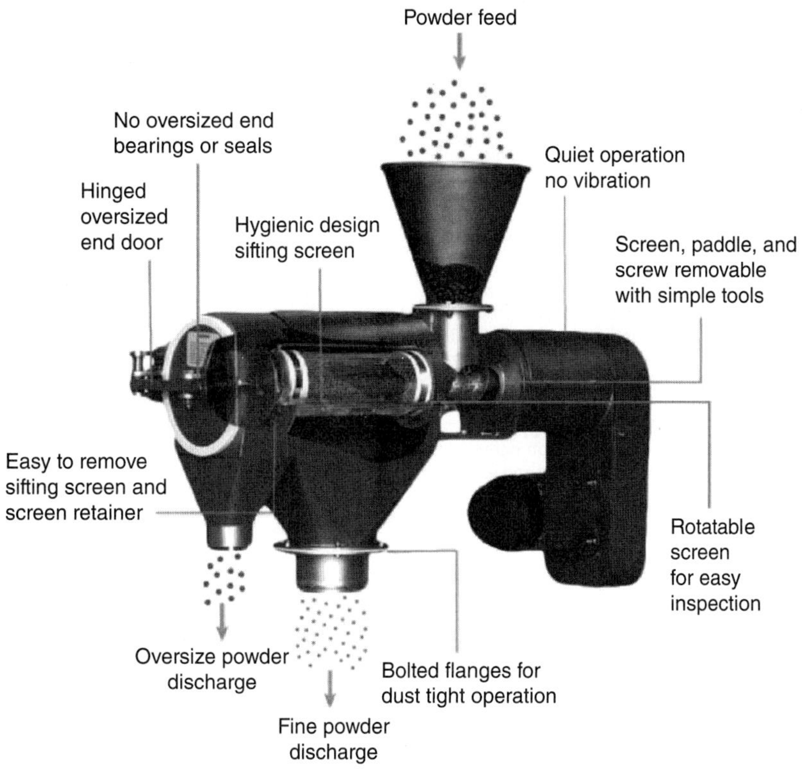

FIG. 21-127 Paddle-driven screener (http://www.kemutecusa.com/products/kek-centrifugal-sifters/).

SIZE REDUCTION

General References: Annual reviews of size reduction, *Ind. Eng. Chem.*, October or November issues, by Work from 1934 to 1965, by Work and Snow in 1966 and 1967, and by Snow in 1968, 1969, and 1970; and in *Powder Technol.* **5:** 351 (1972), and 7 (1973). Snow and Luckie, **10:** 129 (1973), **13:** 33 (1976), **23**(1): 31 (1979). *Chemical Engineering Catalog*, Reinhold, New York, annually. Cremer-Davies, *Chemical Engineering Practice*, vol. 3, *Solid Systems*, Butterworth, London, and Academic, New York, 1957. *Crushing and Grinding: A Bibliography*, Chemical Publishing, New York, 1960. European Symposia on Size Reduction: 1st, Frankfurt, 1962, published in 1962. Rumpf (ed.), *Verlag Chemie*, Düsseldorf; 2d ed., Amsterdam, 1966, published in 1967. Rumpf and Pietsch (eds.), *DECHEMA-Monogr.* **57,** 3d, Cannes, 1971, published in 1972. Rumpf and Schönert (eds.), *DECHEMA-Monogr.* **69.** Gaudin, *Principles of Mineral Dressing*, McGraw-Hill, New York, 1939. International Mineral Processing Congresses: *Recent Developments in Mineral Dressing*, London, 1952, published in 1953, Institution of Mining and Metallurgy; *Progress in Mineral Dressing*, Stockholm, 1957, published in London, 1960, Institution of Mining and Metallurgy; 6th, Cannes, 1962, published in 1965. Roberts (ed.), Pergamon Press, New York; 7th, New York, 1964, published in 1965. Arbiter (ed.), vol. 1: *Technical Papers*, vol. 2: *Milling Methods in the Americas*, Gordon and Breach, New York; 8th, Leningrad, 1968; 9th, Prague, 1970; 10th, London, 1973; 11th, Cagliari, 1975; 12th, São Paulo, 1977. Lowrison, *Crushing and Grinding*, CRC Press, Cleveland, Ohio, 1974. *Pit and Quarry Handbook*, Pit & Quarry Publishing, Chicago, 1968. Richards and Locke, *Text Book of Ore Dressing*, 3d ed., McGraw-Hill, New York, 1940. Rose and Sullivan, *Ball, Tube and Rod Mills*, Chemical Publishing, New York, 1958. Snow, *Bibliography of Size Reduction*, vols. 1 to 9 (an update of the previous bibliography to 1973, including abstracts and index). U.S. Bur. Mines Rep. SO122069, available IIT Research Institute, Chicago, Ill. 60616. Stern, "Guide to Crushing and Grinding Practice," *Chem. Eng.* **69**(25): 129 (1962). Taggart, *Elements of Ore Dressing*, McGraw-Hill, New York, 1951. Since a large part of the literature is in German, the availability of English translations is important. Translation numbers cited in this section refer to translations available through the National Translation Center, Library of Congress, Washington, D.C. Also volumes of selected papers in English translation are available from the Institute for Mechanical Processing Technology, Karlsruhe Technical University, Karlsruhe, Germany.

INTRODUCTION

Industrial Uses of Grinding Grinding operations are critical to many industries, including mining, cement, food, agriculture, pharmaceutical, and chemical. Many solid materials undergo size reduction at some point in their processing cycle. Grinding equipment can be used to reduce the size of a solid material and to intimately mix materials, *usually to produce a solid/liquid dispersion in wet milling operations.*

Some of the common reasons for size reduction are to liberate a desired component for subsequent separation, as in separating ores from gangue; to prepare the material for subsequent chemical reaction, i.e., by enlarging the specific surface area, as in cement manufacture; to meet a size requirement for the quality of the end product, as in fillers or pigments for paints, plastics, agricultural chemicals, etc.; and to prepare wastes for recycling.

Types of Grinding: Particle Fracture versus Deagglomeration Two primary types of size reduction occur in grinding equipment: deagglomeration and particle fracture. In deagglomeration, an aggregate of smaller particles (often with a fractal structure) is size-reduced by breaking clusters of particles off the main aggregate without breaking any of the "primary particles" that form the aggregates. In particle fracture, individual particles are broken rather than simply separating individual particles. Most operations involving particles larger than 10 μm (including materials thought of as rocks and stones) usually involve at least some particle fracture, whereas finer grinding is often mostly deagglomeration. At similar particle scales, deagglomeration requires much less energy than particle fracture. For example, fracture of materials down to a size of 0.1 μm is extremely difficult, whereas deagglomeration of materials in this size range is commonly practiced in several industries, including the automotive paint industry and several electronics industries.

Wet versus Dry Grinding Grinding can occur either wet or dry. Some devices, such as ball mills, can be fed either slurries or dry feeds. In practice, it is found that finer size can be achieved by wet grinding than by dry grinding. In wet grinding by media mills, product sizes of 0.5 nm are attainable with suitable surfactants, and deagglomeration can occur down to much smaller sizes. In dry grinding, the size in ball mills is generally limited by ball coating (Bond and Agthe, *Min. Technol.*, AIME Tech. Publ. 1160, 1940) to about 15 μm. In dry grinding with hammer mills or ring-roller mills, the limiting size is about 10 to 20 μm. Jet mills are generally limited to a mean product size of 2 to 10 μm. However, dense particles can be ground to 2 to 3 μm because of the greater ratio of inertia to aerodynamic drag. Dry processes can sometimes deagglomerate particles down to about 1 μm.

Typical Grinding Circuits There are as many different configurations for grinding processes as there are industries that use grinding equipment; however, many processes use the circuit shown in Fig. 21-128*a*. In this circuit a process stream enters a mill where the particle size is reduced; then, upon exiting the mill, the stream goes to some sort of classification device. There a stream containing the oversized particles is recycled back to the mill, and the product of desired size exits the circuit. Some grinding operations are simply one-pass without any recycler or classifier. For very fine grinding or dispersion (under 1 μm), classifiers are largely unavailable, so processes are either single-pass or recirculated through the mill and tested off-line until a desired particle size is obtained.

The fineness to which a material is ground has a marked effect on its production rate. Figure 21-128*b* shows an example of how the capacity decreases while the specific energy and cost increase as the product is ground finer. Concern about the rising cost of energy has led to publication of a report on this issue (National Materials Advisory Board, *Comminution and Energy Consumption,* Publ. NMAB-364, National Academy Press, Washington, D.C., 1981; available from National Technical Information Service, Springfield, Va.). This has shown that U.S. industries use approximately 32 billion kWh of electric energy per annum in size reduction operations. More than one-half of this energy is consumed in the crushing and grinding of minerals, one-quarter in the production of cement, one-eighth in coal, and one-eighth in agricultural products. In the production of fine chemicals and pharmaceuticals, however, energy consumption is typically not a concern since producing a product with the desired properties is of far greater concern.

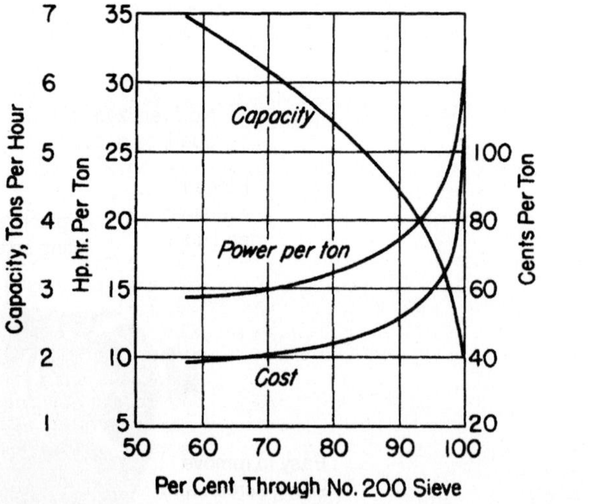

FIG. 21-128*b* Variation in capacity, power, and cost of grinding relative to fineness of product.

THEORETICAL BACKGROUND

Introduction The theoretical background for size reduction is often introduced with particle breakage. It is relatively easy to write down force balances around a particle and make some predictions about how particles might break. Of particular interest in size reduction processes are predictions about the size distribution of particles after breakage and the force/energy required to break particles of a given size, shape, and material.

It has, however, proved difficult to relate theories of particle fracture to properties of interest to the grinding practitioner. This is so, in part, because single-particle testing machines, although they do exist, are expensive and time-consuming to use. To get any useful information, many particles must be tested, and it is unclear that these tests reflect the kind of forces encountered in a given piece of grinding equipment. Even if representative fracture data can be obtained, this information needs to be combined with information on the force distribution and particle mechanics inside a particular grinding device to be useful for scale-up or prediction of the effectiveness of a device. Most of this information (force distribution and particle motion inside devices) has not been studied in detail from either a theoretical or an empirical point of view. However, this is beginning to change with the advent of more powerful computers combined with advances in numerical methods for fluid mechanics and discrete element models.

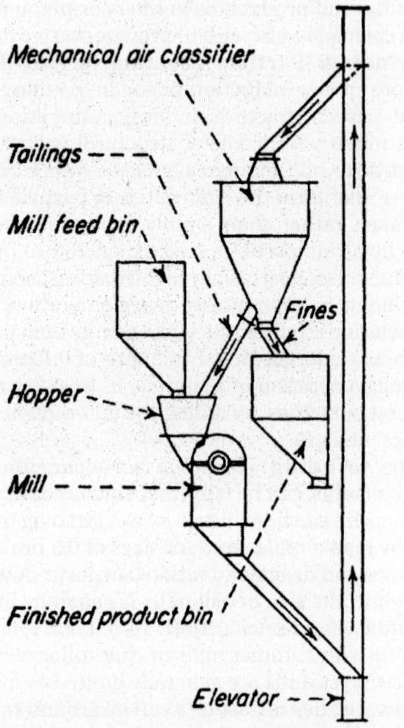

FIG. 21-128*a* Hammer mill in closed circuit with an air classifier.

The practitioner is therefore limited to scale-up and scale-down from testing results of geometrically similar equipment (see the subsection Energy Required and Scale-Up later) and using models that treat the devices as empirical "black boxes" while using a variety of population balance and grind rate theories to keep track of the particle distributions as they go into and out of the mills (see the subsection Modeling of Milling Circuits later).

Also note that the use of multiple mills in series often produces a narrower particle size distribution than a single mill when grinding materials to a much smaller particle size. The rule of thumb is to only grind the material one order of magnitude (that is, 1000 to 100 μm) in a single processing step. This is so because the tendency is to grind the material too fine if you try to do too much in a single pass.

Single-Particle Fracture The key issue in all breakage processes is the creation of a stress field inside the particle that is intense enough to cause breakage. The state of stress and the breakage reaction are affected by many parameters that can be grouped into both particle properties and loading conditions, as shown in Fig. 21-129.

The reaction of a particle to the state of stress is influenced by the material properties, state of stress itself, and presence of microcracks and flaws. Size reduction will start and continue as long as energy is available for the creation of new surface. The stresses provide the required energy and forces necessary for crack growth on the inside and on the surface of the particle.

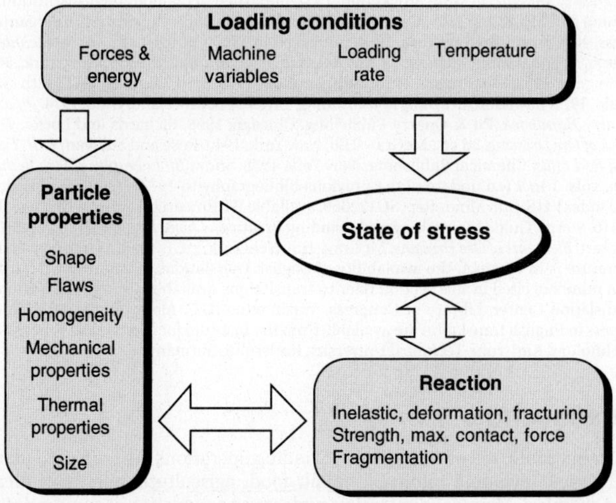

FIG. 21-129 Factors affecting the breakage of a particle. (*After Heiskanen,1995.*)

However, a considerable part of the energy supplied during grinding will be wasted by processes other than particle breakage, such as the production of sound and heat, as well as plastic deformation.

The breakage theory of spheres is a reasonable approximation of what may occur in the size reduction of particles, as most size reduction processes involve roughly spherical particles. An equation for the force required to crush a single particle that is spherical near the contact regions is given by Hertz (Timoschenko and Goodier, *Theory of Elasticity,* 2d ed., McGraw-Hill, New York, 1951). In an experimental and theoretical study of glass spheres, Frank and Lawn [*Proc. R. Soc.* (*London*), **A299**(1458): 291 (1967)] observed the repeated formation of ring cracks as increasing load was applied, causing the circle of contact to widen. Eventually a load is reached at which the ring crack deepens to form a cone crack, and at a sufficient load this propagates across the sphere to cause breakage into fragments. The authors' photographs show how the size of flaws that happen to be encountered at the edge of the circle of contact can result in a distribution of breakage strengths. Thus the mean value of breakage strength depends partly on intrinsic strength and partly on the extent of flaws present. Most industrial solids contain irregularities such as microscopic cracks and weaknesses caused by dislocations, nonstochiometric composition, solid solutions, gas- and liquid-filled voids, or grain boundaries.

Inglis showed that these irregularities play a predominant role in particle breakage as the local stresses σ_i generated at the tips of the crack, as shown in Fig. 21-130, were much higher than the gross applied stress σ_N. The effect is expressed by stress concentration factor k

$$k = \frac{\sigma_i}{\sigma_N} = \frac{l}{r} \tag{21-119}$$

which is a function of the crack length l and the tip radius r.

Griffith found that tensile stresses always occur in the vicinity of crack tips, even when the applied gross stresses are compressive. He also showed that the largest tensile stresses are produced at cracks having a 30° angle to the compressive stress. Thus cracks play a key role in propagation, and their effects greatly overshadow the theoretically calculated values for breakage of spheres or other ideal particles.

ENERGY REQUIRED AND SCALE-UP

Energy Laws Fracture mechanics expresses failure of materials in terms of both stress intensity and fracture toughness, in terms of energy to failure. Due to the difficulty of calculating the stresses on particles in grinding devices, many theoreticians have relied on energy-based theories to connect the performance of grinding devices to the material properties of the material being ground. In these cases, the energy required to break an ensemble of particles can be estimated without making detailed assumptions about the exact stress state of the particles, but rather by calculating the energy required to create fresh surface area with a variety of assumptions.

A variety of energy laws have been proposed. These laws are encompassed in a general differential equation (Walker et al., *Principles of Chemical Engineering,* 3d ed., McGraw-Hill, New York, 1937)

$$dE = -C\, dX/X^n \tag{21-120}$$

where E is the work done, X is the particle size, and C and n are constants. For $n = 1$ the solution is Kick's law (Kick, Das Gasetz der propertionalen Widerstande und seine Anwendung, Leipzig, 1885). The law can be written as

$$E = C \log (X_F/X_P) \tag{21-121}$$

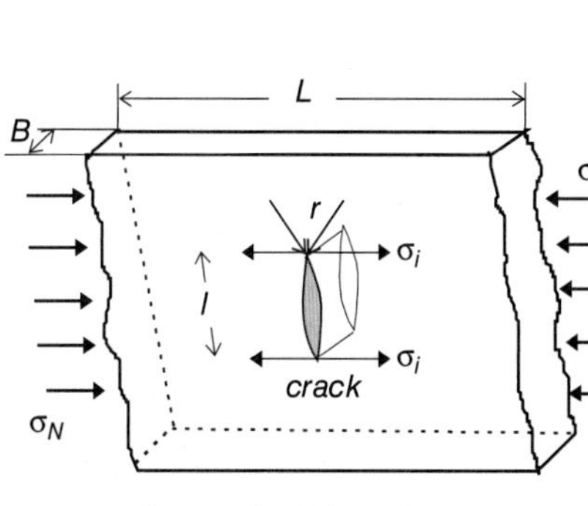

FIG. 21-130 A microcrack in an infinitely large plate.

where X_F is the feed particle size, X_P is the product size, and X_F/X_P is the reduction ratio. For $n > 1$ the solution is

$$E = \left(\frac{C}{n-1}\right)\left(\frac{1}{X_P^{n-1}} - \frac{1}{X_F^{n-1}}\right) \tag{21-122}$$

For $n = 2$ this becomes *Rittinger's law,* which states that the energy is proportional to the new surface produced (Rittinger, *Lehrbuch der Aufbereitungskunde,* Ernst and Korn, Berlin, 1867).

The *Bond law* corresponds to the case in which $n = 1.5$ [Bond, *Trans. Am. Inst. Min. Metall. Pet. Eng.* **193:** 484 (1952)]

$$E = 100E_i\left(\frac{1}{\sqrt{X_P}} - \frac{1}{\sqrt{X_F}}\right) \tag{21-123}$$

where E_i is the *Bond work index,* or work required to reduce a unit weight from a theoretical infinite size to 80 percent passing 100 μm. Extensive data on the work index have made this law useful for rough mill sizing, especially for ball mills. Summary data are given in Table 21-15. The work index may be found experimentally from laboratory crushing and grinding tests or from commercial mill operations. Some rules of thumb for extrapolating the work index to conditions different from those measured are that for dry grinding the index must be increased by a factor of 1.34 over that measured in wet grinding; for open-circuit operations another factor of 1.34 is required over that measured in closed-circuit operations; if the product size X_P is extrapolated below 70 μm, an additional correction factor is $(10.3 + X_P)/1.145X_P$. Also for a jaw or gyratory crusher, the work index may be estimated from

$$E_i = 2.59C_s/\rho_s \tag{21-124}$$

where C_s = impact crushing resistance, (ft·lb)/in of thickness required to break, ρ_s = specific gravity, and E_i is expressed in kWh/ton.

The relation of energy expenditure to the size distribution produced has been thoroughly examined [Arbiter and Bhrany, *Trans. Am. Inst. Min. Metall. Pet. Eng.* **217:** 245–252 (1960); Harris, *Inst. Min. Metall. Trans.* **75**(3): C37 (1966); Holmes, *Trans. Inst. Chem. Eng.* (*London*), **35:** 125–141 (1957); and Kelleher, *Br. Chem. Eng.* **4:** 467–477 (1959); **5:** 773–783 (1960)].

The energy laws have not proved very successful in practice, most likely because only a very small amount of the energy used in milling devices is actually employed for breakage. A great deal of energy input into a mill is used to create noise and heat as well as simply move the material around the device. Although few systematic studies have been done, less (often, much less) than 5 percent of the energy input into a typical grinding device actually goes into breaking the material. The majority of the remaining energy is eventually converted to frictional heat, most of which heats up the product and the mill.

Mill efficiency can be judged in terms of energy input into the device as compared to the particle size achieved for a given material. It is rare that one grinding device will be more than twice as energy-efficient as another device in order to achieve the same particle size for the same material, and there are usually other trade-offs for the more energy-efficient device. In particular, more energy-efficient devices have a tendency to have large, heavy mechanical components that cause great damage to equipment when moved, swung, etc. These, however, tend to be much more costly for the same capacity and harder to maintain than smaller, high-speed devices. For example, for many materials, roll mills are more energy-efficient than hammer mills, but they are also significantly more costly and have higher maintenance costs.

Fine Size Limit (See also Single-Particle Fracture above.) It has long been thought that a limiting size is attainable, and, in fact, it is almost a logical necessity that grinding cannot continue down to the molecular level. Nonetheless, recent results suggest that stirred media mills are capable of grinding many materials down to particle sizes near 100 nm, finer than many predicted limits [see, e.g., S. Mende et al., *Powder Technol.* **132:** 64–73 (2003) or F. Stenger et al., *Chem. Eng. Sci.* **60:** 4557–4565 (2005)]. The requirements to achieve these sizes are high energy input per unit volume, very fine media, a slurry formulated with dispersants designed to prevent agglomeration of the very fine particles, and a great deal of energy and time. With improved technology and technique, finer grinds than ever before are being achieved, at least on the laboratory scale. The energy requirements of these processes are such that it is unlikely that many will be cost-effective except for high-value products. From a practical point of view, if particles much under 1 μm are desired, it is much better to synthesize them close to this size than to grind them down.

Breakage Modes and Grindability Different materials have a greater or lesser ease of grinding, or grindability. In general, soft, brittle materials are easier to grind than hard or ductile materials. Also different types of

TABLE 21-15 Average Work Indices for Various Materials*

Material	No. of tests	Specific gravity	Work index†
All materials tested	2088	—	13.81
Andesite	6	2.84	22.13
Barite	11	4.28	6.24
Basalt	10	2.89	20.41
Bauxite	11	2.38	9.45
Cement clinker	60	3.09	13.49
Cement raw material	87	2.67	10.57
Chrome ore	4	4.06	9.60
Clay	9	2.23	7.10
Clay, calcined	7	2.32	1.43
Coal	10	1.63	11.37
Coke	12	1.51	20.70
Coke, fluid petroleum	2	1.63	38.60
Coke, petroleum	2	1.78	73.80
Copper ore	308	3.02	13.13
Coral	5	2.70	10.16
Diorite	6	2.78	19.40
Dolomite	18	2.82	11.31
Emery	4	3.48	58.18
Feldspar	8	2.59	11.67
Ferrochrome	18	6.75	8.87
Ferromanganese	10	5.91	7.77
Ferrosilicon	15	4.91	12.83
Flint	5	2.65	26.16
Fluorspar	8	2.98	9.76
Gabbro	4	2.83	18.45
Galena	7	5.39	10.19
Garnet	3	3.30	12.37
Glass	5	2.58	3.08
Gneiss	3	2.71	20.13
Gold ore	209	2.86	14.83
Granite	74	2.68	14.39
Graphite	6	1.75	45.03
Gravel	42	2.70	25.17
Gypsum rock	5	2.69	8.16
Ilmenite	7	4.27	13.11
Iron ore	8	3.96	15.44
Hematite	79	3.76	12.68
Hematite—specular	74	3.29	15.40
Oolitic	6	3.32	11.33
Limanite	2	2.53	8.45
Magnetite	83	3.88	10.21
Taconite	66	3.52	14.87
Kyanite	4	3.23	18.87
Lead ore	22	3.44	11.40
Lead-zinc ore	27	3.37	11.35
Limestone	119	2.69	11.61
Limestone for cement	62	2.68	10.18
Manganese ore	15	3.74	12.46
Magnesite, dead burned	1	5.22	16.80
Mica	2	2.89	134.50
Molybdenum	6	2.70	12.97
Nickel ore	11	3.32	11.88
Oil shale	9	1.76	18.10
Phosphate fertilizer	3	2.65	13.03
Phosphate rock	27	2.66	10.13
Potash ore	8	2.37	8.88
Potash salt	3	2.18	8.23
Pumice	4	1.96	11.93
Pyrite ore	4	3.48	8.90
Pyrrhotite ore	3	4.04	9.57
Quartzite	16	2.71	12.18
Quartz	17	2.64	12.77
Rutile ore	5	2.84	12.12
Sandstone	8	2.68	11.53
Shale	13	2.58	16.40
Silica	7	2.71	13.53
Silica sand	17	2.65	16.46
Silicon carbide	7	2.73	26.17
Silver ore	6	2.72	17.30
Sinter	9	3.00	8.77
Slag	12	2.93	15.76
Slag, iron blast furnace	6	2.39	12.16
Slate	5	2.48	13.83
Sodium silicate	3	2.10	13.00
Spodumene ore	7	2.75	13.70
Syenite	3	2.73	14.90
Tile	3	2.59	15.53
Tin ore	9	3.94	10.81
Titanium ore	16	4.23	11.88
Trap rock	49	2.86	21.10
Uranium ore	20	2.70	17.93
Zinc ore	10	3.68	12.42

*Allis-Chalmers Corporation.
†Caution should be used in applying the average work index values listed here to specific installations since individual variations between materials in any classification may be quite large.

grinding equipment apply forces in different ways, and this makes them better suited to particular classes of materials. Figure 21-131 lists the modes of particle loading as they occur in industrial mills. This loading can take place either by slow compression between two planes or by impact against a target. In these cases the force is normal to the plane. If the applied normal forces are too weak to affect the whole of the particle and are restricted to a partial volume at the surface of the particle, the mode is attrition. An alternative way of particle loading is to apply a shear force by moving the loading planes horizontally. The table indicates that compression and impact are used more for coarse grinding, while attrition and abrasion are more common in fine and superfine grinding.

Hard materials (especially Mohs hardness 7 and above) are usually ground by devices designed for abrasion/attrition modes. For example, roll mills would rarely, if ever, be used for grinding of quartz, but media mills of various sorts have been successfully used to grind industrial diamonds. This is so primarily because both compression and high-energy impact modes have substantial contact between the mill and the very hard particles, which causes substantial wear of the device. Many attrition and abrasion devices, however, are designed so that a large component of grinding occurs by impact of particles on one another, rather than impact with the device. Wear still occurs, but it's less dramatic than with other devices.

Ductile materials are an especially difficult problem for most grinding devices. Almost all grinding devices are designed for brittle materials and have some difficulties with ductile materials. However, devices with compression or abrasion modes tend to have the greatest difficulty with these kinds of materials. Mills with a compression mode will tend to flatten and flake these materials. Flaking can also occur in mills with a tangential abrasion mode, but smearing of the material across the surface of the mill is also common. In both cases, particle agglomeration can occur, as opposed to size reduction. Impact and attrition devices tend to do somewhat better with these materials, since their high-speed motion tends to cause more brittle fracture.

Conversely, mills with impact and attrition modes often do poorly with heat-sensitive materials where the materials become ductile as they heat up. Impact and attrition mills cause significant heating at the point of impact, and it is not uncommon for heat-sensitive materials (e.g., plastics) to stick to the device rather than being ground. In the worst cases, cryogenic grinding can be necessary for highly ductile or heat-sensitive materials.

Grindability Methods Laboratory experiments on single particles have been used to correlate grindability. In the past usually it has been assumed that the total energy applied could be related to the grindability whether the energy is applied in a single blow or by repeated dropping of a weight on the sample [Gross and Zimmerly, *Trans. Am. Inst. Min. Metall. Pet. Eng.* **87:** 27, 35 (1930)]. In fact, the results depend on the way in which the force is applied (Axelson, Ph.D. thesis, University of Minnesota, 1949). In spite of this, the results of large mill tests can often be correlated within 25 to 50 percent by a simple test, such as the number of drops of a particular weight needed to reduce a given amount of feed to below a certain mesh size. Two methods having particular application for coal are known as the *ball-mill* and *Hardgrove* methods. In the ball-mill method, the relative amounts of energy necessary to pulverize different coals are determined by placing a weighed sample of coal in a ball mill of a specified size and counting the number of revolutions required to grind the sample so that 80 percent of it will pass through a no. 200 sieve. The grindability index in percent is equal to 50,000, divided by the average of the number of revolutions required by two tests (ASTM designation D-408).

In the Hardgrove method, a prepared sample receives a definite amount of grinding energy in a miniature ball-ring pulverizer. The unknown sample is compared with a coal chosen as having 100 grindability. The Hardgrove grindability index = $13 + 6.93W$, where W is the weight of material passing the no. 200 sieve (see ASTM designation D-409).

Chandler [*Bull. Br. Coal Util. Res. Assoc.* **29**(10): 333 and **29**(11): 371 (1965)] finds no good correlation of grindability measured on 11 coals with roll crushing and attrition, and so these methods should be used with caution.

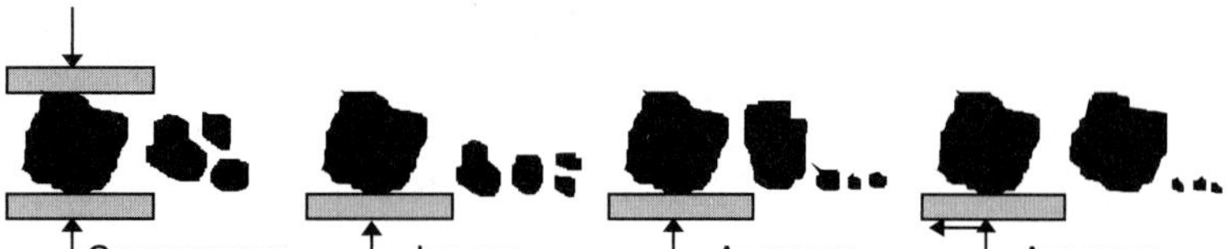

	Compression	Impact	Attrition	Abrasion
Coarse				
crushers	XX			
hammer crusher		XX		
Medium				
roller mills	XX			X
high pressure	XX			X
rolls	XX	XX	XX	X
tumbling mills				
Fine				
vibrating mills	X	XX	XX	
planetary mills	X	XX	X	
hammer mills		XX		
cutter mills		XX	X	
Super fine				
pin mills		XX	XX	
micro impact mills		XX	XX	X
opposed jet mills		XX	XX	X
spiral jet mills		X	XX	X
stirred ball mills			XX	XX

FIG. 21-131 Breakage modes in industrial mills. (*Heiskanen,1995.*)

The Bond grindability method is described in the subsection Capacity and Power Consumption. Manufacturers of various types of mills maintain laboratories in which grindability tests are made to determine the suitability of their machines. When grindability comparisons are made on small equipment of the manufacturers' own class, there is a basis for scale-up to commercial equipment. This is better than relying on a grindability index obtained in a ball mill to estimate the size and capacity of different types such as hammer or jet mills.

OPERATIONAL CONSIDERATIONS

Mill Wear Wear of mill components costs nearly as much as the energy required for comminution—hundreds of millions of dollars a year. The finer stages of comminution result in the greatest wear, because the grinding effort is greatest, as measured by the energy input per unit of feed. Parameters that affect wear fall into three categories: (1) the ore, including hardness, presence of corrosive minerals, and particle size; (2) the mill, including composition, microstructure, and mechanical properties of the material of construction, size of mill, and mill speed; and (3) the environment, including water chemistry and pH, oxygen potential, slurry solids content, and temperature [Moore et al., *Int. J. Mineral Processing* **22:** 313–343 (1988)]. An abrasion index in terms of kilowatthour input per pound of metal lost furnishes a useful indication. In *wet grinding,* a synergy between mechanical wear and corrosion results in higher metal loss than with either mechanism alone [Iwasaki, *Int. J. Mineral Processing* **22:** 345–360 (1988)]. This is due to removal of protective oxide films by abrasion, and by increased corrosion of stressed metal around gouge marks (Moore et al., 1998). Wear rate is higher at lower solids content, since ball coating at high solids protects the balls from wear. This indicates that the mechanism is different from dry grinding. The rate of wear without corrosion can be measured with an inert atmosphere such as nitrogen in the mill. Insertion of marked balls into a ball mill best measures the wear rate at conditions in industrial mills, so long as there is not a galvanic effect due to a different composition of the balls. The mill must be cleared of dissimilar balls before a new composition is tested.

Sulfide ores promote corrosion due to galvanic coupling by a chemical reaction with oxygen present. Increasing the pH generally reduces corrosion. The use of harder materials enhances wear resistance, but this conflicts with achieving adequate ductility to avoid catastrophic brittle failure, so these two effects must be balanced. Wear-resistant materials can be divided into three groups: (1) abrasion-resistant steels, (2) alloyed cast irons, and (3) nonmetallics [see Durman, *Int. J. Mineral Processing* **22:** 381–399 (1988) for a detailed discussion of these].

Cast irons of various sorts are often used for structural parts of large mills such as large ball mills and jaw crushers, while product contact parts such as ball-mill liners and cone crusher mantels are made from a variety of steels.

In many milling applications, mill manufacturers offer a choice of steels for product-contact surfaces (such as mill liner), usually at least one low-alloy "carbon" steel, and higher-alloy stainless steels. The exact alloys vary significantly with mill type. Stainless steels are used in applications where corrosion may occur (many wet grinding operations, but also high-alkali or high-acid minerals), but are more expensive and have lower wear resistance.

Nonmetallic materials include natural rubber, polyurethane, and ceramics. Rubber, owing to its high resilience, is extremely wear-resistant in low-impact abrasion. It is inert to corrosive wear in mill liners, pipe linings, and screens. It is susceptible to cutting abrasion, so that wear increases in the presence of heavy particles, which penetrate, rather than rebound from, the wear surface. Rubber can also swell and soften in solvents. Advantages are its low density, leading to energy savings, ease of installation, and soundproofing qualities. Polyurethane has similar resilient characteristics. Its fluidity at the formation stage makes it suitable for the production of the wearing surface of screens, diaphragms, grates, classifiers, and pump and flotation impellers. The low heat tolerance of elastomers limits their use in dry processing where heat may build up.

Ceramics fill a niche in comminution where metal contamination cannot be tolerated such as pigments, cement, electronic materials, and pharmaceuticals (where any sort of contamination must be minimized). Use of ceramics has greatly increased in recent years, in part due to finer grinding requirements (and therefore higher energy and higher wear) for many industries and in part due to an increased production of electronic materials and pharmaceuticals. Also the technology to produce mill parts from very hard ceramics such as tungsten carbide and yttria-stabilized zirconia has advanced, making larger parts available (although these are often expensive). Ceramic tiles have been used for lining roller mills and chutes and cyclones, where there is a minimum of impact.

Safety The explosion hazard of nonmetallic materials such as sulfur, starch, wood flour, cereal dust, dextrin, coal, pitch, hard rubber, and plastics is often not appreciated [Hartmann and Nagy, *U.S. Bur. Mines Rep. Invest.* **3751** (1944)]. Explosions and fires may be initiated by discharges of static electricity, sparks from flames, hot surfaces, mechanical sparks, and spontaneous combustion. Metal powders also present a hazard because of their *flammability.* Their combustion is favored during grinding operations in which ball, hammer, or ring-roller mills are employed and during which a high grinding temperature may be reached. Many finely divided metal powders in suspension in air are potential *explosion hazards,* and causes for ignition of such dust clouds are numerous [Hartmann and Greenwald, *Min. Metall.* **26:** 331 (1945)]. Concentration of the dust in air and its particle size are important factors that determine explosibility. Below a lower limit of concentration, no explosion can result because the heat of combustion is insufficient to propagate a flame front. Above a maximum limiting concentration, an explosion cannot be produced because insufficient oxygen is available. Note, however, that concentration control in a milling operation is difficult and rarely used to protect the process from a dust explosion. The finer the particles, the more easily ignition is accomplished and the more rapid is the rate of combustion. This is illustrated in Fig. 21-132.

Isolation of the mills, use of nonsparking materials of construction, and magnetic separators to remove foreign magnetic material from the feed are useful *precautions* [Hartman, Nagy, and Brown, *U.S. Bur. Mines Rep. Invest.* **3722** (1943)]. Stainless steel has less sparking tendency than ordinary steel or forgings. Reduction of the oxygen content of air present in grinding systems is a means for preventing dust explosions in equipment [Brown, *U.S. Dep. Agri. Tech. Bull.* **74** (1928)]. The use of *inert gas* has particular adaptation to pulverizers equipped with air classification; flue gas can be used for this purpose, and it is mixed with the air normally present in a system. If inertion is used as the protection method, the limiting oxygen concentration of the material should be measured to determine the appropriate oxygen concentration at which to operate. It is also possible to design the system to contain a dust explosion, to provide the system with explosion venting, or explosion suppression. Various consensus standards on handling combustible dusts are available from the National Fire Protection Association, Quincy, Mass., and should be followed when designing systems handling combustible dusts.

Temperature Stability Many materials are temperature-sensitive and can tolerate temperatures only slightly above room temperature, including many food products, polymers, agricultural chemicals, and pharmaceuticals. This is a particular problem in grinding operations, as grinding inevitably adds heat to the ground material. The two major problems are that either the material will simply be damaged or denatured in some way, such as food products, or the material may melt or soften in the mill, usually causing significant operational problems.

Ways to deal with heat-sensitive materials include choosing a less energy-intensive mill, or running a mill at below optimum energy input. Some mills run naturally cooler than others. For example, jet mills can run cool because they need high gas flow for operation, and this has a significant cooling effect despite the high-energy intensity. Variable-speed drives are commonly used in stirred media mills to control the energy input to heat-sensitive slurries as energy input (and therefore temperature) is a strong function of stirrer speed.

Adding greater cooling capability is often effective, but it can be expensive. Compositions containing fats and waxes are pulverized and blended readily if refrigerated air is introduced into their grinding systems [U.S. Patents 1,739,761

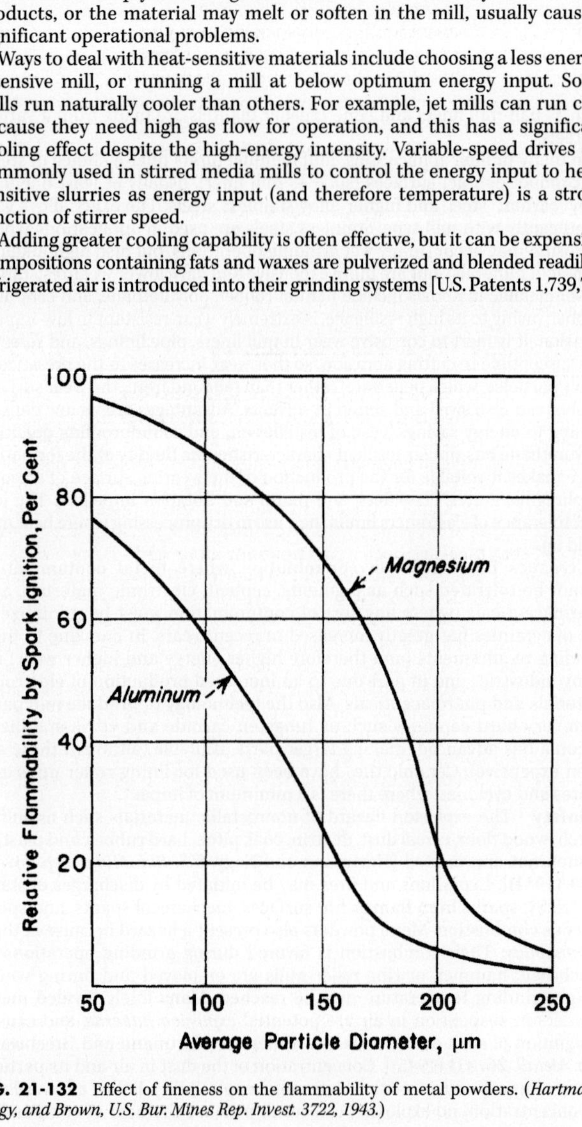

FIG. 21-132 Effect of fineness on the flammability of metal powders. (*Hartmann, Nagy, and Brown, U.S. Bur. Mines Rep. Invest. 3722, 1943.*)

and 2,098,798; see also Hixon, *Chem. Eng. Progress* **87:** 36–44 (May 1991) for flow sheets]. Many mills are also often jacketed, and cooling water is used to remove the excessive heat.

Hygroscopicity Some materials, such as salt, are very hygroscopic; they pick up water from air and deposit on mill surfaces, forming a hard cake. Mills with air classification units may be equipped so that the circulating air can be conditioned by mixing with hot or cold air, gases introduced into the mill, or dehumidification to prepare the air for the grinding of hygroscopic materials. Flow sheets including air dryers are also described by Hixon.

Dispersing Agents and Grinding Aids Grinding aids are helpful under some conditions. For example, surfactants make it possible to ball-mill magnesium in kerosene to 0.5-μm size [Fochtman, Bitten, and Katz, *Ind. Eng. Chem. Prod. Res. Dev.* **2:** 212–216 (1963)]. Without surfactants the size attainable was 3 μm; the rate of grinding was very slow at sizes below this. Also the water in wet grinding may be considered to act as an additive.

Chemical agents that increase the rate of grinding are an attractive prospect since their cost is low. However, despite a voluminous literature on the subject, there is no accepted scientific method to choose such aids; there is not even agreement on the mechanisms by which they work. The subject has been reviewed [Fuerstenau, *KONA Powder and Particle* **13:** 5–17 (1995)]. In wet grinding there are several theories, which have been reviewed [Somasundaran and Lin, *Ind. Eng. Chem. Process Des. Dev.* **11**(3): 321 (1972); Snow, annual reviews, op. cit., 1970–1974; see also Rose, *Ball and Tube Milling,* Constable, London, 1958, pp. 245–249]. Additives can alter the rate of wet ball milling by changing the slurry viscosity or by altering the location of particles with respect to the balls. These effects are discussed in the subsection Tumbling Mills.

In conclusion, there is still no theoretical way to select the most effective additive. Empirical investigation, guided by the principles discussed earlier, is the only recourse. A number of commercially available grinding aids may be tried. Also a kit of 450 surfactants that can be used for systematic trials (Model SU-450, Chem Service Inc., West Chester, Pa.) is available. Numerous experimental studies lead to the conclusion that dry grinding is limited by ball coating and that additives function by reducing the tendency to coat [Bond and Agthe, *Min. Technol.,* AIME Tech. Publ. 1160 (1940)]. Most materials coat if they are ground finely enough, and softer materials coat at larger sizes than hard materials do. The presence of more than a few percent of soft gypsum promotes ball coating in cement-clinker grinding. The presence of a considerable number of coarse particles above 35-mesh inhibits coating. Balls coat more readily as they become scratched. Small amounts of moisture may increase or decrease ball coating. Dry materials also coat. Materials used as grinding aids include solids such as graphite, oleoresinous liquid materials, volatile solids, and vapors. The complex effects of vapors have been extensively studied [Goette and Ziegler, *Z. Ver. Dtsch. Ing.* **98:** 373–376 (1956); and Locher and von Seebach, *Ind. Eng. Chem. Process Des. Dev.* **11**(2): 190 (1972)], but water is the only vapor used in practice. The most effective additive for dry grinding is fumed silica that has been treated with methyl silazane [Tulis, *J. Hazard. Mater.* **4:** 3 (1980)].

Cryogenic Grinding Cryogenic grinding is increasingly becoming a standard option for grinding of rubbers and plastics (especially powder coatings, but also some thermoplastics) as well as heat-sensitive materials such as some pharmaceuticals and chemicals. Many manufacturers of fine-grinding equipment have equipment options for cryogrinding, especially manufacturers of hammer mills and other rotary impact mills.

Cryogrinding adds to operating expenses due to the cost and recovery of liquid nitrogen, but capital cost is a more significant drawback to these systems. Modified mills, special feeders, as well as enhanced air handling and recovery systems are required, and these tend to add significant cost to cryogenic systems. Partly for this reason, there is a healthy toll industry for cryogrinding where specialty equipment can be installed and used for a variety of applications to cover its cost. Many manufacturers of liquid nitrogen have information on cryogrinding applications on their websites.

SIZE REDUCTION COMBINED WITH OTHER OPERATIONS

Size Reduction Combined with Size Classification Grinding systems are batch or continuous in operation (Fig. 21-133). Most large-scale operations are continuous; batch ball or pebble mills are used only when small quantities are to be processed. Batch operation involves a high labor cost for charging and discharging the mill. Continuous operation is accomplished in open or closed circuit, as illustrated in Figs. 21-133 and 21-128*a*. *Operating economy* is the object of closed-circuit grinding with size classifiers. The idea is to remove the material from the mill before all of it is ground, separate the fine product in a classifier, and return the coarse for regrinding with the new feed to the mill. A mill with the fines removed in this way performs much more efficiently. This is so because the material is much less likely to be milled to a size finer than required. Coarse material returned to

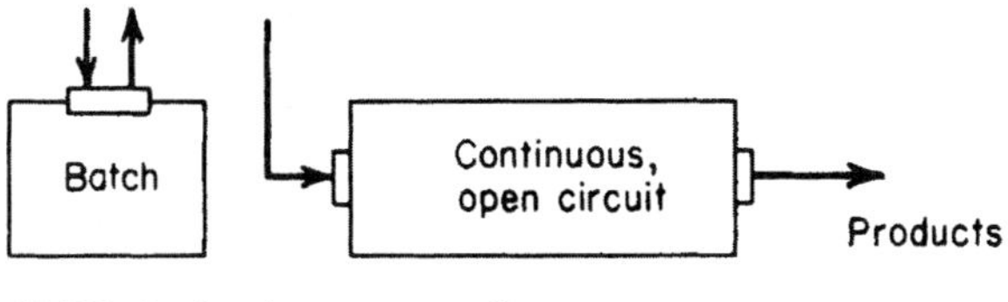

FIG. 21-133 Batch and continuous grinding systems.

a mill by a classifier is known as the *circulating load*; its rate may be from 1 to 10 times the production rate. The ability of the mill to transport material may limit the recycle rate, although mechanical or pneumatic conveyors are often used to recycle the material. Tube mills for use in such circuits may be designed with a smaller length-to-diameter ratio and hence a larger hydraulic gradient for greater flow or with compartments separated by diaphragms with lifters.

Internal Size Classification This plays an essential role in the functioning of machines for dry grinding in the fine size range; particles are retained in the grinding zone until they are as small as required in the finished product; then they are allowed to discharge. By closed-circuit operation the product size distribution is narrower and will have a larger proportion of particles of the desired size. However, making a *product size within narrow limits* (such as between 20 and 40 μm) is often requested but usually is not possible regardless of the grinding circuit used. The reason is that particle breakage is a random process, both as to the probability of breakage of particles and as to the sizes of fragments produced from each breakage event. The narrowest size distribution ideally attainable is one that has a slope of 1.0 when plotted on Gates-Gaudin-Schumann coordinates $[Y = (X/k)^m]$. This can be demonstrated by examining the Gaudin-Meloy size distribution $[Y = 1 - (1 - X/X')^r]$. This is the distribution produced in a mill when particles are cut into pieces of random size, with r cuts per event. The case in which r is large corresponds to a breakage event producing many fines. The case in which r is 1 corresponds to an ideal case such as a knife cutter, in which each particle is cut once per event and the fragments are removed immediately by the classifier. The Meloy distribution with $r = 1$ reduces to the Schumann distribution with a slope of 1.0. Therefore, no practical grinding operation can have a slope greater than 1.0. Slopes typically range from 0.5 to 0.7. The specified product may still be made, but the finer fraction may have to be disposed of in some way. Within these limits, the size distribution of the classifier product depends both on the recycle ratio and on the sharpness of cut of the classifier used.

Size Classification Often only a particular range of product sizes is wanted for a given application. Since the particle breakage process always yields a spectrum of sizes, the product size cannot be directly controlled; however, mill operation can sometimes be varied to produce fewer fines at the expense of producing more coarse particles. By recycling the classified coarse fraction and regrinding it, production of the wanted size range is optimized. Such an arrangement of classifier and mill is called a *mill circuit* and is dealt with further later. More complex systems may include several unit operations such as mixing (Sec. 18), drying (Sec. 16), and agglomerating (see the subsection Size Enlargement). Sound-dampening enclosures may be used to reduce noise from high-speed mills. Chillers, air coolers, and explosion proofing may be added to meet requirements. Weighing and packaging facilities complete the system. Batch ball mills with low ball charges can be used in dry mixing or standardizing of dyes, pigments, colors, and insecticides to incorporate wetting agents and inert extenders. Disk mills, hammer mills, and other high-speed disintegration equipment are useful for final intensive blending of insecticide compositions, earth colors, cosmetic powders, and a variety of other finely divided materials that tend to agglomerate in ribbon and conical blenders. Liquid sprays or gases may be injected into the mill or airstream, for mixing with the material being pulverized to effect chemical reaction or surface treatment.

Other Systems Involving Size Reduction Industrial applications usually involve a number of processing steps combined with size reduction [Hixon, *Chem. Eng. Progress* **87:** 36–44 (May 1991)].

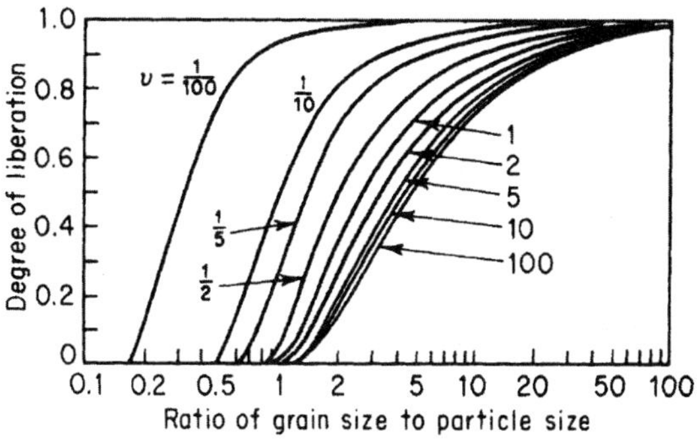

FIG. 21-134 Fraction of mineral B that is liberated as a function of volumetric abundance ratio v of gangue to mineral B (1/grade), and ratio of grain size to particle size of broken fragments (1/fineness). [*Wiegel and Li,* Trans. Soc. Min. Eng.-Am. Inst. Min. Metall. Pet. Eng. ***238:*** *179 (1967).*]

Drying The *drying* of materials often occurs while they are being pulverized in the mill. This occurs due to the high heat generation often seen in grinding operations. These mills are often air-swept to carry away the vapors generated. If the vapors are flammable, appropriate explosion protection must be provided on the mill.

Beneficiation Ball and pebble mills, batch or continuous, offer considerable opportunity for combining a number of *processing steps* that include grinding [Underwood, *Ind. Eng. Chem.* **30:** 905 (1938)]. Mills followed by air classifiers can serve to *separate components of mixtures* because of differences in specific gravity and the component that is pulverized readily. Grinding followed by *froth flotation* has become the beneficiation method most widely used for metallic ores and for nonmetallic minerals such as feldspar. Magnetic separation is the chief means used for upgrading taconite iron ore. Magnetic separators frequently are employed to remove tramp magnetic solids from the feed to high-speed hammer and disk mills.

Liberation Most ores are heterogeneous, and the objective of grinding is to release the valuable mineral component so that it can be separated. Calculations based on a random-breakage model assuming no preferential breakage [Wiegel and Li, *Trans. Am. Inst. Min. Metall. Pet. Eng.* **238:** 179–191 (1967)] agreed, at least in general trends, with plant data on the efficiency of release of mineral grains. Figure 21-134 shows that the desired mineral B can be liberated by coarse grinding when the grade is high so that mineral A becomes a small fraction and mineral B a large fraction of the total volume; mineral B can be liberated only by fine grinding below the grain size, when the grade is low so that there is a small proportion of grains of B. Similar curves, somewhat displaced in size, resulted from a more detailed integral geometry analysis by Barbery [*Minerals Engg.* **5**(2): 123–141 (1992)]. There is at present no way to measure grain size on-line and thus to control liberation. A review of liberation modeling is given by Mehta et al. [*Powder Technol.* **58**(3): 195–209 (1989)]. Many authors have assumed that breakage occurs preferentially along grain boundaries, but there is scant evidence for this. On the contrary, Gorski [*Bull. Acad. Pol. Sci. Ser. Sci. Tech.* **20**(12): 929 (1972)], from analysis of microscope sections, finds an intercrystalline character of comminution of dolomite regardless of the type of crusher used. The liberation of a valuable constituent does not necessarily translate directly to recovery in downstream processes. For example, flotation tends to be more efficient in intermediate sizes than at coarse or fine sizes [McIvor and Finch, *Minerals Engg.* **4**(1): 9–23 (1991)]. For coarser sizes, failure to liberate may be the limitation; finer sizes that are liberated may still be carried through by the water flow. A conclusion is that overgrinding should be avoided by judicious use of size classifiers with recycle grinding.

MODELING AND SIMULATION OF GRINDING PROCESSES

MODELING OF MILLING CIRCUITS

Grinding processes have not benefited as much as some other types of processes from the great increase in computing power and modeling sophistication in the 1990s. Complete simulations of most grinding processes that would be useful to practicing engineers involve breakage mechanics and gas-phase or liquid-phase particle motion coupled in a complex way that is not yet practical to study. However, with the continuing increase of computing power, it is unlikely that this state will continue much longer. Fluid mechanics modeling is well advanced, and the main limitation to modeling many devices is having enough computer power to keep track of a large number of particles as they move and are size-reduced. Traditionally, particle breakage is modeled by using variations of the population balance methodology described below, but more recent models have tended to use discrete element models that track the particles individually. The latter requires greater computing power, but may provide a more realistic way of accounting for particle dynamics in a device.

Computer simulation, based on population balance models [Bass, *Z. Angew. Math. Phys.* **5**(4): 283 (1954)], traces the breakage of each size of particle as a function of grinding time. Furthermore, the simulation models separate the breakage process into two aspects: a breakage rate and a mean fragment size distribution. These are both functions of the size of particle being broken. They usually are not derived from knowledge of the physics of fracture, but are empirical functions fitted to milling data. The following formulation is given in terms of a discrete representation of size distribution; there are comparable equations in integrodifferential form.

BATCH GRINDING

Grinding Rate Function Let w_k = weight fraction of material retained on each screen of a nest of n screens; w_k is related to P_k, the fraction coarser than size X_k, by

$$w_k = (\partial P_k/\partial X_k)\,\Delta X_k \tag{21-125}$$

where ΔX_k is the difference between the openings of screens k and $k+1$. The *grinding rate function* S_u is the rate at which the material of upper size u is selected for breakage in an increment of time, relative to the amount of that size present:

$$dw_u/dt = -S_u w_u \tag{21-126}$$

Breakage Function The *breakage function* $\Delta B_{k,u}$ gives the size distribution of product breakage of size u into all smaller sizes k. Since some fragments from size u are large enough to remain in the range of size u, the term $\Delta B_{u,u}$ is not zero, and

$$\sum_{k=n}^{u} \Delta B_{k,u} = 1 \tag{21-127}$$

The differential equation of batch grinding is deduced from a balance on the material in the size range k. The rate of accumulation of material of size k equals the rate of production from all larger sizes minus the rate of breakage of material of size k:

$$\frac{dw_k}{dt} = \sum_{u=1}^{k} [w_u S_u(t)\,\Delta B_{k,u}] - S_k(t) w_k \tag{21-128}$$

In general, S_u is a function of all the milling variables. Also $\Delta B_{k,u}$ is a function of breakage conditions. If it is assumed that these functions are constant, then relatively simple solutions of the grinding equation are possible, including an analytical solution [Reid, *Chem. Eng. Sci.* **20**(11): 953–963 (1965)] and matrix solutions [Broadbent and Callcott, *J. Inst. Fuel* **29:** 524–539 (1956); **30:** 18–25 (1967); and Meloy and Bergstrom, *7th Int. Min. Proc. Congr. Tech. Pap.*, 1964, pp. 19–31].

Solution of Batch Mill Equations In general, the grinding equation can be solved by numerical methods, e.g., the Euler technique (Austin and Gardner, *1st European Symposium on Size Reduction*, 1962) or the Runge-Kutta technique. The matrix method is a particularly convenient formulation of the Euler technique. Reid's *analytical solution* is useful for calculating the product as a function of time t for a constant feed composition. It is

$$w_{L,k} = \sum_{n=1}^{k} a_{k,n} \exp(-\bar{S}_n\,\Delta t) \tag{21-129}$$

where the subscript L refers to the discharge of the mill, 0 to the entrance, and $\bar{S}_n = 1$ is the "corrected" rate function defined by $\bar{S}_n = 1 - \Delta B_{n,n}$ and then B is normalized with $\Delta B_{n,n} = 0$. The coefficients are

$$a_{k,k} = w_{0k} - \sum_{n=1}^{k-1} a_{k,n} \tag{21-130}$$

and

$$a_{k,n} = \sum_{u=n}^{k-1} \frac{S_u\,\Delta B_{k,u}\,a_{n,u}}{\bar{S}_k - \bar{S}_n} \tag{21-131}$$

The coefficients are evaluated in order since they depend on the coefficients already obtained for larger sizes.

The basic idea behind the *Euler method* is to set the change in w per increment of time as

$$\Delta w_k = (dw_k/dt)\,\Delta t \tag{21-132}$$

where the derivative is evaluated from Eq. (21-129). Equation (21-132) is applied repeatedly for a succession of small time intervals until the desired duration of milling is reached. In the matrix method, a modified rate function is defined $S'_k = S_k\,\Delta t$ as the amount of grinding that occurs in some small time Δt. The result is

$$\mathbf{w}_L = (\mathbf{I} + \mathbf{S'B} - \mathbf{S'})\,\mathbf{w}_F = \mathbf{M}\mathbf{w}_F \tag{21-133}$$

where the quantities $\mathbf{w}$ are vectors, $\mathbf{S'}$ and $\mathbf{B}$ are the matrices of rate and breakage functions, and $\mathbf{I}$ is the unit matrix. This follows because the result obtained by multiplying these matrices is just the sum of products obtained from the Euler method. Equation (21-133) has a physical meaning. The unit matrix times $\mathbf{w}_F$ is simply the amount of feed that is not broken; $\mathbf{S'B}\mathbf{w}_F$ is the amount of feed that is selected and broken into the vector of products; and $\mathbf{S'}\mathbf{w}_F$ is the amount of material that is broken out of its size range and hence must be subtracted from this element of the product. The entire term in parentheses can be considered as a mill matrix $\mathbf{M}$. Thus the milling operation transforms the feed vector to the product vector. Meloy and Bergstrom (*7th Int. Min. Proc. Congr. Tech. Pap.*, 1964, pp. 19–31) pointed out that when Eq. (21-133) is applied over a series of p short-time intervals, the result is

$$\mathbf{w}_L = \mathbf{M}^p\mathbf{w}_F \tag{21-134}$$

Matrix multiplication happens to be cumulative in this special case. It is easy to raise a matrix to a power on a computer since three multiplications give the eighth power, etc. Therefore, the matrix formulation is well adapted to computer use.

CONTINUOUS-MILL SIMULATION

Residence Time Distribution Batch grinding experiments are the simplest type of experiments to produce data on grinding coefficients. But scale-up from batch to continuous mills must take into account the *residence-time distribution* in a continuous mill. This distribution is apparent if a tracer experiment is carried out. For this purpose, background ore is fed continuously, and a pulse of tagged feed is introduced at time t_0. This tagged material appears in the effluent distributed over a period of time, as shown by a typical curve in Fig. 21-135. Because of this distribution some portions are exposed to grinding for longer times than others. Levenspiel (*Chemical Reaction Engineering*, Wiley, New York, 1962) shows several types of residence time distribution that can be observed.

Data on large mills indicate that a curve like that of Fig. 21-135 is typical (Keienberg et al., *3d European Symposium on Size Reduction,* 1972, p. 629). This curve can be accurately expressed as a series of arbitrary functions (Merz and Molerus, *3d European Symposium on Size Reduction,* 1972, p. 607). A good fit is more easily obtained if we choose a function that has the right shape since then only the first two moments are needed. The log-normal probability curve fits most available mill data, as was demonstrated by

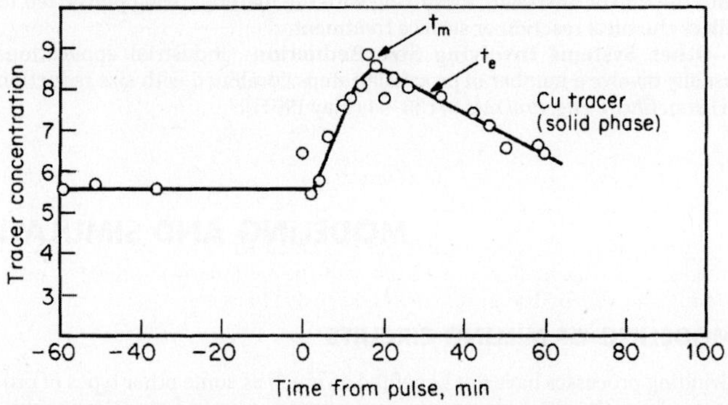

FIG. 21-135 Ore transit through a ball mill. Feed rate is 500 lb/h. (*Courtesy Phelps Dodge Corporation.*)

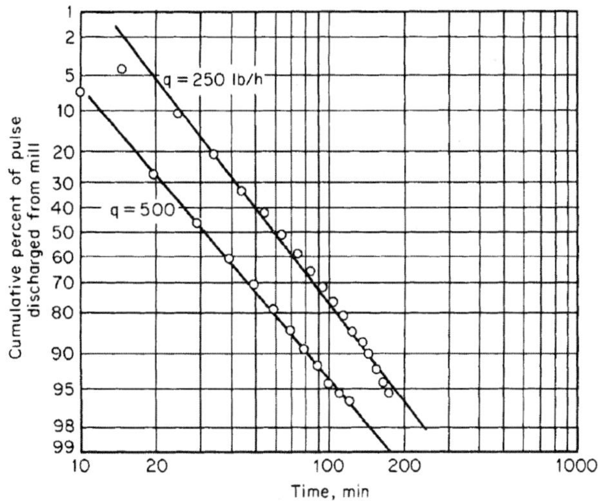

q, lb/h	σ	t_e, min	Dt_e/L^2
250	0.29	39	0.23
500	0.32	18	0.28

FIG. 21-136 Log-normal plot of residence-time distribution in Phelps Dodge mill.

Mori [*Chem. Eng.* (*Japan*) **2**(2): 173 (1964)]. Two examples are shown in Fig. 21-136. The log-normal plot fails only when the mill acts as a nearly perfect mixer.

To measure a residence time distribution, a pulse of tagged feed is inserted into a continuous mill and the effluent is sampled on a schedule. If it is a dry mill, a soluble tracer such as salt or dye may be used, and the samples analyzed conductimetrically or colorimetrically. If it is a wet mill, the tracer must be a solid of similar density to the ore. Materials such as copper concentrate, chrome brick, or barites have been used as tracers and analyzed by X-ray fluorescence. To plot results in log-normal coordinates, the concentration data must first be normalized from the form of Fig. 21-135 to the form of cumulative percent discharged, as in Fig. 21-136. For this, one must either know the total amount of pulse feed or determine it by a simple numerical integration using a computer.

The data are then plotted as in Fig. 21-136, and the coefficients in the log-normal formula of Mori can be read directly from the graph. Here $t_e = t_{50}$ is the time when 50 percent of the pulse has emerged. The standard deviation σ is the time between t_{16} and t_{50} or between t_{50} and t_{84}. Knowing t_e and σ, one can reconstruct the straight line in log-normal coordinates. One can also calculate the vessel dispersion number Dt_e/L^2, which is a measure of the sharpness of the pulse (Levenspiel, *Chemical Reactor Omnibook*, Oregon State University Bookstores Inc., 1979, p. 100.6). This number has been erroneously called the Peclet number by some. Here D is the particle diffusivity. A few available data are summarized (Snow, *International Conference on Particle Technology*, IIT Research Institute, Chicago, 1973, p. 28) for wet mills. Other experiments are presented for dry mills [Hogg et al., *Trans. Am. Inst. Min. Metall. Pet. Eng.* **258:** 194 (1975)]. The most important variables affecting the vessel dispersion number are L/diameter of the mill, ball size, mill speed, scale expressed either as diameter or as throughput, degree of ball filling, and degree of material filling.

Solution for Continuous Milling In the method of Mori [*Chem. Eng.* (*Japan*) **2**(2): 173 (1964)], the residence time distribution is broken up into a number of segments, and the batch-grinding equation is applied to each of them. The resulting size distribution at the mill discharge is

$$\mathbf{w}(L) = \mathbf{w}(t)\,\Delta\varphi \tag{21-135}$$

where $\mathbf{w}(t)$ is a matrix of solutions of the batch equation for the series of times t, with corresponding segments of the cumulative residence time curve. Using the Reid solution, Eq. (21-129), this becomes

$$\mathbf{w}(L) = \mathbf{RZ}\,\Delta\varphi \tag{21-136}$$

since the Reid solution [Eq. (21-129)] can be separated into a matrix $\mathbf{Z}$ of exponentials exp $(-St)$ and another factor R involving only particle sizes. Austin, Klimpel, and Luckie (*Process Engineering of Size Reduction: Ball Milling*, Society of Mining Engineers of AIME, 1984) incorporated into this form a tanks-in-series model for the residence time distribution.

CLOSED-CIRCUIT MILLING

In closed-circuit milling, the tailings from a classifier are mixed with fresh feed and recycled to the mill. Calculations can be based on a material balance and an explicit solution such as Eq. (21-133). Material balances for the normal circuit arrangement (Fig. 21-137) give

$$q = q_F + q_R \tag{21-137}$$

where q = total mill throughput, q_F = rate of feed of new material, and q_R = recycle rate. A material balance on each size gives

$$w_{0,k} = \frac{q_F w_{F,k} + \dfrac{qR}{R}\eta_k w_{L,k}}{q} \tag{21-138}$$

where $w_{0,k}$ = fraction of size k in the mixed feed streams, R = recycle ratio, and η_k = classifier selectivity for size k. With these conditions, a calculation of the transient behavior of the mill can be performed by using any method of solving the milling equation and iterating over intervals of time τ = residence time in the mill. This information is important for evaluating mill circuit control stability and strategies. If the throughput q is controlled to be a constant, as is often done, then τ is constant, and a closed-form matrix solution can be found for the steady state [Callcott, *Trans. Inst. Min. Metall.* **76**(1): C1–11 (1967)]. The resulting flow rates and composition vectors are given in Fig. 21-137. Calcott (*Process Engineering of Size Reduction: Ball Milling*,

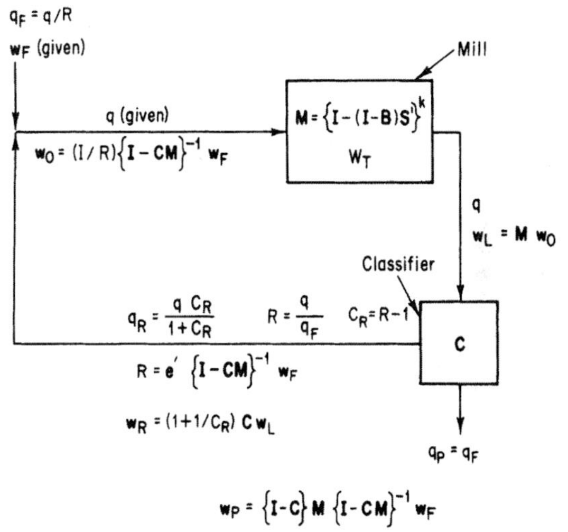

Nomenclature

C_R = circulating load, $R - 1$
C = classifier selectivity matrix, which has classifier selectivity-function values η on diagonal zeros elsewhere
I = identity matrix, which has ones on diagonal, zeros elsewhere
M = mill matrix, which transforms mill-feed-size distribution into mill-product-size distribution
q = flow rate of a material stream
R = recycle ratio q/q_F
w = vector of differential size distribution of a material stream
W_T = holdup, total mass of material in mill

Subscripts:
0 = inlet to mill
F = feed stream
L = mill-discharge stream
P = product stream
R = recycle stream, classifier tailings

FIG. 21-137 Normal closed-circuit continuous grinding system with stream flows and composition matrices, obtained by solving material-balance equations. [*Callcott*, Trans. Inst. Min. Metall. *76*(*1*): *C1-11* (*1967*).]

Society of Mining Engineers of AIME, 1984) gives equations for the reverse-circuit case, in which the feed is classified before it enters the mill. These results can be used to investigate the effects of changes in feed composition on the product. Separate calculations can be made to find the effects of classifier selectivity, mill throughput or recycle, and grindability (rate function) to determine optimum mill-classifier combinations [Lynch, Whiten, and Draper, *Trans. Inst. Min. Metall.* **76**(C169): 179 (1967)]. Equations such as these form the basis for computer codes that are available for modeling mill circuits (Austin, Klimpel, and Luckie, *Process Engineering of Size Reduction: Ball Milling,* Society of Mining Engineers of AIME, 1984).

DATA ON BEHAVIOR OF GRINDING FUNCTIONS

Several breakage functions were suggested early [Gardner and Austin, *1st European Symposium on Size Reduction,* op. cit., 1962, p. 217; Broadbent and Calcott, *J. Inst. Fuel* **29:** 524 (1956); 528 (1956); 18 (1957); **30:** 21 (1957)]. The simple Gates-Gaudin-Schumann equation has been most widely used to fit ball-mill data. For example, this form was assumed by Herbst and Fuerstenau [*Trans. Am. Inst. Min. Metall. Pet. Eng.* **241**(4): 538 (1968)] and Kelsall et al. [*Powder Technol.* **1**(5): 291 (1968); **2**(3): 162 (1968); **3**(3): 170 (1970)]. More recently it has been observed that when the Schumann equation is used, the quantity of coarse fragments cannot be made to agree with the mill product distribution regardless of the choice of rate function. This observation points to the need for a breakage function that has more coarse fragments, such as the function used by Reid and Stewart (Chemica meeting, 1970) and Stewart and Restarick [*Proc. Australas. Inst. Min. Metall.* **239:** 81 (1971)] and shown in Fig. 21-138. This graph can be fitted by a *double Schumann equation*

$$B(X) = A\left(\frac{X}{X_0}\right)^s + (1-A)\left(\frac{X}{X_0}\right)^r \qquad (21\text{-}139)$$

where A is a coefficient less than 1.

In the investigations mentioned earlier, the breakage function was assumed to be normalizable; i.e., the shape was independent of X_0. Austin and Luckie [*Powder Technol.* **5**(5): 267 (1972)] allowed the coefficient A to vary with the size of particle breaking when grinding soft feeds.

Grinding Rate Functions These were determined by tracer experiments in laboratory mills by Kelsall et al. (op. cit.) and in similar work by Szantho and Fuhrmann [*Aufbereit. Tech.* **9**(5): 222 (1968)]. These curves can be fitted by the equation

$$\frac{S}{S_{\max}} = \left(\frac{X}{X_{\max}}\right)^{\alpha} \exp\left(-\frac{X}{X_{\max}}\right) \qquad (21\text{-}140)$$

That a maximum must exist should be apparent from the observation of Coghill and Devaney (U.S. Bur. Mines Tech. Pap., 1937, p. 581) that there is an

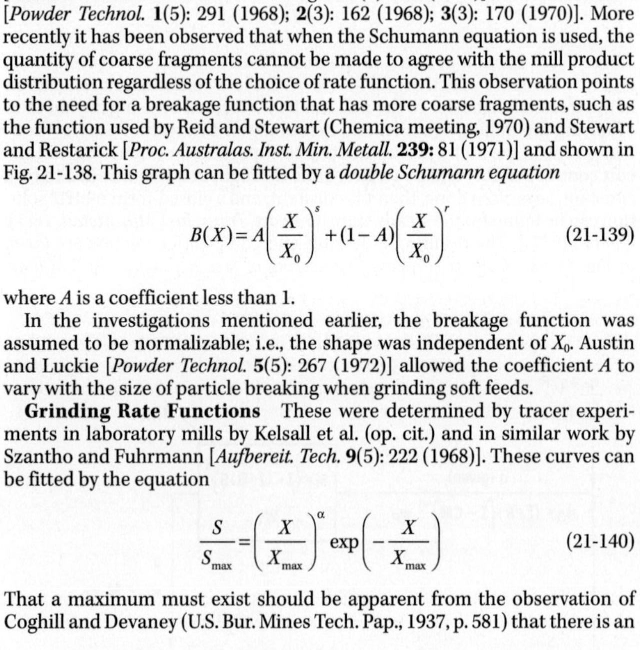

FIG. 21-138 Experimental breakage functions. (*Reid and Stewart, Chemical meeting, 1970.*)

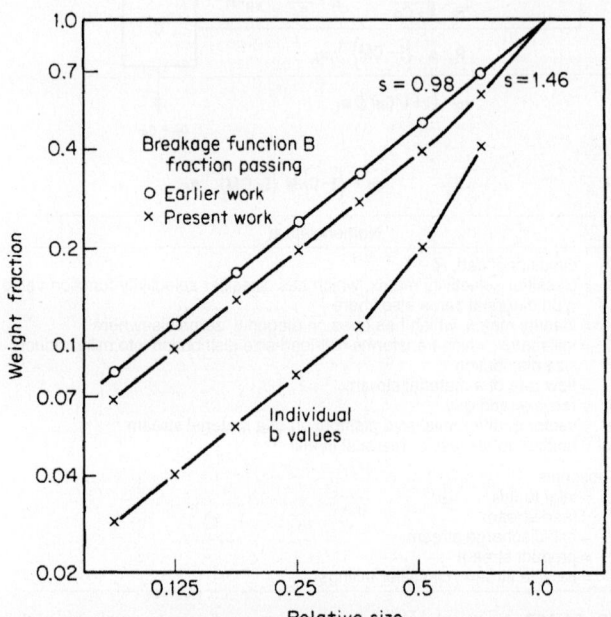

optimum ball size for each feed size. The position of this maximum depends on the ball size. In fact, the feed size for which S is a maximum can be estimated by inverting the formula for optimum ball size given by Coghill and Devaney in the subsection Tumbling Mills.

SCALE-UP AND CONTROL OF GRINDING CIRCUITS

Scale-Up Based on Energy Since large mills are usually sized on the basis of power draft (see subsection Energy Laws), it is appropriate to scale up or convert from batch to continuous data by

$$S(X)_{\text{cont}} = S(X)_{\text{batch}} \frac{(W_T/KW)_{\text{batch}}}{(W_T/KW)_{\text{cont}}} \qquad (21\text{-}141)$$

Usually W_T is not known for continuous mills, but it can be determined from $W_T = t_eQ$, where t_e is determined by a tracer measurement. Equation (21-141) will be valid if the holdup W_T is geometrically similar in the two mills or if operating conditions are in the range in which total production is independent of holdup. Studies of the kinetics of milling [Patat and Mempel, *Chem. Ing. Tech.* **37**(9): 933; (11): 1146; (12): 1259 (1965)] indicate that there is a range of holdup in which this is true. More generally, Austin, Luckie, and Klimpel (loc. cit.) developed empirical relations to predict S as holdup varies. In particular, they observe a slowing of grinding rate when mill filling exceeds ball void volume due to cushioning.

Parameters for Scale-Up Before simulation equations can be used, the parameter matrices **S** and **B** must be back-calculated from experimental data, which turns out to be difficult. One reason is that **S** and **B** occur as a product, so they are to some extent indeterminate; errors in one tend to be compensated by the other. Also the number of parameters is larger than the number of data values from a single size distribution measurement; but this is overcome by using data from grinding tests at a series of grinding times. This should be done anyway, since the empirical parameters should be determined to be valid over the experimental range of grinding times.

It may be easier to fit the parameters by forcing them to follow specified functional forms. In the earliest attempts, it was assumed that the forms should be normalizable (have the same shape regardless of the size being broken). With complex ores containing minerals of different friability, the grinding functions **S** and **B** exhibit complex behavior near the grain size (Choi et al., *Particulate and Multiphase Processes Conference Proceedings* **1:** 903–916). Grinding function **B** is not normalizable with respect to feed size, and **S** does not follow a simple power law.

There are also experimental problems. When a feed size distribution is ground for a short time, there is not enough change in the size distribution in the mill to distinguish between particles being broken into and out of intermediate sizes, unless individual feed size ranges are tagged. Feeding narrow-size fractions alone solves the problem, but changes the milling environment; the presence of fines affects the grinding of coarser sizes. Gupta et al. [*Powder Technol* **28**(1): 97–106 (1981)] ground narrow fractions separately, but subtracted the effect of the first 3 min of grinding, after which the behavior had become steady. Another experimental difficulty arises from the recycle of fines in a closed circuit, which soon "contaminates" the size distribution in the mill; it is better to conduct experiments in open circuit or in batch mills on a laboratory scale.

There are few data demonstrating scale-up of the grinding-rate functions **S** and **B** from pilot- to industrial-scale mills. Weller et al. [*Int. J. Mineral Processing* **22:** 119–147 (1988)] ground chalcopyrite ore in pilot and plant mills and compared predicted parameters with laboratory data of Kelsall [*Electrical Engg. Trans.,* Institution of Engineers Australia, **EE5**(1): 155–169 (1969)] and Austin, Klimpel, and Luckie (*Process Engineering of Size Reduction, Ball Milling,* Society of Mining Engineers, New York, 1984) for quartz. Grinding function **S** has a maximum for a particle size that depends on ball size, which can be expressed as $X_s/X_t = (d_s/d_t)^{2.4}$, where s = scaled-up mill, t = test mill, d = ball size, and X = particle size of maximum rate. Changing ball size also changes the rates according to $\mathbf{S}_s/\mathbf{S}_t = (d_s/d_t)^{0.55}$. These relations shift one rate curve onto another and allow scale-up to a different ball size. Mill diameter also affects rate by a factor $(D_s/D_t)^{0.5}$.

Lynch (*Mineral Crushing and Grinding Circuits, Their Simulation Optimization Design and Control,* Elsevier Scientific Publishing Co., Amsterdam, 1977) and Austin, Klimpel, and Luckie (loc. cit.) developed scale-up factors for ball load, mill filling, and mill speed. In addition, slurry solids content is known to affect the rate, through its effect on slurry rheology. Austin, Klimpel, and Luckie (loc. cit.) present more complete simulation examples and compare them with experimental data to study scale-up and optimization of open and closed circuits, including classifiers such as hydrocyclones and screen bends. Differences in the classifier will affect the rates in a closed circuit. For these reasons scale-up is likely to be uncertain unless conditions in the large mill are as close as possible to those in the test mill.

CRUSHING AND GRINDING EQUIPMENT: DRY GRINDING—IMPACT AND ROLLER MILLS

JAW CRUSHERS

Design and Operation These crushers may be divided into two main groups, the *Blake* (Fig. 21-139), with a movable jaw pivoted at the top, giving greatest movement to the smallest lumps; and the *overhead eccentric,* which is also hinged at the top, but through an *eccentric-driven* shaft that imparts an elliptical motion to the jaw. Both types have a removable crushing plate, usually corrugated, fixed in a vertical position at the front end of a hollow rectangular frame. A similar plate is attached to the swinging movable jaw. The Blake jaw is moved through a knuckle action by the rising and falling of a second lever (pitman) carried by an eccentric shaft. The vertical movement is communicated horizontally to the jaw by double-toggle plates. Because the jaw is pivoted at the top, the throw is greatest at the discharge, preventing choking.

The *overhead eccentric jaw crusher* falls into the second type. These are single-toggle machines. The lower end of the jaw is pulled back against the toggle by a tension rod and spring. The choice between the two types of jaw crushers is generally dictated by the feed characteristics, tonnage, and product requirements (Pryon, *Mineral Processing,* Mining Publications, London, 1960; Wills, *Mineral Processing Technology,* Pergamon, Oxford, 1979). Greater wear caused by the elliptical motion of the overhead eccentric and direct transmittal of shocks to the bearing limit use of this type to readily breakable material. Overhead eccentric crushers are generally preferred for crushing rocks with a hardness equal to or lower than that of limestone. Operating costs of the overhead eccentric are higher for the crushing of hard rocks, but its large reduction ratio is useful for simplified low-tonnage circuits with fewer grinding steps. The double-toggle type of crushers cost about 50 percent more than the similar overhead-eccentric type of crushers.

Comparison of Crushers The jaw crusher can accommodate the same size rocks as a gyratory, with lower capacity and also lower capital and maintenance costs, but similar installation costs. Therefore, they are preferred when the crusher gap is more important than the throughput. Relining the gyratory requires greater effort than for the jaw as well as more space above and below the crusher.

Performance Jaw crushers are applied to the primary crushing of hard materials and are usually followed by other types of crushers. In smaller sizes they are used as single-stage machines. Typical capabilities and specifications are shown in Table 21-16*a*.

GYRATORY CRUSHERS

The development of improved supports and drive mechanisms has allowed gyratory crushers to take over most large hard-ore and mineral-crushing applications. The largest expense of these units is in relining them. Operation is intermittent; so power demand is high, but the total power cost is not great.

Design and Operation The gyratory crusher consists of a cone-shaped pestle oscillating within a larger cone-shaped mortar or bowl. The angles of the cones are such that the width of the passage decreases toward the bottom of the working faces. The pestle consists of a mantle which is free to turn on its spindle. The spindle is oscillated from an eccentric bearing below. Differential motion causing attrition can occur only when pieces are caught simultaneously at the top and bottom of the passage owing to different radii at these points. The circular geometry of the crusher gives a favorably small *nip angle* in the horizontal direction. The nip angle in the vertical direction is less favorable and limits feed acceptance. The vertical nip angle is determined by the shape of the mantle and bowl liner; it is similar to that of a jaw crusher.

Primary crushers have a steep cone angle and a small reduction ratio. *Secondary crushers* have a wider cone angle; this allows the finer product to be spread over a larger passage area and also spreads the wear over a wider area. Wear occurs to the greatest extent in the lower, fine-crushing zone. These features are further extended in cone crushers; therefore, secondary gyratories are much less popular than secondary cone crushers, but they can be used as primaries when quarrying produces suitable feed sizes. The three general types of gyratory crusher are the *suspended-spindle, supported-spindle,* and *fixed-spindle* types. Primary gyratories are designated by the size of feed opening, and secondary or reduction crushers by the diameter of the head in feet and inches. There is a close opening and a wide opening as the mantle gyrates with respect to the concave ring at the outlet end. The close opening is known as the *close setting* or the *closed-side setting,* while the wide opening is known as the *wide-side* or *open-side setting.*

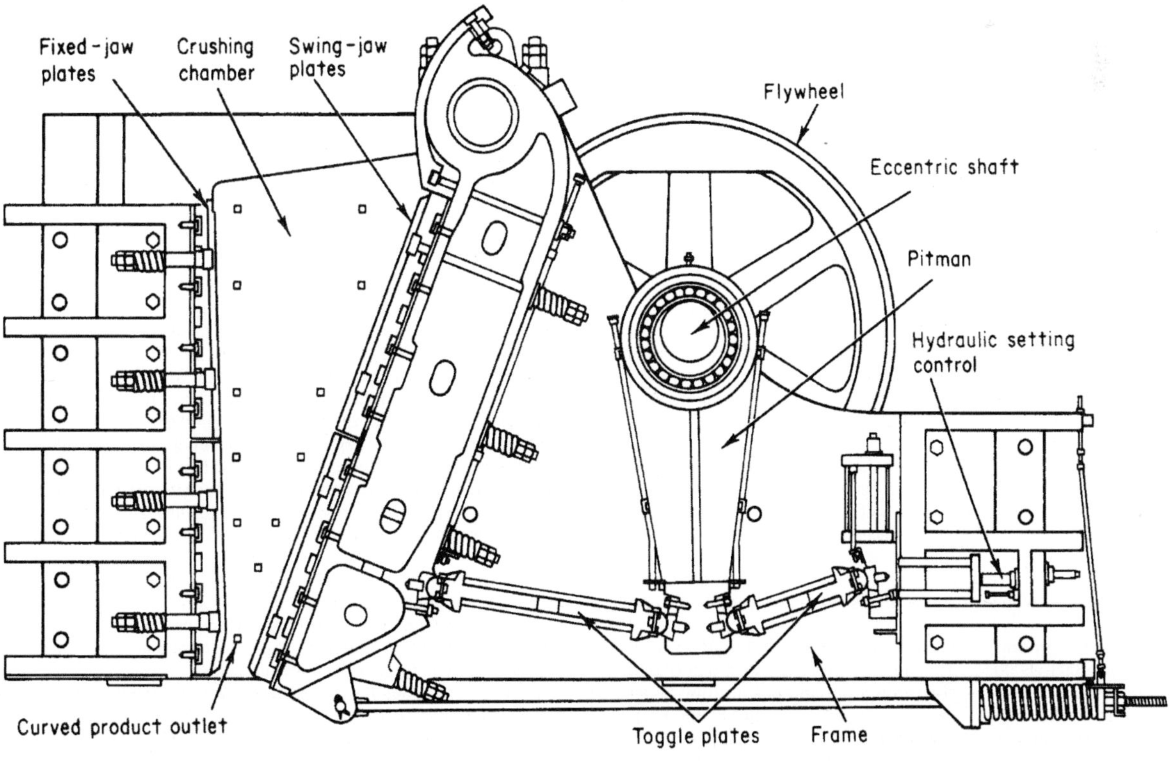

FIG. 21-139 Blake jaw crusher. (*Allis Mineral Systems Grinding Div., Svedala Industries, Inc.*)

TABLE 21-16*a* Performance of Nordberg C Series Eccentric Jaw Crushers

		C95	C105	C0	C100	C05	C110	C125	C10	C15	C10	C200
Feed opening width mm (in)		930 (37)	1060 (42)	800 (32)	1000 (40)	1375 (54)	1100 (44)	1250 (49)	1400 (55)	1400 (55)	1600 (63)	2000 (79)
Feed opening depth mm (in)		580 (23)	700 (28)	510 (20)	760 (30)	760 (30)	850 (34)	950 (37)	1070 (42)	1100 (43)	1200 (47)	1500 (59)
Power kW (HP)		90 (125)	110(150)	75 (100)	110 (150)	160 (200)	160 (200)	160 (200)	200 (250)	200 (300)	250 (350)	400 (500)
Speed (rpm)		330	300	350	260	260	230	220	220	220	220	200
Product size mm (in)	Closed side setting mm (in)	tph (Stph)	tph (Stph)	tph (Stph)	tph (Stph)	tph (Stph)	tph (Stph)	tph (Stph)	tph (Stph)	tph (Stph)	tph (Stph)	tph (Stph)
0–30	20			*								
0–1⅛	¾			*								
0–35	25	*	*	*								
0–1⅜	1	*	*	*								
0–45	30	*	*	*								
0–1¾	1⅛	*	*	*								
0–60	40	*	*	55–75	*	*	*					
0–2⅜	1⅝	*	*	60–80	*	*	*					
0–75	50	*	*	65–95	*	*	*					
0–3	2	*	*	75–100	*	*	*					
0–90	60	105–135	*	80–110	*	*	*					
0–3½	2⅜	115–150	*	90–120	*	*	*					
0–105	70	125–155	135–175	95–135	125–175	210–270	160–220					
0–4⅛	2¾	135–170	150–190	110–145	140–190	230–295	175–240					
0–120	80	140–180	155–195	110–150	145–200	240–300	175–245	*				
0–4¾	3⅛	155–200	170–215	120–165	160–215	260–330	195–270	*				
0–135	90	160–200	175–225	125–175	160–220	260–330	190–275	*				
0–5⅜	3½	175–220	195–245	140–190	175–240	285–360	215–300	*	*	*		
0–150	100	175–225	195–245	140–190	180–250	285–365	215–295	245–335	*	*	*	
0–6	4	195–250	210–270	150–210	200–275	315–400	235–325	270–370	*	*	*	
0–185	125	220–280	245–315	175–245	220–310	345–435	260–360	295–405	325–445	335–465	*	*
0–7	5	240–310	270–345	195–270	245–340	375–480	285–395	325–445	355–490	370–510	*	*
0–225	150	265–335	295–375	210–290	265–365	405–515	310–430	345–475	380–530	395–545	430–610	*
0–9	6	290–370	325–410	230–320	290–400	445–565	340–470	380–525	420–580	435–600	475–670	*
0–260	175	310–390	345–435	245–335	310–430	465–595	350–490	395–545	435–605	455–625	495–695	630–890
0–10	7	340–430	380–480	270–370	340–270	515–650	390–540	435–600	480–665	500–690	545–765	695–980
0–300	200		390–500		355–490	530–670	405–555	445–615	495–685	510–710	560–790	710–1000
0–12	8		430–550		390–535	580–740	445–610	490–675	545–750	565–780	615–870	780–1100
0–340	225							495–685	550–760	570–790	625–880	785–1105
0–13	9							545–750	605–835	630–870	685–965	860–1215
0–375	250							545–755	610–840	630–870	685–965	865–1215
0–15	10							600–830	670–925	695–960	755–1060	950–1340
0–410	275									690–950	745–1055	940–1320
0–16	11									760–1045	820–1160	1030–1455
0–450	300										815–1145	1015–1435
0–18	12										895–1260	1120–1575

*Smaller closed side settings can be often used depending on application and production requirements.
(*From Metso Minerals brochure.*)

Specifications usually are based on closed settings. The setting is adjustable by raising or lowering the mantle.

The length of the crushing stroke greatly affects the capacity and screen analysis of the crushed product. A very short stroke will give a very evenly crushed product but will not give the greatest capacity. A very long stroke will give the greatest capacity, but the product will contain a wider product size distribution.

Performance Crushing occurs through the full cycle in a gyratory crusher, and this produces a higher crushing capacity than a similar-size jaw crusher, which crushes only in the shutting half of the cycle. Gyratory crushers also tend to be easier to operate. They operate most efficiently when they are fully charged, with the main shaft fully buried in charge. Power consumption for gyratory crushers is also lower than that of jaw crushers. These are preferred over jaw crushers when capacities of 818 MT/h (900 tons/h) or higher are required.

Gyratories make a product with open-side settings of 5 to 10 in at discharge rates from 600 to 6000 tons/h, depending on size. Most manufacturers offer a throw from ¼ to 2 in. The throughput and power draw depend on the throw and the hardness of the ore, and on the amount of undersized material in the feed. Removal of undersized material (which can amount to one-third of the feed) by a stationary grizzly can reduce power draw. See Table 21-16*b*.

Gyratory crushers that feature wide-cone angles are called *cone crushers.* These are suitable for secondary crushing, because crushing of fines requires greater work and creates greater wear; the cone shape provides greater working area than primary or jaw crushers for grinding of the finer product. Crusher performance is harmed by sticky material in the feed, more than 10 percent fines in the feed smaller than the crusher setting, excessive feed moisture, feed size segregation, uneven distribution of feed around the circumference, uneven feed control, insufficient capacity of conveyors and closed-circuit screens, extremely hard or tough feed material, and operation at less than recommended speed. Rod mills are sometimes substituted for crushing of tough ore, since they provide more easily replaceable metal for wear.

Control of Crushers The objective of crusher control is usually to maximize crusher throughput at some specified product size, without overloading the crusher. Usually only three variables can be adjusted: feed rate, crusher opening, and feed size in the case of a secondary crusher. Four modes of control for a crusher are as follows:

1. Setting overload control, where the gap setting is fixed except that it opens when overload occurs. A hardness change during high throughput can cause a power overload on the crusher, which control should protect against.
2. Constant power setting control, which maximizes throughput.

TABLE 21-16*b* Performance of Nordberg Superior MK-II Gyratory Crusher [in mtph (stph)]*

				Open side settings of discharge opening—millimeters (inches)										
Size	Feed opening mm (in.)	Pinion RPM	Max. KW (HP)	125 mm (5.0")	140 mm (5½")	150 mm (6")	165 mm (6½")	175 mm (7")	190 mm (7½")	200 mm (8")	215 mm (8½")	230 mm (9")	240 mm (9½")	250 mm (10")
42–65	1065 (42)	600	375 (500)		1635 (1800)	1880 (2075)	2100 (2315)	2320 (2557)						
50–65	1270 (50)	600	375 (500)			2245 (2475)	2625 (2895)	2760 (3040)						
54–75	1370 (54)	600	450 (600)			2555 (2820)	2855 (3145)	3025 (3335)	3215 (3545)	3385 (3735)				
62–75	1575 (62)	600	450 (600)			2575 (2840)	3080 (3395)	3280 (3615)	3660 (4035)	3720 (4205)				
60–89	1525 (60)	600	600 (800)				4100 (4520)	4360 (4805)	4805 (5295)	5006 (5520)	5280 (5820)	5550 (6115)		
60–110	1525 (60)	514	1000 (1400)					5575 (6150)	5845 (6440)	6080 (6705)	6550 (7220)	6910 (7620)	7235 (7975)	7605 (8385)

**From Metso Minerals brochure.*

3. Pressure control, which provides settings that give maximum crusher force and hence throughput.

4. Feeding rate control, for smooth operation. Setting control influences mainly product size and quality, while feed control determines capacity. Flow must also be synchronized with the feed requirements of downstream processes such as ball mills, and improved crusher efficiency can reduce the load on the more costly downstream grinding.

IMPACT BREAKERS

Impact breakers include heavy-duty hammer crushers, rotor impact breakers, and cage mills. They are generally coarse breakers which reduce the size of materials to about 1 mm. Fine hammer mills are described in a following subsection. Not all rocks shatter well by impact. Impact breaking is best suited for the reduction of relatively nonabrasive and low-silica-content materials such as limestone, dolomite, anhydrite, shale, and cement rock, with the most popular application being on limestone. Most of these devices, such as the hammer crusher shown in Fig. 21-140, have top-fed rotors (of various types) and open bottoms through which product discharge occurs. Some hammer crushers have screens or grates.

Hammer Crusher Pivoted hammers are mounted on a horizontal shaft, and crushing takes place by impact between the hammers and breaker plates. Heavy-duty hammer crushers are frequently used in the quarrying industry, for processing municipal solid waste, and to scrap automobiles.

The rotor of these machines is a cylinder to which is affixed a tough steel bar. Breakage can occur against this bar or on rebound from the walls of the device. Free impact breaking is the principle of the rotor breaker, and it does not rely on pinch crushing or attrition grinding between rotor hammers and breaker plates. The result is a high reduction ratio and elimination of secondary and tertiary crushing stages. By adding a screen on a portable mounting, a complete, compact mobile crushing plant of high capacity and efficiency is provided for use in any location.

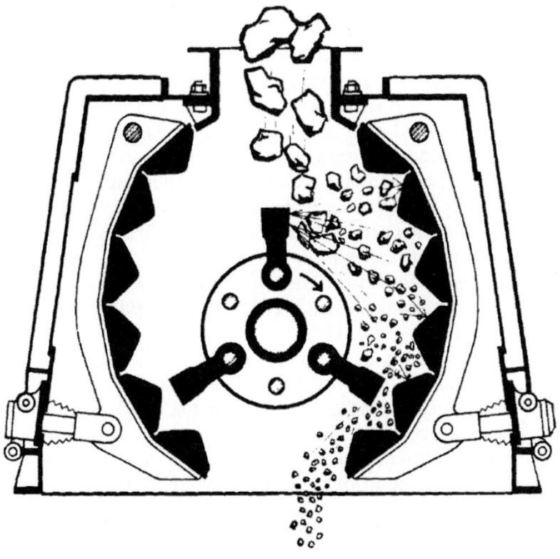

FIG. 21-140 Reversible impactor. (*Pennsylvania Crusher Corp.*)

The ring granulator features a rotor assembly with loose crushing rings, held outwardly by centrifugal force, which chop the feed. It is suitable for highly friable materials that may give excessive fines in an impact mill. For example, bituminous coal is ground to a product below 2 cm (¾ in). They have also been successfully used to grind abrasive quartz to sand size, due to the ease of replacement of the ring impact elements.

Cage Mill In a cage mill, cages of one, two, three, four, six, and eight rows, with bars of special alloy steel, revolving in opposite directions produce a powerful impact action that pulverizes many materials. Cage mills are used for many materials, including quarry rock, phosphate rock, and fertilizer, and for disintegrating clays, colors, press cake, and bones. The advantage of multiple-row cages is the achievement of a greater reduction ratio in a single pass, and these devices can produce products significantly finer than other impactors in many cases, as fine as 325-mesh. These features and the low cost of the mills make them suitable for medium-scale operations where complicated circuits cannot be justified.

Prebreakers Aside from the normal problems of grinding, there are special procedures and equipment for breaking large masses of feed into smaller sizes for further grinding. There is the breaking or shredding of bales, as with rubber, cotton, or hay, in which the compacted mass does not readily come apart. There also is often caking in bags of plastic or hygroscopic materials which were originally fine. Although crushers are sometimes used, the desired size reduction ratio often is not obtainable. Furthermore, a lower capital investment may result through choosing a less rugged device which progressively attacks the large mass to remove only small amounts at a time. Typically, these devices are toothed rotating shafts in casings.

HAMMER MILLS

Operation Hammer mills for fine pulverizing and disintegration are operated at high speeds. The rotor shaft may be vertical or horizontal, but generally is the latter. The shaft carries hammers, sometimes called *beaters.* The hammers may be T-shaped elements, stirrups, bars, or rings fixed or pivoted to the shaft or to disks fixed to the shaft. The grinding action results from *impact* and *attrition* between lumps or particles of the material being ground, the housing, and the grinding elements. A cylindrical screen or grating usually encloses all or part of the rotor. The fineness of product can be regulated by changing the rotor speed, feed rate, or clearance between hammers and grinding plates as well as by changing the number and type of hammers used and the size of discharge openings in the screen.

The *screen* or *grating discharge* for a hammer mill serves as an internal classifier, but its limited area does not permit effective use when small apertures are required. A larger external screen may then be required. The feed must be nonabrasive with a hardness of 1.5 or less. Hammer mills can reduce many materials so that substantially all the product passes a 200-mesh screen.

One of the subtleties of operating a hammer mill is that, in general, screen openings should be sized to be much larger than the desired product size. The screen serves to retain very large particles in the mill, but particles that pass through the screen are usually many times smaller than the screen opening. Thus, changing the screen opening size may strongly affect the coarse end of a product size distribution, but will have limited effect on the median particle size and very little effect at all on the fines. These are most strongly affected by the hammer speed, but the number and type of hammers are also key variables. Screens with very fine openings (500 μm and less) can be used in smaller laboratory mills to produce very fine product, but tend not to be rugged enough for large-scale use. The particle size distribution in hammer mill products tends to be very broad, and in cases where a relatively narrow product size distribution is desired, some sort of grinding circuit with an external classifier is almost always needed.

There are a large number of hammer mill manufacturers. The basic designs are very similar, although there are subtle differences in performance and sturdiness that can lead to varying performance. For example, some machines have lower maximum rotation speeds than others. Less rugged and powerful machines might be fully adequate for vegetable materials (e.g., wood), but not suitable for fine mineral grinding. Occasionally, vendors are particularly experienced in a limited set of products and have designs that are especially suited for these. For relatively common materials, it is usually better to use vendors with practical experience in these materials.

Pin Mills In contrast to peripheral hammers of the rigid or swing types, there is a class of high-speed mills having pin breakers in the grinding circuit. These may be on a rotor with stator pins between circular rows of pins on the rotor disk; or they may be on rotors operating in opposite directions, thereby securing an increased differential speed. There are machines with both vertical and horizontal shafts. In the devices with horizontal shafts, feed is through the top of the mill similar to hammer mills. In devices with vertical shafts, feed is along the shaft, and centrifugal force helps impact the outer ring of pins.

Unlike hammer mills, pin mills do not have screens. Pin mills have a higher energy input per pass than hammer mills and can generally grind softer materials to a finer particle size than hammer mills, while hammer mills perform better on hard or coarse materials. Because pin mills do not have retaining screens, residence time in pin mills is shorter than in hammer mills, and pin mills are therefore more suitable for heat-sensitive materials or cryogrinding.

Universal Mills Several manufacturers are now making "universal mills," which are essentially hammer mill–style devices with fairly narrow chambers that can be either fitted with either a variety of hammer types and screens (although usually only fixed hammers) or set up as a pin mill. These are useful where frequent product changes are made and it is necessary to be able to rapidly change the grind characteristics of the devices, such as small lot manufacturing or grinding research.

Hammer Mills with Internal Air Classifiers A few mills are designed with internal classifiers. These are generally capable of reducing products to particle sizes below 45 μm, with a limit of about 10 μm, depending on the material. A good example of this type of mill is the Hosokawa Mikro-ACM mill, which is a pin mill fitted with an air classifier. There are also devices more like hammer mills, such as the Raymond vertical mill, which do not grind quite as finely as the pin mill–based machine but can handle slightly more abrasive materials.

The *Mikro-ACM pulverizer* is a pin mill with the feed being carried through the rotating pins and recycled through an attached vane classifier. The classifier rotor is separately driven through a speed control which may be adjusted independently of the pin-rotor speed. Oversize particles are carried downward by the internal circulating airstream and are returned to the pin rotor for further reduction. The constant flow of air through the ACM maintains a reasonably low temperature, which makes it ideal for handling heat-sensitive materials, and it is commonly used in the powder coating and pharmaceutical industries for fine grinding.

ROLL CRUSHERS

Once popular for coarse crushing in the minerals industry, these devices long ago lost favor to gyratory and jaw crushers because of their poorer wear characteristics with hard rocks. Roll crushers are still commonly used for grinding of agricultural products such as grains, and for both primary and secondary crushing of coal and other friable rocks such as oil shale and phosphate. They are also commonly used for grinding of superabsorbent polymers and coffee beans due to the ability to produce a very narrow particle size distribution, which is important for product quality. The roll surface is smooth, corrugated, or toothed, depending on the application. *Smooth rolls* tend to wear ring-shaped corrugations that interfere with particle nipping, although some designs provide a mechanism to move one roll from side to side to spread the wear. *Corrugated rolls* give a better bite to the feed, but wear is still a problem. *Toothed rolls* are still practical for rocks of not too high silica content, since the teeth can be regularly resurfaced with hard steel by electric arc welding. Toothed rolls are frequently used for crushing coal and chemicals. For further details, see the 6th Edition of this handbook.

The *capacity* of roll crushers is calculated from the ribbon theory, according to the formula

$$Q = dLs/2.96 \tag{21-142}$$

where Q = capacity, cm^3/min; d = distance between rolls, cm; L = length of rolls, cm; and s = peripheral speed, cm/min. The denominator becomes 1728 in engineering units for Q in cubic feet per minute, d and L in inches, and s in inches per minute. This gives the theoretical capacity and is based on the rolls discharging a continuous, solid uniform ribbon of material. The actual capacity of the crusher depends on the roll diameter, feed irregularities, and hardness and varies between 25 and 75 percent of theoretical capacity.

ROLL PRESS

One of the newer comminution devices, the *roll press,* has achieved significant commercial success, especially in the cement industry. It is used for fine crushing, replacing the function of a coarse ball mill or of tertiary crushers. Unlike ordinary roll crushers, which crush individual particles, the roll press is choke-fed and acts on a thick stream or ribbon of feed. Particles are crushed mostly against other particles, so wear is very low. A roll press can handle a hard rock such as quartz. Energy efficiency is also greater than in ball mills.

The product is in the form of agglomerated slabs. These are broken up in either a ball mill or an impact or hammer mill running at a speed too slow to break individual particles. Some materials may even deagglomerate from the handling that occurs in conveyors. A large proportion of fines is produced, but a fraction of coarse material survives. This makes recycle necessary.

From experiments to grind cement clinker to −80 μm, as compression is increased from 100 to 300 MPa, the required recycle ratio decreases from 4 to 2.8. The energy required per ton of throughput increases from 2.5 to 3.5 kWh/ton. These data are for a 200-mm-diameter pilot-roll press. The status of 150 installations in the cement industry is reviewed [Strasser et al., *Rock Products* **92**(5): 60–72 (1989)]. In cement clinker milling, wear is usually from 0.1 to 0.8 g/ton, and for cement raw materials it is between 0.2 and 1.2 g/ton, whereas it may be 20- to 40-in ball mills.

The size of the largest feed particles should not exceed 0.04 × roll diameter D according to Schoenert ["The Characteristics of Comminution with High Pressure Roller Mills," *KONA Powder and Particle Journal* **9**: 149–158, (1991)]. However, it has been found [Wuestner et al., *Zement-Kalk-Gips* **41**(7): 345–353 (1987); English edition, pp. 207–212] that particles as large as 3 to 4 times the roll gap may be fed to an industrial press.

Machines with up to 2500-kW installed power and 1000 tons/h (900 tons/h) capacity have been installed. The largest presses can supply feed for four or five ball mills. Operating experience (Wuestner et al., loc. cit.) has shown that roll diameters of about 1 m are preferred, as a compromise between production rate and stress on the equipment. The press must be operated choke-fed, with a substantial depth of feed in the hopper; otherwise it will act as an ordinary roll crusher.

ROLL RING-ROLLER MILLS

Roll ring-roller mills (Fig. 21-141) are equipped with rollers that operate against grinding rings. Pressure may be applied with heavy springs or by centrifugal force of the rollers against the ring. Either the ring or the rollers may be stationary. The grinding ring may be in a vertical or horizontal position. Ring-roller mills also are referred to as *ring roll mills* or *roller mills* or *medium-speed mills.* The ball-and-ring and bowl mills are types of ring-roller mill. Ring-roller mills are more energy-efficient than ball mills or hammer mills. The energy to grind coal to 80 percent passing 200-mesh was determined (Luckie and Austin, *Coal Grinding Technology—A Manual for Process Engineers,* Kennedy Van Saun Corporation, Danville, P.A., 1980) as ball mill, 13 hp/ton; hammer mill, 22 hp/ton; roller mill, 9 hp/ton.

Raymond Ring-Roller Mill The Raymond ring-roller mill (Fig. 21-141) is a typical example of a ring-roller mill. The base of the mill carries the grinding ring, rigidly fixed in the base and lying in the horizontal plane. Underneath the grinding ring are tangential air ports through which the air enters the grinding chamber. A vertical shaft driven from below carries the roller journals. Centrifugal force urges the pivoted rollers against the ring. The raw material from the feeder drops between the rolls and ring and is crushed. Both centrifugal air motion and plows move the coarse feed to the nips. The air entrains fines and conveys them up from the grinding zone, providing some classification at this point. An air classifier is also mounted above the grinding zone to return oversize particles. The method of classification used with Raymond mills depends on the fineness desired. If a medium-fine product is required (up to 85 or 90 percent through a no. 100 sieve), a single-cone air classifier is used.

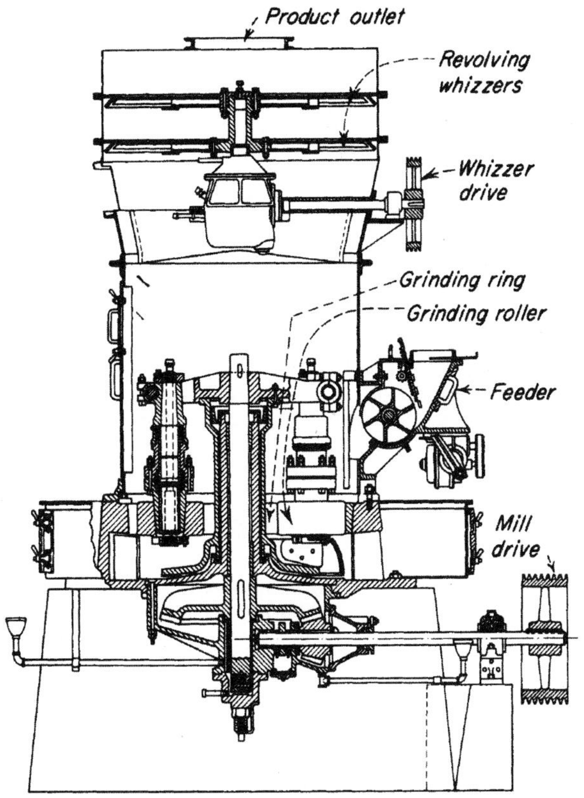

FIG. 21-141 Raymond high-side mill with an internal whizzer classifier. (*ABB Raymond Div., Combustion Engineering Inc.*)

This consists of a housing surrounding the grinding elements with an outlet on top through which the finished product is discharged. This is known as the *low-side mill*. For a finer product and when frequent changes in fineness are required, the whizzer-type classifier is used. This type of mill is known as the *high-side mill*. The Raymond ring-roll mill with internal air classification is used for the large-capacity fine grinding of most of the softer nonmetallic minerals. Materials with a Mohs scale hardness up to and including 5 are handled economically on these units. Typical natural materials handled include barites, bauxite, clay, gypsum, magnesite, phosphate rock, iron oxide pigments, sulfur, talc, graphite, and a host of similar materials. Many of the manufactured pigments and a variety of chemicals are pulverized to high fineness on such units. Included are such materials as calcium phosphates, sodium phosphates, organic insecticides, powdered cornstarch, and many similar materials. When properly operated under suction, these mills are entirely dust-free and automatic.

PAN CRUSHERS

Design and Operation The pan crusher consists of one or more grinding wheels or mullers revolving in a pan; the pan may remain stationary and the mullers be driven, or the pan may be driven while the mullers revolve by friction. The mullers are made of tough alloys such as Ni-Hard. Iron scrapers or plows at a proper angle feed the material under the mullers.

Performance The dry pan is useful for crushing medium-hard and soft materials such as clays, shales, cinders, and soft minerals such as barites. Materials fed should normally be 7.5 cm (3 in) or smaller, and a product able to pass no. 4 to no. 16 sieves can be delivered, depending on the hardness of the material. High reduction ratios with low power and maintenance are features of pan crushers. Production rates can range from 1 to 54 MT/h (1 to 60 tons/h) according to pan size and hardness of material as well as fineness of feed and product.

The *wet pan* is used for developing plasticity or molding qualities in ceramic feed materials. The abrasive and kneading actions of the mullers blend finer particles with the coarser particles as they are crushed [Greaves-Walker, *Am. Refract. Inst. Tech. Bull.* **64** (1937)], which is necessary to achieve a high packing density and an increase in strength of the final ceramic product.

CRUSHING AND GRINDING EQUIPMENT: FLUID-ENERGY OR JET MILLS

DESIGN

Jet milling, also called *fluid-energy grinding*, is an increasingly used process in the chemical industry for processing brittle, heat-sensitive materials into very fine powders with a narrow size distribution. For more than 90 years, jet mills have been built and applied successfully on a semilarge scale in the chemical industry. A number of famous designs are extensively described in a number of patents and publications.

Most such mills are variations on one of the fundamental configurations depicted in Fig. 21-142. The designs differ from one another by the arrangement of the nozzles and the classification section. In the following paragraphs the jet mill types are briefly discussed.

The key feature of jet mills is the conversion of high pressure to kinetic energy. The operating fluid enters the grinding chamber through nozzles placed in the wall. The feed particles brought into the mill through a separate inlet are entrained by expanding jets and accelerated to velocities as high as the velocity of sound. In fact, three collision geometries can be distinguished: (1) interparticle collisions due to turbulence in a free jet; (2) collisions between particles accelerated by opposed jets; and (3) impact of particles on a target. The turbulent nature of the jets causes particles to have differences in velocity and direction. Particle breakage in jet mills is mainly a result of interparticle collisions: wall collisions are generally thought to be of minor importance only, except in mill type D (Fig. 21-142). Fluid-energy-driven mills are a class of impact mills with a considerable degree of attrition due to eccentric and gliding interparticle impacts. The grinding mechanism via mutual collisions means that jet mills operate with virtually no product contact. In other words, the contamination grade is low.

The classification of product leaving the mill depends on a balance between centrifugal forces and drag forces in the flow field around the mill outlet. Mill types A and C create a free vortex at the outlet, while jet mill D makes use of gravity. Type B has an integrated rotor. The final product quality is largely determined by the success of classification.

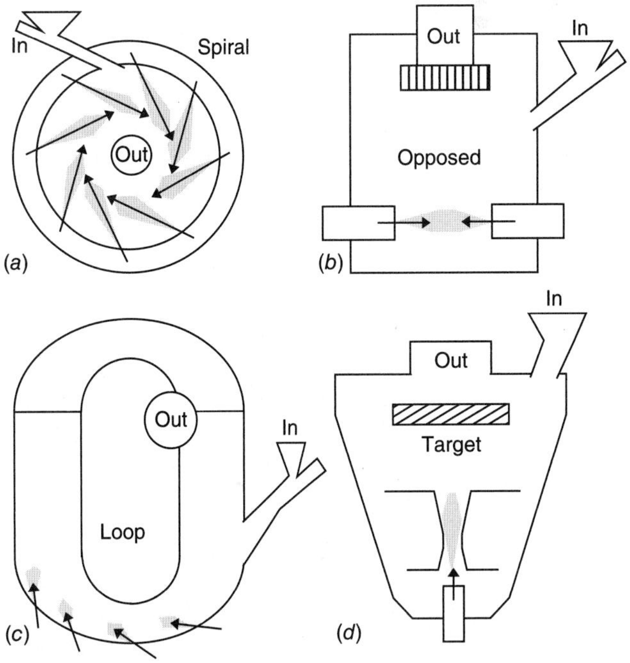

FIG. 21-142 Schematic representation of basic jet mill designs: (*a*) spiral; (*b*) opposed; (*c*) loop; (*d*) target.

TYPES

Spiral Jet Mill The original design of the spiral jet mill, also called a *pancake mill*, is shown in Fig. 21-142. This design was first described by Andrews in 1936 and patented under the name Micronizer. A number of nozzles are placed in the outer wall of the mill through which the grinding medium, a gas or steam, enters the mill.

A spiral jet mill combines both grinding and classification by the same jets. The vortex causes coarse particles of the mill contents to be transferred to the outer zone, as fines can leave through the central outlet. The solid feed is brought into the mill by an air pusher. The outlet is placed in the center of the mill chamber. The working principle of this mill was extensively investigated by Rumpf.

Spiral jet mills are notable for their robust design and compactness. Their direct air operation avoids the need for separate drive units. Another significant argument for the use of jet mills is the lower risk of dust explosions.

Opposed Jet Mill Opposed jet mills are fluid-energy-driven mills that contain two or more jets aligned toward each other (see Fig. 21-142*b*). Different versions are on the market, based on a design patented by Willoughby (1917). In this type of jet mill, opposed gas streams entrain the mill holdup. At the intersection of the jets, the coarse particles hit one another. The grinding air carries the particles upward in a kind of fluidized bed to the classification zone.

Adjustment of the rotor speed allows a direct control of the particle size of the end product. The feed is introduced to the mill by a rotary valve. Drawbacks are the higher cost of investment and maintenance. These types of mills are described by Vogel and Nied.

Other Jet Mill Designs Figure 21-142*d* shows one of the earliest jet mill designs (around 1880), but it is still in use today. In this mill a jet loaded with particles is impacting on an anvil. Consequently, the impact efficiency is high for relatively large particles. Very fine grinding becomes difficult as small particles are decelerated in the stagnant zone in front of the target. Fines are dragged out in an airstream by a fan, as coarse material is recirculated to the jet entry. Points of improvement have included better classification and abrasive-resistant target material. This device is suitable to incorporate as a pregrinder.

The loop mill (Fig. 21-142*c*), also called *Torus mill*, was designed by Kidwell and Stephanoff (1940). The grinding fluid is brought into the grinding section. The fines leave the mill through the classification section.

CRUSHING AND GRINDING EQUIPMENT: WET/DRY GRINDING—MEDIA MILLS

OVERVIEW

Another class of grinding mills is media mills. These mills grind materials primarily through the action of mechanically agitated balls made out of metals (mostly steel) or various ceramics. Different mills use different methods of agitation. Some are more commonly used for dry grinding, others for wet grinding, and still others for both. Types of media mills include tumbling mills, stirred media mills, and vibratory mills.

MEDIA SELECTION

A key to the performance of media mills is the selection of an appropriate grinding medium. Jörg Schwedes and his students have developed correlations which are effective in determining optimal media size for stirred media mills [Kwade et al., *Powder Technol.* **86** (1996); and Becker et al., *Int. J. Miner. Process.* **61** (2001)]. Although these correlations were developed for stirred media mills, the principles developed apply to all media mills.

In this methodology, energy input is broken up into *stress intensity* (SI) and *stress frequency* (SF), defined as

$$SI = (\rho_m - \rho)D_m^3V_t^2$$

$$SF = \omega(D_m/D)^2t$$

where ρ is slurry density, ρ_m is media density, D_m is media diameter, ω is the rotational speed of a rotating mill, D is the rotor diameter of a rotating mill, and V_t is the tip speed of a rotating mill.

Stress intensity is related to the kinetic energy of media beads, and stress frequency is related to the frequency of collisions.

When stress intensity is plotted versus media particle size achieved at constant grinding energy (such as in Fig. 21-143) for limestone, it can be seen that a large number of experimental data can be collapsed onto a single curve. There is a relatively narrow range of stress intensity which gives the smallest particle size, and larger or smaller stress intensities give increasingly larger particle sizes at the same energy input.

This can be explained in physical terms in the following way. For each material, there is a critical stress intensity. If the stress intensity applied during grinding is less than the critical stress intensity, then very little grinding occurs. If the applied stress intensity is much greater than the critical stress intensity, then unnecessary energy is being used in bead collisions, and a greater grinding rate could be obtained by using smaller beads that would collide more frequently. This has a very practical implication for choosing the size and, to some extent, the density of grinding beads. At a constant stirring rate (or tumbling rate or vibration rate), a small range of media sizes give an optimal grinding rate for a given material in a given mill. In practice, most mills are operated using media slightly larger than the optimal size, as changes in feed and media quality can shift the value of the critical stress

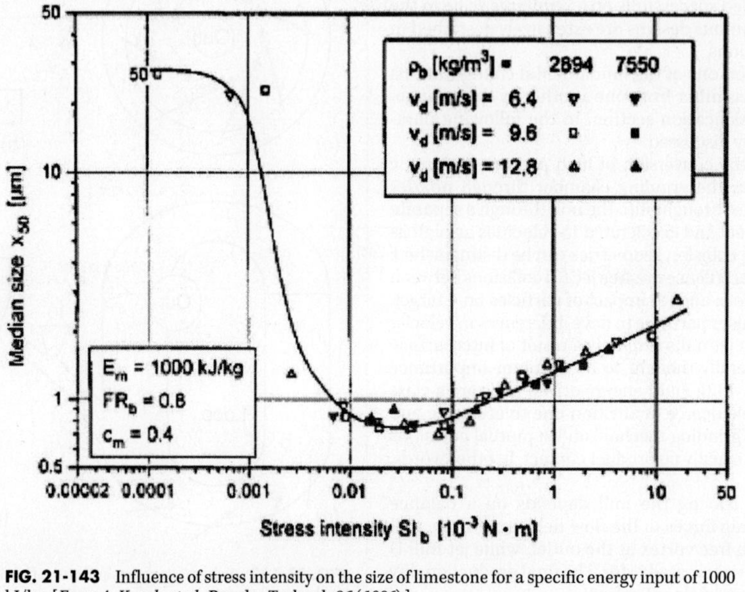

FIG. 21-143 Influence of stress intensity on the size of limestone for a specific energy input of 1000 kJ/kg. [*From A. Kwade et al.,* Powder Technol. ***86*** *(1996).*]

intensity over the lifetime of an industrial process, and the falloff in grinding rate when one is below the critical stress intensity is quite dramatic.

Another important factor when choosing media mills is media and mill wear. Most media mills have fairly rapid rates of media wear, and it is not uncommon to have to replace media monthly or at least add partial loads of media weekly. Media wear will reduce the grind rate of a mill and can cause significant product contamination. Very hard media materials often have low wear rates, but can cause very rapid mill wear. Media with a good balance of properties tend to be specialty ceramics. Commonly used ceramics include glass, specialty sand, alumina, zirconia (although this is higher in mill wear), zirconia-silica composites, and yttria- or ceria-stabilized zirconia. Yttria-stabilized zirconia is particularly wear-resistant but is very expensive. Steel is often used as a medium and has a very good combination of low cost, good wear life, and gentle mill wear if a product can handle slight discoloration and iron content from the medium.

TUMBLING MILLS

Ball, pebble, rod, tube, and compartment mills have a cylindrical or conical shell, rotating on a horizontal axis, and are charged with a grinding medium such as balls of steel, flint, or porcelain or with steel rods. The *ball mill* differs from the tube mill by being short in length; its length, as a rule, is not much different from its diameter (Fig. 21-144). Feed to ball mills can be as large as 2.5 to 4 cm (1 to 1½ in) for very fragile materials, although the top size is generally 1 cm (½ in). Most ball mills operate with a reduction ratio of 20:1 to 200:1. The largest balls are typically 13 cm (5 in) in diameter. The tube mill is generally long in comparison with its diameter, uses smaller balls, and produces a finer product. The compartment mill consists of a cylinder divided into two or more sections by perforated partitions; preliminary grinding takes place at one end and finish grinding at the charge end. These mills have a length-to-diameter ratio in excess of 2 and operate with a reduction ratio of up to 600:1.

Rod mills deliver a more uniform granular product than other revolving mills while minimizing the percentage of fines, which are sometimes detrimental. The *pebble mill* is a tube mill with flint or ceramic pebbles as the grinding medium and may be lined with ceramic or other nonmetallic liners. The *rock-pebble mill* is an autogenous mill in which the medium consists of larger lumps scalped from a preceding step in the grinding flow sheet.

Design The conventional type of *batch mill* consists of a cylindrical steel shell with flat steel-flanged heads. Mill length is equal to or less than the diameter [Coghill, De Vaney, and O'Meara, *Trans. Am. Inst. Min. Metall. Pet. Eng.* **112:** 79 (1934)]. The discharge opening is often opposite the loading manhole and for wet grinding usually is fitted with a valve. One or more vents are provided to release any pressure developed in the mill, to introduce inert gas, or to supply pressure to assist discharge of the mill. In dry grinding, the material is discharged into a hood through a grate over the manhole while the mill rotates. Jackets can be provided for heating and cooling.

Material is fed and discharged through hollow trunnions at opposite ends of *continuous mills.* A grate or diaphragm just inside the discharge end may be employed to regulate the slurry level in wet grinding and thus control retention time. In the case of *air-swept mills,* provision is made for blowing air in at one end and removing the ground material in air suspension at the same end or the other end. Ball mills usually have *liners* which are replaceable when they wear. Both all-rubber liners and rubber liners with metal lifter bars are currently used in large ball mills [McTavish, *Mining Engg.* **42:** 1249–1251 (Nov. 1990)]. Lifters must be at least as high as the ball radius, to key the ball charge and ensure that the balls fall into the toe area of the mill [Powell, *Int. J. Mineral Process.* **31:** 163–193 (1991)]. Special operating problems occur with smooth-lined mills owing to erratic slip of the charge against the wall. At low speeds the charge may *surge* from side to side without actually tumbling; at higher speeds tumbling with *oscillation* occurs. The use of lifters prevents this [Rose, *Proc. Inst. Mech. Eng.* (*London*) **170**(23): 773–780 (1956)].

Pebble mills are frequently lined with nonmetallic materials when iron contamination would harm a product such as a white pigment or cement. Belgian silex (silica) and porcelain block are popular linings. Silica linings and ball media have proved to wear better than other nonmetallic materials. Smaller mills, up to about 50-gal capacity, are made in one piece of ceramic with a cover.

Multicompartment Mills Multicompartment mills feature grinding of coarse feed to finished product in a single operation, wet or dry. The primary grinding compartment carries large grinding balls or rods; one or more secondary compartments carry smaller media for finer grinding.

Operation cascading and *cataracting* are the terms applied to the motion of grinding media. The former applies to the rolling of balls or pebbles from top to bottom of the heap, and the latter refers to the throwing of the balls through the air to the toe of the heap. The criterion by which the ball action in mills of various sizes may be compared is the concept of *critical speed.* It is the theoretical speed at which the centrifugal force on a ball in contact with the mill shell at the height of its path equals the force on it due to gravity:

$$N_c = 42.3/\sqrt{D} \qquad (21\text{-}143)$$

where N_c is the critical speed, r/min, and D is diameter of the mill, m (ft), for a ball diameter that is small with respect to the mill diameter. The numerator becomes 76.6 when D is expressed in feet. *Actual mill speeds* range

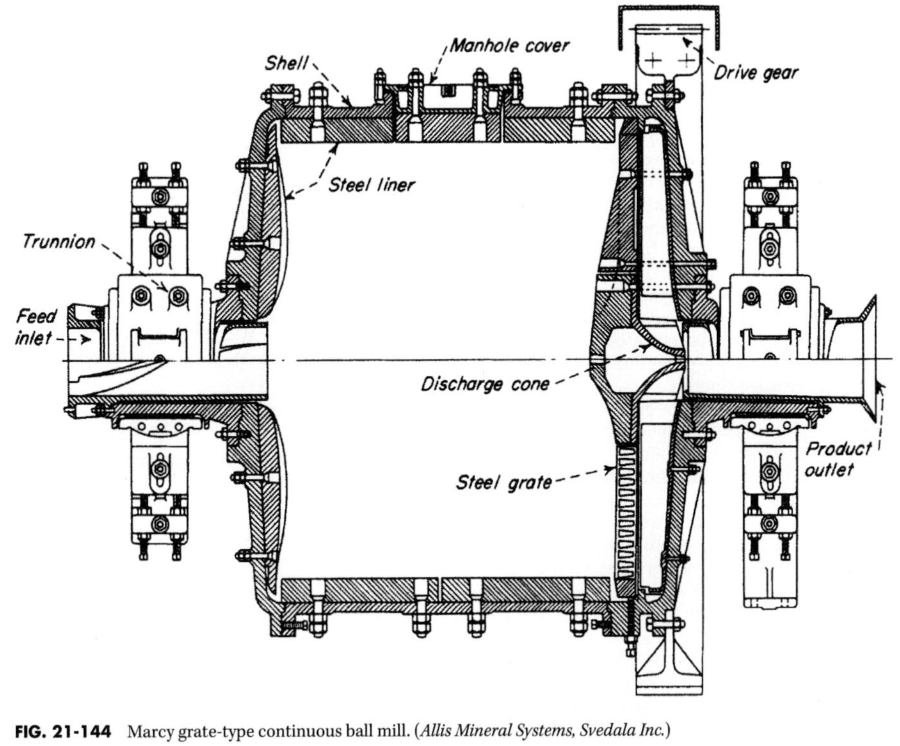

FIG. 21-144 Marcy grate-type continuous ball mill. (*Allis Mineral Systems, Svedala Inc.*)

from 65 to 80 percent of critical. It might be generalized that 65 to 70 percent is required for fine wet grinding in viscous suspension and 70 to 75 percent for fine wet grinding in low-viscosity suspension and for dry grinding of large particles up to 1-cm (½-in) size. Unbaffled mills can run at 105 percent of critical to compensate for slip. The chief factors determining the size of *grinding balls* are fineness of the material being ground and maintenance cost for the ball charge. A coarse feed requires a larger ball than a fine feed. The need for a calculated ball size feed distribution is open to question; however, methods have been proposed for calculating a rationed ball charge [Bond, *Trans. Am. Inst. Min. Metall. Pet. Eng.* **153:** 373 (1943)]. The recommended optimum size of makeup rods and balls is [Bond, *Min. Eng.* **10:** 592–595 (1958)]

$$D_b = \sqrt{\frac{X_p E_i}{K n_r}} \sqrt{\frac{\rho_s}{\sqrt{D}}} \quad (21\text{-}144)$$

where D_b = rod or ball diameter, cm (in); D = mill diameter, m (ft); E_i is the work index of the feed; n_r is speed, percent of critical; ρ_s is feed specific gravity; and K is a constant equal to 214 for rods and 143 for balls. The constant K becomes 300 for rods and 200 for balls when D_b and D are expressed in inches and feet, respectively. This formula gives reasonable results for production-sized mills but not for laboratory mills. The ratio between the recommended ball and rod sizes is 1.23.

Material and Ball Charges The load of a grinding medium can be expressed in terms of the percentage of the volume of the mill that it occupies; i.e., a bulk volume of balls half filling a mill is a 50 percent ball charge. The void space in a static bulk volume of balls is approximately 41 percent. The amount of material in a mill can be expressed conveniently as the ratio of its volume to that of the voids in the ball load. This is known as the *material-to-void ratio.* If the solid material and its suspending medium (water, air, etc.) just fill the ball voids, the *M/V* ratio is 1, for example. Grinding-media loads vary from 20 to 50 percent in practice, and *M/V* ratios are usually near 1.

The material charge of continuous mills, called the *holdup,* cannot be set directly. It is indirectly determined by operating conditions. There is a maximum throughput rate that depends on the shape of the mill, the flow characteristics of the feed, the speed of the mill, and the type of feed and discharge arrangement. Above this rate the holdup increases unstably. The holdup of material in a continuous mill determines the mean residence time and thus the extent of grinding. Gupta et al. [*Int. J. Mineral Process.* **8:** 345–358 (Oct. 1981)] analyzed published experimental data on a 40×40-cm grate discharge laboratory mill and determined that holdup was represented by H_w = (4.020 − 0.176WI)× F_w + (0.040 + 0.01237WI)S_w − (4.970 + 0.395WI), where WI is *Bond work index* based on 100 percent passing a 200-mesh sieve, F_w is the solids feed rate, kg/min, and S_w is weight percent of solids in the feed. This represents experimental data for limestone, feldspar, sulfide ore, and quartz. The influence of WI is believed to be due to its effect on the number of fines present in the mill. Parameters that did not affect H_w are the specific gravity of feed material and feed size over the narrow range studied. Sufficient data were not available to develop a correlation for overflow mills, but the data indicated a linear variation of H_w with F as well. The mean residence time τ (defined as H_w/F) is the most important parameter since it determines the time over which particles are exposed to grinding. Measurements of the water (as opposed to the ore) of several industrial mills (Weller, *Automation in Mining Mineral and Metal Processing,* 3d IFAC Symposium, 1980, pp. 303–309) showed that the maximum mill filling was about 40 percent, and the maximum flow velocity through the mill was 40 m/h. Swaroop et al. [*Powder Technol.* **28:** 253–260 (Mar.–Apr. 1981)] found that the material holdup is higher and the vessel dispersion number $D\tau/L^2$ (see the subsection Continuous-Mill Simulation) is lower in the rod mill than in the ball mill under identical dimensionless conditions. This indicates that the known narrow-product-size distribution from rod mills is partly due to less mixing in the rod mill, in addition to different breakage kinetics.

The holdup in grate-discharge mills depends on the grate openings. Kraft et al. [*Zement-Kalk-Gips Int.* **42**(7): 353–359 (1989); English edition, 237–239] measured the effect of various hole designs in wet milling. They found that slots tangential to the circumference gave higher throughput and therefore lower holdup in the mill. Total hole area had little effect until the feed rate was raised to a critical value (30 m/h in a mill with 0.26-m diameter and 0.6 m long); above this rate the larger area led to lower holdup. The open area is normally specified between 3 and 15 percent, depending on the number of grinding chambers and other conditions. The slots should be 1.5 to 16 mm wide, tapered toward the discharge side by a factor of 1.5 to 2 to prevent blockage by particles.

Dry versus Wet Grinding The choice between wet and dry grinding is generally dictated by the end use of the product. If the presence of liquid with the finished product is not objectionable or the feed is moist or wet, generally wet grinding is preferable to dry grinding, but power consumption, liner wear, and capital costs determine the choice. Other factors that influence the choice are the performance of subsequent dry or wet classification steps, the cost of drying, and the capability of subsequent processing steps for handling a wet product. The net production in wet grinding in the Bond grindability test varies from 145 to 200 percent of that in dry grinding depending on the mesh [Maxson, Cadena, and Bond, *Trans. Am. Int. Min. Metall. Pet. Eng.* **112:** 130–145, 161 (1934)]. Ball mills have a large field of application for wet grinding in closed circuit with size classifiers, which also perform advantageously when wet.

Dry Ball Milling In fine dry grinding, surface forces come into action, causing cushioning and ball coating, resulting in a less efficient use of energy. Grinding media and liner-wear consumption per ton of ground product are lower for a dry grinding system. However, power consumption for dry grinding is about 30 percent larger than for wet grinding. Dry grinding requires the use of dust-collecting equipment.

Wet Ball Milling See Fig. 21-145. The *rheological properties* of the slurry affect the grinding behavior in ball mills. Rheology depends on solids content, particle size, and mineral chemical properties [Kawatra and Eisele, *Int. J. Mineral Process* **22:** 251–259 (1988)]. Above 50 vol% solids, a mineral slurry may become pseudoplastic, i.e., it exhibits a yield value (Austin, Klimpel, and Luckie, *Process Engineering of Size Reduction: Ball Milling,* AIME, 1984). Above the yield value, the grinding rate decreases, and this is believed to be due to adhesion of grinding media to the mill wall, causing centrifuging [Tangsatitkulchai and Austin, *Powder Technol.* **59**(4): 285–293 (1989)]. Maximum power draw and fines production is achieved when the solids content is just below that which produces the critical yield. The solids concentration in a pebble-mill slurry should be high enough to give a slurry viscosity of at least 0.2 Pa · s (200 cP) for best grinding efficiency [Creyke and Webb, *Trans. Br. Ceram. Soc.* **40:** 55 (1941)], but this may have been required to key the charge to the walls of the smooth mill used.

Since viscosity increases with number of fines present, mill performance can often be improved by closed-circuit operation to remove fines. Chemicals such as surfactants allow the solids content to be increased without increasing the yield value of the pseudoplastic slurry, allowing a higher throughput. They may cause foaming problems downstream, however. Increasing temperature lowers the viscosity of water, which controls the viscosity of the slurry under high-shear conditions such as those encountered in the cyclone, but does not greatly affect chemical forces. Slurry viscosity can be most directly controlled by controlling the solids content.

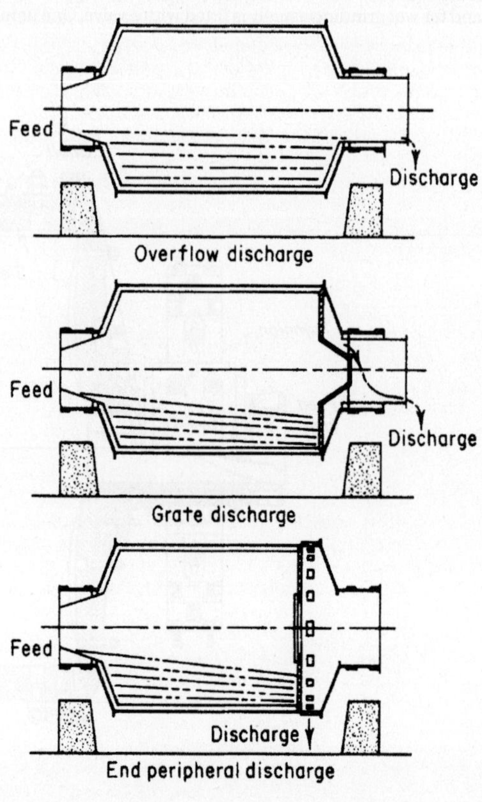

FIG. 21-145 Continuous ball-mill discharge arrangements for wet grinding.

MILL EFFICIENCIES

In summary, controlling factors for cylindrical mills are as follows:

1. Mill speed affects capacity, as well as liner and ball wear, in direct proportion up to 65 to 75 percent of critical speed.
2. Ball charge equal to 35 to 50 percent of the mill volume gives the maximum capacity.
3. Minimum-size balls capable of grinding the feed give maximum efficiency.
4. Bar-type lifters are essential for smooth operation.
5. Material filling equal to ball-void volume is optimum.
6. Higher-circulating loads tend to increase production and decrease the amount of unwanted fine material.
7. Low-level or grate discharge with recycle from a classifier increases grinding capacity over the center or overflow discharge; but liner, grate, and media wear is higher.
8. Ratio of solids to liquids in the mill must be considered on the basis of slurry rheology.

Capacity and Power Consumption One of the methods of mill sizing is based on the observation that the amount of grinding depends on the amount of energy expended, if one assumes comparable good practice of operation in each case. The energy applied to a ball mill is primarily determined by the size of mill and load of balls. Theoretical considerations show the net power to drive a ball mill to be proportional to $D^{2.5}$, but this exponent may be used without modification in comparing two mills only when operating conditions are identical [Gow et al., *Trans. Am. Inst. Min. Metall. Pet. Eng.* **112:** 24 (1934)]. The net power (the gross power draw of the mill minus the power to turn an empty mill) to drive a ball mill was found to be

$$E = [(1.64L - 1)K + 1][(1.64D)^{2.5}E_2] \qquad (21\text{-}145)$$

where L is the inside length of the mill, m (ft); D is the mean inside diameter of the mill, m (ft); E_2 is the net power used by a 0.6- by 0.6-m (2- by 2-ft) laboratory mill under similar operating conditions; and K is 0.9 for mills less than 1.5 m (5 ft) long and 0.85 for mills over 1.5 m long. This formula may be used to scale up pilot milling experiments in which the diameter and length of the mill are changed, but the size of balls and the ball loading as a fraction of mill volume are unchanged. More accurate computer models are now available.

Morrell [*Trans. Instn. Min. Metall.*, Sec. C, **101:** 25–32 (1992)] established equations to predict power draft based on a model of the shape of the rotating ball mass. Photographic observations from laboratory and plant-sized mills, including autogenous, semiautogenous, and ball mills, showed that the shape of the material charge could roughly be represented by angles that gave the position of the toe and shoulder of the charge. The power is determined by the angular speed and the torque to lift the balls. The resulting equations show that power increases rapidly with mill filling up to 35 percent, then varies little between 35 and 50 percent. Also net power is related to mill diameter to an exponent less than 2.5. This agrees with Bond [*Brit. Chem. Engr.* pp. 378–385 (1960)] who stated from plant experience that power increases with diameter to the 2.3 exponent or more for larger mills. Power input increases faster than volume, which varies with diameter squared. The equations can be used to estimate holdup for control of autogenous mills.

STIRRED MEDIA MILLS

Stirred media mills have a wide range of applications. They are often found in minerals processing grinding circuits for grinding in the size range of 5 to 50 μm, and they are the only mill capable of reliably grinding materials to submicrometer sizes. They are very commonly used for grinding and dispersion of dyes, clays, and pigments and are also used for biological cell disruption.

Stirred media mills are also the dominant process equipment used for dispersing fine powders into liquid, e.g., pigment dispersions, and have largely displaced ball mills in these applications. In these applications, they are capable of dispersing powders down to particle sizes below 100 nm effectively and reliably.

Stirred media mills are used almost exclusively for wet grinding. In general, the higher the tip speed of the rotor, the lower the viscosity that can be tolerated by the mill. At high viscosity, very little bead motion occurs. Similarly, mills with lower tip speeds can tolerate the use of larger, heavier media, since gravity will cause additional motion in this case.

Design In stirred mills, a central paddle wheel or disced armature stirs the media at speeds from 100 to 3000 r/min (for some lab units). Stirrer tip speeds vary from 2 m/s for some attritors to 18 m/s for some high-energy mills.

Attritors In the Attritor (Union Process Inc.) a single vertical armature rotates several long radial arms. The rotation speeds are much slower than with other stirred media mills, and the grinding behavior in these mills tends to be more like that in tumbling mills than in other stirred media mills. They can be used for higher-viscosity applications. These are available in batch, continuous, and circulation types.

Vertical Mills Vertical mills are, generally speaking, older designs whose chief advantage is that they are inexpensive. They are vertical chambers of various shapes with a central agitator shaft. The media are stirred by discs or pegs mounted on the shaft. Some mills are open at the top, while others are closed at the top. Most mills have a screen at the top to retain media in the mill.

The big drawback to vertical mills is that they have a limited flow rate range due to the need to have a flow rate high enough to help fluidize the media and low enough to avoid carrying media out of the top of the mill. The higher the viscosity of the slurry in the mill, the more difficult it is to find the optimal flow rate range. Slurries that change viscosity greatly during grinding, such as some high-solid slurries, can be particularly challenging to grind in vertical mills.

Horizontal Media Mills Horizontal media mills are the most common style of mill and are manufactured by a large number of companies. Figure 21-146 illustrates the Drais continuous stirred media mill. The mill has a horizontal chamber with a central shaft. The media are stirred by discs or pegs mounted on the shaft. The advantage of horizontal machines is the elimination of gravity segregation of the feed. The feed slurry is pumped in at one end and discharged at the other where the media are retained by a screen or an array of closely spaced, flat discs. Most are useful for slurries up to about 50 Pa · s (50,000 cP). Also note that slurries with very low viscosities (under 1 Pa · s) can sometimes cause severe mill wear problems. Several manufacturers have mill designs where either the screen rotates or the mill outlet is designed in such a way as to use centrifugal force to keep media off the screen. These mills can use media as fine as 0.1 mm. They also have the highest flow rate capabilities. Hydrodynamically shaped screen cartridges can sometimes accommodate media as fine as 0.1 mm.

Agitator discs are available in several forms: smooth, perforated, eccentric, and pinned. The effect of disc design has received limited study, but pinned discs are usually reserved for highly viscous materials. Cooling water is circulated through a jacket and sometimes through the central shaft. The working speed of disc tips ranges from 5 to 18 m/s regardless of mill size. A series of mills may be used with decreasing media size and increasing rotary speed to achieve desired fine particle size.

Annular Gap Mills Some mills are designed with a large interior rotor that has a narrow gap between the rotor and the inner chamber wall. These annular gap mills generally have higher energy input per unit volume than do the other designs. Media wear tends to be correspondingly higher as well. Despite this, these mills can be recommended for heat-sensitive slurries, because the annular design of the mills allows for a very large heat-transfer surface.

Manufacturers There are many manufacturers of stirred media mills worldwide. Major manufacturers of stirred media mills include Netzsch, Buhler, Drais (now part of Buhler), Premier (now part of SPX), Union Process, and Morehouse Cowles. Many of these manufacturers have devices specifically adapted for specific industries. For example, Buhler has some mills specifically designed to handle higher-viscosity inks, and Premier has a mill designed specifically for milling/flaking of metal powders.

PERFORMANCE OF BEAD MILLS

Variables affecting the milling process are listed below:

- Agitator speed
- Feed rate
- Size of beads
- Bead charge, percent of mill volume
- Feed concentration

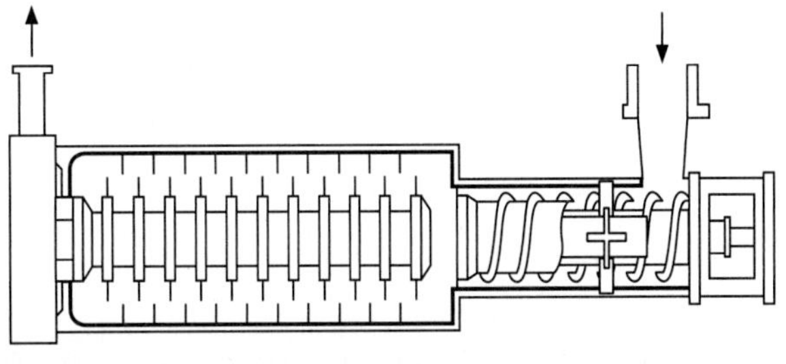

FIG. 21-146 Drais wet grinding and dispersing system (U.S. patent 3,957,210) Draiswerke GmbH. [*Stehr*, International J. Mineral Processing, ***22**(1–4): 431–444 (1988).*]

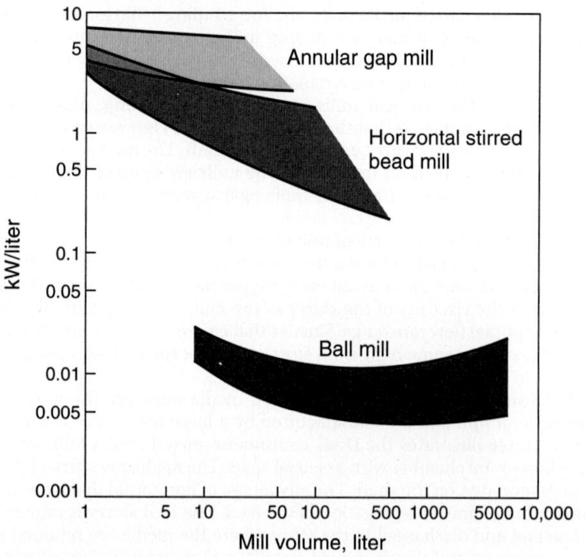

FIG. 21-147 Specific power of bead and ball mills. [*Kolb*, Ceramic Forum International *70*(*5*)*: 212–216* (*1993*).]

Density of beads
Temperature
Design of blades
Shape of mill chamber
Residence time

The availability of more powerful, continuous machines has extended the possible applications to both lower and higher size ranges, from 5- to 200-μm product size, and to a feed size as large as 5 mm. The energy density may be 50 times larger than that in tumbling-ball mills, so that a smaller mill is required (Fig. 21-147). Mills range in size from 1 to 1000 L, with installed power up to 320 kW. Specific power ranges from 10 to 200 or even 2000 kWh/ton, with feed rates usually less than 1 ton. For stirred media mills, an optimum media size is about 20 times greater than the material to be ground. It is possible to relate the Reynolds number to mill power draw in the same way that this is done for rotating mixers (see Fig. 21-148).

In vertical disc-stirred mills, the media should be in a fluidized condition (White, *Media Milling,* Premier Mill Co., 1991). Particles can pack in the bottom if there is not enough stirring action or feed flow, or in the top if flow is too high. These conditions are usually detected by experiment. A study of bead milling [Gao and Forssberg, *Int. J. Mineral Process.* **32**(1–2): 45–59 (1993)] was done in a continuous Drais mill of 6-L capacity having seven 120 × 10-mm horizontal discs. Twenty-seven tests were done with variables at three levels. Dolomite was fed with 2 m^2/g surface area in a slurry ranging from 65 to 75 percent solids by weight, or 39.5 to 51.3 percent by volume. Surface area produced was found to increase linearly with grinding time or specific energy consumption. The variables studied strongly affected the milling rate; two extremes differed by a factor of 10. An optimum bead density for this feed material was 3.7. Evidently the discs of the chosen design could not effectively stir the denser beads. Higher slurry concentration above 70 wt% solids reduced the surface production per unit energy. The power input increased more than proportionally to speed.

Residence Time Distribution Commercially available bead mills have a diameter-to-length ratio ranging from 1:2.5 to 1:3.5. The ratio is expected to affect the *residence time distribution* (RTD). A wide distribution results in overgrinding some feed and undergrinding others. Data from Kula and Schuette [*Biotechnol. Progress* **3**(1): 31–42 (1987)] show that in a Netzsch LME20 mill, the RTD extends from 0.2 to 2.5 times the nominal time, indicating extensive stirring. The RTD is even more important when the objective is to reduce the top size of the product as Stadler et al. [*Chemie-Ingenieur-Technik* **62**(11): 907–915 (1990)] showed, because much of the feed received less than one-half the nominal residence time. A narrow RTD could be achieved by rapidly flowing material through the mill for as many as 10 passes.

VIBRATORY MILLS

The dominant form of industrial vibratory mill is the type with two horizontal tubes, called the *horizontal tube mill.* These tubes are mounted on springs and given a circular vibration by rotation of a counterweight. Many feed flow arrangements are possible, adapting to various applications. Variations include polymer lining to prevent iron contamination, blending of several components, and milling under inert gas and at high and low temperatures.

The vertical vibratory mill has good wear values and a low noise output. It has an unfavorable residence time distribution, since in continuous operation it behaves as a well-stirred vessel. Tube mills are better for continuous operation. The mill volume of the vertical mill cannot be arbitrarily scaled up because the static load of the upper media, especially with steel beads, prevents thorough energy introduction into the lower layers. Larger throughputs can therefore be obtained only by using more mill troughs, as in tube mills. The primary applications of vibratory mills are in fine milling of medium to hard minerals primarily in dry form, producing particle sizes of 1 μm and finer. Throughputs are typically 10 to 20 tons/h. Grinding increases with residence time, active mill volume, energy density and vibration frequency, and media filling and feed charge.

The amount of energy that can be applied limits the tube size to 600 mm, although one design reaches 1000 mm. Larger vibratory amplitudes are more favorable for comminution than higher frequency. The development of larger vibratory mills is unlikely in the near future because of excitation problems. This has led to the use of mills with as many as six grinding tubes.

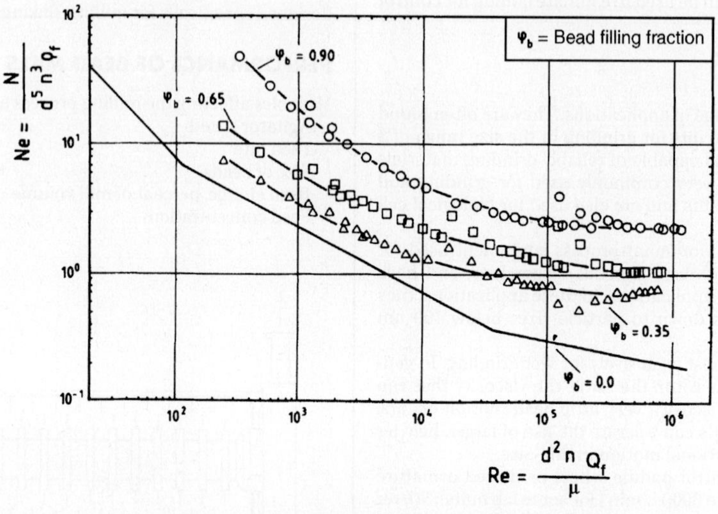

FIG. 21-148 Newton number as a function of Reynolds number for a horizontal stirred bead mill, with fluid alone and with various filling fractions of 1-mm glass beads. [*Weit and Schwedes,* Chemical Engineering and Technology *10*(*6*)*: 398–404* (*1987*).] (N = power input, W; d = stirrer disk diameter, n = stirring speed, 1/s; μ = liquid viscosity, Pa · s; Q_f = feed rate, m^3/s.)

Performance The grinding media diameter should preferably be 10 times that of the feed and should not exceed 100 times the feed diameter. To obtain improved efficiency when reducing size by several orders of magnitude, several stages should be used with different media diameters. As fine grinding proceeds, rheological factors alter the charge ratio, and power requirements may increase. Size availability varies, ranging from 1.3 cm (½ in) down to 325 mesh (44 μm).

Advantages of vibratory mills are (1) simple construction and low capital cost, (2) very fine product size attainable with large reduction ratio in a single pass, (3) good adaptation to many uses, (4) small space and weight requirements, and (5) ease and low cost of maintenance. Disadvantages are (1) limited mill size and throughput, (2) vibration of the support and foundation, and (3) high noise output, especially when run dry. The vibratory-tube mill is also suited to wet milling. In fine wet milling, this narrow residence time distribution lends itself to a simple open circuit with a small throughput. But for tasks of grinding to colloid-size range, the stirred media mill has the advantage.

Residence Time Distribution Höffl [Freiberger, *Forschungshefte A* **750:** 119 pp. (1988)] carried out the first investigations of residence time distribution and grinding on vibratory mills and derived differential equations describing the motion. In vibratory horizontal tube mills, the mean axial transport velocity increases with increasing vibrational velocity, defined as the product $r_s\Omega$, where r_s = amplitude and Ω = frequency. Apparently, the media act as a filter for the feed particles and are opened by vibrations. Nevertheless, good uniformity of transport is obtained, indicated by *vessel dispersion numbers* $D\tau/L^2$ (see the subsection Simulation of Milling Circuits earlier) in the range of 0.06 to 0.08 measured in limestone grinding under conditions where both throughput and vibrational acceleration are optimum.

HICOM MILL

The *Hicom mill* is technically a vertical vibratory mill, but its design allows much higher energy input than do typical vibratory mills. The Hicom mill uses an irregular "nutating" motion to shake the mills, which allows much higher than normal *g* forces. Consequently, smaller media can be used and much higher grinding rates can be achieved. Hicom mill dry grinding performance tends to be competitive with jet mills, a substantial improvement over other vibratory mills. The Hicom mill is primarily used for dry grinding although it can also be used for wet grinding.

PLANETARY BALL MILLS

In planetary ball mills, several ball mill chambers are mounted on a frame in a circular pattern. The balls are all rotated in one direction (clockwise or counterclockwise), and the frame is rotated in the opposite direction, generating substantial centrifugal forces (10 to 50 *g*, depending on the device).

Planetary ball mills are difficult to make at large scale due to mechanical limitations. The largest mills commercially available have volumes in the range of 5 gal. Larger mills have been made, but they have tended to have very significant maintenance difficulties.

DISK ATTRITION MILLS

The disk or attrition mill is a modern counterpart of the early buhrstone mill. Stones are replaced by steel disks mounting interchangeable metal or abrasive grinding plates rotating at higher speeds, thus permitting a much broader range of application. They have a place in the grinding of tough organic materials, such as wood pulp and corn grits. Grinding takes place between the plates, which may operate in a vertical or a horizontal plane. One or both disks may be rotated; if both, then in opposite directions. The assembly, comprising a shaft, disk, and grinding plate, is called a *runner.* Feed material enters a chute near the axis, passes between the grinding plates, and is discharged at the periphery of the disks. The grinding plates are bolted to the disks; the distance between them is adjustable.

DISPERSERS AND EMULSIFIERS

Media Mills and Roll Mills Both media mills and roll mills are commonly used for powder dispersion, especially in the paint and ink industries.

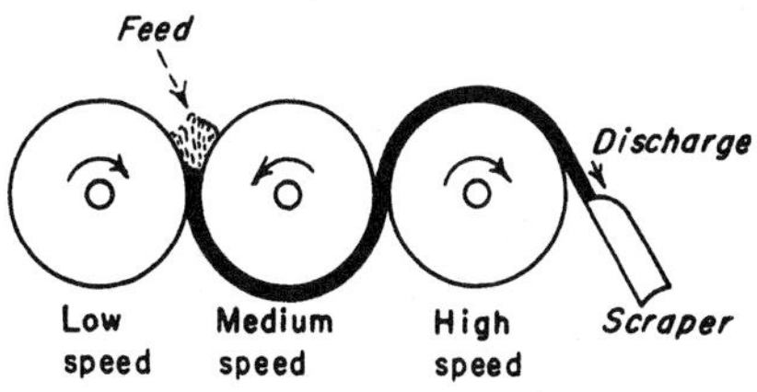

FIG. 21-149 Roller mill for paint grinding.

Media mills used for these operations are essentially the same as described above, although finer media are used than are common in particle-grinding operations (down to 0.2 mm). Often some sort of high-speed mixer is needed to disperse the powder into a liquid before trying to disperse powder in the media mill. Otherwise, large clumps of powder in the slurry can clog the mill.

Paint-grinding roller mills (Fig. 21-149) consist of two to five smooth rollers operating at differential speeds. A paste is fed between the first two rollers (at low speed) and is discharged from the final roller (at high speed) by a scraping blade. The paste passes from the surface of one roller to that of the next because of the differential speed, which also applies shear stress to the film of material passing between the rollers. Roll mills are sometimes heated so that higher-viscosity pastes can be ground and, in some cases, so that solvent can be removed.

Both of these mills can achieve very small particle size dispersion (below 100 nm, if the primary particle size of the powder is small enough). However, formulation with surfactants is absolutely necessary to achieve fine particle dispersions. Otherwise, the particles will simply reagglomerate after leaving the shear field of the machine.

Dispersion and Colloid Mills Colloid mills have a variety of designs, but all have a rotating surface, usually a cone or a disc, with another surface near the rotor that forms a uniform gap (e.g., two discs parallel to each other). The liquid to be emulsified is pumped between the gaps. Sometimes the design allows some pumping action between the rotor and the stator, and some machines of this type resemble centrifugal pumps in design. Colloid mills are relatively easy to clean and can handle materials with viscosity. For this reason, they are very common in the food and cosmetic industries for emulsifying pastes, creams, and lotions.

Pressure Homogenizers These are the wet grinding equivalents to jet mills, but they are used almost exclusively for emulsion and deagglomeration. There are several different styles of these, but all operate by generating pressures between 1000 and 50,000 psi using high-pressure pumps, with all the pressure drop occurring in a very small volume, such as flowing through an expansion valve. Some devices also have liquid jets which impinge on one another, similar to certain kinds of jet mills.

A **high-pressure valve homogenizer** such as the Gaulin and Rannie (APV Gaulin Group) forces the suspension through a narrow orifice. The equipment has two parts: a high-pressure piston pump and a homogenizer valve [Kula and Schuette, *Biotechnol. Progress* **3**(1): 31–42 (1987)]. The pump in production machines may have up to six pistons. The valve opens at a preset or adjustable value, and the suspension is released at high velocity (300 m/s) and impinges on an impact ring. The flow changes direction twice by 90°, resulting in turbulence. There is also a two-stage valve, but it has been shown that it is better to expend all the pressure across a single stage. The temperature of the suspension increases about 2.5°C per 10-MPa pressure drop. Therefore, intermediate cooling is required for multiple passes. Submicrometer-size emulsions can be achieved with jet homogenizers.

Microfluidizer The microfluidizer operates much the same as the valve homogenizers, but has a proprietary interaction chamber rather than an expansion valve. While valve homogenizers often have difficulties with particle slurries due to wear and clogging of the homogenizing valves, microfluidizers are much more robust and are often used in pharmaceutical processing. Interaction chambers for these applications must be made of specialized materials and can be expensive. Slurry particle sizes similar in size to those in media mill operations can be achieved with the microfluidizer.

SIZE ENLARGEMENT

General References: Salman, Hounslow, and Seville (eds.), *Granulation*, Elsevier, 2007; Litster and Ennis, *The Science and Engineering of Granulation Processes*, Kluwer Academic Publishers, 2005; Iveson, Litster, Hapgood, and Ennis, "Nucleation, Growth, and Breakage Phenomena in Agitate Wet Granulation Processes: A Review," *Powder Technology* **117:** 3 (2001); Parikh (ed.), *Handbook of Pharmaceutical Granulation Technology*, 2d ed., Taylor & Francis, 2005; Capes, *Particle Size Enlargement*, Elsevier, New York, 1980; Pietsch, *Size Enlargement by Agglomeration*, Wiley, Chichester, UK, 1992; Pietsch, *Roll Pressing*, Heyden, London, 1976; Çelik (ed.), *Pharmaceutical Compaction Technology*, 2d ed., CRC Press, 2016; Stanley-Wood (ed.), *Enlargement and Compaction of Particulate Solids*, Butterworth & Co. Ltd., 1983; Knepper (ed.), *Agglomeration*, Interscience, New York, 1962; Mead (ed.), *Encyclopedia of Chemical Process Equipment*, Reinhold, New York, 1964.

INTRODUCTION

Size enlargement is any process whereby small particles are agglomerated, compacted, or otherwise brought together into larger, relatively permanent masses in which the original particles can still be distinguished. The term encompasses a variety of unit operations or processing techniques dedicated to particle agglomeration. *Agglomeration* is the formation of aggregates or granules through binding of feed and/or recycled solids. By assembling particles into a single granule, size enlargement processes are a means of particle design.

The feed typically consists of a mixture of solid ingredients, referred to as a *formulation*, which may include an active or key ingredient, binders, diluents, disintegrants, flow aids, surfactants, wetting agents, lubricants, fillers, or end-use aids (e.g., sintering aids, colors or dyes, taste modifiers). The active ingredient is often referred to as the *technical* or *API* (active product ingredient), and it is the end-use ingredient of value, such as a drug substance, fertilizer, pesticide, or a key detergent agent. Size enlargement is performed for a wide variety of reasons such as improved flow, better dissolution, etc. Table 21-17 provides an expanded list of potential reasons for size enlargement.

A wide variety of size enlargement methods are available. A classification of these methods is given in Table 21-18 with key process attributes as well as typical subsequent processing. For example, a primary purpose of wet granulation in the case of pharmaceutical processing is to create free-flowing, nonsegregating blends of ingredients of controlled strength, which may be reproducibly metered in subsequent tableting or for vial- or capsule-filling operations.

Size enlargement processes are employed by a wide range of industries, including pharmaceutical and food processing, consumer products, fertilizer and detergent production, and the mineral processing industries. Agglomeration can be induced in several ways. A solvent or slurry can be atomized onto the bed of particles coating either the particle or granule surfaces. Conversely, the spray drops can form small nuclei that subsequently can agglomerate. The solvent or slurry may contain a suspended or dissolved binder, or a solid binder may be present as one component of the powder feed. Alternatively, the solvent may induce dissolution and recrystallization in the case of soluble particles. Slurries often contain the same particulate matter as the dry feed, and granules may be formed, either completely or partially, as the droplets solidify in flight prior to reaching the particle bed. This method, called *spray drying*, is an extreme case where no further, intended agglomeration takes place after granule formation. Agglomeration may also be induced by heat, which either leads to controlled sintering of the particle bed or induces sintering or partial melting of a binder component of the feed. Agglomerates may also be formed by applying excess pressure to make compacts (e.g., tablets, briquettes, ribbons). A bulk of the discussion below applies to wet agglomeration. Some discussion will also cover dry agglomeration or compaction processes.

TABLE 21-17 Objectives of Size Enlargement

Production of useful structural forms, as in pressing of intricate shapes in powder metallurgy.
Provision of a defined quantity to facilite dispensing and metering, as in agricultural chemical granules or pharmaceutical tablets.
Elimination of dust-handling hazards or losses, as in briquetting of waste fines.
Improved product appearance, or product renewal.
Reduced caking and lump formation, as in granulation of fertilizer.
Improved flow properties, generally defined as enhanced flow rates with improved flow rate uniformity, as in granulation of pharmaceuticals for tableting or ceramics for pressing.
Increased bulk density for storage and tableting feeds.
Creation of nonsegregating blends of powder ingredients with ideally uniform distribution of key ingredients, as in sintering of fines for steel or agricultural chemical or pharmaceutical granules.
Control of solubility, as in instant food products.
Control of porosity and surface-to-volume ratio, as with catalyst supports.
Improvement of heat-transfer characteristics, as in ores or glass for furnace feed.
Remove of particles from liquid, as with polymer additives, which induce clay flocculation.

Reprinted from *Design and Optimization of Granulation and Compaction Processes for Enhanced Product Performance*, Ennis, 2006, with permission of E&G Associates.

MECHANISMS

Binding Mechanisms Solid bridges can form between particles by the sintering of ores, the crystallization of dissolved substances during drying, or the hardening of bonding agents such as glue and resins. Liquid bridges produce cohesion through interfacial forces and capillary suction. The strength of the liquid bridges in a granule depends on the state of saturation. Several states can be distinguished in an assembly of particles held together by a liquid (Fig. 21-150). At the lowest liquid loading, small amounts of liquid are held as discrete lens-shaped rings at the points of contact of the particles. This is the *pendular* state. As the liquid content increases, the rings coalesce and there is a continuous network of liquid interspersed with air. This is the *funicular* state. When all the pore space in the agglomerate is completely filled, the capillary state has been reached. In most cases the capillary state has the highest particle strength. Binder viscosity markedly increases the strength of liquid bound granules. If more liquid is added beyond the capillary state, the granule will become a droplet of powder suspension rather than a granule at all. At this condition, the strength of the agglomerate is merely the strength of the liquid itself, which is naturally very weak. For poorly wetting solids, a pseudo-droplet can form, where air is entrained within the droplet.

Intermolecular and electrostatic forces bond very fine particles without the presence of material bridges. Such bonding is responsible for the tendency of particles less than about 1 μm in diameter to form agglomerates spontaneously under agitation. With larger particles, however, these short-range forces are insufficient to counterbalance the weight or inertia of the particle. High compaction pressures act to plastically flatten interparticle contacts and substantially enhance short-range forces. Mechanical interlocking of particles may occur during the agitation or compression of, for example, fibrous particles, but it is probably only a minor contributor to agglomerate strength in most cases.

Wet Agglomeration Rate Processes In wet agglomeration a liquid is added to an agitated powder bed to promote and allow aggregation. The liquid temporarily combines the feed solids into granules via liquid bridges (described above) until permanent bridges are formed. The mechanisms underlying this process are then strongly tied to the mechanical properties of the liquid, the solids, and their interactions. Broadly speaking, wet agglomeration is composed of three sets of rate processes: (1) wetting and nucleation; (2) consolidation and growth; and (3) attrition and breakage. These processes are described in detail below.

Wetting and Nucleation Here the description of the wetting process follows Hapgood et al. [*AIChE J.* **49:** 350 (2003)]. In essence, wet agglomeration is equivalent to uniform distribution of liquid into a powder bed. The process then begins when liquid droplets are introduced to the powder bed and allowed to disperse. There are two possibilities: (1) the particles are much smaller than the droplets; or (2) the droplets are much smaller than the particles (Fig. 21-151). The first is known as *immersion* or *engulfment nucleation*. If successful, the droplet will completely engulf the particles, making a granule nucleus. The second scenario results in coating or distribution nucleation. Here the droplets decorate the surface of the granule, providing points for future liquid bridges. If the three-phase contact angle is low, the droplet footprints will spread over the surface of the primary particles. Conversely, for higher contact angles, the droplets may remain as small blobs at their original impact points. In either scenario, the granule nucleus is not formed until other primary particles collide with the wetted point of the particle under conditions that result in coalescence. Rate processes of distributed nucleation are then very similar to those of coalescence, which is described later. Therefore, immersion nucleation will be the focus.

The stages of immersion nucleation are shown in Fig. 21-152. First a droplet must be formed via some spray, atomization, or jetting device. Next the droplet impacts onto the surface of the powder bed. After impacting, the droplet will then wet the powder bed by capillary penetration. Finally, once the granule nucleus is formed, it will be subjected to breakage, and therefore liquid distribution, by mechanical agitation. If droplets impact the bed infrequently, each droplet will produce its own granule nucleus. The granule nucleus size distribution will then correlate very well with the spray droplet size distribution. Under these conditions, the spray system

TABLE 21-18 Size Enlargement Methods and Application

Method	Product size (mm)	Granule density	Scale of operation	Additional comments and processing	Typical applications
Tumbling granulators Drums Discs	0.2–20	Moderate	0.5–800 tons/h	Very spherical granules Fluid-bed or rotary kiln drying	Fertilizers, iron and other ores, agricultural chemicals
Mixer-granulators Continuous high-shear (e.g., Shugi mixer) Batch high-shear (e.g., vertical mixer)	0.1–0.5 0.1–2	Low Moderate to high	Up to 50 tons/h Up to 500-kg batch	Handles cohesive materials, both batch and continuous, as well as viscous binders and nonwettable powders Fluid-bed, tray, or vacuum/microwave on-pot drying	Chemicals, detergents, clays, carbon black Pharmaceuticals, ceramics, clays
Fluidized granulators Fluidized beds Spouted beds Wurster coaters	0.1–1	Low (agglomerated) Moderate (layered)	100–900 kg batch 50 tons/h continuous	Flexible, relatively easy to scale, difficult for nonwettable powders and viscous binders, good for coating applications Same vessel drying, air handling requirements	Continuous: fertilizers, inorganic salts, food, detergents Batch: pharmaceuticals, agricultural chemicals, nuclear wastes
Centrifugal granulators	0.3–3	Moderate to high	Up to 200-kg batch	Powder layering and coating applications Fluid-bed or same-pot drying	Pharmaceuticals, agricultural chemicals
Spray methods Spray drying Prilling	0.05–0.2 0.7–2	Low Moderate		Morphology of spray-dried powders can vary widely Same vessel drying	Instant foods, dyes, detergents, ceramics, pharmaceuticals Urea, ammonium nitrate
Pressure compaction Extrusion Roll press Tablet press Molding press Pellet mill	>0.5 >1 10	High to very high	Up to 5 tons/h Up to 50 tons/h Up to 1 ton/h	Very narrow size distributions, very sensitive to powder flow and mechanical properties Often subsequent milling and blending operations	Pharmaceuticals, catalysts, inorganic chemicals, organic chemicals, plastic preforms, metal parts, ceramics, clays, minerals, animal feeds
Thermal processes Sintering	2–50	High to very high	Up to 100 tons/h	Strongest bonding	Ferrous and nonferrous ores, cement clinker, minerals, ceramics
Liquid systems Immiscible wetting in mixers Sol-gel processes Pellet flocculation	<0.3	Low	Up to 10 tons/h	Wet processing based on flocculation properties of particulate feed, subsequent drying	Coal fines, soot, and oil removal from water Metal dicarbide, silica hydrogels Waste sludges and slurries

Reprinted from *Design and Optimization of Granulation and Compaction Processes for Enhanced Product Performance,* Ennis, 2006, with permission of E&G Associates.

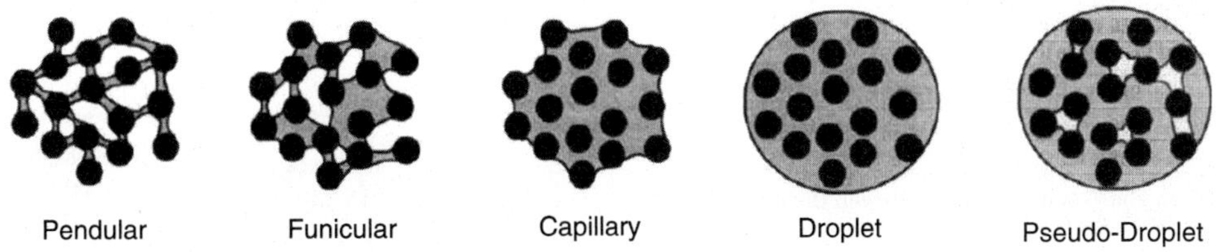

FIG. 21-150 Different states of saturation of liquid-bound granules. (*After Iveson et al,* Powder Technol. *2001, with permission.*)

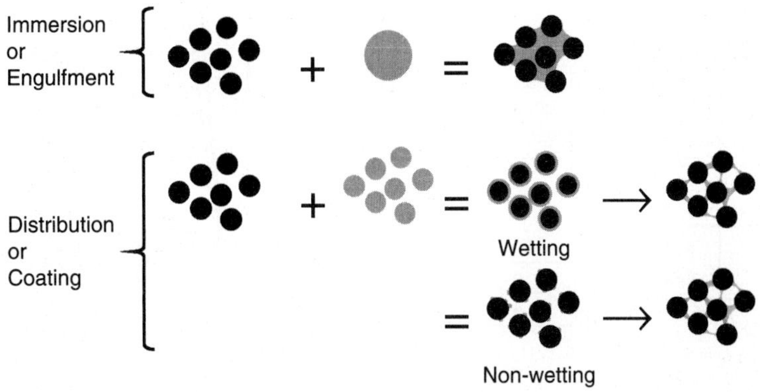

FIG. 21-151 Nucleation mechanisms (black = feed particles, gray = droplets).

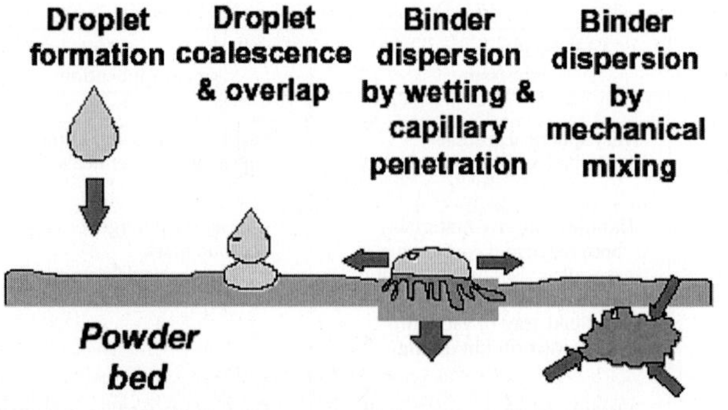

FIG. 21-152 Stages of wetting for fine powder compared to drop size.

(droplet generation device and spray formulation) will control the granulation. This is known as the *drop-controlled regime.* Conversely, if droplets coalesce because their deposition footprints overlap, a large wetted mass will be formed that needs to be broken by agitation. This is known as the *mechanical dispersion* regime.

Figure 21-153 shows a regime map that allows quantitative determination of a granulator's nucleation behavior. On the horizontal axis is the dimensionless spray flux Ψ_a. This quantity equals the rate at which deposited droplet footprint area is generated via the spray, divided by the rate at which powder bed surface area is offered for droplets to deposit. It is calculated using the equation

$$\Psi_a = \frac{\dot{a}}{\dot{A}} = \frac{3\dot{V}}{2\dot{A}d_d} \qquad (21\text{-}146)$$

where $\dot{V}$ is the volume flow rate of liquid from the spray, d_d is the droplet diameter, and $\dot{A}$ is the rate at which area is replenished via agitation below the spray. Area $\dot{A}$ is calculated from the product of the powder bed's surface velocity under the spray and the width of the spray deposition area perpendicular to the direction of the powder bed surface velocity. For values of $\Psi_a < 0.1$, droplets will rarely overlap. Fresh powder bed surface is presented to the spray before the area taken up by droplet footprints is exhausted. For values of $\Psi_a > 1.0$, the droplets will overlap on the powder bed surface and the nucleation will require mechanical dispersion. The vertical axis of Fig. 21-153 represents the dimensionless droplet penetration time τ_p. This quantity equals the rate at which liquid wets the bed via capillary penetration divided by the powder circulation time. The droplet penetration time is calculated via the equations

$$t_p = 1.35\frac{V_d^{2/3}}{\varepsilon_{\text{eff}}^2}\frac{\mu}{R_{\text{eff}}\gamma\cos\theta} \qquad R_{\text{eff}} = \frac{\varphi d_{3,2}}{3}\frac{\varepsilon_{\text{eff}}}{1-\varepsilon_{\text{eff}}} \qquad \varepsilon_{\text{eff}} = \varepsilon_{\text{tap}}(1-\varepsilon+\varepsilon_{\text{tap}}) \qquad (21\text{-}147)$$

where V_d is the droplet volume; μ is the liquid viscosity; γ is the liquid surface tension; θ is the three-phase contact angle between the solid surface, the liquid, and air; R_{eff} is the effective pore radius where ϕ is the powder

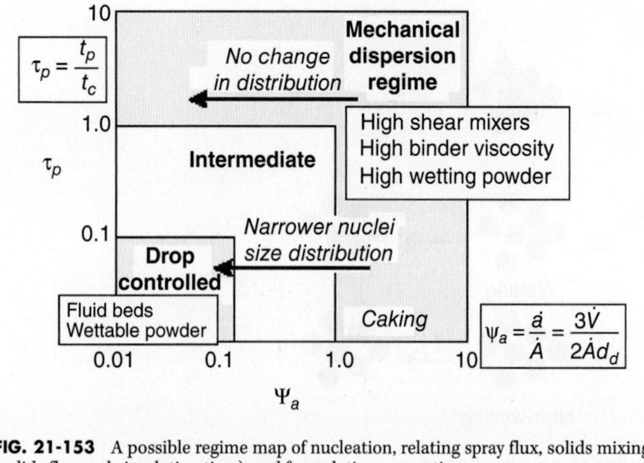

FIG. 21-153 A possible regime map of nucleation, relating spray flux, solids mixing (solids flux and circulation time), and formulation properties.

sphericity, $d_{3,2}$ is the powder's Sauter mean diameter, and ε_{eff} is the effective porosity (porosity excluding macropores); where ε_{tap} is the porosity tapped density state, and ε is the porosity in the granulator. The circulation time is equal to the average time necessary for a material element of flowing powder to leave the spray and then reenter it. It is roughly equal to the path length around the granulator through the spray area, divided by the mean powder bed velocity. For values of $\tau_p < 0.1$, droplets have ample time to wet the powder surface before their position returns to the spray. For values of $\tau_p > 1.0$, droplets will remain on the surface while more droplets are deposited, causing powder bed surface droplet coalescence. Under such conditions liquid must be dispersed mechanically.

There are several implications of the nucleation regime map given in Fig. 21-153 with regard to troubleshooting of wetting and nucleation problems. If drop penetration times are large, then making adjustments to spray may not be sufficient to narrower granule size distributions when remaining in the mechanical regime. Significant changes to wetting and nucleation occur only if changes take the system across a regime boundary. This can occur in an undesirable way if processes are not scaled with due attention to remaining in the drop-controlled regime.

Consolidation and Growth Here the description of the wetting process follows Iveson et al. [*Powder Technology* **117**: 3 (2001)]. When granules come into contact, they either rebound (lose contact) or stick (remain in contact). Rebound can be avoided only if the incoming energy is somehow dissipated. That energy can be dissipated by interparticle friction, plastic mechanical deformation of the granules, or viscous dissipation of the liquid; or it can be stored in capillary bridges of liquid between particles. Based on these mechanisms, it must be concluded that the mechanical properties of the liquid and availability of liquid at the surface both strongly influence the probability of coalescence. They are also both strongly related. At an instant in time, a granule has a fixed amount of solid, liquid, and gas (pore) volume. That liquid may be present within the granule or on the surface. As the granule is consolidated, gas is removed and liquid is squeezed to the surface. Conceptually, a change in state from the left of Fig. 21-153 to the right may be achieved by liquid addition or granule consolidation. That is, liquid can be made available on the surface by either liquid addition or granule consolidation. Hence granule consolidation and growth are strongly related.

Strength and consolidation Consolidation occurs when the imposed stress on a granule exceeds its peak flow yield strength σ_Y. This strength can be characterized by squeezing a puck of wet mass and reporting the force versus displacement. Figure 21-154 shows typical engineering stress versus natural strain curves for 35-μm glass ballotini mixed with 60 Pa·s silicone oil at 70 percent volume saturation for different compression velocities [after Iveson et al., *Powder Technology* **127**: 149 (2002)]. The yield strength of the model granule corresponds to the maximum stress achieved and is dependent on the crosshead velocity. Figure 21-154 shows dimensionless strength

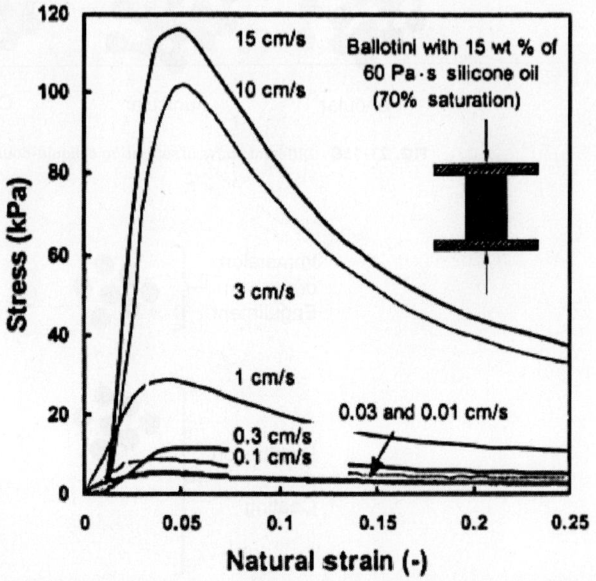

FIG. 21-154 Typical compact stress response for fast compression versus crosshead compression velocity for glass ballotini ($d_{3,2}$ = 35 μm) and compact diameter 20 mm, length 25 mm. [*After Iveson et al.,* Powder Technol. ***127:** 149 (2002), with permission.*]

versus dimensionless speed for several other glass with silicone oil or water systems. The data can be correlated using the relation

$$\frac{\sigma_Y d_{3,2}}{\gamma \cos\theta} = k_1 + k_2 \text{Ca}^n \qquad (21\text{-}148)$$

where σ_Y is the yield strength, $d_{3,2}$ is the particle Sauter mean diameter, γ is the liquid surface tension, θ is the contact angle, and Ca is the capillary number with $\text{Ca} = \mu d_{3,2}\dot{\varepsilon}/\gamma \cos\theta$, where μ is the liquid viscosity, $\dot{\varepsilon}$ is the compressive strain rate, and $k_1 = 5.3$, $k_2 = 280$, and $n = 0.58$ are dimensionless constants. The correlation demonstrates that at low velocity, capillary bonding dominates the granule strength. At higher impact velocity, viscous effects contribute. It is therefore necessary to characterize and conceptualize the wet mass under strain rates relevant to the conditions of actual granulation. Roughly speaking, strength increases with binder viscosity, decreasing particle size, increasing particle size distribution breadth, and increasing surface tension.

Coalescence A simple illustrative model of granule coalescence was developed by Ennis et al. [*Powder Technology* **65**: 257 (1991)]. This model ignores consolidation caused by collisions and treats the granules themselves as rigid. Such a model is conceptually accurate for high-strength or lightly stressed granules (low collision velocity, e.g., fluidized bed). The model is shown diagrammatically in Fig. 21-155. Here the granules are treated as spheres speckled with asperities and covered with a thin liquid film. When the granules collide, energy is dissipated in the viscous film between them. Eventually their surfaces come into contact at the asperities, and they rebound. After rebounding, the granules begin to separate while continuously dissipating energy via the liquid film. If all the inbound collision energy has not been dissipated, the granules separate. Otherwise, they have coalesced. This analysis results in the simple expression for coalescence $\text{St}_v \leq \text{St}_v^*$, where the *viscous Stokes number* St_v and the *critical viscous Stokes number* St_v^* are described, respectively, by

$$\text{St}_v \equiv \frac{4}{9}\frac{\rho_p d_p u}{\mu} \quad \text{and} \quad \text{St}_v^* \equiv \left(1+\frac{1}{e}\right)\ln\left(\frac{h}{h_a}\right) \qquad (21\text{-}149)$$

where ρ_p is the density of the granules, u is the collision velocity, μ is the liquid viscosity, e is the particle-particle coefficient of restitution, h is the liquid layer thickness, h_a is the asperity height, and d_p is the harmonic mean particle diameter of the two colliding particles

$$d_p = \left(\frac{1}{d_1} + \frac{1}{d_2}\right)^{-1} \qquad (21\text{-}150)$$

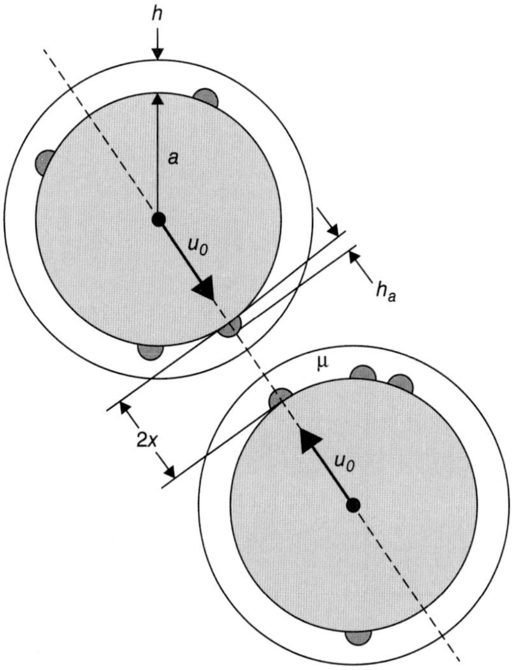

FIG. 21-155 Schematic of two colliding granules each of which is covered by a viscous layer. (After *Ennis et al.,* Powder Technology, *1991, with permission.*)

The viscous Stokes number gives a measure of particle inertia to film viscosity. The collision velocity can be approximated in fluidized beds as $u = 6U_B d_p / D_B \delta^2$ where U_B is the bubble velocity, D_B is the bubble diameter, and δ is the bubble fraction. In drums, pans, and mixers the collision velocity may be approximated by $u = \omega d_p$, ωR, or $(Rg \sin\phi)^{1/2}$, where ω is the angular rotation speed, R is the vessel radius, g is the acceleration of gravity, and ϕ is the angle of inclination. The fact that St_v is proportional to the harmonic mean diameter results in the finer of the two particles dictating the coalescence probability of the collision. Under otherwise identical circumstances, fine-fine collisions and fine-coarse interactions will result in coalescence more frequently than coarse-coarse interactions. According to this model, three regimes of behavior can be identified: $\text{St}_v << \text{St}_v^*$ (noninertial), $\text{St}_v \approx \text{St}_v^*$ (inertial), and $\text{St}_v >> \text{St}_v^*$ (coating).

In the noninertial regime, every collision results in coalescence, provided binder is present. Here the growth rate is independent of the granule kinetic energy, particle size, and binder viscosity (provided other rate processes are constant). Therefore, the growth is controlled by wetting, i.e., the physical distribution and rate of penetration of liquid binder (see the subsection Wetting and Nucleation).

As granules grow in size, St_v increases, leading to localized regions in the process where St_v exceeds St_v^*. In this inertial regime of granulation, the granule size, binder viscosity, and collision velocity determine the proportion of the bed in which granule rebound is possible. Increases in binder viscosity and decreases in agitation intensity increase the extent of granule growth. When the spatial average of St_v exceeds St_v^*, growth is balanced by rebound, leading to the coating regime of granulation. Growth continues only by coating of granules with feed fines. Note that this model only predicts whether coalescence can occur; it does not predict the rate of granulation. The rate of granulation is also dependent on the rate of collision events. However, this model still provides a useful concept for understanding granule coalescence micromechanisms and provides physical limits on maximum permissible granule sizes.

The reasoning of the above model was extended to include deformable granules by Liu et al., *AIChE J.* **46:** 529 (2000). This extended model treats the granules below the liquid layer as elastic, perfectly plastic solids with Young's modulus E and yield strength σ_Y. Therefore, some energy dissipation may also occur during plastic deformation, further limiting rebound (Fig. 21-156). This effect is captured by a second important dimensionless group; the deformation Stokes number $\text{St}_{\text{def}} = \frac{1}{2} m u^2 / d_p^3 \sigma_Y$ where m and d_p are the harmonic mean mass and particle diameter of the colliding granules. Figure 21-157 shows a typical regime map of possible collisions. On the map, type I coalescence corresponds to energy dissipation occurring only in the viscous layer during approach without granule-granule contact. Type II coalescence involves energy dissipation due to plastic deformation directly and more viscous dissipation due to surface flattening. At low St_{def} this model returns results identical to the previous nondeformable model. Unlike the previous model, increased collision velocity does not always increase the likelihood of rebound. In fact, it is possible to transition from type I coalescence to rebound and then to type II by only increasing the collision velocity. Because granule deformation is accounted for, the model can also account for the possibility of liquid squeeze into the contact zone from

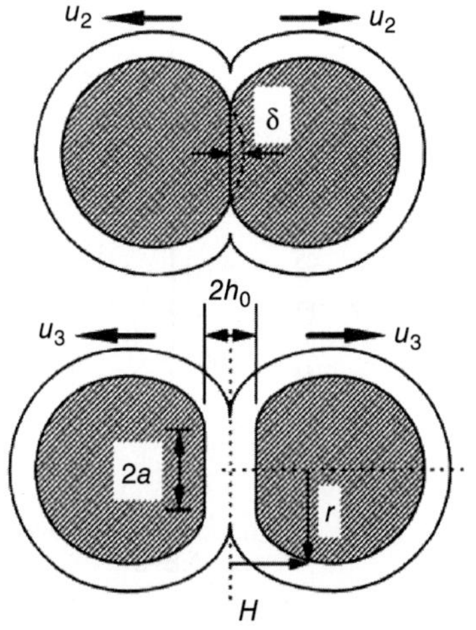

FIG. 21-156 Schematic of two deformable colliding granules each of which is covered by a viscous layer. (*After Liu et al.,* AIChE J. *2004, with permission.*)

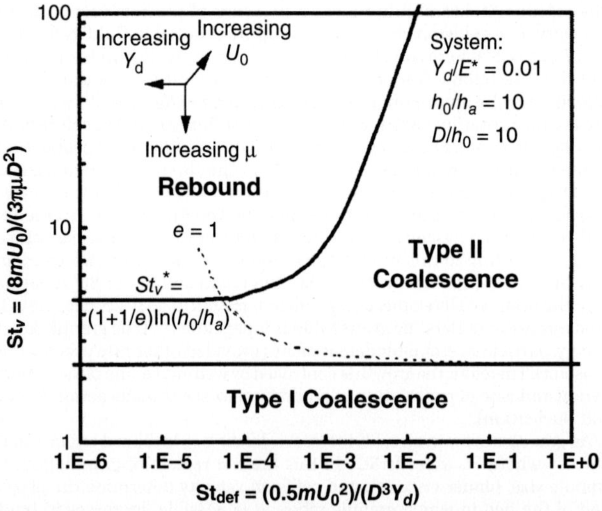

FIG. 21-157 Collision regime map showing regions of coalescence and rebound for surface wet deformable granules. (*After Liu et al.,* AIChE J. *2004, with permission.*)

initially surface dry granules. Figure 21-158 shows a typical regime map of possible surface dry collisions. For low St_{def}, only rebound occurs because no dissipation mechanism exists without plastic deformation. Above a critical value of St_{def}, coalescence is dependent on St_v, similarly to the surface wet case. As with the previous model, this approach has flaws, but is still conceptually useful. It does not account for consolidation or the saturation increase from consolidation, or granule breakage. Also the equations describing the regime boundary interfaces are complex and difficult to comprehend at a glance. However, the model still gives a good framework for comprehending very complicated ideas and how they are interrelated.

Granule Growth Based on extension of the above reasoning, Iveson and Litster [*AIChE J.* **44:** 1510 (1998)] concluded that granule growth must be controlled primarily by the presence of surface liquid and the deformability of granules. As explained above, surface liquid can appear if liquid is added or if pores are eliminated via consolidation. The two relevant dimensionless parameters are then the maximum pore saturation s_{max} and the deformation Stokes number St_{def} (defined above). The *maximum pore saturation* is defined as the volume occupied by liquid in the granules densest state. It can be calculated using the relation

$$s_{max} = w\frac{\rho_s}{\rho_l}\frac{1-\varepsilon_{min}}{\varepsilon_{min}} \qquad (21\text{-}151)$$

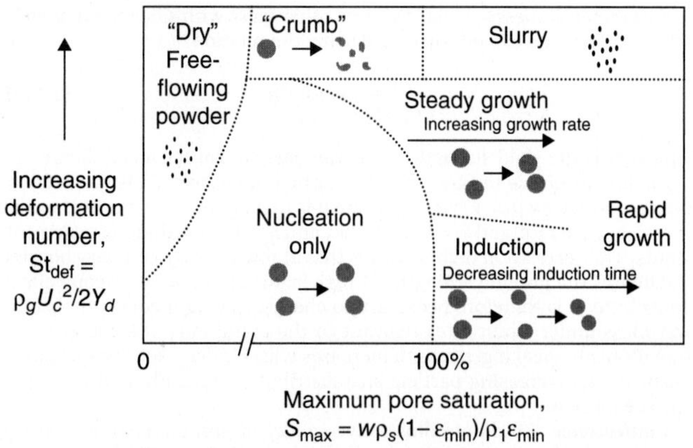

FIG. 21-159 Granule growth regime map. (*Source: Salman et al.,* Granulation, *with permission.*)

where w is the solid-to-liquid mass ratio, ρ_s is the density of the solid particles, ρ_l is the density of the liquid, and ε_{min} is the minimum granule porosity under the given operating conditions. Figure 21-159 shows a regime map of granule growth behavior, using these two dimensionless parameters as coordinates. The vertical axis can be thought of as stress/strength. When that ratio is high, greater consolidation and greater breakage are expected. The horizontal axis is pore saturation. When that ratio is high, more growth is expected, unless the system is overly wet and forms a slurry. Starting at the upper left, the "dry" regime occurs when insufficient moisture is added to agglomerate and the granules are too weak. With stronger granules (or less stress) "nucleation only" may occur. That is, primary particles will coalesce, but there is insufficient moisture to allow for further coalescence. In the "crumb" regime, granules form but are too weak and break immediately, preventing true growth. The center of the regime map describes "steady" growth where liquid is consistently present for growth by continuous squeezing to the surface via consolidation. In this regime, the batch mean particle size will grow at a nominally constant rate. For lower deformation Stokes numbers, "induction" growth may occur. Here the granules are too strong for the liquid to be squeezed to the surface with nearly each collision. Instead no growth happens for some time while liquid is trapped within the granule. Eventually consolidation will allow surface moisture and coalescence will proceed quickly. For obvious reasons, growth in this regime can be hard to control. For either growth regime, if sufficient moisture is added (to the right), "rapid" growth will result. Here the granules are nearly always saturated, so strength is irrelevant because surface moisture is always present. Growth progresses very quickly here, and it is also difficult to control.

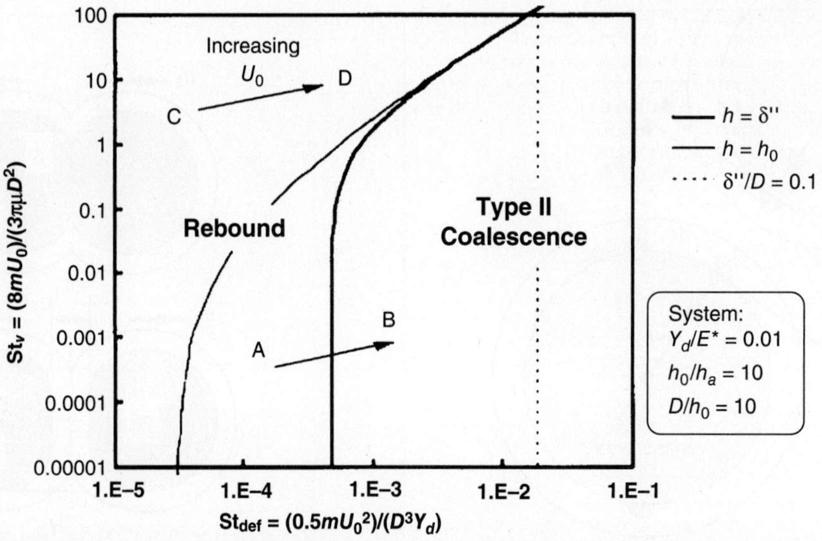

FIG. 21-158 Collision regime map showing regions of coalescence and rebound for surface dry deformable granules. (*After Liu et al.,* AIChE J. *2004, with permission.*)

Note that the vertical axis is not quantitative. Furthermore, a priori knowledge of the granule yield strength and minimum porosity is not possible. This map, and the logic behind it, is not meant to be predictive. Instead it provides a sound descriptive foundation for understanding the complex phenomena at play in a granule growth process. Using these tools allows an engineer to identify what type of behavior is occurring and propose directional changes of process and formulation parameters to recover the desired behavior. Because all parameters within these theories are dimensionless, they can also be used as bases for scale-up.

Very roughly, increasing liquid content tends to move the granulation to the right and slightly up on the regime map. The upward motion is due to granules with higher moisture generally having lower strength. If the primary particle size is decreased, the granulation moves downward and perhaps slightly left. Fine particles produce stronger granules and will have a higher value of ε_{min}. Likewise, wider size distributions and increased liquid surface tension also increase strength, resulting in a similar effect. The influence of viscosity is less obvious. For the most part, increased viscosity increases strength, resulting in movement up and slightly left. However, the surface dissipation is much stronger with greater viscosity, resulting in more successful collisions. Increased equipment speed (e.g., superficial velocity, rotation speed, or agitator speed) generally moves a granulation up on the regime map. This movement is compounded by slight increase in strength due to granule strength viscous effects being rate-dependent.

Attrition and Breakage When one is discussing attrition and breakage, it is important to distinguish between in granulator granule breakage and final product agglomerate breakage. Mechanical degradation of the final product is clearly always a problem. However, size reduction of granules during the granulation may be favorable. To return to the discussion in the subsection Wetting and Nucleation, in the mechanical dispersion regime, breakage is necessary to effectively distribute liquid. For systems with very viscous or poorly wetting liquids, breakage is the rate-controlling process. Furthermore, breakage in the granulator may dictate the equilibrium granule size. The previous subsections have described physics limiting granule size based on what is attainable via coalescence. If, however, granules break at a smaller size than that limit, then a new equilibrium will be found.

As with all breakage phenomena, prediction is difficult (see the subsection Size Reduction). The growth regime map (Fig. 21-159) suggests that mass breakage may be controlled directly by the deformation Stokes number St_{def}. The demarcation between the crumb and growth regimes occurs at a fixed St_{def}. This result is partially substantiated by work from Tardos et al. [*Powder Technology* **94:** 245 (1997)] demonstrating that a critical value $St_{def} \approx 0.2$ defines breakage of granules in a uniform shear field. Extension of this result to granulator use is supported by Smith et al. [*Chemical Engineering Science* **65:** 5651 (2010)] who showed that drop-nucleated granules in high-shear mixers break at a similar magnitude critical St_{def} value.

Compaction Mechanisms In compaction processes, particles are directly forced together by extreme stress. Figure 21-160 conceptually shows the density-stress relationship over a broad range of stresses. At low values of stress, the external stresses first rearrange the particles' microstructure, destroying any macrovoids. Once those voids are filled, the interparticle porosity is decreased until particles themselves must deform elastically to accommodate added strain and densification. Until this point, the increased density may be reversible by decreasing the imposed stress and agitating the powder (i.e., fluffing). However, if the external stress is further increased, the particles will plastically deform or fracture. Plastic deformation results in increased contact area between the particles, increasing the bulk cohesion of the compact. Fracture broadens the size distribution and allows fines to fill the void space between the larger particles, along with increased contact area and contact density for cohesion. With further increases in stress, the compact density will approach the true density of the material. Key steps

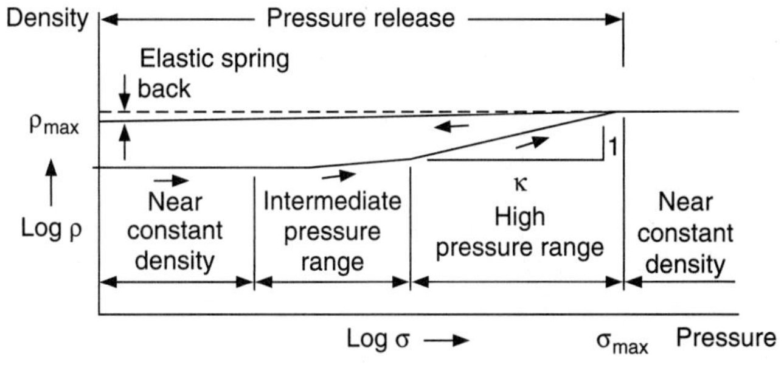

FIG. 21-160 Compressibility diagram of a typical powder illustrating four stages of compaction.

toward performing this densification in an industrial unit operation are (1) powder filling or feeding, (2) stress application, and (3) stress removal and compact ejection in the case of confined compression techniques.

Powder Filling Powder filling and compact weight variability are strongly influenced by bulk density, powder flowability, and powder-tool wall friction (see the subsection Bulk Solids Flow and Hopper Design) as well as any contributing segregation tendencies of the feed. For example, in tableting, the weight of a tablet is equal to the weight of powder allowed to flow into the die. For powders with poor flowability, this weight can be inconsistent. For fixed displacement presses, low fill will also result in low compact density and potentially friable tablets. Likewise, overfilled dies will be overly compacted and may exhibit poor dissolution. For roll-pressing or briquetting, poor flow may also cause inconsistent ribbon or briquette density or may simply be rate-limiting. For guidance on flowability and flow rate prediction, see the subsection Bulk Solids Flow and Hopper Design.

Stress Application and Removal As stress is increased, density also increases, as shown in Fig. 21-160. Here by *stress* we mean the *maximum principal stress*, sometimes called *pressure* in the compaction literature. Many relationships exist for correlating stress and density; these are known as *compaction relations* (see Table 21-19). These relations are useful for predicting stresses necessary to achieve a given compaction state or for inferring density from stress data. In some cases, the fit parameters of the relations can be used to infer individual particle properties. For metal powders the value of K^{-1} is roughly equal to the hardness of the metal powder when the

TABLE 21-19 Common Compaction Relations*

Equation	Authors
$\ln \frac{\rho_i - \rho_i}{\rho_t - \rho_c} = KP_A$	Athy, Shapiro, Heckel, Konopicky, Seelig
$\ln \frac{\rho_c}{\rho_i}\left(\frac{\rho_c - \rho_i}{\rho_t - \rho_c}\right) = KP_A$	Ballhausen
$\ln \frac{\rho_i}{\rho_t}\left(\frac{\rho_t - \rho_i}{\rho_t - \rho_c}\right) = KP_A$	Spencer
$\ln \frac{\rho_c}{\rho_i} = KP_A^a$	Nishihara, Nutting
$\ln \frac{\rho_t - \rho_c}{\rho_t} + K\left(\frac{\rho_c}{\rho_t - \rho_c}\right)^{1/3} = aP_A$	Murray
$\ln \frac{\rho_t}{\rho_c}\left(\frac{\rho_c - \rho_i}{\rho_t - \rho_i}\right) = \ln Ka - (b+c)P_A$	Cooper and Eaton
$\frac{\rho_i}{\rho_c} = 1 - KP_A^a$	Umeya
$\rho_c = KP_A^a$	Jaky
$\rho_c = K(1 - P_A)^a$	Jenike
$\rho_c - \rho_i = KP_A^{1/3}$	Smith
$\rho_c - \rho_i = KP_A^2$	Shaler
$\frac{\rho_c - \rho_i}{\rho_c} = \frac{K \times aP_A}{1 + aP_A}$	Kawakita
$\frac{\rho_t}{\rho_c}\left(\frac{\rho_c - \rho_i}{\rho_t - \rho_i}\right) = \frac{KP_A}{1 + KP_A}$	Aketa
$\frac{1}{\rho_c} = \mathrm{K} - a \ln P_A$	Walker, Bal'shin, Williams, Higuchi, Terzaghi
$\rho_c = K + a \ln P_A$	Gurnham
$\frac{1}{\rho_c} = K - a \ln P_A$	Jones
$\frac{1}{\rho_c} = K - a \ln (P_A - b)$	Mogami
$\frac{\rho_c - \rho_i}{\rho_c} = KP_A\ \rho_i + a\left(\frac{PA}{P_A + b}\right)$	Tanimoto
$\frac{\rho_c - \rho_i}{\rho_c} = \ln (KP_A + b)$	Rieschel

*ρ_t, density of powder; ρ_i, initial apparent density of powder; ρ_c, density of powder applied pressure P_A; K, a, b, and c are constants.

Heckle relation is used. For powders, the parameter a^{-1} correlates well with particle yield strength when the Kawakita relation is used. In all these cases the relations were developed using bulk average stresses and densities. However, powders notoriously do not transmit stress uniformly well. Take, for example, Janssen's analysis of silo stresses (see subsection Bulk Solids Flow and Hopper Design). There the weight of the solids is supported by frictional stresses at the wall. Due to friction, the stresses at extreme depths are independent of depth. Likewise, if the powder weight is supplemented by an imposed stress, much of that stress is supported by friction at the walls. Powder far from the imposed stress will not experience that stress at all due to frictional dissipation. This scenario plays out exactly in compaction processes. In tablet compaction, high-aspect-ratio tablets will achieve lower stresses farther from the moving punch faces. These lower stresses result in lower densities, hence lower compact strength. These effects may be mitigated by application of lubricants such as magnesium stearate. Note that while lubricants decrease friction, they will increase powder cohesion as well. Increased cohesion may worsen powder flowability, leading to the concerns discussed above.

Compaction processes are also typically conducted at strain rates high enough to cause concern for interparticle gas penetration. As the powder is densified, interstitial gas must be evacuated. A powder's gas permeability controls the gas pressure developed within the bulk powder during feeding and compaction. The *permeability* is defined as the ratio of gas velocity to pressure gradient through a powder at a given level of consolidation. High-permeability solids release gas without resistance. Low-permeability powders cannot release the gas, and it becomes entrained as pressurized air pockets. If the developed pressure exceeds the fracture strength of the compact, it may instantly break once the external stress is relieved. If not, the pocket will remain as an inclusion and a stress concentrator, resulting in a weaker product. In addition, as powders discharge from feed hoppers, they undergo expansion during movement, requiring gas flow into the powder. This countercurrent flow of gas impedes the powder discharge. Therefore, low-permeability powders may also cause powder flow concerns, as described above. Permeability generally decreases with decreasing particle size and powder porosity. For general trends refer to Ergun's equation (see Sec. 17).

In addition to entrained gas effects, the powders themselves may exhibit rate-dependent phenomena. At high strain rates, many bulk solids behave viscoelastically or viscoplastically. The concerns of maldistribution of stress or insufficient compaction may need to be reevaluated if the process rates are substantially changed. It is always necessary to conduct process trials at comparable rates to what will be practiced commercially.

Ejection Upon stress removal, the compact expands due to remaining elastic recovery of the powder. This recovery is dependent on the compacted solids' elastic modulus as well as any expansion of remaining entrapped air. This expansion can result in loss of particle bonding and flaw development, which is exacerbated for cases of wide distributions in compact stress due to poor stress transmission. The final step of stress removal involves compact ejection, where any remaining elastic stresses are removed. If elastic recovery is substantial, strain gradients will cause delamination or other fracture of the compact. Note that it is not necessarily elastic recovery in and of itself that causes these problems, but the nonuniform relaxation of stress caused by elastic recovery. Therefore, problems with elastic recovery will be exacerbated by previous problems of nonuniform stress transmission and densification.

AGGLOMERATION EQUIPMENT

Wet Agglomeration Equipment Particle size enlargement equipment can be classified into several groups, with typical objectives summarized in Table 21-17 and advantages and applications summarized in Table 21-18. Terminology is industry-specific. In the following discussion, particle size enlargement in tumbling, mixer, and fluidized-bed granulators is referred to as *granulation*. Granulation processes vary from low to medium levels of applied shear and stress, producing granules of low to medium density. The presence of liquid binder is essential for granule growth and green strength. Granulation includes pelletization or balling as used in the iron ore industry, but does not include the breakdown of compacts by screening as used in some tableting industries. Compaction processes include dry compaction techniques such as roll pressing and tableting. Compaction processes rely on pressure to increase agglomerate density and give sufficient compact strength, either with or without liquid binder, with a resulting high compact density.

In fluidized-bed granulators, the bed of solids is supported and mixed by fluidization gas, generally with simultaneously drying. With small bed agitation intensities and high binder viscosities (due to drying), fluidized-bed granulators can produce one of the lowest-density granules of all processes (with the exception of spray drying). Fluidized and spouted fluidized beds are also used for coating or layering applications from solution or melt feeds, which can produce spherical, densely layered granules. At the other extreme of granulation processes are high-shear mixer-granulators, where mechanical blades and choppers induce binder distribution and growth, producing dense, sometimes irregular granules. Fluidized beds are generally a low-agitation, low-deformation process so that mechanical dispersion of liquid is not possible. As such, they must be operated in the drop-controlled nucleation regime or with dispersion nucleation (see the subsection Wetting and Nucleation). Tumbling granulators such as rotating discs and drums produce spherical granules of low to medium density, and they lie between fluidized bed and mixer in terms of bed agitation intensity and granule density. They have the highest throughput of all granulation processes, often with high recycle ratios. In these devices, spontaneous segregation allows for internal classification and can produce very tight size distributions of uniform spherical granules. Mixer and fluid-bed granules operate as continuous or batch processes.

Tumbling Granulators In tumbling granulators, particles are set in motion by the tumbling action caused by the balance between gravity and centrifugal forces. The most common types of tumbling granulators are drum and inclined disc granulators. Their use is widespread including the iron ore industry (where the process is sometimes called *balling* or *wet pelletization*), fertilizer manufacturing, and agricultural chemicals. Tumbling granulators generally produce granules in the size range of 1 to 20 mm and are not suitable for making granules smaller than 250 μm. Granule density generally falls between that of fluidized-bed and mixer granulators, and it is difficult to produce highly porous agglomerates in tumbling granulators. Tumbling equipment is also suitable for coating large particles, but it is difficult to coat small particles, as growth by coalescence of the seed particles is hard to control. Drum and disc granulators generally operate in continuous feed mode. A key advantage to these systems is the ability to run at large scale. Drums with diameters up to 4 m and throughputs up to 100 tons/h are widely used in the mineral industry.

Disc or pan granulators A disc granulator is shown in Fig. 21-161. It is also referred to as a *pelletizer* in the iron ore industry or a *pan granulator* in the agricultural chemical industry. The equipment consists of a rotating tilted disc or pan with a rim. Solids and fluid agents are continuously added to the disc. A coating of the feed material builds up on the disc, and the thickness of this layer is controlled by scrapers or a plow, which may oscillate mechanically. The surface of the pan may also be lined with expanded metal or an abrasive coating to promote proper lifting and cascading of the particulate bed, although this is generally unnecessary for fine materials. Solids are typically introduced to the disc by either volumetric or gravimetric feeders,

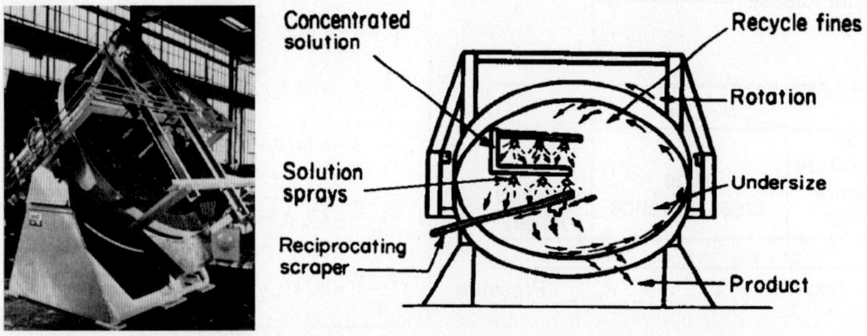

FIG. 21-161 A typical disc granulator. (*Capes,* Particle Size Enlargement, *Elsevier, 1980.*)

preferable at the bottom edge of the rotating granular bed. Gravimetric feeding generally improves granulation performance due to smaller fluctuations in feed rates. Such fluctuations act to disrupt rolling action in the disc and can lead to maldistribution of moisture and local buildup on the disc surface. Wetting fluids that promote growth are generally applied by a series of single-fluid spray nozzles distributed across the face of the bed. Solids feed and spray nozzle locations have a pronounced effect on granulation performance and granule structure.

The required disc rotation speed is given in terms of the critical speed, i.e., the speed at which a single particle is held stationary on the rim of the disc due to centripetal forces. The critical speed N_c is given by

$$N_c = \sqrt{\frac{g \sin \delta}{2\pi^2 D}} \tag{21-152}$$

where g is the gravitational acceleration, δ is the angle of the disc to the horizontal, and D is the disc diameter. The typical operating range for discs is 50 to 75 percent of critical speed, with angles δ of 45° to 55°. This range ensures a good tumbling action. If the speed is too low, sliding will occur. If the speed is too high, particles are thrown off the disc or openings develop in the bed, allowing spray blow-through and uneven buildup on the disc bottom. The choice of proper speed is influenced by flow properties of the feed materials, bed moisture, and pan angle, in addition to granulation performance.

A key feature of disc operation is the inherent size classification (Fig. 21-162). Centripetal forces throw small granules and ungranulated feed high on the disc, whereas large granules remain in the eye and exit as product. In addition, the granular bed generally sits on a bed of ungranulated powder and freshly formed nuclei. Size segregation (see the subsection Solids Mixing) leads to exit of only product granules from the eye at the rim of the disc. This classification effect substantially narrows exit granule size distribution, as compared to drum granulators, and discs typically operate with little or no pellet recycle. Because of this segregation, positioning of the feed and spray nozzles is key in controlling the balance of granulation rate processes and resultant granule structure. Disc granulators produce the narrowest first-pass granule size distribution of all granulation systems, second only to compaction processes of wet extrusion or fluid-bed coating systems. Total holdup and granule residence time distribution vary with changes in operating parameters, which affect the granule motion on the disc. Total holdup (mean residence time) increases with decreasing pan angle, increasing speed, and increasing moisture content. The residence time distribution for a disc lies between the mixing extremes of plug flow and completely mixed, and it can have a marked effect on granule size distribution and structure. Increasing the disc angle narrows the residence time distribution and promotes layered growth. Residence times of 1 to 2 min are common.

Drum granulators Granulation drums are common in the metallurgical and fertilizer industries and are primarily used for very large throughput applications. In contrast to discs, there is no output size classification, and high recycle rates of off-size product are common. As a first approximation, granules can be considered to flow through the drum in plug flow, although backmixing to some extent is common. As illustrated in Fig. 21-163, a granulation drum consists of an inclined cylinder, which may be either open-ended or fitted with annular retaining rings. Either feeds may be premoistened by mixers to form granule nuclei, or liquid may be sprayed onto the tumbling bed via nozzles or distributor pipe systems. Drums are usually tilted longitudinally a few degrees from the horizontal (0 to 10°) to assist flow of granules through the drum. The critical speed for the drum is calculated from Eq. (21-152) with $\delta = 80°$ to 90°. To achieve a cascading, tumbling motion of the load, drums operate at lower fractions of critical speed than discs, typically 30 to 50 percent of N_c. If drum speed is too low, intermittent sliding of the bed will occur with poor tumbling motion; if too high, material will be pinned to the drum wall, increasing the likelihood of bed cataracting and spray blow-through. Scrapers of various designs are often employed to control buildup of the drum wall. Holdup in the drum is between 10 and 20 percent of the drum volume. Drum lengths range from 2 to 5 times the diameter. Power and capacity scale with drum volume. Holdup and mean residence time are controlled by drum length, with difficult systems requiring longer residence times than those that agglomerate readily. Residence times of 1 to 2 min are common.

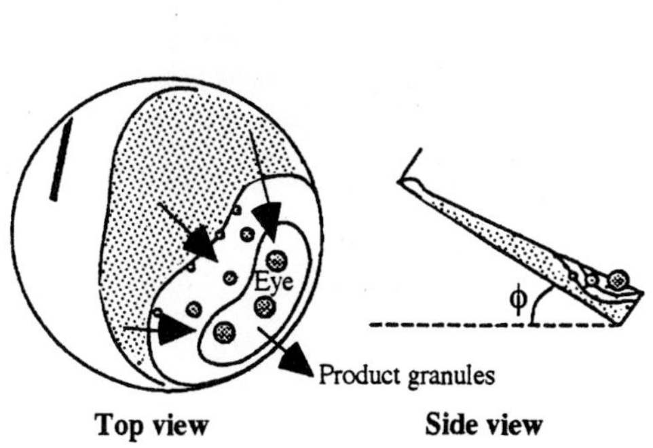

FIG. 21-162 Granule segregation on a disc granulator, illustrating a size classified granular bed sitting on ungranulated feed powder.

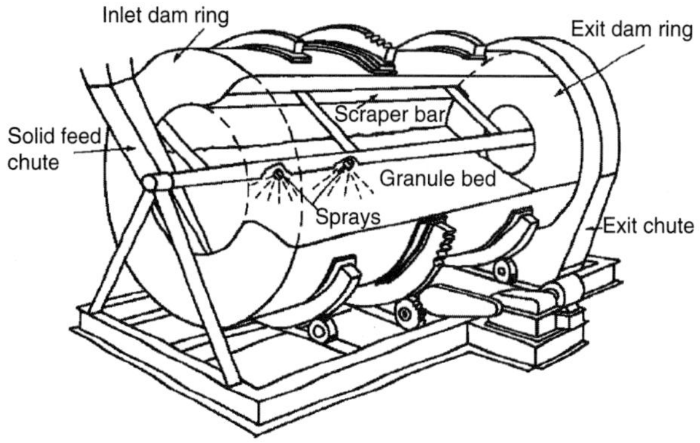

FIG. 21-163 A rolling drum granulator. (*Capes,* Particle Size Enlargement, *Elsevier, 1980.*)

Drum granulation plants often have significant recycle of undersize—and sometimes crushed oversize—granules. Recycle ratios between 2:1 and 5:1 are common in iron ore balling and fertilizer granulation circuits. This large recycle stream has a major effect on circuit operation, stability, and control. A surge of material in the recycle stream affects both the moisture content and the size distribution in the drum. Surging and limiting cycle behavior are common.

Mixer Granulators Mixer granulators contain an agitator to mix particles and liquid and cause granulation. In fact, mixing any wet solid will cause some granulation, even if unintentionally. Mixer granulators have a wide range of applications including ceramics, pharmaceuticals, agrichemicals, and detergents (Table 21-18), and they have the following advantages:

- They can process plastic, sticky materials and can spread viscous binders. That is, they can operate in the mechanical dispersion regime of wetting and the deformable regime of growth (see the subsection Wetting and Nucleation).
- They are less sensitive to operating conditions than tumbling and fluid-bed granulators, although associated with this is less understanding of control and scale-up of granulation mechanisms.
- High-intensity mixers are the only type of granulator that can produce small (<2-mm) high-density granules.

Power and maintenance costs are higher than for tumbling granulators. Outside of high-intensity continuous systems (e.g., the Schugi® in-line mixer), mixers are not feasible for very large throughput applications if substantial growth is required. Granules produced in mixer granulators may not be as spherical as those produced in tumbling granulators, and they are generally denser due to higher agitation intensity. Control of the amount of liquid phase and the intensity and duration of mixing determine agglomerate size and density. Due to greater compaction and kneading action, generally less liquid is required in mixers than in tumbling and fluid-bed granulation. As opposed to tumbling and fluidized-bed granulators, an extremely wide range of mixer granulator equipment is available. The equipment can be divided somewhat arbitrarily into low- and high-shear mixers, although there is considerable overlap in shear rates, and actual growth mechanisms also depend on wet mass rheology in addition to shear rates.

Low-shear mixers Low-speed mixers include (1) ribbon or paddle blenders, (2) planetary mixers, (3) orbiting screw mixers, (4) sigma blade mixers, and (5) double-cone or V blenders, operating with rotation rates or impeller speeds less than 100 rpm. (See Fig. 21-164.) Pug mills and paddle mills as well as ribbon blenders are used for both batch and continuous applications. These devices have horizontal troughs in which rotate central shafts with attached mixing blades of bar, rod, paddle, and other designs. The vessel may be of single- or double-trough design. The rotating blades throw material forward and to the center to achieve a kneading, mixing action. These mills have largely been replaced by tumbling granulators in metallurgical and fertilizer applications, but they are still used as a premixing step for blending very different raw materials, e.g., filter cake with dry powder.

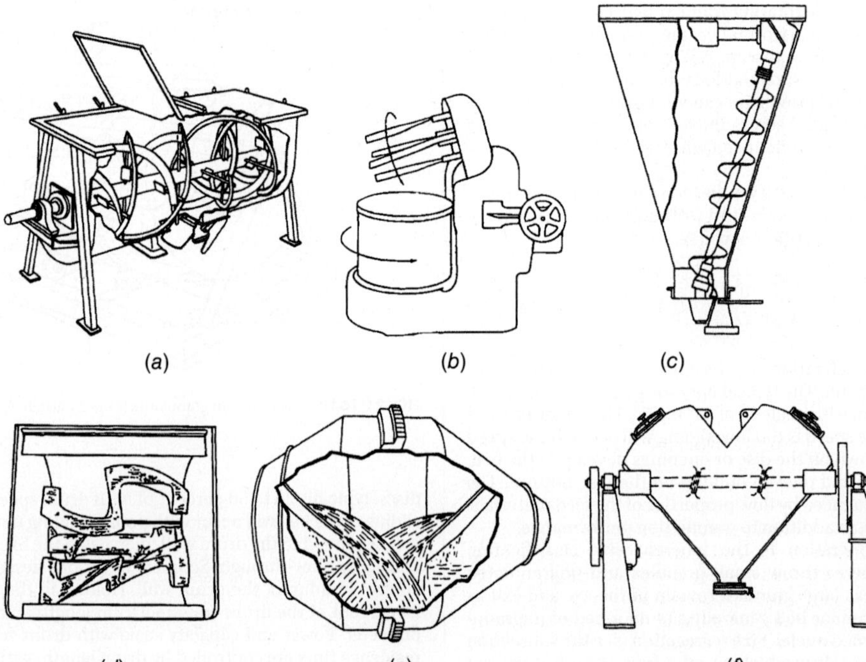

FIG. 21-164 Examples of low-shear mixers used in granulation. (*a*) Ribbon blender; (*b*) planetary mixer; (*c*) orbiting screw mixer; (*d*) sigma blender; (*e*) double-cone blender with baffles; (*f*) V blender with breaker bar. (See also "Solids Mixing.") [(*b*) and (*d*), Chirkot and Propst, in Parikh (ed.), *Handbook of Pharmaceutical Granulation Technology*, 2d ed., Taylor & Francis, 2005.]

Batch planetary mixers are used extensively in the pharmaceutical industry for powder granulation. A typical batch size of 100 to 200 kg has a power input of 10 to 20 kW. Mixing times in these granulators are quite long (20 to 40 min), and many have been replaced with batch high-shear mixers.

High-shear mixers High-speed mixers include continuous shaft mixers and batch high-speed mixers. Continuous shaft mixers have blades or pins rotating at high speed on a central shaft. Both horizontal and vertical shaft designs are available (Figs. 21-165 and 21-166). Examples include the vertical Schugi® mixer and the horizontal pin or peg mixers. These mixers operate at high speed (200 to 3500 rpm) to produce granules of 0.5 to 1.5 mm with a residence time of a few seconds, during which intimate mixing of a sprayed liquid binder and fine cohesive feed powder is achieved. However, little time is available for substantial product growth or densification, and the granulated product is generally fine, irregular, and fluffy with low bulk density. Schugi® and pin mixer capacities may range up to 200 tons/h with power requirements up to 200 kW. Typical plant capacities of peg mixers are 10 to 20 tons/h. Examples of applications include detergents, agricultural chemicals, foodstuffs, clays, ceramics, and carbon black.

Batch high-shear mixer granulators are used extensively in the pharmaceutical industry, where they are valued for their robustness to processing a range of powders as well as their ease of enclosure. Plow-shaped mixers rotate on a horizontal shaft at 60 to 800 rpm, with impeller tip speeds of the order of 10 m/s. Most designs incorporate an off-center high-speed cutter or chopper rotating at much higher speed (500 to 3500 rpm), which breaks down over wet powder mass and limits the maximum granule size. Scale ranges from 10 to 1200 L, with granulation times on the order of 5 to 10 min, which includes both wet massing and granulation stages operating at low and high impeller speed, respectively. Several designs with both vertical and horizontal shafts are available (see Fig. 21-167).

Fluidized Granulators In fluidized granulators (fluidized beds and spouted beds), particles are set in motion by air rather than by mechanical agitation. Applications include fertilizers, industrial chemicals, agricultural chemicals, pharmaceutical granulation, and a range of coating processes. Fluidized granulators produce high-porosity granules by means of agglomeration of powder feeds or high-strength, layered granules due to coating of seed particles or granules by liquid feeds.

Figure 21-168 shows a typical production-size batch fluid-bed granulator. The air handling unit dehumidifies and heats the inlet air. Heated fluidization air enters the processing zone through a distributor, which also supports the particle bed. Liquid binder is sprayed through an air atomizing nozzle located above, in, or below the bed. Bag filters or cyclones are needed to remove dust from the exit air. Other fluidization gases such as nitrogen are also used in place of, or in combination with, air to avoid potential explosion hazards due to fine powders. Continuous fluid-bed granulators are used in the fertilizer, food, and detergent industries. For fertilizer applications, near-size granules are recycled to control the granule size distribution. Dust is not recycled directly, but is first remelted or slurried in the liquid feed.

FIG. 21-165 The Schugi Flexomix® vertical high-shear continuous granulator. (*Courtesy Bepex Corp.*)

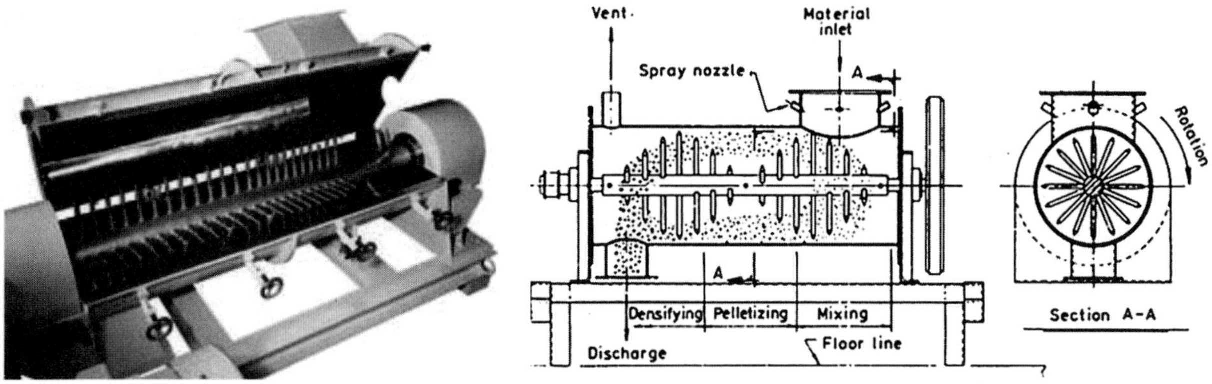

(a) (b)

FIG. 21-166 Examples of horizontal high-shear mixers. (*a*) CB 75 horizontal pin mixer. (*Courtesy Lödige GmbH*). (*b*) Peg granulator. (*Capes, Size Enlargement, 1985.*)

Advantages of fluidized beds over other granulation systems include high volumetric intensity, simultaneous drying and granulation, high heat- and mass-transfer rates, and robustness with respect to operating variables on product quality. Disadvantages include high operating costs with respect to air handling and dust containment and the potential of defluidization due to uncontrolled growth, making them unsuitable generally for very viscous fluid binders or unwettable powders. [See Parikh (ed.), *Handbook of Pharmaceutical Granulation Technology*, 2d ed., CRC Press, 2009 for additional details.]

The hydrodynamics of fluidized beds is covered in detail in Sec. 17. Only aspects specifically related to particle size enlargement are discussed here. Granular products from fluidized beds are generally group B or group D particles under Geldart's powder classification. However, for batch granulation, the bed may initially consist of a group A powder. For granulation, fluidized beds typically operate in the range $1.5U_{mf} < U < 5U_{mf}$, where U_{mf} is the minimum fluidization velocity and U is the operating superficial gas velocity. For batch granulation, the gas velocity may need to be increased significantly during operation to maintain the velocity in this range as the bed particle size increases.

For group B and D particles, nearly all the excess gas velocity $U - U_{mf}$ flows as bubbles through the bed. The flow of bubbles controls particle mixing, attrition, and elutriation. Readers should refer to Sec. 17 for important information and correlations on Geldart's powder classification, minimum fluidization velocity, bubble growth and bed expansion, and elutriation. In summary, however, it is important that mixing, bed turnover, solids flux, bed expansion, shear within the dense phase of the bed, and heat- and mass-transfer control drying scale with fluid-bed excess gas velocity $U-U_{mf}$.

Due to the good mixing and heat transfer properties of fluidized beds, the exit gas temperature is assumed to be the same as the bed temperature, when operating with proper fluidization. Fluidized-bed granulators also act simultaneously as dryers and therefore are subject to the same mass and energy balance limits as dryers: (1) Solvent concentration of the atomized binding fluid in the exit air cannot exceed the saturation value for the solvent in the fluidizing gas at the bed temperature, and (2) the supplied energy in the inlet air must be sufficient to evaporate the solvent and maintain the bed at the desired temperature. Both these limits restrict the maximum rate of liquid feed or binder addition for a given inlet gas velocity and temperature. The liquid feed rate, however, is generally further restricted to avoid excess coalescence.

Compaction Equipment

Tablet Compaction Tableting presses are employed in applications having strict specifications for weight, thickness, hardness, density, and appearance in the agglomerated product. They produce simpler shapes at higher production rates than do molding presses. A single-punch press is one that will take one station of tools consisting of an upper punch, a lower punch, and a die. A rotary press employs a rotating round die table with multiple stations of punches and dies. Older rotary machines are single-sided; that is, there is one fill station and one compression station to produce one tablet per station at every revolution of the rotary head. Modern high-speed rotary presses are double-sided; that is, there are two feed and compression stations to produce two tablets per station at every revolution of the rotary head. For successful tableting, a material must have suitable flow properties to allow it to be fed to the tableting machine. Wet or dry granulation is typically used to improve the flow properties of materials. Dry granulation is simply a precompaction, typically roll pressing or loose slug tableting. It is important to note that if a compaction process is used to improve flowability, the flowable powder must still be compactable in the tableting operation. If the powder is overcompacted prior to tableting, further deformation of the powder will not be possible, and no interparticle bonds will form.

Figure 21-169 illustrates the stations of a typical rotary tablet press of die filling, weight adjustment, compaction, punch unloading, tablet ejection, and tablet knockoff. See the earlier subsection Compaction Mechanisms for discussion of the impact of powder properties on die filling, compaction,

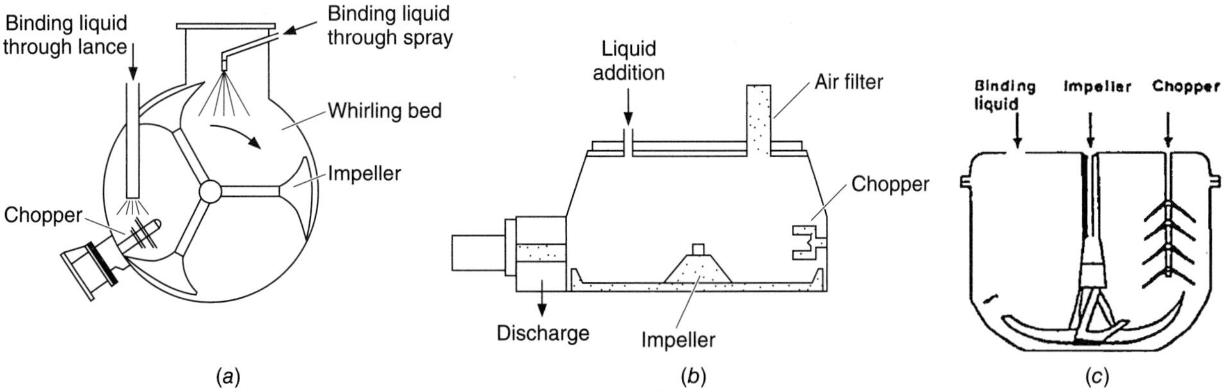

(a) (b) (c)

FIG. 21-167 High-shear mixer granulators for pharmaceutical granule preparation for subsequent tableting. (*a*) Horizontal plough shear, (*b*) vertical bottom-driven shear, and (*c*) vertical top-driven shear.

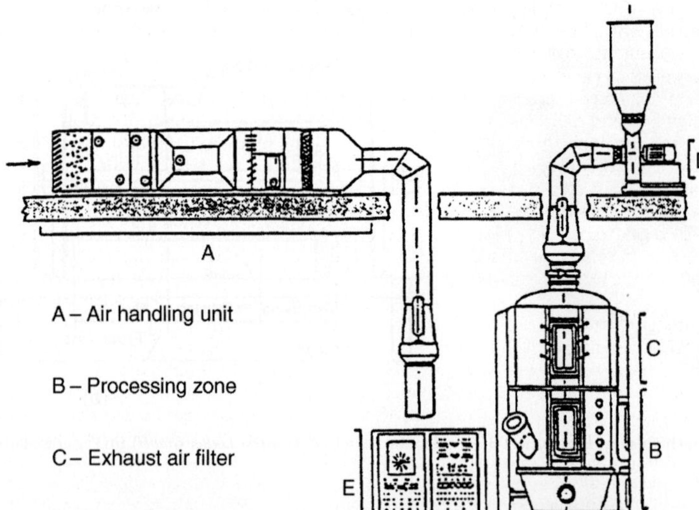

FIG. 21-168 Fluid-bed granulator for batch processing of powder feeds. [*Ghebre-Sellasie (ed.),* Pharmaceutical Pelletization Technology, *Marcel Dekker, 1989.*]

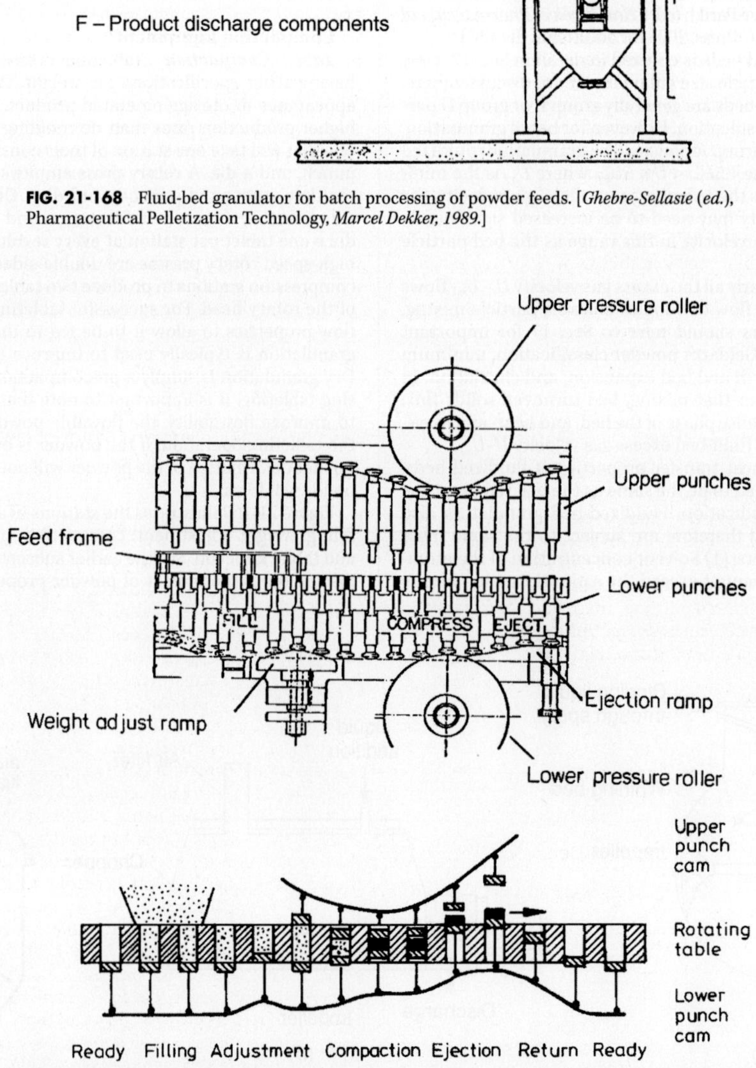

FIG. 21-169 Typical multistation rotary tableting press, indicating stages of tableting for one station. (*Pietsch,* Size Enlargement by Agglomeration, *Wiley, Chichester, 1992.*)

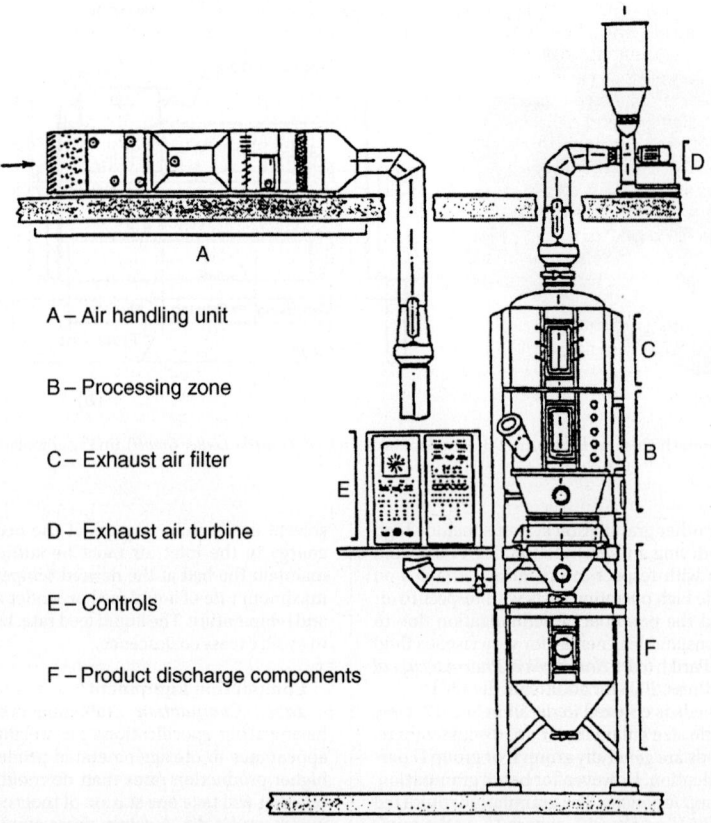

and ejection forces. As discussed above, these stages of compaction improve with increased stress transmission (controlled by lubrication and die geometry), decreased deaeration time (increasing powder permeability and decreasing production rate), increased plastic, permanent deformation, and increased powder flowability (decreasing powder cohesion, increased powder flowability, and increased die diameter and clearances).

Roll Pressing and Briquetting Roll presses compact raw material as it is carried into the gap between two rolls rotating at equal speeds (Fig. 21-170). The size and shape of the agglomerates are determined by the geometry of the roll surfaces. Pockets or indentations in the roll surfaces form briquettes the shape of eggs, pillows, teardrops, or similar forms from a few grams up to 2 kg (5 lb) or more in weight. Smooth or corrugated rolls produce a solid sheet, which can be milled down into the desired particle size on conventional grinding equipment.

Roll presses can produce large quantities of materials at low cost, but the product is less uniform than that from molding or tableting presses. The introduction of the proper quantity of material into each of the rapidly rotating pockets in the rolls is the most difficult problem in the briquetting operation. Various types of feeders have helped to overcome much of this difficulty. The impacting rolls can be either solid or divided into segments. Segmented rolls are preferred for hot briquetting, as the thermal expansion of the equipment can be controlled more easily. Roll presses provide a mechanical advantage in amplifying the feed pressure to some maximum value. This maximum pressure and the roll compaction time control compact density. Generally speaking, as compaction time decreases (e.g., by increasing roll speed), the minimum necessary pressure for quality compacts increases. There may be an upper limit of pressure as well for friable materials or elastic materials prone to delamination.

Pressure increase occurs in two regions of the press (see Fig. 21-170). Above the angle of nip, sliding occurs between the material and roll surface as material is forced into the rolls, with intermediate pressure ranging from 1 to 10 psi. Energy is dissipated primarily through overcoming particle friction and cohesion. Below the angle of nip, no slip at the roll occurs as the powder is compressed into a compact, and pressure may increase up to several thousand pounds per square inch. Both of these intermediate- and high-pressure regions of densification are indicated in the compressibility diagram of Fig. 21-160. The overall performance of the press and its mechanical advantage (i.e., maximum to feed pressure ratio) depend on the mechanical and frictional properties of the powder. For design procedures, see Johanson [*Journal of Applied Mechanics* **32:** 4 (1965)]. The nip angle generally increases with decreasing compressibility, or with increasing roll friction angle and effective angle of friction. The mechanical advantage pressure ratio increases and the time of compaction decreases with decreasing nip angle since the pressure is focused over a smaller roll area. In addition, the mechanical advantage generally increases with increasing compressibility and roll friction.

MODELING

Before any detailed modeling is to be conducted, a simple mass and energy balance of the process should be completed. Also even rough computation of relevant dimensionless groups and comparison to regime maps (see the earlier subsection Consolidation and Growth) can provide sufficient descriptive understanding to guide an engineer toward a successful size enlargement process operation. For wet agglomeration processes, in cases where greater precision is required, or more mathematics is desired, the appropriate tool is *population balance modeling* (PBM). For compaction processes the appropriate tool is *finite element analysis* (FEA).

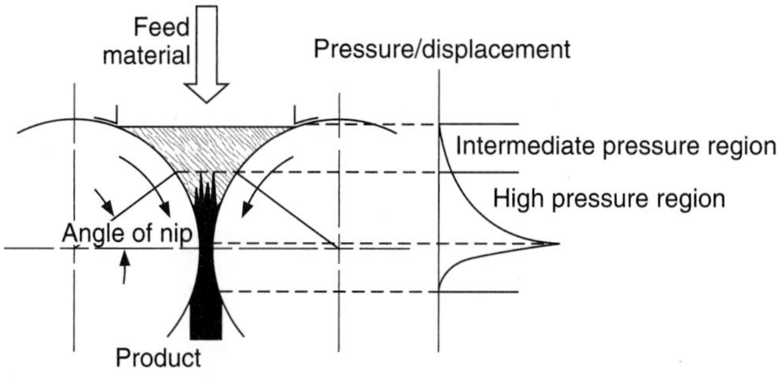

FIG. 21-170 Regions of compression in roll presses. Slippage and particle rearrangement occur above the angle of nip, and powder compaction at high pressure occurs in the nonslip region below the angle of nip.

Population Balance Modeling PBM is a mathematical modeling technique which describes the interacting kinetics of distributed quantities. In size enlargement processes, the most relevant distributed quantity is the granule size. Hence, many PBMs of size enlargement track the time evolution of the particle size distribution. A PBM can be useful for providing (1) a critical evaluation of data to determine the controlling granulation mechanisms, (2) prediction of granule size distributions, (3) sensitivity analysis relating operating conditions to product quality, and (4) full agglomerate circuit simulation. Widespread use of PBM has been limited for a number of reasons. First, the rate expressions for the underlying mechanisms are difficult to predict and many times are unknown. Furthermore, the resulting equations of a PBM are conceptually and mathematically complex and difficult to solve, even numerically (i.e., by multidimensional nonlinear partial integrodifferential equations). However, several commercial process simulation software providers now offer PBM tools that allow the user to develop a PBM without the need for the user to develop the PBM solver. With these advances, it is possible that PBM will gain wider use in the near future.

Here only a brief introduction to population balances is provided. For a more thorough introduction refer to (in order of increasing complexity and mathematical rigor) Randolph and Larson, *Theory of Particulate Processes*, Elsevier, 2012; Hulburt and Katz, *Chemical Engineering Science* **19:** 555 (1964); and Ramkrishna, *Population Balances*, Academic Press, 2000. The PBM is a statement of continuity for particulate systems. It includes a rate expression for each mechanism that changes a particle property (e.g., size, porosity, saturation). The PBM follows the change in the granule size distribution as granules are born, die, grow, and enter or leave the control volume. Those rate phenomena are nucleation, growth, consolidation, and breakage (see the subsection Wet Agglomeration Rate Processes). For mathematical convenience, growth and breakage are separated into continuous and discontinuous transformations. Continuous growth occurs via layering; i.e., larger granules grow continuously from one size to another via coating by smaller particles. Discontinuous growth occurs via coalescence. After a successful coalescence event, two comparably sized granules form a single larger granule. Mathematically the size of the granules discontinuously jumps to a larger size. The relevant analogs for breakage are abrasion and fragmentation. Abrasion occurs when fine particles are chipped from the edges of larger particles such that they continuously shrink in size. Fragmentation occurs when a large particle suffers a breakage event and forms multiple comparably sized fragments. As an example, consider a batch granulator with no breakage. The size distribution of granules is given by the function $n(v, t)$, where $n(v, t)$ is the number of granules with individual granule volume in the range v to $v + dv$ at time t. The time evolution of this size distribution is then given by the equation

$$\begin{aligned}\frac{\partial}{\partial t}n(v,t)+\frac{\partial}{\partial v}[G(v,t)n(v,t)] &= B(v,t)\\ &+\frac{1}{2N(t)}\int_{u=0}^{u=v}\beta(u,v-u,t)n(u,t)n(v-u,t)du \\ &-\frac{1}{N(t)}\int_{u=0}^{u\to\infty}\beta(u,v,t)n(v,t)n(v,t)du\end{aligned} \tag{21-153}$$

where G is the growth rate (volume per time) due to layering, B is the nucleation rate (number per volume per time), N is the total number of particles in the system (number), and β is the coalescence kernel (per number per time). The use of the term *kernel* comes from the integral equation literature. The function $\beta(u, v)$ describes the rate at which granules of volume u and v collide and coalesce. The leftmost term of Eq. (21-153) represents the accumulation of particles of volume v. The next term represents granules growing both out of and into size v due to layering. The derivative arises because this mechanism involves immediately adjacent sizes. The first term on the right represents birth of granules due to nucleation. The next two terms on the far right represent birth and death due to coalescence, respectively. Birth of a granule of volume v occurs when granules of volume u and $u - v$ collide and coalesce. When these two collide, they become one, hence division by 2. Also a granule can only coalesce with another to become size v if both were previously less than size v. Therefore, the integration is over sizes less than v. The last term represents the loss of a granule of size v when it collides with any other size granule, hence integration over all sizes.

Note the left-hand side of Eq. (21-153) is a convective derivative when the velocity along the v coordinate is G. For this reason, interpreting a population balance as a conservation law across a population of particles in a state space of external (physical position) and internal (particle characteristic) coordinates is very convenient for theoretical development. Also Eq. (21-153) was written in terms of granule volume v. However, it may be written in terms of granule size x or expanded to track other internal coordinates as well, e.g., granule liquid and pore volume.

For mathematical forms and parameter-fitting procedures of the various rate laws used in agglomeration processes, refer to Chaudhury et al. ["Mechanistic Modeling of High-Shear and Twin Screw Mixer Granulation Processes," in Pandey and Bharadwaj (eds.), *Predictive Modeling of Pharmaceutical Unit Operations*, Elsevier, 2017, p. 99] and references therein.

Finite Element Analysis *Finite element analysis* (FEA) is a numerical method useful for computing stresses in solids under load and deformation. By their nature, compaction processes are essentially just solids under large strain; therefore, finite element analysis lends itself well to prediction involving these processes. Several commercial FEA software packages exist as a result of the tool's widespread use in mechanical design. The crux of FEA use for compaction is the mathematical description of the material yield and flow known as the *plasticity rule*. For application to powder compaction, the Drucker-Prager Cap (DPC) model is the most accepted plasticity rule. A recent and thorough description of DPC use for powder compaction is given by Cunningham et al. ["Modeling of Powder Compaction with the Drucker-Prager-Cap Model," in Pandey and Bhardwaj (eds.), *Predictive Modeling of Pharmaceutical Unit Operations*, Elsevier, 2017, p. 205]. This approach has been applied to roller compaction and tablet compaction for examining relative density spatial inhomogeneities [e.g., Muliadi et al., *Powder Technology* **237**: 386 (2013); Sinha et al., *Powder Technology* **202**: 46 (2010)]. It has also successfully been applied to predicting tablet debossement feature failures [Swaminathan, S., Hilden, J., Ramey, B., et al., *J. Pharm. Innov.* **11**: 214 (2016)].

FEEDING, METERING, AND DOSING

General References: Wilson, *Feeding Technology for Plastics Processing*, Hanser, 1998; Dhodapkar et al., *Chemical Engineering*, pp. 26–33, Jan. 2006; Fayed and Skocir, *Mechanical Conveyors—Selection and Operation*, Technomic, 1997; Colijn and Carroll, *British Chemical Engineering*, Aug. 1970; Thomson in Fayed and Otten, *Handbook of Powder Science and Technology*, 2d ed., Chapman & Hall, 1997.

INTRODUCTION

Feeders are used in the chemical process industry to control the withdrawal rate of bulk solids from storage silos, bins, and hoppers. Their sizes can range from large feeding devices at the bottom of storage silos to small additive feeders in compounding applications. Feeding technology heavily draws upon the following disciplines (Wilson, *Feeding Technology for Plastics Processing*, Hanser, 1998):

1. *Bulk solids handling:* Characterization, flowability, silo/hopper design, stress and power calculations
2. *Mechanical design:* Hardware design, reliability, fabrication tolerances, bearings, seals, and usability
3. *Electronic hardware and controls:* Weighing technology, PLCs, smart controls, communication, solids flowmeters
4. *Statistics:* Data analysis and processing (sampling, variability, accuracy or linearity, precision or repeatability, control charts)

Feeders should not be confused with dischargers or discharge aids. Discharge aids are designed to assist the gravity discharge of solids out of bins, hoppers, and silos, whereas a feeder is a throttling device that assumes an uninterrupted and consistent incoming flow from the upstream storage container. A feeder is not designed to assist in discharge from the hopper above; however, poor feeder design and operation can negatively impact the flow out of a well-designed mass flow hopper.

Similarly, feeders are not conveyors either, although they share many design attributes of conveyors. Conveyors, such as vibratory, screw, or belt conveyors, are designed to transport solids between two points. They do not control, regulate, or throttle the mass flow rate of the process stream. While there are broad similarities between screw feeders and screw conveyors, their operating characteristics are starkly different.

A well-designed feeder will exhibit the following characteristics:

1. Delivery of consistent and controlled discharge of material over the specified rates (turndown)
2. Uniform withdrawal of material across the outlet cross section of the feed hopper
3. Interface with the feed hopper to minimize stresses and power draw (especially for large installation)
4. Acceptable product attrition rate
5. Delivery of required accuracy (linearity) and precision (repeatability) over the entire range of feed rates
6. Ease of maintenance and acceptable wear and failure rates of key components
7. Ability to interface with existing process control system along with flexibility in programming and troubleshooting
8. Long-term stability of operation

The feed hopper (or storage silo) and the feeding device must be treated as a system where the selection of suitable hopper geometry, feeder, and hopper-feeder interface is intricately related. If the flow out of the hopper is erratic and inconsistent, even the most sophisticated feeder hardware and software technology will not result in acceptable performance. However, a feeder design that does not withdraw product uniformly across the hopper outlet or keep the discharge cross section "fully live" can result in poor performance, especially for fine powders and sticky products. The third leg of this stool is the hopper-feeder interface or transition piece. The design of the transition piece is critical for belt, vibratory, apron, and rotary feeders.

The applications of feeding devices in the process can be broadly classified as (1) those requiring feed rate control and (2) those requiring dosing or batching of a known amount of product. The essential hardware (mechanical design) is common, and the main differences are in the system configurations and the control logic. All the mechanical feeding devices (feeders) operate in volumetric mode where they modulate and control the volume of bulk solids passing through them, hence the term *volumetric feeders*. The most widely used volumetric feeders are screw, belt, vibratory, apron, rotary airlock, plow, and table feeders. Since the bulk density of incoming material can vary, these feeders cannot maintain a constant mass flow rate at the discharge. Gravimetric feeders, which are essentially volumetric feeders with a feedback control logic (see Fig. 21-171), are designed to deliver a specified mass flow rate or totalized weight (batching) regardless of variations in the bulk density. Common gravimetric feeders are *loss-in-weight* (LIW) screw or vibratory, and weigh-belt feeders. In fact, any feeder or conveyor, when coupled with a solids mass flowmeter downstream and with appropriate control logic, can operate as a gravimetric feeding device.

VOLUMETRIC FEEDERS

Eight major types of volumetric feeders are used in industry:

- Screw feeder
- Vibratory feeder
- Belt feeder
- Apron feeder
- Rotary airlock
- Plow feeder
- Table feeders
- Slide-gate

Numerous variants and derivative designs exist in each category.

Most of these feeders will work satisfactorily for free-flowing materials, and the choice may depend on process constraints and economics. However, for difficult-to-handle products, a careful assessment of feeder suitability must be made. As discussed previously, volumetric feeders are at the heart of volumetric and gravimetric feeding technology. Therefore, selection of a suitable feeder plus its design features to match the process conditions and performance metrics is critical. The following factors must be taken into consideration in selecting a volumetric feeder (Wilson, *Feeding Technology for Plastics Processing*, Hanser, 1998; Wahl, *Powder and Bulk Engineering*, May 1985; Dhodapkar et al., *Chemical Engineering*, pp. 26–33, Jan. 2006):

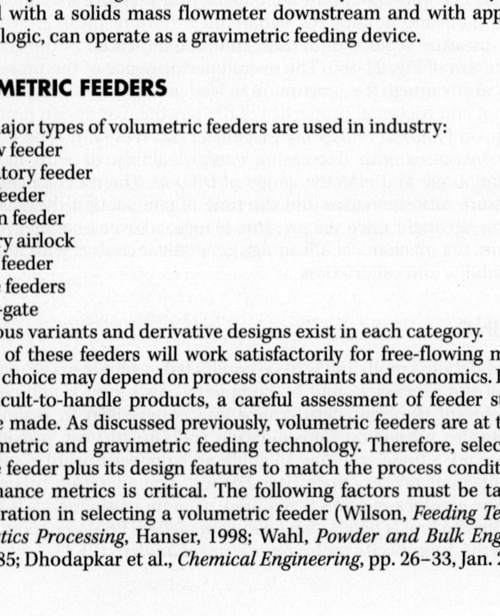

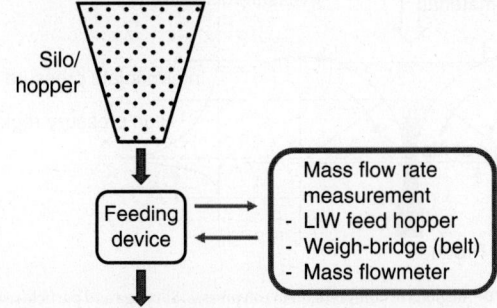

FIG. 21-171 Gravimetric feeder concept.

1. Range of bulk materials—single product or multiple grades—and variability in product due to upstream process conditions
2. Properties of bulk materials at process conditions (temperature, pressure, humidity)
   a. Bulk density, compressibility
   b. Deaeration and flooding characteristics
   c. Particle size and shape; presence of large chunks
   d. Moisture content
   e. Sensitivity to temperature exposure
   f. Friability
   g. Surface properties (sticky, greasy)
   h. Flowability
   i. Dust explosibility
   j. Corrosivity
   k. Toxicity
3. Process requirements
   a. Range of feed rates: maximum, minimum, turndown
   b. Accuracy (linearity), precision (repeatability) in the range
   c. Seal against pressure
   d. Sanitary requirements (e.g., food, pharmaceutical); allowable cross-contamination
   e. Need for dust control and containment
   f. Ability to handle unexpected materials (tramp metal, large chunks)
   g. Material of construction
   h. Inerting
4. Compatibility with the feed hopper or silo outlet shape
5. Available space or headroom constraints
6. Vibrations near installation
7. Process upstream and downstream

Screw Feeders Screw feeders are the most common volumetric feeders in practice today. The essential design of a screw dates back to Archimedes (third century BC) where he used a helical surface around a central cylindrical shaft to pump water. The mechanism to push solids forward remains identical; however, a wide range of designs are now available to handle different types of solids.

They are ideally suited for applications that require a dust-tight enclosure, low available headroom, wide range of particle sizes, and under hoppers and bins with elongated or slot outlets. They are not recommended when the product is friable, sticky, greasy, or highly abrasive. Highly aeratable products with flooding tendency are not compatible with screw feeders. Screw feeders used in additive or compounding technology share similar limitations, some of which are overcome with selection of the twin-screw configuration. These are discussed later in the text.

It is well known from practical experience, and quantified by Jenike's theory (Jenike, "Storage of Solids," Bulletin 123, University of Utah, 1964) that the wedge-shaped hoppers or transition hoppers with elongated slot outlet with aspect ratio (length/width) > 3 have favorable arching dimension as compared to the axisymmetric conical outlet. Also mass flow can be achieved at relatively shallower hopper angles (Schulze, *Powders and Bulk Solids*, Springer, 2008) in wedge-shaped hoppers. A screw feeder is an ideal discharge device for a slot outlet.

While screw conveyors operate at fill efficiencies less than 45 percent, screw feeders operate at fill efficiencies between 85 and 100 percent. The surrounding barrel constrains the solids as they are pushed forward by a smooth wall helical surface. For a screw feeder to function properly, the screw surface should be smooth, the barrel (or trough) surface must be rough, and the product should not adhere to the surface of the screw. Otherwise, the solids will simply roll in place with the shaft (called logging), and the discharge will cease.

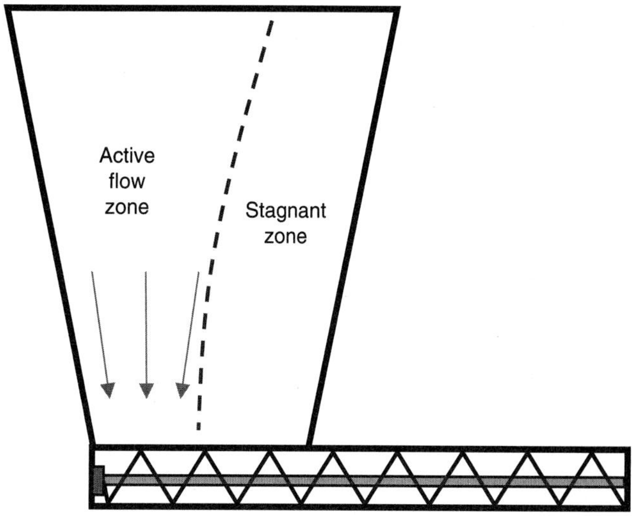

FIG. 21-172 Discharge pattern with constant pitch screw.

When the outlet of the bin and silo is smaller than 1 to 1.5 screw pitches, the product is drawn into the flights such that the entire outlet cross section of bin and silo is "live." However, in typical slot outlets, where the length of the slot far exceeds the screw pitch, the product is preferentially withdrawn from the back of a constant pitch screw (see Fig. 21-172). This creates a stagnant zone within the hopper, which manifests itself as a rathole with cohesive powders or variability in residence time distribution or segregation problems with coarse solids.

An ideal screw design, therefore, is one in which the capacity of the screw increases in the direction of discharge (Marinelli and Carson, *Chemical Engineering Progress*, pp. 47–51, Dec. 1992; Carson, *Powder and Bulk Engineering*, Dec. 1987; Thompson, *Chemical Engineering*, Oct. 30, 1978). It allows the product to be uniformly extracted from the hopper above. Various approaches to achieve uniform drawdown have been evaluated, namely, increasing pitch, tapering flight diameter, tapering the shaft, and combinations thereof. More detailed discussions can be found in the cited references. The optimal screw feeder configuration, as proposed by Carson, has a conical shaft and constant pitch in the beginning, followed by increasing pitch, and then constant pitch in the conveying section downstream of the outlet (Fig. 21-173). It is also recommended that the bin outlet width be equal to the screw diameter. The fabrication tolerances for the screw pitch and the clearance between the

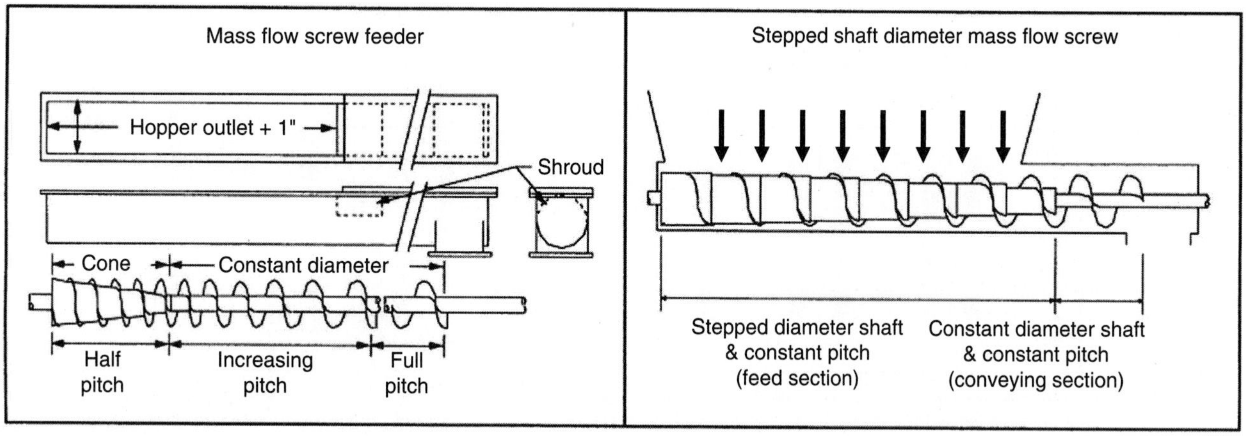

FIG. 21-173 Optimal screw design for mass-flow screw feeder.

screw and the trough are critical design parameters. An optimal *L/D* ratio is between 3 and 6, and the screw should not operate faster than 40 rpm (depending on size).

Screws, by their very design, provide an open helical path for solids flow. Conversely, air from downstream processes can flow back into the feeder if sufficient pressure differential exists. This can cause aeration of powders in the silo/bin, thereby reducing the feeder capacity, and, in the worst case, flooding (uncontrolled discharge) of the powder. To reduce the leakage rate, several modifications to the screw design are commonly made: (1) reduce the pitch of the screw at the discharge end, (2) remove a section of the screw in the middle of the screw, and (3) place a gate or flapper at the discharge end (Bell et al., *Chemical Engineering Progress*, Feb. 2003).

One of the main differences, apart from its size and capacity, between screw feeders installed as discharge devices below a storage silo/bin and unitized feeders used for additive or component feeding (such as in compounding applications) is the presence of a flow-conditioning mechanism in the latter. A combination of small screw size and poor flowability of fine cohesive powders requires a conditioning (or agitation) mechanism. Two approaches are most prevalent: (1) agitator, horizontal/vertical powered independently or coupled with the screw motor, and (2) flexing wall feed hopper. They serve to prevent arching and reduce the bulk density variations in the material. The arching problems in large silos/bins can be addressed by designing for mass flow and keeping the outlet dimension sufficiently large. However, due to the scale of small feeders, these options are often not viable.

Feeder Sizing The unitized feeders must be sized such that the screw rpm is approximately 75 percent of the maximum rpm at the highest rate (Bell et al., *Chemical Engineering Progress*, Feb. 2003). An oversize feeder will deliver pulsating flow at minimum rates. A typical turndown ratio is 10:1; however, higher turndowns can be achieved with special screw designs.

Screw Selection The common choices for screw designs are auger type, spiral type (also called *pigtail* screws) and spiral type with core. Unlike bin/silo discharge screws, where auger screws with progressively increasing capacity (mass flow screws) are common, most of the screws in unitized and additive feeders are simple screws. The handling properties of bulk material dictate the suitability of a screw in an application (see Fig. 21-174). A standard screw has a pitch-to-diameter ratio of 1. Increased-pitch screws offer relatively higher capacity but more pulsatile flows at lower speeds, whereas reduced-pitch screws deliver smoother flow with lower volumetric capacity. To reduce the pulsations in the output, complex single screws with double or triple flight design are available.

As shown in Fig. 21-172, twin-screw (or multiscrew) options are also commercially available. Twin-screw designs deliver a less pulsatile flow and are designed to be self-cleaning. A twin-screw configuration interfaces with a larger outlet opening, hence it is more adept for cohesive powders.

Feed Hopper It is highly recommended that the feeder hopper be designed for mass flow per Jenike's theory.

Testing and Evaluation For a new application, with no prior operational experience in feeding the product, vendor testing is recommended. The vendor's knowledge and experience on similar products can be leveraged if the material in question is free-flowing. However, product with unusual characteristics (e.g., cohesive, sticky, lumpy, stringy, hygroscopic, caking) must be tested for proper selection and sizing. Guidelines for vendor testing can be found in Wilson (*Feeding Technology for Plastics Processing*, Hanser, 1998) and Johnson [*Proceedings of the Powder & Bulk Solids Conference*, pp. 659–668 (1993)]. Obtaining a representative sample for testing is critical. It is also recommended to run longer trials (>1 day) for new materials in critical applications.

Vibratory Feeders Vibratory or oscillatory motion has been applied in bulk solids handling operations for screening, discharge aid, and conveying for a long time. Vibratory conveyors and feeders share the same hardware (trough, excitation mechanism). Both operate on the principle of imparting forward trajectory to the bed of solids on an oscillating trough or pan. A sinusoidal force is applied to the trough with an excitation mechanism. Various excitation mechanisms are available, with electromagnetic drivers and eccentrically loaded electric motors being the most common.

The primary function of a feeder is to control and modulate the feed rate from the silo/hopper outlet, whereas a conveyor is designed to convey the product regardless of the rate of incoming stream. A commonly accepted breakpoint in frequency is 900 vibrations per minute (15 Hz). Conveyors typically operate at lower frequencies (5 to 15 Hz) and higher stroke length (5 to 100 mm), and feeders operate at higher frequencies (15 to 120 Hz) and lower stroke length (1 to 25 mm). The resulting acceleration on bulk material is 1 to 4 *g* for conveyors and 4 to 14 *g* for feeders. Feeders require stronger structural support than conveyors since the reactionary stresses are proportional to stroke but to the square of the frequency.

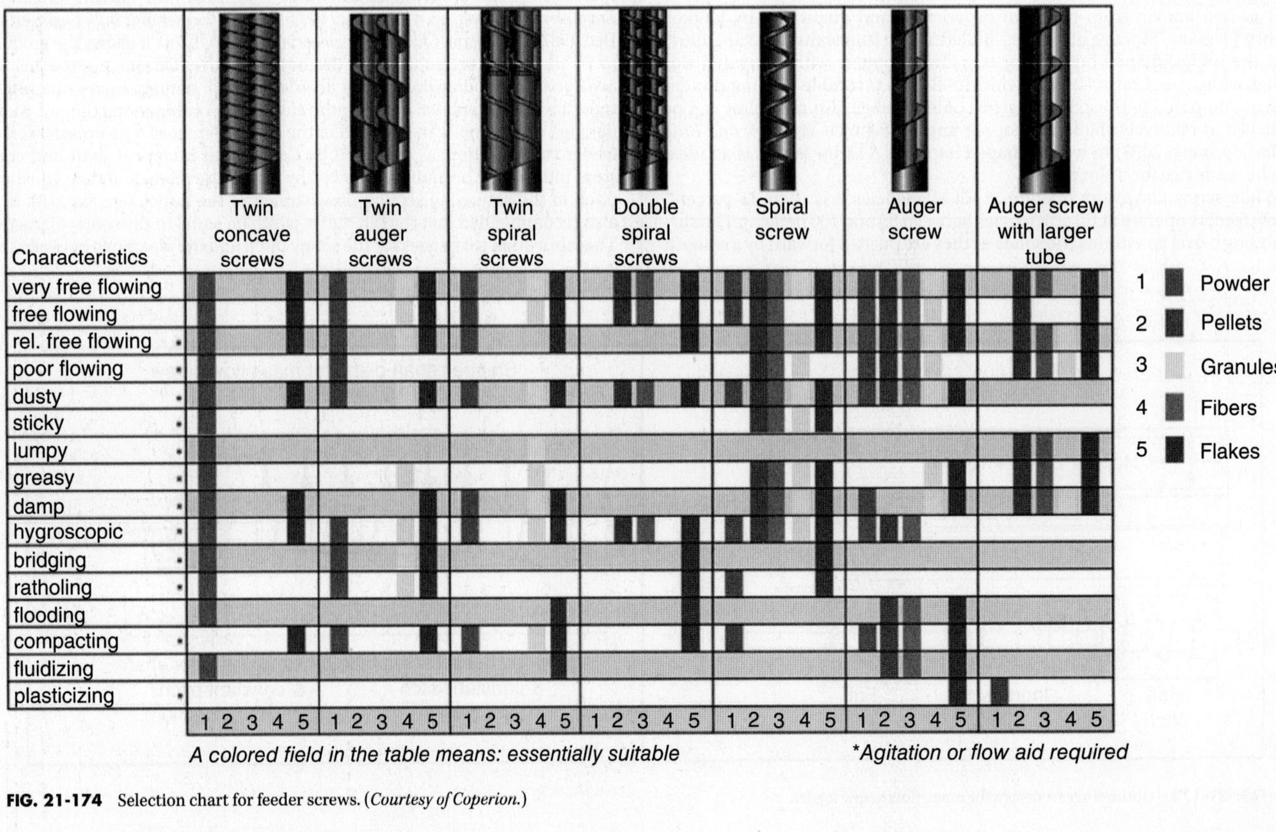

FIG. 21-174 Selection chart for feeder screws. (*Courtesy of Coperion.*)

A vibratory feeder operates by imparting a forwarding motion to the bed of material in a series of throws and catches as the trough vibrates due to the stroke applied at an angle (usually between 20° and 45°). The velocity of the moving bed is a function of stroke, stroke angle, frequency, trough friction, and material properties. For all practical purposes, the velocity can be considered proportional to stroke and frequency. The volumetric rate of the feeder is equal to the product of the cross-sectional area of the bed on the trough and the velocity of the moving bed.

Vibratory feeders are most suitable for friable products and where hygiene, containment, and cleanability are important.

Vibratory Feeder Designs There are three primary types of feeders based on the excitation mechanism: brute force; natural frequency; and two-mass tuned feeders (Fayed and Skocir, *Mechanical Conveyors—Selection and Operation*, Technomic, 1997; Colijn and Carroll, *British Chemical Engineering*, Aug. 1970; Thomson in Fayed and Otten, *Handbook of Powder Science and Technology*, 2d ed., Chapman & Hall, 1997).

The driving mechanism is attached directly to the feeder trough in the brute-force design (Fig. 21-175). The drive imparts force through the center of gravity of the feeder. These feeders are rugged in design, and they are able to withstand shock loads and variable loads. They are considered to be low-cost machines with higher operating cost. The ability to change the feed rate is limited.

The excitation mechanism in natural frequency feeders is attached to the base of the feeder, and the trough acts as an oscillating mass attached to the base with springs. Using the principle of resonance, if the frequency of applied force is equal to the natural frequency of the spring-mass system, then a natural magnification is achieved. To avoid self-destruction due to excessive amplitudes at the resonant frequency (similar to the famous Tacoma, Washington, bridge failure), these feeders are operated at subresonant frequencies (80 to 90 percent of resonant frequency). These feeders can operate smoothly and quietly with minimal energy input and low maintenance (Fayed and Skocir, *Mechanical Conveyors—Selection and Operation*, Technomic, 1997). If the material load on the feeder is increased, it will lower the resonant frequency of the system and bring the vibrating system closer to resonant frequency. Consequently, the intensity of vibrations is amplified, and the feeder cleans itself out. This response is similar to a vibratory conveyor. However, feeders are required to deliver a constant rate regardless of the incoming material load. In that case, a highly detuned and/or damped system with heavy trough is used.

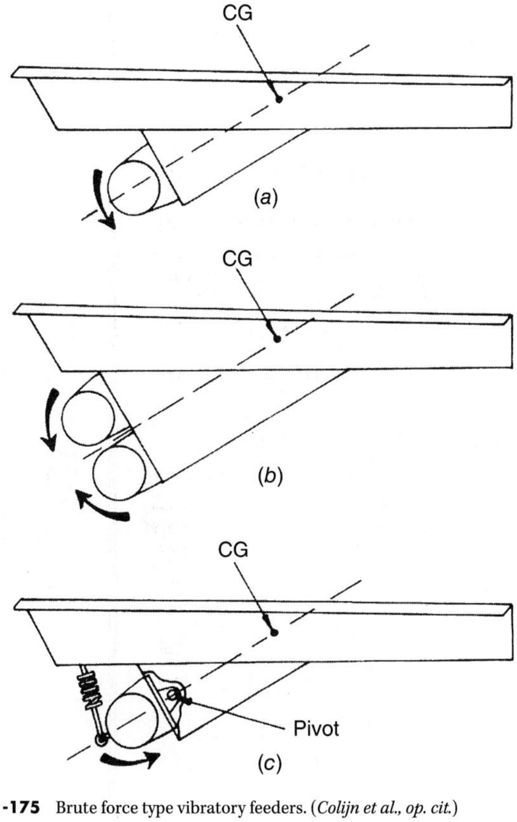

FIG. 21-175 Brute force type vibratory feeders. (*Colijn et al., op. cit.*)

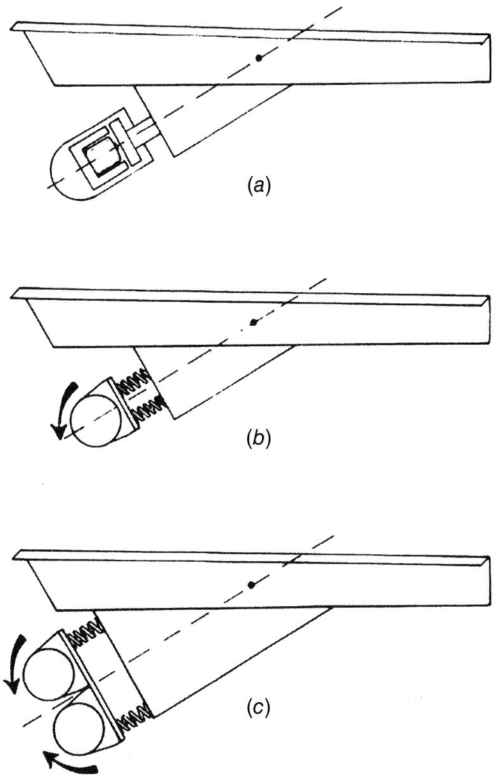

FIG. 21-176 Tuned two-mass type vibratory feeders. (*Colijn et al., op. cit.*)

Tuned two-mass feeders (see Fig. 21-176) are similar to the natural frequency feeders except that the excitation mechanism and the trough constitute a two-mass system joined by springs. This design is, by far, the most common in process industry.

The conveying speed of material on the trough is typically between 0.05 and 0.30 m/s (10 to 60 ft/min).

Declining Feeders The capacity of vibratory feeders is proportional to the frequency and amplitude (1/2 stroke) of the vibrations. Increasing the intensity of vibrations also increases the structural stresses on the feeder and its support, and this will result in more frequent failures. The capacity of vibratory feeders can be increased by about twofold by declining the feeder toward the discharge end. The downward inclination can be up to 20° from the horizontal. In high-capacity declining feeders, the trough length is shorter than the slope generated by the dynamic angle of repose. Increase in capacity comes at the cost of reduced feed rate control [Rademacher, *Feeders and Vibratory Conveyors*, The University of Newcastle Research Associates (TUNRA)—Australia Report, Aug. 1979].

Belt Feeders Belt feeders consist of a rubber or elastomeric belt that extracts the material from the outlet by shearing action. Unlike belt conveyors, belt feeders are relatively flat and short in length. To prevent spillage of material, a skirt is usually provided. The belt speeds may range from 1 to 100 ft/min. For uniform withdrawal across the bin/hopper outlet, a slight taper is provided at the outlet (see Fig. 21-177).

Apron Feeders Apron feeders are used for applications where high impact loads may cause excessive wear of belt feeders. They are more expensive than belt feeders for comparable throughput. See Fig. 21-178.

Rotary Vane/Airlock Feeders Rotary airlock feeders are volumetric feeders that can operate successfully against adverse pressure gradients up to 4 bar. The valve consists of vanes or blades (6 to 12) attached symmetrically to a rotating shaft, called the *rotor assembly*, which is enclosed in a cast and machined housing (see Fig. 21-179). The airlock between the high- and low-pressure sides is maintained by the tight clearances between the rotor and housing. Additional seals are provided to protect the bearings from solids. Numerous variations of rotor designs and housing configurations are available to match the application needs (Fig. 21-180) (Thomson, *Chemical Engineering*, Oct. 30, 1978; Pittman, *Chemical Engineering*, July 1985).

The feed rate can be modulated by changing the speed (rpm) of the rotor (see the Pneumatic Conveying subsection for more details). The fill

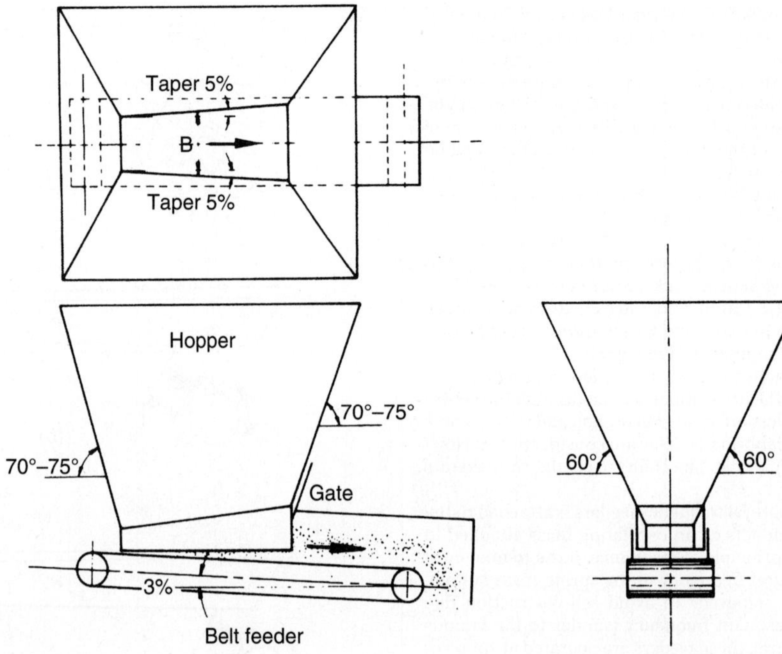

FIG. 21-177 Belt feeder configuration. (*Colijn, op. cit.*)

efficiency of the rotor pockets will depend on material flowability, pressure differential, and rotor speed. The fill efficiency begins to drop at high speeds (typically >30 rpm); therefore, the feed rate becomes nonlinear with speed (Fig. 21-181). At low speeds, especially for large feeders with full pockets, the discharge rate is pulsatile. The discharge rate can be smoothed by selecting a reduced capacity rotor and operating it at higher speeds. Rotary feeders are not recommended for highly abrasive products due to high wear.

Rotary Plow (Plough) Feeders These are used to feed bulk solids from bins with elongated outlet or stockpiles onto conveyors (Fig. 21-182). The bulk solids are discharged from the bin outlet onto a flat shelf, forming a pile due to the angle of repose. A traveling rotary plow with straight or curved blades extracts the material from the pile and discharges to the conveyor below. Linear drag plows can also be used. More details on capacity and power consumption calculations can be found in Colijn, and Carroll, *British Chemical Engineering*, Aug. 1970.

Table Feeders Table feeders have a rotating disk under a hopper which is mounted off-center from the axis of the hopper. The disk has a groove or a series of holes that fill as they traverse the bottom discharge. The material in the groove or the holes is extracted as it emerges from the bottom of the hopper. These feeders are used in compounding applications for precise feeding of highly cohesive additives. See Fig. 21-183.

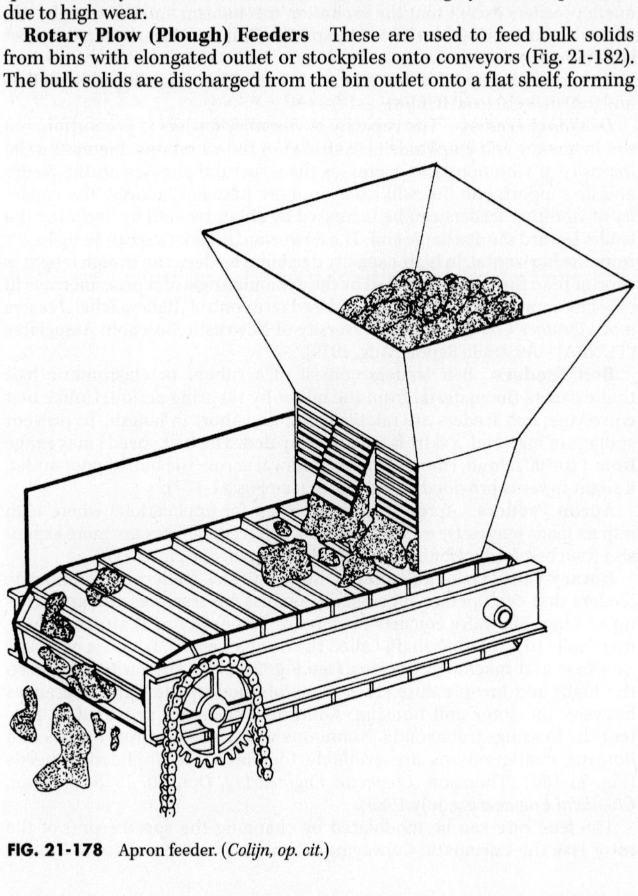

FIG. 21-178 Apron feeder. (*Colijn, op. cit.*)

FIG. 21-179 Typical rotary feeder. (*Courtesy of Coperion.*)

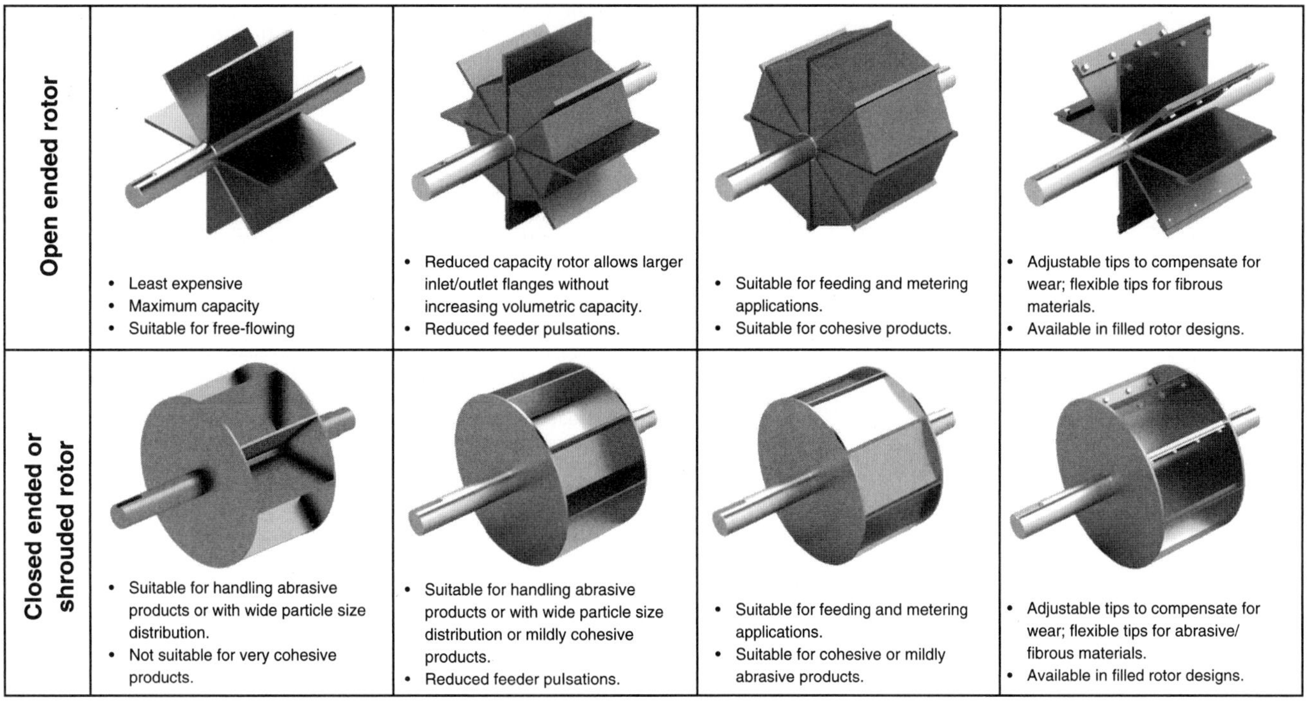

FIG. 21-180 Various vane designs (https://www.younginds.com). (*Courtesy of Young Industries.*)

Slide Gate The discharge rate of coarse and free-flowing materials, such as plastic pellets or grains, out of a slide gate is reasonably predictable with the existing correlations and semitheoretical models (Dhodapkar et al., *Chemical Engineering Progress*, May 2016). The slide gate can be used in conjunction with other gravimetric methods or solids flowmeters to design an accurate feeding system.

GRAVIMETRIC FEEDERS

When feeder accuracy of less than 1 percent is required, gravimetric feeders must be used. A gravimetric feeder is essentially a volumetric feeder with a feedback loop that controls the feeder parameters (typically speed) to achieve the desired set point (Fig. 21-171). Gravimetric feeders can be only as good as the volumetric feeder it is based on. Therefore, selection and specification of a suitable feeder is the key first step.

There are three common approaches for measuring the discharge rate for control input: loss in weight (LIW), weigh bridge (weigh belt), and solids flowmeter.

LIW Feeder Screw feeders are the most common feeding component of a LIW feed system. The feeder consists of a volumetric feeder (e.g., screw), feed hopper, refill valve, weigh scale, and control system for feedback control. The entire assembly is mounted on a weigh scale and isolated from

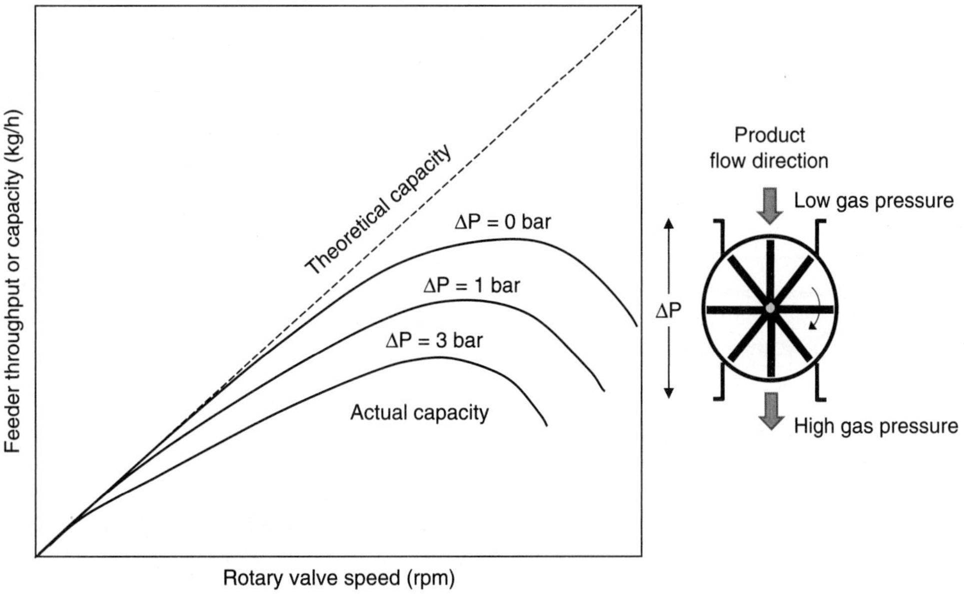

FIG. 21-181 Effect of rotor speed and pressure differential on filling efficiency of rotary feeder.

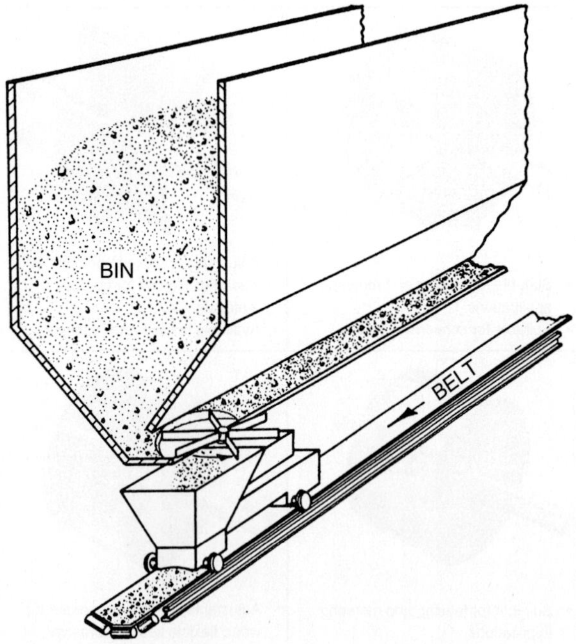

FIG. 21-182 Plow feeder. (*Colijn, op. cit.*)

the upstream and downstream process. The rate of change of weight is equal to the instantaneous feed rate. The control system based on feedback control adjusts the screw speed to maintain a constant discharge (mass) flow rate. The control schemes in the commercial LIW feeders have become very sophisticated with the clever use of noise filtering, data averaging, and fault analysis.

While discharging the feed hopper, the LIW feeder operates in gravimetric mode, as discussed above. However, the hopper must eventually be refilled, and the LIW logic does not apply when the hopper is being emptied and refilled at the same time. The feeder control logic temporarily puts the feeder into "volumetric mode" by keeping the screw rpm constant while the hopper is being refilled. The refill time is usually less than 10 s, and the refill rates should be 10 to 20 times the average feed rate. Additional time must be allowed for deaeration and weight stabilization before the gravimetric mode can be resumed. These delays can be programmed into the control logic. Since the feeder accuracy is unknown and is worse during the volumetric mode, the duration of the volumetric mode must be minimized. Often the size of the feed hopper is increased to reduce the number of refills required. Typical hopper refill rates range from 6 to 10 times per hour.

The vendor of the LIW feeder "assumes" that the reliable discharge will occur once the refill valve is opened. Therefore, reliable design of the supply hopper or storage silo is critical for successful feeding application. Further, insufficient venting of displaced air from the feed hopper will result in excessive refill and stabilization times. See Fig. 21-184.

Weigh-Belt Feeder A belt feeder can be converted to a weigh belt by replacing one of the idlers with a weighing element called the *weigh bridge*. It measures the weight of the material on the belt as it passes over it. Weigh belt feeders (Fig. 21-186) are typically installed directly under storage silos. The feeder-hopper interface must be designed to ensure uniform withdrawal across the outlet. A feed gate is installed to achieve desired loading (layer thickness) on the belt. The belt speed serves as the primary means to modulate and control the feed rate based on the feedback from the weigh bridge.

Unlike LIW feeders, weigh belts do not require a feed hopper, which results in a low headroom requirement. They can be designed to feed against pressure by making the enclosure dust-tight. Belt feeders are susceptible to flooding and spillage when handling fine aeratable powders. A comparison of LIW and gravimetric feeders is given in Table 21-20.

Solids Flowmeters The technologies for measuring the mass flow rate of bulk solids as a granular stream and as a suspension (e.g., pneumatic conveying) have evolved on multiple fronts. The key relevant, widely practiced, and successful mass flowmeters are based on the following concepts:

1. Force measurement due to impact on target plate
2. Measurement of centripetal force on a smooth chute or combination of chutes
3. Coriolis force measurement
4. Microwave technology

Coriolis-based mass flowmeters are the only ones that do not require a field calibration and are agnostic to the bulk properties (density, friction, particle size). However, because of the spinning wheel configuration it cannot be used for friable products.

Performance Metrics for Gravimetric Feeders A good working knowledge of statistical methods is essential to analyze the performance of gravimetric feeders. Some feeder vendors advertise performance

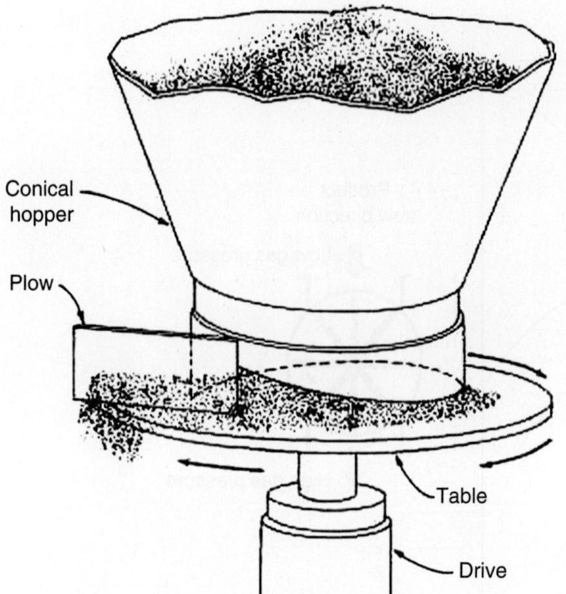

FIG. 21-183 Table feeder.

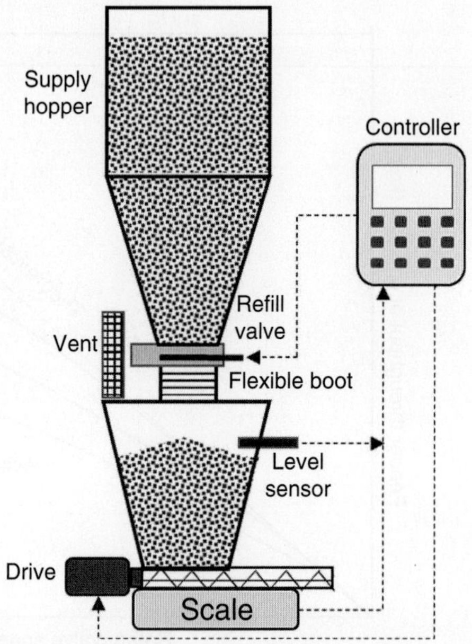

FIG. 21-184 Loss-in-weight feeder configuration.

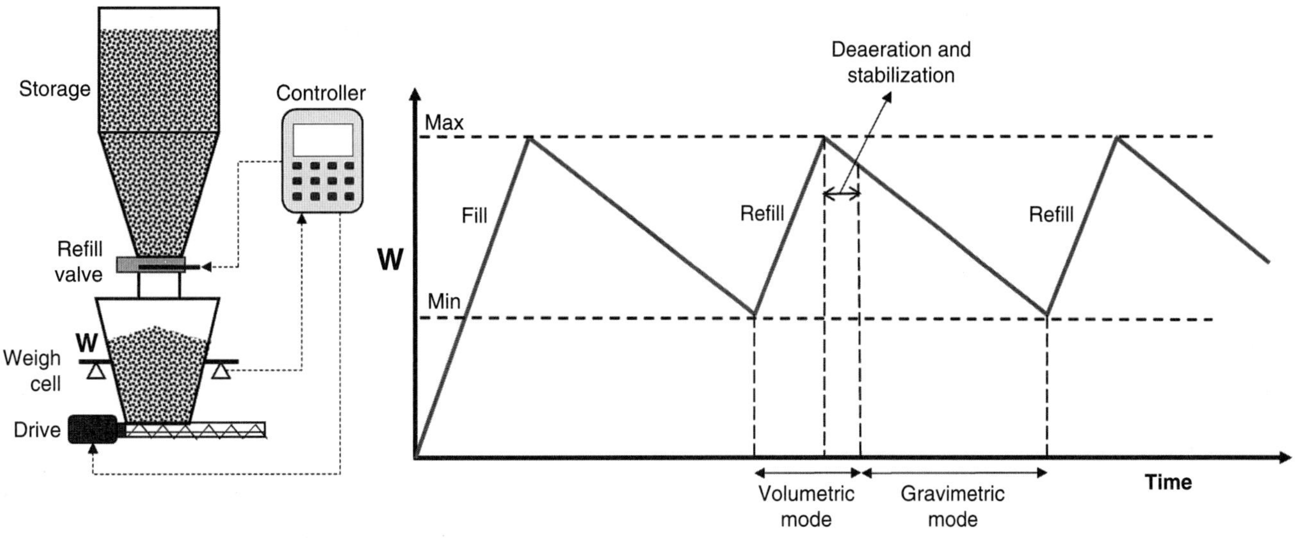

FIG. 21-185 Loss-in-weight feeder operating cycle.

specifications without elaborating the statistical basis, which can lead to wrong equipment selection or project failure due to lack of performance.

The important performance metrics are accuracy or linearity; precision or repeatability; and stability.

1. Accuracy or linearity: This is the measure of how close the feeder's set point is to the actual delivery rate. But quoting a single number without further qualification is misleading. The following additional information must be provided:

- What was percentage of error based on set point or full scale?
- What was the turndown from full-scale setpoint (10:1 or 15:1 or 20:1)?
- How many consecutive samples were taken?
- How many screw revolutions or belt revolutions were there during the sampling time?
- What was the total duration of sampling?
- What was the magnitude of change in weighing system (X percent of scale capacity) during the test?

2. Precision or repeatability: It is the measure of variance in the discharge rate. It is a measure of consistency rather than accuracy. In addition to the qualifiers listed above, it is necessary to specify the confidence level (1σ, 2σ, or 3σ).

3. Stability: It is the measure of performance drift over time due to the quality of hardware, software, and electronics. Control charts can be used to detect unusual changes.

An excellent treatment of these metrics can be found in Wilson (*Feeding Technology for Plastics Processing*, Hanser, 1998). A comprehensive glossary of feeding terms was compiled by Kuchneman (*Powder and Bulk Engineering*, Dec. 2000).

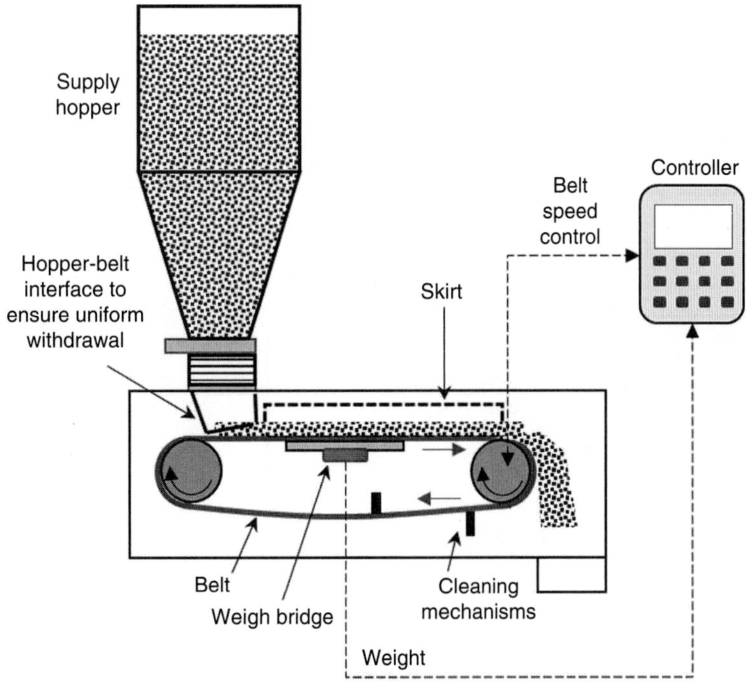

FIG. 21-186 Weigh-belt feeder.

TABLE 21-20 Comparison of LIW and Weigh Belt

Selection factor	Weigh-belt feeder	LIW feeder
Feed rate range	High feed rates at lower capital costs	Covers entire range of feed rates, especially low rates
Range of materials	Limited	Wide
Self-taring capability	No	Yes
Headroom requirement	Low	High
Dust containment	Poor	Good
Scale accessibility	Poor	Good
Metering capability	Yes	No
Suitability for higher pressure	Yes	No

Reprinted with permission from *Powder and Bulk Engineering*, December 1990, "Selecting a gravimetric feeder: Weighbelt or loss-in-weight," Andy Kovats, Control and Metering Limited.

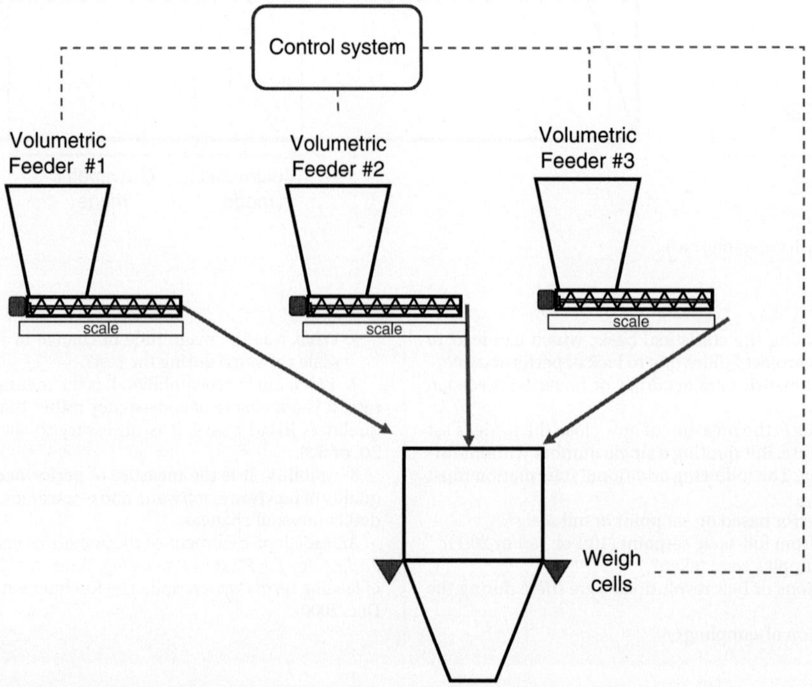

FIG. 21-187 Gain-in-weight batching concept.

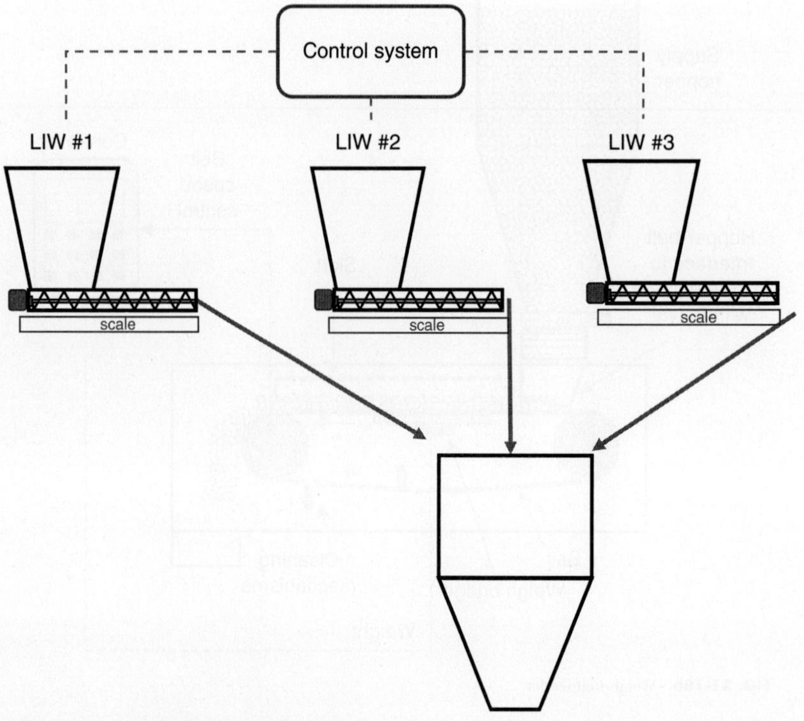

FIG. 21-188 Loss-in-weight batching concept.

Dosing or Batching Application Batch operations, such as addition of components in a batch mixer, require feeders to accurately deliver a specified amount of product to the batch irrespective of the feed rate. This is known as *endpoint control* or *dosing*. It is not uncommon to add over a dozen different components to the mix with compositional contribution ranging from fractions to tens of percent.

There are two approaches to batching operations: (1) gain-in-weight batching; and (2) loss-in-weight batching. Gain-in-weight batching, as shown in Fig. 21-187, uses a simple weigh hopper to measure the amount of component added at any given time. The feeders are programmed to deliver the majority of the dose at high rate and then to switch over to low rate (dribble or trickle mode) to reach the target set point on totalized weight. Each component must be added sequentially. The sensitivity of the scale must be sufficient to measure the addition of minor components with desired accuracy. Because of this sequential approach, the cycle time can be long for a multicomponent mixture. Such systems can be low cost due to minimal hardware requirements.

Loss-in-weight batching, as shown in Fig. 21-188, essentially uses multiple LIW feeders in parallel to add each of the components. The components can be added simultaneously, thereby saving on cycle time. The accuracy is better since each feeder can be customized based on the dosing amount. Simultaneous addition of mixture components can also create a more homogeneous premix which improves the dispersion of minor ingredients. However, these benefits come with increased cost of hardware and headroom requirement.

Section 22

Waste Management

Louis Theodore, Eng.Sc.D. *Consultant, Theodore Tutorials, Professor of Chemical Engineering, Manhattan College; Member, Air and Waste Management Association (Coeditor, Pollution Prevention)*

Paul S. Farber, P.E., M.S. *Principal, P. Farber & Associates, LLC, Willowbrook, Illinois; Member, American Institute of Chemical Engineers, Air & Waste Management Association (Coeditor, Air Pollution Management of Stationary Sources)*

Kenneth N. Weiss, P.E., BCEE, B.Ch.E, M.B.A. *Managing Partner, ERM; Member, Air and Waste Management Association (Introduction to Waste Management and Regulatory Overview)*

Matthew K. Heermann, P.E., B.S. *Consultant—Fossil Power Environmental Technologies, Sargent & Lundy LLC, Chicago, Illinois (Introduction to Waste Management and Regulatory Overview)*

Rita D'Aquino, M.E. *Consultant, Member, American Institute of Chemical Engineers (Pollution Prevention)*

John D. McKenna, Ph.D. *Principal, ETS, Inc.; Member, American Institute of Chemical Engineers, Air and Waste Management Association (Air Pollution Management of Stationary Sources)*

Robert R. Sharp, P.E., Ph.D. *Environmental Consultant; Professor of Environmental Engineering, Manhattan College; Member, American Water Works Association; Water Environment Federation Section Director (Wastewater Management)*

Susan A. Thorneloe, M.S. *U.S. EPA/Office of Research & Development, National Risk Management Research Laboratory; Member, Air and Waste Management Association, International Waste Working Group, the Advisory Board for the International Waste Management and Landfill Symposium (Solid Waste Management); Advisory Board Member of the International Journal "Waste Management" Elsevier and the IWWG Advisory Board*

INTRODUCTION TO WASTE MANAGEMENT AND REGULATORY OVERVIEW

The contributions of Anthony J. Buonicore to the "Waste Management" section in the sixth and later editions of this handbook are acknowledged.

POLLUTION PREVENTION

GREEN CHEMISTRY AND GREEN ENGINEERING

AIR POLLUTION MANAGEMENT OF STATIONARY SOURCES

INDUSTRIAL WASTEWATER MANAGEMENT

MANAGEMENT OF SOLID WASTES

List of Abbreviations

Abbreviation	Definition
3P	Pollution prevention pays
ABS	Alkyl benzene sulfonate
ACC	Annualized capital costs
BACT	Best available control technology
BAT	Best available technology
BCOD	Biodegradable chemical oxygen demand
BCT	Best conventional technology
BOD	Biochemical oxygen demand
BSRT	Biomass solids retention time
BTEX	Benzene, toluene, xylene
CAA	Clean Air Act
CAAA	Clean Air Act Amendments
CCP	Comprehensive costing procedures
CFR	Code of federal regulations
CKD	Cement kiln dust
COD	Chemical oxygen demand
CPI	Chemical process industries
CRF	Capital recovery factor
CTDMPLUS	Complex terrain dispersion model plus algorithms for unstable situations
CWRT	Center for Waste Reduction Technologies
DCF	Direct installation cost factor
DO	Dissolved oxygen
DRE	Destruction and removal efficiency
EBCT	Empty bed contact time
EMAS	Eco-management and audit scheme
EMS	Environmental management system
EPA	Environmental Protection Agency
FML	Flexible membrane liner
GAC/GAX	Granular activated carbon
HAPS	Hazardous air pollutants
HAZWOPER	Hazardous waste operators
HCS	Hauled-container systems
HRT	Reactor hydraulic retention time
HSWA	Hazardous and Solid Waste Act
I-TEF	International toxic equivalency factor
ICF	Indirect installation cost factor
ISO	International Organization for Standardization
LAER	Lowest Achievable Emissions Rate
LCA	Life-cycle assessment
LCC	Life-cycle costing
LOX	Liquid oxygen
MACT	Maximum achievable control technology
MSDS	Material safety data sheets
MSW	Municipal solid waste
MWC	Municipal waste combustors
MWI	Medical waste incinerators
NAAQS	National Ambient Air Quality Standard
NBOD	Nitrogenous biochemical oxygen demand
NIMBY	Not in my back yard
NPDES	National pollutant discharge elimination system
NSPSs	New source performance standards
PCB	Polychlorinated biphenyl
PIES	Pollution prevention information exchange systems
PM	Particulate matter
POTW	Publicly owned treatment work
PPIC	Pollution prevention information clearinghouse
PSD	Prevention of significant deterioration
RACT	Reasonably Available Control Technology
RCRA	Resource Conservation and Recovery Act
RDF	Refuse-derived fuel
SARA	Superfund Amendments and Reauthorization Act
SCR	Selective catalytic reduction
SCS	Stationary-container systems
SE	Strength of treated waste
SMART	Save money and reduce toxics
SO	Strength of untreated waste
SS	Suspended solids
TCC	Total capital cost
TCP	Traditional costing procedures
TGNMO	Total gas nonmethane organics
TOC	Total organic carbon
TSA	Total systems approach
TSCA	Toxic Substances Control Act
TSD	Treatment, storage, and disposal
UASB	Upflow anaerobic sludge blanket
VOC	Volatile organic compound
VOST	Volatile organic sampling train
VSS	Volatile suspended solids
WRAP	Waste reduction always pays
WTE	Waste-to-energy (systems)

General References: United States EPA, *Pollution Prevention Fact Sheet,* Washington, D.C., March 1991; Keoleian, G., and D. Menerey, "Sustainable Development by Design: Review of Life Cycle Design and Related Approaches," *Air & Waste Management* **44** (May 1994); Dupont, R., and K. Ganesan, and L. Theodore, *Pollution Prevention: Sustainability, Industrial Ecology, and Green Engineering,* 2d ed, CRC Press/Taylor & Francis Group, Boca Raton, Fla., 2017; *Getting at the Source,* World Wildlife Fund, Washington, D.C., 1991, p. 7; United States EPA, 1987 *National Biennial RCRA Hazardous Waste Report—Executive Summary,* Washington, D.C., GPO, 1991, p. 10; Theodore, L., *Pollution Prevention: An ETS Theodore Tutorial,* ETS International, Inc., Roanoke, Va., 1993; United States EPA, *The EPA Manual for Waste Minimization Opportunity Assessments,* Cincinnati, Ohio, August 1988; Santoleri, J., J. Reynolds, and L. Theodore, *Introduction to Hazardous Waste Incineration,* 2d ed., Wiley, New York, 2000; ICF Technology Incorporated, *New York State Waste Reduction Guidance Manual,* Alexandria, Va., 1989; Neveril, R. B., *Capital and Operating Costs of Selected Air Pollution Control Systems,* EPA Report 450/5-80-002, Gard, Inc., Niles, Ill., December 1978; Vatavuk, W. M., and R. B. Neveril, "Factors for Estimating Capital and Operating Costs," *Chemical Engineering* (November 3, 1980): 157–162; Vogel, G. A., and E. J. Martin, "Hazardous Waste Incineration," *Chemical Engineering* (September 5, 1983): 143–146 (part 1); Vogel, G. A., and E. J. Martin, "Hazardous Waste Incineration," *Chemical Engineering* (October 17, 1983): 75–78 (part 2); Vogel, G. A., and E. J. Martin, "Estimating Capital Costs of Facility Components," *Chemical Engineering* (November 28, 1983): 87–90; Ulrich, G. D., *A Guide to Chemical Engineering Process Design and Economics,* Wiley-Interscience, New York, 1984; California Department of Health Services, *Economic Implications of Waste Reduction, Recycling, Treatment, and Disposal of Hazardous Wastes: The Fourth Biennial Report,* Sacramento, Calif., 1988, p. 110; Wilcox, J., and L. Theodore, *Engineering and Environmental Ethics,* Wiley, New York, 1998; Theodore, L., *Chemical Engineering: The Essential Reference,* McGraw-Hill, New York, 2014; Abulencia, P., and L. Theodore, *Open-Ended Problems: The Future Chemical Engineering Approach,* Scrivener/Wiley Publishing, Beverly, Mass., 2015; Anastas, P., and T. Williamson, in *Green Chemistry: An Overview, Green Chemistry: Designing Chemistry for the Environment*; ACS Symposium Series 626, ed. P. T. Anastas and T. C. Williamson, American Chemical Society, Washington, D.C., 1995, pp. 1–17; Anastas, P., and J. Warner, *Green Chemistry: Theory and Practice,* Oxford University Press, New York, 1998; Anastas, P., and J. Zimmerman, "Design Through the Twelve Principles of Green Engineering," *Environ. Sci. Technol* **37:** 94A–101A (2003); Pearce, D., A. Mankandya, and E. Barbier, *Blueprint for a Green Economy,* Earthscan, London, 1989; Bishop, P., *Pollution Prevention,* Waveland Press, Prospect Heights, Ill., 2000; United Nations, Report of the World Commission on Environment and Development, General Assembly Resolution 42/187, http://www.un.org/documents/ga/res/42/ares42-187.htm (retrieved Oct. 31, 2007); Graedel, T., and B. Allenby, *Industrial Ecology,* 2d ed., Prentice-Hall, Upper Saddle River, N.J., 2003; Allenby, B., *Industrial Ecology: Policy, Framework and Implementation,* Prentice-Hall, Upper Saddle River, N.J., 1999; Ehlrich, P., and J. Holdren, *Impact of Population Growth,* Scrivener, Washington, D.C., 1971; Tabak, H., et al., "Biodegradability Studies with Organic Priority Pollutant Compounds," United States EPA, MERL, Cincinnati, Ohio, April 1980; Levin, M. A., and M. A. Gealt, *Biotreatment of Industrial and Hazardous Waste,* McGraw-Hill, New York, 1993; Sutton, P. M., and P. N. Mishra, "Biological Fluidized Beds for Water and Wastewater Treatment: A State-of-the-Art Review," WPCF Conference, October 1990; Envirex Equipment Bulletin FB. 200-R2 and private communication, Waukesha, Wis., 1994; Donavan, E. J., Jr., "Evaluation of Three Anaerobic Biological Systems Using Paper Mill Foul Condensate," HydroQual, Inc., EPA, IERL contract 68-03-3074, 1968; Mueller, J. A., K. Subburama, and E. J. Donavan, *Proc. 39th Ind. Waste Conf.,* 599, Ann Arbor, Mich., 1984; ASME Research Committee on Industrial and Municipal Waste—Keeping Society's Options Open—MWCs, A Case Study on Environmental Regulation, 1997; *Chartwell Information,* EBI, Inc., San Diego, June 2004; Wilson, D. G., ed., *Handbook of Solid Waste Management,* Van Nostrand Reinhold, New York, 1997; *Wastes: Engineering Principles and Management Issues,* McGraw-Hill, New York, 1977.

INTRODUCTION TO WASTE MANAGEMENT AND REGULATORY OVERVIEW

In this section, a number of references are made to laws and procedures that have been formulated in the United States with respect to waste management. An engineer handling waste management problems in another country would be well advised to know the specific laws and regulations of that country. Nevertheless, the treatment given here is believed to be useful as a general guide.

MULTIMEDIA APPROACH TO ENVIRONMENTAL REGULATIONS IN THE UNITED STATES

Among the most complex problems to be faced by industry over the past 50 years is the management of natural resource impacts. In the 1970s the engineering profession became acutely aware of its responsibility to society, particularly for the protection of public health and welfare. The decade saw the formation and rapid growth of the U.S. Environmental Protection Agency (EPA) and the passage of federal and state laws governing virtually every aspect of the environment. The end of the decade, however, brought a realization that only the more simplistic problems had been addressed. A limited number of large sources had removed substantial percentages of a few readily definable air pollutants from their emissions. The incremental costs to improve the removal percentages would be significant and would involve increasing numbers of smaller sources, and the health hazards of a host of additional toxic pollutants remained to be quantified and control techniques developed.

Moreover, in the 1970s, air, water, and waste were treated as separate problem areas to be governed by their own statutes and regulations. Toward the latter part of the decade, however, it became obvious that environmental problems were closely interwoven and should be treated in concert. The traditional type of regulation—command and control—had severely restricted compliance options.

The 1980s began with EPA efforts redirected to take advantage of the case-specific knowledge, technical expertise, and imagination of those being regulated. Providing plant engineers with an incentive to find more efficient ways of abating pollution would greatly stimulate innovation in control technology. This is a principal objective, for example, of EPA's "controlled trading" air pollution program, established in the Offsets Policy Interpretative Ruling issued by the EPA in 1976, with statutory foundation given by the Clean Air Act Amendments of 1977. The Clean Air Act Amendments of 1990 expanded the program even more to the control of sulfur oxides under Title IV. In effect, a commodities market on clean air was developed. Efforts to manage greenhouse gas emissions in the 2000s followed this same approach globally, encouraging engineers to find more efficient production methods to reduce resource consumption and energy demands.

The rapidly expanding body of regulation and global environmental concerns presents a formidable challenge to traditional practices of corporate decision making, management, and long-range planning. Those responsible for new plant planning, construction, and operation must take stock of the emerging requirements and develop a fresh approach.

The full impact of the Clean Air Act, the Clean Water Act, the Safe Drinking Water Act, the Resource Conservation and Recovery Act, the Comprehensive Environmental Responsibility, Compensation and Liability (Superfund) Act, and the Toxic Substances Control Act, and now the Clean Power Plan, is still not often generally appreciated. The combination of these requirements, some of which impose conflicting demands or establish differing time schedules, makes the task of obtaining all regulatory approvals very complex, and this has a significant impact on project schedules.

One of the major impacts of environmental regulations is to increase the lead time required to plan and build new plants or to modify existing ones. When new plants generate major environmental complexities, the implications can be profound. The added lead time will vary widely from one case to another, depending on which environmental requirements apply and what difficulties are encountered. For major expansions in any field of heavy industry, however, the delay resulting from environmental requirements could add two to three years. Moreover, there is always the possibility that regulatory approval will be denied. So contingency plans must be developed to meet production needs.

The 1990s saw the emergence of environmental management systems (EMSs) across the globe, including the ISO 14001 environmental management system and the European Union's Eco-Management and Audit Scheme (EMAS). Any EMS is a continual cycle of planning, implementing, reviewing, and improving the processes and actions that an organization undertakes to meet its business and environmental goals. EMSs are built on the Plan, Do, Check, Act model (Fig. 22-1) that leads to continual improvement. Planning includes identifying environmental aspects (and corresponding impacts and compliance obligations) and establishing goals (plan); implementing includes training and operational controls (do); checking includes monitoring and corrective action (check); and reviewing includes progress review and acting to make needed changes to the EMS (act). Organizations that have implemented an EMS sometimes require their major suppliers to become EMS-certified as environmental programs become part of everyday business.

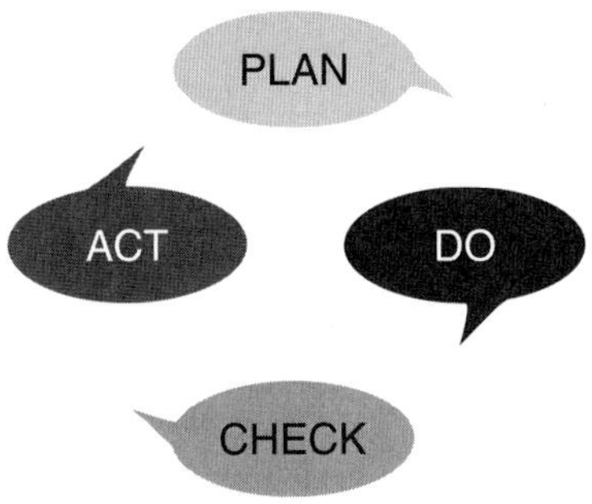

FIG. 22-1 Plan, Do, Check, Act model.

Any company planning a major expansion must concentrate on environmental factors from the outset. Since many environmental approvals require a public hearing, the views of local elected officials and the community at large are extremely important. To an unprecedented degree, the political acceptability of a project can now be crucial. By the 2000s, active stakeholder engagement in projects emerged as a significant force to be reckoned with for any substantial capital project. Every project benefits from a stakeholder engagement plan to support the project schedule and avoid project delays.

Plant Strategies At the plant level, a number of actions should be undertaken to manage the physical, operational, and business impacts associated with environmental issues. These include:

1. Maintaining an accurate source-emission inventory
2. Continually evaluating process operations to identify potential modifications that might reduce or eliminate environmental impacts
3. Ensuring that good housekeeping and strong preventive-maintenance programs exist and are followed
4. Investigating available and emerging pollution control technologies
5. Keeping well informed of the regulations and the directions in which they are moving
6. Working closely with the appropriate regulatory agencies and maintaining open communications to discuss the effects that new regulations may have
7. Keeping the public informed through a good public relations program
8. Building environmental awareness and engagement programs across the employee base to proactively identify risks and opportunities and drive innovation
9. Implementing an EMS

It is unrealistic to expect that at any point in the foreseeable future the authorities will change direction, reduce the effect of regulatory controls, or reestablish the preexisting legal situation in which private companies were free to build major facilities with little or no regulatory restraint.

Corporate Strategic Planning Contingency planning represents an essential component of sound environmental planning for a new plant and other major operational changes. Environmental uncertainties should be specified and related to other contingencies (such as marketing, competitive reactions, politics, and foreign trade) and mapped out in the overall corporate strategy.

Environmental factors should also be incorporated into a company's research and development (R&D) program. Since the planning horizons for new projects may now extend to 5 to 10 years, R&D programs can be designed for specific projects or products. These may require new process modifications, end-of-pipe control technologies, or other adjustments to compliance strategy. It is important to consider both business risks and opportunities associated with environmental issues; R&D programs that address both can create significant value for a company.

Another clear need is to integrate environmental factors into financial planning for major projects. Tradeoff decisions about financing may have to change as the project goes through successive stages of environmental planning and permit negotiations. For example, requirements for the use of more expensive pollution control technology may increase total project costs significantly. A change from end-of-pipe to process modification technology may preclude the use of industrial revenue bond financing under Internal Revenue Service (IRS) rules, but it may reduce operating costs. Regulatory delays can affect assumptions about both the rate of expenditure and inflation factors. Investment, production, environmental, and legal factors are all interrelated, and they can have a major impact on corporate cash flow.

Companies must be thoughtful and proactive in their interactions with local officials and attentive to the management of public opinion. The social responsibility and reputation of companies is an important issue that can significantly affect siting and permitting outcomes. Companies should be strategic about the timing and approach to conducting public hearings and other stakeholder engagement activities. Management must be sensitive to the fact that local officials and other stakeholders who have a voice in the approval process have views and constituencies that go beyond attracting new jobs.

It is also critical to think long term about stakeholder engagement. In general, you want to build goodwill with key stakeholders who can affect your legal and social license to operate; you do not want to be engaging them for the first time when you need their support on an issue. The stakeholder engagement landscape is further complicated by the power of social media.

From all these factors, it is clear that the approval and construction of major new industrial plants or expansions is a far more complicated operation than it has been in the past, even the recent past. Environmental restrictions have prevented the construction of certain facilities in places where earlier they might have been built. In other cases, the acquisition of required approvals may generate a heated technical and political debate that can extend the regulatory process for years.

Often, new requirements may be imposed while a company is seeking approval for a new plant. Thus, companies intending to expand their basic production facilities should anticipate their needs far in advance, prepare to meet the regulatory and social challenges they may eventually confront, and choose locations and design facilities with careful consideration of the environment. The purpose of this section is to help the engineer to meet this environmental regulatory challenge.

UNITED STATES AIR QUALITY LEGISLATION AND REGULATIONS

The Clean Air Act, enacted in 1970 and extensively amended in 1977 and 1990, establishes a comprehensive program for controlling and improving the air quality in the United States through both state and federal regulation. Generally, the EPA is charged with identifying air pollutants that endanger the public health and welfare and with formulating the National Ambient Air Quality Standards (NAAQS). The states are given primary responsibility for assuring air quality within their geographic areas by adopting state implementation plans (SIPs) that specify how the NAAQS will be achieved and maintained. For each NAAQS, a state must adopt a SIP that specifies emission limitations applicable to sources of air pollution and other measures designed to achieve and maintain that NAAQS. These emission limits are the basis for the air quality regulations each state enforces. Among the measures included in the SIP is a preconstruction permit program to ensure that emissions due to new or modified sources of air pollution do not cause or contribute to a violation of a NAAQS.

National Ambient Air Quality Standards The NAAQS are routinely updated. Six pollutants are subject to NAAQS, including sulfur dioxide (SO_2), nitrogen dioxide (NO_2), ozone (O_3), particulate matter in various forms (PM), carbon monoxide (CO), and lead (Pb). The current standards may be found at: https://www.epa.gov/criteria-air-pollutants/naaqs-table.

EPA designates areas of the country as "attainment" or "nonattainment," depending on whether they met the NAAQS for a particular pollutant, or "unclassifiable," if there is insufficient available information to classify an area. The air pollution control requirements included in the SIP for an area depend on the area's designation. As part of the site selection process, it is important to understand if the area is considered attainment or nonattainment because that has a considerable impact on the requirements for obtaining mandatory preconstruction permits and the project schedule. No construction beyond site-clearing activities is allowed prior to receipt of the necessary air quality preconstruction permit.

New Source Performance Standards (NSPSs) In addition to the NAAQS, the Clean Air Act also directs EPA to promulgate categorical emissions standards that set minimum standards of performance for any new or modified emissions sources; known as new source performance standards, these are intended to require the best system of emission reduction. By 2016, EPA had promulgated approximately 100 NSPSs. The current standards can be found in the U.S. Code of Federal Regulations at 40 CFR 60 online at www.ecfr.gov or http://www.gpo.gov/fdsys/pkg/CFR-2011-title40-vol6/xml/CFR-2011-title40-vol6-part60.xml.

National Emission Standards for Hazardous Air Pollutants (NESHAPs) EPA promulgates National Emission Standards for Hazardous Air Pollutants in two different forms. Pollutant-specific emissions standards have been promulgated for asbestos, beryllium, mercury, vinyl chloride, benzene, arsenic, radionuclides, and coke-oven emissions. These standards can be found in the U.S. Code of Federal Regulations at 40 CFR 63 online at www.ecfr.gov and represent the EPA's early attempts to limit emissions of hazardous air pollutants from both new and existing facilities. EPA's approach to regulating HAPs was dramatically changed by the 1990 amendments to the CAA, which designated 189 pollutants as HAPs requiring regulation. This list is subject to periodic updating from time to time. As of 2016, there were 187 HAPs. The current list can be found at https://www.epa.gov/haps/initial-list-hazardous-air-pollutants-modifications and includes many common materials routinely used at industrial sites, such as methanol, toluene, and chlorine. Major sources are defined as any source (new or existing) that emits (after control) 10 tons a year or more of any regulated HAP or 25 tons a year or more of any combination of HAPs. A major source of HAPs must comply with emissions standards known as MACT (Maximum Achievable Control Technology) standards. More than 125 such standards have been promulgated, addressing nearly every variety of industrial source in the United States from surface coating to batch chemical production to primary aluminum production. The current list of categorical MACT standards can be found at https://www3.epa.gov/ttn/atw/mactfnlalph.html. If a proposed new source is not subject to a MACT standard, then the new source must undergo a case-by-case MACT evaluation to identify the appropriate level of control, which is generally the most stringent achieved in practice.

Operating Permits In 1990, an operating permit requirement was added to the Clean Air Act, often referred to as the Title V Operating Permit program. The Title V permit program is administered and enforced primarily by state and local air permitting authorities pursuant to EPA-approved permit programs and subject to EPA oversight. Under the Title V program, all CAA requirements applicable to a particular source must be set forth in a comprehensive permit, often called a Title V permit or an operating permit. Industrial sources must apply for a timely operating permit (typically within one year of start-up) and may continue to operate without a permit until the state permitting authority issues the permit or denies the application. Title V operating permits generally have a five-year life and then must be renewed. Plant owners and operators are required to submit timely and complete operating permit applications and renewal applications, typically 12 months before the expiration of an existing permit. However, sources are free to operate under the existing permit if a permitting authority fails to issue a timely update, provided a timely renewal application was submitted. The Title V permit provides, in a single document, all of a facility's air pollution control obligations along with associated monitoring, recordkeeping, and reporting obligations, and it is a resource for anyone responsible for facility operations.

MAJOR PROJECT PERMITTING

The Clean Air Act and associated federal and state regulations require preconstruction permits authorizing new plant construction or significant modification of existing operations.

Prevention of Significant Deterioration (PSD) Of all the federal laws placing environmental controls on industry (and, in particular, on new plants), perhaps the most confusing and restrictive are the limits imposed for the prevention of significant deterioration (PSD) of air quality. These limits apply to areas of the country that are already cleaner than required by ambient air-quality standards. This regulatory framework evolved from judicial and administrative action under the 1970 Clean Air Act and subsequently was given full statutory foundation by the 1977 Clean Air Act Amendments.

EPA established an area classification scheme to be applied in all such regions. The basic idea was to allow a moderate amount of industrial development but not enough to degrade air quality to a point at which it barely complied with standards. In addition, states were to designate certain areas where pristine air quality was especially desirable. All air-quality areas were categorized as Class I, Class II, or Class III. Class I areas were pristine areas subject to the tightest control. Permanently designated Class I areas included international parks, national wilderness areas, memorial parks exceeding 5000 acres, and national parks exceeding 6000 acres. Although the nature of these areas is such that industrial projects would not be located within them, their Class I status could affect projects in neighboring areas where meteorological conditions might result in the transport of emissions into them. Class II areas were areas of moderate industrial growth. Class III areas were areas of major industrialization. Under EPA regulations promulgated in December 1974, all areas were initially categorized as Class II. States were authorized to reclassify specified areas as Class I or Class III. No area has been recategorized to Class III, so generally, industrial plants are built in Class II areas and must have no impact on Class I areas.

The EPA regulations also established another critical concept known as the *increment.* This was the numerical definition of the amount of additional pollution that may be allowed through the combined effects of all new growth in a locality (see Table 22-1). To ensure that the increments would not be used up hastily, EPA specified that each major new plant must install best available control technology (BACT) to limit emissions. BACT is determined based on a case-by-case engineering analysis and is more stringent than any applicable NSPS.

TABLE 22-1 PSD Air Quality Increments (μ/m^3)

	Class I area	Class II area	Class III area
Sulfur dioxide			
Annual	2	20	40
24 h	5	91	182
3 h	25	512	700
PM_{10}			
Annual	4	17	34
24 h	8	30	60
$PM_{2.5}$			
Annual	1	4	8
24 h	2	9	18
Nitrogen dioxide			
Annual	2.5	25	50

To implement these controls, EPA requires that every new source undergo preconstruction review. The regulations prohibited a company from starting construction on a new source until the review had been completed and provided that, as part of the review procedure, public notice should be given and an opportunity provided for a public hearing on any disputed questions.

Sources Subject to Prevention of Significant Deterioration (PSD) Sources subject to PSD regulations (40 CFR, Sec. 52.21) are major stationary sources and major modifications located in attainment areas and unclassified areas. A major stationary source was defined as any source listed in Table 22-2 with the potential to emit 100 tons per year or more of any pollutant regulated under the Clean Air Act (CAA) or any other source with the potential to emit 250 tons per year or more of any CAA pollutant. The potential to emit is defined as the maximum capacity to emit the pollutant under applicable emission standards and permit conditions (after application of any air pollution control equipment), excluding secondary emissions. A major modification is defined as any physical or operational change of a major stationary source producing a "significant net emissions increase" of any CAA pollutant (see Table 22-3).

As part of the permit review process, ambient air quality monitoring data defining baseline air quality is required to accompany the permit application for all CAA pollutants for which there are NAAQS. If existing data are not available to characterize the site, it may be necessary to collect data for up to one year before completing required environmental permit

TABLE 22-2 Sources Subject to PSD Regulation If Their Potential to Emit Equals or Exceeds 100 Tons per Year

Fossil-fuel-fired steam electric plants of more than 250 million Btu/h heat input
Coal-cleaning plants (with thermal dryers)
Kraft-pulp mills
Portland-cement plants
Primary zinc smelters
Iron and steel mill plants
Primary aluminum-ore-reduction plants
Primary copper smelters
Municipal incinerators capable of charging more than 250 tons of refuse per day
Hydrofluoric, sulfuric, and nitric acid plants
Petroleum refineries
Lime plants
Phosphate-rock-processing plants
Coke-oven batteries
Sulfur-recovery plants
Carbon-black plants (furnace process)
Primary lead smelters
Fuel-conversion plants
Sintering plants
Secondary metal-production plants
Chemical-process plants
Fossil-fuel boilers (or combinations thereof) totaling more than 250 million Btu/h heat input
Petroleum-storage and -transfer units with total storage capacity exceeding 300,000 bbl
Taconite-ore-processing plants
Glass-fiber-processing plants
Charcoal-production plants

TABLE 22-3 Significant Net Emissions Increase

Pollutant	Tons/yr
$Pm_{2.5}$	10
PM_{10}	15
SO_2	40
NO_x	40
VOC	40
CO	100
Lead	0.6
PM	25
Fluorides	3
Sulfuric acid mist	7
Hydrogen sulfide	10
Total reduced sulfur	10
Municipal waste combustor	
Acid gases	40
Metals	15
Organics	3.5×10^{-6}
CFCs (11, 12, 113, 114 & 115)	Any increase
Halons (1211, 1301, 2402)	Any increase
Any emissions increase resulting in a >1 μ/m^3 24-h impact in a Class I area	

applications and assessments. The EPA or the state may accept representative existing monitoring data collected within three years of the permit application to satisfy monitoring requirements.

EPA regulations provide exemptions from BACT and ambient air impact analyses if the emissions increase that results from the modification is accommodated by historical emissions decreases at the plant site occurring within the five years prior the project. This exemption is referred to as the "no significant net increase in emissions" provision of the applicability analysis. A complete PSD review would include a case-by-case determination of the controls required by BACT, an ambient air-impact analysis to determine whether the source might violate applicable increments or air-quality standards; an assessment of the effect on visibility, soils, and vegetation; submission of monitoring data; and full public review. The time required to complete such a review and obtain the permit or license required to commence construction can be anywhere from six months to one year, depending on project complexity, after the application is submitted to the government authorities, and the time required to prepare such an application can be as much as an additional year.

Nonattainment (NA) Those areas of the United States failing to attain compliance with ambient air-quality standards are considered nonattainment areas. New plants may be built in nonattainment areas only if stringent conditions are met. Emissions have to be controlled to the greatest degree possible, and more than equivalent offsetting emission reductions have to be obtained from other sources to assure progress toward achievement of the ambient air-quality standards. Specifically, (1) the new source must be equipped with pollution controls to assure lowest achievable emission rate (LAER), which in no case can be less stringent than any applicable NSPS; (2) all existing sources owned by an applicant in the same state must be in compliance with applicable state implementation plan requirements or be under an approved schedule or an enforcement order to achieve such compliance; (3) the applicant must obtain sufficient emissions offsets to more than make up for the emissions to be generated by the new source (after application of LAER); and (4) the emissions offsets must provide "a positive net air quality benefit in the affected area."

LAER is deliberately a technology-forcing standard of control. The enabling statute stated that LAER must reflect (1) the most stringent emission limitation contained in the implementation plan of any state for such category of sources, unless the applicant can demonstrate that such a limitation is not achievable, or (2) the most stringent limitation achievable in practice within the industrial category, whichever is more stringent. In no event could LAER be less stringent than any applicable NSPS. While the statutory language defining BACT directed that "energy, environmental, and economic impacts and other costs" be taken into account, the comparable provision on LAER does not provide for consideration of economic or related impacts.

For existing pollutant sources in nonattainment areas, reasonable available control technology (RACT) would be required. EPA defines RACT by industrial category.

Minor Source Preconstruction Permits Commonly after the application of air pollution controls, the emissions associated with new facilities or modifications of existing facilities will be less than the emissions thresholds needed to trigger PSD or nonattainment new source review. In such cases, a minor source preconstruction permit is required from the state permitting authority, and new source performance standards (NSPS) apply.

Controlled-Trading Programs The legislation enacted under the Clean Air Act Amendments of 1977 provided the foundation for EPA's controlled-trading program, the essential elements of which include:

- Netting provisions of PSD
- Offsets policy (under nonattainment)
- Banking and brokerage (under nonattainment)

While these different policies vary broadly in form, their objective is essentially the same: to substitute flexible economic-incentive systems for the current rigid, technology-based regulations that specify exactly how companies must comply. These market mechanisms have made regulating easier for EPA and less burdensome and costly for industry.

PSD No-Net-Increase Provisions Under the PSD program, the affected source is the entire plant. This source definition allows a company to determine the most cost-effective way to control pollution any time a plant is modified. Emission increases associated with a new process or production line may be compensated for by historical emission decreases at other parts of the plant. As long as the net emissions increase for the entire site is below the levels identified in Table 22-3, a PSD review is not required.

Offsets Policy Emissions offsets were EPA's first application of the concept that one source could meet its environmental protection obligations by getting another source to assume additional control actions. In nonattainment areas, pollution from a proposed new source, even one that controls its emissions to the lowest possible level, would aggravate existing violations of ambient air-quality standards and trigger the statutory prohibition. The offsets policy provided these new sources with an alternative. The source could proceed with construction plans, provided that:

1. The source would control emissions to the lowest achievable level.
2. Other sources owned by the applicant were in compliance or on an approved compliance schedule.
3. Existing sources were persuaded to reduce emissions by an amount at least equal to the pollution that the new source would add.

Banking and Brokerage Policy EPA's banking policy is aimed at providing companies with incentives to find more offsets. Under the original offset policy, a firm shutting down or modifying a facility could apply the reduction in emissions to new construction elsewhere in the region only if the changes were made simultaneously. However, with banking, a company can deposit the reduction for later use or sale. Such a policy clearly establishes that clean air (or the right to use it) has direct economic value.

REGULATORY DIRECTION

The current direction of regulations and air pollution control efforts is clearly toward significantly reducing the emissions to the environment of a broad range of compounds, including:

1. Volatile organic compounds and other ozone precursors (CO and NO_x)
2. Hazardous air pollutants, including carcinogenic organic emissions and heavy metal emissions
3. Acid rain precursors, including SO_x and NO_x
4. $PM_{2.5}$, which represents very fine particulate matter less than 2.5 microns in aerodynamic diameter. The latest generation of emission standards are addressing concerns related to nitrates and sulfate condensates in the atmosphere referred to as secondary $PM_{2.5}$ formation.

Although it is not possible to predict the future, it is possible to prepare for it and influence it. It is highly recommended that maximum flexibility be designed into new air pollution control systems to allow for increasingly more stringent emission standards for both particulates and gases. Further, it is everyone's responsibility to provide a thorough review of existing and proposed new processes and to make every attempt to identify economical process modifications and/or material substitutions that reduce or, in some cases, eliminate both the emissions to the environment and the overdependency on retrofitted or new end-of-pipe control systems.

UNITED STATES WATER QUALITY LEGISLATION AND REGULATIONS

The Clean Water Act (CWA) The CWA establishes the basic structure for regulating discharges of pollutants into the waters of the United States and establishing quality standards for surface waters. The basis of the CWA was enacted in 1948 and was called the Federal Water Pollution Control Act, but the act was significantly reorganized and expanded in 1972. Clean Water Act became the act's common name with the 1972 amendments. Subsequent amendments have modified certain provisions in the CWA. For example, revisions in 1981 streamlined the municipal wastewater treatment plant construction grants process, and further revisions in 1987 replaced the construction grants program with the State Water Pollution Control Revolving Fund. Other laws have also changed parts of the CWA. For example, Title I of the Great Lakes Critical Programs Act of 1990 incorporated provisions of the Great Lakes Water Quality Agreement between the United States and Canada into the CWA.

The 1972 amendments to the act made it unlawful to discharge any pollutant from a point source into navigable waters unless the discharge is permitted under the National Pollutant Discharge Elimination System (NPDES) permit program. The term *navigable waters* is defined in section 502(7) of the act as waters of the United States, including the territorial seas. Although the jurisdictional scope of the CWA appears to be limited to discharges of pollutants into navigable waters (i.e., waters that are, were, or could be used for interstate or foreign commerce), the Supreme Court has interpreted the term "waters of the United States" expansively to include not just traditional navigable waters, but interstate waters, impoundments of waters, tributaries, and adjacent wetlands.

On June 29, 2015, the U.S. Environmental Protection Agency (EPA) and the U.S. Army Corps of Engineers published in the *Federal Register* a final rule clarifying the definition of waters of the United States (80 FR 37054-37127). In that rule, the agencies defined waters of the United States to include eight categories of jurisdictional waters, including traditional navigable waters, interstate waters, territorial seas, impoundments of jurisdictional waters, tributaries, adjacent waters, and waters with a significant nexus with jurisdictional waters. Waters are considered to have a significant nexus if they significantly affect the chemical, physical, or biological integrity of traditional navigable waters, interstate waters, or the territorial seas.

The term *pollutant* is defined in the CWA to include dredged spoil, solid waste, incinerator residue, sewage, garbage, sewage sludge, munitions, chemical wastes, biological materials, radioactive materials, heat, wrecked or discarded equipment, rock, sand, cellar dirt, and industrial, municipal, and agricultural waste discharged into water. However, courts have also interpreted the definition of *pollutant* broadly to include almost any material, as well as material characteristics including toxicity and acidity.

The CWA gives EPA the authority to establish water quality standards for all contaminants in surface waters, and to implement pollution control programs such as establishing point-source-based effluent limitation guidelines for specific industries (including municipal dischargers, industrial dischargers, and industrial users of municipal treatment works), and establishing effluent limitations for toxic and priority pollutants (applicable to all dischargers).

Source-Based Effluent Limitation Guidelines Effluent guidelines are national wastewater discharge standards that are developed by EPA on an industry-by-industry basis. The guidelines apply to particular point-source categories and pertain to discharges into waters of the United States without regard to the quality of the receiving water. Effluent guidelines have been developed for many point-source categories, including steam-electric power generation, electroplating, petroleum refining, soap and detergent manufacturing, and numerous others. Subcategorized effluent limitation guidelines are codified at 40 CFR, Chapter I, Subchapter N, Parts 401 through 471 of the Code of Federal Regulations (CFR).

Subcategory-specific effluent limitation guidelines set technology-based limitations for specific pollutants at several levels of control, including: BPT (best practicable technology currently available), BAT (best available technology economically achievable), and BCT (best conventional control technology). These terms are defined in CWA section 304(b). Effluent limitations are based on performance of specific technologies and apply to all facilities within a given source category, but the regulations do not require the use of a specific control technology.

New dischargers, which generally refer to any source, the construction of which is commenced after the publication of a proposed discharge standard, must meet source-based effluent limitations derived from new source performance standards (NSPS). NSPS reflect effluent reductions that are achievable based on the "best available demonstrated control technology," and they represent the most stringent controls attainable through the application of best available demonstrated control technology for all pollutants. In establishing NSPS, EPA is directed to take into consideration the cost of achieving the effluent reduction and any non–water quality environmental impacts and energy impacts.

Effluent limitations and NSPS generally apply to discharges made directly into waters of the United States; however, the 40 CFR 122.45(h) and 40 CFR 403.6 rules apply to "internal discharges." Internal discharges are defined in the rule as those where sampling is impracticable or analysis of the stream after blending with other streams is expected to dilute any pollutant to below its applicable detection limit. These internal discharge rules are intended to prevent circumvention of discharge limits by dilution.

Pollutant Types CWA section 304(a)(4) designates the following as conventional pollutants: biochemical oxygen demand (BOD_5), total suspended solids (TSS), fecal coliform, pH, and any additional pollutants EPA defines as conventional. The agency designated oil and grease as an additional conventional pollutant on July 30, 1979 (see 44 FR 44501). Subcategory-specific effluent guidelines generally include discharge limits for conventional pollutants based on control achievable through the implementation of BCT.

In addition to the list of conventional pollutants, two lists have special significance to water quality regulatory programs in the CWA: the Toxic Pollutant List and the Priority Pollutant List.

The Toxic Pollutant List is referenced in section 307(a)(1) of the CWA. The list was first published in the *Federal Register* on January 31, 1978 (43 FR 4108). In a final rule on July 31, 1979, EPA published the list again and added the list to the Code of Federal Regulations at 40 CFR 401.15 (44 FR 44501). EPA removed three pollutants from the list in 1981 after determining that their chemical properties did not justify their inclusion (46 FR 10723).

There are 65 entries in the Toxic Pollutant List; however, many of the entries are not individual compounds (e.g., benzene) but groups of compounds (e.g., haloethers, DDT and metabolites, and chlorinated benzenes). Many of these groups are too large for effective test methods; therefore, EPA published the Priority Pollutants List.

Starting with the list of toxic pollutants, EPA used several criteria to select and prioritize the list of toxic pollutants. The Priority Pollutant List contains 126 chemicals and chemical compounds for which EPA has published analytical test methods. The list was developed to provide more specific definition and test methods for certain toxic pollutants, and it identifies the regulated pollutants by their individual chemical names. The list is published at 40 CFR part 423, Appendix A. Although the list is published as an appendix to part 423 (the Steam Electric Generating Unit Effluent Guidelines), it is not limited in its relevance to that one industrial category. The Priority Pollutant List is an important starting point for EPA to consider, for example, in developing national discharge standards (such as effluent guidelines) or in national permitting programs (such as NPDES).

Water Quality Standards (WQS) These are the standards specifying the conditions to be achieved in a given body of water. They are usually included as maximum numeric limitations for chemicals and chemical compounds included in the Priority Pollutant List, numerical limitations for conditions such as temperature and pH, and general prohibitions against deleterious conditions such as the presence of sludge, floating oil, or toxicity. Water quality standards are promulgated pursuant to CWA section 303, which states that the "standards shall be such as to protect the public health or welfare, enhance the quality of water and serve the purposes of this Act. Such standards shall be established taking into consideration their use and value for public water supplies, propagation of fish and wildlife, recreational purposes, and agricultural, industrial, and other purposes, and also taking into consideration their use and value for navigation."

Section 303(c) of the CWA requires every state to develop water quality standards applicable to all water bodies, or segments of water bodies, that lie within the state. Such standards serve the dual purposes of establishing the water quality goals for a specific water body and serve as the regulatory basis for the establishment of water quality–based treatment controls and strategies beyond the technology-based levels of treatment. Under CWA section 304(a), EPA publishes recommended water quality criteria guidance, which can include concentrations for specific chemicals or levels of parameters in water that protect aquatic life and human health.

EPA established the core of the WQS regulation in a final rule issued in 1983 (54 FR 51400). That rule clarified WQS provisions that had been in place since 1977, and it codified the standards at 40 CFR part 131. To support the WQS regulation, EPA has published a number of guidance documents, including the *Water Quality Standards Handbook* (WQS Handbook, second edition, EPA 823-B-94-005a, August 1994), and the *Technical Support Document for Water Quality–Based Toxics Control* (revised edition, EPA 505/2-90-001, March 1991). EPA has also published a number of WQS criteria tables, including the Aquatic Life Criteria Table, Human Health Criteria Table, and Organoleptic Effects (e.g., taste and odor) Criteria Table.

WQS criteria in these publications and tables are not rules, nor do they automatically become part of a state's water quality standards. However, these criteria are used as guidance to assist states in developing their water quality standards. States may adopt the criteria published by EPA, modify EPA's criteria to reflect site-specific conditions, or adopt different criteria based on other scientifically defensible methods. Once a state has developed its standards, EPA must approve or disapprove them. States are required to review their water quality standards at least once every three years and revise them as necessary.

National Pollutant Discharge Elimination System (NPDES) Permits These permits authorize the discharge pollutants to waters of the United States. CWA section 402 established the NPDES permit program to be administered by EPA or authorized states, territories, or eligible tribes. The NPDES permit program provides two types of permits, individual and general, that may be used to authorize point-source discharges of pollutants.

Individual permits are issued by the state or EPA to a single facility; they include discharge limitations and monitoring requirements specifically tailored to that facility. Issuance of an individual NPDES permit requires submission of a permit application. EPA has developed eight individual permit application forms for applicants seeking coverage under an individual permit. Forms and standards required for an individual permit application are codified in 40 CFR 122.21. Once a facility submits the appropriate application forms, the permitting authority develops a permit for that facility based on the information in the permit application (e.g., type of activity, nature of discharge, and receiving water quality). The authority issues the permit to the facility for a specific time period (not to exceed five years) with a requirement that the facility reapply before the permit expiration date.

General permits are developed by the state or EPA to cover classes or categories of dischargers with similar qualities under a single permit. General permits typically require facilities seeking permit coverage to submit a notice of intent (NOI) to be covered, the contents of which are described in the general permit. General permits may offer a cost-effective option for permitting agencies because of the large number of facilities that can be covered under a single permit.

Under CWA section 402(b), a state or eligible tribe may obtain authorization to administer the NPDES permit program. In order to obtain authorization, the state or eligible tribe must demonstrate to EPA that it has the authorities and resources necessary to implement the permitting program. To date, 46 states and the Virgin Islands have obtained authorization to administer the NPDES permit program.

EPA will review the quality of each permitting program through regularly scheduled Permit and Program Quality Review (PQR) processes. The PQR includes reviews and evaluations of NPDES permit language, NPDES permit fact sheets, calculations, supporting documents in the administrative record, and state permitting program initiatives. Should the PQR result in a significant finding, the EPA may withdraw a state's NPDES permitting authority. Citizens may also petition EPA to withdraw a state's permitting authority. This has happened eight times since 1989, when a concerned citizens group in Kansas petitioned EPA to withdraw the state's permitting authority. None of the petitions have resulted in the withdrawal of a state's NPDES permitting authority.

Effluent Limitations in NPDES Permits These serve as the primary mechanism for controlling discharges of pollutants to receiving waters. When developing effluent limitations for an NPDES permit, a permit writer must consider technology-based effluent limits, as well as discharge limits that are protective of the water quality standards of the receiving water (i.e., water quality–based effluent limits or total maximum daily loads). Two more recent methods for determining effluent limits are whole effluent toxicity and watershed-based permitting. State legislatures may require their state's permitting authority to apply more stringent limits than those recommended by EPA.

The first step in developing effluent limits is deriving the technology-based effluent limitations (TBELs). TBELs include effluent limitations and standards promulgated under section 301 of the CWA (i.e., Effluent Guidelines), new source performance standards promulgated under section 306 of CWA, case-by-case effluent limitations determined under section 402(a)(1) of CWA, or a combination of the three. TBELs in NPDES permits require a minimum level of treatment of pollutants for point-source discharges based on available treatment technologies, while allowing the discharger to use any available control technique to meet the limits. For existing industrial (and other nonmunicipal) facilities, TBELs are derived by using the effluent limitation guidelines or using best professional judgment (BPJ) on a case-by-case basis in the absence of national guidelines and standards.

As previously described, EPA applies a tiered approach when establishing source-specific effluent limits. Effluent limits may be based on the following: the best practicable control technology (BPT) limits based on the performance of currently operating treatment systems; the best available technology economically achievable (BAT) limits based on the best-performing (not necessarily currently operating) treatment technologies that are economically achievable; or the best conventional pollutant control technology (BCT) limits, which are based on a two-part cost-reasonableness test that first compares the cost of pollutant removal by industries in a source category to those of a publicly owned treatment works (POTW) and then evaluates the cost of additional treatment beyond BPT. BCT is typically used to establish discharge limits for conventional pollutants [i.e., biological oxygen demand (BOD_5), fecal coliform, total suspended solids [TSS], pH, and oil and grease].

Once the TBELs are established, the NPDES permit writer determines whether, after application of the TBELs, the discharge will cause, have the reasonable potential to cause, or contribute to an excursion above a state WQS. If the permit writer determines that, notwithstanding application of technology-based controls, the discharge will cause or contribute to an excursion above a state WQS, the permit writer derives effluent limitations necessary to meet the state WQS (referred to as water quality–based effluent limits or WQBELs). WQBELs can include numeric limits for specific chemical constituents in the discharge, as well as discharge limits developed by the permitting authority based on a total maximum daily load (TMDL) of a pollutant discharged into the water body.

TMDLs identify the amount of a specific pollutant or property of a pollutant, from point, nonpoint, and natural background sources, including a margin of safety, that may be discharged into a water body and still ensure that the water body attains the applicable WQS. The permitting authority will allocate a portion of the TMDL to each point source discharging to the water body. The allocations of pollutant loadings to point sources are called wasteload allocations (WLAs) and will be included in the facility's NPDES permit.

Water quality–based effluent limits in NPDES permits, whether WQBELs or TMDLs, must be consistent with the assumptions used to derive the WLA. Unlike source-based limitations, water quality–based effluent limitations can be established regardless of either the cost or the technology required to implement them. At industrial sites, WQBELs and TMDLs will probably be more stringent than the source-based TBELs since they are usually imposed on discharges to an already impaired water body.

Because of the difficulties associated with setting individual water quality–based permit limitations, permitting authorities may develop limits based on whole effluent toxicity (WET). WET describes the aggregate toxic effect of an aqueous sample (e.g., whole effluent wastewater discharge) as measured by an organism's response upon exposure to the sample (e.g., lethality, impaired growth, or impaired reproduction). WET tests replicate the total effect of environmental exposure of aquatic life to toxic pollutants in an effluent without requiring the identification of the specific pollutants. EPA recommends using WET tests in NPDES permits together with requirements based on chemical-specific water quality standards.

Whole effluent toxicity monitoring requirements may be included in NPDES permits to generate a valid baseline of WET data that can be used to determine whether the potential for whole effluent toxicity (i.e., toxicity that would result in an excursion of a state or tribal WQS) has been demonstrated. If reasonable potential has been demonstrated, then a WET limit must be included in future permits. WET test results are then used to determine compliance with NPDES WET permit limits.

Permitting authorities may also take a watershed-based approach to develop discharge limits. Watershed-based permitting is a process that emphasizes addressing all stressors within a watershed, rather than point sources on a discharge-by-discharge basis. Watershed-based permitting can encompass a variety of activities, ranging from synchronizing permits within a basin to developing water quality–based effluent limits using a multiple discharger modeling analysis. EPA believes that watershed-based permitting will provide enhanced environmental outcomes through the integration of NPDES and other water-related programs (e.g., Safe Drinking Water Act source water assessment plans), better targeting of limited state resources, a more streamlined permitting process (e.g., one large public meeting for the watershed rather than numerous small meetings for each permit), and watershed-wide monitoring plans to reduce duplication of samples. EPA also believes that watershed-based permits could be useful in facilitating the trading of pollutant discharge credits similar in nature to the cap-and-trade program for SO_2.

Section 316(a) (published to 40 CFR part 125, subpart H in August 2014) applies to point sources with thermal discharges. Section 316(a) authorizes the NPDES permitting authority to impose alternative effluent limitations for the control of the thermal component of a discharge in lieu of the effluent limits that would otherwise be required under sections 301 or 306 of the CWA. These regulations identify the criteria and process for determining whether an alternative effluent limitation (i.e., a thermal variance from the otherwise applicable effluent limit) may be included in a permit and, if so, what that limit should be.

Cooling Water Intake Structures These structures are located at large industrial facilities and electric power generating facilities, and they are subject to the provisions in section 316(b) of the CWA. The section 316(b) rules, codified at 40 CFR parts 122 and 125, apply to facilities that withdraw water from waters of the United States, have an NPDES permit, have a design intake flow greater than 2 MGD, and use at least 25 percent of the water withdrawn for cooling purposes only. Section 316(b) of the CWA requires that the location, design, construction, and capacity of cooling water intake structures located on a navigable water body must reflect the "best technology available for minimizing adverse environmental impacts." Adverse environmental impacts associated with cooling water intake structures include fish impingement (when fish and other aquatic life are trapped against cooling water intake screens) and entrainment (when aquatic organisms, eggs, and larvae are drawn into a cooling system, through the heat exchanger, and then pumped back out to the source water).

The section 316(b) regulations are unique in that they apply to the intake of water and not the discharge. Power plants and industrial facilities that reduce cooling water to a level commensurate with closed-cycle cooling systems can generally satisfy the section 316(b) requirements using conventional intake technologies. Plants using once-through cooling systems may be required to modify their intake structures to reduce impingement and/or entrainment. A critical design parameter established by EPA to minimize impingement rates is the velocity of the water as it passes through the intake screens, which EPA recommends be kept below 0.5 ft/s. Other technologies that may be required to meet the section 316(b) standards include fish collection and return devices, fine-mesh screens, and variable-speed pumps.

Other Federal Laws Apply to the NPDES Permit Program The National Environmental Policy Act (NEPA), 42 U.S.C. 4321 et seq., was signed into law in 1970. NEPA covers all aspects of environmental protection, not just water quality. NEPA created the Council on Environmental Quality (CEQ) to coordinate federal environmental protection efforts. NEPA also requires that agencies conduct environmental impact reviews [Environmental Assessments (EA) and Environmental Impact Statements (EIA)] for projects significantly affecting the quality of the human environment. Section 511 of the CWA establishes that only EPA-issued permits to dischargers subject to new source performance standards (NSPS) are subject to NEPA's environmental review procedures; however, states may have their own versions of NEPA that may have broader environmental review requirements.

The Coastal Zone Management Act (CZMA), 16 U.S.C. 1451 et seq., was enacted to protect the nation's coastal zone and is implemented through state–federal partnerships. Section 307(c) of CZMA prohibits the issuance of NPDES permits for activities affecting land or water use in coastal zones unless the permit applicant certifies that the proposed activity complies with the state coastal zone management program.

The Oil Pollution Act (OPA), 33 U.S.C. 2701 et seq., was enacted in 1990 and amended section 311 of the CWA. The OPA defined *oil* to include both petroleum and nonpetroleum oils (e.g., vegetable oils) and added reporting and pollution prevention requirements for onshore oil storage facilities. The oil pollution prevention regulation, commonly known as the sheen rule, requires the person in charge of a facility or vessel to report to the federal government discharges that: cause a sheen or discoloration on the surface of discharged water, violate water quality standards found in the facility's NPDES permit, or cause a sludge or emulsion to be deposited beneath the surface of the receiving water or adjoining shorelines. OPA also requires facilities to proactively develop and maintain spill prevention, control, and countermeasures (SPCC) plans and, for worst-case spills, facility response plans (FRPs) that meet the requirements of 40 CFR part 112.

EPA collects environmental information in many forms, such as environmental permit compliance reports, discharge monitoring reports (DMRs), and permit applications for new and existing dischargers. This database is a valuable resource for plant engineers who are responsible for environmental compliance and environmental compliance system design. EPA provides several resources intended to ease access to the database. Two of these resources are the Envirofacts website and WATERS GeoViewer.

The Envirofacts (https://www3.epa.gov/enviro/index.html) website is a one-stop source for environmental information maintained by EPA. Information available through the website is collated from over 20 separate data collection systems, such as the Enforcement and Compliance History Online (ECHO), Permit Compliance System (PCS), and Integrated Compliance Information System (ICIS). The user accesses this vast database by entering a location, or name, of a discharger into the search engine incorporated into the Envirofacts home page.

WATERS GeoViewer (https://www.epa.gov/waterdata/waters-geoviewer) dynamically displays snapshots of EPA Office of Water program data in the form of a map similar to a Google Map or other mapping website. The 2018 version of WATERS GeoViewer depicts the status of NPDES permits for each state; summary information from the Clean Watershed Needs Survey; and water quality assessments. WATERS GeoViewer also contains water-related geographic themes such as watersheds, the national stream network known as the National Hydrography Dataset, state and federal water quality program sample points, and other water-related map layers. WATERS GeoViewer enables you to create customized maps at national and local scales.

Regulatory Direction EPA regularly reviews ELGs. As ELGs come up for review, it is likely that they will become more stringent. For example, the new ELG for the Steam Electric Power Generating Point Source Category, published in 2015, reduced arsenic, mercury, selenium, and nitrate/nitrite discharge limits from flue gas desulfurization systems. This ELG also banned the discharge of pollutants in ash transport water.

In 2014, EPA published a final rule codifying guidance on the use of "sufficiently sensitive" analytical methods for NPDES permit applications and compliance monitoring. This rule takes advantage of increases in the sensitivity of modern analytical methods. As instruments become more sensitive, EPA will likely respond by updating its list of approved analytical methods.

UNITED STATES SOLID WASTE LEGISLATION AND REGULATIONS

Much of the current activity in the field of solid waste management, especially with respect to hazardous wastes and resources recovery, is a direct consequence of recent legislation. Therefore, it is important to review the principal legislation that has affected the entire field of solid waste management.

What follows is a brief review of existing legislation that affects the management of solid wastes. The actual legislation must be consulted for specific detail. Implementation of the legislation is accomplished through regulations adopted by federal, state, and local agencies. Because these regulations are revised continuously, they must be monitored continuously, especially when design and construction work is to be undertaken.

Rivers and Harbors Act, 1899 Passed in 1899, the Rivers and Harbors Act directed the U.S. Army Corps of Engineers to regulate the dumping of debris in navigable waters and adjacent lands.

Solid Waste Disposal Act (SWDA), 1965 Modern solid waste legislation dates from 1965, when the Solid Waste Disposal Act, Title II of Public Law 88-272, was enacted by Congress. The principal intent of this act was to promote the demonstration, construction, and application of solid waste management and resource-recovery systems that preserve and enhance the quality of air, water, and land resources.

National Environmental Policy Act (NEPA), 1969 The National Environmental Policy Act of 1969 was the first federal act that required coordination of federal projects and their impacts with the nation's resources. The act specified the creation of the Council on Environmental Quality in the Executive Office of the President. This body has the authority to force every federal agency to submit to the council an environmental impact statement on every activity or project it may sponsor or over which it has jurisdiction.

Resource Recovery Act, 1970 The Solid Waste Disposal Act of 1965 was amended by Public Law 95-512, the Resources Recovery Act of 1970. This act directed that the emphasis of the national solid waste management program should be shifted from disposal as its primary objective to that of recycling and reuse of recoverable materials in solid wastes or the conversion of wastes to energy.

Resource Conservation and Recovery Act (RCRA), 1976 RCRA is the primary statute governing the regulation of solid and hazardous waste. It completely replaced the Solid Waste Disposal Act of 1965 and supplemented the Resource Recovery Act of 1970; RCRA itself was substantially amended by the Hazardous and Solid Waste Amendments of 1984 (HSWA). The Revisions to the Definition of Solid Waste (DSW) Final Rule, effective July 13, 2015, modifies the EPA's 2008 Definition of Solid Waste rule to ensure that it protects human health and the environment from the mismanagement of hazardous secondary materials intended for recycling, while promoting sustainability through the encouragement of safe and environmentally responsible recycling of such materials. The EPA Hazardous Waste Generator Improvements Rule proposed in 2015 provides a much-needed update to the hazardous waste generator regulations to make the rules easier to understand, facilitate better compliance, provide greater flexibility in how hazardous waste is managed, and close important gaps in the regulations. The principal objectives of RCRA as amended are to:

- Promote the protection of human health and the environment from potential adverse effects of improper solid and hazardous waste management
- Conserve material and energy resources through legitimate waste recycling and recovery
- Reduce or eliminate the generation of hazardous waste as expeditiously as possible

To achieve these objectives, RCRA authorized EPA to regulate the generation, treatment, storage, transportation, and disposal of hazardous wastes. The structure of the national hazardous waste regulatory program envisioned by Congress is laid out in Subtitle C of RCRA (sections 3001 through 3019), which authorized EPA to:

- Promulgate standards governing hazardous waste generation and management
- Promulgate standards for permitting hazardous waste treatment, storage, and disposal facilities
- Inspect hazardous waste management facilities
- Enforce RCRA standards
- Authorize states to manage the RCRA Subtitle C program, in whole or in part, within their respective borders, subject to EPA oversight

Federal RCRA hazardous waste regulations are set forth in 40 CFR parts 260 through 279. The core of the RCRA regulations establishes the cradle-to-grave hazardous waste regulatory program through seven major sets of regulations:

- Identification and listing of regulated hazardous wastes (part 261)
- Standards for generators of hazardous waste (part 262)

- Standards for transporters of hazardous waste (part 263)
- Standards for owners/operators of hazardous waste treatment, storage, and disposal facilities (parts 264, 265, and 267)
- Standards for the management of specific hazardous wastes and specific types of hazardous waste management facilities (part 266)
- Land disposal restriction standards (part 268)
- Requirements for the issuance of permits to hazardous waste facilities (part 270)
- Standards and procedures for authorizing state hazardous waste programs to be operated in lieu of the federal program (part 271)
- Approved state hazardous waste management programs (part 272)
- Standards for universal waste management (part 273)
- Standards for management of used oil (part 279)

EPA, under Section 3006 of RCRA, may authorize a state to administer and enforce a state hazardous waste program in lieu of the federal Subtitle C program. To receive authorization, a state program must:

- Be equivalent to the federal Subtitle C program
- Be consistent with, and no less stringent than, the federal program and other authorized state programs
- Provide adequate enforcement of compliance with Subtitle C requirements

Toxic Substances Control Act (TSCA), 1976 The two major goals of the Toxic Substances Control Act, passed by Congress in 1976, are (1) the acquisition of sufficient information to identify and evaluate potential hazards from chemical substances and (2) the regulation of the manufacture, processing, distribution, use, and disposal of any substance that presents an unreasonable risk of injury to the health of the environment.

Under TSCA, the EPA has issued a ban on the manufacture, processing, and distribution of products containing PCBs. Exporting of PCBs has also been banned. TSCA also required that PCB mixtures containing more than 50 ppm PCBs must be disposed of in an acceptable incinerator or chemical waste landfill. All PCB containers or products containing PCBs had to be clearly marked, and records had to be maintained by the operator of each facility handling at least 45 kilograms of PCBs. These records include PCBs in use in transformers and capacitors, PCBs in transformers and capacitors removed from service, PCBs stored for disposal, and a report on the ultimate disposal of the PCBs. Additionally, PCB transformers must be registered, and records must be kept of hydraulic systems, natural gas pipeline systems, voltage regulators, electromagnetic switches, and ballasts.

TSCA also placed restrictions on the use of chlorofluorocarbons, asbestos, and fully halogenated chlorofluoroalkanes, such as aerosol propellants.

Regulatory Direction There is no doubt that pollution prevention continues to be the regulatory direction. The development of new and more efficient processes and waste minimization technologies will be essential to support this effort.

An area certain to receive regulatory attention in the future is nanotechnology. The applications and implications of nanotechnology not only are changing lives every day, but also are moving so fast that many have not yet grasped their tremendous impact (Theodore, L., and R. Kunz, *Nanotechnology: Environmental Implications and Solutions,* Wiley, Hoboken, N.J., 2002). It is clear that nanotechnology will revolutionize the way industry operates, creating chemical processes and products that are more efficient and less expensive. A primary obstacle to achieving this goal will be the need to control, reduce, and ultimately eliminate the environmental and related problems associated with nanotechnology, or dilemmas that may develop through its misuse.

Completely new legislation and regulatory rulemaking may be necessary for environmental control of nanotechnology. However, in the meantime, one may speculate on how the existing regulatory framework might be applied to the nanotechnology area as this emerging field develops over the next several years.

Commercial applications of nanotechnology are likely to be regulated under the Toxic Substances Control Act (TSCA), which authorized EPA to review and establish limits on the manufacture, processing, distribution, use, and/or disposal of new materials the EPA determines to pose "an unreasonable risk of injury to human health or the environment." The term *chemicals* is defined broadly by TSCA. Unless qualifying for an exemption under the law [R&D (a statutory exemption requiring no further approval by EPA), low-volume production, low environmental releases along with low volume, or plans for limited test marketing], a prospective manufacturer is subject to the full-blown premanufacture notice (PMN) procedure. This requires submittal of said notice, along with toxicity and other data to EPA at least 90 days before commencing production of the chemical substance (Theodore, L., and R. Kunz, *Nanotechnology: Environmental Implications and Solutions,* Wiley, Hoboken, N.J., 2002).

Approval then involves recordkeeping, reporting, and other requirements under the statute. Requirements will differ, depending on whether EPA determines that a particular application constitutes a "significant new use" or a "new chemical substance." EPA can impose limits on production, including an outright ban when it is deemed necessary for adequate protection against "an unreasonable risk of injury to health or the environment." EPA may revisit a chemical's status under TSCA and change the degree or type of regulation when new health and environmental data warrant. EPA is expected to be issuing several new TSCA test rules in 2004 [Bergeson, L. L., "Expect a Busy Year at EPA," *Chem. Processing* 17 (April 14, 2004)]. If the experience with genetically engineered organisms is any indication, there will be a push for EPA to update regulations in the future to reflect changes, advances, and trends in nanotechnology [Bergeson, L. L., "Genetically Engineered Organisms Face Changing Regulations," *Chem. Processing* (March 2004)].

Workplace exposure to chemical substances and the potential for pulmonary toxicity are subject to regulation by the Occupational Safety and Health Administration under the Occupational Safety and Health Act (OSHA), including the requirement that potential hazards be disclosed on material safety data sheets (MSDSs). (An interesting question arises as to whether carbon nanotubes, chemically carbon but with different properties because of their small size and structure, are indeed to be considered the same as or different from carbon black for MSDS purposes.) Both government and private agencies can be expected to develop the requisite threshold limit values (TLVs) for workplace exposure. Also, EPA may once again use TSCA to assert its own jurisdiction, appropriate or not, to minimize exposure in the workplace.

Wastes from a commercial-scale nanotechnology facility would be treated under the Resource Conservation and Recovery Act (RCRA), provided that they meet the criteria for RCRA waste. RCRA requirements could be triggered by a listed manufacturing process or the act's specified hazardous waste characteristics. The type and extent of regulation would depend on how much hazardous waste is generated and whether the wastes generated are treated, stored, or disposed of on-site.

Finally, opponents of nanotechnology, especially, may be able to use NEPA to impede nanotechnology research funded by the U.S. government. A "major federal action significantly affecting the quality of the human environment" is subject to the environmental impact provision under NEPA. (Various states also have environmental impact assessment requirements that could delay or put a stop to construction of nanotechnology facilities.) Time will tell.

POLLUTION PREVENTION

This subsection is drawn in part from Wainwright, B., and L. Theodore, "Pollution Prevention Overview," chap. 1 in *Pollution Prevention: Problems and Solutions*, ed. L. Theodore, R. R. Dupont, and J. Reynolds, Gordon and Breach, New York, 1994.

FURTHER READING: American Society for Testing and Materials, *Standard Guide for Industrial Source Reduction,* draft copy dated June 16, 1992; American Society for Testing and Materials, *Pollution Prevention, Reuse, Recycling and Environmental Efficiency,* ASTM, West Conshohocken, Pa., June 1992; California Department of Health Services, *Economic Implications of Waste Reduction, Recycling, Treatment, and Disposal of Hazardous Wastes: The Fourth Biennial Report,* Sacramento, Calif., 1988; Citizen's Clearinghouse for Hazardous Waste, *Reduction of Hazardous Waste: The Only Serious Management Option,* CCHW, Falls Church, Va., 1986; Congress of the United States, Office of Technology Assessment, *Serious Reduction of Hazardous Waste: For Pollution Prevention and Industrial Efficiency,* GPO, Washington, D.C., 1986; DuPont, R., and K. Ganesan, and L. Theodore, *Pollution Prevention,* 2d ed, CRC Press/Taylor & Francis Group, Boca Raton, Fla., 2017; Friedlander, S., "Pollution Prevention—Implications for Engineering Design, Research, and Education," *Environment* (May 1989): 10; Theodore, L., and R. Kunz, *Nanotechnology: Environmental Implications and Solutions,* Wiley, New York, 2005; Theodore, L., *Pollution Prevention: An ETS Theodore Tutorial,* ETS International, Roanoke, Va., 1994; Theodore, L., *A Citizen's Guide to Pollution Prevention,* East Williston, N.Y., 1993; United States EPA, *Facility Pollution Prevention Guide,* (EPA/600/R-92/088), Washington, D.C., May 1992; United States EPA, "Pollution Prevention Fact Sheets," GPO, Washington, D.C., 1991; United States EPA, *1987 National Biennial RCRA Hazardous Waste Report—Executive Summary,* GPO, Washington, D.C., 1991; United States EPA, Office of Pollution Prevention, *Report on the U.S. Environmental Protection Agency's*

Pollution Prevention Program, GPO, Washington, D.C., 1991; Wilcox, J., and L. Theodore, *Engineering and Environmental Ethics,* Wiley, New York, 1998; World Wildlife Fund, *Getting at the Source—Executive Summary,* 1991.

INTRODUCTION

The amount of waste generated in the United States has reached staggering proportions; according to the United States Environmental Protection Agency, 250 million tons of solid waste alone are generated annually. Although both the Resource Conservation and Recovery Act (RCRA) and the Hazardous and Solid Waste Act (HSWA) encourage businesses to minimize the wastes they generate, the majority of the environmental protection efforts are still centered around treatment and pollution clean-up.

The passage of the Pollution Prevention Act of 1990 has redirected industry's approach to environmental management; pollution prevention has now become the environmental option of the 21st century. Whereas typical waste management strategies concentrate on end-of-pipe pollution control, pollution prevention tries to handle waste at the source (i.e., source reduction). As waste handling and disposal costs increase, the application of pollution prevention measures is becoming more attractive than ever before. Industry is currently exploring the advantages of multimedia waste reduction and developing agendas to *strengthen* environmental design while *lessening* production costs.

There are profound opportunities for both industries and individuals to prevent the generation of waste; indeed, pollution prevention is today primarily stimulated by economics, legislation, liability concerns, and the enhanced environmental benefit of managing waste at the source. The Pollution Prevention Act of 1990 established pollution prevention as a national policy, declaring that "waste should be prevented or reduced at the source wherever feasible, while pollution that cannot be prevented should be recycled in an environmentally safe manner" (United States EPA, *Pollution Prevention Fact Sheet,* Washington, D.C., March 1991). The EPA's policy establishes the following hierarchy of waste management:

1. Source reduction
2. Recycling/reuse
3. Treatment
4. Ultimate disposal

The hierarchy's categories are prioritized to promote the examination of each alternative before the investigation of subsequent options (i.e., the most preferable alternative should be thoroughly evaluated before consideration is given to a less accepted option). Practices that decrease, avoid, or eliminate the generation of waste are considered source reduction and can include the implementation of procedures as simple and economical as good housekeeping. Recycling is the use, reuse, or reclamation of wastes and/or materials that may involve the incorporation of waste recovery techniques (e.g., distillation, filtration). Recycling can be performed at the facility (i.e., on-site) or at an off-site reclamation facility. Treatment involves the destruction or detoxification of wastes into nontoxic or less toxic materials by chemical, biological, or physical methods, or by any combination of these control methods. Disposal has been included in the hierarchy because it is recognized that residual wastes will exist; the EPA's so-called ultimate disposal options include landfilling, land farming, ocean dumping, and deep-well injection. However, the term *ultimate disposal* is a misnomer; it is included here because of its earlier adaptation by the EPA.

Table 22-4 provides a rough timetable demonstrating the United States' approach to waste management. Note how waste management has begun to shift from pollution *control*-driven activities to pollution *prevention* activities.

The application of waste management practices in the United States has recently moved toward securing a new pollution prevention ethic. The performance of pollution prevention assessments and their subsequent implementation will encourage increased activity into methods that will further aid in the reduction of hazardous wastes. One of the most important and propitious consequences of the pollution prevention movement will be the development of life-cycle design and standardized life-cycle cost-accounting procedures. These two consequences are briefly discussed in the two paragraphs that follow. Additional information is provided in a later subsection.

TABLE 22-4 Waste Management Timetable

Prior to 1945	No control
1945–1960	Little control
1960–1970	Some control
1970–1975	Greater control (EPA is founded)
1975–1980	More sophisticated control
1980–1985	Beginning of waste-reduction management
1985–1990	Waste-reduction management
1990–1995	Formal pollution prevention programs (Pollution Prevention Act)
1995–2000	Widespread acceptance of pollution prevention
After 2000	The sky's the limit

The key element of life-cycle design is life-cycle assessment (LCA). LCA is generally envisioned as a process to evaluate the environmental burdens associated with the cradle-to-grave life cycle of a product, process, or activity. A product's life cycle can be roughly described in terms of the following stages:

1. Raw material
2. Bulk material processing
3. Production
4. Manufacturing and assembly
5. Use and service
6. Retirement
7. Disposal

Maintaining an objective process while spanning this life cycle can be difficult given the varying perspective of groups affected by different parts of that cycle. LCA typically does not include any direct or indirect monetary costs or impacts to individual companies or consumers.

Another fundamental goal of life-cycle design is to promote sustainable development at the global, regional, and local levels. There is significant evidence that suggests that current patterns of human and industrial activity on a global scale are not following a sustainable path. Changes to achieve a more sustainable system will require that environmental issues be more effectively addressed in the future. Principles for achieving sustainable development should include [Keoleian, G., and D. Menerey, "Sustainable Development by Design: Review of Life Cycle Design and Related Approaches," *Air & Waste Management* 44 (May 1994)]:

1. *Sustainable resource use (conserving resources, minimizing depletion of nonrenewable resources, using sustainable practices for managing renewable resources).* There can be no product development or economic activity of any kind without available resources. Except for solar energy, the supply of resources is finite. Efficient designs conserve resources while also reducing impacts caused by material extraction and related activities. Depletion of nonrenewable resources and overuse of otherwise renewable resources limits their availability to future generations.

2. *Maintenance of ecosystem structure and function.* This is a principal element of sustainability. Because it is difficult to imagine how human health can be maintained in a degraded, unhealthy natural world, the issue of ecosystem health should be a more fundamental concern. Sustainability requires that the health of all diverse species as well as their interrelated ecological functions be maintained. As only one species in a complex web of ecological interactions, humans cannot separate their success from that of the total system.

3. *Environmental justice.* The issue of environmental justice has come to mean different things to different people. Theodore (Theodore, L., personal notes) has indicated that the subject of environmental justice contains four key elements that are interrelated: environmental racism, environmental health, environmental equity, and environmental politics. (Unlike many environmentalists, Theodore has contended that only the last issue, politics, is a factor in environmental justice.) A major challenge in sustainable development is achieving both intergenerational and intersocietal environmental justice. Overconsuming resources and polluting the planet in such a way that it enjoins future generations from access to reasonable comforts irresponsibly transfer problems to the future in exchange for short-term gains. Beyond this intergenerational conflict, enormous inequities in the distribution of resources continue to exist between developed and less developed countries. Inequities also occur within national boundaries.

Life cycle is a perspective that can consider the true costs of producing products and services by analyzing the costs of potential environmental degradation and energy consumption, as well as more customary costs like capital expenditure and operating expenses. A host of economic and related terms have appeared in the literature. Some of these include total cost assessment, life-cycle costing, and full-cost accounting. Unfortunately, these terms have come to mean different things to different people at different times. In an attempt to remove this ambiguity, the following three economic terms are defined here:

1. *Traditional costing procedure (TCP).* This accounting procedure *only* takes into account capital and operating (including environmental) costs.

2. *Comprehensive costing procedure (CCP).* This procedure includes not only the traditional capital and operating costs but also peripheral costs, such as liability, regulatory expenses, borrowing power, and social considerations.

3. *Life-cycle costing (LCC).* This type of analysis requires that all the traditional costs of project or product systems, from raw material acquisition to end product disposal, be considered.

The TCP approach is relatively simple and can be easily applied to studies involving comparisons of different equipment, different processes, or even parts of processes. CCP has now emerged as the most realistic approach that can be employed in economic project analyses. It is the recommended procedure for pollution prevention studies. The LCC approach is usually applied to the life-cycle analysis (LCA) of a product or service. It has found occasional application in project analysis.

The remainder of this subsection on pollution prevention will provide you with the necessary background to understand the meaning of pollution prevention and its useful implementation. Assessment procedures and the economic benefits derived from managing pollution at the source are discussed, along with methods of cost accounting for pollution prevention. In addition, regulatory and nonregulatory methods to promote pollution prevention and overcome barriers are examined, and ethical considerations are presented. By eliminating waste at the source, everyone can protect the environment by reducing the amount of waste that would otherwise need to be treated or ultimately disposed; accordingly, we will discuss pollution prevention both at home and at work.

POLLUTION PREVENTION HIERARCHY

As discussed in the Introduction, the hierarchy set forth by the EPA in the Pollution Prevention Act establishes an order in which waste management activities should be employed to reduce the quantity of waste generated. The preferred method is source reduction, as indicated in Fig. 22-2. This approach actually precedes traditional waste management by addressing the source of the problem before it occurs.

Although the EPA's policy does not consider recycling or treatment to be actual pollution prevention methods, these methods present an opportunity to reduce the amount of waste that might otherwise be discharged into the environment. Clearly, the definition of pollution prevention and its synonyms (e.g., waste minimization) must be understood to fully appreciate and apply these techniques.

Waste minimization generally considers all of the methods in the EPA hierarchy (except for disposal) appropriate to reduce the volume or quantity of waste requiring disposal (e.g., source reduction). The definition of *source reduction* as applied in the Pollution Prevention Act, however, is "any practice that reduces the amount of any hazardous substance, pollutant, or contaminant entering any waste stream or otherwise released into the environment . . . prior to recycling, treatment or disposal" (United States EPA, *Pollution Prevention Fact Sheet,* Washington, D.C., March 1991). Source reduction reduces the amount of waste generated; it is therefore considered true pollution prevention and has the highest priority in the EPA hierarchy.

Recycling (or reuse) refers to the use (or reuse) of materials that would otherwise be disposed of or treated as a waste product. A good example is a rechargeable battery. Wastes that cannot be directly reused may often be recovered on-site through methods such as distillation. When on-site recovery or reuse is not feasible due to quality specifications or the inability to perform recovery on-site, off-site recovery at a permitted commercial recovery facility is often a possibility. Such management techniques are considered secondary to source reduction and should only be used when pollution cannot be prevented.

The treatment of waste is the third element of the hierarchy and should be used only in the absence of feasible source reduction or recycling opportunities. Waste treatment involves the use of chemical, biological, or physical processes to reduce or eliminate waste material. The incineration of

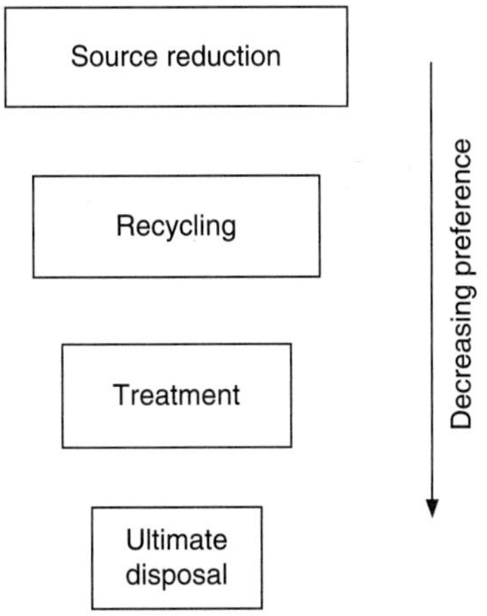

FIG. 22-2 Pollution prevention hierarchy.

wastes is included in this category and is considered "preferable to other treatment methods (i.e., chemical, biological, and physical) because incineration can permanently destroy the hazardous components in waste materials" (Dupont, R., K. Ganesan, and L. Theodore, *Pollution Prevention,* 2nd ed, CRC Press/Taylor & Francis Group, Boca Raton, Fla., 2017). It can also be employed to reduce the volume of waste to be treated.

Several of these pollution prevention elements are used by industry in combination to achieve the greatest waste reduction. Residual wastes that cannot be prevented or otherwise managed are then disposed of only as a last resort.

Figure 22-3 provides a more detailed schematic representation of the two preferred pollution prevention techniques, source reduction and recycling.

MULTIMEDIA ANALYSIS AND LIFE-CYCLE ANALYSIS

Multimedia Analysis In order to properly design and then implement a pollution prevention program, sources of all wastes must be fully understood and evaluated. A multimedia analysis involves a multifaceted approach. It must not only consider one waste stream but all potentially contaminated media (e.g., air, water, land). Past waste management practices have been concerned primarily with treatment. All too often, such methods solve one waste problem by transferring a contaminant from one medium to another (e.g., air stripping); such waste shifting is *not* pollution prevention or waste reduction.

Pollution prevention techniques must be evaluated with a thorough consideration of all media, hence the term multimedia. This approach is a clear departure from previous pollution treatment or control techniques, where it was acceptable to transfer a pollutant from one source to another in order to solve a waste problem. Such strategies merely provide short-term solutions to an ever-increasing problem. As an example, air pollution control equipment prevents or reduces the discharge of waste into the air but at the same time can produce a solid (hazardous) waste problem.

Life-Cycle Analysis The multimedia approach to evaluating a product's waste stream(s) aims to ensure that the treatment of one waste stream does not result in the generation or increase in an another waste output. The environmental impacts of producing a product or service must be evaluated over its entire life cycle. This life-cycle analysis or total systems approach (Theodore, L., personal notes) is crucial to identifying opportunities for improvement. As described earlier, this type of evaluation identifies "energy use, material inputs, and wastes generated during a product's life: from extraction and processing of raw materials to manufacture and transport of a product to the marketplace and finally to use and dispose of the product" (*Getting at the Source,* World Wildlife Fund, Washington, D.C., 1991, p. 7).

During a forum convened by the World Wildlife Fund and the Conservation Foundation in May 1990, various steering committees recommended that a three-part life-cycle model be adopted. This model consists of the following:

1. An inventory of materials and energy used, and environmental releases from all stages in the life of a product or process
2. An analysis of potential environmental effects related to energy use and material resources and environmental releases
3. An analysis of the changes needed to bring about environmental improvements for the product or process under evaluation

Traditional cost analysis often fails to include factors relevant to future damage claims resulting from litigation, the depletion of natural resources, the effects of energy use, and the like. As a result, waste management options such as treatment and disposal may appear better if an overall life-cycle cost analysis is not performed. Environmental costs from cradle to grave must be evaluated together with more conventional production costs to accurately determine genuine production costs. In the future, a total systems approach will most likely involve a more careful evaluation of pollution, energy, and safety issues. For example, if one was to compare the benefits of coal versus oil as a fuel source for an electric power plant, the use of coal might be considered economically favorable. In addition to the cost issues, however, one must be concerned with the environmental effects of coal mining (e.g., transportation and storage before use as a fuel). Society often has a tendency to overlook the fact that there are serious health and safety matters (e.g., miner exposure) that must be considered along with the effects of fugitive emissions. When these effects are weighed alongside standard economic factors, the full cost benefits of coal usage may be eclipsed by environmental costs. Thus, many of the economic benefits associated with pollution prevention are often unrecognized due to inappropriate cost-accounting methods. For this reason, economic considerations are briefly detailed later.

POLLUTION PREVENTION ASSESSMENT PROCEDURES

The first step in establishing a pollution prevention program is to get management commitment. This is necessary because of the need for project structure and control. Management will determine the amount of funding

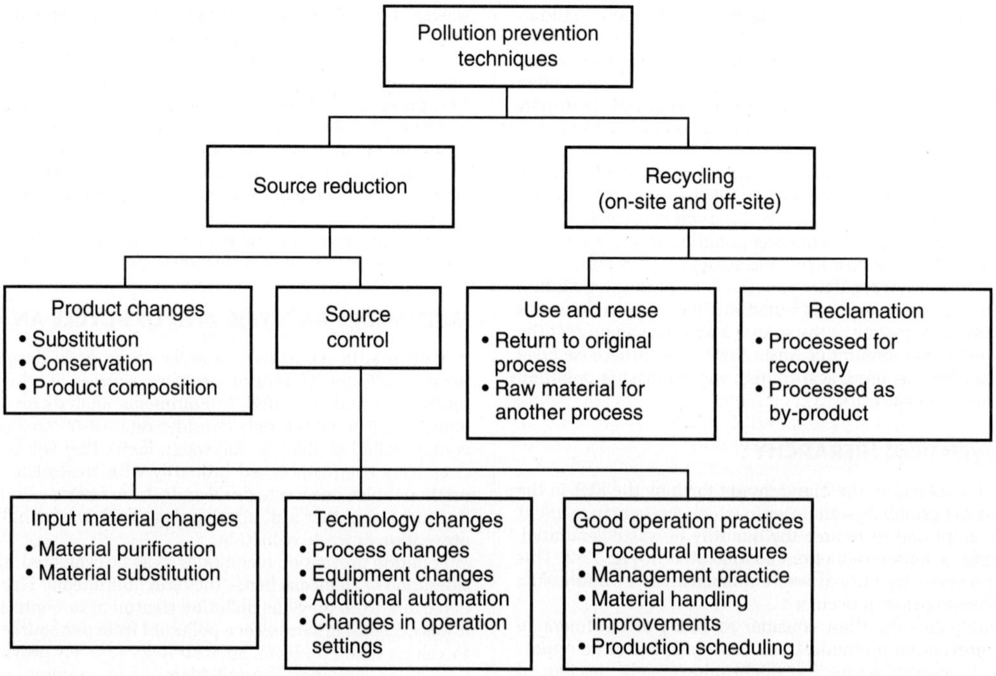

FIG. 22-3 Pollution prevention techniques.

allotted for the program as well as specific program goals. The data collected during the evaluation is then used to develop options for reducing the types and amounts of waste generated. Figure 22-4 depicts a systematic approach that can be used during the procedure. After a particular waste

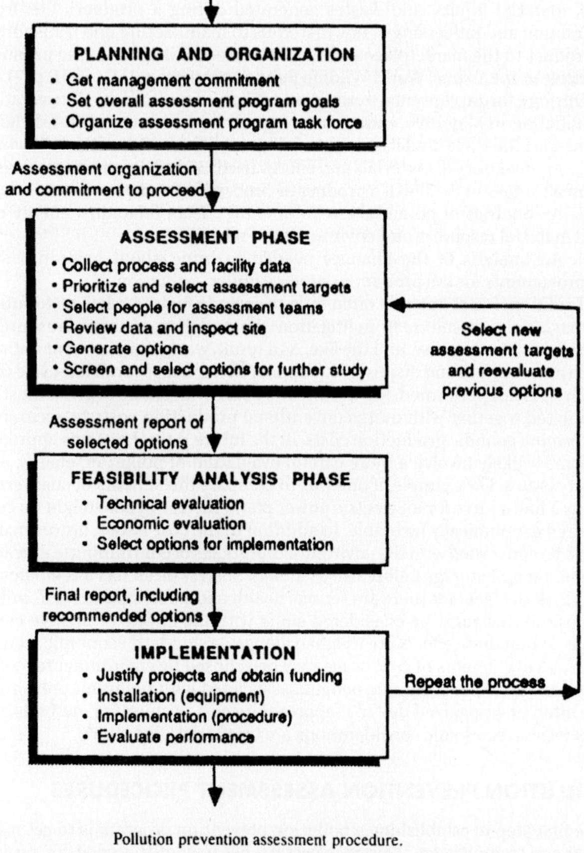

FIG. 22-4 Pollution prevention assessment procedure.

stream or area of concern is identified, feasibility studies are performed involving both economic and technical considerations. Finally, preferred alternatives are implemented. The four phases of the assessment (planning and organization, assessment, feasibility, and implementation) are introduced in the following subsections. Sources of additional information as well as information on industrial programs are also provided.

Planning and Organization The purpose of this phase is to obtain management commitment, to define and develop program goals, and to assemble a project team. Proper planning and organization are crucial to the successful performance of the pollution prevention assessment. Both managers and facility staff play important roles in the assessment procedure by providing the necessary commitment and familiarity with the facility, its processes, and current waste management operations. It is the benefits of the program, including economic advantages, liability reduction, regulatory compliance, and improved public image, that often lead to management support.

Once management has made a commitment to the program and goals have been set, a program task force is established. The selection of a team leader will depend on many factors, including the ability to effectively interface with both the assessment team and management staff.

The task force must be able to identify pollution reduction alternatives and to be aware of inherent obstacles to the process. Such obstacles often arise from anxiety that the program will have a negative effect on product quality or will result in production losses. According to an EPA survey, 30 percent of industry comments centered on concern that product quality would decline if waste minimization techniques were implemented (United States EPA, 1987 *National Biennial RCRA Hazardous Waste Report—Executive Summary*, Washington, D.C., GPO, 1991, p. 10). Therefore, the assessment team, and the team leader in particular, must be ready to respond to these and other concerns [Keoleian, G., and D. Menerey, "Sustainable Development by Design: Review of Life Cycle Design and Related Approaches," *Air & Waste Management* **44** (May 1994)].

Assessment Phase The assessment phase aims to collect data needed to identify and analyze pollution prevention opportunities. Assessment of the facility's waste reduction needs includes the examination of (hazardous) waste streams, process operations, and the identification of techniques that often promise the reduction of waste generation. Information is often derived from observations made during a facility walk-through, interviews with employees (e.g., operators, line workers), and review of site or regulatory records. One professional organization suggests the following sources of information be reviewed, as available (ASTM, Philadelphia, Pa.):

1. Product design criteria
2. Process flow diagrams for all solid waste, wastewater, and air emissions sources
3. Site maps showing the location of all pertinent units (e.g., pollution control devices, points of discharge)

4. Environmental documentation, including: material safety data sheets (MSDSs), military specification data, permits (e.g., NPDES, POTW, RCRA), SARA Title III reports, waste manifests, and any pending permits or application information

5. Economic data, including cost of raw materials; cost of air, wastewater, and (hazardous) waste treatment; waste management operating and maintenance costs; and waste disposal costs

6. Managerial information: environmental policies and procedures; prioritization of waste management concerns; automated or computerized waste management systems; inventory and distribution procedures; maintenance scheduling practices; planned modifications or revisions to existing operations that would affect waste generation activities; and the basis of source reduction decisions and policies

The use of process flow diagrams and material balances are worthwhile ways to quantify losses or emissions and provide essential data to estimate the size and cost of additional equipment, other data to evaluate economic performance, and a baseline for tracking the progress of minimization efforts (Theodore, L., personal notes). Material balances should be applied to individual waste streams or processes and then used to construct an overall balance for the facility. Details on these calculations are available in the literature (Theodore, L., *Pollution Prevention: An ETS Theodore Tutorial*, ETS International, Inc., Roanoke, Va., 1993). In addition, an introduction to this subject is provided in the next subsection.

The data collected are then used to prioritize waste streams and operations for assessment. Each waste stream is assigned a priority based on corporate pollution prevention goals and objectives. Once waste origins are identified and ranked, potential methods to reduce the waste stream are evaluated. The identification of alternatives is generally based on discussions with the facility staff; review of technical literature; and contacts with suppliers, trade organizations, and regulatory agencies.

Alternatives identified during this phase of the assessment are evaluated using screening procedures so as to reduce the number of alternatives requiring further exploration during the feasibility analysis phase. Some criteria used during this screening procedure include cost-effectiveness; implementation time; economic, compliance, safety and liability concerns; waste reduction potential; and whether the technology is proven (Dupont, R., K. Ganesan, and L. Theodore, *Pollution Prevention*, 2d ed, CRC Press/Taylor & Francis Group, Boca Raton, Fla., 2017; Theodore, L., *Pollution Prevention: An ETS Theodore Tutorial*, ETS International, Inc., Roanoke, Va., 1993). Options that meet established criteria are then examined further during the feasibility analysis.

Feasibility Analysis The selection procedure is performed by an evaluation of technical and economic considerations. The technical evaluation determines whether a given option will work as planned. Some typical considerations follow:

1. Safety concerns
2. Product quality impacts or production delays during implementation
3. Labor and/or training requirements
4. Creation of new environmental concerns
5. Waste reduction potential
6. Utility and budget requirements
7. Space and compatibility concerns

If an option proves to be technically ineffective or inappropriate, it is deleted from the list of potential alternatives. Either following or concurrent with the technical evaluation, an economic study is performed, weighing standard measures of profitability such as payback period, investment returns, and net present value. Many of these costs (or, more appropriately, cost savings) may be substantial but are difficult to quantify. (Refer to the subsection Economic Considerations Associated with Pollution Prevention Programs.)

Implementation The findings of the overall assessment are used to demonstrate the technical and economic worthiness of program implementation. Once appropriate funding is obtained, the program is implemented not unlike any other project requiring new procedures or equipment. When preferred waste pollution prevention techniques are identified, they are implemented and should become part of the facility's day-to-day management and operation. Following the program's execution, its performance should be evaluated in order to demonstrate effectiveness, generate data to further refine and augment waste reduction procedures, and maintain management support.

It should be noted that waste reduction, energy conservation, and safety issues are interrelated and often complementary. For example, the reduction in the amount of energy a facility consumes usually results in reduced emissions associated with the generation of power. Energy expenditures associated with the treatment and transport of waste are similarly reduced when the amount of waste generated is lessened; at the same time, worker safety is improved due to reduced exposure to hazardous materials. However, this is not always the case. The addition of air pollution control systems at power plants decreases net power output due to the power consumed by the equipment. This in turn requires more fuel to be combusted for the same power exported to the grid. This additional fuel increases pollution from coal mining, transport, ash disposal, and the like. In extreme cases, very high recovery efficiencies for some pollutants can raise, not lower, overall total emissions. Seventy percent removal might produce a 2 percent loss in power output; 90 percent recovery could lead to a 3 percent loss; 95 percent, a 5 percent loss; 99 percent, a 10 percent loss, and so on. The point to be made is that pollution control is generally not a free lunch.

Sources of Information The successful development and implementation of any pollution prevention program not only depends on a thorough understanding of the facility's operations but also requires an intimate knowledge of current opportunities and advances in the field. In fact, 32 percent of industry respondents to an earlier EPA survey identified the lack of technical information as a major factor delaying or preventing the implementation of a waste-minimization program (United States EPA, 1987 *National Biennial RCRA Hazardous Waste Report—Executive Summary*, Washington, D.C., GPO, 1991, p. 10). One of EPA's positive contributions has been the development of a national Pollution Prevention Information Clearinghouse (PPIC) and the Pollution Prevention Information Exchange System (PIES) to facilitate the exchange of information needed to promote pollution prevention through efficient information transfer [Keoleian, G., and D. Menerey, "Sustainable Development by Design: Review of Life Cycle Design and Related Approaches," *Air & Waste Management* 44 (May 1994)].

PPIC is operated by the EPA's Office of Research and Development and the Office of Pollution Prevention. The clearinghouse is comprised of four elements:

1. Repository, including a hard-copy reference library and collection center and an on-line information retrieval and ordering system.

2. PIES, a computerized conduit to databases and document ordering, accessible via modem and personal computer: (703) 506-1025.

3. PPIC uses the RCRA/Superfund and Small Business Ombudsman Hotlines as well as a PPIC technical assistance line to answer pollution prevention questions, access information in the PPIC, and assist in document ordering and searches. To access PPIC by telephone, call:

RCRA/Superfund Hotline, (800) 242-9346
Small Business Ombudsman Hotline, (800) 368-5888
PPIC Technical Assistance, (703) 821-4800

4. PPIC compiles and disseminates information packets and bulletins, and it initiates networking efforts with other national and international organizations.

In addition, the EPA publishes a newsletter called *Pollution Prevention News* that contains information about EPA news, technologies, program updates, and case studies. The EPA's Risk Reduction Engineering Laboratory and the Center for Environmental Research Information has published several guidance documents, developed in cooperation with the California Department of Health Services. The manuals supplement generic waste reduction information presented in the EPA's *Waste Minimization Opportunity Assessment Manual* (United States EPA, *The EPA Manual for Waste Minimization Opportunity Assessments*, Cincinnati, Ohio, August 1988).

Pollution prevention or waste minimization programs have been established at the state level and as such are good sources of information. Both federal and state agencies are working with universities and research centers and may also provide assistance. For example, the American Institute of Chemical Engineers has established the Center for Waste Reduction Technologies (CWRT), a program based on targeted research, technology transfer, and enhanced education.

Industry Programs A significant pollution prevention resource may very well be found with the "competition." Several large companies have established well-known programs that have successfully incorporated pollution prevention practices into their manufacturing processes. These include, but are not limited to: 3M—Pollution Prevention Pays (3P); Dow Chemical—Waste Reduction Always Pays (WRAP); Chevron—Save Money and Reduce Toxics (SMART); and the General Dynamics Zero Discharge Program.

Smaller companies can benefit from the assistance offered by these larger corporations. Access to information is of major importance when implementing efficient pollution prevention programs. By adopting such programs, industry is affirming pollution prevention's application as a good business practice and not simply a noble effort.

ASSESSMENT PHASE: MATERIAL BALANCE CALCULATIONS

The reader is directed to Dupont, Ganesan, and Theodore (2000) and Santoleri, Reynolds, and Theodore (2000) in the General References for further information.

One of the key elements of the assessment phase of a pollution prevention program involves mass balance equations. These calculations are often referred to as material balances; the calculations are performed via the conservation law for mass. The details of this often-used law are described next.

The conservation law for mass can be applied to any process or system. The general form of the law follows:

$$mass\ in - mass\ out + mass\ generated = mass\ accumulated$$

This equation can be applied to the total mass involved in a process or to a particular species, on either a mole or mass basis. The conservation law for mass can be applied to steady-state or unsteady-state processes and to batch or continuous systems. A steady-state system is one in which there is no change in conditions (e.g., temperature, pressure) or rates of flow with time at any given point in the system; the accumulation term then becomes zero. If there is no chemical reaction, the generation term is zero. All other processes are classified as unsteady state.

To isolate a system for study, the system is separated from the surroundings by a boundary or envelope that may either be real (e.g., a reactor vessel) or imaginary. Mass crossing the boundary and entering the system is part of the mass-in term. The equation may be used for any compound whose quantity does not change by chemical reaction or for any chemical element, regardless of whether it has participated in a chemical reaction. Furthermore, it may be written for one piece of equipment, several pieces of equipment, or around an entire process (i.e., a total material balance).

The conservation of mass law is applied during the performance of pollution prevention assessments. As described earlier, a pollution prevention assessment is a systematic, planned procedure designed to identify ways to reduce or eliminate waste. The assessment should characterize the selected waste streams and processes (ICF Technology Incorporated, *New York State Waste Reduction Guidance Manual*, Alexandria, Va., 1989)—a necessary step if a material balance is to be performed. Some of the data required for the material balance calculation may be collected during the first review of site-specific data; however, in some cases, the information may not be collected until an actual site walk-through is performed.

Simplified mass balances should be developed for each of the important waste-generating operations to identify sources and gain a better understanding of the origins of each waste stream. Since a mass balance is essentially a check to make sure that what goes into a process (the total mass of all raw materials) is what leaves the process (the total mass of the products and by-products), the material balance should be made individually for all components that enter and leave the process. When chemical reactions take place in a system, there may be an advantage to performing "elemental balances" for specific chemical elements in a system. Material balances can help to determine concentrations of waste constituents where analytical test data are limited. They are especially useful when there are points in the production process where it is difficult or uneconomical to collect analytical data.

Mass balance calculations are particularly useful for quantifying fugitive emissions such as evaporative losses. Waste stream data and mass balances will enable one to track flow and characteristics of the waste streams over time. Since in most cases the accumulation equals zero (steady-state operation), it can then be assumed that any buildup is actually leaving the process through fugitive emissions or other means. This will be useful in identifying trends in waste/pollutant generation and will also be critical to measuring the performance of pollution prevention systems. The result of these activities is a catalog of waste streams that provides a description of each waste, including quantities, frequency of discharge, composition, and other important information useful for material balance. Of course, some assumptions or educated estimates will be needed when it is impossible to obtain specific information.

When a material balance is performed in conjunction with a pollution prevention assessment, the amount of waste generated becomes known. The success of the pollution prevention program can therefore be measured by using this information on baseline generation rates (i.e., that rate at which waste is generated without pollution prevention considerations). (Details available from L. Theodore.)

BARRIERS AND INCENTIVES TO POLLUTION PREVENTION

As discussed previously, industry is beginning to realize that there are profound benefits to pollution prevention, including cost effectiveness, reduced liability, enhanced public image, and regulatory compliance. Nevertheless, there are also barriers or disincentives for pollution prevention. This section will briefly outline both barriers and incentives that may need to be confronted or considered during the evaluation of a pollution prevention program.

Barriers to Pollution Prevention ("The Dirty Dozen") There are many reasons why more businesses are not reducing the wastes they generate. The following "dirty dozen" are common disincentives:

1. *Technical limitations.* Given the complexity of present manufacturing processes, waste streams exist that cannot be reduced with current technology. Continued research and development are needed.
2. *Lack of information.* Sometimes the information needed to make a pollution prevention decision may be confidential or difficult to obtain. Sometimes decision makers are simply unaware of the information that is available to help them implement pollution prevention programs.
3. *Consumer preference obstacles.* Consumer preference strongly affects the manner in which a product is produced, packaged, and marketed. If the implementation of a pollution prevention program results in an increase in the cost of a product or decreased convenience or availability, consumers might be reluctant to use it.
4. *Concern over product quality decline.* The use of a less hazardous material in manufacturing may result in decreased product life, durability, or competitiveness.
5. *Economic concerns.* Many companies are unaware of the economic advantages of pollution prevention. Legitimate concerns may include lower profit margins or a lack of funds for the initial capital investment.
6. *Resistance to change.* The unwillingness of many businesses to change is rooted in their reluctance to try technologies that may be unproven or based on a combination of the barriers discussed in this section.
7. *Regulatory barriers.* Existing regulations that have created incentives for the control and containment of wastes are at the same time discouraging the exploration of pollution prevention alternatives. Moreover, since regulatory enforcement is often intermittent, current legislation can weaken waste reduction incentives.
8. *Lack of markets.* The implementation of pollution prevention processes and the production of environmentally friendly products will be of no avail if markets do not exist for such goods. As an example, the recycling of newspaper in the United States has resulted in an overabundance of waste paper without markets prepared to take advantage of this raw material.
9. *Management apathy.* Many managers capable of making decisions to begin pollution prevention activities do not appreciate the potential benefits.
10. *Institutional barriers.* In an organization without a strong infrastructure to support pollution prevention plans, waste reduction programs will be difficult to implement. Similarly, if there is no mechanism in place to hold people accountable for their actions, the success of a pollution prevention program will be limited.
11. *Lack of awareness of pollution prevention advantages.* As mentioned in reason no. 5, decision makers may merely be uninformed of the benefits associated with pollution reduction.
12. *Concern over the dissemination of confidential product information.* If a pollution prevention assessment reveals confidential data about a company's product, there may be fears that the organization will lose a competitive edge.

Pollution Prevention Incentives ("A Baker's Dozen") Various means exist to encourage pollution prevention through regulatory measures, economic incentives, and technical assistance programs. Since the benefits of pollution prevention can surpass prevention barriers, a "baker's dozen" incentives are presented here:

1. *Economic benefits.* The most obvious economic benefits of pollution prevention are the savings that result from the elimination of waste storage, treatment, handling, transport, and disposal. Additional, less tangible economic benefits are realized in terms of decreased liability, lower regulatory compliance costs (e.g., permits), lower legal and insurance costs, and improved process efficiency. Pollution prevention almost always pays for itself, especially when the time required to comply with regulatory standards is considered. Several of these economic benefits are discussed separately in this section.
2. *Regulatory compliance.* Quite simply, when wastes are not generated, compliance issues are not a concern. Waste management costs associated with recordkeeping, reporting, and laboratory analysis are reduced or eliminated. Pollution prevention's proactive approach to waste management will better prepare industry for future regulation of many hazardous substances and wastes that are currently unregulated. Regulations have, and will continue to be, a moving target.
3. *Liability reduction.* Facilities are responsible for their wastes from "cradle to grave." By eliminating or reducing waste generation, future liabilities can also be decreased. Additionally, the need for expensive pollution liability insurance requirements may be abated.
4. *Enhanced public image.* Consumers are interested in purchasing goods that are safer for the environment, and this demand, depending on how they respond, can mean success or failure for many companies. Businesses should therefore be sensitive to consumer demands and use pollution prevention efforts to their utmost advantage by producing goods that are environmentally friendly.
5. *Federal and state grants.* Federal and state grant programs have been developed to strengthen pollution prevention programs initiated by states and private entities. The EPA's Pollution Prevention By and For Small

Business grant program awards grants to small businesses to help them develop and demonstrate new pollution prevention technologies.

6. *Market incentives.* Public demand for environmentally preferred products has generated a market for recycled goods and related products; products can be designed with these environmental characteristics in mind, offering a competitive advantage. In addition, many private and public agencies are beginning to stimulate the market for recycled goods by writing contracts and specifications that call for the use of recycled materials.

7. *Reduced waste-treatment costs.* As discussed in reason no. 5 of the dirty dozen, the increasing costs of traditional end-of-pipe waste management practices are avoided or reduced through the implementation of pollution prevention programs.

8. *Potential tax incentives.* In an effort to promote pollution prevention, taxes may eventually need to be levied to encourage waste generators to consider reduction programs. Conversely, tax breaks could be developed for corporations that use pollution prevention methods.

9. *Decreased worker exposure.* By reducing or eliminating chemical exposures, businesses benefit by lessening the potential for chronic workplace exposure and serious accidents and emergencies. The burdens of medical monitoring programs, personal exposure monitoring, and potential damage claims are also reduced.

10. *Decreased energy consumption.* As mentioned previously, energy conservation methods are often complementary. Less waste generated usually means less energy expended to treat and transport it; this also reduces the pollution that is directly associated with energy consumption.

11. *Increased operating efficiencies.* A potential side effect of pollution prevention is an increase in operating efficiency. A pollution prevention assessment can identify both sources of hazardous wastes *and* losses in process performance. The implementation of a waste reduction program will often rectify such problems through modernization, innovation, and the implementation of good operating practices.

12. *Competitive advantages.* By taking advantage of the many benefits of pollution prevention, businesses can gain a competitive edge.

13. *Reduced negative environmental impacts.* A total systems approach to pollution prevention considers the environmental damage that occurs during raw material procurement and waste disposal. Pollution prevention endeavors therefore help to protect the environment.

ECONOMIC CONSIDERATIONS ASSOCIATED WITH POLLUTION PREVENTION PROGRAMS

The purpose of this subsection is to outline the basic elements of a pollution prevention cost-accounting system that incorporates both traditional and less tangible economic variables. The intent is not to present a detailed discussion of economic analysis but to help identify the more important elements that must be considered to properly quantify pollution prevention options.

The greatest driving force behind any pollution prevention plan is the promise of economic opportunities and cost savings over the long term. Pollution prevention is now recognized as one of the lowest-cost options for waste management. Hence, it is quite important to understand the economics of pollution prevention in order to make informed decisions at both the engineering and management levels. Every engineer should be able to do an economic evaluation of a proposed project. If the project cannot be justified economically after *all* factors—and these will be discussed in more detail—have been taken into account, it should obviously not be pursued. The earlier the evaluation, the fewer resources will be wasted.

Before the true cost or profit of a pollution prevention program can be determined, the economic factors must be identified. The two traditional contributing factors are capital costs and operating costs, but there are other important costs and benefits that must be quantified in a meaningful economic analysis. Table 22-5 demonstrates the evolution of various cost-accounting methods. Although Tables 22-4 (see Introduction) and 22-5 are not directly related, the reader can compare some of the similarities between the two.

The total systems approach (TSA) referenced in Table 22-5 aims to quantify not only the economic aspects of pollution prevention but also the social costs associated with the production of a product or service from cradle to grave—that is, the life-cycle costs. The TSA tries to quantify less tangible benefits, such as the reduced risk derived from not using a hazardous substance. The future is certain to see more emphasis placed on the TSA approach in any pollution prevention program. A utility considering the option of converting from a gas-fired boiler to coal-firing is usually not concerned with the environmental effects of mining, transporting, and storing the coal. Pollution prevention approaches since the mid- to late 1990s have become more sensitive to such issues.

Economic evaluation is usually carried out using standard measures of profitability. Each company and organization has its own economic criteria for selecting projects for implementation. For example, a project can be judged on its payback period. For some companies, if the payback period is more than three years, it is a dead proposal. The approach presented in this subsection represents a preliminary, rather than a detailed, analysis. For smaller facilities with only a few (perhaps simple) processes, the entire pollution prevention assessment procedure might be rather informal. Several obvious pollution prevention options, such as the installation of flow controls and good operating practices, may be implemented with little or no economic evaluation. In such cases, no complicated analyses are needed to demonstrate the advantages of pollution prevention.

TABLE 22-5 Economic Analysis Timetable

Prior to 1945	Capital costs only
1945–1960	Capital and some operating costs
1960–1970	Capital and operating costs
1970–1975	Capital, operating, and some environmental control costs
1975–1980	Capital, operating, and environmental control costs
1980–1985	Capital, operating, and more sophisticated environmental control costs
1985–1990	Capital, operating, and environmental controls, and some life-cycle analysis (total systems approach)
1990–1995	Capital, operating, and environmental control costs and life-cycle analysis (total systems approach)
1995–2000	Widespread acceptance of total systems approach
After 2000	To be detailed at a later date

A proper perspective must be maintained between the magnitude of the potential savings and the amount of manpower required to do the technical and economic feasibility analyses. A short description of the various economic factors—including capital and operating costs and other considerations—follows.

Once identified, the costs and savings are placed into their appropriate categories and quantified for subsequent analysis. Equipment cost is a function of many variables, one of the most significant of which is capacity. Other important variables include operating temperature or pressure conditions and degree of equipment sophistication. Preliminary estimates are often made using simple cost-capacity relationships that are valid when the other variables are confined to a narrow range of values.

The usual technique for determining the capital costs (i.e., total capital costs, which include equipment design, purchase, and installation) for the facility is based on the factored method of establishing direct and indirect installation costs as a function of the known equipment costs. This is basically a modified Lang method, whereby cost factors are applied to known equipment costs [Neveril, R. B., *Capital and Operating Costs of Selected Air Pollution Control Systems*, EPA Report 450/5-80-002, Gard, Inc., Niles, Ill., December 1978; Vatavuk, W. M., and R. B. Neveril, "Factors for Estimating Capital and Operating Costs," *Chemical Engineering* (November 3, 1980): 157–162]. The first step is to obtain from vendors the purchase prices of the primary and auxiliary equipment. The total base price, designated by *X*, which should include instrumentation, control, taxes, freight costs, and so on, serves as the basis for estimating the direct and indirect installation costs. These costs are obtained by multiplying *X* by the cost factors, which are available in the literature. See Neveril (1978), Vatavuk and Neveril (1980), Vogel and Martin (September 5, 1983), Vogel and Martin (October 17, 1983), Vogel and Martin (November 28, 1983), Ulrich (1984), and California Department of Health Services (1988) in General References.

The second step is to estimate the direct installation costs by summing all the cost factors involved, which include piping, insulation, foundation and supports, and so on. The sum of these factors is designated as the DCF (direct installation cost factor). The direct installation costs are then the product of the DCF and *X*. The third step consists of estimating the indirect installation costs. Here all the cost factors for the indirect installation costs (engineering and supervision, start-up, construction fees, and so on) are added; the sum is designated by ICF (indirect installation cost factor). The indirect installation costs are then the product of ICF and *X*. Once the direct and indirect installation costs have been calculated, the total capital cost (TCC) may be evaluated as follows:

$$\text{TCC} = X + (\text{DCF})(X) + (\text{ICF})(X)$$

This can then be converted to annualized capital costs (ACC) with the use of the capital recovery factor (CRF). The CRF can be calculated as follows:

$$\text{CRF} = \frac{(i)(1+i)^n}{(1+i)^n - 1}$$

where n = projected lifetime of the project, years
i = annual interest rate, expressed as a fraction

The annualized capital cost (ACC) is the product of the CRF and TCC and represents the total annualized installed equipment cost distributed over the lifetime of the project. The ACC reflects the cost associated with the initial capital outlay over the depreciable life of the system. Although investment and operating costs can be accounted for in other ways, such as present-worth analysis, the capital recovery method is often preferred because of its simplicity and versatility. This is especially true when comparing somewhat similar systems having different depreciable lives. In such decisions, there are usually other considerations besides economic, but if all other factors are equal, the alternative with the lowest total annualized cost should be the most viable.

Operating costs can vary from site to site since these costs reflect local conditions (e.g., staffing practices, labor, utility costs). Operating costs, like capital costs, may be separated into two categories: direct and indirect costs. Direct costs are those that cover material and labor and are directly involved in operating the facility. These include labor, materials, maintenance and maintenance supplies, replacement parts, wastes, disposal fees, utilities, and laboratory costs. However, the major direct operating costs are usually those associated with labor and materials. Indirect costs are those operating costs associated with, but not directly involved in, operating the facility; costs such as overhead (e.g., building/land leasing and office supplies), administrative fees, property taxes, and insurance fees fall into this category.

The main problem with the traditional type of economic analysis is that it is difficult—in some cases impossible—to quantify some of the not-so-obvious economic merits of a pollution prevention program. Several considerations have recently surfaced as factors that need to be taken into account in any meaningful economic analysis of a pollution prevention effort. What follows is a summary listing of these considerations, most of which have been detailed earlier.

1. Decreased long-term liabilities
2. Regulatory compliance
3. Regulatory recordkeeping
4. Dealings with the EPA
5. Dealings with state and local regulatory bodies
6. Elimination or reduction of fines and penalties
7. Potential tax benefits
8. Customer relations
9. Stockholder support (corporate image)
10. Improved public image
11. Reduced technical support
12. Potential insurance costs and claims
13. Effect on borrowing power
14. Improved mental and physical well-being of employees
15. Reduced health-maintenance costs
16. Employee morale
17. Other process benefits
18. Improved worker safety
19. Avoidance of rising costs of waste treatment and disposal
20. Reduced training costs
21. Reduced emergency response planning

Many proposed pollution prevention programs have been squelched in their early stages because a comprehensive economic analysis was not performed. Until the effects just described are included, the true merits of a pollution prevention program may be clouded by incorrect or incomplete economic accounting. Can something be done by industry to remedy this problem? One approach is to use a modified version of the standard Delphi panel. To estimate these other economic benefits of pollution prevention, several knowledgeable people within and perhaps outside the organization are asked to independently estimate, with explanatory details, these economic benefits. Everyone on the panel is then allowed to independently review all responses. The cycle is then repeated until the group's responses approach convergence.

Finally, pollution prevention measures can provide a company with the opportunity of looking their neighbors in the eye and truthfully saying that all that can reasonably be done to prevent pollution is being done. In effect, the company is doing right by the environment. Is there an economic advantage to this? It is not only a difficult question to answer quantitatively but also a difficult one to answer qualitatively. Industry is left to ponder the answer to this question.

GREEN CHEMISTRY AND GREEN ENGINEERING

See Theodore, L., *Chemical Engineering: The Essential Reference*, McGraw-Hill, New York, 2014; Abulencia, P., and L. Theodore, *Open-Ended Problems: The Future Chemical Engineering Approach,* Scrivener/Wiley Publishing, Beverly, Mass., 2015 in General References.

Activities in the field of green engineering and green chemistry are increasing at a near-exponential rate. In this section we try to familiarize you with these topics by defining and presenting principles of each; future trends are also discussed. It is important that the term *green* should not be considered a new method or type of chemistry or engineering. Rather, it should be incorporated into the way scientists and engineers design for categories that include the environment, manufacturability, disassembly recycling, serviceability, and compliance. Today, the major green element is to search for technology to reduce or eliminate waste from operations and processes, with an important priority of not creating waste in the first place.

Green chemistry, also called *clean chemistry*, refers to that field of chemistry dealing with the synthesis, processing, and use of chemicals that reduce risks to humans and the environment (Anastas, P., and T. Williamson, in *Green Chemistry: An Overview, Green Chemistry: Designing Chemistry for the Environment*; ACS Symposium Series 626, ed. P. T. Anastas and T. C. Williamson, American Chemical Society, Washington, D.C., 1995, pp. 1–17).

It is defined as the invention, design, and application of chemical products and processes to reduce or to eliminate the use and generation of hazardous substances (Anastas, P., and J. Warner, *Green Chemistry: Theory and Practice*, Oxford University Press, New York, 1998).

A baker's dozen principles of green chemistry are provided next [Anastas and Warner (1998)]:

1. *Prevention.* It is better to prevent waste than to treat or clean up waste after it has been generated.

2. *Atom economy.* Synthetic methods should be designed to maximize the incorporation of all materials used in the process through the final product.

3. *Less hazardous chemical syntheses.* Whenever practicable, synthetic methods should be designed to use and generate substances that have little or no toxicity to human health and the environment.

4. *Designing safer chemicals.* Chemical methods should be designed to preserve efficacy of function while minimizing toxicity.

5. *Safer chemicals and auxiliaries.* The use of auxiliary substances (e.g., solvents, separation agents) should be made unnecessary whenever possible and should be innocuous when used.

6. *Design for energy efficiency.* Energy requirements should be recognized for their environmental and economic impacts and should be minimized. Synthetic methods should be conducted at ambient temperature and pressure whenever possible.

7. *Use of renewable feedstocks.* A raw material or feedstock should be renewable rather than depleting, wherever and whenever technically and economically practicable.

8. *Reduce derivatives.* Unnecessary derivatization (blocking group, temporary modification of physicochemical processes, etc.) should be avoided whenever possible because such steps require additional reagents and can generate waste.

9. *Catalysis.* Catalytic reagents (as selective, or discriminating, as possible) are superior to stoichiometric reagents.

10. *Biocatalysis.* Enzymes and antibodies should be used to mediate reagents.

11. *Design for degradation.* Chemical products should be designed in a way that at the end of their function they break down into innocuous degradation products and do not persist in the environment.

12. *Real-time analysis for pollution prevention.* Analytical methods need to be further developed to allow for real-time, in-process monitoring and control prior to the formation of hazardous substances.

13. *Inherently safer chemistry for accident prevention.* Substances and the form of a substance used in a chemical process should be chosen to minimize the potential for chemical accidents, including releases, explosions, and fires.

Green engineering is similar to green chemistry in many respects, as witnessed by the underlying urgency of attention to the environment seen in both sets of principles.

A baker's dozen principles of green engineering are provided next [Anastas, P., and J. Zimmerman, "Design through the Twelve Principles of Green Engineering," *Environ. Sci. Technol.* **37:** 94A–101A (2003)]:

1. *Benign rather than hazardous.* Designers must ensure that all material and energy inputs and outputs are as inherently nonhazardous as possible.

2. *Prevention instead of treatment by recycle or reuse.* It is better to prevent waste than to treat or clean up waste after it is generated.
3. *Design for separation.* Separation and purification operations should be a component of the design framework.
4. *Maximize efficiency.* System components should be designed to maximize mass, energy, and temporal efficiency.
5. *Output-pulled versus input-pushed.* Components, processes, and systems should be output-pulled versus input-pushed through the use of energy and materials.
6. *Conserve complexity.* Energy conservation must also consider entropy. Embedded entropy and complexity must be viewed as an investment when making design choices on recycle, reuse, or beneficial disposition.
7. *Durability rather than immortality.* Targeted durability, not immortality, should be a design goal.
8. *Meet need, minimize excess.* Designing for unneeded capacities or capabilities should be considered flawed design; this includes the use of one-size-fits-all solutions.
9. *Minimize material diversity.* Multicomponent products should be designed for material unification to promote disassembly and value retention (to minimize material diversity).
10. *Integrate material and energy flows.* Design of processes and systems must include the integration of interconnectivity with available energy and material flows.
11. *Design for afterlife.* Performance metrics should include designing for performance in (commercial) afterlife.
12. *Renewable rather than depleting.* Design should be based on renewable and readily available inputs throughout the life cycle.
13. *Engaging communities.* Actively engage communities and stakeholders in the development of engineering solutions.

From the preceding definitions, one might conclude that green engineering is concerned with the design, commercialization, and use of all types of processes and products, whereas green chemistry covers only a very small subset of this—the development of chemical processes and products. Thus, green chemistry may be viewed as a subset of green engineering. It is, in fact, a very broad field, encompassing everything from improving energy efficiency in manufacturing processes to developing plastics from renewable resources.

Green chemistry and green engineering are emerging 21st century issues that come under the larger multifaceted spectrum of sustainable development. Sustainable development represents change in consumption patterns toward environmentally more benign products and a change in investment patterns toward augmenting environmental capital (Pearce, D., A. Mankandya, and E. Barbier, *Blueprint for a Green Economy*, Earthscan, London, 1989). In this respect, sustainable development is feasible. It requires a shift in the balance of the way economic progress is pursued (see also next subsection).

SUSTAINABILITY

To *sustain* is defined by some as to "support without collapse." Discussion of how sustainability should be defined was initiated by the Bruntland Commission. This group was given the mission of creating a "global agenda for change" by the General Assembly of the United Nations in 1984. They defined *sustainable* very broadly (Bishop, P., *Pollution Prevention*, Waveland Press, Prospect Heights, Ill., 2000). "Humanity has the ability to make development sustainable—to ensure that it meets the needs of the present without compromising the ability of future generations to meet their own needs" [United Nations, Report of the World Commission on Environment and Development, General Assembly Resolution 42/187, http://www.un.org/documents/ga/res/42/ares42-187.htm (retrieved March 15. 2018)].

Sustainability involves simultaneous progress in four major areas: (1) human, (2) economic, (3) technological, and (4) environmental. Sustainability requires conservation of resources while minimizing the depletion of nonrenewable resources. However, there can be no product development or economic activity of any kind without available resources. Except for solar and nuclear energy, the supply of resources is finite. Efficient designs conserve resources while also reducing impacts caused by materials extraction and related activities. Depletion of nonrenewable resources and overuse of otherwise renewable resources limits their availability to future generations.

Another principal element of sustainability is maintenance of the structure and function of the ecosystem. Because the health of human population is linked to the health of the natural world, the issue of ecosystem health is a fundamental issue of concern to sustainable development. Thus, sustainability requires that the health of all diverse species as well as their interrelated ecological functions be maintained. As only one species in a complex web of ecological interactions, humans cannot separate their survivability from that of the total ecosystem.

Over the next 50 years (i.e., until 2065–2070 or so), projections suggest that the world's population could increase by 50 percent. Global economic activity is expected to increase by 500 percent. Concurrently, global energy consumption and manufacturing activity may rise to approximately three times current levels. These trends could have serious social, economic, and environmental consequences unless a way can be found to use fewer resources more efficiently. The task ahead in this century is to help shape a sustainable future in a cost-effective manner, recognizing that economic and environmental considerations, supported by innovative science and engineering technology, can work together and promote societal benefits. However, unless humans embrace sustainability, they will ultimately deplete Earth's resources and damage its environment to such an extent that conditions for human existence on the planet will be seriously compromised or even impossible.

As we have stated, sustainable development is feasible. Since *sustainable development* means a change in consumption patterns toward environmentally more benign products, and a change in investment patterns, it will require a shift in the balance of the way economic progress is pursued. Environmental concerns must be properly integrated as these changes occur, and the environment must be viewed as an integral part of human well-being.

In summary, *sustainability* is a term that is used with greater frequency in the environmental community and with sometimes varied meanings. In short, it means the capacity to endure. A more specific definition of sustainability refers to the long-term maintenance of well-being, which has environmental, economic, and social dimensions, and includes the responsible management of resources.

INDUSTRIAL ECOLOGY

The term *industrial ecology* has come to mean different things to different people. At a minimum, one may view it as a spectrum of sustainability as it applies to both industry and ecology. Industrial ecology appears to combine two normally divergent words. It is based on the idea that an industrial system is a part of the natural system and can, in some respects, mimic it. *Biological ecology* is defined as the study of the distribution and abundance of living organisms and the interactions between these organisms and their environment, while industrial ecology may be defined as the study of metabolisms of technological organisms, including their use of resources, their potential environmental impacts, and their interactions with the natural world.

Industrial ecology is an emerging field of study that deals with sustainability. The essence of industrial ecology was also defined in the following manner (Graedel, T., and B. Allenby, *Industrial Ecology*, 2d ed., Prentice-Hall, Upper Saddle River, N.J., 2003).

Industrial ecology is the means by which humanity can deliberately and rationally approach and maintain sustainability, given continued economic, cultural, and technological evolution. The concept requires that an industrial system be viewed not in isolation from its surrounding systems, but in concert with them. It is a systems view in which one seeks to optimize the total materials cycle from virgin material, to finished material, to component, to product, to obsolete product, and to ultimate disposal. Factors to be optimized include resources, energy, and capital. Thus, industrial ecology is industrial and technological in the sense that it focuses on industrial processes and related issues, including the supply and use of materials and energy, adoption of technologies, and study of technological environmental impacts. Although social, cultural, political, and psychological topics arise in an industrial ecology context, they are often regarded as ancillary fields not central to industrial ecology itself (Allenby, B., *Industrial Ecology: Policy, Framework and Implementation*, Prentice-Hall, Upper Saddle River, N.J., 1999).

Industrial ecology's emphasis on industries and technologies can be explained with the key equation of industrial ecology. Originating from the IPAT equation (impact, population, affluence, and technology) (Ehlrich, P., and J. Holdren, *Impact of Population Growth*, Scrivener, Washington, D.C., 1971), this equation expresses the relationship between technology, humanity, and the environment in the following form:

$$\text{Environmental impact} = \text{Population} \times \frac{\text{GDP}}{\text{Person}} \times \frac{\text{Environmental impact}}{\text{Unit of GDP}}$$

The term *GDP* is a country's or region's gross domestic product, the measure of industrial and economic activity (Graedel, T., and B. Allenby, *Industrial Ecology*, 2d ed., Prentice-Hall, Upper Saddle River, N.J., 2003). For this equation, the *population* term, a social and demographic one, has shown a rapid increase in the past several decades and continues to increase. The second term, *per capita GDP*, is an economic indicator of the present population's wealth and living standards. Its general trend is rising as well, although there are wide variations among countries and over time. These trends make it clear that the only hope of keeping environmental interaction in the next few decades at an acceptable level is to reduce the third term, *environmental impacts per unit of GDP*, by more than the product of the increases in the first two terms—a substantial challenge! This third term is mainly technological and is a central focus of industrial ecology.

Industrial ecology has a long list of areas where research and development are needed. The urgent theoretical needs are to develop general theories for industrial ecosystem organizations and functions, and to relate technology more rigorously to sustainability. The tasks are substantial, but carrying them out is likely to provide a crucial framework for society in the next few decades, as we seek to reconcile our use of Earth's resources with the ultimate sustainability of the planet and its inhabitants, human and otherwise.

ETHICAL CONSIDERATIONS

Given the evolutionary nature of pollution prevention, it is evident that as technology changes and progress continues, society's opinion of what is both possible and desirable will also change. Government officials, scientists, and engineers will face new challenges to meet society's needs while also meeting the requirements of changing environmental regulations. We must also pay attention to ethical considerations in pollution prevention policy. While we do not intend to explore the philosophy of ethics or morality in depth, we must explain what we mean when we talk about ethics.

Ethics can simply be defined as the analysis of the rightness or wrongness of an act or actions. According to Dr. Andrew Varga, director of the Philosophical Resources Program at Fordham University, in order to discern the morality of an act, it is customary to look at the act on the basis of four separate elements: its object, motive, circumstances, and consequences (Theodore, L., *Chemical Engineering: The Essential Reference*, McGraw-Hill, New York, 2014). Rooted in this analysis is the belief that if one part of the act is bad, then the overall act itself cannot be considered good. Of course, there are instances where the good effects outweigh the bad, and therefore there may be a reason to permit the evil. The application of this principle is not cut and dried, and it requires decisions to be made on a case-by-case basis. Each decision must be based on an understanding of the interaction between technology and the environment. After all, the decisions made today will have an impact not only on this generation but on many generations to come. If one chooses today not to implement a waste reduction program in order to meet a short-term goal of increased productivity, this might be considered a good decision since it benefits the company and its employees. However, should a major release occur that results in the contamination of a local drinking water source, what is the good?

As an additional example, toxicological studies have indicated that test animals exposed to small quantities of toxic chemicals had better health than control groups that were not exposed. A theory has been developed that says that a low-level exposure to the toxic chemical results in a challenge to the animal to maintain homeostasis; this challenge increases the animal's vigor and, correspondingly, its health. However, larger doses seem to cause an inability to adjust, resulting in negative health effects. Based on this theory, some people believe that absolute pollution reduction might not be necessary.

FUTURE TRENDS

One should keep an open mind when dealing with the types of issues we have discussed and when facing challenges perhaps not yet imagined. Clearly, there is no simple solution or answer to many of the questions. The EPA is currently *attempting* to develop a partnership with government, industry, and educators to produce and distribute pollution prevention educational materials. Given EPA's past history and performance, there are understandable doubts as to whether this program will succeed (Abulencia, P., and L. Theodore, *Open-Ended Problems: The Future Chemical Engineering Approach*, Scrivener/Wiley Publishing, Beverly, Mass., 2015).

Finally, no discussion on pollution prevention would be complete without reference to the activities of the 50-year-old International Organization for Standardization (ISO) and the general subject of nanotechnology (Anastas, P., and T. Williamson, in *Green Chemistry: An Overview, Green Chemistry: Designing Chemistry for the Environment*; ACS Symposium Series 626, ed. P. T. Anastas and T. C. Williamson, American Chemical Society, Washington, D.C., 1995, pp. 1–17). Both are briefly described next.

ISO created Technical Committee 207 (TC 207) to develop new standards for environmental management systems (EMS). The ramifications, especially to the chemical industry, which has become heavily involved in the continued development and application of these standards, will be great. TC 207's activities are being scrutinized closely. The relatively new environmental management set of standards is entitled ISO 14000. With this new standard in place, it is expected that customers will require their product suppliers to be certified by ISO 14000. In addition, some service suppliers who have an impact on the environment are expected to obtain an ISO 14000 certification. Certification will imply to the customer that, when the product or service was prepared, the environment was not significantly damaged in the process. This will effectively require that the life-cycle design mentality discussed earlier be applied to all processes. Implementing the life-cycle design framework will require significant organizational and operational changes in business. To effectively promote the goals of sustainable development, life-cycle designs will have to successfully address cost, performance, and cultural and legal factors.

Nanotechnology is concerned with the world of invisible, miniscule particles that are dominated by forces of physics and chemistry that cannot be applied at the micro, macro, or human scale. These particles have come to be labeled by some as nanomaterials. The new technology allows the engineering of matter by processes that effectively deal with atoms. Interestingly, the classic laws of science operate differently at the nanoscale. Nanoparticles have large surface areas and essentially no inner mass; their surface-to-mass ratio is extremely high, and nanostructures or nanomachines developed from these particles have special properties and exhibit unique behavior. This behavior can significantly affect physical, chemical, electrical, biological, mechanical, and other functional qualities/properties. These qualities can be exploited by applied scientists to engineer revolutionary processes. Potential applications range from chemical products (e.g., parts, plastics, powders, specialty metals) to computer systems to pollution prevention to systems for addressing crime and terrorism. In short, the sky's the limit. As far as the environment is concerned, this new technology can end pollution as we know it today. Nano-level systems might produce products without waste since environmental engineering is built in by design. However, activists have begun to organize against the science, calling for a moratorium on nanotechnology products until the social and environmental risks are better understood. At the time of this writing, no nanoregulations had yet been put into place. Two of the main obstacles to applying nanotechnology will be to show how it can control, reduce, and ultimately eliminate environmental and related problems and to address problems that may develop through the use or misuse of the technology.

The impact of all of this on both the industry and the consumer will be significant. Consumers will have the final say. Once consumers refuse to accept products or services that damage the environment, that industry is either out of business or must change its operations to environmentally acceptable alternatives. With the new standards, customers of certified organizations (in addition to other organizations) will be assured that the products or services they purchase have been produced in accordance with universally accepted standards of environmental management. Organizational claims, which today can be misleading or erroneous, will, under the standards, be backed up by comprehensive and detailed environmental management systems that must withstand the scrutiny of intense audits.

Can all of this be achieved in the near future? If it can, then society need only focus its environmental efforts on educating consumers.

AIR POLLUTION MANAGEMENT OF STATIONARY SOURCES

Further Reading: Billings and Wilder, *Fabric Filter Handbook*, U.S. EPA, NTIS Publ. PB 200-648 (vol. 1), PB 200-649 (vol. 2), PB 200-651 (vol. 3), and PB 200-650 (vol. 4), 1970; Buonicore, "Air Pollution Control," *Chem, Eng.* **87**(13): 81 (June 30, 1980); Buonicore and Theodore, *Industrial Control Equipment for Gaseous Pollutants*, vols. I and II, CRC Press, Boca Raton, Fla., 1975; Calvert, *Scrubber Handbook*, U.S. EPA, NTIS Publ. PB 213-016 (vol. 1) and PB 213-017 (vol. 2), 1972; Danielson, *Air Pollution Engineering Manual*, EPA Publ. AP-40, 1973; Davis, *Air Filtration*, Academic, New York, 1973; Kleet and Galeski, *Flare Systems Study*, EPA-600/2-76-079 (NTIS), 1976; Lund, *Industrial Pollution Control Handbook*, McGraw-Hill, New York, 1971; Oglesby and Nichols, *Manual of Electrostatic Precipitator Technology*, U.S. EPA, NTIS Publ. PB 196-380 (vol. 1), PB 196-381 (vol. 2), PB 196-370 (vol. 3), and PB 198-150 (vol. 4), 1970; *Package Sorption System Study*, EPA-R2-73-202 (NTIS), 1973; Rolke et al., *Afterburner Systems Study*, U.S. EPA, NTIS Publ. PB 212-500, 1972; Slade, *Meteorology and Atomic Energy*, AEC (TID-24190), Oak Ridge, Tenn., 1969; Stern, *Air Pollution*, Academic, New York, 1974; Strauss, *Industrial Gas Cleaning*, Pergamon, New York, 1966; Buonicore and Theodore, *Industrial Control Equipment for Gaseous Pollutants*, vol. 1, 2d ed., CRC Press, Boca

Raton, Fla., 1992; Theodore and Buonicore, *Air Pollution Control Equipment: Selection, Design, Operation, and Maintenance,* Prentice Hall, Englewood Cliffs, N.J., 1982; Treybal, *Mass Transfer Operations,* 3d ed., McGraw-Hill, New York, 1980; Turner, *Workbook of Atmospheric Dispersion Estimates,* U.S. EPA Publ. AP-26, 1970; White, *Industrial Electrostatic Precipitation,* Addison-Wesley, Reading, Mass., 1963; McKenna, Mycock, and Theodore, *Handbook of Air Pollution Control Engineering and Technology,* CRC Press, Boca Raton, Fla., 1995; Theodore and Allen, *Air Pollution Control Equipment,* ETSI Prof. Training Inst., Roanoke, Va., 1993; *Air Pollution Engineering Manual,* 2d ed., Wiley, New York, 2000; *Guideline on Air Quality Models (GAQM),* rev. ed., EPA-450/2-78-027R, July 1986; *Compilation of Air Pollution Emission Factors (AP-42),* 4th ed., U.S. EPA, Research Triangle Park, North Carolina, September, 1985; McKenna and Turner, *Fabric Filter—Baghouses I—Theory, Design, and Selection,* ETSI Prof. Training Inst., Roanoke, Va., 1993; Greiner, *Fabric Filter—Baghouses II—Operation, Maintenance, and Troubleshooting,* ETS Prof. Training Inst., Roanoke, Va., 1993; Yang, Y., and E. Allen, "Biofiltration Control of Hydrogen Sulfide. 2. Kinetics, Biofilter Performance, and Maintenance," *JAWA* **44:** 1315–1321 (1994); McKenna, Turner, and McKenna, Jr., *Fine Particle (2.5 microns) Emissions: Regulations, Measurement, and Control,* Wiley, Hoboken, N.J., 2008.

INTRODUCTION

Air pollutants may be classified into two broad categories: (1) natural and (2) human-made. Natural sources of air pollutants include:

- Windblown dust
- Volcanic ash and gases
- Ozone from lightning and the ozone layer
- Esters and terpenes from vegetation
- Smoke, gases, and fly ash from forest fires
- Pollens and other aeroallergens
- Gases and odors from biologic activities
- Natural radioactivity

Such sources constitute background pollution and that portion of the pollution problem over which control activities can have little, if any, effect.

Human-made sources cover a wide spectrum of chemical and physical activities and are the major contributors to urban air pollution. Air pollutants in the United States pour out from over 10 million vehicles, the refuse of approximately 300 million people, the generation of billions of kilowatts of electricity, and the production of innumerable products demanded by everyday living. Hundreds of millions of tons of air pollutants are generated annually in the United States alone. The six pollutants identified in the Clean Air Act are shown in Table 22-6 for the years 1970 to 2014. Annual emission statistics for these six pollutants are considered major indicators of the U.S. air quality. During the 1970 to 2014 period, the total emissions of the six pollutants declined by 70 percent, while at the same time the gross domestic product increased by 350 percent, the population by 50 percent, and energy consumption by 38 percent. Total emissions in the United States are summarized by source category for the year 2014 in Table 22-7.

Air pollutants may also be classified by origin and state of matter:

1. Origin
   a. Primary. Emitted to the atmosphere from a process
   b. Secondary. Formed in the atmosphere as a result of a chemical reaction
2. State of matter
   a. Gaseous. True gases such as sulfur dioxide, nitrogen oxide, ozone, carbon monoxide; vapors such as gasoline, paint solvent, dry cleaning agents
   b. Particulate. Finely divided solids or liquids; solids such as dust, fumes, and smokes; and liquids such as droplets, mists, fogs, and aerosols

Gaseous Pollutants Gaseous pollutants may be classified as inorganic or organic. Inorganic pollutants consist of:

1. *Sulfur gases.* Sulfur dioxide, sulfur trioxide, hydrogen sulfide
2. *Oxides of carbon.* Carbon monoxide, carbon dioxide
3. *Nitrogen gases.* Nitrous oxide, nitric oxide, nitrogen dioxide, other nitrous oxides
4. *Halogens, halides.* Hydrogen fluoride, hydrogen chloride, chlorine, fluorine, silicon tetrafluoride
5. *Photochemical products.* Ozone, oxidants
6. *Cyanides.* Hydrogen cyanide
7. *Ammonium compounds.* Ammonia
8. *Chlorofluorocarbons.* 1,1,1-trichloro-2,2,2-trifluoroethane; trichlorofluoromethane, dichlorodifluoromethane; chlorodifluoromethane; 1,2-dichloro-1,1,2,2-tetrafluoroethane; chloropentafluoroethane

Organic pollutants consist of:

1. Hydrocarbons
   a. Paraffins. Methane, ethane, octane
   b. Acetylene
   c. Olefins. Ethylene, butadiene
   d. Aromatics. Benzene, toluene, benzpyrene, xylene, styrene
2. Aliphatic oxygenated compounds
   a. Aldehydes. Formaldehyde
   b. Ketones. Acetone, methylethylketone
   c. Organic acids
   d. Alcohols. Methanol, ethanol, isopropanol
   e. Organic halides. Cyanogen chloride bromobenzyl cyanide
   f. Organic sulfides. Dimethyl sulfide
   g. Organic hydroperoxides. Peroxyacetyl nitrite or nitrate (PAN)

The most common gaseous pollutants and their major sources and significance are presented in Table 22-8.

Particulate Pollutants Particulates may be defined as solid or liquid matter whose effective diameter is larger than a molecule but smaller than approximately 100 μm. Particulates dispersed in a gaseous medium are collectively termed an *aerosol.* The terms *smoke, fog, haze,* and *dust* are commonly used to describe particular types of aerosols, depending on the size, shape, and characteristic behavior of the dispersed particles. Aerosols are rather difficult to classify on a scientific basis in terms of their fundamental properties, such as settling rate under the influence of external forces, optical activity, ability to absorb an electrical charge, particle size and structure, surface-to-volume ratio, reaction activity, physiological action, and so on. In general, particle size and settling rate have been the most characteristic properties for many purposes. On the other hand, particles on the order of 1 μm or less settle so slowly that, for all practical purposes, they are regarded as permanent suspensions. Despite possible advantages of scientific classification schemes, the use of popular descriptive terms, such as smoke, dust, and mist, which are essentially based on the mode of formation, appears to be a satisfactory and convenient method of classification. In addition, this approach is so well established and understood that it undoubtedly would be difficult to change.

TABLE 22-6 Six Principal Pollutants' Annual Emission Statistics, Millions of Tons per Year

	1970	1980	1990	2000^a	2005	2010	2014^b
Carbon monoxide (CO)	197.3	177.8	143.6	102.4	80.6	61.6	55.1
Nitrogen oxides (NO_x)^c	26.9	27.1	25.1	22.3	20.3	14.8	12.2
Particulate matter (PM)^d							
PM_{10}	12.2^a	6.2	3.2	3.1	3.6	2.7	2.6
$PM_{2.5}{}^e$	NA	NA	2.3	2.6	2.5	1.9	1.7
Sulfur dioxide (SO_2)	31.2	25.9	23.1	16.3	14.5	7.7	4.9
Volatile organic compounds (VOC)	33.7	30.1	23.1	16.9	15.9	15.0	14.2
Lead^f	0.221	0.074	0.005	0.003	0.003	0.003	0.003
Totals^g	301.5	267.2	218.1	161.0	134.9	101.8	89.0

^{a}In 1985 and 1996 EPA refined its methods for estimating emissions. Between 1970 and 1975, EPA revised its methods for estimating particulate matter emissions.
^{b}The estimates for 2014 are final.
^{c}NO_x estimates prior to 1990 include emissions from fires. Fires would represent a small percentage of the NO_x emissions.
^{d}PM estimates do not include condensable PM, or the majority of $PM_{2.5}$ that is formed in the atmosphere from "precursor" gases such as SO_2 and NO_x.
^{e}EPA has not estimated $PM_{2.5}$ emissions prior to 1990.
^{f}The 1999 estimate for lead is used to represent 2000 and 2003 because lead estimates do not exist for these years.
^{g}$PM_{2.5}$ emissions are not added when calculating the total because they are included in the PM_{10} estimate.
SOURCE: EPA Air Pollutant Emission Trends Data (https://www.epa.gov/air-emissions-inventories/air-pollutant-emissions-trends-data).

TABLE 22-7 2014 National Emissions by Source Category and Pollutant, 1000 Short Tons

Source category	CO	NO_x	PM_{10}	$PM_{2.5}$	SO_2	VOC
Fuel comb. elec. util.	784	1,776	280	205	3,195	41
Fuel comb. industrial	961	1,258	275	223	676	112
Fuel comb. other	2,852	555	424	415	219	476
Chemical & allied product mfg.	167	51	22	17	127	83
Metals processing	766	71	63	48	144	34
Petroleum & related industries	672	685	35	29	119	2,774
Other industrial processes	337	353	766	277	188	329
Solvent utilization	2	1	4	4	0	2,811
Storage & transport	27	20	52	21	9	1,043
Waste disposal & recycling	1,111	83	192	165	17	132
Highway vehicles	22,261	4,489	301	167	22	2,159
Off-highway	14,037	2,669	186	175	77	1,845
Miscellaneous	23,777	399	18,015	4,288	198	5,290
Total	67,756	12,412	20,616	6,033	4,991	17,130
Wildfires	12,701	185	1,326	1,125	96	2,891
Total without wildfires	55,055	12,227	19,290	4,908	4,895	14,238
Miscellaneous without wildfires	11,076	215	16,689	3,163	102	2,399

SOURCE: EPA Air Pollutant Emission Trends Data (https://www.epa.gov/air-emissions-inventories/air-pollutant-emissions-trends-data).

Dust is typically formed by the pulverization or mechanical disintegration of solid matter into particles of smaller size by processes such as grinding, crushing, and drilling. Particle sizes of dust range from a lower limit of about 1 mm up to about 100 or 200 μm and larger. Dust particles are usually irregular in shape, and particle size refers to some average dimension for any given particle. Common examples include fly ash, rock dusts, and ordinary flour. *Smoke* implies a certain degree of optical density and is typically derived from the burning of organic materials such as wood, coal, and tobacco. Smoke particles are very fine, ranging in size from less than 0.01 μm up to 1 μm. They are usually spherical if they have a liquid or tarry composition and irregularly shaped if they have a solid composition. Owing to their very fine particle size, smokes can remain in suspension for long periods of time and can exhibit lively brownian motion.

Fumes are typically formed by processes such as sublimation, condensation, or combustion, generally at relatively high temperatures. They range

TABLE 22-8 Typical Gaseous Pollutants and Their Principal Sources and Significance

Air pollutants	From manufacturing sources such as these	In typical industries	Cause these damaging effects
Alcohols	Used as a solvent in coatings	Surface coatings, printing	Sensory and respiratory irritation
Aldehydes	Results from thermal decomposition of fats, oil, or glycerol; used in some glues and binders	Food processing, light process, wood furniture, chipboard	An irritating odor, suffocating, pungent, choking; not immediately dangerous to life; can become intolerable in a very short time
Ammonia	Used in refrigeration, chemical processes such as dye making, explosives, lacquer, fertilizer	Textiles, chemicals	Corrosive to copper, brass, aluminum, and zinc; high concentration producing chemical burns on wet skin
Aromatics	Used as a solvent in coatings	Surface coatings, printing	Irritation of mucous membranes, narcotic effects; some are carcinogens
Arsine	Any soldering, pickling, etching, or plating process involving metals or acids containing arsenic	Chemical processing, smelting	Breakdown of red cells in blood
Carbon dioxide	Fuel combustion; calcining	Industrial and electric utility boilers, cement and lime production	Greenhouse gas
Carbon monoxide	Fuming of metallic oxides, gas-operated fork trucks	Primary metals; steel and aluminum	Reduction in oxygen-carrying capacity of blood
Chlorine	Manufactured by electrolysis, bleaching cotton and flour; by-product of organic chemicals	Textiles, chemicals	Attacks entire respiratory tract and mucous membrane of eye
Chlorofluorocarbons	Used in refrigeration and production of porous foams; degreasing agent	Refrigeration, plastic foam production, metal fabricating	Attack stratospheric ozone layer; greenhouse gas
Hydrochloric acid	Combustion of coal wastes containing chlorinated plastics	Coal-fired boilers, incinerators	Irritant to eyes and respiratory system
Hydrogen cyanide	From metal plating, blast furnaces, dyestuff works	Metal fabricating, primary metals, textiles	Capable of affecting nerve cells
Hydrogen fluoride	Catalyst in some petroleum refining, etching glass, silicate extraction; by-product in electrolytic production of aluminum	Petroleum, primary metals, aluminum	Strong irritant and corrosive action on all body tissue; damage to citrus plants, effect on teeth and bones of cattle from eating plants
Hydrogen sulfide	Refinery gases, crude oil, sulfur recovery, various chemical industries using sulfur compounds	Petroleum and chemicals; Kraft pulping process	Foul order of rotten eggs; irritating to eyes and respiratory tract; darkening exterior paint
Ketones	Used as a solvent in coatings	Surface coatings, printing	Sensory and respiratory irritation
Lead	Incineration, smelting and casting, transportation	Copper and lead smelting, MSWs	Neurological impairments; kidney, liver, and heart damage.
Nitrogen oxides	High-temperature combustion: metal cleaning, fertilizer, explosives, nitric acid; carbon-arc combustion; manufacture of H_2SO_4	Metal fabrication, heavy chemicals, industrial and electric utility boilers	Irritating gas affecting lungs; vegetation damage
Odors	Slaughtering and rendering animals, tanning animal hides, canning, smoking meats, roasting coffee, brewing beer, processing toiletries	Food processing, allied industries	Objectionable odors
Ozone	Reaction product of VOC and nitrogen oxides	Not produced directly	Irritant to eyes and respiratory system
Phosgenes	Thermal decomposition of chlorinated hydrocarbons, degreasing, manufacture of dyestuffs, pharmaceuticals, organic chemicals	Metal fabrication, heavy chemicals	Damage capable of leading to pulmonary edema, often delayed
Sulfur dioxide	Fuel combustion (coal, oil), smelting and casting, manufacture of paper by sulfite process	Primary metals (ferrous and nonferrous); pulp and paper	Sensory and respiratory irritation, vegetation damage, corrosion, possible adverse effect on health

in particle size from less than 0.1 μm to 1 μm. Similar to smokes, they settle very slowly and exhibit strong brownian motion.

Mists or fogs are typically formed by the condensation of either water or other vapors on suitable nuclei, giving a suspension of small liquid droplets, or by the atomization of liquids. Particle sizes of natural fogs and mists lie between 2 and 200 μm. Droplets larger than 200 μm are more properly classified as drizzle or rain. Many of the important properties of aerosols that depend on particle size are presented in Sec. 17, Fig. 17-40.

When a liquid or solid substance is emitted to the air as particulate matter, its properties and effects may be changed. As a substance is broken up into smaller and smaller particles, more of its surface area is exposed to the air. Under these circumstances, the substance, whatever its chemical composition, tends to combine physically or chemically with other particles or gases in the atmosphere. The resulting combinations are often unpredictable. Very small aerosol particles (from 0.001 to 0.1 μm) can act as condensation nuclei to facilitate the condensation of water vapor, thus promoting the formation of fog and ground mist. Particles less than 2 or 3 μm in size (about half by weight of the particles suspended in urban air) can penetrate the mucous membrane and attract and convey harmful chemicals such as sulfur dioxide. In order to address the special concerns related to the effects of very fine, inhalable particulates, EPA now has ambient standards in place for both PM_{10} and $PM_{2.5}$.

By virtue of the increased surface area of the small aerosol particles and as a result of the adsorption of gas molecules or other such properties that can facilitate chemical reactions, aerosols tend to exhibit greatly enhanced surface activity. Many substances that oxidize slowly in their massive state will oxidize very quickly or may even explode when dispersed as fine particles in the air. Dust explosions, for example, are often caused by the unstable burning or oxidation of combustible particles, brought about by their relatively large specific surfaces. Adsorption and catalytic phenomena can also be extremely important in analyzing and understanding the problems of particulate pollution. The conversion of sulfur dioxide to corrosive sulfuric acid assisted by the catalytic action of iron oxide particles, for example, demonstrates the catalytic nature of certain types of particles in the atmosphere. Finally, aerosols can absorb radiant energy and can rapidly conduct heat to the surrounding gases of the atmosphere. These are gases that ordinarily would be incapable of absorbing radiant energy by themselves. As a result, the air in contact with the aerosols can become much warmer.

Estimating Emissions from Sources Knowledge of the types and rates of emissions is fundamental to the evaluation of any air pollution problem. A comprehensive material balance on the process can often assist in this assessment. Estimates of the rates at which pollutants are discharged from various processes can also be obtained by using published emission factors. See *Compilation of Air Pollution Emission Factors (AP-42),* 5th ed., U.S. EPA, Research Triangle Park, North Carolina, September, 1995, with all succeeding supplements and the EPA Technology Transfer Network's CHIEF (www3.epa.gov/ttn/chief). The emission factor is a statistical average of the rate at which pollutants are emitted from the burning or processing of a given quantity of material or on the basis of some other meaningful parameter. Emission factors are affected by the techniques used in the processing, handling, or burning operations, by the quality of the material used, and by the efficiency of the air pollution control. Since the combination of these factors tends to be unique to a source, emission factors appropriate for one source may not be satisfactory for another source. Hence, care and good judgment must be exercised in identifying appropriate emission factors. If appropriate emission factors cannot be found or if air pollution control equipment is to be designed, specific source sampling should be conducted. The major industrial sources of pollutants, the air contaminants emitted, and typical control techniques are summarized in Table 22-9.

Effects of Air Pollutants

Materials The damage that air pollutants can do to some materials is well known: ozone in photochemical smog cracks rubber, weakens fabrics, and fades dyes; hydrogen sulfide tarnishes silver; smoke dirties laundry; acid aerosols ruin nylon hose. Among the most important effects are discoloration, corrosion, the soiling of goods, and impairment of visibility.

1. *Discoloration.* Many air pollutants accumulate on and discolor buildings. Not only does sooty material blacken buildings, but it can accumulate and become encrusted. This can hide lines and decorations and thereby disfigure structures and reduce their aesthetic appeal. Another common effect is the discoloration of paint by certain acid gases. A good example is the blackening of white paint with a lead base by hydrogen sulfide.

2. *Corrosion.* A more serious effect and one of great economic importance is the corrosive action of acid gases on building materials. Such acids can cause stone surfaces to blister and peel; mortar can be reduced to powder. Metals are also damaged by the corrosive action of some pollutants. Another common effect is the deterioration of tires and other rubber goods. Cracking and apparent "drying" occur when these goods are exposed to ozone and other oxidants.

3. *Soiling of goods.* Clothes, real estate, automobiles, and household goods can easily be soiled by air contaminants, and the more frequent cleaning thus required can become expensive. Also, more frequent cleaning often leads to a shorter life span for materials and to the need to purchase goods more often.

4. *Impairment of visibility.* The impairment of atmospheric visibility (i.e., decreased visual range through a polluted atmosphere) is caused by the scattering of sunlight by particles suspended in the air. It is not a result of sunlight being obscured by materials in the air. Since light scattering, and not obscuration, is the main cause of the reduction in visibility, reduced visibility due to the presence of air pollutants occurs primarily on bright days. On cloudy days or at night there may be no noticeable effect, although the same particulate concentration may exist at these times as on sunny days. Reduction in visibility creates several problems. The most significant are the adverse effects on aircraft, highway, and harbor operations. Reduced visibility can reduce quality of life and also cause adverse aesthetic impressions that can seriously affect tourism and restrict the growth and development of any area. Extreme conditions, such as dust storms or sandstorms, can actually cause physical damage by themselves.

Vegetation Vegetation is more sensitive than animals to many air contaminants, and methods have been developed that use plant response to measure and identify contaminants. The effects of air pollution on vegetation can appear as death, stunted growth, reduced crop yield, and degradation of color. It is interesting to note that in some cases of color damage, such as the silvering of leafy vegetables by oxidants, the plant may still be used as food without any danger to the consumer; however, the consumer usually will not buy such vegetables on aesthetic grounds, so the grower still sustains a loss. Among the pollutants that can harm plants are sulfur dioxide, hydrogen fluoride, and ethylene. Plant damage caused by constituents of photochemical smog has been studied extensively. Damage has been attributed to ozone and peroxyacetyl nitrites, higher aldehydes, and products of the reaction of ozone with olefins. However, none of the cases precisely duplicates all features of the damage observed in the field, and the question remains open to some debate and further study.

Animals Considerable work continues to be performed on the effects of pollutants on animals, including, for a few species, experiments involving mixed pollutants and mixed gas-aerosol systems. In general, such work has shown that mixed pollutants may act in several different ways. They may produce an effect that is additive, amounting to the sum of the effects of each contaminant acting alone; they may produce an effect that is greater than the simply additive (synergistic) or less than the simply additive (antagonistic); or they may produce an effect that differs in some other way from the simply additive.

The mechanism by which an animal can become poisoned in many instances is completely different from that by which humans are affected. As in humans, inhalation is an important route of entry in acute air pollution exposures such as the Meuse Valley and Donora incidents (see the paragraph on Humans that follows). However, probably the most common exposure for herbivorous animals grazing within a zone of pollution will be the ingestion of feed contaminated by air pollutants. In this case, inhalation is of secondary importance.

Air pollutants that present a hazard to livestock, therefore, are those that are taken up by vegetation or deposited on the plants. Only a few pollutants have been observed to cause harm to animals. These include arsenic, fluorides, lead, mercury, and molybdenum.

Humans There seems to be little question that, during many of the more serious episodes, air pollution can have a significant effect on health, especially upon the young, elderly, or people already in ill health. Hundreds of excess deaths have been attributed to incidents in London in 1952, 1956, 1957, and 1962; in Donora, Pennsylvania, in 1948; in New York City in 1953, 1963, and 1966; and Bhopal, India in 1989. Many of the people affected were in failing health, and they were generally suffering from lung conditions. In addition, hundreds of thousands of persons have suffered from serious discomfort and inconvenience, including eye irritation and chest pains, during these and other such incidents. Such acute problems are actually the lesser of the health problems. There is considerable evidence of a chronic threat to human health from air pollution. This evidence ranges from the rapid rise of emphysema as a major health problem, through the identification of carcinogenic compounds in smog, to statistical evidence that people exposed to polluted atmospheres over extended periods of time suffer from a number of ailments and a reduction in their life span. There may even be a significant indirect exposure to air pollution. As we have noted, air pollutants may be deposited onto vegetation or into bodies of water, where they enter the food chain. The impact of such indirect exposures is still under review.

A large body of evidence is available to indicate that atmospheric pollution in varying degrees does affect health adversely. [Amdur, Melvin, and Drinker, "Effect of Inhalation of Sulfur Dioxide by Man," *Lancet* **2:** 758 (1953); Barton, Corn, Gee, Vassallo, and Thomas, "Response of Healthy Men to Inhaled Low

TABLE 22-9 Control Techniques Applicable to Unit Processes at Important Emission Sources

Industry	Process of operation	Air contaminants emitted	Control techniques
Aluminum reduction plants	Materials handling: Buckets and belt Conveyor or pneumatic conveyor	Particulates (dust)	Exhaust systems and baghouse
	Anode and cathode electrode preparation: Cathode (baking) Anode (grinding and blending)	Hydrocarbon emissions from binder Particulates (dust)	Exhaust systems and mechanical collectors
	Baking	Particulates (dust), CO, SO_2, hydrocarbons, and fluorides	High-efficiency cyclone, electrostatic precipitators, scrubbers, catalytic combustion or incinerators, flares, baghouse
	Pot charging	Particulates (dust), CO, HF, SO_2, CF_4, and hydrocarbons	High-efficiency cyclone, baghouse, spray towers, floating-bed scrubber, electrostatic precipitators, chemisorption, wet electrostatic precipitators
	Metal casting	Cl_2, HCl, CO, and particulates (dust)	Exhaust systems and scrubbers
Asphalt plants	Materials handling, storage and classifiers: elevators, chutes, vibrating screens	Particulates (dust)	Wetting; exhaust systems with a scrubber or baghouse
	Drying: rotary oil- or gas-fired	Particulates, SO_2, NO_x, VOC, CO, and smoke	Proper combustion controls, fuel-oil preheating where required; local exhaust system, cyclone and a scrubber or baghouse
	Truck traffic	Dust	Paving, wetting down truck routes
Cement plants	Quarrying: primary crusher, secondary crusher, conveying, storage	Particulates (dust)	Wetting; exhaust systems with fabric filters
	Dry processes: materials handling, air separator (hot-air furnace)	Particulates (dust)	Local exhaust system with mechanical collectors and baghouse
	Grinding	Particulates (dust)	Local exhaust system with cyclones and baghouse
	Pneumatic, conveying and storage	Particulates (dust)	
	Wet process: materials handling, grinding, storage	Wet materials, no dust	
	Kiln operations: rotary kiln	Particulates (dust), CO, SO_x, NO_x, hydrocarbons, aldehydes, ketones	Electrostatic precipitators, acoustic horns and baghouses, scrubber
	Clinker cooling: materials handling	Particulates (dust)	Local exhaust system and electrostatic precipitators or fabric filters
	Grinding and packing, air separator, grinding, pneumatic conveying, materials handling, packaging	Particulates (dust)	Local exhaust system and fabric filters
Coal-preparation plants	Materials handling: conveyors, elevators, chutes	Particulates (dust)	Local exhaust system and cyclones
	Sizing: crushing, screening, classifying	Particulates (dust)	Local exhaust system and cyclones
	Dedusting	Particulates (dust)	Local exhaust system, cyclone precleaners, and baghouse
	Storing coal in piles	Blowing particulates (dust)	Wetting, plastic-spray covering
	Refuse piles	H_2S, particulates, and smoke from burning storage piles	Digging out fire, pumping water onto fire area, blanketing with incombustible material
	Coal drying: rotary, screen, suspension, fluid-bed, cascade	Dust, smoke, particulates, sulfur oxides, H_2S	Exhaust systems with cyclones and fabric filters
Coke plants	By-product-ovens charging	Smoke, particulates (dust)	Pipe-line charging, careful charging techniques, portable hooding and scrubber or baghouses
	Pushing	Smoke, particulates (dust), SO_2	Minimizing green-coke pushing, scrubbers and baghouses
	Quenching	Smoke, particulates (dust and mists), phenols, and ammonia	Baffles and spray tower
	By-product processing	CO, H_2S, methane, ammonia, H_2, phenols, hydrogen cyanide, N_2, benzene, xylene, etc.	Electrostatic precipitator, scrubber, flaring
	Material storage (coal and coke)	Particulates (dust)	Wetting, plastic spray, fire-prevention techniques
Electric utilities (industrial and commercial facilities)	Coal-fired boilers	Particulates SO_x, NO_x, mercury, HCl	Fabric filters precipitators, low NO_x burners, SCR, and SNCR acid gas scrubbers, fuel additives, activated carbon injection
Fertilizer industry (chemical)	Phosphate fertilizers: crushing, grinding, and calcining	Particulates (dust)	Exhaust system, scrubber, cyclone, baghouse
	Hydrolysis of P_2O_5	PH_3, $P_2O_5PO_4$ mist	Scrubbers, flare
	Acidulation and curing	HF, SiF_4	Scrubbers
	Granulation	Particulates (dust) (product recovery)	Exhaust system, scrubber, or baghouse
	Ammoniation	NH_3, NH_4Cl, SiF_4, HF	Cyclone, electrostatic precipitator, baghouse, high-energy scrubber

TABLE 22-9 Control Techniques Applicable to Unit Processes at Important Emission Sources *(Continued)*

Industry	Process of operation	Air contaminants emitted	Control techniques
Fertilizer industry (chemical)	Nitric acid acidulation Superphosphate storage and shipping Ammonium nitrate reactor Prilling tower	NO_x, gaseous fluoride compounds Particulates (dust) NH_3, NO_x NH_4, NO_3	Scrubber, addition of urea Exhaust system, cyclone or baghouse Scrubber Proper operation control, scrubbers
Foundries: Iron	Melting (cupola): Charging Melting Pouring Bottom drop	 Smoke and particulates Smoke and particulates, fume Oil, mist, CO Smoke and particulates	Closed top with exhaust system, CO afterburner, gas-cooling device and scrubbers, baghouse or electrostatic precipitator, wetting to extinguish fire
Brass and bronze	Melting: Charging Melting Pouring	Smoke particulates, oil mist Zinc oxide fume, particulates, smoke Zinc oxide fume, lead oxide fume	Low-zinc-content red brass: use of good combustion controls and slag cover; high-zinc-content brass: use of good combustion controls, local Exhaust system, and baghouse or scrubber
Aluminum	Melting: charging, melting, pouring	Smoke and particulates	Charging clean material (no paint or grease); proper operation required; no air-pollution-control equipment if no fluxes are used and degassing is not required; dirty charge requiring exhaust system with scrubbers and baghouses
Zinc	Melting: Charging	Smoke and particulates	Exhaust system with cyclone and baghouse, charging clean material (no paint or grease)
	Melting	Zinc oxide fume	Careful skimming of dross
	Pouring	Oil mist and hydrocarbons from diecasting machines	Use of low-smoking die-casting lubricants
	Sand-handling shakeout	Particulates (dust), smoke, organic vapors	
	Magnetic pulley, conveyors, and elevators, rotary cooler, screening, crusher-mixer	Particulates (dust)	Exhaust system, cyclone, and baghouse
	Coke-making ovens	Organic acids, aldehydes, smoke, hydrocarbons	Use of binders that will allow ovens to operate at less than 204°C (400°F) or exhaust systems and afterburners
Galvanizing operations	Hot-dip-galvanizing-tank kettle: dipping material into the molten zinc; dusting flux onto the surface of the molten zinc	Fumes, particulates (liquid), vapors: NH_4Cl, ZnO, $ZnCl_2$, Zn, NH_3, oil, and carbon	Close-fitting hoods with high in-draft velocities (in some cases, the hood may not be able to be close to the kettle, so the in-draft velocity must be precipitators
Kraft pulp mills	Digesters: batch and continuous	Mercaptans, methanol (odors)	Condensers and use of lime kiln, boiler, or furnaces as afterburners
	Multiple-effect evaporators	H_2S, other odors	Caustic scrubbing and thermal oxidation of noncondensables
	Recovery furnace	H_2S, mercaptans, organic sulfides, and disulfides	Proper combustion controls for fluctuating load and unrestricted primary and secondary air flow to furnace and dry-bottom electrostatic precipitator; noncontact evaporator
	Weak and strong black-liquor oxidation	H_2S	Packed tower and cyclone
	Smelt tanks	Particulates (mist or dust)	Demisters, venturi, packed tower, or impingement-type scrubbers
	Lime kiln	Particulates (dust), H_2S	Venturi scrubbers
Municipal and industrial incinerators	Single-chamber incinerators	Particulates, smoke, volatiles, CO, SO_x, ammonia, organic acids, aldehydes, NO_x, dioxins, hydrocarbons, odors, HCl, furans	Afterburner, combustion controls
	Multiple-chamber incinerators (retort, inline):	Particulates, smoke, and combustion contaminants	Operating at rated capacity, using auxiliary fuel as specified, and good maintenance, including timely cleanout of ash
	Flue-fed	Particulates, smoke, and combustion contaminants	Use of charging gates and automatic controls for draft; afterburner
	Wood waste	Particulates, smoke, and combustion contaminants	Continuous-feed systems; operation at design load and excess air; cyclones
	Municipal incinerators (50 tons and up per day):	Particulates, smoke, volatiles, CO, ammonia, organic acids, aldehydes, NO_x, furans, hydrocarbons, SO_x,	Preparation of materials, including weighing, grinding, shredding; control of tipping area, furnace design with

TABLE 22-9 Control Techniques Applicable to Unit Processes at Important Emission Sources *(Continued)*

Industry	Process of operation	Air contaminants emitted	Control techniques
Municipal and industrial incinerators	Municipal incinerators (50 tons and up per day): Pathological incinerators Industrial waste	hydrogen chloride, mercury, dioxins and odors, hydrocarbons, HCl, dioxins, furans Particulates, smoke, and combustion contaminants	proper automatic controls; proper start-up techniques; maintenance of design operating temperatures; use of electrostatic precipitators, scrubbers, baghouses, and activated carbon injection; proper ash cleanout Proper charging, acid gas scrubber, baghouse Modified fuel feed, auxiliary fuel and dryer systems, cyclones, scrubbers
Nonferrous smelters, primary:			
Copper	Roasting	SO_2, particulates, fume	Exhaust system, settling chambers, cyclones or scrubbers and electrostatic precipitators for dust and fumes and sulfuric acid plant for SO_2
	Reverberatory furnace	Smoke, particulates, metal oxide fumes, SO_2	Exhaust system, settling chambers, cyclones or scrubbers and electrostatic precipitators for dust and fumes and sulfuric acid plant for SO_2
	Converters: charging, slag skim, pouring, air or oxygen blow	Smoke, fume, SO_2	Exhaust system, settling chambers, cyclones or scrubbers and electrostatic precipitators for dust and fumes and sulfuric acid plant for SO_2
Lead	Sintering	SO_2, particulates, smoke	Exhaust system, cyclones and baghouse or precipitators for dust and fumes, sulfuric acid plant for SO_2
	Blast furnace	SO_2, CO, particulates, lead oxide, zinc oxide	Exhaust system, settling chambers, afterburner and cooling device, cyclone, and baghouse
	Dross reverberatory furnace	SO_2, particulates, fume	Exhaust system, settling chambers, cyclone and cooling device, baghouse
	Refining kettles	SO_2, particulates	Local exhaust system, cooling device, baghouse or precipitator
Cadmium	Roasters, slag, fuming furnaces, deleading kilns	Particulates	Local exhaust system, baghouse or precipitator
Zinc	Roasting	Particulates (dust) and SO_2	Exhaust system, humidifier, cyclone, scrubber, electrostatic precipitator, and acid plant
	Sintering	Particulates (dust) and SO_2	Exhaust system, humidifier, electrostatic precipitator, and acid plant
	Calcining	Zinc oxide fume, particulates, SO_2, CO	Exhaust system, baghouse, scrubber or Acid plant
	Retorts: electric arc		
Nonferrous smelters, secondary	Blast furnaces and cupolas-recovery of metal from scrap and slag	Dust, fumes, particulates, oil vapor, smoke, CO	Exhaust systems, cooling devices, CO burners and baghouses or precipitators
	Reverberatory furnaces	Dust, fumes, particulates, smoke, gaseous fluxing materials	Exhaust systems, and baghouses or precipitators, or venturi scrubbers
	Sweat furnaces	Smoke, particulates, fumes	Precleaning metal and exhaust systems with afterburner and baghouse
	Wire reclamation and autobody burning	Smoke, particulates	Scrubbers and afterburners
Paint and varnish manufacturing	Resin manufacturing: closed reaction vessel	Acrolein, other aldehydes and fatty acids (odors), phthalic anhydride (sublimed)	Exhaust systems with scrubbers and fume burners
	Varnish: cooking-open or closed vessels	Ketones, fatty acids, formic acids, acetic acid, glycerine, acrolein, other aldehydes, phenols and terpenes; from tall oils, hydrogen sulfide, alkyl sulfide, butyl mercaptan, and thiofen (odors)	Exhaust system with scrubbers and fume burners; close-fitting hoods required for open kettles
	Solvent thinning	Olefins, branched-chain aromatics and ketones (odors), solvents	Exhaust system with fume burners
Rendering plants	Feedstock storage and housekeeping	Odors	Quick processing, washdown of all concrete surfaces, paving of dirt roads, proper sewer maintenance, enclosure, packed towers
	Cookers and percolators	SO_2, mercaptans, ammonia, odors	Exhaust system, condenser, scrubber, or incinerator
	Grinding	Particulates (dust)	Exhaust system and scrubber
Roofing plants (asphalt saturators)	Felt or paper saturators: spray section, asphalt tank, wet looper	Asphalt vapors and particulates (liquid)	Exhaust system with high inlet velocity at hoods (3658 m/s [>200 ft/min]) with either scrubbers, baghouses, or two-stage low-voltage electrostatic precipitators

TABLE 22-9 Control Techniques Applicable to Unit Processes at Important Emission Sources *(Continued)*

Industry	Process of operation	Air contaminants emitted	Control techniques
Roofing plants (asphalt saturators)	Crushed rock or other minerals handling	Particulates (dust)	Local exhaust system, cyclone or multiple cyclones
Steel mills	Blast furnaces: charging, pouring	CO, fumes, smoke, particulates (dust)	Good maintenance, seal leaks; use of higher ratio of pelletized or sintered ore; CO burned in waste-heat boilers, stoves, or coke ovens; cyclone, scrubber, and baghouse
	Electric steel furnaces: charging, pouring, oxygen blow	Fumes, smoke, particulates (dust), CO	Segregating dirty scrap; proper hooding, baghouses or electrostatic precipitator
	Open-hearth furnaces: oxygen blow, pouring	Fumes, smoke, SO_x, particulates (dust), CO, NO_x	Proper hooding, settling chambers, waste-heat boiler, baghouse, electrostatic precipitator, and wet scrubber
	Basic oxygen furnaces: oxygen blowing	Fumes, smoke, CO, particulates (dust)	Proper hooding (capturing of emissions and dilute CO), scrubbers, or electrostatic precipitator
	Raw material storage	Particulates (dust)	Wetting or application of plastic spray
	Pelletizing	Particulates (dust)	Proper hooding, cyclone, baghouse
	Sintering	Smoke, particulates (dust), SO_2, NO_x	Proper hooding, cyclones, wet scrubbers, baghouse, or precipitator

Concentrations of Gas-Aerosol Mixtures," *Arch. Environ. Health* **18:** 681 (1969); Bates, Bell, Burnham, Hazucha, and Mantha, "Problems in Studies of Human Exposure to Air Pollutants," *Can. Med. Assoc. J.* **103:** 833 (1970); Ciocco and Thompson, "A Follow-Up of Donora Ten Years After: Methodology and Findings," *Am. J. Public Health* **51:** 155 (1961); Daly, "Air Pollution and Causes of Death," *Br. J. Soc. Med.* **13:** 14 (1959); Jaffe, "The Biological Effect of Photochemical Air Pollutants on Man and Animals," *Am. J. Public Health* **57:** 1269 (1967); New York Academy of Medicine, Committee on Public Health, "Air Pollution and Health," *Bull. N.Y. Acad. Med.* **42:** 588 (1966); Pemberton and Goldberg, "Air Pollution and Bronchitis," *Br. Med. J.* **2:** 567 (1954); Snell and Luchsinger, "Effect of Sulfur Dioxide on Expiratory Flowrates and Total Respiratory Resistance in Normal Human Subjects," *Arch. Environ. Health* **18:** 693 (1969); Speizer and Frank, "A Comparison of Changes in Pulmonary Flow Resistance in Healthy Volunteers Acutely Exposed to SO_2 by Mouth and by Nose," *J. Ind. Med.* **23:** 75 (1966); Stocks, "Cancer and Bronchitis Mortality in Relation to Atmospheric Deposit and Smoke," *Br. Med. J.* **1:** 74 (1959); Toyama, "Air Pollution and Its Health Effects in Japan," *Arch. Environ. Health* **8:** 153 (1963); U.K. Ministry of Health, "Mortality and Morbidity During London Fog of December 1952," Report on Public Health and Medical Subjects No. 95, London, 1954; U.S. Public Health Service, "Air Pollution in Donora, Pa.: Preliminary Report," Public Health Bull. 306 (1949).]

Atmospheric pollution contributes to excesses of death, increased morbidity, and earlier onset of chronic respiratory diseases. There is evidence of a relationship between the intensity of the pollution and the severity of attributable health effects and a consistency of the relationship between these environmental stresses and diseases of the target organs. Air pollutants can both initiate and aggravate a variety of respiratory diseases, including asthma. In fact, the clinical presentation of asthma may be considered an air pollution host-defense disorder brought on by specific airborne irritants: pollens, infectious agents, and gaseous and particulate chemicals. The bronchopulmonary response to these foreign irritants is bronchospasm and hypersecretion; the airways are intermittently and reversibly obstructed.

Air pollutant effects on neural and sensory functions in humans vary widely. Odorous pollutants cause only minor annoyance; yet, if persistent, they can lead to irritation, emotional upset, anorexia, and mental depression. Carbon monoxide can cause death secondary to the depression of the respiratory centers of the central nervous system. Short of death, repeated and prolonged exposure to carbon monoxide can alter sensory protection, temporal perception, and higher mental functions. Lipid-soluble aerosols can enter the body and be absorbed in the lipids of the central nervous system. Once there, their effects may persist long after the initial contact has been removed. Examples of agents of long-term chronic effects are organic phosphate pesticides and aerosols carrying the metals lead, mercury, and cadmium.

The acute toxicological effects of most air contaminants are reasonably well understood, but we are only beginning to understand the effects of exposure to heterogeneous mixtures of gases and particulates at very low concentrations. Two general approaches can be used to study the effects of air contaminants on humans: epidemiology, which tries to associate the effect in large populations with the cause, and laboratory research, which begins with the cause and tries to determine the effects. Ideally, the two methods should complement each other.

EPA recognizes that scientific studies show a link between inhalable PM (alone and in combination with other pollutants in the air) and a series of health effects. While both coarse and fine particles accumulate in the respiratory system and are associated with many adverse health effects, fine particles have been more clearly tied to the most serious health effects. Fine particles have been associated with decreased lung functions, disease, and even premature death. The elderly and children are among those who appear to be at greatest risk (JAWMA Special Issues, *PM2000, Particulate Matter and Health—The Scientific Basis for Regulatory Decision-Making,* specialty conference and exhibition, July and August 2000).

Epidemiology, the more costly of the two approaches to study, requires great care in planning and often suffers from incomplete data and lack of controls. One great advantage, however, is that moral barriers do not limit its application to humans as they do with some kinds of laboratory research. The method is therefore highly useful and has produced considerable information. Laboratory research is less costly than epidemiology, and its results can be checked against controls and verified by experimental repetition.

A SOURCE-CONTROL-PROBLEM STRATEGY

Strategy Control technology is not a total solution if it creates undesirable side effects while meeting its objectives. Air pollution control must be considered in terms of regulatory requirements, total technological systems (equipment and processes), and ecological consequences, such as the problems of recycling or treatment and disposal of collected pollutants and the production of secondary pollutant streams (such as wastewater) that in themselves would require treatment. The 1990 Clean Air Act Amendments (CAAA) have significantly affected the control approach. The CAAA have placed an increased emphasis on control technology by requiring best available control technology (BACT) on new major sources and modifications, and by requiring maximum achievable control technology (MACT) on new and existing major sources of hazardous air pollutants (HAPs).

The control strategy for environmental impact assessment often focuses on five alternatives whose purpose would be the reduction or the elimination of pollutant emissions:

1. Elimination of the operation entirely or in part
2. Modification of the operation or raw materials
3. Relocation of the operation
4. Application of appropriate control technology
5. Combinations of the four alternatives

In light of the relatively high costs often associated with pollution control systems, engineers are directing considerable effort toward process modifications to eliminate as much of the pollution problem as possible at the source. This includes evaluating alternative manufacturing and production techniques, substituting raw materials, and improving process-control methods. Unfortunately, if there is no alternative, the application of the correct pollution control equipment is essential. The equipment must be designed to comply with regulatory emission limitations on a continual basis, interruptions being subject to severe penalty, depending on the circumstances. The continual performance requirement places a heavy emphasis on operation and maintenance practices. The escalating costs of energy, labor, and materials can make operation and maintenance considerations even more important than the original capital cost.

Factors in Control-Equipment Selection In order to solve an air pollution problem, the problem must be defined in detail. A number of factors must be considered before selecting air pollution control equipment. In general, these factors can be grouped into three categories: environmental, engineering, and economic.

Environmental Factors
1. Equipment location
2. Available space
3. Year-round ambient conditions
4. Availability of adequate utilities (power, water, etc.) and ancillary-system facilities (waste treatment and disposal, etc.)
5. Maximum allowable emission (air pollution regulations and codes)
6. Aesthetic considerations (visible steam or water-vapor plume, etc.)
7. Contributions of the air pollution control system to wastewater and land pollution
8. Contribution of the air pollution control system to plant noise levels

Engineering Factors
1. Contaminant characteristics [e.g., physical and chemical properties, concentration, particulate shape and size distribution (in the case of particulates), chemical reactivity, corrosivity, abrasiveness, and toxicity), electrical resistivity (in the case of particulates)].
2. Gas-stream characteristics (e.g., volume flow rate, temperature, pressure, humidity, composition, viscosity, density, reactivity, combustibility, corrosivity, and toxicity).
3. Design and performance characteristics of the particular control system [i.e., size and weight, fractional efficiency curves (in the case of particulates)], mass-transfer or contaminant-destruction capability (in the case of gases or vapors), pressure drop, reliability, turndown capability, power requirements, utility requirements, temperature limitations, maintenance requirements, operating cycles (including start-up and shutdown), and flexibility toward complying with more stringent air pollution codes. Additionally, if the gas stream being treated will vary in volumetric flow rate, temperature, or pressure based on process rate variations or the time of the year (summer versus winter operation), the control system must be designed to comply with emissions limits over these variations.

Economic Factors
1. Capital cost (equipment, installation, engineering, etc.)
2. Operating cost (utilities, labor, maintenance, etc.)
3. Emissions fees
4. Life-cycle cost over the expected equipment lifetime

Comparing Control-Equipment Alternatives The final choice in equipment selection is usually dictated by the equipment capable of achieving compliance with regulatory codes at the lowest uniform annual cost (amortized capital investment plus operation and maintenance costs). To compare specific control-equipment alternatives, knowledge of the particular application and site is essential. A preliminary screening, however, may be performed by reviewing the advantages and disadvantages of each type of air pollution control equipment. General advantages and disadvantages of the most popular types of air pollution equipment for gases and particulates are presented in Tables 22-10 through 22-20. Other activities that must be accomplished before final compliance is achieved are presented in Table 22-21.

In addition to using annualized cost comparisons in evaluating an air pollution control (APC) equipment installation, the impact of the 1990 Clean Air Act Amendments (CAAA) and resulting regulations also must be included in the evaluation. The CAAA prescribes specific pollution control requirements for particular industries and locations. As an example, the CAAA requires that any major stationary source or major modification plan that is subject to prevention of significant deterioration (PSD) requirements must undergo a best available control technology (BACT) analysis. The BACT analysis is done on a case-by-case basis. The process involves evaluating the possible types of air pollution control equipment that could be used for technology and energy and in terms of the environment and economy. The analysis uses a top-down approach that lists all available control technologies in descending order of effectiveness. The most stringent technology is chosen unless it can be demonstrated that, due to technical, energy, environmental, or economic considerations, this type of APC technology is not feasible.

The 1990 CAAA introduced a new level of control for hazardous (toxic) air pollutants (HAPs). As a result, EPA has identified 188 HAPs for regulation. Rather than rely on ambient air quality standards to set acceptable exposures to HAPs, the CAAA requires that EPA promulgate through the end of the decade maximum achievable control technology (MACT) standards for controlling HAPs emitted from specified industries. These standards are based on the level of control established by the best performing 12 percent of industries in each of the categories identified by EPA.

TABLE 22-10 Advantages and Disadvantages of Cyclone Collectors

Advantages
1. Low cost of construction
2. Relatively simple equipment with few maintenance problems
3. Relatively low operating pressure drops (for degree of particulate removal obtained) in the range of approximately 2- to 6-in water column
4. Temperature and pressure limitations imposed only by the materials of construction used
5. Collected material recovered dry for subsequent processing, recycle, or disposal
6. Relatively small space requirements
7. Can be installed as a pre-collector to reduce particle loading to ESPs or fabric filters

Disadvantages
1. Relatively low overall particulate collection efficiencies, especially on particulates below 10 μm in size
2. High efficiencies require high pressure drops (i.e., high energy consumption)
3. Inability to handle tacky materials

TABLE 22-11 Advantages and Disadvantages of Wet Scrubbers

Advantages
1. No secondary dust sources
2. Relatively small space requirements
3. Ability to collect gases as well as particulates (especially "sticky" ones)
4. Ability to handle high-temperature, high-humidity gas streams
5. Capital cost low (if wastewater treatment system not required) unless special alloys required due to system chemistry
6. For some processes, gas stream already at high pressures (so pressure-drop considerations may not be significant)
7. Ability to achieve high collection efficiencies on fine particulates and acid gases (SO_2)—however, at the expense of pressure drop
8. Ability to handle gas streams containing flammable or explosive materials

Disadvantages
1. Possible creation of water-disposal problem
2. Product collected wet requiring dewatering
3. Corrosion problems more severe than with dry systems
4. Steam plume opacity and/or droplet entrainment possibly objectionable
5. Pressure-drop and horsepower requirements possibly high
6. Solids buildup at the wet/dry interface possibly a problem
7. Relatively high maintenance costs
8. Must be protected from freezing
9. Low exit gas temperature reduces exhaust plume dispersion
10. Moist exhaust gas precludes use of most additional controls

DISPERSION FROM STACKS

Stacks discharging to the atmosphere have long been the most common industrial method of disposing of waste gases. The concentrations to which humans, plants, animals, and structures are exposed at ground level can be reduced significantly by emitting the waste gases from a process at great heights. Although tall stacks may be effective in lowering the ground-level concentration of pollutants, they do not in themselves reduce the amount of pollutants released into the atmosphere. However, in certain situations, their use can be the most practical and economical way of dealing with an air pollution problem.

Preliminary Design Considerations To determine the acceptability of a stack as a means of disposing of waste gases, the acceptable ground-level concentration (GLC) of the pollutant or pollutants must be determined. The topography of the area must also be considered so that the stack can be properly located with respect to buildings and hills that might

TABLE 22-12 Advantages and Disadvantages of Dry Scrubbers

Advantages
1. No wet sludge to dispose of
2. Relatively small space requirements
3. Ability to collect acid gases, especially sulfuric acid mist, at high efficiencies
4. Ability to handle high-temperature gas streams
5. Dry exhaust allows addition of fabric filter to control particulates and for additional acid gas/metals removal
6. Exit gas temperature above adiabatic saturation results in "clear stack" emission
7. More effective than wet scrubbers for toxic metals removal

Disadvantages
1. Acid gas control efficiency not as high as with wet scrubber
2. No particulate collection—dry scrubber generates particulate requiring addition of (usually) fabric filter
3. Pressure drop generally higher than wet scrubbers
4. Lime/soda ash reagent more expensive than limestone used in wet scrubbers

TABLE 22-13 Advantages and Disadvantages of Electrostatic Precipitators

Advantages

1. Extremely high particulate (coarse and fine) collection efficiencies attainable (at a relatively low expenditure of energy)
2. Collected material recovered dry for subsequent processing, recycle, or disposal
3. Low pressure drop
4. Designed for continuous operation with minimum maintenance requirements
5. Relatively low operating costs
6. Capable of operation under high pressure (to 150 lbf/in^2) or vacuum conditions
7. Capable of operation at high temperatures [to 704°C(1300°F)] with special materials of construction
8. Relatively large gas flow rates capable of effective handling

Disadvantages

1. High capital cost
2. Performance sensitive to fluctuations in gas-stream conditions (in particular, flows, temperature, particulate and gas composition, and particulate loadings)
3. Certain particulates difficult to collect owing to extremely high- or low-resistivity characteristics
4. Relatively large space requirements required for installation
5. Explosion hazard when treating combustible gases and/or collecting combustible particulates
6. Special precautions required to safeguard personnel from the high voltage
7. Ozone produced by the negatively charged discharge electrode during gas ionization
8. Relatively sophisticated maintenance personnel required
9. Gas ionization may cause dissociation of gas stream constituents and result in creation of toxic byproducts
10. Condensation can result in particulates difficult to remove from plates
11. Not effective in capturing some contaminants that exist as vapors at high temperatures (e.g., heavy metals, dioxins)
12. Not as effective as fabric filters when used in conjunction with flue gas additives for acid gas and/or mercury removal

TABLE 22-14 Advantages and Disadvantages of Fabric-Filter Systems

Advantages

1. Extremely high collection efficiency on both coarse and fine (submicrometer) particles
2. Relatively insensitive to gas-stream fluctuation; efficiency and pressure drop relatively unaffected by large changes in inlet dust loadings for continuously cleaned filters
3. Filter outlet air capable of being recirculated within the plant in many cases (for energy conservation)
4. Collected material recovered dry for subsequent processing, recycle, or disposal
5. No issues with liquid waste disposal, water pollution, or liquid freezing
6. Corrosion and rusting of components usually not problems
7. No hazard of high voltage, simplifying maintenance and repair and permitting collection of flammable dusts
8. Use of selected fibrous or granular filter aids (precoating), permitting the high-efficiency collection of submicrometer smokes and gaseous contaminants
9. Filter collectors available in a large number of configurations, resulting in a range of dimensions and inlet and outlet flange locations to suit installment requirements
10. Combined with dry sorbent injection systems can be used to reduce emissions of acid gases, heavy metals, and toxic organic compounds
11. Relatively simple operation
12. Sizes can range from small bin-vent filters to multiple compartment baghouses treating extremely high gas volumes

Disadvantages

1. Temperatures much in excess of 288°C (550°F) will require special refractory mineral or metallic fabrics that can be very expensive
2. Certain dusts possibly requiring fabric treatments to reduce dust leakage or, in other cases, assist in the removal of the collected dust
3. Concentrations of some dusts in the collector (~50 g/m^3) forming a possible fire or explosion hazard if a spark or flame is admitted by accident; possibility of fabrics burning if readily oxidizable dust is being collected
4. Periodic high maintenance requirement for bag replacement
5. Fabric life possibly shortened at elevated temperatures and in the presence of acid or alkaline particulate or gas constituents
6. Hygroscopic materials, condensation of moisture, or tarry adhesive components possibly can cause crusty caking or plugging of the fabric or requiring special additives
7. Replacement of fabric filter bags will require respiratory protection for maintenance personnel
8. Medium pressure-drop requirements, typically in the range 4- to 10-in water column

TABLE 22-15 Advantages and Disadvantages of Absorption Systems (Packed and Plate Columns)

Advantages

1. Relatively low pressure drop
2. Standardization in fiberglass-reinforced plastic (FRP) construction permitting operation in highly corrosive atmospheres
3. Capable of achieving relatively high mass-transfer efficiencies
4. Increasing the height and/or type of packing or number of plates capable of improving mass transfer without purchasing a new piece of equipment
5. Relatively low capital cost
6. Relatively small space requirements
7. Ability to collect particulates as well as gases
8. Collected substances may be recovered by distillation

Disadvantages

1. Possibility of creating water (or liquid) disposal problem
2. Product collected wet for disposal or reuse
3. Particulates deposition possibly causing plugging of the bed or plates
4. When FRP construction is used, sensitive to temperature
5. Relatively high maintenance costs
6. Must be protected from freezing

TABLE 22-16 Comparison of Plate and Packed Columns

Packed column

1. Lower pressure drop
2. Simpler and cheaper to construct
3. Preferable for liquids with high-foaming tendencies

Plate column

1. Less susceptible to plugging
2. Less weight
3. Less of a problem with channeling
4. Temperature surge resulting in less damage

TABLE 22-17 Advantages and Disadvantages of Adsorption Systems

Advantages

1. Possibility of product recovery
2. Excellent control and response to process changes
3. No chemical-disposal problem when pollutant (product) recovered and returned to process
4. Capability of systems for fully automatic, unattended operation
5. Capability to remove gaseous or vapor contaminants from process streams to extremely low levels

Disadvantages

1. Product recovery possibly requiring a expensive distillation (or extraction) scheme
2. Adsorbent progressively deteriorating in capacity as the number of cycles increases
3. Adsorbent regeneration requiring a steam or vacuum source, possibly off-site
4. Relatively high capital cost
5. Prefiltering of gas stream may be required to remove any particulate capable of plugging the adsorbent bed
6. Cooling of gas stream possibly required to get to the operating temperature [less than 49°C (120°F)]
7. Relatively high steam requirements to desorb high-molecular-weight hydrocarbons
8. Spent adsorbent may be considered a hazardous waste
9. Some contaminants may undergo a violent exothermic reaction with the adsorbent

TABLE 22-18 Advantages and Disadvantages of Combustion Systems

Advantages
1. Simplicity of operation
2. Capability of steam generation or heat recovery in other forms
3. Capability for virtually complete destruction of organic contaminants (>99.99%)

Disadvantages
1. Relatively high operating costs (particularly associated with fuel requirements)
2. Potential for flashback and subsequent explosion hazard
3. Catalyst poisoning (in the case of catalytic incineration)
4. Incomplete combustion, possibly creating potentially worse pollution problems
5. Even complete combustion may produce SO_2, NO_x, and CO_2 depending on fuel characteristics
6. High temperature components and exhaust may be hazardous to maintenance personnel
7. High maintenance requirements—especially if operation is cyclic

TABLE 22-19 Advantages and Disadvantages of Condensers

Advantages
1. Pure product recovery (in the case of indirect-contact condensers)
2. Water used as the coolant in an indirect-contact condenser (i.e., shell-and-tube heat exchanger), not in contact with contaminated gas stream, and is reusable after cooling
3. May be used to produce vacuum to remove contaminants from process
4. Could be used as a precursor to other processes such as adsorption

Disadvantages
1. Relatively low removal efficiency for gaseous contaminants (at concentrations typical of pollution-control applications)
2. Coolant requirements possibly extremely expensive
3. Direct-contact condenser may produce water discharge problems

TABLE 22-20 Advantages and Disadvantages of Selective Catalytic Reduction of Nitrogen Oxides

Advantages
1. Capable of 95% NO_x removal
2. Uses readily available ammonia or urea reagent
3. Exhaust products are N_2 and water

Disadvantages
1. Spent catalyst may be considered hazardous waste
2. Gas stream must be maintained at proper temperature and humidity
3. Use of anhydrous ammonia will require emergency release plan
4. Heavy particulate loadings can damage pore structure of catalyst (plugging)
5. Potential catalyst poisons if metals (such as arsenic) are present in gas stream

introduce a factor of air turbulence into the operation of the stack. Awareness of the meteorological conditions prevalent in the area, such as the prevailing winds, humidity, and rainfall, is also essential. Finally, an accurate knowledge of the constituents of the waste gas and their physical and chemical properties is paramount.

Wind Direction and Speed Wind direction is measured at the height at which the pollutant is released, and the mean direction will indicate the direction of travel of the pollutants. In meteorology, it is conventional to consider the wind direction as the direction from which the wind blows; therefore, a northwest wind will move pollutants to the southeast of the source.

The effect of wind speed is twofold: (1) Wind speed will determine the travel time from a source to a given receptor; and (2) wind speed will affect dilution in the downwind direction. Generally, the concentration of air pollutants downwind from a source is inversely proportional to wind speed.

Wind speed has velocity components in all directions, so there are vertical motions as well as horizontal ones. These random motions of widely different scales and periods are essentially responsible for the movement and diffusion of pollutants about the mean downwind path. These motions can be considered atmospheric turbulence. If the scale of a turbulent motion (i.e., the size of an eddy) is larger than the size of the pollutant plume in its vicinity, the eddy will move that portion of the plume. If an eddy is smaller than the plume, its effect will be to diffuse or spread out the plume. This diffusion caused by eddy motion is widely variable in the atmosphere, but even when the effect of this diffusion is least, it is in the vicinity of three orders of magnitude greater than diffusion by molecular action alone.

Mechanical turbulence is the induced-eddy structure of the atmosphere due to the roughness of the surface over which the air is passing. Therefore, the existence of trees, shrubs, buildings, and terrain features will cause mechanical turbulence. The height and spacing of the elements causing the roughness will affect the turbulence. In general, the higher the roughness elements, the greater the mechanical turbulence. In addition, mechanical turbulence increases as wind speed increases.

Thermal turbulence is turbulence induced by the stability of the atmosphere. When the Earth's surface is heated by the sun's radiation, the lower layer of the atmosphere tends to rise and thermal turbulence becomes greater, especially under conditions of light wind. On clear nights with wind, heat is radiated from the Earth's surface, resulting in the cooling of the ground and the air adjacent to it. This results in extreme stability of the atmosphere near the Earth's surface. Under these conditions, turbulence is at a minimum. Attempts to relate different measures of turbulence of the wind (or stability of the atmosphere) to atmospheric diffusion have been made for some time. The measurement of atmospheric stability by temperature-difference measurements on a tower is frequently used as an indirect measure of turbulence, particularly when climatological estimates of turbulence are desired.

Lapse Rate and Atmospheric Stability Apart from mechanical interference with the steady flow of air caused by buildings and other obstacles, the most important factor that influences the degree of turbulence and hence the speed of diffusion in the lower air is the variation of temperature with height above the ground, referred to as the *lapse rate*. The dry-adiabatic lapse rate (DALR) is the temperature change for a rising parcel of dry air. The dry-adiabatic lapse rate can be approximated as −1°C per 100 m, or $dT/dz = -10^{-2}$ °C/m, or −5.4°F/1000 ft. If the rising air contains water vapor, the cooling due to adiabatic expansion will result in the relative humidity being increased, and saturation may be reached. Further ascent would then lead to condensation of water vapor, and the latent heat thus released would reduce the rate of cooling of the rising air. The buoyancy force on a warm-air parcel is caused by the difference between its density and that of the surrounding air. The perfect-gas law shows that, at a fixed pressure (altitude), the temperature and density of an air parcel are inversely related; temperature is normally used to determine buoyancy because it is easier to measure than density. If the temperature gradient (lapse rate) of the atmosphere is the same as the adiabatic lapse rate, a parcel of air displaced from its original position will expand or contract in such a manner that its density and temperature remain the same as its surroundings. In this case there will be no buoyancy forces on the displaced parcel, and the atmosphere is termed "neutrally stable."

If the atmospheric temperature decreases faster with increasing altitude than the adiabatic lapse rate (superadiabatic), a parcel of air displaced upward will have a higher temperature than the surrounding air. Its density will be lower, giving it a net upward buoyancy force. The opposite situation exists if the parcel of air is displaced downward, and the parcel experiences a downward buoyancy force. Once a parcel of air has started moving up or down, it will continue to do so, causing unstable atmospheric conditions. If the temperature decreases more slowly with increasing altitude than the adiabatic lapse rate, a displaced parcel of air experiences a net restoring force. The buoyancy forces then cause stable atmospheric conditions (see Fig. 22-5).

Strongly stable lapse rates are commonly referred to as *inversions.* The strong stability inhibits mixing across the inversion layer. Normally these conditions of strong stability extend for only several hundred meters vertically. The vertical extent of the inversion is referred to as the *inversion depth.* Two distinct types are observed: the ground-level inversion, caused by radiative cooling of the ground at night, and inversions aloft, occurring between 500 and several thousand meters above the ground (see Fig. 22-6). Some of the more common lapse-rate profiles with the corresponding effect on stack plumes are presented in Fig. 22-7.

From the viewpoint of air pollution, both stable surface layers and low-level inversions are undesirable because they minimize the rate of dilution of contaminants in the atmosphere. Even though the surface layer may be unstable, a low-level inversion will act as a barrier to vertical mixing, and contaminants will accumulate in the surface layer below the inversion. Stable atmospheric conditions tend to be more frequent and longest in persistence in the autumn, but inversions and stable lapse rates are prevalent at all seasons of the year.

Design Calculations For a given stack height, the calculational sequence begins by first estimating the effective height of the emission, employing an applicable plume-rise equation. The maximum GLC may then be determined by using an appropriate atmospheric-diffusion equation. A simple comparison of the calculated GLC for the particular pollutant with the maximum GLC permitted by the local air pollution codes dictates whether the stack is operating satisfactorily. Conversely, with knowledge of

TABLE 22-21 Compliance Activity and Schedule Chart

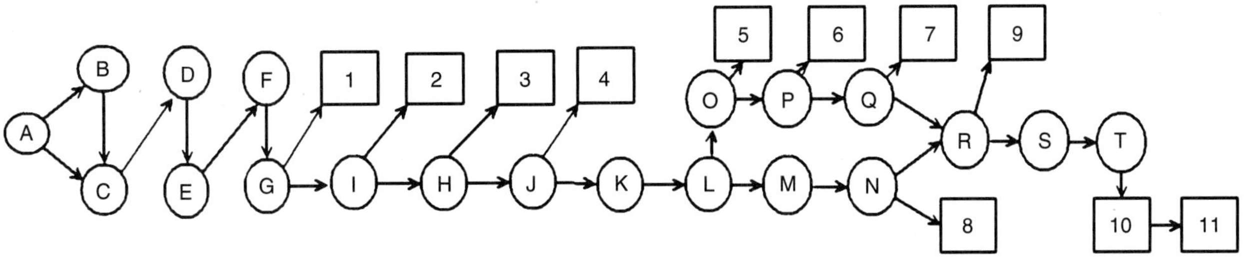

Milestones

1. Date of submittal of final control plan to appropriate agency
2. Date of issuance of equipment/system specifications to suppliers
3. Date of return of proposals from suppliers
4. Date of award of control-device contract
5. Issuance of construction specifications
6. Return of construction bids
7. Award of construction contract
8. Date of initiation of on-site construction or installation of emission-control equipment
9. Date by which on-site construction or installation of emission-control equipment is completed
10. Date by which final compliance is achieved
11. Date of submittal of compliance test results to appropriate agency

Activities

Designation	Activity	Designation	Activity
A-C	Preliminary investigation.	K-L	Review and approval of assembly drawings.
A-B	Source tests, if necessary.	L-M	Vendor prepares fabrication drawings.
C-D	Evaluate control alternatives.	M-N	Fabricate control device.
D-E	Commit funds for total program.	L-O	Prepare engineering drawings.
E-F	Prepare preliminary control plan and compliance schedule for agency.	O-5	Issue construction specifications.
F-G	Agency review and approval.	O-P	Procure construction bids.
G-I	Finalize plans and specifications.	P-Q	Evaluate construction bids.
I-2	Issuance of specifications to suppliers	Q-7	Award construction contract.
I-H	Procure control-device bids.	Q-R	On-site construction.
H-J	Evaluate control-device bids.	N-R	Install control device.
J-4	Award control-device contract.	R-9	Complete construction (system tie-in).
J-K	Vendor prepares assembly drawings.	S-T	Start-up, shakedown, source test
		T-10	Compliance achieved
		10-11	Test results submitted to agency

the maximum acceptable GLC standards, a stack that will satisfy these standards can be properly designed.

Effective Height of an Emission The effective height of an emission rarely corresponds to the physical height of the stack. If the plume is caught in the turbulent wake of the stack or of buildings in the vicinity of the stack, the effluent will be mixed rapidly downward toward the ground. If the plume is emitted free of these turbulent zones, a number of source emission characteristics and meteorological factors influence the rise of the plume. The source emission characteristics include the gas flow rate and the temperature of the effluent at the top of the stack and the diameter of the stack opening. The meteorological factors influencing plume rise include wind speed, air temperature, shear of the wind speed with height, and atmospheric stability. No current theory on plume rise takes into account all of these variables. Most of the equations that have been formulated for computing the effective height of an emission are semi-empirical. When considering any of these plume-rise equations, it is important to evaluate each in terms of the assumptions made and the circumstances existing at the time that the particular correlation was formulated. The formulas generally are not applicable to tall stacks [above 305 m (1000 ft) effective height].

The effective stack height (equivalent to the effective height of the emission) is the sum of the actual stack height, the plume rise due to the exhaust velocity (momentum) of the issuing gases, and the buoyancy rise, which is a function of the temperature of the gases being emitted and the atmospheric conditions.

Some of the more common plume-rise equations have been summarized by Buonicore and Theodore (*Industrial Control Equipment for Gaseous Pollutants,* Vol. 2, CRC Press, Boca Raton, Florida, 1975) and include:

- ASME, *Recommended Guide for the Prediction of the Dispersion of Airborne Effluents,* ASME, New York, 1968.
- Bosanquet-Carey-Halton, *Proc. Inst. Mech. Eng.* (*London*) **162:** 355 (1950).
- Briggs, *Plume Rise,* AEC Critical Review ser., U.S. Atomic Energy Commission, Div. Tech. Inf., 1969.
- Brummage et al., *The Calculation of Atmospheric Dispersion from a Stack,* CONCAWE, The Hague, 1966.
- Carson and Moses, *J. Air Pollut. Control Assoc.* **18:** 454 (1968) and **19:** 862 (1969).
- Canaday, *Int. J. Air Water Pollut.* **4:** 47 (1961).

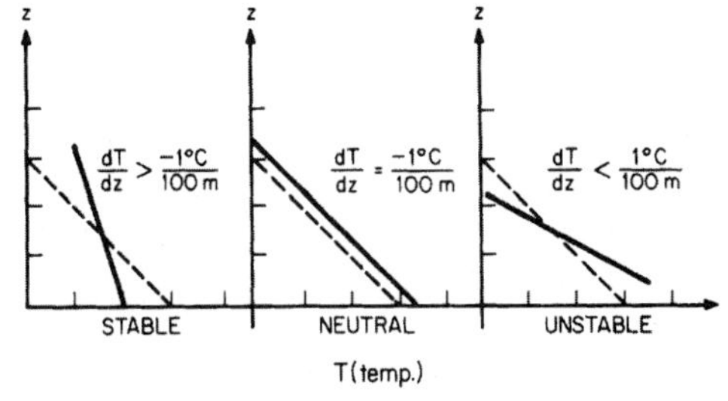

FIG. 22-5 Stability criteria with measured lapse rate.

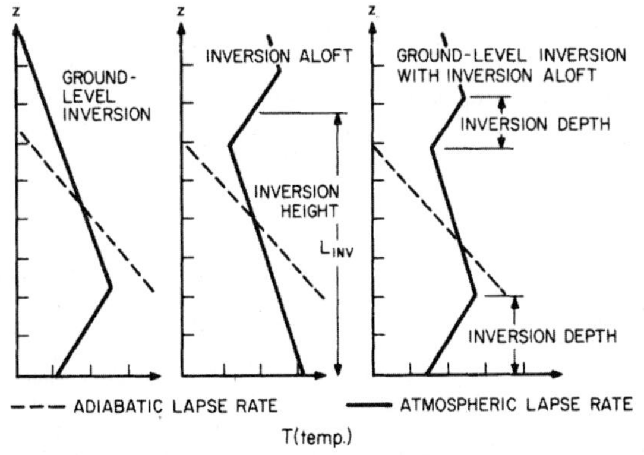

FIG. 22-6 Characteristic lapse rates under inversion conditions.

Temperature gradient | Observation | Description

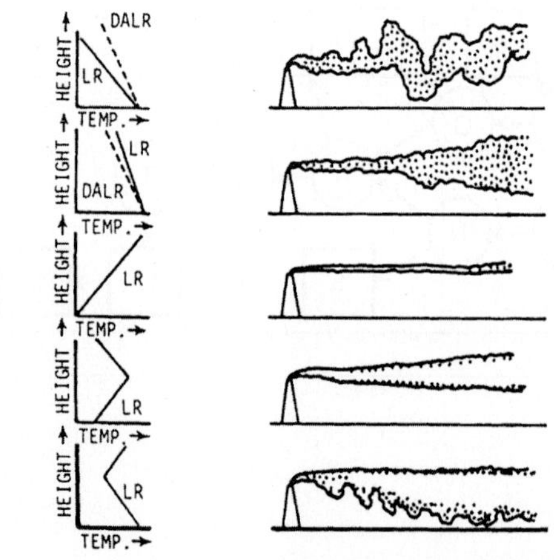

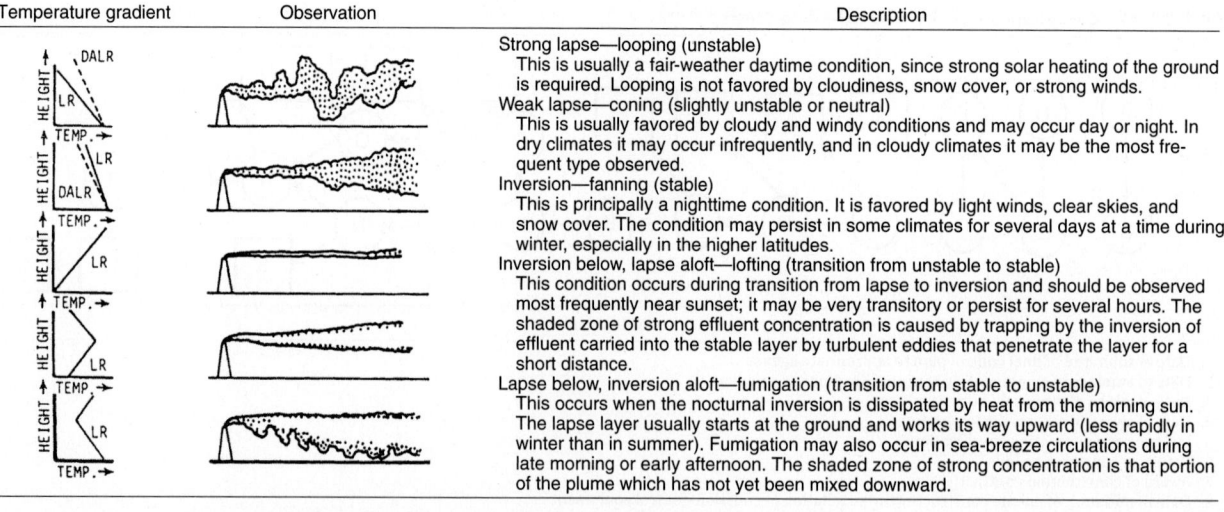

Strong lapse—looping (unstable)
This is usually a fair-weather daytime condition, since strong solar heating of the ground is required. Looping is not favored by cloudiness, snow cover, or strong winds.

Weak lapse—coning (slightly unstable or neutral)
This is usually favored by cloudy and windy conditions and may occur day or night. In dry climates it may occur infrequently, and in cloudy climates it may be the most frequent type observed.

Inversion—fanning (stable)
This is principally a nighttime condition. It is favored by light winds, clear skies, and snow cover. The condition may persist in some climates for several days at a time during winter, especially in the higher latitudes.

Inversion below, lapse aloft—lofting (transition from unstable to stable)
This condition occurs during transition from lapse to inversion and should be observed most frequently near sunset; it may be very transitory or persist for several hours. The shaded zone of strong effluent concentration is caused by trapping by the inversion of effluent carried into the stable layer by turbulent eddies that penetrate the layer for a short distance.

Lapse below, inversion aloft—fumigation (transition from stable to unstable)
This occurs when the nocturnal inversion is dissipated by heat from the morning sun. The lapse layer usually starts at the ground and works its way upward (less rapidly in winter than in summer). Fumigation may also occur in sea-breeze circulations during late morning or early afternoon. The shaded zone of strong concentration is that portion of the plume which has not yet been mixed downward.

FIG. 22-7 Lapse-rate characteristics of atmospheric-diffusion transport of stack emissions.

- Davison-Bryant, *Trans. Conf. Ind. Wastes,* 14th Ann. Meet. Ind. Hyg. Found. Am. **39,** 1949.
- Holland, *Workbook of Atmospheric Dispersion Estimates,* U.S. EPA Publ. AP-26, 1970.
- Lucas, Moore, and Spurr, *Int. J. Air Water Pollut.* **7:** 473 (1963).
- Montgomery et al., *J. Air Pollut. Control Assoc.* **22**(10): 779, TVA (1972).
- Stone and Clarke, *British Experience with Tall Stacks for Air Pollution Control on Large Fossil-Fueled Power Plants,* American Power Conference, Chicago, 1967.
- Stumke, *Staub* **23:** 549 (1963).

See also:

- Briggs, G. A., *Plume Rise Predications: Lectures on Air Pollution and Environmental Impact Analyses,* Workshop Proceedings, American Meteorological Society, Boston, 1975, pp. 59–111.
- Randerson, Darryl, ed., *Plume Rise and Buoyancy Effects, Atmospheric Science and Power Production,* DOE Report DOE/TIC-27601, 1981.

Maximum Ground-Level Concentrations The effective height of an emission having been determined, the next step is to study its path downward by using the appropriate atmospheric-dispersion formula. Some of the more popular atmospheric-dispersion calculational procedures have been summarized by Buonicore and Theodore (1975) and include:

- Bosanquet-Pearson model [Pasquill, *Meteorol. Mag.* **90:** 33, 1063 (1961); and Gifford, *Nucl. Saf.* **2**(4): 47 (1961).]
- Sutton model [*Q.J.R. Meteorol. Soc.* **73:** 257 (1947).]
- TVA Model [Carpenter et al., *J. Air Pollut. Control Assoc.* **21**(8): (1971); and Montgomery et al., *J. Air Pollut. Control Assoc.* **23**(5): 388 (1973).]

See also:

- Bornstein et al., *Simulation of Urban Barrier Effects on Polluted Urban Boundary Layers Using the Three-Dimensional URBMET/TVM Model with Urban Topography,* Air Pollution Proceedings, 1993.
- EPA, *Guidance on the Application of Refined Dispersion Models for Air Toxins Releases,* EPA-450/4-91-007.
- Zannetti, Paolo, *Air Pollution Modeling: Theories, Computational Methods, and Available Software,* Van Nostrand Reinhold, New York, 1990.
- Zannetti, Paolo, *Numerical Simulation Modeling of Air Pollution: An Overview,* Ecological Physical Chemistry, 2d International Workshop, May 1992.

Miscellaneous Effects

Evaporative Cooling When effluent gases are washed to absorb certain constituents before emission, the gases are cooled and become saturated with water vapor. Upon release of the gases, further cooling due to contact with cold surfaces of ductwork or stack is likely. This cooling causes water droplets to condense in the gas stream. Upon release of the gases from the stack, the water droplets evaporate, withdrawing the latent heat of vaporization from the air and cooling the plume. The resulting negative buoyancy reduces the effective stack height. The result may be a plume (with a greater density than that of the ambient atmosphere) that will fall to the ground. If any pollutant remains after scrubbing, its full effect will be felt on the ground in the vicinity of the stack.

Aerodynamic Downwash Should the stack exit velocity be too low as compared with the speed of the crosswind, some of the effluent can be pulled downward by the low pressure on the lee side of the stack. This phenomenon, known as "stack-tip downwash," can be minimized by keeping the exit velocity greater than the mean wind speed (i.e., typically twice the mean wind speed). Another way to minimize stack-tip downwash is to fit the top of the stack with a flat disc that extends for at least one stack diameter outward from the stack.

If it becomes necessary to increase the stack-gas exit velocity to avoid downwash, it may be necessary to remodel the stack exit. A venturi-nozzle design has been found to be the most effective. This design also keeps pressure losses to a minimum.

Building Downwash A review must be conducted for each stack to determine if building downwash effects need to be considered. Atmospheric flow is disrupted by aerodynamic forces in the immediate vicinity of structures or terrain obstacles. The disrupted flow near either building structures or terrain obstacles can both enhance the vertical dispersion of emissions from the source and reduce the effective height of the emissions from the source, resulting in an increase in the maximum GLC.

EPA Air Dispersion Models EPA addresses modeling techniques in the "Guideline on Air Quality Models" [Appendix W (November 2005) of 40 CFR part 51]. The guideline provides a common basis for estimating the air quality concentrations of criteria pollutants used in assessing control strategies and developing emission limits. The continuing development of new air quality models in response to regulatory requirements and the expanded requirements for models to cover ever more complex problems have emphasized the need for periodic review and update of guidance on these techniques. Information on the current status of modeling guidance can always be obtained from EPA's regional offices.

The guideline recommends air quality modeling techniques that should be applied to state implementation plan (SIP) revisions for existing sources and to new source reviews (NSRs), including prevention of significant deterioration (PSD). In addition, the guideline serves to identify, for all interested parties, those techniques and databases that EPA considers acceptable. Dispersion models, while uniquely filling one program need, have become a primary analytical tool in most air quality assessments. Air quality measurements can be used in a complementary manner with dispersion models, with due regard for the strengths and weaknesses of both analysis techniques. The diversity of the nation's topography and climate, and variations in source configurations and operating characteristics, dictate against a strict modeling "cookbook." There is no one model capable of properly addressing all conceivable situations, even within a broad category such as point sources.

The guideline provides a consistent basis for the selection of the most accurate models and databases for use in air quality assessments. There are two levels of sophistication of models. The first level consists of relatively simple estimation techniques that generally use preset, worst-case meteorological conditions to provide conservative estimates of the air quality impact of a specific source, or source category. These are called screening techniques or screening models. The purpose of such techniques is to eliminate the need for more detailed modeling for those sources that clearly will not cause or contribute to ambient concentrations in excess of either the National Ambient Air Quality Standards (NAAQS) or the allowable PSD concentration increments. The second level of models consists of those analytical techniques that provide more detailed treatment of physical and chemical atmospheric processes, require more detailed and precise input data, and provide more specialized concentration estimates. Most of the screening and refined models discussed in the guideline, codes, associated documentation, and other useful information are available for download from EPA's Support Center for Regulatory Air Modeling (SCRAM) web site at http://www.epa.gov/scram001.

CONTROL OF GASEOUS EMISSIONS

There are four chemical-engineering unit operations commonly used for the control of gaseous emissions:

1. *Absorption.* See Sec. 14, Gas Absorption and other Gas-Liquid Operations and Equipment.
2. *Adsorption.* See Sec. 16, Adsorption and Ion Exchange.
3. *Combustion.* See Sec. 24, Energy Resources, Conversion and Utilization.
4. *Condensation.* See Sec. 5 subsection Heat Transfer; Sec. 11, Heat-Transfer Equipment; and the Sec. 12 subsection Evaporative Cooling. For direct-contact condensers, see the Sec. 11 subsection Evaporators, Sec. 12 subsection Evaporative Cooling, and Sec. 11 subsection TEMA Style Shell-and-Tube Heat Exchangers.

These operations, which are routine chemical engineering operations, have been treated extensively in these other sections of this handbook.

There are three additional chemical engineering unit operations that are increasing in use in recent years. They are:

1. Biofiltration
2. Membrane filtration
3. Selective catalytic reduction

and are discussed further in this section.

Absorption The engineering design of gas absorption equipment must be based on a sound application of the principles of diffusion, equilibrium, and mass transfer as developed in Secs. 5 and 14 of this handbook. The main requirement in equipment design is to bring the gas into intimate contact with the liquid; that is, to provide a large interfacial area and a high intensity of interface renewal and to minimize resistance and maximize driving force. This contacting of the phases can be achieved in many different types of equipment, the most important being packed and plate columns. The final choice between them rests with the various criteria that must be met. For example, if the pressure drop through the column is large enough that compression costs become significant, a packed column may be preferable to a plate-type column because of the lower pressure drop.

In most processes involving the absorption of a gaseous pollutant from an effluent gas stream, the gas stream is the processed fluid; hence, its inlet condition (flow rate, composition, and temperature) are usually known. The temperature and composition of the inlet liquid and the composition of the outlet gas are usually specified. The main objectives in the design of an absorption column, then, are the determination of the solvent flow rate and the calculation of the principal dimensions of the equipment (column diameter and height to accomplish the operation). These objectives can be obtained by evaluating, for a selected solvent at a given flow rate, the number of theoretical separation units (stage or plates) and converting them into practical units, column heights, or number of actual plates by means of existing correlations.

The general design procedure consists of a number of steps to be taken into consideration. These include:

1. Solvent selection
2. Equilibrium-data evaluation
3. Estimation of operating data (usually consisting of a mass and energy in which the energy balance decides whether the absorption balance can be considered isothermal or adiabatic)
4. Column selection (should the column selection not be obvious or specified, calculations must be carried out for the different types of columns and the final based on economic considerations)
5. Calculation of column diameter (for packed columns, this is usually based on flooding conditions, and, for plate columns, on the optimum gas velocity or the liquid-handling capacity of the plate)
6. Estimation of column height or number of plates (for packed columns, column height is obtained by multiplying the number of transfer units, obtained from a knowledge of equilibrium and operating data, by the height of a transfer unit; for plate columns, the number of theoretical plates determined from the plot of equilibrium and operating lines is divided by the estimated overall plate efficiency to give the number of actual plates, which in turn allows the column height to be estimated from the plate spacing)
7. Determination of pressure drop through the column (for packed columns, correlations dependent on packing type, column-operating data, and physical properties of the constituents involved are available to estimate the pressure drop through the packing; for plate columns, the pressure drop per plate is obtained and multiplied by the number of plates)

Solvent Selection The choice of a solvent is most important. Often water is used because it is very inexpensive and plentiful, but the following properties must also be considered:

1. *Gas solubility.* A high gas solubility is desired, since this increases the absorption rate and minimizes the quantity of solvent necessary. Generally, solvents of a chemical nature similar to that of the solute to be absorbed will provide good solubility.
2. *Volatility.* A low solvent vapor pressure is desired, since the gas leaving the absorption unit is ordinarily saturated with the solvent, and much may thereby be lost.
3. *Corrosiveness.*
4. *Cost.*
5. *Viscosity.* Low viscosity is preferred for reasons of rapid absorption rates, improved flooding characteristics, lower pressure drops, and good heat-transfer characteristics.
6. *Chemical stability.* The solvent should be chemically stable and, if possible, nonflammable.
7. *Toxicity.*
8. *Low freezing point.* If possible, a low freezing point is favored, since any solidification of the solvent in the column makes the column inoperable.

Equipment The principal types of gas-absorption equipment may be classified as follows:

1. Packed columns (continuous operation)
2. Plate columns (staged operations)
3. Open towers
4. Miscellaneous

Of the four categories, the packed column is by far the most commonly used for the absorption of gaseous pollutants in the chemical and refining industrial sectors. Open towers and tray towers have been predominantly used in the electrical generating industry as flue gas desulfurization (FGD) scrubbers. These FGD scrubbers use limestone or lime and a high liquid-to-gas (L/G) ratio for effective removal of sulfur dioxide from flue gas. Miscellaneous gas-absorption equipment could include acid gas scrubbers that are commonly classified as either wet or dry. In wet scrubber systems, the absorption tower uses a lime-based sorbent liquor that reacts with the acid gases to form a wet/solid by-product. Dry scrubbers can be grouped into three categories: (1) spray dryers; (2) circulating dry scrubbers (CDS); and (3) dry injection. Each of these systems yields a dry product that can be captured with a fabric filter baghouse downstream and thus avoids costly wastewater treatment systems. The baghouse is highly efficient in capturing the particulate emissions, and a portion of the overall acid gas removal has been found to occur within the baghouse. Additional information may be found by referring to the appropriate sections in this handbook and the many excellent texts available, such as McCabe and Smith, *Unit Operations of Chemical Engineering,* 3d ed., McGraw-Hill, New York, 1976; Sherwood and Pigford, *Absorption and Extraction,* 2d ed., McGraw-Hill, New York, 1952; Smith, *Design of Equilibrium Stage Processes,* McGraw-Hill, New York, 1963; Treybal, *Mass Transfer Operations,* 3d ed., McGraw-Hill, New York, 1980; McKenna, Mycock, and Theodore, *Handbook of Air Pollution Control Engineering and Technology,* CRC Press, Boca Raton, Fla., 1995; Theodore and Allen, *Air Pollution Control Equipment,* ETSI Prof. Training Inst., Roanoke, Va., 1993.

Adsorption The design of gas-adsorption equipment is in many ways analogous to the design of gas-absorption equipment, with a solid adsorbent replacing the liquid solvent (see Secs. 16 and 19). Similarity is evident in the material- and energy-balance equations as well as in the methods used to determine the column height. The final choice, as one would expect, rests with the overall process economics.

Selection of Adsorbent Industrial adsorbents are usually capable of adsorbing both organic and inorganic gases and vapors. However, their preferential adsorption characteristics and other physical properties make each of them more or less specific for a particular application. General experience has shown that, for the adsorption of vapors of an organic nature, activated carbon has superior properties, having hydrocarbon-selective properties and high adsorption capacity for such materials. Inorganic adsorbents, such as activated alumina or silica gel, can also be used to adsorb organic materials, but difficulties can arise during regeneration. Activated alumina, silica gel, and molecular sieves will also preferentially adsorb any water vapor with the organic contaminant. At times this may be a considerable drawback in the application of these adsorbents for organic-contaminant removal.

The normal method of regeneration of adsorbents is by use of steam, inert gas (i.e., nitrogen), or other gas streams, and in the majority of cases this can cause at least slight decomposition of the organic compound on the adsorbent. Two difficulties arise: (1) incomplete recovery of the adsorbate, although this may be unimportant; and (2) progressive deterioration in the capacity of the adsorbent as the number of cycles increases because of blocking of the pores from carbon formed by hydrocarbon decomposition.

With activated carbon, a steaming process is used, and the difficulties of regeneration are thereby overcome. This is not feasible with silica gel or activated alumina because of the risk of breakdown of these materials when in contact with liquid water.

In some cases, none of the adsorbents has sufficient retaining capacity for a particular contaminant. In these applications, a large-surface-area adsorbent can be impregnated with an inorganic compound or, in rare cases, with a high-molecular-weight organic compound that can react chemically with the particular contaminant. For example, iodine-impregnated carbons are used for the removal of mercury vapor, and bromine-impregnated carbons are used for ethylene or propylene removal. The action of these impregnates is either catalytic conversion or reaction to a nonobjectionable compound or to a more easily adsorbed compound. For this case, general adsorption theory no longer applies to the overall effects of the process. For example, mercury removal by an iodine-impregnated carbon proceeds more quickly at a higher temperature, and a better overall efficiency can be obtained than in a low-temperature system.

Since adsorption takes place at the interphase boundary, the adsorption surface area becomes an important consideration. Generally, the higher the adsorption surface area, the greater its adsorption capacity. However, the surface area has to be "available" in a particular pore size within the adsorbent. At low partial pressure (or concentration), a surface area in the smallest pores in which the adsorbate can enter is the most efficient. At higher pressures, the larger pores become more important; at very high concentrations, capillary condensation will take place within the pores, and the total micropore volume becomes the limiting factor.

The action of molecular sieves is slightly different from that of other adsorbents in that selectivity is determined more by the pore-size limitations of the particular sieve. In selecting molecular sieves, it is important that the contaminant to be removed be smaller than the available pore size. Hence, it is important that the particular adsorbent not only have an affinity for the contaminant in question but also have sufficient surface area available for adsorption.

Design Data The adsorbent having been selected, the next step is to calculate the quantity of adsorbent required and eventually consider other factors, such as the temperature rise of the gas stream due to adsorption and the useful life of the adsorbent under operating conditions. The sizing and overall design of the adsorption system depend on the properties and characteristics of both the feed gas to be treated and the adsorbent. The following information should be known or available for design purposes:

1. Gas stream
   a. Adsorbate concentration
   b. Temperature
   c. Temperature rise during adsorption
   d. Pressure
   e. Flow rate
   f. Presence of adsorbent contaminant material
2. Adsorbent
   a. Adsorption capacity as used on stream.
   b. Temperature rise during adsorption.
   c. Isothermal or adiabatic operation.
   d. Life, if presence of contaminant material is unavoidable.
   e. Possibility of catalytic effects causing an adverse chemical reaction in the gas stream or the formation of polymerization products on the adsorbent bed, with consequent deterioration.
   f. Bulk density.
   g. Particle size, usually reported as a mean equivalent particle diameter. The dimensions and shape of particles affect both the pressure drops through the adsorbent bed and the diffusion rate into the particles. All things being equal, adsorbent beds consisting of smaller particles, although causing a higher pressure drop, will be more efficient.
   h. Pore data, which are important because they may permit elimination from consideration of adsorbents whose pore diameter will not admit the desired adsorbate molecule.
   i. Hardness, which indicates the care that must be taken in handling adsorbents to prevent the formation of undesirable fines.
   j. Regeneration information.

The design techniques used include both stagewise and continuous-contacting methods and can be applied to batch, continuous, and semicontinuous operations.

Adsorption Phenomena The adsorption process involves three necessary steps. The fluid must first come in contact with the adsorbent, at which time the adsorbate is preferentially or selectively adsorbed on the adsorbent. Next, the fluid must be separated from the adsorbent-adsorbate, and, finally, the adsorbent must be regenerated by removing the adsorbate or by discarding used adsorbent and replacing it with fresh material. Regeneration is performed in a variety of ways, depending on the nature of the adsorbate. Gases or vapors are usually desorbed by either raising the temperature (thermal cycle) or reducing the pressure (pressure cycle). The more popular thermal cycle is accomplished by passing hot gas through the adsorption bed in the direction opposite to the flow during the adsorption cycle. This ensures that the gas passing through the unit during the adsorption cycle always meets the most active adsorbent last and that the adsorbate concentration in the adsorbent at the outlet end of the unit is always maintained at a minimum.

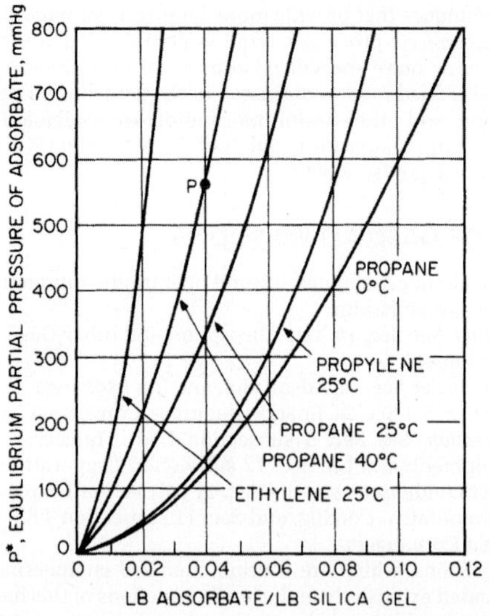

FIG. 22-8 Equilibrium partial pressures for certain organics on silica gel.

In the first step, in which the molecules of the fluid come in contact with the adsorbent, an equilibrium is established between the adsorbed fluid and the fluid remaining in the fluid phase. Figures 22-8 through 22-10 show several experimental equilibrium adsorption isotherms for a number of components adsorbed on various adsorbents. Consider Fig. 22-8, in which the concentration of adsorbed gas on the solid is plotted against the equilibrium partial pressure p^0 of the vapor or gas at constant temperature. At 40°C, for example, pure propane vapor at a pressure of 550 mm Hg is in equilibrium with an adsorbate concentration at point *P* of 0.04 lb adsorbed propane per pound of silica gel. Increasing the pressure of the propane will cause more propane to be adsorbed, while decreasing the pressure of the system at *P* will cause propane to be desorbed from the carbon.

The adsorptive capacity of activated carbon for some common solvent vapors is shown in Table 22-22.

Adsorption-Control Equipment If a gas stream must be treated for a short period, usually only one adsorption unit is necessary, provided, of

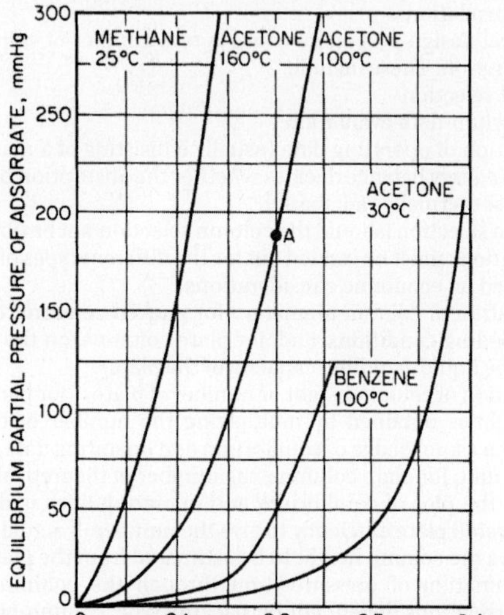

FIG. 22-9 Equilibrium partial pressures for certain organics on carbon.

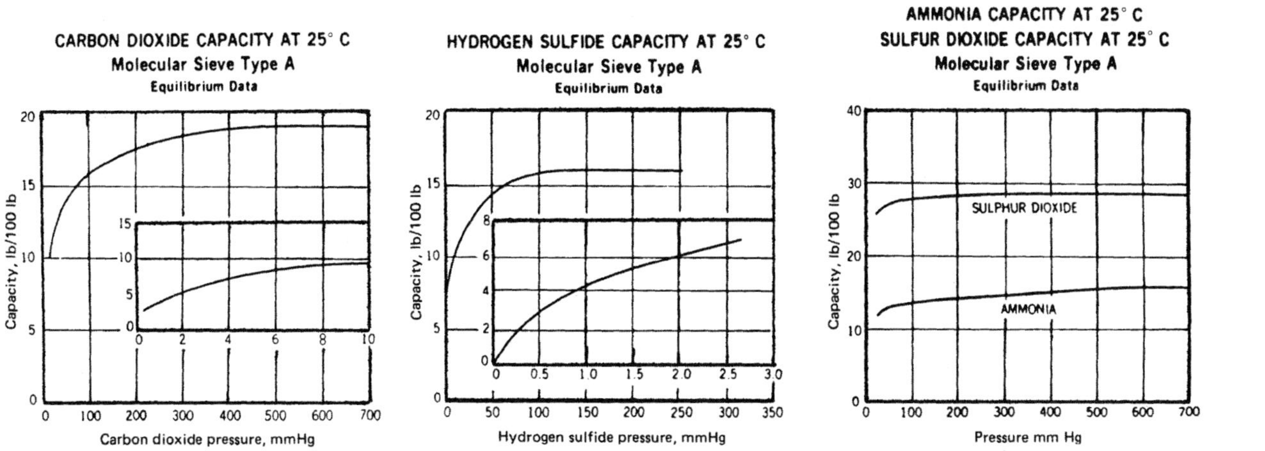

FIG. 22-10 Equilibrium partial pressures for certain gases on molecular sieves. (*A. J. Buonicore and L. Theodore, Industrial Control Equipment for Gaseous Pollutants, vol. I, CRC Press, Boca Raton, Fla., 1975.*)

course, that a sufficient time interval is available between adsorption cycles to permit regeneration. However, this is usually not the case. Since an uninterrupted flow of treated gas is often required, it is necessary to employ one or more units capable of operating in this fashion. The units are designed to handle gas flows without interruption and are characterized by their mode of contact, either staged or continuous. By far the most common type of adsorption system used to remove an objectionable pollutant from a gas stream consists of a number of fixed-bed units operating in such a sequence that the gas flow remains uninterrupted. A two- or three-bed system is usually employed, with one or two beds bypassed for regeneration while one is adsorbing. A typical two-bed system is shown in Fig. 22-11*a*, while a typical three-bed system is shown in Fig. 22-11*b*. The type of system best suited for a particular job is determined from several factors, including the amount and rate of material being adsorbed, the time between cycles, the time required for regeneration, and the cooling time, if required.

Typical of continuous-contact operation for gaseous-pollutant adsorption is the use of a fluidized bed. During steady-state staged-contact operation, the gas flows up through a series of successive fluidized-bed stages, permitting maximum gas-solid contact on each stage. A typical arrangement of this type is shown in Fig. 22-12 for multistage countercurrent adsorption with regeneration. In the upper part of the tower, the particles are contacted countercurrently on perforated trays in relatively shallow beds with the gas stream containing the pollutant, the adsorbent solids moving from tray to tray through downspouts. In the lower part of the tower, the adsorbent is regenerated by similar contact with hot gas, which desorbs and carries off the pollutant. The regenerated adsorbent is then recirculated by an airlift to the top of the tower.

Although the continuous-countercurrent type of operation has found limited application in the removal of gaseous pollutants from process streams (for example, the removal of carbon dioxide and sulfur compounds such as hydrogen sulfide and carbonyl sulfide), by far the most common type of operation in current use is the fixed-bed adsorber. The relatively high cost of continuously transporting solid particles as required in steady-state operations makes fixed-bed adsorption an attractive, economical alternative. If intermittent or batch operation is practical, a simple one-bed system, cycling alternately between the adsorption and regeneration phases, will suffice.

TABLE 22-22 Adsorptive Capacity of Common Solvents on Activated Carbons*

Solvent	Carbon bed weight, %†
Acetone	8
Heptane	6
Isopropyl alcohol	8
Methylene chloride	10
Perchloroethylene	20
Stoddard solvent	2–7
1,1,1-Trichloroethane	12
Trichloroethylene	15
Trichlorotrifluoroethane	8
VM&P naphtha	7

*Assuming steam desorption at 5 to 10 psig.
†For example, 8 lb of acetone adsorbed on 100 lb of activated carbon.

Additional information may be found by referring to the appropriate sections of this handbook. A comprehensive treatment of adsorber design principles is given in Buonicore and Theodore, *Industrial Control Equipment for Gaseous Pollutants,* vol. 1, 2d ed., CRC Press, Boca Raton, Fla., 1992.

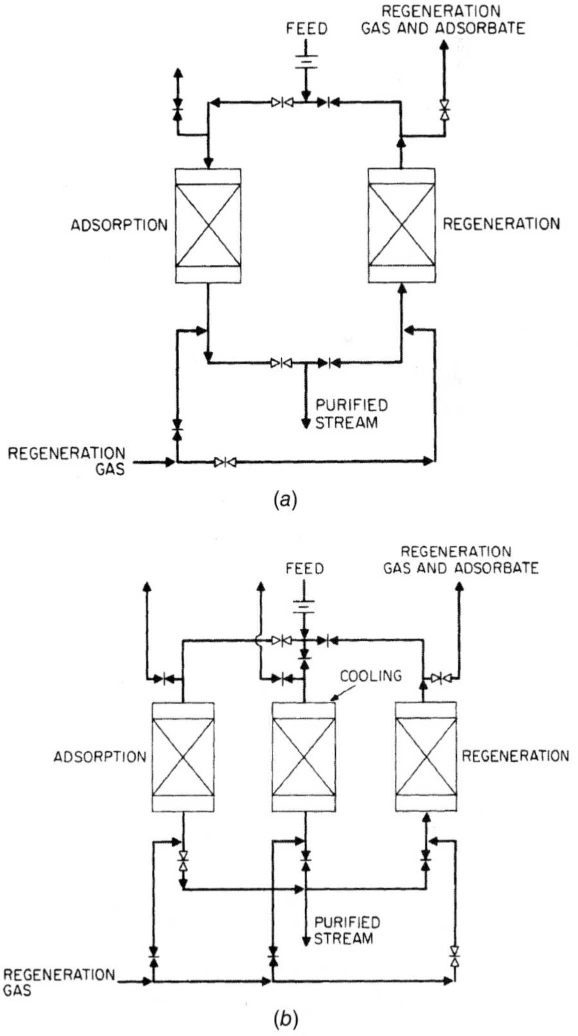

FIG. 22-11 (*a*) Typical two-bed adsorption system. (*b*) Typical three-bed adsorption system.

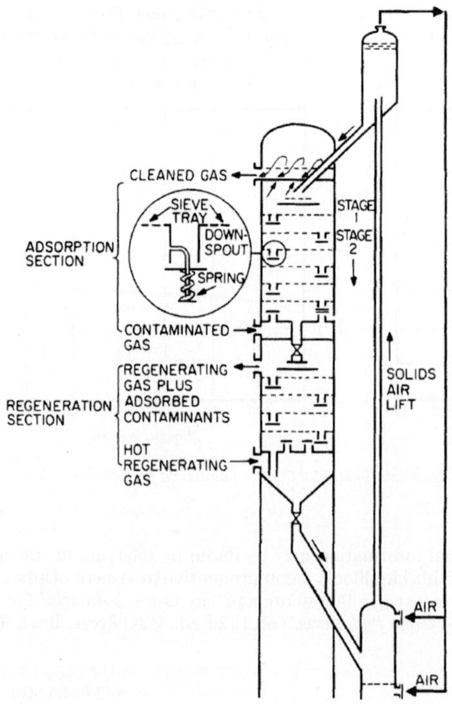

FIG. 22-12 Multistage countercurrent adsorption with regeneration.

Combustion Many organic compounds released from manufacturing operations can be converted to innocuous carbon dioxide and water by rapid oxidation (chemical reaction): combustion. However, combustion of gases containing halides may require the addition of acid gas treatment to the combustor exhaust.

Three rapid oxidation methods are typically used to destroy combustible contaminants: (1) flares (direct-flame combustion), (2) thermal combustors, and (3) catalytic combustors. The thermal and flare methods are characterized by the presence of a flame during combustion. The combustion process is also commonly referred to as afterburning or incineration.

To achieve complete combustion (i.e., the combination of the combustible elements and compounds of a fuel with all the oxygen that they can use), sufficient space, time, and turbulence and a temperature high enough to ignite the constituents must be provided.

The three T's of combustion—time, temperature, and turbulence—govern the speed and completeness of the combustion reaction. For complete combustion, the oxygen must come into intimate contact with the combustible molecule at sufficient temperature and for a sufficient length of time for the reaction to be completed. Incomplete reactions may result in the generation of aldehydes, organic acids, carbon, and carbon monoxide.

Combustion-Control Equipment Combustion-control equipment can be divided into three types: (1) flares, (2) thermal incinerators, (3) catalytic incinerators.

Flares In many industrial operations and particularly in chemical plants and petroleum refineries, large volumes of combustible waste gases are produced. These gases result from undetected leaks in the operating equipment, from upset conditions in the normal operation of a plant in which gases must be vented to avoid dangerously high pressures in operating equipment, from plant start-ups, and from emergency shutdowns. Large quantities of gases may also result from off-specification product or from excess product that cannot be sold. Flows are typically intermittent, with flow rates during major upsets of up to several million cubic feet per hour.

The preferred control method for excess gases and vapors is to recover them in a blowdown recovery system. However, large quantities of gas, especially those during upset and emergency conditions, are difficult to contain and reprocess. In the past, all waste gases were vented directly to the atmosphere. However, widespread venting caused safety and environmental problems, and in practice it is now customary to collect such gases in a closed flare system and to burn them as they are discharged.

Although flares can be used to dispose of excess waste gases, such systems can present additional safety problems. These include the explosion potential, thermal radiation hazards from the flame, and the problem of toxic asphyxiation during flameout. Aside from these safety aspects, there are several other problems associated with flaring that must be dealt with during the design and operation of a flare system. These problems include the formation of smoke, the luminosity of the flame, noise during flame, and the possible emission of by-product air pollutants during flaring.

The heat content of the waste stream to be disposed is another important consideration. The heat content of the waste gas falls into two classes. The gases can either support their own combustion or not. In general, a waste gas with a heating value greater than 7443 kJ/m^3 (200 Btu/ft^3) can be flared successfully. The heating value is based on the lower heating value of the waste gas at the flare. Below 7443 kJ/m^3, enriching the waste gas by injecting another gas with a higher heating value may be necessary. The addition of such a rich gas is called "endothermic flaring." Gases with a heating value as low as 2233 kJ/m^3 (60 Btu/ft^3) have been flared but at a significant fuel demand. It is usually not feasible to flare a gas with a heating value below 3721 kJ/m^3 (100 Btu/ft^3). If the flow of low-Btu gas is continuous, thermal or catalytic incineration can be used to dispose of the gas. For intermittent flows, however, endothermic flaring may be the only possibility.

Although most flares are used to dispose of intermittent waste gases, some continuous flares are in use, but generally only for relatively small volumes of gases. The heating value of large-volume continuous-flow waste gases is usually too valuable to lose in a flare. Vapor recovery or the use of the vapor as a fuel in a process heater is preferred over flaring. Since auxiliary fuel must be added to the gas for it to flare, large, continuous flows of a low-heating-value gas are usually more efficient to burn in a thermal incinerator than in the flame of a flare.

Flares are mostly used for the disposal of hydrocarbons. Waste gases composed of natural gas, propane, ethylene, propylene, butadiene, and butane probably constitute over 95 percent of the material flared. Flares have been used successfully to control malodorous gases such as mercaptans and amines, but care must be taken when flaring these gases. Unless the flare is very efficient and gives good combustion, obnoxious fumes can escape unburned and cause a nuisance.

Flaring of hydrogen sulfide should be avoided because of its toxicity and low odor threshold. In addition, burning relatively small amounts of hydrogen sulfide can create enough sulfur dioxide to cause crop damage or a local nuisance. For gases whose combustion products may cause problems, such as those containing hydrogen sulfide or chlorinated hydrocarbons, flaring is not recommended.

Thermal Incinerators Thermal incinerators or afterburners can be used over a fairly wide but low range of organic vapor concentration. The concentration of the organics in air must be substantially below the lower flammable level (lower explosive limit). As a rule, a factor of four is employed for safety precautions. Reactions are conducted at elevated temperatures to ensure high chemical-reaction rates for the organics. To achieve this temperature, it is necessary to preheat the feed stream using auxiliary energy. Along with the contaminant-laden gas stream, air and fuel are continuously delivered to the incinerator (see Fig. 22-13).

The fuel and contaminants are combusted with air in a firing unit (burner). The burner may use the air in the process-waste stream as the combustion air for the auxiliary fuel, or it may use a separate source of outside air. The products of combustion and the unreacted feed stream are intensely mixed and enter the reaction zone of the unit. The pollutants in the process-gas stream are then reacted at the elevated temperature. Thermal incinerators generally require operating temperatures in the range of 650°C to 980°C (1200°F to 1800°F) for combustion of most organic pollutants (see Table 22-23).

A residence time of 0.2 to 1.0 is often recommended, but this factor is dictated primarily by complex kinetic considerations. The kinetics of hydrocarbon (HC) combustion in the presence of excess oxygen can be simplified into the following first-order rate equation:

$$\frac{d(\text{HCl})}{dt} = -k[\text{HCl}] \tag{22-1}$$

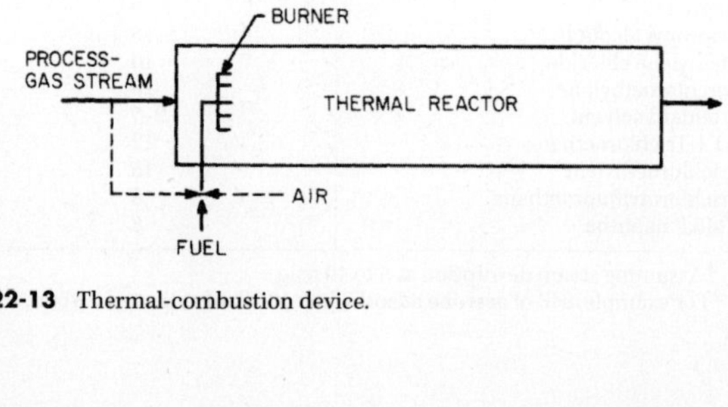

FIG. 22-13 Thermal-combustion device.

TABLE 22-23 Thermal Afterburners: Conditions Required for Satisfactory Performance in Various Abatement Applications

Abatement category	Afterburner residence time, s	Temperature, °F
Hydrocarbon emissions: 90 + % destruction of HC	0.3–0.5	1100–1250*
Hydrocarbons + CO: 90 + % destruction of HC + CO	0.3–0.5	1250–1500
Odor		
50–90% destruction	0.3–0.5	1000–1200
90–99% destruction	0.3–0.5	1100–1300
99 + % destruction	0.3–0.5	1200–1500
Smokes and plumes		
White smoke (liquid mist)		
Plume abatement	0.3–0.5	800–1000†
90 + % destruction of HC + CO	0.3–0.5	1250–1500
Black smoke (soot and combustible particulates)	0.7–1.0	1400–2000

*Temperatures of 1400 to 1500°F (760 to 816°C) may be required if the hydrocarbon has a significant content of any of the following: methane, cellosolve, and substituted aromatics (e.g., toluene and xylenes).

†Operation for plume abatement only is not recommended, since this merely converts a visible hydrocarbon emission into an invisible one and frequently creates a new odor problem because of partial oxidation in the afterburner.

where k = pseudo-first-order rate constant (s^{-1}). If the initial concentration is $C_{A,o}$, the solution of Eq. (22-1) is:

$$\ln\left(\frac{C_A}{C_{A,o}}\right) = -kt \tag{22-2}$$

Equation (22-2) is often used for a kinetic modeling of a burner using mole fractions in the range of 0.15 and 0.001 for oxygen and HC, respectively. The rate constant is generally of the following Arrhenius form:

$$k = Ae^{-E/RT} \tag{22-3}$$

where: A = pre-exponential factor, s– (see Table 2, *Air Pollution Engineering Manual,* Van Nostrand Reinhold, New York, 1992, p. 62)
E = activation energy, cal/gmol (see Table 2, *Air Pollution Engineering Manual,* Van Nostrand Reinhold, New York, 1992, p. 62)
R = universal gas constant, 1.987 cal/gmol °K
T = absolute temperature, °K

The referenced table for A and E is a summary of first-order HC combustion reactions.

The end combustion products are continuously emitted at the outlet of the reactor. The average gas velocity can range from as low as 3 m/s (10 ft/s) to as high as 15 m/s (50 ft/s). These high velocities are required to prevent settling of particulates (if present) and to minimize the dangers of flashback and fire hazards. Space velocity calculations are given in the Incinerator Design and Performance Equations subsection.

The fuel is usually natural gas. The energy liberated by reaction may be directly recovered in the process or indirectly recovered by suitable external heat exchange (see Fig. 22-14).

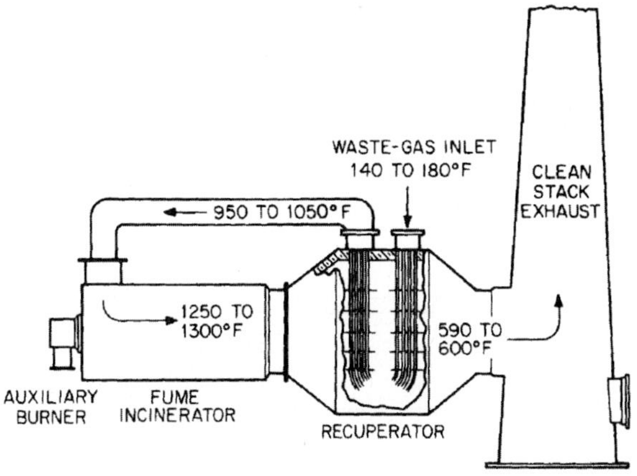

FIG. 22-14 Thermal combustion with energy (heat) recovery.

Because of the high operating temperatures, the unit must be made of metals capable of withstanding this condition. Combustion devices are usually made with an outer steel shell that is lined with refractory material. Refractory-wall thickness is usually in the 0.05- to 0.23-m (2- to 9-in) range, depending on temperature considerations.

Some of the advantages of the thermal incinerators are:
1. Removal of organic gases
2. Removal of submicrometer organic particles
3. Simplicity of construction
4. Small space requirements

Some of the disadvantages are:
1. High operating costs
2. Fire hazards
3. Flashback possibilities

Catalytic Incinerators Catalytic incinerators are an alternative to thermal incinerators. For simple reactions, the effect of the presence of a catalyst is to (1) increase the rate of the reaction, (2) permit the reaction to occur at a lower temperature, and (3) reduce the reactor volume.

In a typical catalytic incinerator for the combustion of organic vapors, the gas stream is delivered to the reactor continuously by a fan at a velocity in the range of 3 to 15 m/s (10 to 30 ft/s), but at a lower temperature, usually in the range of 350°C to 425°C (650°F to 800°F), than the thermal unit. (Design and performance equations used for calculating space velocities are given in the next section.) The gases, which may or may not be preheated, pass through the catalyst bed, where the combustion reaction occurs. The combustion products, which again are made up of water vapor, carbon dioxide, inerts and unreacted vapors, are continuously discharged from the outlet at a higher temperature. Energy savings can again be effected by heat recovery from the exit stream.

Metals in the platinum family are recognized for their ability to promote combustion at low temperatures. Other catalysts include various oxides of copper, chromium, vanadium, nickel, and cobalt. These catalysts are subject to poisoning, particularly from halogens, halogen and sulfur compounds, zinc, arsenic, lead, mercury, and particulates. It is therefore important that catalyst surfaces be clean and active to ensure optimum performance.

Catalysts may be porous pellets, usually cylindrical or spherical in shape, ranging from 0.16 to 1.27 cm ($^1/_{16}$ to ½ in) in diameter. Small sizes are recommended, but the pressure drop through the reactor increases. Among other shapes are honeycombs, ribbons, and wire mesh. Since catalysis is a surface phenomenon, a physical property of these particles is that the internal pore surface is nearly infinitely greater than the outside surface.

The following sequence of steps is involved in the catalytic conversion of reactants to products:
1. Transfer of reactants to and products from the outer catalyst surface
2. Diffusion of reactants and products within the pores of the catalyst
3. Activated adsorption of reactants and the desorption of the products on the active centers of the catalyst
4. Reaction or reactions on active centers on the catalyst surface

At the same time, energy effects arising from chemical reaction can result in the following:
1. Heat transfer to or from active centers to the catalyst-particle surface
2. Heat transfer to and from reactants and products within the catalyst particle
3. Heat transfer to and from moving streams in the reactor
4. Heat transfer from one catalyst particle to another within the reactor
5. Heat transfer to or from the walls of the reactor

Some of the advantages of catalytic incinerators are:
1. Lower fuel requirements as compared with thermal incinerators
2. Lower operating temperatures
3. Minimum insulation requirements
4. Reduced fire hazards
5. Reduced flashback problems

The disadvantages include:
1. Higher initial cost than thermal incinerators
2. Catalyst poisoning
3. Necessity of first removing large particulates
4. Catalyst-regeneration problems
5. Catalyst disposal

Incinerator Design and Performance Equations The key incinerator design and performance calculations are the required fuel usage and physical dimensions of the unit. The following is a general calculation procedure to use in solving for these two parameters, assuming that the process gas stream flow, inlet temperature, combustion temperature, and required residence time are known. The combustion temperature and residence time for thermal incinerators can be estimated using Table 22-23 and Eqs. (22-1) through (22-3).

1. The heat load needed to heat the inlet process gas stream to the incinerator operating temperature is:

$$Q = \Delta H \tag{22-4}$$

2. Correct the heat load for radiant heat losses, *RL:*

$$Q = (1 + RL)(\Delta H);\ RL = \text{fractional basis} \tag{22-5}$$

3. Assuming that natural gas is used to fire the burner with a known heating value of HV_G, calculate the available heat at the operating temperature. A shortcut method usually used for most engineering purposes is:

$$HA_T = (HV_G)\left(\frac{HA_T}{HV_G}\right)_{\text{ref}} \tag{22-6}$$

where the subscript "ref" refers to a reference fuel. For natural gas with a reference HV_G of 1059 Btu/scf, the heat from the combustion using no excess air would be given by:

$$(HA_T)_{\text{ref}} = -0.237(t) + 981;\ T = F \tag{22-7}$$

4. The amount of natural gas needed as fuel (*NG*) is given by:

$$NG = \frac{Q}{HA};\ \text{consistent units} \tag{22-8}$$

5. The resulting volumetric flow rate is the sum of the combustion of the natural gas *q* and the process gas stream *p* at the operating temperature:

$$q_T = q_p + q_c \tag{22-9}$$

A good estimate for q_c is:

$$q_c = (11.5)(NG) \tag{22-10}$$

6. The diameter of the combustion device is given by:

$$S = \frac{q_T}{\upsilon_t} \tag{22-11}$$

where υ_t is defined as the velocity of the gas stream at the incinerator operating temperature.

Condensation Often in air pollution control practice, it becomes necessary to treat an effluent stream consisting of a condensable pollutant vapor and a noncondensable gas. One control method to remove such pollutants from process-gas streams that is often overlooked is condensation. Condensers can be used to collect condensable emissions discharged to the atmosphere, particularly when the vapor concentration is high. Condensers have been used as a pretreatment option when combined with another air pollution control system, such as adsorption. This combination keeps the adsorbent from being quickly overloaded by the pollutant. Condensation is usually accomplished by lowering the temperature of the gaseous stream, although an increase in pressure will produce the same result. The former approach is usually employed by industry, since pressure changes (even small ones) on large volumetric gas-flow rates are often economically prohibitive.

Condensation Equipment There are two basic types of condensers used for control: contact and surface. In contact condensers, the gaseous stream is brought into direct contact with a cooling medium so that the vapors condense and mix with the coolant (see Fig. 22-15). The more widely used system, however, is the surface condenser (or heat exchanger), in which the vapor and the cooling medium are separated by a wall (see Fig. 22-16). Since high removal efficiencies cannot be obtained with low-condensable vapor concentrations, condensers are typically used for pretreatment before some other more efficient control device, such as an incinerator, absorber, or adsorber.

Contact Condensers Spray condensers, jet condensers, and barometric condensers all use water or some other liquid in direct contact with the vapor to be condensed. The temperature approach between the liquid and the vapor is very small, so the efficiency of the condenser is high, but large volumes of the liquid are necessary. If the vapor is soluble in the liquid, the system is essentially an absorptive one. If the vapor is not soluble, the system is a true condenser, in which case the temperature of the vapor must be below the dew point. Direct-contact condensers are seldom used for the removal of organic solvent vapors because the condensate will contain an organic-water mixture that must be separated or treated before disposal. They are, however, the most effective method of removing heat from hot gas streams when the recovery of organics is not a consideration.

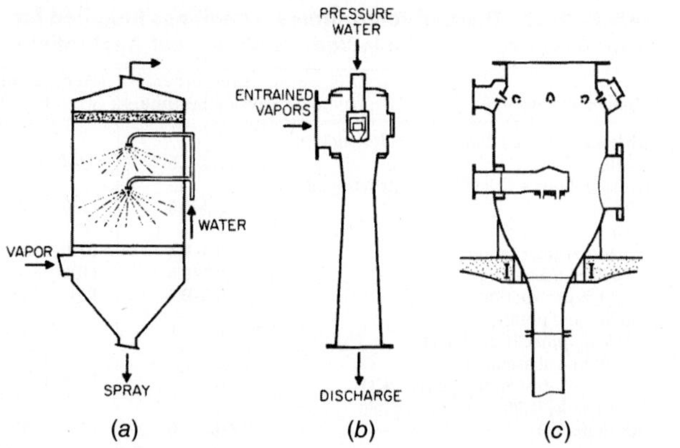

FIG. 22-15 Typical direct-contact condensers. (*a*) Spray chamber. (*b*) Jet. (*c*) Barometric.

In a direct-contact condenser, a stream of water or other cooling liquid is brought into direct contact with the vapor to be condensed. The liquid stream leaving the chamber contains the original cooling liquid plus the condensed substances. The gaseous stream leaving the chamber contains the noncondensable gases and such condensable vapor as did not condense; it is reasonable to assume that the vapors in the exit gas stream are saturated. It is then the temperature of the exit gas stream that determines the collection efficiency of the condenser.

The advantages of contact condensers are that (1) they can be used to produce a vacuum, thereby creating a draft to remove odorous vapors and also reduce boiling points in cookers and vats; (2) they usually are simpler and less expensive than the surface type; and (3) they usually have considerable odor-removing capacity because of the greater condensate dilution (13 lb of 60°F water is required to condense 1 lb of steam at 212°F and cool the condensate to 140°F). The principal disadvantage is the large water requirement. Depending on the nature of the condensate, odor in the wastewater can be offset by using treatment chemicals.

Direct-contact condensers involve the simultaneous transfer of heat and mass. Design procedures available for absorption, humidification, cooling towers, and the like may be applied with some modifications.

Surface Condensers Surface condensers (indirect-contact condensers) are used extensively in the chemical-process industry. They are employed in the air-pollution-equipment industry for recovery, control, and/or removal of trace impurities or contaminants. In the surface type, coolant does not contact the vapor condensate. There are various types of surface condensers, including the shell-and-tube, fin-fan, finned-hairpin, finned-tube-section, and tubular. The use of surface condensers has several advantages. Salable condensate can be recovered. If water is used for coolant, it can be reused, or the condenser may be air-cooled when water is not available. Also, surface condensers require less water and produce 10 to 20 times less condensate. Their disadvantage is that they are usually more expensive and require more maintenance than the contact type.

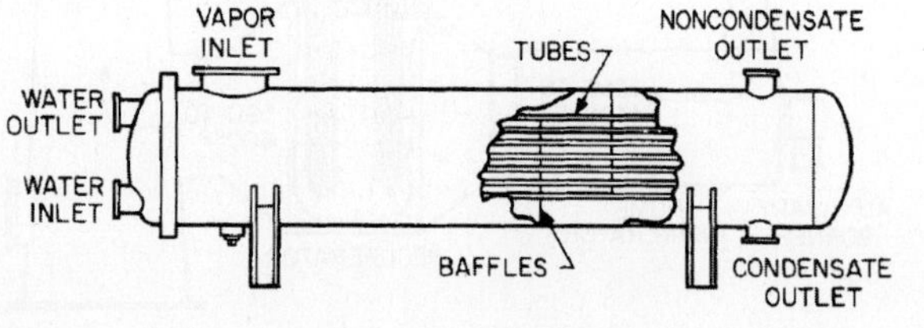

FIG. 22-16 Typical surface condenser (shell-and-tube).

Biological APC Technologies

General References

Biofiltration: Ottengraph, S. S. P., "Exhaust Gas Purification," in *Biotechnology*, vol. 8, ed. H. J. Rehn and G. Reed, VCH Verlagsgesellschaft, Weinheim, Germany, 1986; Ottengraph, S. S. P., "Biological Elimination of Volatile Xenobiotic Compounds in Biofilters," *Bioprocess Eng.* **1:** 61–69 (1986); Smith, F. L., et al., "Development and Demonstration of an Explicit Lumped Parameter Biofilter Model and Design Equation Incorporating Monod Kinetics," *J. Air & Waste Manage. Assoc.* **52**(2): 208–219 (2002); Margulis, L., *Microcosmos*, Summit Books, New York, 1986.

Henry's Law Constants: Solubilities subsection in Sec. 2 of this handbook; DIPPR 911; *ENVIRON*, EPCON International, Houston; "Environmental Simulation Program," OLI Systems, Inc., Morris Plains, N.J. (www.olisystems.com); Mackay, D., et al., "A Critical Review of Henry's Law Constants for Chemicals of Environmental Interest," *J. Phys. Chem. Ref. Data* **10**(4): 1175–1199 (1981); Tse, G., et al., "Infinite Dilution Activity Coefficients and Henry's Law Coefficients of Some Priority Water Pollutants Determined by Relative Gas Chromatographic Method," *Environ. Sci. Technol.* **26:** 2017–2022 (1992); Smith, F. L., and A. H. Harvey, "Avoid Common Pitfalls When Using Henry's Law," *Chem. Eng. Prog.* **103**(9), 2007.

Phase-Equilibrium Data: Shaw, D. G., and A. Maczynski, eds., IUPAC Solubility Data Series, Vol. 81: "Hydrocarbons in Water and Seawater—Revised and Updated," published in 12 parts in *J. Phys. Chem. Ref. Data*, 2005 and 2006; Gmehling, J., et al., "Vapor-Liquid Equilibrium Data Collection: Aqueous-Organic Systems," *DECHEMA Chemistry Data Series*, vol. I, part 1-1d, Schön & Wetzel GmbH, Frankfurt/Main, Germany, 1988.

Bioscrubbers, biotrickling filters, and biofilters are biological APC technologies that use enzymatic catalytic oxidation to completely break down (metabolize) biodegradable air pollutants to H_2O, CO_2, and salts. These pollutants, present in the airstream as vapors, can be either organic or inorganic (e.g., H_2S, NH_3, CO). The oxidation catalysts (enzymes) are produced and maintained within moist, living, active biomass, which operates near ambient temperature and pressure. In practice, these biological APC processes represent a subclass of gas adsorption processes, in which pollutant mass transfer from the gas phase to the liquid phase is enhanced by a fast chemical reaction in the liquid phase, which contains the biomass. Unlike incineration, biodegradation of compounds containing sulfur, nitrogen, and halogens produces benign by-products, biomass, and inorganic ions (SO_4^{-2}, NO_3^{-1}, Cl^{-1}, etc.), which require little or no subsequent treatment before their release into the environment. Biological processes can be particularly desirable because they neither transfer the identified pollutants into another phase (as with carbon adsorption or condensation) nor produce additional, collateral air pollution (such as SO_x, NO_x, CO, CO_2, and particulates) from fuel combustion as does incineration.

Biomass: The Community of Microorganisms* The microorganisms that make up the biomass in these devices are similar to those found in soil and in biological wastewater treatment plants: bacteria, often with smaller populations of fungi, ciliates, and other protists. Bacteria are extremely diverse in their biochemical degradation abilities. They all lack cellular nuclei. Individually, a bacterium has limited metabolic capabilities because a single bacterium has limited quantities of DNA. However, individual bacteria can modify their DNA by exchange and transfer with other bacteria present. Nevertheless, this limited DNA restricts the ability of one kind of bacterium to produce all the different enzymes needed to completely break down all available substrates. Consequently, at steady state, differing bacteria live together in mixed consortia (communities) such that, taken together, these consortia can produce all the enzymes needed to completely break down the available substrates to CO_2, H_2O, salts, and other components.

In bioscrubbers, biotrickling filters, and biofilters, the principal microbial populations metabolize the primary substrates, the biodegradable air pollutants. The secondary and generally smaller microbial populations assist the process of total breakdown by consuming bacterial metabolic intermediates from the primary substrates as well as other waste products produced by the other bacteria. In biotrickling filters and biofilters, the biomass exists within the airstream as a fixed biofilm, a thin layer of microbes growing on an inert attachment surface. An important advantage of an attached biofilm is that the microbes tend to become distributed along chemical gradients by depth within the biofilm. Often, the most rapidly growing microbes are concentrated near the outer surface, where both oxygen and the primary substrate are most abundant, and the slower-growing, secondary microbes are concentrated near the attachment surface. This configuration protects the secondary microbes with an outer layer of expendable microbes. It also provides a nearly infinite mean cell residence time for the often slow-growing but essential secondary microbes, needed for complete breakdown of the pollutants. These bacterial populations are consumed by still other predatory microbial populations, which decompose them, recycling their substances to the consortia. Acting together, these microbial populations finish the complete breakdown of the primary substrates.

*The contributions to this subsection by Prof. Lynn Margulis, University of Massachusetts, Amherst, are gratefully appreciated.

Biofiltration: Theoretical Considerations

Maximum Biofilter Inlet Pollutant Concentration For an aerobic fixed-film device, the local total flux for all pollutants into the biofilm must not exceed the local flux of oxygen into the biofilm, on a stoichiometric basis. Therefore, the maximum acceptable total pollutant concentration is fixed, at the biofilter inlet, by the low maximum oxygen solubility in the biofilm, at the biofilm interface. This oxygen concentration is about 8 mg/L (8 ppm) for air at ambient conditions. If the total pollutant concentration exceeds this maximum limit, aerobic biological conditions cannot be maintained throughout the biofilter, a condition which can result in biofilter failure.

For a single pollutant biofilter application, using the stoichiometric coefficients υ for complete pollutant oxidation, diffusivities *D*, and molecular weights (MW), the maximum acceptable inlet biofilm interface pollutant POL concentration S_{POL} can be estimated from the maximum available biofilm interface oxygen concentration S_{O_2}:

$$S_{POL} = \left[\frac{\nu_{POL}}{\nu_{O_2}}\right]\left[\frac{D_{O_2}}{D_{POL}}\right]\left[\frac{MW_{POL}}{MW_{O_2}}\right]S_{O_2} \qquad (22\text{-}12)$$

Pollutant Solubility At these very low biofilm interface concentrations, biodegradable pollutant solubility in water can be described by using Henry's law:

$$p_i = H_i C_i \qquad (22\text{-}13)$$

where *p* is the partial pressure of the solute *i* in the gas phase; *C* is the concentration of the solute in the aqueous liquid phase; and *H* is the proportionality constant, generally referred to as the Henry's law constant. (Henry's law constants are not true constants but rather functions of temperature.) Experimental data are needed to estimate *H*, however, because pollutant-water mixtures are highly nonideal. The best reference for these solubilities is corroborated data from a computerized data bank that specializes in environmental aqueous systems. Such data banks are especially helpful for calculating solubilities for uncommon pollutants of environmental interest, as well as for compounds that dissociate in water, that is, those that are acidic or basic.

Estimating Henry's Law Constants Henry's law constants can be estimated from extensive data published for solute-water LLE (liquid-liquid equilibria):

$$H_i = \frac{(1-x_{water})P_i^{sat}}{x_i} \qquad (22\text{-}14)$$

where x_{water} is the equilibrium mole fraction of water in the organic phase (often negligible), P_i^{sat} is the vapor pressure of the organic solute, and x_i is the equilibrium mole fraction of the solute in the water phase.

Extensive, reviewed tabulated pollutant data also exist (see subsection Solubilities in Sec. 2 of this handbook). Any single tabulated Henry's law constant *H* is published with its measurement temperature, typically 20°C or 25°C. The constant *H* for a subcritical pollutant (e.g., toluene) is proportional to its pure component vapor pressure P^{sat}, and therefore *H* is an exponential function of temperature *T*. Using Henry's and Raoult's laws, a pollutant's constant *H* for a given ambient design temperature can be estimated by using its tabulated value and its vapor pressure, by extrapolating ≤25°C over the ambient temperature range (4°C < *T* < 50°C), from T_1 to T_2:

$$H_2 = \frac{\gamma_2 P_2^{sat}}{\gamma_1 P_1^{sat}} H_1 \cong \frac{P_2^{sat}}{P_1^{sat}} H_1 \qquad (22\text{-}15a)$$

For a more soluble (or completely miscible) pollutant (e.g., ethanol), a Henry's law constant can be similarly estimated by using its tabulated infinite dilution activity coefficient γ_∞, measured at T_1, by extrapolation using P_2^{sat}, from T_1 to T_2:

$$H_2 = \gamma_{\infty 2} P_2^{sat} \cong \gamma_{\infty 1} P_2^{sat} \qquad (22\text{-}15b)$$

Such estimations for H_2 are single-point extrapolations of nonlinear functions, and therefore they should be used with caution. In both cases, estimation error is introduced due to the nonlinear temperature dependence of the activity coefficient.

Biofilter Kinetics and Design Model These biological systems are complex, so normal simplifying assumptions are approximate at best: biomass homogeneity, independent and uniform multisubstrate kinetics, etc. Nevertheless, for understanding the basics of these bioabsorbers, it is helpful to consider them in terms of a single substrate system, with uniform classic kinetics and mass transfer, and no liquid layer external to the fixed biofilm. Chemical and environmental engineers use different expressions to describe the kinetics of substrate (pollutant) utilization—the shifting order form and the Monod (or Michaelis-Menten) form:

$$\frac{d(S_f)}{dt}=\frac{k_1 S_f}{1+k_2 S_f}=\frac{kX_f S_f}{K_S+S_f} \tag{22-16}$$

where the variables are rate-limiting substrate concentration S_f, first-order reaction rate coefficient $k^1 = k_1$, zero-order rate coefficient $k^0 = k_1/k_2$, maximum utilization rate coefficient for the rate-limiting substrate k, density of the active substrate degraders X_f, and Monod half-velocity coefficient K_S.

Biofilters are chemically enhanced absorbers and are therefore mass transfer limited (see the subsection Absorption with Chemical Reaction in Sec. 14). The magnitudes for the Hatta [= Damkohler II = (Thiele modulus)^2] numbers are quite low, perhaps below 5. Nevertheless, for design simplicity, mass-transfer limitation is generally assumed to be in the liquid phase (the biofilm). For a single-component biofilter, the simplified biofilter model and design equation is

$$\tau=\frac{H}{k_L a_f}\ln\left|\frac{S_{g(\mathrm{IN})}}{S_{g(\mathrm{OUT})}}\right| \tag{22-17}$$

where the variables are the empty-bed residence time τ, a dimensionless variable H, liquid-phase local mass-transfer coefficient k_L, and air–liquid phase effective specific surface a_f. [The units used for the dimensionless H must be the same as those used for S_g (Eq. (22-17).] If the mass-transfer coefficient is for an ideal, single-component, fast first-order reaction system, for a biofilm having no external liquid layer of any significant thickness (i.e., not more than a moist surface), and all other variables are assumed to have constant magnitudes, then

$$k_L=\sqrt{k^1 D_f}=\sqrt{\frac{kX_f D_f}{K_S}} \tag{22-18a}$$

where D_f is the pollutant diffusion coefficient within the biofilm and k_L has a single magnitude, a function of temperature. For a single component following Monod reaction kinetics, k_L is defined as

$$k_L=\sqrt{\frac{2kX_f D_f}{S_i^2}\left(S_i-S_{\min}-K_S\ln\frac{K_S+S_i}{K_S+S_{\min}}\right)} \tag{22-18b}$$

where S_i and $S_{\min}$ are the pollutant biofilm concentrations, at the biofilm interface (maximum) and at the biofilm attachment surface (minimum), respectively. All other variables are assumed to have constant magnitudes throughout the biofilter. (For a "deep biofilm," $S_{\min}$ is assumed to approach the limit of zero.) Therefore, k_L is a function of the local pollutant concentration as well as a function of temperature. Note that Eq. (22-18b) predicts a lower value for k_L than does Eq. (22-18a), until the pollutant concentration approaches the limit of zero, where they are equal.

General Process Description For any APC design, it is difficult to accurately characterize variations of temperature, flow rate, and composition for the polluted source stream, especially during the engineering design of the upstream process source. For biological APC design, we have an imperfect scientific understanding of the behavior of biomass under long-term steady-state operation; more importantly, we lack complete knowledge of how the biomass and media respond to stream variations, especially significant step change variations. This APC field remains a technical art. The magnitude of $k_L a_f$ cannot be calculated with confidence for any media formulation from first principles, especially for multiple pollutants. Consequently, and especially for multiple components, responsible vendors can be expected to recommend field pilot testing for unfamiliar applications as an appropriate expense for achieving a minimum annualized system cost design. What follows is written to help chemical engineers to effectively discuss their project needs with responsible vendors of biological APC technologies and to help them understand what they receive from such vendors.

Most natural, and most artificial, organic compounds can be biodegraded. However, only some biodegradable compounds are candidates for economical biological APC, which typically means a residence time measured in minutes [Eq. (22-17)]. Although $k_L a_f$ is important, its practical range of magnitudes is modest, and candidate suitability for treatment is principally determined by its solubility, coupled with the ratio of the design inlet and outlet concentrations. Note, for instance, that economical designs are achievable for pollutants with low concentrations and low solubility (flavors and fragrances) or for high concentrations and high solubility (ethanol, butyric acid). In practice, compounds such as ethanol, acetone, formaldehyde, acetic acid, and diethylether are good candidates; BTEX, pinenes, and similarly soluble compounds are acceptable candidates; and largely nonpolar hydrocarbons such as alkanes (and highly halogenated compounds in general) are poor candidates. The most expensive media typically can offer the highest values for $k_L a_f$ and are generally justified through reduction of the required EBRT (reduced media volume) needed for a specific application.

An ID fan, which minimizes leakage of polluted air, is normally used to maintain airflow through the process and to overcome the total system pressure drop, typically measured in centimeters of water. The optimum design temperature is determined by the tradeoff between solubility, which decreases with temperature, and biological kinetics, which increases. Another optimization factor is how much heating or cooling is necessary for the inlet stream. Operation is generally controlled to within a narrow range of a couple of degrees. The selected design temperature may range from perhaps 20°C to 45°C, commonly 30°C to 35°C. Particulates or aerosols, including organic liquid droplets, should be removed (perhaps >15 μm). The inlet stream is usually prehumidified to 95 to nearly 100 percent relative humidity. Finally, temperature, particulate, and humidity control is often managed in a single step.

Biofilters Biofilters fall into two categories: soil bed biofilter (open bed) and artificial media biofilters (confined media).

Soil bed biofilters consist of an air distribution system of perforated piping buried under a layer of soil media, typically exposed outdoors. The treated air is emitted directly to atmosphere (Fig. 22-17). The soil media used can consist of a blend of natural soil components and other materials. This is a fixed-film bioreactor, with the biofilm attached to other inert soil materials. Soil media are biologically self-sustaining, hydrophilic, and self-buffering and often last for decades of use. Over a hundred are in operation. Although most soil beds are employed for odor control, any otherwise suitable candidate pollutant should be treatable in a soil bed. Soil bed operating temperature is maintained by inlet air temperature control, and soil biofilters are practical for any climate, including very cold conditions, such as Calgary, Alberta (Canada).

Artificial media biofilters employ a biologically active medium housed in a containment vessel (Fig. 22-18). The support media can consist of natural and synthetic materials, such as compost, peat, bark, or wood chips, and polymer foam, activated carbon, rubber tire waste, or ceramic pellets. The media are typically blended, wetted, and biologically seeded (inoculated) before installation. Inert materials, such as polystyrene beads, may be blended with the media to prevent compaction and pressure drop increase and to maintain biofilm-specific surface. Media moisture content can be a critical operating parameter. Secondary humidification is often used, via water spraying, which can be accurately controlled by weigh cell measurement of the total media bed mass. Media replacement may be required once or twice per decade. Depending on the media and the pollutants, nutrients or pH buffers may be applied with the water spray. Some biofilters are kept

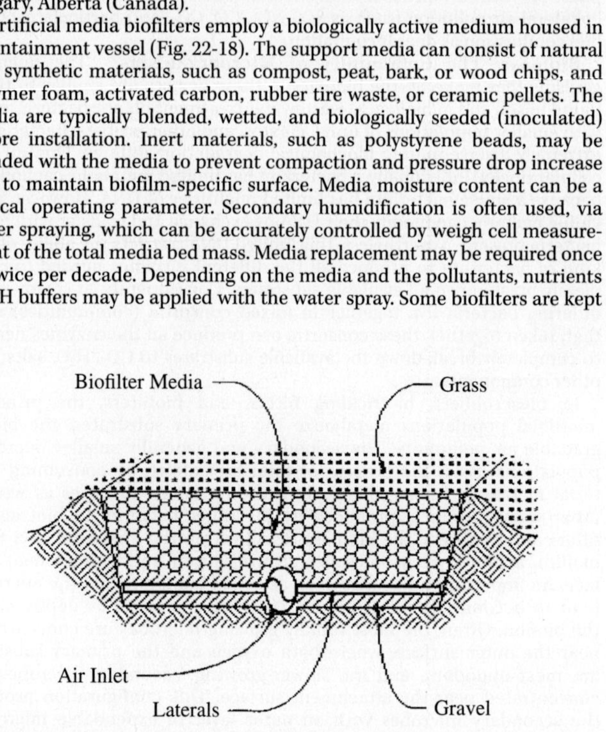

FIG. 22-17 Pretreated contaminated airflows into an inlet manifold, lateral piping, and into a continuous layer of gravel or crushed stone, providing even air distribution across the bottom of the media. Air is biofiltered, flowing upward through the biofilter media, exiting to atmosphere through the media surface or a grass cover. (*Courtesy of Bohn Biofilter, www.bohnbiofilter.com.*)

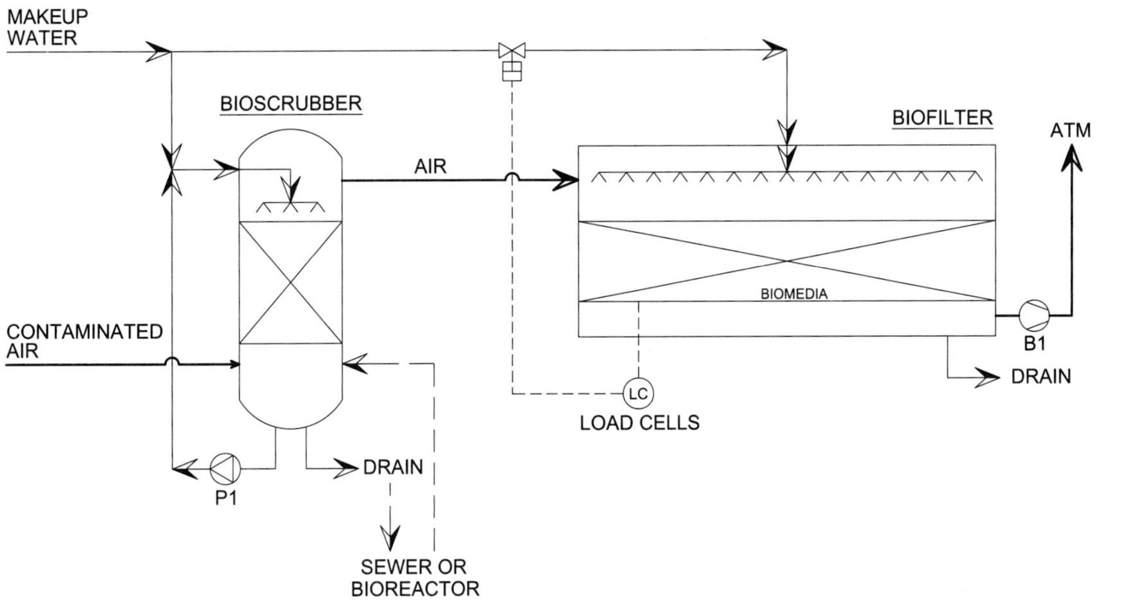

FIG. 22-18 Contaminated air is pretreated (shown: a countercurrent bioscrubber) before passing into the biofilter. Air is biofiltered, flowing downward through the media. Biomedia moisture content is accurately controlled via load-cell control of water sprays on media surface. (*Courtesy of Waterleau Biofilter–Bioton, Bel-Air, www.waterleau.com.*)

wet enough to drip, to purge soluble waste salts. Generally, artificial media biofilters can be designed for higher loadings than can soil bed biofilters. Upflow and downflow designs are practical.

Bioscrubbers Bioscrubbers combine physical absorption with biological degradation, using two separate unit operations. The pollutants are absorbed into water within the scrubber. This water is then directed to an external biological reactor, where the pollutants are degraded. The treated liquid is recycled to the bioscrubber. One configuration employs water and a fixed-film bioreactor, while a second configuration employs MLSS and a slurry reactor. In both cases, separating absorption from reaction permits scrubbing pollutant concentrations far above those that a fixed-film system would survive (due to O_2 limitation). Scrubbing can be designed to attenuate substantial and rapid composition change in the inlet, assisting the performance of a downstream biotreatment step. The scrubber can be multistaged (packed column, Fig. 22-18) or single-staged (a venturi scrubber or spray chamber, Fig. 22-19). Scrubbers with horizontal airflow are also available. Nutrients, pH buffers, and makeup water are added, and biomass and waste salts are purged, as needed. Bioscrubbing is commonly combined with temperature, particulate, and humidity control as feed pretreatment to biotrickling filters and biofilters. A combined multipollutant plot of

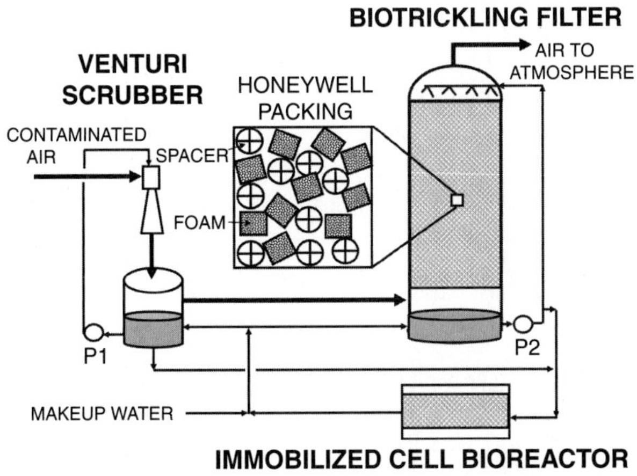

FIG. 22-19 Contaminated air is pretreated (shown: a Venturi scrubber) before passing into the biotrickling filter. Air is biofiltered, flowing downward through the media. (*Courtesy of Honeywell PAI—Biological Air Treatment System, www.honeywellpai.com.*)

$\log H = f(1/T)$ is a helpful design aid in the selection of the optimum stream composition for the scrubber effluent stream.

Biotrickling Filters Biotrickling filters are a hybrid of bioscrubbers and biofilters. A fixed biofilm (often a deep biofilm) is employed, typically attached to a synthetic medium containing blocks of polymer foam, lava rock, sintered ceramics, or other material (Fig. 22-19). Water permeability is preferred for the biofilm support media. Inert materials, such as Pall rings or other distillation-type packing materials, may be blended with the media to prevent compaction and increase of pressure drop and to maintain biofilm-specific surface. Water is recirculated through the filter. Although the water layer forms a resistance to pollutant mass transfer into the fixed biofilm, the water stream also absorbs pollutants that can be removed from the filter for separate biological treatment, as described previously for bioscrubbers. As for a bioscrubber, this hybrid can respond more effectively to rapid changes in the feed, and can survive higher inlet concentrations, than can a biofilter. This feature, in turn, can improve the performance of a downstream biotreatment step. Nutrients, pH buffers, and makeup water are added, and biomass and waste salts are purged, as needed. Upflow, downflow, and horizontal airflow designs are practical.

Use of Activated Carbon Activated carbon (AC) can be used to enhance these processes in at least three ways. Positioned in front of the process, AC can provide pollutant concentration capacitance, reducing concentration peak magnitudes and rapid rate of variation and providing minimum concentrations during periods of no or low source pollutant flow. Positioned behind the process, AC can increase the overall process elimination efficiency, as well as scavenging nonbiologically degradable compounds (e.g., polyhalogenated organics). Mixed with the media, AC is reported to combine these two functions, enhancing biological removal and protecting the system during periods of high pollutant inlet variation. Note that AC positioned in front can present a fire hazard for certain pollutants, such as acetone. AC positioned in back is exposed to very high humidity, which restricts adsorption capacity.

Relative Biofiltration Treatment Costs The true final annualized cost for treating a polluted airstream with any specific APC technology is complex and will be affected by many factors, including project-specific factors. Nevertheless, Fig. 22-20 can provide helpful guidance to the relationships between the probable final costs for treating a polluted airstream using different APC technologies.

Membrane Filtration Membrane systems have been used for several decades by the chemical process industries (CPIs) to separate colloidal and molecular slurries. Membrane filtration was not considered commercially viable pollution control technology until recently. This was due to fouling problems exhibited when handling process streams with high solids content. This changed with the advent of membranes composed of high-flux cellulose acetate. In Europe and Asia, membrane filtration systems have been used for several years, primarily for ethanol dewatering for synthetic fuel plants. Membrane systems were selected

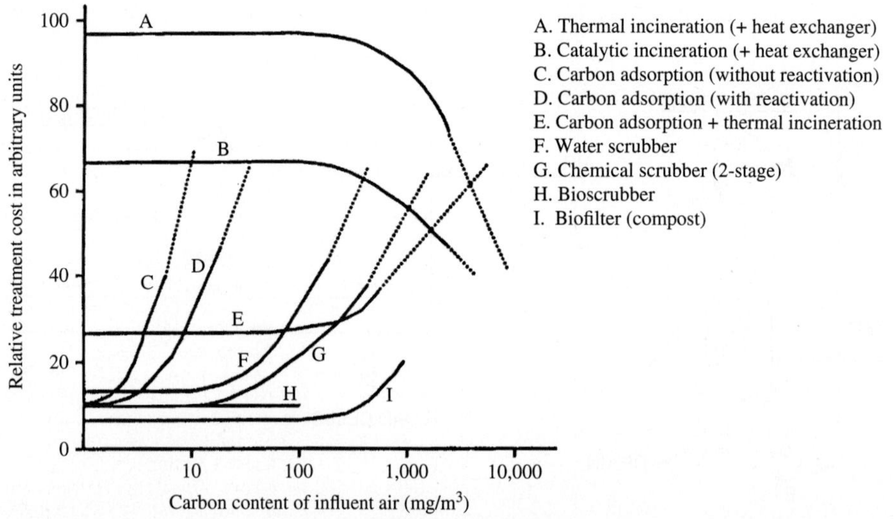

FIG. 22-20 Relative (and approximate) treatment costs for removing VOCs from air [Kosky, K. F., and C. R. Neff (1988). Innovative Biological Degradation System for Petroleum Hydrocarbons Treatment, in *Proceedings of the Conference on Petroleum Hydrocarbons and Organic Chemicals in Ground Water: Prevention, Detection and Restoration,* National Water Well Association, Dublin, Ohio]. Assumption: 40,000 m^3/h at 1982 price level. (Note: Current costs for chemical scrubbing can be two or more times greater than for biofiltration.)

in these cases for their (1) low energy utilization; (2) modular design; (3) low capital costs; (4) low maintenance; and (5) superior separations. In the United States, as EPA regulations become increasingly stringent, there has been a renewed interest in advanced membrane filtration systems. The driving factors are the membrane system's ability to operate in a pollution-free, closed-loop manner with minimum wastewater output. Capital and maintenance costs are low because of the small number of moving parts within membrane systems.

Process Descriptions Selectively permeable membranes have had an increasingly wide range of uses and configurations as the need for more advanced pollution control systems has arisen. There are four major types of membrane systems: (1) pervaporation; (2) reverse osmosis (RO); (3) gas absorption; and (4) gas adsorption. Only membrane pervaporation is commercialized at present.

Membrane Pervaporation Since 1987, membrane pervaporation has become widely accepted in the CPI as an effective means of separation and recovery of liquid-phase process streams. It is most commonly used to dehydrate liquid hydrocarbons to yield a high-purity ethanol, isopropanol, and ethylene glycol product. The method basically consists of a selectively permeable membrane layer separating a liquid feed stream and a gas phase permeate stream as shown in Fig. 22-21. The permeation rate and selectivity is governed by the physicochemical composition of the membrane. Pervaporation differs from reverse osmosis systems in that the permeate rate is not a function of osmotic pressure, since the permeate is maintained at saturation pressure (Anastas, P., and J. Warner, *Green Chemistry: Theory and Practice,* Oxford University Press, New York, 1998).

Three general process groups are commonly used when describing pervaporation: (1) water removal from organics; (2) organic removal from water (solvent recovery); and (3) organic/organic separation. Organic/organic separations are very uncommon and therefore will not be discussed further.

Ethanol Dehydration The membrane-pervaporation process for dehydrating ethanol was first developed by GFT in West Germany in the mid-1970s, with the first commercial units being installed in Brazil and the Philippines. At both sites, the pervaporation unit was coupled to a continuous sugarcane fermentation process that produced ethanol at concentrations up to 96 percent after vacuum pervaporation. The key advantages of the GFT process are (1) no additive chemicals are required; (2) the process is skid-mounted (low capital costs and small footprint); (3) and there is a low energy demand. The low energy requirement is achieved because only a small fraction of the water is actually vaporized, and the required permeation driving force is provided by only a small vacuum pump. The basic ethanol dehydration process schematic is shown in Fig. 22-22. A key advantage

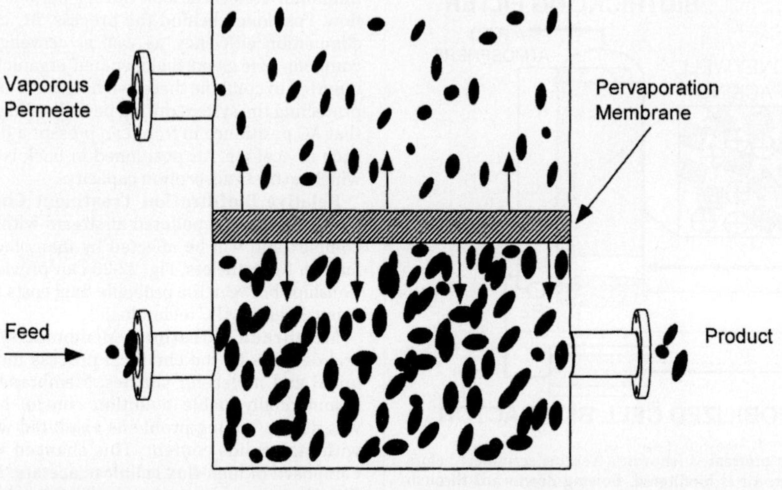

FIG. 22-21 Pervaporation of gas from liquid feed across membrane to vaporous permeate. (*Source: Redrawn from Anastas, P., and J. Warner, Green Chemistry: Theory and Practice, Oxford University Press, New York, 1998*)

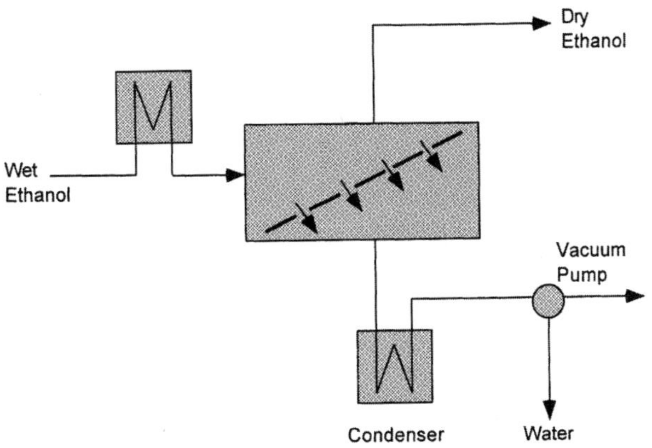

FIG. 22-22 Ethanol dehydration using pervaporation membrane. (*Source: Redrawn from Anastas, P., and J. Warner, Green Chemistry: Theory and Practice, Oxford University Press, New York, 1998.*)

of all pervaporation processes is that vapor-liquid equilibria and possible resulting azeotropic effects are irrelevant [see Fig. 22-23 and Anastas and Warner (1998)].

Solvent Recovery The largest current industrial use of pervaporation is the treatment of mixed organic process streams that have become contaminated with small (10 percent) quantities of water. Pervaporation becomes very attractive when dehydrating streams down to less than 1 percent water. The advantages result from the small operating costs relative to distillation and adsorption. Also, distillation is often impossible, since azeotropes commonly form in multicomponent organic/water mixtures.

Pervaporation occurs in three basic steps:

1. Preferential sorption of chemical species
2. Diffusion of chemical species through the membrane
3. Desorption of chemical species from the membrane

Steps 1 and 2 are controlled by the specific polymer chemistry and its designed interaction with the liquid phase. The last step, consisting of evaporation of the chemical species, is considered to be a fast, nonselective process. Step 2 is the rate-limiting step. The development by the membrane manufacturers of highly selective, highly permeable composite membranes that resist fouling from solids has subsequently been the key to the commercialization of pervaporation systems. The membrane

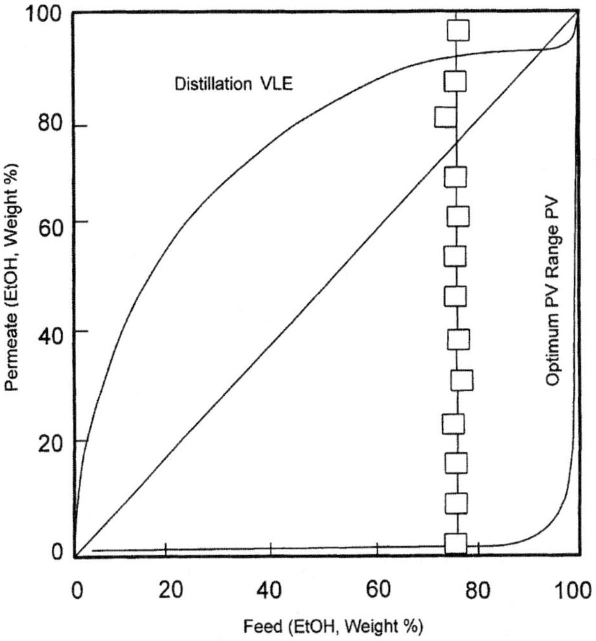

FIG. 22-23 Comparison of two types of pervaporation membranes to distillation of ethanol-water mixtures. (*Source: Redrawn from Anastas, P., and J. Warner, Green Chemistry: Theory and Practice, Oxford University Press, New York, 1998.*)

composition and structure are designed in layers, with each layer fulfilling a specific requirement. Using membrane dehydration as an example, a membrane filter would be composed of a support layer of nonwoven porous polyester below a layer of polyacrylonitrile (PAN) or polysulfone ultrafiltration membrane and a layer of 0.1-mm-thick cross-linked polyacrylate, polyvinyl alcohol (PVA). Other separation membranes generally use the same two sublayers. The top layer is interchanged according to the selectivity desired.

Emerging Membrane Control Technologies The recent improvements in membrane technology have spawned several potentially commercial membrane filtration uses.

Reverse osmosis (RO) membranes. A type of membrane system for treating oily wastewater is currently undergoing commercialization by Bend Research, Inc. The system uses a tube-side feed module that yields high fluxes while being able to handle high-solids-content waste streams [Anastas, P., and J. Zimmerman, "Design through the Twelve Principles of Green Engineering," *Environ. Sci. Technol.* **37:** 94A–101A (2003)]. Another type of reverse osmosis technique is being designed to yield ultrapurified HF recovered from spent etching solutions. It is estimated that 20,000 tons of spent solution is annually generated in the United States and that using membrane RO could save 1 million bbl/yr of oil.

In situ filter membranes. In situ membranes are being fitted into incinerator flue-gas stacks in an attempt to reduce hydrocarbon emissions. Two types of commercially available gas separation membranes are being studied: (1) flat cellulose acetate sheets; and (2) hollow-tube fiber modules made of polyamides.

Vibratory shear-enhanced membranes. The vibratory shear-enhancing process (VSEP) is just starting commercialization by Logic International, Emeryville, Calif. It employs the use of intense sinusoidal shear waves to ensure that the membrane surfaces remain active and clean of solid matter. The application of this technology would be in the purification of wastewater.

Vapor permeation. Vapor permeation is similar to vapor pervaporation except that the feed stream for permeation is a gas. The future commercial viability of this process is based on energy and capital costs savings derived from the feed already being in the vapor phase, as in fractional distillation, so no additional heat input would be required. Its foreseen application areas would be the organics recovery from solvent-laden vapors and pollution treatment. One commercial unit was installed in Germany in 1989 (Pearce, D., A. Mankandya, and E. Barbier, *Blueprint for a Green Economy*, Earthscan, London, 1989).

Selective Catalytic Reduction of Nitrogen Oxides The traditional approach to reducing ambient ozone concentrations has been to reduce VOC emissions, an ozone precursor. In many areas, it has now been recognized that elimination of persistent exceedances of the National Ambient Air Quality Standard for ozone may require more attention to reductions in the other ingredients in ozone formation, nitrogen oxides (NO_x). In such areas, ozone concentrations are controlled by NO_x rather than VOC emissions.

Selective catalytic reduction (SCR) has been used to control NO_x emissions from utility boilers in Europe and Japan for over a decade. Applications of SCR to control process NO_x emissions in the chemical industry are becoming increasingly common. A typical SCR system is shown in Fig. 22-24.

In a chemical industry application, NO_x–laden fumes are preheated by effluent from the catalyst vessel in the feed/effluent heat exchanger and then heated by a gas- or oil-fired heater to over 600°F. In an electric utility application, the flue gas is taken from the boilers' economizer at a temperature of 600°F to 800°F for treatment. A controlled quantity of ammonia is injected into the gas stream before it is passed through a metal oxide, zeolite, or promoted zeolite catalyst bed. The NO_x is reduced to nitrogen and water in the presence of ammonia in accordance with the following exothermic reactions:

$$6NO + 4NH_3 \rightarrow 5N_2 + 6H_2O$$

$$4NO + 4NH_3 + O_2 \rightarrow 4N_2 + 6H_2O$$

$$2NO_2 + 4NH_3 + O_2 \rightarrow 3N_2 + 6H_2O$$

$$NO + NO_2 + 2NH_3 \rightarrow 2N_2 + 3H_2O$$

NO_x analyzers at the preheater inlet and catalyst vessel outlet monitor NO_x concentrations and control the ammonia feed rate. In the chemical/petroleum industry, the effluent gives up much of its heat to the incoming gas in the feed/effluent exchanger and is discharged at about 350°F. In the utility industry, the gas stream exits the SCR reactor and enters into a combustion air preheater and is cooled to 300°F to 400°F before being sent to particulate and acid gas removal equipment.

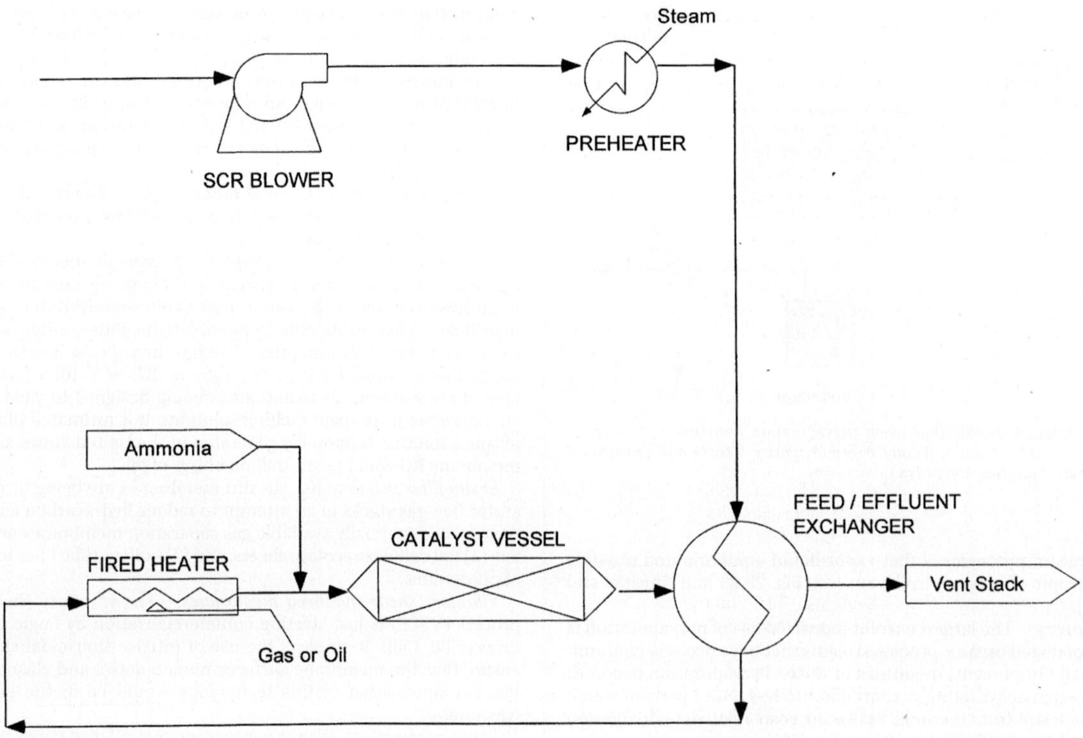

FIG. 22-24 Selective catalytic reduction of nitrogen oxides.

SOURCE CONTROL OF PARTICULATE EMISSIONS

There are four conventional types of equipment used for the control of particulate emissions:

1. Mechanical collectors
2. Wet scrubbers
3. Electrostatic precipitators
4. Fabric filters

Each is discussed in Sec. 17 of this handbook under Gas-Solids Separations. The effectiveness of conventional air pollution control equipment for particulate removal is compared in Fig. 22-25. These fractional efficiency curves indicate that the equipment is least efficient in removing particulates in the 0.1- to 1.0-μm range. For wet scrubbers and fabric filters, the very small particulates (0.1 μm) can be efficiently removed by brownian diffusion. The smaller the particulates, the more intense their brownian motion, and the easier their collection by diffusion forces. Larger particulates (>1 μm) are collected principally by impaction, and removal efficiency increases with particulate size. The minimum in the fractional efficiency curve for scrubbers and filters occurs in the transition range between removal by brownian diffusion and removal by impaction. The introduction of PTFE membranes can be used to greatly improve both the filtration efficiency and the cake release ability of a filter bag and to address the collection of particles in the 0.1 to 1.0 μm range.

A somewhat similar situation exists for electrostatic precipitators (ESPs). Particulates larger than about 1 μm have high mobilities because they are highly charged. Those smaller than a few tenths of a micrometer can achieve moderate mobilities with even a small charge because of aerodynamic slip. A minimum in collection efficiency usually occurs in the transition range between 0.1 and 1.0 μm. The situation is further complicated because not all particulates smaller than about 0.1 μm acquire charges in an ion field. Hence, the efficiency of removal of very small particulates decreases after reaching a maximum in the submicrometer range. Advances in ESP technology, including wide plate spacing (up to 18 in plate-to-plate distance), advanced discharge electrodes providing more electron generation and greater reliability, and high-frequency transformer/rectifier (T/R) sets capable of putting more power into a gas stream, have enabled ESPs to achieve the same outlet dust concentrations previously only seen with fabric filters.

The selection of the optimum type of particulate collection device (e.g., ESP or fabric filter baghouse) is often not obvious without conducting a site-specific economic evaluation. This situation has been brought about by both the recent reductions in the allowable emissions levels and advancements with fabric filter and ESP technologies. Such technoeconomic evaluations can result in

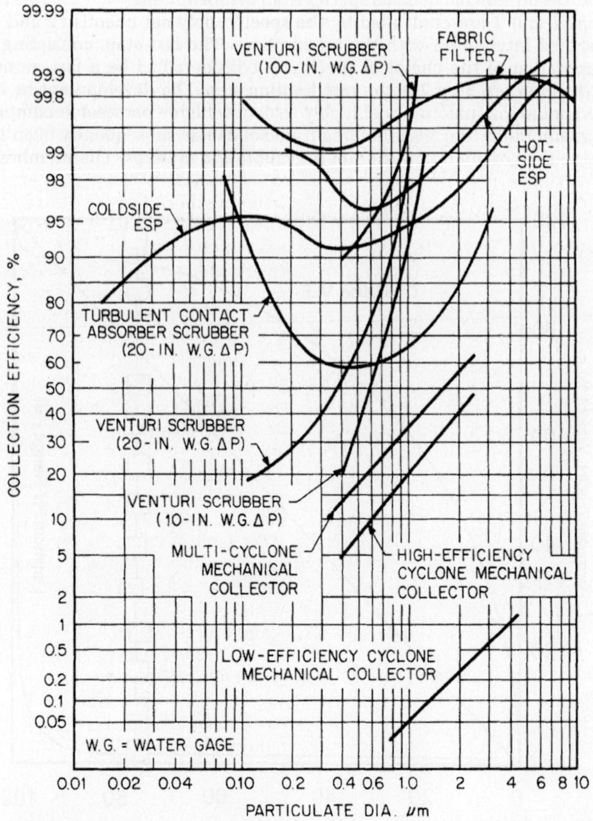

FIG. 22-25 Fractional efficiency curves for conventional air-pollution-control devices. [*Chem. Eng.* 87(13): 83 (June 30, 1980).]

application and even site-specific differences in the final optimum choice (see *Precip Newsletter,* 220, June 1994 and *Fabric Filter Newsletter,* 223, June 1994).

Improvements in existing control technology for fine particulates and the development of advanced techniques are top-priority research goals. Conventional control devices have certain limitations. Precipitators, for example, are limited by the magnitude of charge on the particulate, the electric field, and dust reentrainment. Also, the resistivity of the particulate material may adversely affect both charge and electric field. Advances to overcome resistivity issues and extend the performance of precipitators not limited by resistivity (see Buonicore and Theodore, "Control Technology for Fine Particulate Emissions," DOE Rep. ANL/ECT-5, Argonne National Laboratory, Argonne, Illinois, October 1978) have included flue gas and fuel additives, advanced discharge electrodes, and changes in rapper design. Recent design developments with the potential to improve precipitator performance also include pulse energization, wide plate spacing, and high-frequency T/R [transformer/rectifier] sets [Balakrishnan et al., "Emerging Technologies for Air Pollution Control," *Pollut. Eng.* **11:** 28–32 (Nov. 1979); "Pulse Energization," *Environ. Sci. Technol.* **13**(9): 1044 (1974); Midkaff, "Change in Precipitator Design Expected to Help Plants Meet Clean Air Laws," *Power* **126**(10): 79 (1979); Robert A. Mastropietro, "Impact of ESP Performance on $PM_{2.5}$," European Particulate Control Users Group Meeting, Pisa, Italy, Nov. 7, 2000.]

Fabric filters are limited by physical size and bag-life considerations. Some sacrifices in efficiency might be tolerated if higher air/cloth ratios could be achieved without reducing bag life (improved pulse-jet systems). Improvements in fabric filtration may also be possible by enhancing electrostatic effects that may contribute to rapid formation of a filter cake after cleaning [J. C. Mycock, J. Turner, and J. Farmer, "Baghouse Filtration Products Verification Testing, How It Benefits the Boiler Baghouse Operator," Council of Industrial Boiler Owners Conference, May 6, 2002; C. Jean Bustard et al. "Long Term Evaluation of Activated Carbon Injection for Mercury Control Upstream of a COHPAC Fabric Filter," Air Quality IV, Arlington, Va., Sept. 22, 2003].

Nanofiber filters hold great potential for cost-effective fine particle control. Cartridge filters built with nanotechnology have been reported to improve filter performance. In 2008 the most common types of media for cartridge filters were (1) cellulose, (2) cellulose blended with a synthetic fiber, and (3) cellulose with a nanofiber layer, that is, nanofiber filters. Given the fact that most nanofiber filters are relatively inexpensive, they have become increasingly popular. Their tiny, uniform, pore sizes make these filters more efficient and easier to clean than other types of cartridge filters. Lab tests have shown that nanofiber filters can last up to twice as long as standard cellulose and blended cellulose filters. This is probably a result of nanofiber filters requiring fewer jet pulses per cleaning [McKenna, J., J. Turner, and J. McKenna, Jr., *Fine Particle (2.5 Microns) Emissions: Regulations, Measurement, and Control,* Wiley, New York, 2008].

It appears that a fair amount of research and development is under way all around the world regarding nanofiber filtration with potential for fundamental improvement for fine particle emission control [McKenna, J., J. Turner, and J. McKenna, Jr., *Fine Particle (2.5 Microns) Emissions,* Wiley, New York, 2008]. The EPA is currently sponsoring filter R&D aimed at retrofitting existing baghouses with nanofiber filters to replace conventional bags [*Fabric Filter Newsletter,* Issue 485, The McIlvaine Company, March 2016 and the EPA Newsroom (www.epa.gov/newsreleases/epa-funds-small-businesses-develop-environmental-technologies)].

Wet scrubber technology is limited by scaling and fouling, overall reliability, and energy consumption. The use of supplementary forces acting on particulates to cause them to grow or otherwise be more easily collected at lower pressure drops is being closely investigated. The development of electrostatic and flux-force-condensation scrubbers is a step in this direction, and it has been used for scrubbers that treat gases from hazardous waste incinerators.

The electrostatic effect can be incorporated into wet scrubbing by charging the particulates or the scrubbing-liquor droplets or both. Electrostatic scrubbers may be capable of achieving the same efficiency for fine-particulate removal as is achieved by high-energy scrubbers, but at substantially lower power input. The major drawbacks are increased maintenance of electrical equipment and higher capital cost.

Flux-force-condensation scrubbers combine the effects of flux force (diffusiophoresis and thermophoresis) and water-vapor condensation. These scrubbers contact hot, humid gas with subcooled liquid, or they inject steam into saturated gas, and they have demonstrated that a number of these novel devices can remove fine particulates (see Fig. 22-26). Although limited in terms of commercialization, these systems may find application in many industries.

EMISSIONS MEASUREMENT

Introduction An accurate quantitative analysis of the discharge of pollutants from a process must be determined before the design or selection of control equipment. If the unit is properly engineered by using the emission

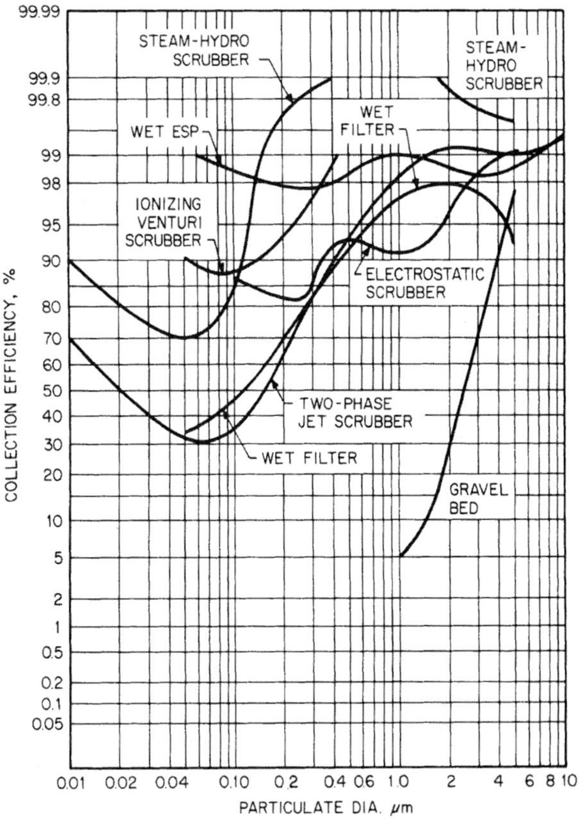

FIG. 22-26 Fractional efficiency curves for novel air-pollution-control devices. [*Chem. Eng.* **87**(13): 85 (June 30, 1980).]

data as input to the control device and the code requirements as maximum-effluent limitations, most pollutants can be successfully controlled.

Sampling is the keystone of source analysis. Sampling methods and tools vary in their complexity according to the specific task; therefore, a degree of both technical knowledge and common sense is needed to design a sampling function. Sampling is done to measure quantities or concentrations of pollutants in effluent gas streams, to measure the efficiency of a pollution-abatement device, to guide the designer of pollution control equipment and facilities, or to appraise contamination from a process or a source. A complete measurement requires a determination of the concentration and contaminant characteristics as well as the associated gas flow. Most statutory limitations require mass rates of emissions; both concentration and volumetric-flow-rate data are therefore required.

The selection of a sampling site and the number of sampling points required are based on attempts to get representative samples. To accomplish this, the sampling site should be at least eight stack or duct diameters downstream and two diameters upstream from any flow disturbance, such as a bend, expansion, contraction, valve, fitting, or visible flame.

Once the sampling location has been decided on, the flue cross section is laid out in a number of equal areas, the center of each being the point where the measurement is to be taken. For rectangular stacks, the cross section is divided into equal areas of the same shape, and the traverse points are located at the center of each equal area, as shown in Fig. 22-27. The ratio of length to width of each elemental area should be selected. For circular stacks, the cross section is divided into equal annular areas, and the traverse points are located at the centroid of each area. The location of the traverse points as a percentage of diameter from the inside wall to the traverse point for circular-stack sampling is given in Table 22-24. The number of traverse points needed on each of two perpendiculars for a particular stack may be estimated from Fig. 22-28.

Once these traverse points have been determined, velocity measurements are made to determine gas flow. The stack-gas velocity is usually determined by means of a pitot tube and differential-pressure gauge. When velocities are very low [less than 3 m/s (10 ft/s)] and when great accuracy is not required, an anemometer may be used. For gases moving in small pipes at relatively high velocities or pressures, orifice-disk meters or venturi meters may be used. These are valuable as continuous or permanent measuring devices.

TABLE 22-24 Location of Traverse Points in Circular Stacks

(Percent of stack diameter from inside wall to traverse point)

	Number of traverse points on a diameter											
Traverse point number on a diameter	2	4	6	8	10	12	14	18	20	22	24
1	14.6	6.7	4.4	3.2	2.6	2.1	1.8	1.4	1.3	1.1	1.1
2	85.4	25.0	14.6	10.5	8.2	6.7	5.7	4.4	3.9	3.5	3.2
3		75.0	29.6	19.4	14.6	11.8	9.9	7.5	6.7	6.0	5.5
4		93.3	70.4	32.3	22.6	17.7	14.6	10.9	9.7	8.7	7.9
5			85.4	67.7	34.2	25.0	20.1	14.6	12.9	11.6	10.5
6			95.6	80.6	65.8	35.6	26.9	18.8	16.5	14.6	13.2
7				89.5	77.4	64.4	36.6	23.6	20.4	18.0	16.1
8				96.8	85.4	75.0	63.4	29.6	25.0	21.8	19.4
9					91.8	82.3	73.1	38.2	30.6	26.2	23.0
10					97.4	88.2	79.9	61.8	38.8	31.5	27.2
11						93.3	85.4	70.4	61.2	39.3	32.3
12						97.9	90.1	76.4	69.4	60.7	39.8
13							94.3	81.2	75.0	68.5	60.2
14							98.2	85.4	79.6	73.8	67.7
15								89.1	83.5	78.5	72.8
16								92.5	87.1	82.0	77.0
17								95.6	90.3	85.4	80.8
18								98.6	93.3	88.4	83.9
19									96.1	91.3	86.8
20									96.7	94.0	89.5
21										96.5	92.1
22										98.9	94.5
23											96.8
24											96.9

Once a flow profile has been established, sampling strategy can be considered. Since sampling collection can be simplified and greatly reduced depending on flow characteristics, it is best to complete the flow-profile measurement before sampling or measuring pollutant concentrations.

Sampling Methodology The following subsections review the methods specified for sampling commonly regulated pollutants as well as sampling for more exotic volatile and semivolatile organic compounds. In all sampling procedures, the main concern is to obtain a representative sample; the U.S. EPA has published reference sampling methods for measuring emissions of specific pollutants so that uniform procedures can be applied in testing to obtain a representative sample. Table 22-25 provides a list of some of the most commonly employed test methods. A complete listing of all EPA test methods can be found by going online to the U.S. EPA Technology Transfer Network Emission Measurement Center at www.epa.gov/ttn/emc/promgate.html. These are the CFR-promulgated test methods published in the *Federal Register* and are the federal government's official legal versions. The test methods reviewed in the following subsections address measuring the emissions of the following pollutants: particulate matter, sulfur dioxide, nitrogen oxides, carbon monoxide, fluorides, hydrogen chloride, total gaseous organics, multiple metals, volatile organic compounds, and semivolatile organic compounds.

Each sampling method requires the use of complex sampling equipment that must be calibrated and operated in accordance with specified reference methods. Additionally, the process or source that is being tested must be operated in a specific manner, usually at rated capacity, under normal procedures.

Velocity and Volumetric Flow Rate The U.S. EPA has published Method 2 as a reference method for determining stack-gas velocity and volumetric flow rate. At several designated sampling points, which represent equal portions of the stack volume (areas in the stack), the velocity and temperature are measured with instrumentation shown in Fig. 22-29.

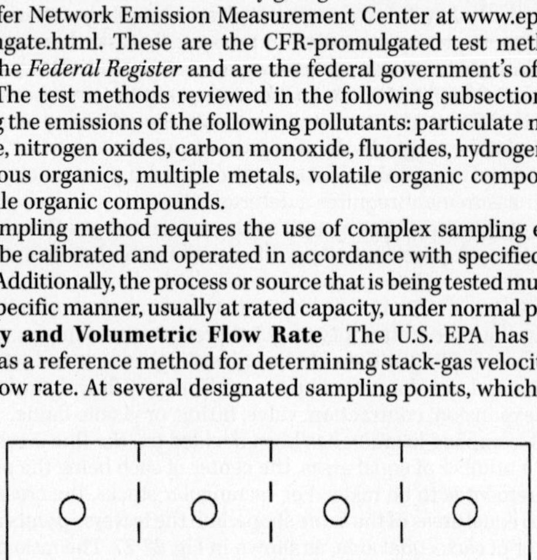

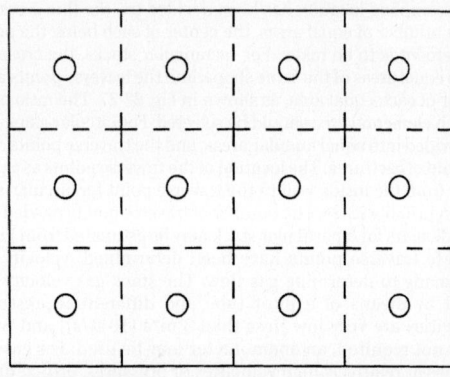

FIG. 22-27 Example showing rectangular stack cross section divided into 12 equal areas, with a traverse point at centroid of each area.

Measurements to determine volumetric flow rate usually require approximately 30 to 60 minutes, depending on the size of the stack being sampled. Since sampling rates depend on stack-gas velocity, a preliminary velocity check is usually made before testing for pollutants to aid in selecting the proper equipment and in determining the approximate sampling rate for the test.

The volumetric flow rate determined by this method is usually within ±10 percent of the true volumetric flow rate. If it is suspected that streamline flow does not exist at the sampling point, then the use of EPA Method 2F ("Determination of Stack Gas Velocity with 3 Dimensional Probe") is recommended for a more accurate measurement.

Sources of Methods and Information For the person trying to obtain current information or to enter into the field of source measurements, several particularly helpful information sources are available. The U.S. EPA methods fall into two groups—those used by EPA's Office of Air Quality Planning and Standards (OAQPS) and those used by EPA's Office of Solid Waste (OSW).

The Emission Measurement Technical Information Center (EMTIC) at Research Triangle Park, N.C., is supported by EPA's Office of Air Quality Planning and Standards. Perhaps the most efficient of several available forms of assistance is the EMTIC Bulletin Board System (BBS). Test methods are included along with announcements, utility programs, miscellaneous documents, and other information. The EMTIC/BBS may be reached on the Internet at https://www.epa.gov/technical-air-pollution-resources. An EMC representative can be found through their support directory at https://www.epa.gov/emc/emc-methods-support. EMC sponsors workshops and training courses jointly with EPA's Air Pollution Training Institute. Training videotapes, a newsletter, and other mailings are also available from EMC at https://www.epa.gov/emc/emc-instructional-material.

An excellent source for information concerning OSW's SW-846 Methods is the Methods Information Communication Exchange (MICE). MICE can be reached on the Internet at mice@lan828.ehsg.saic.com. A telephone call to the MICE line, at 703-821-4690, will put the information seeker in touch with an automated information service or with a live representative. Although the function of MICE is to provide information, they will usually send copies of up to three methods. They will not provide copies of the entire SW-846 Methods Manual. The SW-846 Methods Manual may be obtained on CD-ROM or hard copy from the National Technical Information Service (NTIS). The NTIS order number for the CD-ROM, which includes the third edition and updates 1–3, is PB97-501928INQ. NTIS has a web site at http://www.ntis.gov and may also be reached by telephone at 703-487-4650. SW-846 may also be obtained from the Government Printing Office (GPO). Ordering information for GPO is *Test Methods for Evaluating Solid Waste, Physical/Chemical Methods, SW-846 Manual,* 3d ed., Document No. 955-001-000001. Available from Superintendent of Documents, U.S. Government Printing Office, Washington, D.C., November 1986.

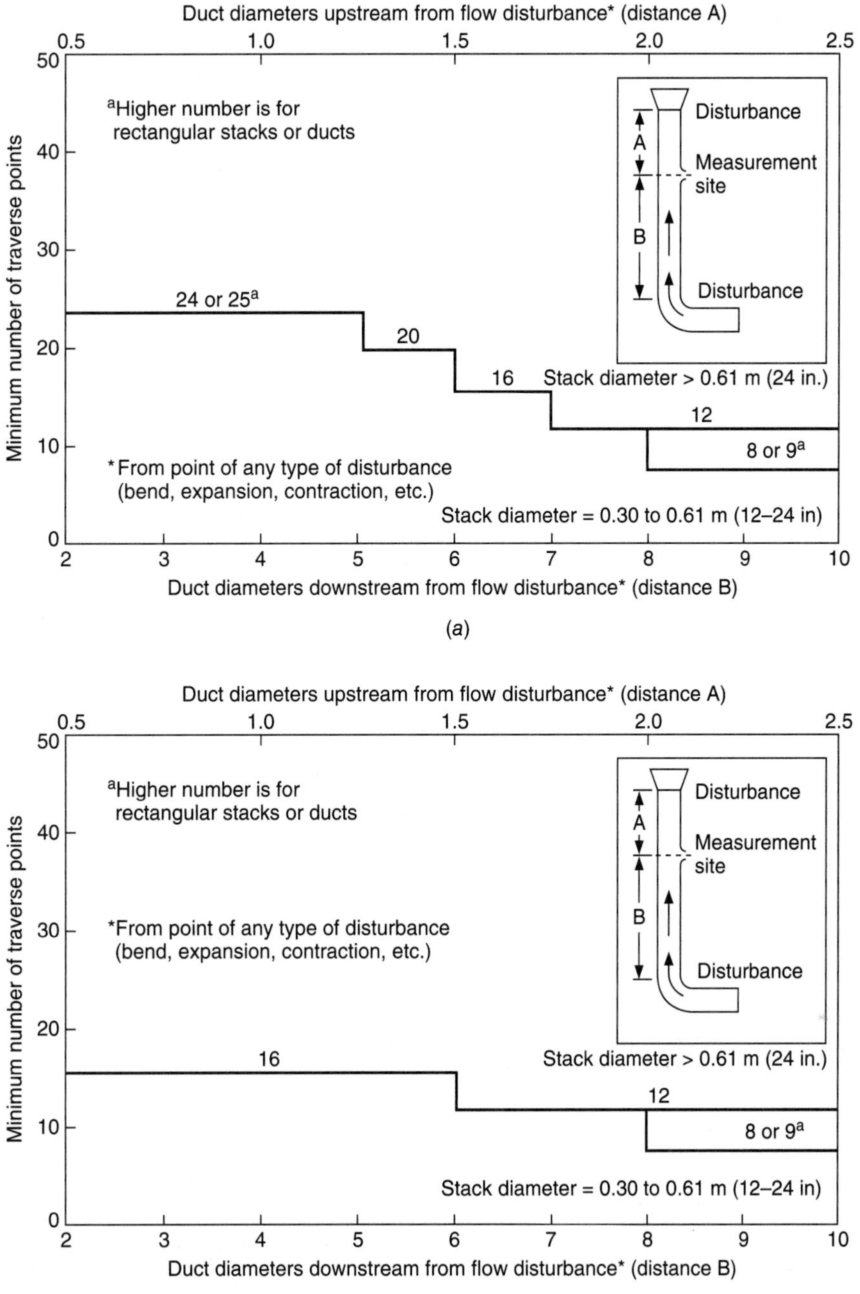

FIG. 22-28 Minimum number of traverse points for (*a*) particulate traverse and (*b*) velocity (nonparticulate) traverse.

The full document is available from the U.S. Government Printing Office, telephone 202-783-3238. GPO also has a web site at http://www.access.gpo.gov.

For more information, or for copies of the California Environmental Protection Agency, Air Resources Board Methods (a.k.a. CARB methods), contact http://www.arb.ca.gov/testmeth/testmeth.htm or telephone the Engineering and Laboratory Branch at 916-263-1630.

EPA reports may be ordered from NTIS at the web site or telephone number previously given.

INDUSTRIAL WASTEWATER MANAGEMENT

Further Reading: Eckenfelder, W. W., *Industrial Water Pollution Control,* 2d ed., McGraw-Hill, New York, 1989; Metcalf & Eddy, Inc., *Wastewater Engineering, Treatment and Resource Recovery,* 5th ed., McGraw-Hill, New York, 2014; Nemerow, N. L., and A. Dasgupta, *Industrial and Hazardous Waste Treatment,* Van Nostrand Reinhold, New York, 1991; Water Environment Federation, *Short Cut Nitrogen Removal: Nitrite Shunt and Deammonification,* WEF, Alexandria, Va., 2015.

INTRODUCTION

All industrial operations produce some wastewaters that must be returned to the environment. Wastewaters can be classified as (1) domestic wastewaters, (2) process wastewaters, and (3) cooling wastewaters. Domestic wastewaters are produced by plant workers, shower facilities, and cafeterias. Process wastewaters result from spills, leaks, product washing, and noncooling

TABLE 22-25 Selected EPA Test Methods

Commonly Used in Stationary Source Compliance Tests

Method	Test parameter(s)
Office of Air Quality Planning and Standards (OAQPS)	
Methods 1–4	Test location, volumetric flow, gas composition, moisture content
Method 5	Particulate matter
Method 6, 6c	Sulfur dioxide
Method 7, 7e	Nitrogen oxides
Method 9	Opacity
Method 10	Carbon monoxide
Method 13	Total fluoride
Method 23	Dioxins and furans
Method 25	VOCs
Method 26	Halogens, halides
Method 29	Metals
Method 30A, 30B	Mercury
Method 201	Particulate PM_{10}
Method 202	Condensible particulate matter
Conditional Method 13	Sulfuric acid vapor or mist
Conditional Method 27	Ammonia
Office of Solid Waste (OSW)	
Method 0010	Modified Method 5, semivolatile organics
Method 0030	Volatile organics

processes. Cooling wastewaters are the result of various cooling processes and can be once-pass systems or multiple-recycle cooling systems. Once-pass cooling systems employ large volumes of cooling waters that are used once and returned to the environment. Multiple-recycle cooling systems have various types of cooling towers to return excess heat to the environment, and they require periodic blowdown to prevent excess buildup of salts.

Domestic wastewaters are generally handled by the normal sanitary-sewerage system to prevent the spread of pathogenic microorganisms that might cause disease. Normally, process wastewaters do not raise the threat of pathogenic microorganisms, but they do have the potential to damage the environment through either direct or indirect chemical reactions. Some process wastes are readily biodegraded and create an immediate oxygen demand. Other process wastes are toxic and represent a direct health hazard to living organisms. Cooling wastewaters are the least dangerous, but they can contain process wastewaters as a result of leaks in the cooling systems. Recycle cooling systems tend to concentrate both inorganic and organic contaminants to a point at which environmental damage can be created. In general, wastewater characteristics such as organics, solids, nitrogen, phosphorous, heavy metals, and whole effluent toxicity (WET) have been the primary focus of wastewater treatment.

Recently, concern for subtle aspects of environmental damage has started to take precedence over other types of damage. We are discovering that the presence of substances in water at concentrations far below those that will produce overt toxicity or excessive reduction of dissolved oxygen levels can have a major impact by changing the balance of organisms in the aquatic ecosystem. This is beginning to have an effect on water quality standards and, consequently, allowable discharges. Unfortunately, we do not yet know enough to create definitive standards for these subtle effects.

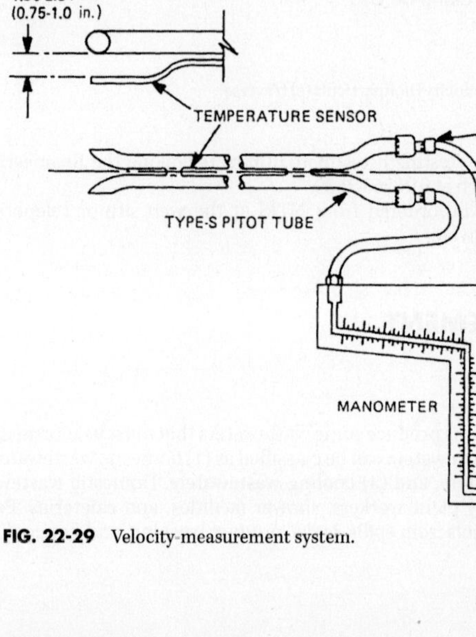

FIG. 22-29 Velocity-measurement system.

Emerging chemicals of concern include pharmaceutically active agents, endocrine disrupters, and low concentrations of heavy metals and metalloids. Over the next decade, we hope to learn enough to make it possible to delineate defensible standards. Another recent concern is that of contaminated stormwater from industrial sites. Federal guidelines to control this wastewater are currently being developed. Most concern is for product spills on plant property outside the production facility and from rain falling on piles of spoils.

UNITED STATES LEGISLATION, REGULATIONS, AND GOVERNMENT AGENCIES

Federal Legislation Public Law 92-500 promulgated in 1972 created the primary framework for management of water pollution in the United States. This act has been amended many times since 1972. Major amendments occurred in 1977 (Public Laws 95-217 and 95-576), 1981 (Public Laws 97-117 and 97-164), and 1987 (Public Law 100-4). This law and its various amendments are referred to as the Clean Water Act. The Clean Water Act addresses a large number of issues of water pollution management. A general review of these issues was presented earlier in this handbook. Primary with respect to control of industrial wastewater is the National Pollutant Discharge Elimination System (NPDES), originally established by PL 92-500. Any municipality or industry that discharges wastewater to the navigable waters of the United States must obtain a discharge permit under the regulations set forth by the NPDES. Under this system, there are three classes of pollutants (conventional pollutants, priority pollutants, and nonconventional/nonpriority pollutants). Conventional pollutants are substances such as biochemical oxygen demand (BOD), suspended solids (SS), pH, oil and grease, and coliforms. Priority pollutants are a list of 129 substances originally set forth in a consent decree between the Environmental Protection Agency and several environmental organizations. This list was incorporated into the 1977 amendments. Most of the substances on this list are organics, but it does include most of the heavy metals. These substances are generally considered to be toxic. However, the toxicity is not absolute; it depends on the concentration. In addition, many of the organics on the list are under appropriate conditions biodegradable. In reality, substances for inclusion on the Priority Pollutant List were chosen on the basis of a risk assessment rather than only a hazard assessment. The third class of pollutants could include any pollutant not in the first two categories. Examples of substances that are currently regulated in the third category are total nitrogen, ammonia, phosphorous, sodium, and chlorine residual. An emerging class of pollutants of concern include pharmaceutical and personal care products (PPCPs) and endocrine disrupting compounds (EDCs). PPCPs and EDCs can be found in both industrial and municipal wastewater; they have garnered increased attention as many eco-toxicological and water quality studies have indicated their ubiquity in natural waters and their potential impacts on receiving water ecosystems.

The overall goals of the Clean Water Act are to restore and maintain water quality in the navigable waters of the United States. The initial standard was to ensure that these waters would be clean enough for recreation (swimmable) and ecologically secure (fishable). Initially this was to be achieved by curtailment of the discharge of pollutants. The regulatory approach was to eventually phase out the discharge of any pollutant into water. Obviously, if no discharge occurs, the waters will be maintained in close to a pristine condition. However, the reduction of pollutant discharge to zero was found to be impractical. Thus, at present, the NPDES has prescribed the limits on discharges as a function of the type of pollutant, the type of industry discharging, and the desired water quality. The specific requirements for each discharger are established in the NPDES permit issued to that discharger. These permits are reviewed every five years and are subject to change at the time of review.

A major tactic that was adopted in the Clean Water Act was to establish uniform technology standards, by class of pollutant and specific industry type, that applied nationwide to all dischargers. Thus, a kraft mill in Oregon would have to meet essentially the same discharge standards as a kraft mill in New York. In establishing these standards, the EPA took into account the state of the art of waste treatment in each industry as well as cost and ecological effectiveness. These discharge standards have been published in the *Federal Register* for more than 30 industrial categories and several hundred subcategories (the Commerce Department Industrial Classification System was used to establish these categories). These have been promulgated over an extensive period of time. Readers should consult the index to this document to find the regulations that apply to a particular industry. Table 22-26 presents the major industrial categories.

Not only are the discharge standards organized on an industry-by-industry basis, but they also differ depending on which of the three classes of pollutants is being regulated. The standards for conventional pollutants are best conventional technology (BCT). The standards for priority pollutants are best available technology (BAT), as are those for nonconventional, nonpriority pollutants. These standards envision that the technology will not be

TABLE 22-26 Industry Categories

1. Adhesives and sealants
2. Aluminum forming
3. Asbestos manufacturing
4. Auto and other laundries
5. Battery manufacturing
6. Coal mining
7. Coil coating
8. Copper forming
9. Electric and electronic components
10. Electroplating
11. Explosives manufacturing
12. Ferroalloys
13. Foundries
14. Gum and wood chemicals
15. Inorganic chemicals manufacturing
16. Iron and steel manufacturing
17. Leather tanning and finishing
18. Mechanical products manufacturing
19. Nonferrous metals manufacturing
20. Ore mining
21. Organic chemicals manufacturing
22. Pesticides
23. Petroleum refining
24. Pharmaceutical preparations
25. Photographic equipment and supplies
26. Plastic and synthetic materials manufacturing
27. Plastic processing
28. Porcelain enamelling
29. Printing and publishing
30. Pulp and paperboard mills
31. Soap and detergent manufacturing
32. Steam electric power plants
33. Textile mills
34. Timber products processing

limited to treatment but may include revision in the industrial processing or reuse of effluents (pollution prevention). They are usually presented as mass of pollutant discharged per unit of product produced.

In some cases, BCT and BAT may not be sufficient to ensure that water quality standards are achieved in a stream segment. Studies have indicated that approximately 10 percent of the stream segments in the United States will have their water quality standards violated, even if all the dischargers to that stream segment meet BCT and BAT regulations. These segments are referred to as water quality limited segments. The NPDES permit for those who discharge into water quality limited segments must provide for pollutant removal in excess of that required by BCT and BAT so that water quality standards (which are jointly established by each state and EPA) are achieved. The stream segments in which water quality standards will be met by the application of BCT and BAT are referred to as effluent quality limited segments.

For industrial discharges that enter municipal sewers, and thus eventually municipal treatment plants, NPDES regulations are nominally the same as for industries that discharge directly to navigable waters—that is, they must meet BCT and BAT standards. These are referred to as categorical industrial pretreatment standards. However, if it can be demonstrated that the municipal treatment plant can remove a pollutant in the industrial waste, a removal credit can be assigned to the permit, thus lowering the requirement on the industrial discharger. The monetary charge to the industry by the municipality for this service is negotiable, but it must be within parameters established by EPA for industrial user charges if the municipality received a federal grant for construction of the treatment plant. A removal credit cannot be assigned if removal of the industrial pollutant makes it difficult for the municipal plant to dispose of sludge. In addition, an industry cannot discharge any substance that can result in physical damage to the municipal sewer or interfere with any aspect of treatment plant performance, even if the substance is not covered by BCT or BAT regulations.

The EPA has established a National Pretreatment Program to address the growing impacts industrial effluents have on publicly owned treatment works (POTWs). These impacts include biological toxicity and inhibition causing biological upsets that can reduce plant efficiency; accumulation of heavy metals in primary and secondary sludge that may affect beneficial reuse applications; increase in whole effluent toxicity (WET), which can affect plant discharge permit requirements; and increasing organic and solid loadings on plants that are reaching their design capacity. The EPA has established two sets of rules that form the basis of the National Pretreatment Program, the categorical pretreatment standards and prohibited discharge standards. The categorical pretreatment standards are industry- or technology-based limitations on pollutant discharges to POTWs promulgated by the U.S. EPA in accordance with section 307 of the Clean Water Act. These standards apply to specific process wastewaters of particular industrial categories (see 40 CFR 403.6 and 40 CFR parts 405 to 471). The prohibited discharge standards forbid certain chemicals and discharges by any sewer system. These specific discharges are described in full in the *Federal Register* and include 126 specific chemicals. The program requires all POTWs with a flow greater than 5 mgd or those receiving significant industrial flow to establish an industrial pretreatment program. States and individual municipalities can develop their own pretreatment programs to address system-specific industrial wastewater impacts. Currently, over 1500 municipalities have established their own pretreatment programs that either meet or exceed the EPA pretreatment standards. Along with setting standards for pretreatment, the EPA identifies best available technologies to economically control industrial effluents. In addition to industrial discharges, the EPA has initiated a program to better control stormwater discharges associated with industrial activity. This program is part of the national NPDES Phase I Stormwater Program.

Environmental Protection Agency President Richard Nixon created the EPA in 1970 to coordinate all environmental pollution control activities at the federal level. The EPA was placed directly under the Office of the President so that it could be more responsive to the political process. In the succeeding decade, the EPA produced a series of federal regulations increasing federal control over all wastewater pollution control activities. In January 1981, President Ronald Reagan reversed the trend of growing federal regulation and began to decrease the role of the federal EPA. However, during the 1980s, an equilibrium was achieved between those who wanted less regulation and those who wanted more. In general, industry and the EPA reached agreement on the optimum level of regulation.

State Water Pollution Control Offices Every state has its own water pollution control office. Some states have reorganized along the lines of the federal EPA with state EPA offices, while others have kept their water pollution control offices within state health departments. Before 1965, each state controlled its own water pollution control programs. Conflicts between states and uneven enforcement of state regulations resulted in the federal government's assuming the leadership role. Unfortunately, conflicts between states shifted to being conflicts between the states and the federal EPA. By 1980, the state water pollution control offices were primarily concerned with handling most of the detailed permit and paperwork for the EPA and in furnishing technical assistance to industries at the local level. In general, most of the details of regulation are carried out by state water pollution control agencies with oversight by EPA.

WASTEWATER CHARACTERISTICS

Wastewater characteristics vary widely from industry to industry. These characteristics affect the treatment techniques chosen for use in meeting discharge requirements. Some general characteristics that should be considered in planning are given in Table 22-27. Because of the large number of pollutant substances, wastewater characteristics are not usually considered on a substance-by-substance basis. Rather, substances with similar polluting effects are grouped into classes of pollutants or characteristics.

Priority Pollutants The greatest recent concern has been for priority pollutants. These materials are treated on an individual-substance basis for regulatory control. Thus, each industry could receive a discharge permit that lists an acceptable level for each priority pollutant. Table 22-28 presents a list of these substances; most are organic, but some inorganics are included. All are considered toxic, but, as indicated previously, there is wide variation in their toxicity. Most of the organics are biologically degradable despite

TABLE 22-27 Wastewater Characteristics

Property	Characteristic	Example	Size or concentration
Solubility	Soluble	Sugar	>100 gm/L
	Insoluble	PCB	<1 mg/L
Stability, biological	Degradable	Sugar	
	Refractory	DDT, metals	
Solids	Dissolved	NaCl	$<10^{-9}$ m
	Colloidal	Carbon	$>10^{-6}$–$<10^{-9}$ m
	Suspended	Bacterium	$>10^{-6}$ m
Organic	Carbon	Alcohol	
Inorganic	Inorganic	Cu^{2+}	
pH	Acidic	HNO_3	
	Neutral	Salt (NaCl)	1–12
	Basic	NaOH	
Temperature	High–low	Cooling	>5°
		Heat exchange	>30°
Toxicity	Biological effect	Heavy metals	Varies
		Priority compounds	
Nutrients	N	NH_3	Varies
	P	PO_4^{3-}	

TABLE 22-28 List of Priority Chemicals*

Compound name
1. Acenaphthene†
2. Acrolein†
3. Acrylonitrile†
4. Benzene†
5. Benzidine†
6. Carbon tetrachloride† (tetrachloromethane)
Chlorinated benzenes (other than dichlorobenzenes)
7. Chlorobenzene
8. 1,2,4-Trichlorobenzene
9. Hexachlorobenzene
Chlorinated ethanes† (including 1,2-dichloroethane, 1,1,1-trichloroethane, and hexachloroethane)
10. 1,2-Dichloroethane
11. 1,1,1-Trichloroethane
12. Hexachloroethane
13. 1,1-Dichloroethane
14. 1,1,2-Trichloroethane
15. 1,1,2,2-Tetrachloroethane
16. Chloroethane (ethyl chloride)
Chloroalkyl ethers† (chloromethyl, chloroethyl, and mixed ethers)
17. Bis(chloromethyl) ether
18. Bis(2-chloroethyl) ether
19. 2-Chloroethyl vinyl ether (mixed)
Chlorinated napthalene†
20. 2-Chloronapthalene
Chlorinated phenols† (other than those listed elsewhere; includes trichlorophenols and chlorinated cresols)
21. 2,4,6-Trichlorophenol
22. *para*-Chloro-*meta*-cresol
23. Chloroform (trichloromethane)†
24. 2-Chlorophenol†
Dichlorobenzenes†
25. 1,2-Dichlorobenzene
26. 1,3-Dichlorobenzene
27. 1,4-Dichlorobenzene
Dichlorobenzidine†
28. 3,3′-Dichlorobenzidine
Dichloroethylenes† (1,1-dichloroethylene and 1,2-dichloroethylene)
29. 1,1-Dichloroethylene
30. 1,2-*trans*-Dichloroethylene
31. 2,4-Dichlorophenol†
Dichloropropane and dichloropropene†
32. 1,2-Dichloropropane
33. 1,2-Dichloropropylene (1,2-dichloropropene)
34. 2,4-Dimethylphenol†
Dinitrotoluene†
35. 2,4-Dinitrotoluene
36. 2,6-Dinitrotoluene
37. 1,2-Diphenylhydrazine†
38. Ethylbenzene†
39. Fluoranthene†
Haloethers† (other than those listed elsewhere)
40. 4-Chlorophenyl phenyl ether
41. 4-Bromophenyl phenyl ether
42. Bis(2-chloroisopropyl) ether
43. Bis(2-chloroethoxy) methane
Halomethanes† (other than those listed elsewhere)
44. Methylene chloride (dichloromethane)
45. Methyl chloride (chloromethane)
46. Methyl bromide (bromomethane)
47. Bromoform (tribromomethane)
48. Dichlorobromomethane
49. Trichlorofluoromethane
50. Dichlorodifluoromethane
51. Chlorodibromomethane
52. Hexachlorobutadiene†
53. Hexachlorocyclopentadiene†
54. Isophorone†
55. Naphthalene†
56. Nitrobenzene†
Nitrophenols† (including 2,4-dinitrophenol and dinitrocresol)
57. 2-Nitrophenol
58. 4-Nitrophenol
59. 2,4-Dinitrophenol†
60. 4,6-Dinitro-*o*-cresol
Nitrosamines†
61. *N*-Nitrosodimethylamine
62. *N*-Nitrosodiphenylamine
63. *N*-Nitrosodi-*n*-propylamine
64. Pentachlorophenol†
65. Phenol†
Phthalate esters†
66. Bis(2-ethylhexyl) phthalate
67. Butyl benzyl phthalate
68. Di-*n*-butyl phthalate
69. Di-*n*-octyl phthalate
70. Diethyl phthalate
71. Dimethyl phthalate
Polynuclear aromatic hydrocarbons (PAH)†
72. Benzo(a)anthracene (1,2-benzanthracene)
73. Benzo(a)pyrene (3,4-benzopyrene)
74. 3,4-Benzofluoranthene
75. Benzo(k)fluoranthene (11,12-benzofluoranthene)
76. Chrysene
77. Acenaphthylene
78. Anthracene
79. Benzo(ghl)perylene (1,12-benzoperylene)
80. Fluorene
81. Phenanthrene
82. Dibenzo(a,h)anthracene (1,2,5,6-dibenzanthracene)
83. Indeno (1,2,3-cd) pyrene (2,3-*o*-phenylenepyrene)
84. Pyrene
85. Tetrachloroethylene†
86. Toluene†
87. Trichloroethylene†
88. Vinyl chloride† (chloroethylene)
Pesticides and metabolites
89. Aldrin†
90. Dieldrin†
91. Chlordane† (technical mixture and metabolites)
DDT and metabolites†
92. 4-4′-DDT
93. 4,4′-DDE (*p,p*′-DDX)
94. 4,4′-DDD (*p,p*′-TDE)
Endosulfan and metabolites†
95. α-Endosulfan-alpha
96. β-Endosulfan-beta
97. Endosulfan sulfate
Endrin and metabolites†
98. Endrin
99. Endrin aldehyde
Heptachlor and metabolites†
100. Heptachlor
101. Heptachlor epoxide
Hexachlorocyclohexane (all isomers)†
102. α-BHC-alpha
103. β-BHC-beta
104. γ-BHC (lindane)-gamma
105. δ-BHC-delta
Polychlorinated biphenyls (PCB)†
106. PCB-1242 (Arochlor 1242)
107. PCB-1254 (Arochlor 1254)
108. PCB-1221 (Arochlor 1221)
109. PCB-1232 (Arochlor 1232)
110. PCB-1248 (Arochlor 1248)
111. PCB-1260 (Arochlor 1260)
112. PCB-1016 (Arochlor 1016)
113. Toxaphene†
114. antimony (total)
115. arsenic (total)
116. asbestos (fibrous)
117. beryllium (total)
118. cadmium (total)
119. chromium (total)
120. copper (total)
121. cyanide (total)
122. lead (total)
123. mercury (total)
124. nickel (total)
125. selenium (total)
126. silver (total)
127. thallium (total)
128. zinc (total)
129. 2,3,7,8-Tetrachlorodibenzo-*p*-dioxin (TCDD)

*Adapted from Eckenfelder, W. W. Jr., *Industrial Water Pollution Control*, 2d ed., McGraw-Hill, New York, 1989.
†Specific compounds and chemical classes as listed in the consent degree.

their toxicity (Allenby, B., *Industrial Ecology: Policy, Framework and Implementation*, Prentice-Hall, Upper Saddle River, N.J., 1999; Ehlrich, P., and J. Holdren, *Impact of Population Growth*, Scrivener, Washington, D.C., 1971). U.S. EPA has collected data on the occurrence of these substances in various industrial wastes and their treatability. A recent trend has been to avoid their use in industrial processing.

Organics The organic composition of industrial wastes varies widely, primarily due to the different raw materials used by each industry. These organics include proteins, carbohydrates, fats and oils, petrochemicals, solvents, pharmaceuticals, small and large molecules, and solids and liquids. Another complication is that a typical industry produces many diverse waste streams. A good practice is to conduct a material balance throughout an entire production facility. This survey should include a flow diagram, the location and sizes of piping, tanks, and flow volumes, and an analysis of each stream. The results of an industrial waste survey for an industry are given in Table 22-29. Noteworthy is the range in waste sources, including organic soap, toilet articles, ABS (alkyl benzene sulfonate), and the relatively clean but hot condenser water that makes up half the plant flow, while the strongest wastes—spent caustic and fly ash—have the lowest flows. See Tables 22-29 and 22-30 for information on the average characteristics of wastes from specific industries.

An important measure of the waste organic strength is the five-day biochemical oxygen demand (BOD_5). Because this test measures the demand for oxygen in the water environment caused by organics released by industry

TABLE 22-29 Industrial Waste Components of a Soap, Detergents, and Toilet Articles Plant

Waste source	Sampling station	COD, mg/L	BOD, mg/L	SS, mg/L	ABS, mg/L	Flow, gal/min
Liquid soap	*D*	1,100	565	195	28	300
Toilet articles	*E*	2,680	1,540	810	69	50
Soap production	*R*	29	16	39	2	30
ABS production	*S*	1,440	380	309	600	110
Powerhouse	*P*	66	10	50	0	550
Condenser	*C*	59	21	24	0	1100
Spent caustic	*B*	30,000	10,000	563	5	2
Tank bottoms	*A*	120,000	150,000	426	20	1.5
Fly ash	*F*			6750		10
Main sewer		450	260	120	37	2150

NOTE: gal/min = 3.78×10^{-3} m^3/min.

SOURCE: Eckenfelder, W. W., *Industrial Water Pollution Control,* 2d ed., McGraw-Hill, New York, 1989.

and municipalities, it has been the primary parameter in determining the strength and effects of a pollutant. This test determines the oxygen demand of a waste exposed to biological organisms (controlled seed) for an incubation period of five days. Usually this demand is caused by degradation of organics according to the following simplified equation, but reduced inorganics in some industries may also cause demand (i.e., Fe^{2+}, S^{2-}, and SO_3^{2-}):

$$\text{Organic waste} + O_2(\text{DO}) \xrightarrow[\text{microbes}]{\text{seed}} CO_2 + H_2O$$

This wet lab test measures the decrease in dissolved oxygen (DO) concentration in five days, which is then related to the sample strength. If the test is extended over 20 days, the BOD_{20} (ultimate BOD) is obtained and corresponds more closely to the chemical oxygen demand (COD) test. The COD test uses strong chemical oxidizing agents with catalysts and heat to oxidize the wastewater and obtain a value that is almost always larger than the 5- and 20-day BOD values. Some organic compounds resists chemical oxidation, resulting in a low COD. A major advantage of the COD test is the completion time of less than three hours, versus five days for the BOD_5 test. Unfortunately, state and federal regulations generally require BOD_5 values, but approximate correlations can be made to allow computation of BOD from COD. A more rapid measure of the organic content of a waste is the instrumental test for total organic carbon (TOC), which takes a few minutes and may be correlated to both COD and BOD for specific wastes. Unfortunately, BOD_5 results are subject to wide statistical variations, and they require close scrutiny and experience to conduct. For municipal wastewaters, BOD_5 is about 67 percent of the ultimate BOD and 40 to 45 percent of the COD, indicating a large amount of nonbiodegradable COD and the continuing need to run BOD as well as COD and TOC. An example of BOD, COD, and TOC relationships for chemical industry wastewater is given in Table 22-30. The concentrations of the wastewaters vary by two orders of magnitude, and the BOD/COD, COD/TOC, and BOD/TOC ratios vary less than twofold. The table indicates that correlation/codification is possible, but care and continual scrutiny must be exercised.

Another technique for organics measurement that overcomes the long period required for the BOD test is the use of continuous respirometry. Here the waste (full-strength rather than diluted as in the standard BOD test) is contacted with biomass in an apparatus that continuously measures the dissolved oxygen consumption. This test determines the ultimate BOD in a few hours if a high level of biomass is used. The test can also yield information on toxicity, the need to develop an acclimated biomass, and required rates of oxygen supply.

In general, low-molecular-weight, water-soluble organics are biodegraded readily. As organic complexity increases, solubility and biodegradability decrease. Soluble organics are metabolized more easily than insoluble organics. Complex carbohydrates, proteins, and fats and oils must be hydrolyzed to simple sugars, aminos, and other organic acids before metabolism. Petrochemicals, pulp and paper, slaughterhouse, brewery, and many other industrial wastes containing complex organics have been satisfactorily treated biologically, but proper testing and evaluation is necessary.

TABLE 22-30 BOD, COD, and TOC Relationships

Type of waste	BOD_5, mg/L	COD, mg/L	TOC, mg/L	BOD_5/COD	COD/TOC	BOD_5/TOC
Chemical	700	1,400	450	0.50	3.12	1.55
Chemical	850	1,900	580	0.45	3.28	1.47
Chemical	8,000	17,500	5,800	0.46	3.02	1.38
Chemical	9,700	15,000	5,500	0.65	2.72	1.76
Chemical	24,000	41,300	9,500	0.58	4.35	2.53
Chemical	60,700	78,000	26,000	0.78	3.00	2.34
Chemical	62,000	143,000	48,140	0.43	2.96	1.28

Adapted from Eckenfelder, W. W. and D. L. Ford, *Water Pollution Control,* Pemberton Press, Austin and New York, 1970.

Inorganics The inorganics in most industrial wastes are the direct result of inorganic compounds in the carriage water. Soft-water sources will have lower inorganics than hard-water or saltwater sources. However, some industrial wastewaters can contain significant quantities of inorganics that result from chemical additions during plant operation. Many food processing wastewaters are high in sodium. While domestic wastewaters have a balance in organics and inorganics, many process wastewaters from industry are deficient in specific inorganic compounds. Biodegradation of organic compounds requires adequate nitrogen, phosphorus, iron, and trace salts. Ammonium salts or nitrate salts can provide the nitrogen, while phosphates supply the phosphorus. Either ferrous or ferric salts or even normal steel corrosion can supply the needed iron. Other trace elements needed for biodegradation are potassium, calcium, magnesium, cobalt, molybdenum, chloride, and sulfur. Carriage water or demineralizer wastewaters or corrosion products can supply the needed trace elements for good metabolism. Occasionally, it is necessary to add specific trace elements or nutrient elements.

pH and Alkalinity Wastewaters should have pH values between 6 and 9 for minimum impact on the environment. Wastewaters with pH values less than 6 will tend to be corrosive as a result of the excess hydrogen ions. On the other hand, raising the pH above 9 will cause some of the metal ions to precipitate as carbonates or as hydroxides at higher pH levels. Alkalinity is important in keeping pH values at the right levels. Bicarbonate alkalinity is the primary buffer in wastewaters. It is important to have adequate alkalinity to neutralize the acid waste components as well as those formed by partial metabolism of organics. Many neutral organics, such as carbohydrates, aldehydes, ketones, and alcohols, are biodegraded through organic acids that must be neutralized by the available alkalinity. If alkalinity is inadequate, sodium carbonate is a better form to add than lime. Lime tends to be hard to control accurately, and it results in high pH levels and precipitation of the calcium that forms part of the alkalinity. In a few instances, sodium bicarbonate may be the best source of alkalinity.

Temperature Most industrial wastes tend to be on the warm side. For the most part, temperature is not a critical issue below 37°C if wastewaters are to receive biological treatment. It is possible to operate thermophilic biological wastewater treatment systems up to 65°C with acclimated microbes. Low-temperature operations in northern climates can result in very low winter temperatures and slow reaction rates for both biological treatment systems and chemical treatment systems. Increased viscosity of wastewaters at low temperatures makes solids separation more difficult. Efforts are generally made to keep operating temperatures between 10°C and 30°C if possible.

Dissolved Oxygen Oxygen is a critical environmental resource in receiving streams and lakes. Aquatic life requires reasonable dissolved oxygen (DO) levels. EPA has set minimum stream DO levels at 5 mg/L during summer operations, when the rate of biological metabolism is at a maximum. It is important that wastewaters have maximum DO levels when they are discharged and a minimum of oxygen-demanding components so that DO remains above 5 mg/L. DO is a poorly soluble gas in water, having a solubility around 9.1 mg/L at 20°C and 101.3-kPa (1-atm) air pressure. As the temperature increases and the pressure decreases with higher elevations above sea level, the solubility of oxygen decreases. Thus, DO is at a minimum when BOD rates are at a maximum. Lowering the temperature yields higher levels of DO saturation, but the biological metabolism rate decreases. Warm-wastewater discharges tend to aggravate the DO situation in receiving waters.

Solids Total solids is the residue remaining from a wastewater dried at 103°C to 105°C. It includes the fractions shown in Fig. 22-30. The first

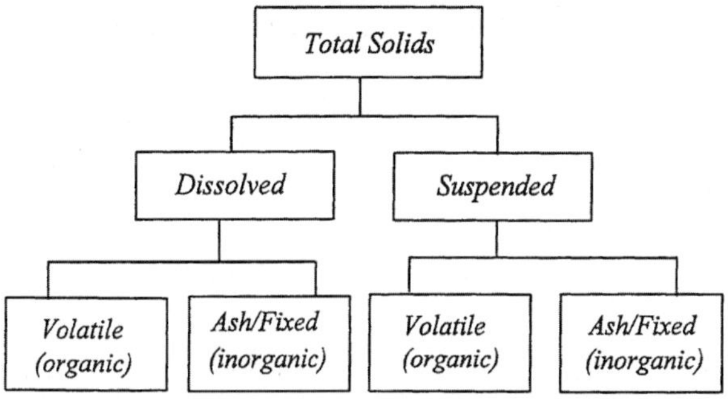

FIG. 22-30 Solids identification. Abbreviations: TS, total solids; SS, suspended solid; D, dissolved; V, volatile.

TABLE 22-31 Solids Variation in Industrial Wastewater

Type of solids	Plating	Pulp and paper
Total	High	High
Dissolved	High	High
Suspended	Low	High
Organic	Low	High
Inorganic	High	High

separation is the portion that passes through a 2-mm filter (dissolved) and those solids captured on the filter (suspended). Combustion at 500°C further separates the solids into volatile and ash (fixed) solids. Although ash and volatile solids do not distinguish inorganic from organic solids exactly, due to loss of inorganics on combustion, the volatile fraction is often used as an approximate representation of the organics present. Another type of solids, settleable solids, refers to solids that settle in an Imhoff cone in one hour. Industrial wastes vary substantially in these types of solids and require individual wastewater treatment process analysis. An example of possible variation is given in Table 22-31.

Nutrients and Eutrophication Nitrogen and phosphorus cause significant problems in the environment and require special attention in industrial wastes. Nitrogen, phosphorus, or both may cause aquatic biological productivity to increase, resulting in low dissolved oxygen and eutrophication of lakes, rivers, estuaries, and marine waters. Table 22-32 shows the primary nutrient forms that cause problems, while the equation for total oxygen demand (TOD) below shows the biological oxidation or oxygen-consuming potential of the most common nitrogen forms.

When organics containing reduced nitrogen are degraded, they usually produce ammonium, which is in equilibrium with ammonia. As the pK for $NH_3 \leftrightarrow NH_4^+$ is 9.3, the ammonium ion is the primary form present in virtually all biological treatment systems because they operate at pH <8.5 and usually in the pH range of 6.5 to 7.5. In aerobic reactions, ammonium is oxidized by nitrifying bacteria (nitrosomonas) to nitrite, and each mg of NH_4^+—N oxidized will require 3.43 mg DO. Further oxidation of nitrite by nitrobacter yields nitrate and uses an additional 1.14 mg of DO, for a total DO consumption of 4.57 mg. Thus, organic and ammonium nitrogen can exert significant biochemical oxygen demand in the water environment. This nitrogen demand is referred to as nitrogenous or NBOD, whereas organic BOD is CBOD (carbonaceous). In the treatment of wastewaters with organics and ammonium, the TOD may have to be satisfied in accordance with the approximate formula:

$$\text{TOD} \cong 1.5\ \text{BOD}_5 + 4.5\ (\text{TKN})$$

where TKN (total Kjeldahl nitrogen) = Organic N + NH_4^+—N.

Phosphorus is not oxidized or reduced biologically, but ortho-P may be formed from organic and poly-P. Ortho-P may be removed by chemical precipitation or biologically with sludges and will be covered in a later section.

While many industrial wastes are so low in nitrogen and phosphorus that these must be added if biologically based treatment is to be used, others contain very high levels of these nutrients. For example, paint production wastes are high in nitrogen, and detergent production wastes are high in phosphorus. Treatment for removal of these nutrients is required in areas where eutrophication is a problem.

Whole Effluent Toxicity (WET) Whole effluent toxicity testing is an important part of the EPA's approach to protecting receiving waters from toxic effluents. The EPA has established 17 analytical methods for testing acute and chronic toxicity of point sources (municipal and industrial wastewater effluents) and their impact on surface water environments. *WET* is a term used to describe the toxic effects imposed on a species or population of aquatic organisms caused by exposure to an effluent. WET is determined analytically by exposing sensitive indigenous organisms to effluents using WET test protocols established by the EPA. WET tests report on the general acute and chronic toxicity of all constituents in a complex effluent and do not represent the toxicity of specific chemicals. To this end, the WET test results could represent the combined toxic effect of heavy metals, un-ionized ammonia, and other toxic constituents in an effluent. The EPA and state regulators have been using WET as an analytical tool for controlling industrial discharges and developing and enforcing industrial pretreatment programs.

TABLE 22-32 Nutrient Forms

Parameter	Example
Organic N	protein
Ammonia	NH_3
Ammonium	NH_4^+
Nitrite	NO_2^-
Nitrate	NO_3^-
Organic P	malathion
Ortho-P	PO_4^{3-}
Poly-P	$(PO_4^{3-})_x$

Oil and Grease These substances are found in many industrial wastes (e.g., meat packing, petrochemical, and soap production). They tend to float on the water surface, blocking oxygen transfer, interfering with recreation, and producing an aesthetically poor appearance in the water. Measurement is by a solvent extraction procedure. Many somewhat different substances will register as oil and grease in this test. Often oil and grease interfere with other treatment operations, so they must be removed during the initial stages of treatment.

Emerging Pollutants of Concern (EPOCs) Emerging pollutants of concern include pharmaceutical and personal care products (PPCPs) and endocrine disrupting compounds (EDC). PPCPs include compounds such as antibiotics, antidepressants, pain medication, fragrances, and other chemicals associated with the production and use of drugs and personal care products. EDCs include natural and synthetic estrogens, as well as industrial chemicals (e.g., dioxins, chlorinated biphenyls, specific pesticides) that mimic estrogens and their interaction with animal endocrine systems. The concentrations of PPCPs and EDCs in municipal wastewaters are typically very low (nano- and pico-gram levels), but they may be higher in some industrial wastewaters. Currently there are no federal regulations for EDCs and PPCPs, other than those that may be on the Priority Pollutant List or those that may result in elevated whole effluent toxicity (WET). However, there is growing evidence that residual PPCPs and EDCs that pass through wastewater treatment plants can affect water ecosystems (e.g., feminization of fish populations). This will likely lead to the regulation of some of these emerging pollutants of concern in the future, especially as various types of water reuse play a larger role in local and regional water supply management.

WASTEWATER TREATMENT

Industrial wastewater contains a vast array of pollutants in soluble, colloidal, and particulate forms, both inorganic and organic. The required effluent standards are diverse, varying with the industrial and pollutant class. Consequently, there can be no standard design for industrial water pollution control. Instead, each site must be designed to achieve optimum performance. However, each of the many proven processes for industrial waste treatment can remove more than one type of pollutant and can be applied to more than one industry. In the subsections that follow, waste treatment processes are discussed from a generalized perspective. In most cases, a combination of several processes is used to achieve the degree of treatment required at the least cost.

Much of the experience and data from wastewater treatment has been gained from municipal treatment plants. Industrial liquid wastes are similar to municipal wastewater, but they can also differ in significant ways. Typical design parameters developed for municipal wastewater operations must not be blindly used for industrial wastewater. As part of the design process, it is best to run laboratory and small pilot tests with the wastewater from a specific plant. It is critical to understand the temporal variations in industrial wastewater strength, flow, and waste components and the effect of these variations on the performance of various treatment processes. Laboratory and pilot studies are often neglected in an effort to minimize costs; instead, designers use the characteristics of wastes from similar plants. This strategy often results in failure, delay, and increased costs. Careful studies on the actual waste at a plant site cannot be overemphasized.

PRETREATMENT

Many industrial wastewater streams should be pretreated before discharge to municipal sewerage systems or even to a central industrial sewerage system. Many POTWs have pretreatment programs that follow the EPA program by regulating specific industries and specific chemical constituents. Product substitution, process alteration, or pretreatment should be considered to address wastewater streams that may fall under the local or federal pretreatment program. Pretreatment may include single- or multiple-unit processes aimed at reducing the overall toxicity and adverse impact of the waste stream on the total treatment system.

Equalization Equalization is one of the most important pretreatment devices. The batch discharge of concentrated wastes is best suited for equalization. It may be important to equalize wastewater flows, wastewater concentrations, or both. Periodic wastewater discharges tend to overload treatment units. Flow equalization tends to level out the hydraulic loads on treatment units. It may or may not level out concentration variations, depending on the extent of mixing within the equalization basin. Mechanical mixing may be adequate if the wastes are purely chemical in their reactivity. Biodegradable wastes normally require aeration mixing so that the microbes are kept aerobic and nuisance odors are prevented. Diffused aeration systems offer better mixing under variable load conditions than mechanical surface aeration equipment. Mixing and oxygen transfer are both important with biodegradable wastewaters. Operation on regular cycles determines the size of the equalization basin. There is no advantage in making the equalization basin any larger than necessary to level out wastewater variations. Industrial operation on a five-day, 40-hour week will normally make a two-day equalization basin as large as needed for continuous operation of the wastewater treatment system under uniform conditions.

Neutralization Acidic or basic wastewaters must be neutralized before discharge. If an industry produces both acidic and basic wastes, these wastes may be mixed together at the proper rates to obtain neutral pH levels. Equalization basins can be used as neutralization basins. When separate chemical neutralization is required, sodium hydroxide is the easiest base material to handle in a liquid form and can be used at various concentrations for in-line neutralization with a minimum of equipment. Yet lime remains the most widely used base for acid neutralization. Limestone is used when reaction rates are slow and considerable time is available for reaction. Sulfuric acid is the primary acid used to neutralize high-pH wastewaters unless calcium sulfate might be precipitated as a result of the neutralization reaction. Hydrochloric acid can be used for neutralization of basic wastes if sulfuric acid is not acceptable. For very weak basic wastewaters, carbon dioxide can be adequate for neutralization.

Grease and Oil Removal Grease and oils tend to form insoluble layers with water as a result of their hydrophobic characteristics. These hydrophobic materials can be easily separated from the water phase by gravity and simple skimming, provided they are not too well mixed with the water before separation. If the oils and greases form emulsions with water as a result of turbulent mixing, the emulsions are difficult to break. Separation of oil and grease should be carried out near the point of their mixing with water. In a few instances, air bubbles can be added to the oil and grease mixtures to separate the hydrophobic materials from the water phase by flotation. Chemicals have also been added to help break the emulsions. American Petroleum Institute (API) separators have been used extensively by the petroleum industry to remove oils from wastewaters. The food industries use grease traps to collect the grease before its discharge. Unfortunately, grease traps are designed for regular cleaning of the trapped grease. Too often they are allowed to fill up and discharge the excess grease into the sewer or are flushed with hot water and steam to fluidize the grease for easy discharge to the sewer. A grease trap should be designed for a specific volume of grease to be collected over specific time periods. Care should be taken to design the trap so that the grease can easily be removed and properly handled. Neglected or poorly designed grease traps are worse than no grease traps at all.

Toxic Substances Recent federal legislation has made it illegal for industries to discharge toxic materials in wastewaters. Each industry is responsible for determining if any of its wastewater components are toxic to the environment and to remove them before the wastewater discharge. The EPA has identified a number of priority pollutants that must be removed and kept under proper control from their origin to their point of ultimate disposal. Major emphasis has recently been placed on heavy metals and on complex organics that have been implicated in possible cancer production. Pretreatment is essential to reduce heavy metals below toxic levels and to prevent the discharge of any toxic organics. Many industries will be required to pretreat waste streams under the local or federal pretreatment program. In the past, these pretreatment programs have focused on heavy metals and toxic organics. More recently they have begun to concentrate on whole effluent toxicity, with future emphasis likely to be on low concentrations of heavy metals and metalloids, pharmaceutical agents, endocrine disrupters, and other emerging pollutants of concern.

PRIMARY TREATMENT

Wastewater treatment is directed toward the removal of pollutants with the least effort. Suspended solids are removed by either physical or chemical separation techniques and are handled as concentrated solids.

Screens Fine screens such as hydroscreens are used to remove moderate-size particles that are not easily compressed under fluid flow. Fine screens are normally used when the quantities of screened particles are large enough to justify the additional units. Mechanically cleaned fine screens have been used for separating large particles. A few industries have used large bar screens to catch large solids that could clog or damage pumps or equipment following the screens.

Grit Chambers Industries with sand or hard, inert particles in their wastewaters have found aerated grit chambers useful for the rapid separation of these inert particles. Aerated grit chambers are relatively small, with total volume based on three-minute retention at maximum flow. Diffused air is normally used to create the mixing pattern shown in Fig. 22-31, with

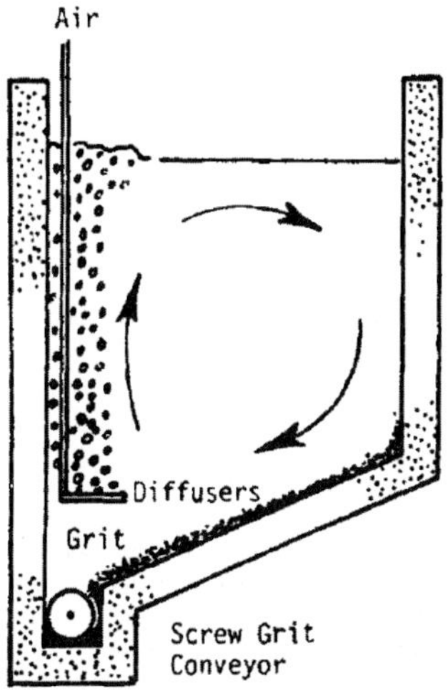

FIG. 22-31 Schematic diagram of an aerated grit chamber.

the heavy, inert particles removed by centrifugal action and friction against the tank walls. The air flow rate is adjusted for the specific particles to be removed. Floatable solids are removed in the aerated grit chamber. It is important to provide for the regular removal of floatable solids from the surface of the grit chamber; otherwise, nuisance conditions will be created. The settled grit is normally removed with a continuous screw and buried in a landfill.

Gravity Sedimentation Slowly settling particles are removed with gravity sedimentation tanks. For the most part, these tanks are designed on the basis of retention time, surface overflow rate, and minimum depth. A sedimentation tank can be rectangular or circular. The important factor affecting its removal efficiency is the hydraulic flow pattern through the tank. The energy contained in the incoming wastewater flow must be dissipated before the solids can settle. The wastewater flow must be distributed properly through the sedimentation volume for maximum settling efficiency. After the solids have settled, the settled effluent should be collected without creating serious hydraulic currents that could adversely affect the sedimentation process. Effluent weirs are placed at the ends of rectangular sedimentation tanks and around the periphery of circular sedimentation tanks to ensure uniform flow out of the tanks. Once the solids have settled, they must be removed from the sedimentation-tank floor by scraping and hydraulic flow. Conventional sedimentation tanks have sludge hoppers to collect the concentrated sludge and to prevent the removal of excess volumes of water with the settled solids. Cross-sectional diagrams of conventional sedimentation tanks are shown in Figs. 22-32 and 22-33.

Design criteria for gravity sedimentation tanks normally provide for two-hour retention based on average flow, with longer retention periods used for light solids or inert solids that do change during their retention in the tank. Sedimentation time should not be too long; otherwise, the solids will compact too densely and will affect solids collection and removal. Organic solids generally will not compact to more than 5 to 10 percent. Inorganic solids will compact up to 20 or 30 percent. Centrifugal sludge pumps can handle solids up to 5 or 6 percent, while positive-displacement sludge pumps can handle solids up to 10 percent. With solids above 10 percent, the sludge tends to lose fluid properties and must be handled as a semisolid rather than a fluid. Circular sedimentation tanks have steel truss boxes with angled sludge scrapers on the lower side. As the sludge scrapers rotate, the solids are pushed toward the sludge hopper for removal on a continuous or semicontinuous basis. The rectangular sedimentation tanks employ

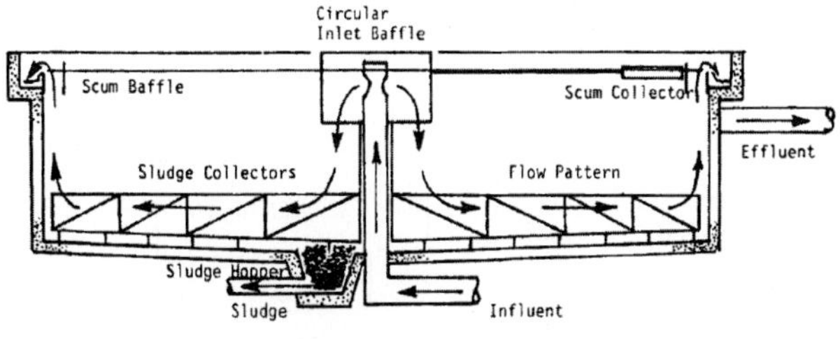

FIG. 22-32 Schematic diagram of a circular sedimentation tank.

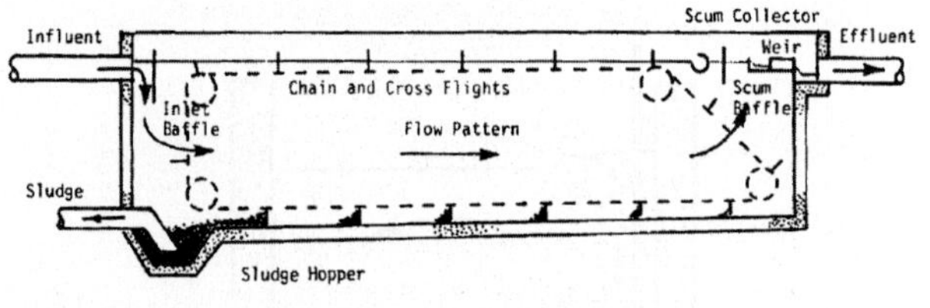

FIG. 22-33 Schematic diagram of a rectangular sedimentation tank.

chain-and-flight sludge collectors or rail-mounted sludge collectors. When floating solids can occur in primary sedimentation tanks, surface skimmers are mounted on the sludge scrapers so that the surface solids are removed at regular intervals.

The surface overflow rate (SOR) for primary sedimentation is normally held close to 40.74 $m^3/(m^2 \cdot day)$ [1000 $gal/(ft^2 \cdot day)$] for average flow rates, depending on the solids characteristics. Lowering the SOR below 40.74 $m^3/(m^2 \cdot day)$ does not produce improved effluent quality in proportion to the reduction in SOR. Generally, the minimum depth of sedimentation tanks is 3.0 m (10 ft), with circular sedimentation tanks having a minimum diameter of 6.0 m (20 ft) and rectangular sedimentation tanks having length-to-width ratios of 5:1. Chain-and-flight limitations generally keep the width of rectangular sedimentation tanks to increments of 6.0 m (20 ft) or less. While hydraulic overflow rates have been limited on the effluent weirs, operating experience has indicated that the recommended limit of 186 $m^3/(m \cdot day)$ [15,000 $gal/(ft \cdot day)$] is lower than necessary for good operation. A circular sedimentation tank with a single-edge weir provides adequate weir length and is easier to adjust than one with a double-sided weir. More problems appear to be created from improper adjustment of the effluent weirs than from improper length.

Chemical Precipitation Lightweight suspended solids and colloidal solids can be removed by chemical precipitation and gravity sedimentation. In effect, the chemical precipitate is used to agglomerate the tiny particles into large particles that settle rapidly in normal sedimentation tanks. Aluminum sulfate, ferric chloride, ferrous sulfate, lime, and polyelectrolytes have been used as coagulants. The choice of coagulant depends on the chemical characteristics of the particles being removed, the pH of the wastewaters, and the cost and availability of the precipitants. While the precipitation reaction results in the removal of the suspended solids, it increases the amount of sludge to be handled. The chemical sludge must be considered along with the characteristics of the original suspended solids in evaluating sludge-processing systems.

Normally, chemical precipitation requires a rapid mixing system and a flocculation system ahead of the sedimentation tank. With a rectangular sedimentation tank, the rapid-mixer and flocculation units are added ahead of the tank. With a circular sedimentation tank, the rapid-mixer and flocculation units are built into the tank. Schematic diagrams of chemical treatment systems are shown in Figs. 22-34 and 22-35. Rapid mixers are designed to provide 30-second retention at average flow with sufficient turbulence to mix the chemicals with the incoming wastewaters. The flocculation units are designed for slow mixing at 20-minute retention. These units are designed to cause the particles to collide and increase in size without excessive shearing. Care must be taken to move the flocculated mixture from the flocculation unit to the sedimentation unit without disrupting the large floc particles.

The parameter used to design rapid mix and flocculation systems is the root mean square velocity gradient G, which is defined by the following equation:

$$G = \left(\frac{P}{VU}\right)^{1/2} \left(\frac{1}{s}\right)$$

where P = power input to the water (ft·lb/s)
V = mixer or flocculator volume $(ft)^3$
U = absolute viscosity of water $(lb \cdot s/ft^2)$

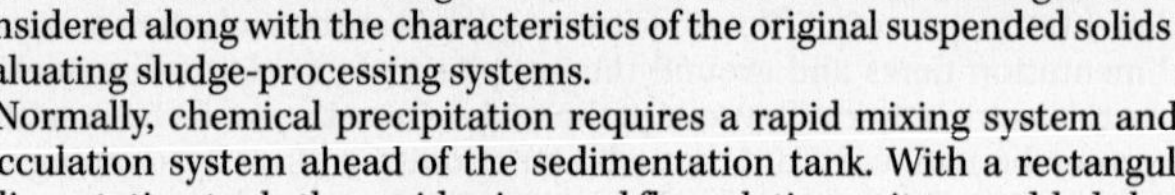

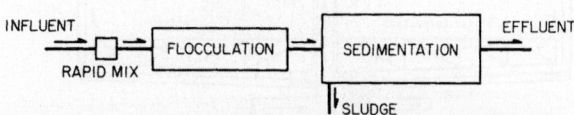

FIG. 22-34 Schematic diagram of a chemical precipitation system for rectangular sedimentation tanks.

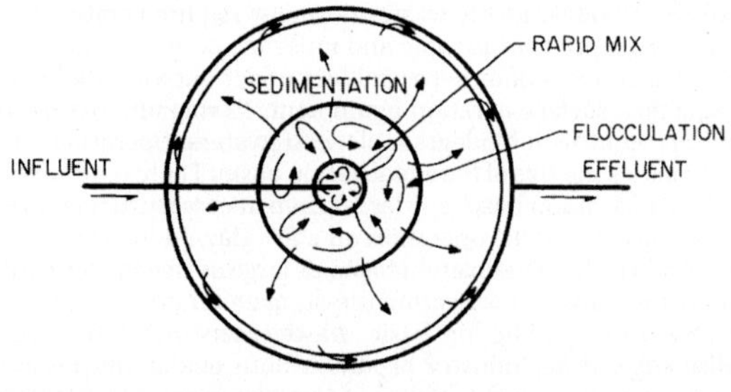

FIG. 22-35 Schematic diagram of a chemical precipitation system for circular sedimentation tanks.

Optimum mixing usually requires a G value of greater than 1000 inverse seconds. Optimum flocculation occurs when G is in the range 10 to 100 inverse seconds.

Chemical precipitation can remove 95 percent of the suspended solids, up to 50 percent of the soluble organics, and the bulk of the heavy metals in a wastewater. Removal of soluble organics is a function of the coagulant chemical, with iron salts yielding the best results and lime the poorest. Metal removal is primarily a function of pH and the ionic state of the metal. Guidance is available from solubility product data.

SECONDARY TREATMENT

Secondary treatment uses processes in which microorganisms, primarily bacteria, stabilize waste components. The mixture of microorganisms is usually referred to as biomass. A portion of the waste is oxidized, releasing energy; the remainder is used as building blocks of protoplasm. The energy released by biomass metabolism is used to produce the new units of protoplasm. Thus, the incentive for the biomass to stabilize waste is that it provides the energy and basic chemical components required for reproduction. The process of biological waste conversion is illustrated by Eq. (22-19).

$$\underset{\text{(electron donor)}}{\text{Waste}} + \text{Biomass} + \underset{\text{acceptor}}{\text{Electron}} \xrightarrow[\substack{\text{Proper}\\ \text{environment}\\ \text{conditions}}]{} \underset{\text{biomass}}{\text{More}} + \begin{array}{c}\text{End products:}\\ \text{Oxidized electron}\\ \text{donor}\\ \text{Reduced electron}\\ \text{acceptor}\end{array} \tag{22-19}$$

As this equation indicates, the waste generally serves as an electron donor, necessitating that an electron acceptor be supplied. A variety of substances can be used as electron acceptors, including molecular oxygen, carbon dioxide, oxidized forms of nitrogen, sulfur, and organic substances. The characteristics of the end products of the reaction are determined by the electron acceptor. Table 22-33 is a list of typical end products as a function of the electron acceptor. In general, the end products of this reaction are at a much lower energy level than the waste components, thus resulting in the release of energy previously mentioned. Although this process is usually used to stabilize organic substances, it can also be used to oxidize inorganics. For example, biomass-mediated oxidation of iron, nitrogen, and sulfur is known to occur in nature and in anthropogenic processes.

Equation (22-19) describes the biomass-mediated reaction and indicates that proper environmental conditions are required for the reaction to take place. These conditions are required by the biomass, not the electron donor or acceptor. The environmental conditions include pH, temperature, nutrients, ionic balance, and so on. In general, biomass can function over a wide pH range, generally from 5 to 9. However, some microbes require a much narrower pH range; for example, effective methane fermentation

TABLE 22-33 Electron Acceptors and End Products for Biological Reactions

Electron acceptors	End product
Molecular oxygen	Water, CO_2, oxidized nitrogen
Oxidized nitrogen	N_2, N_2O, NO, CO_2, H_2O
Oxidized sulfur	H_2S, S, CO_2, H_2O
CO_2, acetic acid, formic acid	CH_4, CO_2, H_2
Complex organics	H_2, simple organics, CO_2, H_2O

requires a pH in the range of 6.5 to 7.5. It is just as important to maintain a relatively constant pH in the process as it is to stay within the acceptable range. Microorganisms can function effectively at the extremes of their pH range, provided they are given the opportunity to acclimate to these conditions. Continual changes in pH are detrimental, even if the organisms are on the average near the middle of their effective pH range. A similar situation prevails for temperature. Most organisms can function well over a broad range of temperatures, but they do not adjust well to frequent fluctuations of even a few degrees. There are three major temperature ranges in which microorganisms function. The psychrophilic range (5°C to 20°C), the mesophilic range (20°C to 45°C), and the thermophilic range (45°C to 70°C). In general, the microbes that function in one of these temperature ranges cannot function efficiently in the other ranges. Because it is generally uneconomical to adjust the temperature of a waste, most processes are operated in the mesophilic range. If the normal temperature of the waste is above or below the mesophilic range, the process will be operated in the psychrophilic or thermophilic range as appropriate. However, occasionally the temperature of the waste is changed to improve performance. For example, some anaerobic treatment processes are operated under thermophilic conditions, even though the waste must be heated to achieve this temperature range. This is carried out in order to speed up the degradation of complex organics or to kill mesophilic pathogens. Any time the biological operation of a process moves away from its optimum or most effective range, be it pH, temperature, nutrients, or what you have, the rate of biological processing falls.

All microorganisms require varying amounts of a large number of nutrients. These are required because they are necessary components of bacterial protoplasm. The nutrients can be divided into three groups: macro, minor, and micro. The macronutrients are those that comprise most of the biomass. These are given by the commonly accepted formula for biomass ($C_{60}H_{87}O_{23}N_{12}P$). The carbon, hydrogen, and oxygen are normally supplied by the waste and water, but the nitrogen and phosphorous must often be added to industrial wastes to ensure that a sufficient amount is present. A good rule is that the mass of nitrogen should be at least 5 percent of the BOD, and the mass of phosphorous should be at least 20 percent of the mass of nitrogen. One of the major operational expenses is the purchase of nitrogen and phosphorous for addition to biologically based treatment processes. The quantities of nitrogen and phosphorous stated here are actually in excess of the minimum amounts needed. The actual amount needed depends on the quantity of excess biomass wasted from the system and the amount of N and P available in the waste. This will be expanded upon later in this subsection. The minor nutrients include the typical inorganic components of water. These are given in Table 22-34. The range of concentrations required in the wastewater for the minor nutrients is 1 to 100 mg/L. The micronutrients include the substances that we normally refer to as trace metals and vitamins. It is interesting to note that the trace metals include virtually all of the toxic heavy metals. This reinforces our earlier point that toxicity is a function of concentration and not an absolute parameter. Whether or not the substances referred to as vitamins will be required depends on the types of microorganisms required to stabilize the waste materials. Many microorganisms can make their own vitamins from the waste components; thus, a supplement is not needed. However, occasionally the addition of an external source of vitamins is essential to the success of a biologically based waste treatment system. In general, the trace nutrients must be present in a waste at a level of a few micrograms per liter.

One aspect of the basic equation for the biological treatment of waste we have not yet mentioned is that biomass appears on both sides of the equation. The only reason that microorganisms function in waste treatment systems is because they can reproduce. Thus, the quantity of biomass in a waste treatment system is higher after the treatment process than before it. This is favorable in that there is a continual production of the organisms required to stabilize the waste. Thus, one of the major reactants is, in effect, available free of charge. However, there is an unfavorable side in that unless some organisms are wasted from the system, an excess level will build up, and the process could choke on organisms. The wasted organisms are referred to as sludge. A major cost component of all biologically based processes is the need to provide for the ultimate disposal of this sludge.

TABLE 22-34 Minor- and Micronutrients Required for Biologically Mediated Reactors

Minor 1–100 mg/L
Sodium, potassium, calcium, magnesium, iron, chloride, sulfate
Micro 1–100 μg/L
Copper, cobalt, nickel, manganese, boron, vanadium, zinc, lead, molybdenum, various organic vitamins, various amino acids

Biologically based treatment processes probably account for most of the treatment systems used for industrial waste management because of their low cost and because most substances are amenable to biological breakdown. However, some substances are difficult to degrade biologically. Unfortunately, it is not possible at present to predict a priori the biodegradability of a specific organic compound; instead, we must depend on experience and testing. The collective experience of the field has been put into compendia by the EPA in a variety of documents. However, these data are primarily qualitative. There have been some attempts to develop a system of prediction of biodegradability based on a number of compound parameters, such as solubility, the presence or absence of certain functional groups, compound polarity, and so on. Unfortunately, none of these systems has advanced to the point where reliable quantitative predictions are possible. Another complication is that some organics that are easily biodegradable at low concentrations exert a toxic effect at high concentrations. Thus, literature data can be confusing. Phenol is a typical compound that shows ease of biodegradation when the concentration is below 500 mg/L but poor biodegradation at higher concentrations. Another factor affecting both biodegradation and toxicity is whether or not a substance is in solution. In general, if a substance is not in solution, it is not available to affect the biomass. Thus, the presence of a waste in substances that can precipitate, complex, or absorb other waste components can have a significant effect on reports of biodegradability or toxicity. A quantitative estimate of toxicity can be obtained in terms of the change in kinetic parameters of a system. These kinetic parameters are discussed next.

Design of Biological Treatment Systems In the past, the design of biologically based waste treatment systems has been derived from rules of thumb. During the past two decades, however, a more fundamental system has been developed, and it is now widely used to design such systems. This system is based on a fundamental understanding of the kinetics and stoichiometry of biological reactions. The system is codified in terms of equations in which four pseudo constants appear. These pseudo constants are k_m, the maximum substrate utilization rate (1/time); K_s, the half maximal velocity concentration (mg/L); Y, the yield coefficient; and b, the endogenous respiration rate (1/time). These are referred to as pseudo constants because, in the mathematical manipulation of the equations in which they appear, they are treated as constants. However, the value of each is a function of the nature of the microbes, the pH, the temperature, and the components of the waste. It is important to remember that if any of these change, the value of the pseudo constants may change as well. The kinetics of biological reactions are described by Eq. (22-20).

$$\frac{dS}{dt}=\frac{k_m SX}{K_s+S} \tag{22-20}$$

where t = time (days)
S = waste concentration (mg/L)
X = biomass concentration (mg/L)
k_m = (mg/L substrate ÷ mg/L biomass) – time

The accumulation or growth of biosolids is given by Eq. (22-21).

$$\frac{dX}{dt}=Y\frac{dS}{dt}-bX \tag{22-21}$$

where X = biomass level (mg/L).

The equations that have been developed for design using these pseudo constants are based on steady-state mass balances of the biomass and the waste components around both the reactor of the system and the device used to separate and recycle microorganisms. Thus, the equations that can be derived will be dependent on the characteristics of the reactor and the separator. It is impossible to present equations here for all the different types of systems. As an illustration, the equations for a common system (a complete-mix stirred tank reactor with recycle) follow.

$$S_e=\frac{K_s[1+\text{BSRT}(b)]}{\text{BSRT}(YK_m-b)-1} \tag{22-22}$$

$$X=\frac{(\text{BSRT})(Y)(S_o-S_e)}{\text{HRT}[1+(b)\text{BSRT}]} \tag{22-23}$$

where S_e = influent waste concentration (mg/L)
S_o = treated waste concentration (mg/L)
HRT = reactor hydraulic retention time

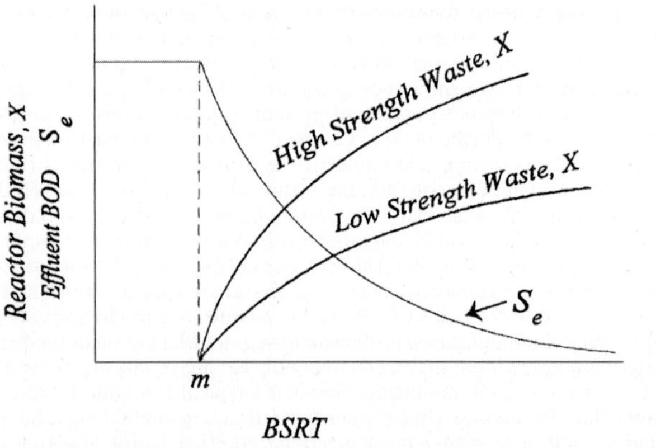

FIG. 22-36 Effect of BSRT on biological treatment process performance. m = minimum BSRT.

Note that these equations predict some unexpected results. The strength of the untreated waste (S_o) has no effect on the strength of the treated waste (S_e). Neither does the size of the reactor (HRT). Rather, a parameter referred to as the biomass solids retention time (BSRT) is the key parameter determining the system performance. This is illustrated in Fig. 22-36. As the BSRT increases, the concentration of untreated waste in the effluent decreases irrespective of the reactor size or the waste strength. However, the reactor size and waste strength have a significant effect on the level of biomass (X) that is maintained in the system at steady state. Since the development of an excess level of biomass in the system can lead to system upset, it is important to be aware of the waste strength and the reactor size. But treatment performance with respect to removal of the waste components is again a function only of BSRT and the value of the pseudo constants. The BSRT also has an effect on the level of biomass in the system and the quantity of excess biomass produced (Fig. 22-37). The latter will determine the quantity of waste sludge that must be dealt with as well as the N and P requirement. The N and P in the biomass removed from the system each day must be replaced. From the formula given previously for biomass, N is 12 percent and P is 2.3 percent by weight in biomass grown under ideal conditions (see Fig. 22-37). Also, note that m, the minimum BSRT, is the minimum time required for the microorganisms to double in mass. Below this minimum, washout occurs, and substrate removal approaches zero.

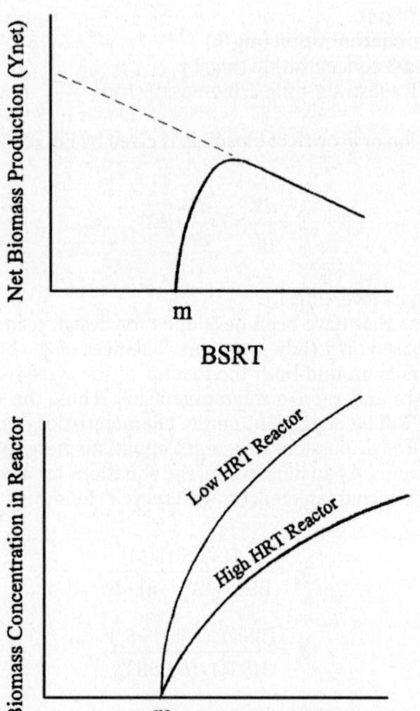

FIG. 22-37 Biomass production.

Although identical results are not obtained for other reactor configurations, the design equations yield similar patterns. The dominant parameters in determining system performance are again the BSRT and the pseudo constants. The latter are not under the control of the design engineer because they are functions of the waste and the microorganisms that developed in the system. The BSRT is the major design parameter under the control of the design engineer. This parameter has been defined as the ratio of the biomass in the reactor to the biomass produced from the waste each day. At steady state, the level of biomass in the system is constant; thus, the biomass produced must equal the biomass wasted. The minimum BSRT that can be used is that which will produce the degree of treatment required. Generally, this minimum value is less than the range of BSRT used in design and operation. This provides not only a safety factor, but, in addition, it is necessary in order to foster the growth of certain favorable groups of microorganisms in the system. Generally, BSRT values in the range of 3 to 15 days are used in most systems, although BSRT values of less than one day for high-rate systems are usually adequate to ensure greater than 95 percent destruction of waste components in all but anaerobic systems. For the latter, a minimum BSRT of eight days is needed for 95 percent destruction. BSRT is controlled by the concentration of biomass (X) in the system and the quantity wasted each day; thus, the treatment plant operator can alter BSRT by altering the rate of biomass wasting.

Reactor Configurations The two major types of biological reactors are suspended growth reactors and fixed-film or biofilm reactors. In the former, the waste and the microorganisms move through the reactor, with the microorganisms constantly suspended in the flow. After exiting the reactor, the suspension flows through a separator, which separates the organisms from the liquid. Some of the organisms are wasted as sludge, while the remainder are returned to the reactor. The supernatant is discharged either to the environment or to other treatment units (Fig. 22-38). The functioning of the separator is very important. Poor performance of the separator will result in high solids in the effluent and a reduction of organisms in the recycle and eventually also in the reactor. The level of the BSRT used in the design of suspended growth systems is set to produce a biomass of the proper level in the system—that is, a level at which waste degradation is rapid but not so high that excess loading on the separator will occur. BSRT also influences the ability of biomass to self-flocculate and thus be removed in the separator (usually a clarifier). A variety of reactor configurations are used in suspended activated sludge systems, each aimed at meeting different effluent criteria (e.g., BOD removal, nutrient removal).

In the fixed-film reactor, the organisms grow on an inert surface that is maintained in the reactor. The inert surface can be granular material, proprietary plastic packing, rotating discs, wood slats, mass-transfer packing, or even a sponge-type material. The reactor can be flooded or have a mixed gas–liquid space (Fig. 22-39). The biomass level on the packing is controlled by hydraulic scour produced as the waste liquid flows through the reactor. There is no need for a separator to ensure that the biomass level in the reactor is maintained. The specific surface area of the packing is the design parameter used to ensure an adequate level of biomass. However, a separator is usually supplied to capture biomass washed from the packing surface by the flow of the waste, thus providing a clarified effluent. As with the suspended growth system, if this biomass escapes the system, it will result in a return to the environment of organic-laden material, thus negating the effectiveness of the waste treatment. As indicated above, biomass level in a fixed-film reactor is maintained by a balance between the rate of growth on the packing (which is a function of the strength of the waste, the yield coefficient, and the BSRT), and the rate of hydraulic flushing by the waste flow. Recycle of treated wastewater is used to control the degree of hydraulic flushing. Thus, typical design parameters for fixed-film reactors include both an organic loading and a hydraulic loading. Details on the typical levels for these will be given in a later subsection.

The reactor concepts just described can be used with any of the electron acceptor systems previously discussed, although some reactor types perform better with specific electron acceptors. For example, suspended

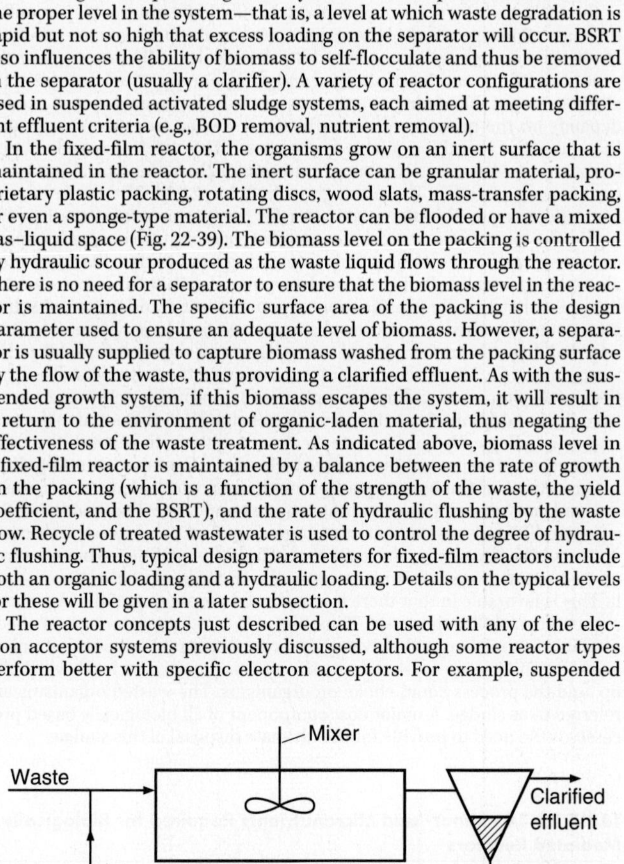

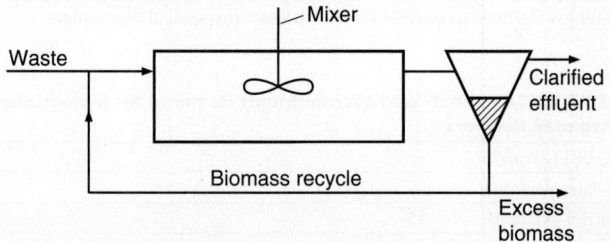

FIG. 22-38 Diagram of a suspended growth system.

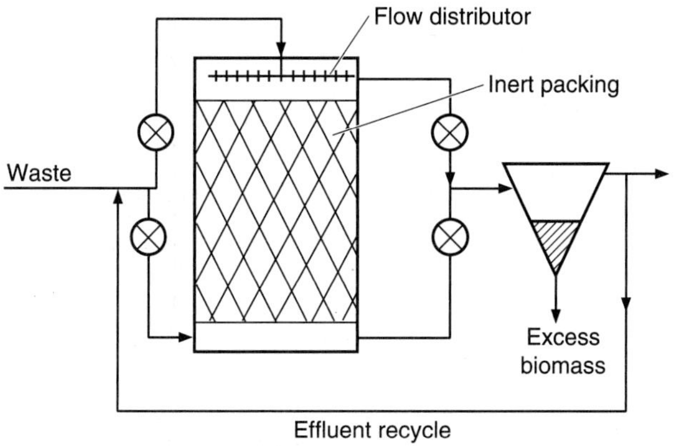

FIG. 22-39 Diagram of a fixed-film system.

growth systems are generally superior to fixed-film systems when molecular oxygen is the electron acceptor because it is easier to supply oxygen to suspended growth systems.

On the other hand, fixed-film systems that use nitrate as the electron acceptor can be superior to activated sludge systems since nitrogen bubbles produced won't affect sludge settling, and oxygen diffusion limitations are not an issue. Different reactor types are better suited for different strengths and types of wastewater. Weak wastewaters are generally easier to treat with fixed-film systems since the appropriate amount of biomass is inherently maintained in the system. Very strong wastes are typically better treated using anaerobic systems to eliminate oxygen limitation and to significantly reduce waste sludge production. In addition, the yield coefficient Y is much lower for anaerobic systems, reducing sludge production and excessive biomass accumulation. Different types and combinations of activated and fixed-film reactor types can be used to perform tertiary treatment on wastewaters and to reclaim hazardous waste streams that are contaminated with recalcitrant chemicals such as PCBs, halogenated aliphatics, and PAHs. Table 22-35 summarizes some of the advantages and disadvantages of various combinations of reactor type, electron acceptor, and waste strength.

Determination of Kinetic and Stoichiometric Pseudo Constants These parameters are most important for predicting the performance of biologically based treatment systems. It would be ideal if tabulations of these were available for various industrial wastes as a function of pH, temperature, and nutrient levels. Unfortunately, little reliable data have been codified. Only certain trends have been established, and these are primarily the result of studies on municipal wastewater. For example, the yield coefficient Y has been shown to be much higher for systems that are aerobic (molecular oxygen as the electron acceptor) than for anaerobic systems (sulfate or carbon dioxide as the electron acceptors). Systems where oxidized nitrogen is the electron acceptor (termed *anoxic*) exhibit yield values intermediate between aerobic and anaerobic systems. The endogenous respiration rate is higher for aerobic and anoxic systems than for anaerobic systems. However, no trends have been established for values of the maximum specific growth rate or the half maximal velocity concentration. Thus, the values applicable to a specific waste must be determined from laboratory studies.

The laboratory studies used small-scale (1- to 5-L) reactors. These are satisfactory because the reaction rates observed are independent of reactor size. Several reactors are operated in parallel on the waste, each at a different BSRT. When steady state is reached after several weeks, data on the biomass level (X) in the system and the untreated waste level in the effluent (usually in terms of BOD or COD) are collected. These data can be plotted for equation forms that will yield linear plots on rectangular coordinates. From the intercepts and the slope of the lines, it is possible to determine values of the four pseudo constants. Table 22-36 presents some available data from the literature on these pseudo constants. Figure 22-40 illustrates the procedure for their determination from the laboratory studies discussed previously.

Activated Sludge This treatment process is the most widely used aerobic suspended growth reactor system. It will consistently produce a high-quality effluent (BOD_5 and SS of 20–30 mg/L). Operational costs are higher than for other secondary treatment processes primarily because of the need to supply molecular oxygen using energy-intensive mechanical aerator- or sparger-type equipment. Removal of soluble organics, colloidal substances, particulates, and inorganics are achieved in this system through a combination of biological metabolism, adsorption, and entrapment in the biological floc. Indeed, many pollutants that are not biologically degradable are removed during activated sludge treatment by adsorption or entrapment by the floc. For example, most heavy metals form hydroxide or carbonate precipitates under the pH conditions maintained in activated sludge, and most organics are easily adsorbed to the surface of the biological floc. A qualitative guide to the latter is provided by the octanol-water partition coefficient of a compound.

All activated sludge systems include a suspended growth reactor in which the wastewater, recycled sludge, and molecular oxygen are mixed. The latter must be dissolved in the water; thus the need for an energy-intensive pure oxygen or air supply system. Usually, air is the source of the molecular oxygen rather than pure oxygen. Energy for mixing the reactor contents is supplied by the aeration equipment. All systems include a separator and pump station for sludge recycle and sludge wasting. The separator is usually a sedimentation tank that is designed to function as both a clarifier and a thickener. Many modifications of the activated sludge process have been developed over the years, and these are described next. Most of these involve differences in the way the reactor is compartmentalized with respect to the introduction of waste, recycle, or oxygen supply.

Modifications The modifications of activated sludge systems offer considerable choice in processes. Some of the most popular modifications of the activated sludge process are illustrated in Fig. 22-41. Conventional activated sludge uses a long, narrow reactor with air supplied along the length of the reactor. The recycle sludge and waste are introduced at the head end of the reactor, producing a zone of high waste to biomass concentration and high oxygen demand. This modification is used for relatively dilute wastes, such as municipal wastewater. Step aeration systems distribute the waste along the length of the aerator, thus reducing the oxygen demand at the head end of the reactor and spreading the oxygen demand more uniformly over the whole reactor. In the complete mix system, the waste and sludge recycle are uniformly distributed over the whole reactor, resulting in the waste load and oxygen demand being uniform in the entire reactor. Complete-mixing activated sludge is the most popular system for industrial wastes because of its ability to absorb shock loads better than other modifications. Contact stabilization is a modification of activated sludge that is best suited to wastewaters having high suspended solids and low soluble organics. Contact stabilization employs a short-term mixing tank to adsorb the suspended solids and metabolize the soluble organics, a sedimentation tank for solids separation, and a reaeration tank for stabilization of the suspended organics. Extended-aeration systems are actually long-term-aeration, completely mixed activated sludge systems. They employ 24- to 48-hour aeration periods and high mixed-liquor suspended solids to provide complete stabilization of the organics and aerobic digestion of the activated sludge in the same aeration tank. The oxidation ditch is a popular form of the extended-aeration system that uses mechanical aeration. Pure-oxygen systems are designed to treat strong industrial wastes in a series of completely mixed units having relatively short contact periods. One of the latest modifications of activated sludge employs powdered activated carbon to adsorb complex organics and to assist in solids separation. Another modification employs a redwood-medium trickling filter ahead of a short-term aeration tank with mixed liquor recycled over the redwood-medium tower to provide heavy microbial growth on the redwood as well as in the aeration tank.

As indicated previously, success with the activated sludge process requires that the biomass have good self-flocculating properties. Significant research effort has been expended to determine the conditions that favor the development of good settling biomass cultures. These have

TABLE 22-35 Favorable (F) and Unfavorable (U) Combinations of Electron Acceptor, Waste Strength, and Reactor Type

Suspended growth reactor		
Electron acceptor	Waste strength	Condition
Aerobic	Low–modest	F
Aerobic	High	U
Anoxic	Low–modest	F
Anoxic	High	U
Anaerobic	Low–modest	U
Anaerobic	High	F
Fixed-film reactor		
Electron acceptor	Waste strength	Condition
Aerobic	Low	F
Aerobic	Modest–high	U
Anoxic	Low–modest	F
Anoxic	High	U
Anaerobic	Low–modest	F
Anaerobic	High	F

TABLE 22-36 Typical of Values for Pseudo Constants

Biomass type	Substrate	k_{max}, mg substrate/mg biomass · day	Y, mg biomass/mg substrate	K_s, mg substrate/liter	b, 1/day	Remarks
Mixed culture	Sewage (COD)	5–10	0.5	50	0.05	Aerobic 20°C
Mixed culture	Glucose (COD)	7.5	0.6	10	0.07–.1	Aerobic 20°C
Mixed culture	Skim milk (COD)	5	—	100	0.05	Aerobic 20°C
Mixed culture	Soybean waste (COD)	12	—	355	0.144	Aerobic 20°C
Methane bacteria	Acetic acid (COD)	8.7	.04	165	0.035	35°C
Anaerobic mixed	Propanoic acid (COD)	7.7	.04	60	0.035	35°C
Anaerobic mixed	Sewage sludge (COD)	6.7	.04	1,800	0.03	35°C
Anoxic	NO_3 as N	0.375	0.8	0.1	0.04	Methanol feed
Aerobic	NH_4^+ as N	5	0.2	1.4	.05	20°C

indicated that nutritional deficiency, levels of dissolved oxygen between 0 and 0.5 mg/L, and pH values below 6.0 will favor the predomination of filamentous biomass. Filamentous organisms settle and compact poorly and thus are difficult to separate from liquid. By avoiding the preceding conditions, and with the application of selector technology, the predomination of filaments in the biomass can be eliminated. A selector is often a short contact (15–30 min) reactor set ahead of the main activated sludge reactor. All of the recycle sludge and all of the waste are routed to the selector. In the selector, either a high rate of aeration (aerobic selector) is used to keep the dissolved oxygen above 2 mg/L, or no aeration occurs (anaerobic selector) so that the dissolved oxygen is zero. Because filamentous organisms are microaerophilic, they cannot predominate when the dissolved oxygen level is either zero or high. The anaerobic selector not only selects in favor of nonfilamentous biomass, but it also fosters luxury uptake of phosphorus, and it is used in systems where phosphorus removal is desired.

Aeration Systems These systems control the design of aeration tanks. Aeration equipment has two major functions: mixing and oxygen transfer. Diffused-aeration equipment uses either a fixed-speed, positive-displacement blower or a high-speed turbine blower for readily adjustable air volumes. Air diffusers can be located along one side of the aeration tank or spread over the entire bottom of the tank. They can be either fine-bubble or coarse-bubble diffusers. Fine-bubble diffusers are more efficient in oxygen transfer but require more extensive air-cleaning equipment to prevent them from clogging as a result of dirty air. Mechanical-surface-aeration equipment is more efficient than diffused-aeration equipment but is not as flexible. Economics has dictated the use of large-power aerators, but tank configuration has tended to favor the use of greater numbers of lower-power aerators. Oxidation ditches use horizontal rotor-type aerators. Mechanical failure caused by wind and wave forces can be a serious problem with mechanical-aeration equipment. Slow-speed mechanical-surface-aeration units present fewer problems than the high-speed mechanical-surface-aeration units. Deep tanks, greater than 3.0 m (10 ft), require draft tubes to ensure proper hydraulic flow through the aeration tank. Short-circuiting is one of the major problems associated with mechanical aeration equipment. Combined mechanical- and diffused-aeration systems have enjoyed some popularity for industrial-waste systems that treat variable organic loads. The mechanical mixers provide the fluid mixing with the diffused aeration varied for different oxygen-transfer rates.

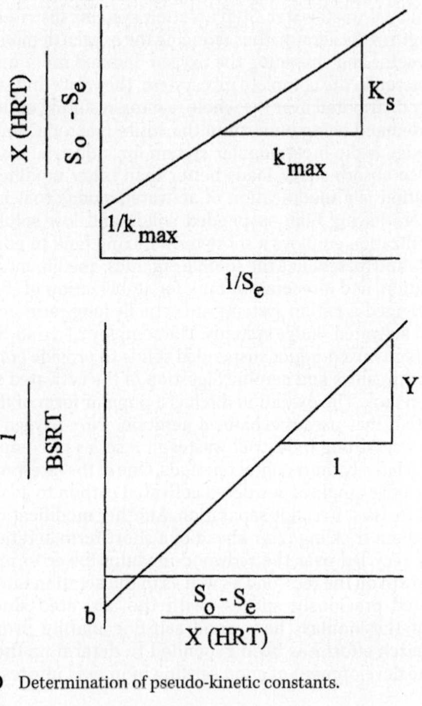

FIG. 22-40 Determination of pseudo-kinetic constants.

Diffused-aeration systems transfer from 20 to 40 mg/(L O_2·h). Combined mechanical- and diffused-aeration systems can transfer up to 65 mg/(L O_2·h), while mechanical-surface aerators can provide up to 90 mg/(L O_2·h). Pure-oxygen systems can provide the highest oxygen-transfer rate, up to 150 mg/(L O_2·h). Aeration equipment must provide sufficient oxygen to meet the peak oxygen demand; otherwise, the system will fail to provide proper treatment. For this reason, the peak oxygen demand and the rate of transfer for the desired equipment determine the size of the aeration tank in terms of retention time. Economics dictates a balance between the size of the aeration tank and the size of the aeration equipment. As the cost of power increases, economics will favor building a larger aeration tank and smaller aerators. It is equally important to examine the hydraulic flow pattern around each aerator to ensure maximum efficiency of oxygen transfer. Improper spacing of aeration equipment can waste energy.

There is no standard aeration-tank shape or size. Aeration tanks can be round, square, or rectangular. Shallow aeration tanks are more difficult to mix than deeper tanks. Yet aeration-tank depths have ranged from 0.6 m (2 ft) to 18 m (60 ft). The oxidation-ditch systems tend to be shallow, while some high-rate diffused-aeration systems have used very deep tanks to provide more efficient oxygen transfer.

Regardless of the aeration equipment used, oxygen-transfer rates must provide from 0.6 to 1.4 kg of oxygen/kg BOD_5 (0.6 to 1.4 lb oxygen/lb BOD_5) stabilized in the aeration tank for carbonaceous-oxygen demand. Nitrogen oxidation can increase oxygen demand at the rate of 4.3 kg (4.3 lb) of oxygen/kg (lb) of ammonia nitrogen oxidized. At low oxygen-transfer rates, more excess activated sludge must be removed from the system than at high oxygen-transfer rates. Here again the economics of sludge handling must be balanced against the cost of oxygen transfer. The quantity of waste activated sludge will depend on wastewater characteristics. The inert suspended solids entering the treatment system must be removed with the excess activated sludge. The soluble organics are stabilized by converting a portion of the organics into suspended solids, producing from 0.3 to 0.8 kg (0.3 to 0.8 lb) of volatile suspended solids/kg (lb) of BOD_5 stabilized. Biodegradable suspended solids in the wastewaters will result in destruction of the original suspended solids and their conversion to a new form. Depending on the chemical characteristics of the biodegradable suspended solids, the conversion factor will range from 0.7 to 1.2 kg (0.7 to 1.2 lb) of microbial solids produced/kg (lb) of suspended solids destroyed. If the suspended solids produced by metabolism are not wasted from the system, they will eventually be discharged in the effluent. While considerable efforts have been directed toward developing activated sludge systems that totally consume the excess solids, no such system has proved to be practical. The concept of total oxidation of excess sludge is fundamentally unsound and should be recognized as such.

A definitive determination of the waste sludge production and the oxygen requirement can be obtained using the pseudo constants referred to previously. The ultimate BOD in the waste will be accounted for by the sum of the oxidation and sludge synthesis [Eq. (22-24)].

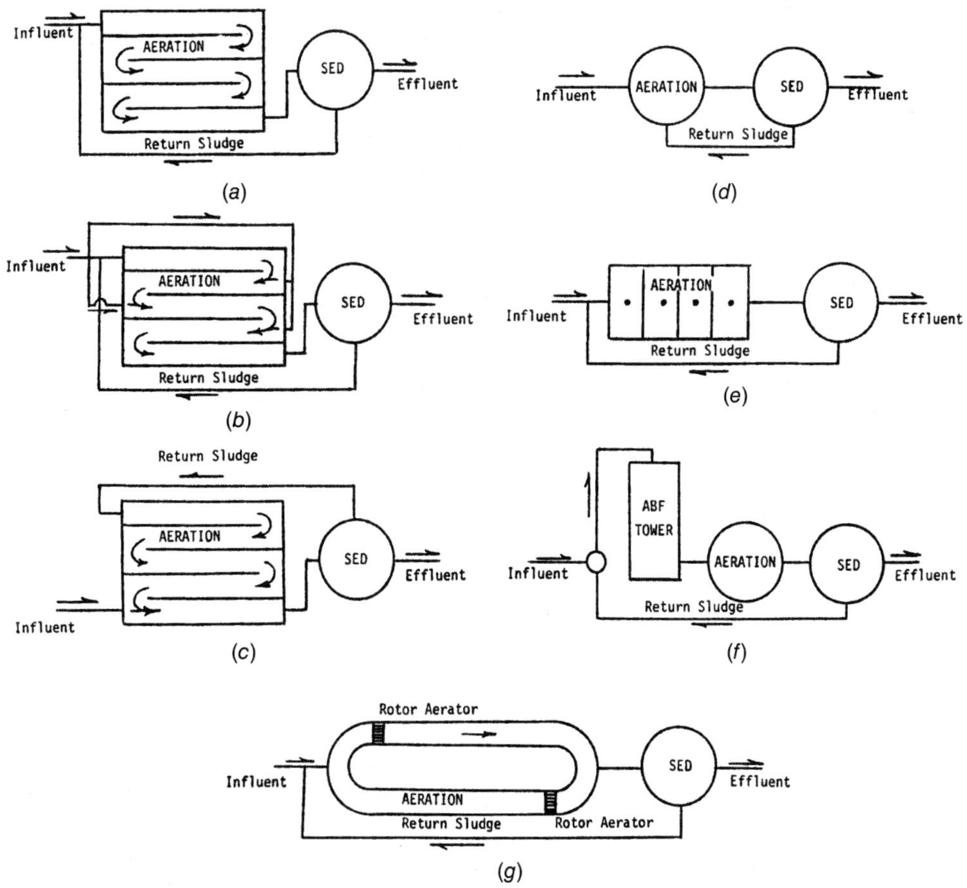

FIG. 22-41 Schematic diagrams of various modifications of the activated sludge process. (*a*) Conventional activated sludge. (*b*) Step aeration. (*c*) Contact stabilization. (*d*) Complete mixing. (*e*) Pure oxygen. (*f*) Activated biofiltration (ABF). (*g*) Oxidation ditch.

Thus:

$$\text{Waste BOD}_U = \text{oxygen required} + \text{BOD}_U \text{ of sludge produced} \qquad (22\text{-}24)$$

where: BOD_U of sludge = $(Y_{\text{net}})(1.42)(\text{Waste BOD}_U)$

$$Y_{\text{net}} = \frac{Y}{1 + b(\text{BSRT})}$$

1.42 = Factor converting biomass to oxygen units; i.e., $\text{BOD}_U \approx \text{COD}$

Sedimentation Tanks These tanks are an integral part of any activated sludge system. It is essential to separate the suspended solids from the treated liquid if a high-quality effluent is to be produced. Circular sedimentation tanks with various types of hydraulic sludge collectors have become the standard secondary sedimentation system. Square tanks have been used with common-wall construction for compact design with multiple tanks. Most secondary sedimentation tanks use center-feed inlets and peripheral-weir outlets. Recently, efforts have been made to employ peripheral inlets with submerged-orifice flow controllers and either center-weir outlets or peripheral-weir outlets adjacent to the peripheral-inlet channel.

Aside from flow control, basic design considerations have centered on surface overflow rates, retention time, and weir overflow rate. Surface overflow rates have been slowly reduced from 33 $m^3/(m^2 \cdot day)$ [800 gal/($ft^2 \cdot day$)] to 24 $m^3/(m^2 \cdot day)$ to 16 $m^3/(m^2 \cdot day)$ [600 gal/($ft^2 \cdot day$) to 400 gal/($ft^2 \cdot day$)] and even to 12 m^3 ($m^2 \cdot day$) [300 gal/($ft^2 \cdot day$)] in some instances, based on average raw-waste flows. Operational results have not demonstrated that lower surface overflow rates improve effluent quality, making 33 $m^3/(m^2 \cdot day)$ [800 gal/($ft^2 \cdot day$)] the design choice in most systems. Retention time has been found to be an important design factor, averaging 2 h on the basis of raw-waste flows. Longer retention periods tend to produce rising sludge problems, while shorter retention periods do not provide for good solids separation with high-return sludge flow rates. Effluent-weir overflow rates have been limited to 186 $m^3/(m \cdot day)$ [15,000 gal/($ft \cdot day$)], with a tendency to reduce the rate to 124 $m^3/(m \cdot day)$ [10,000 gal/($ft \cdot day$)]. Lower effluent-weir overflow rates are obtained by using dual-sided effluent weirs cantilevered from the periphery of the tank. Unfortunately, proper adjustment of dual-side effluent weirs has created more hydraulic problems than the weir overflow rate. Field data have shown that effluent quality is not really affected by weir overflow rates up to 990 $m^3/(m \cdot day)$ [80,000 gal/($ft \cdot day$)] or even 1240 $m^3/(m \cdot day)$ [100,000 gal/($ft \cdot day$)] in a properly designed sedimentation tank. A single peripheral weir, being easy to adjust and keep clean, appears to be optimal for secondary sedimentation tanks from an operational point of view.

Depth tends to be determined from the retention time and the surface overflow rate. As surface overflow rates were reduced, the depth of sedimentation tanks was reduced to keep retention time from being excessive. It was recognized that depth was a valid design parameter and was more critical in some systems than retention time. As mixed-liquor suspended-solids (MLSS) concentrations increase, the depth should also be increased. Minimum sedimentation-tank depths for variable operations should be 3.0 m (10 ft) with depths to 4.5 m (15 ft) if 3000 mg/L MLSS concentrations are to be maintained under variable hydraulic conditions. With MLSS concentrations above 4000 mg/L, the depth of the sedimentation tank should be increased to 6.0 m (20 ft). The key is to keep a definite freeboard over the settled-sludge blanket so that variable hydraulic flows do not lift the solids over the effluent weir.

Scum baffles around the periphery of the sedimentation tank and radial scum collectors are standard equipment to ensure that rising solids or other scum materials are removed as quickly as they form. Hydraulic sludge-collection tubes have replaced the center sludge well, but they have caused a new set of operational problems. These tubes were designed to remove the settled sludge at a faster rate than conventional sludge scrapers. To obtain good hydraulic distribution in the sludge-collection tubes, it was necessary to increase the rate of return sludge flow and decrease the concentration of return sludge. The higher total inflow to the sedimentation tank created increased forces that lifted the settled-solids blanket at the wall, causing loss of excessive suspended solids and lower effluent quality. Operating data tend to favor conventional secondary sedimentation tanks over hydraulic sludge-collection systems. Return-sludge rates normally range from 25 to 50 percent for MLSS concentrations up to 3300 mg/L. Most return-sludge pumps are centrifugal pumps with capacities up to 100 percent raw-waste flow.

Gravity settling can concentrate activated sludge to 10,000 mg/L, but hydraulic sludge-collecting tubes tend to operate best below 8000 mg/L. The excess activated sludge can be wasted either from the return sludge or from a separate waste-sludge hopper near the center of the tank. The low solids concentrations result in large volumes of waste activated sludge in comparison with primary sludge. Unfortunately, the physical characteristics of waste activated sludge prevent significant concentration without the expenditure of considerable energy. Gravity thickening can produce 2 percent solids, while air flotation can produce 4 percent solids concentration. Centrifuges can concentrate activated sludge from 10 to 15 percent solids, but the capture is limited. Vacuum filters can equal the performance of centrifuges if the sludge is chemically conditioned. Filter presses and belt-press filters can produce cakes with 15 to 25 percent solids. It is very important that the excess activated sludge formed in the aeration tanks be wasted on a regular basis; otherwise, effluent quality will deteriorate. Care should be taken to ensure that sludge-thickening systems do not control activated sludge operations. Alternative sludge-handling provisions should be available during maintenance on sludge-thickening equipment. At no time should final sedimentation tanks be used for the storage of sludge beyond that required by daily operational variations.

Anaerobic/Anoxic Activated Sludge The activated sludge concept (i.e., suspended growth reactor) can be used for anaerobic or anoxic systems in which no oxygen or air is added to the reactor. An anoxic activated sludge is used for systems in which the removal of nitrate is a goal or where nitrate is used as the electron acceptor. These systems (denitrification) will be successful if the nitrate is reduced to low levels in the reactor so that nitrate reduction to nitrogen gas does not take place in the clarifier-thickener. Nitrogen gas production in the clarifier will result in the escape of biological solids with the effluent as nitrogen bubbles floating sludge to the clarifier surface. For nitrogen reduction, a source of organics (electron donor) is required. Any inexpensive carbohydrate can be effectively used for nitrate removal. Many systems use methanol as the donor because it is rapidly metabolized; others use the organics in sewage in order to reduce chemical costs. Nitrate reduction is invariably used as part of a system in which organics and nitrogen removal are goals. In such systems, the nitrogen in the waste is first oxidized to nitrate and then reduced to nitrogen gas. Figure 22-42 presents some flow sheets for such systems. Table 22-37 presents some design parameters for nitrate-reduction-activated sludge systems. Anaerobic activated sludge has been used for strong industrial wastes high in degradable organic solids. In these systems, a high rate of gasification takes place in the sludge separator so that a highly clarified effluent is usually not obtained. A vacuum degasifier is incorporated in such systems to reduce solids loss. Such systems have been used primarily with meat packing wastes that are warm, high in BOD, and yield a high level of bicarbonate buffer as a result of ammonia release from protein breakdown. All of these conditions favor anaerobic processing. The use of this process scheme provides a high BSRT (15–30 days), usually required for anaerobic treatment, at a low HRT (1–2 days).

Lagoons Lagoons are low-cost, easy-to-operate wastewater treatment systems capable of producing satisfactory effluents. Nominally, a lagoon is a suspended-growth, no-recycle reactor with a variable degree of mixing. In lagoons in which mechanical or diffused aeration is used, mixing may be sufficient to approach complete mixing (i.e., solids maintained in suspension). In other types of lagoons, most solids settle and remain on the lagoon bottom, but some mixing is achieved as a result of gas production from bacterial metabolism and wind action. Lagoons are categorized as aerobic, facultative, or anaerobic on the basis of degree of aeration. Aerobic lagoons primarily depend on mechanical or diffused air supply. A facultative lagoon is dependent primarily on natural surface aeration and oxygen generated by algal cells. These two lagoon types are relatively shallow to encourage surface aeration and to provide for maximum algal activity. The third type of lagoon is maintained under anaerobic conditions to foster methane fermentation. This system is often covered with floating polystyrene panels to block surface aeration and to help prevent a drop in temperature. Anaerobic lagoons are several meters deeper than the other two types. Lagoon flow schemes can be complex, employing lagoons in series and recycle from downstream to upstream lagoons. The major effect of recycle is to maintain control of the solids. If solids escape the lagoon system, a poor effluent is produced. Periodically controlled solids removal must take place or solids will escape.

Lagoons are, in effect, inexpensive reactors. They are shallow basins either cut below grade or formed by dikes built above grade or a combination of a cut and a dike. The bottom must be lined with an impermeable barrier and the sides protected from wind erosion. These systems are best used where large areas of inexpensive land are available.

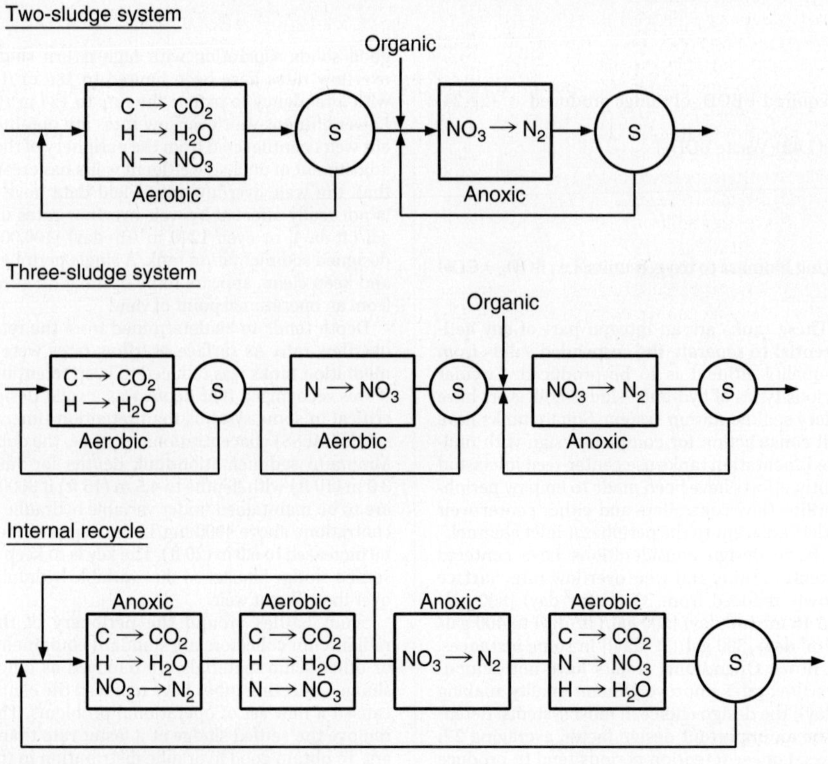

FIG. 22-42 Nitrogen removal systems.

TABLE 22-37 Design Parameter for Nitrogen Removal

System	BSRT, d	HRT, h	X, mg/L	pH
Carbon removal†	2–5	1–3	1000–2000	6.5–8.0
Nitrification†	10–20	0.5–3	1000–2000	7.4–8.6
Denitrification†	1–5	0.2–2	1000–2000	6.5–7.0
Internal recycle*	10–40	8–20	2000–4000	6.5–8.0

*Total for all reactors
†Separate reactors

Facultative Lagoons These lagoons have been designed to use both aerobic and anaerobic reactions. Normally, facultative lagoons consist of two or more cells in series. The settleable solids tend to settle out in the first cell and undergo anaerobic metabolism with the production of organic acids and methane gas, which bubbles out to the atmosphere. Algae at the surface of the lagoon use sunlight for their energy in converting carbon dioxide, water, and ammonium ions into algal protoplasm with the release of oxygen as a waste product. Aerobic bacteria use the oxygen released by the algae to stabilize the soluble and colloidal organics. Thus, the bacteria and algae form a symbiotic relationship as shown in Fig. 22-43. The interesting aspect of facultative lagoons is that the organic matter in the incoming wastewaters is not stabilized but rather is converted to microbial protoplasm, which has a slower rate of oxygen demand. In fact, in some facultative lagoons inorganic compounds in the wastewaters are converted to organic compounds with a total increase in organics within the lagoon system.

Facultative lagoons are designed on the basis of organic load in relationship to the potential sunlight availability. In the northern United States, facultative lagoons are designed on the basis of 2.2 g/(m^2·day) [20 lb BOD_5/(acre·day)]. In the middle part of the United States, the organic load can be increased to 3.4 to 4.5 g/(m^2·day) [30 to 40 lb BOD_5/(acre·day)], while in the southern part the organic load can be increased to 6.7 g/(m^2·day) [60 lb BOD_5/(acre·day)]. The depth of lagoons is normally maintained between 1.0 and 1.7 m (3 and 5 ft). A depth less than 1.0 m (3 ft) encourages the growth of aquatic weeds and permits mosquito breeding. In dry areas, the maximum depth may be increased above 1.7 m (5 ft), depending on evaporation. Most facultative lagoons depend on natural wind action for mixing and should not be placed in screened areas where wind action is blocked.

Effluent quality from facultative lagoons is related primarily to the suspended solids created by living and dead microbes. The long retention period in the lagoons allows the microbes to die off, leaving a small particle that settles slowly. The release of nutrients from the dead microbes permits the algae to survive by recycling the nutrients. Thus, the algae determine the ultimate effluent quality. The use of series ponds with well-designed transfer structures between ponds permits maximum retention of algae within the ponds and the best-quality effluent. Normally the soluble BOD_5 is under 5 or 10 mg/L with a total effluent BOD_5 under 30 mg/L. The effluent suspended solids will vary widely during the different seasons of the year, being a maximum of 70 to 100 mg/L in the summer months and a minimum of 10 to 20 mg/L in the winter months. If suspended-solids removal is essential, chemical precipitation is the best method available at present. Slow sand filters and rock filters have been studied for suspended-solids removal; they work well as long as the effluent suspended solids are relatively low, 40 to 70 mg/L.

Aerated Lagoons These lagoons originated from efforts to control overloaded facultative lagoons. Since the lagoons were deficient in oxygen, additional oxygen was supplied by either mechanical surface aerators or diffused aerators. Mechanical surface aerators were quickly accepted as the primary aerators because they could be quickly added to existing ponds and moved to strategic locations. Unfortunately, the high-speed, floating surface aeration units were not efficient, and large numbers were required for existing lagoons. The problem was simply one of poor mixing in a very shallow lagoon.

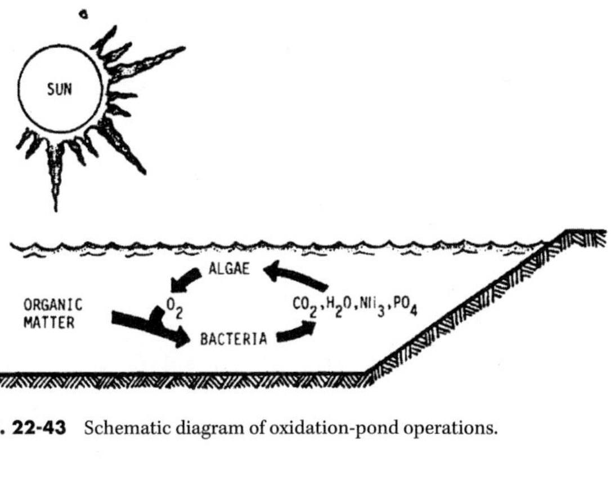

FIG. 22-43 Schematic diagram of oxidation-pond operations.

Eventually, diffused aeration equipment was added to relatively deep lagoons [3.0 to 6.0 m (10 to 20 ft)]. Mixing became the most significant parameter for good oxygen transfer in aerated lagoons. From an economical point of view, it was found that a completely mixed aerated lagoon with 24-hour retention provided the best balance between mixing and oxygen transfer. As the organic load increased, the fluid-retention time also increased. Short-term aeration permitted metabolism of the soluble organics by the bacteria, but time did not permit metabolism of the suspended solids. The suspended solids were combined with the microbial solids produced from metabolism and discharged from the aerated lagoon to a solids-separation pond. Data from the short-term aerated lagoon indicated that 50 percent BOD_5 stabilization occurred, with conversion of the soluble organics to microbial cells. The problem was separation and stabilization of the microbial cells. Short-term sedimentation ponds permitted separation of the solids without significant algae growths but required cleaning at frequent intervals to keep them from filling with solids and flowing into the effluent. Long-term lagoons permitted solids separation and stabilization but also permitted algae to grow and affect effluent quality.

Aerated lagoons were simply dispersed microbial reactors that permitted conversion of the organic components in the wastewaters to microbial solids without stabilization. The residual organics in solution were very low, less than 5 mg/L BOD_5. By adding oxygen and improving mixing, the microbial metabolism reaction was speeded up, but the stabilization of the microbial solids has remained a problem to be solved.

Anaerobic Lagoons These lagoons were developed when a major fraction of the organic contaminants consisted of suspended solids that could be removed easily by gravity sedimentation. The anaerobic lagoons are relatively deep [8.0 to 6.0 m (10 to 20 ft)], with a short fluid-retention time (3 to 5 days) and a high BOD_5 loading rate, up to 3.2 kg/(m^3·day) [200 lb/(1000 ft^3·day)]. Microbial metabolism in the settled-solids layer produces methane and carbon dioxide, which quickly rise to the surface, carrying some of the suspended solids. A scum layer that retards oxygen transfer and release of obnoxious gases is quickly produced in anaerobic lagoons. Mixing with a grinder pump can provide a better environment for metabolism of the suspended solids. The key for anaerobic lagoons is adequate buffer to keep the pH between 6.5 and 8.0. Protein wastes have proved to be the best pollutants to be treated by anaerobic lagoons, with the ammonium ions reacting with carbon dioxide and water to form ammonium bicarbonate as the primary buffer. High-carbohydrate wastes are poor in anaerobic lagoons since they produce organic acids without adequate buffer, making it difficult to maintain a suitable pH for good microbial growth.

Anaerobic lagoons do not produce a high-quality effluent but can reduce the BOD load by 80 to 90 percent with a minimum of effort. Since anaerobic lagoons work best on strong organic wastes, their effluent must be treated by either aerated lagoons or facultative lagoons. An anaerobic lagoon is simply the first stage in the treatment of strong organic wastewaters.

Fixed-Film Reactor Systems A major advantage of fixed-film systems is that a flocculent-type biomass is not necessary as the biomass remains in the reactor attached to inert packing. Biomass does periodically slough off or break away from the packing, usually in large chunks that can be easily removed in a clarifier. On the other hand, the time of contact between the biomass and the waste is much shorter than in suspended growth systems, making it difficult to achieve the same degree of treatment, especially in aerobic systems. Aerobic, anoxic, and anaerobic fixed-film systems are used for waste treatment.

Aerobic systems, including trickling filters and rotating biological contactors (RBC), are operated in a nonflooded mode to ensure adequate oxygen supply. Other aerobic, anoxic, and anaerobic systems employ flooded reactors. The most common systems are packed beds (anaerobic trickling filter) and fluidized or expanded bed systems.

Trickling Filters For years, trickling filters were the mainstay of biological wastewater treatment systems because of their simplicity of design and operation. Trickling filters have been replaced by activated sludge to achieve better effluent quality. Trickling filters are simply fixed-medium biological reactors with the wastewaters being spread over the surface of the fixed media where biofilm microbes grow and remove organics from the wastewater. Oxygen from the air permits aerobic reactions to occur at the surface of the microbial layer, but anaerobic metabolism occurs at the bottom of the microbial layer where oxygen does not penetrate.

Originally, the medium in trickling filters was rock, but rock has largely been replaced by plastic, which provides greater void space per unit of surface area and occupies less volume within the filter. A plastic medium permitted trickling filters to be increased from a medium depth of 1.8 m (6 ft) to one of 4.2 m (14 ft) and even 6.0 m (20 ft). The wastewaters are normally applied by a rotary distributor or a fixed-spray nozzle. The spraying or discharging of wastewaters above the trickling-filter medium permits better distribution over the medium and oxygen transfer before reaching the medium. The effluent from the trickling-filter medium is captured in a claytile underdrain system or in a tank below the plastic medium. It is important that the bottom of the trickling filter be open for air to move quickly through the filter and bring adequate oxygen for the microbial reactions.

If a high-quality effluent is required, trickling filters must be operated at a low hydraulic-loading rate and a low organic-loading rate. Low-rate trickling filters are operated at hydraulic loadings of 2.2×10^{-5} to 4.3×10^{-5} $m^3/(m^2 \cdot s)$ [2 million to 4 million gal/(acre·day)]. High-rate trickling filters are designed for 10.8×10^{-5} to 40.3×10^{-5} $m^3/(m^2 \cdot s)$ [10 million to 40 million gal/(acre·day)] hydraulic loadings and organic loadings up to 1.4 kg/(m^3·day) [90 lb BOD_5/(1000 ft^3·day)]. Plastic-medium trickling filters have been designed to operate at up to 108×10^{-5} $m^3/(m^2 \cdot s)$ [100 million gal/(acre·day)] or even higher, with organic loadings up to 4.8 kg/(m^3·day) [300 lb BOD_5/(1000 ft^3·day)]. Low-rate trickling filters will produce better than 90 percent BOD_5 and suspended-solids reductions, while high-rate trickling filters will produce from 65 to 75 percent BOD_5 reduction. Plastic-medium trickling filters will produce from 59 to 85 percent BOD_5 reduction, depending on the organic-loading rate. It is important to recognize that concentrated industrial wastes will require considerable hydraulic recirculation around the trickling filter to obtain the proper hydraulic-loading rate without excessive organic loads. With high recirculation rates, the organic load is distributed over the entire volume of the trickling filter for maximum organic removal. The short fluid-retention time within the trickling filter is the primary reason for the low treatment efficiency.

Rotating Biological Contactors (RBCs) Rotating biological contactors consist of a series of circular plastic disks, 3.0 to 3.6 m (10 to 12 ft) in diameter, immersed to approximately 40 percent diameter in a contact tank. The RBC disks rotate at 2 to 5 r/min. As the disks travel through the wastewaters, a layer of wastewater adheres to the surface. As the disks travel into the air, the microbes on the disk surface oxidize the organics in the wastewater. Thus, only a small amount of energy is required for aeration. Movement of the disks through the water creates a shear force that controls the thickness of the biofilm on the disks.

Rotating biological contactors have been very popular in treating industrial wastes because of their relatively small size and their low energy requirements. However, mechanical failures, limited treatment efficiency, and the production of odors have reduced their use. Existing RBCs are often covered with plastic shells and equipped with odor control systems, including chemical scrubbers and activated carbon adsorption units. RBC units also work best under uniform organic loads, requiring equalization tanks for many industrial waste streams.

RBCs are designed based on both a hydraulic-loading rate and an organic-loading rate. Normally, hydraulic-loading rates of up to 0.16 $m^3/(m^2 \cdot day)$ [4 gal/(ft^2·day)] of surface area are used, with organic loading rates up to 44 kg/(m^2·day) [9 lb BOD_5/(ft^2·day)]. Treatment efficiency is primarily a function of hydraulic retention time and organic-loading rate. At low organic-loading rates, RBC units will produce nitrified effluent similar to low-rate trickling filters.

Packed-Bed Fixed-Film Systems These systems are essentially anaerobic trickling filters. A wide variety of packed media is used to support biofilm growth and accumulation, ranging in size from 40 mesh granules to 7.5-cm (3-in) stones. Many systems use open structure plastic packing similar to that used in aerobic trickling filters.

The systems using granular media packing are used for anoxic denitrification. They are usually downflow, thus serving the dual function of filtration and denitrification. Contact times are short (EBCT <15 min), but excellent removal is achieved due to the high level of biomass retained in the reactor. Pacing the methanol dose to the varying feed nitrate concentration is crucial. Frequent, short-duration backwash (usually several times per day) is required or the nitrogen bubbles formed will bind the system, causing poor results. Extended backwash every two to three days is required or the system will clog on the biomass growth. Thus, several units in parallel or a large holding tank are needed to compensate for the downtime during backwash. Backwash does not remove all the biomass; a thin film remains, coating the packing. Thus, denitrification begins immediately when the flow is restored.

The systems using the larger packing are used in the treatment of relatively strong, low-suspended-solids industrial waste. These systems are closed columns, usually run in an upflow mode with a gas space at the top. These are operated under anaerobic conditions, with waste conversion to methane and carbon dioxide as the goal. Effluent recycle is often used to help maintain the pH in the inlet zone in the correct range of 6.5 to 7.5 for the methane bacteria. Some wastes require the addition of alkaline material to prevent a pH drop. Sodium bicarbonate is often recommended for pH control because it is easier to handle than lime or sodium hydroxide, and because an overdose of bicarbonate will only raise the pH modestly. An overdose of lime or sodium hydroxide can easily raise the pH above 8.0. Table 22-38 gives some performance data for systems treating industrial wastes. HRTs of one to two days are used, as the buildup of growth on the packing ensures a BSRT of 20 to 50 days. It should be possible to lower the HRT further, but in practice this has not been successful because biomass starts to escape from the system, or plugging occurs. Some escape is due to high gasification rates, and some is due to the fact that anaerobic sludge attaches less tenaciously to packing than aerobic or anoxic sludge. These systems can handle wastes with moderate solids levels. Periodically, solids must be removed from the reactor to prevent plugging of the packing or loss of solids in the effluent.

Biological Fluidized Beds This high-rate process has been used successfully for aerobic, anoxic, and anaerobic treatment of municipal and industrial wastewaters. Many small- and large-scale applications for hazardous waste, contaminated groundwater, nontoxic industrial waste, and municipal wastewater have been reported (Tabak, H., et al., "Biodegradability Studies with Organic Priority Pollutant Compounds," United States EPA, MERL, Cincinnati, Ohio, April 1980; Levin, M. A., and M. A. Gealt, *Biotreatment of Industrial and Hazardous Waste*, McGraw-Hill, New York, 1993). The basic element of the process is a bed of solid carrier particles, such as sand or granular activated carbon, placed in a reactor through which wastewater is passed upflow with sufficient velocity to impart motion or to fluidize the carrier. An active biofilm develops on carrier particles, which removes contaminants from the wastewater through biological and adsorptive mechanisms. Figure 22-44 is a schematic of the process.

The influent wastewater enters the reactor through a pipe manifold and is introduced downflow through nozzles that distribute the flow uniformly at the base of the reactor. Reversing direction at the bottom, the flow fluidizes the carrier when the fluid drag overcomes the buoyant weight of the carrier and its attached biomass layer. During start-up (before much biomass has accumulated), the flow velocity required to achieve fluidization

TABLE 22-38 Anaerobic Process Performance on Industrial Wastewater UASB, Submerged Filter (SF), FBR

Process	Wastewater	Reactor size, MG	COD, g/l	OVL, kg/m^{3}d	%CODr	HRT-d	°C
SF	Rum slops	3.5	80–105	15	71	7.8	35
SF	Modified guar	0.27	9.1	7.5	60	1.0	37
SF	Chemical	1.5	14	11	90	0.7	H
SF	Milk	0.2	3	7.5	60	0.5	32
SF	PMFC	10^{-5}	13.7	23	72	0.6	36
UASB	Potato	0.58	2.5	3	85	0.7	35
UASB	Sugar beet	0.21	3	16	88	—	—
UASB	Brewery	1.16	1.6–2.2S	4.4	83	0.4	30
UASB	Brewery	1.16	2.0–2.4S	8.7	78	0.2	30
UASB	PMFC	10^{-5}	13.7	4–5	87	2.9	36
FBR	PMFC	10^{-5}	13.7	35–48	88	0.4	36
FBR	Soft drink	0.04	3.0	6–7	75	0.5	35
FBR	Chemical	0.04	35	14	95	2.5	35

OVL = organic volumetric load (COD); S = settled effluent; H = heated; PMFC = paper mill foul condensate (5,6).
NOTE: MG = 3785 m^3.

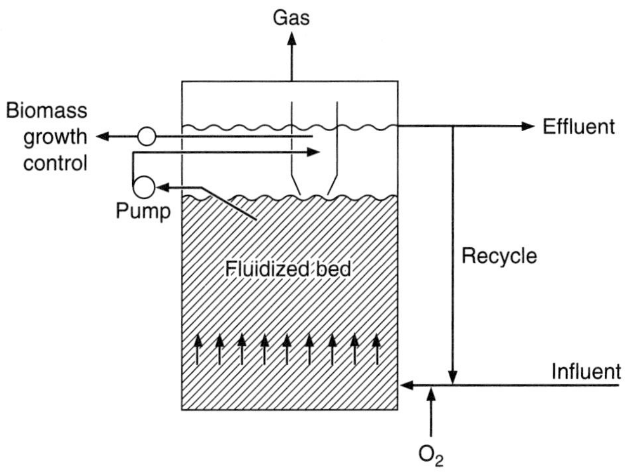

FIG. 22-44 Schematic of fluidized-bed process.

is higher than after the biomass attaches. Recycle of treated effluent is adjusted to achieve the desired degree of fluidization. As biomass accumulates, the particles of coated biomass will separate to a greater extent at constant flow velocity. Thus, as the system ages and more biomass accumulates, the extent of bed expansion increases (the volume of voids increases). This phenomenon prevents clogging of the bed with biomass. Consequently, higher levels of biomass attachment are possible than in other types of fixed-film systems. However, eventually, the degree of bed expansion may become excessive. Reduction of recycle will reduce expansion but may not be feasible because recycle has several purposes (i.e., supply of nutrients, alkalinity, and dilution of waste strength). Control of the expanded bed surface level is automatically accomplished using a sensor that activates a biomass growth control system at a prescribed level and maintains the bed at the proper depth. A pump removes a portion of the attached biomass, separates the biomass and inert carrier by abrasion, and pumps the mixture into a separator. Here the heavy carrier settles back into the fluidized bed, and the sheared biomass is removed from the system by gravity or a second pump. Other growth control designs are also used. Effluent is withdrawn from the supernatant layer above the fluidized bed. The reactor is usually not covered unless it is operating under anaerobic conditions and methane, odorous gases, or other safety precautions are mandated.

When aerobic treatment is to be provided to high concentrations of organics, pure oxygen or hydrogen peroxide may be injected into the wastewater before it enters the reactor. Liquid oxygen (LOX) or pressure swing absorption (PSA) systems have been used to supply oxygen. Air may be used for low DO demands.

In full-scale applications, this process has been found to operate at significantly higher volumetric loading rates for wastewater treatment than other processes. The primary reasons for the very high rates of contaminant removal is the high biologically active surface area available (approximately 1000 ft^2/ft^3 of reactor) and the high concentration of reactor biological solids (8000–40,000 mg/L) that can be maintained (Table 22-39). Because of these atypical high values, designs usually indicate a 200 to 500 percent reduction in reactor volume when compared to other fixed-film and suspended growth treatment processes. Table 22-40 is a list of full-scale commercial applications of the process operated at high wastewater concentrations, including aerobic oxidation of organics, anoxic denitrification, and anaerobic treatment systems.

Of special note is the enhancement to the process when granular activated carbon (GAC) is used as the carrier. Because GAC has adsorptive properties, organic compounds present in potable waters and wastewater at low concentrations, often less than 10 mg/L, are removed by adsorption and subsequently consumed by the biological organisms that grow in the fluidized bed. The BTEX compounds, methylene chloride, chlorobenzene, plastics industry toxic effluent, and many others are removed in this manner. BTEX contamination of groundwater from leaking gasoline storage tanks is a major problem, and 16 full-scale fluidized-bed process applications have been made. Contaminated groundwater is pumped to the ground surface for treatment using the fluidized bed in the aerobic mode. Often the level of BTEX is 1 to 10 mg/L, and about 99 percent is removed in less than 10 minutes' detention time. Installations in operation range in size from 30 to 3000 gpm. The smaller installations are often skid-mounted and may be moved from location to location at a given site.

TABLE 22-39 Process Comparisons

Biological process	MLVSS, mg/L	Surface area, ft^2/ft^3
Activated sludge	1500–3000	—
Pure oxygen activated sludge	2000–5000	—
Suspended growth nitrification	1000–2000	—
Suspended growth denitrification	1500–3000	—
Fluidized bed-CBOD removal	12,000–20,000	800–1200
Fluidized bed-nitrification	8,000–12,000	800–1200
Fluidized bed-denitrification	25,000–40,000	800–1200
Trickling filter-CBOD removal	—	12–50
RBC-CBOD removal	—	30–40

A major advantage of this process over stripping towers and vacuum systems for treating volatile organics (VOCs) is the elimination of effluent gas treatment.

Pilot-plant studies and full-scale plants have shown successful treatment of wastewaters with 5000 to 50,000 mg/L COD from dairy, brewery, other food preparation wastes; paper pulp wastes; deicing fluids; and other hazardous and nonhazardous materials. Design criteria for biological fluidized-bed systems are included in Table 22-41.

A third type of reactor system is the upflow anaerobic sludge blanket (UASB) reactor system. The UASB is a continuous-flow system that maintains a dense, flocculated biomass that is granular. The wastewater passes through this dense sludge blanket to achieve desired treatment efficiency. The UASB, anaerobic filter, and fluidized-bed systems have all been applied successfully to treat a variety of industrial waste streams. Table 22-38 summarizes some of the applications of these treatment technologies.

Design criteria for satisfactory biological fluidized-bed treatment systems include the major parameters given in Table 22-41.

PHYSICAL-CHEMICAL TREATMENT

Processes or unit operations that fall under this classification include adsorption, ion exchange, stripping, chemical oxidation, and membrane separations. All of these are more expensive than biological treatment but are used for the removal of pollutants that are not easily removed by biomass. Often these are used in series with biological treatment; but sometimes they are used as stand-alone processes.

Adsorption This is the most widely used of the physical-chemical treatment processes. It is used primarily for the removal of soluble organics with activated carbon serving as the adsorbent. Most liquid-phase-activated carbon adsorption reactions follow a Freundlich isotherm [Eq. (22-25)].

$$y = kc^{1/m} \qquad (22\text{-}25)$$

where y = adsorbent capacity, mass pollutant/mass carbon
c = concentration of pollutant in waste, mass/volume
k and n = empirical constants

EPA has compiled significant data on values of k and n for environmentally significant pollutants with typical activated carbons. Assuming equilibrium is reached, the isotherm provides the dose of carbon required for treatment. In a concurrent contacting process, the capacity is set by the required effluent concentration. In a countercurrent process, the capacity of the carbon

TABLE 22-40 Full-Scale Commercial Applications of Fluidized Bed Process

Application	Type	Reactor volume, m^3
CBOD—paint—General Motors	Aerobic	108
C,NBOD—sanitary and automotive—GM	Aerobic	165
Chemical—Grindsted Products Denmark	Aerobic	730
Nitrification—fish hatchery—Idaho	Aerobic	820
BTEX, groundwater—Ohio	Aerobic	250
Chemical, Toxicity reduction—Texas	Aerobic	225
Denitrification, Nuclear Fuel, DOE, Fornald, OH	Anoxic	53
CBOD, Denitrification—municipal, Nevada	Anoxic	1450
Petrochemical—Reliance Ind. India	Anaerobic	850
Soft drink, Delaware	Anaerobic	166
Brewery—El Aguila, Spain	Anaerobic	1570

TABLE 22-41 Typical Design Parameters for Fluidized Beds

Waste	Mode	Influent concentration, mg/L	Volumetric load, lb/ft³·d	Hydraulic detention time, hr	Volatile solids (biosolids), g/L
Organic COD	Aerobic	<2000	0.1–0.6	0.5–2.5	12–20
Organic COD	Aerobic	<10	0.02–0.1	0.1–0.5	4–8
NH_3-N	Aerobic	<25	0.05–0.3	0.1–0.5	7–14
NO_3^--N	Anoxic	<5000	0.4–2	0.1–2.4	20–40
Organic COD	Anaerobic	>4000	0.5–4	2–24	20–40

NOTE: $lb/ft^3 \cdot d = 16\ kg/m^3 \cdot d$.

TABLE 22-42 Selective Ion-Exchange Resins

Application	Exchanger type	Composition	Regenerant
Softening	Cation	Polystyrene matrix Sulfonic acid functional groups	NaCl
Heavy metals	Cation	Polystyrene matrix Chelating functional groups	Mineral acids
Chromate	Anion	Polystyrene matrix Tertiary or quaternary ammonium functional groups	Sodium carbonate or alkaline NaCl
Nitrate	Anion	Polystyrene matrix Tributyl ammonium functional group	NaCl

is set by the untreated waste pollutant concentration. Thus countercurrent contacting is preferred.

Activated carbon is available in powdered form (200–400 mesh) and granular form (10–40 mesh). The latter is more expensive but is easier to regenerate and easier to use in a countercurrent contactor. Powdered carbon is applied in well-mixed slurry-type contactors for detention times of several hours, after which separation from the flow occurs by sedimentation. Often coagulation, flocculation, and filtration are required in addition to sedimentation. Because it is difficult to regenerate, powdered carbon is usually discarded after use. Granular carbon is used in column contactors with EBCT of 30 minutes to one hour. Often several contactors are used in series, providing for full countercurrent contact. A single contactor will provide only partial countercurrent contact. When a contactor is exhausted, the carbon is regenerated either by a thermal method or by passing a solvent through the contactor. For waste-treatment applications where a large number of pollutants must be removed but the quantity of each pollutant is small, thermal regeneration is favored. In situations where a single pollutant in large quantity is removed by the carbon, solvent extraction regeneration can be used, especially where the pollutant can be recovered from the solvent and reused. Thermal regeneration is a complex operation. It requires removal of the carbon from the contactor, drainage of free water, transport to a furnace, heating under controlled conditions of temperature, oxygen, time, water vapor partial pressure, quenching, transport back to the reactor, and reloading of the column. Five to 10 percent of the carbon is lost in this regeneration process due to burning and attrition during each regeneration cycle. Multiple hearth, rotary kiln, and fluidized-bed furnaces have all been successfully used for carbon regeneration.

Pretreatment before carbon adsorption is usually for the removal of suspended solids. Often this process is used as tertiary treatment after primary and biological treatment. In either situation, the carbon columns must be designed to provide for backwash. Some solids will escape pretreatment, and biological growth will occur on the carbon, even with extensive pretreatment. Originally carbon treatment was viewed only as applicable for the removal of toxic organics or those that are difficult to degrade biologically. Present practice applies carbon adsorption as a procedure for the removal of all types of organics. Some biological activity will occur in virtually any activated carbon unit, so the design must be adjusted accordingly.

Ion Exchange This process has been used for many years for the treatment of industrial water supplies but not for wastes. However, some new ion-exchange materials have recently been developed that can be used to remove specific pollutants. These new resins are primarily useful for the selective removal of heavy metals, even though the target metals are present at low concentration in a wastewater containing many other inorganics. An ion-exchange process is usually operated with the waste being run downflow through a series of columns containing the appropriate ion-exchange resins. EBCT contact times of 30 min to 1 h are used. Pretreatment for suspended solids and organics removal is practiced as well as pH adjustment.

The capacity of an ion-exchange resin is a function of the type and concentration of regenerant. Because ion-exchange resins are so selective for the target compound, a significant excess of the regenerant must be used. Up to a point, the more regenerant, the greater the capacity of the resin. Unfortunately, this results in waste of most of the regenerant. In fact, the bulk of the operational cost for this process is for regenerant purchase and its disposal; thus, regeneration must be optimized. Regenerants used (selection depends on the ion-exchange resin) include sodium chloride, sodium carbonate, sulfuric acid, sodium hydroxide, and ammonia. Table 22-42 gives information on some of the new highly selective resins. These resins may provide the only practical method for the reduction of heavy metals to the very low levels required by recent EPA regulations.

Stripping Air stripping is applied for the removal of volatile substances from water. Henry's law is the key relationship for use in the design of stripping systems. The minimum gas-to-liquid ratio required for stripping is given by:

$$\frac{G}{L} = \frac{S_{in} - X_{out}}{(H)X_{out}}$$

for a concurrent process and

$$\frac{G}{L} = \frac{X_{in} - X_{out}}{(H)X_{in}}$$

for a countercurrent process where:

G = airflow rate, mass/time
L = waste flow rate, volume/time
X = concentration of pollutant in waste, mass/volume
H = Henry's constant for the pollutant in water, volume/mass

Higher ratios of gas to liquid flow are required than those computed above because mass-transfer limitations must be overcome. Stripping can occur by sparging air into a tank containing the waste. Indeed, stripping of organics from activated sludge tanks is a concern because of the possibility that the public will be exposed to airborne pollutants, including odors. A much more efficient stripping procedure is to use counterflow or cross-flow contact towers. Procedures for the design of these and mass-transfer characteristics of various packings are available elsewhere in this handbook.

Stripping has been successfully and economically employed for the removal of halogenated organics from water and wastes with dispersion of the effluent gas to the atmosphere. However, recent EPA regulations have curtailed this practice. Now removal of these toxic organics from the gas stream is also required. Systems employing activated carbon (prepared for use with gas streams) are used, as are systems to oxidize the organics in the gas stream. However, the cost of cleaning up the gas stream often exceeds the cost of stripping these organics from the water.

Chemical Oxidation Chemical oxidation has gained renewed interest in the water treatment industry due to both innovation and the need to remove more recalcitrant pollutants. The efficiency of this process varies because many side reactions can occur that will consume the oxidant, decreasing its efficacy. In addition, complete oxidation of organics to carbon dioxide and water often will not occur unless oxidants are used in combination or significant overdose is used. Partial oxidation can be used as a pretreatment to biological or carbon adsorption treatment by increasing biodegradability and adsorption affinity. In addition, specific AOP processes can produce a self-renewing chain reaction that can significantly reduce the dose of ozone needed to accomplish oxidation.

Oxidants commonly used include ozone, permanganate, chlorine, chlorine dioxide, and ferrate, often in combination with catalysts. Standard-type mixed reactors are used with contact times of several minutes to an hour.

Advanced Oxidation Processes Advanced oxidation processes (AOPs) use the extremely strong oxidizing power of hydroxyl radicals to oxidize recalcitrant organic compounds to the mineral end products, including CO_2, H_2O, and HCL. The hydroxyl radicals are produced on-site using a combination of traditional oxidation processes, such as ozone/hydrogen peroxide, ultraviolet radiation/ozone, ultraviolet radiation/hydrogen peroxide, Fenton's reagent (ferrous iron and hydrogen peroxide) with ultraviolet radiation, and titanium dioxide/ultraviolet radiation (catalytic oxidation). The hydroxyl radical is a powerful nonselective oxidant; it can effectively and rapidly oxidize most organic compounds. As shown in Table 22-43, the hydroxyl radical has greater oxidizing power than do traditional oxidizing reagents. AOPs can be used to remove overall organic content (COD) or to destroy specific compounds. Table 22-43 shows the relative oxidation power

TABLE 22-43 Relative Oxidation Power of Select Oxidizing Reagents

Oxidizing species	Relative oxidation power
Chlorine	1.0
Hypochlorous acid	1.1
Permanganate	1.24
Hydrogen peroxide	1.31
Ozone	1.52
Atomic oxygen	1.78
Hydroxyl radical	2.05
Titanium dioxide	2.35

of a select number of oxidizing species compared to chlorine. AOPs are most economical for treating industrial and hazardous waste streams that have a low level of recalcitrant contamination (<50 ppm). AOPs are emerging as an efficient and cost-effective treatment technology for specific industrial wastewaters and are likely to have broader applications as the various technologies are proved in the field.

Membrane Processes These processes use a selectively permeable membrane to separate pollutants from water. Most of the membranes are formulated from complex organics that polymerize during membrane preparation. This allows the membrane to be tailored to discriminate by molecular size or by degree of hydrogen bonding potential. Ultrafiltration membranes discriminate by molecular size or weight, while reverse osmosis membranes discriminate by hydrogen-bonding characteristics. The permeability of these membranes is low: from 0.38 to 3.8 m/day (10–100 gal/d/ft^2). The apparatus in which they are used must provide a high surface area per unit volume.

Membrane Bioreactors (MBRs) Membrane bioreactors are a technology that combines biological degradation of waste products with membrane filtration (Fig. 22-45). Typically, MBRs are operated as activated sludge systems with high biomass concentrations (10,000 to 15,000 mg/L MLSS) and long solids retention times (30 to 60 days). The membranes act as a solid–liquid separation to replace secondary clarifiers and granular media polishing filters. With the development of more economical, durable, and efficient membrane components, MBR systems have become feasible options for the aerobic treatment of municipal wastewaters, and potentially for the anaerobic treatment of industrial (i.e., low- to medium-strength) wastewaters. Two different process configurations for MBRs (sidestream or submerged modules) are used within wastewater treatment. The sidestream process is operated with high velocities and high transmembrane pressures (up to 5 bar). This results in high energy consumption and excessive shear forces within the reactor. The process with submerged modules was especially designed for biological wastewater treatment. The modules are submerged in the aerobic activated sludge zone, where flow across the membrane surface is attained by aeration and mixing, resulting in a low flux rate with low transmembrane and vacuum pressures. Submerged membranes typically last longer (8+ years) and require less frequent cleaning and maintenance. With recent advances in membrane materials and process technology, MBRs have become more reliable, durable, and affordable. MBRs can be used for any level of treatment, but they are most cost-effective when high-quality effluent is required.

The submerged MBR system is a small-footprint, single-process unit that can achieve a high-quality standard with low solids production. The membrane systems typically use either micro- or ultramembranes configured as hollow fibers, flat sheets, or tubes, depending on the manufacturer. Experience over the past decade has proved MBRs to be reliable and cost-effective for high-quality effluent applications, and relatively easy to maintain and operate. The key parameters for MBR design and operation are flux rate, transmembrane pressure, fouling rate, and membrane cleaning schedule. MBRs have many possible applications in industrial and advanced municipal wastewater treatment and are becoming the preferred technology for wastewater reuse applications.

INDUSTRIAL REUSE AND RESOURCE RECOVERY

Many industries that use large volumes of freshwater are turning toward industrial reuse to minimize the costs associated with attaining freshwater and disposing of wastewater. The need to minimize water consumption is a function of economics, water shortages caused by drought and overdevelopment, increasingly stringent regulations, and environmental awareness. Industrial reuse water is wastewater that is produced on-site that does not contain sewage and has been adequately treated for use in some part of the industrial operation. Reuse has become more economically feasible as a result of new treatment technologies (MBRs, AOPs, etc.) and rising costs of freshwater and wastewater disposal.

There are many industrial sources and uses of industrial reuse water. Common industrial processes that require large quantities of water, such as evaporative cooling towers and power stations, can be supplied using reclaimed wastewater. Other industrial applications for reuse water include boilers; stack scrubbing; washing of vehicles, buildings, and mechanical parts; and process water. The key to effective industrial reuse is to match the proper treatment processes (chemical, physical, biological) to achieve the desired quality of reuse water. Zero-discharge industrial systems have recently been implemented in the United States and appear to be a growing trend, sustaining water resources and promoting smart industrial development. Some common industries that have implemented water reuse include the paper pulp industry, the pharmaceutical industry, chemical manufacturers, power plants, and refineries.

Increasingly, wastewater treatment plants are being converted to resource recovery facilities, where the emphasis is on both meeting effluent water quality criteria and producing or recovering valuable resources, such as energy, reusable water, and nutrient fertilizer. The main reasons for this are to reduce operating costs, increase sustainability, and provide additional revenue streams. Currently, the primary areas of resource recovery have focused on improved biogas production through optimized anaerobic digestion and codigestion with high-strength organic wastes (fats, oils and grease, food waste, etc.); sludge pretreatment; improved energy recovery, including implementation of combined heat and power (CHP) systems for the use of produced biogas; and nutrient recovery, mainly via the engineered production of struvite from the nutrient-rich wastewater stream produced during the dewatering of anaerobically digested sludge (e.g., dewatering centrate). There is ongoing research aimed at developing new methods for recovering a variety of organics (alcohols, volatile fatty acids, polyhydroxyalkanoates, etc.), inorganics (potassium, ammonia, hydrogen peroxide, etc.), and precious metals from wastewater and sludges.

SLUDGE PROCESSING

Objectives Sludges consist primarily of the solids removed from liquid wastes during their processing. Thus, sludges could contain a wide variety of pollutants and residuals from the application of treatment chemicals, including large organic solids, colloidal organic solids, metal sulfides, heavy-metal hydroxides and carbonates, heavy-metal organic complexes, calcium and magnesium hydroxides, calcium carbonate, precipitated soaps and detergents, and biomass and precipitated phosphates. Because sludge, even after extensive concentration and dewatering, is still greater than 50 percent water by weight, it can also contain soluble pollutants such as ammonia, priority pollutants, and nonbiologically degradable COD.

The general treatment or management of sludge involves stabilization of biodegradable organics, concentration and dewatering, and ultimate disposal of the stabilized and dewatered residue. A large number of individual unit processes and unit operations are used in a sludge-management scheme. Those most frequently used are discussed next. Occasionally, only one of these is needed, but usually several are used in a series arrangement.

Because of the wide variability in sludge characteristics and the variation in the acceptability of treated sludges for ultimate disposal (this is a function of the location and characteristics of the ultimate disposal site), it is impossible to prescribe any particular sludge-management plan.

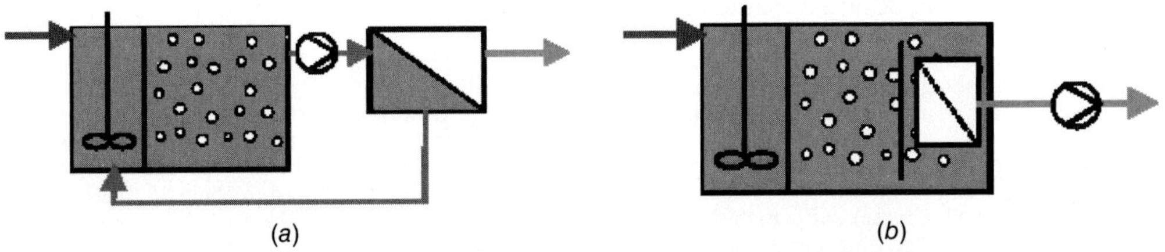

(a) (b)

FIG. 22-45 Membrane bioreactor configurations for wastewater treatment. (*a*) With sidestream module; (*b*) with submerged module.

In the following subsections, the general performance of individual sludge-treatment processes and operations is presented.

Thickening and Flotation Generated sludges are often dilute (1–2 percent solids by weight). In order to reduce the volumetric loading on other processes, the first step in sludge processing is often concentration. The most popular process is gravity thickening, which is carried out in treatment units similar to circular clarifiers. Organic sludges from primary treatment can usually be concentrated to 5 to 8 percent solids. Sludges from secondary treatment can be thickened to 2 percent solids. The potential concentration with completely inorganic sludges is higher (greater than 10 percent solids) except for sludges high in metal hydroxides. Polymers are often used to speed up thickening and increase concentration. Thickening is enhanced by long retention of the solids in the thickening apparatus. However, when biodegradable organics are present, solids retention time must be at a level that will not foster biological activity, lest odors, gas generation, and solids hydrolysis occur. Loading rates on thickeners range from 50 to 122 $kg/m^2/d$ (10–25 $lb/ft^2/d$) for primary sludge to 12 to 45 $kg/m^2/d$ (2.5–9 $lb/ft^2/d$). Solids detention time is 0.5 days in summer to several days in winter.

Gravity Belt Thickening Gravity belt thickening (GBT) is an increasingly popular and effective method of sludge thickening. Gravity belt thickening involves the even distribution of feed sludge across a horizontal moving fabric belt one to three meters wide. Residual water in the thin layer of distributed sludge is allowed to drain through the moving fabric filter via gravity. As the thickened sludge approaches the end of the belt, it is removed from the belt and pumped to downstream sludge treatment processes. Gravity belt thickening typically requires polymer addition to achieve high solids capture (>90 percent). Depending on the type of sludge being treated (primary, waste activated, blended, etc.), GBTs can process approximately 700 to 900 lb of dry solids per hour per meter width of belt or 100 to 200 gallons of sludge per minute per meter of belt. GBTs can achieve a solids concentration between 4 and 8 percent, depending on type and solids concentration of the feed sludge. There are many GBT vendors who offer units for all types of applications and flows.

Flotation Air flotation has been used to concentrate secondary sludges to about 4 percent solids. Incoming sludge is saturated with air at 275 to 350 kPa (40 to 50 psig) before being released in the flotation tank. As the air comes out of solution, the fine bubbles carry the solids to the surface. The air bubbles compact the solids as a floating mass. Normally, the air-to-solids ratio is about 0.01 to 0.05 L/g (0.16–0.8 ft^3/lb). Thickened solids are scraped off the surface, while the effluent is drawn off the middle of the tank and returned to the treatment system. The size of the flotation tanks is determined primarily by the solids-loading rate. A solids loading of 25 to 97 $kg/(m^2 \cdot day)$ [5 to 20 $lb/(ft^2 \cdot day)$] has been found to be adequate. On a flow basis, this generally translates into 0.14 to 2.7 $L/(m^2 \cdot day)$ [0.2 to 4 $gal/(ft^2 \cdot min)$] surface area. Air flotation is enhanced by the addition of polymers, surfactants, and other additives. Flotation has been successfully used with wholly inorganic metal hydroxide sludges. Engineering details on air flotation equipment is available from various equipment manufacturers. Liquid removed during thickening and flotation is usually returned to the head end of the plant.

Stabilization (Anaerobic Digestion, Aerobic Digestion, High Lime Treatment) Sludges high in organics can be stabilized by subjecting them to biological treatment. The most popular system is anaerobic digestion.

Anaerobic Digestion Anaerobic digesters are large, covered tanks with detention times of 30 days, based on the volume of sludge added daily. Digesters are usually heated with an external heat exchanger to 35°C to 37°C to speed the rate of reaction. Mixing is essential to provide good contact between the microbes and the incoming organic solids. Gas mixing and mechanical mixers have been used to provide mixing in the anaerobic digester. Following digestion, the sludge enters a holding tank, which is basically a solids-separation unit and is not normally equipped for either heating or mixing. The supernatant is recycled back to the treatment plant, while the settled sludge is allowed to concentrate to 3 to 6 percent solids before being further processed.

Anaerobic digestion results in the conversion of the biodegradable organics to methane, carbon dioxide, and microbial cells. Because of the energy in the methane, the production of microbial mass is quite low, less than 0.1 kg/kg [0.1 lb volatile suspended solids (VSS)/lb] BCOD metabolized except for carbohydrate wastes. The production of methane is 0.35 m^3/kg (5.6 ft^3/lb) BCOD destroyed. Digester gases range from 50 to 80 percent methane and 20 to 50 percent carbon dioxide, depending on the chemical characteristics of the waste organics being digested. The methane is often used on-site for heat and power generation.

There are three major groups of bacteria that function in anaerobic digestion. The first group hydrolyzes large soluble and nonsoluble organic compounds such as proteins, fats and oils (grease), and carbohydrates, producing smaller water-soluble compounds. These are then degraded by acid-forming bacteria, producing simple volatile organic acids (primarily acetic acid) and hydrogen. The last group (the methane bacteria) split acetic acid to methane and carbon dioxide and produce methane from carbon dioxide and hydrogen. Good operation requires the destruction of the volatile acids as quickly as they are produced. If this does not occur, the volatile acids will build up and depress the pH, which will eventually inhibit the methane bacteria. To prevent this from occurring, the feed of organics to the digester should be as uniform as possible.

If continuous addition of solids is not possible, additions should be made at as short intervals as possible. Alkalinity levels are normally maintained at about 3000 to 5000 mg/L to keep the pH in the range 6.5 to 7.5 as a buffer against variable organic-acid production with varying organic loads. Proteins will produce an adequate buffer, but carbohydrates will require the addition of alkalinity to provide a sufficient buffer. Sodium bicarbonate should be used to supply the buffer.

An anaerobic digester is a no-recycle complete mix reactor. Thus, its performance is independent of organic loading but is controlled by hydraulic retention time (HRT). Based on kinetic theory and values of the pseudo constants for methane bacteria, a minimum HRT of three to four days is required. To provide a safety factor and to compensate for load variation as indicated earlier, HRT is kept in the range 10 to 30 days. Thickening of feed sludge is used to reduce the tank volume required to achieve the long HRT values. When the sludge is high in protein, the alkalinity can increase to greater than 5000 mg/L, and the pH will rise past 7.5. This can result in free ammonia toxicity. To avoid this situation, the pH should be reduced below 7.5 with hydrochloric acid. The use of nitric or sulfuric acid will result in significant operational problems.

Sludge Pretreatment In an effort to reduce sludge production or to increase biogas production by anaerobic digestion, a number of new technologies have been developed to increase the biodegradability of thickened sludge before anaerobic digestion. Most of these processes aim to lyse the bacterial cells present in biological sludges and begin the hydrolysis process, thus increasing the biodegradability of complex organics. Proven processes include thermal hydrolysis (THP) and thermal/chemical hydrolysis (TCHP). THP is similar to a pressure cooker because it uses high pressure and temperature (165°C) to rupture the bacterial cells and make them more readily biodegradable. TCHP systems use the addition of caustic to increase the pH and more moderate temperatures (65°C vs. 165°C) to improve biodegradability. THP processes have been shown to increase digester capacity (200 percent) by increasing organic loading, while improving overall volatile solids destruction (60 percent VSr) and biogas production (15 to 20 percent). In addition, THP treated sludge has improved dewaterability, which can result in up to a 33 percent reduction in dewatering cake mass. Major manufacturers of THP systems include CAMBI, Veolia, Sustec, and Haarslev. THCP systems have shown the ability to achieve similar results to some THP processes using lower temperatures and pressures. As an additional and increasingly important benefit, THP processes and many TCHP processes can be used to achieve Class A biosolids.

Aerobic Digestion Waste activated sludge can be treated more easily in aerobic treatment systems than in anaerobic systems. The sludge has already been partially aerobically digested in the aeration tank. For the most part, only about 25 to 35 percent of the waste activated sludge can be digested. An additional aeration period of 15 to 20 days should be adequate to reduce the residual biodegradable mass to a satisfactory level for dewatering and return to the environment. One of the problems in aerobic digestion is the inability to concentrate the solids to levels greater than 2 percent. A second problem is nitrification. The high protein concentration in the biodegradable solids results in the release of ammonia, which can be oxidized during the long retention period in the aerobic digester. Limiting oxygen supply to the aerobic digester appears to be the best method to handle nitrification and the resulting low pH.

A new concept is to use an on/off air supply cycle. During aeration, nitrates are produced. When the air is shut off, nitrates are reduced to nitrogen gas. This prevents acid buildup and removes nitrogen from the sludge. High power costs for aerobic digestion limit the applicability of this process.

High Lime Treatment This method uses doses of lime sufficient to raise the pH of sludge to 12 or above. As long as the pH is maintained at this level, biological breakdown will not occur. In this sense, the sludge is stable. However, any reduction of pH as a result of contact with CO_2 in the air will allow biological breakdown to begin. Thus, this technique should only be used as temporary treatment until further processing can occur. It is not permanent stabilization like anaerobic or aerobic digestion.

Sludge Dewatering Dewatering is different from concentration in that the latter still leaves a substance with the properties of a liquid. The former produces a product that is essentially a friable solid. When the water content of sludge is reduced to about 70 to 80 percent, it forms a porous solid called sludge cake. There is no free water in the cake because the water is chemically combined with the solids or tightly adsorbed on the internal pores. The operations described next that are used to dewater sludge can be applied at any stage of the sludge management process, but often they follow concentration or biological stabilization. Chemical conditioning is almost always used to aid dewatering.

Lime, alum, and various ferric salts have been used to condition sludge before dewatering. Lime reacts to form calcium carbonate crystals, which act as a solid matrix to hold the sludge particles apart and allow the water to escape during dewatering. Alum and iron salts help displace some of the bound water from hydrophilic organics and form part of the inorganic matrix. Chemical conditioning increases the mass of sludge to be ultimately handled from 10 to 25 percent, depending on the characteristics of the individual sludge. Chemical conditioning can also help remove some of the fine particles by incorporating them into insoluble chemical precipitates. The water (supernatant) removed from the sludge during dewatering is often high in suspended solids and organics. The addition of polymers before, during, or after dewatering will often reduce the level of pollutants in the supernatant.

Centrifugation Both basket and solid-bowl centrifuges have been used to concentrate waste sludges. Field data have shown that it is possible to obtain 10 to 20 percent solids with waste activated sludge, 15 to 30 percent solids with a mixture of primary and waste activated sludge, and up to 30 to 35 percent solids with primary sludge alone. Centrifuges result in 85 to 90 percent solids capture with good operation. The problem is that the centrate contains the fine solids not easily removed. The centrate is normally returned to the treatment process, where it may or may not be removed. Economics do not favor centrifuges unless the sludge cake produced is at least 25 to 30 percent solids. For the most part, centrifuges are designed by equipment manufacturers from field experience. With varying sludge characteristics, centrifuge characteristics will also vary widely.

Vacuum Filtration Vacuum filtration has been the most common method employed in dewatering sludges. Vacuum filters consist of a rotary drum covered with a cloth-filter medium. Various plastic fibers as well as wool have been used for the filter cloth. The filter operates by drawing a vacuum as the drum rotates into chemically conditioned sludge. The vacuum holds a thin layer of sludge, which is dewatered as the drum rotates through the air after leaving the vat. When the drum rotates the cloth to the opposite side of the apparatus, air-pressure jets replace the vacuum, causing the sludge cake to separate from the cloth medium as the cloth moves away from the drum. The cloth travels over a series of rollers, with the sludge being separated by a knife edge and dropping onto a conveyor belt by gravity. The dewatered sludge is moved on the conveyor belt to the next concentration point, while the filter cloth is spray-washed and returned to the drum before entering the sludge vat. Vacuum filters yield the poorest results on waste activated sludge and the best results on primary sludge. Waste activated sludge will concentrate to between 12 and 18 percent solids at a rate of 4.9 to 9.8 kg dry cake/(m^2·h) [1 to 2 lb/(ft^2·h)]. Primary sludge can be dewatered to 25 to 30 percent solids at a rate of 49 kg dry cake/(m^2·h) [10 lb/(ft^2·h)].

Pressure Filtration Pressure filtration has been used increasingly since the early 1970s because of its ability to produce a drier sludge cake. The pressure filters consist of a series of plates and frames separated by a cloth medium. Sludge is forced into the filter under pressure, while the filtrate is drawn off. When maximum pressure is reached, the influent-sludge flow is stopped, and the pressure filter is allowed to discharge the residual filtrate before opening the filter and allowing the filter cake to drop by gravity to a conveyor belt below the filter press. The pressure filter operates at a pressure between 689 and 1380 kPa (100 and 200 psig) and takes 1.5 to 4 h for the pressure cycle. Normally, 20 to 30 min is required to remove the filter cake. The sludge cakes will vary from 20 to 25 percent for waste activated sludge to 50 percent for primary sludge. Chemical conditioning is necessary to obtain good dewatering of the sludges.

Belt-Press Filters The newest filter for handling waste activated sludge is the belt-press filter. The belt press uses a continuous cloth-filter belt. Waste activated sludge is spread over the filter medium, and water is removed initially by gravity. The open belt with the sludge moves into contact with a second moving belt, which squeezes the sludge layer between rollers with ever-increasing pressure. The sludge cake is removed at the end of the filter press by a knife blade, with the sludge dropping by gravity to a conveyor belt. Belt-press filters can produce sludge with 20 to 30 percent solids.

Sand Beds Sand filter beds can be used to dewater either anaerobically or aerobically digested sludges. They work best on relatively small treatment systems located in relatively dry areas. The sand bed consists of coarse gravel graded to fine sand in a series of layers to a depth of 0.45 to 0.6 m (1.5 to 2 ft). The digested sludge is placed over the entire filter surface to a depth of 0.3 m (12 in) and allowed to sit until dry. Free water will drain through the sand bed to an open pipe underdrain system and will be removed from the filter. Air drying will slowly remove the remaining water. The sludge must be cleaned from the bed by hand before adding a second layer of sludge. The sludge layer will drop from an initial thickness of 3 m (12 in) to about 0.006 m (1⁄4 in). An open sand bed can generally handle 49 to 122 kg dry solids/(m^2·yr) [10 to 25 lb/(ft^2·yr)]. Covered sand beds have been used in wet climates as well as in cold climates, but economics does not favor their use.

Separate Centrate Treatment The supernatant (centrate or filtrate) from anaerobic sludge dewatering operations is typically high in nitrogen (500 to 1500 mg/L as N) and phosphorous (100 to 200 mg/L as P). These levels may be elevated if there are significant nutrient-rich industrial loads to the treatment systems (e.g., proteinaceous wastes) or if thermal hydrolysis pretreatment processes are used. Many plants are electing to treat these nutrient-rich centrate streams separately to improve overall total nitrogen removal or to recover nutrients. A typical separate centrate treatment (SCT) process uses return sludge from the main treatment process to carry out nitrification and denitrification in a separate aerobic anoxic reactor. Generally, the goal is to achieve complete nitrification with some degree of denitrification, or total TN removal. The effluent from the SCT process is returned to the main plant biological process, where the nitrified rich SCT sludge serves to seed the main plant with active nitrifier biomass and to provide nitrate that can be removed in anoxic zones in the main plant using COD from the influent wastewater. Due to the high levels of ammonia in centrate, SCT systems often have alkalinity added to maintain a pH above 6.5, and they may also have supplemental carbon (e.g., methanol, glycerol) to enhance denitrification.

Recent developments in the treatment of centrate- and ammonia-rich industrial wastewaters have resulted in the development of shortcut nitrogen processes, including processes that take advantage of the nitrite shunt and deammonification, which is carried out by anaerobic ammonia oxidizing bacteria (anammox). Figure 22-46 shows the three different pathways currently used to biologically remove ammonia nitrogen.

Figure 22-46*a* shows the traditional nitrification/denitrification pathway, along with the required oxygen and carbon demand. Figure 22-46*b* shows the nitrite shunt or simultaneous nitrification/denitrification (SND) pathway. SND converts ammonia nitrogen to nitrite and then uses readily biodegradable COD to convert nitrite to nitrogen gas, bypassing the production of nitrate (nitratation) and the subsequent anoxic conversion back to nitrite (denitratation). This process saves on aeration and supplemental carbon requirements. The process may be achieved through the maintenance of low DO ($<$0.5 mg/L).

Figure 22-46*c* shows the deammonification pathway, which uses anaerobic ammonia oxidizing (anammox) biomass to convert ammonia and nitrite directly to nitrogen gas without the need for any organic carbon. This process uses about a third of the aeration needed for tradition nitrogen removal, but it requires effective suppression of the nitrite oxidizing bacteria (NOB) population to be effective. Suppression of the NOB can be challenging in real-world operations, and the industry is developing effective methods for maintaining deammonification. Deammonification systems also require the selection and retention of slow-growing anammox biomass via preferential settling of granulized anammox biomass or the use of anammox biofilms. Deammonification systems have been implemented throughout the world to treat ammonia-rich industrial wastewater streams and anaerobically digested sludge dewatering streams. At this writing, four companies have patented deammonification technologies, and many groups are working to develop deammonification systems to treat more dilute ammonia streams such as municipal wastewater.

SLUDGE DISPOSAL

Incineration Incineration has been used to reduce the volume of sludge after dewatering. The organic fractions in sludges lend themselves to incineration if the sludge does not have an excessive water content. Multiple-hearth and fluid-bed incinerators have been extensively used for sludge combustion.

A multiple-hearth incinerator consists of several hearths in a vertical cylindrical furnace. The dewatered sludge is added to the top hearth and is slowly pushed through the incinerator, dropping by gravity to the next lower layer until it finally reaches the bottom layer. The top layer is used for drying the sludge with the hot gases from the lower layers. As the temperature of the furnace increases, the organics begin to degrade and undergo combustion. Air is used to add the necessary oxygen and to control the temperature during combustion. It is very important to keep temperatures above 600°C to ensure complete oxidation of the volatile organics. One of the problems with the multiple-hearth incinerator is volatilization of odorous organics during the drying phase before the temperature reaches combustion levels. Even afterburners on the exhaust-gas line may not be adequate for complete oxidation. Air pollution control devices are required on all incinerators to remove fly ash and corrosive gases. The ash from the incinerator must be cooled, collected, and conveyed back to the environment, normally to a sanitary landfill for burial. The residual ash will weigh from 10 to 30 percent of

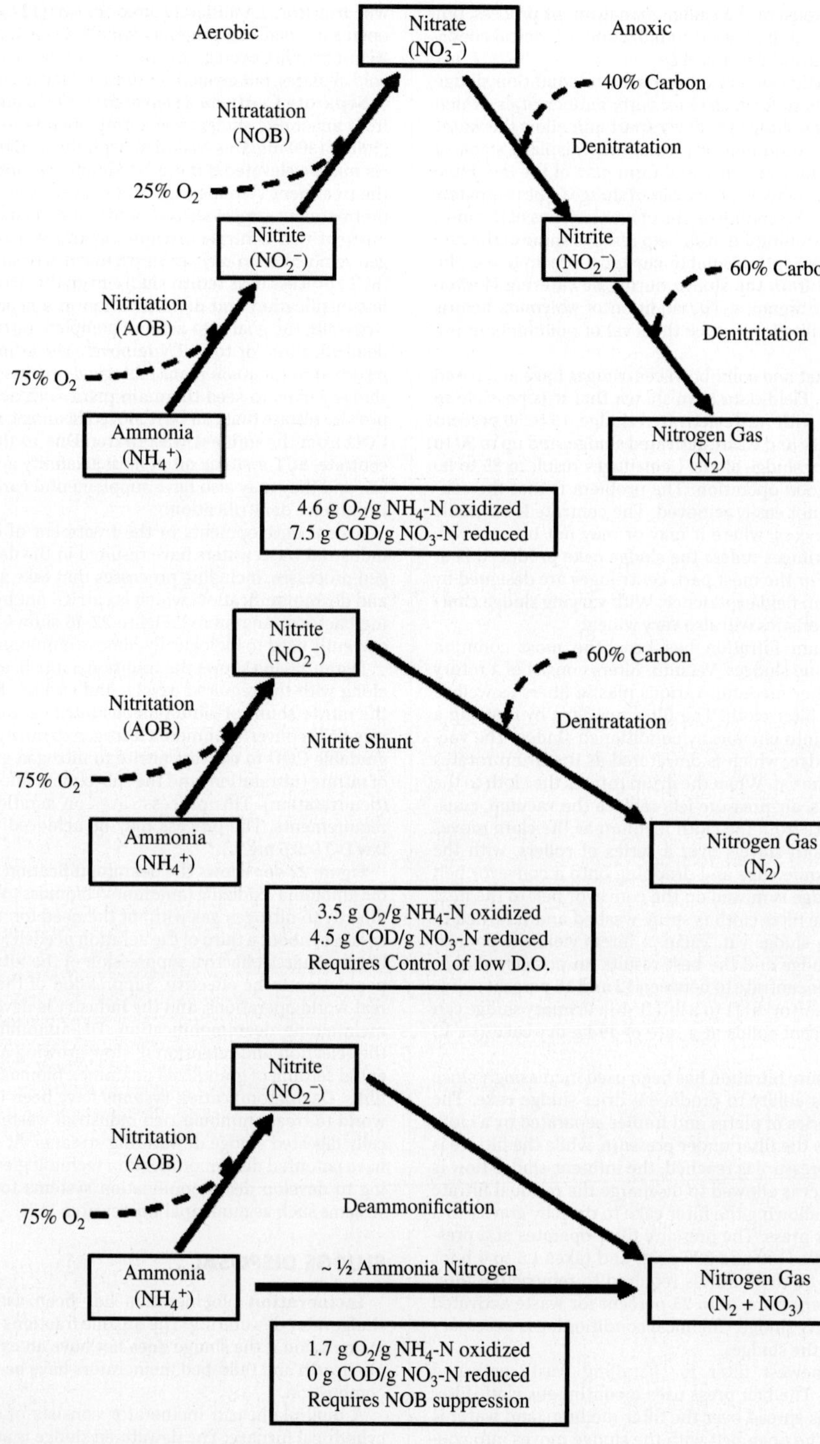

FIG. 22-46 Pathways used to biologically remove ammonia nitrogen.

the original dry weight of the sludge. Supplemental fuels are needed to start the incinerator and to ensure adequate temperatures with sludges containing excessive moisture, such as activated sludge. Heat recovery from wastes is being given more consideration. It is possible to combine the sludges with other wastes to provide a better fuel for the incinerator.

A fluid-bed incinerator uses hot sand as a heat reservoir for dewatering the sludge and combusting the organics. The turbulence created by the incoming air and the sand suspension requires the effluent gases to be treated in a wet scrubber before final discharge. The ash is removed from the scrubber water by a cyclone separator. The scrubber water is normally returned to the treatment process and diluted with the total plant effluent. The ash is normally buried.

Sanitary Landfills Dewatered sludge, either raw or digested, is often buried in a sanitary landfill to minimize the environmental impact. Increased concern over sanitary landfills has made it more difficult simply to bury dewatered sludge. Sanitary landfills must be made secure from leachate and must be monitored regularly to ensure that no environmental damage occurs. The moisture content of most sludges makes them a problem at sanitary landfills designed for solid wastes, requiring separate burial even at the same landfill.

Beneficial Reuse of Biosolids Biosolids are biological treatment sludges that are stabilized by using a variety of methods, including alkaline stabilization, anaerobic digestion, aerobic digestion, composting, or heat drying and pelletization. The disposal of biosolids has typically been done through incineration and landfilling. With the severe decline in the number of landfills and the difficulty with incineration meeting stringent Clean Air Act standards, biosolids are increasingly being land-applied as a beneficial reuse. The specific type of land application is a function of the quality of biosolids being applied. Beneficial reuse of biosolids is regulated by the EPA (Part 503 rules), which has established guidelines and specific criteria for reuse application. The EPA biosolids rules regulate chemical contaminants in the biosolids (metals), pathogen and pathogen indicator content, sludge stabilization methodology, land application site access and setback distances, vector attraction reduction methods, and the crop type and harvesting schedule.

EPA Class A or Class B biosolids can be land-applied. Land application serves two purposes: (1) to provide an inexpensive fertilizer and soil conditioner and (2) to continue treatment of the sludge through exposure to sunlight, plant uptake of metals, and desiccation of pathogens. The benefits of biosolids land application include addition of nutrients (micro and macro) to soil, replacing agricultural dependency on chemical fertilizer; improving soil texture and increasing biological activity to promote root growth; slowly releasing nutrients to reduce off-site transport of nutrients; and lower cost than landfilling or incineration.

The implementation of more stringent air emissions requirements, particularly those stemming from the CAA Amendments of 1990, have resulted in very significant reductions in emissions from the WTE industry (see Table 22-58). The emissions reductions have been achieved both as the result of the closure of outdated facilities and the installation of air pollution control equipment.

Even before 1990, MWC facilities had much lower emissions than the incinerators that were commonplace in many urbanized areas earlier in the century. Today, filterable particulate emissions are, on average, approximately 0.02 pound per ton of municipal solid waste (MSW) burned. Similar dramatic improvements have been made in other emissions as well. Historically, municipal waste combustors were a leading source of dioxin emissions, ranking as the largest single source in the 1980s. In 2012, MWC facilities represented just 0.54 percent of total controlled combustion sources and less than 0.1 percent of total controlled and open burning sources of dioxin [Dwyer, H., and N. J. Themelis, "Inventory of U.S. 2012 Dioxin Emissions to Atmosphere," *Waste Management* **46:** 242–246 (December 2015), http://dx.doi.org/10.1016/j.wasman.2015.08.009].

MANAGEMENT OF SOLID WASTES

INTRODUCTION

In the past 50 years, humans have consumed more resources than in all previous history. How society uses materials is fundamental to many aspects of our economic and environmental future. The rapid rise in material use has led to habitat destruction, biodiversity loss, overly stressed fisheries, and desertification. Using materials from "waste" reduces discards, conserves natural resources and energy, and provides multiple economic benefits. Solid waste is generated and managed throughout a material's life cycle from extraction or harvest of materials and food (e.g., mining, forestry, and agriculture), to production and transport of goods, provision of services, reuse of materials, and, if necessary, disposal. Solid wastes are generated from human and animal activities and from natural and intentional disasters. The U.S. EPA and the European Union have both developed a nonhazardous materials and waste management hierarchy ranking waste management approaches in order from most to least environmentally preferable. The hierarchy emphasizes reducing, reusing, and recycling as key to sustainable materials management, but it recognizes that no single waste management approach is suitable for managing all materials and waste streams in all circumstances.

The management of solid waste has changed significantly over the past 50 years, moving from uncontrolled dumping or burning to complex systems that integrate multiple processes to recover materials or energy. Some parts of the world face environmental degradation and public-health risks from uncollected waste in streets and other public areas, drainage systems clogged with wastes, and contamination of water resources near uncontrolled dump sites. The United States, Europe, and other countries have established programs for solid waste management and are transitioning from "waste" to "materials" management.

Often it is not obvious which processes or technologies will best meet the needs of a community when accounting for differences in existing infrastructure, population density, applicable regulations, energy grid mix, societal or community priorities, and cost. This subsection provides an overview of (1) the transition from "waste" management to more sustainable materials management, (2) process units and activities associated with solid waste management, (3) applicable U.S. legislation, regulations, and policy, (4) different categories of solid wastes as defined by legislation, along with information on properties important to consider in evaluating technology options, (5) statistics on U.S. solid waste generation, (6) descriptions of individual waste management units, technological options, and applicable regulations, and (7) the important role of planning in achieving more sustainable solid waste management and a more resilient community in the aftermath of earthquakes and other natural disasters. In addition, available tools and other information are identified to support those responsible for solid waste management. Not included in this section—but perhaps in future updates—is information on the management of contaminated waste sites in addition to the management of marine debris (which is an issue for many coastal communities).

TRANSITIONING FROM "WASTE" MANAGEMENT TO "MATERIALS" MANAGEMENT

Rather than focusing strictly on end-of-pipe solutions, a life-cycle perspective can be used to evaluate a product's potential impact (Figure 22-47) from materials extraction to end-of-life management. Life-cycle assessment and systems thinking can be used to optimize energy and resource recovery to reduce environmental and other impacts from the use and management of material resources flowing through the economy. Using more ecologically intelligent design is moving us toward cradle-to-cradle thinking [McDonough, W., and M. Braungart, *Cradle to Cradle: Remaking the Way We Make Things*, North Point Press, New York, 2002; Christensen, T., *Solid Waste Technology and Management*, Wiley, New York, 2012]. Resources are available to encourage more sustainable material management, including a report on the understanding of the economic implications of material reuse and recycling (Recycling Economic Information Report, U.S. EPA, 2016). The report shows how the U.S. economy has grown through recycling by creating jobs, building more competitive manufacturing industries, and converting waste materials into valuable raw materials.

Sustainable materials management focuses on impacts throughout a product or material's life cycle. For example, a product can be redesigned

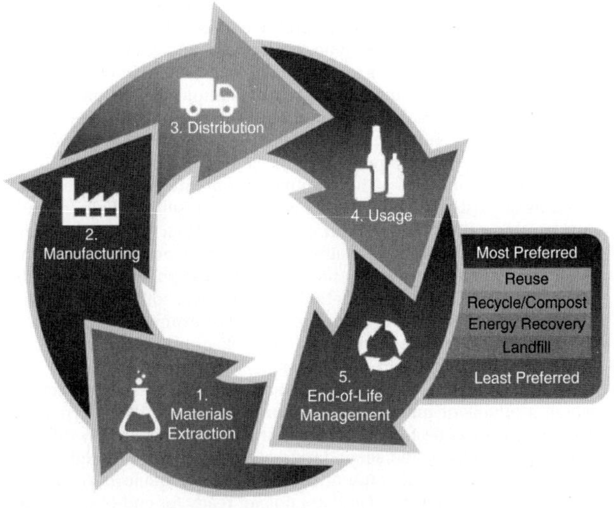

FIG. 22-47 Sustainable material management's life-cycle perspective. (*U.S. Environmental Protection Agency, Sustainable Materials Management. Last updated October 18, 2016. https://www.epa.gov/smm/sustainable-materials-management-basics.*)

to use different, fewer, less toxic, or more durable materials in manufacture. It can be designed so that at the end of its useful life it can be readily disassembled. The manufacturer works to ensure the best use of the product, its maintenance, and return at end-of-life. Further, the manufacturer has a similar relationship with its suppliers, which helps the manufacturer to respond more quickly to changing demands, including environmental impacts along the supply chain. Leading companies have been able to demonstrate the environmental and economic advantages of this approach across the supply chain.

For those working in solid or hazardous waste management, a valuable source of information is the U.S. EPA's *RCRA Orientation Manual* (update released in 2014), which reflects the changing dynamics of the industry in addition to the changes in regulations, guidance, initiatives, and congressional mandates that affect solid waste management. To understand the magnitude of the changes that have occurred since RCRA was enacted, see *25 Years of RCRA: Building on Our Past to Protect Our Future*, published in 2001, which identifies the many changes and major accomplishments in the solid waste management field. For a look ahead, *Sustainable Materials Management: The Road Ahead* suggests a road map for the future based on materials management—fulfilling human needs and prospering, while using fewer materials, reducing toxics, and recovering more of the materials used.

Sustainable design also requires understanding the costs associated with solid waste management [U.S. EPA, *Full Cost Accounting in Action: Case Studies of Six Solid Waste Management Agencies*, EPA 530-R-98-018, Dec 1998]. Assessment of all solid waste management costs, including direct (e.g., operational), indirect (e.g., shared administration), and capital costs (e.g., land, buildings, equipment) as well as environmental costs, will help inform solid waste planning decisions. Sustainable design will minimize the aggregate of all costs. Sustainable management analyses may, at times, identify solutions that contradict the waste hierarchy, but that still offer the lowest environmental impact at a practical cost across a material's or product's life cycle. Often what may make the most environmental and economic sense for a community that is urban or in a population-dense region will be different from what is preferable for more rural or remote regions. There are many other important considerations, such as differences in waste composition, energy grid mix, market fluctuations and market value of recovered materials, and local infrastructure. There is much diversity in the management of solid waste and materials, and local drivers must be considered along with the life-cycle environmental and economic tradeoffs. The next subsection identifies the process units or technologies and activities associated with solid waste or "materials" management.

SOLID WASTE MANAGEMENT PROCESS UNITS AND ACTIVITIES

The process units and activities associated with solid waste management are described next. The process units largely fall within the end-of-life management stage of the product life cycle presented in Fig. 22-47. It is important to understand the individual process units, using a system analysis to calculate the material and energy flows of each process unit. Interactions occur across the activities and should be considered in developing more sustainable management plans. How do options compare, considering both the life-cycle economics and the life-cycle environmental tradeoffs, for food waste and other waste streams in both residential and commercial sectors? We must consider differences in waste composition and in the options for collection, transport, and management. For example, is there on-site or off-site composting? Anaerobic or aerobic digestion? What is the final fate of the products at the end of their useful life? Are they suitable only for daily landfill cover, or can they supply nutrient value for agriculture?

1. *Reduction and reuse.* These activities evaluate potential reductions of discards or waste generation at all points in the supply chain: manufacturing, transportation, distribution, and consumption. Industrial processes can be redesigned to reduce the amount of waste generated. For example, transfer lines between processes can be blown clear pneumatically to drive residual liquid into the batch mix tank, reducing washout. Another form of source reduction is diverting wastes to reuse. For example, containers may be cleaned and reused. Reusable transport packaging will reduce packaging waste at the point of product distribution.

2. *Waste generation.* Waste generation encompasses those activities in which products or materials are no longer valued in their current form and are ready for end-of-life management. Wastes are typically differentiated from by-products and secondary materials, which, although possibly not valued for their initial or intended use, may be valuable as an input to a different product and process and are not yet ready for end-of-life management. As shown in Figure 22-47, products and materials reaching the end of their useful life are either collected for materials recovery, combusted with energy recovery, or disposed through landfilling. Resources from the U.S. EPA for increasing the recovery of materials from MSW can be accessed at https://www.epa.gov/recycle. The quantity of waste has been shown to decrease in response to economic decline, resulting in reduced consumer consumption by extended product use, increased product reuse, and delayed product purchase.

3. *On-site handling, storage, and processing.* This process unit encompasses those activities associated with the handling, storage, and processing of solid wastes at or near the point of generation. On-site storage can be important because of aesthetic considerations, public health, public safety, and economics. For incident waste management, the activities include a staging area for forensics and sorting of materials for either recovery or disposal, depending on the hazards and materials involved.

4. *Collection.* Collection includes the gathering of waste for hauling to the location where the collection vehicle is emptied. Wastes might be hauled to a transfer station, a processing facility, a combustor with or without energy recovery, or a landfill disposal site. All mass burn combustion facilities in the United States recover energy from the burning of waste and remove ferrous and nonferrous materials from the ash. The complexity, timing, equipment, and frequency of collection will vary by the material or waste to be managed, the community or regional population density, air quality concerns, and the distance to be hauled. Conventional wisdom suggests that the best place to capture materials for recovery from MSW is at the point of generation before it is commingled with other wastes. Source-separated recycling is also referred to as dual-stream recycling. This means, for example, keeping the fiber component—paper and cardboard—separate from containers, including glass and plastic containers and cans. Single stream is a method of recycling that allows paper, cardboard, plastic, glass, and metal to be mixed together for pickup. It has been on the increase in residential recycling (or collection) programs. Many states, such as Florida and California, are finding that they can increase recycling rates using commingled or single-stream recycling instead of separation at the point of collection. The tradeoffs between single and dual recycling streams are in the level of material contamination, the quality and quantity of recovered materials, and the processing cost of the collected recyclables. A 2016 report on "state of curbside" provides an analysis of data from 465 curbside recycling programs in the United States. This report found that most collection streams were single stream. The report concludes that successful programs benefit from strong community engagement, and there is a need for more consistency in programs to minimize confusion from those participating [*The 2016 State of Curbside Report*, Prepared by the Recycling Partnership; https://therecyclingpartnership.app.box.com/s/i0wvano7hi3dr3ivqxv689y4zzo583l2; Lakhan, C., "A Comparison of Single and Multi-Stream Recycling Streams in Ontario, Canada," *Resources 2015* **4:** 384–397; doi:10.3390/resources4020384, www.mdpi.com/journal/resources].

5. *Transfer and transport.* Transferring and transporting waste involves two steps: (1) the transfer of wastes from the smaller collection containers to the larger transport equipment and (2) the subsequent transport of the wastes to a transfer station, a processing facility, a combustor with or without energy recovery, or a landfill. Waste transported through a transfer station is usually hauled longer distances to a combustor or landfill. Communities near a coast or river often use barges to haul trash. Rail hauling is also becoming more common for large regional landfills and municipal waste combustors (MWC) [also referred to as waste to energy (WTE)].

6. *Processing and recovery.* The activities for processing and recovery include all the techniques, equipment, and facilities used both to improve the efficiency of the other process components and to recover usable materials, conversion products, or energy from solid wastes. Residues from the recovery process typically go to disposal. If wastes cannot be processed for recovery of materials through recycling or composting, energy recovery through biological and thermal conversion processes or direct combustion is preferable to landfilling if the desire is to maximize energy and resource recovery. However, landfilling may be more desirable if cost is the major driver, although environmental externalities may be important to consider, such as long-term monitoring and maintenance requirements at landfills since emissions may be generated for decades.

7. *Combustion with energy recovery.* Activities associated with the combustion of MSW include unloading solid waste from collection trucks and placing the waste in a storage bunker, and using an overhead crane to mix and subsequently feed the waste into the combustion chamber. Approximately 80 percent of waste combustion facilities are mass burn units. The remaining facilities are refuse-derived fuel (RDF) (~15 percent) and modular units (5 percent) (Energy Recovery Council, 2016 Directory of Waste-to-Energy Facilities, http://energyrecoverycouncil.org/wp-content/uploads/2016/06/ERC-2016-directory.pdf). For both RDF and modular units, there are additional activities different from the mass burn units. Heat released from combustion converts water to steam, which is sent to a turbine generator to produce electricity. Steam may also be exported directly for productive use in combined heat and power (CHP) or steam-only plants. MSW ash is a by-product produced during the combustion process. Captured fly ash

particles fall into hoppers (funnel-shaped receptacles) and are transported by an enclosed conveyor system to the ash discharger. They are then wetted to prevent fugitive dust emissions and mixed with the bottom ash from the grate. The facility transports the ash residue to an enclosed building, where it is processed for removal of recyclable scrap metals. In the United States, MSW combustor ash is mixed and discarded in landfills. European MWC ash is combined, as described above, and managed in an ash monofill, an MSW landfill, or as daily cover in an MSW landfill. In Europe, MWC bottom ash is typically managed separately from the fly ash. The bottom ash is often used in construction or in another beneficial reuse application that meets applicable European Union (EU) regulations.

8. *Disposal through landfilling.* The final set of process activities is disposal. Disposal is the ultimate fate of all solid wastes, whether they are collected and transported to a transfer station or directly to a landfill site, semisolid wastes (sludge) from industrial or municipal treatment plants and air pollution control devices, combustion residue, compost, small-quantity generator hazardous waste, or other substances from various solid waste processing plants. Many U.S. landfills incorporate systems that recover energy from the landfill gas produced by the anaerobic decomposition of biodegradable wastes.

9. *Administrative and other activities supporting solid waste management.* Aspects of solid waste management systems not listed previously include financing, operations, equipment management, personnel, reporting, cost accounting and budgeting, contract administration, ordinances and guidelines, and public communications. Landfill owners and operators are required to provide for the long-term monitoring and maintenance at the landfill to ensure the integrity of the liner, surface and side slope stability, cover material (to minimize cracks and fissures from droughts), and maintenance and monitoring of gas and leachate control technology. Another potential factor affecting waste facilities on the east and southeastern U.S. coasts are building the facilities to withstand potential flooding due to sea level rise or flooding associated with hurricanes or extreme weather events.

UNITED STATES LEGISLATION, REGULATIONS, AND POLICY

It is important to have a working knowledge of environmental laws pertaining to waste management and their related implementing regulations. The Resource Conservation and Recovery Act (RCRA; also known as the Solid Waste Disposal Act), passed in 1976, was established to set up a framework for the proper management of solid and hazardous waste. RCRA establishes criteria for distinguishing hazardous waste from nonhazardous wastes. For hazardous waste, RCRA subtitle C further establishes requirements for waste generators, transporters, and treatment, storage, and disposal facilities (including permitting requirements). RCRA gives states the primary responsibility for developing programs to control nonhazardous waste management, and local regulations may also apply.

Other major legislation affecting the management of solid waste includes the Toxic Substances Control Act (TSCA) and the Clean Air Act (CAA). On June 22, 2016, TSCA was amended to require that new and existing chemicals be evaluated using a risk-based safety standard that includes considerations for vulnerable populations, increased transparency, and chemical information. Primary components of the CAA are Title I, dealing with fugitive emissions monitoring; Title III, reduction of organic hazardous air pollutants; and Title V, operating permits. The 1990 CAA Amendments added requirements for industrial plants to control hazardous air pollutants (HAPs). Most significantly, CAA regulations apply to waste combustion and MSW landfills. CAA regulations may also apply to other facilities that manage waste and materials, including anaerobic digestion, composting, and certain recycling operations (e.g., paper mills, steel foundries).

Under the CAA, EPA has promulgated regulations for both large and small MSW combustors. Based on available data for 1990 to 2005 from the National Emissions Inventories for MSW combustion, total emissions of hazardous air pollutants dropped more than 94 percent in this time period from nearly 58,000 tons in 1990 to about 3300 tons in 2005. Criteria pollutants, such as carbon monoxide, particulate matter, and sulfur dioxide, were also further reduced. Particulate matter was reduced by 96 percent from 1990 to 2005 (i.e., 18,600 to 780 tons per year). Sulfur dioxide emissions for MSW combustion was reduced by 88 percent from 1990 to 2005 (38,300 to 4600 tons per year). Consequently, municipal waste combustors have become a comparatively minor source of combustion-related air pollution as compared to other sources, such as automobiles, trucks, power plants, fireplaces, wood stoves, industrial manufacturing processes, volcanoes, forest fires, and backyard trash burning.

In 2016, CAA regulations were finalized to reduce emissions of gas from MSW landfills (landfills are a major source of methane), along with pollutants that contribute to tropospheric ozone. Recovery and use of methane helps avoid the loss of methane to the atmosphere and conserves fossil fuel. It can be either directly used as boiler fuel or used to generate electricity through combustion in a boiler, turbine, or internal combustion engine. As of 2016, 652 landfill gas to energy projects were in operation in the United States, extracting gas from about 600 landfills (LMOP, https://www.epa.gov/lmop). Within the United States, an outreach program was created to encourage landfill methane recovery and utilization.

TYPES OF SOLID WASTE AND FACTORS IMPORTANT TO ITS MANAGEMENT AND ESTIMATES OF GENERATION

Solid wastes, as noted previously, include all solid or semisolid materials that are no longer considered of sufficient value to be retained. The types and sources of solid wastes, the physical and chemical composition of solid wastes, and typical solid waste generation rates are considered in this subsection.

Types of Solid Waste The term *solid wastes* is all-inclusive and encompasses all sources, types of classifications, compositions, and properties. As a basis for subsequent discussions, it will be helpful to define the various types of solid wastes that are generated. Many definitions of solid waste terms and the classifications are established by federal or state waste management regulations. However, commonly used terms that do not have legal definitions may be used in varying ways, and the use of published data requires considerable care, judgment, and common sense. The following definitions are intended to serve as a guide. [Unless otherwise noted, definitions are adapted from Tchobanoglous, G., and F. Keith, *Handbook of Solid Waste Management*, 2d ed., McGraw-Hill, New York, 2002.]

1. *Solid waste under RCRA regulations.* The term *solid waste*, when used in a regulatory context under RCRA, includes garbage or refuse, sludges, air pollution control residues, and other discarded material resulting from industrial, commercial, mining, and agricultural operations, and from community activities. A solid waste can be a liquid, solid, or contained gaseous material. A solid waste is generated when it is discarded, which can occur if the material is abandoned, is considered inherently waste-like, or is recycled in certain ways.

2. *Municipal solid waste (MSW).* MSW consists of everyday items such as product packaging, yard trimmings, furniture, appliances, tires, food waste, paper products, and consumer electronics. MSW comes from residential, commercial, and institutional sources. Office and cafeteria-type wastes from industrial facilities are included in MSW, but industrial process wastes are excluded. MSW is also called garbage or rubbish [U.S. EPA and Report on the Environment (ROE) Municipal Solid Waste indicator. https://cfpub.epa.gov/roe/indicator_pdf.cfm?i=53. Accessed August 18, 2016].

2a. *Garbage.* An older term that is used to represent easily degradable waste (i.e., putrescible) such as animal, fruit, or vegetable residues resulting from the handling, preparation, cooking, and eating of foods. In warm weather, putrescible wastes will decompose rapidly, posing unique challenges to collect. The materials or waste in this group are already included in the definition of municipal solid waste (as per RCRA). However, this term is used internationally and is therefore defined.

2b. *Rubbish.* Rubbish—another older term not typically used—consists of combustible and noncombustible solid wastes, excluding food wastes or other putrescible materials. Typically, combustible rubbish consists of materials such as paper, cardboard, plastics, textiles, rubber, leather, wood, furniture, and garden trimmings. Noncombustible rubbish consists of items such as glass, crockery, tin cans, aluminum cans, ferrous and other nonferrous metals, dirt, and construction wastes. Again, the types of materials or wastes once referred to as rubbish are already included in RCRA's definition for municipal solid waste.

2c. *MSW generation rates.* Municipal solid waste—as defined by RCRA—refers to materials and products from residential, commercial, institutional, and industrial sources. The MSW generation rate is based on the weight of materials that enter the waste stream before recovery or disposal. Many materials within MSW are recovered for reuse or recycling. A major focus for future recovery is organic waste as industry works to meet the U.S. goal of reducing food loss and waste by 50 percent by 2030. This is important for multiple reasons. First, if food waste is composted, then the resulting material can be used as a soil amendment. Second, if food waste or other rapidly decomposing waste is landfilled, more than likely the methane and other emissions escape to the atmosphere, given that gas collection wells are not installed until three to five years after waste burial [Levis, J. W., and M. A. Barlaz, "Is Biodegradability a Desirable Attribute for Discarded Solid Waste? Perspectives from a National Landfill Greenhouse Gas Inventory Model," *Environ. Sci. Technol.* **45:** 5470–5476 (2011)]. Third, food waste is the largest single stream of materials disposed in MSW, and there are many options for managing food waste that use nutrients from composted food waste to support agriculture and landscaping industries. Once separated and diverted from disposal, food waste is either aerobically composted or processed in an anaerobic digester, which is much more common in Europe than in the United States [Hodge, K. L., J. W. Levis, J. F. DeCarolis, and M. A. Barlaz, "Systematic Evaluation of Industrial, Commercial, and Institutional Food Waste Management Strategies in the United States," *Environ. Sci. Technol.* **50:** 8444–8452 (2016)].

Characterizing the quantity and composition of waste to be managed in a community is critical to being able to develop effective management plans. MSW tonnage (or waste generation) is used to estimate emissions, evaluate management-related policies, and track resource recovery and consumption for city planning and public works budgeting. Currently there are two sources of MSW data, which vary by about 50 percent. The Environmental Research and Education Foundation (EREF) gathered data using a bottom-up, facility-based methodology, whereas the U.S. government statistics are based on a top-down approach using data from industry associations, businesses, and government sources, such as the U.S. Department of Commerce and the U.S. Census Bureau. EREF reported 347 million tons of MSW were generated in 2014, whereas the EPA reported 258 million tons of MSW were generated in 2014. EREF facility survey data found that 64 percent of MSW is managed at landfills, 21 percent recycled, 9 percent managed at waste-to-energy facilities, and 6 percent composted. The EREF report suggests that MSW tonnage is higher and recycling rates are lower than what is reported by the U.S. EPA [EREF (2016) MSW Management in the U.S.: 2010 and 2013, https://erefdn.org/product/municipal-solid-waste-management-u-s-2010-2013/. Accessed 11/4/2016].

3. *Industrial process wastes.* Unwanted material produced during industrial operations or manufacturing processes. These wastes are generated by activities such as manufacturing, agricultural operations, and mining. Many industries are finding successful strategies to reduce discards and build products that use fewer toxics, and they are built so that resources such as fiber and metals can be more readily recovered.

One of the largest waste streams—coal combustion residues (CCR)—is a by-product of the coal-fired electric utility industry. Two separate regulatory determinations were made (in 1993 and in 2000) to exclude large-volume coal combustion waste and the remaining fossil fuel combustion waste from hazardous waste regulations under RCRA subtitle C. As of 2015, 61 million out of 117 million tons of CCR (or 53 percent by mass) were being used to make a range of products from concrete and other cementitious materials, structural fills and embankments, and mining applications (American Coal Ash Association, https://www.acaa-usa.org/Portals/9/Files/PDFs/2015-Survey_Results_Table.pdf). Flue gas desulfurization (FGD) gypsum is also regarded as a CCR and when its use replaces mined gypsum. FGD gypsum is being successfully used to make wallboard.

On April 17, 2015, EPA issued federal regulations establishing requirements for the disposal of residues generated from the combustion of coal at electric utilities and independent power producers. These regulations establish technical requirements for CCR landfills and surface impoundments under subtitle D of RCRA, the nation's primary law for regulating solid waste. The regulations address the risks from coal ash disposal (1) where damage cases found contaminants leaking from land disposal units into groundwater, (2) catastrophic failure of coal ash impoundments, and (3) blowing of contaminants into the air as dust. The 2015 final rule also supports the responsible beneficial use of CCR by distinguishing beneficial use from disposal. There is widespread interest in the use of fly ash and other industrial secondary materials based on a range of benefits. In the planning section, beneficial use of fly ash and other industrial secondary materials will be discussed, along with resources available to help evaluate the potential adverse impacts to human health and the environment from beneficially using secondary materials.

4. *Ashes and residues.* These are the materials remaining from the burning of wood, coal, coke, and other combustible wastes. CCRs have already been discussed under industrial wastes. Ashes and other combustion residues consist of fine powdery materials, cinders, clinkers, and small amounts of burned and partially burned materials.

4a. *Bottom ash.* Heavy, nonairborne residue resulting from combustion in a furnace or incinerator that must be removed mechanically. Bottom ash constitutes the majority of the total ash created by solid waste combustion.

4b. *Fly ash.* Particles that rise with flue gases from combustion rather than sinking to the bottom of the furnace or incinerator. Fly ash represents a small portion of the total ash produced from combustion and can be removed by pollution control equipment. Fly ash from particulate control devices (i.e., electrostatic precipitators or baghouses) and CKD (cement kiln dust) are used in waste stabilization and waste bulking operations. Fly ash—as with coal combustion at power plants—represents residues resulting from air pollution control devices that can vary in complexity depending on the coal type and applicable regulations. To prevent cross-media transfers, it is important to ensure that when fly ash is managed, pollutants captured in the flue gas and transferred to fly ash are not later released to the environment. In the CCR regulations, the potential cross-media transfers were considered for the air, land, and water, with the ultimate fate based on the waste form and environmental conditions the CCR is exposed to over time. [Thorneloe, S., D. Kosson, F. Sanchez, A. Garrabrants, and G. Helms, "Evaluating the Fate of Metals in Air Pollution Control Residues from Coal-Fired Power Plants," *Environmental Science and Technology* **44:** 7351–7356 (2010); Senior, C., S. Thorneloe, B. Khan, and D. Goss, "Fate of Mercury Collected from Air Pollution Control Devices," *Environmental Management* (July 2009): 15–21.]

5. *Construction and demolition wastes (C&D).* C&D wastes are from the construction, renovation, and demolition of buildings, roads, and bridges. These wastes include steel, wood products, bricks, clay, tile, drywall and plaster, asphalt shingles, asphalt, concrete, and portland cement. In addition to building materials, C&D also includes plumbing, heating, and electrical products, lead paint, and asbestos requiring special handling and management.

6. *Special wastes.* The definition of special wastes varies from state to state. Some typical regulated and nonregulated special wastes include cement kiln dust, crude oil and natural gas waste, fossil fuel combustion waste, mining waste, and mineral processing waste. Both federal and state regulations must be met in managing special wastes.

7. *Treatment plant wastes.* Solid and semisolid wastes from water, wastewater, and industrial waste-treatment facilities.

8. *Agricultural wastes.* Wastes and residues resulting from diverse agricultural activities, such as the planting and harvesting of row, field, and tree and vine crops, the production of milk, the production of animals for slaughter, and the operation of feedlots, are collectively called agricultural wastes. Animal feeding operations—or confined animal feedlot operations (CAFOs)—are subject to national pollutant discharge elimination system (NPDES) regulations and can be a source of ammonia and methane emissions, depending on how the waste is managed.

Animal wastes can be valuable sources of plant nutrients (fertilizer), animal nutrients (feedstuffs), and feedstock for energy (methane) generation. Diseased animals are also included as agricultural waste. For any animals killed to eradicate epizootic diseases, guidance is available through the USDA's Animal and Plant Health Inspection Service for response options to make communities more resilient by encouraging immediate action as needed to prevent further loss of animal or vegetable products.

9. *Electronic wastes.* Many materials such as gold, silver, palladium, and rare earth elements can be found in electronic scrap that comes from consumer, business, and government electronic equipment that is near or at the end of its useful life. A National Strategy for Electronics Stewardship was initiated in 2011 to lay the groundwork for improving the design of electronic products and enhancing our management of used or discarded electronics.

10. *Medical waste.* Medical waste is a subset of wastes generated at health care facilities, such as hospitals, physicians' offices, dental practices, blood banks, and veterinary hospitals/clinics, as well as medical research facilities and laboratories. Generally, medical waste is health care waste that may be contaminated by blood, body fluids, or other potentially infectious materials and is often referred to as regulated medical waste.

11. *Homeland security materials and waste.* Homeland security incidents include natural disasters, such as earthquakes, tornados, wildfires, floods, and hurricanes; unintentional human-made disasters, such as oil spills and nuclear power plant accidents; and intentional human-made incidents, such as those generated by a contamination incident using chemical, biological, radiological, or nuclear (CBRN) material. The composition, quantity, toxicity, and complexity of waste and material management will vary by incident type. Response options available for each type of incident are documented as part of the planning process by state and local government in a preincident waste management plan. Waste streams may include animal carcasses contaminated with biological agents, large quantities of vegetative debris and commingled construction and demolition debris, contaminated water resulting from decontamination and cleanup or post-incident precipitation events, and chemical, biological, and radiologically contaminated wastes. Guidance is available for planning and managing waste resulting from homeland security incidents, including: information on important considerations to address, tools to support planning efforts, and waste management options to increase resiliency following an event. [U.S. EPA, Managing Materials and Wastes for Homeland Security Incidents web site, URL: https://www.epa.gov/homeland-security-waste. Accessed August 17, 2016.] A four-tiered waste management hierarchy is used to guide waste management decision making for homeland security waste by reusing and recyling as much material as possible. Prior planning for potential incidents should include identifying opportunities for waste minimization and developing criteria and options for reuse and recycling. Management activities include waste staging, sampling, characterization, packaging, transportation, reuse, recycling, treatment, and disposal.

12. *Radioactive materials.* The regulation of radioactive materials is shared by the U.S. EPA, the Food and Drug Administration, the U.S.

Nuclear Regulatory Commission, and state governments. Because of their potentially hazardous properties, the use of certain radioactive materials is closely regulated to protect the health and safety of the public and the environment. Low-level radioactive waste includes radioactively contaminated protective clothing, tools, filters, rags, medical tubes, and many other items. Waste incidental to reprocessing refers to certain waste by-product that results from reprocessing spent nuclear fuel, which the U.S. Department of Energy (DOE) has distinguished from high-level waste. High-level radioactive waste is irradiated or used nuclear reactor fuel. Uranium mill tailings are the residues remaining after the processing of natural ore to extract uranium and thorium. Because of their highly radioactive fission products, high-level radioactive waste and spent fuel must be handled and stored with care. Since the only way radioactive waste finally becomes harmless is through decay, which for high-level radioactive wastes can take hundreds of thousands of years, the wastes must be stored and finally disposed of in a way that provides adequate protection to the public for a very long time. Currently within the United States, there is no geological repository for permanent disposal of high-level radioactive waste, although decades have been spent evaluating a potential site for long-term disposal. Therefore, this section does not include a discussion on radioactive waste management.

Hazardous Waste RCRA provides a legal definition of the term *hazardous waste*, and the U.S. EPA provides a detailed definition in RCRA regulations 40 CFR parts 260 and 261. A hazardous waste is a waste with properties that make it dangerous or capable of having a harmful effect on human health or the environment if it is mismanaged. Hazardous waste is generated from many sources and may come in many forms, including liquids, solids, gases, and sludges. A waste may be hazardous if it exhibits one or more of the following characteristics: (1) ignitability, (2) corrosivity, (3) reactivity, and (4) toxicity (as determined by the leaching potential of hazardous constituents). A detailed definition of these terms was first published in the *Federal Register* on May 19, 1980. (See 45 FR 33084-33133.) A list of hazardous wastes identified from specific industries, common manufacturing and industrial processes, and discarded commercial chemical products can be found in 40 CFR parts 261.30 through 261.34.

Sources of Wastes Knowledge of the sources and types of solid wastes, along with data on the composition and rates of generation, is basic to the design and operation of the process components associated with the management of solid wastes.

Industrial Waste Sources and types of industrial solid wastes generated by different industry groups are reported in Table 22-44. The expected specific wastes in the table are most readily identifiable.

Hazardous Waste Sources of hazardous waste include industrial manufacturing processes, national and regional laboratories, mining operations, military bases, and small businesses. They may come in many forms, including liquids, solids, gases, and sludges, and household hazardous waste. Hazardous waste is a waste with properties that make it dangerous or capable of having a harmful effect on human health or the environment, as defined by RCRA and its regulations. The identification of amounts and types of hazardous wastes is a concern at each source, with an emphasis on those sources where significant waste quantities are generated. EPA's hazardous waste site provides information on U.S. hazardous waste definitions, management, and regulations.

Spills also generate hazardous wastes and must also be considered. After a spill, the wastes requiring collection and disposal are often significantly greater than the amount of spilled material, especially when an absorbing material, such as straw, sand, or "oil-dry," is used to soak up liquid hazardous wastes. Similarly, the volume and mass of contaminated soil requiring excavation are far greater than the waste volume involved in a spill. Both the adsorbing material and the liquid are classified as hazardous waste.

Properties of Solid Waste Information on the properties of solid wastes is important in evaluating alternative equipment needs, systems, and management programs and plans.

TABLE 22-44 Sources and Types of Industrial Wastes*

Industry	Waste-generating processes	Expected specific wastes
Ordnance and accessories	Manufacturing, assembling	Metals, plastic, rubber, paper, wood, cloth, chemical residues
Food and kindred products	Processing, packaging, shipping	Meats, fats, oils, bones, offal, vegetables, fruits, nuts and shells, cereals
Textile mill products	Weaving, processing, dyeing, shipping	Cloth and filter residues
Apparel and other finished products	Cutting, sewing, sizing, pressing	Cloth, fibers, metals, plastics, rubber
Lumber and wood products	Sawmills, millwork plants, wooden containers, miscellaneous wood products, manufacturing	Scrap wood, shavings, sawdust; in some instances, metals, plastics, fibers, glues, sealers, paints, solvents
Furniture, wood	Manufacture of household and office furniture, partitions, office and store fixtures, mattresses	Those listed under Code 24, in addition, cloth and padding residues
Furniture, metal	Manufacture of household and office furniture, lockers, springs, frames	Metals, plastics, resins, glass, wood, rubber, adhesives, cloth paper
Paper and allied products	Paper manufacture, conversion of paper and paperboard, manufacture of paperboard boxes and containers	Paper and filter residues, chemicals, paper coatings and filters, inks, glues, fasteners
Printing and publishing	Newspaper publishing, printing, lithography, engraving, bookbinding	Paper, newsprint, cardboard, metals, chemicals, cloth, inks, glues
Chemicals and related products	Manufacture and preparation of organic and inorganic chemicals (ranging from drugs and soaps to paints and varnishes and explosives)	Organic and inorganic chemicals, metals, plastics, rubber, glass, oils, paints, solvents, pigments
Petroleum refining and related industries	Manufacture of paving and roofing materials	Asphalt and tars, felts, paper, cloth, fiber
Rubber and miscellaneous plastic products	Manufacture of fabricated rubber and plastic products	Scrap rubber and plastics, lampblack, curing compounds, dyes
Leather and leather products	Leather tanning and finishing, manufacture of leather belting and packaging	Scrap leather, thread, dyes, oils, processing and curing compounds
Stone, clay, and glass products	Manufacture of flat glass, fabrication or forming of glass, manufacture of concrete, gypsum, and plaster products; forming and processing of stone products, abrasives, asbestos, and miscellaneous non-mineral products	Glass, cement, clay, ceramics, gypsum, asbestos, stone, paper, abrasives
Primary metal industries	Melting, casting, forging, drawing, rolling, forming, extruding operations	Ferrous and nonferrous metals scrap, slag, sand, cores, patterns, bonding agents
Fabricated metal products	Manufacture of metal cans, hand tools, general hardware, non-electrical heating apparatus, plumbing fixtures, fabricated structural products, wire, farm machinery and equipment, coating and engraving of metal	Metals, ceramics, sand, slag, scale, coatings, solvents, lubricants, pickling liquors
Machinery (except electrical)	Manufacture of equipment for construction, elevators, moving stairways, conveyors, industrial trucks, trailers, stackers, machine tools, etc.	Slag, sand, cores, metal scrap, wood, plastics, resins, rubber, cloth, paints, solvents, petroleum products
Electrical	Manufacture of electrical equipment, appliances and communication apparatus, machining, drawing, forming, welding, stamping, winding, painting, plating, baking, firing operations	Metal scrap, carbon, glass, exotic metals, rubber, plastics, resins, fibers, cloth residues, PCBs
Transportation equipment	Manufacture of motor vehicles, truck and bus bodies, motor-vehicle parts and accessories, aircraft and parts, ship and boat building, repairing motorcycles and bicycles and parts, etc.	Metal scrap, glass, fiber, wood, rubber, plastics, cloth, paints, solvents, petroleum products
Professional scientific controlling instruments	Manufacture of engineering, laboratory and research instruments and associated equipment	Metals, plastics, resins, glass, wood, rubber, fibers, abrasives
Miscellaneous manufacturing	Manufacture of jewelry, silverware, plate ware, toys, amusements, sporting and athletic goods, costume novelties, buttons, brooms, brushes, signs, advertising displays	Metals, glass, plastics, resin, leather, rubber, composition, bone, cloth, straw, adhesives, paints, solvents

*Std. Industrial Classification Manual, 1972.

Physical Composition Information and data on the physical composition of solid wastes including (1) density of solid wastes, (2) moisture content, and (3) particle size are presented next.

1. *Density.* Typical densities for various wastes as found in containers are reported by source in Table 22-45. Because the densities of solid waste vary markedly with geographic location, season of the year, and length of time in storage, care should be used in selecting typical values.

2. *Moisture content.* The moisture content of solid wastes usually is expressed as the mass of moisture per unit mass of wet or dry material. In the wet-mass method of measurement, the moisture in a sample is expressed as a percentage of the wet mass of the material; in the dry-mass method, it is expressed as a percentage of the dry mass of the material. In equation form, the wet-mass moisture content is expressed as

$$\text{Moisture content (\%)} = \left\{a - \frac{b}{a}\right\} \times 100 \qquad (22\text{-}26)$$

where a = initial mass of sample as delivered and b = mass of sample after drying. Typical data on the moisture content for the solid waste components are given in Table 22-46. The moisture content of industrial wastes can vary greatly.

3. *Particle size.* The material handling properties of solid wastes are dependent on particle size and distribution. This applies as well to feed preparation and air pollution control, which are affected by solid waste particle size and cohesiveness. For wastes such as bulk soils, the amount of fines (from clay and silt) is critical for system design. Cohesiveness, which varies with moisture content, is important for bin and conveyor design.

TABLE 22-45 Typical Density and Moisture-Content Data for Domestic, Commercial, and Industrial Solid Waste

Item	Density, kg/m³		Moisture content, % by mass	
	Range	Typical	Range	Typical
Residential (uncompacted)				
Food wastes (mixed)	130–480	290	50–80	70
Paper	40–130	85	4–10	6
Cardboard	40–80	50	4–8	6
Plastics	40–130	65	1–4	2
Textiles	40–100	65	6–15	10
Rubber	100–200	130	1–4	2
Leather	100–260	160	8–12	10
Garden trimmings	60–225	100	30–80	60
Wood	130–320	240	15–40	20
Glass	160–480	195	1–4	2
Tin cans	50–160	90	2–4	3
Nonferrous metals	65–240	160	2–4	2
Ferrous metals	130–1150	320	2–4	2
Dirt, ashes, etc.	320–1000	480	6–12	8
Ashes	650–830	745	6–12	6
Rubbish (mixed)	90–180	130	5–20	15
Residential (compacted)				
In compactor truck	180–450	300	15–40	20
In landfill (normally compacted)	360–500	450	15–40	30*
In landfill (well-compacted)	590–740	600	15–40	30*
Commercial				
Food wastes (wet)	475–950	535	50–85	75
Appliances	150–200	180	0–5	
Wooden crates	110–160	110	10–30	20
Tree trimmings	100–180	150	20–80	50
Rubbish (combustible)	50–180	120	5–25	15
Rubbish (noncombustible)	180–360	300	5–15	10
Rubbish (mixed)	140–180	160	5–20	12
Construction; demolition; remediation				
Mixed demolition (noncombustible)	1000–1600	1420	2–10	4
Mixed demolition (combustible)	300–400	360	4–15	8
Mixed construction (combustible)	180–360	260	4–15	8
Broken concrete	1200–1800	1540	0–5	—
Contaminated soil	1200–1900	1600	5–25	10
Industrial wastes				
Chemical sludges (wet)	800–1100	1000	75–99	80
Fly ash	700–900	800	2–10	4
Leather scraps	100–250	160	6–15	10
Metal scraps (heavy)	1500–2000	1780	0–5	
Metal scraps (light)	500–900	740	0–5	
Metal scraps (mixed)	700–1500	900	0–5	
Oils, tars, asphalt	800–1000	950	0–5	2
Sawdust	100–350	290	10–40	15
Textile wastes	100–220	180	6–15	10
Wood (mixed)	400–675	500	10–40	20
Agricultural wastes				
Agricultural (mixed)	400–750	560	48–80	50
Fruit wastes (mixed)	250–750	360	60–90	75
Manure (wet)	900–1050	1000	75–96	94
Vegetable wastes (mixed)	200–700	360	50–80	65

*Depends on degree of surface water infiltration.

Adapted from G. Tchobanoglous, H. Theisen, and S. Vigil, *Integrated Solid Waste Management: Engineering Principles and Management Issues*, McGraw-Hill, New York, 1993.

TABLE 22-46 Typical Proximate-Analysis and Energy-Content Data for Components in Domestic, Commercial, and Industrial Solid Waste*

	Proximate analysis, % by mass				Energy content, kJ/kg		
Component	Moisture	Volatile matter	Fixed carbon	Non-combustible	As collected	Dry	Moisture- and ash-free
Food and food products							
Fats	2.0	95.3	2.5	0.2	37,530	38,296	38,374
Food wastes (mixed)	70.0	21.4	3.6	5.0	4,175	13,917	16,700
Fruit wastes	78.7	16.6	4.0	0.7	3,970	18,638	19,271
Meat wastes	38.8	56.4	1.8	3.1	17,730	28,970	30,516
Paper products							
Cardboard	5.2	77.5	12.3	5.0	16,380	17,278	18,240
Magazines	4.1	66.4	7.0	22.5	12,220	12,742	16,648
Newsprint	6.0	81.1	11.5	1.4	18,550	19,734	20,032
Paper (mixed)	10.2	75.9	8.4	5.4	15,815	17,611	18,738
Waxed cartons	3.4	90.9	4.5	1.2	26,345	27,272	27,615
Plastics							
Plastics (mixed)	0.2	95.8	2.0	2.0	32,000	32,064	32,720
Polyethylene	0.2	98.5	<0.1	1.2	43,465	43,552	44,082
Polystyrene	0.2	98.7	0.7	0.5	38,190	38,266	38,216
Polyurethane	0.2	87.1	8.3	4.4	26,060	26,112	27,316
Polyvinyl chloride	0.2	86.9	10.8	2.1	22,690	22,735	23,224
Wood, trees, etc.							
Garden trimmings	60.0	30	9.5	0.5	6,050	15,125	15,316
Green wood	50.0	42.3	7.3	0.4	4,885	9,770	9,848
Hardwood	12.0	75.1	12.4	0.5	17,100	19,432	19,542
Wood (mixed)	20.0	67.9	11.3	0.8	15,444	19,344	19,500
Leather, rubber, textiles, etc.							
Leather (mixed)	10	68.5	12.5	9.0	18,515	20,572	22,858
Rubber (mixed)	1.2	83.9	4.9	9.9	25,330	25,638	28,493
Textiles (mixed)	10	66.0	17.5	6.5	17,445	19,383	20,892
Glass, metals, etc.							
Glass and mineral	2	—	—	96–99+	196†	200	200
Metal, tin cans	5	—	—	94–99+	1,425†	1,500	1,500
Metals, ferrous	2	—	—	96–99+	—	—	—
Metals, nonferrous	2	—	—	94–99+	—	—	—
Miscellaneous							
Office sweepings	3.2	20.5	6.3	70	8,535	8,817	31,847
Multiple wastes	20 (15–40)	53 (30–60)	7 (5–15)	20 (9–30)	10,470	13,090	17,450
Industrial wastes	15 (10–30)	58 (30–60)	7 (5–15)	20 (10–30)	11,630	13,682	17,892

*G. Tchobanoglous, H. Theisen, and S. Vigil, *Integrated Solid Waste Management: Engineering Principles and Management Issues*, McGraw-Hill, New York, 1993.
†Energy content is from coatings, labels, and attached materials.

Chemical Composition Information on the chemical composition of solid wastes is important in determining which options to evaluate for energy and resource recovery since it will vary based on how a material may be recovered for reuse or recycling, composted, combusted (with energy recovery), or landfilled (with or without energy recovery). The chemical composition of buried waste changes over time and affects both leachate generation (and toxicity) and landfill gas emissions. In the United States, less biodegradable waste is being landfilled in response to increasing rates of recycling and composting. In contrast, in Europe, regulations prevent organics from being landfilled due to concerns about methane emissions from landfills (even the best-controlled sites still emit fugitive methane emissions). As a result, in Europe, there is more widespread use of anaerobic digestion and MSW combustion. If solid wastes are to be used as fuel (the use of hazardous waste as a fuel must comply with applicable regulations), the six most important properties to be known are

1. Proximate analysis [AL]
   a. Moisture (loss at 105°C for 1 h)
   b. Volatile matter (additional loss on heating to 950°C)
   c. Ash (residue after burning)
   d. Fixed carbon (remainder)
2. Fusion point of ash
3. Ultimate analysis, percent of C (carbon), H (hydrogen), O (oxygen), N (nitrogen), S (sulfur), and ash
4. Heating value
5. Organic chlorine
6. Organic sulfur

Typical proximate-analysis data for the combustible components of industrial wastes and MSWs are presented in Table 22-46.

Typical data on the inert residue and energy values for solid wastes may be converted to a dry basis by using Eq. (22-27):

$$\frac{\text{kJ}}{\text{kg (dry basis)}} = \frac{\text{kJ}}{\text{kg (as discarded)}}\left(\frac{100}{100-\%\text{ moisture}}\right) \tag{22-27}$$

The corresponding equation on an ash-free basis is

$$\frac{\text{kJ}}{\text{kg (ash-free dry basis)}} = \frac{\text{kJ}}{\text{kg (as discarded)}} \times\left(\frac{100}{100-\%\text{ ash}-\%\text{ moisture}}\right) \tag{22-28}$$

Representative data on the ultimate analysis of typical industrial and municipal-waste components are presented in Table 22-47. If energy values are not available, approximate values can be determined by using the Boie formula or the modified Dulong formula (Eq. 22-29) and the data in Table 22-47 [Nzihou, J. F., et al., "Using Dulong and Vandralek Formulas to Estimate the Calorific Heating Value of a Household Waste Model," *Int. J. Sci. & Engineering Res.* **5**(1): 1878–1883 (January 2014), ISSN 2229-5518].

TABLE 22-47 Typical Ultimate-Analysis Data for Components in Domestic, Commercial, and Industrial Solid Waste*

Components	Percent by mass (dry basis)					
	Carbon	Hydrogen	Oxygen	Nitrogen	Sulfur	Ash
Foods and food products						
Fats	73.0	11.5	14.8	0.4	0.1	0.2
Food wastes (mixed)	48.0	6.4	37.6	2.6	0.4	5.0
Fruit wastes	48.5	6.2	39.5	1.4	0.2	4.2
Meat wastes	59.6	9.4	24.7	1.2	0.2	4.9
Paper products	45.4	6.1	42.1	0.3	0.1	6.0
Cardboard	43.0	5.9	44.8	0.3	0.2	5.0
Magazines	32.9	5.0	38.6	0.1	0.1	23.3
Newsprint	49.1	6.1	43.0	< 0.1	0.2	23.3
Paper (mixed)	43.4	5.8	44.3	0.3	0.2	6.0
Waxed cartons	59.2	9.3	30.1	0.1	.1	1.2
Plastics						
Plastics (mixed)	60.0	7.2	22.8	—	—	10.0
Polyethylene	85.2	14.2	—	< 0.1	< 0.1	0.4
Polystyrene	87.1	8.4	4.0	0.2	—	0.3
Polyurethane†	63.3	6.3	17.6	6.0	< 0.1	4.3
Polyvinyl chloride‡	45.2	5.6	1.6	0.1	0.1	2.0
Wood, trees, etc.						
Garden trimmings	46.0	6.0	38.0	3.4	0.3	6.3
Green timber	50.1	6.4	42.3	0.1	0.1	1.0
Hardwood	49.6	6.1	43.2	0.1	< 0.1	0.9
Wood (mixed)	49.5	6.0	42.7	0.2	< 0.1	1.5
Wood chips (mixed)	48.1	5.8	45.5	0.1	< 0.1	0.4
Glass, metals, etc.						
Glass and mineral	0.5	0.1	0.4	< 0.1	—	98.9
Metals (mixed)	4.5	0.6	4.3	< 0.1	—	90.5
Leather, rubber, textiles						
Leather (mixed)	60.0	8.0	11.6	10.0	0.4	10.0
Rubber (mixed)	69.7	8.7	—	—	1.6	20.0
Textiles (mixed)	48.0	6.4	40.0	2.2	0.2	3.2
Miscellaneous						
Office sweepings	24.3	3.0	4.0	0.5	0.2	68.0
Oils, paints	66.9	9.6	5.2	2.0	—	16.3
Refuse-derived fuel (RAF)	44.7	6.2	38.4	0.7	< 0.1	9.9

*G. Tchobanoglous, H. Theisen, and S. Vigil, *Integrated Solid Waste Management: Engineering Principles and Management Issues*, McGraw-Hill, New York, 1993.
†Organic content is from coatings, labels, and other attached materials.
‡Remainder is chlorine.

$$\text{kJ/kg} = 337C + 1428\left(H - \frac{1}{8}O\right) + 95S \tag{22-29}$$

where C = carbon, percent
H = hydrogen, percent
O = oxygen, percent
S = sulfur, percent

Quantities of Solid Wastes Data and factors to consider in estimating quantities of solid wastes and generation rates are considered briefly in the following paragraphs.

Typical Generation Rates Typical unit waste generation rates from selected commercial and industrial sectors are reported in Table 22-48. Because waste generation practices change rapidly, the presentation of "typical" waste generation data may not be reliable over time or for specific local conditions.

Factors That Affect Generation Rates Factors that influence the quantity of industrial waste and MSW generated include (1) the extent of salvage and recycle operations, (2) community, company, and consumer attitudes, (3) legislation and regulations, and (4) economics. The existence of salvage and recycling operations definitely affects the quantities of wastes collected. Robust domestic and international salvage and recovered materials markets increased the quantity of materials collected. Perhaps the most important factor affecting the generation of certain types of waste is the existence of local, state, and federal regulations for the use and disposal of specific material. In general, the more regulated the waste, the higher the cost for treatment and disposal and therefore the greater the incentive to reduce generation of the waste and to minimize discards.

TABLE 22-48 Solid Waste Generation Rates for Selected Commercial and Industrial Sectors*

Sector	Rate†
Commercial Sector	
Arts, Entertainment, & Recreation	3.08
Durable Wholesale & Trucking	2.99
Education	0.50
Hotels & Lodging	2.14
Medical & Health	0.74
Public Administration	0.39
Restaurants	2.92
Retail Trade - Food & Beverage Stores	6.64
Retail Trade - All Other	2.41
Services - Management, Administrative, Support, & Social	1.44
Services - Professional, Technical, & Financial	2.31
Services - Repair & Personal	1.50
Not Elsewhere Classified	1.20
Industrial Sector	
Manufacturing - Electronic Equipment	0.75
Manufacturing - Food & Nondurable Wholesale	1.85
Manufacturing - All Other	1.50

*Adapted from CalRecycle, 2014 Generator-Based Characterization of Commercial Sector Disposal and Diversion in California. 2015. Available at http://www.calrecycle.ca.gov/Publications/Documents/1543/20151543.pdf.
†All data are reported in tons per employee per year.

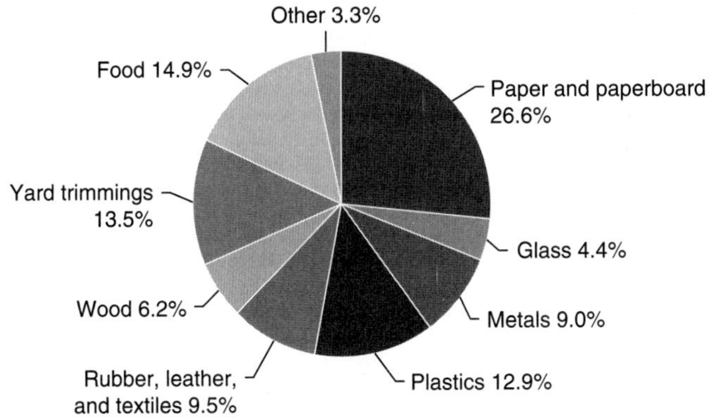

FIG. 22-48 2014 percent of total MSW generation—258 million tons. (*Advancing Sustainable Materials Management: Facts and Figures 2014. U.S. Environmental Protection Agency, Office of Resource Conservation and Recovery, November 2016, EPA 530-R-17-01. https://www.epa.gov/sites/production/files/2016-11/documents/201_smmfactsheet_508.pdf, accessed February 23, 2017.*)

In 2014, U.S. residents, commercial establishments, and institutions produced more than 258 million tons of MSW, which is approximately 4.4 lb of waste per person per day, up from 2.7 lb per person per day in 1960 and 3.7 lb per person per day in 1980. As a percent of total MSW generation, paper and paperboard products account for the largest portion (26 percent), followed by food at 15 percent, yard trimmings at 13 percent, and plastic products at 13 percent of total generation. See Figs. 22-48 and 22-49.

Several waste management practices, such as source reduction, recycling, and composting, divert materials from the waste stream. Source reduction involves altering the design, manufacture, or use and reuse of products and materials to reduce the amount and toxicity of what gets thrown away. A portion of each material category was recycled or composted in 2014. The highest rates of recycling and composting were achieved with paper products, yard trimmings, and metal products. Over 64 percent (44 million tons) of paper and paperboard was recycled in 2014. About 61 percent (21 million tons) of yard trimmings were recovered for composting in 2014. This represents almost a fivefold increase from 1990. Recycling these organic materials alone diverted over 28 percent of MSW from combustion with energy recovery and landfills. In addition, about 7.9 million tons, or about 34 percent, of metals were recovered for recycling. Additional information is available in EPA's report *Advancing Sustainable Materials Management: Facts and Figures 2014*. See Figs. 22-47 and 22-48.

As stated earlier, a hierarchy is used to rank strategies that are considered most environmentally preferable. Source reduction (including reuse) is the most preferred, followed by recycling and composting. About 13 percent of MSW is combusted with energy recovery, and the remaining 53 percent is landfilled. See Fig. 22-50. As discussed earlier, EREF data suggest that MSW tonnage may be higher and recycling rates lower than what is reported by the U.S. EPA.

Source reduction is defined by RCRA as any change in the design, manufacturing, purchase, or use of materials or products (including packaging) to reduce their amount or toxicity before they become MSW. Source reduction helps to reduce waste generation at all points in the supply chain: manufacturing, transportation, distribution, and consumption. Prevention also refers to the reuse of products or materials. Source reduction includes a broad range of activities. Example source reduction activities are included in Table 22-49. Additional information is available in EPA's report *Advancing Sustainable Materials Management: Facts and Figures 2013*, Chapter 3 [U.S. EPA, *Advancing Sustainable Materials Management: Facts and Figures, 2013*, chap. 3, June 2015. https://www.epa.gov/smm/advancing-sustainable-materials-management-facts-and-figures-report].

Recovery through recycling and composting diverted 89 million tons of MSW away from combustion with energy recovery and landfilling in 2014, up from 33 million tons in 1990 and 70 million tons in 2000. See Fig. 22-51.

ON-SITE HANDLING, STORAGE, AND PROCESSING

The handling, storage, and processing of solid wastes are key steps in the solid waste management life cycle.

On-Site Handling On-site handling refers to the activities associated with the handling of solid wastes until they are placed in the containers used for storage before collection. Depending on the type of collection service, handling may also be required to move loaded containers to the collection point and to return the empty containers to the point where they are stored between collections.

Conventional Solid Wastes In most office, commercial, and industrial buildings, solid wastes that accumulate in individual offices or work locations usually are collected in relatively large containers mounted on casters. Once filled, these containers are removed by means of the service elevator, if there is one, and emptied into (1) large storage containers, (2) compactors used in conjunction with the storage containers, (3) stationary compactors

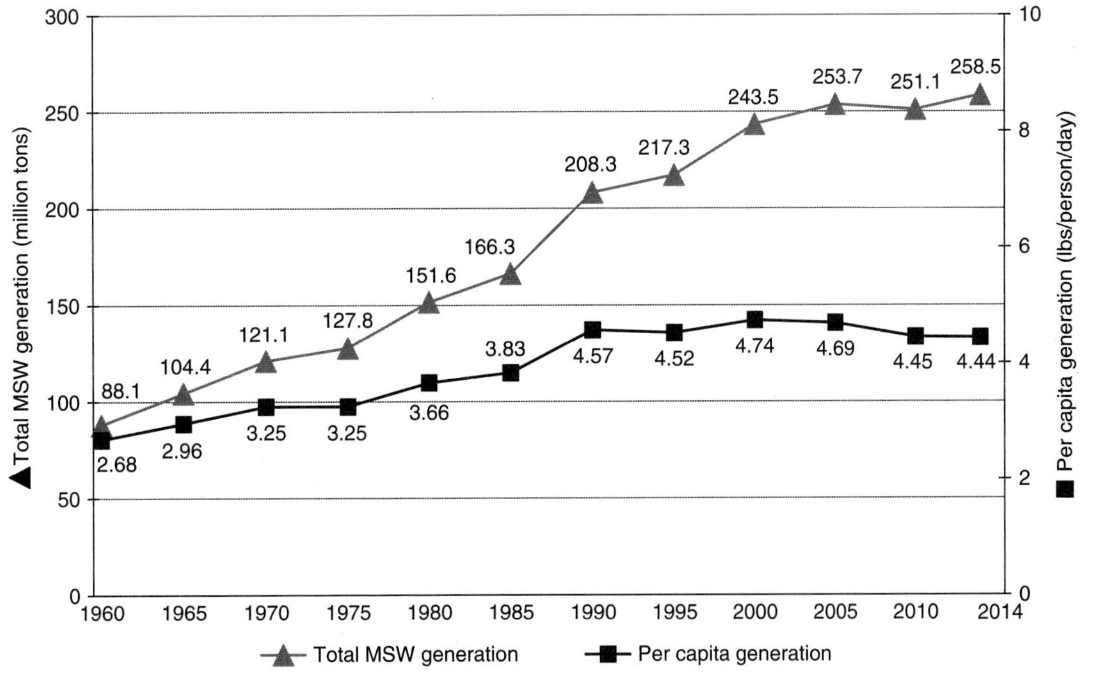

FIG. 22-49 Trends in MSW generation, 1960 to 2013. (*Advancing Sustainable Materials Management: Facts and Figures 2013, U.S. Environmental Protection Agency, Office of Resource Conservation and Recovery, November 2015, EPA 530-R-15-002. https://www.epa.gov/sites/production/files/2015-09/documents/2013_advncng_smm_rpt.pdf, accessed August 15, 2016.*)

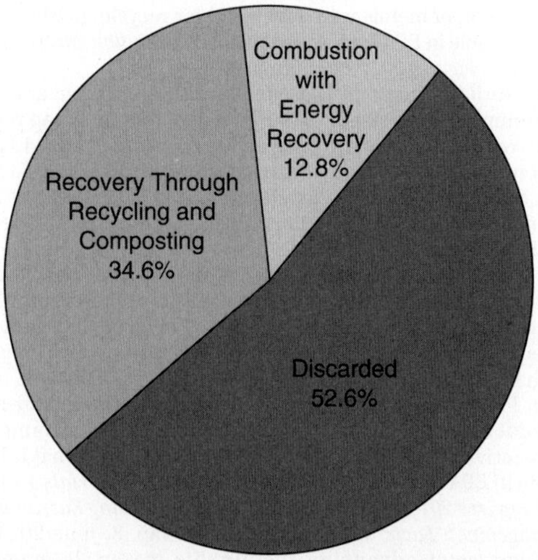

FIG. 22-50 Management of MSW in the United States, 2014. (*Advancing Sustainable Materials Management: Facts and Figures 2014, U.S. Environmental Protection Agency, Office of Resource Conservation and Recovery, June 2014, EPA 530-R-17-01. https://www.epa.gov/sites/production/files/2016-11/documents/201_smmfactsheet_508.pdf, accessed February 23, 2017.*)

that can compress the material into bales or into specially designed containers, or (4) other processing equipment.

Hazardous Waste When hazardous waste is generated, the waste generator is responsible for proper handling. Special containers are usually provided, and trained personnel [OSHA 1910.120 required such workers to have hazardous waste operations (HAZWOPER) training] are responsible for the handling of these wastes. Hazardous wastes include solids, sludges, and liquids; hence, container requirements vary with the form of waste. In addition, household hazardous waste requires appropriate management that does not include pouring them down the drain, on the ground, into storm sewers, or included with household waste.

On-Site Storage Factors that must be considered in the on-site storage of solid wastes include (1) the type of container to be used, (2) the container location, (3) restrictions on how long waste can be temporarily stored, (4) public health and aesthetics, (5) the collection methods to be used, and (conducting) future transport method.

Containers To a large extent, the types and capacities of the containers used depend on the characteristics of the solid wastes to be collected, the collection frequency, and the space available for the placement of containers.

1. *Containers for conventional wastes.* The types and capacities of containers now commonly used for on-site storage of solid wastes are summarized in Table 22-50. The small containers are used in individual offices and workstations. The medium- and large-size containers are used at locations where large volumes are generated.

2. *Containers for hazardous wastes.* On-site storage practices are a function of the types and amounts of hazardous wastes generated and the time period over which waste generation occurs. Usually, when large quantities are generated, special facilities are used that have sufficient capacity to hold wastes accumulated over a period of several days. When only small amounts of hazardous wastes are generated on an intermittent basis, they may be containerized, and limited quantities may be stored for periods covering months. Storage methods, conditions, and timeframes are also regulated under RCRA. General information on the storage containers used for hazardous wastes and the conditions of their use is presented in Table 22-51.

Container Location The location of containers at existing commercial and industrial facilities depends on both the location of available space and service-access conditions. In newer facilities, specific service areas have been included for this purpose. Often, because the containers are not owned by the commercial or industrial activity, the locations and types of containers to be used for on-site storage must be worked out jointly between the industry and the public or private collection agency.

On-Site Processing of Solid Wastes On-site processing methods are used to (1) recover usable materials from solid wastes, (2) reduce the volume, or (3) alter the physical form. The most common on-site processing operations as applied to large commercial and industrial sources include manual sorting, compaction, and incineration. These and other processing operations are considered in the portion of this subsection dealing with processing and resource recovery. Factors that should be considered in the selection of on-site processing equipment are summarized in Table 22-52. Refer to RCRA for requirements for processing hazardous waste on-site.

Materials-Recovery Systems Cardboard, paper, plastics, glass, aluminum, tin, and some ferrous metals are the principal recoverable materials contained in MSW. Most commonly, waste generators segregate these materials from MSW to some degree. After collection, downstream processing and further separation of these materials usually occurs at materials recovery facilities (MRFs). Key activities at MRFs may include opening of bags (bag breaking), the manual separation of bulky items, magnetic field separation, eddy-current separation of nonferrous metals, optical sorting (particularly for plastic streams), screening, densification (compaction), the storage of recovered materials, and the baling of separated materials for transportation. Advanced MRF designs might also incorporate ballistics and gravity separation. A material flow diagram of an example MRF for source-separated materials is presented in Fig. 22-52. [Adapted from Tchobanoglous, G., and F. Keith, *Handbook of Solid Waste Management*, 2d ed., McGraw-Hill, New York, 2002.]

The design and layout of the physical facilities that make up an MRF flow are an important part of the implementation and successful operation of such systems. Important factors that must be considered in the design and layout of such systems include (1) whether recyclables are source-separated, dual stream, or single stream, (2) the types of material to be recovered, (3) the form in which the materials to be recovered will be delivered to the MRF, (4) end-market specifications, (5) the containerization and storage of processed materials for the buyer, (6) ease and economy of operation, and (7) environmental controls. Some companies collect cardboard on site, bale it, and send it directly to a mill or intermediary processing facility. There are also deposit legislation states where designated containers and packaging are redeemed by the public at special collection facilities, depots, or vending machines.

PROCESS AND RESOURCE RECOVERY

This subsection identifies the techniques and methods used to recover materials, conversion of products, and energy production from solid wastes. Generators may not use thermal processes for treating hazardous waste on-site. Topics include (1) processing techniques for solid waste, (2) processing techniques for hazardous wastes, (3) materials-recovery systems, (4) recovery of biological conversion products, (5) thermal processes, and (6) waste-to-energy systems.

Because many of the techniques, especially those associated with the recovery of materials and energy and the processing of solid hazardous wastes, are in a state of flux with respect to application and design criteria, the objective here is only to introduce them. If these techniques are to be considered in the development of waste management systems, current engineering design and performance data must be obtained from consultants, operating records, field tests, equipment manufacturers, and available literature.

TABLE 22-49 Example Source Reduction Practices

Source reduction practice	Activity	Result
Reuse	Reusing products or packaging	Reduce manufacturing and consumer waste
Redesigning products or packaging	Substituting lighter materials	Reduce quantity of materials or toxicity of materials used
	Selecting packaging for protection and longer product life	Reduce loss of perishable products
	Redesign products for longer life	Reduce manufacturing and consumer waste
Right-sizing packaging	Removal of unnecessary layers	Reduced packaging requiring waste management
	Reducing excess packaging	Reduced packaging requiring waste management

EPA. 2015. Advancing Sustainable Materials Management: Facts and figures 2013. US Environmental Protection Agency, Office of Resource Conservation and Recovery. June. EPA530-R-15-002. https://www.epa.gov/sites/production/files/2015-09/documents/2013_advncng_smm_rpt.pdf, accessed 15 August 2016.
Additional information available in EPA's report *Advancing Sustainable Materials Management: Facts and Figures 2013*, Chapter 3.

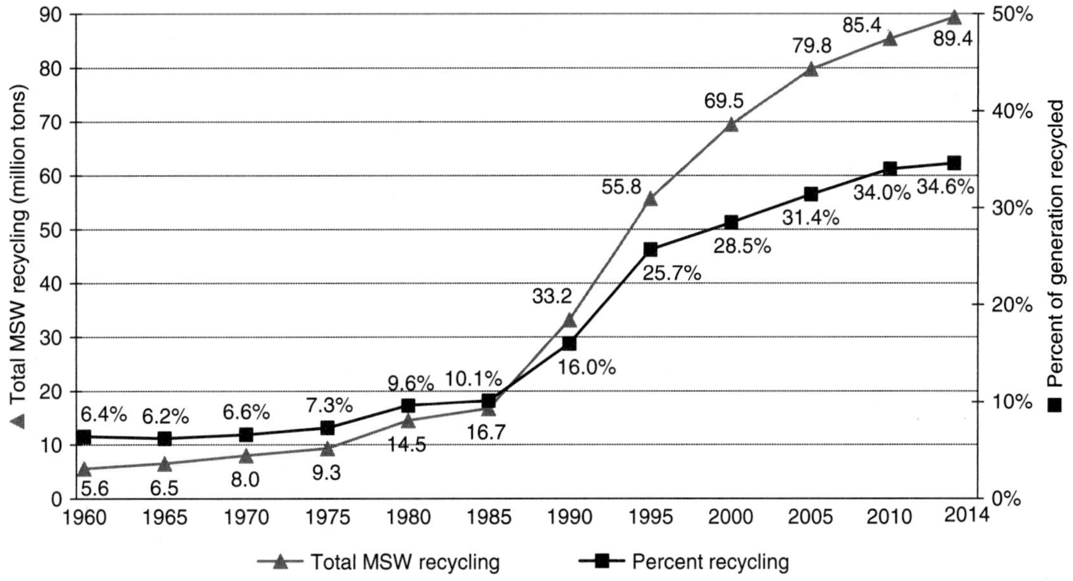

FIG. 22-51 Trends in MSW recovery through recycling and composting, 1960 to 2013. (*Advancing Sustainable Materials Management: Facts and Figures 2013, U.S. Environmental Protection Agency, Office of Resource Conservation and Recovery, June 2015, EPA 530-R-15-002. https://www.epa.gov/sites/production/files/2015-09/documents/2013_advncng_smm_rpt.pdf, accessed August 15, 2016.*)

In any processing (and disposal) scheme, the key item is knowledge of the characteristics of the waste being handled. Without this information, effective processing or treatment is impossible. For this reason, the characteristics of the wastes must be known before they are accepted and hauled to a treatment or disposal site. Proper identification of the waste constituents is the responsibility of the waste generator.

Processing Techniques for Solid Waste Processing techniques used in solid waste management systems (1) improve the efficiency of the systems, (2) recover resources (usable materials), and (3) prepare materials for recovery of conversion products and energy. The more important techniques used for processing solid wastes are summarized in the following list:

- *Component separation.* Component separation is the heart of recycling, and it is required to raise the concentration of the material to the point that it is worth reclaiming and to reduce the amount of contamination to meet specifications. Recycling occurs on a major scale, and for MSW, it accounted for 34.6 percent of the waste stream in 2014.
- *Manual component separation.* The manual separation of solid waste components can be accomplished at the source where solid wastes are generated, at a transfer station, at a centralized processing station, or at the disposal site. Manual sorting at the source of generation is one way to achieve the recovery and reuse of materials. The number and types of components salvaged or sorted (e.g., cardboard and high-quality paper, metals, and wood) depend on the location, the opportunities for recycling, and the end-use market.
- *Automated component separation.* Combined with manual separation, automatic separation techniques reduce overall costs and produce a cleaner recovered product. Various types of equipment are used for automated component separation, including screens for separating materials based on their size, electromagnets and permanent magnets for recovering ferrous metals, air classifiers and flotation systems for separating materials by density, and eddy current systems for separating conducting from nonconducting materials. Advanced systems use sensors and microprocessors for identifying and classifying materials with certain properties and compressed air jets for removing the selected material from the waste stream. Advanced systems can separate material and products by color, size, shape, structural properties, and chemical composition. [Tchobanoglous, G., and F. Keith, *Handbook of Solid Waste Management*, 2d ed., McGraw-Hill, New York, 2002.]

TABLE 22-50 Data on the Types and Sizes of Containers Used for On-Site Storage of Solid Wastes

Container type	Capacity			Dimensions*	
	Unit	Range	Typical	Unit	Typical
Small-capacity					
Plastic or metal (office type)	L	16–40	28	mm	$(180 \times 300)B \times (260 \times 380)T \times 380H$
Plastic or galvanized metal	L	75–150	120	mm	$510D \times 660H$
Barrel, plastic, aluminum, or fiber barrel	L	20–250	120	mm	$510D \times 660H$
Disposable paper bags (standard, leak-resistant, and leakproof)	L	75–210	120	mm	$380W \times 380d \times 1100H$
Disposal plastic bag	L	20–200	170	mm	$460W \times 380d \times 1000H$
Medium-capacity					
Side or top loading	m^3	0.75–9	3	mm	$1830W \times 1070d \times 1650H$
Bulk bags	m^3	0.3–2	1	mm	$1000W \times 1000d \times 1000H$
Large-capacity					
Open top, rolloff (also called debris bags)	m^3	9–38	27	mm	$2440W \times 1830H \times 6100L$
Used with stationary compactor	m^3	15–30	23	mm	$2440W \times 1830H \times 5490L$
Equipped with self-contained compaction mechanism	m^3	15–30	23	mm	$2440W \times 2440H \times 6710L$
Trailer-mounted					
Open top	m^3	15–38	27	mm	$2440W \times 3660H \times 6100L$
Enclosed, equipped with self-contained compaction mechanism	m^3	15–30	27	mm	$2440W \times 3660H \times 7320L$

*B = bottom, T = top, D = diameter, H = height, W = width, L = length, and d = depth.

TABLE 22-51 Typical Data on Containers Used for Storage and Transport of Hazardous Waste

Waste category	Container Type	Container Capacity	Auxiliary equipment and conditions of use
Radioactive substances	Lead encased in concrete	Varies with waste	Isolated storage buildings; high-capacity hoists and lighting equipment; special container markings
	Lined metal drums	210 L	
Corrosive, reactive, and toxic chemicals	Metal drums	210 L	Washing facilities for empty containers, special blending precautions to prevent hazardous reactions; incompatible wastes stored separately
	Plastic drums	210 L, up to 500 L	
	Lined metal drums	210 L	
	Lined and unlined storage tanks	Up to 20 m^3	
Liquids and sludges	Drums	142–3400 L	Drums, hand-trucks, pallets, forklifts
	Trucks	11–30 m^3	Transfer piping, hoses, pumps
	Vacuum tankers	11–15 m^3	Transfer piping, hoses, pumps
Biological wastes	Sealed plastic bags	120 L	Heat sterilization prior to bagging; special heavy-duty bags with hazard warning printed on sides
	Lined boxes, lined metal drums	57 L	
Flammable wastes	Metal drums	210 L	Fume ventilation, temperature control
	Storage tanks	Up to 20 m^3	
Explosives	Shock-absorbing containers	Varies	Temperature control; special container markings

- *Storage and transfer.* When solid wastes are to be processed for material recovery, storage and transfer facilities should be considered an essential part of the processing operation. Important factors in the design of such facilities include (1) the size of the material before and after processing, (2) the density of the material, (3) the angle of repose before and after processing, (4) the abrasive characteristics of the materials, and (5) the moisture content.
- *Mechanical volume reduction.* Mechanical volume reduction, also known as densification or compaction, is perhaps the most important factor in the development and operation of solid waste management systems.

TABLE 22-52 Factors That Should Be Considered in Evaluating On-Site Processing Equipment

Factor	Evaluation
Capabilities	What will the device or mechanism do? Will its use be an improvement over conventional practices?
Reliability	Will the equipment perform its designated functions with little attention beyond preventive maintenance? Has the effectiveness of the equipment been demonstrated in use over a reasonable period of time or merely predicted?
Service	Will servicing capabilities beyond those of the local building maintenance staff be required occasionally? Are properly trained service personnel available through the equipment manufacturer or the local distributor?
Safety of operation	Is the proposed equipment reasonably foolproof so that it may be operated by tenants or building personnel with limited mechanical knowledge or abilities? Does it have adequate safeguards to discourage careless use?
Ease of operation	Is the equipment easy to operate by a tenant or building personnel? Unless functions and actual operations of equipment can be carried out easily, they may be ignored or bypassed by paid personnel and most often by "paying" tenants.
Efficiency	Does the equipment perform efficiently and with a minimum of attention? Under most conditions, equipment that completes an operational cycle each time that it is used should be selected.
Environmental effects	Does the equipment pollute or contaminate the environment? When possible, equipment should reduce environmental pollution presently associated with conventional functions.
Health hazards	Does the device, mechanism, or equipment create or amplify health hazards?
Aesthetics	Do the equipment and its arrangement offend the senses? Every effort should be made to reduce or eliminate offending sights, odors, and noises.
Economics	What are the economics involved? Both first and annual costs must be considered. Future operation and maintenance costs must be assessed carefully. All factors being equal, equipment produced by well-established companies, having a proven history of satisfactory operations should be given appropriate consideration.
Flexibility	Will the equipment or its placement allow future changes to the process to handle wastes with differing characteristics?

*From G. Tchobanouglous, H. Theissen, and R. Eliassen, *Solid Wastes: Engineering Principles and Management Issues*, McGraw-Hill, New York, 1977.

Volume reduction can occur at various stages during the waste management process, including the points of generation, collection, processing, and disposal. At materials recovery facilities (MRFs), the most common forms of compaction are can flatteners, can densifiers (which create aluminum bales/biscuits for easier transport), pelletizers (used for compressing plastics and light MSW), and balers (used for paper and cardboard and, more recently, to densify some metals and plastics). [Tchobanoglous, G., and F. Keith, *Handbook of Solid Waste Management*, 2d ed., McGraw-Hill, New York, 2002.]

- *Chemical volume reduction.* Incineration has been the method commonly used to reduce the volume of wastes chemically. One of the most attractive features of the incineration process is that it can be used to reduce the original volume of combustible solid wastes by 80 to 90 percent. The technology of incineration has advanced, with many mass burn facilities now having two or more combustors with capacities of 1000 tons/day of refuse per unit. However, regulations limiting low volatile and semivolatile metals, acid gases, free chlorine and hydrogen chloride, unburned hydrocarbons, and dioxin and furan emissions have resulted in higher costs and operating complexity.
- *Mechanical size alteration.* The objective of size reduction is to obtain a final product that is reasonably uniform and considerably reduced in size in comparison to its original form. Size reduction does not necessarily imply volume reduction. In some situations, the total volume of the material after size reduction may be greater than the original volume. The two classes of size reduction equipment are distinguished as high-speed impact (hammermills, flail mills, and rotary grinders) and high-torque shear action (counterrotating blades that shred material). [Tchobanoglous, G., and F. Keith, *Handbook of Solid Waste Management*, 2d ed., McGraw-Hill, New York, 2002.] The gain in ease of materials handling must be weighed against the substantial operating costs for size reduction equipment. A substantial market for shearing technology exists in shredding tires to produce fuel for cement kilns and utility boilers. With finer shredding, the size-reduced tire shreds can be used in asphalt pavement or used to make thick, resilient mats for playing fields. However, be aware that concerns have been raised about the safety of recycled tire crumb used in athletic playing fields and playgrounds in the United States. Limited studies have not shown an elevated health risk from playing on fields with tire crumb, but the existing studies do not comprehensively evaluate the concerns about health risks from exposure to tire crumb. Research is ongoing to evaluate the potential pathways in which people may be exposed to tire crumb based on their activities on the fields.
- *Drying and dewatering.* In many solid waste energy-recovery and incineration systems, the shredded light fraction is predried by thermal and mechanical processes to decrease weight. Although energy requirements for drying wastes vary with the application, the required energy input can be estimated by using a value of about 4300 kJ/kg of water evaporated. Drying can often permit increased waste throughput in many treatment systems. For incinerators, it produces more stable combustion and better ash quality.
- *Bulking of liquid wastes.* Liquid wastes are prohibited from being disposed in landfill cells. Wastes with free liquids are mixed with a bulking agent, such as dry sawdust, cement kiln dust, or fly ash. This process is also referred to as liquid waste solidification. However, liquid wastes are being added to "wet" landfill sites to accelerate the decomposition of biodegradable wastes. This is discussed in a later subsection.

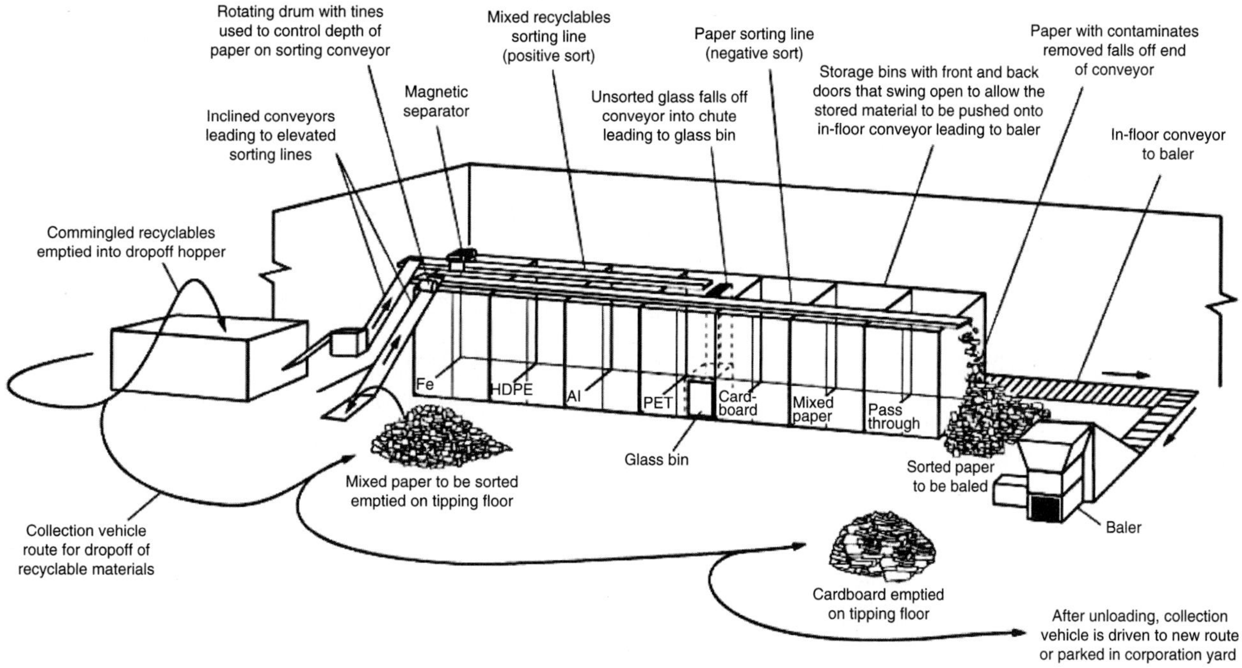

FIG. 22-52 Example: Material flow in an MRF for source-separated materials. *(From G. Tchobanoglous and F. Keith, Handbook of Solid Waste Management, 2d ed., McGraw-Hill, New York, 2002.)*

Processing of Hazardous Waste As with conventional solid wastes, the processing of hazardous wastes is undertaken for three purposes: (1) to recover useful materials, (2) to reduce the amount of wastes that must be disposed in landfills, (3) to prepare the wastes for ultimate disposal, and (4) to recover energy from fuel blending, especially nonhalogenated solvents.

Processing Techniques The processing of hazardous wastes on a batch basis can be accomplished by physical, chemical, thermal, and biological processes. The various individual processes in each category are reported in Table 22-53.

Biological Conversion Processes Biological conversion products that can be derived from wastes include alcohols and a variety of other intermediate organic compounds. The principal biological and thermal processes are reported in Table 22-53. Aerobic composting and anaerobic digestion, the two most highly developed biological processes, are described next.

TABLE 22-53 Biological and Thermal Processes Used for Recovery of Conversion Products from Solid Waste*

Process	Conversion product	Preprocessing required	Comments
Biological			
Aerobic composting	Humus-like material (i.e., compost)	Separation of organic fraction, particle size reduction	Nearly 5000 composting sites in the United States[a]
Anaerobic digestion (low- and high-solids)	Methane gas	Separation of organic fraction, particle size reduction	Approximately 200 AD facilities in the United States[b]
Enzymatic hydrolysis	Glucose from cellulose	Separation of cellulose-containing materials	Pretreatment advancements and emerging biotechnology tools might reduce high costs associated with current techniques[c]
Fermentation	Ethanol, single-cell protein	Separation of organic fraction, particle size reduction	Used in conjunction with the hydrolytic process (i.e., acid or enzymatic hydrolysis to produce glucose)
Thermal			
Combustion with heat recovery	Energy in the form of steam or electricity	None	Markets for steam and electricity required; proved in numerous full-scale applications; air-quality regulations possibly prohibiting use
Thermal desorption	Clean soil used for fill or reclaimed raw materials such as de-oiled metal trimmings	Sizing, blending, dilution	In routine use for Superfund organic contaminated soil
Supplementary fuel firing in boilers	Energy in the form of steam	Shredding, air separation, magnetic separation	If at least capital investment desired, existing air-quality regulations possibly prohibiting use
Gasification	Energy in the form of low-energy gas	Shredding, air separation, magnetic separation	Gasification is relatively obscure in use in the United States for solid waste
Pyrolysis	Energy in the form of gas and oil and char	Shredding, magnetic separation	Technology proved only in pilot applications, but full-scale use has rarely succeeded due to high operating costs and lack of materials for gas, oil, and char only
Hydrolysis	Glucose, furfural	Shredding, air separation	Technology on pilot scale only
Chemical conversion	Oil, gas, cellulose acetate	Shredding, air separation	Technology on pilot scale only

*Adapted from G. Tchobanoglous, H. Theisen, and S. Vigil, *Integrated Solid Waste Management: Engineering Principles and Management Issues*, McGraw-Hill, New York, 1993, page 710.

Additional Sources:

[a]Platt, B., N. Goldstein, C. Coker, S. Brown. July 2014. Institute for Local Self-Reliance (ILSR). *State of Composting in the U.S.: Executive Summary*, page ES-5.

[b]Environmental Research & Education Foundation (EREF). August 2015. *Anaerobic Digestion of Municipal Solid Waste: Report on the State of Practice*, page 1.

[c]Yang, B., Dai, Z., Ding, S-Y., Wyman, C. 2011. *Biofuels.* Enzymatic hydrolysis of cellulosic biomass, pages 421 and 437. http://www.bioenergycenter.org/besc/publications/yang_enzymatic_hydrolysis.pdf.

[CalRecycle, *Life Cycle Assessment and Economic Analysis of Organic Waste Management and Greenhouse Gas Reduction Options* (2009). Available at http://www.calrecycle.ca.gov/climate/Events/LifeCycle/2009/default.htm. As obtained on March 8, 2011.] The recovery of gas from landfills is discussed in the portion of this subsection dealing with landfill disposal.

Aerobic Composting When biodegradable organic materials (e.g., paper, food waste, cotton) are separated from solid wastes and subjected to bacterial decomposition, the end product is called *compost* or *humus*. The entire process, involving both separation and bacterial conversion of the organic solid wastes, is known as *composting*. Decomposition of the organic solid wastes is best accomplished aerobically (i.e., presence of oxygen) [Tchobanoglous, G., and F. Keith, *Handbook of Solid Waste Management*, 2d ed., McGraw-Hill, New York, 2002, pp. 12.4 and 12.26].

Most composting operations involve three basic steps: (1) preprocessing of solid wastes, (2) aerobic decomposition of the organic fraction of solid wastes, and (3) product preparation and marketing [Tchobanoglous, G., H. Theisen, and S. Vigil, *Integrated Solid Waste Management: Engineering Principles and Management Issues*, McGraw-Hill, New York, 1993, p. 684]. Several techniques—such as static piles, turned windrows, passively/actively aerated systems, bioreactors, or in-vessel systems—have been developed to accomplish the decomposition step [Platt, B., N. Goldstein, C. Coker, and S. Brown, Institute for Local Self-Reliance (ILSR). "Executive Summary: State of Composting in the U.S." (July 2014), p. ES-2]. Once the solid wastes have been converted to a humus, they are ready for the third step of product preparation and marketing. This step may include fine grinding, blending with various additives, granulation, bagging, storage, shipping, and, in some cases, direct marketing. The principal design considerations associated with the biological decomposition of prepared solid wastes are presented in Table 22-54.

Anaerobic Digestion Anaerobic digestion (AD) is a series of biological processes in which microorganisms break down biodegradable material in the absence of oxygen, resulting in biogas (i.e., methane) and digestate. AD technologies are generally characterized as "wet" (low-solids, with a moisture content greater than 85 percent) or "dry" (high-solids, with a moisture content of less than 80 percent) [Environmental Research & Education Foundation (EREF), *Anaerobic Digestion of Municipal Solid Waste: Report on the State of Practice*, August 2015, p. 1]. In either technology, there are three basic steps in the conversion process. The first step is to prepare the organic fraction of the solid wastes for anaerobic digestion; it usually includes receiving, sorting, separation, and size reduction. The second step is the addition of moisture and nutrients, blending, pH adjustment, heating of the slurry, and anaerobic digestion in a reactor with continuous flow, in which the contents are well mixed. The third step involves capture, storage, and—if necessary—separation of the gas components. Dewatering and disposal of the digested sludge must also be accomplished; digester effluent may be land-applied and residual solids converted into soil amendment or animal bedding [U.S. Environmental Protection Agency, *Postconsumer Food Diverted Through Donation, Animal Feed, Anaerobic Digestion, and Composting for 2013*, April 2015, p. 6]. Important design considerations are reported in Table 22-55. As of 2016, only 40 to 50 AD facilities (out of more than 2000 AD sites in the United States) are in use for processing food waste (or scraps). In 2015, a collaboration named ReFED was formed by over 30 business, nonprofit, foundation, and government leaders committed to reducing food waste in the United States. This collaboration produced a road map to reduce food waste that identifies both economic and environmental drivers important to reducing the generation of food waste and maximizing the energy and resource recovery from food waste.

Thermal Processes Conversion products that can be derived from solid wastes include heat, gases, a variety of oils, and various related organic compounds. The principal thermal processes that have been used for the recovery of usable conversion products from solid wastes are reported in Table 22-53.

Incineration Solid waste can be combusted with excess oxygen (O_2) in a furnace (e.g., multiple-hearth) strictly to reduce the mass of waste requiring final disposal. Once common as a means of MSW treatment in the United States, its use for nonhazardous solid waste is limited today, with the exception of some smaller on-site installations. Volume reduction is nominally 90 percent, depending on the ash content of the solid waste material and the completeness of combustion.

Combustion with Energy Recovery Solid waste can be combusted with excess oxygen (O_2) in water-wall boilers to recover the resultant heat energy to generate steam and/or electricity in a waste-to-energy facility (WTE). Lower-grade heat energy remaining in combustion gases after primary heat recovery can be used to dry the incoming solid waste fuel or to preheat combustion air or boiler feedwater makeup.

Facilities designed to process MSW can be further divided into mass-burn and refuse-derived-fuel (RDF) units. Mass-burn units are the most prevalent design and employ direct combustion of waste without any preprocessing. In contrast, RDF facilities preprocess the waste through size reduction (shredding), removal of metals, and removal of fines. RDF boilers can be of suspension, stoker, and fluidized-bed designs, while mass-burn boilers are typically of a stoker design, although some smaller facilities use a rotary combustion design. Both types of combustors are field-erected and generally range in size from 100 to over 1000 tons/day of refuse feed per unit (Fig. 22-53). Figure 22-54 provides a generalized schematic of a typical mass-burn WTE facility.

RDF fuels, as well as source-separated waste material (e.g., whole or chipped tires, wood wastes), can also be fired directly in cement kilns to provide heat energy to the calcining process. In addition to heat content, the use of these waste-derived fuels in cement kilns requires careful attention to waste properties, including ash content, chloride content, and the presence of metals that can interfere with the final product quality if not managed appropriately. Waste-derived fuels can also be fired directly in large industrial and utility boilers that are now used for the production of power with

TABLE 22-54 Important Design Considerations for Aerobic Composting Processes*

Item	Comment
Particle size	For optimum results the size of solid wastes should be between 25 and 75 mm (1 and 3 in).
Carbon-to-nitrogen (C/N) ratio	Initial carbon-to-nitrogen ratios (by mass) between 25 and 50 are optimum for aerobic composting. At lower ratios, ammonia is given off. Biological activity is also impeded at lower ratios. At higher ratios, nitrogen may be a limiting nutrient.
Blending and seeding	Composting time can be reduced by seeding with partially decomposed solid wastes to the extent of about 1 to 5 percent by weight. Sewage sludge can also be added to prepared solid wastes. When sludge is added, the final moisture content is the controlling variable.
Moisture content	Moisture content should be in the range between 50 and 60 percent during the composting process. The optimum value appears to be about 55 percent.
Mixing/turning	To prevent drying, caking, and air channeling, material in the process of being composted should be mixed or turned on a regular schedule or as required. Frequency of mixing or turning will depend on the type of composting operation.
Temperature	For best results, temperature should be maintained between 122°F and 131°F (50°C and 55°C) for the first few days and between 131°F and 140°F (55°C and 60°C) for the remainder of the active composting period. If temperature goes beyond 151°F (66°C), biological activity is reduced significantly.
Control of pathogens	If properly conducted, it is possible to kill all the pathogens, weeds, and seeds during the composting process. To do this, the temperature must be maintained between 140°F and 158°F (60°C and 70°C) for 24 h.
Air requirements	The theoretical quantity of oxygen required can be estimated. Air with at least 50 percent of the initial oxygen concentration remaining should reach all parts of the composting material for optimum results, especially in mechanical systems.
pH control	To achieve an optimum aerobic decomposition, pH should remain at 7 to 7.5 range. To minimize the loss of nitrogen in the form of ammonia gas, pH should not rise above about 8.5.
Degree of decomposition	The degree of decomposition can be estimated by measuring the final drop in temperature, degree of self-heating capacity, amount of decomposable and resistant organic matter in the composted material, rises in the redox potential, oxygen uptake, growth of the fungus, *Chaetomium gracilis*, and the starch-iodine test.
Land requirement	The land requirements for a plant with a capacity of 50 ton/d will be 1.5 to 2.0 acres. The land area required for a larger plant will be less on a ton/d basis.

*Adapted from G. Tchobanoglous, H. Theisen, and R.Eliassen, Solid Wastes: Engineering Principles and Management Issues, McGraw-Hill, New York, 1977 and G. Tchobanoglous, H. Theisen, and S. Vigil, *Integrated Solid Waste Management: Engineering Principles and Management Issues*, McGraw-Hill, New York, 1993, page 687.

TABLE 22-55 Important Design Considerations for Anaerobic Digestion of the Organic Fraction of MSW*

Item	Comments	
	Low-solids	High-solids
Size of material	Wastes to be digested should be shredded to a size that will not interfere with the efficient functioning of pumping and mixing operations.	Wastes to be digested should be shredded to a size that will not interfere with the efficient functioning of feeding and discharging mechanisms.
Mixing equipment	To achieve optimum results and to avoid scum buildup, mechanical mixing is recommended.	The mixing equipment will depend on the type of reactor to be used.
Percentage of solid wastes mixed with sludge	Although amounts of waste varying from 50 to 90+ percent have been used, 60 percent appears to be a reasonable compromise.	Depends on the characteristics of the sludge.
Hydraulic and mean cell-residence time (low-solids)/mass retention time (high-solids)	Washout time is in the range of 3 to 4 days. Use 10 to 20 days for design, or base design on results of pilot plant studies.	Use 20 to30 days for design, or base design on results of pilot plant studies.
Loading rate	0.04 to 0.10 lb/ft^3·day (0.6 to 1.6 kg/m^3·day). Not well defined at present time. Significantly higher rates have been reported.	Based on biodegradable volatile solids (BVS): 0.375 to 0.4 lb/ft^3·day (6 to 7 kg/m^3·day). Not well defined at present time. Significantly higher rates have been reported.
Solids concentration	Equal to or less than 8 to 10 percent (4 to 8 percent typical).	Between 20 to 35 percent (22 to 28 percent typical).
Temperature	Between 85°F and 100°F (30°C to 38°C) for mesophilic and between 131°F and 140°F (55°C to 60°C) for thermophilic reactor.	Between 85°F and 100°F (30°C to 38°C) for mesophilic and between 131°F and 140°F (55°C to 60°C) for thermophilic reactor.
Destruction of volatile solid wastes (low-solids) or BVS (high-solids)	Depends on the nature of the waste characteristics. Varies from about 60 to 80 percent; 70 percent can be used for estimating purposes.	Varies from about 90 to 98 + percent depending on the mass retention time and the BVS loading rate.
Total solids destroyed	Varies from 40 to 60 percent, depending on amount of inert material present originally.	Varies depending on the lignin content of the feedstocks.
Gas production	8 to 12 ft^3/lb (0.5 to 0.75 m^3/kg) of volatile solids destroyed (CH_4 = 55 percent; CO_2 = 45 percent).	10 to 16 ft^3/lb (0.625 to 1.0 m^3/kg) of BVS destroyed (CH_4 = 50 percent; CO_2 = 50 percent).

*Adapted from G. Tchobanoglous, H. Theisen, and R. Eliassen, *Solid Wastes: Engineering Principles and Management Issues*, McGraw-Hill, New York, 1977 and G. Tchobanoglous, H. Theisen, and S. Vigil, *Integrated Solid Waste Management: Engineering Principles and Management Issues*, McGraw-Hill, New York, 1993,

pulverized or stoker coal and oil, although the practice is not as common as direct combustion in the specially designed units described previously. Suspension, spreader-stoker, and double-vortex firing systems have been used, depending on the degree of processing.

Trucks deliver MSW to the scale house of the facility, where the truck is weighed and identified. At (1) in Fig. 22-54, the trucks unload waste into a storage pit in an enclosed tipping hall. The forced draft fans (2) for the boiler pull air from the building around the tipping and pit areas, maintaining the building under negative pressure which acts to contain any odors. Overhead cranes (3) are used to mix the waste to facilitate good combustion and load the waste into a feed chute. Hydraulic rams (4) push the waste onto the inclined stoker grate (5) of the furnace and control the rate of fuel addition. Waste is fed to the stoker system that agitates the waste in a systematic fashion to provide continuous drying, volatilization, ignition, and ultimately complete combustion of the waste over a nominal 60 minutes. At the opposite end of the stoker grate from the feed chute, "bottom ash" is removed from the furnace through the ash discharger (6) for further processing.

Combustion air is delivered to the furnace both below and above the stoker system, resulting in complete combustion at temperatures approaching 2000°F. Complete combustion is achieved by having sufficient turbulence and mixing, maintaining a consistent load, and adequate residence time at high combustion temperatures. The conditions within the furnace are controlled through a sophisticated combustion control system operated by a licensed boiler operator. The combustion process is monitored by feedback into the control system, and appropriate adjustments are made by the operator. Data from the facility's continuous emissions monitoring system (CEMS), particularly carbon monoxide (CO) levels, which indicate the completeness of combustion, are closely monitored to ensure proper combustion.

FIG. 22-53 Nine hundred ton per day mass-burn WTE facility, Onondaga County, New York. Covanta.

The heat of combustion generates steam in the boiler (7). At most facilities in the United States, electricity is generated from the steam using a condensing steam turbine driving an electrical generator (8). After providing the in-plant electrical needs, excess energy is distributed to the electrical grid through a step-up transformer and switchgear (9). The turbine exhaust steam is passed through either an air- or water-cooled condenser (10) to return water to the thermal cycle. In an arrangement more commonly used in Europe, some facilities generate both electricity and steam for export to industrial use or to a municipal steam loop used for building heating and cooling. These facilities may be equipped with an extraction turbine designed to capture energy from the boiler outlet steam as it is reduced to a lower temperature and pressure suitable for the steam client's needs.

Air pollution control technology is a combination of reagent addition and mechanical systems to remove acid gases, particulate, hazardous air pollutants, and heavy metals. While specific designs vary from facility to facility, the most common air pollution control technology is described here. Ammonia is injected into the furnace to control NO_x (i.e., nitrogen oxides) through selective noncatalytic reduction (SNCR). Flue gas recirculation and different air control regimes may also be used in conjunction with SNCR to help control NO_x emissions. Newer plants may also incorporate selective catalytic reduction to achieve even lower NO_x emissions levels.

After the flue gas exits the furnace, activated carbon is injected to adsorb mercury and dioxin. The flue gas is then passed through a semidry scrubber (11), also referred to as a spray dryer absorber (SDA). In the scrubber, lime slurry is added to begin neutralizing the acid gases. Water is also injected into the scrubber to control flue gas temperature to control dioxin formation and to ensure the efficacy of the acid gas neutralization reactions. This process typically removes more than 95 percent of sulfur dioxide and hydrochloric acid.

Following the scrubber, the flue gas enters a baghouse (12), in which PM from the flue gas is collected. This not only reduces particulate emissions, but also mercury and organics adsorbed to activated carbon, and metals. The cleaned gas passes through an induced draft fan (13) on its way to the stack (14). Continuous emission monitors check the flue gas for opacity

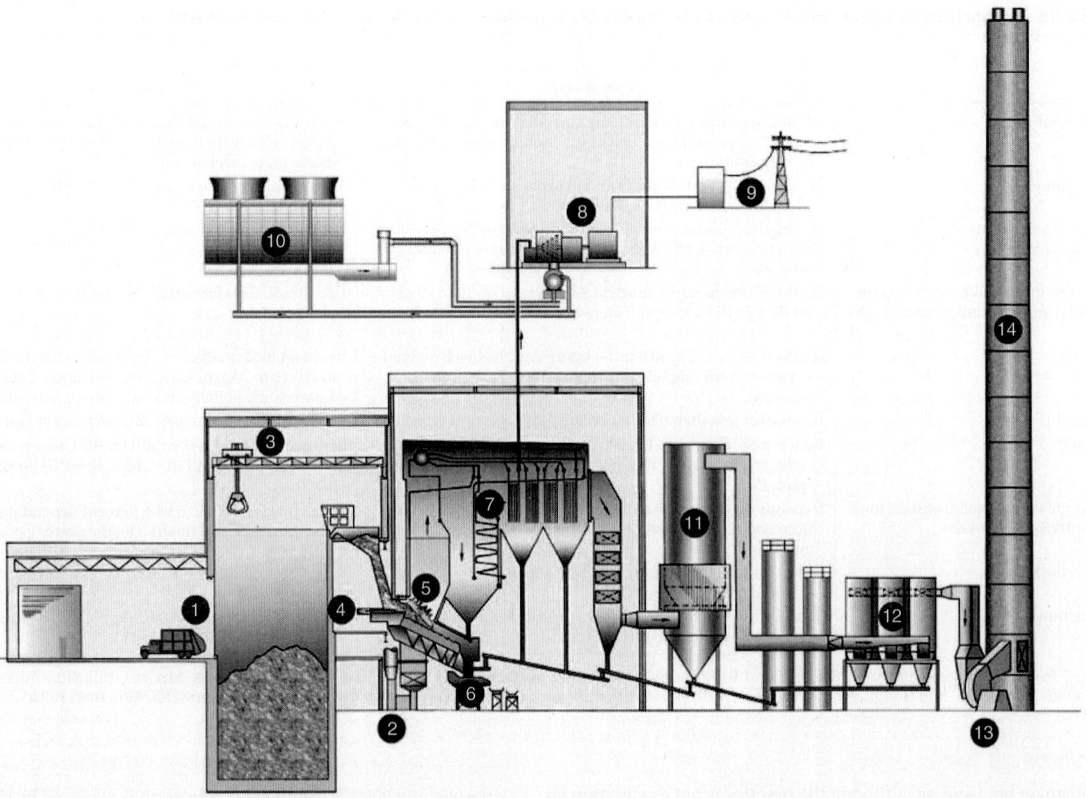

FIG. 22-54 Generalized schematic of a typical mass-burn WTE facility. Covanta.

and pollutant concentrations [typically carbon monoxide (CO), hydrogen chloride (HCl), nitrogen oxides (NO_x), and sulfur dioxide (SO_2)] to assess the performance of the combustion and APC processes and ensure compliance with permit limits.

After the combustion of the waste, ferrous metals are recovered from the ash for recycling by screening and magnetic separation. Nonferrous metals are removed using an eddy current separator, which imparts a temporary charge on the metal, allowing its separation. WTE facility operators are also beginning to use more advanced technologies, including the incorporation of rare-earth magnets into ferrous removal systems, and optical sorting technology to improve metal recoveries.

In the United States, bottom ash from the combustion process is typically combined with fly ash from the scrubber-baghouse air pollution control system. This "combined ash" is about 10 percent of the volume of the initial waste and 25 percent of the initial weight. Part of this weight is water added to control fugitive emissions of ash as it is handled. The combined ash is tested routinely in accordance with the toxicity characteristic leaching procedure (TCLP) to confirm that it is nonhazardous under RCRA. The combined ash is typically landfilled together with MSW, managed in a separate ash monofilll, or used as daily cover at an MSW landfill. Opportunities exist and are being explored for further beneficial reuse as aggregate or as a raw material for cement or asphalt manufacturing.

Gasification The gasification process involves the partial combustion in an oxygen-deficient environment of a carbonaceous or hydrocarbon fuel to generate combustible gas rich in carbon monoxide and hydrogen. A gasifier is similar to an incinerator operating under reducing conditions. Gasification is an endothermic process and requires a heat source, such as syngas combustion, char combustion, or steam. The primary product of gasification, syngas, can be converted into heat, power, fuels, fertilizers, or chemical products. Heat to sustain the process is derived from exothermic reactions, while the combustible components of the low-energy gas are primarily generated by endothermic reactions. The two types of processes are high (can reach up to 1200°C) and low temperature (600–875°C) gasification and plasma gasification. When a gasifier is at atmospheric pressure with air as the partial oxidant, the end products of the gasification process are a low-energy gas typically containing approximately (by volume) 10 percent CO_2, 20 percent CO, 15 percent H_2, and 2 percent CH_4, with the balance being N_2 and a carbon-rich ash. Because of the diluting effect of the nitrogen in the input air, the low-energy gas has an energy content in the range of the 5.2 to 6.0 MJ/m^3 (140 to 160 Btu/ft^3). When pure oxygen is used as the oxidant, a medium-energy gas with an energy content in the range of approximately 12.9 to 13.8 MJ/m^3 (345 to 370 Btu/ft^3) is produced. Gasifiers were in widespread use on coal and wood until natural gas displaced them in the 1930s through the 1950s. Some large coal gasifiers are in use today in the United States and worldwide. Although solid waste could be gasified, there are few known successful commercial operations in the United States specific to the management of solid waste. A comparison was made of the LCA environmental tradeoffs for emerging technologies for MSW (e.g., gasification, pyrolysis, anaerobic digestion) [U.S. EPA, *State of Practice for Emerging Waste Conversion Technologies*, EPA 600/R-12/684, October 2012]. A major issue with evaluating emerging technologies is the lack of comparable data on process by-product emissions and residues. [FL DEP (Florida Department of Environmental Protection, Office of Permitting and Compliance), Emission Sources: NSR/PSD Construction Permits - St. Lucie Plasma Gasification Facility, St. Lucie County, Florida, 2011. Available at: http://www.dep.state.fl.us/Air/emission/construction/geoplasma.htm; Environment and Plastics Industry Council (EPIC), Canadian Plastics Industry Association, *The Gasification of Residual Plastics Derived from Municipal Recycling Facilities*, 2004. Available at: http://www.plastics.ca/articles_merge/gasification_pf.php.]

Pyrolysis Pyrolysis is an endothermic process—also referred to as cracking—using heat to thermally decompose carbon-based material in the complete absence of oxygen. The main products are a mixture of gaseous products, liquid products (typically oils of various kinds and volatility), and solids (char and any metals or minerals that might have been components of the feedstock). In North America, pyrolysis is used on mixed plastics—liquid petroleum-type products predominate—which generally require additional refining. As with gasification, the application of pyrolysis to mixed MSW could generate a gaseous mixture of carbon monoxide (CO) and hydrogen (H_2) called syngas that can be used to generate steam and electricity. The specific products generated, particularly the liquid and solid fractions, and their proportions are likely to vary significantly, depending on reactor design, reaction conditions, and feedstock. Pyrolysis has been tested in pilot plants, and many full-scale demonstration systems have operated. However, few have attained any long-term commercial use. Major issues appear to be a lack of markets for the unstable and acidic pyrolytic oils, the char formed, and the high operating costs.

Depending on the type of reactor used, the physical form of solid wastes to be pyrolyzed can vary from unshredded raw wastes to the finely ground portion of the wastes remaining after two stages of shredding and air classification. Upon heating in an oxygen-free atmosphere, most organic substances can be split via thermal cracking and condensation reactions into gaseous, liquid, and solid fractions. There are three different gasification processes being marketed for use on solid waste. In the first, thermal pyrolysis or cracking, the feedstock is heated at temperatures ranging from 450°C to 900°C in the absence of a catalyst. Thermal cracking typically uses mixed plastics from industrial or municipal sources to yield low-octane liquid and gas products that must be refined to produce usable fuel feedstock. The second process, catalytic pyrolysis or cracking, is similar to thermal pyrolysis, except that a catalyst is used to reduce the process temperature and time. This method has been shown to process a variety of plastics, including high- and low-density polyethylene, polypropylene, and polystyrene to produce liquid and gas products. The third, hydrocracking or hydrogenation, occurs when reacting the feedstock with hydrogen and a catalyst. The process occurs under moderate temperatures and pressures (e.g., 150°C–400°C and 30–100 bar hydrogen) to produce gasoline fuels from various waste feedstocks, including plastics recovered from MSW, plastics mixed with coal, plastics mixed with refinery oils, and scrap tires.

Three major component fractions resulting from the pyrolysis are (1) a gas stream containing primarily hydrogen, methane, carbon monoxide, reduced sulfur compounds, carbon disulfide, and various other gases, depending on the organic characteristics of the material being pyrolyzed; (2) a fraction that consists of a tar or oil stream that is liquid at room temperature and has been found to contain hundreds of chemicals such as acetic acid, acetone, methanol, and phenols; and (3) a char consisting of almost pure carbon plus any inert material that may have entered the process. Distribution of the product fractions varies with the temperature at which the pyrolysis is carried out. Some of these residues may be reused (if approved by a state environmental agency), while others must be disposed of in a landfill. The amount of residual waste produced is about 15 to 20 percent of the overall [plastics] feedstock used in the process.

Determination of Energy Output and Efficiency for Waste to Energy An analysis of the net amount of electrical energy generated per ton from a representative municipal waste combustor is presented in Table 22-56. Based on a typical MSW gross calorific value of 11,600 MJ/tonne, assumed 20 percent moisture in MSW, 70 percent boiler efficiency, 31 percent turbine-generator system efficiency, and 14 percent in-house load, the net power for export is 600 kW/tonne (t). For this example, the overall efficiency is 18.7 percent. New facilities can achieve greater than 20 percent as a result of more efficient boiler and turbine design.

Use of WTE for MSW Management In 2016, there were 77 operating WTE facilities with a total waste capacity of 95,023 tons/day and a gross energy recovery capacity equivalent to 2747 MW [ERC (Energy Recovery Council), 2016 *Directory of Waste-to-Energy Facilities*, http://energyrecoverycouncil.org/wp-content/uploads/2016/06/ERC-2016-directory.pdf]. In 2014, WTE facilities generated 14.3 million MWh of electricity, not including steam exports. The MSW throughput and net electrical generation from the industry has been relatively stable since 2001. In general, WTE facilities are located in heavily populated areas of the country along the east coast, driven by the lack of low-cost land and open space for landfills compared to the midwestern and western states.

Regulations Applicable to WTE Facilities WTE facilities that combust MSW are regulated by federal, state, and local environmental laws, regulations, and permits. Under CAA section 129, many emissions limitations were developed for nine pollutants from municipal waste combustors: particulate matter, sulfur dioxide, hydrogen chloride, oxides of nitrogen, carbon monoxide, lead, cadmium, mercury, and dioxins and furans (hereafter collectively referred to as "dioxins"). The required standards are developed by the U.S. EPA through the maximum achievable control technology (MACT) process defined in section 111 of the CAA. Requirements for existing facilities are established as emissions guidelines that are implemented through state implementation plans (SIPs) in those states that have delegated authority under the CAA. In those states without delegated authority, the requirements are implemented through a federal implementation plan (FIP). The requirements for new facilities are covered under new source performance standards (NSPS).

TABLE 22-56 Representative Energy Output and Efficiency for Typical Municipal Waste Combustor

Item	Value
Gross energy available in waste, MJ/t	11,630
Heat loss from vaporization of water, MJ/t	−1.170
Other boiler heat losses (e.g., radiative, ash, flue gas temperature), 70% boiler eff., HHV basis, MJ/t	−2,320
Turbine-generator system loss, 30% overall system efficiency, MJ/t	−5.620
Gross electrical generation, kWh/t	700
In-house plant load, kWh/t	−100
Net electrical generation, kWh/t	600
Overall efficiency, HHV basis	18.7%

For both new and existing facilities, facility-specific limits and conditions are generally set within the facility's Title V air permit, a permitting mechanism named for the title establishing the program under the CAA. Current emissions guidelines and NSPSs for large (mass-burn waterwall) combustors are shown in Table 22-57. Emissions guidelines and NSPSs for small MWCs are in subparts BBBB and AAAA of 40 CFR 60, respectively. All standards are expressed in concentration units on a dry basis, corrected for 7 percent O_2 and U.S. EPA standard conditions (68°F, 1 atm). Adherence with these standards is demonstrated through either continuous emission monitors or periodic stack testing.

The U.S. EPA is required to review, and revise as appropriate, the standards set for MWCs under sections 111 and 129 of the CAA every five years.

The actual standards for a facility may be lower than the emissions guidelines or NSPS. States with delegated authority under the CAA have the authority to establish more stringent standards. For example, New York State has a lower limit for mercury emissions from large MWCs of 28 μg/dscm @ 7 percent O_2. In addition, the permitting of new major sources, or major modifications of existing sources, of air pollution are subject to prevention of significant deterioration (PSD) requirements of the new source review (NSR) process. Under these requirements, new sources or major modifications to existing sources located in areas that are either unclassified or determined to be in attainment with the National Ambient Air Quality Standards (NAAQS) are required to meet best available control technology (BACT) requirements. BACT review, made on a case-by-case basis in consideration of energy, environmental, and economic impact, can result in significantly lower emissions limits. For example, in the permitting for the West Palm Beach WTE facility that began operating in 2015, the BACT review process resulted in a 24-hour average NO_x limit of 50 ppm, well below the NSPS value of 150 ppm.

Facilities located in nonattainment areas are subject to the most stringent lowest achievable emission rate (LAER) requirements. Under these requirements, economics are not considered. LAER is derived from the most stringent emissions limitation implemented and achieved in practice. In addition, the facility must "offset" its emissions for the pollutant deemed to be in nonattainment by achieving emissions reductions in the vicinity. In addition to air regulations, MWCs are also subject to solid waste permits from working agencies implementing applicable Clean Water Act and RCRA regulations. Many if not all of the process units are subject to regulations under RCRA, CAA, and the Clean Water Act.

Air Emissions from Municipal Waste Combustion The implementation of more stringent air emissions requirements, particularly those stemming from the CAA Amendments of 1990, have resulted in significant reductions in emissions from the WTE industry (Table 22-58). The emissions reductions have been achieved both as the result of closure of outdated facilities and the installation of air pollution control equipment.

Even before 1990, MWC facilities had much lower emissions than the incinerators commonplace in many urbanized areas earlier in the century. Today, filterable particulate emissions are, on average, approximately 0.02 pound per ton of MSW burned. Similar dramatic improvements have been made in other emissions as well. Historically, municipal waste combustors were a leading source of dioxin emissions, ranking as the largest single source in the 1980s. In 2012, MWC facilities represented just 0.54 percent of total controlled combustion sources and less than 0.1 percent of total controlled and open burning sources of dioxin [Dwyer, H., and N. J. Themelis, "Inventory of U.S. 2012 Dioxin Emissions to Atmosphere," *Waste Management* **46**(December): 242–246 (2015). Available from http://dx.doi.org/10.1016/j.wasman.2015.08.009].

Landfill Disposal Disposal on or in the Earth's mantle is the most common method for the long-term handling of (1) solid wastes that are collected and are of no further use, (2) the residual matter remaining after solid wastes have been processed, and (3) the residual matter remaining after the recovery of conversion products or energy has been accomplished. Adhering to federal regulations, modern landfills are designed to minimize releases from landfills and impacts to human health and safety and the environment through proper siting, design, construction, operation, and maintenance.

Landfilling of Solid Waste Landfilling involves the controlled placement of solid wastes into defined cells followed by compaction with heavy equipment. Operation of a landfill requires proper engineered barriers (liner systems), equipment selection, filling sequences, placement methods,

TABLE 22-57 Current Federal Air Emission Standards for Large Mass-Burn Waterwall Municipal Waste Combustors*

Pollutant	Emission guidelines for existing facilities†	New source performance standards‡	Measurement method	Units
Nitrogen oxides (NO_x)	205	150	CEMS	ppmv
Filterable particulate	25	20	Stack test	mg/m^3
Sulfur dioxide (SO_2)	29	30	CEMS	ppmv
Hydrogen chloride (HCl)	29	25	CEMS	ppmv
Carbon monoxide (CO)	100	100	CEMS	ppmv
Mercury (Hg)	50	50	Stack test	$\mu g/m^3$
Cadmium (Cd)	35	10	Stack test	$\mu g/m^3$
Lead (Pb)	400	140	Stack test	$\mu g/m^3$
Total dioxins & furans (PCDD/F)	30	13	Stack test	ng/m^3

*Actual permit standards may vary based on best available control technology and on state and local requirements.
†Standards applicable to units constructed on or before September 20, 1994: see 40 CFR 60, subpart Cb.
‡Standards applicable to units for which construction is commenced after September 20, 1994, or for which reconstruction or modification is commenced after June 19, 1996. See 40 CFR 60, subpart Eb.
Adapted from 60 FR 65415, Dec. 19, 1995, as amended at 62 FR 45119, 45125, Aug. 25, 1997; 71 FR 27333, May 10, 2006, and 60 FR 65419, Dec. 19, 1995, as amended at 62 FR 45121, 45126, Aug. 25, 1997; 71 FR 27336, May 10, 2006.

compaction, and daily cover selection. Daily placement of cover materials on top of recently disposed waste is critical to control fire, odors, and a host of disease factors. Controlling site access through fencing and perimeter monitoring is important to discourage scavenging and other unlawful access. In addition to these immediate safety and health concerns, two major long-term pollutant emission concerns associated with landfilling are leachate and landfill gas formation and migration. Landfill design includes the use of barrier layers for controlling leachate and landfill gas migration.

Nonhazardous waste landfills are regulated under the RCRA subtitle D (solid waste) and hazardous waste landfills under RCRA subtitle C. PCB wastes are regulated under TSCA. Under subtitle D, state and local governments are the primary planning, regulating, and implementing entities for the management of nonhazardous solid waste. Federal regulatory definitions for MSW landfills, industrial landfills, construction and demolition debris landfills, and hazardous waste landfills are shown in Table 22-59.

Important aspects in the implementation of landfills include (1) site selection and design, (2) landfilling methods and operations, (3) occurrence of gases and leachate in landfills, and (4) movement and control of landfill gases and leachate. Landfilling is a large-scale operation. The number of MSW landfills in the United States is decreasing, and the capacity of remaining landfills is rising, as newer landfills are larger and more regional (obtaining waste from different communities). The number of landfills decreased substantially from 1988 to 2014. There were nearly 8000 in 1988, decreasing to 1908 landfills in service nationwide in 2014. Over the long term, the tonnage of MSW landfilled annually has decreased. In 1990, MSW disposed to landfills was 145 million tons, decreasing to 136 million tons in 2014. The tonnage landfilled results from an interaction among generation, recycling, composting, and combustion with energy recovery, which do not necessarily rise and fall at the same rate. Figure 22-55 shows the decrease of landfills that has occurred over the 25-year period from 1988 to 2014. Today's landfills are typically—relative to what was typical in the 1980s—larger and farther away from where the waste is generated.

Site Selection Factors that must be considered in evaluating potential MSW landfill sites are summarized in Table 22-60. Final selection of a disposal site is based on the results of a preliminary site survey, engineering design and cost studies, and an environmental impact assessment. Suitable areas for landfills are identified based on their physical stability, lack of flooding risks, the avoidance of protected ecological habitat, hydrological barriers, and the vulnerability of local surface water bodies and groundwater. When a site is selected for consideration, the "fatal flaw test" is often used to determine its suitability. The fatal flaw test evaluates the proximity of active seismic fault(s), airports, and schools. Additional examples of reasons a candidate site is not selected is that the location is too small, is located near a floodplain, includes wetlands or an endangered species habitat, is found to be a sacred land, or has unsuitable solid conditions. States and local governments play a significant role in siting landfills, and RCRA regulations part 258 subpart B lists additional siting restrictions.

Landfilling Methods and Operations To use the available area at a landfill site effectively, a plan of operation for the placement of solid wastes must be prepared in compliance with subparts C through G of 40 CFR part 258. Various operational methods have been developed primarily on the basis of field experience. The principal methods used for landfilling may be classified as (1) excavated cell, (2) area, and (3) depression. [Berge, N., E. E. Batarseh, and D. R. Reinhart, "Landfilling: Operation and Monitoring," chap. 10.13 in *Solid Waste Technology & Management*, ed. T. H. Christensen, Wiley, New York, 2010, p. 917.] (See Fig. 22-56.)

1. *Excavated cell.* Excavated cells are typically lined with a geomembrane liner, low-permeability clay, or a combination of the two to limit gas and leachate movement before filling with waste. Excavated soil is usually used on-site for daily cover materials. Cells are typically square and up to 1000 feet in width and length with side slopes of 1.5:1 to 2:1 [Tchobanoglous, G., H. Theisen, and S. Vigil, *Integrated Solid Waste Management: Engineering Principles and Management Issues*, McGraw-Hill, New York, 1993, p. 374]. At the end of each day's operation or when the cell is filled, a layer of cover material is placed over the waste. Excavated soil or alternative materials such as mulch, geotextile tarps, or foam may be used.

2. *Area method.* The area method is used when the terrain is unsuitable for the excavation of cells in which to place the solid wastes or the groundwater table is near the surface. The leachate control system is built above grade. The filling operation usually is started by building an earthen levee against which wastes are placed in thin layers and then compacted. Each layer is compacted as the filling progresses, until the thickness of the compacted wastes reaches a height varying from 2 to 3 m (6 to 10 ft). Similar to an excavated cell, a layer of cover material is placed over the completed fill. The cover soil must be hauled in by truck or earthmoving equipment from adjacent land or from borrow-pit areas, or an alternative daily cover material such as reusable geotextile covers are used. Successive lifts are placed on top of one another until the final grade called for in the ultimate development plan is reached. A final layer of cover material is used when the fill reaches the final design height.

3. *Depression method.* At locations where natural or artificial depressions exist, it is often possible to use them effectively for landfilling operations.

TABLE 22-58 Annual Emissions from U.S. MWC Facilities in 1990 and 2005

Pollutant	1990 emissions (tons/year)	2005 emissions (tons/year)	Percent reduction
Dioxins and furans, TEQ basis*	4400	15	99+
Mercury	57	2.3	96
Cadmium	9.6	0.4	96
Lead	170	5.5	97
Particulate matter	18,600	780	96
Hydrochloric acid (HCl)	57,400	3,200	94
Sulfur dioxide (SO_2)	38,300	4,600	88
Nitrogen oxides (NO_x)	64,900	49,500	24

*Dioxin/furan emissions are in units of grams per year toxic equivalent quantity (TEQ), using 1989 NATO toxicity factors; all other pollutant emissions are in units of tons per year.
SOURCE: US EPA, Air Emissions from MSW Combustion Facilities, https://archive.epa.gov/epawaste/nonhaz/municipal/web/html/airem.html (accessed February 13, 2018).

TABLE 22-59 Landfills, Waste Type, and Regulatory Definition

Subtitle D	Example waste type*	Regulatory definition†
Municipal solid waste landfills (MSWFs) CFR part 258	• Municipal solid waste from – Residential – Commercial – Institutional • Industrial nonhazardous process wastes • Construction and demolition debris • Land clearing debris • Nonhazardous sludges • Transportation parts and equipment • Agricultural wastes • Oil and gas wastes • Mining wastes • Auto bodies • Fats, grease, and oils	MSWLF unit means a discrete area of land or an excavation that receives household waste, and that is not a land application unit, surface impoundment, injection well, or waste pile, as those terms are defined in this section. A MSWLF unit also may receive other types of RCRA Subtitle D wastes, such as commercial solid waste, nonhazardous sludge, and industrial solid waste. Such a landfill may be publicly or privately owned. A MSWLF unit may be a new MSWLF unit, an existing MSWLF unit or a lateral expansion. A construction and demolition landfill that receives residential lead-based paint waste and does not receive any other household waste is not a MSWLF unit.
Industrial solid waste landfills CFR part 257 Any landfill other than MSWF, an RCRA subtitle C hazardous waste landfill, or a TSCA landfill	• Industrial nonhazardous process waste • Commercial solid waste Conditionally exempt small-quantity generator waste	Industrial solid waste means solid waste generated by manufacturing or industrial processes that is not a hazardous waste regulated under subtitle C of RCRA. Such waste may include, but is not limited to, waste resulting from the following manufacturing processes: Electric power generation; fertilizer/agricultural chemicals; food and related products/by-products; inorganic chemicals; iron and steel manufacturing; leather and leather products; nonferrous metals manufacturing/foundries; organic chemicals; plastics and resins manufacturing; pulp and paper industry; rubber and miscellaneous plastic products; stone, glass, clay, and concrete products; textile manufacturing; transportation equipment; and water treatment. This term does not include mining waste or oil and gas waste.
	• Construction and demolition (C&D) – Steel – Wood products – Bricks – Clay – Tile – Drywall and plaster – Asphalt shingles – Asphalt concrete – Portland cement.	Construction and demolition (C&D) landfill means a solid waste disposal facility subject to the requirements of subparts A or B of this part that receives construction and demolition waste and does not receive hazardous waste (defined in §261.3 of this chapter) or industrial solid waste (defined in §258.2 of this chapter). Only a C&D landfill that meets the requirements of subpart B of this part may receive conditionally exempt small quantity generator waste (defined in §261.5 of this chapter). A C&D landfill typically receives any one or more of the following types of solid wastes: roadwork material, excavated material, demolition waste, construction/renovation waste, and site clearance waste.
	• Coal combustion residual (CCR)	CCR landfill means an area of land or an excavation that receives CCR and which is not a surface impoundment, an underground injection well, a salt dome formation, a salt bed formation, an underground or surface coal mine, or a cave. For purposes of this subpart, a CCR landfill also includes sand and gravel pits and quarries that receive CCR, CCR piles, and any practice that does not meet the definition of a beneficial use of CCR.
RCRA subtitle C	Waste type*†	Regulatory definition†
Hazardous waste landfill CFR part 264	A solid waste as defined in CFR Title 40 Section 261.2	Landfills authorized under RCRA to accept hazardous waste for disposal. Landfills authorized to accept these wastes must follow very stringent guidelines for their design and operation.

*Adapted from EPA. 2015. *Advancing Sustainable Materials Management: Facts and figures 2013.* US Environmental Protection Agency, Office of Resource Conservation and Recovery. June. EPA530-R-15-002. https://www.epa.gov/sites/production/files/2015-09/documents/2013_advncng_smm_rpt.pdf, accessed August 15, 2016 and EPA. 2016. Industrial and Construction and Demolition (C&D) Landfills. https://www.epa.gov/landfills/industrial-and-construction-and-demolition-cd-landfills, accessed September 1, 2016.
†Code of Federal Regulations Title 40 Section 257.2, 258.2, 261.2.

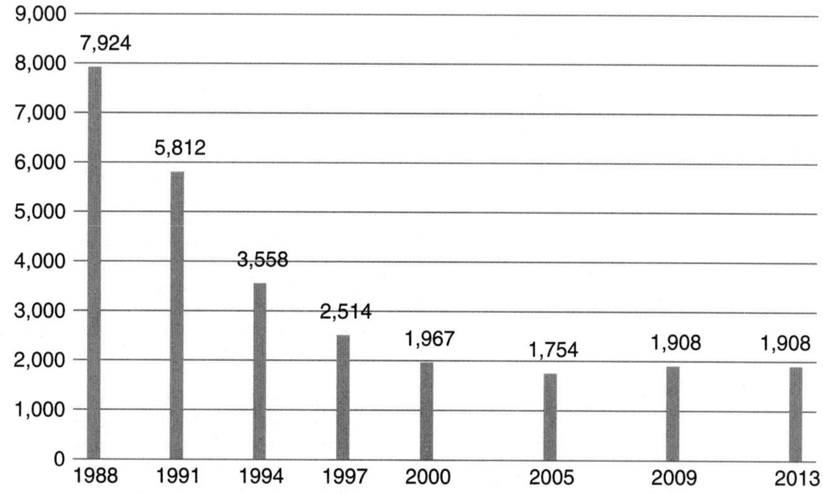

FIG. 22-55 Number of landfills, 1988 through 2013. (*Advancing Sustainable Materials Management: Facts and Figures 2013. U.S. Environmental Protection Agency, Office of Comment [KC3]: AU: Art shows commingled materials. Resource Conservation and Recovery, June 2015, EPA 530-R-15-002. https://www.epa.gov/sites/production/files/2015-09/documents/2013_advncng_smm_rpt.pdf, accessed August 15, 2016.*)

TABLE 22-60 Important Factors in Preliminary Selection of Landfill Sites

Factor	Remarks
Available land area	In selecting potential land disposal sites, it is important to ensure that sufficient land area is available.
Impact of processing and resource recovery	It is important to project the extent of resource-recovery processing activities that are likely to occur in the future and determine their impact on the quantity and condition of the residual materials to be disposed of.
Haul distance	Although minimum haul distances are desirable, other factors must also be considered. These include collection-route location, types of wastes to be hauled, local traffic patterns, and characteristics of the routes to and from the disposal site (condition of the routes, traffic patterns, and access conditions).
Soil conditions and topography	Because it is necessary to provide material for each day's landfill and a final layer of cover after the filling has been completed, data on the amounts and characteristics of the soils in the area must be obtained. Local topography will affect the type of landfill operation to be used, equipment requirements, and the extent of work necessary to make the site usable.
Climatological conditions	Local weather conditions must also be considered in the evaluation of potential sites. Under winter conditions where freezing is severe, landfill cover material must be available in stockpiles when excavation is impractical. Wind and wind patterns must also be considered carefully. To avoid blowing or flying papers, windbreaks must be established.
Surface-water hydrology	The local surface-water hydrology of the area is important in establishing the existing natural drainage and runoff characteristics that must be considered.
Geologic and hydrogeologic conditions	Geologic and hydrogeologic conditions are perhaps the most important factors in establishing the environmental suitability of the area for a landfill site. Data on these factors are required to assess the pollution potential of the proposed site and to establish what must be done to the site to control the movement of leachate or gases from the landfill.
Local environmental conditions	The proximity of both residential and industrial developments is extremely important. Great care must be taken in their operation if they are to be environmentally sound with respect to noise, odor, dust, flying paper, and vector control.
Ultimate uses	Because the ultimate use affects the design and operation of the landfill, this issue must be resolved before the layout and design of the landfill are started.

SOURCE: Adapted from Perry's *Chemical Engineers Handbook*, 8th Edition, McGraw-Hill, New York, NY.

Canyons, ravines, dry borrow pits, and quarries have been used for this purpose. The techniques to place and compact solid wastes in depression landfills vary with the geometry of the site, the characteristics of the cover material, the hydrology and geology of the site, and access to the site. In a canyon site, filling starts at the head end of the canyon and ends at the mouth. The practice prevents the accumulation of water behind the landfill [Tchobanoglous, G., H. Theisen, and S. Vigil, *Integrated Solid Waste Management: Engineering Principles and Management Issues*, McGraw-Hill, New York, 1993, p. 376].

Occurrence of Gases and Leachate in Landfills The following biological, physical, and chemical events occur when solid wastes are placed in a landfill: (1) biological decay of organic materials, either aerobically or anaerobically, with the evolution of gases and liquids; (2) chemical oxidation of waste materials; (3) movement of gases within and from the fill; (4) movement of liquids caused by differential heads; (5) dissolving and leaching of organic and inorganic materials by water and leachate moving through the fill; (6) movement of dissolved material by concentration gradients and osmosis; and (7) uneven settlement caused by consolidation of material into voids.

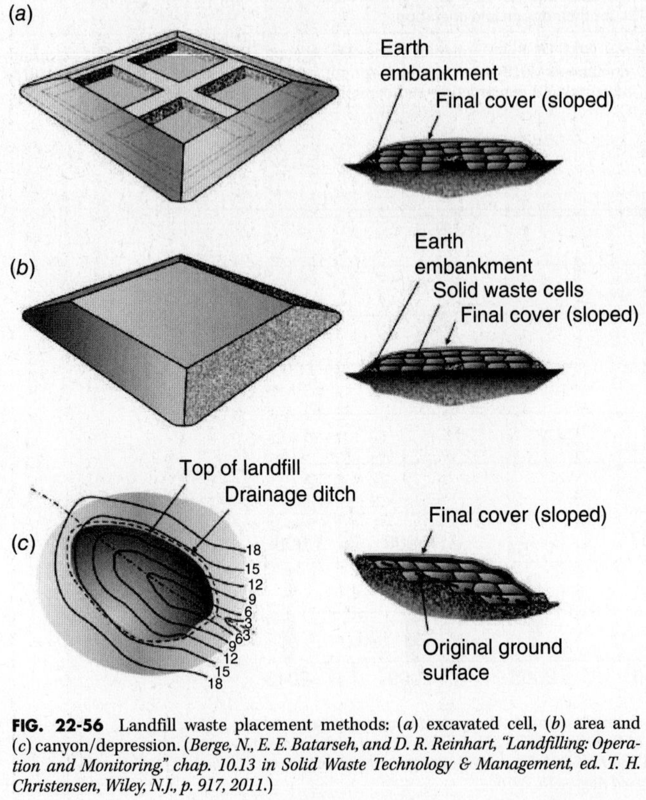

FIG. 22-56 Landfill waste placement methods: (*a*) excavated cell, (*b*) area and (*c*) canyon/depression. (*Berge, N., E. E. Batarseh, and D. R. Reinhart, "Landfilling: Operation and Monitoring," chap. 10.13 in Solid Waste Technology & Management, ed. T. H. Christensen, Wiley, N.J., p. 917, 2011.*)

Bacterial decomposition initially occurs under aerobic conditions because a certain amount of air is trapped within the landfill. However, the oxygen in the trapped air is exhausted within days, and long-term decomposition occurs under anaerobic conditions.

Gases in landfills. Gases found in landfills include carbon dioxide, carbon monoxide, hydrogen, hydrogen sulfide, methane, nitrogen, and oxygen. Carbon dioxide and methane are the principal gases produced from the anaerobic decomposition of the organic solid waste components; nonmethane organic compounds amount to less than 1 percent of landfill gas. Often, atmospheric air dilutes landfill gas as a result of air infiltration. This occurs when gas takes the path of least resistance and leaks through cracks and fissures in the landfill surface cover, wellheads, and leachate wells and sumps. The gas is extracted under negative pressure. Air infiltration through cracks or fissures in the cover or around gas wells is minimized to prevent potential landfill fires. This can be achieved through routine maintenance of the surface and side slope cover and by using leak detection as required by Clean Air Act regulations, and in accordance with appropriate safeguards through automated checks of landfill gas composition in individual wellheads.

The anaerobic conversion of organic compounds is thought to occur in three steps. The first involves the enzyme-mediated transformation (hydrolysis) of higher-weight molecular compounds into compounds suitable for use as a source of energy and cell carbon; the second is associated with the bacterial conversion of the compounds resulting from the first step into identifiable lower-molecular-weight intermediate compounds; and the third step involves the bacterial conversion of the intermediate compounds into simpler end products, such as carbon dioxide (CO_2) and methane (CH_4). The overall anaerobic conversion of organic wastes can be represented with the following equation:

$$C_aH_bO_cN_d \rightarrow nC_wH_xO_yN_z + mCH_4 + sCO_2 + rH_2O + (d - nz)NH_3 \quad (22\text{-}30)$$

where $s = a - nw = m$ and $r = c - ny - 2s$. The term $C_aH_bO_cN_d$ represents the composition on a molar basis of the material present at the start of the process.

The rate of decomposition in landfills, as measured by gas production, reaches a peak within the first two years and then slowly tapers off, continuing in many cases for periods of up to 25 years or more. The total volume of the gases released during anaerobic decomposition can be estimated in a number of ways. If all the organic constituents in the wastes (with the exception of plastics, rubber, and leather) are represented with a generalized formula of the form $C_aH_bO_cN_d$, the total volume of gas can be estimated by using Eq. (22-30) with the assumption of completed conversion to carbon dioxide and methane.

Leachate in landfills. Leachate is formed by the percolation of water through landfills and from water interacting with waste in an anaerobic environment concentrating organic and inorganic constituents. The concentration of leachate constituents peaks within the first two to three years of operation and slowly decreases as the landfill matures; this occurs as organics are removed through washout and waste degradation [Maximova, A., and B. Koumanova, "Study on the Content of Chemicals in Landfill Leachate," in *Chemicals as Intentional and Accidental Global Environmental Threats*, ed. L. Simeonov and E. Chirila, Springer, Netherlands, 2006, pp. 345–356); Qasim, S. R., and W. Chiang, *Sanitary Landfill Leachate: Generation, Control, and Treatment,* Technomic, Lancaster, Pa. 1994]. The concentration of organic compounds follows a decreasing trend over time, whereas the concentration of inorganic constituents tends to vary over time due to adsorption, complexation, precipitation, and dissolution affecting the mobilization of heavy metals within a landfill [Bolyard, Stephanie, "Application of Landfill Treatment Approaches for the Stabilization of Municipal Solid Waste," *Electronic Theses and Dissertations,* Paper 4878 (2016). http://stars.library.ucf.edu/etd/4878].

Metal concentrations in leachate are affected by pH, the presence of organic complexing agents such as humic and fulvic acids, and the presence of inorganic complexing/precipitating agents such as ammonia, carbonates, hydroxides, and chlorides. Studies suggest that a small fraction of these metals in leachate are present as free metal ions. Most metals are associated with organic and inorganic colloidal fractions [Baun, D. L., and T. H. Christensen, "Speciation of Heavy Metals in Landfill Leachate: A Review," *Waste Management & Research* **22**(1): 3–23 (2004). doi: 10.1177/0734242x04042146.] The metals most likely to be found associated with organic colloids are Cd, Cr, Cu, Pb, and Zn. In particular, metals in older leachate form stable complexes with high molecular weight organic components [Calace, N., A. Liberatori, B. M. Petronio, and M. Pietroletti, "Characteristics of Different Molecular Weight Fractions of Organic Matter in Landfill Leachate and Their Role in Soil Sorption of Heavy Metals," *Environmental Pollution* **113**(3): 331–339 (2001), doi: http://dx.doi.org/10.1016/S0269-7491(00)00186-X; Christensen, J. B., D. L. Jensen, and T. H. Christensen, "Effect of Dissolved Organic Carbon on the Mobility of Cadmium, Nickel and Zinc in Leachate Polluted Groundwater," *Water Research* **30**(12): 3037–3049 (1996), doi: http://dx.doi.org/10.1016/S0043-1354(96)00091-7]. The speciation of metals will have a strong impact on which leachate treatment approaches are most effective at removing metals [Baun, D. L., and T. H. Christensen, "Speciation of Heavy Metals in Landfill Leachate: A Review," *Waste Management & Research* **22**(1): 3–23 (2004). doi: 10.1177/0734242x04042146].

Leachate is typically treated at a local wastewater treatment facility. Pollutants that do not biologically degrade may be concentrated in the biosolids and may be released in the wastewater effluent and could later be a source of organic pollution to surface water bodies [Clarke, B. et al., "Investigating Landfill Leachate as a Source of Trace Organic Pollutants," *Chemosphere* **127:** 269–275 (2015)]. In addition to organics, landfill leachate can contain pharmaceuticals and other contaminants of emerging concern [Masoner, J. et al., "Landfill Leachate as a Mirror of Today's Disposable Society: Pharmaceuticals and Other Contaminants of Emerging Concern in Final Leachate from Landfills in the Conterminous United States," *Environ. Toxicol. Chem.* **35**(4): 906–918 (2016)].

Gas and Leachate Movement and Control The landfill gases generated are either collected and combusted or vented to the atmosphere. Most municipal solid waste landfills with a design capacity greater than or equal to 2.5 million metric tons and greater than or equal to 2.5 million cubic meters of waste are subject to federal new source performance standards or emission guidelines. If these subject landfills reach an emissions threshold of 34 tonnes of nonmethane organic compounds (NMOC) per year, they must install a collection and control system within 30 months of reaching the threshold. Additional information on performance standards is available on U.S. EPA's Air Toxics web site. Through a collection and control system, the leachate is removed for treatment and proper disposal.

1. *Gas movement.* Predicting the distance gas can travel through the limited pore spaces in the landfilled waste and cover material is difficult. This migration, either vertically or horizontally, is influenced by concentration, pressure, and permeability. Gases move from areas of higher gas concentration to areas of lower concentrations. Since the concentrations are higher within the landfill, gases will also migrate to areas outside of the landfill. Gas movement is restricted by compacted waste, cover materials, and barriers such as geomembranes, resulting in gas accumulation and increased pressure. The variation in pressure will cause the gas to move from areas of higher pressure to areas of lower pressure. The increased pressure causes subsurface pressures to be higher than atmospheric pressure. Permeability measures the ability of gases to travel through the connected pore spaces. Landfill covers are typically built of low-permeability materials, increasing horizontal movement more than vertical movement. Methane can accumulate below buildings or in other enclosed spaces near a landfill. With proper venting at the landfill source, methane should not pose a public safety problem [Agency for Toxic Substances and Disease Registry (ATSDR), *Landfill Gas Primer: An Overview for Environmental Health Professionals,* chap. 2, November 2001. http://www.atsdr.cdc.gov/hac/landfill/html/intro.html].

Because carbon dioxide is about 1.5 times as dense as air and 2.8 times as dense as methane, it tends to move toward the bottom of a landfill. As a result, the concentration of carbon dioxide in the lower portions of a landfill may be high for years.

Methane is a potent greenhouse gas, which means it is efficient at trapping radiation in the atmosphere, thereby causing a warming effect, and the release of VOCs causes air quality issues such as smog and ozone formation. In the gas of an MSW landfill, VOCs comprise about 39 percent by weight of NMOCs [U.S. EPA, "2.4: Municipal Solid Waste Landfills," in *Compilation of Air Pollutant Emission Factors* (AP-42), November 1998. Table 2.4-2, footnote c. https://www3.epa.gov/ttn/chief/ap42/ch02/final/c02s04.pdf].

2. *Control of gas movement.* Gas movement is controlled by installing vents made of materials that are more permeable than the surrounding soil. Passive systems, more typically installed at smaller landfills, are used to control the emission location to avoid lateral migration to areas adjacent to the landfill. Passive systems can be installed during the waste filling phase (Figure 22-57*a*) or after completion of waste disposal (Figure 22-57*b*). Both systems use a permeable layer of gravel and perforated piping to control the movement of gases within and from the landfill. [Christensen, T. H., *Solid Waste Technology and Management,* vol. II, chap. 10, Wiley, New York, 2011.]

Active landfill gas collection systems connect a series of wells (vertical or horizontal) through common intake header pipes. Vertical wells are typically installed in landfill areas that have reached desired fill height (typically 30 feet or higher) [Townsend, T. G., J. Powell, P. Jain, Q. Xu, T. Tolaymat, and D. Reinhart, *Sustainable Practices for Landfill Design and Operation,* Springer-Verlag, New York, 2015, p. 13.2]. Horizontal gas wells can be installed during the first lift (and subsequent lifts) of waste disposal. Figure 22-58 depicts a landfill gas collection infrastructure. Similar to passive collection systems, active systems require permeable materials, such as gravel and perforated piping. Unlike passive systems, an active system moves the gas from the perforated pipes to a control system through enclosed pipe near the surface

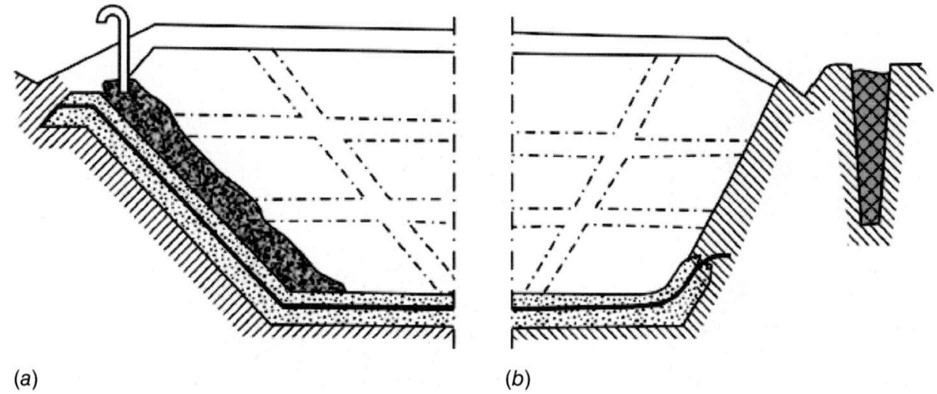

FIG. 22-57 Passive gas venting system installed (*a*) inside the landfill during filling and (*b*) outside after landfilling. (*Willumsen, H., and M. A. Barlaz, "Landfilling: Gas Production, Extraction and Utilization" chap. 10.10 in Solid Waste Technology and Management, ed. T. H. Christensen, Wiley, New York, p. 846.*)

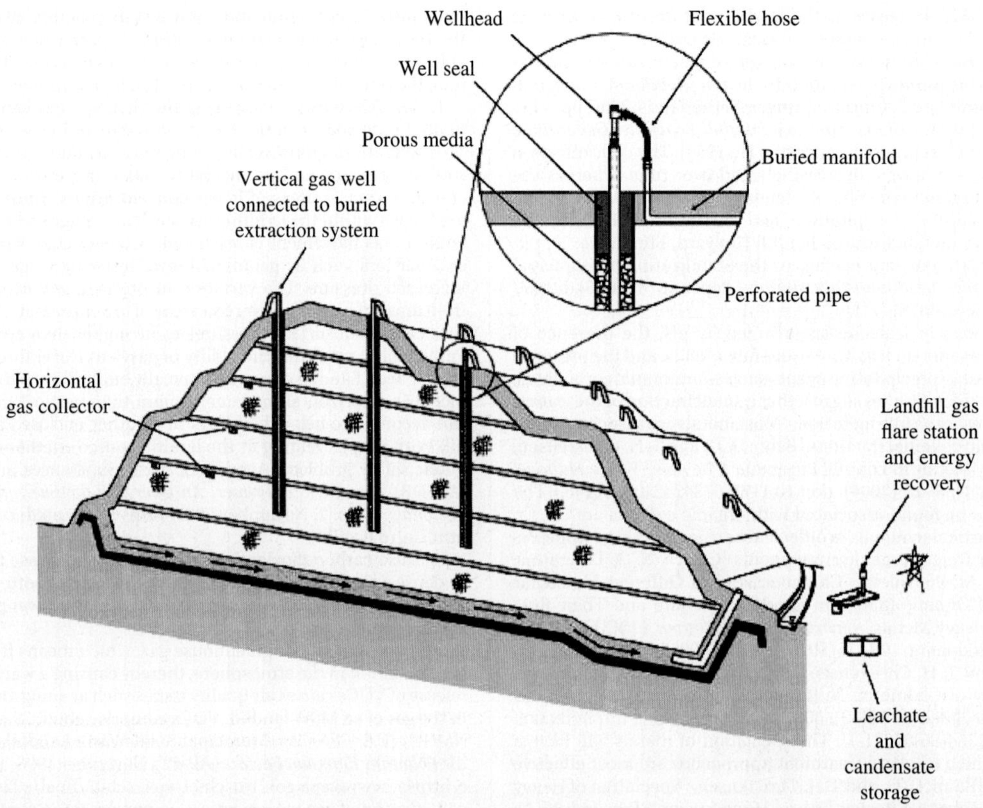

FIG. 22-58 Landfill gas collection system infrastructure. (*Townsend, T. G., J. Powell, P. Jain, Q. Xu, T. Tolaymat, and D. Reinhart, Sustainable Practices for Landfill Design and Operation, chap. 13, "Landfill Gas," Springer Science+Business Media LLC, New York, 2015, p. 282.*)

for collection and management through recovery. Figure 22-59 provides a cross section of a vertical gas well (Figure 22-59*a*) and a horizontal gas well (Figure 22-59*b*).

The movement of landfill gases through adjacent soil formations is controlled by building barriers of materials that are less permeable than the native soil. The two major low-permeability material categories used separately or in combination are earthen materials and geotechnical materials. Earthen materials include clay soils; geotechnical materials, known as geomembranes, are manufactured from plastic polymers, most commonly high-density polyethylene (HDPE). The top barrier, referred to as the cap or cover, is also multilayered and may include compacted soil, a clay layer, a geomembrane, drainage layers (synthetic or natural), and a grassed layer to control erosion. Geomembrane caps that are textured or incorporate artificial grass can be used in place of a grassed layer.

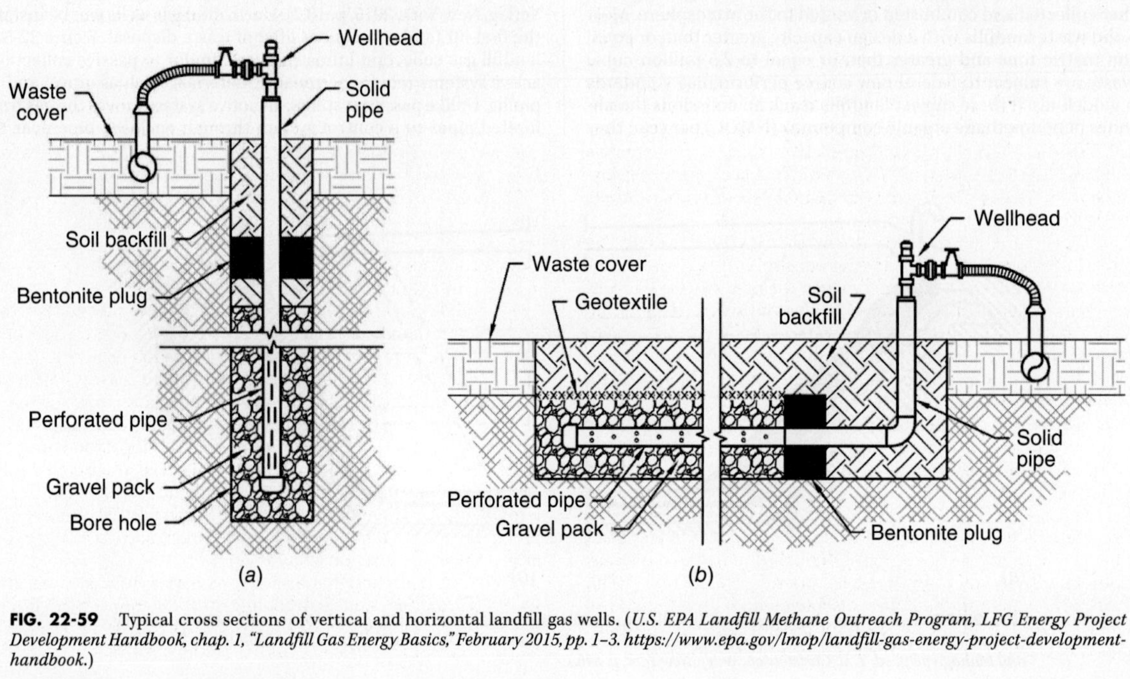

FIG. 22-59 Typical cross sections of vertical and horizontal landfill gas wells. (*U.S. EPA Landfill Methane Outreach Program, LFG Energy Project Development Handbook, chap. 1, "Landfill Gas Energy Basics," February 2015, pp. 1–3. https://www.epa.gov/lmop/landfill-gas-energy-project-development-handbook.*)

3. *Active landfill gas recovery.* As mentioned above, the movement of gases in landfills can be controlled by installing gas recovery wells in closed landfills or completed sections of landfills. Gas recovery systems have been installed in hundreds of landfills in the United States. Vertical gas recovery wells are the most common type, in which the wells are drilled vertically through the waste mass and the wellheads are connected to lateral piping. Another option is putting horizontal collection piping in trenches in the waste mass. Some systems include both types of wells. With either approach, a system of wells is connected to a header, and the gas is extracted from the waste mass using negative pressure controlled by a blower or vacuum induction technology [U.S. EPA, Landfill Methane Outreach Program, "Landfill Gas Energy Basics," chap. 1 in *LFG Energy Project Development Handbook*, February 2015, pp. 1–3. https://www.epa.gov/lmop/landfill-gas-energy-project-development-handbook].

The collected landfill gas can be flared to destroy most of the gas and its constituents. Flare options include candlestick (open) flares or enclosed flares. Alternatively, the methane component of the landfill gas can be used beneficially as an energy source. Landfill gas has a high heating value (HHV) of about 500 British thermal units per standard cubic foot (Btu/scf), about half that of natural gas [U.S. EPA, Landfill Methane Outreach Program, "Landfill Gas Energy Basics," chap. 1 in *LFG Energy Project Development Handbook*, February 2015, pp. 1–2. https://www.epa.gov/lmop/landfill-gas-energy-project-development-handbook].

Following minimal treatment, landfill gas can be used directly as a medium-Btu gas in industrial, commercial, or institutional applications (equipment can include boilers; cement, pottery, or brick kilns; process heaters; and ovens). With additional cleanup, landfill gas can fuel technologies (examples include reciprocating internal combustion engines, gas and steam-fed turbines, and microturbines) to produce electricity for on-site use, sale to the grid, or sale to direct power purchasers. Landfill gas can also be cleaned and concentrated through CO_2 removal to "pipeline quality" for injection into commercial natural gas pipelines, or compression or liquefaction into renewable compressed natural gas or renewable liquefied natural gas [U.S. EPA, Landfill Methane Outreach Program, "Project Technology Options," chap. 3 in *LFG Energy Project Development Handbook*, February 2015. https://www.epa.gov/lmop/landfill-gas-energy-project-development-handbook]. The capital and operation and maintenance costs for landfill gas recovery and beneficial use projects will vary based on many site-specific variables, including the available amount of landfill gas, type of technology, location of the gas or electricity end user, and permit conditions and restrictions [U.S. EPA, Landfill Methane Outreach Program, "Project Economics and Financing," chap. 4 in *LFG Energy Project Development Handbook*, February 2015. https://www.epa.gov/lmop/landfill-gas-energy-project-development-handbook].

4. *Leachate movement.* Precipitation is the principal liquid source of leachate, with additional leachate produced during the decomposition of the landfilled waste. Under normal conditions, leachate is found in the bottom of landfills. From there it must be removed from the landfill through the leachate collection system. Design and operation of the leachate collection system is regulated under RCRA subtitle D, part 258, to minimize the migration of the leachate through the liner barrier systems to adjacent areas, including the underlying soils and groundwater. Moisture flow through a landfill can be estimated by Darcy's laws for saturated flow and conductivity and unsaturated flow. Applying the flow calculations to predictive models, the production of leachate can be estimated for specific site characteristics. Additional discussion of Darcy's law and predictive moisture flow models can be found in Chapter 5 of Townsend, T. G., J. Powell, P. Jain, Q. Xu, T. Tolaymat, and D. Reinhart, *Sustainable Practices for Landfill Design and Operation*, Springer-Verlag, New York, 2015.

5. *Control of leachate movement.* As leachate percolates through the landfill, many of the chemical and biological constituents originally contained in the waste will be removed. Concern over the migration of concentrated environmentally harmful constituents to areas outside of the landfill, in particular the underlying soils and groundwater, led to regulated design criteria including (1) a flexible geomembrane overlaying two feet of compacted clay soil lining the bottom and sides of the landfill, (2) a leachate collection system designed and built to maintain less than a 30-cm depth of leachate on the liner, (3) a leachate collection and removal system that sits on top of the liner system and removes leachate from the landfill for treatment and proper disposal, and (4) continued groundwater monitoring and corrective action requirements.

The goal of the leachate collection system is to remove all leachate from the landfill at a rate sufficient to prevent an unacceptable hydraulic head from occurring at any point over the lining system, thus preventing releases to the environment. There are many components to a collection system, including liners, filters, pumps, manholes, discharge lines, and liquid level monitors. The collection system drainage layer provides horizontal drainage of the leachate to a point of gravitational collection or pumping.

Equally important in controlling the movement of leachate within the landfill is the control of surface-water infiltration. Surface infiltration can be controlled effectively with the use of an impermeable clay layer, membrane liners, an appropriate surface slope (1 to 2 percent), and adequate drainage.

6. *Settlement and structural characteristics of landfills.* The settlement of landfills depends on the initial compaction, characteristics of the wastes, degree of decomposition, and effects of consolidation when the leachate and gases are formed in the landfill. The height of the completed fill will also influence the initial compaction and degree of consolidation.

7. *Potential for landfill fires or elevated temperature events at landfills.* There are a number of processes that may lead to heating events at landfills. The most common is a landfill fire, which may be a result of a hot load of waste buried in the landfill, cigarettes, sparks, or natural phenomena such as lightning strikes. Landfill fires occur at or near the waste surface and require the availability of oxygen. A second type of heating event may result when the gas removed from a gas extraction well exhibits an elevated temperature, which could result from overpumping of the well and subsequent air intrusion. The air reacts with either methane or the buried waste, and heat accumulates. A small percentage of municipal solid waste landfills in North America have reported elevated and increasing temperatures above those normally associated with a hot gas well. These landfills tend to be characterized by a decreasing trend in the ratio of CH_4/CO_2 in individual gas wells or areas of the landfill, increasing leachate strength as quantified by the chemical oxygen demand (COD) and total dissolved solids (TDS), a decreasing leachate pH, the presence of H_2 in the landfill gas, and unusually rapid settlement. These indicators will vary at each landfill due to differences in the location and proximity of leachate collection sumps and gas wells relative to the heating events.

PLANNING

From a sustainability perspective, the generation of waste is an inefficient use of natural resources. Looking ahead, it will become increasingly important to place greater emphasis on sustainable materials management by preventing waste and encouraging reuse. A scarcity of rare earth metals, which are widely used in electronics, is a growing concern to manufacturers. Many are reconsidering the design of their products and evaluating the flow of materials across the product life cycle in order to conserve resources and minimize waste. Resources from the U.S. EPA for increasing the recovery of materials from waste can be accessed at https://www.epa.gov/recycle.

Ideally, solid waste management plans should help encourage more sustainable materials management. However, change is incremental. Communities must evaluate their existing programs and infrastructures, their opportunities to consolidate operations with nearby communities, the fluctuation and market value of materials, local economics, the energy grid mix, and other factors. Efficient and effective solid waste management plans will encourage source reduction and will evaluate total emissions over time. Materials discarded in landfills will produce leachate and gas emissions for decades, while other processes produce emissions that are instantaneous. However, local drivers, especially costs, may limit more sustainable options in favor of disposal through landfilling.

Sustainable materials management represents a paradigm shift in how society views the use of natural resources and the protection of the environment. More sustainable solutions must account for complex interrelationships among processes and competing management objectives. Plans must consider changes that will occur in waste composition in response to source reduction, reuse, and recycling programs.

Regardless of waste type, solid waste management planning can:

- encourage stakeholder communication,
- promote sustainable materials management,
- increase community preparedness, and
- facilitate community resiliency.

The importance of conserving natural resources has led Europe, the United States, and other countries to reevaluate their use of secondary materials and industrial by-products. These materials may be used to fill road embankments, for example. Mined gypsum may be replaced by flue-gas desulfurization (FGD) gypsum from power plants that use wet scrubbers. However, there have been cases where material was used indiscriminately. Therefore, the coal combustion residues (CCR) regulation defined *beneficial use* as the incorporation of an industrial material into a commercial product that (1) provides functional benefit, (2) meets relevant design specifications and performance standards for the proposed use, (3) replaces virgin or raw materials in a product already on the market, and (4) is implemented in an environmentally acceptable manner.

Global competition for limited resources will only intensify as the world population and economies grow. A more productive and less costly use of materials will help our society to remain economically competitive, contribute to our prosperity, and protect the environment in a resource-constrained future. We want to make sure we have enough resources to meet not only today's needs, but those of the future as well. How our society uses materials is fundamental to our economic and environmental future.

The beneficial use of nonhazardous secondary waste materials is managed by state environmental agencies. The relevant state and federal authorities should be consulted to identify all the requirements that would apply to proposed beneficial uses of waste materials. The U.S. EPA developed a methodology to help evaluate the potential adverse impacts to human health and the environment associated with the beneficial use of secondary waste materials. The methodology does not require any specific test methods, only that the data generated are of sufficient quality to support the final conclusions of the evaluation. In 2013, the U.S. EPA applied the methodology to evaluate the two largest encapsulated beneficial uses of coal ash: coal fly ash used as a direct substitute for portland cement in concrete and FGD gypsum used as a replacement for mined gypsum in wallboard. U.S. EPA's evaluation concluded that the beneficial use of encapsulated CCR in concrete and wallboard is appropriate because environmental releases are comparable to or lower than those from analogous non-CCR products, or are at or below relevant regulatory and health-based benchmarks. [U.S. EPA, *Coal Combustion Residual Beneficial Use Evaluation: Fly Ash Concrete and FGD Gypsum Wallboard*, EPA 530-R-14-001, February 2014; van der Sloot, H., D. Kosson, A. C. Garrabrants, and J. Arnold, *The Impact of Coal Combustion Fly Ash Used as a Supplemental Cementitious Material on the Leaching of Constituents from Cements and Concretes*, Prepared for the U.S. EPA/ORD; EPA-600/R-12/704, October 2012; Garrabrants, A. C., D. S. Kosson, R. DeLapp, and H. A. van der Sloot, "Effect of Coal Combustion Fly Ash Use in Concrete on the Mass Transport Release of Constituents of Potential Concern," *Chemosphere* **103:** 131–139 (2014); Kosson, D. S., A. C. Garrabrants, R. DeLapp, and H. A. van der Sloot, "pH-Dependent Leaching of Constituents of Potential Concern from Concrete Materials Containing Coal Combustion Fly Ash, *Chemosphere* **103:** 140–147 (2014).]

As part of the evaluation of the use of fly ash in the production of cementitious materials, the U.S. EPA used the Leaching Environmental Assessment Framework (LEAF) and associated test methods to develop a source term for leaching of inorganic constituents of potential concern (COPC) (e.g., arsenic, boron, chromium, mercury, selenium) to compare to applicable environmental thresholds. A source term is an estimate of constituent release from a material or waste and is used in subsequent fate and transport modeling to evaluate exposure for use in a risk assessment. The results from the assessments provide the data needed to assess whether leaching occurs considering the waste form and the environmental conditions under which the waste is managed.

[Garrabrants, A. C., D. S. Kosson, H. A. van der Sloot, F. Sanchez, and O. Hjelmar, *Background Information for the Leaching Environmental Assessment Framework Test Methods*, EPA/600/R-10/170, December 2010; Thorneloe, S. A., D. S. Kosson, F. Sanchez, A. C. Garrabrants, and G. Helms, "Evaluating the Fate of Metals in Air Pollution Control Residues from Coal-Fired Power Plants," *Environ. Sci. Technol.* **44**(19): 7351–7356 (2010); Kosson, D., F. Sanchez, P. Kariher, L. Turner, D. Delapp, P. Seignette, and S. Thorneloe, *Characterization of Coal Combustion Residues from Electric Utilities—Leaching and Characterization Data*, EPA-600/R-09/151, December 2009; Sanchez, F., D. Kosson, R. Keeney, R. DeLapp, L. Turner, P. Kariher, and S. Thorneloe, *Characterization of Coal Combustion Residues from Electric Utilities Using Wet Scrubbers for Multi-Pollutant Control*, EPA-600/R-08/077, July 2008; Sanchez, F., R. Keeney, D. Kosson, R. Delapp, and S. Thorneloe, *Characterization of Mercury-Enriched Coal Combustion Residues from Electric Utilities Using Enhanced Sorbents for Mercury Control*, EPA-600/R-06/008, February 2006; Sanchez, F., C. H. Mattus, M. I. Morris, and D. S. Kosson, "Use of a New Framework for Evaluating Alternative Treatment Processes for Mercury Contaminated Soils," *Environ. Eng. Sci.* **19**(4): 251–269 (2002); Kosson, D. S., H. A. van der Sloot, F. Sanchez, and A. C. Garrabrants, "An Integrated Framework for Evaluating Leaching in Waste Management and Utilization of Secondary Materials," *Environ. Eng. Sci.* **19**(3): 159–204 (2002).]

Implementation of LEAF for beneficial use applications is considered a significant achievement because it provides standardized methods that can be used in any national, academic, or commercial laboratory to provide consistent and uniform data that can be used to develop more reliable source terms for predicting potential environmental release of pollutants from leaching of constituents of potential concern to ground and surface waters. The LEAF methods have been validated, and guidance for their use was made available in 2017 (https://www.epa.gov/hw-sw846/frequent-questions-about-leaf-methods-and-how-guide) providing examples of LEAF's use to evaluate the beneficial use of coal fly ash and other industrial by-products. Comparable methods for organics are under development (U.S. EPA, *Workshop Report: Considerations for Developing Leaching Test Methods for Semi- and Non-Volatile Organic Compounds*, EPA 600-R-16-057, 2016). The methods are expected to be used in helping facilitate delisting decisions, assessing treatment effectiveness for older contaminated sites (e.g., Superfund sites), and making beneficial use decisions.

The LEAF methods determine the maximum potential to leach and the release rate based on a site-specific assessment that considers differences in the drivers that affect leaching of inorganics such as pH, liquid-to-solid ratio, and waste form. Comparable methods are available in Europe and other countries. The use of LEAF addresses challenges posed by the Science Advisory Board and the National Academy of Science in the appropriate use of leaching tests that provide a realistic understanding of chemical and mass transport fundamentals for contaminant release from complex sources [Sanchez, F., and D. S. Kosson, "Probabilistic Approach for Estimating the Release of Contaminants Under Field Management Scenarios," *Waste Management* **25:** 463–472 (2005)].

Tools are available to help develop solid waste management plans. Several tools are available in Europe that are specific to the local energy grid mix, regulations, and typical process units for material and discards management. In the United States, a decision support tool (DST) is available for use by solid waste planners at state and local levels to analyze and compare MSW management strategies with respect to greenhouse gases (GHGs), cost, energy use and production, and environmental releases to air, land, and water. These strategies can be applied to source reduction, waste collection and transportation, materials recovery facilities, transfer stations, compost facilities, combustion and refuse-derived fuel facilities, and landfills. The tool is referred to as the MSW DST and is used by academics, consultants, the federal government and military, industries, nongovernmental organizations, and state and local governments. In academia, the tool is used to teach across multiple disciplines and curricula, including waste and materials management, industrial ecology, environmental engineering, and life-cycle assessment. Communities are using the tool to benchmark current practices and track environmental improvements (e.g., energy savings and GHG and criteria emissions reductions) achieved by optimizing MSW as a resource.

[Kaplan, P. O., S. R. Ranjithan, and M. A. Barlaz, "Use of Life Cycle Analysis to Support Solid Waste Management Planning for Delaware," *Environ. Sci. Technol.* **43**(5): 1264–1270 (2009); Kaplan, P. O., J. DeCarolis, and S. Thorneloe, "Is It Better to Burn or Bury Waste For Clean Electricity Generation?" *Environ. Sci. Technol.* **43**(6): 1711–1717 (2009); Thorneloe, S. A., K. Weitz, and J. Jambeck, "Application of the U.S. Decision Support Tool for Materials and Waste Management," *Waste Management* **27:** 1006–1020 (2007); Jambeck, J., K. A. Weitz, H. Solo-Gabriele, T. Townsend, and S. Thorneloe, "CCA-Treated Wood Disposed in Landfills and Life-Cycle Trade-Offs with Waste-to-Energy and MSW Landfill Disposal," *Waste Manag.* **27**(8): S21–S28 (2007); Kaplan, P. O., M. A. Barlaz, and S. R. Ranjithan, "A Procedure for Life-Cycle-Based Solid Waste Management with Consideration of Uncertainty," *J. Ind. Ecol.* **8**(4): 155–172 (2004); Weitz K. A., S. A. Thorneloe, S. R. Nishtala, S. Yarkosky, and M. Zannes, "The Impact of Municipal Solid Waste Management on Greenhouse Gas Emissions in the United States," *J. Air Waste Manag. Assoc.* **52:** 1000–1011 (2002).]

The MSW DST can be used to optimize the system given constraints (e.g., determine the most cost-effective strategy for reaching specific policy goals, such as diverting 40 percent landfill waste). The current version of the MSW DST is available through a project web site maintained by Research Triangle Institute, who are stewards of the tool [Thorneloe, S. A., K. Weitz, and J. Jambeck, "Application of the U.S. decision support tool for materials and waste management," *Waste Manag.* **27:** 1006–1020 (2007)]. A second generation of the tool is under development and includes a multivariate optimization algorithm to facilitate dynamic analysis, reflecting the evolution of the MSW composition and management practices over time. Updates have been made to the user interface based on feedback from the current user group, and it includes the ability to answer community-specific questions. Plans also include the use of advanced visualization and interpretation features for use in communicating results with stakeholders using either customized or standardized reporting.

Another tool available is the waste reduction model (WARM) to help solid waste planners and organizations track and voluntarily report greenhouse gas (GHG) emissions reductions from several different waste management practices. WARM calculates estimates of GHG emissions of baseline and alternative U.S. waste management practices—source reduction, recycling, anaerobic digestion, combustion, composting, and landfilling. The tool is web based and is available online.

In addition to planning for MSW management, it is important to plan for potential homeland security incidents or natural disasters with waste streams that typically are not handled by communities or waste management facilities. The U.S. EPA encourages preincident waste management planning and has developed guidance and other resources that identifies: important considerations to address, tools to support planning efforts, and waste management options to increase resiliency following an event. Figure 22-60 depicts the recommended process to guide emergency

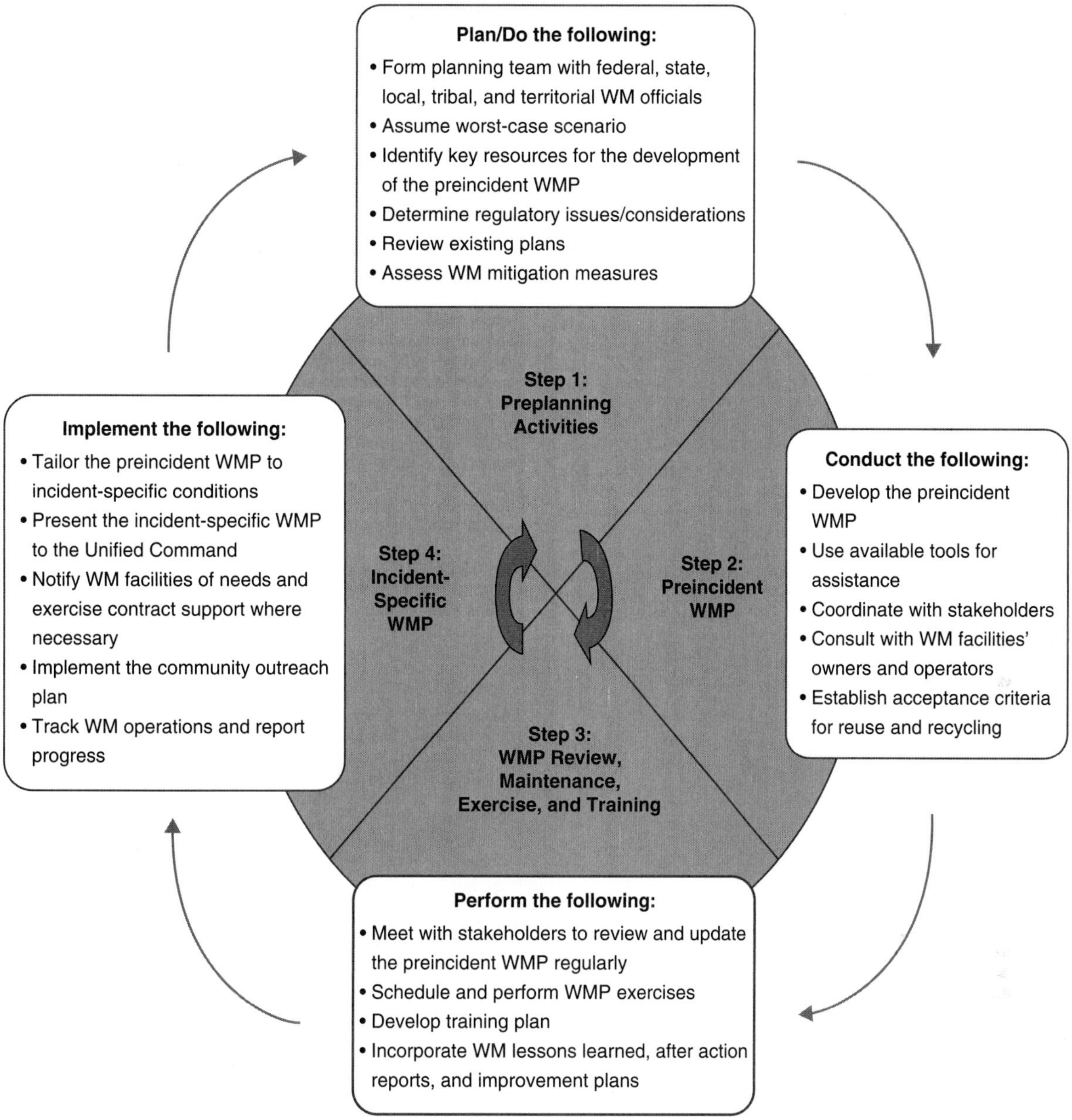
Plan/Do the following:
• Form planning team with federal, state, local, tribal, and territorial WM officials
• Assume worst-case scenario
• Identify key resources for the development of the preincident WMP
• Determine regulatory issues/considerations
• Review existing plans
• Assess WM mitigation measures
Step 1: Preplanning Activities
Conduct the following:
• Develop the preincident WMP
• Use available tools for assistance
• Coordinate with stakeholders
• Consult with WM facilities' owners and operators
• Establish acceptance criteria for reuse and recycling
Step 2: Preincident WMP
Step 3: WMP Review, Maintenance, Exercise, and Training
Perform the following:
• Meet with stakeholders to review and update the preincident WMP regularly
• Schedule and perform WMP exercises
• Develop training plan
• Incorporate WM lessons learned, after action reports, and improvement plans
Step 4: Incident-Specific WMP
Implement the following:
• Tailor the preincident WMP to incident-specific conditions
• Present the incident-specific WMP to the Unified Command
• Notify WM facilities of needs and exercise contract support where necessary
• Implement the community outreach plan
• Track WM operations and report progress

FIG. 22-60 All-hazards waste management planning process. (*U.S. EPA. Waste Management Benefits, Planning and Mitigation Activities for Homeland Security Incidents. Preincident All-Hazards Waste Management Planning Process. https://www.epa.gov/homeland-security-waste/waste-management-benefits-planning-and-mitigation-activities-homeland#planning, accessed November 9, 2016.*)

managers and planners through four steps that cover the initiation, creation, updating, and implementation of a waste management plan. [U.S. EPA, Managing Materials and Wastes for Homeland Security Incidents web site. URL: https://www.epa.gov/homeland-security-waste. Accessed October 24, 2016.] If the incident involves radioactive waste, the Waste Estimation Support Tool (WEST), a GIS-based decision support tool, can be useful for estimating the characteristics, amount, and residual radioactivity of waste generated from remediation and cleanup activities after a radiological incident, including those involving radiological dispersal devices, improvised nuclear devices, and nuclear power plants.

At a minimum, solid waste management plans should consider the hierarchy from nonhazardous to the hazardous, widespread results of a national incident. However, more sustainable approaches to solid waste management can be accomplished through the use of material and energy balances using life-cycle analysis and systems analysis. Using available tools, material and energy flows associated with waste can be calculated as a first step in achieving circular economy. The goal is to identify holistic solutions that optimize resource and energy recovery from solid waste, offset depleting resources, and move us toward more sustainable materials management.

Section 23

Process Safety

Daniel A. Crowl, Ph.D., CCPSC *AIChE/CCPS Staff Consultant; Adjunct Professor, University of Utah; Professor Emeritus of Chemical Engineering, Michigan Technological University; Fellow, American Institute of Chemical Engineers; Fellow, AIChE Center for Chemical Process Safety (Section Coeditor, Process Safety, Introduction)*

Robert W. Johnson, M.S.Ch.E. *President, Unwin Company; Fellow, American Institute of Chemical Engineers (Section Coeditor, Process Safety, Chemical Reactivity, Storage and Handling of Hazardous Materials)*

John Alderman, M.S., P.E., C.S.P. *Managing Partner, Hazard and Risk Analysis, LLC (Electrical Area Classification, Fire Protection Systems)*

Paul Amyotte, Ph.D., P.Eng. *Professor of Chemical Engineering and C.D. Howe Chair in Process Safety, Dalhousie University; Fellow, Chemical Institute of Canada; Fellow, Canadian Academy of Engineering (Dust Explosions)*

Ray Bennett, Ph.D., P.E., CEFEI *Senior Principal Engineer, Baker Engineering and Risk Consultants, Inc.; Member, American Petroleum Institute 752, 753, and 756 (Estimation of Damage Effects)*

Laurence G. Britton, Ph.D. *Process Safety Consultant; Fellow, American Institute of Chemical Engineers; Fellow, Energy Institute; Member, Institute of Physics (U.K.) (Flame Arresters)*

J. Wayne Chastain, B.S., P.E., CCPSC *Engineering Associate, Eastman Chemical Company; Member, American Institute of Chemical Engineers (Layer of Protection Analysis)*

Martin P. Clouthier, M.Sc., P.Eng. *Director, Jensen Hughes Consulting Canada Ltd. (Dust Explosions)*

Michael Davies, Ph.D. *President and CEO, Braunschweiger Flammenfilter GmbH (PROTEGO), Member, American Institute of Chemical Engineers; Member, National Fire Protection Association (Flame Arresters)*

Arthur M. Dowell, III, P.E., B.S. *President, A M Dowell III PLLC; Fellow, American Institute of Chemical Engineers; Senior Member, Instrumentation, Systems and Automation Society (Risk Analysis)*

Hans K. Fauske, D.Sc. *Emeritus President and Regent Advisor, Fauske and Associates, LLC; Fellow, American Institute of Chemical Engineers; Fellow, American Nuclear Society; Member, National Academy of Engineering (Pressure Relief Systems)*

Walter L. Frank, B.S., P.E., CCPSC *President, Frank Risk Solutions, Inc.; AIChE/CCPS Staff Consultant; Fellow, American Institute of Chemical Engineers; Fellow, AIChE Center for Chemical Process Safety (Hazards of Vacuum, Hazards of Inerts)*

Dennis C. Hendershot, M.S. *Process Safety Consultant; Fellow, American Institute of Chemical Engineers (Inherently Safer Design and Related Concepts, Hazard Analysis, Key Procedures)*

Chad V. Mashuga, Ph.D., P.E. *Assistant Professor of Chemical Engineering, Texas A&M University (Flammability, Combustion and Flammability Hazards, Explosions, Vapor Cloud Explosions, Boiling-Liquid Expanding-Vapor Explosions)*

Georges A. Melhem, Ph.D. *President and CEO, IoMosaic; Fellow, American Institute of Chemical Engineers (Emergency Relief Device Effluent Collection and Handling)*

David A. Moore, B.Sc., M.B.A., P.E., C.S.P. *President, AcuTech Consulting Group; Member, ASSE, ASIS, NFPA (Security)*

Thomas H. Pratt, Ph.D., P.E., C.S.P. *Retired; Emeritus Member, NFPA 77 (Static Electricity)*

Richard W. Prugh, M.S., P.E., C.S.P. *Principal Process Safety Consultant, Chilworth Technology, Inc., a Dekra Company; Fellow, American Institute of Chemical Engineers; Member, National Fire Protection Association (Toxicity)*

Thomas O. Spicer III, Ph.D., P.E. *Professor; Maurice E. Barker Chair in Chemical Engineering, Chemical Hazards Research Center Director, Ralph E. Martin Department of Chemical Engineering, University of Arkansas; Fellow, American Institute of Chemical Engineers (Atmospheric Dispersion)*

Angela Summers, Ph.D., P.E. *President, SIS-TECH; Adjunct Professor, Department of Environmental Management, University of Houston–Clear Lake; Fellow, International Society of Automation; Fellow, American Institute of Chemical Engineers; Fellow, AIChE Center for Chemical Process Safety (Safety Instrumented Systems)*

Ronald J. Willey, Ph.D., P.E. *Professor, Department of Chemical Engineering, Northeastern University; Fellow, American Institute of Chemical Engineers (Case Histories)*

John L. Woodward, Ph.D. *Senior Principal Consultant, Baker Engineering and Risk Consultants, Inc.; Fellow, American Institute of Chemical Engineers (Discharge Rates from Punctured Lines and Vessels)*

INTRODUCTION TO PROCESS SAFETY

CASE HISTORIES

HAZARDOUS MATERIALS AND CONDITIONS

INHERENTLY SAFER DESIGN AND RELATED CONCEPTS

PROCESS SAFETY ANALYSIS

CONSEQUENCE MODELS

SAFETY EQUIPMENT, PROCESS DESIGN, AND OPERATION

INTRODUCTION TO PROCESS SAFETY

General References: AIChE/CCPS, *Guidelines for Chemical Process Quantitative Risk Analysis,* 2d ed., American Institute of Chemical Engineers, New York, 2000; AIChE/CCPS, *Guidelines for Hazards Evaluation Procedures,* 3d ed., American Institute of Chemical Engineers, New York, 2008; AIChE/CCPS, *Guidelines for Risk Based Process Safety,* American Institute of Chemical Engineers, New York, 2007; AIChE/CCPS *Layer of Protection Analysis: Simplified Process Risk Assessment,* American Institute of Chemical Engineers, 2001; Crowl and Louvar, *Chemical Process Safety: Fundamentals with Applications,* 3d ed., Prentice-Hall, Englewood Cliffs, N.J., 2012; Mannan, ed., *Lees' Loss Prevention in the Process Industries,* 4th ed., Elsevier, Amsterdam, 2012.

Chemical plant accidents, more precisely called incidents, have resulted in loss of life and injury, damage to the environment, loss of capital equipment and inventory, loss of public confidence in the chemical industry, enactment of new regulations, and permanent changes to the practice of chemical engineering.

The term *loss prevention* can be applied in any industry, but it is widely used in the process industries, where it usually means the same as *process safety.*

Chemical plants, and other industrial facilities, may contain large quantities of hazardous materials. The materials may be hazardous due to toxicity, reactivity, flammability, or explosivity. A chemical plant may also contain large amounts of energy—the energy either is required to process the materials or is contained in the materials themselves. An *incident* occurs when control of this material or energy is lost. An *incident* is defined as an unplanned event leading to undesired consequences. The consequences might include injury to people, damage to the environment, loss of inventory and production, or damage to equipment.

A *hazard* is defined as a chemical or physical condition that has the potential for causing damage to people, property, or the environment (AIChE/CCPS, *Guidelines for Chemical Process Quantitative Risk Analysis,* 3d ed., American Institute of Chemical Engineers, New York, 2008, p. 6). Hazards exist in a chemical plant due to the nature of the materials processed or due to the physical conditions under which the materials are processed, such as high pressure or temperature. These hazards are present most of the time. An initiating event is required to begin the incident process. Once initiated, the incident follows a sequence of steps, called the event sequence, that results in an incident outcome. The consequences of the incident are the resulting effects of the incident. For instance, a rupture in a pipeline due to corrosion (initiating event) results in leakage of a flammable liquid from the process. The liquid evaporates and mixes with air to form a flammable cloud, which finds an ignition source (event sequence), resulting in a fire (incident outcome). The consequences of the incident are considerable fire damage and loss of production.

Risk is defined as a measure of human injury, environmental damage, or economic loss in terms of both the incident likelihood (probability) and the magnitude of the loss or injury (consequence) (AIChE/CCPS, *Guidelines for Chemical Process Quantitative Risk Analysis,* 2d ed., American Institute of Chemical Engineers, New York, 2000, pp. 5–6). It is important that both likelihood and consequence be included in risk. For instance, seat belt use is based on a reduction in the consequences of an incident. However, many people argue against seat belts based on probabilities, which is an incorrect application of the risk concept.

Table 23-1 shows a hierarchy of safety program levels. This hierarchy can be used for process safety, laboratory safety, or any safety program. The lowest level, 0, corresponds to no safety program or even disdain for safety. Level 1 is a reactive program, where incidents cause people to make changes. This safety program will only move forward by a series of incidents. Level 2 is a safety program run by rules and regulations. The problem, of course, is that there are never enough rules and regulations to cover all potential situations. Level 3 has management systems, such as hot work permits, to manage safety. Level 4 contains performance monitoring to drive improvements. Finally, Level 5 is an adapting program, where safety is a core value driving success. One cannot skip levels in the development of a safety program; the program must work through successive levels.

TABLE 23-1 Hierarchy of Safety Programs

Safety level	Name	Description
5	Adapting	Safety is a core value of the organization and a primary driver for successful enterprise
4	Performing	Performance monitoring using metrics to drive continuous improvement
3	Managing	Management systems such as process safety management, lockout/tagout, etc.
2	Complying	Complying with rules and regulations
1	Reacting	Reacting to incidents as they occur
0	Neglecting	Maybe even disdain for safety

Up until the late 1990s, the primary approach to process safety was through hazards identification and risk assessment (HIRA). This procedure is shown in Fig. 23-1. The procedure begins with a complete description of the process. This includes detailed process flow diagrams (PFDs) and piping and instrumentation diagrams (P&IDs), complete specifications on all equipment, maintenance records, operating procedures, and so forth. A hazard identification procedure is then selected (see Hazard Analysis subsection) to identify the hazards and their nature. This is followed by identification of all potential event sequences and potential incidents (scenarios) that can result in loss of control of energy or material. Next is an evaluation of both the consequences and the probability. The consequences are estimated by using source models (to describe the release of material and energy) coupled with a consequence model to describe the incident outcome. The consequence models include dispersion, fire, and explosion modeling. The results of the consequence models are used to estimate the impacts on people, environment, and property. The incident probability is estimated by using fault trees or generic databases for the initial event sequences. Event trees may be used to account for mitigation and post-release incidents. Finally, the risk is estimated by combining the potential consequence for each event with the event frequency and summing over all events.

Once the risk is determined, a decision must be made on risk acceptance. This can be done by comparison to a relative or absolute standard. If the risk is acceptable, then the decision is made to build or operate the process. If the risk is not acceptable, then something must be changed. This could

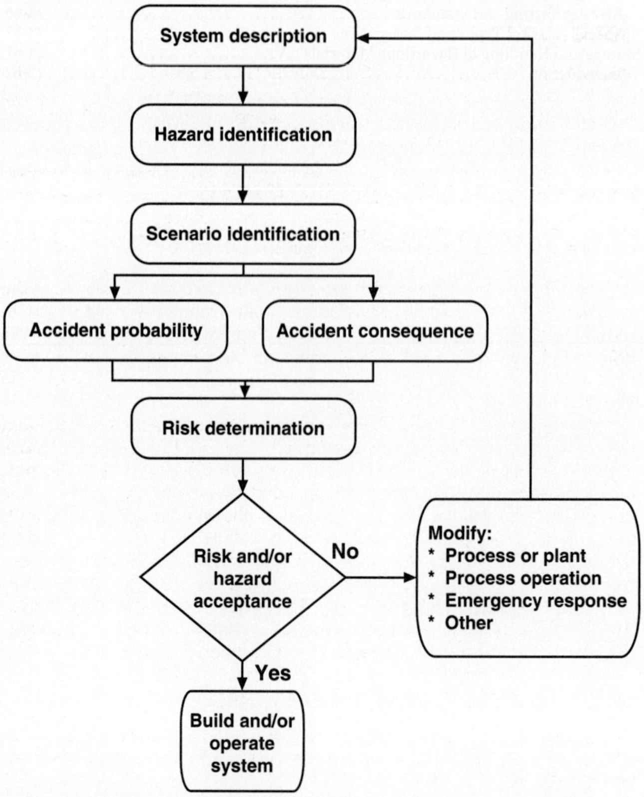

FIG. 23-1 The hazard identification and risk assessment procedure. [*Guidelines for Hazards Evaluation Procedures,* Center for Chemical Process Safety (CCPS) of the American Institute of Chemical Engineers (AIChE); copyright 1985 AIChE and reproduced with permission.]

include the process design, the operation, maintenance, or additional layers of protection that might be added.

The HIRA procedure requires the application of many complex mathematical models to estimate consequences, and it may be beyond the capability of many practitioners.

A more recent procedure is called LOPA, for Layer of Protection Analysis (AIChE/CCPS, *Layer of Protection Analysis: Simplified Process Risk Assessment*, American Institute of Chemical Engineers, 2001). LOPA is an order-of-magnitude quantitative method for analyzing and assessing risk. The approach uses simplified methods to characterize the consequences and estimate the frequencies of incidents. Layers of protection are then added to the process. The protection layers may include inherently safer design, a basic process control system, passive devices such as blast walls, active devices such as relief valves, and even human intervention. The purpose of LOPA is to add layers of protection to reduce the risk to tolerable levels.

Identifying the hazards and the consequences and estimating the risk is only a small part of a comprehensive risk-based process safety management system. In 2007, the American Institute of Chemical Engineers (AIChE) published a new guidelines book on risk-based process safety management (AIChE/CCPS, *Guidelines for Risk Based Process Safety,* American Institute of Chemical Engineers, New York, 2007). This book lists 20 elements of risk-based process safety organized into four major categories. The four categories and 20 elements are shown in Table 23-2.

TABLE 23-2 The 20 Elements of Risk-Based Process Safety Arranged in Four Major Categories

Commit to Process Safety
1. Process safety culture
2. Compliance with standards
3. Process safety competency
4. Workforce involvement
5. Stakeholder outreach
Understand Hazards and Risks
6. Process knowledge management
7. Hazard identification and risk analysis
Manage Risk
8. Operating procedures
9. Safe work practices
10. Asset integrity and reliability
11. Contractor management
12. Training and performance assurance
13. Management of change
14. Operational readiness
15. Conduct of operations
16. Emergency management
Learn from Experience
17. Incident investigation
18. Measurement and metrics
19. Auditing
20. Management review and continuous improvement

AIChE/CCPS, *Guidelines for Risk Based Process Safety*, American Institute of Chemical Engineers, New York, 2007. Reprinted by permission.

CASE HISTORIES

General References Kletz, T. A., *Learning from Accidents*, 3d ed., Gulf Professional Publishing, Houston, 2001; Kletz, T. A., *What Went Wrong? Fifth Edition: Case Histories of Process Plant Disasters and How They Could Have Been Avoided*, Butterworth-Heinemann, Oxford, UK, 2009; Sanders, R. E., *Chemical Process Safety: Learning from Case Histories*, 4th ed., Elsevier Science, Amsterdam, 2015; Urben, P. G., ed., *Bretherick's Handbook of Reactive Chemical Hazards*, 7th ed., Academic Press, New York, 2006.

INTRODUCTION

Summarized here are several major incidents that have occurred over the past 20 years. These incidents were selected based on their diversity and their severity. The last one involved nitrogen asphyxiation, which is the leading cause of death within the chemical process industry.

HYDROCARBON FIRES AND EXPLOSIONS

Explosion and fires at the Texaco refinery, Milford Haven, Wales, July 24, 1994. Reference: HSE Books, Her Majesty's Stationary Office, Norwich, UK 1997.

On July 24, 1994, an explosion followed by a number of fires occurred at the Texaco refinery in Milford Haven, Wales, Great Britain. Prior to this explosion, a severe electrical storm caused plant disturbances that affected many unit operations. With a calculated explosion blast energy equivalent of at least 4 tons of TNT, significant portions of the refinery were damaged. No fatalities occurred.

The root cause was traced back to alarm overload (human factors) and key sensors not working when demanded. Due to a process disturbance, liquid was continuously pumped into a process vessel with a closed outlet valve. The control system indicated that this valve was open. As the unit overfilled, the only means of exit was a relief system designed only for vapor. When the liquid reached the relief system, its momentum ripped the ductwork apart. A massive release of hydrocarbons went into the environment. At the time, operating personnel were responding to 275 alarms, of which 80 percent had high priority. This incident also reminds the practitioner of the value of safety instrumented systems, which can bring out-of-control operation to a safe state without human intervention.

UNCONTROLLED CHEMICAL REACTIONS

T2 Laboratories Inc. reactive chemical explosion, Jacksonville, Fla., December 19, 2007, 4 deaths, 32 injured. Reference: http://www.csb.gov/t2-laboratories-inc-reactive-chemical-explosion/

On December 19, 2007, an explosion and subsequent fire jolted an industrial area north of Jacksonville, Florida. At 1400 lb of TNT equivalent, it was the largest explosion that the U.S. Chemical Safety and Hazard Identification Board (CSB) had investigated up to July 26, 2015. The process involved in the explosion began with a series of chemical reactions that converted metallic sodium and methylcyclopentadiene into a gasoline additive used for boosting octane. The stepwise reaction was completed in a batch reactor, a 2400-gal autoclave with 3-inch-thick walls located in an open outdoor environment, at about 2000 gallons per batch. A control room was 100 feet away. The first step involved placing metallic sodium (in mineral oil) into the reactor along with methylcyclopentadiene reactant and the solvent diethylene glycol dimethyl ether (diglyme). The overall reaction is exothermic and needs to be initiated at 150°C. Once initiated, hydrogen is generated. The hydrogen in this case exited the reactor to the surroundings. The jacketed reactor operated at an internal pressure of about 3 bar. It was fitted with a pressure relief system set to open at 27 bar. This difference is a design flaw. Normally, relief systems should be designed to open at 10 percent above the anticipated normal working pressure conditions within reactors. This may be below the maximum allowable working pressure (MAWP) of the vessel; however, if a vessel is allowed to build up unnecessary pressure to the MAWP before relieving, the risk of an explosion is increased. If the relief had occurred at 5 bar, the vessel would not have ruptured. Neither the chemist nor the chemical engineer were aware of the hazards associated with the solvent diglyme, which can decompose violently at temperatures above 230°C (Urben, P. G., ed., *Bretherick's Handbook of Reactive Chemical Hazards,* 7th ed., Academic Press, New York, 2006). The resultant overpressurization was approximately 10 times the amount expected due to this secondary decomposition.

Another chemical deserving respect is ammonium nitrate. Two major incidents involving ammonium nitrate occurred in 2013 and 2015. On April 17, 2013, twelve emergency responders and three members of the public lost their lives as a result of an explosion at the West Fertilizer Company, located in West, Texas (https://www.csb.gov/west-fertilizer-explosion-and-fire-/). Several findings are relevant. The emergency responders had trained for dealing with an anhydrous ammonia scenario at the plant. They did not train for the more serious hazard related to storage of ammonium nitrate fertilizer. A fire broke out in a wooden structure used to store ammonium nitrate. Under the right conditions, such as exposure to high temperature along with a shock wave source, fertilizer grade ammonium nitrate (FGAN) can detonate, creating a significant explosion. The proximity of homes and schools to the storage facility further exacerbated the impacts of this incident.

A more severe ammonium nitrate explosion occurred in Tianjin, China, on August 12, 2015, that killed 165 people and injured over 800. For details, see http://www.nytimes.com/2016/02/06/world/asia/tianjin-explosions-were-result-of-mismanagement-china-finds.html?_r=0.

DUST EXPLOSIONS

Dust explosion February 7, 2008. Reference: CSB investigations, http://www.csb.gov/imperial-sugar-company-dust-explosion-and-fire/

On February 7, 2008, a huge dust explosion and fire at an Imperial Sugar processing plant in Port Wentworth, Georgia, killed 14 workers and injured 38 others. Massive amounts of combustible sugar dust had accumulated around the packaging area. The sugar granules were transported by conveyors and bucket elevators. Dust particles fell to the sides, accumulating on supporting structures and flat surfaces. A modification was made where one of the conveyor belts was covered by steel panels. This environment created an airborne dust concentration within the ignitable region along with an ignition source (an overheated bearing was suspected) that created the first explosion. A series of severe secondary explosions followed. Housekeeping (dust control) is critical for plants that deal with solids handling.

TOXIC RELEASES

Ruptured chlorine transfer hose. Reference: http://www.csb.gov/csb-safety-advisory-calls-for-validation-of-chlorine-transfer-hoses/

On August 14, 2002, a one-inch chlorine transfer hose, which was used in a railcar offloading operation in Missouri, catastrophically ruptured and initiated a sequence of events that led to the release of 48,000 lb of chlorine gas into neighboring areas. The root cause was that the wrong material was used to make the ruptured hose. The distributor fabricated transfer hoses with Schedule 80 Monel 400 end fittings that contained high-density polyethylene spiral guards. Two new hoses were put into service on June 15, 2002. One of the hoses failed when chlorine interacted with the polyethylene.

Most plastics react chemically with chlorine because of their hydrocarbon structural makeup. This reactivity is avoided if fluorine atoms are substituted into the hydrocarbon molecule. The Chlorine Institute recommends that hoses constructed with such an inner lining have a structural layer braid of polyvinylidene fluoride (PVDF) monofilament material or a structural braid of Hastelloy C-276. An underlying lesson here is material compatibility. Material compatibility tables exist for engineers to consult, including in other sections of this handbook. A second lesson is to manage changes and to complete a hazard analysis for any change from design intentions.

NITROGEN ASPHYXIATION

Incident at Union Carbide Corporation, Hahnville, Louisiana, March 27, 1998. Reference: http://www.csb.gov/union-carbide-corp-nitrogen-asphyxiation-incident/

On March 27, 1998, two workers at a Union Carbide Corporation plant in Hahnville, Louisiana, were overcome by nitrogen gas while performing a black light inspection at an open end of a 48-in pipe. One employee was killed and an independent contractor was seriously injured due to nitrogen asphyxiation. Nitrogen was being injected into a nearby reactor to prevent contamination of a catalyst by oxygen and related materials. The nitrogen also flowed through some of the piping systems connected to the reactors. No warning sign was posted on the pipe opening identifying it as a confined space. Nor was there a warning that the pipe contained potentially hazardous nitrogen.

Nitrogen asphyxiation has been a significant cause of occupational fatalities of plant personnel in the past 30 years. Such incidents occurred in a variety of facilities, including industrial plants, laboratories, and medical facilities. Almost half involve contractors. The Occupational Safety and Health Administration (OSHA) requires employers to maintain workplace oxygen at levels between 19.5 and 23.5 percent. The human body is adversely affected by lower concentrations. An atmosphere of only 4 to 6 percent oxygen causes the victim to fall into a coma in less than 40 seconds. Oxygen must be administered within minutes to offer a chance of survival. Failure to detect an oxygen-deficient (nitrogen-enriched) atmosphere has been a significant factor in several incidents. One of the first safety lessons taught to the author was in a paper mill in 1974: if a person went down in a confined space, *do not enter*, notify others immediately, and await people who were qualified with proper equipment to enter the confined space.

HAZARDOUS MATERIALS AND CONDITIONS

FLAMMABILITY

Nomenclature

K_G	Deflagration index for gases (bar · m/s)
K_{St}	Deflagration index for dusts (bar · m/s)
LFL	Lower flammability limit (vol% fuel in air)
LOC	Limiting oxygen concentration
n	Number of combustible species
P	Pressure
T	Temperature (°C)
t	Time (s)
UFL	Upper flammability limit (vol% fuel in air)
V	Vessel volume (m^3)
y_i	Mole fraction of component i on a combustible basis
z	Stoichiometric coefficient for oxygen
ΔH_c	Net heat of combustion (kcal/mol)

General References: Crowl, *Understanding Explosions*, American Institute of Chemical Engineers, New York, 2003; Crowl and Louvar, *Chemical Process Safety: Fundamentals with Applications*, 3d ed., Prentice-Hall, Upper Saddle River, N.J., 2011, Chaps. 6 and 7; Eckoff, *Dust Explosions in the Process Industries*, 3d ed., Elsevier Science, Burlington, Mass., 2003; Kinney and Graham, *Explosive Shocks in Air*, 2d ed., Springer-Verlag, New York, 1985; Lewis and von Elbe, *Combustion, Flames and Explosions of Gases*, 3d ed., Academic Press, New York, 1987; Mannan, *Lees' Loss Prevention in the Process Industries*, 4th ed., Elsevier, Amsterdam, 2012, Chap. 16: Fire, Chap. 17: Explosion.

Introduction Fire and explosions in chemical plants and refineries are rare, but when they do occur, they are very dramatic.

Accident statistics have shown that fires and explosions represent 97 percent of the largest accidents in the chemical industry (J. Coco, ed., *Large Property Damage Losses in the Hydrocarbon-Chemical Industry: A Thirty Year Review*, J. H. Marsh and McLennan, New York, 1997).

Prevention of fires and explosions requires

1. An understanding of the fundamentals of fires and explosions
2. Proper experimental characterization of flammable and explosive materials
3. Proper application of these concepts in the plant environment

The technology does exist to handle and process flammable and explosive materials safely, and to mitigate the effects of an explosion. The challenges to this problem are as follows:

1. Combustion behavior varies widely and is dependent on a wide range of parameters.
2. There is an incomplete fundamental understanding of fires and explosions. Predictive methods are still under development.
3. Fire and explosion properties are not fundamentally based and are an artifact of a particular experimental apparatus and procedure.
4. High-quality data from a standardized apparatus that produces consistent results are lacking.
5. The application of these concepts in a plant environment is difficult.

The Fire Triangle The fire triangle is shown in Fig. 23-2. It shows that a fire will result if fuel, oxidant, and an ignition source are present. In reality, the fuel and oxidant must be within certain concentration ranges, and the ignition source must be robust enough to initiate the fire. The fire triangle applies to gases, liquids, and solids. Liquids are volatized and solids decompose prior to combustion in the vapor phase. For dusts arising from solid materials, the

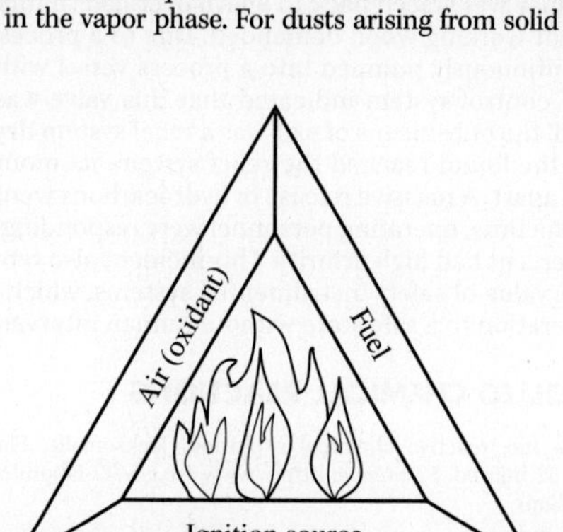

FIG. 23-2 The fire triangle showing the requirement for combustion of gases and vapors. [D. A. Crowl, *Understanding Explosions*, Center for Chemical Process Safety (CCPS) of the American Institute of Chemical Engineers (AIChE); copyright 2003 AIChE and reproduced with permission.]

particle size, distribution, and suspension in the gas are also important parameters in the combustion—these are sometimes included in the fire triangle.

The usual oxidizer in the fire triangle is oxygen from air. However, gases such as fluorine and chlorine, liquids such as peroxides and chlorates, and solids such as ammonium nitrate and some metals can serve as oxidizers. Exothermic decomposition without oxygen is also possible—for example, with ethylene oxide or acetylene.

Ignition arises from a wide variety of sources, including static electricity, hot surfaces, sparks, open flames, and electric circuits. Ignition sources are elusive and difficult to eliminate entirely, although efforts should always be made to reduce them.

If any side of the fire triangle is removed, a fire will not result. In the past, the most common method of fire control was the elimination of ignition sources. However, experience has shown that this is not robust enough. Current fire control prevention methods continue with the elimination of ignition sources while focusing efforts more strongly on preventing flammable mixtures.

Definition of Terms The following are terms necessary to characterize fires and explosions (Crowl and Louvar, *Chemical Process Safety: Fundamentals with Applications,* 3d ed., Prentice-Hall, Upper Saddle River, N.J., 2011, pp. 227–229).

Autoignition temperature (AIT) This is a fixed temperature above which adequate energy is available in the environment to provide an ignition source.

Boiling-liquid expanding-vapor explosion (BLEVE) A BLEVE occurs if a vessel that contains a liquid at a temperature above its atmospheric pressure boiling point ruptures. The subsequent BLEVE is the explosive vaporization of a large fraction of the vessel contents, possibly followed by combustion or explosion of the vaporized cloud if it is combustible. This type of explosion occurs when an external fire heats the contents of a tank of volatile material. As the tank contents heat, the vapor pressure of the liquid within the tank increases, and the tank's structural integrity is reduced because of the heating. If the tank ruptures, the hot liquid volatilizes explosively.

Combustion or fire Combustion or fire is a chemical reaction in which a substance combines with an oxidant and releases energy. Part of the energy released is used to sustain the reaction.

Confined explosion This explosion occurs within a vessel or a building.

Deflagration In this explosion, the reaction front moves at a speed less than the speed of sound in the unreacted medium.

Detonation In this explosion, the reaction front moves at a speed greater than the speed of sound in the unreacted medium.

Dust explosion This explosion results from the rapid combustion of fine solid particles. Many solid materials (including common metals such as iron and aluminum) and natural materials (including sugar, grain dust, and sawdust) become flammable when reduced to a fine powder and suspended in air.

Explosion An explosion is a rapid expansion of gases resulting in a rapidly moving pressure or shock wave. The expansion can be mechanical (by means of a sudden rupture of a pressurized vessel), or it can be the result of a rapid chemical reaction. Explosion damage is caused by the pressure or shock wave.

Fire point The fire point is the lowest temperature at which a vapor above a liquid will continue to burn once ignited; the fire point temperature is higher than the flash point.

Flammability limits Vapor–air mixtures will ignite and burn only over a well-specified range of compositions. The mixture will not burn when the composition is lower than the lower flammable limit (LFL); the mixture is too lean for combustion. The mixture is also not combustible when the composition is too rich, that is, when it is above the upper flammable limit (UFL). A mixture is flammable only when the composition is between the LFL and the UFL. Commonly used units are volume percent of fuel (percentage of fuel plus air).

Lower explosion limit (LEL) and upper explosion limit (UEL) are used interchangeably with LFL and UFL.

Flash point (FP) The flash point of a liquid is the lowest temperature at which it gives off enough vapor to form an ignitable mixture with air. At the flash point, the vapor will burn but only briefly; inadequate vapor is produced to maintain combustion. The flash point generally increases with increasing pressure.

There are several different experimental methods used to determine flash points. Each method produces a somewhat different value. The two most commonly used methods are open cup and closed cup, depending on the physical configuration of the experimental equipment. The open-cup flash point is a few degrees higher than the closed-cup flash point.

Ignition Ignition of a flammable mixture may be caused by a flammable mixture coming into contact with a source of ignition with sufficient energy or by the gas reaching a temperature high enough to cause the gas to autoignite.

Mechanical explosion A mechanical explosion results from the sudden failure of a vessel containing high-pressure, nonreactive gas.

Minimum ignition energy This is the minimum energy input required to initiate combustion.

Overpressure The pressure over ambient that results from an explosion.

Shock wave This is an abrupt pressure wave moving through a gas. A shock wave in open air is followed by a strong wind; the combined shock wave and wind is called a blast wave. The pressure increase in the shock wave is so rapid that the process is mostly adiabatic.

Unconfined explosion Unconfined explosions occur in the open. This type of explosion is usually the result of a flammable gas spill. The gas is dispersed and mixed with air until it comes in contact with an ignition source. Unconfined explosions are rarer than confined explosions because the explosive material is often diluted below the LFL by wind dispersion. These explosions are destructive because large quantities of gas and large areas are often involved.

Figure 23-3 is a plot of concentration versus temperature, and it shows how several of these definitions are related. The exponential curve in Fig. 23-3 represents the saturation vapor pressure curve for the liquid material. Typically, the UFL increases and the LFL decreases with temperature. The LFL theoretically intersects the saturation vapor pressure curve at the flash point, although experimental data are not always consistent. The autoignition temperature is actually the lowest temperature of an autoignition region. The behavior of the autoignition region and the flammability limits at higher temperatures are not well understood.

The flash point and flammability limits are not fundamental properties but are defined only by the specific experimental apparatus and procedure used.

Table 2-153 provides flammability data for a number of compounds.

Combustion and Flammability Hazards

Vapor Mixtures Often, flammability data are required for vapor mixtures. The flammability limits for the mixture can be estimated using LeChatelier's rule [LeChatelier, "Estimation of Firedamp by Flammability Limits," *Ann. Mines* (1891), ser. 8, **19:** 388–395, with translation in *Process Saf. Prog.* **23**(3): 172 (2004)].

$$\mathrm{LFL}_{\mathrm{mix}} = \frac{1}{\sum_{i=1}^{n} y_i/\mathrm{LFL}_i} \qquad (23\text{-}1)$$

where LFL_i = lower flammability limit for component i (in vol%)
y_i = mole fraction of component i on a combustible basis
n = number of combustible species

Note that Eq. (23-1) is only applied to the combustible species, and the mole fraction is computed using only the combustible species.

LeChatelier's rule is empirically derived and is not universally applicable. Mashuga and Crowl ["Derivation of LeChatelier's Mixing Rule for Flammable Limits," *Process Saf. Prog.* **19**(2): 112–118 (2000)] determined that the following assumptions are present in LeChatelier's rule:

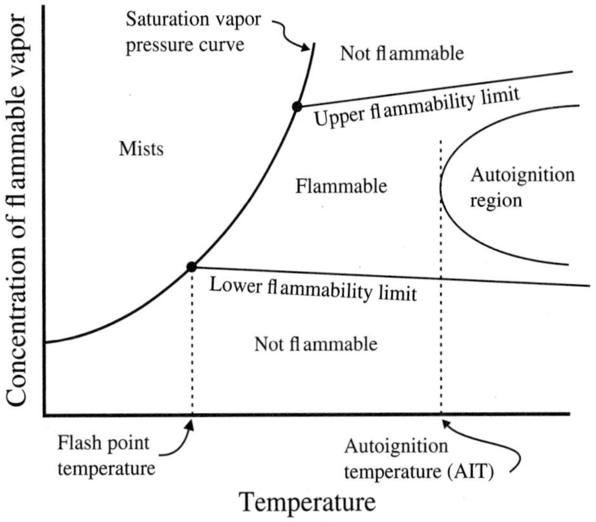

FIG. 23-3 The relationship between the various flammability properties. (D. A. Crowl and J. F. Louvar, *Chemical Process Safety: Fundamentals with Applications,* 2d ed., © 2002. Adapted by permission of Pearson Education, Inc., Upper Saddle River, N.J.)

1. The product heat capacities are constant.
2. The number of moles of gas is constant.
3. The combustion kinetics of the pure species is independent of and unchanged by the presence of other combustible species.
4. The adiabatic temperature rise at the flammability limit is the same for all species.

These assumptions were found to be reasonably valid at the LFL and less so at the UFL, even though Eq. (23-1) can be written for the UFL.

Liquid Mixtures Flash point temperatures for liquid mixtures can be estimated if only one component is flammable and the flash point temperature of the flammable component is known. In this case, the flash point temperature is estimated by determining the temperature at which the vapor pressure of the flammable component in the mixture is equal to the pure component vapor pressure at its flash point. Estimation of flash point temperatures for mixtures of several flammable components can be done by a similar procedure, but it is recommended that the flash point temperatures be measured experimentally.

Flammability Limit Dependence on Temperature In general, as the temperature increases, the flammability range widens, that is, the LFL decreases and the UFL increases. Zabetakis et al. (Zabetakis, Lambiris, and Scott, "Flame Temperatures of Limit Mixtures," *7th Symposium on Combustion,* Butterworths, London, 1959) derived the following empirical equations, which are approximate for many hydrocarbons:

$$\mathrm{LFT}_T = \mathrm{LFL}_{25} - \frac{0.75}{\Delta H_c}(T-25)$$
$$\mathrm{UFL}_T = \mathrm{UFL}_{25} + \frac{0.75}{\Delta H_c}(T-25) \tag{23-2}$$

where ΔH_c is the net heat of combustion (kcal/mol) and T is the temperature (°C).

Flammability Limit Dependence on Pressure Pressure has little effect on the LFL except at very low pressures (<50 mmHg absolute) where flames do not propagate (Zabetakis, "Fire and Explosion Hazards at Temperature and Pressure Extremes," *AIChE Inst. Chem. Engr. Symp.* ser. 2, American Institute of Chemical Engineers, New York, 1965, pp. 99–104).

The UFL increases as the pressure is increased. A very approximate equation for the change in UFL with pressure is available for some hydrocarbon gases (Zabetakis, 1965):

$$\mathrm{UFL}_P = \mathrm{UFL} + 20.6(\log P + 1) \tag{23-3}$$

where P is the pressure (megapascals absolute) and UFL is the upper flammability limit (vol% fuel in air at 1 atm).

Estimating Flammability Limits There are a number of very approximate methods available to estimate flammability limits. However, for critical safety values, experimental determination as close as possible to actual process conditions is always recommended.

Jones ["Inflammation Limits and Their Practical Application in Hazardous Industrial Operations," *Chem. Rev.* **22**(1): 1–26 (1938)] found that for many hydrocarbon vapors, the LFL and UFL can be estimated from the stoichiometric concentration of fuel:

$$\mathrm{LFL} = 0.55C_{st}$$
$$\mathrm{UFL} = 3.50C_{st} \tag{23-4}$$

where C_{st} is the stoichiometric fuel concentration (vol% fuel in air).

For a stoichiometric combustion equation of the form

$$(1)\, C_mH_xO_y + z\, O_2 \rightarrow m\, CO_2 + (x/2)\, H_2O \tag{23-5}$$

it follows that

$$z = m + \frac{x}{y} - \frac{y}{2} \tag{23-6}$$

and furthermore that

$$C_{st} = \frac{100}{1 + z/0.21} \tag{23-7}$$

Equation (23-7) can be used with (23-4) to estimate the LFL and UFL.

Suzuki ["Empirical Relationship between Lower Flammability Limits and Standard Enthalpies of Combustion of Organic Compounds," *Fire and Materials* **18**: 333–336 (1994); Suzuki and Koide, "Correlation between Upper Flammability Limits and Thermochemical Properties of Organic Compounds," *Fire and Materials* **18**: 393–397 (1994)] provides more detailed correlations for the UFL and LFL in terms of the heat of combustion.

Flammability limits can also be estimated by using calculated adiabatic flame temperatures and a chemical equilibrium program [Mashuga and Crowl, "Flammability Zone Prediction Using Calculated Adiabatic Flame Temperatures," *Process Saf. Prog.* **18**(3): 127–134 (1999)].

Limiting Oxygen Concentration (LOC) Below the limiting oxygen concentration it is not possible to support combustion, independent of the fuel concentration. The LOC is expressed in units of volume percent oxygen. The LOC is dependent on the pressure, the temperature, and the inert gas. Table 23-3 lists a number of LOCs, and it shows the change in LOC if carbon dioxide is the inert gas instead of nitrogen.

The LOC can be estimated for many hydrocarbons from

$$\mathrm{LOC} = z\,\mathrm{LFL} \tag{23-8}$$

where z is the stoichiometric coefficient for oxygen [see Eq. (23-5)] and LFL is the lower flammability limit.

Flammability Diagram Figure 23-4 shows a typical flammability diagram. Point A shows how the scales are oriented—at any point on the diagram, the concentrations must add up to 100 percent. Point A is 60 percent fuel, 20 percent oxygen, and 20 percent nitrogen. The air line represents all possible combinations of fuel and air—it intersects the nitrogen axis at 79 percent nitrogen, which is the percentage in air. The stoichiometric line represents all stoichiometric combinations of fuel and oxygen. If the combustion reaction is written according to Eq. (23-5), then the intersection of the stoichiometric line with the oxygen axis is given by

$$100\left(\frac{z}{1+z}\right) \tag{23-9}$$

The LFL and UFL are drawn on the air line from the fuel axis values.

TABLE 23-3 Limiting Oxygen Concentrations (Volume Percent Oxygen Concentrations above Which Combustion Can Occur)

Gas or vapor	N_2/Air	CO_2/Air
Methane	12	14.5
Ethane	11	13.5
Propane	11.5	14.5
n-Butane	12	14.5
Isobutane	12	15
n-Pentane	12	14.5
Isopentane	12	14.5
n-Hexane	12	14.5
n-Heptane	11.5	14.5
Ethylene	10	11.5
Propylene	11.5	14
1-Butene	11.5	14
Isobutylene	12	15
Butadiene	10.5	13
3-Methyl-1-butene	11.5	14
Benzene	11.4	14
Toluene	9.5	—
Styrene	9.0	—
Cyclopropane	11.5	14
Gasoline		
(73/100)	12	15
(100/130)	12	15
(115/145)	12	14.5
Kerosene	10 (150°C)	13 (150°C)
JP-1 fuel	10.5 (150°C)	14 (150°C)
Natural gas	12	14.5
Acetone	11.5	14
t-Butanol	NA	16.5 (150°C)
Carbon disulfide	5	7.5
Carbon monoxide	5.5	5.5
Ethanol	10.5	13
Ethyl ether	10.5	13
Hydrogen	5	5.2
Hydrogen sulfide	7.5	11.5
Isobutyl formate	12.5	15
Methanol	10	12
Methyl acetate	11	13.5

Data from NFPA 68, *Venting of Deflagrations* (Quincy, Mass.: National Fire Protection Association, 1994).

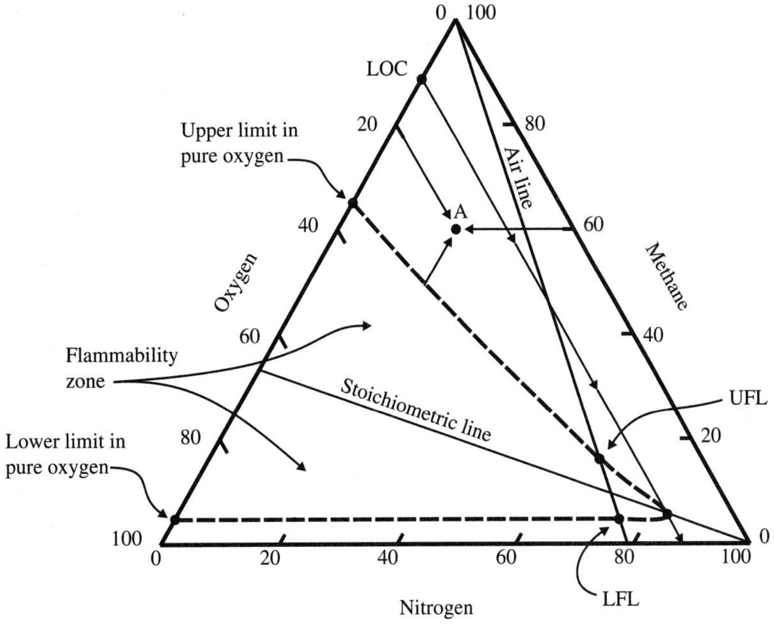

FIG. 23-4 Flammability diagram for methane at an initial temperature and pressure of 25°C and 1 atm. [C. V. Mashuga and D. A. Crowl, "Application of the Flammability Diagram for Evaluation of Fire and Explosion Hazards of Flammable Vapors," *Process Safety Progress,* vol. 17, no. 3; copyright 1998 American Institute of Chemical Engineers (AIChE) and reproduced with permission.]

A typical flammability zone for most hydrocarbon vapors is shown in Fig. 23-4. Any concentration within the flammability zone is defined as flammable.

The LOC is found where the lowest constant oxygen concentration intersects the nose of the flammability zone, as shown in Fig. 23-4.

Crowl (*Understanding Explosions*, American Institute of Chemical Engineers, New York, 2003, App. A) derived a number of rules for using flammability diagrams:

1. If two gas mixtures *R* and *S* are combined, the resulting mixture composition lies on a line connecting points *R* and *S* on the flammability diagram. The location of the final mixture on the straight line depends on the relative moles in the mixtures combined: If mixture *S* has more moles, the final mixture point will lie closer to point *S*. This is identical to the lever rule used for phase diagrams.

2. If a mixture *R* is continuously diluted with mixture *S*, the mixture composition follows along the straight line between points *R* and *S* on the flammability diagram. As the dilution continues, the mixture composition moves closer and closer to point *S*. Eventually, at infinite dilution the mixture composition is at point *S*.

3. For systems having composition points that fall on a straight line passing through an apex corresponding to one pure component, the other two components are present in a fixed ratio along the entire line length.

Figure 23-5 shows how nitrogen can be used to avoid the flammable zone when taking a vessel off-line for maintenance. In this case, nitrogen enters the vessel until a concentration is reached at point *S*. Then air can enter, arriving at point *R*. Figure 23-6 shows the reverse procedure when bringing a vessel on-line. Starting at point *A*, nitrogen is added until point *S* is reached. Then fuel is pumped in until point *R* is reached. In both cases, the flammable zone is avoided.

A complete flammability diagram requires hundreds of tests in a combustion sphere [Mashuga and Crowl, "Application of the Flammability Diagram for the Evaluation of Fire and Explosion Hazards of Flammable Vapors," *Process Saf. Prog.* **17**(3): 176–183 (1998)]. However, an approximate diagram can be drawn by using the LFL, UFL, LOC, and flammability limits in pure oxygen. The following procedure is used:

1. Draw the flammability limits in air as points on the air line, using the fuel axis values.

2. Draw the flammability limits in pure oxygen as points on the oxygen scale, using the fuel axis values. Table 23-4 provides a number of values for the flammability limits in pure oxygen. These are drawn on the oxygen axis using the fuel axis concentrations.

3. Use Eq. (23-9) to draw a point on the oxygen axis, and then draw the stoichiometric line from this point to the 100 percent nitrogen apex.

4. Locate the LOC on the oxygen axis. Draw a line parallel to the fuel axis until it intersects the stoichiometric line. Draw a point at the intersection.

5. Connect the points to estimate the flammability zone.

In reality, not all the data are available, so a reduced form of the preceding procedure is used to draw a partial diagram (Crowl, *Understanding Explosions*, American Institute of Chemical Engineers, New York, 2003, p. 27).

Ignition Sources and Energy Table 23-5 provides a list of the ignition sources for major fires. As seen in Table 23-5, ignition sources are very common and cannot be used as the only method of fire prevention.

The minimum ignition energy (MIE) is the minimum energy input required to initiate combustion. All flammable materials (gases, dusts, and aerosols) have an MIE. The MIE depends on the species, concentration, particle or droplet size, pressure, and temperature. A few MIEs are provided in Table 23-6. In general, experimental data indicate that

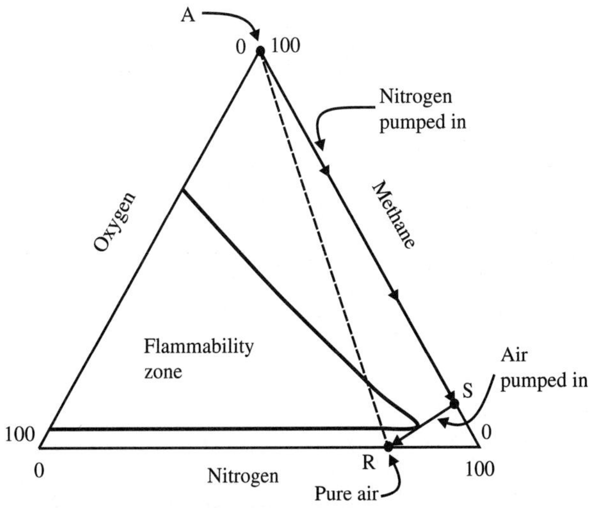

FIG. 23-5 A procedure for avoiding the flammability zone for taking a vessel out of service. [D. A. Crowl, *Understanding Explosions,* Center for Chemical Process Safety (CCPS) of the American Institute of Chemical Engineers (AIChE); copyright 2003 AIChE and reproduced with permission.]

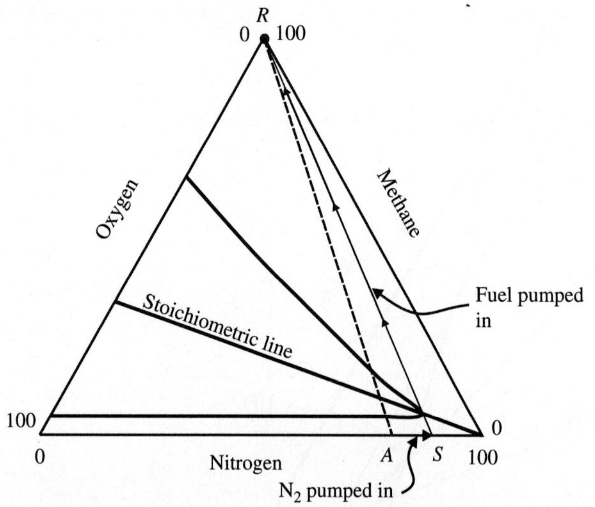

FIG. 23-6 A procedure for avoiding the flammability zone for placing a vessel into service. [D. A. Crowl, *Understanding Explosions,* Center for Chemical Process Safety (CCPS) of the American Institute of Chemical Engineers (AIChE); copyright 2003 AIChE and reproduced with permission.]

1. The MIE increases with increasing pressure.
2. The MIE for dusts is, in general, at energy levels somewhat higher than that of combustible gases.
3. An increase in nitrogen concentration increases the MIE.

Most hydrocarbon vapors have an MIE of about 0.25 mJ. This is very low—a static spark that you can feel is greater than about 20 mJ. Dusts typically have MIEs of about 10 mJ. In both the vapor and dust cases, wide variability in the values is expected.

Aerosols and Mists The flammability behavior of vapors is affected by the presence of liquid droplets in the form of aerosols or mists. Aerosols are liquid droplets or solid particles of a size small enough to remain suspended in air for prolonged periods. Mists are suspended liquid droplets produced by the condensation of vapor into liquid or by the breaking up of liquid into a dispersed state by splashing, spraying, or atomizing.

For liquid droplets with diameters less than 0.01 mm, the LFL is virtually the same as the substance in vapor form. For mechanically formed mists with drop diameters between 0.001 and 0.2 mm, the LFL decreases as the drop diameter increases. In experiments with larger drop diameters, the LFL was less than one-tenth of the vapor LFL. Flame speed increases with decreasing drop diameter up to a point, then decreases. See Lin et al. [Lin, Chen, Mashuga, Mannan, "Improved Electrospray Design for Aerosol Generation and Flame Propagation Analysis," *J. Loss Prev.* **38**(3): 148–155 (2015)]. Thus, suspended droplets have a profound effect on flammability.

Explosions

Introduction Gas explosions depend on many parameters, including temperature, pressure, gas composition, ignition source, geometry of surroundings, turbulence in the gas, mixing, time before ignition, and so forth. Thus, gas explosions are difficult to characterize and predict.

TABLE 23-4 Flammability Limits in Pure Oxygen

Compound	Formula	Limits of flammability in pure oxygen: Lower	Upper
Hydrogen	H_2	4.0	94
Carbon monoxide*	CO	15.5	94
Ammonia	NH_3	15.0	79
Methane	CH_4	5.1	61
Ethane	C_2H_6	3.0	66
Ethylene	C_2H_4	3.0	80
Propylene	C_3H_6	2.1	53
Cyclopropane	C_3H_6	2.5	60
Diethyl ether	$C_4H_{10}O$	2.0	82
Divinyl ether	C_4H_6O	1.8	85

*The limits are insensitive to p_{H_2O} above a few millimeters of mercury.

Data from B. Lewis and G. von Elbe, *Combustion, Flames and Explosions of Gases* (New York: Harcourt Brace Jovanovich, 1987).

TABLE 23-5 Ignition Sources of Major Fires*

Electrical (wiring of motors)	23%
Smoking	18
Friction (bearings or broken parts)	10
Overheated materials (abnormally high temperatures)	8
Hot surfaces (heat from boilers, lamps, etc.)	7
Burner flames (improper use of torches, etc.)	7
Combustion sparks (sparks and embers)	5
Spontaneous ignition (rubbish, etc.)	4
Cutting and welding (sparks, arcs, heat, etc.)	4
Exposure (fires jumping into new areas)	3
Incendiarism (fires maliciously set)	3
Mechanical sparks (grinders, crushers, etc.)	2
Molten substances (hot spills)	2
Chemical action (processes not in control)	1
Static sparks (release of accumulated energy)	1
Lightning (where lightning rods are not used)	1
Miscellaneous	1

**Accident Prevention Manual for Industrial Operations* (Chicago: National Safety Council, 1974).

An explosion occurs when energy is released into the gas phase in a very short time, typically milliseconds or less. If the energy is released into the gas phase, the energy causes the gases to expand very rapidly, forcing back the surrounding gas and initiating a pressure wave that moves rapidly outward from the blast origin. The pressure wave contains energy that causes damage to the surroundings. A prediction of the damage effects from an explosion requires a clear understanding of how this pressure wave behaves.

Detonation and Deflagration The difference between a detonation and a deflagration depends on how fast the pressure wave moves out from the blast origin. If the pressure wave moves at a speed less than the speed of sound in the ambient gas, then a deflagration results. If the pressure wave moves at a speed greater than the speed of sound in the ambient gas, then a detonation results.

For ideal gases, the speed of sound is a function of temperature and molecular weight only. For air at 20°C, the speed of sound is 344 m/s (1129 ft/s).

For a detonation, the reaction front moves faster than the speed of sound, pushing the pressure wave or shock front immediately ahead of it. For a deflagration, the reaction front moves at a speed less than the speed of sound, resulting in a pressure wave that moves at the speed of sound, moving away from the reaction front. A noticeable difference is found in the resulting pressure-time or pressure-distance plots.

The difference in behavior between a detonation and a deflagration results in a significant difference in the damage. For a deflagration, the damage is usually localized. However, for a detonation, the damage is more widespread.

TABLE 23-6 Minimum Ignition Energy (MIE) for Selected Gases

Chemical	Minimum ignition energy, mJ
Acetylene	0.020
Benzene	0.225
1,3-Butadiene	0.125
n-Butane	0.260
Cyclohexane	0.223
Cyclopropane	0.180
Ethane	0.240
Ethene	0.124
Ethylacetate	0.480
Ethylene oxide	0.062
n-Heptane	0.240
Hexane	0.248
Hydrogen	0.018
Methane	0.280
Methanol	0.140
Methyl acetylene	0.120
Methyl ethyl ketone	0.280
n-Pentane	0.220
2-Pentane	0.180
Propane	0.250

Data from I. Glassman, *Combustion,* 3d ed. (New York: Academic Press, 1996).

For high explosives, such as TNT, detonations are the normal result. However, for flammable vapors, deflagrations are more common.

Confined Explosions A confined explosion occurs in a building or process. Empirical studies on deflagrations (Tang and Baker, "A New Set of Blast Curves from Vapor Cloud Explosions," 33d Loss Prevention Symposium, AIChE, 1999; Mercx, van Wees, and Opschoor, "Current Research at TNO on Vapour Cloud Explosion Modeling," *Plant/Operations Progress,* October 1993) have shown that the behavior of the explosion is highly dependent on the degree of confinement. Confinement may be due to process equipment, buildings, storage vessels, foliage, and anything else that impedes the expansion of the reaction front.

These studies have found that increased confinement leads to flame acceleration and increased damage. The flame acceleration is caused by increased turbulence that stretches and tears the flame front, resulting in a larger flame front surface and an increased combustion rate. The turbulence is caused by two phenomena. First, the unburned gases are pushed and accelerated by the combustion products behind the reaction front. Second, turbulence is caused by the interaction of the gases with obstacles. The increased combustion rate results in additional turbulence and additional acceleration, providing a feedback mechanism for even more turbulence.

Characterizing Explosive Behavior for Vapors and Dusts Figure 23-7 is a schematic of a device used to characterize explosive vapors. This vessel is typically 4 to 20 L. It includes a gas handling and mixing system (not shown), an igniter to initiate the combustion, and a high-speed pressure transducer capable of measuring the pressure changes at the millisecond level.

The igniter can be of several types, including a fuse wire, spark, or chemical ignition system. A typical energy for ignition is 10 J, although gases can be ignited at much lower levels.

The gas is metered into the chamber to provide a mixture of a known composition. At a specified time, the igniter is activated, and data are collected from the pressure transducer.

A typical pressure time plot is shown in Fig. 23-8. After ignition, the pressure increases rapidly, reaches a peak, and then diminishes as the reaction products are consumed and the gases are quenched and cooled by the vessel wall.

The experiment is repeated over a range of concentrations. A plot of the maximum pressure versus fuel concentration is used to determine the flammability limits, as shown in Fig. 23-9. A pressure increase of 7 percent over initial ambient pressure is used to define the flammability limits (ASTM E918-83, *Standard Procedure for Determining Limits of Flammability of Chemicals at Elevated Temperature and Pressure,* ASTM International, West Conshohocken, Pa., 2011).

Figure 23-10 shows a device used to characterize the combustion of dusts. In this case, the dusts are initially contained in a small sample holder external to the vessel, and the dust is dispersed in pneumatically just prior to ignition. A typical pressure–time curve for the dust apparatus is shown in Fig. 23-11. The vessel is initially at a pressure less than atmospheric, but the pressure increases to atmospheric after injection dispersion. After dispersion, a delay time occurs prior to ignition. The results are highly dependent on the delay time.

Two parameters are used to characterize the combustion for both the vapor and dust cases. The first is the maximum pressure during the combustion process, and the second is the maximum rate of pressure increase. Empirical studies have shown that a deflagration index can be computed from the maximum rate of pressure increase:

$$K_G \text{ or } K_{St} = \left(\frac{dP}{dt}\right)_{max} V^{1/3} \tag{23-10}$$

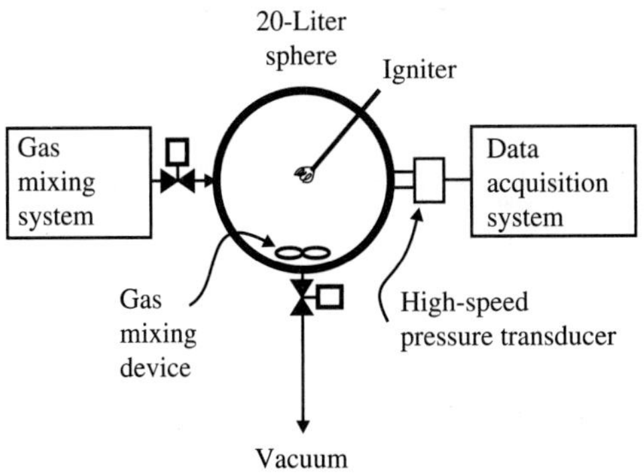

FIG. 23-7 An apparatus for collecting explosion data for gases and vapors. [D. A. Crowl, *Understanding Explosions,* Center for Chemical Process Safety (CCPS) of the American Institute of Chemical Engineers (AIChE); copyright 2003 AIChE and reproduced with permission.]

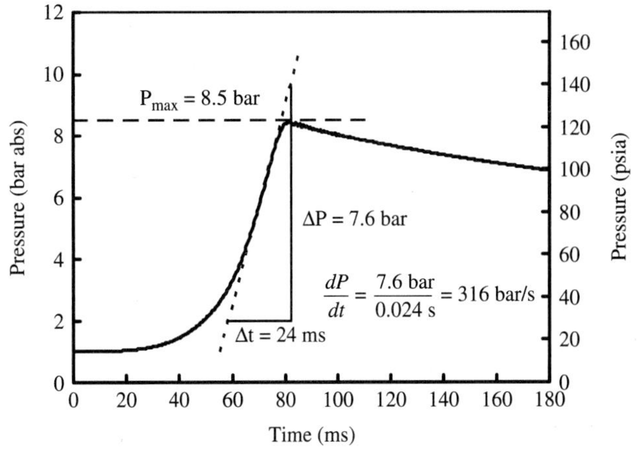

FIG. 23-8 Typical pressure versus time data obtained from gas explosion apparatus shown in Fig. 23-7. (Daniel A. Crowl and Joseph F. Louvar, *Chemical Process Safety: Fundamentals with Applications,* 2d ed., © 2002. Adapted by permission of Pearson Education, Inc., Upper Saddle River, N.J.)

where K_G = deflagration index for gases (bar·m/s)
K_{St} = deflagration index for dusts (bar·m/s)
P = pressure (bar)
t = time (s)
V = is the vessel volume (m^3)

The higher the value of the deflagration index, the more robust the combustion. Table 23-7 contains combustion data for gases, while Table 23-8 contains combustion data for dusts.

Vapor Cloud Explosions A vapor cloud explosion (VCE) occurs when a large quantity of flammable material is released, is mixed with enough air to form a flammable mixture, and is ignited. Damage from a VCE is due mostly to the overpressure, but significant damage to equipment and personnel may occur due to thermal radiation from the resulting fireball.

A VCE requires several conditions to occur (*Estimating the Flammable Mass of a Vapor Cloud,* American Institute of Chemical Engineers, New York, 1999):

1. The released material must be flammable.
2. A cloud of sufficient size must form prior to ignition.
3. The released material must mix with an adequate quantity of air to produce a sufficient mass in the flammable range.
4. The speed of the flame propagation must accelerate as the vapor cloud burns. This acceleration can be due to turbulence, as discussed in the section on confined explosions. Without this acceleration, only a flash fire will result.

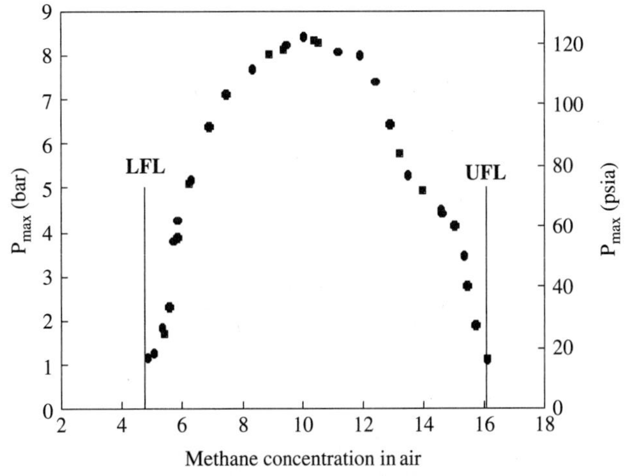

FIG. 23-9 Maximum pressure as a function of volume percent concentration for methane in air in a 20-L test sphere. The initial temperature and pressure are 25°C and 1 atm. The stoichiometric concentration is 9.51 percent methane. [C. V. Mashuga and D. A. Crowl, "Application of the Flammability Diagram for Evaluation of Fire and Explosion Hazards of Flammable Vapors," *Process Safety Progress,* vol. 17, no. 3; copyright 1998 American Institute of Chemical Engineers (AIChE) and reproduced with permission.]

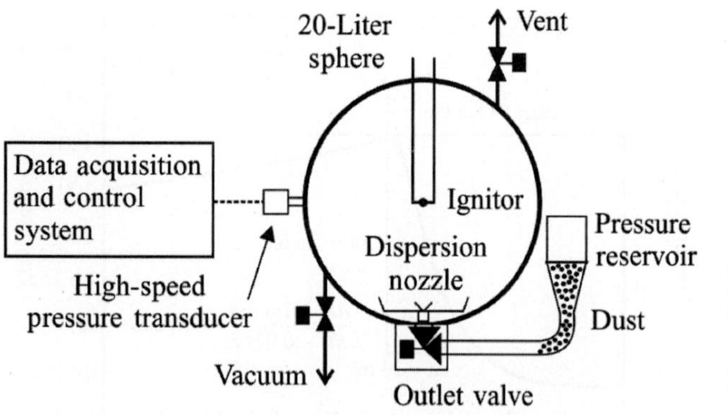

FIG. 23-10 An apparatus for collecting explosion data for dusts.

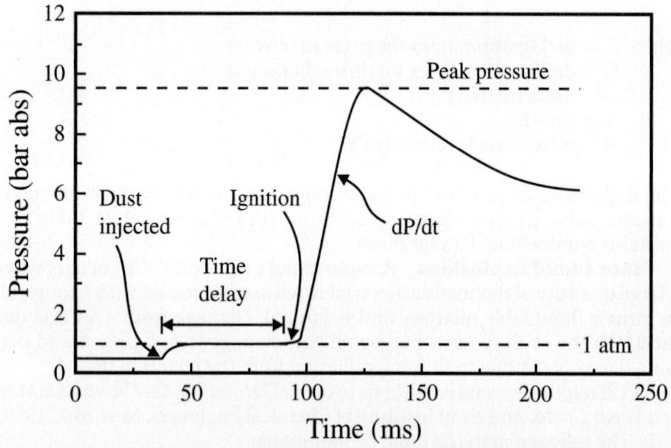

FIG. 23-11 Pressure data from dust explosion device. (Daniel A. Crowl and Joseph F. Louvar, *Chemical Process Safety: Fundamentals with Applications,* 2d ed., © 2002. Adapted by permission of Pearson Education, Inc., Upper Saddle River, N.J.)

Most VCEs involving flammable liquids or gases result only in a deflagration—detonations are unlikely. As the confinement of the vapor cloud increases, due to congestion from process equipment, the flame accelerates, and higher overpressures are achieved. The higher overpressures may approach the severity of a detonation.

Four methods are available to estimate the damage from a VCE: TNT equivalency, TNO Multi-Energy method, Baker-Strehlow-Tang method, and computational fluid dynamics. The TNT equivalency method is discussed in the Estimation of Damage Effects subsection. The other methods are discussed elsewhere (*Guidelines for Evaluating the Characteristics of Vapor Cloud Explosions, Flash Fires and BLEVES,* American Institute of Chemical Engineers, New York, 1994; *Guidelines for Chemical Process Quantitative Risk Analysis,* 2d ed., American Institute of Chemical Engineers, New York, 2000).

Boiling-Liquid Expanding-Vapor Explosions A boiling-liquid expanding-vapor explosion, commonly called a BLEVE (pronounced blevee), occurs when a vessel containing a liquid stored at a temperature above its normal boiling point fails catastrophically. After failure, a fraction of the liquid flashes almost instantaneously into vapor. Damage may be caused, in part, by the rapid expansion of the vapor and fragments from the failing vessel. The liquid may be water.

The most damaging BLEVE occurs when a vessel contains a flammable liquid stored at a temperature above its normal boiling point. The vessel walls below the liquid level are maintained at a low temperature due to the rapid heat transfer to the liquid. However, the vessel walls exposed to the fire above the liquid level will heat rapidly due to the much lower heat transfer to the vapor. The vessel wall temperature will increase to a point where the strength of the vessel wall is significantly reduced. The vessel wall will fail catastrophically, resulting in the flashing of a large quantity of flammable liquid into vapor. Since a fire is already present, the resulting vapor cloud will ignite almost immediately. Overpressures from the vessel failure may result, but most of the damage is caused by radiation from the resulting large fireball.

DUST EXPLOSIONS

Nomenclature

$(dP/dt)_{max}$	Maximum rate of pressure rise in a constant-volume explosion, bar/s
K_{St}	Size- or volume-normalized (standardized) maximum rate of pressure rise in a constant-volume explosion, bar·m/s
P_{des}	Maximum design pressure of an enclosure, bar(g)
P_{max}	Maximum explosion pressure in a constant-volume explosion, bar(g)
P_{red}	Maximum reduced pressure in a vented enclosure, bar(g)
P_{stat}	Static burst pressure of a relief vent, bar(g)
LIT	Minimum ignition temperature of a dust layer or dust deposit, °C
LOC	Limiting oxygen concentration in the atmosphere for flame propagation in a dust cloud, vol%
MEC	Minimum explosible dust concentration, g/m^3
MIE	Minimum ignition energy of a dust cloud (electric spark), mJ
MIT	Minimum ignition temperature of a dust cloud, °C

TABLE 23-7 Maximum Pressures and Deflagration Indices for a Number of Gases and Vapors

Chemical	Maximum pressure P_{max}, barg			Deflagration index K_G, bar·m/s		
	NFPA 68 1997	Bartknecht 1993	Senecal 1998	NFPA 68 1997	Bartknecht 1993	Senecal 1998
Acetylene	10.6			109		
Ammonia	5.4			10		
Butane	8.0	8.0		92	92	
Carbon disulfide	6.4			105		
Diethyl ether	8.1			115		
Ethane	7.8	7.8	7.4	106	106	78
Ethyl alcohol	7.0			78		
Ethyl benzene	6.6	7.4		94	96	
Ethylene			8.0			171
Hydrogen	6.9	6.8	6.5	659	550	638
Hydrogen sulfide	7.4			45		
Isobutane			7.4			67
Methane	7.05	7.1	6.7	64	55	46
Methyl alcohol		7.5	7.2		75	94
Methylene chloride	5.0			5		
Pentane	7.65	7.8		104	104	
Propane	7.9	7.9	7.2	96	100	76
Toluene		7.8			94	

Data selected from:
NFPA 68: Venting of Deflagrations (Quincy, Mass.: National Fire Protection Association, 1997).
W. Bartknecht, *Explosionsschutz: Grundlagen und Anwendung* (New York: Springer-Verlag, 1993).
J. A. Senecal and P. A. Beaulieu, "K_G: Data and Analysis," *31st Loss Prevention Symposium* (New York: American Institute of Chemical Engineers, 1997).

TABLE 23-8 Combustion Data for Dust Clouds*

Dust	Median particle size, μm	Minimum explosive dust conc., g/m³	P_{max}, barg	K_{St}, bar·m/s	Minimum ignition energy, mJ
Cotton, Wood, Peat					
Cotton	44	100	7.2	24	—
Cellulose	51	60	9.3	66	250
Wood dust	33	—	—	—	100
Wood dust	80	—	—	—	7
Paper dust	<10	—	5.7	18	—
Feed, Food					
Dextrose	80	60	4.3	18	—
Fructose	200	125	6.4	27	180
Fructose	400	—	—	—	>4000
Wheat grain dust	80	60	9.3	112	—
Milk powder	165	60	8.1	90	75
Rice flour	—	60	7.4	57	>100
Wheat flour	50	—	—	—	540
Milk sugar	10	60	8.3	75	14
Coal, Coal Products					
Activated carbon	18	60	8.8	44	—
Bituminous coal	<10	—	9.0	55	—
Plastics, Resins, Rubber					
Polyacrylamide	10	250	5.9	12	—
Polyester	<10	—	10.1	194	—
Polyethylene	72	—	7.5	67	—
Polyethylene	280	—	6.2	20	—
Polypropylene	25	30	8.4	101	—
Polypropylene	162	200	7.7	38	—
Polystyrene (copolymer)	155	30	8.4	110	—
Polystyrene (hard foam)	760	—	8.4	23	—
Polyurethane	3	<30	7.8	156	—
Intermediate Products, Auxiliary Materials					
Adipinic acid	<10	60	8.0	97	—
Naphthalene	95	15	8.5	178	<1
Salicylic acid	—	30	—	—	—
Other Technical, Chemical Products					
Organic dyestuff (blue)	<10	—	9.0	73	—
Organic dyestuff (red)	<10	50	11.2	249	—
Organic dyestuff (red)	52	60	9.8	237	—
Metals, Alloys					
Aluminum powder	<10	60	11.2	515	—
Aluminum powder	22	30	11.5	110	—
Bronze powder	18	750	4.1	31	—
Iron (from dry filter)	12	500	5.2	50	—
Magnesium	28	30	17.5	508	—
Magnesium	240	500	7.0	12	—
Silicon	<10	125	10.2	126	54
Zinc (dust from collector)	<10	250	6.7	125	—
Other Inorganic Products					
Graphite (99.5% C)	7	<30	5.9	71	—
Sulfur	20	30	6.8	151	—
Toner	<10	60	8.9	196	4

St Classes for Dusts

Deflagration index K_{St}, bar·m/s	St class
0	St-0
1–200	St-1
200–300	St-2
>300	St-3

*Data selected from R. K. Eckoff, *Dust Explosions in the Process Industries* (Oxford, England: Butterworth-Heinemann, 1997).

General References: Amyotte, *An Introduction to Dust Explosions. Understanding the Myths and Realities of Dust Explosions for a Safer Workplace*, Elsevier/Butterworth-Heinemann, Waltham, Mass. (2013); Bartknecht, *Dust Explosions. Course, Prevention, Protection*, Springer-Verlag, Berlin (1989); Barton, ed., *Dust Explosion Prevention and Protection: A Practical Guide*, Institution of Chemical Engineers, Rugby (2002); Center for Chemical Process Safety, *Guidelines for Safe Handling of Powders and Bulk Solids*, Wiley, Hoboken, N.J. (2004); Eckhoff, *Dust Explosions in the Process Industries*, 3rd ed., Elsevier/Gulf Professional Publishing, Boston, Mass. (2003); Frank, Rodgers, and Colonna, *NFPA Guide to Combustible Dusts*, National Fire Protection Association, Quincy, Mass. (2012); Mannan, *Lees' Loss Prevention in the Process Industries*, 4th ed., vol. 2, chap. 17, Elsevier/Butterworth-Heinemann, Oxford, UK (2012); Zalosh, *Dust Explosions*, Chapter 70 in *SFPE Handbook of Fire Protection Engineering*, ed. Hurley, Springer, New York (2016).

Incidents Dust explosions arise in industrial scenarios where combustible dusts are stored, transported, processed, or otherwise handled. Injury and loss of life, damage to facilities, business interruption, and environmental degradation can result from dust explosions involving a wide range of materials including coal, grain, foodstuffs, metals, paper, pharmaceuticals, plastics, resins, sulfur, textiles, rubber, and wood (among many others). Devastating dust explosions have occurred worldwide for centuries. Recent major dust explosions investigated by the U.S. Chemical Safety Board (*www.csb.gov*) have involved powdered polyethylene, phenolic resin, aluminum, and sugar.

Fundamentals Dust explosions are classified as chemical rather than physical explosions due to the occurrence of combustion and flame propagation through the reaction mass. Most dust explosions involve deflagrations, in which the flame travels at less than the speed of sound in the unreacted medium and is preceded by a pressure or blast wave. It is the subsequent bursting or rupturing (due to internal pressure development) of the enclosure in which the deflagration occurs that produces explosion effects. The necessary conditions for a dust explosion are well expressed by the *explosion pentagon* consisting of fuel, oxidant, ignition source, mixing (or dispersion), and confinement.

Combustible dust is defined as finely divided combustible particulate solid material that presents a flash fire or explosion hazard when suspended in air or a process-specific oxidizing medium over a range of concentrations, regardless of particle size, shape, or chemical composition (NFPA 654, *Standard for the Prevention of Fire and Dust Explosions from the Manufacturing, Processing, and Handling of Combustible Particulate Solids*, National Fire Protection Association, Quincy, Mass., 2017). This definition highlights two critical features of combustible dusts: (1) There is no universal particle-size boundary above which all combustible dusts become nonexplosible. Finer sizes of a given combustible dust are, however, more readily ignitable and yield more severe overpressure consequences than coarser sizes. This effect is caused by the increase in specific surface area that accompanies a reduction in particle dimensions. (2) Dusts having shapes other than spherical (e.g., flake or cylindrical particles) can also present a dust explosion hazard.

Other parameters, both physical (e.g., moisture content) and chemical (e.g., heat of combustion), also play a role in ease of ignition and the ensuing overpressure development. Dust composition determines the actual combustion mode, with the majority of dust deflagrations occurring as homogeneous gas-phase reactions of evolved volatile matter and oxidant. Hybrid mixtures of combustible dust and flammable gas or vapor are a unique concern in the process industries that require particular attention.

Air contains nominally 21 percent oxygen by volume, thus ensuring that the oxidant component of the dust explosion pentagon is readily available. While an oxidant must be present for combustion to occur, its removal does not need to be complete to prevent the occurrence of a dust explosion (as subsequently described with respect to the concept of a limiting oxygen concentration).

Combustible dust ignition sources are pervasive in industry and include: (1) flames and direct heat, (2) hot work (e.g., welding and cutting), (3) incandescent material (e.g., smoldering particles), (4) hot surfaces (e.g., overheated bearings), (5) electrostatic discharges (such as may result from material handling or inadequate grounding and bonding), (6) electrical sparks (such as may be caused by switching operations), (7) friction sparks and hot spots (caused by rubbing between solids and friction-induced heating, respectively), (8) impact sparks (ignition by surface heating resulting from metal-on-metal impact), (9) self-heating (spontaneous combustion), (10) lightning, and (11) shock waves. As illustrated by this list, many dust ignition sources involve either an energetic spark or elevated temperature of a boundary surface.

Solid-phase dust particles are strongly influenced by gravity; further, an essential prerequisite for a dust explosion is the existence of a dust/oxidant suspension. These points highlight the importance of the mixing criterion of the explosion pentagon, which can be satisfied by either dust cloud formation inside process equipment or dispersion of hazardous accumulations of dust in the workplace. How long a dust cloud remains in suspension depends on a number of factors, including the size, shape, and specific surface area of the particles, the moisture content and density of the dust, and possible agglomeration of particles pre- and post-dispersion. Some degree of turbulence will always exist in a dust cloud; while highly turbulent dust clouds are harder to ignite, they yield more violent explosions.

Confinement allows for the development of explosion overpressure. The degree of confinement does not, however, need to be total (or complete) for a dust explosion to arise. While an unconfined dust explosion is expected to be a rare occurrence, workplace congestion can create a high blockage ratio and flow obstruction leading to turbulence generation in the unburned dust cloud and enhanced explosion violence. In the case of limited or no confinement, the result may be a flash fire with the potential for secondary fires or explosions.

Hazard Identification Flash fire and explosion hazards can thus be present in equipment used to process combustible particulate solids, or in areas where there is hazardous accumulation of fugitive dust. For most industrial applications, the recommended approach to identifying these hazards is to conduct a systematic and documented dust hazard analysis (DHA), which encompasses material, equipment, and building hazard evaluations. The objectives of a dust hazard analysis are to: (1) identify combustible dust fire and explosion hazards, (2) evaluate existing explosion protection measures compared to good engineering practice as exemplified by contemporary engineering standards and guidelines, and (3) where practicable, develop solutions to close gaps between existing protective measures and good engineering practice.

Material explosibility parameters should be determined for dust samples representative of the stage of the process or location being evaluated. Each component of the process or material handling system is considered, and areas outside equipment are evaluated for hazardous accumulation of fugitive dust. An example and a list of the minimum criteria for an acceptable dust hazard analysis are provided in NFPA 652, *Standard on the Fundamentals of Combustible Dust*, National Fire Protection Association, Quincy, Mass., 2016.

There are many recognized hazard evaluation methodologies (e.g., failure modes and effects analysis, hazard and operability study, what-if scenario analysis) that can be used to perform a dust hazard analysis. However, the vast majority of deflagration hazards exist in common industrial processes (e.g., spray drying and material transfer systems) and equipment (e.g., dust collectors, spray dryers, and blenders), for which the associated fire and explosion hazards are well understood, and proven protection schemes have been established. Therefore, a comparative gap analysis using normative engineering standards as a benchmark for industry best practice is usually the most effective means of hazard identification. NFPA offers many occupancy-specific standards related to facilities handling or processing combustible particulate solids that can be used for this purpose. The most widely applicable standard in this regard is NFPA 654, *Standard for the Prevention of Fire and Dust Explosions from the Manufacturing, Processing, and Handling of Combustible Particulate Solids*, National Fire Protection Association, Quincy, Mass., 2017.

Explosibility Characteristics Table 23-9 gives a listing of key dust explosibility parameters, relevant methodologies for their determination, and examples of why the determination of such parameters is important from an industrial perspective. The first three entries in Table 23-9 [P_{max}, $(dP/dt)_{max}$, and K_{St}] relate material hazards to the dust explosion risk component of consequence severity (explosion violence); knowledge

TABLE 23-9 Important Dust Explosibility Parameters

Parameter	Example test methodology	Example industrial applications
P_{max}	ASTM E1226-12a, *Standard Test Method for Explosibility of Dust Clouds*, ASTM International, West Conshohocken, Penn. (2012)	Venting Suppression Containment Isolation Partial inerting
$(dP/dt)_{max}$	ASTM E1226-12a, *Standard Test Method for Explosibility of Dust Clouds*, ASTM International, West Conshohocken, Penn. (2012)	As per P_{max}
K_{St}	ASTM E1226-12a, *Standard Test Method for Explosibility of Dust Clouds*, ASTM International, West Conshohocken, Penn. (2012)	As per P_{max}
MEC	ASTM E1515-14, *Standard Test Method for Minimum Explosible Concentration of Combustible Dusts*, ASTM International, West Conshohocken, Penn. (2014)	Control of dust concentrations
MIE	ASTM E2019-03(2013), *Standard Test Method for Minimum Ignition Energy of a Dust Cloud in Air*, ASTM International, West Conshohocken, Penn. (2013)	Control of ignition sources (e.g., grounding and bonding)
MIT	ASTM E1491-06(2012), *Standard Test Method for Minimum Autoignition Temperature of Dust Clouds*, ASTM International, West Conshohocken, Penn. (2012)	Control of process and surface temperatures (dust clouds)
LIT	ASTM E2021-15, *Standard Test Method for Hot-Surface Ignition Temperature of Dust Layers*, ASTM International, West Conshohocken, Penn. (2015)	Control of process and surface temperatures (dust layers)
LOC	ASTM E2931-13, *Standard Test Method for Limiting Oxygen (Oxidant) Concentration of Combustible Dust Clouds*, ASTM International, West Conshohocken, Penn. (2013)	Inerting (with gaseous diluents)

Adapted with permission from Amyotte and Eckhoff, "Dust Explosion Causation, Prevention and Mitigation: An Overview," *Journal of Chemical Health and Safety* **17**(1): 15–28 (2010).

of these parameters is therefore required for dust explosion mitigation. [K_{St} is the product of $(dP/dt)_{max}$ and the cube root of the volume of the standardized vessel in which $(dP/dt)_{max}$ was determined in accordance with a standardized test procedure.] The remaining entries in Table 23-9 are associated with dust material hazards in relation to explosion likelihood (ignition sensitivity) and are thus important in terms of dust explosion prevention and risk assessment. Additional parameters may also be relevant; for example, powder conductivity (as indicated by volume resistivity) gives insight into the potential for ignition by bulking brush (or cone) electrostatic discharge.

It is critical to recognize that the explosibility characteristics shown in Table 23-9 are not intrinsic material properties. Values of these parameters are strongly dependent on material characteristics (e.g., moisture content, particle size distribution, and shape) and experimental conditions (e.g., apparatus dimensions, turbulence level, and ignition energy). However, when determined by qualified personnel using standardized test procedures and standardized equipment, they provide the best available information for understanding material hazards and explosion risk as described in the previous paragraph. Dust explosion testing standards exist worldwide; they include various normative protocols developed by organizations such as ASTM International (shown in Table 23-9), IEC (International Electrotechnical Commission), and ISO (International Organization for Standardization). In a similar manner, and in addition to the NFPA standards referenced here, other documentation giving risk control measures for combustible dusts is available, including European EN standards and ATEX directives, and VDI guidelines (Germany).

Typical values of the parameters in Table 23-9 can be found in various databases and the general references cited here. Because of the aforementioned dependence on dust morphology and chemical composition, caution must be exercised when using literature values of explosibility parameters for any type of hazard identification or risk assessment. Textbooks and databases can be helpful as indicators of dust explosibility, but they should not be seen as a substitute for test data on the actual material being processed. With these points in mind, one can generally expect that P_{max} for combustible dusts will be on the order of 8 to 10 bar(g), while K_{St} for most dusts is <200 bar·m/s. [Again, these are generalizations only for the sake of illustration; P_{max} for some metal dusts, such as iron, is only 3 to 4 bar(g), while fine aluminum dust can have K_{St} values >300 bar·m/s.] MEC for many combustible dusts is <100 g/m^3, and MIE values can be in the low-mJ range meaning these materials are susceptible to ignition by electrostatic discharge.

Equipment Hazards An equipment explosion hazard is deemed to exist when there is potential for all components of the explosion pentagon to occur at the same time. Loss history is replete with examples of equipment dust explosions involving dust collection systems, grinders, pulverizers, silos, conveying systems, dryers, mixers, blenders, and many other types of equipment. Approximately half of industrial dust explosions involve dust collectors. Deflagrations in one piece of equipment can propagate to connected equipment or process areas, thus initiating secondary events with potentially devastating consequences (as described later).

When evaluating whether there is an equipment explosion hazard, it is important to consider not just normal modes of operation, but also potential upset scenarios that could lead to the necessary conditions for a dust explosion. It is also important to note that some components of the explosion pentagon may be present only to a limited extent—for example, a low-reactivity fuel or low degree of confinement. As previously mentioned, the ignition of a dust/air suspension in this case could produce a potentially fatal flash fire.

Explosion Prevention In principle, then, dust explosion prevention can be achieved by eliminating at least one of the components of the explosion pentagon. In many cases, though (particularly for existing processes), it is impractical to eliminate the fuel. An explosion hazard can still be eliminated, however, by changing the dust properties. For example, application of the moderation principle of inherent safety by increasing the milled particle size distribution for a material such as polyethylene could render the dust nonexplosible. Depending on the process by which metallic dust is generated, the particle surface may be oxidized to the extent that the dust is no longer explosible. Combustible dust can also be inerted by adding sodium bicarbonate or other solid inertant; however, the explosion severity of a particular dust may be increased if the quantity of inertant is inadequate. It is also possible in certain applications to maintain the quantity of fuel available for combustion below the MEC. NFPA 69 (*Standard on Explosion Prevention Systems*, National Fire Protection Association, Quincy, Mass., 2014) provides guidance on deflagration prevention by combustible concentration reduction.

Explosion prevention can also be achieved by operating below the LOC of the fuel, which for combustible dust suspensions typically lies in the range of 8 to 15 vol%. NFPA 69 further provides design guidelines for deflagration prevention by oxidant concentration reduction. Nitrogen is commonly used as a diluent; other diluents such as rare gases, carbon dioxide, steam, or flue gas may be used.

Ignition source control is usually employed as a layer of protection in preventing a dust explosion, rather than as a primary basis for safety. This is especially the case for dusts with MIE values less than 1 J because certain types of electrostatic discharge can reach this energy level. In some carefully evaluated situations for specific types of equipment involving dusts with relatively high MIEs, it is possible to use ignition source control as a basis for safety. A formalized and documented hazard analysis performed by a qualified person should be undertaken in such cases. Due to the ubiquity of ignition sources outside equipment boundaries, ignition source control in general plant areas is not a defensible basis for safety. In cases where equipment fugitive dust emission cannot be controlled and effective housekeeping practices cannot be achieved (e.g., milling rooms), deflagration venting and control of ignition sources (e.g., by using classified electrical equipment) are commonly used safeguards. Information on the proper selection of electrical equipment in combustible dust atmospheres is given in NFPA 499, *Recommended Practice for the Classification of Combustible Dusts and of Hazardous (Classified) Locations for Electrical Installations in Chemical Process Areas*, National Fire Protection Association, Quincy, Mass., 2017.

Explosion Mitigation The effects and potential consequences of a dust deflagration are typically mitigated by one or more of three methods (Table 23-9): (1) explosion protection by deflagration venting, (2) deflagration control by suppression, and (3) deflagration control by pressure containment. Dust deflagrations can propagate between connected equipment; this can be prevented by active or passive isolation.

Deflagration venting works on the principle of relieving pressure that develops in a vessel or compartment before damaging overpressures can be achieved. As shown in Fig. 23-12, explosion vents are designed to open at relatively low pressures (well below P_{des}). Design guidelines for deflagration venting are provided in NFPA 68, *Standard on Explosion Protection by Deflagration Venting*, National Fire Protection Association, Quincy, Mass., 2013. The required size of the vent is influenced by the explosibility parameters of the material (P_{max} and K_{St}), as well as relevant design features of the protected enclosure (P_{red}) and the vent itself (P_{stat}). Larger vent areas are needed for more reactive dusts. Other factors that influence the required vent size are vent panel mass, length-to-diameter ratio of the equipment, and elevated operating pressure. For equipment with high turbulence levels, increased vent area may also be required.

An important drawback with deflagration venting is the ensuing fireball and ejection of burning material, which may initiate fires in other places. The thermal, pressure, and projectile hazards associated with a vented deflagration may cause worker fatalities or serious injury. For this reason, deflagration venting is usually only feasible for equipment located outdoors, in areas where access is restricted. For equipment located indoors, vent ducts can be employed to direct a vented deflagration outdoors; the use of vent ducts will require an increased vent area. Design considerations for vent ducts are also addressed in NFPA 68, *Standard on Explosion Protection by Deflagration Venting*, National Fire Protection Association, Quincy, Mass., 2013.

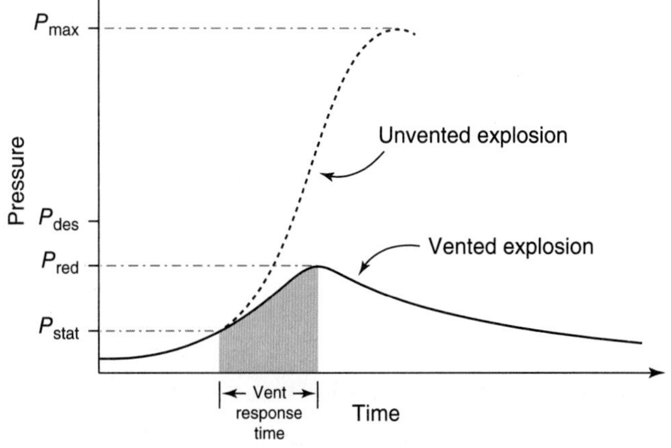

FIG. 23-12 Illustration of pressure rise during vented and unvented explosions. [Adapted with permission from Pekalski, Zevenbergen, Lemkowitz and Pasman, "A Review of Explosion Prevention and Protection Systems Suitable as Ultimate Layer of Protection in Chemical Process Installations," *Process Safety and Environmental Protection*, **83:** 1–17 (2005).]

The length and volume of a vent duct are limiting factors. The effects of a vented deflagration may also be mitigated by using a flame arresting and particulate retention device. Such devices have been tested for specific applications and are not suitable for all types of dust.

Deflagration control by suppression has the advantage of avoiding the development of a fireball. This method involves early detection of dust cloud ignition and rapid injection of a suppressant such as sodium bicarbonate. The protected equipment must withstand the increased pressure associated with the injection of the suppressant. There are also ongoing preventive maintenance and inspection requirements. There are challenges with this application for small volumes and highly reactive dusts such as aluminum.

Various types of equipment are conducive to deflagration control by pressure containment. While it is usually impractical to employ pressure containment for larger equipment such as silos, dust collectors, or spray dryers, this method is suitable for robust process equipment such as grinders. Equipment can be designed as explosion-pressure-resistant (i.e., the equipment is designed to withstand the expected explosion pressure without permanent deformation) or explosion-pressure-shock resistant (i.e., the equipment is designed to withstand the expected explosion pressure without rupturing, but allowing permanent deformation). A disadvantage for the latter is that deformed equipment will need to be replaced.

Secondary Explosion Hazards Explosible dust clouds are optically thick and have airborne dust concentrations orders of magnitude higher than those acceptable by occupational hygiene standards. These facts help to explain why *primary* dust explosions generally occur inside process units while *secondary* explosions (or flash fires) occur due to entrainment of dust layers by the blast waves arising from a primary explosion. The primary event might be a dust or gas explosion originating in a process unit, or it could be any disturbance energetic enough to disperse combustible dust layered on the floor and other surfaces (such as improper cleaning practices using compressed air). This domino sequence of primary and secondary explosions is illustrated in Fig. 23-13.

The dust layer thickness needed to support an explosion following layer dispersion is on the order of only a millimeter or less. In general, a hazard assessment should be performed if settled dust forms an opaque layer on surfaces; for example, if the underlying surface color cannot be discerned or if visible markings such as footprints can be left in the dust. There is thus a strong need for effective removal of dust accumulations from all exposed surfaces at all heights by means of a regularly scheduled housekeeping program. It should be noted that housekeeping, while absolutely essential as a preventive measure, is not as effective as the inherently safer approach of designing building surfaces to prevent dust accumulation, and designing equipment to contain dust so that it does not escape and does not have to be cleaned up in the first place. Avoidance of dust layer formation, and removing those deposits that do accumulate, has the added benefit of eliminating the fuel source for dust layer fires. Finally, such actions enable compliance with the various NFPA standards that stipulate allowable thicknesses of combustible dust layers (either qualitatively or quantitatively) for different materials and industries: NFPA 61, *Standard for the Prevention of Fires and Dust Explosions in Agricultural and Food Processing Facilities,* National Fire Protection Association, Quincy, Mass., 2017; NFPA 484, *Standard for Combustible Metals,* National Fire Protection Association, Quincy, Mass., 2015; NFPA 655, *Standard for Prevention of Sulfur Fires and Explosions,* National Fire Protection Association, Quincy, Mass., 2012; NFPA 664, *Standard for the Prevention of Fires and Explosions in Wood Processing and Woodworking Facilities,* National Fire Protection Association, Quincy, Mass., 2017.

STATIC ELECTRICITY

Nomenclature

A	area, m^2
C	capacitance, F
d	diameter, m
I	current, A
L	length, m
Q	charge, C
R	resistance, Ω
V	electric potential, V
v	velocity, m/s
W	energy, J
ε	dielectric constant, dimensionless
ε_0	permittivity of a vacuum, 8.845×10^{-12} F/m
κ	conductivity, S/m
λ	surface resistivity, Ω
ρ	volume resistivity, Ωm
τ	time constant, s

General References: API RP 2003, *Protection Against Ignitions Arising Out of Static, Lightning and Stray Currents,* 7th ed., American Petroleum Institute, Washington D.C., 2008; British Standard BS 5958, *Control of Undesirable Static Electricity,* British Standards Institution, London, 1991; Britton, L. G., "Avoiding Static Ignition Hazards in Chemical Operations," AIChE-CCPS, 1999; ESCIS, "Static Electricity: Rules for Plant Safety," *Plant/Operations Progress* **7**(1): 1 (1988); NFPA 77, *Recommended Practice on Static Electricity,* 2014 Edition, National Fire Protection Association, Quincy, Mass.; Pratt, T. H., "Electrostatic Ignitions of Fires and Explosions," AIChE-CCPS, 2000; Walmsley, H. L., "Avoidance of Electrostatic Hazards in the Petroleum Industry," *J. Electrostatics* **27**(1 and 2): 1–200 (1992).

Definitions

Bonding Providing a low-resistance conductor, $<10^6$ Ω, between conductive objects so that they are at the same electrical potential.

Combustible Capable of undergoing combustion; used here to apply to mists and dusts.

Conductive Having the ability to allow the passage of an electric charge; see Table 23-11.

Conductivity An intrinsic property of a material that governs its ability to transmit an electric charge; the reciprocal of resistivity. Conventionally, conductivity, κ in S/m (or usually pS/m, picosiemens per meter) is used for bulk liquids, and surface conductivity, in Ω, is used for solid surfaces.

Flammable Capable of burning in air (or oxygen) when mixed in suitable proportions.

Grounding (earthing) Providing a low electrical resistance between conductive objects and ground (earth), $<10^6$ Ω, so that the objects are at the same electrical potential as the earth.

Incendive Having the ability to ignite a combustion reaction in a flammable material.

Insulative Having the ability to resist the flow of an electric charge; see Table 23-11.

Minimum ignition energy, MIE The minimum energy in a capacitive discharge required to ignite an optimum concentration of the material in question.

Resistivity An intrinsic property of a material that governs its ability to resist the transmission of an electric charge; the reciprocal of conductivity. Conventionally, resistivity, ρ in Ωm, is used for bulk solids.

Introduction All matter is made up of atoms that contain negatively charged electrons and positively charged protons. In their basic state, atoms are electrically neutral; but when an electron is lost or gained, the atom

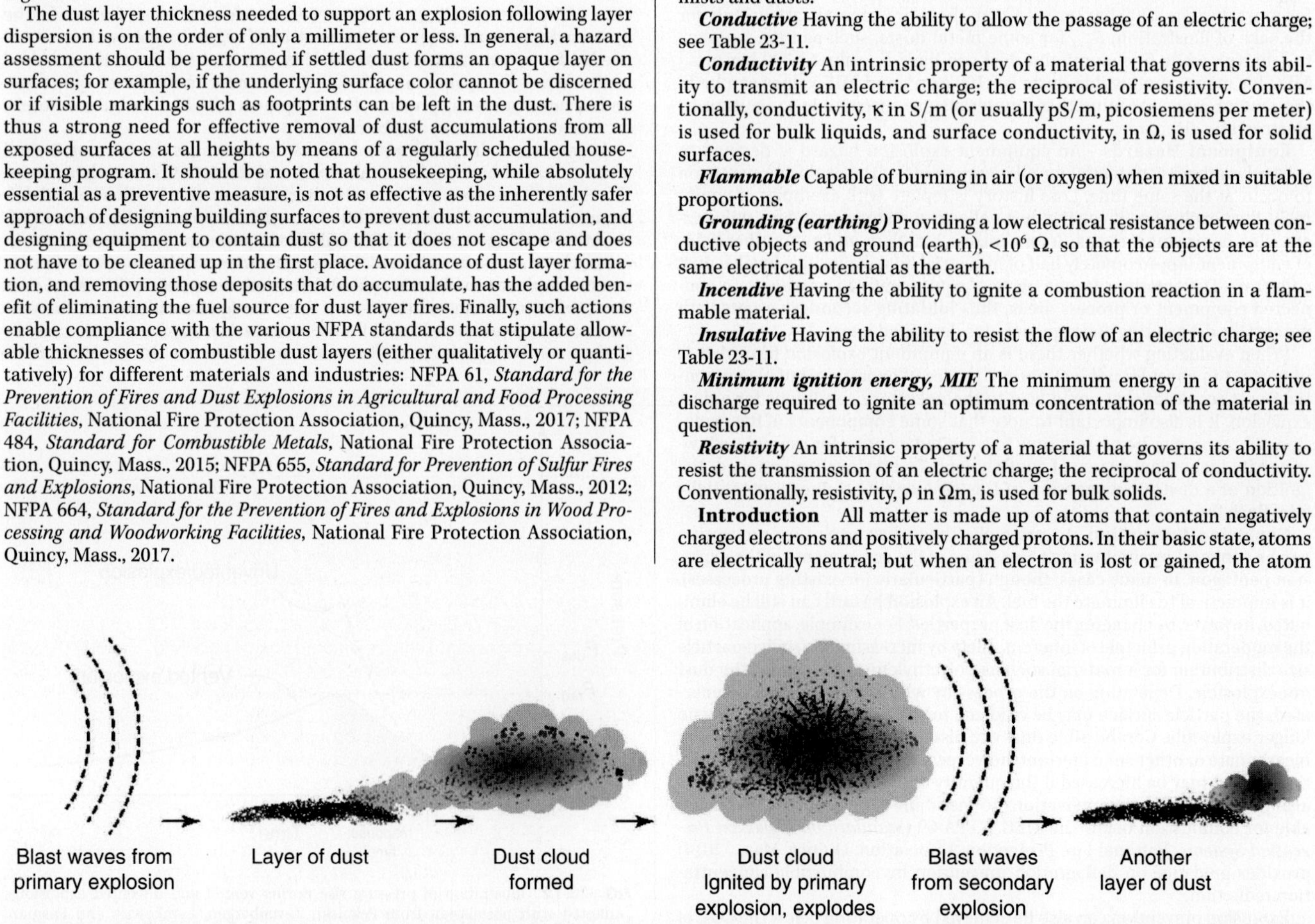

FIG. 23-13 Secondary dust explosion resulting from primary explosion. [Adapted with permission from Abbasi and Abbasi, "Dust Explosions—Cases, Causes, Consequences, and Control," *Journal of Hazardous Materials,* **140:** 7–44 (2007).]

TABLE 23-10 Terms for Resistivity of Solids

Term	Volume resistivity, ρ Ωm	Surface resistivity, λΩ
Conductive	$\rho < 10^2$	$\lambda < 10^5$
Dissipative	$10^2 < \rho < 10^9$	$10^5 < \lambda < 10^{12}$
Insulative	$\rho > 10^9$	$\lambda > 10^{12}$

Data from Electrostatic *Discharge Control Handbook*, EOS/ESD Association, Rome, N.Y., 1994.

becomes an ion. An electrostatic charge is a result of a quantity of ions of the same polarity being accumulated at the same place at the same time. An excess of electrons results in a negative charge and a deficiency of electrons results in a positive charge; like charges repel each other and unlike charges attract each other. An accumulation of charge results in an electric field that will apply a force, either repulsive or attractive, on any electric charge within that field. When like charges accumulate on or in a material, they will repel each other and subsequently either dissipate by finding their way to ground, recombine with their opposite counterpart by moving through the material, or find their way into some sort of spark. The time it takes for a charge to dissipate depends on the conductivity (or resistivity) of the material upon which the charge is held (Tables 23-10 and 23-11). Since all materials have some conductivity, it is a misnomer to use the term *nonconductor*, but it is commonly used to characterize a material that can pass only a very little charge. Traditionally, *conductivity* is used to characterize liquids and *resistivity* is used to characterize solids. Volume resistivity, ρ, is an intrinsic property of a material and is determined by measuring the resistance, R, of a sample of the subject material of a known constant cross-sectional area, A, and a known length, L. Surface resistivity, λ, of a material is determined by measuring the surface resistance across a square of the subject material where L_1 and L_2 are the lengths of the edges of the square. The units for λ are ohms, but they are sometimes termed "ohms per square." Conductivity, κ, is the reciprocal of resistivity. Since the dissipation of a static charge is exponential, the relaxation of the charge is characterized by the relaxation time constant, τ, which is the time required for the charge to dissipate to 0.368 (1/e) of its original charge. The relationships for these quantities are as follows:

$$\rho = RA/L, \quad \lambda = R(L_1/L_2), \quad \kappa = 1/\rho, \quad \text{and} \quad \tau = \rho\varepsilon\varepsilon_0 \qquad (23\text{-}11)$$

Anywhere charges accumulate, attention must be given to the possibility of having electrostatic ignitions of flammable atmospheres. In order for ignitions to occur, four conditions must exist: (1) a means of separating charge, (2) a means of accumulating the separated charge and maintaining an electrical potential, (3) a discharge of all (or part) of the accumulated charge into a spark, and (4) an ignitable mixture where spark discharge occurs. It follows then that ignitions can be avoided by eliminating any one (or more) of the four requirements.

Electrostatic Charging—Charge Separation When a solid or liquid is at rest, the atoms arrange themselves into their lowest energy state where molecular polarization usually occurs. Usually the electrons will either be at the surface or tucked down beneath the surface; that is, the molecules arrange themselves with either their positive or negative ends pointing up. At solid–solid interfaces, the charges can align themselves such that one charge is preferentially on one surface and the opposite charge is on the other. Likewise, this sort of charge separation occurs at liquid–solid, liquid–gas, and two-phase liquid interfaces.

TABLE 23-11 Conductivity of Example Liquids (See NFPA 77)

Liquid	Conductivity K, pS/m	Dielectric constant ε	Time constant τ, s
Mercury	1.06×10^{18}	—	$<10^{-6}$
Salt water	5×10^{12}	—	$<10^{-6}$
Ethylene glycol	1.16×10^{8}	37.7	2.9×10^{-6}
Acetone	6×10^{6}	20.7	3×10^{-5}
Water (extremely pure)	4.3×10^{6}	80.4	1.7×10^{-4}
Aniline	2.4×10^{6}	6.89	2.5×10^{-5}
Ethyl alcohol	1.35×10^{5}	24.55	1.6×10^{-3}
Acetic acid	1.12×10^{6}	6.15	4.9×10^{-5}
Methylene chloride	4300	8.93	1.8×10^{-2}
Diethyl ether	30	4.6	1.4
Toluene	1	2.38	21
Benzene (pure)	5×10^{-3}	2.3	~100
Hexane (pure)	1×10^{-5}	1.90	~100

Data from NFPA 77, *Recommended Practice on Static Electricity*, 2014 Edition, National Fire Protection Association, Quincy, Mass.

Charge Accumulation Once charge separation is established at the interfaces, movement can easily gather the charges en masse and result in the accumulation of a lot of charge in one place, along with its concomitant electric field.

Pure, standard-state gases do not accumulate static charges, and seldom, if ever, do they contribute to electrostatic scenarios. However, solids, liquids, mists, and dusts can accumulate charge. Isolated conductors are the most hazardous accumulators of static charges because they can give up all their energy into a spark discharge which, if large enough, will ignite flammable and combustible materials. Insulative solids can accumulate charges that can be potentially hazardous, depending on conditions. Any time powders are moved about in process equipment, there will be contact charging both between particles and on the surface of the equipment. Charge separation and accumulation therefore occur in operations such as sieving, pouring, chuting, grinding, micronizing, drying, fluidization, and pneumatic or mechanical transport. Charges can also accumulate on insulative solid surfaces by energetic processes such as rubbing, separating films, depositing a charged dust or mist, or moving a material through a pipe. The pneumatic transport of an insulative solid into an isolated or insulative receptacle is one of the larger generators or accumulators of charge in chemical processes.

Experience has shown on many occasions that the addition of a powder or granular material from a plastic bag into an open container having ignitable vapors can result in the ignition of the vapors (or dusts). In these cases, the solid material has previously been moving about within the bag during normal handling and has created a double layer of charge at the bag/product interface—there is a positive charge on the solid and a negative charge on the bag, or vice versa. In these cases, the charge density can be very high because there is little electric field away from the layer. The positive and negative charges are separated, but they are adjacent to one another, so they result in a neutral entity. When the material is dumped into the container, these entities are rapidly separated en masse, creating a large surface charge on the surface of the bag so that an incendive brush discharge can result. This mechanism also occurs in the use of flexible intermediate bulk containers (FIBCs); therefore, the proper type of A, B, C, or D FIBC must be used where applicable to prevent ignitions of vapors or dusts.

A double layer of charge is normally present on the free surface of a liquid where charges of one polarity gather at the surface and charges of the opposite polarity gather at a layer just below. When the liquid is moved across a solid surface, separation will occur by the removal of some charge from the top layer. The outer charges will tend to stick to the solid, while the inner, opposite charge will tend to move along with the liquid. For example, when a liquid is pumped through a pipe, such charge separation occurs in the double layer, and a current of ions will move along with the liquid. This phenomenon is termed the "streaming current," which can be as high as several tens of microamperes.

Liquids with a conductivity between 50 pS/m and 10^4 pS/m have been termed "semiconductive" (NFPA 77); this term applies to those liquids that can act as accumulators when charging rates are extremely high or when they are electrically isolated. Spark discharges are possible from the more conductive of these liquids. Streaming currents do not occur in liquids of high conductivity, but they are a ubiquitous concern for liquids with a conductivity less than 50 pS/m that have longer relaxation times (see Table 23-11) and can hold a charge for significant periods of time. The equations for the time constant, τ, work reasonably well for liquids having conductivities down to approximately 0.1 pS/m, but they do not apply for lower conductivities where dissipation is hyperbolic rather than exponential. Antistatic additives can be used to increase the conductivity of liquids and obviate some of the potential hazards in operations where low-conductivity liquids are used. However, where antistatic additives can be leached out in other operations, they should be used with caution.

When the double layer of a free surface of a liquid is disturbed to form a mist, charge separation and accumulation will occur, and the resulting mist can be highly charged. This can occur in both conductive and insulative liquids: splashing, bubbling, stirring, splash loading, hoses, the wind blowing, shower heads, Niagara Falls, etc.

Charge Induction Charge induction requires two objects: a primary object that has already been charged and a secondary object that responds to the electric field created by the primary object. That is, the electric field from the primary object causes charge separation to occur within the body of the secondary object. The objects can be either conductive or insulative, but the primary concern in electrostatic hazards is when the primary object is insulative and the secondary object is conductive. For example, consider the case where a flammable, conductive liquid is in an insulative plastic container and the outside of the container is charged by rubbing. The outer charge will induce itself onto the inner conductive liquid, and if it is poured into a grounded, conductive receptacle, a spark may occur that could ignite the liquid. Consider the opposite case where a charged, insulative liquid is poured into an isolated, conductive container. The charge will induce itself onto the container, and a spark may occur if the container touches ground.

Electrostatic Discharge By far the most common discharges in chemical processes are the *capacitive spark discharges* that occur between two conductive bodies at a different electrical potential. The flow of electrons in the spark is concentrated in a channel of high-energy ions, which, if large enough, can ignite flammable materials. Spark discharges occur when the electric field in the gap between the two conductors exceeds the dielectric breakdown of the air in the gap. When breakdown in the air begins, a channel of ionized gases is formed that has a low resistance such that the electrical energy stored in the conductor can freely flow into the spark. The critical electric field strength in air of some three million volts per meter can be exceeded by increasing the voltage across the gap or by closing the gap—for example, by touching a finger to a doorknob.

If there is enough energy in the spark, a flammable vapor, mist, or dust will ignite; if there is not enough energy, ignition will not occur. A minimum amount of energy is needed to cause the ignition of a given configuration of vapor, mist, or dust; this is called the minimum ignition energy, MIE. Spark discharges can occur from ungrounded conductive objects in the workplace: buckets, carts, cans, drums, tanks, process equipment, and humans, to name a few. When such items are involved, the energy, W, dissipated into the spark can be estimated if two of the three following parameters of capacitance, C, charge, Q, and voltage, V, are known (or reasonably estimated) by the following relations:

$$C = Q/V \quad \text{and} \quad W = CV^2/2 = QV/2 = Q^2/2C \tag{23-12}$$

The MIEs of organic vapors are between 0.01 and 1 mJ. Mists and dusts have higher MIEs and are typically in the range of 1 to 5000 mJ.

Corona discharge is a one-electrode diffuse discharge where the strength of an electric field exceeds the breakdown potential of air. Corona discharge commonly occurs at sharp points or edges of grounded conductors. Corona discharges are not incendive to ordinary vapors. *Brush discharge* is a higher-energy form of corona discharge that occurs at grounded conductors having rounded surfaces. Brush discharges can be incendive to vapors but not dusts. *Propagating brush discharge* is an energetic discharge that occurs when a grounded electrode approaches a highly charged insulative surface that has a conductive backing—for example, a plastic film on a metal surface. Propagating brush discharges can have energies of several joules and are therefore incendive to vapors and dusts. *Bulking brush discharge* occurs on the surface of a powder as it is being bulked into a container; it can have an effective energy of 10 to 25 mJ. Bulking brush discharges are therefore incendive to vapors and some dusts.

Bonding and Grounding By far the easiest and most effective way to prevent the accumulation of charge on conductive objects is to maintain proper bonding and grounding of all conductive process equipment. In order to maintain potentials below hazardous levels in conductive process equipment, a resistance to ground of less than $10^6\ \Omega$ is quite sufficient to preclude incendive sparks. When process equipment is installed (or maintenance performed) where conductive paths to ground are to be depended upon to maintain a path to ground, a resistance of less than 10 Ω should be confirmed for all of the equipment. The notion here is that if the resistance to ground is less than 10 Ω initially, then a resistance of less than $10^6\ \Omega$ will be maintained over the lifetime of the equipment. If resistance is not less than 10 Ω, the equipment owner/operator needs to find out why and have it fixed. In cases where two or more pieces of equipment must be bonded together so there can be no potential difference between them, clamps and wires may be used. Such equipment should be robust enough to withstand the rigors of the workplace, and the clamps must be strong enough to penetrate through coatings of paint and rust. Flexible hoses made of nonconductive material that incorporate conductive stiffening wires are often overlooked and left isolated. These should be installed to ensure that the wires are in good contact with conductive, grounded equipment. Likewise, any wires in boots or socks used for gravity transfer operations must also be grounded.

Human Electrostatic Discharge The human body is a conductor of electricity and should be treated as such because it may become isolated so that it can accumulate a charge, which can then lead to the ignition of a vapor, mist, or dust by spark discharge. Common ways of accumulating charge are by walking over an insulative surface, by rubbing an insulative surface, by rising from an insulative chair, or by similar contact or separation actions. Also, a charge can be accumulated on a person when a charged, insulative liquid is emptied from a conductive container. The rubbing together of clothes while moving around the workplace separates charge so that a positive charge will be on one layer but an equal and opposite charge will be on the adjacent layer. Since there is no net charge, sparking will not occur; but if some clothing is removed, a net charge will remain on the person; this can then discharge into an incendive spark. Charges on a person can be dissipated by the use of conductive flooring and electrostatic dissipative footwear; resistance to earth should be between $10^6\ \Omega$ and $10^9\ \Omega$.

Humidity When two insulative materials are contacted and then separated, the amount of charge accumulated on the separated faces will depend on the surface conductivity of the materials being separated. The higher the surface conductivities of the materials being separated, the easier it is for the charges to recombine as the separation occurs; conversely, the lower the surface conductivity of the materials, the more difficult it is for the charges to recombine in situ. Surface conductivity can vary by orders of magnitude with the amount of moisture adsorbed on the surface; this, in turn, varies with humidity. The classic example of charge separation is walking across a carpet when the humidity is low and getting a spark when touching a doorknob. This sort of charge accumulation is surface conductivity dependent; therefore, charge accumulation can sometimes be reduced, or even eliminated, by increasing the humidity. However, one must be careful to increase the humidity where the electrostatic charge is generated, not the ambient humidity outside of the process equipment.

CHEMICAL REACTIVITY

GENERAL REFERENCES: ASTM E1231, *Standard Practice for Calculation of Hazard Potential Figures of Merit for Thermally Unstable Materials*, ASTM International, West Conshohocken, Penn., 2015; ASTM E2012, *Standard Guide for the Preparation of a Binary Chemical Compatibility Chart*, ASTM International, West Conshohocken, Penn., 2012; Barton and Rogers, eds., *Chemical Reaction Hazards: A Guide to Safety*, 2d ed., Gulf Publishing Co., Houston, Tex., 1997; Beever, "Scaling Rules for Prediction of Thermal Runaway," *Intl. Symp. on Runaway Reactions*, CCPS-AIChE, Cambridge, Mass., 1989; Bowes, "A General Approach to the Prediction and Control of Potential Runaway Reaction," *Inst. Chem. Eng., Symp. Ser. 68*, 1981; Bowes, *Self-Heating: Evaluating and Controlling the Hazards*, Elsevier, Amsterdam, 1984; *Bretherick's Reactive Chemical Hazards Database*, version 3.0, Butterworth-Heinemann, Oxford, UK, 2000; Center for Chemical Process Safety (CCPS), Chemical Reactivity Worksheet, version 4.0, www.aiche.org/ccps/resources/chemical-reactivity-worksheet-40, 2016; CCPS, *Guidelines for Chemical Reactivity Evaluation and Application to Process Design*, AIChE, New York, 1995; CCPS, *Guidelines for Safe Storage and Handling of Reactive Materials*, AIChE, New York, 1995; CCPS, Safety Alert, "Reactive Material Hazards: What You Need To Know," AIChE, New York, 2001; Davies, *Organic Peroxides*, Butterworths, London, 1961; ESCIS, *Thermal Process Safety: Data, Assessment Criteria, Measures*, Safety Series, Booklet 8, Expert Commission for Safety in the Swiss Chemical Industry, Basel, Switzerland, April 1993; Fenlon, "Calorimetric Methods Used in the Assessment of Thermally Unstable or Reactive Chemicals," *Intl. Symp. for the Prevention of Major Chemical Accidents*, AIChE, 1987; Fisher et al., "Emergency Relief System Design Using DIERS Technology: The Design Institute for Emergency Relief Systems (DIERS) Project Manual," AIChE, New York, 1992; Frank-Kamenetskii, *Diffusion and Heat Transfer in Chemical Kinetics*, Plenum Press, New York, 1969; Frurip et al., "A Review of Chemical Compatibility Issues," *AIChE Loss Prevention Symp.*, Houston, Tex., 1997; Grewer, *Thermal Hazards of Chemical Reactions*, 2d ed., Elsevier, London, 2015; Hazard Investigation, "Improving Reactive Hazard Management," Report No. 2001-01-H, U.S. Chemical Safety and Hazard Investigation Board, Washington, D.C., October 2002; Hendershot, "A Checklist for Inherently Safer Chemical Reaction Process Design and Operation," *Intl. Symp. on Risk, Reliability and Security*, CCPS-AIChE, New York, October 2002; "How to Prevent Runaway Reactions, Case Study: Phenol-Formaldehyde Reaction Hazards," EPA 550-F99-004, U.S. Environmental Protection Agency, Washington, D.C., August 1999; HSE, *Designing and Operating Safe Chemical Reaction Processes*, U.K. Health and Safety Executive, 2000; *International Conference on Managing Chemical Reactivity Hazards and High Energy Release Events*, CCPS-AIChE, New York, 2003; *International Symp. on Runaway Reactions and Pressure Relief Design*, DIERS-AIChE, New York, 1995; Johnson et al., *Essential Practices for Managing Chemical Reactivity Hazards*, CCPS-AIChE, New York, 2003; Kohlbrand, "Reactive Chemical Screening for Pilot Plant Safety," *Chem. Eng. Prog.* **81**(4): 52 (1985); Kohlbrand, "The Relationship Between Theory and Testing in the Evaluation of Reactive Chemical Hazards," *Intl. Symp. for the Prevention of Major Chemical Accidents*, AIChE, New York, 1987; Leggett, "Chemical Reaction Hazard Identification and Evaluation: Taking the First Steps," *AIChE Loss Prevention Symp.*, New Orleans, March 2002; Lewis, *Sax's Dangerous Properties of Industrial Materials*, 12th ed., Wiley, New York, 2012; Mannan, *Lees' Loss Prevention in the Process Industries*, 4th ed., Butterworth-Heinemann, Oxford, UK, 2012; Mosley et al., "Screen Reactive Chemical Hazards Early in Process Development," *Chem. Eng. Prog.* **96**(11): 51–65 (2000); NFPA 49, "Hazardous Chemicals Data," and NFPA 491, "Hazardous Chemical Reactions," in *Fire Protection Guide to Hazardous Materials*, 14th ed., National Fire Protection Association, Quincy, Mass., 2010; NFPA 400, "Hazardous Materials Code," National Fire Protection Association, Quincy, Mass., 2016; Pohanish and Greene, *Wiley Guide to Chemical Incompatibilities*, 3rd ed., Wiley, New York, 2009; *Second Intl. Symp. on Runaway Reactions, Pressure Relief Design and Effluent Handling*, DIERS-AIChE, New York, 1998; Stoessel, *Thermal Safety of Chemical Processes: Risk Assessment and Process Design*, Wiley-VCH, Veinheim, 2008; Stull, "Linking Thermodynamics and Kinetics to Predict Real Chemical Hazards," *Loss Prevention 7*, AIChE, 1976, p. 67; Thomas, "Self Heating and Thermal Ignition—A Guide to Its Theory and Application," ASTM STP 502, pp. 56–82, ASTM International, 1972; Urben, ed., *Bretherick's Handbook of Reactive Chemical Hazards*, 7th ed., Academic Press, Oxford, UK, 2007; Yoshida et al., *Safety of Reactive Chemicals and Pyrotechnics*, Elsevier Science, London, 1995.

Introduction *Chemical reactivity* is the tendency of substances to undergo chemical change. A *chemical reactivity hazard* is a situation with the potential for an uncontrolled chemical reaction that can result directly or indirectly in serious harm to people, property, and/or the environment. A chemical reaction can get out of control whenever the reaction environment is not able to safely absorb the energy and products released by the reaction. The possibility of such situations should be anticipated not only in the reaction step of chemical processes but also in storage, mixing, physical processing, purification, waste treatment, environmental control systems, and any other areas where reactive materials are handled or reactive interactions are possible.

The main business of most chemical companies is to manufacture products by means of controlled chemical reactions. The reactivity that makes chemicals useful can also make them hazardous. Chemical reactions are usually carried out without mishap, but sometimes they get out of control because of problems such as the wrong or contaminated raw material being used, changed operating conditions, excess material additions, unanticipated time delays, failed equipment, incompatible materials of construction, loss of a critical utility, or loss of temperature control. Such mishaps can be worse if the chemistry under both normal and abnormal conditions is not fully understood. Therefore, it is essential that chemical process designers and operators understand the nature of the reactive materials and the chemistry involved under both normal and foreseeable abnormal situations, and what it takes to control intended reactions and avoid unintended reactions throughout the entire life cycle of a process facility.

Life Cycle Considerations

Considering Chemical Reactivity during Process Development Decisions made at the early development stages of a process facility, including conceptual and research phases, will largely determine the nature and magnitude of the chemical reactivity hazards that will need to be contained and controlled throughout the life cycle of the facility. For this reason, chemical reactivity hazards should be considered from the outset of process development, including creative thinking about feasible alternatives to the use of reactive materials or to the employment of highly energetic reactive systems. What may seem reasonable to the research chemist—handling materials in very small quantities—will have vast implications for the design and ongoing operation of a full-scale facility that must safely control the intended chemical reactions and avoid unintended reactions throughout the facility's lifetime.

Mosley et al. (2000) describe a chemistry hazard and operability (CHAZOP) analysis approach, similar to a HAZOP study but applied at the early development stages of a new process. Guide words are used to identify possible deviations from the intended process chemistry and to prompt discussion on their significance.

Many companies designate a particular person or position as the "owner" of the process chemistry; this responsibility is likely to change as the life cycle progresses from development to design to commissioning and ongoing operation. Data on the hazardous properties of the chemical reactions to be used and the materials to be handled should be assembled into a formal documentation package. Screening tests (described later in this subsection) may also need to be performed early in the development process to identify the consequences of abnormal reactions and of deviations such as exceeding the normal reaction temperature. This documentation package will then form part of the information base on which safeguards can be developed to control chemical reactivity hazards.

Considering Inherently Safer Approaches Specific to Reactivity Hazards The basic concepts of inherently safer plants, and the general strategies for making a facility inherently safer, are detailed in the later subsection on Inherently Safer Design and Related Concepts. Strategies that focus on chemical reactivity hazards, and steps to conduct a review of these strategies, are highlighted in that section.

Instead of choosing to receive and store a highly reactive raw material, it may be possible to use a less hazardous material that is one step farther along in the formulation or synthesis chain. Alternatively, a decision may be made to generate the material on demand and eliminate all or most storage and handling of the material. Many reactive materials can be handled in dilute solutions, dissolved in less hazardous solvents, or otherwise handled under inherently safer conditions. (For some reactive materials, such as benzoyl peroxide, handling as a dilute paste or solution is essential to the safe handling of the material.)

Inherently safer facilities with respect to chemical reactivity hazards must focus on the magnitude of stored chemical energy, the kinetics of how fast the energy could be released, and the possible reaction products that may themselves have hazardous properties such as toxicity or flammability.

With respect to kinetics, a slower reaction might be considered at first glance to be inherently safer than a rapid reaction. This may indeed be the case, if the energy and products of the slower reaction can always be dissipated safely without causing harm or loss. However, this is often not the case, for two important reasons. First, regardless of the speed of the reaction, the same potential chemical energy is still thermodynamically present, and it may be available under abnormal conditions, such as an external fire or the introduction of a catalytic contaminant. Second, a slower reaction may allow unreacted material to accumulate. Hence, faster reactions are generally more desirable, as discussed in the general reaction considerations that follow.

Finally, with respect to reaction products, a chemical reaction that does not generate hazardous reaction products or by-products is inherently safer than one that does. However, one must consider more than the obvious hazardous reaction products. The generation of any kind of noncondensible gases can cause a vessel rupture due to internal overpressurization if it is not adequately vented or relieved.

The following is a typical agenda for an inherent safety review at the concept or development stage of a new facility involving reactivity hazards (Johnson et al., *Essential Practices for Managing Chemical Reactivity Hazards*, AIChE, New York, 2003):

1. Review what is known of the chemical reactivity hazards (as well as other hazards) that will need to be contained and controlled in the proposed process. This existing level of knowledge might come from past experience, suppliers, literature reviews, incident reports, or other sources.
2. Based on the level of knowledge of chemical reactivity hazards, determine if additional screening of reactivity hazards is necessary. Having reactive functional groups might indicate the need to perform literature searches, access databases, or run differential scanning calorimetry.
3. Discuss possible process alternatives and their relative hazards, including discussions on such topics as alternative solvents and possible incompatibilities to avoid.
4. Brainstorm and discuss possible ways to reduce or eliminate the hazards.
5. Obtain consensus on significant unknowns that will need to be addressed.
6. Document the review, including attendees, scope, approach, and decisions.
7. Assign follow-up items with responsibilities, goal completion dates, and a closure mechanism such as reconvening after a designated number of weeks.

Scale-Up Considerations A key consideration when scaling up a reactive process, such as from a pilot plant to a full-scale facility, is to ensure adequate heat removal for normal or abnormal exothermic reactions. Heat generation is proportional to *volume* (mass) in a reactive system, whereas heat removal is only proportional to *area* (surface area) at best. Even though the reaction temperature can be easily controlled in the laboratory, this does not mean that it can be adequately controlled in a plant-scale reactor. Increasing the size of a reactor—or of another process or storage vessel where, for example, polymerization or slow degradation can occur—without adequately considering heat transfer can have disastrous effects. The careful design of the agitation or recirculation system is likewise important when scaling up, and the combined effects on the design of the emergency relief system must be taken into account.

Scale-up can also have a significant effect on the basic process control system and safety systems in a reactive process. In particular, a larger process will likely require more sensors at different locations in the process to be able to rapidly detect the onset of out-of-control situations. Consideration should be given to the impact of higher-temperature gradients in plant-scale equipment compared to a laboratory or pilot plant reactor (Hendershot, "A Checklist for Inherently Safer Chemical Reaction Process Design and Operation," *Intl. Symp. on Risk, Reliability and Security*, CCPS-AIChE, New York, October 2002).

Designing Processes for Control of Intended Chemical Reactions

General Considerations The following should be taken into account whenever one is designing or operating a chemical process that involves intended chemical reactions [Hendershot (2002)]. CCPS (*Guidelines for Process Safety in Batch Reaction Systems*, AIChE, New York, 1999) also describes many key issues and process safety practices to consider that are oriented toward the design and operation of batch reaction systems.

- *Know the heat of reaction for the intended and other potential chemical reactions.*
- *Calculate the maximum adiabatic temperature for the reaction mixture (i.e., if the reaction goes to completion and all of the heat of reaction goes toward heating up the remaining products of reaction).*
- *Determine the stability of all components of the reaction mixture at the maximum adiabatic reaction temperature.* This might be done through literature searching, supplier contacts, or experimentation.
- *Understand the stability of the reaction mixture at the maximum adiabatic reaction temperature.* Are there any chemical reactions, other than the intended reaction, that can occur at the maximum adiabatic reaction temperature? Consider possible decomposition reactions, particularly those that generate gaseous products.
- *Determine the heat addition and heat removal capabilities of the reactor.* Do not forget to consider the reactor agitator as a source of energy—about 2550 Btu/(h·hp).

- *Identify potential reaction contaminants.* In particular, consider possible contaminants that are ubiquitous in a plant environment, such as air, water, rust, oil, and grease. Think about the possible catalytic effects of trace metal ions such as sodium, calcium, and others commonly present in process water.
- *Consider the impact of possible deviations from intended reactant charges and operating conditions.* For example, is a double charge of one of the reactants a possible deviation, and, if so, what is the impact?
- *Identify all heat sources connected to the reaction vessel and determine their maximum temperature.*
- *Determine the minimum temperature to which the reactor cooling sources could cool the reaction mixture.*
- *Understand the rate of all chemical reactions.* Thermal hazard calorimetry testing can provide useful kinetic data.
- *Consider possible vapor-phase reactions.* These might include combustion reactions, other vapor-phase reactions such as the reaction of organic vapors with a chlorine atmosphere, and vapor-phase decomposition of materials such as ethylene oxide or organic peroxide.
- *Understand the hazards of the products of both intended and unintended reactions.*
- *Rapid reactions are desirable.* In general, it is preferable to have chemical reactions occur immediately when the reactants come into contact.
- *Avoid batch processes in which all the potential chemical energy is present in the system at the start of the reaction step.*
- *Avoid using control of reaction mixture temperature as a means for limiting the reaction rate.*
- *Avoid feeding a material to a reactor at a higher temperature than the boiling point of the reactor contents.* This can cause rapid boiling of the reactor contents and vapor generation.

Exothermic Reactions and "Runaway Reactions" The term *runaway reaction* is often improperly used to refer to any uncontrolled chemical reaction. As properly used, it refers to loss of control of a kinetically limited, exothermic reaction that proceeds at a stable, controlled rate under normal conditions with adequate removal of the heat of reaction (Fig. 23-14). When the situation changes so that the heat of reaction is not adequately removed, the excess heat increases the temperature of the reaction mass, which in turn increases the reaction rate and thus the rate of heat release as an exponential function of reaction temperature. If not limited by some means such as (1) the limiting reactant being exhausted, (2) a solvent removing the heat of reaction by boiling off, or (3) quenching or inhibiting the reaction, this "bootstrap" situation can result in an exponential temperature rise that can reach as high as hundreds of Celsius degrees per minute. The resulting temperature increase, generation of gaseous reaction products, and/or boil-off of evaporated liquid can easily exceed a pressure or thermal limit of the containment system if not adequately relieved. The elevated temperatures may also initiate a secondary or side reaction that is even more rapid or energetic.

This runaway situation can be understood by comparing Fig. 23-14 with Fig. 23-15, which has two new lines added, for two possible upset conditions in a process with a cooling coil or other heat exchanger being used to absorb the heat of an exothermic reaction. The temperature of the cooling medium might increase (shift from line 1 to line 2), or the heat-transfer coefficient might decrease, such as by heat exchanger fouling (shift from line 1 to line 3). When one of these shifts gets past point T_{NR} (temperature of no return), then the heat removal line no longer crosses the heat generation line and, as a result, stable operation is no longer possible. The heat of reaction causes the system temperature to increase, which further increases the rate of heat generation, which further increases the system temperature, and so on.

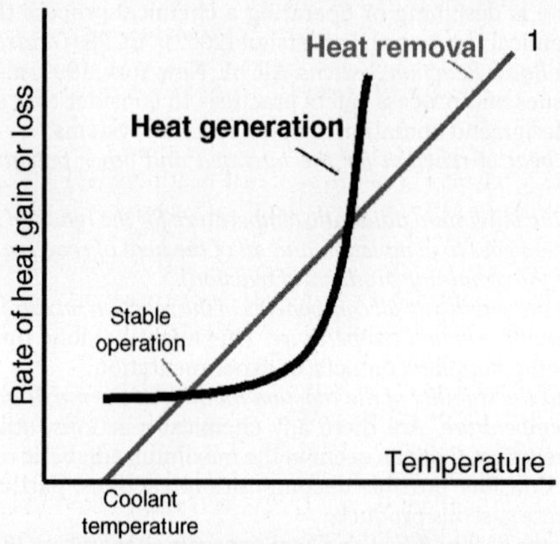

FIG. 23-14 For stable operation, all heat generated by an exothermic reaction is transferred to the surroundings, by whatever means (*conduction, evaporation, etc.*).

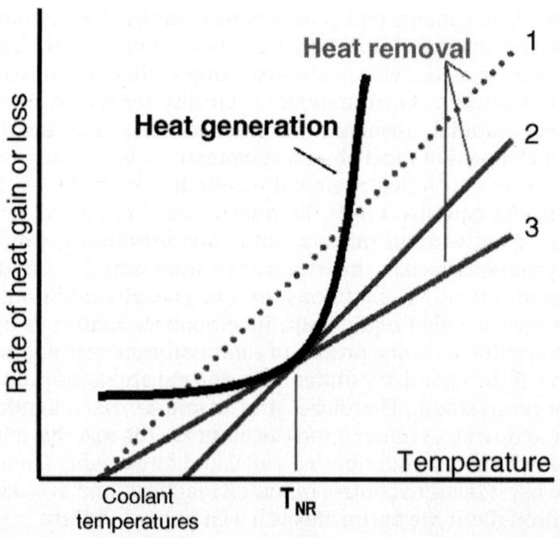

FIG. 23-15 For an exothermic reaction system with heat removal, such as to a vessel jacket and cooling coil, the limit of stable operation is reached as the reaction temperature increases to T_{NR} (temperature of no return), beyond which the rate of heat generation, which increases exponentially with increasing temperature, exceeds the capability of the system to remove the heat of reaction (see text).

Many possible abnormal situations can initiate a runaway reaction. These include:

- A loss of flow of cooling medium to or from the reactor
- An increase in the temperature of the cooling medium
- A general increase in the temperature of the storage or process configuration, such as might be caused by an extreme ambient condition or a loss of refrigeration
- Abnormal heat addition to the reactive material or mixture, such as might be caused by an external fire or the injection of steam into a vessel jacket or directly into the material or mixture
- Intentional heating of a vessel containing thermally sensitive material, due to a lack of recognition of the runaway hazard or other reason
- Gradual fouling of the heat exchange surfaces to the point that maximum coolant flow is no longer sufficient to remove the heat of reaction
- Loss of agitation or circulation of the reactant mass or other reduction in the heat-transfer coefficient, or contact with the heat exchange surface
- Insulation of the system, resulting in less heat dissipation
- Addition of a contaminant or excess catalyst that would increase the reaction rate
- Excess or rapid addition of a limiting reactant
- Adding reactants at too low a process temperature, then increasing the temperature
- Delay or interruption in vessel agitation, allowing unreacted material to accumulate, then starting or restarting agitation
- Delayed addition of catalyst
- Blockage of a vapor line or other means of increasing the system pressure
- Loss of a moderating diluent or solvent
- Inadequate inhibitor concentration in a storage container, or inadequate mixing of the inhibitor (including as a result of the freezing of the material)
- Transfer of the reactive material or mixture to a location not capable of removing the heat of reaction

As can be seen from the preceding list, runaway reactions do not occur by a single mechanism. They can take place not only in reactors but also in raw material and product storage containers, vessels, purification systems, and anywhere else exothermic reactive systems and self-reacting materials (described later in this subsection) are involved.

Historical perspective. An analysis of thermal runaways in the United Kingdom (Barton, J. A., and P. F. Nolan, "Incidents in the Chemical Industry Due to Thermal-Runaway Chemical Reactions," *Hazards X: Process Safety in Fine and Specialty Chemical Plants, IChemE series 115*, 1989, pp. 3–18) indicated that such incidents occur because of the following general causes:

- Inadequate understanding of the process chemistry and thermochemistry
- Inadequate design for heat removal
- Inadequate control systems and safety systems
- Inadequate operational procedures, including training

Semibatch reactions. The inherently safer way to operate exothermic reaction processes is to determine a temperature at which the reaction occurs very rapidly [Hendershot (2002)]. The reactor can be operated at this temperature while feeding one or more of the reactants gradually to limit the potential energy contained in the reactor. This type of gradual addition process is often called *semibatch.* It is desirable to have a physical way to restrict the rate at which the limiting reactant is added—for example, a metering pump, a small feed line, or a restricting orifice. Ideally, the limiting reactant should react immediately, or very quickly, when it is charged. The reactant feed can be stopped if necessary if there is any kind of a failure (e.g., loss of cooling, power failure, loss of agitation), and the reactor will contain little or no potential chemical energy from unreacted material. Some means to confirm actual reaction of the limiting reagent is also desirable. A direct measurement is best, but indirect methods such as monitoring the demand for cooling from an exothermic batch reactor can also be effective.

Design of Emergency Relief and Effluent Treatment Systems Containment systems are only rarely designed with a high enough pressure and temperature rating to fully contain a runaway reaction. For this reason, overpressure protection is critically important as a last line of defense against loss events that can result from runaway reactions. The parts of this section on Pressure Relief Systems and on Emergency Relief Device Effluent Collection and Handling address design basis selection, relief calculations, and effluent treatment system configurations for reactive system overpressure protection.

Endothermic Reactions An endothermic reaction process is generally easier to bring to a safe state if an out-of-control situation is detected. Discontinuing heat input is usually the primary line of defense to stop the operation. In this regard, the endothermic reaction is inherently safer than an exothermic reaction.

The following should especially be taken into account:

- The final product of an endothermic chemical reaction has a greater energy content than the starting materials. For this reason, materials with net positive heats of formation are often termed *endothermic compounds.* (For example, most high explosives are endothermic compounds.) This energy content can potentially be released in an uncontrolled manner if enough energy is again added to the material, such as by heating it to a decomposition temperature.
- Likewise, if control of an endothermic reaction process is lost, such as by a heating control valve opening too far or by a steam leak directly into the reaction mass, a degradation reaction or other secondary or side reaction may be initiated that can be exothermic and can lead to a thermal runaway.
- Some endothermic compounds can gradually degrade, decompose, become more concentrated, or become sensitized over time.

Designing Facilities for Avoidance of Unintended Reactions

General Considerations The following general design and operational considerations for avoiding unintended chemical reactions are summarized from a CCPS Safety Alert (*Reactive Material Hazards: What You Need To Know*, AIChE, New York, 2001):

- Train all personnel to be aware of reactivity hazards and incompatibilities and to know maximum storage temperatures and quantities.
- Design storage and handling equipment with compatible materials of construction.
- Avoid heating coils, space heaters, and all other heat sources for thermally sensitive materials.
- Avoid confinement when possible; otherwise, provide adequate emergency relief protection.
- Avoid the possibility of pumping a liquid reactive material against a closed or plugged line.
- Locate storage areas away from operating areas in secured and monitored locations.
- Monitor material and building temperatures where feasible with high-temperature alarms.
- Clearly label and identify all reactive materials and what must be avoided (e.g., heat, water).
- Positively segregate and separate incompatible materials, using dedicated equipment if possible.
- Use dedicated fittings and connections to avoid unloading a material to the wrong tank.
- Rotate inventories for materials that can degrade or react over time.
- Pay close attention to housekeeping and fire prevention around storage and handling areas.

Identifying Potential Reactions The U.S. Chemical Safety and Hazard Investigation Board's Hazard Investigation "Improving Reactive Hazard Management" (Washington, D.C., 2002) highlighted the importance of identifying chemical reactivity hazards, following an examination of 167 previous reactive incidents. CCPS has published a preliminary screening methodology for identifying where reactive hazards are likely to exist [Johnson et al. (2003)]. The flowchart for the preliminary screening methodology is shown in Fig. 23-16.

The following paragraphs break down the types of reactive materials and reactive interactions that an engineer may need to address in the design of a chemical process or other facility such as a warehouse where reactive materials are handled [Johnson et al. (2003)]. These can be considered to be in three larger categories:

- Self-reactive substances (polymerizing, decomposing, rearranging)
- Substances that are reactive with ubiquitous substances such as air (spontaneously combustible/pyrophoric, peroxide forming), water (water reactive), or ordinary combustibles (oxidizers)
- Incompatible materials

Polymerizing, Decomposing, and Rearranging Substances Most of these substances are stable under normal conditions or with an added inhibitor, but they can energetically self-react with the input of thermal, mechanical, or another form of energy sufficient to overcome their activation energy barriers (see Sec. 7, Reaction Kinetics, under Theoretical Methods). The rate of self-reaction can vary from imperceptibly slow to violently explosive, and it is likely to accelerate if the reaction is exothermic or self-catalytic.

The tendency of a material such as acrylic acid or styrene to *polymerize* is usually recognized. The chemical supplier's safety data sheet (SDS) can be checked and the supplier contacted to learn whether hazardous polymerization might be expected. A less energetic means of self-reaction is by molecular *rearrangement*, such as by isomerizing, tautomering, disproportionating, or condensing.

The *decomposition* of some materials into smaller, more stable molecules can be initiated by mechanical shock alone; these materials are known as *shock-sensitive.* Many commercially important chemicals are *thermally sensitive* and decompose with the addition of heat. For storage situations, the critical temperature at which the thermal energy is sufficient to start an uncontrolled reaction in a particular storage configuration for a specified time is known as the *self-accelerating decomposition temperature* (SADT).

Decomposing materials are sometimes referred to as *unstable.* They generally have a positive heat of formation, so energy will be released when the decomposition reaction occurs. Self-reactive materials can often be recognized by the presence of certain chemical structures that tend to confer reactivity. These include

- Carbon–carbon double bonds not in benzene rings (e.g., ethylene, styrene)
- Carbon–carbon triple bonds (e.g., acetylene)
- Nitrogen-containing compounds (NO_2 groups, adjacent N atoms, etc.)
- Oxygen–oxygen bonds (peroxides, hydroperoxides, ozonides)
- Ring compounds with only three or four atoms (e.g., ethylene oxide)
- Metal- and halogen-containing complexes (metal fulminates, halites, halates, etc.)

A more complete list is given in CCPS's *Guidelines for Safe Storage and Handling of Reactive Materials* (AIChE, New York, 1995), and specific compounds can be investigated in resources such as *Bretherick's Handbook* (Urben, ed., *Bretherick's Handbook of Reactive Chemical Hazards,* 7th ed., Academic Press, Oxford, UK, 2007).

General considerations for avoiding unintended reactions with self-reacting substances include:

- Knowing the mechanisms and boundaries of what will initiate a self-reaction
- Maintaining diluents or inhibitors to extend the boundaries where feasible and avoiding the mechanisms (such as shock and overtemperature) that would initiate the self-reaction
- Having reliable controls and last-resort safety systems in place to detect and deal with an incipient out-of-control condition

Specific design considerations for a few substances, including acrylic acid, styrene, organic peroxides, ethylene oxide, and 1,3-butadiene, are given in CCPS' *Guidelines for Safe Storage and Handling of Reactive Materials* (AIChE, New York, 1995) on the basis of an industry-practice survey. Industry user groups distribute detailed information for other substances. These include methacrylic acid and methacrylate esters (http://www.mpausa.org) and ethylene oxide (https://www.americanchemistry.com/ProductsTechnology/Ethylene-Oxide/EO-Product-Stewardship-Manual-3rd-edition/default.aspx).

Spontaneously Combustible and Pyrophoric Substances *Spontaneously combustible* substances will readily react with the oxygen in the atmosphere, igniting and burning even without an ignition source. Ignition may be immediate, or it may result from a self-heating process that may take minutes or hours. As a result, some spontaneously combustible substances are known as *self-heating* materials.

Pyrophoric materials ignite spontaneously on short exposure to air under ordinary ambient conditions. Some materials that are considered pyrophoric require a minimum relative humidity in the atmosphere for spontaneous ignition to occur. The potential of pyrophoric materials to exhibit this behavior is usually well known due to the extreme care required for their safe handling.

Pyrophoric and other spontaneously combustible substances will generally be identified as such on their product literature, safety data sheets (SDSs), or International Chemical Safety Cards (ICSCs). If transported, these substances should be identified as DOT/UN Hazard Class 4.2 materials for shipping purposes and labeled as spontaneously combustible. For pyrophoric substances, the NFPA 704 diamond for container or vessel labeling has a red (top) quadrant with a rating of 4, indicating the highest severity of

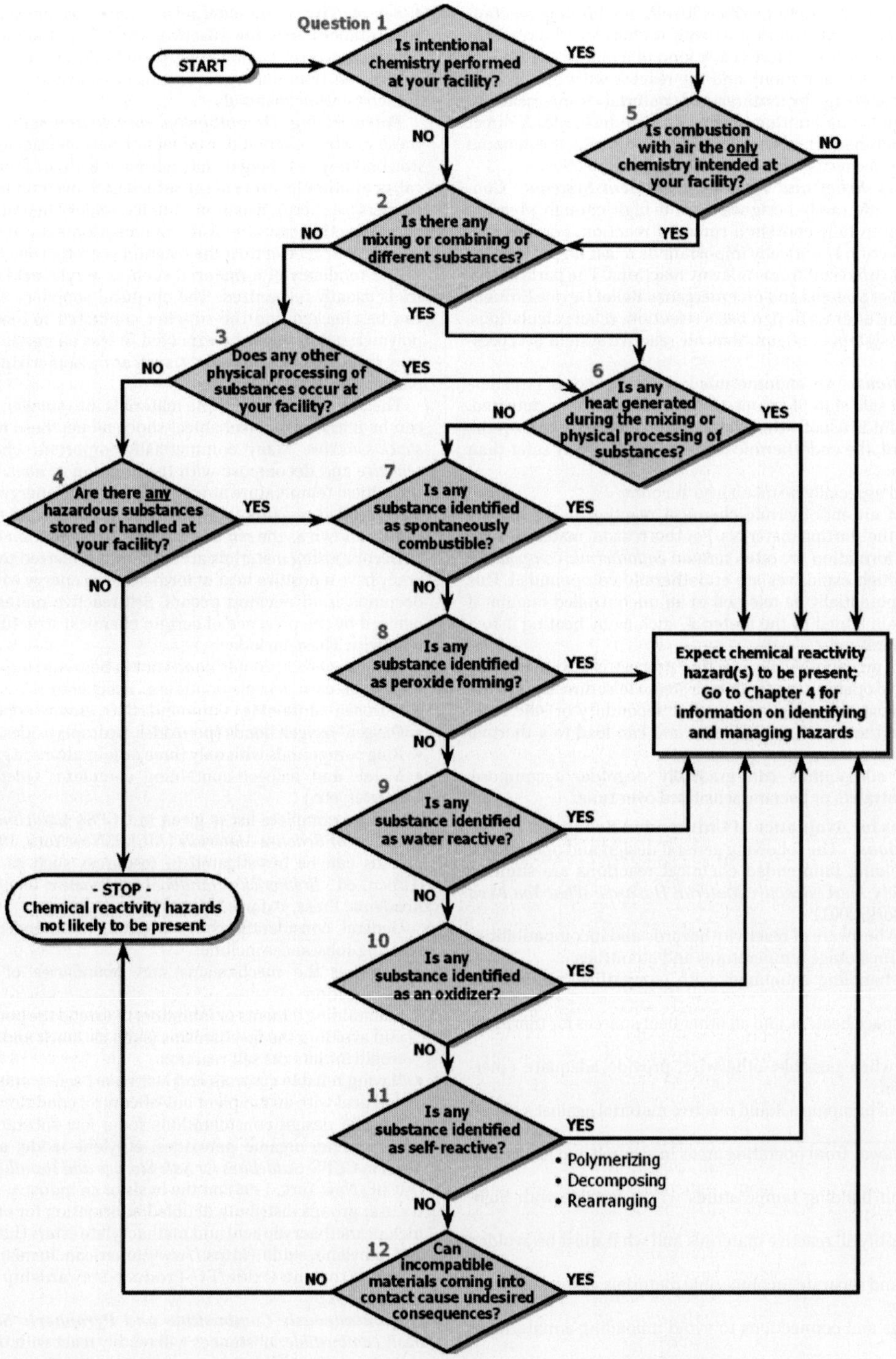

FIG. 23-16 CCPS preliminary screening for chemical reactivity hazards. [From Johnson et al., *Essential Practices for Managing Chemical Reactivity Hazards*, CCPS-AIChE, New York, 2003; copyright AIChE; reproduced with permission.]

flammability hazard (NFPA 704, *Standard System for the Identification of the Hazards of Materials for Emergency Response*, National Fire Protection Association, Quincy, Mass., 2012). Note that pyrophoric materials often exhibit one or more other reactivity hazards as well, such as water reactivity.

A scenario that has resulted in many fires and explosions in petroleum refineries involves iron sulfide. An impure, pyrophoric sulfide forms when streams containing hydrogen sulfide or other volatile sulfur compounds are processed in ferrous equipment. Oxidation of moist iron sulfide is highly exothermic. Opening iron-sulfide-containing equipment without adequate purging can result in rapid self-heating and ignition of the iron sulfide, which can then ignite other residual flammable gases or liquids in the equipment.

Many scenarios involving spontaneous combustion involve a combination of materials exposed to sufficient air, often in an insulating situation that prevents heat generated by a slow oxidation reaction from dissipating, which results in a self-heating situation.

Lists of pyrophoric materials that include less common chemicals, including metals, can be found in volume 2 of *Bretherick's Handbook* [Urben (2007)]. Other spontaneously combustible substances are tabulated by their proper shipping names and UN/NA numbers in the U.S. Dept. of Transportation regulation 49 CFR 172.101. Possible causes of uncontrolled reactions associated with pyrophoric and other spontaneously combustible materials are listed in Johnson et al. (2003).

Peroxide Formers *Peroxide formers* will react with oxygen in the atmosphere to form unstable peroxides, which in turn might explosively decompose if concentrated. Peroxide formation, or peroxidation, usually happens slowly over time, when a peroxide-forming liquid is stored with limited access to air.

Substances that are peroxide formers will often have an inhibitor or stabilizer added to prevent peroxidation. They are often not easily identifiable as peroxide formers by using SDSs or ICSCs. Rather, they are often identified by another characteristic, such as flammability, for storage and shipping purposes. Examples of peroxide formers include 1,3-butadiene, 1,1-dichloroethylene, isopropyl ether, and alkali metals. Johnson et al. (2003) tabulate other chemical structures susceptible to peroxide formation.

The total exclusion of air from vessels and equipment containing peroxide formers, and the establishment and observing of strict shelf life limitations, are basic strategies for managing peroxide-forming hazards.

Water-Reactive Substances *Water-reactive substances* will chemically react with water, particularly at normal ambient conditions. For fire protection purposes, a material is considered water-reactive if a gas or at least 30 cal/g (126 kJ/kg) of heat is generated when it is mixed with water (NFPA 704, *Standard System for the Identification of the Hazards of Materials for Emergency Response*, National Fire Protection Association, Quincy, Mass. 2012), using a two-drop mixing calorimeter.

Water reactivity can be hazardous by one or more of several mechanisms. The heat of reaction can cause thermal burns, ignite combustible materials, or initiate other chemical reactions. Flammable, corrosive, or toxic gases are often formed as reaction products. The violence of some reactions may disperse hazardous materials. Even slow reactions can generate sufficient heat and off-gases to overpressurize and rupture a closed container.

Water-reactive substances will almost always be identified as such on their SDSs or ICSCs. They may be identified as DOT/UN Hazard Class 4.3 materials for shipping purposes and labeled as dangerous when wet. However, some water-reactive materials are classified otherwise. Acetic anhydride is designated Class 8; it may also be identified as a combustible liquid.

The total exclusion of water from vessels and equipment containing water-reactive substances and the maintenance of the primary containment integrity over time are obvious design and operational considerations when handling water-reactive substances. Drying of equipment prior to start-up and careful design of provisions for cleaning and purging of equipment are also essential.

Oxidizers and Organic Peroxides An *oxidizer* is any material that readily yields oxygen or other oxidizing gas, or that readily reacts to promote or initiate combustion of combustible materials and can undergo a vigorous, self-sustained decomposition due to contamination or heat exposure (NFPA 400, *Hazardous Materials Code*, National Fire Protection Association, Quincy, Mass., 2016). Thus, most oxidizers can be thought of as being reactive with ordinary combustible liquids or solids, which are commonly used as process, packaging, general use, or structural materials. They can also react with many reducing chemicals, and they are often self-reactive with the organic constituent of the organic peroxide molecule.

Oxidizers will nearly always be identified as such on their SDSs or ICSCs. They may be identified as DOT/UN Hazard Class 5.1 materials for shipping purposes and labeled as oxidizers. However, some oxidizers are classified otherwise.

Volume 2 of *Bretherick's Handbook* [Urben (2007)] lists many structures and individual chemical compounds having oxidizing properties. NFPA 400 (2016) can be consulted for typical organic peroxide formulations. Note, however, that some organic peroxide formulations burn with even less intensity than ordinary combustibles and present no chemical reactivity hazard. NFPA 400 (2016) contains safety provisions for the storage of liquid and solid oxidizers and of organic peroxide formulations.

Incompatible Materials In this context, *incompatible* refers to two or more materials not being able to be combined with each other without undesired consequences. ASTM E2012 (ASTM International, West Conshohocken, Penn.) gives a method for preparing a binary compatibility chart for identifying incompatibilities. The Chemical Reactivity Worksheet (CCPS-AIChE, www.aiche.org/ccps/resources/chemical-reactivity-worksheet-40, 2016) uses a group compatibility method to predict the results of mixing any binary combination of over 5000 specific chemicals, including many common mixtures and solutions. Materials to be considered include not only raw materials and products but also by-products, waste products, cleaning solutions, normal and possible abnormal materials of construction, possible contaminants and degradation products, material that could be left in the process from a previous batch or cleanout, and materials in interconnected piping, heat-transfer systems, waste collection systems, or collocated storage.

The essence of the ASTM E2012 approach is to determine incompatibility *scenarios* that could foreseeably occur by examining all possible binary combinations. It may be necessary to review a process by using a systematic method such as a process hazard analysis (PHA) to identify all incompatibility scenarios that have a significant likelihood of occurrence and severity of consequences. The same review can then be used to evaluate whether adequate safeguards exist or whether further risk reduction is warranted.

Where the consequences of combining two or more materials under given conditions of temperature, confinement, etc., are unknown and cannot be predicted with certainty, testing may need to be performed to screen for potential incompatibilities. Two common test methods used for this purpose are differential scanning calorimetry and mixing cell calorimetry (described later in this section).

Design considerations to avoid contact between incompatible materials include the following: total exclusion of an incompatible substance from the facility; quality control and sampling of incoming materials; approval procedures for bringing new chemicals and materials of construction on-site; dedicated fittings and unloading spots; vessel, piping, and container labeling; dedicated or segregated storage; segregated diking, drainage, and vent systems; quality control of materials of construction (both initial construction and ongoing maintenance and modifications); sealless pumps, double tube sheets, and other means of excluding seal fluid, heat-transfer fluid, and other utility substances; positive isolation of interconnections by physical disconnects, blinding, or double block and bleed valves; avoidance of manifolds with flexible connections; use of compatible purge gases, cleaning solutions, heat-transfer fluids, insulation, fire-extinguishing and suppression agents whenever possible; and removal of unused materials from the site. These design considerations will always need to be accompanied by procedure training, hazard awareness, and operating discipline to be effective on an ongoing basis.

Designing Preventive Safeguards to Handle Uncontrolled Reactions Last-resort layers of protection are designed into many reactive chemical storage and handling operations as final lines of defense to avert a loss event such as an explosion or a hazardous material release, if the operation exceeds safe operating limits and it is not possible to regain control by using the operation's normal control mechanisms. Summarized here are some typical last-resort preventive safeguards (CCPS, *Guidelines for Safe Storage and Handling of Reactive Materials,* 1995, chap. 5). These safeguards are likely to be considered *safety instrumented systems* (SISs); they should be selected, designed, and maintained accordingly. (See the later subsection on Safety Instrumented Systems.) These last-resort systems require some amount of time to detect an incipient or out-of-control condition and bring the reactive system to a safe state; hence, they would not normally apply to situations such as decomposition of a shock-sensitive material.

Inhibitor Injection Inhibitor injection systems are primarily used with materials known to polymerize. If the material begins to polymerize in an uncontrolled manner, then injection of an inhibitor may be able to interrupt the polymerization reaction before a runaway reaction occurs and design limits are exceeded. The type of inhibitor needed will depend on the nature of the polymerization reaction. For example, a free-radical scavenger such as hydroquinone is likely to be used as an inhibitor for a material that reacts by free-radical polymerization. The inhibitor is often the same inhibitor that is used to maintain stability during normal transportation and storage, but injected in a much larger quantity. Injection of an inhibitor to kill an out-of-control reaction during operation or before a batch is completed is often called a *short-stop* system.

An inhibitor injection system needs to be carefully designed and maintained in order to be a reliable last-resort preventive safeguard. Since the inhibitor injection system is on standby and may not ever be actually used, the potency of the inhibitor and the functionality of the system components will need to be effectively tested on a periodic basis without excessive disruption of normal operations. CCPS's *Guidelines for Engineering Design for Process Safety, 2nd Edition* (AIChE, New York, 2012) discusses testing of continuous-process safety systems. This functional testing is important not only for the checking of the inhibitor supply and the proper operation of the delivery system, but also for testing the means of detecting an out-of-control situation and verifying successful control.

Quench Systems Quench systems in some form can be used for nearly all types of reactive systems. A quench system involves the addition of flooding quantities of water or other quenching medium to the reactive material. The quenching medium might be a subcooled material such as liquid nitrogen or dry ice in special applications.

The means by which a quench system works depends on the nature of the reactive material. As an example, for water-reactive materials, a quench system will destroy the material in a last-resort situation and generally form less-hazardous products, and, at the same time, the excess quenching medium will absorb much of the heat of reaction. Most quench systems are designed to both cool down and dilute a material that may be reacting uncontrollably; the quenching medium may also interfere with the chemical reaction or deactivate a catalyst.

Dump Systems For an inhibitor injection or quench system, the inhibitor or quenching medium is transferred from an external supply to the reactive material. In a dump system, the reactive material is transferred from the storage/handling facility to a safer location that is the same size or, more commonly, larger than the normal capacity of the vessel or process being protected.

This allows depressurizing and deinventorying of the reacting mass from the process in an out-of-control situation, such as an incipient runaway reaction.

Depressuring Systems A last-resort depressurizing system can be added to a reactive system to vent off excessive pressure buildup in a tank vessel in a controlled manner before reaching the relief valve or rupture disk set pressure. Such a depressurizing system typically consists of a remotely actuated vent valve connected to the vapor space of the vessel, with the venting discharge directed to a scrubber, flare, or other treatment system of adequate capacity. The system can be designed to be actuated either manually, by a control room or field operator, or by detection of high pressure and/or high temperature in the vessel.

Reactive Hazard Reviews and Process Hazard Analyses Reactive hazards can be evaluated using reviews of all new processes and of all existing processes on a periodic basis. Reviews should include:

1. Review of process chemistry, including reactions, side reactions, heat of reaction, potential pressure buildup, and characteristics of intermediate streams
2. Review of reactive chemicals test data for evidence of flammability characteristics, exotherms, shock sensitivity, and other evidence of instability
3. Review of planned operation of process, especially the possibility of upsets, modes of failure, unexpected delays, redundancy of equipment and instrumentation, critical instruments and controls, and worst-credible-case scenarios
4. Review of actual operational experience and abnormal situations, including equipment failures, process upsets, protective system functioning (or not functioning), unexplained exotherms, near misses, and actual loss-of-containment events.

These reviews either can be in addition to, or combined with, periodic process hazard analyses (PHAs) by using methods such as hazard and operability (HAZOP) studies and what-if analyses. The latter should consciously focus on identifying scenarios in which intended reactions could get out of control and unintended reactions could be initiated. One means of accomplishing this as part of a HAZOP study has been to include *chemical reaction* as one of the parameters to be investigated for each study node. Johnson and Unwin ("Addressing Chemical Reactivity Hazards in Process Hazard Analysis," *Proc. Intl. Conf. and Workshop on Managing Chemical Reactivity Hazards and High Energy Release Events,* CCPS-AIChE, New York, 2003) describe other PHA-related approaches for studying chemical reactivity hazards.

Worst-Case Thinking At every point in the operation, the process designer should conceive of worst-case combinations of circumstances that could realistically exist during the facility lifetime, such as loss of cooling water, power failure, wrong combination or amount of reactants, wrong valve position, plugged lines, instrument failure, loss of compressed air, air leakage, loss of agitation, deadheaded pumps, raw material impurities, and receipt of wrong materials. An engineering evaluation should then be made of the worst-case consequences, with the goal that the plant will be safe even if the worst case occurs. The previous discussion of calculating the maximum adiabatic temperature rise, then considering what might happen if it is realized, is an example of this type of analysis. A HAZOP study could be used to help identify abnormal situations and worst-case consequences.

Reactivity Testing Much of the data needed for the design of facilities with reactivity hazards involve the determination of thermal stability, including:

1. The possibility of an exothermic reaction occurring
2. The temperature at which measurable exothermic reactions start
3. The rate of reaction as a function of temperature
4. The heat generated per unit mass of material

In many cases, data on the increase of pressure during a reaction are also required, especially for vent sizing, and on the composition of the product gases.

The term *detected onset temperature* T_{onset} is used in two contexts:

1. In a testing context, it refers to the first detection of exothermic activity on the thermogram. The differential scanning calorimeter (DSC) has a typical scan rate of 10°C/min, whereas the accelerating rate calorimeter (ARC®) may have a sensitivity of 0.02°C/min. Consequently, the temperature at which thermal activity is detected by the DSC can be as much as 50°C different from ARC data.

2. The second context is the process reactor. There is a potential for a runaway if the net heat gain of the system exceeds its total heat loss capability. A self-heating rate of 3°C/day is not unusual for a monomer storage tank in the early stages of a runaway. This corresponds to 0.002°C/min, which is below ARC detection limits.

Sources of Reactivity Data Several important sources of reactivity data are described in the following paragraphs.

Calculations Potential energy that can be released by a chemical system can often be predicted by thermodynamic calculations. If there is little energy, the reaction still may be hazardous if gaseous products are produced. Kinetic data are usually not available in this way. Thermodynamic calculations should be backed up by actual tests.

Differential Scanning Calorimetry Sample and inert reference materials are heated in such a way that the temperatures are always equal. Onset-of-reaction temperatures reported by the DSC are higher than the true onset temperatures, so the test is mainly a screening test.

Differential Thermal Analysis (DTA) A sample and inert reference material are heated at a controlled rate in a single heating block. This test is basically qualitative and can be used for identifying exothermic reactions. Like the DSC, it is also a screening test. Reported temperatures are not reliable enough for quantitative conclusions. If an exothermic reaction is observed, it is advisable to conduct tests in the ARC.

Mixing Cell Calorimetry (MCC) The MCC provides information about the instantaneous temperature rise that results from the mixing of two compounds. Together, DSC and MCC provide a reliable overview of the thermal events that may occur in a process.

Accelerating Rate Calorimetry (ARC) This equipment determines the self-heating rate of a chemical under near-adiabatic conditions. It usually gives a conservative estimate of the conditions for, and consequences of, a runaway reaction. Pressure and rate data from the ARC may sometimes be used for pressure vessel emergency relief design. Activation energy, heat of reaction, and approximate reaction order can usually be determined. For multiphase reactions, agitation can be provided. Non-stirred ARC runs may give answers that do not adequately duplicate plant results when there are reactants that may settle out or that require mixing for the reaction to be carried out (DeHaven and Dietsche, "Catalyst Explosion: A Case History," *Plant/Oper. Prog.,* April 1990).

Vent Sizing Package (VSP2™) The VSP is an extension of ARC technology. The VSP2 is a bench-scale apparatus for characterizing runaway chemical reactions. It makes possible the sizing of pressure relief systems with less engineering expertise than is required with the ARC or other methods.

Advanced Reactive System Screening Tool (ARSST™) The ARSST measures sample temperature and pressure within a sample containment vessel. The ARSST determines the potential for runaway reactions and measures the rate of temperature and pressure rise (for gassy reactions) to allow determinations of the energy and gas release rates. This information can be combined with simplified methods to assess reactor safety relief venting requirements.

Shock Sensitivity Shock-sensitive materials react exothermically when subjected to a pressure pulse. Materials that do not show an exotherm on a DSC or DTA are presumed not to be shock-sensitive. Testing methods include:

- *Drop weight test* A weight is dropped on a sample in a metal cup. The test measures the susceptibility of a chemical to decompose explosively when subjected to impact. This test should be applied to any materials known or suspected to contain unstable atomic groupings.
- *Confinement cap test* Detonability of a material is determined by using a blasting cap.
- *Adiabatic compression test* High pressure is rapidly applied to a liquid in a U-shaped metal tube. Bubbles of hot compressed gas driven into the liquid may cause explosive decomposition of the liquid. This test is intended to simulate water hammer and sloshing effects in transportation, such as humping of railway tank cars. It is very severe and gives worst-case results.

Obtaining test data for designing a facility with significant reactivity hazards requires familiarity with a range of test equipment and a substantial amount of experience in the interpretation of test results.

TOXICITY

Nomenclature

A', B', C'	Antoine equation values, mm Hg, $\log_{10}$
B	Breathing rate, m^3/h
C_{MG/M^3}	Concentration of the gas or vapor, mg/m^3
C_N	Relative concentration of a gas or vapor, vol%
C_{PPM}	Concentration of the gas or vapor, ppmv (parts per million by volume)
D	Oral or ingestion dose, mg/kg of body weight
L	Concentration of gas or vapor for a specific injury, ppmv
M	Molecular weight of the gas or vapor, g/g-mol
t	Duration of exposure, min
T	Temperature, °C

General References: Crowl and Louvar, *Chemical Process Safety—Fundamentals with Applications*, 3d ed., chaps. 2 and 3, Prentice-Hall, Upper Saddle River, N.J., 2011; U.S. Department of Labor, Occupational Safety and Health Standards, *Process Safety Management of Highly Hazardous Chemicals*, Title 29 of the Code of Federal Regulations, 29 CFR 1910.119(d)(1)(i), (e)(3)(vii), and (f)(1)(iii)(A), 1992; U.S. Environmental

Protection Agency, *Risk Management Programs for Chemical Accidental Release Prevention*, Title 40 of the Code of Federal Regulations, 40 CFR 68.15(b)(3), 15(c), 24(c)(7), and 26(b)(1), 1993.

Introduction This section is primarily devoted to the short-term (acute) hazards of inhaling toxic gases, vapors, and dusts, together with methods for controlling the hazards. Also included is a brief discussion of skin-contact hazards, since inhalation and skin-contact hazards are encountered in chemical plants, with ingestion and injection of toxic materials being much less common hazards. Several types of toxic materials are *not* included in this discussion: (1) the typically chronic (low-concentration, long-term exposure) hazards of mutagens, teratogens, and carcinogens, (2) drugs, viruses, and other biological toxins, and (3) warfare agents such as nerve gases. Also excluded are discussions of aquatic and soil toxicity, with the degradation that is provided by atmospheric oxygen and ozone, by microbial action, and by sunlight. The Internet can provide sources for discussions of these other types of toxicity hazards, and control of such hazards.

Many natural and artificial substances are toxic to humans (and animals). Liquids and solids can be ingested, or exposure can be through the skin, eyes, or other external passages into the body. Where these substances are gaseous or volatile, toxic effects can result from inhalation. As a result of accidents and tests, it has been discovered that some of these substances are more toxic than others. Quantification of the degree of hazard has become important in devising appropriate measures for containing these substances, for providing healthful breathing atmospheres through ventilation or scrubbing, and in protecting personnel involved in cleanup of accidental spills with various types of breathing apparatus.

History of Toxic-Material Releases There have been relatively few releases of toxic materials that have affected the public, but some notable incidents that involved acute exposures are: (1) the truck-overturn release of ammonia in Houston, Texas, on May 11, 1976; (2) the runaway-reaction release of tetrachlorodibenzoparadioxin in Seveso, Italy, on July 9, 1976; (3) the tank car puncture release of chlorine in Mississauga, Ontario, Canada, on November 10, 1979; (4) the train derailment release of chlorine in Montanas, Mexico, on August 1, 1981; (5) the runaway-reaction release of methyl isocyanate in Bhopal, India, on December 3, 1984; (6) a release of liquid ammonia in Jonova, Lithuania, on March 20, 1989; (7) the rupture of a chlorine hose in Festus, Missouri, on August 14, 2002; (8) the release of allyl chloride in Dalton, Georgia, on April 12, 2004; (9) the rupture of a derailed chlorine tank car in Graniteville, South Carolina, on January 6, 2005; and (10) the poisoning caused by methyl bromide fumigant in the U.S. Virgin Islands on March 20, 2015.

Recent incidents that involved employee exposures to toxic gases and vapors were: (1) release of reaction-generated hydrogen sulfide in Pennington, Alabama, on January 16, 2002; (2) release of phosgene in Belle, West Virginia, on January 23, 2010; and (3) release of methyl mercaptan in LaPorte, Texas, on November 14, 2014. Also noteworthy were nitrogen-asphyxiation fatalities in Hahnville, Louisiana, on March 27, 1998; and in Delaware City, Delaware, on November 5, 2005.

Evaluation of Toxicity Hazards The present Process Safety Management standard of the Occupational Safety and Health Act (OSHA) requires "toxicity information" and "a qualitative evaluation of a range of the possible safety and health effects of failure of controls on employees in the workplace." Similarly, the Risk Management Programs for Chemical Accidental Release Prevention (RMP) standard of the Environmental Protection Agency's Clean Air Act Amendments requires toxicity information, with a qualitative evaluation of a range of the possible safety and health effects of failure of controls on public health and the environment, and with analysis of the off-site consequences of the worst-case release scenario and the other more-likely significant accidental-release scenarios.

The *Registry of Toxic Effects of Chemical Substances* (Accelrys, Inc., 2016) lists 25 different routes for intake of toxic substances by animals (and, therefore, humans). However, the most likely hazard route in chemical plants would be inhalation, followed by eye and skin contact, with oral (ingestion) and subcutaneous (needle-stick) less likely. Among the types of toxic substances are teratogens, mutagens, carcinogens, biological toxins such as venoms, viral toxins, toxic pollens and other natural materials such as poison ivy and poisonous pitches, chemical-warfare agents, and substances that can result in aquatic toxicity. Several references are concerned with the wide range of toxicity hazards (Bingham and Cohrssen, *Patty's Toxicology*, 6th ed., Wiley, New York, 2012; Stine and Brown, *Principles of Toxicology*, 3d ed., CRC Press, Boca Raton, Fla., 2015). This section is concerned primarily with inhalation toxicity.

The containment of toxic materials is the primary line of defense against toxicity hazards. The secondary line of defense—against releases of toxic materials from containment—is protection of personnel from inhalation of toxic gases or vapors. Determining the appropriate degree of protection requires knowledge of the concentrations that are hazardous and estimations of the hazardous concentrations that could occur. Thus, to perform safety and health evaluations, *quantitative* knowledge of the effects of exposure to toxic materials is needed. Data are available from several resources, as are discussed next.

Toxicity-Hazards Data Toxicity data for many of the more common substances are readily available (American Conference of Governmental Industrial Hygienists, *Threshold Limit Values for Chemical Substances and Physical Agents,* ACGIH, Cincinnati, Ohio, 2016; National Institute for Occupational Safety and Health, *Pocket Guide to Chemical Hazards,* CDC/NIOSH, Atlanta, Ga., 2016). Additional toxicity-hazards data are available for the general population, for specific locations, for different durations of exposure, and for different effects of exposure. To aid in the interpretation of inhalation-toxicity data, the following definitions of abbreviations and acronyms are provided. The Internet provides further information about definitions and the sources of toxicity data.

AEGL-1, -2, -3 The *acute exposure guideline level* (concentration) is (1) the airborne concentration of a substance above which it is predicted that the general population, including susceptible individuals, could experience notable discomfort, irritation, or certain asymptomatic nonsensory effects, and the effects are not disabling and are transient and reversible upon cessation of exposure; (2) the airborne concentration of a substance above which it is predicted that the general population, including susceptible individuals, could experience irreversible or other serious, long-lasting adverse health effects or an impaired ability to escape; or (3) the airborne concentration of a substance above which it is predicted that the general population, including susceptible individuals, could experience life-threatening health effects or death.

ATC Values of *acute toxicity concentration* are published as part of the New Jersey Toxic Catastrophe Prevention Act, for the protection of the public.

ATE Values of *acute toxicity estimate* are derived from tests on animals with the objective of meeting the requirement of the "harmonized" criteria for toxicity, as set by the IOMC (InterOrganization program for the sound Management of Chemicals) and the CG/HCCS (Coordinating Group for the Harmonization of Chemical Classification Systems).

DTL A *dangerous toxic load* is the dose (usually concentration times duration) that would give a specified adverse toxic effect.

ERPG-1, -2, -3 The *emergency response planning guidelines* are for one-hour exposures, (1) no effects other than mild transient adverse health effects or perception of a clearly defined objectionable odor; (2) no irreversible or other serious health effects or symptoms that could impair a person's ability to take protective action; (3) no life-threatening health effects.

HEC The *human equivalent concentration* is obtained by comparing the concentrations that affect humans to the concentrations that give a similar effect in rats (or other animals), to enable translation of rat data to human toxicity limits.

IDLH The *immediately dangerous to life and health* inhalation conditions that pose an immediate threat to life or health, including the hindering of the ability of a worker to escape, or causing severe eye or respiratory irritation, disorientation, or incoordination, as a consequence of a 30-minute exposure.

LCLo The *lethal concentration low* is the lowest concentration in air, other than LC50, which has caused death in humans or animals.

LC50 The *lethal concentration 50* is the concentration of a substance in air, exposure to which for a specified length of time is expected to cause the death of 50 percent of an entire defined experimental animal population.

LDLo The *lethal dose low* is the lowest dose, other than LD50 of a substance introduced by any route, other than inhalation, over a given period of time reported to have caused death in humans or animals.

LD50 The *lethal dose 50* is the dose of a substance that is expected to cause the death of 50 percent of an entire defined experimental animal population, by any route other than inhalation.

MAC *Maximum allowable concentration,* regardless of period of exposure.

MAK *Maximum allowable concentration* (MAK in German) is the maximum permissible concentration of a substance as a gas, vapor, or aerosol in the air at the workplace that does not normally affect worker health or cause unreasonable nuisance even with repeated and long-term exposures. Tables of MAK values are also referred to as biological tolerance values (BAT in German).

MEL *Maximum exposure limit* (concentration) for a long-term (8-hour) or short-term (15-minute) exposure.

NOAEL/NOAEC A limiting concentration below which there is *no observable adverse effect.*

OES *Occupational exposure standard* is the standard at which there is no indication of risk to health.

PEL The *permissible exposure limit* is the limit which must not be exceeded during any eight-hour work shift of a 40-hour workweek, as enforced by

OSHA. A table of "annotated" PELs was published by OSHA 2006, comparing California PELs, NIOSH RELs, and ACGIH TLVs (threshold limit value).

REL A *recommended exposure limit*, as based on TLV, TLV-STEL, or TLV-C.

SLOD (DTL) The *dangerous toxic load* that would give a *significant likelihood of death* is defined as the dose (usually concentration times duration) that would cause mortality to 50 percent of an exposed population, and is typically equal to the LC_{50} (lethal concentration to 50 percent of the exposed population for a four-hour exposure).

SLOT (DTL) The *dangerous toxic load* for the *specified level of toxicity* is defined as the dose (usually concentration times duration) that would cause severe distress to almost everyone in the (land use) area, with a substantial fraction of the exposed population requiring medical attention, with some people being seriously injured, or requiring prolonged treatment, and with highly susceptible people possibly being killed.

TLV-C The airborne concentration that should not be exceeded during any part of the working exposure.

TLV-STEL The *threshold limit value–short-term exposure limit* is the airborne concentration to which it is believed that workers can be exposed continuously for a short period of time without suffering from (1) irritation, (2) chronic or irreversible tissue damage, (3) dose-rate-dependent toxic effects, or (4) narcosis of sufficient degree to increase the likelihood of accidental injury, impaired self-rescue, or materially reduced work efficiency.

TLV-TWA *The threshold limit value–time-weighted average* is the airborne concentration of a chemical substance to which it is believed that nearly all workers may be repeatedly exposed, day after day, over a working lifetime, without adverse health effects, as time-weighted average for eight-hour exposures.

WEELC Ceiling limits of *workplace environmental exposure levels* should not be exceeded at any time during the workday.

WEEL8 Values of *workplace environmental exposure levels* are expressed as time-weighted average (TWA) concentrations for eight-hour exposures. Worker excursion exposure levels may exceed three times the pertinent WEEL for no more than a total of 30 minutes during an eight-hour workday.

Because many of the toxicity values for the preceding criteria were derived from a common database, the values for the same or similar effects and for the same duration of exposure may also be similar.

Quantitative Evaluation of Toxicity Hazards The concentration limits for gases and vapors typically are given in terms of parts per million by volume (ppm, or ppmv), and the concentration limits for dusts and aerosols are given in terms of milligrams per cubic meter (mg/m^3). Since toxicity data are sometimes presented in terms of mg/m^3 (or mg/L), the ppm concentrations can be converted to mg/m^3 concentrations using the following relationship, for normal temperatures and pressures:

$$C_{MG/M^3} = C_{PPM}\,(M/0.0244)\,(1000/1{,}000{,}000) = 0.041\,C_{PPM}\,M \quad \text{mg/m}^3 \tag{23-13}$$

where M is the molecular weight of the gas or vapor, and 0.0244 is the volume of one gram-molecular weight of air at 25°C and 14.7 pounds per square inch, absolute (psia), in cubic meters. Concentrations in terms of mg/m^3 are useful when evaluating the toxicity hazards from releases from process equipment (in terms of kilograms per second), or if there is a combination of toxic gas, vapor, and/or dust.

The toxicity hazard of a combination of inhalable toxic substances can be evaluated through use of the Le Chatelier principle (Mannan, *Lees' Loss Prevention in the Process Industries*, 3d ed., vol. 2, chap. 18 and section 25.3, Elsevier, Oxford, UK, 2005, pp. 25/7–25/11), provided that the effects are merely additive (as contrasted with negatively or positively synergistic). The relationship is as follows:

$$L_{MIX} = [C_1/L_1] + [C_2/L_2] + \ldots \tag{23-14}$$

where C is the concentration of the gas or vapor of interest in the mixture of gases or vapors (on an air-free basis); the values of L are the lower flammable limits for the Le Chatelier flammability equation, or the values of the toxicity limits (such as the LC50 concentrations); and L_{MIX} is the effective flammable or toxic limit of the mixture.

To facilitate the use of these data, several types of graphical and "probit" equation methods are available (Mannan, *Lees' Loss Prevention in the Process Industries*, 3d ed., vol. 1, p. 9/68, and vol. 2, p. 18/56, Elsevier, Oxford, UK, 2005).

The effects of inhalation exposures to toxic materials are strongly dependent on both the concentration of the material in air and the duration of the exposure. For short-term exposures, the dose received is the product of the concentration and the duration, as given by the "Haber Law" (Haber, *Funf Vortrage aus den Jahren 1920–1923*, Springer-Verlag, Berlin, 1924; Fleming and D'Alonzo, *Modern Occupational Medicine*, Lea & Febiger, Philadelphia, 1960, p. 78)

$$D = C_{PPM}\,t \quad \text{ppm-minutes} \tag{23-15}$$

where D is the inhaled dose of the toxic gas or vapor, in ppm-minutes.

However, for long-term exposures, there are metabolic processes in the human body that can result in detoxification. An exponent can be applied to the concentration or to the duration so that a good relationship can be between (1) the concentration that results in a pertinent physiological effect and (2) the duration of exposure to that concentration (Elkins, *The Chemistry of Industrial Toxicology*, 2d ed., Wiley, New York, 1959, p. 242).

There is a rather strong relationship between inhalation toxicity and oral/ingestion toxicity. "When only oral data are available, the approach used in occupational hygiene is known as the 'Stockinger-Woodward' approach, which assumes an 80-kilogram human breathes at a rate of 1.25 m^3 per hour, with 100 percent absorption of the inhaled material" (Knight and Thomas, *Practical Guide to Chemical Safety Testing*, Smithers Rapra Publishing, Shrewsbury, UK, 2003, pp. 149–150 and table 7.2). Based on these assumptions, a relationship can be developed to obtain an approximate value for the inhalation dose of a toxic vapor:

$$D_{INH} = C_{PPM}\,t = 2000\,D_{MG/KG}/[M\,B] \quad \text{ppm-minutes} \tag{23-16}$$

In Eq. (23-16), D_{INH} is the inhalation dose (in ppm-minutes) that would cause the injury of interest, $D_{MG/KG}$ is the ingestion dose of the toxic liquid that causes that injury (in milligrams per kilogram of body weight), M is the molecular weight of the vapor, B is the breathing rate (about 0.02 cubic meter per minute), C_{PPM} is the concentration of the toxic vapor in air (in ppm), and t is the duration of exposure to the toxic vapor (in minutes). This relationship could be rearranged to give an approximate oral dose if the hazardous inhalation concentration is known.

A special class of "toxic" materials is the asphyxiants. They include nitrogen—which has been termed "the most dangerous gas" (Kletz, *What Went Wrong—Case Histories of Process Plant Disasters*, Gulf, Houston, Tex., 1985, pp. 145–149); carbon dioxide; several fluorocarbons and chlorofluorocarbons; the noble gases such as helium, neon, and argon; and flammable gases such as propane and ethylene. See the Hazards of Inerts subsection for more information on asphyxiation hazards.

Another special class of toxic materials is the organophosphate compounds, which are found in many insecticides and in some herbicides. These compounds can have severe neurotoxic and reproductive effects (Marrs, Maynard, and Sidell, *Chemical Warfare Agents*, Wiley, New York, 1996, pp. 83–96 and 115–133), so exceptional precautions should be taken when handling organic compounds containing phosphorus.

Worst-Case Evaluation To ensure that adequate precautions have been taken to counteract the accidental release of toxic liquids, the worst-case concentration of toxic vapors above the liquid should be determined. The relationship between temperature and vapor concentration is determined by a combination of the Antoine equation and Dalton's law:

$$C_{PPM} = (10^6/760)\,10^{A' - (B'/(T + C'))} \quad \text{parts per million, by volume} \tag{23-17}$$

Values for A′, B′, and C′ are available in the open literature (Speight, *Lange's Handbook of Chemistry*, 16th ed., McGraw-Hill, New York, 2005, p. 2.297), or other relationships between vapor pressure and temperature could be used (Yaws, *Chemical Properties Handbook*, McGraw-Hill, New York, 1999, p. 159).

Safeguards Against Toxicity Hazards The best protection against toxicity hazards is complete containment of hazardous materials within processing equipment. Where complete containment of toxic liquids or gases is impractical, exhaust ventilation (preferably to a scrubber) can limit or eliminate exposure to toxic materials. The exhaust ventilation rate (velocity or volumetric rate) may be calculable for a worst-case rate of toxic gas release or for a worst-case spill of toxic liquid (U.S. Environmental Protection Agency, *Risk Management Program Guidance for Offsite Consequence Analysis*, Appendix D, Equation D1, 2006). However, tests to determine concentrations in air usually would be needed to determine the required rate of local exhaust ventilation for dusty processes and for fugitive releases of gases (American Conference of Governmental Industrial Hygienists, *Industrial Ventilation*, 28th ed., chap. 7, ACGIH, Cincinnati, Ohio, 2013).

If containment and exhaust ventilation are not considered adequate, cartridge respirators or self-contained breathing apparatus can provide protection against inhalation (and, in some cases, ingestion) of toxic materials. In 1994, the Occupational Safety and Health Standards were amended to require that "the employer shall assess the workplace to determine if hazards are present, or are likely to be present, which necessitate the use of personal protective equipment (PPE). If such hazards are present, or likely to be present, the employer shall select, and have each affected employee use, the types of PPE that will protect the affected employee from the hazards identified in the hazard assessment" [Occupational Safety and Health Standards, 29 CFR

TABLE 23-12 Assigned Protection Factors

Type of respirator	Quarter mask	Half mask	Full facepiece	Helmet/hood	Loose-fitting facepiece
Air-purifying respirator	5	10	50	....	....
Powered air-purifying respirator (PAPR)	....	50	1,000	25/1,000	25
Supplied-air respirator (SAR) or airline respirator					
• Demand mode	....	10	50	....	....
• Continuous flow mode	....	50	1,000	25/1,000	25
• Pressure-demand or other positive-pressure mode	....	50	1,000	....	....
Self-contained breathing apparatus (SCBA)					
• Demand mode	....	10	50	50	....
• Pressure-demand or other positive-pressure mode (e.g., open/closed circuit)	....	....	10,000	10,000	....

Occupational Safety and Health Standards, 29 CFR 1910.134 and Appendices A, B, and C to 1910.134, and Appendix B to Subpart I, 2011.

1910.132(d), 2011]. Subsequently—on November 22, 2006—OSHA issued a Table of Assigned Protection Factors (Table 23-12). The APF represents the factor by which the concentration of a contaminant in the atmosphere outside the mask would be reduced to a concentration inside the mask.

The OSHA hazard assessment would aid in determining the type of breathing protection that would be appropriate for the toxicity hazard. Guidelines for appropriate use of breathing protection are given in this OSHA standard.

Evaluation of Ingestion Hazards Toxicity data are available for the acute (single-dose) ingestion/oral toxicity of many toxic materials (Accelrys, Inc., *Registry of Toxic Effects of Chemical Substances,* 2016; Lewis, *Sax's Dangerous Properties of Industrial Materials,* 12th ed., Wiley, New York, 2012). However, very few data are available for prolonged ingestion or periodic doses of toxic materials. It is likely that metabolic processes would operate to increase the total burden required for toxic effects for such chronic exposures, except for some materials (such as mercury and lead) that apparently can accumulate in the body. In the absence of better skin-contact data, it might be appropriate to use parenteral or subcutaneous injection data for a worst-case exposure (e.g., through a cut in the skin). However, the use of intravenous or intraperitoneal data might overstate the skin exposure toxicity of a material.

The primary route for ingestion of toxic materials (especially dusts, mists, and vapors) is by the swallowing of mucus and saliva that have absorbed these materials during breathing. Cilia in the nose and esophagus (windpipe) sweep foreign materials that have been embedded or absorbed by these fluids toward the pharynx, where the contaminated fluid is swallowed (Guyton and Hall, *Textbook of Medical Physiology*, 13th ed., Elsevier, Philadelphia, 2016, pp. 504, 817, and 819).

Evaluation of Skin-Contact Hazards An anecdotal description of a fatal skin exposure to fluoroacetic acid can be used to estimate a mass-transfer coefficient of about 15 $\mu g/(min \cdot cm^2)$ for this molten material [Peters, R. A., et al., "Subacute Fluoroacetate Poisoning," *J. Occup. Med.* **23**(2): 112 (1981)]. In this incident, the exposed area was about 0.15 m^2, the duration of exposure was about one minute, and the subcutaneous lethal dose is about 280 micrograms per kilogram. Similar mass-transfer coefficients could be derived for other materials having skin absorption hazards, if the surface area exposed, the duration of exposure, and the injurious-dose data were available.

The U.S. Department of Transportation has developed guidance for the corrosivity of chemical substances (U.S. DOT, *Shippers—General Requirements for Shipments and Packagings,* 49 CFR 173.136 and 49 CFR 173.137, *Assignment of Packing Group,* 2011). The following definitions apply for Class 8 corrosive materials:

Packing Group 1—Great Danger: Full thickness destruction of human skin (exposure time, 3 min or less; observation time, 60 min)

Packing Group 2—Medium Danger: Full thickness destruction of human skin (exposure time, 3 to 60 min; observation time, 14 days)

Packing Group 3—Minor Danger: Full thickness destruction of human skin (exposure time, 1 to 4 h; observation time, 14 days).

These quantitative criteria could be used to estimate the effects of corrosive exposures of other durations.

Another effect of skin-contact toxicity is dermatitis. This can be caused by "physical" agents, such as detergents and solvents that remove the natural oils from the skin and thereby render the skin susceptible to materials that ordinarily do not affect the skin. Dermatitis also can be caused by desiccants and water-reactive chemicals that remove moisture from the skin, generating heat and causing burns. Other causes are oxidizers; protein precipitants; allergic or anaphylactic proteins; friction, pressure, and trauma; thermal and electromagnetic radiation; biological agents; and plant poisons (National Safety Council, *Fundamentals of Industrial Hygiene,* 6th ed., Itasca, Ill., 2012).

Dermatitis can be prevented or controlled by containment of skin-contact hazards and the use of tools to avoid contact (engineering controls) or by the wearing of protective clothing, including gloves and eye and face protection, and good personal hygiene, including hand and face washing (administrative controls).

Conclusions Toxicity data are available for many thousands of solid, liquid, and gaseous chemicals and other materials. The data for inhalation toxicity provide guidance for concentration and duration limits, for protection of the public, chemical plant employees, and emergency response personnel. Similar data for ingestion and skin contact with toxic materials are not as readily available. Investigation into toxic effects is continuing so that toxic materials can be handled safely.

OTHER HAZARDS

Hazards of Vacuum

Introduction Some types of equipment (e.g., storage tanks) often have a relatively low resistance to the damage that can be caused by internal vacuum. The low vacuum rating for such equipment and the amount of damage that can result are often surprising, and potentially costly, lessons learned by plant engineers and operators.

Equipment Limitations A robust internal pressure rating for a piece of equipment is no guarantee that it will withstand an appreciable vacuum. Industry loss experience includes failures of vessels with design pressure ratings in excess of 25 psig (Sanders, "Victims of Vacuum," *Proceedings of the 27th Annual Loss Prevention Symposium,* AIChE, New York, 1993). Low-pressure storage tanks are particularly fragile. For example, an atmospheric fixed-roof storage tank may only withstand a vacuum of 2.5 mbar (0.036 psi or 1 in water) (BP, *Hazards of Trapped Pressure and Vacuum,* 3rd ed., IChemE, Rugby, UK, 2009). Jacketed vessels can be particularly vulnerable to internal vacuum, since the operating pressure of the heat-transfer medium in the jacket adds to the differential pressure that would otherwise exist between the atmosphere and the vessel interior.

While many pressure vessels may withstand a significant vacuum, design calculations are required to confirm this. Unless specifically rated for vacuum service, equipment should be assumed to be subject to damage by vacuum. When equipment is procured, consideration should be given to including the vacuum rating in the pressure vessel calculations and code stamp. In many instances, the additional cost of doing so will be an insignificant fraction of the total procurement cost for the vessel (see the subsection Protective Measures for Equipment).

Consequences of Vacuum Damage Vessels, tank trucks, or railcars can be dimpled by partial collapse or, more significantly, crushed like used beverage cans. Fortunately, equipment damaged by underpressurization does not fail explosively, as might occur with overpressurized equipment. Nevertheless, the loss of containment of equipment contents is a real risk, as damage to the vessel or to the piping connected to the vessel could result in significant releases of toxic, flammable, or otherwise hazardous materials. Alternatively, vacuum within equipment could lead to ingress of air into inerted or fuel-rich systems, creating a fire or explosion hazard within the equipment.

Common Causes of Equipment Underpressurization Equipment can be exposed to excessive vacuum due to an unanticipated mechanism creating a vacuum or the failure or inadequate design of protective systems provided to mitigate the hazard. A common scenario involves the pumping, draining, or siphoning of liquid from a tank that has no, or inadequate, venting capacity and thus cannot allow the entry of air at a rate sufficient to backfill behind the dropping liquid level. Vacuums can also be created when a blower, fan, compressor, or jet ejector removes gases from equipment. Other mechanisms for generating a vacuum, which have been demonstrated by industry experience, include the following:

- *Condensation of vapors or cooling of hot gases.* Steam is commonly used to clean vessels and, less frequently, to create an inert atmosphere inside of equipment. Steam condensing inside a closed vessel can create a significant vacuum, and vessels (e.g., railcars) have collapsed when all vessel inlets were closed after steam cleaning. The sudden cooling of a storage tank by a thunderstorm can create a vacuum when gases in the vessel headspace cool or vapors of volatile liquids condense. The American Petroleum Institute (API, *Venting Atmospheric and Low-Pressure Storage Tanks,* Standard 2000, Washington, D.C., 2014) provides guidance for in-breathing requirements as a function of tank capacity to protect against this latter scenario.
- *Absorption of a gas in a liquid.* A vessel collapsed when ammonia vapor from the headspace dissolved in water within the vessel (Mannan, S., *Lees' Loss Prevention in the Process Industries,* vol. 3, 4th ed., Elsevier, Oxford, UK, 2012, p. 2600).
- *Chemical reactions that remove gases from the headspace.* The corrosion of the interior of a steel vessel, especially if the vessel has been chemically cleaned, can consume and remove a significant quantity of the oxygen from the vessel atmosphere. Other chemical reactions (e.g., ammonia reacting with hydrogen chloride to form ammonium chloride) can reduce the amount of gas or vapor in the vessel.

Prudent design requires that equipment be protected from credible underpressurization scenarios. Equipment damage can result when such protections are omitted, improperly sized, incorrectly designed or installed, or inadequately maintained. Common failures include the following:

- *Failure to consider appropriate challenges when determining the required relief capacity* (e.g., maximum rates of liquid withdrawal or cooling of vessel contents). Credible contingencies (e.g., thunderstorm cooling a vessel during steam-out) should be considered.
- *Inadequate capacity, or failure, of vessel blanketing systems.* Inert gas supplies are often piped to vessels to maintain a reduced-oxygen atmosphere during liquid withdrawal. Coincident high demand for inert elsewhere, closure of a valve, or depletion of the supply could result in the failure to prevent a vacuum.
- *Operating errors.* Many vessel collapses have resulted from closing or failing to open a valve in a vent line. For this reason, valves in vacuum relief lines should be avoided, and they may be prohibited by some design codes.
- *Maintenance errors.* One common error is the failure to remove an isolation blind in a vent line when returning a vessel to service. Even a thin sheet of plastic placed over an open nozzle may be sufficient to allow a vessel-damaging vacuum to be produced (BP, *Hazards of Trapped Pressure and Vacuum,* 3rd ed., IChemE, Rugby, UK, 2009).
- *Inappropriate modifications.* A hose was temporarily connected to a tank vent line and then submerged in a drum of liquid. Only a few inches of submergence were required to effectively block the vent. A vacuum was created when liquid was drained from the vessel, collapsing the vessel (Mannan, S., *Lees' Loss Prevention in the Process Industries,* vol. 2, 4th ed., Elsevier, Oxford, UK, 2012, p. 1829).
- *Failure of process controllers.* Control failures can either initiate events (e.g., increase the speed of an exhauster) or disable protections (e.g., reduce the rate of supply of inert gas to a vessel).
- *Plugging of vent lines or devices.* Process materials can occlude vents. Monomers requiring an inhibitor to prevent polymerization can evaporate from a tank and then condense in the vent line, free of the inhibitor. Waxes and other high-melting-point materials can solidify upon cooling in the vent system, dusts can accumulate, and water vapor can condense to form liquid seals in low points of vent lines or freeze in the winter. Such scenarios are a particular problem in cases where flame arrestors, screens, or other devices introduce small apertures in the vent flow path. Plugging of vent lines by animal or insect nests is not uncommon.
- *Inadequate or incorrect maintenance.* Mechanical devices such as vacuum breakers and flame arrestors require routine maintenance attention to ensure that they provide their intended protective function. Incorrect maintenance (e.g., changing the vacuum breaker set pressure) could defeat the intended protection.

Lees (Mannan, S., *Lees' Loss Prevention in the Process Industries,* vol. 3, 4th ed., Elsevier, Oxford, UK, 2012, p. 2600), BP (*Hazards of Trapped Pressure and Vacuum,* IChemE, Rugby, UK, 2009), and Kletz (*What Went Wrong?—Case Histories of Process Plant Disaster,* Gulf Publishing Company, Houston, Tex., 1989) include additional case histories providing valuable lessons about how equipment failures and human errors can combine to inflict vacuum damage.

Protective Measures for Equipment If equipment is subject to a vacuum, the inherently safer alternative would be to design the equipment to withstand a full vacuum. While this may not be economically feasible for large storage tanks, the incremental cost for smaller vessels may not be prohibitive, particularly when traded off against the capital and continued operating and maintenance costs of some alternatives (e.g., protective instrumentation systems). The incremental fabrication cost of providing a suitable vacuum rating can be less than 10 percent for vessels of up to 3000 gal nominal capacity and having a 15 psig pressure rating [Wintner, "Check the Vacuum Rating of Your Tanks," *Chemical Engineering* (February 1991), pp. 157–159].

Careful process hazards analysis may show that a particular vessel need not be designed to withstand a full vacuum (e.g., if the maximum attainable vacuum is limited by the performance characteristics of an exhauster). Whatever the vacuum rating, rated vessels must be periodically inspected to ensure that internal or external corrosion has not diminished the vessel strength.

Reliable protections against excessive vacuum should be provided whenever equipment cannot withstand the vacuums that can credibly be achieved. In some low-risk situations, protections may consist of administrative controls implemented by adequately trained personnel. Where the risk of damage is higher or where design standards or codes require, engineered protections should be implemented.

Where process, safety, and environmental considerations permit, vacuum protection may be provided by properly sized ever-open vents. Alternatively, active protective devices and systems are required. Vacuum breaker valves designed to open and admit air at a predetermined vacuum in the vessel are commonly used on storage tanks, but they may not be suitable for some applications involving flammable liquids. Inert gas blanketing systems may be used if adequate capacity and reliability can be ensured. Where the source of the vacuum can be deenergized or isolated, suitably reliable instrumented systems (e.g., interlocks) can be provided.

API (*Venting Atmospheric and Low-Pressure Storage Tanks,* Standard 2000, Washington, D.C., 2014) provides guidance for vacuum protection of low-pressure storage tanks. Where vacuum relief devices are provided, they should communicate directly with the vapor space in the vessel and should be installed so that they cannot be sealed off by the liquid contents in the vessel. Valves should be avoided in the inlets or outlets of vacuum relief devices unless the valves are reliably car-sealed or locked open, or excess relief capacity is provided (e.g., via multiple-way valves).

Hazards of Inerts

Introduction The use of inert gases to displace oxygen from equipment atmospheres in order to prevent combustion and, perhaps, consequent explosions has been described in the subsection Flammability. Other applications for inerting exist, including preventing (1) corrosion or other deterioration of out-of-service equipment, (2) degradation of oxygen-sensitive products, or (3) exothermic reactions with air- or water-reactive materials. While the risk of asphyxiation in an oxygen-deficient environment is the most frequently recognized concern, other hazards such as toxicity, temperature and pressure extremes, and chemical incompatibilities also must be considered.

Sources of Inerts The most commonly used inert gases are N_2 and CO_2, but other gases and vapors such as argon (Ar), helium (He), steam, and exhaust gases from combustion devices are also used. The choice of the most appropriate inert for a given application must be based upon factors such as cost, availability, reliability of supply, effectiveness, and compatibility with process streams (Cunliff, "Avoiding Explosions by Means of Inerting Systems," *IChemE Symposium Series no. 148,* Rugby, UK, 2001; Grossel and Zalosh, *Guidelines for Safe Handling of Powders and Bulk Solids,* CCPS-AIChE, Cambridge, Mass., 2004).

Traditionally, inerts have been obtained from sources such as high-pressure gas cylinders or tube trailers or through the evaporation of cryogenic liquids from bulk tanks. Other sources of inerts include the following (NFPA 69, *Standard on Explosion Prevention Systems,* National Fire Protection Association, Quincy, Mass., 2014; FM Global, Loss Prevention Data Sheet 7-59, *Inerting and Purging of Tanks, Process Vessels, and Equipment,* Johnston, R.I., 2000):

- On-site cryogenic air separation plants
- Gas generators burning or catalytically oxidizing a hydrocarbon to produce an oxygen-deficient product gas
- Nitrogen produced by the air oxidation of ammonia
- Nitrogen produced by removal of oxygen from air using pressure swing adsorption (PSA) or membrane separation units

Inert gas streams generated on site should be carefully monitored to ensure detection of O_2 contamination of the product gas in the event of equipment failure or operational upset.

Asphyxiation and Toxicity Hazards An asphyxiant is a chemical (either a gas or a vapor) that can cause death or unconsciousness by suffocation (BP, *Hazards of Nitrogen and Catalyst Handling,* IChemE, Rugby, UK, 2009). A simple asphyxiant is a chemical, such as N_2, He, or Ar, whose effects are caused by the displacement of O_2 in air, reducing the O_2 concentration below its normal value of approximately 21 vol%. The physiological effects of oxygen concentration reduction by simple asphyxiants are illustrated in Table 23-13 (Air Products, Dangers of Oxygen-Deficient Atmospheres, Safetygram 17, Allentown, Penn., 2014).

The physiological processes leading to death from hypoxia (i.e., insufficient supply of oxygen to the body tissues) are described by Air Products (Air Products, *Dangers of Oxygen-Deficient Atmospheres,* Safetygram 17,

TABLE 23-13 Effects of Breathing Oxygen-Deficient Atmospheres

Oxygen content of air, vol %	Signs and symptoms of persons at rest
19.5–23.5	OSHA-prescribed limits on oxygen content for breathing gases for atmosphere-supplying respirators.
19	Some adverse physiological effects occur, but they may not be noticeable.
15–19	Impaired thinking and attention. Increased pulse and breathing rate. Reduced coordination. Decreased ability to work strenuously. Reduced physical and intellectual performance without awareness.
12–15	Poor judgment. Faulty coordination. Abnormal fatigue upon exertion. Emotional upset.
10–12	Very poor judgment and coordination. Impaired respiration that may cause permanent heart damage. Possibility of fainting within a few minutes without warning. Nausea and vomiting.
<10	Inability to move. Fainting almost immediate. Loss of consciousness. Convulsions. Death.

SOURCE: *Air Products, Dangers of Oxygen-Deficient Atmospheres,* Safetygram 17, Allentown, PA, 2014 and OSHA, 29 CFR 1910.134, *Respiratory Protection Standard.*

Allentown, Penn., 2014). In an oxygen-free environment, loss of consciousness occurs within seconds after the first breath, followed by death within 2 to 4 min. A person exposed to an oxygen-deficient environment may not recognize the warning signs and may not be able to reason or take protective action before unconsciousness occurs. Victims removed from an O_2-deficient atmosphere require resuscitation through the administration of O_2 to prevent death [U.S. Chemical Safety and Hazard Investigation Board (CSB), *Hazards of Nitrogen Asphyxiation,* Safety Bulletin no. 2003-10-B, 2003]. In this safety bulletin, the CSB identified 80 nitrogen asphyxiation deaths and 50 injuries occurring in 85 incidents between 1992 and 2002.

Physical exertion increases oxygen demand and may result in oxygen deficiency symptoms at higher oxygen concentrations, and individuals in poor health may be less tolerant of reduced oxygen concentrations (Air Products, *Dangers of Oxygen-Deficient Atmospheres,* Safetygram 17, Allentown, Penn., 2014). The guidance in Table 23-13 assumes a sea-level location and should be applied cautiously for facilities at significant altitudes; however, OSHA's Respiratory Protection Standard accepts 19.5 vol% as a safe lower O_2 concentration up to an altitude of 8000 ft (OSHA, 29 CFR 1910.134, *Respiratory Protection Standard*).

A chemical asphyxiant works by interfering with the body's ability to absorb or transport O_2 to the tissues. A relevant example is CO, which can be present in inert gas streams produced by combustion. Exposure to CO concentrations of approximately 400 ppm can be life threatening for a three-hour exposure, and 12,800 ppm can cause death after 1 to 3 min (Air Products, *Carbon Monoxide,* Safetygram 19, Allentown, Penn., 2014).

While CO_2 is typically called a simple asphyxiant (like N_2), data indicate that, at higher concentrations, CO_2 has an effect more significant than that attributable solely to the corresponding decrease in O_2 concentration. At a concentration of 15 vol% CO_2, unconsciousness occurs in less than a minute (Air Products, *Carbon Dioxide,* Safetygram 18, Allentown, Penn., 2014). Diluting air with CO_2 to establish a concentration of 15 vol% CO_2 would leave an O_2 concentration of about 18 vol% (21 vol% × 0.85). As Table 23-13 indicates, for the simple asphyxiant effect, unconsciousness is not likely to result until the O_2 concentration has been reduced below about 10 vol%. Air Products provides an explanation of the physiological effects of CO_2 (Air Products, *Carbon Dioxide,* Safetygram 18, Allentown, Penn., 2014).

Injuries and fatalities from asphyxiation are often associated with personnel entry into inerted equipment or enclosures. Safety requirements for confined space access are provided by OSHA (29 CFR 1910.146, *Confined Space Entry Standard*), the American National Standards Institute (ANSI, Z117.1, *Safety Requirements for Confined Spaces,* 2009), Hodson ("Safe Entry into Confined Spaces," *Handbook of Chemical Health and Safety,* American Chemical Society, Washington, D.C., 2001), and BP (*Hazards of Nitrogen and Catalyst Handling,* IChemE, Rugby, UK, 2009). OSHA has established 19.5 vol% as the minimum safe oxygen concentration for confined space entry without supplemental oxygen supply (see Table 23-13). Note that OSHA imposes a safe upper limit on O_2 concentration of 23.5 vol% to protect against the enhanced flammability hazards associated with O_2-enriched atmospheres.

Physical Hazards A variety of physical hazards are presented by the various inerts in common usage.

High temperature. The high-temperature off-gases from combustion-based sources of inerts typically must be quenched before use. Water scrubbing, in addition to reducing the temperature, can remove soot and sulfur compounds (which could react with moisture to form corrosive acids) present in the off-gas. The humidity of the resultant gas stream may make it unsuitable for inerting applications where moisture cannot be tolerated.

Use of steam as an inert requires that equipment be maintained at an elevated temperature to limit condensation that would lower the inert concentration. FM Global (Loss Prevention Data Sheet 7-59, *Inerting and Purging of Tanks, Process Vessels, and Equipment,* Johnston, R.I., 2000) recommends a minimum temperature of 160°F. The use of steam may not be appropriate in systems where brittle materials (such as cast iron) may be stressed by thermal expansion and in systems where materials of construction (e.g., plastics) may be weakened or damaged by high temperatures. Protection for personnel to prevent thermal burns from equipment may be required.

In addition, some equipment or equipment supports may not have the strength to support a significant load of condensate, and provisions must be made for removal of condensate from the inerted equipment.

Low temperature. The atmospheric boiling points for N_2, CO_2, He, and Ar are −196°C, −79°C, −269°C, and −186°C, respectively. The potential for cryogenic burns must be addressed in operating and maintenance procedures and in specifying personal protective equipment requirements.

Cryogenic temperatures can cause embrittlement of some materials of construction (e.g., carbon steel) and must be considered in the design of inert gas delivery systems. Controls should be provided to ensure that operational upsets do not allow the migration of cryogenic liquids to piping or equipment not designed to withstand such low temperatures.

The potential for the condensation and fractional distillation of air on the outside of equipment containing cryogenic liquids with boiling points less than that of O_2 must be considered. For example, because N_2 boils at a lower temperature than O_2 (−196°C versus −183°C), air can condense on the outside of piping that carries liquid nitrogen. The liquid that drops off the piping will be enriched in O_2 and can pose an enhanced fire or explosion risk in the vicinity of the equipment.

High pressure. Cryogenic liquids produce large volumes of gas upon evaporation (for example, one volume of liquid N_2 produces 694 volumes of gas at 20°C) (Air Products, *Safe Handling of Cryogenic Liquids,* Safetygram 16, Allentown, Penn., 2014). Containers such as transport and storage vessels must be provided with overpressure relief to address this hazard. An additional concern is the hydrostatic pressure that can be produced if cryogenic liquids are trapped in a liquid-full system. Without a vapor space to allow liquid expansion, extremely high pressures can be produced; accordingly, pressure relief devices must be installed in sections of equipment where cryogenic liquids might become trapped between closed valves.

Given the large liquid-to-gas expansion ratio, consideration should be given to limiting the quantity of cryogenic liquid stored inside tight enclosures or buildings that could become pressurized. The asphyxiation hazard associated with inert gases was addressed previously.

Portable containers of high-pressure inert gases can operate at pressures of thousands of pounds per square inch. Suitable precautions are required to protect containers and associated regulators and piping from damage. Refer to CGA (CGA, *Safe Handling of Compressed Gases in Containers,* Publication P-1, 2015; CGA, *Precautions for Connecting Compressed Gas Containers to Systems,* Publication SB-10, 2009) and Air Products (Air Products, *Handling, Storage, and Use of Compressed Gas Cylinders,* Safetygram 10, Allentown, Penn., 2015) for guidance.

Air Products (Air Products, *Product Migration of Liquefied Compressed Gases in Manifolded Systems,* Safetygram 38, Allentown, Penn., 2015) provides precautionary guidance with respect to manifolding of liquid-containing cylinders. A temperature difference of only a few degrees between cylinders can cause gas from the warmer cylinder to migrate through the manifold to the cooler cylinder, where it could condense and potentially fill the cylinder. A liquid-filled cylinder could rupture if it was subsequently valved closed.

Static electricity. The use of high-pressure CO_2 for inerting poses a concern for potential static electricity hazards. CO_2 converts directly to a solid if the liquid is depressurized below 61 psig (Air Products, *Carbon Dioxide,* Safetygram 18, Allentown, Penn., 2014). Consequently, discharge of liquid CO_2 produces CO_2 "snow" that, when moving at a high velocity, can generate static electric charge. Incendive sparks (5 to 15 mJ at 10 to 20 kV) have been reported (Urben, *Bretherick's Handbook of Reactive Chemical Hazards,* 6th ed., Butterworth-Heinemann, Oxford, UK, 1999).

Obscuring clouds. As releases of cryogenic liquids evaporate, the resulting cold vapors condense moisture in the surrounding air, which can create a highly visible fog. It should be noted that the fog does not define the extent of the vapor cloud but, rather, the extent of the cloud that is cold enough to condense the moisture in the air. Concentrations of inert vapors, perhaps insufficient to support respiration, can extend beyond the visible cloud. Fog clouds can obscure emergency evacuation pathways and hamper access by emergency responders (Air Products, Safe Handling of *Cryogenic Liquids,* Safetygram 16, Allentown, Penn., 2014).

Bulk storage. It is not uncommon for facilities to store large quantities of cryogenic liquids for inerting purposes. Often, the storage tanks and related equipment (such as evaporators) are owned and maintained by the cryogenic supplier. Larger inventories increase the risk in the event of a release. It is important that a dialogue be maintained between the supplier and the user regarding respective operating and emergency response roles.

Chemical Incompatibility Hazards While N_2 and CO_2 may act as inerts with respect to many combustion reactions, they are far from being chemically inert. Only the noble gases (e.g., Ar and He) can, for practical purposes, be regarded as true inerts. Frank ("Inerting for Explosion Prevention," *Proceedings of the 38th Annual Loss Prevention Symposium,* AIChE, New York, 2004) lists a number of incompatibilities for N_2, CO_2, and CO (which can be present in gas streams from combustion-based inert gas generators). Notable incompatibilities for N_2 are lithium metal and titanium metal (which is reported to burn in N_2). CO_2 is incompatible with many metals (e.g., aluminum and the alkali metals), bases, and amines. It also forms carbonic acid in water, which can corrode some materials. CO is a strong reducing agent and is incompatible with oxidizers, potassium, sodium, some aluminum compounds, and certain metal oxides. Trace metals and residual organic compounds may contaminate gas streams from combustion-based inert gas generators, posing a variety of potential incompatibility and product quality concerns.

Hazards of Over-Inerting Certain polymerization inhibitors added to stabilize monomers require a small concentration of dissolved O_2 to be effective (NFPA 69, *Standard on Explosion Prevention Systems,* National Fire Protection Association, Quincy, Mass., 2014). For example, methyl acrylate and ethyl acrylate are commonly stabilized with hydroquinone monomethyl ether. Industry guidance recommends a minimum concentration of 5 vol% O_2 in the atmosphere above the acrylate to prevent polymerization (European Basic Acrylic Monomer Manufacturers Association, *Safe Handling and Storage of Acrylic Esters,* 2003).

INHERENTLY SAFER DESIGN AND RELATED CONCEPTS

General References: CCPS, *Inherently Safer Chemical Processes: A Life Cycle Approach.* 2d ed., AIChE, New York, 2009; CCPS, *Guidelines for Engineering Design for Process Safety,* 2nd ed., AIChE, New York, 2012; *Guidelines for Hazard Evaluation Procedures,* 3d ed., AIChE, New York, 2008; *Final Report: Definition for Inherently Safer Technology in Production, Transportation, Storage, and Use,* U.S. Department of Homeland Security, Science & Technology, Chemical Security Analysis Center, Aberdeen Proving Ground, Md., July 19, 2010; Kletz, T. A., "What You Don't Have, Can't Leak," *Chemistry and Industry* (May 6, 1978), pp. 287-292; Kletz, T. A., and P. Amyotte, *Process Plants: A Handbook for Inherently Safer Design,* 2d ed., Boca Raton, Fla., CRC Press, 2010; National Research Council, *The Use and Storage of Methyl Isocyanate (MIC) at Bayer CropScience,* The National Academies Press, Washington, D.C., 2012.

Introduction For many years, the usual procedure in a plant design was to identify the hazards by one of the systematic techniques described in the Hazard Analysis subsection or by waiting until an accident occurred, and then adding protective equipment to control the hazard or to protect people from the consequences of a loss event. This protective equipment is often complex and expensive, and it requires regular testing and maintenance. It often interferes with the smooth operation of the plant and is sometimes bypassed. Gradually, industry came to realize that, whenever possible, plants should be designed to be user-friendly and able to withstand human error and equipment failure without serious effects on safety (and output and efficiency). When handling flammable, explosive, toxic, or corrosive materials, only very low failure rates of people and equipment are tolerable. These low rates may be impossible or impracticable to achieve consistently for long periods.

The most effective way of designing user-friendly plants is to avoid, when possible, large inventories of hazardous materials in process or storage. "What you don't have, can't leak" [Kletz, T. A., *Chemistry and Industry* (May 6, 1978), pp. 287-292]. This sounds obvious, but until the explosion at Flixborough, UK, in 1974, little systematic thought was given to ways of reducing inventories. Industry simply designed a plant and accepted whatever inventory the design required, confident that it could be kept under control. The Flixborough explosion weakened that confidence, and the 1984 toxic gas release in Bhopal, India, almost destroyed that confidence. Plants in which we avoid a hazard, for example, by reducing inventories or avoiding hazardous reactions are usually called inherently safer. In recent years, the U.S. Chemical Safety Board (CSB) has often highlighted the potential benefits of inherently safer design in many of its incident reports and the resulting recommendations.

Some examples of inherently safer and more user-friendly plants are summarized next (Kletz, T. A., and P. Amyotte, *Process Plants: A Handbook for Inherently Safer Design,* 2d ed., CRC Press, Boca Raton, Fla., 2010; CCPS, *Inherently Safer Chemical Processes: A Life Cycle Approach.* 2d ed., Wiley, Hoboken, N.J., 2009).

Intensification or Minimization One approach is to use such a small amount of hazardous material or energy that it does not matter even if that material or energy is released. For example, in the 1984 Bhopal toxic gas release, the material that leaked (methyl isocyanate, MIC) and killed over 2000 people was an intermediate that it was not essential to store. After 1984, many companies had reduced their stocks of other highly toxic raw materials and intermediates (CCPS, *Inherently Safer Chemical Processes: A Life Cycle Approach.* 2d ed., Wiley, Hoboken, N.J., 2009). By 2012, storage and transport of large amounts of MIC, and nearly all presence of MIC as an in-process intermediate, had been eliminated in the United States [National Research Council, *The Use and Storage of Methyl Isocyanate (MIC) at Bayer CropScience,* The National Academies Press, Washington, D.C., 2012]. Intensification is the preferred route to inherently safer designs as the plants, being smaller, may also be cheaper (Bell, N. A. R., "Major Loss Prevention in the Process Industries," *IChemE Symposium Series* No. 34, IChemE, Rugby, Warwickshire, UK, 1971, p. 50). Designing and operating a facility closer to a zero energy state is also inherently safer, such as developing chemistry or improved catalysts that allow operating at 20 psig instead of 2000 psig.

Substitution If intensification is not possible, then an alternative is to use a safer material in place of a hazardous one. Thus it is possible to replace flammable solvents, refrigerants, and heat transfer media with nonflammable or less flammable materials or other hazardous products with safer ones. For example, replacement of oil-based paints with aqueous latex paints eliminates the fire hazard of flammable solvents, as well as the toxicity of some of those solvents, throughout the supply and use chain, from the initial manufacture of the paint, through distribution, warehousing, retail sale, final use by the consumer, and waste disposal of leftover paint. Processes that use hazardous raw materials or intermediates may be replaced by processes that do not use them.

Attenuation or Moderation Another alternative to intensification is attenuation (moderation), or using a hazardous material under less hazardous conditions. Thus large quantities of liquefied chlorine, ammonia, and petroleum gas can be stored as refrigerated liquids at atmospheric pressure instead of under pressure at ambient temperature. Dyestuffs that form explosive dusts can be handled as slurries. Other combustible dusts can be handled as granules or pellets rather than as fine powders. Aqueous solutions of neutralizing acids or bases, such as ammonia, hydrogen chloride, and others, can often be used in place of anhydrous materials. Benzoyl peroxide, a common polymerization initiator, is much more stable when used as a water-wetted solid rather than as a dry powder.

Limitation of Effects of Failures Effects of failures can be limited by equipment design or change in reaction conditions, rather than by adding protective equipment. For example:

- Heating media such as steam or hot oil should not be hotter than the temperature at which the materials being heated are liable to ignite spontaneously, decompose, or otherwise react uncontrollably.
- Tubular reactors may be safer than stirred tank reactors as the inventory is usually lower.
- Vapor-phase reactors are safer than liquid-phase ones as the mass flow rate through a hole of a given size is much less.
- A small, deep diked area around a storage tank is safer than a large, shallow one as the evaporation rate is lower and the area of any fire from spilled material is smaller (CCPS, *Inherently Safer Chemical Processes: A Life Cycle Approach,* 2d ed., Wiley, Hoboken, N.J., 2009).
- Changing the order of operations, reducing the temperature, or changing other parameters can prevent many runaway reactions.
- Reducing the frequency of hazardous operations, such as sampling or maintenance, reduces exposure to hazards.

Simplification Simpler plants are friendlier than complex plants because they provide fewer opportunities for error and less equipment that might fail. Some of the reasons for complication in plant design are:

- The need to control hazards. If an ISD strategy can be implemented, there is less need for protective equipment, and plants will be simpler.
- A desire for flexibility. Multistream plants with many crossovers and valves have many leak points, and errors in valve settings are easily made.
- Lavish provision of installed spares with accompanying isolation and changeover valves.
- Failure to identify hazards until late in design. By this time it may be impossible to avoid the hazard, and complex equipment must be added to control it.

Knock-On Effects (Domino Effects) Plants should be designed so that those incidents that do occur do not produce knock-on or domino effects. Some examples include:

- Provide generous spacing between process units to prevent the spread of an incident from one area to another. Some distance recommendations can be found in industry and other fire protection guidelines and standards, as well as from many insurance companies.

- Locate equipment that is liable to leak outside so that leaks of flammable gases and vapors are dispersed by natural ventilation. Indoors, a few tens of kilograms are sufficient to create a flammable vapor cloud in the building, with the potential for an explosion. If leaks of nonflammable toxic gases are liable to occur, it may be safer to locate the plant indoors in a containment building equipped with vapor detectors and appropriate treatment of ventilation air before discharge to the atmosphere.
- Constructing storage tanks so the roof-wall weld will fail before the base-wall weld, preventing spillage of the contents if the tank is overpressurized and fails. In general, in designing equipment, consider the way in which it is likely to fail and, when possible, locate or design the equipment to minimize the consequences.

Making Incorrect Assembly Impossible Plants should be designed so that incorrect assembly is difficult or impossible. For example, compressor valves should be designed so that the inlet and exit valves cannot be interchanged, and hose connections of different types or sizes should be used for compressed air and nitrogen.

Making Status Clear It should be possible to see at a glance if equipment has been assembled or installed incorrectly or whether it is in the open or shut position. For example:

- Check valves should be marked so that installation the wrong way is obvious. It should not be necessary to look for a faint arrow hardly visible beneath a layer of dirt.
- Use gate valves with rising spindles instead of valves with nonrising spindles because it is easy to see whether they are open or shut. Ball valves and cocks are friendlier if the handles cannot be replaced in the wrong position.
- The position of a figure-8 plate is apparent at a glance. If a slip plate is used, the projecting tag should be readily visible, even if the line is insulated.

Error Tolerance Whenever possible, equipment should tolerate poor installation or operation without failure. Expansion loops in pipework are more tolerant of poor installation than are expansion joints (bellows). Fixed pipes, or articulated arms, if flexibility is necessary, are more robust than hoses.

Ease of Control Processes with a flat response to a disturbance are more tolerant of upsets than those with a steep response. Processes in which a rise of temperature decreases the rate of reaction are more easily controlled than those in which the rate of reaction increases as temperature increases, but this not common in chemistry practiced in industry. However, there are a few examples of processes in which a rise in temperature reduces the rate of reaction. For example, in the manufacture of peroxides, water is removed by a dehydrating agent. If magnesium sulfate is used as the agent, a rise in temperature causes release of water by the agent, diluting the reactants and stopping the reaction [Gerritsen, H. G., and C. M. Van't Land, *I&EC Process Design* **24:** 893 (1985)].

Software In some programmable electronic systems, errors are much easier to detect and correct than others. Accidentally pressing the wrong key or touching the wrong place on a touch screen should never produce serious consequences. If we use the term *software* in the wider sense to cover all procedures, as distinct from hardware or equipment, some software is much friendlier. For example, when writing any kind of procedure—operating, maintenance, or any other activity—a good goal is to write a procedure where the *correct* way to do the operation, the *safe* way, and the *easiest* way are all the same. People will not be tempted to find shortcuts in a procedure if the correct procedure is easy.

U.S. Department of Homeland Security (USDHS) Definition of Inherently Safer Design The USDHS and CCPS developed a definition of "inherently safer technology" or IST [*Final Report: Definition for Inherently Safer Technology in Production, Transportation, Storage, and Use*, U.S. Department of Homeland Security, Science & Technology, Chemical Security Analysis Center, Aberdeen Proving Ground, Md., July 19, 2010]. IST is of potential interest to USDHS because, if the hazard associated with a process facility can be eliminated or significantly reduced, the facility may become a less attractive target for a terrorist. An abstracted definition of IST follows (the complete definition can be found in the USDHS report):

> Inherently Safer Technology (IST), also known as Inherently Safer Design (ISD), permanently eliminates or reduces hazards to avoid or reduce the consequences of incidents. IST is a philosophy, applied to the design and operation life cycle. IST considers options, including eliminating or reducing a hazard, substituting less hazardous material, less hazardous process conditions, and design to reduce the potential for, or consequences of, failure.
>
> ISTs are relative. A technology can only be described as inherently safer when compared to a different technology, including a description of the hazard or set of hazards being considered, their location, and the potentially affected population.
>
> ISTs are based on an informed decision process. Because an option may be inherently safer with regard to some hazards and inherently less safe with regard to others, decisions about the optimum strategy for managing risks from all hazards are required. The decision process must consider the entire life cycle, the full spectrum of hazards and risks, and the potential for transfer of risk from one impacted population to another.

Actions Needed for the Design of Inherently Safer and User-Friendly Plants Everybody who is responsible for the design and operation of process plants needs to be aware that there are opportunities for improving the friendliness of the plants they design, manage, operate, and maintain. These people will participate in a variety of activities which identify hazards—for example, design and safety reviews, process hazard analysis studies, job safety analyses, incident investigations, management of change reviews, writing operating and maintenance procedures, a "walk around" plant safety inspection, and many others. When a hazard is identified using one of these techniques, the participants in the activity should challenge themselves to identify opportunities to eliminate the hazard before accepting the existence of the hazard and identifying ways to manage it. Once a hazard has been identified, ask the following questions:

- Can the hazard be eliminated?
- If not, can the magnitude of the hazard be significantly reduced?
- Do the alternatives identified by answering the first two questions increase the magnitude of any other hazards or create new hazards? If so, consider all hazards in selecting the best alternative.
- What technical and management systems are required to manage the hazards that inevitably will remain? (All too often, engineers go directly to this question and do not challenge the presence of the hazard, but rather immediately focus on how to manage the hazard.)

To achieve many of the changes suggested here, it is necessary to carry out much more critical examination of alternatives during the early stages of design than has been customary in most companies. However, it is never too late, and there have been significant applications of inherently safer concepts in old plants that have operated for many years. The hazard analysis discussions later in this chapter describe techniques for identifying hazards at all stages in a process life cycle. When a new plant is needed, it is wanted as soon as possible, and there is no time to develop inherently safer designs. When designing a new plant, be conscious of all the improvements that could have been made if there was more time. These improvements should be noted and work on their feasibility started, ready for the plant after next.

Organizations have generally used two approaches to incorporating inherent safety considerations into the design and operation of process plants: instituting stand-alone inherently safer design reviews, and incorporating inherently safer design consideration into existing process safety reviews (discussed in the Hazard Analysis section). CCPS (*Inherently Safer Chemical Processes: A Life Cycle Approach*, 2d ed., Wiley, Hoboken, N.J., 2009) and Section 3.7 of *Guidelines for Hazard Evaluation Procedures*, 3d ed. (AIChE, New York, 2008) provide guidance on how inherently safer design can be incorporated into process hazard analysis. CCPS (*Guidelines for Engineering Design for Process Safety*, 2nd ed., AIChE, New York, 2012) provides a series of checklists, including specific suggestions for inherently safer designs for many types of common process industry equipment.

PROCESS SAFETY ANALYSIS

HAZARD ANALYSIS

General References: Alaimo, *Handbook of Chemical Health and Safety*, Oxford University Press, New York, 2001; CCPS, *Guidelines for Engineering Design for Process Safety*, 2d ed., AIChE, New York, 2012; CCPS, *Guidelines for Hazard Evaluation Procedures*, 3d ed., AIChE, New York, 2008 (and 2d ed., 1992); CCPS, *Guidelines for Risk Based Process Safety*, AIChE, New York, 2007; Center for Chemical Process Safety (CCPS), *Guidelines for Chemical Process Quantitative Risk Analysis*, 2d ed., AIChE, New York, 2000; Council of the European Union, Directive 2012/18/EU, *Control of Major Accident Hazards Involving Dangerous Substances*, July 4, 2012; Crawley and Tyler, *HAZOP: Guide to Best Practice*, 3d ed., Elsevier, Amsterdam, 2015; Crowl and Louvar, *Chemical Process Safety: Fundamentals with Applications*, 3d ed., Prentice-Hall, Englewood Cliffs, N.J., 2011; Dowell, "Managing the PHA Team," *Process Safety Progress* **13**(1) (1994, January); *Dow's Chemical Exposure Index Guide*, AIChE, New York, 1998; *Dow's Fire & Explosion Index Hazard Classification Guide*, 7th ed., AIChE, New York, 1994; Harris, *Patty's Industrial Hygiene*, 6th ed., Wiley, New York, 2010; IEC 60812, *Analysis Techniques* *for System Reliability—Procedure for Failure Mode and Effects Analysis (FMEA)*, 2d ed., International Electrotechnical Commission, Geneva, 2006; IEC 61025, *Fault Tree Analysis (FTA)*, 2d ed., International Electrotechnical Commission, Geneva, 2006; IEC 61882, *Hazard and Operability Studies (HAZOP Studies)—Application Guide*, 2d ed., International Electrotechnical Commission, Geneva, 2016; Kletz, *HAZOP and HAZAN*, 4th ed., Institution of Chemical Engineers, Rugby, Warwickshire, UK, 2006; Knowlton, *Hazard and Operability Studies*, Chemetics International Co., Ltd., Vancouver, British Columbia, February 1989; Mannan, *Lees' Loss Prevention in the Process Industries*, 4th ed., Butterworth-Heinemann, London, 2012; Skelton, *Process Safety Analysis: An Introduction*, Gulf Publishing, Houston, Tex., 1997; Tweeddale, *Managing Risk and Reliability of Process Plants*, Gulf Professional, Houston, Tex., 2003; U.S. EPA, *Risk Management Programs for Chemical Accidental Release Prevention Requirements*, 40 CFR Part 68; U.S. OSHA, *Process Safety Management of Highly Hazardous Chemicals, Explosives, and Blasting Agents*, 29 CFR 1910.119; Wells, *Hazard Identification and Risk Assessment*, IChemE, Rugby, Warwickshire, UK, 1996.

Introduction The meaning of *hazard* is often confused with *risk*. Hazard is defined as the inherent potential of a material or activity to harm people, property, or the environment. Hazard does not have a probability component. There are differences in terminology on the meaning of risk in the published literature that can lead to confusion. Risk has been defined in various ways. In this edition of the handbook, risk is defined as "a measure of human injury, environmental damage, or economic loss in terms of both the incident likelihood and magnitude of the injury, damage, or loss" (CCPS, *Guidelines for Chemical Process Quantitative Risk Analysis*, 2d ed., AIChE, New York, 2000, pp. 5–6). Risk analysis implies a probability of something occurring. The term *hazard identification and risk analysis* (HIRA) has been defined to include all activities required to identify hazards, to identify potential loss event scenarios, and to evaluate the risks of a facility throughout its life cycle, to ensure that risks to employees, the public, or the environment are properly managed (CCPS, *Guidelines for Risk Based Process Safety*, AIChE, New York, 2007, Chapter 9).

Definition of Terms Following are some definitions that are useful in understanding the components of hazards.

Accidental release Defined in the United States Environmental Protection Agency's Risk Management Program (RMP) regulation as an unanticipated emission of a regulated substance or other extremely hazardous substance into the ambient air from a stationary source (U.S. EPA, *Risk Management Programs for Chemical Accidental Release Prevention Requirements*, 40 CFR Part 68).

Acute hazard The potential for injury or damage to occur as a result of an instantaneous or short-duration exposure to the effects of an incident.

Cause-consequence A method for illustrating the possible outcomes arising from the logical combination of selected input events or states. A combination of fault tree and event tree models.

Chemical Exposure Index (CEI) A method, developed by the Dow Chemical Company, used to identify and rank the relative acute health hazards associated with potential chemical releases. The CEI is calculated from five factors: a measure of toxicity; the quantity of volatile material available for a release; the distance to each area of concern; the molecular weight of the material being evaluated; and process variables that can affect the conditions of a release such as temperature, pressure, and reactivity.

Chronic hazard The potential for injury or damage to occur as a result of prolonged exposure to an undesirable condition.

Consequence Result of a specific event. In the context of qualitative hazard evaluation procedures, the consequences are the effects following from the initiating cause, with the consequence description taken through to the loss event and sometimes to the loss event impacts. In the context of quantitative risk analyses, the consequence refers to the physical effects of the loss event usually involving a fire, explosion, or release of toxic or corrosive material.

Consequence analysis The analysis of the effects of incident outcome cases independent of frequency or probability.

Domino effect An incident that starts in one piece of equipment and affects other nearby items, such as vessels containing hazardous materials, by thermal blast or fragment impact. This can lead to escalation of consequences or frequency of occurrence. This is also known as a knock-on effect.

Event An occurrence involving the process caused by equipment performance or human action or by an occurrence external to the process.

Event tree A logic model that graphically portrays the combinations of events and circumstances in an incident sequence.

Failure modes and effects analysis (FMEA) A systematic, tabular method for evaluating and documenting the effects of known types of component failures.

Fault tree A logic model that graphically portrays the combinations of failures that can lead to a specific main failure or incident of interest (Top event).

Fire and Explosion Index (FEI) A method developed by the Dow Chemical Company for ranking the relative potential fire and explosion effect radius and property damage or business interruption impacts associated with a process. Analysts calculate various hazard and exposure factors using material characteristics and process data.

Hazard An inherent physical or chemical characteristic that has the potential for causing harm to people, the environment, or property.

Hazard and operability (HAZOP) study A scenario-based hazard evaluation procedure in which a team uses a series of guide words to identify possible deviations from the intended design or operation of a process, then examines the potential consequences of the deviations and the adequacy of existing safeguards.

Hazard identification and risk analysis (HIRA) A collective term that encompasses all activities involved in identifying hazards and evaluating risk at facilities, throughout their life cycle, to make certain that risks to employees, the public, or the environment are consistently controlled within the organization's risk tolerance.

Incident An unplanned event or sequence of events that either resulted in or had the potential to result in adverse impacts.

Loss event Point of time in an abnormal situation when an irreversible physical event occurs that has the potential for loss and harm impacts. Examples include release of a hazardous material, ignition of flammable vapors or ignitable dust cloud, and overpressurization rupture of a tank or vessel. An incident might involve more than one loss event, such as a flammable liquid spill (first loss event) followed by the ignition of a flash fire and pool fire (second loss event) that heats up an adjacent vessel and its contents to the point of rupture (third loss event).

Process hazard analysis (PHA) An organized effort to identify and evaluate hazards associated with processes and operations to enable their control. This review normally involves the use of qualitative techniques to identify and assess the significance of hazards. Conclusions and appropriate recommendations are developed. Occasionally, quantitative methods are used to help prioritize risk reduction.

Safeguard Any device, system, or action that would likely interrupt the chain of events following an initiating cause or that would mitigate loss event impacts.

What-if analysis A scenario-based hazard evaluation procedure using a brainstorming approach in which typically a team that includes one or more persons familiar with the subject process asks questions or voices concerns about what could go wrong, what consequences could ensue, and whether the existing safeguards are adequate.

What-if/checklist analysis A what-if analysis that uses some form of checklist or other listing of broad categories of concern to structure the what-if questioning.

Process Hazard Analysis Regulations In the United States, the Occupational Safety and Health Administration (OSHA) rule for Process Safety Management (PSM) of Highly Toxic Hazardous Chemicals, 29 CFR 1910.119, part (e), requires an initial PHA and an update every five years for processes that handle listed chemicals or contain over 10,000 pounds (4356 kg) of flammable material. The PHA must be done by a team, must include employees such as operators and mechanics, and must have at least one person skilled in the methodology employed. Suggested methodologies for PHA are listed in Table 23-14. The PHA must consider hazards listed in the PSM Rule, part (e), including information from previous incidents with potential for catastrophic consequences, engineering and administrative controls and consequences of their failure, facility siting, and human factors. The consequences of failure of controls must be considered.

As required by the Clean Air Act Amendments of 1990, the U.S. Environmental Protection Agency (EPA) mandates a Risk Management Program (RMP) for listed substances (*Risk Management Programs for Chemical Accidental Release Prevention Requirements*, 40 CFR Part 68). RMP requires:

1. A hazard assessment that details the potential effects of an accidental release, an accidental release history of the last five years, and an evaluation of worst-case and alternative accidental releases;

2. A prevention program that includes safety precautions and maintenance, monitoring, and employee training measures; and

TABLE 23-14 Process Hazard Analysis Methods Listed in the United States

- What-if
- Checklist
- What-if/checklist
- Hazard and operability study (HAZOP)
- Failure modes and effects analysis (FMEA)
- Fault tree analysis (FTA)
- An appropriate equivalent methodology

OSHA Process Safety Management Standard (29 CFR 1910.119).

3. An emergency program that spells out emergency health care, employee training measures, and procedures for informing the public and response agencies, should an accidental release occur.

Most countries also have regulations analogous to the U.S. regulations. For example, the European Union issued the "Seveso III" Directive in 2012 (replacing the 1982 and 1996 directives) which requires all member states to implement regulations for the control of major accident hazards. Also, in addition to the U.S. government requirements, many state and local governments have implemented regulations requiring process hazard analysis and risk management.

Hazard Identification and Analysis Tools The hazard and risk assessment tools used vary with the stage of the project from the early design stage to plant operations. Many techniques are available. In the following discussion, they will be categorized as hazard identification and analysis tools, hazard ranking methods, and logic model methods. Reviews done early in projects often result in easier, more effective changes.

Safety, Health, Environmental, and Loss Prevention Reviews Most process industry companies have specific internal protocols defining these reviews, which may have different names or descriptions in different organizations. In most organizations, these reviews are conducted at various stages in the process life cycle, from initial process conceptualization, through laboratory development, scale-up, plant design, start-up, operation, modification, and shutdown. The scope and focus of the review will be different at different stages in development, with reviews early in process development focusing on major hazards and strategies for managing the risks. As the process and plant become clearly defined, the reviews will focus more on details of the design and operation. The purpose of the reviews is to have an independent (from the development, design, or operation team) evaluation of the process and layout from safety, industrial hygiene, environmental, and loss prevention points of view. It is often desirable to combine these reviews to improve the efficiency of the use of time for the reviewers (CCPS, *Guidelines for Hazard Evaluation Procedures,* 3d ed., AIChE, New York, 2008, chaps. 4.1, 4.2, 14, 17, 18).

Checklists Checklists are simple means of applying experience to designs or situations to ensure that the features appearing in the list are not overlooked. Checklists tend to be general and may not be appropriate to a specific situation. They may not handle the novel design or unusual process adequately (CCPS, *Guidelines for Hazard Evaluation Procedures,* 3d ed., AIChE, New York, 2008, chaps. 4.4, 17, appendix A). The CCPS *Guidelines for Engineering Design for Process Safety,* 2d ed. (AIChE, New York, 2012) provides a useful set of checklists for common chemical processing equipment.

What-If Analysis At each process step, what-if questions are formulated and answered to evaluate the effects of component failures or procedural errors. This technique relies on the experience level of the questioner. What-if methods are often used in conjunction with checklists (CCPS, *Guidelines for Hazard Evaluation Procedures,* 3d ed., AIChE, New York, 2008, chaps. 5.2, 22), and this methodology is referred to as a what-if-checklist review. The discussions in the HAZOP section that follows relate to review team composition (expertise, size, and leadership), the potential use of qualitative risk-ranking tools, documentation, and follow-up on actions; these also apply to what-if and what-if-checklist reviews.

Failure Mode and Effect Analysis (FMEA) This is a systematic study of the causes of failures and their effects. All causes or modes of failure are considered for each element of a system, and then all possible outcomes or effects are recorded (CCPS, *Guidelines for Hazard Evaluation Procedures,* 3d ed., AIChE, New York, 2008, chaps. 5.4, 21).

Industrial Hygiene Reviews These reviews evaluate the potential of a process to cause harm to the health of people. The review normally deals with chronic effects of exposure to chemicals and other harmful agents (e.g., noise, heat, repetitive motion) in the workplace. Chapter 44 of *Patty's Industrial Hygiene* (2010) reviews occupational health and safety management systems.

Facilities Reviews There are many kinds of facilities reviews that are useful in detecting and preventing process safety problems. They include pre-start-up safety reviews (PSSR—before the plant, equipment, or procedure modification operates), new-plant reviews (the plant has started, but is still new), reviews of existing plants (safety, technology, and operations audits and reviews), management reviews, critical instrument reviews, and hazardous materials transportation reviews.

Hazard and Operability Study (HAZOP) HAZOP is a formal hazard identification and evaluation procedure designed to identify hazards to people, process plants, and the environment (CCPS, *Guidelines for Hazard Evaluation Procedures,* 3d ed., AIChE, New York, 2008, Chapters 5.3, 15, 19; CCPS, *Guidelines for Chemical Process Quantitative Risk Analysis,* 2d ed., AIChE, New York, 2000, pp. 583–587). The technique aims to stimulate, in a systematic way, the imagination of designers and people who operate plants or equipment to identify potential hazards. HAZOP studies assume that a hazard or operating problem can arise when there is a deviation from the design or operating intention. Actions to control identified hazard or operational scenarios can then be taken before a real incident occurs. The primary goal in performing a HAZOP is to identify, not to analyze or quantify the potential incidents and loss events in a process. The end product of a HAZOP is a list of concerns and recommendations for the prevention of problems, not an analysis of the occurrence, frequency, overall effects, and the definite solution. A HAZOP study is most cost effective when done during plant design; it is easier and cheaper to change a design than to modify an existing plant. However, HAZOP is a valuable process hazard analysis tool at any stage in the life cycle of a plant. These studies make use of the combined experience and training of a group of knowledgeable people in a structured setting. Some key HAZOP terms follow.

- ***Intention*** How the part or process is expected to operate.
- ***Guide words*** Simple words used to qualify the intention in order to guide and stimulate creative thinking and so discover deviations; Table 23-15 describes commonly used guide words.
- ***Deviations*** Departures from the intention discovered by the systematic application of guide words.
- ***Causes*** Reasons that deviations might occur.
- ***Consequences*** Results of deviations if they occur.
- ***Safeguards*** Prevention, mitigation, and control features that already exist in the plant, or that are already incorporated into a new design.
- ***Actions*** Prevention, mitigation, and control features that do not currently exist and are recommended by the HAZOP team; actions may also include recommendations for additional study if the HAZOP team does not have sufficient information, or time to understand a concern sufficiently to make a specific recommendation.

The guide words are used in conjunction with the process intentions to generate possible deviations from the intended operation (see Table 23-15). Some examples of deviations that might be generated in the course of a HAZOP study include:

- No flow
- Reverse flow
- Less flow
- Increased (more) temperature
- Decreased (less) pressure
- Reverse flow from a reactor to a feed tank
- Composition change (more or less)
- Sampling (other)
- Corrosion/erosion (other)

The HAZOP team then determines the specific causes of each deviation. For example, no flow of a particular material in a specified pipe might include causes such as a manual valve improperly closed, a stopped pump, or a pipe plugged with solids. The HAZOP team then determines the consequences of the deviation for each cause and qualitatively decides the magnitude of the hazard. The team identifies any existing safeguards in the plant or design and qualitatively judges whether they are adequate. If the team determines that additional safeguards are required, it may recommend specific actions. The team may determine that the issue requires more detailed study than can be accommodated in the time frame of a HAZOP meeting, and it may recommend more extensive evaluation to determine if further action is needed and what that action should be.

TABLE 23-15 Some Hazop Guide Words Used in Conjunction with Process Parameters

Guide word	Meanings	Comments
No, Not, None	Complete negation of design intentions	No part of intention is achieved and nothing else occurs
More	Quantitative increases	Quantities and relevant physical properties such as flow rates, heat, pressure
Less	Quantitative decreases of any relevant physical parameters	Same as above
As well as	Qualitative increase	All design and operating intentions are achieved as well as some additional activity
Part of	A qualitative decrease	Some parts of the intention are achieved, others are not
Reverse	Logical opposite of intention	Activities such as reverse flow or chemical reaction, or poison instead of antidote
Other than	Complete substitution	No part of intention is achieved; something quite different happens

SOURCE: Knowlton, 1989.

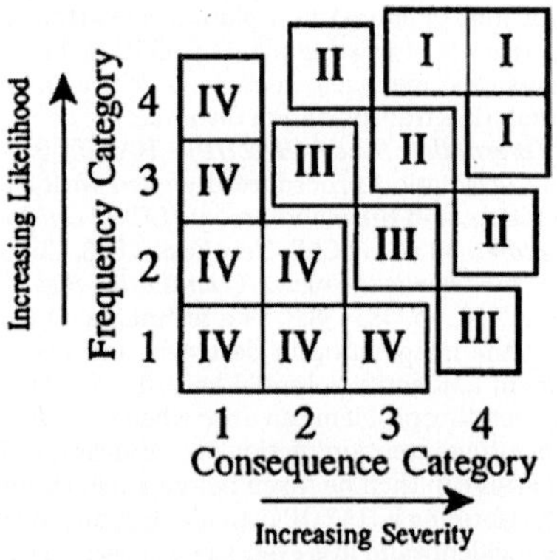

FIG. 23-17 Example qualitative risk matrix. (*From* Guidelines for Hazard Evaluation Procedures, Second Edition with Worked Examples, *AIChE, New York, 1992, reprinted with permission.*)

Many HAZOP studies incorporate a qualitative evaluation of risk to help the team to evaluate the adequacy of existing safeguards and the need for additional safeguards. The CCPS *Guidelines for Hazard Evaluation Procedures,* 3d ed., Chapter 7 (AIChE, New York, 2008), shows how a true order-of-magnitude quantitative evaluation of risk can be incorporated into HAZOP studies. This involves constructing a risk matrix, such as the one shown in Fig. 23-17. The team determines, based on its knowledge of the plant, experience, and engineering judgment, which of the several consequence and likelihood categories in the risk matrix best describe the deviation-cause-consequence under consideration. Scenarios with high consequence and high frequency represent a large risk; those with low likelihood and consequence are of low risk. An organization can use the matrix to establish guidelines for which of the boxes in the risk matrix require action.

HAZOP studies may be done for batch as well as continuous processes (CCPS, *Guidelines for Hazard Evaluation Procedures,* 3d ed., AIChE, New York, 2008, chap. 9.1). For a continuous process, the working document is usually a set of flow sheets or piping and instrumentation diagrams (P&IDs). Batch processes have another dimension, time. Time is usually not significant with a continuous process that is operating smoothly, although start-up and shutdown must also be considered when the continuous process will resemble a batch process. For batch processes the working documents consist not only of the flowsheets or P&IDs but also the operating procedures. One way to incorporate this fourth dimension is to use guide words associated with time, such as those described in Table 23-16.

HAZOP studies involve team members, at least some of whom have had experience in the plant design to be studied. These team members apply their expertise to achieve the aims of HAZOP. There are three overall aims to which any HAZOP study should be addressed:

1. Use deviations as the starting point to identify as many scenarios as possible that describe what can go wrong, and evaluate how likely each scenario is to occur and how severe the consequences could be.

2. Decide whether action is required, and identify ways in which the problem can be solved.

3. Identify cases in which a decision cannot be made immediately, and decide what information or action is required.

The team leader is a key to the success of a HAZOP study and should have adequate training for the job. Proper planning is important to success. The leader is actually a facilitator (a discussion leader and one who keeps the meetings on track) whose facilitating skills are just as important as technical knowledge. The leader outlines the boundaries of the study and ensures that the design intention is clearly understood. The leader applies guide words and encourages the team to discuss causes, consequences, and possible remedial actions for each deviation. Prolonged discussions of how a problem may be solved should be avoided. Ideally, the team leader should be accompanied by a scribe or recorder, freeing the leader for full-time facilitating. The scribe should take notes in detail for full recording of as much of the meeting as is necessary to document a thorough discussion of potential loss event scenarios, and to capture the intent of actions and recommendations. Many companies offer computer tools to help facilitate and document HAZOP studies.

TABLE 23-16 HAZOP Guide Words Associated with Time

Guide word	Meaning
No time	Step(s) missed
More time	Step does not occur when it should
Less time	Step occurs before previous step is finished
Wrong time	Flow or other activity occurs when it should not

SOURCE: Knowlton, 1989.

Team size is important. Having fewer than three contribution members, excluding the scribe and leader, will probably reduce team effectiveness. A team size of five to eight, including the leader and scribe, is probably optimum. The time required for HAZOP studies is significant and may vary widely, depending on the complexity of the system and how the process is broken down into study nodes. Some organizations divide the process into small study nodes, perhaps individual lines. In this case, each node may take on the order of an hour to review. Other organizations divide the process into larger study nodes, perhaps a single major piece of process equipment such as a distillation column, extractor, or reactor, along with all of the piping and support equipment associated with that major equipment. In this case, the time will be much longer for each node (although there will be fewer study nodes), the review for each requiring up to a full working day or more, depending on complexity. It should be recognized that the time required for HAZOP studies may not really be additional time for the project as a whole. If started early enough in the design, the HAZOP may actually save time on the project by identifying safety and operational concerns early in the design process, avoiding rework if these issues are identified later. It also should make possible smoother start-ups and make the process or plant safer and easier to operate, which will more than pay back the cost of the HAZOP study during the life of the plant.

Documentation of the discussions during a HAZOP or other PHA, and of response to all recommended actions, is important for many reasons, including:

- Making sure that there is a timely and appropriate response to all recommended actions
- Ensuring that people charged with responding to the recommendations understand the safety concern that led to the recommendation
- Allowing management to monitor the status of all recommendations and provide adequate resources for follow-up
- Facilitating the use of the HAZOP for other PSM activities such as management of change, incident investigation, development of training and procedures, and periodic revalidation of the HAZOP.

The key to good HAZOP documentation is to do it right away while the information is fresh. All HAZOP discussions concerning significant hazards should be documented, even if the discussion does not result in a recommendation because the hazard is determined to be adequately managed by existing safeguards. This demonstrates that the HAZOP has discussed the hazard and evaluated the safeguards, and it highlights existing safeguards that must be maintained to ensure safe operation. Table 23-17 is an example of a HAZOP documentation worksheet. Following the HAZOP study, there should be a process to ensure that all actions are properly acted upon, that this is done in a timely manner, and that the response has been fully documented. Initial HAZOP planning should establish the management follow-up procedure that will be used to track responses to recommendations (CCPS, *Guidelines for Hazard Evaluation Procedures,* 3d ed., AIChE, New York, 2008, chap. 8).

Hazard Ranking Methods Hazard ranking methods (CCPS, *Guidelines for Hazard Evaluation Procedures,* 3d ed., AIChE, New York, 2008, chaps. 4.3, 20) allow the analyst to compare the hazards of several processes, plants, or activities. They can be used to compare alternative chemical process routes, plant designs, plant siting options, or other design choices. Hazard ranking methods can also be used for prioritizing facilities for additional risk management studies. They generally produce a numerical score for the process being evaluated. The scores generally do not have units and are only meaningful relative to each other in the context of the hazard index being used. Some of the more commonly used hazard ranking methods are:

- National Fire Protection Association (NFPA) Standard System for Identification of Health, Flammability, Reactivity, and Related Hazards (NFPA 704, 2012)
- Dow Fire and Explosion Index (*Dow's Fire & Explosion Index Hazard Classification Guide,* 7th ed., AIChE, New York, 1994)
- Dow Chemical Exposure Index (*Dow's Chemical Exposure Index Guide,* AIChE, New York, 1998).

Consequence-Based Ranking Systems Release consequence modeling can be used to rank potential chemical hazards based on the hazard distance resulting from a specified loss of containment. For example, the U.S. EPA's RMP regulations (*Risk Management Programs for Chemical Accidental Release Prevention Requirements,* 40 CFR Part 68) require consequence modeling for a predefined worst-case scenario: release of the entire contents of the largest container of a material in 10 min. EPA provides lookup tables and software (RMP*Comp) to assist in estimating the hazard distances for materials covered by the RMP regulations.

Logic Model Methods The following tools are most commonly used in quantitative risk analysis, but they can also be useful qualitatively to understand the combinations of events that can cause a loss event. The logic models can also be useful in understanding how protective systems affect various potential accident scenarios. These methods will be thoroughly discussed in the Risk Analysis subsection. Also, hazard identification and

TABLE 23-17 Sample HAZOP Study Results

P&ID No.: E-250, Revision D
Meeting Date: ##/##/##

Team: Mr. Smart, Mr. Associate, Ms. Piper, Mr. Stedman, Mr. Volt (all from the ABC Anywhere Plant)

1.0 Line — Air supply line to incinerator (Intention: Supply 15,000 scfm air to incinerator at ambient temperature and 3 in. WC)

Item	Deviation	Cause	Consequences	Safeguards	Actions
1.1.1	No Air Flow to Incinerator	Air fan #1 fails off or fan #1 power off	Loss of combustion. Release out the stack	• Redundant fan on standby with autostart • Incinerator shutdowns on low-low air pressure PSLL-1, low temperature, or flameout	1, 2
			Incinerator explosion	• Redundant fan on standby with autostart • Incinerator shutdowns on low-low air pressure PSLL-1, low temperature, or flameout • Accumulation, then reignition of gas+air required for damaging pressure to develop	1, 2
1.1.2		FCV-1 fails closed, FT1 fails high signal, or FT-2 fails low signal	Loss of combustion. Release out the stack	• Mechanical stop on FCV-1 • Incinerator shutdown on low-low air pressure PSLL-1, low temperature, or flameout	2
			Incinerator explosion	• Mechanical stop on FCV-1 • Incinerator shutdown on low-low air pressure PSLL-1, low temperature, or flameout • Accumulation, then reignition of gas+air required for damaging pressure to develop	2
1.1.3		Loss of electric power	Loss of combustion. Release out the stack	• Incinerator automatic shutdown on loss of electric power	
			Incinerator explosion	• Incinerator automatic shutdown on loss of electric power • Accumulation, then reignition of gas+air required for damaging pressure to develop	
1.1.4		Plugged air screen	Loss of combustion. Release out the stack	• Air screen cleaned weekly • Incinerator shutdown on low-low air pressure PSLL-1, low temperature, or flameout	2
			Incinerator explosion	• Air screen cleaned weekly • Incinerator shutdown on low-low air pressure PSLL-1, low temperature, or flameout • Accumulation, then reignition of gas+air required for damaging pressure to develop	2

SOURCE: CCPS, *Guidelines for Hazard Evaluation Procedures, Third Edition*, AIChE, New York, 2008, reprinted with permission.

evaluation tools discussed in this section are valuable precursors to a quantitative risk analysis (QRA) or a layer of protection analysis (LOPA). Generally a QRA or LOPA quantifies the risk of hazard scenarios that have been identified by using tools such as those discussed above.

- Fault tree analysis (CCPS, *Guidelines for Hazard Evaluation Procedures,* 3d ed., AIChE, New York, 2008, chaps. 5.5, 16)
- Event tree analysis (CCPS, *Guidelines for Hazard Evaluation Procedures,* 3d ed., AIChE, New York, 2008, chaps. 5.6, 16)
- Cause-consequence diagram (CCPS, *Guidelines for Hazard Evaluation Procedures,* 3d ed., AIChE, New York, 2008, chap. 5.7)

RISK ANALYSIS

General References: Arendt, S., "Management of Quantitative Risk Assessment in the Chemical Process Industry," *Plant/Operations Progress* **9**(4) (1990, October); Center for Chemical Process Safety (CCPS), *Guidelines for Chemical Process Quantitative Risk Analysis,* 2d ed., AIChE, New York, 1999; CCPS, *Guidelines for Chemical Transportation Risk Analysis,* 2d ed., AIChE, New York, 2008; EFCE, *Risk Assessment in the Process Industries,* European Federation of Chemical Engineering, Publication Series no. 45, 1985; Mannan, S., ed., *Lees' Loss Prevention in the Process Industries,* 4th ed., Butterworth-Heinemann, London, 2012; World Bank, *Manual of Industrial Hazard Assessment Techniques,* Office of Environmental and Scientific Affairs, World Bank, Washington, D.C., 1985.

Frequency Estimation References: Billington and Allan, *Reliability Evaluation of Engineering Systems: Concepts and Techniques,* Plenum Press, New York, 1983; CCPS, *Guidelines for Process Equipment Reliability Data,* AIChE, New York, 1989; Fussell, Powers, and Bennetts, "Fault Trees: A State of the Art Discussion," *IEEE Transactions on Reliability,* IEEE, New York, 1974; Roberts, N. H., et al., *Fault Tree Handbook,* NUREG-0492, Washington, D.C.; *OREDA: Offshore Reliability Data Handbook,* OREDA, 2002; Swain and Guttmann, *Handbook of Human Reliability Analysis with Emphasis on Nuclear Power Plant Applications,* NUREG/CR-1278, USNRC, Washington, D.C., 1983.

Risk Estimation References: Baybutt, "Uncertainty in Risk Analysis," Conference on Mathematics in Major Accident Risk Assessment, Oxford University, UK, 1986; Considine, *The Assessment of Individual and Societal Risks,* SRD Report R-310, Safety and Reliability Directorate, UKAEA, Warrington, UK, 1984; Rasmussen, *Reactor Safety Study: An Assessment of Accident Risk in U.S. Commercial Nuclear Power Plants,* WASH-1400 NUREG 75/014, Washington, D.C., 1975; Rijnmond Public Authority, *A Risk Analysis of 6 Potentially Hazardous Industrial Objects in the Rijnmond Area—A Pilot Study,* D. Reidel, Boston, 1982; U.K. Health and Safety Executive, *Canvey—An Investigation of Potential Hazards from the Operations in the Canvey Island/Thurrock Area,* HMSO, London, 1978.

Risk Criteria and Risk Decision Making: Ale, "The Implementation of an External Safety Policy in the Netherlands," *Intl. Conference on Hazard Identification and Risk Analysis, Human Factors and Reliability in Process Safety,* AIChE, New York, 1992, pp. 173–183; CCPS, *Guidelines for Developing Quantitative Safety Risk Criteria,* AIChE, New York, 2009; Gibson, "Hazard Analysis and Risk Criteria," *Chemical Engineering Progress* (November 1980): 46–50.

Introduction The previous sections dealt with techniques for qualitative hazard analysis only. This section addresses the quantitative methodologies available to analyze and estimate risk, which is a function of both the consequences of an incident and its frequency. The application of these methodologies in most instances is not trivial. A significant allocation of resources is necessary. Therefore, a selection process or risk prioritization process is advised before considering a risk analysis study.

Important definitions are as follows.

Accepted risk The risk is considered tolerable for a given activity by those responsible for managing or regulating the operation of a hazardous facility. The term *acceptable risk* has often been used, but this inevitably raises the question, "Acceptable to whom?" Tweeddale (*Managing Risk and Reliability of Process Plants,* Gulf Professional, Houston, Tex., 2003, p. 70) suggests that *accepted risk* is a better term because it makes it clear that the risk has been accepted by those responsible for the decisions on how to build, operate, and regulate the facility.

Event sequence A specific, unplanned sequence of events composed of initiating events and intermediate events that may lead to an incident.

Frequency The rate at which observed or predicted events occur.

Incident outcome The physical outcome of an incident; e.g., a leak of a flammable and toxic gas could result in a jet fire, a vapor cloud explosion, a vapor cloud fire, or a toxic cloud.

Probability The likelihood of the occurrence of events, the values of which range from 0 to 1.

Probability analysis Way to evaluate the likelihood of an event occurring. By using failure rate data for equipment, piping, instruments, and fault tree techniques, the frequency (number of events per unit time) can be quantitatively estimated.

Probit model A mathematical model of dosage and response in which the dependent variable (response) is a probit number that is related through a statistical function directly to a probability.

Quantitative risk assessment (QRA) The systematic development of numerical estimates of the expected frequency and consequence of potential accidents associated with a facility or an operation. Using consequence and probability analyses and other factors such as population density and expected weather conditions, QRA estimates the fatality rate for a given set of events.

Risk A measure of economic loss or human injury in terms of both incident likelihood (frequency) and the magnitude of the loss or injury (consequence).

Risk analysis The development of an estimate of risk based on engineering evaluation and mathematical techniques for combining estimates of incident consequences and frequencies. Incidents in the context of the discussion in this chapter are acute events that involve loss of containment of material or energy.

Risk assessment The process by which results of a risk analysis are used to make decisions, either through a relative ranking of risk reduction strategies or through comparison with risk targets. The terms *risk analysis* and *risk assessment* are often used interchangeably in the literature.

A typical hazard identification process, such as a HAZOP study, is sometimes used as a starting point for the selection of potential major risks for risk analysis. Other selection or screening processes can also be applied. However major risks are chosen, a HAZOP study is a good starting point to develop information for the risk analysis study. A major risk may qualify for risk analysis if the magnitude of the incident is potentially quite large (high potential consequence) or if the frequency of a severe event is judged to be high (high potential frequency) or both. A flowchart that describes a possible process for risk analysis is shown in Fig. 23-18.

The components of a risk analysis involve the estimation of the frequency of an event, an estimation of the consequences, and the selection and generation of the estimate of risk itself.

A risk analysis can have a variety of potential goals:

1. To screen or bracket a number of risks in order to prioritize them for possible future study
2. To estimate risk to employees
3. To estimate risk to the public
4. To estimate financial risk
5. To evaluate a range of risk reduction measures
6. To meet legal or regulatory requirements
7. To assist in emergency planning

The scope of a study required to satisfy these goals will be dependent upon the extent of the risk, the depth of the study required, and the level of resources available (mathematical models and tools and skilled people to perform the study and any internal or external constraints).

A risk analysis can be applied to fixed facilities or transportation movements, although much of the attention today still centers on the former. In a fixed-facility risk analysis, QRA can aid risk management decisions with respect to

1. Chemical processes
2. Process equipment
3. Operating procedures
4. Chemical inventories
5. Storage conditions

In a transportation risk analysis (TRA), the risk parameters are more extensive, but more restrictive in some ways. Examples of risk parameters that could be considered are

1. Alternate modes of transport
2. Routes
3. Travel restrictions
4. Shipment size
5. Shipping conditions (e.g., pressure, temperature)
6. Container design
7. Unit size (e.g., bulk versus drums)

The objective of a risk analysis is to reduce the level of risk wherever practical. Much of the benefit of a risk analysis comes from the discipline that it imposes and the detailed understanding of the major contributors of the risk that follows. There is general agreement that if risks can be identified and analyzed, then measures for risk reduction can be effectively selected.

The expertise required to perform a risk analysis is substantial. Although various software programs are available to calculate the frequency of events or their consequences, or even risk estimates, engineering judgment and experience are still very much needed to produce meaningful results. And although professional courses are available in this subject area, there is a significant learning curve required, not only for engineers to become practiced risk analysts, but also for management to be able to understand and interpret the results. For these reasons, it may be helpful to use a consultant organization in this field.

LOPA (Layer of Protection Analysis) is a simplified order-of-magnitude quantitative risk analysis method for process safety risk assessment. See LOPA details later in this subsection.

The analysis of a risk—that is, its estimation—leads to the assessment of that risk and the decision-making processes of selecting the appropriate level of risk reduction. In most studies, this is an iterative process of risk analysis and risk assessment until the risk is reduced to some specified level. The question of how safe is safe enough has to be addressed either implicitly or explicitly in the decision-making process. This subject is discussed in detail later in this section.

Frequency Estimation There are two primary sources for estimates of incident frequencies. These are historical records and the application of fault tree analysis and related techniques, and they are not necessarily applied independently. Specific historical data can sometimes be usefully applied as a check on frequency estimates of various subevents of a fault tree, for example.

The use of historical data provides the most straightforward approach to the generation of incident frequency estimates but is subject to the applicability and the adequacy of the records. Care should be exercised in extracting data from long periods of the historical record over which design or operating standards or measurement criteria may have changed.

An estimate of the total population from which the incident information has been obtained is important and may be difficult to obtain.

Fault tree analysis and other related event frequency estimation techniques, such as event tree analysis, play a crucial role in the risk analysis process. Fault trees are logic diagrams that depict how components and systems can fail. The undesired event becomes the top event, and subsequent subevents, and eventually basic causes, are then developed and connected through logic gates. The fault tree is completed when all basic causes, including equipment failures and human errors, form the base of the tree. Practitioners have developed general rules for construction, but no specific rules for which events or gates to use. The construction of a fault tree is still more of an art than a science. Although a number of attempts have been made to automate the construction of fault trees from process flow diagrams or piping instrumentation diagrams, these attempts have been largely unsuccessful [Andow, P. K., "Difficulties in Fault Tree Synthesis for Process Plant," *IEEE Transactions on Reliability R-29*(1): 2 (1980)].

Once the fault tree is constructed, quantitative failure rate and probability data must be obtained for all basic causes. A number of equipment failure rate databases are available for general use. However, specific equipment failure rate data is generally lacking and, therefore, data estimation and reduction techniques must be applied to generic databases to help compensate for this shortcoming. Accuracy and applicability of data will always be a concern, but experienced practitioners can generally obtain useful results from quantifying fault trees.

Human error probabilities can also be estimated using methodologies and techniques originally developed in the nuclear industry. A number of different models are available (Swain, "Comparative Evaluation of Methods for Human Reliability Analysis," GRS Project RS 688, 1988). This estimation process should be done with great care, as many factors can affect the reliability of the estimates. Methodologies using expert opinion to obtain failure rate and probability estimates have also been used where there is sparse or inappropriate data.

In some instances, plant-specific information relating to frequencies of subevents (e.g., a release from a relief device) can be compared against results derived from the quantitative fault tree analysis, starting with basic component failure rate data.

An example of a fault tree logic diagram using AND and OR gate logic is shown in Fig. 23-19.

The logical structure of a fault tree can be described in terms of Boolean algebraic equations. Some specific prerequisites to the application of this methodology are:

- Equipment states are binary (working or failed).
- Transition from one state to another is instantaneous.
- Component failures are statistically independent.
- The failure rate and repair rate are consistent for each equipment item.
- After repair, the component is returned to the working state.

Minimal cut set analysis is a mathematical technique for developing and providing probability estimates for the combinations of basic component failures and/or human error probabilities that are necessary and sufficient to result in the occurrence of the top event.

A number of software programs are available to perform these calculations, given the basic failure data and fault tree logic diagram (CCPS, *Guidelines for Chemical Process Quantitative Risk Analysis,* 2d ed., AIChE, New York, 1999). Other less well known approaches to quantifying fault tree event frequencies are being practiced, which result in gate-by-gate calculations using discrete-state, continuous-time, Markov models [Doelp et al., "Quantitative Fault Tree Analysis, Gate-by-Gate Method," *Plant Operations Progress* **4**(3): 227–238, (1984)].

Identification and quantitative estimation of common-cause failures are general problems in fault tree analysis. Boolean approaches are generally better suited to handle common-cause failures mathematically. The basic assumption is that failures are completely independent events, but in reality

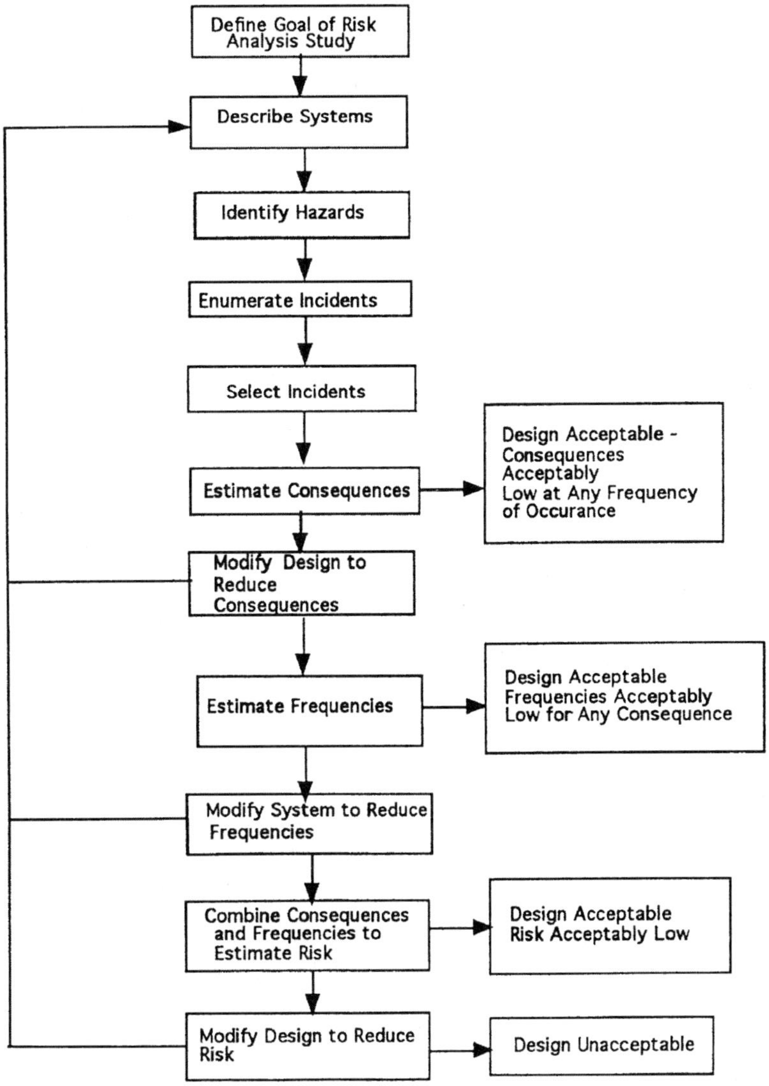

FIG. 23-18 One version of a risk analysis process. (*CCPS,* Guidelines for Developing Quantitative Safety Risk Criteria, *AIChE, New York, 2009, p. 25, by permission.*)

dependencies will exist, and these are categorized as common-cause failures (CCFs). Both qualitative and quantitative techniques can be applied to identify and assess CCFs. An excellent overview of CCF is available (CCPS, *Guidelines for Chemical Process Quantitative Risk Analysis,* 2d ed., AIChE, New York, 1999).

Event tree analysis is another useful frequency estimation technique for risk analysis. It is a bottom-up logic diagram, which starts with an identifiable event. Branches are then generated, which lead to specific chronologically based outcomes with defined probabilities. Event tree analysis can provide a logic bridge from the fault tree top event, such as a flammable release, into specific incident outcomes (e.g., no ignition, flash fire, or vapor cloud explosion). Probabilities for each limb in the event tree diagram are assigned and, when multiplied by the starting frequency, produce frequencies at each node point for all of the various incident outcome states. The probabilities for all of the limbs at any given level of the event tree must sum to 1.0. Event trees are generally very helpful in the generation of a final risk estimate. Figure 23-20 shows an event tree for a release of LPG (CCPS, *Guidelines for Chemical Process Quantitative Risk Analysis,* 2d ed., AIChE, New York, 1999).

Consequence Estimation Given that an incident (release of material or energy) has been defined, the consequence of the release of material or energy and the ultimate effects on the receptors of the risk (for example, injury or fatality to people, damage to environment, or financial loss) must be estimated. Consequence estimation evaluates the sequence of the release to the point of its ultimate impact on the receptors. Consequence analysis for releases is described later in this section.

The output of these frequency and consequence calculation processes is one or more pairs of incidents or incident outcome case frequencies paired with their effects (consequence or impact).

Risk Estimation There are a number of risk measures that can be estimated. The specific risk measures chosen are generally related to the study objective and the depth of study, along with any preferences or requirements established by the decision makers. Generally, risk measures can be broken down into three categories: risk indices, individual risk measures, and societal risk measures.

Risk indices are usually single-number estimates, which may be used to compare one risk with another or used in an absolute sense compared to a specific target. For risks to employees, the *fatal accident rate* (FAR) is a commonly applied measure. The FAR is a single-number index, which is the expected number of fatalities from a specific event based on 10^8 exposure hours. For workers in a chemical plant, the FAR would be calculated for a specific event as

$$\text{FAR} = \frac{10^8}{8760} EP\tau \tag{23-18}$$

where FAR = expected number of fatalities from a specific event based on 10^8 exposure hours
E = frequency of event, yr^{-1}
P = probability of being killed by the event at a specific location
τ = fraction of time spent at specific location

However, the worker in a chemical plant could be exposed to other potential events that might result in a fatality. In this case, the overall FAR for the worker would be

$$\mathrm{FAR} = \frac{10^8}{8760}\sum_i E_i \sum_j P_{ij}\tau_j \tag{23-19}$$

Each subscript i is a specific event, and each subscript j is a specific location. References are available that provide FAR estimates for various occupations, modes of transportation, and other activities [Kletz, T. A., "The Risk Equations—What Risk Should We Run?" *New Scientist* (May 12, 1977): 320–325].

Figure 23-21 is an example of an individual risk contour plot, which shows the expected frequency of an event causing a specified level of harm at a specified location, regardless of whether anyone is present at that location to suffer that level of harm.

The total individual risk at each point is equal to the sum of the individual risks at that point from all incident outcome cases.

$$\mathrm{IR}_{x,y} = \sum_{i=1}^{n} \mathrm{IR}_{x,y,i} \tag{23-20}$$

where $\mathrm{IR}_{x,y}$ = total individual risk of fatality at geographic location x,y
$\mathrm{IR}_{x,y,i}$ = individual risk of fatality at geographic location x,y from incident outcome case i
n = total number of incident outcome cases

A common form of societal risk measure is an *F-N* curve, which is normally presented as a cumulative distribution plot of frequency F versus number of fatalities N. An example of this type of measure is shown in Fig. 23-22.

Any individual point on the curve is obtained by summing the frequencies of all events resulting in that number of fatalities or greater. The slope of the curve and the maximum number of fatalities are two key indicators of the degree of risk.

Risk Criteria Once a risk estimate is prepared, alternatives to reduce the risk can be determined, but one always faces the challenge of how low a risk level is low enough. Different countries in the world have established numerical criteria, primarily as a tool to address the Seveso regulations [Ale, "The Implementation of an External Safety Policy in the Netherlands," *Intl. Conference on Hazard Identification and Risk Analysis, Human Factors and Reliability in Process Safety,* AIChE, New York, 1992, pp. 173–183 and HSE (2001)]. In addition, over the years various chemical companies have established numerical targets [Gibson, "The Use of Risk Criteria in the Chemical Industry," *Trans. IChemE* **71**(Pt. B): 117–123 (1993); Helmers and Schaller, "Calculated Process Risks and Health Management," *Plant/Operations Progress* **1**(3): 190–194 (1982); Renshaw, "A Major Accident Prevention Program," *Plant/Operations Progress* **9**(3): 194–197, (1990)]. Pikaar and Seaman provide a good survey paper on the use of risk guidelines by both governments and operating companies (Pikaar and Seaman, *A Review of Risk Control,* Zoetermeer, Ministerie VROM, Netherlands, 1995). Guidance is now available from industry experts for developing risk criteria (CCPS, *Guidelines for Developing Quantitative Safety Risk Criteria,* AIChE, New York, 2009).

Risk Decision Making Risk criteria represent the first step in risk decision making. Efforts must be made to reduce the risk at least to a "tolerable"

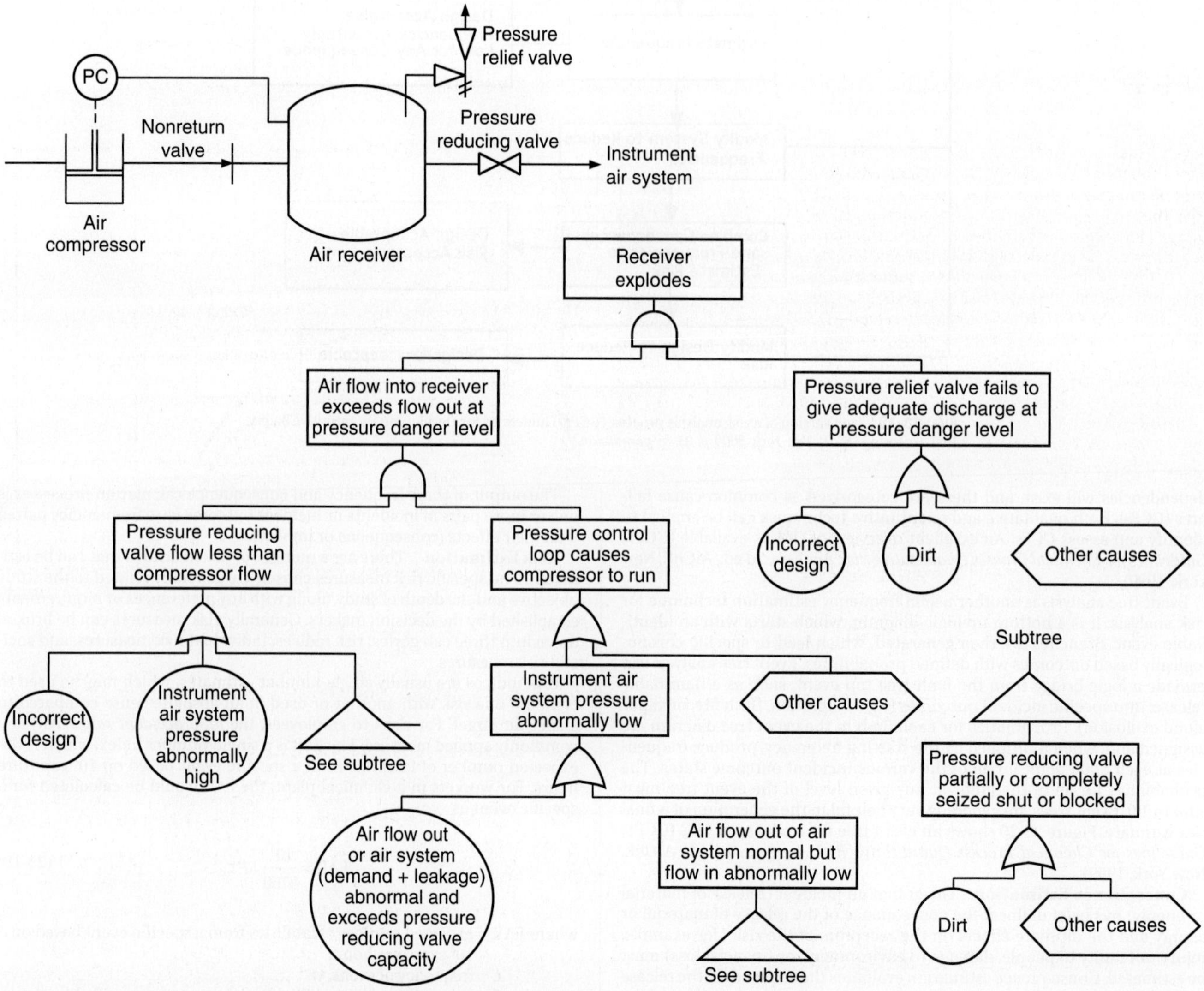

FIG. 23-19 Process drawing and fault tree for explosion of an air receiver. (*F. P. Lees,* Loss Prevention in the Process Industries, *Vol 1, 1980, pp. 200, 201, by permission.*)

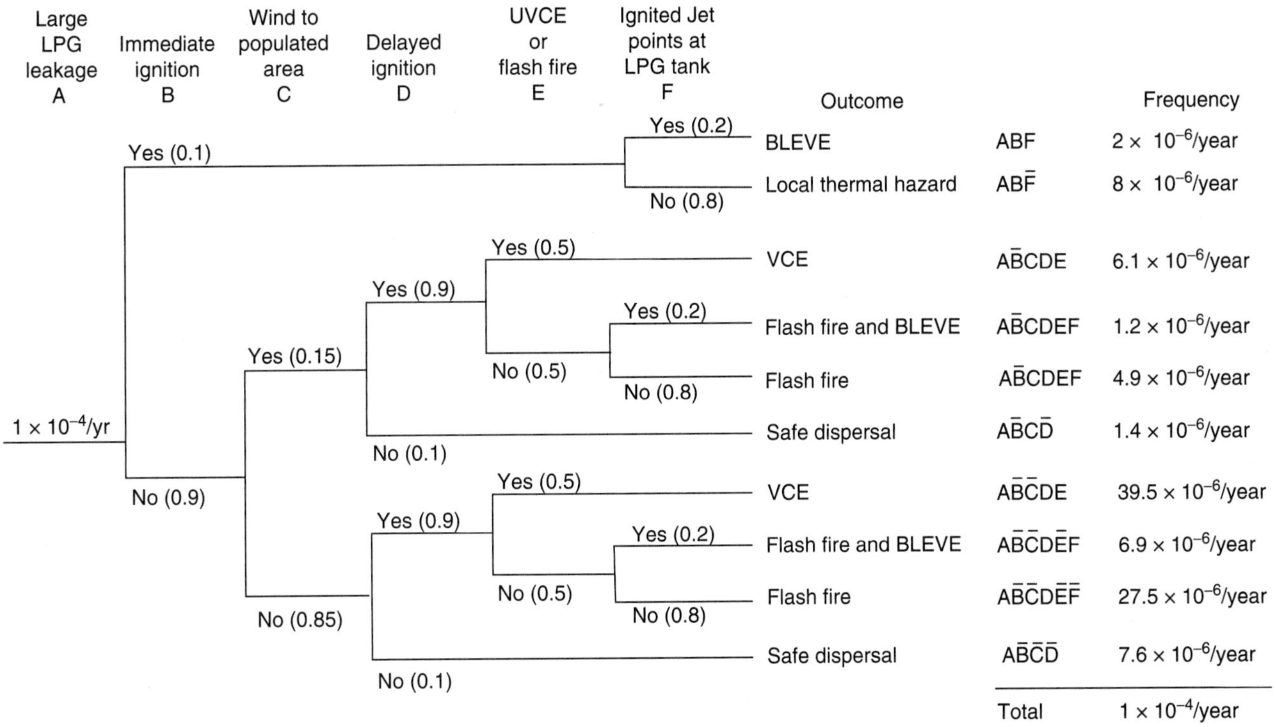

FIG. 23-20 Sample event tree for a release of LPG. (*CCPS, Guidelines for Chemical Process Quantitative Risk Analysis, 2d Ed., AIChE, New York, 1999, p. 329, by permission.*)

level, which may be dictated by a government or an operating company. Once the tolerable level is reached, then additional risk reductions should still be evaluated. A number of criteria also provide guidance on a level of risk that requires no further risk reduction. Between these two levels is a range of risk where risk reduction options need to be evaluated. How far to continue reducing risk below the tolerable level is up to each company and is a function of costs and benefits. At some point, additional risk reductions have little benefit. The "as low as reasonably practicable" (ALARP) principle is described by CCPS (*Guidelines for Developing Quantitative Safety Risk Criteria,* AIChE, New York, 2009, pp. 44–46). Hamm and Schwartz summarize some strategies for consideration (Hamm and Schwartz, "Issues and Strategies in Risk Decision Making," *Intl. Process Safety Management Conference and Workshop,* San Francisco, AIChE, New York, 1993, pp. 351–371). CCPS provides an introduction to a number of decision analysis tools that could be applied (CCPS, *Tools for Making Acute Risk Decisions with Chemical Process Safety Implications,* AIChE, New York, 1994).

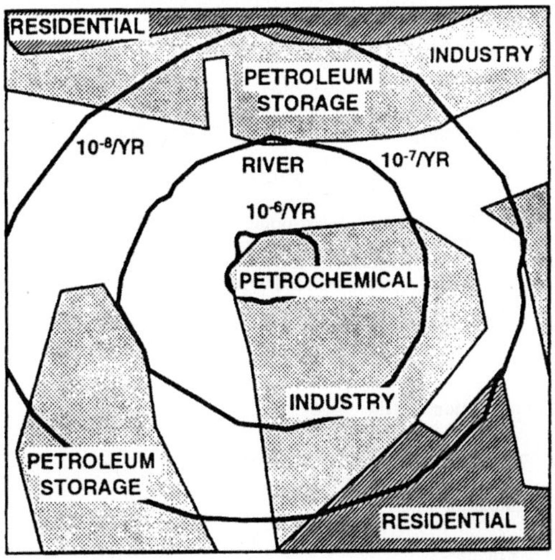

FIG. 23-21 Example of an individual risk contour plot. (*CCPS,* Guidelines for Developing Quantitative Safety Risk Criteria, *AIChE, New York, 2009, p. 32, by permission.*)

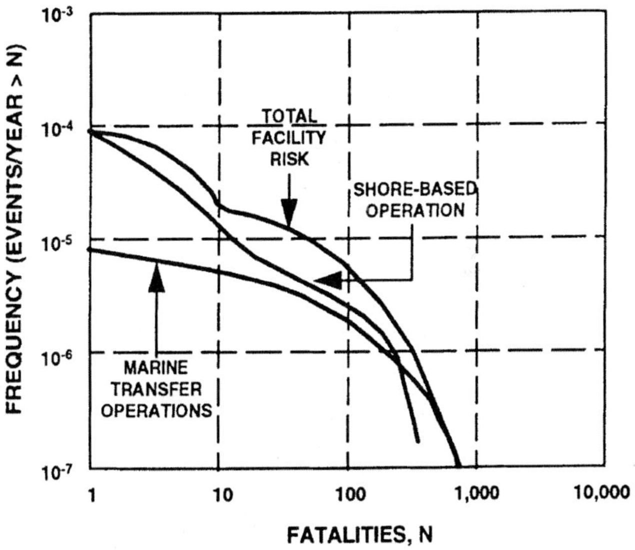

FIG. 23-22 Example of a societal risk *F-N* curve. (*CCPS,* Guidelines for Developing Quantitative Safety Risk Criteria, *AIChE, New York, 2009, p. 36, by permission.*)

LAYER OF PROTECTION ANALYSIS

General References Center for Chemical Process Safety, *Layer of Protection Analysis: Simplified Process Risk Assessment,* AIChE, New York, 2001; Center for Chemical Process Safety, *Guidelines for Initiating Events and Independent Protection Layers in Layer of Protection Analysis,* AIChE, New York, 2015; Center for Chemical Process Safety, *Guidelines for Enabling Conditions and Conditional Modifiers in Layer of Protection Analysis,* AIChE, New York, 2014.

Layer of protection analysis (LOPA) is a simplified single-scenario quantitative risk analysis methodology typically used as an extension of qualitative risk analysis. LOPA is applied to estimate the frequency of a specified scenario using strict rules for the factors included in the analysis to try to ensure that the estimate is conservative. For a specified single cause-consequence scenario, LOPA combines the initiating event frequency with the probability of failure of independent protection layers (IPLs) to predict a frequency. The resulting frequency can be either compared to predefined risk criteria or used to compare two or more scenarios in a relative risk analysis. LOPA remains simpler than more rigorous techniques by restricting the level of detail of the analysis to order-of-magnitude estimates of frequency and probability of failure on demand, as well as requiring that all factors used in the analysis are independent, eliminating the need for estimating the impact of dependency.

Initiating Events Each LOPA is an analysis of a single cause-consequence scenario. The cause of the scenario in a LOPA is an initiating event, which can be any fault or failure in a process that starts an event sequence leading to an undesired consequence. Initiating events typically fall into one of four categories: control system failures, human errors, mechanical failures, and external events. Control system failures include any aspect of a control loop that can fail, including failure of a control valve open, closed, or unresponsive, failure of a sensor or transmitter, or failure of the basic process control system (BPCS) logic solver itself. Human errors encompass failures of commission and omission and include errors in the control room or in the field. Mechanical failures include failures of primary containment, including pump seals, gaskets, piping and vessels, as well as failures of pumps, compressors, agitators, and other mechanical equipment. External events include loss of distributed services such as power or cooling water, external impacts from vehicles or cranes, and severe weather events. Table 23-18 shows typical initiating events used in a LOPA, along with prevalent values for the initiating failure frequency.

Independent Protection Layers Independent protection layers (IPLs) are safeguards in the process that either prevent or mitigate the scenario being evaluated. IPLs meet specific criteria specified in the LOPA methodology; they must be independent, effective, and audited. Each IPL used in a scenario must be independent of all other factors that are used in the analysis, including all other IPLs and the initiating event. As an example, using multiple safety instrumented systems (SISs) that are all implemented through a single logic solver would not be allowed in a LOPA scenario, since the SISs are not independent. The effectiveness of an IPL is based both on its ability to respond quickly enough to prevent the consequences of the event and on its ability to address the scenario. An example of an ineffective IPL is a relief valve that is sized for some overpressure scenarios but not for the case under consideration. An IPL must be audited or validated periodically to ensure that the safeguard is maintained and offers the appropriate level of risk reduction. For example, flame arresters are often installed in piping to prevent flame propagation; however, they can only be validated as IPLs if a program is in place to inspect them and to make sure they are working properly. Example IPLs are shown in Table 23-19.

Enabling Conditions Enabling conditions are additional factors in a LOPA that are neither failures nor protection layers, but are conditions that must be present for the event being analyzed to occur. Enabling conditions reflect operating phases or other conditions that are present infrequently but are required for the event to escalate to the consequence of concern. Enabling conditions are dimensionless probabilities that must be independent of the initiating event and any IPLs in order to be used in a LOPA. An example of an enabling condition would be a plant that normally operates using water as the solvent and does not pose a flammability hazard but is occasionally cleaned using a flammable solvent. A release resulting in a fire can only occur during the cleaning operation, which is a rare condition for the facility. The fraction of time the operation is in the cleaning mode could be used as an enabling condition in the LOPA.

Risk Criteria LOPA typically uses one of two types of endpoints for the analysis: loss of primary or secondary containment or ultimate consequences such as fatalities, environmental impact, damage to facilities, business interruption, cost of repairs, or lost profit. When loss of primary or secondary containment is the endpoint for the analysis, the risk criteria indicating the tolerable frequency of an event is normally based on the type, condition, and quantity of material that could be released. The type of release includes its flammability and toxicity. The condition is an indication of its temperature as compared to its boiling point and flash point. Risk criteria based on ultimate consequences will establish tolerable frequencies for each category of consequence to be evaluated using the LOPA methodology.

The risk criteria for LOPA are single-scenario risk criteria, as compared to individual risk or societal risk criteria, which are aggregate risk criteria.

TABLE 23-18 Example Initiating Events and Initiating Event Frequencies for LOPA

Initiating event	Initiating event frequency (failures per year)
Basic process control system (BPCS) failure	10^{-1}
Cooling water failure	10^{-1}
Pressure regulator failure	10^{-1}
Pump seal leak	1

TABLE 23-19 Example Independent Protection Layers and Probabilities of Failure on Demand for LOPA

Independent protection layer	Probability of failure on demand (PFD)
Safety instrumented system (PFD dependent on safety integrity level)	0.1–0.001
Spring-operated pressure relief valve in clean service	0.01
Rupture disk in clean service	0.01

Whether single-scenario risk criteria are based on loss of containment or ultimate consequences, a single tolerable frequency for each consequence may be established, or a range may be established indicating a broadly acceptable risk and a not-to-exceed risk, with the area between requiring consideration of additional controls.

When risk criteria are based on loss of containment, the factors already mentioned are all that are needed to conduct a LOPA. Multiplication of the initiating event frequency, any enabling condition that is applicable, and the IPLs for the scenario yield a predicted frequency for the event that can be compared to the risk criteria for the scenario.

Conditional Modifiers When ultimate consequences are the basis for the risk criteria for LOPA, conditional modifiers are often used in the analysis. Conditional modifiers are probabilities included in LOPA that reflect uncertainties in the event reaching the impact terms of interest that are not failures of safeguards. Common conditional modifiers are probability of ignition, probability of personnel presence, probability of injury or fatality, and probability of equipment damage or business interruption. When the endpoint of a LOPA is in impact terms, conditional modifiers may be used but are not required. Conditional modifiers must be independent of all other factors that are included in a LOPA, and they must be maintained over time to be effective. When conditional modifiers are used in a LOPA, they are additional probabilities that are included in the analysis to determine the predicted frequency of the scenario for comparison with the target frequency for the event from the risk criteria.

LOPA Results A LOPA gives a predicted frequency for a scenario using a strictly defined, conservative methodology. The comparison of the predicted frequency to the tolerable frequency of the scenario from the risk criteria indicates the adequacy of the existing controls and the additional risk reduction reliability needed if the controls are deemed inadequate. The recommended risk reduction can be used to determine the safety integrity level of any safety instrumented system applied to the scenario or the reliability of any other controls that might be recommended.

LOPA Example A vessel is filled by a pump that can exceed the maximum allowable working pressure (MAWP) of the vessel if level control failure causes the vessel to be overfilled. Exceeding the MAWP can result in vessel rupture, release of the contents, and fire or explosion. The consequence analysis of this scenario indicates that if personnel are in the area, significant injuries could occur, and the risk criterion applied is selected to be no more than one event in 10^5 years for this example. A properly sized relief system is in place to prevent the consequence of concern. The only conditional modifier deemed to apply was the probability of personnel presence. An example LOPA for this scenario is shown in Table 23-20.

In this example scenario, the LOPA indicates that an additional IPL would be recommended, and the probability of failure on demand of that IPL should be 0.1 or less to meet the specified risk criterion.

TABLE 23-20 Example LOPA for Overfill and Overpressure of Vessel

Factor	Description	Probability	Frequency (per year)
Consequence description	Failure of vessel due to overfill caused by level control failure		
Risk criterion	Severe injuries		10^{-5}
Initiating event	BPCS level control loop failure		10^{-1}
Independent protection layers	Relief valve	0.01	
Conditional modifiers	Probability of personnel presence	0.1	
Predicted frequency of event			10^{-4}
Additional risk reduction needed to meet risk criterion		0.1	

CONSEQUENCE MODELS

DISCHARGE RATES FROM PUNCTURED LINES AND VESSELS

Nomenclature

a	Coefficient in nonequilibrium factor (Eq. 23-75)
A	Cross-sectional area perpendicular to flow, m^2
C_D	Discharge coefficient
C_{DG}	Discharge coefficient for gas flow
C_{DL}	Discharge coefficient for liquid flow
C_p	Heat capacity at constant pressure, J/(kg·K)
C_v	Heat capacity at constant volume, J/(kg·K)
D	Pipe diameter, m
D_T	Tank diameter, m
f	Fanning friction factor
F_I	Pipe inclination factor defined by Eq. (23-32)
g	Gravitational acceleration, m/s^2
G	Mass flux, kg/(m^2·s)
H	Specific enthalpy, J/kg
H_{GL}	Heat of vaporization ($H_G - H_L$) at saturation, J/kg
K	Slip velocity, u_G/u_L
K_e	Number of velocity heads for fittings, expansions, contractions, and bends
L	Length of pipe, m
m	Mass discharged, kg
MF	Momentum flux or velocity heads, N/m^2 or Pascal
N	$4fL/D + K_e$, pipe resistance or nonequilibrium compressibility factor
P	Pressure, Pa (N/m^2)
Q'	Q'/w, Heat transfer rate, W
Q	Q'/w, Heat transfer rate per mass discharge rate, J/kg
R	Gas constant, J/(kgmol·K)
Re	Reynolds number, dimensionless, $D\rho u/\mu$
S	Specific entropy, J/(kg·K)
t	Time, s
T	Temperature, K
u	Velocity, m/s
v	Specific volume, m^3/kg
v_{GL}	Specific volume difference, $v_G - v_L$, m^3/kg
w	Mass discharge rate, kg/s
x	Vapor quality, kg vapor/kg mixture or mole fraction of liquid components
y	Mole fraction of vapor components
X_m	Lockhart Martinelli parameter
z	Vertical distance, m
Greek letters	
α	Volume fraction vapor, m^3 vapor/m^3 mixture
ε	Dimensionless specific volume, v/v_0
γ	Heat capacity ratio, C_p/C_v
η	Pressure ratio, P/P_0
φ^2	Two-phase multiplier, pressure drop for two-phase flow divided by pressure drop for single-phase flow
μ	Single or two-phase viscosity, Pa·s
θ	Overall inclination angle of pipe to the horizontal from source to break, degrees
ρ	Density, kg/m^3
ω	Slope of dimensionless specific volume to reciprocal dimensionless pressure, defined by Eqs. (23-51) and (23-52)
Subscripts	
a	Ambient
b	Backpressure
bub	Bubble point, pressure and saturation temperature when first vapor bubbles appear
c	η_{cn}/η_s, ratio of choked pressure to saturation pressure
ch	Choked
C	Condensable components (the contaminant)
d	Discharge
dew	Dew point, pressure and saturation temperature when first (or last) liquid occurs
diff	Differential form
e	Equivalent, for two-phase specific volume with slip
g,G	Gas or vapor
H	Homogeneous flow (slip velocity ratio of unity)
int	Integral form
I	Inert or padding gas component
L	Liquid
max	Maximum
N	Nonequilibrium
ori	Orifice
p	Pipe flow
s	Saturation (or bubble point)
sol	Solids
S	Constant specific entropy
v	Vapor pressure
0	Initial stagnation conditions
1	Point at which backpressure from pipe is felt after entrance from tank
2	Plane at vena contracta or at pipe puncture
*	Dimensionless

General References: Cheremisinoff and Gupta, eds., *Handbook of Fluids in Motion,* Ann Arbor Science, Ann Arbor, Mich., 1983; Chisholm, *Two-Phase Flow in Pipelines and Heat Exchangers,* George Godwin, New York, in association with the Institution of Chemical Engineers, 1983; Fisher et al., *Emergency Relief System Design Using DIERS Technology,* AIChE, New York, 1992; Levenspiel, "The Discharge of Gases from a Reservoir Through a Pipe," *AIChE J.* **23**(3): 402–403, 1977; Woodward, "Discharge Rates Through Holes in Process Vessels and Piping," in Fthenakis, ed., *Prevention and Control of Accidental Releases of Hazardous Gases,* Van Nostrand Reinhold, New York, 1993, pp. 94–159.

Overview Modeling the consequences of accidental releases of hazardous materials begins with the calculation of discharge rates. In the most general case, the discharged material is made up of a volatile flashing liquid and vapor along with noncondensable gases and solid particles. Most of the model development is given in terms of two-phase, vapor plus liquid flow, but it can be readily extended to three phases with vapor, liquid, plus suspended solids flow. The solids contribute primarily to the density and heat capacity of the mixture, contributing energy to the flashing and to the subsequent dispersion.

Types of Discharge Hazardous accidental releases can occur from vessels, pipelines, reactors, and distillation columns. The more common accidental events are breaks in a vessel or its associated piping. Figure 23-23 illustrates this. For a vessel, the piping attachments can be from a dip leg or from the bottom or top of the vessel. A puncture in the vapor space of the vessel or a break in a top-attachment pipe, initially at least, discharges vapor plus the padding gas used to maintain vessel pressure. This padding gas can be air for an atmospheric vessel. The discharged vapor can cool upon expansion and condense liquids when the pressure in the jet drops to the dew point pressure P_{dew}.

A puncture in the liquid space of the vessel or a break in the bottom-attachment or dip-leg pipe, initially, at least, discharges liquid plus any solids present without any noncondensable components. The liquid can begin to flash when the pressure drops to the bubble point pressure P_{bub}. If the liquid is extremely volatile, it could totally evaporate when the pressure drops below the dew point, producing vapor plus solids. The initial mass vapor fraction x_0 is zero, as is the initial volume fraction α_0.

For a puncture, break, or pressure relief valve (PRV) opening from a reactor or distillation column, there may be no clear-cut level distinguishing the liquid and vapor phases. That is, the system is initially mixed. In this case, noncondensable gases, condensable vapors, and liquid plus solids are initially discharged. The value of α_0 is nonzero and less than unity, reflecting the contributions of the gases and vapors.

For a blowdown calculation, the conditions change. With a tank vapor space release, as the pressure decreases, the liquid could reach its bubble point and begin to flash. Vapor bubbles in the liquid generate a liquid swell, so a frothy, two-phase interface rises and could drive out all the vapor (along with noncondensables) and begin two-phase flashing flow of liquid without noncondensables. Thus, the discharge calculation would also change from a vapor discharge to a two-phase flashing liquid discharge.

The energy and momentum balance equations are drawn across planes at points 0, 1, and 2, as illustrated for a general case in Fig. 23-24.

Energy Balance Method for Orifice Discharge Solutions can be found for the discharge rate by solving for the energy balance or the momentum balance. The energy balance solution is quite simple and general, but it is sensitive to inaccuracies in physical properties

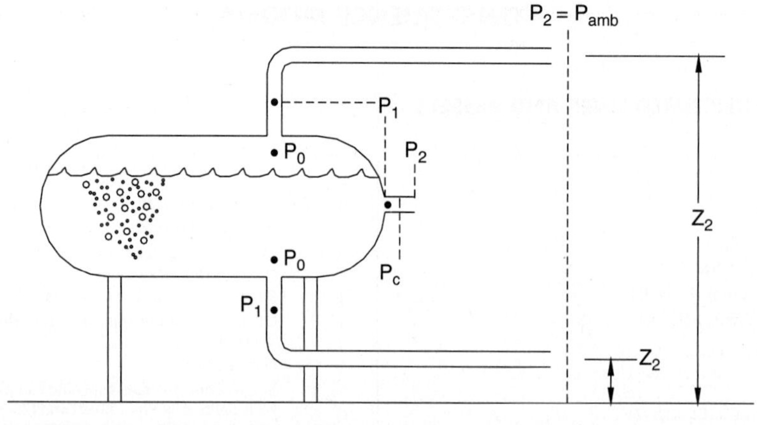

FIG. 23-23 Types of accidental discharge and pressure evaluation points.

correlations. The following equations apply to orifice flow. A separate momentum balance solution is applied to pipe flow to find the pressure losses. The balances are written across an orifice from a plane 0 inside the tank at stagnant conditions (i.e., far enough away from the orifice that the velocity inside the tank is negligible) to a plane at the backpressure P_b at point 2.

$$H_0 = H_2 + \frac{1}{2}(G^2 v_e^2)_2 + Q \tag{23-21}$$

Rarely, if ever, is the heat-transfer term Q nonnegligible.

Perform an isentropic expansion. That is, as the pressure decreases from P_0 to the backpressure P_b (usually ambient pressure P_a), select intermediate values of pressure P_1. At each P_1 find the temperature T_{S1} that keeps S constant. Solve for the vapor fraction x, using the entropy balance between planes 0 and 1:

$$S_0 = xS_{G1} + (1 - x - x_{sol})S_{L1} + x_{sol}S_{sol} \tag{23-22}$$

Since the mass fraction of solids is constant, the vapor flash fraction is

$$x = \frac{S_0 - [(1 - x_{sol})S_{L1} - x_{sol}S_{sol}]_{T_{S1}}}{(S_{G1} - S_{L1})_{T_{S1}}} \tag{23-23}$$

Find the enthalpy H_2 at T_{S1} from physical property correlations. Velocity is found solving Eq. (23-21) for homogeneous flow (with $Q = 0$):

$$Gv_e = u = \sqrt{2(H_0 - H_2)} \tag{23-24}$$

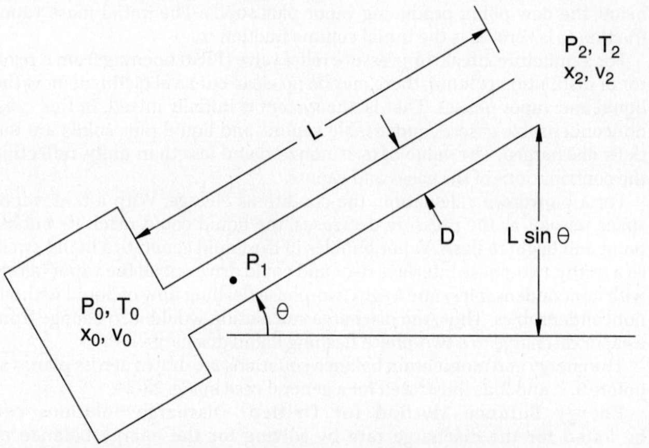

FIG. 23-24 Configuration modeled for pipe flow with elevation change.

Find the phase densities ρ_G and ρ_L from an equation of state at P_1, T_{S1} and density ρ as the reciprocal of the homogeneous specific volume v [the weighted average $xv_G + (1 - x)v_L$]. Calculate the mass flux from

$$G = u\rho = \frac{u}{\upsilon} \tag{23-25}$$

Search with values of P_1 until G is maximized. The choke pressure, P_{ch}, is the value of P_1 that produces a maximum value of mass flux G_{max}. The discharge rate w is given from the mass flux, a discharge coefficient C_D, and the orifice cross-sectional area A as

$$w = C_D A G_{max} \tag{23-26}$$

The coefficient C_D varies with the smoothness of the puncture and the vapor/gas fraction [see Bragg, S., "Effect of Compressibility on the Discharge Coefficient of Orifices and Convergent Nozzles," *J. Mech. Eng. Sci.* **2:** 35–44 (1960); Jobson, "On the Flow of Compressible Fluids Through Orifices," *Proc. Instn. Mech. Engrs.* **169**(37): 767–776 (1955); "The Two-Phase Critical Flow of One-Component Mixtures in Nozzles, Orifices, and Short Tubes," *Trans. ASME J. Heat Transfer* **93**(5): 179–187 (1976); Morris, S. "Two-Phase Discharge Coefficients for Nozzles and Orifices," Report to Energy Technology Unit, Dept. of Mech. Eng. Heriot-Watt Univ. Edinburgh, UK, 1989].

This approach is illustrated for two-phase flashing flow with a multicomponent mixture (mole fractions 0.477 allyl alcohol, 0.3404 allyl chloride, 0.1826 dodecane). Figure 23-25 plots the flash curve showing that as pressure decreases, flashing begins at the bubble point of 7.22 bar(a) (104.7 psia) and increases to 30.8 mass percent. At the same time, temperature follows the saturation curve down. Figure 23-26 plots the two-phase density,

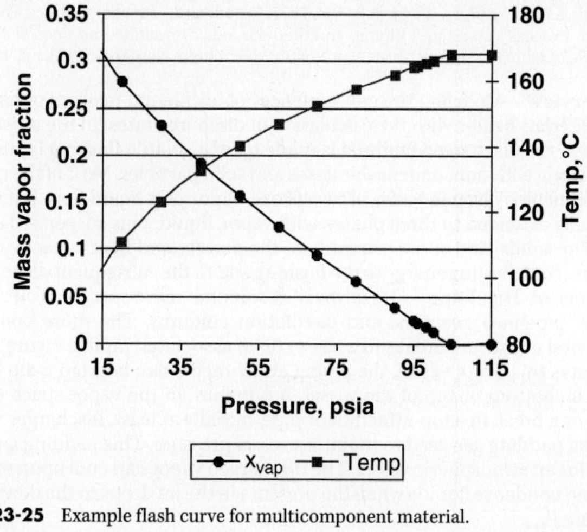

FIG. 23-25 Example flash curve for multicomponent material.

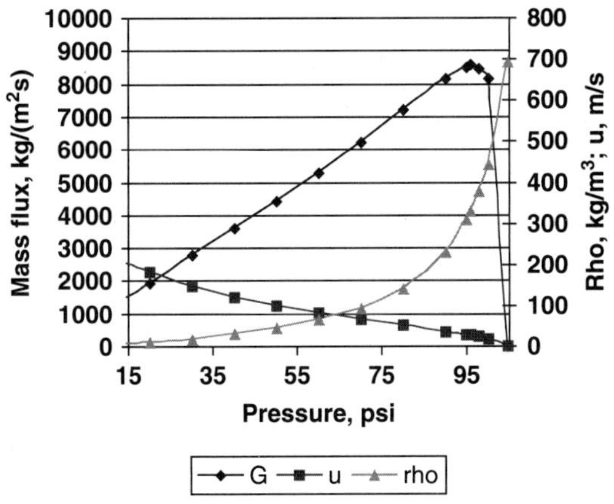

FIG. 23-26 Example of using the energy balance to find the orifice discharge flux.

the velocity, and the product of these, the mass flux G. In the plot G is a maximum at 8586 kg/(m$^2\cdot$s) at the choke pressure just below the bubble point. This is so because the two-phase density drops quickly as flashing occurs at reduced pressures, while the velocity increases more slowly. With a discharge coefficient of 0.61 and a 2-in orifice (area of 0.002027 m^2) the discharge rate w is 10.6 kg/s.

Momentum Balance in Dimensionless Variables For pipe flow, it is necessary to solve the momentum balance. The momentum balance is simplified by using the following dimensionless variables:

$$\text{Pressure ratio: } \eta = \frac{P}{P_0}$$

$$\text{Mass flux ratio: } G_* = \frac{G}{\sqrt{P_0\rho_0}} \qquad (23\text{-}27)$$

$$\text{Specific volume ratio: } \varepsilon = \frac{\upsilon}{\upsilon_0}$$

The discharge relationships are derived by solving the differential momentum balance over a tank plus pipe:

$$\upsilon\, dP + G^2\upsilon\, d\upsilon + \left(4f\frac{dz}{D} + K_e\right)\frac{1}{2}G^2\upsilon_L^2\,\phi_L^2 + g\sin\theta\, dz = 0 \qquad (23\text{-}28)$$

The terms represent, respectively, the effect of pressure gradient, acceleration, line friction, and potential energy (static head). The effect of fittings, bends, and entrance effects, etc., is included in the term K_e correlated as a number of effective "velocity heads." The inclination angle θ is the angle to the horizontal from the elevation of the pipe connection to the vessel to the discharge point. The term ϕ_L^2 is the two-phase multiplier that corrects the liquid-phase friction pressure loss to a two-phase pressure loss [see Lockhart and Martinelli, "Proposed Correlation of Data for Isothermal Two-Phase, Two-Component Flow in Pipes," *Chem. Eng. Prog.* **45**(1): 39–48 (1949)]. Equation (23-28) is written in units of pressure/density.

In dimensionless variables, the momentum balance is

$$\varepsilon\, d\eta + G_*^2\,\varepsilon\, d\varepsilon + N\frac{1}{2}G_*^2\,\varepsilon^2\phi_L^2 + \frac{g\sin\theta\, dz}{P_0\upsilon_0} = 0 \qquad (23\text{-}29)$$

where N collects the friction loss terms in terms of the number of velocity heads, adding the corrector K_e, as

$$N = 4f_L\frac{dz}{D} + K_e \qquad (23\text{-}30)$$

The momentum balance for homogeneous flow can be factored to a form that enables integration as

$$-N = \frac{G_{*p}^2\varepsilon\, d\varepsilon + \varepsilon\, d\eta}{\frac{1}{2}G_{*p}^2\varepsilon^2\varphi_L^2 + F_I} \qquad (23\text{-}31)$$

by defining a pipe inclination factor F_I that represents the ratio of potential energy to the flow energy. It is defined as

$$F_I = \frac{gD\sin\theta}{4f_{LO}P_0\upsilon_0} \qquad (23\text{-}32)$$

where F_I is positive for upflow, negative for downflow, and zero for horizontal flow. This contributes to making downflows higher and upflows lower than horizontal flow.

Analytical Solutions for Orifice and Pipe Flow Equation (23-31) can be solved analytically for pipe breaks and tank punctures for the following cases:

- Subcooled liquid flow
- Adiabatic expansion of ideal gases
- Flashing liquid flow without noncondensable gases ($\alpha_0 = 0$)
- Subcooled liquid mixed with noncondensable gases ($\alpha_0 > 0$) (frozen flow)
- Flashing liquid mixed with noncondensable gases (hybrid flow)

The solution for frozen flow and hybrid flow is given by Leung and Epstein ["Flashing Two-Phase Flow Including the Effects of Noncondensable Gases," *ASME Trans. J. Heat Transfer* **113**(1): 269 (1991)].

Orifice Discharge for Gas Flow The analytic solution for discharge through an orifice of an ideal gas is derived by invoking the equation of state for adiabatic expansion of an ideal gas:

$$\frac{P}{P_0} = \left(\frac{\rho}{\rho_0}\right)^{\gamma} = \left(\frac{T}{T_0}\right)^{\gamma/(\gamma-1)} \qquad (23\text{-}33)$$

where

$$\gamma = \frac{C_P}{C_V} \qquad (23\text{-}34)$$

The solution is

$$G_{*g}^2 = \frac{2\gamma}{\gamma-1}\eta_2^{2/\gamma}\left(1 - \eta_2^{(\gamma-1)/\gamma}\right) \qquad (23\text{-}35)$$

This solution applies for both subsonic and choked flow. If the flow is choked, the exit pressure ratio η_2 is replaced by the choked pressure ratio h_{ch}, given by

$$\eta_{ch} = \left(\frac{2}{\gamma+1}\right)^{\gamma/(\gamma-1)} \qquad (23\text{-}36)$$

Equation (23-36) must be evaluated to test for choked flow in any event. When η_{ch} from Eq. (23-36) is substituted for η_2 into Eq. (23-35), the general solution reduces to the choked flow solution (where G is G_g):

$$G_{*g}^2 = \gamma\left(\frac{2}{\gamma+1}\right)^{(\gamma+1)/(\gamma-1)} \qquad (23\text{-}37)$$

The discharge rate is found by using Eq. (23-37) with Eqs. (23-27) and (23-26).

Blowdown of Gas or Liquid Discharge through Orifice An analytic solution is available for blowdown (time-dependent discharge) of an ideal gas from a tank. [Woodward and Mudan, "Liquid and Gas Discharge Rates Through Holes in Process Vessels," *J. Loss Prevention* **4**(3): 161–165 (1991)]. The time-varying mass of gas in the tank m_T is the product of the tank volume V_T and the density ρ:

$$m_T(t) = V_T\rho(t) \qquad (23\text{-}38)$$

Differentiating, solving for dt, and integrating gives

$$t=-V_T\int_{P_0}^{P}\frac{d\rho}{w} \tag{23-39}$$

where w is the time-varying discharge rate, dm_T/dt. Typically, all but a small fraction of the mass in a tank is discharged at sonic flow, so a sonic flow solution is most useful. Transforming P and ρ to temperature T, using Eq. (23-33) allows Eq. (23-39) to be integrated to give a solution in terms of the initial discharge rate w_0 and the initial tank mass m_{T0}:

$$w(t)=w_0[F(t)]^{(\gamma+1)/(\gamma-1)} \tag{23-40}$$

where

$$F(t)=(1+At)^{-1} \tag{23-41}$$

$$A=\frac{w_0\,(\gamma+1)}{2\,m_{T0}} \tag{23-42}$$

Liquid blowdown is treated by Lee and Sommerfeld [Lee and Sommerfeld, "Maximum Leakage Times Through Puncture Holes for Process Vessels of Various Shapes," *J. Hazardous Mat.* **38**(1): 27–40 (1994); Lee and Sommerfeld, "Safe Drainage or Leakage Considerations and Geometry in the Design of Process Vessels," *Trans. IChemE* **72**(part B): 88–89 (1994)].

Pipe and Orifice Flow for Subcooled Liquids Since liquids are essentially incompressible, ε is constant at ε_0, and $d\varepsilon$ is zero in Eq. (23-31). Recognizing that η_0 and ε_0 are unity, we see that integration gives

$$-N\left(\frac{1}{2}G_{*p}^2\,\varepsilon_0^2+F_I\right)=\varepsilon_0\int_{\eta_0}^{\eta_2}d\eta=\eta_2-\eta_0$$

or

$$G_{*p}^2=2\left(\frac{1-\eta_2}{N}-F_I\right) \tag{23-43}$$

The solution for orifice flow is a special case with N equal to unity (entrance losses only) and F_I equal to zero, giving

$$G_{*\text{ori}}=\sqrt{2(1-\eta_2)} \tag{23-44}$$

Numerical Solution for Orifice Flow With orifice flow, the last two terms of the momentum balance (line resistance and potential energy change) are negligible. The momentum balance, Eq. (23-29), reduces to

$$G_*^2\,\varepsilon\,d\varepsilon=-\varepsilon d\eta \tag{23-45}$$

This equation can be treated in differential form and in integral form. In differential form it becomes

$$G_{*\text{diff}}=-\left(\frac{d\varepsilon}{d\eta}\right)^{-1/2}\quad\text{or}\quad G_{\text{diff}}=-\left(\frac{d\upsilon}{dP}\right)^{-1/2} \tag{23-46}$$

For the integral form, express Eq. (23-45) as

$$\frac{1}{2}G_*^2\,d\varepsilon^2=-\varepsilon d\eta$$

to obtain

$$G_*d\varepsilon=(-2\varepsilon\,d\eta)^{1/2} \tag{23-47}$$

In general, the limits of integration are from ε_0, η_0 to an arbitrary final point ε, η (recognizing that $\varepsilon_S=\varepsilon_0=1$). However, this method works better using an indefinite integration of $d\varepsilon$:

$$G_{*\text{int}}=\frac{1}{\varepsilon}\left(2\int_{\eta}^{\eta_0}\varepsilon\,d\eta\right)^{1/2} \tag{23-48}$$

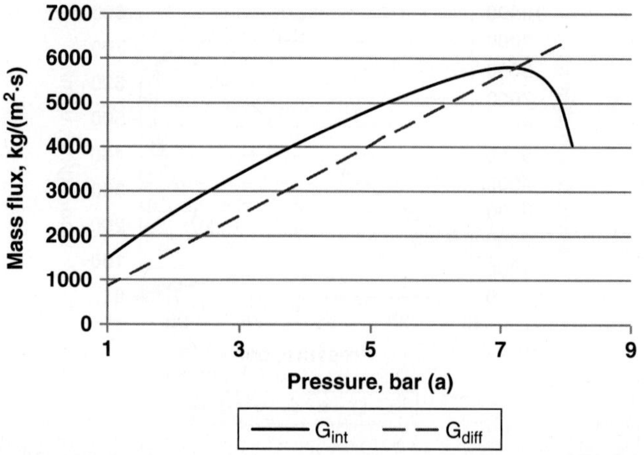

FIG. 23-27 Numerical solution of the momentum balance for orifice flow using Eqs. (23-46) and (23-49).

or in dimensional form

$$G_{\text{int}}=\frac{1}{\upsilon}\left(2\int_{P}^{P_0}\upsilon dP\right)^{1/2} \tag{23-49}$$

Equations (23-46) and (23-49) are readily evaluated at intermediate pressure points P_1 in the range $P_a<P_1<P_0$, giving two curves for G versus P, which cross at the choke point. This is also the point of the correct mass flux as illustrated in Fig. 23-27 for the 40:30:30 mixture of propane, isobutene, n-butane. In this example, the intersection occurs at $P=7.222$ bar(a), giving $G=5792.6$ kg/(m²s).

The disadvantage of this solution is that it is sensitive to inaccuracies in the physical properties correlations used to evaluate the flash fraction and specific volumes. The inaccuracies can be readily apparent when the curves are not smooth. The DIPPR properties are broadly used [Daubert and Danner, "Physical and Thermodynamic Properties of Pure Chemicals, Data Compilation," DIPPR (Design Inst. of Phys. Props. Res.), AIChE, New York, 1989]. These can be compared to the accurate values of the STRAPP program [Ely and Huber, "NIST Thermophysical Properties of Hydrocarbon Mixtures Database Version 1.0 Users' Guide, U.S. Dept. of Commerce, National Institute of Standards and Technology (NIST), Gaithersburg, Md., February 1990].

Omega Method Model for Compressible Flows The factored momentum balance, Eq. (23-31), can be analytically integrated after first relating the dimensionless specific volume ε to the dimensionless pressure ratio η. A method to do this, designated the *omega method,* was suggested by Leung, "A Generalized Correlation for One-Component Homogeneous Equilibrium Flashing Choked Flow," *AIChE J.* **32**(10): 1743–1746 (1986).

$$\varepsilon=\begin{cases}\omega\left(\dfrac{\eta_s}{\eta}-1\right)+1 & \text{if }\dfrac{\eta_s}{\eta}>1\\[2ex] 1.0 & \text{if }\dfrac{\eta_s}{\eta}\le 1\end{cases} \tag{23-50}$$

Equation (23-50) represents a linear relationship between the two- or three-phase specific volume and reciprocal pressure (υ versus P^{-1} or ε versus η^{-1}) beginning at the bubble point P_S, where η is η_s. For single components, ω_s is found by using the Clapeyron equation to give

$$\omega_s=\frac{C_{pL}T_0P_S}{\upsilon_{L0}}\left[\frac{\upsilon_{GL0}(P_S)}{H_{GL0}(P_S)}\right]^2 \tag{23-51}$$

This value of ω can be called the saturation value or ω_s, since it applies only with flashing liquids (i.e., in the flashing region with pressures less than the bubble point, as seen in Fig. 23-25). Alternately, use the slope of the ε versus η^{-1} curve between the bubble point and a second, lower pressure at ε_2, η_2 to evaluate ω_s, or

$$\omega_s=\frac{\varepsilon_2-1}{\eta_S/\eta_2-1}-1 \tag{23-52}$$

This averages over a wider pressure range than Eq. (23-51) and is more accurate, especially for multicomponent mixtures [Woodward, "An Amended Method for Calculating Omega for an Homogeneous Equilibrium Model of Predicting Discharge Rates", *J. Loss Prev. in Process Ind.* **8**(5): 253–259 (1995)].

When noncondensable gases are present, the definition of ω is generalized to

$$\omega = \alpha_0 + (1 - \alpha_0)\,\omega_s \tag{23-53}$$

where α_0 is the void fraction of noncondensables calculated from the mass fraction and the specific volume of noncondensables (accounting for temperature and pressure),

$$\alpha_0 = x_{\upsilon 0}\upsilon_{\upsilon 0}/\upsilon_0 \tag{23-54}$$

Homogeneous Equilibrium Omega Method for Orifice and Horizontal Pipe Flow The homogeneous equilibrium model (HEM) solution is obtained by substituting Eq. (23-50) for ε and integrating the momentum balance with respect to η [Leung and Grolmes, "The Discharge of Two-Phase Flashing Flow in a Horizontal Duct," *AIChE J.* **33**(3): 524–527 (1987); also errata, **34**(6): 1030 (1988); Leung, "Two-Phase Flow Discharge in Nozzles and Pipes—A Unified Approach," *J. Loss Prevention Process Ind.* **3**(27): 27–32 (1990a); Leung and Ciolek, "Flashing Flow Discharge of Initially Subcooled Liquid in Pipes," *ASME Trans. J. Fluids Eng.* **116**(3) (1994)].

For pipe flow, two solutions are needed, the orifice flow solution giving $G_{*\text{ori}}$ at the pipe break and the pipe flow solution giving G_{*p}. Plane 1 in Fig. 23-24 is at the pressure ratio η_1, where the pipe pressure balance begins and the inlet orifice pressure loss ends. The final pipe solution finds η_1 so that

$$G_{*\text{ori}} = G_{*p} \tag{23-55}$$

The integration over both pressure spans must be conducted over the subcooled region and the flashing region. These are the high and low subcooling cases that set the endpoints for integration. With high subcooling, the bubble point pressure ratio η_s will fall in the integration span ($\eta_s > \eta_1$), giving flashing in the orifice (case 1). In the low subcooling case, flashing occurs in the pipe ($\eta_s < \eta_1$, case 2). These two options are illustrated in Fig. 23-28.

Over the subcooled region from η_0 to η_s, $d\varepsilon$ is zero and ε is constant at unity, ε_0. Leung (1990a) plots solutions for G_* and η_c as functions of η_s and α_0 [also Leung, "Similarity Between Flashing and Non-Flashing Two-Phase Flow," *AIChE J.* **36**(5): 797 (1990b); Leung, "Size Safety Relief Valves for Flashing Liquids," *Chem. Eng. Prog.* **88**(2): 70–75 (1992); Leung and Epstein, "A Generalized Correlation for Two-Phase Non-Flashing Homogeneous Choked Flow," *ASME J. Heat Transfer* **112**(2) (1990)]. The solution is given first for horizontal flow, with the flow inclination factor $F_I = 0$. The solutions for the two cases are as follows:

Case 1: Flashing in Orifice

$$G_{*\text{ori}}^2 = \frac{2\{(1-\eta_s)+[\omega\eta_s \ln(\eta_s/\eta_{\text{ch}})-(\omega-1)(\eta_s-\eta_{\text{ch}})]\}}{\varepsilon_{\text{ch}}^2} \tag{23-56}$$

$$G_{*p}^2 = 2\left\{\frac{\dfrac{\eta_1-\eta_2}{1-\omega}+\dfrac{\omega\eta_s}{(1-\omega)^2}\ln\dfrac{\eta_2\,\varepsilon_2}{\eta_1\,\varepsilon_1}}{N+2\ln\left(\dfrac{\varepsilon_2}{\varepsilon_1}\right)}\right\} \tag{23-57}$$

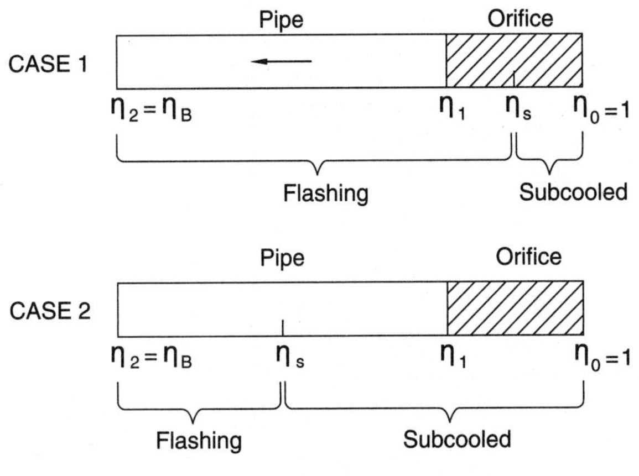

FIG. 23-28 Options for integration range of HEM omega method.

Case 2: Low Subcooled Region, Flashing in the Pipe

$$G_{*\text{ori}}^2 = 2(\eta_0-\eta_1) \tag{23-58}$$

$$G_{*p}^2 = 2\left[\frac{\eta_1-\eta_s+\dfrac{\eta_s-\eta_2}{1-\omega}+\dfrac{\eta_s-\eta_2}{1-\omega}+\dfrac{\omega\eta_s}{(1-\omega)^2}\ln\left(\dfrac{\eta_2\varepsilon_2}{\eta_s\varepsilon_S}\right)}{N+2\ln\left(\dfrac{\varepsilon_2}{\varepsilon_S}\right)}\right] \tag{23-59}$$

The choke point is found by the usual maximization relationship:

$$\left(\frac{dG_*}{d\eta_2}\right)_{\eta_2=\eta_{\text{ch}}} = 0 \tag{23-60}$$

This gives an implicit equation in the choked pressure ratio η_c (η_{ch}/η_s):

If $\omega \le 2$:

$$\eta_c^2+\omega(\omega-2)(1-\eta_c)^2+2\omega^2\ln\eta_c+2\omega^2(1-\eta_c)=0 \tag{23-61}$$

An explicit equation provides an adequate approximate solution for larger values of ω:

For $\omega \ge 2$:

$$\eta_c = 0.55 + 0.217\ln\omega - 0.046(\ln\omega)^2 + 0.004(\ln\omega)^3 \tag{23-62}$$

For orifice flow, the definition of choked flow in terms of the backpressure ratio η_b is

- Choked compressible: $\eta_b \le \eta_{\text{ch}}$ set $\eta = \eta_{\text{ch}}$
- Subsonic compressible: $\eta_b > \eta_{\text{ch}}$ set $\eta = \eta_b$
- Subcooled liquid: always subsonic

The omega method HEM solution for orifice flow is plotted in Fig. 23-29. The solution for flashing liquids without noncondensables is to the right of $\omega = 1$, and the solution for frozen flow with subcooled liquids plus noncondensables is to the left. The omega method HEM solution for horizontal pipe flow is plotted in Fig. 23-30 as the ratio of pipe mass flux to the mass flux for a perfect nozzle, $G_{*p}/G_{*\text{ori}}$. Leung (1990a) points out this ratio can be recognized as equivalent to a discharge coefficient, C_D. When $\omega = 1$ and $\alpha_0 = 0$, Fig. 23-30 is in agreement with the relationship for classical incompressible flow

$$C_D = (1 + 4fL/D)^{-1/2} \tag{23-63}$$

and Fig. 23-30 with $\omega = 1$ or $\alpha_0 = 1$ gives the same result as isothermal pipe flow with $k = 1$ for a gas.

HEM for Inclined Pipe, Discharge If a pipe leak occurs at an elevation above or below the pump or source tank, the elevation change can be idealized between the source and break elevations, as shown in Fig. 23-24. That is, elevation changes can be treated as an inclined pipe with a nonzero inclination factor F_I. This is an approximation to the actual piping isometrics, but it is often an adequate approximation.

The HEM solution provided by Leung and Epstein ["The Discharge of Two-Phase Flashing Flow from an Inclined Duct," *ASME Trans. J. of Heat Transfer* **112**(2): (1990)] is implicit in G_{*p}:

Case 1: High Subcooling, Flashing in Orifice

Use Eq. (23-56) for $G_{*\text{ori}}$:

$$N+\ln\left\{\frac{X(\eta_2)}{X(\eta_1)}\left[\frac{\eta_1}{\eta_2}\right]^2\right\}=\frac{1-\omega}{c}(\eta_1-\eta_2)$$
$$+\frac{b(1-\omega)-c\omega\eta_s}{2c^2}\ln\frac{X(\eta_2)}{X(\eta_1)} \tag{23-64}$$
$$+\frac{bc\,\omega\eta_s-(1-\omega)(b^2-2ac)}{2c^2}[I_0(\eta_2)-I_0(\eta_1)]$$

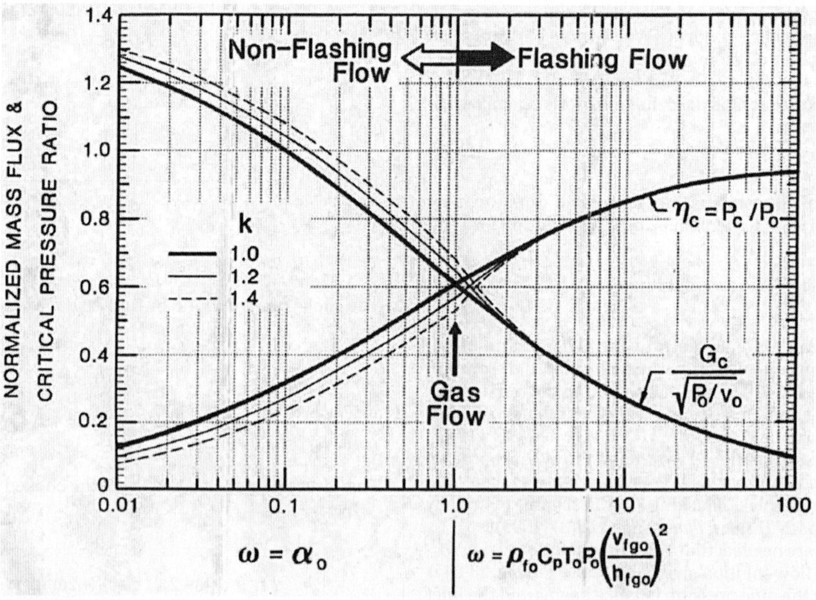

FIG. 23-29 Omega method solution for orifice flow of flashing liquids and for noncondensible gas plus subcooled liquids.

Case 2: Low Subcooling, Flashing in Pipe
Use η_b for η_1 in Eq. (23-58) for $G_{*\mathrm{ori}}$:

$$N+\ln\left\{\left[\frac{X(\eta_2)}{X(\eta_S)}\right]\left[\frac{\eta_S}{\eta_2}\right]^2\right\}=\frac{\eta_1-\eta_s}{\frac{1}{2}G_*^2+F_l}+\frac{1-\omega}{c}(\eta_s-\eta_2)$$
$$+\frac{b(1-\omega)-c\omega\eta_s}{2c^2}\ln\frac{X(\eta_2)}{X(\eta_s)}$$
$$+\frac{bc\,\omega\eta_s-(1-\omega)(b^2-2ac)}{2c^2}[I_0(\eta_2)-I_0(\eta_S)] \qquad (23\text{-}65)$$

where

$$a=\frac{1}{2}G_*^2\,\omega^2\eta_S^2 \qquad (23\text{-}66)$$

$$b=\frac{1}{2}G_*^2\,2\omega(1-\omega)\eta_s \qquad (23\text{-}67)$$

$$c=\frac{1}{2}G_*^2(1-\omega)^2+F_l \qquad (23\text{-}68)$$

$$q=4ac-b^2 \qquad (23\text{-}69)$$

$$X(\eta)=a+b\eta+c\eta^2 \qquad (23\text{-}70)$$

$$I_0(\eta)=\int\frac{d\eta}{X(\eta)} \qquad (23\text{-}71)$$

If $q>0$, $F_l>0$, upflow:

$$I_0(\eta)=\frac{2}{\sqrt{q}}\tan^{-1}\frac{2c\eta+b}{\sqrt{q}} \qquad (23\text{-}72)$$

If $q<0$, $F_l<0$, downflow:

$$I_0(\eta)=\ln\frac{2c\eta+b-\sqrt{-q}}{2c\eta+b+\sqrt{-q}} \qquad (23\text{-}73)$$

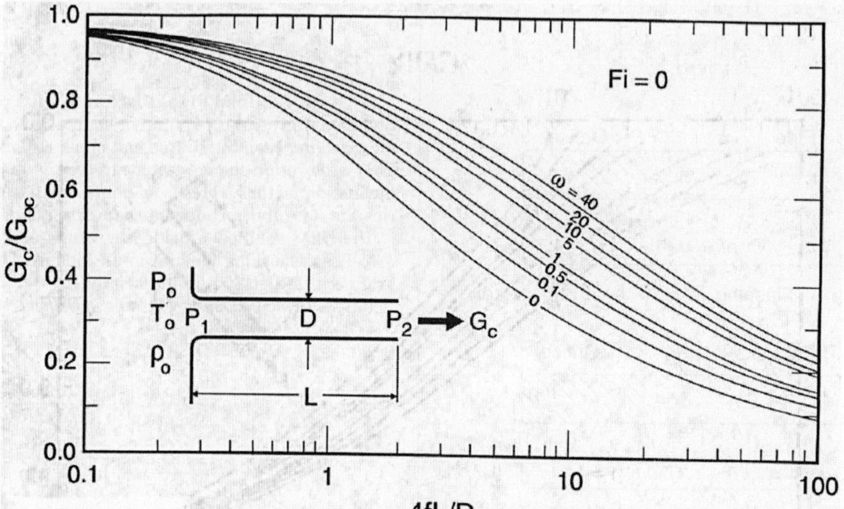

FIG. 23-30 Omega method solution for flashing liquid horizontal pipe flow.

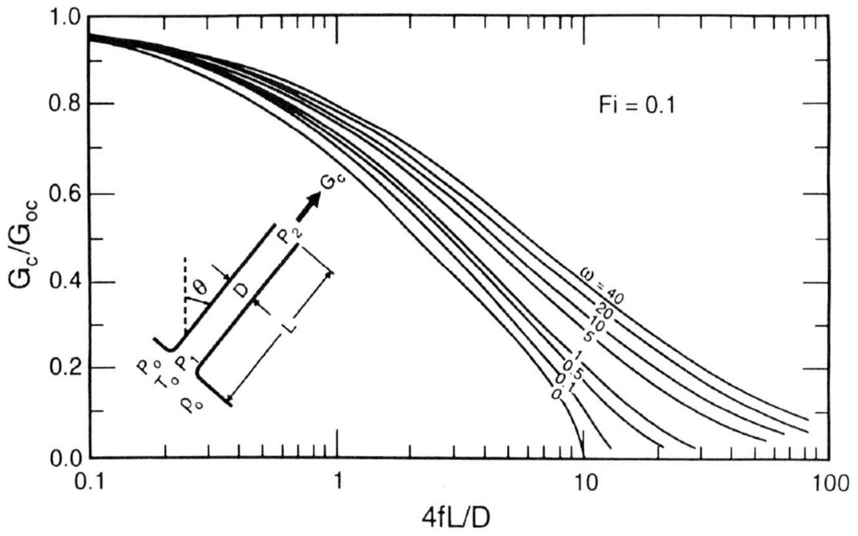

FIG. 23-31 Omega method HEM solution for inclined pipe flow at $F_I = 0.1$.

If q = 0, F_I = 0, horizontal flow:

$$I_0(\eta) = \frac{-1}{(1-\omega)\varepsilon\eta} \tag{23-74}$$

The omega method HEM solution for inclined pipe flow is illustrated in Fig. 23-31 for $F_I = 0.10$.

Nonequilibrium Extension of Omega Method The omega method HEM tends to produce discharge rates that are low, particularly for short pipes. To correct this deficiency, Diener and Schmidt proposed a modification they term the *nonequilibrium compressibility factor N*, defined by

$$N = \left[x_0 + C_{PL0} T_0 P_0 \left(\frac{\upsilon_{GL0}}{H_{GL0}^2}\right) \ln\left[\frac{1}{\eta_c}\right]\right]^a \tag{23-75}$$

with

$$a = \begin{cases} 0.6 & \text{for orifices, control valves, short nozzles} \\ 0.4 & \text{for pressure relief valves, high} - \text{lift control valves} \\ 0 & \text{for long nozzles, office with large area ratio} \end{cases}$$

The choked pressure ratio η_c is found by using Eqs. (23-61) and (23-62). The nonequilibrium compressibility factor N is used to modify ω:

$$\omega = \alpha_0 + \omega_S N \tag{23-76}$$

[Diener and Schmidt, "Sizing of Throttling Device for Gas/Liquid Two-Phase Flow, Part 1: Safety Valves, Part 2: Control Valves, Orifices, and Nozzles," *J. Hazardous Mat.* **23**(4): 335–344 (2004); **24**(1): 29–37 (2005)].

Differences Between Subcooled and Saturated Discharge for Horizontal Pipes Data by Uchida and Narai (1966) in Figs. 23-32 and 23-33 illustrate the substantial differences between subcooled and saturated-liquid discharge rates [Uchida and Narai, "Discharge of Saturated Water Through Pipes and Orifices," *Proc. 3d Intl. Heat Transfer Conf., ASME,* Chicago **5:** 1–12 (1966)]. Discharge rates decrease with increasing pipe length in both cases, but the drop in discharge rate is much more pronounced with saturated liquids. This is so because the flashed vapor effectively chokes the flow and decreases the two-phase density.

Accuracy of Discharge Rate Predictions Model verification is difficult because agreement with data varies substantially from one set of experimental data to another. The accuracy of the HEM and the NEM correction is illustrated in Fig. 23-34 and compared with data for saturated water by Uchida and Narai (1966). The NEM correction using the power coefficient a of 0.6 decreases omega and increases the predicted discharge rates so that good predictions are obtained for pipe lengths greater than 0.5 m.

The accuracy of the energy balance method for discharge of flashing liquids through orifices and horizontal pipes is illustrated in Figs. 23-35 and 23-36. The ratio of predicted to observed mass flux is plotted for saturated water data by Uchida and Narai and by Sozzi and Sutherland (Sozzi and Sutherland, "Critical Flow of Saturated and Subcooled Water at High Pressure," General Electric Co. Report no. NEDO-13418, July 1975; also *ASME Symp. on Non-Equilibrium Two-Phase Flows,* 1975). Orifice flow rates are underpredicted by about the same factor with the energy balance method and with the NEM. Discharge predictions for short (0.2-m) pipes are overpredicted by the energy balance method. In this region, the assumption of homogeneous equilibrium is not justified. A model that takes slip velocity into account may improve predictions for short pipes.

Experimental data on discharge rates for two-phase flow are also given by Graham, "The Flow of Air-Water Mixtures Through Nozzles," National Engineering Laboratories (NEL) Report no. 308, East Kilbride, Glasgow, 1967; Simpson, Rooney, and Grattan, "Two-Phase Flow Through Gate Valves and Orifice Plates," Int. Conf. on the Physical Modeling of Multi-Phase Flow, Coventry, UK, April 19–20, 1983; Tangren, Dodge, and Siefert, "Compressibility Effects in Two-Phase Flow," *J. Applied Phys.* **20:** 637–645 (1949); Van den Akker, Snoey, and Spoelstra, "Discharges of Pressurized Liquefied Gases Through Apertures and Pipes," *I. Chem. E. Symposium Ser.* (London) **80:** E23–35 (1983); Watson, Vaughan, and McFarlane, "Two-Phase Pressure Drop with a Sharp-Edged Orifice," National Engineering Laboratories (NEL) Report no. 290, East Kilbride, Glasgow, 1967.

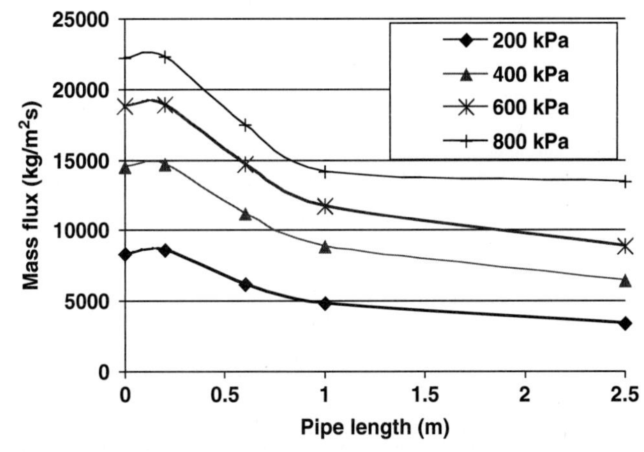

FIG. 23-32 Subcooled water data by Uchida and Narai (1966).

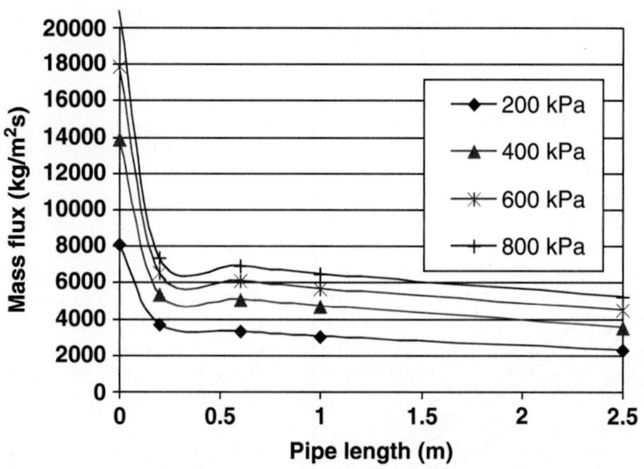

FIG. 23-33 Saturated water data by Uchida and Narai (1966).

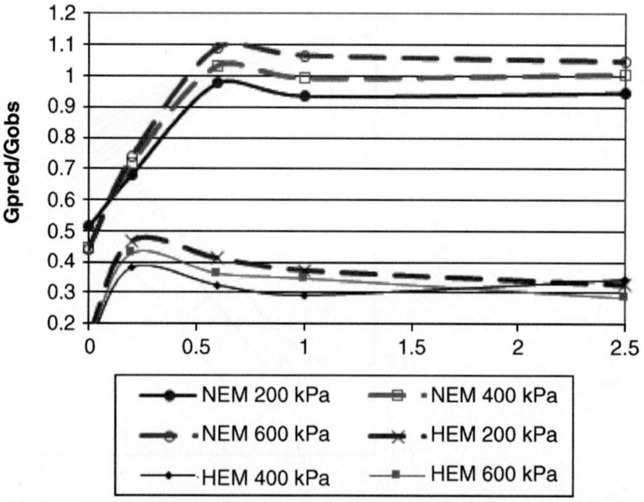

FIG. 23-34 Accuracy of omega method HEM and NEM correction compared with saturated water data by Uchida and Narai (1966).

ATMOSPHERIC DISPERSION

Nomenclature

$\langle C \rangle_1$	(Ensemble) time-averaged concentration for averaging time t_1, mass per volume (kg/m³)
d_0	Displacement height, length (m)
E	Airborne contaminant mass rate in a plume, mass per time (kg/s)
E_t	Total airborne contaminant mass in a puff, mass (kg)
g	Acceleration due to gravity, length squared per time (m/s²)
k	Von Karman's constant (typically 0.4)
L	Monin-Obukhov length, length (m)
p	Index reflecting decrease in time-averaged concentration with averaging time
t_d	Along wind (x direction) dispersion time scale, time (s)
t_h	Time scale for contaminant to pose a particular hazard, time (s)
t_s	Source time scale, length of time for contaminant to become airborne, time (s)
t_t	Time scale for a contaminant cloud to reach hazard endpoint distance or sensor location, time (s)
u	(Characteristic) wind speed at elevation z, length per time (m/s)
u_r	Wind speed at reference height z_r, length per time (m/s)
u_*	Friction velocity, length per time (m/s)
x_v	Virtual source distance upwind of real source, length (m)
x	(Downwind) distance in wind direction, length (m)
x_h	Hazard endpoint distance, length (m)
z_o	Surface roughness, length (m)
z_r	Reference height, length (m)
	Greek symbols
α	Monin-Obukhov length coefficient
ε	Height of ground covering, length (m)
ρ_s	Source contaminant density, mass per volume (kg/m³)
$\sigma_x, \sigma_y, \sigma_z$	Dispersion coefficients in x, y, z directions, length (m)
σ_θ	Wind direction standard deviation, angle
	Subscripts
x	Along wind direction
y	Lateral (crosswind) direction
z	Vertical direction

GENERAL REFERENCES: Arya, *Air Pollution Meteorology and Dispersion*, Oxford University Press, New York, 1999; Britter and McQuaid, *Workbook on the Dispersion of Dense Gases*, Health and Safety Executive Report 17/1988, Sheffield, UK, 1988; De Visscher, *Air Dispersion Modeling—Foundations and Applications*, Wiley, New York, 2014; Mannan, *Lees' Loss Prevention in the Process Industries*, 4th ed., chap. 15, Elsevier Butterworth-Heinemann, Oxford, UK, 2012; Panofsky and Dutton, *Atmospheric Turbulence*, Wiley, New York, 1984; Pasquill and Smith, *Atmospheric Diffusion*, 3d ed., Ellis Horwood Limited, Chichester, UK, 1983; *Seinfeld, Atmospheric Chemistry and Physics of Air Pollution*, chaps. 12–15, Wiley, New York, 1986; Turner, *Workbook of Atmospheric Dispersion Estimates*, U.S. Department of Health, Education, and Welfare, Washington, D.C., 1970.

Introduction Atmospheric dispersion models predict the dilution of an airborne contaminant after its release (and depressurization) to the atmosphere. The discussion in this section focuses on episodic releases representing acute biological, flammable, or toxic hazards. Physical and mathematical dispersion models discussed here are typically used for forensic purposes (comparison with actual events such as experiments, field trials, or accidents); regulatory purposes (estimating the consequences of releases of toxic or flammable materials for siting requirements); or planning purposes (operations or emergency response preparedness). Release consequences may be used to estimate the risk to a facility and workforce or the surrounding population. All such estimates can be used to identify appropriate emergency response or mitigation measures and the priority with which such measures should be considered. (While the principles discussed here are used in emergency response, models for emergency response are more specialized, for example, typically requiring real-time meteorological input parameters.)

Parameters Affecting Atmospheric Dispersion The parameters important to atmospheric dispersion can be divided roughly into three categories: contaminant source; atmospheric and terrain properties; and contaminant interaction with the atmosphere.

Contaminant Source The contaminant source includes such factors as:

1. *Rate or total amount of contaminant that becomes airborne.* For a continuous contaminant release of sufficient duration, the contaminant concentration at a fixed downwind distance is roughly proportional to the rate at which the contaminant becomes airborne. Likewise, the contaminant concentration at a fixed downwind distance is roughly proportional to the amount of contaminant released if the release is instantaneous. However, materials released from containment may not immediately become airborne. For example, a liquid stored at atmospheric pressure below its boiling point will form a liquid pool when released. The rate at which the contaminant becomes airborne will depend on the heat-transfer rate to the

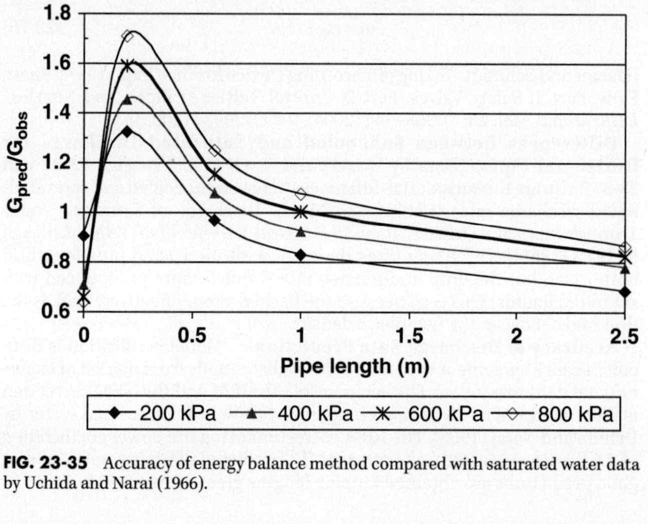

FIG. 23-35 Accuracy of energy balance method compared with saturated water data by Uchida and Narai (1966).

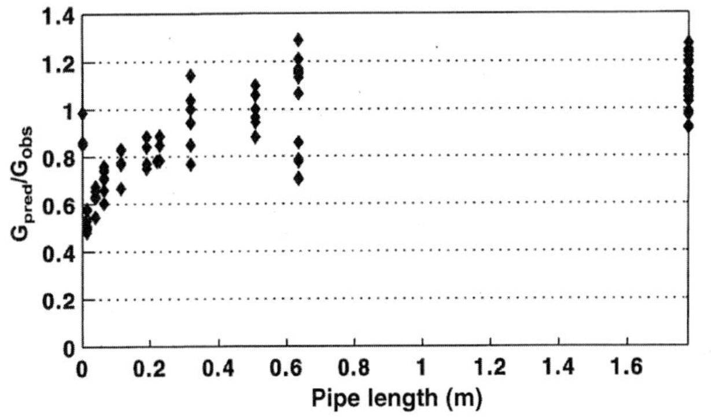

FIG. 23-36 Accuracy of energy balance method for flashing liquid discharge through orifices and horizontal pipes.

liquid pool and the mass-transfer rate from the liquid pool to air. However, the same liquid stored at a temperature above its boiling point can form an aerosol when depressurized so that the liquid phase is airborne. If the aerosol droplet size formed in the release is too large to remain suspended, the liquid phase will fall to the ground (rainout), and the resulting liquid pool evaporation will again be dictated by heat- and mass-transfer considerations. In addition to liquids stored above their boiling point, aerosols can be formed from pressurized liquids such as hydraulic fluids. Aerosols that do not rain out can create particularly hazardous conditions because the suspended liquid phase results in much denser contaminant clouds (more hazardous material per unit volume than the same material in the gas phase).

2. *Release momentum.* For jet releases, the amount of air entrained in an unobstructed jet is proportional to the jet velocity. Depending on the orientation of the jet relative to nearby obstructions, the momentum of a jet can be dissipated without significant air entrainment. The degree of initial air entrainment can be an important determinant of the hazard extent, particularly for flammable hazards. It would be conservative to assume the source momentum is dissipated without air dilution, but this could underestimate the initial air dilution. Explosive releases are high-momentum, instantaneous releases. For explosive releases, a rough first approximation is to assume that the mass of contaminant in the explosion is mixed with 10 times that mass of air.

3. *Release buoyancy.* Contaminant release buoyancy is determined by the initial density of the airborne contaminant (due to temperature, molecular weight, and whether or not an aerosol is present). Lighter-than-air contaminants will rise and be more readily dispersed. Denser-than-air contaminants will tend to stay near ground level, and the atmospheric dispersion of such materials can be importantly determined by (negative) buoyancy effects. The degree of importance is typically quantified by a Richardson number (proportional to the density difference with air and the quantity or rate of contaminant released and inversely proportional to the square of the wind speed); the higher the Richardson number, the more important the effect of negative buoyancy on the dispersion. Denser-than-air contaminants can displace the atmospheric flow field and move in a direction influenced more by terrain than wind. By contrast, contaminants that do not perturb the atmospheric flow field and simply follow the airflow are termed *passive* contaminants.

4. *Other source conditions.* The source height and the source area are also important source characteristics. The greatest impact is typically associated with ground-level sources, so elevated sources are not considered in the model discussion that follows.

Atmospheric and Terrain Parameters In addition to terrain and basic atmospheric parameters listed here, many other meteorological effects can be important under some circumstances (e.g., inversion layers) but are beyond the scope of this introduction.

1. *General terrain characteristics.* General terrain characteristics influence the level of atmospheric turbulence and the shape of the vertical wind speed profile, parameterized with the surface roughness and displacement height. In a similar fashion to pipe roughness, the surface roughness z_0 is roughly proportional to the height of the (uniform) ground covering ε ($z_0 \sim 0.05\varepsilon$ to 0.15ε) and can be used to infer the amount of vertical turbulence (mixing) in the atmospheric flow. Obstacles enhance vertical turbulence above the obstacle height (or ε in the case of ground covering). The ground covering height that determines z_0 should not be larger than the depth of the contaminant cloud being modeled. Obstacles can also displace the atmospheric flow; the displacement height d_0 is the height below which the (average) ambient wind speed is negligible. If the displacement height is negligible (so that flow blockage is not important), contaminant concentration at a fixed downwind distance decreases with increased surface roughness (an order-of-magnitude increase in surface roughness decreases the concentration by roughly a factor of 3).

2. *Wind direction.* Wind direction is the most important determinant of the *location* of hazard zones, with notable exceptions involving high-momentum releases or releases where buoyancy is important (particularly for denser-than-air contaminants involving terrain effects such as valleys and slopes, especially under low-wind-speed conditions). Wind direction variability is much larger in the lateral (crosswind) direction than in the vertical direction (typical of flat-plate boundary layers) and is measured with the standard deviation σ_θ, which is a function of atmospheric stability. From a practical point of view, visual indicators of wind direction (such as a wind sock) can be helpful in predicting the direction a contaminant cloud will move, but as discussed above, denser-than-air contaminants can displace the atmospheric flow field and move in a direction influenced more by terrain or obstacles than wind.

3. *Atmospheric stability.* Atmospheric stability is often characterized by Pasquill stability class (ranging from A through F). Atmospheric stability classes are broad classifications that are used to further characterize the continuum of atmospheric turbulence available to the dispersion process. Neutral atmospheric stability (class D) occurs most often and indicates that the vertical momentum flux is not influenced by the vertical heat flux in the lowest atmospheric layers. Unstable atmospheric conditions (instability increasing from classes C through A) indicate the atmospheric vertical heat flux is enhancing the vertical mixing of the atmosphere, such as when ground surface heating due to insolation (incident solar radiation) enhances vertical atmospheric mixing (turbulence). Stable atmospheric conditions (stability increasing from classes E through F) indicate the atmospheric vertical heat flux is suppressing the vertical mixing of the atmosphere (reduced dilution), such as when ground surface cooling at night suppresses vertical mixing (turbulence). All atmospheric dispersion models rely in some way on measured atmospheric parameters, and these parameters are often correlated on the basis of Pasquill stability class. Consequently, atmospheric dispersion model results should be viewed as only representative of the atmospheric conditions present in a real or hypothetical release. Atmospheric stability class is an important determinant of concentration at a fixed downwind distance in dispersion models. For passive contaminants, concentration at a fixed downwind distance can be roughly an order of magnitude higher for F stability than for D stability. Table 23-21 illustrates the relationship between atmospheric stability class and other meteorological conditions. Determination of Pasquill stability class can be made with detailed measurements such as those discussed by Golder ["Relations among Stability Parameters in the Surface Layer," *Boundary Layer Meteorol.* **3:** 47–58 (1972)]. Note that the worst-case stability conditions (smallest rate of dilution of a contaminant cloud) are for D stability in daytime and F stability at night (the period from one hour before sunset to one hour after dawn). (Pasquill stability class G is used by some for conditions more stable than F stability.)

4. *Wind speed.* Wind speed is the most important determinant of the *magnitude* of atmospheric turbulence available to the dispersion/dilution process. The rate of air dilution is driven by the available turbulent kinetic energy (from the atmosphere or the release itself), and the ambient turbulent kinetic energy is roughly proportional to the square of the wind speed. Because wind speed varies with height above the ground, the reference wind speed u_r must be specified at a particular height z_r (typically 10 m, provided local flow obstructions are not important at this height). For a continuous contaminant release, the contaminant concentration at a fixed downwind distance is (roughly) inversely proportional to the wind speed. Denser-than-air effects are less important at higher wind speeds. The wind speed profile is affected by atmospheric stability, surface roughness z_0, and displacement height d_0 (the height below which the wind speed is negligible). The vertical profile of the wind speed is logarithmic, and for neutral and stable conditions

$$u = \frac{u_*}{k}\left[\ln\left(\frac{z-d_0}{z_0}\right) + \alpha\left(\frac{z}{L}\right)\right] \tag{23-77}$$

when $z \gg z_0$ and d_0; k is von Karman's constant (typically 0.4), u_* is the friction velocity (u_*^2 characterizes the atmospheric mixing rate), L is the Monin-Obukhov length (a measure of stability class), and α is the Monin-Obukhov length coefficient ($\alpha = 5.2$ for neutral to stable conditions) (De Visscher, *Air Dispersion Modeling—Foundations and Applications*, Wiley, New York, 2014). For D stability, $L = \infty$ so $z/L = 0$, and the second term in brackets is zero. Since Eq. (23-77) must also hold for u_r and z_r, u_* is proportional to u_r (all other things being equal), so reduction of the wind speed by a factor of 2 reduces the atmospheric mixing by a factor of 4.

TABLE 23-21 Typical Atmospheric Stability Classes in Terms of Wind Speed, Insolation, and State of the Sky

Surface wind speed m/s	Insolation			Night	
	Strong	Moderate	Slight	Thinly overcast or ≥4/8 low cloud	≤3/8 cloud
<2	A	A–B	B	—	—
2–3	A–B	B	C	E	F
3–5	B	B–C	C	D	E
5–6	C	C–D	D	D	D
>6	C	D	D	D	D

For A–B, take the average of values for A and B, etc. Pasquill and Smith relate insolation to conditions in England. Seinfeld (*Atmospheric Chemistry and Physics of Air Pollution*, Wiley, New York, 1986) classifies insolation greater than 700 W/m^2 as strong, less than 350 W/m^2 as slight, and between these limits as moderate. Night refers to the period from 1 h before sunset to 1 h after dawn. The neutral class D should be used, regardless of wind speed, for overcast conditions during day or night and for all sky conditions during the hour preceding or following the night period.

SOURCE: Pasquill and Smith, *Atmospheric Diffusion*, 3d ed., Ellis Horwood Limited, Chichester, U.K., 1983.

5. *Obstacles or flow obstructions.* In contrast to general terrain characteristics, obstacles have length scales much larger than the depth or width of the contaminant cloud or the height of the characteristic wind speed, and they can increase or decrease contaminant concentration, depending on location. Obstacles can increase concentration by delaying the dispersal of the contaminant cloud; for example, inside a dike, contaminant concentration is higher, but downwind of the dike, concentrations can be lower. The downwind side of an obstacle can temporarily trap portions of a contaminant cloud.

Contaminant Interaction with the Atmosphere Contaminant interaction with the atmosphere is important for several reasons:

1. There are chemical reactions between the released contaminant and ambient (humid) air or surfaces. If the released contaminant reacts, any reacted material can no longer be considered airborne if the reaction products are not hazardous, and so chemical reactions effectively reduce the rate or amount of airborne contaminant. Some reactions can be characterized as dry or wet deposition.

2. Phase changes are typically associated with the evaporation of any suspended liquid phase in an aerosol release. As air is mixed with an aerosol, equilibrium constraints cause additional evaporation of the liquid phase, which reduces the temperature of the liquid phase (and the vapor phase if thermal equilibrium is maintained).

3. Ground-to-contaminant cloud heat transfer acts to warm the contaminant cloud if the cloud temperature is lower than ambient. Ground-to-cloud heat transfer can be important for cold clouds at ground level because the buoyancy of the contaminant cloud can be significantly reduced for cold clouds with contaminant molecular weight less than that of air. At higher wind speeds, heat transfer is by forced convection, and even though such conditions produce higher heat-transfer coefficients than do low-wind-speed conditions, heat transfer is typically more important at low-wind-speed conditions because of two effects: (1) at low wind speeds, the amount of air entrainment is reduced, so the heat-transfer driving force is larger; and (2) at low wind speeds, the contact time between the contaminant cloud and the ground is longer.

Atmospheric Dispersion Models Atmospheric dispersion models generally fall into the categories discussed next. Regardless of the modeling approach taken, it should be verified that the appropriate physical phenomena are being modeled; the model should be validated by comparison with relevant data (at field and laboratory scale). The choice of modeling techniques may be influenced by the expected distance to the level of concern.

1. *Physical or wind tunnel models* Wind tunnel models have long been used to study the atmospheric flow around structures such as buildings and bridges to predict pressure loading and local velocities. Wind tunnel measurement of contaminant concentrations for release scenarios can be used to estimate hazard zones. However, wind tunnel models are generally considered incapable of simultaneously scaling mechanical turbulence and thermally induced turbulence (a verification issue). Wind tunnel experiments can be very useful for validating mathematical models.

2. *Empirical models* Empirical models rely on the correlation of atmospheric dispersion data for characteristic release types. Two examples of empirically based models are the Pasquill-Gifford model (for passive contaminants) and the Britter-McQuaid model (for denser-than-air contaminants), both of which are described in this subsection. Empirical models can be useful for the validation of other mathematical models, but they are limited to the characteristic release scenarios considered in the correlation. Selected empirical models will be discussed in greater detail because they can provide a reasonable first approximation of the hazard extent for many release scenarios, and they can be used as screening tools to indicate which release scenarios are most important to consider with more sophisticated approaches.

3. *First principle mathematical models* These models solve the basic conservation equations for mass and momentum in their form as partial differential equations (PDEs), along with some method of turbulence closure and appropriate initial and boundary conditions. Such models have become more common with the steady increase in computing power and sophistication of numerical algorithms. However, many potential problems must be addressed. In the verification process, the PDEs being solved must adequately represent the physics of the dispersion process, especially for processes such as ground-to-cloud heat transfer, phase changes for condensed phases, and chemical reactions. Also, turbulence closure methods (and associated boundary and initial conditions) must be appropriate for the dispersion processes present, especially for denser-than-air contaminants. Regardless of the algorithm for solving the PDEs, any solution must demonstrate resolution independence (i.e., the numerical solution must be independent of grid spacing or time step). Finally, models should be validated against relevant information for the scenario considered. Despite decreased computational costs, such models still require a significant investment for simulating release scenarios.

4. *Simplified mathematical models* These models typically begin with the basic conservation equations of the first principle models, but they make simplifying assumptions (typically related to similarity theory) to reduce the problem to the solution of (simultaneous) ordinary differential equations. In the verification process, such models must also address the relevant physical phenomena, and they must be validated for the application being considered. Typically, such models are easily solved on a computer with less user interaction than that needed to solve PDEs. Simplified mathematical models may also be used as screening tools to identify the most important release scenarios; however, other modeling approaches should be considered only if they address and have been validated for the important aspects of the scenario under consideration.

All mathematical models predict (ensemble) time-averaged cloud behavior for a particular set of release conditions. To illustrate in very broad terms, consider a set of trials (field experiments) with continuous contaminant releases (as plumes) that are conducted under identical atmospheric conditions. Suppose that you could measure the concentration on the plume centerline at a given downwind distance with a reasonably fast concentration sensor. Owing to the turbulent nature of the atmosphere and the dispersion process, the measured concentration on the plume centerline at a given downwind location will differ for each trial. If these measurements are averaged during the period for which the contaminant is present, the average of measurements will not change after a sufficient number of trials; this is an ensemble average. This ensemble average reflects the instantaneous concentration, provided the averaging time of the sensor is sufficiently fast. If one considers the difference between any one measured data set and this ensemble average, the measurements will show peak concentrations higher than the average (mean). Peak-to-mean concentration values depend on many factors, but for many purposes, a peak-to-mean concentration ratio of 2 can be assumed. For this hypothetical example, the concentrations were assumed to be measured on the plume centerline. However, due to variation in the wind direction related to large-scale atmospheric turbulence, the centerline of a passive contaminant plume does not remain at the same ground location. This is illustrated in Fig. 23-37 for a ground-level release on flat, unobstructed terrain. This effect is termed *plume meander.* (Note that denser-than-air contaminant plumes would exhibit less of this effect because such plumes may actually displace the atmospheric flow field.) For a fixed ground-level location, the concentration sensor will be at various locations within the plume (potentially moving in and out of the plume). For a constant average wind direction, concentrations at a fixed ground-level location will again approach an (ensemble) average over a sufficiently long averaging time (10-min averaging time has proved to be standard). In this case, concentrations are (ensemble-averaged) 10-min concentrations. (Other properties of the concentration distribution are considered in the literature.) For the effect of plume meander, the relationship between concentrations for various averaging times for ground-level plumes can be approximated as

$$\frac{\langle C\rangle_2}{\langle C\rangle_1} = \left(\frac{t_1}{t_2}\right)^p \qquad (23\text{-}78)$$

where $\langle C\rangle_1$ and $\langle C\rangle_2$ are (ensemble) time-averaged concentrations with averaging times t_1 and t_2, and p is an index typically taken to be around 0.2 for passive plumes. In addition to other more sophisticated approaches, some dispersion models adjust the dispersion coefficients to account for the effect

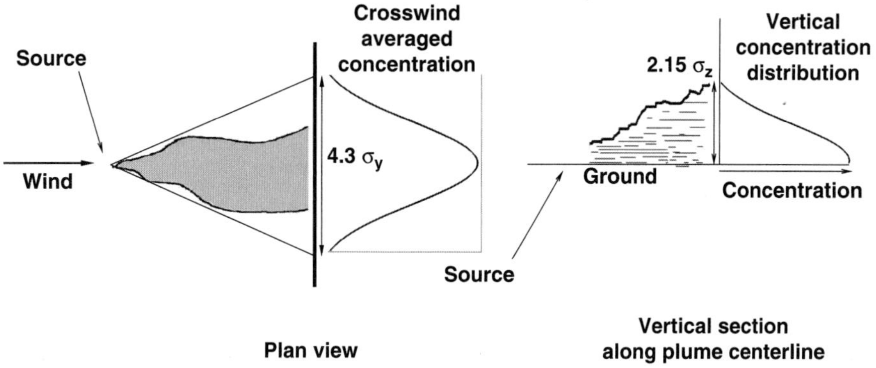

FIG. 23-37 Schematic representation of time-averaged distribution and spread for a continuous plume. σ_y and σ_z are the statistical measures of crosswind and vertical dimensions; $4.3\sigma_y$ is the width corresponding to a concentration 0.1 of the central value when the distribution is of gaussian form (a corresponding cloud height is $2.15\sigma_z$). (*Redrawn from Pasquill and Smith,* Atmospheric Diffusion, *3d ed., Ellis Horwood Limited, Chichester, U.K., 1983*).

of plume meander. Equation (23-78) shows that concentration decreases as averaging time increases. Based on a comparison between passive puff and (10-min) plume coefficients, the averaging time associated with puff coefficients is about 20 s, which is the smallest value that should properly be used in Eq. (23-78). Wind (plume) meander does not change puff dispersion coefficients or concentrations (but does change the puff's location). Note that this is only a rough illustration of the processes that are discussed in much greater detail in the literature.

Basic Scenario Time Scales There are several competing time scales that are important:

1. The source time scale t_s describes the length of time for the contaminant to become airborne; the source time scale is also limited by the inventory of contaminant.

2. The hazard endpoint time scale t_h describes the length of time required for the contaminant to pose a hazard. There are many different time scales associated with various toxicity levels (e.g., TLV-C ceiling limit values are never to be exceeded, TLV-STEL values are not to be exceeded in a 15-min period). Time scales associated with flammability hazards reflect the maximum local concentration (and also typically include peak-to-mean concentration ratios), and for reasons previously discussed they are considered representative of dispersion model averaging times of around 20 s.

3. The travel time t_t is the time required for a contaminant cloud to reach the endpoint distance x_e (or sensor location). As a first approximation, $t_t = x_e/(u_r/2)$ for ground-level clouds; for elevated releases, $t_t = x_h/u$ where u is the characteristic wind speed.

Many models typically represent releases based on plume ($t_s >> t_t$) or puff ($t_s << t_t$) behavior. Other time scale restrictions are considered for the models discussed here.

Scenario Development and Simulation The typical procedure for assessing the consequences of an airborne contaminant release is as follows:

1. Identify release scenarios by which containment can be lost along with the hazards of that scenario. Hazards may differ, depending on the physical state of the released contaminant and the circumstances of the scenario. Hazards arise from the properties of the released material, such as biological agents, toxic materials, or flammable materials; flammable or reactive materials may pose an explosion hazard, depending on their reactivity and the degree to which the air-contaminant mixture is confined. Scenarios include a description of the applicable atmospheric conditions (which may be dictated by regulatory requirements).

2. Develop an appropriate source model to define the source description (see previous subsection) for each scenario.

3. Use an appropriate atmospheric dispersion model to assess the consequences of each scenario. For screening purposes, less costly atmospheric dispersion models can be used to identify the most important scenarios. More expensive modeling procedures can be applied to the most important scenarios, provided such procedures are more appropriate and accurate. Screening methods may also be useful in considering the validity of more complicated models.

4. Determine if the resulting consequences or risk to people and property is acceptable. For unacceptable scenarios, mitigation measures should be applied, such as those discussed by Prugh and Johnson (*Guidelines for Vapor Release Mitigation*, AIChE, New York, 1988), and amended scenarios should be reassessed. If mitigation measures cannot sufficiently reduce the consequences or risk, the appropriate business and ethical decision would be to discontinue such operations.

Passive Contaminant (Pasquill-Gifford) Dispersion Models The Gaussian dispersion model is based on the assumption of a passive contaminant release (i.e., the contaminant cloud moves at the wind speed immediately after release). Based on the theoretical model of a passive contaminant, the spatial distribution would have a Gaussian distribution with characteristic length scales. Using extensive observations of (steady-state) plumes, the Pasquill-Gifford dispersion model correlates characteristic vertical and lateral length scales (or dispersion coefficients σ_z and σ_y, respectively) with atmospheric stability class as a function of distance; other correlations have also been proposed for the plume dispersion coefficients taking other effects into account, such as the effect of surface roughness z_0. Less extensive observations of instantaneously released puffs have been used to characterize the length scales of puffs (with the additional length scale σ_x to characterize dispersion in the along-wind direction). Although the Pasquill-Gifford approach provides for the prediction of the concentration distribution, the following discussion is limited to the maximum predicted concentration since this is most important for hazard assessment purposes.

Pasquill-Gifford Plume Model At a given downwind distance x, the maximum (average) concentration for a (continuous) passive plume from a point source is

$$\langle C \rangle_1 = \frac{E}{\pi \sigma_y \sigma_z u} \tag{23-79}$$

where E is the mass rate at which the contaminant becomes airborne and u is the characteristic wind speed (typically taken to be u_r). Pasquill-Gifford plume dispersion coefficients as a function of downwind distance and atmospheric stability are available from many sources [Seinfeld, *Atmospheric Chemistry and Physics of Air Pollution*, Wiley, New York, 1986; Mannan, *Lees' Loss Prevention in the Process Industries*, 4th ed., chap. 15, Elsevier Butterworth-Heinemann, Oxford, UK, 2012; Griffiths, "Errors in the Use of the Briggs Parameterization for Atmospheric Dispersion Coefficients," *Atmos. Environ.* **28**(17): 2861–2865 (1994)]. Passive dispersion coefficients are typically not provided for distances less than 100 m or greater than a few kilometers because predicted concentrations outside this range must be viewed with some caution (e.g., meteorological conditions may not persist over such large time scales, and at such long distances, large-scale meteorological and terrain features may dictate plume behavior in ways not accounted for by this simple approach).

Note that the predicted values of σ_y and σ_z are sensitive to the specification of atmospheric stability. Between D and F stability classes, σ_y for D stability is roughly 1.5 times greater than for F stability, and σ_z is roughly 2 to 3 times greater. Since $\langle C \rangle_1$ is inversely proportional to $\sigma_y\sigma_z$, the predicted $\langle C \rangle_1$ for F stability is roughly 3 to 5 times greater than for D stability.

Pasquill-Gifford Puff Model At a given downwind distance x, the maximum (average) concentration for a (instantaneous) passive puff from a point source is

$$\langle C \rangle = \frac{2E_t}{(2\pi)^{3/2} \sigma_x \sigma_y \sigma_z} \tag{23-80}$$

where E_t is the total contaminant mass that becomes airborne. Pasquill-Gifford puff dispersion coefficients as a function of downwind distance (or travel time) and atmospheric stability are available from many sources (e.g., De Visscher, *Air Dispersion Modeling—Foundations and Applications*, Wiley, New York, 2014); σ_x can be approximated by σ_y in puff models. As for passive plumes, note that passive puff dispersion coefficients are not provided for distances of less than 100 m (where near-source effects will be important), and predicting concentrations for distances longer than a few kilometers must be viewed with some caution for the reasons cited previously.

Note that the predicted values of σ_y and σ_z are sensitive to the specification of atmospheric stability for puffs as well. Between D and F stability classes, σ_y for D stability is roughly three times greater than for F stability, and σ_z is roughly seven times greater. Since $\langle C \rangle$ is inversely proportional to $\sigma_y^2\sigma_z$, the predicted $\langle C \rangle$ for F stability is roughly 60 times greater than for D stability.

Choosing Between Passive Puff or Plume Models Beyond the general distinction that we have discussed between plumes and puffs, additional guidance on choosing between plume and puff behavior is based on an along-wind dispersion time scale given by $t_d = 2\sigma_x/u_r$, where σ_x is evaluated at the endpoint distance x_e. Hanna and Franzese [*J. Appl. Meteorol.* **39:** 1700–1714 (2000)] used several data sets to determine $\sigma_x \approx 0.1t_t\, u$, so $t_d = 0.2t_t$. The release can be modeled as a puff if $t_s < t_d = 0.2t_t$ and as a plume if $t_s - t_t > 2.5t_d = 0.5t_t$. For $0.2t_t \le t_s \le 1.5t_t$, the predicted concentration is typically considered the larger of the puff and plume predicted concentrations. Note that (10-min average) plume model concentrations should be corrected for averaging time using Eq. (23-78) if the hazards under consideration have $t_h < 10$ min.

Denser-than-Air Contaminant Dispersion Models Releases of denser-than-air contaminants are typically modeled poorly with passive dispersion models, mainly because of the passive model assumption that the contaminant cloud immediately moves at the ambient wind speed. Britter and McQuaid (*Workbook on the Dispersion of Dense Gases*, Health and Safety Executive Report 17/1988, Sheffield, UK, 1988) proposed a correlation for estimating the dispersion of denser-than-air contaminants from area sources for plume and puff releases. Their objective was to produce correlations that predicted the distance to a given concentration level within a factor of 2. Their analysis identified the dominant independent variables as: density of released contaminant after depressurization to atmospheric pressure ρ_s; volumetric rate E/ρ_s (or total volume E_t/ρ_s) of contaminant released; characteristic wind speed u_r (typically taken to be at 10-m elevation z_r); and characteristic source dimension D_s. Based at least in part on the fact that currently available field test data for denser-than-air contaminants do not clearly indicate the importance of these parameters, Britter and McQuaid ignored the effect of surface roughness, atmospheric stability, and exact source dimensions. (Many models for denser-than-air behavior indicate these parameters to be important.) Other effects were not included in the analysis, including dilution due to source momentum and condensation of ambient humidity; such effects may be of crucial importance for contaminants that have a molecular weight less than that of air, such as liquefied natural gas (LNG), ammonia, and hydrogen fluoride. Consult the literature for model details.

Recommended Procedure for Screening Estimates The recommended procedure for making concentration estimates at a specified downwind distance with the simplified models discussed here is as follows:

1. For a given release scenario, estimate the state of the released contaminant after it has depressurized and become airborne (including any initial dilution). The initial mole fraction of hazardous components will be applied to the final reported concentrations and hazardous endpoint concentrations throughout the process. If source momentum is important (as in a jet release or for plume rise), other models are available that can address these considerations. Disregarding the dilution due to source momentum will likely result in higher concentrations downwind, but not always.
2. Consider the important time scales involved, and decide whether a puff or plume model is indicated. If this choice is unclear, assume a plume release.
3. Determine whether denser-than-air behavior is important.
4. When denser-than-air effects are important, use the Britter-McQuaid (plume or puff) models. Otherwise, assume the release is passive and use the Pasquill-Gifford (plume or puff) models.
5. Adjust values for the averaging time correction for plume predictions with Eq. (23-78). If the hazard time scale t_h is different from the model averaging time scale (10 min for plumes), then the predicted concentration should be adjusted to t_h, but only if $t_s \ge t_h$; if $t_s < t_h$, then adjust the predicted concentration to t_s. For flammable hazard considerations ($t_h \approx 0$), it may also be appropriate to take into account that the peak concentration is roughly twice the (ensemble) predicted mean concentration.
6. For plume predictions, confirm that plume behavior applies by consideration of appropriate time scales. If plume behavior is not justified, revise the calculations with the puff model and recheck the dispersion time scale. Report the appropriate concentration or distance.

ESTIMATION OF DAMAGE EFFECTS

Nomenclature

I	Impulse
F	Applied force
P_{so}	Side on or incident overpressure
R	Maximum resistance of a structure
T	Natural period
t_d	Duration of positive phase
VCE	Vapor cloud explosion

GENERAL REFERENCES: American Society of Civil Engineers, *Design of Blast Resistant Buildings in Petrochemical Facilities*, 2d ed., ASCE, Reston, Va., 2010; Glasstone, Samuel, *The Effects of Nuclear Weapons*, 3d ed., U.S. Government Printing Office, Washington, D.C., 1977; International Association of Oil and Gas Producers, *Risk Assessment Directory—Vulnerability of Plant/Structure*, Report No. 434-15, March 2010; U.S. Department of Defense, Explosives Safety Board, *Technical Paper 14—Approved Methods and Algorithms for DoD Risk-Based Explosives Siting Rev. 4*, Alexandria, Va., November 2008.

Types of Explosions and Effects on Blast Loads Buildings and equipment may be subject to several types of explosions, including:

- *Vapor cloud explosions.* These occur when a flammable cloud is ignited. The intensity of the explosion is a function of the quantity and reactivity of the fuel, the amount of congestion within the cloud, and the confinement of the cloud. VCEs may be detonations or deflagrations, with deflagrations being more common in industrial applications. The blast wave from a detonation will have an instantaneous rise to the peak overpressure. Deflagrations may have a more gradual rise to peak pressure and may be centrally peaked. Both deflagrations and detonations are typically treated as shock waves when estimating the damage to structures at processing sites. The actual wave shape is typically accounted for in assessing structural components on off-shore platforms and vessels.
- *Condensed-phase explosions/other uncontrolled chemical reactions.* Explosions of this type can result in blast pressures that are much higher than most VCEs. Condensed-phase explosions are typically detonations, and uncontrolled chemical reactions may be modeled as detonations.
- *Boiling liquid expanding vapor explosions (BLEVEs)/pressure-volume ruptures/physical explosions.* The blast loading from these types of explosions are typically modeled as detonations.

Regardless of the type of explosion, the following information is required to estimate the damage to a building, vehicle, or piece of equipment:

- *Overpressure.* Overpressure is indicative of the force applied to a structure. The overpressure generated by the blast wave when it arrives at the target location is referred to as the side-on overpressure, the static overpressure, or the incident overpressure. This pressure is the pressure that is applied to surfaces that are parallel to the direction of the blast wave, such as the ground, flat roofs, and the side and leeward walls of a building. The loads on the blastward side of a building will be increased by a reflection factor. The value of the reflection factor varies with both the orientation of the building and the magnitude of the incident pressure and the blast wave shape. It is important to understand whether the pressure resulting from a calculation is the incident or the applied pressure. The ASCE (American Society of Civil Engineers, *Design of Blast Resistant Buildings in Petrochemical Facilities*, 2d ed., ASCE, Reston, Va., 2010) provides an excellent simplified guide for determining the loads on a building once an incident pressure is provided.
- *Impulse.* Impulse is indicative of the energy applied to a surface, and it is determined by integrating the pressure-time curve. For most simplified approaches, the loading is assumed to be a triangular pulse, so the impulse is given by:

$$I = \tfrac{1}{2} P_{so}\, t_d \tag{23-81}$$

where I = impulse
P_{so} = peak side-on pressure
t_d = duration of the positive phase

- *Dynamic pressure.* Dynamic pressure results from the movement of the air within the blast wave. While being a relatively small component of the total load on a building or other large, blocky target, the dynamic pressure does provide a substantial portion of the loading on drag-sensitive targets such as poles, tower supports, and pipe racks. A good description of the simplified methods available for estimating the drag loads on various objects is provided in Glasstone (Glasstone, S., *The Effects of Nuclear Weapons*, 3d ed., U.S. Government Printing Office, Washington, D.C., 1977). The reason for requiring both the overpressure and the impulse is illustrated by the following two examples.

Many older references [U.S. Department of the Interior, Office of Oil and Gas, *Minimizing Damage to Refineries from Nuclear Attack, Natural and Other Disasters*, Washington, D.C., 1970; see also Glasstone (1977)] provide estimates of structural damage in terms of overpressure alone. These references tend to be conservative since much of the original information resulted from research in support of nuclear weapons development and testing. The nuclear weapons produced very long-duration blast loads and are thus far more damaging than an industrial explosion that produces the same peak overpressure with a substantially shorter duration.

Since the terrorist attacks of 2001, many blast-resistant products, such as windows and doors, have been developed to meet U.S. government specifications. These products are developed to resist the short-duration loads that result from small (<100 kg TNT) explosions, but they are offered with just a peak pressure designation (such as a 4 psi window). A window designed to resist an applied load of 4 psi with an impulse of 28 psi-ms (a common government specification) is not likely to be suitable for use at a refinery where a 2 psi incident load with a duration of 50 ms produces an applied (reflected) pressure of 4 psi and an impulse of 100 psi-ms.

Multiple combinations of pressure and impulse can cause the same damage to a structure. These combinations are graphically represented by iso-damage curves. The generic shape of an iso-damage curve for a single building component is shown in Fig. 23-38. The axes of the curve have been normalized to the components maximum resistance (R) and natural period (T). An iso-damage curve may be developed from a theoretical basis or may be represented by an empirical fit to test data or sophisticated calculations. The curve shown illustrates that two dissimilar blast loads may produce the same amount of damage. For example, a high pressure ($F/R = 10$) and low impulse loading ($I/RT = 1$) are predicted to produce the same damage as a low-pressure ($F/R = 1$) blast wave with significantly more impulse ($I/RT = 10$). The impulse asymptote identifies the minimum impulse required to produce the level of damage represented by the curve. The pressure asymptote likewise identifies the minimum pressure required to produce the damage represented by the curve. Older references that typically only correlate damage with pressure are in effect identifying the pressure asymptote.

If the blast load's duration (t_d) is less than 30 percent of the natural period of a building component, the loading is considered impulsive. If t_d is greater than three times the component's natural period, then the load is considered quasi-static. Durations between these limits are considered to be in the dynamic region (the knee of the iso-damage curve). Individual building components may have substantially different natural periods. A light, flexible roof (such as steel decking supported by open web steel joists) may have a natural period much greater than the masonry wall of a building. Building iso-damage curves can be developed using component iso-damage curves; the development of a building iso-damage curve is beyond the scope of this discussion.

The need to accurately assess the exact pressure and duration of a predicted blast load depends on whether a component or structure is in the impulsive, dynamic, or quasi-static regime. The influence of the various load characteristics is provided in Table 23-22.

Regardless of the source of the explosion, the potential for damage to a building or other structure will decrease as the distance from the explosion increases. One unique aspect of a VCE is that a building may be within the explosion cloud. When a building is within the explosion, the use of simplified methods is questionable, and a more reasonable approach is to rely on computational fluid dynamic codes (CFD) such as FLACS (Health and Safety Executive, *Explosion Loading on Topsides Equipment: Part 1—Treatment of Explosion Loads, Response Analysis and Design*, Offshore Technology Report—OTO 1999 046, March 2000; and Health and Safety Executive, *Explosion Loading on Topsides Equipment: Part 2—Determination of Explosion Loading on Offshore Equipment Using FLACS*, Offshore Technology Report—OTO 1999 047, March 2000).

Alternative Approaches to Estimating Structural Damage There are many methods for calculating structural damage from explosions. The choice of method depends on the purposes of the estimate.

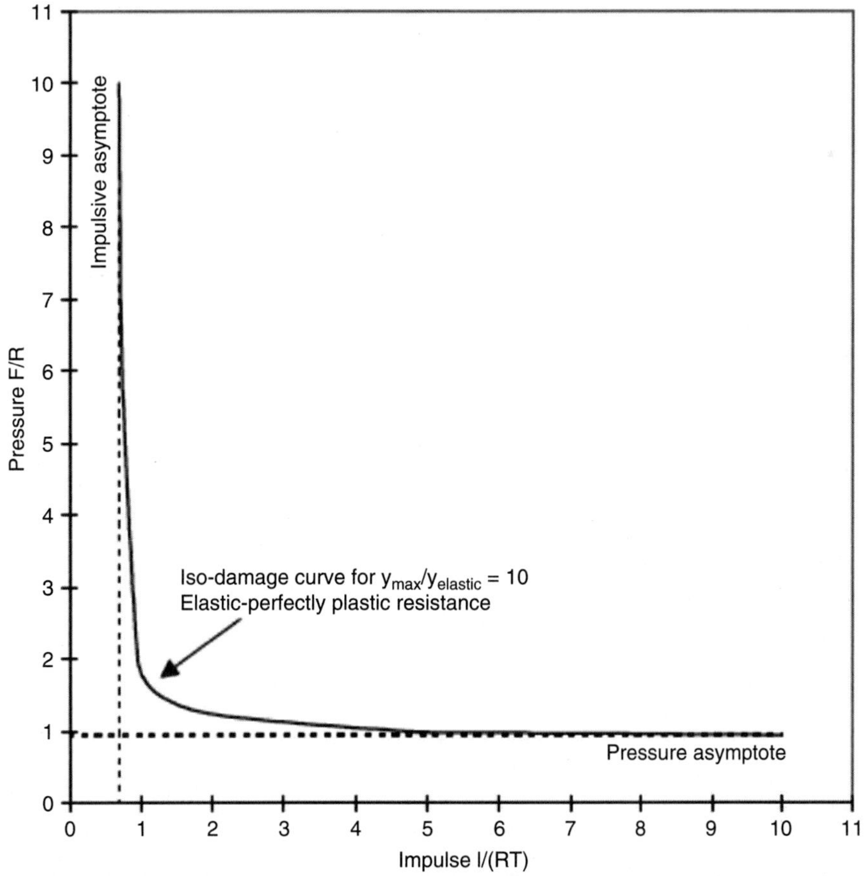

FIG. 23-38 Normalized iso-damage curve.

TABLE 23-22 Regimes of Dynamic Response

	Impulsive ($t_d/T < 0.3$)	Dynamic ($0.3 < t_d/T < 3$)	Quasi-static ($t_d/T < 3$)
Peak load	Preserving the exact peak pressure is not critical.	Preserve peak value—the response is sensitive to increases or decreases in peak load for a smooth pressure pulse.	
Duration	Preserving the exact load duration is not critical.	Preserve load duration since in this range it is close to the natural period of the structure. Even slight changes may affect response.	Not critical if response is elastic, but it is critical when response is plastic.
Impulse	Accurate representation of the impulse is critical.	Accurate representation of the impulse is important.	Accurate representation of the impulse is not important.
Rise time	Preserving rise time is not important.	Preserving rise time is important; ignoring it can significantly affect response.	

SOURCE: International Association of Oil and Gas Producers, *Risk Assessment Directory—Vulnerability of Plant/Structure*, Report No. 434-15, March 2010, Table 2.11, page 11.

The ASCE (American Society of Civil Engineers, *Design of Blast Resistant Buildings in Petrochemical Facilities*, 2d ed., ASCE, Reston, Va., 2010), the Process Industry Practices (*STC 01018 PIP Blast Resistant Building Design Criteria*, Austin, Tex., 2006), the U.S. Department of Defense (U.S. Army Corps of Engineers, *Single Degree of Freedom Structural Response Limits for Antiterrorism Design*, PDC-TR 06-08, October 20, 2006), NORSOK (NORSOK Standard N-004, *Design of Steel Structures Rev. 2*, October 2004), and the ISCE (UK Institute of Structural Civil Engineers, *Blast Effects on Buildings*, 2d ed., Thomas Telford Publishers, London, 2009) have published guidance documents that are readily available and are listed in Table 23-23. These references provide guidance for the calculation of the response of various structures to blast loads. Guidance is provided for the analysis of reinforced concrete, reinforced and unreinforced masonry, and other types of construction.

Damage Estimates in Support of Facility Siting Studies If the estimate is being performed as a part of a facility siting study in accordance with American Petroleum Institute (API) Recommended Practice (RP) 752 (American Petroleum Institute, Recommended Practice 752—*Management of Hazards Associated with Process Plant Permanent Buildings*, 3d ed., December 2009), it will have to address a wide variety of potential explosion scenarios. API RP 752 requires that the method consider both the peak overpressure and impulse. Thus simple damage look-up tables are not appropriate for this purpose. The use of building damage levels (BDLs) is a common building siting evaluation criterion. Building damage increases as the severity of the blast load increases, and it may be represented as a continuous or discrete function. When the discrete approach is used, BDLs are categorized into a number of damage states ranging from minimal damage to collapse [Baker, Q. A., et al., "Explosion Risk and Structural Damage Assessment Code (ERASDAC)," 30th DoD Explosive Safety Seminar, Department of Defense Explosive Safety Board, Arlington, Va., 2002]. An open source of BDL curves is available from the DDESB. The DDESB curves predict the "percentage of damage" rather than discrete damage states. Examples of percent damage as a function of pressure and impulse for a ductile and a brittle structure are shown in Fig. 23-39 and Fig. 23-40, respectively. The curve to the upper right is the 100 percent damage curve, and the curve on the lower left is the 0.1 percent damage curve. The limitations of this approach are that it does not readily allow the identification of the type of damage that has occurred or which building components may be governing the percentage of damage to the structure.

Typical discrete BDLs used in the process industry are shown in Table 23-24. One advantage of this approach is that the nature of the damage is indicated by the damage description. Iso-damage curves expressed as pressure-impulse (P-i) diagrams serve to define the boundaries between the damage states when discrete BDLs are used. An illustration of the pressure-impulse curves may be presented as upper bounds on the lower damage state as shown in Fig. 23-41.

Calculating Damage or Response in Support of Design Activities BDLs should not be used as a design tool. The response of the building components should be calculated using dynamic structural analysis methods that are appropriate for the level of response that can be tolerated. The most common approach is to perform a dynamic structural analysis using either single- or two-degree-of freedom calculations. The details for performing these calculations and typical design criteria are discussed in the sources shown in Table 23-23.

Calculation of Damage or Response in Support of Incident Investigation When an explosion occurs at a processing facility, the damaged buildings and structures may serve as damage indicators. The damage to the structures can be used to make an estimate of the blast pressures and impulses at various locations on the plant, and this information is used to support or disqualify different explosion scenarios. The use of tabulated values of structural damage as a function of pressure may be useful in the early stages of such an investigation. A fairly comprehensive compilation of these values is provided in Table 23-25. Note that the correlations in the table are for very long duration (0.2 to 2 s) blast loads, and the structural damage to houses is mitigated by the blowout of windows and doors. More detailed analyses may be warranted in later stages of an investigation to improve estimates of blast loads causing the observed damage.

Equipment Damage Modes Equipment may fail in two different modes:

- *Failure to operate.* This failure mode is applicable to pumps, valves, motors, electrical components, controls, and other pieces of equipment that are expected to operate in some manner. The explosion damages the

TABLE 23-23 Recent Publications in Blast Resistant Design

Publishing organization	Title	Summary
U.S. Army Corps of Engineers, Protective Design Center	*Single Degree of Freedom Structural Response Limits for Antiterrorism Design*	Published in 2008 and includes direct correlation between building damage, component damage, and numerical limits on component response. Comprehensive in that it addresses reinforced and unreinforced masonry, steel, concrete, prestressed concrete, and wood components. No direct correlation between extent of damage and numerical values of occupant vulnerability.
U.S. Department of Defense, Explosives Safety Board (DDESB)	Technical Paper 14 - *Approved Methods and Algorithms for DoD Risk-Based Explosives Siting*	Provides P-i damage curves for a number of building types and components. Provides occupant vulnerability for each building type as a function of building damage. Building damage is defined as a continuous function.
American Society of Civil Engineers	*Design of Blast Resistant Buildings in Petrochemical Facilities*	Published in 1997 and updated in 2010. Provides good overall discussion of issues, including loadings and limits on component responses. Does not address building damage or occupant vulnerability per se.
NORSOK	*Design of Steel Structures*	Published in 2004, and while intended primarily for offshore structures, provides discussion of response limits and design charts that address the shape of the loading function as well as the presence of membrane action. Only addresses steel components.
Construction Industries Institute	PIP STC 01018 *Blast Resistant Building Design Criteria*	Published in 2006, this document provides information on design and analysis approaches as well as numerical values limiting the deformation of structural components.
UK Institution of Structural Civil Engineers	*Blast Effects on Buildings*, 2d ed.	Published in 2009 and provides guidance on the design of buildings to resist both high explosives and detonations and deflagrations due to industrial, vapor cloud, and dust explosions.

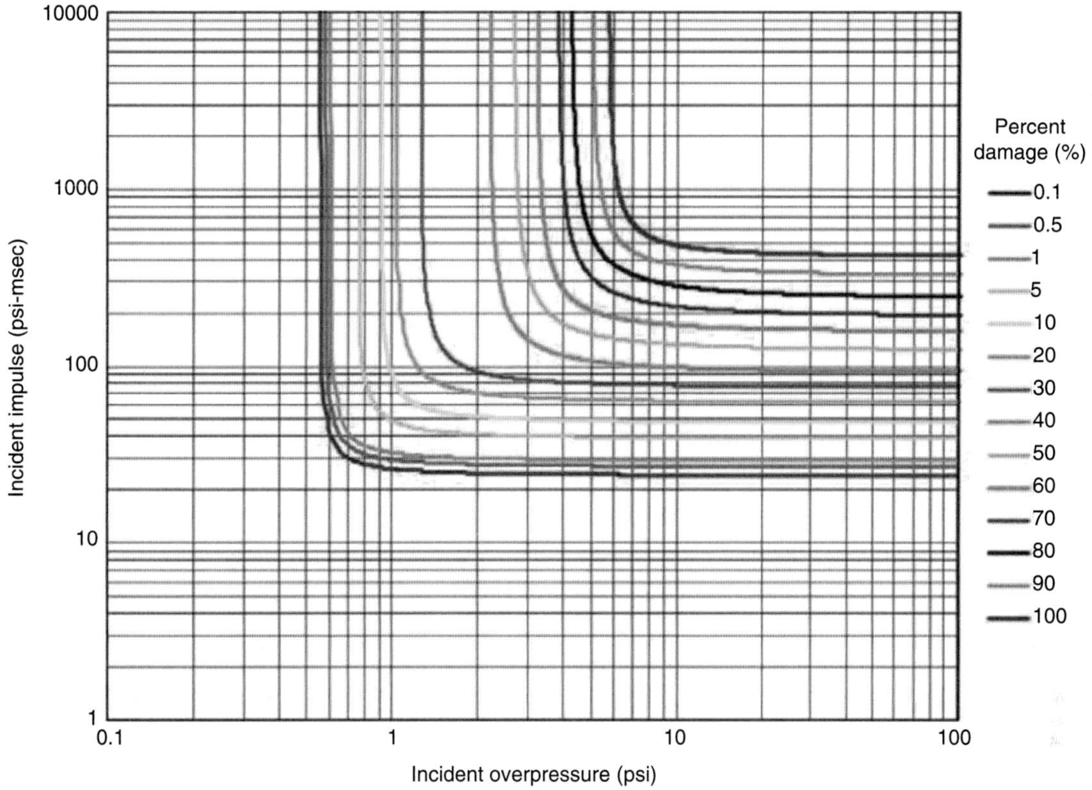

FIG. 23-39 DDESB BDL curves for a building with ductile properties.

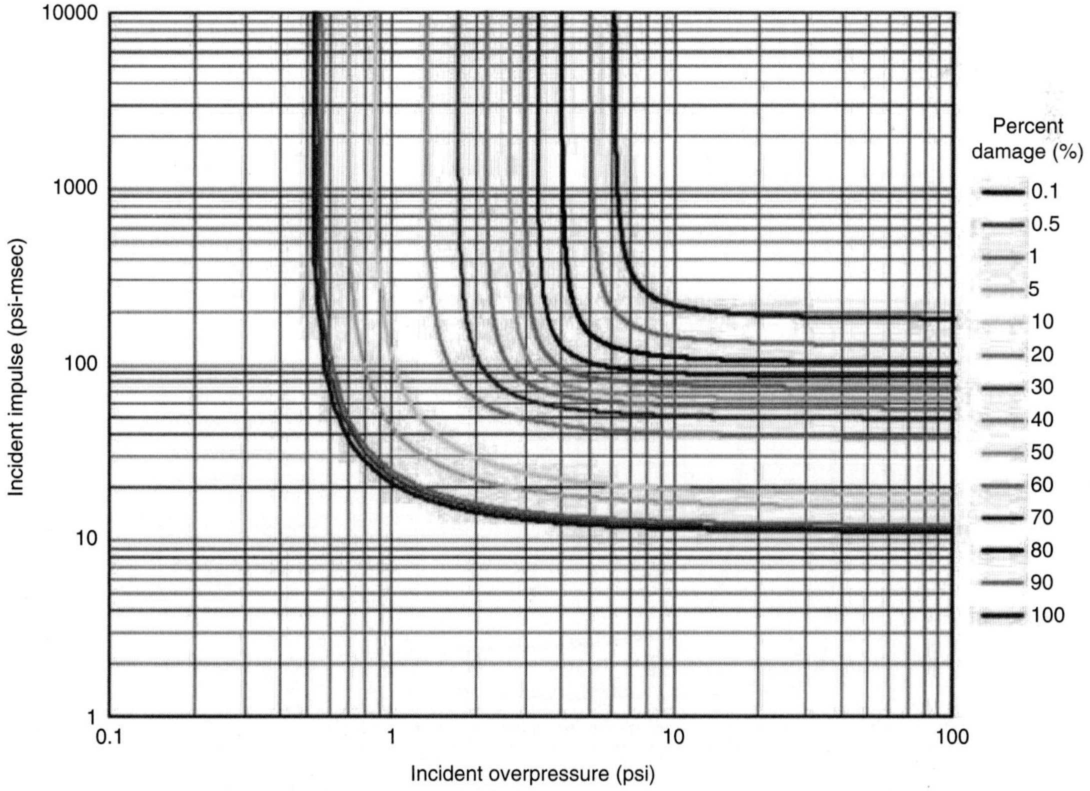

FIG. 23-40 DDESB BDL curves for a building with brittle properties.

TABLE 23-24 Typical Industry Building Damage Level Descriptions (Baker, 2002)

Building damage level (BDL)	BDL name	Damage description
1	Minor	Onset of visible damage to reflected wall of building.
2A	Light	Reflected wall components sustain permanent damage requiring replacement. Other walls and roof have visible damage that is generally repairable.
2B	Moderate	Reflected wall components are collapsed or very severely damaged. Other walls and roof have permanent damage requiring replacement.
3	Major	Reflected wall has collapsed. Other walls and roof have substantial plastic deformation that may be approaching incipient collapse.
4	Collapse	Complete failure of the building roof and a substantial area of walls.

equipment either by physically deforming it or by producing an internal mechanical shock that disrupts its operation.

- *Failure to contain.* This failure mode applies to pipes, vessels, and tanks as well as to operating equipment that contains fluids or gases. The damage is a result of deformations of the equipment caused by the explosion. Leaks may occur at flanges and connections at lower blast loads than required to produce significant deformations in a pipe or vessel.

Data on the shock tolerance of equipment may be available from vendors. Alternatively, the U.S. Department of Energy publishes some generic equipment tolerance data (*Seismic Evaluation Procedure for Equipment in U.S. Department of Energy Facilities*, DOE/EH-0545, and available online at https://ehss.energy.gov/au/seismic/seismic.pdf).

Calculating the deformation of equipment requires a detailed structural assessment of the equipment and its mountings. Acceptance criteria are not widely published, but they have been determined for some types of equipment as part of seismic evaluations.

Potential Injuries to Occupants The risk to building occupants is commonly referred to as occupant vulnerability (OV). OV is essentially the risk of death or life-threatening injury should the explosion occur. However, there is some variation in the definition between organizations, so the analyst should be careful that the calculations match the definition.

Personnel may be injured by explosions in four different manners:

- *Direct blast injuries.* The overpressure damages the lungs or other soft organs. This injury mechanism is rare in VCEs unless the person is within the exploding cloud. However, it is more likely to occur when a condensed-phase explosion occurs.
- *Translational injuries.* The person is thrown against a hard object and is injured by the impact.

TABLE 23-25 Damage Estimates for Common Structures Based on Overpressure

Pressure		
Psig	kPa	Damage
0.02	0.14	Annoying noise (137 dB if of low frequency 10–15 Hz).
0.03	0.21	Occasional breaking of large glass windows already under strain.
0.04	0.28	Loud noise (143 dB), sonic boom, glass failure.
0.1	0.69	Breakage of small windows under strain.
0.15	1.03	Typical pressure for glass breakage.
0.3	2.07	"Safe distance" (probability 0.945 of no serious damage to typical brick buildings below this value); projectile limit; some damage to house ceilings; 10% window glass broken.
0.4	2.76	Limited minor structural damage.
0.5–1.0	3.4–6.9	Large and small windows usually shattered; occasional damage to window frames.
0.7	4.8	Minor damage to house structures.
1.0	6.9	Partial demolition of houses, made uninhabitable.
1.0–2.0	6.9–13.8	Corrugated asbestos shattered; corrugated steel or aluminum panels, fastenings fail, followed by buckling; wood panels (standard housing fastenings) fail, panels blown in.
1.3	9.0	Steel frame of clad building slightly distorted.
2	13.8	Partial collapse of walls and roofs of houses.
2.0–3.0	13.8–20.7	Concrete or cinder block walls, not reinforced, shattered.
2.3	15.8	Lower limit of serious structural damage.
2.5	17.2	50% destruction of brickwork of houses.
3	20.7	Heavy machines (3000 lb) in industrial building suffer little damage; steel frame building distorted and pulled away from foundations.
3.0–4.0	20.7–27.6	Frameless, self-framing steel panel building demolished; rupture of oil storage tanks.
4	27.6	Cladding of light industrial buildings ruptured.
5	34.5	Wooden utility poles snapped; tall hydraulic press (40,000 lb) in building slightly damaged.
5.0–7.0	34.5–48.2	Nearly total collapse of houses.
7	48.2	Loaded lighter-weight (British) train wagons overturned.
7.0–8.0	48.2–55.1	Brick panels, 8–12 inch thick, not reinforced, fail by shearing or flexure.
9	62	Loaded train boxcars completely demolished.
10	68.9	Probable total destruction of buildings; heavy machine tools (7000 lb) moved and badly damaged, very heavy machine tools (12,000 lb) survive.

SOURCE: International Association of Oil and Gas Producers, *Risk Assessment Directory—Vulnerability of Plant/Structure*, Report No. 434-15, March 2010, Table 2.8, page 8.

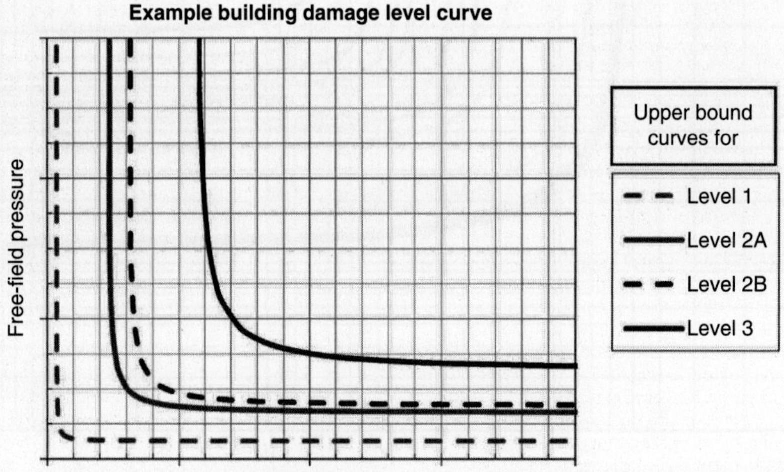

FIG. 23-41 Illustration of discrete state BDL curves.

- *Fragment injuries.* The person is struck by fragments from the explosion source. This is the primary injury mechanism for military weapons and may be a significant cause of injuries from a bursting pipe or vessel.
- *Debris impact injuries.* This is the most common cause of injuries to personnel within buildings at industrial sites. Buildings may collapse under blast loadings that would be tolerable for personnel in the open, but the collapsing building or failing wall results in severe or fatal injuries to personnel.

The DDESB provides injury models that are correlated with building damage levels. A model widely used in the processing industries is provided in Baker, Q.A., et al., "Explosion Risk and Structural Damage Assessment Code (ERASDAC)," *30th DoD Explosive Safety Seminar,* Department of Defense Explosive Safety Board, Arlington, Va., 2002.

SAFETY EQUIPMENT, PROCESS DESIGN, AND OPERATION

PRESSURE RELIEF SYSTEMS

Nomenclature

A	Vent area, m^2
A_{vessel}	Vessel cross-sectional area, m^2
C	Specific heat, J/kg-K
C_D	Discharge coefficient
G	Mass flux, kg/m^2-s
h_{fg}	Latent heat of vaporization, J/kg
j_g	Superficial velocity, m/s
k	Isentropic coefficient
m	Sample mass in test apparatus, kg
M_w	Molecular weight, kg/kgmole
P	Pressure, psia
$\dot{P}$	Rate of pressure rise, psi/min
q	Heat input rate, J/s
R	Gas constant, 8314 Pa-m^3/K-kgmole
T	Absolute temperature, K
$\dot{T}$	Rate of temperature rise, °C/min
u_∞	Bubble rise velocity, m/s
v	Gas volume in test apparatus, m^3
v_f, v_g	Specific volumes of liquid and gas, m^3/kg
V	Vessel volume or reactant volume, m^3
W_{req}	Required relief rate, kg/s
x_o	Vapor or gas quality
	Greek letters
α	Void fraction
β	Volumetric expansion coefficient, °C^{-1}
λ	Latent heat of evaporation, J/kg
ρ_f, ρ_g	Liquid density and gas density, kg/m^3

General References: Center for Chemical Process Safety (CCPS), *Guidelines for Pressure Relief and Effluent Handling Systems,* American Institute of Chemical Engineers, New York, 1998; DIERS Project Manual, *Emergency Relief System Design Using DIERS Technology,* American Institute of Chemical Engineers, New York, 1992; Fauske, "Determine Two-Phase Flows During Releases," *Chem. Eng. Prog.,* February 1999; Fauske, "Properly Size Vents for Nonreactive and Reactive Chemicals," *Chem. Eng. Prog.,* February 2000; Fauske, "A Practical Approach to Capacity Certification," *Chem. Process.,* February 2003; Fauske, "Revisiting DIERS Two-Phase Methodology for Reactive Systems Twenty Years Later," *Process Saf. Prog.,* September 2006. See more references under Codes, Standards, and Guidelines subsection.

Introduction All process designs should try to arrive at an inherently safe facility, that is, one from which a worst-case event cannot cause injury to personnel, damage to equipment, or harm to the environment. Incorporating safety features that are intrinsic (built-in) rather than extrinsic (added-on) to the basic design, together with the use of high-integrity equipment and piping, provide the first lines of defense against the dramatic, often catastrophic, effects of an overpressure and subsequent rupture. In recent years, many companies have incorporated the principles of depressuring or instrumental shutdown of key equipment as a means to control a release and avoid the actuation of pressure relief devices. This minimizes the probability of failure of the device because, once activated, the device may no longer be dependable. Since maintenance of relief devices can be sporadic, this redundancy provides yet another layer of safety. However, regardless of the number of lines of defense and depressuring systems in place, overpressure protection must still be provided. Emergency pressure relief systems are intended to provide the last line of protection and thus must be designed for high reliability, even though they will have to function infrequently.

Self-actuated pressure relief systems must be designed to limit the pressure rise that can occur as a result of overcompressing, overfilling, or overheating either an inert or a chemically reactive medium in a closed system. Pressure generation is usually the result of either expansion of a single-phase medium (by material addition or heating) or a shift of the phase equilibrium in a multiphase medium (as a result of composition or temperature changes, particularly in the case of a reactive system). These mechanisms of pressure generation differ from what is commonly referred to as explosion venting. Events such as dust explosions and flammable vapor deflagrations propagate nonuniformly from a point of initiation, generating pressure or shock waves. Such venting problems are not included in these discussions.

Relief System Terminology Refer to API-RP520 Part I for complete terminology.

Pressure relief valve (PRV) A pressure relief device designed to open and relieve excess pressure and to reclose after normal conditions have been restored. PRV is a generic term applied to *relief valve* (set up for liquid flow), *safety valve* (set up for gas or vapor flow), and *safety relief valve* (set up for either liquid or compressible flow).

Rupture disk A nonreclosing pressure relief device actuated by static differential pressure and designed to function by the bursting of a pressure-containing disk.

Set pressure The inlet gauge pressure at which a PRV will start to open (or a rupture disk will burst) under service conditions of temperature and back-pressure.

Backpressure is the pressure existing at the outlet of a relief device. The value under no-flow conditions is superimposed backpressure. The value under flowing conditions consists of both superimposed backpressure and built-up pressure due to piping pressure drop.

Blowdown The reduction in flowing pressure below the set point required for a PRV to close.

Overpressure A pressure increase above the set point during relief flow, usually expressed as a percentage of the differential set pressure.

Relieving pressure is the set pressure plus the overpressure.

Maximum allowable working pressure (MAWP) The maximum allowed pressure at the top of the vessel in its normal operating position at the operating temperature specified for that pressure.

Design pressure The design pressure is used to determine the minimum thickness of a vessel component and may be used in place of MAWP where the latter has not been established. It is equal to or less than the MAWP. It is the pressure specified on the equipment purchase order.

Accumulation The rise of pressure above the MAWP of the protected system, usually expressed as a percentage of the gauge MAWP. Maximum allowable accumulations are established by applicable codes for emergency operating and fire contingencies.

Codes, Standards, and Guidelines Industry practice is to conform to the applicable regulations, codes, and recommended practices. In many cases, these will provide different guidelines. A suggested approach would be to review all applicable codes, standards, and recommended practices prior to choosing a design basis. The Design Institute for Emergency Relief Systems (DIERS) was established by AIChE to address sizing aspects of relief systems for two-phase, vapor–liquid flashing flow regimes. The DIERS Project Manual (*Emergency Relief System Design Using DIERS Technology,* American Institute of Chemical Engineers, New York, 1992) and the CCPS Guidelines (*Guidelines for Pressure Relief and Effluent Handling Systems,* American Institute of Chemical Engineers, New York, 1998) are continually being updated to further the general knowledge in emergency relief system design as illustrated in the ninth edition.

NFPA 30 and API Standard 2000 provide guidance for the design of overpressure protection involving storage tanks that operate at or near atmospheric pressure. In particular, NFPA 30 focuses on flammability issues, while API 2000 addresses both pressure and vacuum requirements. The ASME code (sections I and VIII) and API RP 520 are the primary references for pressure relief device sizing requirements.

Designers of emergency pressure relief systems should be familiar with the following list of regulations, codes of practice, and industry standards and guidelines in the United States.

API RP 520. *Sizing, Selection, and Installation of Pressure-Relieving Devices in Refineries.* Part I, *Sizing and Selection*, 9th ed., July 2014, and part II, *Installation*, 6th ed., December 1994. American Petroleum Institute, Washington, D.C.

API RP 521, 1997. *Guide for Pressure-Relieving and Depressuring Systems*, 4th ed., American Petroleum Institute, Washington, D.C.

API STD 526, 1995. *Flanged Steel Pressure Relief Valves*, 4th ed., American Petroleum Institute, Washington, D.C.

API STD 2000. *Venting Atmospheric and Low-Pressure Storage Tanks, Nonrefrigerated and Refrigerated*, 7th ed., American Petroleum Institute, Washington, D.C., March 2014.

API RP 2001, 1984. *Fire Protection in Refineries*, American Petroleum Institute, Washington, D.C.

ASME, 2001. *Boiler and Pressure Vessel Code*, sec. I, *Power Boilers*, and sec. VIII, *Pressure Vessels*, American Society of Mechanical Engineers, New York.

ASME, 1988. *Performance Test Code PTC-25, Safety and Relief Valves.* American Society of Mechanical Engineers, New York.

CCPS, 1993. *Engineering Design for Process Safety.* American Institute of Chemical Engineers, New York.

CCPS, 1998. *Guidelines for Pressure Relief and Effluent Handling Systems*, American Institute of Chemical Engineers, New York.

DIERS, 1992. *Emergency Relief System Design Using DIERS Technology, DIERS Project Manual*, American Institute of Chemical Engineers, New York.

National Board of Boiler and Pressure Vessel Inspectors, 2004. *Pressure Relieving Device Certifications (Red Book NB-18)*, National Board of Boiler and Pressure Vessel Inspectors, Columbus, Ohio.

NFPA 30. *Flammable and Combustible Liquids Code*, National Fire Protection Association, Quincy, Mass., 2014 edition.

OSHA 1910.106, 2005. *Flammable and Combustible Liquids*, Regulations (Standards – 29 CFR), U.S. Dept. of Labor, Occupational Safety and Health Administration, Washington, D.C.

Relief System Design The most difficult part of designing an adequate emergency pressure relief system is determining the emergency events (credible design scenarios) for which to design. The difficulty arises primarily because the identification of credible design scenarios usually involves highly subjective judgments, which are often influenced by economic situations. Unfortunately, there exists no universally accepted list of credible design scenarios. Relief systems must be designed for the credible chain of events that results in the most severe venting requirements (worst credible scenario). Credibility is judged primarily by the number and the time frame of causative failures required to generate the postulated emergency. Only totally independent equipment or human failures should be considered when judging credibility. A failure resulting from another failure is an effect, rather than an independent causative factor. A suggested guideline for assessing credibility as a function of the number and time frames of independent causative events is:

- Any single failure is *credible.*
- Two or more simultaneous failures are *not credible.*
- Two events in sequence are *credible.*
- Three or more events in sequence are *not credible.*

The first step in scenario selection is to identify all the credible emergencies using the preceding guidelines (or a similar set). This is perhaps best accomplished by identifying all the possible sources of pressure and vacuum. Table 23-26 lists a number of commonly existing pressure and vacuum sources.

Fire The main consequence of fire exposure is heat input causing thermal expansion, vaporization, or thermally induced runaway reaction/decomposition resulting in a pressure rise. An additional result of fire exposure is the possibility of overheating the wall of the equipment in the vapor space where the wall is not cooled by the liquid. In this case, the vessel wall may fail due to the high temperature, even though the relief system is operating. Hence API RP-521 recommends vapor depressuring facilities for high-pressure services (greater than 17 bar or 250 psig). Guidelines for estimating the heat input from a fire are found in API recommended practices, NFPA 30 (for bulk storage tanks), OSHA 1910.106, and corporate engineering standards. In determining the heat input from fire exposure, NFPA allows credit for application of water spray to a vessel; API allows no such credit.

Pressure vessels (including heat exchangers and air coolers) in a plant handling flammable fluids are subject to potential exposure to external fire. A vessel or group of vessels that could be exposed to a pool fire must be protected by pressure relief device(s). Additional protection to reduce the device relief load can be provided by insulation, water spray, drainage, or remote-controlled depressuring devices. Plant layout should consider spacing requirements, such as those set forth by NFPA, API, Industrial Risk Insurers, or Factory Mutual, and must include accessibility for fire-fighting personnel and equipment. Several pieces of equipment located adjacent to each other that cannot be isolated by shutoff valves can be protected by a common relief device, providing the interconnecting piping is large enough to handle the required relief load, and the relief set pressure is no higher than the minimum MAWP of these pieces of equipment.

Operational Failures A number of scenarios of various operational failures may result in the generation of overpressure conditions:

- *Blocked outlet.* Operation or maintenance errors (especially following a plant turnaround) can block the outlet of a liquid or vapor stream from a piece of process equipment, resulting in an overpressure condition.
- *Opening a manual valve.* Manual valves that are normally closed to isolate two or more pieces of equipment or process streams can be inadvertently opened, causing the release of a high-pressure stream or resulting in vacuum conditions. Other effects may include the development of critical flows, flashing of liquids, or the generation of a runaway chemical reaction.
- *Cooling water failure.* The loss of cooling water is one of the more commonly encountered causes of overpressurization. Two examples of the critical consequences of this event are the loss of condensing duty in column overhead systems and the loss of cooling for compressor seals and lube oil systems. Different scenarios should be considered for this event, depending on whether the failure affects a single piece of equipment (or process unit) or is plantwide.
- *Power failure.* The loss of power will shut down all motor-driven rotating equipment, including pumps, compressors, air coolers, and vessel agitators. As with cooling water failure, power failure can have a negative cascading effect on other equipment and systems throughout the plant.
- *Instrument air failure.* The consequences of the loss of instrument air should be evaluated in conjunction with the failure mode of the control valve actuators. It should not be assumed that the correct air failure response will occur on these control valves, as some valves may stick in their last operating position.
- *Thermal expansion.* Equipment and pipelines that are liquid-full under normal operating conditions are subject to hydraulic expansion if the temperature increases. Common sources of heat that can result in high pressures due to thermal expansion include solar radiation, steam or other heated tracing, heating coils, and heat transfer from other pieces of equipment.
- *Vacuum.* Vacuum conditions in process equipment can develop due to a wide variety of situations, including:
  - Instrument malfunction
  - Draining or removing liquid with venting
  - Shutting off purge steam without pressuring with noncondensible vapors
  - Extreme cold ambient temperatures resulting in subatmospheric vapor pressures
  - Water addition to vessels that have been steam-purged

If vacuum conditions can develop, then either the equipment must be designed for vacuum conditions or a vacuum relief system must be installed.

Equipment Failure Most equipment failures that can lead to overpressure situations involve the rupture or break of internal tubes inside heat exchangers and other vessels and the failure of valves and regulators. Heat exchangers and other vessels should be protected with a relief system of sufficient capacity to avoid overpressure in case of internal failure. Characterization of the types of failure and the design of the relief system are left to the discretion of the designer. API RP 521 presents guidance in determining these requirements, including criteria for deciding when a full tube rupture is likely. In cases involving the failure of control valves and regulators, it is important to evaluate both the fail-open and fail-closed positions.

Runaway Reactions Runaway temperature and pressure in process vessels can occur as a result of many factors, including loss of cooling, feed or

TABLE 23-26 Common Sources of Pressure and Vacuum

Heat related
- Fire
- Out-of-control heaters and coolers
- Ambient temperature changes
- Runaway chemical reactions

Equipment and systems
- Pumps and compressors
- Heaters and coolers
- Vaporizers and condensers
- Vent manifold interconnections
- Utility headers (steam, air, water, etc.)

Physical changes
- Gas absorption (e.g., HCl in water)
- Thermal expansion
- Vapor condensation

TABLE 23-27 Summary of Device Characteristics

	Reclosing devices		Nonreclosing devices
	Relief valves	Disk-valve combinations	Rupture disks
Fluid above normal boiling point	+	+	–
Toxic fluids	+	+	–
Corrosive fluids	–	+	+
Cost	–	–	+
Minimum pipe size	–	–	+
Testing and maintenance	–	–	+
Won't fatigue and fail low	+	+	–
Opens quickly and fully	–	–	+

NOTE: + indicates advantageous
– indicates disadvantageous

quench failure, excessive feed rates or temperatures, contaminants, catalyst problems, and agitation failure. Of major concern is the high rate of energy release in runaway reactions or the formation of gaseous products, which generally cause a rapid pressure rise in the equipment. In order to properly assess these effects, the reaction kinetics must either be known or obtained experimentally. In general, a lower relief set pressure (much below the equipment MAWP) is desirable for these runaway reaction systems in order to relieve the system at a lower reaction rate since most reactions are Arrhenius in behavior.

Pressure Relief Devices The most common method of overpressure protection is through the use of safety relief valves or rupture disks, which discharge into a containment vessel, a disposal system, or directly to the atmosphere. Table 23-27 summarizes some of the device characteristics and the advantages.

Safety Relief Valves (SRVs) Conventional safety relief valves are used in systems where built-up backpressures typically do not exceed 10 percent of the set pressure. The spring setting of the valve is reduced by the amount of superimposed backpressure expected. Higher built-up backpressures can result in a complete loss of continuous valve relief capacity. The designer must examine the effects of other relieving devices connected to a common header on the performance of each valve. Some mechanical considerations of conventional relief valves are presented in the ASME code; however, the manufacturer should be consulted for specific details.

Balanced safety relief valves may be used in systems where built-up or superimposed backpressure is high or variable. In general, the capacity of a balanced valve is not significantly affected by backpressures below 30 percent of set pressure. Most manufacturers recommend keeping the backpressure on balanced valves below 45 to 50 percent of the set pressure. Consult API-526 and valve manufacturers for the maximum outlet pressure limit for bellows-type SRVs.

For both conventional and balanced SRVs, the inlet pressure loss, including the mounting nozzle entrance loss, rupture disk flow resistance, and inlet pipe friction, is recommended to stay below 3 percent of the differential set pressure or else valve instability may occur, resulting in degraded relief capacity.

Pilot-Operated Relief Valves In a pilot-operated relief valve, the main valve is combined with and controlled by a smaller, self-actuating pressure relief valve. The pilot is a spring-loaded valve that senses the process pressure and opens the main valve by lowering the pressure on the top of an unbalanced piston, diaphragm, or bellows of the main valve. Once the process pressure is lowered to the blowdown pressure, the pilot closes the main valve by permitting the pressure in the top of the main valve to increase. Pilot-operated relief valves are commonly used in clean, low-pressure services and in services where a large relieving area at high set pressures is required. The set pressure of this type of valve can be close to the operating pressure. Pilot-operated valves are often chosen when operating pressures are within 5 percent of set pressures and a close-tolerance valve is needed.

Rupture Disks A rupture disk is a nonreclosing device designed to function by the bursting of a pressure-retaining disk. This assembly consists of a thin, circular membrane usually made of metal, plastic, or graphite that is firmly clamped in a disk holder. When the process reaches the bursting pressure of the disk, the disk ruptures and releases the pressure. Rupture disks can be installed alone or in combination with other types of devices. Once blown, rupture disks do not reseat; thus, the entire contents of the upstream process equipment will be vented. Rupture disks of nonfragmented type are commonly used in series (upstream) with a safety relief valve to prevent corrosive fluids from contacting the metal parts of the valve. In addition, this combination is a reclosing system.

The burst tolerances of rupture disks are typically about ± 5 percent for set pressures above 2.76 barg (40 psig). Consult API-520-I (section 2.3.6.1) on the proper selection and burst setting of the rupture disks.

Pressure-Vacuum Relief Valves For applications involving atmospheric and low-pressure storage tanks, pressure-vacuum relief valves (PVRVs) are used to provide pressure relief. Such devices are not included in the scope of ASME section VIII and thus are not code certified; hence they are used mostly in noncoded vessels (<1.03 barg or 15 psig design). These units combine both a pressure and a vacuum relief valve into a single assembly that mounts on a nozzle on top of the tank, and they are usually sized to handle the normal in-breathing and out-breathing requirements.

Sizing of Pressure Relief Systems A critical point in design is determining whether or not the relief system must be sized for single-phase or two-phase relief flow. Two-phase flow often occurs during a runaway, but it can also occur in nonreactive systems such as vessels with gas spargers, vessels experiencing high heat input rates, or systems containing known foaming agents such as latex. The so-called drift flux methodology [Zuber and Findlay, "Average Volumetric Concentration in Two-Phase Flow Systems," *Trans. ASME, J. Heat Transf.* **87:** 453–468 (1965); Wallis, *One-Dimensional Two-Phase Flow*, McGraw-Hill, New York, 1969] has been extended and applied to both a volumetric heating case (uniform vapor generation throughout the liquid) and a wall heating case (vaporization occurring only at vessel wall) in the DIERS (Design Institute for Emergency Relief Systems) study (DIERS 1992). The DIERS methodology is important as a means of addressing situations, such as two-phase flow, not covered adequately by current ASME and API methods. The recent CCPS guidelines (CCPS 1998) are the best source of updated information on these methods. For a top-vented vessel, the important mechanism for the liquid carryover resulting in two-phase relief is "boil-over." The vessel hydrodynamic model based on drift flux formulation is used to estimate the quality (i.e., vapor mass fraction) entering the vent system. The churn-turbulent regime vessel model is generally reserved for nonfoaming and nonviscous liquids. This regime would yield the highest degree of vapor–liquid disengagement. The bubbly regime vessel model is generally applied to foamy liquids and viscous systems. The bubbly regime would yield only limited disengagement in the vessel.

Required Relief Rate The required relief rate is the vent rate W (kg) required to remove the volume being generated within the protected equipment when the equipment is at its highest allowed pressure:

$$W_{\text{req}} = \frac{\text{Net Volume Generation Rate}}{\text{Specific Volume of Vent Stream}} \tag{23-82}$$

For steady-state design scenarios, the required vent rate, once determined, provides the capacity information that is required to properly size the relief device and associated piping. For transient situations (e.g., two-phase venting of a runaway reactor), the required vent rate would require the simultaneous solution of the applicable material and energy balances on the equipment together with the in-vessel hydrodynamic model. Special cases yielding simplified solutions are given next. For clarity, nonreactive systems and reactive systems are presented separately.

Nonreactive Systems

Constant Flow into Protected Equipment (Blocked Outlet) For the steady-state design scenario with a constant flow of fluid W_{in} (kg/s) from a pressure source that is above the maximum allowed pressure in the protected equipment, volume is being generated within the equipment at a rate of $W_{\text{in}}/\rho_{\text{in}}$ where ρ_{in} (kg/m^3) is the incoming fluid density evaluated at the maximum allowed pressure. Denoting ρ_{out} (kg/m^3) as the vent stream fluid density, Eq. (23-82) then yields the required vent rate:

$$W_{\text{req}} = (W_{\text{in}}/\rho_{\text{in}})\rho_{\text{out}} \tag{23-83}$$

Constant Heat Input into Protected Equipment—Thermal Relief If the addition of heat to the equipment does not cause the fluid to boil, then the volume generation rate is the thermal expansion rate of the fluid:

$$W_{\text{req}} = q\beta/C_p \tag{23-84}$$

where q is the heat input rate (J/s), β is the coefficient of volumetric expansion at constant pressure (°C^{-1}), and C_p is the constant-pressure specific heat (J/kg°C).

The properties are evaluated at maximum allowed pressure conditions. For liquids, β can typically be evaluated from the specific volume change over a 5°C temperature increment. For ideal gases, Eq. (23-84) becomes

$$W_{\text{req}} = q/(C_pT) \tag{23-85}$$

where T is the absolute temperature (K).

Boiling-Liquid Relief (Nonuniform Flow Regime) If the fluid is at its boiling point, then volume is generated through the phase change that occurs upon vaporization. For *nonfoamy fluids*, vents sized for all-vapor relief are adequate, even if some initial two-phase venting is predicted (Fauske 2006). The required vent rate based on the volumetric rate balance criterion of Eq. (23-82) is given for a single-component fluid as

$$W_{\text{req}} = \frac{q}{h_{fg}}\left(\frac{v_{fg}}{v_g}\right) \tag{23-86}$$

where h_{fg} is the latent heat of vaporization (J/kg), v_g is the vapor specific volume (m^3/kg), v_{fg} is the specific volume increase upon vaporization = $v_g - v_f$ (m^3/kg), and v_f is the liquid specific volume (m^3/kg), and the required minimum vent area $A_{\min}$ (m^2) is given as

$$A_{\min} = W_{\text{req}}/G \tag{23-87}$$

where G ($kg \cdot m^{-2} \cdot s^{-1}$) is the discharge mass flux. As an example for highly subcritical conditions applicable to low-pressure API tanks, the discharge mass flux is given by

$$G = (2\Delta P \rho_v)^{1/2} \tag{23-88}$$

where ΔP (Pa) is the available pressure drop and ρ_v (kg/m^{-3}) is the vapor density.

In the case of a foamy system, ρ_v in Eq. (23-88) is replaced by

$$\rho = \rho_l(1-\alpha) + \rho_v\alpha \tag{23-89}$$

where ρ (kg/m^{-3}) is the two-phase density, ρ_l (kg/m^{-3}) is the liquid density, and α is the foam void fraction. With a typical value of $\alpha \approx 0.99$, for systems of interest the required vent area $A_{\min}$ for foamy conditions increases by a factor of 2.

Reactive Systems Following the AIChE DIERS methodology, three types of reactive systems are usually distinguished for venting character, including vapor, gassy, and hybrid. The venting character and corresponding relief area requirement are easily determined using the DIERS calorimetry methodology, including the Vent Sizing Package 2 (VSP2™) and Advanced Reactive System Screening Tool (ARSST™) commercialized by Fauske & Associates, LLC.

For the *vapor system* (i.e., total pressure is equal to the vapor pressure), the principal parameter determining the vent size requirement is the rate of temperature rise $\dot{T}$ (°C/$\min^{-1}$), measured at the relief set pressure P (psia). Since, for the vapor system, the reaction is entirely tempered by the latent heat of vaporization, the lowest practical relief set pressure (which is well below the maximum allowable pressure) results in the smallest relief area requirement.

For the *gassy system*, in the absence of any tempering (i.e., the total pressure is equal to the noncondensible gas pressure), the principal parameter determining the vent size requirement is the measured maximum rate of pressure rise $\dot{P}_{\max}$ (psi/$\min^{-1}$). In this case, the smallest vent size requirement is obtained by considering the maximum allowable pressure P.

For the *hybrid system*, with both gas production and vaporization occurring simultaneously (i.e., the total pressure is equal to the sum of the gas partial pressure and the vapor pressure), both the rate of temperature rise $\dot{T}$ (°C/$\min^{-1}$) and the rate of pressure rise $\dot{P}$ (psi/$\min^{-1}$) are needed to determine the proper vent size for a specified venting pressure.

Vapor Systems Vent Sizing In order to eliminate oversizing and the potential for valve instability (chatter), the smallest vent size obtained by considering all vapor venting evaluated at the selected relief set pressure well below MAWP is recommended. In this case, if two-phase flow should occur (may be likely), only a modest overpressure will occur before a turnaround in pressure (Fauske 2006).

The required relief rate for all vapor venting is given by

$$W_{\text{req}} = \frac{V(1-\alpha_o)\rho_l c\dot{T}}{\lambda} \tag{23-90}$$

where V (m^3) is the vessel volume, $(1-\alpha_o)$ is the initial fill fraction, ρ_l (kg/m^{-3}) is the liquid density, c ($J \cdot kg^{-1} \cdot K^{-1}$) is the liquid specific heat, $\dot{T}$ (K/s^{-1}) is the rate of temperature rise corresponding to the relief set pressure ($\dot{T}$ obtained from the VSP2 calorimeter is directly scalable), and λ (J/kg^{-1}) is the latent heat of evaporation. In case of vapor critical flow, the minimum vent area $A_{\min}$ is given by Eq. (23-87), resulting in

$$A_{\min}/V = \frac{(1-\alpha_o)\rho_l c\dot{T}}{0.67C_D\lambda P}\left[\frac{RT}{M_{w,v}}\right]^{1/2} \tag{23-91}$$

where R (8314 Pa-m^3/K-kg mole), T (K) is the vapor temperature corresponding to the relief set pressure P (Pa), $M_{w,v}$ (kg/kg mole) is the vapor molecular weight, and C_D is the discharge coefficient.

In case necessary physical properties are lacking under the conditions of the emergency scenario (more often the case than not), the following simple design method requiring no physical properties can be used that is consistent with all relevant experimental data (Fauske 2006)

$$A/V = \frac{C}{P\left(1+\dfrac{1.98\times10^3}{P^{1.75}}\right)^{0.286}}\left(\frac{\dot{T}}{C_D}\right) \tag{23-92}$$

where A (m^2) is the vent area, V (m^3) is the volume of reactant, P (psig) is the gauge relief set pressure, $C = 3.5 \times 10^{-3}$ for churn turbulent flow, and $C = 7.0 \times 10^{-3}$ for bubbly or foamy system. Equation (23-92) is a combined expression of the subcritical and critical vapor flow expressions of A/V using corresponding water physical properties (Fauske 2000).

As an example, considering the DIERS large-scale 2.2 m^3 styrene-ethylbenzene runaway test, A/V (m^{-1}) = 2.08×10^{-3}, P = 64.8 psig, and $\dot{T}$ (°C/$\min^{-1}$) = 21.6, compares to

$$A/V = \frac{3.5\times10^{-3}}{64.8\left(1+\dfrac{1.98\times10^3}{64.8^{1.75}}\right)^{0.286}}\left(\frac{21.6}{0.45}\right) = 2.03\times10^{-3}\ \text{m}^{-1}$$

Two-phase flow occurred upon relief opening, resulting in a modest overpressure based on absolute pressure of 24 percent.

Gassy Systems Vent Sizing The required relief rate for all gas venting at peak reactive condition is given by

$$W_{\text{req}} = \frac{V(1-\alpha)\rho_l}{m_t}v\frac{\dot{P}}{P}\rho_g \tag{23-93}$$

and considering gas critical flow, the vent area A (m^2) is given by

$$A/V = \frac{1}{0/61C_D}\frac{(1-\alpha)\rho_l v}{m_t}\frac{\dot{P}}{P}\left(\frac{M_{w,g}}{RT}\right)^{1/2} \tag{23-94}$$

where m_t (kg) is the sample mass in the test apparatus, v (m^3) is the volume occupied by gas in the test apparatus, $\dot{P}$ (Pa/s) is the maximum pressure rise rate measured in the test apparatus, $M_{w,g}$ (kg/kgmole) is the gas molecular weight, and α is the average void fraction. This illustrates the benefit from two-phase flow during the period before turnaround in peak venting pressure P corresponding to complete disengagement evaluated from

$$\frac{V(1-\alpha)\rho_l}{m_t}\frac{v\dot{P}}{P} = \frac{2\alpha}{1-1.5\alpha}u_\infty A_{\text{vessel}} \tag{23-95}$$

where u_∞ is the bubble rise velocity (~0.2 m s^{-1}) and A_{vessel} is the cross-sectional area of vessel (m^2).

Based on the preceding equations (Fauske 2000) and considering the ARSST parameters of $m_t = 0.01$ kg, $v = 3.5 \times 10^{-4}$ m^3, and setting $M_{w,g} = 44$ kg/kgmole and combining critical and subcritical expressions of A/V, leads to the following simple design method for gassy systems:

$$A/V = \frac{3.5\times10^{-3}}{P\left[1+\dfrac{1.98\times10^3}{P^{1.75}}\right]^{0.286}}\dot{P}(1-\alpha) \tag{23-96}$$

which is consistent with relevant available experimental data.

A 58-gallon test vessel with an effective vent diameter of 10.9 inches filled with neat dicumyl peroxide ($A/V = 0.27$) is vented safely with a peak pressure of about 5 psig. Applying Eq. (23-95) with $V = 0.22$ m^3, $A_{\text{vessel}} = 0.26$ m^2,

$\rho_l = 1000$ kg m^{-3}, and $u_\infty = 0.2$ m/s^{-1}, along with RSST information of $m_t = 0.01$ kg, $v = 3.5 \times 10^{-4}$ m^3, $\dot{P} = 71.7$ psi/s^{-1}, and a peak relief pressure $P = 20$ psia, results in $\bar{\alpha} \approx 0.66$, and from Eq. (23-96),

$$A/V = \frac{3.5\times 10^{-3}}{5\left[1+\dfrac{1.98\times 10^{3}}{5^{1.75}}\right]^{0.286}} \times 4300(1-0.66) = 0.26 \text{ m}^{-1}$$

which is consistent with the experiment value of 0.27 m^{-1}.

A full-scale 460-gallon test vessel with a vent diameter of 22.5 inches filled with neat dicumyl peroxide ($A/V = 0.14$) is vented safely with a peak relief pressure of 20 psig. Again, applying Eq. (23-95) for these conditions results in $\bar{\alpha} \approx 0.66$, and applying Eq. (23-96),

$$A/V = \frac{3.5\times 10^{-3}}{20\left[1+\dfrac{1.98\times 10^{3}}{20^{1.75}}\right]^{0.286}} \times 4300(1-0.66) = 0.13 \text{ m}^{-1}$$

which is consistent with the experiment value of 0.14 m^{-1}.

Hybrid Systems Vent Sizing The required relief rate for all vapor-gas venting at tempering is given by

$$W_{\text{req}} = \frac{V(1-\alpha_o)\rho_l c\dot{T}}{\lambda} + \frac{V(1-\alpha_o)\rho_l v}{m_t}\left(\frac{\dot{P}}{P}\right)\rho_g \tag{23-97}$$

and considering vapor-gas critical flow, the vent area A (m^2) is given by

$$A/V = \frac{(1-\alpha_o)\rho_l c\dot{T}}{0.61 C_D \lambda P}\left[\frac{RT}{M_{w,v}}\right]^{1/2} + \frac{(1-\alpha_o)\rho_l v}{0.61 C_D m_t}\left(\frac{\dot{P}}{P}\right)\left[\frac{M_{w,g}}{RT}\right]^{1/2} \tag{23-98}$$

and the following derivations for vapor and gassy systems can then be simplified to the Fauske generalized vent sizing equation

$$A/V = \frac{3.5\times 10^{-3}}{P\left[1+\dfrac{1.98\times 10^{3}}{P^{1.75}}\right]^{0.286}}(\dot{T}+\dot{P}) \tag{23-99}$$

which is applicable to vapor systems ($\dot{P} = 0$), gassy systems ($\dot{T} = 0$), and hybrid systems ($\dot{T} + \dot{P}$) as illustrated next (Fauske 2006).

An experiment with a contaminated 200-kg, 50 percent H_2O_2 solution is used here to compare the result with Eq. (23-99). The runaway reaction tempered and vented safely with a value of $A/V = 2.6 \times 10^{-2}$ m^{-1} and a resulting relief pressure of about 1 psig. The self-heat rate at tempering equaled 55°C/min^{-1}, and the corresponding rate of pressure rise was 14 psi/min^{-1} (from a 10 g sample in the ARSST), and it results in

$$A/V = \frac{3.5\times 10^{-3}}{1\left[1+\dfrac{1.98\times 10^{3}}{1^{1.75}}\right]^{0.286}}(55+14) = 2.75\times 10^{-2} \text{ m}^{-1}$$

which compares to the experimental value of 2.6×10^{-2} m^{-1}.

Pressure Relief Valve (PRV)

Two-Phase Flow Evaluation for Vapor Systems While the required vent area is determined considering all vapor venting [Eq. (23-91)] in a top-reacted vessel, two-phase flow will occur at the opening of the valve when

$$\bar{\alpha} > \alpha_o \tag{23-100}$$

where $\bar{\alpha}$ is the average void fraction due to liquid swell and α_o is the initial vessel void fraction. Here the controlling parameters are the superficial velocity j_g (m/s^{-1}) set equal to the vessel superficial vapor velocity corresponding to all vapor flow through the valve

$$j_g = \frac{W_{\text{req}}[\text{Eq. (23-90)}]}{\rho_v A_{\text{vessel}}} \tag{23-101}$$

and the appropriate flow regime, such as churn turbulent, determines the minimum value of $\bar{\alpha}$

$$j_g/u_\infty = \frac{2\bar{\alpha}}{1-1.5\bar{\alpha}} \tag{23-102}$$

If criterion 19 [Eq. (23-100)] is satisfied, resulting in two-phase flow, which may be the case for mechanically stirred systems or systems experiencing depressurization, the valve vapor quality, x_o, and the resulting two-phase mass flow rate, G (kg · m^{-2} · s^{-1}), can be estimated from the "coupling" equation (also referred to as the DIERS coupling equation) proposed by H. K. Fauske et al., "Emergency Relief System—Sizing and Scaleup," *Plant/Operations Prog.*, January 1983:

$$x_o G A_{\text{vent}} = j_g \rho_v A_{\text{vessel}} \tag{23-103}$$

and the valve two-phase flashing flow methodology proposed by Fauske (1999)

$$G = \left[\frac{1-x_o}{G_{ERM}^2} + \frac{x_o}{C_{D,v}^2 G_v^2}\right]^{1/2} \tag{23-104}$$

where

$$G_{ERM} = \frac{\lambda}{v_{fg}}(TC)^{1/2} \tag{23-105}$$

and

$$G_V = P\left[\frac{M_{w,v}}{TR}\right]^{1/2}\left[k\left(\frac{1}{k+1}\right)^{(k+1)/(k-1)}\right]^{1/2} \tag{23-106}$$

and $C_{D,v}$ is the valve vendor's recommended discharge coefficient for all vapor flow. Detailed comparisons with experimental data are provided by Fauske (1999 and 2003).

This subsection on Pressure Relief Systems will conclude with some recommendations for how to prevent PRV instability (chatter).

Stable PRV Operation Typical causes of chatter (instability and potential valve damage) include:

- Excessive inlet pressure loss (3 percent rule).
- Excessive backpressure (10 percent rule).
- Oversized valve.
- Natural oversizing. There are a finite number of standard valve-nozzle sizes to choose from, and the calculated area (A) cannot be expected to correspond exactly to one of these sizes.

In practice, a 10 percent safety factor is automatically applied to the calculated area, and then the standard-size nozzle area that is the closest to the resulting value on the high side is selected. This may lead to potential oversizing of more than 50 percent.

Assuring Stable PRV Operation

Fire–Nonreactive Systems

- Base PRV nozzle size on MAWP. Applying current practice, this is the relief set pressure. It provides the smallest PRV size.
- To avoid unstable PSV operation (chatter), use the preceding base PRV size, but set the actual relief set pressure equal to the practical value below MAWP.

Following this procedure, the PRV is initially undersized, resulting in the pressure continuing to rise upon PRV opening, but the resulting overpressure will never reach MAWP. This approach has the following advantages:

- Assure smallest PRV size.
- Eliminate concerns about satisfying the 3 percent and 10 percent rules, or if these rules eliminate unstable operation or damaging chatter, eliminating the other effects such as acoustic effects and pipe vibration and oversized PRVs.
- Given a finite fire duration, it keeps the pressure below MAWP due to early fire heat removal by latent heat of evaporation. This may rule out the onset of exothermicity, in case it should run away at elevated temperature.

Reactive Vapor Systems

- Select PRV with lowest practical relief set pressure (<< MAWP) and base the sizing upon all vapor venting.

Following this procedure, the pressure will continue to rise following PRV opening due to the occurrence of two-phase flow. The allowance of significant overpressure will have the same benefits as those discussed for the fire–nonreactive systems, including the smallest PRV size. In summary, undersize the valve at a relief set pressure sufficiently below MAWP to assure PRV stability.

Reactive Gassy Systems

- Use rupture discs set at the lowest practical pressure (<< MAWP). This will provide early reactant loss and the smallest rupture disc size evaluated at MAWP. Also note that the use of PRV has the potential to lead to chattering and damage to the PRV independent of the PRV set pressure.

Based on Experience

- Plant people are reluctant to give up the practice of setting PRV set pressures as the design pressure—especially for the fire case.
- The only thing that moves such people is to see guidance in standards like API 520 and 521.
- Our view of setting relief activation below the design pressure is always good and does not violate any standards.

EMERGENCY RELIEF DEVICE EFFLUENT COLLECTION AND HANDLING

General References: AIChE/CCPS, *Guidelines for Pressure Relief and Effluent Handling Systems,* American Institute of Chemical Engineers, New York, 1998.

Reasons for Containment The emergency relief effluent (discharge) from a relief system of a reactor or a process vessel will often need treatment before it can be vented directly to atmosphere. Typical relief system discharges contain flammable or toxic materials and can create fire and explosion hazards as well as toxicity hazards, both on-site and off-site. There may also be an environmental impact, which can result in additional cleanup and remediation costs.

In particular, a two-phase discharge can lead to more hazard than an all-vapor discharge. A two-phase discharge can lead to the formation of aerosols and heavier-than-air vapor clouds. More mass is airborne from a two-phase discharge than from an all-vapor discharge. As a result, the potential hazard footprints are larger. Rainout of toxic and flammable material (liquid droplets) can occur near the release point on-site or downwind from the release point off-site.

Effluent Handling Strategies Emergency relief effluent can consist of (a) all-vapor flow, (b) vapor-liquid flow, and (c) vapor-liquid-solid flow. One of four effluent handling strategies can be used, depending on the relief effluent type, phase, and hazard nature:

Direct Discharge to Atmosphere This is typically adequate for an all-vapor discharge via a tall vent stack. The required vent stack height and diameter for proper dispersion are determined based on a specific design flow rate in order to prevent potential toxicity, flammability, and explosion impacts. The presence of nearby buildings and other structures where people are working should be taken into consideration in any dispersion analysis that is performed to determine the required vent stack height.

Flaring and Incineration This is also recommended for all-vapor flow because the presence of liquid droplets or mists will decrease the incineration efficiency. The purpose of incineration/flaring is to turn chemicals into less hazardous combustion products. For example, the efficient combustion of CH compounds will lead to the formation of carbon monoxide, carbon dioxide, and water. Both elevated flare stacks and ground flares can be used or considered.

Partial Containment and Separation The purpose of this effluent handling strategy is to first separate and contain the liquid and then to pass on the vapor for further treatment and handling. Typical separation equipment includes:

Horizontal separators (gravity)
Vertical separators (gravity)
Cyclone separators
Quench tanks (open)
Vent tanks (open)
Impingement separators
Emergency scrubbers (absorbers)

Total Containment Under this type of effluent handling scheme, a containment vessel with a large enough volume is used to collect the relief system discharge. The pressure in the discharging vessel is decreased because of the large volume available in the containment vessel. This additional containment volume can be added either in the same vessel or by using an external containment vessel, that is, a catch or dump tank. The types of containment vessels that can be used include:

All-vapor vent tanks (closed). This is mostly applicable to relief system effluents where large volumes of noncondensible gas or vapor are produced, so-called gassy systems. The presence of a large enough vent containment volume can reduce the pressure in the discharging vessel so that venting to atmosphere is avoided. If it is impractical to provide a large enough vent containment tank, a vent containment tank with a smaller volume can be used so that a smaller rate of vapor can be vented to atmosphere in a safe manner.

Quench tanks (closed). This is desirable for relief system effluents containing primarily condensible vapors, so-called vapor systems. For vapor systems, the pressure is driven by the liquid temperature, and by cooling the liquid using a quench fluid, the pressure can be decreased. The quench fluid will either cool the effluent temperature below the onset temperature of a hazardous runaway reaction or it will slow the reaction enough to prevent two-phase flow. The incoming flow is mixed with the quench tank liquid using a sparger. A good sparger design is essential for proper mixing and optimal quench tank performance.

Combinations of effluent handling strategies are often considered, as shown in Fig. 23-42. Most of the separation equipment will work well for low-viscosity and nonfoamy materials, while quench tanks and cyclone separators can work well for high-viscosity and foamy materials. Figure 23-43 illustrates a selection strategy for effluent handling equipment that is recommended by the Center for Chemical Process Safety (CCPS) *Guidelines for Pressure Relief and Effluent Handling Systems.* It can be seen from this figure that CCPS suggests directing emergency releases to some form of containment or mitigation on a widespread basis.

Emergency relief vents on vessels handling highly hazardous chemicals or subject to potential runaway reactions should be directed to a catch tank designed to perform the required vapor/liquid separation; where conditions warrant, the provisions of CCPS recommendations regarding quench tanks and scrubbing can be employed as shown in Fig. 23-43.

Additional Considerations The selection and design of effluent handling and collection systems should meet recognized and generally accepted good engineering practice (RAGAGEP). Effluent handling systems where runaway reactions are present must address the potential for continuing chemical reactions within the effluent handling system as well as in the reactor or vessel that is the source of the effluent. Effluent handling systems should be designed to effect separation and containment for an initial relief that is typically liquid rich at the start of the relief transient and that can become vapor rich during or toward the end of the relief transient. The pressure design rating of containment vessels should be able to withstand a deflagration from any flammable vapors that accumulate in the system. Where quench tanks are used, extreme ambient conditions should be considered; for example, a quench fluid or solvent should not freeze. Dispersion, thermal radiation, and vapor cloud explosion potential and assessments should be considered where partial containment and separation systems are used and where flammable or toxic materials are emitted from stacks or flares. Structural and mechanical considerations include the adequacy of structural support for dynamic and static loads, and the proper material selection for the chemical service and for the temperature extremes caused by fires, runaway reactions, or cold temperatures resulting from depressurizations and condensation of light ends. More detailed information about these additional factors can be found in the AIChE/CCPS *Guidelines for Pressure Relief and Effluent Handling Systems,* American Institute of Chemical Engineers, New York, 1998.

FLAME ARRESTERS

General References: Brandes, E., and T. Redeker, Maximum Experimental Safe Gap of Binary and Ternary Mixtures, Fourth International Symposium on Hazards, Prevention and Mitigation of Industrial Explosions (ISHPMIE), Bourges, France, October 2002, pp. 207–213; Britton, L. G., "Using Maximum Experimental Safe Gap to Select Flame Arresters," *Process Safety Progress* **19**(3) (2000); Britton, L. G., "Using Heats of Oxidation to Evaluate Flammability Hazards," *Process Safety Progress* **21**(1) (2002); Center for Chemical Process Safety, "Deflagration and Detonation Flame Arresters," chap 13 in *Guidelines for Engineering Design for Process Safety*, American Institute of Chemical Engineers, New York, 1993; Center for Chemical Process Safety, "Effluent Disposal Systems," chap 15 in *Guidelines for Engineering Design for Process Safety*, American Institute of Chemical Engineers, New York, 1993; Davies, M., and T. Heidermann, Investigation of Common Application Failures Proven by Life Field Testing of Endurance Burning Tested End-of-Line Flame Arresters, 37th Loss Prevention Symposium, American Institute of Chemical Engineers, 2003; Davies, M., and T. Heidermann, "Protect Your Process with the Proper Flame Arresters," *CEP*, December 2013, pp. 16–22; Förster, H., and C. Kersten, Investigation of Deflagrations and Detonations in Pipes and Flame Arresters by High Speed Framing, Fourth International Symposium on Hazards, Prevention and Mitigation of Industrial Explosions (ISHPMIE), Bourges, France, October 21–25, 2002; Grossel, S. S., *Deflagration and Detonation Flame Arresters*, AIChE-CCPS Concept Book, New York, 2002; Howard, W. B., "Flame Arresters and Flashback Preventers," *Plant/Operations Progress* **1**(4) (1982); Howard, W. B., "Precautions in Selection, Installation and Use of Flame Arresters," *Chem. Eng. Prog.*, April 1992; International Standard ISO 16852, *Flame Arrester Performance Requirements, Test Methods and Limits of Use*, International Organization for Standardization, Geneva, 2008; Lapp and Vickers, International Data Exchange Symposium on Flame Arresters and Arrestment Technology, Banff, Alberta, October 1992; Lunn, G. A., "The Maximum Experimental Safe Gap: The Effect of Oxygen Enrichment and the Influence of Reaction Kinetics," *Journal of Hazardous Materials* **8:** 261–270 (1984); NFPA 69, *Standard on Explosion Prevention Systems*, 2014; Piotrowski, T., "Specification of Flame Arresting Devices for Manifolded Low Pressure Storage Tanks," *Plant/Operations Progress* **10**(2) (1991); Roussakis and Lapp, A Comprehensive Test Method for In-Line Flame

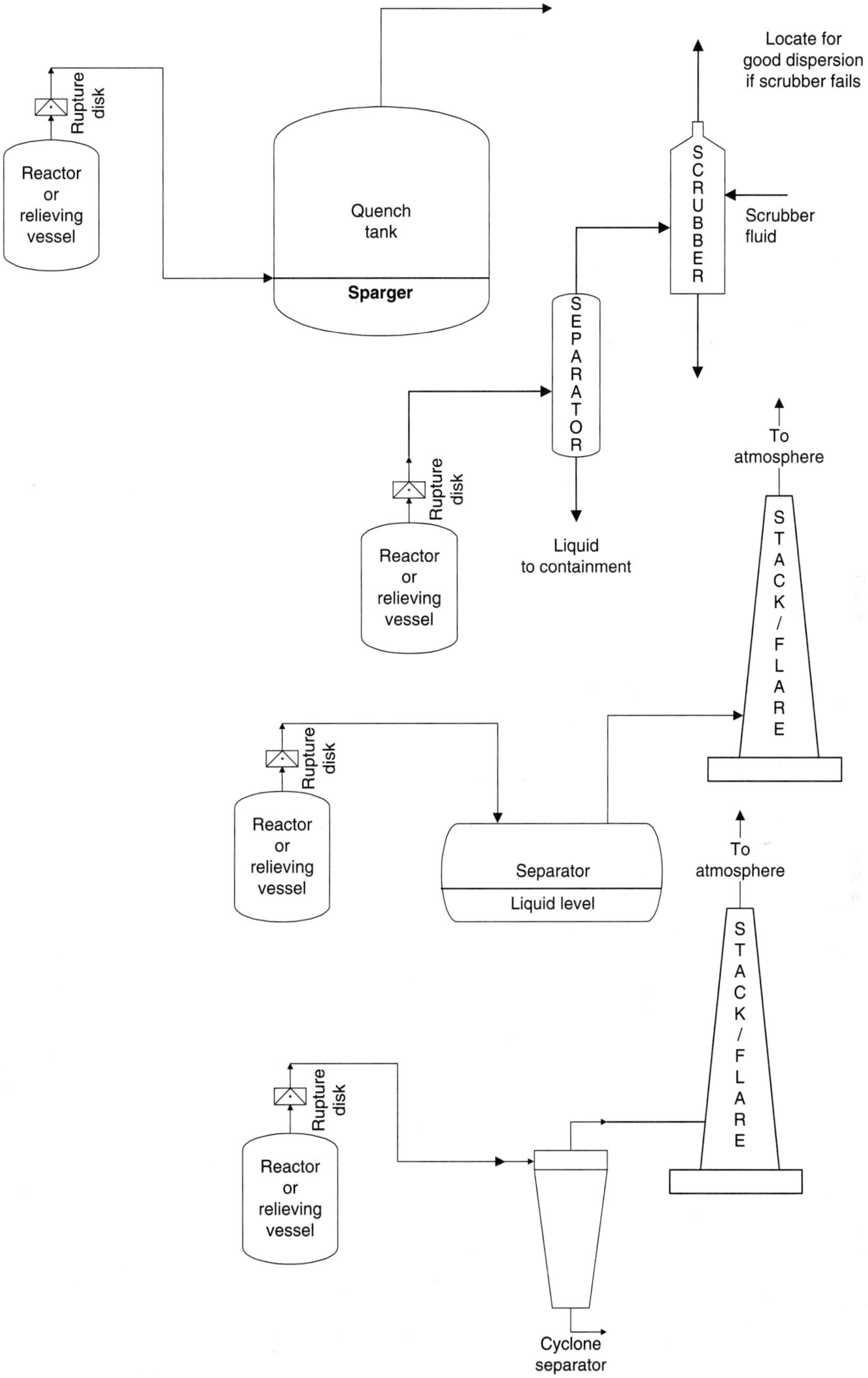

FIG. 23-42 Typical effluent handling system configurations.

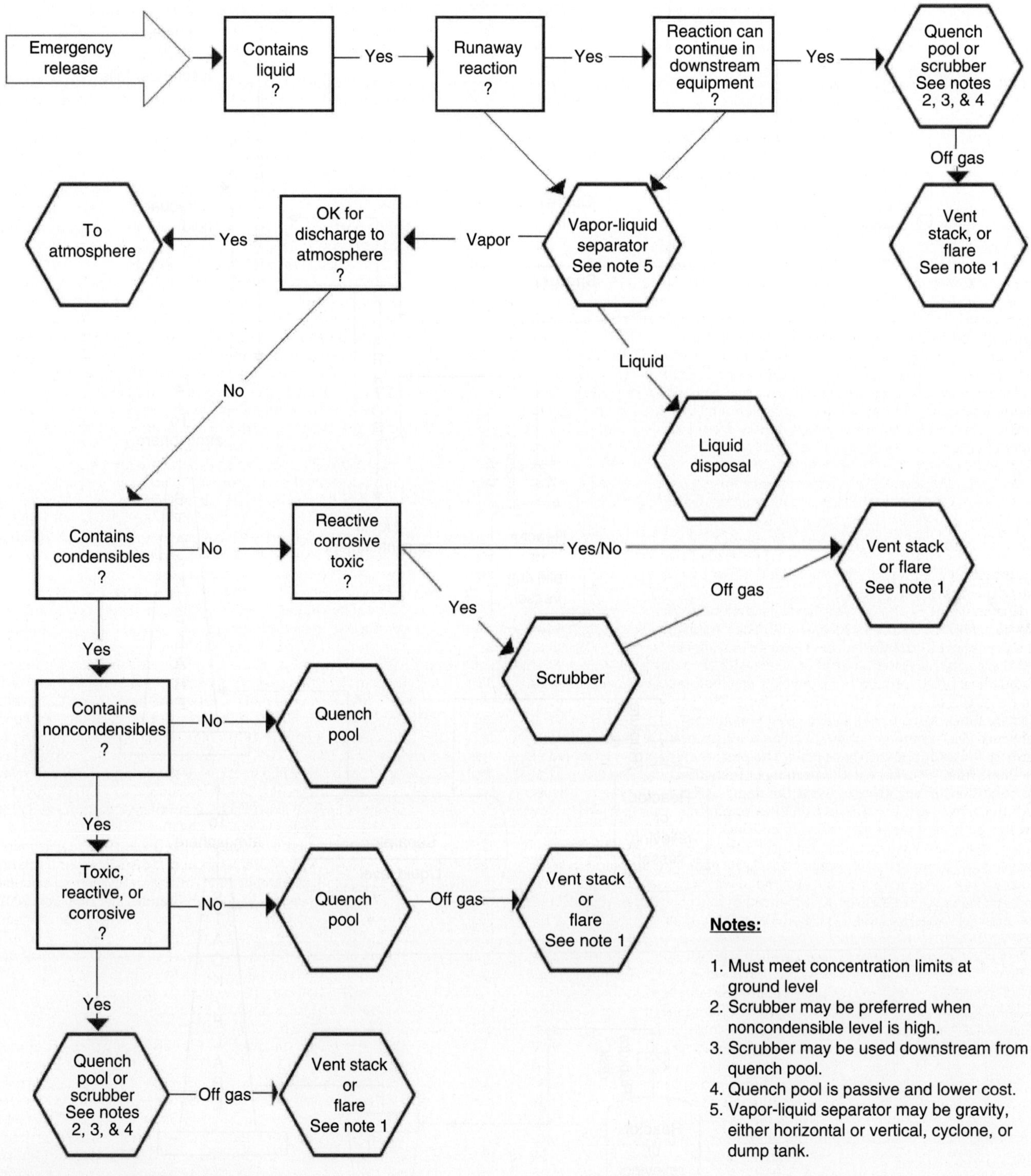

FIG. 23-43 Recommended strategy for the selection of effluent handling systems. (AIChE/CCPS, *Guidelines for Pressure Relief and Effluent Handling Systems,* 1st ed., American Institute of Chemical Engineers, New York, 1998.)

Arresters, *Plant/Operations Progress* **10**(2) (1991); Thibault, P., L. Britton, and F. Zhang, "Deflagration and Detonation of Ethylene Oxide Vapor in Pipelines," *Process Safety Progress* **19**(3) (2000).

General Considerations In 2002, the Center for Chemical Process Safety (CCPS) of the American Institute of Chemical Engineers (AIChE) published a *Concept Book* on this topic (Grossel 2002 in General References). The book was intended to expand and update the coverage given in Center for Chemical Process Safety, "Deflagration and Detonation Flame Arresters," (1993); see General References.

According to ISO 16852 (2008), a flame arrester is a device fitted to the opening of an enclosure or to the connecting pipe work of a system of enclosures and whose intended function is to allow flow but to prevent the transmission of flame.

Some preliminary considerations for arrester selection and placement are:

1. Identify the at-risk equipment and the potential ignition sources in the piping system to determine where arresters should be placed and what general type (deflagration or detonation, unidirectional or bidirectional) is needed.

2. Determine the worst-case gas mixture combustion characteristics, system pressure, system temperature, and permissible pressure drop across the arrester, to help select the most appropriate element design. Not only does element design affect pressure drop, but also the rate of blockage due to particle impact, liquid condensation, and chemical reaction (e.g., polymerization) can make some designs impractical, even if in-service and out-of-service arresters are provided in parallel.

3. The possibility of a stationary flame residing on the arrester element surface should be evaluated, and the need for additional safeguards, should such an event occur, should be evaluated (see the subsection Endurance Burn)

4. Consider the limitations any construction materials when exposed to reactive or corrosive stream components.

5. The upset conditions shall not exceed the test conditions at which the arrester was certified. Under certain upset conditions, such as a high-pressure excursion or elevated oxygen concentration (Lunn 1984), there may be no flame arrester available for the task.

6. Consider the type and location of the arrester with respect to ease of maintenance, particularly for large in-line arresters.

To maximize effectiveness, attention should be given to proper selection, application, and maintenance of the device. Since arresters may fail on demand, it is good engineering practice to conduct layers of protection analysis to determine what additional mitigation may be required (Förster, H., "Flame Arresters—The New Standard and Its Consequences," *Proceedings of the International ESMG Symposium, Part 2: Industrial Explosion Protection,* Nürnberg, Germany, March 27–29, 2001). For marine vapor control systems in the United States, flame arrester applications are regulated by the U.S. Coast Guard. For other applications, alternative test protocols and procedures have been developed by different agencies.

Combustion: Deflagrations and Detonations A deflagration is a combustion wave propagating at less than the speed of sound as measured in the unburned gas immediately ahead of the flame front. Flame speed relative to the unburned gas is typically 10 to 100 m/s, although owing to the expansion of hot gas behind the flame, several hundred meters per second may be achieved relative to the pipe wall. The combustion wave propagates by a process of heat transfer and species diffusion across the flame front, and there is no coupling in time or space with the weak shock front generated ahead of it. Deflagrations typically generate maximum pressures in the range of 2 to 12 times the initial pressure. The pressure peak coincides with the flame front, although a marked pressure rise precedes it; thus the unburned gas is compressed as the deflagration proceeds, depending on the flame speed and vent paths available. The precompression of gas ahead of the flame front (also known as cascading or pressure piling) establishes the gas conditions in the arrester when the flame enters it, and hence affects both the arrestment process and the maximum pressure generated in the arrester body. It is most important to consider the influences of pressure, temperature, and oxygen concentration on the safe gap [safe gap at standard conditions = maximum experimental safe gap]. The safe gap is a good indicator of how difficult it is to stop a flame with a flame arrester. In addition, the safe gap of mixtures is influenced by the chemical interaction of the different gases [Brandes and Redeker (2002); Lunn (1984) in General References].

As the deflagration flame travels through piping, its speed increases due to flow-induced turbulence and compressive heating of the unburned gas ahead of the flame front. Turbulence is especially enhanced by flow obstructions such as valves, elbows, and tees. Once the flame speed has attained the order of 100 m/s, a *deflagration-to-detonation transition* (DDT) can occur. The travel distance for this to occur is referred to as the *run-up* distance for detonation. This distance varies with the gas mixture sensitivity and increases with pipe diameter. It is often difficult to estimate the run-up distance at which a DDT may occur in a piping system, and when to specify either a deflagration or a detonation flame arrester. However, some indication of this can be obtained from Förster and Kersten (2002). The ISO standard 16852 (2008) limits the use of in-line deflagration arresters to $L/D = 50$ for group D and C applications and $L/D = 30$ for group B. Tabulated run-up distances are generally for straight pipe runs, and DDT can occur for smaller distances in pipe systems containing flow obstructions. At the instant of transition, a transient state of *overdriven detonation* is achieved, and it persists for a distance of a few pipe diameters. Overdriven detonations propagate at speeds greater than the speed of sound (as measured in the burned gas immediately behind the flame front), and side-on pressure ratios (at the pipe wall) in the range of 50 to 100 have been measured. The peak pressure is variable, depending on the amount of precompression during deflagration.

After the abnormally high velocities and pressures associated with DDT have decayed, a state of stable detonation is attained. A detonation is a combustion-driven shock wave propagating at the speed of sound as measured in the burned gas immediately behind the flame front. Since the speed of sound in this hot gas is much larger than that in the unburned gas or the ambient air, and the flame front speed is augmented by the burned gas velocity, stable detonations propagate at supersonic velocities relative to an external fixed point. A typical velocity for a stoichiometric gas mixture is 1800 m/s. The wave is sustained by chemical energy released by shock compression and ignition of the unreacted gas. The flame front is coupled in space and time with the shock front, with no significant pressure rise ahead of the shock front.

Due to the complexity of deflagration and detonation, flame arrester performance must be demonstrated by realistic testing. Such testing has demonstrated that arresters capable of stopping even overdriven detonations may fail under high-pressure deflagration test conditions. Users should request a detailed test report from the flame arrester manufacturer, and only an approved arrester tested to a recognized test standard by an independent third party (e.g., notified body) should be installed.

Combustion: Gas Characteristics and Sensitivity Combustion thermodynamic calculations allow the determination of peak deflagration and detonation pressures, plus stable detonation velocity. Other relevant gas characteristics are entirely experimental. The sensitivity to detonation depends on the detonable range and fundamental burning velocity, although no specific correlations or measures of sensitivity exist based on fundamental properties. National Electrical Code (NEC) groups are commonly used to rank gases for the purposes of flame arrester selection. By this method, group A gases (acetylene) are considered most sensitive to detonation (and most difficult to arrest), while group D gases (such as saturated hydrocarbons) are considered least sensitive and easiest to arrest. As formerly applied, successful testing of an arrester using one gas in an NEC electrical group was assumed to mean that the arrester would be suitable for all other gases in that group. As currently applied in various codes, a representative test gas from each group (such as hydrogen in group B, ethylene in group C, or propane in group D) is typically used for arrester certification, and its ability to arrest a different gas or gas mixture is deduced by comparing the respective maximum experimental safe gaps.

Experimental Safe Gap The arrester is assumed to be suitable for any single gas or mixture having an MESG [maximum experimental safe gap (MESG) is a standardized measurement of how easily a gas flame will pass through a narrow gap bordered by heat-absorbing metal] greater than or equal to that of the representative test gas. Although the arrester may be directly tested using the gas or gas mixture of interest, this is rarely done owing to cost. The MESG comparison procedure is currently applied to both deflagration and detonation arresters. Specific testing protocols are specified by the codes used. Since MESGs provide an independent ranking for arrester selection, reference to NEC groups is not essential. Nevertheless, arresters continue to be described in terms of the NEC group of the representative test gas used for certification. For example, an arrester might be described as a group C (ethylene MESG ≥ 0.65 mm) type. A detailed discussion of the use of MESGs is given in Britton (2000).

There have been no systematic studies proving that arrester performance can be directly correlated with MESG, especially if the MESG for a mixture is estimated by using the Le Chatelier rule. An alternative way to calculate the MESG and safe gap of mixtures is given in Brandes and Redeker (2002). Furthermore, major revisions to the MESG test method have caused many historic MESG values to increase significantly; the current test method minimizes compression of the unburned gas mixture and may therefore underestimate the likelihood of arrester failure via autoignition.

Corrosion Consideration should be given to possible corrosion since this may weaken the structure, increase the pressure drop, and decrease the effectiveness of the element (Grossel 2002).

Directionality To select an arrester for any service, the potential sources of ignition must be established. If the arrester will encounter a flame arriving only from one direction, a *unidirectional* arrester can be used. If a flame may arrive from either direction, a *bidirectional* arrester is needed.

Endurance Burn Under certain conditions, a successfully arrested flame may stabilize on the unprotected side of an arrester element. Should this condition not be corrected, the arrester may fail. Temperature sensors may be incorporated at the arrester to indicate a stabilized flame condition and either alarm or initiate appropriate action, such as valve closure. It is very important to install an endurance burning flame arrester in the same way as tested to avoid malfunction (Davies and Heidermann 2003).

Installation Flame arresters are distinguished in accordance with the installation (in-line, end-of-line, in equipment). Davies and Heidermann (2003) give additional information.

Maintenance NFPA 69, *Standard on Explosion Prevention Systems* (2014) provides good guidance on the maintenance of flame arresters.

Monitoring The differential pressure across the arrester element can be monitored to determine the possible need for cleaning. The pressure taps must not create a flame path around the arrester. It can be important to provide temperature sensors, such as resistance thermometers, at the arrester to detect flame arrival and endurance burning.

Operating Temperature, Pressure, and Oxygen Concentration Arresters are certified to some specified set of maximum temperature, pressure, and oxygen concentration.

Pressure Drop Flow resistance depends on flame arrester design and on a time-dependent fouling factor due to corrosion, or accumulation of liquids, particles, or polymers, depending on the system involved. A fouling factor (20 percent or greater) might be estimated for sizing the flame arrester. It is important that certified flow curves for the arrester be used rather than calculated curves since the latter can be highly optimistic. Measures to prevent condensation or polymerization (e.g., heat tracing, choice of arrester material) should be applied.

Deflagration Arresters The two types of deflagration arresters normally encountered are end-of-line deflagration arresters (Figs. 23-44 and 23-45) and in-line deflagration arresters (International Standard ISO 16852 2008).

End-of-Line Arresters These vent directly to atmosphere or are equipped with a pressure vacuum relief valve or a short piece of pipe. The allowed length of pipe connected to an in-line deflagration arrester must be established by proper testing with the appropriate gas mixture and the pipe diameter involved. Turbulence-promoting irregularities in the flow (bends, tees, elbows, valves, etc.) cannot be used unless testing has addressed the exact geometry. If atmospheric tanks are equipped with flame arresters combined with vents, fouling or blockage by extraneous material can inhibit gas flow to the degree that the tank can be damaged by vacuum or overpressure. Whether or not flame arresters are used, proper inspection and maintenance of these vent systems is required. A number of explosions have occurred due to misapplication of end-of-line or tank vent deflagration arresters where detonation arresters or in-line deflagration arresters should have been used.

Detonation and Other In-Line Arresters If the point of ignition is remote from the arrester location, the arrester is an in-line type, such as might be situated in a vapor collection system connecting several tanks (Fig. 23-46). ISO 16852 and NFPA 69 provide guidance on the different types of in-line flame arresters and their limits of use. In some cases, in-line arresters need to stop deflagrations only. However, in such cases it must be demonstrated that detonations cannot occur in the pipework system.

Detonation arresters are typically used in conjunction with other measures to decrease the risk of flame propagation. For example, in vapor control systems the vapor is often enriched, diluted, or inerted, with appropriate instrumentation and control (Center for Chemical Process Safety, "Effluent Disposal Systems," 1993). In cases where ignition sources are present or predictable (such as most vapor destruction systems), the detonation arrester is used as a last-resort method anticipating possible failure of vapor composition control. Where vent collection systems have several vapor/oxidant sources, stream compositions can be highly variable, and this can be additionally complicated when upset conditions are considered. It is often cost-effective to perform hazard analyses, such as HAZOP or fault tree analysis, to determine whether such vent streams can enter the flammable region and, if so, what composition corresponds to the worst credible case. Such an analysis is also suitable to assess alternatives to arresters.

Effect of Pipe Diameter Changes Arrester performance can be impaired by local changes in pipe diameter. It was shown that a minimum distance of 120 pipe diameters should be allowed between the arrester and any increase in pipe diameter; otherwise, a marked reduction in maximum allowable operating pressure would occur (Lapp and Vickers 1992). As a rule, arresters should be mounted in piping either equal to or smaller than the nominal size of the arrester.

Arrester Testing and Standards Flame arresters shall be tested for their specific application considering the process conditions and the intention of use (e.g., end of line, in-line, detonation, deflagration, atmospheric explosion, short-time burning).

ISO 16852 (2008) is the most up to date and comprehensive flame arrester standard for performance requirements, test methods, and limits for use. A wide range of tests are detailed, depending on the intended use of the flame arrester, and they include end-of-line and in-line deflagration tests as well as detonation tests. Other standards are discussed in the General Reference section (Grossel 2002).

In addition, flow measurement is required to generate flow curves for the flame arresters, and constructional requirements are detailed along with the requirements for pressure and leak testing.

Deflagration Arrester Testing For end-of-line, tank vent, and in-line deflagration flame arresters, ISO 16852 (2008) requires manufacturers to provide users with data for flow capacity at operating pressures, proof of success during an endurance burn or continuous flame test as required, and evidence of flashback test results (for end-of-line arresters) or explosion test results (for in-line or tank vent arrester applications).

Endurance burn testing generally implies that the ignited gas mixture and flow rate are adjusted to give the worst-case heating (based on temperature observations on the protected side of the element surface), that the burn continues for an unlimited duration, and flame penetration does not occur. For short-time burning flame testing, a gas mixture and flow rate are established at specified conditions, and the mixture burns on the flame arrester for a specified duration. The endurance burn test is usually a more severe test than the short-time burn. Davies and Heidermann (2003) give additional information.

Flashback tests incorporate a flame arrester on top of a tank with a large plastic bag surrounding the flame arrester. A specific gas mixture (e.g., propane, ethylene, or hydrogen at the most sensitive composition in air) flows through and fills the tank and the bag. Deflagration flames initiated in the bag must not pass through the flame arrester into the protected vessel.

ISO 16852 (2008) requires in-line deflagration flame arresters to be tested in a closed piping system with a run-up distance from ignition of not more than 50 pipe diameters when tested with propane or ethylene and not more than 30 pipe diameters when tested with hydrogen.

Whatever the application, a user should be aware that not all test procedures are the same, or of the same severity, or use the same rating designations. Therefore, it is important to review the test procedure and determine whether the procedure used is applicable to the intended installation and potential hazard the flame arrester is meant to prevent.

Detonation Arrester Testing Detonation arresters are extensively tested for proof of performance against deflagrations, detonations, and, if required, short-time or endurance burns. The test gas must be selected to have either the same or a lower MESG than the gas in question. Typical MESG benchmark gases are stoichiometric mixtures of propane, hexane,

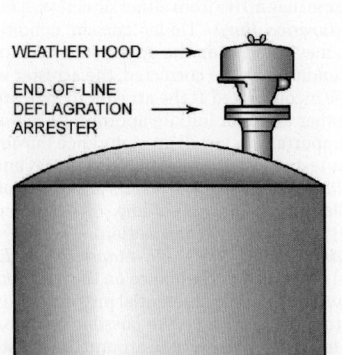

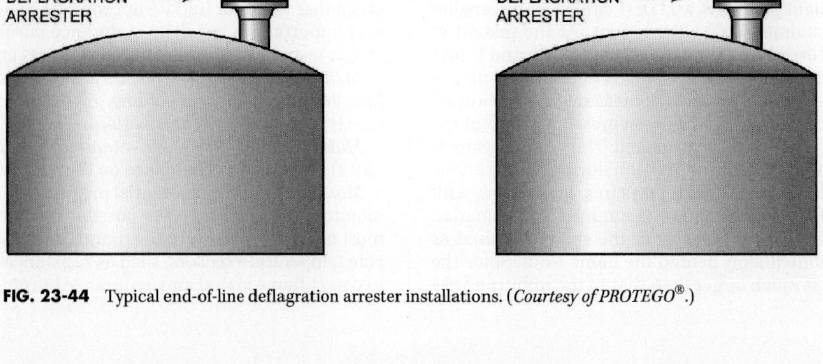

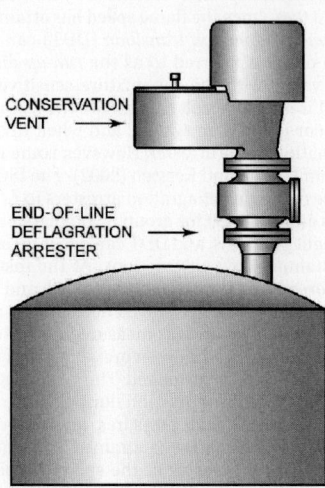

FIG. 23-44 Typical end-of-line deflagration arrester installations. (*Courtesy of PROTEGO®.*)

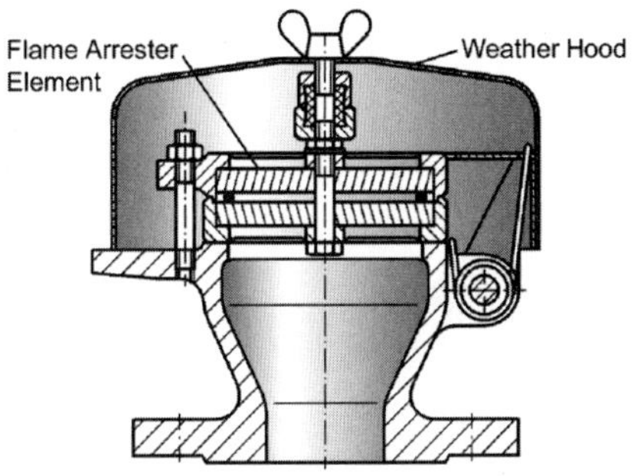

FIG. 23-45 Typical end-of-line deflagration arrester design. (*Courtesy of PROTEGO®.*)

or gasoline in air to represent group D gases having an MESG equal to or greater than 0.9 mm, and ethylene in air to represent group C gases with an MESG no less than 0.65 mm. Commercially available arresters are typically certified for use with one or another of these benchmark gas types. An ethylene-type arrester is selected if the gas in question has an MESG of less than 0.9 mm but not less than 0.65 mm. The ISO 16852 (2008) standard introduces four distinct test procedures for detonation arresters, depending on the intended use.

Type 1 Unstable detonation with restriction
Type 2 Unstable detonation without restriction
Type 3 Stable detonation with restriction
Type 4 Stable detonation without restriction

Testing is conducted in a fully closed system, allowing the detonation arrester to be tested for the required operating pressure. Additional testing is also required if the operating temperature exceeds 140°F (60°C). Five detonation tests are required along with five deflagration tests with a short run-up distance and five deflagration tests with a longer run-up distance. There are well-defined values for the required detonation pressure to be achieved for the detonation testing. For detonation arresters tested with restriction, both the deflagration and detonation tests are carried out with the restriction in place.

Additional testing is carried out for short-time burning (up to 30 minutes) or endurance burning as required.

This test does not use propane for group D gases but use hexane instead due to its lower autoignition temperatures. For group C tests, ethylene can be used for all test stages.

Care must be taken when applying the MESG method (Center for Chemical Process Safety, "Deflagration and Detonation Flame Arresters," 1993; Britton 2000; Brandes and Redeker 2002). The user has the option to request additional tests to address such concerns and may wish to test actual stream compositions rather than simulate them on the basis of MESG values.

ISO 16852 (2008) allows users to select a detonation arrester best suited to the intended application.

For installations governed by the USCG in Appendix A of 33 CFR, part 154 (*Marine Vapor Control Systems*), the USCG test procedures must be followed.

Grossel (2002) discusses differences between the requirements of disparate agencies.

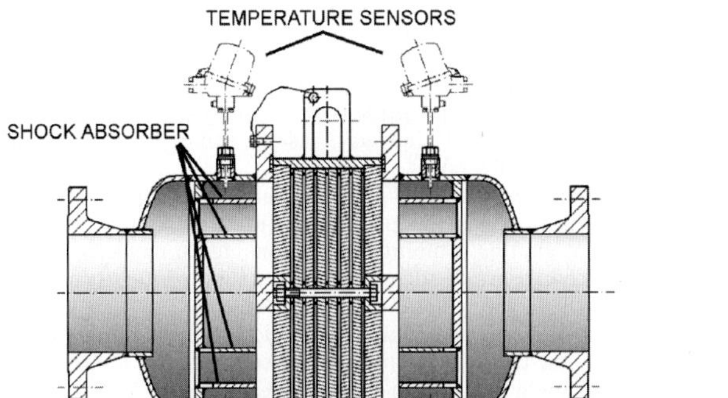

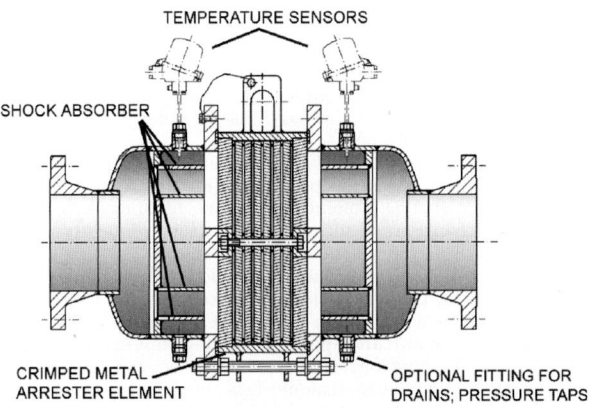

FIG. 23-46 Typical detonation arrester design (crimped ribbon type). (*Courtesy of PROTEGO®.*)

Special Arrester Types and Alternatives

Hydraulic (Liquid Seal) Flame Arresters These are most commonly used in large-pipe-diameter systems where fixed-element flame arresters are either cost prohibitive or otherwise impractical (e.g., very corrosive gas or where the gas contains solid particles that would quickly plug a conventional arrester element). These arresters contain a liquid, usually water-based, to provide a flame barrier. Figure 23-47 shows one design. Realistic tests are needed to ensure performance, as described in ISO 16852 (2008). Note that hydraulic flame arresters may fail at high flow rates, producing a sufficiently high concentration of gas bubbles to allow the transmission of flame. This is distinct from the more obvious failure mode caused by failure to maintain adequate liquid level.

Alternatives to Arresters Alternatives to the use of flame arresters are mentioned in NFPA 69 (2014) and include fast-acting isolation valves, vapor suppression systems, velocity-type devices in which gas velocity is designed to exceed flashback velocity, and control of the flammable mixture. The latter alternative often involves reduction of oxygen concentration to less than the limiting oxygen concentration (LOC) of the gas stream.

STORAGE AND HANDLING OF HAZARDOUS MATERIALS

GENERAL REFERENCES: API Standard 620, *Design and Construction of Large, Welded, Low-Pressure Storage Tanks,* American Petroleum Institute, Washington, D.C.; AP402, *Air Pollution Engineering Manual,* 2d ed., U.S. Environmental Protection Agency, Office of Air Quality Planning and Standards, 1973; ASME, *Process Piping: The Complete Guide to ASME B31.3,* 3d ed., American Society of Mechanical Engineers, New York, 2009; ASME, *ASME Boiler and Pressure Vessel Code; ASME Code for Pressure Piping; ASME General and Safety Standards; ASME Performance Test Codes,* American Society of Mechanical Engineers, New York; *Chemical Exposure Index,* AIChE, New York, 1998; CGA, *Handbook of Compressed Gases,* 5th ed., Compressed Gas Association, Chantilly, Va., 2013; CCPS, *Guidelines for Engineering Design for Process Safety,* 2d ed., CCPS-AIChE, New York, 2012; CCPS, *Guidelines for Facility Siting and Layout,* CCPS-AIChE, New York, 2003; CCPS, *Guidelines for Process Safety in Batch Reaction Systems,* CCPS-AIChE, New York, 1999; CCPS, *Guidelines for Safe Storage and Handling of High Toxic Hazard Materials,* CCPS-AIChE, New York, 1988; CCPS, *Guidelines for Safe Storage and Handling of Reactive Materials,* CCPS-AIChE, New York, 1995; CCPS, *Guidelines for Asset Integrity Management,* CCPS-AIChE, New York, 2017; Grossel and Crowl, eds., *Handbook of Highly Toxic Materials Handling and Management,* Marcel Dekker, New York, 1995; Kletz, "Friendly Plants," *Chem. Eng. Prog.,* July 1989; Kletz and Amyotte, *Process Plants: A Handbook for Inherently Safer Design,* 2d ed., CRC Press, Boca Raton, Fla., 2010; Kohan, *Pressure Vessel Systems: A User's Guide to Safe Operations and Maintenance,* McGraw-Hill, New York, 1987; Mannan, *Lees' Loss Prevention in the Process Industries,* 4th ed., Butterworth-Heinemann, Oxford, UK, 2012; *Ventsorb for Industrial Air Purification,* Bulletin 23-56c.

Introduction Most chemicals handled in the process industries have physical, chemical, and toxicological hazards to a greater or lesser degree. This requires that these hazards be contained and controlled throughout the entire life cycle of the facility to avoid loss, injury, and environmental damage. The provisions that will be necessary to contain and control the hazards will vary significantly depending on the chemicals and process conditions required.

Established Practices Codes, standards, regulatory requirements, industry guidelines, recommended practices, and supplier specifications have all developed over the years to embody the collective experience of industry and its stakeholders in the safe handling of specific materials. These should be the engineer's first resource in seeking to design a new facility.

The *ASME Boiler and Pressure Vessel Code,* section VIII (American Society of Mechanical Engineers, New York), is the standard resource for the design, fabrication, installation, and testing of storage tanks and process vessels rated as pressure vessels—that is, those designed to operate at an internal or external gauge pressure above 15 psi (103 kPa). ASME B31.3 is a basic resource for process piping systems.

Examples of established practices and other resources—some of which pertain to the safe storage and handling of specific hazardous chemicals, classes of chemicals, or facilities—include those listed in Table 23-28 from the publications of two U.S. organizations, NFPA and the Compressed Gas Association (CGA). Other organizations that may have pertinent standards include the International Standards Organization (ISO), the American National Standards Institute (ANSI), ASTM International (Conshohocken, Pa; www.astm.org), and other well-established national standards such as British Standards and Deutsches Institut für Normung e.V. (DIN) standards. Local codes and regulations should be checked for applicability, and the latest version should always be used when employing established practices.

Basic Design Strategies The storage and handling of hazardous materials involve risks that can be reduced to very low levels by good planning, design, maintenance, operation, and management practices. Facilities that handle hazardous materials typically represent a variety of risks, ranging from small leaks, which require prompt attention, to large releases,

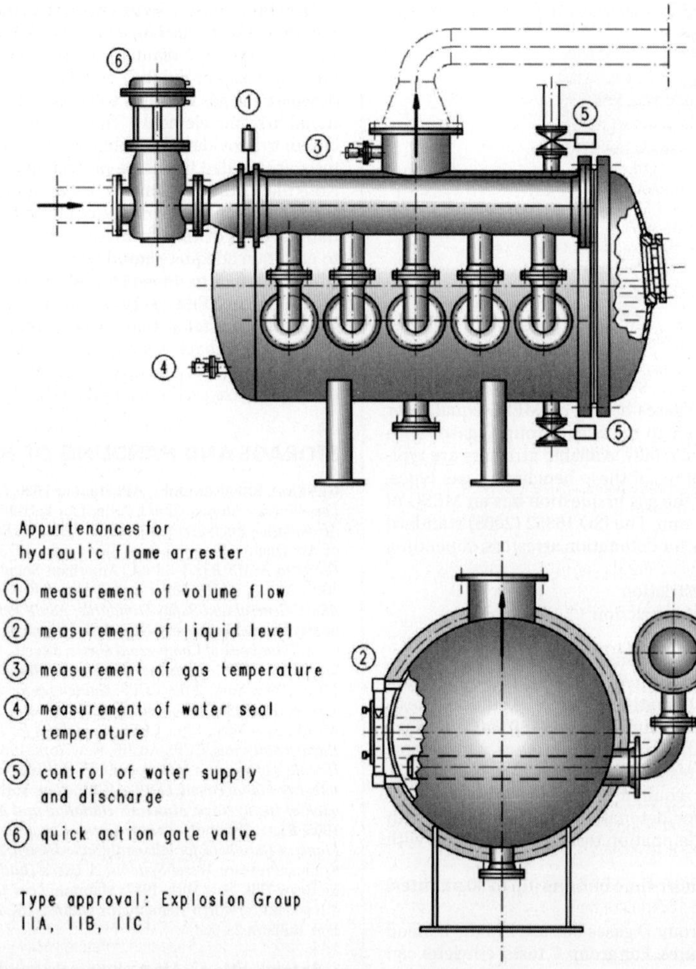

FIG. 23-47 Tested and approved hydraulic (liquid seal) flame arrester. (*Courtesy of PROTEGO®*.)

which are extremely rare in well-managed facilities but which have the potential for widespread impact (CCPS, *Guidelines for Safe Storage and Handling of High Toxic Hazard Materials,* AIChE, New York, 1988). It is essential that good techniques be developed for identifying significant hazards and providing adequate preventive and mitigative safeguards. Hazards can be identified and evaluated by using approaches discussed in the subsection on Process Safety Analysis.

Loss of containment due to mechanical failure or improper operation is a major cause of chemical process incidents. The design of storage and piping systems should be based on minimizing the likelihood of loss of containment, with the accompanying release of hazardous materials, and on limiting the amount of the release. An effective emergency response program that can reduce the impacts of a release should be available.

Thus, the basic design strategy for storing and handling hazardous materials can be summarized as follows, with reference to other parts of this section in parentheses:

1. Understand the hazardous properties of the materials to be stored and handled (Flammability; Reactivity; Toxicity; Other Hazards), as well as the physical hazards associated with the expected process design.
2. Reduce or eliminate the underlying hazards as much as feasible (Inherently Safer Design).
3. Evaluate the potential consequences of major and minor loss-of-containment events and other emergency situations involving hazardous materials and energies, and take this into account during site selection and facility layout and when evaluating the adequacy of personnel, public, and environmental protection (Discharge Rates from Punctured Lines and Vessels; Atmospheric Dispersion; Estimation of Damage Effects).
4. Design and build a robust and well-protected primary containment system following codes, standards, regulations, and other established practices (Security).
5. Design and implement a reliable and fault-tolerant basic process control system to ensure that the design limitations of the primary containment system are not exceeded.
6. Include provisions for detecting abnormal process conditions and bringing the process to a safe state before an emergency situation occurs (Safety Instrumented Systems).
7. Design, install, and maintain reliable and effective emergency relief systems, as well as mitigation systems such as secondary containment, deluge, and suppression systems, to reduce the severity of consequences in the event an emergency situation does occur (Pressure Relief Systems; Emergency Relief Device Effluent Collection and Handling).
8. Evaluate the risks associated with the process and its safety systems taken as a whole, including consideration of people, property, business, and the environment, that could be affected by loss events; determine whether the risks have been adequately reduced (Hazard Analysis; Risk Analysis; Discharge Rates from Punctured Lines and Vessels; Atmospheric Dispersion; Estimation of Damage Effects).
9. Take human factors into account in the design and implementation of the control system and the facility procedures (Key Procedures).
10. Ensure that staffing, training, inspections, tests, maintenance, and management of change are all adequate to maintain the integrity of the system throughout the facility's lifetime (Key Procedures).

Designers and operating companies will address these items in different ways, according to their established procedures. The steps that are addressed elsewhere in this section are not repeated here.

Site Selection, Layout, and Spacing Facility siting decisions that will have critical, far-reaching implications are made very early in a new facility's life cycle, or in the early planning stages of a site expansion project. The degree of public and regulatory involvement in this decision-making process, as well as the extent of prescriptive requirements and established practices in this area, vary considerably among countries, regions, and

TABLE 23-28 Examples of Established Practices Related to Storage and Handling of Hazardous Materials

Designation	Title
National Fire Protection Association (Quincy, Mass.; *www.nfpa.org*)	
NFPA 30	Flammable and Combustible Liquids Code
NFPA 30B	Code for the Manufacture and Storage of Aerosol Products
NFPA 36	Standard for Solvent Extraction Plants
NFPA 45	Standard on Fire Protection for Laboratories Using Chemicals
NFPA 53	Recommended Practice on Materials, Equipment and Systems Used in Oxygen-Enriched Atmospheres
NFPA 55	Standard for the Storage, Use, and Handling of Compressed Gases and Cryogenic Fluids in Portable and Stationary Containers, Cylinders, and Tanks
NFPA 58	Liquefied Petroleum Gas Code
NFPA 59A	Standard for the Production, Storage, and Handling of Liquefied Natural Gas (LNG)
NFPA 68	Guide for Venting of Deflagrations
NFPA 69	Standard on Explosion Prevention System
NFPA 318	Standard for the Protection of Semiconductor Fabrication Facilities
NFPA 326	Standard for the Safeguarding of Tanks and Containers for Entry, Cleaning, or Repair
NFPA 329	Recommended Practice for Handling Releases of Flammable and Combustible Liquids and Gases
NFPA 400	Hazardous Chemical Code
NFPA 430	Code for the Storage of Liquid and Solid Oxidizers
NFPA 432	Code for the Storage of Organic Peroxide Formulations
NFPA 434	Code for the Storage of Pesticides
NFPA 484	Standard for Combustible Metals, Metal Powders, and Metal Dusts
NFPA 490	Code for the Storage of Ammonium Nitrate
NFPA 495	Explosive Materials Code
NFPA 497	Recommended Practice for the Classification of Flammable Liquids, Gases, or Vapors and of Hazardous (Classified) Locations for Electrical Installations in Chemical Process Areas
NFPA 499	Recommended Practice for the Classification of Combustible Dusts and of Hazardous (Classified) Locations for Electrical Installations in Chemical Process Areas
NFPA 654	Standard for the Prevention of Fire and Dust Explosions from the Manufacturing, Processing, and Handling of Combustible Particulate Solids
NFPA 655	Standard for Prevention of Sulfur Fires and Explosions
NFPA 704	Standard System for the Identification of the Hazards of Materials for Emergency Response
Compressed Gas Association (Chantilly, Va.; *www.cganet.com*)	
CGA G-1	Acetylene
CGA G-2	Anhydrous Ammonia
CGA G-3	Sulfur Dioxide
CGA G-4	Oxygen
CGA G-5	Hydrogen
CGA G-6	Carbon Dioxide
CGA G-8.1	Standard for Nitrous Oxide Systems at Consumer Sites
CGA G-12	Hydrogen Sulfide
CGA G-14	Code of Practice for Nitrogen Trifluoride (EIGA Doc. 92/03)
CGA P-1	Safe Handling of Compressed Gases in Containers
CGA P-8	Safe Practices Guide for Cryogenic Air Separation Plants
CGA P-9	The Inert Gases: Argon, Nitrogen, and Helium
CGA P-12	Safe Handling of Cryogenic Liquids
CGA P-16	Recommended Procedures for Nitrogen Purging of Tank Cars
CGA P-32	Safe Storage and Handling of Silane and Silane Mixtures
CGA P-34	Safe Handling of Ozone-Containing Mixtures Including the Installation and Operation of Ozone-Generating Equipment
CGA S-1.1	Pressure Relief Device Standards—Part 1—Cylinders for Compressed Gases
CGA S-1.2	Pressure Relief Device Standards—Part 2—Cargo and Portable Tanks for Compressed Gases
CGA S-1.3	Pressure Relief Device Standards—Part 3—Stationary Storage Containers for Compressed Gases

NOTE: Always check the latest edition when using established practices.

companies. Insurance carriers are also generally involved in the process, particularly with regard to fire protection considerations.

From the perspective of process safety, key considerations with respect to site selection, layout, and spacing can be summarized as

- Where on-site personnel (including contractors and visitors), critical equipment, the surrounding public, and sensitive environmental receptors are located with respect to hazardous materials and processes.
- Whether the design and construction of control rooms and other occupied structures, as well as detection, warning, and emergency response provisions, will provide adequate protection in the event of a major fire, explosion, or toxic release event.

Recommended distances for the spacing of buildings and equipment for fire protection were issued as GAP 2.5.2, *Plant Layout and Spacing for Oil and Chemical Plants* (HSB Industrial Risk Insurers, Hartford, Conn., 2001). These are referenced in "Typical Spacing Tables" included as Appendix A of the CCPS *Guidelines for Facility Siting and Layout* (AIChE, New York, 2003).

Other resources pertaining to siting and layout include

- *Dow's Fire & Explosion Index Hazard Classification Guide,* 7th ed. (AIChE, New York, 1994), which gives an empirical radius of exposure and damage factor based on the quantity and characteristics of the material being stored and handled.
- API RP 752, *Management of Hazards Associated with Location of Process Plant Buildings,* 3d ed. (American Petroleum Institute, Washington, D.C., 2009), which gives consequence-based and risk-based approaches to evaluating protection afforded by occupied structures.

Storage

Storage Facilities Dating back to at least 1974, when a vapor cloud explosion in Flixborough, UK, claimed 28 lives and destroyed an entire chemical plant (Mannan, *Lees' Loss Prevention in the Process Industries,* 4th ed., Butterworth-Heinemann, Oxford, UK, 2012), a major emphasis in the safe storage and handling of hazardous materials has been to reduce hazardous material inventories. Inventory reduction can be accomplished not only by using fewer and smaller storage tanks and vessels but also by eliminating any nonessential intermediate storage vessels and batch process weigh tanks and generating hazardous materials on demand when feasible. Note, however, that the reduction of inventory may require more frequent and smaller shipments and improved inventory management.

There may be more chances for errors in connecting and reconnecting with small shipments. Quantitative risk analysis of storage facilities has revealed solutions that may run counter to intuition [Schaller, *Plant/Oper. Prog.* **9**(1), 1990]. For example, reducing inventories in tanks of hazardous materials does little to reduce risk in situations where most of the exposure arises from the number and extent of valves, nozzles, and lines connecting the tank. Removing tanks from service altogether, on the other hand, generally helps. A large pressure vessel may offer greater safety than several small pressure vessels of the same aggregate capacity because there are fewer associated nozzles and lines. In addition, a large pressure vessel is inherently more robust, or it can economically be made more robust by deliberate overdesign than can a number of small vessels of the same design pressure. On the other hand, if the larger vessel has larger connecting lines, the relative risk may be greater if release rates through the larger lines increase the risk more than the inherently greater strength of the vessel reduces it. In transporting hazardous materials, maintaining tank car integrity in a derailment is often the most important line of defense.

Safer Storage Conditions The hazards associated with storage facilities can often be reduced significantly by changing storage conditions. The primary objective is to reduce the driving force available to transport the hazardous material into the atmosphere in case of a leak [Hendershot, "Alternatives for Reducing the Risks of Hazardous Material Storage Facilities," *Environ. Prog.* **7:** 180ff (1988, August)]. Some methods to accomplish this follow.

Dilution Dilution of a low-boiling hazardous material reduces the hazard in two ways:

1. The vapor pressure is reduced. This has a significant effect compared to the rate of release of a material that would boil at less than ambient temperature. It may be possible to store an aqueous solution at or near atmospheric pressure, such as by storing aqueous ammonium hydroxide instead of anhydrous ammonia.

2. In the event of a spill, the atmospheric concentration of the hazardous material will be reduced, resulting in a smaller hazard downwind of the spill.

Refrigeration Loss of containment of a liquefied gas under pressure and ambient temperature causes immediate flashing of a large proportion of the liquified gas. This is followed by slower evaporation of the residue. The hazard from a gas under pressure is normally much less in terms of the amount of material stored, but the physical energy released if a confined explosion occurs at high pressure is large.

Refrigerated storage of hazardous materials that are stored at or below their atmospheric boiling points mitigates the consequences of containment loss in three ways:

1. The rate of release, in the event of loss of containment, will be reduced because of the lower vapor pressure in the event of a leak.

2. Material stored at a reduced temperature has little or no superheat, and there will be little or no flashing to vapor in case of a leak. Vaporization will be mainly determined by liquid evaporation from the surface of the spilled liquid, which depends on weather and surface conditions and on pool size.

3. The amount of material released to the atmosphere will be further reduced because liquid entrainment from the two-phase flashing jet resulting from a leak will be reduced or eliminated. Refrigerated storage is most effective in mitigating storage facility risk if the material is refrigerated when received.

The economics of storing liquefied gases are such that it is usually attractive to use pressure storage for small quantities, pressure or semirefrigerated storage for medium to large quantities, and fully refrigerated storage for very large quantities. Quantitative guidelines can be found in Mannan (*Lees' Loss Prevention in the Process Industries,* 4th ed., Butterworth-Heinemann, Oxford, UK, 2012).

It is generally considered that there is a greater hazard in storing large quantities of liquefied gas under pressure than at low temperatures and low pressures. The trend is toward replacing pressure storage with refrigerated low-pressure storage for large inventories. However, it is necessary to consider the risk of the entire system, including the refrigeration system, and not just the storage vessel. The consequences of failure of the refrigeration system must be considered. Each case should be carefully evaluated on its own merits.

Preventing Leaks and Spills from Accumulating Under Tanks or Equipment Around storage and process equipment, it is a good idea to design dikes that will not allow toxic and flammable materials to accumulate around the bottom of tanks or equipment in case of a spill. If liquid is spilled and ignited inside a dike where there are storage tanks or process equipment, the fire may be continuously supplied with fuel, and the consequences can be severe. It is usually much better to direct possible spills and leaks to an area away from the tank or equipment and provide a firewall to shield the equipment from most of the flames if a fire occurs. Figure 23-48 shows a diking design for directing leaks and spills to an area away from tanks and equipment.

The surface area of a spill should be minimized for hazardous materials that have a significant vapor pressure at ambient conditions, such as acrylonitrile or chlorine. This will make it easier and more practical to collect vapor from a spill or to suppress vapor release with foam or by other means. This may require a deeper nondrained dike area than normal or some other design that will minimize surface area, in order to contain the required volume. It is usually not desirable to cover a diked area to restrict loss of vapor if the spill consists of a flammable or combustible material.

Minimal Use of Underground Tanks The U.S. Environmental Protection Agency's (EPA) Office of Underground Storage Tanks defines underground tanks as those with 10 percent or more of their volume, including piping, underground. An aboveground tank that does not have more than 10 percent of its volume (including piping) underground is excluded from the underground tank regulations. Note, however, that a 5000-gal tank sitting wholly atop the ground but having 1400 ft of 3-in buried pipe or 350 ft of 6-in buried pipe is considered an underground storage tank.

At one time, burying tanks was recommended because it minimized the need for a fire protection system, dikes, and distance separation. At many companies, this is no longer considered good practice. Mounding, or

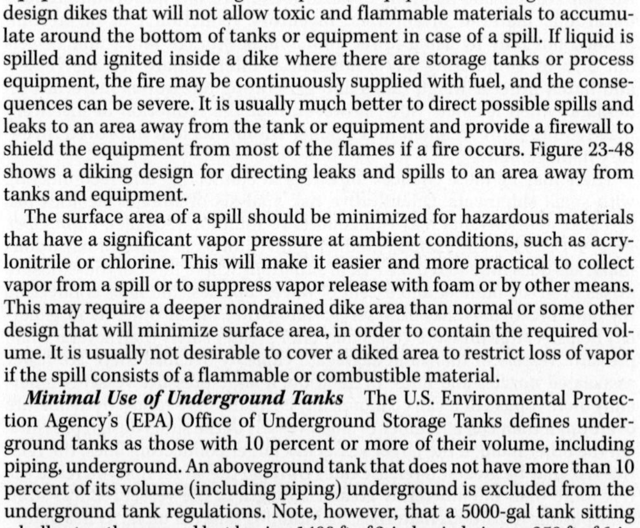

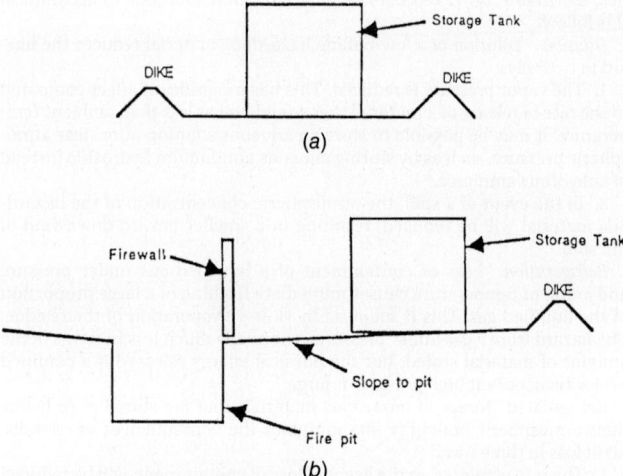

FIG. 23-48 Methods of diking for flammable liquids: (*a*) Traditional diking method allows leaks to accumulate around the tank. In case of fire, the tank will be exposed to flames that can be supplied by fuel from the tank and will be hard to control. (*b*) In the more desirable method, leaks are directed away from the tank. In case of fire, the tank will be shielded from most flames and fire will be easier to fight. (*From Englund, in* Advances in Chemical Engineering, *vol. 15, Academic Press, San Diego, 1990, pp. 73–135, by permission.*)

burying tanks above grade, has most of the same problems as burying tanks below ground and is usually not recommended.

Problems with buried tanks include

- Difficulty in monitoring interior and exterior corrosion (shell thickness)
- Difficulty in detecting leaks
- Difficulty of repairing a tank if the surrounding earth is saturated with chemicals from a leak
- Potential contamination of groundwater due to leakage.

Government regulations concerning buried tanks have become stricter. This is a result of the many leaking tanks identified as causing adverse environmental and human health problems.

Design of Tanks, Piping, and Pumps Six basic tank designs are used for the storage of organic liquids: (1) fixed-roof, (2) external floating-roof, (3) internal floating-roof, (4) variable vapor space, (5) low-pressure, and (6) high-pressure tanks. The first four tank designs listed are not generally considered suitable for high toxic hazard materials.

Low-Pressure Tanks (up to 15 psig) Low-pressure storage tanks for high toxic hazard materials should meet, as a minimum, American Petroleum Institute (API) Standard 620, *Recommended Rules for the Design and Construction of Large Welded, Low-Pressure Storage Tanks* (American Petroleum Institute, Washington, D.C.). This standard covers aboveground tanks designed for all gauge pressures less than or equal to 15 psi (103 kPa) and metal temperatures less than or equal to 250°F (121°C). There are no specific requirements in API Standard 620 for high toxic hazard materials.

API Standard 650, *Welded Steel Tanks for Oil Storage* (American Petroleum Institute, Washington, D.C.), has limited applicability to the storage of high toxic hazard materials because it prohibits refrigerated service and limits gauge pressures to 2.5 psi (17 kPa) and only if designed for certain conditions. Most API 650 tanks have a working pressure approaching atmospheric pressure, and hence their pressure-relieving devices must generally vent directly to the atmosphere. This standard's safety factors and welding controls are less stringent than those required by API Standard 620.

Horizontal and vertical cylindrical tanks are used to store highly toxic liquids and other hazardous materials at atmospheric pressure. Horizontal, vertical, and spherical tanks are used for refrigerated liquefied gases that are stored at atmospheric pressure. The design pressure of tanks for atmospheric pressure and low-pressure storage at ambient temperature should not be less than 100 percent of the vapor pressure of the material at the maximum design temperature. The maximum design metal temperature to be used takes into consideration the maximum temperature of material entering the tank; the maximum ambient temperature, including solar radiation effects; and the maximum temperature attainable by expected or reasonably foreseeable abnormal chemical reactions.

Since discharges of vapors from highly hazardous materials cannot simply be released to the atmosphere, the use of a weak seam roof is not normally acceptable. It is best that tanks in low-pressure hazardous service be designed and stamped for a gage pressure of 15 psi (103 kPa) to provide maximum safety, and pressure relief systems must be provided to vent relieved overpressure to equipment that can collect, contain, and treat the effluent.

The minimum design temperature should be the lowest temperature to which the tank will be subjected, taking into consideration the minimum temperature of material entering the tank, the minimum temperature to which the material may be autorefrigerated by rapid evaporation of low-boiling liquids or mechanically refrigerated, the minimum ambient temperature of the area where the tank is located, and any possible temperature reduction by endothermic physical processes or chemical reactions involving the stored material. API Standard 620 provides for installations in areas where the lowest recorded one-day mean temperature is 50°F (10°C).

While either rupture disks or relief valves are allowed on storage tanks by code, rupture disks by themselves should not be used on tanks for the storage of toxic or other highly hazardous materials because they do not close after opening, which can lead to a continuing, uncontrolled release of hazardous material to the atmosphere.

API Standard 620 requires a combined pneumatic hydrotest at 125 percent of design tank loading. In tanks designed for low-density liquids, the upper portion is not fully tested. For highly hazardous materials, consideration should be given for hydrotesting at the maximum specified design liquid level. It may be required that the lower shell thickness be increased to withstand a full head of water and that the foundation be designed such that it can support a tank full of water, or the density of the liquid if it is greater than the density of water. Testing in this manner not only tests the containment capability of the tank, but also provides an overload test for the tank and the foundation similar to the overload test for pressure vessels. API Standard 620 also requires radiography.

Proper preparation of the subgrade and grade is extremely important for tanks that are to rest directly on grade. Precautions should be taken to prevent ground freezing under refrigerated tanks because this can cause the ground to heave and damage the foundation or the tank. Designing for free air circulation under the tank is a method for passive protection from ground freezing.

Steels lose their ductility at low temperatures and can become subject to brittle failure. There are specific requirements for metals to be used for refrigerated storage tanks in API Standard 620, Annexes Q and R.

Corrosive chemicals and external exposure can cause tank failure. Materials of construction should be chosen so that they are compatible with the chemicals and exposure involved. Welding reduces the corrosion resistance of many alloys, leading to localized attack at the heat-affected zones. This may be prevented by the use of the proper alloys and weld materials, in some cases combined with annealing heat treatment.

External corrosion can occur under insulation, especially if the weather barrier is not maintained or if the tank is operating at conditions at which condensation is likely. This form of attack is hidden and may be unnoticed for a long time. Inspection holes and plugs should be installed in the insulation to monitor possible corrosion under the insulation.

Pressure Vessels (above 15 psig) The design of vessels above a gauge pressure of 15 psi (103 kPa) falls within the scope of the *ASME Boiler and Pressure Vessel Code*, section VIII, "Pressure Vessels, Division I" (American Society of Mechanical Engineers, New York), and should be designated as lethal service if required. *Lethal service* means containing substances that are "poisonous gases or liquids of such a nature that a very small amount of the gas or vapor of the liquid mixed or unmixed with air is dangerous to life when inhaled. This class includes substances which are stored under pressure or may generate a pressure if stored in a closed vessel." This is similar to, but not the same as, the same definition as that for Category M fluid service of the *ASME Pressure Piping Code* (see below). Pressure vessels for the storage of highly hazardous materials should be designed in accordance with requirements of the ASME code even if the vessels could be exempted because of high pressure or size. The code requires that the corrosion allowance be adequate to compensate for the more or less uniform corrosion expected to take place during the life of the vessel and that it not weaken the vessel below design strength.

Venting and Drainage Low-pressure storage tanks are particularly susceptible to damage if good venting practices are not followed. A vent that does not function properly at all times may cause damage to the tank from pressure that is too high or too low. Vapors should go to a collection system, if necessary, to contain toxic and hazardous vents.

Piping Piping systems for toxic fluids fall within chapter VIII of the *ASME Pressure Piping Code*, "Piping for Category M Fluid Service" (American Society of Mechanical Engineers, New York). Category M fluid service is defined as "fluid service in which the potential for personnel exposure is judged to be significant and in which a single exposure to a small quantity of a toxic fluid, caused by leakage, can produce serious irreversible harm to persons on breathing or bodily contact, even when prompt restorative measures are taken."

Piping systems in hazardous service should meet the requirements for both Category M fluid service and for "severe cyclic conditions." Piping systems should be subjected to a flexibility analysis, and if they are found to be too rigid, flexibility should be added. Severe vibration pulsations should be eliminated. Expansion bellows, flexible connections, and glass equipment should be avoided. Pipelines should be designed with the minimum number of joints, fittings, and valves. Joints should be flanged or butt-welded. Threaded joints should not be used.

Instrumentation Instrument systems are an essential part of the safe design and operation of systems for storing and handling hazardous materials. They are key elements of systems to eliminate the threat of conditions that could result in loss of containment. They are also used for early detection of releases so that mitigating action can be taken before these releases result in serious effects on workers or the public or on the environment.

Pumps and Gaskets The most common maintenance problem with centrifugal pumps is with the seals. Mechanical seal problems account for most of the pump repairs in a chemical plant, with bearing failures a distant second. The absence of an external motor (on canned pumps) and a seal is appealing to those experienced with mechanical seal pumps.

Sealless pumps are very popular and are widely used in the chemical industry. These pumps are manufactured in two basic types: canned motor and magnetic drive. Magnetic-drive pumps have thicker "cans," which hold in the process fluid, and the clearances between the internal rotor and can are greater than in canned-motor pumps. This permits greater bearing wear before the rotor starts wearing through the can. Many magnetic-drive pump designs now have incorporated a safety clearance, which uses a rub ring or a wear ring to support the rotating member in the event of excessive bearing wear or failure. This design feature prevents the rotating member (outer magnet holder or internal rotating shaft assembly) from accidentally rupturing the can, and it provides a temporary bearing surface until the problem bearings can be replaced. Because most magnetic-drive pumps use permanent magnets for both the internal and external rotors, there is less heat transferred to the pumped fluid than with canned-motor pumps. Some canned-motor pumps have fully pressure-rated outer shells that enclose the canned motor; others do not. With magnetic-drive pumps, containment of leakage through the can to the outer shell can be a problem. Even though the shell may be thick and capable of holding high pressures, there is often an elastomeric lip seal on the outer magnetic rotor shaft with little pressure containment capability.

Canned-motor pumps typically have a clearance between the rotor and the containment shell or can, which separates the fluid from the stator, of only 0.008 to 0.010 in (0.20 to 0.25 mm). The can needs to be thin to allow magnetic flux to flow to the rotor. It is typically 0.010 to 0.015 in (0.25 to 0.38 mm) thick and made of Hastelloy. The rotor can wear through the can very rapidly if the rotor bearing wears enough to cause the rotor to move slightly and begin to rub against the can. The can may rupture, causing uncontrollable loss of the fluid being pumped.

It should not be assumed that just because there is no seal, sealless pumps are always safer than pumps with seals, even with the advanced technology now available in sealless pumps. Use sealless pumps with considerable caution when handling hazardous or flammable liquids.

Sealless pumps rely on the process fluid to lubricate the bearings. If the wear rate of the bearings in the fluid being handled is not known, the bearings can wear unexpectedly, causing rupture of the can.

Running a sealless pump dry can cause complete failure. If there is cavitation in the pump, hydraulic balancing in the pump no longer functions, and excessive wear can occur, leading to failure of the can. The most common problem with sealless pumps is bearing failure, which occurs either by flashing the fluid in the magnet area because of a drop in flow below minimum flow or by flashing in the impeller eye as it leaves the magnet area. It is estimated that 90 percent of conventional canned-motor pump failures are the result of dry running. Canned pumps are available that can be operated dry for as long as 48 h, according to manufacturer claims.

It is especially important to avoid deadheading a sealless pump. Deadheaded sealless pumps can cause overheating. The bearings may be damaged, and the pump may be overpressurized. The pump and piping systems should be designed to avoid dead spots when pumping monomers. Monomers in dead spots may polymerize and plug the pump. There are minimum flow requirements for sealless pumps. It is recommended that a recirculation system be used to provide internal pump flow whenever the pump operates. Inlet line filters are recommended, but care must be taken not to cause excessive pressure drop on the suction side. Typical inlet filters use sieve openings of 0.0059 in (0.15 mm).

A sealless pump that is not properly operated can rupture with potentially serious results. The can may fail if valves on both sides of the pump are closed and the fluid in the pump expands, due to either heating up from a cold condition or starting the pump. If the pump is run dry, the bearings can be ruined. The pump can heat up and be damaged if there is insufficient flow to take away heat from the windings. Sealless pumps, especially canned-motor pumps, produce a significant amount of heat, since nearly all the electric energy lost in the system is absorbed by the fluid being pumped. *If this heat cannot be properly dissipated, the fluid will heat up with possibly severe consequences.* Considerable care must be used when installing a sealless pump to be sure that improper operations cannot occur.

The instrumentation recommended for sealless pumps may seem somewhat excessive. However, sealless pumps are expensive, and they can be made to last for a long time, compared to conventional centrifugal pumps where seals may need to be changed often. Most failures of sealless pumps are caused by running them dry and damaging the bearings. Close monitoring of temperature is necessary in sealless pumps. Three temperature sensors (resistance temperature devices, or RTDs) are recommended: (1) in the internal fluid circulation loop, (2) in the magnet, or shroud, area, and (3) in the pump case area.

It is important that sealless pumps be flooded with liquid before starting, to avoid damage to bearings from imbalance or overheating. Entrained gases in the suction can cause immediate imbalance problems and lead to internal bearing damage. Some type of liquid sensor is recommended. Sealless pumps must not be operated deadheaded (pump liquid full with inlet and/or outlet valves closed). Properly installed and maintained, sealless pumps, both canned-motor and magnetic-drive, offer an economical and safe way to minimize hazards and leaks of hazardous liquids.

Loss-of-Containment Causes Table 23-29 indicates four basic ways in which containment can be lost. These cause categories can be used both as a checklist of considerations during the design process and as a starting point for evaluating the adequacy of safeguards as part of a process hazard and risk analysis.

Maintaining the Mechanical Integrity of the Primary Containment System The second main category in the tabulation of

TABLE 23-29 Summary of Loss-of-Containment Causes in the Chemical Industry

I. Containment lost via an "open-end" route to atmosphere
- A. Due to genuine process relief or dumping requirements
- B. Due to maloperation of equipment in service; e.g., spurious relief valve operation
- C. Due to operator error; e.g., drain or vent valve left open, misrouting of materials, tank overfilled, unit opened up under pressure

II. Containment failure under design operating conditions due to imperfections in the equipment
- A. Imperfections arising prior to commissioning and not detected before start-up
- B. Imperfections due to equipment deterioration in service and not detected before the effect becomes significant
- C. Imperfections arising from routine maintenance or minor modifications not carried out correctly, e.g., poor workmanship, wrong materials

III. Containment failure under design operating conditions due to external causes
- A. Impact damage, such as by cranes, road vehicles, excavators, machinery associated with the process
- B. Damage by confined explosions due to accumulation and ignition of flammable mixtures arising from small process leaks, e.g., flammable gas buildup in analyzer houses, in enclosed drains, around submerged tanks
- C. Settlement of structural supports due to geologic or climatic factors or failure of structural supports due to corrosion, etc.
- D. Damage to tank trucks, railcars, containers, etc., during transport of materials on- or off-site
- E. Fire exposure
- F. Blast effects from a nearby explosion (unconfined vapor cloud explosion, bursting vessel, etc.), such as blast overpressure, projectiles, structural damage
- G. Natural events (acts of God) such as windstorms, earthquakes, floods, lightning

IV. Containment failure due to deviations in plant conditions beyond design limits
- A. Overpressurizing of equipment
- B. Underpressurizing of non-vacuum-rated equipment
- C. High metal temperature (causing loss of strength)
- D. Low metal temperature (causing cold embrittlement and overstressing)
- E. Wrong process materials or abnormal impurities (causing accelerated corrosion, chemical attack of seals or gaskets, stress corrosion cracking, embrittlement, etc.)

SOURCE: Summarized from Appendix A of Prugh and Johnson, *Guidelines for Vapor Release Mitigation*, CCPS-AIChE, New York, 1988.

loss-of-containment causes pertains to containment failure under design operating conditions due to imperfections in the equipment. This group of causes is the main focus of a facility's asset integrity management (AIM) program (CCPS, *Guidelines for Asset Integrity Management*, AIChE, New York 2017), also known as a mechanical integrity (MI) program. The AIM program should also detect other imperfections such as previous periods of operating outside design limits, or improper process materials or impurities that cause accelerated corrosion, chemical attack of seals or gaskets, stress corrosion cracking, embrittlement, etc. AIM programs include quality management of the initial construction and of maintenance materials; routine preventive maintenance activities; regular inspections and nondestructive testing (NDT) of vessels, tanks, and piping to detect corrosion, pitting, erosion, cracking, creep, etc.; functional testing of standby equipment, including alarms, safety instrumented systems, and emergency relief systems; and correcting problems that are identified while using, inspecting, testing, or maintaining the equipment and instrumentation. CCPS (*Guidelines for Asset Integrity Management*, AIChE, New York, 2017) provides guidance on developing and implementing an asset integrity management program.

Release Detection and Mitigation *Mitigation* means reducing the severity of consequences of a loss event such as a major release, fire, and/or explosion. The choice of mitigation strategies will depend on the nature of the hazardous materials and energies that can be released and the degree to which risk reduction is needed to ensure people, property, and the environment are adequately protected. The latter will be affected by the proximity of populations and sensitive environments to the facility. An unstaffed remote natural gas facility will obviously not warrant the same mitigation measures as a facility using large quantities of high-toxic-hazard materials with other industry or residences nearby.

To be effective, a mitigation strategy will need to be capable of:
- Detecting either an incipient or an actual loss event
- Deciding on and initiating the proper course of action to mitigate the situation
- Reducing the severity of consequences at the source, in transit, and/or at the receptor locations
- Preventing domino effects that could have even more severe consequences

Each of these steps might be performed by direct action of operations or emergency response personnel and/or by automatic systems. An example of the latter might be an array of toxic or flammable gas detectors that might trip an emergency shutdown system that closes remotely actuated block valves and vents off the process pressure to a flare if two adjacent sensors read above a predetermined vapor concentration.

Mitigation measures can also be *passive* safeguards, meaning that they require no human intervention and no engineered sensing and actuation system to work. Examples of passive mitigation measures are secondary containment systems, blast-resistant and fire-resistant structures, insulated or low-heat-capacity spill surfaces to reduce the rate of evaporation, and an increased distance between the hazardous materials and energies and the sensitive receptors.

SAFETY INSTRUMENTED SYSTEMS

REFERENCES: *Guidelines for Safe Automation of Chemical Processes*, American Institute of Chemical Engineers, New York, 2017; IEC 61511, *Functional Safety: Safety Instrumented Systems for the Process Industry Sector*, International Electrotechnical Commission, Geneva, Switzerland, 2016; ISA TR84.00.04, *Guidelines for the Implementation of ANSI/ISA 84.00.01-2004* (IEC 61511), International Society of Automation, Research Triangle Park, N.C., 2015; ISA TR84.00.02, *Safety Instrumented Functions (SIF)—Safety Integrity Level (SIL) Evaluation Techniques*, International Society of Automation, Research Triangle Park, N.C., 2015; ISA TR84.00.03, *Mechanical Integrity of Safety Instrumented Systems*, International Society of Automation, Research Triangle Park, N.C., 2012.

GENERAL REFERENCES: *Guidelines for Safe and Reliable Instrumented Protective Systems*, American Institute of Chemical Engineers, New York, 2007; *Guidelines for Hazard Evaluation Procedures, Third Edition with Worked Examples*, American Institute of Chemical Engineers, New York, 2008; *Guidelines for Initiating Events and Independent Protection Layers in Layers of Protection Analysis*, American Institute of Chemical Engineers, New York, 2014; IEC 62443, *Security for Industrial Automation and Control Systems*, Parts 1–3, International Electrotechnical Commission, Geneva, Switzerland, 2009–2013.

Glossary

Process control system System that responds to input signals from the process, its associated equipment, other programmable systems, and/or an operator and generates output signals, causing the process and its associated equipment to operate in the desired manner.

Compensating measures Planned means for managing process risk during periods of process operation with known faults or problems that increase risk.

Core attribute Fundamental underlying property of a protection layer. The core attributes are independence, functionality, integrity, reliability, auditability, management of change, and access security.

Independent protection layer An IPL is a device, system, or action that is capable of preventing a hazardous event from proceeding to the undesired consequence regardless of the initiating cause (or its consequences) or the failure of any other protection layer.

Safety instrumented function (SIF) A task performed by a safety instrumented system that takes the process to a specified safe state when abnormal operation is detected.

Safety instrumented system (SIS) Any combination of separate and independent devices (sensors, logic solvers, final elements, and support systems) designed and managed to achieve a specified safety integrity level. An SIS may implement one or more safety instrumented functions.

Safety integrity level (SIL) Discrete level (one of four categories) used to specify the likelihood that a safety instrumented function will perform its required task under all operational states within a specified time.

Introduction The chemical industry relies on instrumented systems to support the safe execution of many process tasks. A process control system is used to maintain unit operations within prescribed normal operating limits. Operators supervise the unit operation by monitoring process variables and, when necessary, taking action using instrumented systems. The SIS detects the existence of unacceptable process conditions and takes action on the process to achieve or maintain a safe state of the process. In the past, these systems have also been called emergency shutdown systems, safety interlock systems, and safety critical systems.

In 2004, the European Committee for Electrotechnical Standardization (CENELEC) and the American National Standards Institute (ANSI) recognized IEC 61511 as a consensus standard for the implementation of SISs in the process industry. IEC 61511 covers the complete life cycle of the SIS and documents the requirements for its assessment, design, operation, maintenance, testing, and management. The standard establishes a numerical benchmark for the SIS's performance known as the safety integrity level (SIL) and provides requirements on how to design and manage the SIS to achieve the target SIL. IEC 61511 was updated in 2016 with greater emphasis on the operation and maintenance activities and tracking metrics needed to sustain SIS performance long-term.

The chemical industry has also found significant benefits to plant productivity and operability when SIS work processes are used to design and manage other instrumented protective systems (IPSs), such as those that mitigate potential economic and business losses. The CCPS's *Guidelines for Safe and Reliable Instrumented Protective Systems* (2007) discusses the activities and quality control measures needed to design for safety, reliability, operability, and maintainability. CCPS's *Guidelines for Safe Automation of Chemical Processes* (2017) addresses design and management practices needed to safely apply automation systems in control and safety applications.

Hazard and Risk Analysis Identify hazardous events as early as possible in the design of process equipment so that measures can be taken to reduce or eliminate potential loss events. Inherently safer design strategies, such as minimize, substitute, moderate, and simplify, should be implemented to make the process operation more resilient to disturbances and to reduce the need for protection layers.

When it is no longer practical to reduce the risk further by process design modification, protection layers are used to mitigate the remaining process risk. IPLs must meet the necessary rigor associated with the seven core attributes. CCPS's *Guidelines for Initiating Event and Independent Protection Layers* builds upon traditional risk analysis to define specific design and management practices that are needed to sustain the protection in the long term.

There are two critical activities to be completed during the risk assessment phase. First, the safety functions (i.e., those functions that detect and respond to hazardous events) are identified by using an accepted hazard and risk analysis (H&RA) methodology. Second, each safety function is allocated to a protection layer that is designed and managed to achieve the required risk reduction.

An H&RA involves a review of the process design and its control, operation, and maintenance practices. The review is conducted by a multidisciplinary team with expertise in the design and operation of the process unit. The team uses a systematic screening process to determine how deviations from normal operation lead to hazardous events. A security risk assessment is also conducted to identify the needed countermeasures and security provisions, including the need for additional automatic monitoring and reporting. IEC 62443 provides guidance for cyber security of industrial control and safety systems. The H&RA identifies areas where the process risk is too high, requiring the implementation of safety functions. The team's objective is to reduce the risk to below the owner/operator's risk criteria.

Process risk is defined by the frequency of the occurrence and the potential consequence severity of the hazardous event. The initiating causes (e.g., single causes or multiple causes and conditions) are identified for each hazardous event, and their frequency of occurrence is estimated. The consequence severity is the logical conclusion to the propagation of the hazardous event if no protection layers are implemented as barriers to the event.

The gap between the process risk and the owner/operator's risk criteria establishes the requirements for risk reduction. The risk gap can be managed by a single safety function or by multiple functions allocated to protection layers. The team defines the risk reduction that must be provided by each safety function and allocates the safety function to a protection layer that is designed and managed to achieve the allocated risk reduction.

When the safety function is allocated to the SIS, it is a safety instrumented function (SIF). The risk reduction allocated to the SIF defines its target SIL. This target is related to the SIF probability of failure on demand (PFD), e.g., SIL 1 (PFD range: 0.01 to 0.1), SIL 2 (PFD range: 0.001 to 0.01), SIL 3 (PFD range: 0.0001 to 0.001), and SIL 4 (PFD range: 0.00001 to 0.0001).

The identification of safety functions continues until the process risk associated with the hazard is reduced to meet the risk criteria. When there is insufficient risk reduction provided, the team makes recommendations for process design changes (e.g., inherently safer design), improvement to existing functions, or the implementation of new functions. These recommendations are generally prioritized based on the magnitude of the gap between the mitigated process risk (the risk considering the presence of existing functions) and the risk criteria.

Process Requirements Specification The process requirements specification is typically developed by process engineering based on the list of functions recommended by the various risk analyses performed on the unit operation and functions required by industry codes and standards. The process engineer determines the specific needs of the operations group and defines the best overall strategy for addressing the hazardous events during each planned operating mode given the operator staffing and their capability. The process requirements specification documents the result of a rationalization process that seeks to optimize the SIS functionality, reliability, and integrity using the inherently safer practices as recommended by CCPS (2016). The process requirements specification should define:

- Instrument justification and alarm rationalization basis
- The safe state, including safety and nonsafety actions
- The process safety time available for SIS response
- The reliability requirements necessary to achieve desired process unit availability
- The operability requirements for each process operating mode, such as start-up, reduced rates, degraded operation, maintenance, and shutdown
- The process conditions when it is safe to do SIS testing
- Why, when, and how each function in the SIS takes action
- Why, when, and how the operator should interact with the SIS, including any manual operation or bypass modes
- The compensating measures needed to ensure safe operation while any part of the SIS is degraded or disabled

Safety Requirements Specification There is often quite a bit of give and take between the process requirements specification and the safety requirements specification in the early stages of the project. For example, the ideal process measurement may not be practical in the existing installation. At all times, it should be recognized that the goal is to prevent the hazardous event from propagating to a loss event. The design documentation should establish a clear connection between the hazardous events and the functions implemented by the SIS to achieve the risk reduction strategy. Process simulations should verify that the SIS response times are fast enough to prevent propagation.

The SIS design phase focuses on achieving the target SIL through careful selection of the devices (e.g., user approved for safety), redundancy, on-line diagnostics, and frequent inspections and proof tests. ISA TR84.00.04-2015 gives extensive guidance on how to practically implement IEC 61511.

Application-specific standards by organizations such as the American Society of Mechanical Engineers (ASME), the American Petroleum Institute (API), and the National Fire Protection Association (NFPA) provide additional requirements and guidance for specialized applications like overpressure protection, overfill prevention, and burner management systems.

The SIS uses dedicated devices, including process sensors that detect the hazardous event, a logic solver that decides what to do, and final elements that take action on the process. Often, a single logic solver implements multiple functions, so the potential for common-cause failures between functions should be considered during design, since these failures can result in a significantly higher potential for the loss event to occur than assumed in the risk analysis.

The SIS is normally designed to fail-safe on loss of power and takes action only when the process demands that it do so. These demands often occur when safe operating limits are exceeded due to process control system failures. The SIS is generally separate and independent of the process control system, including its hardware, software, and user interfaces, such as operator, maintenance, and engineering interfaces.

Systematic errors can occur anywhere in the design and implementation process or during the operational life of an SIS device. These errors put the SIS on the path to failure in spite of the design elements incorporated to achieve robust hardware and software systems. Systematic errors are minimized using work processes that address potential human errors in the SIS design and management (e.g., programming errors or hardware specification errors).

Random hardware failure can occur throughout the device life as components age in the operating environment of the process unit. These failures can cause a device to fail dangerously, that is, it cannot perform as required. The failure rate of each device is estimated by examining its dangerous failure modes and their frequency of occurrence. This failure rate is used to calculate the PFD of the SIS considering its specific devices, redundancy, diagnostics, common-cause failure potential, and proof test interval. The PFD is then compared to the target SIL to determine whether the design is adequate. ISA TR84.00.02-2015 provides guidance on the verification of the SIL using quantitative and qualitative techniques.

Engineering, Installation, Commissioning, and Validation (EICV) This phase involves the physical realization of the SRS. The bulk of the work in this phase is not a process engineering effort. Detailed engineering, installation, and commissioning is generally an I&E function. However, this is where the assumptions and requirements developed by the process engineer are put into practice and validated.

Validation is performed as part of a site acceptance test (SAT). Validation involves a complete end-to-end test to demonstrate that the SIS works as installed in the operating environment. It proves that SIS devices execute the logic according to the specification and ensures that the SIS and its devices interact as intended with other systems, such as the process control system and operator interface. The SAT also provides an opportunity for a first-pass validation of the procedures developed for the operating basis (see next subsection).

A functional safety assessment is conducted prior to start-up to ensure that the SIS design and construction are complete. This assessment provides information for the unit pre-start-up safety review, which is used to verify that the process unit is ready for safe start-up. All documentation is formally updated to as-built status, incorporating any modifications made since the last formal revision. Once the functional safety assessment has approved the SIS for process unit start-up, formal management of change is followed to evaluate proposed modifications to the SIS or its associated documentation. Any deviation from the approved SRS is reviewed and approved by appropriate parties prior to change implementation.

Operating Basis The resources and skills of plant operations are considered during engineering design, since the equipment must be safely monitored and operated by these personnel. They must be trained on the new SIS and on their responsibilities prior to the SIS being turned over to operations and maintenance personnel. Consideration should be given to the content and depth of the information that must be communicated to various personnel. This is especially important as the responsibility for the SIS transitions from the project team to operations and maintenance control.

The process engineer is responsible for defining the content of SIS operating procedures, which should cover SIS-specific information (e.g., set points, SIS actions, and the hazard that is being prevented with SIS), the correct use of bypasses and resets, the operator response to SIS alarms and trips, when to execute a manual shutdown, and provisions for operation with detected faults (e.g., compensating measures). These procedures, along with analogous ones developed by maintenance and reliability engineering for maintenance activities, make up the backbone of the operating basis.

ISA TR84.00.03-2012 provides guidance on the types of procedures needed to sustain the SIS equipment in the "as good as new" condition. Since a device can fail at any time during its life, periodic proof tests are performed to demonstrate the functionality of the SIS. Proof tests are covered by operation and maintenance procedures that ensure that the test is done correctly, consistently, and safely and that the device is returned to a fully operational state after test. Each test serves as an opportunity for personnel to see the SIS in action and to validate the procedures associated with its operation. Failures found during testing need to be investigated using root cause analysis, and corrective actions should be taken to prevent reoccurrence.

Periodic assessment and audit of process demand records, process safety events, test results, organization competency, and management system documents help ensure that the SIS performance remains consistent with the H&RA assumptions and the process requirements specification. Quality assurance of the installed system requires procedures to collect performance data to support prior use analysis and continued use of the equipment, to verify that the in-service performance remains consistent with the claimed SIL, and to ensure that action is taken to correct deficiencies.

SECURITY

General References: ANSI/API Standard 780, *Security Risk Assessment Methodology for the Petroleum and Petrochemical Industries,* Washington D.C., May 2013; CCPS, *Guidelines for Analyzing and Managing the Security Vulnerabilities of Fixed Chemical Sites,* American Institute of Chemical Engineers, New York, 2002; *GIE Security Risk Assessment Methodology,* Gas Infrastructure Europe, Brussels, 2015; ISO 31000, *Risk Management—Principles and Guidelines on Implementation,* International Organization for Standardization, Geneva, 2009; National Institute of Justice, *Chemical Facility Vulnerability Assessment Methodology,* Washington, D.C., July 2002; Synthetic Organic Chemical Manufacturers Association (SOCMA), *Manual on Chemical Site Security Vulnerability Analysis Methodology and Model,* Washington, D.C., 2002.

Introduction Prior to the terror attack on the United States on September 11, 2001, commonly known as 9/11, chemical process safety activities focused mainly on accidental release risks and excluded most considerations of intentional releases. Security was provided mostly for lesser threats than such extreme acts of violence, and terrorism was generally not provided for except in high-security areas of the world. Exceptions to this included general concerns for sabotage. This was due to a perception that these risks were managed adequately, and that the threat of a terrorist attack, particularly on chemical manufacturing facilities or transportation systems, was remote.

Following 9/11, it has become increasingly apparent that the threat of intentional harm to infrastructure, especially where hazardous materials are manufactured, stored, processed, or transported, had to be considered a credible concern. Security for the chemical industry has taken on increased emphasis as a result, and such organizations as the American Institute of Chemical Engineers recognized the paradigm shift and published guidelines on analyzing these threats (CCPS, *Guidelines for Managing and Analyzing the Security Vulnerabilities of Fixed Chemical Sites,* AIChE, New York, 2002). The concerns of international terrorism have spread to many countries around the world, and addressing this concern is now a permanent part of the requirements of the chemical engineering profession. Chemical engineers now must include *chemical process security* as a critical element of the management of a process facility.

Chemical process security management has as its objectives:

1. To minimize the risk of harm to the public or employees from intentional acts against a process facility, and
2. To protect the assets (including employees) of the process facility to maintain the ongoing integrity of the operation and to preserve the value of the investment.

Process security and process safety have many parallels and use many common programs and systems for achieving their ends. Process security management requires a systems approach to develop a comprehensive security program, which shares many common elements with process safety management.

Chemical process security includes, but goes beyond, traditional *physical security.* Physical security includes such considerations as guards, barriers, surveillance equipment, and other physical system considerations. Physical security is an element of chemical process security, but physical security alone is not always adequate to address the new challenges of security against extreme acts of violence, such as terrorism. Effective chemical process security must also consider the integration of broader process elements, including technology, chemical usage and quantities, procedures, administrative controls, training, and cyber interface with those traditional physical security elements.

The chemical engineer has an opportunity to influence these considerations in all stages of a process life cycle, including concept, engineering, construction, commissioning, operations, modification, and decommissioning. Security issues that are recognized in the concept and design phases of a project allow for cost-effective considerations that can eliminate or greatly minimize security risks. For example, if a buffer zone can be provided between the public areas and a plant fence, and then again between a plant fence and critical process equipment, those two zones can effectively provide such benefits as

- ***Detection zone(s)***, given they are free of obstacles and have sufficient depth to allow for adversaries to be detected while attempting unlawful entry
- ***Standoff zone(s)***, given they have sufficient depth to keep adversaries from using explosives or standoff weapons effectively from the perimeter
- ***Delay zone(s)*** allowing intervening force the time to respond or time for operators to take evasive action before an adversary reaches a target following detection

A chemical engineer may have a choice of inherent safety variables, such as quantity stored or process temperatures and pressures, or process safety measures such as emergency isolation valves or containment systems, all of which may greatly reduce the vulnerabilities or the consequences of intentional loss. These are in addition to traditional security measures that include physical security, background checks, administrative controls, access controls, or other protective measures. For a more complete discussion of the options, refer to the CCPS document *Guidelines for Analyzing and Managing the Security Vulnerabilities of Fixed Chemical Sites* (AIChE, New York, 2002) and other references [*Counterterrorism and Contingency Planning Guide*, special publication from *Security Management Magazine* and American Society for Industrial Security, 2001; Dalton, *Security Management: Business Strategies for Success*, Butterworth-Heinemann Publishing, Newton, Mass., 1995; Walsh and Healy, eds., *Protection of Assets Manual*, Merritt Co., Santa Monica, Calif. (four-volume loose-leaf reference manual, updated monthly)].

Threats of Concern Terrorist acts can be the most problematic to defend against since they may be more extreme or malevolent than other crimes focused on monetary gains or outcomes with less malicious intent. Terrorists often use military-style tactics not usually provided for in base chemical facility design. Chemical facility security must be considered in context with local and national homeland security law enforcement activities, as well as with emergency response capabilities. There is a practical limit to the ability of a chemical site to prevent or mitigate a terrorist act. Above a certain level of threat, the facility needs to rely on law enforcement and government protective services to provide enhanced physical security against extreme acts of intentional harm. The security posture must be both threat-based and risk-based; therefore, the continued application of extremely robust security measures is not usually necessary.

The acts of concern for terrorism can be generally defined as involving the four motives shown in Table 23-30.

Other adversaries that must be considered applicable include those capable and interested in perpetrating a full spectrum of security acts. These may include outside parties or insiders or a combination of the two working in collusion.

The threats that are applicable and the adversaries that may be culpable are characterized to understand their capabilities, intent, and therefore potential targets and tactics. The targets and acts of interest to various

TABLE 23-30 Example List of Candidates to Be Considered as Critical Assets (ANSI/API Standard 780 SRA Methodology)

Security motives of concern	Candidate critical assets threatened
Loss of containment, damage, or injury	• The public, employees, contractors and visitors • Process equipment handling hazardous chemicals, including processes, pipelines, and storage tanks. Marine vessels and facilities, pipelines, and other transportation systems
Theft	• Hazardous chemicals processed, stored, manufactured, or transported • Metering stations, process control and inventory management systems • Critical business information from telecommunications and information management systems, including Internet-accessible assets • Important economic assets ranging from intellectual property to physical assets
Contamination	• Raw material, intermediates, catalysts, products, processes, storage tanks, and pipelines • Critical business or process data
Degradation of assets	• Processes containing hazardous chemicals • Business image and community reputation • Utilities (electric power, steam, water, natural gas, and specialty gases) • Telecommunications systems • Business systems
Other security events (determined to be relevant)	• Corporate identity and reputation and related value • Personnel • Critical data • Operational integrity • Records

Adapted from ANSI/API Standard 780, *Security Risk Assessment Methodology for the Petroleum and Petrochemical Industries*, Washington, D.C., 2013.

adversaries will vary with the group. For example, a terrorist may be interested in destroying a process through violent means, such as by the use of an explosive device. An activist or disgruntled insider may be interested only in a nonviolent protest or in causing some limited physical damage, but not in harming the environment, coworkers, or the public in the process. The various adversaries and strategies of interest form the basis of the security risk assessment (SRA), which is the foundation of a chemical process security management system specific to address the anticipated threats.

Overall Objectives of Terrorism Terrorists try to accomplish their goals by creating fear and uncertainty in the target population through the use of violent acts. The underlying goals include fundamentalist objectives, such as purity of religion or idealistic goals, but they may include power struggles, such as trying to overthrow a government, or reparations, such as revenge for past actions. The reason for a chemical plant being targeted may be that it serves an adversary of the terrorist (economic or military significance) or that it can be weaponized to cause third-party harm (health and safety consequences from intentional release of hazardous materials).

Security Vulnerability Assessment/Security Risk Assessment A security vulnerability assessment (SVA) or security risk assessment (SRA) is intended to identify security vulnerabilities and risks and protection gaps that can be exploited by the adversary from a wide range of threats ranging from vandalism to terrorism. With the recognition of threats, consequences, and vulnerabilities, the risk of security events can be evaluated, and a security management system can be organized to cost-effectively mitigate those risks through business-prudent countermeasures.

Security risk assessment methodologies include the following five steps:

1. *Characterization.* Characterize the facility or operation to understand what critical assets need to be secured, their importance, and their infrastructure dependencies and interdependencies.

2. *Threat assessment.* Identify and characterize threats against those assets, and evaluate the assets in terms of the attractiveness of the targets to each threat and the consequences if they are damaged, compromised, or stolen.

3. *Vulnerability assessment.* Identify potential security vulnerabilities that enhance the probability that the threat will successfully accomplish the act.

4. *Risk evaluation.* Determine the risk represented by these events or conditions by determining the likelihood of a successful event and the maximum credible consequences of an event if it were to occur; rank the risk of the event occurring and, if it is determined to exceed risk guidelines, make recommendations for lowering the risk.

5. *Risk treatment.* Identify and evaluate risk mitigation options (both net risk reduction and benefit/cost analyses) and reassess the risk to ensure adequate countermeasures are being applied. Evaluate the appropriate response capabilities for security events and the ability of the operation or facility to adjust its operations to meet its goals in recovering from the incident.

SRA and SVA Methodologies There are several SRA and SVA techniques and methods available to the industry, all of which share common elements. The following is a list of some available methodologies published by various governments, private, and trade and professional organizations. Some are merely chapters or sections of documents that address security or risk assessment/risk management in broader terms. Some are SRA or SVA publications by themselves. Some of these methods are complete, systematic analytical techniques, and others are mere checklists.

- American Institute of Chemical Engineers, Center for Chemical Process Safety: *Guidelines for Analyzing and Managing the Security Vulnerabilities of Fixed Chemical Sites* (AIChE, New York, 2002)
- ANSI/API Standard 780: *Security Risk Assessment Methodology for the Petroleum and Petrochemical Industries* (Washington, D.C., 2013)
- Gas Infrastructure Europe: *GIE Security Risk Assessment Methodology* (Brussels, 2015)
- National Institute of Justice: *Chemical Facility Vulnerability Assessment Methodology* (Sandia VAM, Washington, D.C., 2002)
- Synthetic Organic Chemical Manufacturers Association: *Manual on Chemical Site Security Vulnerability Analysis Methodology and Model* (Washington, D.C., 2002).

One approach to conducting an SRA is shown in Fig. 23-49. This methodology was published by the ANSI/API, Standard 780, *Security Risk Assessment Methodology for the Petroleum and Petrochemical Industries* (Washington, D.C., 2013). The ANSI/API Standard 780 SRA is founded on a risk-based approach to managing petroleum and petrochemical facility security. To begin the process, companies usually perform an enterprise-level screening methodology to sort out significant risks among multiple sites and to determine priorities for analysis and implementation of any recommended changes.

Overview of the Security Risk Assessment Methodology The SRA is conducted in several stages, with five steps.

Step 1: Asset Characterization The asset characterization includes analyzing information that describes the technical details of the facility assets to support the analysis, identifying the pathways or access points, potential critical assets, identifying the hazards and consequences of concern, and supporting infrastructure and identifying existing elements of protection. For the SRA, the *Consequences (C)* represent the degree of event severity in terms of short- and long-term injury, damage, or business interruption costs that would reasonably result were an adversary able to successfully engage critical assets, with the understanding that malevolent acts may yield more severe effects than expected from accidental events. Table 23-31 provides adjectival references for qualitative Definitions of Consequence Severity of the Undesired Event used to plot severity on a scale of 1 to 5.

Step 2: Threat Assessment and Asset Attractiveness The consideration of possible threats includes internal threats, external threats, and internally assisted threats (i.e., collusion between internal and external threats). The evaluation and subsequent selection of the specific threats includes higher-order (terrorist) threats as well as lesser facility-specific threats. For the SRA, the *Threat (T)* is an unwanted incident perpetrated by threats to attack or sabotage critical assets; to include the capability, intent, and methods targeted at the intentional exploitation of vulnerability to create an undesired consequence. Threat ranking is expressed in factors of:

- Credible existence of threat for the location of the assets
- Intelligence about the threat, including general history of events
- Suspected intent or motivation
- Intelligence about the threat specific to the location being analyzed
- Assessed capability and ability of the adversary to execute the act
- Degree of interest the adversary is perceived to have in attacking or diverting assets or systems (the Attractiveness of the asset to the adversary), given targeting choices at the facility if an attack or theft were to occur

Step 2 also determines the target attractiveness of each asset from the adversary's perspective as identified in the threat assessment. Various factors must be considered to determine the attractiveness to the adversary type to make a ranking judgment of target attractiveness. *Attractiveness (A)* is a factor that modifies the Threat estimate to result in the likelihood of the security event for a specific act or against a specific asset. This factor can be evaluated as a composite estimate based on such factors as the perceived value of a target to the Threat, and the Threat's choice of targets to avoid discovery and to maximize the probability of success. This may be related to a conditional probability between 0.0 and 1.0 in increments of 0.2 for each

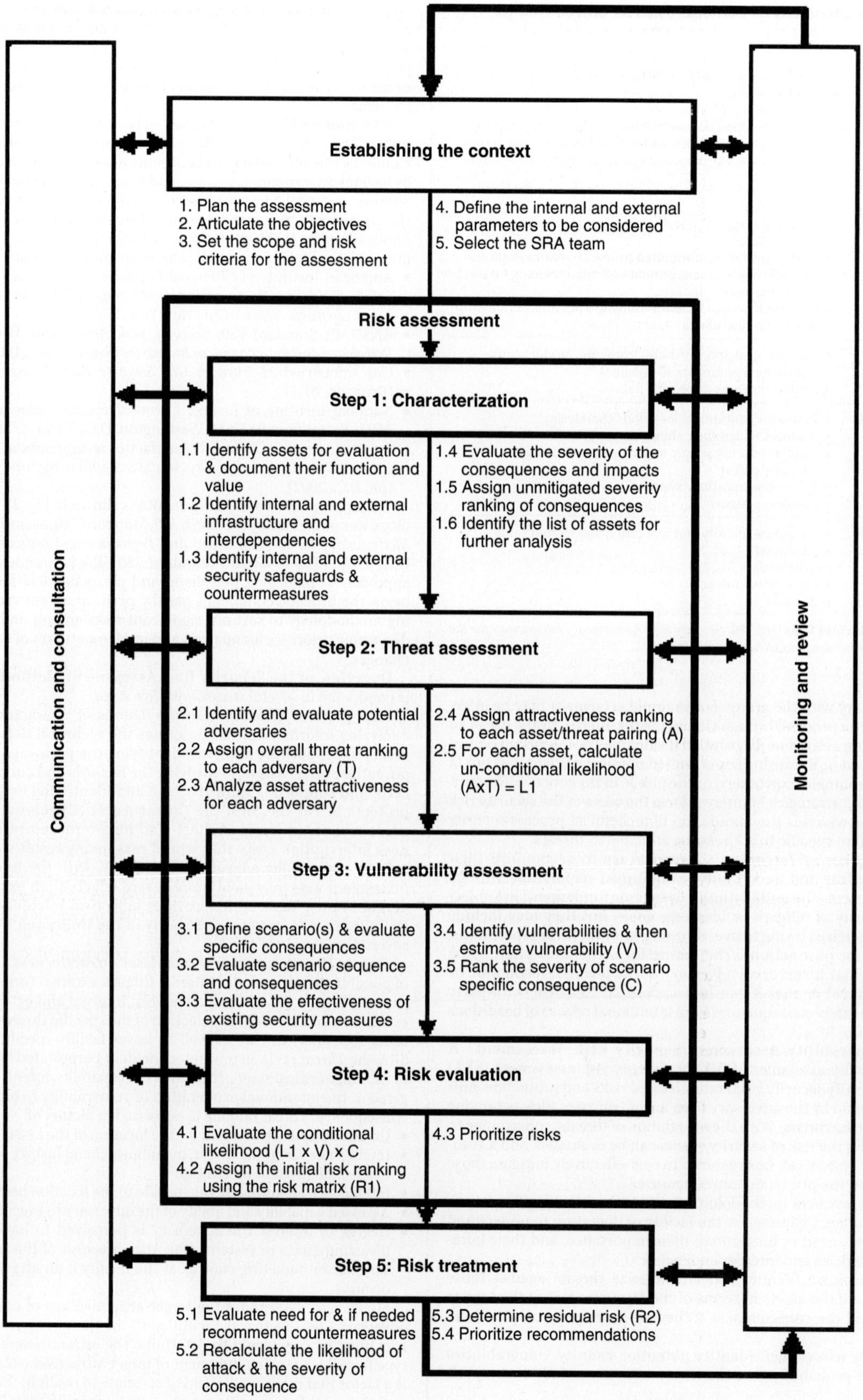

FIG. 23-49 One approach to conducting an SRA. (*Adapted from ANSI/API Standard 780, Security Risk Assessment Methodology for the Petroleum and Petrochemical Industries, Washington, D.C., 2013.*)

TABLE 23-31 Example Definitions of Consequences of the Event (ANSI/API Standard 780 SRA Methodology)

Description	Ranking
A. Possibility of minor injury on-site; no fatalities or injuries anticipated off-site. B. No environmental impacts C. Up to $X loss in property damage D. Very short term (up to X weeks) business interruption/expense E. Very low or no impact or loss of reputation or business viability; mentioned in local press	1
A. On-site injuries that are not widespread but only in the vicinity of the incident location; no fatalities or injuries anticipated off-site B. Minor environmental impacts to immediate incident site area only, less than X year(s) to recover C. $X to $X loss in property damage D. Short term (>X week to Y months) business interruption/expense E. Low loss of reputation or business viability; query by regulatory agency; significant local press coverage	2
A. Possibility of widespread on-site serious injuries; no fatalities or injuries anticipated off-site B. Environmental impact on-site and/or minor off-site impact, Y year(s) to recover C. Over $X to $X loss in property damage D. Medium term (Y to Z months) business interruption/expense E. Medium loss of reputation or business viability; attention of regulatory agencies; national press coverage	3
A. Possibility of X to Y on-site fatalities; possibility of off-site injuries B. Very large environmental impact on-site and/or large off-site impact, between Y and Z years to recover C. Over $X to $X loss in property damage D. Long-term (X to Y years) business interruption/expense E. High loss of reputation or business viability; prosecution by regulator; extensive national press coverage	4
A. Possibility of any off-site fatalities from large-scale toxic or flammable release; possibility of multiple on-site fatalities B. Major environmental impact on-site and/or off-site (e.g., large-scale toxic contamination of public waterway), more than XX years/poor chance of recovery C. Over $X loss in property damage D. Very long-term (>X years) business interruption/expense; large-scale disrupttion to the national economy, public or private operations; loss of critical data E. Very high loss of reputation or business viability; international press coverage	5

Adapted from ANSI/API Standard 780, *Security Risk Assessment Methodology for the Petroleum and Petrochemical Industries*, Washington, D.C., 2013

of the five levels as an additional means of relating to the Attractiveness estimate.

Step 3: Vulnerability Analysis The vulnerability analysis includes the relative pairing of each target asset and threat to identify potential vulnerabilities related to process security events. This involves the identification of existing countermeasures and their level of effectiveness in reducing those vulnerabilities. The degree of vulnerability of each valued asset and threat pairing is evaluated by the formulation of security-related scenarios to conduct the security vulnerability analysis, including the assignment of risk rankings to those security-related scenarios.

During the vulnerability analysis, each target is paired with the applicable Threat(s) using threat ranking criteria. Some targets located together in close proximity may be grouped together based on the potential collateral effects that could reasonably be expected if the adjacent target was attacked, and on the security countermeasures that would have to be breached in order for the attack to be successful. The existing safeguards that would function in an attack on the target and the existing security countermeasures for that target/Threat combination are also identified, where Threat is the likelihood of an adversary to be interested in causing harm, its ability to cause harm, and the likelihood of success based on its capabilities. Attractiveness is the probability that the adversary will target that particular asset, and vulnerability is a probability that the adversary's attack against a given target will be successful.

The *Likelihood (L)* is an estimate of the probability or frequency that a given act will result in a given consequence. It is both a function of the Threat seeking out the asset and attempting the act as well as the successful execution of the act to achieve the Threat's goals. The combination of the two factors Threat (T) and Attractiveness (A) produce a surrogate estimate for the Likelihood of the act (L_1) where $T \times A = L_1$ for each scenario, which is either a probability of the event or a frequency over a given period of time.

Step 4: Risk Assessment and Risk Treatment The risk assessment determines the relative degree of risk in terms of the expected effect on each critical asset as a function of consequence and probability of occurrence. Using the assets identified during step 1 (Asset Characterization), the risks are prioritized based on the likelihood of a successful attack, which is a function of the threats assessed under step 2 and the degree of vulnerability identified under step 3.

The level of effectiveness of the existing safeguards and countermeasures is then examined to determine what (if any) security vulnerabilities exist for each target or set of targets, to include physical security, cyber security, administrative controls, and other safeguards. The objective is to identify measures that protect the entire facility or an individual critical asset to determine the effectiveness of the protection. The likelihood of an adversary successfully exploiting some vulnerability and an estimate of the resulting sabotage damage are then determined. Based on this assessment, judgments are made on the degree of risk and the need for and the value of implementing additional countermeasures.

The *Vulnerability (V)* is any weakness that can be exploited by one or more threats to gain unauthorized access and subsequent destruction, degradation, or theft of an asset. Vulnerability is a surrogate for the Likelihood of expected success (L_2) for each scenario ($L_2 = V$).

Possible consequences of the postulated security events are assessed, and a risk ranking scale is used to estimate the degree of severity. These scenario deliberations provide a basis for prioritizing the results of the vulnerability analysis and for evaluating the need for recommendations. Each credible security scenario (e.g., each combination of target and threat) is assigned a risk ranking based on the severity of the undesired event that assesses the exposures to casualties, replacement cost, environmental consequences, business interruption, and company reputation. Five levels of increasing severity are applied to each of the five categories, providing a reference to determine the most credible severity in each category, where consequence severity ranking is the worst credible outcome across all of the categories.

The level of risk of the adversary exploiting the asset is determined in terms of the existing security countermeasures, as a function of consequence and probability of occurrence, to determine the relative degree of risk in terms of the expected effect on each critical asset or pathway. This analysis is a function of the consequences or impacts to the critical functions from the disruption or loss of the critical asset, and the likelihood of a successful attack as a function of the threat or adversary and the degree of vulnerability of the asset. The scenarios are then risk-ranked using a risk ranking matrix, such as the example in Fig. 23-50, plotting each scenario based on its likelihood and consequences, to categorize the assets into discrete levels of risk for the application of appropriate countermeasures. Hence the overall likelihood calculation is expressed as ($L_1 \times V$) on one axis and plotted with the severity of consequence on the other axis.

Step 5: Proposed Countermeasures Strategy and Recommendation Prioritization Based on the vulnerabilities identified and the risk that the layers of protection are breached, appropriate enhancements to the security countermeasures are recommended. Countermeasure options are identified to further reduce vulnerabilities following the security doctrines of deter, detect, delay, respond, mitigate, and possibly prevent. Based on the consequences and the likelihood that the existing layers of protection were breached, appropriate business-prudent enhancements to the security countermeasures are recommended.

			Likelihood (L)				
Consequences (C)			VL	L	M	H	VH
			1	2	3	4	5
	VH	5	3	4	4	5	5
	H	4	2	3	4	4	5
	M	3	2	2	3	4	4
	L	2	1	2	2	3	4
	VL	1	1	1	2	2	3

FIG. 23-50 Example risk matrix for security risk assessment.

The evaluation of potentially applicable countermeasures also considers other factors, including:

- Reduced probability of successful attack
- Degree of risk reduction by the options
- Reliability and maintainability of the options
- Capabilities and effectiveness of mitigation options
- Costs of mitigation options

The SRA is the cornerstone for integrating chemical security management and process safety management strategies into a comprehensive process safety and security strategy. Security risk reduction opportunities during the process life cycle are explained, as well as various process risk management strategies (including inherent safety) that are applicable.

Defining the Risk to Be Managed For the purposes of the SRA, the definition of risk is an expression of the likelihood that a defined threat will target and successfully attack a specific security vulnerability of a particular target or combination of targets to cause a given set of consequences. This can be expressed as Security Risk (R_S) as a function of Consequences, Vulnerability, and Threat, or

$$R_S = f(C, V, T) \tag{23-107}$$

where C = Direct and indirect consequence of a successful act against an asset
V = Vulnerability of the asset to the act
T = Threat associated with the act
R_S = The likelihood of a successful act against an asset, assuming both the likelihood of the act occurring (L_1) and the likelihood of success (L_2) causing a given set of Consequences. Therefore, R_S = a function of (C, L_1, L_2) or $R_S = C, (A \times T), V$ (ANSI/API Standard 780 *Security Risk Assessment Methodology for the Petroleum and Petrochemical Industries*, Washington, D.C., 2013, pp. 11–12).

This is contrasted with the usual accidental risk definitions. The risk variables to be analyzed are defined as shown in ANSI/API Standard 780, *Security Risk Assessment Methodology for the Petroleum and Petrochemical Industries* (Washington, D.C., 2013, p. 31). A challenge for security risk analysis is that the accurate prediction of the frequency and location of terrorist acts is not considered credible. Thus, the analyst has a choice of assuming a frequency of a certain attack based on the threat assessment or assuming the attack frequency is continuous, thereby focusing solely on the conditional likelihood of success of the adversary who attempts an attack. While the latter approach provides a baseline for making decisions about vulnerability, it does not fully answer the question of costs versus benefits of any countermeasures. Certain crimes other than terrorism may be more predictable or frequent, allowing for statistical analysis to help frame the risks and justify countermeasure expenditures. Due to this limitation, the factor of attractiveness is considered along with consequences, threat, and vulnerability, to determine the priorities for and design of security measures for the industry.

Security Strategies A basic premise is that not all security risks can be completely prevented. Appropriate strategies for managing security can vary widely depending on the circumstances, including the type of facility and the threats facing the facility. As a result, it is difficult to prescribe security measures that apply to all facilities in all industries. Instead, it is suggested to use the SRA as a means of identifying, analyzing, and reducing vulnerabilities. The specific situations must be evaluated individually by local management using best judgment of applicable practices. Appropriate security risk management decisions must be made commensurate with the risks. This flexible method recognizes that there is not a uniform approach to security in the chemical process industry, and that resources are best applied to mitigate higher-risk situations.

Security strategies for the process industries are generally based on the application of five key concepts against each threat: deterrence, detection, delay, response, and recovery.

A complete security design includes these concepts in layers of protection or a defense in depth arrangement. The most critical assets should be placed in the center of conceptual concentric levels of increasingly more stringent security measures. In the concept of rings of protection, the spatial relationship between the location of the target asset and the location of the physical countermeasures is often important.

In the case of malicious acts, the layers or rings of protection must be particularly robust because the adversaries are intentionally trying to breach the protective features, and they can be counted on to use whatever means are available to succeed. This could include explosives or other initiating events that result in widespread common-cause failures. Some particularly motivated adversaries might commit suicide while trying to breach the security layers of protection.

Countermeasures and Security Risk Management Concepts Countermeasures are actions taken to reduce or eliminate one or more vulnerabilities. Countermeasures include hardware, technical systems, software, interdiction response, procedures, and administrative controls.

Security risk reduction at a site can include the following strategies:

- Physical security
- Cyber security
- Crisis management and emergency response plans
- Policies and procedures
- Information security
- Intelligence
- Inherent safety

Security Management System A comprehensive process security management system must include management program elements that integrate and work in concert with other management systems to control security risks. The 13 management practices shown in Table 23-32 is an example of a management system developed by the American Chemistry Council.

The purpose of a security management system is to ensure the ongoing, integrated, and systematic application of security principles and programs to protect personnel and assets in a dynamic security environment to ensure the continuity of the operation and supporting or dependent infrastructure. Traditional industrial facility security management tended to focus on protection of persons and property from crime (e.g., theft of property, workplace violence) and crime prevention, response, and investigation. While these are still elements of facility security, a management system allows the incorporation of broader security concerns relating to intentional attack on fixed assets, by higher-order adversaries. To develop and implement a *security* management system not only provides a more thorough, dynamic, risk-based, and proactive approach, but also allows security management to be integrated into a facility's overall environmental, health, and safety management systems.

The American Chemistry Council's Responsible Care® Security Code is designed to encourage continuous improvement in security performance by using a risk-based approach to identify, assess, and address vulnerabilities;

TABLE 23-32 American Chemistry Council's Responsible Care® Security Code Process Security Management System*

1.	Leadership commitment
2.	Analysis of threats, vulnerabilities, and consequences
3.	Implementation of security measures
4.	Information and cyber security
5.	Documentation
6.	Training, drills, and guidance
7.	Communications, dialogue, and information exchange
8.	Response to security threats
9.	Response to security incidents
10.	Audits
11.	Third-party verification
12.	Management of change
13.	Continuous improvement

**Site Security Guidelines for the U.S. Chemical Industry,* American Chemistry Council, October 2001.

prevent or mitigate incidents; enhance training and response capabilities; and maintain and improve relationships with key stakeholders. As a condition of membership in the council, each member company must implement the Security Code for facilities, transportation and value chain, and cyber security.

ELECTRICAL AREA CLASSIFICATION

GENERAL REFERENCES: API Recommended Practice 500, *Classification of Locations for Electrical Classification at Petroleum Facilities Classified as Class I, Division 1 and Division 2,* American Petroleum Institute, Washington D.C., 2012; Canadian Electrical Code CSA C22.1, Canadian Standard Association, Toronto, Canada, 2015; IEC 60079-10, *Electrical Apparatus for Explosive Gas Atmospheres,* part 10, "Classification of Hazardous Areas," International Electrotechnical Commission, Geneva, Switzerland, 2002; NFPA 70, National Electric Code, Chapter 5, Article 500 and 505, National Fire Protection Association (NFPA), Quincy, Mass., 2014; NFPA 70B, *Recommended Practice for Electrical Equipment Maintenance,* NFPA, Quincy, Mass., 2016; NFPA 496, *Purged and Pressurized Enclosures for Electrical Equipment,* NFPA, Quincy, Mass., 2017; NFPA 497, *Recommended Practice for the Classification of Flammable Liquids, Gases or Vapors in Chemical Process Areas,* NFPA, Quincy, Mass., 2017; NFPA 499, *Recommended Practice for the Classification of Combustible Dusts in Chemical Process Areas,* NFPA, Quincy, Mass., 2017.

Background Flammable liquids, vapors, and gases, combustible dusts, and ignitable fibers are hazardous materials found in many processing facilities. Electrical equipment has the potential to act as an ignition source for these hazardous materials. The concept of the electrical classification area and the divisions within the area was created to provide a graded measure of the potential for an ignition event.

A hazardous (classified) location is a space containing any of the following:

- An atmosphere in which an ignitable concentration of flammable gas, vapor, or dust is present or might occasionally be present.
- Electrical equipment on which combustible dust might accumulate and interfere with heat dissipation from the equipment.
- Surfaces (especially horizontal surfaces) that contain easily ignitable fibers.

Once the Class, Division (Zone), and Group have been defined, electrical equipment for the location can be identified. This is more important when installing new electrical equipment in an existing classified area than in a new installation.

Electrical Classification Areas are classified according to the properties of the material being used and the surrounding atmosphere. Elements that affect area classifications may include the availability of flammable or explosive material, operating temperature and pressure, flash points, autoignition temperature, vapor density of the material, resistivity of dust or fibers, dust layer ignition temperature, open or sealed conduit, and ventilation. Each room, section, or area must be considered individually in determining its classification. Normal activities such as draining liquids, disconnecting hose, drumming, and sampling can affect the electrical classification. The overall classification of the area should also be considered, such as a control building within a process unit: although the process unit may be electrically classified, the control building could be purged, making it nonclassified.

In recent years, electrical area classification has become more focused on the potential of release and distance from the release point than the potential for a flammable vapor/air mixture. Therefore, equipment types are often mixed inside buildings or units, instead of all being explosion-proof.

Classification is a complex issue, but the main factors to consider are:

- Process conditions
- Materials being used
- Where the material is used
- The probability of the presence of a gas
- The quantity and duration of hazardous vapor
- The amount of ventilation
- The nature of the gas: lighter or heavier than air

Electrical Classification System in the United States In the United States and other parts of the world, hazardous (classified) locations are typically designated by Class, Division, and Group.

Classes Three distinct classes of hazardous (classified) locations have been established:

- *Class I areas* are characterized as areas containing flammable vapors escaping from a flammable or heated combustible liquid and areas containing flammable gases.
- *Class II areas* contain combustible dust suspended in air or combustible dust accumulations that can interfere with heat dissipation from electrical equipment or can be ignited by that equipment.
- *Class III areas* contain accumulations of fibers.

Divisions Each classified area is divided into either Division 1 or 2. For a Class I or Class II area,

- Division 1 is likely to contain the hazardous condition during normal operations or frequently because of maintenance and repair.
- Division 2 is likely to contain the hazardous condition only under abnormal circumstances, such as process upset or equipment failure.

TABLE 23-33 Typical Materials and Groups

Group	Material
A	Acetylene
B	Acrolein, butadiene, ethylene oxide, propylene oxide, hydrogen
C	Ethylene, ethyl ether, hydrogen sulfide
D	Acetone, acrylonitrile, ammonia, benzene, butane, ethane, ethanol, hexanes, heptanes, isoprene, methane (natural gas), methanol, petroleum naphtha, propane, propylene, styrene, toluene, vinyl acetate, vinyl chloride, xylenes
E	Aluminum, magnesium, titanium, zinc, bronze, chromium, tin, cadmium
F	Carbonaceous dusts, carbon, charcoal, coal
G	Alfalfa, cocoa, coffee, corn, corn starch, rice, sugar, wheat

For Class III areas, the division classification is based on whether the area is used for processing or storage. A Class III manufacturing area may be Division 1, while a Class III warehouse is likely to be Division 2.

Groups Equipment protective features also depend on the degree or severity of a hazard to which equipment is exposed. Class I hazardous materials are typically placed into one of four groups, depending on their physical properties and characteristics. Dusts, which are Class II materials, are similarly grouped by degree of hazard. Typical materials and their Groups are shown in Table 23-33.

Zone Alternative Method NFPA 70, Article 505 (NFPA 70, National Electric Code, Chapter 5, Article 505, National Fire Protection Association, Quincy, Mass.) allows the use of Zones instead of Divisions, as shown in Table 23-34. This method is also used in other areas of the world.

European Hazardous Area The European Union (EU) ATEX Directive 94/9/EC covers electrical and mechanical equipment and protective systems, which may be used in potentially explosive atmospheres (flammable gases, vapors or dusts). Table 23-35 shows the relationship between the category and the expected zone of use.

Equipment Certification Once a hazardous location has been classified, appropriate electrical equipment should be chosen for that area. In general, equipment must be approved for use in a classified area. Listed equipment for hazardous (classified) areas is marked to show the code-specified environments where it can be safely used. These markings often include the maximum surface temperature of the equipment under normal operating conditions.

Most general-purpose electrical equipment has an unacceptably high probability of igniting flammable concentrations of vapors, dusts, and fibers. Consequently, a range of electrical equipment has been developed to minimize this probability by means of special design and construction features. This includes limiting the surface temperature of the equipment, minimizing the potential for sparking, and controlling vapor travel into or out of an electrical enclosure, including upset and cleanout or turnaround conditions.

While explosion-proof and dust-ignition-proof enclosures are most often used in hazardous areas, there are other National Electrical Manufacturers Association (NEMA) type enclosures for electrical equipment located in nonhazardous areas. When selecting heat-generating electrical equipment for a hazardous (classified) location, its hottest external-surface operating temperature should be compared to the ignition temperature of the surrounding gas, vapor, or dust. Lowering the ignition temperature for organic dusts that

TABLE 23-34 Zone Definitions

Zone	Definition
0	Where flammable gases or vapors are present continuously or for long periods of time
1	Where flammable gases or vapors are present under normal operating conditions or exist frequently under repair or maintenance conditions
2	Where flammable gases or vapors are present only during abnormal conditions
21	Where a combustible dust, such as a cloud, is likely to occur during normal operation in sufficient quantity to be capable of producing an explosive concentration of combustible or ignitable dust mixture in air
22	Combustible dust clouds may occur infrequently and may persist for only a short time, or the accumulation or layers of combustible dust may be present under abnormal conditions and give rise to ignitable mixtures of dust in air. Following an abnormal condition, if the removal of dust accumulations or layers cannot be assured, then the area is classified as Zone 21

TABLE 23-35 EU Categories and Zones

Category	Zone	Definition
1	0	An area in which an explosive gas atmosphere is present continuously or for long periods of time.
2	1	An area in which an explosive gas atmosphere is likely to occur in normal operations.
3	2	An area in which an explosive gas atmosphere is not likely to occur in normal operation and, if it occurs, will only exist for a short time.

dehydrate or carbonize should be considered. Care should be taken to ensure that these special features of the equipment match the flammability and ignition characteristics of the materials to which it is likely to be exposed.

FIRE PROTECTION

General References: API Recommended Practice 2218, *Fireproofing Practices in Petroleum and Petrochemical Processing Plants*, American Petroleum Institute, Washington, D.C., 2013; CCPS, *Guidelines for Fire Protection in Chemical, Petrochemical, and Hydrocarbon Processing Facilities*, AIChE, New York, 2003; CCPS, *Guidelines for Engineering Design for Process Safety*, 2d ed, AIChE, New York, 2012; *Fire Protection Handbook*, 20th ed., National Fire Protection Association, Quincy, Mass., 2008.

Introduction to Fire Protection Potential fires in process facilities include vessel and equipment fires (internal or external), ground-level pool fires, multilevel and three-dimensional fires resulting from spills or releases at elevated levels, liquid or gas jet fires from leaks, gas fires from vaporizing liquefied gas releases, or combinations of these. The selection of appropriate fire protection for a specific type of facility or item of equipment should be based on the life-cycle stage of the facility and the results of a fire hazard analysis. Typically, the protection features available include one or more of the following:

- Elimination of the hazard and its resulting scenarios
- Prevention (reduction of probability) of its occurrence
- Detection and control
- Mitigation of its consequences
- Emergency response

The elimination of a fire hazard may be the ideal solution, but it is often not possible. In general, the optimum level of fire protection is achieved by selecting from the other appropriate prevention and mitigation options. The higher the performance availability (or the lower the probability of failure on demand) of each selected fire protection feature, the more effective the overall fire protection system. The generally preferred approach to improve effectiveness is to select a combination of *passive* and *active* fire protection features. All fire protection features (passive and active) require periodic inspection and maintenance.

Process-related fires are a principal concern in many processing facilities. Other fires that can occur in specific areas within a facility include fires involving:

- Solid material, e.g., fires involving wood, paper, dust, plastic, etc.
- Electrical equipment, e.g., transformer fires
- Oxygen, e.g., systems for oxygen addition
- Combustible metals, e.g., sodium
- Pyrophoric materials, e.g., aluminum alkyl

Passive Fire Protection Passive fire protection involves systems that require no human intervention or automatic operation. Examples are spacing between units and storage tanks, relief systems, fireproofing, and automatic isolation and depressuring systems.

Structural Fireproofing Fireproofing is a fire-resistant material or system that is applied to a surface to delay heat transfer to that surface. Fireproofing protects against intense and prolonged heat exposure that can cause the weakening of steel and the eventual collapse of unprotected equipment, vessels, and supports, which can lead to the spread of burning liquids and substantial loss of property. The primary purpose is to improve the capability of equipment and structures to maintain their integrity until the fire is extinguished by either stopping the fuel source or active fire protection methods.

The value of fireproofing is realized during the early stages of a fire when efforts are primarily directed at shutting down units, isolating fuel flow to the fire, actuating fixed suppression equipment, and setting up portable firefighting equipment. During this critical period, if nonfireproofed equipment and pipe supports fail due to fire-related heat exposure, they could collapse and cause gasket failures, line breaks, and equipment failures, resulting in expansion of the fire. Fireproofing may be applied to control or power wiring to allow the operation of emergency isolation valves, venting vessels, or water spray systems during a fire.

TABLE 23-36 Advantages and Limitations of Various Extinguishing Agents

Agent	Type of extinguishment	Advantages	Limitations
Water	Cooling Smothering Dilution Exposure	Available Very low cost	Not for Class C electrical fires; freezes at 32°F (0°C); reactive with some material, e.g., sodium, magnesium; cannot extinguish low-flash-point materials
Foam	Smothering	Best for Class B pool fires (two-dimensional fires)	Not for electrical fires; foam blanket may break up; not applicable for LPG
CO_2	Smothering	Nonreactive No residue Class C fires	Reduces O_2 level; toxic to people (asphyxiant); not applicable for oxidizers
Dry chemical	Chain breaking	Class B and C fires	Fire reflash if not completely extinguished or hot surfaces are present (especially flammable/combustible liquids)
Clean agent	Chain breaking Inerting	Good for Class A, B, and C fires	Not for outdoors; may produce toxic gases

Active Fire Protection Active fire protection systems are typically installed to provide the desired mitigation by either automatic or manual activation.

Firefighting Agents There are different types of fires and many different firefighting agents for combating them. An understanding of how these different types of firefighting agents are used in fire protection is essential because their effectiveness can vary widely when applied to different types of fires. Table 23-36 highlights the advantages and limitations of the various extinguishing agents.

Fire Water Systems Normally, fire water demands range between 7600 and 38,000 liters per minute (2000 and 10,000 gpm). The design capacity of the fire water system should be at a minimum four hours of continuous operation of the largest fire water demand. The fire water demand capacity is based on a number of factors, including:

- Sources of water available
- Reliability of make-up water supply
- Potential for escalation to other areas of the facility
- Isolation philosophy and the ability to depressurize high-pressure units

The reliability of the fire water supply should be such that the loss of any one source does not result in a loss of more than 50 percent of the flow requirements of the system.

A facility drawing water from a stream or lake should either have independent locations from which to draw or have a back-up supply from the city system or a private well. Care should be taken when using potable and nonpotable sources so that cross-contamination does not occur.

Fire water pumping capacity (flow rate) should be sufficient to provide the required amount of water at required pressure to the fire areas having the greatest demand. At least 50 percent of the pumping capacity should be from diesel-driven pumps. Fire water systems should be designed in accordance with NFPA 20 and 22 (National Fire Protection Association, Quincy, Mass.).

Water Spray Systems The term *water spray* refers to the use of water discharged from nozzles having a predetermined pattern, droplet size, velocity, and density. Fixed water spray systems are most commonly used to protect flammable liquid and gas vessels, piping and equipment, process structures and equipment, electrical equipment, oil switches, rotating electrical machinery, and openings through which conveyors pass. Water spray systems should be designed in accordance with NFPA 15 (National Fire Protection Association, Quincy, Mass.).

Water Mist Systems A water mist system is a proprietary fire protection system using very fine water sprays. The very small water droplets allow the water mist to control or extinguish fires by cooling the flame and fire plume, oxygen displacement by water vapor, and radiant heat attenuation. Water mist systems are intended for rapid suppression of fires using water discharged into completely enclosed, limited-volume spaces. Water mist systems are desirable for spaces where the amount of water that can be stored or discharged is limited. Water mist systems should be designed in accordance with NFPA 750 (National Fire Protection Association, Quincy, Mass.).

Foam Systems Foam is primarily used to extinguish two-dimensional surface fires involving liquids that are lighter than water. Foams may be used to insulate and protect against exposure to radiant heat. They also prevent the ignition of flammable liquids that are inadvertently exposed to the air (typically due to a spill) by separating them from air by spreading foam completely over the exposed surface. However, progressive foam breakdown can render the protective foam coating useless; thus, frequent reapplication may be necessary. Foam systems should be designed in accordance with NFPA 11 and NFPA 16 (National Fire Protection Association, Quincy, Mass.).

Carbon Dioxide Systems Fixed CO_2 systems may be total flooding or local application systems. CO_2 systems should not be used where personnel may be present because of the asphyxiation hazard. Total flooding carbon dioxide systems may be used where there is a permanent enclosure around the area or equipment that is adequate to enable the required concentration of CO_2 to be maintained. Carbon dioxide systems should be designed in accordance with NFPA 12 (National Fire Protection Association, Quincy, Mass.).

Dry Chemical Systems Dry chemicals are recognized for their unusual efficiency in extinguishing two-dimensional fires involving flammable liquids. Fast extinguishing action is achieved, provided the agent engulfs the fire without interruption of the application. The finely divided powder acts with a chain-breaking reaction by inhibiting the oxidation process within the flame itself. If there is risk of reignition from embers or hot surfaces, these ignition sources should be quenched or cooled with water and secured with foam, or the source of fuel should be shut off before attempting extinguishment. Dry chemical systems should be designed in accordance with NFPA 17 (National Fire Protection Association, Quincy, Mass.).

Clean Agent Systems Clean agents are electrically nonconductive, volatile, or gaseous fire extinguishing agents that do not leave a residue. Clean agents fall within two categories: halocarbons and inert gases. Typical halocarbons include hydrofluorocarbons (HFCs), hydrochlorofluorocarbons (HCFCs), perfluorocarbons (PFCs or FCs), and fluoroiodocarbons (FICs). Typical inert gases include argon, nitrogen, or combinations of these agents. Inert gas systems should not be used where personnel may be present because of the asphyxiation hazard. Clean agent fire extinguishing systems are used primarily to protect enclosures. Clean agent systems should be designed in accordance with NFPA 2001 (National Fire Protection Association, Quincy, Mass.).

KEY PROCEDURES

General References: CCPS, *Guidelines for Risk Based Process Safety*, American Institute of Chemical Engineers, New York, 2007; Kletz, *Learning from Accidents*, 3d ed, Gulf Professional, Boston, 2001; Mannan, ed., *Lees' Loss Prevention in the Process Industries*, 4th ed., Butterworth-Heinemann, London, 2012; U.S. Environmental Protection Agency, *Risk Management Programs for Chemical Accidental Release Prevention Requirements*, 40 CFR part 68; U.S. Occupational Safety and Health Administration, *Process Safety Management of Highly Hazardous Chemicals, Explosives, and Blasting Agents*, 29 CFR 1910.119.

Process Safety Management A process safety management (PSM) system encompasses all of the management processes and work practices required to manage all process risks associated with the hazards in a process plant. In the United States, the Occupational Safety and Health Administration (OSHA) PSM Standard (29 CFR 1910.119) and the Environmental Protection Agency (EPA) Risk Management Program (RMP) Rule (40 CFR Part 68) establish minimum PSM requirements for covered processes. Other countries, as well as some state and local jurisdictions, also have PSM requirements. The Center for Chemical Process Safety (CCPS) published *Guidelines for Risk Based Process Safety* [American Institute of Chemical Engineers (AIChE), New York, 2007], outlining the components of a modern PSM program. Risk-based process safety, or RBPS, includes the provisions of OSHA PSM and EPA RMP, and it identifies a total of 20 elements, some of which are not included in the OSHA and EPA regulations. The 20 elements of RBPS are grouped into four "pillars" as follows:

- ***Commit to Process Safety***
  - *Process Safety Culture.* The combination of group values and behaviors that determine the manner in which process safety is managed
  - *Compliance with Standards.* Identifying codes, standards, and good practices relevant to process operations and maintaining systems to ensure proper compliance with requirements
  - *Process Safety Competency.* Developing and constantly improving a thorough knowledge of the process technology, making that knowledge available to everybody who needs it throughout the organization, and always applying that knowledge to understand process risks and how to properly manage them
  - *Workforce Involvement.* Actively involving the right people at all levels of the organization, particularly including front-line operating and maintenance workers, in all process safety activities to ensure complete understanding of how the process actually behaves
  - *Stakeholder Outreach.* Actively involving all stakeholders; for example, all employees, contractors, stockholders or owners, neighbors and the community, nearby businesses and plants, government and regulators at all levels, outside emergency response organizations, outside medical providers, schools, and anybody else who has a stake or can be affected by operations
- ***Understand Hazards and Risk***
  - *Process Knowledge Management.* Systems to document, update, and maintain all information required to understand and manage the hazards associated with the materials, equipment, and chemical and physical processes of the plant, and to make it accessible to people who need it, at the time they need it
  - *Hazard Identification and Risk Analysis.* Procedures to identify process hazards, potential loss event scenarios, and what must be done to properly manage those risks (see earlier discussions in this Process Safety Section on Inherently Safer Design, Hazard Analysis, and Risk Analysis)
- ***Manage Risk***
  - *Operating Procedures.* Detailed written instructions on how the process is to be started up, operated, and shut down, including critical process safety control parameters and operating limits, what to do if they are exceeded, and emergency procedures
  - *Safe Work Practices.* Procedures related to testing, inspecting, calibrating, maintaining, and repairing equipment or other process facilities; these are generally the activities which will require a work permit, such as confined space entry, hot work, lockout/tagout, line breaking, trenching and excavation, and overhead crane lifts
  - *Asset Integrity and Reliability.* Systems to ensure that process equipment is designed, installed, inspected, tested, and maintained so that the equipment robustly contains hazardous materials and energies, performs as required when needed, and has deficiencies corrected promptly
  - *Contractor Management.* Systems to ensure good communication between the company and any contractor engaged in any work, including qualifying of contractors (in terms of capability to do the task safely); to ensure that contractors understand company safety policies, safe work practices, workplace hazards, emergency response, and anything else they need to know to work safely; and to ensure that contractors inform the company of any hazards that may arise from their work activities
  - *Training and Performance Assurance.* Systems to train all employees so they are qualified to do their jobs, and testing to ensure that the training has been understood and that the employees can do the job correctly
  - *Management of Change.* Systems to identify changes in equipment, chemicals, other materials, procedures, control systems and software, and organization, and to categorize the identified changes so that an appropriate review can be conducted; this includes appropriate risk assessment and hazard review procedures for all categories of changes, approval processes, training of people on the changes, validation of correct implementation, and documentation of the changes
  - *Operational Readiness.* Procedures to ensure that all equipment is ready for use before it is started up; this includes a pre–start-up safety review (PSSR) for new plants, or anything that has been subjected to a management of change review
  - *Conduct of Operations.* Ensuring that operation of the plant and all process safety management systems is consistent with the intent as described in the appropriate documentation; in a few words, "say what you are going to do" and "do what you say" on all activities related to process safety
  - *Emergency Management.* Emergency response plans and procedures at all levels; for example, unit response to a process upset, spill and leak response procedures, firefighting procedures, evacuation or shelter-in-place procedures, and coordination of emergency response with outside agencies
- ***Learn from Experience***
  - *Incident Investigation.* Systems to identify, report, and appropriately investigate process safety incidents and near misses, to implement process improvements to prevent recurrence, and to share the lessons with others
  - *Measurements and Metrics.* Quantitative measures to monitor the health and performance of PSM systems
  - *Auditing.* Periodic review by qualified auditors of PSM systems, including internal self-audits by local management and plant, corporate, or outside audits
  - *Management Review and Continuous Improvement.* Ongoing review of all process safety management systems and visible, active involvement and communication of the importance of PSM by management at all levels of a company

Many of the elements of RBPS have been discussed in previous sections. Additional discussion of some of them follows.

Safety Culture The safety culture of an organization has been recognized as critical to the success of a process safety management program. For example, Hopkins (*Lessons from Longford: The Esso Gas Plant Explosion*, CCH Australia Limited, Sydney, Australia, 2000) discusses the safety culture aspects of the causes of an explosion at a gas processing plant in Longford, Australia. Hopkins (*Failure to Learn: The BP Texas City Refinery Disaster*, CCH Australia Limited, Sydney, Australia, 2008) and the final report of the BP North American Refineries Independent Safety Panel (Baker et al., *The Report of the BP U.S. Refineries Independent Safety Review Panel*, 2007, available at www.csb.gov) identify many safety culture aspects of the March 2005 refinery explosion in Texas City, Texas. In *Guidelines for Risk Based Process Safety* (AIChE, New York, 2007), CCPS includes an extensive discussion of process safety culture and identifies some aspects of a weak or a sound safety culture (Table 23-37).

Process Safety Competency and Process Knowledge Management (Institutional Memory) Institutional memory is an important aspect of process safety competency (CCPS, *Guidelines for Defining Process Safety Competency Requirements*, AIChE, New York, 2015). Process safety incidents (loss events) do not occur because we do not know how to prevent them, but often because we do not use the information that is available. The recommendations made after a loss event happens, or identified in some kind of safety review or process hazard analysis, are forgotten when the people involved have left the plant. The procedures they introduced are allowed to lapse, the equipment they installed is no longer used, and a similar loss event happens again. The following actions can prevent or reduce this loss of information:

- Include a note on the reason why as part of every instruction, code, and standard as well as accounts of loss events that would not have occurred if the instruction, code, or standard had been followed.
- Describe old loss events as well as recent ones in safety bulletins and newsletters, and discuss them at safety meetings.
- Follow up at regular intervals (e.g., during audits) to see that the recommendations made after loss events are being followed, in design as well as operations.
- Make sure that recommendations for changes in design are acceptable to the design organization.
- Remember that the first step down the road to a loss event is taken when someone turns a blind eye to a missing blind or any other deviation from intended operation.
- For each process unit, keep a memory book, a folder of reports on past loss events (paper or computer files) that is compulsory reading for new people working in the unit at all levels—operating, maintenance, technical, and management—and that others dip into from time to time. It should include relevant reports from other companies but should not include personal safety incident descriptions such as cuts and bruises.
- Never remove equipment before knowing why it was installed, and do not modify equipment without understanding the original design basis. Never abandon a procedure without knowing why it was adopted.
- When people are moving to other jobs in the company or leaving it, make sure that the remaining employees at all levels have adequate knowledge and experience, and make an effort to capture the knowledge of the people who are leaving.
- Include important loss events of the past in the training of company employees. The training should include accounts of loss events that demonstrate the need for codes, standards, or instructions. Ask audience members to say what they think should be done to prevent similar loss events from happening again. More will be remembered after a discussion than after a lecture, and audience members are more likely to be convinced when they have worked out for themselves the actions that should be taken (Kletz, *Lessons from Disaster—How Organizations Have No Memory and Accidents Recur*, Institution of Chemical Engineers, Rugby, United Kingdom, 1993, chap. 10). Suitable loss events for discussion can be found in books of case histories and from online resources, but loss events that have happened in the past in the facility where people work will have the greatest impact.
- There are many databases of loss events, reports of major investigations from various government agencies throughout the world (for example, the investigation reports and videos from the U.S. Chemical Safety and Hazard Investigation Board at www.csb.gov), as well as books of case histories, but they tend to be little used. Better retrieval systems are needed so that people can find details of past loss events, in their own and other companies, and can understand the recommendations made after the loss event.

TABLE 23-37 Some Characteristics of Weak and Sound Safety Cultures

Weak culture	Sound culture
Little value to process safety	Process safety a core value, integrated into all operations
Poor sense of vulnerability	Clear understanding of risk and how to control it
Minimal resources for risk control	Appropriate resources based on understanding of risks
Signals of safety problems overlooked	Emphasis on learning from past experience
Poor performance accepted and normalized	Effort to continuously improve performance
Relies on management to identify and manage risk	All employees are involved in identifying hazards and managing risk

Summarized from *Guidelines for Risk Based Process Safety*, AIChE, New York, 2007, Table 3.1

Operating Procedures Safety by design should always be our aim, but it may be impossible or too expensive, and then we have to rely on procedures (CCPS, *Guidelines for Writing Effective Operating and Maintenance Procedures*, AIChE, New York, 1996). Key features of all procedures are as follows:

- They should be as simple as possible and use simple language, so as to help the reader rather than protect the writer.
- People who are actually doing the work should be involved in writing procedures. They may be aware of important details in how the job is done that are not obvious to those who do not have direct experience actually doing the activity.
- Procedures must include safe operating limits for critical process safety operating parameters (temperature, pressure, flow rate, or whatever is important for the specific process), the consequences of violating these safe operating limits, and the actions required if the safe operating limits are exceeded.
- Procedures should be explained to and discussed with those who will have to carry them out, not just sent to them electronically or by distributing a paper copy to them.
- Regular checks and audits should be made to confirm that the procedures are being carried out correctly.

Safe Work Practices Safe work practices are important for safely performing nonroutine work activities. Nonroutine can be defined as "any activity that is not fully described in an operating procedure" (CCPS, *Guidelines for Risk Based Process Safety*, AIChE, New York, 2007). Some examples include confined space entry, hot work, lockout/tagout, opening vessels or pipes, excavation, and other maintenance and construction activities. A common example of a specific safe work practice is preparation of equipment for maintenance. An essential feature of this procedure is a permit-to-work system. The operating team members prepare the equipment and prepare a permit describing the work to be done, the preparation carried out, the remaining hazards, and the precautions necessary. The permit is then accepted by the person or group who will carry out the work and is returned when the work is complete. The permit system reduces the chance that hazards will be overlooked, lists ways of controlling them, and informs those doing the job what precautions they should take. The system should cover such matters as who is authorized to issue and accept work permits, the training all personnel involved should receive, and the period of time for which the permits are valid. It should also cover the following:

- ***Isolation of the Equipment Under Maintenance*** Poor or missing isolation has been the cause of many serious loss events. Do not rely on valves except for quick, low-risk jobs. Use blinds, or disconnecting and blanking, unless the job is so quick that blinding or disconnection would take as long and be as hazardous as the main job. Valves used for isolation, including isolation while fitting blinds or disconnecting, should be locked shut following the facility's lockout/tagout procedures. Blinds should be made to the same standard (pressure rating and material of construction) as the plant. Plants should be designed so that blinds can be inserted without difficulty; that is, there should be sufficient flexibility in the pipework, or a slip-ring figure 8 plate (spectacle blind) should be used. Electricity should be isolated by locking out the power supply, or by removal of fuses. Do not leave the fuses lying around for anyone to replace. Always try out electrical equipment after isolating or defusing to check that the correct power sources have been isolated or the right fuses have been removed.
- ***Identification of Equipment*** Many loss events have occurred because maintenance workers opened up the wrong equipment. Equipment that is under repair should be numbered or labeled unambiguously. Temporary labels should be used if there are no permanent ones. Pointing out the correct equipment is not sufficient; the equipment needs to be specifically identified in the permit documentation.

- ***Freeing from Hazardous Materials*** Equipment that is to be repaired should be freed as far as possible from hazardous materials. Gases can be removed by sweeping out with nitrogen (if the gases are flammable) or air, water-soluble liquids by washing with water, and oils by steaming. Some materials such as heavy oils and materials that polymerize are very difficult or impossible to remove completely. Tests should be carried out to make sure that the concentration of any remaining hazardous material is below an agreed-upon level.
- ***Put Machinery in the Lowest Energy State*** For example, the forks of forklift trucks should be lowered, and springs should not be compressed or extended. Do not work under heavy suspended loads. For some machinery, the lowest energy state is less obvious, and people who are experts in the design and operation of the machinery should be consulted.
- ***Handover*** Permits should be handed over (and returned when the job is complete) person-to-person. They should not be left on the table for people to sign when they come in to work.
- ***Change of Intent*** If there is a change in the work to be done, the permit should be returned and a new one issued (Crowl and Grossel, eds., *Handbook of Toxic Materials Handling and Management*, Marcel Dekker, 1994, chap. 12).

Inspection, Testing, and Preventive Maintenance of Protective Equipment The inspection, testing, and preventive maintenance of plant safety equipment is an important subset of the RBPS element Asset Integrity and Reliability (CCPS, *Guidelines for Asset Integrity Management*, AIChE, New York, 2017), which includes all activities required to ensure that all plant equipment is:

- Properly designed
- Fit for use for the intended service
- Correctly installed
- Inspected, tested, and maintained throughout its service life, and any identified deficiencies corrected, so the equipment remains fit for use, is working properly, and is available for use when needed

This section will discuss some specific aspects related to protective equipment in a plant. All protective equipment should be scheduled for regular preventive maintenance, inspection, and testing. This is particularly important if failure is latent (hidden); for example, it is not known if an interlock, alarm, or relief valve is in working order unless it is tested. The frequency of testing or inspection depends on the failure rate and the length of time considered tolerable if it fails. Relief valves fail at a rate of about 0.01 per year on the average, and testing every 1 or 2 years is often adequate. However, service conditions such as corrosion, potential polymerization, fouling with solids, and other environmental factors can reduce reliability. Protective systems based on instruments, such as trips, alarms, and safety instrumented systems, may fail more often, and more frequent testing may be necessary. Frequency of maintenance and testing should be based on data on the reliability of all of the equipment in the protective system, in the actual plant service and environment. Historical data from the plant where the system is installed or in a similar service, from past maintenance and testing records, provides the best information for establishing maintenance and testing schedules. Pressure systems (vessels and pipework) in noncorrosive duties can go for many years between inspections, but in some duties, they may need to be inspected annually or even more often.

All protective equipment should be designed so that it can be tested or inspected safely, without disrupting plant operation, and provision should be made to provide protection for the plant while the protective equipment is being tested or maintained. Audits should include a check that the tests are carried out and the results acted upon. The supervisor or engineer responsible should be reminded when a test or inspection is due, and senior managers should be informed if a test has not been carried out by the due date. Test and inspection schedules should include specific procedures for the test, and the results of the tests must be documented. Test procedures should include all components of the protective system. Test results should be reviewed for failure patterns that might indicate recurring issues, and the frequency of failure should be compared to the reliability assumed in the original design. If equipment is found to have a higher failure rate than the designer assumed, the protective system is probably not as reliable as expected when it was designed, and it may need to be improved to provide the needed protection. Test results should be displayed for all to see, such as on a board in the control room.

Tests should be like real life. For example, a high-temperature trip failed to work despite regular testing. It was removed from its case before testing, so the test did not disclose that the pointer rubbed against the case. This prevented it from indicating a high temperature.

Operators sometimes regard tests and inspections as a nuisance, interfering with the smooth operation of the plant. Operator training should emphasize that protective equipment is there for their protection, and they should "own" it.

Management of Change Many loss events have occurred when modifications to the plant had unforeseen and unsafe side effects (Sanders, *Chemical Process Safety: Learning from Case Histories*, 4th ed, Butterworth-Heinemann, London, 2015). No such modifications should therefore be made until they have been authorized by a professionally qualified person who has made a systematic attempt to identify and assess the consequences of the proposal, by hazard and operability study or a similar technique (CCPS, *Guidelines for the Management of Change for Process Safety*, AIChE, New York, 2008). When the modification is complete, the person who authorized it should inspect it to make sure that the design intention has been followed and that it "looks right." What does not look right is usually wrong and should at least be checked.

Unauthorized modifications are particularly liable to occur:

- During start-ups, as changes may be necessary to get the plant on line.
- During maintenance, as the maintenance workers may be tempted to improve the plant as well as to repair it. They may suggest modifications, but they should put the plant back as it was unless a change has been authorized.
- When the modification is cheap and no financial authorization is necessary. Many seemingly trivial modifications have had tragic results.
- When the modification is temporary. Twenty-eight people were killed in an explosion resulting from the failure of a temporary modification at Flixborough, UK (Mannan, ed., *Lees' Loss Prevention in the Process Industries*, 4th ed, Butterworth-Heinemann, London, 2012, Appendix A1; Kletz, *Learning from Accidents*, 3d ed, Gulf Professional, Boston, 2001, chap. 8).
- When one modification leads to another, and then another [Kletz, "Modification Chains," *Plant/Operations Progress* **5**(3): 136 (1986)].
- When organizations are changed, especially when staffing is reduced. Such changes should be studied as thoroughly as changes in equipment or processes (CCPS, *Guidelines for Managing Process Safety Risks during Organizational Change*, AIChE, New York, 2013).

Incident Investigation and Human Error

Incident Investigation Although most companies investigate loss events (and many investigate dangerous incidents in which no one was injured or there was no significant loss), these investigations are often superficial, and the organization fails to learn all of the lessons for which it has paid the high price of an incident. The collection of evidence is usually adequate, but often only superficial conclusions are drawn from it. Identifying the causes of a loss event is like peeling an onion. The outer layers deal with the immediate technical causes and triggering events, while the inner layers deal with ways of avoiding the hazard and with the underlying weaknesses in management systems (Kletz, *Learning from Accidents*, 3d ed, Gulf Professional, Boston, 2001; CCPS, *Guidelines for Investigating Chemical Process Incidents*, 2d ed, AIChE, New York, 2003).

Dealing with the immediate technical causes of a leak will prevent another leak for the same reason. If the plant can be modified to use so little of the hazardous material that leaks do not matter, or to use safer material, as discussed in the Inherently Safer Design subsection, all significant leaks of that hazardous material can be eliminated. If the management system and plant design can be improved, many more incidents will be prevented.

Other points to watch when drawing conclusions from the facts include:

1. Avoid the temptation to list causes for which little or nothing can be done. For example, a source of ignition should not be listed as the primary cause of a fire or explosion, as leaks of flammable gases are liable to ignite even though all known sources of ignition are eliminated. The real cause is whatever led to the formation of a flammable mixture of gas or vapor and air. The removal of known sources of ignition should, however, be included in the recommendations. Similarly, human error should not be listed as a cause. The investigation should address the root causes of the human error.

2. Do not produce a long list of recommendations without any indication of the relative contributions they will make to the reduction of risk or without any comparison of costs and benefits. Resources are not unlimited (and resources are not just monetary, but also the time and expertise of the engineers and other personnel who have to implement the recommendation). The more that is spent on reducing one hazard, the less there is left to spend on reducing others.

3. A specific person should be made responsible for carrying out each agreed-upon recommendation, and a completion date scheduled for each. Otherwise, the actions may not be implemented, and the loss event may happen again.

4. Avoid the temptation to overreact after a loss event by installing an excessive amount of protective equipment or adopting complex procedures that are unlikely to be followed after a few years have passed. Sometimes the incident report on a loss event will call for more protective equipment, even though the protective equipment that was available was not used.

Sometimes a loss event occurs because complex procedures were not followed, and the report recommends extra procedures. It would be better to find out why the original equipment was not used or why the original procedures were not followed.

5. Remember that few, if any, loss events have a single cause. In most cases, many people had an opportunity to prevent it, from the chemist who developed the process to the operator who closed the wrong valve. Figure 23-51 shows by example the opportunities that were available to prevent a fire or minimize the consequences of an apparently simple incident: an expansion joint (bellows) was incorrectly installed in a pipeline so that it was distorted. After some months it leaked, and a passing vehicle ignited the escaping vapor. Damage was extensive because, to save on costs, the surrounding equipment had not been fire-protected.

The fitter who installed the expansion joint incorrectly could have prevented the fire. So could the person who was responsible for training and supervising the fitter; so could the designers if they had not specified an expansion joint, had carried out a HAZOP study, or had consulted experts; so could the author of the company's design standards, and those responsible for training designers, those responsible for inspecting workmanship, and anyone who kept his or her eyes open while walking around the plant.

6. When you are reading an incident report, look for the things that are not said. For example, a gland leak on a liquefied flammable gas pump caught fire and caused considerable damage. The report drew attention to the congested layout, the amount of redundant equipment in the area, the fact that a gearbox casing had been made of aluminum which melted, and several other unsatisfactory features. It did not stress that there had been a number of gland leaks on this pump over the years, that reliable glands are available for liquefied gases at ambient temperatures, and therefore there was no need to have tolerated a leaky pump on this service.

7. At one time most loss events were said to be due to human error, and in a sense they all are. If someone—a designer, manager, operator, or maintenance worker—had done something differently, the loss event would not have occurred. However, the term *human error* is not very helpful; different types of error require quite different actions to prevent their happening again. Classification of errors as discussed in the next subsection is recommended because it helps us to understand the type of action needed to prevent a repeat incident.

Human Error Human error is often a factor in process safety incidents (Kletz, *An Engineer's View of Human Error*, 3d ed, Taylor and Francis,

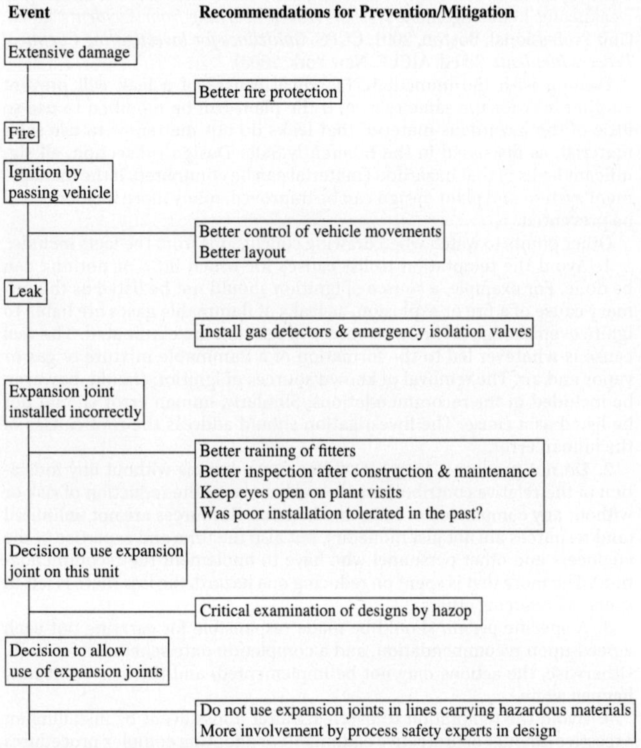

FIG. 23-51 An example of the many ways by which an accident could have been prevented.

New York, 2001; CCPS, *Guidelines for Preventing Human Error in Process Safety*, AIChE, New York, 2004; CCPS, *Human Factors Methods for Improving Performance in the Process Industries*, AIChE, New York, 2006). However, human error should never be considered a root cause for an incident investigation. The investigation should identify why the human error occurred.

- Some errors, called *mistakes*, are due to poor training or instructions—someone did not know what to do. It is a management responsibility to provide good training and instructions and to avoid instructions that are designed to protect the writer rather than help the reader. However the instructions are written, problems will arise that are not covered, and so people, particularly operators, should be trained to be flexible and able to diagnose and handle unforeseen situations. If the instructions are hard to follow, can the job be simplified?
- Some deviations, called *violations* or *noncompliance*, occur because someone knew what to do but made a decision not to do it. As discussed in the subsection on Inherently Safer Design, an objective should be to simplify the job; if the correct method is difficult, an incorrect one will be used. Explain the reasons for the instructions, carry out checks from time to time to see that instructions are being followed, and do not turn a blind eye if they are not. Some violations make the job easier, and many are made by people who think they have found a better way of doing the job. If instructions are incorrect, they will not be followed because people understand that the provided instructions do not work. In some cases, not following incorrect instructions may even prevent a loss event. This may encourage a culture where not following instructions becomes acceptable. The methods of behavioral science can be used to reduce violations. Specially trained members of the workforce keep their eyes open and tactfully draw the attention of fellow workers to violations such as failure to wear protective clothing.
- Some deviations occur because the job is beyond the physical or mental ability of the person asked to do it, sometimes beyond anyone's ability. In this case, the plant design should be improved.
- The most common type of human error is a momentary slip or lapse of attention. This happens to everyone from time to time and cannot be prevented by telling people to be more careful, telling them to keep their minds on the job, or better training. In fact, slips and lapses of attention are most likely to occur when people are well trained. They put themselves on autopilot and carry out the task without continually monitoring progress, though they may check it from time to time. These errors are more likely to occur when people are distracted or stressed. To avoid slips and lapses of attention, the plant design or method of working should be changed in order to remove opportunities for error, to minimize the consequences of an error, or to provide opportunities for recovery. Whenever possible, design user-friendly plants (see Inherently Safer Design subsection) that can withstand errors (and equipment failures) without suffering serious effects on safety (and output and efficiency). It is more effective to change the performance of equipment than to try to change the behavior of people.

When a loss event report says that the event was due to human error, the writer usually means an error by an operator or other frontline worker. But designers and managers also make errors, and design procedures and management systems must be designed to minimize the potential for such errors to result in a loss event.

Key Performance Indicators In recent years, there has been significant progress in identifying appropriate metrics for monitoring process safety performance and the health of PSM systems (CCPS, *Guidelines for Process Safety Metrics*, AIChE, New York, 2009; CCPS, *Guidelines for Integrating Management Systems and Metrics to Improve Process Safety Performance*, AIChE, New York, 2016). The process industries have recognized that traditional safety measures such as injury rates and lost time incidents do not provide a good measure of process safety performance. For example, a release of a large amount of flammable gas occurs, say a leak of 5 tons of propane gas, but since the wind happens to be blowing in a certain direction, the flammable vapor cloud does not ignite. It blows away and is safely dissipated. This event will have no impact on traditional safety measures because nobody was injured. It may be reported as an environmental release, but providentially there were no safety consequences this time. However, this is clearly a significant process safety event, and there could have been multiple fatalities and injuries, and financial losses of millions of dollars. Many companies and industry associations, such as the American Petroleum Institute (API) (ANSI/API Recommended Practice 754, *Process Safety Performance Indicators for the Refining and Petrochemical Industries*, 2d ed, Washington D.C., 2016) and the American Chemistry Council (ACC), are beginning to use and report a variety of new measures of process safety performance. This is an evolving area of development, and, with more experience with the newly developed metrics, changes and improvements will continue.

CCPS's document *Process Safety Leading and Lagging Metrics* (AIChE, New York, 2011, http://www.aiche.org/sites/default/files/docs/pages/metrics%20english%20updated.pdf) identifies three types of metrics:

- *Lagging metrics*, a retrospective set of metrics that are based on incidents that meet the threshold of severity that should be reported as part of an industrywide process safety metric.
- *Leading metrics*, a forward-looking set of metrics that indicate the performance of the key work processes, operating discipline, or layers of protection that prevent incidents.
- *Near miss* and other internal lagging metrics that describe less severe incidents (below the threshold for inclusion in an industry lagging metric) or unsafe conditions that activated one or more layers of protection. They are generally considered to be good indicators of conditions that could ultimately lead to a more severe incident.

CCPS conducted a survey to determine what metrics were being used by participating companies (CCPS, *Process Safety Leading Indicators Industry Survey*, AIChE, New York, 2013, http://www.aiche.org/sites/default/files/docs/pages/8404_leading.web_v2.pdf). Table 23-38 summarizes some of the most commonly used leading process safety metrics from that survey.

TABLE 23-38 Some Commonly Reported Leading Process Safety Metrics from the 2013 CCPS Process Safety Metrics Survey

Number of past due and/or having approved extension of audit action items
Number of past due and/or having approved extension of PHA action items
Number of outstanding incident investigation action items closed
Demands on safety systems
Incident (PSI) or loss of primary containment (LOPC)
Training for process safety management (PSM) critical positions
Procedures current and accurate
Activation of a safety instrumented system (SIS)
Safe operating limit excursions
Activation of mechanical shutdown system

Energy Resources, Conversion, and Utilization

Shabbir Ahmed, Ph.D. *Chemical Engineer, Chemical Sciences and Engineering Division, Argonne National Laboratory (Section Editor)*

David K. Schmalzer, Ph.D., P.E. *Argonne National Laboratory (Retired), Member, American Chemical Society, American Institute of Chemical Engineers (Resources and Reserves, Liquid Petroleum Fuels)*

Dirk T. Van Essendelft, Ph.D. *Chemical Engineer, National Energy Technology Laboratory, U.S. Department of Energy (Coal)*

Lawrence J. Shadle, Ph.D. *Mechanical Engineer, National Energy Technology Laboratory, U.S. Department of Energy (Coke)*

Nicholas S. Siefert, Ph.D., P.E. *Mechanical Engineer, National Energy Technology Laboratory, U.S. Department of Energy (Other Solid Fuels)*

Dirk Link, Ph.D. *Chemist, National Energy Technology Laboratory, U.S. Department of Energy (Nonpetroleum Liquid Fuels)*

George A. Richards, Ph.D. *Mechanical Engineer, National Energy Technology Laboratory, U.S. Department of Energy (Natural Gas, Liquefied Petroleum Gas, Other Gaseous Fuels)*

Yongkoo Seol, Ph.D. *Geologist, National Energy Technology Laboratory, U.S. Department of Energy (Natural Gas)*

Dushyant Shekhawat, Ph.D., P.E. *Chemical Engineer, National Energy Technology Laboratory, U.S. Department of Energy (Natural Gas, Fuel and Energy Costs)*

Daniel J. Soeder, M.S. *Director, Energy Resources Initiative, South Dakota School of Mines & Technology (Gaseous Fuels)*

Massood Ramezan, Ph.D., P.E. *Sr. Technical Advisor, KeyLogic Systems, Inc. (Coal Conversion)*

Gary J. Stiegel, P.E., M.S. *Technology Manager (Retired), National Energy Technology Laboratory, U.S. Department of Energy (Coal Conversion)*

Peter J. Loftus, D. Phil. *Chief Scientist, Primaira LLC, Member, American Society of Mechanical Engineers (Heat Generation)*

Ian C. Kemp, M.A. (Cantab) *Scientific Leader, GlaxoSmithKline; Fellow, Institution of Chemical Engineers; Associate Member, Institution of Mechanical Engineers (Pinch Analysis)*

(Francis) Lee Smith, Ph.D. *Principal, Wilcrest Consulting Associates, LLC, Katy, Texas; Partner and General Manager, Albutran USA, LLC, Katy, Texas (Energy Recovery)*

Joseph D. Smith, Ph.D. *Professor of Chemical and Biochemical Engineering, Missouri University of Science and Technology (Thermal Energy Conversion and Utilization)*

Dennis W. Dees, Ph.D. *Senior Electrochemical Engineer, Chemical Sciences and Engineering Division, Argonne National Laboratory (Electrochemical Energy Storage)*

INTRODUCTION

FUELS

HEAT GENERATION

PINCH ANALYSIS

ENERGY RECOVERY

THERMAL ENERGY CONVERSION AND UTILIZATION

Acronyms

Acronym	Definition
AFBC	atmospheric fluidized-bed combustion
AGC-21	Advanced Gas Conversion process
BGL	British Gas and Lurgi process
COE	cost of electricity
COED	Char Oil Energy Development process
CSR	coke strength after reaction
DOE	U.S. Department of Energy
EDS	Exxon Donor Solvent process
FBC	fluidized-bed combustion
FSI	free swelling index
HAO	hydrogenated anthracene oil
HGI	Hardgrove Grindability Index
HHV	higher heating value
HPO	hydrogenated phenanthrene oil
HRI	Hydrocarbon Research, Inc.
IGCC	integrated gasification combined-cycle
KRW	Kellogg-Rust-Westinghouse process
LHV	lower heating value

Acronym	Definition
LNG	liquefied natural gas
MTG	methanol-to-gasoline process
OTFT	once-through Fischer-Tropsch process
PC	pulverized coal
PFBC	pressurized fluidized-bed combustion
POTW	publicly owned treatment works
Quad	10^{15} Btu
RO	reverse osmosis
SASOL	South African operation of synthetic fuels plants
SMDS	Shell Middle Distillate Synthesis process
SNG	synthetic natural gas
SRC	solvent-refined coal
TDS	total dissolved solids
TES	thermal energy storage
TIC	turbine inlet cooling
TSS	total suspended solids
VM	volatile matter

INTRODUCTION

General References: Loftness, *Energy Handbook*, 2d ed., Van Nostrand Reinhold, New York, 1984; Energy Information Administration, *Emissions of Greenhouse Gases in the United States 2003*, U.S. Dept. of Energy, DOE/EIA-0573 (2004); Howes and Fainberg, eds., *The Energy Source Book*, American Institute of Physics, New York, 1991; Johansson, Kelly, Reddy, and Williams, eds., Burnham, exec. ed., *Renewable Energy—Sources for Fuels and Electricity*, Island Press, Washington, 1993; Turner, *Energy Management Handbook*, 5th ed., The Fairmont Press, Lilburn, Ga., 2004; National Energy Policy, National Energy Policy Development Group, Washington, May 2001.

Energy is usually defined as the capacity to do work. Nature provides us with many sources of energy, some difficult to use efficiently (e.g., solar radiation and wind energy), others more concentrated or energy dense and therefore easier to use (e.g., fossil fuels). Energy sources can be classified also as *renewable* (solar and nonsolar) and *nonrenewable*. Renewable energy resources are derived in a number of ways: gravitational forces of the sun and moon, which create the tides; the rotation of the earth combined with solar energy, which generates the currents in the ocean and the winds; the decay of radioactive minerals and the interior heat of the earth, which provide geothermal energy; photosynthetic production of organic matter; and the direct radiation of the sun. These energy sources are called renewable because they are either continuously replenished or, for all practical purposes, are inexhaustible.

Nonrenewable energy sources include the fossil fuels (natural gas, petroleum, shale oil, coal, and peat) as well as uranium. Fossil fuels are both energy dense and widespread, and much of the world's industrial, utility, and transportation sectors rely on the energy contained in them. Concerns over global warming notwithstanding, fossil fuels will remain the dominant fuel form for the foreseeable future. This is so for two reasons: (1) the development and deployment of new technologies able to use renewable energy sources, such as solar, wind, and biomass, are uneconomic at present, in most part owing to the diffuse or intermittent nature of the sources; and (2) concerns persist about the storage or disposal of spent nuclear fuel and nuclear proliferation.

Fossil fuels, therefore, remain the focus of this section; their principal use is in the generation of heat and electricity in the industrial, utility, and commercial sectors, and in the generation of shaft power in transportation. The material in this section deals primarily with the conversion of the chemical energy in fossil fuels to heat and electricity. Material from *Perry's Chemical Engineers' Handbook*, 8th ed., Sec. 24, has been updated and condensed. Three new technology areas have been added: Shale Gas, which has dramatically changed the fuel landscape, Pinch Analysis, to cover an important segment of energy analysis, and Energy Storage, to reflect a fast-growing field that is aimed at balancing the demand-supply asynchronicity and emission-free power at the point of use.

FUELS

RESOURCES AND RESERVES

Proven worldwide energy resources are large. The largest remaining known reserves of crude oil, used mainly for producing transportation fuels, are located in the Middle East and in the former Soviet Union. Large reserves of natural gas exist in the former Soviet Union and the Middle East. The combined technologies of directional drilling and hydraulic fracturing have led to major increases in natural gas reserves and production in the United States, particularly in the "shale" deposits. Coal is the most abundant fuel on earth and a major fuel for electricity in the United States and Asia. Annual world consumption of energy is still currently less than 1 percent of combined world reserves of fossil fuels. The resources and reserves of the principal fossil fuels in the United States—coal, petroleum, and natural gas—follow.

	ZJ*	
Fuel	Proven reserves	Total estimated resource
Coal	12.6	96.2
Petroleum	0.14	1.35
Natural gas	0.33	2.40

ZJ* = 10^{21} J. To convert to 10^{18} Btu multiply by 0.948

The energy content of fossil fuels in commonly measured quantities is as follows:

	Energy content	
Bituminous and anthracite coal	30.2 MJ/kg	26×10^6 Btu/U.S. ton
Lignite and subbituminous coal	23.2 MJ/kg	20×10^6 Btu/U.S. ton
Crude oil	38.5 MJ/L	5.8×10^6 Btu/bbl
Natural-gas liquids	25.2 MJ/L	3.8×10^6 Btu/bbl
Natural gas	38.4 MJ/m^3	1032 Btu/ft^3

1 bbl = 42 U.S. gal = 159 L = 0.159 m^3

SOLID FUELS

General References: *Annual Book of ASTM Standards,* sec. 5, ASTM International, West Conshohocken, Pa., 2016; *European (EN) Standards on Solid Biofuels,* EN14961 series (*Fuel Specification and Classes*), 6 parts, EN 15234 series (*Fuel quality assurance*), 6 parts, 2011; Lowry and Elliott, eds., *Chemistry of Coal Utilization,* Wiley, New York, 2d suppl. vol., 1981; Van Krevelen, *Coal,* 3d ed., Elsevier Science, Amsterdam, 1993;.

Coal, Coal Blends, and Modeling Traditionally, solid fuels usually meant coal. However, in today's carbon-constrained world, engineers must think of solid fuels as a continuum from biomass all the way through high anthracitic coals as well as their derivative products. This continuum is easily understood through the van Krevelen diagram (Fig. 24-1), which shows the continuous transition of solid fuels from biomass to anthracite. While coal remains an important part of today's energy portfolio, biomass is becoming an important fuel because of its carbon-neutral content. Coal is formed when biomass is immobilized and subsequently buried for long periods of time and exposed to heat and pressure, which cause it to lose oxygen and hydrogen on its way to forming condensed aromatic carbon chains. This process is known as "coalification" and is discussed in detail in "Coal-Bearing Depositional Systems" [by Diessel, Springer, Berlin (1992)].

Classification As coal is formed, the physical properties change, and it can be grouped into four categories according to ASTM D388-15; see Table 24-1 for classification of coals by rank.

The classifications are either based on the gross calorific value on a moist, mineral-matter-free (mmf) basis for lower ranking coals or on the values of the fixed carbon and volatile matter from a proximate analysis on a dry mmf basis. It is worth noting that typical proximate analysis is often reported on an as-received basis, dry basis, or dry-ash-free (daf) basis. All of these bases are different than mineral-matter-free (mmf) basis, which contains a correction to the fraction of the ash for the presence of sulfides in the coal that are not reflected in the analytical numbers. These bases and the proper conversions can be found in ASTM D3180. The recommended course of action is to convert all analytical values to an as-received basis and then use the Parr formulas to calculate the appropriate coal rank values (as outlined in ASTM D388). The methods outlined in ASTM D388 are commonly used in cases of high importance such as litigation. However, in daily practice it is much more common to compare fuels on a daf basis (which is usually close to the dry mmf basis), and is frequently reported in literature.

Biomass is much more difficult to classify because the physical characteristics can vary much more, depending on the source and how the biomass is handled. The European Committee for Standardization has published several standards for biomass fuels. Most notable are EN 14961 series for classification and specification and EN 15234 for quality assurance for solid biofuels. While these standards are used in most European Union (EU) member countries, they have yet to be officially adopted in most other countries around the world.

Solid Fuel Composition Solid fuel composition is typically measured through ultimate and proximate analysis. An ultimate analysis is performed following ASTM D3176 and is used to determine the mass fractions of ash, carbon, hydrogen, nitrogen, sulfur, and oxygen (by difference). These values are intrinsic to the fuel sample and do not depend on thermal treatments. A proximate analysis is performed by either ASTM D7582 or ASTM D3172 and is used to determine the mass fractions of moisture, volatile matter, fixed carbon, and ash. The D3172 standard was developed as a series of separate tests, and the D7582 standard was developed to do the analysis via a single, thermo-gravimetric analysis (TGA). The TGA method is becoming more popular because it is less labor intensive and less expensive. In this method, a sample is loaded into the TGA and subjected to a standardized heating profile. A drying step reveals the amount of both surface and capillary adsorbed water. A ramp of 30 K/min (54°F/min) to 1223 K (1742°F) in an inert environment reveals the volatile matter content. Finally, an oxidation step reveals the ash content. The fixed carbon is determined by difference. The volatile matter is important because it drives flame characteristics and has a relatively high heating value compared to the fixed carbon, which is especially valuable in low-rank coals and biomass. The fixed carbon is important as a slower-burning fuel, as are derivative solid products such as coke, activated carbon, and biochar. Even though these standards were developed specifically for coal and coke, it is common to use them for most any kind of solid, carbon-based fuels.

Modeling Fuels The ASTM tests for proximate analysis are useful for comparing coals but are not reflective of the high heating rates experienced in many combustion processes [>10,000 K/s (> 17500°F/s)]. As heating rate increases, the volatile yield tends to increase and the average molecular weight tends to shift toward heavier compounds such as tars, while the fixed carbon content decreases. Slower heating rates allow more of the heavy volatiles to crack into secondary carbons and light gases, which is useful when tailoring solid fuel–derived products. Liquid yield can be increased by fast pyrolysis, whereas solid product yields are increased by slow pyrolysis.

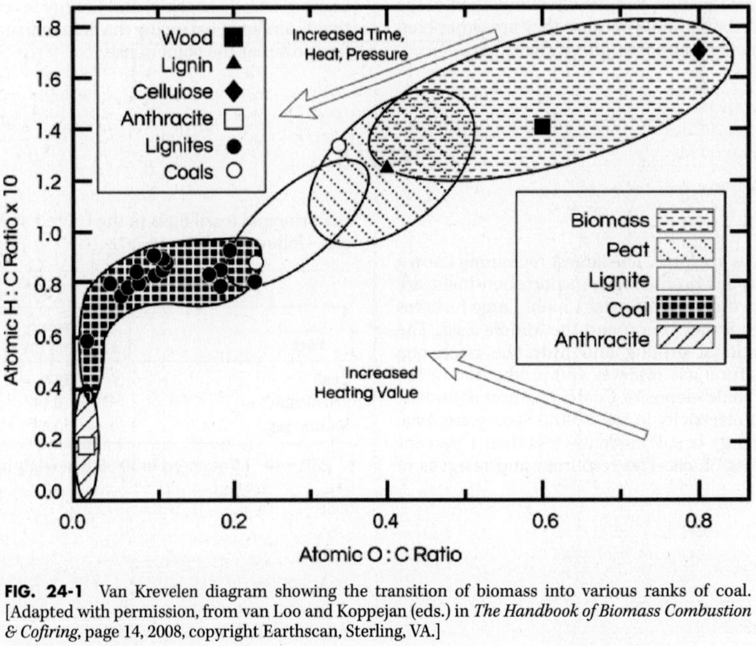

FIG. 24-1 Van Krevelen diagram showing the transition of biomass into various ranks of coal. [Adapted with permission, from van Loo and Koppejan (eds.) in *The Handbook of Biomass Combustion & Cofiring,* page 14, 2008, copyright Earthscan, Sterling, VA.]

TABLE 24-1 Classification of Coals by Rank[*]

Class/group	Fixed carbon limits (dry, mineral-matter-free basis), %: Equal or greater than	Fixed carbon limits: Less than	Volatile matter limits (dry, mineral-matter-free basis), %: Greater than	Volatile matter limits: Equal or less than	Gross calorific value limits (moist, mineral-matter-free basis)[†] MJ/kg: Equal or greater than	MJ/kg: Less than	Btu/lb: Equal or greater than	Btu/lb: Less than	Agglomerating character
Anthracitic:									
Meta-anthracite	98	—	—	2	—	—	—	—	
Anthracite	92	98	2	8	—	—	—	—	Nonagglomerating
Semianthracite[‡]	86	92	8	14	—	—	—	—	
Bituminous:									
Low-volatile bituminous coal	78	86	14	22	—	—	—	—	
Medium-volatile bituminous coal	69	78	22	31	—	—	—	—	
High-volatile A bituminous coal	—	69	31	—	32.6[§]	—	14,000[§]	—	Commonly agglomerating[¶]
High-volatile B bituminous coal	—	—	—	—	30.2[§]	32.6	13,000[§]	14,000	
High-volatile C bituminous coal	—	—	—	—	26.7	30.2	11,500	13,000	
					24.4	26.7	10,500	11,500	Agglomerating
Subbituminous:									
Subbituminous A coal	—	—	—	—	24.4	26.7	10,500	11,500	
Subbituminous B coal	—	—	—	—	22.1	24.4	9,500	10,500	
Subbituminous C coal	—	—	—	—	19.3	22.1	8,300	9,500	Nonagglomerating
Lignitic:									
Lignite A	—	—	—	—	14.7	19.3	6,300	8,300	
Lignite B	—	—	—	—	—	14.7	—	6,300	

Adapted, with permission, from D388-99, Standard Classification of Coals by Rank; copyright ASTM International, 100 Barr Harbor Drive, West Conshohocken, PA 19428.

[*]This classification does not apply to certain coals, as discussed in source.

[†]*Moist* refers to coal containing its natural inherent moisture but not including visible water on the surface of the coal.

[‡]If agglomerating, classify in low-volatile group of the bituminous class.

[§]Coals having 69 percent or more fixed carbon on the dry, mineral-matter-free basis shall be classified according to fixed carbon, regardless of gross calorific value.

[¶]It is recognized that there may be nonagglomerating varieties in these groups of the bituminous class and that there are notable exceptions in the high-volatile C bituminous group.

Several good tools have been developed to estimate the products of pyrolysis: the Chemical Percolation Devolatilization model by Sandia National Laboratory and University of Utah [Fletcher et al., "Chemical Percolation Model for Devolatilization. 3. Direct Use of Carbon-13 NMR Data to Predict Effects of Coal Type," *Energy Fuels* **6**(4): 414–431 (1992)]; the Functional-Group, Depolymerization, Vaporization, Cross-Linking model by Advanced Fuel Research [Solomon et al., "General Model of Coal Devolatilization," *Energy Fuels* **2:** 405–422 (1988)]; and PC Coal Lab by Niksa Energy Associates [Niksa S., PC Coal Lab.]. These software tools use empirical methods along with internal models to predict the yields and compositions of the fixed carbons and volatile gases based on a prescribed heating profile. Finally, the Carbonaceous Chemistry for Computational Modeling (C3M) software by the National Energy Technology Laboratory (NETL) was developed to interface these complex models with other computational modeling tools, such as multiphase computational fluid dynamics simulators [van Essendelft et al., "Advanced Chemical Surrogate Model Development within C3M for CFD Modeling, Part 1: Methodology Development for Coal Pyrolysis," *Ind. Eng. Chem. Res.* **53:** 7780–7796 (2014)]. Together, these tools make a user-friendly environment for engineers and scientists to predict solid fuel behavior under a wide variety of conditions. C3M also incorporates other important chemistry rates for fuel conversion modeling and spans the use of biomass, coal, and coke.

Properties/Characteristics of Solid Fuels *Heating value* is typically measured using an oxygen-bomb calorimeter following the procedure outlined in ASTM D5865. The reported values from these tests are the higher heating value (HHV), which includes the heat evolved when any water present condenses. The lower heating value (LHV) is related to the HHV by simply subtracting the weight of water from weight of fuel burned (W) multiplied by the latent heat of vaporization at standard conditions (K), which is 2395 kJ/kg (1030 Btu/lb). This results in the heat evolved when the water present does not condense.

$$\text{LHV} = \text{HHV} - K \cdot W \tag{24-1}$$

The mass of combustion water can be determined from the percentage of hydrogen in the ultimate analysis.

$$\text{LHV} = \text{HHV} - \%\text{H} \cdot 214 \text{ kJ/kg} \tag{24-2}$$

$$\text{LHV} = \text{HHV} - \%\text{H} \cdot 92.0 \text{ Btu/lb} \tag{24-3}$$

Here, %H is the percentage of hydrogen from the ultimate analysis on a moisture laden as-determined basis.

It is not always feasible to analyze every sample of interest, and it is valuable to be able to estimate the heating value from the ultimate analysis. There are several popular equations for estimating these properties, most notably the Boie and Dulong equations. However, Mason and Gandhi at the Institute of Gas Technology demonstrated a more accurate equation that reduces error across all coal ranks [Mason and Gandhi, "Formulas for Calculating the Heating Value of Coal and Coal Char: Development, Tests, and Uses," *Fuel Processing Tech.* **7:** 11–22 (1983)]. The HHV on a dry basis can be estimated as

$$\text{HHV} = 340.95\,\%\text{C} + 1323.0\,\%\text{H} + 68.4\,\%\text{S} - 15.3\,\%\text{A} - 119.9\,(\%\text{O} + \%\text{N})\text{ kJ/kg} \tag{24-4}$$

$$\text{HHV} = 146.58\,\%\text{C} + 568.79\,\%\text{H} + 29.4\,\%\text{S} - 6.58\,\%\text{A} - 51.55\,(\%\text{O} + \%\text{N})\text{ Btu/lb} \tag{24-5}$$

Here, %C, %H, %O, %N, %S, and %A are the mass percentages from the ultimate analysis for carbon, hydrogen, oxygen, nitrogen, sulfur, and ash on a dry basis. The National Renewable Energy Laboratory (NREL) reviewed and published a similar correlation for biomass and liquid fuels (Gaur, S., and T. B. Reed, *An Atlas of Thermal Data for Biomass and Other Fuels*, NREL, Golden, Colo., 2012).

$$\text{HHV} = 349.1\,\%\text{C} + 1178.3\,\%\text{H} + 100.5\,\%\text{S} - 103.4\,\%\text{O} - 15.1\,\%\text{N} - 21.1\,\%\text{A kJ/kg} \tag{24-6}$$

$$\text{HHV} = 150.1\,\%\text{C} + 506.58\,\%\text{H} + 43.21\,\%\text{S} - 44.45\,\%\text{O} - 6.49\,\%\text{N} - 9.07\,\%\text{A Btu/lb} \tag{24-7}$$

The Mason and Gandhi equation was tested across 775 samples of coal covering all ranks with an average standard deviation of 300.1 kJ/kg (129 Btu/lb). The NREL model was tested across 119 samples of various biomass, liquid fuels, and other solid fuels with an average error of 337 kJ/kg (145 Btu/lb). For the best accuracy, it is recommended that the Mason and Gandhi equation be used for coal and the NREL correlation be used for biomass and liquid fuels.

Heat of formation values are often calculated when modeling fuel reactions. This is done by energy balance around the complete combustion equation assuming all carbon ends up as carbon dioxide, all hydrogen ends up as liquid water, all sulfur ends up as SO_2, and all nitrogen ends up as N_2, and that any oxygen is consumed in the reaction. The energy balance simplifies to the following [van Essendelft et al., "Advanced Chemistry Surrogate

Model Development within C3m for CFD Modeling, Part 1: Methodology Development for Coal Pyrolysis," *Ind. Eng. Chem. Res.* **53:** 7780–7796 (2014)]:

$$\Delta\hat{H}_{f\_\text{fuel}} = -HHV + \frac{\%C}{100}\Delta\hat{H}_{f\_\text{CO2}(g)} + \frac{\%S}{100}\Delta\hat{H}_{f\_\text{SO2}(g)} + \frac{\%H}{2*100}\Delta\hat{H}_{f\_\text{H2O}(l)} \tag{24-8}$$

Here $\Delta\hat{H}_f$ denotes the heat of formation on a mass basis, and any following species denote the values for those species, and %C, %H, and %S denote the mass percentages of carbon, sulfur, and hydrogen from the ultimate analysis. Units should be maintained in a consistent way.

The *specific heat* of a variety of carbon-containing compounds can be estimated using a two-parameter Einstein equation over the range of 273 to 1073 K (32°F to 1472°F) [Merric, "Mathematical Models of the Thermal Decomposition of Coal: 2. Specific Heats and Heats of Reaction," *Fuel* **62:** 540–546 (1983)].

$$C_{p\_FCVM} = \frac{R}{Aw_{\text{fuel}}}\left[f\left(\frac{380}{T}\right) + 2f\left(\frac{1800}{T}\right)\right] \tag{24-9}$$

$$f(x) = \frac{\exp(x)}{((\exp(x)-1)/x)^2} \tag{24-10}$$

Here, R is the universal gas constant, Aw is the average atomic weight, and T is the temperature in Kelvin. The equation will yield the mass normalized specific heat for the solid fuel on a dry, ash-free basis. The units are set by the choice of units for the gas constant and atomic weight. The average atomic weight can be calculated from the ultimate analysis and the known values for the atomic weights of carbon, hydrogen, oxygen, nitrogen, sulfur.

$$Aw_{\text{fuel}} = \frac{C+H+O+N+S}{\frac{C}{Aw_C}+\frac{H}{Aw_H}+\frac{O}{Aw_O}+\frac{N}{Aw_N}+\frac{S}{Aw_S}} \tag{24-11}$$

For most coals this value should be in the range of 7 to 12 kg/kmol, and it may be less for biomass. The specific heat of ash can be taken as a simple linear form [Merric, "Mathematical Models of the Thermal Decomposition of Coal: 2. Specific Heats and Heats of Reaction," *Fuel* **62:** 540–546 (1983)].

$$C_{p\_\text{ash}} = 594 + 0.586T\left(\frac{J}{\text{kg}\cdot\text{K}}\right) \tag{24-12}$$

The specific heat of the moisture can be assumed to be the same as the specific heat of liquid water (4186 J/kg/K). The bulk specific heat of the fuel can be calculated from the weight percentages of the ultimate analysis on an as-received basis.

$$C_{p\_\text{bulk}} = \frac{w_{\text{H2O}}C_{p\_\text{H2O}} + (w_{FC}+w_{VM})C_{p\_FCVM} + (w_{\text{ash}}+w_{\text{sulf}})C_{p\_\text{ash}}}{100} \tag{24-13}$$

Sulfur is an important consideration for solid fuel use due to concerns about its effects on acid rain and human health. Most biomass is very low in sulfur (generally below 0.2 percent on a dry basis), and most derivative products are also low sulfur unless exposed to sulfur-containing compounds. Coal can contain considerable amounts of sulfur, up to 7 percent. Depending on the environment during coalification, coal can pick up sulfur from the environment. The sulfur ends up bound in one of three states: as organically bound, as iron pyrite inclusions, or as mineral sulfate inclusions. The pyrite and mineral inclusions can be removed by size reduction (liberation) and washing using standard coal processing techniques. However, 20 to 80 percent of the sulfur can be organically bound. There are no known cost-effective methods to remove organically bound sulfur. When oxidized, all these sulfur sources produce sulfur dioxide (SO_2) emissions, which contribute to acid rain and a variety of human health issues. In the United States, eastern coals generally have a higher sulfur content than western coals, as can be seen in Table 24-2.

Due to EPA regulations and the cost of sulfur abatement equipment, many power producers in the East are importing low-sulfur coal from the West or from international sources.

Mercury in solid fuels, due to its highly toxic nature, has become a growing concern. Mercury content in solid fuels is quite low. For biomass the mercury contents range from 14 to 72 µg/kg [Friedli et al., "Mercury in Smoke from Biomass Fires," *Geophy. Res. Lett.* **28:** 3223–3226 (2001)].

TABLE 24-2 Estimate of Recoverable U.S. Coal Reserves by Sulfur Ranges and Regions (Pg)

	Low sulfur*	Medium sulfur*	High sulfur*	Total
Appalachia	10.6	18.4	21.1	50.2
Interior	0.7	9.1	52.6	62.4
West	79.6	49.5	8.0	137.1
U.S. total	90.9	77.0	81.7	249.6

*Sulfur content ranges defined in units of kg·S/kJ (lb·S/MBtu) as follows: low sulfur, < 0.29 (< 0.60); medium sulfur, 0.29 to 0.80 (0.61 to 1.67); high sulfur, > 0.80 (>1.67).

Coals typically range from 70 up to 240 µg/kg (Tewalt et al., *Mercury in U.S. Coal—Abundance, Distribution, and Modes of Occurrence*, Fact Sheet 095-01, U.S. Geological Survey, Reston, Va., 2001). To date, no emissions standards have been set and upheld as law within the United States. The ease with which mercury can be captured largely depends on the form it takes during combustion. Ionic forms of mercury are much easier to capture with existing technology than elemental forms. Coals with chlorine contents of 0.05 percent or higher tend to produce ionic mercury forms. The coal type, combustion conditions, and flue gas cleanup equipment can have a dramatic impact on the emissions of mercury, ranging from 10 to 90 percent capture.

Ash and slag are by-products of burning solid fuels; any noncombustible material in a solid fuel ends up as ash in fuel conversion processes. Ash can fuse to form sintered and fused masses that can impede material flow or heat transfer and erode and corrode high-temperature materials. The transition of free ash to slag is known as ash fusibility. The fusibility is characterized through four temperatures—initial deformation, softening, hemispherical, and fluid temperature—and is determined by ASTM D1857/D1857M. In general, solid fuel conversion equipment is operated either below the ash fusion temperature [~<1373 K (<2000 °F)] to avoid any risk of sintering or well above the fusion temperature such that the slag becomes fluid. Ash and slag are complex mixtures of a wide variety of possible mineral phases. The ash fusion temperature is dominated by the lowest melting phases as well as their abundance in the ash. The most accurate predictions for ash fusion temperature have come from computer learning algorithms like neural networks and support vector machines. If equipment operates above the ash fusion temperature, such as slagging gasifiers, it is valuable to predict slag viscosity. The multiphase properties of slag make the prediction of slag viscosity difficult, and slag is often a non-newtonian fluid. Most modern slag viscosity models are a variant of the Urbain model [Urbain, "Viscosity Estimation of Slags," *Steel Res.* **58:** 111–116 (1987)]. The constituents of biomass ash generally act as fluxing agents for slag and tend to lower the ash fusion temperature, which is an important consideration in designing fuel conversion systems because the ash fusion temperature is usually a limiting factor in design. Ash from nonslagging processes is typically referred to as fly ash, which is a valuable by-product for cement manufacturers.

Free swelling index (FSI) is a measure of a solid fuel's propensity to swell during heating. This is important in applications with fixed and fluid beds and is important in the production of coke. The test is done according to ASTM D720/D720M. An FSI of 4 or more is usually considered troublesome for fixed/fluid beds.

Grindability of solid fuels is important because size reduction is needed for the liberation of contaminants, handling of the solid, and increasing reaction kinetics. The grindability of coal is determined using the Hardgrove Grindability Index (HGI), which has been standardized in ASTM D409/D409M. The HGI is a relative reference to a standard coal whose value is arbitrarily set to 100. The higher the HGI, the easier it is to grind. The FSI and HGI of a large number of coals has been compiled by the Bureau of Mines (*Coal Conversion Systems Technical Data Book*, Institute of Gas Technology, Springfield, Va., 1978). The grindability of biomass is quite different from coal, and no standards have yet been determined. It is well known that raw biomass is a visco-plastic, highly anisotropic material that needs high shear rates to fracture. Thus a high-energy hammer mill is usually needed to work with raw biomass. Low-temperature thermal treatments such as torrefaction are gaining popularity to improve the friability of biomass such that it can be used together with coal in existing material handling equipment. However, work is still needed to standardize mill capacity formulations and associated energy consumption so that engineers can properly size and design equipment.

Bulk density of solid fuels depends on the true density (which is most affected by moisture content), the size and shape of the solid fragments, and the settling of the material. Biomass bulk densities can vary substantially depending on the type of biomass, how the material is handled, and any processing done (palletization, drying, heat treatments, etc.). The following are typical ranges for coals:

	kg/m³	lb/ft³
Anthracite	800–930	50–58
Bituminous	670–910	42–57
Lignite	640–860	40–54

Size stability is the tendency of a solid fuel to withstand breakage during handling and shipping. Size stability is measured by ASTM D440, in which the fuel is dropped at a certain height onto a steel plate and the size distribution before and after the test are compared. A complimentary test is friability, which is the tendency to break during handling. Friability is measured by ASTM D441, which is very similar to ASTM D440, but a standard tumbler is used.

Coke Coke is the solid, infusible material remaining after the carbonization of coal, coal pitch, petroleum residues, and other carbonaceous materials. The quality and yield of coke depend on the feedstock and the heat treatment conditions. Different coke materials are often identified by their feedstock unless it is from coal (e.g., petroleum coke), the manufacturing process (e.g., vertical slot oven coke), the end use (e.g., blast furnace coke), or even the particle size distribution (e.g., coke breeze). Carbonization of coal into coke involves a complex set of physical and chemical changes, including depolymerization, cracking, polymerization, and condensation, resulting in devolatilization, softening, swelling, and finally, resolidification (van Krevelen *Coal: Typology, Physics, Chemistry, Constitution*, 2d ed., Elsevier, New York, 1993).

Coke Production Process Description Crushed coal is heated slowly in a coke oven in a deep fixed bed so that the volatiles produced have an opportunity to soften and partially dissolve the coal's macromolecular network. This is thought to create a metaplast, which then repolymerizes, solidifies, and anneals after prolonged exposure to high temperatures for 20 to 40 hours. The material is quenched with water, producing a solid product consisting primarily of nonvolatile carbon fused together into strong, hard, yet porous material—coke.

Not all coals produce carbonized solid products suitable for coke applications. Tests such as the free swelling index (FSI), Gray King, or Roga index are used to evaluate the potential for a coal to carbonize into a consolidated coke. Medium and high-volatile bituminous coals with high vitrinite content are preferred for metallurgical applications because coals in this rank range can produce high-strength coals with low friability (Sundholm et al., "Manufacture of Metallurgical Coke and Recovery of Coal Chemicals," in *Ironmaking*, AISE Steel Foundation, Pittsburgh, Pa., 1999, pp. 381–546). These bituminous coals exhibit the greatest swelling and fluidity upon heating in an inert environment as measured by Dilatometer and Giesler plastometer techniques. It is important that coals used for metallurgical coke production have low concentrations of inert materials. This includes both the mineral matter that produces ash upon combustion and the inertinite macerals, which are fossilized remnants of charred or microbiologically carbonized plant detritus. The size of the coke particles depends on the content of these inert materials; the higher the inert content, the smaller the coke materials produced [Schapiro et al., "Recent Developments in Coal Petrography," *Blast Furnace, Coke Oven, and Raw Materials Committee Proc.* **20:** 89–112 (1961)].

Coals are often blended in order to achieve all of the characteristics desired for the application using the coke. Coke properties, structure, and reactivity are dependent on the maximum processing temperature, heating rate, residence time, and pressure. The specific influence of these process conditions varies for different feedstocks. However, the coke strength after reaction with CO_2 (CSR) is a useful screening parameter to predict the behavior of coal blends [Valia, "Prediction of Coke Strength after Reaction with CO_2 from Coal Analyses at Inland Steel Company," *Iron and Steel Soc. Trans.* **11:** 55–65 (1990)]. The CSR is a linearly additive property such that coke produced from blending several coals together can be predicted from a weighted average of their individual CSR values. Subsequent research developed models to estimate the CSR knowing the heat treatment conditions and fluidity temperature range and estimating a catalytic index from the coal's ash content, ash composition, and sulfur content [Marsh and Walker, "The Effects of Impregnation of Alkali Salts upon Carbonization Properties," *Fuel Proc. Tech.* **2:** 61–75 (1979)].

Metallurgical Coke The single most important characteristic of metallurgical coke is its strength to withstand breakage and abrasion during handling and its use in the blast furnace. Standard tests performed in a ball mill apparatus that are used to characterize these coke properties are the stability index for breakage and the hardness index for abrasion [ASTM D409]. Essentially all coal-derived coke in the United States is produced at temperatures between 1175 K (1655°F) and 1425 K (2105°F) for metallurgical applications. This represents nearly 5 percent of the total bituminous coal consumed in the United States (Energy Information Administration Annual Coal Report, 2016, https://www.eia.gov/coal annual/pdf/acr.pdf). About 90 percent of this type of coke is made in slot-type by-product recovery ovens, and the rest is made in heat recovery ovens. Blast furnaces use about 90 percent of the production; the rest is mainly for use in foundries and gasification plants.

TABLE 24-3 Chemical and Physical Properties of Cokes Used in the United States

Property	Metallurgical coke	Petroleum delayed coke	Petroleum fluid coke
Volatile matter, wt% dry	0.5–1.0	8–18	3.7–7.0
Sulfur content, wt% dry	0.6–1.0	—	1.5–10.0
Ash content, wt% dry	8–12	0.05–1.6	0.1–2.8
True density, g/cm³	0.8–0.99	1.28–1.42	1.5–1.6
Grindability index	60–68	40–60	20–30
Coke strength after reaction (CSR), % of original	55–65	—	—

Metallurgical coke quality is assessed by evaluating coke strength after reaction with CO_2 (CSR); vitrinite, volatile matter (VM), ash, moisture, sulfur, and phosphorous contents; and fluidity. The maximum level of CSR is rank dependent, maximizing at vitrinite reflectance between 1 and 1.4 percent in the range of high- and medium-volatile bituminous coals. A coke reactivity index (CRI) has been adopted by ASTM D5341-93a, which correlates very loosely to the earlier CSR [Diez et al., "Coal for Metallurgical Coke Production: Prediction of Coke Quality and Future Requirements for Coke Making," *Int. J. Coal Geo.* **50:** 389–412 (2002)]. The ranges of chemical composition and physical attributes of metallurgical coke used in the United States are compared to petroleum-derived counterparts in Table 24-3. The yields of typical by-products from these high-temperature coking ovens are: 72.1 percent coke, 17.1 percent gas, 5.0 percent tar, 4.4 percent water, 1.2 percent light oil, and 0.3 percent ammonia by weight of coal.

Foundry coke is designed for use in metal-refining cupolas and is of larger particle size, and produced at lower temperatures, than blast furnace coke, though for a longer period of time. Foundry coke has specifications not required of blast furnace coke, including: volatile matter, sulfur, and ash must be less than 1.0, 0.7, and 8 wt% on a dry basis, and the size should exceed 100 mm (4 in).

Low- and medium-temperature cokes (773 to 1023 K or 932°F to 1382°F) are no longer produced in the United States to any significant extent; however, there continues to be some interest in low-temperature carbonization as a source of by-products, including hydrocarbon liquids and gases to substitute for petroleum and natural-gas resources and specialty chemicals such as fertilizer.

Pitch coke is produced using the distillate bottoms from coal tars. Pitch coke has about 1.0 percent volatile matter and significantly lower ash and sulfur contents than those produced directly from coal, with less than 1.0 percent wt. ash and 0.5 percent wt. sulfur as-received.

Petroleum coke is made using the process residuals (e.g., vacuum distillate bottoms) from petroleum refining. It is prepared from one of two distinctly different processes: delayed coking and fluid coking. Delayed coke is produced by heating a gas oil or heavier feedstock to 755 to 810 K (900°F to 1000°F) and spraying it into a large vertical cylinder, where cracking and polymerization reactions occur. Fluid coke is made in a fluidized-bed reactor, where preheated feed is sprayed onto a fluidized bed of coke particles. The fluid coke product is continuously withdrawn by size classifiers in the solids recycle loop of the circulating reactor system. Petroleum coke contains many of the impurities from its feedstock; thus, the sulfur content is usually high, and appreciable quantities of vanadium may be present. Pitch and petroleum feedstocks are preferred over coal when mineral matter content is deleterious to the end application. Most petroleum coke is used for fuel, but some premium delayed coke known as "needle coke" is used to make anodes for the aluminum industry. That coke is first calcined to less than 0.5 percent volatiles at 1573 to 1673 K (2372°F to 2552°F) before it is used to make anodes.

Other Solid Fuels *Coal char* is, generically, the nonagglomerated, nonfusible residue from the thermal treatment of coal; however, it is more specifically the solid residue from low- or medium-temperature carbonization processes. Char can be used as a fuel or a carbon source. Chars have compositions between those of coal and coke. For example, the volatile matter, sulfur content, and heating values of the chars are often lower, and the ash content is often higher, than those of the original coal.

Peat is partially decomposed plant matter that has accumulated in a water-saturated environment. It is the precursor of coal but is not classified as coal. Peat is used extensively as a fuel primarily in Ireland and the former Soviet Union, but in the United States its main use is in horticulture and agriculture. Although analyses of peat vary widely, a typical high-grade peat has 90 percent water, 3 percent fixed carbon, 5 percent volatile matter, 1.5 percent ash, and 0.10 percent sulfur. The moisture-free heating value is approximately 20.9 MJ/kg (9000 Btu/lb).

Wood, wood scraps, bark, and wood product plant waste streams are major elements of the biomass industry. In 1991, about 1.7 EJ [1.6×10^{15} Btu (quads)] of energy were obtained from wood and wood wastes, representing about 60 percent of the total biomass-derived energy in the United States. Typical higher heating values are 20 MJ/kg (8600 Btu/lb) for oven-dried hardwood and 20.9 MJ/kg (9000 Btu/lb) for oven-dried softwood. These values are accurate enough for most engineering purposes. U.S. Department of Agriculture Handbook 72 (revised 1974) gives the specific gravity of the important softwoods and hardwoods, which is useful if heating value on a volume basis is needed.

Charcoal is the residue from the pyrolysis of wood. It absorbs moisture readily, often containing as much as 10 to 15 percent water. In addition, it usually contains about 2 to 3 percent ash on a weight basis. The higher heating value of charcoal is about 27.9 to 30.2 MJ/kg (12,000 to 13,000 Btu/lb).

Bagasse is the solid residue remaining after sugarcane has been crushed by pressure rolls. It usually contains from 40 to 50 percent water. The dry bagasse has a heating value between 18.6 and 20.9 MJ/kg (8000 to 9000 Btu/lb).

Corn stover is the residue remaining after corn has been harvested, such as the leaves, the stalks, and sometimes the husk and cob. The dry corn stover has a higher heating value around 18 MJ/kg DAF (7800 Btu/lb). While it is crucial that some of the corn stover should remain in the field in order to replace nutrients and organic carbon, there may be benefits, such as increased yields, when some of the corn stover is not placed back into the soil, which means that some corn stover can be sold for external use either as combustion fuel or as a chemical feedstock.

Municipal solid waste (MSW), which results in the generation of large quantities of solid wastes, is a significant feature of most societies. In the United States over the last few decades, the rate has been about 2 kg (4.4 lb) per capita per day. In 2006 in the United States, the EPA found that the breakdown of MSW was the following: 33.9 percent paper/paperboard, 12.9 percent yard trimmings, 12.4 percent food scraps, 11.7 percent plastics, 7.6 percent metals, 5.5 percent wood, 5.3 percent glass, 4.7 percent textiles, 2.6 percent rubber/leather, and 2.3 percent other materials.

The fuel value of municipal solid wastes is usually sufficient to enable self-supporting combustion, leaving only the incombustible residue and reducing by 90 percent the volume of waste consigned to landfill. The heat released by the combustion of waste can be recovered and utilized, or the MSW can be used as feedstock in gasifiers and converted into synthetic natural gas or liquid fuels.

The *nonrenewable portions of MSW* are those waste materials that ultimately derive from oil and natural gas, such as plastics, rubber, and tires. For example, tire-derived fuel (TDF), which is produced by shredding and processing waste tires and which has a heating value of 30.2 to 37.2 MJ/kg (13,000 to 16,000 Btu/lb), is an important fuel for use in cement kilns and as a supplement to coal in steam raising.

LIQUID FUELS

Liquid Petroleum Fuels The discussion here focuses on burner fuels rather than transportation fuels. There is overlap, particularly for fuels in the distillate or "gas oil" range. This is particularly evident in the Northeast, where the No. 2 heating oil sulfur specification is being phased down to match the ultralow sulfur diesel (ULSD), 15 ppm S. This phase-in is illustrated in Fig. 24-2.

Other factors such as the Tier 3 gasoline specifications, which will phase down average gasoline sulfur content to 10 ppm, will affect refining and distribution. Crude oils vary widely in composition and quality, which affects refining and blending. As many as one-quarter to one-half of the molecules in crude may contain sulfur atoms, and some contain nitrogen, oxygen, vanadium, nickel, or arsenic. Desulfurization, hydrogenation, cracking (to lower molecular weight), and other refining processes are performed on selected fractions before they are blended and marketed as fuels.

Specifications The American Society for Testing and Materials has developed specifications (*Annual Book of ASTM Standards,* Conshohocken, Pa., updated annually) that are widely used to classify fuels. Table 24-4 shows fuels covered by ASTM D 396-14, Standard Specification for Fuel Oils. D 396 omits kerosene (low-sulfur, clean-burning No. 1 fuels for lamps and freestanding flue less domestic heaters), which is covered separately by ASTM D 3699-13b. As noted previously, fuel sulfur specifications are becoming stricter, which is not fully reflected in the current ASTM D396 document.

In drawing contracts and making acceptance tests, refer to the pertinent ASTM standards.

ASTM Standards contain specifications (classifications) and test methods for burner fuels (D 396), motor and aviation gasoline (D 4814-14a and D 910-13a), diesel fuels (D 975-14a), and aviation and gas-turbine fuels (D 1655-13a and D 2880-13b). ASTM D 4057-12 contains procedures for sampling bulk oil in tanks, barges, and other vessels.

Fuel specifications from different sources may differ in test limits on such features as sulfur content and density, but the same general categories are recognized worldwide: kerosene-type vaporizing fuel, distillate (or gas oil) for atomizing burners, and more viscous blends and residuals for commerce and heavy industry.

Foreign specifications are generally available from the American National Standards Institute, New York; United States federal specifications, at Naval Publications and Forms, Philadelphia. The International Association for Stability, Handling and Use of Liquid Fuels maintains a web site (www.iash.net) with extensive references to fuel standards (subscription required).

Equipment manufacturers and large-volume users often write fuel specifications to suit particular equipment, operating conditions, and economics. Nonstandard test procedures and restrictive test limits should be avoided; they reduce the availability of fuel and increase its cost.

Bunker-fuel specifications for merchant vessels were described by ASTM D 2069, *Standard Specification for Marine Fuels*, which was withdrawn in 2003. Specifications under ASTM D-396-14 or foreign specifications may be substituted as appropriate.

Chemical and Physical Properties Petroleum fuels contain paraffins, isoparaffins, naphthenes, and aromatics, plus organic sulfur, oxygen, and nitrogen compounds that were not removed by refining. Olefins are absent or negligible except when created by severe refining. Vacuum-tower distillate with a final boiling point equivalent to 730 to 840 K (850°F to 1050°F) at atmospheric pressure may contain from 0.1 to 0.5 ppm vanadium and nickel, but these metal-bearing compounds do not distill into No. 1 and 2 fuel oils.

Black, viscous residuum directly from the still at 472 K (390°F) or higher serves as fuel in nearby furnaces or may be cooled and blended to make

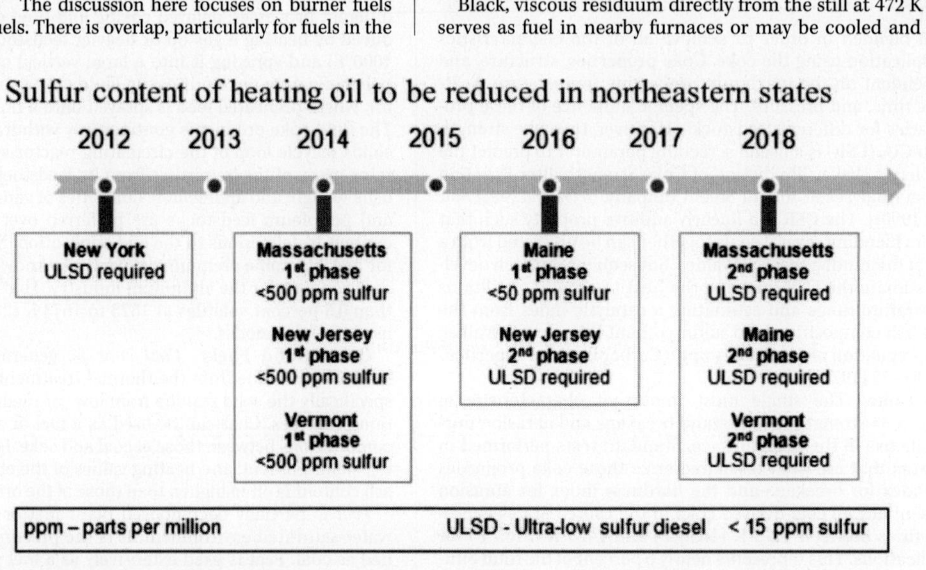

FIG. 24-2 Phase-in schedule for Ultra Low Sulfur Diesel (ULSD) in the northeastern states. (*Today in Energy*, U.S. Energy Information Administration, April 18, 2012.)

TABLE 24-4 Detailed Requirements for Fuel Oils[a]

Property	ASTM test method[b]	No. 1 low sulfur[c]	No. 1[c]	No. 2 low sulfur[c]	No. 2[c]	Grade no. 4 (light)[c]	No. 4	No. 5 (light)	No. 5 (heavy)	No. 6
Flash point, °C, min	D 93—Proc. A	38	38	38	38	38	—	—	—	—
	D 93—Proc. B	—	—	—	—	—	55	55	55	60
Water and sediment, % vol, max	D 2709	0.05	0.05	0.05	0.05	—	—	—	—	—
	D 95 + D 473	—	—	—	—	(0.50)[d]	(0.50)[d]	(1.00)[d]	(1.00)[d]	(2.00)[d]
Distillation temperature, °C	D 86									
10% volume recovered, max		215	215	—	—	—	—	—	—	—
90% volume recovered, min		—	—	282	282	—	—	—	—	—
90% volume recovered, max		288	288	338	338	—	—	—	—	—
Kinematic viscosity at 40°C, mm^2/s	D 445									
Min		1.3	1.3	1.9	1.9	1.9	>5.5	—	—	—
Max		2.1	2.1	3.4	3.4	5.5	24.0[e]	—	—	—
Kinematic viscosity at 100°C, mm^2/s	D 445									
Min		—	—	—	—	—	—	5.0	9.0	15.0
Max		—	—	—	—	—	—	8.9[e]	14.9[e]	50.0[e]
Ramsbottom carbon residue on 10% distillation residue, % mass, max	D 524	0.15	0.15	0.35	0.35	—	—	—	—	—
Ash, % mass, max	D 482	—	—	—	—	0.05	0.10	0.15	0.15	—
Sulfur, % mass max[f]	D 129	—	0.50	—	0.50	—	—	—	—	—
	D 2622	0.05		0.05						
Copper strip corrosion rating, max, 3 h at 50°C	D 130	No. 3	No. 3	No. 3	No. 3	—	—	—	—	—
Density at 15°C, kg/m^3	D 1298									
Min		—	—	—	—	>876[g]	—	—	—	—
Max		850	850	876	876	—	—	—	—	—
Pour point, °C, max[h]	D 97	−18	−18	−6	−6	−6	−6	—	—	[i]

Adapted, with permission, from D 396-06, Standard Specification for Fuel Oils; copyright ASTM International, 100 Barr Harbor Drive, West Conshohocken, PA 19428.

[a]It is the intent of these classifications that failure to meet any requirement of a given grade does not automatically place an oil in the next lower grade unless in fact it meets all requirements of the lower grade. However, to meet special operating conditions, modifications of individual limiting requirements may be agreed upon among the purchaser, seller, and manufacturer.

[b]The test methods indicated are the approved referee methods. Other acceptable methods are indicated in Sections 2 and 5.1 of ASTM D 396.

[c]Under U.S. regulations, Grades No. 1, No. 1 Low Sulfur, No. 2, No. 2 Low Sulfur, and No. 4 (Light) are required by 40 CFR Part 80 to contain a sufficient amount of the dye Solvent Red 164 so its presence is visually apparent. At or beyond terminal storage tanks, they are required by 26 CFR Part 48 to contain the dye Solvent Red 164 at a concentration spectrally equivalent to 3.9 lb per thousand barrels of the solid dye standard Solvent Red 26.

[d]The amount of water by distillation by Test Method D 95 plus the sediment by extraction by Test method D 473 shall not exceed the value shown in the table. For Grade No. 6 fuel oil, the amount of sediment by extraction shall not exceed 0.50 mass percent, and a deduction in quantity shall be made for all water and sediment in excess of 1.0 mass percent.

[e]Where low sulfur fuel oil is required, fuel oil falling in the viscosity range of a lower numbered grade down to and including No. 4 can be supplied by agreement between the purchaser and supplier. The viscosity range of the initial shipment shall be identified, and advance notice shall be required when changing from one viscosity range to another. This notice shall be in sufficient time to permit the user to make the necessary adjustments.

[f]Other sulfur limits may apply in selected areas in the United States and in other countries.

[g]This limit ensures a minimum heating value and also prevents misrepresentation and misapplication of this product as Grade No. 2.

[h]Lower or higher pour points can be specified whenever required by conditions of storage or use. When a pour point less than −18°C is specified, the minimum viscosity at 40°C for Grade No. 2 shall be 1.7 mm^2/s and the minimum 90% recovered temperature shall be waived.

[i]Where low sulfur fuel oil is required, Grade No. 6 fuel oil will be classified as Low Pour (−15°C max) or High Pour (no max). Low Pour fuel oil should be used unless tanks and lines are heated.

commercial fuels. Diluted with 5 to 20 percent distillate, the blend is No. 6 fuel oil. With 20 to 50 percent distillate, it becomes No. 4 and No. 5 fuel oils for commercial use, as in schools and apartment houses. Distillate-residual blends also serve as diesel fuel in large stationary and marine engines. However, distillates with inadequate solvent power will precipitate asphaltenes and other high-molecular-weight colloids from *visbroken* (severely heated) residuals. A blotter test, ASTM D 4740-04(2014), will detect sludge in pilot blends. Tests employing centrifuges, filtration (D 4870-09), and microscopic examination have also been used.

No. 6 fuel oil contains from 10 to 500 ppm vanadium and nickel in complex organic molecules, principally porphyrins. These cannot be removed economically, except incidentally during severe hydrodesulfurization (Amero, Silver, and Yanik, *Hydrodesulfurized Residual Oils as Gas Turbine Fuels*, ASME Pap. 75-WA/GT-8). Salt, sand, rust, and dirt may also be present, giving No. 6 a typical ash content of 0.01 to 0.5 percent by weight.

Ultimate analyses of some typical fuels are shown in Table 24-5. The sulfur content of fuels will often be lower than shown in the table, reflecting the growing regulatory constraints on sulfur content.

The hydrogen content of petroleum fuels can be calculated from density with the following formula, with an accuracy of about 1 percent for petroleum liquids that contain no sulfur, water, or ash:

$$H = 26 - 15s \tag{24-14}$$

where H = percent hydrogen and s = relative density at 15°C (with respect to water), also referred to as specific gravity.

Relative density is usually determined at ambient temperature with specialized hydrometers. In the United States these hydrometers commonly are graduated in an arbitrary scale termed *degrees API*. This scale relates

TABLE 24-5 Typical Ultimate Analyses of Petroleum Fuels

Composition, %	No. 1 fuel oil (41.5° API)	No. 2 fuel oil (33° API)	No. 4 fuel oil (23.2° API)	Low sulfur, No. 6 F.O. (12.6° API)	High sulfur, No. 6 (15.5° API)
Carbon	86.4	87.3	86.47	87.26	84.67
Hydrogen	13.6	12.6	11.65	10.49	11.02
Oxygen	0.01	0.04	0.27	0.64	0.38
Nitrogen	0.003	0.006	0.24	0.28	0.18
Sulfur	0.09	0.22	1.35	0.84	3.97
Ash	<0.01	<0.01	0.02	0.04	0.02
C/H Ratio	6.35	6.93	7.42	8.31	7.62

NOTE: The C/H ratio is a weight ratio.

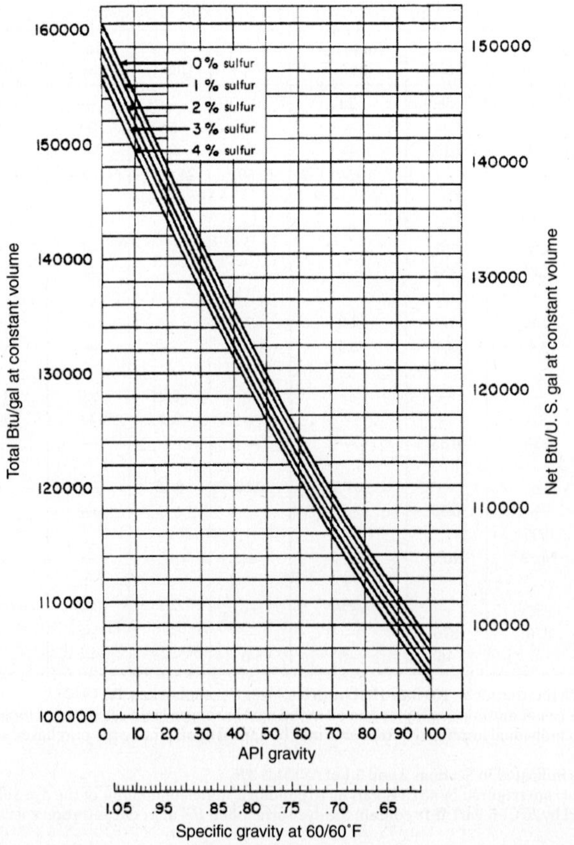

FIG. 24-3 Heat of combustion of petroleum fuels. To convert Btu/U.S. gal to kJ/m³, multiply by 278.7.

inversely to relative density s (at 60°F) as follows (see also the abscissa scale of Fig. 24-3):

$$\text{Degrees API} = \frac{141.5}{s} - 131.5 \tag{24-15}$$

For practical engineering purposes, relative density at 15°C (288 K), widely used in countries outside the United States, is considered equivalent to specific gravity at 60°F (288.6 K). With the adoption of SI units, the American Petroleum Institute favors absolute density at 288 K instead of degrees API.

The hydrogen content, heat of combustion, specific heat, and thermal conductivity data herein were abstracted from Bureau of Standards Miscellaneous Publication 97, *Thermal Properties of Petroleum Products.* These data are widely used, although other correlations have appeared, notably that by Linden and Othmer [*Chem. Eng.* **54**(4, 5) (1947)].

Heat of combustion can be estimated within 1 percent from the relative density of the fuel by using Fig. 24-3. Corrections for water and sediment must be applied for residual fuels, but they are insignificant for clean distillates.

Pour point ranges from 213 K (−80°F) for some kerosene-type jet fuels to 319 K (115°F) for waxy No. 6 fuel oils. *Cloud point* (which is not measured on opaque fuels) is typically 3 to 8 K higher than pour point unless the pour point has been depressed by additives. Typical petroleum fuels are practically newtonian liquids between the cloud point and the boiling point and at pressures below 6.9 MPa (1000 psia).

Fuel systems for No. 1 (kerosene) and No. 2 fuel oil (diesel, home heating oil) are not heated. Systems for No. 6 fuel oil are usually designed to preheat the fuel to 300 to 320 K (90°F to 120°F) to reduce viscosity for handling and to 350 to 370 K (165°F to 200°F) to reduce viscosity further for proper atomization. No. 5 fuel oil may also be heated, but preheating is usually not required for No. 4. (See Table 24-4.) Steam or electric heating is employed as dictated by economics, climatic conditions, length of storage time, and frequency of use. Pressure relief arrangements are recommended on sections of heated pipelines when fuel could be inadvertently trapped between valves.

The *kinematic viscosity* of a typical No. 6 fuel oil declines from 5000 mm²/s (0.054 ft²/s) at 298 K (77°F) to about 700 mm²/s (0.0075 ft²/s) and 50 mm²/s (0.000538 ft²/s) on heating to 323 K (122°F) and 373 K (212°F), respectively. Viscosity of 1000 mm²/s or less is required for manageable pumping. Proper boiler atomization requires a viscosity between 15 and 65 mm²/s.

Thermal expansion of petroleum fuels can be estimated as volume change per unit volume per degree. ASTM-IP Petroleum Measurement Tables [ASTM D 1250-08(2013) IP 200] are used for volume corrections in commercial transactions.

Heat capacity (specific heat) of petroleum liquids between 0°C and 205°C (32°F and 400°F), having a relative density of 0.75 to 0.96 at 15°C (60°F), can be calculated within 2 to 4 percent of the experimental values from the following equations:

$$c = \frac{1.685 + (0.039 \times {}^\circ\text{C})}{\sqrt{s}} \tag{24-16}$$

$$c' = \frac{0.388 + (0.00045 \times {}^\circ\text{F})}{\sqrt{s}} \tag{24-17}$$

where c is heat capacity, kJ/(kg·°C) or kJ/(kg·K), and c' is heat capacity, Btu/(lb·°F). Heat capacity varies with temperature, and the arithmetic average of the values at the initial and final temperatures can be used for calculations relating to the heating or cooling of oil.

The *thermal conductivity* of liquid petroleum products is given in Fig. 24-4. Thermal conductivities for asphalt and paraffin wax in their

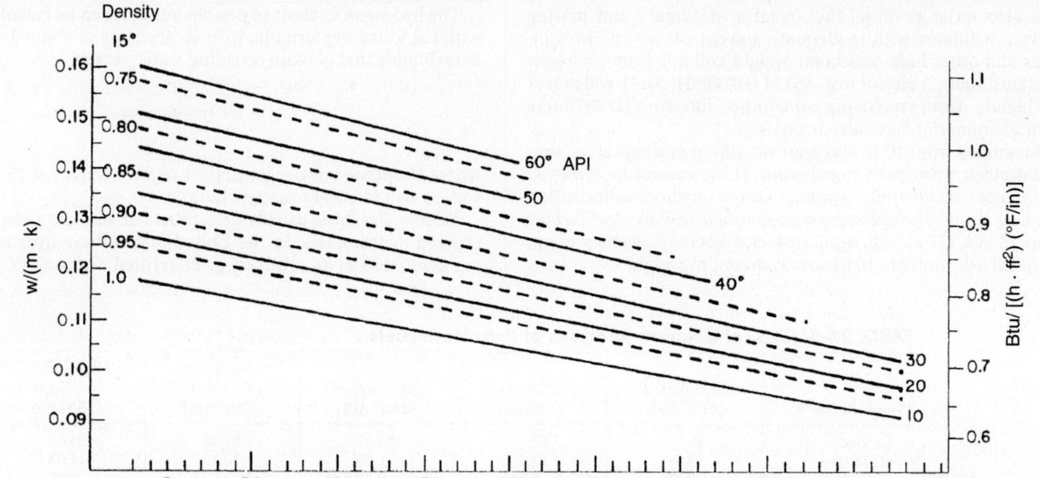

FIG. 24-4 Thermal conductivity of petroleum liquids. The solid lines refer to density expressed as degrees API; the broken lines refer to relative density at 288 K (15°C). [K = (°F + 459.7)/1.8.]

solid states are 0.17 and 0.23 W/(m·K), respectively, for temperatures above 273 K (32°F) [1.2 and 1.6 Btu/(h·ft^2)(°F/in)].

Commercial Considerations Fuels are sold in gallons and in multiples of the 42-gal barrel (0.159 m^3) in the United States, while a weight basis is used in other parts of the world. Transactions exceeding about 20 to 40 m^3 (5000 to 10,000 U.S. gal) usually involve volume corrections to 288 K (60°F) for accounting purposes.

Receipts of tank-car quantities or larger are usually checked for gravity, appearance, and flash point to confirm product identification and absence of contamination. Testing for sulfur content has become routine as sulfur specifications have become more stringent.

Safety considerations design and the location of storage tanks, vents, piping, and connections are specified by state fire marshals, underwriters' codes, and local ordinances. In NFPA 30, *Flammable and Combustible Liquids Code, 2015* (published by the National Fire Protection Association, Quincy, Mass.), liquid petroleum fuels are placed in Class I through Class III B based on their flash point, boiling point, and vapor pressure.

NFPA 30 details the design features and safe placement of handling equipment for flammable and combustible liquids.

Nonpetroleum Liquid Fuels *Tar sands* (also called *oil sands*) are a mixture of clay, sand, and bitumen, which is a heavy, viscous, asphalt-like oil. Traditional methods for mining and recovery of the desired product bitumen include strip mining followed by extraction using hot water or steam, or in situ recovery of heavy oil using methods that heat the bitumen below ground to promote its flow and recovery. Hot water is used to physically separate the bitumen from the sand for further processing. In situ methods include the injection of steam or solvents deep into the ground. Once recovered, the thick bitumen is then processed into various liquid fractions, such as naphtha, light oils, and heavy oils. Canada produces the world's largest amount of syncrude from oil sands, with Canadian syncrude production increasing from 0.27 million m^3/d (1.7 MB/d) in 2010 to over 0.32 million m^3/d (2 MB/d) in 2014.

Oil shale is a sedimentary rock that contains kerogen, which can be released as petroleum-like liquids upon heating of the rock through a process called retorting. The shale is either mined followed by surface retorting, or processed in situ to recover the liquids without removing the shale. The retorting process involves heating to 755 to 810 K (900°F to 1000°F), where pyrolysis converts the kerogen into a syncrude. Catalytic processing with hydrogen is then used to remove impurities and create a product that can be processed using conventional petroleum refining techniques to create gasoline, diesel fuel, and other hydrocarbons.

Coal-Derived Fuels Liquid fuels can be obtained from coal using a variety of processes, most of which fall under either direct liquefaction or indirect liquefaction. Direct liquefaction typically involves the treatment of coal with high temperatures and pressures along with solvents to create liquids which are highly aromatic in nature. This liquid must then undergo various degrees of hydrotreatment to increase the hydrogen-to-carbon ratio and improve the quality of the crude. Indirect liquefaction involves the gasification of the coal into a gaseous mixture containing carbon monoxide and hydrogen, among many other gaseous impurities. The carbon monoxide and hydrogen, known as "synthesis gas" or syngas, can then be reacted over an appropriate catalyst to form hydrocarbon chains of various lengths. The two catalyst systems most often employed for this process, referred to as the Fischer-Tropsch process (or F-T process), are iron-based or cobalt-based. The hydrocarbons produced have chain lengths ranging from C1 (methane) to C40+ (wax). Depending on the desired properties of the final hydrocarbon material, a significant amount of cracking, isomerization, and hydrotreating may be required. In addition to the production of hydrocarbon transportation fuels, gasification also supports the conversion of coal to other substitute fuels, including methanol and dimethyl ether (DME). DME can be produced directly from syngas, or from the conversion of methanol (via dehydration reaction). DME, with a chemical formula of CH_3-O-CH_3, is a clean-burning material and has been used as a substitute for liquefied petroleum gas (LPG). DME can also be liquefied and, with appropriate modifications to the fuel system, used as an alternative liquid transportation fuel for diesel-type applications. Methanol can also be converted into gasoline using a patented process (Mobil's MTG process) based on zeolite catalyst ZSM-5.

Renewable Fuels Efforts to curb the environmental impact of carbon dioxide emissions have led to efforts to use renewable feedstocks, such as biomass materials, in liquid fuel production schemes. Renewable resources can be combined with coal in a gasification scheme to produce liquid fuels, similar to the process mentioned previously. Renewable resources can also be used as a feedstock for the production of other liquid fuels, including bioethanol, biodiesel, or other liquid biofuels.

Bioethanol is produced by the fermentation of sugars produced from the hydrolysis of cellulose in energy crops, such as corn and sugarcane, as well as cellulosic biomass, such as corn stover and switchgrass. Bioethanol is often used as a blend component with gasoline, and its oxygenating properties promote cleaner burning and lower emissions.

Biodiesel is a mixture of fatty acid methyl esters (FAMEs) that can be used as a substitute for some transportation fuels. Biodiesel is an oxygenated mixture that is produced using vegetable oils, seed oils, or animal fats. Catalyzed reactions of these oils and fats with methanol or ethanol produce long-chain FAMEs.

Renewable fuels have also been produced using algae and algal oils. Once the algal biomass material has been harvested and dewatered (which are the most energy-intensive steps of the process), the internal oils can be extracted and thermocatalytically converted into either a chemically compatible biofuel (FAMEs as produced above) or a chemically transparent hydrocarbon fuel.

GASEOUS FUELS

Characteristics of Gaseous Fuels Gaseous fuels have a variety of properties that affect their use as fuels. Important characteristics of a few common gaseous fuels are listed in Table 24-6. Natural gas properties are listed as typical ranges because the composition of natural gas varies. A brief discussion of some of the properties is presented next.

Two *heating values* are in common use—the lower heating value (LHV) and the higher heating value (HHV). The difference between them is that the HHV assumes that product water is condensed to a liquid, whereas the LHV assumes product water is a vapor, so they differ by the heat of condensation of product water. This difference should be kept in mind when evaluating the use or economics of fuel supplies. For example, power system efficiency is referenced to LHV for gas-fueled engines but to HHV for solid-fueled boilers, which sometimes creates confusion when comparing the two efficiencies. Stoichiometric air/fuel ratios, on both a mass and volume basis, are also listed in Table 24-6. Together with the heating value, these ratios can be used to estimate the quantities of fuel and air needed to generate a given quantity of heat, providing an initial estimate of the size of air- and fuel-handling equipment.

The *Wobbe index* is defined as the heating value divided by the square root of the specific gravity (relative to air) of the fuel ($HHV/\sqrt{S.G.}$). Two chemically different fuels that have the same Wobbe index will produce the same heat output in a burner designed to operate at a given pressure, making the Wobbe index a key parameter for evaluating fuel interchangeability. Pipelined natural gas in the United States has a specified range of Wobbe index to ensure that commercial and residential burners produce constant heat in devices with fuel pressure regulation (typical of many installations).

Adiabatic flame temperature is the calculated temperature reached by combustion of a gaseous fuel, without heat loss, assuming a stoichiometric oxygen level and ambient initial temperature. It provides a theoretical maximum upper temperature from fuel combustion in ambient air. The actual, achieved flame temperature depends on the heat loss, completeness of combustion, and air supply conditions (temperature, humidity, and amount of excess air). Adiabatic flame temperature can be computed for fuel and oxidant mixtures with arbitrary oxidant conditions (air or oxygen) using equilibrium software codes such as Cantera (http://cantera.org/docs/sphinx/html/index.html) and NASA Glen Research center's CEA (Chemical Equilibrium with Applications).

Flame speed is a measured property of each fuel and should be considered a descriptor of how "fast" a flame can propagate in a fuel. The actual flame speed in a given device depends on many factors (i.e., temperature, fuel/air ratio, turbulence level), and the listed speeds are shown to emphasize the maximum. It is widely known that hydrogen has a very high flame speed compared to methane, and factors unique to hydrogen combustion must be considered in devices that use premixed hydrogen flames.

The *thermophysical properties* of most of the pure fuels are listed in other sections of this handbook. For complex mixtures such as natural gas, properties can be calculated using various computer codes, including REFPROP software developed by the National Institute of Standards and Technology (NIST).

Sources of Gaseous Fuels Natural gas is the most common gaseous fuel and is found in underground reservoirs across the globe. Natural gas composition depends on the source and varies widely from deposit to deposit. Natural gas can contain a variety of nonhydrocarbons, such as hydrogen sulfide, carbon dioxide, nitrogen, and water vapor. These constituents (with the possible exception of nitrogen) are removed (typically via amine scrubbing or, more recently, membrane separation) in a processing plant. A typical composition, before the removal of heavier hydrocarbons, is in the range (by volume) as follows: CH_4 (80–99%); C_2H_6 (2–9%); C_3H_8 (0.5–3%); and C_4H_{10} (0.1–1%). Pentane, hexane, heptane, octane, and some inerts (CO_2, N_2, and He) may also be present in natural gas. Gas containing appreciable quantities of higher hydrocarbons is usually referred to

TABLE 24-6 Important Characteristics of Common Gaseous Fuels

	Natural gas (typical)	Methane	Ethane	Propane	Hydrogen	Carbon monoxide
Lower heating value, MJ/kg (Btu/lb)	38–50 (16,500–21,500)	50.0 (21,500)	47.5 (20,420)	46.4 (19,950)	120.1 (51,630)	10.11 (4350)
Higher heating value, MJ/Kg (Btu/lb)	45–55 (19,500–23,500)	55.5 (23,860)	51.9 (22,310)	50.4 (21,670)	142.1 (61,090)	10.11 (4350)
Lower heating value, MJ/m^3 (Btu/ft^3)	29–38 (778–1020)	33.9 (910)	60.4 (1621)	86.5 (2322)	10.2 (274)	12.0 (322)
Higher heating value, MJ/m^3 (Btu/ft^3)	33–42 (886–1127)	37.3 (1001)	66.0 (1771)	94.0 (2523)	12.11 (325)	12.0 (322)
Stoichiometric air/fuel ratio (mass basis)*	15 – 19	17.2	16.1	15.7	34.3	2.47
Stoichiometric air/fuel ratio (volume basis)*	9 – 12	9.55	16.7	23.9	2.39	2.39
Adiabatic flame temperature, K (°F)	2213 (3524)	2191 (3484)	2222 (3540)	2240 (3572)	2318 (3713)	2223 (3542)
Flame speed, m/s (ft/s)	Depends on composition	0.45 (1.48)	0.48 (1.57)	0.46 (1.51)	2.83 (9.28)	0.52 (1.71)
Specific gravity†	0.55–0.65	0.554	1.038	1.522	0.070	0.967
Wobbe Index $MJ/m^3/\sqrt{S.G.}$ ($Btu/ft^3/\sqrt{S.G.}$)	41–53 (1100–1422)	50.1 (1345)	64.8 (1739)	76.2 (2045)	45.8 (1229)	12.2 (327)

*For 15% excess air.
†Relative to air.
Source: Reed, Richard J., ed. *North American Combustion Handbook: A Basic Reference on the Art and Science of Industrial Heating with Gaseous and Liquid Fuels.* 3rd ed. Vol. 1. Cleveland, OH: North American Mfg., 1986; Baukal, Charles E., and Robert Schwartz, eds., *The John Zink Combustion Handbook.* Boca Raton, FL: CRC Press, 2001.

as "wet" gas because the heavier gases can be condensed to a liquid. The processing of natural gas also includes the removal of higher hydrocarbons; liquefied propane and butane are referred to as liquefied petroleum gas (LPG).

Hydrocarbon gas liquids refer to the gases produced along with oil or gas. These hydrocarbons exist as a gas at ambient pressure and temperature but are liquid at higher pressure. Butane and propane are common examples. Most of the propane produced in the United States comes from natural gas processing plants. The term *associated gas* usually refers to gases produced during oil production, and *refinery gas* includes those gases separated during oil refining.

Methane hydrates have been considered a potential unconventional gas resource due to the vast volume of methane gas [about 4.5×10^{16} m^3 ($\sim 1.6 \times 10^{18}$ ft^3)] preserved in deep ocean sediment and permafrost. Gas hydrate is naturally occurring clathrate compounds consisting of water molecules and gaseous guest molecules, mostly methane. While several drilling-based production technologies (chemical injection, thermal stimulation, gas exchange) are explored, reservoir depressurization has been generally accepted and demonstrated to be a relatively simple and effective method of gas production. Sand-hosted deep-water hydrate reservoir provides for the most feasible gas production with its warm temperatures and high intrinsic permeability. Along with operational and economic uncertainties, there are environmental risks, including seafloor subsidence, gas leaks, and brackish water production, associated with gas production from methane hydrate reservoirs.

Hydrogen does not occur naturally; it is primarily produced by steam reforming of natural gas. As explained elsewhere in this section, hydrogen can also be produced by gasifying solid fuels (like coal or biomass) with oxygen to produce syngas ($CO+H_2$), extending the H_2 yield with the water-gas shift ($CO + H_2O \rightarrow CO_2 + H_2$), and then purifying it, typically by pressure swing adsorption.

Gaseous fuel is also produced by the anaerobic digestion of landfill wastes, farm wastes, or wastewater treatment facilities. The gas is typically a combination of methane and CO_2 and can be used for on-site power generation and heating.

Shale Gas Development Natural gas production in the United States has transitioned from mainly conventional to largely unconventional "tight gas" and shale over the last decade, as shown in Fig. 24-5, with the trend likely to continue in the future. For many years, the bulk of commercial oil and gas was produced from "conventional" underground reservoirs using standard vertical well-drilling technologies. The production of economic quantities of oil and gas from "unconventional" reservoirs requires special engineering techniques, such as horizontal drilling and hydraulic fracturing. Unconventional resources tend to be less concentrated, but total quantities are often very large, sometimes extending throughout almost the entire volume of the formation.

Filling a conventional reservoir with oil and natural gas is a complex process requiring a source rock containing organic material, thermal maturation from geothermal heat applied over geologic time in the absence of oxygen, a porous and permeable reservoir rock, a trap and seal on the reservoir rock to contain the hydrocarbons, and finally, a migration pathway for hydrocarbons from the source rock to reach the reservoir rock (Selley, R. C., and S. A. Sonneberg, *Elements of Petroleum Geology*, 3d ed., Academic Press, Cambridge, Mass., 2014). Unconventional reservoirs like gas shales contain hydrocarbons generated in place from thermally mature organic material that was deposited with the shale. Unconventional gas does not require the concentration or "beneficiation" of hydrocarbons like those in a conventional reservoir. Instead, the gas is produced directly from the source rock. This is a new concept in petroleum geology: the source rock is also the reservoir rock. The U.S. Geological Survey refers to gas shales as "continuous resources," to distinguish them from conventional reservoirs restricted to traditional traps and seals (Charpentier, R. R., and T. A. Cook, *USGS Methodology for Assessing Continuous Petroleum Resources*, U.S. Geological Survey Open File Report 2011-1167, 2011). A well can be drilled and stimulated almost anywhere within a continuous resource using the proper production technology and can be expected to produce economical amounts of hydrocarbons. The amount of recoverable gas in U.S. shale formations vastly exceeds the amount of recoverable gas remaining in conventional reservoirs.

The production rate per unit volume of shale is low, so the key is bringing a wellbore into contact with a large volume of rock. Unfortunately, most organic-rich shales tend to be only a few hundred feet (a few tens of meters) thick, significantly limiting the volume contacted by traditional vertical wells.

By the 1990s, big oil companies working offshore in extremely deep water had greatly improved directional drilling technology by adding downhole motors, steerable bits, and better inertial navigation and telemetry. George P. Mitchell applied this drilling technology to the Barnett Shale and found that horizontal wells could extend to distances of 3000 to 5000 feet (1 to 1.5 km) and remain in the target shale their entire length. The directional control allowed multiple wells to be drilled from a single location. The long horizontal boreholes, known as laterals, were then stimulated with a series of hydraulic fractures, instead of the single hydraulic fracture treatments used in vertical wells (Fig. 24-6).

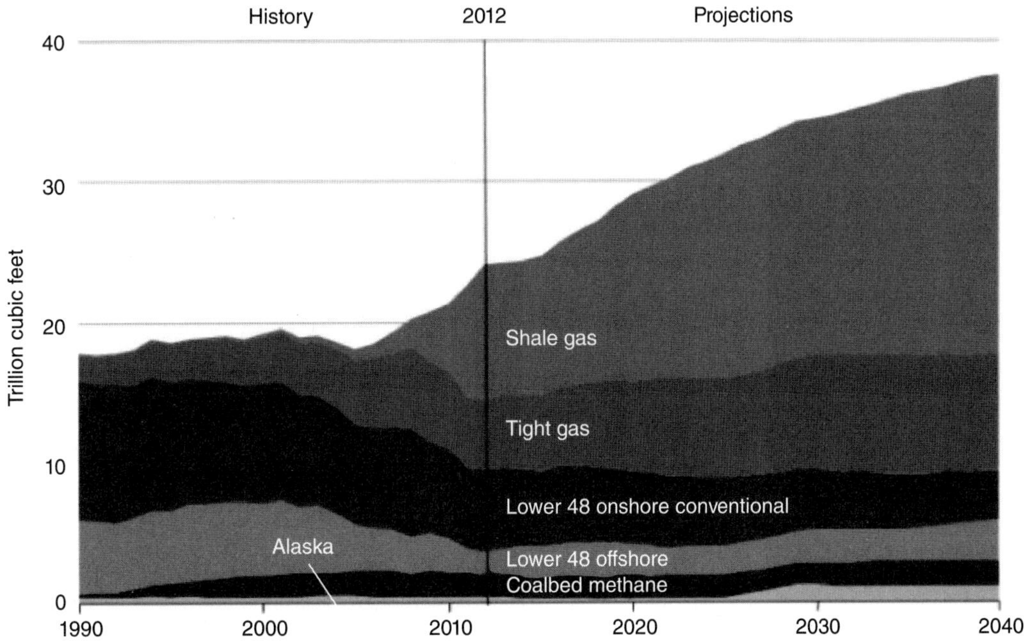

FIG. 24-5 Shale gas and other tight gas as a percentage of total gas resources in the United States since 1990, projected to 2040. Source: U.S. Energy Information Administration.

Hydraulic fracturing was invented in 1947 by Floyd Farris of Stanolind Oil and Gas Company in Kansas [Montgomery, C. T., and M. B. Smith, "Hydraulic Fracturing: History of an Enduring Technology," *Journal of Petroleum Technology* **62**(12): 26–32 (2010)]. The technology uses hydraulic pressure applied from the surface to overcome rock strength and create a high-permeability fracture into the reservoir in a direction perpendicular to the wellbore. Sand or other fine materials are emplaced in the new fracture to act as "proppant" that holds it open after the pressure has been released. Mitchell Energy was able to stage multiple hydraulic fractures at 500 to 1000 ft intervals (150 to 300 m) along the horizontal wellbore, contacting a much larger volume of shale than possible with a vertical well.

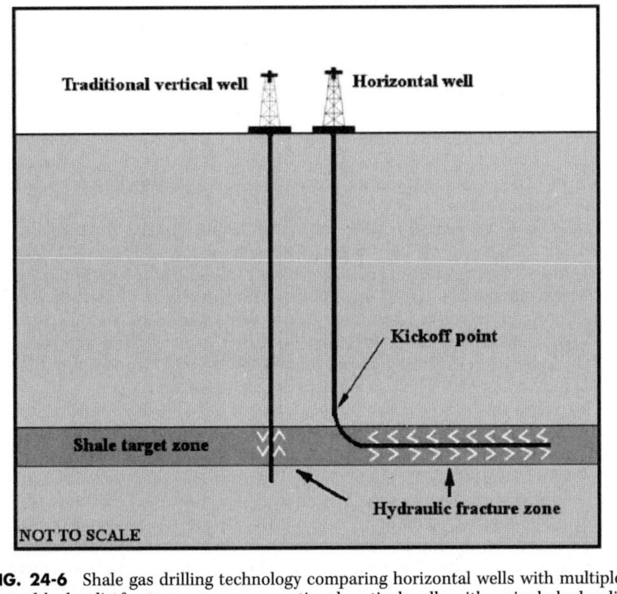

FIG. 24-6 Shale gas drilling technology comparing horizontal wells with multiple, staged hydraulic fractures versus conventional vertical wells with a single hydraulic fracture. The horizontal wells contact much more formation volume. Modified from Soeder (2012).

A number of different fluids and methods can be used for a frac. Chemical information posted on the Frac Focus web site (http://fracfocus.org/) indicates that the materials generally consist mostly of water, proppant sand, and a fraction of a percent of chemical additives (Fig. 24-7).

A horizontal shale gas well may use 1.2 to 1.9 million liters (300,000–500,000 gallons) of water for each individual hydraulic fracturing episode, or "stage," within a lateral. For a typical lateral length of 5000 feet (1.5 km), the stimulation design would call for ten separate frac stages that would require a total of 12 to 19 million liters or 3 to 5 million gallons of fluid. Shale fracs tend to use less proppant sand than other kinds of fracs to maintain open fractures after pressure is released. So-called "light sand" fracs are more effective on shale, and they also minimize the use of viscous gels like guar gum to carry in proppant. Hydraulic fracturing in shale creates natural rough spots or "asperities" on fracture walls that help prop open the fracture when pressure is released. Nevertheless, because of the high volume of hydraulic fractures in shale, even light sand fracs use a lot of sand. Concerns have been raised about the damage to landscapes from mining this sand, which must be composed of evenly sized, well-rounded, and high compressive strength quartz grains to work well as a proppant. The Jordan Sandstone in Wisconsin has been extensively mined as frac sand (Parsen, M., and J. Zambito, *Frac Sand in Wisconsin*, Fact Sheet 05, Wisconsin Geological and Natural History Survey, Madison, Wis., 2014, http://wgnhs.uwex.edu/pubs/fs05/).

One of the most common chemicals added to a frac consists of polyacrylamide, a friction reducer that creates an extremely slippery fluid known as "slickwater." Slickwater plays two roles: to reduce friction losses and offset pressure reduction as the frac fluid is pumped to the formation from the surface down a long string of narrow production casing, and to help the fluid penetrate and open up existing natural fractures in the formation. Other chemicals commonly used include hydrochloric acid to clean the perforations in the production casing, ethylene glycol compounds for corrosion resistance, a scale inhibitor such as phosphoric acid, and a biocide.

The biocide is probably the most hazardous material in use. Biocides come in two general types: oxidizing and nonoxidizing [Kahrilas, G. A., J. Blotevogel, P. S. Stewart, and T. Borch, "Biocides in Hydraulic Fracturing Fluids: A Critical Review of Their Usage, Mobility, Degradation, and Toxicity," *Environmental Science & Technology* **49**(1): 16–32 (2015)]. Oxidizing biocides (such as bleach and peroxide) attack microbes, but they also corrode equipment and damage rock formations. Most hydraulic fracturing operations use nonoxidizing biocides. These fall into two classes: lytic biocides that act by dissolving the cell walls of bacteria, and electrophilic biocides that act by binding themselves to bacterial cell walls (Kahrilas et al. 2015).

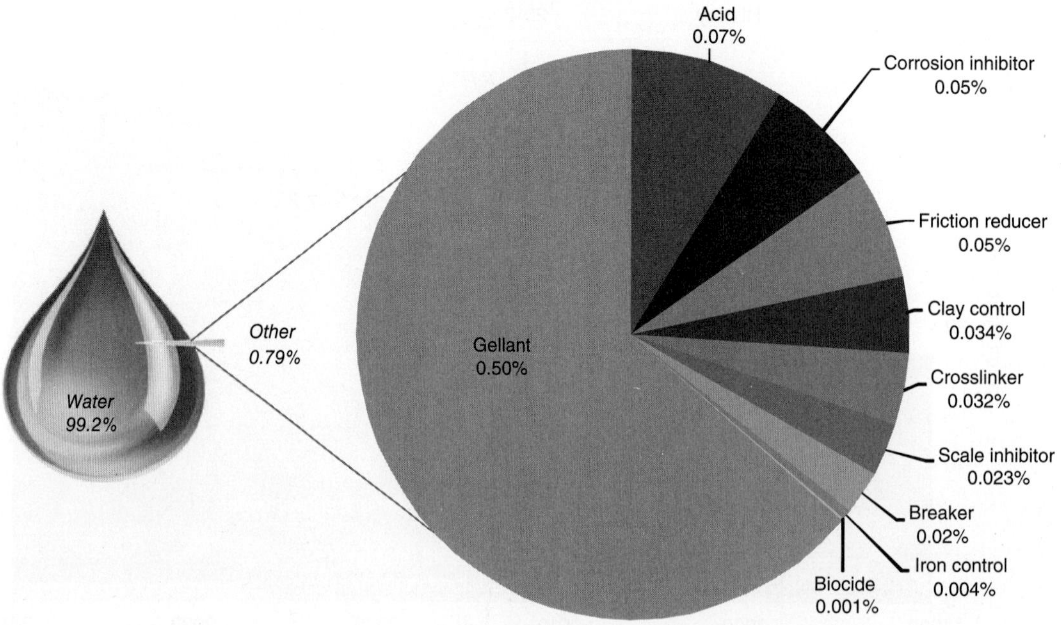

FIG. 24-7 Typical chemical additives by volume in hydraulic fracture fluids used on shale. Source: Ground Water Protection Council, 2012, FracFocus records: January 1, 2011 through February 27, 2012: accessed online at http://www.fracfocus.org. Modified from Soeder et al. (2014).

Glutaraldehyde is a common biocide used in hydraulic fracturing. Others include tetrakis hydroxymethyl-phosphonium sulfate and quaternary ammonium chloride (source: FracFocus web site: http://fracfocus.org/). Biocides are necessary because if not repressed, bacteria introduced downhole with the frac fluids can consume organic and sulfate compounds, creating hydrogen sulfide gas (H_2S) as a by-product. The H_2S causes the production gas to be "sour" and corrosive. It must be removed before the gas can be sold to a pipeline. H_2S is also toxic. Alternatives to biocides, such as treatment with ultraviolet light and other options, have been tried, but they have not been found to be as economical (Kahrilas et al. 2015).

Methods of hydraulic fracturing have evolved on shale plays. After the fluids are pumped down a well to fracture the rock, some portion (known as "flowback") returns to the surface with the gas. The flowback starts out fresh but becomes increasingly salty over time, with a total dissolved solids (TDS) content six to ten times greater than seawater. In the early days, these produced fluids were typically disposed of through local publicly owned treatment works (POTW), which did little to remove the TDS. Current water management practice recycles the flowback into the next frac and disposes of residual liquid waste down Class II underground injection control (UIC) wells.

Operators from the Gulf Coast entering shale gas plays like the Marcellus were convinced early on that unless they were careful about the composition of the frac water being injected, it might cause smectite clays to swell or shrink, resulting in plugged-up flowpaths or collapsed borehole walls. Because of these concerns about swelling clays, service companies obtained drinking-quality water from municipalities for hydraulic fracturing, used it once, and then disposed of the produced water. A fresh supply of high-quality water was then brought in for the next frac job. Not only was this environmentally and economically unsustainable, it was also unnecessary. The Marcellus Shale has been buried so deeply and for so long under tremendous heat and pressure that most of the swelling clays have been recrystallized into more stable clay minerals like chlorite and illite, which are essentially nonswelling. Much lower quality (and cheaper) water sources can be used for a Marcellus Shale frac. Operators switched from tap water to using untreated raw water from streams or effluent from POTWs for frac water.

In 2011, after an appeal by the Pennsylvania Department of Environmental Protection to stop disposing of produced water in POTWs, the industry began recycling produced water into the next frac. Recycling is less stressful to streams and aquatic ecosystems, and like other widely used environmental practices it also has some tangible economic benefits. Recycling provides savings on transportation costs because the water is already at the well site. It also provides significant savings in disposal costs, which have increased fivefold in Pennsylvania over the past few years for high-TDS waters.

Recycling flowback comes with several caveats. The water used in a hydraulic fracture treatment has to be essentially free of suspended solids, such as sediment. Total suspended solids (TSS) will plug up pores and microfractures if they are allowed to persist into the next frac job. Most of the on-site treatments of produced water are designed to remove the TSS. Techniques include advanced filtration systems, additives to clump or flocculate the clays, centrifuge-like settling processes, and other methods. The TSS filtration techniques currently in use at drill sites allow nearly all of the lower-salinity produced water to be recycled [Maloney, K. O., and D. A. Yoxtheimer, "Production and Disposal of Waste Materials from Gas and Oil Extraction from the Marcellus Shale Play in Pennsylvania," *Environmental Practice* **14:** 278–287 (2012)].

Hydraulic fracture water slated for reuse can contain moderate concentrations of TDS, but if the amount of dissolved solids gets too high, it will interfere with recycling when certain metals, like sodium or calcium, reach critical concentrations. Too much sodium or calcium in the water inhibits the performance of polyacrylamide and other additives. The current processes for treating high-TDS waters on-site are inefficient, with low throughput volumes.

The combination of directional drilling and staged hydraulic fracturing technology has enabled shale gas operators to produce economical amounts of gas, leading to the extensive development of shale gas resources in the United States.

Shale Gas Processing The hydrocarbons present in source rocks like shale are controlled by two factors: kerogen type and thermal maturity. Kerogen is organic material largely derived from dead plants deposited with the source rock sediment. It is classified into three types, based on origin. Type I kerogen is derived primarily from freshwater algae, and tends to be waxy. Type II kerogen is derived from lipid-rich marine algae, and tends to be oily. Type III kerogen is derived from cellulose-rich, woody land plants, and tends to be coaly. There is a fourth type that is inert.

The three active kerogens break down under thermal maturation to create different hydrocarbon compounds. At low thermal maturity, microbial action in the sediments creates biogenic gas from all three kerogen types, which is essentially pure methane with traces of carbon dioxide. At higher thermal maturity, in the so-called oil window, the type I and type II kerogens generate petroleum. The type III "gas prone" kerogen produces "thermogenic" methane in the oil window, which is isotopically different from the biogenic type. Thermal maturity beyond the oil window takes the type I and type II kerogen into the "wet gas" zone, where both types produce methane and light hydrocarbons such as ethane, butane, propane,

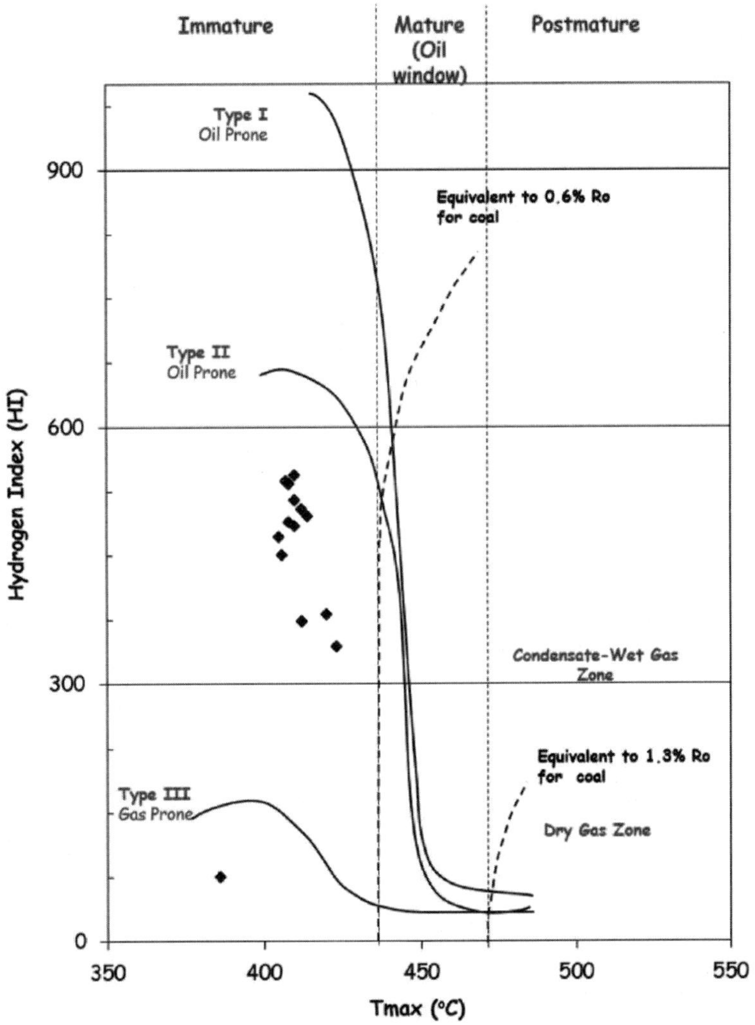

FIG. 24-8 Thermal maturity diagram. Hydrogen index defines the kerogen type, and T_{max} indicates the highest temperature reached over geologic time. The paths show how the kerogen changes as it "cooks" while trending toward greater thermal maturity. Equivalent T_{max} for vitrinite reflectance (Ro) from coal is also shown. The data points plotted are for Niobrara Shale samples from South Dakota; these are oil prone, but too immature to make much oil. Modified from Soeder et al. (2015).

and hexane. These are often referred to as natural gas liquids (NGLs) or condensate because they have a tendency to be produced from a well in vapor form, and then condense into liquids at the lower temperatures and pressures on the surface. At still higher thermal maturity, all of the kerogen types produce "dry gas," composed of the simplest hydrocarbon, methane. Other gases that may occur with the hydrocarbons include carbon dioxide, nitrogen, and hydrogen sulfide, along with inert gases such as helium and argon. The inert gases are contributed by crustal processes in the Earth, and they often provide clues to the origin and age of the natural gas. A thermal maturity diagram with the trajectories of the three kerogen types responding to temperature is shown in Fig. 24-8.

Natural gas transmission pipelines require produced gas to meet certain standards before they will accept it. These include min and max BTU values, maximum water vapor, CO_2, N_2, and H_2S contents, and pressure. BTU values are based on dry gas, and the inclusion of condensates, such as propane, will make the BTU value too high. More importantly, propane, butane, and other NGLs can be sold separately at a far higher price than dry gas. Therefore, many gas well operators have processing plants to recover NGLs and otherwise condition the produced gas before placing it in a pipeline.

Shale gas development has changed the economics of natural gas in the United States. Production of dry gas increased greatly between 2008 and 2014 (see Fig. 24-5), with little increase in natural gas utilization. This caused the wellhead price to drop from around \$11 per million BTU in 2008 to about \$2 per million BTU in 2014.

Because of the low prices for dry natural gas, many developers have been concentrating on the NGL and shale oil plays, where the economics are better. A number of gas shales, including the Marcellus in Pennsylvania, the Utica in Ohio, the Niobrara in Colorado, and the Eagle Ford in Texas contain dry gas in one part of the formation, and natural gas liquids in another, depending on burial depth and geologic history. The Niobrara Formation, for example, is buried quite deeply in the Denver-Julesburg Basin, under as much as 11,000 feet of rock. The thermal maturation at this burial depth led to the generation of significant amounts of condensate and even oil along with the natural gas. The rock is much shallower to the east in Nebraska and South Dakota, where it has never been buried deeper than 3000 feet, and in some cases much less. In these areas, the Niobrara produces only biogenic methane gas (Sonnenberg, S. A., "The Niobrara Petroleum System: A New Resource Play in the Rocky Mountain Region," in *Revisiting and Revitalizing the Niobrara in the Central Rockies*, ed. J. E. Estes-Jackson and D. S. Anderson, Rocky Mountain Association of Geologists, Denver, Colo., 2011, pp. 13–32). Most of the focus for drilling and production of the Niobrara has been in the deeper basins like the Denver-Julesburg or Powder River Basin in Wyoming.

Gas processing plants are located within these plays to remove NGLs from the production stream before it enters the pipeline. For example, Whiting Petroleum Company operates the Red Tail Gas Plant in the Pawnee National Grassland about 15 miles north of Raymer, Colorado. This plant is located between the Niobrara production wells and the transmission pipeline. It is designed to remove propane, butane, ethane, and other NGLs

TABLE 24-7 Ranges of Natural Gas Composition, in Percent, for 1280 Analyses from the Central Appalachian Basin from Colón-Román and Ruppert, 2014

Gas	No. of analyses	Range, %	No. of analyses with "0" gas	No. of analyses labeled "trace"
Methane	1267	1.2–99.4	3	0
Ethane	1262	0.01–64.7	4	0
Propane	1120	0.001–16.2	19	23
n-butane	993	0.0010–5.6	205	25
i-butane	854	0.0012–3.1	221	57
n-pentane	987	0.0019–2.0000	285	100
i-pentane	602	0.0049–1.4	254	113
Cyclohexane	791	0.1–0.7	213	401
Hexane and Hexane plus	794	0.0011–1.5	245	44
Nitrogen	1218	0.1–78.1	22	3
Oxygen	435	0.0056–21.35	407	212
Argon	59	0.003–1.9	270	502
Hydrogen	300	0.0021–2.27	451	94
Hydrogen sulfide	2	0.0260–0.51	957	1
Carbon dioxide	830	0.001–83.53	188	209
Helium	1236	0.0099–1.080	4	0

NOTE: There was no attempt to remove results that were reported as zero in the original reports. However, the user should consider using the zero values with caution because it is not known whether the gas was analyzed and the component was not detected or whether no attempt was made to analyze the component.

from the production stream using a series of cooling towers and condensers. Additional treatment loops dewater the gas and adjust CO_2 content before the gas enters a compressor and then the pipeline.

The bulk of production from gas shales is methane, with minor amounts of condensate. However, even seemingly trivial amounts can be quite valuable. Marcellus Shale gas production in the Appalachian Basin is primarily dry gas, although a condensate "sweet spot" occurs in far western Pennsylvania and the northern panhandle of West Virginia, where the shale is shallow enough to contain NGLs but still thick enough and deep enough to be worth drilling. The most significant liquid produced from the Marcellus in this location is ethane, an important feedstock to the plastics industry, where it is a major component of polyethylene. The development of the Marcellus Shale resulted in the establishment of several new plastics manufacturing plants near Charleston, W.Va., and Monaca, Pa. Prior to the establishment of these plants, the ethane was shipped to plants on the Gulf Coast for processing. These were the first new plastics plants established in the Ohio Valley since the 1980s.

Table 24-7 shows the results of a USGS assessment of gas composition in the Appalachian Basin by Colón-Román and Ruppert based on the analysis of 1280 gas samples (Colón-Román, Y. A., and L. F. Ruppert, *Central Appalachian Basin Natural Gas Database—Distribution, Composition, and Origin of Natural Gases*, U.S. Geological Survey Open-File Report 2014–1207, 2014, http://dx.doi.org/10.3133/ofr20141207). The values cover quite a large range. Developing a better understanding of the geologic controls on the generation of NGLs will improve the predictability of this important resource.

The most prominent and best-known of tight oil plays is the Bakken Shale, in the Williston Basin of North Dakota. The Bakken produces both light petroleum and natural gas from a limestone unit sandwiched in between two organic-rich shales. Although considered an unconventional play, the slightly more porous and permeable limestone acts as a natural gathering system to collect the oil from the tighter shales and direct it to a well. This system is relatively inefficient, and Bakken producers estimate that they recover approximately 6 percent of the oil in place from the shale. Even this small recovery, which leaves 94 percent of the resource behind in the ground, has made North Dakota the second largest oil-producing state in the United States, second only to Texas, according to the U.S. Energy Information Administration in 2014.

REFERENCES FOR THIS SECTION: Charpentier, R. R., and T. A. Cook, *USGS Methodology for Assessing Continuous Petroleum Resources*, U.S. Geological Survey Open File Report 2011–1167, 2011; Colón-Román, Y. A., and L. F. Ruppert, *Central Appalachian Basin Natural Gas Database—Distribution, Composition, and Origin of Natural Gases*, U.S. Geological Survey Open-File Report 2014–1207, 2014, http://dx.doi.org/10.3133/ofr20141207; Kahrilas, G. A., J. Blotevogel, P. S. Stewart, and T. Borch, "Biocides in Hydraulic Fracturing Fluids: A Critical Review of Their Usage, Mobility, Degradation, and Toxicity," *Environmental Science & Technology* **49**(1): 16–32 (2015); Maloney, K. O., and D. A. Yoxtheimer, "Production and Disposal of Waste Materials from Gas and Oil Extraction from the Marcellus Shale Play in Pennsylvania," *Environmental Practice* **14**: 278–287 (2012); Montgomery, C. T., and M. B. Smith, "Hydraulic Fracturing: History of an Enduring Technology," *Journal of Petroleum Technology* **62**(12): 26–32 (2010); Parsen, M., and J. Zambito, *Frac Sand in Wisconsin*, Fact Sheet 05, Wisconsin Geological and Natural History Survey, Madison, Wis., 2014, http://wgnhs.uwex.edu/pubs/fs05/; Selley, R., *Elements of Petroleum Geology*, 3d ed., Academic Press, Cambridge, Mass., 2014; Soeder, D. J., S. Sharma, N. Pekney, L. Hopkinson, R. Dilmore, B. Kutchko, B. Stewart, K. Carter, A. Hakala, and R. Capo, "An Approach for Assessing Engineering Risk from Shale Gas Wells in the United States," *International Journal of Coal Geology* **126**: 4–19 (2014); Soeder, D. J., J. F. Sawyer, A. Freye, and S. Singh, *Assessment of Hydrocarbon Potential in the Niobrara Formation, Rosebud Sioux Reservation, South Dakota*, Paper 2153622, Proceedings of Unconventional Resources Technology Conference (URTeC), San Antonio, Tex., July 20–22, 2015, DOI 10.15530/urtec-2015-2153622; Soeder, D. J., and, W. M. Kappel, "Water Resources and Natural Gas Production from the Marcellus Shale," U.S. Geological Survey Fact Sheet 2009–3032, 2009; Sonnenberg, S. A., "The Niobrara Petroleum System: A New Resource Play in the Rocky Mountain Region," in *Revisiting and Revitalizing the Niobrara in the Central Rockies*, ed. J. E. Estes-Jackson and D. S. Anderson, Rocky Mountain Association of Geologists, Denver, Colo., 2011, pp. 13–32.

Gaseous Fuel Distribution and End Use Global natural gas consumption in 2014 was 3393 billion cubic meters (*BP Statistical Review of World Energy*, June 2016, 65th ed.) and 756 billion cubic meters in the United States (EIA data). Natural gas is distributed in long-distance pipelines at pressures of up to 1750 psig in pipelines that can be as large as 107 cm (42 inches) in diameter. Compressor stations are typically situated at 50- to 100-mile intervals along the pipelines and use gas turbines, reciprocating engines, or electric motors to drive compressors. Electric power generation consumes 35 percent of the natural gas used in the United States, with industry (27%), residential (17%), commercial (12%) (nonmanufacturing, hotels, retail, etc.), and gas production and distribution (9%) consuming the rest. About 0.1 percent is used to fuel vehicles.

The large amount of natural gas used for heating creates a seasonal demand. Approximately 6 billion cubic feet of natural gas storage capacity (underground storage typically consisting of depleted oil or gas reservoirs, but also sometimes other geology) is available in the United States to meet this demand.

Natural gas can be cooled and liquefied, creating readily transportable liquefied natural gas (LNG). Today, LNG is transported by sea in refrigerated tankers and stored in tanks at terminals, where it is regasified for pipeline distribution. Regasification requires the addition of heat, which can change the composition of LNG being added to the pipeline (the lightest hydrocarbons will evaporate first), and it must be carefully controlled.

Gaseous fuels are also important chemical feedstocks. Figure 24-9 provides a brief summary of value-added chemicals that can be produced from natural gas. Methane in general, the main constituent of natural gas, is a highly stable molecule and requires extreme reaction conditions (e.g., high temperature and oxidative environment) for its chemical conversion. Therefore, not many direct methane conversion routes are commercialized yet. Mainly methane is converted to synthesis gas (carbon monoxide and hydrogen) using the steam reforming process. The resulting synthesis gas serves as a building block for a wide range of chemicals (see Fig. 24-9). The main chemical reactions of natural gas components are listed in Table 24-8.

Hydrogen (produced from steam reforming of natural gas) is critical to the production of ammonia via the Haber process, which supplies most of the world's fertilizer. Hydrogen is also used in various oil refining processes, including sulfur removal by hydrodesulfurization and the production of lighter fractions by hydrocracking. Another important chemical feedstock found in natural gas is ethane, which can be separated and thermally "cracked" to produce ethylene, an important precursor for plastics production.

Storage challenges restrict the use of gaseous fuels for transportation. However, both natural gas and propane vehicles are available and being used, especially in fleets, where a refueling station is part of the fleet. According to the DOE's Office of Energy Efficiency and Renewable Energy Alternative Fuels Data Center, there are approximately 150,000 natural gas vehicles in the United States comprising roughly 1 percent of the 15.2 million worldwide fleet in 2015. Hydrogen is used to power a small number of fuel cell vehicles.

FUEL AND ENERGY COSTS

Fuel costs vary widely both geographically and temporally. Oil and gas markets have been highly volatile in recent years, while steam coal markets have not. Much combustion equipment is designed for a specific fuel, limiting the potential for fuel switching to take advantage of price trends. The costs given in Table 24-9 are U.S. averages not necessarily applicable to a specific location; they do provide fuel cost trends.

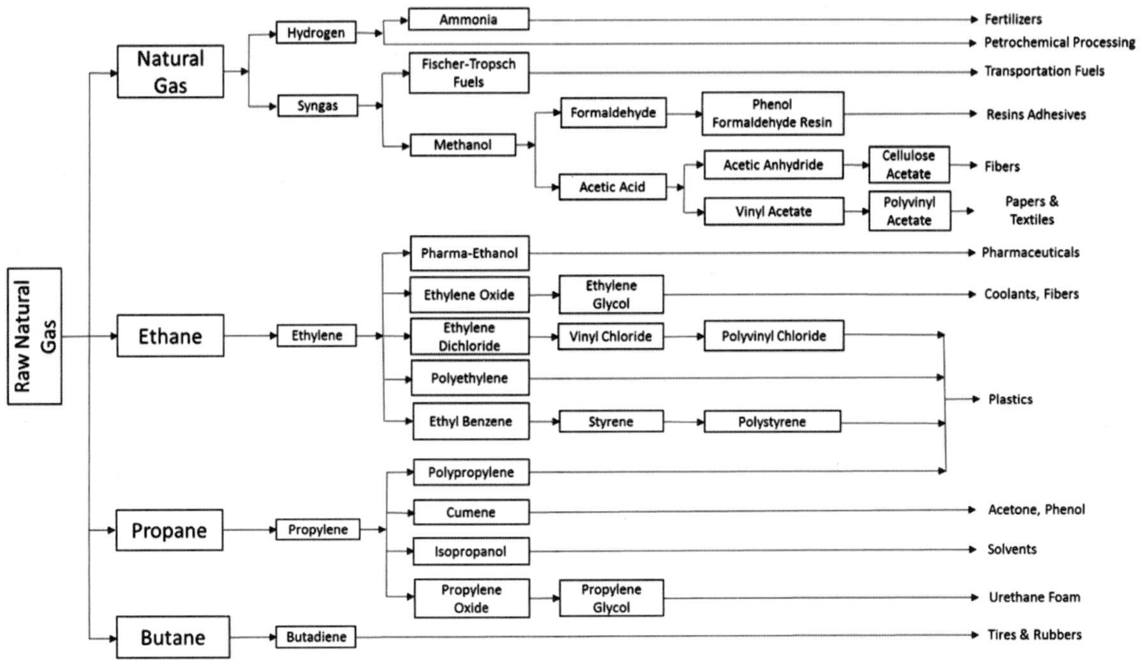

FIG. 24-9 Value-added chemicals from natural gas.

COAL CONVERSION

Coal is the most abundant fossil fuel and is projected to remain as a major source of energy throughout the world for the foreseeable future. However, because liquids and gases are more desirable than solid fuels, technologies have been, and continue to be, developed to economically convert coal into liquid and gaseous fuels. Figure 24-10 is a simplified depiction of the different routes from coal to clean gaseous and liquid fuels.

COAL GASIFICATION

General References: Bell, Towler, and Fan, *Coal Gasification and Its Applications*, Elsevier, New York, 2010; *Coal Gasification Guidebook: Status, Applications, and Technologies*, Electric Power Research Institute, EPRI TR-102034, Palo Alto, Calif., 1993; Fuel Gasification Symp., 152d American Chemical Society Mtg., September 1966; Higman and Tam, "Advances in Coal Gasification, Hydrogenation, and Gas Treating for the Production of Chemicals and Fuels," *Chemical Reviews*, (2013, October 21); Higman and van der Burgt, *Gasification*, 2d ed., Elsevier, New York, 2008; Kent, ed., *Riegel's Handbook of Industrial Chemisty* 10th ed., chap. 17, 2003; Stiegel and Ramezan, "The Case for Gasification," *EM Magazine*, December 2004, pp. 27–33; Kristiansen, *Understanding Coal Gasification*, IEA Coal Research, March 1996; Liu, Song, and Subramani, *Hydrogen and Syngas Production and Purification Technologies*, Wiley, New York, 2010; Lowry, ed., *Chemistry of Coal Utilization*, suppl. vol., Wiley, New York, 1963, and 2d suppl. vol., ed. Elliot, 1981; Speight, *Gasification of Unconventional Feedstocks*, Gulf Professional Publishing, New York, 2014.

Background The advantages of gaseous fuels have resulted in an increased demand for gas and have led to the invention of advanced processes for coal gasification. Converting coal to combustible gas has been practiced commercially since the early 19th century. Chapter 17 of Riegel's *Handbook of Industrial Chemistry* (Tenth Edition) provides a good summary of the early history of coal gasification. Coal-derived gas was distributed in urban areas of the United States for residential and commercial uses until its displacement by lower-cost natural gas, starting in the 1940s. At about that time, the development of oxygen-based gasification processes was initiated. An early elevated-pressure gasification process, developed by Lurgi Kohle u Mineralöltechnik GmbH, is still in use. The compositions of gases produced by several more advanced gasification processes are listed in Table 24-10.

Theoretical Considerations The chemistry of coal gasification can be approximated by assuming coal is only carbon and considering the most important reactions involved (see Table 24-11). Reaction (24-18), the combustion of carbon with oxygen, which can be assumed to go to completion, is highly exothermic and supplies most of the thermal energy for the other gasification reactions. The oxygen used in the gasifier may come from direct feeding of air or may be high-purity oxygen from an air separation unit. Endothermic reactions (24-20) and (24-21), which represent the conversion of carbon to combustible gases, are driven by the heat energy supplied by reaction (24-18).

TABLE 24-8 Chemical Reactions of Natural Gas Components

Process	Reaction	Reaction heat* kJ/mole (Btu/lbmole)
Combustion	$CH_4 + 2O_2 => CO_2 + 2H_2O$	−802.5 (−345,000)
Steam reforming	$CH_4 + H_2O => CO + 3H_2$	205.7 (88,500)
Partial oxidation	$CH_4 + 0.5O_2 => CO + 2H_2$	−35.9 (−15,400)
Dry reforming	$CH_4 + CO_2 => 2CO + 2H_2$	247.0 (106,200)
Decomposition	$CH_4 => C(s) + 2H_2$	91.1 (39,200)
Oxidative coupling	$CH_4 + 0.5O_2 => 0.5C_2H_4 + H_2O$	−141.1 (−60,700)
Selective partial oxidation	$CH_4 + 0.5O_2 => CH_3OH$	−126.4 (−54,300)
Dehydroaromatization	$6CH_4 => C_6H_6 + 9H_2$	88.7 (38,100)
Ethane cracking	$C_2H_6 => C_2H_4 + H_2$	137.1 (58,900)
Propane cracking	$C_3H_8 => C_3H_6 + H_2$	124.2 (53,400)
Butane cracking to 1,3 butadiene	$C_4H_{10} => C_4H_6 + H_2$	235.7 (101,300)

*Calculated based on all the reactant and product species in gaseous phase except carbon (solid) in decomposition reaction.

TABLE 24-9 Time-Price Relationships for Fossil Fuels

Year	Coal, \$/Mg (\$/U.S. ton)	Wellhead natural gas \$/1000 m³(\$/1000 scf)	Crude oil, domestic first purchase price \$/m³ (\$/bbl)
1975	21.33 (19.35)	15.54 (0.44)	48.24 (7.67)
1985	27.78 (25.20)	88.64 (2.51)	151.52 (24.09)
1995	20.76 (18.83)	54.74 (1.55)	91.96 (14.62)
2005	26.00 (23.59)	258.86 (7.33)	316.25 (50.28)
2011	40.69 (36.91)	139.50 (3.95)	602.12 (95.73)
2016	43.68 (39.63)	92.52 (2.62)	317.76 (50.52)

SOURCE: *Annual Energy Review 2011*, DOE/EIA-0384, September 2012, Tables 7.9, 6.7, and 5.18, respectively; 2016 prices from daily prices posted on June 9, 2016 on the EIA web site.

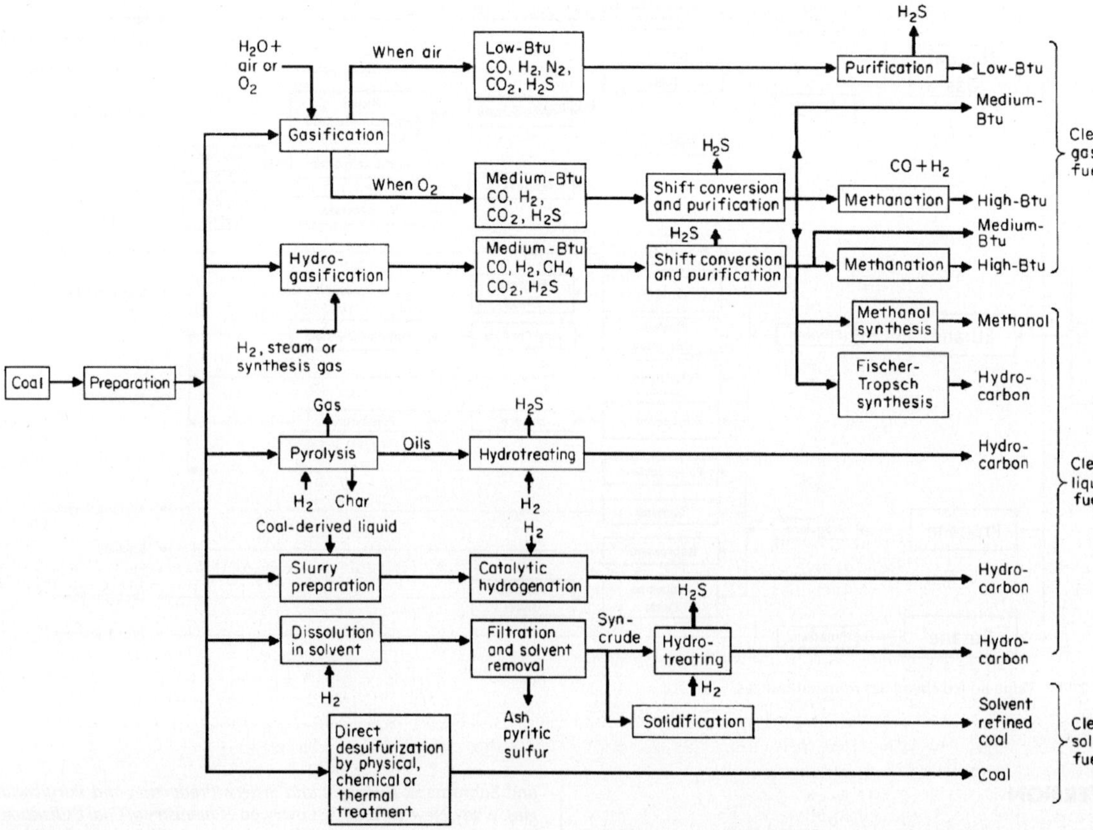

FIG. 24.10 The production of clean fuels from coal.

TABLE 24-10 Coal-Derived Gas Compositions

Gasifier technology	Sasol/Lurgi[a]	Texaco/GE Energy[b]	BGL[c]	E-Gas/CB&I	Shell/Uhde[d]	Siemens	Transport Gasifier[e]	Transport Gasifier[f]
Type of bed	Moving	Entrained	Moving	Entrained	Entrained	Entrained	Circulating	Circulating
Coal feed form	Dry coal	Coal slurry	Dry coal	Coal slurry	Dry coal	Dry coal	Dry coal	Dry coal
Coal type	Illinois #6	Illinois #6	Illinois #6	Illinois #6	Illinois #5	Illinois #6	Low rank	Low rank
Oxidant	Oxygen	Oxygen	Oxygen	Oxygen	Oxygen	Oxygen	Air	Oxygen
Pressure, MPa (psia)	0.101 (14.7)	4.22 (612)	2.82 (409)	2.86 (415)	2.46 (357)	4.2 (609)	4.51 (654)	4.02 (583)
Ash form	Slag	Slag	Slag	Slag	Slag	Slag	Powdery	Powdery
Composition, vol%								
H_2								
CO	52.2	30.3	26.4	33.5	26.7	25.7	11.2	27.1
CO_2	29.5	39.6	45.8	44.9	63.1	66.9	17.9	34.3
CH_4	5.6	10.8	2.9	16.0	1.5	2.0	8.6	16.2
Other hydrocarbons	4.4	0.1	3.8	1.8	0.03	0.03	2.7	5.3
H_2S	0.3	—	0.2	—	—	—	—	
COS	0.9	1.0	1.0	1.0	1.3	1.6	0.3	0.29
N_2 + Ar	0.04	0.02	0.1	0.1	0.1	0.2	0.02	0.02
H_2O	1.5	1.6	3.3	2.5	5.2	0.7	51.1	1.2
NH_3 + HCN	5.1	16.5	16.3	—	2.0	2.9	7.8	15.2
HCl	0.5	0.1	0.2	0.2	0.02	0.02	0.33	0.4
H_2S:COS	—	0.02	0.03	0.03	0.03	0.02	0.01	—
	20:1	42:1	11:1	10:1	9:1	10:1	19:1	15:1

[a]Rath, "Status of Gasification Demonstration Plants," *Proc. 2d Annu. Fuel Cells Contract Review Mtg.*, DOE/METC-9090/6112, p. 91.
[b]*Coal Gasification Guidebook: Status, Applications, and Technologies*, Electric Power Research Institute, EPRI TR-102034, 1993. pp. 5–28.
[c]*Coal Gasification Guidebook: Status, Applications, and Technologies*, Electric Power Research Institute, EPRI TR-102034, 1993. pp. 5–58.
[d]*Coal Gasification Guidebook: Status, Applications, and Technologies*, Electric Power Research Institute, EPRI TR-102034, 1993. pp. 5–48.
[e]*Kemper County IGCC™ Project – Detailed Design Report*, Southern Company Services, Inc., NETL DE-FC26-06NT42391, 2014, Appendix A.
[f]*Cost and Performance Baseline for Fossil Energy Plants, Volume 3a, DOE/NETL-2010/1399 Report*, May 2011.

TABLE 24-11 Chemical Reactions in Coal Gasification

Reaction	Reaction heat kJ/(kg · mol) (HHV)	Process	Number
Solid–gas reaction			
$C + O_2 \rightarrow CO_2$	+393,790	Combustion	(24-18)
$C + 2H_2 \rightarrow CH_4$	+74,900	Hydrogasification	(24-19)
$C + H_2O \rightarrow CO + H_2$	−175,440	Steam-carbon	(24-20)
$C + CO_2 \rightarrow 2CO$	−172,580	Boudouard	(24-21)
Gas-phase reaction			
$CO + H_2O \rightarrow H_2 + CO_2$	−2860	Water-gas shift	(24-22)
$CO + 3H_2O \rightarrow CH_4 + H_2$	+250,340	Methanation	(24-23)
$CH_2 \rightarrow \left(1 - \frac{X}{4}\right)C + \left(\frac{X}{4}\right)CH_4$		Pyrolysis	(24-24)
$CH_2 + mH_2 \rightarrow \left[1 - \left(\frac{X+2m}{4}\right)\right]C + \left(\frac{X+2m}{4}\right)CH_4$		Hydropyrolysis	(24-25)

Hydrogen and carbon monoxide produced by the gasification reaction react with each other and with carbon. The hydrogenation of carbon to produce methane, reaction (24-19), is exothermic and contributes heat energy. Similarly, the methanation of CO, reaction (24-23), can also contribute heat energy. These reactions are affected by the water-gas-shift reaction (24-2), the equilibrium of which controls the extent of reactions (24-20) and (24-21).

Several authors have shown (Gumz, *Gas Producers and Blast Furnaces*, Wiley, New York, 1950; Elliott and von Fredersdorff, *Chemistry of Coal Utilization*, 2d suppl. vol., ed. Lowry, Wiley, New York, 1963) that there are three fundamental gasification reactions: the Boudouard reaction [(24-21) in Table 24-11], the heterogeneous water-gas-shift reaction (24-22), and the hydrogasification reaction (24-19). The equilibrium constants for these reactions are sufficient to calculate equilibrium for all the reactions listed. Unfortunately, it is not possible to calculate accurate gas composition by using reactions (24-18) to (24-23). One reason is that not all reactions may be in equilibrium. Another reason is that other reactions are taking place. Since gasification of coal always involves elevated temperatures, thermal decomposition (pyrolysis) takes place as coal enters the gasification zone of the reactor. Reaction (24-24) treats coal as a hydrocarbon and postulates its thermal disintegration to produce carbon (coke) and methane. Reaction (24-25) illustrates the stoichiometry of hydrogasifying part of the carbon to produce methane.

These reactions can be used to estimate the effect of changes in operating parameters on gas composition. As temperature increases, endothermic reactions are favored over exothermic reactions. Methane production will decrease, and CO production will be favored as reactions are shifted in the direction in which heat absorption takes place. An increase in pressure favors reactions in which the number of moles of products is less than the number of moles of reactants. At higher pressure, the production of CO_2 and CH_4 will be favored.

It is generally believed that oxygen reacts completely in a very short distance from the point at which it is mixed or comes in contact with coal or char. The heat evolved acts to pyrolyze the coal, and the char formed then reacts with carbon dioxide, steam, and other gases formed by combustion and pyrolysis. The assumption made in Table 24-11 that the solid reactant is carbon is probably close to correct, but the type of char formed affects the kinetics of gas–solid reactions. The overall reaction is probably rate controlled below 1273 K (1832°F). Above this temperature, pore diffusion has an overriding effect, and at very high temperatures surface-film diffusion probably controls. For many gasification processes, the reactivity of the char is quite important and can depend on feed coal characteristics, the method of heating, the rate of heating, and particle-gas dynamics.

The importance of these concepts can be illustrated by the extent to which the pyrolysis reactions contribute to gas production. In a moving bed gasifier, the particle is heated through several distinct thermal zones. In the initial heat-up zone, coal carbonization or devolatilization dominates. In the successively hotter zones, char devolatilization, char gasification, and fixed carbon combustion are the dominant processes. About 17 percent of total gas production occurs during coal devolatilization, and about 23 percent is produced during char devolatilization. The balance is produced during char gasification and combustion.

Gasifier Types and Characteristics The four main types of gasifier reactors, moving bed, entrained bed, fluidized bed, and transport bed as shown in Fig. 24-11, are all in commercial use. The moving bed is sometimes referred to as a "fixed" bed because the coal bed is kept at a constant height. These gasifiers can differ in many ways: size, type of coal fed, feed and product flow rates, residence time, and reaction temperature. Gas compositions from the gasifiers discussed are listed in Table 24-10.

In a *moving bed gasifier*, depending on the temperature at the base of the coal bed, the ash can either be dry or in the form of molten slag. If excess steam is added, the temperature can be kept below the ash fusion point, in which case the coal bed rests on a rotating grate that allows the dry ash to fall through for removal. To reduce steam usage, a slagging bottom gasifier was developed by British Gas and Lurgi (BGL) in which the ash is allowed to melt and drain off through a slag tap. This gasifier has over twice the capacity per unit of cross-section area of the dry-bottom gasifier. The BGL technology is currently licensed by Envirotherm and Zemag Clean Energy Technology.

In a *fluidized-bed gasifier*, the problem of coal agglomeration is eliminated by a fluidized-bed gasifier developed by GTI. The U-Gas technology employs an agglomerating-ash fluidized-bed gasifier in which crushed limestone can be injected with the coal for sulfur capture. Char and ash that exit the gasifier with the product gas are recycled to the hot agglomerating and jetting zone, where temperatures are high enough to pyrolyze fresh coal introduced at that point, gasify the char, and soften the ash particles. The ash particles stick together and fall to the base of the gasifier, where they are cooled and removed. The agglomerating fluid-bed gasifier can use either air or oxygen. Pressurized operation has several advantages: slightly higher methane formation, resulting in higher heating value of the gas; increased heat from the methanation reactions, which reduces the amount of oxygen needed; reduced heat losses through the wall and, consequently, improved efficiency; and higher capacity. The U-Gas technology is being licensed by GTI's partner Synthesis Energy Systems.

A *dry-fed entrained flow gasifier*, for which a primary example of an oxygen-blown, dry-feed, entrained-flow gasifier is the Shell gasifier. An advantage of Shell coal gasification technology is its ability to process a range of coals, with a wide variety of coals (from brown coal to anthracite) having been successfully tested. As with other entrained-flow gasifiers, disadvantages of the Shell process include a high oxygen requirement and a high waste heat recovery duty. However, the ability to feed dry coal reduces the oxygen requirement below that of single-stage entrained-flow gasifiers that use a slurry feed and makes the Shell gasifier somewhat more efficient. The penalty for this small efficiency improvement is a more complex coal-feeding system. Uhde, Siemens, MHI, Shell, and Thermal Power Research Institute (TPRI) are marketing this technology.

For *slurry-fed entrained flow gasifiers*, three commercial examples are the GE gasifier (formerly Texaco), the CB&I E-Gas gasifier (formerly ConocoPhillips), and the East China University of Science and Technology (ECUST). The Texaco gasifier is similar to the Siemens gasifier, except that the coal is fed as a slurry. Reactor exit gas is cooled either by direct water injection or by a radiative cooler directly below the reactor. ECUST offers an opposed multiburner (OMB) slurry-fed gasifier that is similar to the GE gasifier except that the coal slurry is fed through four burners positioned at 90° intervals around the top of the gasifier, forming two pairs of opposed burners. The E-Gas gasifier differs from other systems in that it uses a two-stage reactor. The bulk of the feed slurry and all of the oxygen are sent to the first (horizontal) stage, where the coal is gasified. Hot gas flows into the second (vertical) stage, where the remainder of the coal slurry is injected. Hot fuel gas is cooled in a fired-tube boiler fuel gas cooler.

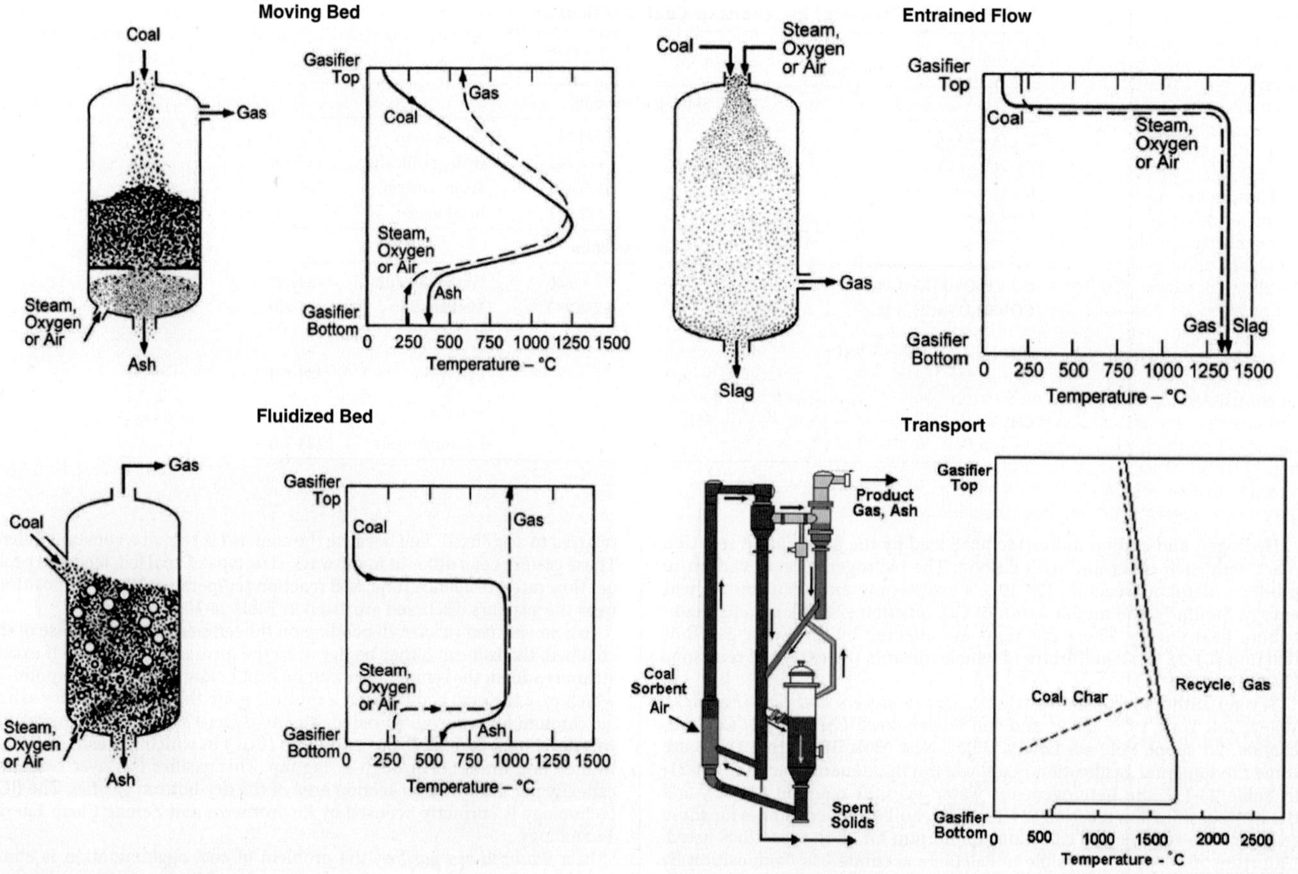

FIG. 24-11 Gasifier types and temperature profiles: (*a*) moving bed (dry ash), (*b*) entrained flow, (*c*) fluidized bed, (*d*) transport.

The *transport bed gasifier*, also known as the TRIG™ gasifier, was developed by Southern Company and KBR and is currently being deployed at Mississippi Power's Plant Radcliff IGCC plant in Kemper County, Mississippi. The transport gasifier is an advanced pressurized circulating fluidized-bed gasifier that is based on KBR's extensive experience in fluidized catalytic cracking. The gasifier operates at moderate temperatures (815°C–1065°C) and pressures up to 600 psig, achieving high carbon conversion while mitigating the formation of tars and oils. Because of its low operating temperature, short residence time, and dry feed injection, the gasifier exhibits high cold gas efficiency, is very amenable to gasifying high ash- and/or high moisture-containing coals, and works especially well on highly reactive low-rank coals. The gasifier is capable of operating using either air or pure oxygen as the oxidant, depending on the application, such as power generation or chemicals production.

Gasification-Based Liquid Fuels and Chemicals Liquid fuels and chemicals from gasification-based synthesis gas are described in the coal liquefaction section following this section. While the downstream areas of power system and indirect liquefaction plants will differ markedly, the gasification sections will be quite similar and are described in this section.

Gasification-Based Power Systems An important driving force for coal gasification process development is the environmental superiority of gasification-based power generation systems, generally referred to as integrated gasification combined-cycle (IGCC) power production (Fig. 24-12). Coal is crushed prior to being fed to a reactor, where it is gasified through contact with steam and air or oxygen. Partial oxidation produces the high-temperature [1033 to 2255 K (1400°F to 3600°F), depending on the type of gasifier] reducing environment necessary for gasification. The product fuel gas passes through heat recovery and cleanup, where particulates (dust), sulfur, and other impurities (e.g., mercury and arsenic) are removed. After cleanup, the fuel gas is composed primarily of hydrogen and carbon oxides. At this point, the fuel gas is either burned with compressed air and expanded through a gas turbine to generate electricity and releasing all carbon as carbon dioxide, or it can undergo water gas shift to produce a fuel containing only hydrogen and carbon dioxide. The carbon dioxide is separated from the hydrogen and sequestered, while the hydrogen is burned in a gas turbine to generate electricity without carbon dioxide emissions. In either case, heat is recovered from the turbine's hot exhaust gas to produce steam (at subcritical conditions), which is expanded in a steam turbine for additional electric power generation.

All four basic gasifier types could be incorporated into IGCC plant designs, although to date entrained-flow gasifiers have actually been deployed while the transport gasifier is currently undergoing commissioning. With each gasifier type, the oxidant can be air or oxygen, depending upon the

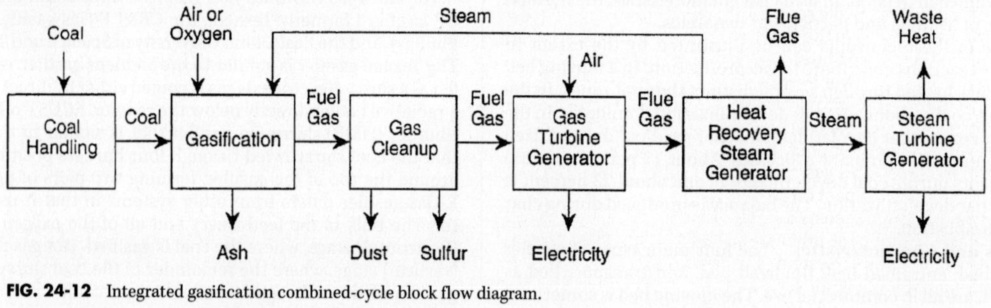

FIG. 24-12 Integrated gasification combined-cycle block flow diagram.

application, and the coal can be fed dry (all gasifier types) or in a slurry (entrained bed only). The composition of the fuel gas as well as its pressure and temperature are determined by the design of the gasifier, the gas cleanup system, and the intended application.

There are several features of IGCC power systems that contribute to their improved thermal efficiency and environmental superiority compared to a conventional pulverized-coal fired power plant. First, the mass flow rate of gas from an oxygen-fired gasifier is about one-fourth that from a combustor, and it operates substoichiometrically, while the combustor is air-blown and operates with excess air. Because of the elevated operating pressure and the lack of nitrogen dilution, the volumetric gas flow to the sulfur removal system is actually only 0.5 percent to 1 percent of the volumetric flow to the flue gas desulfurization unit; this lowers the capital cost of the gas cleanup system.

Furthermore, the sulfur in coal-derived fuel gas is mainly present as hydrogen sulfide, which is more easily recovered than the sulfur dioxide in flue gas. Not only can hydrogen sulfide be easily converted to elemental sulfur, a more valuable by-product than the calcium sulfate produced when lime is used to remove sulfur dioxide from flue gas, but also neither lime nor limestone is required. Nitrogen in the coal is largely converted to nitrogen gas or ammonia in the gasifier, which is easily removed by water washing, thus reducing nitrogen oxide emissions when the fuel gas is burned. Carbon dioxide can also be scrubbed from the fuel gas, and even if further reductions in carbon emissions are required, the carbon monoxide in the fuel gas can be converted to hydrogen and carbon dioxide before CO_2 removal. Finally, it has been estimated that the cost of mercury removal from an IGCC system would be only about one-tenth the cost for a conventional power plant.

Another advantage is that the IGCC system generates electricity by both combustion (Brayton cycle) and steam (Rankine cycle) turbines. The inclusion of the Brayton topping cycle improves efficiency compared to a conventional power plant's Rankine cycle–only generating system. Typically, about two-thirds of the power generated comes from the Brayton cycle and one-third from the Rankine cycle. The efficiency of the gas turbine is a function of its firing temperature within the turbine combustor, and efforts are underway to develop gas turbines with significantly higher firing temperatures (i.e., >3000°F) than those deployed in today's IGCC plants.

Current Status There have been substantial advancements in the development of gasification-based power systems during the last two decades. Programs are in place in the United States to support demonstration projects and for conducting research to improve the efficiency, cost effectiveness, and environmental performance of IGCC power generation. Areas of research that are likely to produce significant improvements in coal gasification technology are the gasification reactor, air separation (oxygen production), fuel gas cleanup (removal of sulfur, mercury, particulates, and other pollutants), and advanced turbines and fuel cells.

Gasification technology is being widely used throughout the world. A study conducted in 2014 indicated that there were 272 commercial operating gasification projects with a total of 686 gasifiers worldwide. Total capacity of those projects in operation was 116,000 MW (thermal) with another 109,000 MW (thermal) in various stages of development. As we discuss later, in addition to producing fuel gas for power production, synthesis gas production by gasification is also the first step in the indirect liquefaction of coal and the production of chemicals. Furthermore, gasification of carbonaceous, hydrogen-containing fuels is an effective method of thermal hydrogen production and is considered to be a key technology in the transition to a hydrogen economy. Therefore, the possibility exists for the coproduction of electric power and liquid fuels and chemicals while sequestering carbon dioxide. Such an option could allow a gasifier in an IGCC system to operate at full capacity at all times, producing fuel gas at times of peak power demand and a mix of fuel gas and synthesis gas at other times.

In 2005, there were four coal-based IGCC power plants in operation in the world: Tampa Electric in Polk County, Florida, based on a Texaco (now GE) gasifier; Wabash repowering project in Indiana, based on E-Gas (now CB&I E-Gas); and Buggenum in The Netherlands and Puertollano in Spain, both based on Shell gasifiers (Holt, N. A. H., "IGCC Technology—Status, Opportunities, and Issues," *EM Magazine*, December 2004, pp. 18–26). Today, the Buggenum facility has been shut down, while several others are either in operation or under construction (i.e., Vresova in the Czech Republic, Duke Edwardsport and Southern Kemper in the United States, and GreenGen in China).

Cost of Gasification-Based Power Systems Comparing power options is complicated by the many different parameters that must be considered in making a cost determination: coal cost; coal properties, including sulfur and moisture contents; ambient temperature; elevation; degree of process integration; gas turbine model; and gas cleanup method. These and many other factors have a significant impact on cost.

While a comparison of absolute costs among different power systems is difficult, the costs of the component units are usually within given ranges. For an oxygen-blown IGCC power system, the breakdown of the capital cost for the four component units is: air separation plant (10 to 15 percent), gasifier including gas cleanup (30 to 40 percent), combined-cycle power unit (40 to 45 percent), and balance of plant (5 to 10 percent). The breakdown of the cost of electricity is: capital charge (52 to 56 percent), operating and maintenance (14 to 17 percent), and fuel (28 to 32 percent). One of the main challenges to the development and deployment of IGCC power plants has been the much higher capital cost compared to that of a natural gas–fired generating unit, thereby negating the fuel cost savings. The costs of recent IGCC plants have come in well above initial estimates for a number of reasons. However, the capital costs of IGCC power plants are expected to decline considerably as more of these facilities are built, standard designs are developed, and economies of scale are realized.

COAL LIQUEFACTION

General References: Anderson, *The Fischer-Tropsch Synthesis*, Academic Press, New York, 1984; Dry, "The Fischer-Tropsch Synthesis," *Catalysis Science and Technology*, vol. 1, Springer-Verlag, New York, 1981; Kent, ed., "Coal Technology," chap. 17 in *Riegel's Handbook of Industrial Chemistry*, 10th ed., Kluwer Academic/Plenum Publishers, New York, 2003; Lowry, ed., *Chemistry of Coal Utilization*, suppl. vol., Wiley, New York, 1963, and 2d suppl. vol., Elliott, ed., 1981; Rao, Stiegel, Cinquegrane, and Srivastava, "Iron-based Catalyst for Slurry-Phase Fischer-Tropsch Process: Technology Review," *Fuel Processing Technology* **30:** 83–151 (1992); Sheldon, *Chemicals from Synthesis Gas*, D. Reidel Publishing Co., Dordrecht, Netherlands, 1983; Srivastava, McIlvried, Gray, Tomlinson, and Klunder, American Chemical Society Fuel Chemistry Division Preprints, Chicago, 1995; Wender, "Reactions of Synthesis Gas," *Fuel Processing Technology* **48:** 189–297 (1996); Wu and Storch, "Hydrogenation of Coal and Tar," *U.S. Bur. Mines Bull.* **633**, 1968.

Background Coal liquefaction denotes the process of converting solid coal into a liquid fuel. The primary objective of any coal liquefaction process is to increase the hydrogen-to-carbon atomic ratio. For a typical bituminous coal, this ratio is about 0.8, while for light petroleum it is about 1.8. A secondary objective is to remove sulfur, nitrogen, oxygen, and ash so as to produce a nearly pure hydrocarbon. There are several ways to accomplish liquefaction: (1) pyrolysis, (2) direct hydrogenation of the coal at elevated temperature and pressure, (3) hydrogenation of coal slurried in a solvent, and (4) gasification of coal to produce synthesis gas (a mixture of hydrogen and carbon monoxide, also referred to as syngas) followed by the use of Fischer-Tropsch (F-T) or other chemistry to produce liquid products. The first three of these approaches are generally referred to as direct liquefaction, in that the coal is directly converted to a liquid. The fourth approach is termed indirect liquefaction because the coal is first converted to an intermediate product.

Pyrolysis In pyrolysis, coal is heated in the absence of oxygen to drive off volatile components, leaving behind a solid residue enriched in carbon, known as char or coke. Most coal pyrolysis operations are for the purpose of producing metallurgical coke, with the liquids produced being considered only as a by-product. However, some small-scale work has been done to maximize liquids production by heating the coal at carefully controlled conditions of temperature versus time, usually in several stages. Although capable of producing a significant liquid yield, this approach has two major drawbacks. First, the liquids produced are of low quality and require significant upgrading to convert them to salable products. Second, a large fraction of the original heating value of the coal remains in the char, which must be profitably marketed to make the pyrolysis process economically feasible. Two processes that reached a high state of development were the COED process developed by FMC Corporation, which used a series of fluidized beds operating at successively higher temperatures, and the TOSCOAL process, which used a horizontal rotating kiln.

Direct Hydrogenation In direct hydrogenation, pulverized coal is contacted with hydrogen at carefully controlled conditions of temperature and pressure. The hydrogen reacts with the coal, converting it into gaseous and liquid products. In some cases, the coal is impregnated with a catalyst before being introduced into the reactor. Again, small-scale experiments have been successfully conducted. However, the major difficulty with this approach is scaleup to commercial size. Significant technical problems exist in feeding a large volume of powdered coal (powdered coal is necessary to provide a large surface area for reaction) into a reactor at high pressure, heating it to the desired temperature, and then quenching the products.

Direct Liquefaction Figure 24-13 presents a simplified process flow diagram of a typical direct coal liquefaction plant using coal slurry hydrogenation. Coal is ground and slurried with a process-derived solvent, mixed with a hydrogen-rich gas stream, preheated, and sent to a one- or two-stage liquefaction reactor system. In the reactor(s), the organic fraction of the coal dissolves in the solvent, and the dissolved fragments react with hydrogen to form liquid and gaseous products. Sulfur in the coal is converted to hydrogen sulfide, nitrogen is converted to ammonia, and oxygen is converted to water.

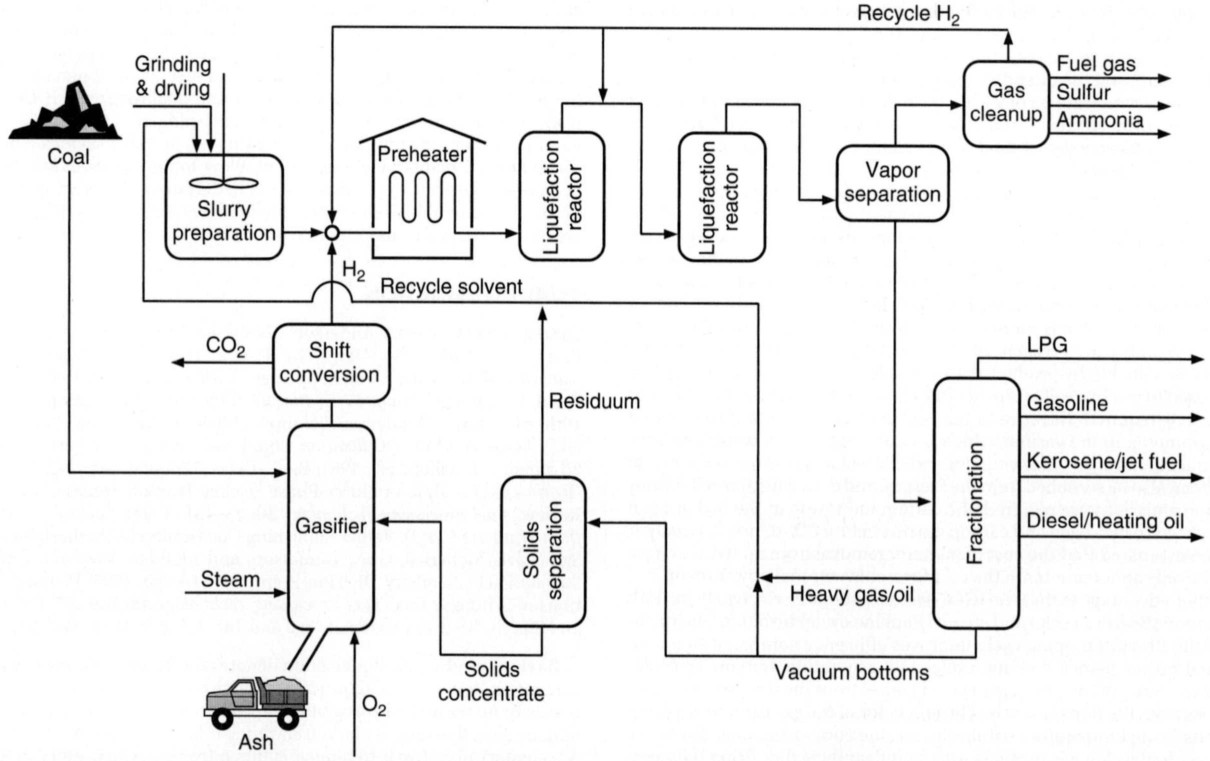

FIG. 24-13 Direct liquefaction of coal.

The reactor products go to vapor/liquid separation. The gas is cleaned and, after removal of a purge stream to prevent buildup of inerts, mixed with fresh hydrogen and recycled. The liquid is sent to fractionation for recovery of distillates. Heavy gasoil is recycled as process solvent, and vacuum bottoms are gasified for hydrogen production. Ash from the gasifier is sent to disposal. Heavy direct liquefaction products contain polynuclear aromatics and are potentially carcinogenic. However, this problem can be avoided by recycling to extinction all material boiling above the desired product end point.

Direct Liquefaction Kinetics Hydrogenation of coal in a slurry is a complex process, the mechanism of which is not fully understood. It is generally believed that coal first decomposes in the solvent to form free radicals, which are then stabilized by the extraction of hydrogen from hydroaromatic solvent molecules, such as tetralin. If the solvent does not have sufficient hydrogen transfer capability, the free radicals can recombine (undergo retrograde reactions) to form heavy, nonliquid molecules. A greatly simplified model of the liquefaction process is shown here.

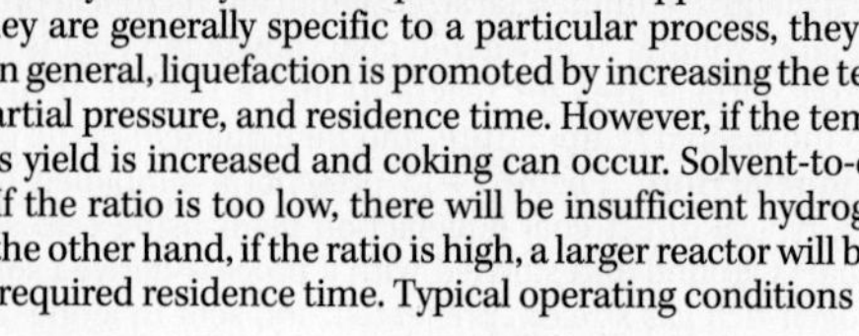

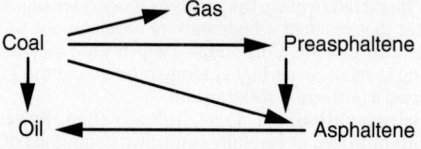

Many factors affect the rate and extent of coal liquefaction, including temperature, hydrogen partial pressure, residence time, coal type and analysis, solvent properties, solvent-to-coal ratio, ash composition, and the presence or absence of a catalyst. Many kinetic expressions have appeared in the literature, but since they are generally specific to a particular process, they will not be listed here. In general, liquefaction is promoted by increasing the temperature, hydrogen partial pressure, and residence time. However, if the temperature is too high, gas yield is increased and coking can occur. Solvent-to-coal ratio is important. If the ratio is too low, there will be insufficient hydrogen transfer activity; on the other hand, if the ratio is high, a larger reactor will be needed to provide the required residence time. Typical operating conditions are:

Temperature	670–730 K (750°F–850°F)
Pressure	10.3–20.7 MPa (1500–3000 psia)
Residence time	0.5–1 h
Solvent-to-coal ratio	1.5–2 kg/kg (1.5–2 lb/lb)

For the most highly developed processes, moisture- and ash-free (maf) coal conversion can be as high as 90 to 95 percent with a C_4+ distillate yield of 60 to 75 wt% and a hydrogen consumption of 5 to 7 wt%. When an external catalyst is used, it is typically some combination of cobalt, nickel, and molybdenum on a solid acid support, such as silica alumina. In slurry hydrogenation processes, catalyst life is typically fairly short because of the large number of potential catalyst poisons present in the system.

Several variations of the slurry hydrogenation process, depicted in Fig. 24-13 and discussed next, were tested at pilot plant scale. Table 24-12 presents typical operating conditions and yields for these processes.

The *SRC-I process*, developed by the Pittsburg & Midway Coal Mining Co. in the early 1960s, was not really a liquefaction process; rather, it was designed to produce a solid fuel for utility applications. Only enough liquid was produced to keep the process in solvent balance. The bottoms product was subjected to filtration or solvent extraction to remove ash and then solidified to produce a low-ash, low-sulfur substitute for coal. However, the value of the product was not high enough to make this process economically viable.

The *SRC-II process*, developed by Gulf Oil Corp., was an improved version of the SRC-I process, designed to produce more valuable liquid products, rather than a solid. The major difference between SRC-II and the earlier process was the recycle of a portion of the fractionator bottoms to the slurry feed tank. This increased the ash content of the reactor feed. This ash, particularly the iron pyrites in the ash, acted as a catalyst and improved product yield and quality.

The *Exxon Donor Solvent (EDS) process*, developed by the Exxon Research and Engineering Co., differed from the typical process in that, before being recycled, the solvent was hydrogenated in a fixed-bed reactor using a hydrotreating catalyst, such as cobalt or nickel molybdate. Exxon found that the use of this hydrogen donor solvent with carefully controlled properties improved process performance. Exxon developed a solvent index, based on solvent properties, which correlated with solvent effectiveness.

The *H-Coal process*, based on H-Oil technology, was developed by Hydrocarbon Research, Inc. (HRI). The heart of the process was a three-phase, ebullated-bed reactor in which catalyst pellets were fluidized by the upward flow of slurry and gas through the reactor. The reactor contained an internal tube for recirculating the reaction mixture to the bottom of the catalyst bed. Catalyst activity in the reactor was maintained by the withdrawal of small quantities of spent catalyst and the addition of fresh catalyst. The addition of a catalyst to the reactor is the main feature that distinguishes the H-Coal process from the typical process.

TABLE 24-12 Direct Liquefaction Process Conditions and Product Yields

Developer	Gulf[a]	Gulf	Exxon	HRI	SCS,[b] EPRI, Amoco	HTI
Process	SRC-I	SRC-II	EDS	H-Coal	Two-stage	Two-stage
Coal type	Kentucky 9 & 14	Illinois No. 6	Illinois No. 6	Illinois No. 6	Illinois No. 6	Illinois No. 6
Operating conditions						
Nominal reactor residence time, h	0.5	0.97	0.67			
Coal space velocity per stage, kg/(h·m^3) (lb/[h·ft^3])				530 (33.1)	825[c] (51.7)	310 (19.4)
Temperature, K (°F)	724 (842)	730 (855)	722 (840)	726 (847)	700 (800)	692 (787)
Total pressure, MPa (psia)	10.3 (1500)	13.4 (1950)	10.3 (1500)			19.2 (2790)
H_2 partial pressure, MPa (psia)	9.7 (1410)	12.6 (1830)		12.6 (1827)	18.3 (2660)	
Catalyst type	Coal minerals	Coal minerals	Coal minerals	Supported catalyst (Co/Mo)	AKZO-AO-60 (Ni/Mo)	AKZO-AO-60 (Ni/Mo)
Catalyst replacement rate, kg/kg (lb/U.S. ton) mf coal					1.5×10^{-3} (3.0)	2.3×10^{-3} (4.5)
Product yields, wt % maf coal						
H_2	−2.4	−5.0	−4.3	−5.9	−6.0	−7.2
H_2O	—	—	12.2[d]	8.3	9.7	9.8
H_2S, CO_x, NH_3	—	9.6	4.2[e]	5.0	5.2	5.2
C_1-C_3	3.7[f]	13.9[f]	7.3	11.3	6.5	5.6
C_4^+ distillate	13.5[g]	46.8[g]	38.8	53.1	65.6	73.3
Bottoms[h]	68.4	30.7	41.8	28.2	19.0	13.3
Unreacted coal[i]	5.4	4.6	—	6.4	7.0	5.0
Distillate end point, K (°F)	727 (850)	727 (850)	911 (1180)	797 (975)	797 (975)	524 (975)

[a]In partnership with Pittsburg & Midway Coal Mining Co.
[b]Southern Company Services, Inc., prime contractor for Wilsonville Facility.
[c]Coal space velocity is based on settled catalyst volume.
[d]CO_x is included.
[e]CO_x is excluded.
[f]C_4 is included.
[g]C_4 is excluded.
[h]Unreacted coal is included.
[i]"Unreacted coal" is actually insoluble organic matter remaining after reaction.

Two-stage liquefaction is an advanced process concept that provides higher yields of better-quality products by carrying out the coal dissolution and the hydrogenation/hydrocracking steps in separate reactors whose conditions are optimized for the reaction that is occurring. Either or both reactors may be catalytic. Slurry catalysts have been tested, in addition to the more conventional supported catalysts, as a means of simplifying reactor design and removing process constraints. The U.S. Department of Energy and its private sector collaborators, Hydrocarbon Technologies, Inc., and others, have advanced the development of the two-stage direct liquefaction process to commercialization status during the last two decades. Coal-derived product quality has been improved dramatically (less than 50 ppm nitrogen content, for example) through the addition of in-line fixed-bed hydrotreating of the product stream.

In *coal-oil coprocessing*, coal is slurried in petroleum residuum rather than in recycle solvent, and both the coal and petroleum components are converted into high-quality fuels in the slurry reactor. This variation offers the potential for significant cost reduction by eliminating or reducing recycle streams. More importantly, fresh hydrogen requirements are reduced because the petroleum feedstock has a higher initial hydrogen content than coal. As a result, plant capital investment is reduced substantially. Other carbonaceous materials, such as municipal waste, plastics, cellulosics, and used motor oils might also serve as co-feedstocks with coal in this technology.

Commercial Operations The world's only commercial-scale direct coal liquefaction plant, located in the Inner Mongolia Autonomous Region of China, was dedicated in 2004. Start-up of the plant began in December 2008. The first train of the first phase of the Shenhua Direct Coal Liquefaction Plant liquefies 2,100,000 Mg/a (2,315,000 ton/yr) of coal from the Shangwan Mine in the Shenhua coal field of Inner Mongolia. The plant uses a combination of technologies developed in the United States, Japan, and Germany with modifications and enhancements developed in China. The process uses a two-stage reactor system and includes an in-line hydrotreater; it produces 591,900 Mg/a (652,460 ton/yr) of diesel, 174,500 Mg/a (192,350 ton/yr) of naphtha, 70,500 Mg/a (77,710 ton/yr) of LPG, and 8300 Mg/a (9150 ton/yr) of ammonia. When completed, the plant is expected to include 10 trains producing approximately 10,000,000 Mg/a (11,000,000 ton/yr) of oil products.

Indirect Liquefaction Unlike the processes just described, indirect liquefaction is not limited to coal but may be performed using any carbonaceous feed, such as natural gas, petroleum residues, petroleum coke, coal, and biomass. Figure 24-14 presents a simplified process flow diagram for a typical indirect liquefaction process using coal as the feedstock. The syngas is produced in a gasifier (see the description of coal gasifiers earlier in this section), which partially combusts the coal or other feed at high temperature [1500–1750 K (2200°F–2700°F)] and moderate pressure [2–4 MPa (300–600 psia)] with a mixture of oxygen (or air) and steam. In addition to H_2 and CO, the raw synthesis gas contains other constituents, such as CO_2, H_2S, NH_3, N_2, H_2O, and CH_4, as well as particulates and, with some gasifiers, tars.

The syngas leaving the gasifier is cooled and passed through particulate removal equipment. Following this, depending on the requirements of the syngas conversion process, it may be necessary to adjust the H_2/CO ratio. Modern high-efficiency gasifiers typically produce syngas with a H_2/CO molar ratio between 0.45 and 0.7, which is lower than the stoichiometric ratio of about 2 for Fisher-Tropsch (F-T) synthesis or methanol production. Some F-T catalysts, particularly iron catalysts, have water gas shift conversion activity and permit operation with a low H_2/CO ratio—see reaction (24-29). Others, such as cobalt catalysts, possess little shift activity and require adjustment of the H_2/CO ratio before the syngas enters the synthesis reactor.

After shift conversion (if required), acid gases (CO_2 and H_2S) are scrubbed from the synthesis gas. A guard chamber is sometimes used to remove the last traces of H_2S, since F-T catalysts are generally very sensitive to sulfur poisoning. The cleaned gas is sent to the synthesis reactor, where it is converted at moderate temperature and pressure, typically 498 to 613 K (435°F–645°F) and 1.5 to 6.1 MPa (220–880 psia). Products, whose composition depends on operating conditions, the catalyst employed, and the reactor design, include saturated hydrocarbons (mainly straight-chain paraffins from methane through n-C_{50} and higher), oxygenates (methanol, higher alcohols, ethers), and olefins.

Fischer-Tropsch synthesis is the best known technology for producing hydrocarbons from synthesis gas. This technology had its beginning in Germany in 1902 when Sabatier and Senderens hydrogenated CO to methane using a nickel catalyst. In 1926 Fischer and Tropsch were awarded a patent for the discovery of a catalytic technique to convert syngas to liquid hydrocarbons similar to petroleum.

The *basic reactions* in the F-T synthesis are:

Paraffin formation:

$$(2n+1)H_2 + nCO \rightarrow C_nH_{2n+2} + nH_2O \tag{24-26}$$

Olefin formation:

$$2nH_2 + nCO \rightarrow C_nH_{2n} + nH_2O \tag{24-27}$$

Alcohol formation:

$$2nH_2 + nCO \rightarrow C_nH_{2n+1}OH + (n-1)H_2O \tag{24-28}$$

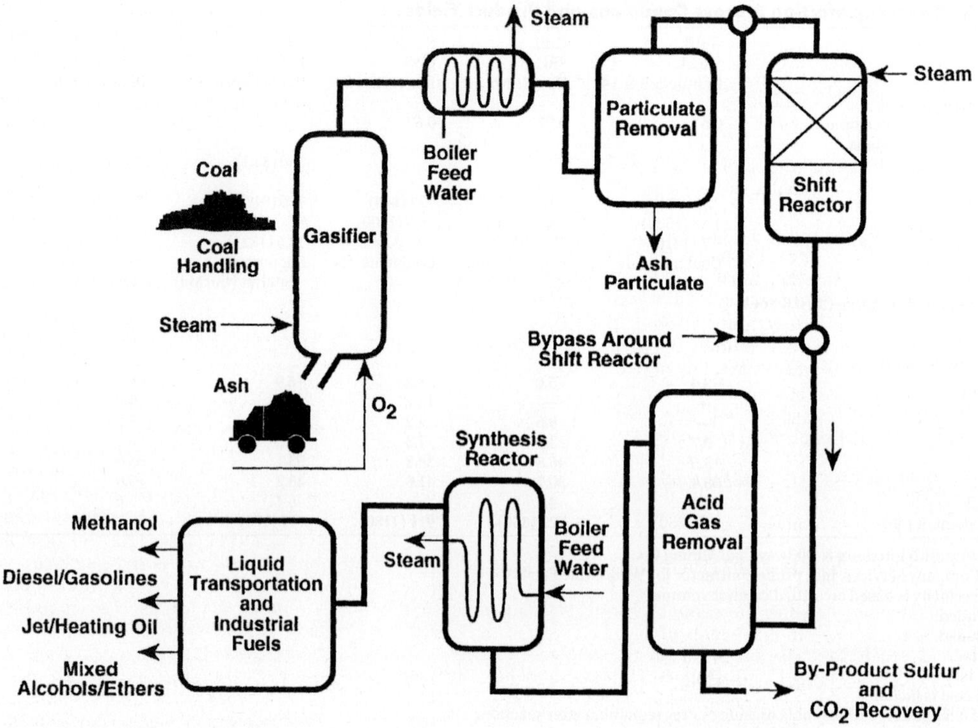

FIG. 24-14 Indirect liquefaction of coal.

Other reactions that may occur during F-T synthesis, depending on the catalyst employed and the reaction conditions used, are:

Water-gas shift:

$$CO + H_2O \leftrightarrow CO_2 + H_2 \tag{24-29}$$

Boudouard reaction:

$$2CO \rightarrow C(s) + CO_2 \tag{24-30}$$

Surface carbonaceous deposit formation:

$$(x + y/2)H_2 + xCO \rightarrow C_xH_y + xH_2O \tag{24-31}$$

Catalyst oxidation-reduction:

$$xM + yH_2O \leftrightarrow M_xO_y + yH_2 \tag{24-32}$$

$$xM + yCO_2 \leftrightarrow M_xO_y + yCO \tag{24-33}$$

Bulk carbide formation:

$$xM + yC \rightarrow M_xC_y \tag{24-34}$$

In the preceding equations, M represents a catalytic metal atom.

The production of hydrocarbons using traditional F-T catalysts is governed by chain growth (polymerization) kinetics. The theoretical equation describing the distribution of hydrocarbon products, commonly referred to as the Anderson-Schulz-Flory (ASF) equation, is:

$$\log(W_n/n) = n \log \alpha + \log[(1 - \alpha)^2/\alpha] \tag{24-35}$$

where W_n is the weight fraction of products with carbon number n, and α is the chain growth probability, that is, the probability that a carbon chain on the catalyst surface will grow by adding another carbon atom rather than desorb from the catalyst surface and terminate. In deriving Eq. (24-35), α is assumed to be independent of chain length. However, α is dependent on temperature, pressure, H_2/CO ratio, and catalyst composition. As α increases, the average carbon number of the product also increases. When α equals 0, methane is the only product formed. As α approaches 1, the product becomes predominantly wax. In practice, α is not really independent of chain length. Methane production, particularly with cobalt catalysts, is typically higher than predicted; and C_2 yield is often lower. Some investigators have found a significant deviation from the ASF distribution for higher carbon number products and have proposed a dual alpha mechanism to explain their results.

Figure 24-15 provides a graphical representation of Eq. (24-36) showing the weight fraction of various products as a function of α. This figure shows that there is a particular α that will maximize the yield of any desired product, such as gasoline or diesel fuel. Based on the ASF equation, the weight fraction of material between carbon numbers m and n inclusive is given by:

$$W_{mn} = m\alpha^{m-1} - (m-1)\alpha^m - (n+1)\alpha^n + n\alpha^{n+1} \tag{24-36}$$

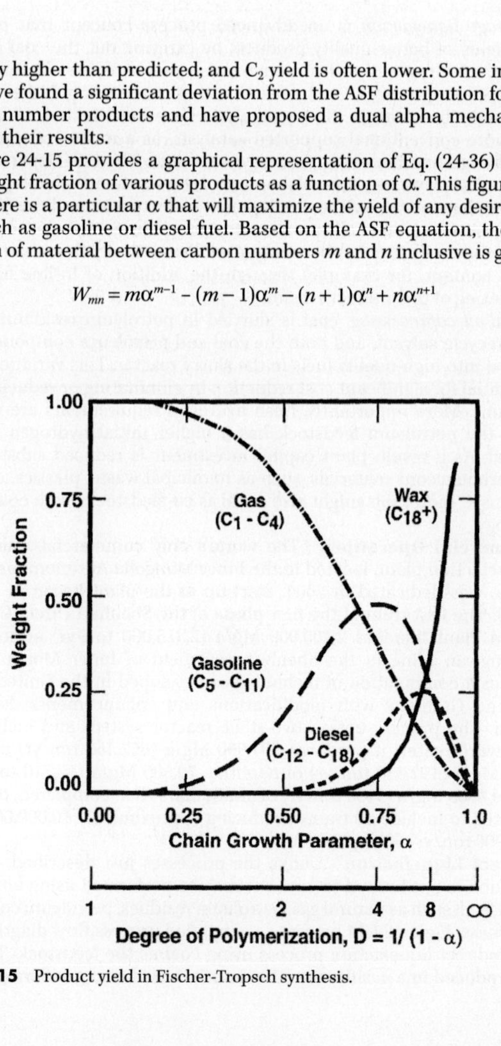

FIG. 24-15 Product yield in Fischer-Tropsch synthesis.

The α to maximize the yield of the carbon number range from m to n is given by:

$$\alpha_{opt} = \left(\frac{m^2 - m}{n^2 + n}\right)^{\frac{1}{n-m+1}} \tag{24-37}$$

For *F-T catalysts,* the patent literature is replete with recipes for the production of the catalysts, with most formulations being based on iron, cobalt, or ruthenium, typically with the addition of some promoter(s). Nickel is sometimes listed as an F-T catalyst, but nickel has too much hydrogenation activity and produces mainly methane. In practice, because of the cost of ruthenium, commercial plants use either cobalt-based or iron-based catalysts. Cobalt is usually deposited on a refractory oxide support, such as alumina, silica, titania, or zirconia. Iron is typically not supported, and it may be prepared by precipitation. Of the large number of promoters listed in patents, rhenium or one of the noble metals may be used to promote cobalt, and iron is often promoted with potassium.

F-T Reactor Design The F-T reaction is highly exothermic and, for hydrogen-rich syngas, it can be symbolically represented by

$$CO + 2H_2 \rightarrow -CH_2- + H_2O \quad \Delta H = -165 \text{ kJ/mol of } -CH_2- \tag{24-38}$$

For CO-rich syngas, the overall reaction is:

$$2CO + H_2 \rightarrow -CH_2- + CO_2 \quad \Delta H = -207 \text{ kJ/mol of } -CH_2- \tag{24-39}$$

Because of the high heat release, adequate heat removal from the reactor is critical. High temperatures result in high yields of methane, as well as coking and sintering of the catalyst. Three types of reactors (tubular fixed bed, fluidized bed, and slurry) provide good temperature control, and all three types are being used for synthesis gas conversion. Some newer reactor designs using microchannels and microfibrous entrapped catalysts are under development and offer improved temperature control. The first plants used tubular or plate-type fixed-bed reactors. Later, SASOL, in South Africa, used fluidized-bed reactors, and most recently, slurry reactors have come into use. Fluid-bed reactors are limited to the production of products that are volatile at reactor operating conditions. Nonvolatile products accumulate on the catalyst and destroy its fluidizing properties.

F-T reactor operations can be classified into two categories: high-temperature, 613 K (645°F), or low-temperature, 494 to 544 K (430°F–520°F). The Synthol and fixed fluidized-bed reactors developed by SASOL are typical of high-temperature operation. Using an iron-based catalyst, these reactors produce a very good gasoline having high olefinicity and a low boiling range. The olefin fraction can readily be oligomerized to produce diesel fuel. Low-temperature operation, typical of fixed-bed and slurry reactors, produces a much more paraffinic and straight-chain product. The chain growth parameter can be tailored to give the desired product selectivity. The primary diesel fraction, as well as the diesel-range product from hydrocracking of the wax, is an excellent diesel fuel.

Chemicals from Syngas A wide range of products can be produced from syngas. These include such chemicals as methanol, ethanol, isobutanol, dimethyl ether, dimethyl carbonate, and many other chemicals. Typical methanol-producing reactions are:

$$CO + 2H_2 \rightarrow CH_3OH \tag{24-40}$$

$$CO_2 + 3H_2 \rightarrow CH_3OH + H_2O \tag{24-41}$$

Once methanol is produced, it can be converted into an extensive range of materials. The following reactions illustrate some of the chemicals of major importance that can be made from methanol. Among these are dimethyl ether, acetic acid, methyl acetate, acetic anhydride, vinyl acetate, formaldehyde, and methyl tertiary butyl ether (MTBE).

$$2CH_3OH \rightarrow CH_3OCH_3 + H_2O \tag{24-42}$$

$$CH_3OH + CO \rightarrow CH_3COOH \tag{24-43}$$

$$CH_3COOH + CH_3OH \rightarrow CH_3COOCH_3 + H_2O \tag{24-44}$$

$$CH_3COOCH_3 + CO \rightarrow CH_3COOCOCH_3 \tag{24-45}$$

$$CH_3COOH + CH_2{=}CH_2 + \tfrac{1}{2}O_2 \rightarrow CH_2{=}CHCOCH_3 + H_2O \tag{24-46}$$

$$CH_3OH + \tfrac{1}{2}O_2 \rightarrow CH_2O + H_2O \tag{24-47}$$

$$CH_3OH + (CH_3)_2C{=}CH_2 \rightarrow CH_3OC(CH_3)_3 \tag{24-48}$$

Reaction (24-42) can occur in parallel with the methanol-producing reactions, thereby overcoming the equilibrium limitation on methanol formation. Higher alcohols can also be formed, as illustrated by reaction (24-28), which can generate either linear or branched alcohols, depending on the catalyst used and the operating conditions. The production of methyl acetate, reaction (24-44), from synthesis gas is currently being practiced commercially. Following methanol synthesis, half the methanol is reacted with carbon monoxide to form acetic acid, which is reacted with the rest of the methanol to form methyl acetate.

Methyl acrylate and methyl methacrylate, which are critical to the production of polyesters, plastics, latexes, and synthetic lubricants, can also be produced. For example, methyl methacrylate can be produced from propylene and methanol:

$$C_3H_6 + \tfrac{1}{2}O_2 + CO + CH_3OH \rightarrow CH_2{=}C(CH_3)COOCH_3 + H_2O \tag{24-49}$$

Commercial Operations The only commercial indirect coal liquefaction plants for the production of transportation fuels are currently operated by SASOL in South Africa. Construction of the original plant in Sasolburg was begun in 1950, and operations began in 1955. This plant employed both fixed-bed (Arge) and entrained-bed (Synthol) reactors. The fixed-bed reactors have been converted to natural gas, and the Synthol reactors have been replaced by advanced fixed fluidized-bed reactors. Two additional plants, located in Secunda, that employ dry-ash Lurgi Mark IV coal gasifiers and entrained-bed (Synthol) reactors for synthesis gas conversion were constructed with startups in 1980 and 1983. In addition to producing a significant fraction of South Africa's transportation fuel requirements, these plants produce more than 120 other products from coal. A new indirect coal liquefaction plant using Siemens gasification technology is currently under construction by Shenhua Ningxia Coal Group in Yinchuan, China.

SASOL and others, including Exxon, Statoil, Air Products and Chemicals, Inc., and the U.S. Department of Energy, have engaged in the development of slurry bubble column reactors for F-T and oxygenate synthesis. SASOL commissioned a five-meter-diameter slurry reactor in 1993, which doubled the wax capacity of the SASOL I facility. The development work on slurry reactors shows that they have several advantages over competing reactor designs: (1) excellent heat transfer capability resulting in nearly isothermal reactor operations, (2) high catalyst and reactor productivity, (3) ease of catalyst addition and withdrawal, (4) simple construction, and (5) the ability to process hydrogen-lean synthesis gas successfully. Because of the small particle size of the catalyst used in slurry reactors, effective separation of catalyst from the products can be difficult but is crucial to successful operation.

The United States has two commercial facilities that convert coal to fuels or chemicals via a syngas intermediate. The Great Plains Synfuels Plant, located in Beulah, North Dakota, and operated by Dakota Gasification Company (DGC), produces synthetic natural gas (SNG) from North Dakota lignite. Fourteen Lurgi dry-ash gasifiers in the plant convert approximately 15,400 Mg/d (17,000 U.S. ton/day) of lignite into syngas, which is methanated to about 4.7×10^6 Nm^3 (166×10^6 std ft^3) of pipeline-quality gas. Aromatic naphtha and tar oil are also produced in the gasification section. The plant operates at 120 percent of its original design capacity. In addition to SNG, a wide assortment of other products are produced and sold (anhydrous ammonia, ammonium sulfate, phenol, cresylic acid, naphtha, krypton and xenon gases, liquid nitrogen, and carbon dioxide). POSCO is operating a 500,000 tons/year coal to SNG plant using E-Gas gasification technology at their Gwangyang Steel Works in South Korea. Several coal to SNG plants are in various stages of development in China.

Eastman Chemical Company has operated a coal-to-methanol plant in Kingsport, Tennessee, since 1983. Although originally designed to process 34 Mg/h (37 U.S. ton/h) of coal, the two GE (formerly Texaco) gasifiers (one is a backup) process 49 Mg/h (547 U.S. ton/h) of coal to synthesis gas. Using ICI methanol technology, the synthesis gas is converted to methanol, which is an intermediate in the production of methyl acetate and acetic acid. The plant produces about 225,000 Mg/a (250,000 U.S. ton/yr) of acetic anhydride. As part of the DOE Clean Coal Technology Program, Air Products and Chemicals, Inc., and Eastman Chemical Company constructed and operated a 9.8 Mg/h (260 U.S. ton/d) slurry-phase reactor for the conversion of synthesis gas to methanol.

Despite the success of SASOL, most of the commercial interest in Fischer-Tropsch synthesis technology is based on natural gas, as this represents a way to bring remote gas deposits to market using conventional tankers. In 1985, Mobil commercialized its methanol-to-gasoline (MTG) technology in New Zealand, natural gas being the feedstock. This fixed-bed process converted synthesis gas to 4000 Mg/d (4400 U.S. ton/day) of methanol; the methanol could then be converted to 2290 m^3/d (14,400 bbl/d) of gasoline. Owing to economic factors, the plant was used primarily for the production of methanol.

Shell Gas B.V. constructed a 2337 m^3/d (14,700 bbl/d) F-T plant in Malaysia that started operations in 1994. The Shell Middle Distillate Synthesis (SMDS) process uses natural gas as the feedstock to fixed-bed reactors containing cobalt-based catalyst. The heavy hydrocarbons from the F-T reactors are converted to distillate fuels by hydrocracking and hydroisomerization. The quality of the products is very high, the diesel fuel having a cetane number in excess of 75 with no sulfur.

For a number of years, the largest F-T facility based on natural gas was the Mossgas plant located in Mossel Bay, South Africa. Natural gas is converted to synthesis gas in a two-stage reformer and subsequently converted to hydrocarbons by SASOL's Synthol technology. The plant, commissioned in 1992, has a capacity of 7155 m^3/d (45,000 bbl/d). Sasol subsequently built two additional gas-to-liquids plants. The 34,000 bbl/d Oryx plant in Ras Laffan Industrial City in Qatar began operations in 2007, while their 33,000 bbl/d Escravos project in Nigeria began operations in 2014. The largest gas-to-liquids plant in operation today was built by Shell in Ras Laffan Industrial City in Qatar. This plant employs Shell gasification technology and produces 140,000 bbl/d of liquids. Commercial operations began in 2011. Today, there are a number of both small- and large-scale gas-to-liquid projects under evaluation to capitalize on flared gas in oil fields and the abundant supply of natural gas produced from shale formations.

HEAT GENERATION

General References: *Application of FBC for Power Generation,* Electric Power Research Institute, EPRI PR-101816, Palo Alto, Calif., 1993; Basu and Fraser, *Circulating Fluidized Bed Boilers: Design and Operations,* Butterworth and Heinemann, Boston, 1991; Baukal, ed., *The John Zink Hamworthy Combustion Handbook,* 2d ed., CRC Press, Boca Raton, Fla., 2013; Boyen, *Thermal Energy Recovery,* 2d ed., Wiley, New York, 1980; Cuenca and Anthony, eds., *Pressurized Fluidized Bed Combustion,* Springer, Berlin, 1995; *Proceedings of International FBC Conference(s),* ASME, New York, 1991–2005; Reed, R. J., *North American Combustion Handbook,* 3d ed., vols. I and II, North American Manufacturing Company, Cleveland, Ohio, 1996; Singer, ed., *Combustion: Fossil Power Systems,* 4th ed., Combustion Engineering, Windsor, Conn., 1993; Stultz and Kitto, eds., *Steam: Its Generation and Use,* 42nd ed., Babcock and Wilcox, Charlotte, N.C., 2015.

COMBUSTION BACKGROUND

Theoretical Oxygen and Air for Combustion The amount of oxidant (oxygen or air) just sufficient to burn the carbon, hydrogen, and sulfur in a fuel to carbon dioxide, water vapor, and sulfur dioxide is the *theoretical* or *stoichiometric oxygen* or *air* requirement. The chemical equation for complete combustion of a fuel is

$$C_xH_yO_zS_w + \left(\frac{4x+y-2z+4w}{4}\right)O_2 = x\,CO_2 + \left(\frac{y}{2}\right)H_2O + w\,SO_2 \qquad (24\text{-}50)$$

x, y, z, and w being the number of atoms of carbon, hydrogen, oxygen, and sulfur, respectively, in the fuel. For example, 1 mol of methane (CH_4) requires 2 mol of oxygen for complete combustion to 1 mol of carbon dioxide and 2 mol of water. If air is the oxidant, each mol of oxygen is accompanied by 3.76 mol of nitrogen.

The volume of theoretical oxygen (at 0.101 MPa and 298 K) needed to burn any fuel can be calculated from the ultimate analysis of the fuel as follows:

$$24.45\left(\frac{C}{12}+\frac{H}{4}-\frac{O}{32}+\frac{S}{32}\right) = m^3 O_2/\text{kg fuel} \qquad (24\text{-}51)$$

where C, H, O, and S are the decimal weights of these elements in 1 kg of fuel. (To convert to ft^3 per lb of fuel, multiply by 16.02.) The mass of oxygen (in kg) required can be obtained by multiplying the volume by 1.31. The volume of theoretical air can be obtained by using a coefficient of 116.4 in Eq. (24-51) in place of 24.45.

Figure 24-16 gives the theoretical air requirements for a variety of combustible materials on the basis of fuel higher heating value (HHV). If only the fuel lower heating value is known, the HHV can be calculated from Eq. (24-1).

Excess Air for Combustion More than the theoretical amount of air is necessary in practice to achieve complete combustion. This excess air is expressed as a percentage of the theoretical air amount. The *equivalence ratio* is defined as the ratio of the actual fuel/air ratio to the stoichiometric fuel/air ratio. Equivalence ratio values less than 1.0 correspond to fuel-*lean* mixtures. Conversely, values greater than 1.0 correspond to fuel-*rich* mixtures.

Products of Combustion For lean mixtures, the *products of combustion* (POC) of a sulfur-free fuel consist of carbon dioxide, water vapor, nitrogen, oxygen, and possible small amounts of carbon monoxide and unburned hydrocarbon species. Figure 24-17 shows the effect of fuel/air ratio on the flue gas composition resulting from the combustion of natural gas. In the case of solid and liquid fuels, the POC may also include solid residues containing ash and unburned carbon particles.

Equilibrium combustion product compositions and properties may be readily calculated using thermochemical computer codes that minimize the Gibbs free energy and use thermodynamic databases containing polynomial curve-fits of physical properties. Two widely used versions are those developed at NASA Lewis (Gordon and McBride, NASA SP-273, 1971, now available online at http://www.grc.nasa.gov/WWW/CEAWeb) and at Stanford University (Reynolds, *STANJAN Chemical Equilibrium Solver,* Stanford University, 1987).

Flame Temperature The heat released by the chemical reaction of fuel and oxidant heats the POC. Heat is transferred from the POC, primarily by radiation and convection, to the surroundings, and the resulting temperature in the reaction zone is the flame temperature. If there is no heat transfer to the surroundings, the flame temperature equals the theoretical, or adiabatic, flame temperature.

Figure 24-18 shows the available heat in the products of combustion for various common fuels. The available heat is the total heat released during combustion minus the flue-gas heat loss (including the heat of vaporization of any water formed in the POC).

Flammability Limits There are both upper (or rich) and lower (or lean) limits of flammability of fuel/air or fuel/oxygen mixtures. Outside these limits, a self-sustaining flame cannot form. Flammability limits for common fuels are listed in Table 24-13.

Flame Speed Flame speed is defined as the velocity, relative to the unburned gas, at which an adiabatic flame propagates normal to itself through a homogeneous gas mixture. It is related to the combustion reaction rate and is important in determining burner flashback and blow-off limits. In a premixed burner, the flame can *flash back* through the flameholder and ignite the mixture upstream of the burner head if the mixture velocity at the flameholder is lower than the flame speed. Conversely, if the mixture velocity is significantly higher than the flame speed, the flame may not stay attached to the flameholder and is said to *blow off.* Flame speed is strongly dependent on fuel/air ratio, passing from nearly zero at the lean limit of flammability through a maximum and back to near zero at the rich limit of flammability. Maximum flame speeds for common fuels are provided in Table 24-13.

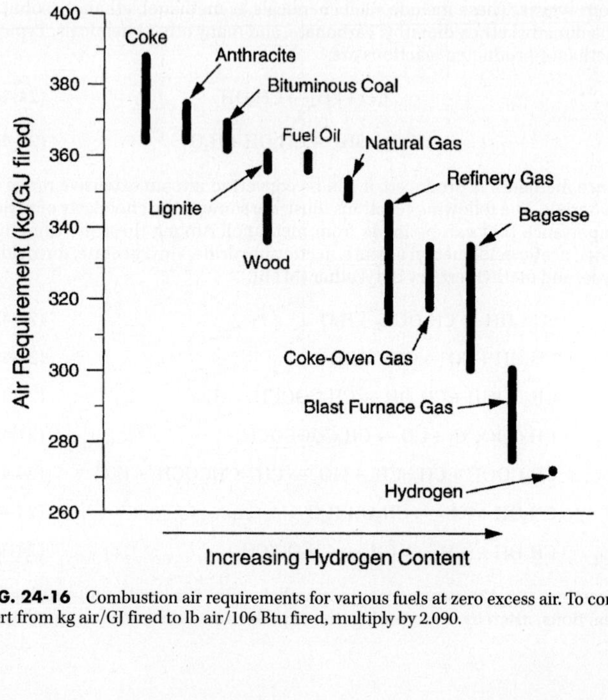

FIG. 24-16 Combustion air requirements for various fuels at zero excess air. To convert from kg air/GJ fired to lb air/106 Btu fired, multiply by 2.090.

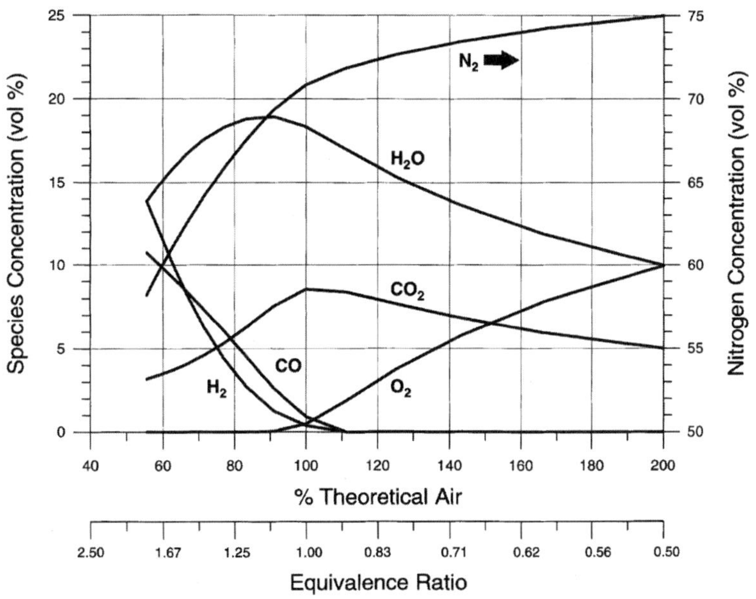

FIG. 24-17 Effect of fuel/air ratio on flue gas composition for a typical U.S. natural gas containing 93.9 percent CH_4, 3.2 percent C_2H_6, 0.7 percent C_3H_8, 0.4 percent C_4H_{10}, 1.5 percent N_2, and 1.1 percent CO_2 by volume.

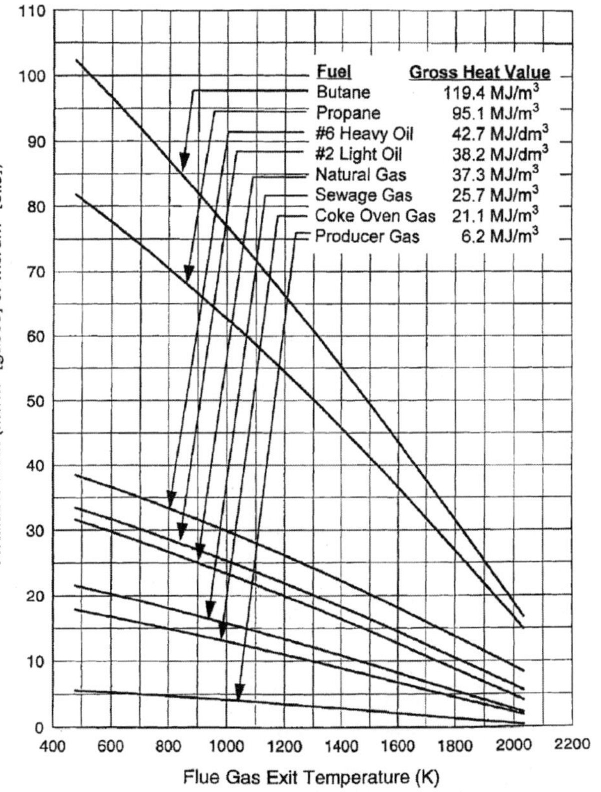

FIG. 24-18 Available heats for some typical fuels. The fuels are identified by their gross (or higher) heating values. All available heat figures are based on complete combustion and fuel and air initial temperature of 288 K (60°F). To convert from MJ/Nm³ to Btu/ft³, multiply by 26.84. To convert from MJ/dm³ to Btu/gal, multiply by 3588.

Pollutant Formation and Control in Flames Key combustion-generated air pollutants include nitrogen oxides (NO_x), sulfur oxides (principally SO_2), particulate matter, carbon monoxide, and unburned hydrocarbons.

Nitrogen Oxides These can form by three reaction paths, each having unique characteristics (see Fig. 24-19), and are responsible for the formation of NO_x during combustion processes: (1) *thermal* NO_x, which is formed by the combination of atmospheric nitrogen and oxygen at high temperatures; (2) *fuel* NO_x, which is formed from the oxidation of fuel-bound nitrogen; and (3) *prompt* NO_x, which is formed by the reaction of fuel-derived hydrocarbon fragments with atmospheric nitrogen. [NO_x is used to refer to $NO + NO_2$. NO is the primary form in combustion products (typically 95 percent of total NO_x). NO is subsequently oxidized to NO_2 in the atmosphere.]

Thermal NO_x formation is described by the Zeldovich mechanism:

$$N_2 + O \rightleftharpoons NO + N \tag{24-52}$$

$$N + O_2 \rightleftharpoons NO + O \tag{24-53}$$

$$N + OH \rightleftharpoons NO + H \tag{24-54}$$

The first of these reactions is the rate-limiting step. Assuming that O and O_2 are in partial equilibrium, the NO formation rate can be expressed as follows:

$$\frac{d[NO]}{dt} = A[N_2][O_2]^{1/2} \exp\left(\frac{-E}{RT}\right) \tag{24-55}$$

As indicated, the rate of NO formation increases exponentially with temperature, and, of course, oxygen and nitrogen must be available for thermal NO_x to form. Thus, thermal NO_x formation is rapid in high-temperature lean zones of flames.

Fuel-bound nitrogen (FBN) is the major source of NO_x emissions from combustion of nitrogen-bearing fuels such as heavy oils, coal, and coke. Under the reducing conditions surrounding the burning droplet or particle, the FBN is converted to fixed nitrogen species such as HCN and NH_3. These, in turn, are readily oxidized to form NO if they reach the lean zone of the flame. Between 20 and 80 percent of the bound nitrogen is typically converted to NO_x, depending on the design of the combustion equipment. With prolonged exposure (order of 100 ms) to high temperature and reducing conditions,

TABLE 24-13 Combustion Characteristics of Various Fuels*

Fuel	Minimum ignition temp., K/°F	Calculated flame temperature,† K/°F in air	Calculated flame temperature,† K/°F in O_2	Flammability limits, % fuel gas by volume in air: lower	Flammability limits, % fuel gas by volume in air: upper	Maximum flame velocity, m/s and ft/s: in air	Maximum flame velocity, m/s and ft/s: in O_2	% theoretical air for max. flame velocity
Acetylene, C_2H_2	578/581	2905/4770	3383/5630	2.5	81.0	2.67/8.75	—	83
Blast furnace gas	—	1727/2650	—	35.0	73.5	—	—	—
Butane, commercial	753/896	2246/3583	—	1.86	8.41	0.87/2.85	—	—
Butane, n-C_4H_{10}	678/761	2246/3583	—	1.86	8.41	0.40/1.3	—	97
Carbon monoxide, CO	882/1128	2223/3542	—	12.5	74.2	0.52/1.7	—	55
Carbureted water gas	—	2311/3700	3061/5050	6.4	37.7	0.66/2.15	—	90
Coke oven gas	—	2261/3610	—	4.4	34.0	0.70/2.30	—	90
Ethane, C_2H_4	745/882	2222/3540	—	3.0	12.5	0.48/1.56	—	98
Gasoline	553/536	—	—	1.4	7.6	—	—	—
Hydrogen, H_2	845/1062	2318/4010	3247/5385	4.0	74.2	2.83/9.3	—	57
Hydrogen sulfide, H_2S	565/558	—	—	4.3	45.5	—	—	—
Mapp gas, (allene) C_3H_4	728/850	—	3200/5301	3.4	10.8	—	4.69/15.4	—
Methane, CH_4	905/1170	2191/3484	—	5.0	15.0	0.45/1.48	4.50/14.76	90
Methanol, CH_3OH	658/725	2177/3460	—	6.7	36.0	—	0.49/1.6	—
Natural gas	—	2214/3525	2916/4790	4.3	15.0	0.30/1.00	4.63/15.2	100
Producer gas	—	1927/3010	—	17.0	73.7	0.26/0.85	—	90
Propane, C_3H_8	739/871	2240/3573	3105/5130	2.1	10.1	0.46/1.52	3.72/12.2	94
Propane, commercial	773/932	2240/3573	—	2.37	9.50	0.85/2.78	—	—
Propylene, C_3H_6	—	—	3166/5240	—	—	—	—	—
Town gas (brown coal)	643/700	2318/3710	—	4.8	31.0	—	—	—

*For combustion with air at standard temperature and pressure. These flame temperatures are calculated for 100 percent theoretical air, disassociation considered. Data from *Gas Engineers Handbook*, Industrial Press, New York, 1965.

†Flame temperatures are theoretical—calculated for stoichiometric ratio, dissociation considered.

however, these fixed nitrogen species may be converted to molecular nitrogen, thus avoiding the NO formation path.

Prompt NO_x refers to a mechanism that can lead to small quantities of NO_x. Hydrocarbon fragments (such as C, CH, CH_2) may react with atmospheric nitrogen under fuel-rich conditions to yield fixed nitrogen species such as NH, HCN, H_2CN, and CN. These, in turn, can be oxidized to NO in the lean zone of the flame. In most flames, especially those from nitrogen-containing fuels, the prompt mechanism is responsible for only a small fraction of the total NO_x. Its control is important only when attempting to reach the lowest possible emissions.

NO_x emission control is preferable to minimize NO_x formation through control of the mixing, combustion, and heat-transfer processes rather than through postcombustion techniques such as selective catalytic reduction. Four techniques for doing so, illustrated in Fig. 24-20, are air staging, fuel staging, flue-gas recirculation, and lean premixing.

Air staging in the introduction of combustion air can control NO_x emissions from all fuel types. The combustion air stream is split to create a fuel-rich primary zone and a fuel-lean secondary zone. The rich primary zone converts fuel-bound nitrogen to molecular nitrogen and suppresses thermal NO_x. Heat is removed prior to the addition of the secondary combustion air. The resulting lower flame temperatures [below 1810 K (2800°F)] under lean conditions reduce the rate of formation of thermal NO_x. This technique has been widely applied to furnaces and boilers, and it is the preferred approach for burning liquid and solid fuels. Staged-air burners are typically capable of reducing NO_x emissions by 30 to 60 percent, relative to uncontrolled levels. Air staging can also be accomplished by the use of overfire air systems in boilers.

Fuel staging in the introduction of fuel is an effective approach for controlling NO_x emissions when burning gaseous fuels. The first combustion stage is very lean, resulting in low thermal and prompt NO_x. Heat is removed

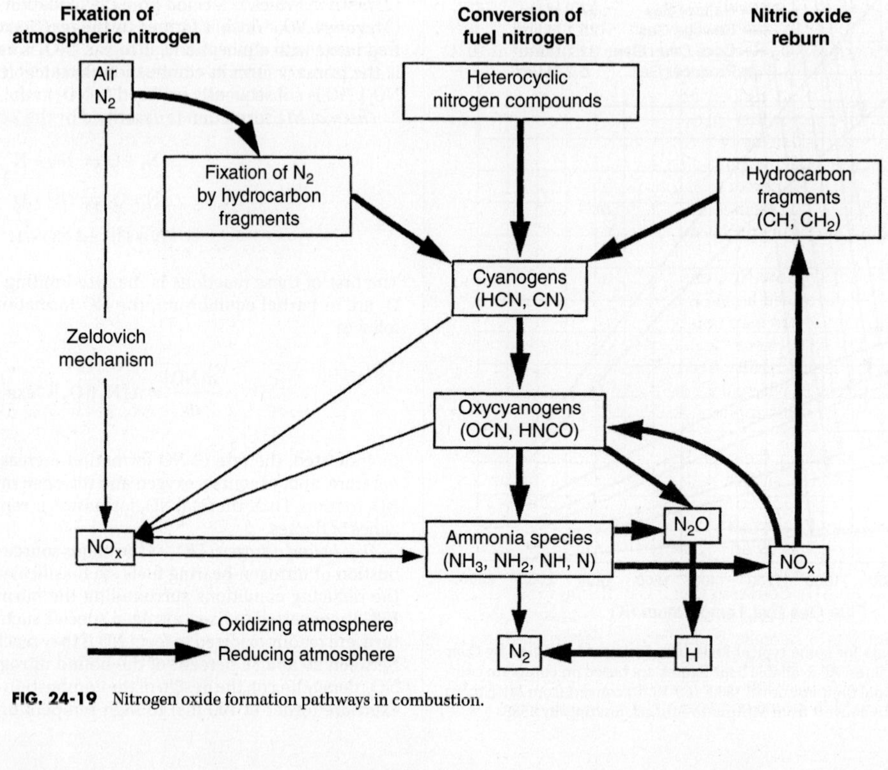

FIG. 24-19 Nitrogen oxide formation pathways in combustion.

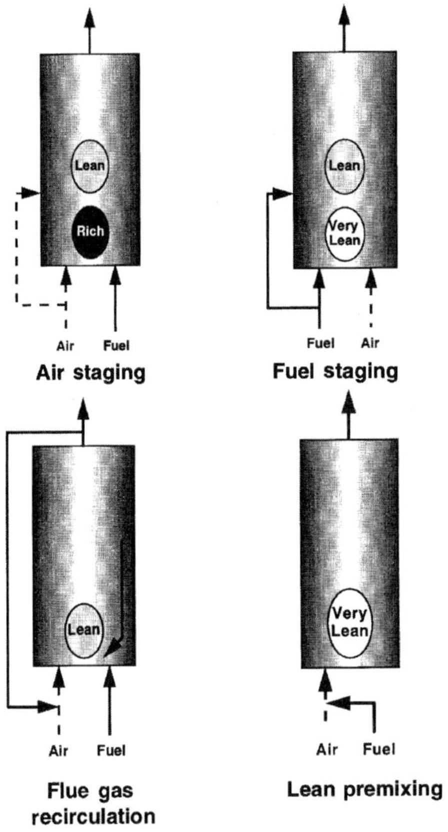

FIG. 24-20 Combustion modifications for NO_x control.

prior to the injection of the secondary fuel. The secondary fuel entrains flue gas prior to reacting, further reducing flame temperatures. In addition, NO_x reduction through reburning reactions may occur in the staged jets. This technique is the favored approach for refinery- and chemical plant–fired heaters using gaseous fuels. Staged-fuel burners are typically capable of reducing NO_x emissions by 40 to 70 percent, relative to uncontrolled levels.

Flue gas recirculation, alone or in combination with other modifications, can significantly reduce thermal NO_x. Recirculated flue gas is a diluent that reduces flame temperatures. External and internal recirculation paths have been applied: internal recirculation can be accomplished by jet entrainment using either combustion air or fuel jet energy; external recirculation requires a fan or a jet pump (driven by the combustion air). When combined with staged-air or staged-fuel methods, NO_x emissions from gas-fired burners can be reduced by 50 to 90 percent. In some applications, external flue-gas recirculation can decrease thermal efficiency. Condensation in the recirculation loop can cause operating problems and increase maintenance requirements.

Lean premixing to obtain very low NO_x emissions can be achieved by premixing gaseous fuels (or vaporized liquid fuels) with air and reacting at high excess air. The uniform and very lean conditions in such systems favor very low thermal and prompt NO_x. However, achieving such low emissions requires operating near the lean stability limit. This is an attractive NO_x control approach for gas turbines, where operation at high excess air does not incur an efficiency penalty. In this application, NO_x emissions have been reduced by 75 to 95 percent.

Sulfur Oxides Sulfur occurs in fuels as inorganic minerals (primarily pyrite, FeS_2), organic structures, sulfate salts, and elemental sulfur. Sulfur contents range from parts per million in pipeline natural gas, to a few tenths of a percent in diesel and light fuel oils, to 0.5 to 5 percent in heavy fuel oils and coals. Sulfur compounds are pyrolized during the volatilization phase of oil and coal combustion and react in the gas phase to form predominantly SO_2 and some SO_3. Conversion of fuel sulfur to these oxides is generally high (85 to 90 percent) and is relatively independent of combustion conditions. From 1 to 4 percent of the SO_2 is further oxidized to SO_3, which is highly reactive and extremely hygroscopic. It combines with water to form sulfuric acid aerosol, which can increase the visibility of stack plumes. It also elevates the dew point of water so that, to avoid back-end condensation and resulting corrosion, the flue-gas discharge temperature must be raised to about 420 K (300°F), reducing heat recovery and thermal efficiency. This reaction is enhanced by the presence of fine particles, which serve as condensation nuclei. Some coals may contain ash with substantial alkali content. During the combustion of these fuels, the alkali may react to form condensed phase compounds (such as sulfates), thereby reducing the amount of sulfur emitted as oxides. Reductions in SO_2 emissions may be achieved either by removing sulfur from the fuel before and/or during combustion, or by post-combustion flue-gas desulfurization (wet scrubbing using limestone slurry, for example).

Combustion-Related Particulate Emissions These may consist of one or more of the following types, depending on the fuel.

Mineral matter derived from ash constituents of liquid and solid fuels can vaporize and condense as sub-micron-size aerosols. Larger mineral matter fragments are formed from mineral inclusions that melt and resolidify downstream.

Sulfate particles formed in the gas phase can condense. In addition, sulfate can become bound to metals and can be adsorbed on unburned carbon particles.

Unburned carbon includes unburned char, coke, cenospheres, and soot.

Particles of char are produced as a normal intermediate product in the combustion of solid fuels. Following initial particle heating and devolatilization, the remaining solid particle is termed *char.* Char oxidation requires considerably longer periods (ranging from 30 ms to over 1 s, depending on particle size and temperature) than the other phases of solid fuel combustion. The fraction of char remaining after the combustion zone depends on the combustion conditions as well as the char reactivity.

Cenospheres are formed during heavy oil combustion. In the early stages of combustion, the oil particle is rapidly heated and evolves volatile species, which react in the gas phase. Toward the end of the volatile-loss phase, the generation of gas declines rapidly, and the droplet (at this point, a highly viscous mass) solidifies into a porous coke particle known as a *cenosphere.* This is called *initial coke.* For the heaviest oils, the initial coke particle diameter may be 20 percent larger than the initial droplet diameter. For lighter residual oils, it may be only one-third of the original droplet diameter. After a short interval, the initial coke undergoes contraction to form *final coke.* Final coke diameter is about 80 percent of the initial droplet diameter for the heaviest oils. At this time the temperature of the particle is approximately 1070 to 1270 K (1470°F to 1830°F). Following coke formation, the coke particles burn out in the lean zone, but the heterogeneous oxidation proceeds slowly. Final unburned carbon levels depend on a balance between the amount of coke formed and the fraction burned out. Coke formation tends to correlate with fuel properties such as asphaltene content, C:H ratio, or Conradson carbon residue. Coke burnout depends on combustion conditions and coke reactivity. Coke reactivity is influenced by the presence of combustion catalysts (e.g., vanadium) in the cenospheres.

Formation of soot is a gas-phase phenomenon that occurs in hot, fuel-rich zones. Soot occurs as fine particles (0.02 to 0.2 μm), often agglomerated into filaments or chains that can be several millimeters long. Factors that increase soot formation rates include high C:H ratio, high temperature, very rich conditions, and long residence times at these conditions. Pyrolysis of fuel molecules leads to soot precursors such as acetylene and higher analogs and various polyaromatic hydrocarbons. These condense to form very small (<2 nm) particles. The bulk of solid-phase material is generated by *surface growth*—the attachment of gas-phase species to the surface of the particles and their incorporation into the particulate phase. Another growth mechanism is *coagulation,* in which particles collide and coalesce. Soot particle formation and growth are typically followed by soot oxidation to form CO and CO_2. Eventual soot emission from a flame depends on the relative balance between the soot-formation and oxidation reactions.

Carbon Monoxide This is a key intermediate in the oxidation of all hydrocarbons. In a well-adjusted combustion system, essentially all the CO is oxidized to CO_2, and the final emission of CO is very low indeed (a few parts per million). However, in systems that have low temperature zones (for example, where a flame impinges on a wall or a furnace load) or are in poor adjustment (for example, an individual burner fuel/air ratio out of balance in a multiburner installation or a misdirected fuel jet that allows fuel to bypass the main flame), CO emissions can be significant. The primary method of CO control is good combustion system design and practice.

Unburned Hydrocarbon Species These may be emitted from hydrocarbon flames. In general, there are two classes of unburned hydrocarbons: (1) small molecules that are the intermediate products of combustion (for example, formaldehyde) and (2) larger molecules that are formed by pyrosynthesis in hot, fuel-rich zones within flames, e.g., benzene, toluene, xylene, and various polycyclic aromatic hydrocarbons (PAHs). Many of these species are listed as hazardous air pollutants (HAPs) in Title III of the Clean Air Act Amendments of 1990 and are therefore of particular concern. In a well-adjusted combustion system, the emission of HAPs is extremely low (typically, parts per trillion to parts per billion). However, the emission of certain HAPs may be of concern in poorly designed or maladjusted systems.

COMBUSTION OF SOLID FUELS

There are three basic modes of burning solid fuels, each identified with a furnace design specific for that mode: in suspension, in a bed at rest* on a grate (fuel-bed firing), or in a fluidized bed. Although many variations of these generic modes and furnace designs have been devised, the fundamental characteristics of equipment and procedure remain intact. They will be described briefly.

Suspension Firing Suspension firing of pulverized coal (PC) is commoner than fuel-bed or fluidized-bed firing of coarse coal in the United States. This mode of firing affords higher steam-generation capacity, is independent of the caking characteristics of the coal, and responds quickly to load changes. Pulverized coal firing accounts for approximately 55 percent of the power generated by electric utilities in the United States. It is rarely used on boilers of less than 45.4 Mg/h (100,000 lb/h) steam capacity because its economic advantage decreases with size.

A simplified model of PC combustion includes the following sequence of events: (1) on entering the furnace, a PC particle is heated rapidly, driving off the volatile components and leaving a char particle; (2) the volatile components burn independently of the coal particle; and (3) on completion of volatiles combustion, the remaining char particle burns. While this simple sequence may be generally correct, PC combustion is an extremely complex process involving many interrelated physical and chemical processes.

Devolatilization The volatiles produced during rapid heating of coal can include H_2, CH_4, CO, CO_2, and C_2-C_4 hydrocarbons, as well as tars, other organic compounds, and reduced sulfur and nitrogen species. The yield of these various fractions is a function of both heating rate and final particle temperature. The resulting char particle may be larger in diameter than the parent coal particle, owing to swelling produced by volatiles ejection. The particle density also decreases.

Char oxidation dominates the time required for complete burnout of a coal particle. The heterogeneous reactions responsible for char oxidation are much slower than the devolatilization process and gas-phase reaction of the volatiles. Char burnout may require from 30 ms to over 1 s, depending on combustion conditions (oxygen level, temperature), and char particle size and reactivity. Char reactivity depends on parent coal type. The rate-limiting step in char burnout can be chemical reaction or gaseous diffusion. At low temperatures or for very large particles, chemical reaction is the rate-limiting step. At higher temperatures, boundary-layer diffusion of reactants and products is the rate-limiting step.

Pulverized-Coal Furnaces In designing and sizing PC furnaces, particular attention must be given to the following fuel-ash properties:

- Ash fusion temperatures, including the spread between initial deformation temperature and fluid temperature
- Ratio of basic (calcium, sodium, potassium) to acidic (iron, silicon, aluminum) ash constituents, and specifically iron-to-calcium ratio
- Ash content
- Ash friability

These characteristics influence furnace plan area, furnace volume, and burning zone size required to maintain steam production capacity for a given fuel grade or quality.

Coal properties influence pulverizer capacity and the sizing of the air heater and other heat-recovery sections of a steam generator. Furnace size and heat-release rates are designed to control slagging characteristics. Consequently, heat-release rates in terms of the ratio of net heat input to plan area range from 4.4 MW/m^2 [1.4×10^6 (Btu/h·ft^2)] for severely slagging coals to 6.6 MW/m^2 [2.1×10^6 Btu/(h·ft^2)] for low-slagging fuels.

The various burner and furnace configurations for PC firing are shown schematically in Fig. 24-21. The U-shaped flame, designated as fantail vertical firing (Fig. 24-21*a*), was developed initially for pulverized coal before the advent of water-cooled furnace walls. Because a large percentage of the total combustion air is withheld from the fuel stream until it projects well down into the furnace, this type of firing is well suited for solid fuels that are difficult to ignite, such as those with less than 15 percent volatile matter. Although this configuration is no longer used in central-station power plants, it may find favor again if low-volatile chars from coal-conversion processes are used for steam generation or process heating.

Modern central stations use the other burner-furnace configurations shown in Fig. 24-21, in which the coal and air are mixed rapidly in and close to the burner. The primary air, used to transport the pulverized coal to the burner, comprises 10 to 20 percent of the total combustion air. The secondary air comprises the remainder of the total air and mixes in or near the burner with the primary air and coal. The velocity of the mixture leaving the burner must be high enough to prevent flashback in the primary air-coal piping. In practice, this velocity is maintained at about 31 m/s (100 ft/s).

In tangential firing (Fig. 24-21*b*), the burners are arranged in vertical banks at each corner of a square (or nearly square) furnace and directed toward an imaginary circle in the center of the furnace. This results in the formation of a large vortex with its axis on the vertical centerline. The burners consist of an arrangement of slots one above the other, admitting, through alternate slots, primary air/fuel mixture and secondary air. It is possible to tilt the burners upward or downward, the maximum inclination to the horizontal being 30°, enabling the operator to selectively use in-furnace heat-absorbing surfaces, especially the superheater.

The circular burner shown in Fig. 24-22 is widely used in horizontally fired furnaces and is capable of firing coal, oil, or gas in capacities as high as 174 GJ/h (1.65×10^8 Btu/h). In such burners the air is often swirled to create a zone of reverse flow immediately downstream of the burner centerline, which provides for combustion stability.

Low-NO_x burners are designed to delay and control the mixing of coal and air in the main combustion zone. A typical low-NO_x air-staged burner is illustrated in Fig. 24-23. This combustion approach can reduce NO_x emissions from coal burning by 40 to 50 percent. Because of the reduced flame temperature and delayed mixing in a low-NO_x burner, unburned carbon emissions may increase in some applications and for some coals. Overfire air is another technique for staging the combustion air to control NO_x emissions when burning coal in suspension-firing systems. Overfire air ports are installed above the top level of burners on wall- and tangential-fired boilers. The use of overfire air can reduce NO_x emissions by 20 to 30 percent. Reburn is an NO_x control strategy that involves diverting a portion of the fuel from the burners to a second combustion zone (reburn zone) above the main burners. Completion air is added above the reburn zone to complete fuel burnout. The reburn fuel can be natural gas, oil, or pulverized coal, though natural gas is used in most applications. In this approach, the stoichiometry in the reburn zone is controlled to be slightly rich (equivalence ratio of ~1.15), under which conditions a portion (50 to 60 percent) of the NO_x is converted to molecular nitrogen.

The *pulverizer* is the heart of any solid-fuel suspension-firing system. Air is used to dry the coal, transport it through the pulverizer, classify it, and transport it to the burner, where the transport air provides part of the air for combustion. The pulverizers themselves are classified according to whether they are under positive or negative pressure and whether they operate at slow, medium, or high speed.

Pulverization occurs by impact, attrition, or crushing. The capacity of a pulverizer depends on the grindability of the coal and the fineness desired, as shown by Fig. 24-24. Capacity can also be seriously reduced by excessive moisture in the coal, but it can be restored by increasing the temperature of the primary air. Figure 24-25 indicates the temperatures needed. For PC boilers, the coal size usually is 65 to 80 percent through a 200-mesh screen, which is equivalent to 74 μm.

Cyclone Furnaces In *cyclone firing* (Fig. 24-21*d*) the coal is not pulverized but is crushed to 4-mesh (4.76-mm) size and admitted tangentially with primary air to a horizontal cylindrical chamber, called a *cyclone furnace,* which is connected peripherally to a boiler furnace. Secondary air also is admitted, so that almost all of the coal burns within the chamber. The combustion gas then flows into the boiler furnace. In the cyclone furnace, finer coal particles burn in suspension, and the coarser ones are thrown centrifugally to the chamber wall, where most of them are captured in a sticky wall coating of molten slag. The secondary air, admitted tangentially along the top of the cyclone furnace, sweeps the slag-captured particles and completes their combustion. A typical firing rate is about 18.6 GJ/(h·m^3) [500,000 Btu/(h·ft^3)]. The slag drains continuously into the boiler furnace and thence into a quenching tank. Figure 24-26 shows a cyclone furnace schematically.

Fuel-Bed Firing Fuel-bed firing is accomplished with mechanical stokers, which are designed to achieve continuous or intermittent fuel feed, fuel ignition, proper distribution of the combustion air, free release of the gaseous combustion products, and continuous or intermittent disposal of the unburned residue. These aims are met with two classes of stokers, distinguished by the direction of fuel feed to the bed: underfeed and overfeed. Overfeed stokers are represented by two types, distinguished by the relative directions of fuel and air flow (and also by the manner of fuel feed): crossfeed, also termed mass-burning, and spreader. The principles of these three methods of fuel-bed firing are illustrated schematically in Fig. 24-27.

Underfeed Firing Both fuel and air have the same relative direction in the underfeed stoker, which is built in single-retort and multiple-retort designs. In the *single-retort,* side-dump stoker, a ram pushes coal into the retort toward the end of the stoker and upward toward the tuyere blocks, where air is admitted to the bed. This type of stoker will handle most bituminous coals and anthracite, preferably in the size range 19 to 50 mm

*The burning fuel bed may be moved slowly through the furnace by the vibrating action of the grate or by being carried on a traveling grate.

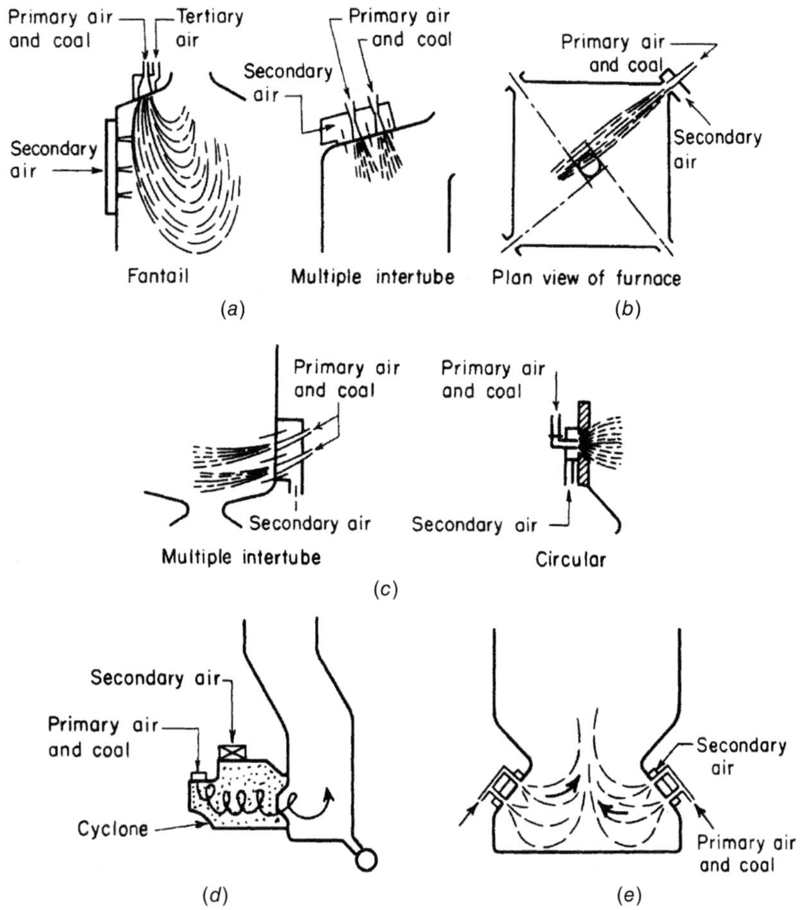

FIG. 24-21 Burner and furnace configurations for pulverized-coal firing: (*a*) vertical firing; (*b*) tangential firing; (*c*) horizontal firing; (*d*) cyclone firing; (*e*) opposed-inclined firing.

(¾ to 2 in) and no more than 50 percent through a 6-mm (¼-in) screen. Overfire air or steam jets are often used in the bridgewall at the end of the stoker to promote turbulence.

In the *multiple-retort stoker,* rams feed coal to the top of sloping grates between banks of tuyeres. Auxiliary small sloping rams perform the same function as the pusher rods in the single retort. Air is admitted along the top of the banks of tuyeres, and on the largest units the tuyeres themselves are given a slight reciprocating action to agitate the bed further. This type of stoker operates best with caking coals having a relatively high ash-softening temperature. Coal sizing is up to 50 mm (2 in) with 30 to 50 percent through a 6-mm (¼-in) screen.

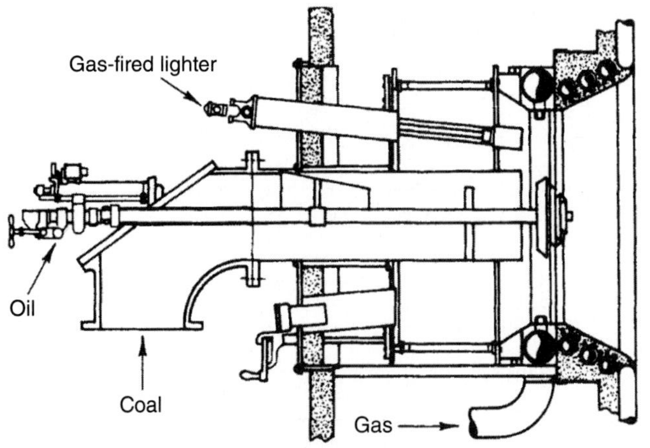

FIG. 24-22 Circular burner for pulverized coal, oil, or gas. (From *Marks' Standard Handbook for Mechanical Engineers,* 8th ed., McGraw-Hill, New York, 1978.)

Overfeed Firing: Crossfeed (Mass-Burning) Stokers Crossfeed stokers are also termed mass-burning stokers because the fuel is dumped by gravity from a hopper onto one end of a moving grate, which carries it into the furnace and down its length. Because of this feature, crossfeed stokers are commonly called *traveling-grate* stokers. The grate may be either of two designs: *bar grate* or *chain grate.* Alternatively, the burning fuel bed may be conveyed by a vibratory motion of the stoker (*vibrating-grate*).

The fuel flows at right angles to the airflow. Only a small amount of air is fed at the front of the stoker, to keep the fuel mixture rich, but as the coal moves toward the middle of the furnace, the amount of air is increased, and most of the coal is burned by the time it gets halfway down the length of the grate. Fuel-bed depth varies from 100 to 200 mm (4 to 8 in), depending on the fuel, which can be coke breeze, anthracite, or any noncaking bituminous coal.

Overfeed Firing: Spreader Stokers Spreader stokers burn coal (or other fuel) by propelling it into the furnace. A portion of the coal burns in suspension (the percentage depending on the coal fineness), while the rest burns on a grate. In most units, coal is pushed off a plate under the storage hopper onto revolving paddles (either overthrow or underthrow) which distribute the coal on the grate (Fig. 24-27*c*). The angle and speed of the paddles control coal distribution. The largest coal particles travel the farthest, while the smallest ones become partially consumed during their trajectory and fall on the forward half of the grate. The grate may be stationary or traveling. The fuel and air flow in opposite directions.

Some spreaders use air to transport the coal to the furnace and distribute it, while others use mechanical means to transport the coal to a series of pneumatic jets.

The performance of spreader stokers is affected by changes in coal sizing. The equipment can distribute a wide range of fuel sizes, but it distributes each particle on the basis of size and weight. Normal size specifications call for 19-mm (¾-in) nut and slack with not more than 30 percent less than 6.4 mm (¼ in).

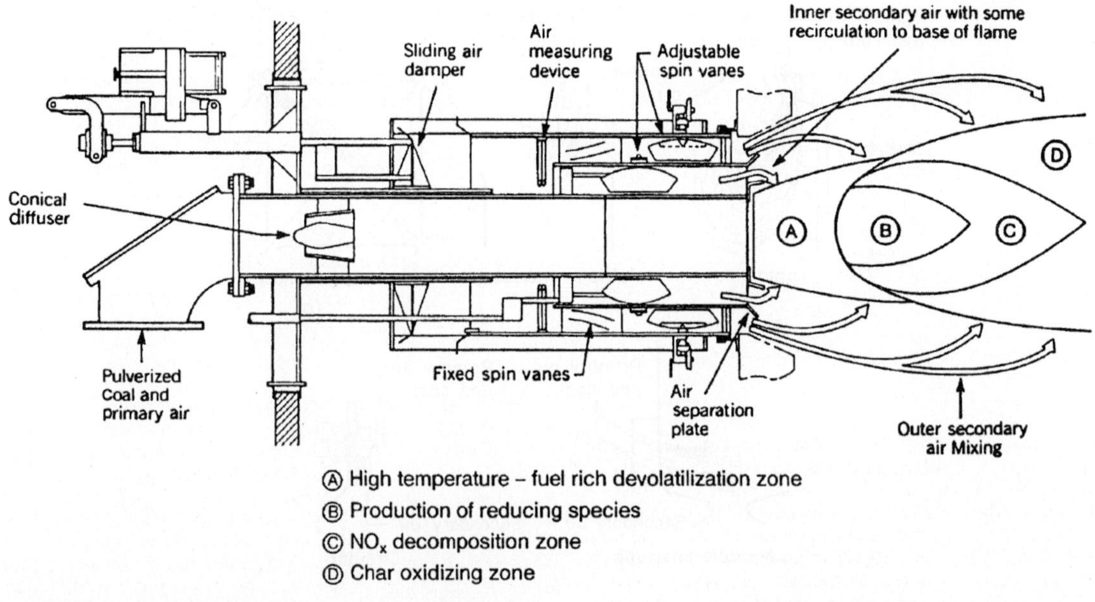

FIG. 24-23 Low-NO_x pulverized coal burner. (Babcock & Wilcox Co.)

Typically, approximately 30 to 50 percent of the coal is burned in suspension. If excessive fines are present, more coal particles will be carried out of the furnace and burned in suspension, and very little ash will be available to provide a protective cover for the grate surface. On the other hand, if sufficient fines are not present, not all the fuel will be burned on the grate, resulting in derating of the unit and excessive dumping of live coals to the ash hopper.

Excess air is usually 30 to 40 percent for stationary and dumping grates, while traveling grates are operated with from 22 to 30 percent excess air. Preheated air can be supplied for all types of grates, but the temperature is usually limited to 395 to 422 K (250°F to 300°F) to prevent excessive slagging of the fuel bed.

Overfire air nozzles are located in the front wall underneath the spreaders and in the rear wall from 0.3 to 0.9 m (1 to 3 ft) above the grate level. These nozzles use air directly from a fan or inspirate air with steam to provide turbulence above the grate for most effective mixing of fuel and air. They supply about 15 percent of the total combustion air.

Comparison of Suspension and Fuel-Bed Firing A major factor to consider when comparing a stoker-fired boiler with a PC boiler is the reduction in efficiency due to carbon loss. The carbon content of the ash passing out of a spreader stoker furnace varies from 30 to 50 percent. Overall efficiency of the stoker can be increased by reburning the ash: it is returned to the stoker grate by gravity or a pneumatic feed system. A continuous-ash-discharge spreader-stoker-fired unit will typically have a carbon loss of 4 to 8 percent, depending on the amount of ash reinjection. A properly designed PC boiler, on the other hand, can maintain an efficiency loss due to unburned carbon of less than 0.4 percent.

A difference between these firing methods may also be manifested in the initial fuel cost. For efficient operation of a spreader-stoker-fired boiler, the

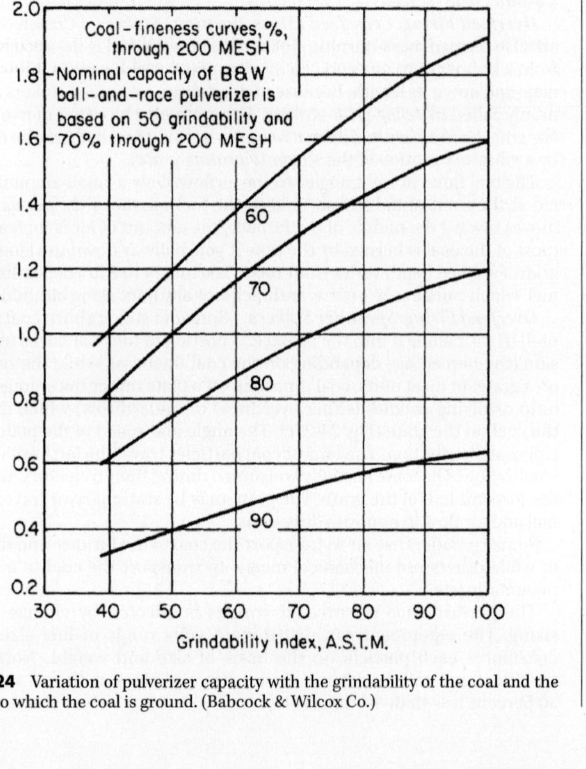

FIG. 24-24 Variation of pulverizer capacity with the grindability of the coal and the fineness to which the coal is ground. (Babcock & Wilcox Co.)

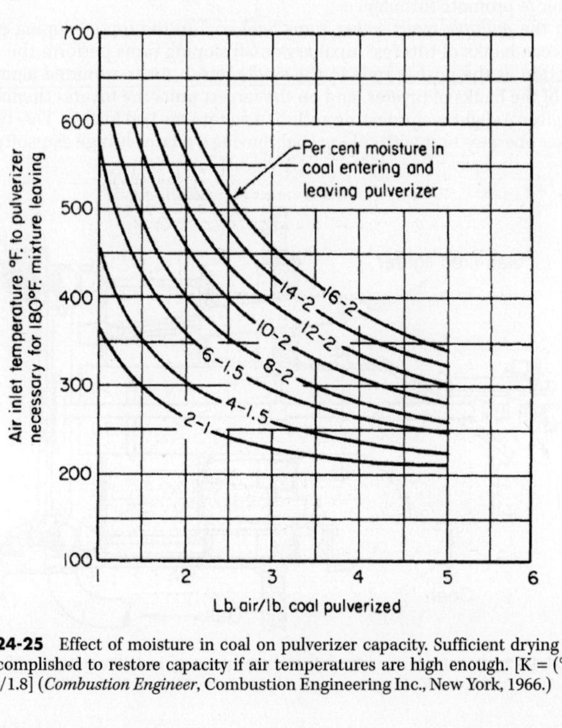

FIG. 24-25 Effect of moisture in coal on pulverizer capacity. Sufficient drying can be accomplished to restore capacity if air temperatures are high enough. [K = (°F + 459.7)/1.8] (*Combustion Engineer*, Combustion Engineering Inc., New York, 1966.)

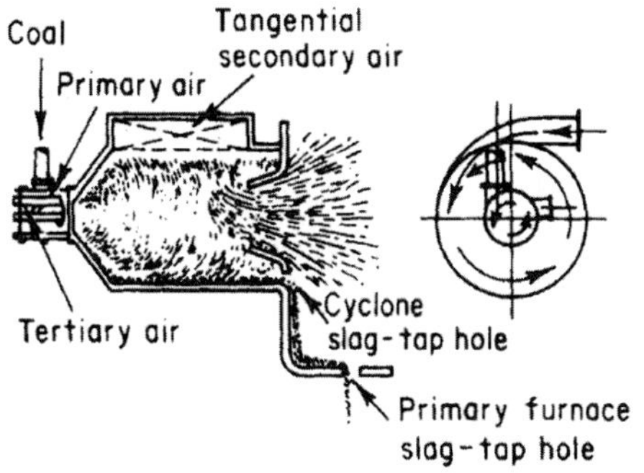

FIG. 24-26 Cyclone furnace. (From *Marks' Standard Handbook for Mechanical Engineers*, 9th ed., McGraw-Hill, New York, 1987.)

coal must consist of a proper mixture of coarse and fine particles. Normally, double-screened coal is purchased because less expensive run-of-mine coal does not provide the optimum balance of coarse and fine material.

An advantage of a stoker-fired furnace is its easy adaptability to firing almost any unsized solid fuels. Bark, bagasse, or refuse can normally be fired on a stoker to supplement the coal with a minimum amount of additional equipment. Thus, such supplementary waste fuels may be able to contribute a higher percentage of the total heat input in a stoker-fired furnace than in a PC furnace without expensive equipment modifications.

Fluidized-Bed Combustion The principles of gas–solid fluidization and their application to the chemical process industry are treated in Sec. 17. Their general application to combustion is reviewed briefly here, and their more specific application to fluidized-bed boilers is discussed later in this section.

In fluidized-bed combustion (FBC), fuel is burned in a bed of particles supported in an agitated state by an upward flow of air introduced by an air distributor. The bed particles may be sand or ash derived from the fuel, or spent sorbent (limestone or dolomite) if in-bed sulfur capture is included

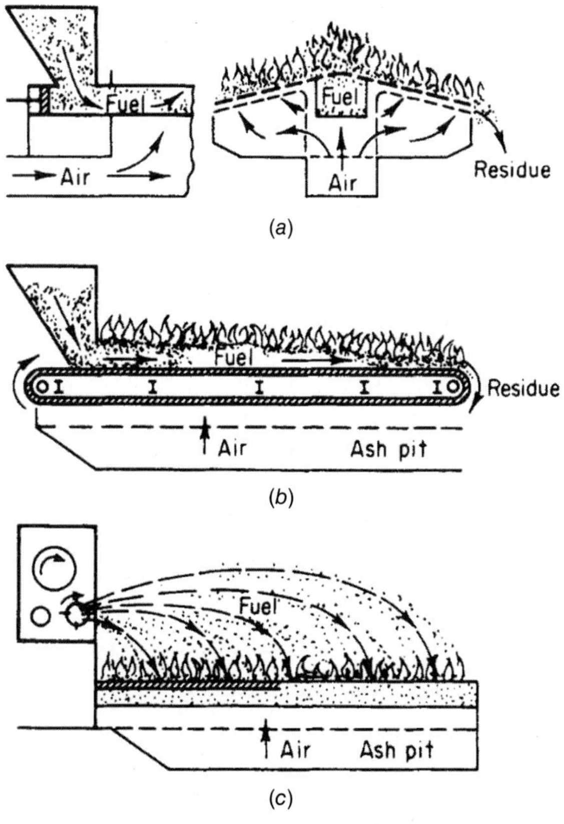

FIG. 24-27 Basic types of mechanical stokers: (*a*) underfeed; (*b*) crossfeed; (*c*) overfeed (spreader stoker).

in the design. The fluidizing action promotes good solids mixing and gas–solids contacting that allow high combustion efficiency to be achieved at temperatures significantly lower than those of a pulverized coal (PC) furnace [typically 1116 K (1550°F) compared to over 1589 K (2400°F)]. These lower temperatures also result in reduced slagging and fouling problems and significantly lower NO_x formation. This latter benefit, in conjunction with in-bed sulfur capture capability, constitutes one of the great advantages of fluidized-bed combustion: in situ pollution control. Having this control built into the furnace reduces the extent of back-end cleanup required for a given emissions standard.

There are two types of FBC units being sold commercially, *bubbling* and *circulating*, both operating at atmospheric pressure. Typical superficial fluidizing velocities are as follows.

Bubbling FBC	1.5 to 2.7 m/s (5 to 9 ft/s)
Circulating FBC	3.7 to 7.3 m/s (12 to 24 ft/s)

Pressurized FBC (PFBC) designs, bubbling and circulating, have been under development since the 1970s, but all work has now been curtailed. Work on an advanced PFBC design, incorporating partial coal gasification and char combustion, has also been curtailed. Several commercial bubbling PFBC units were built in the 1990s, covering the range 80 to 340 MW_e. However, no new PFBC plants have been announced since 1996, and consequently the technology is dropped from this edition of the handbook.

Bubbling Beds A large proportion of the noncombustible feedstock—sand, ash, or sorbent-derived material—remains in the combustor forming the bed. Bed depth is maintained by draining off excess material. Most of the gas in excess of that required for minimum fluidization appears as bubbles (voids), and these carry particles upward in their wake, resulting in the rapid vertical mixing of the bed material that promotes even temperatures and good gas–solids contacting.

Early bubbling FBC units were designed to burn coal, and the heat released was removed by heat transfer to in-bed tubes and/or to the water-wall tubes used to enclose the furnace. These surfaces experienced high rates of metal loss through the combined effects of erosion and abrasion. Protective measures such as plasma-sprayed coatings and metal fins to disrupt the solids flow pattern were used. These were effective for only short periods before requiring replacement, and so maintenance requirements were high.

The two largest bubbling-bed FBC units built, 160 and 350 MW_e, entered into service in the late 1980s burning bituminous coal. However, for coal applications, the overall performance of circulating fluidized-bed units has proved superior, and bubbling-bed designs of this size are no longer marketed. The bubbling technology is now mainly used for burning high-moisture biomass fuels for which in-bed heat-transfer surfaces are not required. These units are usually designed for generating capacities below 120 MW_e, and typical fuels include wood wastes, paper mill and sewage sludges, and peat.

The boilers are of water-wall construction, with the walls in the bed region lined with refractory to protect against erosion and abrasion damage. The heat released in the bed is removed from the flue gas by the exposed water walls and by convection pass tubing set in the gas path after the boiler. The coal boilers included multiclones to collect elutriated carbon and recycle it back to the boiler to increase combustion efficiency. However, because biomass is more reactive, recycle is not required.

To accommodate the wide variations in fuel moisture content while maintaining close bed temperature control, two control strategies are used.

- Operate the bed substoichiometrically so that bed temperature varies with the fluidizing air (primary combustion air). There is an equal and opposite change in the air fed above the bed (secondary combustion air) to complete the combustion process, the total air remaining constant. As the fuel moisture increases, the primary airflow rate increases, increasing the fluidizing velocity and releasing more heat in the bed to evaporate the additional water and keep the bed temperature constant. The bed operates less substoichiometrically, and in-bed combustion efficiency increases. At the same time the secondary air is reduced, and as there is less above-bed combustion, the upper furnace temperature falls. The control action is reversed when the fuel moisture content decreases. The split of primary to secondary airflow ranges from above 80:20 for high-moisture fuels to below 50:50 for low-moisture fuels.
- The preceding strategy results in variations in fluidizing velocity that might produce poor fluidization at some operating conditions with a resulting reduction in combustion efficiency. To maintain fluidizing velocity constant, recirculated flue gas is mixed with the primary combustion air to lower its oxygen content while keeping the combined fluidization gas flow rate constant. Air may be fed above the bed to complete combustion, but this is controlled independently of the air passing to the bed. As the fuel moisture increases, less flue gas is recirculated, the oxygen content

of the primary combustion air and in-bed combustion both increase, and secondary air decreases. The control action is reversed when the fuel moisture content decreases.

Circulating Beds These fluidized beds operate at higher velocities, and virtually all the solids are elutriated from the furnace. The majority of the elutriated solids, still at combustion temperature, are captured by reverse-flow cyclone(s) and recirculated to the foot of the combustor. The foot of the combustor is a potentially very erosive region because it contains large particles not elutriated from the bed, and they are being fluidized at high velocity. Consequently, the lower reaches of the combustor do not contain heat-transfer tubes, and the water walls are protected with refractory. Some combustors have experienced damage at the interface between the water walls and the refractory, and measures similar to those employed in bubbling beds have been used to protect the tubes in this region.

The circulating-bed design is the leading FBC technology and is the only design currently offered at sizes above 120 MW_e. Units as large as 320 MW_e have been built incorporating subcritical steam conditions with reheat, and supercritical designs of up to 600 MW_e are available. The fuels used in these designs are coals or opportunity fuels such as petroleum coke, although one 240-MW_e unit is designed to operate solely on biomass. Circulating FBC boilers as small as 25 MW_e have been built to fire either coal or biomass. However, as the size gets smaller, the circulating FBC becomes less competitive for biomass firing, and as indicated earlier, the bubbling FBC design is favored for this application.

The heat released in the furnace is removed from the flue gas by the exposed upper water-wall tubes. As the units increase in size, more heat-transfer surface is required than is provided by the walls. Surface can be added by wrapping horizontal tubing over the walls of the upper furnace, or by added wing walls, sections of water wall extending short distances into the furnace enclosure. In some designs, tubes are extended across the upper furnace where, although the fluidizing velocity is still high, the erosion potential is low because the solids are finer and their concentration is lower. In some designs, heat is removed from the recirculated solids by passing them through a bubbling-bed heat exchanger before returning them to the furnace. The potential for erosion or abrasion of the tubes in this bed is low because the recirculating solids are mostly less than 500 μm and are well fluidized at velocities of only around 0.6 m/s (2 ft/s).

The combustion air is introduced at two levels, 60 to 70 percent being introduced through the distributor and 30 to 40 percent above the bed. This staged entry results in the lower reaches operating substoichiometrically, which helps to reduce NO_x emissions but tends to reduce the fluidizing velocity at the base of the combustor. To compensate for this and to increase solids mixing by increasing the local gas velocity, the portion of the combustor below the secondary air entry points is tapered.

Fuel Flexibility An advantage of FBC designs is fuel flexibility: A single unit can burn a wider range of fuels than a PC furnace, thus offering owners an improved bargaining position to negotiate lower fuel prices. Among the fuels fired are bituminous and subbituminous coals, anthracite culm, lignite, petroleum coke, refuse-derived fuel, biomass, industrial and sewage sludges, and shredded tires. But fuel flexibility can be achieved only if the unit is designed for the range of fuels intended to be burned. For example, to maintain the same firing rate, a feed system designed for a certain fuel must be capable of feeding a lower calorific fuel at a higher rate. Similarly, to maintain the same degree of sulfur capture, feeders must be capable of delivering sorbent over a range of rates matching the sulfur contents of the fuels likely to be fed. This increase in operating flexibility increases the capital cost, offsetting at least partially the economic benefits of reduced fuel pricing.

Sulfur Emissions Sulfur present in a fuel is released as SO_2, a known contributor to acid rain deposition. By adding limestone or dolomite to a fluidized bed, much of this can be captured as calcium sulfate, a dry nonhazardous solid. As limestone usually contains over 40 percent calcium, compared to only 20 percent in dolomite, it is the preferred sorbent, resulting in lower transportation costs for the raw mineral and the resulting ash product. Moreover, the high magnesium content of the dolomite makes the ash unsuitable for some building applications and so reduces its potential for utilization. Whatever sorbent is selected, for economic reasons it is usually from a source local to the FBC plant. If more than one sorbent is available, plant trials are needed to determine the one most suitable, as results from laboratory-scale reactivity assessments are unreliable.

At atmospheric pressure, calcium carbonate almost completely calcines to free lime, and it is this that captures the sulfur dioxide. As the free lime is not completely sulfated, the resulting sorbent ash is very alkaline, consisting primarily of $CaSO_4$ and CaO, with small amounts of $CaCO_3$.

$$CaCO_3 \rightleftharpoons CaO + CO_2 \tag{24-56}$$

$$CaO + SO_2 + [O] \rightleftharpoons CaSO_4 \tag{24-57}$$

The sulfation reaction has an optimum at a mean bed temperature of around 1116 K (1550°F). Units are usually designed to operate at this temperature, but changes in operating conditions to achieve power output and accommodate changes in fuel composition often result in the plant operating at above or below the optimum temperature. To compensate for the ensuing reduction in sulfur capture, more sorbent is fed to the bed.

Achieving very high levels of sulfur control (~98%) with only in-bed sulfur capture requires Ca/S molar ratios as high as 3 (3 mol of calcium in the sorbent for each mole of sulfur in the coal). For a 3 percent sulfur coal and a sorbent with 38 percent calcium, the sorbent-to-coal weight ratio is 0.3. To reduce the sorbent demand for the same sulfur capture efficiency, a back-end dry scrubber is used to remove additional SO_2 from the flue gas. This has been demonstrated to lower the Ca/S molar ratio to 1.9, a sorbent-to-coal weight ratio of 0.19. For a coal with 12 percent ash, this 37 percent reduction in sorbent feed rate reduces the amount of ash sent to disposal by 17 percent. This reduction in solids handling requirements improves process economics primarily by lowering the operating costs of the plant. The capital cost remains roughly the same, the cost of the dry scrubber being mainly offset by the reduction in solids handling equipment.

The sorbent used in the dry scrubber can be either lime or reactivated ash captured by the baghouse. The baghouse ash contains unreacted sorbent particles with a surface coating of calcium sulfate that prevents SO_2 reaching the CaO at their core. Hydrating these solids causes the material to swell and crack, thus exposing the unreacted CaO, which when recycled back to the boiler is available to react with additional SO_2. As lime is more expensive than limestone, using the baghouse ash is the preferred approach. It also eliminates the need to provide separate handling facilities for lime.

Nitrogen Oxide Emissions FBC units achieve excellent combustion and sulfur emission performance at relatively modest combustion temperatures in the range of 1060 to 1172 K (1450°F to 1650°F). At these temperatures, no atmospheric nitrogen is converted to NO_x. The prime variables influencing NO_x formation are excess air, mean bed temperature, nitrogen content of the fuel, and Ca/S molar ratio. With respect to the latter, high sorbent feed rates increase the free lime content, and this catalyzes NO_2 formation. Typical NO_x emissions, consisting of around 90 percent NO and 10 percent NO_2, are in the range 86 to 129 mg/MJ (0.2 to 0.3 lb/MBtu). These values have been reduced to as low as 21 mg/MJ (0.05 lb/MBtu) by injecting ammonia into the boiler freeboard to promote selective noncatalytic reduction (SNCR) reactions. This is a less costly approach than the selective catalytic reduction (SCR) units required for PC plants.

Because the operating temperature is lower, FBC units release more N_2O than do PC units. Nitrous oxide is a greenhouse gas that absorbs 270 times more heat per molecule than carbon dioxide and as such is likely to come under increased scrutiny in the future. The emissions at full load from coal-fired units are around 65 mg/MJ (0.15 lb/MBtu), but these increase as load is reduced and furnace temperature falls. Measurements from biomass-fired FBCs have not been made. Combustion processes do not contribute greatly to current U.S. N_2O emissions; agriculture and motor vehicles account for 86 percent of the total.

Particulate Emissions To meet environmental regulations, FBC boilers use a back-end particulate collector, such as a baghouse or an electrostatic precipitator (ESP). Compared to PC units, because of its sulfur sorbent content, the ash from a circulating CFB has higher resistivity and is finer because the flue gas path contains cyclones. Both factors result in a reduction in ESP collection efficiency, although lowering the gas velocity and providing additional collection fields can compensate for this, albeit at increased capital cost. Bubbling FBC designs fired on biomass do not normally feed sulfur sorbents and do not include dust collectors in the flue gas path. For these applications suitably designed ESPs are often included. In general, however, baghouses are the preferred particulate collection devices.

FBC ash is irregular, whereas PC ash, because it melts at the elevated operating temperatures, is spherical. This difference in shape influences baghouse design in three ways: (1) FBC ash does not flow from the collection hoppers as readily, and special attention has to be given to their design; (2) FBC ash forms a stronger cake, requiring more frequent and more robust cleaning mechanisms, e.g., shake-deflate and pulse-jet technologies; and (3) this more robust action in conjunction with the more abrasive, irregular particles results in filter bags being more prone to failure in FBC systems. Careful selection of bag materials (synthetic felts generally perform best) and good installation and maintenance practices minimize the latter problem.

Mercury Emissions Under the provisions of the Mercury and Toxic Air Standards, first proposed by the U.S. Environmental Protection Agency (EPA) in 2011 and upheld by the U.S. Supreme Court in 2016, power plants will be required to reduce mercury emissions by approximately 90 percent relative to 2011 levels, generally over a four-year period.

Tests on operating fluidized beds show that inherent mercury capture from the flue gas for bituminous coals is over 90 percent, but for lignites capture it is closer to 50 percent. The reasons for this difference are not

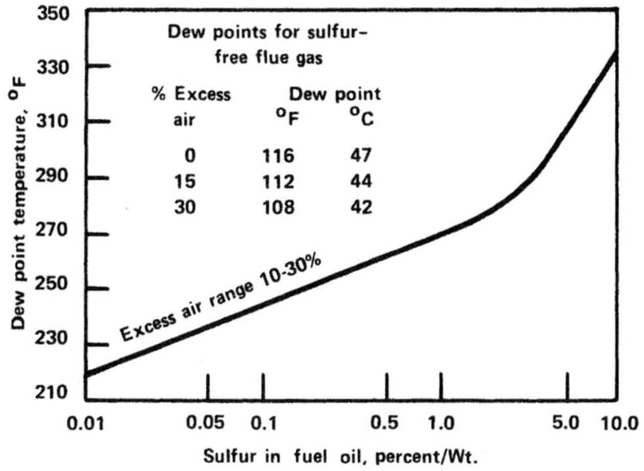

FIG. 24-28 Maximum flue-gas dew point versus percent of sulfur in typical oil fuels. (K = [°F + 459.7]/1.8.)

fully understood, but they could be explained by the higher reactivity of lignite, resulting in lower unburned carbon carryover. Carbon is known to adsorb elemental and ionic mercury, and a low carbon content will result in reduced mercury capture. The combustion efficiency for less reactive bituminous coal is lower, resulting in more carbon carryover and possibly accounting for the higher mercury capture. Recent tests have shown that halide-impregnated activated carbons injected into flue gas streams can capture over 90 percent of the mercury present. So the low inherent mercury capture associated with lignite can be compensated for by activated carbon injection ahead of the dust collection device used.

COMBUSTION OF LIQUID FUELS

Oil is typically burned as a suspension of droplets generated by atomizing the fuel. As the droplets pass from the atomizer into the flame zone, they are heated both by radiation from the flame and by convection from the hot gases that surround them, and the lighter fuel components vaporize. The vapors mix with surrounding air and ignite. Depending on the fuel type, the fuel droplet may be completely vaporized or it may be partially vaporized, leaving a residual char or coke particle.

Fuel oils can contain a significant amount of sulfur: in the case of high-sulfur No. 6, it may be as much as 4 percent. SO_2 is the principal product of sulfur combustion with stoichiometric or leaner fuel/air mixtures, but with the excess air customarily used for satisfactory combustion, SO_3 can form and then condense as sulfuric acid at temperatures higher than the normally expected dew point. Thus air preheaters and other heat recovery equipment in the flue-gas stream can be endangered. Figure 24-28 shows the maximum safe upper limits for dew points in the stacks of furnaces burning sulfur-containing oil and emitting unscrubbed flue gas.

Atomizers Atomization is the process of breaking up a continuous liquid phase into discrete droplets. Figure 24-29 shows the idealized

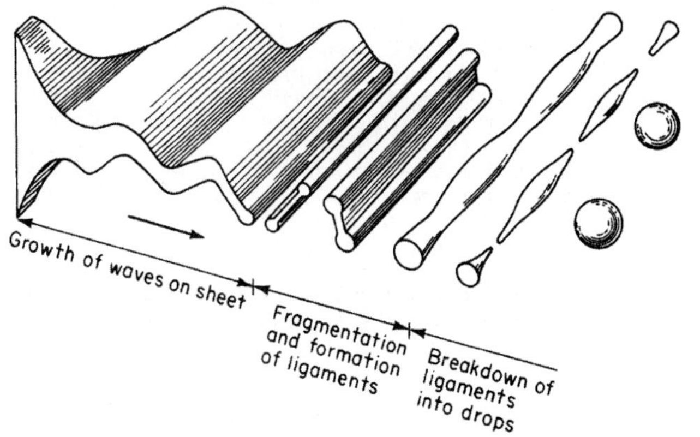

FIG. 24-29 Idealized process of drop formation by breakup of a liquid sheet. (After Dombrowski and Johns, *Chem. Eng. Sci.* **18:** 203, 1963.)

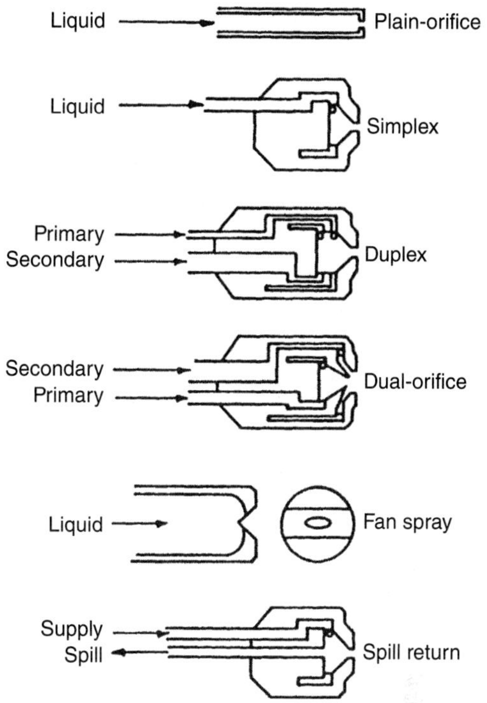

FIG. 24-30a Common types of atomizers: pressure atomizers. (From Lefebvre, *Atomization and Sprays*, Hemisphere, New York, 1989. Reproduced with permission. All rights reserved.)

process by which the surface area of a liquid sheet is increased until it forms droplets. Atomizers may be classified into two broad groups (see Fig. 24-30), pressure atomizers, in which fuel oil is injected at high pressure, and twin-fluid atomizers, in which fuel oil is injected at moderate pressure and a compressible fluid (steam or air) assists in the atomization process. Low oil viscosity (less than 15 mm^2/s) is required for effective atomization (i.e., small droplet size). Light oils, such as No. 2 fuel oil, may be atomized at ambient temperature. However, heavy oils must be heated to produce the

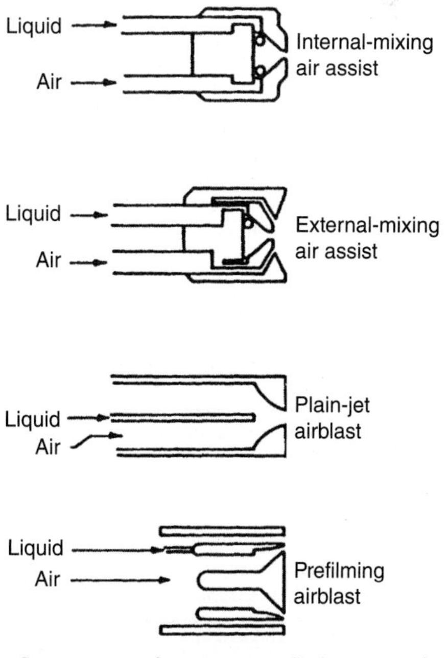

FIG. 24-30b Common types of atomizers: twin-fluid atomizers. (From Lefebvre, *Atomization and Sprays*, Hemisphere, New York, 1989. Reproduced with permission. All rights reserved.)

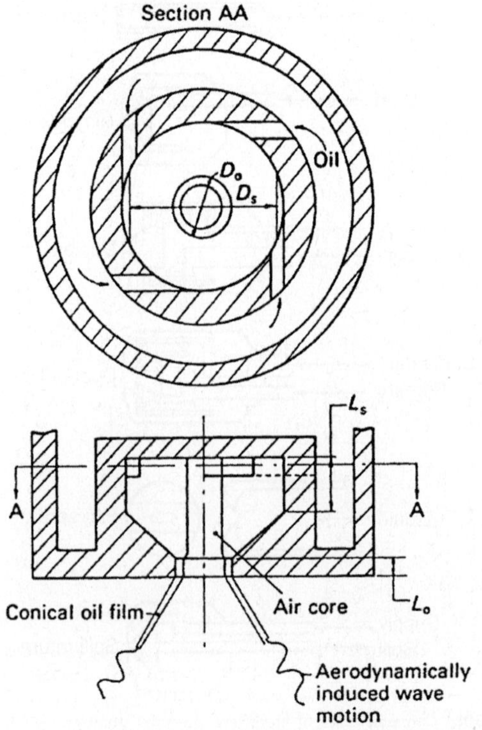

FIG. 24-31 Swirl pressure-jet atomizer. (From Lefebvre, *Atomization and Sprays*, Hemisphere, New York, 1989. Reproduced with permission. All rights reserved.)

desired viscosity. Required preheats vary from approximately 373 K (212°F) for No. 6 oil to 623 K (480°F) for vacuum bottoms.

Pressure Atomizers The most common type of pressure atomizer is the swirl type (Fig. 24-31). Entering a small cup through tangential orifices, the oil swirls at high velocity. The outlet forms a dam around the open end of the cup, and the oil spills over the dam in the form of a thin conical sheet, which subsequently breaks up into thin filaments and then droplets. Depending on the fuel viscosity, operating pressures range from 0.69 to 6.9 MPa (100 to 1000 psia), and the attainable turndown ratio is approximately 4:1. Pressure atomization is most effective for lighter fuel oils.

Twin-Fluid Atomizers In a twin-fluid atomizer, the fuel stream is exposed to a stream of air or steam flowing at high velocity. In the internal-mixing configuration (Fig. 24-32), the liquid and gas mix inside the nozzle before discharging through the outlet orifice. In the external-mixing nozzle, the oil stream is impacted by the high-velocity gas stream outside the nozzle. The internal type requires lower flows of secondary fluid. In industrial combustion systems, steam is the preferred atomizing medium for these nozzles. In gas turbines, compressed air is more readily available. Maximum oil pressure is about 0.69 MPa (100 psia), with the steam or air pressure being maintained about 0.14 to 0.28 MPa (20 to 40 psia) in excess of the oil pressure. The mass flow of atomizing fluid varies from 5 to 30 percent of the fuel flow rate and represents

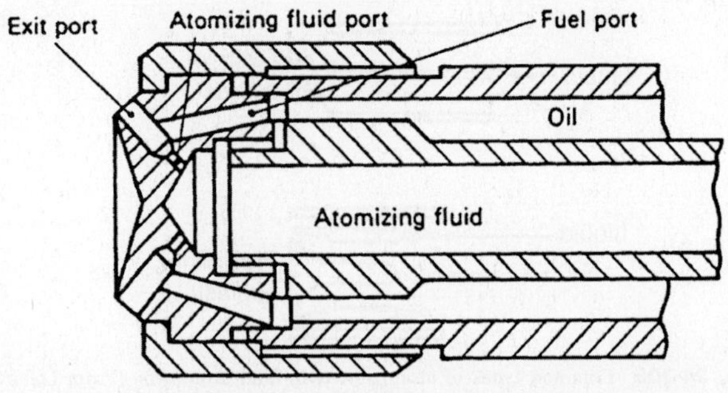

FIG. 24-32 Y-jet twin-fluid atomizer. (From Lefebvre, *Atomization and Sprays*, Hemisphere, New York, 1989. Reproduced with permission. All rights reserved.)

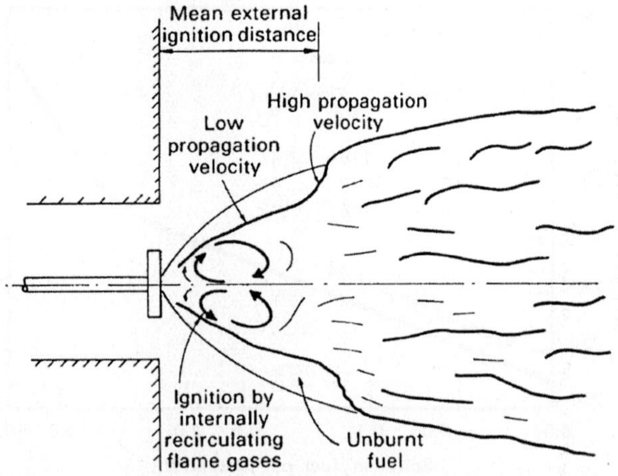

FIG. 24-33 Structure of typical oil flame. (From Lawn, *Principles of Combustion Engineering for Boilers*, Academic Press, London, 1987. Reprinted with permission.)

only a modest energy consumption. Turndown performance is better than for pressure atomizers and may be as high as 20:1.

A well-designed atomizer will generate a cloud of droplets with a mean size of about 30 to 40 μm and a top size of about 100 μm for light oils such as No. 2 fuel oil. Mean and top sizes are somewhat larger than this for heavier fuel oils.

Oil Burners The structure of an oil flame is shown in Fig. 24-33, and Fig. 24-34 illustrates a conventional circular oil burner for use in boilers. A combination of stabilization techniques is used, typically including swirl. It is important to match the droplet trajectories to the combustion aerodynamics of a given burner to ensure stable ignition and good turndown performance.

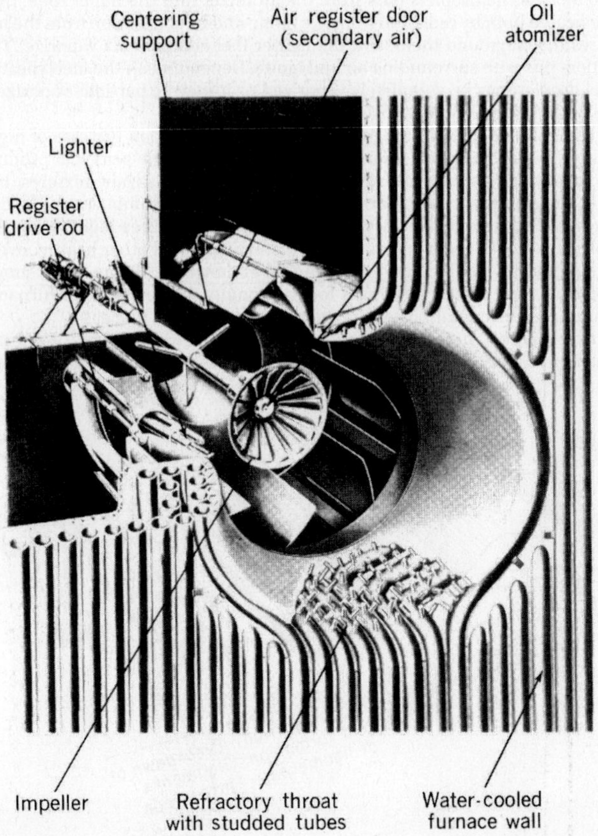

FIG. 24-34 Circular register burner with water-cooled throat for oil firing. (Babcock & Wilcox Co.)

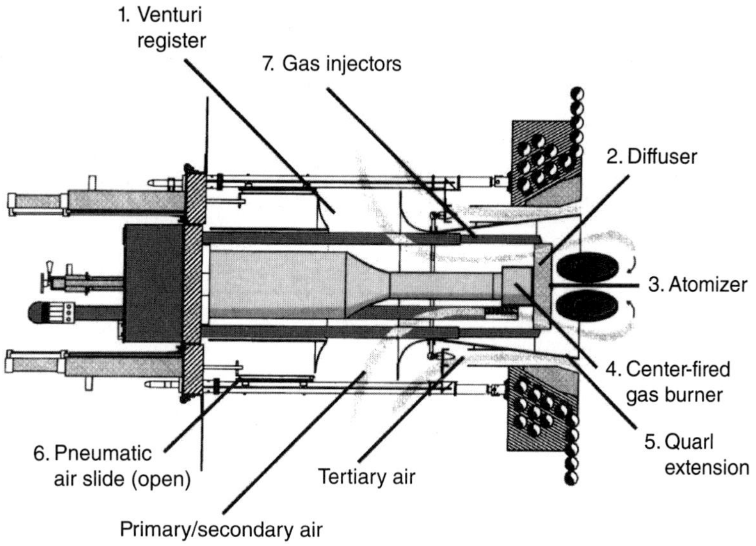

FIG. 24-35 Low-NO_x combination oil/gas forced-draft boiler burner. (Todd Combustion, Inc.)

Many oil burners are designed as combination gas/oil burners. An example of a modern low-NO_x oil/gas forced-draft burner is shown in Fig. 24-35. This is an air-staged design, with the air divided into primary, secondary, and tertiary streams. An air-staged natural draft process heater oil/gas burner is illustrated in Fig. 24-36.

Emissions of unburned carbon (primarily coke cenospheres) may be reduced by (1) achieving smaller average fuel droplet size (e.g., by heating the fuel to lower its viscosity or by optimizing the atomizer geometry), (2) increasing the combustion air preheat temperature, or (3) firing oils with high vanadium content (vanadium appears to catalyze the burnout of coke).

COMBUSTION OF GASEOUS FUELS

Combustion of gas takes place in two ways, depending upon when gas and air are mixed. When gas and air are mixed before ignition, as in a Bunsen

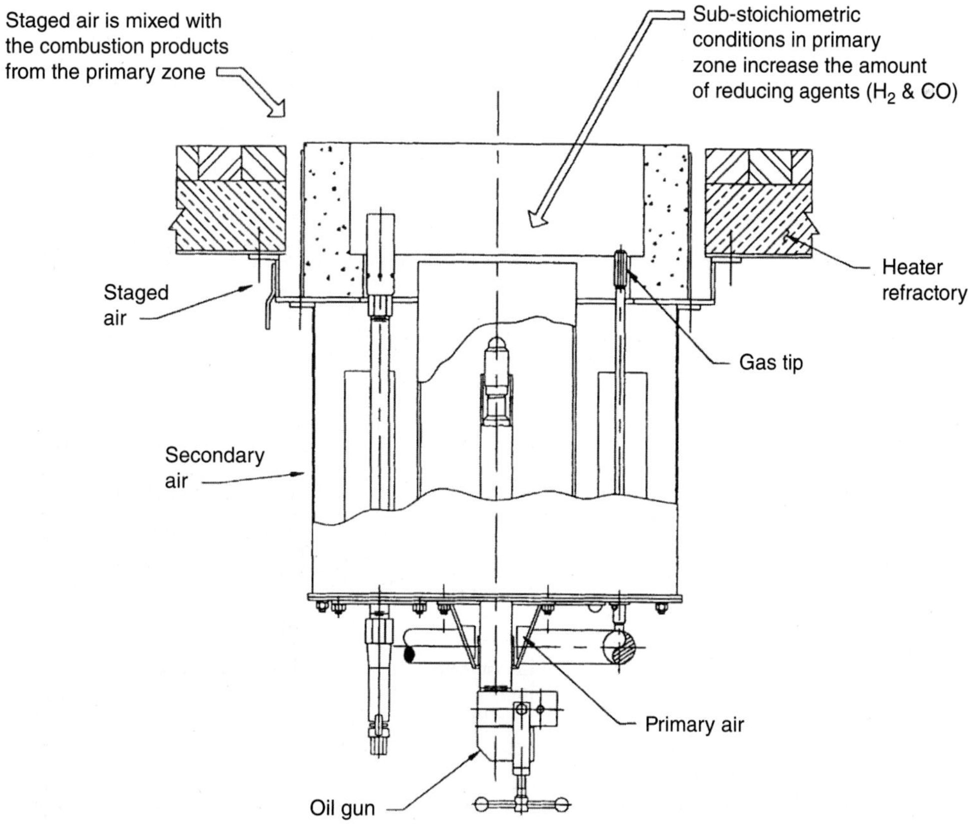

FIG. 24-36 Air-staged natural-draft combination oil/gas burner. (Callidus Technologies, Inc.)

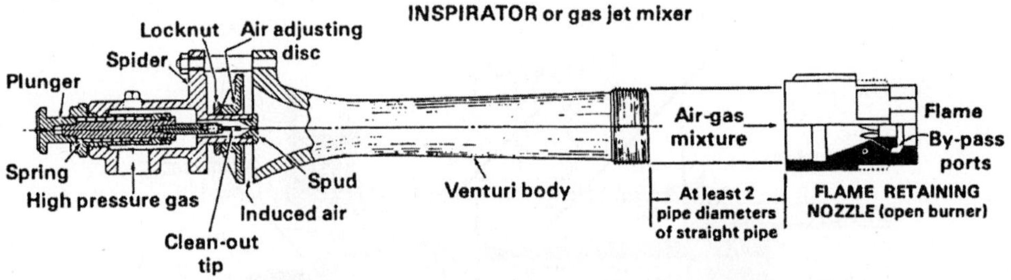

FIG. 24-37 Inspirator (gas-jet) mixer feeding a large port premix nozzle of the flame retention type. High-velocity gas emerging from the spud entrains and mixes with air induced in proportion to the gas flow. The mixture velocity is reduced and pressure is recovered in the venturi section. (From *North American Combustion Handbook*, 3d ed., North American Manufacturing Company, Cleveland, 1996.)

burner, burning proceeds by hydroxylation. The hydrocarbons and oxygen form hydroxylated compounds that become aldehydes; the addition of heat and additional oxygen breaks down the aldehydes to H_2, CO, CO_2, and H_2O. Inasmuch as carbon is converted to aldehydes in the initial stages of mixing, no soot can be developed even if the flame is quenched.

Cracking occurs when oxygen is added to hydrocarbons after they have been heated, decomposing the hydrocarbons into carbon and hydrogen, which, when combined with sufficient oxygen, form CO_2 and H_2O. Soot and carbon black are formed if insufficient oxygen is present or if the combustion process is arrested before completion.

Gas Burners Gas burners may be classified as premixed or nonpremixed. Many types of flame stabilizers are employed in gas burners. Bluff body, swirl, and combinations thereof are the predominant stabilization mechanisms.

Fully Premixed Burners A fully premixed burner includes a section for completely mixing the fuel and air upstream of the burner. The burner proper consists essentially of a flame holder. The porting that admits the mixture to the combustion chamber is designed to produce a fairly high velocity through a large number of orifices to avoid the possibility of the flame flashing back through the flame holder and igniting the mixture upstream of the burner.

Surface combustion devices are designed for fully premixing the gaseous fuel and air and burning it on a porous radiant surface. The close coupling of the combustion process with the burner surface results in low flame temperatures and, consequently, low NO_x formation. Surface materials can include ceramic fibers, reticulated ceramics, and metal alloy mats. This approach allows the burner shape to be customized to match the heat transfer profile with the application.

Partially Premixed Burners These burners have a premixing section in which a mixture that is flammable but overall fuel-rich is generated. Secondary combustion air is then supplied around the flame holder. The fuel gas may be used to aspirate the combustion air or vice versa, the former being the more common. Examples of both are provided in Figs. 24-37 and 24-38.

Nozzle-Mix Burners The most widely used industrial gas burners are of the nozzle-mix type. The air and fuel gas are separated until they are rapidly mixed and reacted after leaving the ports. These burners allow a wide range of fuel/air ratios, a wide variety of flame shapes, and multifuel firing capabilities. They can be used to generate special atmospheres by firing at very rich conditions (50 percent excess fuel) or very lean conditions (1000 percent excess air). By changing nozzle shape and degree of swirl, the flame profile and mixing rates can be varied widely, from a rapid-mixing short flame ($L/D = 1$), to a conventional flame ($L/D = 5$ to 10), to a slow-mixing long flame ($L/D = 20$ to 50).

Staged Burners As was pointed out earlier under Pollutant Formation and Control in Flames, the proper staging of fuel or air in the combustion process is one technique for minimizing NO_x emissions. Gas burners that achieve such staging are available.

Air-staged burners Low-NO_x air-staged burners for firing gas (or oil) are shown in Fig. 24-34. A high-performance, low-NO_x burner for high-temperature furnaces is shown in Fig. 24-39. In this design, both air-staging and external flue-gas recirculation are used to achieve extremely low levels of NO_x emissions (approximately 90 percent lower than conventional burners). The flue gas is recirculated by a jet pump driven by the primary combustion air.

Fuel-staged burners Fuel-staged burner use is the preferred combustion approach for NO_x control because gaseous fuels typically contain little or no fixed nitrogen. Figure 24-40 illustrates a fuel-staged natural draft refinery process heater burner. The fuel is split into primary (30 to 40 percent) and secondary (60 to 70 percent) streams. Furnace gas may be internally recirculated by the primary gas jets for additional NO_x control. NO_x reductions of 80 to 90 percent have been achieved by staging fuel combustion.

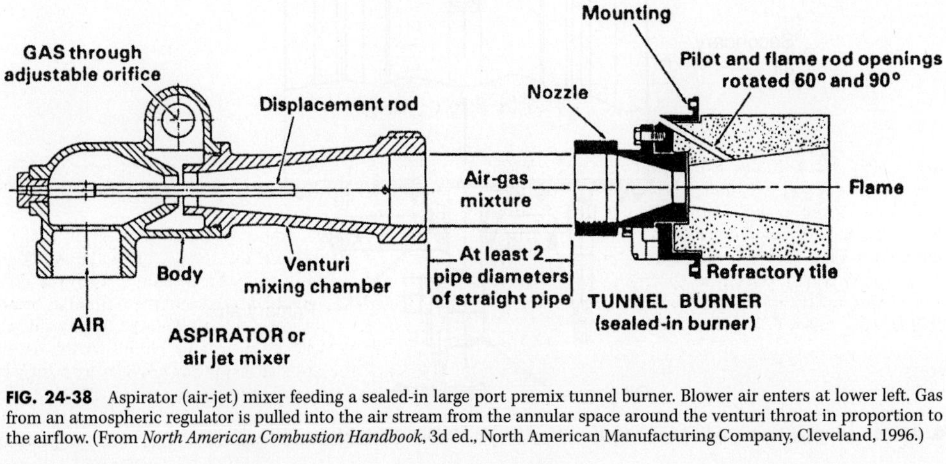

FIG. 24-38 Aspirator (air-jet) mixer feeding a sealed-in large port premix tunnel burner. Blower air enters at lower left. Gas from an atmospheric regulator is pulled into the air stream from the annular space around the venturi throat in proportion to the airflow. (From *North American Combustion Handbook*, 3d ed., North American Manufacturing Company, Cleveland, 1996.)

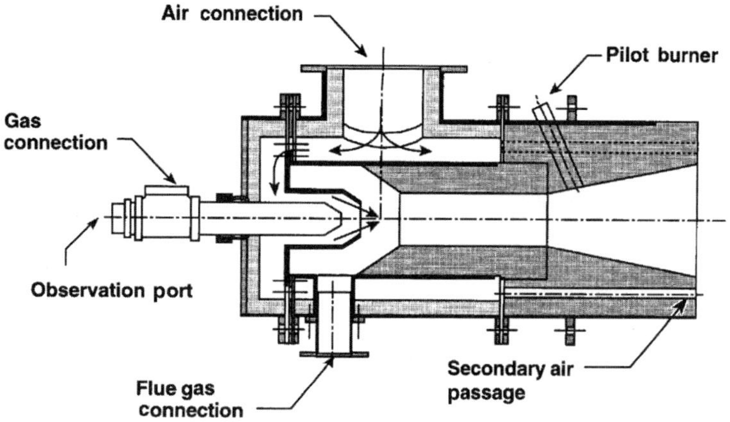

FIG. 24-39 Low-NO_x burner with air-staging and flue-gas recirculation for use in high-temperature furnaces. (Hauck Manufacturing Company. Developed and patented by the Gas Research Institute.)

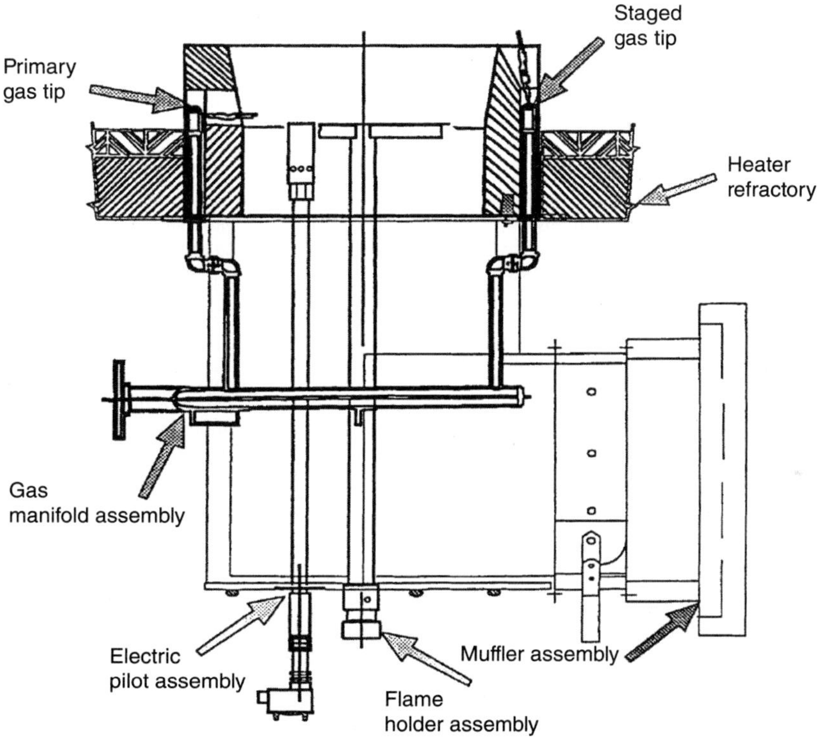

FIG. 24-40 Low-NO_x fuel-staged burner for a natural draft refinery process heater. (Callidus Technologies, Inc.)

PINCH ANALYSIS

General References: ESDU Data Items 87030, 89001, and 90027, *Process Integration and Pinch Technology*, available from ESDU International plc, London, 1987, 1989, 1990; Kemp, *Pinch Analysis and Process Integration—User Guide to Process Integration for the Efficient Use of Energy*, 2d ed., Butterworth Heinemann, Elsevier, Oxford, UK, and New York, 2007; Linnhoff, Townsend, et al., *User Guide to Process Integration for the Efficient Use of Energy*, Institution of Chemical Engineers, Rugby, UK, 1982; Smith, *Chemical Process Design and Integration*, Wiley, New York, 2005.

BASIC PRINCIPLES

Pinch analysis (also known as pinch technology or process integration) helps to set rigorous energy targets for a plant, process, or site by assessing minimum required energy consumption and identifying energy-saving opportunities. The overall methodology is described in detail in the General References.

Pinch analysis examines the process flows that require or release heat. These are categorized into hot streams (which give up heat as they cool down or condense) and cold streams (which require heat). All the heating requirements for cold streams could be fulfilled by hot utilities (e.g., steam, hot water, furnace gases), and likewise the cooling needs could be met by cold utilities (e.g., cooling water, chilled water, or refrigeration). However, heat can be recovered from hot streams at a higher temperature to cold streams at a lower temperature; see Fig. 24-41. All heat exchange reduces both hot and cold utility use, and hence reduces fuel and power use and emissions. Pinch analysis allows rigorous *energy targets* to be calculated for

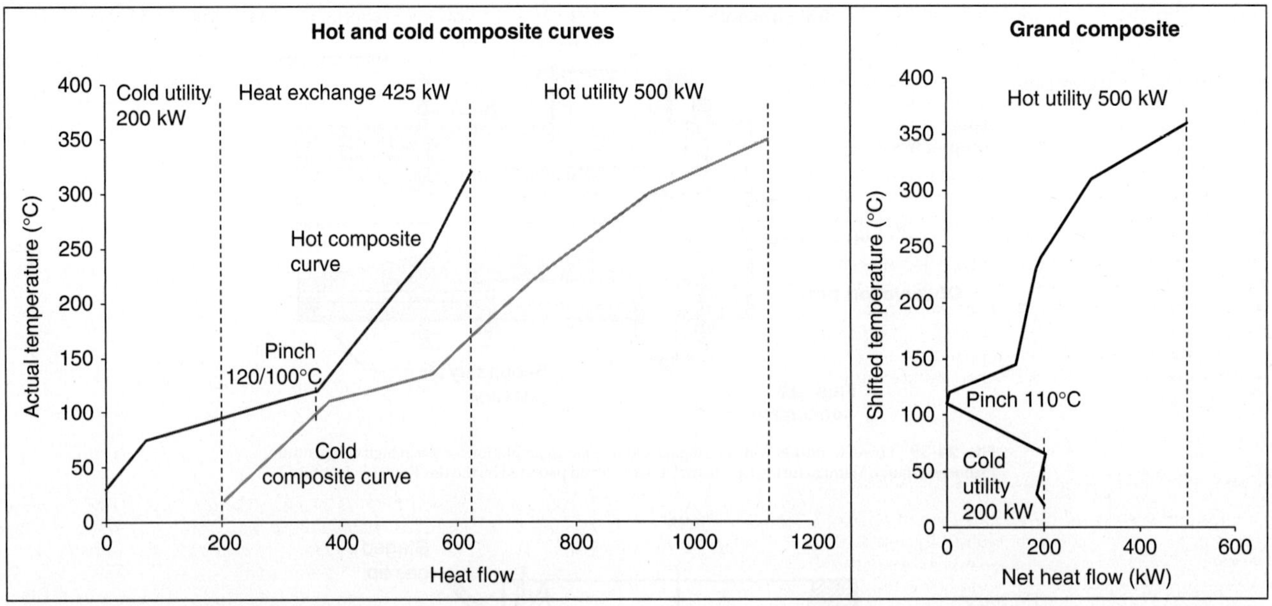

FIG. 24-41 Composite and grand composite curves for a typical liquid-phase process.

how much heat exchange is possible, and hence the minimum possible levels of hot and cold utility use.

Most processes have a *pinch temperature.* Above this temperature they have a net heat requirement; below the pinch, there is net waste heat rejection. Heating below the pinch, cooling above the pinch, or heat exchange across the pinch all incur an energy penalty. Conversely, heat pumps only achieve a real energy saving if they work backwards across the pinch, upgrading useless below-pinch waste heat to useful above-pinch heat. Hence, a pinch analysis of a system is an important prerequisite of any energy-saving project, to ensure that it will achieve its aims.

Streams are characterized by their temperature and heat load (kW), the latter being calculated as

$$Q_{\text{stream}} = WC_p(T_{\text{in}} - T_{\text{out}}) \tag{24-58}$$

The hot and cold streams can be effectively represented on a temperature-heat load diagram, as shown in Fig. 24-41 for a typical liquid-phase process. Where there are multiple hot and cold streams, their heat loads can be summed together to produce *composite curves.* The hot composite curve is the sum of the heat loads of all the hot streams over the temperature ranges where each one exists. Likewise, the cold composite curve is the sum of the heat loads of all the cold streams. A minimum temperature difference for heat exchange, $\Delta T_{\min}$, must be selected, and it has been chosen as 20°C. This gives the vertical distance between the curves. The point of closest vertical approach is the pinch; in Fig. 24-42, this is at a temperature of 100°C for the cold streams and 120°C for the hot streams. The region of overlap between the hot and cold composite curves shows the opportunity for heat exchange, recovering heat from hot to cold streams. The remaining heating and cooling in the nonoverlapping region must be supplied by heating or cooling utilities.

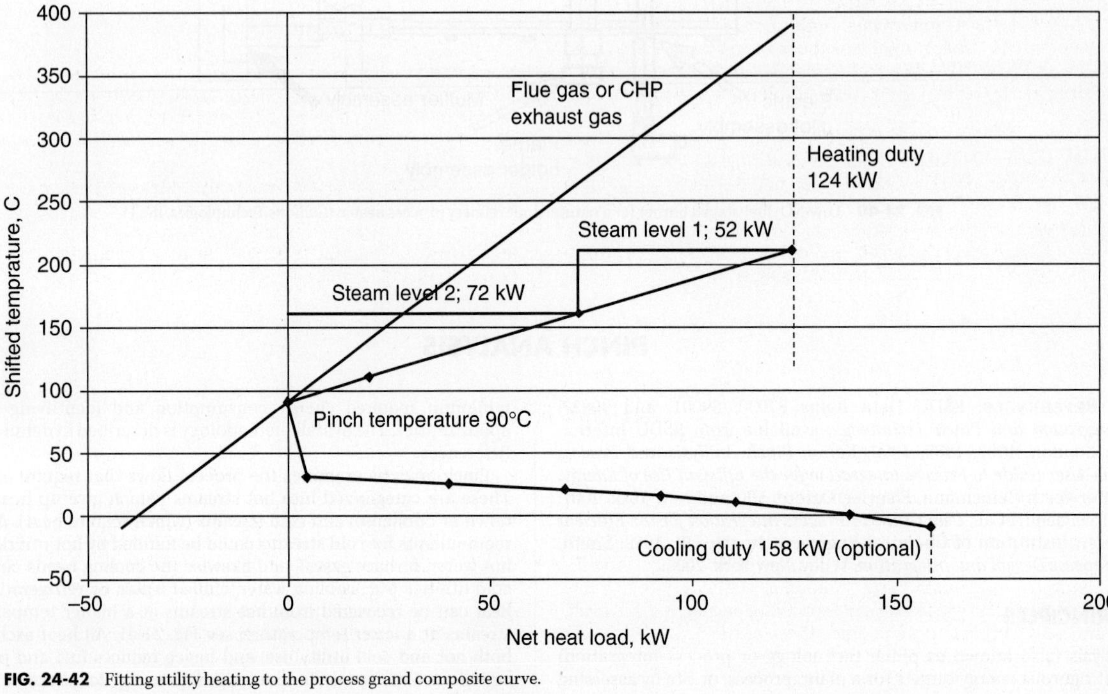

FIG. 24-42 Fitting utility heating to the process grand composite curve.

To obtain the net requirements for hot or cold utility (external supply of heating or cooling) at any specific temperature, the total heat required for the cold streams is subtracted from that available from the hot streams at any temperature. This gives the *grand composite curve* (GCC), shown on the right-hand side of Fig. 24-41. Above the pinch, hot utility is required; below the pinch, cold utility is needed. To allow for the minimum temperature difference for heat exchange ΔT_{min}, hot stream temperatures must be reduced by half this amount (10°C) and the cold stream temperatures increased, to give "shifted temperatures." The grand composite curve shows the exact location of the pinch more clearly than the composite curves.

A heat recovery system or *heat exchanger network* can then be designed to try to achieve the energy targets. It must be remembered that pinch analysis targets give the maximum feasible heat recovery and that some aspects of this, particularly heat exchangers with small loads, may be uneconomical. Very often, there is a capital–energy tradeoff where some potential heat recovery is sacrificed to give a cheaper, simpler project with a better economic rate of return.

THE APPROPRIATE PLACEMENT PRINCIPLE

The appropriate placement principle for a unit operation or a utility states that, to minimize energy use, it should ideally be placed so that it releases all its heat above the pinch temperature and above the GCC of the process, or receives all its heat below the pinch and below the GCC. It applies to:

- Heating and cooling utilities (e.g., steam, hot furnace or flue gases, cooling water)
- Unit operations with large heat loads, especially those including evaporation and condensation at fixed temperatures, such as distillation columns and evaporators

This can reveal opportunities for *process change* to maximize heat recovery. For example, if a distillation column or evaporator releases a large amount of latent heat just below the GCC of the remaining process, a small increase in operating pressure can increase the release temperature so that this can heat the rest of the process. Alternatively, the background process may be changed to receive heat at a slightly lower temperature.

PINCH ANALYSIS AND UTILITY SYSTEMS

The grand composite curve also helps in optimizing the configuration of the utility systems that supply the heating and cooling requirements. The operating line of the hot utility system needs to lie entirely above the process GCC, and the cold utility system below it. If heat is provided by condensing steam, this plots as a horizontal line at the condensation temperature; multiple steam levels may be used. Alternatively, heat may be supplied from a hot gas stream (air or flue gas); this releases sensible heat over a range of temperatures and plots as a sloping line. Both methods are illustrated in Fig. 24-42. The net cost of supplying the heat can be substantially reduced by using a cogeneration (CHP) system; the exhaust from a gas turbine or a reciprocating engine, or steam from a steam turbine, are suitable, and the GCC is vital to optimize the CHP system for the given plant.

Below the pinch, heat is usually rejected to cooling water. However, if the pinch temperature is high enough, it can be used to raise steam, or heat boiler feedwater or air. Low-temperature heat may be used for building space heating. Similarly, if a gas or diesel engine is used as a cogeneration system, the high-temperature exhaust heat can be used for process heating duties or to raise steam, while the jacket cooling water (typically at 60°C to 80°C and below the process pinch) can be used for space heating.

TABLE 24-14 Carbon Footprint for Fuels and Electric Power for Selected Countries

Fossil fuels	Natural gas	0.184
	Fuel oil	0.25
	Diesel oil	0.25
	Coal	0.342
Electricity	France	0.088
	Germany	0.458
	UK	0.541
	USA	0.613
	China	0.836
	India	0.924
	Australia	0.953
Cogeneration	Gas engine	0.27

Data from National Energy Foundation, UK, 2010. All figures in kg CO_2 kWh^{-1}.

HEAT PUMPS

Heat pumps are systems that use a small amount of high-grade energy (e.g., mechanical or electrical power, high-pressure steam) to upgrade a larger amount of low-grade heat to a higher temperature. Examples include:

- Closed-cycle heat pumps
- Refrigeration systems
- Absorption heat pumps and heat transformers/splitters
- Mechanical vapor recompression
- Thermal vapor recompression (thermocompressors, ejectors)

The economics of heat pumps are best where a low-cost power source is available, such as hydroelectric power.

The appropriate placement principle again applies to heat pumps, but in this case, as the heat is released at a higher temperature than the cooling, the heat pump should be placed so that the heat is released above the GCC, and any heat absorbed from the process should come below the GCC. In other words, the heat pump should work backwards across the pinch. If the heat pump is not working across the pinch, then one of the associated heat duties could be fulfilled by heat exchange instead, usually at significantly lower cost.

Many processes are unsuitable for heat pumping; for example, in Fig. 24-42, to upgrade any significant amount of heat, it must be lifted by about 70°C on the GCC, plus the ΔT_{min} across two sets of heat exchangers, and this temperature lift is too high to be economical. However, for systems where there are large net heat loads just above and below the pinch, with a low temperature lift, heat pumping may be worthwhile. Examples include food and agricultural dryers with a large air recycle, large swimming pool facilities, and mechanical or thermal vapor recompression on distillation columns and evaporators (see Sec. 12 and Sec. 13).

CARBON FOOTPRINT

As well as saving process energy, heat recovery leads to a reduction in carbon footprint (greenhouse gas emissions, due to carbon dioxide). Carbon footprint may also be reduced by switching to a different energy source. The CO_2 saving from a project depends on the source of the heat, as shown in Table 24-14.

The low figure for French electricity is because it is predominantly generated by nuclear power; countries with extensive hydroelectric generation will have similar values. Extensive use of renewables (e.g., wind power) also lowers carbon footprint. Note also the major reduction from cogenerating power compared to standalone stations using fossil fuel.

ENERGY RECOVERY

Most processing energy enters and then leaves the process as energy, separate from the product. The energy enters as electricity, steam, fossil fuels, etc., and then leaves, released to the environment as heat, through "coolers," hot combustion flue gases, waste heat, etc. Recovering heat to be used elsewhere in the process is important to increase process efficiency and minimize cost. Minimizing the total annualized costs for this flow of energy through the process is a complex engineering task in itself, separate from classic process design. Since these costs include the costs for getting energy into and out of the process, they should be evaluated together, as elements integrated within a larger system. Such a holistic system evaluation impacts how the overall project will be designed (utilities supply, reaction and separations design, pinch analyses, 3D process layout, plot plan, etc.). Therefore, evaluation and selection of the process energy technology system should be performed at the start of the project design cycle, during technology selection VIP (see the subsection Value-Improving Practices in Sec. 9), when the potential to influence project costs exists at its maximum value.

Following the 1970s energy crisis, enhanced technology systems have been developed which can significantly reduce the annualized costs for process energy. Several of these technologies are presented next, because they are broadly applicable, have a rapid payback, and can make a significant reduction in overall annualized energy costs. Wet surface air coolers (WSACs), an evaporative cooling technology, are presented in Sec. 12.

ECONOMIZERS

General References: Ball, D., et al., *Condensing Heat Exchanger Systems for Oil-Fired Residential/Commercial Furnaces and Boilers: Phase I and II*, US DOE BNL-51617, 1982; Butcher, T. A., N. Park, and W. Litzke, "Condensing Economizers: Thermal Performance and Particulate Removal Efficiencies," in ASME *Two Phase Flow and Heat Transfer, HTD*, vol. 197, 1992 (for U.S. DOE reports see http://www.osti.gov/energycitations); Huijbregts, W. M. M., and R. G. I. Leferink, "Latest Advances in the Understanding of Acid Dewpoint Corrosion: Corrosion and Stress Corrosion Cracking in Combustion Gas Condensates," *Anti-Corrosion Methods and Materials* **51**(3): 173–188 (2004), http://www.hbscc.nl/; Lahtvee, T., and O. Schaus, *Study of Materials to Resist Corrosion in Condensing Gas-Fired Furnaces*, Final Report to Gas Research Institute, GRI-80/0133, February 1982; Okkes, A. G., "Get Acid Dew Points of Flue Gas," *Hydrocarbon Processing*, July 1987, pp. 53–55; Razgaitis, R., et al., *Condensing Heat Exchanger Systems for Residential/Commercial Furnaces: Phase III*, US DOE BNL-51770, 1984; Razgaitis, R., et al., *Condensing Heat Exchanger Systems for Residential/Commercial Furnaces and Boilers: Phase IV*, BNL-51943, 1985.

Economizers improve boiler thermal efficiency by recovering heat from the combustion flue gases exhausted from the steam boiler section. The recovered heat is used to heat colder streams (heat sinks) before ultimate discharge of the waste gas to atmosphere. This recovered heat displaces the need to burn additional fuel to heat these same streams.

Normally, after being heated, these streams are used in the boiler area [deaerator feedwater, cold return condensate, boiler feedwater, reverse osmosis (RO) feedwater] or in the combustion chamber (air preheat). However, economizers can be used to recover and supply heat elsewhere, such as hot process water or hot utility water, especially as used in the food processing and pulp/paper industries. Additionally, recovered flue gas waste heat can be used indirectly; for example, remote process streams can be heated locally with hot steam condensate, and then the cooled return steam condensate can be reheated in the flue gas economizer. An extension of these concepts is provided by the application of using hot water to vaporize LNG: hot-water-based liquid is used to vaporize the process stream [LNG: stored near 122 K (−151°C), returning near 273 K (0°C)] to the hot water heater. Before entering the hot water heater, the cooled stream recovers flue gas waste heat in a condensing economizer.

Acid Dew Point For fossil fuels, the acid dew point temperature is that temperature at which the actual mixed acid vapor pressure equals the mixed acid vapor saturation pressure. The mixed acid dew point can be approximated by the sulfuric acid dew point (Fig. 24-43). It can be described as a function of the SO_3 and water content of the flue gas (Huijbregts). These concentrations result from the sulfur, hydrogen, and free water content of the fuel; the relative humidity of the air; and the amount of excess air used. Using the equation of Verhoff, where T is degrees K and P is mm Hg (see Okkes 1987):

$$T_{\text{dew}}(SO_3) = 1000/[2.276 - 0.0294\ln(P_{H2O}) - 0.0858\ln(P_{SO3}) + 0.0062\ln(P_{H2O}\,P_{SO3})] \tag{24-59}$$

The corrosiveness of flue gas condensate is further complicated by the presence of other components (Cl^{-1}, NO_3^{-1}, etc.). The sources of these components can be either the fuel or the combustion air (salt, ammonia, Freon, chlorine, chlorinated VOCs, etc.), usually producing a more corrosive condensate.

Water Dew Point For flue gas, the water dew point is that temperature at which the actual water vapor pressure equals the water saturation vapor pressure. Cooling the flue gas below this temperature will result in the formation of liquid water [or ice, below 273 K (0°C)]. For example, burning natural gas with 3 percent excess oxygen (15 percent excess air), the flue gas water dew point would be about 330 K (56.7°C) (Fig. 24-44).

Boiler Thermal Efficiency Traditionally, boiler thermal efficiency is calculated as Q_{OUT}/Q_{IN}, where Q_{IN} is the LHV (lower heating value) of the fuel. A rule of thumb for economizers is that boiler efficiency increases by about 1 percent for every 22°C (40°F) drop in temperature of the dry flue gas. These two statements do not reveal the considerable quantity of *additional heat*, available to be recovered through *condensation of the water vapor* in the flue gas, which is lost to atmosphere with hot flue gas. Based on fuel HHV (higher heating value), the total latent heat loss can be substantial: an additional 9.6 percent (natural gas), 8.0 percent (propane), 6.5 percent (heating oil).

Conventional Economizers Conventional economizers can be constructed from relatively inexpensive materials, such as low-alloy carbon steels, if they will be operated dry on the gas side, with flue-gas-side metal temperatures above the acid dew point. This practice is done to protect the economizer from corrosion, caused by the acidic flue gas condensate. Conventional economizers can also be constructed from more expensive materials, and they can be operated below the acid dew point but above the water dew point. This practice permits greater heat recovery, but with a generally lower payback. A compromise practice for operation below the acid dew point is to use less expensive but less corrosion-resistant materials, accepting an accelerated rate of corrosion, and periodically replacing the damaged heat-transfer surfaces when needed. Nevertheless, when high-sulfur fuel is burned (such as oil or coal), typically the water inlet feed to a conventional economizer is preheated to a temperature above the anticipated acid dew point.

Condensing Economizers Flue gas condensing waste heat economizers are designed to operate *below the flue gas water dew point*. This temperature can range from about 316 K (43°C) to 333 K (60°C), depending on the amount of hydrogen and water in the fuel, the amount of excess combustion air used, and the humidity of the air. [Higher flue gas water dew points can be encountered for other industrial applications, such as product driers, fryers (food processing), and waste water incinerators.] Such economizers recover *flue gas sensible heat* as well as *water vapor latent heat* from the hot flue gas. Fuel consumption is reduced in proportion to the efficiency increase.

Condensing economizers are constructed from inexpensive, but durable, corrosion-resistant materials. Extensive materials testing has been performed for operation in this service, including for coal combustion

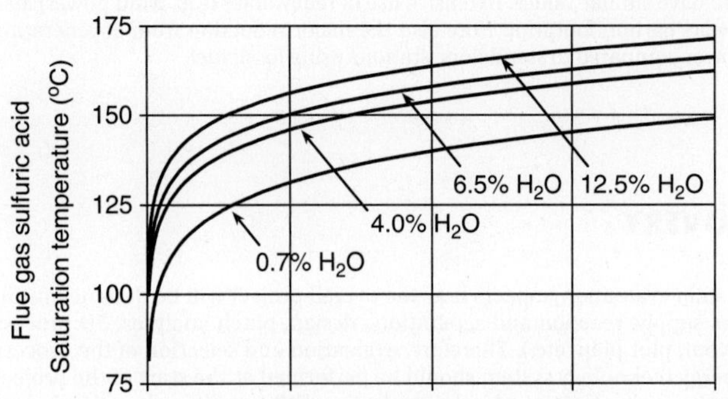

FIG. 24-43 Calculated sulfuric acid dew points, as a function of SO_3 content, for various flue gas water vapor concentrations. (Courtesy W. M. M. Huijbregts, 2004.)

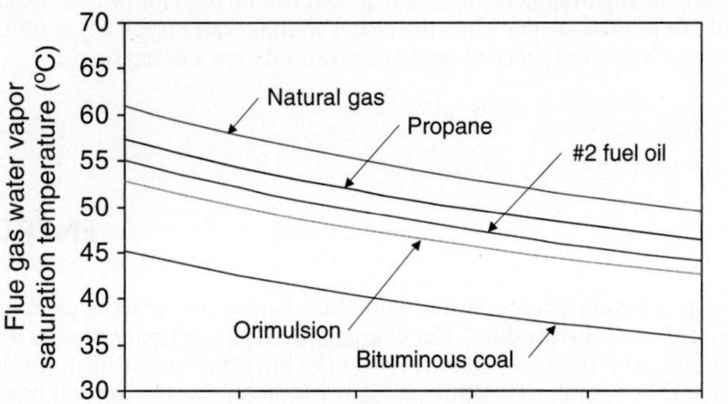

FIG. 24-44 Calculated flue gas water vapor dew points, for different fuel types, as a function of excess air [Orimulsion (28.3 percent water), Pittsburgh Seam 8 (5 percent water)]. (Courtesy T. A. Butcher, US DOE; www.bnl.gov.)

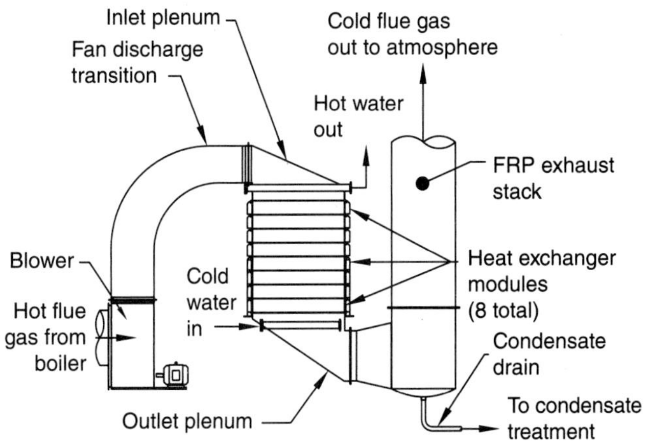

FIG. 24-45 Standard equipment arrangement, flue gas condensing economizer waste heat recovery system (flow: left to right). The ID fan draws hot flue gas from the boiler, propelling it into the top of the condensing economizer. (Courtesy CHX Condensing Heat Exchanger Co.; www.chxheat.com.)

(Lahtvee and Schaus 1982; Ball 1982; Razgaitis et al. 1984, 1985; Butcher, Park, and Litzke 1992). The metallurgy for the tube-side liquid is determined by the liquid chemistry requirements (usually water-based liquid); 304 stainless steel is typical.

For gas-side materials, one available technology employs Teflon-covered metal tubing and Teflon tube sheets. This technology is often operated across both the acid and water dew points, and can accept inlet gas temperatures to 533 K (260°C). Typical applications may achieve a cold-end ΔT below 45°C (80°F), improve the boiler thermal efficiency by about 10 percent (LHV basis), and have a simple payback of 2 to 3 years, based on *fuel avoidance* (Figs. 24-45 and 24-46).

A second technology employs metallic finned tubing, extruded over the water tubing. Aluminum 1000 series fins are preferred, for heat-transfer reasons in natural gas applications, but stainless steel (or other material) fins are used for higher temperatures and/or more corrosive flue gas. This second technology both is less expensive and has better heat transfer (per ft^2). Consequently, for the same payback, the cold-end approach can be lower and the water outlet temperature and the boiler efficiency improvement higher. Flue gas condensate from the combustion of natural gas typically has a pH of around 4.3, and aluminum fins are suitable. For more acidic (or erosive) flue gas conditions, other metallurgy (Incoloy® 825 and Hastelloy®), or a Hersite or equivalent coating, may be used to prevent corrosion damage (Fig. 24-47).

Flue gas condensates at different temperatures, compositions, and relative corrosivity condenses and exists at different positions within the condensing economizer. These positions are not fixed in space or time but move back and forth in response to changing load conditions in either stream. Condensing economizers are typically equipped with water spray nozzles for periodic washdown of the flue gas side, to be used (infrequently) for natural gas combustion, but more frequently for services having heavier pollutant loading, such as oil or coal. Over 200 such heat exchangers have been installed, some in service for more than 20 years (2005). This technology is suitable for heat recovery applications of any magnitude.

Several environmental benefits are gained through the use of this condensing technology. Burning less fuel proportionally reduces *collateral combustion emissions* (NO_x, SO_x, CO, CO_2 and particulates, including PM-2.5). Additionally, *flue gas pollutant removal* occurs in the condensing economizer, as has been extensively investigated, characterized, and modeled by the U.S. DOE (Butcher, Park, and Litzke 1992), including applications burning coal and Orimulsion. Typically, the condensate will contain most (by mass, >90 percent) of the highly dissociated inorganic matter (H_2SO_4, HCl, HNO_3, HNO_2, NH_3, salts, etc.) and the larger-diameter particulates (>10 μm) and a lower but substantial fraction (>60 percent) of the smaller-diameter particulates (<5 μm). Such gross pollutant removal can be a cost-effective first stage for a traditional air pollution control system by reducing the volume of the flue gas to be treated (water content and temperature) and by reducing the concentration of the pollutants in the flue gas to be treated downstream. Unlike spray quenching, such indirect quench-cooling

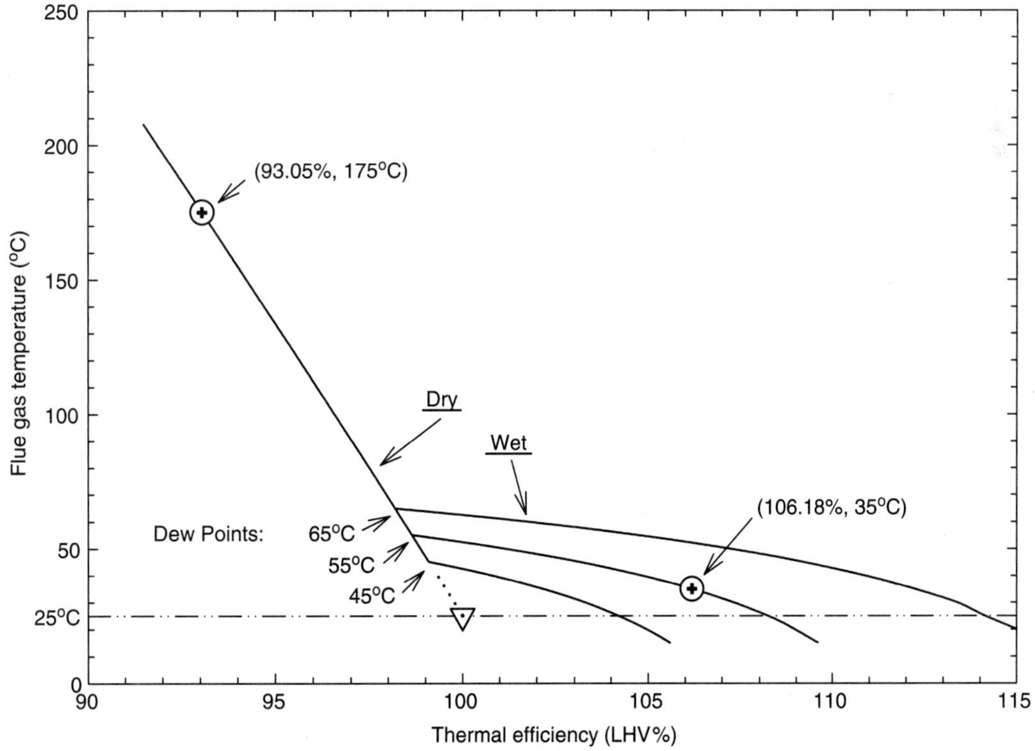

FIG. 24-46 Characteristic curves for boiler thermal efficiency as a function of flue gas effluent temperature and flue gas water dew points. Based on the LHV of a fuel, and stoichiometric reaction, 100 percent efficiency would be achieved if sufficient combustion heat were recovered and removed, so that the temperature of the effluent flue gas was reduced to 25°C. For a flue gas with a 55°C dew point, recovering additional heat via condensation by cooling from 175°C to 35°C (as shown) would increase the overall efficiency by more than 13 percent. (Courtesy Combustion & Energy Systems, Ltd.; www.condexenergy.com.)

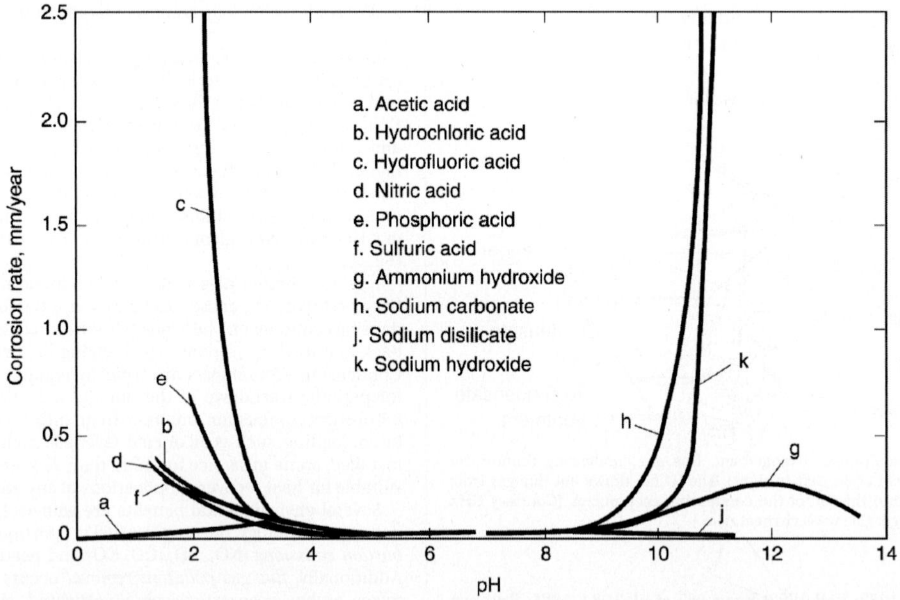

FIG. 24-47 Effect of corrosion on 1100-H14 aluminum alloy by various chemical solutions. Observe the minimal corrosion in the pH range from 4.0 to 9.0. The low corrosion rates in acetic acid, nitric acid, and ammonium hydroxide demonstrate that the nature of the individual ions in solution is more important than the degree of acidity or alkalinity. (With permission from ASM International; www.asminternational.org. Courtesy of Combustion & Energy Systems, Ltd.; www.condexenergy.com.)

segregates the flue gas pollutants from the cooling system fluid, generating a much smaller, more concentrated wastewater stream for subsequent waste treatment. Although pollutant removal percentages are high, they are functions of the specific real-time mass- and heat-transfer conditions within the economizer. Condensate treatment from boiler economizers normally is neutralization, often using the boiler blowdown, and release to the sewer and sewage treatment.

REGENERATORS

Storage of heat is a temporary operation since perfect thermal insulators are unknown; thus, heat is absorbed in solids or liquids as sensible or latent heat to be released later at designated times and conditions. The collection and release of heat can be achieved in two modes: on a batch basis, as in the checkerbrick regenerator for blast furnaces, or on a continuous basis, as in the Ljungstrom air heater.

Checkerbrick Regenerators Preheating combustion air in open-hearth furnaces, ingot-soaking pits, glass-melting tanks, by-product coke ovens, heat-treating furnaces, and the like has been universally carried out in regenerators constructed of fireclay, chrome, or silica bricks of various shapes. Although many geometric arrangements have been used in practice, the so-called basketweave design has been adopted in most applications.

Blast-Furnace Stoves Blast-furnace stoves are used to preheat the air that is blown into a blast furnace. A typical blast furnace, producing 1500 Mg (1650 U.S. ton) of pig iron per day, will be blown with 47.2 m^3/s (100,000 std ft^3/min) of atmospheric air preheated to temperatures ranging in normal practice from 755 to 922 K (900°F to 1200°F). A set of four stoves is usually provided, each consisting of a vertical steel cylinder 7.3 m (24 ft) in diameter, 33 m (108 ft) high, topped with a spherical dome. Characteristic plan and elevation sections of a stove are shown in Fig. 24-48. The interior comprises three regions: in the cylindrical portion, (1) a side combustion chamber, lens-shaped in cross section, bounded by a segment of the stove wall and a mirror-image bridgewall separating it from (2) the chamber of the cylinder that is filled with heat-absorbing checkerbrick, and (3) the capping dome, which constitutes the open passage between the two chambers.

The heat-exchanging surface in each stove is just under 11,500 m^2 (124,000 ft^2). In operation, each stove is carried through a two-step, 4-h cycle. In a 3-h *on-gas* step, the checkerbricks in a stove are heated by the combustion of blast-furnace gas. In the alternating *on-wind* 1-h step, they are cooled by the passage of cold air through the stove. At any given time, three stoves are simultaneously on gas, while a single stove is on wind.

At the start of an on-wind step, about one-half of the air, entering at 366 K (200°F), passes through the checkerbricks, the other half being bypassed around the stove through the cold-blast mixer valve. The gas passing through the stove exhausts at 1366 K (2000°F). Mixing this with the unheated air produces a blast temperature of 811 K (1000°F). The temperature of the heated air from the stove falls steadily throughout the on-wind step. The fraction of total air volume bypassed through the mixer valve is continually decreased by progressively closing this valve, its operation being automatically regulated to maintain the exit gas temperature

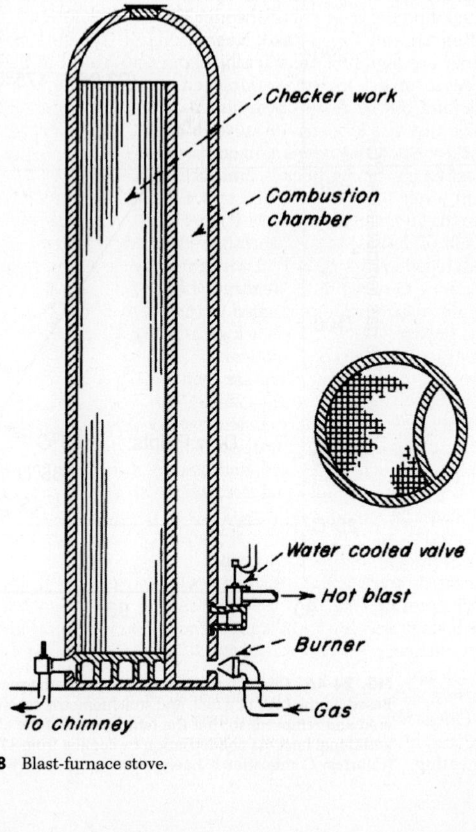

FIG. 24-48 Blast-furnace stove.

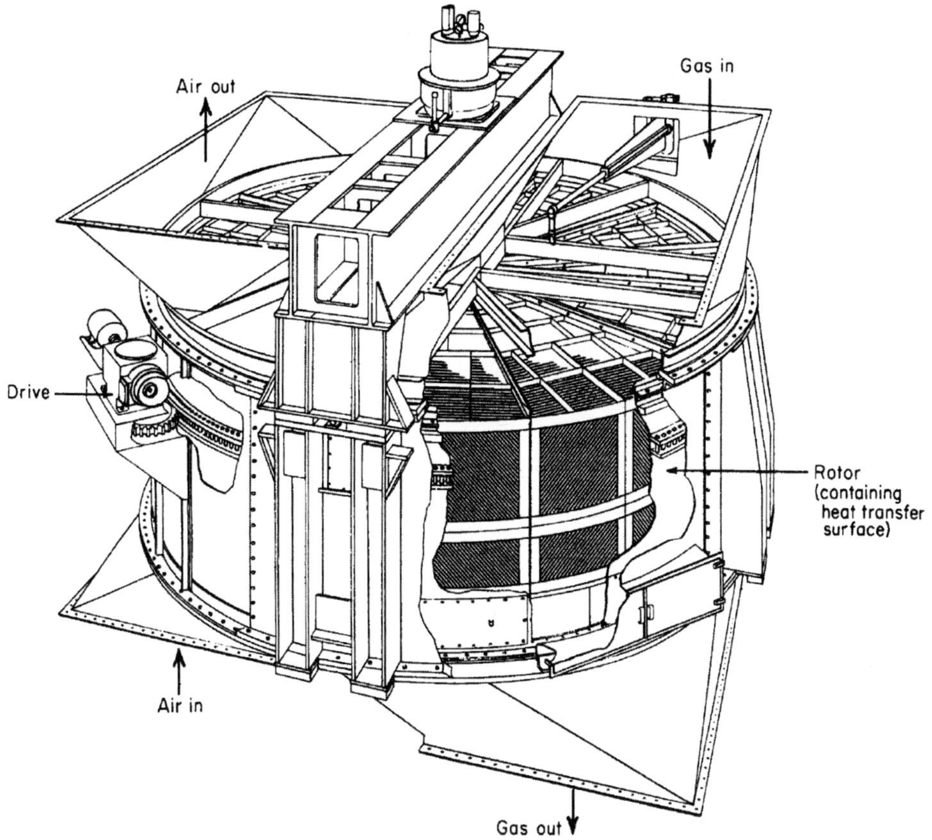

FIG. 24-49 Ljungstrom air heater.

at 811 K. At the end of 1 h of on-wind operation, the cold-blast mixer valve is closed, sending the entire blast through the checkerbricks.

Open-Hearth and Glass-Tank Regenerators These contain checkerbricks that are modified considerably from those used in blast-furnace stoves because of the higher working temperatures, more drastic thermal shock, and dirtier gases encountered. Larger bricks form flue cross sections five times as large as the stove flues, and the percentage of voids in the checkerbricks is 51 percent, in contrast to 32 percent voids in stoves. The vertical height of the flues is limited by the elevation of the furnace above plant level. Short flues from 3 to 4.9 m (10 to 16 ft) are common in contrast to the 26- to 29-m (85- to 95-ft) flue lengths in blast-furnace stoves.

As a result of the larger flues and the restricted surface area per unit of gas passed, regenerators employed with this type of furnace exhibit much lower efficiency than would be realized with smaller flues. In view of the large amount of iron oxide contained in open-hearth exhaust gas and the alkali fume present in glass-tank stack gases, however, smaller checkerbrick dimensions are considered impractical.

Ljungstrom Heaters A familiar continuous regenerative-type air heater is the Ljungstrom heater (Fig. 24-49). The heater assembly consists of a slow-moving rotor embedded between two peripheral housings separated from one another by a central partition. Through one side of the partition, a stream of hot gas is being cooled, and through the other side, a stream of cold gas is being heated. Radial and circumferential seals sliding on the rotor limit the leakage between the streams. The rotor is divided into sectors, each of which is tightly packed with metal plates and wires that promote high heat-transfer rates at low pressure drop.

These heaters are available with rotors up to 6 m (20 ft) in diameter. Gas temperatures up to 1255 K (1800°F) can be accommodated. Gas face velocity is usually around 2.5 m/s (500 ft/min). The rotor height depends on service, efficiency, and operating conditions but usually is between 0.2 and 0.91 m (8 and 36 in). Rotors are driven by small motors with rotor speed up to 20 r/min. Heater effectiveness can be as high as 85 to 90 percent heat recovery. Lungstrom-type heaters are used in power-plant boilers and also in the process industries for heat recovery and for air-conditioning and building heating.

Regenerative Burners In these systems, a compact heat storage regenerator (containing ceramic balls, for example) is incorporated into the burner. Operating in pairs, one burner fires while the other exhausts: combustion air is preheated in the regenerator of the firing burner, and furnace gas gives up heat to the regenerator in the exhausting burner (see Fig. 24-50). Burner operations are switched periodically. Such systems can yield combustion air preheats between 933 K (1220°F) and 1525 K (2282°F) for furnace temperatures between 1073 K (1472°F) and 1723 K (2642°F), respectively. Corresponding fuel savings compared to cold-air firing will vary approximately from 30 to 70 percent.

Miscellaneous Systems Many other systems have been proposed for transferring heat regeneratively, including the use of high-temperature liquids and fluidized beds for direct contact with gases, but other problems that limit industrial application are encountered. These systems are covered by methods described in other sections of this handbook.

RECUPERATORS

Regenerators are by nature intermittent or cycling devices, although, as set forth previously, the Ljunstrom design avoids interruption of the fluid stream by cycling the heat-retrieval reservoir between the hot and cold fluid streams. Truly continuous counterparts of regenerators exist, however, and they are called *recuperators.*

The simplest configuration for a recuperative heat exchanger is the metallic radiation recuperator. An inner tube carries the hot exhaust gases and an outer tube carries the combustion air. The bulk of the heat transfer from the hot gases to the surface of the inner tube is by radiation, whereas that from the inner tube to the cold combustion air is predominantly by convection.

Shell-and-tube heat exchangers may also be used as recuperators; convective heat transfer dominates in these recuperators. For applications involving higher temperatures, ceramic recuperators have been developed. These devices can allow operation at up to 1823 K (2822°F) on the gas side and over 1093 K (1508°F) on the air side. Early ceramic recuperators were built of furnace brick and cement, but the repeated thermal cycling caused

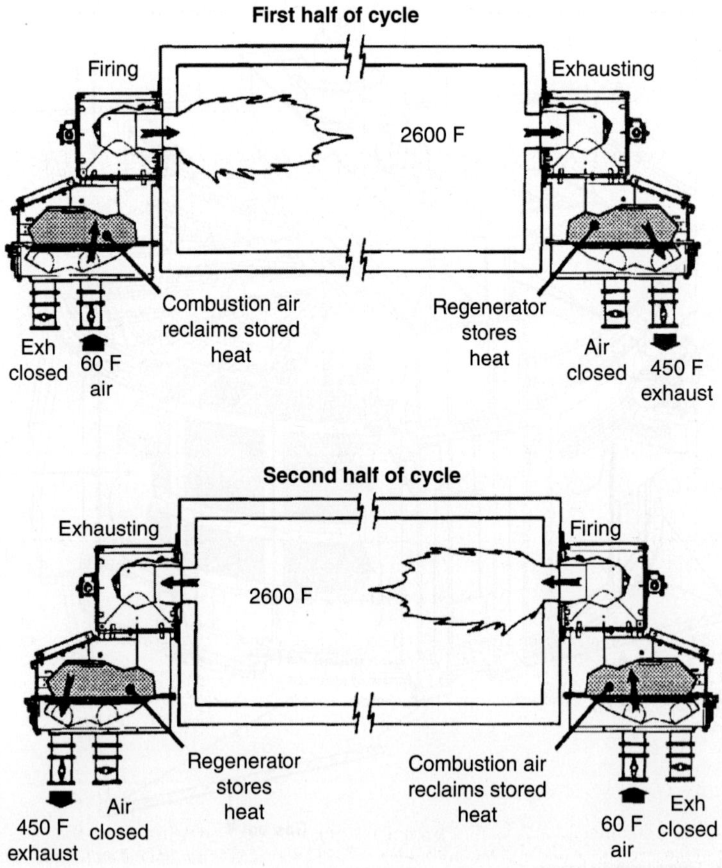

FIG. 24-50 Schematic of a regenerative burner system. (North American Manufacturing Co.)

cracking and rapid deterioration of the recuperator. Later designs have used various approaches to overcome the problems of leakage and cracking. In one configuration, for example, silicon carbide tubes carry the combustion air through the waste gas, and flexible seals are used in the air headers. In this manner, the seals are maintained at comparatively low temperatures and the leakage rate can be reduced to a few percent of the total flow.

TURBINE INLET (AIR) COOLING

GENERAL REFERENCES: See "Bibliography" at http://www.turbineinletcooling.org/resources.

Turbine inlet cooling (TIC) can increase gas turbine (GT) power output on a hot day by 10 to 30 percent, while improving (reducing) the turbine heat rate (kJ/kW$_e$) by as much as 5 percent. By increasing the air compressor inlet air density, turbine inlet air cooling is the most cost-effective method for increasing turbine gross power output for fixed-altitude GTs.

GTs are constant-volume machines, such that a fixed-speed GT air compressor section draws a nearly constant volumetric flow of inlet air, independent of ambient air conditions. Air density drops with increased altitude (reduced barometric pressure), increased ambient temperature, and, to a lesser degree, increased water content (specific humidity).

Increasing the air density increases the GT inlet air mass flow. For a given stoichiometric fuel/air ratio and a given combustion temperature, increased air mass flow allows increased fuel flow, resulting in increased GT power output. Additionally, compressor efficiency increases with decreased air temperatures, resulting in less parasitic compressor shaft work consumed and greater net turbine power output. Therefore, TIC increases net incremental power output faster than incremental fuel consumption, resulting in improved overall fuel efficiency (reduced heat rate); see Fig. 24-51.

Evaporative Technologies These TIC technologies cool the inlet air by vaporizing water in direct contact with the inlet air, using the latent heat of vaporization of water. These technologies either spray water directly into the airstream as a fog or mist or evaporate the water from fixed wetted media placed in the airstream.

Performance of these technologies is typically an 85 percent approach to the ambient wet-bulb temperature (T_{WB}) from the ambient dry-bulb temperature (T_{DB}), limited to a 3°F approach to the ambient T_{WB}. For locations with high T_{DB} and low T_{WB}, the cooling benefit can be significant. For locations with humid conditions, the relative cooling achievable by evaporative technologies can be very limited. Additionally, on a given day, the relative changes in hourly T_{DB} and T_{WB} will result in variable GT peak power output.

Evaporative TIC technologies have the minimum installed and operating costs. The operating electric parasitic load is generally considered insignificant. Water chemistry management is crucial to prevent maintenance problems caused by dissolved or suspended matter in the water, fouling either the evaporation media or the turbines themselves.

Refrigeration Technologies These TIC technologies indirectly cool the inlet air, typically using either chilled water or a vaporizing refrigerant, in a finned coil heat exchanger placed in the incoming airstream. Refrigeration technologies can achieve and maintain much lower air compressor inlet temperatures, largely independent of changing ambient conditions (Fig. 24-52). Best design practices (ice formation/consequent GT damage) have restricted the OEM recommended minimum inlet air cooling temperatures to the 5°C to 10°C (41°F to 50°F) range. Refrigeration can be supplied by either absorption or mechanical chillers. Absorption chillers employ either LiBr or ammonia absorption. Electrically driven centrifugal packaged chiller systems, using chilled water as a secondary refrigerant, account for the majority of refrigeration machine types. R-123, R-134a, and R-717 are the most common refrigerants offered by vendors. Several installations have

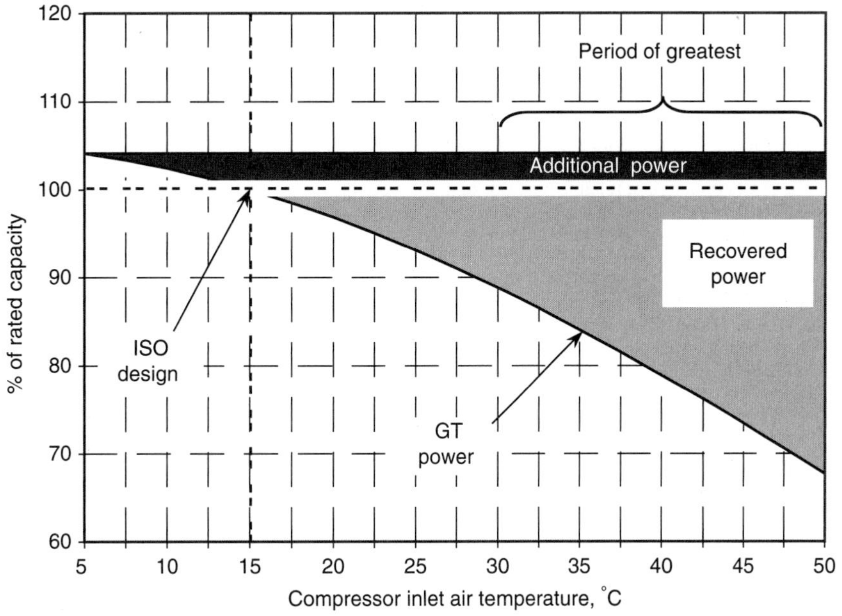

FIG. 24-51 Effect of inlet air ambient temperature on the power output of a typical GT. If ambient air at 95°F (35°C) were cooled to 50°F (10°C), the gross GT power output would be increased by approximately 22 percent, and the gross heat rate improved by 3.7 percent. Operated at its ISO conditions [15°C (59°F) at sea level], GT rated performance is 100 percent. (Turbine Air Systems; www.tas.com.)

used direct refrigeration through R-22 or ammonia (R-717) vaporization in the TIC air coil, with refrigerant condensation in a wet surface air cooler (WSAC) or an air-cooled condenser. Site-specific sources of cooling (the heat sink), such as LNG vaporization, have also been used. The water condensate generated by cooling inlet air below the ambient dew point can be employed as makeup water for a cooling tower or a WSAC.

While refrigeration technologies can provide a near-constant GT inlet air temperature, and thus a nearly constant GT power output, refrigeration TIC systems are significantly more expensive than evaporative systems. However, the incremental net power output from the refrigeration system, when measured on a unit cost basis (\$ per kW_e), costs much less than the unit cost of the GT itself, thereby decreasing the overall unit cost of the GT plant. Refrigeration technologies require a significant energy input to operate. In the case of electric-driven chillers, the parasitic electric load can consume 10 percent to 15 percent of the GT gross incremental increase in power. Therefore, the net overall plant incremental output, after allowance for the refrigeration system parasitic load, is used for evaluating the economics of these systems.

Thermal Energy Storage (TES) A TES system consists of an insulated cold storage tank and a chiller, generating ice or chilled water, which is accumulated in the tank. A TES system is typically installed when TIC is

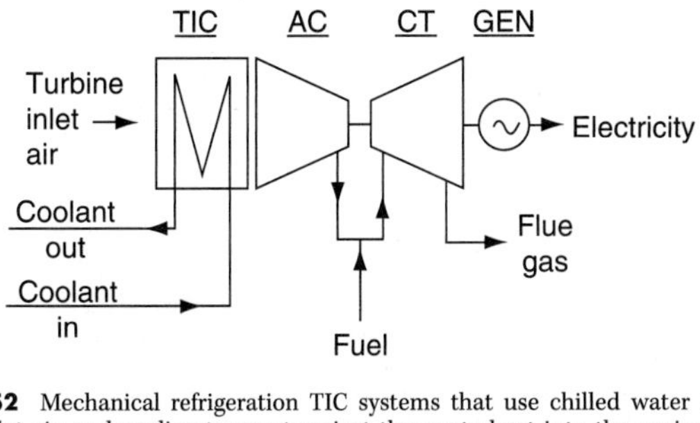

FIG. 24-52 Mechanical refrigeration TIC systems that use chilled water to cool turbine inlet air, and cooling towers to reject the waste heat into the environment, account for the majority of refrigeration TIC systems sold.

required for only a limited number of operating hours per day. The principal TES benefit is that the *net* electrical peak output of the power plant is maximized by shifting the parasitic electric power loss (chiller power demand) from the period of peak demand (midday) to the off-peak hours (night). A second benefit is a reduced overall capital cost for a smaller chiller (usually half-sized) because the smaller chiller is operated either continuously (partial storage) or over a nighttime period (full storage). A third benefit is the reduced operating costs, stemming from the ability to use lower-value off-peak power for the chilling load. A fourth benefit is that the chiller will typically perform better during the nighttime off-peak period, due to lower ambient temperatures, reducing the energy required for refrigerant compression. A fifth benefit is the ability to increase or decrease generation capacity in a short time period, when called upon, to stabilize the transmission system or grid the plant is tied into. However, a TES site footprint may need to be larger to include the TES storage tank.

Summary Evaporative technologies have the lowest capital and operating costs, but their benefit is limited by ambient conditions. They are most effective where the peak T_{WB} is lowest, especially at times of peak demand, as are found in hot, arid regions. Refrigeration technologies can achieve and maintain constant lower temperatures, independent of ambient conditions. However, they have higher initial capital and operating costs. Although refrigeration technologies are favored for use in warm, humid regions, where both the peak daytime T_{WB} and air conditioning demand are highest, they have been selected and installed in more northern temperate latitudes (Canada) and arid climates as well. TES and other hybrids, which combine various technologies, are used to minimize the overall annualized costs for TIC systems. The unit costs for TIC technologies are substantially lower than are the unit costs for new GTs. Therefore, the installation of a new TIC technology to an existing GT system may offer the lowest-cost choice for increasing the output capability of that older system. For new installations, the inclusion of an optimal TIC technology with the new GT should result in minimizing the annualized cost for output power. Typical unit costs for TIC options range from \$25 (evaporative) to \$200 (refrigeration) per kW_e of incremental power output versus \$300 to \$500 per kW_e for a new single-cycle GT. Selection of the most suitable TIC option for any given application (new or retrofit) requires a careful evaluation of (1) the local power requirements and revenues for power generation, (2) the local meteorological conditions, and (3) the capabilities and costs of different TIC technology options offered by the various TIC vendors (see www.turbineinletcooling.org).

THERMAL ENERGY CONVERSION AND UTILIZATION

BOILERS

Steam generators are designed to produce steam for process needs along with electric power generation, or solely for electric power generation. In each case, the goal is to provide the most efficient and reliable boiler design for the least cost. Many factors influence the selection of the type of steam generator and its design, and some are addressed later in discussions of industrial and utility boilers.

Figure 24-53 shows the chief operating characteristics of a range of boilers, from small-scale heating systems to large-scale utility boilers. In the industrial market, boilers have been designed to burn a wide range of fuels and operate at pressures up to about 12 MPa (~1800 psia) and steaming rates extending to about 450,000 kg/h (~1,000,000 lb/h). High-capacity shop-assembled boilers (package boilers) range in capacity from about 4550 kg/h (~10,000 lb/h) to about 270,000 kg/h (600,000 lb/h). These units are designed for operation at pressures up to about 11 MPa (~1650 psia) and about 780 K (~950°F). Figure 24-54 shows a gas- or liquid-fuel-fired unit. While most shop-assembled boilers are gas- or oil-fired, designs are also available to burn pulverized coal. For comparison, a field-erected coal-fired industrial boiler is shown in Fig. 24-55.

Boilers designed for service in electric power utility systems operate at both subcritical-pressure and supercritical-pressure steam conditions. Subcritical-pressure boilers range in design pressures up to about 19 MPa (2700 psia), with steaming capacities up to about 3000 Mg/h (6,500,000 lb/h). Supercritical-pressure boilers have been designed to operate at pressures up to 35 MPa (5000 psia). The 24 MPa (3500 psia) cycle has been firmly established in the utility industry, and boilers with steaming capacities up to 4200 Mg/h (9,300,000 lb/h) and superheat and reheat temperatures of 814 K (1005°F) are in service.

With the drive to reduce greenhouse gas emissions from coal-fired systems, newer boiler designs referred to as Advanced Ultra-Supercritical (A-USC) boilers have been installed and are operating in China, Europe, and the United States. These units (see for example Fig. 24-56) operate at temperatures up to 1030 K (1400°F) and pressures up to 30 MPa (4364 psia), with main steam capacities up to 1830 Mg/h (4,035,000 lb/h) [Weitzel, et al., *Advanced Ultra-Supercritical Power Plant (700 to 760C) Design for Indian Coal.* Babcock & Wilcox Power Generation Group, 2012. BR-1884]. These units generate the same power using approximately 11 percent less coal (11% less CO_2 produced compared to the supercritical design), and they rely on advanced materials to withstand the high temperatures and pressures in the boiler tubes.

The furnace of a large coal-fired steam generator absorbs half of the heat released, so the gas temperature leaving the furnace is about 1376 K (2000°F), depending on the slagging behavior of the fuel.

Boiler Design Issues Boiler design involves the interaction of many variables: water-steam circulation, fuel characteristics, firing systems and heat input, and heat transfer. The furnace enclosure is one of the most critical components of a steam generator and must be conservatively designed to assure high boiler availability. The furnace configuration and its size are determined by combustion requirements, fuel characteristics, emission standards for gaseous effluents and particulate matter, and the need to provide a uniform gas flow and temperature entering the convection zone to minimize ash deposits and excessive superheater metal temperatures. Discussion of some of these factors follows.

Circulation and Heat Transfer Circulation, as applied to a steam generator, is the movement of water or steam or a mixture of both through heated tubes. The circulation objective is to absorb as much heat as possible from the fireside through the tube metal at a rate that assures sufficient cooling of the furnace-wall tubes during all operating conditions, with an adequate margin of reserve for transient upsets. Adequate circulation prevents excessive metal temperatures or temperature differentials that would cause failures due to overstressing, overheating, or corrosion.

The rate of heat transfer from the tubes to the fluid depends primarily on turbulence and the magnitude of the heat flux itself. Turbulence is a function of fluid flow velocity and tube surface roughness. Turbulence has been achieved by designing for high flow velocities, which ensure that nucleate boiling takes place along the inner tube surface. If sufficient turbulent flow does not exist, *departure from nucleate boiling* (DNB) occurs when a thin film of steam on the tube surface forms that impedes heat transfer and results in tube overheating and possible tube failure (rupture). DNB is especially significant in A-USC boilers that use advanced materials in their boiler tubes and are more susceptible to tube failure.

Satisfactory performance is obtained with tubes having helical ribs on the inside surface to produce a swirling flow inside the tubes. The resulting centrifugal action forces the water droplets toward the inner tube surface and prevents the formation of a steam film. The internally rifled tube maintains nucleate boiling at much higher steam temperature and pressure and with much lower flow velocities than are needed in smooth tubes. In modern practice, the most important criterion in drum boilers is the prevention of conditions that favor DNB.

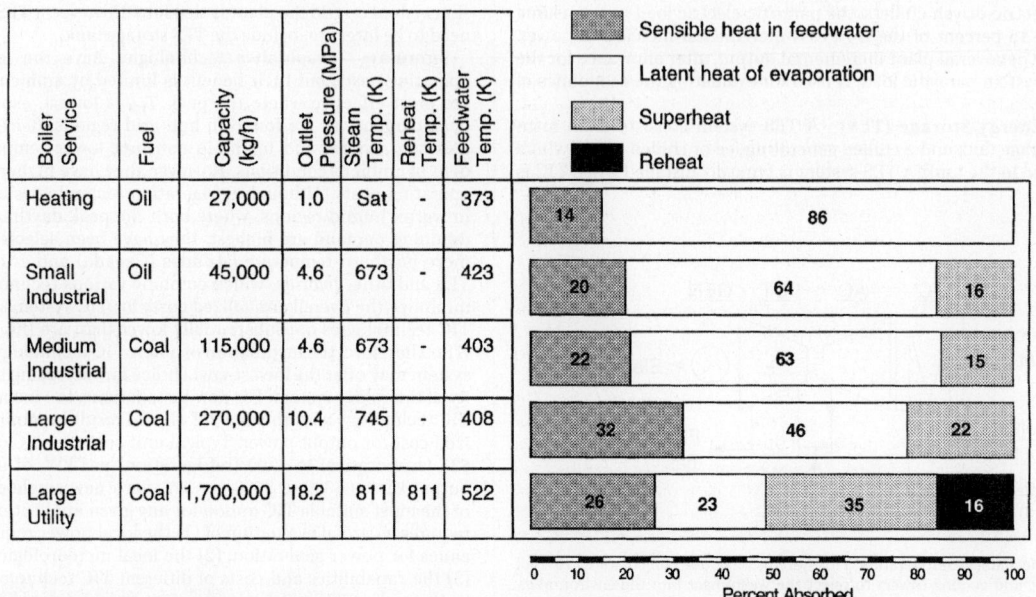

Boiler Service	Fuel	Capacity (kg/h)	Outlet Pressure (MPa)	Steam Temp. (K)	Reheat Temp. (K)	Feedwater Temp. (K)
Heating	Oil	27,000	1.0	Sat	-	373
Small Industrial	Oil	45,000	4.6	673	-	423
Medium Industrial	Coal	115,000	4.6	673	-	403
Large Industrial	Coal	270,000	10.4	745	-	408
Large Utility	Coal	1,700,000	18.2	811	811	522

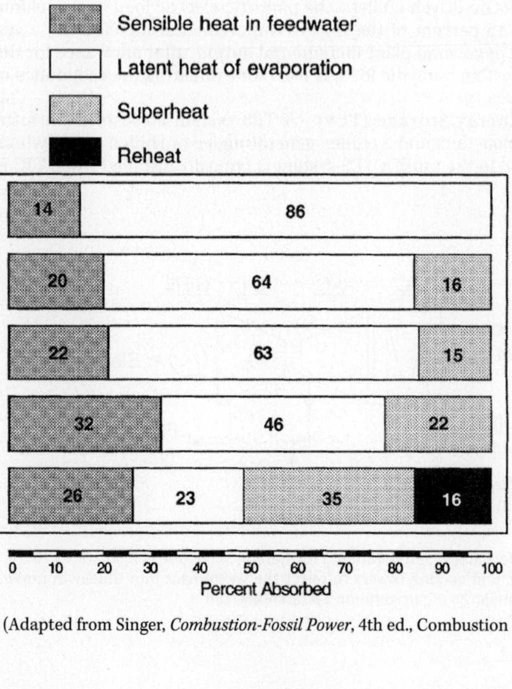

FIG. 24-53 Heat absorption distribution for various types of boilers. (Adapted from Singer, *Combustion-Fossil Power*, 4th ed., Combustion Engineering, Inc., Windsor CT, 1991.)

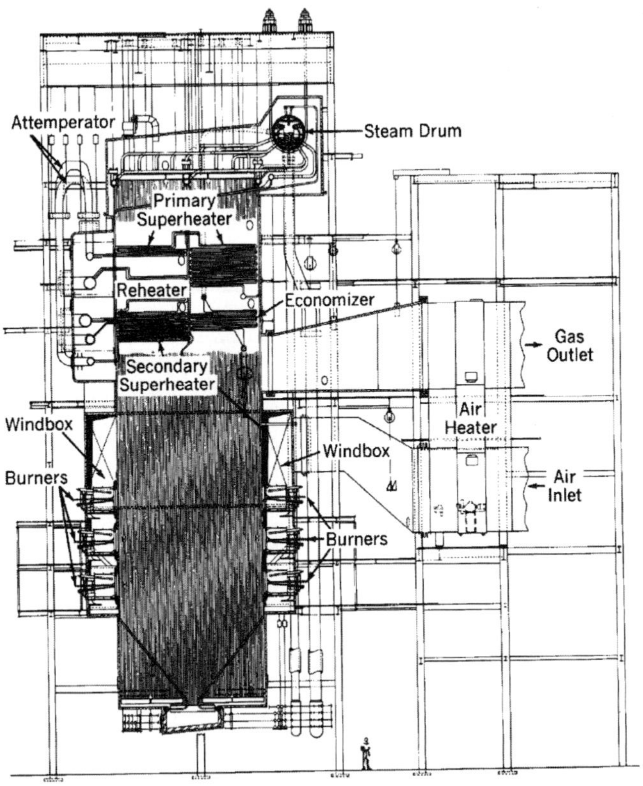

FIG. 24-54 Shop-assembled radiant boiler for natural gas or oil. (Babcock & Wilcox Co.)

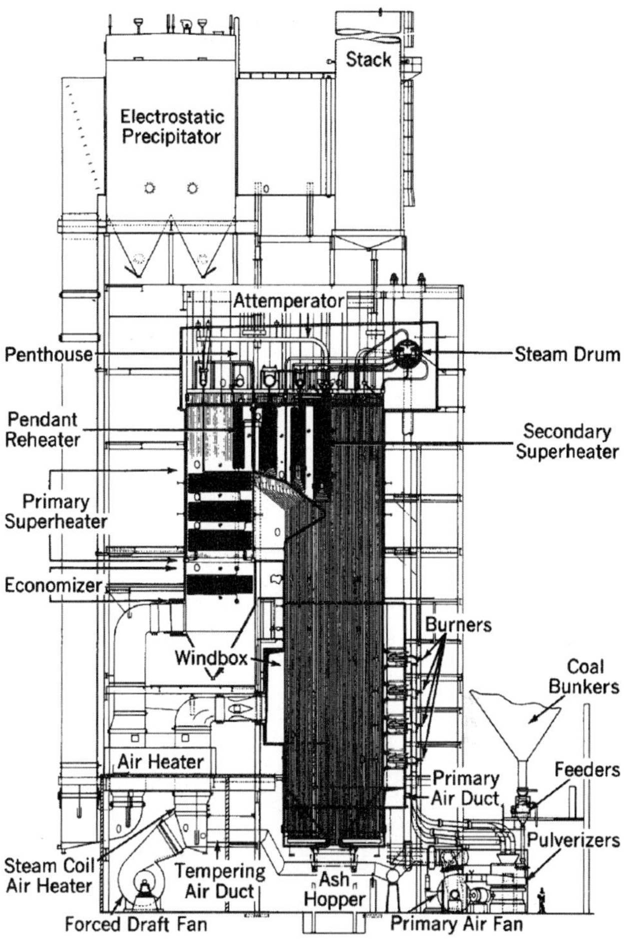

FIG. 24-55 Field-erected radiant boiler for pulverized coal. (Babcock & Wilcox Co.)

UTILITY STEAM GENERATORS

Steam Generator Circulation Systems Circulation systems for utility applications are generally classified as natural circulation and forced (pump-assisted) circulation in drum-type boilers, and as once-through flow in subcritical- and supercritical-pressure boilers.

Natural circulation in a boiler circulation loop depends only on the difference between the mean fluid/water density in the downcomers and the mean density of the steam/water mixture in the heated furnace water walls. The actual circulating pressure head is the difference between the total gravity head in the downcomer and the integrated gravity heads in the upcoming legs of the loop containing the heated tubes. The circulating head must balance the sum of the losses due to friction, shock, and acceleration throughout the loop.

In a once-through system, the feedwater entering the unit absorbs heat until it is completely vaporized into steam. The total mass flow through the water wall tubes equals the feedwater flow and, during normal operation, the total steam flow. As only steam leaves the boiler, there is no need for a steam drum.

Fuel Characteristics Fuel choice has a major impact on boiler design and sizing. Because of the heat transfer resistance offered by ash deposits on the furnace chamber walls in a coal-fired boiler, the mean absorbed heat flux is lower than in gas- or oil-fired boilers, so a greater surface area must be provided to maintain the same total boiler heat duty.

In addition, coal characteristics play a major role in the design and operation of a coal-fired boiler. Coals with low volatile-matter content (see Table 24-1) require higher ignition temperatures, while coals with less than 12 to 14 percent volatile matter may require supplementary fuel to stabilize ignition. Generally, western U.S. coals are more reactive than other coals and, consequently, are easier to ignite, but because of high moisture content they also require higher air temperatures to the mills to dry the coal and to achieve proper pulverization, critical for efficient combustion inside the furnace. Extremely high-ash coal with various minerals composed of alkaline earth metals (sodium, calcium, potassium) also present ignition and surface fouling problems. Specific minerals in the ash and the quantity of ash significantly influence furnace sizing and operation. Accordingly, a detailed review of coal characteristics is critical to determining how fuel quality may affect boiler design and operation.

Superheaters and Reheaters A superheater raises the temperature of the steam generated above the saturation level. An important function is to minimize moisture in the last stages of a turbine to avoid blade erosion. With continued increase of evaporation temperatures and pressures, however, a point is reached at which the available superheat temperature is insufficient to prevent excessive moisture from forming in the low-pressure turbine stages. This condition is resolved by removing the vapor for reheat at intermediate pressure in the boiler and returning it to the turbine for continued expansion to condenser pressure. The thermodynamic cycle using this modification of the Rankine cycle is called the *reheat cycle.*

Economizers Economizers improve boiler efficiency by extracting a portion of available heat from the furnace exhaust gases and transferring it to feedwater, which enters the steam generator at a temperature appreciably lower than the saturation-steam temperature.

Air preheaters also recover additional heat from the flue gas to preheat air fed to the mills and combustion air fed to the coal burners.

Industrial Boilers Industrial boilers are steam generators that provide power, steam, or both to an industrial plant, in contrast to a utility boiler in a steam power plant. A common configuration consists of a stationary water-tube boiler in which some of the steam is generated in a convection-section tube bank (also termed a *boiler bank*). In the original industrial boilers, almost all boiling occurred in that section, but now most industrial steam generators with 180 Mg/h (397,000 lb/h) and greater capacity are radiant boilers. The boiler steam pressure and temperature and feedwater temperature determine the fraction of total heat absorbed in the boiler bank. For a typical coal-fired boiler producing about 90 Mg/h (198,000 lb/h):

% Total steam in boiler bank	Boiler pressure		Steam temperature		Feedwater temperature	
	MPa	Psia	K	°F	K	°F
45	1.4	200	460	369	389	241
30	4.1	600	672	750	389	241
15	10.3	1500	783	950	450	351
10	12.4	1800	811	1000	450	351

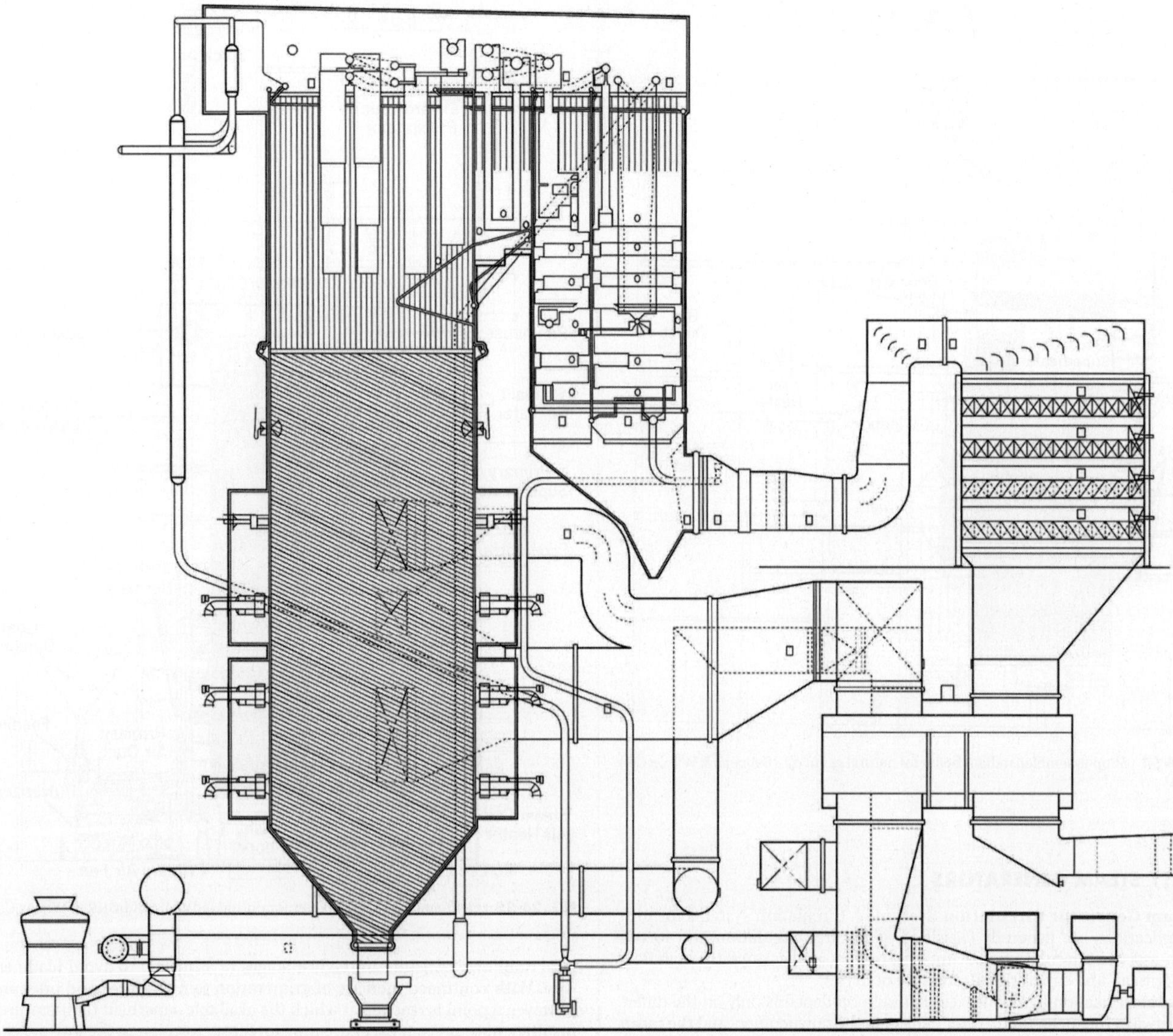

FIG. 24-56 Variable pressure 858 MW boiler for pulverized coal firing. Design pressure 3800 psig (26.2 MPa); superheat and reheat steam temperatures 1085°F (585°C); capacity 6,246,000 lb steam/h (787 kg steam/s). (Babcock & Wilcox Co.)

Higher pressures require thicker plate for operation, which increases the cost of the boiler. Thus, it is normally not economical to use a boiler bank for heat absorption at pressures above 11 MPa (1550 psia).

Industrial boilers are used for many applications, from large power-generating units with sophisticated control systems to maximize efficiency, to small low-pressure units for space or process heating, which emphasizes simplicity and low capital cost. Although their usual primary function is to provide thermal energy as steam, in some cases, steam generation is of secondary importance as a process objective. Examples include a chemical recovery unit used in the Kraft paper process to reform black liquor into white liquor, a carbon monoxide boiler used in an oil refinery to reduce CO emissions from the FCC, or a gas-cooling waste-heat boiler in an open-hearth furnace. It is not unusual for an industrial boiler to serve a multiplicity of functions. For example, a chemical-recovery boiler that converts black liquor into useful chemicals also generates process steam, while a bark-burning unit recovers heat from an otherwise waste by-product, and a waste vent incinerator treats hazardous vapors while producing process steam.

Industrial boilers burn oil, gas, coal, biofuels, and a vast range of product or waste fuels, some of which are shown in Tables 24-15 and 24-16. With the advent of fracking, natural gas has become the principal fuel of choice,

TABLE 24-15 Solid-Waste Fuels Burned in Industrial Boilers

Waste	HHV, kJ/kg*
Bagasse	8374–11,630
Furfural residue	11,630–13,956
Bark	9304–11,630
General wood wastes	10,467–18,608
Coffee grounds	11,397–15,119
Nut hulls	16,282–18,608
Rich hulls	12,095–15,119
Corncobs	18,608–19,306
Rubber scrap	26,749–45,822
Leather	27,912–45,822
Cork scrap	27,912–30,238
Paraffin	39,077
Cellophane plastics	27,912
Polyvinyl chloride	40,705
Vinyl scrap	40,705
Sludges	4652–27,912
Paper wastes	13,695–18,608

*To convert kilojoules per kilogram to British thermal units per pound, multiply by 4.299×10^{-1}.

TABLE 24-16 Fuel Consumption in Boilers in Various Industries

	Annual energy consumption (PJ/a)				
Industry	Total	Residual fuel oil	Distillate fuel oil	Natural gas	Coal
Chemicals	539	—	—	451	88
Food	323	8	6	237	72
Paper	212	29	2	143	38
Petroleum and coal products	169	6	2	159	2
Transportation equipment	56	—	1	48	7
Plastics and rubber products	54	5	1	48	—
Primary metals	40	—	—	40	—
Beverage and tobacco products	34	—	—	29	5
Computer and electronic products	29	1	1	27	—
Wood products	18	1	1	15	1
Textile mills	15	—	1	—	14
Total	1,489	50	15	1,197	227

SOURCE: *2002 Manufacturing Energy Consumption Survey,* Energy Information Administration, U.S. Dept. of Energy.

accounting for approximately 80 percent of all the energy fired in industrial boilers across a range of manufacturing industries (Table 24-16). Coal is the second most prevalent fuel, accounting for about 15 percent of the energy fired. Waste fuels, however, are becoming increasingly important.

An excellent discussion of industrial boilers is presented in Section III: "Applications of Steam" in B&W's 42nd edition of their book *Steam/Its Generation and Use.*

Design Criteria Industrial-boiler designs are tailored to the fuels and firing systems involved. Some of the more important design criteria include:

- Furnace heat-release rates, both W/m^3 and W/m^2 of effective projected radiant surface [$Btu/(h \cdot ft^3)$ and $Btu/(h \cdot ft^2)$]
- Heat release on grates
- Flue-gas velocities through tube banks
- Tube spacings

Table 24-17 gives typical values or ranges of these criteria for gas, oil, and coal-fired industrial boilers. The furnace release rates are important since they establish maximum local absorption rates within safe limits. They also have a bearing on the completeness of combustion and therefore on boiler thermal efficiency and particulate emissions. Limiting heat release on grates (in stoker firing) minimizes carbon loss, controls smoke emissions, and avoids excessive fly ash formation and release.

Limits on flue-gas velocities for gas- or oil-fired industrial boilers are usually determined by the need to limit draft loss. For coal firing, design gas velocities are established to minimize fouling and plugging of tube banks in high-temperature zones and erosion in low-temperature zones.

TABLE 24-17 Typical Design Parameters for Industrial Boilers

Furnace	Heat-release rate, W/m^{2*} of EPRS†
Natural gas-fired	630,800
Oil-fired	551,900–630,800
Coal: pulverized coal	220,780–378,480
Spreader stoker	252,320–410,020

Stoker, coal-fired	Grate heat-release rate, W/m^2
Continuous-discharge spreader	2,050,000–2,207,800
Dump-grade spreader	1,419,300–1,734,700
Overfeed traveling grate	1,261,000–1,734,700

Flue-gas velocity: type Fuel-fired	Single-pass Boiler, m/s	Baffled Boiler, m/s	Baffled Economizer, m/s
Gas or distillate oil	30.5	30.5	30.5
Residual oil	30.5	22.9	30.5
Coal (not lignite)			
Low-ash	19.8–21.3	15.2	15.2–18.3
High-ash	15.2	NA‡	12.2–15.2

*To convert watts per square meter to British thermal units per hour-foot, multiply by 0.317.
†Effective projected radiant surface.
‡Not available.

Convection tube spacing is important when the fuel is residual oil or coal, especially coal with ash having low fusion temperature and high fouling/slagging tendencies. The amount of ash and, even more important, specific ash characteristics must be specified for design.

Natural-circulation and convection boiler banks are the basic design features on which a line of standard industrial boilers has been developed to accommodate the diverse steam, water, and fuel requirements of the industrial market.

Figure 24-57 shows the amount of energy available for power generation in a fire-tube boiler, an industrial boiler, and subcritical- and supercritical-pressure boilers, respectively. Condensing losses decrease substantially, and regeneration of air and feedwater becomes increasingly important in the most advanced central-station boilers.

The boiler designer must proportion heat-absorbing and heat-recovery surfaces in a way to make the best use of heat released by the fuel. Water walls, superheaters, and reheaters are exposed to convection and radiant heat, whereas convection heat transfer predominates in air preheaters and economizers. The relative amounts of these surfaces vary with the size and operating conditions of the specific boiler.

Package Boilers In a fire-tube boiler, hot combustion products flow through tubes immersed in boiler water, transferring heat to it. In a water-tube boiler, combustion heat is transferred to water flowing through tubes that line the furnace walls and boiler passages. The greater safety of water-tube boilers has long been recognized, and they have generally superseded fire-tube configurations except for small package boiler designs. Fire-tube package boilers range from a few hundred to about 18,000 kg/h (40,000 lb/h) steaming capacity. A fire-tube boiler is illustrated in Figs. 24-58 and 24-59. Water-tube package boilers range from a few hundred to 270,000 kg/h (600,000 lb/h) steaming capacity. A water-tube package boiler is illustrated in Fig. 24-60. Most water-tube package boilers use natural circulation and are designed for pressurized firing. The most significant advantage of shop-assembled or package boilers is the cost benefit associated with the use of standard designs and parts.

Package boilers can be shipped complete with fuel-burning equipment, controls, and boiler trim. Larger packaged boilers may be shipped in sections, however, and a shop-assembled boiler with a capacity greater than about 109,000 kg/h (240,000 lb/h) is deliverable only by barge. (For a more detailed discussion of shop-assembled boilers, see Singer, *Combustion-Fossil Power,* 4th ed., Combustion Engineering, Inc., Windsor Conn., 1991, pp. 8.36–8.42.)

Fluidized-Bed Boilers Although the furnace of a fluid-bed boiler has a unique design, the system as a whole consists mainly of standard equipment items adapted to suit process requirements. The systems for feedstock preparation and feeding (biomass, coal, sorbent, and sand), ash removal, and ash disposal are very similar to those found in PC boiler plants, the main difference being that the top size of the material being handled is greater. The water-wall boiler enclosure and convection pass tubing are also like the designs found in PC boilers. The fluidized-bed plant includes particulate removal equipment such as cyclones, baghouses, and ESPs, with designs like those found in other solids-handling process plants.

Bubbling FBCs A simplified schematic of a bubbling FBC plant is presented in Fig. 24-61. This design is used primarily for burning biomass, although coal may be cofired to maintain heat release rates when feeding high-moisture feedstocks such as sludges.

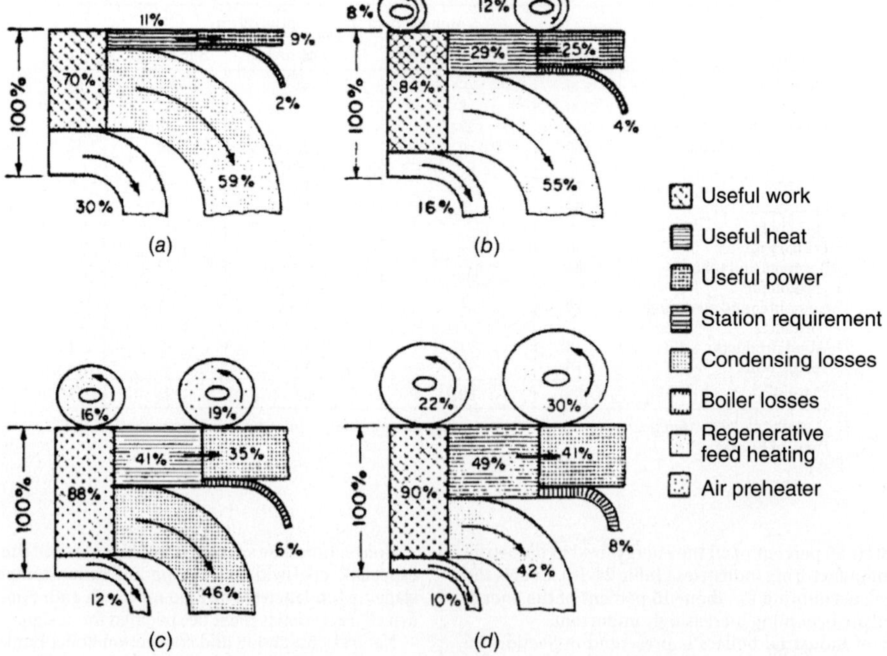

FIG. 24-57 Sankey diagrams for various types of boilers: (*a*) fire-tube boiler; (*b*) industrial boiler; (*c*) subcritical boiler; (*d*) supercritical boiler.

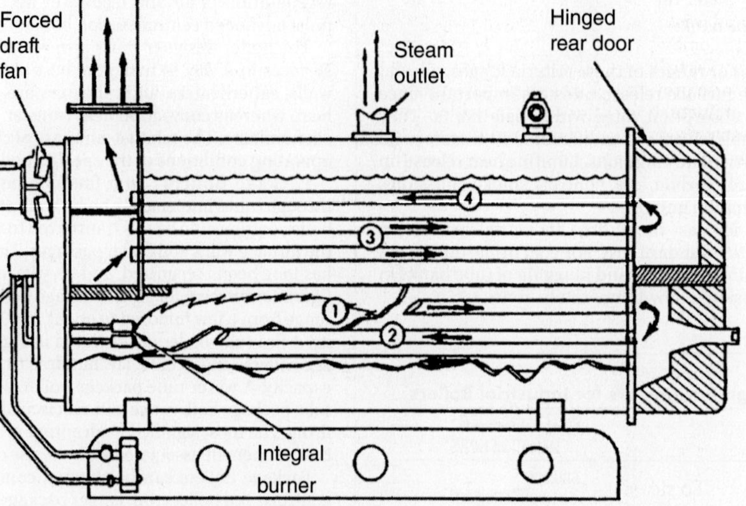

FIG. 24-58 Four-pass packaged fire-tube boiler. Circled numbers indicate passes. (From Cleaver Brooks, Inc. Reproduced from *Gas Engineer's Handbook*, Industrial Press, New York, 1965, with permission.)

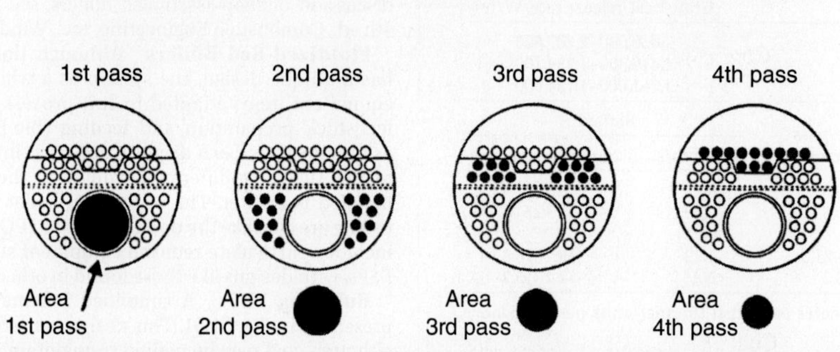

FIG. 24-59 Location and relative size of each of the four passes of the flue gas through a fire-tube boiler. (From Cleaver Brooks, Inc. Reproduced from *Gas Engineer's Handbook*, Industrial Press, New York, 1965, with permission.)

FIG. 24-60 D-type shop-assembled water-tube boiler. (Combustion Engineering Inc.)

The biomass is fed overbed through multiple feed chutes using air jets to help distribute the fuel over the surface of the bed. Variable-speed screw conveyors are commonly used to meter the fuel feed rate and control steam output. Feedstocks such as bark and waste wood are chipped to a top size of 25 mm (1 in) to promote complete combustion. The bed usually consists of sand around 1 m (3 ft) deep. This serves to homogenize the fluidized bed and increase combustion efficiency. It also provides a heat sink to help maintain bed temperature during periods of fluctuating fuel moisture content.

Biomass fuels, especially forestry products and waste wood, can contain oversized material such as rocks and other debris. If this material accumulates in the fluid bed, fluidization deteriorates, resulting in poor fuel mixing that reduces combustion efficiency and lowers thermal efficiency. To avoid this mal-operation, units typically employ an open-bottom floor design that allows easy solids removal while still ensuring even distribution of the fluidizing air. The fluidizing air (primary combustion air) is fed into large-bore pipes extending across the width of the furnace and enters the boiler through a series of nozzles set in the pipes. The pipes are spaced to allow material to drain from all areas of the bed.

Bed temperature control strategies discussed earlier result in the bed operating sub-stoichiometrically. This helps lower NO_x formation but

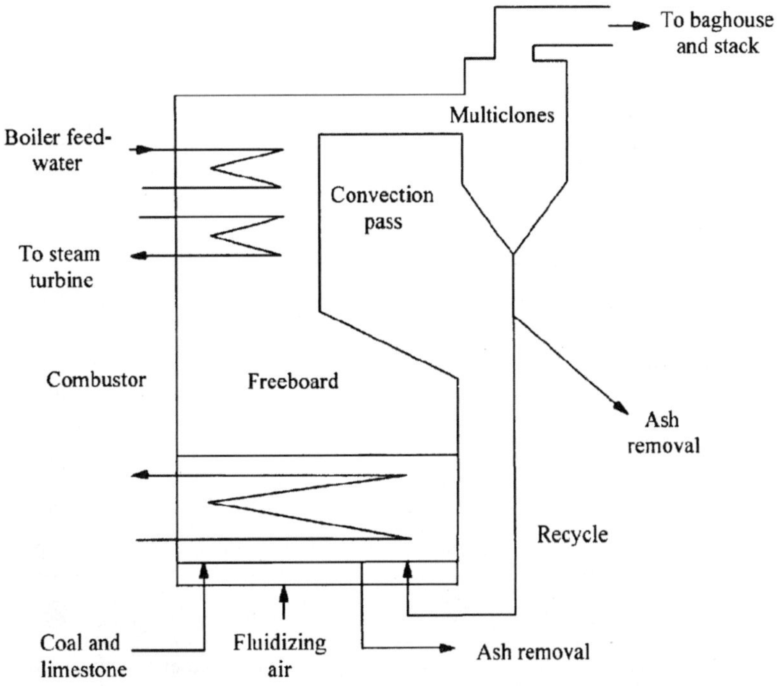

FIG. 24-61 Simplified flow diagram for bubbling AFBC.

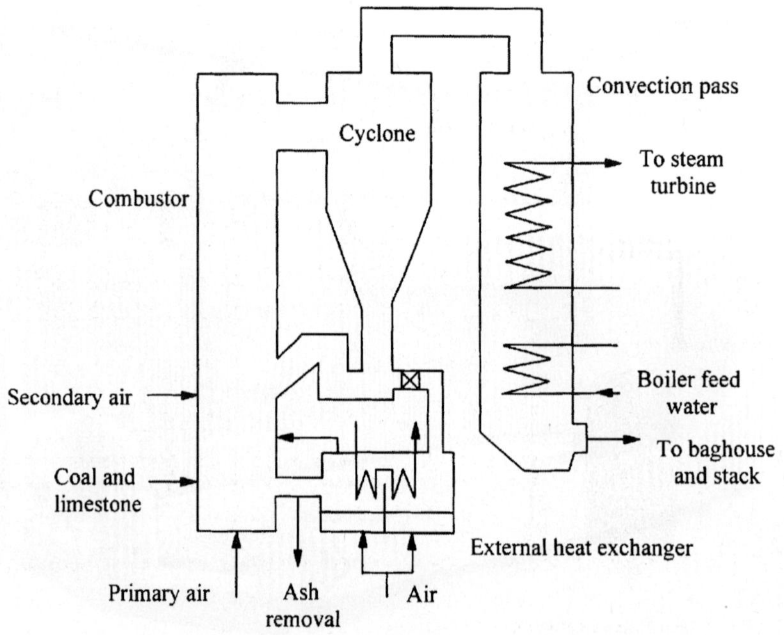

FIG. 24-62 Simplified flow diagram for circulating AFBC.

increases the amount of CO formed and the amount of unburned fuel leaving the bed. To complete the combustion process, secondary air is introduced through multiple inlet ports above the bubbling bed, with the upper furnace height designed to achieve 2- to 3-s flue gas residence time. The velocity of the secondary air should be sufficient to penetrate across the furnace and facilitate good mixing with the flue gas leaving the bed surface. Although the unburned carbon leaving the furnace is normally low, CO levels can still be relatively high at 43 to 128 mg/MJ (0.10 to 0.25 lb/MBtu), which may be a problem in nonattainment areas.

The main reason to use an FBC is that they operate at lower combustion temperatures than normal industrial boilers, which avoids the generation of thermal NO_x during the combustion process.

Circulating FBCs The circulating FBC is more widely deployed than the bubbling-bed design and at far larger sizes. A simplified schematic for a design with an external heat exchanger is presented in Fig. 24-62. As of 2005 there were over 20 units in the range of 200 to 320 MW_e worldwide, and a similar number were in the planning stages.

The most widely used fuels are low-grade coals with high ash and moisture contents, making them unsuitable for use in traditional PC plants. The coal is crushed to a top size of 12 mm (1/2 in), without drying, and fed by gravity into the lower refractory-lined portion of the boiler. The feed points are close to the pressure balance point, and the low backpressure greatly reduces the sealing requirements. In some designs the coal is also introduced into the cyclone ash return lines to simplify fuel feeding and utilize some of the waste heat contained in the ash to dry the feed coal. The larger units average a feed point for every 23 m^2 (250 ft^2) of freeboard cross section. The highly turbulent bed facilitates good mixing of the coal across the combustor. The sorbent is prepared to a top size of 1 mm (0.04 in) and dried so that it can be pneumatically conveyed to the combustor also using one feed point for every 23 m^2 (250 ft^2) of freeboard cross section. Load control is achieved primarily by reducing the coal feed rate, with a corresponding reduction in airflow, to reduce the heat release rate. As the heat-transfer area remains unchanged, the furnace temperature falls.

Almost all the particulate matter leaving the boiler is collected by cyclones and recycled to the base of the unit. The number of cyclones varies in different design concepts, but for the larger units, one cyclone is provided for 100 to 130 MW_e of generating capacity, providing one recycle point for 210 to 290 m^2 (720 to 950 ft^2) of freeboard cross section. The collected particulate matter is returned against a backpressure of 0.02 MPa (3 psi), through a J-valve. The recycle ratio can be as high as 40:1, corresponding to a relatively long mean particle residence time and accounting for the high performance of circulating units. For bituminous coal combustion, efficiency is close to 99 percent, but like the biomass-fired bubbling beds, CO levels are relatively high at around 43 mg/MJ (0.1 lb/MBtu).

Pressurized Circulating FBC A pressurized circulating fluidized bed (PCFB) combustor can use lower quality (high sulfur, high ash) coal and blends of coal and biomass with high thermal efficiencies (~50%). Due to increased operating pressure, this system has a smaller footprint than a traditional FBC (see Fig. 24-63) (PFBC Environmental Energy Technology, Inc., *PFBC: Competitive Clean Coal Power Utilizing Pressurized Fluidized Bed Combined-Cycle Technology* [Yamada et al., "Coal Combustion Power Generation Technology," *Bulletin of the Japan Institute of Energy*, 82(11): pp. 822–829 (2003)].

Though not common, these units have found application in Europe as evidenced by a demonstration plant installed in Cottbus, Germany, for carbon capture. This plant was shut down due to problems with feeding coal into the pressurized combustor. However, these units provide better control of carbon emissions for a wider range of fuel quality.

PROCESS HEATING EQUIPMENT

Many major energy-intensive industries depend on direct-fired or indirect-fired equipment for drying, heating, calcining, melting, or chemical processing. This subsection discusses both direct- and indirect-fired equipment, with the greater emphasis on indirect firing for the process industries.

Direct-Fired Equipment Direct-fired combustion equipment transfers heat by bringing the flame or the products of combustion into direct contact with the process stream. Common examples are rotary kilns, open-hearth furnaces, and submerged-combustion evaporators. Table 24-18 gives the average energy consumption rates for various industries and processes that use direct heat. Section 12 of this handbook describes and illustrates rotary dryers, rotary kilns, and hearth furnaces. Forging, heat treating, and metal milling furnaces are discussed by Mawhinney (*Marks' Standard Handbook for Mechanical Engineers,* 9th ed., McGraw-Hill, New York, 1987, pp. 7.47–7.52). Other direct-fired furnaces are described later in this section.

Indirect-Fired Equipment (Fired Heaters) Indirect-fired combustion equipment (fired heaters) transfers heat released by the combustion of fuels into an open space (furnace) across either a metallic or refractory wall separating the flame and products of combustion from the process stream. Examples of indirect-fired combustion equipment include heat exchangers (discussed in Sec. 11), steam boilers, fired heaters, muffle furnaces, and melting pots. Steam boilers have been treated earlier in this section, and a subsequent subsection on industrial furnaces will include muffle furnaces.

Fired heaters differ from other indirect-fired processing equipment in that the process stream is heated by passage through a coil or tubebank located along the walls and roof of the furnace. Process fluid temperatures range between 478 K (400°F) in crude heaters and 1172 K (1650°F) in ethylene synthesis heaters. High process temperatures are achieved by burning hydrocarbon fuels with air to produce flame temperatures up to 2200 K (3500°F) at near-stoichiometric conditions. To ensure complete combustion and limit NO_x and CO emissions, fired heaters normally operate with at least 10 percent excess air and flame temperatures around 1366 K (2000°F). Combustion data for hydrocarbon fuels is provided by Hougen, Watson, and Ragatz (*Chemical*

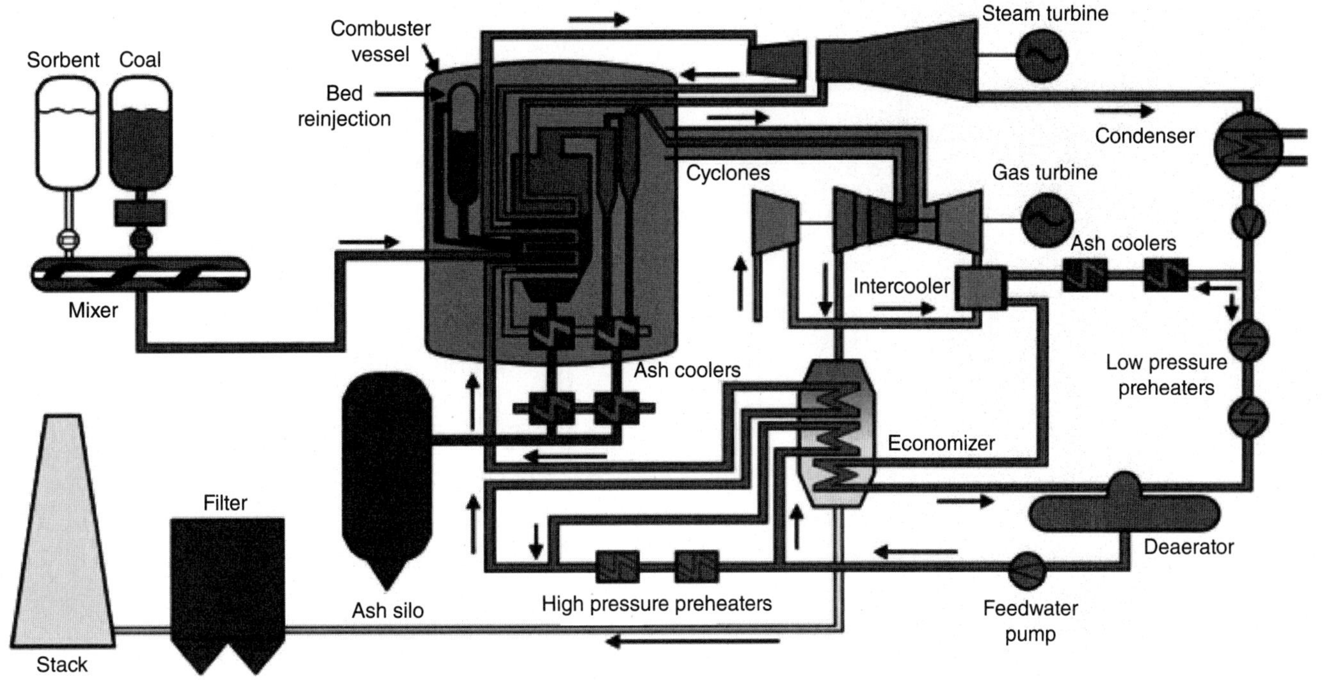

FIG 24-63 Diagram of a pressurized circulating fluidized bed combustion system. (PFBC Environmental Energy Technology, Inc.)

Process Principles, vol. I, Wiley, New York, 1954, p. 409) and in *Marks' Standard Handbook for Mechanical Engineers,* 8th ed. (McGraw-Hill, New York, 1978, p. 4.57). Fired heaters are classified by function and by coil design.

Column reboilers heat and partially vaporize a recirculating stream from a fractionating column. The outlet temperature of a reboiler stream is typically 477 to 546 K (400°F to 550°F).

Fractionator-feed preheaters partially vaporize charge stock from an upstream unfired preheater en route to a fractionating column. A typical refinery application includes a crude feed to an atmospheric column that enters the fired heater as a liquid at 505 K (450°F) and leaves at 644 K (700°F), having become 60 percent vaporized.

Reactor-feed-stream preheaters heat the reactant stream(s) for a high-temperature chemical reaction. The feed stream may be single-phase/single-component [example: steam being superheated from 644 to 1089 K (700°F to 1500°F) for styrene-manufacture reactors]; single-phase/multicomponent [example: preheating the feed to a catalytic reformer, a mixture of hydrocarbon vapors and recycle hydrogen, from 700 to 811 K (800°F to 1000°F) under pressure as high as 4.1 MPa (600 psia)]; or multiphase/multicomponent [example: a mixture of hydrogen gas and liquid hydrocarbon heated from 644 to 727 K (700°F to 850°F) at about 20 MPa (3000 psia) before it enters a hydrocracker].

Heat-transfer-fluid heaters maintain the temperature of a circulating liquid heating medium (e.g., a paraffinic hydrocarbon mixture, a Dowtherm, or a molten salt) at a level that may exceed 673 K (750°F).

Viscous-liquid heaters lower the viscosity of very heavy oils to pumpable levels.

Fired reactors contain tubes or coils in which an endothermic reaction within a stream of reactants occurs. Examples include steam/hydrocarbon reformers, catalyst-filled tubes in a combustion chamber, and pyrolyzers, coils in which alkanes (from ethane to gas oil) are cracked to olefins; in both types of reactors the temperature is maintained up to 1172 K (1650°F).

Coil Design Indirect-fired equipment is conventionally classified by tube orientation: vertical and horizontal. Although there are many variations of each of these two principal configurations, they all are embraced within seven major types, as follows.

A *simple vertical cylindrical heater* (see Fig. 24-64) has vertical tubes arrayed along the walls of a combustion chamber fired vertically from the floor. This type of heater does not include a convection section, and it is inexpensive. It has a small footprint but low efficiency, and it is usually selected for small-duty applications [0.5 to 21 GJ/h (0.5 to 20 × 10^6 Btu/h)].

Vertical cylindrical, cross-tube convection heaters are like the preceding type except for a horizontal convective tube bank above the combustion chamber. The design is economical with a high efficiency, and it is usually selected for higher-duty applications [11 to 210 GJ/h (10 to 200 × 10^6 Btu/h)].

The *arbor (wicket)* heater is a substantially vertical design in which the radiant tubes are inverted U's connecting the inlet and outlet terminal manifolds in parallel. An overhead crossflow convection bank is usually included. This type of design is good for heating large gas flows with low pressure drop. Typical duties are 53 to 106 GJ/h (50 to 100 × 10^6 Btu/h).

In the *vertical-tube, single-row, double-fired heater,* a single row of vertical tubes is arrayed along the center plane of the radiant section that is fired from both sides (Fig. 24-65). Usually this type of heater has an overhead horizontal convection bank. Although it is the most expensive of the fired heater designs, it provides the most uniform heat transfer to the tubes. Duties range from 21 to 132 GJ/h (20 to 125 × 10^6 Btu/h) per cell (twin-cell designs are not unusual).

Horizontal-tube cabin heaters position the tubes of the radiant-section coil horizontally along the walls and the slanting roof for the length of the cabin-shaped enclosure. The convection tube bank is placed horizontally above the combustion chamber. It may be fired from the floor, the side walls,

TABLE 24-18 Average Energy Consumption for Various Industries Using Direct Heat

Industry	Product/process	Energy consumption per unit of product GJ/Mg	10^6 Btu/U.S. ton
Paper	Kraft process	20.9	18.0
	Integrated plant/paper*	34.2	29.5
	Integrated plant/paperboard*	18.8	16.2
Glass	Flat glass	17.3	14.9
	Container glass	18.1	15.6
	Pressed/blown	31.6	27.2
Clay/ceramics	Portland cement	4.6	4.0
	Lime	5.5	4.7
	Mineral wool	42.7	36.8
Steel	Blast furnace and steel mills	20.7	17.8
Nonferrous metals	Primary copper	34.2	29.5
	Secondary copper	4.6	4.0
	Primary lead	25.1	21.6
	Secondary lead	0.8	0.7
	Primary zinc	69.6	60.0
	Secondary zinc	5.0	4.3
	Primary aluminum	78.9	68.0
	Secondary aluminum	5.2	4.5

SOURCE: *Manufacturing Consumption of Energy,* Energy Information Administration, U.S. Dept. of Energy, 1991.

*Mixture of direct and indirect firing.

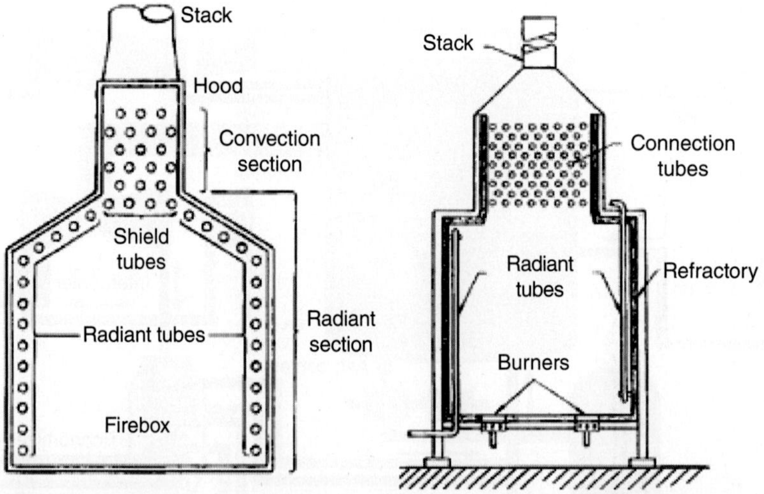

FIG. 24-64 Fired heaters used to heat process fluids by indirect heating: (*a*) box or cabin heater, (*b*) vertical cylindrical heater. (W. R. P. J. R. F. James R. Couper, "Section 8.11 Heat Transfer and Heat Exchangers: Fired Heaters," in *Chemical Process Equipment: Selection and Design*, 3rd ed., Elsevier, New York, 2012, pp. 211–219.)

or the end walls. As in the case of its vertical cylindrical counterpart, its economical design and high efficiency make it the most popular horizontal-tube heater. Duties are 11 to 105 GJ/h (10 to 100 10^6 Btu).

In the *horizontal-tube box heater with side-mounted convection tube bank*, the radiant-section tubes run horizontally along the walls and the flat roof of the box-shaped heater, but the convection section is placed in a box of its own beside the radiant section. Firing is horizontal from the end walls. The design of this heater results in a relatively expensive unit justified mainly by its ability to burn low-grade, high-ash fuel oil. Duties are 53 to 210 GJ/h (50 to 200 × 10^6 Btu/h).

Vertical cylindrical helical coil heaters are hybrid designs that are classified as vertical heaters, but their in-tube characteristics are like those of horizontal heaters. There is no convection section. In addition to the advantages of simple vertical cylindrical heaters, helical coil heaters are easy to drain. They are limited to small-duty applications: 5 to 21 GJ/h (5 to 20 × 10^6 Btu/h).

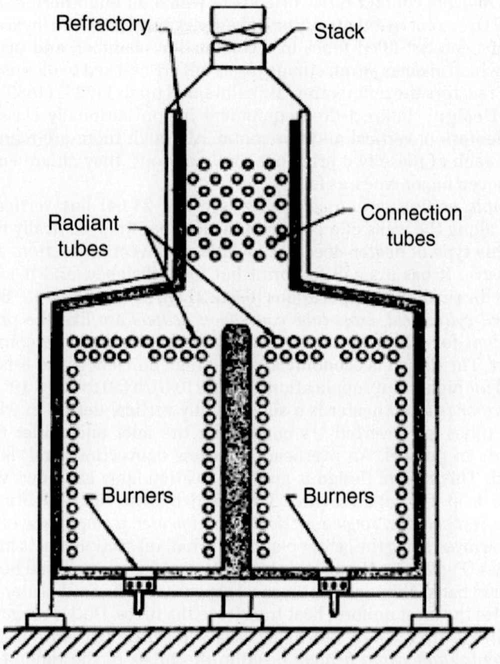

FIG. 24-65 Double-fired cabin (box) heater. (W. R. P. J. R. F. James R. Couper, "Section 8.11 Heat Transfer and Heat Exchangers: Fired Heaters," in *Chemical Process Equipment: Selection and Design*, 3rd ed., Elsevier, New York, 2012, pp. 211–219.)

Schematic elevation sections of a vertical cylindrical, cross-tube convection heater; a horizontal-tube cabin heater; and a vertical cylindrical, helical-coil heater are shown in Fig. 24-66. The seven basic designs and some variations of them are pictured and described in Berman [*Chem. Eng.* **85:** 98–104 (1978)] and by R. K. Johnson [*Combustion* **50**(5): 10–16, (1978)].

The design of both radiant and convection sections of fired heaters, along with some equipment descriptions and operating suggestions, are discussed by Berman in *Encyclopedia of Chemical Processing and Design* (McKetta, ed., vol. 22, Marcel Dekker, New York, 1985, pp. 31–69). He also treats construction materials, mechanical features, and operating points in three other *Chemical Engineering* articles [all in vol. 85 (1978): no. 17, July 31, pp. 87–96; no. 18, August 14, pp. 129–140; and no. 20, Sept. 11, pp. 165–169].

INDUSTRIAL FURNACES

Industrial furnaces serve the manufacturing sector and can be divided into two groups. Boiler furnaces are the larger group and are used solely to generate steam, as discussed earlier in the subsection on industrial boilers. Furnaces of the other group are classified by: (1) source of heat (fuel combustion or electricity), (2) function (heating without change of phase or with melting), (3) process cycle (batch or continuous), (4) mode of heat application (direct or indirect), and (5) atmosphere in furnace (air, protective, or reactive, including vacuum). Each will be discussed briefly.

Source of Heat Industrial furnaces are either fuel-fired or electric, and the first decision that a prospective furnace user must make is between these two. Although electric furnaces are uniquely suited to some applications in the chemical industry (manufacture of silicon carbide, calcium carbide, and graphite, for example), their principal use is in the metallurgical and metal-treatment industries. Generally, the choice between electric and fuel-fired is economic or custom-application dictated, because most tasks that can be done in one can be done equally well in the other. Except for an occasional passing reference, electric furnaces are not considered further, but the interested reader will find detailed reviews in *Kirk-Othmer Encyclopedia of Chemical Technology* (4th ed., vol. 12, articles by Cotchen, Sommer, and Walton, pp. 228–265, Wiley, New York, 1994) and in *Marks' Standard Handbook for Mechanical Engineers* (9th ed., article by Lewis, pp. 7.59–7.68, McGraw-Hill, New York, 1987).

Function and Process Cycle Industrial furnaces are enclosures in which process material is heated, dried, melted, or reacted. Melting is considered a special category because of the peculiar difficulties associated with a solid feed, a hot liquid product, and a two-phase mixture during processing; it is customary, therefore, to classify furnaces as heating or melting.

Melting Furnaces: The Glass Furnace Most melting furnaces, electric or fuel-fired, are found in the metals-processing industry, but a notable exception is the glass furnace. Like most melting furnaces, a glass furnace requires highly radiative flames to promote heat transfer to the feed charge, and it employs regenerators to conserve heat from the high-temperature process [greater than 1813K (2300°F)].

A typical side-port continuous regenerative glass furnace is shown in Fig. 24-67. Side-port furnaces with burners mounted on both sides of the

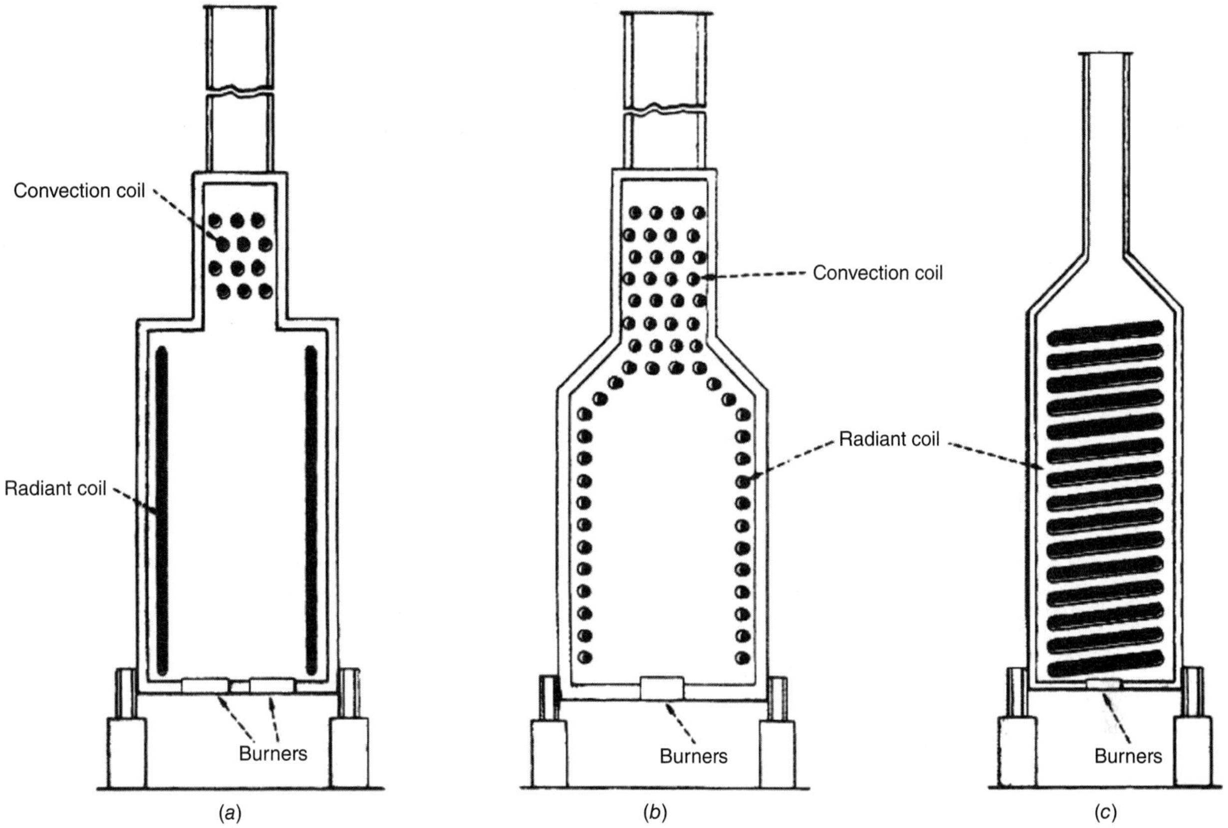

FIG. 24-66 Representative types of fired heaters: (*a*) vertical-tube cylindrical with cross-flow-convection section; (*b*) horizontal-tube cabin; (*c*) vertical cylindrical, helical coil. (From Berman, *Chem. Eng.* **85:** 98–104, June 19, 1978.)

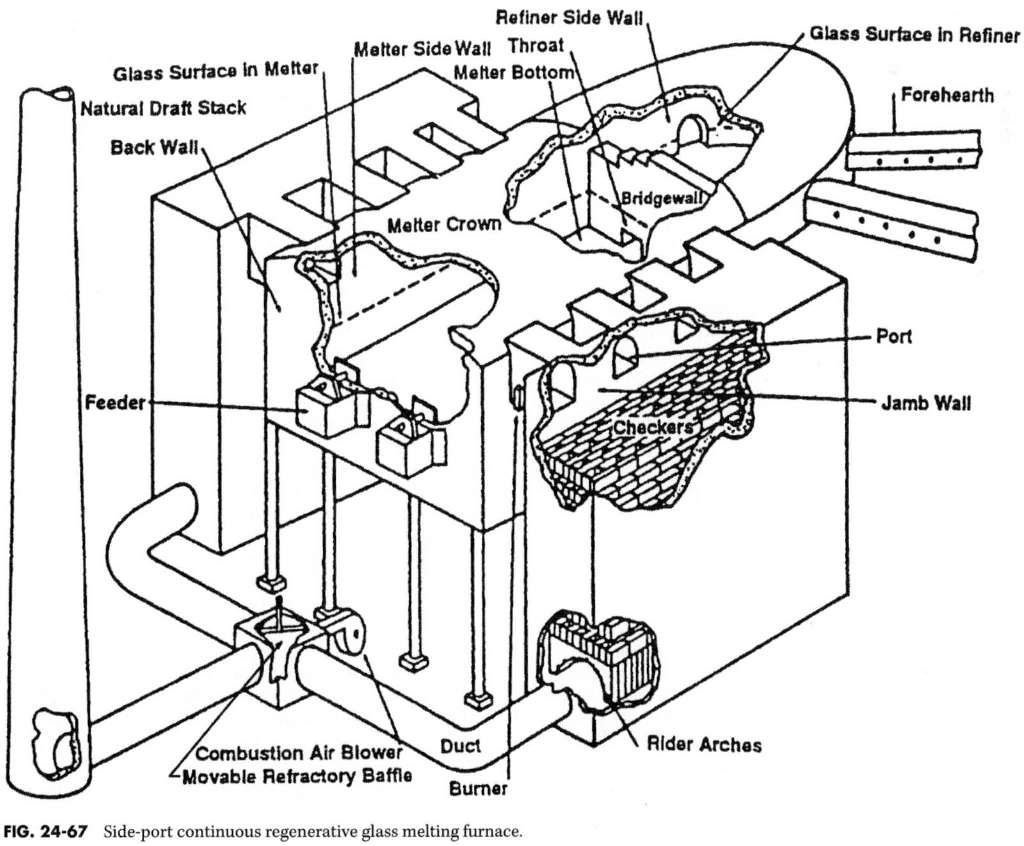

FIG. 24-67 Side-port continuous regenerative glass melting furnace.

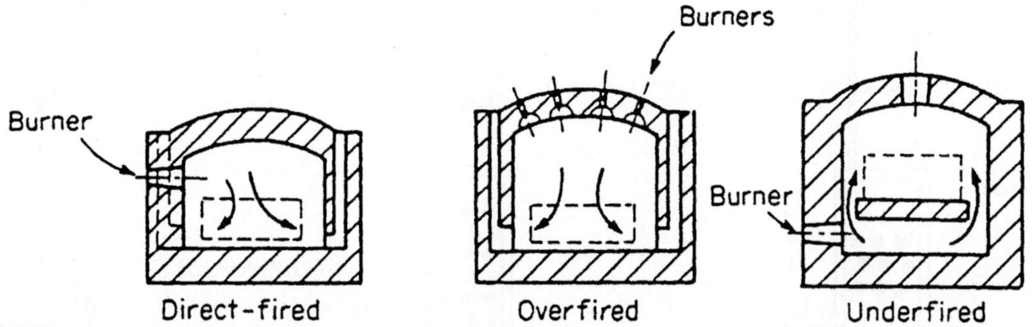

FIG. 24-68 Methods of firing direct-heated furnaces. (From *Marks' Standard Handbook for Mechanical Engineers*, 9th ed., McGraw-Hill, New York, 1987. Reproduced with permission.)

furnace and the sides firing alternately are used in the flat and container glass industries. Refractory-lined flues are used to recover energy from the hot flue gas, with high-temperature flue gas leaving the furnace heating a refractory section called a *checker*. After the checker has reached the desired temperature, gas flow is reversed, and the firing switches to the other side of the furnace. Combustion air is preheated by the hot checker to temperatures up to 1533 K (2300°F). Cycling airflow from one checker to the other is reversed approximately every 15 to 30 minutes.

The glass melt is generally 1 to 2 m (3 to 6 ft) deep, the depth being limited by the need for proper heat transfer to the melt. Container glass furnaces are typically 6 to 9 m (20 to 30 ft) wide and 6 to 12 m (20 to 40 ft) long. Flat glass furnaces tend to be longer, typically over 30 m (100 ft) in length, to ensure complete reaction of the batch ingredients and to improve glass quality (fewer bubbles). Flat glass furnaces typically have a melting capacity of 450 to 630 Mg/day (500 to 750 U.S. ton/d), compared to a maximum of 540 Mg/day (600 U.S. ton/d) for container and pressed or blown glass furnaces.

Though the stoichiometric chemical energy requirement for glassmaking is only some 2.3 GJ/Mg (2×10^6 Btu/U.S. ton) of glass, the inherently low thermal efficiency of regenerative furnaces means that, in practice, at least 7 GJ/Mg (6×10^6 Btu/U.S. ton) is required. Of this total, some 40 percent goes to batch heating and the required heat of reactions, 30 percent is lost through the furnace structure, and 30 percent is lost through the stack. The smaller furnaces used in pressed/blown glass melting are less efficient, so energy consumption may be as high as 17.4 TJ/Mg (15×10^6 Btu/U.S. ton). Industrial furnaces may be operated in batch or continuous mode.

Batch Furnaces This type of furnace is employed mainly for the heat treatment of metals and for the drying and calcination of ceramic articles. In the chemical process industry, batch furnaces may be used for the same purposes as batch-tray and truck dryers when the drying or process temperature exceeds 600 K (620°F). They are employed also for small-batch calcinations, thermal decompositions, and other chemical reactions which, on a larger scale, are performed in rotary kilns, hearth furnaces, and shaft furnaces.

Continuous Furnaces These furnaces may be used for the same general purposes as are the batch type, but usually not on a small scale. The process material may be carried through the furnace by a moving conveyor (chain, belt, roller), or it may be pushed through on idle rollers, the motion being sustained by an external pusher operating on successively entering cars or trays, each pushing the one ahead along the entire length of the furnace and through the exit flame curtains or doors.

FURNACE ATMOSPHERE AND MODE OF HEATING

Direct Heating Industrial furnaces may be directly or indirectly heated, and they may be filled with air or a protective atmosphere, or they may be under a vacuum. Direct heating is accomplished by the hot combustion gases being inside the furnace and therefore in direct contact with the process material. Thus, the material is heated by radiation and convection from the hot gas and by reradiation from the heated refractory walls of the chamber. Three styles of direct firing are illustrated in Fig. 24-68. *Simple direct firing* is used increasingly because of its simplicity and because of improved burners. The *overhead* design allows the roof burners to be so placed as to provide optimum temperature distribution in the chamber. *Underfiring* offers the advantage of the charge's being protected from the flame. The maximum temperature in these direct-heated furnaces is limited to about 1255 K (1800°F) to avoid prohibitively shortening the life of the refractories in the furnace.

Indirect Heating If the process material cannot tolerate exposure to the combustion gas or if a vacuum or an atmosphere other than air is needed in the furnace chamber, indirect firing must be used. This is accomplished in a muffle* furnace or a radiant-tube furnace (tubes carrying the hot combustion gas run through the furnace).

Atmosphere Protective atmosphere within the furnace chamber may be essential, especially in the heat treatment of metal parts. Mawhinney (in *Marks' Standard Handbook for Mechanical Engineers*, 9th ed., McGraw-Hill, New York, 1987, p. 752) lists pure hydrogen, dissociated ammonia (a hydrogen/nitrogen mixture), and six other protective reducing gases with their compositions (mixtures of hydrogen, nitrogen, carbon monoxide, carbon dioxide, and sometimes methane) that are codified for and by the metals-treatment industry. In general, any other gas or vapor that is compatible with the temperature and the lining material of the furnace can be provided in an indirect-fired furnace, or the furnace can be evacuated.

COGENERATION

Cogeneration is an energy conversion process wherein heat from a fuel is simultaneously converted to useful thermal energy (e.g., process steam) and electric energy. The need for either form can be the primary incentive for cogeneration, but there must be opportunity for economic captive use or sale of the other. In a chemical plant, the process and other heating steam is likely to be the primary product; in a public utility plant, electricity is the usual primary product.

Thus, a cogeneration system is designed from one of two perspectives: (1) It may be sized to meet the process heat and other steam needs of a plant or community of industrial and institutional users, so that the electric power is treated as a by-product that must be either used on-site or sold, or (2) it may be sized to meet electric power demand, and the rejected heat used to supply needs at or near the production site. The latter approach is more common if a utility owns the system; the former if a chemical plant is the owner.

Industrial use of cogeneration leads to small, dispersed electric-power-generation installations to reduce or eliminate reliance on large central power plants. Because of the relatively short distances over which thermal energy can be economically transported, process-heat generation is characteristically an on-site process, with or without cogeneration.

Cogeneration systems may not match the varying power and heat demands at all times for most applications. Thus, an industrial cogeneration system's output often must be supplemented by separate on-site heat generation or by the purchase of utility-supplied electric power. If the on-site electric power demand is relatively low, an alternative option is to match the cogeneration system to the heat load and contract for the sale of excess electricity to the local utility grid.

Improved economic performance through fuel saving is normally the main incentive for cogeneration. Since all heat-engine-based electric power systems reject heat to the environment, that rejected heat can often be reclaimed and used to meet all or part of the local thermal energy needs. Using rejected heat usually has no effect on the amount of primary fuel used, yet it often leads to saving all or part of the fuel otherwise required for the thermal-energy generation process. Heat engines also require a high-temperature thermal input, usually receiving the working fluid directly from a heating source, but in some situations, they can obtain their required

*A *muffle* is an impenetrable ceramic or metal barrier between the firing chamber and the interior of the furnace. It heats the process charge by radiation and furnace atmosphere convection.

thermal input as rejected heat from a higher-temperature process. In the former case, the cogeneration process employs a heat-engine topping cycle; in the latter case, a bottoming cycle is used.

The choice of fuel for a cogeneration system is determined by the primary heat-engine cycle. Closed-cycle power systems that are externally fired—the steam turbine, either an indirectly fired open-cycle gas turbine or a closed-cycle gas turbine—can use virtually any fuel that can be burned in a safe and environmentally acceptable manner: coal, municipal solid waste, biomass, and industrial wastes are burnable with closed power systems. Internal combustion engines, on the other hand, including open-cycle gas turbines, are restricted to fuels that have combustion characteristics compatible with the engine type and that yield combustion products clean enough to pass through the engine without damaging it. In addition to natural gas, butane, and the conventional petroleum-derived liquid fuels, refined liquid and gaseous fuels derived from shale, coal, or biomass are fuels that fall in this category. Direct-coal-fired internal combustion engines have been an experimental reality for decades but are not yet a practical reality technologically or economically, due mainly to solids-handling issues.

There are at least three broad classes of application for topping-cycle cogeneration systems:

- Utilities or municipal power systems supplying electric power and low-grade heat [e.g., 422 K (300°F)] for local district heating systems,
- Large residential, commercial, or institutional complexes requiring space heat, hot water, and electricity, and
- Large industrial operations with on-site needs for electricity and heat in the form of process steam, direct heat, or space heat.

Typical Systems All cogeneration systems involve the operation of a heat engine for the production of mechanical work which, in nearly all cases, is used to drive an electric generator. The most common heat-engine types appropriate for topping-cycle cogeneration systems are:

- Steam turbines (backpressure and extraction configurations)
- Open-cycle (combustion) gas turbines
- Indirectly fired gas turbines: open cycles and closed cycles
- Diesel engines

Each heat-engine type has unique characteristics, making it better suited for some cogeneration applications than for others. For example, engine types can be characterized by:

- Power-to-heat ratio at design point
- Efficiency at design point
- Capacity range
- Power-to-heat-ratio variability
- Off-design (part-load) efficiency
- Multifuel capability

The major heat-engine types are described in terms of these characteristics in Table 24-19.

ENERGY STORAGE

The asynchronicity of power supply and demand cycles require the typical power plant (coal, natural gas, or nuclear) to ramp down to less efficient rates or to dump the excess energy during lean demand hours, and to fire up expensive and inefficient peaker plants during peak demands. Other applications, such as portable, remote power and transportation sectors, require grid-independent power. Advances in storage technologies are now prompting investments in R&D or production processes that can not only help bridge the supply and demand schedules, but also enable the transport of energy from generation points to population centers with high consumption.

The energy source can be electricity at a power plant, solar energy conversions (photovoltaic, photoelectochemical, or thermal), electric power from wind turbines, or another source. Two energy storage media that are drawing investments are chemical energy storage (hydrogen or synthetic liquid fuels such as methanol and dimethyl ether) and electrochemical energy storage (batteries).

A key advantage of hydrogen (see Table 24-20) as a storage medium is that it can be stored as a compressed gas or liquid and subsequently used to generate clean electric power with fuel cell systems, stationary or automotive, with water as the only by-product at the point of use. Wider acceptance of this option will depend on a multitude of factors, and it is being led by the deployment of fuel cell vehicles, its supporting infrastructure, and the cost-effectiveness of this pathway.

Hydrogen is primarily produced and consumed as an industrial chemical in petroleum refining ($3 \cdot 10^9$ ft^3/day in U.S. refineries, 2016) (https://www.eia.gov/dnav/pet/hist/LeafHandler.ashx?n=PET&s=8_NA_8PH_NUS_6&f=A) and in the chemical industry (e.g., fertilizers) via the steam methane reforming reaction. The conversion from various hydrocarbon fuels can be generically represented by the following reaction:

$$\begin{aligned} C_nH_mO_p + x(O_2 + 3.76N_2) + (2n - 2x - p)H_2O(l) \\ = nCO_2 + (2n - 2x - p + m/2)H_2 + 3.76xN_2 \end{aligned}$$

where x is the oxygen-to-fuel molar ratio [*Int. J. of Hydrogen Energy* **26:** 291–301 (2001)].

Newer processes being developed for the production of hydrogen as an energy storage medium include water electrolysis, where the electricity generated at power plants or photovoltaic cells or wind turbines is used to electrochemically split the water; photoelectrochemical processes for water splitting [http://energy.gov/eere/fuelcells/hydrogen-production-photoelectrochemical-water-splitting]; thermochemical conversion, where concentrated solar rays are used to split water over a catalyst at temperatures exceeding 1000°C; and other processes. Table 24-21 lists some of the properties of hydrogen.

The most significant energy storage technology development is that of the lithium ion battery. Initially driven by the demand for portable power and remote and uninterruptible power needs, it is now progressing toward larger energy demands, such as the powering of hybrid and electric vehicles and eventually to load leveling in stationary applications.

ELECTROCHEMICAL ENERGY STORAGE

General Battery Descriptors A battery in its essence is a box, can, case, pouch, or other container with two terminals that is built to deliver electrical energy. There are two broad classifications: primary and secondary or rechargeable. Primary batteries are made to deliver their energy once, while rechargeable batteries are at least partially reversible and, as such, are capable of electrical energy storage. The discussion here is limited to rechargeable batteries. Fundamentally, batteries are one or more electrochemical cells made of a positive electrode, a negative electrode, and an electrolyte joining them. (Newman, J., and K. E. Thomas-Alyea, *Electrochemical Systems*, 3d ed., Wiley Interscience, New York, 2004; Rahn, C. D., and C-Y Wang, *Battery Systems Engineering*, Wiley, New York, 2013; Reddy, T. B., *Linden's Handbook of Batteries*, 4th ed., McGraw-Hill, New York, 2002; Root, M., *The TAB Battery Book: An In-Depth Guide to Construction, Design, and Use*, McGraw-Hill, New York, 2011; Vincent, C. A., and B. Scrosati, *Modern Batteries*, 2nd ed., Wiley, New York, 1997; West, A. C., *Electrochemistry and Electrochemical Engineering: An Introduction*, Columbia University, New York, 2012.

TABLE 24-19 Cogeneration Characteristics for Heat Engines

Engine type	Size range, MWe/unit	Efficiency at design point	Part-load efficiency	Multifuel capability	Maximum temperature of recoverable heat, °F (°C)*	Recoverable heat, Btu/kWh†	Typical power-to-heat ratio
Steam turbine							
Extraction-condensing type	30–300	0.25–0.30	Fair	Excellent	200 (93)–600 (315)‡	11,000–35,000	0.1–0.3
Backpressure type	20–200	0.20–0.25	Fair	Excellent	200 (93)–600 (315)‡	17,000–70,000	0.05–0.2
Combustion gas turbines	10–100	0.25–0.30	Poor	Poor	1000 (538)–1200 (649)	3000–11,000	0.3–0.45
Indirectly fired gas turbines							
Open-cycle turbines	10–85	0.25–0.30	Poor	Good	700 (371)–900 (482)	3500–8500	0.4–1.0
Closed-cycle turbines	5–350	0.25–0.30	Excellent	Good	700 (371)–900 (482)	3500–8500	0.4–1.0
Diesel engines	0.05–25	0.35–0.40	Good	Fair to poor	500 (260)–700 (371)	4000–6000	0.6–0.85

*°C + 273 = K.
†1 Btu = 1055 J.
‡Saturated steam.

TABLE 24-20 Attractive Features and Challenges of Hydrogen

Advantages	Challenges
• High energy content per gram • Produces water only on oxidation • Fuel cells convert chemical energy to electric energy with high efficiency • 50% electric only • 95%* with CHP • Transportable as gas or liquid	• Low energy content per volume • Highly flammable • Invisible flame • Light gas, difficult to compress and liquefy • Cost of production, compression, and delivery is still high • Needs specific safety features in enclosed applications

*https://fuelcellsworks.com/archives/2015/09/23/ene-farm-installed-120000-residential-fuel-cell-units/.

To fully understand battery operation, one must consider its chemistry and design. However, batteries are generally measured by a number of electrical properties.

Batteries deliver a current (I) in amperes at a terminal voltage (V) in volts. The relationship between current and voltage is generally quite complicated, but it is often represented by the resistive equation (24-60).

$$V = U - IR \tag{24-60}$$

The open-circuit voltage (U) is determined by thermodynamics, which is set by chemistry, cell configuration (i.e., number of cells in series), and state of charge (SOC). Likewise, the battery internal resistance (R) in ohms is a function of chemistry, cell and battery design, and SOC. In addition, because all the battery kinetic and transport phenomena are lumped into R, it is dependent on cycling history (i.e., integrated effects of current, temperature, and time). The SOC of a battery is defined as the fraction of capacity, usually expressed as a percentage, remaining in the battery (a fully charged battery will be at 100 percent SOC). Capacity (C) in ampere-hours is the quantity of electrons the battery can deliver. Alternatively, one can define a state of discharge (SOD = 1 − SOC). While U obviously also depends on temperature, entropic effects are usually small in batteries compared to the impact of temperature on R. Finally, Eq. (24-60) works for the battery charging or discharging by changing the sign of the current (i.e., discharge current is positive).

Typically, a battery is operated between two cutoff voltages (V_{max} and V_{min}). V_{max} is also referred to as top-of-charge voltage. Similarly, V_{min} is often called bottom-of-discharge voltage. These limits are generally set by the battery stability range. Almost all batteries are thermodynamically unstable (that is, they have side reactions that degrade performance or life). The capacity of the battery is then determined by discharging the battery at a fixed current from V_{max} to V_{min}. Obviously, the capacity can vary with applied current and the current can vary by orders of magnitude, depending on battery size. Typically, a battery size independent discharge or charge rate is defined, which is referred to as the C-rate as given in Eq. (24-61).

$$\text{C-rate} = C/(\text{hours for full discharge or charge}) \tag{24-61}$$

As an example, a C-rate of $C/2$ would fully discharge or charge the battery in 2 hours. Likewise, a C-rate of $2C$ would fully discharge or charge the battery in 30 minutes. The current is then determined by dividing the capacity by the discharge or charge time. While the concept of a C-rate eliminates the battery size dependence, the actual measured capacity will still depend, to some extent, on C-rate. To eliminate this dependence, the capacity is sometimes fixed at a set C-rate (e.g. a $C_1/2$ current is the $C/1$ capacity divided by 2 hours).

TABLE 24-21 Properties of Hydrogen

Autoignition temperature, °C	500–585
Density (NTP), g/cm^3	0.084
Diffusion coefficient (NTP), cm^2/s	0.61
Enthalpy (NTP), kJ/kg	3858
Flame temperature in air, °C	2045
Flammable range (air, vol. %)	4–75
Specific heat, C_p (NTP), J/g-K	14.3
Specific heat, C_v (NTP), J/g-K	10.2
Thermal conductivity (NTP), W/m-K	0.18
Viscosity (NTP), g/cm-s	8.8×10^{-5}
NTP = 20°C, 1 atm.	

Hydrogen Analysis Resource Center, http://hydrogen.pnl.gov/hydrogen-data/hydrogen-properties.

Ultimately, batteries are designed to store and deliver electrical energy and power. Common measures for a battery are the amount of energy and power per unit mass (e.g., specific energy, watt·hours/kilogram) or volume (e.g., energy density, watt·hours/liter), which is generally defined under set conditions (C-rate, SOC, etc.). For energy, battery capacity is both a fundamental design parameter and is one of the most easily measurable indicators of the battery health or remaining life. Further, the battery manufacturer typically establishes a rated capacity under conditions similar to its applications. The rated capacity is usually less than the actual capacity to allow for aging or the application not being able to fully use the capacity, so-called usable capacity.

Similarly, for power, R determines the reversibility of the battery and is both a fundamental design parameter and one of the most easily measurable indicators of the battery's health or remaining life. Consideration of energy efficiency and Joule heating effects suggests minimizing R. However, minimizing R tends to add weight and volume to the battery at added cost. Depending on battery size, R can vary by orders of magnitude. As an alternative to R for a design parameter, the ratio of battery terminal voltage to open circuit voltage (V/U) at a fixed current or power (e.g., maximum rated power or current) can be used. V/U has the advantage of being independent of battery size. Maximum battery power occurs at V/U equal to approximately half, but generally the battery is operated at or above 0.8.

R tends to increase as the battery ages and thus is an important indicator of battery health. While there are many ways to measure R, a common easily implemented technique is to apply a constant current step or pulse to the battery for a fixed amount of time, usually seconds to tens of seconds. R is then determined by the change between initial and final voltages and currents according to Eq. (24-62).

$$R = \frac{\Delta V}{\Delta I} \tag{24-62}$$

Note that this technique can be applied to any step change in current or power. Also, it is at best an approximate method because the SOC of the battery is changing with time. In addition, since R is usually dependent on time, when this technique is used, R is often reported with a time, sometimes as a subscript, indicating the interval of time (e.g., a 10-second pulse would be R_{10s}).

The concept of efficiency is often used to analyze a battery. While there are many that can be applied, two very important ones are current and energy. Both are defined based on each full cycle—for example, the discharge energy of a given cycle divided by the previous charge energy. The energy efficiency is an indicator of the battery's ability to effectively store and deliver electrical energy. The current efficiency—the discharge capacity of a given cycle divided by the previous charge capacity—is an indicator of the level of any side reactions within the battery that could drain capacity from the battery.

Common Battery Technologies Even when we limit our discussion to rechargeable technologies, there are many more types of batteries than we can consider here. Therefore, this discussion will be limited to the most common battery technologies. In general, batteries are classified by the solvent used in the electrolyte, either aqueous or nonaqueous. The cell voltage of aqueous batteries is limited by the decomposition reaction of water, 1.23 volts. Somewhat wider voltage ranges can be obtained by some combination of adjusting the pH of the electrolyte, using additives to suppress the decomposition reactions, having electrode materials that inhibit the kinetics of the decomposition reactions, and adding a gas recombination device. Nonaqueous electrolytes are typically made with an inorganic salt dissolved in one or more organic solvents and they are known to be able to operate at above five volts, but they also suffer from side reactions. In general, the side reactions tend to be more prevalent on the negative electrode; they form reaction products on the active materials that create a solid electrolyte interface (originally called interphase) or SEI. For a battery technology to have adequate life and be viable, all the side reactions must be limited or controlled.

The type of electrode electrochemical reaction can also be classified into two categories, either intercalation or conversion. In conversion reactions, the electrolyte ions react with the active materials (the main electrode material responsible for the energy storage capacity) to form new compounds and phases. Electrodes based on conversion reactions are known to have some of the highest specific capacities (in mAh/g of active material). Historically, the phase growth changes during cycling have come at a cost. The total volume and specific morphology of electrode charge and discharge products can change with SOC and C-rate, which often limits life. Intercalation electrodes are typified by the reactive species diffusing into and out of the active material structure during cycling. The diffusing specie can form a solid solution with the active material or can form new phases within the same particulate structure. Ideally, there are only minimal volume changes associated with intercalation reactions. Typically, solid solutions give a sloping open-circuit voltage curve. The fact that a phase change can occur in both types of electrode reactions suggests that the line between them can get blurred.

By definition, the positive electrode is a cathode during discharge and an anode during charge. By convention, the positive electrode is generally referred to as the cathode, and conversely the negative electrode is called the anode.

Lead-acid batteries were invented more than a century and half ago, and considering the improvements in battery technologies, it is amazing that they are still being used. This is both a testament to the battery and a statement on the importance of cost. The electrolyte is aqueous sulfuric acid, and the electrode conversion reactions are:

Positive electrode:

$$PbO_2(s)+4H^+(aq)+SO_4^{-2}(aq)+2e^- \rightleftharpoons PbSO_4(s)+2H_2O(l)$$

Negative electrode: $Pb(s)+SO_4^{-2}(aq) \rightleftharpoons PbSO_4(s)+2e^-$

Lead-acid batteries have a number of notable qualities. The lead makes them heavy, but it is offset somewhat by the fact that both reactions are multivalent (two electrons are liberated for each lead ion) and that it is a nominal 2-volt system. Also, both electrodes are based on lead compounds that discharge to the same relatively insoluble product. Finally, the electrolyte changes pH and density with the SOC, which is used as an SOC indicator. The specific energy of this technology is rather modest (typically less than 50 Wh/kg), as is its cycle life.

Nickel-metal hydride is part of a series of battery technologies (e.g., nickel-zinc, nickel-hydrogen, and nickel-cadmium) that use the nickel oxyhydroxide positive intercalation electrode.

Positive electrode: $NiO(OH)(s)+H_2O(l)+e^- \rightleftharpoons Ni(OH)_2(s)+OH^-(aq)$

Considered by many the best electrode ever developed, this electrode can be cycled thousands of times. The active material changes little as protons move into and out of the structure. An aqueous hydroxide-based electrolyte is used with a metal hydride negative intercalation electrode.

Negative electrode: $MH(s)+OH^-(aq) \rightleftharpoons M(s)+H_2O(l)+e^-$

M represents a metal alloy structure designed to absorb hydrogen. The nominal voltage for this system is 1.3 volts. The specific energy of this technology is very good for an aqueous system, approaching 100 Wh/kg, as well as having good cycle life.

Lithium-ion batteries are actually a class of battery technologies that are typified by a nonaqueous electrolyte (e.g., $LiPF_6$ dissolved in an organic carbonate-based solvent mixture) with lithium ions that are shuttled back and forth between electrodes during cycling. Nearly all common lithium-ion battery technologies use graphite as an intercalation negative electrode.

Negative electrode: $Li_xC_6(s) \leftrightharpoons Li_{x-\delta}C_6(s)+\delta Li^+(\text{non-aq})+\delta e^-$

The lithium concentration in the graphite varies continuously during cycling through a series of phases, or stages, from initially zero to LiC_6. Graphite has a high specific capacity and operates near lithium potential. While there are a number of positive electrode materials, the most common are the layered materials based on lithium transition metal oxides (e.g., cobalt and nickel). Of these, lithium cobalt oxide ($LiCoO_2$ or LCO) is the most common intercalation positive electrode.

Positive electrode: $Li_yCoO_2(s)+\delta Li^+(\text{non-aq})+\delta e^- \leftrightharpoons Li_{y+\delta}CoO_2(s)$

Almost all consumer electronics applications use LCO/graphite lithium-ion cells, which operate nominally at 4 volts. The high voltage, compared to aqueous systems, gives lithium-ion cells a significant advantage in energy density. However, the organic-based electrolyte and other carbonaceous materials, combined with a sensitivity to overcharging (charging the battery past V_{max}), makes them susceptible to thermal runaway and fires.

The cost of cobalt usually pushes manufacturers to other positive electrode alternatives for larger battery formats. One of these is lithium nickel oxide ($LiNiO_2$), which also operates nominally at 4 volts. Another is a lithium iron phosphate olivine material ($LiFePO_4$), which operates nominally at 3.5 volts. Finally, a relatively new class of positive electrode materials is in some electric vehicle applications. The layered-layered lithium transition metal oxides have a general formula of $xLi_2MnO_3\bullet(1-x)LiMO_2$, where M is some combination of Mn, Ni, and/or Co. These are complex nanostructured materials, which operate nominally at 4 volts. It should be noted here that all these electrode materials are typically modified to improve stability and increase performance. In general, lithium-ion battery technologies have good cycle lives and the highest specific energy of all commercial batteries, with cells well over 200 Wh/kg.

Design Considerations The battery's application tends to dictate the battery design, which would include establishing an operating voltage range, total energy and power needed, cycle and calendar life, and possibly available space and permissible mass. Of course, the cost of the battery would also come into consideration. Because it is more expedient and economical to use off-the-shelf batteries, device designers often adapt their designs to available batteries. Assuming a specific application exists that would justify the battery design and development process, then the first step would be to establish the battery technology of choice. In general, the energy density or specific energy requirements would dictate the battery technology, but it can also be controlled by cost or battery life needs.

Once established, the next step would be to determine the number of cells in series, which can easily be determined from the cell's nominal voltage and the application's voltage range requirements. Here is where the lithium-ion battery technology has a real advantage: for a given application, fewer than half the cells are needed compared to an aqueous electrolyte–based battery technology. Unless off-the-shelf cells are preferred, from a cost perspective there is generally no real reason to have cells in parallel. For safety or manufacturing reasons, battery packs are sometimes designed with cells in parallel. However, there is an added cost for additional cells.

Assuming all cells in the battery are the same, which is generally the case, the energy requirement and capacity of each cell is easily established. The total capacity of the cell is determined by the amount of active materials. As mentioned previously, the active materials can be characterized by their specific capacity. The theoretical specific capacity is determined by the electrochemical reaction. A more useful, practical specific capacity can most easily be established through small cell studies. Typically, the battery is designed to be limited by one electrode or the other (i.e., if there is more capacity in the negative electrode, then the cell is positive limited), which is usually dictated by the battery's stability limitations. The negative-to-positive capacity ratio, N:P, is set as close to 1 as possible to guarantee stability and not waste active materials. Knowing the total battery capacity and N:P, the amount of active materials is then established.

The fundamental electrode design for most of today's technologies is a planar geometry. This has several advantages. It allows the electrodes to be easily stacked in a parallel alternating fashion. The electrodes are double thick in that half the current of each electrode goes to the adjacent electrodes. Further, the planar geometry helps minimize any nonuniformity in the current distribution. Finally, it allows for easier scale-up. Small-area developmental cells can be constructed and tested in a similar manner to the large cells to confirm performance. The larger-area cells can then be designed based on the developmental cell specifications. Of course, the process can be more complicated, although most battery technologies are developed with this type of method.

Each electrode is formed onto a current collector, which is designed to distribute electronic current through the electrodes, while minimizing resistive losses. Current collectors are typically made of metal that is stable in the electrode environment. The current collector design can be relatively simple, as for thin electrodes (0.02 to 0.2 mm) that typically use a foil or an expanded metal mesh as the current collector. Thin electrodes are usually coated or extruded directly onto both sides of the current collector foil or mesh. Thicker electrodes usually need a more substantial current collector capable of supporting the electrode's weight.

Electrode active materials are generally powders to increase the electrochemically active area and to reduce the solid-state diffusion lengths. The electrodes have a porous structure, formed on the current collector, to allow electrolyte access to all the particles. Typical porosities are about a third to limit tortuosity effects and not waste electrolyte or space. While some electrodes are sintered or pressed, a polymer, referred to as a binder, is often used to hold the electrode together, and sometimes a carbon additive is used to enhance the electrode's electronic conductivity if the active materials are not very conductive.

A separator layer is placed between adjacent electrodes to prevent them from making electrical contact. The thin separator sheet, on the order of 0.01 mm, is typically a micro-porous polymer (e.g., polyethylene or polypropylene) or sometimes a cellulose-based film. Obviously, it is critical that the separator have adequate strength and that the electrolyte wet the separator as well as the electrodes. In some cell technologies, a gelled or plasticized electrolyte is used that may allow for the elimination of the separator. In these cases, a liquid electrolyte is blended with a polymer, which can add stability and more rigidity to the cell. Unfortunately, the ionic transport properties of the electrolyte also suffer.

The minimum total cell area is established by the battery power (P) requirements and is highly dependent on the cell's area specific resistance (ASI) and U. The cell ASI is an area-independent impedance that can be obtained from R by multiplying by the cell area (A). Implicit in this calculation is the assumption that the cell current distribution is uniform across the cell area. Because of thermal effects, current collector resistance, and other reasons, this assumption is almost never absolutely true, but it is often approximated. The ASI is generally obtained from small developmental cell studies. The minimum cell area can then be calculated by Eq. (24-63), where N is the number of cells in the series string [Eroglu, D., S. Ha, and K. G. Gallagher, "Fraction of the Theoretical Specific Energy Achieved on Pack Level for Hypothetical Battery Chemistries," *Journal of Power Sources* **267:** 14 (2014)].

$$A = \frac{(\mathrm{ASI})P}{N(U)^2\left[\frac{V}{U}\right]\left(1-\left[\frac{V}{U}\right]\right)} \tag{24-63}$$

In general, the ASI is dependent on electrode thickness, which requires the process for determining the minimum cell area to be iterative. However, for many battery technologies, the ASI changes little with thickness over a wide range, which helps make the process easier [Gallagher, K. G., P. A. Nelson, and D. W. Dees, "Simplified Calculation of the Area Specific Impedance for Battery Design," *Journal of Power Sources* **196:** 2289 (2011)]. Establishing the minimum cell area also defines the maximum positive and negative electrode thicknesses. In general, the maximum electrode thickness is limited by the electrolyte's ability to transport the ions from one electrode to the other. This requirement for most lithium-ion battery applications forces the technology into a thin electrode format. The thin electrode format allows the individual layers to be coated, then stacked (prismatic design) or wound into cells using high-speed manufacturing techniques. Aqueous electrolyte batteries with proton or hydroxide ion conduction can be much thicker, requiring much less electrode area.

Defining the total cell area and component thicknesses does little to define the shape of the battery. This is generally a complex optimization process that has to consider many factors, including available space, manufacturing capability, number of cells, cell and battery housing designs (e.g., can, pouch, box, case), cell and battery ancillary hardware (e.g., current collector tabs and feedthroughs, cell interconnects, cooling, battery management system electronics) and, of course, cost.

Materials of Construction

Lindell R. Hurst, Jr., M.S., P.E. *Senior Materials and Corrosion Engineer, Shell Global Solutions (US) Inc. Retired, Registered Professional Metallurgical Engineer (Alabama, Ohio, North Dakota) (Section Coeditor, Corrosion Prevention)*

Edward R. Naylor, B.S., M.S. *Senior Materials Engineering Associate, AkzoNobel; Certified API 510, 570, 653 and Fixed Equipment Source Inspector (Section Coeditor, Corrosion Prevention)*

Emory A. Ford, Ph.D. *Associate Director, Materials Technology Institute, Chief Scientist and Director of Research, Lyondell/Bassel Retired, Fellow Materials Technology Institute (Section Coeditor, Corrosion Prevention)*

Sheldon W. Dean, Jr., ScD, P.E. *President, Dean Corrosion Technology, Inc.; Fellow, Air Products and Chemicals, Inc., Retired; Fellow, ASTM; Fellow, NACE; Fellow, AIChE; Fellow, Materials Technology Institute (Corrosion Fundamentals, Corrosion Prevention)*

Eugene L. Liening, M.S., P.E. *Manufacturing & Engineering Technology Fellow, The Dow Chemical Company Retired; Fellow, Materials Technology Institute; Registered Professional Metallurgical Engineer (Michigan) (Corrosion Testing)*

Vinay P. Deodeshmukh, Ph.D. *Sr. Applications Development Manager – High Temperature and Corrosion Resistant Alloys, Haynes International Inc. (Corrosion Fundamentals, High-Temperature Corrosion, Nickel Alloys)*

Pradip R. Khaladkar, M.S., P.E. *Principal Consultant, Materials Engineering Group, Dupont Company (Retired), Registered Professional Engineer (Delaware), Fellow, Materials Technology Institute, St. Louis (Nonmetallic Materials)*

Kevin L. Ganschow, B.S., P.E. *Senior Staff Materials Engineer, Chevron Corporation; Registered Professional Mechanical Engineer (California) (Ferritic Steels)*

James D. Fritz, Ph.D. *Consultant, NACE International certified Material Selection Design Specialist; Member of the Metallic Materials and Materials Joining Subcommittees of the ASME Bioprocessing Equipment Standard, the Ferrous Specifications Subcommittee of the ASME Boiler & Pressure Vessel Code, and ASM International (Stainless Steels)*

Richard C. Sutherlin, B.S., P.E. *Richard Sutherlin, PE, Consulting, LLC; Registered Professional Metallurgical Engineer (Oregon) (Reactive Metals)*

Paul E. Manning, Ph.D. *Director CRA Marketing and Business Development, Haynes International (Nickel Alloys)*

INTRODUCTION

CORROSION FUNDAMENTALS

CORROSION PREVENTION

CORROSION-TESTING METHODS

ALLOY DESIGNATIONS

FERRITIC STEELS

STAINLESS STEELS

NICKEL ALLOYS

REACTIVE METALS

OTHER METALS AND ALLOYS

LOW-TEMPERATURE AND CRYOGENIC MATERIALS

NONMETALLIC MATERIALS FOR CORROSION CONTROL

General references: The following books are general references that are useful to a process engineer in understanding corrosion science and engineering and in choosing appropriate materials of construction: Baboian, R., ed., *NACE Corrosion Engineer's Reference Book,* 3d ed., NACE International, Houston, Tex., 2002; Birks, N., G. H. Meier, and F. S. Pettit, *Introduction to the High-Temperature Oxidation of Metals,* 2d ed., Cambridge University Press, Cambridge, UK, 2006; *Corrosion in the Petrochemical Industry,* 2d ed., ASM International, Materials Park, Ohio, 2015; Cramer, S., and B. Covino, Jr., eds., *ASM Handbook,* vol. 13A: *Corrosion: Fundamentals, Testing, and Protection,* ASM International, Materials Park, Ohio, 2003; Cramer, S., and B. Covino, Jr., eds., *ASM Handbook,* vol. 13B: *Corrosion: Materials,* ASM International, Materials Park, Ohio, 2005; Cramer, S., and B. Covino, Jr., eds., *ASM Handbook,* vol. 13C: *Corrosion: Environments and Industries: Materials,* ASM International, Materials Park, Ohio, 2006; Dillon, C., *Materials Selection for the Chemical Process Industry,* 2d ed., McGraw-Hill, New York, 2004; Fontana, M., and P. Greene, *Corrosion Engineering,* 2d ed., McGraw-Hill, New York, 1978; Jones, D., *Principles and Prevention of Corrosion,* 2d ed., Prentice Hall, Englewood Cliffs, N.J., 1992; Lai, G. Y., *High-Temperature Corrosion and Materials Applications,* ASM International, Materials Park, Ohio, 2007; Moniz, B. J., and W. I. Pollock, eds., *Process Industries Corrosion—The Theory and Practice,* NACE International, Houston, Tex., 1986; Uhlig, H., and R. Revie, *Corrosion and Corrosion Control—An Introduction to Corrosion Science and Engineering,* 3d ed., Wiley, New York, 1985.

INTRODUCTION

The selection of materials of construction for the equipment and facilities to produce chemicals is a core competency of chemical engineering. The chemical products desired cannot be manufactured without considering the selection of the optimum materials of construction for safe, economical manufacture and required product quality. Pressure containment and leak prevention are important to maintaining public confidence and the industry's right to operate. This section will introduce corrosion science and fundamentals, corrosion prevention, and common materials of construction used in the process industries. The purpose is to introduce the chemical engineer to the materials he or she will need to successfully operate process facilities safely and economically. Process engineers should engage with trained corrosion and materials engineering professionals and subject matter experts to choose the appropriate materials of construction and to maintain the physical assets of the facilities for which they are responsible.

CORROSION FUNDAMENTALS

INTRODUCTION

Corrosion has been defined as "the deterioration of a material, usually a metal, that results from a chemical or electrochemical reaction with its environment" (ASTM G193, "NACE/ASTM Standard Terminology and Acronyms Related to Corrosion," *Annual Book of ASTM Standards, 2014,* Vol. 03.02, ASTM International, West Conshohocken, Pa., 2014). Corrosion damage can show up as subsurface attack or cracking, so the traditional understanding does not cover the contemporary usage. In addition, the modern definition includes the deterioration of nonmetals, which will be covered in a later subsection. Most corrosion processes involve the chemical oxidation of metals, that is, the metal involved loses electrons as it is converted to corrosion products.

Practically all metals in use today are produced by a smelting process that reduces the ore and purifies the metal so that it can be converted into useful products. Corrosion is simply a reversal of the smelting process. Most metals are not thermodynamically stable in atmospheric oxygen, so corrosion returns the metal to a more stable form. The purpose of corrosion control is therefore to slow this process to the point where it does not interfere with the useful function of the metal item.

Although metals will react with liquids and gases, most corrosion processes that occur at temperatures below about 350°C result from exposure to a liquid phase. At higher temperatures, the reaction rate of metals with gases becomes high enough to cause significant damage. These processes are known as oxidation reactions rather than corrosion.

COST OF CORROSION

There have been many studies over the years to try to determine the cost of corrosion. The results of these calculations show that this cost is in the range of 2 to 4 percent of the GDP (gross domestic product). The use of existing technology could have prevented costs that were about 15 percent of the total costs of corrosion [*Materials Performance* **5**(6): 6 (1995)].

However, these estimates miss a very important effect of corrosion damage. Failures resulting from corrosion can cause injuries and fatalities, as well as environmental damage from loss of containment. Corrosion control is a key element in assuring the safety and reliability of plants and equipment.

CORROSION BY LIQUIDS

Corrosion of metals at lower temperatures usually results from exposure to liquids. The liquid phase does not have to be continuous in time or space. For example, atmospheric corrosion occurs because of occasional exposure of metal surfaces to moisture from dew or rain. However, not all liquids are corrosive. Nonpolar liquids, liquids with a dielectric constant less than 10, are not usually corrosive to most metals (Dean, S. W., *Materials Selector for Hazardous Chemicals, Organic Solvents,* Publication MS-8, Materials Technology Institute, St. Louis, Mo., 2011, p. 9). In the case of polar liquids, many metals and alloys resist corrosion because of a layer of oxidized metal on their surfaces that shields the metals from the liquid. In some cases, the layer is an oxide resulting from the exposure of the metal to air. In other cases, the layer is a corrosion product from the liquid. If the corrosion product is not soluble in the liquid, but forms a continuous adherent layer, the corrosion rate will generally be low or negligible. Polar liquids are much more likely to dissolve corrosion products and metal oxides, thereby allowing the corrosion to continue.

ELECTROCHEMISTRY OF CORROSION

Electrochemical corrosion theory is based on the understanding that all corrosion processes involve two or more separate electrochemical reactions: the anodic reactions and the cathodic reactions. The total of the electrons liberated by the cathodic reactions must exactly equal the electrons consumed by the anodic reactions. The anodic reactions are oxidations in which metallic atoms lose electrons and become positively charged ions. An example is shown in Eq. (25-1):

$$Fe \rightarrow Fe^{++} + 2e \tag{25-1}$$

The electrons shown on the right side of Eq. (25-1) are retained in the metal, while the ferrous ions dissolve into the solution and diffuse away. This reaction is balanced by a reduction reaction, for example, the reduction of molecular oxygen dissolved in the solution, as shown in Eq. (25-2):

$$O_2 + 2H_2O + 4e \rightarrow 4OH^- \tag{25-2}$$

These two reactions occur simultaneously, so there is no accumulation or depletion of electrons in the metal.

Electrode Potential The Gibbs free energy associated with an electrochemical reaction is related to the potential of the reaction through

Eq. (25-3) (Moore, W. J., *Physical Chemistry*, 2d ed., Prentice-Hall, Englewood Cliffs, N.J., 1955, p. 73):

$$\Delta G = -nEF \tag{25-3}$$

where ΔG is the Gibbs free energy of the reaction,
n is the number of electrons transferred,
E is the potential of the reaction,
F is Faraday's constant (96,500 coulombs per gram equivalent).

The realization that the potential of a cell in which a metal surface in contact with an ionic solution coupled to a reference electrode could be used to measure the free energy of the metal surface thus became the basis for electrochemical corrosion theory. The reference electrode that has been adopted as the standard is the platinized platinum/hydrogen gas/hydrogen ion electrode at 1 atm gas pressure, 1 M H^+ activity (concentration), and 25°C, SHE. Unfortunately, this reference electrode is difficult to manage in experimental work, so several other reference electrodes are used. Table 25-1 provides information on converting between the various electrodes currently in use.

It should also be noted that the sign convention used in displaying electrode potentials is the Gibbs Stockholm convention in which increasingly positive potentials represent increasingly oxidizing conditions (ASTM G3, "Standard Practice for Conventions Applicable to Electrochemical Measurements in Corrosion Testing," in *Annual Book of ASTM Standards, 2014*, vol. 03.02, ASTM International, West Conshohocken, Pa., 2014).

TABLE 25-1 Potentials of Reference Electrodes versus the Standard Hydrogen Electrode, SHE, 25°C

Name	Electrode	Designation	Potential vs. SHE (V)
Calomel		SCE	0.241
Silver/silver chloride, 0.1 M	Ag/AgCl/0.1 M KCl	0.1 SSC	0.286
Silver/silver chloride 1 M	Ag/AgCl/ 1.0 M KCl	1.0 SSC	0.235
Silver/silver chloride sat.	Ag/AgCl/Sat. KCl	Sat SSC	0.198
Copper/copper sulfate	Cu/ Sat. $CuSO_4$	CCS	0.3
Mercury/mercury sulfate	Hg/ Hg_2SO_4/H_2SO_4	—	0.616

ASTM G3, "Standard Practice for Conventions Applicable to Electrochemical Measurements in Corrosion Testing," *Annual Book of ASTM Standards*, 2014, Vol. 03.02 (ASTM International, West Conshohocken, Pa., 2014).

Potential-pH Diagrams Marcel Pourbaix used thermodynamic theory to construct diagrams in which the electrode potentials of the metallic elements in aqueous solutions are plotted against the pH of the solution showing where various compounds and ions of the metal would be stable at 25°C (77°) (Veleva, L., and R. D. Kane, "Thermodynamics of Atmospheric Corrosion and the Use of Pourbaix Diagrams," in *ASM Handbook*, vol. 13A, *Corrosion: Fundamentals, Testing, and Protection*, ASM International, Materials Park, Ohio, 2003). An example of one of these diagrams is shown in Fig. 25-1. The electrode potential of the metal against the standard hydrogen reference electrode is shown on the ordinate axis, while the

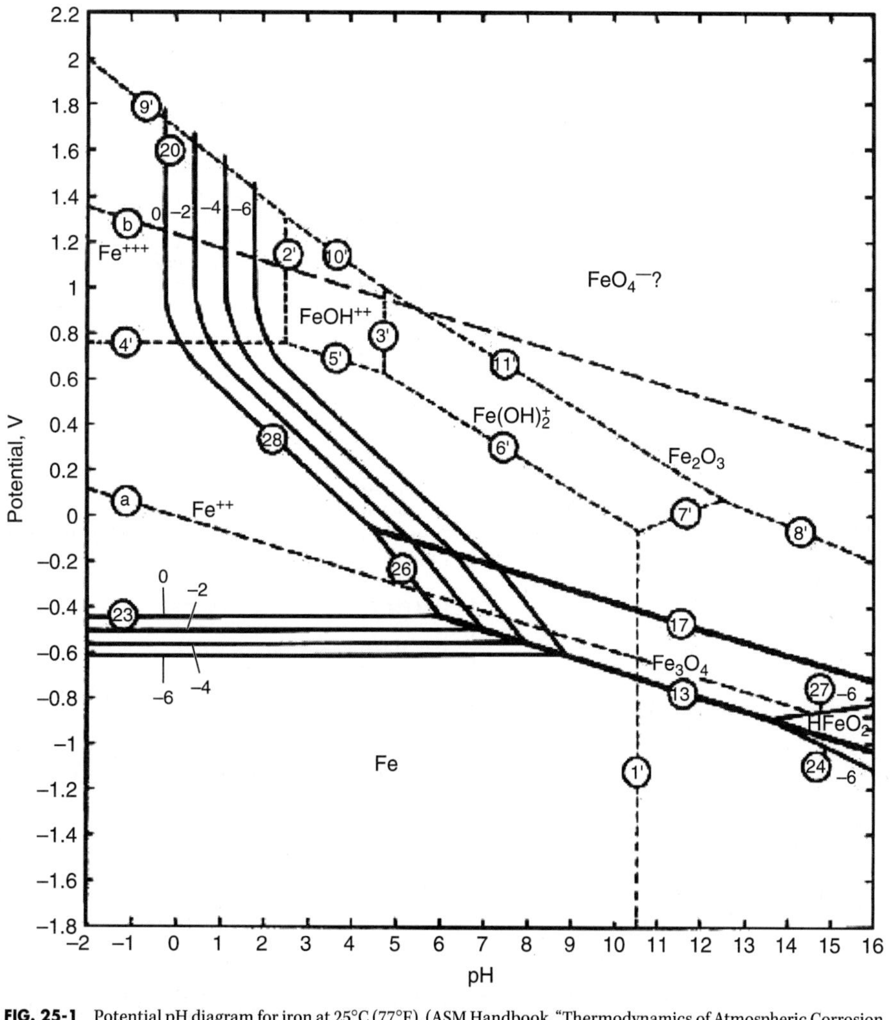

FIG. 25-1 Potential pH diagram for iron at 25°C (77°F). (ASM Handbook, "Thermodynamics of Atmospheric Corrosion and the Use of Pourbaix Diagrams," *Volume 13A Corrosion: Fundamentals, Testing, and Protection*. Reprinted with permission of ASM International. All rights reserved. www.asminternational.org.)

solution pH is shown on the abscissa axis. The horizontal lines on the left side of the diagram show the potential of iron in equilibrium with the concentration of ferrous ions in solution. The regions where iron oxides and hydroxides are stable are also shown. These diagrams have been very helpful in enabling us to understand the reactions that occur during corrosion. However, they must be modified for alloys.

Polarization When current is passed through a metal surface in an ionic solution, the electrode potential is changed. This change is denoted as polarization. J. Tafel discovered that the electrode potential of a metal surface changed as the current density applied to the surface varied [Tafel, J., *Z. Physik. Chem.* **54:** 614 (1905)]. In the case where only one electrochemical reaction was occurring, the relationship between potential, E, and current density followed the expression shown in Eq. (25-4):

$$E = a + b \log i \tag{25-4}$$

where i is the current density, and a and b are constants.

The b constant is known as the Tafel slope, and its value can, in some cases, be predicted from reaction rate theory (Glasstone, S., K. J. Laidler, and H. Eyring, *The Theory of Rate Processes*, McGraw Hill, New York, 1944, pp. 552–599). This expression is also known as Tafel kinetics, and it applies to electrochemical reactions where the rate is limited by the charge transfer process.

In the case of most corrosion processes, both the anodic and cathodic processes are occurring simultaneously at the same overall electron transfer rate. In this case, if a small potential change, for example, <10 mV, is applied to the surface, the current density does not follow the relationship shown in Eq. (25-4), but instead it appears to be linearly related to the applied potential change. A typical polarization diagram is shown in Fig. 25-2 for carbon steel in sulfuric acid (ASTM G82, "Standard Guide for Development and Use of a Galvanic Series for Predicting Galvanic Corrosion Performance," *Annual Book of ASTM Standards, 2014*, vol. 03.02, ASTM International, West Conshohocken, Pa., 2014). With larger potential changes, such as >50 mV, the Tafel behavior begins to appear because one of the reactions becomes dominant. In the low polarization region, the ratio of potential change to

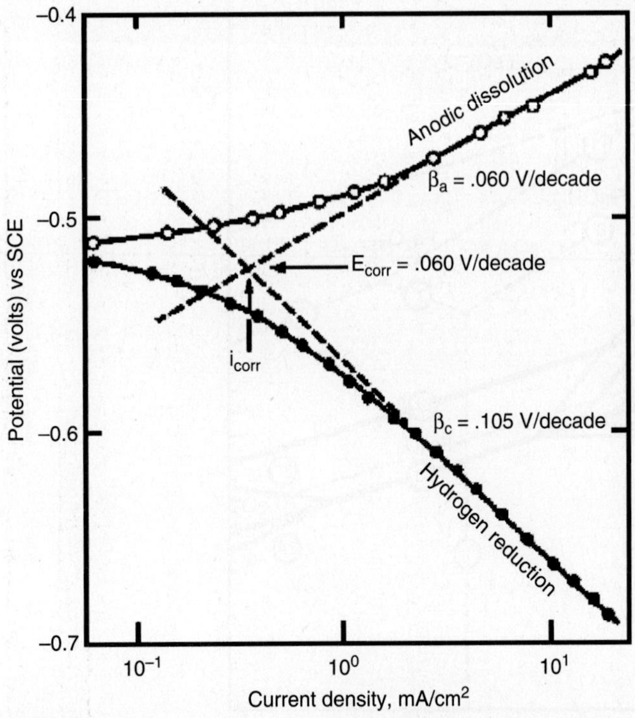

FIG. 25-2 Polarization diagram for carbon steel in deaerated 0.52 N sulfuric acid showing Tafel behavior. (Reprinted with permission, from G82-98(2014) Standard Guide for Development and Use of a Galvanic Series for Predicting Galvanic Corrosion Performance, copyright ASTM International, 100 Barr Harbor Drive, West Conshohocken, PA 19428. A copy of the complete standard may be obtained from ASTM, www.astm.org.)

current density is known as the polarization resistance, R_p, and it is related to the corrosion rate of the metal surface through Eq. (25-5),

$$i_{corr} = B/R_p \tag{25-5}$$

where i_{corr} is the corrosion rate of the metal expressed as a current density.
B is a constant known as the Stern-Geary constant.

The i_{corr} relationship to corrosion rate is given by Eq. (25-6), which is based on Faraday's law (ASTM G102, "Standard Practice for Calculation of Corrosion Rates and Related Information from Electrochemical Measurements," *Annual Book of ASTM Standards, 2014*, vol. 03.02, ASTM International, West Conshohocken, Pa., 2014).

$$CR = (K_1 \cdot i_{corr} \cdot EW)/\rho \tag{25-6}$$

where CR is the corrosion rate expressed as penetration per unit time,
K_1 is a constant,
EW is the equivalent weight of the metal or alloy,
ρ is the density of the metal.

The value of K_1 is determined by the units selected for the other terms of the equation.

Passivity When some alloys are polarized in strong acid solutions, it is possible to observe the development of passivity. Passivity is a corrosion-resistant condition that occurs with many metals and alloys in oxidizing conditions. Stainless steels are typical examples of alloys exhibiting passivity. The polarization diagram of Type 430 (UNS S43000) stainless steel in deaerated 1 N sulfuric acid at 30°C (86°F) is shown in Fig. 25-3 (ASTM G5, "Standard Reference Test Method for Making Potentiodynamic Anodic Polarization Measurements," *Annual Book of ASTM Standards, 2014*, vol. 03.02, ASTM International, West Conshohocken, Pa., 2014). In this case the corrosion rate of the metal is very high in the potential range of –0.3 to –0.5 V versus the SCE reference electrode. However, as the potential increases from this range, the current density, which is proportional to the corrosion rate, decreases by four decades. Surface examinations of the steel have shown that this decrease is caused by the development of a thin, invisible layer of a mixed chrome–iron oxide layer. This mixed oxide layer is almost insoluble in the acid, and this low solubility results in the corrosion resistance of the stainless steel. This phenomenon is responsible for the corrosion resistance of many alloys containing chromium.

GENERAL CORROSION

Corrosion that occurs uniformly over a metal surface is known as general corrosion. Although the rate can vary from point to point on the surface, the variations are relatively small, and they do not persist for extended time periods. As a result, the overall loss tends to be uniform. This type of corrosion is observed in many situations, especially in the corrosion of carbon steel in the atmosphere. In cases where the corrosion rate is modest, engineers may accommodate for the loss of metal over time by adding a corrosion allowance to the required thickness of metal used.

LOCALIZED CORROSION

Pitting Pitting is a form of localized corrosion that occurs in passive metals and metals with thin corrosion product layers (Frankel, G. S., "Pitting Corrosion," in *ASM Handbook*, vol. 13A, *Corrosion: Fundamentals, Testing, and Protection*, ed. S. D. Cramer and B. S. Covino, Jr., ASM International, Metals Park, Ohio, 2003, p. 236). It may occur because of impurities in the metal surface or a variety of other irregularities on the surface that cause anodic attack to localize. A major problem with pitting is that the environment tends to stabilize the localization of the attack so that the pits continue to grow. Figure 25-4 shows common morphologies of pit penetrations as seen in sections through the pit (ASTM G46, "Standard Guide for Examination and Evaluation of Pitting Corrosion," *Annual Book of ASTM Standards, 2014*, vol. 03.02, ASTM International, West Conshohocken, Pa., 2014). Pits usually occur randomly and may result in penetration with loss of containment before any significant loss of strength. Environments containing chloride or other halide ions are particularly aggressive in causing pitting.

Crevice Corrosion Another form of localized corrosion is crevice corrosion. This type of attack is similar to pitting in that the crevice permits the environmental changes that lead to accelerated anodic attack. Chloride

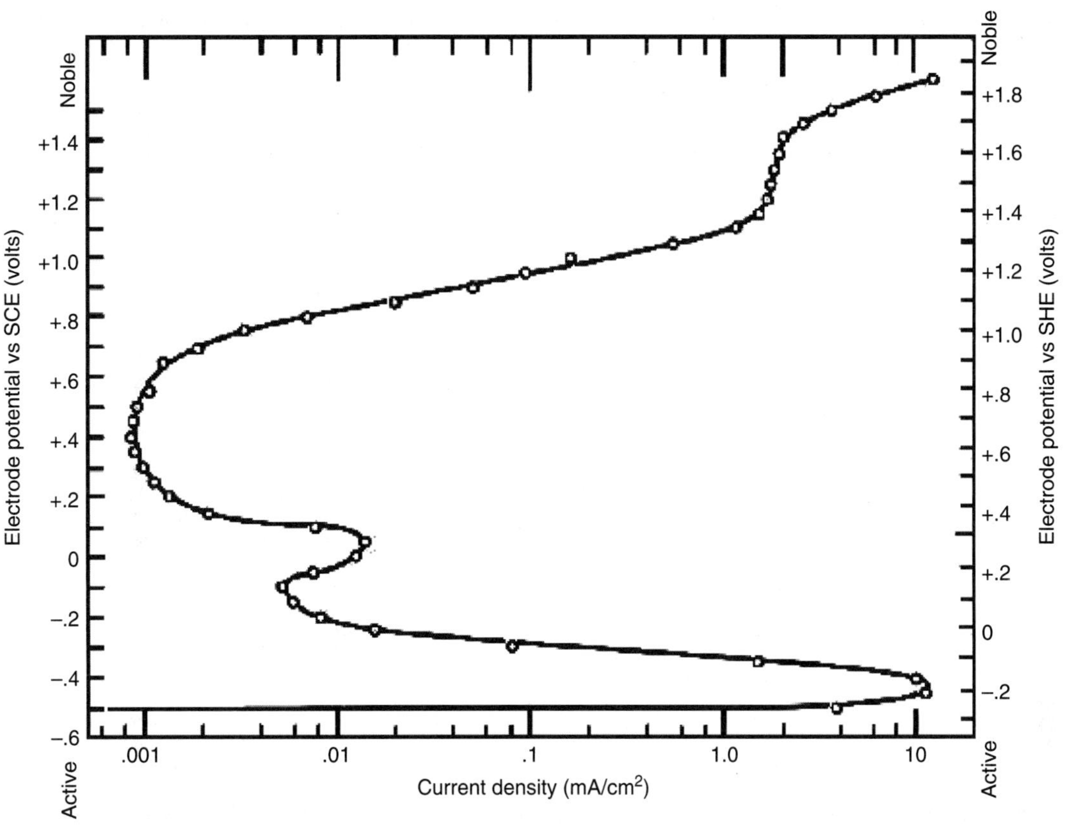

FIG. 25-3 Anodic polarization diagram for type 430 stainless steel in 1 N sulfuric acid at 30°C. (Reprinted with permission, from G3-14 Standard Practice for Conventions Applicable to Electrochemical Measurements in Corrosion Testing, copyright ASTM International, 100 Barr Harbor Drive, West Conshohocken, PA 19428. A copy of the complete standard may be obtained from ASTM, www.astm.org.)

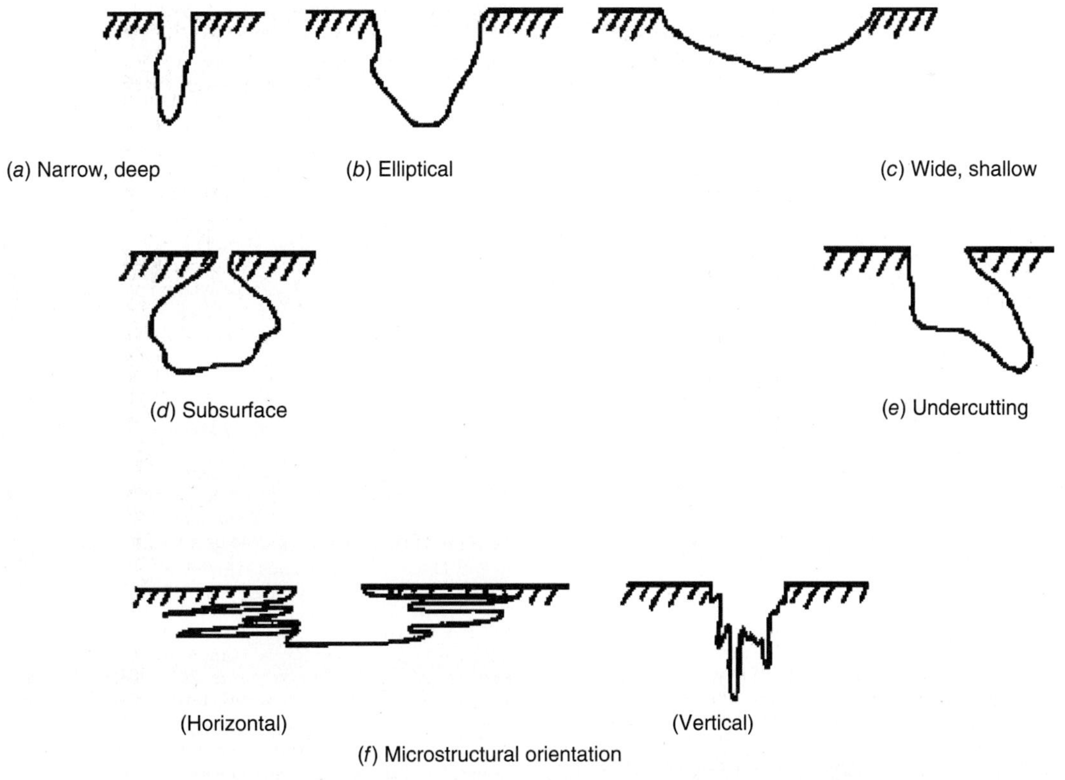

FIG. 25-4 Typical profiles of pits as seen in sections through pits in metals. [Reprinted with permission from G46-94(2013) Standard Guide for Examination and Evaluation of Pitting Corrosion, copyright ASTM International, 100 Barr Harbor Drive, West Conshohocken, PA 19428. A copy of the complete standard may be obtained from ASTM, www.astm.org.]

ions are often involved in the development of the aggressive environment, which is acidic. Crevice corrosion will develop in conditions that are less aggressive than those that cause pitting.

Intergranular Corrosion A very detrimental form of localized corrosion is known as intergranular corrosion. Practically all metals are crystalline, and the crystals that comprise the item are known as grains. Intergranular corrosion results from compositional variations at the grain boundaries that cause these areas to be more susceptible to corrosive attack. The resulting corrosion damage will allow the grains to become loose and wash away. As a result, the penetration removes a large fraction of the metal with only a small fraction being oxidized. The problem was originally observed with higher-carbon stainless steel alloys that had been improperly heat treated or had been welded (Fritz, J. D., "Effects of Metallurgical Variables on the Corrosion of Stainless Steels," in *ASM Handbook*, vol. 13A, *Corrosion: Fundamentals, Testing, and Protection*, ed. S. D. Cramer and B. S. Covino, Jr., ASM International, Metals Park, Ohio, 2003, p. 267). The condition of being susceptible to intergranular corrosion is known as sensitization.

Dealloying Corrosion Dealloying corrosion is another form of localized attack that occurs with some metals. Copper-based alloys are most known for this form of attack (Corcoran, S. G., "Effects of Dealloying Corrosion on Dealloying Corrosion," in *ASM Handbook*, vol. 13A, *Corrosion: Fundamentals, Testing, and Protection*, ed. S. D. Cramer and B. S. Covino, Jr., ASM International, Metals Park, Ohio, 2003, p. 287). Copper-zinc alloys (brasses) are particularly prone to this problem. In that case, the corrosion is termed dezincification. The mechanism is still being debated, but the result is that the zinc in the alloy is lost, and the remaining copper retains the same volume, but has no strength. This attack can occur in spots on a surface, or it can cover the entire surface. Other types of dealloying corrosion include denickelification and dealuminification of certain copper alloys, and graphitization of gray cast irons.

Galvanic Corrosion This form of attack occurs when two different alloys are in contact with each other in a corrosive environment (Baboian, R., "Galvanic Corrosion," in *ASM Handbook*, vol. 13A, *Corrosion: Fundamentals, Testing, and Protection*, ed. S. D. Cramer and B. S. Covino, Jr., ASM International, Metals Park, Ohio, 2003, p. 210). This problem is also known as bimetallic corrosion. In this situation, one of the alloys will corrode at an accelerated rate, while the other suffers little or no damage. This problem is most prevalent in liquids that have a significant electrical conductivity such as seawater. The accelerated corrosion is accompanied by the passage of electrical current through the junction where the two alloys are connected.

Velocity Effects Accelerated corrosion also occurs in areas exposed to high liquid velocities (Aylor, D., and B. Phull, "Evaluation Erosion Corrosion, Cavitation and Impingement," in *ASM Handbook*, vol. 13A, *Corrosion: Fundamentals, Testing, and Protection*, ed. S. D. Cramer and B. S. Covino, Jr., ASM International, Metals Park, Ohio, 2003, p. 639). This problem has also been known as erosion corrosion, flow-assisted corrosion, flow-sensitive corrosion, and a variety of similar terms. There are three different classes of flow-accelerated corrosion. The first class occurs when only the velocity of the liquid and the turbulence it produces cause the damage. The second class covers cases where solid particles entrained in the liquid strike the metal surface and damage the protective corrosion product layer, thereby accelerating the corrosion. The third class involves bubbles or voids in the liquid that strike the surface. In the case of voids, the collapse of the voids causes localized damage and disrupts the corrosion product layer. This case is known as cavitation corrosion. Rising bubbles can also cause enhanced corrosion. This problem has been observed with carbon steel in concentrated sulfuric acid and is known as hydrogen grooving [Dean, S. W., Jr., and G. D. Grab, *Materials Performance* **25**(7): 48–52 (1986)].

In the case of a liquid flowing into a tube or pipe, the corrosion damage tends to be greatest at the inlet end, bends, elbows, or obstacles in the flow path. In these cases, the damage often appears as horseshoe-shaped grooves in the metal. Generally, the cause of the enhanced corrosion rate is the higher mass transfer rate that results from the higher velocity and turbulence it produces.

ENVIRONMENTALLY ASSISTED CRACKING

Corrosion can cause or accelerate the cracking of metals when they are under tensile stress. Four different mechanisms have been observed: stress corrosion cracking (SCC), hydrogen embrittlement (HE), liquid metal embrittlement (LME), and corrosion fatigue (CF).

Stress Corrosion Cracking This type of cracking occurs with alloys in specific environments and limited conditions. It occurs locally in regions where high tensile stresses exist, and it results in penetration or rupture of the metal part.

Experimental studies showed that the cracking processes were related to corrosion. When corrosion was prevented through cathodic protection, cracking did not happen. The cracks followed specific directions in the metal related to the metallurgy of the alloy. For example, cracking in brass tends to be intergranular, that is, it follows grain boundaries in the metal. Figure 25-5 shows a section through an SCC crack in brass (ASTM G186, "Standard Test Method for Determining Whether Gas-Leak-Detector Fluid Solutions Can Cause Stress Corrosion Cracking of Brass Alloys," *Annual Book of ASTM Standards, 2014*, vol. 03.02, ASTM International, West Conshohocken, Pa., 2014). In austenitic stainless steels in chloride environments, the cracks are transgranular, but they follow specific crystallographic directions through the grain. This requires the cracks to reinitiate every time they encounter a grain boundary. In some alloys, both transgranular and intergranular cracks are found. In all of these situations, the cracks tend to be highly branched, propagating in many directions through the metal.

In structures, the most likely locations for cracks are around welds and cold-formed areas. The stresses in these areas are termed *residual stresses*, and they are generally much higher than externally applied stresses.

SCC requires four components (Dean, S. W., *Laboratory Corrosion Testing of Metals and Alloys*, Materials Technology Institute, St. Louis, Mo., 2015, p.129). The alloy must be susceptible to SCC in a corrosive environment. The environment must have the composition and temperature that will allow SCC to proceed. The tensile stress must be high enough for SCC to initiate. The metal item must be in the environment long enough for the cracking process to initiate. As with pitting and crevice corrosion, an induction time is usually observed before cracking begins.

Hydrogen Embrittlement Hydrogen embrittlement (HE) is caused by the presence of atomic hydrogen in the metal, resulting in the development of cracks, or a reduction of ductility and toughness. Although corrosion can generate atomic hydrogen as a corrosion product, it is not the only source of atomic hydrogen that can cause HE. Electroplating, cathodic protection, and high-pressure hydrogen gas are other sources. HE cracks are intergranular and exhibit branching. The cracks are believed to be caused by decohesion at these grain boundaries. As a general rule, alloys become susceptible to HE when their tensile strength exceeds a critical level. Because hardness and tensile strength are correlated, the critical value is often reported as a hardness value.

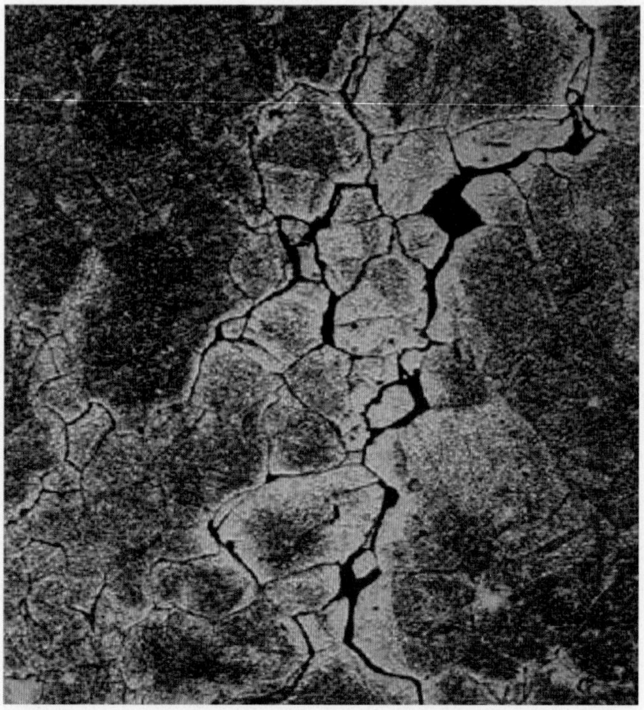

FIG. 25-5 Section through an SCC crack in brass showing intergranular cracking. [Reprinted with permission from G186-05(2011) Standard Test Method for Determining Whether Gas-Leak Detector Fluid Solutions Can Cause Stress Corrosion Cracking of Brass Alloys, copyright ASTM International, 100 Barr Harbor Drive, West Conshohocken, PA 19428. A copy of the complete standard may be obtained from ASTM, www.astm.org.]

In petroleum production and refining, and in chemical processing, HE has been a particular concern when hydrogen sulfide, H_2S, hydrogen cyanide, HCN, or anhydrous hydrogen fluoride, HF, is present. In the case of iron- and nickel-based alloys, the critical hardness is Rockwell C 22 (NACE MR01-75, "Sulfide Stress Cracking Resistant Metallic Materials for Oilfield Equipment," NACE International, Houston, Tex.). This value was determined for H_2S-containing environments, but in less aggressive environments a higher critical hardness may apply. For example, in atmospheric exposures, the value for steel is about Rockwell C 35. It has been found that in such service there is a threshold stress value that is necessary to initiate HE. Stress-relieving and postweld heat treatments are regularly used to minimize the probability of HE for vessels and equipment, especially in H_2S service.

Liquid Metal Embrittlement This form of cracking does not require corrosion for the cracks to develop (Kohlman, D. G., "Liquid Metal Induced Embrittlement," in *ASM Handbook*, vol. 13A, *Corrosion: Fundamentals, Testing, and Protection*, ed. S. D. Cramer and B. S. Covino, Jr., ASM International, Metals Park, Ohio, 2003, p. 381). At temperatures below 350°C (662°F) mercury is the most common agent causing this type of cracking. Copper, nickel, and aluminum-based alloys are susceptible to failures from mercury.

Austenitic stainless steels are susceptible to LME in molten zinc and molten copper. Failures have occurred in fires when molten zinc dripped from galvanized steel onto the stainless steel surface. Molten copper has also caused failures when the copper contacted the hot stainless during welding operations.

Corrosion Fatigue This form of cracking requires a stress that varies or cycles (Phull, B., "Evaluation Corrosion Fatigue," in *ASM Handbook*, vol. 13A, *Corrosion: Fundamentals, Testing, and Protection*, ed. S. D. Cramer and B. S. Covino, Jr., ASM International, Metals Park, Ohio, 2003, p. 625). Crack growth occurs only while the stress is increasing. What distinguishes CF from conventional fatigue is that corrosion is necessary for it to progress, and the stress level at which it occurs is lower than the threshold for conventional fatigue.

OTHER TYPES OF CORROSION

Microbial Influenced Corrosion Metals in exposed waters—including seawater, cooling water, and natural freshwater—and soils will often be covered with biological growth. In the case of natural waters, this is known as slime or fouling. In some cases, the biological layer that develops will cause accelerated corrosion (Dexter, S. C., "Microbiologically Influenced Corrosion," in *ASM Handbook*, vol. 13A, *Corrosion: Fundamentals, Testing, and Protection*, ed. S. D. Cramer and B. S. Covino, Jr., ASM International, Metals Park, Ohio, 2003, p. 398). The corrosion often takes the form of deep pits in the metal. Carbon and stainless steels are susceptible to this form of attack. Sulfate-reducing bacteria are often present in these cases, and the pits often contain iron sulfide in the corrosion products within the pits if these bacteria are present.

Fretting Corrosion Fretting corrosion occurs when metal surfaces rub against each other (Glaeser, W., and I. G. Wright, "Forms of Mechanically Assisted Degradation," in *ASM Handbook*, vol. 13A, *Corrosion: Fundamentals, Testing, and Protection*, ed. S. D. Cramer and B. S. Covino, Jr., ASM International, Metals Park, Ohio, 2003, p. 324). The mechanism is believed to be a result of many regions of the contacting surfaces being welded together, and then the welds are broken off to create tiny balls of rolled-up metal and oxide. This type of attack is not corrosion in the traditional sense, but it is usually included in discussions of corrosion damage.

HIGH-TEMPERATURE GASEOUS CORROSION

Overview All high-temperature alloys and coatings corrode to various extents when exposed to a variety of high-temperature engineering environments. Oxidation is the most commonly observed high-temperature corrosion reaction, and it forms the basis for other high-temperature degradation processes. In mixed gaseous environments containing gases such as H_2S, CO, Cl_2, and NH_3, or in the presence of molten deposits, high-temperature degradation is initiated from more than one type of oxidant. As a consequence, high-temperature corrosion reactions, such as sulfidation, carburization, and nitridation, occur in combination with oxidation. However, the nature of corrosion products and the extent of corrosion varies significantly from alloy to alloy. Ideally, field testing is recommended to understand corrosion behavior of the alloys in mixed gaseous environments and when molten deposits are present on the alloy surface. In most cases, high-temperature corrosion data generated in the lab is used to compare and rank the alloy performances. In this section, different forms of high-temperature corrosion reactions that occur in industries are briefly explained. A special emphasis is given to the corrosion mechanisms found in the chemical and petrochemical process industries.

Oxidation In most industrial processes, oxidation takes place in air or combustion environments. The combustion environment often consists of free molecular oxygen that dictates the oxidation behavior of the alloys; however, sometimes stoichiometric and substoichiometric conditions prevail. In such "reducing" environments, partial pressure of oxygen PO_2 is rather low, and the oxidation reaction is rather sluggish and dependent on CO/CO_2 or H_2/H_2O ratios. In the former case, if secondary oxidant (e.g., sulfur) is present, then it often dictates the corrosion kinetics. The thermodynamic consideration for the formation of oxide is described using the following reaction:

$$\mathrm{M} + \tfrac{1}{2}\,\mathrm{O_2} = \mathrm{MO} \tag{25-7}$$

The oxidation takes place when PO_2 in the environment is greater than PO_2 in equilibrium with the oxide. The PO_2 in equilibrium with oxide for Eq. (25-7) is determined from the standard free energy of formation, as shown in Eq. (25-8), where activities of M and MO are considered to be unity.

$$K_p = \exp\left(-\frac{\Delta G^\circ}{RT}\right) = \frac{1}{PO_2^{1/2}} \tag{25-8}$$

K_p is a reaction rate constant in equilibrium. The plots for free energy of formation for various oxides as a function of temperature (i.e., the Ellingham diagram) shown in Fig. 25-6 are quite useful in determining the stabilities of various oxides and PO_2 in equilibrium with oxide (Shifler, D., "Factors Affecting High-Temperature Corrosion and Materials Properties," in *Metals Handbook*, vol. 13A, *Corrosion: Fundamentals, Testing, and Protection*, ed. S. D. Cramer and B. S. Covino, Jr., ASM International, Metals Park, Ohio, 2003). In reducing environments when PO_2 is controlled by either CO/CO_2 or H_2/H_2O ratios, the reactions rate constant is related to ΔG° in Eq. (25-9):

$$K_p = \exp\left(-\frac{\Delta G^\circ}{RT}\right) = \frac{PCO\ PO_2^{1/2}}{PCO_2} \quad \text{or} \quad \frac{PH_2\ PO_2^{1/2}}{PH_2O} \tag{25-9}$$

If equilibrium ratios PCO/PCO_2 or PH_2/PH_2O are known, then PO_2 can be found from Eq. (25-9). The Ellingham diagram also reports CO/CO_2 or H_2/H_2O ratios for the corresponding oxidation equilibria, which could be used to determine equilibrium PO_2.

The amount of oxidation attack in the alloys can be measured by oxidation kinetics and the amount of metal consumed. The oxidation kinetics follows three principal laws—linear, parabolic, and logarithmic. The linear kinetics shown in Eq. (25-10) occurs when the scale is porous and the reaction is controlled by a gas phase mass transfer. The oxidation reactions follow parabolic kinetics for compact and adherent scales, which are commonly controlled by diffusion of mass transfer through scales. Most of the oxides follow parabolic oxidation rate laws at higher temperatures, and absolute rates shown in Fig. 25-7 are useful in comparing the oxidation kinetics of metals and alloys. The oxides with the highest defect concentrations indicate high rate constants. High-temperature alloys, when exposed to elevated temperatures, follow linear kinetics due to rapid scale growth during transient oxidation, but once a protective scale is established, the oxidation reaction becomes diffusion controlled and follows parabolic behavior. In the case of thin film oxidation, particularly at lower temperatures, scale formation follows inverse or cubic logarithmic rate laws, as shown in Eqs. (25-10) and (25-11).

$$x = k_l t \tag{25-10}$$

$$x^2 = 2k_p t \tag{25-11}$$

Here k_l and k_p are linear and parabolic rate constants, respectively, and x is oxide thickness. The engineering alloys and coatings are designed such that they can establish thin, slow-growing, and adherent oxide scales for high-temperature corrosion protection. It is seen from Fig. 25-8 that the alloys are designed to form either alumina, chromia, or silica scales due to rather low rate constants. It is also seen that even within these scales, depending

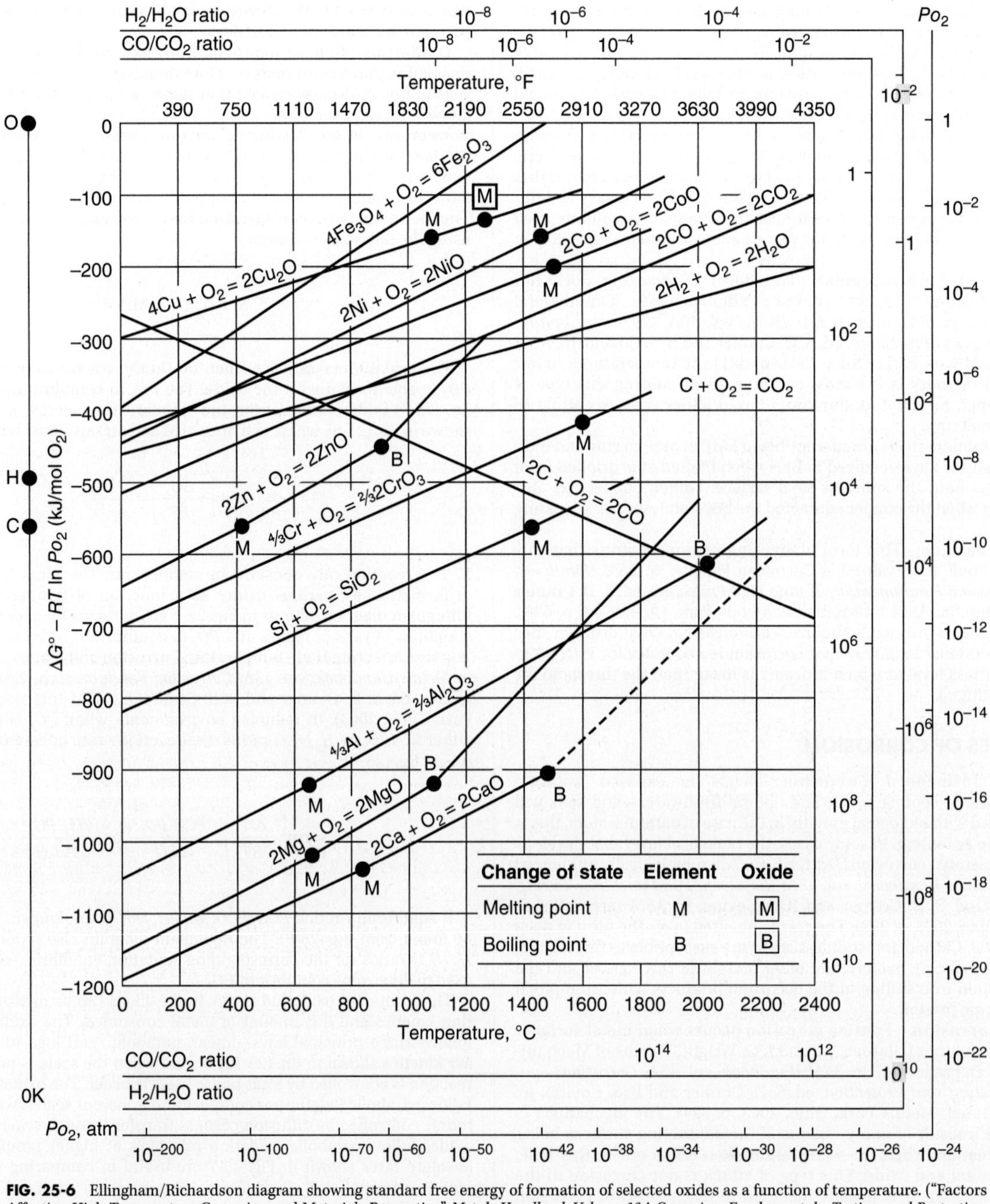

FIG. 25-6 Ellingham/Richardson diagram showing standard free energy of formation of selected oxides as a function of temperature. ("Factors Affecting High-Temperature Corrosion and Materials Properties," *Metals Handbook Volume 13A Corrosion: Fundamentals. Testing, and Protection.* Reprinted with permission of ASM International. All rights reserved. www.asminternational.org.)

upon the alloy, parabolic rate constants vary, but they typically fall within the bands represented.

The scale developed on the alloy surface acts as a diffusion barrier; thus adherent scale formation is extremely important for long-term oxidation or corrosion protection. The adherence of the oxide scale can be determined by the ratio of the molar volumes of the metal to the oxide formed on it, known as the Pilling-Bedford ratio (PBR). When there is no large difference in the molar volumes (PBR ratio is equal or nearly equal to 1), the scale adheres to the metal. Under cyclic conditions, however, oxides still spall due to differences in the coefficients of thermal expansion between the oxide and the underlying alloy substrate. Thus, when the bare metal is exposed to an oxidizing environment, if the alloy cannot reestablish protective or semiprotective scale, nonprotective scale formation occurs, eventually resulting in increased oxidation rates. The reactive elements, such as La, Zr, Y, and Hf, are added to further improve the scale adhesion characteristics of the alloy. A number of external factors affect the oxidation behavior of high-temperature alloys and coatings; these are primarily thermal cyclic conditions, long-term exposures, flowing atmospheres, and scale volatilization in the presence of water vapor. In general, alumina-forming alloys offer better oxidation resistance than chromia-forming alloys, while Ni-base alloys perform significantly better than Fe- and Co-base alloys. A number of references are available that report oxidation behavior and performance characteristics of engineering alloys and coatings used in the industries (Young, D. J., "High Temperature Oxidation and Corrosion of Metals," Elsevier Corrosion Series, 2008, p. 185; Birks, N., G. H. Meier, and F. S. Pettit, *Introduction to the High-Temperature Oxidation of Metals*, 2d ed., Cambridge University Press, Cambridge, UK, 2006, p. 101; Lai, G. Y., *High Temperature Corrosion and Materials Application*, 2d ed., Cambridge University Press, Cambridge, UK, 2006, p. 15).

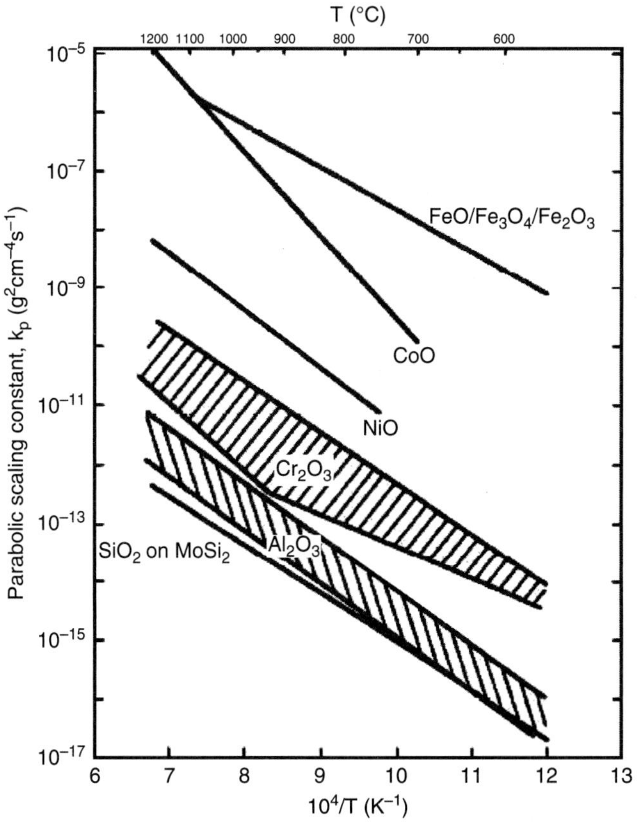

FIG. 25-7 Parabolic rate constants for selected oxides showing scatter bands reported in literature for chromia- and alumina-forming alloys. (Gleeson, B., *High-Temperature Corrosion of Metallic Alloys and Coatings. Corrosion and Environmental Degradation,* Vol. II, p. 179, ed. Schutze, M., Reprinted with permission from Wiley-VCH, Weinheim, Germany, 2000.)

Corrosion in Carbonaceous Environment—Carburization and Metal Dusting In many industrial processes, the combustion of fossil fuels creates exhaust gases containing H_2, CH_x, CO, and CO_2 mixtures, while a number of processes in the chemical and petrochemical industry involve reducing conditions to produce mixture of CO and H_2 via steam methane re-forming of natural gas. The presence of stable carbon as a solid over a wide temperature range from various oxidants CO, CO_2, and CH_x becomes the source of carburization or metal dusting.

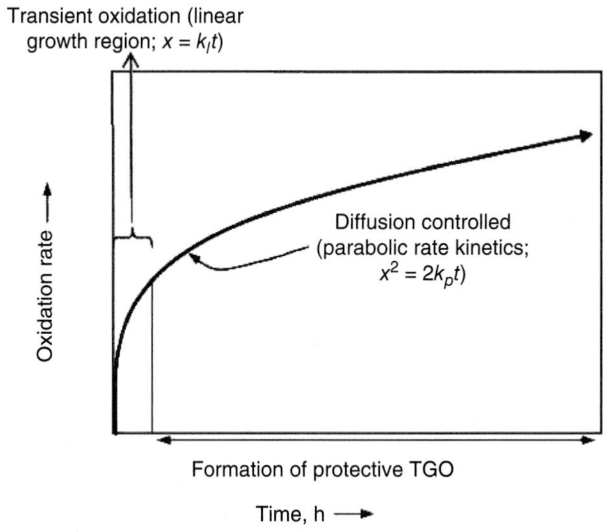

FIG. 25-8 Oxidation kinetics showing linear and parabolic behavior of thermal grown oxide formation in high-temperature alloys.

For thermodynamic consideration, when carbon activity is ≤1, but high enough to stabilize carbides, carburization prevails, while if carbon activity is supersaturated >>1, carbon becomes stable as deposits known as coking. The catalysis process of carbon deposition by the alloy often leads to catastrophic disintegration of the metal surface, resulting in continued metal wastage rates.

Depending on the process conditions, a combination of the following reactions occurs:

$$CO + H_2 = H_2O + C \text{ (CO reduction reaction)} \tag{25-12}$$

$$2CO = CO_2 + C \text{ (Boudouard reaction)} \tag{25-13}$$

$$CH_4 = 2H_2 + C \text{ (Hydrocarbon thermal cracking)} \tag{25-14}$$

The hydrocarbon cracking reaction produces high carbon activity with increased temperature and often controls the carburization reaction at high temperatures [above about 750°C (1382°F)]; however, CO reduction and Boudouard reactions generate high carbon activity even at relatively low temperatures [~450°C to 700°C (1292°F)] and takes time to equilibrate in this temperature range. Consequently, supersaturated gas produces high a_c, resulting in metal dusting reactions. The carbon activity for the preceding reactions is calculated as in Eq. (25-15):

$$a_c = \frac{k_p^{12} PH_2\ PCO}{PH_2O};\ a_c = \frac{k_p^{13} p_{CO}^2}{PCO_2};\ a_c = \frac{k_p^{14} PCH_4}{P^2 H_2} \tag{25-15}$$

where k_p^{12}, k_p^{13}, k_p^{14} are reaction rate constants for Eqs. (25-12), (25-13), and (25-14), respectively, and they are usually calculated from the Gibbs free energy of formation, ΔG°.

Carburization results in the formation of intergranular and intragranular carbides throughout the matrix. The carburization front is dictated by a_c, the local carbon activity, and if a_c is high enough, such that carbides of the elements present in the alloy can be formed, the front moves internally within the alloy matrix. The carbide formation often results in embrittlement and can severely affect the mechanical properties of the alloy. In general, Fe-base alloys are more susceptible to carburization than Ni-base alloys, and the addition of nickel improves the carburization resistance of the alloys by lowering the diffusivity and solubility of carbon (Young, D. J., *High Temperature Corrosion and Materials Application,* Elsevier Corrosion Series, 2008, p. 397). Moreover, minor additions of elements, such as molybdenum, niobium, tungsten, and reactive elements, help improve carburization resistance (Young 2008, p. 397; Lai, G. Y., *High Temperature Corrosion and Materials Application,* 2d ed., Cambridge University Press, Cambridge, UK, 2006, p. 54). The additions of Si and Al also help improve carburization and metal dusting resistance by forming dense scale that slows down carbon ingress through the scale. Additionally, elements such as copper and tin that are noncatalytic to carbon also minimize metal dusting attack in the alloys (Young 2008, p. 397).

Corrosion in Sulfur-Bearing Environment—Sulfidation and Hot Corrosion Sulfur is a strong corrodent and can cause catastrophic corrosion when it is present as an impurity in fuel and feedstocks. In certain industrial processes containing excess air or free molecular oxygen, fuel or feedstock reacts completely to convert sulfur into SO_2 or SO_3 gas mixtures. Such an environment is considered oxidizing-sulfidizing in nature. It is less aggressive for the most high-temperature alloys up to approximately 900°C (1652°F) compared to other sulfidation reactions unless molten deposits are also present. Sulfidation in an O_2–SO_2/SO_3 environment is not dictated by the formation of low-temperature molten sulfide eutectics such as Ni–Ni_3S_2 [T_{melt} = 635°C (1175°F)] unlike in other forms of sulfidation attack (Birks, N., G. H. Meier, F. S. Pettit, *Introduction to the High-Temperature Oxidation of Metals,* 2d ed., Cambridge University Press, Cambridge, UK, 2006, p. 165). Notwithstanding, oxidizing-sulfidizing corrosion can become a catastrophic degradation mechanism in the presence of small amounts of salt vapors. Sulfidizing-oxidizing environment in the presence of salt vapors of alkali sulfates and chlorides can cause the destruction of protective thermally grown oxide (TGO). This form of corrosion attack is defined as "hot corrosion/coal-ash corrosion/oil-ash corrosion," depending on the types of salt deposits and the mixed gaseous environment. There are two types of hot corrosion observed in industrial environments: type I and type II. Typically, type I hot corrosion occurs above the melting temperature of salt deposits [800°C to 1000°C (1472°F to 1832°F)], while type II hot corrosion occurs below the melting temperature of external salt deposits [600°C to 750°C (1112°F to 1382°F)] (Birks, N., G. H. Meier, F. S. Pettit, *Introduction* *to the High-Temperature Oxidation of Metals*, 2d ed., Cambridge University Press, Cambridge, UK, 2006, p. 205). The molten deposits in type II hot corrosion are established due to sufficient PSO_3 present in the system from the reaction between salt deposits and reaction product (e.g., Na_2SO_4–$NiSO_4$ or Na_2SO_4–$CoSO_4$) (Birks et al. 2006, p. 205).

In many environmental conditions, sulfur in the system is converted to H_2S due to a lack of free molecular oxygen in gaseous mixtures containing CO, CH_4, and H_2. Such conditions are typically called reducing or nonoxidizing conditions. In such environments, oxidation and sulfidation occur simultaneously, and the types of scales and precipitates formed depend on the partial pressures of oxygen and sulfur. The phase stability diagrams as shown in Fig. 25-9 for an Ni-S-O system are quite helpful in understanding degradation modes and resulting corrosion products. Sulfidation occurs because of the formation of internal sulfides beneath external oxidation/corrosion products. The common alloying elements that react to form various sulfides are Cr, Fe, Ni, Co, and Mn. While considering the thermodynamic stabilities of various sulfides, it is also important to compare their rates of formation. For instance, the growth rate for chromium sulfides is four to five orders of magnitude faster than that for chromium oxide. Another important consideration unlike oxidation is the formation of molten eutectics (e.g., Ni–Ni_3S_2), which further results in catastrophic degradation of the alloys.

Ni-base alloys are in general resistant to oxidizing-sulfidizing and hot corrosive environments, but they are susceptible to catastrophic sulfidation in reducing environments due to the formation of low-melting-temperature sulfides, while Co-base alloys offer superior performance in reducing environments. Co-base alloys offer rather inferior type II hot corrosion resistance compared to Ni-base alloys. Higher amounts of chromium additions help in lowering sulfidation attack in the alloys along with additions of Al and Si (Young, D. J., *High Temperature Oxidation and Corrosion of Metals*, Elsevier Corrosion Series, 2008, p. 361; Birks, N., G. H. Meier, and F. S. Pettit, *Introduction to the High-Temperature Oxidation of Metals*, 2d ed., Cambridge University Press, Cambridge, UK, 2006, p. 163; Lai, G. Y., *High Temperature Corrosion and Materials Application*, 2d ed., Cambridge University Press, Cambridge, UK, 2006, p. 130). Minor elemental additions of Nb and Zr, and Ti also lower sulfidation kinetics. (Lai 2006, p. 141).

Corrosion by Halogens Corrosion in environments containing chlorine and fluorine is commonly encountered, while corrosion by the presence of bromides and iodides is occasionally found in industrial environments. The elements present in high-temperature alloys react with halogens to form low-melting-temperature metal halides, while some elements form halides that even vaporize in temperature ranges where high-temperature alloys are used. The vaporization of metal halides due to high vapor pressures could cause catastrophic corrosion in high-temperature alloys. Ni-base alloys are generally more resistant in chlorine- and fluorine-bearing environments due to the formation of $FeCl_x$ volatile vapor species, and high chromium, aluminum, and silicon contents help in the corrosion resistance. In reducing fluorine-bearing environments, however, Ni-Mo–based systems seem to offer the best resistance, but the trend reverses when oxygen is present in the system. In O_2 and chlorine/fluorine environments, Mo and W form highly volatile oxychlorides and oxyfluorides, and the use of high refractory-containing alloys should be avoided (Lai 2006, p. 94).

Nitridation Nitridation typically occurs in ammonia or nitrogen-bearing environments. When alloys are exposed to highly oxidizing environments, nitridation is less severe unless the protective scale cannot offer resistance, especially in thermal cycling conditions. For instance, in combustion environments involving thermal cycling, most alloys offer oxidation resistance up to certain point, but after long-term exposures they suffer internal nitridation. Nitrogen gas, N_2, is not as severe as ammonia, and nitridation in nitrogen environments often occurs at temperatures

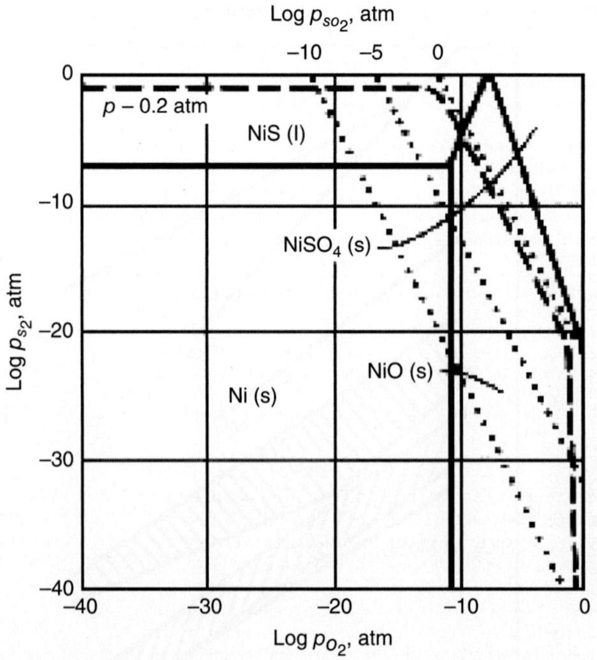

FIG. 25-9 Phase stability diagram for nickel-sulfur-oxygen system at 1250 K (1790°F). ("Factors Affecting High-Temperature Corrosion and Materials Properties," *Metals Handbook Volume 13A Corrosion: Fundamentals, Testing, and Protection.* Reprinted with permission of ASM International. All rights reserved. www.asminternational.org.)

above 1000°C (1832°F), while alloys may suffer nitridation in ammonia at a relatively low temperature range. Nitridation in industrial atmospheres takes place due to dissociation of nitrogen or ammonia, subsequent adsorption followed by absorption of [N] into the metal. Consequently, the metal becomes supersaturated with [N], thus precipitating nitrides of the elements present in the metal. Nitrides of some of the common elements observed in high-temperature alloys include Fe_2N, Cr_2N, MoN, TiN, AlN, and NbN. Ni-based alloys are more resistant than Fe-base alloys, while alumina-forming alloys that can establish alumina scales in reducing environments are generally more resistant than chromia-formers (Lai, G. Y., *High Temperature Corrosion and Materials Application*, 2d ed., Cambridge University Press, Cambridge, UK, 2006).

Corrosion by Molten Salt or Molten Metal This form of degradation is observed when metal is completely immersed in a molten environment, either in molten salt or molten metal. In such an environment, an external gaseous mixture, even if present, does not directly interact with metal. Mixed gaseous environments still affect corrosion properties, but by the reaction with molten medium rather than by its direct interaction with the alloy surface. In this form of corrosion, molten medium reacts with elements present in the alloy and results in leaching of the most reactive elements from the alloy substrate into the molten medium. The resistance of the alloys depends on whether the molten medium is oxidizing or reducing, and the performance of the alloys varies significantly from one medium to another. Some commonly found molten media are molten nitrates, chlorides, carbonates, zinc, and sodium.

CORROSION PREVENTION

MATERIALS SELECTION

Introduction The first step in determining the materials of construction for a project is an evaluation of the environments that will be present to determine if they are corrosive to the materials commonly used. As a general rule, carbon steel is probably the material of choice for projects if corrosion is not a concern, and the maximum temperature will not exceed 425°C (800°F). Above this temperature, a variety of alloy steels and other materials may be required to provide sufficient long-term integrity. Most organic chemicals are not corrosive to steel, and gases and vapors are also not corrosive to steels in most cases. However, acids, molten salts, and aqueous liquids are corrosive to steels, and these environments would normally require resistant materials. Usually the engineering and construction firm engaged to build the project will have specific materials recommendations for the various units in the project. However, it is usually helpful to consult authoritative sources to become familiar with the performance of materials that will be used in the project. Table 25-2 shows several sources that may be helpful. It must be noted that literature references often do not provide details on materials of construction, and the guidance found in these sources may not answer many questions that arise.

TABLE 25-2 Sources of Corrosion Resistance Information

Name	Source	Environment	Reference
Corrosion Survey Database (COR SUR)	NACE	Chemicals	1
MTI Materials Selector (MS) Series	MTI	Industrial chemicals	2
Materials Selector for the Chemical Process Industries	MTI	Industrial chemicals	2
Handbook of Stainless Steels	McGraw-Hill	Chemicals	3
Corrosion Resistance Handbook	Wm. Andrew	Chemicals	4
Chemical Resistance of Plastics and Elastomers	Wm. Andrew	Chemicals	4

NOTE: Vendors usually have good information on materials they supply.
References:
1. NACE International, The Corrosion Society, International, Copyright © 2002.
2. Materials Technology Institute, 1001 Craig Road, Suite 490, St. Louis, MO 63146.
3. Peckner, D., and I. M. Bernstein. Handbook Of Stainless Steels. McGraw-Hill, New York, 1977.
4. Elsevier B.V. Registered office: Radarweg 29, 1043 NX Amsterdam, The Netherlands.

Strategy After information has been obtained on the corrosion rates that can be expected in an environment, it is necessary to determine the strategy for dealing with the possible corrosion that may occur with various materials. In order to judge the corrosion rates that may be found for different materials, see Table 25-3. This table provides guidelines for making decisions on how to deal with the corrosion that may be encountered.

Generally, corrosion-resistant alloys cost more than carbon steel, so the decision to use a corrosion-resistant material will add cost to the project. However, there are options that may reduce the cost while still obtaining the benefit of corrosion-resistant materials. In addition, in some cases, it may be possible to deal with the corrosion loss by increasing the thickness of metal. This additional metal is known as a corrosion allowance, and its cost is minor compared to the cost of many corrosion-resistant alloys. Corrosion allowances are generally used when the corrosion rate is low or moderate and there are no issues of pitting or cracking. In addition, when a corrosion allowance is specified, there must be a program to carry out routine inspections to monitor the loss of metal so that action can be taken before the corrosion loss exceeds the corrosion allowance. As a general rule, corrosion allowances are in the range of 3 to 10 mm (0.125 to 0.4 in)

Another possibility is to investigate the possibility of cladding or lining the vessel or piping with a corrosion-resistant material so that the carbon steel carries the structural load while the layer of corrosion-resistant material provides the protection from corrosion. This approach is effective and has been widely used. The use of cladding or overlaying adds cost beyond the base cost of the materials, so it is seldom economical to use this approach if the total thickness of metal is less than 25 mm (1.0 in). In these cases, it usually is more economical to use solid corrosion-resistant material construction. The subject of cladding will be discussed in more detail later.

The issue of environmentally assisted cracking must be considered and understood with any project. This form of corrosion cannot be solved with a corrosion allowance. Cladding and other barrier layers can be effective, but minor penetrations through the layer can result in catastrophic failures, so it is very important to set up procedures for monitoring the protective layer performance when this form of failure is a possibility. In particular, hydrogen sulfide, hydrogen fluoride, and hydrogen cyanide are very aggressive with carbon steels, and chlorides are aggressive with austenitic stainless steels (NACE Standard MR0175, "Metals for Sulfide Stress Cracking and Stress Corrosion Cracking Resistance in Oilfield Environments," NACE International, Houston, Tex., 1975). Also, concentrated sodium hydroxide will cause cracking with many alloys. When dealing with these chemicals, special precautions are required.

TABLE 25-3 Corrosion Rate Classification

CR µm/yr	CR mpy	Designation	Preventive measures
CR < 25	CR < 1.0	Noncorrosive	1
25 ≤ CR < 50	1 ≤ CR < 2	Slightly corrosive	2
50 ≤ CR < 400	2 ≤ CR < 16	Moderately corrosive	3
400 ≤ CR < 825	16 ≤ CR < 32	Corrosive	4
≥ 825	CR ≥ 32	Severely corrosive	4

Preventive Measures:
1. No preventive measures except in cases where corrosion product contamination is not allowed.
2. Corrosion allowance or protective coatings used.
3. Corrosion allowance, protective coating, cladding, lining, or corrosion-resistant alloy used.
4. Protective cladding, lining, or corrosion-resistant alloy used.

DESIGN

General Process equipment is usually designed to deliver maximum performance at minimum cost, but corrosion concerns may not be reflected in the design decisions. As a result, problems may arise in various areas unless the design is modified to mitigate the corrosion possibility. The following discussion covers several types of equipment with corrosion concerns that have been observed.

Heat Exchangers Heat exchangers are designed to either heat or cool a process fluid. Corrosion issues have been observed with both types of heat exchangers. In many systems, steam is used to heat the process stream. One common issue that occurs in these systems is that the heat exchanger is oversized so that only a portion of the surface area is required to heat the fluid to the required temperature. As a result, condensate fills the bottom of the unit, so the remaining surface area exposed to the steam heats the process fluid. As a result, the process fluid will be overheated, and this can cause corrosive attack in some systems. Vertical exchangers are less susceptible to this problem, but horizontal exchangers can experience severe corrosion on the upper tubes. In exchangers where the process fluid is being vaporized on the shell side of the exchanger, it is important to maintain the liquid level in the shell above the highest tube to prevent dry boiling on the upper tubes. This has been the cause of corrosion and cracking in these tubes.

The decision as to which fluids will be routed through the shell side versus the tube side of a heat exchanger is usually based on pressure drop or fluid considerations. In most cases, these decisions do not affect the corrosion performance of the exchanger, but occasionally the decision becomes important. For example, in a boiler using horizontal tubes with the boiler water on the tube side being heated by flue gas, severe corrosion was observed on the bottoms of some of the tubes. The corrosion was identified as caustic gouging, and it was attributed to poor flow distribution of the boiler water resulting in having the tubes almost dry at the exit. The design in this case was inadequate for the service conditions, with the boiling occurring on the tube side and poor flow control from the manifold feeding the tubes.

Another example of a design issue occurred in a nitric acid plant where the cooler-condenser hot end tube sheet suffered serious corrosion from the condensation of nitric acid on the inlet end surface. This problem occurred because the corrosion products that accumulated there were not washed away by the condensing acid, and the development of a high concentration of chrome six ions aggravated the corrosion. This was another example of horizontal tubes in the exchanger, but with condensing rather than boiling. Insulating the cooling water side of the tube sheet prevented the condensation and minimized the corrosion damage.

Heat exchangers used for cooling and condensing process fluids are less susceptible to corrosion issues, but problems can occur when cooling waters are overheated. In particular, stainless steels are susceptible to pitting and stress corrosion cracking in waters with significant chloride contents. In both of these mechanisms it has been observed that there is a critical temperature above which the problem becomes severe. Another problem that has been observed is that horizontal condensers can have condensate block the lower tubes in the exchanger if the outlet is not designed properly to allow free flow. In this case, the condensate will accumulate until the pressure drop becomes large enough to force the accumulated liquid out in a slug. This slug can damage downstream equipment through a fatigue mechanism [Dean, S. W., "Chloride SSC of Stainless Steel? No – Cyclic Strain Cracking!" *Materials Performance* **39**(9): 78–87 (2000)].

Fired heaters also present design challenges from a corrosion perspective. In systems where steam is present in the process fluid, cool spots that allow condensation to occur above the heated zone will drip, and the thermal shock from this dripping can cause thermal fatigue cracking of components in the furnace. In such cases the use of insulation to eliminate cool spots solved the problem.

Distillation and Absorption Columns These units are often subject to corrosion damage because they are involved with separating and concentrating components that may be corrosive. A cursory examination of such a process would suggest that the most corrosive areas should be at the bottom where it is hottest, or perhaps at the top. However, this is seldom observed. In many cases where water is present, a volatile component in the system is trapped in the column because it ionizes at lower temperatures, thereby becoming less volatile, and recombines at higher temperatures, increasing its volatility. Hydrogen chloride and carbon dioxide both have been observed to exhibit this behavior. As a result, the location where the compound concentrates causes corrosion to occur at a high rate. Several approaches have been used to deal with this problem. In the case of carbon dioxide accumulation, the addition of sodium hydroxide to the process stream was successful in removing the carbon dioxide from the system. In the case of hydrogen chloride, the use of a bypass that removed some of the liquid from the point where it concentrated was successful. Another approach is to upgrade the materials of construction with an alloy that resists the offending contamination in the region where the concentration occurs.

Tankage Although tankage is generally not considered a critical area for corrosion control, there are cases where careful design is required to assure good performance. In flat-bottom storage tanks, the tank bottoms often suffer corrosion damage both from the inside and from the soil side. With organic liquids that are less dense than water, a water layer can develop on the tank bottom from condensation, pressure testing, or minor impurities in the liquid. This layer can become very corrosive over time because in most situations the outlet nozzle for the tank is above the bottom. Corrosion from the soil side is a concern if the sand layer below the tank becomes filled with water and does not drain properly. Cathodic protection systems are now used to prevent soil-side corrosion. Sulfuric acid tanks are a special case, and their design incorporates many important features to minimize corrosion damage (NACE Standard SP02-94, "Design, Fabrication, and Inspection of Fresh Sulfuric Acid Tanks, 2006" and NACE Standard SP02-05, "Design, Fabrication, and Inspection of Tanks for Storage of Petroleum Refining Alkylation Unit Spent Sulfuric Acid at Ambient Temperature, 2015" NACE International, Houston, Tex.).

Piping Although piping standards cover most of the issues involved with piping designs, there are some details that are overlooked in many cases. One important detail is the elimination of dead legs in piping systems. Dead legs are sometimes included to allow for future expansion or changes to the flow system. In other cases they are included to facilitate start-up or shutdown operations where it may be necessary to introduce steam or other material. These areas are locations where corrosive components can accumulate and cause damage without attracting much attention. From a design perspective, if the dead leg cannot be eliminated, an effort should be made to prevent liquids from accumulating in it. This can be accomplished by having the entry point at the top of the pipe rather than at the bottom. Also, placing the valve as close to the junction point as possible can minimize the quantity of liquid that accumulates there.

Bypass lines around heat exchangers are sometimes used to control the fluid temperature exiting the exchanger. The problem in this case is that the mixing point where the bypassed fluid encounters the fluid from the exchanger can be subjected to severe thermal stresses from minor flow variations in the streams. Rather than using a tee or wye connection, it may be better to have a concentric connection so that the mixing occurs over a larger area and does not affect the pipe walls.

ALTERING PROCESS/ENVIRONMENT

Introduction Changing the environment or the process is sometimes the best way to deal with corrosion issues. However, it requires knowledge of the corrosion mechanism that occurs in the environment in question. As a general rule, changes to the environment appear to be impossible in most cases. The ability to find some flexibility in the conditions will often pay dividends in terms of being able to use less expensive materials. Some examples are discussed next.

Passivation Promotors Stainless steels and other alloys with significant chromium contents owe their corrosion resistance to an oxide passive film that forms spontaneously in the presence of oxidizing substances in the environment. This film is insoluble in many aqueous and acid environments, but its ability to form is dependent on the concentration of oxidizing substances in the environment. In cases where the natural formation of this film is marginal, the corrosion resistance of the alloy can be improved by the addition of small amounts of an oxidizing component. This approach has been used with stainless steels in concentrated sulfuric acid. In this case, the addition of small amounts of hydrogen peroxide or nitric acid was successful in reducing the corrosion rate into the noncorrosive region. The addition of air is commonly used in urea reactors to keep the stainless steel in the passive region. In another case, the addition of a small concentration of hydrogen peroxide was successful in reducing the corrosion of stainless steel pipes and tubing in the condenser from a sulfuric acid concentrator.

In the case of steels, passivation is also possible, but it requires a higher level of oxidizer and pH control. Nitrite additions have been used for this purpose. In addition, the condensate from air compressors is usually not corrosive to steels because of its high oxygen content.

Removal of Oxidizers In many aqueous environments, the presence of dissolved oxygen is the driving force for the corrosion of steel. In these systems, the removal of dissolved oxygen can reduce the corrosion rate to a low level. This approach is commonly used in boiler feedwater preparation. In the case of high-pressure boilers, it is particularly important to keep the dissolved oxygen at a very low level, depending on the pressure required. In this case, both mechanical deaeration and chemical scavengers are used. These systems are designed to maintain the boiler tubing in good operating condition and to minimize pitting that will occur if the oxygen level becomes too high.

Scale Enhancers In the case of cooling water, oxygen removal is not practical when cooling towers operate by the evaporation of water. In these cases, corrosion control is achieved by the addition of chemicals that promote the formation of a protective scale on steel and copper alloy tubes and pipes. These chemicals usually include phosphates or molybdates, which are effective in maintaining the system pH and minimizing corrosion of the metals in the system. In the past, chromates were used for this purpose, but they are no longer permitted in most applications. Maintaining the system pH is important in order to keep the pH in a range where the solubility of iron in the liquid is very low.

Cracking Preventives In the cases of anhydrous ammonia and several of the lower molecular weight alcohols, the addition of about 0.1 percent water has been found to be effective in preventing the stress corrosion cracking of carbon steels. Although in some cases this is not possible because of product specifications, where it can be practiced it has been found to be a practical solution to a serious problem (Dean, S. W., *Materials Selector for Hazardous Chemicals, Organic Solvents, Publication MS-8*, Materials Technology Institute, St. Louis, Mo., 2011, p. 9).

Removal of Aggressive Compounds Because minor concentrations of specific compounds can cause corrosion damage, the removal of these compounds is another approach to controlling corrosion. The removal of oxygen from boiler feedwater was mentioned earlier. Chloride removal systems have been used to prepare water for use in stainless steel equipment to reduce the possibility of chloride stress corrosion cracking. Another example is the use of a copper removal system for cooling water used in aluminum exchangers. Aluminum alloys are susceptible to pitting damage in cooling water, but if trace copper contamination is removed from the water, the aluminum will give better performance. In this case, an aluminum foil contactor was placed in the cooling water inlet to the exchanger, and it functioned by replacing copper ions with aluminum ions in the water.

Process Changes Sometimes process changes can be made to prevent or minimize corrosion problems. For example, in distillation columns with external reboilers, the boiling process may concentrate corrosive species on the tube surfaces to the point where they cause rapid corrosion. This is especially true for horizontal tube units. In this case, it might be possible to increase the pressure in the exchanger so that boiling does not occur on the tubes. The vaporization can occur when the fluid passes through a pressure let-down valve entering the column bottom. Another example is the case of a centrifugal pump where the minimum net static head is maintained at the pump inlet to prevent cavitation that will damage the impeller.

When cooling water can be corrosive to exchangers, some possibilities to prevent problems include using fin-fan air exchangers or using a closed-loop inhibited system between the critical exchanger and the cooling tower system.

Another example occurred in a nitration process where sulfuric acid was used to remove the water generated by the reaction, thereby allowing the reaction to proceed to completion. The problem in this case was that sulfuric acid is very corrosive to stainless steels in the concentration range from 80 to 91 percent. As a result, it is conventional to use a sulfuric acid concentrator that concentrates the recycled acid to 93 percent. By concentrating the acid to 89 percent and adding a small concentration of nitric acid, it was possible to eliminate the difficult steps required to reach 93 percent, and also to handle the 89 percent acid with conventional stainless steel alloys. In all of these cases the key is understanding the corrosion mechanisms that occur and taking steps to reduce or eliminate the driving force or to enhance the protective surface film.

BARRIERS TO PREVENT CORROSION

A common technique of corrosion prevention is to place a barrier between the corrosive environment and the underlying alloy that would be corroded by the environment. Barriers can be either nonmetals or an alloy that is more corrosion resistant than the alloy it is protecting.

Paints, Coatings, and Linings Paints, coatings, and linings are systems of pigments with natural or synthetic resins that form a protective film by drying, oxidizing, or polymerizing (Dillon, C. P., *Corrosion Control in the Chemical Process Industries*, 2d ed., Materials Technology Institute, St. Louis, Mo., 1994, p. 346). There are many different types of systems, and the more common ones will be discussed.

The terms *paint* and *coating* are often used interchangeably. For the purpose of this section, *paint* will be used. Paints are applied to the exterior of equipment to prevent corrosion by the external environment. They are a very important means of preventing corrosion under insulation (CUI) because insulation can often act as a water trap that can create very corrosive conditions. A lining is a system applied to the interior of process equipment.

The correct system selection and application is critical to achieving the desired level of corrosion prevention. All paints and linings require some level of surface preparation. If the preparation is inadequate, proper adhesion of the system to the underlying alloy won't be achieved, and premature failure will occur. Common methods of surface preparation are sand or grit blasting, water blasting, and hand or tool cleaning. Different systems require different levels of preparation. There are three common methods of classifying surface preparation: the National Association of Corrosion Engineers (NACE), the Steel Structures Painting Council (SSPC), and the Swedish Pictorial Standards (SA). A commercially available paint or coating system will state what level of surface preparation is required.

When the paint or lining is applied, the first coat is referred to as the primer coat. Some paints/linings will require more coats. Additional coats may be of the same type or a different type. For example, epoxies are often topcoated with a urethane because the urethane has greater ultraviolet light resistance. Two- and three-coat systems are commonly used. A rule of thumb is that for a particular environment, more coats will last longer and provide better corrosion resistance.

Common paints and linings are:

1. *Inorganic zinc.* This is zinc particles in a silicate binder. This is widely used on carbon steel. The zinc particles provide galvanic protection to the steel. It should never be used on stainless steel. In a fire situation, the zinc could melt and cause liquid metal embrittlement of the stainless steel.

2. *Epoxies.* These are formulated from polyphenol and epichlorinhydrin and require a catalyst for curing. They have good corrosion resistance to a wide variety of conditions (Dillon, C. P., *Corrosion Control in the Chemical Process Industries*, 2d ed., Materials Technology Institute, St. Louis, Mo., 1994, p. 349).

3. *Epoxy mastics.* These paints are very useful for maintenance painting because they don't require the same level of surface preparation and are often used over old, still adherent paint systems. Be aware that if the underlying old system has lost its adherence, the new epoxy mastic will not be properly adhered to the underlying metal.

4. *Silicones.* Silicones have very high temperature limits and are normally a niche product for high-temperature applications such as furnace stacks.

5. *Urethanes.* Urethanes are available in a wide variety of colors and have excellent gloss retention and weathering resistance. They are commonly used as topcoats for aesthetic reasons.

6. *Alkyds.* Alkyds do not have good chemical resistance, but they are inexpensive and easy to apply. They have limited use for protecting process equipment but are often used for buildings, both interiors and exteriors.

The best chance of getting a good paint application is during initial fabrication. Once equipment is in service, it may be difficult to repair or replace a paint system once it starts to fail. Restrictions on such surface preparation techniques as grit blasting often preclude getting the desired level of surface preparation. If the equipment is insulated, removal and reinstallation of the insulation will significantly increase repainting costs. Money spent up front on a superior system will often pay for itself in the life-cycle cost of the equipment.

Linings are used on the internal surfaces of process equipment. This is an area that requires great care. Linings will almost invariably have pinholes, referred to as holidays, that allow the process contents to contact the alloy. Even if a lining is pinhole free on initial application, flaws in the lining can develop once the equipment is in service. It is not good practice to rely on a lining to protect equipment from highly corrosive process conditions. Any flaw in the lining can lead to very rapid failure. The most common use of linings is to either slow corrosion in mildly corrosive systems or to protect the process contents from iron contamination that could result from even low levels of corrosion.

Paints and linings can be applied by several methods, including spraying, rolling, and brushing. For large areas, spraying is almost always the economic choice. Paint systems will come with instructions for application. After application, inspection is often needed to ensure quality. This will include a thickness check to be sure the paint is neither too thin nor too thick. For large or critical jobs, a National Association of Corrosion Engineers–qualified coating inspector can be employed to check quality.

A coating system that can be used in lieu of painting is thermal spray aluminum. In this process, a spray of molten aluminum droplets is applied to a steel substrate. This method is considerably more expensive than conventional painting, but it results in a superior corrosion barrier and a much longer life. The life-cycle costs may make this an attractive option. It can be used on both bare and insulated steel.

Galvanizing Galvanizing is the deposition of a zinc coating on a ferritic, normally carbon steel, substrate. The zinc serves two corrosion prevention purposes. The zinc will react with the atmosphere to form a zinc oxide corrosion product that will protect the steel by forming a barrier. Also, zinc is active to steel in the galvanic series and will provide galvanic protection. Galvanizing is done by two primary processes. In hot dip galvanizing, the steel component is placed in a bath of molten zinc. Zinc will solidify on the surface. In electrogalvanizing, zinc ions are electrodeposited on the steel.

Galvanizing is commonly used on structural steel, pipe, fasteners, and other components that will be directly exposed to the atmosphere. Galvanized steel can be difficult to paint due to problems in getting adequate surface preparation. If painting on galvanized steel is necessary, care must be taken, and consulting with a painting subject matter expert is recommended.

Metal Barriers Metal barriers are most commonly used when equipment is constructed from a ferritic steel such as carbon steel with a corrosion-resistant alloy (CRA) placed between it and the process environment. The use of a metal barrier has the advantage of using lower-cost ferritic steel as the underlying substrate, which provides most of the structural strength. The higher-cost CRA is normally much thinner than the steel. This provides for significant cost advantages over the use of solid CRA. The cost advantage increases with increasing thickness of the equipment.

The types of metal barriers can be classified by the construction technique.

1. Loose lining
2. Roll cladding
3. Explosive cladding
4. Weld overlay

Loose Lining Loose linings are thin sheets of a corrosion-resistant alloy placed over the underlying substrate. There are different ways to do this, but a common technique is to attach the sheets to the substrate by fillet welds. As more sheets are laid down, they are overlapped and seal-welded to each other. It is critical that all welds attaching sheets are leak free. Haynes International publication No. H-2010, *Fabrication of HASTELLOY® Corrosion-Resistant Alloys*, shows this technique in detail (Haynes H-2010, *Fabrication of HASTELLOY® Corrosion-Resistant Alloys*, Haynes International, Kokomo, Ind., 2003, p. 18-24).

Good welding technique is critical. A weld flaw may allow process contents to get behind the corrosion-resistant sheets and corrode the steel substrate. This type of problem can be very difficult to detect by inspection, and the first indication of a problem can be a loss of process containment.

Roll Bonding Roll bonding is a technique where a CRA is metallurgically bonded to a ferritic steel by placing the CRA on the ferritic steel and then pressing the two alloys together between rolls at high temperature and high pressure. The critical element for plate quality is the bond between the steel and the CRA. This bond can be measured by the shear strength of the bond and the extent of any unbonded areas between the plates. ASTM A264 covers the purchase requirements for chromium-nickel steel-clad plates. These plates can be fabricated into welded pipe.

Explosion Bonding Explosion bonding creates a metallurgical bond between a CRA and a ferritic steel by placing the CRA on top of the steel, covering it with an explosive charge, and then detonating the charge. The compressive forces created by the detonation create a solid state weld bonding the CRA to the steel. For certain metals such as zirconium, explosion bonding is the only technically viable method.

Weld Overlay Weld overlay, as its name implies, is placing a CRA on a ferritic steel by using a welding process to lay down layers of weld metal on the steel substrate.

Roll bonding, explosion bonding, and weld overlay are often considered competing processes. When these techniques are used in equipment fabrication, quality control of the resulting plate is very important. Standards in common usage to insure quality are:

ASTM A263 Standard Specification for Stainless Chromium Steel-Clad Plate

ASTM A264 Standard Specification for Stainless Chromium-Nickel Steel-Clad Plate

ASTM A265 Standard Specification for Nickel and Nickel-Base Alloy-Clad Steel Plate

ASTM B432 Standard Specification for Copper and Copper Alloy Clad Steel Plate

ASTM B898 Standard Specification for Reactive and Refractory Metal Clad Plate

Nonmetallic Barriers In addition to the metallic barriers discussed, several types of nonmetallic barriers are used in the process industries. These include:

1. Glass linings
2. Rubber linings
3. Refractory linings

Glass-Lined Equipment Glass-lined equipment is fabricated by bonding a layer of glass to a steel substrate. Glass thickness is normally 0.22 cm (0.085 in) or less. Almost all types of equipment are available, including reactors, tanks, pipe, valves, and fittings, subject to size limitations. Glass is resistant to almost all acids except hydrofluoric. Glass-lined equipment has wide usage where high purity is required and where even low corrosion rates can contaminate the product.

Glass is brittle by nature, and any chipping that allows the process to contact the steel substrate may result in a failure by corrosion. Glass-lined equipment is repairable by the use of metal patches. Tantalum is commonly used. However, other materials such as stainless steel can be used, as long as the material has acceptable corrosion resistance to the process environment. The key is to detect the failure of the glass before the failure of the steel occurs. Fault detectors are commercially available that use electrical techniques to determine when a flaw has occurred that results in contact between the process and the steel.

Rubber-Lined Equipment Rubber is a generic term that covers several different elastomeric compounds. Several have good to excellent resistance to certain chemicals and can be used as linings for process equipment. For example, natural rubber is widely used to line hydrochloric acid storage tanks. For specific applications, a subject matter expert should be consulted.

Refractory-Lined Equipment Refractory is normally used in high-temperature applications such as waste heat boilers. When used as an internal lining, what is called the cold wall effect must be considered. The refractory will insulate the steel, causing it to be at a considerably lower temperature. Refractory can crack, and if hot gases reach the steel and condense, corrosion can occur, especially if the hot gases create acid conditions on condensation.

CATHODIC PROTECTION

Cathodic protection (CP) is a corrosion prevention technique that protects equipment by using electrical circuits to force the equipment into a cathodic state. The two basic types are sacrificial anode and impressed current. The most common uses of cathodic protection are on equipment in contact with the soil, including buried pipelines and the bottom sides of large storage tanks.

With sacrificial anodes, the galvanic series is used to electrically connect steel to a less noble metal such as zinc or magnesium. Figure 25-10 is a basic description of a sacrificial anode system.

The steel becomes the cathode and is protected by the anode, which corrodes preferentially. Anodes will be consumed as they corrode. Because of this consumption, they need regular testing to ensure that they are providing adequate levels of protection. Once the level falls below accepted industry standards, replacement is required.

With an impressed current system, an external power supply is used with an inert anode. The normal arrangement is to use an alternating current supply off an available electric supply system. A rectifier converts the alternating current to direct current, with the negative lead to the equipment to be protected and the positive lead to the inert anode, normally graphite. This supplies electrons to the protected equipment, making it the cathode. Figure 25-11 is an example of an impressed current system.

The choice between using sacrificial anodes and impressed current systems is driven by installation and maintenance costs. In remote areas where a local power supply is not available, sacrificial anodes are normally used. However, solar cells used with some type of battery storage have been used.

Paints or other nonmetallic barriers such as a polyethylene wrap are used on the protected equipment. The purpose is to reduce any exposed area. This decreases the amount of protection needed in terms of electrical load. Cathodic protection systems need to be engineered and designed for the specific application. Only subject matter experts in this field should be used for the design, installation, and maintenance of CP systems.

ANODIC PROTECTION

Anodic protection is the opposite of cathodic protection in that it makes the equipment be protected by the anode. This only works with metals or alloys that exhibit active/passive behavior (see the subsection on Passivity under Corrosion Fundamentals). The goal is to use an electrical potential to move the equipment into the passive state. The most common usage is on sulfuric acid storage tanks to reduce iron contamination and on 316L stainless steel coolers used in sulfuric acid production (Dillon, C. P., *Corrosion Control in the Chemical Process Industries*, 2d ed., Materials Technology Institute, St. Louis, Mo., 1994, p. 379). With these systems, design and operation are very critical because an error can result in accelerated corrosion.

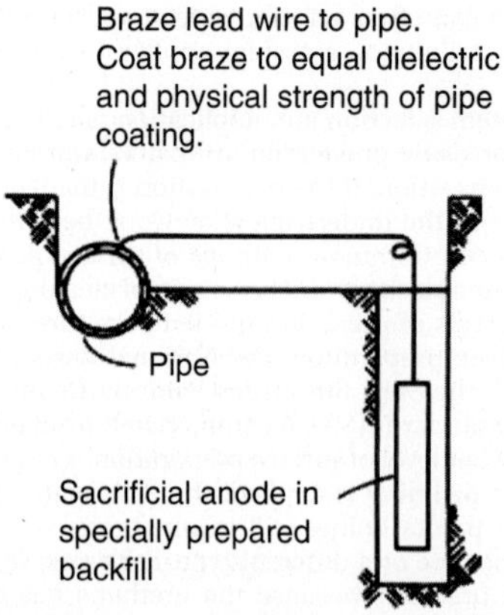

FIG. 25-10 Sacrificial anode cathodic protection system. ("Corrosion in Petroleum Production Operations – Types of Cathodic Protection Systems," *Metals Handbook Volume 13C Corrosion: Environments and Industries.* Reprinted with permission of ASM International. All rights reserved. www.asminternational.org.)

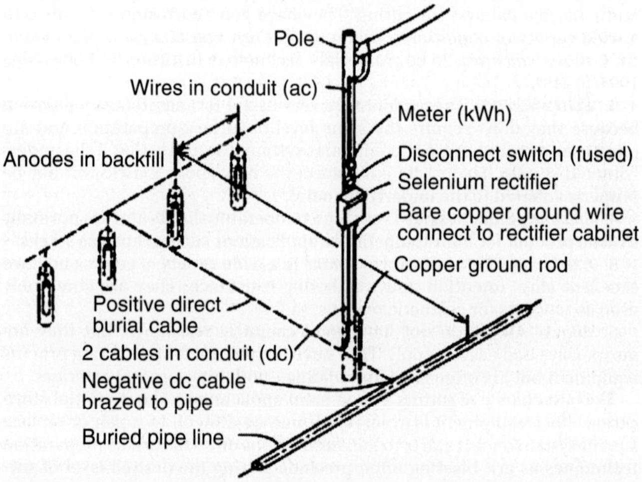

FIG. 25-11 Impressed current cathodic protection system. ("Corrosion in Petroleum Production Operations – Types of Cathodic Protection Systems" *Metals Handbook Volume 13C Corrosion: Environments and Industries.* Reprinted with permission of ASM International. All rights reserved. www.asminternational.org.)

CORROSION-TESTING METHODS*

The purpose of corrosion testing is to identify the optimum choices for process equipment in terms of materials of construction, design, and corrosion-control measures. *Optimum* here means that which comprises the best combination of cost, lifetime, safety, and reliability.

In many cases, information about the corrosion resistance of a material in a specific environment is not available and must be derived experimentally. There is no standard way to evaluate an alloy in an environment. Even if the chemistry of the operating plant environment is duplicated in the laboratory, a variety of other factors affect the result. These include, for example, velocity, hot and cold wall effects, crevices, chemical changes in the fluid during the test, stress levels of the equipment, contamination with products of corrosion, trace impurities, and dissolved gases. Also, the rate of the

*[Includes information from papers by Oliver W. Siebert, John G. Stoecker II, and Ann Van Orden, courtesy of NACE International; Oliver W. Siebert and John R. Scully, courtesy of ASTM; John R. Scully and Robert G. Kelly, courtesy of ASM; and Metal Samples Company, Division of Alabama Specialty Products Company, Munford, Alabama.]

corrosion reaction itself may change with time. Even with these considerations, immersion testing remains the most widely used method for selecting materials of construction.

Corrosion testing may be done either in the field—in either full-scale operation plants or smaller pilot plants—or in laboratories using simulated environments.

Both field and laboratory corrosion tests use specimens (coupons) or electronic corrosion sensors that are installed in the process. The most common electronic corrosion sensors are based on electrical resistance measurements or electrochemical response. Laboratory corrosion tests are generally regarded as producing less reliable results than field tests because the laboratory environment is simulated, and so they may not accurately or completely reproduce the many factors of an operating plant that affect corrosion behavior. Field corrosion tests are conducted in the actual process, and they are better able to account for the variety of potential influences.

CORROSION TESTING: LABORATORY TESTS

Metals and alloys do not respond alike to the many factors that are involved in corrosion. Consequently, it is impractical to establish any universal standard laboratory procedures for corrosion testing. However, some details of laboratory testing need careful attention in order to produce useful results. This section reviews important aspects of laboratory corrosion testing. A comprehensive treatment of this subject is given in MTI's *Laboratory Corrosion Testing of Metals and Alloys,* The Materials Technology Institute, St. Louis, Mo., 2015 (www.mti-global.org).

Coupon Immersion Test The coupon total-immersion test is a relatively simple, nonaccelerated corrosion test method that uses metal test specimens (coupons). The total-immersion test serves quite well to eliminate materials that obviously cannot be used. Further selection among those materials of potential use can be based on a knowledge of the properties of the materials concerned and possibly more sophisticated corrosion tests.

The National Association of Corrosion Engineers (NACE) TMO169-95 "Standard Laboratory Corrosion Testing of Metals for the Process Industries," and ASTM G31, "Recommended Practice for Laboratory Immersion Corrosion Testing of Metals," are the general guides for immersion testing. Small pieces of the candidate metal are exposed to the medium, and the loss of mass of the metal is measured for a given period of time. The mass loss is usually converted mathematically to a corrosion (penetration) rate. A disadvantage of this method is the assumed average-time weight loss. The corrosion rate could be high initially and then decrease with time. In other cases, the rate of corrosion might increase very gradually with time. The description that follows is based on these standards.

Test Piece* The size and the shape of specimens will vary with the purpose of the test, nature of the material, and apparatus used. A large surface-to-mass ratio and a small ratio of edge area to total area are desirable. These ratios can be achieved through the use of rectangular or circular specimens of minimum thickness. Circular specimens should be from sheet stock, not bar stock, to minimize the exposed end grain.

Shapes of typical commercially available test specimens are shown in Fig. 25-12. Other desired shapes are available in various sizes and can be obtained in a variety of shapes, materials of construction, and surface finish as needed.

All specimens should be measured carefully to permit accurate calculation of the exposed areas. An area calculation accurate to plus or minus 1 percent is usually adequate. A consistent surface finish provides more consistent results. Specimens should be degreased by scrubbing with bleach-free scouring powder, followed by thorough rinsing in water and in a suitable solvent, and air-dried. For relatively soft metals such as aluminum, magnesium, and copper, scrubbing with abrasive powder is not always needed, and it can mar the surface of the specimen. The use of towels for drying may introduce an error through contamination of the specimens with grease or lint. The dried specimen should be weighed on an analytic balance. After final preparation of the specimens, they should be stored in a desiccator until exposure if they are not used immediately.

Apparatus A common test apparatus consists of a kettle or flask of suitable size [usually 500 to 5000 mL (16.9 to 169.1 ounces)], a reflux condenser with atmospheric seal, a sparger for controlling atmosphere or aeration, a thermowell and temperature-regulating device, a heating device (mantle, hot plate, or bath), and a specimen-support system. If agitation is required, the apparatus can be modified to accept a suitable stirring mechanism, such as a magnetic stirrer. A typical resin-flask setup for this type of test is shown in Fig. 25-13. Open-beaker tests should *not* be used because of evaporation and contamination.

*Coupons and racks/holders as well as availability information are courtesy of Metal Samples, Munford, Alabama.

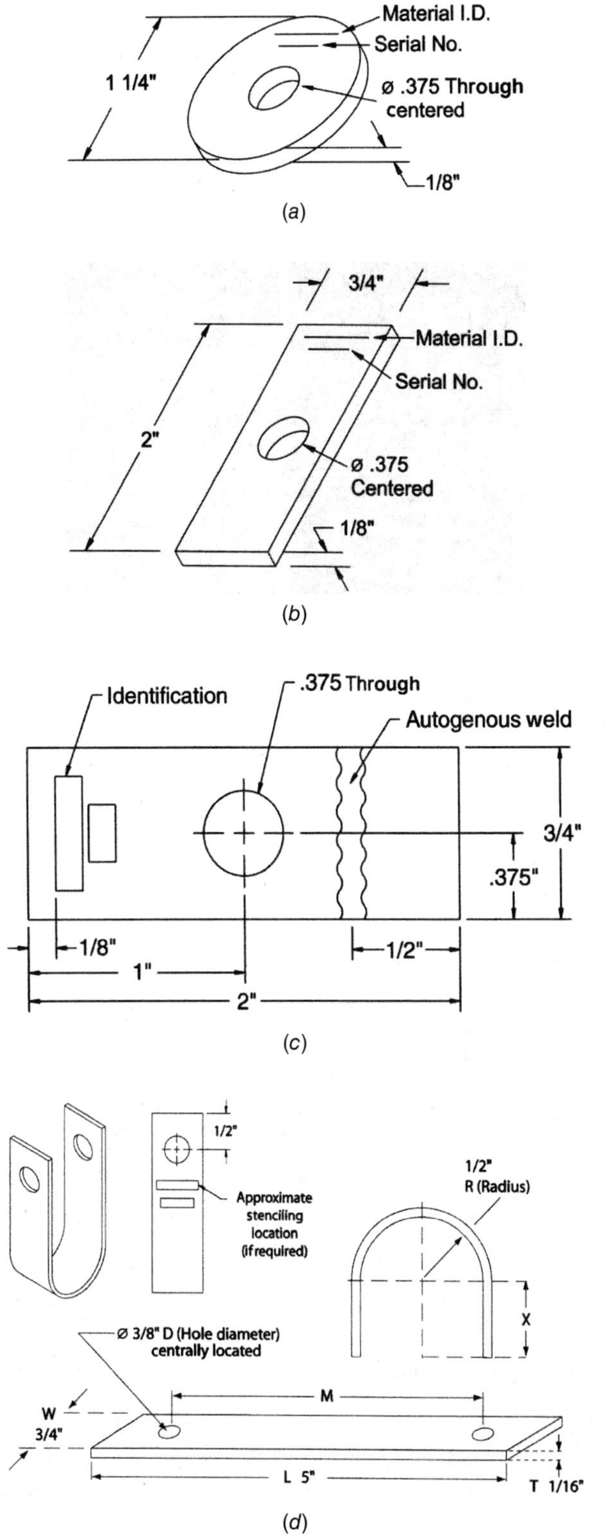

FIG. 25-12 Typical commercially available test coupons: (*a*) circular; (*b*) rectangular; (*c*) welded rectangular; (*d*) horseshoe stressed.

In more complex tests, provisions might be needed for continuous flow or replenishment of the corrosive liquid while simultaneously maintaining a controlled atmosphere.

Heat flux apparatus for testing materials for heat-transfer applications is shown in Fig. 25-14. Here the sample is at a higher temperature than the bulk solution.

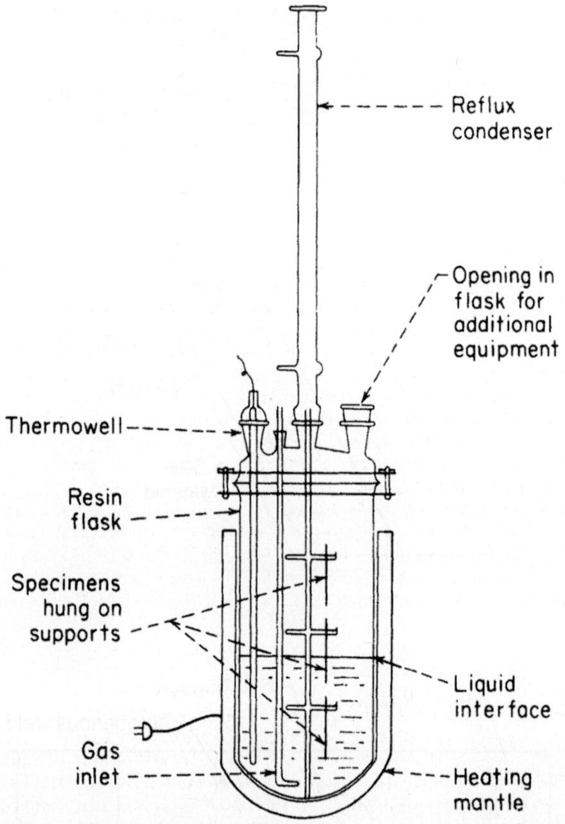

FIG. 25-13 Laboratory-equipment arrangement for corrosion testing. (*Based on NACE Standard TMO169-95.*)

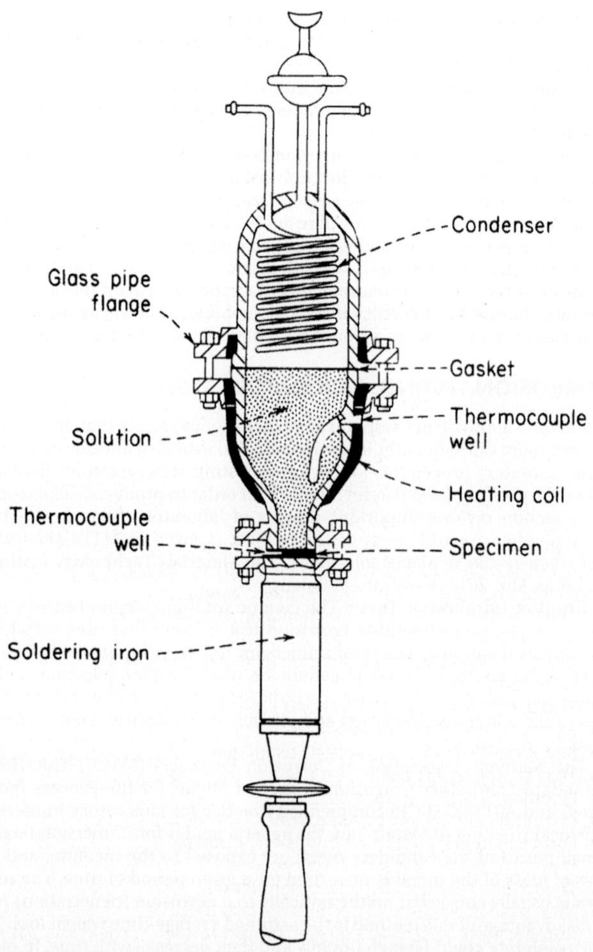

FIG. 25-14 Laboratory setup for the corrosion testing of heat-transfer materials.

A variety of important factors influence corrosion in a laboratory test, and applicable factors must be reported. These may include solution composition (including changes during the test), temperature, oxygen concentration or aeration/deaeration, rate of flow, pH, ratio of solution volume to specimen area, and any other important characteristics of the test environment.

Method of Supporting Specimens The supporting device and container should not be affected by or cause contamination of the test solution. The method of supporting specimens will vary with the apparatus used for conducting the test but should be designed to physically isolate the specimens from each other and any metallic container or supporting device used with the apparatus.

The shape and form of the specimen support should assure free contact of the specimen with the test solution, the liquid/vapor interface, and/or the vapor phase, as desired and as shown in Fig. 25-13. Some common supports are glass or ceramic rods, glass saddles, glass hooks, fluorocarbon plastic strings, and various insulated or coated metallic supports.

Duration of Test The duration of any test will be determined by the nature and purpose of the test. A procedure for evaluating the effect of time on corrosion of the metal and also on the corrosiveness of the environment in laboratory tests has been presented by Wachter and Treseder [*Chem. Eng. Prog.* 315–326 (June 1947)]. This technique is called the planned-interval test. Other procedures that require the removal of solid corrosion products between exposure periods will not accurately measure the normal changes of corrosion with time.

Materials that experience severe corrosion generally do not need lengthy tests to obtain accurate corrosion rates. Although this assumption is valid in many cases, there are exceptions. For example, lead exposed to sulfuric acid corrodes at an extremely high rate at first while building a protective film; then the rate decreases considerably, so further corrosion is negligible. The phenomenon of forming a protective film is observed with many corrosion-resistant materials, and therefore short tests on such materials may indicate high corrosion rates that would be misleading.

Short-time tests also can give misleading results on alloys that form passive films, such as stainless steels. With borderline conditions, a prolonged test may be needed to permit breakdown of the passive film and subsequently more rapid attack. Consequently, tests run for long periods are considerably more realistic than those conducted for short durations. This statement must be qualified by stating that corrosion should not proceed to the point at which the original specimen size or the exposed area is drastically reduced or the metal is perforated.

If anticipated corrosion rates are moderate or low, Eq. (25-16) gives a suggested test duration:

$$\text{Duration of test, h} = \frac{78{,}740}{\text{corrosion rate, mm/yr}} = \frac{2000}{\text{corrosion rate, mils/yr}} \qquad (25\text{-}16)$$

Cleaning Specimens after Test Before specimens are cleaned, their appearance should be observed and recorded. Locations of deposits, variations in types of deposits, and variations in corrosion products are extremely important in evaluating localized corrosion such as pitting and concentration-cell attack.

Cleaning specimens after the test is a vital step in the corrosion-test procedure and, if not done properly, can give rise to misleading test results. Generally, the cleaning procedure should remove all corrosion products from specimens with a minimum removal of sound metal. Set rules cannot be applied to cleaning because procedures will vary with the type of metal being cleaned and the degree of adherence of corrosion products.

Mechanical cleaning is the most common method; it includes scrubbing, scraping, brushing, mechanical shocking, and ultrasonic procedures. Scrubbing with a bristle brush and a mild abrasive is a widely used method. Others are used principally as supplements to remove heavily encrusted corrosion products before scrubbing. Other methods of cleaning that are sometimes used include chemical cleaning and electrolytic cleaning. Care should be used to avoid the removal of sound metal. If removal of sound metal may occur, then that effect must be quantified and the weight loss should be corrected accordingly.

Evaluation of Results After the specimens have been cleaned and reweighed, they should be examined carefully. Localized attack such as pits, crevice corrosion, stress-accelerated corrosion, cracking, or intergranular corrosion should be measured for depth and the area affected.

The depth of localized corrosion should be reported for the actual test period and not interpolated or extrapolated to an annual rate. The rate of initiation or propagation of pits is seldom uniform. The size, shape, and distribution of pits should be noted. A distinction should be made between those occurring underneath the supporting devices (concentration cells) and those on the surfaces that were freely exposed to the test solution. The specimen may be subjected to simple bending tests to determine whether any embrittlement has occurred.

If localized corrosion is not present or is recorded separately in the report, the corrosion rate (penetration rate) can be calculated as shown in Eq. (25-17)

$$\frac{\text{Weight loss}\times 534}{(\text{Area})(\text{time})(\text{metal density})} = \text{mils/yr (mpy)}$$
$$\frac{\text{Weight loss}\times 13.56}{(\text{Area})(\text{time})(\text{metal density})} = \text{mm/yr (mmpy)} \qquad (25\text{-}17)$$

Weight loss is in mg, area is in in^2 of metal surface exposed, time is in hours exposed, and density is in g/cm^3. Densities for alloys can be obtained from the producers or from various metal handbooks.

The following checklist is a recommended guide for reporting all important information and data:

- Corrosive media and concentration (changes during test)
- Volume of test solution
- Temperature (maximum, minimum, and average)
- Aeration (describe conditions or technique)
- Agitation (describe conditions or technique)
- Type of apparatus used for test
- Duration of each test (start, finish)
- Chemical composition or trade name of metals tested
- Form and metallurgical conditions of specimens
- Exact size, shape, and area of specimens
- Treatment used to prepare specimens for test
- Number of specimens of each material tested and whether specimens were tested separately or which specimens were tested in the same container
- Method used to clean specimens after exposure and the extent of any error expected by this treatment
- Actual weight losses for each specimen
- Evaluation of attack if other than general, such as crevice corrosion under support rod, pit depth and distribution, and results of microscopic examination or bend tests
- Corrosion rates for each specimen expressed as millimeters (mils) per year

Laboratory tests are typically not able to simulate every plant condition that affects corrosion. Some conditions to consider when applying laboratory test results to actual plant use include galvanic effects, concentration cells, cold and hot wall effects, and minor constituents in the corrosive environment such as impurities and contaminants.

Electrical Resistance Corrosion Sensors The measurement of corrosion by electrical resistance is performed by measuring the change in resistance of a thin metallic wire or strip as its cross section decreases from the loss of metal by corrosion. Commercial devices are available that convert that measurement to a corrosion rate and compensate for the effect of temperature on electrical resistance. Advantages of this method include that these devices can measure corrosion over short periods of time (hours or days) and that the environment does not have to be an electrolyte or liquid, so measurements are possible in environments such as concrete and corrosive gases. Disadvantages of the technique include that localized corrosion (pitting, crevice corrosion, galvanic, stress corrosion cracking, fatigue, and so forth) is not reliably measured, scatter in the readings can be caused by wide temperature fluctuations or strain on the sensor (as from flow and turbulence), and corrosion products or deposits that are electrically conductive will detrimentally affect the readings.

Linear Polarization This method measures a characteristic of an electrochemical corrosion system called polarization resistance. A relationship exists between polarization resistance and the instantaneous corrosion rate of a freely corroding alloy. The polarization resistance is determined by measuring the amount of applied current needed to change the corrosion potential of the freely corroding specimen by about 10 mV. The slope of the curves thus generated is directly related to the corrosion rate by Faraday's law. Several commercial instruments are available that are used for linear polarization measurements, for both laboratory and field use. The corrosion rate measured is the instantaneous rate, and measurements can be quickly repeated to provide a quasi–real-time view of corrosion. This method is not sensitive to localized corrosion.

Potentiodynamic Polarization The activity of pitting, crevice corrosion, and stress-corrosion cracking is strongly dependent on the corrosion potential (i.e., the potential difference between the corroding metal and a suitable reference electrode). Commercially available instruments (potentiostats) can control the corrosion potential at a desired value and measure the current required to maintain it at that value. A plot of that current over a range of potentials is called a polarization diagram. By using proper experimental techniques, it is possible to define approximate ranges of corrosion potential in which pitting, crevice corrosion, and stress-corrosion cracking will or will not occur. With properly designed electrode sensors, these techniques can be used in the field as well as in the laboratory.

The potentiostat has a three-electrode system: a reference electrode such as a saturated calomel electrode (SCE), a platinum counter (auxiliary) electrode, and a working electrode made of the alloy of interest (Fig. 25-15). Current passes between the counter and working electrodes to maintain the desired potential between the reference and working electrodes. The potentiostat controls that potential, either holding it constant, stepping it, or scanning it anodically or cathodically at some linear rate.

For a given metal/environment system, the potentiostat provides a plot of the current resulting from changes in potential. This is typically presented as a plot of log current density versus potential, or an Evans diagram. A typical active/passive metal anodic polarization curve is seen in Fig. 25-16, generally showing the regions of active corrosion, passivity, and a transpassive region.

Scan Rates Sweeping a range of potentials in the anodic (more electropositive) direction of a potentiodynamic polarization curve at a high scan rate of about 60 V/h (high from the perspective of the corrosion engineer, slow from the perspective of a physical chemist) can indicate regions where intense anodic activity is likely. Second, for otherwise identical conditions, sweeping at a relatively slow rate of potential change of about 1 V/h will indicate regions likely to be relatively inactive. The rapid sweep of the potential range has the object of minimizing film formation, so the currents observed relate to relatively film-free or thin-film conditions. The object of the slow sweep rate experiment is to allow time for filming to occur. A zero scan rate provides the opportunity for maximum stability of the metal surface, but at high electropositive potentials, the environment could be affected or changed. A rapid scan rate compromises the steady-state nature of the metal surface but better maintains the stability of the environment. Whenever possible, corrosion tests should be conducted using multiple techniques: potentiodynamic polarization at various scan rates, crevice, stress, velocity, and so forth. An evaluation of these several results, on a holistic basis, can greatly reduce or temper their individual limitations.

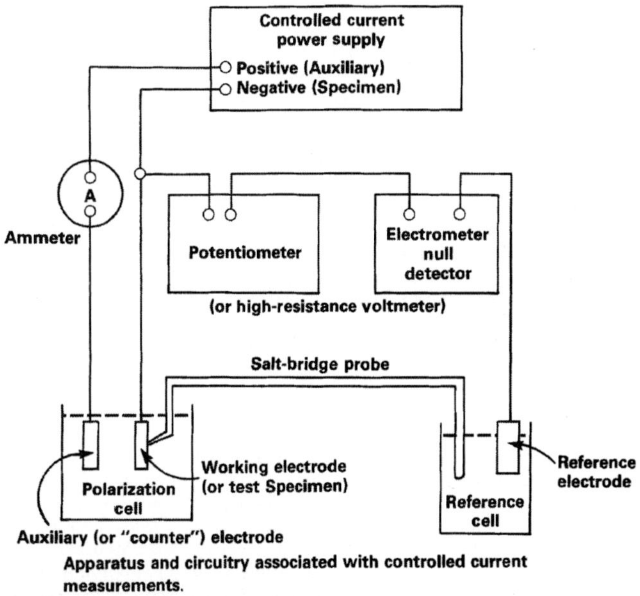

FIG. 25-15 The potentiostat apparatus and circuitry associated with controlled potential measurements of polarization curves.

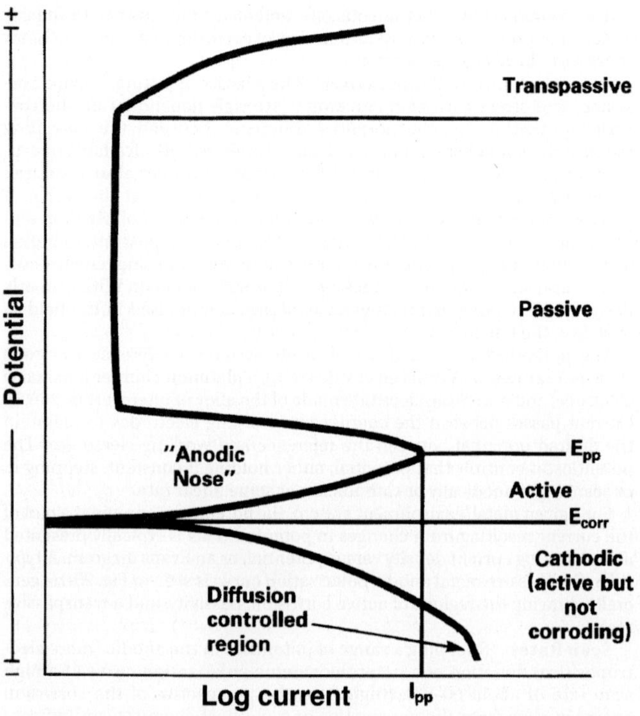

FIG. 25-16 Typical electrochemical polarization curve for an active/passive alloy (with cathodic trace) showing active, passive, and transpassive regions and other important features. (Note: E_p = primary passive potential, E_{corr} = freely corroding potential.)

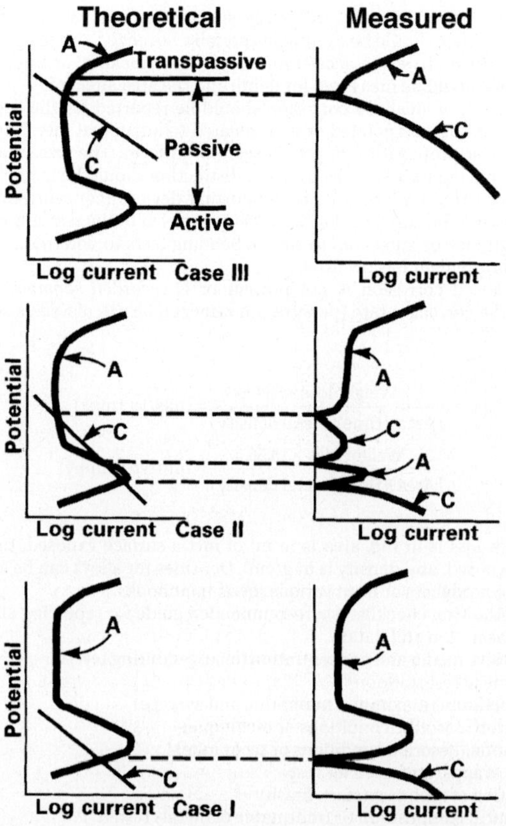

FIG. 25-17 Six possible types of behavior for an active/passive alloy in a corrosive environment.

Slow-Scan Technique In ASTM G5, "Polarization Practice for Standard Reference Method for Making Potentiostatic and Potentiodynamic Anodic Polarization Measurements," all oxygen in the test solution is purged with hydrogen for a minimum of 0.5 hours before introducing the specimen. The test material is then allowed to reach a steady state of equilibrium (open circuit corrosion potential, E_{corr}) with the test medium before the potential scan is conducted. Starting the evaluation of a basically passive alloy that is already in its "stable" condition precludes any detailed study of how the metal reaches that protected state (the normal intersection of the theoretical anodic and cathodic curves is recorded as a zero applied current on the ASTM diagram of potentiostatic potential versus applied current). These intersections between the anodic and cathodic polarization curves are where the total oxidation rate equals the total reduction rate (ASTM G3, "Recommended Practice for Conventions Applicable to Electrochemical Measurements in Corrosion Testing").

Three general reaction types compare the activation-control reduction processes. In Fig. 25-17, in Case I the single reversible corrosion potential (anode/cathode intersection) is in the active region. A wide range of corrosion rates is possible. In Case II the cathodic curve intersects the anodic curve at three potentials, one active and two passive. If the middle active/passive intersection is not stable, the lower and upper intersections indicate the possibility of very high corrosion rates. In Case III, corrosion is in the stable, passive region, and the alloys generally passivate spontaneously and exhibit low corrosion rates. Most investigators report that the ASTM method is effective for studying Case I systems. An alloy-medium system exhibiting Case II and III conditions generally cannot be evaluated by this conventional ASTM method.

The potentiodynamic polarization electrochemical technique can be used to study and interpret corrosion phenomena. It may also furnish useful information on film breakdown or repair.

Those wanting a more detailed review of the subject of electrochemistry (and/or corrosion testing using electrochemistry) are directed to additional reference material from the following: Siebert, O. W., and J. G. Stoecker, "Materials of Construction," in *Perry's Chemical Engineers' Handbook*, 7th ed., sec. 28, McGraw-Hill, New York, 1997, pp. 28-11 to 28-20; Stoecker, J. G., O. W. Siebert, and P. E. Morris, "Practical Applications of Potentiodynamic Polarization Curves in Materials Selection," *Materials Performance* **22**(11): 13–22 (1983); Siebert, O. W., "Correlation of Laboratory Electrochemical Investigations with Field Application of Anodic Protection," *Materials Performance* **20**(2): 38–43 (1981); and the assorted historical literature of Stern, Geary, Evans, Sudbury, Riggs, Pourbaix, and Edeleanu, and other studies referred to in their publications.

Crevice Corrosion Prediction The most common type of localized corrosion is the occluded mode crevice corrosion. Pitting can, in effect, be considered a self-formed crevice. A crevice must be wide enough to permit liquid entry, but sufficiently narrow to maintain a stagnant zone. It is nearly impossible to build equipment without mechanical crevices. On a micro level, scratches can be sufficient crevices to initiate or propagate corrosion in some metal/environment systems. The conditions in a crevice can, with time, become a different and much more aggressive environment than those on a nearby clean, open surface. Crevices may also be created by factors foreign to the original system design, such as deposits, corrosion products, and so forth. In many studies, it is important to know or to be able to evaluate the crevice corrosion sensitivity of a metal to a specific environment and to be able to monitor a system for predictive maintenance.

A common method to test for crevice corrosion is by the immersion test technique, creating a crevice by clamping two metal test specimens together, or clamping a metal specimen in contact with an inert plastic or ceramic material.

Velocity* For corrosion to occur, an environment must be brought into contact with the metal surface, and the metal atoms or ions must be allowed to be transported away. Therefore, the rate of transport of the environment with respect to a metal surface is a major factor in the corrosion system. Changes in velocity may increase or decrease attack, depending on its effect involved. A varying quantity of dissolved gas may be brought in contact with the metal, or velocity changes may alter diffusion or transfer of ions by changing the thickness of the boundary layer at the surface. The boundary layer, which is not stagnant, moves except where it touches the surface. Many metals depend on the development of a protective surface for their corrosion resistance. This may consist of an oxide film, a corrosion product, an adsorbed film of gas, or other surface phenomena. The removal

*See review paper by David C. Silverman, courtesy of NACE International.

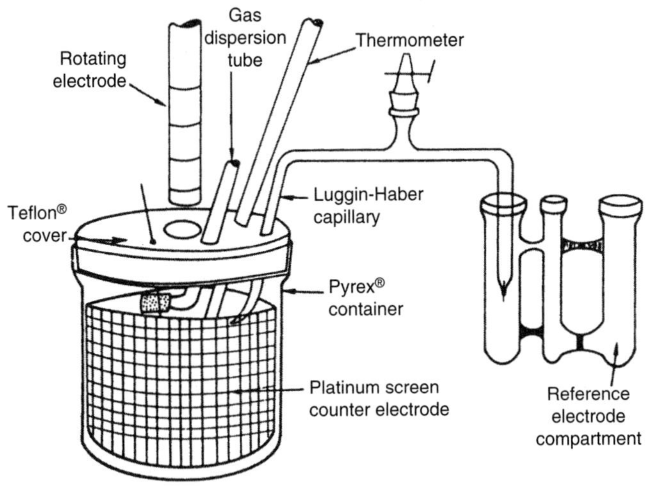

FIG. 25-18 Rotating cylinder electrode apparatus.

of these surfaces by the effect of the fluid velocity exposes fresh metal, and as a result, the corrosion reaction may proceed at an increasing rate. In these systems, corrosion might be minimal until a so-called critical velocity is attained where the protective surface is damaged or removed and the velocity is too high for a stable film to re-form. Above this critical velocity, the corrosion may increase rapidly.

The NACE Landrum Wheel velocity test, originally TM0270-72, is typical of several mechanical-action immersion test methods to evaluate the effects of corrosion. Unfortunately, these laboratory simulation techniques did not consider the fluid mechanics of the environment or metal interface, and service experience very seldom supports the test predictions. A rotating cylinder within a cylinder electrode test system has been developed that operates under a defined hydrodynamics relationship (Figs. 25-18 and 25-19). The assumption is that if the rotating electrode operates at a shear stress comparable to that in plant geometry, the mechanism in the plant geometry may be modeled in the laboratory. Once the mechanism is defined, the appropriate relationship between fluid flow rate and corrosion rate in the plant equipment as defined by the mechanism can be used to predict the expected corrosion rate. If fluid velocity does affect the corrosion rate, the degree of mass-transfer control, if that is the controlling mechanism (as opposed to activation control), can be estimated. Conventional potentiodynamic polarization scans are conducted as described previously. In other cases, the corrosion potential can be monitored at a constant velocity until steady state is attained. While the value of the final corrosion potential is virtually independent of velocity, the time to reach steady state may be dependent on velocity. The mass-transfer control of the corrosion potential can be proportional to the velocity raised to its appropriate exponent. The rate of breakdown of a passive film is velocity-sensitive. To review a very detailed and much needed refinement of the information and application of the rotating electrode technique as used for evaluation of the effect of velocity on corrosion, the reader is directed to a study by David C. Silverman, "The Rotating Cylinder Electrode for Examining Velocity-Sensitive Corrosion—A Review," *Corrosion* **60**(11): 1003–1023 (2004).

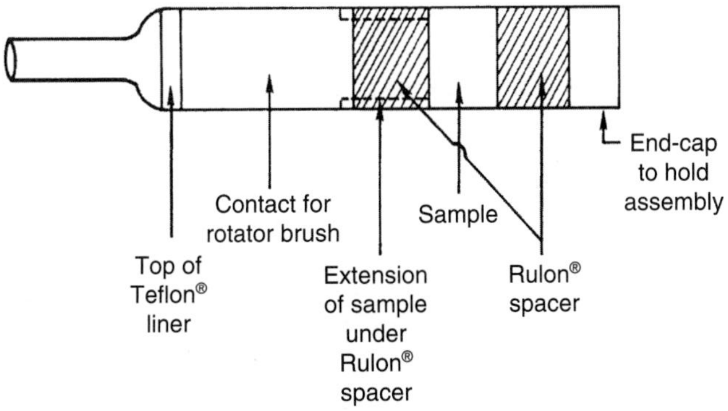

FIG. 25-19 Inner rotating cylinder used in laboratory apparatus of Fig. 25-18.

Environmental Cracking The problem of environmental cracking of metals and their alloys is very important. Stress corrosion cracking (SCC) is cracking caused by the combined effects of stress and corrosion. Stress corrosion cracking is an incompletely understood corrosion phenomenon. Much research activity (aimed mostly at mechanisms) plus practical experience have led to the development of empirical guidelines, but they contain a large element of uncertainty. No single test method has proven to be reproducible enough for known crack-causing environment/alloy systems to justify a high confidence level.

Most SCC testing is done using immersion tests with stressed metal specimens. There are a variety of methods of stressing the specimens, including bending, welding, compressed C-rings, and constant extension of tensile bars (slow strain rate). Prestressed samples, such as those shown in Fig. 25-20, have been used for laboratory and field SCC testing. The variable observed is "time to failure or visible cracking." Such tests do not provide acceleration of failure.

SCC often shows a fairly long induction period (months to years), so such tests must be conducted for long periods before reliable conclusions can be drawn. Several months' duration is not unusual.

In the constant-strain method, the specimen is stretched or bent to a fixed position at the start of the test. The most common shape of the specimens used for constant-strain testing is the U-bend, hairpin, or horseshoe type. A bolt is placed through holes in the legs of the specimen, and it is loaded by tightening a nut on the bolt. In some cases, the stress may be reduced during the test as a result of creep. In the constant-load test, the specimen is supported horizontally at each end and is loaded vertically downward at one or two points and has maximum stress over a substantial length or area of the specimen. The load applied is a predetermined, fixed dead weight. Specimens used in either of these tests may be precracked to assign a stress level or a desired location for fracture to occur or both, as is used in fracture mechanics studies. These tensile-stressed specimens are then exposed to the environment of study.

Slow Strain-Rate Test In the slow strain-rate test (SSRT), a tension specimen is slowly loaded in a test frame to failure under prescribed test conditions. The normal test extension rates are from 2.54×10^{-7} to 2.54×10^{-10} m/s (10^{-5} to 10^{-8} in/s). Failure times are usually 1 to 10 days. The failure mode will be either SCC or tensile overload, sometimes accelerated by corrosion. An advantage of the SSRT, compared to constant-strain tests, is that the protective surface film is ruptured mechanically during the test, thus giving SCC an opportunity to progress. It is common for the potential to be monitored during the SSRT. The strain rates that best generate SCC in various alloys are reported in the literature. The SSRT is generally considered a severe test for SCC. A disadvantage of this test is that indications of failure are not generally observed until the tension specimen is plastically stressed, sometimes significantly, above the yield strength of the metal. Such high-stressed conditions can be an order of magnitude higher than the intended operating stress conditions. Another disadvantage is that crack initiation must occur fairly rapidly in order to have crack growth sufficient to be detected using the SSRT. The occurrence of SCC in alloys requiring long initiation times may go undetected.

Conjunctive Use of Slow- and Rapid-Scan Polarization Potentiodynamic polarization curves can be used to predict SCC-sensitive potential ranges. The technique involves conducting both slow- and rapid-scan sweeps in the anodic direction of a range of potentials. Comparison of the two curves indicates potential ranges within which high anodic activity in the film-free condition reduces to insignificant activity when the time requirements for film formation are met. This indicates heightened susceptibility to SCC. Some SCC theories predict these domains of SCC behavior to be between the primary passive potential and the onset of passivity. This technique helps identify those SCC potential ranges.

Electrochemical Impedance Spectroscopy (EIS) and AC Impedance* Many direct-current test techniques assess the overall corrosion process occurring at a metal surface but treat the metal/solution interface as if it were a pure resistor. Problems of accuracy and reproducibility often encountered in the application of direct-current methods have led to the use of electrochemical impedance spectroscopy (EIS).

Electrode surfaces in electrolytes generally have a surface charge that is balanced by an ion accumulation in the adjacent solution, thus making the system electrically neutral. The first component is a double layer created by a charge difference between the electrode surface and the adjacent molecular layer in the fluid. Electrode surfaces may behave at any given frequency as a network of resistive and capacitive elements from which electrical impedance may be measured and analyzed.

*Excerpted from papers by Oliver W. Siebert, courtesy of NACE International and ASTM.

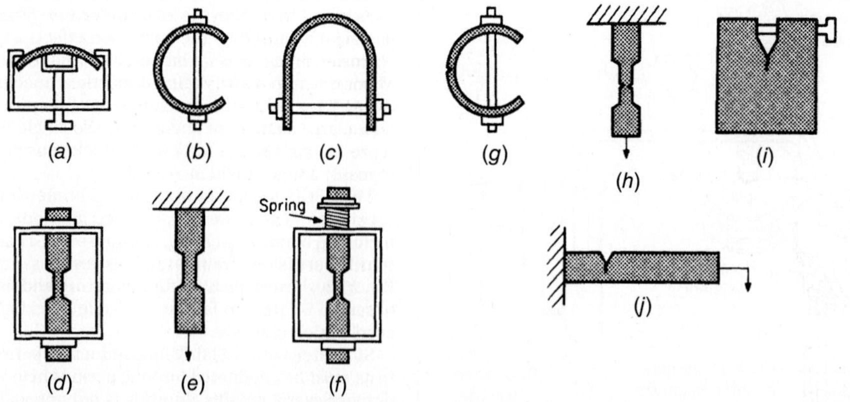

FIG. 25-20 Specimens for stress-corrosion tests. (*a*) Bent beam. (*b*) C ring. (*c*) U bend. (*d*) Tensile. (*e*) Tensile. (*f*) Tensile. (*g*) Notched C ring. (*h*) Notched tensile. (*i*) Precracked, wedge. Open-loading type. (*j*) Precracked, cantilever beam. [Chem. Eng. *78:* 159 (*Sept. 20, 1971*).]

The application of an impressed alternating current on a metal specimen can generate information on the state of the surface of the specimen. The corrosion behavior of the surface of an electrode is related to how that surface responds to this electrochemical circuit. The AC impedance technique involves the application of a small sinusoidal voltage across this circuit. The frequency of that alternating signal is varied, and the voltage and current responses of the system are measured.

A method of analyzing electrical impedance of a corrosion system uses so-called white-noise analysis by the fast Fourier transform technique (FFT). The entire spectrum is derived from one signal. The impedance components thus generated are plotted on either a Nyquist (real versus imaginary) or Bode (log real versus log frequency plus log phase angle versus log frequency) plot. These data are analyzed by computer to determine the polarization resistance, and thus the corrosion rate if Tafel slopes are known. EIS measurements are also used with coated specimens to better understand how the coating's dielectric properties change with time.

Other Electrochemical Test Techniques A summary of electrochemical test techniques can be found in Chapter 7, "Electrochemical Tests," 2005 *ASTM Manual* 20, by John R. Scully. He presents theories associated with the mechanisms of corrosion, and he describes common test methods used to predict corrosion. These include methods based on concentration polarization effects, frequency modulation methods, electronic noise resistance, and the scratch-repassivation method for local corrosion.

CORROSION TESTING: PLANT TESTS

It is not always practical or convenient to investigate corrosion problems in the laboratory. It may be difficult to discover the conditions of service and reproduce them exactly. This is especially true with process streams that change with time, such as those that may occur in batch processes, evaporation, distillation, polymerization, sulfonation, synthesis, and processes using recycled mother liquors. Laboratory tests may also experience contamination of the test solution by corrosion products, or depletion of a corrosive component, which significantly affects test results.

In such cases, it is usually preferable to carry out the corrosion-testing program by exposing specimens in operating equipment under actual conditions of service. Such in-plant testing has the additional advantages that it is possible to test a large number of specimens at the same time and that little technical supervision is required.

Test Specimens In carrying out plant tests, it is necessary to install the test specimens so that they will not come into contact with other metals and alloys. This avoids having their normal behavior disturbed by galvanic effects. It is also desirable to protect the specimens from mechanical damage.

There is no standard size or shape for corrosion-test specimens. They usually weigh from 10 to 50 g (0.35 to 1.76 ounces) and preferably have a large surface-to-mass ratio. Disks 40 mm (1½ in) in diameter by 3.2 mm (⅛ in) thick and similarly dimensioned square and rectangular specimens are the most common. Surface preparation varies with the aim of the test, but machine grinding of surfaces or polishing with a 120 grit is common. Samples should not have sheared edges, should be clean (no heat-treatment scale remaining unless this is specifically part of the test), and should be identified by stamping. See Fig. 25-21 for a typical plant test assembly.

The choice of materials for the specimen holder is important. Materials must be durable enough to ensure satisfactory completion of the test. It is good practice to select very resistant materials for the test assembly. Common insulating materials used are resistant plastics such as nylon and PTFE.

The method of supporting the specimen holder during the test is important. The holder must be located so as to cover the conditions of exposure to be studied. It may have to be submerged, or exposed only to the vapors, or located at liquid level, or holders may be called for at all three locations. Various means are used to support the holders in liquids or in vapors. The simplest is to suspend the holder by means of a heavy wire or light metal chain. Holders have been strung between heating coils, clamped to agitator shafts, welded to evaporator tube sheets, and so on.

For tests in pipelines of 75-mm (3-in) diameter or larger, a spool holder such as that shown in Fig. 25-22 has been used. This frame is designed so that it may be placed in a pipeline in any position without permitting the disk specimens to touch the wall of the pipe. As with the strip-type holder, this assembly does not materially interfere with the fluid through the pipe, and it permits the study of corrosion effects prevailing in the pipeline.

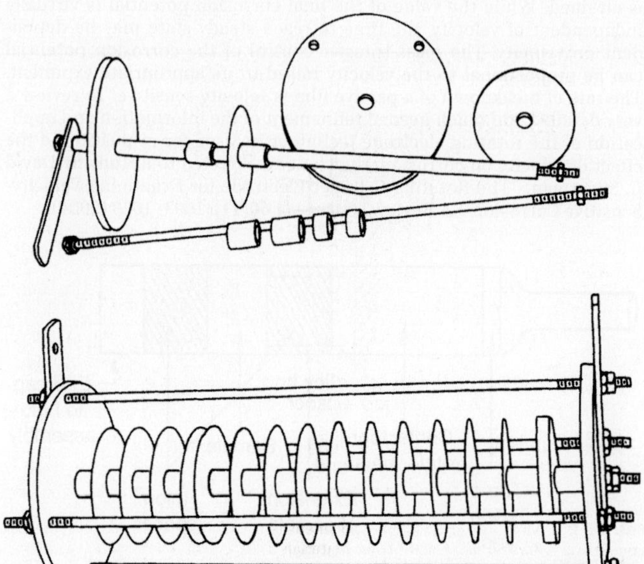

FIG. 25-21 Assembly of a corrosion-test spool and specimens. (*Mantell, ed.,* Engineering Materials Handbook, *McGraw-Hill, New York, 1958.*)

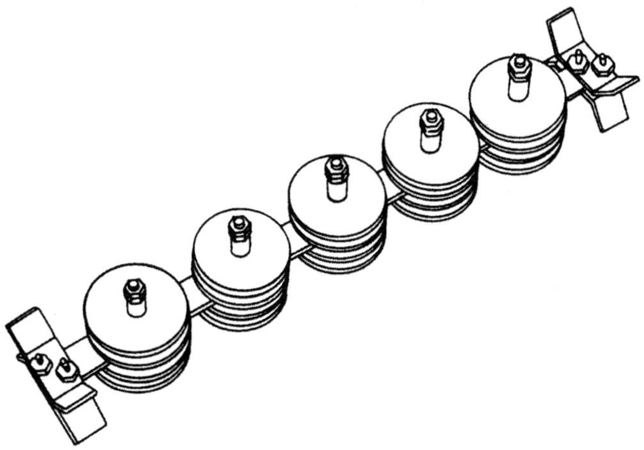

FIG. 25-22 Spool-type specimen holder for use in a 3-in-diameter or larger pipe. (*Mantell, ed.,* Engineering Materials Handbook, *McGraw-Hill, New York, 1958.*)

Another way to study corrosion in pipelines is to install in the line short sections of pipe of the materials to be tested. If possible, these test sections should be electrically isolated from each other and from the rest of the piping system.

It is occasionally desirable to expose corrosion-test specimens in operating equipment without the use of specimen holders of the type described. This can be done by attaching specimens directly to some part of the operating equipment and by providing the necessary isolation against galvanic effects as shown in Fig. 25-23.

Test Results The methods of cleaning specimens and evaluating results after plant corrosion tests are identical to those described earlier for laboratory tests.

On-Line Corrosion Monitoring* On-line corrosion monitoring is used to provide in-situ, real-time information about corrosion in an operating plant. The most common devices are based on either electrical resistance or linear polarization resistance (LPR) electrochemical measurements. These devices are particularly useful to detect changes in process conditions (upsets) that render equipment susceptible to corrosion.

Electrical resistance sensors measure changes in the electrical resistance of a metal element as it gets thinner from corrosion. As the metal gets thinner, its resistance increases. The most common metal elements used in plant monitoring are thin cylinders with welded ends. These sensors are only quasi-real-time, and they may require days or weeks to accumulate enough corrosion to indicate a rate. The higher the corrosion rate, the shorter the time required. These sensors are affected by temperature variations and strain effects (e.g., from flow or vibration), both of which affect electrical resistance.

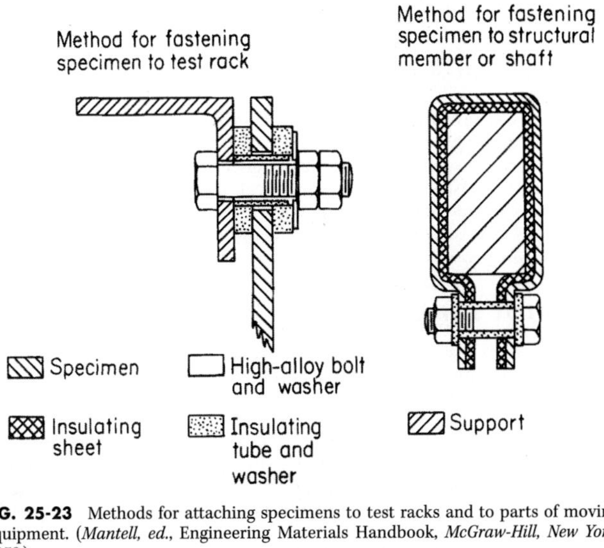

FIG. 25-23 Methods for attaching specimens to test racks and to parts of moving equipment. (*Mantell, ed.,* Engineering Materials Handbook, *McGraw-Hill, New York, 1958.*)

*Excerpted from papers by Oliver W. Siebert, courtesy of NACE International and ASTM.

Linear polarization resistance (LPR) sensors provide instantaneous corrosion rate measurements, and they work on the principle outlined in the ASTM guide on making polarization resistance measurements (G59, "Standard Practice for Conducting Potentiodynamic Polarization Resistance Measurements"). LPR sensors measure the current density required to create a small shift (usually 5 to 10 mV) in the corrosion potential of the test electrode, which provides an indication of the corrosion rate.

Unique Uses of On-Line Corrosion Monitoring The major use of corrosion monitoring is to measure the corrosion rate in the plant or the field. As noted previously, on-line corrosion monitoring can also be used to detect process upsets that change the corrosivity of the process. Monitoring can also be used to optimize the chemistry and level of corrosion inhibitors used. Too little inhibitor allows unacceptable levels of corrosion, and too much inhibitor increases costs without adequate benefit. Optimizing inhibitor use in terms of concentration, location in the process, and method of addition can be helped by the use of carefully placed corrosion sensors. Corrosion sensors may also be used to monitor scale formation, and some sensors claim to provide information about localized corrosion.

Other Types of Sensors Other types of sensors are used to measure corrosion, or to measure environmental characteristics known to influence corrosion.

The electrochemical techniques of EIS and polarization testing we have discussed as laboratory methods may also be adapted for in-plant use. Electrochemical noise monitoring is another method that may be used in the plant. This method uses naturally occurring fluctuations in current and potential of a metal to provide information on the reaction kinetics at the surface and the corrosion rate.

Ultrasonic thickness monitoring may be permanently mounted on the outside of process equipment to provide real-time indications of metal loss. Such devices provide measurements for small areas, or multiple transducers may be ganged to provide measurements for larger areas.

Some sensors do not measure corrosion directly, but instead yield measurements that are useful to indicate a change in corrosive conditions. Examples include:

Pressure probes. Pressure monitors or transducers may be of use in corrosion monitoring in environments where a buildup of gases such as hydrogen or H_2S may contribute to corrosion.

Gas probes. The hydrogen patch probe is mounted on the outside of a pipe or vessel, and it measures the amount of corrosion-generated hydrogen coming through a metal wall.

pH probes. Monitoring pH may aid in the early detection of corrosion. The acidity or alkalinity of the environment is often a controlling parameter for corrosion, and it may be controllable.

Ion probes. Measuring metal ions in solution with specific ion electrodes may provide a direct indication of corrosion, and measuring other types of ions (chloride for example) may allow one to infer a change in corrosive conditions.

Considerations for Measuring Corrosion Rate with Coupon Specimens Corrosion rates may vary during testing, as described in the discussion about laboratory testing. Because the rate obtained from coupon testing is averaged over time, one must be aware that actual rates may have varied during the time of testing.

The frequency of sampling is important. Generally, measurements made over longer times are more valid. This is especially true for low corrosion rates, under 1 mil/yr, mpy (0.001 in/yr). When corrosion rates are this low, longer times should be used.

A variety of factors may throw off these rates, as outlined in ASTM G31, "Standard Practice for Laboratory Immersion Corrosion Testing of Metals." Coupon-type tests cannot be correlated with changing plant conditions that may dramatically affect process equipment lifetimes. Other methods must be used if more frequent measurements are desired or correlation with plant conditions is necessary.

A plot of mass loss versus time can provide information about changes in the conditions under which the test has been run. One example of such a plot comes from the ASTM Standard G96, "Standard Guide for Online Monitoring of Corrosion in Plant Equipment (Electrical and Electrochemical Methods.)" As mentioned previously, weight loss measurements are not appropriate for the measurement of localized corrosion, such as pitting and crevice corrosion. Accurate corrosion penetration rates for localized corrosion require direct measurement on the coupon. Instruments commonly used for this include pit-depth gages, calibrated microscopes, and profile meters based on laser or optical principles.

Heat Flux Tests Removable tube test heat exchangers can be useful in the field for monitoring heat flux (corrosion) conditions, per NACE TM0286-94 (similar to laboratory test, Fig. 25-14).

ALLOY DESIGNATIONS

The Unified Numbering System (UNS) has become the most widely accepted system for identifying metals and alloys. The UNS is built around a series of 18 primary alphanumeric designations, as shown in Table 25-4. ASTM E527 is the standard practice for numbering metals and alloys in the UNS system.

Each metal and alloy to receive a UNS designation is placed in one of the primary systems and then given a unique five-digit identifier. In most cases a UNS specification characterizes the chemical composition of the metal or alloy. It does not include other criteria, such as mechanical properties, heat treatment, or form.

Prior to the adoption of UNS, there was no universally recognized system for numbering alloys. Different organizations such as the Aluminum Association (AA), Copper Development Association (CDA), and American Iron and Steel Institute (AISI) had their own systems. In many cases, UNS has incorporated these commonly known numbers from those systems. For example, AISI 304 stainless steel has the UNS number S30400.

In addition to the UNS system, other organizations generate materials standards of importance to the chemical process industries. Some of these are:

1. American National Standards Institute (ANSI), formerly American Standards Association (ASA). ANSI promulgates the piping codes used in the chemical-process industries.
2. American Society of Mechanical Engineers (ASME). This society generates the Boiler and Pressure Vessel Codes.

TABLE 25-4 UNS Numbering System

Axxxxx	Aluminum and aluminum alloys	Lxxxxx	Low-melting-point metals and alloys
Cxxxxx	Copper and copper alloys	Mxxxxx	Miscellaneous nonferrous metals and alloys
Dxxxxx	Specified mechanical propertied steels	Nxxxxx	Nickel and nickel alloys
Exxxxx	Rare earth and rare earth-like alloys	Pxxxxx	Precious metals and alloys
Fxxxxx	Cast irons	Rxxxxx	Reactive and refractory metals and alloys
Gxxxxx	AISI and SAE carbon and alloy steels	Sxxxxx	Stainless steels, valve steels, superalloys
Hxxxxx	AISI H-steels	Txxxxx	Tool steels
Jxxxxx	Cast steels	Wxxxxx	Welding filler metals
Kxxxxx	Miscellaneous steels and ferrous alloys	Zxxxxx	Zinc and zinc alloys

3. American Society for Testing and Materials (ASTM). This society generates specifications for most of the materials used in the ANSI Piping Codes and the ASME Boiler and Pressure Vessel Codes.
4. International Organization for Standardization (ISO). This organization is engaged in generating standards for worldwide use.

ASTM and ASME have incorporated UNS designations into their specifications.

FERRITIC STEELS

CARBON STEEL

Carbon steel is the most common material used in industry due to its relative low cost, high strength, favorable properties, and general application in many chemical processing services. Carbon steel has predictable behavior, can be used at elevated temperatures up to 540°C (1000°F), and seldom fails in an unexpected fashion. Carbon steel's excellent ductility permits mechanical manipulation into various shapes; its formability coupled with ease of welding allows different component parts to be joined into one continuous piece of equipment, such as a pressure vessel or a long run of piping.

Conventional carbon steel contains mostly iron with small amounts of carbon (a few tenths of a percent) that strengthen the material and create its excellent mechanical properties. The grades most commonly used in the chemical process industries have tensile strengths in the 345- to 485-MPa (50,000 to 70,000 lbf/in^2) range. Manipulating the alloy elements, different heat treatments, and cold working creates higher-strength steel with other property enhancements or weaknesses.

Carbon steel has routine applications in processing organic chemicals found in the refining and chemical industries. Refining process plants are often constructed almost entirely of conventional carbon steel, with limited applications of other alloys where high temperatures, corrosion, or metallurgical damage preclude steel's use.

Carbon steel has limited applications above 425°C (800°F) due to a metallurgical transformation called graphitization, a breakdown of iron carbides after long periods of high-temperature exposure. Oxidation also increases to unacceptable levels above 510°C (950°F) when carbon steel is exposed to air (typical furnace conditions). With temperatures above 425°C (800°F), low-alloy steels (discussed next) become the first alternative material considered.

Carbon steel can tolerate neutral and basic conditions very well. Industry available data can provide predicted corrosion rates with exposure to different chemicals or conditions. Less predictable corrosion occurs in applications where organic chemical streams with trace amounts of water are being processed; water will drop out in low spots and may be acidic or permit scaling that accelerates corrosion in these areas. Carbon steel is generally the most economic alternative in chemical exposure applications with corrosion rates less than 0.12 mm/yr (5 mils/yr). Typically, steel equipment has a designed corrosion allowance that permits some corrosion to occur while still giving the equipment an acceptable design life (nominally 20 years).

For highly corrosive environments, the wetted interior of a carbon steel vessel can be internally alloy clad or weld overlaid to protect the underlying carbon steel. Roll/explosion-bond cladding or weld overlay uses highly corrosion-resistant alloys such as stainless steels, a high nickel alloy, titanium, or zirconium. In many cases, a base carbon steel vessel using cladding or weld overlay provides more favorable equipment properties or fabrication costs than a solid corrosion-resistant alloy.

Carbon steel often experiences corrosion under installed thermal insulation, which is used to retain heat or protect personnel. Several techniques minimize corrosion under insulation, including organic coatings (paint) or spraying the surface with TSA (thermal sprayed aluminum).

CAST IRONS

Cast irons contain mostly iron with a few percent of carbon, and they cost less than carbon steel. Cast irons have very low ductility (they are brittle), and they can fail in a less predictable fashion than carbon steel. There are several types of cast iron, including gray iron, ductile iron, white iron, and malleable iron. Ductile iron is the most common grade used in the chemical industry because of its superior mechanical properties. Cast irons have limited use in the chemical, petrochemical, or refining industries. Some applications include pump parts, such as impellers and pump casings. Cast iron has low corrosion rates in low-temperature water, which enables it to be used in low-pressure water services.

LOW-ALLOY STEELS

Low-alloy steels contain one or more alloying agents to provide mechanical and corrosion-resistant properties superior to those of carbon steel. These alloying elements allow for the production of a large variety of standard and proprietary grades. Nickel increases toughness and improves low-temperature properties and corrosion resistance. Chromium and silicon improve hardness, abrasion resistance, corrosion resistance, and resistance to oxidation. Molybdenum provides strength at elevated temperatures. The addition of small amounts of alloying materials greatly improves corrosion resistance to atmospheric environments but has little effect against liquid corrosives. The alloying elements produce a tight, dense, adherent rust film; however, in acid or alkaline conditions, these minor alloying elements only create corrosion resistance equivalent to that of carbon steel.

The most common low-alloy steels used in chemical processing and refining contain up to 9 percent chromium and up to 1 percent molybdenum, referred to as low-chrome or chrome-moly (Cr-Mo steels). The primary reasons to use a low-alloy steel instead of carbon steel include the following:

- Improved high-temperature strength, useful up to at least 649°C (1200°F).
- Improved resistance to sulfidation corrosion in the absence of high-pressure hydrogen. Carbon steel is limited to about 260°C to 290°C (500°F to 550°F). Typically, at least 5 percent Cr is needed to improve sulfidation resistance. API RP 939-C gives information on avoiding sulfidation corrosion.
- Improved resistance to high-temperature hydrogen attack; carbon steel is limited to as little as 232°C (450°F), depending on the hydrogen partial pressure. API RP 941 has information to set practical limits on the use of carbon and low-alloy steels in hydrogen service at elevated temperatures and pressures.

Most welded low-alloy steels have hard welds and weld heat affected zones (HAZs), which can crack following fabrication or when put into service. Certain process environments promote cracking. Post-welding heat treatment (PWHT), generally prescribed by fabrication codes (ASME/ANSI), will reduce hardness, improve fracture toughness, and provide more resistance to process environment cracking. Low-alloy steels carry a significant cost premium compared to carbon steel due to the additional alloying elements and the required PWHT during fabrication.

The higher strength of low-alloy steels permits process equipment designs with a reduced or lower minimum thickness. Thinner wall thickness reduces the amount of welding required, reduces equipment weight, and translates into cost savings. Newer low-alloy steels with small additions of vanadium allow an even greater reduction in wall thickness. However, vanadium-enhanced low-alloy steels have more stringent fabrication requirements, which add complication but reduce overall costs relative to nonenhanced low-alloy steels.

STAINLESS STEELS

Stainless steel is typically defined as a ferrous alloy containing at least 10 percent (weight percent) chromium. Since their discovery in the early part of the 20th century, stainless steels have become primary construction materials for process vessels, heat exchangers, storage tanks, and process piping. This growth has been driven by the need for corrosion-resistant materials in the chemical, petrochemical, power generation, pulp and paper, pharmaceutical, and food processing industries.

Today there are literally hundreds of different stainless steel alloys or "grades," and the choice of a grade for any specific application is based on the required level of corrosion resistance, strength, and toughness as well as other factors such as fabrication requirements, availability, and cost. Stainless steels are divided into groups or families based on their crystal structure and strengthening mechanism. The five primary stainless steel families are austenitic, ferritic, duplex, martensitic, and precipitation-hardened stainless steels. For a more complete listing of commonly specified grades used for wrought plate, sheet, and strip, see the following ASTM standards:

- A240, Standard Specification for Chromium and Chromium-Nickel Stainless Steel Plate, Sheet, and Strip for Pressure Vessels and General Applications
- A666, Standard Specification for Annealed or Cold-Worked Austenitic Stainless Steel Sheet, Strip, Plate, and Bar
- A693, Standard Specification for Precipitation-Hardening Stainless Steel and Heat Resistant Plate Sheet, and Strip

AUSTENITIC STAINLESS STEELS

Austenitic stainless steels are by far the largest and most widely used of all the stainless steels. This group of stainless steels is characterized by having a microstructure that is either entirely or predominately austenite phase, which is a nonmagnetic phase that has a face-center cubic crystal structure. Some of the austenitic stainless steels also contain a small amount of ferrite phase, which is a ferromagnetic phase with a body-centered crystal structure. The compositions of the austenitic stainless steels are adjusted to produce the desired microstructure or properties. This is done by maintaining a balance between the austenite stabilizing elements (Ni, C, Mn, and N) and the ferrite stabilizing elements (Cr, Mo, Si, and Nb).

As a family, the austenitic stainless steels tend to have a relatively low yield strength, high work hardening rate, high tensile strength, good ductility, and excellent low-temperature toughness. They also tend to be more susceptible to chloride stress corrosion cracking than are the ferritic and duplex stainless steels. Austenitic stainless steels have good weldabilty and are readily fabricated into complex shapes. This family of stainless steels cannot be hardened or strengthened by heat treatment, but it can be strengthened by cold forming or work hardening (see ASTM A666). The austenitic family of stainless steels can be further divided into subgroups, depending on their alloying content.

300 Series Austenitic Stainless Steel The most widely used of all stainless steels are the AISI 300 series austenitic grades, which include standard grades such as 304 (S30400), 304L (S30403), 316 (31600), and 316L (S31603). Some of the more commonly specified 300 series grades and their chemical compositions are summarized in Table 25-5. The microstructure of these standard grades consists primarily of austenite phase with a small amount of ferrite phase, typically in the range of 3 to 8 vol%. These grades are used widely across all industry sectors and are available in a wide variety of product forms, including plate, sheet, strip, tubing, piping, forgings, and bar.

The 304 and 304L grades are often termed 18-8 stainless steels, which refers to the nominal 18 percent (wt%) chromium and 8 percent nickel content. The 316 and 316L grades have just over 16 percent Cr, a minimum Ni content of 10 percent, and a 2 to 3 percent Mo addition, which improves the resistance to localized corrosion, especially by chlorides and other halides. The L refers to low carbon content (≤0.030%) which is desirable for welded applications due to the improved resistance to chromium carbide precipitation. In fact, because the composition of an L grade is a subset of the regular grade, many 300 series stainless steels are promoted as being "dual-certified," that is, meeting the requirements for both 304 and 304L, as

TABLE 25-5 List of Commonly Used 300 Series Austenitic Stainless Steels*

Common name	UNS number	C	Mn	Si	Cr	Ni	Mo	N	Other
301	S30100	0.15	2	1	16–18	6.0–8.0	...	0.1	
301L	S30103	0.03	2	1	16–18	6.0–8.0	...	0.2	
302	S30200	0.15	2	0.75	17–19	8.0–10.0	...	0.1	
304	S30400	0.07	2	0.75	17.5–19.5	8.0–10.5	...	0.1	
304L	S30403	0.03	2	0.75	17.5–19.5	8.0–12.0	...	0.1	
305	S30500	0.12	2	0.75	17.0–19.0	10.5–13.0	...	...	
309S	S30908	0.08	2	0.75	22.0–24.0	12.0–15.0	...	...	
310S	S31008	0.08	2	1.5	24.0–26.0	19.0–22.0	...	...	
321	S32100	0.08	2	0.75	17–19	9.0–12.0	...	0.1	Ti 5 × (C + N) min, 0.70 max
347	S34700	0.08	2	0.75	17–19	9.0–13.0	...	...	Nb 10 × (C) min, 1.0 max
316	S31600	0.08	2	0.75	16–18	10.0–14.0	2.0–3.0	0.1	
316L	S31603	0.03	2	0.75	16–18	10.0–14.0	2.0–3.0	0.1	
317	S31700	0.08	2	0.75	18–20	11.0–15.0	3.0–4.0	0.1	
317L	S31703	0.03	2	0.75	18–20	11.0–15.0	3.0–4.0	0.1	
317LM	S31725	0.03	2	0.75	18–20	13.5–17.5	4.0–5.0	0.2	
317LMN	S31726	0.03	2	0.75	17–20	13.5–17.5	4.0–5.0	0.10–0.20	

*Single values are maximum values unless otherwise noted.

TABLE 25-6 List of Commonly Used 200 Series Austenitic Stainless Steels*

Common name	UNS number	C	Mn	Si	Cr	Ni	Mo	N	Other
0201	S20100	0.15	5.5–7.5	1	16–18	3.5–5.5	...	0.25	
201L	S20103	0.03	5.5–7.5	0.75	16–18	3.5–5.5	...	0.25	
201LN	S20153	0.03	6.4–7.5	0.75	16–17.5	4.0–5.0	...	0.10–0.25	Cu 1.0 max
202	S20200	0.15	7.5–10.0	1	17–19	4.0–6.0	...	0.25	
Nitronic® 30	S20400	0.03	7.0–9.0	1	15–17	1.5–3.0	1.5–3.0	0.15–0.30	
XM-19 Nitronic® 50	S20910	0.06	4.0–6.0	0.75	20.5–23.5	11.5–13.5	1.5–3.0	0.20–0.40	
XM-17	S21600	0.08	7.5–9.0	0.75	17.5–22.0	5.0–7.0	2.0–3.0	0.25–0.50	
XM-18	S21603	0.03	7.5–9.0	0.75	17.5–22.0	5.0–7.0	2.0–3.0	0.25–0.50	
Nitronic® 60	(S21800)	0.1	7.0–9.0	3.5–4.5	16–18	8.0–9.0	...	0.08–0.18	
XM-11 Nitronic® 40	(S21904)	0.04	8.0–10.0	0.75	19–21.5	5.5–7.5	...	0.15–0.40	
XM-29 Nitronic® 33	(S24000)	0.08	11.5–14.5	0.75	17–19	2.3–3.7	...	0.20–0.40	

*Single values are maximum values unless otherwise noted.

an example. Chromium carbide precipitation can also be minimized by the addition of the stabilizing elements titanium, niobium, or tantalum, which are added to the stabilized grades, type 321 and 347. The 317L (S31703), 317LM (S31725), and 317LMN (S31726) grades contain progressively higher levels of Cr, Ni, and Mo, which provide increased corrosion resistance and slightly higher strength levels than the 316L grade.

200 Series Stainless Steel The AISI 200 series austenitic stainless steels contain a lower level of Ni and higher levels of Mn and N than the 300 series stainless steels (see Table 25-6). This subgroup of stainless steels tends to have higher strength levels and larger strain hardening coefficients than the 300 series grades. Because of the lower nickel contents, the 200 series grades are often considered a lower-cost alternative to the 300 series stainless steels.

The most widely used grades in this category are the 201 (S20100) and 201L (S20103) grades, which have a corrosion resistance that approaches that of the 304 and 304L grades. The 201/201L grades are used in a wide range of applications, including structural members, hose clamps, piston rings, roofing, and transit cars. The 201LN™ (S20153) is a modified 201 grade that has a very stable austenite phase at low temperatures and is ideally suited for cryogenic tanks and vessels down to −196°C (−320°F).

Other commonly specified 200 series stainless steels include the Nitronic 30® (S20400), Nitronic 60® (S21800), XM-11 (S21904), XM-17 (S21600), and XM-18 (S21603). The Nitronic 30® (S20400) grade is a nitrogen strengthened grade that has good resistance to abrasion and metal-to-metal wear. The Nitronic 60® (S21800) grade is known for having outstanding resistance to galling and is often specified for fasteners, pins, and bushings. The XM-11 (S21904) grade has a corrosion resistance slightly better than types 304 and 304L and is twice as strong based on yield strength. The XM-17 (S21600) and the low-carbon equivalent XM-18 (S21603) stainless steels have a corrosion resistance that approaches the 316/316L grades, and the XM-19 (S20910) grade's resistance is similar to type 317L.

High-Performance Austenitic Stainless Steel The more highly alloyed austenitic grades are often termed high-performance or super austenitic grades. These steels have increased levels of Cr, Ni, Mo, and occasionally N and are designed to provide increased resistance to aggressive environments, such as stronger acids and bases and higher chloride-bearing environments such as seawater, brackish water, and brines (see Table 25-7). A Cu addition used in the 904L (N08904) and alloy 20 (N08020) stainless steels provides improved resistance to reducing acids. As a group, the high-performance grades tend to have higher strength levels and greater resistance to stress corrosion cracking than the 300 and 200 series stainless steels.

The 6 percent Mo super austenitic stainless steels, which include the grades AL-6XN® (N08367), 254 SMO® (S31254), and 25-6Mo (N08926), contain approximately 20 percent Cr, 6 percent Mo, and 0.20 percent N. These grades show very good resistance to various chemical environments and can resist long-term exposures to seawater. In many environments, 6 percent Mo grades have a corrosion resistance that is just below that of the Ni/Cr/Mo alloys such as N06625.

The most corrosion-resistant austenitic stainless steels are the 7 percent Mo grades 654 SMO® (S32654) and 27-7MO (S31277), which have a resistance in many environments that approaches that of the C276 (N10276) Ni/Cr/Mo alloy.

FERRITIC STAINLESS STEELS

The ferritic family of stainless steels is the AISI 400 series and is the second most widely used group of stainless steels. This family is characterized by having a microstructure that consists of ferrite phase and possibly small amounts of carbides and nitrides. Because the ferrite phase is ferromagnetic, this group of stainless steels has a magnetic attraction similar to carbon steels.

Ferritic stainless steels tend to have higher strength and much better resistance to chloride stress corrosion cracking than the austenitic grades; however, they do have reduced formability and weldability. The ferritic grades cannot be hardened or strengthened by heat treatment. They have limited toughness compared to other types of stainless steels. The toughness can be further reduced by a large grain size and thicker cross section. Because of the toughness limitations, the ferritic grades are only produced as thinner sheet and strip products. It is rare to find products thicker than 0.150 inches, which restricts their use to tubing and other thin-gage applications.

The ferritic family includes stainless steels with a wide range of corrosion resistance (see Table 25-8). The most widely used ferritic grade is type 430, which has a corrosion resistance just below that of the 304 austenitic stainless steel. On the low end of corrosion resistance is type 409, which is widely used in automotive exhaust systems, and on the upper end are the super ferritic stainless steels such as SEA-CURE® (S44660) and AL 29-4C® (S44735), which have a corrosion resistance in many environments that is equal to or slightly better than the 6 percent Mo super austenitic grades. Because of their very low Ni content, ferritic stainless steels are less costly on a weight basis than the austenitic grades, with a similar corrosion resistance.

Ferritic grades that contain more than about 12 percent Cr are susceptible to a loss of ductility due to alpha prime precipitation that occurs when

TABLE 25-7 List of Commonly Used High-Performance Austenitic Stainless Steels*

Common name	UNS number	C	Mn	Si	Cr	Ni	Mo	N	Cu	Other
Alloy 20 20Cb-3	N08020	0.07	2	1	19.0–21.0	32–38	2.0–3.0	...	3.0–4.0	Nb 8×C min., 1.00 max.
Alloy 28	N08028	0.03	2.5	1	26.0–28.0	29.5–32.5	3.0–4.0	...	0.6–1.4	
3127 hMo Alloy 31	N08031	0.015	2	0.3	26.0–28.0	30–32	6.0–7.0	0.15–0.25	1.0–1.4	
AL-6XN®	N08367	0.03	2	1	20.0–22.0	23.5–25.5	6.0–7.0	0.18–0.25	0.75	
904L	N08904	0.02	2	1	19.0–23.0	23.0–28.0	4.0–5.0	0.1	1.0–2.	
25-6MO, 1925 hMo	N08926	0.02	2	0.5	19.0–21.0	24.0–26.0	6.0–7.0	0.15–0.25	0.5–1.5	
254 SMO®	S31254	0.02	1	0.8	19.5–20.5	17.5–18.5	6.0–6.5	0.18–0.25	0.5–1.0	
UR 66	S31266	0.03	2.0–4.0	1	23.0–25.0	21.0–24.0	5.2–6.2	0.35–0.60	1.0–2.5	W 1.5–2.5
27-7MO	S31277	0.02	3	0.5	20.5–23.0	26.0–28.0	6.5–8.0	0.30–0.40	0.5–1.5	
254N	S32053	0.03	1	1	22.0–24.0	24.0–26.0	5.0–6.0	0.17–0.22	...	
654 SMO®	S32654	0.02	2.0–4.0	0.5	24.0–25.0	21.0–23.0	7.0–8.0	0.45–0.55	0.3–0.6	
4565	S34565	0.03	5.0–7.0	1	23.0–25.0	16.0–18.0	4.0–5.0	0.40–0.60	...	Nb 0.10

*Single values are maximum values unless otherwise noted.

TABLE 25-8 List of Commonly Used Ferritic Stainless Steels*

Common name	UNS number	C	Mn	Si	Cr	Ni	Mo	N	Other
403	S40300	0.15	1	0.5	11.5–13.0	0.6	...	...	
405	S40500	0.08	1	1	11.5–14.5	0.6	...	...	Al 0.10–0.30
409	S40910	0.03	1	1	10.5–11.7	0.5	...	0.03	Ti 6 × (C = N) min., 0.50 max., Nb 0.17
430	S43000	0.12	1	1	16.0–18.0	0.75	...	...	
439	S43035	0.03	1	1	17.0–19.0	0.5	...	0.03	Ti [0.2 + 4(C + N)] min., 1.10 max., Al 0.15
434	S43400	0.12	1	1	16.0–18.0	...	0.75–1.25	...	
436	S43600	0.12	1	1	16.0–18.0	...	0.75–1.25	...	Nb 5 × C min., 0.80 max.
444	S44400	0.025	1	1	17.5–19.5	1	1.75–2.50	0.035	(Ti + Nb) [0.20 + 4(C = N)]min., 0.80 max.
XM-27 Sea-Cure®	S44660	0.03	1	1	25.0–28.0	1.0–1.35	3.0–4.0	0.04	(Ti + Nb) 0.20–1.0, Ti + Nb 6 × (C = N) min.
AL29-4C®	S44735	0.03	1	1	28.0–30.0	1	3.6–4.2	0.045	(Ti + Nb) 0.20–1.00, (Ti + Nb) 6 × (C + N) min.

*Single values are maximum values unless otherwise noted.

heated to temperatures of 315°C to 525°C (600°F to 950°F). Because of this, the higher alloyed ferritic grades typically have a maximum service temperature of 315°C (600°F).

DUPLEX STAINLESS STEELS

Duplex stainless steel (DSS) is called "duplex" because it has a two-phase microstructure consisting of ferrite and austenite. Wrought DSS products typically have an austenite/ferrite phase balance of 50 to 55 vol% austenite (45 to 50 percent ferrite). The duplex microstructure gives this family of stainless steels a desirable combination of properties, including relatively high strength, good toughness, and improved chloride stress corrosion cracking resistance compared to the 300 series austenitic stainless steels.

Because the levels of Ni and Mo are reduced compared to an austenitic grade with similar corrosion resistance, the duplex grades are often less expensive than their austenitic counterparts. Since the mid-1990s, the use of DSS in North America has grown substantially, and many new duplex grades have been developed. Although the development of new duplex grades provides users with more material options, the lack of availability of all product forms for many of the newer grades can be an issue.

Because of the high yield strength of the duplex grades, the required section thickness for tanks and pressure vessels can often be reduced with DSS construction, resulting in substantial cost savings for heavy-wall or high-pressure constructions. In order to avoid precipitation of the undesirable alpha prime in the ferrite phase, most DSSs have a maximum service temperature of 315°C (600°F). As with the other stainless steel families, DSSs have a range of corrosion resistance and can be divided into subgroups depending on alloying content (see Table 25-9).

Lean Duplex Stainless Steels The lean duplex grades such as LDX 2101® (S32101), UR 2202 (S32202), ATI 2102® (S82011), and 2304 (S32304) are characterized by having relatively low levels of Ni and Mo. To compensate for the reduced Ni content, which is a strong austenite former, these grades have increased levels of N and Mn, which provide an acceptable austenite/ferrite phase balance. As a group, the lean grades have high strength, particularly yield strength, and exhibit a pitting and crevice corrosion resistance that is similar to or just below that of the 316L austenitic grade. Because of their relatively high yield strength, the lean DSSs are well suited for tank construction and structural applications.

Standard Duplex Stainless Steels The standard DSSs typically contain 21 to 25 percent Cr, 2 to 3 percent Mo, and 0.15 percent N. Grades within this category are used widely across all industry sectors, and type 2205 (S32205) has become the most widely used of all of the duplex grades. The 2205 grade has a yield strength that is more than twice that of type 316 and a pitting and crevice corrosion resistance that is similar to that of type 904L austenitic stainless steel. Applications for standard DSSs include tanks, piping, process vessels, and structural applications.

Super Duplex Stainless Steels The super DSSs typically contain 25 percent Cr, 3.5 to 4.0 percent Mo, and 0.25 to 0.27 percent N. The pitting and crevice corrosion resistance of the super DSS are essentially equivalent to the 6 percent Mo super austenitic stainless steels. In North America, the most widely used super DSS grades are the 2507 (S32750) and Zeron 100 (S32760) grades. Common applications include piping, heat exchangers, tanks, and process vessels used for chemical processing and marine applications.

Hyper Duplex Stainless Steels The hyper DSSs, SAF 3207® (S33207) and SAF 2707® (S32707), are the most highly alloyed of the duplex family. They contain 26 to 30 percent Cr, 3.5 to 5.0 percent Mo, and 0.30 to 0.50 percent N. The hyper DSSs are designed for more aggressive acid and chloride-containing environments. The hyper grades are produced in limited quantities and are currently only available as seamless tubing.

MARTENSITIC STAINLESS STEELS

The martensitic stainless steels also belong to the AISI 400 series but differ from the ferritic grades by having the ability to be strengthened by heat treatment. This family of stainless steels is characterized by having a microstructure that consists predominately of the martensite phase and possibly lesser

TABLE 25-9 List of Commonly Used Duplex Stainless Steels*

Common name	UNS number	C	Mn	Si	Cr	Ni	Mo	N	Cu	Other
					Lean duplex stainless steels					
LDX 2101® 2101	S32101	0.04	4.0–6.0	1	21.0–22.0	1.35–1.70	0.1–0.8	0.20–0.25	0.1–0.8	
URANUS® 2202 2202	S32202	0.03	2	1	21.5–24.0	1.00–2.80	0.45	0.18–0.26	...	
2304	S32304	0.03	2.5	1	21.5–24.5	3.0–5.5	0.05–0.6			
ATI 2102® 2102	S82011	0.03	2.0–3.0	1	20.5–23.5	1.0–2.0	0.1–1.0	0.15–0.27	0.50–0.60	
					Standard duplex stainless steels					
2205	S31803	0.03	2	1	21.0–23.0	4.5–6.5	2.5–3.5	0.08–0.20	...	
2205	S32205	0.03	2	1	22.0–23.0	4.5–6.5	3.0–3.5	0.14–0.20	...	
255	S32550	0.04	1.5	1	24.0–27.0	4.5–6.5	2.9–3.9	0.10–0.25	1.5–2.5	
					Super duplex stainless steels					
2507	S32750	0.03	1.2	0.8	24.0–26.0	6.0–8.0	3.0–5.0	0.24–0.32	0.5	
ZERON 100 Z100	S32760	0.03	1	1	24.0–26.0	6.0–8.0	3.0–4.0	0.20–0.30	0.50–1.00	W 0.5–1.0
					Hyper duplex stainless steels					
SAF 2707 HD 2707	S32707	0.03	1.5	0.5	26.0–29.0	5.5–9.5	4.0–5.0	0.30–0.50	1	Co 0.5–2.0
	S33207	0.03	1.5	0.8	29.0–33.0	6.0–9. 0	3.0–5.0	0.40–0.60	1	

*Single values are maximum values unless otherwise noted.

amounts of secondary phases such as ferrite, austenite, and carbides. At elevated temperatures of approximately 1070°C (1900°F), this family of stainless steels has an austenitic structure that can be transformed into a highly strained body-centered tetragonal phase, martensite, when cooled to room temperature. In order to achieve a suitable combination of high strength (high hardness) and toughness, steels hardened by the martensite transformation must be tempered at lower temperatures, 93°C to 704°C (200°F to 1300°F). The martensitic grades are typically alloyed with 10.5 to 18 percent Cr and lesser amounts of elements such as Ni and Mo. Common grades include 410 (S41000), 416 (S41600), 420 (S42000), 440A (S44002), and 440C (S44004). This family of stainless steels is ferromagnetic, similar to carbon steel (see Table 25-10).

Martensitic stainless steels have high strength, good wear resistance, low toughness, and a relatively high ductile-to-brittle transition temperature. They are very difficult to weld, and they typically require a post-weld heat treatment. Because of this, the martensitic grades are often restricted to nonwelded applications.

The martensitic grades do not have very high Cr levels, and some of the Cr that is present is tied up as carbides. This results in a relatively low corrosion resistance, and the martensitic stainless steels are less resistant than the standard 304/304L austenitic grades. Because of their limited toughness and corrosion resistance, the martensitic grades are only specified when high strength is required. Common applications include turbine blades, cutlery, furnace parts, fasteners, and shafts.

PRECIPITATION HARDENING STAINLESS STEEL

The precipitation hardening (PH) stainless steels belong to the AISI 600 series, and like the martensitic stainless steels, they can be strengthened by heat treatment. The defining characteristic of this family of stainless steels is that they all rely on a precipitation mechanism for some or all of their strengthening. An age-hardening heat treatment is used to produce fine intermetallic precipitates that provide increased strength. The PH stainless steels can be subdivided into martensitic, semi-austenitic, and fully austenitic grades. Because of their higher Cr levels, the PH grades have better corrosion resistance than the 400 series martensitic grades, and they are used for high-strength applications that require more corrosion resistance. Common applications for the PH grades include fasteners, aircraft fittings, shafts, gears, bellows, and jet engine parts. Some of the more commonly specified PH grades and their chemical compositions are summarized in Table 25-11.

Martensitic PH Stainless Steels This group of stainless steels includes the 17-4PH® (S17400), 15-5 PH® (S15500), 13-8 PH® (S13800), Custom 450® (S45000), and Custom 455® (S45500) grades. With the martensitic PH grades, an untempered martensitic microstructure is produced during cooling from the solution anneal temperature. A subsequent age-hardening heat treatment between 482°C and 621°C (900°F and 1150°F) increases the strength due to the precipitation and tempers the martensite structure for improved toughness and ductility. The martensitic PH grades are well suited for bar, wire, and forging applications.

Semi-Austenitic PH Stainless Steels The semi-austenitic grades have a balanced composition that results in a solution-annealed microstructure that is predominately austenite phase. This more ductile structure can then be formed using standard techniques. After fabrication, a thermal treatment is used to transform the austenite to martensite, followed by an aging heat treatment, which increases the strength due to a precipitation mechanism. Common semi-austenitic PH grades include 17-7PH® (S17700), 15-7 PH® (S15700), AM-350® (S35000), and AM 355® (S35500) grades. The semi-austenitic PH grades are used primarily for flat-rolled applications and applications that require extensive forming.

Austenitic PH Stainless Steels The A286 (S66286) grade is the only PH grade that remains primarily austenitic after the aging heat treatment. The A286 grade is hardened after solution annealing by a one-step aging heat treatment to precipitate gamma prime phase. The A286 grade is used in applications that require high strength and good corrosion resistance at temperatures up to 704°C (1300°F). Common uses include jet and automotive engine components, gas turbine blades, high-temperature fasteners, and springs.

HIGH-TEMPERATURE STAINLESS STEELS

Alloys that are designed primarily for higher-temperature applications are often termed high-temperature or heat-resistant alloys. Although not formally defined, a high-temperature application for stainless steels is an environment

TABLE 25-10 List of Commonly Used Martensitic Stainless Steels*

Common name	UNS number	C	Mn	Si	Cr	Ni	Mo	Other
403	S40300	0.15	1	0.5	11.5–13.0	0.6	...	
410	S41000	0.08–0.15	1	1	11.5–13.5	0.75	...	
XM-30	S41040	0.18	1	1	11.0–13.0	...	...	Nb 0.05–0.30
414	S41400	0.15	1	1	11.5–13.5	0.75	1.25–2.50	
415	S41500	0.05	0.50–1.00	0.6	11.5–14.0	3.5–5.5	0.50–1.00	
416	S41600	0.15	1.25	1	12.0–14.0	...	0.6	S 0.15 min.
420	S42000	0.15 min	1	1	12.0–14.0	...	...	
431	S43100	0.2	1	1	15.0–17.0	1.25–2.50	...	
444A	S44402	0.60–0.75	1	1	16.0–18.0	...	0.75	
444B	S44403	0.75–0.95	1	1	16.0–18.0	...	0.75	
444C	S44404	0.95–1.20	1	1	16.0–18.0	...	0.75	

*Single values are maximum values unless otherwise noted.

TABLE 25-11 List of Commonly Used Precipitation Hardening Stainless Steels*

Common name	UNS number	C	Mn	Si	Cr	Ni	Mo	Other
Martensitic PH stainless steels								
PH 13-8 Mo	S13800	0.05	0.2	0.1	12.25–13.25	7.5–8.5	2.00–2.50	Al 0.90–1.35, N 0.10
15-5 PH	S15500	0.07	1	1	14.0–15.5	3.5–5.5	...	Cu 2.5–4.5, (Nb + Ta) 0.15–0.45
17-4 PH	S17400	0.07	1	1	15.0–17.5	3.0–5.0	...	Cu 3.0–5.0, (Nb + Ta) 0.15–0.45
Custom 450	S45000	0.05	1	1	14.0–16.0	5.0–7.0	0.50–1.00	Cu 1.24–1.75, Nb 8 × C min.
Custom 455	S45500	0.05	0.5	0.5	11.0–12.5	7.5–9.5	0.5	Ti 0.9–1.4, Cu 1.5–2.5, (Nb + Ta) 0.1–0.5
Semi-austenitic PH stainless steels								
PH 15-7 Mo	S15700	0.09	1	1	14.0–16.0	6.50–7.75	2.00–3.00	Al 0.75–1.50
17-7 PH	S17700	0.09	1	1	16.0–18.0	6.50–7.75	...	Al 0.75–1.50
AM-350	S35000	0.07–0.11	0.50–1.25	0.5	16.0–17.0	4.0–5.0	2.50–3.25	N 0.07–0.13
AM-355	S35500	0.10–0.15	0.50–1.25	0.5	15.0–16.0	4.0–5.0	2.50–3.25	N 0.07–0.13
Austenitic PH stainless steels								
A-286	S66286	0.08	2	1	13.5–16.0	24.0–27.0	1.0–1.5	Ti 1.90–2.35, Al 0.35 max, V 0.1–0.5, B 0.003–0.010

*Single values are maximum values unless otherwise noted.

TABLE 25-12 List of Commonly Used High-Temperature Stainless Steels*

Common name	UNS number	C	Mn	Si	Cr	Ni	Mo	N	Other
				Wrought high-temperature stainless steels					
304H	S30409	0.04–0.1	2	0.75	18–20	8–10.5	...	...	
153 MA®	S30415	0.04–0.06	0.8	1.0–2.0	18–19	10-Sep	...	0.12–0.18	Ce 0.03–0.08
316H	S31609	0.04–0.1	2	0.75	16–18	14-Oct	2.0–3.0	...	
321H	S32109	0.04–0.1	2	0.75	17–19	12-Sep	...	...	Ti 4 × (C + N) min, 0.70 max
347H	S34709	0.04–0.1	2	0.75	17–19	13-Sep	...	...	Cb 8 × (C) min, 1.0 max
348H	S34809	0.04–0.1	2	0.75	17–19	13-Sep	...	...	(Cb + Ta) 8 × (C) min, 1.0 max, Ta 0.10, Co 0.20
253 MA®	S30815	0.05–0.1	0.8	1.4–2.0	20–22	12-Oct	...	...	N 0.14–0.20, Ce 0.03–0.08
309H	S30909	0.04–0.1	2	0.75	22–24	15-Dec	...	...	
310H	S31009	0.04–0.1	2	0.75	24–26	19–22	...	...	
N08810	800H	0.05–0.10	1.5	1	19–23	30–35	...	...	Cu 0.75 min, Fe 39.5 min., Al 0.15–0.60, Ti 0.15–0.60
405	S40500	0.08	1	1	11.5–14.5	0.6	...	...	Al 0.1–0.3
439	S43035	0.03	1	1	17–19	0.5	...	0.03	Ti [0.2++4(C + N)] min, 1.10 max; Al 0.15
446	S44600	0.2	1.5	1	23–27	0.75	...	0.2	
				Cast high-temperature stainless steels					
HC	J92605	0.5	1	2	26–30	4	0.5		
HD	J93005	0.5	1.5	2	26–30	7-Apr	0.5		
HE	J93403	0.2–0.5	2	2	26–30	11-Aug	0.5		
HF	J92603	0.2–0.4	2	2	18–23	12-Aug	0.5		
HH	J93503	0.2–0.5	2	2	24–28	14-Nov	0.5		
HI	J94003	0.2–0.5	2	2	26–30	14–18	0.5		
HK	J94224	0.2–0.6	2	2	24–28	18–22	0.5		
HL	N08604	0.2–0.6	2	2	28–32	18–22	0.5		
HN	J94213	0.2–0.5	2	2	19–23	23–27	0.5		
HP	N08705	0.35–0.75	2	2.5	24–28	33–37	0.5		
HT	N08605	0.35–0.75	2	2.5	15–19	33–37	0.5		
HU	N08004	0.35–0.75	2	2.5	17–21	37–41	0.5		

*Single values are maximum values unless otherwise noted.

that exposes construction materials to temperatures of 500°C (950°F) or higher. Heat-resistant stainless steels are often considered as a separate group, and a list of some of the more common heat-resistant stainless steels is given in Table 25-12. Because many of the commonly specified heat-resistant grades are only available as castings, Table 25-12 also includes cast grades.

High-temperature environments place special demands on construction materials, including the need for good metallurgical stability and sufficiently high strength and creep resistance. In addition, many high-temperature environments can be very corrosive to construction materials. Common modes of high-temperature attack include:

- Oxidation
- Carburization and metal dusting
- Nitridation
- Halogen corrosion
- Sulfidation
- Molten salt corrosion
- Molten metal attack
- Corrosion by ash/salt deposits

Common applications for high-temperature stainless steels are industrial furnaces, turbochargers, gas turbines, power-plants, and equipment used to produce petrochemical products, chemicals, cement, and glass. When choosing a heat-resistant stainless steel, the required high-temperature mechanical properties and the resistance to high-temperature corrosion must be considered.

NICKEL ALLOYS

Nickel alloys are vital to the process industries. This is because their environmental resistance and attainable high-temperature strengths are generally much greater than those of the austenitic stainless steels. Furthermore, they are relatively easy to form and weld into complex components.

Advantages of nickel as an alloy base include its reasonable price, its inherent nobility (resistance to liquid and gaseous environments), and the fact that beneficial elements, such as chromium, are highly soluble in nickel. It also has an ideal atomic structure (face-centered cubic, or gamma phase) throughout its solid range; this imparts excellent ductility (ability to deform under stress prior to fracturing). The nickel alloys can be divided into two major groups, as follows:

Those designed primarily for use at temperatures below 500°C (932°F) in corrosive liquids.

Those designed for use at temperatures above 500°C (932°F) in corrosive gases, or just very hot air, which causes oxidation of metallic materials.

These are more commonly known as the corrosion-resistant nickel alloys and the high-temperature nickel alloys. For the former, the main concern is their resistance to the environment. For the latter, there are additional concerns, such as:

1. The stability of the structure at high operating temperatures
2. Material strength at high operating temperatures
3. Creep (slow, progressive, plastic deformation of the material, even at low stresses)

CORROSION-RESISTANT NICKEL ALLOYS

Introduction Nickel alloys are extremely important to the process industries, being significantly more resistant to aqueous corrosion than the stainless steels. Commercially pure nickel, for example, is the material of choice for sodium hydroxide, while the nickel-copper (Ni-Cu) alloys are outstanding in hydrofluoric acid. The nickel-molybdenum (Ni-Mo) alloys are exceptional in perhaps the two most important industrial chemicals: hydrochloric and sulfuric acids. The nickel-chromium-molybdenum (Ni-Cr-Mo) alloys offer great versatility, being resistant to a wide range of chemicals, including chlorides, which are notorious for causing pitting, crevice attack, and stress corrosion cracking of lesser materials.

The role of chromium in the corrosion-resistant nickel alloys is to enable passivation (i.e., the formation of protective oxide or hydroxide films). Copper, which is closer to the noble end of the electromotive series than nickel, enhances its resistance to corrosion in the absence of passive films. Molybdenum, which is a potent solid solution strengthening agent in nickel, greatly enhances the corrosion resistance of nickel in so-called reducing

TABLE 25-13 Nickel Alloy Families

Type	Primary attributes
Commercially pure nickel	Resistant to caustic alkalis and foodstuffs, useful electrical and magnetic properties
Nickel-copper (Ni-Cu)	Resistant to seawater and hydrofluoric acid
Nickel-molybdenum (Ni-Mo)	Resistant to pure, reducing acids
Nickel-chromium (Ni-Cr)	Resistant to oxidizing acids, resistant to chloride-induced stress corrosion cracking
Nickel-chromium-molybdenum (Ni-Cr-Mo)	Resistant to both oxidizing and reducing acids, extremely resistant to chloride-induced pitting, crevice corrosion, and stress corrosion cracking
Nickel-chromium-iron (Ni-Cr-Fe)	Lower-cost alternate to Ni-Cr
Nickel-iron-chromium (Ni-Fe-Cr)	Even lower-cost alternate to Ni-Cr

acids (i.e., those that induce hydrogen evolution at cathodic sites) and, when combined with chromium, contributes to the integrity of passive films, especially their ability to withstand chlorides.

There are several nickel alloy families, grouped according to their major (10 wt% or greater) alloying elements, as shown in Table 25-13.

Commercially Pure Nickels Numerous materials are sold under the guise of commercially pure nickel. Most contain in excess of 99 percent nickel, and most contain small, elemental additions to control specific properties. Several are used in the electronics industry, where the electrical and magnetic properties of pure nickel are used to advantage.

From a corrosion standpoint, the commercially pure nickels are important for two reasons. First, they have outstanding resistance to the caustic alkalis (caustic soda and caustic potash) over wide ranges of concentration and temperature. Second, they are very easy to form into complex shapes and have inherent resistance to mild corrosives; thus they are suitable for food processing equipment.

The commercially pure nickel most widely used in the chemical process industries is nickel 200. For applications above approximately 300°C (572°F), nickel 201 is generally preferred. This has a lower carbon content and is thus resistant to graphitization; it also has higher creep resistance.

Two of the commercially pure nickels (alloys 300 and 301), are age-hardenable. The former was designed for electronic applications. The latter was designed as an alternate to nickel 200, for applications requiring high strength. The applications of alloy 301 include plastic extrusion press parts, glass molds, and corrosion-resistant springs.

Ni-Cu Alloys Nickel and copper, neighbors in the Periodic Table, share the same atomic structure (fcc). Moreover, this structure is retained in all mixtures of the two elements, at all temperatures in the solid range. This has given rise to several commercially important nickel-copper and copper-nickel alloys. The nickel-copper alloys contain approximately 30 to 45 wt% copper. The primary attributes of the Ni-Cu alloys are their resistance to seawater, brackish water, and hydrofluoric acid. They also have moderate resistance to other nonoxidizing acids. They withstand cavitation erosion and are therefore ideally suited to applications in flowing water, such as propellers and pumps.

Ni-Mo Alloys The Ni-Mo, or B-type, alloys' chief benefit is their high resistance to pure hydrochloric and sulfuric acids over large ranges of concentration and temperature. They also resist pure hydrobromic acid, some concentrations of hydrofluoric acid, food-grade phosphoric acid, acid chlorides, and other nonoxidizing halide salt solutions. The primary limitation of the Ni-Mo alloys is that they cannot tolerate either oxidizing acids, such as nitric, or acids that contain oxidizing species. Such species include oxygen, hydrogen peroxide, chlorine, bromine, ferric ions, and cupric ions.

The most widely used wrought material in this family is B-3 alloy, which was developed in the early 1990s. B-3 alloy has deliberate additions of chromium and iron, which provide the material with more structural stability than its predecessors. The most widely used cast Ni-Mo alloy is N-7M.

Ni-Cr Alloys Early experiments involving the addition of chromium to nickel not only resulted in materials resistant to oxidizing acids, but also paved the way for the development of a wide range of oxidation-resistant high-temperature alloys. Indeed, some of the first Ni-Cr materials were, and still are, used for heating elements in domestic appliances. It is not surprising, therefore, that the most common corrosion-resistant Ni-Cr materials, namely alloys 600 and 625, were designed primarily for use at high temperatures. The primary corrosion-related attributes of alloy 600 include excellent resistance to sodium hydroxide and good resistance to stress corrosion cracking (SCC) relative to many stainless steels.

Alloy 625 was developed in the 1950s for use as steam-line piping in supercritical steam power plants. To impart solid solution strength, significant additions of molybdenum and niobium were used. During the development, it was discovered that niobium induces a very effective hardening precipitate, known as gamma double-prime, and for some time the development of alloy 625 was delayed to allow the development of a gamma double-prime strengthened superalloy.

Ni-Cr-Mo Alloys The chief attribute of the Ni-Cr-Mo (C-type) alloys is their versatility. This stems from the fact that they contain high levels of both chromium and molybdenum. Chromium induces passivation in oxidizing acids, as it does in the stainless steels. Molybdenum provides resistance to reducing acids, in particular hydrochloric acid.

Hydrochloric acid and the chloride salts are the chemicals most responsible for the commercial success of the Ni-Cr-Mo alloys. These compounds are encountered throughout the chemical process industries, and they can be very damaging to the stainless steels. The Ni-Cr-Mo alloys are particularly resistant to pitting, crevice attack, and stress corrosion cracking, insidious and unpredictable forms of corrosion caused by chlorides. Just as the Ni-Cr-Mo alloys resist hydrochloric acid and associated salts, they also resist the corresponding compounds of bromine and fluorine. Indeed, the Ni-Cr-Mo alloys are among the few metallic materials that withstand warm hydrofluoric acid. Among their other attributes, the Ni-Cr-Mo alloys resist sulfuric and phosphoric acids, various high-temperature organic acids, and certain concentrations of caustic soda and caustic potash.

The first Ni-Cr-Mo material, alloy C, was introduced in the early 1930s in cast form. Wrought products of this alloy became available in the 1940s, notably with a lower carbon limit. Post-weld annealing was necessary with alloy C due to the fact that, prior to the 1960s, there was no way to minimize the carbon and silicon contents during melting. With the advent of argon-oxygen decarburization (AOD) during melting, however, low-carbon, low-silicon versions became reality, the first being C-276 alloy. For applications requiring even greater resistance to weld heat-affected zone sensitization, C-4 alloy was introduced in the 1970s. To increase its thermal stability, additions of iron and tungsten were omitted.

The chromium content of C-276 and C-4 alloys is approximately 16 wt%. By the mid-1980s, it was realized that many industrial environments contain oxidizing impurities, and that higher chromium contents might be generally beneficial to the versatility of the nickel-chromium-molybdenum alloys. Thus, C-22 alloy, 686 alloy, alloy 59, and C-2000 alloy, introduced in the 1980s and 1990s, contain 21 to 23 percent chromium. C-2000 alloy is unique among the Ni-Cr-Mo materials in having a deliberate copper addition, to enhance its resistance to sulfuric acid. The original cast alloy C is still made today, under the guise of its AFS designation, CW-12MW, although newer cast materials (CW-2M and CX-2MW) with enhanced properties are also widely used.

Due to the needs of the process industries for such versatile materials, the Ni-Cr-Mo system is still an active field of development. Recent advances include an age-hardenable version of C-22 alloy (C-22HS alloy) and a high-molybdenum version (HYBRID-BC1 alloy).

Ni-Cr-Fe Alloys The nickel-chromium-iron alloys were originally designed to fill the performance gap between the high-molybdenum stainless steels and the nickel-chromium-molybdenum alloys. Thus, they possess good resistance to chloride-induced phenomena, such as pitting, crevice corrosion, and stress corrosion cracking, and they exhibit moderate resistance to the halogen acids, in particular hydrochloric.

The first low-carbon, wrought material in this family, G-3 alloy, which was introduced in the 1970s, quickly became established in two major applications. First, cold-reduced tubes of G-3 alloy became standard for moderately sour oil and gas wells. Second, G-3 alloy found use in evaporators used to concentrate "wet process" phosphoric acid in the agrichemical industry.

The success of G-3 alloy in these applications led, in the 1980s, to the introduction of G-30 and G-50 alloys. G-30 alloy was a high-chromium variant with enhanced resistance to phosphoric acid, and G-50 alloy was a high-molybdenum variant with enhanced resistance to stress corrosion cracking in elevated-temperature environments that contained hydrogen sulfide, as found in oil and gas wells. Copper was added to G-3 and G-30 alloys to enhance their resistance to sulfuric acid. It should be mentioned that a material developed in the 2000s, G-35 alloy, has superseded G-30 alloy for "wet process" phosphoric acid applications. However, it falls within the Ni-Cr category and contains 33 wt% chromium.

Ni-Fe-Cr Alloys Alloy 825 and its age-hardenable cousin alloy 925 contain significantly more iron than chromium and thus constitute a separate grouping. With iron contents of 30 and 28 wt%, respectively, they are compositionally close to the high-nickel, austenitic stainless steels. The negative aspects of such high iron contents are reduced solubilities of key elements, such as molybdenum, and reduced resistance to environmental cracking. On the other hand, high iron contents reduce the cost of making the alloys. One of the key additions to alloy 825 is copper. As in other materials, it provides enhanced resistance to sulfuric acid.

Table 25-14 shows the nominal composition of several nickel-based alloys.

TABLE 25-14 Common Nickel-Based Alloys

Type	Alloy	UNS	Ni	Fe	Cr	Mo	W	Cu	Mn	Other
Ni	200	N02200	Bal.	0.2	—	—	—	0.1	0.2	Si 0.2, C 0.08
Ni-Cu	400	N04400	Bal.	1.2	—	—	—	31.5	1	Si 0.2, C 0.2
Ni-Mo	B-3	N10675	Bal.	1.5	1.5	28.5	3*	0.2*	3*	Si 0.1*, C 0.01*, Al 0.5*
Ni-Cr	600	N06600	Bal.	8	15.5	—	—	0.2	0.5	Si 0.2, C 0.08
Ni-Cr	625	N06625	Bal.	2.5	21.5	9	—	—	0.2	Si 0.2, C 0.05, Al 0.2, Ti 0.2, Nb + Ta 3.6
Ni-Cr-Mo	C-22	N06022	Bal.	3	22	13	3	0.5*	0.5*	Si 0.08*, C 0.01*, V 0.35*
Ni-Cr-Mo	C-276	N10276	Bal.	5	16	16	4	0.5*	1*	Si 0.08*, C 0.01*, V 0.35*
Ni-Cr-Mo	C-2000	N06200	Bal.	3*	23	16	—	1.6	0.5*	Si 0.08*, C 0.01*, Al 0.5*
Ni-Cr-Mo	59	N06059	Bal.	1.5*	23	16	—	—	—	
Ni-Cr-Mo	686	N06686	Bal.	2*	20.5	16	3.9	—	—	
Ni-Cr-Fe	G-30	N06030	Bal.	15	30	5.5	2.5	2	1.5*	Si 0.8*, C 0.03*, Co 5*, Nb 0.8
Ni-Fe-Cr	825	N08825	Bal.	30	21.5	3	—	2.2	0.5	Si 0.2, C 0.03, Al 0.1, Ti 0.9
Ni-Cr-Mo	CW-12MW	N30002	Bal.	6	16.5	17	4.5	—	1*	C 0.12*
Ni-Cr-Mo	CW-2M	N26455	Bal.	2	16	16	1	—	1*	C 0.02*
Ni-Cr-Mo	CX-2MW (UNS N26022)	N26022	Bal.	4	21	13	3	—	1*	C 0.02*

*Maximum composition.

HIGH-TEMPERATURE NICKEL ALLOYS

High-temperature nickel-base alloys have found widespread use in the chemical and petrochemical process industries, primarily for their high-temperature strength and environmental resistance in hot gaseous atmospheres. Among the mechanical properties, tensile strength is important, but in most cases creep/stress rupture strength dictates the stress allowable for the alloy at higher temperatures. Most of the high-temperature components are subjected to thermal cyclic exposures and loading; thus they are also required to be resistant to thermal fatigue and low cycle fatigue (LCF). In addition, various other properties, such as thermal stability, susceptibility to stress relaxation cracking, fabricability, and repairability are quite relevant from an alloy selection standpoint. It is noteworthy that the alloy properties and their behaviors vary significantly across the temperature ranges. Therefore, the alloy is often selected based on its unique characteristics and properties in the temperature range of interest along with its availability and cost. The properties of the alloy may still vary due to other factors, including grain size changes, heat-to-heat compositional variations, "rich" or "lean" compositions, property variations from differences in the product forms (plate, bar, sheet, foil, pipe, and tube) and most importantly, homogeneity and the presence of impurities within the alloy.

The high-temperature strength in nickel-base alloys is imparted by solid solution strengthening from alloying elements such as Mo, Ta, and W, or precipitation hardening. Most of the high-temperature alloys rely on the formation of a chromia (Cr_2O_3) scale for high-temperature corrosion protection. However, in certain unique high-temperature corrosive environments, alumina (Al_2O_3) or silica (SiO_2) scale-forming alloys are also used.

The nominal compositions of various wrought and cast nickel-base alloys used in the chemical process industries for high-temperature applications are given in Table 25-15. The solid solution strengthened chromia-forming alloys can be broadly grouped according to their primary alloying elements as follows: Ni-Fe-Cr, Ni-Cr-Fe, and Ni-Cr-Mo/W, while additional nickel-base alloys that do not fall within the preceding classes can be categorized as high-temperature corrosion-resistant alloys, precipitation-strengthened alloys, and cast alloys.

Ni-Fe-Cr Alloys The examples that can be grouped into this category are alloys 800/800H/800HT®, RA-330®, HR-120®. In general, these alloys are considered an upgrade over austenitic stainless steels, while they are also used as low-cost alternatives to high nickel–containing alloys. These alloys offer optimum high-temperature strength and resistance to oxidation, carburization, and other types of high-temperature degradation processes. Alloy 800H is a refinement of alloy 800 by the addition of higher carbon, while 800HT is an improvement over 800H, offering better creep strength by virtue of a higher Al + Ti content. In contrast, RA-330 is a Ni-Fe-Cr alloy modified with silicon addition to offer better corrosion resistance at high temperatures.

Ni-Cr-Fe Alloys The alloys grouped into this category are alloys 600, 601, and X, all of which have relatively low Fe contents compared to the previous class. These exhibit good high-temperature strength and corrosion resistance. They are also readily available and have excellent fabricability. Alloy 601 offers higher strength and better corrosion resistance than alloy 600 due to the addition of Al and a higher Cr content (Table 25-15) (INCONEL® 600 and 601 Tech Brochures, www.specialmetals.com; HASTELLOY® X Tech Brochure, www.haynesintl.com).

Ni-Cr-Mo/W Alloys Alloys such as 625, 617, and 230® can be grouped into this class since they exhibit superior properties compared to those of the other groups. They are used for outstanding high-temperature strength, corrosion resistance, thermal fatigue resistance, and fabricability. However, none of the alloys in this class requires a separate precipitation hardening treatment to impart additional strength. Among the three alloys, alloys 230 and 617 exhibit higher creep and stress rupture strength than alloy 625 above 760°C (HAYNES® 230® Tech Brochure, www.haynesintl.com), while alloy 617 offers better stress rupture strength and alloy 230 better creep strength. In addition, an issue with alloy 625 is that it is susceptible to a loss in room-temperature ductility after long-term thermal exposures in the temperature range 760°C to 871°C (1400°F to 1598°F) (Radavich, J. F., and A. Fort, *Effects of Long-Term Exposures in Alloy 625 at 1200°F, 1400°F,*

TABLE 25-15 Common High-Temperature Nickel-Base Alloys

Alloy	UNS	Type	Ni	Cr	Co	Fe	Mo	W	Ti	Al	Nb	Other
800	N08800	High-temperature corrosion alloy	Bal.	21	—	39.5	—	—	0.3	0.3	—	C 0.10 max, Al + Ti 0.3-1.20
Alloy X	N06002	High-temperature corrosion alloy	Bal.	22	1.5	18	9	0.6	—	—	—	C 0.10
602CA®	N06025	High-temperature corrosion alloy	Bal.	25	—	9	—	—	0.15	2.1	—	C 0.20, Y 0.05-0.12
718	N07718	Precipitation-strengthened superalloy	Bal.	19	—	18.5	3	—	0.9	0.5	5.1 (Nb + Ta)	C 0.04, Si 0.2, Mn 0.2
230	N06230	Solid solution strengthened superalloy	Bal.	22	5*	3*	2	14	—	0.3	—	C 0.1, La 0.02, Si 0.4, Mn 0.5
HR-160	N122160	High-temperature corrosion alloy	Bal.	28	29	2*	1*	1*	0.5	0.4*	—	C 0.05, Si 2.75, Mn 0.5
HK-40	J94204	Casting	20	25	—	—	0.5	0.5	—	—	—	C 0.35-0.45
HP Mod.	N28701	Casting	35	25	—	—	—	0.5	—	—	1	

*Maximum composition.

and 1600°F, Superalloys 718, 625, 706 and Various Derivatives, ed. E. A. Loria, The Minerals, Metals & Materials Society, 1994). In terms of oxidation resistance, alloy 230 offers superior oxidation resistance among the chromia-formers (Deodeshmukh, V., and S. Srivastava, "Static Oxidation Data for High Temperature Alloys," Haynes International, unpublished results, 2012).

Other High-Temperature Alloys The nickel-base alloys resistant to certain types of high-temperature corrosion often contain elements such as Al and Si along with Cr. The examples include alumina-forming alloys 214® and 602CA® for oxidation and carburization resistance, and alloy HR-160®, which forms a subscale of silica scale, for sulfidation resistance.

High-Temperature Cast Alloys High-temperature nickel castings are important, particularly for furnace tubes. These tubes are centrifugally cast. For many years the HK-40, 25Cr/20Ni alloys were the most common. These represented an improvement over wrought alloys because of superior tensile strengths. In recent years, the HP modified 25Cr/35Ni/Nb grades have been introduced to the commercial market. These have superior stress rupture properties, allowing for the use of thinner tubes with better heat transfer. The HP-modified grades are often marketed under trade names and can have significantly different compositions and properties. ASTM 608 is the specification for high-temperature nickel alloy centrifugally cast tubes.

REACTIVE METALS

CORROSION RESISTANCE OF REACTIVE AND REFRACTORY METALS

Reactive and refractory metals have very unique properties compared to some of the more common materials like carbon and stainless steels. For corrosion resistance, these materials are similar in some aspects, but they behave much differently in various environments. Titanium and niobium prefer oxidizing media but will have difficulty in alkaline environments. Zirconium can be used in both strong acids and strong caustics. Tantalum can be used in strong acids, but caution should be exercised when placing tantalum in strong, hot caustics. Ti, Zr, Nb, and Ta are generally very resistant to organic compounds. In organic media, some water is needed to maintain passivity, but the amount will depend on the alloy used. Generally, titanium and niobium are used up to the boiling point in lower-concentration acids, whereas zirconium and tantalum can be used in many environments above the boiling point. All of these materials are very susceptible to attack in hydrofluoric acid and fluoride-containing solutions. All of these materials have excellent resistance to chloride media. Of the four materials, titanium is one of the materials most used in chloride solutions. All of these alloys have good resistance to crevice corrosion, with the exception of some titanium commercially pure (CP) grades, which will suffer crevice corrosion under certain conditions. This susceptibility can be alleviated by the addition of Pd or Ru, or by the addition of Ni and Mo (e.g., titanium grade 12).

PHYSICAL AND MECHANICAL PROPERTIES

Both titanium and zirconium are hexagonal, close-packed materials. Niobium and tantalum are body-centered cubic materials. All four of these materials have high melting points, and their densities vary from the low-density titanium metal at 0.163 g/cm^3 (4.51 lb/in^3) to the very high-density tantalum metal of 0.600 g/cm^3 (16.6 lb/in^3).

All of the reactive and refractory metals are protected by stable, adherent oxide films. Because these materials are very reactive, they will re-form the oxide film instantaneously (in the presence of water or oxygen) if the oxide layer is damaged. The oxide film that forms will depend on the environment to which the metal is exposed. Table 25-16 shows the oxide compositions formed on these metals. Some of these materials form multiple oxide films (Ti and Nb) in various types of media (either reducing or oxidizing). Their corrosion resistance will be best when the most stable oxide is formed on the surface.

TABLE 25-16 Comparing Oxide Forms on Titanium, Zirconium, Niobium, and Tantalum

Metal	Possible oxides	Most stable and corrosion resistant oxide	Conditions favoring formation of most corrosion resistant oxide
Ti	TiO Ti_2O_3 TiO_2 (Ti_3O_5 rare)	TiO_2	Ti's film is often a mixture of three oxides. The relative amounts of each form depends on the oxidizing power of the media. The amount of protective TiO_2 present decreases with increasing reducing power of the media. Ti's oxide film is, therefore, conditional.
Zr	ZrO_2	ZrO_2	Zr always forms ZrO_2 on its surface, even under very reducing conditions. Zr's protective oxide film is, therefore, very reliable.
Nb	NbO NbO_2 Nb_2O_5	Nb_2O_5	Like Ti, Nb will form lower oxides under reducing conditions. Nb's oxide film is, therefore, conditional.
Ta	Ta_2O_5	Ta_2O_5	Ta's oxide film is likely to be Ta_2O_5 even in reducing media. Ta's oxide film is, therefore, reliable.

HYDROGEN EMBRITTLEMENT

As the term suggests, hydrogen embrittlement is a degradation process related to the absorption of hydrogen and a resulting loss of ductility. Although these materials are very reactive, they have differences in their susceptibility for hydrogen pickup. All of these materials (Ti, Zr, Nb, Ta) will have a tendency to absorb hydrogen under certain conditions with Zr having the lowest solubility for hydrogen and lower tendency for hydrogen absorption. Because these materials are very reactive, caution should be exercised when they are galvanically connected to other dissimilar metals in corrosive media. For tantalum, it has been shown that very small deposits of platinum (over-voltage element) were effective in preventing hydrogen pickup. These deposits or "spots" of platinum can be applied either by platinum spot welding or by the electroplating processes.

TITANIUM

Introduction Titanium is a lightweight, corrosion-resistant material. There are a number of titanium alloys available for use in chemical applications. The chemical processing alloy grades range from the very low-strength, high-ductility Ti Grade 1 alloy to the very high-strength Ti Grade 38 alloy. Many of these alloys are approved for use in ASME Boiler and Pressure Vessel Code construction. The common CP grades are variations of Grades 1 and 2. The CP grades have good corrosion properties in many applications, but these properties will improve in the more severe corrosive media with the addition of Pd, Ru, or Ni and Mo, as in the case of Ti Grade 12. While these alloying elements will help increase titanium's resistance to the higher concentrations of stronger acids, it will also enable the alloy to resist crevice corrosion. Table 25-17 shows those alloys and alloy compositions approved for use in ASME Boiler and Pressure Vessel Code construction.

TABLE 25-17 Titanium Alloys Approved for Use in ASME Boiler and Pressure Code Construction

Alloy grade	UNS	Common name	Composition
1	R50250	CP Ti	Unalloyed Ti
2	R50400	CP Ti	Unalloyed Ti
2H	R50400	CP Ti	Unalloyed Ti
3	R50550	CP Ti	Unalloyed Ti
7	R52400	Ti-Pd	Grade 2 + 0.15 Pd
7H	R52400	Ti-Pd	Grade 2 + 0.15 Pd
9	R56320	Ti325	3Al-2.5V
11	R52250	Ti-Pd	Grade 1 + 0.15 Pd
12	R53400	Ti-Code 12	Grade 2 + 0.3Mo-0.8Ni
16	R52402	Ti-Pd Lean	Grade 2 + 0.05 Pd
16H	R52402	Ti-Pd Lean	Grade 2 + 0.05 Pd
17	R52252	Ti-Pd Lean	Grade 2 + 0.05 Pd
26	R52404	Ti-Ru	Grade 2 + 0.1Ru
26H	R52404	Ti-Ru	Grade 2 + 0.1Ru
27	R52254	Ti-Ru	Grade 2 + 0.1Ru
28	R56323	Ti-3-2.5 Ru	Grade 2 + 0.1Ru
38	R54250	Ti425	Ti 4Al-2.5V-1.5Fe

Crevice Corrosion Titanium can experience crevice corrosion when exposed to hot chlorides, bromides, iodides, fluorides, or sulfate-containing solutions. Crevice corrosion in titanium will generally occur in chloride solutions where the temperature is greater than 70°C (180°F) or the pH is less than 10. This type of localized corrosion may occur in tight crevices, such as under gaskets, deposits, or scales, or in certain metal joints. One way to effectively prevent this type of corrosion in titanium is to use grades with Pd or Ru additions and alloys with Ni and Mo (Ti Grade 12).

MAJOR CHEMICAL PROCESSING APPLICATIONS FOR TITANIUM

Water and Seawater One of the largest chemical applications for titanium is in seawater and chloride applications. Titanium has excellent corrosion resistance in water and seawater, even in excess of 315°C (600°F). When the temperature exceeds 75°C (170°F) in a high chloride (seawater) environment, the possibility of crevice corrosion can exist. (See preceding discussion of Crevice Corrosion.)

Oxidizing Media (Peroxides, Chlorine) Titanium has good corrosion resistance in oxidizing media and oxidizing acids, such as chromic acid, wet chlorine, chlorites, hypochlorites, and perchlorates. The corrosion resistance of titanium in these media can, however, depend on the oxidizing ions (acid purity) in solution. Titanium should never be used in dry chlorine since this can cause a rapid attack and possible ignition.

Oxidizing Acids (Nitric, Chromic, Perchloric) Titanium also has good resistance to the oxidizing acids. For example, titanium has good resistance to nitric acid when the temperature is below boiling and there are certain metal species (Ti, Fe, Cr, Si) or metal ions (Pt or Pd) present in the solution. Titanium should not, however, be used in red fuming nitric acid since a pyrophoric reaction may occur with a potentially explosive effect.

Reducing Media (Hydrochloric, Sulfuric, Phosphoric) Titanium is used in some reducing media in the lower concentrations below the boiling point as shown in Figs. 25-24 and 25-25. Titanium is not very resistant to strong reducing acids because these media will cause a breakdown of the oxide film. In these media, the titanium alloys with the Pd or Ru additions or with Ni/Mo alloying elements will have better corrosion resistance than the unalloyed CP grades.

Alkaline Media (Sodium Hydroxide, Potassium Hydroxide, Ammonium Hydroxide) Titanium alloys are generally resistant to alkaline media at the lower concentrations but will begin to absorb hydrogen at the higher concentrations and temperatures. In these conditions the titanium may suffer hydrogen embrittlement when the temperature exceeds 80°C (175°F) and where the pH is greater than 12.

Organics and Organic Acids (Acetic, Citric, Formic, Lactic) Titanium generally has good resistance to organic media and organic acids.

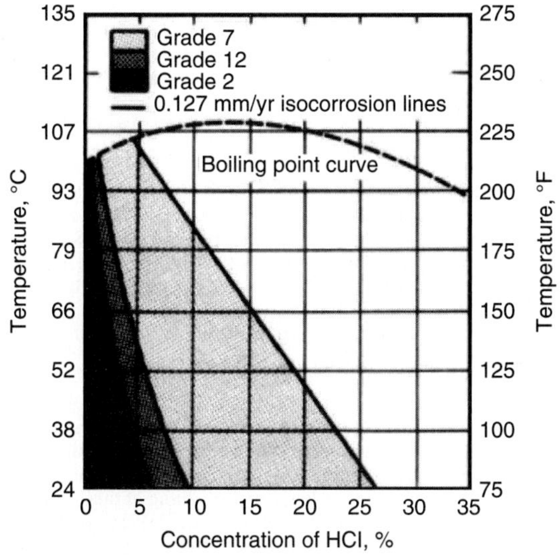

FIG. 25-24 Isocorrosion curve of titanium alloys in hydrochloric acid; 0.13 mm/yr (5 mpy) lines. ("Corrosion of Titanium and Titanium Alloys," *Metals Handbook Volume 13B Corrosion: Environments and Industries.* Reprinted with permission of ASM International. All rights reserved. www.asminternational.org.)

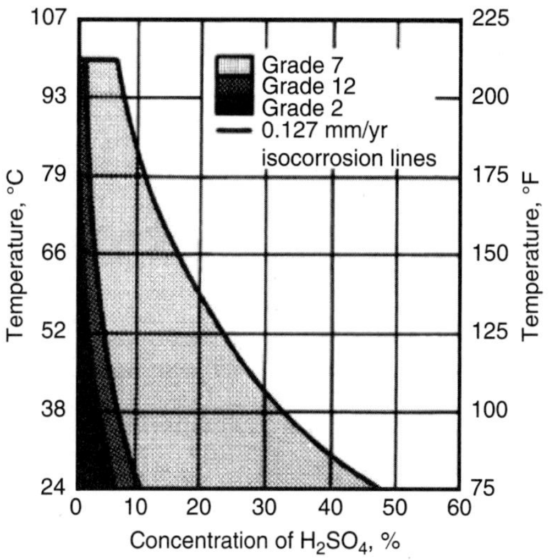

FIG. 25-25 Isocorrosion curve of titanium alloys in sulfuric acid; 0.13 mm/yr (5 mpy) lines. ("Corrosion of Titanium and Titanium Alloys," *Metals Handbook Volume 13B Corrosion: Environments and Industries.* Reprinted with permission of ASM International. All rights reserved. www.asminternational.org.)

The presence of oxidizing species or certain metal ions and aeration will improve the corrosion resistance of titanium in these media.

Fabrication of Titanium Titanium can be fabricated into solid and clad equipment. It can be machined and formed using conventional equipment. Titanium can be welded using inert gas welding processes such as gas tungsten arc (GTAW), gas metal arc (GMAW) and plasma arc welding (PAW).

ZIRCONIUM

Zirconium is a highly corrosion-resistant material used in many chemical environments. Zirconium is unique in the sense that it exhibits excellent resistance to both strong acids and strong alkaline media, including the mineral acids such as nitric, sulfuric, hydrochloric, formic, and acetic. Zirconium for chemical processing applications is generally available in three grades: Zr700 for explosion cladding use, Zr702 for all types of chemical equipment, and Zr705 for pumps, valves, and fasteners.

MAJOR CHEMICAL PROCESSING APPLICATIONS FOR ZIRCONIUM

Water and Seawater Zirconium has good resistance in all types of water, seawater, brine, and chloride environments, even at elevated temperatures to approximately 300°C (572°F).

Oxidizing Media (e.g., Nitric) Zirconium will have excellent corrosion resistance in oxidizing media such as nitric acid. Zirconium is used to the full range of concentrations, even at the higher temperatures as shown in Fig. 25-26. At concentrations above 70 percent, zirconium will have the tendency to stress corrosion crack. This susceptibility can be eliminated by the use of a stress-relief anneal heat treatment on the final equipment.

Reducing Media (Hydrochloric, Sulfuric, Phosphoric) Zirconium has excellent resistance to sulfuric acid at the low concentrations to well over the boiling point and up to about 70 percent above the boiling temperature, as shown in Fig. 25-27. Zirconium is one of the few materials capable of handling sulfuric acid above the boiling temperature. In sulfuric acid concentrations and temperatures exceeding the "weld limit line," the welds must be heat treated, or else preferential attack on the welds will occur. In hydrochloric acid, zirconium also has good corrosion resistance through the full range of 0 to 36 percent acid at temperatures exceeding the boiling point, as shown in Fig. 25-28. The presence of oxidizing impurities in the HCl will, however, cause zirconium to suffer from localized pitting and stress corrosion cracking. Laboratory studies as well as commercial applications have shown that the use of acid pickling followed by a stress-relief heat treat will reduce or eliminate this tendency for localized corrosion. Zirconium also has good resistance to phosphoric acid, but the presence of the fluoride ion (caused by the ores used in producing the phosphoric acid) will increase the corrosion rate of the zirconium significantly.

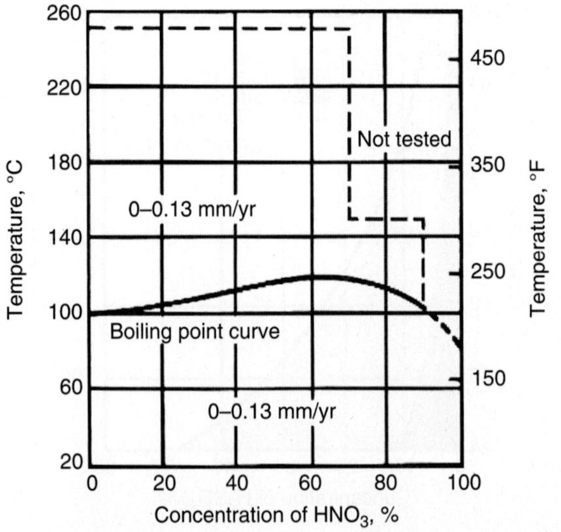

FIG. 25-26 Isocorrosion curve of zirconium in nitric acid. ("Corrosion of Zirconium and Zirconium Alloys," *Metals Handbook Volume 13B Corrosion: Environments and Industries*. Reprinted with permission of ASM International. All rights reserved. www.asminternational.org.)

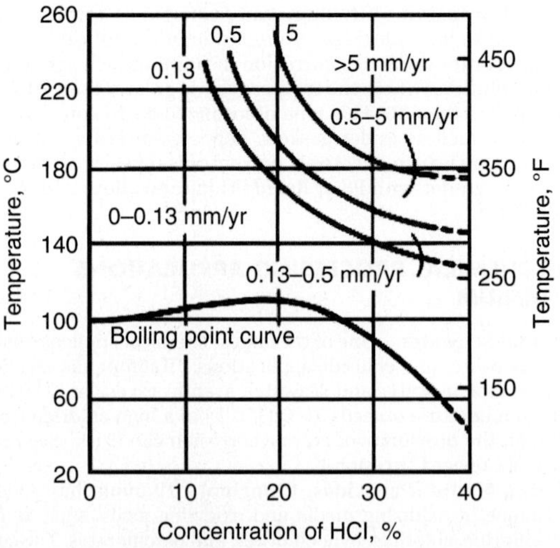

FIG. 25-28 Isocorrosion curve of zirconium in hydrochloric acid. ("Corrosion of Zirconium and Zirconium Alloys," *Metals Handbook Volume 13B Corrosion: Materials*. Reprinted with permission of ASM International. All rights reserved. www.asminternational.org.)

Organic Solutions (Acetic, Formic) Zirconium has excellent corrosion resistance to all organic media. One of the largest chemical applications for zirconium is its use in acetic acid environments where zirconium will exhibit very low corrosion rates up to 260°C (500°F). Zirconium will even handle the purer acetic acid environments if the water content is maintained to at least 600 ppm. Zirconium also has excellent resistance to formic acid environments.

Alkaline Media (Sodium Hydroxide, Potassium Hydroxide, Aluminum Hydroxide, and Ammonium Hydroxide) Zirconium has excellent resistance to most caustics even above the boiling temperature.

Urea (Ammonium Carbamate Media) Zirconium has excellent resistance to the urea synthesis reaction at temperatures exceeding 220°C (428°F) and in the presence of ammonium carbamate, which is very corrosive to many other materials. This application was one of the earliest chemical processing applications for zirconium.

Fabrication of Zirconium Zirconium can be fabricated and welded into heat exchangers, columns, piping systems, pressure vessels, valves, and pumps for the chemical process industry. Because zirconium is a reactive metal, it is sensitive to contamination during welding and thereby limited to inert gas welding processes such as gas tungsten arc (GTAW) and plasma arc welding (PAW).

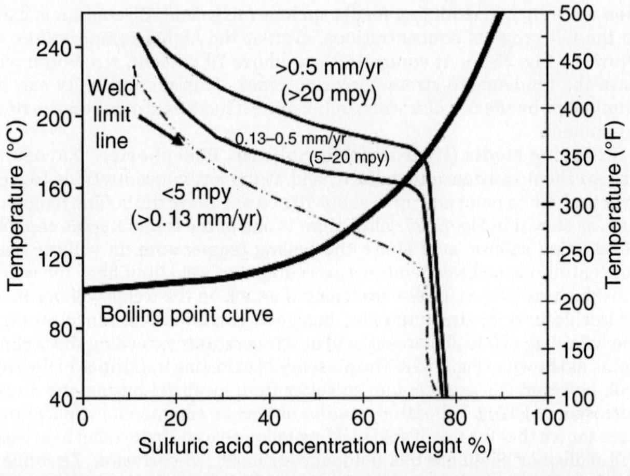

FIG. 25-27 Isocorrosion curve of zirconium in sulfuric acid. (Reprinted with permission from *ASTM STP728 Industrial Applications of Titanium and Zirconium*. Copyright ASTM International, 100 Barr Harbor Drive, West Conshohocken, PA 19428.)

TANTALUM

Tantalum alloys for chemical processing applications include pure unalloyed tantalum (UNS R05200) and Ta2.5W (UNS R05252) and, on a limited basis, Ta40Nb (UNS R05240). Tantalum is virtually immune to corrosion, including resistance to both strong oxidizing and highly reducing media, but Ta and its alloys are susceptible to hydrogen pickup and embrittlement in strong, hot alkaline media.

MAJOR CHEMICAL PROCESSING APPLICATIONS FOR TANTALUM

Oxidizing Media (Nitric, Chromic, Perchloric, Chlorine, Conc. Sulfuric) Tantalum has excellent resistance to even severely oxidizing environments such as concentrated nitric acid well above boiling, chromic acid, and chlorine environments.

Reducing Media (Hydrochloric, Sulfuric, Hydrobromic, Phosphoric, Formic, Oxalic) Tantalum resists most reducing media even well above the boiling temperatures.

Tantalum is also very resistant to hydrochloric acid to temperatures approaching 200°C (392°F). As the temperature exceeds 200°C (392°F), care should be taken due to tantalum's hydrogen absorption susceptibility at the higher temperatures (Fansteel published Corrosion Data Survey on Tantalum, "Hydrogen Embrittlement of Tantalum," Fansteel, Inc., North Chicago, Ill., May 8, 1972, pp. 123–133).

Tantalum is resistant to sulfuric acid through the full range of concentrations and temperatures below 150°C (302°F). Tantalum can even be used in concentrated 98 percent sulfuric acid as high as 200°C (392°F). Although there are a number of materials that could be used below sulfuric acid's boiling point, tantalum is one of the only choices (along with zirconium) above the boiling point, as shown in Fig. 25-29 (Sutherlin, R. C, "Zirconium and Zirconium Alloys for Use in Sulfuric Acid Applications," presented at the 2003 ACHEMA Conference, Frankfurt am Main, Germany, May 19–24, 2003).

Tantalum can also be used in strong bromine and hydrobromic acid environments.

Fabrication of Tantalum Tantalum, like niobium, is used as thin tubing or as a liner (cladding) in chemical equipment. Although tantalum is not approved as an ASME Boiler and Pressure Vessel Code material, it can be used in ASME BPV code equipment as a corrosion-resistant liner within the pressure envelope of an approved ASME BPV material. Tantalum is extremely ductile and can be easily formed. Welding of tantalum is limited to the use of gas tungsten arc (GTAW), plasma arc (PAW), and electron beam (EBW) welding. This metal is extremely sensitive to contamination during welding and should only be welded by fabricators experienced in refractory metals. It is used in lined piping systems as a corrosion barrier, in thermowells to protect thermocouples, and as rupture disks in pressure-relief devices.

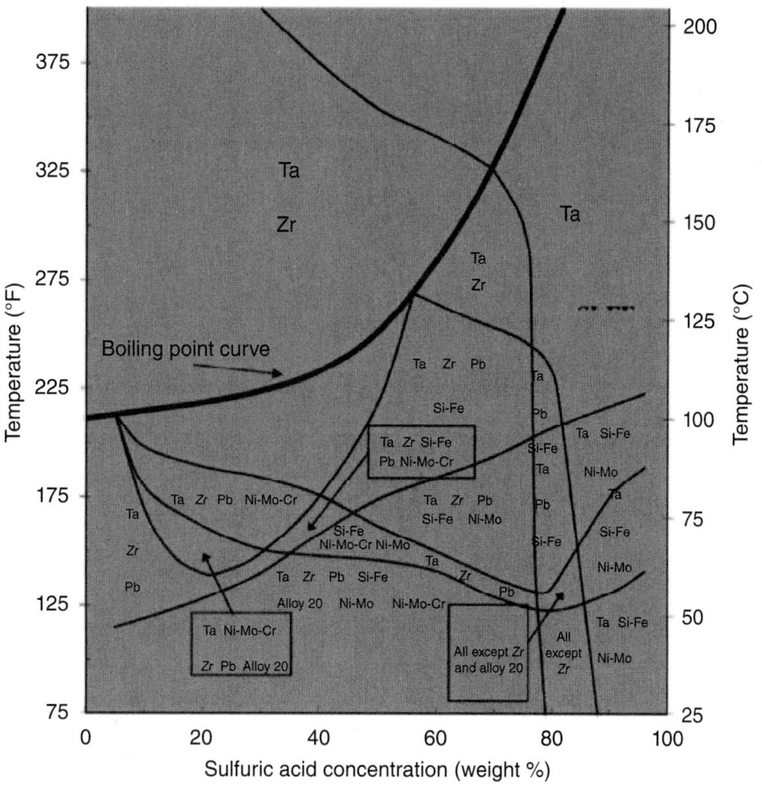

FIG. 25-29 Isocorrosion curve of various materials for use in sulfuric acid, Note: Materials shown in this figure will exhibit a corrosion rate of less than 0.5 mm/yr (20 mpy) except for Zr and Ta, which have a corrosion rate of less than 0.125 mm/yr (5 mpy). (Reprinted with permission of ATI Specialty Alloys and Components.)

NIOBIUM

Niobium resists a wide variety of chemical media, including mineral acids, organic media, and most salt solutions. Niobium is resistant to most oxidizing acids such as nitric acid, but it may be subject to hydrogen embrittlement in reducing acids. Like titanium and tantalum, niobium may be subject to embrittlement in certain alkaline media.

The niobium alloys used in chemical processing applications include pure niobium (Grades 1 and 2) and Nb1Zr.

MAJOR CHEMICAL PROCESSING APPLICATIONS FOR NIOBIUM

Niobium has good resistance to most organic and mineral acids at temperatures to 100°C (212°F). Like titanium, niobium exhibits better corrosion resistance if oxidizing impurities are present in the corrosive environment. Niobium will, however, suffer an increased corrosion attack in acids if the fluoride ion is present. Niobium exhibits especially good resistance in chromic acid and bromine environments.

Niobium will be embrittled in alkaline solutions at even the lower concentrations and higher temperatures or the higher concentrations and ambient temperatures.

Fabrication of Niobium Niobium can be used in chemical equipment such as heat exchangers and lined vessels. Niobium's fabrication methods and welding processes include welding techniques similar to those used for tantalum equipment.

OTHER METALS AND ALLOYS

The metals and alloys listed previously are by no means the only construction materials used in the chemical process industries. Some additional metals and alloys are described next.

ALUMINUM ALLOYS

Aluminum alloys have some specific applications. Although not normally considered a corrosion-resistant alloy, certain grades of aluminum are used to store and handle concentrated (>93%) nitric acid at ambient temperature. The most commonly used grades are A91100, A93003, A95052, and A95454. It is important to use low-silicon welding rods during fabrication or repair (Dillon, C. P., *Corrosion Control in the Chemical Process Industries*, 2d ed., Materials Technology Institute, St. Louis, Mo., 1994, p. 228). Aluminum is also very important in the fabrication of heat exchangers in gas liquefaction and evaporation units. It has very good low-temperature impact properties, making it useful for cryogenic services.

Aluminum is an amphoteric material, meaning it can be corroded by both acids and bases. It is rapidly attacked by caustic and most acids except as noted with nitric acid.

There are a large number of aluminum alloys categorized by composition, heat treatment, and mechanical working. Some of the more common aluminum alloys used in the process industries are shown in Table 25-18.

COPPER ALLOYS

The most important copper alloys used in the process industries are brasses (Cu-Zn alloys) and copper-nickel alloys. Some of the common copper-based alloys are shown in Table 25-19. The main use has been for heat exchanger tubes. Cu-Ni alloys are widely used in seawater exchangers.

TABLE 25-18 Some Commonly Used Aluminum Alloys

		Composition (nominal)						
Common name	UNS	Al	Si	Cu	Fe	Zn	Mn	Other
1100	A91100	99.0 min		0.5–0.20		0.10 max	0.50 max	Si + Fe 1.0 max
3003	A93003	Rem.	0.6 max	0.5–0.20	0.7 max	0.10 max	1.0–1.5	
5052	A95052	Rem.	0.25 max	0.10 max	0.4 max	0.10 max	0.10 max	Mg 2.2–2.8, Cr 0.15–0.25
5454	A95454		0.25 max	0.10 max	0.4 max	0.25 max	0.50–1.0	Mg 2.4–3.0, Cr 0.05–0.20, Ti 0.20

TABLE 25-19 Some Commonly Used Copper Alloys

		Composition (nominal)					
Common name	UNS	Cu	Zn	Ni	Sn	Fe (max)	Other
Red brass	C23000	85	15			0.05 max	Pb 0.06 max.
Admiralty brass	C44300	72	27		0.9–1.2	0.06 max	Pb 0.07 max.
90-10 Copper-nickel	C70600	86	1.0 max	10		1.0–1.8	Pb 0.05 max., Mn 1.0 max.
85-15 Copper-nickel	C72200	83		15		0.5–1.0	Cr 0.3–0.7, Mn 0.4–0.9
70-30 Copper-nickel	C71500	68	1.0 max	30		0.4–0.7	Pb 0.05 max., Mn 1.0 max.

TABLE 25-20 Some Commonly Used High-Cobalt Alloys

		Composition (nominal)									
Common name	UNS	Ni	Co	Fe	Cr	Mo	W	Mn	Si	C	La	Other
HAYNES 556	R30556	20	18	31	22	3	2.5	1	0.4	0.1	0.02	Ta 0.6, Al 0.1, N 0.20, Zr 0.02
HAYNES 188	R30188	22	39	3*	22	—	14	1.25*	0.35	0.1	0.03	B 0.015*
HAYNES 25	R30605	10	51	3*	20	—	15	1.5*	0.4	0.1	—	—
HAYNES HR-160	N12160	37	29	2*	28	1*	1*	0.5	2.75	0.05	—	Nb 1*
ULTIMET	R31233	9	54	3	26	5	2	0.8	0.3	0.06	—	N 0.08

*Maximum.

Brasses are susceptible to dezincification attack. All copper alloys can be attacked by ammonia and amines, depending on concentration and the presence of oxygen.

COBALT ALLOYS

Alloys with a high cobalt content are important in high-temperature applications, particularly when sulfidation attack is a concern. In addition, high cobalt-containing alloys are also used for corrosion and wear resistance applications. Alloys 556® (UNS R30556) and HR-160® (UNS N12160) are widely used because of their resistance to sulfidation, while ULTIMET® alloy offers excellent aqueous corrosion and wear resistance, along with resistance to weld-related cracking. Other high-cobalt alloys HAYNES® 188 (UNS R30188) and 25 (UNS R30605) combine excellent high-temperature strength with good resistance to oxidizing and sulfidizing environments up to 980°C (1796°F) for prolonged exposures. Alloy 25 also offers excellent resistance to metal galling. Some common cobalt-containing alloys are shown in Table 25-20.

LEAD

Lead was once commonly used because of its good resistance to sulfuric acid. Its use has dramatically decreased because of its toxicity.

LOW-TEMPERATURE AND CRYOGENIC MATERIALS

Selecting metals and alloys for low-temperature and cryogenic applications is a complex subject that involves many considerations, including special material testing and fabrication issues. It should only be done by a subject matter expert familiar with the applicable codes and standards. The information that follows is a short introduction to the subject.

As temperatures fall, carbon and low-alloy steels show a sharp reduction in ductility and resistance to fracture. This reduction can occur over a narrow temperature range. The temperature at which this reduction occurs varies by grade and heat. In general, this becomes a consideration whenever the temperature falls below 24°C (75°F). The generally accepted design and fabrication codes take this reduction of properties into account.

There are special grades of carbon steel for lower temperatures. These include ASTM A516 normalized for plate, ASTM A333 for pipe (includes both carbon and low-alloy steels), and ASTM A334 for tube (includes both carbon and low-alloy steels).

The addition of nickel to carbon steels as an alloying element can lower the allowable design temperature at which adequate physical properties are retained. These grades include ASTM A203, ASTM A353, and ASTM A645 for plate, ASTM A333 for pipe, ASTM A334 for tubes, ASTM A320 for bolting, ASTM A350, ASTM 420, and ASTM A522 for flanges and fittings. The specified nickel additions are 2¼, 3½, 5, 8, and 9 percent. As the nickel content is increased, the allowable temperature decreases.

Austenitic stainless steels do not undergo the sudden decrease in ductility and resistance to fracture that carbon and other ferritic steels experience. For this reason they are often used for very low temperatures, including cryogenic applications. Type 304 stainless steel is probably the most common grade used for cryogenic applications.

Although not normally considered a structural material, aluminum and some aluminum alloys have important applications in cryogenic processes as heat transfer materials. Aluminum is used in heat exchangers to transfer heat between different liquid streams in processes such as liquefied natural gas processing and gas separation units.

NONMETALLIC MATERIALS FOR CORROSION CONTROL

INTRODUCTION

Nonmetallic materials can be classified as inorganic nonmetallic, such as refractories and glass, and organic nonmetallic, such as plastics, also known as polymers, polymerics, or polymer-based materials.

INORGANIC NONMETALLICS

Glass and Glassed Steel Glass is an inorganic product of fusion that is cooled to a rigid condition without crystallizing. With unique properties compared with metals, these materials require special considerations in their design and use.

Glass has excellent resistance to all acids except hydrofluoric and hot, concentrated phosphoric acid (H_3PO_4). It is also subject to attack by hot alkaline solutions. Glass is particularly suitable for piping when transparency is desirable.

The chief drawback of glass is its fragility and brittleness, and it is also subject to damage by thermal shock. However, glass armored with epoxy-polyester fiberglass offers some protection against breakage. Similarly, glassed steel combines the corrosion resistance of glass with the working strength of steel on the outside. Glass linings are resistant to all concentrations of hydrochloric acid to 120°C (250°F), to dilute concentrations of sulfuric to the boiling point, to concentrated sulfuric to 230°C (450°F), and to all concentrations of nitric acid to the boiling point.

Porcelain and Stoneware Porcelain and stoneware materials are about as resistant to acids and chemicals as glass but with greater strength, a property offset by a greater potential for thermal shock. Porcelain enamels are used to coat steel, but the enamel has slightly inferior chemical resistance. Some refractory coatings, capable of taking very high temperatures, are also available.

Brick Construction Brick-lined construction can be used for many severely corrosive conditions under which high alloys would fail. Brick linings can be installed over metal, concrete, and fiberglass structures. Acid-resistant bricks are made from carbon, red shale, or acid-resistant refractory materials. Red-shale brick is not used above 175°C (350°F) because of spalling. Acid-resistant refractories can be used up to 870°C (1600°F).

A number of cement materials are used with brick. Standard are polymer resin, silicate, and sulfur-based materials. The most widely used resins are furane, vinyl ester, phenolic, polyester, and epoxies. Carbon-filled furanes and phenolics are good against nonoxidizing acids, salts, and solvents. Silicates and silica-filled resins should not be used in hydrofluoric or fluorosilicic acid applications. Sulfur-based cements are limited to 93°C (200°F), while resins can be used to about 180°C (350°F). Silicate-based cements are available for service temperatures up to 1000°C (1830°F).

Brick porosity, which can be as high as 20 percent, requires an intermediate lining of lead, asphalt, rubber, or plastic. This membrane functions as the primary barrier to protect the substrate from corrosion damage. The brick lining provides thermal and mechanical protection for the membrane. The membrane system also allows for the differential thermal expansion between the brick lining and supporting substrate. The design of brick linings exposed to higher operating pressure should take into account chemical expansion in addition to thermal expansion.

Cement and Concrete Concrete is an aggregate of inert reinforcing particles in an amorphous matrix of hardened cement paste. Concrete made of Portland cement has limited resistance to acids and bases and will fail mechanically following absorption of crystal-forming solutions such as brines and various organics. Concretes made of corrosion-resistant cements (such as calcium aluminate) or polymer resins can be selected for specific chemical exposures.

Soil Clay is the primary construction material for settling basins and waste treatment evaporation ponds. Since there is no single type of clay even within a given geographic area, the shrinkage, porosity, absorption characteristics, and chemical resistance must be checked for each application. Geotextiles can be incorporated into basin and pond clay construction to improve the performance of the structure.

ORGANIC NONNMETALLIC MATERIALS

These materials are based on long-chain organic molecules shaped into forms that can be used in chemical handling applications. A full description of these materials is done in reference to:

1. Generic type, that is, the organic chemical nature. Examples are polymer families such as styrenics, vinyls, and fluoropolymers.

2. Mechanical properties: rigid [modulus >690 MPa (>100,000 psi)], semirigid [modulus between 69 MPa and 690 MPa (10,000 to 100,000 psi)], and nonrigid [modulus <69 MPa (10,000 psi)].

3. Thermal processing: thermoplastic (materials that can remelted and reprocessed) or thermosetting (materials that cannot be remelted and reprocessed due to post-molding cross linking or curing).

The word *elastomers* refers to mechanical property basis; such materials are nonrigid and have the unique feature of high elongation and high recovery. Composites are a special class of materials that are made by using the polymer as a continuous phase and fibers or particulates as a discrete phase. Various combinations are possible, depending on the final mechanical properties desired. Examples are wood and fiber-reinforced plastic. Polymeric materials have an inherent limitation in that their maximum service temperature for chemical handling is about 204°C (400°F). They also have lower mechanical properties, such as tensile, elongation, stiffness, and impact. They often are heterogeneous and anisotropic, that is, they have different properties in different directions. Hence their use is limited to linings of metallic housings or self-supporting structures at low temperatures. They are also workmanship sensitive since the final material of construction is often formed during fabrication. Most importantly, they are viscoelastic, that is, they creep readily, making their performance more time and temperature sensitive than that of metals. There are fewer nondestructive testing techniques for in-service condition assessment. Yet there are many applications where only polymeric materials can be used due to their outstanding chemical resistance and favorable strength-to-weight ratio. Various manufacturing techniques such as injection molding, extrusion, blow molding, transfer molding, contact molding, and compression molding are used to manufacture finished parts.

COMMONLY USED THERMOPLASTIC MATERIALS

General use thermoplastics (in roughly ascending order of performance and cost) include the following:

Polyethylene (PE) (-CH_2-CH_2-)n Various grades such as linear low density, high density, and ultrahigh molecular weight are available, although high-density grade is the most popular, particularly the cross-linked type. Commonly used forms are piping and rotomolded storage tanks. PE has excellent resistance to mineral acids up to 43°C (110°F). PEs have limited resistance to organic solvents. Among the polymer materials, urethanes, nylons, and polyethylene have the best abrasion resistance, but only PE is commonly used for chemical handling.

Polyvinyl Chloride (PVC) (-CH_2-CHCl-)n- A polymer of vinyl chloride monomer is used in either plasticized or unplasticized form. A chlorinated version (CPVC) is also used for higher temperatures and chemical resistance. Common forms are piping and sheet linings for vessels and columns. It has good resistance to sulfuric acid and limited resistance to organic solvents.

Polypropylene (PP) (-CH_2-$CHCH_3$-)n In terms of chemical and temperature resistance, PP is superior to PE and PVC. It is used as self-supporting pipe or as lining for metallic piping. It is also used as vessel linings in sheet form.

Mid-Performance Partially Fluorinated Polymers The presence of fluorine significantly increases the chemical and temperature resistance of polymers. They are used as linings for piping, valves, and vessels. These include the following materials:

Polyvinyledene difluoride (PVDF) (-CH_2-CF_0-)n Trademark: Kynar by Arkema. Maximum use temperature 121°C (250°F).

Ethylene Chloro-tri-fluoroethylene (ECTFE) (-CH_2-CH_2-CF_2-CFCl-)n Trademark Halar of Ausimont. Maximum use temperature 135°C (275°F).

Ethylene tetrafluroethylene (ETFE) (-CH_2-CH_2-CF_2-CF_2-)n Trademark Tefzel of Chemours. Maximum use temperature 149°C (300°F).

High-Performance Fully Fluorinated Polymers Chemical and temperature resistance is higher than that of the partially fluorinated polymers. Among all polymer-based materials, these have the highest chemical and temperature resistance. Because their mechanical properties are lower than the partially fluorinated, they are used principally as linings for vessels, piping, valves, hoses, and expansion joints. These materials include the following:

Polytetrafluoroethylene (PTFE) (-CF_2-CF_2-)n PTFE was the first grade of fluoropolymer discovered. This polymer is processed by the powder metallurgy technique of compaction and sintering due to its very high melt viscosity. This is also the most commonly used material for lining pipes, hoses, expansion joints, and gaskets. Its maximum service temperature is 260°C (500°F). One trade name is Teflon PTFE.

Fluorinated Ethylene Propylene (FEP) (-CF_2-CF_2-CF_2-$CFCF_3$-)n This grade is melt processible and therefore can be easily welded. It is extruded and injection- and transfer-molded but with a lower service temperature, 204°C (400°F), than that of PTFE. A common example is Teflon FEP by Dupont.

PerfluoroAlkoxy (PFA) The highest-performing material of this class. Melt processible, and the maximum service temperature is 260°C (500°F). An example is Teflon FEP by Dupont.

COMMONLY USED THERMOSETTING MATERIALS

Introduction Used with glass, carbon, or polyester reinforcement, thermosetting resins are used for fiber-reinforced plastic (FRP) vessels and piping. FRP is excellent for storing and transporting acids such hydrochloric, sulfuric, and phosphoric. It is also used for hydrofluoric acid up to 5 percent concentrations, but it has limited resistance to organic solvents. The addition of SiC_4 in the corrosion barrier can improve the abrasion resistance. Typical FRP laminates consist of a corrosion barrier of 2.54 to 5.08 mm (0.1 to 0.2 in) made up of a resin-rich layer or layers of veil (with 90 percent resin) followed by two to four layers of chopped strand mat. The veils can be of glass, carbon, or polyester. Together the corrosion barrier has 70 percent resin and 30 percent glass. The corrosion barrier is followed by structural layers of either the hand layup construction, which consists of alternate layers of mats and woven roving, or the filament-wound construction, which uses continuous strand roving. The structural thickness typically has 30 percent resin. These resins are also used with fibers, fabric, or particulate fillers as coatings. All resins are used as copolymers with styrene as a reactive diluent.

BisA Fumerate Polyester Resins for FRP The earliest resins used for FRP vessels and piping with good chemical resistance in general service. An example is Atlac 382 resin by Reichold.

Chlorendic Anhydride–Based Polyester Resins Chlorine-containing polyester resins used for handling chlorine-containing chemicals. Example is Hetron 192 by Ashland Company. Used as copolymers with styrene.

Epoxy Vinyl Ester and Novolac Epoxy Vinyl Ester Resins Currently these classes of resin are the most commonly used materials for FRP applications. They are always used as copolymers with styrene. Brominated versions of these resins are also available for fire retardancy. Novolac epoxies have better temperature and solvent resistance than the straight vinyl esters. Examples are the Derakane family of resins by Ashland Company.

Furan A polymer of furfural alcohol, Furan has excellent organic solvent resistance in addition to resistance to acids. Due to changing environmental regulations, the corrosive nature of the resin catalyst, and greater difficulty with handling during fabrication, Furan is used less often than the vinyl esters.

Epoxy Resins These are more brittle than the vinyl esters but have excellent chemical resistance in certain applications. They tend to be stronger and hence are used in high-pressure piping, particularly for downhole applications.

COMMONLY USED ELASTOMERS (THERMOSETTING RUBBERS)

Introduction Elastomers or rubbers are used as linings for vessels, valves, hoses, and expansion joints. They are used for seals and gaskets in very large quantities. Elastomers are broadly classified as natural and synthetic rubbers. They are generally more economical than other options, easy to manufacture, and more forgiving in terms of installation and repair. Natural rubbers (Cis-1,4-polyisoprene) are made of the rubber latex produced by rubber trees, *Hevea brasiliensis*. Synthetic rubbers are a product of the petrochemical industry. In addition to the natural rubbers in their soft, semihard, and hard forms, the following are used for chemical handling:

1. Butyl or chlorobutyl rubbers—poly(isobutylene-co-isoprene)
2. BUNA N or nitrile rubber—poly(butadiene-co-acrylonitrile)
3. EPDM (ethylene propylene diene monomer), also known as EPM
4. Chloroprene—chlorinated polyisoprene or polychloroprene, popularly known as neoprene
5. Chlorosulfonated polyethylene (Hypalon®)

Fluorine-containing elastomers such as FKM (e.g., Viton®) or FFKM (e.g., Kalrez®) are used for high-temperature and highly corrosive applications, principally as seals and gaskets.

LIFE CYCLE OF EQUIPMENT

Life cycle deals with materials selection, design, fabrication, shipping, operation, maintenance, and replacement issues.

STORAGE TANKS, REACTORS, TRANSPORTATION EQUIPMENT: FRP

Material Selection Material selection involves choosing the appropriate resin and reinforcement for both the corrosion barrier and the structural layers and a cure system per ASTM C581 or the new MTI (Materials Technology Institute) guidelines, which involve coupons exposed to a medium at the application temperature. The cross sections of these coupons are studied for the type and extent of damage. Using this information, the life of the laminate can be estimated.

Design, Construction, Inspection, and Transportation FRP is a multilayered construction of corrosion barrier and structural thickness built by contact-molded (also known as hand layup) or filament-wound techniques. Using a thermosetting resin and curing with either methyl ethyl ketone peroxide (MEKP), benzoyl peroxide (BPO), or cumene hydroperoxide (CHP), the structure is completed. For vessels, ASME RTP-1 is a widely accepted industry code for design, fabrication, inspection, and transport in North America, and EN13121 is a comparable code in Europe. The rest of the world follows one of them. For pressures higher than 0.10 MPa (14.7 psig), ASME Section X is used. Large-diameter field-erected or field-fabricated tanks are less common, but new guidelines are available from the Materials Technology Institute (Begsjo, P., and S. Rohmhild, *Accelerated Testing of FRP*, Materials Technology Institute, St. Louis, Mo., 2014).

Maintenance (In-Service Inspection, Condition Assessment, Fitness for Service, NDT, and Destructive Techniques) of FRP In-service inspection consists primarily of visual examination. Damage observed is blisters, cracking (gross and fine), delamination, fiber prominence, chemical attack, and abrasion. Acoustic emission testing is often used to determine overall structural integrity. The frequency of in-service inspection is determined by the nature of service and the consequences of failure. Repair of damaged FRP is possible, but test patches are recommended to ensure success. Destructive testing in the form of a cutout is possible, but it should be considered a last resort. With proper interpretation of this information, remaining life estimates can be made for the corrosion barrier, which is effectively the remaining life of the unit itself. If the cross-sectional examination indicates damage to the structural layer, retained mechanical properties should be determined in case the unit needs to be taken out of service immediately.

After all the testing is completed, the path forward is one of operate as is, re-rate, repair/resurface, or replace. Re-rating is usually for storage vessels; it is an option if it is acceptable to operate the unit at a reduced capacity. Resurfacing is usually costly, so replacing the unit is often economically justified, and it eliminates uncertainty.

VESSELS WITH LININGS

Vessel linings are classified in three ways.

1. Chemical nature
2. Thickness
3. Application method—sheet, spray, or trowel

The most commonly used classification is the thickness. Thick linings are defined as those greater than 0.635 mm (0.025 in). Usually they are 1.01 mm (0.040 in) and higher. Thick linings are used where the corrosion rate of carbon steel is greater than 10 mils per year. Thin linings have thicknesses less than 0.635 mm (0.025 in) and are used for situations where the corrosion rate of carbon steel is less than 0.254 mm (0.010 in) per year, or where the corrosion is localized, such as pitting or crevice corrosion. Thin linings are also used for nonstick applications or for product purity.

Selection of Lining Testing for chemical compatibility is done in two steps, screening and application specific. Screening is done by exposing simple coupons in liquid and vapor phases and tracking the changes in mechanical properties over time. Criteria such as those shown in Table 25-21 are used. High-weighted-value candidates are selected for further evaluation.

EVALUATION OF SCREENING TESTING BY COUPON IMMERSION

After screening, the candidate materials are subjected to a one-sided test (Atlas cell test), ASTM C868. Postexposure evaluation is done by visual examination and by a peel-pull test for loss of adhesion to and condition of the substrate. Since linings are permeable, particularly the fluoropolymer ones, permeation testing is carried out. Permeation is a complicated topic, and engineering decisions are usually made using the rules of thumb shown in Table 25-22. Other factors such as shop versus field, repairability, and NDT also play a role in lining selection.

Design and Fabrication of Vessels to Be Lined Vessels need to be designed and fabricated in such a way that coatings and linings can be effectively applied or installed. This usually means no sharp corners or crevices.

TABLE 25-21 Evaluation of Screening Testing by Coupon Immersion

Weighted value	Weight* change	Volume* change	Hardness change (units)	Mechanical† property retained	Visual‡ observed change	BTT (min)	Permeation rate (μg/cm²/min)
10	0–0.25	0–2.5	0–2	> = 97	No change	≤ 1 = 0	≤ 0.9
9	> 0.25–0.5	> 2.5–5.0	> 2–4	94 < 97		> 1–2 = 2 > 1 ≤ 2	
8	> 0.5–0.75	> 5.0–10.0	> 4–6	90 < 94		> 2–5 = 3 > 2 ≤ 5	> 0.9–9
7	> 0.75–1.0	> 10.0–20.0	> 6–9	85 < 90	Slightly discolored; slightly bleached	> 5–10 = 4 > 5 ≤ 10	
6	> 1.0–1.5	> 20.0–30.0	> 8–12	80 < 85	Discolored; yellow, agent; flexible	> 10–30 = 5 > 10 ≤ 30	> 9–90
5	> 1.5–2.0	> 30.0–40.0	> 12–15	75 < 80	Possible; stress crack agent; flexible; possible oxidizing agent; slightly crazed	> 30–120 = 6 > 30 ≤ 120	
4	> 2.0–3.0	> 40.0–50.0	> 15–18	70 < 75	Distored; wraped; softened; slight swelling; blistered; known stress crack agent	> 120–240 = 7 > 120 ≤ 240	> 90–900
3	> 3.0–4.0	> 50.0–70.0	> 18–21	60 < 70	Cracking; crazing; plasticize, oxidizer; softened; swelling; surface hardened	> 240–480 = 9 > 240 ≤ 480	
2	> 4.0–6.0	> 60.9–90.0	> 21–25	50 < 80	Severe distortion; oxidizer; and plastixizer, deterioratedd	> 480–960 = 9 > 480 ≤ 960	> 900–9000
1	> 6.0	> 90.0	> 25	> 0 < 50	Decomposed	> 960 = 10 > 960	
				0	Solvent dissolved, disintegrated		> 9000

*All values are given as percent change from original.

†Percent mechanical properties retained include tensile strength, elongation, modulus, flexural strength, and impact. If the percent retention is greater than 100 percent, a value of 200 minus the percent property retained is used in the calculations.

‡Due to the variety of information of this type reported, this information can be used only as a guideline.

Welded joints also need to be smooth and in some cases ground flush. Nozzle sizes cannot be less than 50.8 mm (2 in) with projections of no more than 152.4 mm (6 in). NACE standard SP0178 should be followed. Surface preparation for most nonmetallic linings is white metal blasting according to Steel Structure Paint Council standard SSPC SP 5.

Spray-Applied Thin Coatings Thermosetting coatings of epoxy, phenolic, and phenolic-modified epoxy are the most commonly used. They are multicoat systems averaging about 0.31 to 0.41 mm (0.012 to 0.016 in) of dry film thickness.

Spray, Brush, or Trowel-Applied Thick Linings Vinyl ester and furan resins are used with chopped fibers, fabric, or particulate fillers for a combination of corrosion control and strength. Broadly they are classified as flake glass linings or mat linings. A high-voltage spark tester is used to ensure continuity.

Elastomeric (Rubber) Sheet Lining These are typically 6.35 mm (0.25 in) thick and are used for handling and transporting hazardous mineral acids such as hydrochloric, hydrofluoric, and phosphoric acids and abrasive slurries. Most commonly used elastomeric linings are the natural rubbers—soft, semihard, or hard, depending on the temperature and chemical resistance required. Selection of rubbers is done by screening test by coupon immersion and application-specific testing. The interpretation of coupon data requires a great deal of skill and experience. Table 25-23 shows basic guidelines.

Rubber linings are applied as uncured sheets and are vulcanized (cured) in the autoclave (in the shop) or with steam in the field. The level of cure is indicated by Shore A or D scales of Durometer hardness. Integrity of the final lining is tested by a high-voltage spark test. Degradation of rubber in service occurs by blistering, chemical attack (hardening or softening), separation of laps, cracking (micro or gross), and mechanical damage such as abrasion or tears. Damaged rubber can be repaired relatively easily with patching done by chemical cure rather than steam vulcanizing.

TABLE 25-22 Permeation Rules

Permeation variables		
Factor	**Change**	**Effect on permeation**
Permeant concentration	+	+
Temperature	+	+
Pressure	+	+
Permeant/polymer chemical similarity	+	+
Voids in polymer	+	+
Permeant size/shape	+	−
Polymer thickness	+	−
Polymer crystallinity	+	−
Polymer chain stiffness	+	−
Polymer interchain forces	+	−

Key to Table 25-22: As the concentration of the permeant goes up (+), the permeation rate goes up (+).

Thermoplastic Linings PVC, CPVC, polypropylene, and fluoropolymers are used as linings in a variety of ways. For PVC, CPVC, and polypropylene, the most common form is sheet linings directly "wall-papered" on the inside of a metallic vessel. The sheets need to be welded to each other, so welding technology is an important part of this type of lining. Fluoropolymer linings such as PFA, PTFE, and FEP are installed in one of five ways, as shown in Table 25-24. The choice depends on such factors as vessel geometry, complexity, size, shop or field, and cost. In loose linings, bonded linings, and dual-laminate structures, welding of the fluoropolymers is involved. The types of welding procedures used are mostly flow fusion and hot gas hand welding. The quality of welding is determined by a short-term tensile test as described by ASTM C 1147 as well as German Welding Institute (DVS) standards in the qualification stage. High voltage (15 kV to 25 kV) is used to detect pinholes or breaks in the weld seam. In-service condition assessment is done mostly by visual inspection and by destructive peel test if required. Spark testing of in-service lining should be avoided as far as possible, but if necessary the voltage should be reduced to 50 percent of the original setting. Typical degradation modes of fluoropolymer liners are discoloration, blistering, cracking, separation of welds, and occasionally environmental stress cracking. Direct

TABLE 25-23 Interpretation of Immersion Testing of Elastomers Coupons

	Root cause of property damage				
	A	B	C	D	E
Property change	Deleterious process medium absorption	Extraction of compound ingredients	Medium attack on the filler system	Attack and degradation of cross-links and/or the polymer backbone	
Hardness	Decrease	Increase	Usually decreases	Increase (hard/brittle)	Decrease (soft/gummy)
Mass and volume	Increase	Decrease	Increase	*(Often)* Increase	Increase
Tensile strength at break	Decrease	Increase *(or decrease)*	Decrease	Decrease	Decrease
Modulus	Decrease	Increase *(or decrease)*	Decrease	Increase	Decrease
Elongation at break	Increase	*(Usually)* Decease	Increase	Decrease	Increase or decrease *(often increase)*

TABLE 25-24 Fluoropolymer Lining Options

Lining technology	Materials available	Fabrication	Design	Size limitation
Adhesively bonded fabric backed sheets to carbon steel	All fluoropolymers possible. Most commonly used: PVDF, ETFE, FEP, and PFA. Thickness range from 60 mils to 120 mils	Welding, vacuum bagging, use of adhesive	Pressure determined by ASME code, shop and field, full vacuum up to 120 F, max temp determined by adhesive and permeation consideration	None
Loose lining of sheets	All fluoropolymers possible. Most commonly used: ETFE, modified PTFE, ETFE, FEP, and PFA. Thickness range from 60 mils to 120 mils	Welding, thermoforming	Very limited vacuum handling ability. Pressure determined by ASME code. Max temperature determined by permeation consideration	Determined by body flange. Usually 3 ft diameter by 6 ft
Dual laminate (fluoropolymers inside FRP)	Same as adhesively bonded	Liner fabricated first by welding and FRP then laminated on top	Very limited vacuum handling ability. Pressure determined by ASME code. Max temperature determined by permeation consideration	14 ft diameter max
Rotolining	ETFE and PFA (two-coat system). Thickness range from 90 mils to 250 mils	Liner rotationally molded in a CS vessel	Pressure determined by ASME code. Vacuum limit not determined but expected to be somewhat higher	6 ft diameter by 22 ft
Spray and baked coatings -Electrostatic powder coating -Liquid dispersin coating	ETFE, FEP, PFA (with or without wire mesh reinforcement), PVDF (with or without fabric reinforcement), ECTFE	Multicoat application of dry powder, or liquid dispersion and baked	Pressure determined by ASME code. Vacuum limit not determined but expected to be somewhat higher	20 in diameter by 40 ft

chemical attack is rare, but permeation is a bane of fluoropolymers. It cannot be eliminated, but it can be mitigated by following some rules shown in Table 25-24. Repairs to fluoropolymer linings can be difficult because polymer morphology could change due to absorption and permeation. "Cold" welding—that is, welding very close to the melting temperature—slow-speed welding, and using nitrogen can help in the repair procedures.

Piping Most commonly used plastic piping is custom-made FRP and PTFE-lined steel. PP and PVDF-lined piping are also used for less corrosive or lower-hazard services. Commodity or machine-made FRP piping is usually not recommended for hazardous service. Rubber-lined piping and solid thermoplastic piping such as PE, PP, and PVC/CPVC are rarely used in North America but are still rather common in other parts of the world. ASME B31.3 is the governing code for materials, design, fabrication, inspection, and joining of pressure piping. ASTM F1545 is used for all lined piping for material, construction, and more importantly for qualifying the producer. For in-service inspection of FRP piping, internal boroscopic examination is the only possible nondestructive examination technique, and it requires the pipe circuit to be clean and out of operation. Often, removing sections of piping and destructively testing them on a statistical sampling basis provides valuable information. A very large percentage of FRP failures are at the wrapped joints, particularly those made in the field. For the PTFE or PVDF-lined piping, the failures are often at the flanged joints. Collapse of the liners due to temperature and pressure cycling is a common failure mode. Like FRP pipe, lined pipe systems must be shut down and clean to allow for proper inspection.

Valves Valves with thermoplastic liners are very common. PTFE, modified PTFE, PFA, FEP, ETFE, and PVDF are the most commonly used materials. Ball, plug, butterfly, diaphragm, and clamp valves use these materials for body liners, ball or plug covers, and diaphragms.

Expansion Joints Solid PTFE expansion joints are common, although some companies choose not to use them for safety reasons. Unlike vessels, piping, and valves, PTFE expansion joints have no metallic housings. PTFE-lined elastomeric expansion joints are a better option.

Hoses Many polymeric material components are in chemical hoses. Inner liners are typically of PE, PTFE, ETFE, PVDF, or elastomers such as natural rubber, synthetic butyl, or chlorobutyl rubber, EPDM, and chloroprene. Carbon-filled PTFE is also used for static charge dissipation where required. The overbraiding can be metallic, such as stainless steel or high-nickel alloys, or polymeric, such as PVDF. Sometimes an elastomeric cover is also used for additional handling protection. End connections can be flanged or couplings. Transfer hoses are typically 25.4 to 50.8 mm (1 or 2 in) with lengths of 6.1 m (20 ft). Transfer hoses should not be used as permanently installed flexible piping. Most operations replace the hoses at six-month to one-year frequencies.

Seals and Gaskets All elastomers are used as gaskets. Usually elastomeric gaskets are full-faced. PTFE (filled, nonfilled, or expanded) is the most commonly used nonelastomeric material. PTFE is also used in the semimetallic forms, such as envelope or spiral-wound configurations. The selection of gaskets involves two considerations, chemical compatibility (determined by coupon testing) and compression set (at the application temperature). Proper joint assembly techniques, proper gasket preload, and achieving recommended torque are vital. Gaskets are changed every time the joint is disassembled. O-rings are commonly used for severe service applications such as valves and mechanical seals. Other cross sections such as T- and U-rings are also sometimes used.

Column Internals ETFE and PP are the most commonly used column packing materials. They can be produced in all the common forms, such as saddles or rings. The most common form of failure is embrittlement.

Index